On the Cover

The cover depicts an extraordinary sight of coordinated movement—or murmuration—of a flock of starlings in response to a potential predator (a hawk). Individual birds coordinate their flight by sending, receiving, and processing signals from their immediate neighbors, although the exact processes that allow a flock of starlings to respond in almost perfect unison are still unknown. Murmuration is one of the most visible and beautiful examples of a critical transition, in which small perturbations to a system result in major transformations. Such critical transitions are important in many areas of biology, from the actions of individual proteins and cells to the behavior of flocks of birds.

A stunning video of the event is at
Life11e.com/cover

Life

The Science of Biology

ELEVENTH EDITION

Life

The Science of Biology

ELEVENTH EDITION

 SINAUER ASSOCIATES

 MACMILLAN

David Sadava
The Claremont Colleges

David M. Hillis
University of Texas, Austin

H. Craig Heller
Stanford University

Sally D. Hacker
Oregon State University

The Cover
See the inside front cover for an explanation of the murmuration event shown on the cover of the book. Go to **Life11e.com/cover** to see a video of this stunning event. Photo courtesy of Giangiorgio Crisponi; video © Dylan Winter.

The Frontispiece
Schooling in fish is highly synchronized and (like murmuration in starlings) is sometimes a defense against a predator, in this case a blacktop reef shark (*Carcharhinus melanopterus*) at Baa Atoll, Maldives. Photo © Frank Krahmer/ Getty Images.

Life: The Science of Biology, Eleventh Edition
Copyright © 2017 by Oxford University Press
Sinauer Associates is an imprint of Oxford University Press
This book may not be reproduced in whole or in part without permission.

Address Editorial Correspondence To:
Sinauer Associates, Inc., P.O. Box 407, Sunderland, MA 01375-0407 U.S.A.

Address Orders To:
MPS/W. H. Freeman & Co., Order Department, 16365 James Madison Highway, U. S. Route 15, Gordonsville, VA 22942 U.S.A.

Examination Copy Information: 1-800-446-8923

Library of Congress Cataloging-in-Publication Data
Names: Sadava, David E., author.
Title: Life: The science of biology / David Sadava, The Claremont Colleges, David
    M. Hillis, University of Texas, H. Craig Heller, Stanford University,
    Sally D. Hacker, Oregon State University.
Description: Eleventh edition. | Sunderland, MA : Sinauer Associates, Inc.,
    2016. | Includes index.
Identifiers: LCCN 2016032791 | ISBN 9781319010164
Subjects: LCSH: Biology--Textbooks.
Classification: LCC QH308.2 .L565 2016 | DDC 570--dc23
LC record available at https://lccn.loc.gov/2016032791

Second Printing June 2017
Printed in United States of America

Contents in BRIEF

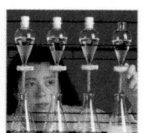
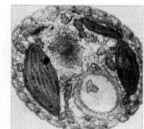
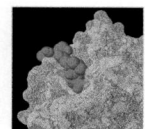

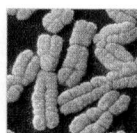
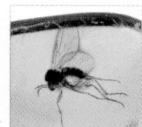

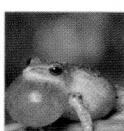

Life The Science of Biology

A Note from the Authors

Biology inspires wonder. The image of the flock of starlings on our cover shows an astonishing behavior known as murmuration—mesmerizing patterns that emerge out of thousands of birds flying in unison. This fascinating behavior is an example of what motivates biologists to study the complexities of life.

Biology is dynamic. It is constantly changing as new insights lead to new ideas and new tools to test those ideas. Think of the use of drones and satellites to photograph penguin populations in Antarctica. Think of the developments in biological imaging and computation. Think of genome sequencing and its effect on our understanding of everything from human diseases to the tree of life. Think of the use of ice cores to understand past global climate.

Biology is a system. Biological systems are made up of different levels of organization—from molecules to ecosystems—that are interconnected and complex. Biologists are beginning to use integrated approaches to understand the complex properties of biological systems.

Biology is life. We face many challenges as humans including emerging diseases, feeding people in a sustainable way, population growth, degradation of natural systems, and climate change. We understand that humans are integrally connected to and dependent on all life on Earth.

Our goal—and challenge—in writing *Life* is to engage students in all these aspects of biology by motivating learning through active discovery. We focus on key concepts and contemporary examples that provide a foundation for further study. We have consulted and collaborated with faculty, students, and experts in the fields of biology and education. As you will see in the next several pages, in this, our Eleventh Edition, we engage students by explaining how biology affects their daily lives and how new knowledge is discovered. To help students master the concepts and principles, we introduce a number of new active learning activities.

Please read on.

David Sadava   David Hillis   Craig Heller   Sally Hacker

With roots 10 editions deep, the Eleventh Edition blends carefully reviewed content and an emphasis on experiments with a contemporary approach to active learning. Throughout, we focus on three driving themes:

Life is ENGAGING

We want students to come away from the introductory biology course with a sense of wonder and an ability to pursue biological questions. To that end, we keep the writing lively, the examples modern and exciting, and the emphasis on how we know what we know.

To help students engage with the course, an **Investigating Life** thread weaves through each chapter. Each individual part of the thread appeals to students, inspires them to ask questions, and keeps them wanting to know, "What next?" The beauty of the thread is that it gives students an authentic scientific experience beyond just reading about someone's research results.

Life is ACTIVE

With the help of the Advisory Board, we've developed a series of pedagogical features that integrate tools for instructors with those for student independent study. Students work their way toward real understanding of biological principles. We don't just ask them to memorize information—we invite them to actively participate in the process of discovery.

For instructors wanting to make active learning a part of their classroom, we've created an **Active Learning Guide**, with start-up material, ideas, and a complete guide to all the *Life* activities that are ready to add to your in-class repertoire.

Life is FOCUSED ON SKILLS

We want students to feel comfortable as they learn to manipulate and interpret data. The various types of problem-solving exercises in *Life* help them develop practical, analytical, and quantitative skills—skills that will benefit them whether they continue in biology or not. Exercises vary in skill level, type, and approach, with answers in the back of the book.

The response to our **Work with the Data** boxes has been so consistently positive that we now include at least one in every chapter. This feature asks students to analyze data from original scientific experiments, and includes questions in the text plus, now, a corresponding set of alternative exercises online in LaunchPad. We also provide an appendix, **Making Sense of Data: A Statistics Primer**, to help students prepare for the quantitative work they'll be doing.

Experience *Life* through LAUNCHPAD

With this edition's **LaunchPad**, *Life* is more than ever a truly integrated text/media resource. LaunchPad gives students everything they need to prepare for class and exams, while giving instructors everything they need to set up a course, customize the content, craft presentations, assign homework, assess students, and guide the progress of individuals and the class as a whole.

in  LaunchPad

"What makes this book stand out **most among the competition** is the way in which there is a continued sense of reinforcement for novice learners."

—Claudette Davis, George Mason University

Life is ENGAGING

Investigating Life

The **Investigating Life** thread begins every chapter with a fascinating real-life story emerging from actual research. The story concludes with an **Opening Question** that is explored throughout the chapter's Investigating Life thread. By the end of the chapter, students should be able to understand the answer to the Opening Question.

NEW

> investigating**life**

How Aspirin Works

Despite suffering from the "ague," the Reverend Edward Stone went walking in the English countryside. Feverish, tired, with aching muscles and joints, he came across a willow tree. Although apparently unaware that many ancient healers used willow bark extracts to reduce fever, the clergyman knew of the tradition of natural remedies for various diseases. The willow reminded him of the bitter extracts from the bark of South American trees then being sold (at high prices) to treat fevers.
found it
oms.
nd it into
plained
ults of
and's
salicylic
rld. The

licylic
w ge-
years
synthe-
gh the
acidity

irritated the digestive system. In the late 1890s, th chemical company Bayer synthesized a milder ye fective form, acetylsalicylic acid, which it markete The new medicine's success launched Bayer to w nence as a pharmaceutical company, a position it today.

In the 1960s and 1970s, aspirin use declined when other pain-reducing medications became able. But over this same time, clinical studies re use for aspirin: it is an effective anticoagulant, s vent heart attacks and strokes caused by blood many people take a daily low dose of aspirin as against clotting disorders.

Fever, joint pain, headache, blood clots: Wha symptoms have in common? They all are media acid products called prostaglandins and molecu from them. Salicylic acid blocks the synthesis o prostaglandin. The biochemical mechanism by works was described in 1971. As we will see, a ing of this mechanism requires an understandin and enzyme function—two subjects of this cha

Q&A How do anti-inflammatory drugs work as enzyme inhibitors?

NEW

> investigating**life**   **How Do Anti-Inflammatory Drugs Work as Enzyme Inhibitors?**

experiment

Original Papers: Vane, J. R. 1971. Inhibition of prostaglandin synthesis as a mechanism of action of aspirin-like drugs. *Nature* 231: 232–235.

Smith, J. B. and A. L Wells. 1971. Aspirin selectively inhibits prostaglandin production in human platelets. *Nature* 231: 235–238.

The opening story in this chapter describes how willow bark, the source of what became aspirin, is a centuries-old remedy for pain and inflammation. Until the late twentieth century, it was assumed that aspirin acted directly on the nervous system. Working at England's Royal College of Surgeons, a group led by John Vane showed that aspirin acts instead as an inhibitor of the enzymatic reaction that produces prostaglandin (PG), a fatty acid derivative important in inflammation.

HYPOTHESIS▶ Aspirin acts as an anti-inflammatory drug by inhibiting an enzyme that catalyzes prostaglandin synthesis.

METHOD

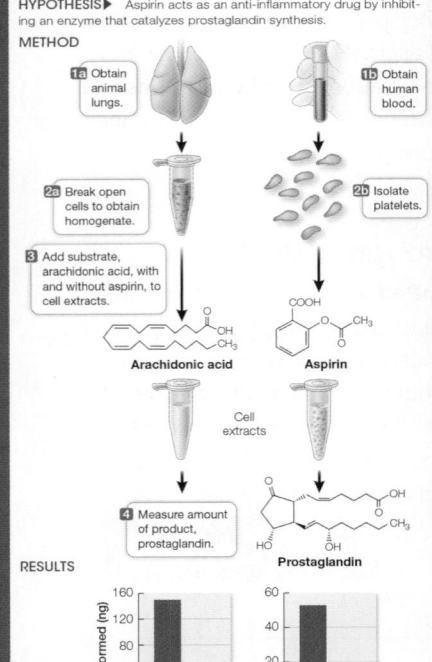

1a Obtain animal lungs.

1b Obtain human blood.

2a Break open cells to obtain homogenate.

2b Isolate platelets.

3 Add substrate, arachidonic acid, with and without aspirin, to cell extracts.

Arachidonic acid   Aspirin

Cell extracts

4 Measure amount of product, prostaglandin.

Prostaglandin

RESULTS

PG formed (ng) — Aspirin concentration (μg/mL)

CONCLUSION▶ In both animal and human cells, aspirin blocks the synthesis of the pro-inflammatory molecule, prostaglandin, in test tube experiments.

work with the data

For his discovery of the mechanism by which aspirin relieves pain, John Vane was awarded the Nobel Prize and knighted by Queen Elizabeth II. Key to the experiments was the assumption that enzyme activity and mechanism are the same outside the organism as they are inside. In the lab, if given its substrate and the same set of environmental conditions as in the cytoplasm, an enzyme will catalyze production of its product.

QUESTIONS▶

1. In the first set of experiments, lung tissue from guinea pigs was broken up to form a cell-free extract (homogenate). The substrate, arachidonic acid, was added to the extract, and after 30 min the amount of prostaglandin (PG) was measured. The results are shown in **Table A**. Plot PG synthesis versus aspirin concentration. What can you conclude?

2. A similar set of experiments was performed on human platelets (membrane-bound cell fragments derived from white blood cells). These cells are known to produce PG by the same enzyme mechanism as lung tissue under certain environmental conditions. The results are shown in **Table B**. What can you conclude from these results and about the generalizability of your conclusion from Question 1?

Table A

Aspirin concentration (μg/mL)	PG synthesis (ng)
0	220
1	172
2	136
10	99
50	33
80	0

Table B

Aspirin concentration (μg/mL)	PG synthesis (ng)
0	53
0.01	48
0.1	35
1	18
10	7

3. In a third set of experiments, platelets were isolated from human volunteers and their ability to make PG was measured (no aspirin present). Then these people were given a clinically effective dose of aspirin, and after a short period their platelets were isolated and PG synthesis measured (again, no aspirin in the test tube experiments). The results from three people are shown in **Table C**. How do these data either reinforce or refute your conclusion from Questions 1 and 2?

Table C

Individual	Prostaglandin synthesis (ng)	
	Before aspirin	After aspirin
1	160	16
2	108	5
3	103	20

A similar **work with the data** exercise may be assigned in **LaunchPad**.

Experiment & Work with the Data

The chapter continually returns to the opening story and question: the **Experiment** describes the original research behind the story and the **Work with the Data** gives students the opportunity to analyze research results for themselves. Work with the Data includes questions in the book and an additional corresponding exercise online and assignable in LaunchPad.

Life is ENGAGING

The **Investigating Life** thread provides a consistent pathway through each chapter that engages students from the first page (**Opening Story**), through the body of the chapter (**Experiment, Work with the Data**), to the last page (**Future Directions**). Through this emphasis on real experiments and data analysis, these learning tools show students how the concepts they're learning apply to the real world.

NEW

investigatinglife

Q&A How do anti-inflammatory drugs work as enzyme inhibitors?

Go into a drugstore or watch ads on TV, and it becomes obvious that there are many drugs that alleviate pain. Like aspirin, many of them target the prostaglandin pathway for inflammation. John Vane's research and later work showed that aspirin irreversibly inhibits the enzyme COX. But it turns out that there are two forms of COX, called COX-1 and COX-2, and there are several kinds of prostaglandins. COX-1 catalyzes the production of protective prostaglandins that are involved with blood clotting (which is why people take aspirin to prevent future heart attacks due to blood clots) and the integrity of the lining of the stomach. COX-2 catalyzes production of prostaglandins involved with inflammation and associated pain. Aspirin inhibits both COX-1 and COX-2. So it is not surprising that people are advised to go easy on aspirin: it may block pain, but its use can lead to stomach upsets and an inability to clot blood after a wound. Knowledge of aspirin's limitations has led to a search for inhibitors of COX-2 only. Most of them are competitive inhibitors. How would this affect how often you need to take the drug? Look up one of these drugs and read the description of how it works.

Future directions

As details at the atomic level of the binding of enzymes to substrates are worked out, biologists use the data to try to predict which substrates might bind to which enzymes. An important consideration in binding is ΔG; the less positive this parameter, the more likely binding will occur. Factors such as ionic attractions and van der Waals forces contribute to the ΔG of binding. These factors are also important in the binding of proteins to nonsubstrate molecules, including other proteins and RNA. Indeed, it is becoming apparent that in general a protein in the cell does not exist in isolation, but is bound to something(s) else. If the protein is an enzyme, its substrate binding must be its most preferred partner. Understanding all the possible molecular interactions will result in a deeper understanding of what really goes on at the chemical level inside the cell.

Answer to Opening Question and Future Directions

As the final piece of the Investigating Life thread, the **Opening Question**, which students have been considering throughout the whole chapter, is restated and the answer is explained in full detail. The final **Future Directions** section explores new questions and research opportunities that the chapter content leads to, setting the stage for students to consider how the textbook material matters in their lives.

Media Clip 32.5: Komodo Dragons Bring Down Prey

Textbook Reference: Key Concept 32.4 Life on Land Contributed to Vertebrate Diver

in LaunchPad

Media Links

Throughout the chapter, **Media Links** prompt students to go online for video clips, animations, activities, and simulations that correspond to what they're reading. All of these items can be assigned in LaunchPad, and most are accompanied by assessments.

IN-BOOK PROMPT:

Most lizards are insectivores, although some are herbivores and a few prey on other vertebrates. Most lizards walk on four limbs (**Figure 32.21B**), although limblessness has evolved repeatedly among the lizards, especially in burrowing and grassland species. The largest lizard is the predaceous Komodo dragon of the East Indies, which grows as long as 3 meters and can weigh more than 150 kilograms.

▶ Media Clip 32.5 **Komodo Dragons Bring Down Prey**
www.Life11e.com/mc32.5

Life is ACTIVE

We have developed a series of learning tools under the guidance of an Advisory Board of 19 experts in science education and pedagogy. These features are woven into the chapters so seamlessly that **active learning** is the natural result. All of these features are aligned to the new Learning Outcomes, giving students a clear expectation of what they need to learn.

Focus Your Learning

Each chapter is now organized around Key Concepts. **Focus Your Learning** statements zero in on the essential ideas in each Key Concept. These statements guide students in their reading and prepare them for the upcoming Recaps and Apply What You've Learned exercise.

Recaps with Learning Outcomes

Recaps summarize each section and include related **Learning Outcomes** and thought-provoking questions (with answers in the back of the book). Learning Outcomes are aligned with the section's Focus Your Learning statements, and they tell students how to show that they've mastered the material. The questions that follow are aligned with and reinforce the Learning Outcomes. This alignment helps students self-assess whether they should review the section or continue reading.

NEW

key concept 21.2 Phylogeny Can Be Reconstructed from Traits of Organisms

To illustrate how a phylogenetic tree is constructed, consider the eight vertebrate animals listed in **Table 21.1**: lamprey, perch, salamander, lizard, crocodile, pigeon, mouse, and chimpanzee. We will initially assume that any given derived trait arose only once during the evolution of these animals (that is, there has been no convergent evolution), and that no derived traits were lost from any of the descendant groups (there has been no evolutionary reversal). For simplicity, we have selected traits that are either present (+) or absent (–).

focus your learning

- Modern phylogenetic methods employ the principle of parsimony and mathematical models (when appropriate) to analyze morphological, developmental, paleontological, behavioral, and molecular data.

NEW

21.2 recap

Phylogenetic trees can be constructed by using the parsimony principle to find the simplest explanation for phylogenetic data. Maximum likelihood methods incorporate more explicit mathematical models of evolutionary change to reconstruct evolutionary history.

learning outcomes

You should be able to:

- Analyze a phylogenetic tree to identify synapomorphies, homoplasies, and relationships among taxa.
- Reconstruct a phylogenetic tree from a data matrix of characters.

The matrix at right supplies data for seven land plants and an outgroup: an aquatic plant known as a stonewort. Each trait is scored as either present (+) or absent (–) in each of the plants. Use this data matrix to reconstruct the phylogeny of land plants and answer the questions that follow.

Taxon	Protected embryos	True roots	Persistently green sporophyte	Vascular cells	Stomata	Megaphylls (true leaves)	Seeds
Stonewort (outgroup)	–	–	–	–	–	–	–
Liverwort	+	–	–	–	–	–	–
Pine tree	+	+	+	+	+	+	+
Bracken fern	+	+	+	+	+	+	–
Club moss	+	+	+	+	+	–	–
Sphagnum moss	+	–	–	–	+	–	–
Hornwort	+	–	+	–	+	–	–
Sunflower	+	+	+	+	+	+	+

1. Which two of these taxa are most closely related?
2. Plants that produce seeds are known as seed plants. What is the sister group to the seed plants among these taxa?
3. Which two traits evolved along the same branch of your reconstructed phylogeny?
4. Are there any homoplasies in your reconstructed phylogeny?

Life is ACTIVE

Simulations

Forty interactive **simulations** put the controls in students' hands and give them the opportunity to learn by doing. Topics include:

- Lipid Bilayer Composition
- Genetic Drift
- Visual Receptive Fields
- Island Biogeography
- Population Growth

Life, Eleventh Edition features a significantly increased number of simulations compared to the previous edition.

Please log in to LaunchPad to review the entire library of simulations.

Apply What You've Learned

Chapters conclude with an exercise that challenges students to put their newly acquired knowledge into action. The exercise presents a summary of research that is relevant to the chapter material followed by questions that, like Recap questions, align to the relevant Focus Your Learning statements, which are listed at the start of the exercise. Answers are in the back of the book.

ACTIVITY 21.2: PHYLOGENY AND MOLECULAR EVOLUTION SIMULATION

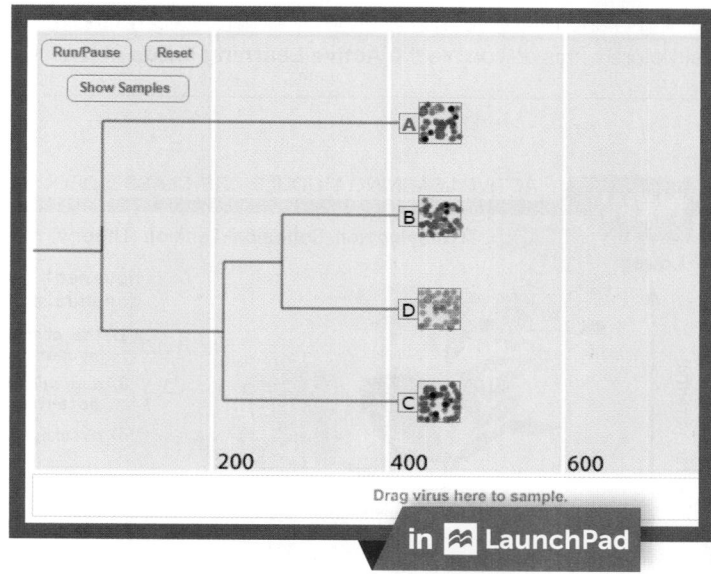

in 📖 LaunchPad

NEW

▶ Apply What You've Learned

Review

35.1 Plants receiving insufficient quantities of essential elements can be evaluated by observing symptomatic changes in physical features and/or growth.

35.1 Experiments using hydroponic growth conditions have allowed scientists to determine the essential elements plants need for growth.

Original Paper: Brown, P. H., R. M. Welch and E. E. Cary. 1987. Nickel: A micronutrient essential for higher plants. *Plant Physiology* 85: 801–803.

Both barley and wheat are cereal crops used for feeding humans and livestock, and the two have many similarities. Because of its characteristics when fermented, however, barley is favored over wheat for beer making. When you consider how much beer is brewed around the world, it's no wonder that researchers have spent considerable effort looking for ways to optimize barley growth.

Optimizing growth of any crop can be accomplished if the crop's nutritional requirements are known. In the case of barley, one research group investigated the element nickel as a possible essential micronutrient. Earlier studies where a single generation of plants was grown without Ni indicated that it was not a mineral requirement. They began by creating a stock growth solution containing extremely low amounts of nickel ions (less than 30 nanograms per liter [ng/L]). Then they divided this stock solution into three containers and added nickel sulfate ($NiSO_4$) to two of these. This step produced supplemented concentrations of 0, 0.6, and 1.0 micromolar (μM) nickel sulfate in the three final

Next, one portion of the seeds from each group was analyzed for nickel content. The other portion from each group was placed in nickel-free growth medium and tested for germination. The graph below shows seed germination success plotted against nickel content of the same seeds.

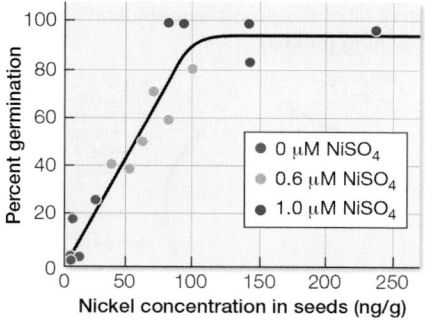

Questions

1. The researchers concluded that nickel is an essential micronutrient of barley. Do you agree with their conclusion? Explain your answer.

2. Analyze the data about the mass and number of seeds produced by the third-generation barley plants. What conclusion can you draw from these data? Is this consistent with your answer to Question 1 above? Explain.

Life is ACTIVE

Encouraging students to be active when they work with the book on their own is just the beginning. We are committed to active learning in any setting, including in the classroom, for any teacher who wants to try it. Perhaps the most powerful new tools for instructors in this edition are the **Active Learning Guide** and accompanying set of **Active Learning Modules** and **Exercises**.

NEW | ACTIVE LEARNING MODULE PRE-CLASS VIDEO

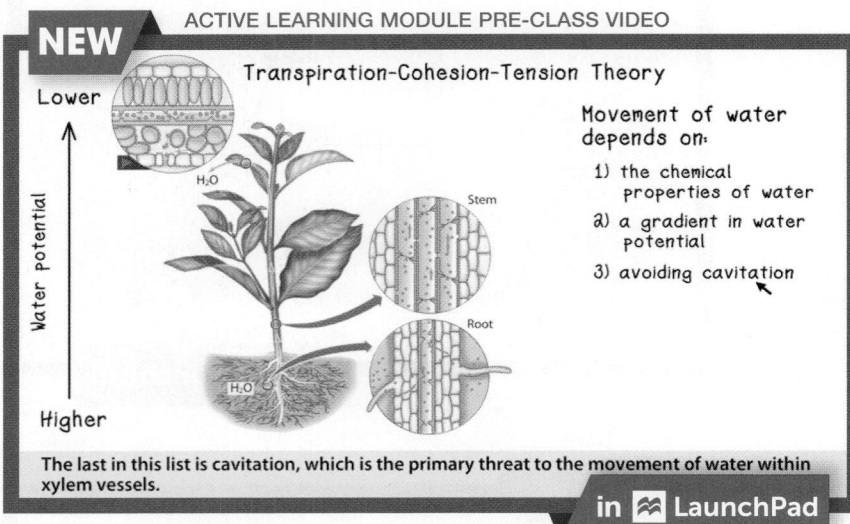

in ☰ LaunchPad

Active Learning Modules

The new **Active Learning Modules** provide everything an instructor needs to successfully implement an active approach to teaching key topics. Each module includes many resources, including:

- Pre-class video specifically created for the module
- Pre-quiz and post-quiz
- Handout for in-class work
- Detailed in-class exercise
- Detailed instructor's guide

These modules are easy to implement and are a great way to add more active learning to the classroom.

NEW | ACTIVE LEARNING MODULE IN-CLASS EXERCISE SLIDES

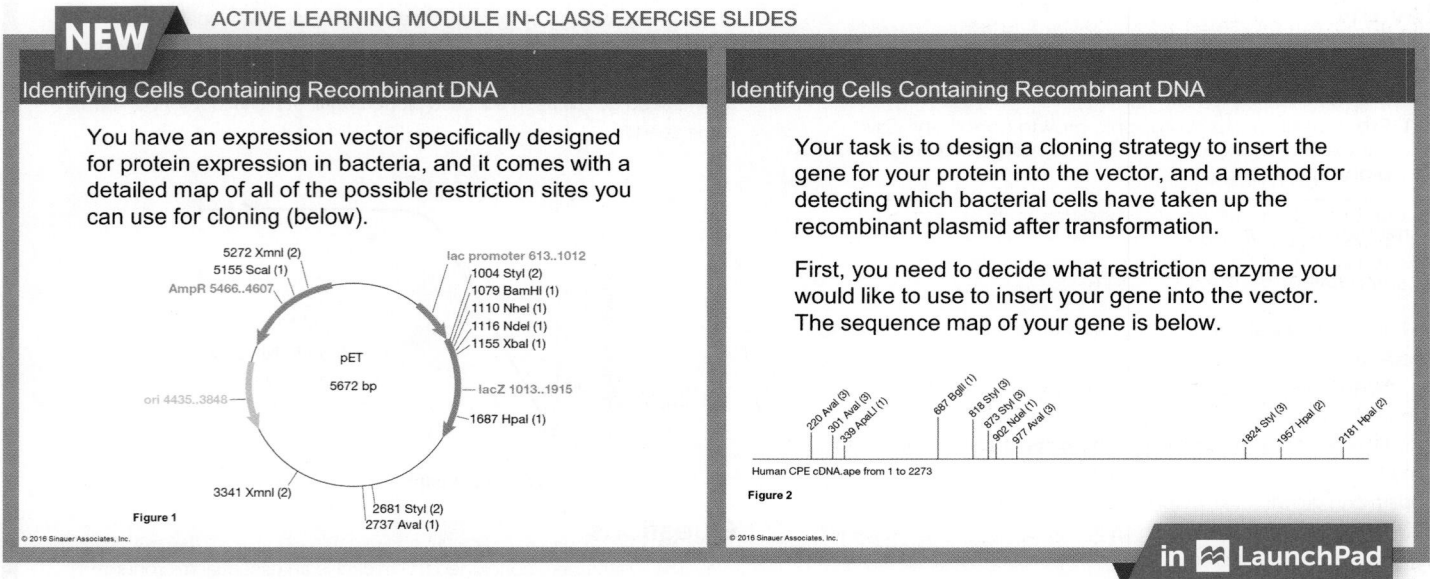

in ☰ LaunchPad

"A lot of books try to sell their active learning focus, but this one actually follows through in a meaningful way."

—Carly Jordan, George Washington University

Life is ACTIVE

Active Learning Guide

The **Active Learning Guide** provides extensive resources and support for implementing active learning techniques in the classroom. This all-new guide provides instructors with a thorough introduction to the concepts, techniques, and benefits of active learning, along with chapter-by-chapter guidance on teaching using the many active learning resources we provide for *Life*, Eleventh Edition.

Part 1: Introduction to Active Learning

- What is active learning?
- Designing your course for active learning
- Using active learning in the classroom
- How to implement *Life*'s resources

Part 2: Chapter-by-Chapter Guidance

Each Active Learning Guide chapter contains:

- Overview of the textbook chapter
- The student resources available in LaunchPad for the chapter, with descriptions
- Specific concepts/content that are particularly well-suited to active learning techniques
- The chapter's Active Learning Exercises and how to integrate them into class sessions
- The relevant Active Learning Modules (if applicable) and how to integrate them into the course
- Suggestions for additional active learning opportunities
- Links to specific additional online resources

Part 3: Appendices

- Guide to Using Learning Standards in Teaching and Course Development
- Overview of Bloom's Taxonomy

"This textbook works well with any teaching style....
It is a great tool to help instructors transform their
classroom into an active learning environment."
—Robert Osuna, University at Albany, State University of New York

"The active learning modules included with this
text take the 'fear factor' out of active learning."
—Candice Damiani, University of Pittsburgh

Life is FOCUSED ON SKILLS

Life has long been recognized for bringing quantitative analysis and critical thinking to the forefront of the introductory biology course. The new edition allows for more hands-on experience than ever with data manipulation skills, both in the book and in LaunchPad.

Work with the Data

In the **Work with the Data** exercises, students analyze the results of an original scientific experiment, then work through a series of questions. In response to overwhelmingly positive feedback, the new edition includes at least one Work with the Data box in each chapter.

Online Companion Exercises

For additional practice, every Work with the Data exercise has a corresponding **Online Companion** in LaunchPad, addressing the same skills as its in-book equivalent.

work with the data

Donald Levin of the University of Texas proposed that *Phlox drummondii* has red flowers only in locations where it is sympatric with pink-flowered *P. cuspidata* because having red flowers decreases interspecific hybridization. To test this hypothesis, Levin introduced equal numbers of red- and pink-flowered *P. drummondii* individuals into an area with many pink-flowered *P. cuspidata*. At the end of the flowering season, he assessed the genetic composition of the seeds produced by *P. drummondii*. The results are shown in the table below.

Morph (flower color)	Number of seeds (progeny)		
	P. drummondii	Hybrid	Total
Red	181 (87%)	27 (13%)	208
Pink	86 (62%)	53 (38%)	139

QUESTIONS▶

1. Check the 95% confidence intervals for the proportion of hybrid seeds in red- and pink-flowered *P. drummondii* in the graph in the Results of the experiment. There are many websites available for calculating confidence intervals; a good one is the Vassar College statistical computation site, VassarStats.net. You can go to this site and select "Proportions" from the left-hand menu, then select "The Confidence Interval of a Proportion." What are the numerical values of the 95% confidence intervals?

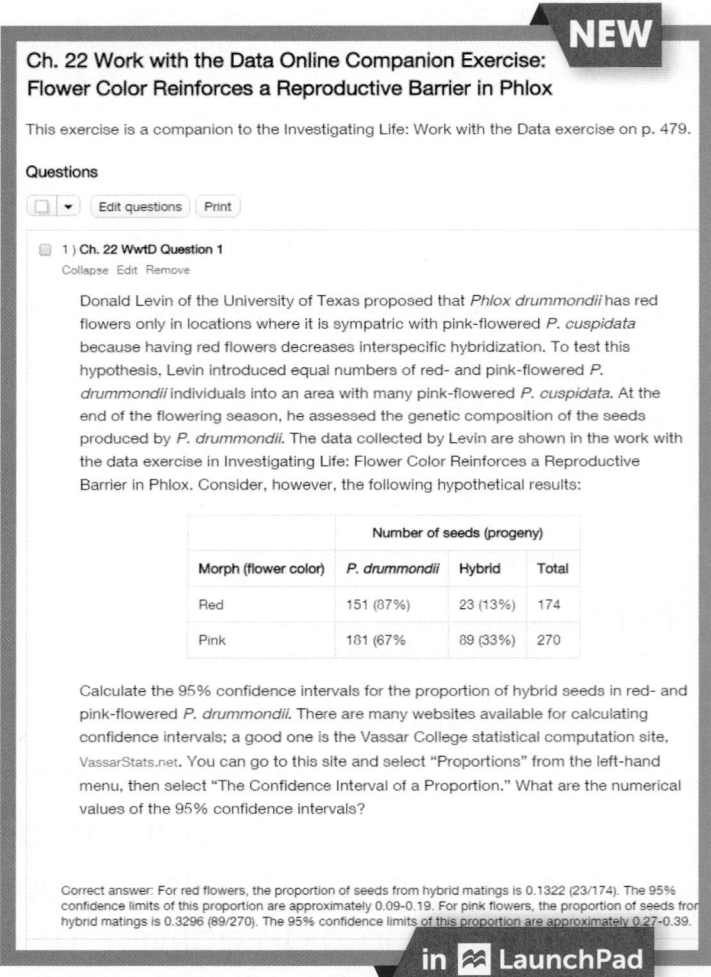

NEW

Ch. 22 Work with the Data Online Companion Exercise: Flower Color Reinforces a Reproductive Barrier in Phlox

This exercise is a companion to the Investigating Life: Work with the Data exercise on p. 479.

Questions

Edit questions   Print

1) **Ch. 22 WwtD Question 1**
Collapse Edit Remove

Donald Levin of the University of Texas proposed that *Phlox drummondii* has red flowers only in locations where it is sympatric with pink-flowered *P. cuspidata* because having red flowers decreases interspecific hybridization. To test this hypothesis, Levin introduced equal numbers of red- and pink-flowered *P. drummondii* individuals into an area with many pink-flowered *P. cuspidata*. At the end of the flowering season, he assessed the genetic composition of the seeds produced by *P. drummondii*. The data collected by Levin are shown in the work with the data exercise in Investigating Life: Flower Color Reinforces a Reproductive Barrier in Phlox. Consider, however, the following hypothetical results:

Morph (flower color)	Number of seeds (progeny)		
	P. drummondii	Hybrid	Total
Red	151 (87%)	23 (13%)	174
Pink	181 (67%)	89 (33%)	270

Calculate the 95% confidence intervals for the proportion of hybrid seeds in red- and pink-flowered *P. drummondii*. There are many websites available for calculating confidence intervals; a good one is the Vassar College statistical computation site, VassarStats.net. You can go to this site and select "Proportions" from the left-hand menu, then select "The Confidence Interval of a Proportion." What are the numerical values of the 95% confidence intervals?

Correct answer: For red flowers, the proportion of seeds from hybrid matings is 0.1322 (23/174). The 95% confidence limits of this proportion are approximately 0.09–0.19. For pink flowers, the proportion of seeds from hybrid matings is 0.3296 (89/270). The 95% confidence limits of this proportion are approximately 0.27–0.39.

in 🌊 **LaunchPad**

"I especially like the "Work with the Data" exercises. Too often this sort of critical thinking is left to upper level courses."

— Susan Reigler, Indiana University Southeast

Life is FOCUSED ON SKILLS

Apply What You've Learned

Exploring the same types of problems that scientists typically investigate, the **Apply What You've Learned** exercises allow students to hone both critical thinking and data analysis skills in fascinating contexts. Answers appear in the back of the book.

▶ Apply What You've Learned

Review

22.4 Reproductive isolation is reinforced in sympatry through selection for prezygotic mechanisms that prevent hybridization, which results from postzygotic mechanisms that reduce the fitness of hybrids.

Evolutionary biologists are particularly interested in reinforcement's role in speciation because this process directly connects natural selection to the origin of species.

Different *Drosophila* species vary in the concentrations of different hydrocarbons on their cuticles. These hydrocarbons often serve as reproductive signals, with females preferring to mate with males that have the suite of hydrocarbons from the females' own species over those of other species. As they often are reproductive signals, the hydrocarbons may be the target of the selection involved in reinforcement.

Two species of *Drosophila*, *D. serrata* and *D. birchii*, are found on Australia's east coast and overlap in range. *D. serrata* popula-

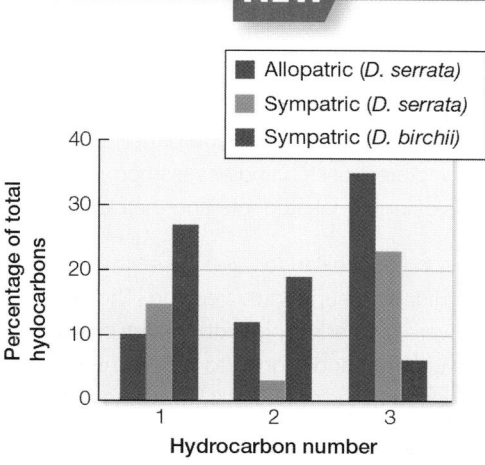

Questions

1. Based on the data, which hydrocarbon(s) show a pattern that is consistent with reinforcement? Explain your answer.

2. Suppose the allopatric populations of *D. serrata* were taken

---

How Does Statistics Help Us Understand the Natural World?

Statistics is essential to scientific discovery. Most biological studies involve five basic steps, each of which requires statistics:

- **Step 1: Choose an Experimental Design**
  Clearly define the scientific question and the methods necessary to tackle the question.

- **Step 2: Collect Data**
  Gather information about the natural world through observations and experiments.

- **Step 3: Organize and Visualize the Data**
  Use tables, graphs, and other useful representations to gain intuition about the data.

- **Step 4: Summarize the Data**
  Summarize the data with a few key statistical calculations.

- **Step 5: Make Inferences from the Data**
  Use statistical methods to draw general conclusions from the data about the world and the ways it works.

Step 1: Choose an Experimental Design

We make observations and conduct experiments to gain knowledge about the world. Scientists come up with scientific ideas based on

Making Sense of Data: A Statistics Primer

This primer (an appendix in the text and also in LaunchPad), lays the proper groundwork for understanding statistics and data, providing helpful student support for all of the quantitative exercises in the new edition.

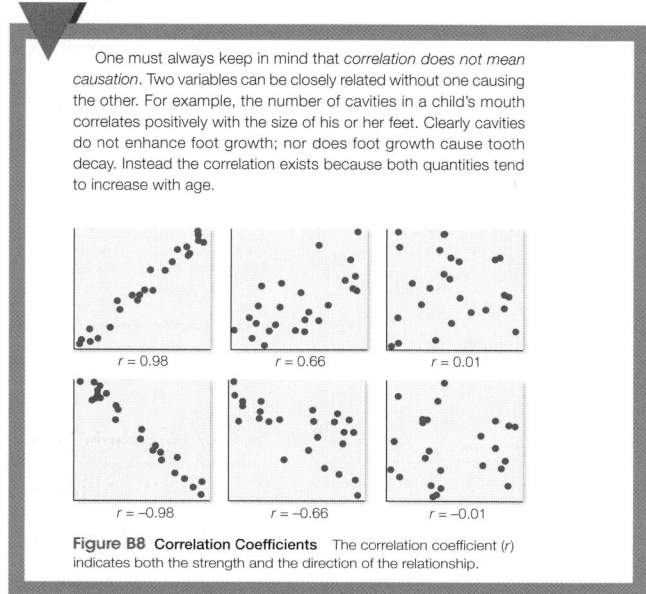

One must always keep in mind that *correlation does not mean causation*. Two variables can be closely related without one causing the other. For example, the number of cavities in a child's mouth correlates positively with the size of his or her feet. Clearly cavities do not enhance foot growth; nor does foot growth cause tooth decay. Instead the correlation exists because both quantities tend to increase with age.

Figure B8 Correlation Coefficients   The correlation coefficient (r) indicates both the strength and the direction of the relationship.

Life in LAUNCHPAD

LAUNCHPADWORKS.COM

Built to address the biggest classroom issues instructors face, LaunchPad gives students everything they need to prepare for class and exams, while giving instructors everything they need to quickly set up a course, shape the content to their syllabus, craft presentations and lectures, assign and assess homework, and guide the progress of individual students and the class as a whole.

All of the resources in LaunchPad for *Life*, Eleventh Edition have been created specifically to support this edition. The **Activities** and **Animations** were created using textbook art; all quiz and test questions are updated and aligned to the Eleventh Edition; everything is referenced to the textbook's Key Concepts; and substantial new active learning resources support the use of the book in an active learning context.

Searchable Question Bank

LaunchPad's enhanced question bank functionality makes it easier to search for and select questions from any of the *Life* question banks (Test Bank, Learning-Curve, Summative Quizzes). With the new filtering functionality, instructors can get the precise mix of questions they want by filtering for:

- Learning Outcome (new for the Eleventh Edition)
- Focus Your Learning statement
- Key Concept
- Bloom's Level
- Difficulty (for LearningCurve questions)
- Source

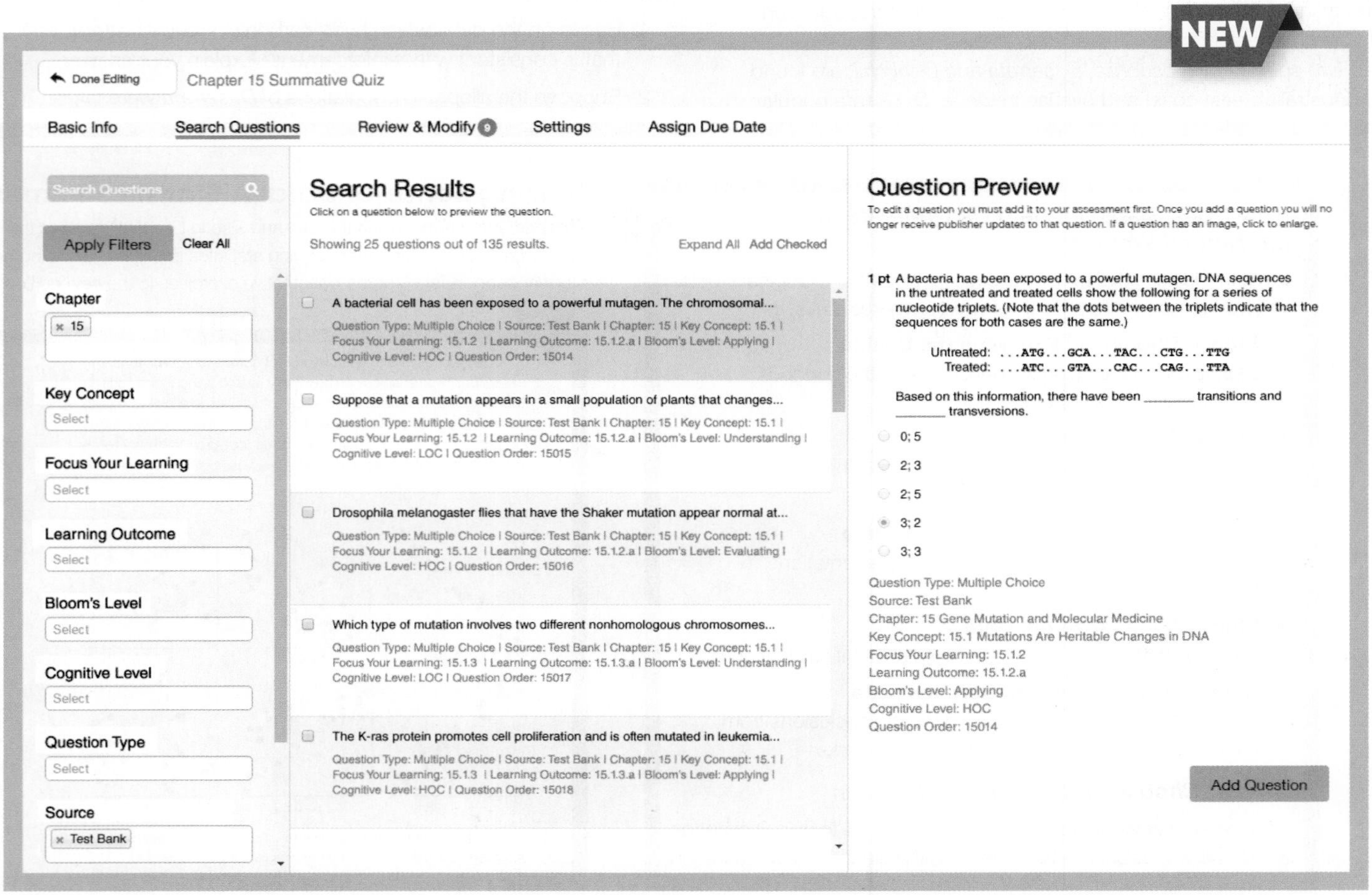

NEW

← Done Editing   Chapter 15 Summative Quiz

Basic Info   **Search Questions**   Review & Modify ●   Settings   Assign Due Date

Search Questions 🔍

Apply Filters   Clear All

Chapter
× 15

Key Concept
Select

Focus Your Learning
Select

Learning Outcome
Select

Bloom's Level
Select

Cognitive Level
Select

Question Type
Select

Source
× Test Bank

Search Results
Click on a question below to preview the question.

Showing 25 questions out of 135 results.   Expand All   Add Checked

☐ A bacterial cell has been exposed to a powerful mutagen. The chromosomal...
Question Type: Multiple Choice | Source: Test Bank | Chapter: 15 | Key Concept: 15.1 | Focus Your Learning: 15.1.2 | Learning Outcome: 15.1.2.a | Bloom's Level: Applying | Cognitive Level: HOC | Question Order: 15014

☐ Suppose that a mutation appears in a small population of plants that changes...
Question Type: Multiple Choice | Source: Test Bank | Chapter: 15 | Key Concept: 15.1 | Focus Your Learning: 15.1.2 | Learning Outcome: 15.1.2.a | Bloom's Level: Understanding | Cognitive Level: LOC | Question Order: 15015

☐ Drosophila melanogaster flies that have the Shaker mutation appear normal at...
Question Type: Multiple Choice | Source: Test Bank | Chapter: 15 | Key Concept: 15.1 | Focus Your Learning: 15.1.2 | Learning Outcome: 15.1.2.a | Bloom's Level: Evaluating | Cognitive Level: HOC | Question Order: 15016

☐ Which type of mutation involves two different nonhomologous chromosomes...
Question Type: Multiple Choice | Source: Test Bank | Chapter: 15 | Key Concept: 15.1 | Focus Your Learning: 15.1.3 | Learning Outcome: 15.1.3.a | Bloom's Level: Understanding | Cognitive Level: LOC | Question Order: 15017

☐ The K-ras protein promotes cell proliferation and is often mutated in leukemia...
Question Type: Multiple Choice | Source: Test Bank | Chapter: 15 | Key Concept: 15.1 | Focus Your Learning: 15.1.3 | Learning Outcome: 15.1.3.a | Bloom's Level: Applying | Cognitive Level: HOC | Question Order: 15018

Question Preview
To edit a question you must add it to your assessment first. Once you add a question you will no longer receive publisher updates to that question. If a question has an image, click to enlarge.

1 pt A bacteria has been exposed to a powerful mutagen. DNA sequences in the untreated and treated cells show the following for a series of nucleotide triplets. (Note that the dots between the triplets indicate that the sequences for both cases are the same.)

Untreated: ...ATG...GCA...TAC...CTG...TTG
Treated: ...ATC...GTA...CAC...CAG...TTA

Based on this information, there have been _____ transitions and _____ transversions.

○ 0; 5
○ 2; 3
○ 2; 5
● 3; 2
○ 3; 3

Question Type: Multiple Choice
Source: Test Bank
Chapter: 15 Gene Mutation and Molecular Medicine
Key Concept: 15.1 Mutations Are Heritable Changes in DNA
Focus Your Learning: 15.1.2
Learning Outcome: 15.1.2.a
Bloom's Level: Applying
Cognitive Level: HOC
Question Order: 15014

Add Question

Life in LAUNCHPAD

NEW RELEASE!

LearningCurve
macmillan learning

LearningCurve adaptive quizzing gives each student individualized question sets and feedback based on their correct and incorrect responses. All the questions link back to the eBook to encourage students to read the book in preparation for classtime and exams.

Updated Interface

LearningCurve's new streamlined user experience makes features such as the Study Plan, eBook, and Hints more prominent. Also new—all questions now feature text hints to guide the student's thinking.

Instructor Customization

LearningCurve organizes chapter quizzes by Key Concept and makes it easy for instructors to hide questions on concepts they are not covering. With this, each student's quizzes can focus on the exact content being taught.

NEW

LearningCurve

Chapter 10 Photosynthesis: Energy from Sunlight

Target score progress

You have: **226** points   Target: **750**

LearningCurve tips for success

Your personalized study plan

▶ Key Concept 10.1 Photosynthesis Uses Light to Make Carbohydrate
48% accuracy

▶ Key Concept 10.2 Photosynthesis Converts Light Energy into Chemical Energy
78% accuracy

▶ Key Concept 10.3 Chemical Energy Trapped in Photosynthesis Is Used to Synthesize Carbohydrates
33% accuracy

Study plan suggestions

▤ eBook: 10.3 Chemical Energy Trapped in Photosynthesis Is Used to Synthesize Carbohydrates

☟ Activity 10.1 The Calvin Cycle

☟ Animated Tutorial 10.3 Tracing the Pathway of CO_2

▶ Key Concept 10.4 Plants Have Adapted Photosynthesis to Environmental Conditions
33% accuracy

▶ Key Concept 10.5 Photosynthesis Is an Integral Part of Plant Metabolism
100% accuracy

NEW

Target Score Progress: 🏠 **Back to study plan**

Hint

What can local changes in membrane potential generate?   ✕

Why is it important for a neuron to maintain a membrane potential instead of allowing ions to reach equilibrium?

○ **The membrane potential maintains fluid balance in the cell.**

○ **The membrane potential allows the cell to respond to a stimulus.**

○ **If the ions were in equilibrium, the cell would swell and burst.**

○ **Maintaining equilibrium would require too much energy.**

○ **The cell must maintain a greater intracellular concentration of positive ions to maintain its cytoskeleton.**

Need help on this question?

📄 **Read the ebook page on this topic**
(no penalty)

Get a hint
(fewer points)

Show answer
(no points)

Life and ASSESSMENT

Life's assessment resources have long been appreciated for their accuracy, clarity, and ease-of-use. For this edition, we have enriched our assessment program to offer instructors more of exactly what they want. Along with updating and revising all assessment banks and adding new questions at higher cognitive levels, we have fully aligned all in-book and online content for the Eleventh Edition to the new Learning Outcomes.

Learning Outcomes

The new **Learning Outcomes** in *Life* are the core of the book's learning standards (Key Concepts, Focus Your Learning statements, and Learning Outcomes). Each chapter in this edition is structured around these standards. In the book, Learning Outcomes are aligned to Focus Your Learning statements, which are in turn aligned with Key Concepts. Learning Outcomes appear in the Recaps, where they show students what they should be able to do after reading a Key Concept section. Questions in the Recaps allow students to test themselves on those Learning Outcomes (and answers are in the back of the book).

Online, all Eleventh Edition assessment questions and resources are aligned to a more extensive set of Learning Outcomes available for instructors. This alignment across the book and all assessment pieces provides instructors a concrete way of assessing students on their mastery of the important material in each chapter.

Summative Quizzes

Each chapter has a Summative Quiz composed of 20 questions spanning the chapter's Key Concepts. Quizzes are pre-built and ready to assign in Launch-Pad. Summative Quizzes are completely customizable—instructors can add, revise, or remove questions to match their course content.

LearningCurve
macmillan learning

For each chapter, LearningCurve offers an extensive set of questions that are distinct from those in the Test Bank and Summative Quizzes. Questions are organized around Key Concepts, aligned to Learning Outcomes, and ranked by difficulty. Thus, students can master the material at a manageable pace, facing difficult questions only after answering easy and moderate questions correctly.

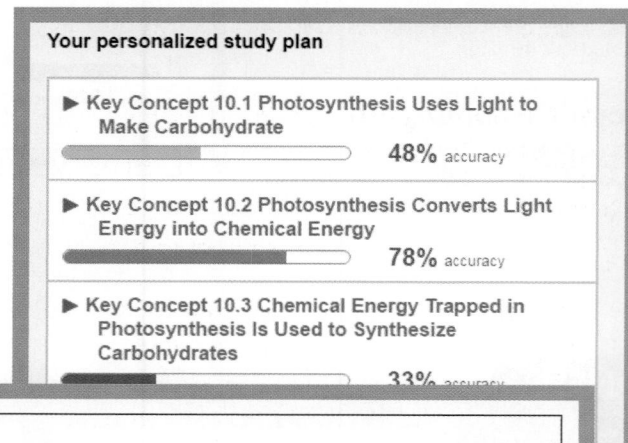

Your personalized study plan

▶ **Key Concept 10.1 Photosynthesis Uses Light to Make Carbohydrate**
  48% accuracy

▶ **Key Concept 10.2 Photosynthesis Converts Light Energy into Chemical Energy**
  78% accuracy

▶ **Key Concept 10.3 Chemical Energy Trapped in Photosynthesis Is Used to Synthesize Carbohydrates**
  33% accuracy

NEW

Learning Standards for Chapter 8: Energy, Enzymes, and Metabolism

KEY CONCEPT	SUBHEAD	FOCUS YOUR LEARNING STATEMENT	LEARNING OUTCOME
8.1 Physical Principles Underlie Biological Energy Transformations	There are two basic types of energy	8.1.1 Energy is the capacity for change in biological systems and takes one of two forms: potential or kinetic.	8.1.1.a Describe the forms of energy in biological systems.
			8.1.1.b Differentiate between potential energy and kinetic energy.
	There are two basic types of metabolism	8.1.2 Energy changes in living things often involve chemical reactions.	8.1.2.a Explain how to determine whether a reaction is anabolic or catabolic.
			8.1.2.b Compare and contrast anabolic and catabolic reactions.
		8.1.3 Chemical transformations in living systems are governed by the laws of thermodynamics.	8.1.3.a Identify the meaning of the term "thermodynamics."
	The first law of thermodynamics: Energy is neither created nor destroyed	8.1.4 The first law of thermodynamics states that energy is neither created nor destroyed.	8.1.4.a Explain how the first law of thermodynamics relates to biological systems.
	The second law of thermodynamics: Disorder tends to increase	8.1.5 The second law of thermodynamics states that disorder is constantly increasing in the universe.	8.1.5.a Compare and contrast the first and second laws of thermodynamics.
			8.1.5.b Explain the concept of free energy.
			8.1.5.c Use the equation $\Delta G = \Delta H - T\Delta S$ to describe biological systems.
			8.1.5.d Apply the second law of thermodynamics to biological systems.
	Chemical reactions release or consume energy	8.1.6 Chemical reactions in biological systems are either exergonic or endergonic.	8.1.6.a Differentiate between exergonic and endergonic reactions.
	Chemical equilibrium and free energy are related	8.1.7 Chemical reactions in biological systems are often reversible.	8.1.7.a Use an example to explain the concept of chemical equilibrium.
			8.1.7.b Predict the direction of reaction for a system at equilibrium when conditions are changed.
			8.1.7.c Describe the change in free energy for a reaction given information about its point of equilibrium.

SAMPLE FROM FULL SET OF LEARNING OUTCOMES FOR CHAPTER 8

Life and ASSESSMENT

Test Bank

The *Life* **Test Bank** has been significantly revised for the Eleventh Edition, to meet four goals:

- Update content to match the revised textbook chapters
- Align questions to the new Learning Outcomes
- Increase the number of questions with higher cognitive (Bloom's) levels
- Develop a new section of short answer questions to allow for open-ended questions and problems that don't fit the multiple choice or fill-in-the-blank framework

All questions are available in LaunchPad and are easily searchable by Chapter, Key Concept, Focus Your Learning statement, Learning Outcome, and Bloom's level using LaunchPad's new question bank tools.

Assessment in the Book and in LaunchPad

FEATURE	LAUNCHPAD ASSESSMENT RESOURCES
Key Concepts	• All quiz and test questions are tagged by Key Concept and searchable in LaunchPad • LearningCurve study plans and instructor reports are organized around the Key Concepts
Focus Your Learning Statements	• Questions in Recaps for students • All quiz and test questions are tagged by Focus Your Learning statement and are searchable in LaunchPad
Learning Outcomes	All Recap, quiz, and test questions are aligned with Learning Outcomes and are searchable in LaunchPad
Work with the Data	Companion Work with the Data exercises in LaunchPad
Summative Quizzes	Each chapter includes a pre-built Summative Quiz in LaunchPad
Active Learning Modules	All new Active Learning Modules include a pre-quiz, and post-quiz in LaunchPad, plus additional assessment questions
Simulations	Companion quizzes assignable in LaunchPad
Animations	Pre-built quizzes in LaunchPad
Activities	All activities are now assignable and report on completion to the LaunchPad gradebook
Test Bank	An extensive test bank is provided for each chapter, in a variety of formats: LaunchPad question banks, Word documents, and in the Diploma test-creation program (software included)

Life is VISUAL

The art in *Life*, Eleventh Edition surpasses the art in previous editions. The authors and artists have examined each and every figure in the text, refining them for clarity, readability, and beauty. Many figures now include a question.

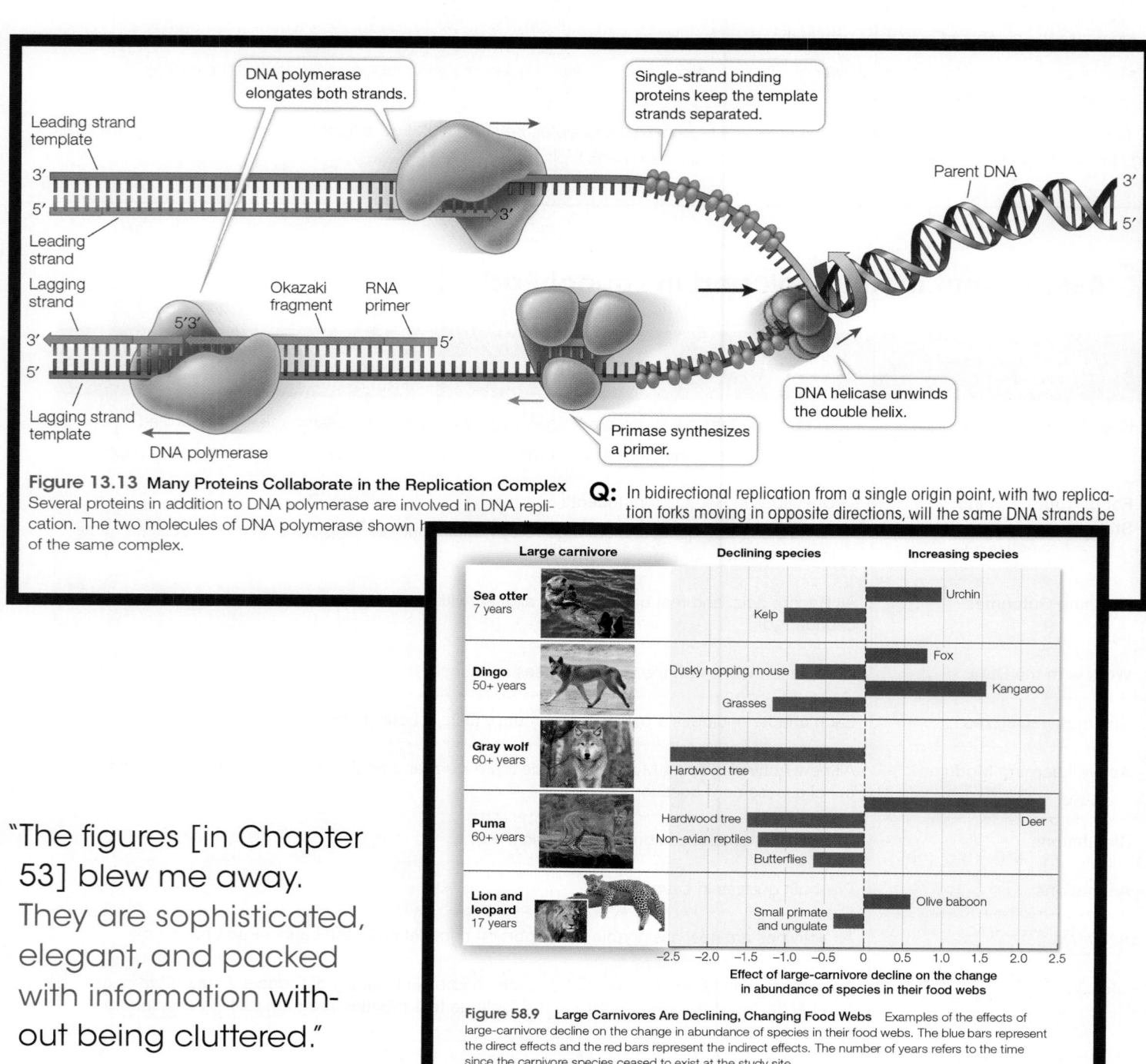

Figure 13.13 Many Proteins Collaborate in the Replication Complex Several proteins in addition to DNA polymerase are involved in DNA replication. The two molecules of DNA polymerase shown h of the same complex.

Q: In bidirectional replication from a single origin point, with two replication forks moving in opposite directions, will the same DNA strands be

Figure 58.9   Large Carnivores Are Declining, Changing Food Webs Examples of the effects of large-carnivore decline on the change in abundance of species in their food webs. The blue bars represent the direct effects and the red bars represent the indirect effects. The number of years refers to the time since the carnivore species ceased to exist at the study site.

"The figures [in Chapter 53] blew me away. They are sophisticated, elegant, and packed with information without being cluttered."

— Jill DeVito, University of Texas, Arlington

Focus Figures

Throughout the new edition, students will find specially created figures that take them through an essential concept. These illustrations let students know that a particular concept is absolutely essential, needs their full attention, and gives them a more interactive experience with the material.

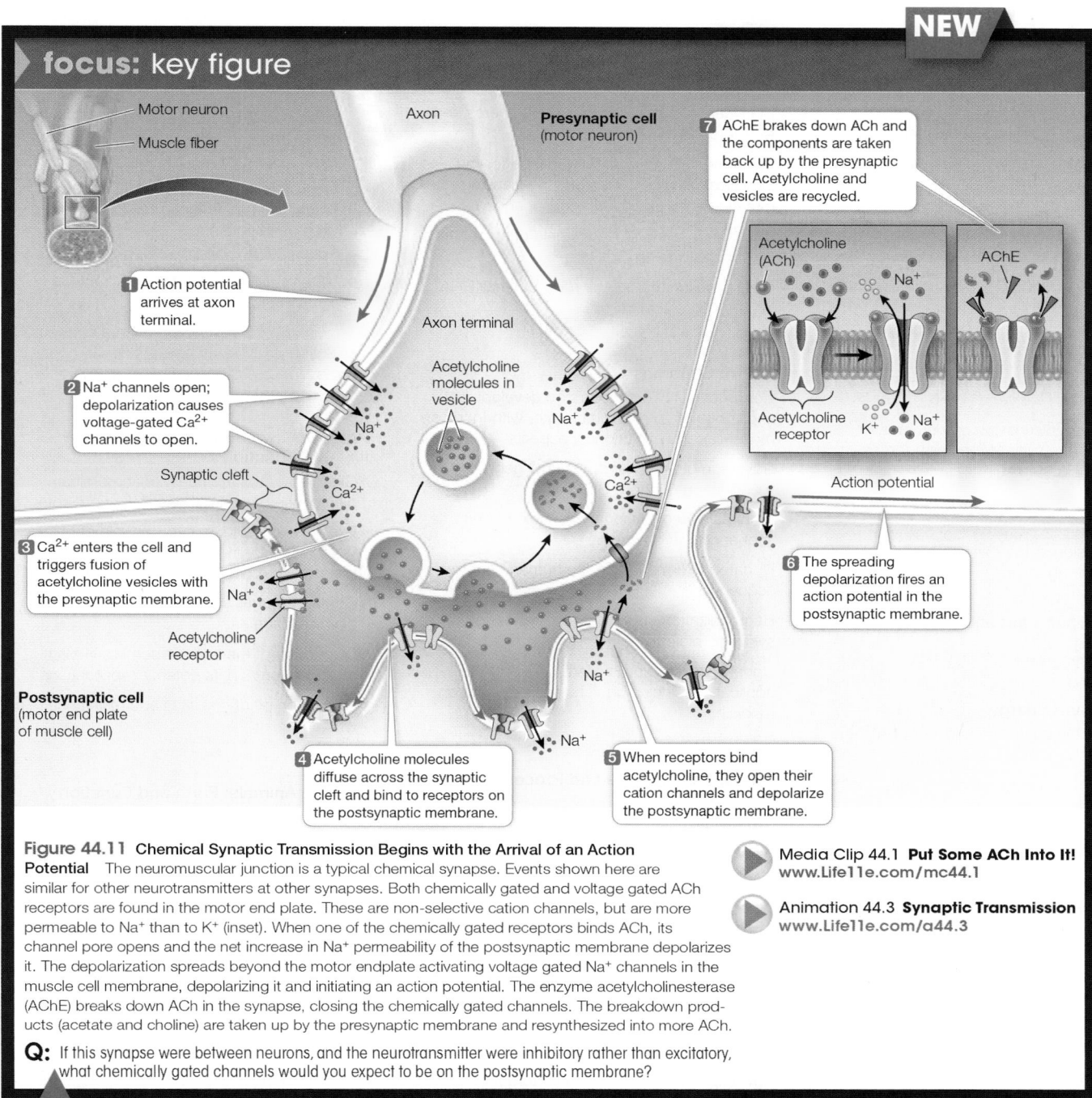

> **focus: key figure**
>
> NEW
>
> Motor neuron
> Muscle fiber
>
> Axon
>
> **Presynaptic cell** (motor neuron)
>
> **7** AChE brakes down ACh and the components are taken back up by the presynaptic cell. Acetylcholine and vesicles are recycled.
>
> Acetylcholine (ACh)
> Na^+
> AChE
> Acetylcholine receptor
> K^+ Na^+
>
> **1** Action potential arrives at axon terminal.
>
> Axon terminal
>
> Acetylcholine molecules in vesicle
>
> **2** Na^+ channels open; depolarization causes voltage-gated Ca^{2+} channels to open.
> Na^+
> Na^+
>
> Synaptic cleft
> Ca^{2+}
> Ca^{2+}
>
> Action potential
>
> **3** Ca^{2+} enters the cell and triggers fusion of acetylcholine vesicles with the presynaptic membrane.
> Na^+
>
> **6** The spreading depolarization fires an action potential in the postsynaptic membrane.
>
> Acetylcholine receptor
> Na^+
>
> **Postsynaptic cell** (motor end plate of muscle cell)
>
> Na^+
>
> **4** Acetylcholine molecules diffuse across the synaptic cleft and bind to receptors on the postsynaptic membrane.
>
> **5** When receptors bind acetylcholine, they open their cation channels and depolarize the postsynaptic membrane.

Figure 44.11 Chemical Synaptic Transmission Begins with the Arrival of an Action Potential The neuromuscular junction is a typical chemical synapse. Events shown here are similar for other neurotransmitters at other synapses. Both chemically gated and voltage gated ACh receptors are found in the motor end plate. These are non-selective cation channels, but are more permeable to Na^+ than to K^+ (inset). When one of the chemically gated receptors binds ACh, its channel pore opens and the net increase in Na^+ permeability of the postsynaptic membrane depolarizes it. The depolarization spreads beyond the motor endplate activating voltage gated Na^+ channels in the muscle cell membrane, depolarizing it and initiating an action potential. The enzyme acetylcholinesterase (AChE) breaks down ACh in the synapse, closing the chemically gated channels. The breakdown products (acetate and choline) are taken up by the presynaptic membrane and resynthesized into more ACh.

Q: If this synapse were between neurons, and the neurotransmitter were inhibitory rather than excitatory, what chemically gated channels would you expect to be on the postsynaptic membrane?

Media Clip 44.1 **Put Some ACh Into It!** www.Life11e.com/mc44.1

Animation 44.3 **Synaptic Transmission** www.Life11e.com/a44.3

Each of these figures concludes with a **thought-provoking question** that can lead to further exploration as the student is working through the text—another way of encouraging students to be active learners. Answers are in the back of the book.

Life is CURRENT

Life, Eleventh Edition Content Updates

PART 1: The Science of Life and Its Chemical Basis

- Streamlined and more focused description of electron orbitals
- Additional examples of functional groups and roles in biological molecules
- Additional examples of protein functions
- Reorganization and clearer presentation of the levels of protein structure
- Updated information on the search for evidence of conditions for life outside of Earth

PART 2: Cells

- Discovery of a new organelle, the tannosome, that encloses pigments
- New discussion of substructure and instability of microtubules
- Reorganization of discussion of membrane structure to feature connections with the cytoskeleton inside the cell
- Overview of membrane permeability of various substances
- Further discussion of the roles of receptor agonists and antagonists in human behavior and metabolism
- Clarification of the targets and roles of protein kinases

PART 3: Cells and Energy

- Clarification of the chemistry and role of ATP in energy metabolism
- New discussion of ribozymes along with protein enzymes
- Description of the mechanism of aspirin as an example of enzyme inhibition
- New description of feedback inhibition of the purine pathways and its role in arthritis
- Clearer description of electron transport and ATP synthesis, and their role in fat metabolism
- Clarification of the structure and function of photosynthetic reaction centers

PART 4: Genes and Heredity

- Revised and clarified discussion of DNA packaging into chromosomes
- More emphasis on human genetics (e.g., red hair)

- More examples of problem solving in genetics to clarify concepts
- Clarification of the aspects of DNA structure that allow specific interactions with other molecules
- Clarification of the functions of telomeres and telomerases
- Experimental and genetic evidence for the identification of a human disease-causing mutation and gene (e.g., breast cancer)
- More emphasis on the new field of epigenetics (e.g., behavior)
- Clearer description of the roles of small RNAs in gene regulation

PART 5: Genomes

- Consolidated two chapters on development and evolution to a single chapter with focus on molecular and evolutionary aspects
- Coverage of a new and powerful gene modification method: CRISPR
- Emphasis on evolutionary aspects of genome sequencing (e.g., cats)
- Updated information on the human genome sequence
- Reorganization and update of discussion of synthetic biology
- Updated discussion of stem cells and their applications, including veterinary and human medicine

PART 6: The Patterns and Processes of Evolution

- New Investigating Life thread about the evolution of echolocation, and ways moths have evolved to avoid detection by echolocation
- Updated information about the evolution of Darwin's finches, as an example of the relationship between speciation and geographic isolation
- Added studies about the evolution of flu viruses, cross-species transfers, and why the 1918 flu epidemic was so severe
- New experiments about the factors that influence the extinction of populations

PART 7: The Evolution of Diversity

- Updated studies that clarify the evolution of eukaryotes within the archaeans
- New experiments and information about the communication among bacteria through quorum sensing
- Added experiments and applications about biofuel production from algae
- New experiments and problems about the application of fungal diversity to study environmental change
- Updated information about the phylogeny of animals, and what it tells us about the evolution of nervous systems
- New information and experiments about the evolution of viviparity, and the advantages and disadvantages of viviparity versus oviparity

PART 8: Flowering Plants: Form and Function

- Clear illustrated summary of anatomical differences between monocots and dicots
- Experimental evidence for the role of plant vascular tissue in transporting substances in the stem (e.g., cyanogenic glycosides)
- Greater emphasis on the role of mycorrhizae in plant nutrition
- Discussion of the role of apomixis in plant reproduction and its potential applications
- Update on plant diseases and resistance (e.g., wheat rust)

PART 9: Animals: Form and Function

- Description of a previously unrecognized mammalian adaptation for heat loss and the development of a new technology that amplifies that adaptation to improve physical performance
- Graphic description of the neuromuscular coupling extending to the control of Ca^{2+} release from the sarcoplasmic reticulum
- Investigation of the use of RNAi to prevent hypertrophic cardiomyopathy
- Fascinating investigations of the capacity of seals to spend most of their time underwater and not breathing
- Exceedingly clear presentations of the mechanisms of membrane potentials, action potentials, conduction of action potentials, and synaptic transmission

Life is CURRENT

- Use of pit viper infrared sensory system to investigate transient receptor potential proteins
- Exciting story of being able to see what a rat thinks—at least in terms of the memory of place

PART 10: Ecology
- More attention is given to the importance of scale and multiple levels of organization in ecology. Special emphasis has been placed on global change and the role of humans as part of Earth's systems
- Inclusion of more quantitative and conceptual material with a focus on visualization (through figures) and active learning (through Work with the Data questions)

- Examples have been diversified; fewer insect examples and more marine, invasive species, and endangered species examples
- New content on the role of ecology in human society and how it differs from environmental science and environmentalism
- New content on the effects of Earth's topography on physical processes such as temperature inversion, ocean upwelling, and rain shadow effects
- Major revision and reorganization of population growth models and examples, including new material on life table analysis
- New content on disturbance and succession including the concept of alternative stable states

- New content on nutrient cycling at ecosystem and global scales, including decomposition, nutrient transformations, and the causes of climate change
- New content on the mechanisms behind species extinctions including the concept of extinction vortex
- New content on the causes and consequences of biodiversity loss including pollution, overharvesting, invasive species, emerging diseases, and the effects of climate change

Updates to the Media and Assessment Resources

Activities
- A new format for most activities, designed to better engage the student
- New functionality that reports to the LaunchPad gradebook upon activity completion

Simulations
- Twelve new simulations on topics such as Macromolecules, Active and Passive Transport, and Natural Selection
- Each simulation is now accompanied by an assignable quiz in LaunchPad

Animations
- A new video format that incorporates captions directly into the video

Work with the Data
- Each in-book Work with the Data exercise has an online companion exercise in LaunchPad, designed to give students practice with the same analytical skills

Active Learning Exercises
- Extensively revised and reformatted

New Active Learning Guide
- A thorough primer on active learning in biology
- Includes sections such as:
  - What is active learning?
  - Designing your course for active learning
  - Using active learning in the classroom
  - How to implement *Life* 11e resources
- Chapter-by-chapter ideas, suggestions, and resources

New Active Learning Modules
- Robust modules designed to provide all the support an instructor needs to teach selected topics in an active learning setting
- Each module includes a detailed instructor's guide, a pre-quiz, a pre-class video, an in-class exercise, a post-quiz, and assessment questions

Test Bank
- All questions aligned to new Learning Outcomes
- A new set of short answer questions in each chapter
- Additional questions at higher Bloom's Levels

LearningCurve
- Expanded question banks for each chapter
- New fill-in-the-blank questions
- Hints added to all questions

Acclaim for Life

"This textbook fits nicely with the current focus on active learning and the AAAS Vision and Change report."

—Michael Baltzley, Western Oregon University

"In most cases, I found the text to be exceptionally well-written and clear…The author(s) were particularly good at explaining 'big' concepts… The writing/presentation style of the author(s) was very readable…. In particular, the author(s) use a number of 'real-world' examples to highlight the information in the text. I found that this made the text and information very relatable and easier to follow/understand."

—Mike Shaughnessy, Northeastern State University

"The authors and publishing team are to be commended for producing a set of products that promise to promote active learning. The products of Life 11e contain numerous, novel features that distinguish them from similar products."

—David Marcey, California Lutheran University

"Opening with an intriguing story with promise of future investigation is the way to go. Many freshmen, introductory biology courses have a large number of students interested in medicine/medical fields. When "real" scenarios can be used to supplement the material within the chapter and help a student make those applicable connections, the chapter is much more interesting and meaningful."

—Melissa Reedy, University of Illinois, Urbana-Champaign

"Without question, my favorite element was the open-ended concept questions included in the section recaps. These questions are well-written, thought provoking, not easily answered or looked up, and require an understanding of the concepts rather than a memorization of flash card terms."

—Brad Mehrtens, University of Illinois, Urbana-Champaign

"I love that [Connect the Concepts] makes a direct connection to earlier text/figures; this emphasizes and encourages the integrative, rather than linear, process of learning biology."

—Tess Killpack, Wellesley College

"I really like the extensive use of examples and the figures, recaps, problems, etc. I think this makes the chapter much more accessible to students and gives them a built-in way to test their understanding beyond simple memorization questions. These features are what make this chapter and the other ecology chapters in this text that I have reviewed stand out from other texts."

—Amy Downing, Ohio Wesleyan University

"I did find the chapter engaging specifically because real case studies were used as part of the conceptual framework. This was one of the major points that biology educators were calling for at the Vision and Change forum a few years ago."

—Andrew David, Clarkson University

Authors

DAVID HILLIS   SALLY HACKER   CRAIG HELLER   DAVID SADAVA

DAVID SADAVA is the Pritzker Family Foundation Professor of Biology, Emeritus at the Keck Science Center of Claremont McKenna, Pitzer, and Scripps, three of The Claremont Colleges. In addition, he is Adjunct Professor of Cancer Cell Biology at the City of Hope Medical Center in Duarte, California. Twice winner of the Huntoon Award for superior teaching, Dr. Sadava has taught courses on introductory biology, biotechnology, biochemistry, cell biology, molecular biology, plant biology, and cancer biology. In addition to *Life: The Science of Biology* and *Principles of Life*, he is the author or coauthor of books on cell biology and on plants, genes, and crop biotechnology. His research has resulted in many papers coauthored with his students, on topics ranging from plant biochemistry to pharmacology of narcotic analgesics to human genetic diseases. For the past 20 years, he has investigated multidrug resistance in human small-cell lung cancer with a view to understanding and overcoming this clinical challenge. At the City of Hope, his current work focuses on new anti-cancer agents from plants. He is the featured lecturer in "Understanding Genetics: DNA, Genes and their Real-World Applications" and "What Science Knows About Cancer" video courses for The Great Courses series.

DAVID M. HILLIS is the Alfred W. Roark Centennial Professor in Integrative Biology and the Director of the Dean's Scholars Program at the University of Texas at Austin, where he also has directed the School of Biological Sciences and the Center for Computational Biology and Bioinformatics. Dr. Hillis has taught courses in introductory biology, genetics, evolution, systematics, and biodiversity. He has been elected to the National Academy of Sciences and the American Academy of Arts and Sciences, awarded a John D. and Catherine T. MacArthur fellowship, and has served as President of the Society for the Study of Evolution and of the Society of Systematic Biologists. He served on the National Research Council committee that wrote the report *BIO 2010: Transforming Undergraduate Biology Education for Research Biologists*, and he serves on the Executive Committee of the National Academies Scientific Teaching Alliance. His research interests span much of evolutionary biology, including experimental studies of viral evolution, empirical studies of natural molecular evolution, applications of phylogenetics, analyses of biodiversity, and evolutionary modeling. He is particularly interested in teaching and research about the practical applications of evolutionary biology.

H. CRAIG HELLER is the Lorry I. Lokey/Business Wire Professor in Biological Sciences and Human Biology at Stanford University. He has taught neurobiology and physiology in the core biology courses at Stanford since 1972 and served as Director of the Program in Human Biology, Chairman of the Biology Department, and Associate Dean of Research. Dr. Heller is a fellow of the American Association for the Advancement of Science and a recipient of the Walter J. Gores Award for excellence in teaching and the Kenneth Cuthbertson Award for Exceptional Service to Stanford University. His research is on the neurobiology of sleep and circadian rhythms, mammalian hibernation, the regulation of body temperature, the physiology of human performance, and the neurobiology of learning and memory. He has done research on many species and problems ranging from sleeping kangaroo rats, diving seals, hibernating bears and squirrels, photoperiodic hamsters, and exercising athletes. Dr. Heller has extended his enthusiasm for promoting active learning via the development of a two-year curriculum in human biology for the middle grades and through the production of Virtual Labs—interactive computer-based modules to teach physiology.

SALLY D. HACKER is Professor at Oregon State University where she has been a faculty member since 2004. She has taught courses in introductory ecology, community ecology, invasion biology, field ecology, and marine biology. She was awarded the Murray F. Buell Award by the Ecological Society of America and the Young Investigator Prize by the American Society of Naturalists. Dr. Hacker's research explores the structure, function, and services of natural and managed ecosystems under varying contexts of species interactions and global change. She has conducted research with plants and animals in rocky intertidal, salt marsh, seagrass, and coastal dune ecosystems. Her work has most recently focused on the protective role of dune ecosystems in mitigating coastal vulnerability due to climate change. In addition to the textbooks *Life: The Science of Biology* and *Ecology* (Sinauer Associates), she is author or coauthor on numerous articles and book chapters exploring community ecology, species interactions, marine invasions, and ecosystem services important to coastal management. She is particularly interested in promoting active and experiential learning for students interested in ecology and field-emersion experiences.

Acknowledgments

While our names are on the cover and we take ultimate responsibility for the contents of this book, we relied on many people for help. On the editorial side, first and foremost was the Advisory Board, whose names follow these paragraphs. These biologists, from institutions large and small that are research- and teaching-oriented, were with us every step of the way. At the outset, as we planned the very different approaches that you can see in this edition, they provided invaluable critiques to our proposals, with many of the advisors having experiences in active teaching environments that are different from ours. Then as we wrote the chapters and designed the illustrations that shaped the reality of our proposals, the Board again provided very useful evaluations. Finally, with their varied experiences in using online material, the members of the Advisory Board were an important source of feedback as we developed the active learning and assessment materials associated with the book. We couldn't have done it as well without them.

On the following pages, you can see three lists of chapter reviewers totaling more than 400 individuals. First were between edition reviewers, who carefully looked at the previous edition of *Life* and told us what we could do better, leave out, or add in the new edition. Then, accuracy reviewers, many of them experts in their field, examined each of our chapters for accuracy and not only spotted errors of fact and concepts, but made suggestions for better presentation. Finally, with our emphasis on active learning and pedagogy, we asked a group of experienced experts in biology education to critique our pedagogical features and give us suggestions for improvement. We are grateful to each and every one of these reviewers.

The team we worked intimately with at Sinauer Associates over many months consisted of a number of old friends and some new ones. In the former category are David McIntyre, who still manages to find fresh and exquisite photographs that illuminate our concepts; Jason Dirks, who directs a team of talented media and supplements editors and writers; Johannah Walkowicz, who again recruited the multitude of reviewers listed here; Joan Gemme, who designed the cover and skillfully arranged the words and images on every page, and, along with Joanne Delphia, designed the interior to provide a fresh look; Chris Small, whose behind the scenes skill at coordinating all facets of art and production remain invaluable; Dean Scudder, who continues to bring a wealth of insight to how best to tell our story; Azelie Fortier, whose interactions with the Advisory Board and potential adopters have informed the development of our new active learning features; and Liz Pierson, the most insightful copyeditor we know.

New to the *Life* team were the editors at Sinauer Associates, Danna Lockwood and Chelsea Holabird, who worked creatively (and tirelessly) to assure that each of our chapters lived up to the challenging goals we set for this Eleventh Edition. Carol Pritchard-Martinez came back after a hiatus to provide superb guidance as our developmental editor. Also, Carrie Mailler provided critical knowledge about our foray into the world of pedagogy and learning objectives.

Macmillan Learning has worked aggressively to bring *Life* to a broader audience. Executive Marketing Manager Will Moore, Executive Editor Andy Dunaway, and Editor Beth Cole (who was our liaison in the early stages of this edition), along with the regional specialists, regional managers, and experienced sales force, are effective ambassadors and skillful transmitters of the strong new features and unique strengths of this new edition. We depend upon their experience and energy to keep us in touch with how *Life* is perceived by its users. Many thanks also to Chris Efstratiou and the whole Macmillan LaunchPad production team.

Finally and above all, we thank Andy Sinauer. Andy is the ideal publisher for his authors. He loves biology, is committed to education, and sets high standards. Throughout the preparation of this new edition, he has been not just our editor but also our guiding spirit.

DAVID SADAVA

DAVID HILLIS

CRAIG HELLER

SALLY HACKER

Reviewers and Contributors

for the Eleventh Edition

Advisory Board

Lori Ann Biederman, Iowa State University

Candice Damiani, Unversity of Pittsburgh

Claudette Davis, George Mason University, Fairfax

Kamal Dulai, University of California, Merced

Ellen S. Goldey, Florida Atlantic University

Christopher Gregg, Louisiana State University

Carly Jordan, The George Washington University

Brenda Leicht, University of Iowa

Kristen Lennon, Hagerstown Community College

Jay Mager, Ohio Northern University

David J. Marcey, California Lutheran University

Robert Osuna, University at Albany, State University of New York

Pamela A. Pape-Lindstrom, Everett Community College

Roger Persell, Hunter College, City University of New York

Anna Powolny, Spelman College

Mitchell Singer, University of California, Davis

Kathryn Spilios, Boston University

Sarah Texel, Towson University

Catherine Ueckert, Northern Arizona University

Between Edition Reviewers

Elizabeth Binney, Western Washington University

Catherine Black, Idaho State University

Andrew Blaustein, Oregon State University

Katie Boes, College of Wooster

Michelle Boone, Miami University of Ohio

Christopher Brown, Georgia Gwinnett College

Theodore Burk, Creighton University

Cheryl Burrell, Wake Forest University

Jean Cardinale, Alfred University

Shelton Charles, Forsythe Technical Community College

Debora Christensen, Drake University

John Cigliano, Cedar Crest College

Claudette Davis, George Mason University

Beverly Clendening, Hofstra University

Catherine Craker, Ohio Northern University

Noelle Cutter, Molloy College

Dean Tolan, Boston University

Amy Downing, Ohio Wesleyan University

William Eldred, Boston University

David Eldridge, Baylor University

Julie Emerson, Amherst College

Stuart Feinstein, University of California, Santa Barbara

Kenneth Filchak, University of Notre Dame

Teresa Fischer, Indian River State College

Pam Freeman, The College of St. Scholastica

Theresa Grove, Valdosta State University

Kathy Hafer, Washington University in St. Louis

Kevin Hamed, Virginia Highlands Community College

Laura L. Hernandez, University of Wisconsin, Madison

Mark Holland, Salisbury University

Lara Hutson, State University of New York at Buffalo

Danielle Ignace, Smith College

David Kittlesen, University of Virginia

Drew Kohlhorst, Georgia State University

Stephen Kolomyjec, Ohio Northern University

Kristen Lennon, Hagerstown Community College

Eric Liebold, Salisbury University

Robert Lowery, Indian River State College

Jay Mager, Ohio Northern University

Stephanie Marin, California State Polytechnic University, Pomona

Elizabeth McCain, Muhlenberg College

Andy McCall, Denison University

Clara Moore, Franklin and Marshall College

Richard Moore, Miami University of Ohio

Ross Nehm, Stony Brook University, State University of New York

Leah Okumura, Wellesley College

Nancy Pencoe, University of West Georgia

Uwe Pott, University of Wisconsin, Green Bay

David Puthoff, Frostburg State University

Diviya Ray, Massachusetts Institute of Technology

Brian Ring, Valdosta State University

Mary K. Ritke, University of Indianapolis

George Robinson, University at Albany, State University of New York

Jason Rothman, California State Polytechnic University, Pomona

Patricia Rugaber, College of Coastal Georgia

Mark Sarvary, Cornell University

Robert Savage, Williams College

Mark Schneegurt, Wichita State University

Fayla Schwartz, Everett Community College

Wayne Silver, Wake Forest University

Neelima Sinha, University of California, Davis

Joel Snodgrass, Towson University

Sita Somara, Forsythe Technical Community College

Kathryn Spilios, Boston University

Cynthia Surmacz, Bloomsburg University

Steven Swoap, Williams College

Edward Tall, Seton Hall University

Mark Taylor, Baylor University

Douglas Thrower, University of California, Santa Barbara

Catherine Ueckert, Northern Arizona University

Laura Vallier, Hofstra University

John Walker, Appalachian State University

Heping Zhou, Seton Hall University

Accuracy Reviewers

Cristian Aguilar, Azusa Pacific University

C. A. Allen, Blinn University

Tracey Anderson, University of Minnesota, Morris

Caleb Bailey, Brigham Young University, Idaho

Michael Balsai, Temple University

Michael Baltzley, Western Oregon University

Adebiyi Banjoko, Chandler Gilbert Community College

Chad Barber, California Lutheran University

Todd Barkman, Western Michigan University

Alexandra Bely, University of Maryland

Michael Berger, Washington State University, Vancouver

Karen K. Bernd, Davidson College

Stephanie Bingham, Barry University

Elizabeth Binney, Western Washington University

Kristopher Blee, California State University, Chico

Ed Bobich, California State Polytechnic University, Pomona

Robert Bohanan, University of Wisconsin

Nancy Boury, Iowa State University

Melissa Bowlin, University of Michigan, Dearborn

Sarah Boyer, Macalester College

Forrest Michael Brem, University of Memphis

Matthew W. Brewer, Georgia State University

Mirjana Brockett, Georgia Institute of Technology

Heather Bruns, Ball State University

Jill Buettner, Richland College

Steve Burnett, Clayton State University

Cheryl Burrell, Forsyth Technical Community College

Steve Bush, Coastal Carolina University

Madhavi Chakravadhanula, Arizona State University

Eric Chambers, Valdosta State University

Shelton Charles, Forsyth Technical Community College

Charles Chen, Azusa Pacific University

Vasanta Chivukula, Atlanta Metro College

Lynn Christenson, Vassar College

Dennis O. Clegg, University of California, Santa Barbara

Craig Clifford, Northeastern State University

Liz Co, Boston University

Reid Compton, University of Maryland

Stephanie Conant, University of Detroit, Mercy

Vikki Connaughton, American University

Cindy Corbitt, University of Louisville

Brett Couch, University of British Columbia

Alistair Cullum, Creighton University

Noelle Cutter, Molloy College

Mary Ellen Czesak, Vassar College

David Dalton, Reed College

Deborah Dardis, Southeastern Louisiana University

Doug Darnowski, Indiana University, Southeast

Farahad Dastoor, University Maine

Andrew David, Clarkson University

Frank David, University of California, Santa Barbara

William Davis, Washington State University

Lewis Deaton, University Louisiana, Lafayette

Kelly Decker, California State University, East Bay

John Demastes, University of Northern Iowa

Fred Deneke, Cosumnes River College

Jill DeVito, University of Texas, Arlington

Sam Donovan, University of Pittsburgh

Amy Downing, Ohio Wesleyan University

Matthew Draud, Armstrong State University

Robert Drewell, Clark University

Kelli Duncan, Vassar College

Erin Easlon, University of California, Davis

John Elder, Valdosta State University

Jean Everett, College of Charleston

Gordon Fain, University of California, Los Angeles

David Fankhauser, University of Cincinnati, Clermont

Melinda Faulkner, Bradley University

Kenneth Filchak, University of Notre Dame

Samuel Flaxman, University of Colorado, Boulder

Amy Frary, Mount Holyoke College

Aaren Freeman, Adelphi University

Mark Garcia, Collin College

Marina Gerson, California State University, Stanislaus

Raffaella Ghittoni, University of Southern California

Brian Gibbens, University of Minnesota

J. Phil Gibson, University of Oklahoma

William Gilliland, DePaul University

Jaime Grace, Bradley University

Matthew Graham, Eastern Connecticut State University

Robert Grammer, Belmont University

Mike Grosbeck, Brigham Young University, Idaho

Theresa Grove, Valdosta State University

Lisa Grubisha, University Wisconsin, Green Bay

Ehren Haderlie, Brigham Young University, Idaho

Rebecca Hale, University of North Carolina at Asheville

Ben Hanelt, University of New Mexico

Amy Hark, Muhlenberg College

Phillip Harris, University of Alabama, Tuscaloosa

Valerie Haywood, Case Western Reserve University

Christiane Healey, University of Massachusetts, Amherst

Tracy Heath, Iowa State University

Heather Heck, University of Maine

Robert Hegna, Palm Beach Atlantic University

Amy Helms, Collin College

Kelly Howe, University of New Mexico

David Hudson, Atlanta Metropolitan State College

Myra Hughey, Vassar College

Ryan Huish, The University of Virginia's College at Wise

Cynthia Hutton, Northland Pioneer College

Chris Ivey, California State University, Chico

Erika Iyengar, Muhlenberg University

Rachel Johnson, University of Minnesota, Morris

Keith Johnson, Bradley University

Scott Johnson, Towson University

Russell Johnson, Colby College

Jennifer Jost, Bradley University

Jacob Kagey, University of Detroit, Mercy

Stephanie Kamel, University North Carolina, Wilmington

Lori Kayes, Oregon State University

David Keller, California State University, Chico

Diane Kelly, University of Massachusetts

Todd Kelson, Brigham Young University Idaho

Kyeorda Kemp, Northeastern State University

Tess Killpack, Wellesley College

Scott Kimball, Baker University

Lara Kingeter, Tarrant County College Northeast Campus

Bridgette Kirkpatrick, Collin College

David Kittlesen, University of Virginia

David Knochel, University of Colorado, Denver

William Kristan, California State University, San Marco

Allan Larson, Washington University in St. Louis

Kristin Latham, Western Oregon University

Kari Lavalli, Boston University

Mark Lazzaro, College of Charleston

Kay Lee-Fruman, California State University, Long Beach

Jeff Leips, University Maryland, Baltimore County

Kristen Lennon, Hagerstown Community College

Eric Liebl, Denison University

Brianna Lindh, Willamette University

Rob Loeb, Pennsylvania State University

John Logsdon, The University of Iowa

Janet Loxterman, Idaho State University

Charles Mallery, University of Miami, Coral Gables

Robert Mans, Armstong State University

Barry Margulies, Towson University

Linda Mayerhofer, University at Albany, State University of New York

Bryant McAllister, The University of Iowa

Lori McGrew, Belmont University

Kyle McQuade, Colorado Mesa University

Mark Meade, Jacksonville State University

Brad Mehrtens, University of Illinois at Urbana-Champaign

Jennifer Metzler, Ball State University

Rick Miller, Formerly at Southeastern Louisiana University

Rich Moore, Miami University, Ohio

Patricia Moore, University of Georgia

Sherri Morris, Bradley University

Barbara Musolf, Clayton State University

Ross Nehm, Stony Brook University, State University of New York

Kim Nelson, Pennsylvania State University

Katie Nemeth, College of St. Scholastica

John Niedzwiecki, Belmont University

Ben Normark, University of Massachusetts, Amherst

Eric Norstrom, DePaul University

Richard Noyes, University of Central Arkansas

Matthew Nusnbaum, Georgia State University

Leah Okumura, Wellesley College

Jonelle Orridge, Broward College

Jennifer Osterhage, University of Kentucky

Kristen Patterson, University of Texas at Austin

James Patton, Vanderbilt University

Paulette Peckol, Smith College

Patricia Peroni, Davidson College
Teresa Petrino, Barry University
Uwe Pott, University of Wisconsin-Green Bay
Dan Potter, University of California, Davis
Daniel Potts, Buffalo State University
Nick Pullen, University Northern Colorado
Nick Ragsdale, Belmont University
Claire Varian Ramos, Colorado State University, Pueblo
Flona Redway, Barry University
Melissa Reedy, University of Illinois Urbana-Champaign
Mark Reedy, Creighton University
Susan Reigler, Indiana University Southeast
Laurel Roberts, University of Pittsburgh
George Robinson, University at Albany, State University of New York
Kevin Roe, Iowa State University
Corey Roelke, University Texas at Arlington
Kristi Curry Rogers, Macalester College
Thomas Rosburg, Drake University
Mike Rosenzweig, Virginia Tech
Jason Rothman, California State Polytechnic University, Pomona
Ben Rowley, University of Central Arkansas
Jodi Rymer, College of the Holy Cross
Robert Savage, Williams College
Brenda Schoffstall, Barry University
Paul Schulter, University of Nevada, Las Vegas
J. Reid Schwebach, George Mason University
Erik Scully, Towson University
Jeanne Serb, Iowa State University
Alan Shabel, University of California. Berkeley
Mike Shaughnessy, Northeastern State University
Mark Sherrard, University of Northern Iowa
Ann Showalter, Clayton State University
Cahleen Shrier, Azusa Pacific University
Wayne Silver, Wake Forest University
Anu Singh-Cundy, Western Washington University
Diviya Sinha, Massachusetts Institute of Technology
Bill Smith, Wake Forest University
Nancy Solomon, Miami University, Ohio
Jay Stachowicz, University of California, Davis
Maria Stanko, New Jersey Institute of Technology
Nathan Staples, Canada College
Karl Sternberg, Western New England University
Tara Turley Stoulig, Southeastern Louisiana University
M. Elizabeth Stroupe, Florida State University
Ken Sweat, Arizona State University
Beckie Symula, University of Mississippi
Nicoladie Tam, University of North Texas
Amina Tassa, Collin College

Casey terHorst, California State University, Northridge
Conrad Toepfer, Brescia University
Dean Tolan, Boston University
W. Robert Trentham, Carson Newman University
Claudia Uhde-Stone, California State University, East Bay
Hemayet Ullah, Howard University
Ximena Valderrama, Ramapo College of New Jersey
Kenneth van Golen, University of Delaware
Doug Van Hoewyk, Coastal Carolina University
John VandenBrooks, Midwestern University
Meenakshi Vijayaraghavan, Tulane University
Janice Voltzow, University of Scranton
Catherine Vrentas, Frostburg State University
Heather Waye, University Minnesota, Morris
Allison Welch, College of Charleston
Gerard White, College of Coastal Georgia
Mary White, Southern Louisiana University
Lisa Whitenack, Allegheny College
Judith Williams, University of San Diego
Daniel Williams, Coastal Carolina University
Gay Williamson, Mississippi State University
Sarah Wojiski, Massachusetts College of Pharmacy and Health Sciences
Weimin Zhong, Yale University

Pedagogy Reviewers

Lori Biederman, Iowa State University
Jennifer Butler, Willamette University
Liz Co, Boston University
Julie Cronk, Columbus State Community College
Deborah Dardis, Southeastern Louisiana University
Farahad Dastoor, University of Maine
Claudette Davis, George Mason University
Tod Duncan, University of Colorado, Denver
Miles Dean Engell, North Carolina State University
Ginger Fisher, University of Northern Colorado
Marina Gerson, California State University, Stanislaus
Valerie Haywood, Case Western Reserve University
Ryan Huish, The University of Virginia's College at Wise
Jamie Jensen, Brigham Young University
Rachel Johnson, University of Minnesota, Morris
Dave Kittlesen, University of Virginia
Kari Lavalli, Boston University
Jeanne Lawless, Binghampton University, State University of New York
Brenda Leicht, University of Iowa

Kristen Lennon, Hagerstown Community College
Carolyn Martineau, DePaul University
Dina Newman, Rochester Institute of Technology
Robert Osuna, University at Albany, State University of New York
Mike Rosenzweig, Virginia Tech
Mitchell Singer, University of California, Davis
Alicia Slater, Stetson University
Kathryn Spilios, Boston University
Ximena Valderrama Ramapo, The College of New Jersey
Catherine Wilcoxson Ueckert, Northern Arizona University
Mary Williams, American Society of Plant Biology, Features Editor, *The Plant Cell*
Robert Wise, University Wisconsin, Osh Kosh

Media and Supplements Contributors

Edward Awad, Vanier College
Susan Lopez Bailey, science writer
Cissy Ballen, Cornell University
Joseph Bruseo, Holyoke Community College
Kenneth Filchak, University of Notre Dame
Donna Francis, University of Massachusetts, Amherst
Richard Gill, Brigham Young University
Richard Gonzalez-Diaz, Seminole State College
Adam Hall, Smith College
Carol Hand, science writer
Phillip Harris, University of Alabama
Margaret Hill, science writer
Norman Johnson, University of Massachusetts Amherst
Carly Jordan, George Washington University
Jessica Kalagher, science writer
Brenda Leicht, University of Iowa
Laurie Beth Leonelli, Sarah Lawrence College
John Lepri, University of North Carolina at Greensboro
Betty McGuire, Cornell University
Penny Molyneux, Smith College
Kristine Nowak, Kennesaw State University
Scott Nowak, Kennesaw State University
Diviya Ray, Massachusetts Institute of Technology
Meredith Safford, Johns Hopkins University
Jessica Santangelo, Hofstra University
John Townsend-Mehler, Montana State University
Mary Tyler, University of Maine
Danielle Webster, University of Central Florida
Robert Wise, University of Wisconsin, Oshkosh
Timothy Wakefield, John Brown University

Media and Supplements

to accompany *Life: The Science of Biology*, Eleventh Edition

 **LaunchPad**

LaunchPad is the easy-to-use online platform that integrates the e-Book, all of the student and instructor media resources, and assessment functions into one clean, unified interface. For more about LaunchPad, see pages xvi–xix and visit **macmillanlearning.com/LaunchPad**.

Student Resources

ACTIVITIES. A range of interactive activities helps students learn and review key facts and concepts through labeling diagrams, identifying steps in processes, and working with simulations of biological systems and processes.

ANIMATIONS. In-depth animations present complex topics in a clear, easy-to-follow format. Each is accompanied by a brief quiz.

MEDIA CLIPS. These engaging video clips depict fascinating examples of some of the many organisms, processes, and phenomena discussed in the textbook.

BIONEWS FROM SCIENTIFIC AMERICAN. BioNews makes it easy for instructors to bring the dynamic nature of the biological sciences and up-to-the-minute currency into their course via an automatically updated news feed in LaunchPad.

WORK WITH THE DATA. Online companions to the Work with the Data exercises in the textbook.

FLASHCARDS. A convenient way for students to learn and review the extensive terminology of introductory biology.

TREE OF LIFE. An interactive version of the Tree of Life from Appendix A, with links to a wealth of information on each group listed.

MATH FOR LIFE. A collection of mathematical shortcuts and references to help students with the quantitative skills they need in the biology laboratory.

SUMMATIVE QUIZZES. Pre-built quizzes that assess overall student understanding of each chapter.

LEARNINGCURVE. A powerful adaptive quizzing system with a game-like format that engages students. See pages xvii–xviii for details.

Instructor Resources

TEXTBOOK FIGURES AND TABLES. Every figure and table from the textbook (including all photos and all unnumbered figures) is provided in both JPEG and PowerPoint formats, in multiple versions, including whole, reformatted, and unlabeled.

POWERPOINT RESOURCES. For each chapter of the textbook, several different PowerPoint presentations are available, providing instructors the flexibility to build presentations in the manner that best suits their needs:

- Textbook Figures and Tables
- Lecture Presentations
- Layered Art Figures
- Active Learning Exercises (see below)
- Active Learning Modules (see below)

VIDEOS. Featuring many new videos for the Eleventh Edition, the wide-ranging collection of video segments helps demonstrate the complexity and beauty of life.

ACTIVE LEARNING GUIDE. New for the Eleventh Edition, this guide provides an excellent overview of active learning techniques in biology, along with chapter-specific support. See page xiii for details.

ACTIVE LEARNING EXERCISES. Set up for easy integration into lectures, these short exercises pose questions or problems for the class to discuss or solve. Each also includes a multiple-choice element, for easy use with clicker systems.

ACTIVE LEARNING MODULES. New for the Eleventh Edition, these comprehensive modules provide instructors everything they need to teach selected key topics using an active learning approach. See page xii for details.

INSTRUCTOR'S MANUAL. The Instructor's Manual includes the following sections for each chapter: *Overview, What's New, Chapter Opener Resources* (Discussion Questions and Online Resources), *Chapter Outline*, and *Key Terms*.

LEARNING OUTCOMES. The new Learning Outcomes provide instructors using *Life*, Eleventh Edition a thorough set of learning standards, to which all of the assessment resources are aligned. See pages x and xviii for details.

TEST BANK. The *Life* Test Bank includes over 7,000 questions, all referenced to specific Key Concepts, Learning Outcomes, and Bloom's Levels. Each chapter includes a wide range of multiple-choice, fill-in-the-blank, and short-answer questions.

COMPUTERIZED TEST BANK. (CD, ISBN 978-1-319-08581-0) The entire Test Bank, plus the Summative Quizzes and LearningCurve questions are all included in Blackboard's easy-to-use Diploma program (software included). Designed for both novices and advanced users, Diploma allows instructors to quickly and easily create or edit questions and create quizzes or exams.

COURSE MANAGEMENT SYSTEM SUPPORT. As a service for *Life* adopters using Blackboard, Moodle, Desire2Learn, or other course management systems, full electronic course packs are available.

Additional Resources

Increase participation. Confirm understanding. Measure performance.

iclicker is an award-winning student response solution that allows you to easily transform your classroom into an active learning environment. For more information, visit **iclicker.com**.

LabPartner

Macmillan Learning Lab Solutions

Macmillan Learning Lab Solutions unites the same brands you know and trust—W. H. Freeman, Hayden-McNeil, Sapling Learning, and Late Nite Labs—and provides the resources to build a seamless, comprehensive lab experience for your students. For more information, visit **macmillanlearning.com/LabSolutions**.

CatchUp Math & Stats for the Life Sciences

Michael Harris, Gordon Taylor, and Jacquelyn Taylor
(ISBN 978-1-4292-0557-3)
Presented in brief, accessible units, this primer will help students quickly brush up on the quantitative skills they need to succeed in biology.

A Student Handbook for Writing in Biology, Fifth Edition

Karin Knisely (ISBN 978-1-3191-2181-5)
This book provides practical advice to students who are learning to write according to the conventions in biology, using the standards of journal publication as a model.

BioStats Basics: A Student Handbook

James L. Gould and Grant F. Gould (ISBN 978-0-7167-3416-1)
Engaging and informal, *BioStats Basics* provides introductory-level biology students with a practical, accessible introduction to statistical methodology.

Inquiry Biology: A Laboratory Manual, Volumes 1 and 2

Mary Tyler, Ryan W. Cowan, and Jennifer L. Lockhart (Volume 1 ISBN 978-1-4292-9288-7; Volume 2 ISBN 978-1-4292-9289-4)
This introductory biology laboratory manual is inquiry-based—instructing in the process of science by allowing students to ask their own questions, gather background information, formulate hypotheses, design and carry out experiments, collect and analyze data, and formulate conclusions.

Hayden-McNeil Life Sciences Lab Notebook

(ISBN 978-1-4292-3055-1)
This durable, high-quality carbonless laboratory notebook allows students to hand in originals *or* copies of their lab work, not entire composition books. The Lab Notebook contains Hayden-McNeil's unique white paper carbonless copies and biology-specific reference materials.

The Scientific Teaching Book Series

The Series is a collection of practical guides, intended for all science, technology, engineering and mathematics (STEM) faculty who teach undergraduate and graduate students in these disciplines. The purpose of these books is to help faculty become more successful in all aspects of teaching and learning science, including classroom instruction, mentoring students, and professional development. Authored by well-known science educators, the Series provides concise descriptions of best practices and how to implement them in the classroom, the laboratory, or the department. For readers interested in the research results on which these best practices are based, the books also provide a gateway to the key educational literature. The Series includes:

Scientific Teaching
Jo Handelsman, Sarah Miller, and Christine Pfund
(ISBN 978-1-4292-0188-9)

Transformations: Approaches to College Science Teaching
Deborah Allen and Kimberly Tanner (ISBN 978-1-4292-5335-2)

Entering Research: A Facilitator's Manual
Janet L. Branchaw, Christine Pfund, and Raelyn Rediske
(ISBN 978-1-4292-5857-9)

Discipline-Based Science Education Research: A Scientist's Guide
Stephanie J. Slater, Timothy F. Slater, and Janelle M. Bailey
(ISBN 978-1-4292-6586-7)

Assessment in the College Classroom
Clarissa Dirks, Mary Pat Wenderoth, and Michelle Withers
(ISBN 978-1-4292-8197-3)

Contents

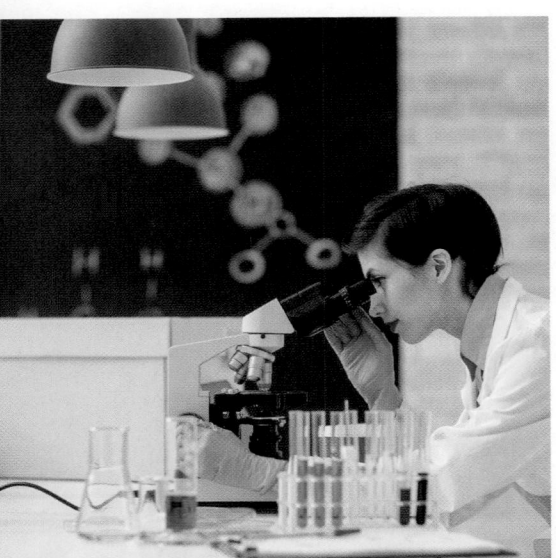

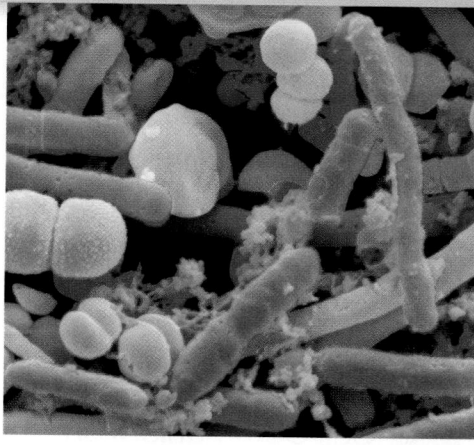

PART TWO   Cells

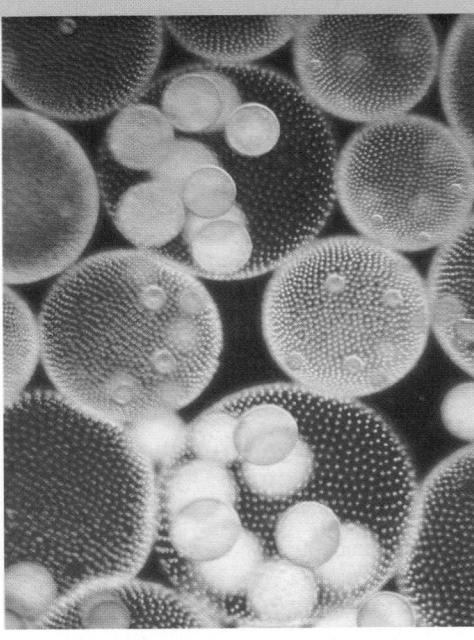

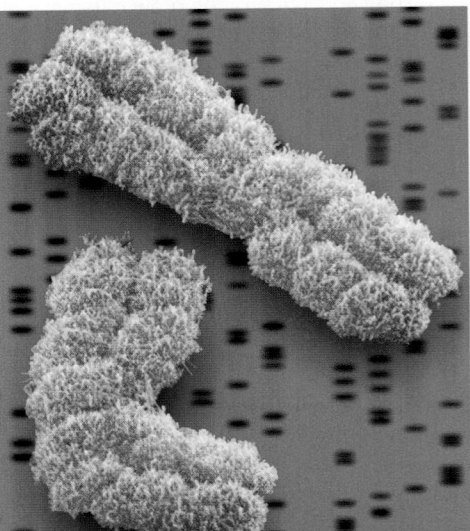

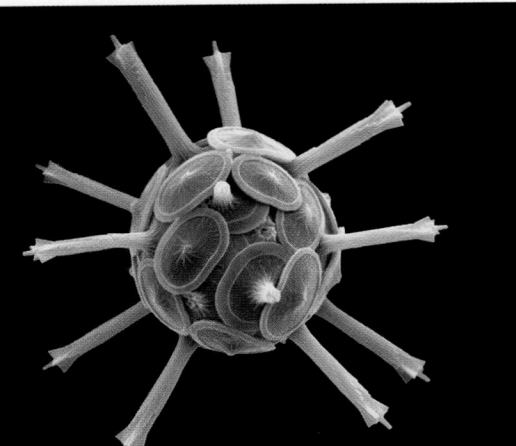

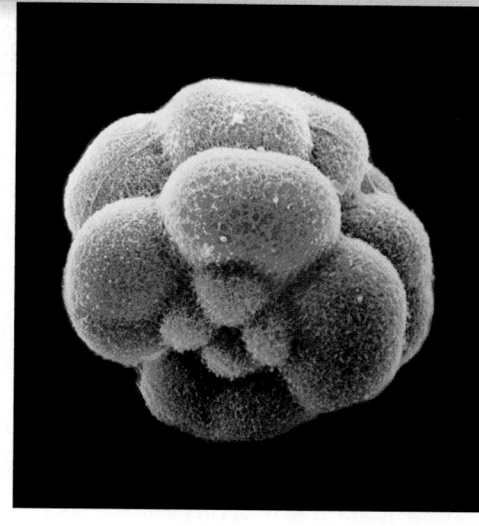

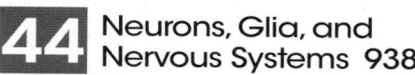

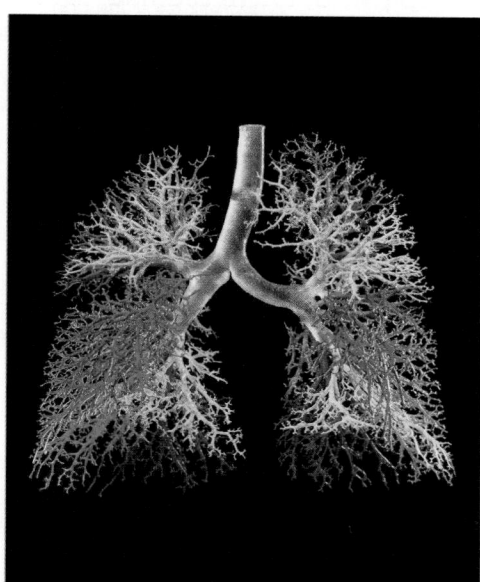

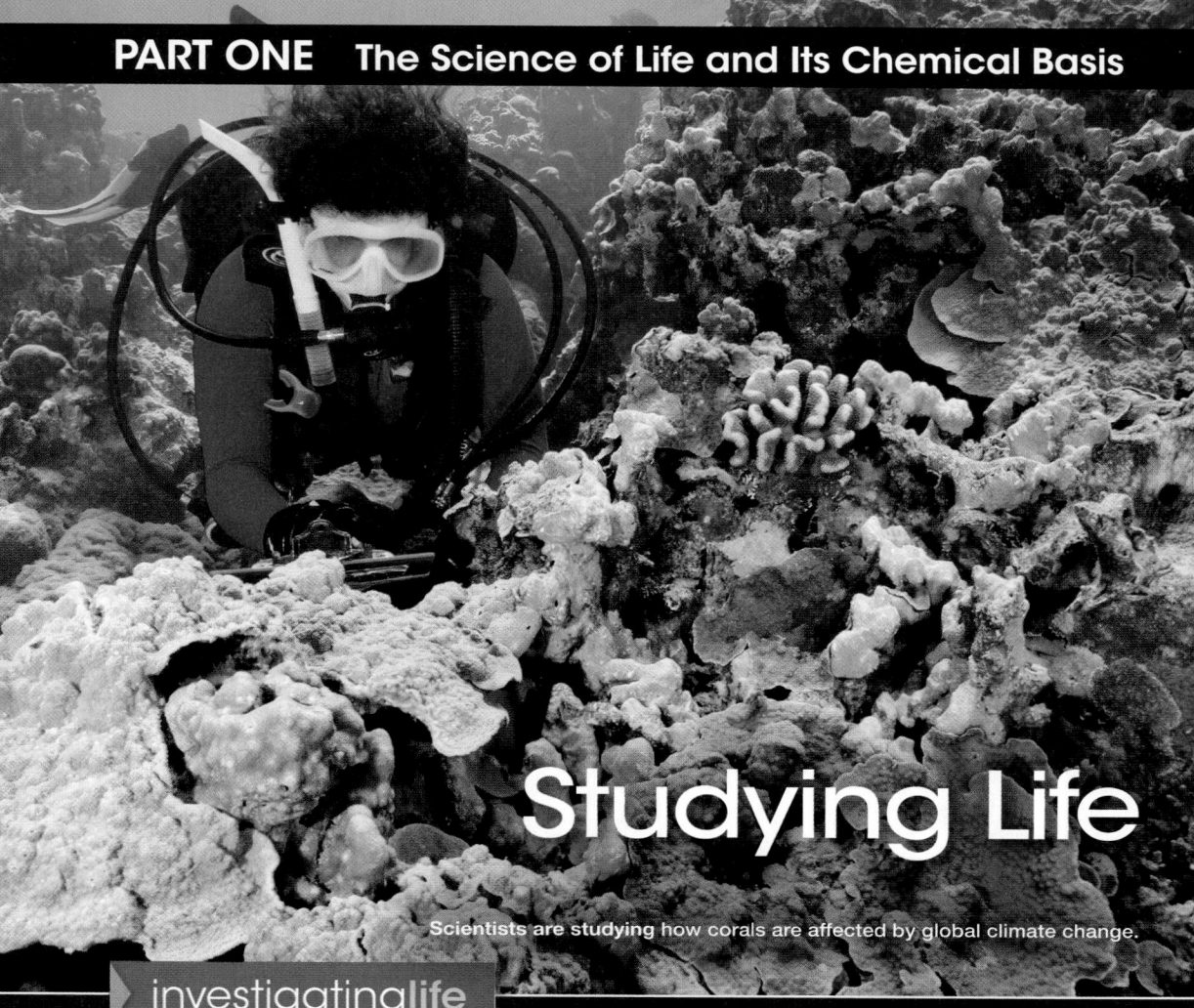

1

Studying Life

Scientists are studying how corals are affected by global climate change.

investigatinglife

Corals in Hot Water

Coral reefs support the largest diversity of life in the oceans. They provide fisheries and storm protection for about a billion people, and are a magnificent source of natural beauty. But coral reefs are endangered. Over the past 20 years about half of the world's reef-building corals have been destroyed by rising water temperature and other factors. High temperatures disrupt a fascinating aspect of coral biology. Corals are animals, yet most corals have algae (dinoflagellates) growing within their cells. Dinoflagellates use the energy of sunlight to produce carbohydrates. Corals provide a home for the dinoflagellates, which in turn provide nutrients for the corals. When high temperature impairs the dinoflagellates, the corals eject them—a process called bleaching. Then, without nutrients from the dinoflagellates, the corals die, unless they can take in new dinoflagellates that are more resistant to the higher temperatures.

Understanding the effects of heat on corals was the motivation of graduate student Rachael Bay and her colleagues working with Professor Steve Palumbi at Stanford University's Hopkins Marine Station. While studying corals in small back-reef pools in American Samoa, they observed that during low tides some pools reached higher temperatures than others. The researchers predicted that corals in the warmer pools had mechanisms enabling them to resist bleaching. To test their prediction, the researchers brought corals into the laboratory, subjected them to temperature fluctuations, and showed that the corals from the warm pools were more resistant to bleaching. They also transplanted corals between the different temperature environments in nature. As a result of these experiments, Rachael and her colleagues proposed that two different processes contribute to the different abilities of the corals to survive heat stress. First, corals growing in the warmer pools could have genetic traits contributing to heat resistance, and second, individual corals may have the same genetic make-up but differ in their ability to adjust to their environment by changing the expression levels of certain genes. More knowledge of the mechanisms of heat stress and heat resistance in corals could lead to new strategies to decrease their losses as their environments change.

Q&A How might experiments on heat stress in corals be used to predict the response of

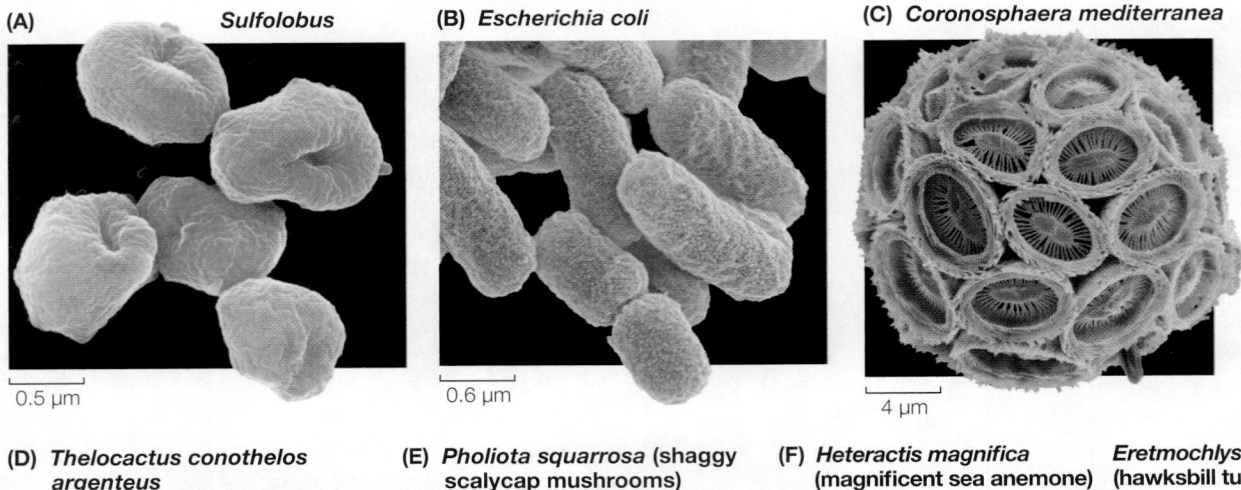

(A) *Sulfolobus*

0.5 μm

(B) *Escherichia coli*

0.6 μm

(C) *Coronosphaera mediterranea*

4 μm

(D) *Thelocactus conothelos argenteus*

(E) *Pholiota squarrosa* (shaggy scalycap mushrooms)

(F) *Heteractis magnifica* (magnificent sea anemone)   *Eretmochlys imbricata* (hawksbill turtle)

Chaetodon punctatofasciatus (spotband butterflyfish)

Figure 1.1 The Many Faces of Life   The processes of evolution have led to the millions of diverse organisms living on Earth today. Prokaryotic archaea **(A)** and bacteria **(B)** are all single-celled organisms, as described in Chapter 25. **(C)** Many protists are unicellular but, as discussed in Chapter 26, their cell structures are more complex than those of the prokaryotes. This protist has manufactured "plates" of calcium carbonate that surround and protect its single cell. (D–F) Most of the visible life on Earth is multicellular. Chapters 27 and 28 cover the green plants **(D)**. The other broad groups of multicellular organisms are the fungi **(E)**, discussed in Chapter 29, and the animals **(F)**, covered in Chapters 30–32.

key concept 1.1 Living Organisms Share Similarities and a Common Origin

Intuitively we all know what life is, but try to define it; it isn't easy. You can easily designate the things around you as living or nonliving, but what are the essential differences? We call the living things organisms. In contrast to nonliving things, organisms sustain and renew themselves. The loss of the ability to sustain and renew means the loss of life, and organisms that die become part of the nonliving world. **Biology** is the scientific study of organisms, both living and after death (e.g., the study of fossils) with the goal of discovering and understanding the diversity and the complex processes that make up life.

focus your learning
- Major characteristics are shared among all living things.
- Living organisms have influenced the history of the planet Earth.
- Biological populations change over time.

Life on our planet is quite diverse (**Figure 1.1**), yet its many diverse forms share common features. What characteristics do organisms share that distinguish them from the nonliving world? Most organisms:

- are composed of a common set of chemical compounds: mainly carbohydrates, fatty acids, nucleic acids, and amino acids.

- are made up of cells.
- use molecules obtained from the environment to synthesize new biological molecules.
- extract energy from the environment and use it to do work.
- contain genetic information—genomes—that enables them to develop, maintain themselves, function, and reproduce.
- use a universal molecular code to build proteins from their genomic information.
- regulate their internal environments.
- exist in populations that evolve over time.

How do you think all organisms came to have these similarities? If life had multiple origins, we would not expect to see such striking similarities in chemical composition, cell structure, cell functions, and genetic codes across the living world. Instead, these common characteristics logically lead to the conclusion that all life has a common ancestry, and that the diverse organisms alive today all originated from one life form. Organisms from a separate origin of life—say, on another planet—might be similar in superficial ways to life on Earth, but they would not have the same genetic code, chemical composition, or cellular structures and functions that we see widely shared among living organisms on Earth. All evidence points to a common origin of life on our planet about 4 billion years ago.

Some forms of life may not display all characteristics listed above all of the time. For example, the seed of a desert plant may go for many years without extracting energy from the environment, converting molecules, regulating its internal environment, or reproducing; yet the seed is alive. Viruses present a special case as well. Viruses are not composed of cells and cannot carry out physiological functions on their own. Viruses depend on the cells of host organisms to carry out these functions for them. Yet viruses contain genetic information, and their populations evolve over time, as we know from witnessing changes in the flu viruses each flu season. Even though viruses are not independent cellular organisms, their existence depends on cells, and it is highly probable that viruses evolved from cellular life forms. Thus most biologists consider viruses to be a part of life.

As you go through this book, you will explore details of the common characteristics of life, how these characteristics arose, and how they work together so that organisms survive and reproduce. Because organisms do not all survive and reproduce with equal success, you will see again and again that through differential survival and reproduction populations of organisms evolve and become adapted to Earth's many environments. The processes of evolution have generated the enormous diversity of life on Earth, and evolution is a central theme of biology.

Life arose from non-life via chemical evolution

Earth formed between 4.6 and 4.5 billion years ago, but it was not a place hospitable to life. The cooling of Earth, the formation of surface water, and the evolution of the first life forms took about 600 million years. If we picture the 4.6-billion-year history of Earth as a 30-day month, life first appeared some time around the end of the first week (**Figure 1.2**).

The young Earth's atmosphere, oceans, and climate were very different than they are today, but experiments simulating those conditions have confirmed that the generation of complex molecules through random physical association of chemicals is possible, even inevitable. The critical step for the evolution of life was the appearance of **nucleic acids**—molecules that could reproduce themselves and serve as templates for the synthesis of **proteins**, large molecules with complex but stable shapes. The variation in the shapes of these proteins enabled them to participate in increasing numbers and kinds of chemical reactions with other molecules. These subjects are covered in Part One of this book.

Cellular structure evolved in the common ancestor of life

An important step in the evolution of life was the enclosure of complex proteins and other biological molecules by membranes that

Figure 1.2 Life's Timeline   Depicting the 4.6 billion years of Earth's history on the scale of a 30-day month provides a sense of the immensity of evolutionary time.

(A) Liposomes

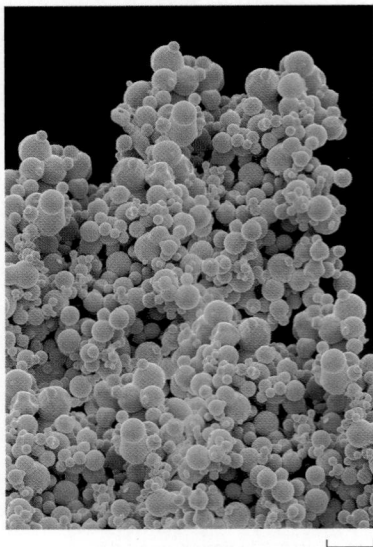

(B) Prokaryotic cells

Cell membrane

200 nm

(C) Eukaryotic cells

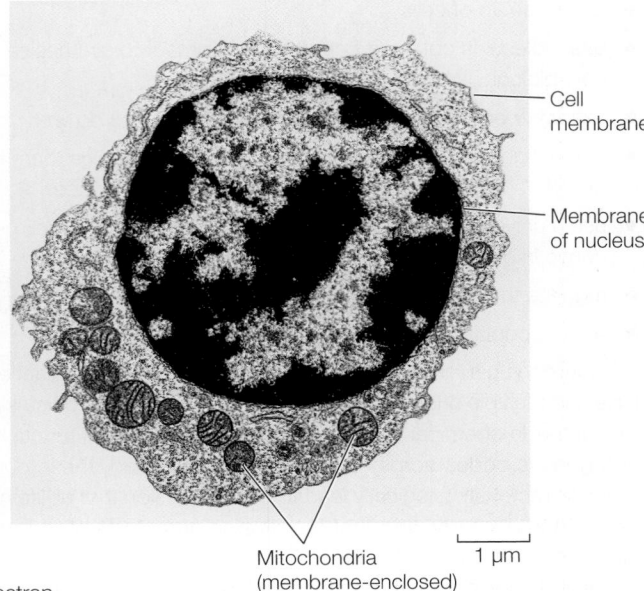

Cell membrane

Membrane of nucleus

Mitochondria (membrane-enclosed)

1 µm

Figure 1.3 Cells Are Building Blocks for Life   These photographs were taken with electron microscopes (see Figure 5.3) and enhanced with added color to highlight details. **(A)** Liposomes are tiny round bubbles known as vesicles and are made out of the same material as a cell membrane. Liposomes can be filled with drugs and used to deliver drugs to cells. **(B)** Two prokaryotic cells of an *Enterococcus* bacterium that lives in the human digestive system. Prokaryotes are unicellular organisms with genetic and biochemical material enclosed inside a single membrane. **(C)** A human white blood cell (lymphocyte) represents one of the many specialized cell types that make up a multicellular eukaryote. Multiple membranes within the cell-enclosing outer membrane segregate the different biochemical processes of eukaryotic cells.

contained them in a compact internal environment separate from the surrounding (external) environment. Fatty acid molecules played a critical role in membrane evolution because these molecules do not dissolve in water. Think of shaking an oil and vinegar salad dressing. The oil breaks up into small droplets, but the droplets do not dissolve in the vinegar and they rapidly coalesce. Similarly, fatty acids can form membranous films on the surface of water. When these films are agitated, they can form spherical structures called **liposomes**. Such structures are now used for delivering drugs to cells (**Figure 1.3A**). In a primordial ocean, such membranous structures could have enveloped assemblages of complex biological molecules. The origin of an internal environment that concentrated reactants and products of chemical reactions led to the first cells with the ability to replicate themselves—the evolution of the first cellular organisms.

For billions of years all the organisms existing on Earth were unicellular and were enclosed by a single outer membrane. Such organisms, like the bacteria that are abundant on you, in you, and all around you, are called **prokaryotes** (**Figure 1.3B**). Two main groups diverged early in life's history: the **Bacteria** and **Archaea**. The third major category of life on earth, the **Eukarya**, arose billions of years later from one group of archaea. In addition to its outer membrane, a eukaryotic cell (**Figure 1.3C**) has internal membranes that enclose specialized internal compartments called organelles. The organelle that gives eukaryotes their name is the **nucleus**—the organelle that contains the cell's genetic information. The word "eukaryote" comes from two Greek words meaning "a true kernel." Other organelles carry out specific functions such as synthesizing biological molecules or providing energy.

How might eukaryotes have arisen from prokaryotes? Infoldings of the prokaryote cell membrane could have formed internal compartments—the organelles—that isolated cell functions from each other, resulting in greater integration or efficiency of cell functions. Another possibility, similar to that seen in the relationship between corals and dinoflagellates described at the beginning of this chapter, is that close, interdependent relationships may have developed between different prokaryote cells, leading to a merger of sorts. Suppose a prokaryote good at converting energy was engulfed (but not digested) by a prokaryote good at synthesizing biological molecules. Each would supply a valuable service for the other, but now one would be an organelle inside the other. The structure of prokaryote and eukaryote cells, their membranes, and their evolution are the subjects of Part Two.

Single-celled organisms were the only forms of life for over half of the history of life on Earth (see Figure 1.2). However, at some point the cells of some eukaryotes didn't separate after cell division, instead they remained attached to each other. Such colonial aggregations of cells made it possible for some of the associated cells to specialize in certain functions, such as reproduction, while other cells specialized in other functions, such as absorbing nutrients or motility. This **cellular specialization** enabled multicellular eukaryotes to increase in size and become more efficient at gathering resources and adapting to specific environments.

Photosynthesis allows some organisms to capture energy from the sun

Living cells require energy to function. The earliest prokaryotes supplied their energy needs—their **metabolism**—by taking in small molecules from their environment, breaking the chemical bonds of these molecules, and using the energy released from those chemical bonds to do cellular work. Many modern prokaryotes still function this way, and they function very successfully. But about 2.5 billion years ago, the emergence of **photosynthesis** changed the nature of life on Earth.

Photosynthesis transforms the energy of sunlight into a form of *chemical energy that can be used to do work such as the synthesis of large molecules. These large molecules can then be used to build cell structures or can be broken down to provide metabolic energy. Photosynthesis is the basis of much of life on Earth today because its energy-capturing processes provide food for other organisms. Early photosynthetic cells were probably similar to present-day prokaryotes called **cyanobacteria** (**Figure 1.4**). Over time, photosynthetic prokaryotes became so abundant that vast quantities of oxygen gas (O_2), which is a by-product of photosynthesis, began to accumulate in the atmosphere.

---

*connect the concepts The pathways that harvest chemical energy to do all the kinds of biological work necessary to support metabolism are presented in Chapter 9.

During the early history of prokaryotic life, there was no O_2 in Earth's atmosphere. In fact, O_2 was toxic to many of the prokaryotes living at that time, and its buildup in the atmosphere resulted in a huge mass extinction. But those organisms that could tolerate O_2 proliferated. Atmospheric O_2 opened up vast new avenues of evolution because **aerobic metabolism**—a biochemical process that uses O_2 to extract energy from nutrient molecules—is far more efficient than **anaerobic metabolism** (which does not use O_2). Most organisms today use aerobic metabolism.

Oxygen in the atmosphere also made it possible for life to move onto land. For most of life's history, UV radiation falling on Earth's surface was so intense that it destroyed any organism that was not shielded by water. The atmospheric accumulation of photosynthetically generated O_2 over a period of more than 2 billion years gradually produced a thick layer of ozone (O_3) in the upper atmosphere. By about 500 million years ago, the ozone layer was sufficiently dense and absorbed enough of the sun's UV radiation to make it possible for organisms to leave the protection of the water and live on land.

Biological information is stored in a genetic code common to all organisms

Before there was a science of biology, humans recognized that offspring resemble parents, and this fact was used by plant and animal breeders to produce variants with desirable qualities. However, not until the famous plant breeding experiments of the Austrian monk Gregor Mendel in the mid-1800s was it demonstrated that inherited traits existed in discrete units (**Figure 1.5**). These discrete units of inheritance were termed **genes** in the early 1900s, giving rise to the science of **genetics**. An enormous body of information about traits of organisms represented by single genes was created before the chemical structure of a gene was known, and hence before the nature of the information that is passed from parent to offspring was known. That mystery was solved in the mid-1900s with the discovery that the molecule deoxyribonucleic acid, **DNA**, is the genetic information that specifies what an organism will look like and how it will function. This "blueprint" for the existence of each individual organism is contained in the sum total of all the DNA

(A) Cyanobacteria

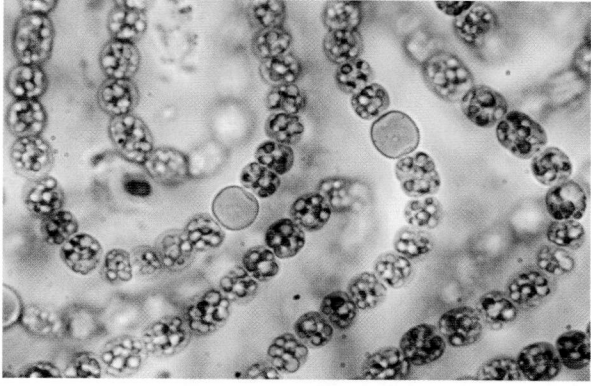

5 μm

(B) Fossilized stromatolite

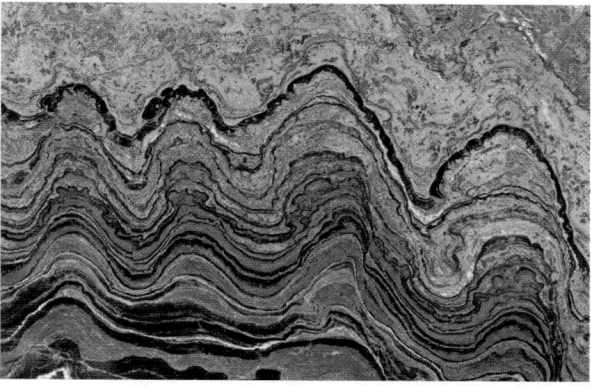

0.5 cm

(C) Living stromatolite

Stromatolites form as small grains of sediment are cemented together by communities of microorganisms, especially **cyanobacteria**.

Figure 1.4 Photosynthetic Organisms Changed Earth's Atmosphere (A) Cyanobacteria are aquatic and photosynthetic: they live in the water and can manufacture their own food. Although they are quite small, they often grow in colonies large enough to see. **(B)** Colonies of photosynthetic cyanobacteria and other microorganisms produced structures called stromatolites that were preserved in the ancient fossil record. This section of fossilized stromatolite reveals layers representing centuries of growth. **(C)** Living stromatolites can still be found in appropriate environments.

molecules contained in each of the organism's cells—its **genome**. DNA molecules are long sequences of four different subunits called **nucleotides**. Genes are specific segments of DNA that encode the

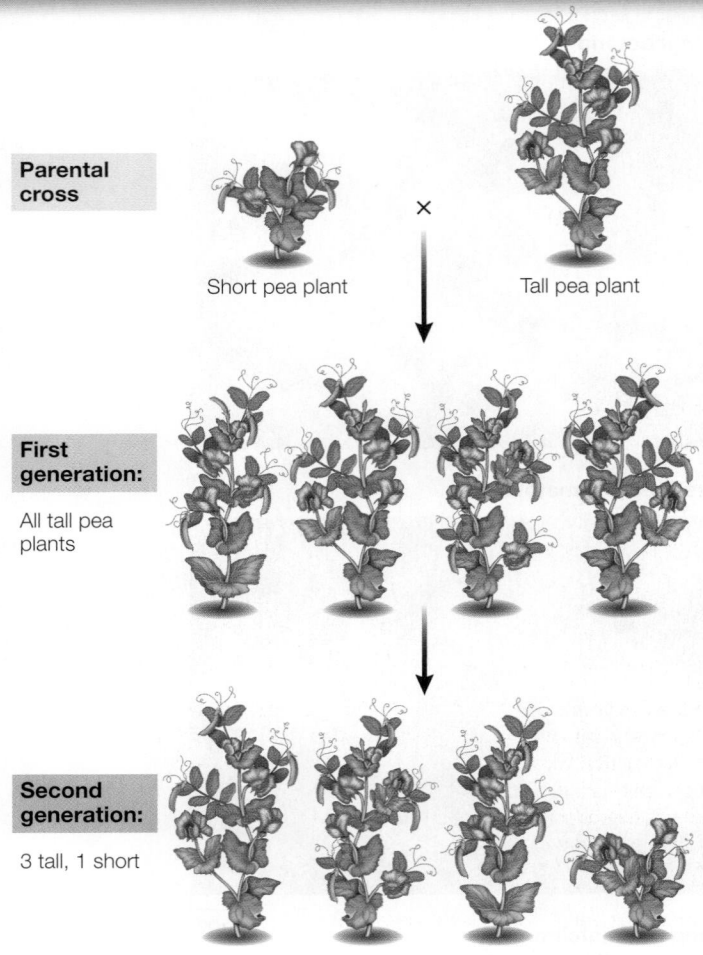

Parental cross

Short pea plant   ×   Tall pea plant

First generation:

All tall pea plants

Second generation:

3 tall, 1 short

Figure 1.5 Genetics Developed as the Study of Discrete Heritable Traits Gregor Mendel discovered the basics of genetics through breeding experiments with pea plants. When he crossed tall plants with short plants, all of the offspring were tall. Then when he crossed these offspring plants, he got one short plant for every three tall plants. From these experiments Mendel concluded that there was a tall factor and a short factor. Each plant inherited one factor from each parent, and the tall factor was dominant.

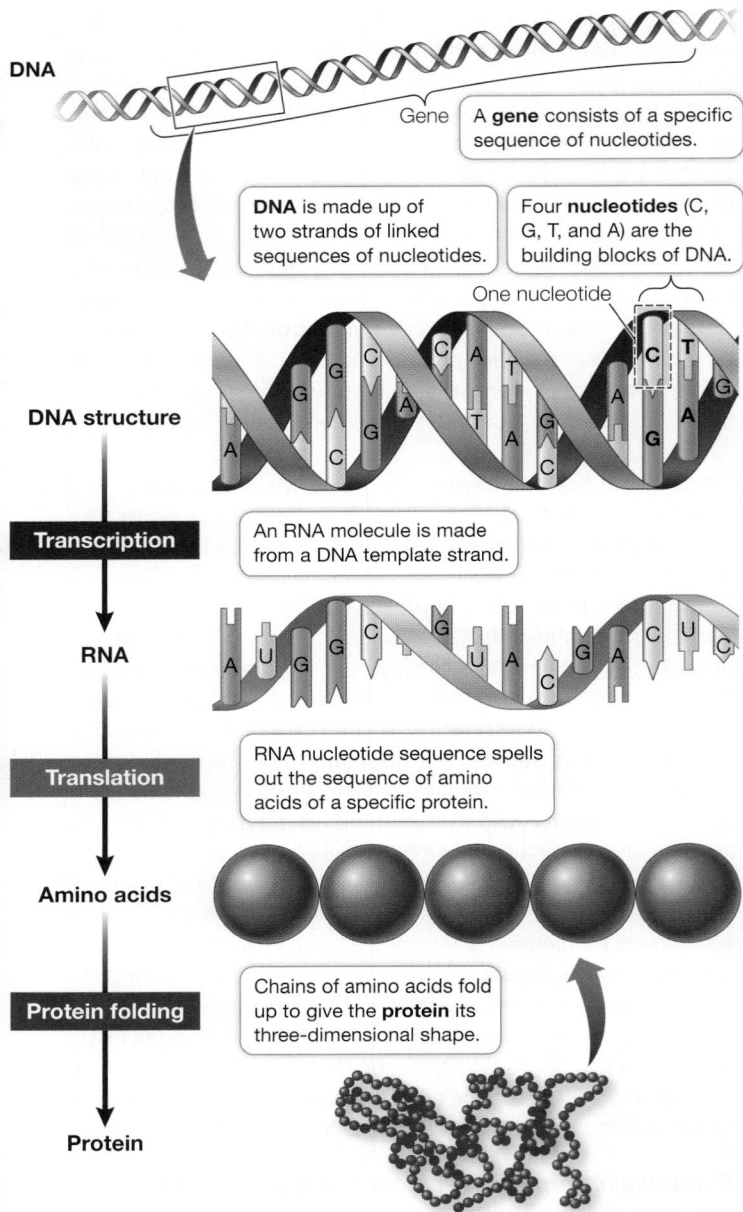

DNA

Gene A **gene** consists of a specific sequence of nucleotides.

DNA is made up of two strands of linked sequences of nucleotides.

Four **nucleotides** (C, G, T, and A) are the building blocks of DNA.

One nucleotide

DNA structure

Transcription An RNA molecule is made from a DNA template strand.

RNA

Translation RNA nucleotide sequence spells out the sequence of amino acids of a specific protein.

Amino acids

Protein folding Chains of amino acids fold up to give the **protein** its three-dimensional shape.

Protein

Figure 1.6 DNA Is Life's Blueprint The instructions for life are contained in the sequences of nucleotides in DNA molecules. Specific DNA nucleotide sequences compose genes. The average length of a single human gene is 16,000 nucleotides. The information in each gene provides the cell with the information it needs to manufacture molecules of a specific protein.

information the cell uses to build proteins. Therefore each gene is defined by a specific sequence of the four nucleotides. The genetic code spells out how sequences of nucleotides are *translated into sequences of amino acids, which are the building blocks of proteins. This translation process involves first **transcribing** some of the DNA information of a gene into the structure of another smaller molecule called **ribonucleic acid** (**RNA**) (**Figure 1.6**). RNA serves as the template for synthesis of a protein. Protein molecules govern the chemical reactions within cells and form much of an organism's structure.

***connect the concepts** How the structure of DNA codes for and is translated into proteins is discussed in Concept 4.1 and in more detail in Chapters 13 and 14.

By analogy with a book, the nucleotides of DNA are like the letters of an alphabet, and protein molecules are the words. Combinations of proteins that form structures and control specific biochemical processes are the sentences and paragraphs. The structures and processes that are organized into different complex systems with specific tasks (such as digestion or movement) are the chapters of the book, and the complete book is the organism. If you were to write out your own genome using four letters to represent the four nucleotides, you would write more than 3 billion letters. Using the size type you are reading now, your genome would fill about 1,000 books the size of this one. The processes of evolution are the authors and editors of all the books in the library of life.

Even though all the cells of a multicellular organism contain essentially the same genome, different cells of the organism have different functions and form different structures—contractile proteins form in muscle cells, hemoglobin in red blood cells, digestive enzymes in gut cells, and so on. We thus surmise that different types of cells in an organism express different parts of the genome. How cells control gene expression to enable a complex organism to develop and function is a major focus of current biological research.

The genome of an organism consists of thousands of genes. This entire genome must be replicated as new cells are produced. However, the replication process is not perfect, and a few errors, known as **mutations**, are likely to occur when the genome is replicated. Mutations occur spontaneously and can be induced by environmental factors such as chemicals and radiation. Most mutations are either harmful or have no effect, but occasionally a mutation improves the functioning of the organism under the environmental conditions it encounters.

The discovery of the structure of DNA by James Watson and Francis Crick in the middle of the twentieth century, and the subsequent elucidation of the mechanisms by which this material encodes and transmits information, transformed biology. These crucial discoveries are detailed in Parts Four and Five of this book.

Populations of all living organisms evolve

A **population** is a group of individuals of the same type of organism that interbreed. The processes of **evolution** result in changes in the genetic makeup of populations through time. Evolution is the major unifying principle of biology. Charles Darwin compiled factual evidence for evolution in his 1859 book *On the Origin of Species*. Darwin argued that differential survival and reproduction among individuals in a population, which he termed *natural selection, could account for much of the evolution of biodiversity.

---

*connect the concepts   The story of Darwin's formulation of evolution by natural selection is covered in Key Concept 20.1.

Although Darwin proposed that all organisms are descended from a common ancestor and therefore are related to one another, he did not have knowledge of the mechanisms of genetic inheritance. Even so, he understood the process of selective breeding of plants and animals to obtain breeds with desired traits. Darwin himself was a pigeon breeder and knew that he could produce pigeons with different colorations, feather patterns, or tail shapes by selecting mates (see Figure 20.5). The steps in such selective breeding can be described as follows:

(A) *Dyscophus guineti*

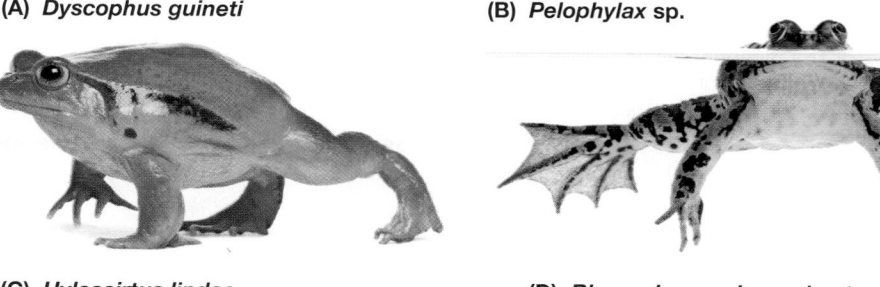

(B) *Pelophylax* sp.

(C) *Hyloscirtus lindae*

(D) *Rhacophorus nigropalmatus*

Figure 1.7 **Adaptations to the Environment**   The limbs of frogs show adaptations to the different environments of each species. **(A)** This terrestrial frog walks across the ground using its short legs and peglike digits (toes). **(B)** Webbed rear feet are evident in this highly aquatic species of frog. **(C)** This arboreal species has toe pads, which are adaptations for climbing. **(D)** A different arboreal species has extended webbing between the toes, which increases surface area and allows the frog to glide from tree to tree.

- Observe trait variation in a population.
- Selectively breed pairs based on a desired trait to produce that trait in their offspring.
- Repeat such crosses in subsequent generations to amplify the trait.

Darwin realized that if humans could select for specific traits in domesticated plants and animals, the same process could operate in nature; hence the term "natural selection" as opposed to artificial (human-imposed) selection. He reasoned that the reproductive capacity of plants and animals, if unchecked, would result in unlimited growth of populations, but we do not observe such growth in nature. In most species, only a small percentage of an individual's offspring survive to reproduce. Thus any trait that confers even a small increase in the probability that its possessor will survive and reproduce increases in the population across generations.

Because organisms with certain traits survive and reproduce best under specific sets of conditions, natural selection leads to **adaptations**: structural, physiological, or behavioral traits that enhance an organism's chances of survival and reproduction in its environment (**Figure 1.7**). The opening story of this chapter described experiments aimed at understanding adaptations that might enable corals to survive in a warmer environment. In addition to natural selection, evolutionary processes such as sexual selection (selection due to mate choice) and genetic drift (the random fluctuation of gene frequencies in a population due to chance events) contribute to biodiversity. These processes operating over evolutionary history have contributed to the remarkable diversity of life on Earth. Part Six describes how natural selection and other evolutionary processes are supported and

Figure 1.8 The Blueprint of Life   Using technologies that reveal the sequence of nucleotides that make up the genome of an organism, we can detect small alterations that are responsible for specific traits. The *EDAR* gene is involved in hair development, and a single nucleotide substitution results in a single amino acid difference between the EDAR proteins of humans with Asian origins and those with African or European origins.

East Asian TCCACGTACAACTCTGAGAAGGCTGTTGTGAAAACGTGGCGCCACCTCGCC
African   TCCACGTACAACTCTGAGAAGGCTGCTGTGAAAACGTGGCGCCACCTCGCC

explained by the massive body of genetic and molecular genetic knowledge mostly produced over the past hundred years.

Over most of the history of biology, the diversity of life was described and catalogued based on structural characteristics of organisms. With the rapid development of molecular genetic tools in recent years, it is now possible to obtain the DNA sequences of parts or all of the genomes of organisms. Now it is possible to compare organisms on the basis of their DNA blueprints, which has given rise to the field of **genomics**. For example, a structural feature that differentiates humans with African or European origins versus East Asian origins is hair. Europeans have fine hair that can be curly, as is the hair of Africans. East Asians have thick hair that is straight. This difference is due to a gene called *EDAR* and to a single nucleotide substitution in that gene resulting in a single amino acid change in the protein product of that gene (**Figure 1.8**).

The ability to sequence the billions of letters in genetic codes generates enormous amounts of data, and to manage and analyze that data, the field of **bioinformatics** has grown in parallel with genomics. Genomics has enormous applications throughout biology, from evolutionary studies to investigations of human health. Advances in genomics and bioinformatics make current initiatives in precision medicine possible.

Biologists trace the evolutionary tree of life

Populations geographically isolated from one another evolve differences. As populations diverge from one another, individuals from one population become less likely to reproduce successfully with individuals of the other populations. Eventually these differences between the populations become so great that they are considered to be different species. Thus species that share a fairly recent evolutionary history are generally more similar

to each other than are species that share an ancestor in the more distant past. By identifying, analyzing, and quantifying similarities and differences among species, biologists can construct **phylogenetic trees** that portray the evolutionary histories of the different groups of organisms. As an example, we can show the evolutionary relationships among humans and our closest living relatives in a branching diagram that shows how they diverged from a common ancestor. In this tree, the living species are shown at the tips of the branches, and the branches show when these groups diverged from one another along a time scale. Our convention in this book will be to place phylogenetic trees on their sides, with the oldest lineages on the left and the most recent to the right.

Tens of millions of species exist on Earth today; many times that number lived in the past but are now extinct. Biologists give

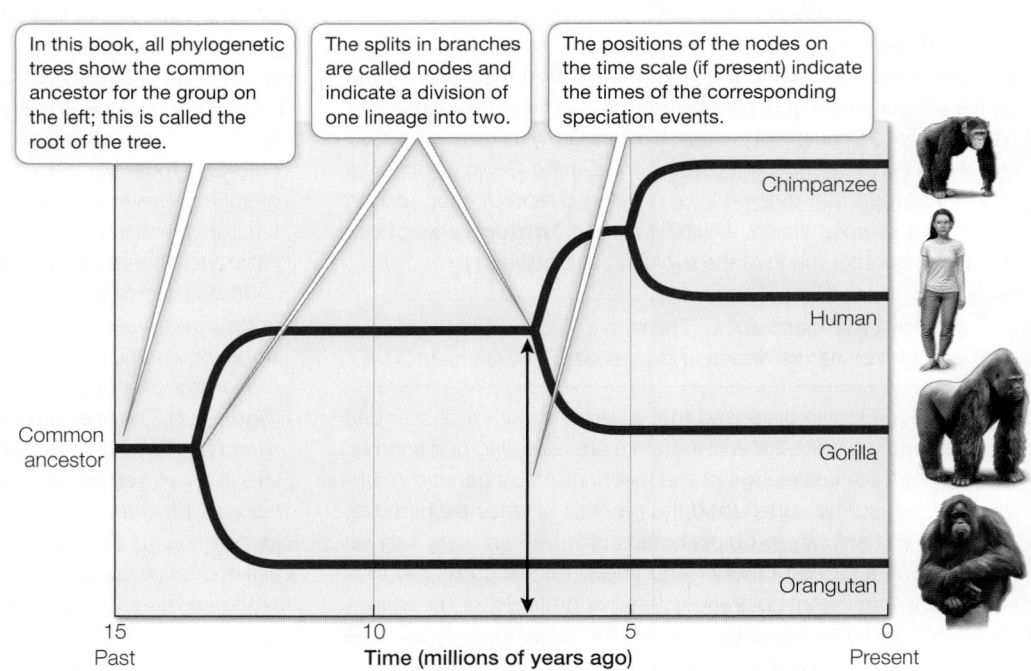

In this book, all phylogenetic trees show the common ancestor for the group on the left; this is called the root of the tree.

The splits in branches are called nodes and indicate a division of one lineage into two.

The positions of the nodes on the time scale (if present) indicate the times of the corresponding speciation events.

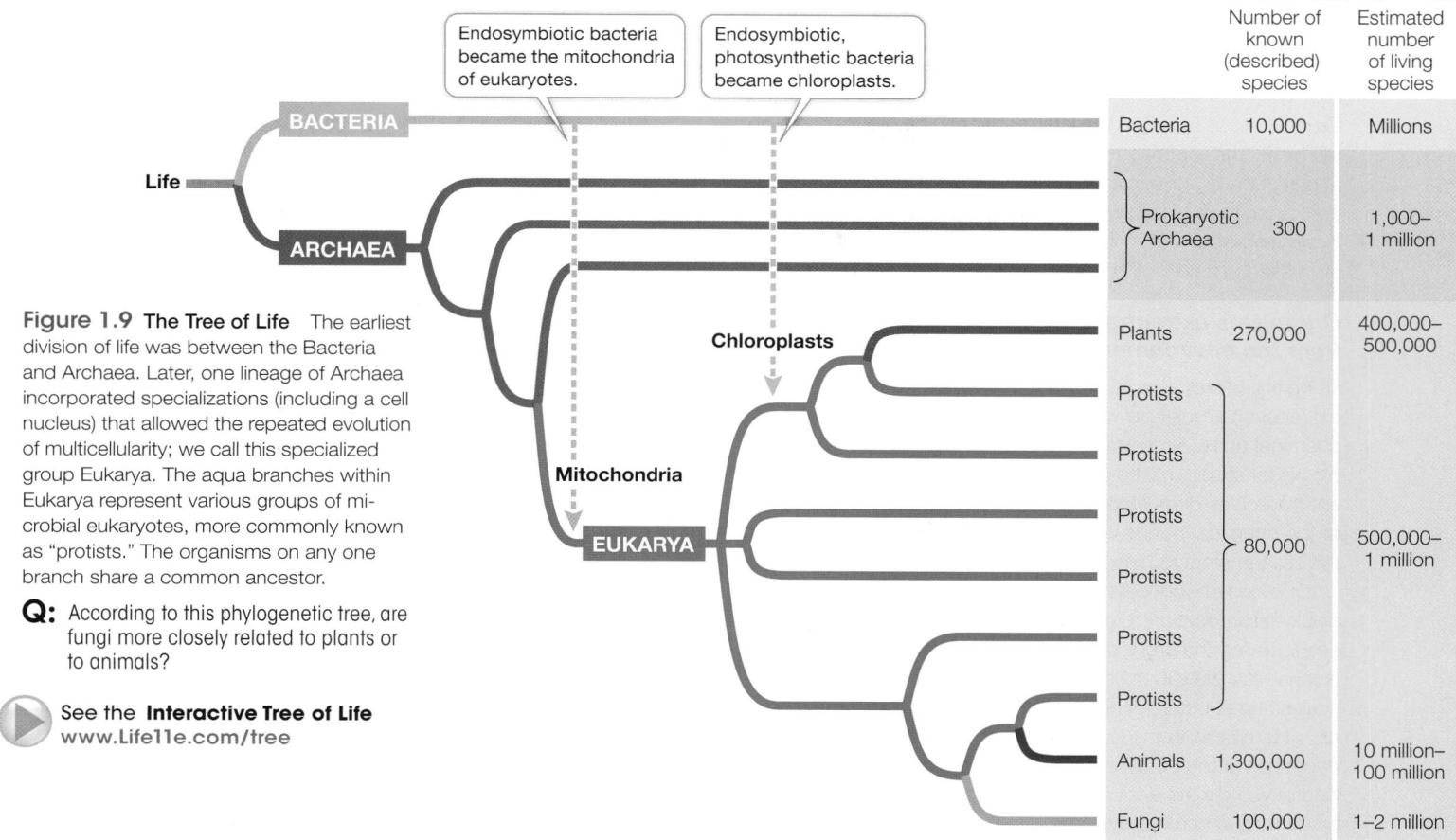

Figure 1.9 The Tree of Life The earliest division of life was between the Bacteria and Archaea. Later, one lineage of Archaea incorporated specializations (including a cell nucleus) that allowed the repeated evolution of multicellularity; we call this specialized group Eukarya. The aqua branches within Eukarya represent various groups of microbial eukaryotes, more commonly known as "protists." The organisms on any one branch share a common ancestor.

Q: According to this phylogenetic tree, are fungi more closely related to plants or to animals?

See the **Interactive Tree of Life**
www.Life11e.com/tree

each of these species a distinctive scientific name formed from two Latinized names—a **binomial**. The first part of the name identifies the species' **genus** (plural *genera*)—a group of species that share a recent common ancestor. The second part of the scientific name identifies a particular species within the genus. For example, the scientific name for modern-day humans is *Homo sapiens*: *Homo* is our genus, *sapiens* refers to our species. *Homo* is Latin for "man," and *sapiens* is from the Latin word for "wise" or "rational." The binomial for our close but extinct relatives, the Neanderthals, is *Homo neanderthalensis*. Note that it is conventional to write the binomial in italics with the genus capitalized but the second part not capitalized.

Much of biology is based on comparisons among species. Meaningful interpretations of those comparisons require an understanding of evolutionary relationships of the respective species. Our ability to reconstruct evolutionary relationships has been greatly enhanced in recent decades by gene sequencing. Genomic sequence analysis and other molecular techniques have enabled biologists to augment evolutionary knowledge based on the fossil record with a vast array of molecular evidence. The result is the ongoing compilation of phylogenetic trees that document and diagram evolutionary relationships. The broadest categories of the tree of life are shown in **Figure 1.9**, and they will be surveyed in more detail in Part Seven. (The tree is expanded in Appendix A, and you can also explore the tree interactively online.)

Although many details remain to be clarified, the broad outlines of the tree of life have been determined. Its branching patterns are based on a rich array of evidence from fossils, structures, metabolic processes, behavior, and molecular analyses of genomes. Recall that the primary division of life—into Archaea and Bacteria—occurred among the early single-celled prokaryotes. Members of these two groups differ so fundamentally that they are believed to have separated into distinct evolutionary lineages early in the history of life. One of the lineages of Archaea incorporated a nucleus and internal, membrane-bound organelles and gave rise to the major group Eukarya. Plants, fungi, and animals are examples of familiar multicellular eukaryotes that evolved independently, from different groups of the unicellular eukaryotes known as **protists**. We know that plants, fungi, and animals had independent origins of multicellularity because each of these three groups is most closely related to different groups of protists, as can be seen from the branching pattern of Figure 1.9.

Cellular differentiation and specialization underlie multicellular life

Single-celled organisms must provide for all of their own needs, but the cells of multicellular organisms can evolve specializations to carry out certain functions exclusively and efficiently because they can depend on other cells to carry out other functions. Thus the cells of a multicellular organism can have different developmental fates. Similar cell types can develop together into **tissues** that accomplish tasks that a single cell cannot. For example, muscle cells develop cellular mechanisms for generating force. A single muscle cell cannot generate much force, but many cells of a muscle tissue can work together to generate considerable force, and working with structural tissues such as bones, they can produce large movements. Different tissue types develop together into **organs** that accomplish specific functions. The heart, brain, and stomach are each composed of several types of tissues, as are the roots, stems, and leaves of plants. Organs whose functions are interrelated can be grouped into **organ systems**; the esophagus, stomach, and intestines, for example, are all part of the digestive system. The hierarchy of biological organization from atom to organism is shown in **Figure 1.10A**. The biology of the two major groups of multicellular organisms, plants and animals, is discussed in detail in Parts Eight and Nine, respectively.

Organisms extract energy and raw materials from the environment

Living organisms acquire nutrients from the environment. Biochemical reactions break down complex nutrient molecules into smaller chemical units. Some of those smaller units are used as building blocks for structures the organism requires. The breakdown of nutrient molecules does more than supply raw materials. The breaking of the chemical bonds of nutrient molecules releases energy that the cell transfers to high-energy molecules that it uses to do work.

One kind of work cells do is mechanical—moving molecules from one cellular location to another, moving whole cells or tissues, or even moving the organism itself (**Figure 1.11A**). The cell also does biochemical work in the building, or synthesis, of new complex molecules and structures from smaller chemical units. For example, we are all familiar with the fact that carbohydrates eaten today may be deposited in the body as fat tomorrow (**Figure 1.11B**). Still another kind of work is the electrical work that is the essence of information processing in nervous systems.

The many biochemical reactions that take place in cells are integrally linked in that the products of one reaction are the raw materials of the next. These complex networks of reactions must be integrated and precisely controlled; when they are not, the result is malfunction and disease.

 Media Clip 1.1 Leaping Lemurs
www.Life11e.com/mc1.1

Living organisms must regulate their internal environment

If different specialized cells, tissues, and organs provide for different needs of the multicellular organism, how are these specific services shared by the cells of the whole organism? All of the cells of the body share an **internal environment** that is made up of extracellular fluids. Cells derive their nutrients from these extracellular fluids, and they deposit their wastes into the extracellular fluid. This internal environment serves the needs of all cells of the body, and therefore its physical and chemical composition must be maintained within a narrow range of physiological conditions that support survival and function. The maintenance of this narrow range of conditions is known as **homeostasis**. A relatively stable internal environment means that cells can function efficiently even when external conditions could not support the lives of individual cells.

***Homeostasis** requires that the activities of the cells and systems of the body be regulated. Regulation requires information—information about internal conditions, external conditions, and what is optimal. Thus organisms must have **sensory mechanisms** to monitor conditions, **effector mechanisms** to alter those conditions, and **signaling mechanisms** to integrate information and enable communication between sensors and effectors. The major informational systems of animals—nervous, hormonal, and immune—use chemical and electrical signals to process information.

***connect the concepts**   The regulation of body temperature is an important example of homeostasis that is discussed in Chapter 39.

The concept of homeostasis also applies to the intracellular environment. Both unicellular and multicellular organisms must regulate the composition of their intracellular environments within a range that allows those cells to survive and function. Individual cells regulate these properties through actions of their membranes and, in the case of eukaryotes, their organelles. Thus self-regulation to maintain a more or less constant internal environment is a general attribute of all life.

Living organisms interact

Organisms do not live in isolation. Besides the internal hierarchy of the individual organism, there is also an external hierarchy of the biological world (**Figure 1.10B**). The populations of all the species that live and interact in a defined area are called a **community**.

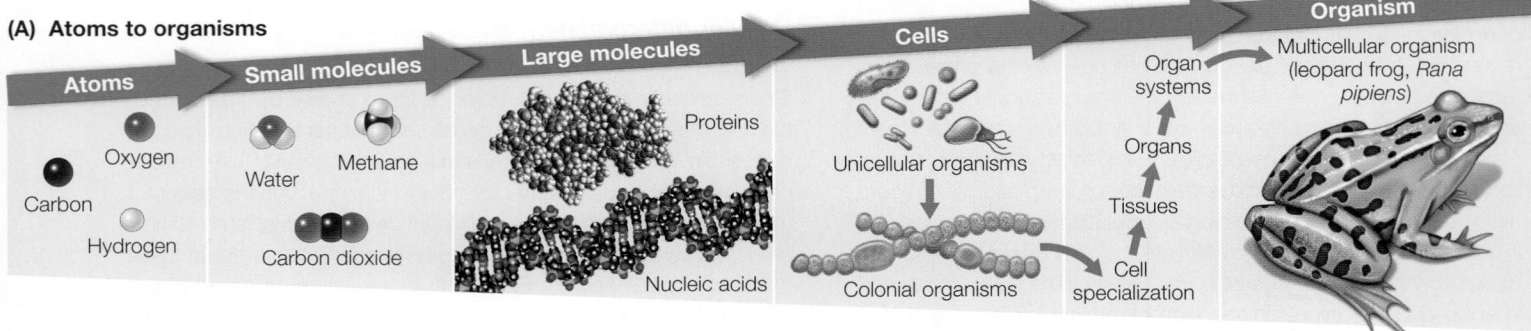

(A) Atoms to organisms

Organism

Cells

Large molecules

Small molecules

Atoms

Carbon
Oxygen
Hydrogen

Water
Methane
Carbon dioxide

Proteins
Nucleic acids

Unicellular organisms
Colonial organisms

Organ systems
Organs
Tissues
Cell specialization

Multicellular organism (leopard frog, *Rana pipiens*)

Figure 1.10 Biology Is Studied at Many Levels of Organization   **(A)** Life's properties emerge when DNA and other molecules are organized in cells, which form building blocks for organisms. **(B)** Organisms exist in populations and interact with other populations to form communities, which interact with the physical environment to make up the many ecosystems of the biosphere.

(A) *Coracias garrulus*

(B) *Spermophilus parryi*

Figure 1.11 Energy Can Be Used Immediately or Stored
(A) Animal cells break down food molecules and use the energy contained in the chemical bonds of those molecules to do mechanical work, such as running, flying, and jumping. This image is a European roller flying with mammal prey in its bill. **(B)** The cells of this Arctic ground squirrel have broken down the complex carbohydrates in the plants it consumed and converted those molecules into fats. The fats are stored in the animal's body to provide an energy supply for the cold months.

Communities together with their abiotic, or physical, environment constitute an **ecosystem**.

Individuals in a population interact in many different ways. Animals eat plants and other animals and compete with other species for food and other resources. Some animals prevent other individuals of their own species from exploiting a resource, be it food, nesting sites, or mates. Animals may also cooperate with members of their own species, forming social units such as a termite colony or a flock of birds. Such interactions have resulted in the evolution of social behaviors such as communication and courtship displays.

Plants also interact with their external biotic and abiotic environment. Plants living on land depend on partnerships with fungi, bacteria, and animals. Some of these partnerships are necessary to obtain nutrients, some to produce fertile seeds, and still others to disperse seeds. Plants compete with each other for light and water and have ongoing evolutionary interactions with the animals that eat them. Through time, many adaptations such as thorns and toxins have evolved that protect plants from predation. Other adaptations such as flowers and fruits help attract the animals that assist in plant reproduction. The interactions of populations of plant and animal species in a community are major evolutionary forces that produce specialized adaptations.

The major ecosystems of Earth that cover broad geographic areas with distinguishing physical features and communities of organisms are known as **biomes**. Examples of biomes include Arctic tundra, coral reefs, and tropical rainforests. All of the biomes on our planet make up the **biosphere**. The ways in which species interact with one another and with their environment in populations, communities, and ecosystems is the subject of ecology, covered in Part Ten of this book.

(B) Organisms to ecosystems

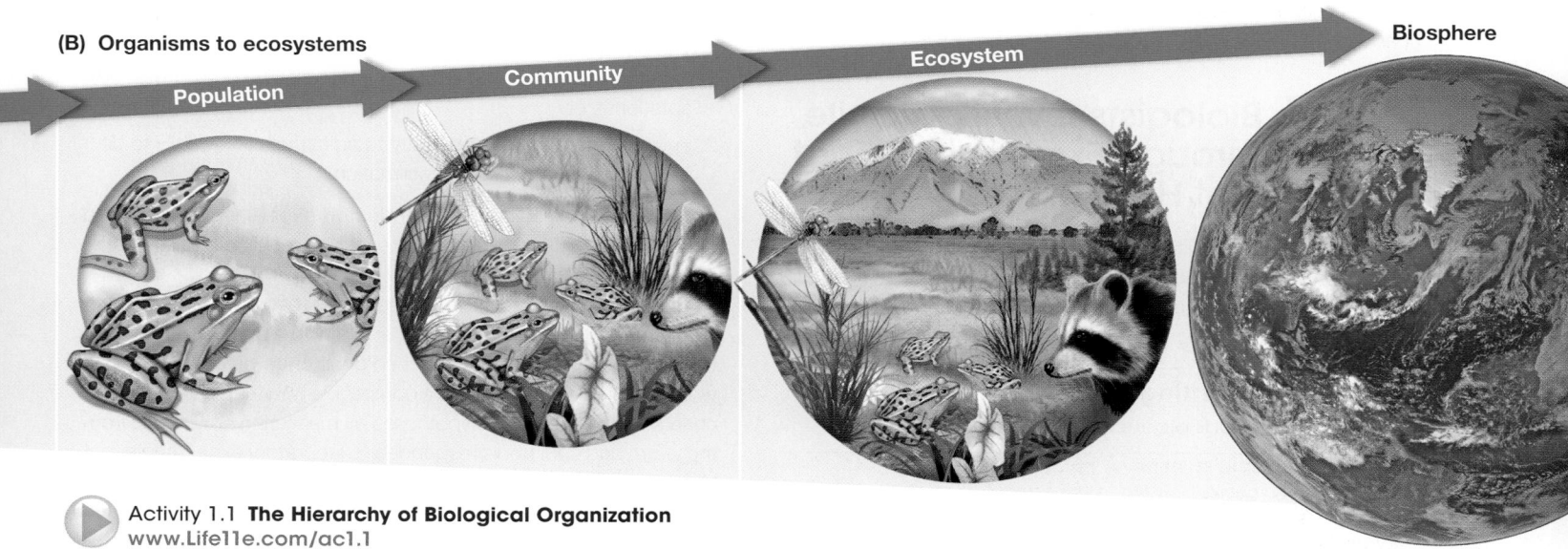

Population   Community   Ecosystem   Biosphere

Activity 1.1 **The Hierarchy of Biological Organization**
www.Life11e.com/ac1.1

▶1.1 recap

All organisms are related by common descent from a single origin of life. They contain genetic information that encodes how they look and how they function. They also reproduce, extract energy from their environment, use energy to do biological work, synthesize complex molecules to build biological structures, regulate their internal environment, and interact with one another. Evolution has produced an enormous diversity of organisms that make up communities and ecosystems. Genomics helps scientists establish the evolutionary relationships between organisms.

learning outcomes

You should be able to:

- Explain how photosynthesis influenced the history of life on Earth.
- Describe evidence that supports the common origin of all life on Earth.
- Explain how natural selection can produce evolutionary change.
- Explain how scientists classify organisms according to evolutionary relationships.

1. How were atmospheric oxygen levels related to the colonization of terrestrial environments?

2. If we discovered life on another planet, how could we tell if it had a separate origin from life on Earth?

3. Assume a population of fish becomes isolated in a dark cave environment, with no access to light from the surface. What major change might you expect in the population over time, and why?

4. What information do biologists use to construct phylogenetic trees?

The preceding section outlined the major features of life—features that we will cover in depth in subsequent chapters of this book. Before going into the details of what we know about life, however, it is important to understand how scientists obtain information and how they use that information to broaden our understanding of life and to put that understanding to practical use.

▶key concept

1.2 Biologists Investigate Life through Experiments That Test Hypotheses

Scientific investigations are based on observation, experimentation, data analysis, and logic. Scientists use many different tools and methods in making observations, collecting data, experimenting, analyzing data, and applying logic.

focus your learning

- Scientific methods produce biological knowledge.

Observing and quantifying are important skills

Biologists observe the world around them. Our ability to observe has been greatly enhanced by technologies such as electron microscopes, rapid genome sequencing, magnetic resonance imaging, and global positioning satellites. These technologies allow us to observe everything from the distribution of molecules in the body to the movement of animals across continents and oceans.

Observation is a basic tool of biology, but as scientists we must be able to quantify the information, or **data**, we collect and observe. Whether we are testing a new drug or mapping the migrations of whales, it is essential to apply mathematical and statistical methods to the data we collect. In our opening discussion of the responses of corals to temperature, it was essential for Rachael to measure temperatures accurately and to have a physical measure of bleaching so she could compare the responses of her corals to different conditions and conclude that the differences were more than just random variation.

Scientific methods combine observation, experimentation, and logic

Textbooks often describe "*the* scientific method," as if there were a single, simple flow chart that all scientists follow. This is an oversimplification. Although flow charts such as the one shown in **Figure 1.12** incorporate much of what scientists do, you should not conclude that scientists necessarily progress through the steps of the process in one prescribed, linear order.

Observations lead to questions. To answer those questions, scientists make additional observations, formulate possible answers, and do experiments to test those possibilities. This investigative approach traditionally has five steps: (1) making observations; (2) asking questions; (3) forming hypotheses, which are tentative answers to the questions; (4) making predictions based on the hypotheses; and (5) testing the predictions by making additional observations or conducting experiments.

After posing a question, a scientist often uses **inductive logic** to propose a tentative answer. Inductive logic involves taking observations or facts and formulating a new proposition that is compatible with those observations or facts. Such a tentative proposition is a **hypothesis** (plural *hypotheses*). For example, in the opener to this chapter, you learned that Rachael Bay observed corals growing in pools that reached temperatures known to kill corals. She formulated two hypotheses:

1. The populations of corals in the warm pools have evolved genetic differences that enable them to survive heat stress (adaptations), or

2. Individual corals have the ability to adjust physiologically to different thermal conditions (acclimation).

Rachael and her colleagues conducted experiments and made observations to test these hypotheses. How do scientists design experiments to test hypotheses?

The next step in the scientific method is to apply a different form of logic—**deductive logic**—that starts with a statement believed to be true (the hypothesis) and then goes on to predict what facts would also have to be true to be compatible with that statement. Rachael's team tested their hypotheses by measuring how corals from the warm and cool pools responded to laboratory experiments that simulated the heat-stress conditions that occurred sometimes in the warm pools. They predicted that the cool-pool corals would bleach at higher rates under the experimental conditions than the warm-pool corals would. That experiment revealed differences between

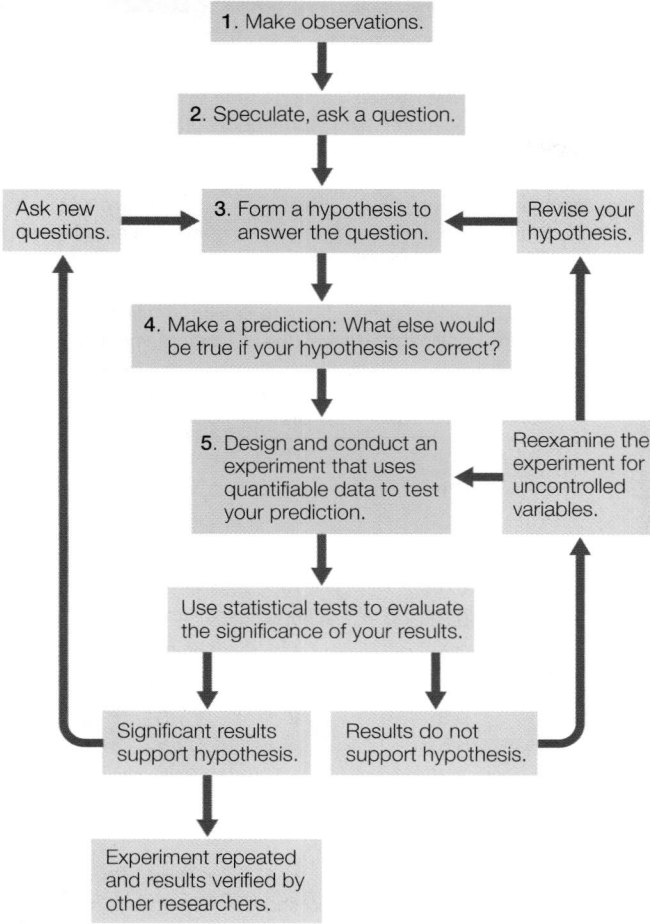

1. Make observations.

2. Speculate, ask a question.

Ask new questions.

3. Form a hypothesis to answer the question.

Revise your hypothesis.

4. Make a prediction: What else would be true if your hypothesis is correct?

5. Design and conduct an experiment that uses quantifiable data to test your prediction.

Reexamine the experiment for uncontrolled variables.

Use statistical tests to evaluate the significance of your results.

Significant results support hypothesis.

Results do not support hypothesis.

Experiment repeated and results verified by other researchers.

Figure 1.12 Scientific Methodology The process of observation, speculation, hypothesis, prediction, and experimentation is a cornerstone of modern science, although scientists may initiate their research at several different points. Answers gleaned through experimentation lead to new questions, more hypotheses, further experiments, and expanding knowledge.

the two populations, but did not reveal whether the differences were a result of long-term evolutionary changes in the populations (adaptations) or short-term acclimation of individuals to their current environments. The researchers therefore looked for genetic differences in the corals living in the warm and cool pools (adaptations). They also looked for differences in gene expression patterns when the corals were transplanted to different environments (acclimation). Did the two populations have fundamental genetic differences, or did the environments they lived in stimulate different patterns of gene expression and therefore physiological acclimation?

 Animation 1.1 Using Scientific Methodology
www.Life11e.com/a1.1

Good experiments have the potential to falsify hypotheses

Once predictions are made from a hypothesis, experiments can be designed to test those predictions. The most informative experiments are those that have the ability to show that the prediction is wrong. If the prediction is wrong, the hypothesis must be questioned, modified, or rejected.

There are two general types of experiments, both of which compare data from different groups or samples:

1. **Controlled experiments** manipulate one factor of interest while holding other variables constant as a means of testing the influence of the manipulated variable.

2. **Comparative experiments** compare data gathered from different populations that differ in multiple, unknown ways.

Rachael Bay and her colleagues conducted both types of experiments, to reveal genetic differences between the coral populations as well as their abilities to acclimate to different temperature environments.

In a controlled experiment, we predict on the basis of our hypothesis that some critical factor, or variable, has an effect on the phenomenon we are investigating. We devise some method to manipulate *only that variable* in an "experimental" group and compare the resulting data with data from an unmanipulated "control" group. If the predicted difference occurs, we then apply statistical tests to ascertain the probability that the manipulation caused the difference (as opposed to the difference being the result of random chance). **Investigating Life: Corals in Hot Water** describes a controlled experiment examining the effect of pool of origin on the response of corals to heat stress done by Rachael Bay and other members of the Palumbi laboratory. A good controlled experiment is not easy to design because biological variables are so interrelated that it is sometimes difficult to alter just one. In this experiment, the controlled variable was whether the experimental corals came from the cool-pool or the warm-pool environment. They were then subjected to the same thermal stresses and their responses measured as extent of bleaching. The control groups were corals not exposed to the thermal stress protocol.

A comparative experiment starts with the prediction that there will be a difference between samples or groups based on the hypothesis. Unlike in a controlled experiment, we cannot control all the variables in a comparative experiment; often we cannot even identify all the variables involved. We are simply gathering and comparing data from different sample groups. Rachael and her colleagues did a comparative experiment in addition to the controlled experiment shown in Investigating Life: Corals in Hot Water. They analyzed and compared the genetic makeup of corals taken from the warm and cool pools.

Statistical methods are essential scientific tools

Whether we do comparative or controlled experiments, we have to decide whether there is a real difference between the samples, individuals, groups, or populations in the study. How do we decide whether a measured difference is enough to support or falsify a hypothesis? In other words, how do we decide in an unbiased, objective way that the measured difference is significant?

Significance is measured with statistical methods. Random variation is almost always present in any set of measurements. Statistical tests calculate the probability that the differences observed in an experiment could be due to random variation. The results of statistical tests are therefore probabilities. A statistical test starts with a **null hypothesis**—the premise that observed differences are the result of random variation that arises from drawing two finite samples from

experiment

Original Paper: Palumbi, S. R., D. J. Barshis, N. Traylor-Knowles and R. A. Bay. 2014. Mechanisms of reef coral resistance to future climate change. *Science* 344: 895–898.

Rachael Bay and her colleagues did a controlled experiment by bringing corals from the warm and the cool tidal pools into the laboratory and then subjecting them to cycles of heat stress that simulated the conditions that sometimes occurred in the warm tidal pools. The outcome measure was bleaching—the loss of photosynthetic symbionts (dinoflagellates) from the coral bodies.

HYPOTHESIS ▶ Heat stress results in coral bleaching, but corals from warm pools are less subject to bleaching under heat stress than are corals from cool pools.

METHOD

1. Transplant corals from warm and cool pools into laboratory aquaria maintained at a safe temperature.
2. Subject experimental corals to heat stress (the experimental condition) by simulating the daily temperature cycle in the warm pools.
3. Measure coral bleaching (the dependent variable) as the ratio of chlorophyll (the photosynthetic pigment of the symbionts) in the corals exposed to heat stress versus in the controls in non-stressed conditions. A ratio of less than 1.0 indicates higher bleaching in the heat-stressed corals than in the controls.

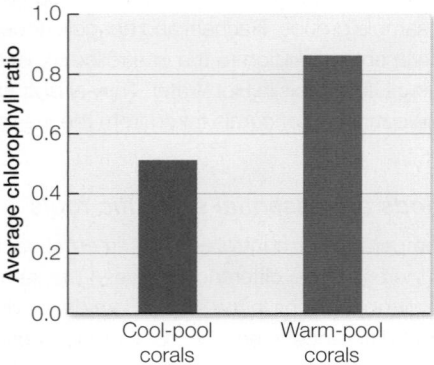

RESULTS Corals from both cool pools and warm pools exhibited higher bleaching under heat stress than controls did, but the cool-pool corals showed higher average levels of bleaching than the warm-pool corals did. (You will calculate the statistical significance of this difference in the following work with the data exercise.)

CONCLUSION ▶ The warm-pool corals showed lower effects of heat stress than the cool-pool corals did, although both populations showed increased bleaching under heat stress. The difference in the populations could be explained either by genetic adaptation to heat stress in the populations of warm-pool corals or by physiological acclimatization of individuals exposed to warm-pool conditions, or by some combination of the two.

work with the data

The bar graph in the Results section of the experiment shows that both cool-pool corals and warm-pool corals showed increased levels of bleaching under heat stress, relative to controls. It also shows that the cool-pool corals showed greater relative bleaching under heat stress than did the warm-pool corals. But how do we know that these differences are greater than would be expected if the variation within and between samples represents nothing more than random variation? Here we will use some simple tests to see if the differences measured are significant (unlikely due to random variation).

Average chlorophyll ratio	Cool-pool corals	Warm-pool corals
0.0	0	0
0.1	0	0
0.2	1	0
0.3	1	0
0.4	4	0
0.5	7	0
0.6	1	0
0.7	1	1
0.8	2	3
0.9	0	3
1.0	0	1

QUESTIONS ▶

1. Prepare a bar graph based on these data. Display the chlorophyll ratios on the *x* axis, and show the number of corals that exhibited each ratio on the *y* axis.
2. First, let's consider if there is a significant effect of heat stress in each of the two groups. At the outset we assume that there is no effect of heat stress in either group—this is called our null hypothesis (H_0). We use statistical tests to decide if we can reject the null hypothesis and conclude that there *is* an effect of heat stress. If the null hypothesis is true, then we would expect that any differences in chlorophyll levels between control and experimental populations are due to

the same population. When quantified observations, or data, are collected, statistical methods are applied to those data to see if there is sufficient evidence to reject the null hypothesis.

More specifically, statistical methods tell us the probability of obtaining the same results by chance even if the null hypothesis were true. *We need to eliminate, insofar as possible, the chance that any differences showing up in the data are merely the result of random variation in the samples tested.* Scientists generally conclude that the differences they measure are significant if statistical tests show that the probability of error (that is, the probability that a difference as large as the one observed could be obtained by mere chance) is 5 percent or lower. They often use more stringent criteria for rejecting the null hypothesis, however, random variation, and so we would expect the chlorophyll ratios to be greater than 1.0 as often as they are less than 1.0. In other words, the expected distribution of values < 1 and > 1 is like the expected distribution of heads and tails in a coin toss. Let's first consider the cool-pool corals. There are 17 samples, and all 17 had chlorophyll ratios < 1.0. The probability of this happening at random is the same as getting 17 heads in 17 flips of a fair coin (by a "fair coin" we mean the probability of heads in one flip = 0.5). Using this criterion, calculate the probability of the null hypothesis (H_0: no effect of heat stress on bleaching) for each of the two populations.

3. Now let's consider if there is a significant difference in the level of bleaching in the cool-pool versus the warm-pool corals. Our null hypothesis in this case is that the levels of bleaching are the same in both groups. If the null hypothesis is true, then we would expect (on average) the same distribution of chlorophyll ratios in the samples of both groups. From the Results section in the experiment, we see that the average chlorophyll ratio observed in the cool-pool samples is 0.5, and the average chlorophyll ratio observed in the warm-pool samples is 0.85. Would we expect a difference this great if the distributions of the two samples were drawn from the same underlying distribution of values? To find out, write each of the 25 observed chlorophyll ratios (from both groups) on an index card. Thoroughly mix and shuffle the cards to randomize their order, and then deal them into two groups of the same size as the cool-pool samples (17) and warm-pool samples (8). Calculate the average ratios in these randomized samples. Repeat this procedure or combine the results of your randomization test with those of other students in your class. How often do you see a difference as great as the observed difference in the two samples (0.85–0.5, or a difference of 0.35)?

4. What do these results suggest about the possible response of coral populations in different environments to ocean warming?

A similar **work with the data** exercise may be assigned in LaunchPad.

depending on the consequences of accepting a wrong hypothesis. Appendix B of this book offers a short primer on statistical methods to which you can refer as you analyze data that will be presented throughout the text.

Discoveries in biology can be generalized

All life is related by descent from a common ancestor, shares a genetic code, and consists of similar biochemical building blocks. Therefore knowledge gained from investigations of one type of organism can, with thought and care, be generalized to other organisms. Biologists use **model systems** for research, knowing that they can extend their findings from such systems to other organisms. For example, our basic understanding of the chemical reactions in cells came from research on bacteria but is applicable to all cells, including those of humans. Similarly, the biochemistry of photosynthesis—the process by which all green plants use sunlight to produce biological molecules—was worked out largely from experiments on *Chlorella*, a unicellular green alga. Much of what we know about the genes that control plant development is the result of work on *Arabidopsis thaliana*, a relative of the mustard plant. Knowledge about how animals, including humans, develop has come from work on sea urchins, frogs, chickens, roundworms, mice, and fruit flies. Being able to generalize from model systems is a powerful tool in biology.

Not all forms of inquiry are scientific

Science is a unique human endeavor that has certain standards of practice. Other areas of scholarship share with science the practice of making observations and asking questions, but scientists are distinguished by what they do with their observations, how they frame their questions, and how they determine their answers. Quantifiable data, subjected to appropriate statistical analysis, are critical in evaluating hypotheses. The Investigating Life boxes and the Work with the Data exercises you will find throughout this book are intended to reinforce this way of thinking. *Scientific observation, hypothesis generation, and experimental testing constitute the most powerful approach humans have devised for learning about the world and how it works.*

Scientific explanations for natural processes are objective and reliable because *a hypothesis must be testable* and *a hypothesis must have the potential of being rejected* by direct observations and experiments. Scientists must clearly describe the methods they use to test hypotheses so that other scientists can repeat their results. Not all experiments are repeated, but surprising or controversial results are always subjected to independent verification. Scientists worldwide share this process of testing and rejecting hypotheses, contributing to a common body of scientific knowledge.

If you understand the methods of science, you can distinguish science from non-science. Art, music, and literature all contribute to the quality of human life, but they are not science. They do not use scientific methods to establish what is fact. Religion is not science, although religions have historically attempted to explain natural events ranging from unusual weather patterns to crop failures to human diseases. Most such phenomena that at one time were mysterious can now be explained in terms of scientific principles. Fundamental tenets of religious faith, such as the existence of a supreme deity or deities, cannot be confirmed or refuted by experimentation and are thus outside the realm of science.

The power of science derives from strict objectivity and absolute dependence on evidence based on *reproducible and quantifiable observations*. A religious or spiritual explanation of a natural phenomenon may be coherent and satisfying for the person holding that view, but it is not testable and therefore it is not science. To invoke a supernatural explanation (such as a "creator" or "intelligent designer" with no known bounds) is to depart from the world of science. Science does not necessarily say that religious beliefs are wrong; they are simply not part of the world of science and are untestable using scientific methods.

Science describes how the world works. It is silent on the question of how the world "ought to be." Many scientific advances that contribute to human welfare also raise major ethical issues. Recent developments in genetics and developmental biology may enable us to select the sex of our children, to use stem cells to repair our bodies, and to modify the human genome. Although scientific knowledge may enable us to do these things, science cannot tell us whether or not we *should* do so, or if we choose to do them, how we should regulate them. Such issues are as crucial to human society as the science itself, and a responsible scientist does not lose sight of these questions or neglect the contributions of the humanities and social sciences in attempting to come to grips with them.

1.2 recap

Scientific methods of inquiry start with the formulation of hypotheses based on observations and data. Comparative and controlled experiments are carried out to test hypotheses. Statistics are essential for drawing conclusions from the data obtained in experiments. Because organisms can be compared through their evolutionary relationships, the results of experiments can frequently be generalized to apply to other species.

learning outcomes

You should be able to:

- Explain the roles of inductive logic and deductive logic in scientific inquiry.
- Describe how a controlled experiment is designed.
- Differentiate between controlled experiments and comparative experiments.
- Explain why model organisms are used in biological research.

1. What is the relationship between a hypothesis and an experiment?
2. What is controlled in a controlled experiment?
3. What features characterize questions that can be answered using a comparative approach rather than through a controlled experiment?
4. Why is it often possible to generalize the results of biological research on one species to very different species?

The vast body of scientific knowledge accumulated over centuries of human civilization allows us to understand and manipulate aspects of the natural world in ways that no other species can. These abilities present us with challenges, opportunities, and above all, responsibilities.

key concept

1.3 Understanding Biology Is Important for Health, Well-Being, and Public-Policy Decisions

Human beings exist in and depend on a world of living organisms. The oxygen we breathe is produced by the photosynthetic activity of countless billions of individual organisms. The food that fuels our bodies comes from the tissues of other living organisms. The fuels that power our cars, trucks, and planes are carbon molecules produced by living organisms—mostly millions of years ago. Inside and out, our bodies are covered in complex communities of living unicellular organisms. Some help us be healthy, but others can invade our bodies and cause mild to serious diseases, or even death. These interactions with other species are not limited to humans. Ecosystems depend on thousands of complex interactions among the millions of species that inhabit Earth. Understanding biological principles is essential for us to lead healthy productive lives and to be good stewards of our planet.

focus your learning

- Biology contributes many things to society.

Modern agriculture depends on biology

Agriculture represents some of the earliest human applications of biological principles. Beginning about 10,000 years ago, humans began to cultivate and harvest grains. It is probable that even in those early times farmers selected the most productive or otherwise favorable plants and animals to use as seed stock for propagation, and over generations farmers continued and refined these practices. By 1960, the worldwide yield of food grains reached a billion tons, but human population growth outpaced agricultural productivity and starvation was common in developing countries. Catastrophic famine was predicted for the 1970s. In response to those predictions, developed countries initiated huge efforts to produce genetically improved, higher-yielding food grain varieties. These highly successful research programs resulted in what has been called the Green Revolution, and by the year 2000 the worldwide yield of food grains, which make up about 50 percent of the calories in the human diet, reached 2 billion tons. Increasing knowledge of plant and animal biology has transformed agriculture in many ways and has resulted in huge boosts in food production.

A new threat to agriculture and the world food supply is global climate change. Some regions are becoming wetter and others drier, and those changes affect traditional agricultures. An example is rice, which normally grows in shallow water in rice paddies (**Figure 1.13**). But, common rice varieties cannot survive complete submergence for more than a few days, and are therefore killed by flooding caused by too much rain. In India and Bangladesh alone, annual flooding can destroy about 4 million tons of rice, enough to feed 30 million people. In recent years plant geneticists Pam Ronald, David Mackill, and their colleagues at the University of California at Davis have discovered a single gene from an ancient variety of rice that when genetically engineered into the current rice varieties makes them capable of surviving total submergence for many days. Advances such as this may help save the benefits of the Green Revolution from the new problems generated by global climate change.

Over the past few decades, detailed knowledge of the genomes of many domestic species and the development of technology for directly recombining genes have allowed biologists to develop new breeds and strains of animals, plants, and fungi of agricultural interest. New strains of crop plants are being developed that are resistant to pests or tolerant of drought or, in the case of rice, flooding. Moreover, understanding evolutionary theory allows biologists to devise

Figure 1.13 A Green Revolution Advances in agricultural plant genetics have vastly increased yields and nutritional value of crops such as the food grains that make up about 50 percent of calories in the human diet. New molecular genetic technologies make even greater advances possible. The strains of rice on the right have been modified to increase resistance to prolonged flooding; the rice plants on the left are unmodified, and have died from over-exposure to flood conditions.

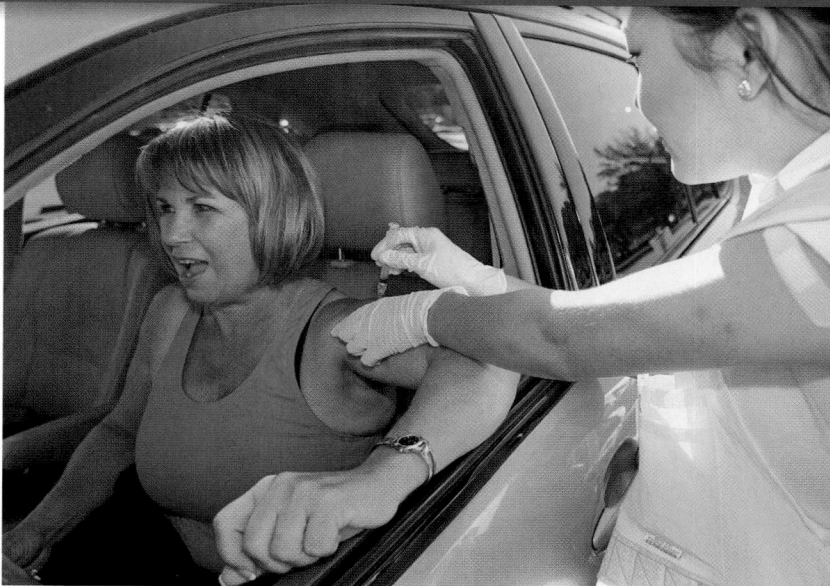

Figure 1.14 Medical Applications of Biology Improve Human Health Vaccination to prevent disease is a biologically based medical practice that began in the eighteenth century. Today evolutionary biology and genomics provide the basis for constant updates to vaccines that protect humans from virus-borne diseases such as flu and measles. In the developed world, vaccinations are so commonplace that some are offered on a "drive-through" basis.

strategies for the application of pesticides that minimize the evolution of pest resistance. And better understanding of plant–fungus relationships results in better plant health and higher productivity. These are just a few of the many ways that biology continues to inform and improve agricultural practice.

Biology is the basis of medical practice

People have speculated about the causes of diseases and searched for methods to combat them since ancient times. Long before the microbial causes of many diseases were known, people recognized that infections could be passed from one person to another, and the isolation of infected persons has been practiced as long as written records have been available.

Modern biological research informs us about how living organisms work, and about why they develop the problems and infections that we call disease. In addition to diseases caused by infectious organisms, we now know that many diseases are genetic—meaning that variants of genes in our genomes cause particular problems in the way we function. Developing appropriate treatments or cures for diseases depends on understanding the origin, basis, and effects of these diseases, as well as understanding the consequences of any changes we make. For example, the recent resurgence of tuberculosis is the result of the evolution of bacteria that are resistant to antibiotics. Dealing with future tuberculosis epidemics will require understanding aspects of molecular biology, physiology, microbial ecology, and evolution—in other words, many of the general principles of modern biology.

Many of the microbial organisms that are periodically epidemic in human populations have short generation times and high mutation rates. For example, we need yearly vaccines for flu because of the high rate of evolution of influenza viruses, the causative agent of flu. Evolutionary principles help us understand how influenza viruses are changing, and can even help us predict which strains of influenza virus are likely to lead to future flu epidemics. This medical understanding—which combines an application of molecular biology, evolutionary theory, and basic principles of ecology—allows medical researchers to develop effective vaccines and other strategies for the control of major epidemics (**Figure 1.14**).

Biology can inform public policy

Thanks to the deciphering of genomes and our newfound ability to manipulate them, vast new possibilities now exist for controlling human diseases and increasing agricultural productivity—but these capabilities raise ethical and policy issues. How much and in what ways should we tinker with the genes of humans and other species? Does it matter whether the genomes of our crop plants and domesticated animals are changed by traditional methods of controlled breeding or by the biotechnology of gene transfer? What rules should govern the release of genetically modified organisms into the environment? Science alone cannot provide all the answers, but wise policy decisions must be based on accurate scientific information.

Biologists are increasingly called on to advise government agencies concerning the laws, rules, and regulations by which society deals with the increasing number of challenges that have a biological basis. As an example of the value of scientific knowledge for the assessment and formulation of public policy, consider a management problem. Scientists and fishermen have long known that Atlantic bluefin tuna (*Thunnus thynnus*) have a western breeding ground in the Gulf of Mexico and an eastern breeding ground in the Mediterranean Sea (**Figure 1.15**). Overfishing led to declining numbers of bluefin tuna, especially in the western-breeding populations, to the point of these populations being endangered.

Initially it was assumed by scientists, fishermen, and policy makers alike that the eastern and western populations had geographically

(A)

Figure 1.15 Bluefin Tuna Do Not Recognize Boundaries
(A) Marine biologist Barbara Block attaches computerized data-recording tracking tags to a live bluefin tuna before returning it to the Atlantic Ocean, where its travels will be monitored. **(B)** Tracking data from eastern (Mediterranean) and western (Gulf of Mexico) breeding populations of bluefin tuna. At one time commercial fishing regulatory agencies assumed that bluefins from western- and eastern-breeding populations fed on their respective sides of the Atlantic. To speed recovery of the endangered western population, fishing quotas were created for each side of the mid-Atlantic Ocean (dashed line).

Q: Based on Dr. Block's data, does it seem likely that fishing quotas based on the mid-Atlantic boundary line would protect the western-breeding population of bluefin tuna?

(B)

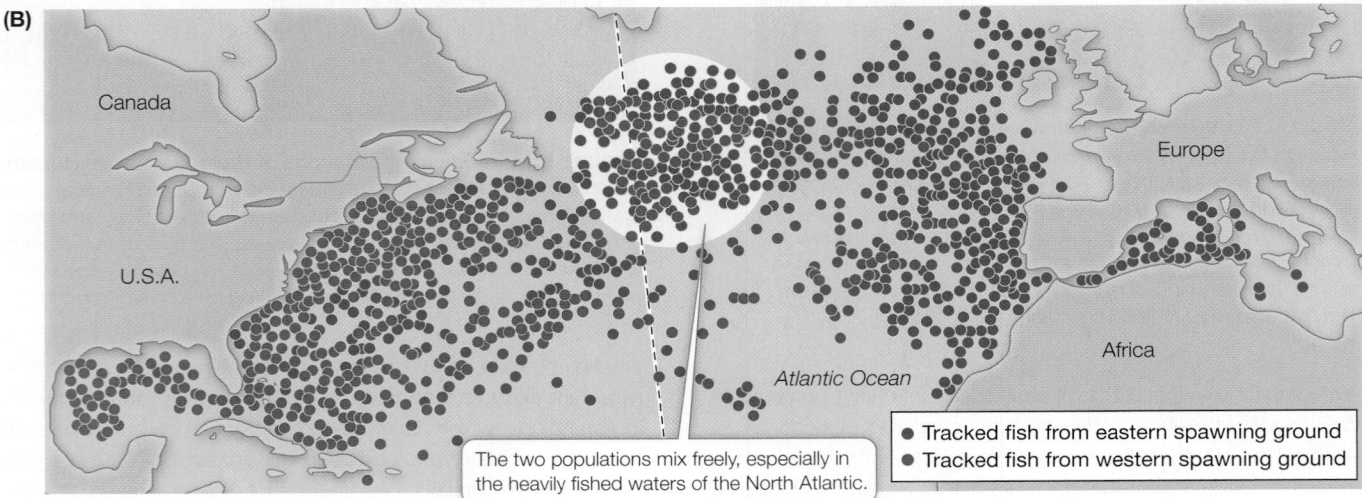

The two populations mix freely, especially in the heavily fished waters of the North Atlantic.

● Tracked fish from eastern spawning ground
● Tracked fish from western spawning ground

separate feeding grounds as well as separate breeding grounds. Acting on this assumption, an international commission drew a line down the middle of the Atlantic Ocean and established strict fishing quotas on the western side of the line, with the intent of allowing the western population to recover. Modern tracking data, however, revealed that in fact the eastern and western bluefin populations mix freely on their feeding grounds across the entire North Atlantic—a swath of ocean that includes the most heavily fished waters in the world. Tuna caught on the eastern side of the line could just as likely be from the western-breeding population as the eastern; thus the established policy could not achieve its intended goal.

Scientific knowledge can make significant contributions to the formulation of wise public policy, but economic and political factors frequently outweigh scientific recommendations. It is particularly lamentable when solid scientific evidence is discounted or ignored because it is inconvenient. An example is the rejection by many policy makers of strong evidence for and scientific consensus on global climate change.

Biology is crucial for understanding ecosystems

Earth has been changing since its formation and continues to change. Human activity, however, is resulting in an unprecedented *rate* of change in the world's ecosystems. For example, the mining and consumption of fossil fuels is releasing massive quantities of

carbon dioxide into Earth's atmosphere. This anthropogenic (human-generated) increase in atmospheric carbon dioxide is largely responsible for the rapid rate of global climate warming recorded over the last 50 years (**Figure 1.16**).

Our use of natural resources is putting stress on the ability of Earth's ecosystems to continue to produce the goods and services on which our society depends. Human activities are changing global climates at an unprecedented rate and are leading to the extinctions of large numbers of species (such as the reef-building corals featured in this chapter). The modern, warmer world is also experiencing the spread of new diseases and the resurgence of old ones. Biological knowledge is vital for determining the causes of these changes and for devising policies to deal with them.

Biodiversity helps us understand, enjoy, and appreciate our world

Beyond issues of policy and pragmatism lies the human "need to know." Most people are fascinated by the richness and diversity of life and want to know more about organisms, how they function, and how they interact (**Figure 1.17**). For the biologist, biodiversity offers an inexhaustible source of new questions and opportunities for new investigations and experiments. New discoveries and greater knowledge generate questions no one thought to ask before. Science is a never-ending quest.

(A) 1941

Riggs Glacier

Muir Glacier

(B) 2004

Riggs Glacier

Figure 1.16 A Warmer World   Earth's climate has been steadily warming for the last 150 years. The rate of this warming trend has also steadily increased, resulting in the rapid melting of polar ice caps, glaciers, and alpine (mountaintop) snow and ice. **(A, B)** These photographs shows the effects of 63 years of climate change on two ancient, longstanding glaciers in Alaska. Over that time, Muir Glacier retreated some 7 kilometers and can no longer be seen from the original vantage point. Understanding how biological populations respond to such change requires integration of biological principles from molecular biology to ecosystem ecology.

Figure 1.17 Exploring Life on Earth   These biologists are collecting insects in the top boughs of a spruce tree in the Carmanah Valley of Vancouver, Canada. Biologists estimate that the number of species discovered to date is only a small percentage of the species that inhabit Earth. To fill this gap in our knowledge, biologists around the world are applying thorough sampling techniques and new genetic tools to document and understand the Earth's biodiversity.

Apart from scientific endeavors, biodiversity enriches our lives. You may be an avid birdwatcher, or enjoy gardening, or seek out particular species if you hunt or fish. Some people observe or collect butterflies, or mushrooms, or other groups of plants and animals. Displays of spring wildflowers bring out throngs of human viewers. Hiking and camping in natural areas full of diverse species are activities enjoyed by millions. Learning about biology greatly increases our enjoyment of the world around us.

1.3 recap

Biology informs us about the structure, processes, and interactions of the living organisms that make up our world. Informed decisions about food and energy production, health, and our environment depend on biological knowledge. Biology also addresses the human need to understand the world around us, and helps us appreciate the diverse planet we call home.

learning outcomes

You should be able to:

- Explain the relevance of biology to people's daily lives.
- Apply biological knowledge to answer a question that pertains to everyday life.

1. Describe an example of how modern biology is applied to agriculture.
2. Discuss an example of a biological problem that is directly related to global climate change.
3. Why are some antibiotics no longer as effective for treating bacterial diseases as they were when these drugs were introduced?

This chapter has provided a brief roadmap of the rest of the book. Thinking about the principles outlined here may help you clarify and make sense of the pages of detailed description to come.

investigatinglife

How might experiments on heat stress in corals be used to predict the response of corals to global warming?

How can we put studies like the one conducted by Rachael and her colleagues to use? One way is to use the information to make predictions. As we've noted in this chapter, abundant evidence shows that Earth's climate is warming as a result of increased levels of CO_2 in our atmosphere. Can corals survive this global warming, or will they all lose their photosynthetic dinoflagellates and die? The experiments and observations done by Rachael's team suggest that higher temperatures will lead to increased coral bleaching, but that some short-term acclimation of individuals can occur. If global warming is slowed sufficiently, longer-term processes of evolution will act on the surviving corals that have genetic adaptations that increase their tolerance to warmer waters. This selection process should result in genetic changes in populations, which will allow more corals to survive. How well acclimation and adaptation can work to ensure the survival of corals depends on many factors, including the average life span of individual corals, the rate of global warming, and the availability of alternative heat-resistant dinoflagellates. The basic research conducted by Rachael's team is a first, but critical, step in predicting the future response of corals to global warming.

Future directions

As described, the corals in Rachael's study have certain species of dinoflagellates growing within their cells. The corals provide a suitable environment for the dinoflagellates and the dinoflagellates provide nutrients for the corals. Such a mutually beneficial relationship is called **symbiosis**. The bleaching response to high temperature implies dysfunction in the dinoflagellates, leading to their expulsion from the corals. Understanding thermal tolerance of the corals therefore requires greater understanding of genetic changes and acclimation of the symbiotic dinoflagellates. Other studies have shown that some corals contain multiple strains of dinoflagellates, and these strains change in abundance when the corals are grown at different temperatures, perhaps contributing to thermal tolerance. There are many species and strains of dinoflagellates, but coral species usually are associated with specific species. Future research could focus on thermal tolerance of dinoflagellates to identify genes that confer thermal tolerance. Additional research could be to find ways of transferring those genes into other species of dinoflagellates or to find ways of infecting corals endangered by rising temperatures with strains of temperature-tolerant symbionts.

Chapter Summary 1

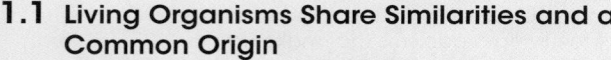

1.1 Living Organisms Share Similarities and a Common Origin

- **Biology** is the scientific study of organisms, including their characteristics, evolution, functions, and interactions.

- All living organisms are related to one another through common descent. Shared features among organisms support the concept of a common ancestry for all life on Earth. These features include: a common set of chemical building blocks, a nearly universal genetic code, sequence similarities across fundamental genes, and similarities in cellular structure and function.

- Cells evolved early in the history of life. **Cellular specialization** allowed multicellular organisms to increase in size and diversity. Review Figure 1.2

- **Photosynthesis** provided a means of capturing energy directly from sunlight, and over time changed Earth's atmosphere.

- The instructions for a cell are contained in its **genome**, which consists of **DNA** molecules made up of sequences of **nucleotides**. Specific segments of DNA called **genes** contain the information the cell uses to make **proteins** through processes of transcription and translation. Review Figure 1.6

- **Evolution**—change in the genetic makeup of biological **populations** through time—is a fundamental principle of life. Populations evolve by means of several different processes, including **natural selection**, which is responsible for the diversity of **adaptations** found in living organisms.

- Biologists use fossils, anatomical similarities and differences, and molecular comparisons of genomes to reconstruct the history of life. Life is divided into two major lineages—**Bacteria**, **Archaea**. One lineage of Archaea became specialized as **Eukarya**, which includes all multicellular organisms. Review Figure 1.9

- Life can be studied at different levels of organization within a biological hierarchy. The specialized cells of multicellular organisms are organized into **tissues**, **organs**, and **organ systems**.

- Living organisms, whether unicellular or multicellular, must regulate their internal environment to maintain **homeostasis**, the range of physical conditions necessary for their survival and function.

- Individual organisms form populations and interact with other organisms of their own and other species. The populations that live and interact in a defined area form a **community**, and communities together with their abiotic (nonliving) environment constitute an **ecosystem**. Review Figure 1.10, Activity 1.1

1.2 Biologists Investigate Life through Experiments That Test Hypotheses

- Scientific methods combine observation, the gathering of information (**data**), experimentation, and the application of logic to study the natural world. Many scientific investigations involve five steps: making observations, asking questions, forming hypotheses, making predictions, and testing those predictions. Review Figure 1.12

- **Hypotheses** are tentative explanations of observations or answers to questions. Predictions made on the basis of a hypothesis are tested with additional observations and experiments. The two general types of experiments are **comparative** and **controlled**. Review Investigating Life: Corals in Hot Water, Animation 1.1

- Quantifiable data are critical in evaluating hypotheses. Statistical methods are applied to quantitative data to establish whether the differences observed could be the result of chance. These methods start with the **null hypothesis** that there are no differences apart from random variation. See Appendix B

- Biological knowledge obtained from a **model system** may be generalized to other species.

▶ 1.3 Understanding Biology Is Important for Health, Well-Being, and Public-Policy Decisions

- Application of biological knowledge is responsible for vastly increased agricultural production.

- Understanding and treating human disease requires integrating a wide range of biological principles, from molecular biology to cell biology, physiology, evolution, and ecology.

- Biologists are often called on to advise government agencies on the solutions to important problems that have a biological component.

- Biology is increasingly important for understanding how organisms interact in a rapidly changing world.

- Biology helps us understand, appreciate, and enjoy the diverse living world.

> Go to **LearningCurve** (in **LaunchPad**) for dynamic quizzing that helps you solidify your understanding of this chapter. **LearningCurve** adapts to your responses, giving you the practice you need to master each key concept.

▶ Apply What You've Learned

Review

1.1 Living organisms have influenced the history of the planet Earth.

1.1 Major characteristics are shared among all living things.

1.1 Biological populations change over time.

1.2 Scientific methods produce biological knowledge.

1.3 Biology contributes many things to society.

Original Paper: Lindsey, H. A., J. Gallie, S. Taylor, and B. Kerr. 2013. Evolutionary rescue from extinction is contingent on a lower rate of environmental change. *Nature* 494: 463–467.

As many as half of the antibiotics that are used in human medicine may be incorrectly prescribed or completely unnecessary. Antibiotics are also given to food animals in an effort to prevent disease. A general overuse of these drugs has contributed to a significant increase in drug-resistant pathogens. If you were raising animals for food, what evidence could you use to decide if it is better to treat the animals routinely with a low dose of antibiotics or wait until symptoms occur and then administer a high dose?

Researchers used serial transfer to propagate 1,255 populations of the bacterium *Escherichia coli* in an environment in which concentrations of the antibiotic rifampicin were increased over time. Each

population began in liquid medium containing no antibiotic and ended at the same maximum concentration of rifampicin.

During the serial transfers, populations received one of three treatments. Populations designated as "Sudden" were exposed to the maximum rifampicin concentration immediately after their first transfer and continued to receive this maximum concentration in each subsequent transfer. Populations designated as "Moderate" were exposed to moderately increasing concentrations of rifampicin; the maximum concentration of rifampicin was reached halfway through the transfers. Populations designated as "Gradual" were exposed to more gradual increases of rifampicin; the maximum concentration of rifampicin was reached at the very end of the transfers.

Figure A summarizes the treatment conditions and **Figure B** shows the percentage of populations in each treatment that survived the experiment. The number above each bar shows how many of the initial 1,255 populations underwent each type of treatment.

Questions

1. What do the results of this investigation show about the effects of sudden versus gradual environmental changes on the evolution of biological populations? Explain how these results help explain extinctions and adaptations of populations in changing environments.

2. In what ways does this investigation model events that shaped the history of life on Earth?

3. Explain what makes it reasonable for us to infer that similar results would be observed if this investigation were repeated using a plant, insect, or mammal population in a changing environment.

4. Identify this investigation as either a comparative or a controlled study. Design an experiment of the other type of study that tests the hypothesis that organisms can develop adaptations to changing surroundings. Explain your reasoning.

5. Suppose a research organization has requested funding for long-term research to develop a continuous pipeline of new antibiotics over the next 50 years. Could the organization use results from this investigation to support their request? Explain.

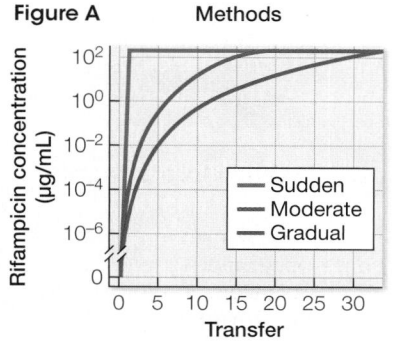

Figure A Methods

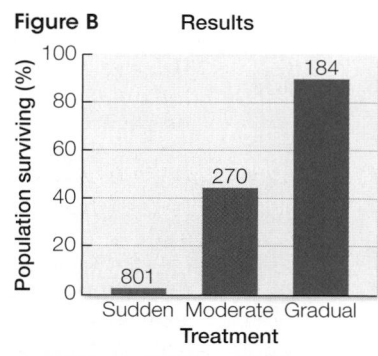

Figure B Results

2

investigatinglife

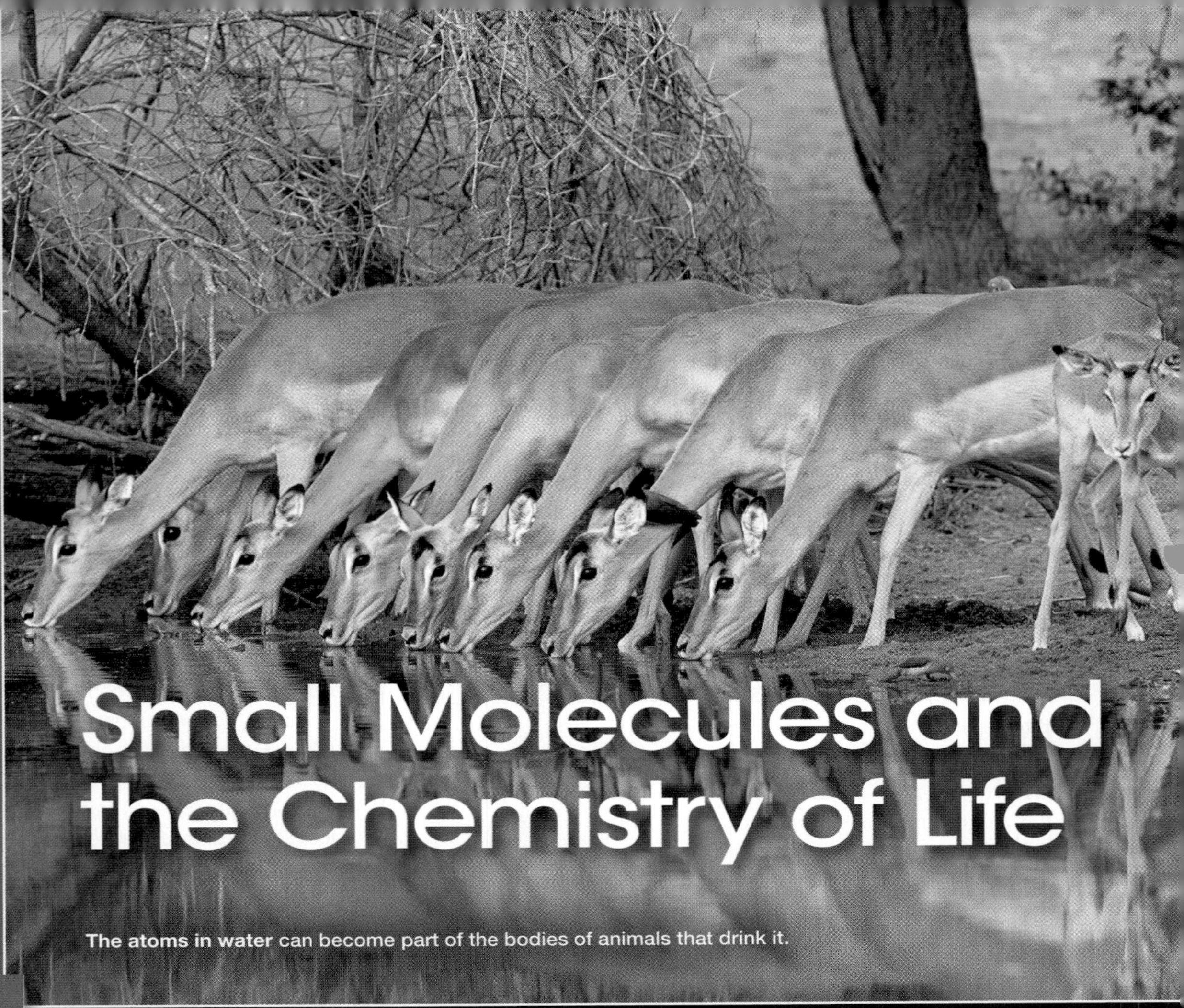

Small Molecules and the Chemistry of Life

The atoms in water can become part of the bodies of animals that drink it.

Tracking a Dinosaur

"You are what you eat—and that applies to teeth" is a modification of a famous saying about body chemistry. One of the important atoms in organisms is oxygen (O), which is part of water (H_2O). Oxygen has two naturally occurring variants called isotopes; they have the same chemical properties but different weights because their nuclei have different numbers of neutrons. Both isotopes of O are incorporated into the bodies of animals that consume the isotopes in water and food.

The hard surface of teeth, called enamel, is made up largely of calcium phosphate, which contains oxygen atoms. The iso-topic composition of the oxygen in enamel varies depending on where an animal was living when the enamel was made. When water evaporates from the ocean, it forms clouds that move inland and release rain. Water with the heavier isotope of O is heavier, and tends to fall more readily than water containing the lighter isotope. As a result, regions of the world that are closer to the ocean receive rain containing more heavy water than do regions farther away, and these differences are reflected in the bodies of animals that dwell in these regions.

This property has been used to reveal an astounding fact about dinosaurs that lived in the great basins of southwestern North America about 150 million years ago. *Camarasaurus* was big—up to 25 meters long and weighing up to 45 metric tons. Henry Ficke from Colorado College analyzed the oxygen isotopes in the enamel of *Camarasaurus* fossils and found two kinds of teeth: Some had the heavy oxygen content typical of rains and rocks in the basin region. But others, surprisingly, had a lower proportion of heavy oxygen, indicating that these animals had lived at higher elevations 300 km to the west. This indicated for the first time that dinosaurs migrated a long way from west to east. The reason for this migration is not clear. *Camarasaurus* ate a plant-based diet, and perhaps the migration was directed at finding food.

Many biological phenomena are analyzed and understood in the context of chemistry. You will learn, for example, that adding even one atom to a large molecule can change its character and function. Analyzing the isotopes in dinosaur teeth is but one example of how chemical analysis has led to advances in biological science.

 What insights into biological systems have

key concept 2.1 Atomic Structure Explains the Properties of Matter

All matter is composed of **atoms**. Atoms are tiny—more than a trillion (10^{12}) of them could make up the period at the end of this sentence. Atoms and their component particles have volume and mass, which are characteristics of all matter. **Mass** is a measure of the quantity of matter present; the greater the mass, the greater the quantity of matter. Atoms also have electric charges.

focus your learning

- Six elements (carbon, hydrogen, oxygen, nitrogen, phosphorus, and sulfur) make up most of the matter found in living organisms.
- Reactive atoms tend to gain or lose electrons from their valence (outer) shell in order to achieve a full valence shell, because a full valence shell confers stability.
- The structure of an atom determines its properties.

What are atoms?

Each atom consists of a dense, positively charged **nucleus**, around which one or more negatively charged **electrons** move (**Figure 2.1**). The nucleus contains one or more positively charged **protons** and may contain one or more **neutrons** with no electric charge.

The mass of a proton serves as a standard unit of measure called the **dalton** (named after the English chemist John Dalton). A single proton or neutron has a mass of about 1 dalton (Da), which is 1.7×10^{-24} grams (0.00000000000000000000000017 g), but an electron is even tinier at 9×10^{-28} g (0.0005 Da). Because the mass of an electron is negligible compared with the mass of a proton or a neutron, the contribution of electrons to the mass of an atom can usually be ignored when measurements and calculations are made. It is electrons, however, that determine how atoms will combine with other atoms to form stable associations.

Each proton has a positive electric charge, defined as +1 unit of charge. An electron has a negative charge equal and opposite to that of a proton (−1). The neutron, as its name suggests,

is electrically neutral, so its charge is 0. Charges that are different (+/−) attract each other, whereas charges that are alike (+/+, −/−) repel each other. Generally, atoms are electrically neutral because the number of electrons in an atom equals the number of protons.

An element consists of only one kind of atom

An **element** is a pure substance that contains only one kind of atom. The element hydrogen consists only of hydrogen atoms; the element iron consists only of iron atoms. The atoms of each element have certain characteristics or properties that distinguish them from the atoms of other elements. These physical and chemical properties (how they interact with other atoms) depend on the numbers of particles the atoms contain.

There are 94 elements in nature, and at least another 24 have been made in physics laboratories. About 98 percent of the tissue of every living organism (except in skeletons) is composed of just six elements:

Carbon (symbol C)	Hydrogen (H)	Nitrogen (N)
Oxygen (O)	Phosphorus (P)	Sulfur (S)

The biological roles of these elements will be our major concern in this book, but other elements are found in living organisms as well. Sodium and potassium, for example, are essential for nerve function; calcium can act as a biological signal; iodine is a component of a vital hormone; and magnesium is bound to chlorophyll in plants.

Each element has a unique number of protons

An element differs from other elements by the number of protons in the nucleus of each of its atoms; the number of protons is designated the **atomic number**. This atomic number is unique to each element and does not change. The atomic number of helium is 2, and an atom of helium always has two protons; the atomic number of oxygen is 8, and an atom of oxygen always has eight protons. Since the number of protons (and electrons) determines how an element behaves in chemical reactions, it is possible to arrange the elements in a table such that those with similar chemical properties are grouped together. This is the familiar **periodic table** you see in **Figure 2.2**.

▶ Media Clip 2.1 **The Elements Song**
www.Life11e.com/mc2.1

Along with a definitive number of protons, every element except hydrogen has one or more neutrons in its nucleus. The **mass number** of an atom is the total number of protons and neutrons in its nucleus. The nucleus of a carbon atom contains six protons and six neutrons and has a mass number of 12. Oxygen has eight protons and eight neutrons and has a mass number of 16. Since the mass of an electron is negligible, the mass number is essentially the mass of the atom in daltons.

By convention, we often print the symbol for an element with the atomic number at the lower left and the mass number at the upper left, both immediately preceding the symbol. Thus hydrogen, carbon, and oxygen can be written as $^{1}_{1}H$, $^{12}_{6}C$, and $^{16}_{8}O$, respectively.

The number of neutrons differs among isotopes

In some elements, the number of neutrons in the atomic nucleus may vary. Different **isotopes** of the same element have the same number of protons but different numbers of neutrons, as you saw

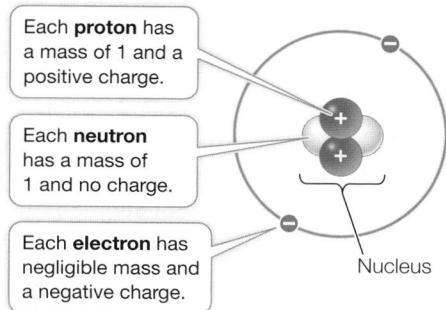

Each **proton** has a mass of 1 and a positive charge.

Each **neutron** has a mass of 1 and no charge.

Each **electron** has negligible mass and a negative charge.

Nucleus

Figure 2.1 The Structure of an Atom An atom has a nucleus composed of neutrons and protons surrounded by a cloud of rapidly moving electrons. Although the nucleus accounts for virtually all of the atomic weight, it occupies only 1/10,000 of the atom's volume. This type of representation, which is called a Bohr model, depicts individual electrons moving in "orbits" at set distances from the nucleus.

Figure 2.2 The Periodic Table The periodic table groups all the elements according to their physical and chemical properties. Elements 1–94 occur in nature; elements with atomic numbers greater than 94 were created in the laboratory.

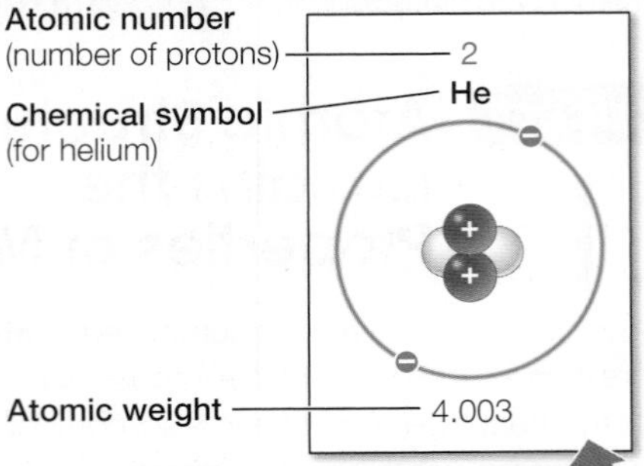

Atomic number (number of protons) ── 2
Chemical symbol (for helium) ── He
Atomic weight ── 4.003

The six elements highlighted in yellow make up 98% of the tissue of most living organisms (except for skeletons).

Elements in the same vertical columns have similar properties in terms of chemical reactions because they have the same number of electrons in their outermost shell.

Elements highlighted in orange are present in small amounts in many organisms.

1 H 1.0079																	2 He 4.003
3 Li 6.941	4 Be 9.012											5 B 10.81	6 C 12.011	7 N 14.007	8 O 15.999	9 F 18.998	10 Ne 20.179
11 Na 22.990	12 Mg 24.305											13 Al 26.982	14 Si 28.086	15 P 30.974	16 S 32.06	17 Cl 35.453	18 Ar 39.948
19 K 39.098	20 Ca 40.08	21 Sc 44.956	22 Ti 47.88	23 V 50.942	24 Cr 51.996	25 Mn 54.938	26 Fe 55.847	27 Co 58.933	28 Ni 58.69	29 Cu 63.546	30 Zn 65.38	31 Ga 69.72	32 Ge 72.59	33 As 74.922	34 Se 78.96	35 Br 79.909	36 Kr 83.80
37 Rb 85.4778	38 Sr 87.62	39 Y 88.906	40 Zr 91.22	41 Nb 92.906	42 Mo 95.94	43 Tc (99)	44 Ru 101.07	45 Rh 102.906	46 Pd 106.4	47 Ag 107.870	48 Cd 112.41	49 In 114.82	50 Sn 118.69	51 Sb 121.75	52 Te 127.60	53 I 126.904	54 Xe 131.30
55 Cs 132.905	56 Ba 137.34		72 Hf 178.49	73 Ta 180.948	74 W 183.85	75 Re 186.207	76 Os 190.2	77 Ir 192.2	78 Pt 195.08	79 Au 196.967	80 Hg 200.59	81 Tl 204.37	82 Pb 207.19	83 Bi 208.980	84 Po (209)	85 At (210)	86 Rn (222)
87 Fr (223)	88 Ra 226.025		104 Rf (261)	105 Db (262)	106 Sg (266)	107 Bh (264)	108 Hs (269)	109 Mt (268)	110 Ds (269)	111 Rg (272)	112 Cn (277)	113 Uut	114 Fl (285)	115 Uup (289)	116 Lv (293)	117 Uus	118 Uuo (293)

Masses in parentheses indicate unstable elements that decay rapidly to form other elements.

Lanthanide series	57 La 138.906	58 Ce 140.12	59 Pr 140.9077	60 Nd 144.24	61 Pm (145)	62 Sm 150.36	63 Eu 151.96	64 Gd 157.25	65 Tb 158.924	66 Dy 162.50	67 Ho 164.930	68 Er 167.26	69 Tm 168.934	70 Yb 173.04	71 Lu 174.97
Actinide series	89 Ac 227.028	90 Th 232.038	91 Pa 231.0359	92 U 238.02	93 Np 237.0482	94 Pu (244)	95 Am (243)	96 Cm (247)	97 Bk (247)	98 Cf (251)	99 Es (252)	100 Fm (257)	101 Md (258)	102 No (259)	103 Lr (260)

in the story opening this chapter. Many elements have several isotopes. Generally, isotopes are formed when atoms combine and/or release particles (decay). The isotopes of hydrogen shown here have special names, but for most elements their isotopes do not have special names.

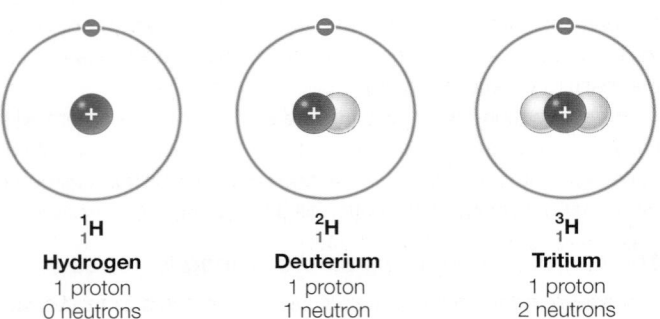

$^{1}_{1}H$
Hydrogen
1 proton
0 neutrons

$^{2}_{1}H$
Deuterium
1 proton
1 neutron

$^{3}_{1}H$
Tritium
1 proton
2 neutrons

The natural isotopes of carbon, for example, are ^{12}C (six neutrons in the nucleus), ^{13}C (seven neutrons), and ^{14}C (eight neutrons). Note that all three (called carbon-12, carbon-13, and carbon-14) have six protons, so they are all carbon. Most carbon atoms are ^{12}C, about 1.1 percent are ^{13}C, and a tiny fraction are ^{14}C. The carbon atoms that make up complex biological molecules are mostly ^{12}C, but some are ^{13}C. The ratio of ^{13}C:^{12}C varies with location and can be used to identify biological samples whose origin is not known (**Investigating Life: Determining Beef Source in Big Macs Using Isotope Analysis**). All carbon isotopes have virtually the same chemical reactivity, which is an important property for their use in experimental biology and medicine.

An element's **atomic weight** (or relative atomic mass) is equivalent to the average of the mass numbers of a representative sample of atoms of that element, with all the isotopes in their normally occurring proportions. More precisely, an element's atomic weight is defined as the ratio of the average mass per atom of the element

experiment

Original Paper: Martinelli, L., G. Nardoto, L. Chesson, F. Rinaldi, J. P. H.B. Ometto, T. Cerling and J. Ehleringer. 2011. Worldwide stable carbon and nitrogen isotopes of Big Mac patties: An example of a truly "glocal" food. *Food Chemistry* 127: 1712–1718.

The Big Mac hamburger can be purchased in 35,000 outlets in 120 countries. The recipe for the Big Mac is virtually the same everywhere; it is claimed that the burger has the same quality meat and nutritional quality everywhere it is served. To investigate this claim, Lesley Chesson and

James Ehleringer at the University of Utah compared the ratio of ^{13}C:^{12}C in Big Mac patties around the world to determine whether the meat had a local or a common source. Beef cattle eat plant food, and carbon atoms from this food end up in meat. In some types of plants, the ratio of ^{13}C:^{12}C is higher than in others. Depending on which type of plants the cattle consume, the beef will also have a different ^{13}C:^{12}C ratio.

HYPOTHESIS▶　Beef in Big Macs served in different countries comes from the same source.

METHOD

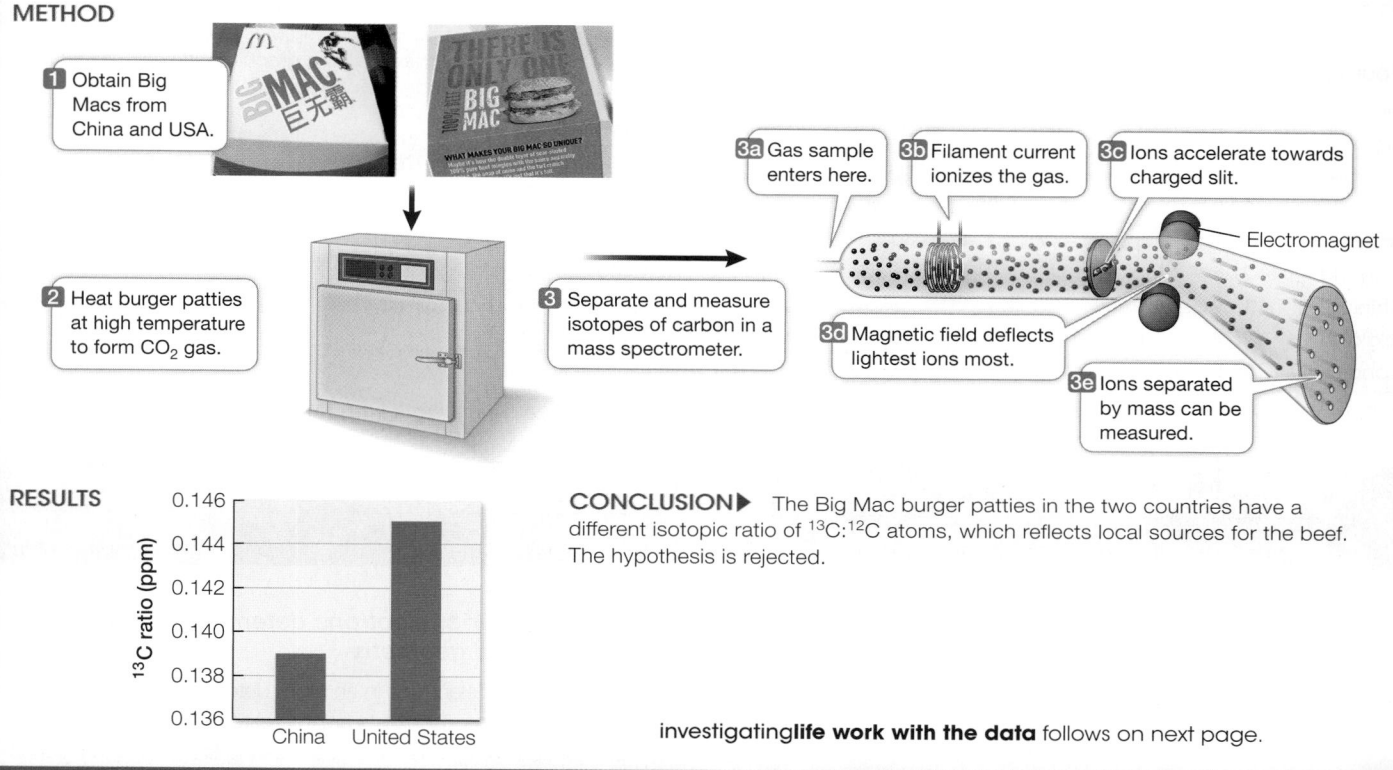

1 Obtain Big Macs from China and USA.

2 Heat burger patties at high temperature to form CO_2 gas.

3 Separate and measure isotopes of carbon in a mass spectrometer.

3a Gas sample enters here.

3b Filament current ionizes the gas.

3c Ions accelerate towards charged slit.

Electromagnet

3d Magnetic field deflects lightest ions most.

3e Ions separated by mass can be measured.

RESULTS

^{13}C ratio (ppm)

China　United States

CONCLUSION▶　The Big Mac burger patties in the two countries have a different isotopic ratio of ^{13}C:^{12}C atoms, which reflects local sources for the beef. The hypothesis is rejected.

*investigating***life work with the data** follows on next page.

---

to 1/12 of the mass of an atom of ^{12}C. Because it is a ratio, atomic weight is a dimensionless physical quantity—it is not expressed in units. The atomic weight of hydrogen, taking into account all of its isotopes and their typical abundances, is 1.00794. This number is fractional because it is the average of the contributing masses of all of the isotopes. This definition implies that in any given sample of hydrogen atoms of a particular element found on Earth, the average composition of isotopes will be constant. But as you saw in the opening to this chapter, that is not necessarily so. Some water has more of the heavy O isotopes. So chemists are now listing atomic weights as ranges, for example, H: 1.00784–1.00811.

Most isotopes are stable. However, some, called **radioisotopes**, are unstable and spontaneously give off energy in the form of α (alpha), β (beta), or γ (gamma) radiation from the atomic nucleus. Known as **radioactive decay**, this release of energy transforms the original atom. The type of transformation varies depending on the radioisotope, but some transformations result

in a different number of protons, so that the original atom becomes a different element.

Scientists can detect released radiation from ***radioisotopes**. For instance, if you feed an earthworm food containing a radioisotope, you can follow the worm's path through the soil by using a detector called a Geiger counter. Most atoms in living organisms are organized into stable associations called **molecules**. If a radioisotope is incorporated into a molecule, it acts as a tag or label, allowing a researcher or physician to track the molecule in an experiment or in the body (**Figure 2.3**).

---

*connect the concepts　Radioisotopes are also used to analyze biochemical pathways (see Key Concept 10.3) and to date fossils (see Key Concept 24.1).

Although radioisotopes are useful in research and in medicine, even a low dose of the radiation they emit has the potential to

work with the data

The experiment compared the meat in Big Macs served in 25 different countries by measuring the ratio of stable carbon isotopes ($^{13}C{:}^{12}C$) in samples of ten beef patties from each country. The hypothesis was based on claims that the beef used in Big Macs comes from a single source, and that the $^{13}C{:}^{12}C$ ratio in patties found everywhere should thus be the same. **Table A** shows the average the $^{13}C{:}^{12}C$ ratio for each of the 25 countries. A higher number means more of the rare, stable isotope ^{13}C than of the much more common ^{12}C.

Table A

Country	$^{13}C{:}^{12}C$ ratio[a]	Country	$^{13}C{:}^{12}C$ ratio	Country	$^{13}C{:}^{12}C$ ratio
All	15.8	Germany	21.7	Portugal	20.7
Argentina	17.2	Hungary	22.0	Scotland	25.2
Australia	19.4	Indonesia	19.5	Slovakia	21.0
Austria	22.0	Israel	20.4	South Africa	13.0
Brazil	11.1	Japan	11.8	Spain	21.1
Canada	21.6	Malaysia	21.5	Sweden	23.2
China	13.9	Mexico	13.9	Turkey	20.5
England	25.4	Netherlands	20.7	USA	14.5
France	21.8	Paraguay	12.1	Uruguay	16.7

[a]Measured in parts per million atoms (× 100).

QUESTIONS ▶

1. Do the average local ratios differ from the 25-country average ratio? What statistical test would you perform to show that the differences between the individual country averages and the all-countries average are significant? (Refer to Appendix B.)

2. The countries from which the hamburgers were analyzed are in different locations on Earth. One way to compare locations is by geographic latitude. You may recall that in latitude, 0 degrees defines the equator and 90 degrees defines the poles. **Table B** shows average $^{13}C{:}^{12}C$ ratios of Big Mac beef patties from different latitudes. Plot the data on a graph of ratio versus latitude. What can you conclude from these data?

Table B

Latitude	$^{13}C{:}^{12}C$ ratio[a]
20°S–20°N	11.4
20°–40°[b]	15.3
40°–60°[b]	21.8

[a]Average value measured in parts per million atoms (× 100).

[b] 20°S–40°S and 20°N–40°N; 40°S–60°S and 40°N–60°N.

A similar **work with the data** exercise may be assigned in **LaunchPad**.

---

damage molecules and cells. However, these damaging effects are sometimes used to our advantage; for example, the radiation from ^{60}Co (cobalt-60) is used in medicine to kill cancer cells.

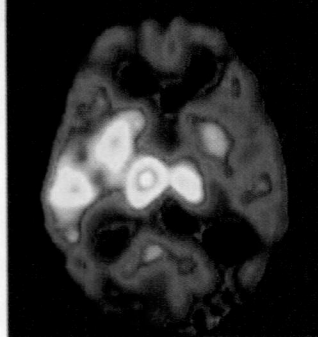 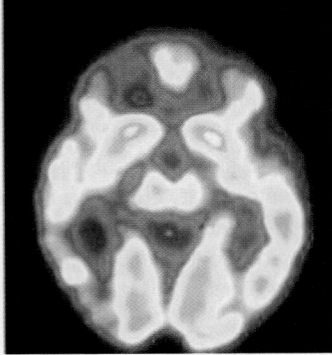

(A) Depressed   **(B) Not depressed**

Figure 2.3 Tagging the Brain   In these images from live persons, a radioactively labeled sugar is used to detect differences between the brain activity of a depressed person **(A)** and that of a person who is not depressed **(B)**. The more active a brain region is, the more sugar it takes up (shown as orange areas). The brain of the depressed person (left) shows less activity than the brain of the person who is not depressed.

Q: How could this technique be used in the clinic?

The behavior of electrons determines chemical bonding and geometry

The number of electrons in an atom determines how it will combine with other atoms. Biologists are interested in how chemical changes take place in living cells. When considering atoms, they are concerned primarily with electrons because the behavior of electrons explains how chemical reactions occur. Chemical reactions alter the atomic compositions of substances and thus alter their properties. Reactions usually involve changes in the distribution of electrons between atoms.

The location of a given electron in an atom at any given time is impossible to determine. We can only describe a volume of space within the atom where the electron is likely to be. The region of space where the electron is found at least 90 percent of the time is the electron's **orbital**. Orbitals have characteristic shapes and orientations, and a given orbital can be occupied by a maximum of two electrons. Thus any atom larger than helium (atomic number 2) must have electrons in two or more orbitals. As we move from lighter to heavier atoms in the periodic table, the orbitals are filled in a specific sequence, in a series of what are known as **electron shells**, or energy levels, around the nucleus.

- *First shell*: The innermost electron shell consists of just one orbital. A hydrogen atom ($_1H$) has one electron in its first shell; helium ($_2He$) has two. Atoms of all other elements have two or more shells to accommodate orbitals for additional electrons.

- *Second shell*: The second shell contains four orbitals and hence holds up to eight electrons.

● *Additional shells*: Elements with more than ten electrons have three or more electron shells. The farther a shell is from the nucleus, the higher the energy level is for an electron occupying that shell.

 Activity 2.1 Electron Orbitals
www.Life11e/ac2.1

The *s* orbitals fill with electrons first, and their electrons have the lowest energy level. Subsequent shells have different numbers of orbitals, but the outermost shells usually hold only eight electrons. In any atom, the outermost electron shell (the **valence shell**) determines how the atom combines with other atoms—that is, how the atom behaves chemically. When a valence shell with four orbitals contains eight electrons, there are no unpaired electrons and the atom is stable—it is least likely to react with other atoms (**Figure 2.4**). Examples of chemically stable elements are helium, neon, and argon. By contrast, atoms that have one or more unpaired electrons in their outer shells are capable of reacting with other atoms.

Atoms with unpaired electrons (i.e., partially filled orbitals) in their outermost electron shells are unstable and will undergo reactions in order to fill their outermost shells. *Reactive atoms can attain stability either by sharing electrons with other atoms or by losing or gaining one or more electrons.* In either case, the atoms involved are bonded together into stable associations called molecules. The tendency of atoms to form stable molecules so that they have eight electrons in their outermost shells is known as the octet rule. Many atoms in biologically important molecules—for example, carbon (C) and

nitrogen (N)—follow this rule. An important exception is hydrogen (H), which attains stability when two electrons occupy its single shell (consisting of just one *s* orbital).

▶ 2.1 recap

Living organisms are composed of the same set of chemical elements as the rest of the universe. An atom consists of a nucleus of protons and neutrons and a characteristic configuration of electrons in orbitals around the nucleus. An atom's chemical properties are determined by its atomic structure.

learning outcomes

You should be able to:

● Compare the elements found in living tissue with elements found in nonliving matter.

● Explain the role that the outermost electron shell plays in determining how an atom may combine with other atoms.

● Explain how elements can be grouped according to their chemical properties in a periodic fashion.

(continued)

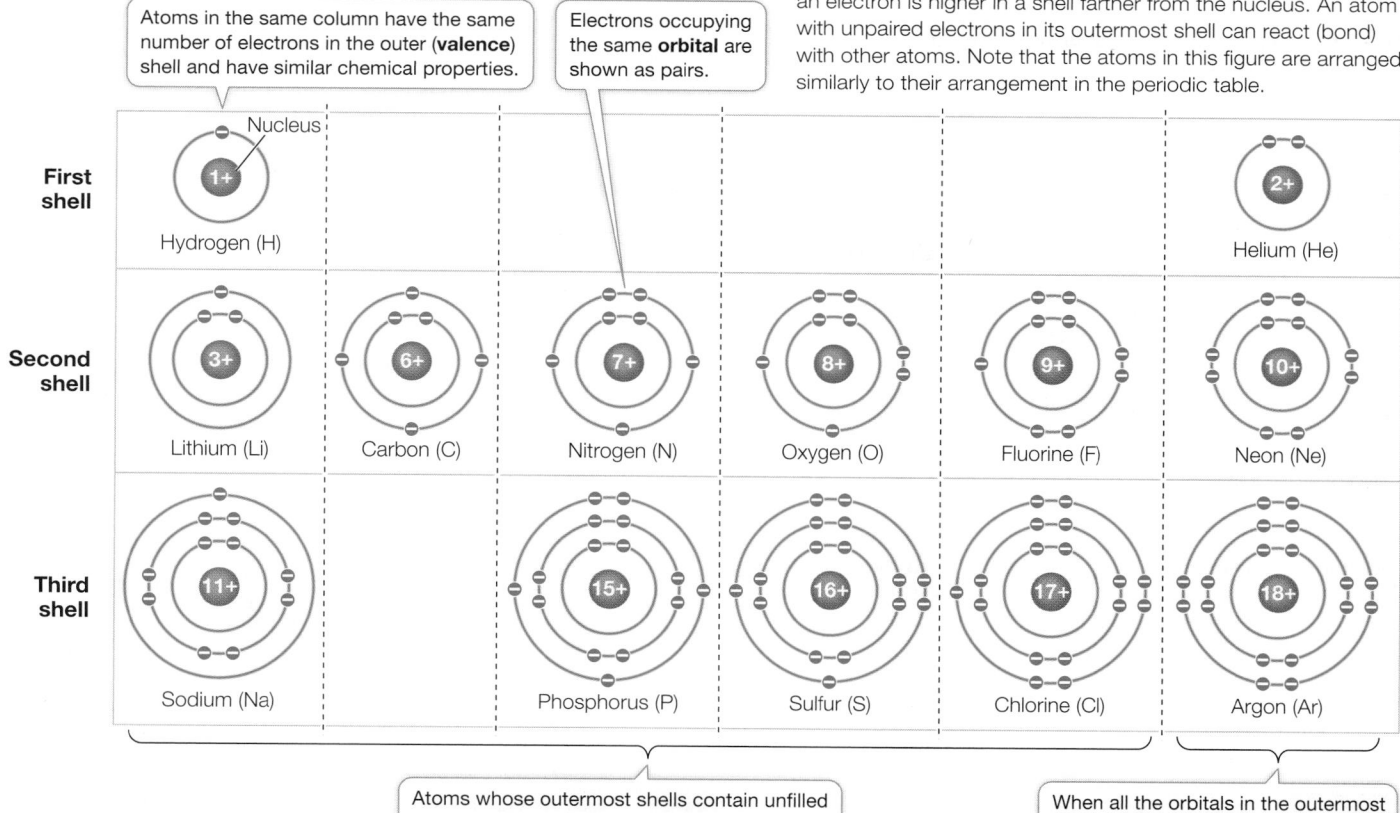

Figure 2.4 Electron Shells Determine the Reactivity of Atoms Each shell can hold a specific maximum number of electrons. Going out from the nucleus, each shell must be filled before electrons can occupy the next shell. The energy level of an electron is higher in a shell farther from the nucleus. An atom with unpaired electrons in its outermost shell can react (bond) with other atoms. Note that the atoms in this figure are arranged similarly to their arrangement in the periodic table.

Atoms in the same column have the same number of electrons in the outer (**valence**) shell and have similar chemical properties.

Electrons occupying the same **orbital** are shown as pairs.

First shell: Hydrogen (H) · Helium (He)

Second shell: Lithium (Li) · Carbon (C) · Nitrogen (N) · Oxygen (O) · Fluorine (F) · Neon (Ne)

Third shell: Sodium (Na) · Phosphorus (P) · Sulfur (S) · Chlorine (Cl) · Argon (Ar)

Atoms whose outermost shells contain unfilled orbitals (unpaired electrons) are **reactive**.

When all the orbitals in the outermost shell are filled, the atom is **stable**.

2.1 recap (continued)

1. Describe how the configuration of the valence shell influences the placement of an element in the periodic table.
2. How does bonding help a reactive atom achieve stability?
3. Would you expect the elemental composition of Earth's crust to be the same as that of the human body? Explain your answer.

You have been introduced to the individual players on the biochemical stage—the atoms. We have shown how the number of unpaired electrons in an atom's valence shell drives its "quest for stability." Next we will describe the different types of chemical bonds that can lead to stability—joining atoms together into molecular structures with hosts of different properties.

key concept 2.2 Atoms Bond to Form Molecules

A **chemical bond** is an attractive force that links two atoms together in a molecule. There are several kinds of chemical bonds (**Table 2.1**). In this section we will begin with covalent bonds, the strong bonds that result from the sharing of electrons. Next we will examine ionic attractions, which form when an atom gains or loses one or more electrons to achieve stability. We will then consider other, weaker kinds of interactions, including hydrogen bonds.

 Animation 2.1 **Chemical Bond Formation**
www.Life11e.com/a2.1

focus your learning

- Covalent bonds are very stable and break only with a large input of energy.
- Polar covalent bonds occur when two atoms share bonding electrons in an unequal fashion, whereas nonpolar covalent bonds occur when the bonding electrons are shared equally.
- Hydrophilic interactions occur between polar molecules, and hydrophobic interactions occur between nonpolar molecules.
- van der Waals forces are weak, noncovalent attractive forces that occur between any two atoms.

Covalent bonds consist of shared pairs of electrons

A **covalent bond** forms when two atoms attain stable electron numbers in their outermost shells by sharing one or more pairs of electrons. Consider two hydrogen atoms coming close to one another, each with an unpaired electron in its single shell (**Figure 2.5**). When the electrons pair up, a stable association is formed, and this links the two hydrogen atoms in a covalent bond, forming the molecule H_2.

A **compound** is a pure substance made up of two or more different elements bonded together in a fixed ratio. Chemical symbols identify the different elements in a compound, and subscript numbers indicate how many atoms of each element are present (e.g., H_2O has two atoms of hydrogen bonded to a single oxygen atom). Every compound has a **molecular weight** that is the sum of the atomic weights of all atoms in the molecule. Looking at the periodic table in Figure 2.2, you can calculate the molecular weight of water to be 18.01. (But remember that this value comes from the average atomic weights of hydrogen and oxygen; the molecular weight of the heavy water in our opening story is higher because it

table 2.1 Chemical Bonds and Interactions

Name	Basis of interaction	Structure	Bond energy[a]
Covalent bond	Sharing of electron pairs		50–110
Ionic attraction	Attraction of opposite charges		3–7
Hydrogen bond	Electrical attraction between a covalently bonded H atom and an electronegative atom		3–7
Hydrophobic interaction	Interaction of nonpolar substances in the presence of polar substances (especially water)		1–2
van der Waals interaction	Interaction of electrons of nonpolar substances		1

[a]Bond energy is the amount of energy in kcal/mol needed to separate two bonded or interacting atoms under physiological conditions.

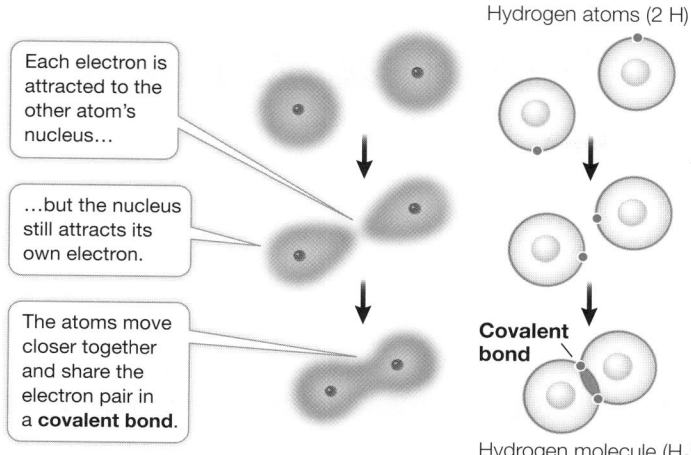

Hydrogen atoms (2 H)

Each electron is attracted to the other atom's nucleus...

...but the nucleus still attracts its own electron.

The atoms move closer together and share the electron pair in a **covalent bond**.

Covalent bond

Hydrogen molecule (H_2)

Figure 2.5 Electrons Are Shared in Covalent Bonds Two hydrogen atoms can combine to form a hydrogen molecule. A covalent bond forms when the electron orbitals of the two atoms overlap in an energetically stable manner.

is formed from heavier isotopes.) Molecules that make up living organisms can have molecular weights of up to half a billion.

How are the covalent bonds formed in a molecule of methane gas (CH_4)? The carbon atom has six electrons: two electrons fill its inner shell, and four unpaired electrons travel in its outer shell. Because its outer shell can hold up to eight electrons, carbon can share electrons with up to four other atoms—*it can form four covalent bonds* (**Focus: Key Figure 2.6A**). When an atom of carbon reacts with four hydrogen atoms, methane forms. Thanks to electron sharing, the outer shell of methane's carbon atom is now filled with eight electrons, a stable configuration. The outer shell of each of the four hydrogen atoms is also filled. Four covalent bonds—four shared electron pairs—hold methane together. **Focus: Key Figure 2.6B** shows several different ways to represent the molecular structure of methane. **Table 2.2** shows the covalent bonding capacities of some biologically significant elements.

STRENGTH AND STABILITY Covalent bonds are very strong, meaning that it takes a lot of energy to break them. At temperatures where life exists, the covalent bonds of biological molecules are quite stable, as are their three-dimensional structures. However, this stability does not preclude change, as you will see.

ORIENTATION For a given pair of elements—for example, carbon bonded to hydrogen—the length of the covalent bond is always the same. And for a given atom within a molecule, the angle of each of its covalent bonds, with respect to the other bonds, is generally the same. This is true regardless of the type of larger molecule that contains the atom. For example, the four filled orbitals around the carbon atom in methane are always distributed in space so that the bonded hydrogen atoms point to the corners of a regular tetrahedron, with carbon in the center (see Figure 2.6B). Even when carbon is bonded to four atoms other than hydrogen, this three-dimensional orientation is more or less maintained. The orientation of covalent bonds in space gives the molecules their three-dimensional geometry,

focus: key figure

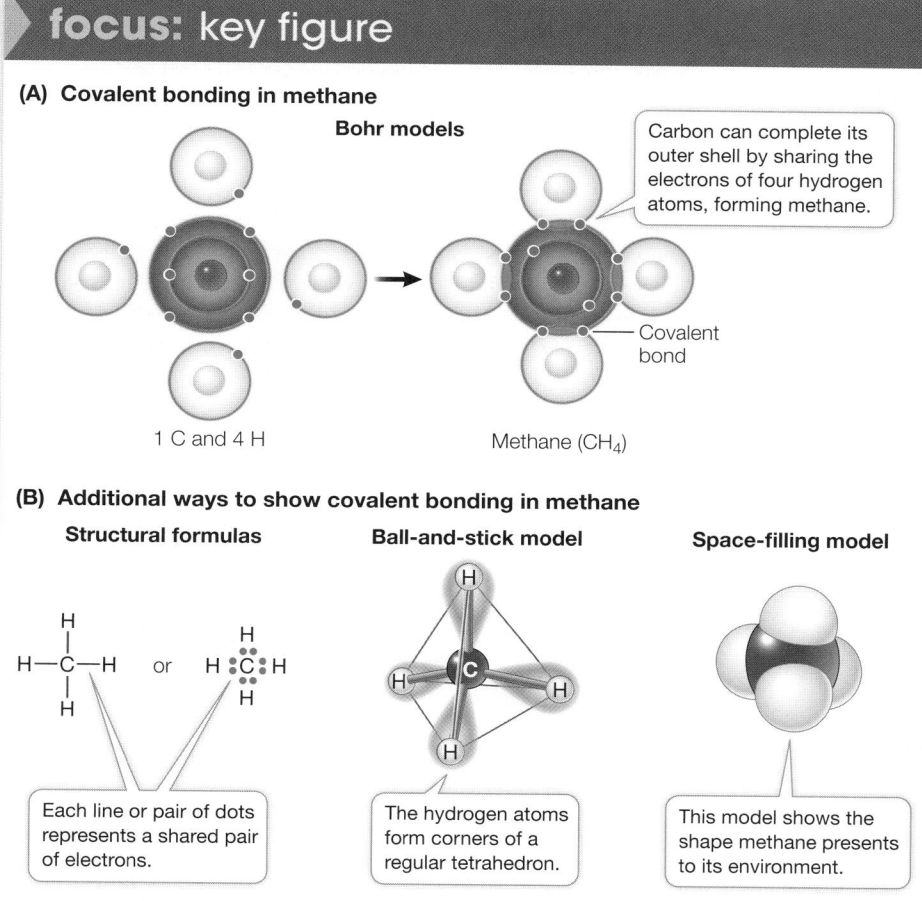

(A) Covalent bonding in methane

Bohr models

Carbon can complete its outer shell by sharing the electrons of four hydrogen atoms, forming methane.

1 C and 4 H

Covalent bond

Methane (CH_4)

(B) Additional ways to show covalent bonding in methane

Structural formulas

Ball-and-stick model

Space-filling model

Each line or pair of dots represents a shared pair of electrons.

The hydrogen atoms form corners of a regular tetrahedron.

This model shows the shape methane presents to its environment.

Figure 2.6 Covalent Bonding Can Form Compounds (A) Bohr models showing the formation of covalent bonds in methane, whose molecular formula is CH_4. Electrons are shown in shells around the nucleus. (B) Three additional ways of representing the structure of methane. In a structural formula, a covalent bond can be indicated with a single line or a shared pair of electron dots. The ball-and-stick model and the space-filling model show the spatial orientations of the bonds. The space-filling model indicates the overall shape and surface of the molecule. In the chapters that follow, different conventions will be used to depict molecules. Bear in mind that these are models to illustrate certain properties, not accurate portrayals of how atoms would actually appear.

Q: Draw the four representations for carbon dioxide.

table **2.2** Covalent Bonding Capabilities of Some Biologically Important Elements	
Element	Usual number of covalent bonds
Hydrogen (H)	1
Oxygen (O)	2
Sulfur (S)	2
Nitrogen (N)	3
Carbon (C)	4
Phosphorus (P)	5

table **2.3** Some Electronegativities	
Element	Electronegativity
Oxygen (O)	3.5
Chlorine (Cl)	3.1
Nitrogen (N)	3.0
Carbon (C)	2.5
Phosphorus (P)	2.1
Hydrogen (H)	2.1
Sodium (Na)	0.9
Potassium (K)	0.8

and the shapes of molecules contribute to their biological functions (see Key Concept 3.1).

Even though the orientations of bonds around each atom are fairly stable, the shapes of molecules can change. Think of a single covalent bond as an axle around which the two atoms, along with their other bonded atoms, can rotate.

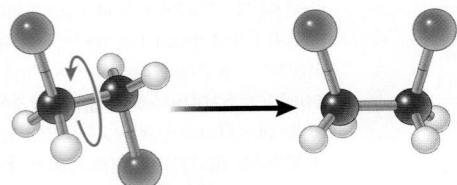

Dichloroethane

Bond rotation has enormous implications for the large molecules that make up living tissues. Since long chains of atoms (especially carbons) can rotate freely, there are many possibilities for the arrangement of atoms within the chain. This allows molecules to alter their structures, for example, to fit other molecules.

MULTIPLE COVALENT BONDS Two atoms can share more than one pair of electrons, forming multiple covalent bonds. These can be represented by lines between the chemical symbols for the linked atoms:

- A single bond involves the sharing of a single pair of electrons (for example, H—H or C—H).
- A double bond involves the sharing of four electrons (two pairs) (C=C).
- Triple bonds—six shared electrons—are rare, but there is one in nitrogen gas (N≡N), which is the major component of the air we breathe.

UNEQUAL SHARING OF ELECTRONS If two atoms of the same element are covalently bonded, there is an equal sharing of the pair(s) of electrons in their outermost shells. However, when the two atoms are of different elements, the sharing is not necessarily equal. One nucleus may exert a greater attractive force on the electron pair than the other nucleus, so that the pair tends to be closer to that atom.

The attractive force that an atomic nucleus exerts on electrons is called its **electronegativity**. The electronegativity of an atom depends roughly on how many positive charges it has (atoms with more protons are more positive and thus more attractive to electrons) and on the distance between the nucleus and the electrons

in the outer (valence) shell (the closer the electrons, the greater the electronegative pull). **Table 2.3** shows the electronegativities (which are calculated to produce dimensionless quantities) of some elements important in biological systems. Note that oxygen (O) is very electronegative; in fact, oxygen is the second most electronegative element (after fluorine). Many organisms exploit the negativity of oxygen; moving electrons between C and O atoms powers living systems, as you will see in many subsequent examples in this text.

If two atoms are close to one another in electronegativity, they will share electrons equally in what is called a **nonpolar covalent bond**. Two oxygen atoms, for example, each with an electronegativity of 3.5, will share electrons equally. So will two hydrogen atoms (each with an electronegativity of 2.1). But when hydrogen bonds with oxygen to form water, the electrons involved are unequally shared; they tend to be nearer to the oxygen nucleus because it is more electronegative than hydrogen. When electrons are drawn to one nucleus more than to the other, the result is a **polar covalent bond** (Figure 2.7).

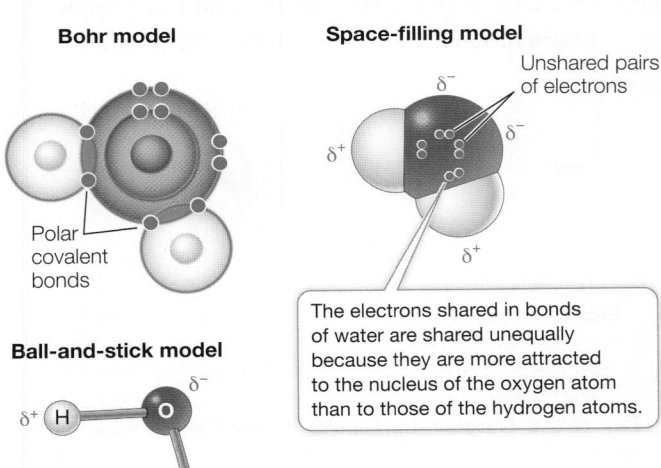

The electrons shared in bonds of water are shared unequally because they are more attracted to the nucleus of the oxygen atom than to those of the hydrogen atoms.

Figure 2.7 Water's Covalent Bonds Are Polar These three representations all illustrate polar covalent bonding in water (H_2O). When atoms with different electronegativities, such as oxygen and hydrogen, form a covalent bond, the electrons are drawn to one nucleus more than to the other. A molecule held together by such a polar covalent bond has partial (δ^+ and δ^-) charges at different surfaces. In water, the shared electrons are displaced toward the oxygen atom's nucleus.

Because of this unequal sharing of electrons, the oxygen end of the hydrogen–oxygen bond has a slightly negative charge (symbolized by δ^- and spoken of as "delta negative," meaning a partial unit of charge), and the hydrogen end has a slightly positive charge (δ^+). The bond is **polar** because these opposite charges are separated at the two ends, or poles, of the bond. The partial charges that result from polar covalent bonds produce polar molecules or polar regions of large molecules. Polar bonds within molecules greatly influence the interactions they have with other polar molecules. Water (H_2O) is a polar compound, and this polarity has significant effects on its physical properties and chemical reactivity, as you will see in later chapters.

Ionic attractions form by electrical attraction

When one interacting atom is much more electronegative than the other, a complete transfer of one or more electrons may take place. Consider sodium (electronegativity 0.9) and chlorine (electronegativity 3.1). A sodium atom has only one electron in its outermost shell; this condition is unstable. A chlorine atom has seven electrons in its outermost shell—another unstable condition. Since the electronegativity of chlorine is so much greater than that of sodium, any electrons involved in bonding will tend to transfer completely from sodium's outermost shell to that of chlorine (**Figure 2.8**). This reaction between sodium and chlorine makes the resulting atoms more

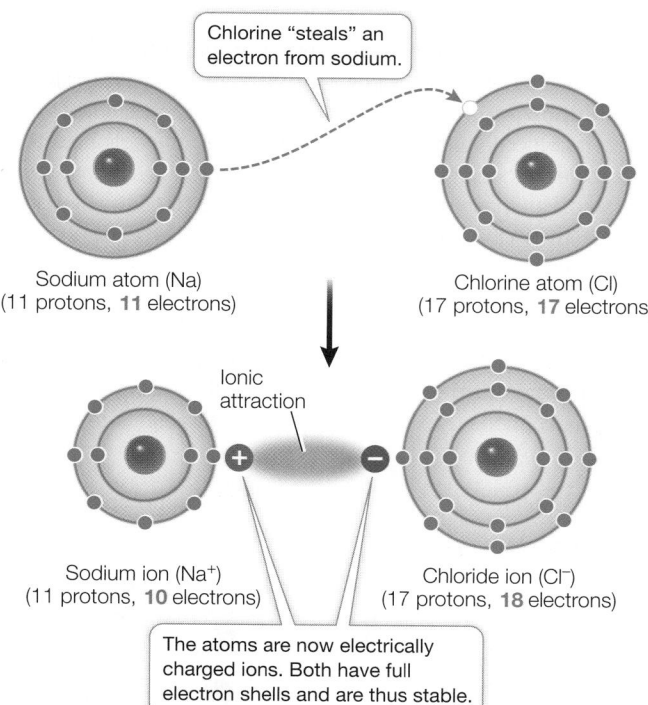

Chlorine "steals" an electron from sodium.

Sodium atom (Na)
(11 protons, **11** electrons)

Chlorine atom (Cl)
(17 protons, **17** electrons)

Ionic attraction

Sodium ion (Na$^+$)
(11 protons, **10** electrons)

Chloride ion (Cl$^-$)
(17 protons, **18** electrons)

The atoms are now electrically charged ions. Both have full electron shells and are thus stable.

Figure 2.8 Formation of Sodium and Chloride Ions When a sodium atom reacts with a chlorine atom, the more electronegative chlorine fills its outermost shell by "stealing" an electron from the sodium. In so doing, the chlorine atom becomes a negatively charged chloride ion (Cl$^-$). With one less electron, the sodium atom becomes a positively charged sodium ion (Na$^+$).

Q: What ions are formed when calcium ion reacts with chlorine? See the periodic table in Figure 2.2.

stable because they both have eight fully paired electrons in their outer shells. The result is two ions.

Ions are electrically charged particles that form when atoms gain or lose one or more electrons:

- The sodium ion (Na$^+$) in our example has a +1 unit of charge because it has one less electron than it has protons. The outermost electron shell of the sodium ion is full, with eight electrons, so the ion is stable. Positively charged ions are called **cations**. Other biologically important cations include Ca^{2+}, H$^+$, Mg^{2+}, and K$^+$.

- The chloride ion (Cl$^-$) has a −1 unit of charge because it has one more electron than it has protons. This additional electron gives Cl$^-$ a stable outermost shell with eight electrons. Negatively charged ions are called **anions**.

Some elements can form ions with multiple charges by losing or gaining more than one electron. Examples are Ca^{2+} (the calcium ion, a calcium atom that has lost two electrons) and Mg^{2+} (the magnesium ion). Two biologically important elements can each yield more than one stable ion. Iron yields Fe^{2+} (the ferrous ion) and Fe^{3+} (the ferric ion), and copper yields Cu$^+$ (the cuprous ion) and Cu^{2+} (the cupric ion). Groups of covalently bonded atoms that carry an electric charge are called **complex ions**; examples include NH$_4^+$ (the ammonium ion), SO$_4^{2-}$ (the sulfate ion), and PO$_4^{3-}$ (the phosphate ion). Once formed, ions are usually stable and no more electrons are lost or gained.

Ionic attractions are bonds formed as a result of the electrical attraction between ions bearing opposite charges. Ions can form bonds that result in stable solid compounds, which we call salts. You're familiar with sodium chloride (NaCl)—table salt; its cations (Na$^+$) and anions (Cl$^-$) are held together by ionic attractions. In solids, the attractions are strong because the ions are close together. However, when ions are dispersed in water, the distances between them can be large, greatly reducing the strength of the attraction. Under the conditions in living cells, an ionic attraction is less strong than a covalent bond (see Table 2.1).

Not surprisingly, ions can interact with polar molecules, since both are charged. This interaction results when a solid salt such as NaCl dissolves in water. Water molecules surround the individual ions, separating them (**Figure 2.9**). The negatively charged chloride ions attract the positive poles of the water molecules, while the positively charged sodium ions attract the negative poles of the water molecules. This special property of water (its polarity) is one reason it is such a good biological solvent (see Key Concept 2.4).

Hydrogen bonds may form within or between molecules with polar covalent bonds

In liquid water, the negatively charged oxygen (δ^-) atom of one water molecule is attracted to the positively charged hydrogen (δ^+) atoms of another water molecule (**Figure 2.10A**). The bond resulting from this attraction is called a **hydrogen bond**. Later in this chapter you'll see how hydrogen bonding between water molecules contributes to many of the properties that make water so important for living systems. Hydrogen bonds are not restricted to water molecules. Such a bond can also form between a strongly electronegative atom in one molecule and a hydrogen atom that is involved in a polar

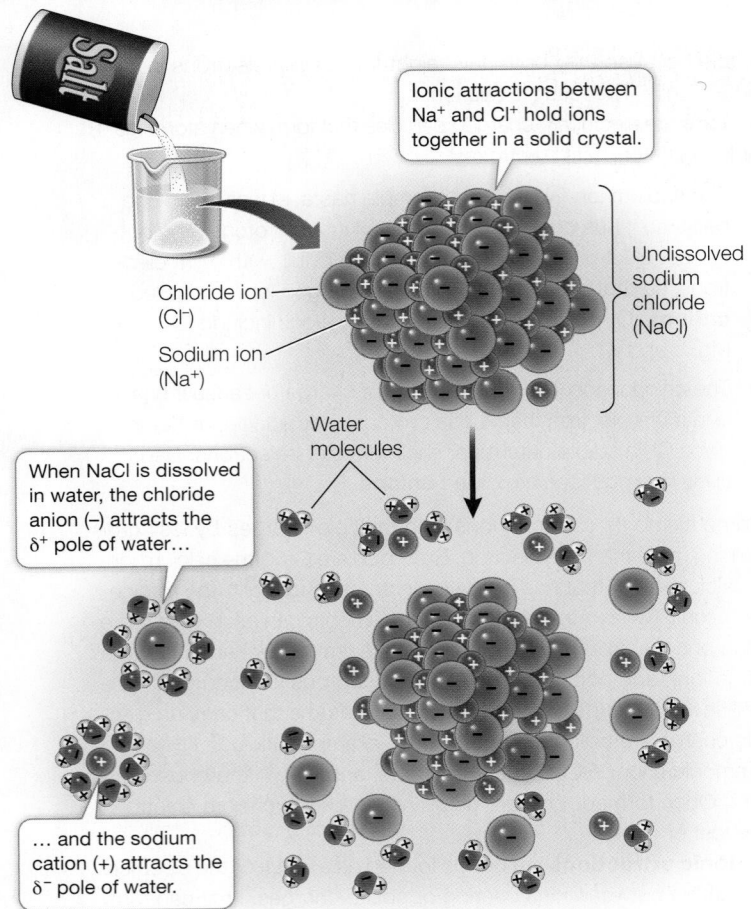

Ionic attractions between Na⁺ and Cl⁺ hold ions together in a solid crystal.

Undissolved sodium chloride (NaCl)

Chloride ion (Cl⁻)

Sodium ion (Na⁺)

Water molecules

When NaCl is dissolved in water, the chloride anion (–) attracts the δ⁺ pole of water…

… and the sodium cation (+) attracts the δ⁻ pole of water.

Figure 2.9 Water Molecules Surround Ions   When an ionic solid dissolves in water, polar water molecules cluster around the cations and anions, preventing them from reassociating.

Q: What happens at the chemical and physical levels when a salt solution evaporates?

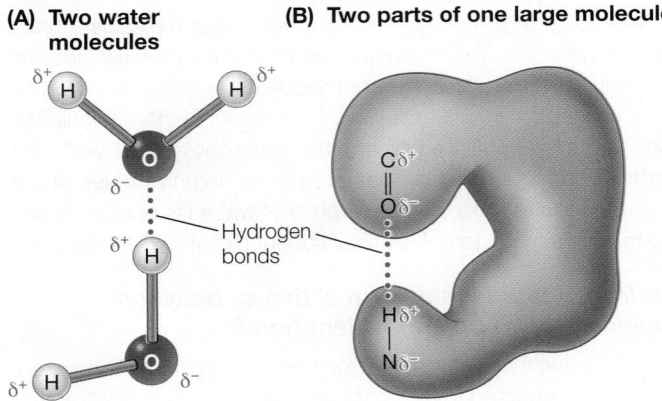

(A) Two water molecules

(B) Two parts of one large molecule

δ^+ H   H δ^+

O

δ^-

δ^+ H

Hydrogen bonds

δ^+ H

O

δ^+ H δ^-

C δ^+

O δ^-

H δ^+

N δ^-

Figure 2.10 Hydrogen Bonds Can Form between or within Molecules   **(A)** A hydrogen bond between two molecules is an attraction between a negative charge on one molecule and the positive charge on a hydrogen atom of the second molecule. **(B)** Hydrogen bonds can form between different parts of the same large molecule.

Q: What happens to a large molecule if increasing heat is applied? Which bonds or interactions are affected?

covalent bond in another molecule, or another part of the same molecule (**Figure 2.10B**).

A hydrogen bond is weaker than most ionic attractions because its formation is due to partial charges (δ^+ and δ^-). It is much weaker than a covalent bond between a hydrogen atom and an oxygen atom (see Table 2.1). Although individual hydrogen bonds are weak, there can be many of them within a single molecule or between two molecules. In these cases, the hydrogen bonds together have considerable strength and can greatly influence the structure and properties of substances. For example, hydrogen bonds play important roles in determining and maintaining the three-dimensional shapes of giant molecules such as DNA and proteins (see Key Concept 3.2).

Hydrophobic interactions bring together nonpolar molecules

Just as water molecules can interact with one another through hydrogen bonds, any molecule that is polar can interact with other polar molecules through the weak (δ^+ to δ^-) attractions of hydrogen bonds. If a polar molecule interacts with water in this way, it is called **hydrophilic** ("water-loving") (**Figure 2.11A**).

Nonpolar molecules, in contrast, tend to interact with other nonpolar molecules. For example, carbon (electronegativity 2.5) forms nonpolar bonds with hydrogen (electronegativity 2.1), and molecules containing only hydrogen and carbon atoms—called **hydrocarbon molecules**—are nonpolar. In water these molecules tend to aggregate with one another rather than with the polar water molecules. Therefore nonpolar molecules are known as **hydrophobic** ("water-hating"), and the interactions between them are called **hydrophobic interactions** (**Figure 2.11B**). Of course, hydrophobic substances do not really "hate" water; they can form weak interactions with it, since the electronegativities of carbon and hydrogen are not exactly the same. But these interactions are far weaker than the hydrogen bonds between the water molecules (see Table 2.1), so the nonpolar substances tend to aggregate.

van der Waals forces involve contacts between atoms

The interactions between nonpolar substances are enhanced by **van der Waals forces**, which occur when the atoms of two molecules are in close proximity. These brief interactions result from

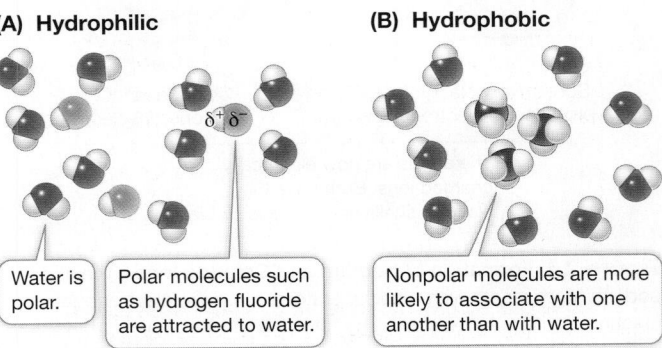

(A) Hydrophilic

(B) Hydrophobic

δ^+ δ^-

Water is polar.

Polar molecules such as hydrogen fluoride are attracted to water.

Nonpolar molecules are more likely to associate with one another than with water.

Figure 2.11 Hydrophilic and Hydrophobic   **(A)** Molecules with polar covalent bonds are attracted to polar water (they are hydrophilic). **(B)** Molecules with nonpolar covalent bonds show greater attraction to one another than to water (they are hydrophobic).

random variations in the electron distribution in one molecule, which can create opposite charge distributions in the adjacent molecule. So there will be a weak, temporary δ^+ to δ^- attraction. van der Waals forces occur in polar and nonpolar molecules. In the latter case, they can induce cohesion between molecules that do not form ionic attractions of hydrogen bonds. Although a single van der Waals interaction is brief and weak, the sum of many such interactions over the entire span of a large nonpolar molecule can result in substantial attraction. This is important when hydrophobic regions of different molecules such as an enzyme and a substrate come together (see Chapter 8).

2.2 recap

Some atoms form strong covalent bonds with other atoms by sharing one or more pairs of electrons. Unequal sharing of electrons produces polarity. Other atoms become ions by losing or gaining electrons, and they interact with other ions or polar molecules. Weak forces such as hydrogen bonds also act to attract atoms to one another.

learning outcomes

You should be able to:
- Explain why covalent bonds are so strong.
- Identify examples of polar covalent and nonpolar covalent bonds.
- Distinguish between hydrophilic and hydrophobic interactions using examples.
- Identify instances in which van der Waals forces are important.

1. Why is a covalent bond stronger than an ionic attraction when a molecule is in water?

2. Consider the following pairs of bonded atoms:

C—H   C＝O   O—P   C—C

For each pair, use Tables 2.2 and 2.3 to determine:
a. whether the bond is polar or nonpolar
b. if the bond is polar, which end is δ^-
c. whether the atom pair contributes hydrophilic or hydrophobic properties to a molecule

3. Geckos are lizards that are amazing climbers. A gecko can climb up a glass surface and stick to it with a single toe. Professor Kellar Autumn at Lewis and Clark College and his students and collaborators have shown that each toe of a gecko has millions of micrometer-sized hairs, and that each hair splits into hundreds of 200-nanometer tips that provide intimate contact with a surface. Careful measurements show that a million of these tips could easily support the animal, but it has far more. The toes stick well on hydrophilic and hydrophobic surfaces. Bending the hairs allows the gecko to detach. What kind of noncovalent force is involved in gecko sticking?

The bonding of atoms into molecules is not necessarily a permanent affair. The dynamic aspect of life involves constant change, even at the molecular level. In the next section we will examine how molecules interact with one another—how they break up, how they find new partners, and what the consequences of those changes can be.

key concept 2.3 Atoms Change Partners in Chemical Reactions

A hallmark of life is that it is dynamic. Things don't sit still, especially at the chemical level. Atoms bonded to other atoms can leave a binding relationship and find a new partner. Bear in mind that this does not happen all the time: your skin is a biological tissue, and its atoms form a stable structure. But change is common, especially among the atoms and molecules dissolved in water. A **chemical reaction** occurs when moving atoms collide with sufficient energy to combine or to change their bonding partners.

focus your learning
- Chemical reactions obey the laws of conservation of energy and conservation of matter.
- Chemical reactions are accompanied by changes in energy.

Consider the combustion reaction that takes place in the flame of a propane stove. When propane (C_3H_8) reacts with oxygen gas (O_2), the carbon atoms become bonded to oxygen atoms instead of hydrogen atoms, and the hydrogen atoms become bonded to oxygen instead of carbon (**Figure 2.12**). As the covalently bonded atoms change partners, the composition of the matter changes; propane and oxygen gas become carbon dioxide and water. This chemical reaction can be represented by the equation

$$C_3H_8 + 5\,O_2 \rightarrow 3\,CO_2 + 4\,H_2O + Energy$$

Reactants → Products

In this equation, propane and oxygen are the **reactants**, and carbon dioxide and water are the **products**. This example is an

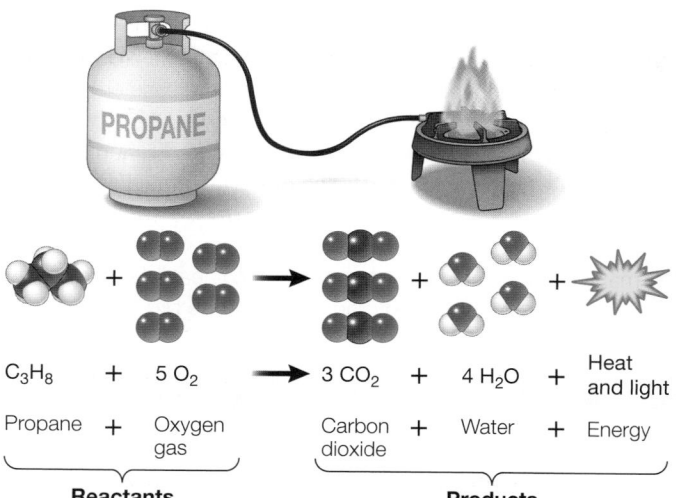

C_3H_8	+	5 O_2	→	3 CO_2	+	4 H_2O	+	Heat and light
Propane	+	Oxygen gas		Carbon dioxide	+	Water	+	Energy

Reactants   **Products**

Figure 2.12 Bonding Partners and Energy May Change in a Chemical Reaction One molecule of propane (a gas used for cooking) from this burner reacts with five molecules of oxygen gas to give three molecules of carbon dioxide and four molecules of water. This reaction releases energy in the form of heat and light.

oxidation-reduction (redox) reaction—a special type of chemical reaction that involves a transfer of electrons between two species:

- The electron acceptor, or oxidizing agent, gains electrons and is said to be *reduced* in a chemical reaction.
- The electron donor, or reducing agent, loses electrons and is said to be is *oxidized* in a chemical reaction.

Can you identify the oxidizing and reducing agents in our example? Because electrons and protons (i.e., hydrogen atoms) are transferred *from* propane, propane is the reducing agent and oxygen is the oxidizing agent, which in this example forms water upon accepting the hydrogens. You will see many examples of redox reactions involving electron/proton transfer in later chapters.

The products of a chemical reaction can have very different properties from the reactants. The reaction shown in Figure 2.12 is said to be *complete*: all the propane and oxygen are used up in forming the two products. The arrow symbolizes the direction of the chemical reaction. The numbers preceding the molecular formulas indicate how many molecules are used or produced.

Note that in this and all other chemical reactions, *matter is neither created nor destroyed*. The total number of carbon atoms on the left side of the equation (3) equals the total number of carbon atoms on the right (3). In other words, the equation is *balanced*. However, there is another aspect of this reaction: the heat and light of the stove's flame reveal that the reaction between propane and oxygen releases a great deal of energy.

Energy is defined as the capacity to do work, but in the context of chemical reactions, it can be thought of as the capacity for change. Chemical reactions do not create or destroy energy, but *changes in the form of energy* usually accompany chemical reactions.

In the reaction between propane and oxygen, a large amount of heat energy is released. This energy was present in another form, called potential chemical energy, in the covalent bonds within the propane and oxygen gas molecules. Not all reactions release energy; indeed, many chemical reactions require that energy be supplied from the environment. Some of this energy is then stored as potential chemical energy in the bonds formed in the products. We will see in future chapters how reactions that release energy and reactions that require energy are often coupled.

Many chemical reactions take place in living cells, and some of these have a lot in common with the oxidation–reduction reaction we saw in the combustion of propane. In cells, the reactants are different (they may be sugars or fats), and the reactions proceed by many intermediate steps that permit the released energy to be harvested and put to use by the cells. But the products are the same: carbon dioxide and water.

> **2.3 recap**

In a chemical reaction, a set of reactants is converted to a set of products with different chemical compositions. A chemical reaction involves breaking old bonds and making new ones. A reaction may release energy or require its input.

learning outcomes

You should be able to:

- Demonstrate an example to show that chemical reactions obey the law of conservation of matter.

> **2.3 recap**

- Explain how energy is involved in a chemical reaction.

1. Using the example of the chemical equation $C_6H_{12}O_6 + O_2 \rightarrow CO_2 + H_2O$, insert numbers so that the numbers of atoms on both sides of the equation are balanced.
2. Use an example to explain how the form of energy can change during a chemical reaction.

We will return to chemical reactions and how they occur in living systems in Chapter 8, 9, and 10, focusing particularly on energy transformations that fuel biological processes. First, however, we will examine the unique properties of the substance in which most biochemical reactions take place: water.

> **key concept** **Water Is Critical for Life**
> **2.4**

Your body is more than 70 percent water by weight, excluding the minerals contained in bones. Water is the dominant component of virtually all living organisms, and most biochemical reactions take place in this watery, or aqueous, environment. Water is an unusual substance with unusual properties. Under conditions on Earth, water exists in solid, liquid, and gas forms, all of which have relevance to living systems. Water allows chemical reactions to occur inside living organisms, and it is necessary for the formation of certain biological structures. In this section we will explore how the structure and interactions of water molecules make water essential to life.

focus your learning

- Biochemical reactions within cells occur in an aqueous environment.
- Moles are used to quantify substances in biological fluids.
- The acid–base properties of water and other compounds in cells allow reversible changes that affect biological functions.

Water has a unique structure and special properties

The molecule H_2O has unique chemical features. As you have already learned, water is a polar molecule that can form hydrogen bonds. The four pairs of electrons in the outer shell of the oxygen atom repel one another, giving the water molecule a tetrahedral shape:

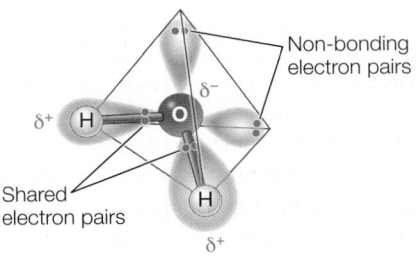

The chemical features of water explain some of its interesting properties, such as the ability of ice to float, the melting and freezing

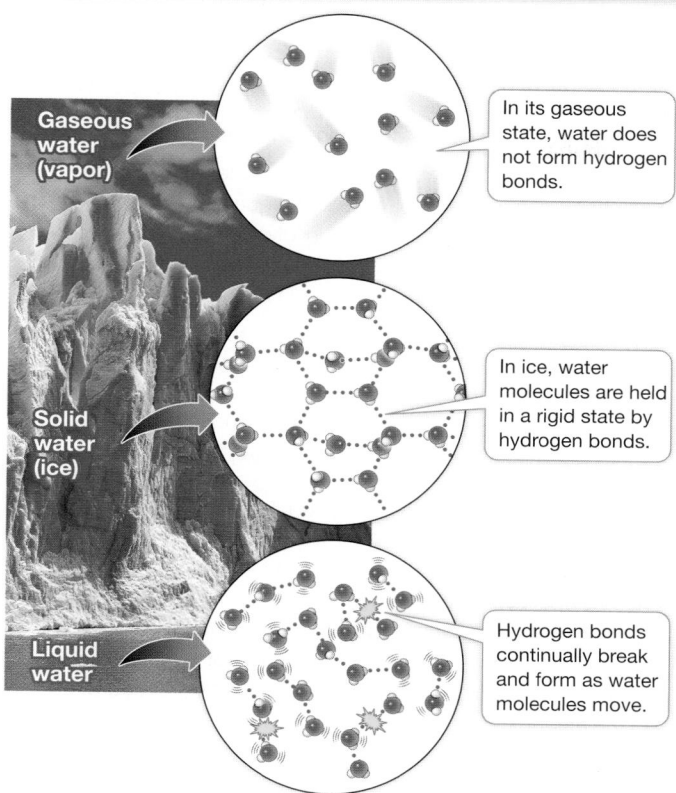

In its gaseous state, water does not form hydrogen bonds.

In ice, water molecules are held in a rigid state by hydrogen bonds.

Hydrogen bonds continually break and form as water molecules move.

Gaseous water (vapor)

Solid water (ice)

Liquid water

Figure 2.13 Hydrogen Bonding and the Properties of Water
Hydrogen bonding occurs between the molecules of water in both its liquid and solid states. Ice is more structured but less dense than liquid water, which is why ice floats. Water forms a gas when its hydrogen bonds are broken and the molecules move farther apart.

and five times that of sand. This is why when you are on the beach as the sun sets, the sand cools much more quickly than the water.

Because it takes a lot of energy to change the physical state of water, water temperatures in oceans and other large bodies of water are remarkably constant throughout the year. The temperature changes of coastal land masses are in turn moderated by large bodies of water. Indeed, water helps minimize variations in atmospheric temperature across the planet. Water's ability to moderate heat helps it function as an insulator, helping prevent an organism's body temperature from rising during a sunny day.

Water also has a high **heat of vaporization**, which means that a lot of heat is required to change water from its liquid to its gaseous state (the process of evaporation). Once again, much of the heat energy is used to break the many hydrogen bonds between the water molecules. This heat must be absorbed from the environment in contact with the water. ***Evaporation** thus has a cooling effect on the environment—whether a leaf, a forest, or an entire land mass. This effect explains why sweating cools the human body: as sweat evaporates from the skin, it uses up some of the adjacent body heat (**Figure 2.14A**).

***connect the concepts**   Living systems use the evaporation of water, which disrupts hydrogen bonds, to dissipate excess heat that would otherwise cause problems. See Key Concept 38.3 for examples in plants, and Key Concepts 39.3–39.5 for examples in animals.

---

temperatures of water, the ability of water to store heat, the formation of water droplets, water's ability to dissolve many substances, and its inability to dissolve many others.

ICE FLOATS   In water's solid state (ice), individual water molecules are held in place by hydrogen bonds. Each molecule is bonded to four other molecules in a rigid, crystalline structure (**Figure 2.13**). Although the molecules are held firmly in place, they are farther apart from one another than they are in liquid water, where the molecules are moving about. In ice, there are cavities between individual water molecules. In other words, solid water is less dense than liquid water, which is why ice floats.

Think of the biological consequences if ice were to sink in water. A pond would freeze from the bottom up, becoming a solid block of ice in winter and killing most of the organisms living there. Once the whole pond was frozen, its temperature could drop well below the freezing point of water. But because ice floats, it forms an insulating layer on the top of the pond, and reduces heat flow to the cold air above. Thus fish, plants, and other organisms in the pond are not subjected to temperatures lower than 0°C, which is the freezing point of pure water.

MELTING, FREEZING, AND HEAT CAPACITY   Compared with many other substances that have molecules of similar size, ice requires a great deal of heat energy to melt. The amount of heat energy required to raise the temperature of 1 gram of a substance by 1°C is called its **specific heat**. Water has a relatively high specific heat because so many hydrogen bonds connecting the water molecules in ice must be broken to change water from solid to liquid. In the opposite process—freezing—a great deal of energy is released to the environment. We say water has a high heat capacity. For example, water has twice the specific heat of ethyl alcohol

COHESION AND SURFACE TENSION   In liquid water, individual molecules are able to move about. The hydrogen bonds between the molecules continually form and break (see Figure 2.14). Chemists estimate that this occurs about a trillion times a minute for a single water molecule!

At any given time, a water molecule forms on average 3.4 hydrogen bonds with other water molecules. These hydrogen bonds explain the cohesive strength of liquid water. This cohesive strength, or **cohesion**, is defined as the capacity of water molecules to resist coming apart from one another when placed under tension. Water's ***cohesive strength** permits narrow columns of liquid water to move from the roots to the leaves of tall trees. When water evaporates from the leaves, the entire column moves upward in response to the pull of the molecules at the top (**Figure 2.14B**). A related property is **adhesion**, the attraction of water molecules to other molecules of a different type. For example, when you put a straw into a cup with water, it "climbs" up the straw so the column is higher than the level in the cup. This adhesive behavior of water—sticking to the sides of the straw—reflects the adhesion of water forming the column.

***connect the concepts**   As described in Key Concept 34.2, the transpiration–cohesion–tension mechanism accounts for the movement of water from roots to leaves. Because of the cohesion between water molecules, water is pulled upward in continuous columns.

(A)

High heat of vaporization: Sweating uses evaporation of water to cool the body.

(B)

Cohesion: Water's cohesive strength helps it to flow from the roots to the leaves in a tree.

(C)

Surface tension: Water molecules stick to one another and help prevent this wolf spider from sinking.

Figure 2.14 Water in Biology   These three properties of water make it beneficial to organisms.

The surface of liquid water exposed to the air is difficult to puncture because the water molecules at the surface are hydrogen-bonded to other water molecules below them. This surface tension of water permits a container to be filled slightly above its rim without overflowing, and it permits spiders to walk on the surface of a pond (**Figure 2.14C**).

The reactions of life take place in aqueous solutions

A **solution** is produced when a substance (the **solute**) is dissolved in a liquid (the **solvent**). If the solvent is water, then the solution is called an aqueous solution. As you know by now, water is polar. Because many important molecules in biological systems are polar, they readily dissolve in water. Being soluble doesn't mean that the molecules lose their identity and properties. They can still react, and indeed many important biochemical reactions occur in aqueous solutions.

Biologists who are interested in the biochemical reactions within cells identify the reactants and products and determine their amounts using two different types of analyses:

1. *Qualitative analyses* focus on identifying the substances involved in chemical reactions. For example, a qualitative analysis would be used to investigate the steps involved and the products formed during respiration, when carbon-containing compounds are broken down to release energy in living tissues.

2. *Quantitative analyses* measure concentrations or amounts of substances. For example, a biochemist would use a quantitative analysis to measure how much of a certain product is formed in a chemical reaction. What follows is a brief introduction to some of the quantitative chemical terms you will see in this book.

Fundamental to quantitative thinking in chemistry and biology is the concept of the mole. A **mole** is the amount of a substance (in grams) that is numerically equal to its molecular weight. So a mole of hydrogen gas (H_2) weighs 2 g, a mole of sodium ion (Na^+) weighs 23 g, and a mole of table sugar ($C_{12}H_{22}O_{11}$) weighs about 342 g.

Quantitative analyses do not yield counts of molecules. Because the amount of a substance in 1 mole is directly related to its molecular weight, it follows that the number of molecules in 1 mole is constant for all substances. So 1 mole of salt contains the same number of molecules as 1 mole of table sugar. This constant number of molecules in a mole is called **Avogadro's number**, and it is 6.02×10^{23} molecules per mole. Chemists work with moles of substances (which can be weighed in the laboratory) instead of actual molecules, which are too numerous to be counted. Consider 34.2 g (just over 1 ounce) of table sugar, $C_{12}H_{22}O_{11}$. This is one-tenth of a mole, or one-tenth of Avogadro's number: 6.02×10^{22} molecules.

A chemist can dissolve 1 mole of table sugar (342 g) in water to make 1 liter of solution, knowing that the mole contains 6.02×10^{23} individual sugar molecules. This solution—1 mole of a substance dissolved in water to make 1 liter—is called a 1 molar (1 M) solution. When a physician injects a certain volume and molar concentration of a drug into the bloodstream of a patient, a rough calculation can be made of the actual number of drug molecules that will interact with the patient's cells. As you know, the dose is important.

The many molecules dissolved in the water of living tissues are not present at concentrations anywhere near 1 molar. Most are in the micromolar (millionths of a mole per liter of solution; μM) to millimolar (thousandths of a mole per liter; mM) range. Some, such as hormone molecules, are even less concentrated than that. While these molarities seem to indicate very low concentrations, remember that even a 1 μM solution has 6.02×10^{17} molecules of the solute per liter.

Aqueous solutions may be acidic or basic

When some substances dissolve in water, they release hydrogen ions (H^+), which are actually single, positively charged protons. Hydrogen ions can interact with other molecules and change their properties. For example, the protons in "acid rain" can damage plants, and you probably have experienced the excess of hydrogen ions that we call "acid indigestion."

Here we will examine the properties of **acids** (defined as substances that release H^+) and **bases** (defined as substances that accept H^+). We will distinguish between strong and weak acids and bases, and provide a quantitative means for stating the concentration of H^+ in solutions: the pH scale.

ACIDS RELEASE H^+

When hydrochloric acid (HCl) is added to water, it dissolves, releasing the ions H^+ and Cl^-:

$$HCl \rightarrow H^+ + Cl^-$$

Because its H^+ concentration has increased, the solution is acidic.

Acids are substances that *release* H^+ ions in solution. HCl is an acid, as is H_2SO_4 (sulfuric acid). One molecule of sulfuric acid will ionize to yield two H^+ and one SO_4^{2-}. Biological compounds that contain —COOH (called a carboxyl group) are acids because the carboxyl group can ionize to —COO$^-$, releasing H^+:

$$-COOH \rightarrow -COO^- + H^+$$

Acids that fully ionize in solution, such as HCl and H_2SO_4, are called strong acids. However, not all acids ionize fully in water. For example, if acetic acid (CH_3COOH) is added to water, some of it will dissociate into two ions (CH_3COO^- and H^+), but some of the original acetic acid will remain as well. Because the reaction is not complete, acetic acid is a weak acid.

BASES ACCEPT H^+

Bases are substances that *accept* H^+ in solution. As with acids, there are strong and weak bases. If NaOH (sodium hydroxide) is added to water, it dissolves and ionizes, releasing OH$^-$ and Na$^+$ ions:

$$NaOH \rightarrow Na^+ + OH^-$$

Because OH$^-$ absorbs H^+ to form water, such a solution is basic. This reaction is complete, and so NaOH is a strong base.

Weak bases include the bicarbonate ion (HCO_3^-), which can accept an H^+ ion and become carbonic acid (H_2CO_3), and ammonia (NH_3), which can accept H^+ and become an ammonium ion (NH_4^+). Biological compounds that contain —NH_2 (the amino group) are also bases because —NH_2 accepts H^+:

$$-NH_2 + H^+ \rightarrow -NH_3^+$$

ACID–BASE REACTIONS MAY BE REVERSIBLE

When acetic acid is dissolved in water, two reactions happen. First, the acetic acid forms its ions:

$$CH_3COOH \rightarrow CH_3COO^- + H^+$$

Then, once the ions are formed, some of them re-form acetic acid:

$$CH_3COO^- + H^+ \rightarrow CH_3COOH$$

This pair of reactions is reversible. A **reversible reaction** can proceed in either direction—left to right or right to left—depending on the relative starting concentrations of the reactants and products. The formula for a reversible reaction can be written using a double arrow:

$$CH_3COOH \sim CH_3COO^- + H^+$$

In terms of acids and bases, there are two types of reactions, depending on the extent of the reversibility:

1. The ionization of strong acids and bases in water is virtually irreversible.

2. The ionization of weak acids and bases in water is somewhat reversible.

WATER IS A WEAK ACID AND A WEAK BASE

The water molecule has a slight but significant tendency to ionize into a hydroxide ion (OH$^-$) and a hydrogen ion (H^+). Actually, two water molecules participate in this reaction. One of the two molecules "captures" a hydrogen ion from the other molecule, forming a hydroxide ion and a hydronium ion:

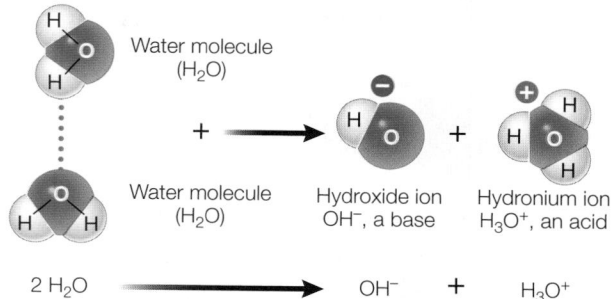

The hydronium ion is, in effect, a hydrogen ion bound to a water molecule. For simplicity, biochemists tend to use a modified representation of the ionization of water:

$$H_2O \rightarrow H^+ + OH^-$$

The ionization of water is important to all living creatures. This fact may seem surprising, since only about 1 water molecule in 500 million is ionized at any given time. But this is less surprising if we focus on the abundance of water in living systems, and the reactive nature of the H^+ ions produced by ionization.

pH: HYDROGEN ION CONCENTRATION

As we have seen, compounds can be either acids or bases, and thus solutions can be either acidic or basic. We can measure how acidic or basic a solution is by measuring its concentration of H^+ in moles per liter (its molarity; see p. 36). Here are some examples:

- Pure water has a H^+ concentration of 10^{-7} M.

- A 1 M HCl solution has a H^+ concentration of 1 M (recall that all the HCl dissociates into its ions).

- A 1 M NaOH solution has a H^+ concentration of 10^{-14} M.

This is a very wide range of numbers to work with—think about the decimals! It is easier to work with the logarithm of the H^+ concentration, because logarithms compress this range: the $\log_{10}$ of 100, for example, is 2; and the $\log_{10}$ of 0.01 is −2. Because most H^+ concentrations in living systems are less than 1 M, their $\log_{10}$ values are negative. For convenience, we convert these negative numbers into positive ones by using the *negative* of the logarithm of the H^+ molar concentration. This number is called the **pH** of the solution:

$$pH = -\log(H^+) \text{ where } (H^+) \text{ is the concentration of free } H^+ \text{ in solution}$$

Since the H^+ concentration of pure water is 10^{-7} M, its pH is $-\log(10^{-7})$ = −(−7), or 7. A smaller negative logarithm means a larger number. In practical terms, a lower pH means a higher H^+ concentration, or greater acidity. In 1 M HCl, the H^+ concentration is 1 M, so the pH is the negative logarithm of 1 ($-\log 10^0$), or 0. The pH of 1 M NaOH is the negative logarithm of 10^{-14}, or 14.

A solution with a pH of less than 7 is acidic—it contains more H^+ ions than OH$^-$ ions. A solution with a pH of 7 is referred to as neutral,

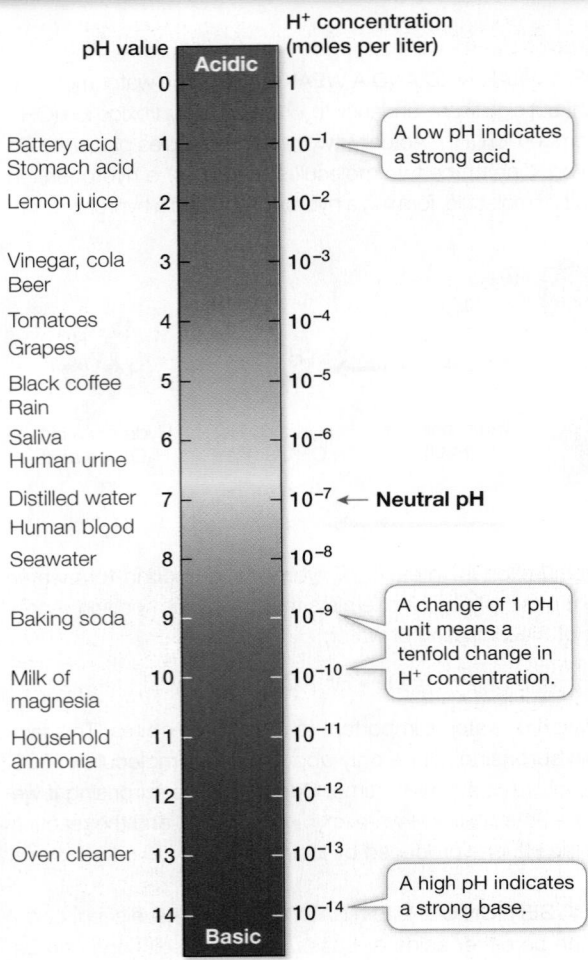

Figure 2.15 pH Values of Some Familiar Substances

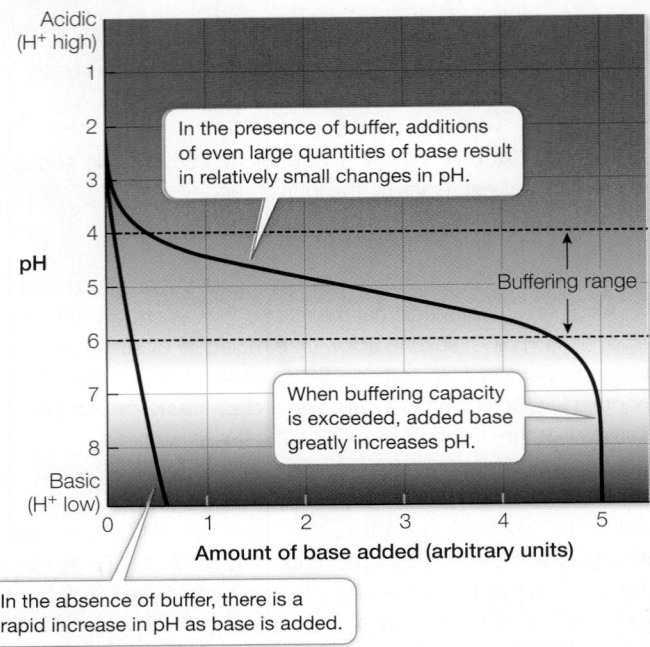

Figure 2.16 Buffers Minimize Changes in pH When base is added to a solution, the pH of the solution increases. Without a buffer, the change is large and the slope of the pH graph is steep. In the presence of a buffer, however, the slope within the buffering range is shallow.

and a solution with a pH value greater than 7 is basic. **Figure 2.15** shows the pH values of some common substances.

Why is this discussion of pH so relevant to biology? Many reactions involve the transfer of an ion or charged group from one molecule to another, and the presence of positive or negative ions in the environment can greatly influence the rates of such reactions. Furthermore, pH can influence the shapes of molecules. Many biologically important molecules contain charged groups (e.g., —COO⁻) that can interact with the polar regions of water, and these interactions influence the way such molecules fold up into three-dimensional shapes. If these charged groups combine with H⁺ or other ions in their environment to form uncharged groups (e.g., —COOH; see above), they will have a reduced tendency to interact with water. These uncharged (hydrophobic) groups might induce the molecule to fold up differently so that they are no longer in contact with the watery environment. Since the three-dimensional structures of biological molecules greatly affect the way they function, organisms do all they can to minimize changes in the pH of their cells and tissues. An important way to do this is with buffers.

BUFFERS The maintenance of internal constancy—homeostasis—is a hallmark of all living things and extends to pH. If biological molecules lose or gain H⁺ ions, their properties can change, thus upsetting homeostasis. For example, people who exercise strenuously produce lactic acid, which ionizes to produce H⁺. Internal constancy within tissues can be achieved with buffers: solutions that

maintain a relatively constant pH even when substantial amounts of acid or base are added. How does this work?

A **buffer** is a mixture of a weak acid and its corresponding base, or a weak base and its corresponding acid. For example, a weak acid is carbonic acid (H_2CO_3), and its corresponding base is the bicarbonate ion (HCO_3^-). If another acid is added to a solution containing this mixture (a buffered solution), not all the H⁺ ions from the acid remain in solution. Instead, many of them combine with the bicarbonate ions to produce more carbonic acid:

$$HCO_3^- + H^+ \rightleftharpoons H_2CO_3$$

This reaction uses up some of the H⁺ ions in the solution and decreases the acidifying effect of the added acid. If a base is added, the reaction essentially reverses. Some of the carbonic acid ionizes to produce bicarbonate ions and more H⁺, which counteracts some of the added base. The buffer minimizes the effect that an added acid or base has on pH. The carbonic acid/bicarbonate buffering system is present in the blood, where it is important for preventing significant changes in pH that could disrupt the ability of the blood to carry vital oxygen to tissues. A given amount of acid or base causes a smaller pH change in a buffered solution than in a non-buffered one (**Figure 2.16**).

Buffers illustrate an important chemical principle of reversible reactions, called the **law of mass action**. Addition of a reactant on one side of a reversible system drives the reaction in the direction that uses up that compound. In the case of buffers, addition of an acid drives the reaction in one direction; addition of a base drives the reaction in the other direction.

You may occasionally use a buffer to relieve indigestion. The lining of the stomach constantly secretes hydrochloric acid, making the stomach contents acidic. But excessive stomach acid can cause discomfort. We can relieve this discomfort by ingesting a salt such as $NaHCO_3$ (sodium bicarbonate), which acts as a buffer.

2.4 recap

Most of the chemistry of life occurs in water, which has unique properties that make it an ideal medium for supporting life. These properties include hydrogen bonding between molecules, high specific heat, and cohesion between molecules. Aqueous solutions can be acidic or basic, depending on the concentration of hydrogen ions. The cells and tissues of organisms are buffered, however, because changes in pH can change the properties of biological molecules.

learning outcomes

You should be able to:

- Explain why water's solvent properties are important in understanding events taking place inside cells.
- Perform quantitative analyses of biochemical compounds using the mole concept.

- Predict and explain changes in pH caused by changes in concentration of weak acids or bases in living tissue.

1. What is a solution, and why do we call water "the medium of life"?
2. A substance in a cell has a concentration of 0.00000001 molar. What does this mean in terms of the number of molecules in a cell that has a volume of 0.000001 liters?
3. In human tissues, CO_2 is formed as a by-product of oxidation of more complex molecules. CO_2 enters the blood-stream and in red blood cells is removed from the blood by the reaction $CO_2 + H_2O \rightarrow HCO_3^- + H^+$. What is the consequence of this reaction on the pH of the blood? What chemical reaction would restore the pH of the blood?

investigatinglife

What insights into biological systems have been discovered through isotope analysis?

The opening investigation of dinosaurs (oxygen atoms) and the investigating life experiment of hamburgers (carbon atoms) are examples of identifying the life history of organisms through isotopic analysis. Recently, this has been applied to humans as well. Hair is a living tissue, with a lot of O and H atoms derived from water in the diet. Ecologist Jim Ehleringer and chemist Thure Cerling have shown that like the dinosaurs, the isotopic ratios of O and H in hair reflect the geographical region where a person drank the water. This can be useful evidence in a forensic investigation, placing a person at a location. Isotopic ratios are also used to identify where plants are grown, because their tissues also have unique ratios. For example, ratios have revealed the geographical origin of poppies used to make heroin. The mass spectrometer, a chemical analysis instrument that detects isotopes, is fast becoming an important tool for biologists.

Future directions

Tracking the isotopes of oxygen and hydrogen in precipitation is useful in understanding patterns of climate change. Water evaporates in warmer regions at the tropical latitudes on Earth and moves toward the cooler poles. As an air mass moves from a warmer to a cooler region, water vapor condenses and is removed as precipitation. The heavy isotopes of H and O tend to fall as precipitation more readily than the lighter isotopes, so as the water vapor moves toward the poles, it becomes enriched in the lighter isotopes. The ratio of heavy to light isotopes that reach the poles depends on the climate—the cooler the climate, the lower the ratio, because more water precipitates as it moves toward the poles, depleting more of the heavier isotopes. Analyses of polar ice cores show that heavy-to-light isotope ratios vary over geological time scales. This has allowed scientists to reconstruct climate change in the past, and to relate it to fossil organisms that lived at those times. Isotope ratios may be useful in monitoring climate change that is happening today.

Chapter Summary 2

2.1 Atomic Structure Explains the Properties of Matter

- Matter is composed of atoms. Each **atom** consists of a positively charged **nucleus** made up of **protons** and **neutrons**, surrounded by **electrons** bearing negative charges. **Review Figure 2.1**
- The number of protons in the nucleus defines an **element**. There are many elements in the universe, but only a few of them make up the bulk of living organisms: C, H, O, P, N, and S. **Review Figure 2.2**
- **Isotopes** of an element differ in their numbers of neutrons. **Radioisotopes** are radioactive, emitting radiation as they break down.
- Electrons are distributed in **electron shells**, which are volumes of space defined by specific numbers of orbitals. Each **orbital** contains a maximum of two electrons. **Review Figure 2.4, Activity 2.1**

- In losing, gaining, or sharing electrons to become more stable, an atom can combine with other atoms to form a **molecule**.

2.2 Atoms Bond to Form Molecules

- A **chemical bond** is an attractive force that links two atoms together in a molecule. **Review Table 2.1, Animation 2.1**
- A **compound** is a substance made up of molecules with two or more different atoms bonded together in a fixed ratio, such as water (H_2O).
- **Covalent bonds** are strong bonds formed when two atoms share one or more pairs of electrons. **Review Figure 2.5, Focus: Key Figure 2.6**
- When two atoms of unequal electronegativity bond with each other, a **polar covalent bond** is formed. The two ends, or poles, of the bond have partial charges (δ^+ or δ^-). **Review Figure 2.7**

(continued)

Chapter Summary 2 (continued)

- An **ion** is an electrically charged particle that forms when an atom gains or loses one or more electrons in order to form a more stable electron configuration. **Anions** and **cations** are negatively and positively charged ions, respectively. Different charges attract each other, and like charges repel each other.
- **Ionic attractions** occur between oppositely charged ions. Ionic attractions are strong in solids (salts) but weaken when the ions are separated from one another in solution. Review **Figure 2.9**
- A **hydrogen bond** is a weak electrical attraction that forms between a δ^+ hydrogen atom in one molecule and a δ^- atom in another molecule (or in another part of the same, large molecule). Hydrogen bonds are abundant in water. Review **Figure 2.10**
- Nonpolar molecules interact very little with polar molecules, including water. Nonpolar molecules are attracted to one another by very weak bonds called **van der Waals forces**.

2.3 Atoms Change Partners in Chemical Reactions

- In **chemical reactions**, atoms combine or change their bonding partners. **Reactants** are converted into **products**.
- Some chemical reactions release **energy** as one of their products; other reactions can occur only if energy is provided to the reactants.
- Neither matter nor energy is created or destroyed in a chemical reaction, but both change form. Review **Figure 2.12**

- In organisms, chemical reactions take place in multiple steps so that released energy can be harvested for cellular activities.

2.4 Water Is Critical for Life

- Water's molecular structure and its capacity to form hydrogen bonds give it unique properties that are significant for life. Review **Figure 2.13**
- The high **specific heat** of water means that water gains or loses a great deal of heat when it changes state. Water's high **heat of vaporization** ensures effective cooling when water evaporates.
- The **cohesion** of water molecules refers to their capacity to resist coming apart from one another. Hydrogen bonding between the water molecules plays an essential role in this property.
- A **solution** is produced when a solid substance (the **solute**) dissolves in a liquid (the **solvent**). Water is the critically important solvent for life. The ability of molecules to lose or accept protons (H^+) is important in biological structure and function.

Go to **LearningCurve** (in **LaunchPad**) for dynamic quizzing that helps you solidify your understanding of this chapter. **LearningCurve** adapts to your responses, giving you the practice you need to master each key concept.

Apply What You've Learned

Review

2.4 The acid–base properties of water and other compounds in cells allow reversible changes that affect biological functions.

Scientists are developing methods to sample a person's breath and test it for higher-than-normal levels of compounds that contain ketones. The buildup of these compounds in the body can indicate type I diabetes. This work could lead to development of a portable device that will allow earlier diagnosis of diabetes, especially in children.

Insulin is a hormone that stimulates cells in the body to take up glucose from the blood. Cells use glucose as a source of energy for normal cell function. In type I diabetes, insulin is not produced, and glucose does not enter cells from the blood. Lacking an incoming supply of glucose, cells begin breaking down fats as a source of energy. When fats break down, several ketone-containing compounds are produced and they can be detected by a breath analysis device. A ketone has the general structure RCOR', where R and R' are the same or different groups of atoms.

One of the ketone-containing products of fat breakdown is acetoacetic acid, which has the structure below. Acetoacetic acid is very harmful when it builds up in high concentration in the body.

$$H_3C - \underset{\underset{O}{\|}}{C} - CH_2 - \underset{\underset{O}{\|}}{C} - OH$$

Questions

1. In water, acetoacetic acid can ionize to form protons and acetoacetate ions. Write the chemical structures for this reaction.

2. Now write out the reversible chemical reaction for the ionization of carbonic acid in blood, which can act as a buffer. Would the pH of blood change if a small amount of acetoacetic acid produced normally in tissues were added to the blood? Explain.

3. What would happen to the pH of the blood if more and more acetoacetic acid built up, as in a case of untreated diabetes? Explain, using what you found in Question 2 above. What would happen to the body's ability to maintain a constant blood pH? How does this explain why people with this condition become severely ill?

4. Acetone is one of the compounds produced from fat breakdown and expelled in the breath. It is targeted for detection by a breath-analysis device. Its chemical structure is shown below. Analyze the structure to determine whether it would be expected to affect the carbonic acid or bicarbonate buffer system in the blood.

$$H_3C - \underset{\underset{O}{\|}}{C} - CH_3$$

5. Suppose scientists succeed in developing a device that detects ketone buildup in the breath. If a person tested positive for higher-than-normal ketones with the breath test, what are some blood tests that could be done to follow up? Suggest three or four different blood tests, other than acetoacetic acid or ketone levels, that could be used to detect the existence of insulin insufficiency. Describe the outcome of each test in the case of a positive result.

Go to **LaunchPad** for the eBook, LearningCurve, animations, activities, flashcards, and additional resources and assignments.

3

Proteins, Carbohydrates, and Lipids

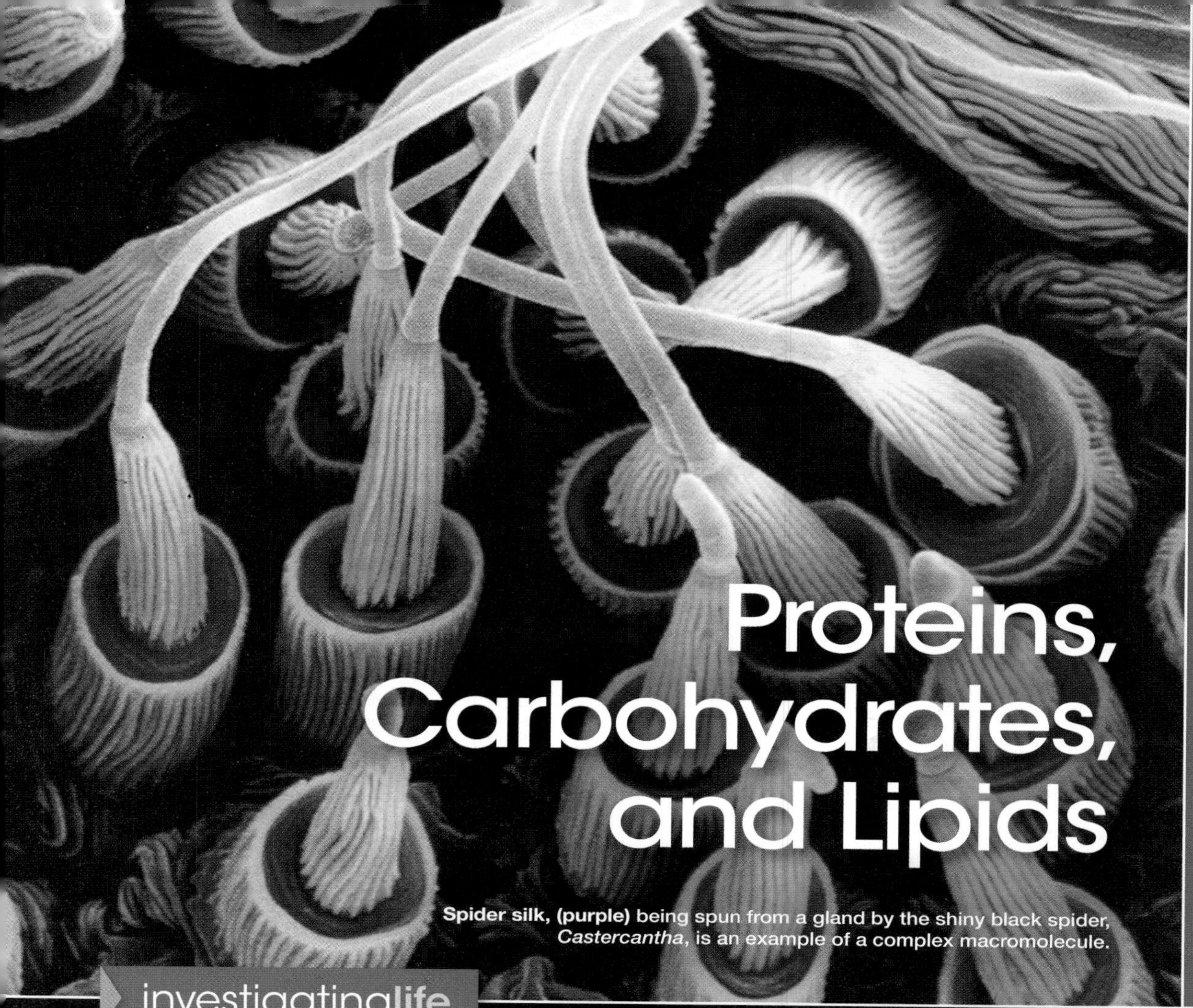

Spider silk, (purple) being spun from a gland by the shiny black spider, *Castercantha*, is an example of a complex macromolecule.

key concepts

3.1 Macromolecules Characterize Living Things

3.2 The Function of a Protein Depends on Its Three-Dimensional Structure

3.3 Simple Sugars Are the Basic Structural Unit of Carbohydrates

3.4 Lipids Are Defined by Their Solubility Rather Than by Chemical Structure

investigatinglife

Weaving a Web

A spider web is an amazing structure. It is not only beautiful to look at, but an architectural wonder serving as the spider's home, its mating place, and its way to capture food. Consider a fly that chances to interact with a spider web. The fibers of the web must slow down the fly, but they cannot break, so they need to stretch to dissipate the energy of the fly's movement. The fibers holding the fly cannot stretch too much; they must be strong enough to hold the web in place. Web fibers are far thinner than a human hair, yet they are very strong. The fibers can also be long; for example, the Darwin's bark spider makes strands up to 25 meters long.

Chemical investigations reveal that spider silk is made of variations on a single type of large molecule called a protein. Proteins are polymers: long chains of individual smaller units called amino acids. The proteins in spider silks have characteristic structures and amino acid compositions that determine particular functions. Proteins in the stretchy web fibers have amino acids that allow them to curl into spirals, and these spirals can slip along one another to change the fiber's length. Another kind of spider silk is the dragline silk, which is less stretchy and used to construct the outline of the web, its spokes, and the lifeline of the spider. The proteins in these strong fibers are made up of amino acids that cause the proteins to fold into flat sheets with ratchets, so that parallel sheets can fit together like Lego blocks. This arrangement makes these fibers hard to pull apart.

Just how strong is spider silk? A popular series of films features the fictional character Spider-Man. He often uses his web to ensnare evildoers and protect the innocent. In one scene, he stops a train by using ten ropes made of spider silk, each about 1 centimeter thick. Scientists who study spider silk have estimated the number of silk fibers in each rope and calculated that this scene would indeed be possible. All of this from covalent bonds holding long chains of amino acids together and weaker forces between the chains! Given the amazing strength shown by these proteins, it is not surprising that there is a lot of interest in using spider silk for human activities.

Q What are practical uses for spider silk?

key concept

3.1 Macromolecules Characterize Living Things

Four kinds of molecules are characteristic of living things: proteins, carbohydrates, lipids, and nucleic acids. With the exception of the lipids, these biological molecules are **polymers** (*poly*, "many," + *mer*, "unit") constructed by the covalent bonding of smaller molecules called **monomers**. Each kind of biological molecule is made up of monomers with similar chemical structures:

- Proteins are formed from different combinations of 20 amino acids, all of which share chemical similarities.
- Carbohydrates can form giant molecules by linking together chemically similar sugar monomers (monosaccharides) to form polysaccharides.
- Nucleic acids are formed from four kinds of nucleotide monomers linked together in long chains.
- Lipids also form large structures from a limited set of smaller molecules, but in this case noncovalent forces maintain the interactions between the lipid monomers that are held together by covalent bonds.

Polymers containing thousands or more atoms are called **macromolecules**. The proteins, carbohydrates, and nucleic acids of living systems certainly fall into this category. Although large lipid structures are not polymers in the strictest sense, it is convenient to treat them as macromolecules (see Key Concept 3.4). Green plants have the living world's most abundant protein (rubisco; see Chapter 10), most abundant carbohydrate (cellulose in plant cell walls), and most abundant lipid (monogalactosyl diglyceride in leaves).

focus your learning

- Isomers are molecules having the same composition but different structures.
- Monomers are chemically linked via condensation reactions to form polymers.

Chemical groupings determine the structures of macromolecules

Certain small groups of atoms, called **functional groups**, occur frequently in biological molecules (**Figure 3.1**). Each functional group has specific chemical properties, and when it is attached to a larger molecule, it confers those properties on the larger molecule. One of these properties is polarity. Looking at the structures in Figure 3.1, can you determine which functional groups are the most polar? (Hint: look for C—O, N—H, and P—O bonds.) The consistent chemical

Figure 3.1 Some Functional Groups Important to Living Systems   Highlighted here are the eight functional groups most commonly found in biologically important molecules. "R" is a variable chemical grouping.

Functional group	Class of compounds and an example	Properties
Hydroxyl 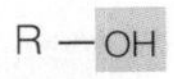 R—OH	**Alcohols** 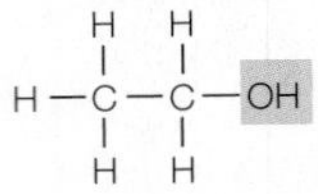 (Ethanol)	Polar. Forms hydrogen bonds with water to help dissolve molecules. Enables linkage to other molecules by condensation.
Aldehyde 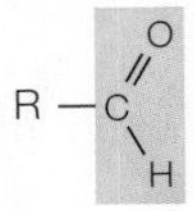	**Aldehydes** 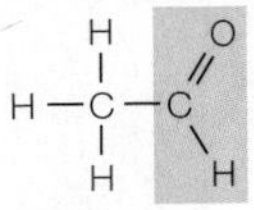 (Acetaldehyde)	Polar. C=O group is very reactive. Important in building molecules and in energy-releasing reactions.
Keto 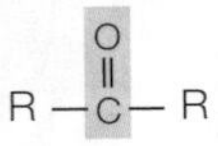	**Ketones** 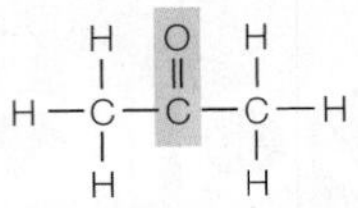 (Acetone)	Polar. C=O group is important in carbohydrates and in energy reactions.
Carboxyl 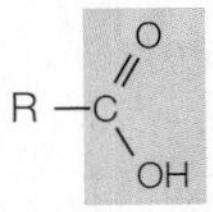	**Carboxylic acids** 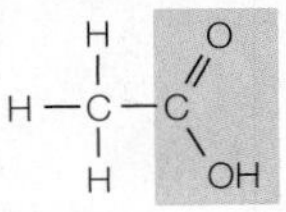 (Acetic acid)	Charged; acidic. Ionizes in living tissues to form —COO⁻ and H⁺. Enters into condensation reactions by giving up —OH. Some carboxylic acids important in energy-releasing reactions.
Amino 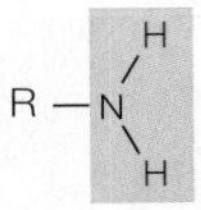	**Amines** 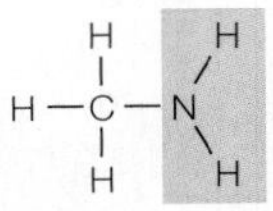 (Methylamine)	Charged; basic. Accepts H⁺ in living tissues to form —NH₃⁺. Enters into condensation reactions by giving up H⁺.
Phosphate 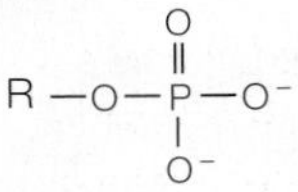	**Organic phosphates** 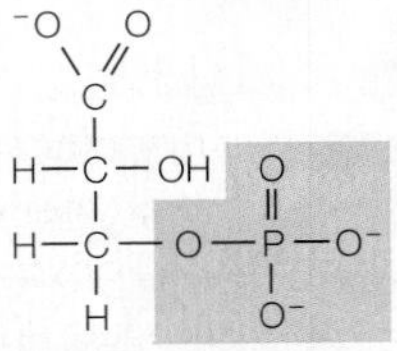 (3-Phosphoglycerate)	Charged; acidic. Enters into condensation reactions by giving up —OH. When bonded to another phosphate, hydrolysis releases much energy.
Sulfhydryl 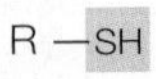 R—SH	**Thiols** 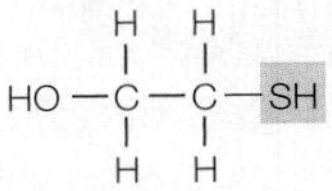 (Mercaptoethanol)	By giving up H, two —SH groups can react to form a disulfide bridge, thus stabilizing protein structure.
Methyl 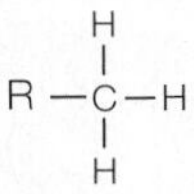	**Alkyl** 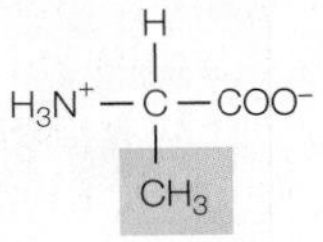 (Alanine)	Nonpolar. Important in interacting with other nonpolar molecules and in energy transfer.

Activity 3.1 **Functional Groups**
www.Life11e.com/ac3.1

Animation 3.1 **Proteins, Carbohydrates, and Lipids**
www.Life11e.com/a3.1

(A) Structural isomers

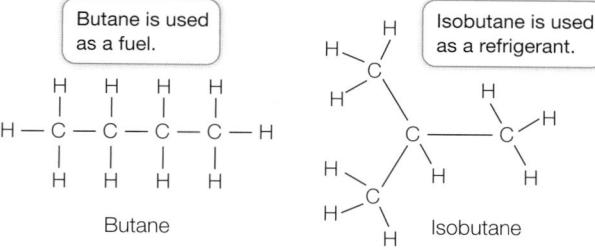

Butane is used as a fuel.

Isobutane is used as a refrigerant.

Butane

Isobutane

(B) *cis-trans* isomers

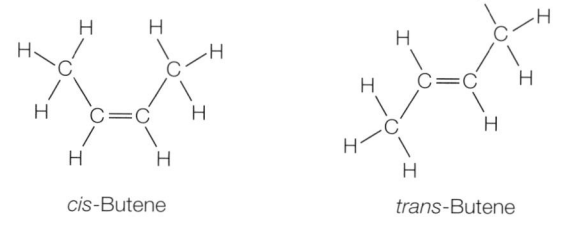

cis-Butene

trans-Butene

(C) Optical isomers

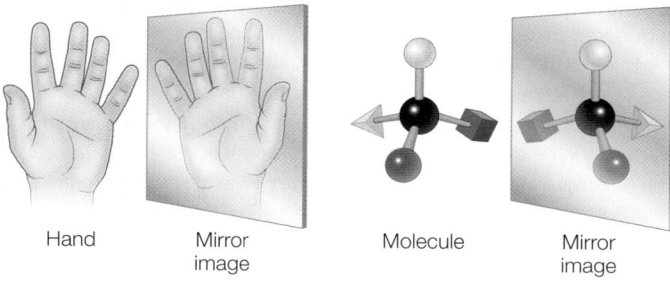

Hand

Mirror image

Molecule

Mirror image

Figure 3.2 Isomers Isomers have the same chemical formula, but the atoms are arranged differently. Pairs of isomers often have different chemical properties and functions.

behavior of functional groups helps us understand the properties of the molecules that contain them.

Macromolecules have many different functional groups. A single large protein may contain nonpolar, polar, and charged functional groups, each of which gives different specific properties to local sites on the macromolecule. As we will see, sometimes these different groups interact within the same macromolecule. They help determine the shape of the macromolecule as well as how it interacts with other macromolecules and with smaller molecules.

Using the same atoms, molecules can differ from one another because their functional groups can be arranged differently. **Isomers** are molecules that have the same ("iso") chemical formula—the same kinds and numbers of atoms—but with the atoms arranged differently. Of the different kinds of isomers, we will consider three: structural isomers, *cis-trans* isomers, and optical isomers.

1. **Structural isomers** differ in how their atoms are joined together. Consider two simple molecules, each composed of four carbon and ten hydrogen atoms bonded covalently, both with the formula C_4H_{10}. These atoms can be linked in two different ways, resulting in different molecules (**Figure 3.2A**).

2. ***cis-trans* isomers** typically involve a double bond between two carbon atoms, where the carbons share two pairs of electrons. When the remaining two bonds of each of these carbons

are to two different atoms or groups of atoms (e.g., a hydrogen and a methyl group; **Figure 3.2B**), these can be oriented on the same side or different sides of the double-bonded molecule. If the different atoms or groups of atoms are on the same side, the double bond is called *cis*; if they are on opposite sides, the bond is *trans*. These molecules can have very different properties.

3. **Optical isomers** occur when a carbon atom has four different atoms or groups of atoms attached to it. This pattern allows for two different ways of making the attachments, each the mirror image of the other (**Figure 3.2C**). Such a carbon atom is called an asymmetric carbon, and the two resulting molecules are optical isomers of one another. You can envision your right and left hands as optical isomers. Just as a glove is specific for a particular hand, some biochemical molecules that can interact with one optical isomer of a carbon compound are unable to "fit" the other.

The structures of macromolecules reflect their functions

The four kinds of biological macromolecules are present in roughly the same proportions in all living organisms (**Figure 3.3**). Also, a protein that has a certain structure, and therefore function, in an apple tree probably has a similar structure and function in a human being, because the protein's chemistry is the same wherever it is found. This impressive biochemical unity reflects the evolution of all life from a common ancestor, by descent with modification. An important advantage of biochemical unity is that some organisms can acquire needed raw materials by eating other organisms. When you eat an apple, the molecules you take in include carbohydrates, lipids, and proteins that can be broken down and rebuilt into the varieties of those molecules needed by humans.

Each kind of macromolecule performs one or more functions such as energy storage, structural support, catalysis (speeding up of chemical reactions), transport of other molecules, regulation of other molecules, defense, movement, or information storage. These roles are not necessarily exclusive; for example, both carbohydrates and proteins can play structural roles, supporting and protecting tissues and organs. However, only the nucleic acids specialize in information storage and transmission. These macromolecules function as

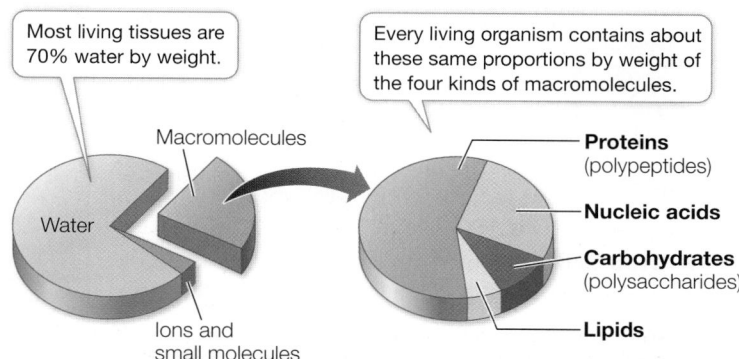

Most living tissues are 70% water by weight.

Every living organism contains about these same proportions by weight of the four kinds of macromolecules.

Macromolecules

Water

Ions and small molecules

Proteins (polypeptides)

Nucleic acids

Carbohydrates (polysaccharides)

Lipids

Figure 3.3 Substances Found in Living Tissues The substances shown here make up the nonmineral components of living tissues (bone would be an example of a mineral component).

experiment

Original Paper: Teule, F., Y.-G. Miao, B.-H. Sohn, Y.-S. Kim, J. Hull, M. J. Fraser, R. V. Lewis and D. L. Jarvis. 2012. Silkworms transformed with chimeric silkworm/spider silk genes spin composite silk fibers with improved mechanical properties. *Proceedings of the National Academy of Sciences USA*. 109: 923–928.

Spider silk is one of the strongest materials known, but it is very hard to harvest in quantity. Industrial biologists are therefore using genetic engineering to produce the protein fibers. The scientists have coaxed silkworms, insect larvae that make silk fibers used in clothing, to make a much stronger composite of silkworm and spider silk.

HYPOTHESIS▶   Genetically engineered silkworms can produce silk with physical properties like those of silk made by spiders.

METHOD

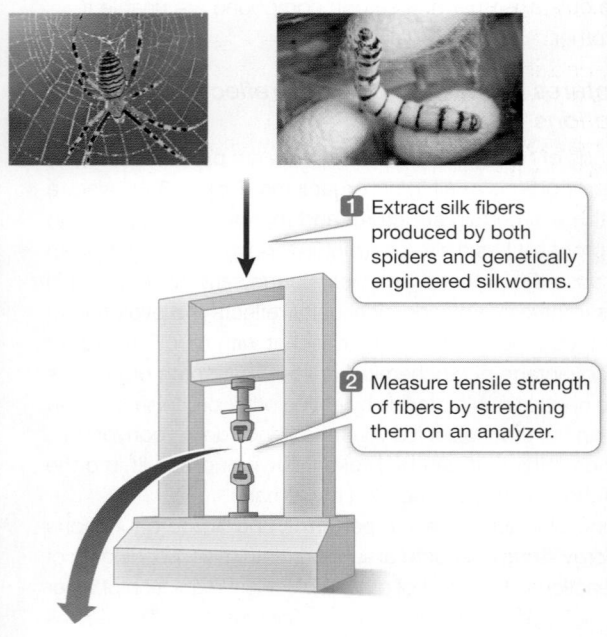

1 Extract silk fibers produced by both spiders and genetically engineered silkworms.

2 Measure tensile strength of fibers by stretching them on an analyzer.

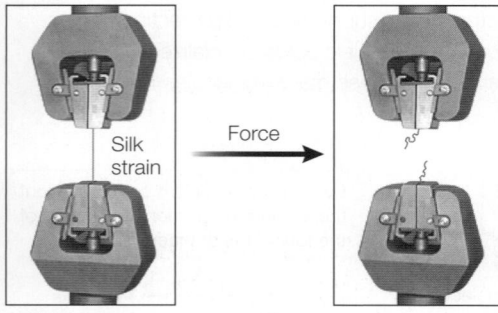

Silk strain

Force

RESULTS   Both silks had identical physical properties of stress (force needed to break) and strain (stretching).

CONCLUSION▶   Silkworms can make composite silk fiber with the same properties as those of the native spider silk protein.

work with the data

Spider silk is a protein-based material that is very strong because of its secondary structure. Because of its potential uses by humans, it is desirable to obtain large amounts of these fibers. Spiders spin out the silk fibers for their web, but the amounts are insufficient for human use. Silkworms make abundant silk fibers that are used in textiles but are much weaker than spider silk. An international team of scientists led by Randy Lewis at the University of Wyoming genetically engineered silkworms to make composite silkworm–spider silk fibers in large quantities. They then tested the properties of these fibers and compared them with those of native spider silk.

QUESTIONS▶

1. To evaluate the properties of the protein fibers, researchers stretched them and measured the forces involved as well as the resulting lengthening of the fibers. You can liken this to stretching a rubber band until it breaks. Stress is the force required to break the fiber; it is measured in millipascals (mPa; 1 Pa is the force in newtons per unit area in m^2). Strain is a measure of how much the fiber lengthens (as a percentage of its original length). **Table A** compares the results from silk from spiders and from genetically engineered silkworms. Did the genetically engineered silkworms make fibers that were similar to native spider silk?

2. Why was thickness (diameter) reported in the data?

3. What statistical test would you perform to determine if the measured properties were significantly different (see Appendix B)?

4. The scientists measured the properties composite of spider silk compared with those of Kevlar (a synthetic fiber with many uses, including in bicycle tires, racing sails, and body armor) and steel. The results are shown in **Table B**. What can you conclude about the spider silk?

Table A

Silk	Diameter (μm)	Max. strain (%)
Silkworms	21.8 ± 1.6	22.0 ± 5.8
Silkworms, gen. engineer	21.1 ± 1.4	31.8 ± 5.2
Spiders	8.1 ± 0.4	19.7 ± 4.8

Table B

Material	Energy required to break filament (joules/kg)	Weight (g/mm³)
Spider silk	120,000	1.3
Kevlar	40,000	1.4
Steel	3,500	7.84

A similar **work with the data** exercise may be assigned in **LaunchPad**.

hereditary material, carrying the traits of both species and individuals from generation to generation.

The sequence and chemical properties of the chain of monomers in a macromolecule determine its three-dimensional shape and function. Some macromolecules fold into compact forms with surface features that make them water-soluble and capable of intimate

(A) Condensation reactions produce water.

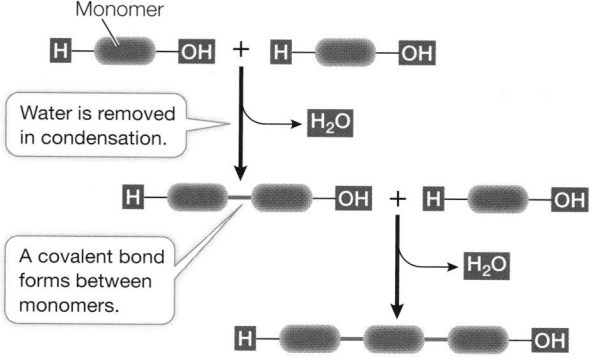

Monomer

Water is removed in condensation.

A covalent bond forms between monomers.

(B) Hydrolysis reactions consume water.

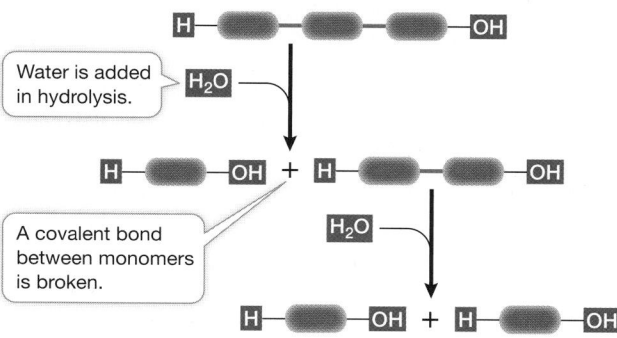

Water is added in hydrolysis.

A covalent bond between monomers is broken.

Figure 3.4 Condensation and Hydrolysis of Polymers
(A) Condensation reactions link monomers into polymers and produce water. **(B)** Hydrolysis reactions break polymers into individual monomers and consume water.

interactions with other molecules. Some proteins and carbohydrates form long, fibrous structures (such as those found in spider silk or hair) that provide strength and rigidity. The specific structure of a macromolecule determines its function in a given environment, regardless of its origin. For example, composite spider silk fiber made by another organism (e.g., the silkworm, a moth larva) has the same properties as spider silk made by a spider (**Investigating Life: Making Spider Silk**).

Most macromolecules are formed by condensation and broken down by hydrolysis

Polymers are formed from monomers by a series of **condensation reactions** (in this case called dehydration reactions; both terms refer to the loss of water). Condensation reactions result in the formation of covalent bonds between monomers. A molecule of water is released with each covalent bond formed (**Figure 3.4A**). The condensation reactions that produce the different kinds of polymers differ in detail, but in all cases polymers form only if water molecules are removed and energy is added to the system. In living systems, specific energy-rich molecules supply the necessary energy.

The reverse of a condensation reaction is a **hydrolysis reaction** (*hydro*, "water," + *lysis*, "break"). Hydrolysis reactions result in the breakdown of polymers into their component monomers. Water reacts with the covalent bonds that link the polymer together. For each covalent bond that is broken, a water molecule splits into

two ions (H^+ and OH^-), each of which becomes part of one of the products (**Figure 3.4B**).

3.1 recap

The four kinds of large molecules that distinguish living tissues are proteins, lipids, carbohydrates, and nucleic acids. Most are polymers: chains of linked monomers. Very large polymers are called macromolecules. Biological molecules carry out a variety of life-sustaining functions.

learning outcomes

You should be able to:
- Distinguish between structural, *cis-trans*, and optical isomers.
- Give an example of two molecules that have the same functional groups but different properties, and explain the reason for the differences.
- Sketch the chemical structures of two biological monomers, and show how a condensation reaction produces a covalent bond between the two.

1. What are the differences between structural, *cis-trans*, and optical isomers?
2. Examine the isomers mannose and galactose below. What makes them isomers of each other? Which functional groups do these carbohydrates contain, and what properties do these functional groups give to the molecules?

α-Mannose   α-Galactose

3. Below are the general structures of three monomers in a biological polymer where A, B, and C are numerous atoms. Draw the reaction that links them together to form a polymer ABC.

$$H-A-OH \qquad H-B-OH \qquad H-C-OH$$

The four types of macromolecules can be seen as the building blocks of life. We will cover the unique properties of the nucleic acids in Chapter 4. The remainder of this chapter will describe the structures and functions of the proteins, carbohydrates, and lipids.

key concept 3.2 The Function of a Protein Depends on Its Three-Dimensional Structure

Proteins have very diverse roles. You'll learn about the many functions of proteins (**Table 3.1**) in virtually every chapter of this book. **Proteins** are polymers made up of 20 amino acids in different proportions and sequences. Proteins range in size, from small ones such as the human hormone insulin, which has 51 amino acids and a molecular weight of 5,733, to huge molecules such as

table 3.1 Proteins and Their Functions

Category	Function
Enzymes	Catalyze (speed up) biochemical reactions
Structural proteins	Provide physical stability and movement
Defensive proteins	Recognize and respond to nonself substances (e.g., antibodies)
Signaling proteins	Control physiological processes (e.g., hormones)
Receptor proteins	Receive and respond to chemical signals
Membrane transporters	Regulate passage of substances across cellular membranes
Storage proteins	Store amino acids for later use
Transport proteins	Bind and carry substances within the organism
Gene regulatory proteins	Determine the rate of expression of a gene
Motor proteins	Cause movement of structures in the cell

the muscle protein titin, with 26,926 amino acids and a molecular weight of 2,993,451. Proteins consist of one or more **polypeptide chains**—unbranched (linear) polymers of covalently linked amino acids. Variation in the sequences of amino acids in the polypeptide chains allows for the vast diversity in protein structure and function. Each chain folds into a particular three-dimensional shape that is specified by the sequence of amino acids present in the chain.

focus your learning

- A protein's tertiary structure describes its three-dimensional shape and is stabilized by hydrogen bonds, hydrophobic interactions, ionic attractions, and in some proteins, disulfide bonds.
- Exposed surface groups on a protein provide both shapes and chemical groups that can interact specifically with other molecules or ions.
- The forces that stabilize a protein's secondary, tertiary, and quaternary structures make the protein vulnerable to disruption by environmental factors.

Monomers of proteins link together to make the macromolecule

Each **amino acid** has both a carboxyl functional group and an amino functional group (see Figure 3.1) attached to the same carbon atom, called the α (alpha) carbon. Also attached to the α carbon atom are a hydrogen atom and a **side chain**, or **R group**, designated by the letter R.

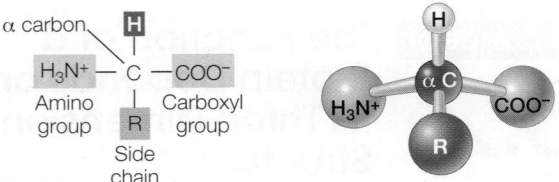

The α carbon in this example is asymmetrical because it is bonded to four different atoms or groups of atoms. Therefore amino acids can exist as optical isomers called D-amino acids and L-amino acids. D and L are abbreviations of the Latin terms for right (*dextro*) and left (*levo*). Only L-amino acids (with the configuration shown above) are commonly found in the proteins of most organisms, and their presence is an important chemical "signature" of life.

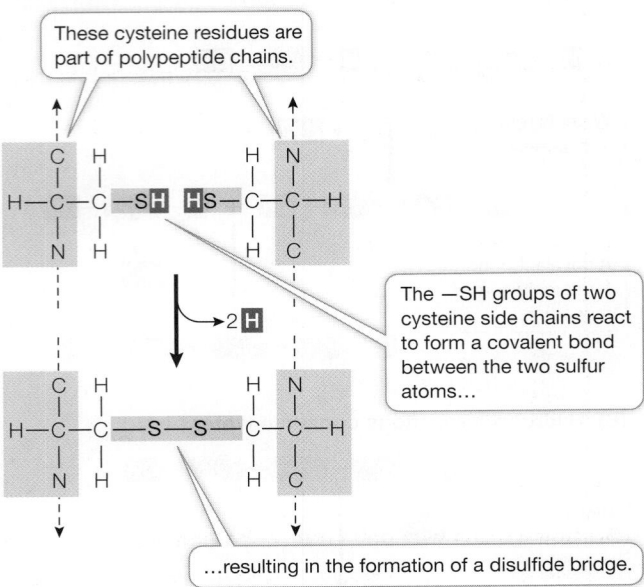

Figure 3.5 A Disulfide Bridge Two cysteine molecules in a polypeptide chain can form a disulfide bridge (—S—S—) by oxidation (removal of H atoms).

At the pH levels typically found in cells (usually about pH 7), both the carboxyl and amino groups of amino acids are ionized: the carboxyl group has lost a hydrogen ion:

$$-COOH \rightarrow -COO^- + H^+$$

and the amino group has gained a hydrogen ion:

$$-NH_2 + H^+ \rightarrow -NH_3^+$$

Thus amino acids are simultaneously acids and bases.

 **Activity 3.2 Features of Amino Acids**
www.Life11e.com/ac3.2

The side chains (or R groups) of amino acids contain functional groups that are important in determining the three-dimensional structure and thus the function of the protein. As **Table 3.2** shows, the 20 amino acids found in living organisms are grouped and distinguished by their side chains:

- Five amino acids have electrically charged (ionized) side chains at pH levels typical of living cells. These side chains attract water (are hydrophilic) and attract oppositely charged ions of all sorts.
- Five amino acids have polar side chains. They are also hydrophilic and attract other polar or charged molecules.
- Seven amino acids have side chains that are nonpolar and thus hydrophobic. In the watery environment of the cell, these hydrophobic groups may cluster together in the interior of the protein.

Three amino acids—cysteine, glycine, and proline—are special cases, although the side chains of the latter two are generally hydrophobic.

- The cysteine side chain, which has a terminal —SH group, can react with another cysteine side chain in an oxidation reaction to form a covalent bond (**Figure 3.5**). Such a bond, called a **disulfide bridge** or disulfide bond (—S—S—), helps determine how a polypeptide chain folds.

table 3.2   The Twenty Amino Acids

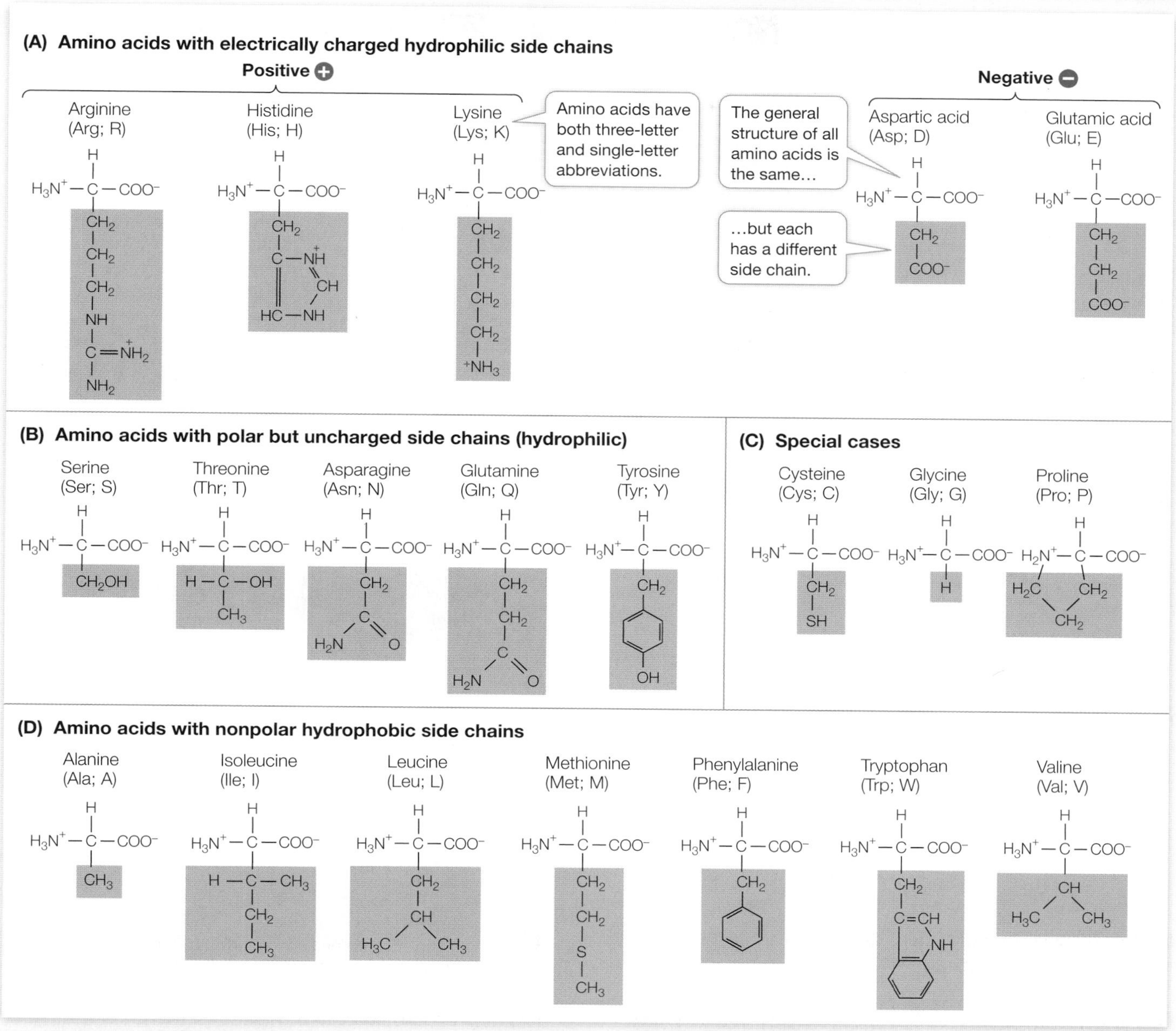

(A) Amino acids with electrically charged hydrophilic side chains

Positive ⊕   Negative ⊖

Arginine (Arg; R)   Histidine (His; H)   Lysine (Lys; K)

Amino acids have both three-letter and single-letter abbreviations.

The general structure of all amino acids is the same...

Aspartic acid (Asp; D)   Glutamic acid (Glu; E)

...but each has a different side chain.

(B) Amino acids with polar but uncharged side chains (hydrophilic)

Serine (Ser; S)   Threonine (Thr; T)   Asparagine (Asn; N)   Glutamine (Gln; Q)   Tyrosine (Tyr; Y)

(C) Special cases

Cysteine (Cys; C)   Glycine (Gly; G)   Proline (Pro; P)

(D) Amino acids with nonpolar hydrophobic side chains

Alanine (Ala; A)   Isoleucine (Ile; I)   Leucine (Leu; L)   Methionine (Met; M)   Phenylalanine (Phe; F)   Tryptophan (Trp; W)   Valine (Val; V)

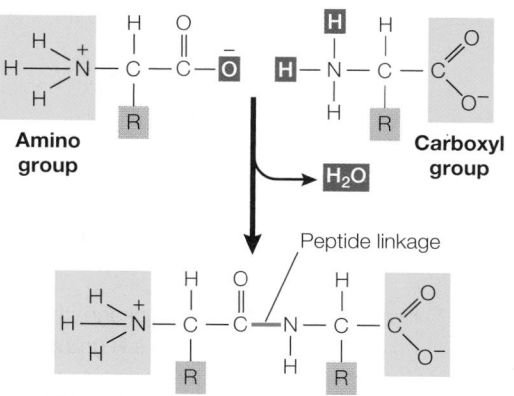

Amino group   Carboxyl group

H₂O

Peptide linkage

N terminus ⟶ C terminus

- The glycine side chain consists of a single hydrogen atom. It is small enough to fit into tight corners in the interiors of protein molecules where larger side chains could not fit.

- Proline possesses a modified amino group that lacks a hydrogen atom and instead forms a covalent bond with the hydrocarbon side chain, resulting in a ring structure. This limits both its hydrogen-bonding ability and its ability to rotate around the α carbon. So proline is often found where a protein bends or loops.

Figure 3.6 Formation of Peptide Linkages   In living things, the dehydration reaction leading to a peptide linkage (also called a peptide bond) has many intermediate steps, but the reactants and products are the same as those shown in this simplified diagram.

focus: key figure

Level	Description	Stabilized by	Example
(A) Primary	Amino acid monomers are joined, forming polypeptide chains.	Peptide bonds	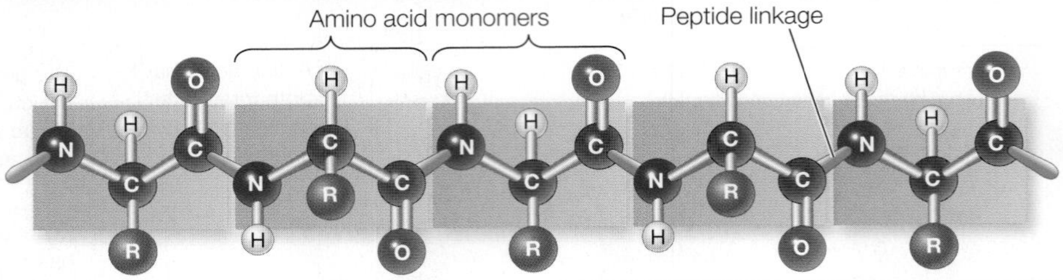 Amino acid monomers Peptide linkage
(B) Secondary	Polypeptide chains may form α helices or β pleated sheets.	Hydrogen bonds	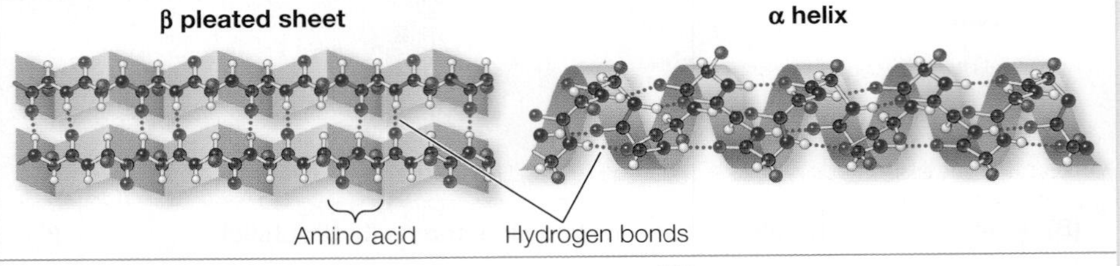 β pleated sheet α helix Amino acid Hydrogen bonds
(C) Tertiary	Polypeptides fold, forming specific shapes.	Hydrogen bonds; disulfide bridges; hydrophobic interactions	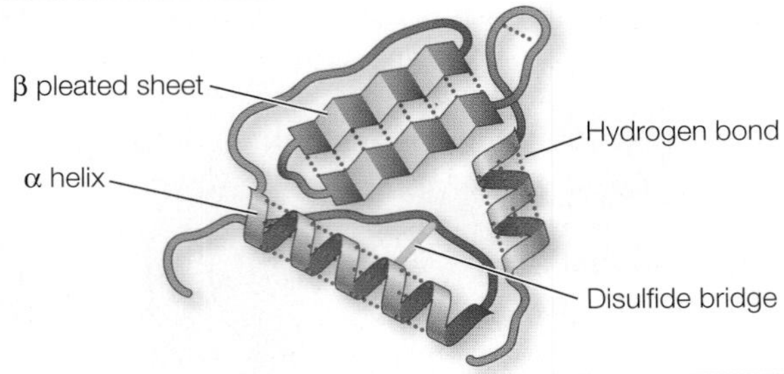 β pleated sheet α helix Hydrogen bond Disulfide bridge
(D) Quaternary	Two or more polypeptides assemble to form larger protein molecules.	Hydrogen bonds; disulfide bridges; hydrophobic interactions; ionic interactions	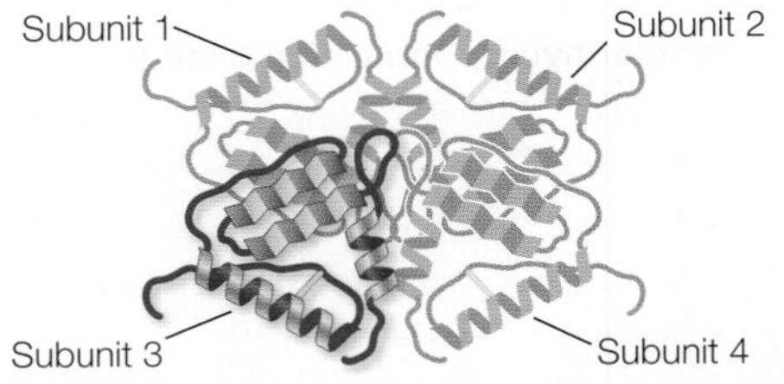 Subunit 1 Subunit 2 Subunit 3 Subunit 4

Figure 3.7 The Four Levels of Protein Structure   Secondary, tertiary, and quaternary structure all arise from the primary structure of the protein.

Q: If a protein is gently heated to break hydrogen bonds, what level(s) of structure will be unaffected, and why?

Peptide linkages form the backbone of a protein

Linking amino acids involves a reaction between carboxyl and amino groups attached to the α carbon. The carboxyl group of one amino acid reacts with the amino group of another, undergoing a condensation reaction that forms a **peptide linkage** (also called a peptide bond). **Figure 3.6** depicts this reaction.

Just as a sentence begins with a capital letter and ends with a period, polypeptide chains have a beginning and an end. The "capital letter" marking the beginning of a polypeptide is the amino group of the first amino acid added to the chain and is known as the N terminus. The "period" is the carboxyl group of the last amino acid added; this is the C terminus.

Two characteristics of the peptide bond are especially important in the three-dimensional structures of proteins:

1. In the C—N linkage, the adjacent α carbons (α C—C—N—α C) are not free to rotate fully, which limits the folding of the polypeptide chain.

2. The oxygen bound to the carbon (C=O) in the carboxyl group carries a slight negative charge (δ^-), whereas the hydrogen bound to the nitrogen (N—H) in the amino group is slightly positive (δ^+). This asymmetry of charge favors hydrogen bonding within the protein molecule itself and between molecules. These bonds contribute to the structures and functions of many proteins.

In addition to these characteristics of the peptide linkage, the particular sequence of amino acids—with their various R groups—in the polypeptide chain also plays a vital role in determining a protein's structure (**Focus Key Figure 3.7A**) and function.

The primary structure of a protein is its amino acid sequence

The precise sequence of amino acids in a polypeptide chain held together by peptide bonds constitutes the **primary structure** of a protein (see Figure 3.7A). The backbone of the polypeptide chain consists of the repeating sequence $-N-C-C-$ made up of the N atom from the amino group, the α C atom, and the C atom from the carboxyl group in each amino acid.

The single-letter abbreviations for amino acids (see Table 3.2) are used to record the amino acid sequence of a protein. Here, for example, are the first 20 amino acids (out of a total of 124) in the protein ribonuclease from a cow:

KETAAAKFERQHMDSSTSAA

The theoretical number of different proteins is enormous. Since there are 20 different amino acids, there could be $20 \times 20 = 400$ distinct dipeptides (two linked amino acids) and $20 \times 20 \times 20 = 8,000$ different tripeptides (three linked amino acids). Imagine this process of multiplying by 20 extended to a protein made up of 100 amino acids (which would be considered a small protein). There could be 20^{100} (that's approximately 10^{130}) such small proteins, each with its own distinctive primary structure. How large is the number 20^{100}? Physicists tell us there aren't that many electrons in the entire universe!

The sequence of amino acids in the polypeptide chain determines its final shape. The properties associated with each functional group in the side chains of the amino acids (see Table 3.2) determine how the protein can twist and fold, thus adopting a specific stable structure that distinguishes it from every other protein.

The secondary structure of a protein requires hydrogen bonding

A protein's **secondary structure** consists of regular, repeated spatial patterns in different regions of a polypeptide chain. There are two basic types of secondary structure, both determined by hydrogen bonding between the amino acids that make up the primary structure: the α helix and the β pleated sheet.

THE ALPHA HELIX   The α (**alpha**) **helix** is a right-handed coil that turns in the same direction as a standard wood screw (see Figure 3.7B and **Figure 3.8**). The R groups extend outward from the peptide backbone of the helix. The coiling results from hydrogen bonds that form between the δ^+ hydrogen of the N—H of one amino acid and the δ^- oxygen of the C=O of another. When this pattern of hydrogen bonding is established repeatedly over a segment of the protein, it stabilizes the coil.

THE BETA PLEATED SHEET   A β (**beta**) **pleated sheet** is formed from two or more polypeptide chains that are almost completely extended and aligned. The sheet is stabilized by hydrogen bonds between the N—H groups on one chain and the C=O groups on the other (see Figure 3.7B). A β pleated sheet may form between separate polypeptide chains or between different regions of a single polypeptide chain that is bent back on itself. The ratcheted, stacked sheets in dragline spider silks (described at the beginning of the chapter and in Investigating Life: Making Spider Silk) are made up of β pleated sheets. Many proteins contain regions of both α helix and β pleated sheet in the same polypeptide chain.

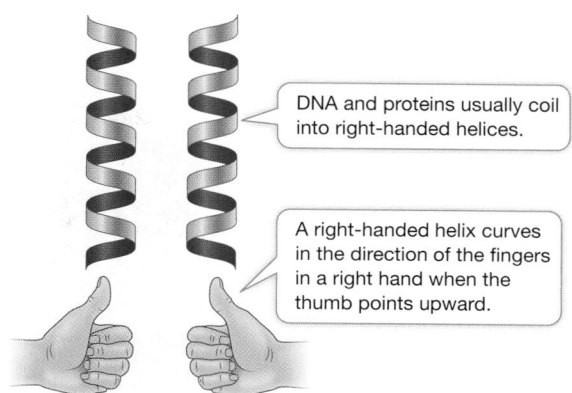

DNA and proteins usually coil into right-handed helices.

A right-handed helix curves in the direction of the fingers in a right hand when the thumb points upward.

Figure 3.8 **Left- and Right-Handed Helices**   A protein will often have one or more right-handed helices as part of its secondary structure.

The tertiary structure of a protein is formed by bending and folding

In many proteins, the polypeptide chain is bent at specific sites and then folded back and forth, resulting in the **tertiary structure** of the protein (see Figure 3.7C). Although α helices and β pleated sheets contribute to the tertiary structure, usually only portions of the macromolecule have these secondary structures, and large regions consist of tertiary structure unique to a particular protein. For example, the proteins found in stretchy spider silks (see the opening story) have repeated amino acid sequences that cause the proteins to fold into spirals. Tertiary structure is a macromolecule's definitive three-dimensional shape, often including a buried interior as well as a surface that is exposed to the environment.

The protein's exposed outer surfaces present functional groups capable of interacting with other molecules in the cell. These molecules might be other macromolecules, including proteins, nucleic acids, carbohydrates, and lipid structures, or smaller chemical substances.

You saw that hydrogen bonding between the N—H and C=O groups within and between chains is responsible for secondary structure. For tertiary structure, the interactions between R groups—the amino acid side chains—and between R groups and the environment are key. Recall the various strong and weak interactions between atoms described in Key Concept 2.2. Here is how these interactions are involved in determining and maintaining tertiary structure:

- Covalent disulfide bridges can form between specific cysteine side chains (see Figure 3.5), holding a folded polypeptide in place.

- Hydrogen bonds between side chains also stabilize folds in proteins.

- Hydrophobic side chains can aggregate together in the interior of the protein, away from water, folding the polypeptide in the process. Close interactions between the hydrophobic side chains are stabilized by van der Waals forces.

- Ionic attractions can form between positively and negatively charged side chains, forming salt bridges between amino acids. Salt bridges can be near the surfaces of polypeptides or buried deep within a protein, away from water. These interactions

(A) Space-filling model

Figure 3.9 Three Representations of Lysozyme Different molecular representations of a protein emphasize different aspects of its tertiary structure: surface features, sites of bends and folds, or sites where alpha or beta structures predominate. These three representations of lysozyme are similarly oriented.

Q: Can you identify regions of the protein that are hydrophilic? Hydrophobic?

 Media Clip 3.1 **Protein Structures in 3D**
www.Life11e.com/mc3.1

A realistic depiction of lysozyme shows dense packing of its atoms.

(B) Stick model

β pleated sheet

α helix

(C) Ribbon model

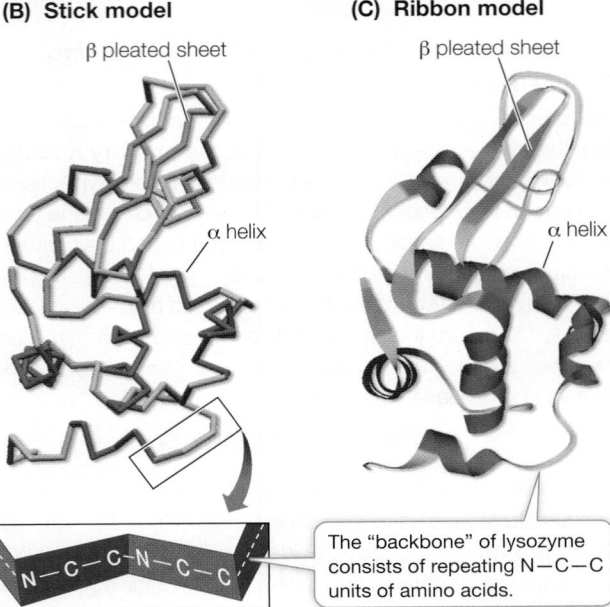

β pleated sheet

α helix

N—C—C—N—C—C

The "backbone" of lysozyme consists of repeating N—C—C units of amino acids.

occur between positively and negatively charged amino acids, for example glutamic acid (which has a negatively charged R group) and arginine (which is positively charged) (see Table 3.2):

Arg ～～C

NH₂   O

C ～～ Glu

+NH₂   ⁻O

A protein folds into its final shape in a way that maximizes all the interactions noted and minimizes inappropriate interactions, such as two positively charged residues (a term identifying monomers in a polymer) being near one another, or a hydrophobic residue being near water. A complete description of a protein's tertiary structure would specify the location of every atom in the molecule in three-dimensional space relative to all the other atoms. **Figure 3.9** shows three ways of modeling the structure of the protein lysozyme. Each way has its uses. The space-filling model might be used to study how other molecules interact with specific sites and R groups on the protein's surface. The stick model emphasizes the sites where bends occur, resulting in folds in the polypeptide chain. The ribbon model, perhaps the most widely used, shows the different types of secondary structure and how they fold into the tertiary structure.

Remember that both secondary and tertiary structure derive from primary structure. If a protein is heated slowly and moderately, the heat energy will disrupt only the weak interactions, causing the secondary and tertiary structure to break down. The protein is then said to be **denatured**. A comparison of native (untreated) and denatured proteins shows major differences:

● Native proteins are compact; denatured proteins have a larger volume.

● Native proteins exist in one, preferred shape; denatured proteins can take many shapes.

● Native proteins have hydrogen bonds that stabilize the structure internally; denatured proteins have hydrogen bonds on the exterior, to water.

You can't "unboil" an egg after it has been hard-boiled; the egg proteins are irreversibly denatured. Amazingly, in some cases a protein can return to its normal tertiary structure when it cools, demonstrating that all the information needed to specify the unique shape of a protein is contained in its primary structure. This was first shown (using chemicals instead of heat to denature the protein) by biochemist Christian Anfinsen for the protein ribonuclease (**Figure 3.10**).

The quaternary structure of a protein consists of subunits

Many functional proteins contain two or more polypeptide chains, called subunits, each of them folded into its own unique tertiary structure. The protein's **quaternary structure** results from the ways in which these subunits bind together and interact (see Figure 3.7D).

Hemoglobin has four polypeptide chains interacting to form quaternary structure (**Figure 3.11**). Hydrophobic interactions, van der Waals forces, hydrogen bonds, and ionic attractions all help hold the four subunits together to form a hemoglobin molecule. However, the weak nature of these forces permits small changes in the quaternary structure to aid the protein's function—which is to carry oxygen in red blood cells. As hemoglobin binds one O_2 molecule, the four subunits shift their relative positions slightly, changing the quaternary structure. Ionic attractions are broken, exposing buried side chains that enhance the binding of additional O_2 molecules. The quaternary structure changes back when hemoglobin releases its O_2 molecules to the cells of the body.

Shape and surface chemistry contribute to protein function

The shapes of proteins allow specific sites on their exposed surfaces to bind noncovalently to other molecules, which may be large or small. The binding is usually very specific because only certain compatible chemical groups will bind to one another. The specificity of protein binding depends on two general properties of the protein: its shape, and the chemistry of its exposed surface groups.

experiment

Figure 3.10A Primary Structure Specifies Tertiary Structure

Original Papers: Anfinsen, C. B., E. Haber, M. Sela and F. White, Jr. 1961. The kinetics of formation of native ribonuclease during oxidation of the reduced polypeptide chain. *Proceedings of the National Academy of Sciences USA* 47: 1309–1314.

White, Jr., F. 1961. Regeneration of native secondary and tertiary structures by air oxidation of reduced ribonuclease. *Journal of Biological Chemistry* 236: 1353–1360.

Using the protein ribonuclease, Christian Anfinsen showed that proteins spontaneously fold into functionally correct three-dimensional configurations. As long as the primary structure is not disrupted, the information for correct folding (under the right conditions) is retained.

HYPOTHESIS▶ Under controlled conditions that simulate the normal cellular environment, a denatured protein can refold into a functional three-dimensional structure.

METHOD Chemically denature a functional ribonuclease so that only its primary structure (i.e., an unfolded polypeptide chain) remains. Once denaturation is complete, remove the disruptive chemicals.

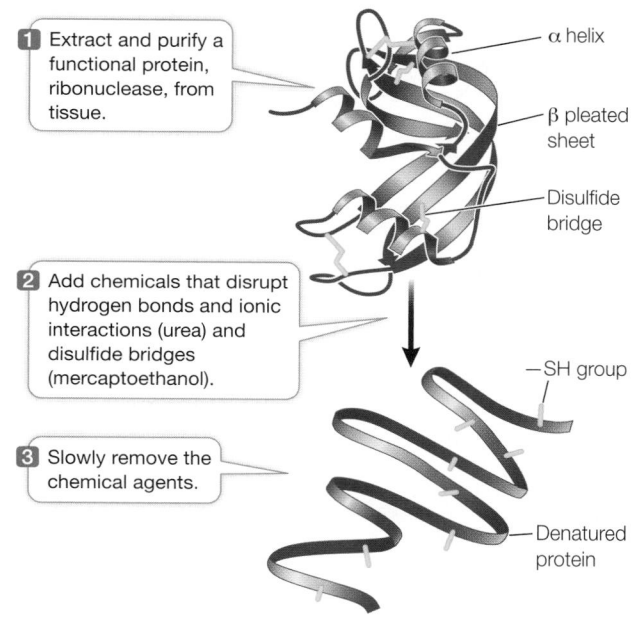

1 Extract and purify a functional protein, ribonuclease, from tissue.

— α helix

— β pleated sheet

— Disulfide bridge

2 Add chemicals that disrupt hydrogen bonds and ionic interactions (urea) and disulfide bridges (mercaptoethanol).

—SH group

3 Slowly remove the chemical agents.

— Denatured protein

RESULTS When the disruptive agents are removed, three-dimensional structure is restored and the protein once again is functional.

CONCLUSION▶ In normal cellular conditions, the primary structure of a protein specifies how it folds into a functional, three-dimensional structure.

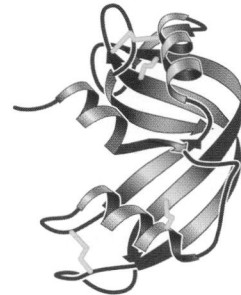

Figure 3.10B **work with the data** follows on next page.

(A) Ribbon model

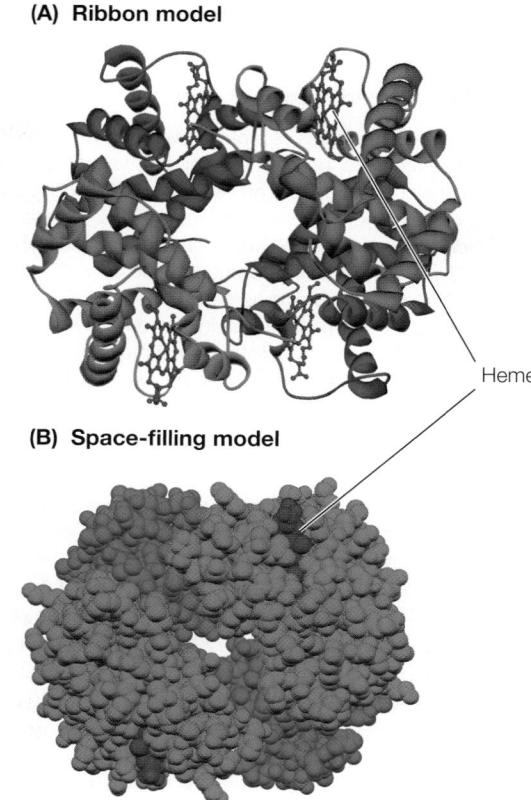

Heme

(B) Space-filling model

Figure 3.11 **Quaternary Structure of a Protein** Hemoglobin consists of four folded polypeptide subunits that assemble themselves into the quaternary structure represented by the ribbon model **(A)** and space-filling model **(B)**. In both graphic representations, each type of subunit is a different color (α subunits are blue and β subunits are green). The heme groups (red) contain iron and are the oxygen-carrying sites.

- *Shape*. When a small molecule collides with and binds to a much larger protein, it is like a baseball being caught by a catcher's mitt: the mitt has a shape that binds to the ball and fits around it. Just as a hockey puck or a ping-pong ball does not fit a baseball catcher's mitt, a given molecule will not bind to a protein unless there is a general "fit" between their three-dimensional shapes.

- *Chemistry*. The exposed R groups on the surface of a protein permit chemical interactions with other substances (**Figure 3.12**). Three types of interactions may be involved: ionic, hydrophobic, or hydrogen bonding. Many important functions of proteins involve interactions between surface R groups and other molecules.

Environmental conditions affect protein structure

Because they are determined by weak forces, the three-dimensional structures of proteins are influenced by environmental conditions. Conditions that would not break covalent bonds can disrupt the weaker, noncovalent interactions that determine secondary, tertiary, and quaternary structure. Such alterations may affect a protein's shape and thus its function. Various conditions can alter the weak, noncovalent interactions:

work with the data

Figure 3.10B Primary Structure Specifies Tertiary Structure

Original Papers: Anfinsen, C. B. et al. 1961; White, Jr., F. 1961.

After the tertiary structures of proteins were shown to be highly specific, the question arose as to how the order of amino acids determined the three-dimensional structure. The second protein whose structure was determined was ribonuclease A (RNase A). This enzyme was readily available from cow pancreases at slaughterhouses and, because it works in the highly acidic environment of the cow stomach, was stable compared with most proteins and easy to purify. RNase A has 124 amino acids. Among these are eight cysteine residues, which form four disulfide bridges. Were these covalent links between cysteines essential for the three-dimensional structure of RNase A? As outlined in Figure 3.10A, Christian Anfinsen and his colleagues set out to answer this question.

QUESTIONS▶

1. Initially, the disulfide bonds (S—S) in RNase A were eliminated because the sulfur atoms in cysteine residues were all reduced (—SH). At time zero, reoxidation began; and at various times, the amount of S—S bond re-formation and the activity of the enzyme were measured by chemical methods. The data are shown in **Figure A**.

   At what time did disulfide bonds begin to form? At what time did enzyme activity begin to appear? Explain the difference between these times.

2. The three-dimensional structure of RNase A was examined by ultraviolet spectroscopy. In this technique, the protein was exposed to different wavelengths of ultraviolet light (measured in nanometers) and the amount of light absorbed by the protein at each wavelength was measured (E). The results are plotted in **Figure B**.

   Look carefully at the plots. What are the differences between the peak absorbances of native (untreated) and reduced (denatured) RNase A? What happened when reduced RNase A was reoxidized (renatured)? What can you conclude about the structure of RNase A from these experiments?

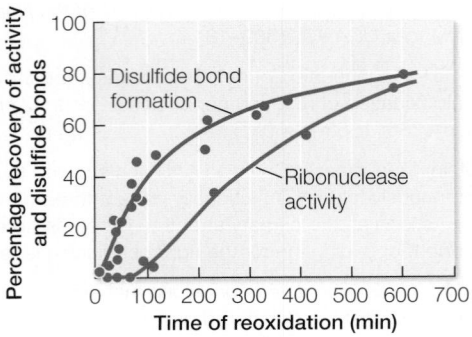

Figure A

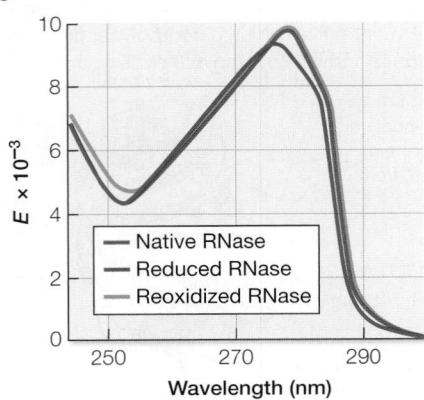

Figure B

A similar **work with the data** exercise may be assigned in **LaunchPad**.

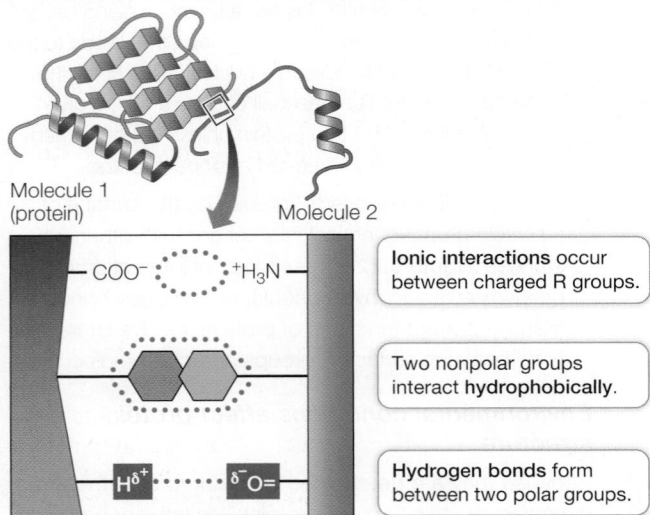

Figure 3.12 Noncovalent Interactions between Proteins and Other Molecules Noncovalent interactions (see p. 51) allow a protein (red) to bind tightly to another molecule (green) with specific properties. Noncovalent interactions also allow regions within the same protein to interact with one another.

Q: Why are these interactions sensitive to heat? (Hint: see Table 2.1.)

- Increases in temperature cause more rapid molecular movements and thus can break hydrogen bonds and hydrophobic interactions.

- Changes in pH can change the pattern of ionization of exposed carboxyl and amino groups in the R groups of amino acids, thus disrupting the pattern of ionic attractions and repulsions.

- High concentrations of polar substances such as urea can disrupt the hydrogen bonding that is crucial to protein structure. Urea was used in the experiment on reversible protein denaturation shown in Figure 3.10.

- Nonpolar substances may also disrupt normal protein structure in cases where hydrophobic interactions are essential to maintain the structure.

Protein shapes can change

As we saw in the case of hemoglobin, which undergoes subtle shape changes when it binds oxygen, the shapes of proteins can change as a result of their interactions with other molecules. Proteins can also change shape if they undergo covalent modifications.

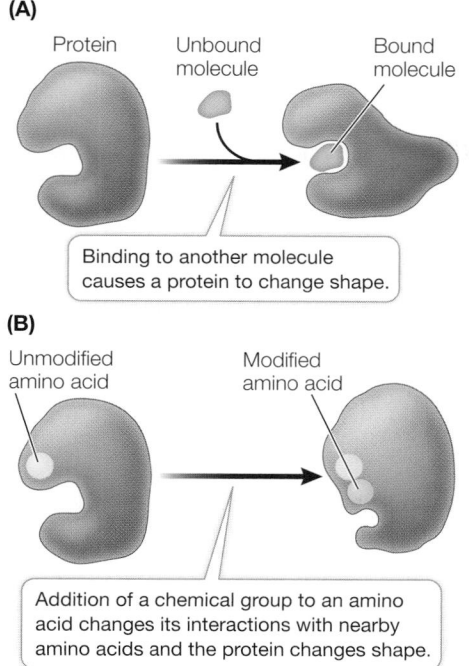

(A)

Protein   Unbound   Bound
               molecule   molecule

Binding to another molecule
causes a protein to change shape.

(B)

Unmodified   Modified
amino acid   amino acid

Addition of a chemical group to an amino
acid changes its interactions with nearby
amino acids and the protein changes shape.

Figure 3.13 Protein Structure Can Change Proteins can change their tertiary structure when they bind to other molecules **(A)** or are modified chemically **(B)**.

- *Proteins interact with other molecules*. Proteins do not exist in isolation. In fact, if a biochemist "goes fishing" with a particular protein, by attaching the protein to a chemical "hook" and inserting it into cells, the protein will often be attached to something else when it is "reeled in." These molecular interactions are reminiscent of the interactions that make up quaternary structure (see above). If a polypeptide comes into contact with another molecule, R groups on its surface may form weak interactions (e.g., hydrophobic, van der Waals) with groups on the surface of the other molecule. This may disrupt some of the interactions between R groups within the polypeptide, causing it to undergo a change in shape (**Figure 3.13A**). You will see many instances of this in the coming chapters.

- *Proteins undergo* ***covalent modifications**. After it is made, the structure of a protein can be modified by the covalent bonding of a chemical group to the side chain of one or more of its amino acids. The chemical modification of just one amino acid can alter the shape and function of a protein. An example is the addition of a charged phosphate group to a relatively nonpolar R group. This can cause the amino acid to become more hydrophilic and to move to the outer surface of the protein, altering the shape of the protein in the region near the amino acid (**Figure 3.13B**).

---

*connect the concepts Covalent modifications of proteins with subsequent changes in shape and function underlie many biological processes, ranging from signaling within the cell (see Key Concept 7.3) to the action of plant hormones on growth (see Key Concept 37.2).

Molecular chaperones help shape proteins

Within a living cell, a polypeptide chain is sometimes in danger of binding the wrong substance. There are two major situations when this can occur:

1. *Just after a protein is made*. When a protein has not yet folded completely, it can present a surface that binds the wrong molecule.

2. *Following denaturation*. Certain conditions, such as moderate heat, can cause some proteins in a living cell to denature without killing the organism. Before the protein can refold, it may present a surface that binds the wrong molecule. In these cases, the inappropriate binding may be irreversible.

Many cells have a special class of proteins, called **chaperones**, that protect the three-dimensional structures of other proteins. They bind to their partner proteins just as they are being made and also when they become denatured. Like the chaperones at a high school dance, chaperone proteins prevent inappropriate interactions and enhance appropriate ones. Typically, a chaperone protein has a cagelike structure that pulls in a polypeptide, causes it to fold into the correct shape, and then releases it (**Figure 3.14**). Tumors make chaperone proteins, possibly to stabilize

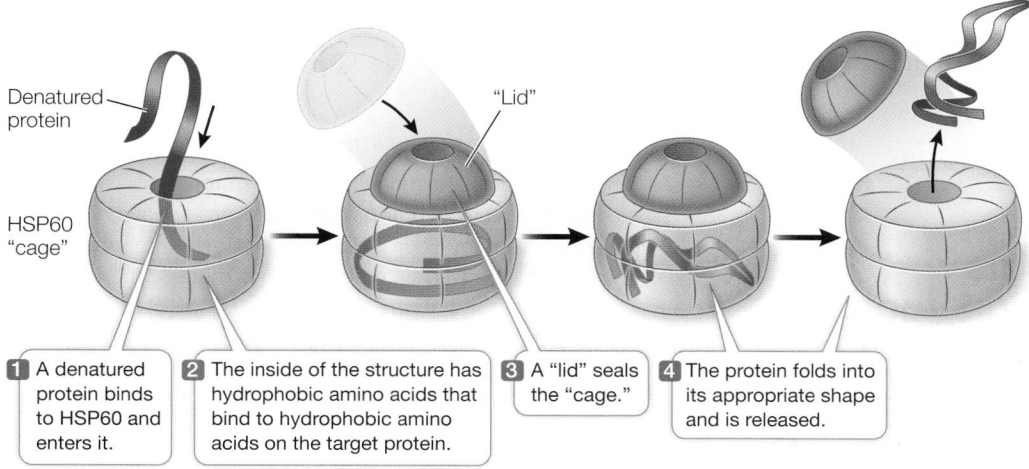

Denatured
protein

HSP60
"cage"

"Lid"

1 A denatured protein binds to HSP60 and enters it.

2 The inside of the structure has hydrophobic amino acids that bind to hydrophobic amino acids on the target protein.

3 A "lid" seals the "cage."

4 The protein folds into its appropriate shape and is released.

Figure 3.14 Molecular Chaperones Protect Proteins from Inappropriate Binding Chaperone proteins surround new or denatured proteins and prevent them from binding to the wrong substances. Heat shock proteins such as HSP60, shown here, make up one class of chaperone proteins.

Q: Why are heat shock proteins important for a cell?

proteins important in the cancer process, and so chaperone-inhibiting drugs are being designed for use in chemotherapy. In some clinical situations, treatment with these inhibitors results in the inappropriate folding of proteins in tumor cells, causing the tumors to stop growing.

3.2 recap

Proteins are polymers of amino acids. The sequence of amino acids in a protein is its primary structure. Secondary, tertiary, and quaternary structures arise through interactions among the amino acids. A protein's three-dimensional shape and exposed chemical groups establish its binding specificity for other substances.

learning outcomes

You should be able to:

- Predict how a protein's tertiary structure will be affected by a change in an amino acid in its primary structure.
- Analyze data to determine why an amino acid change in a protein leads to a change in the protein's shape.
- Predict the effects of various environmental factors on a protein's structure.

1. Suppose that, in a given protein, one lysine is replaced by aspartic acid (see Table 3.2). Does this change occur in the primary structure or in the secondary structure? How might it result in a change in tertiary structure? In quaternary structure?

2. A single amino acid change in a protein can change its shape. For example, the amino acid glycine may normally reside at a certain position in a protein. If glycine is replaced with either glutamic acid or arginine, the protein shape near that amino acid changes significantly. There are two possible explanations for this:

   a. A small amino acid at that position in the polypeptide is necessary for normal shape.

   b. An uncharged amino acid is necessary for normal shape.

   Further amino acid substitutions are done to distinguish between these possibilities. Replacing glycine with serine or alanine results in normal shape, but replacing glycine with valine changes the shape. Which of the two possible explanations is supported by the observations? Explain your answer.

3. Human hair is composed of the protein keratin. At the hair salon, two techniques are used to modify the three-dimensional shape of hair. Styling involves heat, and a perm involves cleaving and re-forming disulfide bonds. How would you investigate these phenomena in terms of protein structure?

key concept 3.3 Simple Sugars Are the Basic Structural Unit of Carbohydrates

Carbohydrates make up a large group of molecules that all have similar atomic compositions but differ greatly in size, chemical properties, and biological functions. Carbohydrates usually have the general formula $(C_1H_2O_1)_n$, (n stands for a number), which makes them appear as "hydrates of carbon" (a hydrate refers to water), hence their name. However, carbohydrates are not really "hydrates" because the water molecules are not intact. Rather, the linked carbon atoms are bonded with hydrogen atoms (—H) and hydroxyl groups (—OH), the components of water.

Carbohydrates have four major biochemical roles:

1. They are a source of stored energy that can be released in a form usable by organisms.

2. They are used to transport stored energy within complex organisms.

3. They serve as carbon skeletons that can be rearranged to form new molecules.

4. They form extracellular assemblies such as cell walls that provide structure to organisms.

Some carbohydrates are relatively small, with molecular weights of less than 100. Others are true macromolecules, with molecular weights in the hundreds of thousands.

focus your learning

- Carbohydrates function primarily in storing and transporting chemical energy and as sources of carbon for building new macromolecules.
- Polysaccharides of glucose all provide energy storage and structural functions but vary in branching patterns and type of glycosidic linkages between glucose units.

There are four categories of biologically important carbohydrate defined by the number of monomers:

1. **Monosaccharides** (*mono*, "one," + *saccharide*, "sugar"), such as glucose, are simple sugars. They are the monomers from which the larger carbohydrates are constructed.

2. **Disaccharides** (*di*, "two") consist of two monosaccharides linked together by covalent bonds. The most familiar is sucrose, which is made up of covalently bonded glucose and fructose molecules.

3. **Oligosaccharides** (*oligo*, "several") are made up of several (3–20) monosaccharides.

4. **Polysaccharides** (*poly*, "many"), such as starch, glycogen, and cellulose, are polymers made up of hundreds or thousands of monosaccharides.

Monosaccharides are simple sugars

All living cells contain the monosaccharide **glucose**; it is the "blood sugar" used to store and transport energy in humans. Cells use glucose as an energy source, breaking it down through a series of reactions that converts stored energy to more usable chemical energy and produce carbon dioxide; this is a cellular form of the combustion reaction you saw in Key Concept 2.3.

Glucose exists in straight chains and in ring forms. The ring forms predominate in virtually all biological circumstances because they are more stable in water. There are two versions of the glucose ring, called α- and β-glucose, which differ only in the orientation of the

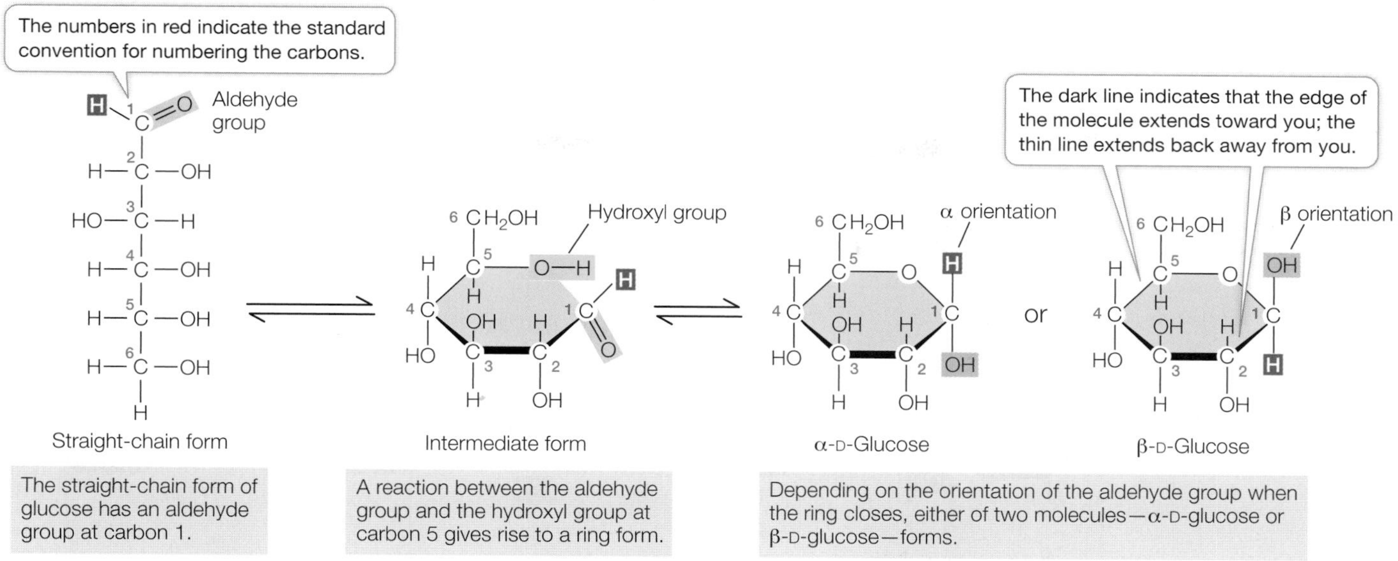

The numbers in red indicate the standard convention for numbering the carbons.

The dark line indicates that the edge of the molecule extends toward you; the thin line extends back away from you.

The straight-chain form of glucose has an aldehyde group at carbon 1.

A reaction between the aldehyde group and the hydroxyl group at carbon 5 gives rise to a ring form.

Depending on the orientation of the aldehyde group when the ring closes, either of two molecules—α-D-glucose or β-D-glucose—forms.

Figure 3.15 From One Form of Glucose to the Other All glucose molecules have the formula $C_6H_{12}O_6$, but their structures vary. When dissolved in water, the α and β "ring" forms of glucose interconvert. The convention used here for numbering the carbon atoms is standard in biochemistry.

Activity 3.3 **Forms of Glucose**
www/Life11e.com/ac3.3

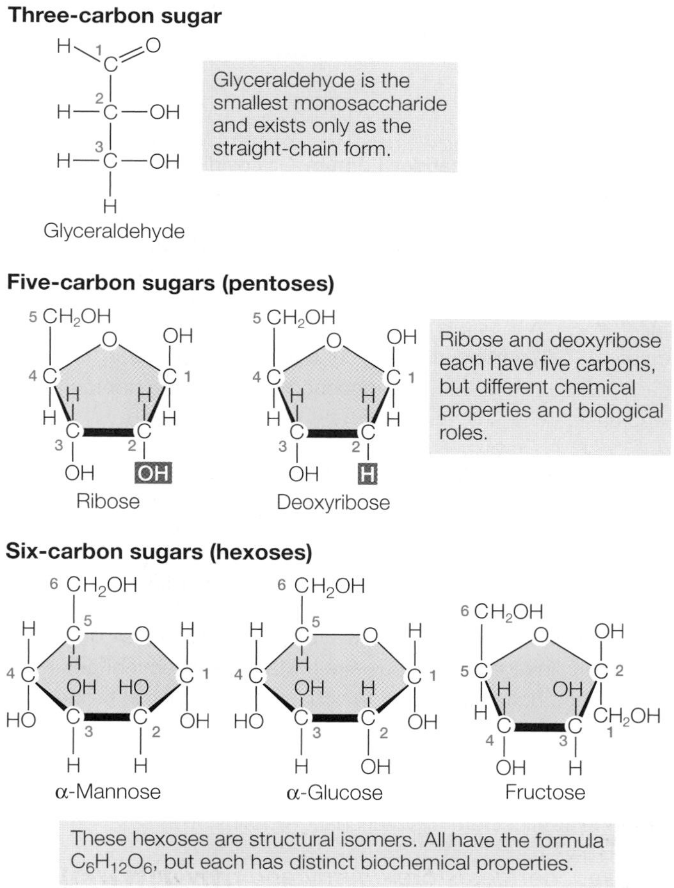

—H and —OH groups attached to carbon 1 (**Figure 3.15**). The α and β forms interconvert and exist in equilibrium when dissolved in water.

Different monosaccharides contain different numbers of carbons. Some monosaccharides are structural isomers, with the same kinds and numbers of atoms but in different arrangements (**Figure 3.16**). Such seemingly small structural changes can significantly alter their properties. Most of the monosaccharides in living systems belong to the D (right-handed) series of optical isomers.

- **Pentoses** (*pente*, "five") are five-carbon sugars. Two pentoses are of particular biological importance: the backbones of the nucleic acids RNA and DNA contain ribose and deoxyribose, respectively (see Key Concept 4.1). These two pentoses are not isomers of each other; rather, one oxygen atom is missing from carbon 2 in deoxyribose (*de-*, "absent"). The absence of this oxygen atom is an important distinction between RNA and DNA.

- The **hexoses** (*hex*, "six") shown in Figures 3.15 and 3.16 are a group of structural isomers with the formula $C_6H_{12}O_6$. Common hexoses are glucose, fructose (so named because it was first found in fruits), mannose, and galactose.

Glycosidic linkages bond monosaccharides

The disaccharides, oligosaccharides, and polysaccharides are all constructed from monosaccharides that are covalently bonded together by condensation reactions that form **glycosidic linkages** (**Figure 3.17**). A single glycosidic linkage between two monosaccharides forms a disaccharide. For example, sucrose—common table sugar (the real stuff, not artificial sweetener) in the human diet and the major disaccharide in plants—is formed from a glucose and a fructose molecule.

The disaccharides maltose and cellobiose are made from two glucose molecules (see Figure 3.17). Maltose and cellobiose are structural isomers, both having the formula $C_{12}H_{22}O_{11}$. However, they have different chemical properties and are recognized by

Three-carbon sugar

Glyceraldehyde is the smallest monosaccharide and exists only as the straight-chain form.

Glyceraldehyde

Five-carbon sugars (pentoses)

Ribose and deoxyribose each have five carbons, but different chemical properties and biological roles.

Ribose   Deoxyribose

Six-carbon sugars (hexoses)

α-Mannose   α-Glucose   Fructose

These hexoses are structural isomers. All have the formula $C_6H_{12}O_6$, but each has distinct biochemical properties.

Figure 3.16 Monosaccharides Are Simple Sugars Monosaccharides are made up of varying numbers of carbons. Some hexoses are structural isomers that have the same kind and number of atoms, but the atoms are arranged differently. Fructose, for example, is a hexose but forms a five-membered ring like the pentoses.

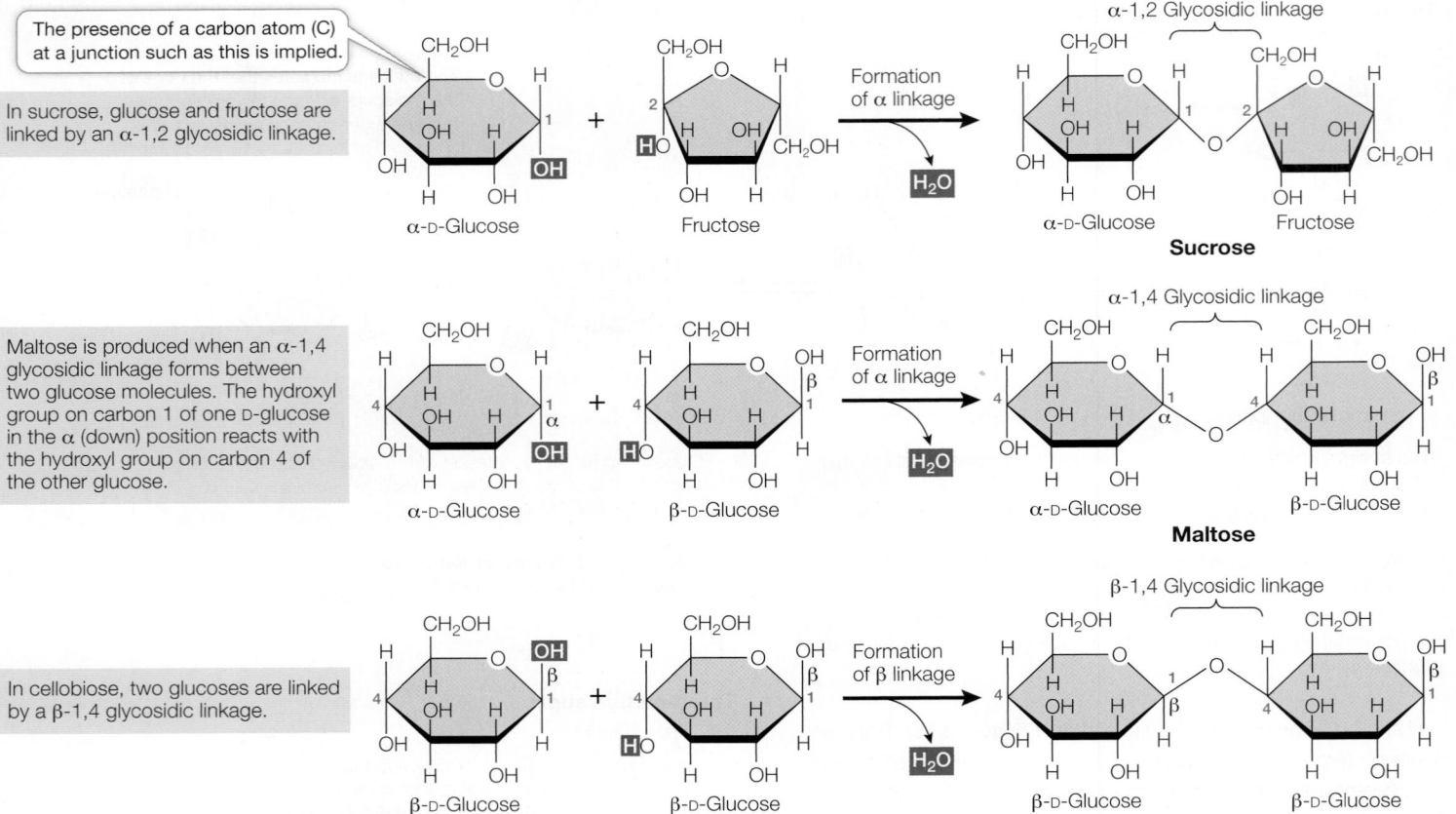

The presence of a carbon atom (C) at a junction such as this is implied.

In sucrose, glucose and fructose are linked by an α-1,2 glycosidic linkage.

Maltose is produced when an α-1,4 glycosidic linkage forms between two glucose molecules. The hydroxyl group on carbon 1 of one D-glucose in the α (down) position reacts with the hydroxyl group on carbon 4 of the other glucose.

In cellobiose, two glucoses are linked by a β-1,4 glycosidic linkage.

Figure 3.17 Disaccharides Form by Glycosidic Linkages
Glycosidic linkages between two monosaccharides can create many different disaccharides. The particular disaccharide formed depends on which monosaccharides are linked, on the site of linkage (i.e., which carbon atoms are involved), and on the form (α or β) of the linkage.

different *enzymes in biological tissues. For example, maltose can be hydrolyzed into its monosaccharides in the human body, whereas cellobiose cannot.

---

*connect the concepts Enzymes are an important class of proteins that may change shape when they come into contact with a reactant in a biochemical reaction. Each enzyme is specific for the reactant it binds. See Key Concept 8.4.

Oligosaccharides contain several monosaccharides bound by glycosidic linkages at various sites. Many oligosaccharides have additional functional groups, which give them special properties. Oligosaccharides are often covalently bonded to proteins and lipids on the outer cell surface, where they serve as recognition signals. The different human blood groups (e.g., the ABO blood types) get their specificities from oligosaccharide chains.

Polysaccharides store energy and provide structural materials

Polysaccharides are large (sometimes gigantic) polymers of monosaccharides connected by glycosidic linkages (**Figure 3.18**). In contrast to polypeptides, polysaccharides are not necessarily linear chains of monomers. Each monomer unit has several sites that are capable of forming glycosidic linkages, and thus branched molecules are possible.

STARCH **Starches** comprise a family of large molecules with similar structures. While all starches are polysaccharides of glucose with α-glycosidic linkages (α–1,4 and α–1,6 glycosidic bonds; see Figure 3.18A), the different starches can be distinguished by the amount of branching that occurs at carbons 1 and 6 (see Figure 3.18B). Starch is the principal energy storage compound of plants. Some plant starches, such as amylose, are unbranched; others are moderately branched (for example, amylopectin). Starch readily binds water; if you're a cook you know this. However, when water is removed, hydrogen bonds form between the unbranched polysaccharide chains, which then aggregate. Large starch aggregates called starch grains can be observed in the storage tissues of plant seeds (see Figure 3.18C). These aggregates are broken up when starch is heated, breaking the hydrogen bonds. The starch becomes less solid and crystalline and water is absorbed, making the starch even more amorphous. This is what happens in baking with wheat flour, and is what gives bread its texture. Next time you eat some bread, think of hydrogen bonds!

GLYCOGEN **Glycogen** is a water-insoluble, highly branched polymer of glucose. It is used to store glucose in the liver and muscles and is thus an energy storage compound for animals, as starch is for plants. Both glycogen and starch are readily hydrolyzed into glucose monomers, which in turn can be broken down to liberate their stored energy.

But if it is glucose that is needed for fuel, why store it in the form of glycogen? The reason is that 1,000 glucose molecules would exert 1,000 times the osmotic pressure of a single glycogen molecule, causing water to enter cells where glucose is stored

(A) Molecular structure

Cellulose

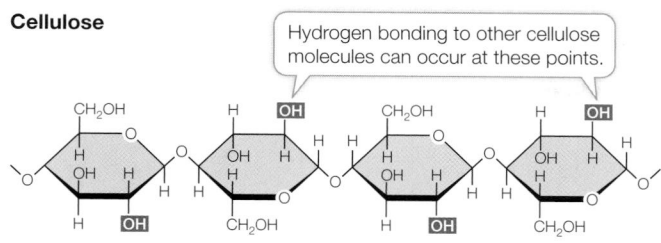

Hydrogen bonding to other cellulose molecules can occur at these points.

Starch and glycogen

Branching occurs here.

Cellulose is an unbranched polymer of glucose with β-1,4 glycosidic linkages that are chemically very stable.

Glycogen and starch are polymers of glucose with α-1,4 glycosidic linkages. α-1,6 Glycosidic linkages produce branching at carbon 6.

(B) Macromolecular structure

Linear (cellulose)

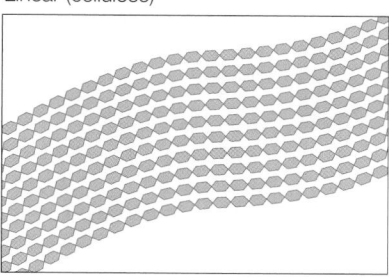

Parallel cellulose molecules form hydrogen bonds, resulting in thin fibrils.

Branched (starch)

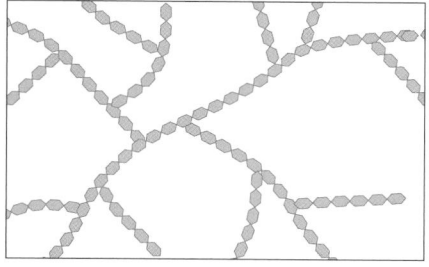

Branching limits the number of hydrogen bonds that can form in starch molecules, making starch less compact than cellulose.

Highly branched (glycogen)

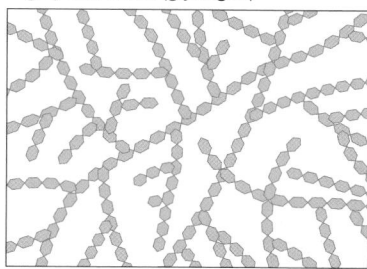

The high amount of branching in glycogen makes its solid deposits more compact than starch.

(C) Polysaccharides in cells

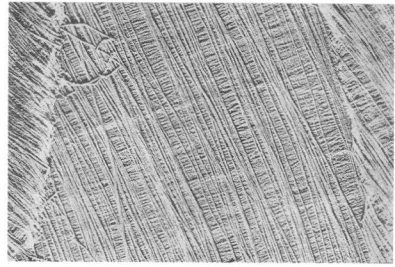

Layers of cellulose fibrils, as seen in this scanning electron micrograph, give plant cell walls great strength.

Within these potato cells, starch deposits (colored red in this scanning electron micrograph) have a granular shape.

The dark clumps in this electron micrograph are glycogen deposits.

Figure 3.18 Representative Polysaccharides Cellulose, starch, and glycogen have different levels of branching and compaction of the polysaccharides.

(see Key Concept 6.3). If it were not for polysaccharides, many organisms would expend a lot of energy expelling excess water from their cells.

CELLULOSE   As the predominant component of plant cell walls, **cellulose** is by far the most abundant organic compound on Earth. Like starch and glycogen, cellulose is a polysaccharide of glucose, but its individual monosaccharides are connected by β- rather than by α-glycosidic linkages. Starch is easily degraded by the actions of chemicals or enzymes. Cellulose, however, is chemically more stable because of its β-glycosidic linkages.

Thus whereas starch is easily broken down to supply glucose for energy-producing reactions, cellulose is an excellent structural material that can withstand harsh environmental conditions without substantial change.

Chemically modified carbohydrates contain additional functional groups

Some carbohydrates are chemically modified by the addition of functional groups such as phosphate, amino, or *N*-acetyl groups (**Figure 3.19**). For example, a phosphate group may be added to one or more of the —OH sites (see Figure 3.19A). Some of the resulting sugar phosphates, such as fructose 1,6-bisphosphate, are important intermediates in cellular energy reactions, which you'll see in Chapter 9.

(A) Sugar phosphate

Fructose 1,6-bisphosphate is involved in the reactions that liberate energy from glucose. (The numbers in its name refer to the carbon sites of phosphate bonding; *bis-* indicates that two phosphates are present.)

Figure 3.19 Chemically Modified Carbohydrates Added functional groups can modify the form and properties of a carbohydrate.

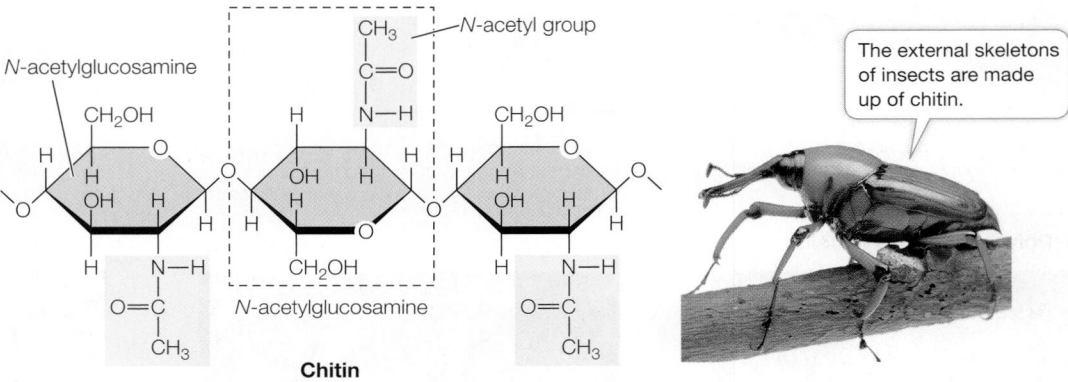

Phosphate groups

Fructose

Fructose 1,6-bisphosphate

(B) Amino sugars

The monosaccharides glucosamine and galactosamine are amino sugars with an amino group in place of a hydroxyl group.

Glucosamine   Amino group   **Galactosamine**

Galactosamine is an important component of cartilage, a connective tissue in vertebrates.

(C) Chitin

Chitin is a polymer of *N*-acetylglucosamine; *N*-acetyl groups provide additional sites for hydrogen bonding between the polymers.

N-acetylglucosamine

N-acetyl group

N-acetylglucosamine

Chitin

The external skeletons of insects are made up of chitin.

When an amino group (NH_2) is substituted for an —OH group, amino sugars, such as glucosamine and galactosamine, are produced (see Figure 3.19B). These compounds are important in the extracellular matrix (see Key Concept 5.4), where they form parts of glycoproteins, which are molecules involved in keeping tissues together. Galactosamine is a major component of cartilage, the material that forms caps on the ends of bones and stiffens the ears and nose. A derivative of glucosamine is present in the polymer chitin (see Figure 3.19C), the principal structural polysaccharide in the external skeletons of insects and many crustaceans (such as crabs and lobsters), and a component of the cell walls of fungi.

▶ 3.3 recap

Carbohydrates are composed of carbon, hydrogen, and oxygen and have the general formula $(C_1H_2O_1)_n$. They provide energy and structure to cells and are precursors of numerous important biological molecules. Monosaccharide monomers can be connected by glycosidic linkages to form disaccharides, oligosaccharides, and polysaccharides.

▶ 3.3 recap

learning outcomes

You should be able to:
- Identify the primary functions of carbohydrates in living organisms.
- Explain why polysaccharides are useful structures for energy storage.
- Use the chemical structures of polysaccharides to explain their properties.

1. What qualities of the polysaccharides starch and glycogen make them useful for energy storage?
2. After looking at the cellulose molecule in Figure 3.18A, can you see why a large number of hydrogen bonds are present in the linear structure of cellulose shown in Figure 3.18B? Why is this structure so strong?

You have seen how amino acid monomers form protein polymers and how sugar monomers form the polymers of carbohydrates. Now we will look at the lipids, which are unique among the four classes of large biological molecules in that they are not, strictly speaking, polymers.

3.4 Lipids Are Defined by Their Solubility Rather Than by Chemical Structure

Lipids—colloquially called fats—are hydrocarbons that are insoluble in water because of their many nonpolar covalent bonds. As you saw in Key Concept 2.2, nonpolar hydrocarbon molecules are hydrophobic and preferentially aggregate together, away from water, which is polar. When nonpolar hydrocarbons are sufficiently close to one another, weak but additive van der Waals forces help hold them together. The huge macromolecular aggregations that can form are not polymers in a strict chemical sense, because the individual lipid molecules are not covalently bonded. With this understanding, it is still useful to consider aggregations of individual lipids as a different sort of polymer.

focus your learning

- Triglycerides are simple lipids composed of glycerol bonded to three fatty acids via ester linkages.
- Because phospholipids are amphipathic, they are able to aggregate to form phospholipid bilayers, which are useful in creating membrane structures.
- Carotenoids, steroids, certain vitamins, and waxes are classified as lipids and have a variety of functions, depending on their chemical structures.

There are several different types of lipids, and they play a number of roles in living organisms:

- Fats and oils store energy.
- Phospholipids play important structural roles in cell membranes.
- Carotenoids and chlorophylls help plants capture light energy.
- Steroids and modified fatty acids play regulatory roles as hormones and vitamins.
- Fat in animal bodies serves as thermal insulation.
- A lipid coating around nerves provides electrical insulation.
- Oil or wax on the surfaces of skin, fur, feathers, and leaves repels water and prevents excessive evaporation of water from terrestrial animals and plants.

Fats and oils are triglycerides

Chemically, fats and oils are **triglycerides**, also known as simple lipids. Triglycerides that are solid at room temperature (around 20°C) are called **fats**; those that are liquid at room temperature are called **oils**. Triglycerides are composed of two types of building blocks: fatty acids and glycerol. **Glycerol** is a small molecule with three hydroxyl (—OH) groups (thus it is an alcohol). A **fatty acid** is made up of a long nonpolar hydrocarbon chain and an acidic polar carboxyl

group (—COOH). These chains are very hydrophobic because of their abundant C—H and C—C bonds, which have similar *electronegativity values and are therefore nonpolar.

---

connect the concepts As discussed in Key Concept 2.2, electronegativity is a measure of the attraction an atomic nucleus exerts on electrons in a covalent bond. When one interacting atom is much more electronegative than the other, a complete transfer of one or more electrons may take place.

A triglyceride contains three fatty acid molecules and one molecule of glycerol. Making a triglyceride involves three condensation (dehydration) reactions. In each reaction, the carboxyl group of a fatty acid bonds with a hydroxyl group of glycerol, resulting in a covalent bond called an **ester linkage** and the release of a water molecule (**Figure 3.20**). The three fatty acids in a triglyceride molecule need not all have the same hydrocarbon chain length or structure; some may be saturated fatty acids, whereas others may be unsaturated:

- In **saturated fatty acids**, all the bonds between the carbon atoms in the hydrocarbon chain are single bonds—there are no double bonds. That is, all the bonds are saturated with hydrogen atoms (**Figure 3.21A**). These fatty acid molecules are relatively straight, and they pack together tightly, like pencils in a box.

- In **unsaturated fatty acids**, the hydrocarbon chain contains one or more double bonds. Linoleic acid is an example of a polyunsaturated fatty acid that has two double bonds near the middle of the hydrocarbon chain, causing kinks in the molecule (**Figure 3.21B**). Such kinks prevent the unsaturated fat molecules from packing together tightly.

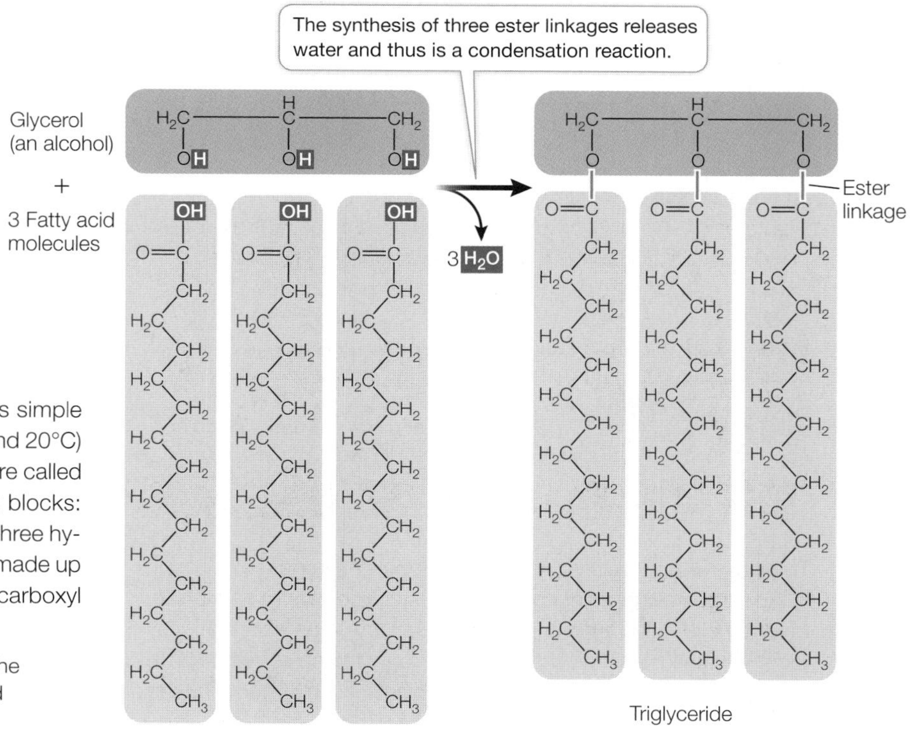

The synthesis of three ester linkages releases water and thus is a condensation reaction.

Glycerol (an alcohol)

+ 3 Fatty acid molecules

3 H_2O

Ester linkage

Triglyceride

Figure 3.20 Synthesis of a Triglyceride In living things, the reaction that forms a triglyceride is more complex, but the end result is the same as shown here.

(A) Palmitic acid

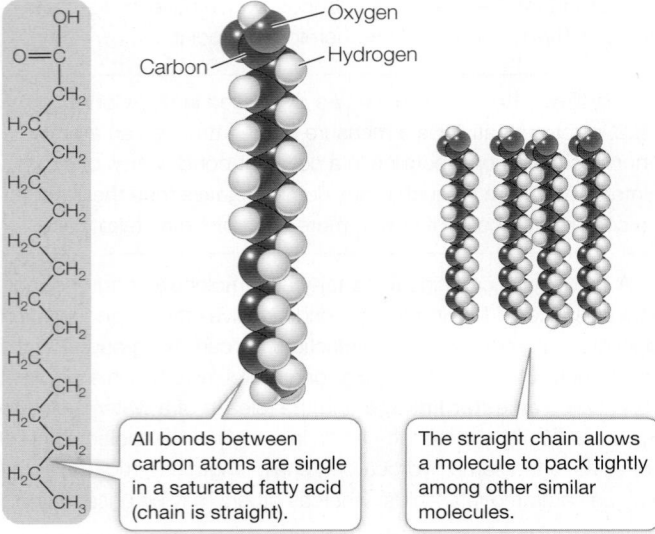

All bonds between carbon atoms are single in a saturated fatty acid (chain is straight).

The straight chain allows a molecule to pack tightly among other similar molecules.

(B) Linoleic acid

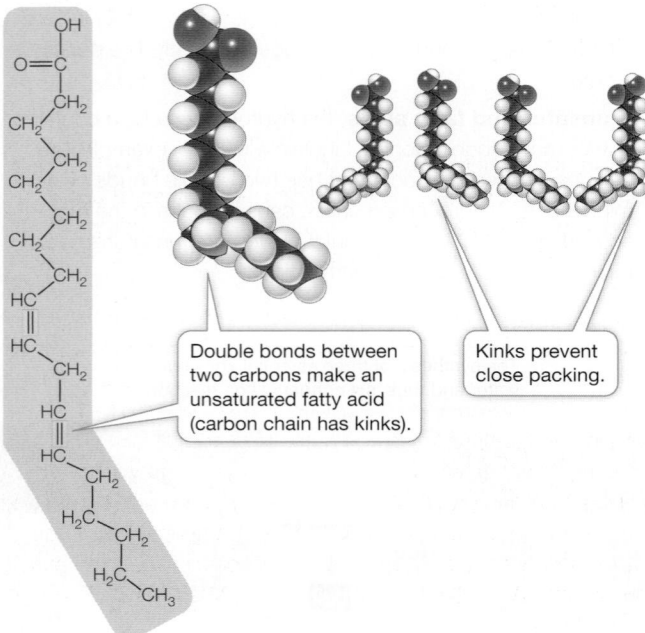

Double bonds between two carbons make an unsaturated fatty acid (carbon chain has kinks).

Kinks prevent close packing.

Figure 3.21 Saturated and Unsaturated Fatty Acids **(A)** The straight hydrocarbon chain of a saturated fatty acid allows the molecule to pack tightly with other, similar molecules. **(B)** In unsaturated fatty acids, kinks in the chain prevent close packing. The color convention in the models shown here (gray, H; red, O; black, C) is commonly used.

The kinks in fatty acid molecules are important in determining the fluidity and melting points of lipids. The triglycerides of animal fats tend to have many long-chain saturated fatty acids packed tightly together; these fats are usually solids at room temperature and have high melting points. The triglycerides of plants, such as corn oil, tend to have short or unsaturated fatty acids. Because of their kinks, these fatty acids pack together poorly and have low melting points, and these triglycerides are usually liquids at room temperature.

Unsaturated   Saturated

Fatty acids are excellent storehouses for chemical energy. As you will see in Chapter 9, when the C—H bond is broken, it releases significant energy that an organism can use for its own purposes, such as movement or building up other complex molecules.

Phospholipids form biological membranes

We have mentioned the hydrophobic nature of the many C—C and C—H bonds in fatty acids. But what about the carboxyl functional group at the end of the molecule? When it ionizes and forms COO⁻, it is strongly hydrophilic. So a fatty acid is a molecule with a hydrophilic end and a long hydrophobic "tail." It has two opposing chemical properties—part hydrophobic and part hydrophilic; the technical term for this is **amphipathic**.

Like triglycerides, **phospholipids** contain fatty acids bound to glycerol by ester linkages. In phospholipids, however, any one of several phosphate-containing compounds replaces the first or third fatty acid, giving phospholipids amphipathic properties (**Figure 3.22A**). The phosphate functional group has a negative electric charge, so this portion of the molecule is hydrophilic, attracting polar water molecules. But the two fatty acids are hydrophobic, so they tend to avoid water and aggregate together or with other hydrophobic substances.

In an aqueous environment, phospholipids line up in such a way that the nonpolar, hydrophobic "tails" pack tightly together and the phosphate-containing "heads" face outward, where they interact with water. The phospholipids can form a **bilayer**: a sheet two molecules thick, with water excluded from the core (**Figure 3.22B**). Biological membranes have this kind of **phospholipid bilayer** structure, and we will devote Chapter 6 to their biological functions.

Some lipids have roles in energy conversion, regulation, and protection

In the paragraphs above we focused on triglycerides and phospholipids—lipids that are involved in energy storage and cell structure. However, there are other nonpolar and amphipathic lipids that have different structures and roles.

CAROTENOIDS   The carotenoids are a family of light-absorbing pigments found in plants and animals. Beta-carotene (β-carotene) is one of the pigments that traps light energy in leaves during photosynthesis. In humans, a molecule of β-carotene can be broken

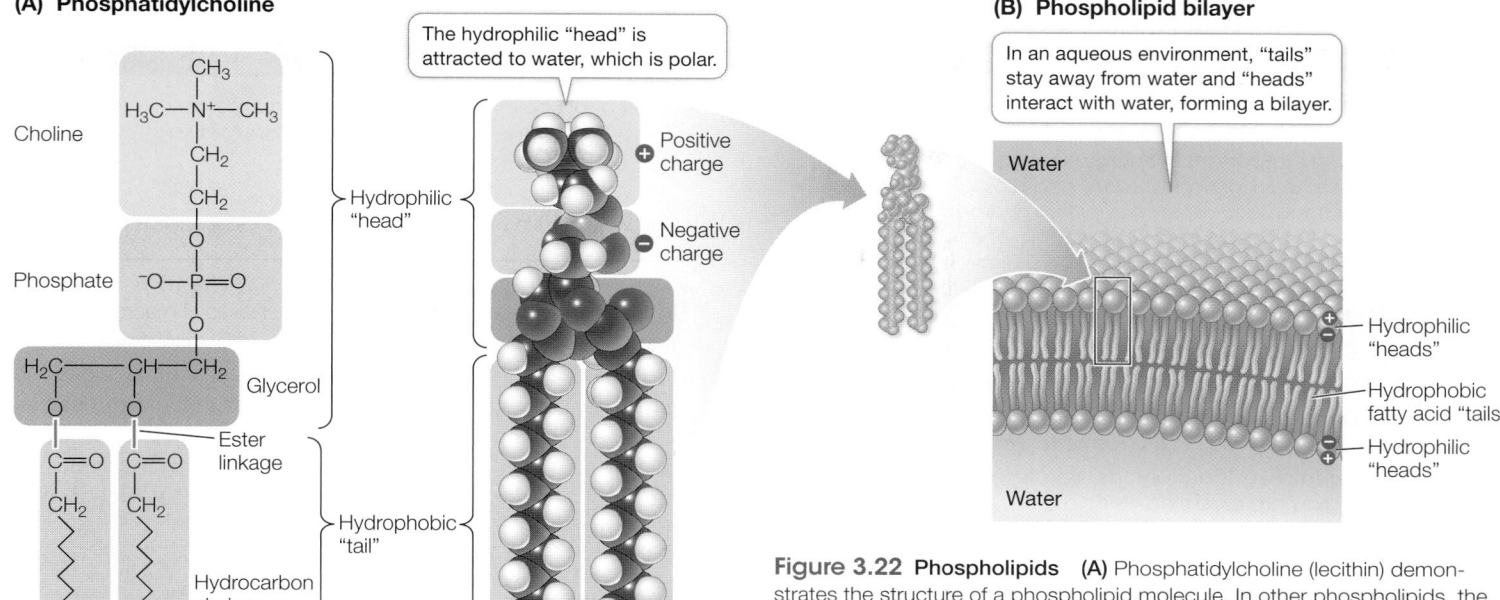

(A) Phosphatidylcholine

Choline

Phosphate

Glycerol

Ester linkage

Hydrocarbon chains

Hydrophilic "head"

Hydrophobic "tail"

The hydrophilic "head" is attracted to water, which is polar.

Positive charge

Negative charge

The hydrophobic "tails" are not attracted to water.

(B) Phospholipid bilayer

In an aqueous environment, "tails" stay away from water and "heads" interact with water, forming a bilayer.

Water

Hydrophilic "heads"

Hydrophobic fatty acid "tails"

Hydrophilic "heads"

Water

Figure 3.22 Phospholipids **(A)** Phosphatidylcholine (lecithin) demonstrates the structure of a phospholipid molecule. In other phospholipids, the amino acid serine, the sugar alcohol inositol, or other compounds replace choline. **(B)** In an aqueous environment, hydrophobic interactions bring the "tails" of phospholipids together in the interior of a bilayer. The hydrophilic "heads" face outward on both sides of the bilayer, where they interact with the surrounding water molecules.

down into two vitamin A molecules. Vitamin A is used to make the pigment *cis*-retinal, which is required for vision.

β-Carotene

Vitamin A

Carotenoids are responsible for the colors of carrots, tomatoes, pumpkins, egg yolks, and butter. The brilliant yellows and oranges of autumn leaves are also from carotenoids.

STEROIDS The steroids are a family of organic compounds whose multiple rings are linked through shared carbons. The steroid cholesterol is an important constituent of membranes, helping maintain membrane integrity (see Key Concept 6.1).

Cholesterol

Other steroids function as hormones: chemical signals that carry messages from one part of the body to another (see Chapter 40). Cholesterol is synthesized in the liver and is the starting material for making steroid hormones such as testosterone and estrogen.

VITAMINS Vitamins are small molecules that are not synthesized by the human body or in some cases are synthesized in inadequate amounts and so must be acquired from the diet (see Chapter 50). For example, vitamin A is formed from the β-carotene found in green and yellow vegetables (see above). In humans, a deficiency of vitamin A leads to dry skin, eyes, and internal body surfaces, retarded growth and development, and night blindness, which is a diagnostic symptom for the deficiency. You'll see in Chapter 18 how plant biologists are producing crops that contain more vitamin A to alleviate this condition. Vitamins D, E, and K are also lipids.

WAXES Birds and mammals have glands in their skins that secrete waxy coatings onto their hair or feathers. These coatings repel water and help keep the hair and feathers pliable. The shiny leaves of plants such as holly, familiar during winter holidays, also have waxy coatings. Waxy coatings on plants can help them retain water and exclude pathogens. Bees make their honeycombs out of wax. Waxes are substances that are hydrophobic and plastic, or malleable, at room temperature. Each wax molecule consists of a saturated, long-chain fatty acid and a saturated, long-chain alcohol joined by an ester linkage. The result is a very long molecule with 40–60 CH_2 groups.

3.4 recap

Lipids include both hydrophobic and amphipathic molecules that are largely composed of carbon and hydrogen. They are important in energy storage, light absorption, intercellular communication, and biological structures. A phospholipid is composed of two hydrophobic fatty acids linked to glycerol and a hydrophilic phosphate group. Cell membranes contain phospholipid bilayers.

learning outcomes
You should be able to:

- Compare and contrast triglycerides in fats and oils.
- Describe how the polar and nonpolar natures of phospholipids result in the formation of a membrane bilayer.
- Explain why carotenoids, steroids, certain vitamins, and waxes are classified as lipids.

3.4 recap

1. If phospholipids were placed in a nonpolar solvent, would a membrane-type bilayer structure form? Explain your answer, using a diagram.
2. What is the difference between fats and oils?
3. Why are steroids and some vitamins classified as lipids?

In this chapter we discussed three of the classes of macromolecules that are characteristic of living organisms. All organisms are made up of these three types of macromolecules, a fact that shows the wonderful biochemical unity of life. This unity in turns implies that all life has a common origin (see Key Concept 1.1). Essential to this origin was a fourth class of macromolecule, the nucleic acids. In the next chapter we will turn to the related topics of nucleic acids and the origin of life.

investigatinglife

Q&A What are practical uses for spider silk?

Spider silk protein has many glycine and alanine resides, which are hydrophobic and cause the protein to fold into a fibrous shape. More than 1,000 individual polypeptide chains hydrogen bond and hydrophobically interact with one another to form long silk fibers. Because of the strength arising from its protein structure, spider silk is much desired for human uses. Using genetic engineering, biologists have been able to harvest composite silk fibers produced by silkworms in industrial quantities. Now that this composite silkworm–spider silk is available, numerous applications (functions related to structure) can be contemplated. For example, a recent study compared surgical sutures made from spider silk with widely used sutures made from synthetic fibers to hold together injured tendons. The spider silk sutures were far superior, holding their

strength over many cycles of tendon movement during healing. Another application on the horizon is bulletproof vests: the energy required to break a spider silk filament is at least 100 times greater than the energy of a typical bullet in impact.

Future directions

Perhaps the most widespread use for spider silk may be in textiles. By definition, textiles are flexible materials made of natural or artificial fibers (threads) that are woven together. The clothes you are wearing are textiles, perhaps made of cotton fibers (made up of cellulose) or polyester fibers (an artificial polymer) or a combination. Of course, silk fibers from silkworms have a long history in textiles, particularly in Asia. The addition of spider silk fibers, or combined spider–silkworm silk to existing textiles, holds great promise to increase the strength and in some cases the flexibility of the textile.

Chapter Summary 3

▶ **3.1 Macromolecules Characterize Living Things**

See Animation 3.1

- **Macromolecules** are **polymers** constructed by the formation of covalent bonds between smaller molecules called **monomers**. Macromolecules in living organisms include polysaccharides, proteins, and nucleic acids. Organizationally, large lipid structures may also be considered like macromolecules.

- **Functional groups** are small groups of atoms that are consistently found together in a variety of different macromolecules. Functional groups have particular chemical properties that they confer on any larger molecule of which they are a part. **Review Figure 3.1, Activity 3.1**

- Structural, *cis-trans*, and **optical isomers** have the same kinds and numbers of atoms but differ in their structures and properties. **Review Figure 3.2**

- The many functions of macromolecules are directly related to their three-dimensional shapes, which in turn result from the sequences and chemical properties of their monomers.

- Monomers are joined by **condensation reactions**, which release a molecule of water for each bond formed. **Hydrolysis reactions** use water to break polymers into monomers. **Review Figure 3.4**

▶ **3.2 The Function of a Protein Depends on Its Three-Dimensional Structure**

- The functions of **proteins** include support, protection (e.g., skin surface), catalysis, transport, defense, regulation, movement, signaling, and storage. **Review Table 3.1**

- Proteins consist of one or more **polypeptide chains**, which are polymers of **amino acids**. Four atoms or groups are attached to a central carbon atom: a hydrogen atom, an amino group, a carboxyl group, and a variable R group. The particular properties of each amino acid depend on its **side chain**, or **R group**, which may be charged, polar, or hydrophobic. Review Table 3.2, Activity 3.2

- **Peptide linkages**, also called peptide bonds, covalently link amino acids into polypeptide chains. These bonds form by condensation reactions between the carboxyl and amino groups. Review Figure 3.6

- The **primary structure** of a protein is the sequence of amino acids in the chain. This chain is folded into a **secondary structure**, which in different parts of the protein may form an **α helix** or a **β pleated sheet**. Review Focus: Key Figure 3.7A–C

- **Disulfide bridges** and noncovalent interactions between amino acids cause polypeptide chains to fold into three-dimensional **tertiary structures**. These bridges as well as noncovalent interactions allow multiple polypeptide chains to form **quaternary structures**. Review Figure Focus: Key Figure 3.7C and D

- Heat, alterations in pH, and certain chemicals can all result in a protein becoming **denatured**. This involves the loss of tertiary and/or secondary structure as well as biological function. Review Figure 3.10

- The specific shape and structure of a protein allow it to bind noncovalently to other molecules. In addition, amino acids may be modified by the covalent bonding of chemical groups to their side chains. Such binding may result in a protein changing its shape. Review Figures 3.12, 3.13

- **Chaperone proteins** enhance correct protein folding and prevent inappropriate binding to other molecules. Review Figure 3.14

▶ 3.3 Simple Sugars Are the Basic Structural Unit of Carbohydrates

- **Carbohydrates** contain carbon bonded to hydrogen and oxygen atoms and have the general formula $(C_1H_2O_1)_n$.

- **Monosaccharides** are the monomers that make up larger carbohydrates. **Hexoses** such as **glucose** are six-carbon monosaccharides; **pentoses** have five carbons. Review Figure 3.16, Activity 3.3

- **Glycosidic linkages**, which have either an α or a β orientation in space, are covalent bonds between monosaccharides. Two linked monosaccharides are called a **disaccharide**; larger units are **oligosaccharides** and **polysaccharides**. Review Figure 3.17

- **Starch** is a polymer of glucose that stores energy in plants, and **glycogen** is an analogous polymer in animals. They can be easily broken down to release stored energy. Review Figure 3.18

- **Cellulose** is a very stable glucose polymer and is the principal structural component of plant cell walls.

▶ 3.4 Lipids Are Defined by Their Solubility Rather Than by Chemical Structure

- **Lipids** are hydrocarbons that are insoluble in water because of their many nonpolar covalent bonds. They play roles in energy storage, membrane structure, light harvesting, intercellular communication, and protection.

- **Fats** and **oils** are **triglycerides**. A triglyceride is composed of three **fatty acids** covalently bonded to a molecule of glycerol by **ester linkages**. Review Figure 3.20

- A **saturated fatty acid** has a hydrocarbon chain with no double bonds. These molecules can pack together tightly. The hydrocarbon chain of an **unsaturated fatty acid** has one or more double bonds that bend the chain, preventing close packing. Review Figure 3.21

- A **phospholipid** has a hydrophobic hydrocarbon "tail" and a hydrophilic phosphate "head"; that is, it is **amphipathic**. In water, the interactions of the tails and heads of phospholipids generate a **phospholipid bilayer**. The heads are directed outward, where they interact with the surrounding water. The tails are packed together in the interior of the bilayer, away from water. Review Figure 3.22

- Other lipids include vitamins A, D, E, and K, steroids, and plant pigments such as carotenoids.

See Activity 3.4 to review macromolecule structures.

Go to **LearningCurve** (in **LaunchPad**) for dynamic quizzing that helps you solidify your understanding of this chapter. **LearningCurve** adapts to your responses, giving you the practice you need to master each key concept.

▶ Apply What You've Learned

Review

3.2 A protein's tertiary structure describes its three-dimensional shape and is stabilized by hydrogen bonds, hydrophobic interactions, ionic attractions, and in some proteins, disulfide bonds.

3.2 Exposed surface groups on a protein provide both shapes and chemical groups that can interact specifically with other molecules or ions.

Original Paper: Conlon, J. M. 2001. Evolution of the insulin molecule: insights into structure-activity and phylogenetic relationships. *Peptides* 22: 1183–1193.

Dogs, like people, tend to have more health problems as they age. About one in every 100 dogs that reach age 12 loses the ability to

synthesize insulin, a hormone in the body that regulates cell uptake of glucose from the blood. Veterinarians prescribe injections of porcine insulin for treatment. Porcine insulin is extracted from pigs' blood and has the same amino acid sequence as canine (dog) insulin. **Figure A** shows the structure of canine insulin, which is composed of two polypeptides, an A chain and a B chain, linked by disulfide bridges. **Figure B** shows how these chains fold to form secondary and tertiary structures. Labels A2, A3, A19, B23, and B24 indicate three amino acids on the A chain and two on the B chain thought to be directly involved in the binding of insulin to target proteins on body cell surfaces.

Since insulin is a protein found in all vertebrate animals, scientists have compared the structures of insulin from different vertebrate species. The table shows a comparison of a few amino acids in the A chain and B chain of insulins from a wide range of vertebrates.

(continued)

Figure A

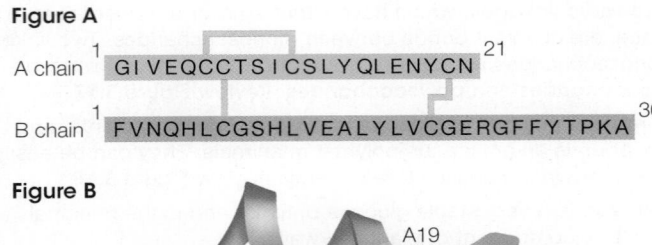

A chain ^{1} GIVEQCCTSICSLYQLENYCN ^{21}

B chain ^{1} FVNQHLCGSHLVEALYLVCGERGFFYTPKA ^{30}

Figure B

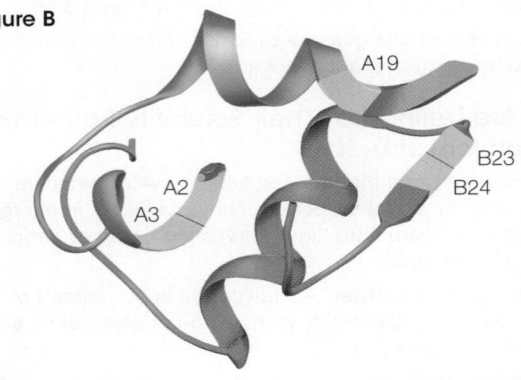

Questions

1. What types of forces are involved in the binding of insulin to its target? What would be true about the positions of the amino acids A2, A3, A19, B23, and B24 that would allow them to be involved in binding?

2. What evidence shown in the table supports the hypothesis that amino acids A2, A3, A19, B23, and B24 are important in insulin's binding activity?

3. What conclusion can you draw about the finding that there are amino acid variations in some positions across vertebrate species? What could account for the finding that only certain amino acids are found at those positions and not at a wide range of amino acids?

4. Develop a hypothesis about the importance of cysteine residues in canine insulin. How could you test your hypothesis?

Species	A1	A2	A3	A4	A5	A19	A21	B12	B16	B23	B24	B25	B26
Canine	Gly	Ile	Val	Glu	Gln	Tyr	Asn	Val	Tyr	Gly	Phe	Phe	Tyr
Guinea pig	—	—	—	Asp	—	—	—	—	—	—	—	—	—
Casiragua	—	—	—	Asp	—	—	—	—	—	—	—	Tyr	Arg
Porcupine	—	—	—	Asp	—	—	—	—	—	—	—	—	—
Cuis	—	—	—	Asp	—	—	—	—	—	—	—	—	Ser
Iguana	—	—	—	—	Gln	—	—	—	—	—	—	Tyr	—
Rattlesnake	—	—	—	—	—	—	—	—	Phe	—	—	Tyr	—
Caecilian	—	—	—	—	Lys	—	—	—	—	—	—	—	—
Wood frog	—	—	—	—	—	—	Ser	—	—	—	—	—	—
Surinam toad	—	—	—	—	—	—	—	—	His	—	—	—	—
Cod	—	—	—	Asp	—	—	—	—	—	—	—	—	—
Eel	—	—	—	—	—	—	—	—	—	—	—	—	Phe
Tilapia	—	—	—	—	Glu	—	—	—	—	—	—	—	—
Bowfin	—	—	—	—	—	—	—	—	Phe	—	—	—	—
Spiny dogfish	—	—	—	—	His	—	—	—	—	—	—	Tyr	—
Hammerhead shark	—	—	—	Asp	His	—	—	—	—	—	—	Tyr	—
Striped gudgeon fish	—	—	—	—	Lys	—	—	—	—	—	—	Tyr	—

Source: Conlon, J. M. 2001. Evolution of the insulin molecule: Insights into structure-activity and phylogenetic relationships. *Peptides* 22: 1183–1193.

Note: A dash indicates the amino acid is the same as in the canine.

Go to **LaunchPad** for the eBook, LearningCurve, animations, activities, flashcards, and additional resources and assignments.

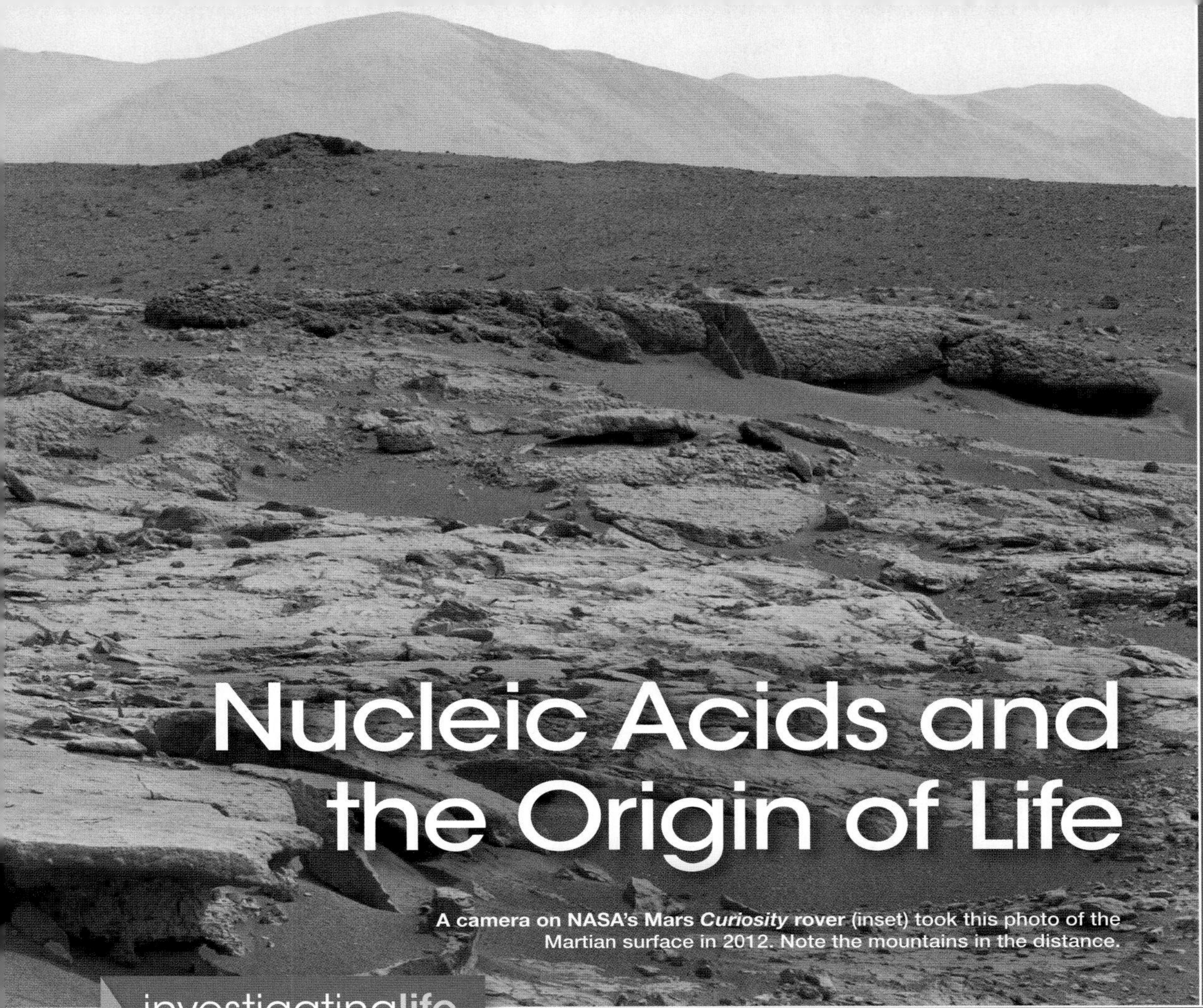

4

key concepts

4.1 Nucleic Acid Structures Reflect Their Functions

4.2 The Small Molecules of Life Originated on Primitive Earth

4.3 The Large Molecules of Life Originated from Small Molecules

4.4 Cells Originated from Their Molecular Building Blocks

Nucleic Acids and the Origin of Life

A camera on NASA's Mars *Curiosity* rover (inset) took this photo of the Martian surface in 2012. Note the mountains in the distance.

investigatinglife

Looking for Life

On Earth you can find organisms and the chemical signatures of life everywhere—even deep under the sea, inside volcanoes, buried in ice, and floating in the air. Determining life's origin is difficult because (with few exceptions) the simple organisms that first existed left no fossils. On Mars, however, things might be different.

Mars has fascinated scientists (and science-fiction writers) since its discovery by astronomers more than 3,000 years ago. A cold planet, Mars has a geology that has changed little over several billion years, and so any evidence of life may be preserved. By the late nineteenth century, distinguished professors of astronomy firmly believed there was life on Mars, based on observations through telescopes that showed polar ice caps that changed with the seasons, darker areas thought to be liquid water, and even markings that resembled canals.

Better telescopes, chemical analyses based on light emission (spectroscopy), orbiting satellites, and landers controlled from Earth have laid to rest many of the initial ideas about life on Mars. But the challenge of finding present or past life on Mars remains. The search revolves around finding organisms, evidence of a current or past environment where life could exist, and chemical signatures of present or past life.

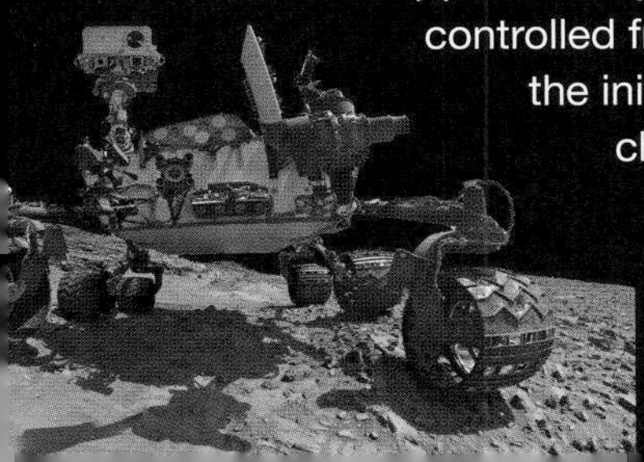

Unlike Earth, Mars now lacks a magnetic field to protect against cosmic radiation arriving from space. The surface of Mars is assaulted with very high levels of radiation, making it inhospitable to life. So any current life would have to be below the surface, where radiation is much lower.

Since life as we know it requires water, determining whether Mars now has or ever had water has been a central question. It is now established that water exists on the surface of Mars—at the poles, frozen, and in water vapor in its atmosphere. There is even evidence that trickles of water may exist on the surface, kept liquid in the bitter cold by a high salt concentration. The discovery of methane (CH_4) in the atmosphere of Mars has generated additional excitement and speculation. Might this methane be a by-product of an organic reaction? The search for evidence of the chemical elements (C, H, O, P, N, and S) of life introduced in Chapter 2 continues.

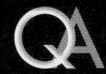

QA Can we find evidence of life on Mars?

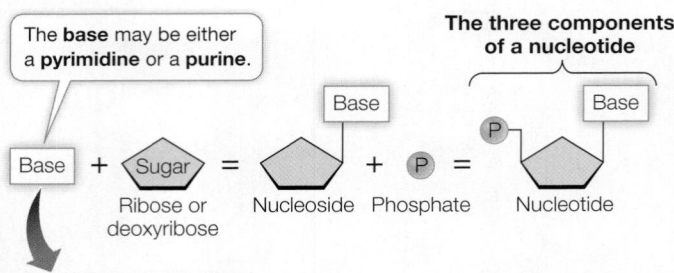

key concept

4.1 Nucleic Acid Structures Reflect Their Functions

Nucleic acids contain life's genetic code. From medicine to evolution, from agriculture to forensics, the properties of nucleic acids affect our lives every day. It is with nucleic acids that the concept of "information" entered the biological vocabulary. Nucleic acids are uniquely capable of coding for and transmitting biological information.

focus your learning

- The structures of DNA and RNA enable their functions in storing and transferring genetic information.
- Base pairing between nucleotides in DNA and RNA provides the structure needed for transfer of genetic information.
- Base sequences of nucleotides in DNA provide the chemical diversity needed for storage of genetic information.
- Nucleotides other than those found in DNA and RNA have diverse functions within the cell.

Nucleic acids are informational macromolecules

Nucleic acids are polymers specialized for the storage, transmission, and use of genetic information. There are two types of nucleic acids: **DNA** (**deoxyribonucleic acid**) and **RNA** (**ribonucleic acid**). Although you have heard of it in countless pronouncements as a metaphor for the essence of something ("It's in her DNA"), DNA is first and foremost a macromolecule. Its structure encodes hereditary information and passes it from generation to generation. Through RNA intermediates, the information encoded in DNA specifies the amino acid sequences of proteins and controls the expression (that is, the synthesis) of other RNAs. In Chapter 11 you will learn how, during cell division and reproduction, information is preserved from the DNA of the parental cell when both the cell and its DNA are duplicated to form two new daughter cells. Chapter 13 will describe how information flows from DNA to RNA to the proteins that ultimately carry out many of life's functions. Here we focus on the chemistry of the nucleic acids, revealing how their structures reflect their functions.

Nucleic acids are polymers composed of monomers called nucleotides. A **nucleotide** has three components: a nitrogen-containing **base**, a pentose sugar, and one to three phosphate groups (**Figure 4.1**). Molecules consisting of a pentose sugar and a nitrogenous base—but no phosphate group—are called **nucleosides**. The nucleotides that make up nucleic acids contain just one phosphate group—they are nucleoside *mono*phosphates.

The bases of the nucleic acids take one of two chemical forms: a six-membered single-ring structure called a **pyrimidine**, or a fused double-ring structure called a **purine** (see Figure 4.1). In DNA, the pentose sugar is **deoxyribose**, which differs from the **ribose** found in RNA by the absence of one oxygen atom (see Figure 3.16).

A nucleic acid forms when nucleotide monomers are added to an existing chain one at a time. The pentose sugar in the last nucleotide of the existing chain and the phosphate on the new nucleotide undergo a condensation reaction (meaning H_2O is removed; see Figure 3.4), and the resulting bond is called a **phosphodiester linkage** (**Figure 4.2**). The phosphate on the incoming nucleotide is attached to the 5′ carbon atom of its sugar, and the linkage occurs between

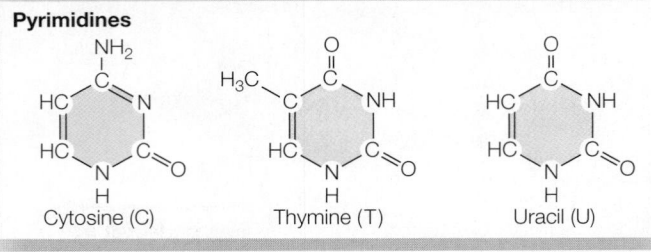

The three components of a nucleotide

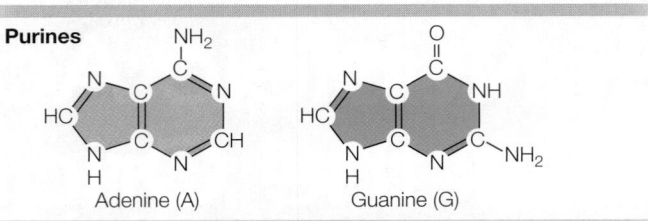

The two groups of bases

Figure 4.1 Nucleotide Chemistry Nucleotide monomers, each consisting of a base, a sugar, and a phosphate group, are the building blocks of DNA and RNA polymers. The bases fall into two groups: the pyrimidines and the purines.

 Activity 4.1 Nucleic Acid Building Blocks
www.Life11e.com/ac4.1

 Animation 4.1 Nucleic Acids
www.Life11e.com/a4.1

it and the 3′ carbon on the last sugar of the existing chain. Because each nucleotide is added to the 3′ carbon of the last sugar, nucleic acids are said to *elongate in the 5′-to-3′ direction*.

As with carbohydrates (see Key Concept 3.3), nucleic acids can range in size. Oligonucleotides are relatively short, with about 20 nucleotide monomers, whereas polynucleotides can be much longer.

- Oligonucleotides include RNA molecules that function as "primers" to begin the duplication of DNA; RNA molecules that regulate the expression of genes; and synthetic DNA molecules used for amplifying and analyzing other, longer nucleotide sequences.

- Polynucleotides, more commonly referred to as nucleic acids, include DNA and some RNA. Polynucleotides can be very long, and indeed are the longest polymers in the living world. Some DNA molecules in humans contain hundreds of millions of nucleotides.

Base pairing occurs in both DNA and RNA

DNA and RNA differ somewhat in their sugar groups, bases, and strand structure (**Table 4.1**). Four bases are found in DNA: **adenine** (**A**), **cytosine** (**C**), **guanine** (**G**), and **thymine** (**T**). RNA is also made up of four different monomers, but its nucleotides include **uracil** (**U**) instead of thymine.

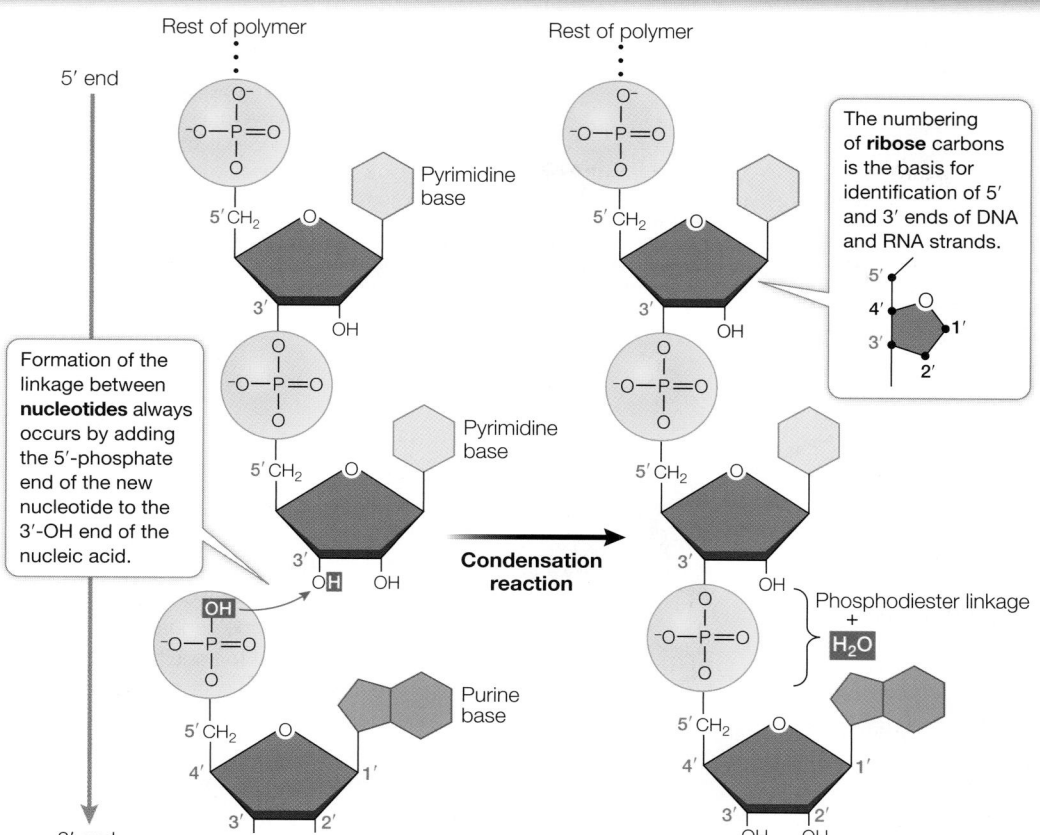

Figure 4.2 Linking Nucleotides Together   Growth of a nucleic acid (RNA in this figure) from its monomers occurs in the 5' (phosphate) to 3' (hydroxyl) direction. Note that the added nucleotide is not initially a monophosphate but actually a triphosphate. The details of this process will be described in Chapter 13.

The numbering of **ribose** carbons is the basis for identification of 5' and 3' ends of DNA and RNA strands.

Formation of the linkage between **nucleotides** always occurs by adding the 5'-phosphate end of the new nucleotide to the 3'-OH end of the nucleic acid.

a considerable force of attraction, which can bind together two polynucleotide strands, or a single strand that folds back onto itself. This attraction is not as strong as a covalent bond, however. This means that individual base pairs are relatively easy to break with a modest input of energy. As you will see, the breaking and making of hydrogen bonds in nucleic acids is vital to their role in living systems.

RNA   Even though RNA is generally single-stranded (**Figure 4.3A**), base pairing can occur between different regions of the molecule. Portions of the single-stranded RNA molecule can fold back and pair with one another (**Figure 4.3B**). Thus complementary hydrogen bonding between ribonucleotides plays an important role in determining the three-dimensional shapes of some RNA molecules. Complementary base pairing can also take place between ribonucleotides and deoxyribonucleotides. Adenine in an RNA strand can pair either with uracil (in another RNA strand) or with thymine (in a DNA strand). Similarly, an adenine in DNA can pair either with thymine (in the complementary DNA strand) or with uracil (in RNA).

DNA   Usually, ***DNA** is double-stranded; that is, it consists of two separate polynucleotide strands of the same length that are held together by hydrogen bonds between base pairs (**Focus: Key Figure 4.4A**). In contrast to RNA's diversity in three-dimensional structure, DNA is remarkably uniform. The A-T and G-C base pairs are about

A key to understanding the structure and function of nucleic acids is the principle of **complementary base pairing**. In DNA, thymine and adenine pair (T-A), and cytosine and guanine pair (C-G). In RNA, the base pairs are A-U and C-G.

Thymine   T ::::: A   **Adenine**

Hydrogen bond

Cytosine   C ::::: G   **Guanine**

Base pairs are held together primarily by hydrogen bonds. As you can see, there are polar C═O and N—H covalent bonds in the bases; these can form hydrogen bonds between the δ^- on an oxygen or nitrogen of one base and the δ^+ on a hydrogen of another base.

Individual hydrogen bonds are relatively weak, but there are so many of them in a DNA or RNA molecule that collectively they provide

table **4.1** Distinguishing RNA from DNA				
Nucleic acid	Sugar	Bases	Name of nucleoside	Strands
RNA	Ribose	Adenine	Adenosine	One
		Cytosine	Cytidine	
		Guanine	Guanosine	
		Uracil	Uridine	
DNA	Deoxyribose	Adenine	Deoxyadenosine	Two
		Cytosine	Deoxycytidine	
		Guanine	Deoxyguanosine	
		Thymine	Deoxythymidine	

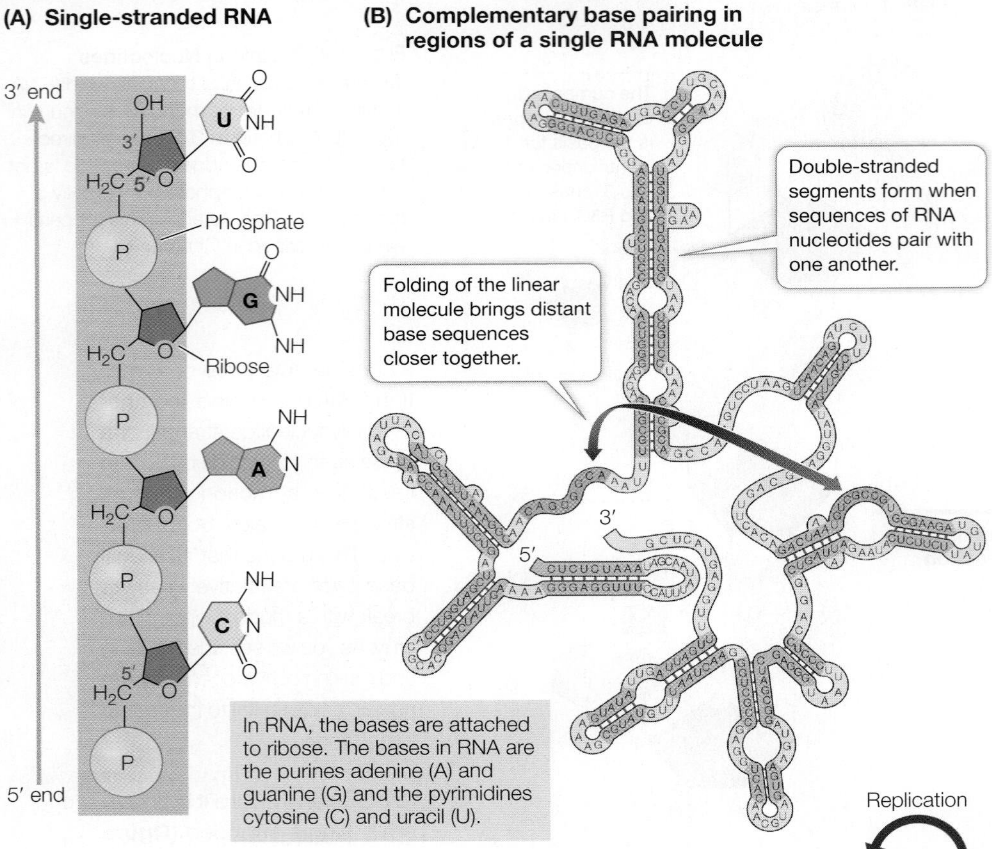

(A) Single-stranded RNA

3' end

Phosphate

Ribose

In RNA, the bases are attached to ribose. The bases in RNA are the purines adenine (A) and guanine (G) and the pyrimidines cytosine (C) and uracil (U).

5' end

(B) Complementary base pairing in regions of a single RNA molecule

Folding of the linear molecule brings distant base sequences closer together.

Double-stranded segments form when sequences of RNA nucleotides pair with one another.

Figure 4.3 RNA **(A)** RNA is usually a single strand. **(B)** When a single-stranded RNA folds back on itself, hydrogen bonds between complementary sequences can stabilize it into a three-dimensional shape with complex surface characteristics.

Q: What would happen if a folded RNA molecule were heated? Recall the effect of heat on hydrogen bonds.

the same size (each is a purine paired with a pyrimidine), and the two polynucleotide strands form a "ladder" that twists into a **double helix (Focus: Key Figure 4.4B)**. The sugar–phosphate groups form the sides of the ladder, and the bases with their hydrogen bonds form the "rungs" on the inside. DNA carries genetic information in its sequence of base pairs rather than in its three-dimensional structure. The key differences among DNA molecules are their different nucleotide base sequences.

> ***connect the concepts*** As explained in Key Concept 13.2, the two strands of a DNA molecule fit together perfectly because they run in opposite directions—that is, they are antiparallel to one another. Equal distance between strands is maintained because a purine on one strand is always found opposite a pyrimidine on the other.

 Activity 4.2 DNA Structure
www.Life11e.com/ac4.2

DNA carries information and is expressed through RNA

DNA is an informational molecule. The information is encoded in the sequence of bases carried in its strands. For example, the information encoded in the sequence TCAGCA is different from the information in the sequence CCAGCA. DNA transmits information in two ways:

1. DNA can be reproduced exactly. This process, called **DNA replication**, is done by polymerization using an existing strand as a base-pairing template.

2. Certain DNA sequences can be copied into RNA, in a process called **transcription**. The nucleotide sequence in the RNA can then be used to specify a sequence of amino acids in a polypeptide chain, in the process called **translation**. The overall process of transcription and translation is called **gene expression**.

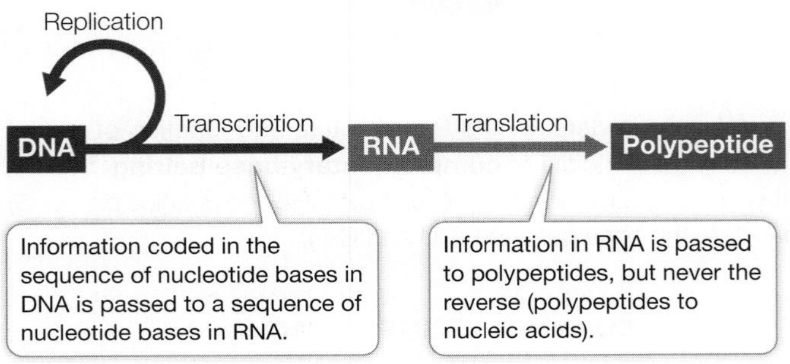

Replication

| DNA | Transcription → | RNA | Translation → | Polypeptide |

Information coded in the sequence of nucleotide bases in DNA is passed to a sequence of nucleotide bases in RNA.

Information in RNA is passed to polypeptides, but never the reverse (polypeptides to nucleic acids).

The details of these important processes will be described in later chapters, but it is important to realize two things at this point:

1. *DNA replication and transcription depend on the base-pairing properties of nucleic acids*. Recall that the hydrogen-bonded base pairs are A-T and G-C in DNA and A-U and G-C in RNA. Consider, for example, this double-stranded DNA region:

   5'-TCAGCA-3'

   3'-AGTCGT-5'

   Transcription of the lower strand will result in a single strand of RNA with the sequence 5'-UCAGCA-3'. Can you figure out the sequence that the top strand would produce?

2. *DNA replication usually involves the entire DNA molecule*. Since DNA holds essential information, it must be replicated completely and accurately so that each new cell or new organism receives a complete set of DNA from its parent (**Figure 4.5A**). The complete set of DNA in a living organism is called its **genome**. However, not all of the information in the genome is needed at all times and in all tissues, and only small sections of the DNA are transcribed into RNA molecules. The sequences of DNA that are transcribed into RNA are called **genes** (**Figure 4.5B**).

focus: key figure

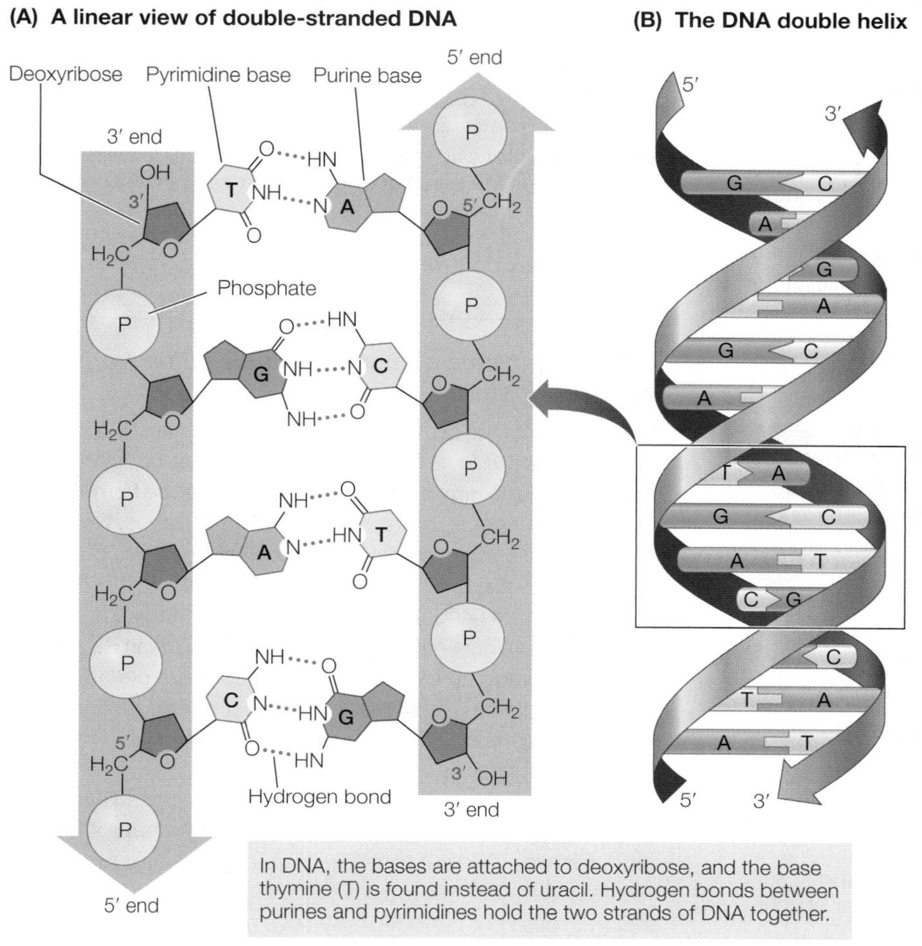

(A) A linear view of double-stranded DNA

Deoxyribose Pyrimidine base Purine base

3′ end

OH

3′

H₂C O

T ⋯⋯ A
O⋯⋯HN
NH⋯⋯N

Phosphate

P

H₂C O

G ⋯⋯ C
O⋯⋯HN
NH⋯⋯N
NH⋯⋯O

P

H₂C O

A ⋯⋯ T
NH⋯⋯O
N⋯⋯HN
O

P

H₂C O

C ⋯⋯ G
NH⋯⋯O
N⋯⋯HN
O⋯⋯HN

5′

H₂C O

Hydrogen bond

P

5′ end

P
5′CH₂ O

P
CH₂ O

P
CH₂ O

P
CH₂ O

3′ OH
3′ end

5′ end

In DNA, the bases are attached to deoxyribose, and the base thymine (T) is found instead of uracil. Hydrogen bonds between purines and pyrimidines hold the two strands of DNA together.

(B) The DNA double helix

5′

3′

G—C
A
G
A
G—C
A
T—A
G—C
A—T
C—G
C
T—A
A—T

5′   3′

Figure 4.4 DNA **(A)** DNA usually consists of two strands running in opposite directions that are held together by hydrogen bonds between purines and pyrimidines on the two strands. **(B)** The two strands in DNA are coiled in a right-handed double helix.

Q: For replication or transcription, DNA must "unwind" to expose bases. What bonds must be broken for this to occur?

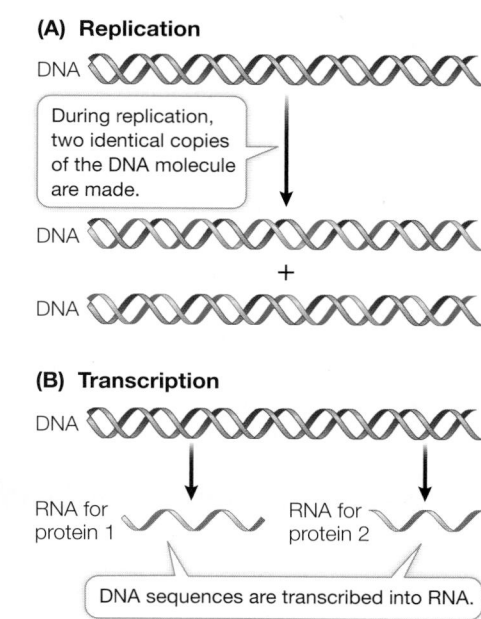

(A) Replication

DNA

During replication, two identical copies of the DNA molecule are made.

DNA

+

DNA

(B) Transcription

DNA

RNA for protein 1   RNA for protein 2

DNA sequences are transcribed into RNA.

Figure 4.5 DNA Replication and Transcription DNA is usually completely replicated **(A)** but only partially transcribed **(B)**. RNA transcripts are produced from genes that code for specific proteins. Transcription of different genes occurs at different times and, in multicellular organisms, in different cells of the body.

Q: What do you think determines whether a particular cell transcribes a sequence of DNA into RNA?

In humans, the gene that encodes the major protein in hair (keratin) is expressed in skin cells that produce hair. The genetic information in the keratin-encoding gene is transcribed into RNA and then translated into a keratin polypeptide. In other tissues such as the muscles, the keratin gene is not transcribed, but other genes are—for example, the genes that encode proteins present in muscles but not in skin or hair. These genes are turned on and off by the control of gene expression, which we will describe in Chapter 16.

The DNA base sequence reveals evolutionary relationships

DNA carries hereditary information from one generation to the next, gradually accumulating changes in its base sequences over long periods of time. A series of DNA molecules stretches back through the lineage of every organism to the beginning of biological evolution on Earth, about 3.8 billion years ago. Therefore closely related living species have more similar base sequences than do species that are more distantly related. The same is true for closely related versus distantly related individuals within a species. The details of how scientists use this information will be covered in Chapter 24.

Improvements in methods for DNA sequencing and analysis by computer programs have enabled scientists to determine the entire DNA base sequences—the genome—of many organisms, including humans, whose genome contains about 3 billion base pairs. These studies have confirmed many of the evolutionary relationships that had been inferred previously from more traditional comparisons of body structure, biochemistry, and physiology. For example, traditional comparisons had indicated that the closest living relative of humans (*Homo sapiens*) is the chimpanzee (genus *Pan*). In fact, the chimpanzee genome shares more than 98 percent of its DNA base sequence with the human genome. Increasingly, scientists turn to DNA analyses to elucidate evolutionary relationships when other comparisons are not possible or are not conclusive. For example, DNA studies revealed a close relationship between starlings and mockingbirds that was not expected on the basis of their anatomy or behavior.

Nucleotides have other important roles

Nucleotides are more than just the building blocks of nucleic acids. As we will describe in later chapters, there are several nucleotides (or modified nucleotides) with other functions:

- ATP (adenosine triphosphate) acts as an agent of energy transfer in many biochemical reactions (see Key Concept 8.2).

- GTP (guanosine triphosphate) serves as an energy source, especially in protein synthesis. It also plays a role in the transfer of information from the environment to cells (see Key Concept 7.2).

- cAMP (cyclic adenosine monophosphate) is a special nucleotide with an additional bond between the sugar and the phosphate group. It is essential in many processes, including the actions of hormones and the transmission of information by the nervous system (see Key Concept 7.3).

- Nucleotides play roles as carriers in the synthesis and breakdown of carbohydrates and lipids.

▶ 4.1 recap

The nucleic acids DNA and RNA are polymers made up of nucleotide monomers. The sequence of nucleotides in DNA carries the information that is used by RNA to specify primary protein structure. The genetic information in DNA is passed from generation to generation and can be used to understand evolutionary relationships.

learning outcomes

You should be able to:

- Explain how purines and pyrimidines differ.
- Explain how biological information is encoded in DNA.
- Identify cellular functions of nucleotides other than information processing.

1. What are the key differences between purines and pyrimidines and how does this relate to the structure of DNA?

2. How can DNA molecules be very diverse, even though they appear to be structurally similar?

3. Single-stranded nucleic acids about 25 bases long are called aptamers. Because many sequences are possible, particular aptamers can be used to bind to specific targets, either as drugs to a protein target or for quantitating a small molecule in tissues or fluids. How can aptamers have such diversity and specificity for binding?

We have seen that the nucleic acids RNA and DNA carry the blueprint of life, and that the inheritance of these macromolecules reaches back to the beginning of evolutionary time. But when, where, and how did nucleic acids arise on Earth? How did the building blocks of life, such as amino acids and sugars, originally arise?

▶ key concept The Small Molecules of Life Originated on Primitive Earth

4.2

You learned in Chapter 2 that living things are composed of the same elements as the inanimate universe (e.g., C, H, O, P, N, and S). But usually the arrangements of these atoms into molecules are unique in biological systems. You will not find biological molecules such as proteins in nonliving matter unless they came from a once-living organism.

focus your learning

- Experimental evidence was used to refute the idea that life could arise from nonliving materials.
- Water did not always exist on Earth, and its presence on Earth was necessary before life could arise.
- Experimental evidence supports the theory that life evolved as the result of chemical and physical changes on early Earth.
- Evidence discovered in meteorites and on Mars raises questions about whether life evolved on bodies other than Earth.

Living organisms do not repeatedly come from inanimate nature

We can't know for certain how life on Earth began. But one thing is sure: life (or at least life as we know it) is not constantly being restarted. That is, **spontaneous generation** of life from inanimate nature is not happening repeatedly before our eyes. Now and in the past, all life has come from life that existed before. But people, including scientists, did not always believe this.

Many cultures and religions have suggested that life can arise repeatedly from nonliving matter. During the European Renaissance (from the fourteenth to seventeenth centuries, a period that witnessed the birth of modern science), most people thought that at least some forms of life arose repeatedly and directly from inanimate or decaying matter by spontaneous generation. People thought, for example, that mice arose from sweaty clothes placed in dim light; that frogs sprang directly from moist soil; and that rotting meat produced flies. One scientist who doubted these assumptions was the Italian physician and poet Francesco Redi. Redi hypothesized that flies arose not by some mysterious transformation of decaying meat, but from other flies that laid their eggs on the meat. In 1668, Redi performed a scientific experiment—a relatively new concept at the time—to test his hypothesis. He set out three jars containing chunks of meat:

1. One jar contained meat exposed to both air and flies.

2. A second jar was covered with a fine cloth so that the meat was exposed to air but not to flies.

3. The third jar was sealed with a lid so the meat was exposed to neither air nor flies.

Jar 1: No lid   Jar 2:   Jar 3: Lid
Fine cloth cover

As he had hypothesized, Redi found maggots, which then hatched into flies, only in the first jar. This finding demonstrated that maggots could occur only where flies were present before. The idea that a complex organism like a fly could appear spontaneously from a nonliving substance in the meat, or from "something in the air," was laid to rest. Well, perhaps not quite to rest.

In the 1660s, the newly invented microscope revealed a vast biological world that had never been seen before. Virtually every environment on Earth was found to be teeming with tiny organisms. Some scientists believed these organisms arose spontaneously from their rich chemical environment, by the action of a "life force." But experiments in the nineteenth century by the great French scientist Louis Pasteur showed that microorganisms can arise only from other microorganisms, and that an environment without life remains lifeless (**Figure 4.6**).

 Animation 4.2 **Pasteur's Experiment**
www.Life11e.com/a4.2

Pasteur's and Redi's experiments showed that living organisms cannot arise from nonliving materials *under the conditions that exist on Earth now*. But their experiments did not prove that spontaneous generation never occurred. Eons ago, conditions on Earth and in the atmosphere above it were vastly different than they are today. Indeed, conditions similar to those found on primitive Earth may have existed, or may exist now, on other bodies in our solar system and elsewhere.

Life began in water

As we emphasized in Chapter 2, water is an essential component of life as we know it. This is why there was great excitement when remotely controlled spacecraft sent from Earth detected water ice on Mars. Astronomers believe our solar system began forming about 4.6 billion years ago, when a star exploded and collapsed to form the sun and about 500 bodies called planetesimals. These planetesimals collided with one another to form the inner planets, including Earth and Mars. The first chemical signatures indicating the presence of life on Earth are about 4 billion years old. So it took 600 million years for the chemical conditions on Earth to become just right for life. Key among those conditions was the presence of water.

Ancient Earth probably had a lot of water high in its atmosphere. But the new planet was hot, and the water remained in vapor form and dissipated into space. As Earth cooled, it became possible for water to condense on the planet's surface—but where did that water come from? One current view is that comets (loose agglomerations of dust and ice that have orbited the sun since the planets formed) struck Earth and Mars repeatedly, bringing to those planets not only water but also other chemical components of life, such as nitrogen and carbon-containing molecules.

As the planets cooled and chemicals from their crusts dissolved in the water, simple chemical reactions would have taken place. Some of these reactions might have led to life, but impacts by large comets and rocky meteorites released enough energy to heat the developing oceans almost to boiling, thus destroying any early life that might have existed. On Earth, these large impacts eventually subsided, and around 3.8 billion years ago, life gained a foothold. There has been life on Earth ever since.

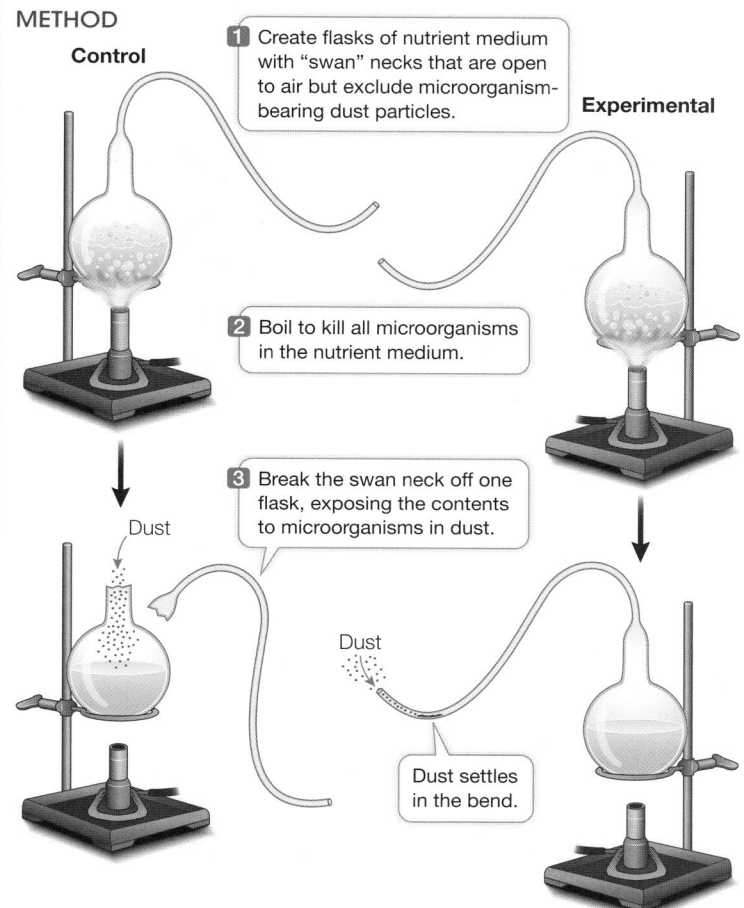

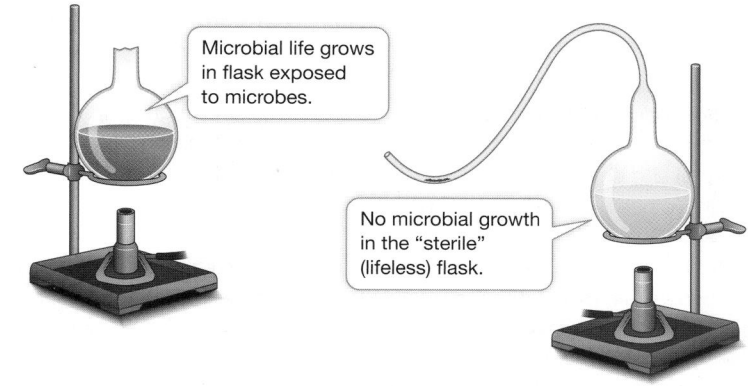

experiment

Figure 4.6 **Disproving the Spontaneous Generation of Life**

Source: Pasteur gave a talk on his research at the "Sorbonne Scientific Soirée" on April 7, 1864. This talk has been translated into English: rc.usf.edu/~levineat/pasteur.pdf.

Previous experiments disproving the spontaneous generation of larger organisms were called into question when microorganisms were discovered. Louis Pasteur's classic experiments disproved the spontaneous generation of microorganisms.

HYPOTHESIS▶ Microorganisms come only from other microorganisms and cannot arise by spontaneous generation.

METHOD

Control

1 Create flasks of nutrient medium with "swan" necks that are open to air but exclude microorganism-bearing dust particles.

Experimental

2 Boil to kill all microorganisms in the nutrient medium.

Dust

3 Break the swan neck off one flask, exposing the contents to microorganisms in dust.

Dust

Dust settles in the bend.

RESULTS Microbial life grows only in the flasks exposed to microorganisms. There is no "spontaneous generation" of life in the sterile flask.

Microbial life grows in flask exposed to microbes.

No microbial growth in the "sterile" (lifeless) flask.

CONCLUSION▶ Living organisms arise from preexisting living organisms.

experiment

Figure 4.7   Could Biological Molecules Have Been Formed from Chemicals Present in Earth's Early Atmosphere?

Original Papers: Miller, S. L. 1953. A production of amino acids under possible primitive earth conditions. *Science* 117: 528–519.

Miller, S. L. and H. C. Urey. 1959. Organic compound synthesis on the primitive earth. *Science* 130: 245–251.

With an increased understanding of the atmospheric conditions that existed on primitive Earth, the researchers devised an experiment to see if these conditions could lead to the formation of organic molecules.

HYPOTHESIS▶   Organic chemical compounds can be generated under conditions similar to those that existed in the atmosphere of primitive Earth.

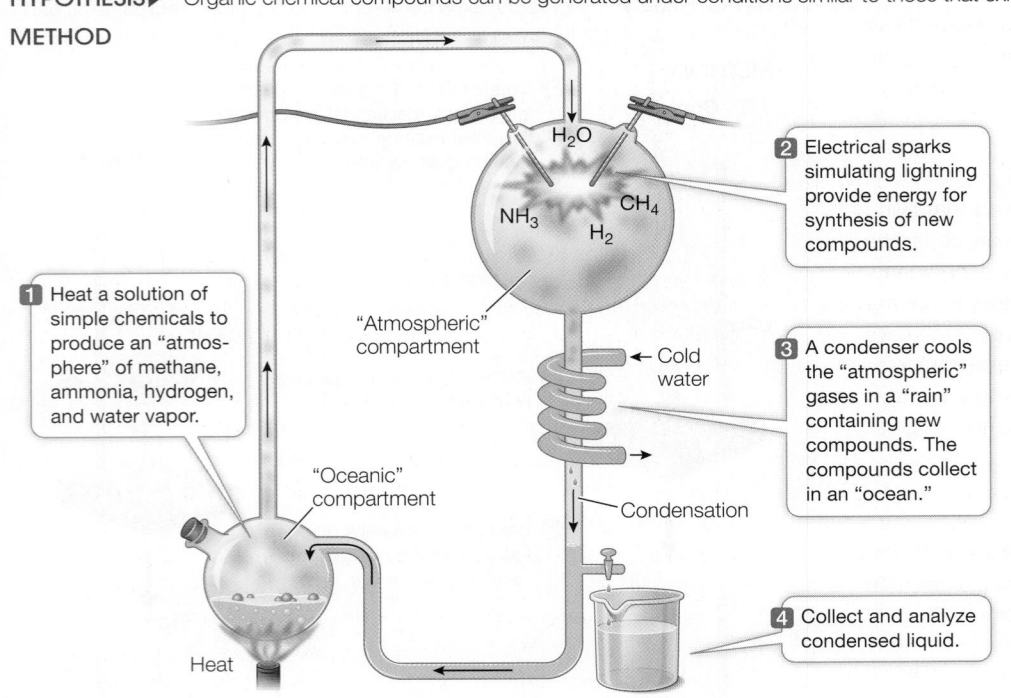

METHOD

RESULTS

1 Heat a solution of simple chemicals to produce an "atmosphere" of methane, ammonia, hydrogen, and water vapor.

"Atmospheric" compartment

2 Electrical sparks simulating lightning provide energy for synthesis of new compounds.

3 A condenser cools the "atmospheric" gases in a "rain" containing new compounds. The compounds collect in an "ocean."

Cold water

Condensation

"Oceanic" compartment

Heat

4 Collect and analyze condensed liquid.

Reactions in the condensed liquid eventually formed organic chemical compounds, including amino acids.

CONCLUSION▶   The chemical building blocks of life could have been generated in the probable atmosphere of early Earth.

Several models have been proposed to explain the origin of life on Earth. The next sections will discuss two alternative theories: that life arose on Earth through chemical evolution, or that life came from outside Earth.

Prebiotic synthesis experiments model early Earth

One theory for the origin of life on Earth, **chemical evolution**, holds that conditions on primitive Earth led to the formation of simple molecules such as monomers (see Key Concept 3.1), and that these molecules led to the formation of life forms. Scientists have sought to reconstruct those primitive conditions, both physically (by varying temperature) and chemically (by re-creating the mixes of elements that may have been present).

HOT CHEMISTRY   In oxygenated water, some trace metals such as molybdenum and rhenium are soluble, and their presence in sediments under oceans and lakes is directly proportional to the amount of oxygen gas (O_2) that was present in and above the water at the times the rocks were formed. Measurements of dated sedimentary cores indicate that none of these rare metals was present prior to 2.5 billion years ago. This and other lines of evidence suggest that there

was little O_2 in Earth's early atmosphere. Oxygen gas is thought to have accumulated about 2.5 billion years ago as the by-product of photosynthesis by single-celled life forms; today 21 percent of our atmosphere is O_2.

In the 1950s Stanley Miller and Harold Urey at the University of Chicago set up an experimental "atmosphere" containing the gases they thought were present in Earth's early atmosphere: hydrogen gas, ammonia, methane gas, and water vapor. They passed an electrical spark through these gases to simulate lightning, a source of energy to drive chemical reactions. Then they cooled the system so the gases would condense and collect in a watery solution, or "ocean" (**Figure 4.7**). After a week of continuous operation, the system contained numerous organic molecules, including a variety of amino acids—the building blocks of proteins.

 Animation 4.3 Synthesis of Prebiotic Molecules
www.Life11e.com/a4.3

COLD CHEMISTRY   Stanley Miller also performed a long-term experiment in which the electrical spark was not used. In 1972 he filled test tubes with ammonia gas, water vapor, and cyanide

(HCN), another molecule that is thought to have formed on primitive Earth. After checking that there were no contaminating substances or organisms that might confound the results, he sealed the tubes and cooled them to −78°C, the temperature of the ice that covers Europa, one of Jupiter's moons. Opening the tubes 27 years later, Miller found amino acids and nucleotide bases. Apparently, pockets of liquid water within the ice had allowed high concentrations of the starting materials to accumulate, thereby speeding up chemical reactions. The important conclusion is that the cold water within ice on ancient Earth, and other celestial bodies such as Mars, Europa, and Enceladus (one of Saturn's moons; satellite photos have revealed geysers of liquid water coming from its interior), may have provided environments for the prebiotic synthesis of molecules required for the subsequent formation of simple living systems.

The results of these experiments were profoundly important in giving weight to speculations about the chemical origin of life on Earth and elsewhere in the universe. Decades of experimental work and critical evaluation followed Miller and Urey's original experiments. In science, an experiment and its results must be repeatable and be reinterpreted and refined as more knowledge accumulates. For example, ideas about Earth's original atmosphere have changed. There is abundant evidence indicating that major volcanic eruptions occurred 4 billion years ago; these would have released carbon dioxide (CO_2), nitrogen (N_2), hydrogen sulfide (H_2S), and sulfur dioxide (SO_2) into the atmosphere. Experiments using these gases in addition to the ones in the original Miller–Urey experiment have produced a more diverse list of organic products:

- All five bases that are present in DNA and RNA (i.e., A, T, C, G, and U)
- All of the 20 amino acids used in protein synthesis
- Many three- to six-carbon sugars
- Certain fatty acids
- Vitamin B_6 (pantothenic acid, a component of coenzyme A)
- Nicotinamide (part of NAD, which is involved in energy metabolism)
- Carboxylic acids such as succinic and lactic acids (also involved in energy metabolism)

Life may have come from outside Earth

In 1969 a remarkable event led to the discovery that a meteorite from space carried molecules that were characteristic of life on Earth. On September 28 of that year, fragments of a meteorite fell around the town of Murchison, Australia. Using gloves to avoid Earth-derived contamination, scientists immediately shaved off tiny pieces of the rock, put them in test tubes, and extracted them in water (**Figure 4.8**). They found several of the molecules that are unique to life, including purines, pyrimidines, sugars, and ten amino acids.

 Media Clip 4.1 DNA Building Blocks from Space
www.Life11e.com/mc4.1

Were these molecules truly brought from space as part of the meteorite, or did they get there after the rock landed on Earth? There are a number of reasons to believe the molecules were not Earthly contaminants:

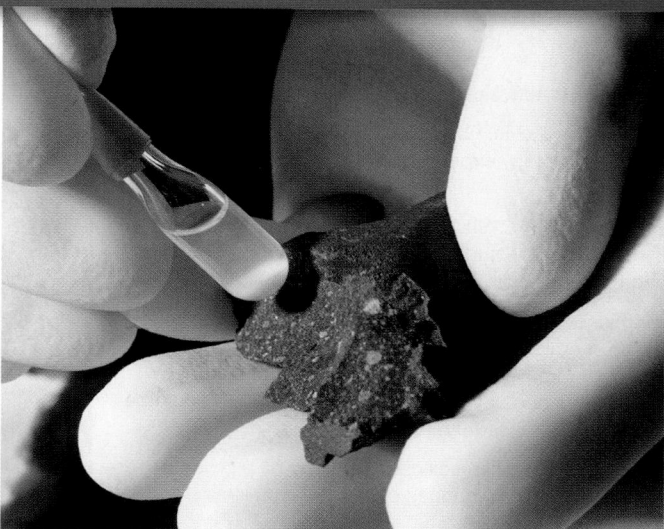

Figure 4.8 The Murchison Meteorite   Pieces from a fragment of the meteorite that landed in Australia in 1969 were put into test tubes with water. Soluble molecules present in the rock—including amino acids, nucleotide bases, and sugars—dissolved in the water.

- The scientists took great care to avoid contamination. They used gloves and sterile instruments, took pieces from below the rock's surface, and did their work very soon after the meteorite landed (they hoped before organisms from Earth could contaminate the samples).
- Amino acids in most living organisms on Earth are L-amino acids; that is, the amino acids of living things are found in only one of the two possible optical isomeric forms (see Figure 3.2). The amino acids in the meteorite, however, were a mixture of L- and D-isomers, with a slight majority of the L form. Thus the amino acids in the meteorite were not likely to have come from a living organism on Earth.
- In the story that opened Chapter 2, we described how the ratio of isotopes in a living organism reflects the ratio of the same isotopes in the environment where the organism lives. The isotope ratios for carbon and hydrogen in the sugars from the meteorite were different from the ratios of those elements found on Earth.

More than 90 meteorites from Mars have been recovered on Earth. Many show signs of water—for example, in minerals such as carbonates that are precipitated from aqueous solution. Some also contain organic molecules that are the chemical signatures of life. While the presence of such molecules suggests that these rocks once harbored life, it does not prove that there were living organisms in the rocks when they landed on Earth. Many scientists find it hard to believe that an organism could survive thousands of years of traveling through space in a meteorite, followed by intense heat as the meteorite passed through Earth's atmosphere. But there is evidence that the heat at the centers of some meteorites may not have been extreme. If this were the case, then a long interplanetary trip by living organisms might have been possible.

Life does not have to reach Earth from space to prove it exists elsewhere. There is an ongoing search for life in celestial bodies. For example, as mentioned in the opening story of this chapter, both orbiting satellites and vehicles probing the landscape of Mars search for conditions suitable for present or past life on Mars. Perhaps the most dramatic attempt to discover life on Mars came in

experiment

Original Papers: Levin, G. V. and O. Straat. 1976. Viking labeled release biology experiment: Interim results. *Science* 194: 1322–1329.

Ponnamperuma, C., A. Shimoyama, M. Yamada, T. Hobo and R. Pal. 1977. Possible surface reactions on Mars: Implications for Viking biology results. *Science* 197: 455–457.

The *Viking* landers on Mars scooped up some of the soil and tested it for chemical interconversions consistent with life. Gilbert Levin and colleagues designed experiments to show remotely, at a distance of millions of miles, that there could be life on Mars at the present time.

HYPOTHESIS▶   Martian soil can be tested by a probe on Mars to show chemical changes consistent with life.

METHOD

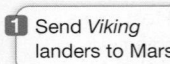

1 Send *Viking* landers to Mars.

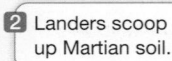

2 Landers scoop up Martian soil.

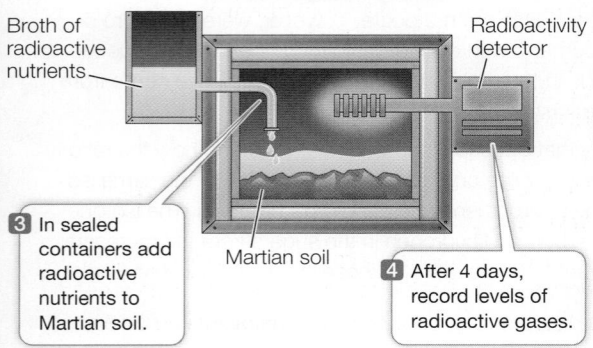

Broth of radioactive nutrients

Radioactivity detector

3 In sealed containers add radioactive nutrients to Martian soil.

Martian soil

4 After 4 days, record levels of radioactive gases.

RESULTS

Radioactive gas detected (counts per minute) after 4 Martian days

Control (no soil)	Martian sample 1	Martian sample 2
500	9,500	12,000

CONCLUSION▶   Martian soil shows chemical changes consistent with life.

work with the data

While it's challenging to design experiments in the lab here on Earth, imagine the challenge of designing experiments done remotely on Mars! A team of biologists and chemists led by Levin and Straat sent instruments to Mars to test for the presence of life by measuring the emission of gases after nutrients were supplied to soils possibly containing Martian organisms. A scoop was used to obtain a 0.2 mL sample of soil which was then placed in a sealed container. The mixture was warmed to about 18°C (much higher than the surrounding Martian environment). Radioactive nutrients (formate, glycolate, glycine, alanine, and lactate) with their carbon atoms labeled in ^{14}C were added. At intervals after nutrient injection, a radioactive detector was employed to test for the release of ^{14}C radioactive gases.

QUESTIONS▶

1. Table A shows data from two of the experiments. Plot the data as ^{14}C radioactive gas versus Martian days after nutrient addition. What can you conclude? Laboratory experiments prior to the two Viking departures showed that the total radioactivity that could be released if all of the nutrients were completely converted to gases was 257,000 counts per minute (cpm). Calculate the fraction of this that appeared in the Martian sample and comment on the result.

2. Table A shows data from Martian soil that had been heated for 3 h at 160°C before nutrient addition. Plot these data on the same graph you made for Question 1. What types of molecules are destroyed by heat, and how does this occur? How do these data affect your conclusions from question 1?

3. Back on Earth, Ponnamperuma and colleagues simulated Martian soil by using a major component of Martian soil called hematite (Fe_2O_3) and added the nutrients in the same way as had been done in the Mars experiment. Their results, including a preheated experiment, are shown in Table B. How do these results affect the conclusions you drew after considering Questions 1 and 2?

Table A

	^{14}C Gases Detected (counts per minute)		
Time (days)	Experiment 1	Experiment 2	Preheated soil
Initial	185	1,100	655
0.2	3,000	5,500	540
0.5	4,800	7,200	500
1.0	6,200	9,500	525
2.0	7,000	11,300	590
3.0	7,600	11,800	610
4.0	8,000	12,000	620

Table B

	^{14}C Gases Detected (counts per minute)	
	Hematite + ^{14}C nutrients	Hematite without nutrients
Experiment	10,140	150
Preheated 160°C	308	107

A similar **work with the data** exercise may be assigned in **LaunchPad**.

1976, when NASA celebrated the 200th anniversary of the political independence of the United States by landing two stationary probes, or landers, called *Viking 1* and *Viking 2* on the planet. **Investigating Life: Can We Find Evidence of Life on Mars?** describes how scientists on Earth directed onboard instruments to evaluate soil on Mars for signs of life.

A distinctive aspect of living systems is their ability to use molecules from the environment to extract chemical energy for growth. During this process, waste products are released. You are familiar with this: you breathe in oxygen gas (O_2), take in nutrients in food, and breathe out carbon dioxide (CO_2). Other organisms do not use O_2 but take in other molecules. In the Mars experiments, scientists directed the Viking landers to scoop up some Martian soil and then exposed it to seven organic molecules as nutrients, all of which can be formed in Miller–Urey experiments (e.g., amino acids; see Figure 4.7). The seven molecules were labeled with a radioisotope of carbon (^{14}C). After a period of exposure, a detector was used to measure any radioactive gases (possibly CO_2) produced.

The results of some of the experiments were astonishing: radioactive gas was detected in some experiments, and some scientists concluded that there might be life on Mars. The fact that there was little radioactive gas produced when the soil was preheated to 160°C to kill living organisms gave credence to this conclusion. Later, simulations in the lab on Earth showed that radioactive gas could could be formed in this experiment with certain kinds of soils, without organisms. So there is an alternative, nonbiological explanation for the Mars data. Nevertheless, some biologists still adhere to the biological explanation for the Mars data, and the quest for life on Mars continues.

4.2 recap

Life does not arise repeatedly through spontaneous generation, but comes from preexisting life. Water is an essential ingredient for the emergence of life. Chemical synthesis experiments provide support for the idea that life's simple molecules formed in the prebiotic environment on Earth. Meteorites that have landed on Earth provide some evidence for an extraterrestrial origin of life, as do experiments with landers on Mars showing possible chemical interconversions that may reflect life.

learning outcomes

You should be able to:

- Explain how Redi's and Pasteur's experiments disproved spontaneous generation.
- Describe conditions on early Earth and theories about how these conditions affected the origin of life on Earth.
- Justify the conditions used by Miller and Urey in their experiments.

1. What conditions existing on Earth today might preclude the origin of life from the prebiotic molecules Miller and Urey used?

2. The interpretation of Pasteur's experiment (see Figure 4.6) depended on the inactivation of microorganisms by heat. We now know of microorganisms that can survive extremely high temperatures (see Chapter 26). Does this change the interpretation of Pasteur's experiment? What experiments would you do to inactivate such microbes?

4.2 recap

3. The Miller–Urey experiment (see Figure 4.7) showed that it was possible for amino acids to be formed from gases that were hypothesized to have been in Earth's early atmosphere. These amino acids were dissolved in water. Knowing what you do about the polymerization of amino acids into proteins (see Figure 3.6), how would you set up an experiment to show that proteins can form under the conditions of early Earth?

Chemistry experiments modeling the conditions of ancient Earth provide clues about the origins of the monomers (such as amino acids) that make up the polymers (such as proteins) that characterize life. How did these polymers develop?

key concept 4.3 The Large Molecules of Life Originated from Small Molecules

The Miller–Urey experiment and others that followed provide a plausible scenario for the formation of the building blocks of life under conditions that prevailed on primitive Earth. The next step in forming and supporting a general theory on the origin of life would be an explanation for how polymers formed from these monomers.

focus your learning

- Evidence from simulation experiments supports the theory that biological polymers formed from chemical reactions taking place in early Earth conditions.
- Scientists hypothesize that the presence of catalysts would have been necessary during the evolution of biological polymers.

Complex molecules could be formed from simpler ones on primitive Earth

Scientists have used a number of model systems to try to simulate conditions under which the first biological polymers might have been made. Each of these systems is based on several observations and speculations:

- Solid mineral surfaces, such as powderlike clays, have large surface areas. Scientists speculate that the silicates in clay may have catalyzed (sped up) the condensation reactions that resulted in organic polymers.

- Hydrothermal vents deep in the ocean, where hot water emerges from beneath Earth's crust, lack oxygen gas and contain metals such as iron and nickel. In laboratory experiments, these metals have been shown to catalyze the polymerization of amino acids in the absence of oxygen.

- In hot pools at the edges of oceans, evaporation may have concentrated monomers to the point where polymerization was favored (the "primordial soup" hypothesis).

In whatever ways the earliest stages of chemical evolution occurred, they resulted in the emergence of monomers and polymers that have probably remained unchanged in their general structures and functions for several billion years.

RNA may have been the first biological catalyst

Many chemical changes occur in living organisms: an example is DNA replication, which we described earlier in this chapter as the process by which a molecule of DNA is duplicated (see Figure 4.5). Many other chemical changes occur in living systems, often involving the hydrolysis or synthesis of macromolecules or conversions among small molecules. As you will see in Chapter 8, these chemical changes can occur spontaneously in aqueous solutions like those that exist in biological systems, but most would occur extremely slowly. Catalysts—molecules that speed up biochemical conversions—solve this problem. So a key to the origin of life is the appearance of catalysts.

Today, the main catalysts in organisms are proteins called enzymes. The many shapes of proteins allow them to bind to diverse substances in solution and speed up chemical reactions. But we know that proteins are made based on information in nucleic acids. So we have a chicken-or-egg problem: If proteins are needed for life, nucleic acids must have appeared first, so that the proteins could be made. But if nucleic acids appeared before proteins, proteins could not have been the first catalysts. Could nucleic acids be catalysts, in addition to their role as blueprints for protein synthesis? The answer is yes.

Like a protein, the three-dimensional structure of a folded RNA molecule presents a unique surface to the external environment (see Figure 4.3). The surfaces of RNA molecules can be every bit as specific as those of proteins. The three-dimensional shapes and other chemical properties of certain RNA molecules allow them to function as catalysts. Catalytic RNAs, called **ribozymes**, can speed up reactions involving their own nucleotides as well as other cellular substances. Although in retrospect it is not too surprising, the discovery of catalytic RNAs was a surprise to a community of biologists who were convinced that all biological catalysts were proteins (enzymes). It took almost a decade for the work of the scientists involved, Thomas Cech and Sidney Altman, to be fully accepted by other scientists, but when it was, the two were awarded the Nobel Prize.

Given that RNA can be both informational (in its nucleotide sequence) and catalytic (because of its ability to form unique three-dimensional shapes), it has been hypothesized that early life consisted of an "RNA world"—a world before DNA. It is thought that when RNA was first made, it could have acted as a catalyst for its own replication as well as for the synthesis of proteins. DNA could eventually have evolved from RNA (**Figure 4.9**). Several lines of evidence support this scenario:

- In living organisms today, the formation of peptide linkages (see Figure 3.6) is catalyzed by ribozymes.

- In certain viruses called retroviruses, there is an enzyme called reverse transcriptase that catalyzes the synthesis of DNA from RNA.

- When a short, naturally occurring RNA molecule is added to a mixture of nucleotides, RNA polymers are formed at a rate 7 million times greater than the formation of polymers without the added RNA. This indicates that the added RNA is a catalyst, not just a template.

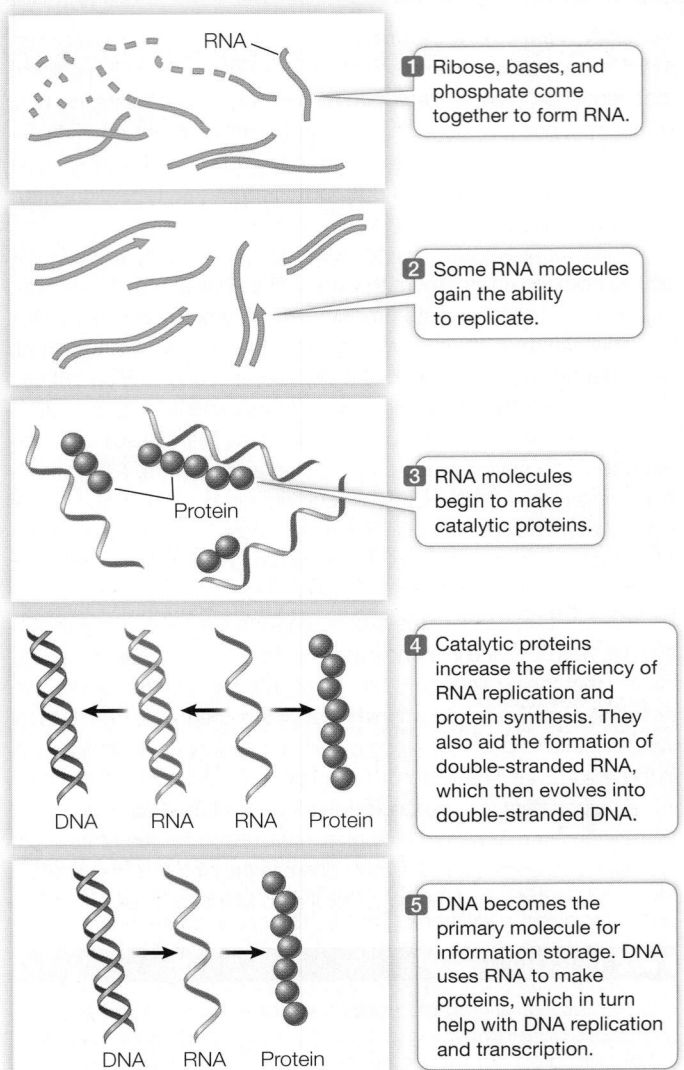

1 Ribose, bases, and phosphate come together to form RNA.

2 Some RNA molecules gain the ability to replicate.

3 RNA molecules begin to make catalytic proteins.

Protein

4 Catalytic proteins increase the efficiency of RNA replication and protein synthesis. They also aid the formation of double-stranded RNA, which then evolves into double-stranded DNA.

DNA   RNA   RNA   Protein

5 DNA becomes the primary molecule for information storage. DNA uses RNA to make proteins, which in turn help with DNA replication and transcription.

DNA   RNA   Protein

Figure 4.9 The "RNA World" Hypothesis This view postulates that in a world before DNA, RNA alone was both the blueprint for protein synthesis and a catalyst for its own replication. Eventually, the information storage molecules of DNA could have evolved from RNA.

- An artificial ribozyme has been developed that can catalyze the assembly of short RNAs into a longer molecule that is an exact copy of itself. This may be how nucleic acid replication evolved.

4.3 recap

The formation of the large polymers that are characteristic of life may have occurred on the surfaces of clay particles, near hydrothermal vents, or in hot pools at the edges of oceans. RNA may have been the first genetic material and catalyst.

learning outcomes

You should be able to:

- Justify the need for catalysts in the origin of life.
- Describe the proposed role for RNA in the formation of long-chain polymers.

4.3 recap

- Provide evidence supporting RNA as a catalyst in prebiotic chemical reactions that gave rise to other biological polymers.

1. Why was the ability to both encode information and catalyze reactions important for the origin of life?
2. Why was the discovery of ribozymes important for the development of the "RNA world" hypothesis?
3. In living organisms, the catalyst for the formation of the peptide bond is an RNA that does not have an informational role. How does this relate to the "RNA world" hypothesis?

The discovery of mechanisms for the formation of small and large molecules is essential to answering questions about the origin of life on Earth. But we also need to understand how organized living systems formed. Such systems display the characteristic properties of life, including reproduction, energy processing, and responsiveness to the environment. These are properties of cells, whose origin we will explore in the next section.

key concept 4.4 Cells Originated from Their Molecular Building Blocks

As you have seen from many of the theories on the origin of life, the evolution of biochemistry occurred under localized conditions. That is, the chemical reactions of life could not occur in a dilute aqueous environment with the molecular participants far apart. There had to be a compartment of some sort that brought together and concentrated the compounds involved in these events. Biologists have proposed that initially this compartment may have simply been a tiny droplet of water on the surface of a rock. But another major event in the origin of life was necessary: the evolution of the cell membrane.

focus your learning

- Membrane development was critical for the evolution of the cell as the smallest unit of life.
- Scientists have discovered fossils of structures that may have been early cells.

Life as we know it is separated from the environment within structurally defined units called **cells**. The internal contents of a cell are separated from the nonbiological environment by a special barrier—a **membrane**. The membrane is not just a barrier; it regulates what goes into and out of the cell, as we will describe in Chapter 6. This role of the surface membrane is very important because it permits the interior of the cell to maintain a chemical composition that is different from that of its external environment.

How did the first cells with membranes come into existence?

Jack Szostak and his colleagues at Harvard University built a laboratory model that gives insights into the origin of cells. To do this, they first put fatty acids (which can be made in prebiotic experiments) into

(A) Hypothetical formation of a protocell

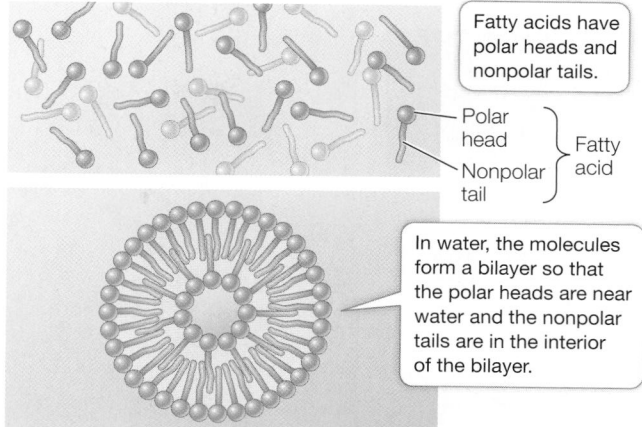

Fatty acids have polar heads and nonpolar tails.

Polar head — Nonpolar tail } Fatty acid

In water, the molecules form a bilayer so that the polar heads are near water and the nonpolar tails are in the interior of the bilayer.

(B) Model of a protocell

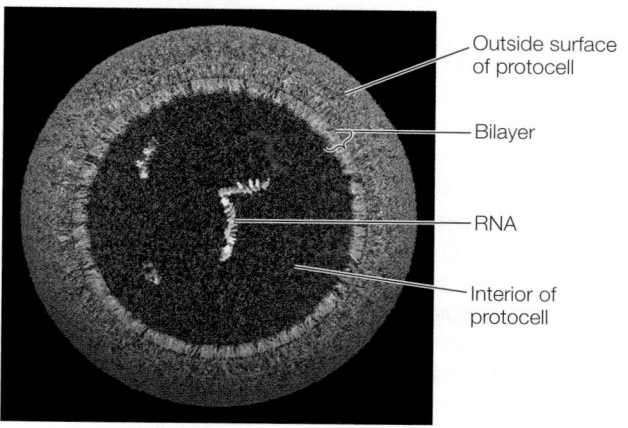

Outside surface of protocell

Bilayer

RNA

Interior of protocell

Figure 4.10 Protocells **(A)** In a series of experiments, Jack Szostak and his colleagues mixed fatty acid molecules in water. The molecules formed spherical structures called protocells, with water surrounded by bilayers of fatty acids. **(B)** A model of a protocell. A portion of the "membrane" has been cut away to reveal the inside of the protocell and the membrane's bilayer structure. Nutrients and nucleotides pass through the "membrane" and enter the protocell, where they copy an already present RNA template. The new copies of RNA remain in the protocell.

water. Recall from Chapter 3 that fatty acids are amphipathic: they have a hydrophilic polar head and a long, nonpolar tail that is hydrophobic (see Figure 3.22). When placed in water, fatty acids will arrange themselves in a round "huddle" much like a football team: the hydrophilic heads point outward to interact with the aqueous environment, and the fatty acid tails point inward, away from the water molecules.

What if some water becomes trapped in the interior of this "huddle"? Now the layer of hydrophobic fatty acid tails is in water, which is an unstable situation. To stabilize this structure, a second layer of fatty acids forms. This **lipid bilayer** has the polar heads of the fatty acids facing both outward and inward, because they are attracted to the polar water molecules present on each side of the double layer. The nonpolar tails form the interior of the bilayer (**Figure 4.10**). These prebiotic, water-filled structures, defined by a lipid bilayer membrane, very much resemble living cells. Scientists refer to these compartments as **protocells**. Examining their properties revealed that:

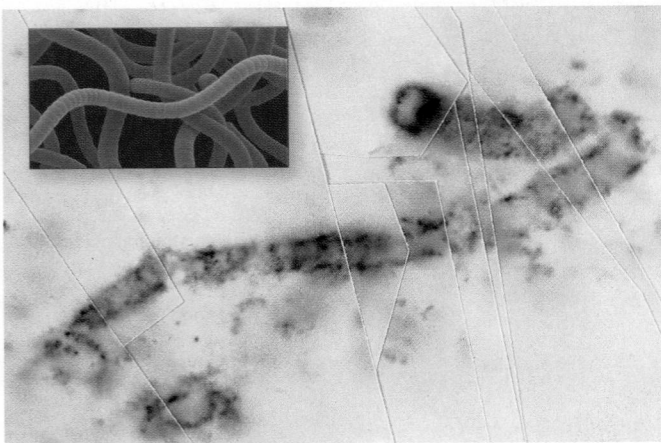

Figure 4.11 The Earliest Cells? This fossil from Western Australia is 3.5 billion years old. Its form is similar to that of modern filamentous cyanobacteria (inset).

- Large molecules such as DNA and RNA could not pass through the bilayer to enter the protocells, but small molecules such as sugars and individual nucleotides could.

- Nucleic acids inside the protocells could replicate using the nucleotides from outside. When the investigators placed a short nucleic acid strand capable of self-replication inside protocells and added nucleotides to the watery environment outside, the nucleotides crossed the barrier, entered the protocells, and became incorporated into new polynucleotide chains. This replication, which can occur without protein catalysis, may have been the first step toward cell reproduction.

Were these protocells truly cells, and was the lipid bilayer produced in these experiments a true cell membrane? Certainly not. The protocells could not fully reproduce, nor could they carry out all the metabolic reactions that take place in modern cells. The simple lipid bilayer had few of the sophisticated functions of modern cell membranes. Nevertheless, the *protocell may be a reasonable facsimile of a cell as it evolved billions of years ago:

- It can act as an organized system of parts, with substances interacting and reacting, in some cases catalytically.

- It includes an interior that is distinct from the exterior environment.

- It is capable of limited replication.

These are all fundamental characteristics of living cells.

---

*connect the concepts Protocells more closely resemble bacteria than plant or animal cells. The evolution of the latter two cell types is believed to have occurred in a series of steps, as explained in Key Concept 26.1.

Some ancient cells left a fossil imprint

In the 1990s scientists made a rare find: a formation of ancient rocks in Australia that had remained relatively unchanged since the rocks first formed 3.5 billion years ago. In one of these rock samples, geologist J. William Schopf of the University of California, Los Angeles, saw chains and clumps of what looked tantalizingly like contemporary cyanobacteria, or "blue-green" bacteria (**Figure 4.11**). Cyanobacteria are believed to have been among the first organisms because they can perform photosynthesis, converting CO_2 and water into carbohydrates. Schopf needed to prove that the chains were once alive, not just the results of simple chemical reactions. He and his colleagues looked for chemical evidence of photosynthesis in the rock samples.

The use of carbon dioxide in photosynthesis is a hallmark of life and leaves a unique chemical signature—a specific ratio of carbon isotopes ($^{13}C{:}^{12}C$) in the resulting carbohydrates. Schopf showed that the Australian material had this isotope signature. Furthermore, microscopic examination of the chains revealed *internal* substructures that are characteristic of living systems and were not likely to be the result of simple chemical reactions. Schopf's evidence suggests that the Australian sample is indeed the remains of a truly ancient living organism.

In 2011 a different team of scientists, working about 32 km from Schopf's discovery, found similar-looking microfossil structures in sandstone rocks that were about 3.4 billion years old. In this case, a chemical analysis of the rocks indicated that these cells used sulfur instead of oxygen in the series of cellular reactions that release chemical energy.

Taking geological, chemical, and biological evidence into account, it is plausible that it took about 500 million to a billion years from the formation of Earth until the appearance of the first cells (**Figure 4.12**). Life has been cellular ever since. In the next chapter we will begin our study of cell structure and function.

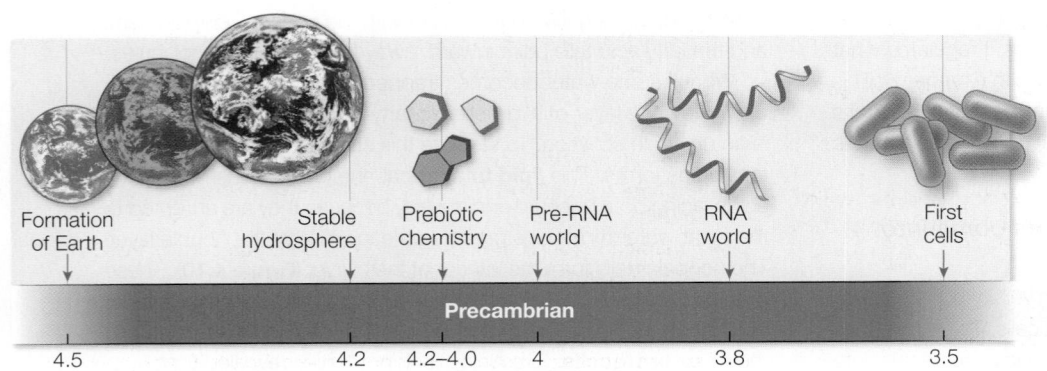

Formation of Earth		Stable hydrosphere	Prebiotic chemistry	Pre-RNA world	RNA world	First cells

Precambrian

| 4.5 | | 4.2 | 4.2–4.0 | 4 | 3.8 | 3.5 |

Billions of years ago

Figure 4.12 The Origin of Life This highly simplified time line gives a sense of the major events that culminated in the origin of life more than 3.5 billion years ago.

4.4 recap

The chemical reactions that preceded living organisms probably occurred in specialized compartments, such as water droplets on the surfaces of minerals. Life as we know it did not begin until the emergence of cells. Protocells made in the laboratory have some of the properties of modern cells. Cell-like structures fossilized in ancient rocks date early cells to about 3.5 billion years ago.

learning outcomes

You should be able to:

- Justify the reasoning behind the hypothesis that a membrane was critical in the evolution of life on Earth.

- Describe the evidence suggesting that microfossils may be the remains of early cells.

___

1. Why is the cell membrane important to the evolution of living organisms?

2. If you wanted to find evidence for the existence of cells more than 3 billion years ago, what would you look for?

investigatinglife

Can we find evidence of life on Mars?

In 2014, NASA declared that finding evidence for current or past life on Mars is a primary goal. At this writing, the *Curiosity* rover is still there, doing experiments and analyses. Later rovers are planned for early in the coming decade. In addition, the European Space Agency, collaborating with Russia, is planning to launch several "ExoMars" rovers in the middle of the decade, which will drill 2 m into the surface of Mars to get cores to analyze them for chemical traces of life below the surface. Most exciting, although uncertain, is a Mars Sample Return Mission, in which soil samples will be brought back to Earth for careful experimentation and analysis in laboratories, just as samples from the moon were brought back more than 40 years ago. If you had a Martian soil sample here on Earth, what analyses and experiments would you perform?

Future directions

A daring space mission is in the final planning stages that may provide key information on the origin of life on Earth. There are more than 600,000 known asteroids, most of them rocks from the early solar system orbiting the sun between Mars and Jupiter. But a few thousand have been kicked out of the typical location and are relatively near Earth, some getting closer than our moon. One, called Bennu, was discovered in 1999 and has been well studied. It is just the right size and composition to contain organic molecules. In 2018, a probe called OSIRIS-REx will arrive at the asteroid, spend about a year near it, and collect a sample of material and then return it to Earth in 2023 for analysis. This sample of the early solar system will provide scientists with a window onto not just what the asteroid is made of, but perhaps the earliest molecules that could have led to life.

Chapter Summary 4

▶ 4.1 Nucleic Acid Structures Reflect Their Function

- The unique functions of the **nucleic acids**—**DNA** and **RNA**—are information storage and transfer. DNA is the hereditary material that passes genetic information from one generation to the next, and RNA uses that information to specify the structures of proteins.

- Nucleic acids are polymers of nucleotides. A **nucleotide** consists of a phosphate group, a pentose sugar (**ribose** in RNA and **deoxyribose** in DNA), and a nitrogen-containing **base**. Review Figure 4.1, Animation 4.1, Activity 4.1

- In DNA, the nucleotide bases are **adenine (A)**, **guanine (G)**, **cytosine (C)**, and **thymine (T)**. **Uracil (U)** replaces thymine in RNA. C, T, and U have single-ring structures and are **pyrimidines**. A and G have double-ring structures and are **purines**.

- The nucleotides in DNA and RNA are joined by **phosphodiester linkages** involving the sugar of one nucleotide and the phosphate of the next, forming a nucleic acid polymer. Review Figure 4.2, Activity 4.2

- **Complementary base pairing** due to hydrogen bonds between A and T, A and U, and G and C occurs in nucleic acids. In RNA, the hydrogen bonds result in a folded molecule. In DNA, the hydrogen bonds connect two strands into a double helix. Review Figure 4.3, Focus: Key Figure 4.4

- The information content of DNA and RNA resides in their base sequences.

- The information in DNA is expressed as RNA in **transcription**. RNA can then specify the amino acid sequence of a protein in **translation**. Review Figure 4.5

▶ 4.2 The Small Molecules of Life Originated on Primitive Earth

- The historical belief that life can originate repeatedly by **spontaneous generation** was disproven experimentally. Review Figure 4.6, Animation 4.2

- A prerequisite for life is the presence of water.

- One hypothesis for the origin of life on Earth is **chemical evolution**: the idea that organic molecules were formed on Earth before life began.

- Chemical experiments modeling the prebiotic conditions on Earth support the idea of chemical evolution. Review Figure 4.7, Animation 4.3

- Some meteorites that have landed on Earth contain organic molecules, suggesting that life might have originated extraterrestrially.

(continued)

Chapter Summary 4 (continued)

- Experiments with lands on Mars yielded some data consistent with the existence of life on that planet. **Review Investigating Life: Can We Find Evidence of Life on Mars?**

▶ 4.3 The Large Molecules of Life Originated from Small Molecules

- Chemical evolution may have led to the polymerization of small molecules into polymers. This may have occurred on the surfaces of clay particles, in hydrothermal vents, or in hot pools at the edges of oceans.
- A catalyst speeds up a chemical reaction. Today most catalysts are proteins, but some RNA molecules can function as both catalysts and information molecules. A catalytic RNA is called a **ribozyme**.
- The existence of ribozymes supports the idea of an "RNA world"— a world before DNA. On early Earth, RNA may have acted as a catalyst for its own replication as well as for the synthesis of proteins. DNA could eventually have evolved from RNA. **Review Figure 4.9**
- In support of the "RNA world" hypothesis, an artificial self-replicating ribozyme was developed in the laboratory.

▶ 4.4 Cells Originated from Their Molecular Building Blocks

- A key to the emergence of living cells was the prebiotic generation of compartments enclosed by **membranes**. Such enclosed compartments permitted the generation and maintenance of internal chemical conditions that were different from those in the exterior environment.
- In the laboratory, fatty acids assemble into **protocells** that have some of the characteristics of cells. **Review Figure 4.10**
- Ancient rocks (3.5 billion years old) have been found with imprints that are probably fossils of early cells.

> Go to **LearningCurve** (in **LaunchPad**) for dynamic quizzing that helps you solidify your understanding of this chapter. **LearningCurve** adapts to your responses, giving you the practice you need to master each key concept.

▶ Apply What You've Learned

Review

4.1 The structures of the polynucleotides DNA and RNA enable their functions in storing and transfering genetic information.

4.1 Base pairing between nucleotides in the polynucleotides DNA and RNA provides the structure needed for transfer of genetic information.

4.1 Base sequences of nucleotides in DNA provide the chemical diversity needed for storage of genetic information.

4.1 Nucleotides other than those found in DNA and RNA have diverse functions within the cell.

Original Paper: Chargaff, E. 1950. Chemical specificity of nucleic acids and mechanisms for their enzymatic degradation. *Experientia.* 6: 201–240.

Nucleic acids show structural similarity across organisms, whether you're looking at DNA from a bacterium, a wheat plant, or a human. If you run chemical analyses on DNA extracted from several organisms, you always find a one-to-one molar ratio of phosphate groups to deoxyribose groups. This one-to-one ratio results from the repeating backbone structure of DNA, which is constructed from many nucleotide monomers polymerized together. All organisms share this structural similarity.

You know, however, that DNA from different organisms is not identical. After all, DNA carries biological information specific to each organism. What can you learn about DNA if you analyze base composition across several species? What can you learn about RNA? The tables below provide data for analyzing these questions.

Questions

1. Calculate the purine-to-pyrimidine ratio for each DNA data set. What pattern do you observe? What does this pattern indicate about DNA structure?

2. Calculate the purine-to-pyrimidine ratio for each RNA data set. What pattern do you observe? What does this pattern indicate about RNA structure?

3. What is the significance of any difference in patterns you found for DNA and RNA as you answered Questions 1 and 2?

4. Calculate the combined AT content and combined GC content in the DNA of each organism listed in the table. How does DNA from different organisms compare with respect to this calculation?

5. Identify two organisms having similar AT content and GC content in their DNA from your answer to Question 4. Explain how these organisms can share this similarity yet have completely different genetic makeups.

Organism and tissue	DNA base composition (%)			
	Adenine	Guanine	Cytosine	Thymine
Herring sperm	27.8	22.2	22.6	27.5
Rat bone marrow	28.6	21.4	21.5	28.4
Human sperm	30.7	19.3	18.8	31.2
Escherichia coli	26.0	24.9	25.2	23.9
Yeast	31.3	18.7	17.1	32.9

Organism and tissue	RNA base composition (%)			
	Adenine	Guanine	Cytosine	Uracil
Rat liver	19.2	28.5	27.5	24.8
Carp muscle	16.4	34.4	31.1	18.1
Yeast	25.1	30.2	20.1	24.6
Rabbit liver	19.7	26.8	25.8	27.6
Cat brain	21.6	31.8	26.0	20.6

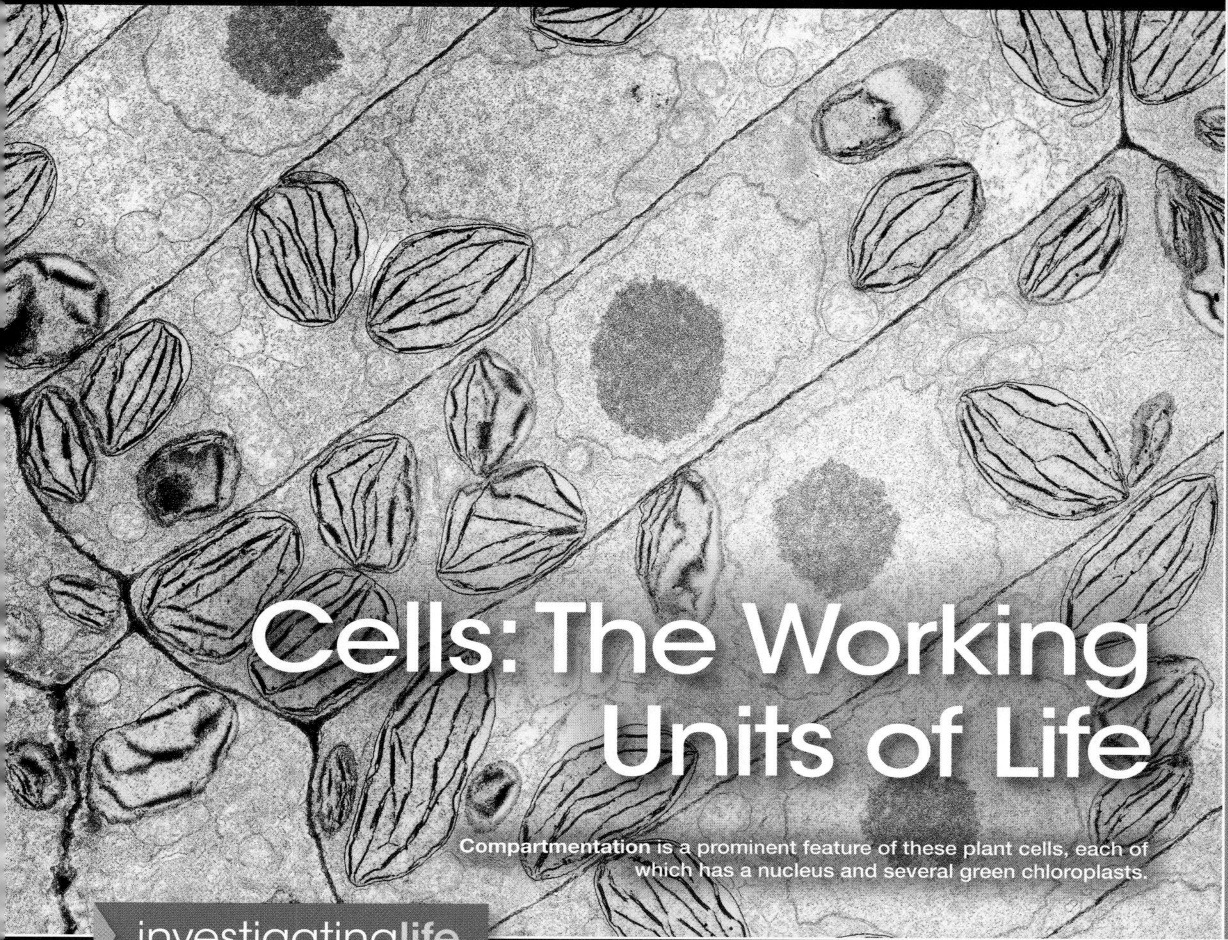

5

Cells: The Working Units of Life

Compartmentation is a prominent feature of these plant cells, each of which has a nucleus and several green chloroplasts.

key concepts

- **5.1** Cells Are the Fundamental Units of Life
- **5.2** Prokaryotic Cells Are the Simplest Cells
- **5.3** Eukaryotic Cells Contain Organelles
- **5.4** Extracellular Structures Have Important Roles
- **5.5** Eukaryotic Cells Evolved in Several Steps

investigatinglife

Natural Sunscreen

Sunlight plays a key role in life, providing the energy that is transformed into stored chemical energy in green plants through the process of photosynthesis. When we eat food, the energy stored in its chemical bonds ultimately came from the sun. But sunlight has a nasty side effect: part of it, the highly energetic ultraviolet light, damages the genetic material, DNA. Even without knowing this chemistry, some people take measures to minimize exposure to excess sunlight, either by staying in the shade or using chemical sunscreens. But there is also a biological mechanism that has evolved to reduce the sun's damage to organisms. In most animals, dark brown or black chemical pigments called melanins are made inside cells, the basic units of life. Melanins absorb ultraviolet light, thereby protecting DNA.

In humans, melanin is made in certain skin cells called melanocytes. Humans first evolved in Africa, and the intense sunlight near the equator acted as an evolutionary selective agent for the development of large numbers of melanocytes. As some of these dark-skinned people migrated to more northern regions where the sun is not as intense, the selection pressure for melanocytes was reduced, and over millennia, genetic changes leading to fewer melanocytes persisted in offspring, resulting in lighter skin. Exposure to intense sunlight in these people results in the production of more melanin, the familiar tanning response.

Melanocytes are specialized cells. You find them in the skin but not typically in internal organs. We will return to the process of how different cells have different functions in later chapters. But even inside the melanocyte, there is specialization. Melanin is made inside a special cell compartment called the melanosome. These compartments can be expelled from melanocytes and transferred to other skin cells, so they too become loaded with pigment. This occurs spontaneously in dark-skinned people, while in paler people it is stimulated by exposure to intense sunlight.

The observation that the chemistry of life takes place within cells, and in some cases within specialized compartments in them, is a key concept of biological science.

QA Are pigments compartmentalized in all types of cells?

key concept

5.1

Cells Are the Fundamental Units of Life

Chapter 1 introduced some of the characteristics of life: chemical complexity, growth and reproduction, the ability to refashion substances from the environment, and the ability to move specific substances into and out of the organism. Cells have all of these characteristics. Just as atoms are the building blocks of chemistry, cells are the building blocks of life.

---

focus your learning

- Cell theory states that the cell is the smallest unit of life, that cells make up all living organisms, and that cells come from existing cells.
- Microscopy allows visual examination of cells.
- Membranes provide a structural role in a cell and also allow the cell to maintain homeostasis and communicate with other cells.

What is the cell theory?

The **cell theory** is an important unifying principle of biology. There are three components of the cell theory:

1. Cells are the fundamental units of life.
2. All living organisms are composed of cells.
3. All cells come from preexisting cells.

To the original cell theory, first stated in 1838, should be added:

4. Modern cells evolved from a common ancestor.

Cells contain water and other small and large molecules, which we examined in Chapters 2–4. Each cell contains at least 10,000 different types of molecules, most of them present in many copies. Cells use these molecules to transform matter and energy, to respond to their environments, and to reproduce. The cell theory has three important implications:

1. Studying cell biology is in some sense the same as studying life. The principles that underlie the functions of the single cell of a bacterium are similar to those governing the approximately 60 trillion human cells of your body.

2. Life is continuous. All the cells in your body came from a single cell, a fertilized egg (zygote). That zygote came from the fusion of two cells, a sperm and an egg, from your parents. The cells of your parents' bodies were all derived from their parents, and so on back through generations and evolution to the initial cell.

3. The origin of life on Earth was marked by the origin of the first cells (see Chapter 4).

Even the largest creatures on Earth are composed of cells, but the cells themselves are usually too small for the naked eye to see. Why are cells so small?

Cell size is limited by the surface area-to-volume ratio

Most cells are tiny. In 1665 Robert Hooke estimated that in 1 square inch of cork, which he examined under his magnifying lens, there were 1,259,712,000 cells! The diameters of cells range from about 1 to 100 micrometers (μm). There are some exceptions: the eggs of birds are large single cells, and individual cells of several types of algae and bacteria are large enough to be viewed with the unaided eye (**Figure 5.1**).

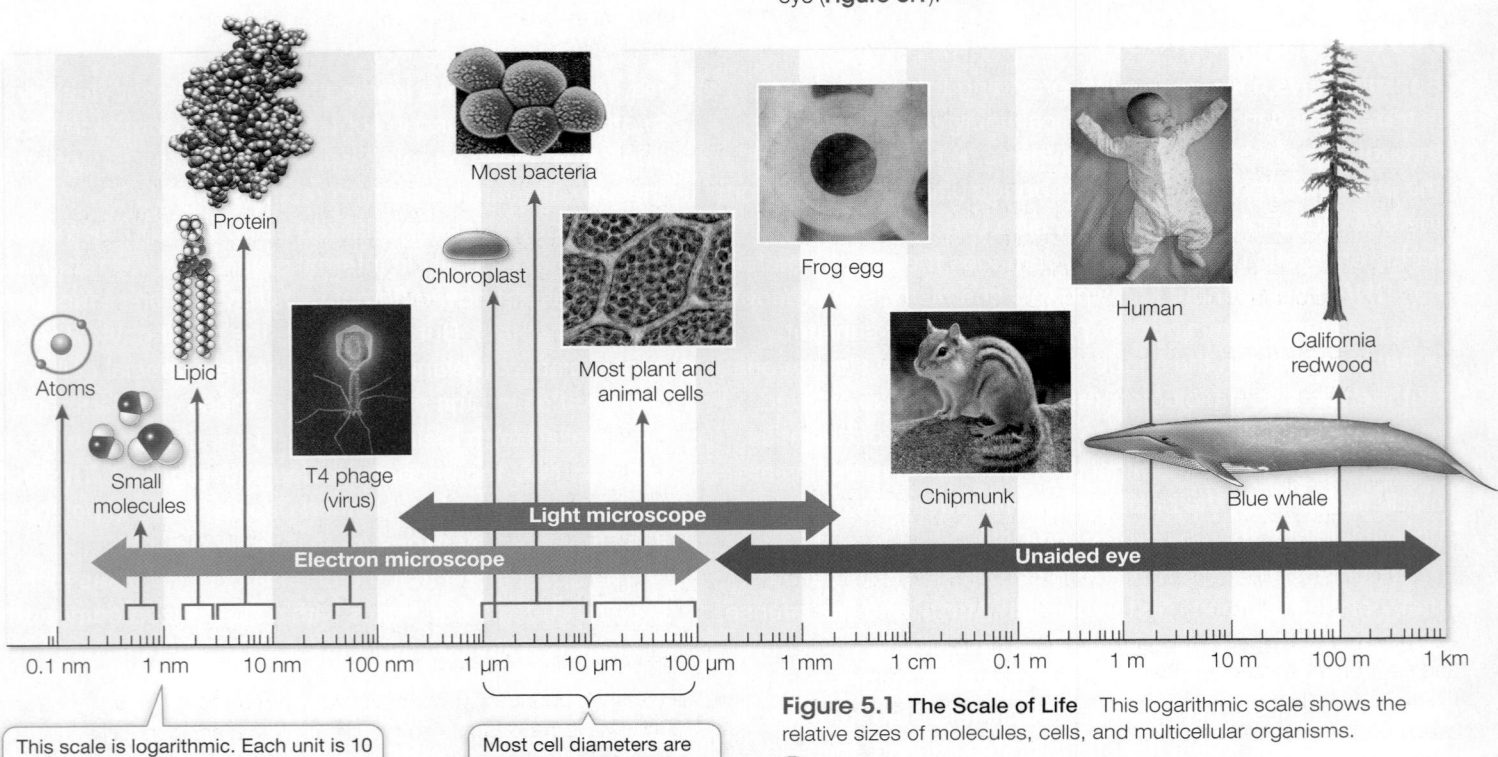

This scale is logarithmic. Each unit is 10 times bigger than the previous unit.

Most cell diameters are in the range of 1–100 μm.

Figure 5.1 **The Scale of Life**   This logarithmic scale shows the relative sizes of molecules, cells, and multicellular organisms.

Activity 5.1 **The Scale of Life**
www.Life11e.com/ac5.1

Diameter	2 μm	20 μm	200 μm
Surface area $4\pi r^2$	12.6 μm²	1,260 μm²	126,000 μm²
Volume $\frac{4}{3}\pi r^3$	4.2 μm³	4,200 μm³	4,200,000 μm³
Surface area-to-volume ratio	3.0:1	0.3:1	0.003:1

Figure 5.2 Why Cells Are Small   As an object grows larger, its volume increases more rapidly than its surface area. Cells must maintain a large surface area-to-volume ratio in order to function. This explains why large organisms are composed of many small cells rather than a few huge ones.

Small cell size is a practical necessity arising from the change in the **surface area-to-volume ratio** of any object as it increases in size. As an object increases in volume, its surface area also increases, but not at the same rate (**Figure 5.2**). This phenomenon has great biological significance. To appreciate this point, let's assume that the amount of chemical activity carried out by a cell is proportional to its volume. The surface area of the cell determines the amount of substances that can enter it from the outside environment, and the amount of waste products that can exit to the environment.

As a living cell grows larger, its chemical activities, and thus its need for resources and its rate of waste production, increase faster than its surface area. (The surface area, being two-dimensional, increases in proportion to the square of the radius, whereas the volume, being three-dimensional, increases much more—in proportion to the cube of the radius.) In addition, substances must move from one site to another within the cell; the smaller the cell, the more easily this is accomplished. This explains why large organisms must consist of many small cells: cells must be small in volume in order to maintain a large enough surface area-to-volume ratio and an ideal internal volume. The large surface area represented by the many small cells of a multicellular organism enables it to carry out the many different functions required for survival.

Microscopes reveal the features of cells

Most cells are too small to be seen with the unaided eye. Microscopes allow cells and details within them to be seen. The property that allows detail to be seen is called resolution. Formally defined, resolution is the minimum distance two objects can be apart and still be seen as two objects. Resolution for the human eye is about 0.2 mm (200 μm). Most cells are much smaller than 200 μm and thus are invisible to the human eye. Microscopes magnify and increase resolution so that cells and their internal structures can be seen clearly (**Figure 5.3**).

There are two basic types of microscopes—light microscopes and electron microscopes—that use different forms of radiation (see Figure 5.3). While the resolution is better in electron microscopy, only dead cells are visualized because they must be prepared in a vacuum. Light microscopes, by contrast, can be used to visualize living cells (for example, by phase-contrast microscopy; see Figure 5.3).

Before we look at cell structure, it is useful to consider the many uses of microscopy. Here is an example: an entire branch of medicine, pathology, makes use of many different methods of microscopy to aid in the analysis of cells and the diagnosis of diseases. For instance, a surgeon might remove from a body some tissue suspected of being cancerous. The pathologist might:

- examine the tissue quickly by phase-contrast microscopy to determine the size, shape, and spread of the cells;
- stain the tissue with a general dye and examine it by bright-field microscopy to bring out features such as the shapes of the nuclei, or cell division characteristics;
- examine the tissue under the transmission electron microscope to observe internal structures such as the mitochondria or the chromatin (these are described in Key Concept 5.3);
- stain the tissue with a specific dye and examine it by microscopy for the presence of proteins that are diagnostic of particular cancers. The results can influence the choice of therapy.

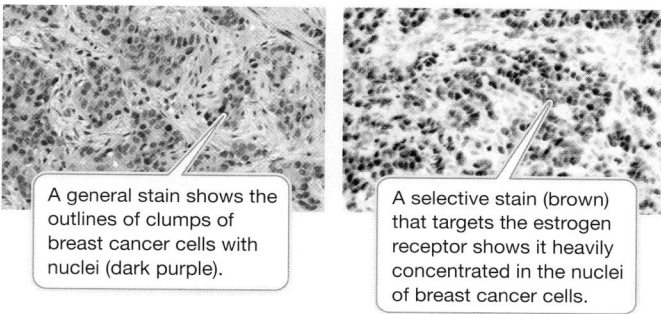

A general stain shows the outlines of clumps of breast cancer cells with nuclei (dark purple).

A selective stain (brown) that targets the estrogen receptor shows it heavily concentrated in the nuclei of breast cancer cells.

The cell membrane forms an outer boundary of every cell

While light microscopy shows many types of cells and even some structure inside them, the **cell membrane** is best observed with an electron microscope. This very thin structure forms the outer boundary of the cytoplasm of every cell, and it has similar thickness and molecular structure in all cells. You will learn about membranes in more detail in Chapter 6. For now, keep in mind that (with the exception of some species of archaea discussed in Chapter 26) the cell membrane consists of a phospholipid bilayer (see Key Concept 3.4), and that a variety of proteins are embedded within the bilayer.

The membrane has several important roles:

- The cell membrane acts as a selectively permeable barrier, preventing some substances from crossing it while permitting other substances to enter and leave the cell.
- By regulating transport across it, the cell membrane allows the cell to maintain a more or less constant internal environment. The maintenance of a constant internal environment (known as homeostasis) is a key characteristic of life and will be discussed in detail in Chapter 40.
- As the cell's boundary with the outside environment, the cell membrane is important in communicating with adjacent cells and receiving signals from the environment. You will learn about this in Chapter 7.

research tools

Figure 5.3 Looking at Cells The six images on this page show some techniques used in light microscopy. The three images on the following page were created using electron microscopes. All of these images are of a particular type of cultured cell known as HeLa cells.

Note that the images in most cases are flat, two-dimensional views. As you look at images of cells, keep in mind that they are three-dimensional structures.

 Activity 5.2 Know Your Techniques
www.Life11e.com/ac5.2

Light microscope

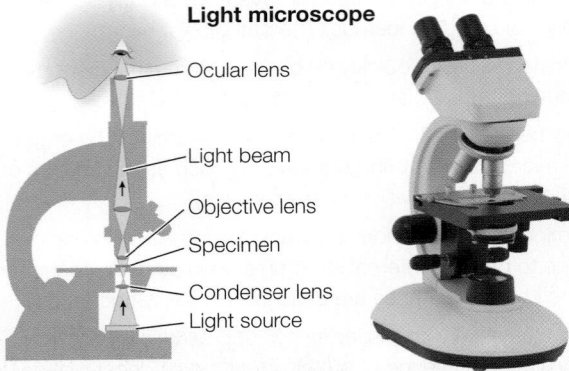

- Ocular lens
- Light beam
- Objective lens
- Specimen
- Condenser lens
- Light source

In a *light microscope*, glass lenses and visible light are used to form an image. The resolution is about 0.2 μm, which is 1,000 times greater than that of the human eye. Light microscopy allows visualization of cell sizes and shapes and some internal cell structures. Internal structures are hard to see under visible light, so cells are often chemically treated and stained with various dyes to make certain structures stand out by increasing contrast.

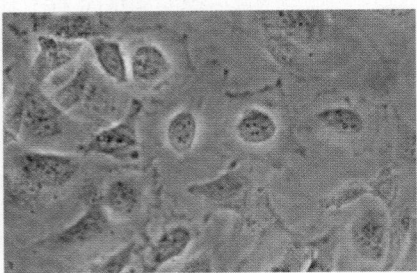

30 μm

In **bright-field microscopy**, light passes directly through these human cells. Unless natural pigments are present, there is little contrast and details are not distinguished.

30 μm

In **phase-contrast microscopy**, contrast in the image is increased by emphasizing differences in refractive index (the capacity to bend light), thereby enhancing light and dark regions in the cell.

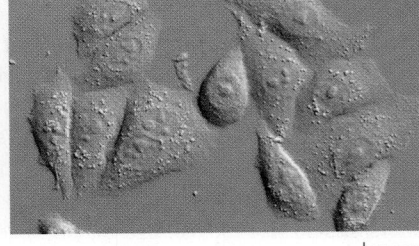

30 μm

Differential interference-contrast microscopy uses two beams of polarized light. The combined images look as if the cell is casting a shadow on one side.

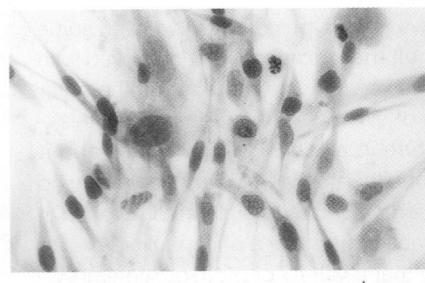

30 μm

In **stained bright-field microscopy**, a stain enhances contrast and reveals details not otherwise visible. Stains differ greatly in their chemistry and their capacity to bind to cell materials, so many choices are available.

20 μm

In **fluorescence microscopy**, a natural substance in the cell or a fluorescent dye that binds to a specific cell material is stimulated by a beam of light, and the longer-wavelength fluorescent light is observed coming directly from the dye.

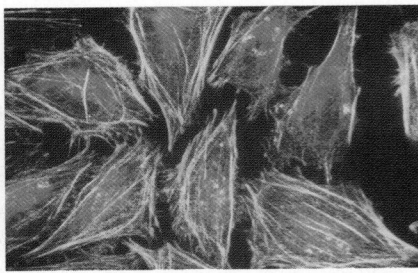

20 μm

Confocal microscopy uses fluorescent materials but adds a system of focusing both the stimulating and emitted light so that a single plane through the cell is seen. The result is a sharper two-dimensional image than with standard fluorescence microscopy.

- The cell membrane often has proteins protruding from it that are responsible for binding and adhering to adjacent cells. Thus the cell membrane plays an important structural role and contributes to cell shape.

Cells may be classified as either prokaryotic or eukaryotic

As you saw in Key Concept 1.1, biologists classify all living things into three domains: Archaea, Bacteria, and Eukarya. The organisms in Archaea and Bacteria are collectively called **prokaryotes**, and

research tools (continued)

Transmission electron microscope

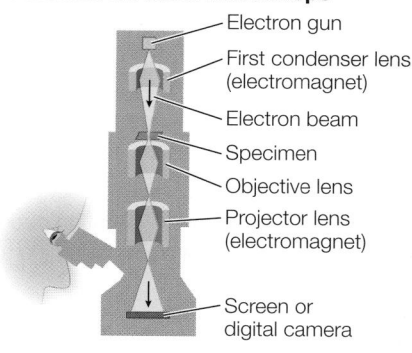

- Electron gun
- First condenser lens (electromagnet)
- Electron beam
- Specimen
- Objective lens
- Projector lens (electromagnet)
- Screen or digital camera

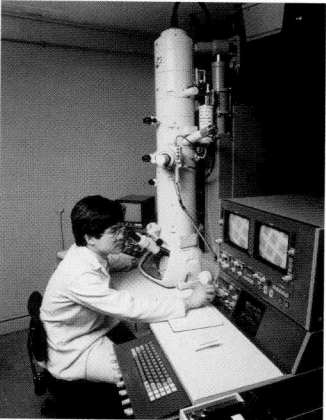

In an *electron microscope*, electromagnets are used to focus an electron beam, much as a light microscope uses glass lenses to focus a beam of light. Since we cannot see electrons, the electron microscope directs them through a vacuum at a fluorescent screen or digital camera to create a visible image. The resolution of electron microscopes is about 2 nm, which is about 100,000 times greater than that of the human eye. This resolution permits the details of many subcellular structures to be distinguished.

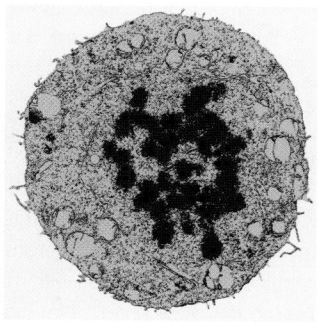

10 μm

In **transmission electron microscopy** (TEM), a beam of electrons is focused on the object by magnets. Objects appear darker if they absorb the electrons. If the electrons pass through they are detected on a fluorescent screen.

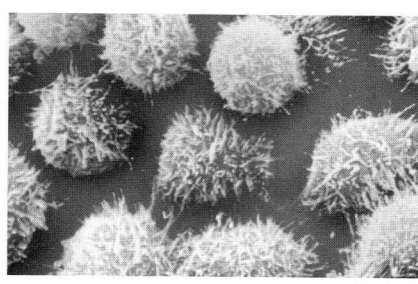

20 μm

Scanning electron microscopy (SEM) directs electrons to the surface of the sample, where they cause other electrons to be emitted. These electrons are viewed on a screen. The three-dimensional surface of the sample can be visualized.

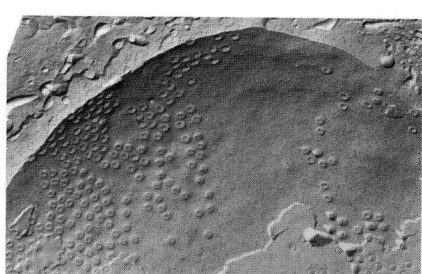

0.1 μm

In **freeze-fracture microscopy**, cells are frozen and then a knife is used to crack them open. The crack often passes through the interior of cell and internal membranes. The "bumps" that appear are usually large proteins or aggregates embedded in the interior of the membrane.

---

they have in common a prokaryotic cell organization. A prokaryotic cell does not typically have membrane-enclosed internal compartments; in particular, it does not have a nucleus. The first cells were probably similar in organization to those of modern prokaryotes.

Eukaryotic cell organization is found in members of the domain Eukarya (**eukaryotes**), which includes the protists, plants, fungi, and animals. In contrast to prokaryotic cells, eukaryotic cells contain membrane-enclosed compartments called **organelles**. The most noticeable organelle is the cell **nucleus**, where most of the cell's DNA is located and where gene expression begins:

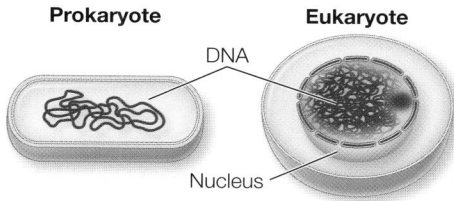

Prokaryote   **Eukaryote**

DNA

Nucleus

Just as a cell is an enclosed compartment, separating its contents from the surrounding environment, so each organelle provides a compartment that separates molecules and biochemical reactions

from the rest of the cell. This "division of labor" provides possibilities for regulation and efficiency that were important in the evolution of complex organisms and helps explain the complexity of eukaryotic cells relative to prokaryotic cells.

5.1 recap

The cell theory is a unifying principle of biology. Cell size is limited in order to maintain a high surface area-to-volume ratio. Both prokaryotic and eukaryotic cells are enclosed within a cell membrane, but prokaryotic cells lack the membrane-enclosed organelles that are found in eukaryotic cells.

learning outcomes

You should be able to:

- List the three tenets of the original cell theory.
- Explain the importance of membranes in cell function.

1. How does cell biology embody all the principles of life?
2. What is the importance of the cell membrane and the membranes that surround organelles?

As we mentioned in this section, there are two structural themes in cell architecture: prokaryotic and eukaryotic. We will now turn to the organization of prokaryotic cells.

key concept 5.2 Prokaryotic Cells Are the Simplest Cells

With diameters or lengths in the range of 1–10 μm, prokaryotic cells are generally smaller than eukaryotic cells, whose diameters are usually in the range of 10–100 μm. Each individual prokaryote is a single cell, but many types of prokaryotes form chains or small clusters of cells, and some occur in large clusters containing hundreds of cells. In this section we will first consider the features shared by cells in the domains Bacteria and Archaea. Then we will examine structural features that are found in some, but not all, prokaryotes.

focus your learning

- All prokaryotic cells contain a cell membrane, nucleoid with DNA, cytoplasm, and ribosomes.

What are the features of prokaryotic cells?

All *prokaryotic cells have the same basic structure (**Figure 5.4**):

- The cell membrane encloses the cell, regulating the traffic of materials into and out of the cell, and separating its interior from the external environment.

- The **nucleoid** is a region in the cell where the DNA is located. As we described in Key Concept 4.1, DNA is the hereditary material that controls cell growth, maintenance, and reproduction.

- The rest of the material enclosed in the cell membrane is called the **cytoplasm** (cytosol).

- **Ribosomes** are complexes of RNA and proteins in the cytoplasm that are about 25 nanometers (nm) in diameter. They can be visualized only with the electron microscope. They are the sites of protein synthesis, where information coded for in nucleic acids directs the sequential linking of amino acids to form proteins.

---

*connect the concepts   In terms of sheer numbers, prokaryotes are the most successful organisms on Earth. As we examine prokaryotic cells in this section, bear in mind that there are vast numbers of prokaryotic species, and that bacteria and archaea are distinguished in numerous ways. These differences, and the vast diversity of organisms in these two domains, will be the subject of Chapter 26.

The cytoplasm is not static. Rather, the substances in this environment are in constant motion. For example, a typical protein moves around the entire cell within a minute, and it collides with many other molecules along the way. This motion helps ensure that biochemical reactions proceed at rates sufficient to meet the needs of the cell. Although they are structurally less complex than eukaryotic cells, prokaryotic cells are functionally complex, carrying out thousands of biochemical reactions.

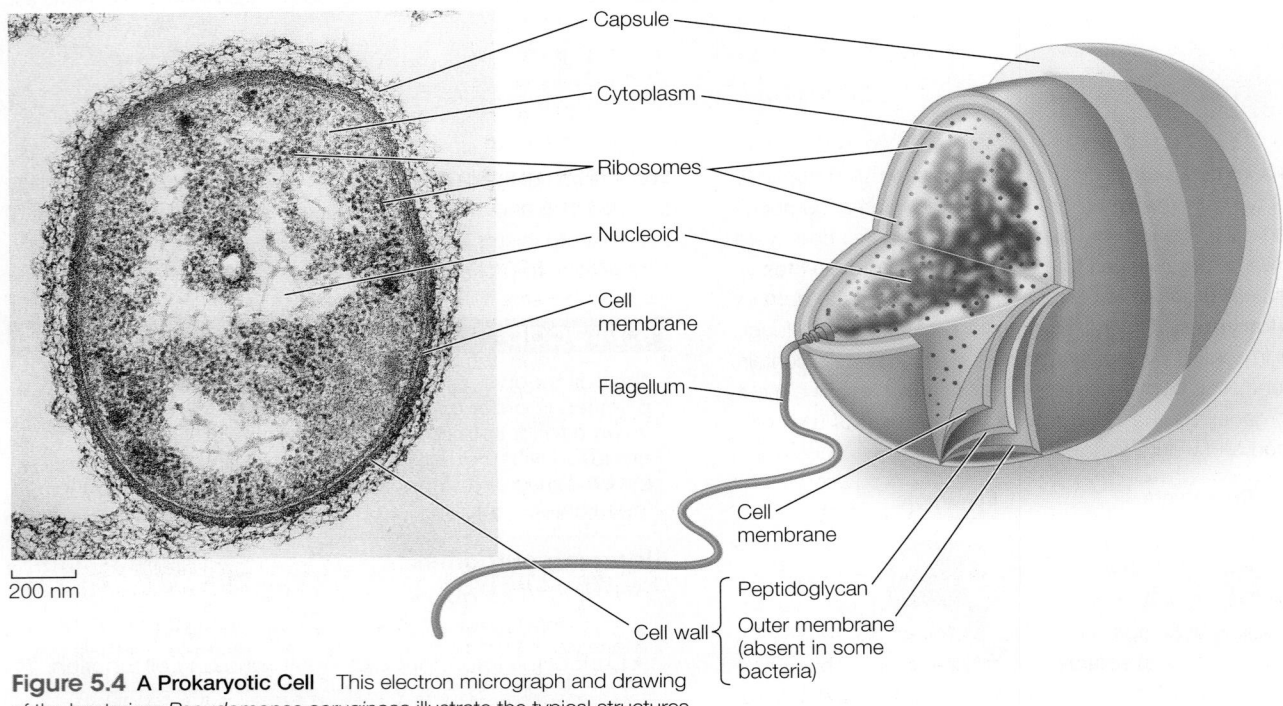

Capsule
Cytoplasm
Ribosomes
Nucleoid
Cell membrane
Flagellum
Cell membrane
Peptidoglycan
Cell wall
Outer membrane (absent in some bacteria)

200 nm

Figure 5.4 A Prokaryotic Cell   This electron micrograph and drawing of the bacterium *Pseudomonas aeruginosa* illustrate the typical structures shared by all prokaryotic cells. In addition, this bacterium has a protective outer membrane that is not present in all prokaryotes. The flagellum and capsule are also structures found in some, but not all, prokaryotic cells.

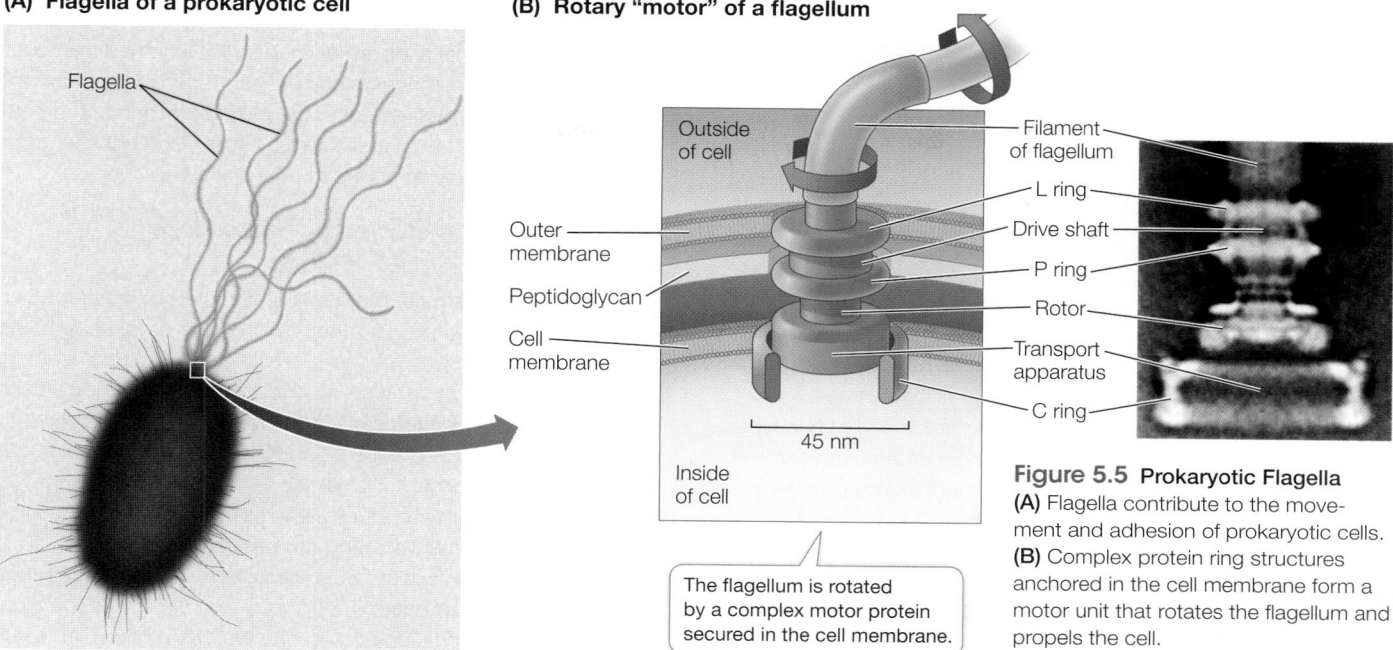

(A) Flagella of a prokaryotic cell

Flagella

(B) Rotary "motor" of a flagellum

Outside of cell

Outer membrane

Peptidoglycan

Cell membrane

Inside of cell

45 nm

Filament of flagellum

L ring

Drive shaft

P ring

Rotor

Transport apparatus

C ring

The flagellum is rotated by a complex motor protein secured in the cell membrane.

Figure 5.5 Prokaryotic Flagella **(A)** Flagella contribute to the movement and adhesion of prokaryotic cells. **(B)** Complex protein ring structures anchored in the cell membrane form a motor unit that rotates the flagellum and propels the cell.

Specialized features are found in some prokaryotes

As they evolved, some prokaryotes developed specialized structures that gave them a selective advantage: cells with these structures were better able to survive and reproduce in particular environments than cells lacking them.

CELL WALLS   Most prokaryotes have a **cell wall** located outside the cell membrane. The rigid wall supports the cell and determines its shape. The cell walls of most bacteria, but not archaea, contain peptidoglycan, a polymer of amino sugars that is linked at regular intervals to short peptides. Cross-linking among these peptides results in a single giant molecule around the entire cell. In some bacteria, another layer, the **outer membrane** (a polysaccharide-rich phospholipid membrane), encloses the peptidoglycan layer (see Figure 5.4). Unlike the cell membrane, this outer membrane is not a major barrier to the movement of molecules across it.

Enclosing the cell wall in some bacteria is a slimy layer composed mostly of polysaccharides and referred to as a **capsule**. In some cases these capsules protect the bacteria from attack by white blood cells in the animals they infect. Capsules also help keep the cells from drying out, and sometimes they help bacteria attach to other cells.

INTERNAL MEMBRANES   Some groups of bacteria—including the cyanobacteria—carry out photosynthesis: they use energy from the sun to convert carbon dioxide and water into carbohydrates. These bacteria have an **internal membrane** system that contains molecules needed for photosynthesis. The development of photosynthesis, which requires membranes, was an important event in the early evolution of life on Earth. Other prokaryotes have internal membrane folds that are attached to the cell membrane. These folds may function in cell division or in various energy-releasing reactions.

FLAGELLA AND PILI   Some prokaryotes swim by using appendages called **flagella**, which sometimes look like tiny corkscrews (**Figure**

5.5A). In bacteria, the filament of the flagellum is made of a protein called flagellin. (As you will see in Key Concept 5.3, the flagella of eukaryotes are quite different in structure, but similar in function.) A complex motor protein spins the bacterial flagellum on its axis like a propeller, driving the cell along. The motor protein is anchored to the cell membrane and, in some bacteria, to the outer membrane of the cell wall (**Figure 5.5B**). We know that the flagella cause the motion of cells because if they are removed, the cells do not move.

Pili are structures made of protein that project from the surfaces of some types of bacterial cells. These hairlike structures are shorter than flagella and are used for adherence. Conjugative pili (sex pili) help bacteria join to one another to exchange genetic material. Fimbriae are composed of the same proteins as pili but are shorter, and help cells adhere to surfaces such as animal cells, for food and protection.

CYTOSKELETON   The **cytoskeleton** is the collective name for protein filaments that play roles in cell division, cell movement, and in maintaining the shapes of cells. One such protein forms a ring structure that constricts during cell division, whereas another forms helical structures that extend down the lengths of rod-shaped cells, helping maintain their shapes. In the past it was thought that only eukaryotic cells had cytoskeletons (see Key Concept 5.3), but more recently, biologists have recognized that cytoskeletal components are also widely distributed among prokaryotes.

5.2 recap

Prokaryotic cells share basic features, including the cell membrane and a region called the nucleoid. The rest of the material enclosed by the cell membrane, including the ribosomes, is called the cytoplasm. Other features, such as cell walls, internal membranes, and flagella, are present in some but not all prokaryotes.

(continued)

As we have already noted, the prokaryotic cell is one of two types of cell recognized in cell biology. The other is the eukaryotic cell. Eukaryotic cells are more structurally and functionally complex than prokaryotic cells.

key concept 5.3 Eukaryotic Cells Contain Organelles

You are probably familiar with several types of cells in your body, such as blood cells, skin cells, and brain cells. While these cells have distinctive structures and functions, they and other eukaryotic cells share many features, and it is these common features that we will discuss in this section.

focus your learning

- Eukaryotic cells differ from prokaryotic cells in that they contain membrane-bound structures known as organelles.
- Each organelle carries out a different specific function that, when combined with others, allows the cell to function as a whole.

Activity 5.3 **Eukaryotic Cell Tour**
www.Life11e.com/ac5.3

Eukaryotic cells generally have lengths or diameters about ten times greater than those of prokaryotes. Like prokaryotic cells, eukaryotic cells have a cell membrane, cytoplasm, and ribosomes. Other structures in eukaryotic cells analogous to those in prokaryotes include a cytoskeleton composed of protein fibers and, outside the cell membrane, an extracellular matrix. But as you learned earlier in this chapter, eukaryotic cells also have compartments within the cytoplasm that are separated by membranes.

Compartmentalization is important to eukaryotic cell function

The membranous compartments of eukaryotic cells are called organelles. All eukaryotic cells have many organelles and structures in common; the most obvious is the cell nucleus. But they also have some differences. For example, many plant cells have chloroplasts that perform photosynthesis. Each type of organelle has a specific role: some organelles have been characterized as factories that make specific products, whereas others are like power plants that take in energy in one form and convert it to a more useful

form. These functional roles are defined by the chemical reactions that occur within the organelles. Ion concentrations differ from one organelle to another. For instance, the pH (a reflection of the concentration on H^+) is 7.4 inside the nucleus, but 4.5 (a concentration 1,000 times higher) inside another organelle, the lysosome.

Organelles can be studied by microscopy or isolated for chemical analysis

Cell organelles and structures were first detected by light and then by electron microscopy. The functions of the organelles could sometimes be inferred by observations and experiments, leading, for example, to the hypothesis (later confirmed) that the nucleus contained the genetic material. Later, the use of stains targeted to specific macromolecules allowed cell biologists to determine the chemical compositions of organelles.

Another way to analyze cells is to take them apart in a process called differential centrifugation, or cell fractionation. This process permits cell organelles and other cytoplasmic structures to be separated from each other and examined using chemical methods. Cell fractionation begins with the destruction of the cell membrane, which allows the cytoplasmic components to flow out into a test tube. The various organelles can then be separated from one another on the basis of size or density (**Figure 5.6**). Biochemical analyses can then be done on the isolated organelles. Microscopy and cell fractionation have complemented each other, giving us a more complete picture of the composition and function of each organelle and structure.

Microscopy of eukaryotic cells has revealed that many of the organelles are similar in appearance in each cell type (**Figure 5.7**). By comparing Figures 5.7 and 5.4 you can see some of the prominent differences between eukaryotic cells and prokaryotic cells.

Ribosomes are factories for protein synthesis

The *ribosomes of prokaryotes and eukaryotes are similar in that both types consist of two different-sized subunits. Eukaryotic ribosomes are somewhat larger than those of prokaryotes, but the structure of prokaryotic ribosomes is better understood. Chemically, ribosomes consist of a special type of RNA called ribosomal RNA (rRNA). Ribosomes also contain more than 50 different protein molecules, which fit together noncovalently, employing forces for protein interaction such as hydrophobic interactions (see Figure 3.12). Additionally, some of the proteins bind noncovalently to the rRNA.

*connect the concepts   As discussed in Key Concept 14.5, the ribosome is the molecular workbench where mRNA is translated into protein.

In prokaryotic cells, ribosomes generally float freely in the cytoplasm. In eukaryotic cells they are found in multiple places: floating freely in the cytoplasm or attached to the surface of the endoplasmic reticulum (a membrane-bound organelle; see below), inside the mitochondria, and inside the chloroplasts in plant cells. In each of these locations, the ribosomes are molecular factories where proteins are synthesized. Although they seem small in comparison with the cells that contain them, by molecular standards ribosomes are huge complexes (about 25 nm in diameter), made up of several dozen different molecules.

Because they lack membranes and are not a compartment, ribosomes are not classified as organelles. Rather, like the cyto-skeleton (see below) they are called cell structures.

The nucleus contains most of the genetic information

As we discussed in Chapter 4, hereditary information is stored in the sequence of nucleotides in DNA molecules. Most of the DNA in eukaryotic cells resides in the nucleus (see Figure 5.7). Information encoded in the DNA is translated into proteins at the ribosomes (a process we will describe in detail in Chapter 14).

Most cells have a single nucleus, which is usually the most prominent organelle. The nucleus of a typical animal cell is approximately 5 μm in diameter—substantially larger than many entire prokaryotic cells (**Figure 5.8A**). The nucleus has several functions in the cell:

- It is the location of most of the cell's DNA and the site of DNA replication.
- It is the site where gene transcription is turned on or off.
- A region within the nucleus, the **nucleolus**, is where ribosomes begin to be assembled from RNA and proteins.

The contents of the nucleus, aside from the nucleolus, are referred to as the nucleoplasm. Similar to the cytoplasm, the nucleoplasm consists of the liquid content of the nucleus and the insoluble molecules suspended within it.

The nucleus is surrounded by an integrated structure composed of two membranes, called the **nuclear envelope**. This structure separates the genetic material from the cytoplasm. Functionally, it separates DNA transcription (which occurs in the nucleus) from translation (which occurs in the cytoplasm) (see Key Concept 4.1). The two membranes of the nuclear envelope are perforated by thousands of nuclear pores (**Figure 5.8B**), each measuring approximately 9 nm in diameter, which connect the nucleoplasm with the cytoplasm.

The pores act as "traffic cops," allowing some molecules to pass into and out of the nucleus and blocking others. This allows the nucleus to regulate its information-processing functions. Small substances, including ions and other molecules with molecular weights of less than 10,000 daltons, move through the pores. But larger molecules, such as most proteins that are made in the cytoplasm, can't get through. In some cases, however, there is a special transport mechanism to get such proteins into the nucleus; you'll learn about this in Chapter 14.

Inside the nucleus, DNA is combined with proteins to form a fibrous complex called **chromatin**. Chromatin occurs in the form of exceedingly long, thin threads called **chromosomes**. Different eukaryotic organisms have different numbers of chromosomes (ranging from two in one kind of Australian ant to hundreds in some plants). Prior to cell division, the chromatin becomes tightly

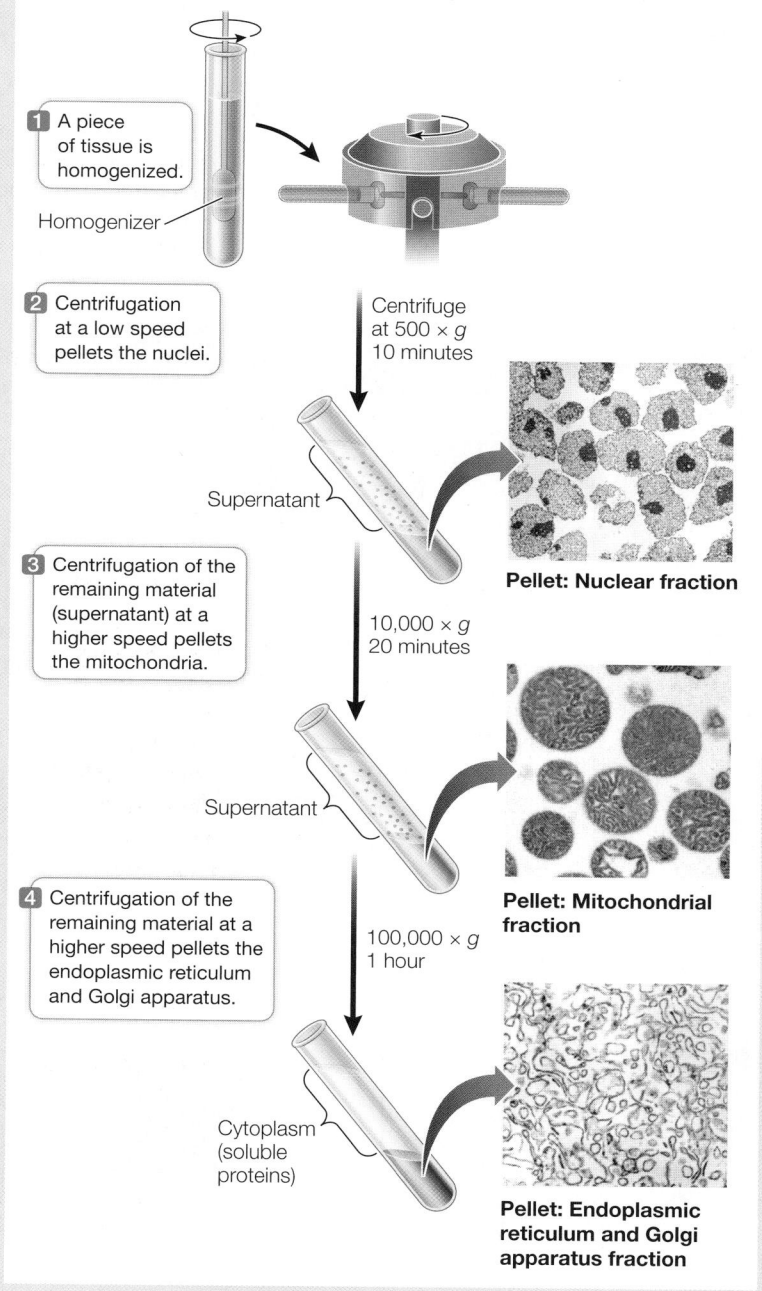

research tools

Figure 5.6 Cell Fractionation Organelles can be separated from one another after cells are broken open and their contents suspended in an aqueous medium. The medium is placed in a tube and spun in a centrifuge, which rotates about an axis at high speed. Centrifugal forces (measured in multiples of gravity, $\times g$) cause particles to sediment (form a pellet) at the bottom of the tube, which may be collected for biochemical study. Heavier particles sediment at lower speeds (lower centrifugal forces) than lighter particles. By adjusting the speed of centrifugation, researchers can separate and partially purify cellular organelles and large particles such as ribosomes.

1 A piece of tissue is homogenized.

Homogenizer

2 Centrifugation at a low speed pellets the nuclei.

Centrifuge at 500 $\times g$ 10 minutes

Supernatant

Pellet: Nuclear fraction

3 Centrifugation of the remaining material (supernatant) at a higher speed pellets the mitochondria.

10,000 $\times g$ 20 minutes

Supernatant

Pellet: Mitochondrial fraction

4 Centrifugation of the remaining material at a higher speed pellets the endoplasmic reticulum and Golgi apparatus.

100,000 $\times g$ 1 hour

Cytoplasm (soluble proteins)

Pellet: Endoplasmic reticulum and Golgi apparatus fraction

compacted and condensed so that the individual chromosomes are visible under a light microscope. This facilitates distribution of the DNA during cell division (**Figure 5.8C**).

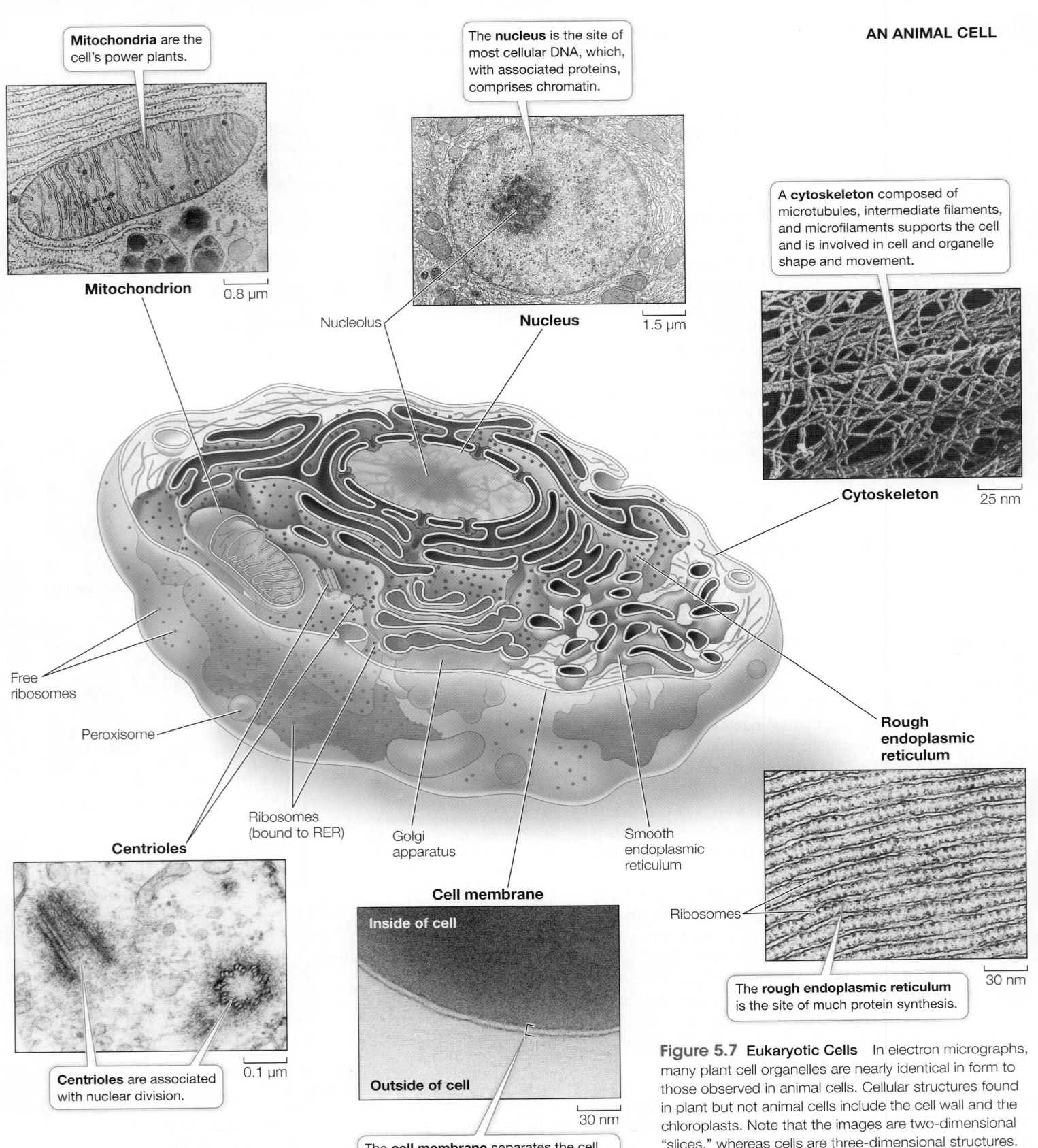

AN ANIMAL CELL

Mitochondria are the cell's power plants.

Mitochondrion   0.8 µm

The **nucleus** is the site of most cellular DNA, which, with associated proteins, comprises chromatin.

Nucleolus

Nucleus   1.5 µm

A **cytoskeleton** composed of microtubules, intermediate filaments, and microfilaments supports the cell and is involved in cell and organelle shape and movement.

Cytoskeleton   25 nm

Free ribosomes

Peroxisome

Centrioles

Ribosomes (bound to RER)

Golgi apparatus

Smooth endoplasmic reticulum

Rough endoplasmic reticulum

Centrioles are associated with nuclear division.   0.1 µm

Cell membrane

Inside of cell

Outside of cell   30 nm

The **cell membrane** separates the cell from its environment and regulates traffic of materials into and out of the cell.

Ribosomes   30 nm

The **rough endoplasmic reticulum** is the site of much protein synthesis.

Figure 5.7 Eukaryotic Cells   In electron micrographs, many plant cell organelles are nearly identical in form to those observed in animal cells. Cellular structures found in plant but not animal cells include the cell wall and the chloroplasts. Note that the images are two-dimensional "slices," whereas cells are three-dimensional structures.

▶ Media Clip 5.1 **The Inner Life of a Cell**
www.Life11e.com/mc5.1

A PLANT CELL

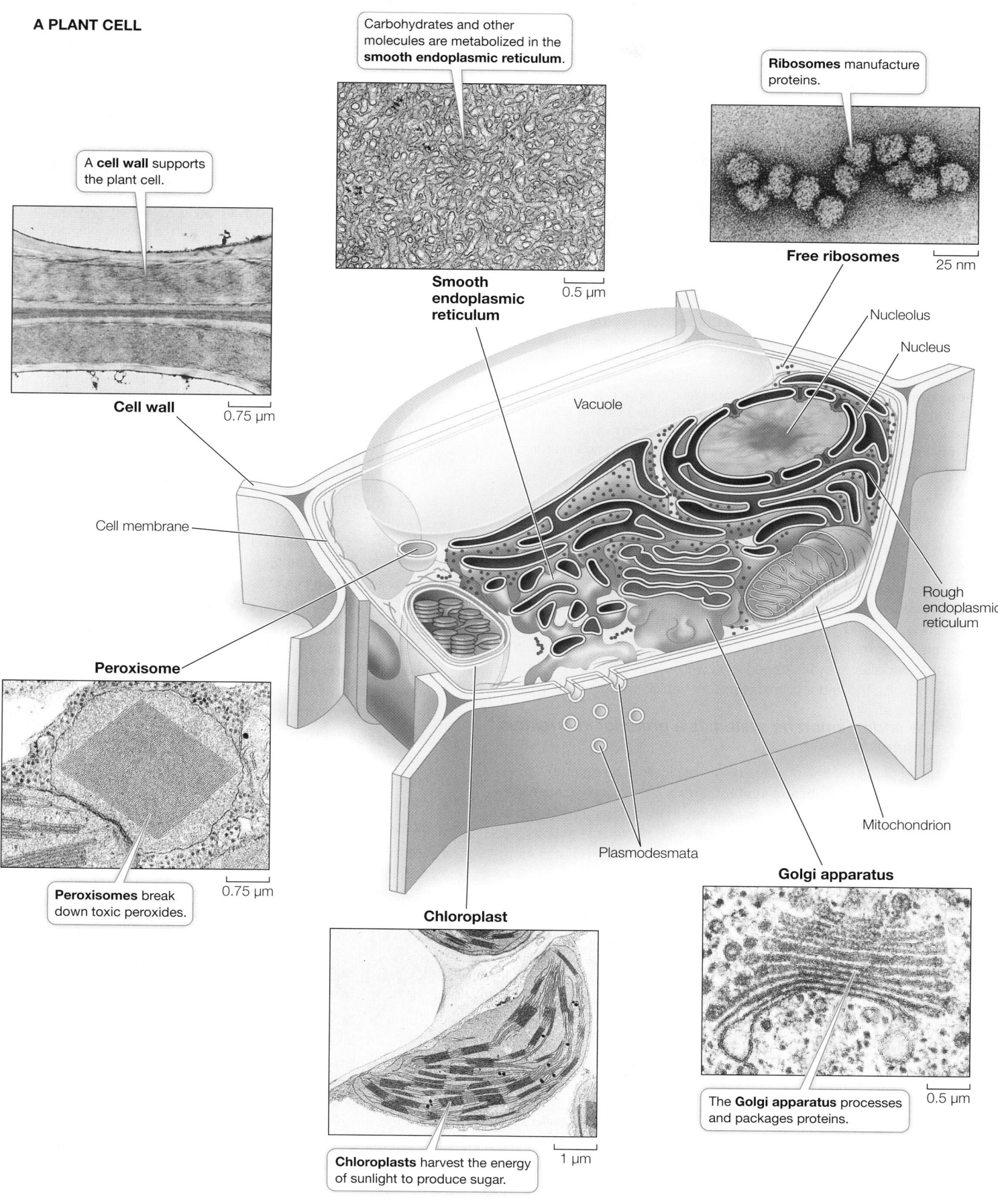

A **cell wall** supports the plant cell.

Cell wall
0.75 μm

Carbohydrates and other molecules are metabolized in the **smooth endoplasmic reticulum**.

Smooth endoplasmic reticulum
0.5 μm

Ribosomes manufacture proteins.

Free ribosomes
25 nm

Nucleolus

Nucleus

Vacuole

Cell membrane

Rough endoplasmic reticulum

Peroxisome

Peroxisomes break down toxic peroxides.
0.75 μm

Plasmodesmata

Mitochondrion

Chloroplast

Chloroplasts harvest the energy of sunlight to produce sugar.
1 μm

Golgi apparatus

The **Golgi apparatus** processes and packages proteins.
0.5 μm

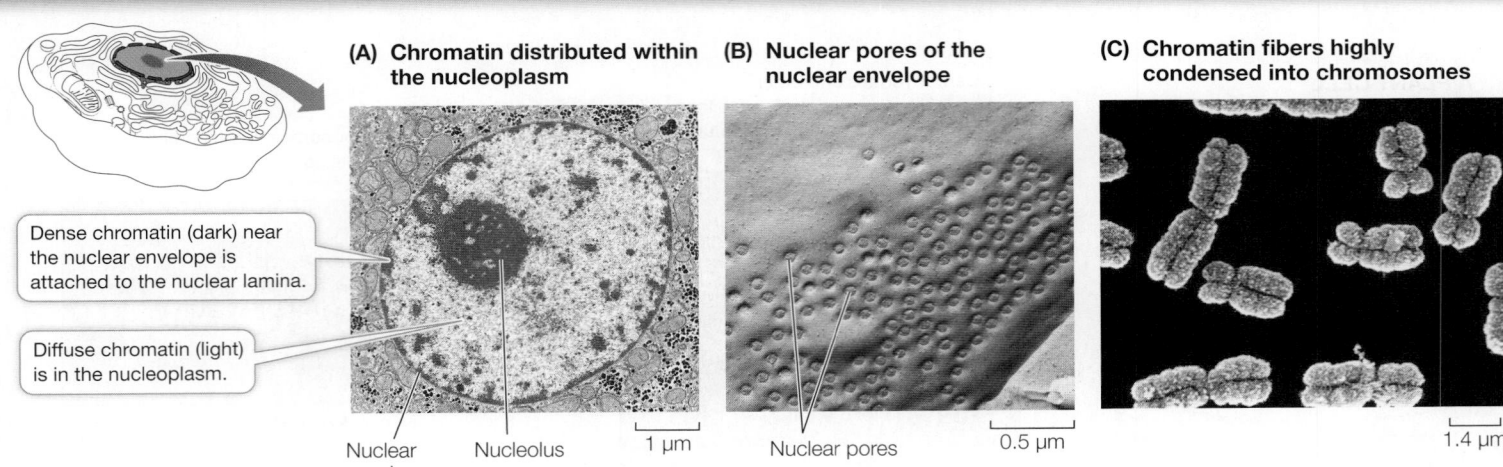

(A) Chromatin distributed within the nucleoplasm

(B) Nuclear pores of the nuclear envelope

(C) Chromatin fibers highly condensed into chromosomes

Dense chromatin (dark) near the nuclear envelope is attached to the nuclear lamina.

Diffuse chromatin (light) is in the nucleoplasm.

Nuclear envelope   Nucleolus   1 µm

Nuclear pores   0.5 µm

1.4 µm

Figure 5.8 The Nucleus, Chromatin, and Chromosomes **(A)** Chromatin consists of nuclear DNA and the proteins associated with it. When the cell is not dividing, the chromatin is dispersed throughout the nucleus. This two-dimensional image was made using a transmission electron microscope. **(B)** The nuclear envelope has many pores that regulate traffic of large molecules such as RNA and proteins in and out of the nucleus. **(C)** The chromatin in dividing cells becomes highly condensed, so that the individual chromosomes can be seen. This three-dimensional image of isolated metaphase chromosomes was produced using a scanning electron microscope.

At the inside edge of the nucleus, the chromatin is attached to a protein meshwork, called the nuclear lamina, which is formed by the polymerization of proteins called lamins into long, thin structures called intermediate filaments (see below). The nuclear lamina maintains the shape of the nucleus by its attachment to both the chromatin and the nuclear envelope. On the outside of the nucleus, the outer membrane of the nuclear envelope folds outward into the cytoplasm and is continuous with the membrane of another organelle, the endoplasmic reticulum, which we will discuss next.

The endomembrane system is a group of interrelated organelles

Much of the volume of some eukaryotic cells is taken up by an extensive **endomembrane system**. This is an interconnected system of membrane-enclosed compartments that are sometimes flattened into sheets and sometimes have other characteristic shapes (see Figure 5.7). The endomembrane system includes the cell membrane, the nuclear envelope, the endoplasmic reticulum, the Golgi apparatus, and lysosomes, which are derived from the Golgi apparatus. Tiny, membrane-surrounded droplets called vesicles shuttle substances between the various components of the endomembrane system (**Focus: Key Figure 5.9**). In drawings and electron microscope pictures, this system appears static, fixed in space and time. But these depictions are just snapshots; in the living cell, the membranes and the materials they contain are in constant motion. Membrane components have been observed to shift from one organelle to another within the endomembrane system. Thus all these membranes must be functionally related.

ENDOPLASMIC RETICULUM   Electron micrographs of eukaryotic cells reveal networks of interconnected membranes branching throughout the cytoplasm, forming tubes and flattened sacs. These membranes are collectively called the **endoplasmic reticulum**, or **ER**. The interior compartment of the ER, referred to as the lumen, is separate and distinct from the surrounding cytoplasm (see Figure 5.9). The ER can enclose up to 10 percent of the interior volume of the cell, and its folds result in a surface area many times greater than that of the cell membrane. There are two types of endoplasmic reticulum, the so-called rough and smooth.

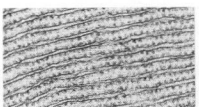 **Rough endoplasmic reticulum (RER)** is called "rough" because of the many ribosomes attached to the outer surface of the membrane, giving it a "rough" appearance in electron microscopy (see Figure 5.7). The bound ribosomes are actively involved in protein synthesis, but that is not the entire story:

- The RER receives into its lumen certain newly synthesized proteins (including proteins destined for lysosomes, the cell membrane, and export from the cell), segregating them away from the cytoplasm. The RER also participates in transporting these proteins to other locations in the cell.

- While inside the RER, proteins can be chemically modified to alter their functions and to "tag" them for delivery to specific cellular destinations.

- Proteins are shipped to destinations elsewhere in the cell enclosed within vesicles that pinch off from the RER.

- Most membrane-bound proteins are made in the RER.

A protein enters the lumen of the RER through a pore as it is synthesized. As with a protein passing through a nuclear pore, there is a special transport mechanism to get proteins into the ER, which you'll learn about in Chapter 14. Once in the lumen of the RER, proteins undergo several changes, including the formation of disulfide bridges and folding into their tertiary structures (see Figure 3.5).

Some proteins are covalently linked to carbohydrate groups in the RER, thus becoming glycoproteins. In the case of proteins directed to the lysosomes, the carbohydrate groups are part of an "addressing" system that ensures that the right proteins are directed to those organelles. This addressing system is very important because the enzymes within the lysosomes are some of the most destructive the cell makes. Were they not properly addressed and contained, they could destroy the cell.

focus: key figure

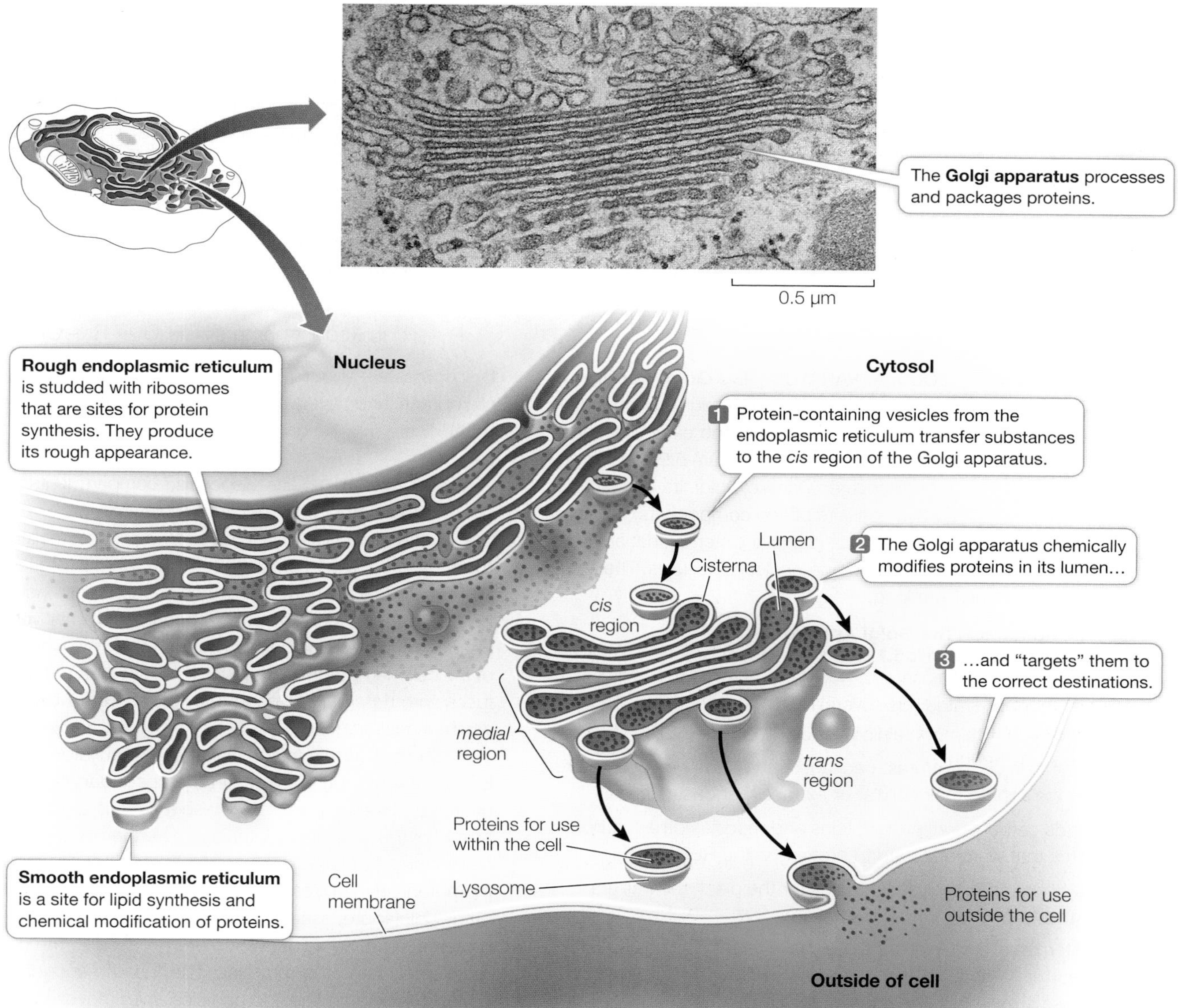

The **Golgi apparatus** processes and packages proteins.

0.5 μm

Rough endoplasmic reticulum is studded with ribosomes that are sites for protein synthesis. They produce its rough appearance.

Nucleus

Cytosol

1 Protein-containing vesicles from the endoplasmic reticulum transfer substances to the *cis* region of the Golgi apparatus.

Lumen

2 The Golgi apparatus chemically modifies proteins in its lumen...

Cisterna

cis region

3 ...and "targets" them to the correct destinations.

medial region

trans region

Proteins for use within the cell

Smooth endoplasmic reticulum is a site for lipid synthesis and chemical modification of proteins.

Cell membrane

Lysosome

Proteins for use outside the cell

Outside of cell

Figure 5.9 The Endomembrane System Membranes of the nucleus, endoplasmic reticulum, and Golgi apparatus form a network connected by vesicles.

Q: What processes in the endomembrane system are mediated by vesicles?

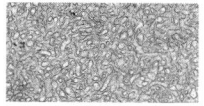 The **smooth endoplasmic reticulum (SER)** lacks ribosomes and is more tubular (and less like flattened sacs) than the RER, but it shows continuity with portions of the RER (see Figure 5.9). Certain proteins that are synthesized in the RER are chemically modified within the lumen of the ***SER**. The SER has four other important roles:

1. It is responsible for the chemical modification of small molecules taken in by the cell that may be toxic to the cell. These modifications make the targeted molecules more polar, so they are more water-soluble and easily removed.

2. It is the site for glycogen degradation in animal cells. We will discuss this important process in Chapter 9.

3. It is the site where lipids and steroids are synthesized, as well as some polysaccharides in plant cells.

4. It stores calcium ions, which when released trigger a number of cell responses.

---

*connect the concepts   Regulating muscle contraction is one of the important roles played by the calcium ions stored by the SER. See Key Concept 47.1.

Cells that synthesize a lot of protein for export are usually packed with RER. Examples include glandular cells that secrete digestive enzymes and white blood cells that secrete antibodies. In contrast, cells that carry out less protein synthesis (such as storage cells) contain less RER. Liver cells, which modify molecules (including toxins) that enter the body from the digestive system, have abundant SER.

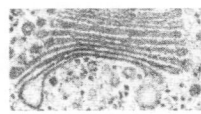 **GOLGI APPARATUS**   The **Golgi apparatus** (or Golgi complex, or just Golgi) is another part of the diverse, dynamic, and extensive endomembrane system (see Figure 5.9). This structure was named after its discoverer, Camillo Golgi. Its appearance varies, but it almost always consists of two components: flattened membranous sacs called cisternae (singular cisterna) that are piled up like saucers, and small membrane-enclosed vesicles. The entire apparatus is about 1 μm long.

 Animation 5.1 **The Golgi Apparatus**
www.Life11e.com/a5.1

The Golgi apparatus has several roles:

- It receives protein-containing vesicles from the RER.
- It modifies, concentrates, packages, and sorts proteins before they are sent to their cellular or extracellular destinations.
- It adds carbohydrates to proteins and modifies other carbohydrates that were attached to proteins in the RER.
- It is where some polysaccharides for the plant cell wall are synthesized.

The cisternae of the Golgi apparatus have three functionally distinct regions:

1. The *cis* region lies nearest to the nucleus or a patch of RER.
2. The *trans* region lies closest to the cell membrane.
3. The *medial* region lies in between (see Figure 5.9).

The terms *cis*, *trans*, and *medial* derive from Latin words meaning, respectively, "on the same side," "on the opposite side," and "in the middle." These three parts of the Golgi apparatus contain different enzymes and perform different functions.

Protein-containing vesicles from the RER fuse with the *cis* membrane of the Golgi apparatus, releasing their cargo into the lumen of the Golgi apparatus cisterna. Other vesicles may move between the cisternae, transporting proteins, and it appears that some proteins move from one cisterna to the next through tiny channels. Vesicles budding off from the *trans* region carry their contents away from the Golgi apparatus. These vesicles go to the cell membrane or to the lysosome, another organelle in the endomembrane system.

ER → *cis*-Golgi → *medial*-Golgi → *trans*-Golgi ⟨ cell membrane / lysosome

How do the vesicles in this system recognize their targets for fusion? In other words, when a vesicle buds off from the ER, how does it "know" to fuse with the *cis*-Golgi membrane? The answer lies in proteins on both membranes appropriately called SNARES. Pairs of SNARES are complementary—that is, they bind to one another like a lock and key. So the SNARE on the ER surface that forms a vesicle can bind to a SNARE on the *cis*-Golgi.

LYSOSOMES   The **primary lysosomes** originate from the Golgi apparatus. They contain digestive enzymes and are the sites where macromolecules—proteins, polysaccharides, nucleic acids, and lipids—are hydrolyzed into their monomers (see Figure 3.4).

$$R_1—R_2 \text{ (linked monomers)} + H_2O \rightarrow R_1—OH + R_2—H$$

Lysosomes are about 1 μm in diameter, are surrounded by a single membrane, and have a densely staining, featureless interior (**Figure 5.10**). There may be dozens of lysosomes in a cell, depending on its needs.

Lysosomes are cellular "disposals," compartments for the breakdown of food, other cells, or foreign objects that are taken up by the cell. These materials enter the cell by a process called **phagocytosis** (*phago*, "eat," + *cytosis*, "cellular"). In this process, a pocket forms in the cell membrane and then deepens and encloses material from outside the cell. The pocket becomes a small vesicle called a phagosome, containing food or other material, which breaks free of the cell membrane to move into the cytoplasm. The phagosome fuses with a primary lysosome to form a **secondary lysosome**, where digestion occurs.

The effect of this fusion is rather like children devouring Halloween candy: the enzymes in the secondary lysosome quickly hydrolyze the food particles. These reactions are enhanced by the acidity of the lysosome's interior, where the pH is lower than in the surrounding cytoplasm. The products of digestion pass through the membrane of the lysosome, providing energy and raw materials for other cellular processes. For example, a protease might hydrolyze proteins to amino acids, which are released to the cytoplasm for reassembly into other proteins. The "used" secondary lysosome, now containing undigested particles, then moves to the cell membrane, fuses with it, and releases the undigested contents to the environment, a process called **exocytosis**.

Phagocytes (see Key Concept 41.1) are specialized cells that have an essential role in taking up and breaking down materials; they are found in nearly all animals and many protists. You will encounter them and their activities again at many places in this book, but at this point one example suffices: in the human liver and spleen, phagocytes digest approximately 10 billion aged or damaged blood cells each day! The digestion products are then used to make new cells to replace those that are digested.

Lysosomes are active even in cells that do not perform phagocytosis. Cells are dynamic systems; some cell components are continually being broken down and replaced by new ones. The programmed destruction of cell components is called **autophagy**, and lysosomes are where the cell breaks down its own materials. With the proper signal, lysosomes can engulf entire organelles, hydrolyzing their constituents.

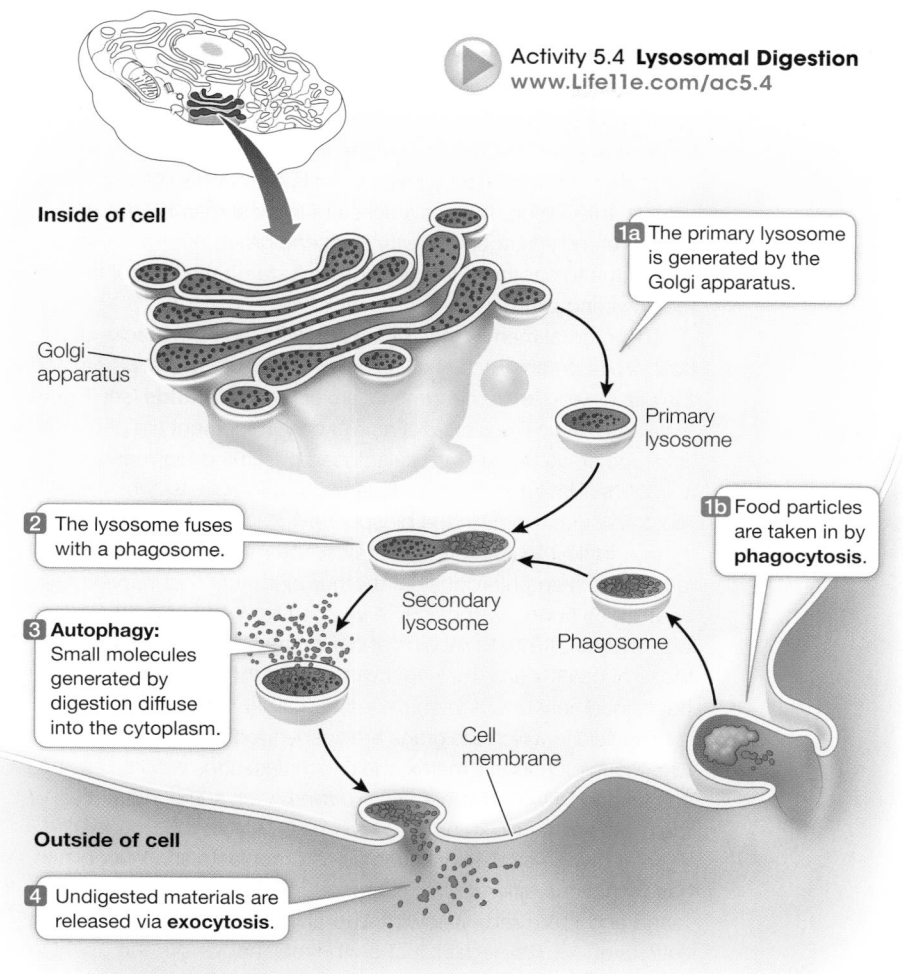

Activity 5.4 **Lysosomal Digestion**
www.Life11e.com/ac5.4

Inside of cell

1a The primary lysosome is generated by the Golgi apparatus.

Golgi apparatus

Primary lysosome

1b Food particles are taken in by **phagocytosis**.

2 The lysosome fuses with a phagosome.

Secondary lysosome

Phagosome

3 Autophagy: Small molecules generated by digestion diffuse into the cytoplasm.

Cell membrane

Outside of cell

4 Undigested materials are released via **exocytosis**.

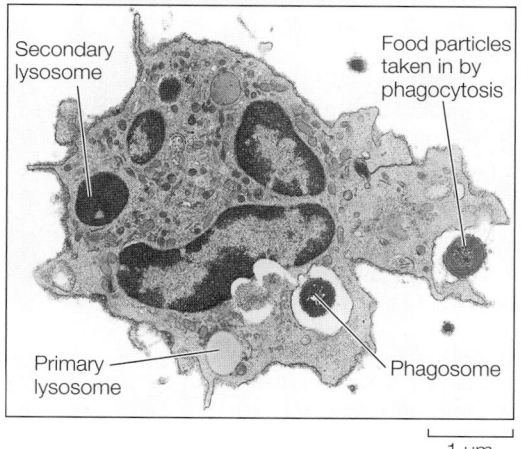

Secondary lysosome

Food particles taken in by phagocytosis

Primary lysosome

Phagosome

1 μm

Figure 5.10 Lysosomes Isolate Digestive Enzymes from the Cytoplasm Lysosomes are sites for the hydrolysis of material taken into the cell by phagocytosis. Discarded components exit the cell by exocytosis.

How important is autophagy? An entire class of human diseases called lysosomal storage diseases are caused by the failure of lysosomes to digest specific cellular components; these diseases are invariably harmful or fatal. An example is Tay-Sachs disease, in which a particular lipid called a ganglioside is not broken down in the lysosomes and instead accumulates in brain cells. In the most common form of this disease, a baby starts exhibiting neurological symptoms and becomes blind, deaf, and unable to swallow after 6 months of age. Death occurs before age 4.

Plant cells do not appear to contain lysosomes, but the vacuole of a plant cell (which we will describe below) may function in an equivalent capacity because it, like lysosomes, contains many digestive enzymes.

Some organelles transform energy

A cell requires energy to make the molecules it needs for activities such as growth, reproduction, responsiveness, and movement. Energy is harvested from fuel molecules in the mitochondria (found in all eukaryotic cells) and from sunlight in the chloroplasts of plant cells. In contrast, energy transformations in prokaryotic cells are associated with enzymes attached to the inner surface of the cell membrane or to extensions of the cell membrane that protrude into the cytoplasm.

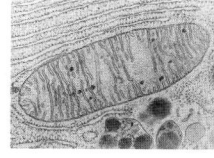

MITOCHONDRIA In eukaryotic cells, the breakdown of energy-rich molecules such as glucose begins in the cytoplasm. The molecules that result from this partial degradation enter the **mitochondria** (singular *mitochondrion*), whose primary function is to harvest the chemical energy of those molecules in a form that the cell can use, namely the energy-rich molecule ATP (adenosine triphosphate) (see Key Concept 8.2). The production of ATP in the mitochondria, using fuel molecules and molecular oxygen (O_2), is called **cellular respiration**.

Typical mitochondria are somewhat less than 1.5 μm in diameter and are 2–8 μm in length—about the size of many bacteria. They can reproduce and divide independently of the central nucleus. The number of mitochondria per cell ranges from one large organelle in some unicellular protists to a few hundred thousand in large egg cells. An average human liver cell contains more than 1,000 mitochondria. Cells that are active in movement and growth require the most chemical energy, and these tend to have the most mitochondria per unit of volume.

Mitochondria have two membranes. The outer membrane is smooth and protective, and it offers little resistance to the movement of substances into and out of the organelle. Immediately inside the outer membrane is an inner membrane, which folds inward in many places and thus has a surface area much greater than that of the outer membrane (**Figure 5.11**). The folds tend to be quite regular, giving rise to shelflike structures called cristae. The inner membrane exerts much more control over what enters and leaves the space it encloses than does the outer membrane. Embedded in the inner

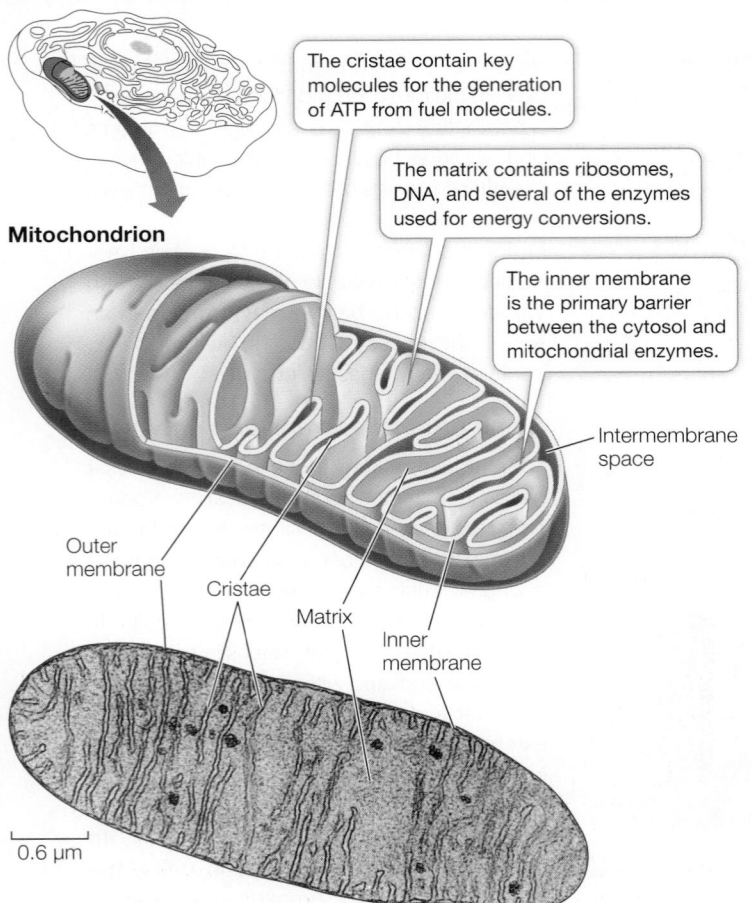

The cristae contain key molecules for the generation of ATP from fuel molecules.

The matrix contains ribosomes, DNA, and several of the enzymes used for energy conversions.

The inner membrane is the primary barrier between the cytosol and mitochondrial enzymes.

Mitochondrion

Intermembrane space

Outer membrane

Cristae

Matrix

Inner membrane

0.6 μm

Figure 5.11 A Mitochondrion Converts Energy from Fuel Molecules into ATP The electron micrograph is a two-dimensional slice through a three-dimensional organelle. As the drawing emphasizes, the cristae are extensions of the inner mitochondrial membrane.

Q: What kinds of human cells would you expect to have a lot of mitochondria?

mitochondrial membrane are many large protein complexes that participate in cellular respiration.

The space enclosed by the inner membrane is referred to as the mitochondrial matrix. In addition to many enzymes, the matrix contains ribosomes and DNA that are used to make some of the proteins needed for cellular respiration. As we will discuss later in this chapter, it is likely that this DNA is the remnant of a larger, complete chromosome from a prokaryote that may have been the mitochondrion's progenitor. In Chapter 9 we will discuss how the different parts of the mitochondrion work together in cellular respiration.

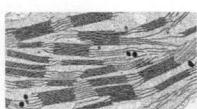

PLASTIDS One class of organelles—the plastids—is present only in the cells of plants and certain protists. Like mitochondria, plastids can divide independently of the cell nucleus and probably evolved from independent prokaryotes. There are several types of plastids, with different functions.

Just as melanosomes are a compartment for pigment in animal cells (see the chapter opening), so are **chloroplasts**, which contain the green pigment chlorophyll and are the sites of photosynthesis (**Figure 5.12**). In photosynthesis, light energy is converted into the

chemical energy of bonds between atoms. The molecules formed by photosynthesis provide food for the photosynthetic organism and for other organisms that eat it. Directly or indirectly, photosynthesis is the energy source for most of the living world.

Like a mitochondrion, a chloroplast is surrounded by two membranes. In addition, there is a series of internal membranes whose structure and arrangement vary from one group of photosynthetic organisms to another. Here we concentrate on the chloroplasts of the flowering plants.

The internal membranes of chloroplasts look like stacks of flat, hollow pita bread. Each stack is called a granum (plural grana) and the pita bread–like compartments are called **thylakoids** (see Figure 5.12). Thylakoid lipids are distinctive: only 10 percent are phospholipids, whereas the rest are galactose-substituted diglycerides and sulfolipids. Because of the abundance of chloroplasts, these are the most abundant lipids in the biosphere.

In addition to lipids and proteins, the membranes of the thylakoids contain chlorophyll and other pigments that harvest light energy for photosynthesis (we will see how they do this in Key Concept 10.2). The thylakoids of one granum may be connected to those of other grana, making the interior of the chloroplast a highly developed network of membranes, much like the ER.

The fluid in which the grana are suspended is called the stroma. Like the mitochondrial matrix, the chloroplast stroma contains ribosomes and DNA, which are used to synthesize some, but not all, of the proteins that make up the chloroplast.

Other types of plastids, such as chromoplasts and leucoplasts, have functions different from those of chloroplasts. Chromoplasts make and store red, yellow, and orange pigments, especially in flowers and fruits—again, just as the melanocytes in the opening story store pigment.

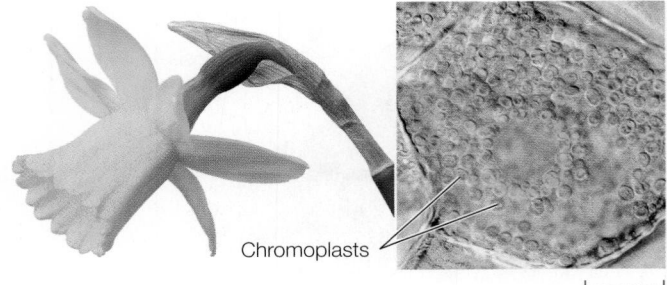

Chromoplasts

20 μm

Leucoplasts are storage organelles that do not contain pigments. An amyloplast is a leucoplast that stores starch.

Leucoplast

Starch grains

1 μm

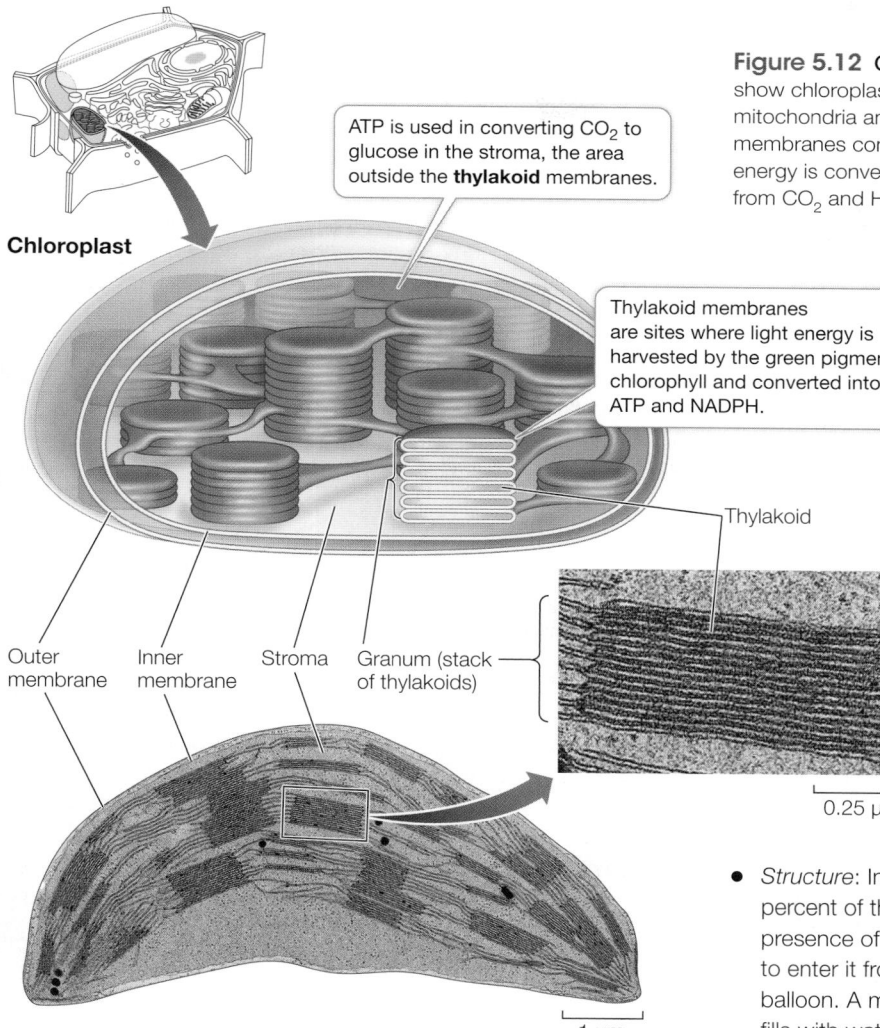

Chloroplast

ATP is used in converting CO_2 to glucose in the stroma, the area outside the **thylakoid** membranes.

Thylakoid membranes are sites where light energy is harvested by the green pigment chlorophyll and converted into ATP and NADPH.

Thylakoid

Outer membrane | Inner membrane | Stroma | Granum (stack of thylakoids)

0.25 μm

1 μm

Figure 5.12 Chloroplasts Feed the World   The electron micrographs show chloroplasts from a leaf of corn. Chloroplasts are large compared with mitochondria and contain extensive networks of thylakoid membranes. These membranes contain the green pigment chlorophyll, and are the sites where light energy is converted into chemical energy for the synthesis of carbohydrates from CO_2 and H_2O.

As with lysosomes, there are rare inherited diseases in humans that involve peroxisomes. In Zellweger syndrome there is a defect in peroxisome assembly, and affected infants are born without peroxisomes. As you can imagine, a consequence of this is the accumulation of toxic peroxides, and the infants seldom live beyond 1 year of age.

Glyoxysomes are similar to peroxisomes and are found only in plants. They are most abundant in young plants and are the locations where stored lipids are converted into carbohydrates for transport to growing cells.

Vacuoles occur in many eukaryotic cells but particularly those of plants, fungi, and protists. They arise from the ER or Golgi apparatus. Plant vacuoles (**Figure 5.13**) have several functions:

- *Structure*: In many plant cells, vacuoles take up more than 90 percent of the cell volume and grow as the cell grows. The presence of dissolved substances in the vacuole causes water to enter it from the cytoplasm, making the vacuole swell like a balloon. A mature plant cell does not swell when the vacuole fills with water, since it has a rigid cell wall. Instead, it stiffens from the increase in water pressure (called turgor), and this supports the plant body (see Figure 6.10).

- *Reproduction*: Vacuoles contain some of the pigments (especially blue and pink ones) in the petals and fruits of flowering plants. These pigments—the anthocyanins—are visual cues that help

There are several other membrane-enclosed organelles

There are several other organelles whose boundary membranes separate their specialized chemical reactions and contents from the rest of the cytoplasm: peroxisomes, glyoxysomes, and vacuoles.

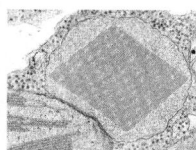

Peroxisomes are organelles that accumulate toxic peroxides, such as hydrogen peroxide (H_2O_2), that occur as by-products of some biochemical reactions. These peroxides are safely broken down inside the peroxisomes without mixing with other parts of the cell.

$$RH_2 + O_2 \rightarrow R + H_2O_2 \text{ (cellular reactions)}$$

$$2\,H_2O_2 \rightarrow 2\,H_2O + O_2 \text{ (inside peroxisome)}$$

Peroxisomes are small organelles, about 0.2–1.7 μm in diameter. They have a single membrane and a granular interior containing specialized enzymes. Peroxisomes are found in at least some of the cells of almost every eukaryotic species.

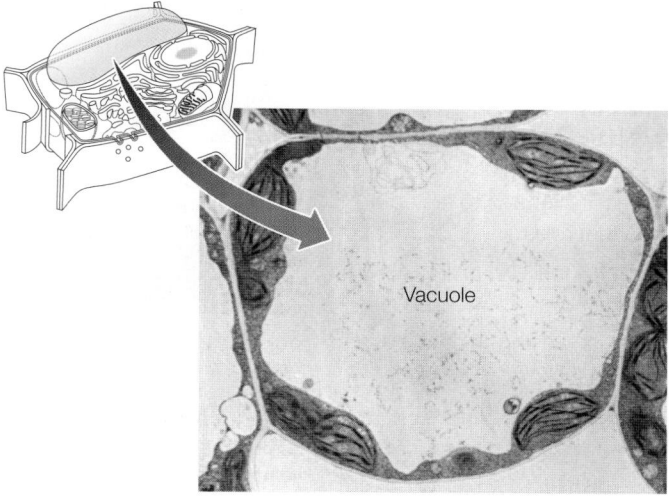

Vacuole

4 μm

Figure 5.13 Vacuoles in Plant Cells Are Usually Large
The large central vacuole in this cell is typical of mature plant cells.

attract animals that assist in pollination or seed dispersal.

- *Digestion*: In some plants, the vacuoles in seeds contain enzymes that hydrolyze stored proteins into monomers. During seed germination, the monomers are used as food by the developing plant seedlings.

- *Storage*: In addition to storing water, vacuoles store toxic molecules and waste products. Because they are poisonous or distasteful, these stored materials deter some animals from eating the plants and may thus contribute to plant defense and survival. An example is tannins, a class of polyphenols that are made in many plants and may be familiar to you as the aftertaste in tea and red wine. Their bitter taste deters predators, and their dark color acts like animal melanin to protect the plant from ultraviolet light damage. Tannins are made in the chloroplast and end up in the vacuole. A newly discovered organelle, appropriately named the tannosome, contains tannins (**Investigating Life: Discovering a New Organelle, the Tannosome**).

So far we have discussed numerous membrane-enclosed organelles. Now we will turn to a group of cytoplasmic structures without membranes.

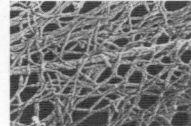

The cytoskeleton is important in cell structure and movement

Cell shapes can sometimes change, and there appear to be rapid movements of structures inside cells. These phenomena involve a meshwork of rods within cells that can be seen in electron microscopy. Experimentation showed that this meshwork—called the cytoskeleton—fills several important roles:

- It supports the cell and maintains its shape.

- It holds cell organelles and other particles in position within the cell.

- It moves organelles and other particles around in the cell.

- It is involved with movements of the cytoplasm, called cytoplasmic streaming.

- It interacts with extracellular structures, helping anchor the cell in place.

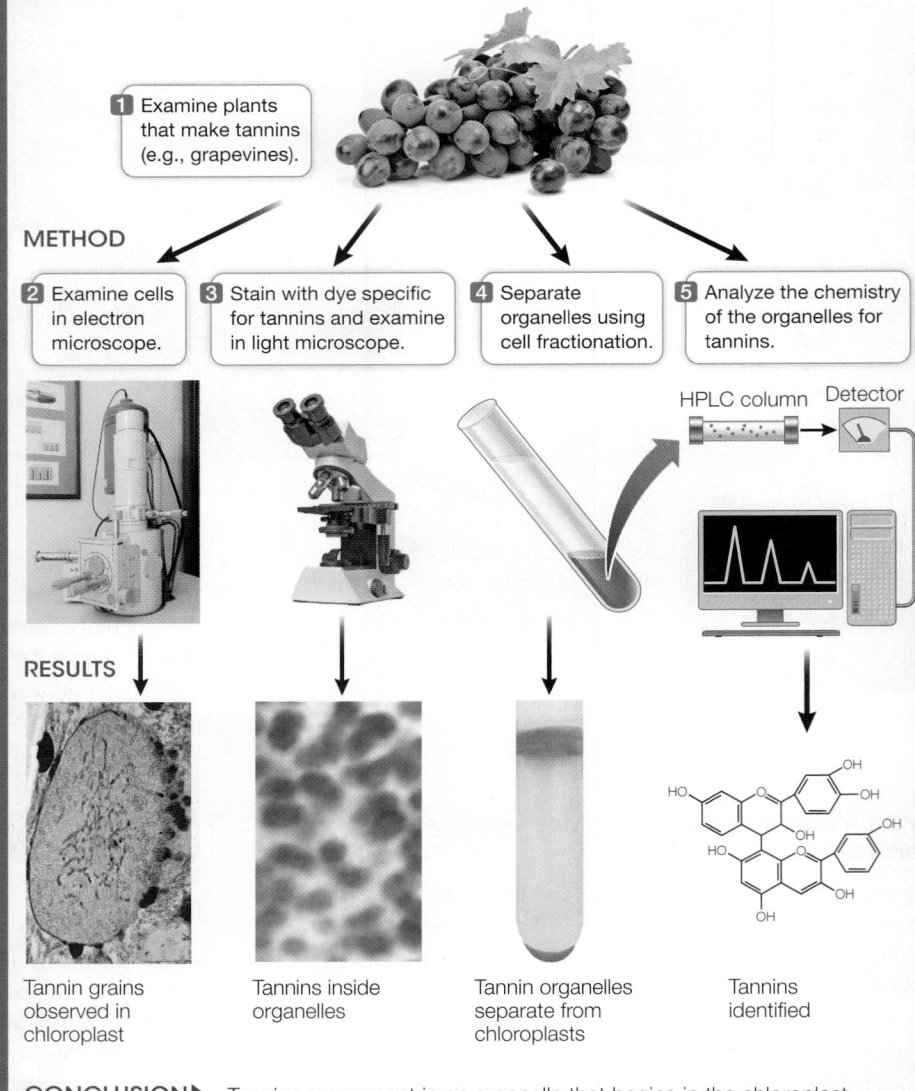

> **investigatinglife**   **Discovering a New Organelle, the Tannosome**
>
> **experiment**
>
> **Original Paper:** Brillouet, J.-M. et al. 2013. The tannosome is an organelle forming condensed tannins in the chlorophyllous organs of *Tracheophyta*. *Annals of Botany* 112: 1003–1014.
>
> Through a series of investigations involving microscopy, cell fractionation, and chemical analysis, biologists discovered that plant tannins are compartmentalized in a heretofore undiscovered structure, the tannosome.
>
> **QUESTION ▶**   Are pigments compartmentalized in all types of cells?
>
> 1 Examine plants that make tannins (e.g., grapevines).
>
> **METHOD**
>
> 2 Examine cells in electron microscope.
> 3 Stain with dye specific for tannins and examine in light microscope.
> 4 Separate organelles using cell fractionation.
> 5 Analyze the chemistry of the organelles for tannins.
>
> HPLC column   Detector
>
> **RESULTS**
>
> Tannin grains observed in chloroplast
> Tannins inside organelles
> Tannin organelles separate from chloroplasts
> Tannins identified
>
> **CONCLUSION ▶**   Tannins are present in an organelle that begins in the chloroplast.

There are three components of the eukaryotic cytoskeleton: microfilaments (smallest diameter), intermediate filaments, and microtubules (largest diameter). These filaments have very different functions.

MICROFILAMENTS   **Microfilaments** can exist as single filaments, in bundles, or in networks. They are about 7 nm in diameter and up to several micrometers long. Microfilaments have two major roles:

The team of biologists and chemists (Brillouet et al., 2013) at the French National Institute for Agricultural Research, led by Geneviève Conejero, sought to investigate the cellular basis of tannin formation. It was well known from electron micrographs that these pigments accumulate in vacuoles, but how they got there was not established. The team used a combination of microscopy (light and electron) (see Figure 5.3), cell fractionation (see Figure 5.6), and chemical analysis to show that tannins are formed in discrete vesicles, which the researchers named tannosomes.

QUESTIONS▶

1. Plant organs rich in tannins were examined by electron microscopy. Representative photographs are shown in **Figure A**. What do these images indicate about the origin and final resting spot of tannins in the cell? Based on what you know of vesicle trafficking in the cell, how do you think the tannins get from the chloroplast to the vacuole?

Figure A

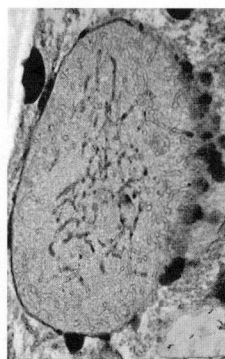 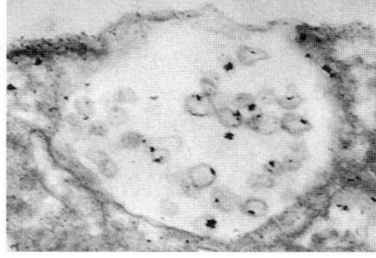

Chloroplast with tannin granules

Vacuole with tannin vesicles

2. The scientists used specific dyes targeted to chlorophyll and tannins and examined slices of plant tissues by light microscopy. **Figure B** shows the results. What can you conclude about the subcellular origin of tannins, and how does your answer relate to your answer to Question 1?

Figure B

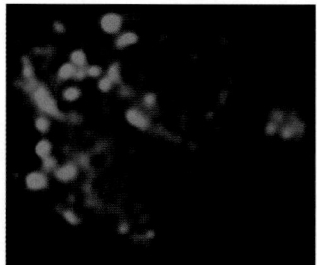 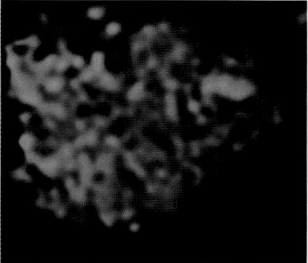

Staining of plant cell for chlorophyll

Staining of plant cell for tannins

3. Cell fractionation was used to try to isolate tannin-containing structures separate from chloroplasts. In a very dense (72%), syrupy sucrose medium, fractionated cell contents showed two membrane-bound fractions (see the experiment). Both were examined by microscopy for tannin staining, and the contents were analyzed for chlorophyll and tannins. The results are shown in the table. What can you conclude about the tannin compartment?

	Upper fraction	Lower fraction
Staining for tannins	No	Yes
Staining for chlorophyll	Yes	Yes
Chemical analysis for tannins	No	Yes

A similar **work with the data** exercise may be assigned in **LaunchPad**.

---

1. They help the entire cell or parts of the cell move.
2. They determine and stabilize cell shape.

Microfilaments are assembled from monomers of **actin**, a protein that exists in several forms and has many functions, especially in animals. The actin found in microfilaments (which are also known as *****actin filaments**) has distinct ends, designated "plus" and "minus." These ends permit actin monomers to interact with one another to form long, double helical chains (**Figure 5.14A**). Within cells, the polymerization of actin into microfilaments is reversible, and the microfilaments can disappear from cells by breaking down into monomers of free actin. Special actin-binding proteins mediate these processes.

> ***** connect the concepts** In animal muscle cells, actin filaments are associated with another protein, the "motor protein" myosin, and the interactions of these two proteins account for the contraction of muscles, as described in Key Concept 47.1.

In non-muscle cells, actin filaments are associated with localized changes in cell shape. For example, microfilaments are involved in the flowing movement of the cytoplasm called cytoplasmic streaming, in amoeboid movement, and in the "pinching" contractions that divide an animal cell into two daughter cells. Microfilaments are also involved in the formation of cellular extensions called pseudopodia (*pseudo*, "false," + *podia*, "feet") that enable some cells to move (**Figure 5.15**). As you will see in Key Concept 41.3, cells of the immune system must move toward other cells during the immune response.

In some cell types, microfilaments form a meshwork just inside the cell membrane. Actin-binding proteins then cross-link the microfilaments to form a rigid netlike structure that supports the cell. For example, microfilaments support the tiny microvilli that line the human intestine, giving it a larger surface area through which to absorb nutrients (**Figure 5.16**).

INTERMEDIATE FILAMENTS   There are at least 50 different kinds of **intermediate filaments**, many of them specific to a few cell

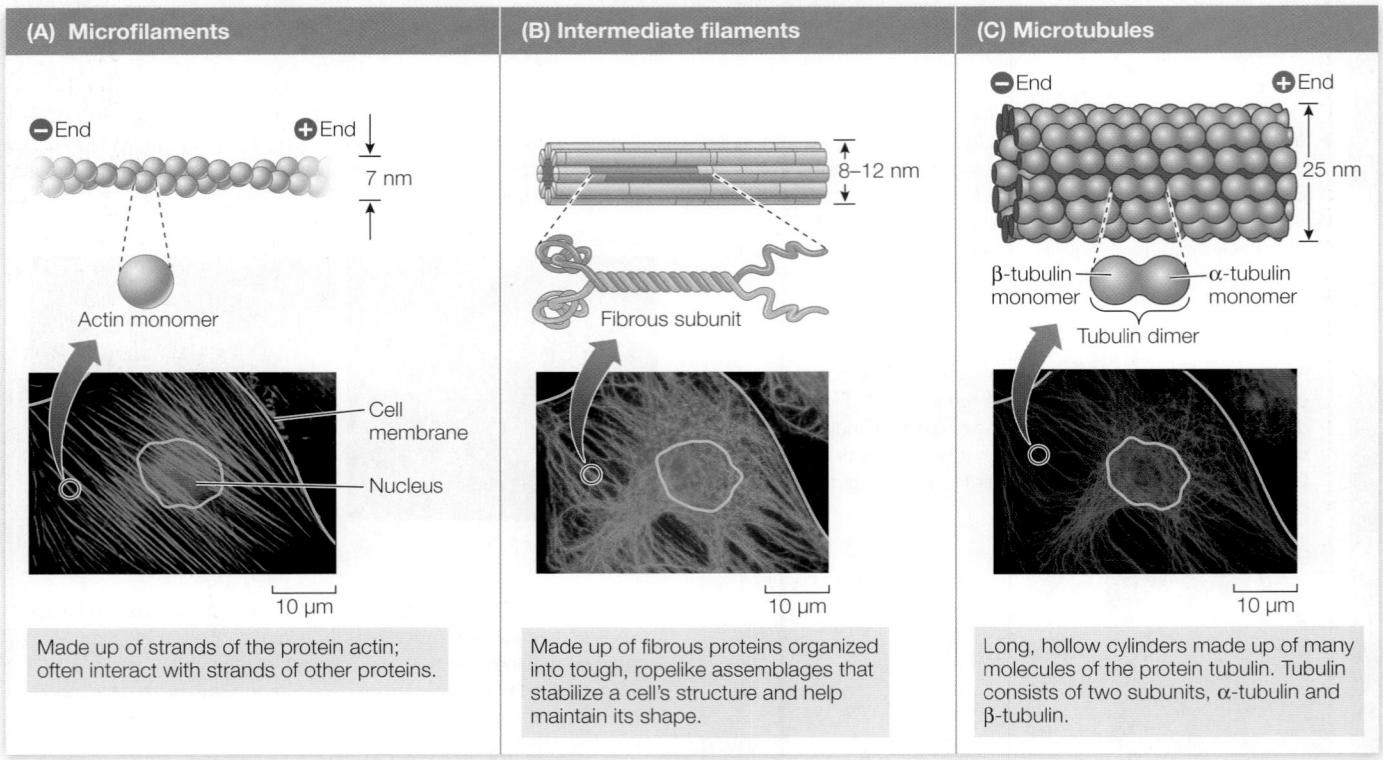

(A) Microfilaments

Made up of strands of the protein actin; often interact with strands of other proteins.

(B) Intermediate filaments

Made up of fibrous proteins organized into tough, ropelike assemblages that stabilize a cell's structure and help maintain its shape.

(C) Microtubules

Long, hollow cylinders made up of many molecules of the protein tubulin. Tubulin consists of two subunits, α-tubulin and β-tubulin.

Figure 5.14 The Cytoskeleton Three highly visible and important structural components of the cytoskeleton are shown here in detail. The photographs are all of the same cell, treated with different fluorescent antibodies that detect microfilaments **(A)**, intermediate filaments **(B)**, or microtubules **(C)**. These structures maintain and reinforce cell shape and contribute to cell movement. The position of the cell's nucleus is near the center of the photos.

types. They generally fall into six molecular classes (based on amino acid sequence) that share the same general structure. One of these classes consists of fibrous proteins of the keratin family, which also includes the proteins that make up hair and fingernails. Intermediate filaments are tough, ropelike protein structures 8–12 nm in diameter **(Figure 5.14B)**. They are more permanent than the other two types in that they do not continually form and re-form, as the microtubules and microfilaments do.

Intermediate filaments have two major structural functions:

1. They anchor cell structures in place. In some cells, intermediate filaments radiate from the nuclear envelope and help maintain the positions of the nucleus and other organelles in the cell. The lamins of the nuclear lamina are intermediate filaments. Other kinds of intermediate filaments help hold in place the complex apparatus of microfilaments in the microvilli of intestinal cells (see Figure 5.16).

2. They resist tension. For example, they maintain rigidity in body surface tissues by stretching through the cytoplasm and connecting specialized membrane structures called desmosomes (see Figure 6.7).

MICROTUBULES The largest-diameter components of the cytoskeletal system, **microtubules**, are long, hollow, unbranched cylinders about 25 nm in diameter and up to several micrometers long. Microtubules have two roles in the cell:

1. They form a rigid internal skeleton for some cells.

2. They act as a framework along which motor proteins can move structures within the cell.

Microtubules are assembled from dimers–two similar molecules in a complex–of the protein **tubulin**, called α-tubulin and β-tubulin. Thirteen chains of tubulin doublets surround the central cavity of the microtubule (**Figure 5.14C**; see also Figure 5.17B).

$$\alpha\text{-tubulin} + \beta\text{-tubulin} \rightarrow \text{doublet} \sim \text{polymer of doublets}$$
$$(\text{microtubule})$$

As in microfilaments, the two ends of a microtubule are different: one is designated the "plus" end and the other the "minus" end. Tubulin dimers can be rapidly added or subtracted, mainly at the plus end, lengthening or shortening the microtubule.

Many microtubules appear to radiate from a region of the cell called the microtubule organizing center. Tubulin polymerization results in a rigid structure, and tubulin depolymerization leads to its collapse.

$$\text{Microtubule} \leftrightarrow \text{tubulin doublets}$$

The capacity to change length rapidly makes microtubules dynamic structures: they are readily adapted for new purposes in the cell. For example, by disassembly and reassembly, microtubules can move to new parts of the cell and assemble new structures needed for cell division. Microtubules from all eukaryotes have this dynamic property, indicating that it is evolutionarily advantageous over a static, unchanging structure.

In plants, microtubules act as a framework for the assembly of cellulose and help orient the cellulose fibers of the cell wall (see Figure 3.18). Electron micrographs of plants frequently show microtubules lying just inside the cell membranes of cells that are forming

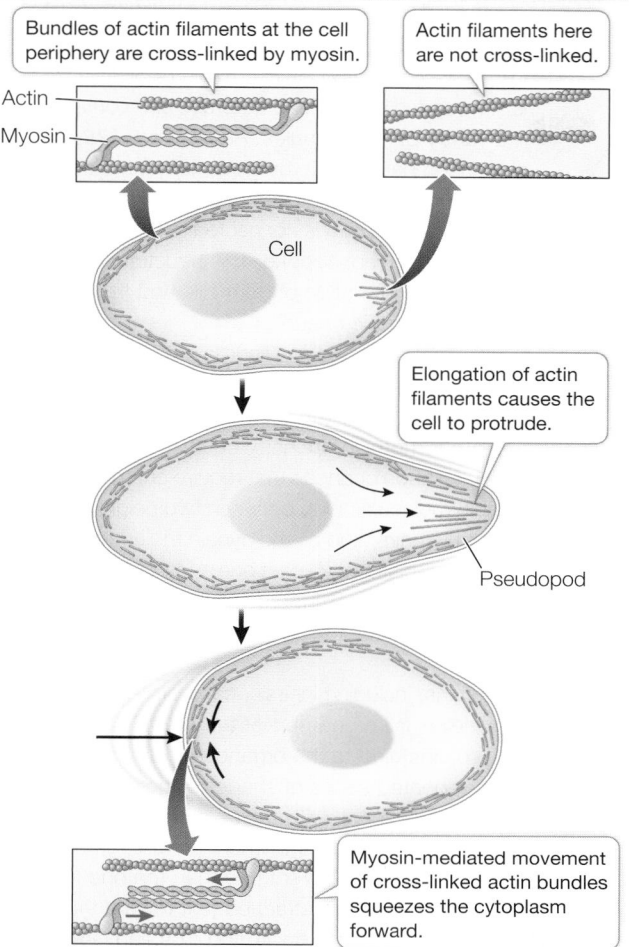

Bundles of actin filaments at the cell periphery are cross-linked by myosin.

Actin filaments here are not cross-linked.

Actin

Myosin

Cell

Elongation of actin filaments causes the cell to protrude.

Pseudopod

Myosin-mediated movement of cross-linked actin bundles squeezes the cytoplasm forward.

Figure 5.15 Microfilaments and Cell Movements Microfilaments mediate the movement of whole cells (as illustrated here for amoeboid movement), as well as the movement of cytoplasm within a cell.

or extending their cell walls. If the orientation of these microtubules is altered experimentally, it leads to a similar change in the cell wall and a new shape for the cell.

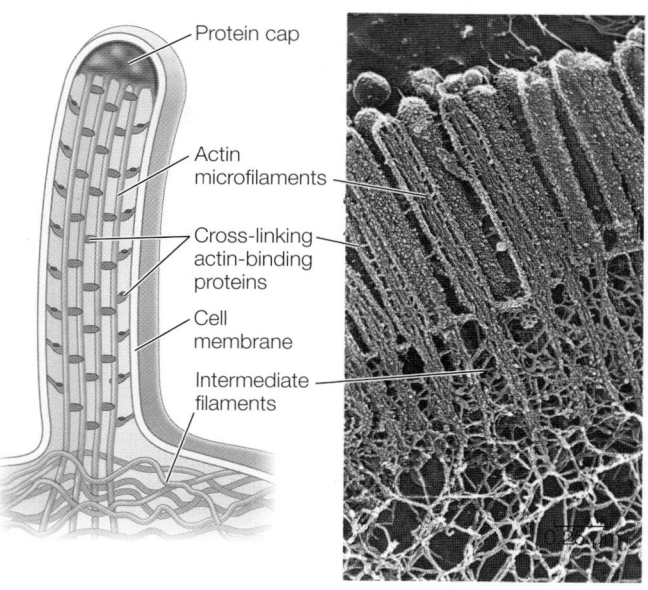

Protein cap

Actin microfilaments

Cross-linking actin-binding proteins

Cell membrane

Intermediate filaments

Figure 5.16 Microfilaments for Support Cells that line the intestine are folded into tiny projections called microvilli, which are supported by microfilaments. The microfilaments interact with intermediate filaments at the base of each microvillus. Microvilli increase the surface area of a cell, facilitating its absorption of small molecules.

Microtubules serve as tracks for **motor proteins**, specialized molecules that use cellular energy to change their shapes and move. Motor proteins bind to and move along the microtubules, carrying materials from one part of the cell to another. Microtubules are also essential in distributing chromosomes to daughter cells during cell division. Because of this, drugs such as vincristine and taxol, which disrupt microtubule dynamics, also disrupt cell division. These drugs are useful for treating cancer, where cell division is excessive.

CILIA AND FLAGELLA Microtubules and their associated proteins line the interior of certain movable appendages on eukaryotic cells: the **cilia (Figure 5.17A)** and flagella. Many cells have one or the

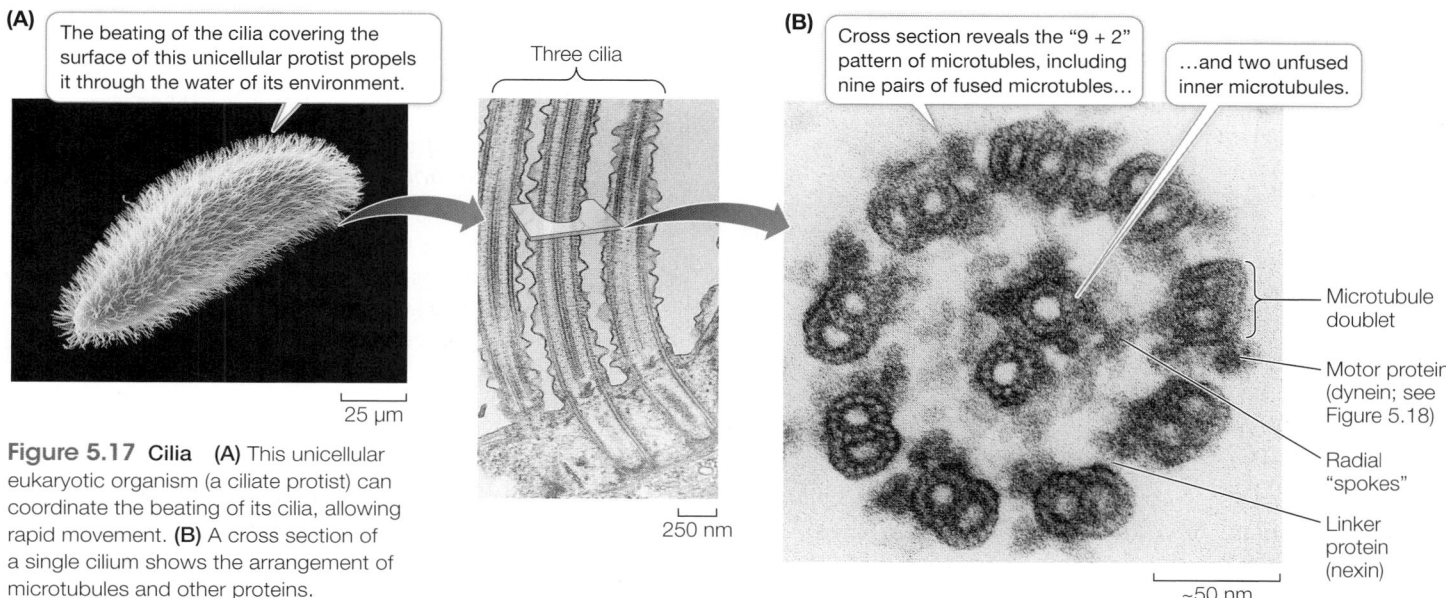

(A)

The beating of the cilia covering the surface of this unicellular protist propels it through the water of its environment.

Three cilia

25 µm

Figure 5.17 Cilia (A) This unicellular eukaryotic organism (a ciliate protist) can coordinate the beating of its cilia, allowing rapid movement. **(B)** A cross section of a single cilium shows the arrangement of microtubules and other proteins.

250 nm

(B)

Cross section reveals the "9 + 2" pattern of microtubules, including nine pairs of fused microtubules...

...and two unfused inner microtubules.

Microtubule doublet

Motor protein (dynein; see Figure 5.18)

Radial "spokes"

Linker protein (nexin)

~50 nm

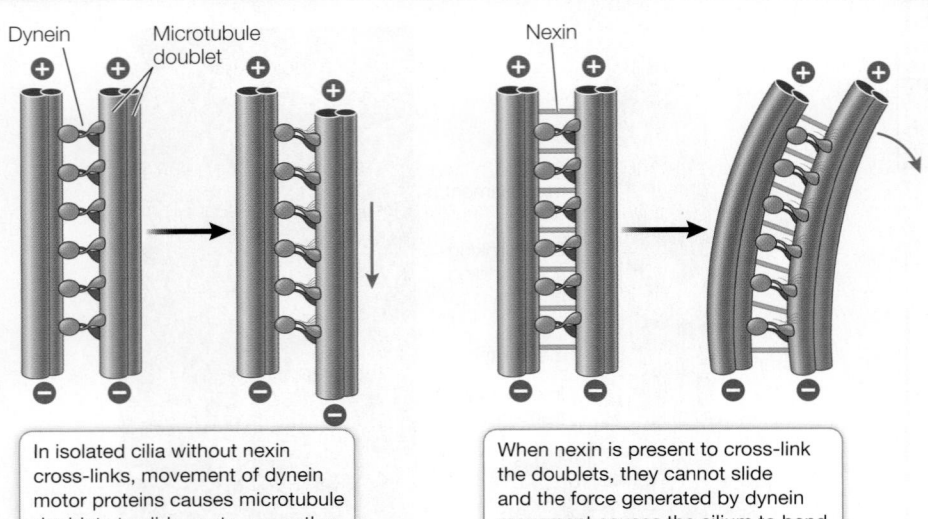

In isolated cilia without nexin cross-links, movement of dynein motor proteins causes microtubule doublets to slide past one another.

When nexin is present to cross-link the doublets, they cannot slide and the force generated by dynein movement causes the cilium to bend.

Figure 5.18 Motor Protein Moves Microtubules in Cilia and Flagella
A motor protein, dynein, causes microtubule doublets to slide past one another. In a flagellum or cilium, anchorage of the microtubule doublets to one another results in bending.

Q: In some people, the nexin links are absent and microtubule doublets are not attached to one another. What do you think the consequences of this would be?

other of these appendages, which form from projections of the cell membrane:

- Cilia are only 0.25 μm in length. They occur by the hundreds on individual cells and move stiffly to either propel the cell (for example, in protists) or to move fluid over a stationary cell (as in the human respiratory system).

- Flagella are longer—100 to 200 μm—and occur singly or in pairs. They can push or pull a cell through its aqueous environment.

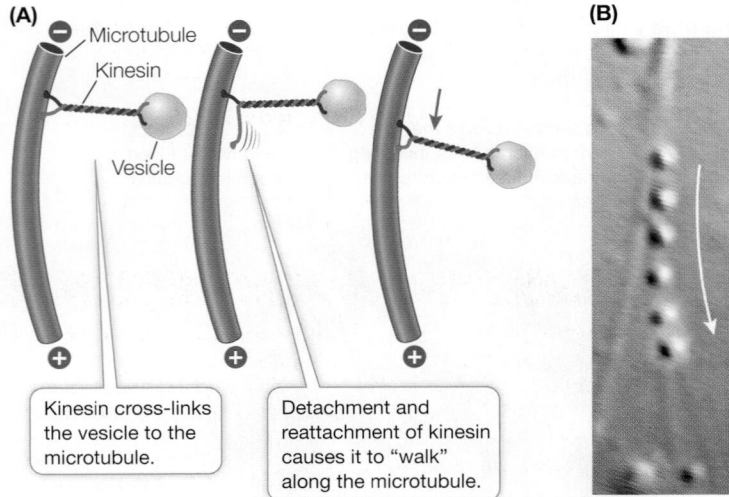

Kinesin cross-links the vesicle to the microtubule.

Detachment and reattachment of kinesin causes it to "walk" along the microtubule.

Figure 5.19 A Motor Protein Pulls Vesicles along Microtubules
(A) Kinesin delivers vesicles to various parts of the cell by moving along microtubule "railroad tracks." **(B)** A vesicle is pulled by kinesin along a microtubule in the protist *Dictyostelium*. The time sequence, with half-second intervals, is shown by the color changes from purple to blue.

In cross section, a typical cilium or eukaryotic flagellum is surrounded by the cell membrane and contains a "9 + 2" array of microtubules. As **Figure 5.17B** shows, nine fused pairs of microtubules form an outer cylinder, and one pair of unfused microtubules runs up the center. Each doublet is connected to the center of the structure by a radial spoke. This structure is essential to the bending motion of both cilia and flagella. How does this bending occur?

The motion of cilia and flagella results from the sliding of the microtubule doublets past one another. This sliding is driven by the motor protein dynein, which, like other motor proteins, works by undergoing reversible shape changes that require chemical energy. Dynein molecules bind between two neighboring microtubule pairs, and as the dynein molecules change shape, the pairs move past one another (**Figure 5.18**). Another protein, nexin, cross-links the pairs and appears to limit how far the doublets can slide. This causes the cilium or flagellum to bend.

Other motor proteins, including kinesin, carry protein-laden vesicles or other organelles from one part of the cell to another (**Figure 5.19**). These proteins bind to the organelle and "walk" it along a microtubule by a repeated series of shape changes. Recall that ***microtubules** are directional, with a plus end where there is a net addition of tubulin and a minus end where there is removal. Cytoplasmic dynein (which has a different role than the one found in cilia and flagella) moves organelles attached to it toward the minus end, whereas kinesin moves them toward the plus end (see Figure 5.14). The ability of microtubules to carry vesicles is used by some fish and amphibians to avoid predators. Motor proteins carrying pigment vesicles aggregate them in the center of the cell in response to environmental or nervous system stimuli. This makes the animal appear lighter. When the stimuli are removed, other motor proteins move the vesicles along microtubules to a more even distribution, restoring a darker hue.

---

*connect the concepts Perhaps the most dramatic role of microtubules is in the process of chromosome movement during cell division. See Key Concept 11.3.

Biologists can manipulate living systems to establish cause and effect

How do we know that the structural fibers of the cytoskeleton can achieve all these dynamic functions? We can observe an individual structure under the microscope, and we can observe the functions of living cells that contain that structure. These observations may suggest that the structure carries out a particular function, but mere correlation does not show cause and effect. For example, light microscopy of living cells reveals that the cytoplasm is actively streaming around the cell, and that cytoplasm flows into an extended portion of an amoeboid cell during movement. The observed presence of cytoskeletal components *suggests, but does not prove,* their role in this process. Science seeks to show the specific links that relate one process, A, to a function, B. In cell

experiment

Figure 5.20A The Role of Microfilaments in Cell Movement—Showing Cause and Effect in Biology

Original Paper: Pollard, T. D and R. R. Weihing. 1974. Actin and myosin in cell movement. *CRC Critical Reviews of Biochemistry* 2: 1–65.

After a test tube demonstration that the drug cytochalasin B prevented microfilament formation from monomeric precursors, the question was asked: Will the drug work like this in living cells and inhibit cell movement in *Amoeba*? Complementary experiments showed that the drug did not poison other cellular processes.

HYPOTHESIS▶ Amoeboid cell movements are caused by the cytoskeleton.

METHOD

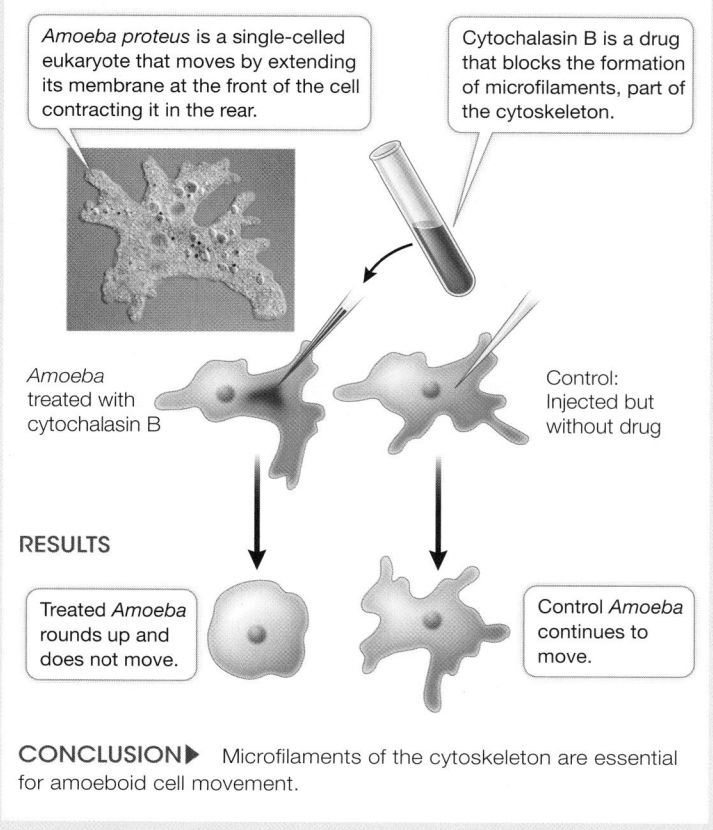

Amoeba proteus is a single-celled eukaryote that moves by extending its membrane at the front of the cell contracting it in the rear.

Cytochalasin B is a drug that blocks the formation of microfilaments, part of the cytoskeleton.

Amoeba treated with cytochalasin B

Control: Injected but without drug

RESULTS

Treated *Amoeba* rounds up and does not move.

Control *Amoeba* continues to move.

CONCLUSION▶ Microfilaments of the cytoskeleton are essential for amoeboid cell movement.

work with the data

Figure 5.20B The Role of Microfilaments in Cell Movement—Showing Cause and Effect in Biology

Original Paper: Pollard, T. D. and R. R. Weihing. 1974.

In a search for natural molecules that have effects on cells—particularly cancer cells—a team of chemists and biologists at Imperial Chemical Industries examined extracts of the fungus *Helminthosporium dematiodeium*. When the extracts appeared to inhibit cell division, the scientists purified the active ingredient and called it cytochalasin B (from the Greek *cyto*, "cell," and *chalasis*, "dislocation"). Remarkably, application of cytochalasin B to dividing cells blocked the division of the cytoplasm but not division of the nucleus, so the result was a binucleate cell. In addition, the drug inhibited cell movement and phagocytosis. These dynamic processes were both hypothesized to involve cytoplasmic microfilaments (actin filaments); the experiment that tested this hypothesis is outlined in Figure 5.20A.

Several important controls were done to validate the conclusions of the experiment. The experiment was repeated in the presence of the following drugs: cycloheximide, which inhibits new protein synthesis; dinitrophenol, which inhibits new ATP formation (energy); and colchicine, which inhibits the polymerization of microtubules. The results are shown in the table.

Condition	Rounded cells (%)
No drug	3
Cytochalasin B	95
Colchicine	4
Cycloheximide	3
Cycloheximide + cytochalasin B	94
Dinitrophenol	5
Dinitrophenol + cytochalasin B	85

QUESTIONS▶

1. Explain the reasoning behind each experiment. Why were these controls important?

2. Interpret the results of each experiment. What can you conclude about movements in *Amoeba* and the cytoskeleton?

A similar **work with the data** exercise may be assigned in **LaunchPad**.

---

biology, two approaches are often used to show that a structure or process A causes function B:

1. *Inhibition*: Use a drug that inhibits A and see if B still occurs. If it does not, then A is probably a causative factor for B. **Figure 5.20** shows an experiment with such a drug (an inhibitor) that demonstrates cause and effect in the case of the cytoskeleton and cell movement.

2. *Mutation*: Examine a cell that lacks the gene (or genes) for A and see if B still occurs. If it does not, then A is probably a causative factor for B. Part Four of this book will describe many experiments using this genetic approach.

▶ 5.3 recap

The hallmark of eukaryotic cells is compartmentalization. Membrane-enclosed organelles process information, transform energy, form internal compartments for transporting proteins, and carry out intracellular digestion. An internal cytoskeleton plays several structural roles.

learning outcomes

You should be able to:

• Distinguish between prokaryotes and eukaryotes.

• Compare the sizes, structures, typical cell numbers, and roles of subcellular structures.

(continued)

5.3 recap (continued)

- Trace the path of a molecule as it moves between organelles in a eukaryotic cell.

1. What are the major differences between prokaryotic cells and eukaryotic cell?

2. Through how many membranes would a molecule have to pass in moving (a) from the interior (stroma) of a chloroplast to the interior (matrix) of a mitochondrion? (b) From the interior of a lysosome to the outside of a cell? (c) From one ribosome to another?

3. The drug vincristine is used to treat many cancers. It appears to work by causing microtubules to depolymerize. Vincristine use has many side effects, including loss of dividing cells and nerve problems. Explain why this might be so.

4. The movements of newly synthesized proteins can be followed through cells using a "pulse–chase" experiment. During synthesis, proteins are tagged with a radioactive isotope (the "pulse"), then the cells are allowed to process the proteins for varying periods of time. The locations of the radioactive proteins are then determined by isolating cell organelles and quantifying their levels of radioactivity. What would be the expected pathway for (a) a lysosomal enzyme and (b) a protein that is released from the cell?

All cells interact with their environments. Many eukaryotic cells are parts of multicellular organisms and must closely coordinate their activities with other cells. The cell membrane plays a crucial role in these interactions, but other structures outside that membrane are involved as well.

key concept 5.4 Extracellular Structures Have Important Roles

Although the cell membrane is the functional barrier between the inside and the outside of a cell, many structures are produced by cells and secreted to the outside of the cell membrane, where they play essential roles in protecting, supporting, or attaching cells to each other. Because they are outside the cell membrane, these structures are said to be extracellular. The peptidoglycan cell wall of bacteria is an example of an extracellular structure (see Figure 5.4). In eukaryotes, other extracellular structures—the cell walls of plants and the extracellular matrices found between the cells of animals—play similar roles. Each of these structures is made up of two main components: a prominent fibrous macromolecule and a gel-like medium in which the fibers are embedded.

focus your learning

- Extracellular structures are important for protection, support, and interacting with other cells.

What is the plant cell wall?

The plant cell wall is a semirigid structure outside the cell membrane (**Figure 5.21**). We will consider the structure and role of the cell wall in more detail in Chapter 33. For now, we note that it consists of cellulose fibers (see Figure 3.18) embedded in other complex polysaccharides and proteins. The plant cell wall has three major roles:

1. It provides support for the cell and plant by remaining rigid. Yet it is flexible enough that it can allow the plant to bend in the wind, for example.

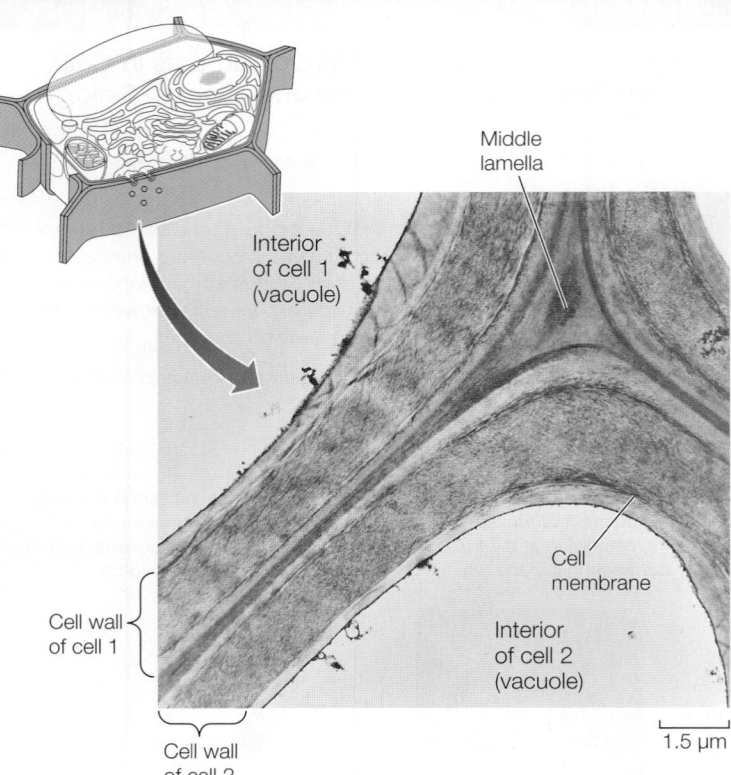

Figure 5.21 The Plant Cell Wall   The semirigid cell wall provides support for plant cells. It is composed of cellulose fibrils embedded in a matrix of polysaccharides and proteins.

2. It acts as a barrier to infection by fungi and other organisms that can cause plant diseases.

3. It contributes to plant form by growing as the plant cells expand.

In some cells, such as those in a leaf, the cell wall is porous to allow the passage of molecules into and out of the cell. In other cells, such as those of the plant's vascular system (which transports water and small molecules between organs), the wall is not porous.

Because of their thick cell walls, plant cells viewed under a light microscope appear to be entirely isolated from one another. But electron microscopy reveals that this is not the case. The cytoplasms of adjacent plant cells are connected by numerous cell membrane–lined channels called **plasmodesmata**, which are about 20–40 nm in diameter and extend through the cell walls (see Figure 5.7). Plasmodesmata permit the diffusion of water, ions, small molecules, RNA, and proteins between connected cells, allowing for the use of these substances far from their sites of synthesis. Plasmodesmata also enable communication between adjacent cells.

The extracellular matrix supports tissue functions in animals

Animal cells lack the semirigid wall that is characteristic of plant cells, but many animal cells are surrounded by, or in contact with, an **extracellular matrix**. This matrix is composed of three types of molecules: fibrous proteins such as **collagen** (the most abundant protein in mammals, constituting over 25 percent of the protein in the human body); a matrix of glycoproteins termed **proteoglycans**, consisting primarily of sugars; and a third group of proteins that link

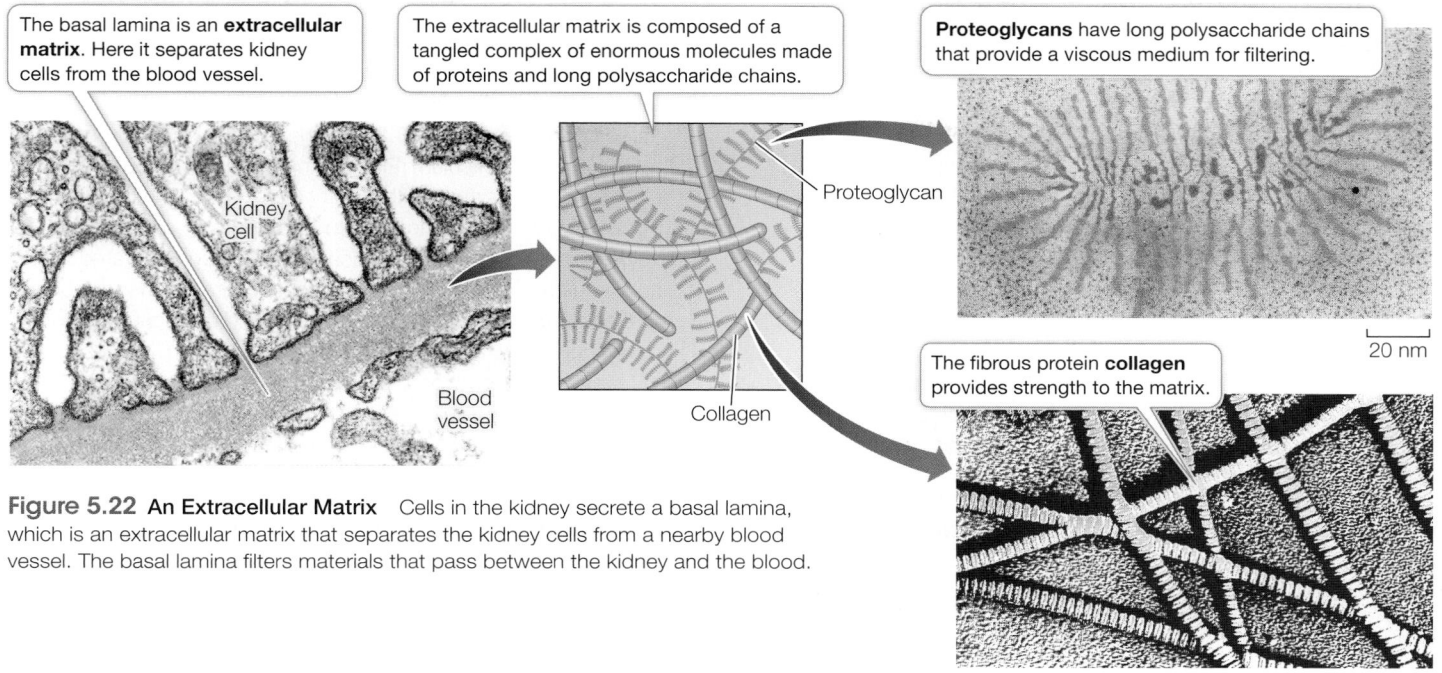

The basal lamina is an **extracellular matrix**. Here it separates kidney cells from the blood vessel.

The extracellular matrix is composed of a tangled complex of enormous molecules made of proteins and long polysaccharide chains.

Proteoglycans have long polysaccharide chains that provide a viscous medium for filtering.

The fibrous protein **collagen** provides strength to the matrix.

Kidney cell

Proteoglycan

Blood vessel

Collagen

20 nm

100 nm

Figure 5.22 **An Extracellular Matrix** Cells in the kidney secrete a basal lamina, which is an extracellular matrix that separates the kidney cells from a nearby blood vessel. The basal lamina filters materials that pass between the kidney and the blood.

the fibrous proteins and the gel-like proteoglycan matrix together (**Figure 5.22**). These proteins and proteoglycans are secreted, along with other substances that are specific to certain body tissues, by cells that are present in or near the matrix.

The functions of the extracellular matrix are many:

- It holds cells together in tissues. In Chapter 6 we will see how there is an intercellular "glue" that is involved in both cell recognition and adhesion.

- It contributes to the physical properties of cartilage, skin, and other tissues. For example, the mineral component of bone is laid down on an organized extracellular matrix.

- It helps filter materials passing between different tissues. This is especially important in the kidney.

- It helps orient cell movements during embryonic development and during tissue repair.

- It plays a role in chemical signaling from one cell to another. Proteins connect the cell's cell membrane to the extracellular matrix. These proteins (for example, integrin) span the cell membrane and are involved with transmitting signals to the interior of the cell. This allows communication between the extracellular matrix and the cytoplasm of the cell.

5.4 recap

Extracellular structures are produced by cells and secreted outside the cell membrane. Most consist of a fibrous component in a gel-like medium.

learning outcomes

You should be able to:

- Explain structure–function relationships in cell walls.
- Compare and contrast plant and animal cell extracellular matrices.
- Analyze changes to cells involving extracellular components.

5.4 recap

1. When a cancer spreads from the initial site to another location in the body, initial events are the detachment of the cancer cells from its surrounding extra cellular matrix and digestion of the matrix as the cell migrates. How might these events occur?

2. How does the structure of a cell wall contribute to its function?

3. How could a biologist use chemical analysis to differentiate between a plant cell and an animal cell?

We have now discussed the structures and some functions of prokaryotic and eukaryotic cells. Both exemplify the cell theory, showing that cells are the basic units of life and of biological continuity. Much of the rest of this part of the book will deal with these two aspects of cells. There is abundant evidence that the simpler prokaryotic cells are more ancient than eukaryotic cells, and that the first cells were probably prokaryotic. We will now turn to the next step in cellular evolution, the origin of eukaryotic cells.

key concept

5.5

Eukaryotic Cells Evolved in Several Steps

Life on Earth was entirely prokaryotic for about 2 billion years—from the time when prokaryotic cells first appeared until about 1.5 billion years ago, when evidence of eukaryotic cells appears in ancient rocks. The advent of compartmentalization—the hallmark of eukaryotes—was a major event in the history of life. It permitted many more biochemical functions to exist in the same cell than had previously been possible. Compared with a typical eukaryote, a single prokaryotic cell is often biochemically specialized, limited in the resources it can use and the functions it can perform.

focus your learning

- Biologists theorize that the endomembrane system and the cell nucleus originated as the result of infoldings of the cell membrane.
- The theory of endosymbiosis proposes that organelles arose when cells engulfed other cells, and this arrangement led to symbiotic relationships.

What is the origin of compartmentalization? We will describe the evolution of eukaryotic organelles in more detail in Key Concept 25.2. Here we outline two major themes in this process.

Internal membranes and the nuclear envelope probably came from the cell membrane

We noted earlier in this chapter that some bacteria contain internal membranes. How could these arise? In electron micrographs, the internal membranes of prokaryotes often appear to be inward folds of the cell membrane. This observation has led to a theory that the endomembrane system and the cell nucleus originated by related processes (**Figure 5.23A**). The close relationship between the endoplasmic reticulum (ER) and the nuclear envelope in today's eukaryotes is consistent with this theory.

A bacterium with enclosed compartments would have several evolutionary advantages. Chemicals could be concentrated in particular regions of the cell, allowing chemical reactions to proceed more efficiently. A biochemical process could be segregated within an organelle with, for example, a different pH from the rest of the cell, creating more favorable conditions for that process. Finally, gene transcription could be separated from translation, providing more opportunities for separate control of these steps in gene expression.

Some organelles arose by endosymbiosis

Symbiosis means "living together," and often refers to two organisms that coexist, each one supplying something that the

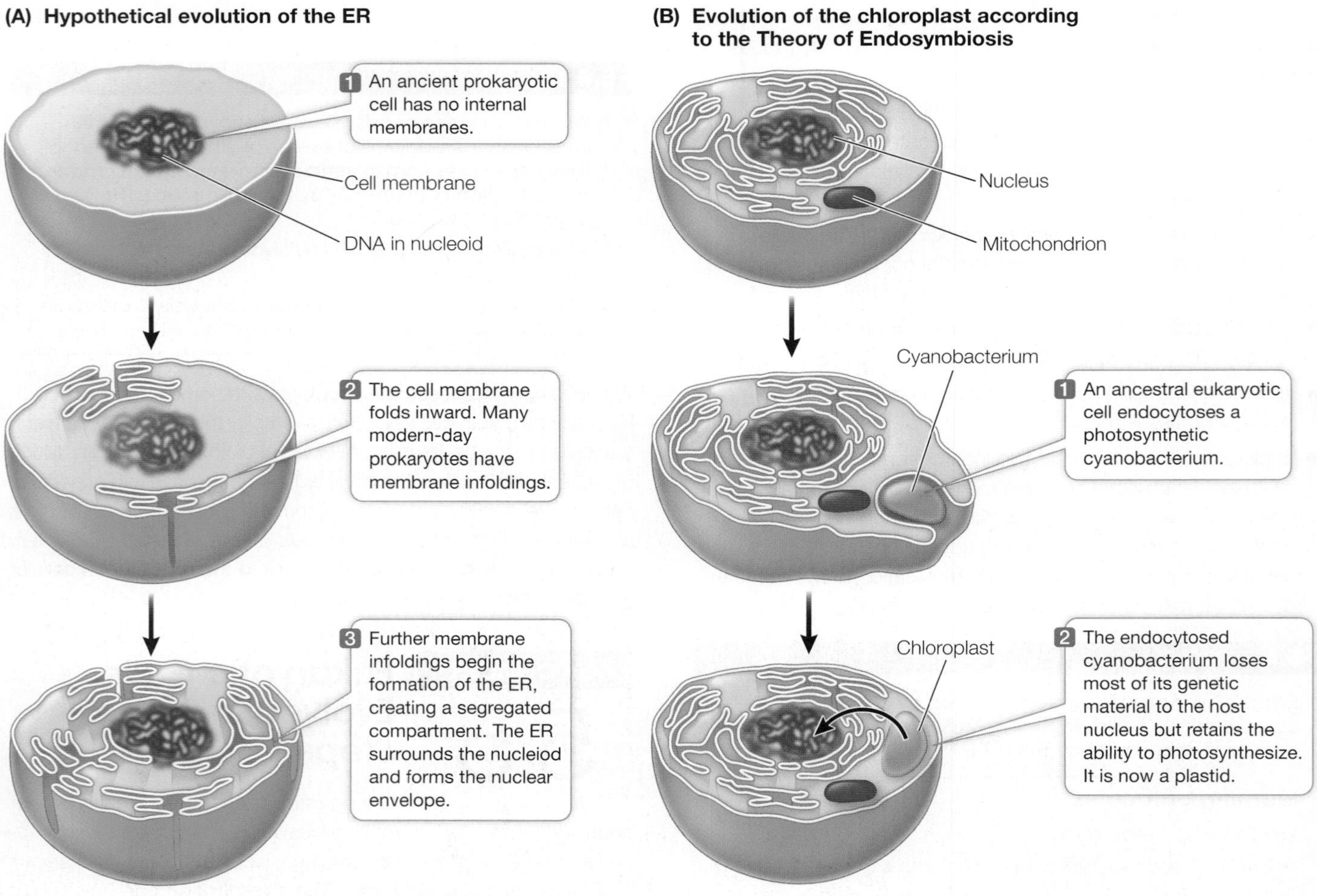

(A) Hypothetical evolution of the ER

1 An ancient prokaryotic cell has no internal membranes.

Cell membrane

DNA in nucleoid

2 The cell membrane folds inward. Many modern-day prokaryotes have membrane infoldings.

3 Further membrane infoldings begin the formation of the ER, creating a segregated compartment. The ER surrounds the nucleiod and forms the nuclear envelope.

(B) Evolution of the chloroplast according to the Theory of Endosymbiosis

Nucleus

Mitochondrion

Cyanobacterium

1 An ancestral eukaryotic cell endocytoses a photosynthetic cyanobacterium.

Chloroplast

2 The endocytosed cyanobacterium loses most of its genetic material to the host nucleus but retains the ability to photosynthesize. It is now a plastid.

Figure 5.23 The Origins of Organelles **(A)** The endomembrane system and nuclear envelope may have been formed by infolding and then fusion of the cell membrane. **(B)** The endosymbiosis theory proposes that some organelles may be descended from prokaryotes that were engulfed by other, larger cells.

other needs. With the **theory of endosymbiosis** biologists have proposed that some organelles—the mitochondria and the plastids—arose not by an infolding of the cell membrane but by one cell ingesting (but not digesting) another cell, giving rise to a symbiotic relationship. Eventually, the ingested cell lost its autonomy and some of its functions. In addition, many of the ingested cell's genes were transferred to the host's DNA. Mitochondria and plastids in today's eukaryotic cells are the remnants of these symbionts, retaining some specialized functions that benefit their host cells.

Consider the case of the plastid. About 2.5 billion years ago some prokaryotes (the cyanobacteria) developed photosynthesis. The emergence of these prokaryotes was a key event in the evolution of complex organisms because they increased the O_2 concentration in Earth's atmosphere (see Key Concept 25.2).

According to the endosymbiosis theory, photosynthetic prokaryotes also provided the precursor of the modern-day plastid. Cells without cell walls can engulf relatively large particles by phagocytosis (see Figure 5.10). In some cases, such as that of phagocytes in the human immune system, the engulfed particle can be an entire cell, such as a bacterium. Plastids may have arisen by a similar event involving an ancestral eukaryote and a cyanobacterium (**Figure 5.23B**).

5.5 recap

Eukaryotic cells arose long after prokaryotic cells. Some organelles may have evolved by infolding of the cell membrane, whereas others probably evolved by endosymbiosis.

learning outcomes

You should be able to:

- Explain how the cell membrane could have been involved in the evolution of the endomembrane system in eukaryotes.
- Relate endosymbiotic theory to observations of modern cells.

1. How could membrane infolding in a prokaryotic cell lead to the formation of the endomembrane system?
2. When they are removed from a cell, organelles such as mitochondria no longer function as they did inside the cell. What aspect of endosymbiotic theory explains this observation?

In this chapter we presented an overview of the components of cells, with some ideas about their structures, functions, and origins. As you now embark on the study of major cellular processes, keep in mind that cellular components do not exist in isolation: they are part of a dynamic, interacting system. In Chapter 6 you will see that the cell membrane is far from a passive barrier, but instead is a multifunctional system that connects the inside of the cell with its extracellular environment.

▶ investigating life

Q&A   Are pigments compartmentalized in all types of cells?

In this chapter, you learned of several situations in which pigments are concentrated in discrete organelles: in humans, melanin is packaged in melanosomes in human skin cells; in plants, tannins are packaged in tannosomes, chlorophyll is packaged in thylakoids, other pigments are contained in chromoplasts, and anthocyanins that may attract pollinators and seed dispersers are packaged in the vacuoles in petals and fruits. You learned, too, of the remarkable ability of motor proteins to move pigment granules into the center of cells in fish and amphibians, to lighten their skin color as a defensive mechanism.

Many biological traits and adaptations are linked to pigments in organelles. For example, the distribution of melanosomes is related to eye color, genetically determined errors in the formation of melanosomes can lead to albinism, and a lack of protection against ultraviolet light. In plants, the formation of tannosomes is associated with plant defenses (see Chapter 38). Understanding the pathway by which tannosomes are formed could have beneficial applications in agriculture. For example, tannins are important to the taste of red wine and manipulating tannin concentrations can result in new flavors.

Future directions

The most common form of human cancer is skin cancer. Most skin cancers are easily treated, but a rare type, melanoma, can be lethal. As its name implies, melanoma occurs when melanocytes in the skin multiply excessively. This can happen if melanocytes are exposed to ultraviolet light from the sun in excess of the ability of their melanin pigment to absorb it. As the tumor develops, it often becomes resistant to the drugs used to treat it. Recent research indicates that one way that drug resistance develops is that the drugs get sucked up by melanosomes and stay in the compartments, preventing them from getting to their targets elsewhere in the cell. Biologists are trying to develop drugs that target melanosomes to prevent this drug sequestration.

Chapter Summary 5

5.1 Cells Are the Fundamental Units of Life

- The **cell theory** is the unifying theory of cell biology. All living things are composed of cells, and all cells come from preexisting cells.

- A cell is small and maintains a large **surface area-to-volume ratio**. This allows it to exchange adequate quantities of materials with its environment. **Review Figures 5.1, 5.2, Activity 5.1**

- Cell structures can be studied with light and electron microscopes. **Review Figure 5.3, Activity 5.2**

- All cells are enclosed by a selectively permeable **cell membrane** that separates their contents from the external environment.

- Whereas certain biochemical processes, molecules, and structures are shared by all kinds of cells, there are two categories of organisms—**prokaryotes** and **eukaryotes**—that can be distinguished by characteristic cell structures.

- Eukaryotic cells are generally larger and more complex than prokaryotic cells. They contain membrane-bound **organelles**, including the **nucleus**.

5.2 Prokaryotic Cells Are the Simplest Cells

- Prokaryotic cells have no internal compartments but have a **nucleoid** region containing DNA. The rest of the material inside the cell membrane, including the **ribosomes**, is called the **cytoplasm** (or cytosol). Some prokaryotes have additional protective structures, including a **cell wall**, an **outer membrane**, and a **capsule**. **Review Figure 5.4**

- Some prokaryotes have folded **internal membranes** such as those used in photosynthesis, and some have **flagella** or **pili** for motility or attachment. **Review Figure 5.5**

- Filamentous proteins in the cytoplasm make up the **cytoskeleton**, which assists in cell division and the maintenance of cell shape.

5.3 Eukaryotic Cells Contain Organelles

See Activity 5.3

- Eukaryotic cells are larger than prokaryotic cells and contain many membrane-enclosed **organelles**. The membranes that envelop organelles ensure compartmentalization of their functions. **Review Figure 5.7**

- The **nucleus** contains most of the cell's DNA and participates in the control of protein synthesis. The DNA and the proteins associated with it form a material called **chromatin**. Each long, thin DNA molecule occurs in a discrete chromatin structure called a **chromosome**. **Review Figure 5.8**

- Within the nucleus is the **nucleolus**, where ribosome assembly begins. The **ribosomes** function as sites of protein synthesis.

- The **endomembrane system**—consisting of the cell membrane, the nuclear envelope, the **endoplasmic reticulum**, the **Golgi apparatus**, and **lysosomes**—is a series of interrelated compartments enclosed by membranes. It segregates proteins and modifies them. **Lysosomes** contain many digestive enzymes. **Review Focus: Key Figure 5.9, Figure 5.10, Activity 5.4, Animation 5.1**

- **Mitochondria** and **chloroplasts** are organelles that process energy. Mitochondria are present in all eukaryotic organisms and contain the enzymes needed for **cellular respiration**. The cells of photosynthetic eukaryotes contain chloroplasts that harvest light energy for photosynthesis. **Review Figures 5.11, 5.12**

- Large **vacuoles** are present in many plant cells. A vacuole consists of a membrane-enclosed compartment full of water and dissolved substances.

- The **microfilaments**, **intermediate filaments**, and **microtubules** of the cytoskeleton provide the cell with shape, strength, and movement. **Review Figures 5.14, 5.15, 5.16**

- **Motor proteins** use cellular energy to change shape and move. They drive the bending movements of **cilia** and **flagella**, and transport organelles along microtubules within the cell. **Review Figures 5.18, 5.19**

5.4 Extracellular Structures Have Important Roles

- The plant cell wall consists principally of cellulose. Cell walls are pierced by **plasmodesmata** that join the cytoplasms of adjacent cells.

- In animal cells, the **extracellular matrix** consists of different kinds of proteins, including **collagen** and **proteoglycans**. **Review Figure 5.22**

5.5 Eukaryotic Cells Evolved in Several Steps

- Infoldings of the cell membrane could have led to the formation of some membrane-enclosed organelles, such as the endomembrane system and the nucleus. **Review Figure 5.23A**

- **Symbiosis** means "living together." The **theory of endosymbiosis** states that mitochondria and chloroplasts originated when larger cells engulfed, but did not digest, smaller cells. Mutual benefits permitted this symbiotic relationship to be maintained, allowing the smaller cells to evolve into the eukaryotic organelles observed today. **Review Figure 5.23B**

Go to **LearningCurve** (in **LaunchPad**) for dynamic quizzing that helps you solidify your understanding of this chapter. **LearningCurve** adapts to your responses, giving you the practice you need to master each key concept.

Apply What You've Learned

Review

5.1 Membranes provide a structural role in a cell and also allow the cell to maintain homeostasis and communicate with other cells.

5.1 Microscopy allows visual examination of cells.

5.3 Each organelle carries out a different specific function that, when combined with others, allows the cell to function as a whole.

Original Paper: Hirschberg, K. et al. 1998. Kinetic analysis of secretory protein traffic and characterization of Golgi to plasma membrane transport intermediates in living cells. *The Journal of Cell Biology* 143: 1485–1503.

Some proteins naturally emit light. One example is green fluorescent protein found in the jellyfish *Aequorea victoria*, which emits a green glow from the rim of its circular body. Scientists are not sure what biological role green fluorescent protein plays in the jellyfish, though one hypothesis is that the light scares off predators. While research on this question is ongoing, green fluorescent protein has been put to work as a useful biological tool. The gene encoding the protein has been isolated and small portions of it linked to genes encoding proteins from other organisms. Expression of these fused genes produces "fusion proteins" that retain their normal biological functions but have fluorescent tags that allow the proteins to be tracked.

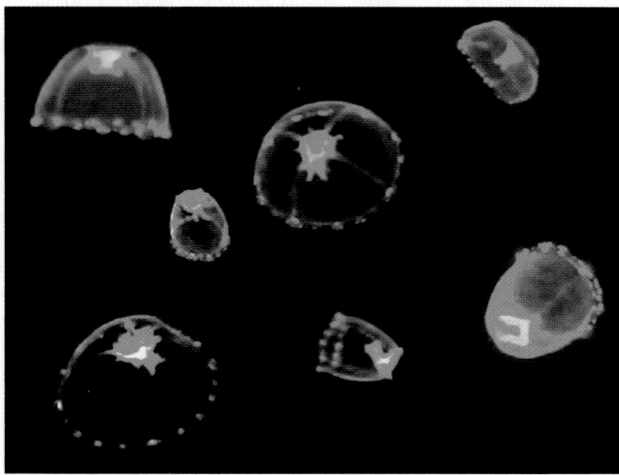

Researchers followed such a fusion protein through a mammalian cell using microscopy. They fused genetic sequences from green fluorescent protein to the gene encoding a protein from a virus. When this virus infects a cell, it directs the cell to make this protein and insert it into the cell membrane. The membrane is used to package new

viruses. By choosing this protein, the researchers identified a nice system for following a protein's movements through a cell. They used a microscope outfitted with fluorescence imaging equipment to observe changes after infecting the cell with virus. The table below summarizes the results.

Time (min)	Relative fluorescence intensity			
	ER	Golgi apparatus	Cell membrane	Total
0	0.95	0.05	0.00	1.00
20	0.64	0.28	0.08	1.00
40	0.38	0.39	0.23	1.00
60	0.17	0.38	0.44	0.99
80	0.05	0.28	0.65	0.98
100	0.00	0.25	0.70	0.95
150	0.00	0.05	0.77	0.82
200	0.00	0.00	0.75	0.75

Questions

1. Create a graph plotting the fluorescence intensity versus time for each category. Use the results to describe the pathway taken by the protein.

2. Suppose that the mammalian cells had been given a molecule that inhibited protein synthesis at the same time they were inoculated with the virus carrying the gene for the fusion protein. How would this have affected the results of the experiment?

3. Another way to track the tagged protein is cell fractionation. If cells are infected with the virus and at various times ruptured and cell fractions isolated by centrifugation, the fluorescent protein can be identified in specific fractions at each time point. Would the researchers have learned anything different using this method? Explain.

4. Green fluorescent protein is not the only fluorescent protein that biologists have used as a tag for following proteins. How do you think researchers could use fluorescent tagging to follow two different proteins simultaneously in the same cell? Could the researchers use this to show how two proteins end up in different locations?

5. Create a graph or table to predict the results of an experiment in which researchers follow the fates of two proteins after they are made on ribosomes in the rough ER: a secretory protein destined to be secreted by the cell and a peroxidase destined for the peroxisome.

Go to **LaunchPad** for the eBook, LearningCurve, animations, activities, flashcards, and additional resources and assignments.

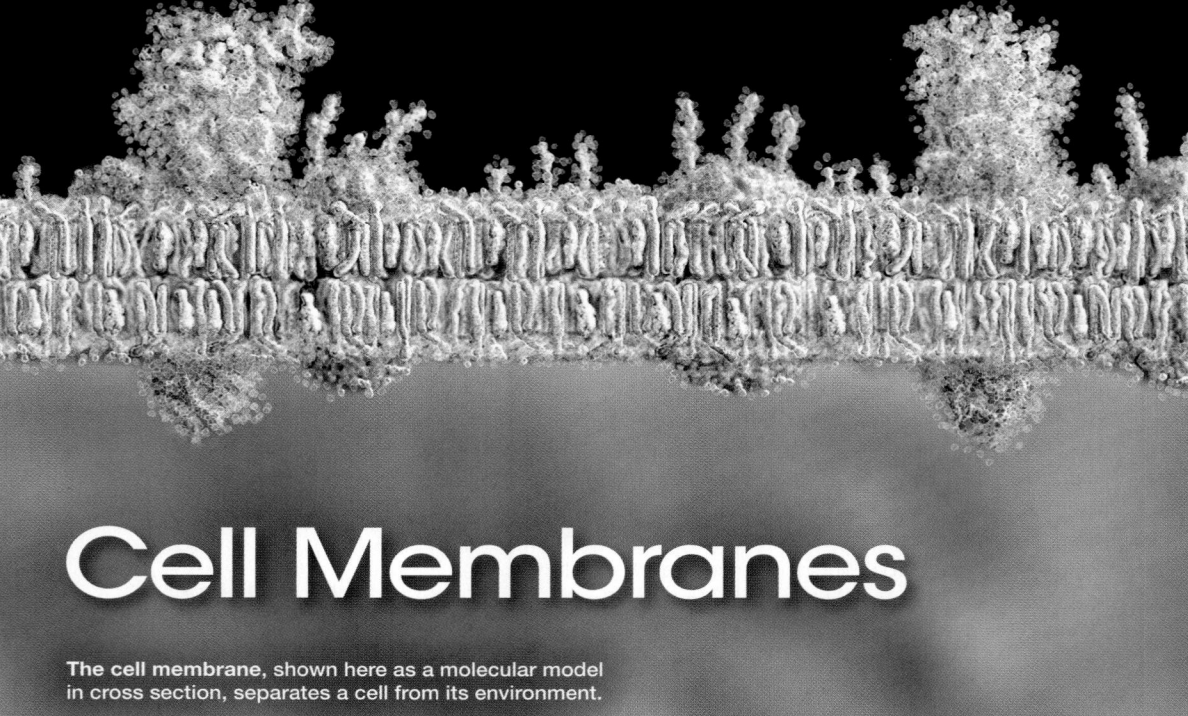

Cell Membranes

The cell membrane, shown here as a molecular model in cross section, separates a cell from its environment.

▶ investigating life

Sweating and Membranes

During World War II, Winston Churchill told the British Parliament, "I have nothing to offer but blood, toil, tears and sweat." He may not have known that the last two, tears and sweat, are transported across cell membranes inside vesicles. As you saw in Chapter 2, sweating is a way to reduce body heat by using excess heat to evaporate water. At peak activity, we may lose as much as 2 liters of water an hour, and if you know about the German air attacks on London during the war, you know that the people indeed toiled hard and must have sweated a lot.

The sweat glands lie just below the surface of the skin.

They are essentially cell-lined tubes surrounded by extracellular fluid. When sweating is triggered, these tubes fill with water and dissolved substances. To get from the extracellular fluid into the tube, water must go through the cells that line the tube.

A hallmark of living cells is the ability to regulate what enters and leaves the cytoplasm. This

is a function of the cell membrane, a hydrophobic lipid bilayer with associated proteins. Because it is insoluble in the aqueous environment both inside and outside cells, the membrane is a physical barrier. But it is also a functional barrier. Whereas water is polar, the interior of the membrane is nonpolar—so water has a natural tendency to avoid the membrane. The rate of movement of water across a lipid bilayer is modest. When you engage in normal activities such as reading this book, the cell membranes enclosing the cells lining the sweat glands do not allow much water to enter or leave. But when you exercise vigorously, tiny membrane-enclosed vesicles inside the cells fill with water and some dissolved salts. In a process called exocytosis, these vesicles fuse with the cell membrane and release their watery contents (sweat) into the tubes. From there the sweat flows to the surface of the skin and evaporates.

Vesicles are not the only way to get polar water across a nonpolar membrane. In the mammalian kidney, and in plant roots, stems, and leaves, special pores called aquaporins occur in the cell membrane. Water can flow through them readily because the proteins lining the channel have a hydrophilic inner surface.

 What is the importance of aquaporin

key concept
6.1

Biological Membranes Are Lipid–Protein Bilayers

The physical organization and functioning of all biological membranes depend on their constituents: lipids, proteins, and carbohydrates. You are already familiar with these molecules from Chapter 3 and with membranes that enclose cells and organelles from Chapter 5. Lipids establish the physical integrity of the membrane and create a barrier to the rapid passage of hydrophilic materials such as water and ions. In addition, the phospholipid bilayer serves as a lipid "lake" in which a variety of proteins "float" (**Focus: Key Figure 6.1**). This general design is known as the **fluid mosaic model**. It is *mosaic* because it is made up of many discrete components, and *fluid* because the components can move freely.

focus your learning

- A lipid bilayer forms as the result of two kinds of interactions: hydrophobic interactions between nonpolar tails of phospholipids, and hydrophilic interactions between polar heads of phospholipids and water molecules.
- Proteins associated with membranes may be embedded in the lipid bilayer or associated with exposed parts of the membrane.
- Membranes are dynamic structures that undergo constant change.
- The degree of a membrane's fluidity is influenced by lipid composition and temperature.

The fluid mosaic model depicts proteins as noncovalently embedded in the phospholipid bilayer by their hydrophobic regions (or domains) or tethered to lipids inserted into the membrane. Proteins may span the membrane or may be bound on the surface. Their hydrophilic regions are exposed to the watery conditions on either side of the bilayer. Membrane proteins have several functions, including moving materials through the membrane and receiving chemical signals from the cell's external environment. Each membrane has a set of proteins suitable for the specialized functions of the cell or organelle it surrounds.

The carbohydrates associated with membranes are attached either to the lipids or to protein molecules. In cell membranes, carbohydrates are located on the outside of the cell, where they may interact with substances in the external environment. Like some of the membrane proteins, carbohydrates are crucial in recognizing specific molecules, such as those on the surfaces of adjacent cells.

Although the fluid mosaic model is largely valid for membrane structure, it does not say much about membrane composition. As you read about the various molecules in membranes in the next sections, keep in mind that some membranes have more protein than lipids, others are lipid-rich, others have significant amounts of cholesterol or other sterols, and still others are rich in carbohydrates.

Lipids form the hydrophobic core of the membrane

The lipids in biological membranes are usually *phospholipids. Recall from Key Concept 2.2 that some compounds are hydrophilic ("water-loving") and others are hydrophobic ("water-hating"), and from Key Concept 3.4 that a phospholipid molecule has regions of both kinds:

1. *Hydrophilic regions*: The phosphate-containing "head" of the phospholipid is electrically charged and therefore associates with polar water molecules.
2. *Hydrophobic regions*: The long, nonpolar fatty acid "tails" of the phospholipid associate with other nonpolar materials; they do not dissolve in water or associate with hydrophilic substances.

---

*connect the concepts   The straight chains of fatty acids allow phospholipids to pack closely together. Review the molecular structure and properties of phospholipids in Key Concept 3.4.

The chemical properties of phospholipids are such that when phospholipids coexist with water, they form a bilayer, with the fatty acid "tails" of the two layers interacting with each other and the polar "heads" facing the outside aqueous environment (**Figure 6.2**). The thickness of a biological membrane is about 8 nanometers (0.008 μm), which is twice the length of a typical phospholipid—another indication that the membrane consists of a lipid bilayer. To give you an idea of the scale, a typical sheet of paper is about 8,000 times thicker than this.

All biological membranes have a similar structure, but they differ in the kinds of proteins and lipids they contain. Membranes from different cells or organelles may differ greatly in their lipid composition. Phospholipids can differ in terms of fatty acid chain length (number of carbon atoms), degree of unsaturation (number of double bonds) in the fatty acids, and the polar groups present (see Chapter 3). The saturated chains allow close packing of fatty acids in the bilayer, whereas the "kinks" in unsaturated fatty acids (see Figure 3.21) make for a less dense, more fluid packing.

Up to 25 percent of the lipid content of an animal cell's cell membrane may be the steroid cholesterol (see Key Concept 3.4). Cholesterol preferentially associates with saturated fatty acids. When present, cholesterol is important for membrane integrity; the cholesterol in your membranes is not hazardous to your health.

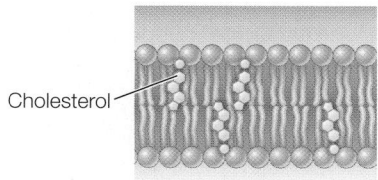

Cholesterol

The fatty acids of the phospholipids make the hydrophobic interior of the membrane somewhat fluid—about as fluid as lightweight olive oil. This fluidity permits some molecules to move laterally within the plane of the membrane.

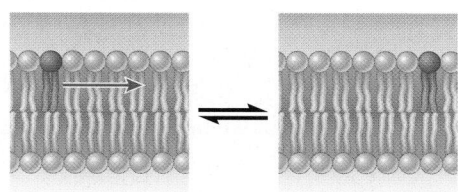

focus: key figure

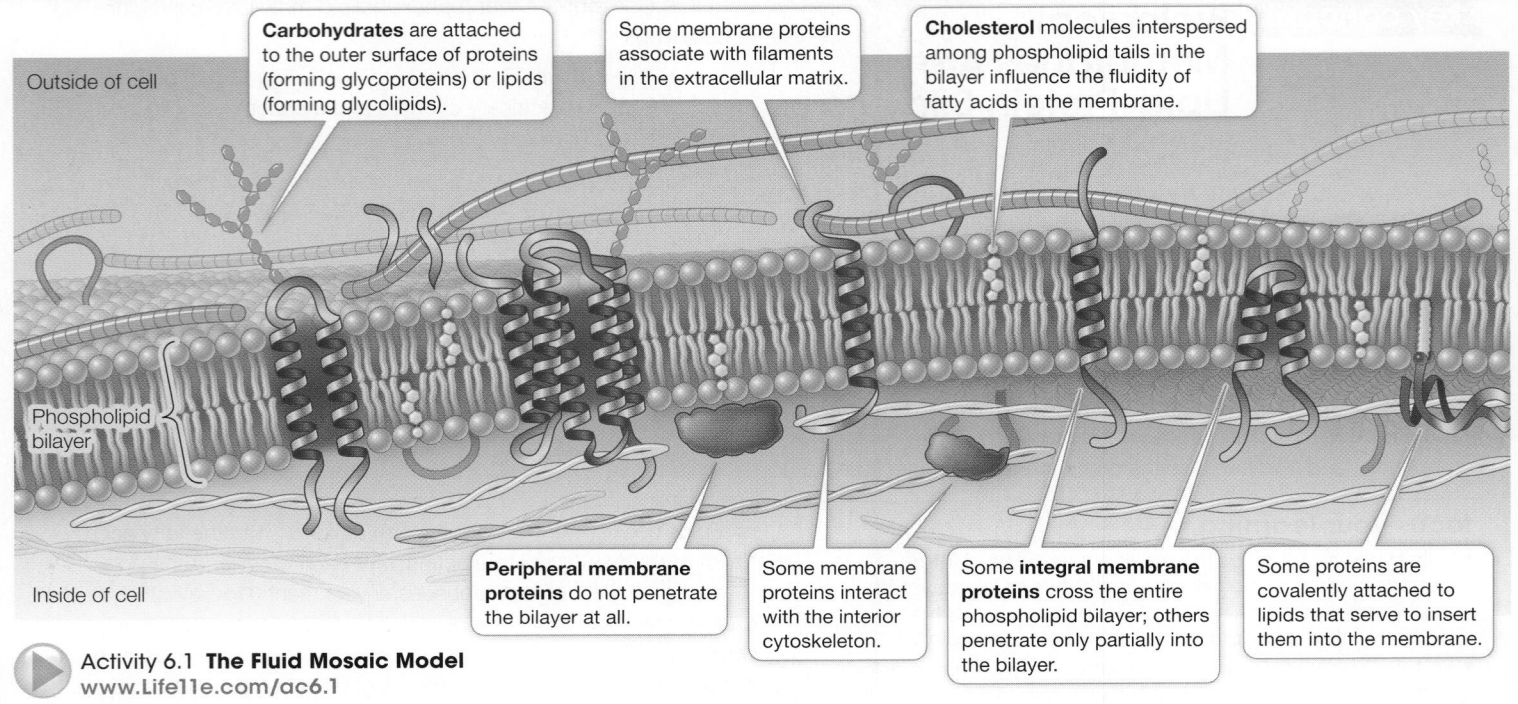

Carbohydrates are attached to the outer surface of proteins (forming glycoproteins) or lipids (forming glycolipids).

Some membrane proteins associate with filaments in the extracellular matrix.

Cholesterol molecules interspersed among phospholipid tails in the bilayer influence the fluidity of fatty acids in the membrane.

Outside of cell

Phospholipid bilayer

Inside of cell

Peripheral membrane proteins do not penetrate the bilayer at all.

Some membrane proteins interact with the interior cytoskeleton.

Some integral membrane proteins cross the entire phospholipid bilayer; others penetrate only partially into the bilayer.

Some proteins are covalently attached to lipids that serve to insert them into the membrane.

Activity 6.1 **The Fluid Mosaic Model**
www.Life11e.com/ac6.1

Figure 6.1 The Fluid Mosaic Model   The general molecular structure of a biological membrane is a continuous phospholipid bilayer that has proteins embedded in it or associated with it.

Q: What kinds of chemical interactions hold some membrane proteins embedded in the membrane and others on the membrane surface?

A given phospholipid molecule in the cell membrane can travel from one end of the cell to the other in a little more than a second! However, a phospholipid molecule in one half of the bilayer is unlikely to spontaneously flip over to the other side. For that to happen, the polar part of the molecule would have to move through the hydrophobic interior of the membrane. Since spontaneous phospholipid flip-flops are rare, the inner and outer halves of the bilayer may be quite different in the kinds of phospholipids they contain.

Membrane fluidity is affected by several factors, two of which are particularly important:

3. *Lipid composition*: Cholesterol and long-chain, saturated fatty acids pack tightly beside one another, with little room for movement. This close packing results in less-fluid membranes. A membrane with shorter-chain fatty acids, unsaturated fatty acids, or less cholesterol is more fluid.

4. *Temperature*: Because molecules move more slowly and fluidity decreases at reduced temperatures, cellular processes that take place within the membrane may slow down or stop under cold conditions in organisms that cannot keep their bodies warm. To address this problem, in some organisms the lipid composition of their membranes changes when they get cold, replacing saturated with unsaturated fatty acids and using fatty acids with shorter tails. These changes play a role in the survival of plants, bacteria, and hibernating animals during the winter.

Activity 6.2 **Lipid Bilayer Composition Simulation**
www.Life11e.com/ac6.2

Membrane proteins are asymmetrically distributed

All biological membranes contain proteins. Typically, cell membranes have 1 protein molecule for every 25 lipid molecules. This ratio varies depending on membrane function. In the inner membrane of the mitochondrion, which is specialized for energy processing, there is 1 protein for every 15 lipids. However, myelin—a modified cell membrane that encloses portions of some neurons (nerve cells) and acts as an electrical insulator—has only 1 protein for every 70 lipids.

Membrane proteins are very diverse. In fact, about one-fourth of the protein-coding genes in the eukaryotic genome encode membrane proteins. There are two general types of membrane proteins: integral proteins and peripheral proteins.

Integral membrane proteins are at least partly embedded in the phospholipid bilayer (see Figure 6.1). Like phospholipids, these proteins have both hydrophilic and hydrophobic regions (domains) (**Figure 6.3**).

1. *Hydrophilic domains*: Stretches of amino acids with hydrophilic side chains (R groups; see Table 3.2) give certain regions of the protein a polar character. These hydrophilic domains interact with water and stick out into the aqueous environment inside or outside the cell.

2. *Hydrophobic domains*: Stretches of amino acids with hydrophobic side chains give other regions of the protein a nonpolar character. These domains interact with the fatty acids in the interior of the phospholipid bilayer, away from water. Some proteins contain covalently linked lipids such as fatty acid chains

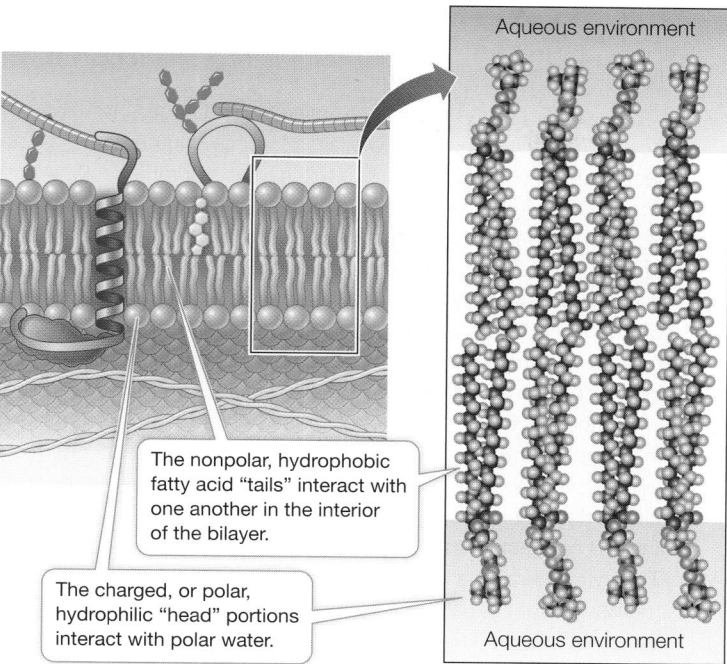

The nonpolar, hydrophobic fatty acid "tails" interact with one another in the interior of the bilayer.

The charged, or polar, hydrophilic "head" portions interact with polar water.

Figure 6.2 Phospholipid Bilayer   The phospholipid bilayer separates two aqueous regions. The eight phospholipid molecules shown on the right represent a small cross section of a membrane bilayer.

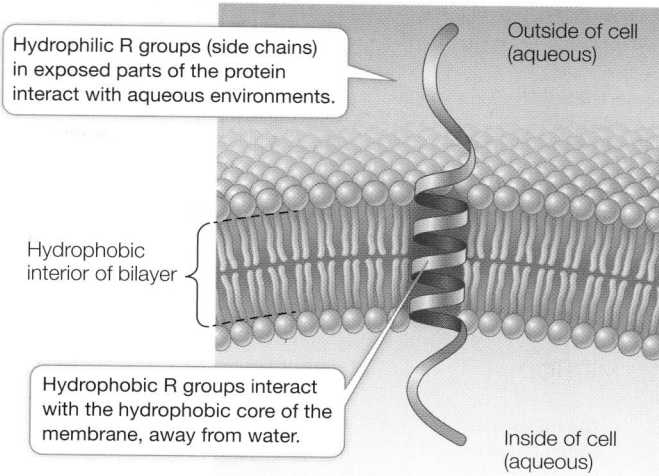

Hydrophilic R groups (side chains) in exposed parts of the protein interact with aqueous environments.

Hydrophobic interior of bilayer

Hydrophobic R groups interact with the hydrophobic core of the membrane, away from water.

Outside of cell (aqueous)

Inside of cell (aqueous)

Figure 6.3 Interactions of Integral Membrane Proteins   An integral membrane protein is held in the membrane by the distribution of the hydrophilic and hydrophobic side chains on its amino acids. The hydrophilic parts of the protein extend into the aqueous cell exterior and the internal cytoplasm. The hydrophobic side chains interact with the hydrophobic lipid core of the membrane.

that do the job of inserting into the membrane (instead of the protein having hydrophobic amino acid regions).

Peripheral membrane proteins lack exposed hydrophobic groups and are not embedded in the bilayer. Instead, they have polar or charged regions that interact with exposed parts of integral membrane proteins, or with the polar heads of phospholipid molecules (see Figure 6.1).

A special preparation method for electron microscopy, called freeze-fracturing, reveals proteins that are embedded in the phospholipid bilayers of cellular membranes (**Figure 6.4**). When the two lipid leaflets (or layers) that make up the bilayer are separated, the proteins can be seen as bumps that protrude from the interior of each membrane. The bumps are not observed when artificial bilayers of pure lipid are freeze-fractured.

Membrane proteins and lipids generally interact only noncovalently. The polar ends of proteins can interact with the polar ends of lipids, and the nonpolar regions of both molecules can interact hydrophobically. As mentioned above, however, some membrane

research tools

Figure 6.4 Membrane Proteins Revealed by the Freeze-Fracture Technique   This HeLa cell (a human cell) membrane was first frozen to immobilize the lipids and proteins, and then fractured so that the bilayer was split open.

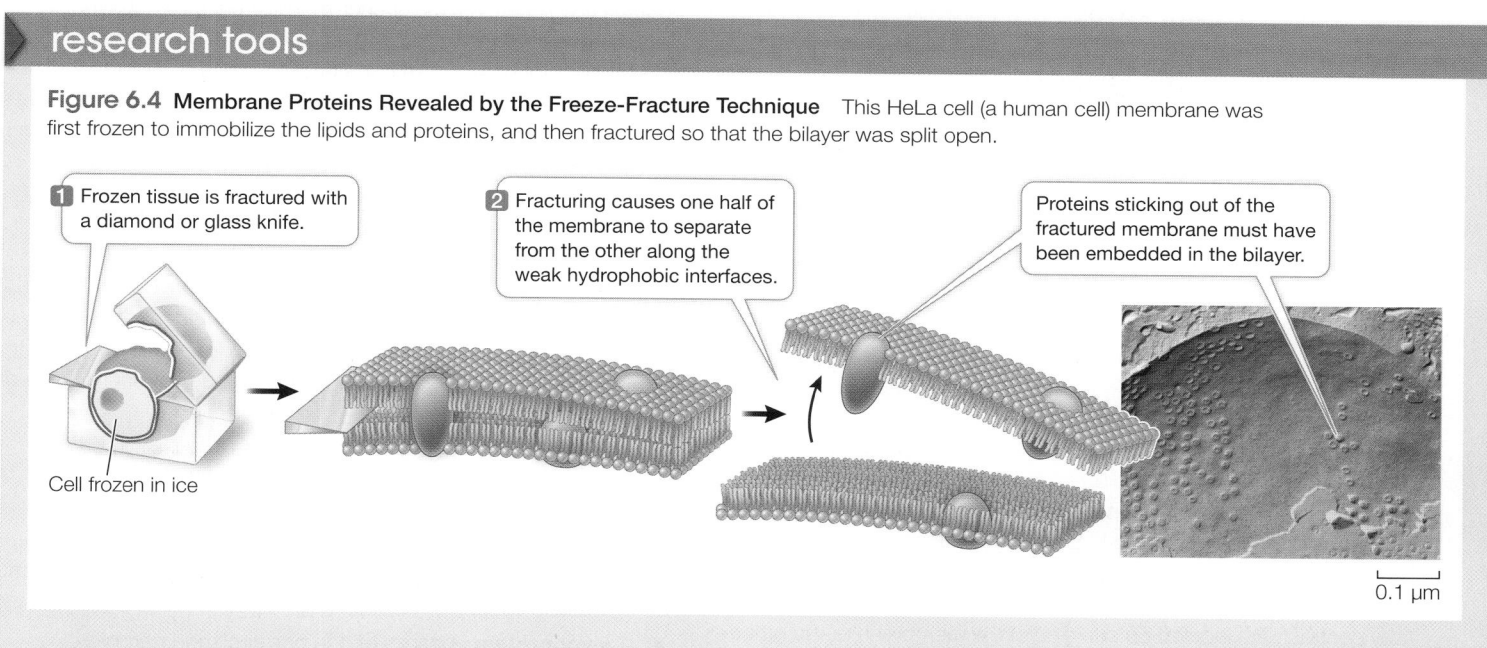

1 Frozen tissue is fractured with a diamond or glass knife.

2 Fracturing causes one half of the membrane to separate from the other along the weak hydrophobic interfaces.

Proteins sticking out of the fractured membrane must have been embedded in the bilayer.

Cell frozen in ice

0.1 μm

experiment

Figure 6.5   Rapid Diffusion of Membrane Proteins

Original Paper: Frye, L. D. and M. Edidin. 1970. The rapid intermixing of cell surface antigens after formation of mouse-human heterokaryons. *Journal of Cell Science* 7: 319–335.

Two animal cells can be fused together in the laboratory, forming a single large cell (heterokaryon). This phenomenon was used to test whether membrane proteins can diffuse independently in the plane of the cell membrane.

HYPOTHESIS▶ Proteins embedded in a membrane can diffuse freely within the membrane.

METHOD

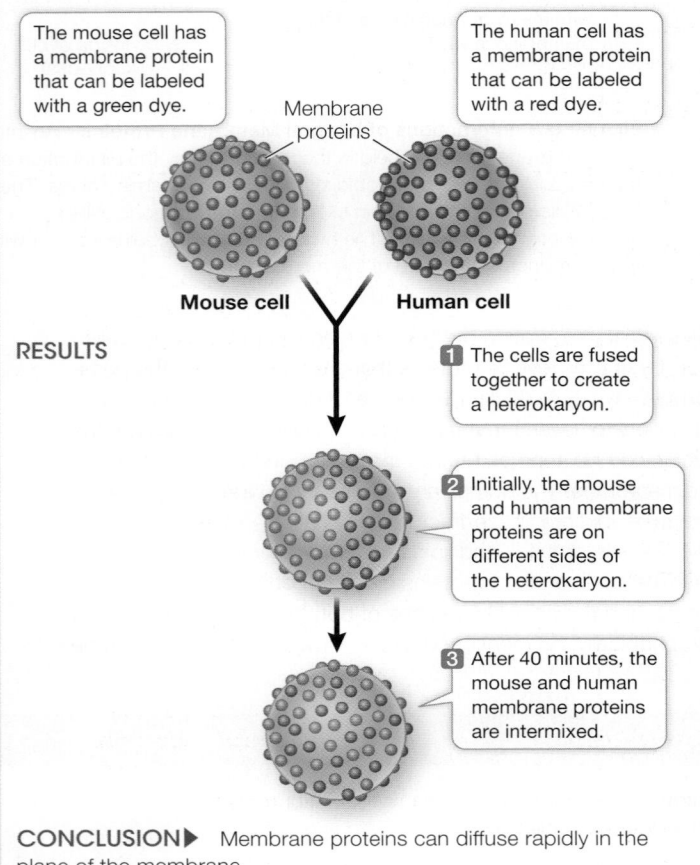

The mouse cell has a membrane protein that can be labeled with a green dye.

The human cell has a membrane protein that can be labeled with a red dye.

Membrane proteins

Mouse cell   **Human cell**

RESULTS

1 The cells are fused together to create a heterokaryon.

2 Initially, the mouse and human membrane proteins are on different sides of the heterokaryon.

3 After 40 minutes, the mouse and human membrane proteins are intermixed.

CONCLUSION▶ Membrane proteins can diffuse rapidly in the plane of the membrane.

---

proteins have hydrophobic lipid components covalently attached to them that allow these proteins to tether themselves to the phospholipid bilayer.

Proteins are asymmetrically distributed on the inner and outer surfaces of membranes. An integral protein that extends all the way through the phospholipid bilayer and protrudes on both sides is known as a **transmembrane protein**. In addition to one or more **transmembrane domains** that extend through the bilayer, such a protein may have domains with other specific functions on the inner and outer sides of the membrane. Peripheral membrane proteins, by contrast, are located on one side of the membrane or the other. This asymmetrical arrangement of membrane proteins gives the two

surfaces of the membrane different properties. As you will soon see, these differences have great functional significance.

Like lipids, some membrane proteins move around relatively freely within the phospholipid bilayer. Experiments that involve the technique of cell fusion illustrate this migration dramatically. When two cells are fused, a single continuous membrane forms and surrounds both cells, and some proteins from each cell distribute themselves uniformly around this membrane (**Figure 6.5**).

Although some proteins are free to migrate in the membrane, others are not, but rather appear to be "anchored" to a specific region of the membrane. These membrane regions are like an enclosure of animals at the zoo: the animals are free to move around within the fenced area but not outside it. An example is the protein in the cell membrane of a muscle cell that recognizes a chemical signal from a neuron. This protein is normally found only at the specific region where the neuron meets the muscle cell. Proteins inside the cell can restrict the movement of proteins within a membrane. The cytoskeleton has components just below the inner face of the membrane that are attached to membrane proteins protruding into the cytoplasm. The stability of the cytoskeletal components may thus restrict movement of attached membrane proteins.

Membranes are constantly changing

Membranes in eukaryotic cells are constantly forming, transforming from one type to another, fusing with one another, and breaking down. As we discussed in Chapter 5, fragments of membrane move, in the form of vesicles, from the endoplasmic reticulum (ER) to the Golgi apparatus, and from the Golgi apparatus to the cell membrane (see Figure 5.9). Secondary lysosomes form when primary lysosomes from the Golgi apparatus fuse with phagosomes from the cell membrane (see Figure 5.10).

Because all membranes appear similar under the electron microscope, and because they interconvert readily, we might expect all subcellular membranes to be chemically identical. However, that is not the case: there are major chemical differences among the membranes of even a single eukaryotic cell. Membranes are changed chemically when they form parts of certain organelles. In the Golgi apparatus, for example, the membranes of the *cis* face closely resemble those of the ER in chemical composition, but those of the *trans* face are more similar to the cell membrane.

Cell membrane carbohydrates are recognition sites

In addition to lipids and proteins, the cell membrane contains carbohydrates (see Figure 6.1). The carbohydrates are located on the outer surface of the cell membrane and serve as recognition sites for other cells and molecules, as you will see in Key Concept 6.2.

Membrane-associated carbohydrates may be covalently bonded to lipids or to proteins:

- A **glycolipid** consists of a carbohydrate covalently bonded to a lipid. Extending out from the cell surface, the carbohydrate may serve as a recognition signal for interactions between cells. For example, the carbohydrates on some glycolipids change when cells become cancerous. This change may allow white blood cells to target cancer cells for destruction.

- A **glycoprotein** consists of one or more short carbohydrate chains covalently bonded to a protein. The bound carbohydrates are oligosaccharides, usually not exceeding 15 monosaccharide units in length (see Key Concept 3.3). A proteoglycan (see Key Concept 5.4) is a more heavily glycosylated protein: it has more carbohydrate molecules attached to it, and the carbohydrate chains are often longer than they are in glycoproteins. The carbohydrates of glycoproteins and proteoglycans often function in cell recognition and adhesion.

The "alphabet" of monosaccharides on the outer surfaces of membranes can generate a large diversity of messages. Recall from Key Concept 3.3 that monosaccharides are simple carbohydrates, often containing five or six carbons in a ring structure, which can bond with one another in various configurations. They may form linear or branched oligosaccharides with many different three-dimensional shapes. An oligosaccharide of a specific shape on one cell can bind to a complementary shape on an adjacent cell. This binding is the basis of cell–cell adhesion.

6.1 recap

The fluid mosaic model applies to the cell membrane and the membranes of organelles. In the aqueous environment of the cell, phospholipids spontaneously assemble into a bilayer. The cell membrane is considered fluid because the phospholipids and some of the proteins that compose it can move within their monolayer. An integral membrane protein has both hydrophilic and hydrophobic domains, which affect its position and function in the membrane. Peripheral membrane proteins are bound to one side or the other of the membrane. Carbohydrates that attach to lipids and proteins on the outside of the membrane serve as recognition and adhesion sites for adjacent cells.

learning outcomes

You should be able to:

- Explain how the hydrophobic and hydrophilic regions of phospholipids are involved in membrane formation.
- Differentiate between integral proteins and peripheral proteins.
- Design an experiment to compare the properties of membranes in different types of cells.
- Compare and contrast information provided by freeze-fracturing techniques and cell fusion experiments.

1. How do the hydrophobic and hydrophilic regions of phospholipids form a membrane bilayer?

2. What differentiates an integral protein from a peripheral protein?

3. What information about membranes is derived from freeze-fracturing and cell fusion experiments?

4. When a normal lung cell becomes a lung cancer cell, there are several important changes in cell membrane properties. How would you investigate the observation that the cancer cell membrane is more fluid, with more rapid diffusion in the plane of the membrane of both lipids and proteins?

Now that you understand the structure of biological membranes, let's see how their components function. In the next section we'll focus on the membrane that surrounds individual cells: the cell membrane. We'll then look at how the cell membrane allows individual cells to be grouped together into multicellular systems of tissues.

key concept 6.2 The Cell Membrane Is Important in Cell Adhesion and Recognition

Often the cells of multicellular organisms exist in specialized groups with similar functions, called tissues. Your body has about 60 trillion cells organized into various kinds of tissues—such as muscle, nerve, and epithelium. Two processes allow cells to arrange themselves in groups:

1. **Cell recognition**, in which one cell specifically recognizes and binds to another cell of a certain type

2. **Cell adhesion**, in which the connection between the two cells is strengthened

Both processes involve the cell membrane. One way to study these processes is to break down a tissue into its individual cells and then allow them to adhere to one another again. This type of experiment is most easily done in relatively simple organisms, such as sponges, which provide good models for studying processes that also occur in the complex tissues of larger species.

focus your learning

- Cell adhesion and cell recognition are specific and depend on protein and carbohydrate molecules present on the cell membrane.
- After two cells recognize and bind to one another, they can contribute additional material to form stable cell junctions that enhance protective, structural, or communication functions in the organism.

Sponges are multicellular marine animals that have only a few distinct cell layers (see Key Concept 30.5). The cells of a sponge adhere to one another but can be separated mechanically by passing the animal several times through a fine wire screen (**Figure 6.6**). Through this process, what was a single animal becomes hundreds of individual cells suspended in seawater. If such cells are stirred gently for a few hours, cell recognition occurs: the cells bump into and recognize one another, sticking together in the same shape and tissue organization as the original sponge. This recognition is species-specific; if disaggregated sponge cells from two different species are placed in the same container and shaken, individual cells will stick only to other cells of the same species. Two different sponges form, just like the ones at the start of the experiment.

Such tissue-specific and species-specific cell recognition and cell adhesion are essential to the formation and maintenance of tissues in multicellular organisms. Think of your own body. What keeps muscle cells bound to muscle cells and skin to skin? Specific cell adhesion is so obvious a characteristic of complex organisms that it is easy to overlook. You will see many examples of specific cell adhesion throughout this book; here we describe its general principles. As you will see, cell recognition and cell adhesion depend on cell membrane proteins.

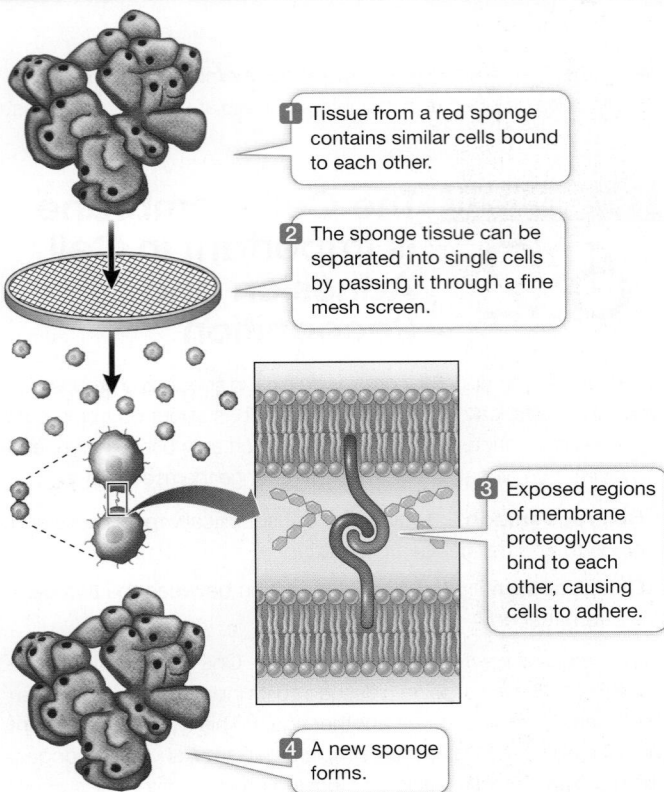

1. Tissue from a red sponge contains similar cells bound to each other.

2. The sponge tissue can be separated into single cells by passing it through a fine mesh screen.

3. Exposed regions of membrane proteoglycans bind to each other, causing cells to adhere.

4. A new sponge forms.

Figure 6.6 Cell Recognition and Adhesion   In most cases (including the aggregation of animal cells into tissues), the binding between molecules is homotypic (same to same).

Cell recognition and adhesion involve proteins and carbohydrates at the cell surface

The molecules responsible for cell recognition and adhesion in sponges are proteoglycans (often 80% carbohydrate by molecular weight) that carry two kinds of carbohydrates. One kind is relatively small and binds to membrane components, keeping the proteoglycan attached to the cell. The other kind of carbohydrate is a larger, sulfated polysaccharide. If the sulfated polysaccharide from a particular species of sponge is purified and attached to cellulose beads, the beads will aggregate together or with sponge cells—but only with cells of the same species from which the polysaccharide was purified. This demonstrates that the sulfated polysaccharide is responsible for both the specific recognition and adhesion of the sponge cells.

Cell adhesion can result from interactions between the carbohydrates that are parts of glycolipids, glycoproteins, or proteoglycans—as is the case in sponge cells. In other cases, a carbohydrate on one cell interacts with a membrane protein on another cell. Or two proteins can interact directly. As we described in Key Concept 3.2, a protein not only has a specific shape, it also has specific chemical groups exposed on its surface where they can interact with other substances, including other proteins. Both of these features allow binding to other specific molecules. Cell adhesion occurs in all kinds of multicellular organisms. In plants, cell adhesion may be mediated by both integral membrane proteins and specific carbohydrates in the cell walls.

In most cases, the binding of cells in a tissue is **homotypic**; that is, the same molecule sticks out of both cells, and the exposed surfaces fit together and bind to each other. This is a factor that keeps skin cells together in a sheet of cells. But **heterotypic** binding (between different molecules on different cells) also occurs. In this case, chemical groups on different surface molecules have an affinity for one another. For example, when the mammalian sperm meets the egg, different proteins on the two types of cells have complementary binding surfaces. Similarly, some algae form male and female reproductive cells (analogous to sperm and eggs) that have flagella to propel them toward each other. Male and female cells can recognize each other by heterotypic glycoproteins on their flagella.

Three types of cell junctions connect adjacent cells

In a complex multicellular organism, cell recognition molecules allow specific types of cells to bind to one another. Often, after the initial binding, both cells contribute material to form additional membrane structures that connect them to one another. These specialized structures, called **cell junctions**, are most evident in electron micrographs of epithelial tissues, which are layers of cells that line body cavities or cover body surfaces. These surfaces often receive stresses or must retain their contents under pressure, so it is particularly important that their cells adhere tightly. We will examine three types of cell junctions that enable animal cells to seal intercellular spaces, reinforce attachments to one another, and communicate with each other. Tight junctions, desmosomes, and gap junctions, respectively, perform these three functions (**Figure 6.7**).

1. **Tight junctions** prevent substances from moving through the spaces between cells. For example, cells lining the bladder have tight junctions so urine cannot leak out into the body cavity. Another important function of tight junctions is to maintain distinct faces of a cell within a tissue by restricting the migration of membrane proteins over the cell surface from one face to the other. This may result in a certain function (e.g., endocytosis) being confined to one region of the cell surface.

2. **Desmosomes** hold neighboring cells firmly together, acting like spot welds or rivets. Materials can still move around in the extracellular matrix. This provides mechanical stability for tissues such as skin that receive physical stress.

3. **Gap junctions** are channels that run between membrane pores in adjacent cells, allowing substances to pass between cells. In the heart, for example, gap junctions allow the rapid spread of electric current (mediated by ions) so the heart muscle cells beat in unison.

Cell membranes adhere to the extracellular matrix

In Key Concept 5.4 you learned about the extracellular matrix of animal cells, which is composed of collagen protein arranged in fibers in a gelatinous matrix of proteoglycans. The attachment of a cell to the extracellular matrix is important in maintaining the integrity of a tissue. In addition, some cells can detach from their neighbors, move, and attach to other cells; this is often mediated by interactions with the extracellular matrix.

A transmembrane protein called **integrin** often mediates the attachment of epithelial cells to the extracellular matrix (**Figure 6.8A**). More than 24 different integrins have been described in human cells.

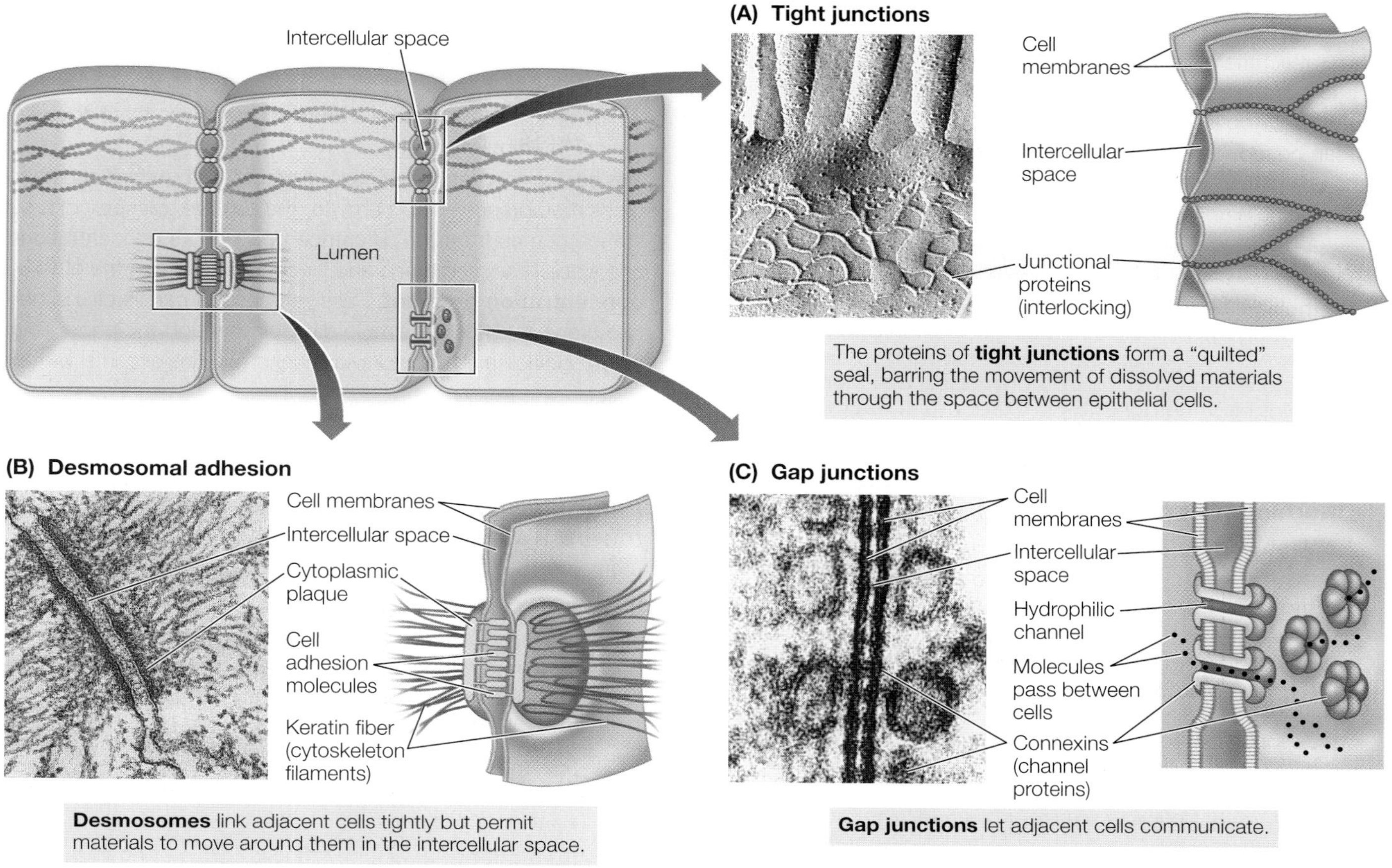

(A) Tight junctions

Intercellular space

Lumen

Cell membranes

Intercellular space

Junctional proteins (interlocking)

The proteins of **tight junctions** form a "quilted" seal, barring the movement of dissolved materials through the space between epithelial cells.

(B) Desmosomal adhesion

Cell membranes

Intercellular space

Cytoplasmic plaque

Cell adhesion molecules

Keratin fiber (cytoskeleton filaments)

Desmosomes link adjacent cells tightly but permit materials to move around them in the intercellular space.

(C) Gap junctions

Cell membranes

Intercellular space

Hydrophilic channel

Molecules pass between cells

Connexins (channel proteins)

Gap junctions let adjacent cells communicate.

Figure 6.7 **Junctions Link Animal Cells Together**   Tight junctions **(A)** and desmosomes **(B)** are abundant in epithelial tissues. Gap junctions **(C)** are also found in some muscle and nerve tissues, in which rapid communication between cells is important. Although all three junction types are shown in the cell at the top left, all three are not necessarily seen at the same time in actual cells.

▶ **Activity 6.3 Animal Cell Junctions**
www.Life11e.com/ac6.3

All of them bind to a protein in the extracellular matrix outside the cell, and to actin filaments, which are part of the cytoskeleton, inside the cell. So in addition to adhesion, integrin has a role in maintaining cell structure via its interaction with the cytoskeleton.

The binding of integrin to the extracellular matrix is noncovalent and reversible. When a cell moves its location within a tissue or organism, one side of the cell detaches from the extracellular matrix while the other side extends in the direction of movement, forming

(A)

Inside of cell

Actin

Integrin has binding sites for the cell cytoskeleton and for the extracellular matrix; the cell is bound to the matrix.

Integrin

Extracellular matrix

Outside of cell

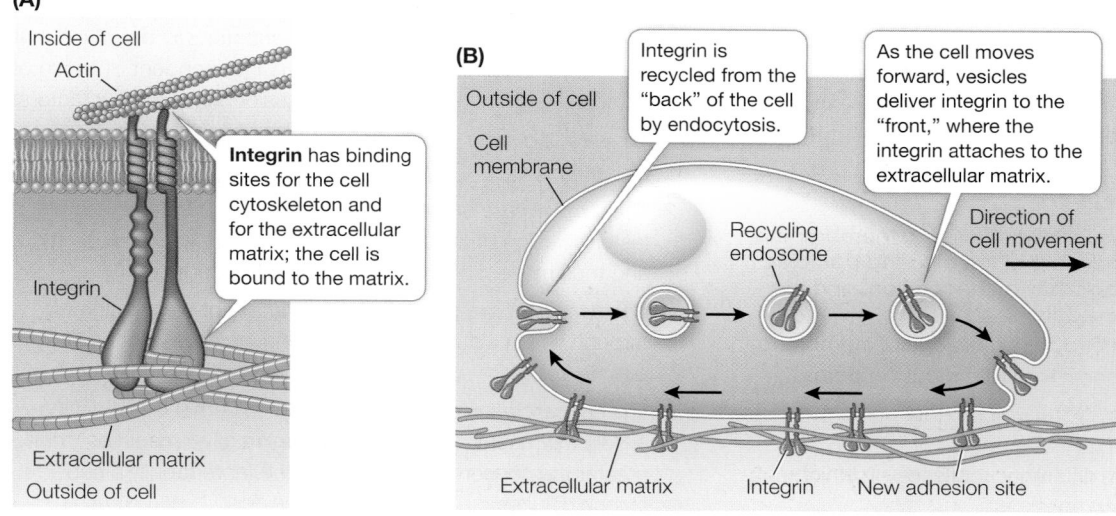

(B)

Outside of cell

Cell membrane

Integrin is recycled from the "back" of the cell by endocytosis.

As the cell moves forward, vesicles deliver integrin to the "front," where the integrin attaches to the extracellular matrix.

Recycling endosome

Direction of cell movement

Extracellular matrix   Integrin   New adhesion site

Figure 6.8 **Integrins and the Extracellular Matrix**   **(A)** Integrins mediate the attachment of cells to the extracellular matrix. **(B)** Cell movements are mediated by integrin attachment.

new attachments in that direction (**Figure 6.8B**). The integrin at the "back" of the cell (away from the direction of movement) is brought into the cytoplasm by endocytosis (see Key Concept 6.5) so that it can be recycled and used for new attachments at the "front" of the cell. These events are important for cell movement within the developing embryo and for the spread of cancer cells.

6.2 recap

In multicellular organisms, cells arrange themselves into tissues via the processes of cell recognition and cell adhesion. These processes are mediated by membrane-associated proteins and carbohydrates. Cell membrane proteins also interact with the extracellular matrix. Cell junctions assist in strengthening tissues and allow cells to communicate with one another.

learning outcomes

You should be able to:

- Design investigations to analyze the function of proteins and carbohydrates in cell adhesion and cell recognition.
- Compare and contrast the three types of cell junctions found in animals.

1. Describe an experiment showing a role for carbohydrates in cell recognition and adhesion.
2. To which of the three types of cell junctions are plasmodesmata most similar (see Figure 5.7B)?

You have seen how the cell membrane and molecules associated with it facilitate binding between cells and the maintenance of cell adhesion. We'll turn now to another major function of membranes: regulating the substances that enter or leave a cell or organelle.

key concept 6.3 Substances Can Cross Membranes by Passive Processes

As you have already learned, cellular membranes have many functions, and control of the cell's internal composition is one of the most important. Biological membranes allow some substances, but not others, to pass through them. This characteristic of membranes is called **selective permeability**. Selective permeability allows the membrane to determine what substances enter or leave a cell or organelle.

focus your learning

- The rate of diffusion of a substance across a membrane is influenced by the size and mass of the diffusing particles, the temperature and the density of the solution, and the size of the concentration gradient.
- Osmosis is diffusion of water across a biological membrane.
- Proteins that act as channels facilitate diffusion across membranes.

There are two fundamentally different processes by which substances cross biological membranes:

1. The processes of **passive transport** do not require the input of chemical energy to drive them.
2. The processes of **active transport** are driven by chemical energy (metabolic energy).

This section focuses on the passive processes by which substances cross membranes. The energy for the passive transport of a substance comes from the difference between its concentration on one side of the membrane and its concentration on the other—its **concentration gradient**. Passive transport can involve either of two types of diffusion: simple diffusion through the phospholipid bilayer, or facilitated diffusion via channel proteins or carrier proteins.

 Animation 6.1 **Passive Transport**
www.Life11e.com/a6.1

Diffusion is the process of random movement toward a state of equilibrium

In a solution, there is a tendency for all of the components to be evenly distributed. You can see this when a drop of food coloring falls into a glass of water. Initially the pigment molecules are very concentrated, but they will move about at random, slowly spreading until the intensity of the color is eventually the same throughout the glass.

A solution in which the solute particles are uniformly distributed is said to be at equilibrium because there will be no future net change in their concentration. Being at equilibrium does not mean that the particles have stopped moving; it just means that they are moving in such a way that their overall distribution does not change.

Diffusion is the process of random movement toward a state of equilibrium. Although the motion of each individual particle is absolutely random, the net movement of particles is directional until equilibrium is reached. Diffusion is thus a net movement from regions of greater concentration to regions of lesser concentration. In a solution with many different substances dissolved in it (solutes), the diffusion of each solute is independent of those of the others. How fast a substance diffuses depends on four factors:

1. The *size and mass* of the molecules or ions: smaller molecules diffuse faster.
2. The *temperature* of the solution: higher temperatures lead to faster diffusion because ions or molecules have more energy, and thus move more rapidly, at higher temperatures.
3. The *density of the solution*: As the density of the solution through which a substance is diffusing increases, the diffusion rate decreases.
4. The *concentration gradient* in the system—that is, the change in solute concentration with distance in a given direction: the greater the concentration gradient, the more rapidly a substance diffuses.

In our example shown above, the gradient is the concentration of food coloring in the droplet when first dropped into the water as compared with the concentration of food coloring in the water on the edge of the glass. When the food coloring is first dropped into the water, there is a high (or steep) concentration gradient. As time passes and the coloring diffuses through the solution, the concentration gradient slowly decreases.

DIFFUSION WITHIN CELLS AND TISSUES   In a small volume such as that inside a cell, solutes distribute themselves rapidly by diffusion. Small molecules and ions may move from one end of a cell to another in a millisecond (10^{-3} s, or one-thousandth of a second). However, the usefulness of diffusion as a transport mechanism declines drastically as distances become greater. In the absence of mechanical stirring, diffusion across more than a centimeter may take an hour or more, and diffusion across meters may take years! Diffusion would not be adequate to distribute materials over the length of a human body, much less that of a larger organism. But within our cells or across layers of one or two cells, diffusion is rapid enough to distribute small molecules and ions almost instantaneously. This is not so when dehydration occurs and the cytoplasm becomes denser as water is lost. Nerve cells in humans are very sensitive to this; when the body becomes dehydrated, unconsciousness may occur because substances important to nerve-cell function diffuse more slowly.

DIFFUSION ACROSS MEMBRANES   In a solution without barriers, all the solutes diffuse at rates determined by temperature, their physical properties, their concentration gradients, and the density of the solution. If a biological membrane divides the solution into separate compartments, then the movement of the different solutes can be affected by the properties of the membrane. The membrane is said to be permeable to solutes that can cross it more or less easily, but impermeable to substances that cannot move across it.

Molecules to which the membrane is impermeable remain in separate compartments, and their concentrations may be different on the two sides of the membrane. Molecules to which the membrane is permeable diffuse from one compartment to the other until their concentrations are equal on both sides of the membrane, and equilibrium is reached. After that point, individual molecules will continue to pass through the membrane, but *there will be no net change in concentration*.

Simple diffusion takes place through the phospholipid bilayer

In **simple diffusion**, small molecules pass through the phospholipid bilayer of the membrane. A molecule that is itself hydrophobic, and is therefore soluble in lipids, enters the membrane readily and is able to pass through it. The more lipid-soluble the molecule is, the more rapidly it diffuses through the membrane bilayer. This statement holds true over a wide range of molecular weights.

By contrast, electrically charged or polar molecules, such as amino acids, sugars, and ions, do not pass readily through a membrane, for two reasons. First, such charged or polar molecules are not very soluble in the hydrophobic interior of the bilayer. Second, such substances form many hydrogen bonds with water and ions

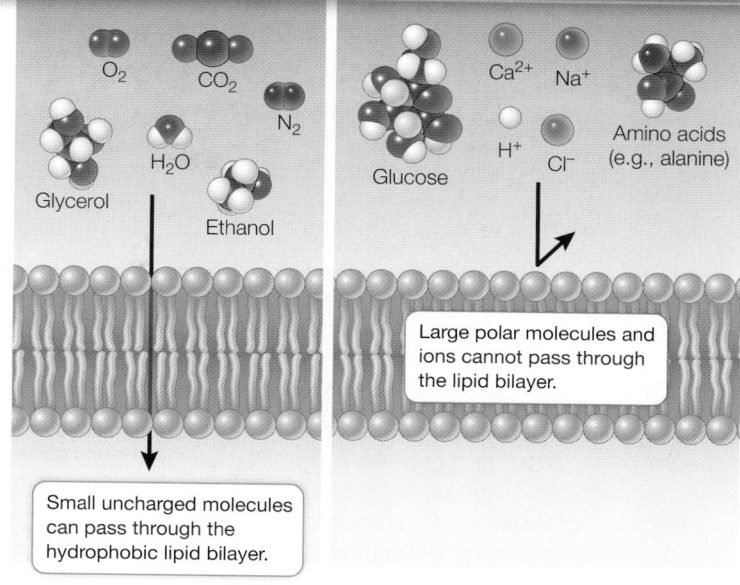

Figure 6.9 Permeability of Phospholipid Bilayers   Small uncharged molecules can diffuse through the membrane, but ions and large polar molecules cannot.

in the aqueous environment, be it the cytoplasm or the cell exterior. The multiplicity of these hydrogen bonds prevents the substances from moving into the hydrophobic interior of the membrane. **Figure 6.9** summarizes these phenomena.

Osmosis is the diffusion of water across membranes

Water molecules pass through specialized channels in membranes (see the opening story) by a diffusion process called *****osmosis**. This completely passive process uses no metabolic energy and depends on the relative concentrations of the water molecules on each side of the membrane. It is important that you remember that in a particular solution, the higher the total solute concentration, the lower the concentration of water molecules. A membrane may allow water but not solutes to pass across it, and in that case, water will diffuse across the membrane toward the side with the higher solute (lower water) concentration.

*****connect the concepts**   Osmosis plays key roles in plant physiology (e.g., in the root; see Key Concept 34.1) and in animal physiology (e.g., in the kidney; see Key Concept 51.1).

Three terms are used to compare the solute concentrations of two solutions separated by a membrane:

1. **Isotonic** solutions have equal solute concentrations (**Figure 6.10A**).
2. A **hypotonic** solution has a lower solute concentration than the other solution with which it is being compared (**Figure 6.10B**).
3. A **hypertonic** solution has a higher solute concentration than the other solution with which it is being compared (**Figure 6.10C**).

In general, when we discuss tonicity in cells, we refer to the outside solution in comparison to the inside of the cell. Therefore a solution that has a higher concentration of solutes than the concentration inside the cell is hypertonic. To reach equilibrium, water would

Figure 6.10 Osmosis Can Modify the Shapes of Cells In a solution that is isotonic with the cytoplasm **(A)**, a plant or animal cell maintains a consistent, characteristic shape because there is no net movement of water into or out of the cell. In these models, it is assumed that the solutes do not move across the membrane. In a solution that is hypotonic to the cytoplasm **(B)**, water enters the cell. An environment that is hypertonic to the cytoplasm **(C)** results in water leaving the cell.

Q: If you overfertilize the soil of a potted plant, the plant often wilts. Why?

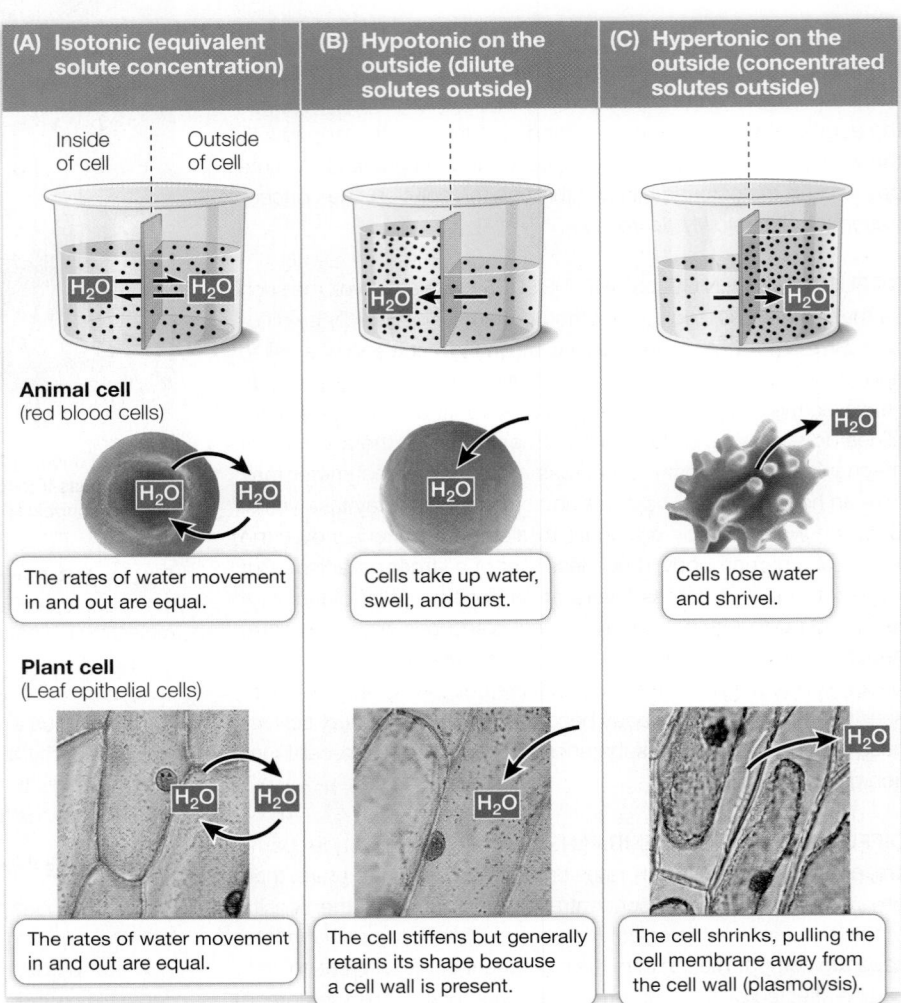

(A) Isotonic (equivalent solute concentration)	(B) Hypotonic on the outside (dilute solutes outside)	(C) Hypertonic on the outside (concentrated solutes outside)

Animal cell (red blood cells)

| The rates of water movement in and out are equal. | Cells take up water, swell, and burst. | Cells lose water and shrivel. |

Plant cell (Leaf epithelial cells)

| The rates of water movement in and out are equal. | The cell stiffens but generally retains its shape because a cell wall is present. | The cell shrinks, pulling the cell membrane away from the cell wall (plasmolysis). |

move out of the cell into the surrounding solution. Water moves from a hypotonic solution across a membrane to a hypertonic solution.

When we say that "water moves," bear in mind that we are referring to the net movement of water. Since it is so abundant, water is constantly moving through protein channels across the cell membrane into and out of cells. What concerns us here is whether the overall movement is greater in one direction or the other.

The concentration of solutes in the environment determines the direction of **osmosis** in all animal cells. A red blood cell takes up water from a solution that is hypotonic to the cell's contents. The cell bursts because its cell membrane cannot withstand the pressure created by the water entry and the resultant swelling. Conversely, the cell shrinks if the solution surrounding it is hypertonic to its contents. The integrity of red and white blood cells is absolutely dependent on the maintenance of a constant solute concentration in the blood plasma: the plasma must be isotonic to the blood cells if the cells are not to burst or shrink. Regulation of the solute concentration of body fluids is thus an important process for organisms without cell walls. Aquatic invertebrates usually have internal solute concentrations that match those of their aqueous environment. However, fish often have a very different internal environment: for example, the environment around a fish living in a freshwater stream is hypotonic to the body fluids inside the fish. Considerable chemical energy must be used to maintain this imbalance.

Unlike animal cells, the cells of plants, archaea, bacteria, fungi, and some protists have cell walls outside of their cell membrane that limit their volumes and keep them from bursting. Cells with sturdy walls take up a limited amount of water, and in so doing they build up internal pressure against the cell wall, which prevents further water from entering. This pressure within the cell is called **turgor pressure**. Turgor pressure keeps plants upright (and lettuce crisp) and is the driving force for the enlargement of plant cells. It is a normal and essential component of plant growth. If enough water leaves the cells, turgor pressure drops and the plant wilts.

Turgor pressure reaches about 100 pounds per square inch (0.7 kg/cm²)—several times greater than the pressure in automobile tires. This pressure is so great that the cells would change shape and detach from one another were it not for adhesive molecules in the plant cell walls and the plasmodesmata linking the cell cytoplasms (see Key Concept 5.4).

Diffusion may be aided by channel proteins

As we saw earlier, polar or charged substances such as water, amino acids, sugars, and ions do not readily diffuse across membranes. But they can cross the hydrophobic phospholipid bilayer passively (that is, without the input of energy) in one of two ways, depending on the substance:

1. **Channel proteins** are integral membrane proteins that form channels across the membrane through which certain substances can pass. Note that these are not channels in the widely used sense of the word (meaning open to the environment like a canal) but tunnels (surrounded by the membrane protein).

2. **Carrier proteins** bind substances and speed up their diffusion through the phospholipid bilayer.

Diffusion that is aided by channel proteins or carrier proteins is called **facilitated diffusion**. The substances diffuse according to their concentration gradients, but their diffusion is facilitated by protein channels or carriers. Although these proteins enhance the rate of diffusion, facilitated diffusion, like simple diffusion, does not require any input of energy.

ION CHANNELS   The best-studied channel proteins are the **ion channels**. As you will see in later chapters, the movement of ions across membranes is important in many biological processes such as respiration within the mitochondria, the electrical activity of the nervous system, and the opening of pores in leaves to allow gas exchange with the environment. Several types of *****ion channel** have been identified—each specific for a particular ion. All show the same basic structure of a hydrophilic pore that allows a particular ion to move through its center.

   *****connect the concepts**   Ion channels play an important role in the excitability of nervous tissues. See Key Concept 45.2.

   Just as a fence may have a gate that can be opened or closed, most ion channels are gated: they can be opened or closed to ion passage (**Figure 6.11**). A **gated channel** opens when a stimulus causes a change in the three-dimensional shape of the channel. In some cases, this stimulus is the binding of a chemical signal, or **ligand** (see Figure 6.11). Channels controlled in this way are called ligand-gated channels. Other channels are mechanically gated and open in response to physical stimuli such as sound waves in the ear. A voltage-gated channel is stimulated to open or close by a change in the voltage (electric charge difference) across the membrane.

AQUAPORINS FOR WATER   As you saw in the opening story of this chapter, water can cross membranes through protein channels called **aquaporins**. These channels function as a cellular plumbing system for moving water. The aquaporin channel is highly specific. Water molecules move in single file through the channel, which excludes ions so that the electrical properties of the cell are maintained. Aquaporins were first identified when a protein from red blood cell membranes was inserted into frog oocytes (immature egg cells). The membranes of these cells are normally impermeable to water, but the membranes of the cells treated with aquaporins became much more permeable (**Investigating Life: Aquaporins Increase Membrane Permeability to Water**).

Carrier proteins aid diffusion by binding substances

As we mentioned earlier, another type of facilitated diffusion involves the binding of the transported substance to a membrane protein called a **carrier protein**. Like channel proteins, carrier proteins facilitate the passive diffusion of substances into or out of cells or organelles. Carrier proteins transport polar molecules such as sugars and amino acids.

   Glucose is the major energy source for most cells, and living systems require a great deal of it. Glucose is polar and cannot readily diffuse across membranes. Eukaryotic cell membranes contain a carrier protein—the glucose transporter—that facilitates glucose uptake into the cell. Binding of glucose to a specific three-dimensional site on one side of the transporter protein causes the protein

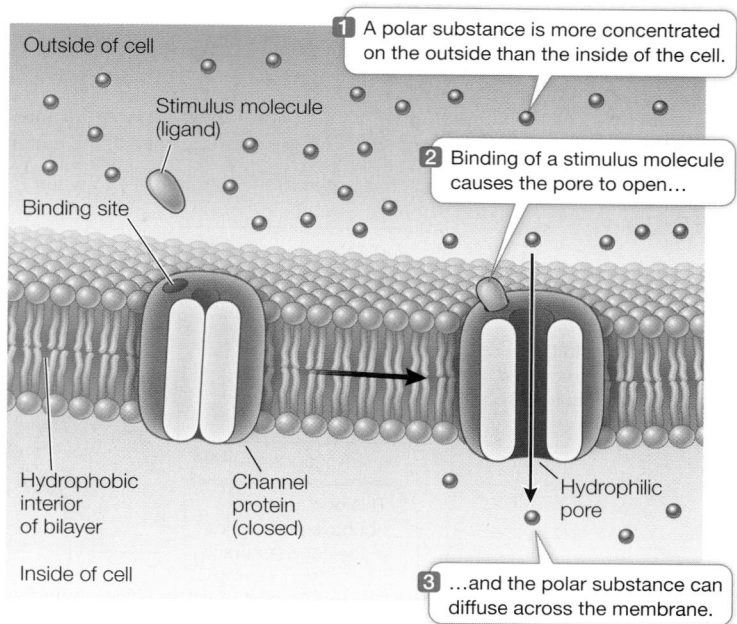

Figure 6.11 A Gated Channel Protein Opens in Response to a Stimulus   The channel protein has a pore of polar amino acids and water. It is anchored in the hydrophobic bilayer by its outer coating of nonpolar R groups. The protein changes its three-dimensional shape when a stimulus molecule (ligand) binds to it, opening the pore so that specific hydrophilic substances can pass through. Other channels open in response to an electric potential (voltage) or mechanical stimuli.

to change its shape and release glucose on the other side of the membrane (**Figure 6.12A**). Since glucose is either broken down or otherwise removed almost as soon as it enters a cell, there is almost always a strong concentration gradient favoring glucose entry (that is, a higher concentration outside the cell than inside).

   Transport by carrier proteins is different from simple diffusion. In simple diffusion, the rate of movement depends on the concentration gradient across the membrane. This is also true for carrier-mediated transport, up to a point. In carrier-mediated transport, as the concentration gradient increases, the diffusion rate also increases, but its *rate* of increase slows, and a point is reached at which the diffusion rate becomes constant. At this point, the facilitated diffusion system is said to be *saturated* (**Figure 6.12B**). This is explained by the fact that a particular cell has a specific number of carrier protein molecules in its cell membrane.

   The rate of diffusion reaches a maximum when all the carrier molecules are fully loaded with solute molecules. Think of waiting for the elevator on the ground floor of a hotel with 50 other people. You can't all get in the elevator (carrier) at once, so the rate of transport (say, ten people at a time) is at its maximum, and the transport system is "saturated." As a consequence, cells that require large amounts of energy, such as muscle cells, have high concentrations of glucose transporters in their membranes so that the maximum rate of facilitated diffusion is greater. Likewise, the human brain has high glucose needs, and the blood vessels that nourish it have high concentrations of glucose transporters.

experiment

Original Paper: Preston, G. M., T. P. Carroll, W. B. Gugino and P. Agre. 1992. Appearance of water channels in *Xenopus* oocytes expressing red cell CHIP28 protein. *Science* 256: 385–387.

A protein was isolated from the membranes of cells in which water diffuses rapidly across the membranes. When the protein was inserted into oocytes, which do not normally have it, the water permeability of the oocytes was greatly increased.

HYPOTHESIS▶ Aquaporin increases membrane permeability to water.

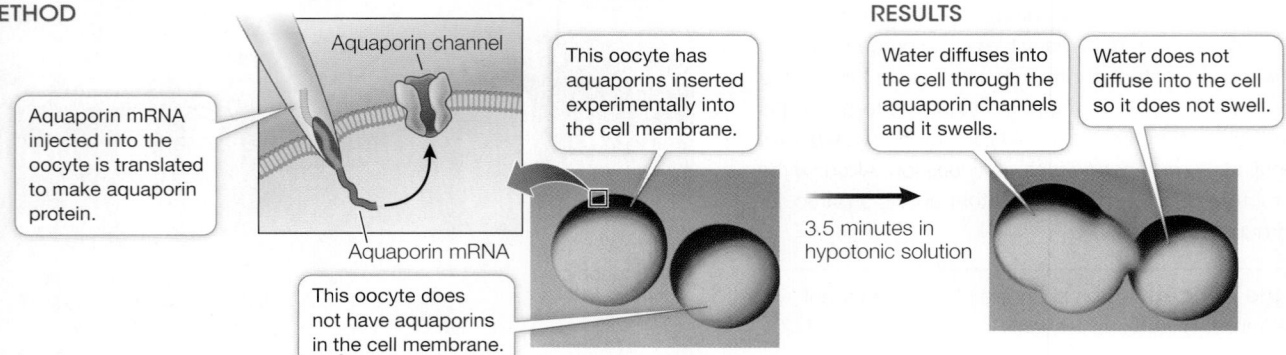

METHOD

Aquaporin channel

Aquaporin mRNA injected into the oocyte is translated to make aquaporin protein.

Aquaporin mRNA

This oocyte does not have aquaporins in the cell membrane.

This oocyte has aquaporins inserted experimentally into the cell membrane.

RESULTS

Water diffuses into the cell through the aquaporin channels and it swells.

Water does not diffuse into the cell so it does not swell.

3.5 minutes in hypotonic solution

CONCLUSION▶ Aquaporin increases the rate of water diffusion across the cell membrane.

work with the data

Although diffusion can account for limited water movement across cell membranes, simple diffusion seemed unlikely to explain the considerable water movement in kidney and red blood cells. By chance, Peter Agre and his colleagues found a major membrane protein, CHIP28, shared by these two cell types, and hypothesized that it was responsible for cell membrane water transport. So they did a "what if" experiment, taking the mRNA for CHIP28 and injecting it into frog oocytes that do not normally make the protein.

QUESTIONS▶

1. Frog oocytes in isotonic liquid were injected with either a small amount of water (control) or with CHIP28 mRNA in a small amount of water. The oocytes were then transferred to a hypotonic medium, and changes in relative cell volume were measured by microscopy. **Table A** shows the results.

Table A

	Cell volume	
Time (min)	CHIP28 mRNA	Water only (control)
0	1.0	1.0
0.5	1.05	1.0
1	1.15	1.02
1.5	1.23	—
2	1.32	1.02
2.5	1.36	—
3	1.41	1.02
4	(burst)	1.03

Plot the data on a graph of relative cell volume versus time after injection. How do the mRNA and control eggs compare? What explains the increase in the volume of the mRNA-injected oocytes? What explains the situation with both cells at 4 minutes after injection?

2. Water permeability (P_f) was calculated by the rate of osmotic swelling. **Table B** shows the results when increasing amounts of CHIP28 mRNA were injected.

Table B

Amount of mRNA (ng)	P_f (cm/sec × 10^{-4})[a]
0	13.7 (3.3)
0.1	50.0 (10.1)
0.5	112 (29.2)
2.0	175 (38.4)
10.0	221 (14.8)

[a] Numbers in parentheses indicate +/− standard deviation.

What can you conclude from these data? What statistical test (see Appendix B) would you do to show the significance of your conclusion?

3. To further investigate the role of CHIP28 in water transport, oocytes were given a known inhibitor of protein-mediated water transport, mercuric chloride, and then tested for water permeability with CHIP28 mRNA. Some of the eggs were also treated with mercaptoethanol, a molecule that overcomes the inhibition of water transport by mercuric chloride. The question was, would the inhibitor actually block water transport mediated by CHIP28, and would mercaptoethanol restore transport? The results are shown in **Table C**.

Table C

mRNA injected	Mercuric chloride	Mercapto-ethanol	P_f (cm/sec × 10^{-4})
None	None	None	27.9
None	Yes	None	20.3
None	Yes	Yes	25.4
CHIP28	None	None	210
CHIP28	Yes	None	80.7
CHIP28	Yes	Yes	188

What can you conclude about the molecular nature of water transport mediated by CHIP28 mRNA? What data support your conclusion? Explain all the controls that were done.

A similar **work with the data** exercise may be assigned in **LaunchPad**.

(A) Glucose uptake by the glucose transporter

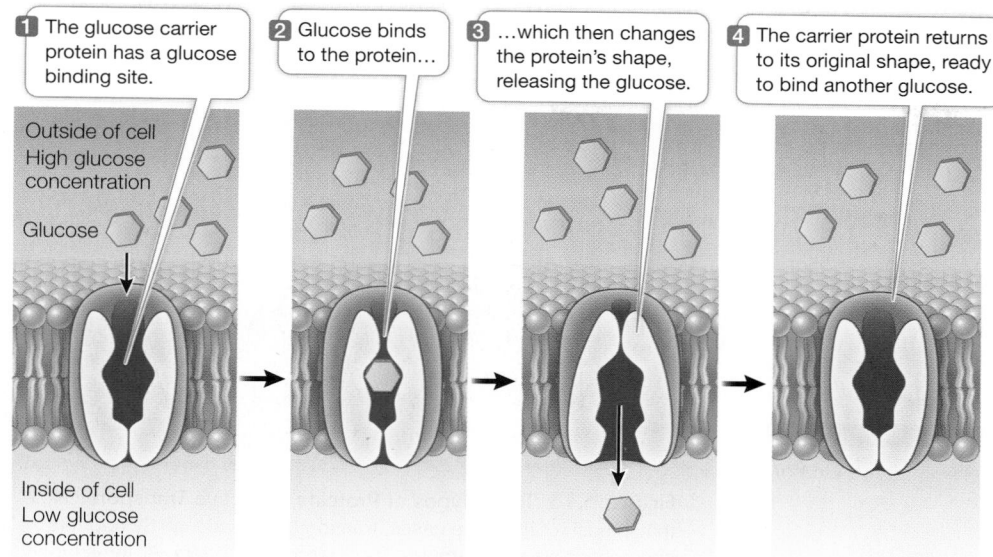

1. The glucose carrier protein has a glucose binding site.

2. Glucose binds to the protein...

3. ...which then changes the protein's shape, releasing the glucose.

4. The carrier protein returns to its original shape, ready to bind another glucose.

Outside of cell
High glucose concentration

Glucose

Inside of cell
Low glucose concentration

(B) Gradual saturation of glucose transporters

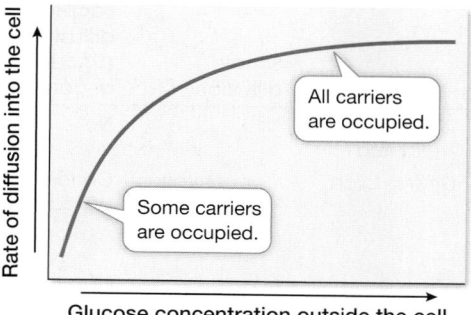

All carriers are occupied.

Some carriers are occupied.

Rate of diffusion into the cell

Glucose concentration outside the cell

Activity 6.4 **Membrane Transport Simulation**
www.Life11e.com/ac6.4

Figure 6.12 **A Carrier Protein Facilitates Diffusion** The glucose transporter is a carrier protein that allows glucose to enter the cell at a faster rate than would be possible by simple diffusion. **(A)** The transporter binds to glucose, brings it into the membrane interior, then changes shape, releasing glucose into the cell cytoplasm. **(B)** The graph shows the rate of glucose entry via a carrier versus the concentration of glucose outside the cell. As the glucose concentration increases, the rate of diffusion increases until the point at which all the available transporters are being used (the system is saturated).

6.3 recap

Diffusion is the movement of ions or molecules from a region of greater concentration to a region of lesser concentration. Osmosis is the diffusion of water through a selectively permeable cell membrane. Channel proteins and carrier proteins can facilitate the diffusion of charged and polar substances, including water, across cell membranes.

learning outcomes

You should be able to:

- Predict rates of diffusion of substances under different conditions.
- Predict the effect of different environmental conditions on the movement of water into and out of living cells.
- Explain the role of channel proteins in facilitating diffusion across membranes.

1. What properties determine whether, and how fast, a substance will diffuse across a membrane?

2. What is the effect on red blood cells if the kidney removes too many ions from the blood and makes it hypotonic to the cells?

3. Cells in the nervous system may have a higher concentration of K^+ ions inside the cell than outside. What is the role of the cell membrane in maintaining this imbalance? How can K^+ membrane channels eliminate this imbalance?

The process of diffusion tends to equalize the concentrations of substances outside and inside cells. However, one hallmark of a living cell is that it can have an internal composition quite different from that of its environment. To achieve this, a cell must sometimes move substances against their concentration gradients. This process requires an input of energy and is known as active transport.

key concept

6.4 Active Transport across Membranes Requires Energy

In many biological situations, the concentration of a particular ion or small molecule inside the cell differs from that outside the cell. Such imbalances are maintained by a protein in the cell membrane that moves the substance against its concentration and/or electrical gradient. Movement of molecules or ions against a gradient is called **active transport** because it requires an input of energy. Often the energy source is adenosine triphosphate (ATP), which has chemical energy stored in its terminal phosphate bond. In eukaryotes, ATP is produced in the mitochondria. Energy is released in a hydrolysis reaction that breaks the terminal phosphate bond of ATP, converting it to adenosine diphosphate (ADP). We will detail how ATP provides energy to cells in Key Concept 8.2.

The differences between diffusion and active transport are summarized in **Table 6.1**.

focus your learning

- Active transport requires energy to move a substance against its concentration gradient.
- Three kinds of membrane proteins are involved in active transport: uniporters, symporters, and antiporters.
- Primary active transport uses ATP hydrolysis directly to provide the energy for transport, whereas secondary active transport uses an ion concentration gradient that was established by ATP hydrolysis.

table 6.1 Membrane Transport Mechanisms

	Simple diffusion	Facilitated diffusion (through channel or carrier)	Active transport
Cellular energy required?	No	No	Yes
Driving force	Concentration gradient	Concentration gradient	ATP hydrolysis (against concentration gradient)
Membrane protein required?	No	Yes	Yes
Specificity	No	Yes	Yes

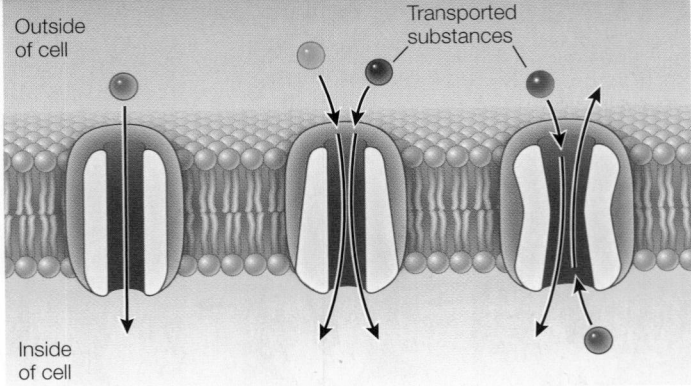

Uniporter transports one substance in one direction.

Symporter transports two different substances in the same direction.

Antiporter transports two different substances in opposite directions.

Outside of cell

Transported substances

Inside of cell

Figure 6.13 Three Types of Proteins for Active Transport Note that in each of the three cases, transport is directional. Symporters and antiporters are examples of coupled transporters. All three types of transporters are coupled to energy sources in order to move substances against their concentration gradients.

 Animation 6.2 **Active Transport**
www.Life11e.com/a6.2

Active transport is directional

In diffusion, ions or molecules can move down their concentration gradients in either direction across the cell membrane. In contrast, active transport is directional, and moves a substance either into or out of the cell or organelle, depending on need. There are three kinds of membrane proteins that carry out active transport (**Figure 6.13**):

1. A **uniporter** moves a single substance in one direction. For example, a calcium-binding protein found in the cell membrane and the ER membranes of many cells actively transports Ca^{2+} to locations where it is more highly concentrated, either outside the cell or inside the lumen of the ER.

2. A **symporter** moves two substances in the same direction. For example, a symporter in the cells that line the intestine must bind Na^+ in addition to an amino acid in order to absorb amino acids from the intestine.

3. An **antiporter** moves two substances in opposite directions, one into the cell (or organelle) and the other out of the cell (or organelle). For example, many cells have a sodium–potassium pump that moves Na^+ out of the cell and K^+ into it.

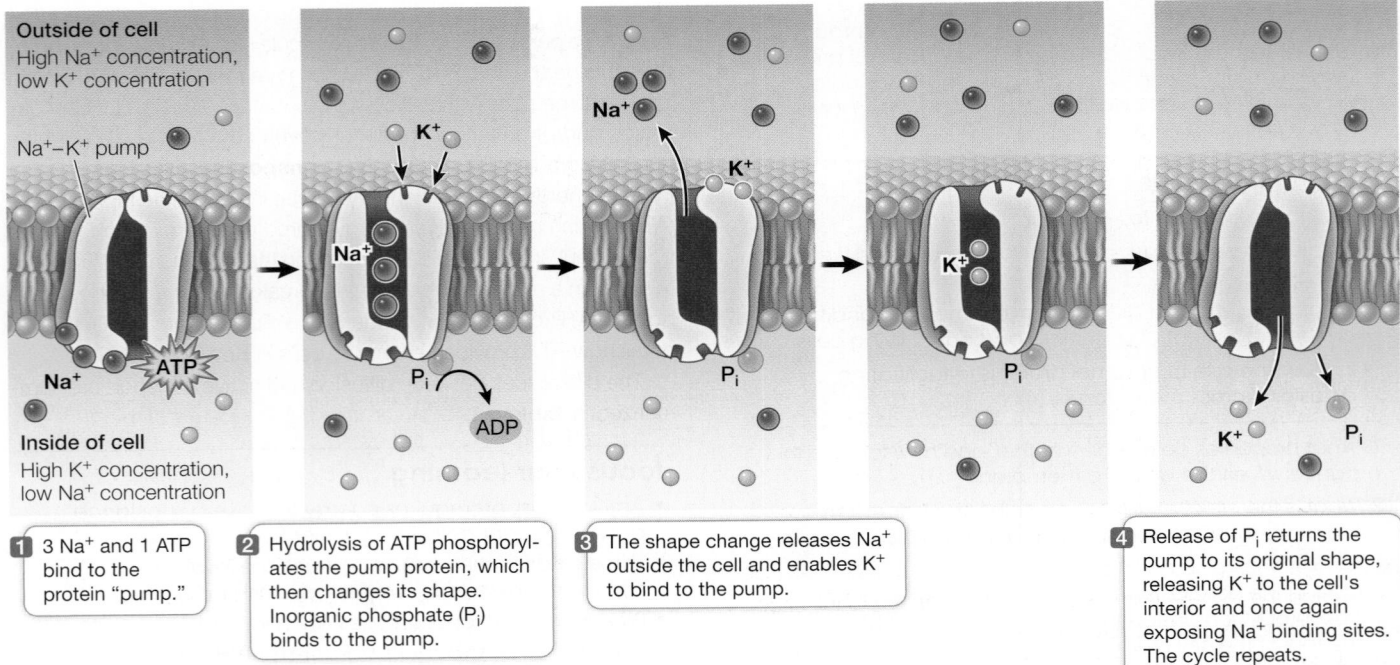

1 3 Na^+ and 1 ATP bind to the protein "pump."

2 Hydrolysis of ATP phosphorylates the pump protein, which then changes its shape. Inorganic phosphate (P_i) binds to the pump.

3 The shape change releases Na^+ outside the cell and enables K^+ to bind to the pump.

4 Release of P_i returns the pump to its original shape, releasing K^+ to the cell's interior and once again exposing Na^+ binding sites. The cycle repeats.

Figure 6.14 Primary Active Transport: The Sodium–Potassium Pump In active transport, energy is used to move a solute against its concentration gradient. Here, energy from ATP is used to move Na^+ and K^+ against their concentration gradients.

Symporters and antiporters are also known as **coupled transporters** because they move two substances at once.

Different energy sources distinguish different active transport systems

There are two basic types of active transport:

1. **Primary active transport** involves the direct hydrolysis of ATP, which provides the energy required for transport.

2. **Secondary active transport** does not use ATP directly. Instead, its energy is supplied by an ion concentration gradient established by primary (ATP-driven) active transport.

In primary active transport, energy released by the hydrolysis of ATP drives the movement of specific ions against their concentration gradients. For example, we mentioned earlier that concentrations of potassium ions (K^+) inside a cell are often much higher than in the fluid bathing the cell. However, the concentration of sodium ions (Na^+) is often much higher outside the cell. A protein in the cell membrane pumps Na^+ out of the cell and K^+ into the cell against these concentration gradients, ensuring that the gradients are maintained (**Figure 6.14**). This **sodium–potassium (Na^+–K^+) pump** is found in all animal cells. The pump is an integral membrane glycoprotein. It breaks down a molecule of ATP to ADP and a free phosphate ion (P_i) and uses the energy released to bring two K^+ ions into the cell and export three Na^+ ions. The Na^+–K^+ pump is thus an antiporter because it moves two substances in different directions.

An important drug used to treat a weak heart inhibits the Na^+–K^+ pump. Digitalis, purified from the foxglove plant, blocks the release of phosphate ions (P_i) from the protein (see Figure 6.14, step 4). This "freezes" the shape of the pump, and it can no longer bind Na^+. So Na^+ accumulates inside heart cells. This ultimately leads to a greater force of contraction of heart muscle, which aids a patient whose heart is not pumping well.

In secondary active transport, the movement of a substance against its concentration gradient is accomplished using energy "regained" by letting ions move across the membrane with their concentration gradients. Think of it this way: A large dam may prevent a river from flowing. The accumulated water represents energy, and when the water is allowed to flow past the dam, the energy it releases can be used to drive turbines that make electricity. It's similar for secondary active transport. Consider, for example, the absorption of glucose into the bloodstream from the digestive tract. Once the Na^+–K^+ pump establishes a concentration gradient of sodium ions, the passive diffusion of some Na^+ back into the cell can provide energy for the secondary active transport of glucose into the cell (**Figure 6.15**). Secondary active transport aids in the uptake of amino acids and sugars, which are essential raw materials for cell maintenance and growth. Both types of coupled transport proteins—symporters and antiporters—are used for secondary active transport.

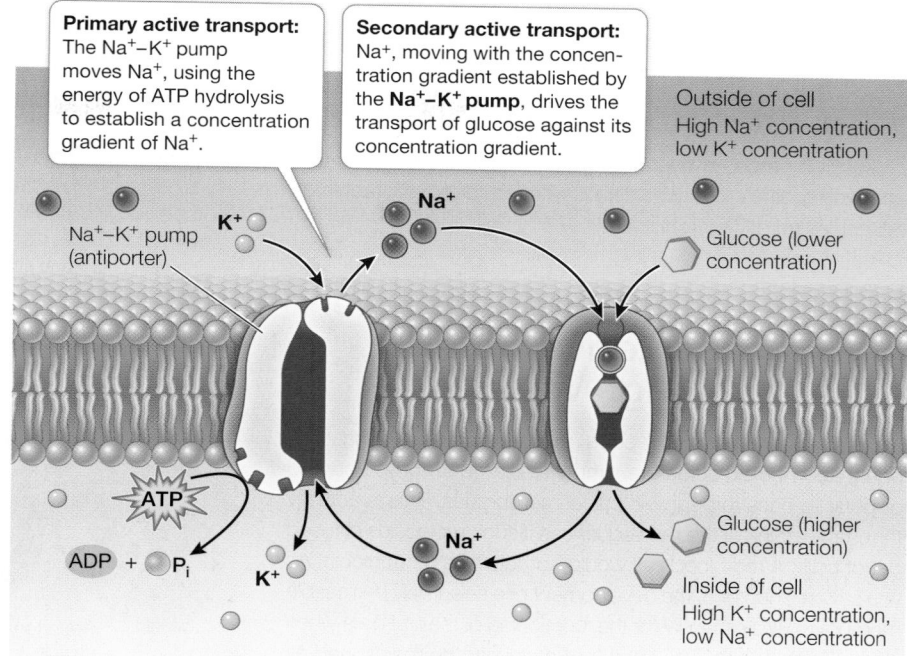

Primary active transport: The Na^+–K^+ pump moves Na^+, using the energy of ATP hydrolysis to establish a concentration gradient of Na^+.

Secondary active transport: Na^+, moving with the concentration gradient established by the **Na^+–K^+ pump**, drives the transport of glucose against its concentration gradient.

Outside of cell
High Na^+ concentration, low K^+ concentration

Na^+–K^+ pump (antiporter)

Glucose (lower concentration)

ATP

ADP + P_i

Glucose (higher concentration)

Inside of cell
High K^+ concentration, low Na^+ concentration

Figure 6.15 Secondary Active Transport The Na^+ concentration gradient established by primary active transport by the sodium–potassium pump (left) powers the secondary active transport of glucose (right). A symporter protein couples the movement of glucose across the membrane against its concentration gradient to the passive movement of Na^+ into the cell.

Q: If the Na^+–K^+ pump is blocked by a drug, how will intercellular concentrations of Na^+ and glucose be affected?

6.4 recap

Active transport across a membrane is directional and requires an input of energy to move substances against their concentration gradients. Active transport allows a cell to maintain small molecules and ions at concentrations very different from those in the surrounding environment.

learning outcomes

You should be able to:

- Compare and contrast diffusion and active transport.
- Explain why active transport requires energy.
- Describe and give examples of the different types of proteins that carry out active transport.
- Differentiate between primary and secondary active transport.

1. Why is energy required for active transport?
2. Why is the sodium–potassium (Na^+–K^+) pump classified as an antiporter?
3. What are the differences between primary active transport and secondary active transport?
4. What are the similarities and differences between active transport and facilitated diffusion?

We have examined a number of passive and active ways in which ions and small molecules can enter and leave cells. But what about large molecules such as proteins? Many proteins are so large that they diffuse very slowly, and their bulk makes it difficult for them to pass through the phospholipid bilayer. A completely different mechanism is needed to move intact large molecules across membranes.

key concept 6.5 Large Molecules Enter and Leave a Cell through Vesicles

Macromolecules such as proteins, polysaccharides, and nucleic acids are simply too large and too charged or polar to pass through biological membranes. This is actually fortunate—think of the consequences if such molecules diffused out of cells: A red blood cell would not retain its hemoglobin! As you saw in Chapter 5, the development of a selectively permeable membrane was essential for the functioning of the first cells when life on Earth began. The interior of a cell can be maintained as a separate compartment with a different composition from that of the exterior environment, which is subject to abrupt changes. However, cells must sometimes take up or secrete (release to the external environment) intact large molecules. In Key Concept 5.3 we described phagocytosis, the mechanism by which solid particles can be brought into the cell by means of vesicles that pinch off from the cell membrane. The general terms for the mechanisms by which substances enter and leave the cell via membrane vesicles are endocytosis and exocytosis.

focus your learning

- Three types of endocytosis occur in cells.
- Cells take in specific molecules from the environment through receptor-mediated endocytosis.
- Exocytosis is the process by which substances are secreted by a cell.

 Animation 6.3 Endocytosis and Exocytosis www.Life11e.com/a6.3

Macromolecules and particles enter the cell by endocytosis

Endocytosis is a general term for a group of processes that bring small molecules, macromolecules, large particles, and even small cells into the eukaryotic cell (**Figure 6.16A**). We described an example of endocytosis earlier in the chapter, involving integrins (see Figure 6.8). Generally, there are three types of endocytosis: phagocytosis, pinocytosis, and receptor-mediated endocytosis. In all three, the cell membrane invaginates (folds inward), forming a small pocket around materials from the environment. The pocket deepens, forming a vesicle. This vesicle separates from the cell membrane and migrates with its contents to the cell's interior.

1. In **phagocytosis** ("cellular eating"), part of the cell membrane engulfs large particles or even entire cells. Unicellular protists use phagocytosis for feeding, and some white blood cells use

(A) Endocytosis

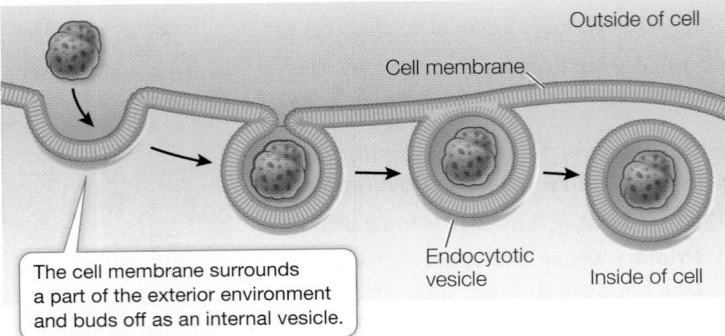

The cell membrane surrounds a part of the exterior environment and buds off as an internal vesicle.

(B) Exocytosis

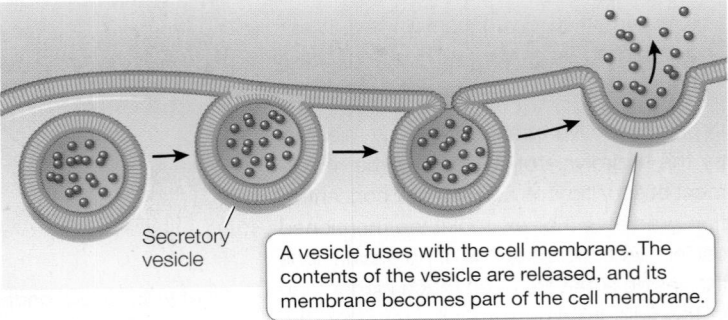

A vesicle fuses with the cell membrane. The contents of the vesicle are released, and its membrane becomes part of the cell membrane.

Figure 6.16 Endocytosis and Exocytosis Endocytosis **(A)** and exocytosis **(B)** are used by eukaryotic cells to take up and release fluids, large molecules, and particles. Smaller cells, such as invading bacteria, can be taken up by endocytosis.

 Media Clip 6.1 An Amoeba Eats by Phagocytosis www.Life11e.com/mc6.1

phagocytosis to defend the body by engulfing foreign cells and substances. The food vacuole or phagosome that forms usually fuses with a lysosome, where its contents are digested (see Figure 5.10).

2. In **pinocytosis** ("cellular drinking"), vesicles also form. However, these vesicles are smaller, and the process operates to bring fluids and dissolved substances into the cell. Like phagocytosis, pinocytosis can be relatively nonspecific regarding what it brings into the cell. For example, pinocytosis occurs constantly in the endothelium, the single layer of cells that separates a tiny blood capillary from the surrounding tissue. Pinocytosis allows cells of the endothelium to rapidly acquire fluids and dissolved solutes from the blood.

3. In **receptor-mediated endocytosis**, molecules at the cell surface recognize and trigger the uptake of specific materials.

Let's take a closer look at this last process.

Receptor-mediated endocytosis is highly specific

Receptor-mediated endocytosis is used by animal cells to capture specific macromolecules from the cell's environment. This process depends on **receptor proteins**, which are proteins that can bind to specific molecules within the cell or in the cell's external environment. In receptor-mediated endocytosis, the receptors are integral membrane proteins located at particular regions on the extracellular

Figure 6.17 Receptor-Mediated Endocytosis The receptor proteins in a coated pit bind specific macromolecules, which are then carried into the cell by a coated vesicle.

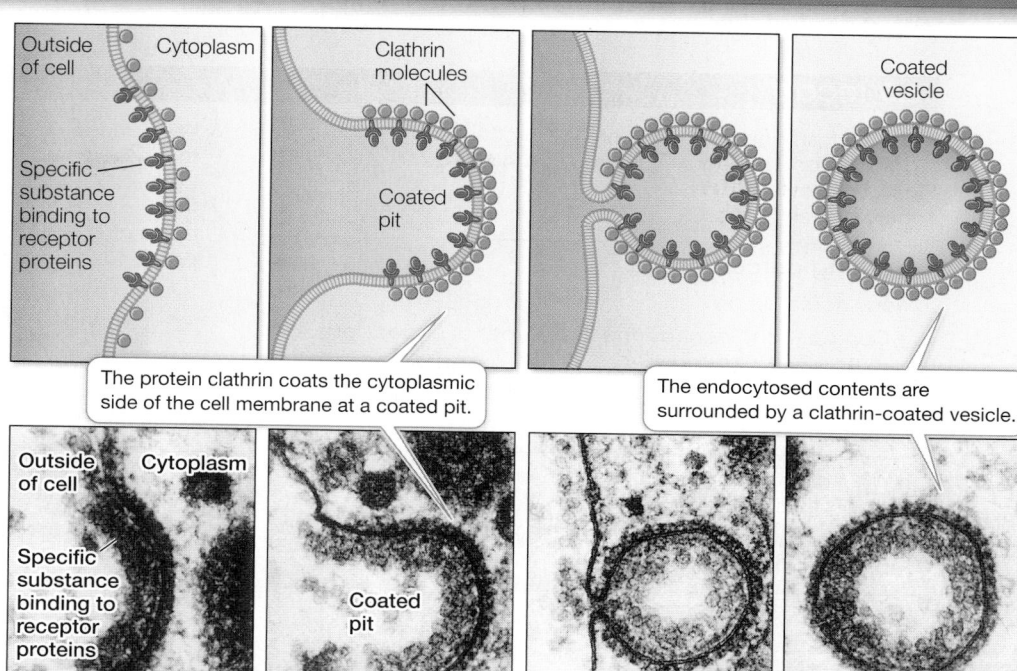

The protein clathrin coats the cytoplasmic side of the cell membrane at a coated pit.

The endocytosed contents are surrounded by a clathrin-coated vesicle.

surface of the cell membrane. These membrane regions are called coated pits because they form slight depressions in the cell membrane and their cytoplasmic surfaces are coated by other proteins, such as clathrin. The uptake process is similar to that in phagocytosis.

When a receptor protein binds to its specific ligand (in this case, the macromolecule to be taken into the cell), its coated pit invaginates and forms a coated vesicle around the bound macromolecule. The clathrin molecules strengthen and stabilize the vesicle, which carries the macromolecule away from the cell membrane and into the cytoplasm (**Figure 6.17**). Once inside, the vesicle loses its clathrin coat and may fuse with a lysosome, where the engulfed material is digested (by the hydrolysis of polymers to monomers) and the products are released into the cytoplasm. Because of its specificity for particular macromolecules, receptor-mediated endocytosis is an efficient method of taking up substances that may exist at low concentrations in the cell's environment.

Receptor-mediated endocytosis is the method by which cholesterol is taken up by most mammalian cells. Water-insoluble cholesterol and triglycerides are packaged by liver cells into lipoprotein particles. Most of the cholesterol is packaged into a type of lipoprotein particle called low-density lipoprotein, or LDL, which is circulated via the bloodstream. When a particular cell requires cholesterol, it produces specific LDL receptors, which are inserted into the cell membrane in clathrin-coated pits. Binding of LDLs to the receptor proteins triggers the uptake of the LDLs via receptor-mediated endocytosis. Within the resulting vesicle, the LDL particles are freed from the receptors. The receptors segregate to a region that buds off and forms a new vesicle, which is recycled to the cell membrane. The freed LDL particles remain in the original vesicle, which fuses with a lysosome. There, the LDLs are digested and the cholesterol made available for cell use.

In healthy individuals, the liver takes up unused LDLs for recycling. People with the inherited disease familial hypercholesterolemia have a deficient LDL receptor in their livers. This prevents receptor-mediated endocytosis of LDLs, resulting in dangerously high levels of cholesterol in the blood. The cholesterol builds up in the arteries that nourish the heart and causes heart attacks. In extreme cases where only the deficient receptor is present, children and teenagers can have severe cardiovascular disease.

Exocytosis moves materials out of the cell

Exocytosis is the process by which materials packaged in vesicles are secreted from a cell when the vesicle membrane fuses with the cell membrane (**Figure 6.16B**). This fusing allows the vesicle contents to be released into the environment, and the vesicle membrane is smoothly incorporated into the cell membrane. In another form of exocytosis, the vesicle touches the cell membrane and a pore forms, releasing the vesicle's contents. There is no membrane fusion in this process, termed "kiss and run." We saw an example of exocytosis in describing sweat glands at the start of the chapter.

Table 6.2 summarizes examples of endocytosis and exocytosis.

table 6.2 Endocytosis and Exocytosis	
Type of process	**Example**
Endocytosis	
Receptor-mediated endocytosis	Specific uptake of large molecules (e.g., LDL)
Pinocytosis	Nonspecific uptake of extracellular fluid (e.g., fluids and dissolved substances from blood)
Phagocytosis	Nonspecific uptake of large undissolved particles (e.g., invading bacteria by cells of the immune system)
Exocytosis	
Release of large molecules	Vesicle fusion with cell membrane (e.g., digestive enzymes in the pancreas)
Release of small molecules	Vesicle fusion with cell membrane (e.g., neurotransmitters at the synapse)

6.5 recap

Endocytosis and exocytosis are the processes by which large particles and molecules are transported into and out of the cell. Endocytosis may be mediated by a receptor protein in the cell membrane.

learning outcomes

You should be able to:

- Compare and contrast the three types of endocytosis.
- Describe the process of receptor-mediated endocytosis.
- Provide examples of processes that involve endocytosis and/or exocytosis.

1. What are the differences between phagocytosis and pinocytosis?
2. Describe receptor-mediated endocytosis and give an example of a process involving this phenomenon.

3. Key Concept 27.2 describes the diatoms, which are protists that have complex glassy structures in their cell walls (see Figure 27.8 and the photo below of eight diatom cells inside their ornate cell walls). These structures form within the Golgi apparatus. How do they reach the cell wall without having to pass through a membrane?

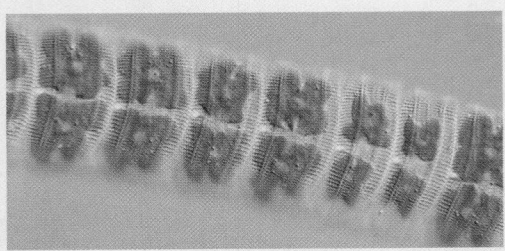

---

investigatinglife

Q&A What is the importance of aquaporin membrane channels?

There are a dozen genes encoding aquaporins in humans, each channel having its particular location in the body. Genetic deficiency resulting in a lack of one of the aquaporins can have significant consequences. For example, one of the aquaporins occurs in the part of the kidney where water is removed from the developing urine and returned to the blood. When this aquaporin is absent, there is more water in urine and the person urinates more and more often; bedwetting is a common result. In plants, aquaporins occur in the cell membrane and tonoplast (membrane that encloses the vacuole). In the cell membranes of root cells, aquaporins form channels for water on its way from the soil to the interior tissues of the root, from which it is transported to the rest of the plant. In the vacuole, aquaporin channels are important in the organelle swelling, resulting in turgor pressure, which is important in plant organ structure and plant cell expansion. All of these aquaporins, and those of other organisms, share a common structure that spans the cell membrane and has a channel through which water molecules pass in single file.

Future directions

A fascinating role for aquaporins is not just in the transport of water across membranes, but in the accompanying transport of substances dissolved in it. For example, in plants boron is an important nutrient, and it does not diffuse efficiently across membranes. Soil water containing boron ions passes readily into plants through aquaporin channels. Even gases such as CO_2 may pass in and out of cells via aquaporins while dissolved in water. Efforts are under way to insert aquaporins into synthetic membranes used for industrial applications. Because aquaporins allow only water and low concentrations of ions to pass through, such membranes could be used to purify contaminated fresh water, or to desalinate water, which as you know has very high ion concentrations. Aquaporins are expressed in skin, where they play a role in keeping the skin moist. This is also a goal of some cosmetics (moisturizers) and has led to a search for, and patenting of, small molecules that stimulate aquaporins. These molecules are components of some widely marketed skin creams.

Chapter Summary 6

6.1 Biological Membranes Are Lipid–Protein Bilayers

- Biological membranes consist of lipids, proteins, and carbohydrates. The **fluid mosaic model** of membrane structure describes a phospholipid bilayer in which proteins can move about within the plane of the membrane. **Review Activity 6.1**
- The two layers of a membrane may have different properties because of their different lipid compositions. Animal cell membranes may contain high concentrations of cholesterol (up to 25%). **Review Activity 6.2**

- The properties of membranes also depend on the **integral membrane proteins** and **peripheral membrane proteins** associated with them. Some proteins, called **transmembrane proteins**, span the membrane. **Review Focus: Key Figure 6.1**
- Carbohydrates, attached to proteins in **glycoproteins** or to phospholipids in **glycolipids**, project from the external surface of the cell membrane and function as recognition signals.
- Membranes are not static structures, but are constantly forming, exchanging components, and breaking down.

6.2 Cell Membrane Is Important in Cell Adhesion and Recognition

- In order for cells to assemble into tissues, they must recognize and adhere to one another. **Cell recognition** and **cell adhesion** depend on membrane-associated proteins and carbohydrates. **Review Figure 6.6**

- Adhesion can involve binding between identical (**homotypic**) or different (**heterotypic**) molecules on adjacent cells.

- **Cell junctions** connect adjacent cells. In some animal cells, **tight junctions** prevent the passage of molecules through the inter-cellular spaces between cells, and they restrict the migration of membrane proteins over the cell surface. **Desmosomes** cause cells to adhere firmly to one another. **Gap junctions** provide channels for communication between adjacent cells. **Review Figure 6.7, Activity 6.3**

- **Integrins** mediate the attachment of animal cells to the extracellular matrix and to each other. Detachment and recycling of integrins allow cells to move. **Review Figure 6.8**

6.3 Substances Can Cross Membranes by Passive Processes

See Animation 6.1

- Membranes exhibit **selective permeability**, regulating which substances pass through them. Substances can cross the membrane by either **passive transport**, which requires no input of chemical energy, or by **active transport**, which uses chemical energy. **Review Figure 6.9**

- **Diffusion** is the movement of a solute from a region of higher concentration to a region of lower concentration. Equilibrium is reached when there is no further net change in concentration.

- In **osmosis**, water diffuses across a membrane from a region of higher water concentration to a region of lower water concentration.

- In an **isotonic** environment, total solute concentrations on both sides of the cell membrane are equal. If the solution surrounding a cell is **hypotonic** to the cell interior, more water enters the cell than leaves it, causing it to swell. In plant cells, this contributes to **turgor pressure**. In a **hypertonic** solution, more water leaves the cell than enters it, causing it to shrivel. **Review Figure 6.10**

- A substance can diffuse passively across a membrane by either **simple diffusion** or **facilitated diffusion**, via a **channel protein** or a carrier protein.

- **Ion channels** are membrane proteins that allow the rapid facilitated diffusion of ions through membranes. **Gated channels** can be opened or closed by chemical **ligands**, or changes in membrane voltage, or mechanical stimuli. **Review Figure 6.11**

- **Aquaporins** are water channels. **Review Investigating Life: Aquaporins Increase Membrane Permeability to Water**

- **Carrier proteins** bind to polar molecules such as sugars and amino acids and transport them across the membrane. The maximum rate of this type of facilitated diffusion is limited by the number of carrier (transporter) proteins in the membrane. **Review Figure 6.12, Activity 6.4**

6.4 Active Transport across Membranes Requires Energy

See Animation 6.2

- **Active transport** requires the use of chemical energy to move substances across membranes against their concentration or electrical gradients. Active transport proteins may be **uniporters**, **symporters**, or **antiporters**. **Review Figure 6.13**

- In **primary active transport**, energy from the hydrolysis of ATP is used to move ions into or out of cells. The **sodium-potassium pump** is an important example. **Review Figure 6.14**

- **Secondary active transport** couples the passive movement of one substance down its concentration gradient to the movement of another substance against its concentration gradient. Energy from ATP is used indirectly to establish the concentration gradient that results in the movement of the first substance. **Review Figure 6.15**

6.5 Large Molecules Enter and Leave a Cell through Vesicles

See Animation 6.3

- **Endocytosis** is the transport of small molecules, macromolecules, large particles, and small cells into eukaryotic cells via invagination of the cell membrane and the formation of vesicles. **Phagocytosis** and **pinocytosis** are types of endocytosis. **Review Figure 6.16A**

- In **exocytosis**, materials in vesicles are secreted from the cell when the vesicles fuse with the cell membrane. **Review Figure 6.16B**

- In **receptor-mediated endocytosis**, a specific **receptor protein** on the cell membrane binds to a particular macromolecule. **Review Figure 6.17**

> Go to **LearningCurve** (in **LaunchPad**) for dynamic quizzing that helps you solidify your understanding of this chapter. **LearningCurve** adapts to your responses, giving you the practice you need to master each key concept.

▶ Apply What You've Learned

Review

6.1 The degree of a membrane's fluidity is influenced by lipid composition and temperature.

Original Paper: Cossins, A. R. and C. L. Prosser. 1978. Evolutionary adaptation of membranes to temperature. *Proceedings of the National Academy of Sciences USA.* 75: 2040–2043.

Fish are poikilotherms, animals having body temperatures that change with their environment. Depending on where you go in the world, you can find fish at all extremes of body temperature. Desert pupfish native to desert springs in California and Mexico tolerate temperatures as high as 42°C. At the other extreme is the Arctic sculpin, which can tolerate water temperatures as low as −2°C. Goldfish live at more moderate temperatures ranging from 5°C to 25°C.

A research group saw this range as an interesting opportunity to run a comparative study of the fluidity and composition of cell membranes. They included rats and hamsters in the study, because rats maintain a constant body temperature of 37°C and represent the high end of the temperature range. The researchers kept each animal at a specific temperature for several days. Arctic sculpin were held at 0°C. One group of goldfish was held at 5°C, and another was held at 25°C. Desert pupfish were held at 34°C, and rats were held at normal room temperature, 21°C. The same neuronal cells were removed from each animal and membranes isolated from these cells. This gave the researchers a set of cell membranes representing environmental temperatures ranging from 0°C to 37°C.

The researchers then added a fluorescent molecule to each of the membranes and incubated them at 20°C. They measured fluorescence and plotted these data against the animal's body temperature, as shown in the graph at right. (Each point represents a single animal.) Fluorescence is inversely related to the movement of the molecules within the membranes. A higher value means less movement of the fluorescent probe and therefore a less fluid membrane.

The table below shows the ratio of saturated to unsaturated fatty acids in the phospholipid phosphatidyl choline for various species in the study.

Ratio of saturated to unsaturated fatty acids				
Arctic sculpin 0°C	Goldfish 5°C	Goldfish 25°C	Desert pupfish 34°C	Rat 37°C
0.593	0.659	0.817	0.990	1.218

Questions

1. What do the data in the graph indicate about the relative fluidity of cell membranes in different species when measured under the same temperature conditions? Is a trend apparent? Explain.

2. Temperature affects membrane fluidity. According to the data, what else influences membrane fluidity?

3. Considering your answer to Question 2, explain how and why this factor affects membrane fluidity.

4. Suppose you isolated cell membranes from an animal and measured the fluorescence of a membrane probe in the same way that these researchers did. If the fluorescence was found to be 0.27, and assuming that the movement of the fluorescent probe is only affected by fatty acid saturation/unsaturation, what can you predict about the ratio of saturated to unsaturated fatty acids in ethanolamine phospholipids present in membranes of this animal? Express your answer to the hundredths place.

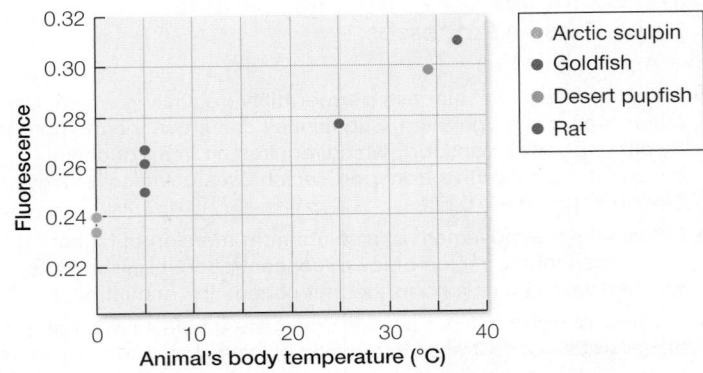

Go to **LaunchPad** for the eBook, LearningCurve, animations, activities, flashcards, and additional resources and assignments.

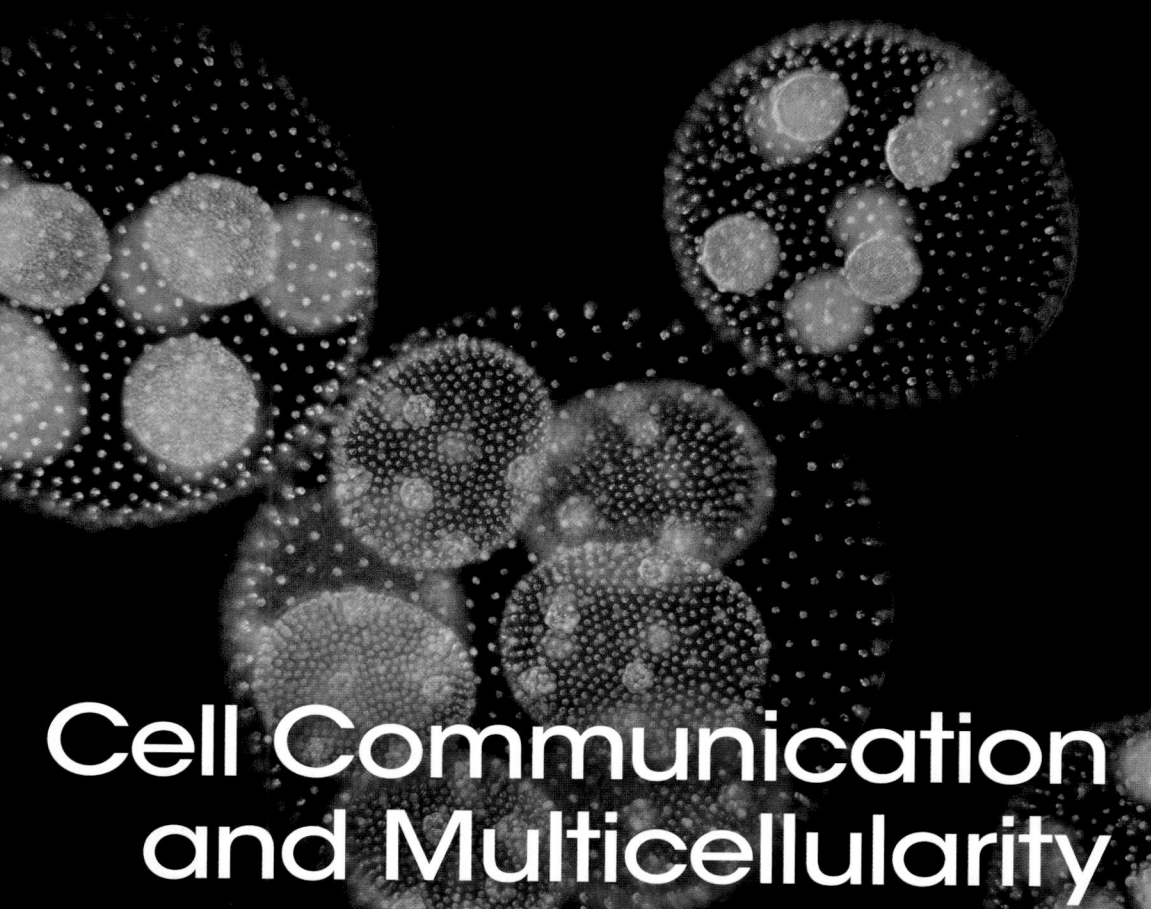

Cell Communication and Multicellularity

7

key concepts

7.1 Signals and Signaling Affect Cell Function

7.2 Receptors Bind Signals to Initiate a Cellular Response

7.3 The Response to a Signal Spreads through the Cell

7.4 Cells Change in Response to Signals in Several Ways

7.5 Adjacent Cells in a Multicellular Organism Can Communicate Directly

The green alga, *Volvox*, has thousands of cells, interacting with one another and specializing.

investigating life

A Signal for Bonding

Prairie voles (*Microtus ochrogaster*) are small rodents that live in temperate climates, where they dig tunnels in fields. When a male prairie vole encounters a female, the female enters her reproductive cycle and mating (which can take hours) often ensues. After mating the pair stays together, building a nest and raising their pups. The bond between the two voles is so strong that they stay together for life. Contrast this behavior with that of the montane vole (*M. montanus*), which is closely related to the prairie vole and lives in the hills not far away. In this species, mating is quick, and afterward the pair separates. The male looks for new mates and the female abandons her young soon after they are born.

The reason for these dramatic behavioral differences lies in the brains of these two species. When prairie voles mate, the brains of both males and females release specific peptides consisting of nine amino acids. In females, the peptide is oxytocin; in males, it is vasopressin. The peptides circulate in the bloodstream and reach all tissues in the body, but they bind to only a few cell types. These cells have surface proteins called receptors, to which the peptides specifically

The binding of peptide to receptor causes the receptor, which extends across the cell membrane, to change shape. Within the cytoplasm, this change sets off a series of events that ultimately result in changes in behavior. The receptors for oxytocin and vasopressin in prairie voles are most concentrated in the regions of the brain that are responsible for behaviors such as bonding and caring for young. In montane voles, there are far fewer receptors for these peptides, and as a result, fewer bonding and caring behaviors. Clearly, oxytocin and vasopressin are signals that induce these behaviors.

Intercellular signaling is a hallmark of multicellular organisms. A cell's response to a signal molecule takes place in three sequential steps. First, the signal binds to a receptor in the cell, often embedded in the outside surface of the cell membrane. Second, signal binding conveys a message to the cell. Third, the cell changes its activity in response to the signal. In a multicellular organism, these steps lead to changes in that organism's functioning, with signaling allowing different types of cells and tissues to work together.

7.1 Signals and Signaling Affect Cell Function

Both prokaryotic and eukaryotic cells process information from their environments. This information can be in the form of a physical stimulus, such as the light reaching your eyes as you read this book, or chemicals that bathe a cell, such as the carbohydrate lactose in the solution around bacteria. A signal may come from outside the organism, such as the scent of a female moth seeking a mate in the dark, or from a neighboring cell within the organism, such as the liver, where signals from other organs cause liver cells to regulate their uptake or release of glucose.

focus your learning

- Chemical signals that target cells can be classified according to signal source and mode of signal delivery.
- A signal transduction pathway includes a signal, a receptor, and a response
- Not all cells respond to a signal, because some lack the ability to receive the signal.

The presence of a signal does not always lead to a response; a cell does not respond to every signal, just as you are not paying close attention to every stimulus in your environment as you read this. To respond to a signal, a cell must have a specific receptor that can detect it and a way to use that information to influence cellular processes. A **signal transduction pathway** is a sequence of molecular events and chemical reactions that lead to a cell's response to a signal. Signal transduction pathways vary greatly in their details, but every such pathway involves a signal, a receptor, and a response. This section will provide a brief overview of signal transduction. We'll consider receptors in Key Concept 7.2, and other aspects of signal transduction in Key Concepts 7.3 and 7.4.

Cells receive several types of signals

The environment is full of signals. For example, our sense organs allow us to respond to light (a physical signal), or odors and tastes (chemical signals). Bacteria and protists can respond to small chemical changes in their surroundings. Plants respond to light as a signal as well as an energy source, for example, by growing toward the source of light. A cell that is deep inside a large multicellular organism and far away from the exterior environment receives signals from neighboring cells and the surrounding extracellular fluids. In multicellular organisms, chemical signals are often made in one part of the body and arrive at target cells by local diffusion or by circulation in the blood or the plant vascular system. Chemical cell signals are usually present in tiny concentrations (as low as 10^{-10} M) (see Chapter 2 for an explanation of molar concentrations) and differ in their sources and mode of delivery (**Figure 7.1**):

- **Autocrine** signals diffuse to and affect the cells that make them. For example, many tumor cells reproduce uncontrollably because they both make, and respond to, signals that stimulate cell division.

- **Juxtacrine** signals affect only cells right next to and in contact with the cell producing the signal. This type of signaling is especially common during development, when cells are in groups and changing to become specialized.

- **Paracrine** signals diffuse to and affect nearby cells. An example occurs in inflammation when the skin is cut. Signals from skin cells are sent to nearby blood cells to aid in healing (see Key Concept 41.2).

- Signals that travel through the circulatory systems of animals or the vascular systems of plants are generally called **hormones**.

A signal transduction pathway involves a signal, a receptor, and responses: Overview

As you saw in the opening story, the elements of a signal transduction pathway are a signal, a receptor, and a response (**Figure 7.2**). For the information from a signal to be transmitted to a cell,

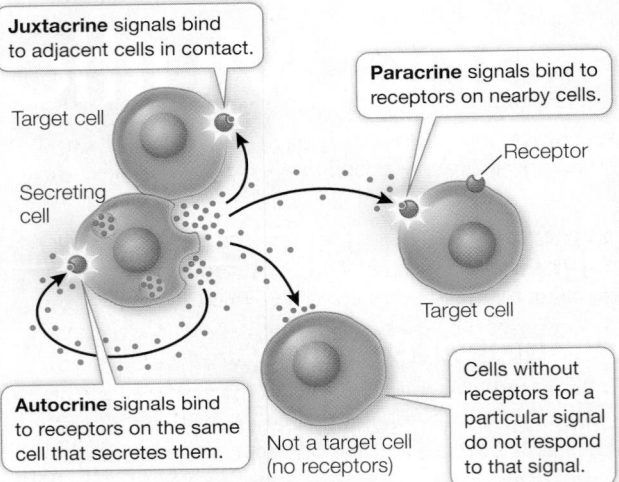

(A) Cell signaling on the local level

Juxtacrine signals bind to adjacent cells in contact.

Target cell

Paracrine signals bind to receptors on nearby cells.

Receptor

Secreting cell

Target cell

Autocrine signals bind to receptors on the same cell that secretes them.

Not a target cell (no receptors)

Cells without receptors for a particular signal do not respond to that signal.

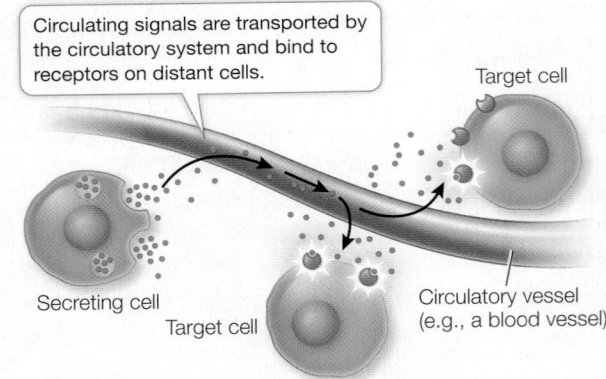

(B) Cell signaling from a distance

Circulating signals are transported by the circulatory system and bind to receptors on distant cells.

Target cell

Secreting cell

Target cell

Circulatory vessel (e.g., a blood vessel)

Figure 7.1 Chemical Signaling Systems (A) A signal molecule can diffuse to and act on the cell that produces it, an adjacent cell, or a nearby cell. (B) Hormones are chemical signals that act on distant cells and must be transported by the organism's circulatory system.

Activity 7.1 Chemical Signaling Systems
www.Life11e.com/ac7.1

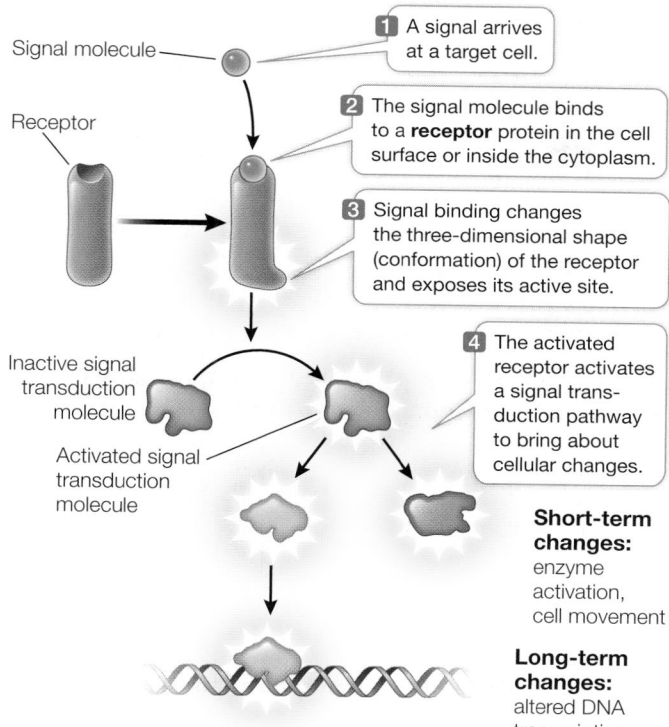

1 A signal arrives at a target cell.

Signal molecule

Receptor

2 The signal molecule binds to a **receptor** protein in the cell surface or inside the cytoplasm.

3 Signal binding changes the three-dimensional shape (conformation) of the receptor and exposes its active site.

Inactive signal transduction molecule

Activated signal transduction molecule

4 The activated receptor activates a signal transduction pathway to bring about cellular changes.

Short-term changes: enzyme activation, cell movement

Long-term changes: altered DNA transcription

Figure 7.2 A Signal Transduction Pathway   This general pathway is common to many cells and situations. The ultimate effects on the cell are either short-term or long-term molecular changes, or both.

the target cell must be able to receive the signal and respond to it. This is the job of receptors. All cells may be exposed to a chemical signal, but most cells in the organism may not be able to respond to it. *Only cells with the appropriate receptors can respond.*

The response can involve enzymes, which catalyze biochemical reactions, and transcription factors, which are proteins that turn the expression of particular genes on and off. An important feature of signal transduction is that the activities of specific enzymes and transcription factors are regulated: they are either activated or inactivated to bring about cellular changes (see Figure 7.2). For example, an enzyme may be activated by the addition of a phosphate group (phosphorylation) to a particular site on the protein, thereby changing the enzyme's shape (see Figure 3.13B) and exposing its active site. The activity of a protein can also be regulated by mechanisms that control its location in the cell. For example, a transcription factor located in the cytoplasm is inactive because it is separated from the genetic material in the nucleus; a signal transduction pathway may result in the transport of the factor to the nucleus, where it can affect gene expression.

In this chapter we'll consider signal transduction pathways in isolation from one another. In life, however, signal transduction pathways are often interconnected. A great deal of **crosstalk**—interactions between different signal transduction pathways—is common. For example, a single activated protein (receptor or enzyme) might activate enzymes or transcription factors in multiple pathways, leading to multiple responses to a single stimulus. Multiple signal transduction pathways might converge on a single transcription factor, allowing the transcription of a single gene to be adjusted in response to several different signals. Crosstalk can also result in the activation of one pathway and the inhibition of another. The phenomenon of crosstalk inside the cell is analogous to the "crosstalk" that occurs

at the level of the whole body. For example, in your limbs you have opposing muscles. When you bend your elbow, you contract one set of muscles and relax the opposing muscles, so that your arm will bend. Because of crosstalk, biologists often refer to "signaling networks" rather than signal transduction pathways, reflecting the high degree of complexity in cellular signaling.

7.1 recap

Cells are constantly exposed to molecular signals that can come from the external environment or from within the body of a multicellular organism. To respond to a signal, the cell must have a specific receptor that detects the signal and activates some type of cellular response. Signal transduction pathways involve regulation of enzymes and transcription factors, and crosstalk often occurs between pathways.

learning outcomes

You should be able to:

- Use the source and distribution mechanism of a signal to identify the type.
- Give an example of how a particular type of signal (autocrine, juxtacrine, paracrine, or hormone) functions in an organism.
- Explain how specificity is achieved in a signal transduction pathway.
- Compare cells found in the same organism that can and cannot respond to a specific chemical signal.

1. Under what conditions would autocrine signaling be most advantageous for a cell?
2. How are hormones delivered to target cell receptors?
3. All cells have the potential to respond to all signals in an organism. What makes signaling specific?

The general features of signal transduction pathways described in this section will recur in more detail throughout the chapter. First, let's consider more closely the nature of the receptors that bind signal molecules.

key concept 7.2   Receptors Bind Signals to Initiate a Cellular Response

Any given cell in a multicellular organism is bombarded with many signals. However, it responds to only some of them, because no cell makes receptors for all signals. A **receptor** protein recognizes its signal very specifically, in much the same way that a membrane transport protein recognizes and binds to the substance it transports. This specificity ensures that only those cells that make a specific receptor will respond to a given signal.

focus your learning

- Cells make specific receptor proteins that recognize only the signals to which they respond.
- The binding of a chemical signal (called a ligand) with its receptor is reversible and is measured by a dissociation constant.

(continued)

• Intracellular receptors are located inside the cell, where they interact with physical signals such as light or with chemical signals that diffuse across the cell membrane.

Receptors that recognize chemical signals have specific binding sites

A **ligand** is a specific chemical signal molecule that fits into a three-dimensional site on its protein receptor (**Figure 7.3A**). Binding of the signaling ligand causes the receptor protein to change its three-dimensional shape, and that conformational change initiates a cellular response. The ligand does not contribute further to this response. In fact, the ligand is usually not changed; its role is purely to "knock on the door."

The sensitivity of a cell to a signal is determined in part by the affinity of the cell's receptors for the signal ligand—the likelihood that the receptor will bind to the ligand at any given ligand concentration. Receptors (R) bind to their ligands (L) according to chemistry's law of mass action. This means that the binding is reversible:

$$R + L \rightleftharpoons RL \tag{7.1}$$

For most ligand–receptor complexes (RL), binding is favored. Reversibility is important, however, because if the ligand were never released, the receptor would be continuously stimulated and the cell would never stop responding.

As with any reversible chemical reaction, the binding and dissociation processes each have a rate constant, here designated k_1 and k_2:

$$\text{Binding: } R + L \xrightarrow{k_1} RL \tag{7.2}$$

$$\text{Dissociation: } RL \xrightarrow{k_2} R + L \tag{7.3}$$

A rate constant relates the rate of a reaction to the concentration(s) of the reactant(s):

$$\text{Rate of binding} = k_1[R][L] \tag{7.4}$$

$$\text{Rate of dissociation} = k_2[RL] \tag{7.5}$$

where "[ ]" indicates the concentration of the substance inside the brackets. Binding of a receptor to a ligand is reversible, and when equilibrium is reached the rate of binding equals the rate of dissociation:

$$k_1[R][L] = k_2[RL] \tag{7.6}$$

If this is rearranged, we get:

$$\frac{[R][L]}{[RL]} = \frac{k_2}{k_1} = K_D \tag{7.7}$$

K_D, the **dissociation constant**, is a measure of the affinity of the receptor for its ligand. The lower the K_D, the higher the affinity of the ligand for its receptor. Some receptors have very low K_D values, which allows them to bind their ligands at very low ligand concentrations; other receptors have higher K_D values and need more ligand to be present to set off their signal transduction pathways.

An entire field of biology and medicine—called pharmacology—is devoted to the study of drugs. These molecules are usually thought of as synthetic, but some are natural substances, such as caffeine (see Figure 7.3B). Drugs function as ligands that bind specific receptors. In the discovery and design of new drugs, it is helpful to know the specific receptor that the drug will bind, because then it is possible to determine the K_D value of its binding. This is one factor that can be

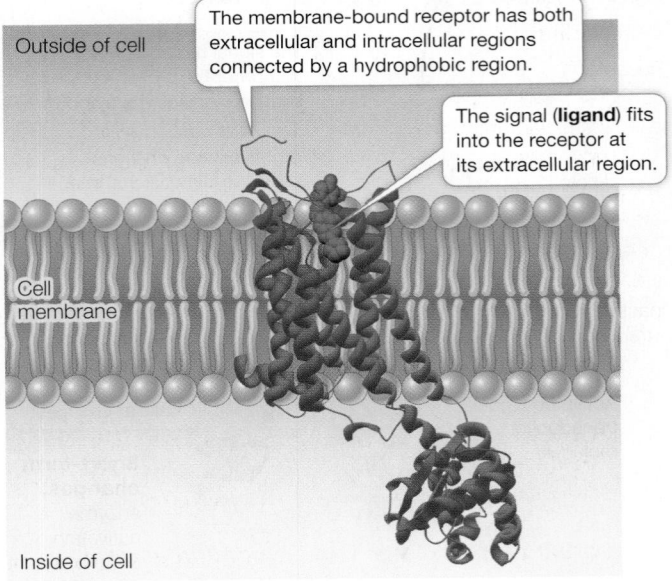

(A) The signal, adenosine, binding to its receptor

Outside of cell

The membrane-bound receptor has both extracellular and intracellular regions connected by a hydrophobic region.

The signal (**ligand**) fits into the receptor at its extracellular region.

Cell membrane

Inside of cell

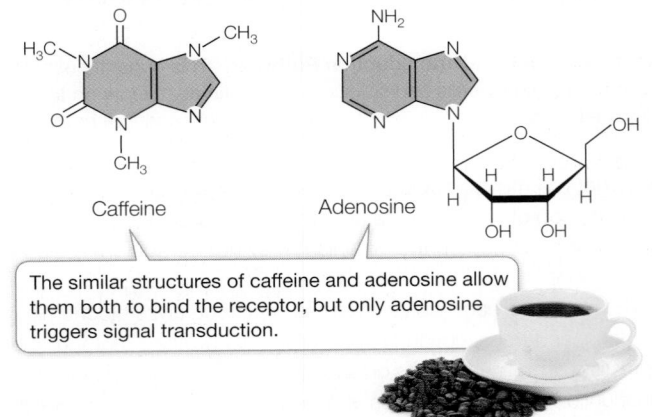

(B) Chemical similarity between caffeine and adenosine

Caffeine   Adenosine

The similar structures of caffeine and adenosine allow them both to bind the receptor, but only adenosine triggers signal transduction.

Figure 7.3 A Signal and Its Receptor   (A) The adenosine 2A receptor occurs in the human brain, where it is involved in inhibiting active wakefulness. **(B)** Adenosine is the normal (endogenous) ligand for the receptor. Caffeine has a structure similar to that of adenosine and can act as an antagonist that binds the receptor and prevents its normal functioning.

Q: Is the binding of adenosine and caffeine covalent or noncovalent? Explain your answer.

taken into consideration when determining dosage levels. Of course, many drugs have side effects, and these are also dosage-dependent.

What happens when a ligand binds to a receptor? When we discussed small molecules binding to proteins in Key Concept 3.2, we described how proteins often change shape when bound. This is exactly what happens to receptors. The change in the receptor's shape may expose a previously hidden group of amino acids on the protein that participate in a biochemical activity. This activity may be the binding of another molecule, such as another protein (e.g., a G protein, discussed below), or a substrate for an enzyme.

Instead of the ligand, other chemicals that resemble it can bind to the receptor. **Agonists** are chemicals that set a receptor into signal transduction mode just as the ligand does. In contrast, inhibitors, or **antagonists**, bind to the receptor and "freeze" it in place, preventing the real ligand from binding, but do not set off signal transduction. Agonists and antagonists can be natural, or we can make them in the lab once we know the details of receptor–ligand binding.

Many substances that alter human behavior bind to specific receptors in the brain and prevent the binding of the receptors' specific ligands. One example is caffeine, which is probably the world's most widely consumed stimulant. In the brain, the nucleoside adenosine acts as a ligand that binds to a receptor on nerve cells, initiating a signal transduction pathway that reduces brain activity, especially feelings of active wakefulness. Because caffeine has a molecular structure similar to that of adenosine, it also binds to the adenosine receptor (**Figure 7.3B**). But in this case binding does not initiate a signal transduction pathway. Rather, it "ties up" the receptor, preventing adenosine binding and thereby allowing continued nerve cell activity and an active feeling.

Receptors can be classified by location and function

There are many kinds of chemical signals. Some ligands are hydrophobic (nonpolar) and can diffuse through membranes, whereas others cannot. Physical signals such as light also vary in their ability to penetrate cells and tissues. Correspondingly, a receptor can be classified by its location in the cell, which largely depends on the nature of its signal (**Figure 7.4**):

- *Membrane receptors*: Large or polar ligands cannot cross the lipid bilayer. Insulin, for example, is a protein hormone that cannot diffuse through the cell membrane; instead, it binds to a transmembrane receptor with an extracellular binding domain.

- *Intracellular receptors*: Small or nonpolar ligands can diffuse across the nonpolar phospholipid bilayer of the cell membrane and enter the cell. The hormone estrogen, for example, is a lipid-soluble steroid (see in-text art p. 61) that can diffuse across the cell membrane; it binds to a receptor inside the cell.

Light of certain wavelengths can penetrate the cells in a plant leaf quite easily, and many types of light receptors in plants are also intracellular.

In complex eukaryotes such as mammals and higher plants, there are three well-studied categories of cell membrane receptors that are grouped according to their functions: ion channels, protein kinase receptors, and G protein-coupled receptors.

ION CHANNELS As you saw in Key Concept 6.3, the cell membranes of many types of cells have gated **ion channels** that allow ions such as Na^+, K^+, Ca^{2+}, or Cl^- to enter or leave the cell. The gate-opening mechanism is an alteration in the three-dimensional shape of the channel protein upon interaction with a signal; thus these proteins function as ***receptors**. Each type of ion channel responds to a specific signal, including sensory stimuli such as light, sound, and electric charge differences across the cell membrane, as well as chemical ligands such as hormones and neurotransmitters.

***connect the concepts** Ion channels are key to the functioning of the nervous system. An example is the connection between nerve and muscle described in Key Concepts 44.3 and 47.1.

The acetylcholine receptor, located in the cell membrane of skeletal muscle cells, is an example of an ion channel. This protein is a sodium channel that binds the ligand acetylcholine, which is a neurotransmitter—a chemical signal released from nerve cells (**Figure 7.5**). When two molecules of acetylcholine bind to the channel, it opens for about a thousandth of a second. That is enough time for Na^+, which is more concentrated outside the cell than inside, to rush into the cell, moving in response to both concentration and electric potential gradients. The change in Na^+ concentration in the cell initiates a series of events that result in muscle contraction.

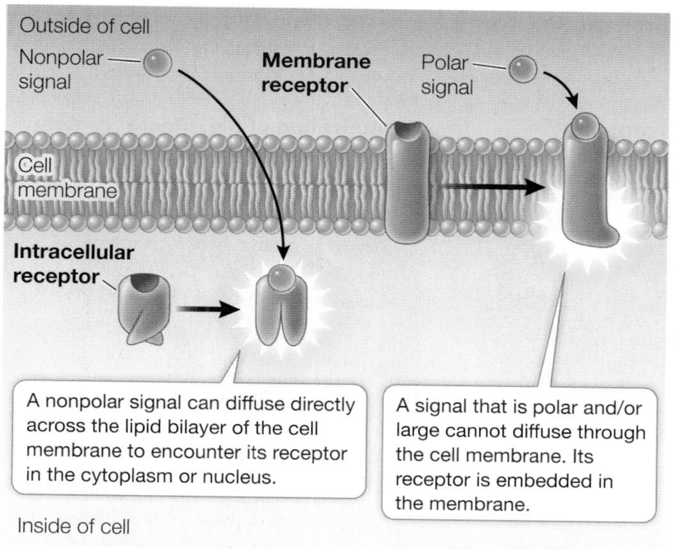

Outside of cell

Nonpolar signal — **Membrane receptor** — Polar signal

Cell membrane

Intracellular receptor

A nonpolar signal can diffuse directly across the lipid bilayer of the cell membrane to encounter its receptor in the cytoplasm or nucleus.

A signal that is polar and/or large cannot diffuse through the cell membrane. Its receptor is embedded in the membrane.

Inside of cell

Figure 7.4 Two Locations for Receptors Receptors can be located inside the cell (in the cytoplasm or nucleus) or in the cell membrane.

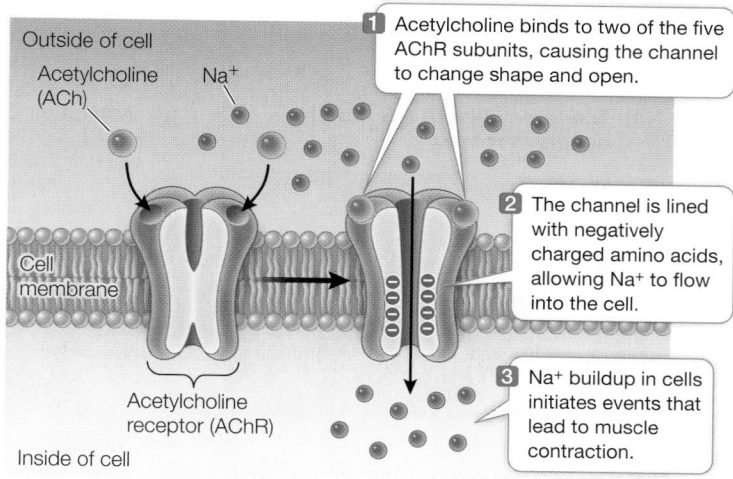

Outside of cell

Acetylcholine (ACh) Na^+

1 Acetylcholine binds to two of the five AChR subunits, causing the channel to change shape and open.

Cell membrane

2 The channel is lined with negatively charged amino acids, allowing Na^+ to flow into the cell.

Acetylcholine receptor (AChR)

Inside of cell

3 Na^+ buildup in cells initiates events that lead to muscle contraction.

Figure 7.5 A Gated Ion Channel The acetylcholine receptor (AChR) is a ligand-gated ion channel for sodium ions. It is made up of five polypeptide subunits. When acetylcholine molecules (ACh) bind to two of the subunits, the gate opens and Na^+ flows into the cell. This channel helps regulate membrane polarity.

Figure 7.6 A Protein Kinase Receptor The hormone insulin binds to a receptor on the outside surface of the cell and initiates a response.

PROTEIN KINASE RECEPTORS   Some eukaryotic receptor proteins, called **protein kinases,** catalyze the phosphorylation (adding phosphate) of themselves or other proteins, thus changing their shapes and therefore their functions.

$$\text{Target protein} + \text{ATP} \xrightarrow{\text{Protein kinase}} \text{Protein}-\text{P} + \text{ADP}$$
(altered shape and function)

Phosphorylation is a reaction that is especially important in biology. Of the estimated 21,000 genes that code for proteins in humans, more than 500 of them encode protein kinases. When you think of all the functions that make up a person, this is indeed an impressive number. The three amino acids that are phosphorylated are therefore worth knowing:

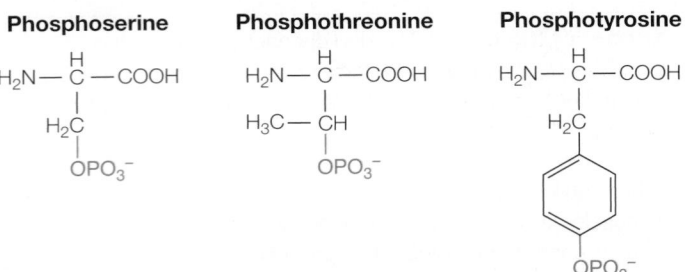

Phosphoserine   **Phosphothreonine**   **Phosphotyrosine**

The receptor for insulin is an example of a protein kinase receptor. Insulin is a protein hormone made by the pancreas. Its receptor has two copies each of two different polypeptide subunits called α and β (**Figure 7.6**). When insulin binds to the receptor, the receptor becomes activated and is able to phosphorylate itself and certain cytoplasmic proteins that are appropriately called insulin-response substrates. These proteins then initiate many cellular responses, including the insertion of glucose transporters (see Figure 6.12) into the cell membrane.

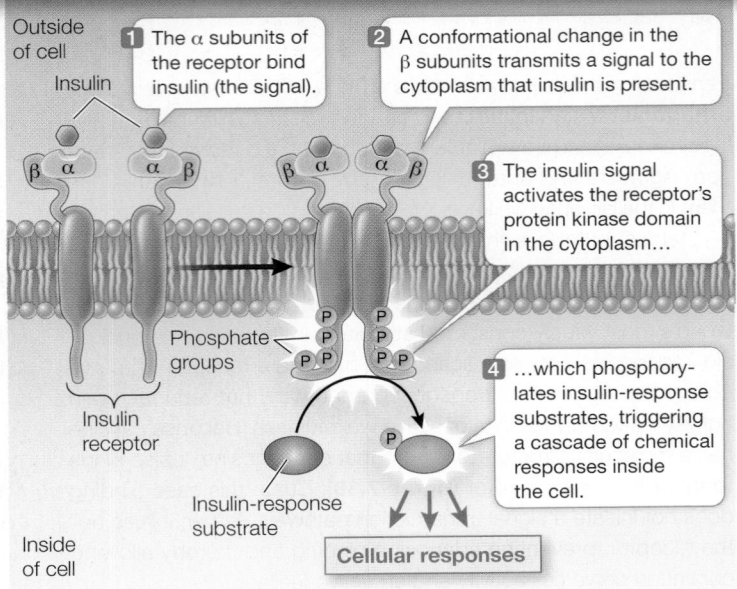

1 The α subunits of the receptor bind insulin (the signal).

2 A conformational change in the β subunits transmits a signal to the cytoplasm that insulin is present.

3 The insulin signal activates the receptor's protein kinase domain in the cytoplasm...

4 ...which phosphorylates insulin-response substrates, triggering a cascade of chemical responses inside the cell.

G PROTEIN-COUPLED RECEPTORS   A third category of eukaryotic cell membrane receptors is the **G protein-coupled receptors**, also referred to by the more impressive sounding name seven-transmembrane domain receptors. These receptors have many roles, including light detection in the mammalian retina (photoreceptors), detection of odors (olfactory receptors), and regulation of mood and behavior (such as mating in mammals and even single-celled yeasts). The receptors that bind the hormones oxytocin and vasopressin, which affect mating behavior in voles (see the opening story), are G protein-coupled receptors.

The seven-transmembrane domains of the receptor protein pass through the phospholipid bilayer and are separated by short loops that extend either outside or inside the cell. Ligand binding on the extracellular side of the receptor changes the shape of its cytoplasmic region, exposing a site that binds to a mobile membrane protein called a **G protein**. The G protein is partially inserted into the lipid bilayer and partially exposed on the cytoplasmic surface of the membrane.

Many G proteins have three polypeptide subunits and can bind three different types of molecules (**Figure 7.7A**):

(A) The three units of a G protein-coupled receptor

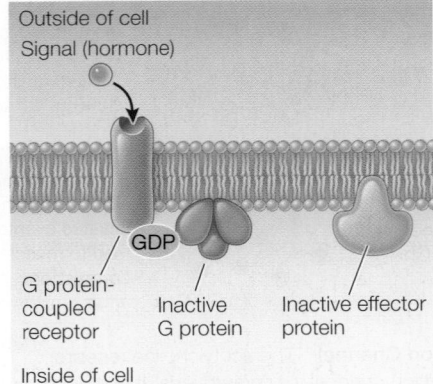

(B) Activation of the G protein

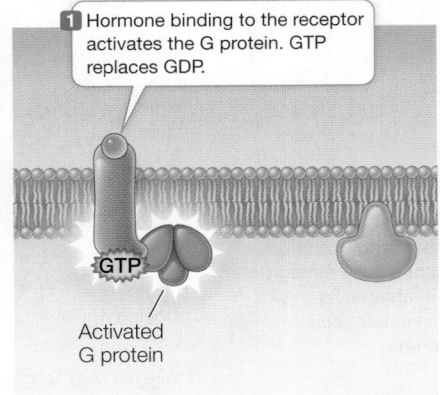

1 Hormone binding to the receptor activates the G protein. GTP replaces GDP.

(C) Activation of the effector protein

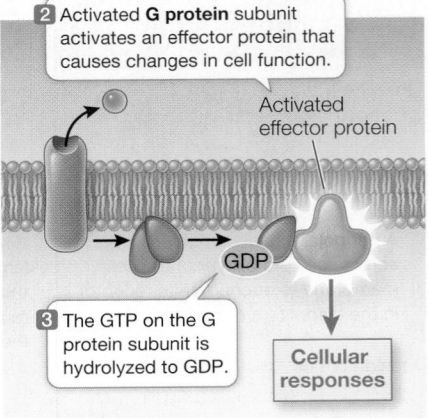

2 Activated **G protein** subunit activates an effector protein that causes changes in cell function.

3 The GTP on the G protein subunit is hydrolyzed to GDP.

Figure 7.7 A G Protein-Coupled Receptor The G protein is an intermediary between the receptor and its effector.

Animation 7.1 A Signal Transduction Pathway www.Life11e.com/a7.1

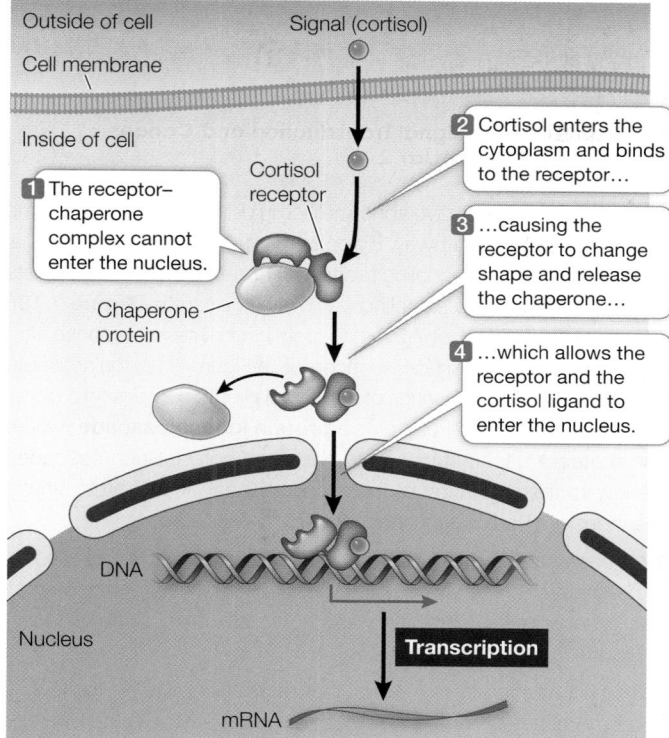

Figure 7.8 An Intracellular Receptor The receptor for cortisol is in the cytoplasm, bound to a chaperone protein that is released when cortisol binds to the receptor.

they are activated; after binding their ligands, these transcription factors move to the nucleus where they bind to DNA and alter the expression of specific genes. A typical example is the receptor for the steroid hormone cortisol. This receptor is normally bound to a chaperone protein that blocks it from entering the nucleus. Binding of the hormone causes the receptor to change its shape so that the chaperone is released (**Figure 7.8**). This release allows the receptor to enter the nucleus, where it affects DNA transcription. Another group of intracellular receptors is always located in the nucleus, and their ligands must enter the nucleus before binding.

> ### 7.2 recap
>
> Receptors are proteins that bind, or are changed by, specific ligands or physical signals. The changed receptor initiates a response in the cell. These receptors are located in the cell membrane or inside the cell.

learning outcomes

You should be able to:

- Explain how and why different cells respond to different chemical signals.
- Using equations, describe the chemical kinetics of receptor–ligand interactions.
- Analyze dissociation constants as a measure of binding affinity.
- Compare and contrast cell membrane receptors and intracellular receptors.

1. What is the dissociation constant K_D, and what does it tell us about the likelihood that a signal will produce a cellular response?
2. What is the chemical nature and importance of specificity in the binding of a receptor to its particular ligand?
3. What are the differences between receptors located in the cytoplasm and those at the cell membrane?

Now that we have discussed signals and receptors, let's examine the characteristics of the molecules (transducers) that mediate the cellular response.

1. The receptor
2. GDP and GTP (guanosine diphosphate and triphosphate, respectively; these are nucleoside phosphates like ADP and ATP)
3. An effector protein (see next paragraph)

When the G protein binds to an activated receptor protein, GDP is exchanged for GTP (**Figure 7.7B**). At the same time, the ligand is usually released from the extracellular side of the receptor. GTP binding causes a conformational change in the G protein. The GTP-bound subunit then separates from the rest of the G protein, diffusing in the plane of the phospholipid bilayer until it encounters an **effector protein** to which it can bind. An effector protein is just what its name implies: it causes an effect in the cell. The binding of the GTP-bearing G protein subunit activates the effector—which may be an enzyme or an ion channel—thereby causing changes in cell function (**Figure 7.7C**).

After activation of the effector protein, the GTP bound to the G protein is hydrolyzed to GDP. The now inactive G protein subunit separates from the effector protein and diffuses in the membrane to collide with and bind to the other two G protein subunits. When the three components of the G protein are reassembled, the protein is capable of binding again to an activated receptor. After binding, the activated receptor exchanges the GDP on the G protein for a GTP, and the cycle begins again.

Intracellular receptors are located in the cytoplasm or the nucleus

Intracellular receptors are located inside the cell and respond to physical signals such as light (e.g., some photoreceptors in plants) or chemical signals that can diffuse across the cell membrane (e.g., steroid hormones in animals). Many intracellular receptors are transcription factors. Some are located in the cytoplasm until

> ### key concept
> ## 7.3 The Response to a Signal Spreads through the Cell

As you have seen, there are different kinds of signals and receptors. Not surprisingly, the ways that signals are transduced, and the resulting cellular responses, also vary. Some signal transduction pathways are quite simple and direct, whereas others involve multiple steps. As we mentioned in Key Concept 7.1, signal transduction pathways can involve enzymes and transcription factors. In addition, **second messengers** can diffuse throughout the cytoplasm and mediate further steps in pathways.

focus your learning

- A signal transduction cascade transmits and amplifies a signal inside a cell.
- Second–messenger molecules can transmit and amplify signals within a cell.
- Signal transduction is regulated in a cell by various mechanisms.

In many cases, a signal can initiate a cascade (chain) of events, in which proteins interact with other proteins, which interact with still other proteins until the final responses are achieved. Through such a cascade, an initial signal can be both amplified and distributed to cause several different responses in the target cell. In this section we will examine the kinds of molecules that transduce signals and look at several different signal transduction pathways.

The cell amplifies its response to ligand binding

Scientists worked out the signal transduction pathway for one growth factor by studying a cell that went wrong. Many human bladder cancers contain an abnormal form of a protein called ras (so named because a similar protein was previously isolated from a *rat* sarcoma tumor). Investigations of these bladder cancers showed that ras was a G protein. Recall that G proteins function as switches with "on" and "off" states. In the "off" state, G protein is bound to GDP, whereas in the "on" state the protein becomes bound to GTP. The abnormal form of the ras protein under study was always

(A) Ras function in a normal cell

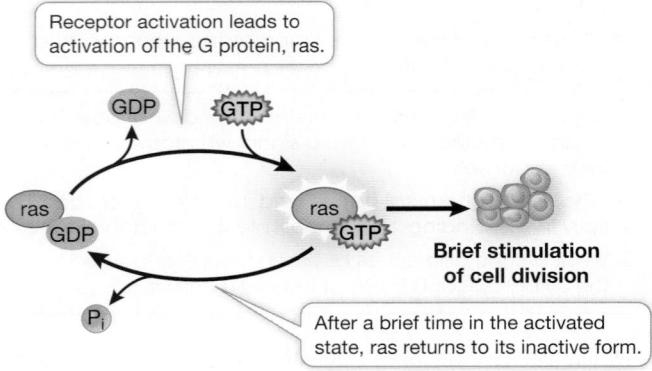

(B) Abnormal ras function in a cancer cell

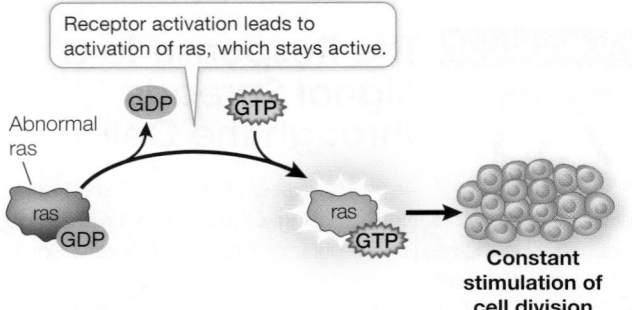

Figure 7.9 Signal Transduction and Cancer **(A)** Ras is a G protein that regulates cell division. **(B)** In some tumors, the ras protein is permanently active, resulting in uncontrolled cell division.

active because it was permanently bound to GTP and thus caused continuous cell division (**Figure 7.9**).

 Animation 7.2 Signal Transduction and Cancer
www.Life11e.com/a7.2

Other cancer cells have abnormalities in different parts of the same signal transduction pathway. By comparing the defects in abnormal cells with the normal signaling process in non-cancer cells, biologists worked out the entire signaling pathway (**Focus: Key Figure 7.10**). Recall from Key Concept 7.2 that G protein mediates a response after activation by the protein kinase receptor. In Figure 7.10, the activated G protein ras initiates a series of additional events comprising a signal transduction pathway known as a **protein kinase cascade**, where one protein kinase activates the next, and so on. Such cascades are key to the regulation of many cellular activities. Protein kinase cascades are useful signal transducers for four reasons:

1. At each step in the cascade of events, the signal is *amplified*, because each newly activated protein kinase is an enzyme that can catalyze the phosphorylation of many target proteins (see Figure 7.10, steps 5 and 6).

2. The information from a signal that originally arrived at the cell membrane is *communicated* to the nucleus, where the expression of multiple genes is often modified.

3. The multitude of steps provides some *specificity* to the process.

4. Different target proteins at each step in the cascade can provide *variation* in the response.

Second messengers can amplify signals between receptors and target molecules

Often there is a small molecule intermediary between the activated receptor and the cascade of events that ensues. Earl Sutherland and his colleagues at Case Western Reserve University discovered one such molecule when they were investigating the activation of the liver enzyme glycogen phosphorylase by the hormone epinephrine. The hormone is released when an animal faces life-threatening conditions and needs energy fast for the fight-or-flight response (see Figure 40.3). Glycogen phosphorylase catalyzes the breakdown of glycogen stored in the liver so that the resulting glucose molecules can be released to the blood. The enzyme is present in the liver cell cytoplasm but is inactive unless the liver cells are exposed to epinephrine:

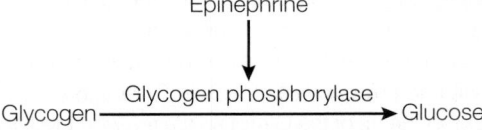

The researchers found that epinephrine could activate glycogen phosphorylase in liver cells that had been broken open, but only if the entire cell contents, including cell membrane fragments, were present. Under these conditions epinephrine was bound to the cell membrane fragments (the location of its receptor), but the active phosphorylase was present in the solution. Adding epinephrine to just cytoplasm with inactive phosphorylase did not result in activation. The researchers hypothesized that there must be a "second messenger" that transmits the epinephrine signal (epinephrine being the "first messenger"). The experiments confirmed the existence of a second messenger, later identified as **cyclic AMP (cAMP)**, which

focus: key figure

Outside of cell

Growth factor

1 A growth factor binds its receptor...

Figure 7.10 A Protein Kinase Cascade In a protein kinase cascade, a series of proteins are sequentially activated.

Q: Sorafenib is a drug that was designed to inhibit raf, which is very active in kidney cancer. How does this drug work on the protein kinase cascade?

ras GDP
Inactive

2 ...which phosphorylates itself.

3 The activated receptor initiates a series of events that allow ras to bind GTP and become activated.

ras GTP
Active raf

ras raf

4 Activated ras binds and activates raf.

MEK MEK P

5 Activated raf is a protein kinase that phosphorylates many molecules of MEK.

MAPK MAPK P

6 Activated MEK is a protein kinase that phosphorylates many molecules of MAP kinase.

Inactive kinase

Activated kinase

Amplification

Amplification

7 MAP kinase, when activated by phosphorylation, can enter the nucleus.

MAPK P

Cellular responses (including stimulation of cell division)

Inside of cell

Nucleus

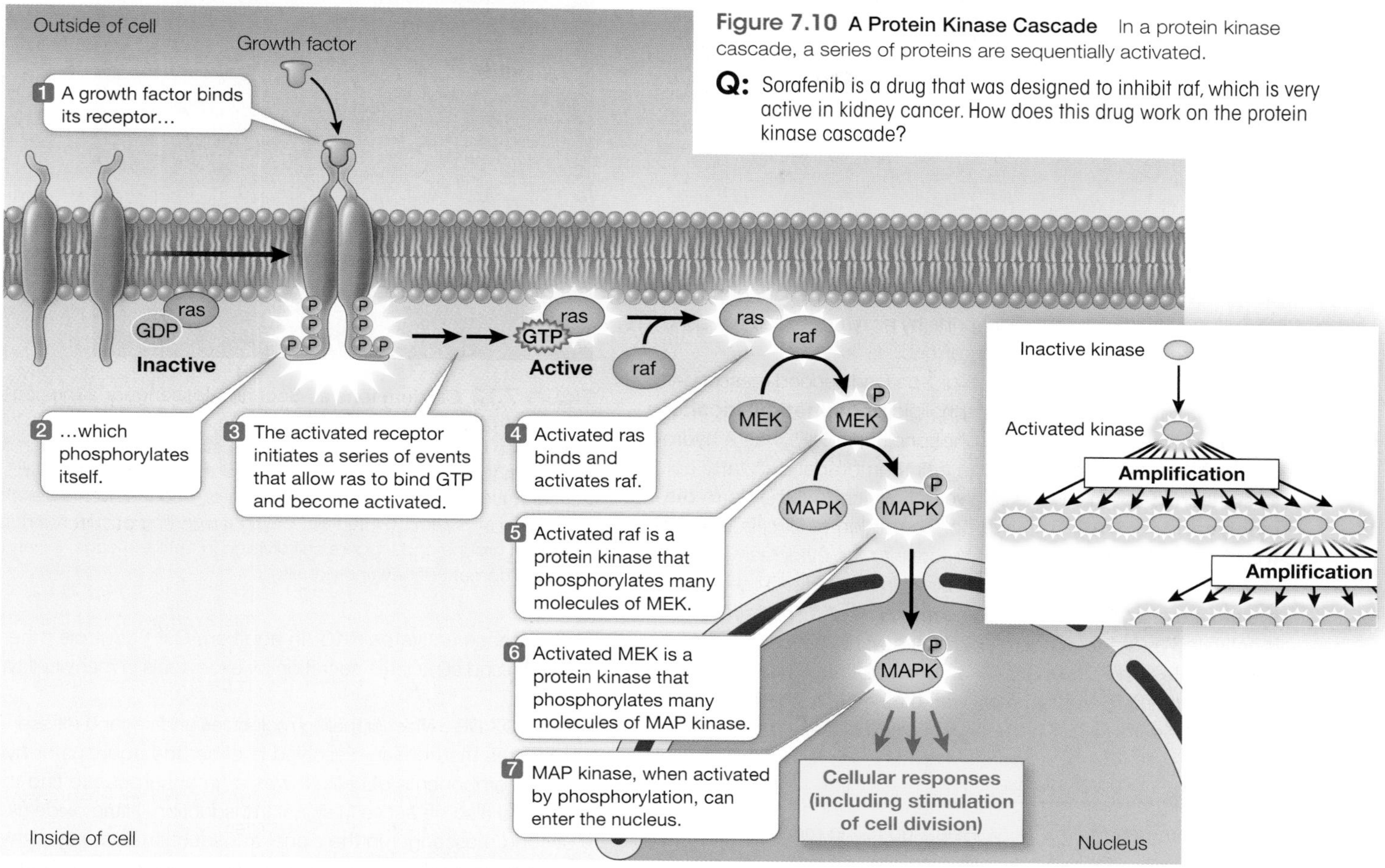

is produced from ATP by the enzyme adenylyl cyclase (**Figure 7.11**). Adenylyl cyclase is activated via a G protein-coupled epinephrine receptor (see the first steps in Figure 7.15).

In contrast to the specificity of receptor binding, second messengers such as cAMP allow a cell to respond to a single event at the cell membrane with many events inside the cell. Thus second messengers serve to rapidly amplify and distribute the signal—for example, binding of a single epinephrine molecule leads to the production of many molecules of cAMP, which then activate many enzyme targets by binding to them noncovalently. In the case of epinephrine and the liver cell, glycogen phosphorylase is just one of several enzymes that are activated.

Second messengers are often involved in crosstalk between different signaling pathways. Activation of the epinephrine receptor is not the only way for a cell to produce cAMP; and as noted, there are multiple targets of cAMP in the cell, and these targets are parts of other pathways.

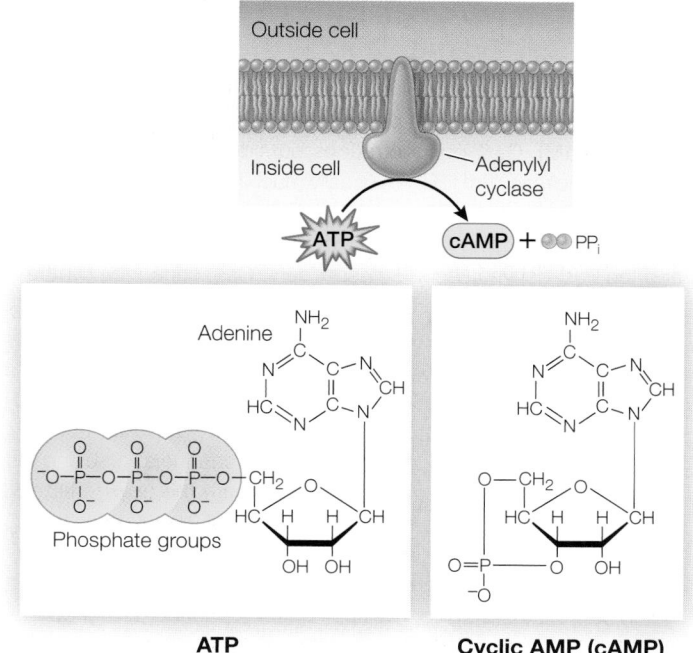

Figure 7.11 The Formation of Cyclic AMP The formation of cAMP from ATP is catalyzed by adenylyl cyclase, an enzyme that is activated by G proteins.

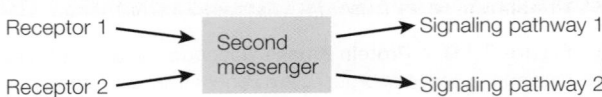

Several other classes of second messengers have since been identified, including lipid-derived second messengers, calcium ions, and nitric oxide.

LIPID-DERIVED SECOND MESSENGERS   In addition to their role as structural components of the cell membrane, phospholipids are also involved in signal transduction. When certain phospholipids are hydrolyzed into their component parts by enzymes called phospholipases, second messengers are formed.

The best-studied examples of lipid-derived second messengers come from hydrolysis of the phospholipid **phosphatidyl inositol-bisphosphate** (**PIP$_2$**). Like all phospholipids, PIP$_2$ has a hydrophobic portion embedded in the cell membrane: two fatty acid tails attached to a molecule of glycerol, which together form **diacylglycerol**, or **DAG**. The hydrophilic portion of PIP$_2$ is **inositol trisphosphate**, or **IP$_3$**, which projects into the cytoplasm.

As with cAMP, the receptors involved in this second-messenger system are often G protein-coupled receptors. A G protein subunit is activated by the receptor, then diffuses within the cell membrane and activates phospholipase C, an enzyme that is also located in the membrane. This enzyme cleaves off the IP$_3$ from PIP$_2$, leaving the diacylglycerol (DAG) in the phospholipid bilayer:

$$\text{PIP}_2 \xrightarrow{\text{Phospholipase C}} \text{IP}_3 \quad + \quad \text{DAG}$$

| In membrane | Released to cytoplasm | In membrane |

IP$_3$ and DAG are both second messengers; they have different modes of action that build on each other to activate protein kinase C (PKC). PKC refers to a family of protein kinases that can phosphorylate a wide variety of target proteins, leading to a multiplicity of cellular responses that vary depending on the tissue or cell type.

CALCIUM IONS   Calcium ions (Ca^{2+}) are scarce inside most cells, which have cytosolic Ca^{2+} concentrations of only about 0.1 micromolar (μM). Ca^{2+} concentrations outside cells and within the endoplasmic reticulum (ER) are usually much higher. Active transport proteins in the cell and ER membranes maintain this concentration difference by pumping Ca^{2+} out of the cytosol. In contrast to cAMP and the lipid-derived second messengers, Ca^{2+} cannot be synthesized to increase the intracellular Ca^{2+} concentration. Instead, Ca^{2+} ion levels are regulated via the opening and closing of ion channels and the action of membrane pumps.

Many signals can cause calcium channels to open, including IP$_3$. The entry of a sperm into an egg is a very important signal that causes a massive opening of calcium channels, resulting in numerous and dramatic changes that prepare the now-fertilized egg for cell division and development (**Figure 7.12**). Whatever the initial signal that causes calcium channels to open, their opening results in a dramatic increase in cytosolic Ca^{2+} concentration, which can increase up to 100-fold within a fraction of a second.

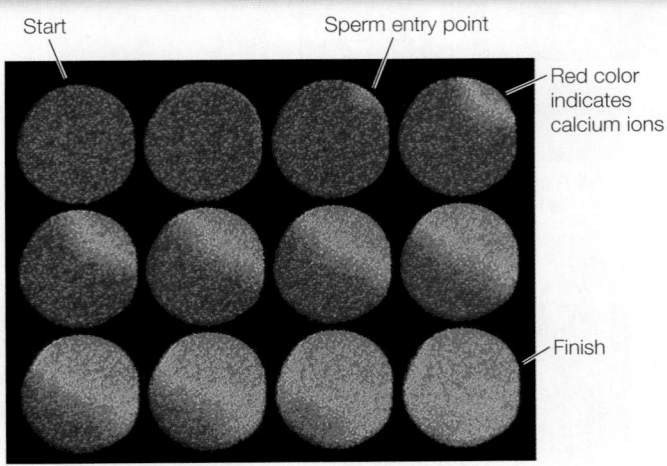

Figure 7.12 Calcium Ions as Second Messengers   The concentration of Ca^{2+} can be measured using a dye that fluoresces when it binds the ion. Here, fertilization in a sea star egg causes a rush of Ca^{2+} from the environment into the cytoplasm. Areas of high Ca^{2+} concentration are indicated by the red color, and the events are photographed at 5-second intervals. Calcium signaling occurs in virtually all animal groups and triggers cell division in fertilized eggs, initiating the development of new individuals.

This increase activates PKC. In addition, Ca^{2+} controls other ion channels and stimulates secretion by exocytosis in many cell types.

NITRIC OXIDE   Most signaling molecules and second messengers are solutes that remain dissolved in either the aqueous or hydrophobic components of cells. It was a great surprise to find that a gas could also be active in signal transduction. Nitric oxide (NO) is a second messenger in the signal transduction pathway between the neurotransmitter acetylcholine (see Key Concept 7.2) and the relaxation of smooth muscles lining blood vessels, which allows more blood flow (**Figure 7.13**). In the body, NO is made from the amino acid arginine by the enzyme NO synthase. When the acetylcholine receptor on the surface of an endothelial cell is activated, IP$_3$ is released from the membrane (via the pathway shown in Figure 7.13), causing a calcium channel in the ER membrane to open and a subsequent increase in cytosolic Ca^{2+}. The Ca^{2+} then activates NO synthase to produce NO. NO is chemically very unstable, readily reacting with oxygen gas as well as other small molecules. Although NO diffuses readily, it does not get far. Conveniently, the endothelial cells are close to the underlying smooth muscle cells, where NO activates an enzyme called guanylyl cyclase (a close relative of adenylyl cyclase). This enzyme catalyzes the formation of cyclic GMP (cGMP): yet another second messenger that contributes to the relaxation of muscle cells.

The discovery of NO as a participant in signal transduction explained the action of nitroglycerin, a drug that has been used for more than a century to treat angina, the chest pain caused by insufficient blood flow to the heart. Nitroglycerin releases NO, which results in relaxation of the blood vessels and increased blood flow. The drug sildenafil (Viagra) was developed to treat angina via the NO signal transduction pathway but was only modestly useful for that purpose. However, men taking it reported more pronounced penile erections. During sexual stimulation, NO acts as a signal, causing an increase in cGMP and a subsequent relaxation of the smooth muscles

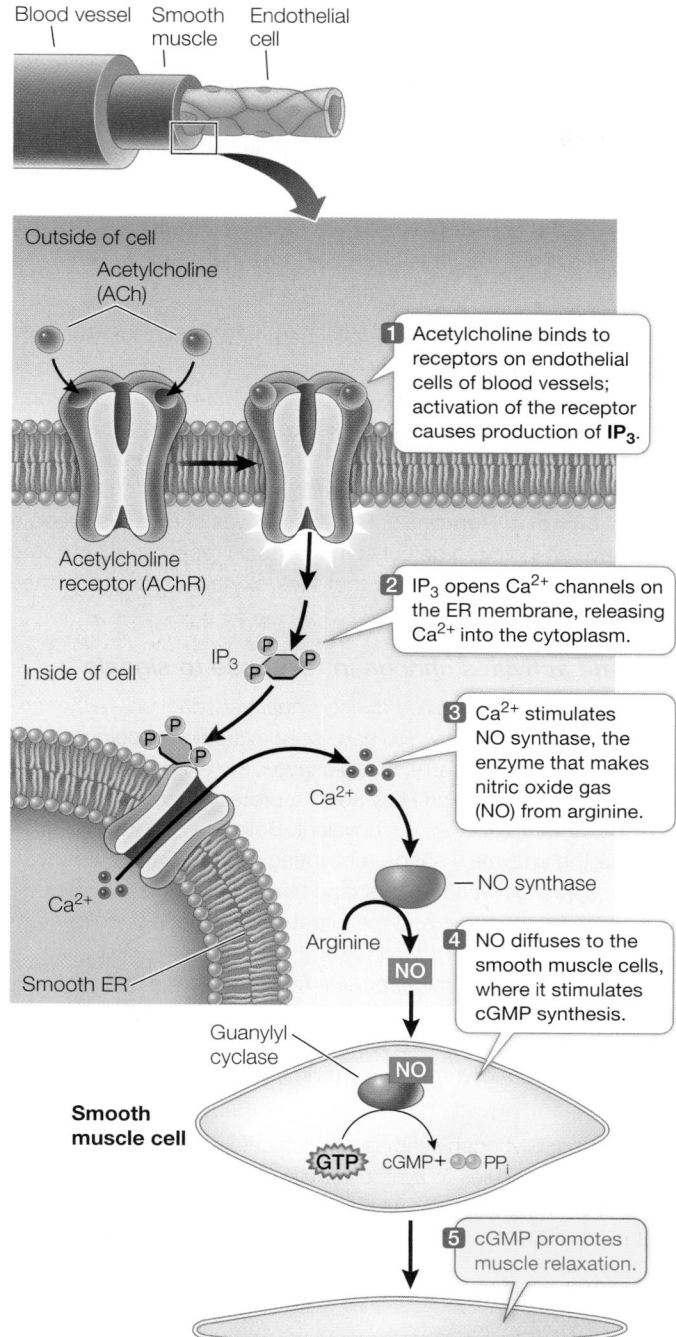

Figure 7.13 **Nitric Oxide in Signal Transduction**   Nitric oxide (NO) is an unstable gas, which nevertheless serves as a mediator between a signal, acetylcholine (ACh), and its effect: the relaxation of smooth muscles.

surrounding the arteries in the corpus cavernosum of the penis. As a result of this signal, the penis fills with blood, producing an erection. Sildenafil acts by inhibiting an enzyme (a phosphodiesterase) that breaks down cGMP—resulting in more cGMP and better erections.

Signal transduction is highly regulated

Cells can regulate the activity of a molecule involved in signal transduction. The concentration of NO, which breaks down quickly, can

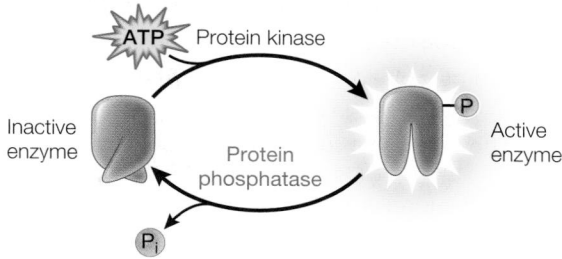

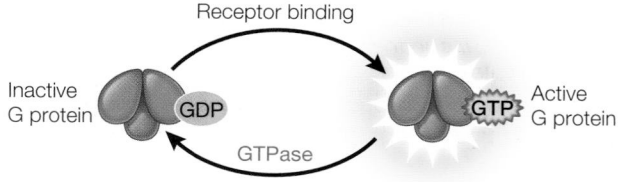

Figure 7.14 **Regulation of Signal Transduction**   Some signals lead to the production of active transducers such as **(A)** protein kinases, **(B)** G proteins, and **(C)** cAMP. Other enzymes (shown in red) inactivate or remove these transducers.

be regulated only by how much of it is made. By contrast, membrane pumps and ion channels regulate the cytosolic concentration of Ca^{2+}, as you have seen. To regulate protein kinase cascades, G proteins, and cAMP, there are enzymes that inactivate the activated transducer (**Figure 7.14**).

The balance between the activities of enzymes that activate and inactivate transducers determines the ultimate cellular response to a signal. Cells can alter this balance in several ways:

- *Synthesis or breakdown of the enzymes*. For example, synthesis of adenylyl cyclase (which synthesizes cAMP) and breakdown of phosphodiesterase (which breaks down cAMP) would tilt the balance in favor of more cAMP in the cell.

- *Activation or inhibition of the enzymes by other molecules*. An example is the inhibition of phosphodiesterase by sildenafil.

Because cell signaling is so important in diseases such as cancer, a search is under way for new drugs that can modulate the activities of enzymes that participate in signal transduction pathways.

▶ 7.3 recap

Signal transduction is the series of steps between the binding of a signal to a receptor and the ultimate cellular response. A protein kinase cascade amplifies a signal through a series of protein phosphorylation reactions. In many cases, a second messenger amplifies and distributes the downstream effects of the signal. The activity of transducer molecules is regulated.

(continued)

learning outcomes

You should be able to:

- Describe how a protein cascade amplifies a signal.
- Explain how the same second-messenger molecule can be used in many different signal transduction pathways.
- Explain how signal transduction pathways are regulated in a cell.

1. How can an activated G protein such as ras amplify a response to ligand binding?
2. Cyclic AMP is a second messenger in many different responses. How can the same messenger act in different ways in different cells?
3. How are signal transduction cascades regulated?

You have seen how the binding of a signal to its receptor initiates the response of a cell to the signal, and how signal transduction pathways amplify the signal and distribute its effects to numerous targets in the cell. In the next section we will look at the third step in the signal transduction process, the actual effects of the signal on cell function.

key concept 7.4 Cells Change in Response to Signals in Several Ways

The effects of a signal on cell function take three primary forms: the opening of ion channels, changes in the activities of enzymes, or differential gene expression. These events set the cell on a path for further and sometimes dramatic changes in form and function.

focus your learning

- A cell can respond to signals by opening or closing ion channels.
- A cell can respond to signals by altering gene transcription.

Ion channels respond to signals by opening or closing

You have seen that ion channels can function as receptors in cell signaling (see Figure 7.5). In Chapters 44 and 45, you will learn how signaling affects ion channels in the central nervous system and in senses, where stimulation of a receptor by light, sound, or touch can lead to opening of an ion channel. A different type of ion channel response lies within some cells. For example, signal transduction using the IP_3–DAG pathway can result in the opening of Ca^{2+} channels in the ER membrane.

Oxytocin, which we described in the opening of this chapter, combines the two ion channel responses. Oxytocin binds to a G protein-coupled receptor (see Figure 7.7), which is expressed in brain tissue as well as muscles involved in birth and lactation.

Activation of the receptor results in signal transduction through the IP_3–DAG pathway, releasing Ca^{2+} into the cytoplasm via ion channels. In the brain, the effect of Ca^{2+} is to indirectly stimulate nerve cell activity by the opening of ion channels for Na^+, as you will learn in Chapter 44. To summarize:

$$\text{Oxytocin} \rightarrow \text{Receptor} \rightarrow \text{G protein activation} \rightarrow$$
$$IP_3 \text{ signal transduction} \rightarrow Ca^{2+} \text{ channel opening} \rightarrow$$
$$Na^+ \text{ channel opening}$$

You saw in the opening of this chapter that the behavioral effect of oxytocin in voles involves bonding between them. Does this also happen in people? Neuroscientist Paul Zak at the Center for Neuroeconomics Studies thinks so. He has done experiments with human volunteers and shown that oxytocin signaling is important in trusting behavior (**Investigating Life: Is Oxytocin a "Trust" Signal in Humans?**). Recently, it was shown that people who have an inherited defect in the oxytocin receptor act in some ways like the montane voles that lack oxytocin receptors: they are less trusting.

Enzyme activities change in response to signals

Enzymes are often modified during signal transduction—either covalently or noncovalently. You have seen examples of both types of protein modifications earlier in this chapter. For example, addition of a phosphate group to an enzyme by a protein kinase is a covalent change; cAMP binding is noncovalent. Both types of modifications change the enzyme's shape, activating or inhibiting its function. In the case of activation, the shape change exposes a previously inaccessible active site, and the target enzyme goes on to perform a new cellular role.

The G protein-mediated protein kinase cascade that is stimulated by epinephrine in liver cells results in the activation by cAMP of a key signaling molecule, protein kinase A. In turn, protein kinase A phosphorylates two other *enzymes, with opposite effects:

1. *Inhibition*: Glycogen synthase, which catalyzes the joining of glucose molecules to synthesize the energy-storing molecule glycogen, is inactivated when a phosphate group is added to it by protein kinase A. Thus the epinephrine signal *prevents glucose from being stored* in the form of glycogen (**Figure 7.15, step 1**).

2. *Activation*: Phosphorylase kinase is activated when a phosphate group is added to it. It is part of a protein kinase cascade that ultimately leads to the activation of glycogen phosphorylase, another key enzyme in glucose metabolism. This enzyme results in the *liberation of glucose molecules* from glycogen (**Figure 7.15, steps 2 and 3**).

---

*connect the concepts   Enzyme regulation determines the rate of chemical transformations in cells, and thus cell functions. See Key Concept 8.5.

The amplification of the signal in this pathway is impressive; as detailed in Figure 7.15, each molecule of epinephrine that arrives

experiment

Original Paper: Zak, P., R. Kurzband and W. T. Matzner. 2005. Oxytocin is associated with human trustworthiness. *Hormones and Behavior* 48: 522–527.

Cell signaling in the brain leads to changes in behavior. This experiment investigates whether the peptide oxytocin correlates with the development of trust between humans.

HYPOTHESIS▶ The signal oxytocin is involved in trust between humans.

METHOD

1 156 people given $10 to participate.

2 People randomly assigned to pairs who do not meet.

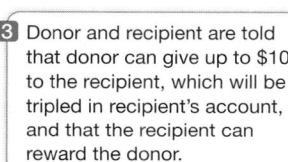

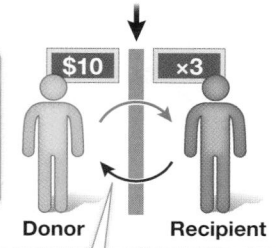

Donor   **Recipient**

3 Donor and recipient are told that donor can give up to $10 to the recipient, which will be tripled in recipient's account, and that the recipient can reward the donor.

4 Establish Trust: Donor hopes recipient will reward donor out of the tripled account more than donor gave. Trustworthiness is seen when recipient feels good about donor's generosity and gives back to donor.

5 Establish a Control Group: In the control group donors are told at random exactly what to give, and recipients are aware of this. So there was no donor intention to be generous.

6 Measure oxytocin levels in blood of donors in both experimental and control groups immediately after the donor decides what to give the recipient.

RESULTS

	Experimental (donor intent)	Control (no intent)
Amount given by donor	$5.52	$5.63
Amount returned by recipient	$6.96	$3.53
Oxytocin level in recipient (pg/ml)	278	198

CONCLUSIONS▶

1. People (recipients) who trust another person (donor, experimental) will be more trustworthy than people who do not have that feeling.
2. Oxytocin levels are higher in people who are trustworthy.

work with the data

Original Paper: Kosfeld, M., M. Heinrichs, P. Zak, U. Fischbaker and E. Fehr. 2005. Oxytocin increases trust in humans. *Nature* 435: 673–676.

Paul Zak used an economics experiment to investigate trust and found that when people felt that a stranger had trusted them with money, the recipients not only reciprocated by giving the donor some money back, but also had elevated levels of the intercellular signaling molecule oxytocin. As in many experiments, although these data were suggestive of a relationship, correlation does not necessarily mean causation. So Zak and his colleagues tried a direct experiment in which subjects were given oxytocin before the experiment, to determine if oxytocin induces greater trust.

QUESTIONS▶

1. In the experiment, donors were divided into control and experimental groups and paired with recipients. All donors were given 12 money units and asked to donate to a paired recipient, whom they did not know or see. Donors knew that whatever amount they donated would be tripled in the recipient's account (up to 36 units) and trusted that the recipient would generously give some money back to the donor. The experimental donor group was given nasal spray containing oxytocin just before the experiment; the control donor group received a nasal spray without oxytocin. The table shows the results.

	Average number of units (SD) sent to recipient
Experimental group (with oxytocin spray)	9.6 (2.8)
Control (no oxytocin spray)	8.1 (3.1)

What can you conclude from these data about oxytocin and trust in the donors? What statistical test would you use to show that the two averages are significantly different?

2. The experiment described in Question 1 was repeated, but this time both donor groups were told exactly how much to give the recipient. The results are shown in the table.

	Average number of units (SD) sent to recipient
Experimental group (with oxytocin spray)	7.5 (3.3)
Control (no oxytocin spray)	7.5 (3.4)

What can you conclude about the effect of oxytocin in this case, and what does it indicate about the trust relationship?

A similar **work with the data** exercise may be assigned in **LaunchPad**.

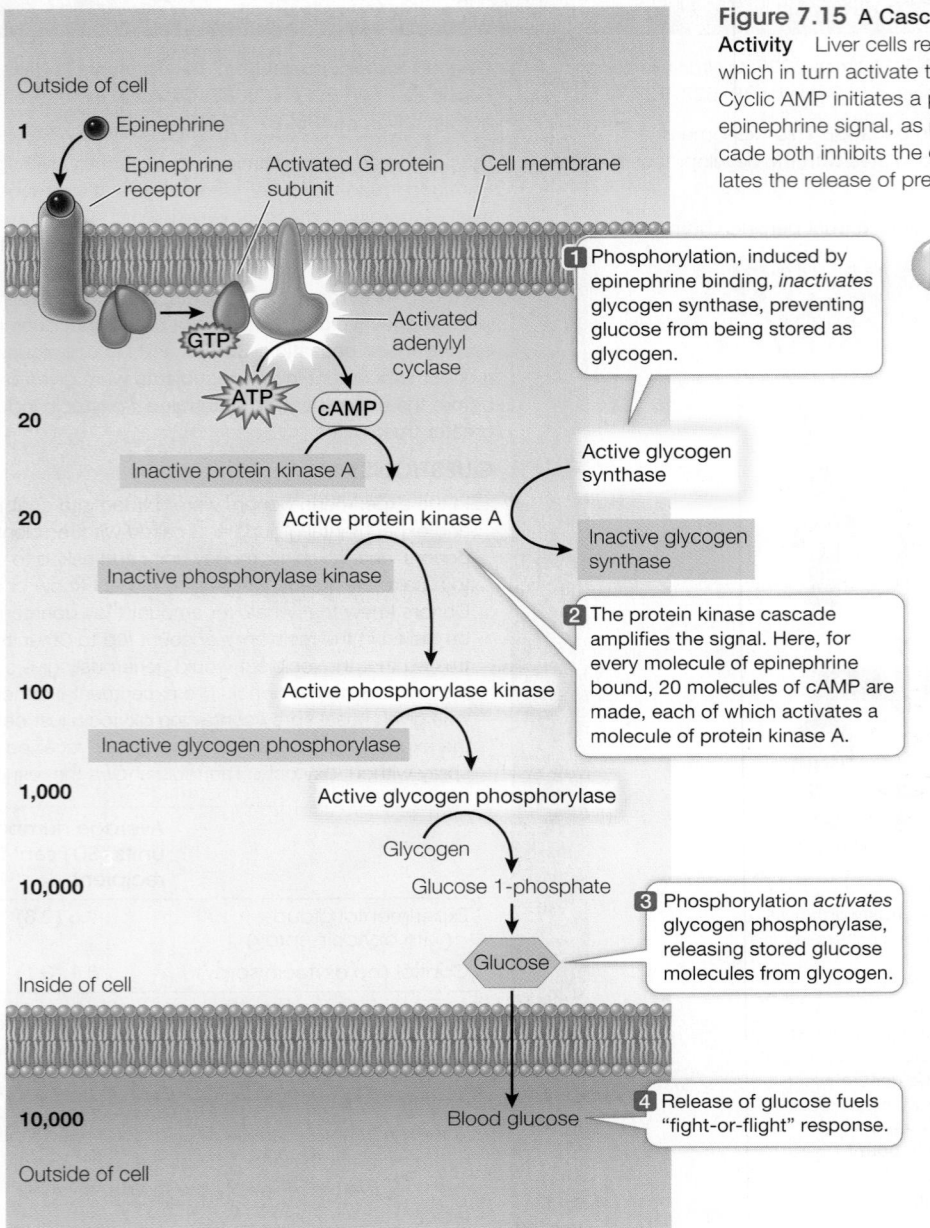

Figure 7.15 A Cascade of Reactions Leads to Altered Enzyme Activity Liver cells respond to epinephrine by activating G proteins, which in turn activate the synthesis of the second messenger cAMP. Cyclic AMP initiates a protein kinase cascade, greatly amplifying the epinephrine signal, as indicated by the bold, black numbers. The cascade both inhibits the conversion of glucose to glycogen and stimulates the release of previously stored glucose.

Activity 7.2 Cell Signaling and Amplification Simulation
www.Life11e.com/ac7.2

1 Phosphorylation, induced by epinephrine binding, *inactivates* glycogen synthase, preventing glucose from being stored as glycogen.

2 The protein kinase cascade amplifies the signal. Here, for every molecule of epinephrine bound, 20 molecules of cAMP are made, each of which activates a molecule of protein kinase A.

3 Phosphorylation *activates* glycogen phosphorylase, releasing stored glucose molecules from glycogen.

4 Release of glucose fuels "fight-or-flight" response.

Signals can initiate DNA transcription

You learned in Key Concept 4.1 that the genetic material, DNA, is transcribed into RNA, which is then translated into a protein whose amino acid sequence is specified by the original DNA sequence. Proteins are important in all cellular functions, so a key way to regulate specific functions in a cell is to regulate which proteins are made, and therefore which DNA sequences are transcribed.

*Signal transduction plays an important role in determining which DNA sequences are transcribed. Common targets of signal transduction are proteins called transcription factors, which bind to specific DNA sequences in the cell nucleus and activate or inactivate transcription of the adjacent DNA regions. For example, the ras signaling pathway (see Figure 7.10) ends in the nucleus. The final protein kinase in the ras signaling cascade, MAPK (mitogen-activated protein kinase; a mitogen is a type of signal that stimulates cell division), enters the nucleus and phosphorylates a protein that stimulates the expression of several genes involved in cell proliferation.

*connect the concepts In this chapter you've learned about signaling pathways that occur in animal cells. Signal transduction pathways play equally important roles in other organisms, including plants, as we will explore extensively in Chapters 36 and 38.

at the cell membrane ultimately results in the release of 10,000 molecules of glucose into the bloodstream:

1	molecule of epinephrine bound to the membrane leads to
20	molecules of cAMP, which activate
20	molecules of protein kinase A, which activate
100	molecules of phosphorylase kinase, which activate
1,000	molecules of glycogen phosphorylase, which produce
10,000	molecules of glucose 1-phosphate, which produce
10,000	molecules of blood glucose

7.4 recap

Cells respond to signal transduction by opening membrane channels, activating or inactivating enzymes, and stimulating or inhibiting gene transcription.

learning outcomes

You should be able to:

- Describe the signal transduction pathway stimulated by oxytocin.
- Compare various mechanisms that cells use to respond to signals.

1. How does oxytocin signaling involve several signaling pathways?
2. Which method(s) of cell response to signaling are rapid and which are slower? Explain.

We have described how signals from a cell's environment can influence the cell. But the environment of a cell in a multicellular organism is more than the extracellular medium—it includes neighboring cells as well. In the next section we'll see how specialized junctions between cells allow them to pass signals from one to another.

key concept 7.5 Adjacent Cells in a Multicellular Organism Can Communicate Directly

The hallmark of multicellular organisms is their ability to have specialized functions in subsets of cells within their bodies. How do these cells communicate with one another so that they can work together for the good of the entire organism? As you learned in Key Concept 7.1, some intercellular signals travel through the circulatory system to reach their target cells. But cells also have more direct ways of communicating. Cells that are packed together within a tissue can communicate directly with their neighbors via specialized intercellular junctions: gap junctions in animals (see Figure 6.7) and plasmodesmata in plants (see Figure 5.7).

focus your learning

- Animal cells can communicate directly with other cells via gap junctions.
- Development of intercellular communication was a key step in the evolution of multicellularity.

How do animal cells directly communicate?

Gap junctions are channels between adjacent cells that occur in many animals, occupying up to 25 percent of the area of the cell membrane (**Figure 7.16A**). Gap junctions traverse the narrow space between the cell membranes of two adjacent cells (the "gap") by means of channel structures called **connexons**. The walls of a connexon are composed of six subunits of the integral membrane protein connexin. In adjacent cells, two connexons come together to form a gap junction that links the cytoplasms of the two cells.

(A) Gap junctions in animal cells

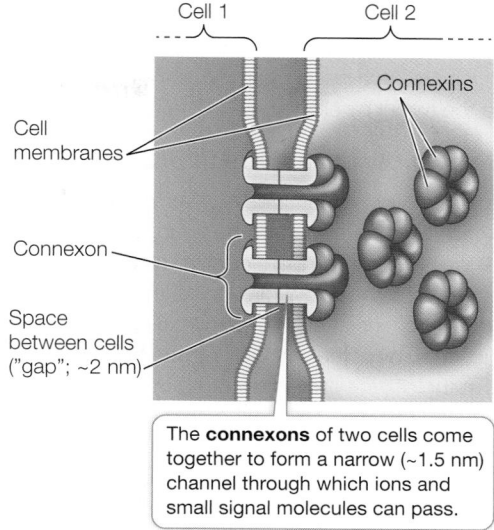

The **connexons** of two cells come together to form a narrow (~1.5 nm) channel through which ions and small signal molecules can pass.

(B) Plasmodesmata in plant cells

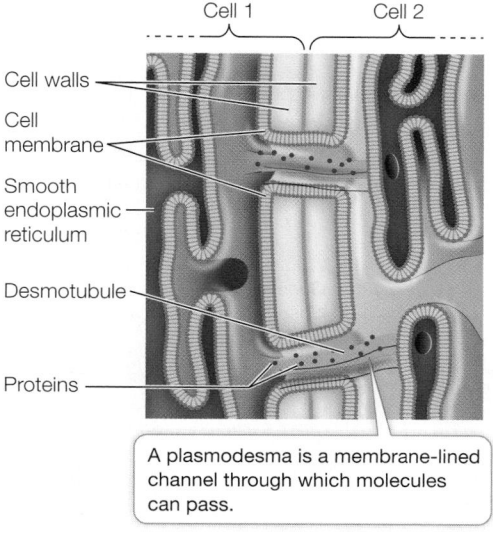

A plasmodesma is a membrane-lined channel through which molecules can pass.

Figure 7.16 Communicating Junctions **(A)** An animal cell may contain hundreds of gap junctions connecting it to neighboring cells. The pores of gap junctions allow small molecules to pass from cell to cell, ensuring similar concentrations of important signaling molecules in adjacent cells so that the cells can coordinate their activities. **(B)** Plasmodesmata connect plant cells. The desmotubule, derived from the smooth endoplasmic reticulum, fills up most of the space inside a plasmodesma, leaving a tiny gap through which small metabolites and ions can pass.

There may be hundreds of these channels between a cell and its neighbors. The channel pores are about 1.5 nanometers (nm) in diameter—too narrow for the passage of large molecules but adequate for passage of small molecules and ions.

Here is an example of gap junctions at work: In the lens of the mammalian eye, only the cells at the periphery are close enough to the blood supply for adequate diffusion of nutrients and wastes. But because lens cells are connected by large numbers of gap junctions, material can diffuse between them rapidly and efficiently.

Hormones and second messengers can move through gap junctions. Sometimes just a few cells in a tissue have the receptor for a particular signal; in such cases, gap junctions allow a coordinated response to the signal by all the cells in the tissue.

Plant cells communicate through plasmodesmata

Instead of gap junctions, plants have **plasmodesmata** (singular *plasmodesma*), which are membrane-lined tunnels that traverse the thick cell walls separating plant cells from one another. A typical plant cell has several thousand plasmodesmata. Plasmodesmata differ from gap junctions in one fundamental way: unlike gap junctions, in which the wall of the channel is made of integral membrane proteins from the adjacent cell membranes, plasmodesmata are lined by the fused cell membranes themselves.

The diameter of a plasmodesma is about 6 nm, far larger than a gap junction channel. But the actual space available for diffusion is about the same—1.5 nm. Examination of the interior of the plasmodesma by transmission electron microscopy reveals that a tubule called the **desmotubule**, apparently derived from the endoplasmic reticulum, fills up most of the opening of the plasmodesma (**Figure 7.16B**). Typically, only small metabolites and ions can move between plant cells.

Plasmodesmata are vital in plants because their circulatory transport system, the vascular system, lacks the tiny vessels (capillaries) that many animals have for bringing gases and nutrients to every cell. For instance, simple diffusion from plant cell to cell across cell membranes is inadequate to account for the movement of a plant hormone from the site of production to the site of action. Instead, plants rely on more rapid diffusion through plasmodesmata to ensure that all cells of a tissue respond to a signal at the same time. There are cases in which larger molecules or particles can pass between cells via plasmodesmata. For example, some viruses can move through plasmodesmata by using "movement proteins" to assist their passage.

Modern organisms provide clues about the evolution of cell–cell interactions and multicellularity

Multicellular organisms can indeed be multicellular—a human has about 60 trillion eukaryotic cells and many more prokaryotic ones. But it's not just cells that make up a human or a rose plant—it is groups of cells specialized into tissues and then groups of tissues in organs (e.g., the nerve cells in the brain or the petals in a flower) that have specific roles. As you will learn in Chapter 19, the embryonic development of tissues and organs takes place in several steps:

- Migration of cells to form a group
- Organization of cells in the cluster
- Specialization of cells into tissues
- Connections between cells

Even though single-celled organisms continue to be highly successful on Earth, over time complex multicellular organisms evolved, along with their division of biological labor among specialized cells. The transition from single-celled to multicellular life took a long time.

Indeed, while there is evidence that single-celled organisms arose about 500 million to a billion years after the formation of Earth (see Chapter 4), the first evidence of true multicellular organisms dates from more than a billion years later. Multicellularity probably arose several times.

Studying the evolutionary origin of multicellularity is a challenge because it happened so long ago. The closest unicellular relatives of most modern animals and plants probably existed hundreds of millions of years ago. The transition from single-celled to multicellular organisms may have occurred in several steps:

- Aggregation of cells into a cluster
- Intercellular communication within the cluster
- Specialization of some cells within the cluster
- Organization of specialized cells into groups (tissues)

Does this list look familiar? (See above for the origin of organs in the embryo.)

A key event would have been the evolution of intercellular communication, which is necessary to coordinate the activities of different cells within a multicellular organism.

We can visualize how the evolution of multicellularity might have occurred by looking at the "Volvocine line" of aquatic green algae (Chlorophyta). These plants range from single cells to complex multicellular organisms with differentiated cell clusters (**Figure 7.17**). Included in this range are a single-celled organism (*Chlamydomonas*); an

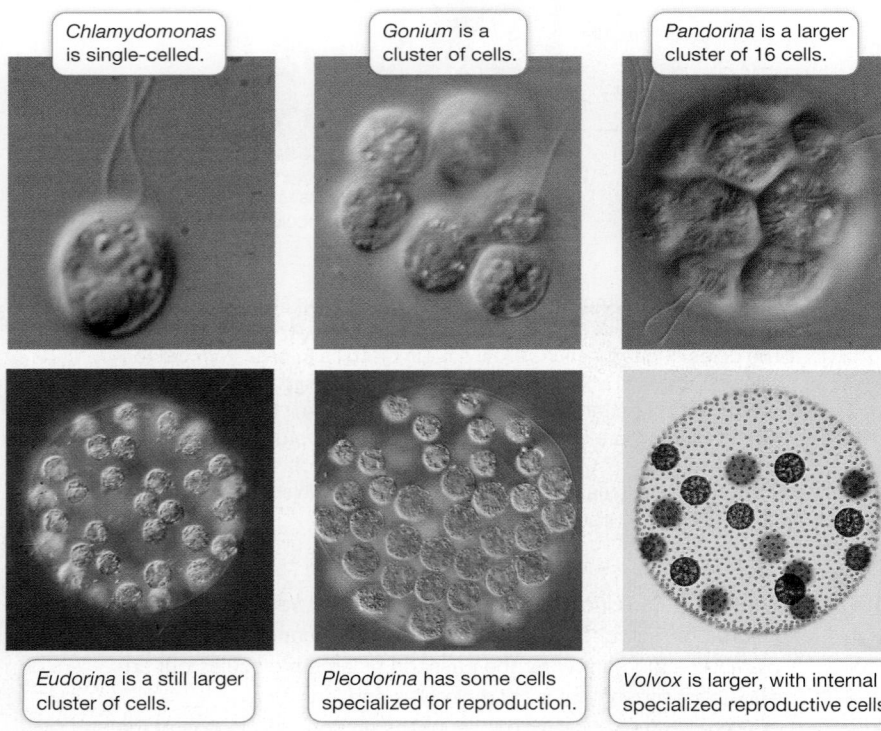

Chlamydomonas is single-celled.

Gonium is a cluster of cells.

Pandorina is a larger cluster of 16 cells.

Eudorina is a still larger cluster of cells.

Pleodorina has some cells specialized for reproduction.

Volvox is larger, with internal specialized reproductive cells.

Figure 7.17 Multicellularity   The evolution of intercellular interactions in a multicellular organism can be inferred from these green algae.

Q: Why was the evolution of direct communication important for tissue formation?

organism that occurs in small cell clusters (*Gonium*); species with larger cell clusters (*Pandorina* and *Eudorina*); a colony of somatic and reproductive cells (*Pleodorina*); and a larger, 1,000-celled alga with somatic and reproductive cells organized into separate tissues (*Volvox*).

Chlamydomonas is the single-celled member of this group. It has two cellular phases: a swimming phase, when the cells have flagella and move about, and a non-swimming phase, when the flagella are reabsorbed (disaggregated) and the cell undergoes cell division (reproduction). Compare this with *Volvox*: most of the cells of this multicellular, spherical organism are on the surface; the beating of their flagella gives the organism a rolling motion as it swims toward light, where it can perform photosynthesis. But some *Volvox* cells are larger and located inside the sphere. These cells are specialized for reproduction: they lose their flagella and then divide to form offspring.

 Media Clip 7.1 Social Amoebas Aggregate on Cue
www.Life11e.com/mc7.1

The separation of somatic and reproductive functions in *Volvox* is possible because of a key intercellular signaling mechanism that coordinates the activities of the separate tissues within the organism. *Volvox* has a gene whose protein product is produced by the outer, motile cells and travels to the reproductive cells, causing them to lose their flagella and divide. This gene is not active in species such as *Gonium* and *Pandorina*, which show cell aggregation but no cell specialization.

7.5 recap

Cells can communicate with their neighbors through specialized cell junctions. In animals, these structures are gap junctions; in plants, they are plasmodesmata. The evolution of intercellular communication and tissue formation can be inferred from existing organisms, such as certain related green algae.

learning outcomes

- Explain the biological advantages of gap junctions in animals.
- Design investigations to answer questions about intercellular communication.

1. List the characteristics of direct communication by cell junctions and chemical signaling between cells. What are the advantages of direct communication over chemical signaling?

2. How does the Volvocine line of green algae show possible steps in the evolution of cell communication and tissue formation?

3. The tiny invertebrate *Hydra* has an apical region with tentacles and a long, slender body. *Hydra* can reproduce asexually when cells on the body wall differentiate and form a bud, which then breaks off as a new organism. Buds form only at certain distances from the apex, leading to the idea that the apex releases a signal molecule that diffuses down the body and, at high concentrations (i.e., near the apex), inhibits bud formation. *Hydra* lacks a circulatory system, so this inhibitor must diffuse from cell to cell. If you have an antibody that binds to connexons and plugs up the gap junctions, how would you test the hypothesis that *Hydra's* inhibitory factor passes through these junctions?

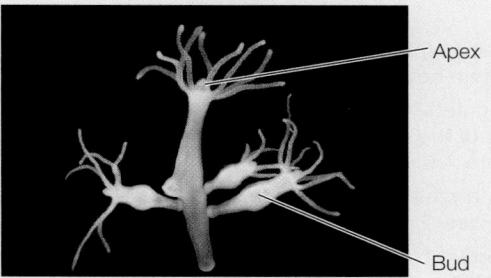

Apex

Bud

investigatinglife

 Is oxytocin the "trust" signal in humans?

Paul Zak's experiments showing the involvement of oxytocin in the trust involved in money transfers between people opened up a new field of study called neuroeconomics. In 1759 the philosopher Adam Smith wrote that social and economic interactions between people are driven by what he called "fellow feeling." Now, 250 years later, the oxytocin studies validate this and have put it into the language of cell signaling. What other things besides economic generosity release oxytocin? Zak and others have shown that many pleasant activities, even massage and hugs, release oxytocin and induce positive behaviors as a result. Oxytocin is released during sexual activity in humans, and the release results in bonding behaviors just as it does in voles. Of course, oxytocin is now available for purchase. An ad proclaims "We 100% guarantee that Oxytocin Spray will enrich your social life or your money back."

Future directions

In this chapter you have seen how a natural substance such as caffeine acts on receptors to alter cell function. As signal transduction pathways have been described for important processes such as the regulation of cell division, it has become possible to perturb intermediate steps in the pathway between signal and cell effects. In the signaling that stimulates cells to reproduce, two important steps are the focus of intense interest. One is at the cell membrane and involves the ras GTPase that converts GTP bound to ras to GDP that can be inactive in cancer cells (see Figure 7.9). Drugs that target this GTPase make it more active, so there is less active ras in the cell and the signal transduction pathway for cell division does not get under way. Another area of intense interest is an intermediate step in the phospholipase C pathway (see Figure 7.14) that involves the protein kinase mTOR. Several drugs that target mTOR inactivate it, resulting in less cell division in tumor cells and cells involved in

Chapter Summary 7

7.1 Signals and Signaling Affect Cell Function

- Cells receive many signals from the environment and from other cells. Chemical signals are often at very low concentrations. **Autocrine** signals affect the cells that make them; **juxtacrine** signals affect adjacent cells; **paracrine** signals diffuse to and affect nearby cells; and **hormones** are carried through the circulatory systems of animals or the vascular systems of plants. Review Figure 7.1, Activity 7.1

- A **signal transduction pathway** involves the interaction of a signal molecule with a receptor; the transduction of the signal via a series of steps within the cell; and effects on the function of the cell. Review Figure 7.2

- Signal transduction pathways involve regulation of enzymes and transcription factors. A great deal of **crosstalk** occurs between pathways.

7.2 Receptors Bind Signals to Initiate a Cellular Response

- Cells respond to signals only if they have specific **receptor** proteins that can recognize those signals.

- Binding of a signal **ligand** to its receptor obeys the chemical law of mass action. A key measurement of the strength of binding is the **dissociation constant** (K_D).

- Depending on the nature of its signal or ligand, a receptor may be located in the cell membrane or inside the target cell. Review Figure 7.4

- Receptors located in the cell membrane include **ion channels**, **protein kinases**, and **G protein-coupled receptors**.

- Ion channel receptors are "gated." The gate "opens" when the three-dimensional structure of the channel protein is altered by ligand binding. Review Figure 7.5

- Protein kinase receptors catalyze the phosphorylation of themselves or other proteins. Review Figure 7.6

- A **G protein** has three important binding sites, which bind a G protein-coupled receptor, GDP or GTP, and an **effector protein**. A G protein can either activate or inhibit an effector protein. Review Figure 7.7, Animation 7.1

- **Intracellular receptors** include certain photoreceptors in plants and steroid hormone receptors in animals. A lipid-soluble ligand such as a steroid hormone may enter the cytoplasm or the nucleus before binding. Many intracellular receptors are transcription factors. Review Figure 7.8

7.3 The Response to a Signal Spreads through the Cell

- A **protein kinase cascade** amplifies the response to receptor binding. Review Focus: Key Figure 7.10, Animation 7.2

- Second messengers include **cyclic AMP (cAMP)**, **inositol trisphosphate (IP$_3$)**, **diacylglycerol (DAG)**, and calcium ions. IP$_3$ and DAG are derived from the phospholipid **phosphatidyl inositol-bisphosphate (PIP$_2$)**. Review Figures 7.11, 7.12

- The gas nitric oxide (NO) is involved in signal transduction in human smooth muscle cells. Review Figure 7.13

- Signal transduction can be regulated in several ways. The balance between activating and inactivating the molecules involved determines the ultimate cellular response to a signal. Review Figure 7.14

7.4 Cells Change in Response to Signals in Several Ways

- The cellular responses to signals may include the opening of ion channels, the alteration of enzyme activities, or changes in gene expression.

- Activated enzymes may activate other enzymes in a signal transduction pathway, leading to impressive amplification of a signal. Review Figure 7.15, Activity 7.2

- Protein kinases covalently add phosphate groups to target proteins; cAMP binds target proteins noncovalently. Both kinds of binding change the target protein's shape to expose or hide a region involved in function.

7.5 Adjacent Cells in a Multicellular Organism Can Communicate Directly

- Many adjacent animal cells can communicate with one another directly through small pores in their cell membranes called **gap junctions**. Protein structures called **connexons** form thin channels between two adjacent cells through which small signal molecules and ions can pass. Review Figure 7.16A

- Plant cells are connected by somewhat larger pores called **plasmodesmata**, which traverse both cell membranes and cell walls. The **desmotubule** narrows the opening of the plasmodesma. Review Figure 7.16B

- The evolution of cell communication and tissue formation can be inferred from existing organisms, such as certain green algae. Review Figure 7.17

See Activity 7.3 for a concept review of this chapter.

Go to **LearningCurve** (in **LaunchPad**) for dynamic quizzing that helps you solidify your understanding of this chapter. **LearningCurve** adapts to your responses, giving you the practice you need to master each key concept.

Apply What You've Learned

Review

7.1 A signal transduction pathway includes a signal, a receptor, and a response.

7.3 A signal transduction cascade transmits and amplifies a signal inside a cell.

Original Papers: Forst, S., J. Delgado and M. Inouye. 1989. Phosphorylation of OmpR by the osmosensor EnvZ modulates expression of the *ompF* and *ompC* genes in *Escherichia coli*. *Proceedings of the National Academy of Sciences USA* 86: 6052–6056.

Cai, S. J. and M. Inouye. 2002. EnvZ-OmpR interaction and osmoregulation in *Escherichia coli*. *Journal of Biological Chemistry* 277: 24155–24161.

Escherichia coli (**Figure A**) is a bacterium that lives in the human intestine, where it contributes to digestion by breaking down food molecules. You might think that these bacteria have a stable environment, but these microorganisms face a periodic threat. During digestion in the intestine, the solute concentrations around the bacteria fluctuate and at times exceed the solute concentration inside the cells. The cells must respond to this hypertonic environment to prevent desiccation due to water loss.

Figure A

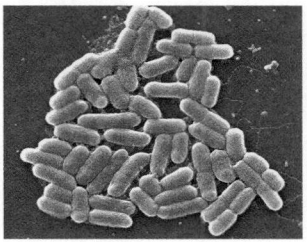

To investigate how the bacterial cells manage this problem, researchers grew *E. coli* in two solutions. One solution was hypertonic, with a solute concentration higher than that inside the cell. The other solution was isotonic, with a solute concentration similar to that inside the cell. The researchers isolated membrane proteins from the cells and separated them according to size, using polyacrylamide gel electrophoresis. After staining the proteins on the gel, one of the membrane proteins, OmpC, showed differences depending on the environmental conditions used to grow the cells. **Figure B** shows these results. The amount of stain is proportional to the amount of protein in the gel.

Figure B

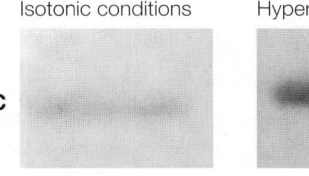

Isotonic conditions   Hypertonic conditions

OmpC

The researchers suspected that a signal transduction pathway was involved in *E. coli's* response to a hypertonic environment. They investigated whether any proteins became phosphorylated by a protein kinase as part of such a pathway. Interesting results were noted with respect to two proteins: a membrane protein (EnvZ) and a protein found in the cytoplasm (OmpR). After purifying EnvZ and OmpR from *E. coli*, each protein was incubated separately with ^{32}P-radiolabeled ATP. This was done to examine protein phosphorylation: proteins that are phosphorylated in the presence of ^{32}P-ATP become labeled ^{32}P. **Table A** summarizes the results of this experiment.

Table A

	EnvZ + ^{32}P-ATP	OmpR + ^{32}P-ATP
^{32}P label incorporated into protein	High	None

Next, the researchers tried mixing ^{32}P-radiolabeled EnvZ with OmpR. After mixing the two proteins, samples were removed from the mixture at 0, 2, 5, and 10 minutes. **Table B** summarizes the results.

Table B

	Time after mixing ^{32}P-EnvZ + OmpR (min)			
	0	2	5	10
^{32}P label present in EnvZ	High	Medium	Low	None
^{32}P label present in OmpR	None	Low	Medium	High

Finally, the researchers measured the quantities of EnvZ and OmpR in *E. coli* cells after they were grown under different conditions. Values in **Table C** are numbers of molecules per cell.

Table C

	Nutrient broth (isotonic)	Nutrient broth + 20% sucrose (hypertonic)
EnvZ	63	113
OmpR	2,043	3,525

Questions

1. What evidence shows that bacterial cells respond in their membrane protein composition to changes in environmental solute concentration?

2. What do the data in Tables A and B indicate about the roles of EnvZ and OmpR? Write two equations to illustrate your answer.

3. What evidence suggests that signal amplification occurs in this signal transduction system?

4. Draw a sketch to illustrate a possible mechanism for how this signal transduction pathway responds to a hypertonic environment. Your sketch should account for all of the data provided.

Go to **LaunchPad** for the eBook, LearningCurve, animations, activities, flashcards, and additional resources and assignments.

8

PART THREE Cells and Energy

Energy, Enzymes, and Metabolism

Energy transformations are a hallmark of life.

▶ investigating life

How Aspirin Works

Despite suffering from the "ague," the Reverend Edward Stone went walking in the English countryside. Feverish, tired, with aching muscles and joints, he came across a willow tree. Although apparently unaware that many ancient healers used willow bark extracts to reduce fever, the clergyman knew of the tradition of natural remedies for various diseases. The willow reminded him of the bitter extracts from the bark of South American trees then being sold (at high prices) to treat fevers. Removing some willow bark, Stone sucked on it and found it did indeed taste bitter—and that it relieved his symptoms.

Later he gathered a pound of willow bark and ground it into a powder, which he gave to about 50 people who complained of pain; all said they felt better. Stone reported the results of this "clinical test" in a letter to the Royal Society, England's most respected scientific body. Stone had discovered salicylic acid, the basis of the most widely used drug in the world. The date of his letter (which still exists) was April 25, 1763.

The chemical structure of salicylic acid (named for *Salix*, the willow genus) was worked out about 70 years later, and soon chemists could synthesize it in the laboratory. Although the compound alleviated pain, its acidity irritated the digestive system. In the late 1890s, the German chemical company Bayer synthesized a milder yet equally effective form, acetylsalicylic acid, which it marketed as aspirin. The new medicine's success launched Bayer to world prominence as a pharmaceutical company, a position it maintains today.

In the 1960s and 1970s, aspirin use declined somewhat when other pain-reducing medications became widely available. But over this same time, clinical studies revealed a new use for aspirin: it is an effective anticoagulant, shown to prevent heart attacks and strokes caused by blood clots. Today many people take a daily low dose of aspirin as a preventive against clotting disorders.

Fever, joint pain, headache, blood clots: What do these symptoms have in common? They all are mediated by fatty acid products called prostaglandins and molecules derived from them. Salicylic acid blocks the synthesis of the primary prostaglandin. The biochemical mechanism by which aspirin works was described in 1971. As we will see, an understanding of this mechanism requires an understanding of protein and enzyme function—two subjects of this chapter.

Q How do anti-inflammatory drugs work as enzyme inhibitors?

key concept 8.1 Physical Principles Underlie Biological Energy Transformations

A **chemical reaction** occurs when atoms have sufficient energy to combine or change their bonding partners. Consider the hydrolysis of the disaccharide sucrose to its component monomers, glucose and fructose (see p. 56 for the chemical structures of these sugars). We can express this reaction by a chemical equation:

$$\text{Sucrose} + H_2O \rightarrow \text{glucose} + \text{fructose}$$
$$(C_{12}H_{22}O_{11}) \qquad (C_6H_{12}O_6) \ (C_6H_{12}O_6)$$

In this equation, sucrose and water are the **reactants**, and glucose and fructose are the **products**. During the reaction, some of the bonds in sucrose and water are broken and new bonds are formed, resulting in products with chemical properties that are very different from those of the reactants. The sum total of all the chemical reactions occurring in a biological system at a given time is called **metabolism**. Metabolic reactions involve energy changes; for example, the energy contained in the chemical bonds of sucrose (reactants) is greater than the energy in the bonds of the two products, glucose and fructose.

focus your learning

- The second law of thermodynamics states that disorder is constantly increasing in the universe.
- Chemical reactions in biological systems are either exergonic or endergonic.

Physicists define **energy** as the capacity to do work, which occurs when a force operates on an object over a distance. In biochemistry, it is more useful to think of energy as *the capacity for change*. In biochemical reactions, energy changes are usually associated with changes in the chemical compositions and properties of molecules.

There are two basic types of energy

Energy comes in many forms: chemical, electrical, heat, light, and mechanical (**Table 8.1**). But all forms of energy can be considered as one of two basic types (**Figure 8.1**):

table 8.1 Energy in Biology

Form of energy	Example in biology
Chemical: Stored in bonds	Chemical energy stored in covalent bonds is released during the hydrolysis of polymers
Electrical: Separation of charges	Electrical gradients across cell membranes help drive the movement of ions through channels
Heat: Transfer due to temperature difference	Heat can be released by chemical reactions, and this can alter the internal temperature of an organism
Light: Electromagnetic radiation stored as photons	Light energy is captured by pigments in the eye and by plant pigments in photosynthesis
Mechanical: Energy of motion	Mechanical energy is used in muscle movements and movements within cells

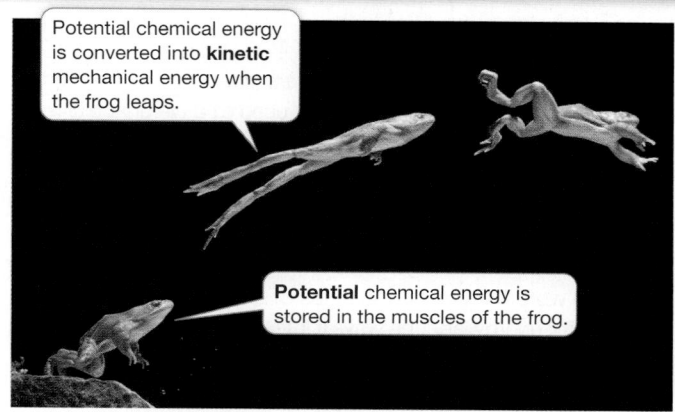

Potential chemical energy is converted into **kinetic** mechanical energy when the frog leaps.

Potential chemical energy is stored in the muscles of the frog.

Figure 8.1 Energy Conversions and Work A leaping frog illustrates both the conversion between potential and kinetic energy and the conversion of energy from one form (chemical) to another (mechanical).

1. **Potential energy** is the energy of state or position—that is, stored energy. It can be stored in many forms: in chemical bonds, as a concentration gradient, or even as an electric charge imbalance.
2. **Kinetic energy** is the energy of movement—that is, the type of energy that does work, that makes things change. For example, heat causes molecular motions and can even break chemical bonds.

Potential energy can be converted into kinetic energy and vice versa, and the form that the energy takes can also be converted. Think of reading this book: light energy is converted to chemical energy in your eyes, and then is converted to electrical energy in the nerve cells that carry messages to your brain. When you decide to turn a page, the electrical and chemical energy of nerves and muscles are converted to kinetic energy for movement of your hand and arm.

There are two basic types of metabolism

Energy changes in living systems usually occur as chemical changes, in which energy is stored in, or released from, chemical bonds.

Anabolic reactions (collectively anabolism) link smaller molecules to form larger, more complex molecules (for example, the synthesis of sucrose from glucose and fructose). Anabolic reactions require an input of energy. Energy is captured in the chemical bonds that are formed (for example, the glycosidic bond between the two monosaccharides). This captured energy is stored in the chemical bonds as potential energy:

$$\text{Glucose} + \text{fructose} + \text{energy} \rightarrow \text{Sucrose}$$

Catabolic reactions (collectively catabolism) break down larger, more complex molecules into smaller ones and often release the energy stored in the chemical bonds. For example, when sucrose is hydrolyzed, energy is released. In a biological system the released energy may be recaptured in new chemical bonds, or it may be used as kinetic energy—moving atoms, molecules, cells, or the whole organism:

$$\text{Sucrose} + H_2O \rightarrow \text{Glucose} + \text{fructose} + \text{energy}$$

Catabolic and anabolic reactions are often linked. The energy released in catabolic reactions is often used to drive anabolic reactions—that is, to do biological work. For example, the energy released by the breakdown of glucose (catabolism) is used to drive anabolic reactions such as the synthesis of triglycerides. This is why you can accumulate fat if you eat food in excess of your energy requirements.

The **laws of thermodynamics** (*thermo*, "energy," + *dynamics*, "change") were derived from studies of the fundamental physical properties of energy, and the ways it interacts with matter. The laws apply to all matter and all energy transformations in the universe. Their application to living systems helps us understand how organisms and cells harvest and transform energy to sustain life.

The first law of thermodynamics: Energy is neither created nor destroyed

The first law of thermodynamics states that in any energy conversion, energy is neither created nor destroyed. In other words, during any conversion of energy, the total energy in the system under consideration before and after the conversion is the same (**Figure 8.2A**). As you will see in the next two chapters, the potential energy present in the chemical bonds of carbohydrates and lipids can be converted to potential energy in the form of adenosine triphosphate (ATP). This can then be converted into kinetic energy to do mechanical work (such as in muscle contractions) or biochemical work (such as protein synthesis).

The second law of thermodynamics: Disorder tends to increase

Although energy cannot be created or destroyed, the second law of thermodynamics states that when energy is converted from one form to another, some of that energy becomes unavailable for doing work (**Figure 8.2B**). In other words, no physical process or chemical reaction is 100 percent efficient; some of the released energy is lost to a form associated with disorder. Think of disorder as a kind of randomness that is due to the thermal motion of particles; this energy is of such a low value and so dispersed that it is unusable. **Entropy** is a measure of the disorder in a system.

It takes energy to impose order on a system. Unless energy is applied to a system, it will be randomly arranged or disordered. The second law applies to all energy transformations, but we will focus here on chemical reactions in living systems.

NOT ALL ENERGY CAN BE USED   In any system, the total energy includes the usable energy that can do work and the unusable energy that is lost to disorder:

$$\text{Total energy} = \text{usable energy} + \text{unusable energy}$$

In biological systems, the total energy is called **enthalpy** (*H*). The usable energy that can do work is called **free energy** (*G*). Free energy is what cells require for all the chemical reactions involved in growth, cell division, and maintenance. The unusable energy is represented by entropy (*S*) multiplied by the absolute temperature (*T*). Thus we can rewrite the word equation above more precisely as:

$$H = G + TS \qquad (8.1)$$

Because we are interested in usable energy, we rearrange Equation 8.1:

$$G = H - TS \qquad (8.2)$$

(A)

The First Law of Thermodynamics
The total amount of energy before a transformation equals the total amount after a transformation. No new energy is created, and no energy is lost.

Energy transformation

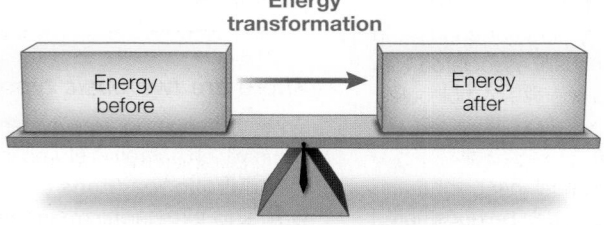

(B)

The Second Law of Thermodynamics
Although a transformation does not change the total amount of energy within a closed system (one that is not exchanging matter or energy with the surroundings), after any transformation the amount of energy available to do work is always less than the original amount of energy.

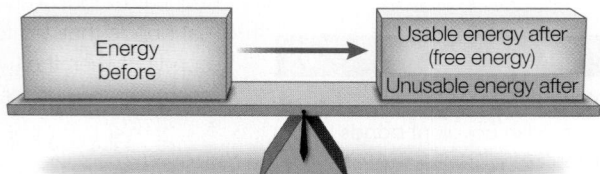

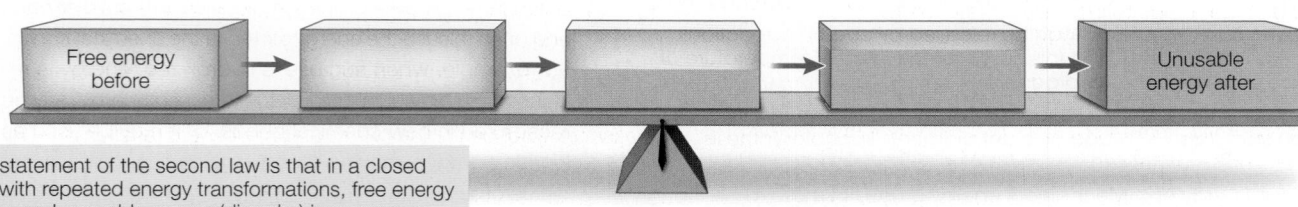

Another statement of the second law is that in a closed system, with repeated energy transformations, free energy decreases and unusable energy (disorder) increases—a phenomenon known as the increase in **entropy**.

Figure 8.2 The Laws of Thermodynamics   (A) The first law states that energy cannot be created or destroyed. **(B)** The second law states that after energy transformations, some energy becomes unavailable to do work.

Although we cannot measure G, H, or S absolutely, we can determine the change in each at a constant temperature. Such energy changes are measured in calories (cal) or joules (J).[1] A change (in energy, or another quantity) is represented by the Greek letter delta (Δ). The change in free energy (ΔG) of any chemical reaction is equal to the difference in free energy between the products and the reactants:

$$\Delta G_{reaction} = G_{products} - G_{reactants} \qquad (8.3)$$

A change in free energy can be either positive or negative; that is, the free energy of the products can be more or less than the free energy of the reactants. If the products have more free energy than the reactants, then there must have been some input of energy into the reaction. (Remember that energy cannot be created, so some energy must have been added from an external source.)

At a constant temperature ΔG is defined in terms of the change in total energy (ΔH) and the change in entropy (ΔS):

$$\Delta G = \Delta H - T\Delta S \qquad (8.4)$$

Equation 8.4 tells us whether free energy is released or required by a chemical reaction:

- If ΔG is negative ($\Delta G < 0$), free energy is released.
- If ΔG is positive ($\Delta G > 0$), free energy is required.

If the necessary free energy is not available, the reaction does not occur. The sign and magnitude of ΔG depend on the two factors on the right side of the equation:

1. ΔH: In a chemical reaction, ΔH is the total amount of energy added to the system ($\Delta H > 0$) or released ($\Delta H < 0$).

2. ΔS: Depending on the sign and magnitude of ΔS, the entire term, $T\Delta S$, may be negative or positive, large or small. In other words, in living systems at a constant temperature (no change in T), the magnitude and sign of ΔG can depend a lot on changes in entropy.

If a chemical reaction increases entropy, its products are more disordered or random than its reactants. If there are more products than reactants, as in the hydrolysis of a protein to its amino acids, the products have considerable freedom to move around. The disorder in a solution of amino acids will be large compared with that in the protein, in which peptide bonds and other forces prevent free movement. So in hydrolysis, the change in entropy (ΔS) will be positive. Conversely, if there are fewer products and they are more restrained in their movements than the reactants (as for amino acids being joined in a protein), ΔS will be negative.

DISORDER TENDS TO INCREASE   The second law of thermodynamics also predicts that as a result of transformations involving energy, disorder tends to increase; some energy is always lost to random thermal motion (entropy). Chemical changes, physical changes, and biological processes all tend to increase entropy (see Figure 8.2B), and this explains why some reactions proceed in one direction rather than another.

How does the second law apply to organisms? Consider the human body, with its highly organized tissues and organs composed of large, complex molecules. You might think that this order and complexity are in conflict with the second law—after all, entropy, or disorder, should be maximized—but they are not for two reasons:

1. *Getting ordered is coupled to the generation of disorder.* Making 1 kg of a human body (soft tissues, not bones) requires the catabolism of about 10 kg of highly ordered biological materials (our food), which are converted into CO_2, H_2O, and other simple molecules. So this process creates far more disorder (more energy is lost to entropy in the small molecules) than the amount of order (total energy; enthalpy) stored in large molecules in 1 kg of a person.

2. *Life requires a constant input of energy to maintain order.* Without this energy, the complex structures of living systems would break down. Because energy is used to generate and maintain order, there is no conflict with the second law of thermodynamics.

Having seen that the laws of thermodynamics apply to organisms, let's see how these laws apply to biochemical reactions inside the cell.

Chemical reactions release or consume energy

As you saw earlier, anabolic reactions link simple molecules to form more complex molecules, so they tend to increase complexity (order) in the cell. By contrast, catabolic reactions break down complex molecules into simpler ones, so they tend to decrease complexity (generate disorder).

- Catabolic reactions may break down an ordered reactant into smaller, more randomly distributed products. Reactions that release free energy ($-\Delta G$) are called **exergonic** reactions (**Figure 8.3A**). For example:

  Complex molecules → free energy + small molecules

- Anabolic reactions may make a single product (a highly ordered substance) out of many smaller reactants (less ordered). Reactions that require or consume free energy ($+\Delta G$) are called **endergonic** reactions (**Figure 8.3B**). For example:

  Free energy + small molecules → complex molecules

In principle, chemical reactions are reversible and can run both forward and backward. For example, if compound A can be converted into compound B (A → B), then B, in principle, can be converted into A (B → A), although *the concentrations of A and B determine which of these directions will be favored.* You can think of the overall reaction as resulting from competition between the forward and reverse reactions (A ⇌ B). According to the law of mass action, increasing the concentration of A makes the forward reaction (A → B) happen more often than the reverse reaction, just as increasing the concentration of B favors the reverse reaction (B → A).

There are concentrations of A and B at which the forward and reverse reactions take place at the same rate. At these concentrations,

---

[1]A calorie is the amount of heat energy needed to raise the temperature of 1 gram of pure water from 14.5°C to 15.5°C. In the SI system, energy is measured in joules (J); 1 J = 0.239 cal; conversely, 1 cal = 4.184 J. Thus, for example, 486 cal = 2,033 J, or 2.033 kJ. Although they are defined here in terms of heat, the calorie and the joule are measures of mechanical, electrical, or chemical energy. When you compare data on energy, always compare joules with joules and calories with calories.

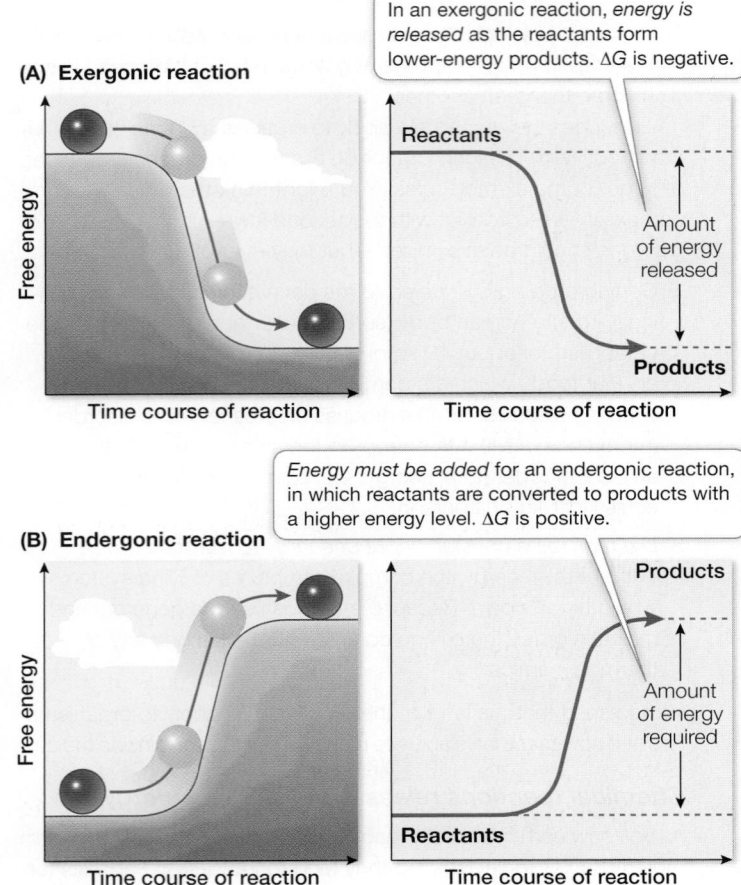

(A) Exergonic reaction

In an exergonic reaction, *energy is released* as the reactants form lower-energy products. ΔG is negative.

Reactants

Free energy

Time course of reaction

Amount of energy released

Products

Time course of reaction

Energy must be added for an endergonic reaction, in which reactants are converted to products with a higher energy level. ΔG is positive.

(B) Endergonic reaction

Products

Free energy

Time course of reaction

Amount of energy required

Reactants

Time course of reaction

Figure 8.3 Exergonic and Endergonic Reactions **(A)** In an exergonic reaction, the reactants behave like a ball rolling down a hill, and energy is released. **(B)** A ball will not roll uphill by itself. Driving an endergonic reaction, like moving a ball uphill, requires the addition of free energy.

no further net change in the system is observable, although individual molecules are still forming and breaking apart. This balance between forward and reverse reactions is known as **chemical equilibrium**. Chemical equilibrium is a state of no net change, and a state in which ΔG = 0.

Chemical equilibrium and free energy are related

Every chemical reaction proceeds to a certain extent, but not necessarily to completion (all reactants converted into products). Each reaction has a specific equilibrium point, which is related to the relative free energy content of the reactants and products. To understand the principle of equilibrium, let's consider the following example, which takes place inside most cells but which we can do in the lab. This is the interconversion of glucose 1-phosphate and glucose 6-phosphate, a rearrangement of a phosphate group from one position on the ring of carbon atoms to another:

Glucose 1-phosphate ⇌ glucose 6-phosphate

Imagine that we start out with an aqueous solution of glucose 1-phosphate that has a concentration of 0.02 M. (M stands for molar concentration; see Key Concept 2.4.) The solution is maintained

under constant environmental conditions (25°C and pH 7). As the reaction proceeds to equilibrium, the concentration of the product, glucose 6-phosphate, rises from 0 M to 0.019 M, while the concentration of the reactant, glucose 1-phosphate, falls from 0.02 M to 0.001 M, at which point, equilibrium is reached (**Figure 8.4**). *At equilibrium, the reverse reaction, from glucose 6-phosphate to glucose 1-phosphate, progresses at the same rate as the forward reaction.*

At equilibrium, then, this reaction has a product-to-reactant ratio of 19:1 (0.019/0.001), so the forward reaction has gone 95 percent of the way to completion ("to the right," as written above). This result is obtained every time the experiment is run under the same conditions.

The change in free energy (ΔG) for any reaction is related directly to its point of equilibrium. The further toward completion the point of equilibrium lies, the more free energy is released. In an exergonic reaction, ΔG is a negative number. The total value of ΔG also depends on the beginning concentrations of the reactants and products and other conditions such as temperature, pressure, and pH of the solution. Biochemists often calculate ΔG using standard laboratory conditions: 25°C, one atmosphere pressure, one molar (1 M) concentrations of the solutes, and pH 7. The **standard free energy change** calculated using these conditions is designated ΔG⁰. In our example of the conversion of glucose 1-phosphate to glucose 6-phosphate, ΔG⁰ = –1.7 kcal/mol, or –7.1 kJ/mol.

A large, positive ΔG for a reaction means that it does not proceed to the right (A → B). If the concentration of B is initially high relative to that of A, such a reaction runs "to the left" (A ← B), and at equilibrium nearly all of B is converted into A. A ΔG value near zero is characteristic of a readily

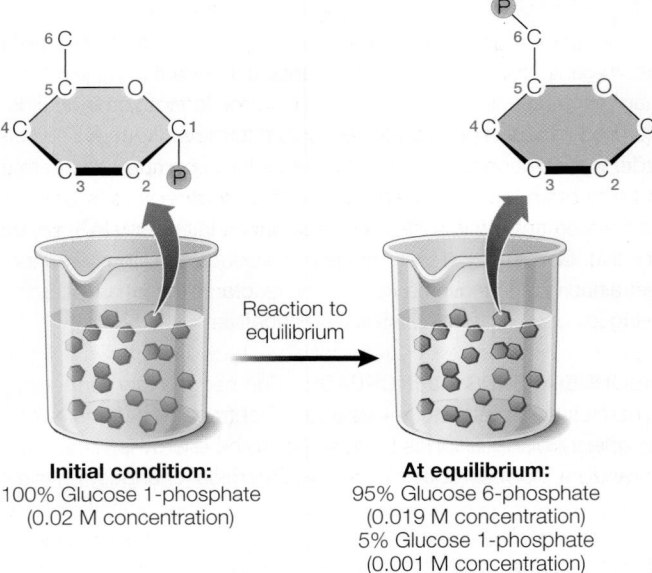

Reaction to equilibrium

Initial condition:
100% Glucose 1-phosphate
(0.02 M concentration)

At equilibrium:
95% Glucose 6-phosphate
(0.019 M concentration)
5% Glucose 1-phosphate
(0.001 M concentration)

Figure 8.4 Chemical Reactions Run to Equilibrium No matter what quantities of glucose 1-phosphate and glucose 6-phosphate are dissolved in water, when equilibrium is attained, there will always be 95 percent glucose 6-phosphate and 5 percent glucose 1-phosphate.

reversible reaction: reactants and products have almost the same free energies.

In Chapters 9 and 10 you will learn the biochemical conversions that extract energy from food and light. In turn, this energy is used to synthesize carbohydrates, lipids, and proteins. All of the chemical reactions carried out by living organisms are governed by the principles of thermodynamics and equilibrium.

▶ 8.1 recap

Two laws of thermodynamics govern energy transformations in biological systems. A biochemical reaction can release or consume energy, and it may not run to completion, but instead end up at a point of equilibrium.

learning outcomes

You should be able to:

- Apply the second law of thermodynamics to biological systems.
- Differentiate between exergonic and endergonic reactions.

1. What is the difference between endergonic and exergonic reactions, and what is the importance of positive and negative ΔG?
2. What makes it possible for endergonic reactions to proceed in organisms?

The principles of thermodynamics that we have been discussing apply to all energy transformations in the universe, so they are very powerful and useful. Now let's apply them to reactions in cells that involve the currency of biological energy, ATP.

▶ key concept 8.2 ATP Plays a Key Role in Biochemical Energetics

Cells use adenosine triphosphate (ATP) for the capture and transfer of the free energy they require to do chemical work. ATP operates as a kind of "energy currency." Just as it is more efficient and convenient for you to trade money for a lunch in a café than to trade your actual labor, it's useful for cells to have a single currency for transferring energy between different reactions and cell processes. So some of the free energy that is released by exergonic reactions is captured in the formation of ATP from adenosine diphosphate (ADP) and inorganic phosphate (HPO_4^{2-}, which is commonly abbreviated to P_i). The ATP can then be hydrolyzed in the cell to release free energy to drive endergonic reactions. (In some reactions, guanosine triphosphate [GTP] is used as the energy transfer molecule instead of ATP, but we will focus on ATP here.)

focus your learning

- ATP releases usable energy for a cell when hydrolyzed to ADP and P_i.
- Through coupling reactions, ATP drives endergonic reactions using energy derived from exergonic reactions.

ATP has another important role in the cell beyond its use as an energy currency: it can be converted into a building block for nucleic acids (see Chapter 4). The structure of ATP is similar to that of other nucleoside triphosphates, but two things about ATP make it especially useful to cells:

1. ATP releases a relatively large amount of energy when hydrolyzed to ADP and P_i.
2. ATP can phosphorylate (donate a phosphate group to) many different molecules, which gain some of the energy that was stored in the ATP.

ATP hydrolysis releases energy

An ATP molecule consists of the nitrogenous base adenine bonded to ribose (a sugar), which is attached to a sequence of three phosphate groups (**Figure 8.5A**). The hydrolysis of a molecule of ATP yields free energy, as well as ADP and an inorganic phosphate ion (P_i). Thus:

$$ATP + H_2O \rightarrow ADP + P_i + \text{free energy}$$

The important property of this reaction is that it is exergonic, releasing free energy. Under standard laboratory conditions, the change in free energy (ΔG) for this reaction is about −7.3 kcal/mol (−30 kJ/mol). However, under cellular conditions ΔG can be as much as −14 kcal/mol. We give both values here because you will encounter them both and you should be aware of their origins. Both are correct, but in different conditions.

A molecule of ATP can be hydrolyzed either to ADP and P_i, or to adenosine monophosphate (AMP) and a pyrophosphate ion ($P_2O_7^{4-}$; commonly abbreviated as PP_i). Two characteristics of ATP account for the free energy released by the loss of one or two of its phosphate groups:

1. Because phosphate groups are negatively charged and so repel each other, it takes energy to get two phosphates near enough to each other to make the covalent bond that links them together. Some of this energy is stored as potential energy in the P~O bonds between the phosphates in ATP (the wavy line indicates a high-energy bond).
2. The free energy of this P~O bond (called a phosphoric acid anhydride bond) is much higher than the energy of the O—H bond that forms as a result of hydrolysis. So it is the lower free energy state of the system that releases energy.

The structure of ATP

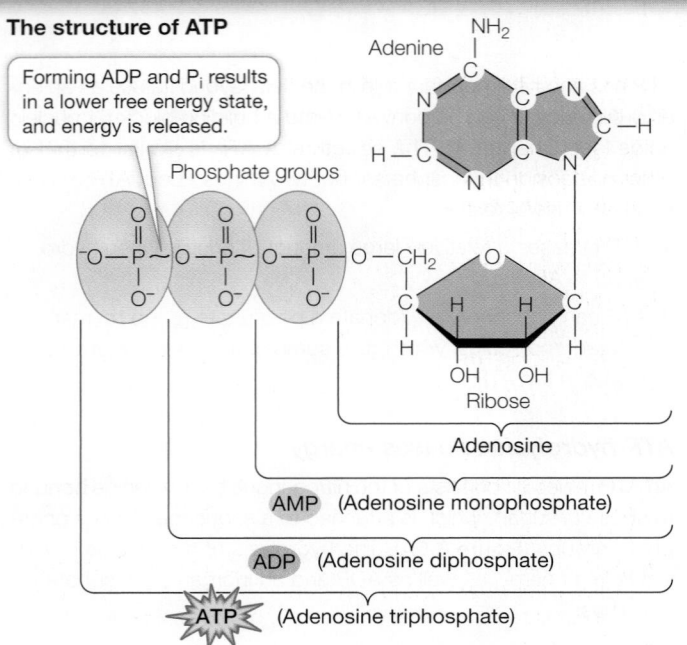

Forming ADP and P_i results in a lower free energy state, and energy is released.

Adenine

Phosphate groups

AMP (Adenosine monophosphate)

ADP (Adenosine diphosphate)

ATP (Adenosine triphosphate)

Ribose

Adenosine

Figure 8.5 ATP   ATP is richer in energy than its relatives ADP and AMP.

 Activity 8.1 ATP and Coupled Reactions
www.Life11e.com/ac8.1

Cells use the energy released by ATP hydrolysis to fuel endergonic reactions (such as the biosynthesis of complex molecules), for active transport, for movement, and even for the production of light (bioluminescence). Another interesting example of the use of ATP involves converting its chemical energy into light energy.

 Media Clip 8.1 Bioluminescence in the Deep Sea
www.Life11e.com/mc8.1

ATP couples exergonic and endergonic reactions

As you have just seen, the hydrolysis of ATP is exergonic and yields ADP, P_i, and more free energy (or AMP, PP_i, and more free energy). The reverse reaction, the formation of ATP from ADP and P_i, is endergonic and consumes as much free energy as is released by the hydrolysis of ATP:

$$ADP + P_i + free\ energy \rightarrow ATP + H_2O$$

Many exergonic reactions in the cell can provide the energy to convert ADP into ATP. For eukaryotes and many prokaryotes, the most important of these reactions is cellular respiration, in which some of the energy released from fuel molecules is captured in ATP. The formation and hydrolysis of ATP constitute what might be called an "energy-coupling cycle," in which ADP picks up energy from exergonic reactions to become ATP, which then donates energy to endergonic reactions. ATP is the common component of these reactions and is the agent of coupling, as illustrated in **Figure 8.6**.

Coupling of exergonic and endergonic reactions is very common in metabolism. Free energy is captured

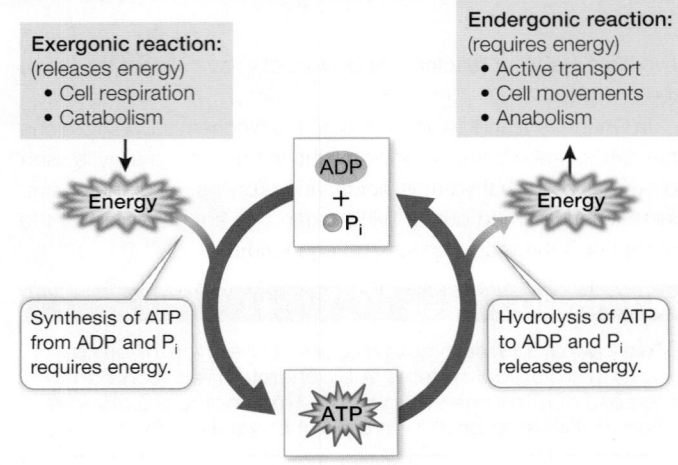

Figure 8.6 Coupling of Reactions   Exergonic cellular reactions release the energy needed to make ATP from ADP. The energy released from the conversion of ATP back to ADP can be used to fuel endergonic reactions.

and retained in the P~O bonds of ATP. ATP then diffuses to another place in the cell, where its hydrolysis releases the free energy to drive an endergonic reaction. For example, the formation of glucose 6-phosphate from glucose (**Figure 8.7**), which has a positive ΔG (is endergonic), will not proceed without the input of free energy from ATP hydrolysis, which has a negative ΔG (is exergonic). The overall ΔG for the coupled reactions (when the two ΔGs are added together) is negative. Hence the reactions proceed exergonically when they are coupled by occurring at the same time and place, and glucose

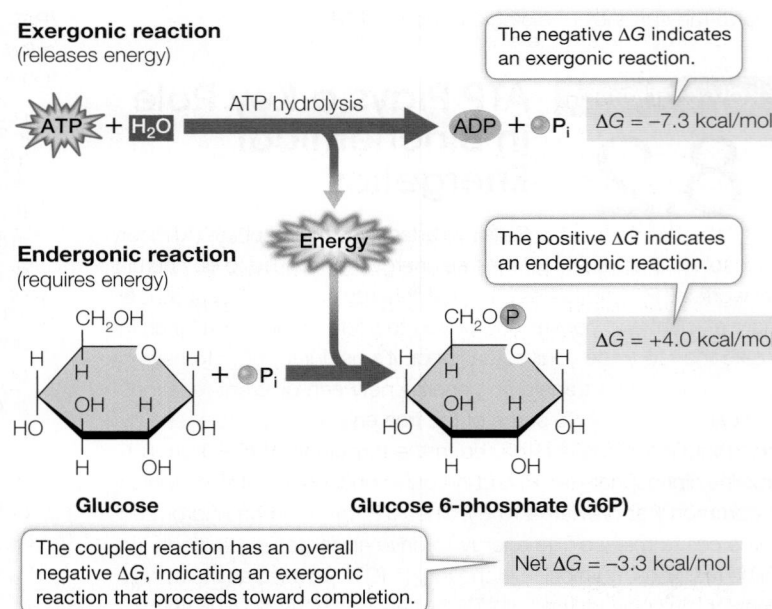

Figure 8.7 Coupling of ATP Hydrolysis to an Endergonic Reaction   The addition of phosphate derived from the hydrolysis of ATP to glucose forms the molecule glucose 6-phosphate (in a reaction catalyzed by hexokinase). ATP hydrolysis is exergonic and the energy released drives the second reaction, which is endergonic.

Q: What would be the magnitude of the ΔG of a different conversion that could drive the synthesis of ATP?

6-phosphate is synthesized. As you will see in Chapter 9, this is the initial reaction in the catabolism of glucose.

An active cell requires the production of millions of molecules of ATP per second to drive its biochemical machinery. From previous chapters, you may already be familiar with some of the activities in the cell that require energy from the hydrolysis of ATP:

- Active transport across a membrane (see Figure 6.14)
- Condensation reactions that use enzymes to form polymers (see Figure 3.4A)
- Modifications of cell signaling proteins by protein kinases (see Figure 7.15)
- Motor proteins that move vesicles along microtubules (see Figure 5.19)

An ATP molecule is typically consumed within a second of its formation. At rest, an average person produces and hydrolyzes about 40 kg of ATP per day—as much as some people weigh. This means that each ATP molecule undergoes about 10,000 cycles of synthesis and hydrolysis every day!

8.2 recap

ATP is the "energy currency" of cells. Potential energy is stored in the P~O bonds of ATP. Some of the free energy released by exergonic reactions can be captured in the form of ATP. This energy can then be released by ATP hydrolysis and used to drive endergonic reactions.

learning outcomes

You should be able to:

- Explain the characteristics of ATP that account for the high free energy released during its hydrolysis to form ADP and P_i.
- Analyze evidence to identify reactions as exergonic or endergonic.
- Predict biochemical transformations that require coupling to ATP hydrolysis.

1. What chemical properties allow ATP to store energy?
2. The following is a biochemical reaction that involves ATP:

   Phosphoenolpyruvate → pyruvate ($\Delta G = +10$ kcal/mol)

   a. Is the reaction exergonic or endergonic?
   b. Is ATP hydrolyzed or formed during the reaction?

ATP is synthesized and used up very rapidly. But these biochemical reactions—and nearly all the others that take place inside a cell—could not proceed so rapidly without the help of enzymes.

key concept
8.3 Enzymes Speed Up Biochemical Transformations

When we know the change in free energy (ΔG) of a reaction, we know where the equilibrium point of the reaction lies: the more negative the ΔG value is, the further the reaction proceeds to products and toward completion. However, ΔG tells us nothing about the *rate* of a reaction—how fast it moves toward equilibrium. When we measure reaction rates of biochemical reactions in the lab, it quickly becomes apparent that in terms of the rapid needs for the chemistry of life, reaction rates are very slow—in fact, so slow that the cells would not survive without a way to speed up the reactions. That is the role of catalysts: substances that speed up reactions without themselves being permanently altered. A catalyst does not cause a reaction to occur that would not proceed without it, *but merely increases the rate of the reaction*, allowing equilibrium to be approached more rapidly. This is an important point: *No catalyst makes a reaction occur that cannot otherwise occur*.

focus your learning

- Catalysts speed up the rate of a reaction but do not allow the occurrence of a reaction that would not otherwise take place.
- Most enzymes are protein catalysts, and all enzymes have active sites where specific reactants called substrates bind and react to form products that diffuse away.
- Enzymes do not alter the equilibrium point or free energy change in a reaction.

Most biological catalysts are proteins called enzymes. A biological catalyst is a molecular framework or scaffold within which chemical catalysis takes place. The enzyme binds the reactants and sometimes participates in the reaction itself; however, such participation does not permanently change the enzyme. That's the hallmark of any catalyst: It ends up in exactly the same chemical condition after a reaction as before it.

In this section we will discuss the energy barrier that controls the rate of a chemical reaction. Then we will focus on the roles of enzymes: how they interact with specific reactants, how they lower the energy barrier, and how they permit reactions to proceed more quickly.

To speed up a reaction, an energy barrier must be overcome

An exergonic reaction may release free energy, but without a catalyst it will take place very slowly because there is an energy barrier between reactants and products. Think about the hydrolysis of sucrose, which we described in Key Concept 8.1:

$$\text{Sucrose} + H_2O \rightarrow \text{glucose} + \text{fructose}$$

In your body, this reaction is part of digestion. Even if water is abundant, the sucrose molecule will only very rarely bind the H atom and the —OH group of water at the appropriate locations to break the covalent bond between glucose and fructose *unless there is an input of energy to initiate the reaction*. Such an input of energy will place the sucrose into a reactive mode called the **transition state**. The energy input required for sucrose to reach this state is called the **activation energy** (E_a). Here's a more familiar example that can show you the ideas of activation energy and transition state:

$$\text{Fireworks} + O_2 \rightarrow CO_2 + H_2O + \text{energy (heat and light)}$$

Fireworks don't start exploding until you give them a spark: that's the activation energy. It excites the molecules in the fireworks so they will react with oxygen in the air. Once the transition state is reached, the explosive reaction occurs.

focus: key figure

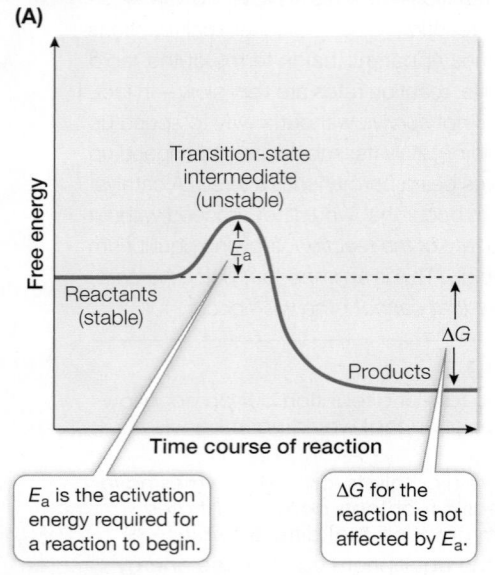

(A)

Transition-state intermediate (unstable)

E_a

Reactants (stable)

Products

ΔG

Free energy

Time course of reaction

E_a is the activation energy required for a reaction to begin.

ΔG for the reaction is not affected by E_a.

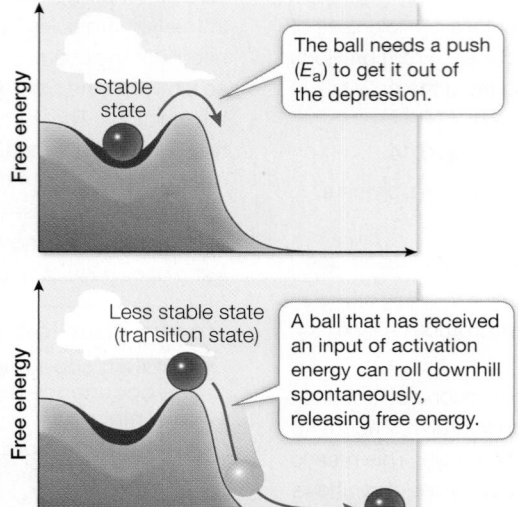

(B)

Free energy

Stable state

The ball needs a push (E_a) to get it out of the depression.

Free energy

Less stable state (transition state)

A ball that has received an input of activation energy can roll downhill spontaneously, releasing free energy.

Figure 8.8 Activation Energy Initiates Reactions **(A)** In any chemical reaction, an initial stable state must become less stable before change is possible. **(B)** A ball on a hillside provides a physical analogy for the biochemical principle graphed in (A).

Q: Would activation energy be required for an endergonic reaction?

Generally, exergonic reactions proceed only after the reactants are pushed over the energy barrier by some added energy. The energy barrier thus represents the amount of energy needed to start the reaction, the activation energy (**Focus: Key Figure 8.8A**). Recall the ball rolling down the hill in Figure 8.3A. The ball has a lot of potential energy at the top of the hill. However, if it is stuck in a small depression, it will not roll down the hill, even though that action is exergonic. To start the ball rolling, a small amount of energy (activation energy) is needed to push it out of the depression (**Focus: Key Figure 8.8B**). In a chemical reaction, the activation energy is the energy needed to change the reactants into unstable molecular forms called transition-state intermediates.

Transition-state intermediates have higher free energies than either the reactants or the products. Their bonds may be stretched and therefore unstable. Although the amount of activation energy needed for different reactions varies, it is often small compared with the change in free energy of the overall reaction. The activation energy put in to start a reaction is recovered during the ensuing "downhill" phase of the reaction, so it is not a part of the net free energy change, ΔG (see Figure 8.8A).

Where does activation energy come from? If reactants are at the human body temperature (37°C), they are moving around. A few are moving fast enough that their kinetic energy can overcome the energy barrier, enter the transition state, and react. So the reaction takes place—but very slowly. If the system is heated, all the reactant molecules move faster and have more kinetic energy, and the reaction speeds up. You have probably used this method in the chemistry laboratory or the kitchen.

However, adding heat to increase the average kinetic energy of the molecules so they can react would not work in living systems. Such a nonspecific approach would accelerate all chemical reactions, including destructive ones such as the denaturation of proteins (see Chapter 3). A more effective way to speed up a

reaction in a living system is to lower the energy barrier by bringing the reactants close together. In living cells, enzymes accomplish this task.

Enzymes bind specific reactants at their active sites

Catalysts increase the rates of chemical reactions. Most nonbiological catalysts are nonspecific. For example, powdered platinum catalyzes virtually any reaction in which molecular hydrogen (H_2) is a reactant. In contrast, most biological catalysts are highly specific. An *enzyme usually recognizes and binds to only one or a few closely related reactants, yielding specific products.

---

*connect the concepts   To understand enzymes and their activity, you must first understand the structures and chemical properties of proteins. See Key Concept 3.2.

In an enzyme-catalyzed reaction, the reactants are called **substrates**. Substrate molecules bind to a particular place on the enzyme protein called the **active site**, where catalysis takes place (**Figure 8.9**). The specificity of an enzyme in speeding up a particular reaction results from the exact three-dimensional shape and structure of its active site, into which only a narrow range of substrates can fit. Other molecules—with different shapes, different functional groups, and different properties—cannot fit properly and bind to the active site. This specificity is comparable to the specific binding of a membrane transport protein or receptor protein to its specific ligand, as described in Chapters 6 and 7.

The names of enzymes often reflect their functions and end with the suffix "ase." For example, the enzyme sucrase catalyzes the hydrolysis of sucrose, which we have referred to above:

$$\text{Sucrose} + H_2O \xrightarrow{\text{Sucrase}} \text{glucose} + \text{fructose}$$

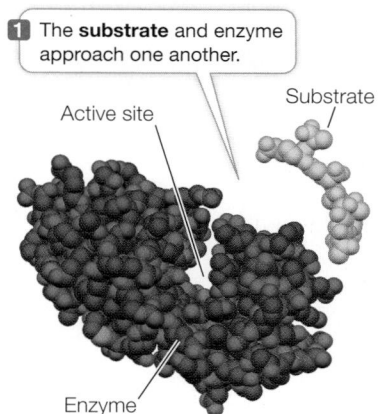

1 The **substrate** and enzyme approach one another.

Active site

Substrate

Enzyme

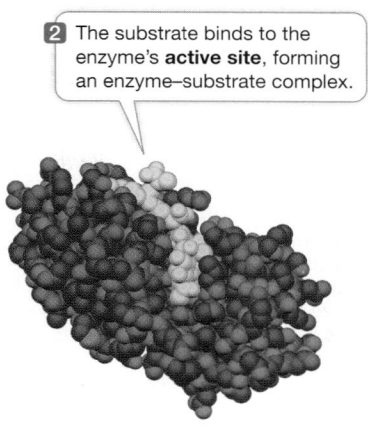

2 The substrate binds to the enzyme's **active site**, forming an enzyme–substrate complex.

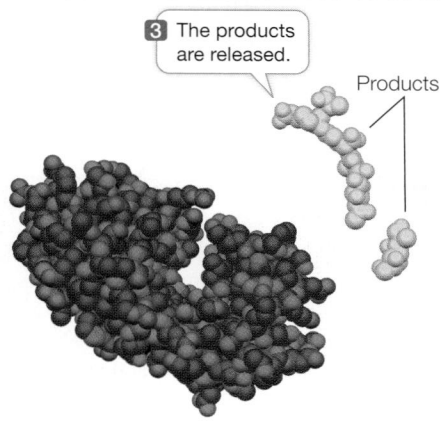

3 The products are released.

Products

Figure 8.9 Enzyme and Substrate   A reaction involving an enzyme is illustrated by lysozyme. Lysozyme catalyzes breakage of bonds in the peptidoglycans of bacterial cell walls. (See Key Concept 5.2 for a description of peptidoglycans.)

Thousands of enzymes have been identified, each catalyzing the transformation of its particular substrate(s). In general, enzymes fall into six categories:

1. *Oxidoreductases* transfer electrons between molecules, especially in energy metabolism. An example of an enzyme that catalyzes this reaction is oxidoreductase:

$$A– + B \rightarrow A + B–$$

You will see other examples in the next two chapters.

2. *Transferases* transfer groups of atoms (functional groups) (see Figure 3.1) between molecules. The basic reaction can be shown as:

$$AX + B \rightarrow A + BX$$

where A = the donor, B = the acceptor, and X = the functional group. For example, aminotransferases transfer $–NH_2$ groups between molecules, linking carbohydrates and amino acids.

3. *Hydrolases* add water to covalent bonds to break down molecules. An example (shown above) is sucrase, which hydrolyzes sucrose to its constituent monosaccharides.

4. *Lyases* catalyze the breaking of various chemical bonds by means other than hydrolysis and oxidation, often forming a new double bond or a new ring structure. For example, a lyase catalyzes the cleavage of the phosphorus–oxygen bond of ATP:

$$ATP \rightarrow cAMP + PP_i$$

5. *Isomerases* move functional groups from one place to another within a molecule, forming an isomer with the same atoms, differently bonded. For example, glucose and fructose are isomers of $C_6H_{12}O_6$ (see Figure 3.17).

6. *Ligases* join two molecules together. You will see examples of ligases when you learn about DNA replication in Chapter 13.

A special note on language: When biologists refer to an enzyme, they usually use a verb for its action. For example, as noted above (#6), we used the words "ligases join…." It is important for you to realize that this wording is shorthand for what really happens: ligases catalyze the joining. Enzymes don't "do" the reaction; they catalyze it.

When a substrate binds to the active site of an enzyme, the result is called an **enzyme–substrate complex** (**ES**) (see Figure 8.9). The ***chemical forces** that hold the complex together may include hydrogen bonding, electrical attraction, or temporary covalent bonding. The enzyme–substrate complex gives rise to product and free enzyme:

$$E + S \rightarrow ES \rightarrow E + P$$

where E is the enzyme, S is the substrate, P is the product, and ES is the enzyme–substrate complex. The free enzyme (E) is in the same chemical form at the end of the reaction as at the beginning. While bound to the substrate, it may change chemically, but by the end of the reaction it has been restored to its initial form and is ready to bind more substrate.

---

*connect the concepts   Binding of a substrate to an enzyme involves not just a "fit" but also noncovalent interactions between the enzyme and substrate. See Key Concept 3.2 for an outline of noncovalent interactions.

The relationship between enzyme and substrate may seem familiar; it is analogous to the binding of a receptor and ligand that we described in Chapter 7. In that instance, we described the affinity of a receptor for its ligand with the dissociation constant (K_D). The lower the K_D, the tighter the binding. For enzymes and their substrates, K_D values are often in the range 10^{-5} to 10^{-6} M, which favors the formation of the ES. In practical terms, the K_D values of the ES are such that the binding between enzyme and substrate is somewhat reversible; the substrate can be released before the reaction. However, many enzyme-catalyzed reactions will proceed because the ES is short-lived and the product(s) form quickly.

Enzymes lower the energy barrier but do not affect equilibrium

Reactants that are bound to the enzyme, forming an enzyme–substrate complex, require less activation energy than do the transition-state intermediates in the corresponding uncatalyzed reaction (**Figure 8.10**). Thus the enzyme lowers the energy barrier for the reaction—it offers the reaction an easier path, speeding it up. When an enzyme lowers the energy barrier, both the forward and the reverse reactions speed up, so the enzyme-catalyzed reaction proceeds toward equilibrium more rapidly than the uncatalyzed reaction. *The final equilibrium is the same with or without the enzyme.* Similarly, adding an enzyme to

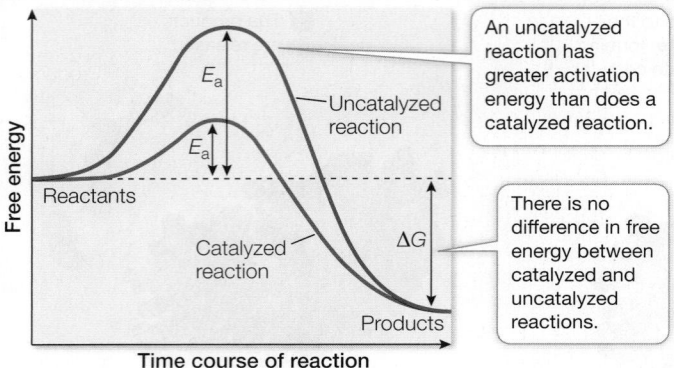

Figure 8.10 Enzymes Lower the Energy Barrier   Although the activation energy is lower in an enzyme-catalyzed reaction than in an uncatalyzed reaction, the energy released is the same with or without catalysis. In other words, E_a is lower but ΔG is unchanged. Lower activation energy means the reaction will take place at a faster rate.

 Activity 8.2 Free Energy Changes
www.Life11e.com/ac8.2

a reaction does not change the difference in free energy (ΔG) between the reactants and the products (see Figure 8.10).

Enzymes can change the rate of a reaction substantially. For example, if a particular protein that has arginine as its terminal amino acid just sits in solution, the protein molecules tend toward disorder and the terminal peptide bonds break, releasing the arginine residues (ΔS increases). Without an enzyme, this is a very slow reaction—it takes about 7 years for half of the protein molecules to undergo the reaction. But, with the enzyme carboxypeptidase A catalyzing the reaction, half the arginines are released in less than a second! Rate enhancement by enzymes varies from 1 million times to an amazing 10^{26} by sulfate monoesterases. The consequence of catalysis for living cells is not difficult to imagine. Such increased reaction rates make new realities possible.

▶ 8.3 recap

A chemical reaction requires a "push" over the energy barrier to get started. An enzyme reduces the activation energy needed to start a reaction by binding the reactants (substrates), which speeds up the reaction. Biological catalysts are highly specific; only a narrow range of substrates can fit into the enzyme's active site, so generally only one specific reaction can be catalyzed by a given enzyme. Enzymes affect only the *rates* at which reactions proceed—they do not affect the equilibrium or free energy associated with a given reaction.

learning outcomes

You should be able to:

- Describe a catalyst.
- Describe how an enzyme–substrate complex forms.
- Describe the relationship between an enzyme and the equilibrium point of the reaction it catalyzes.

1. Explain how the structure of an enzyme makes that enzyme specific.
2. What is the relationship between an enzyme and the equilibrium point of a reaction?

▶ 8.3 recap

3. When potatoes are peeled, the enzyme polyphenol oxidase causes discoloration by catalyzing the oxidation of certain molecules, using O_2 as a substrate. Explain these observations:

   a. Browning is reduced in potatoes that are peeled under water and kept there.

   b. Potatoes that have been boiled at 100°C and then cooled do not brown when they are sliced.

Now that you have a general understanding of the structures, functions, and specificities of enzymes, let's look more closely at how they work.

▶ key concept

8.4 Enzymes Bring Substrates Together so Reactions Readily Occur

During and after the formation of the enzyme–substrate complex, chemical interactions occur. These interactions contribute directly to the breaking of old bonds and the formation of new ones. In catalyzing a reaction, an enzyme may use one or more strategies.

focus your learning

- Several different mechanisms enable enzymes to speed up chemical reactions.
- An enzyme may undergo a change in shape called induced fit as the result of binding its substrate.
- Many enzymes require additional chemical partners to catalyze a chemical reaction.
- The concentration of substrate influences the rate of an enzyme-catalyzed reaction.

Enzymes can orient substrates

When free in solution, substrates are moving from place to place randomly while at the same time vibrating, rotating, and tumbling around. They may not have the proper orientation to interact when they collide. Part of the activation energy needed to start a reaction is used to bring together specific atoms so that bonds can form (**Figure 8.11A**). For example, if acetyl CoA and oxaloacetate are to form citrate (a step in the metabolism of glucose; see Key Concept 9.2), the two substrates must be oriented so that the carbon atom of the methyl group of acetyl CoA can form a covalent bond with the carbon atom of the carbonyl group of oxaloacetate. The active site of the enzyme citrate synthase has just the right shape to bind these two molecules so that these atoms are adjacent.

Enzymes can induce strain in the substrate

Once a substrate has bound to its active site, an enzyme can cause bonds in the substrate to stretch, putting it in an unstable transition state (**Figure 8.11B**). For example, lysozyme is a protective enzyme abundant in tears and saliva that destroys invading bacteria by cleaving peptidoglycans in their cell walls. Lysozyme's active site "stretches" the bonds between the glycan monomers, rendering the bonds unstable and more reactive to lysozyme's other substrate, water.

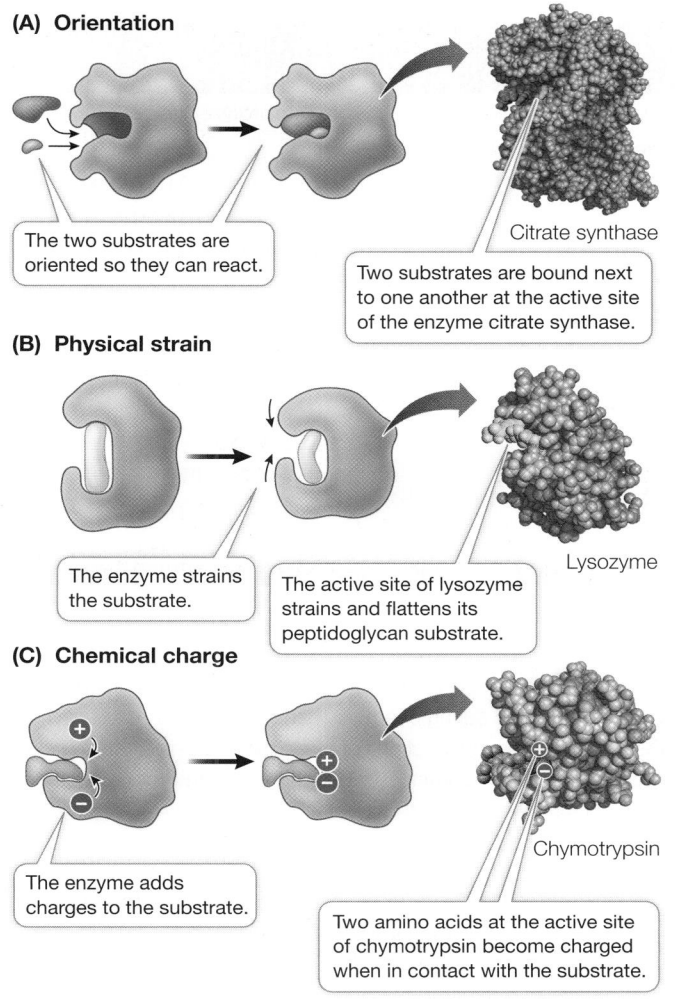

(A) Orientation

The two substrates are oriented so they can react.

Two substrates are bound next to one another at the active site of the enzyme citrate synthase.

Citrate synthase

(B) Physical strain

The enzyme strains the substrate.

The active site of lysozyme strains and flattens its peptidoglycan substrate.

Lysozyme

(C) Chemical charge

The enzyme adds charges to the substrate.

Two amino acids at the active site of chymotrypsin become charged when in contact with the substrate.

Chymotrypsin

Figure 8.11 Life at the Active Site   Enzymes use several mechanisms to cause their substrates to enter the transition state: **(A)** orientation, **(B)** physical strain, and **(C)** chemical charge.

Enzymes can temporarily add chemical groups to substrates

The side chains (R groups) of an enzyme's amino acids may be direct participants in making the enzyme's substrates more chemically reactive (**Figure 8.11C**).

- In *acid–base catalysis*, the acidic or basic side chains of the amino acids in the active site transfer H^+ to or from the substrate, destabilizing a covalent bond in the substrate, and permitting it to break.

- In *covalent catalysis*, a functional group in a side chain forms a temporary covalent bond with a portion of the substrate.

- In *metal ion catalysis*, metal ions such as copper, iron, and manganese, which are often firmly bound to side chains of enzymes, can lose or gain electrons without detaching from the enzymes. This ability makes them important participants in oxidation–reduction reactions, which involve the loss or gain of electrons.

Molecular structure determines enzyme function

Many enzymes are much larger than their substrates. An enzyme is typically a protein containing hundreds of amino acids. It may consist of a single folded polypeptide chain or of several subunits (see Key Concept 3.2). Its substrate is generally a small molecule or a small part of a large molecule. The active site of the enzyme is usually quite small, not more than 6 to 12 amino acids. You may be asking yourself:

- What features of the enzyme's active site allow it to recognize and bind the substrate?
- What is the role of the rest of the protein?

THE ACTIVE SITE IS SPECIFIC TO THE SUBSTRATE(S)   The remarkable ability of an enzyme to select exactly the right substrate(s) depends on a precise interlocking of molecular shapes and interactions of chemical groups at the active site. The binding of a substrate to the active site depends on the same kinds of forces that maintain the tertiary structure of the enzyme: hydrogen bonds, the attraction and repulsion of electrically charged groups, and hydrophobic interactions. The specific fit of the substrate in the active site of lysozyme is illustrated in Figures 8.9 and 8.11B.

AN ENZYME CHANGES SHAPE WHEN IT BINDS A SUBSTRATE   Just as a membrane receptor protein may undergo precise changes in conformation upon binding to its ligand (see Chapter 7), some enzymes change their shapes when they bind their substrate(s). These shape changes, which are called **induced fit**, alter the shape of the active site(s) of the enzyme.

An example of induced fit can be seen in the enzyme hexokinase (see Figure 8.7), which catalyzes the reaction

$$\text{Glucose} + \text{ATP} \rightarrow \text{glucose 6-phosphate} + \text{ADP}$$

Induced fit brings reactive side chains from the hexokinase active site into alignment with the substrates (**Figure 8.12**), facilitating its catalytic mechanisms. Equally important, the folding of hexokinase to fit around the substrates (glucose and ATP) excludes water

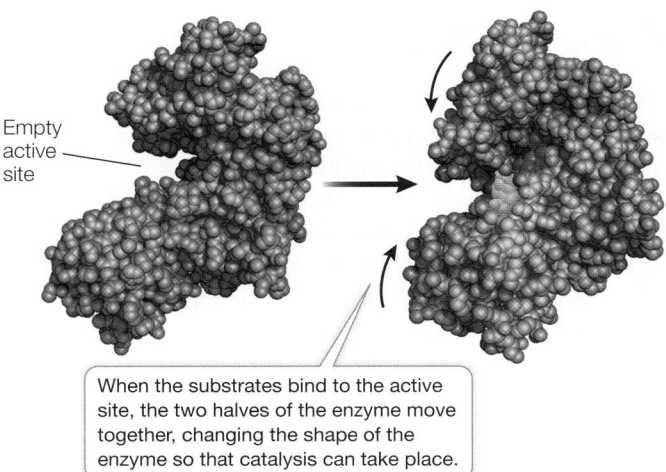

Empty active site

When the substrates bind to the active site, the two halves of the enzyme move together, changing the shape of the enzyme so that catalysis can take place.

Figure 8.12 Some Enzymes Change Shape When Substrate Binds to Them   Shape changes result in an induced fit between enzyme and substrate, improving the catalytic ability of the enzyme. Induced fit can be observed in the enzyme hexokinase, seen here with and without its substrates, glucose (red) and ATP (yellow).

Q: Do covalent bonds in hexokinase break when it changes shape? Explain.

from the active site. This is essential, because if water were present, the ATP could be hydrolyzed to ADP and P_i. But since water is absent, the transfer of a phosphate from ATP to glucose is favored.

Induced fit at least partly indicates what the role(s) of the rest of the protein may be:

● It provides a framework so that the amino acids of the active site are properly positioned in relation to the substrate(s).

● It participates in significant changes in protein shape and structure that result in induced fit.

● It provides binding sites for regulatory molecules (see Key Concept 8.5).

Not all enzymes are proteins

For most of the twentieth century, biologists thought that all enzymes were proteins. After all, they reasoned, only proteins have the diversity of three-dimensional structures and functional groups to bind to substrates and catalyze their transformations (review Key Concept 3.2). Then it was discovered that some RNA molecules can act as enzymes by catalyzing changes in substrates, in most cases other RNA molecules. You will learn about RNA splicing in Key Concept 14.4. Briefly, this involves a cell making a larger RNA that gets cut and spliced to remove internal region(s). The splicing reaction is catalyzed by a ribozyme. Another important ribozyme occurs at the ribosome, where an RNA molecule catalyzes the formation of a peptide bond between amino acids (see Key Concept 14.5).

In retrospect, it is not too surprising that some RNA molecules can act as biological catalysts (**ribozymes**). As you saw in Figure 4.3, most RNAs are single strands of nucleotides that can fold back on themselves and form a three-dimensional structure by hydrogen bonding. A substrate can fit into the RNA structure, and functional groups on the RNA (such as $-NH_2$ and $C=O$) can be involved in catalysis.

Some enzymes require other molecules in order to function

Many enzymes require the presence of nonprotein chemical "partners" in order to function (**Table 8.2**):

● *Prosthetic groups* are distinct, non–amino acid atoms or molecular groupings that are permanently bound to their enzymes. An example is flavin adenine dinucleotide (FAD), which is bound to succinate dehydrogenase, an important enzyme in cellular respiration (see Key Concept 9.3).

● *Inorganic cofactors* include ions such as copper, zinc, and iron that are permanently bound to certain enzymes. For example, the enzyme alcohol dehydrogenase contains the cofactor zinc.

● A *coenzyme* is a nonprotein carbon-containing molecule that is required for the action of one or more enzymes. It is usually relatively small compared with the enzyme to which it binds.

A coenzyme moves from enzyme to enzyme, adding or removing chemical groups from the substrate. A coenzyme is like a substrate in that it does not permanently bind to the enzyme: it binds to the active site, changes chemically during the reaction, and then separates from the enzyme to participate in other reactions. There is actually no clear distinction between the functions of coenzymes and some substrates. For example, ATP and ADP have been described as

table **8.2** Some Examples of Nonprotein "Partners" of Enzymes	
Type of molecule	Role in catalyzed reactions
Prosthetic groups	
Heme	Binds ions, O_2, and electrons
FAD	Carries electrons/protons
Retinal	Converts light energy
Inorganic cofactors	
Iron (Fe^{2+} or Fe^{3+})	Oxidation/reduction
Copper (Cu^+ or Cu^{2+})	Oxidation/reduction
Zinc (Zn^{2+})	Stabilizes DNA binding structure
Coenzymes	
Biotin	Carries $-COO^-$
Coenzyme A	Carries $-CO-CH_3$
NAD	Carries electrons/protons
ATP	Provides/extracts energy

coenzymes, even though they are really substrates that gain or lose phosphate groups during chemical reactions. The term "coenzyme" was coined before the functions of these molecules were fully understood. Biochemists continue to use the term, and for consistency with the field, we use it in this book.

The substrate concentration affects the reaction rate

For a reaction of the type A → B, the rate of the uncatalyzed reaction is directly proportional to the concentration of A. The higher the concentration of substrate, the faster the rate of the reaction. The appropriate enzyme not only speeds up the reaction; it also changes the shape of a plot of rate versus substrate concentration (**Figure 8.13**). For a given concentration of enzyme, the rate of the enzyme-catalyzed reaction initially increases as the substrate concentration increases from zero, but then it levels off. At some point, further increases in the substrate concentration

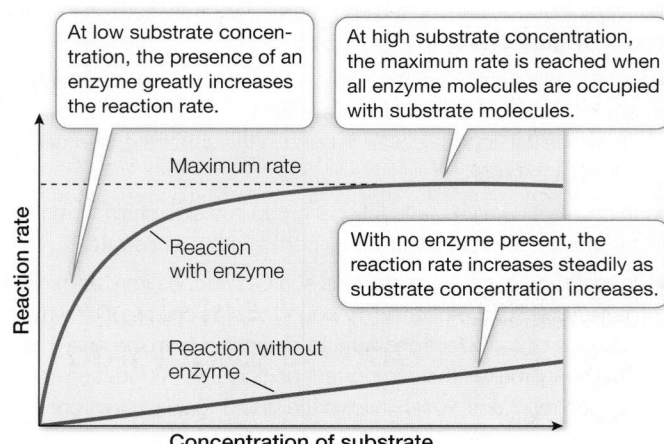

At low substrate concentration, the presence of an enzyme greatly increases the reaction rate.

At high substrate concentration, the maximum rate is reached when all enzyme molecules are occupied with substrate molecules.

Maximum rate

Reaction with enzyme

With no enzyme present, the reaction rate increases steadily as substrate concentration increases.

Reaction without enzyme

Reaction rate

Concentration of substrate

Figure 8.13 Catalyzed Reactions Reach a Maximum Rate
Because there is usually less enzyme than substrate present, the reaction rate levels off when the enzyme becomes saturated.

do not significantly increase the reaction rate—the maximum rate has been reached.

The concentration of an enzyme is usually much lower than that of its substrate (e.g., in cells expressing the enzyme sucrase, there are far more molecules of sucrose than of the enzyme). As more and more substrate molecules bind to the available enzyme molecules, there is a saturation phenomenon like the one that occurs in facilitated diffusion (see Figure 6.12). When all the enzyme molecules are bound with substrate molecules, the enzyme is working as fast as it can—at its maximum rate. Nothing is gained by adding more substrate, because no free enzyme molecules are left to act as catalysts. Under these conditions the active sites are said to be saturated.

The maximum rate of a catalyzed reaction can be used to measure how efficient the enzyme is. The turnover number is the maximum number of substrate molecules that one enzyme molecule can convert to product per unit of time. This number ranges from 1 molecule every 2 seconds for lysozyme to an amazing 40 million molecules per second for the liver enzyme catalase.

8.4 recap

Enzymes orient their substrates to bring together specific atoms so that bonds can form. An enzyme can participate in the reaction it catalyzes by temporarily changing shape or destabilizing the enzyme–substrate complex. Some enzymes require prosthetic groups, inorganic cofactors, or coenzymes in order to function.

learning outcomes

You should be able to:

- Give examples of mechanisms for enzyme catalysis.
- Describe coenzymes and cofactors and how they function.
- Predict changes in the rate of an enzyme-catalyzed reaction as factors in the reaction are varied.

1. The chemical composition of an enzyme before and after a catalyzed reaction is the same, but during the reaction an enzyme can change. How and why does this happen?
2. Compare enzymes and coenzymes with regard to chemical structure and function.
3. Plot reaction rate versus substrate concentration, and explain the shape of the curve.

You've seen in this section how individual enzymes work on their substrates. However, enzymes inside organisms don't operate in isolation—there may be thousands of different enzymes within a given cell. Let's see how all these different enzymes work together in a complex organism.

key concept
8.5
Enzyme Activities Can Be Regulated

The biochemical reactions in cells operate within metabolic pathways in which the product of one reaction is a reactant for the next. These pathways do not exist in isolation, but interact extensively, and each reaction in each pathway is catalyzed by a specific enzyme.

focus your learning

- The cell carries out large numbers of metabolic reactions that are interconnected and regulated by the enzymes that catalyze them.
- Small molecules can regulate enzyme activity by binding to and activating or inhibiting an enzyme.
- Allosteric regulation results from the binding of regulator molecules to sites on an enzyme other than the active site.
- Environmental factors such as pH and temperature affect enzyme activity.

Within a cell or organism, the presence and activity of enzymes determine the "flow" of chemicals through different metabolic pathways. The amount of enzyme activity, in turn, can be controlled in two ways:

1. *Regulation of gene expression.* The gene for an enzyme protein may be expressed more or less, resulting in more or fewer enzyme molecules in the cell. You learned in Chapter 7 that some signal transduction pathways result in changes in gene expression, and that often the genes that are switched on or off encode enzymes.

2. *Regulation of enzyme activity.* The enzyme may change its shape to hide its active site from the substrate, or a regulator molecule may prevent the substrate from reaching the active site. Blocking the activity of a single enzyme can affect an entire metabolic pathway.

The flow of molecules through interacting metabolic pathways can be studied, but this process quickly becomes complicated, because each pathway influences the others. Computer algorithms are used to model these pathways and show how they mesh in an interdependent system (**Figure 8.14**). Such models can help predict what will happen if the concentration of one molecule or another is altered. This new field of biology is called **systems biology**, and it has numerous applications.

In this section we will investigate the roles of enzymes in organizing and regulating metabolic pathways. We will also examine how the environment—particularly temperature and pH—affects enzyme activity.

Activity 8.3 **System Simulation**
www.Life11e.com/ac8.3

Enzymes can be regulated by inhibitors

Various chemical inhibitors can bind to enzymes, slowing down the rates of the reactions they catalyze. Some inhibitors occur naturally in cells; others are artificial. Naturally occurring inhibitors regulate metabolism; artificial ones can be used to treat disease, to kill pests, or to study how enzymes work. In some cases the inhibitor binds the enzyme irreversibly, and the enzyme becomes permanently inactivated. In other cases the inhibitor has reversible effects; it can separate from the enzyme, allowing the enzyme to function fully as before. The removal of a natural reversible inhibitor increases an enzyme's rate of catalysis.

IRREVERSIBLE INHIBITION If an inhibitor covalently binds to certain side chains at the active site of an enzyme, it will permanently inactivate the enzyme by destroying its capacity to interact with its

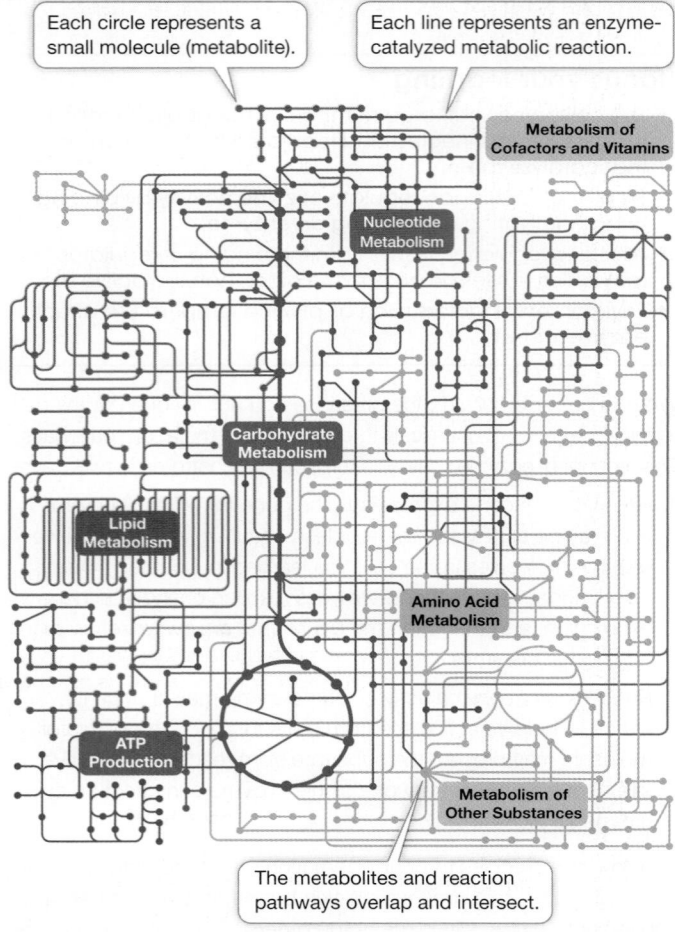

Each circle represents a small molecule (metabolite).

Each line represents an enzyme-catalyzed metabolic reaction.

Metabolism of Cofactors and Vitamins

Nucleotide Metabolism

Carbohydrate Metabolism

Lipid Metabolism

Amino Acid Metabolism

ATP Production

Metabolism of Other Substances

The metabolites and reaction pathways overlap and intersect.

Figure 8.14 Metabolic Pathways   The complex interactions of metabolic pathways can be modeled by the tools of systems biology. In cells, the main elements controlling these pathways are enzymes.

normal substrate. An example of an irreversible inhibitor is aspirin, which was described in the opening of this chapter. Aspirin (acetylsalicylic acid) binds to a target enzyme called cyclooxygenase (COX) near its active site. When it binds, aspirin releases its acetyl

group ($-CH_2CH_3$) and this group of atoms is transferred to bond covalently to serine at its polar hydroxyl group (**Figure 8.15**). The group of atoms now "coating" serine sticks out into a channel in the protein and prevents the substrate, arachidonic acid, from reaching the active site. So prostaglandin is no longer produced and its stimulation of inflammation and pain are blocked. The elucidation of the mechanism of aspirin's action as an inhibitor of a pathway involved in inflammation was a landmark in modern pharmacology, the study of drugs (**Investigating Life: How Do Anti-Inflammatory Drugs Work as Enzyme Inhibitors?**).

REVERSIBLE INHIBITION   In some cases an inhibitor is similar enough to a particular enzyme's natural substrate to bind noncovalently to its active site, yet different enough that the enzyme catalyzes no chemical reaction. While such a molecule is bound to the enzyme, the natural substrate cannot enter the active site and the enzyme is unable to function. Such a molecule is called a **competitive inhibitor** because it competes with the natural substrate for the active site (**Figure 8.16A and B**). In this case, the degree of inhibition depends on the relative concentrations of the substrate and the inhibitor: if the inhibitor concentration is higher, it is more likely to bind the active site of the enzyme than the substrate, and vice versa. The inhibition is reversible because if the concentration of substrate is increased or if the concentration of inhibitor is reduced, the substrate is more likely to bind, and the enzyme is active again.

An example of a competitive inhibitor is the drug methotrexate. An important coenzyme in the formation of purines (components of nucleic acids) is tetrahydrofolate, which is formed from dihydrofolate in a reaction catalyzed by dihydrofolate reductase (DHFR):

$$\text{Dihydrofolate} \xrightarrow{\text{DHFR}} \text{tetrahydrofolate}$$

When cancer cells reproduce, they need to replicate their DNA, and so they need to produce purines. This makes DHFR an ideal target for an anticancer drug. A team led by Sidney Farber at Harvard Medical School first showed that an analog of dihydrofolate could

Figure 8.15 Irreversible Inhibition   Aspirin transfers an acetyl group, which covalently bonds to a serine amino acid near the active site of its target enzyme. This permanent change prevents the substrate from binding.

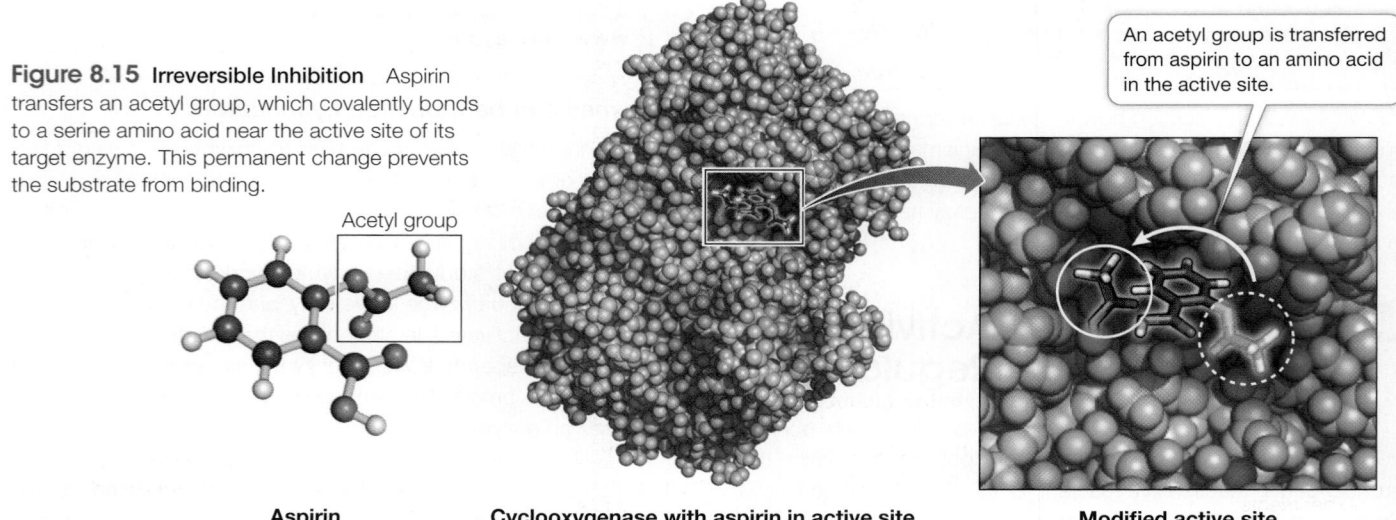

Acetyl group

An acetyl group is transferred from aspirin to an amino acid in the active site.

Aspirin   **Cyclooxygenase with aspirin in active site**   **Modified active site**

experiment

Original Papers: Vane, J. R. 1971. Inhibition of prostaglandin synthesis as a mechanism of action of aspirin-like drugs. *Nature* 231: 232–235.

Smith, J. B. and A. L Wells. 1971. Aspirin selectively inhibits prosta-glandin production in human platelets. *Nature* 231: 235–238.

The opening story in this chapter describes how willow bark, the source of what became aspirin, is a centuries-old remedy for pain and inflammation. Until the late twentieth century, it was assumed that aspirin acted directly on the nervous system. Working at England's Royal College of Surgeons, a group led by John Vane showed that aspirin acts instead as an inhibitor of the enzymatic reaction that produces prostaglandin (PG), a fatty acid derivative important in inflammation.

HYPOTHESIS▶   Aspirin acts as an anti-inflammatory drug by inhibiting an enzyme that catalyzes prostaglandin synthesis.

METHOD

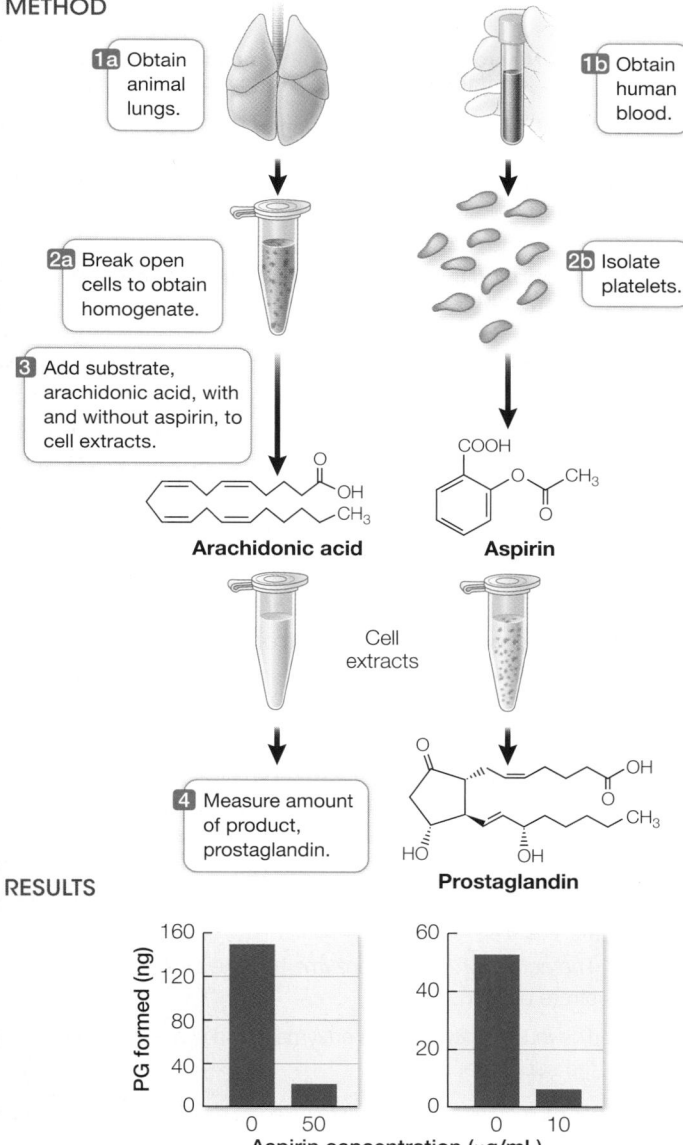

1a Obtain animal lungs.

1b Obtain human blood.

2a Break open cells to obtain homogenate.

2b Isolate platelets.

3 Add substrate, arachidonic acid, with and without aspirin, to cell extracts.

Arachidonic acid

Aspirin

Cell extracts

4 Measure amount of product, prostaglandin.

Prostaglandin

RESULTS

PG formed (ng)

Aspirin concentration (μg/mL)

CONCLUSION▶   In both animal and human cells, aspirin blocks the synthesis of the pro-inflammatory molecule, prostaglandin, in test tube experiments.

work with the data

For his discovery of the mechanism by which aspirin relieves pain, John Vane was awarded the Nobel Prize and knighted by Queen Elizabeth II. Key to the experiments was the assumption that enzyme activity and mechanism are the same outside the organism as they are inside. In the lab, if given its substrate and the same set of environmental conditions as in the cytoplasm, an enzyme will catalyze production of its product.

QUESTIONS▶

1. In the first set of experiments, lung tissue from guinea pigs was broken up to form a cell-free extract (homogenate). The substrate, arachidonic acid, was added to the extract, and after 30 min the amount of prostaglandin (PG) was measured. The results are shown in **Table A**. Plot PG synthesis versus aspirin concentration. What can you conclude?

2. A similar set of experiments was performed on human platelets (membrane-bound cell fragments derived from white blood cells). These cells are known to produce PG by the same enzyme mechanism as lung tissue under certain environmental conditions. The results are shown in **Table B**. What can you conclude from these results and about the generalizability of your conclusion from Question 1?

Table A

Aspirin concentration (μg/mL)	PG synthesis (ng)
0	220
1	172
2	136
10	99
50	33
80	0

Table B

Aspirin concentration (μg/mL)	PG synthesis (ng)
0	53
0.01	48
0.1	35
1	18
10	7

3. In a third set of experiments, platelets were isolated from human volunteers and their ability to make PG was measured (no aspirin present). Then these people were given a clinically effective dose of aspirin, and after a short period their platelets were isolated and PG synthesis measured (again, no aspirin in the test tube experiments). The results from three people are shown in **Table C**. How do these data either reinforce or refute your conclusion from Questions 1 and 2?

Table C

	Prostaglandin synthesis (ng)	
Individual	Before aspirin	After aspirin
1	160	16
2	108	5
3	103	20

A similar **work with the data** exercise may be assigned in **LaunchPad**.

(A) Normal enzyme–substrate binding

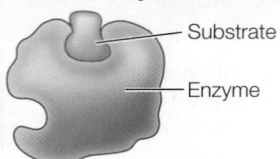

Substrate

Enzyme

Normal substrate binding to enzyme active site.

(B) Competitive inhibition

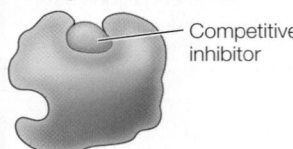

Competitive inhibitor

A **competitive inhibitor** binds to the active site, preventing substrate binding.

(C) Uncompetitive inhibition

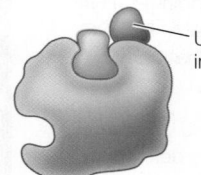

Uncompetitive inhibitor

An **uncompetitive inhibitor** binds to the enzyme–substrate complex, preventing release of products.

(D) Noncompetitive inhibition

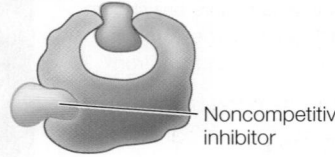

Noncompetitive inhibitor

A **noncompetitive inhibitor** binds at a site other than the active site, changing enzyme structure so that normal substrate binding cannot occur.

Figure 8.16 Reversible Inhibition

 Animation 8.1 **Enzyme Catalysis**
www.Life11e.com/a8.1

treat leukemia. The drug used by Farber (aminopterin) has since been replaced by a similar analog, methotrexate:

Dihydrofolate

Methotrexate

This successful drug is used to treat inflammatory diseases such as psoriasis and rheumatoid arthritis, as well as cancer.

An **uncompetitive inhibitor** (**Figure 8.16C**) binds to the enzyme–substrate complex, preventing the complex from releasing

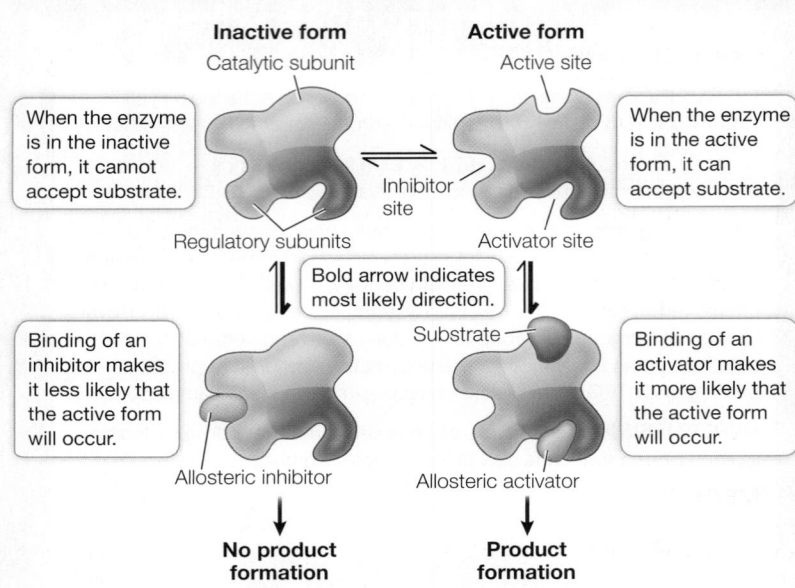

Inactive form
Catalytic subunit

Active form
Active site

When the enzyme is in the inactive form, it cannot accept substrate.

Inhibitor site

When the enzyme is in the active form, it can accept substrate.

Regulatory subunits

Activator site

Bold arrow indicates most likely direction.

Binding of an inhibitor makes it less likely that the active form will occur.

Substrate

Binding of an activator makes it more likely that the active form will occur.

Allosteric inhibitor

Allosteric activator

No product formation

Product formation

Figure 8.17 Allosteric Regulation of Enzymes   Active and inactive forms of an enzyme can be interconverted, depending on the binding of effector molecules at sites other than the active site. Binding an inhibitor stabilizes the inactive form, and binding an activator stabilizes the active form.

 Animation 8.2 **Allosteric Regulation of Enzymes**
www.Life11e.com/a8.2

products. Unlike competitive inhibition, uncompetitive inhibition cannot be overcome by adding more substrate.

A **noncompetitive inhibitor** binds to an unbound enzyme at a site distinct from the active site. Binding causes a change in the shape of the enzyme that alters its activity (**Figure 8.16D**). The active site may no longer bind the substrate, or if it does, the rate of product formation may be reduced. Like competitive inhibitors, noncompetitive inhibitors can become unbound, so their effects are reversible.

Allosteric enzymes are controlled via changes in shape

The change in enzyme shape that is due to noncompetitive inhibitor binding is an example of allostery (*allo*, "different," + *stereos*, "shape"). **Allosteric regulation** occurs when an effector molecule binds to a site other than the active site of an enzyme, *inducing the enzyme to change its shape*. The change in shape alters the affinity of the active site for the substrate, and so the rate of the reaction is changed.

Often, an enzyme will exist in the cell in more than one possible shape (**Figure 8.17**):

- The *active form* of the enzyme has the proper shape for substrate binding.
- The *inactive form* of the enzyme has a shape that cannot bind the substrate.

Other molecules, collectively referred to as effectors, can influence which form the enzyme takes:

- Binding of an inhibitor to a site other than the active site can stabilize the inactive form of the enzyme, making it less likely to convert to the active form.

- The active form can be stabilized by the binding of an activator to another site on the enzyme.

Like substrate binding, the binding of inhibitors and activators to their regulatory sites (also called allosteric sites) is highly specific. Most (but not all) enzymes that are allosterically regulated are proteins with quaternary structure; that is, they are made up of multiple polypeptide subunits. The polypeptide that has the active site is called the catalytic subunit. The allosteric sites are often located on different polypeptides, called the regulatory subunits (see Figure 8.17).

Some enzymes have multiple subunits containing active sites, and the binding of substrate to one of the active sites causes allosteric effects. When substrate binds to one subunit, there is a slight change in protein structure that influences the adjacent subunit. The slight change to the second subunit makes its active site more likely to bind to the substrate. So the reaction speeds up as the sites become sequentially activated.

As a result, an allosteric enzyme with multiple active sites and a nonallosteric enzyme with a single active site differ greatly in their reaction rates when the substrate concentration is low. Graphs of reaction rates plotted against substrate concentrations show this relationship. For a nonallosteric enzyme, the plot is hyperbolic:

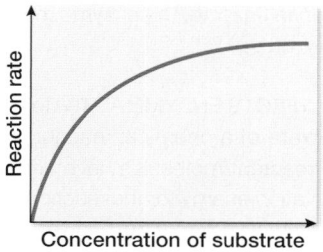

The reaction rate first increases sharply with increasing substrate concentration, then tapers off to a constant maximum rate as the supply of enzyme becomes saturated.

For a multisubunit allosteric enzyme, the graph looks different, having a sigmoid (S-shaped) appearance:

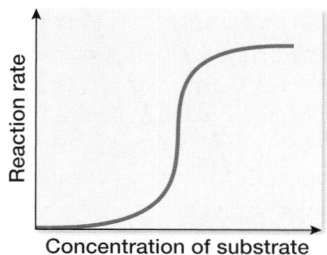

At low substrate concentrations, the reaction rate increases only gradually as substrate concentration increases. After the substrate binds to the first active site of the enzyme (the slowly increasing part of the curve), there is a change in the enzyme's quaternary structure such that the other sites become more likely to bind substrate, so the reaction speeds up (the rapidly increasing part of the curve). Once all sites are saturated with substrate, the reaction rate reaches a plateau (the upper, flat part of the curve). Within a certain range, the reaction rate is extremely sensitive to relatively small changes in substrate concentration. In addition, allosteric enzymes are very sensitive to low concentrations of inhibitors. Because of this sensitivity, allosteric enzymes are important in regulating entire metabolic pathways.

Allosteric effects regulate many metabolic pathways

Metabolic pathways typically involve a starting material, various intermediate products, and an end product that is used for some purpose by the cell. In each pathway there are a number of reactions, each forming an intermediate product and each catalyzed by a different *enzyme. The first step in a pathway is called the commitment step, meaning that once this enzyme-catalyzed reaction occurs, the "ball is rolling," and the other reactions happen in sequence, leading to the end product. But what if the cell has no requirement for that product—for example, if that product is available from its environment in adequate amounts? It would be energetically wasteful for the cell to continue making something it does not need.

---

*connect the concepts   The number of molecules of a particular enzyme present in a cell at a given time is controlled in large part by the synthesis of that protein. See Key Concept 16.1 for a discussion of the regulation of protein synthesis.

One way to avoid this problem is to shut down the metabolic pathway by having the final product inhibit the enzyme that catalyzes the commitment step (**Figure 8.18**). Often this inhibition occurs allosterically. When the end product is present at a high concentration, some of it binds to an allosteric site on the commitment step enzyme, thereby causing it to become inactive. Thus the final product acts as a noncompetitive inhibitor (described earlier in this section) of the first enzyme in the pathway. This mechanism is known as feedback inhibition or end-product inhibition. We will describe many other examples of such inhibition in later chapters.

Many enzymes are regulated through reversible phosphorylation

Many enzymes involved in signal transduction are regulated via reversible phosphorylation (see Figure 7.14A). An enzyme can be activated by a protein kinase, which adds a phosphate from ATP to one or more specific amino acids. This results in a change in the shape of the enzyme, making it active. Such activation is reversible because another enzyme called a protein phosphatase can catalyze the hydrolysis and removal of phosphate groups, so that the enzyme becomes inactive again. In addition to the enzymes involved in signal transduction, many other enzymes and proteins in the cell (such as ion channels) are regulated via reversible phosphorylation. Reflecting the important role of protein phosphorylation in cell functions, the human genome contains more than 500 protein kinase genes: about 2 percent of all the protein-coding genes we have.

Enzymes are affected by their environment

Enzymes enable cells to perform chemical reactions and carry out complex processes rapidly without using the extremes of temperature and pH employed by chemists in the laboratory. However, because of their three-dimensional structures and the chemistry of the side chains in their active sites, enzymes (and their substrates) are

(A) Feedback inhibition

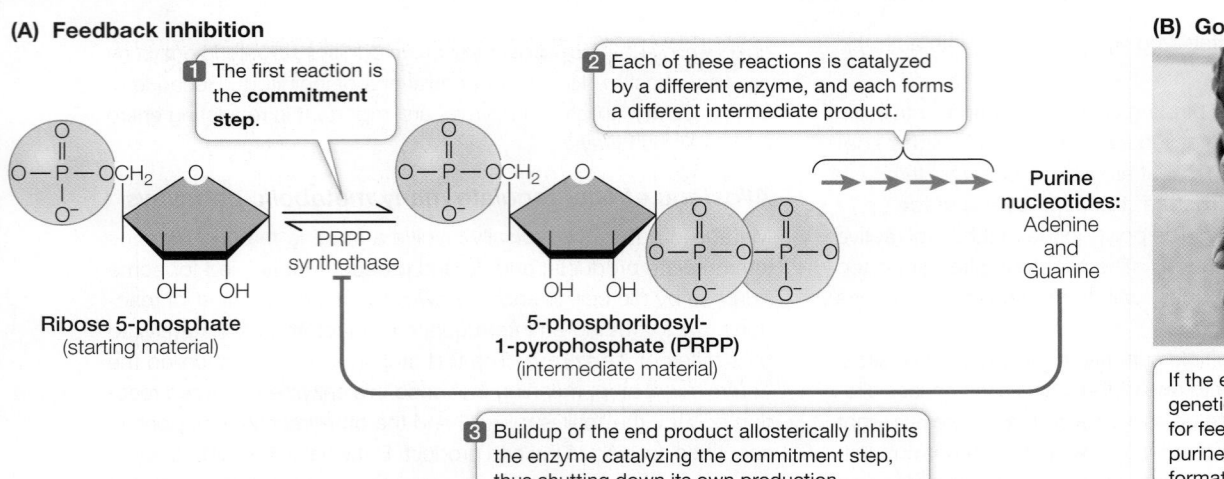

1 The first reaction is the **commitment step**.

2 Each of these reactions is catalyzed by a different enzyme, and each forms a different intermediate product.

Purine nucleotides: Adenine and Guanine

Ribose 5-phosphate (starting material)

PRPP synthethase

5-phosphoribosyl-1-pyrophosphate (PRPP) (intermediate material)

3 Buildup of the end product allosterically inhibits the enzyme catalyzing the commitment step, thus shutting down its own production.

(B) Gout

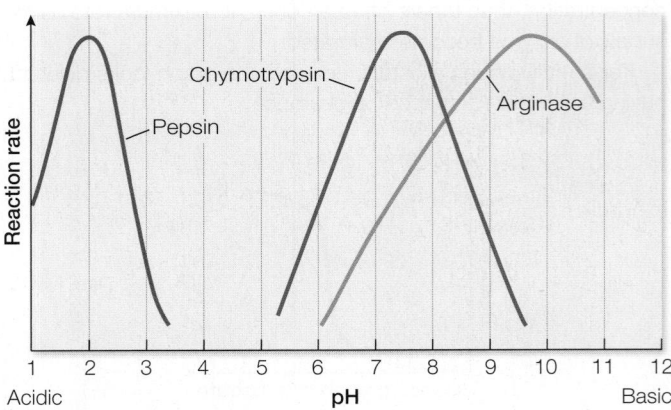

If the enzyme PRPP synthetase is genetically unable to bind purines for feedback inhibition, excess purines accumulate and lead to the formation of crystals in joints (gout).

Figure 8.18 Feedback Inhibition of Metabolic Pathways
(A) The first reaction in a metabolic pathway is referred to as the commitment step. It is often catalyzed by an enzyme that can be allosterically inhibited by the end product of the pathway. The specific pathway shown here is the synthesis of purine nucleotides (ATP, GTP) from ribose 5-phosphate in humans. **(B)** Gout, an accumulation of crystals in joints caused by an excess of purines, can occur if there is a failure of feedback inhibition.

highly sensitive to changes in temperature and pH. In Key Concept 3.2 we described the general effects of these environmental factors on proteins. Here we will examine their effects on enzyme function (which, of course, depends on enzyme structure and chemistry).

pH AFFECTS ENZYME ACTIVITY   The rates of most enzyme-catalyzed reactions depend on the pH of the solution in which they occur. While the water inside cells is generally at a neutral pH of 7, the presence of acids, bases, and buffers can alter this. Each enzyme is most active at a particular pH; its activity decreases as the solution is made more acidic or more basic than the ideal (optimal) pH (**Figure 8.19**). As an example, consider the human digestive system (see Key Concept 50.3). The pH inside the human stomach is highly acidic, around pH 1.5. However, many enzymes that hydrolyze macromolecules in the intestines, such as proteases, have pH optima in the neutral range. So when food enters the small intestine, a buffer (bicarbonate) is secreted into the intestine to raise the pH to 6.5. This allows the hydrolytic enzymes to be active and digest the food.

An important factor in the effect of pH on enzyme function is ionization of the carboxyl, amino, and other groups on either the substrate or the enzyme. In neutral or basic solutions, carboxyl groups (—COOH) release H^+ to become negatively charged carboxylate groups (—COO⁻). However, in neutral or acidic solutions, amino groups (—NH₂) accept H^+ to become positively charged —NH₃⁺ groups (see the discussion of acids and bases in Key Concept 2.4). Thus in a neutral solution, an amino group is electrically attracted to a carboxyl group on another molecule or another part of the same molecule, because both groups are ionized and have opposite charges. If the pH changes, however, the ionization of these groups may change. For example, at a low pH (high H^+ concentration, such as the stomach contents where the enzyme pepsin is active), the excess H^+ may react with —COO⁻ to form —COOH. If this happens, the group is no longer negatively charged and can

no longer interact with positively charged groups in the protein, so the folding of the protein may be altered. If such a change occurs at the active site of an enzyme, the enzyme may no longer be able to bind to its substrate.

TEMPERATURE AFFECTS ENZYME ACTIVITY   In general, warming increases the rate of a chemical reaction because a greater proportion of the reactant molecules have enough kinetic energy to provide the activation energy for the reaction. Enzyme-catalyzed reactions are no different (**Figure 8.20**). However, temperatures that

Figure 8.19 pH Affects Enzyme Activity   An enzyme catalyzes its reaction at a maximum rate. The activity curve for each enzyme peaks at its optimal pH. For example, pepsin is active in the acidic environment of the stomach, whereas chymotrypsin is active in the small intestine.

Q: Inside the cell, the pH of the cytoplasm is typically 7.2, but inside the lysosome the pH is 4.8. A protease, which catalyzes the hydrolysis of proteins, is active within the lysosome but is inactive in the cytoplasm. How can this occur?

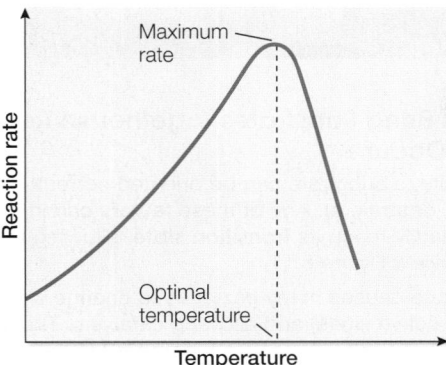

Figure 8.20 Temperature Affects Enzyme Activity   Each enzyme is most active at a particular optimal temperature. At higher temperatures the enzyme becomes denatured and inactive; this explains why the activity curve falls off abruptly at temperatures above the optimum.

are too high inactivate enzymes, because at high temperatures enzyme molecules vibrate and twist so rapidly that some of their noncovalent bonds break. When an enzyme's tertiary structure is changed by heat it loses its function. Some enzymes denature at temperatures only slightly above that of the human body, but a few are stable even at the boiling point (or freezing point) of water. All enzymes, however, have an optimal temperature for activity.

In general, enzymes adapted to warm temperatures do not denature at those temperatures, because their tertiary structures are held together largely by covalent bonds and disulfide bridges, instead of the more heat-sensitive weak chemical interactions. Most enzymes in humans are more stable at high temperatures than are those of the bacteria that infect us, so that a moderate fever tends to denature bacterial enzymes, but not our own.

8.5 recap

The rates of most enzyme-catalyzed reactions are affected by interacting molecules (such as inhibitors and activators) and by environmental factors (such as temperature and pH). Reversible phosphorylation is another important mechanism for regulating enzyme activity.

learning outcomes

You should be able to:

- Explain how metabolic pathways can shift in different directions.
- Distinguish between different types of inhibitors.
- Apply the concept of allosteric regulation to explain experimental results.

1. What is feedback inhibition? How might the reactions shown in Figure 8.18A fit into a systems diagram like the one shown in Figure 8.14?
2. Consider an enzyme that is subject to allosteric regulation. If a competitive inhibitor (not an allosteric inhibitor) is added to a solution containing such an enzyme, the ratio of enzyme molecules in the active form to those in the inactive form increases. Explain this observation.
3. In humans, hydrogen peroxide (H_2O_2) is a dangerous toxin produced as a by-product of several metabolic pathways. The accumulation of H_2O_2 is prevented by its conversion to harmless H_2O, a reaction catalyzed by the appropriately named enzyme catalase. Air pollutants can inhibit this enzyme and leave individuals susceptible to tissue damage by H_2O_2. How would you investigate whether catalase has an allosteric or a nonallosteric mechanism and whether the pollutants are acting as competitive or noncompetitive inhibitors?

 investigatinglife

How do anti-inflammatory drugs work as enzyme inhibitors?

Go into a drugstore or watch ads on TV, and it becomes obvious that there are many drugs that alleviate pain. Like aspirin, many of them target the prostaglandin pathway for inflammation. John Vane's research and later work showed that aspirin irreversibly inhibits the enzyme COX. But it turns out that there are two forms of COX, called COX-1 and COX-2, and there are several kinds of prostaglandins. COX-1 catalyzes the production of protective prostaglandins that are involved with blood clotting (which is why people take aspirin to prevent future heart attacks due to blood clots) and the integrity of the lining of the stomach. COX-2 catalyzes production of prostaglandins involved with inflammation and associated pain. Aspirin inhibits both COX-1 and COX-2. So it is not surprising that people are advised to go easy on aspirin: it may block pain, but its use can lead to stomach upsets and an inability to clot blood after a wound. Knowledge of aspirin's limitations has led to a search for inhibitors of COX-2 only. Most of them are competitive inhibitors. How would this affect how often you need to take the drug? Look up one of these drugs and read the description of how it works.

Future directions

As details at the atomic level of the binding of enzymes to substrates are worked out, biologists use the data to try to predict which substrates might bind to which enzymes. An important consideration in binding is ΔG; the less positive this parameter, the more likely binding will occur. Factors such as ionic attractions and van der Waals forces contribute to the ΔG of binding. These factors are also important in the binding of proteins to nonsubstrate molecules, including other proteins and RNA. Indeed, it is becoming apparent that in general a protein in the cell does not exist in isolation, but is bound to something(s) else. If the protein is an enzyme, its substrate binding must be its most preferred partner. Understanding all the possible molecular interactions will result in a deeper understanding of what really goes on at the chemical level inside the cell.

Chapter Summary 8

8.1 Physical Principles Underlie Biological Energy Transformations

- **Energy** is the capacity to do work. In a biological system, the usable energy is called **free energy (G)**. The unusable energy is **entropy (S)**, a measure of the disorder in the system.

- **Potential energy** is the energy of state or position; it includes the energy stored in chemical bonds. **Kinetic energy** is the energy of motion; it is the type of energy that can do work.

- The **laws of thermodynamics** apply to living organisms. The first law states that energy cannot be created or destroyed. The second law states that energy transformations decrease the amount of energy available to do work (free energy) and increase disorder. **Review Figure 8.2**

- The change in free energy (ΔG) of a reaction determines its **chemical equilibrium**, the point at which the forward and reverse reactions proceed at the same rate.

- An **exergonic** reaction releases free energy and has a negative ΔG. An **endergonic** reaction consumes or requires free energy and has a positive ΔG. Endergonic reactions proceed only if free energy is provided. **Review Figure 8.3**

- **Metabolism** is the sum of all the biochemical (metabolic) reactions occurring in an organism at a given time. **Catabolic reactions** are associated with the breakdown of complex molecules and release energy (are exergonic). **Anabolic reactions** build complexity in the cell and are endergonic.

8.2 ATP Plays a Key Role in Biochemical Energetics

- Adenosine triphosphate (ATP) serves as an energy currency in cells. Hydrolysis of ATP releases a relatively large amount of free energy.

- The ATP cycle couples exergonic and endergonic reactions, harvesting free energy from exergonic reactions, and providing free energy for endergonic reactions. **Review Figure 8.6, Activity 8.1**

8.3 Enzymes Speed Up Biochemical Transformations

- The rate of a chemical reaction is independent of ΔG but is determined by the energy barrier. **Review Focus: Key Figure 8.8**

- Enzymes are protein catalysts that affect the rates of biological reactions by lowering the energy barrier, supplying the **activation energy (E_a)** needed to initiate reactions. **Review Figure 8.10, Activity 8.2**

- A **substrate** binds to the enzyme's active site—the site of catalysis—forming an **enzyme–substrate complex (ES)**. Enzymes are highly specific for their substrates. **Review Figure 8.9**

- Enzymes can be classified as to the type of chemical reaction catalyzed.

8.4 Enzymes Bring Substrates Together so Reactions Readily Occur

- At the active site, a substrate can be oriented correctly, chemically modified, or strained. Any of these factors can induce the substrate to readily reach its **transition state**, allowing the reaction to proceed. **Review Figure 8.11**

- Binding substrate causes many enzymes to change shape, exposing their active site(s) and allowing catalysis. The change in enzyme shape caused by substrate binding is known as **induced fit**. **Review Figure 8.12**

- Some enzymes require other substances, known as cofactors, to carry out catalysis. Prosthetic groups are permanently bound to enzymes; coenzymes are not. A coenzyme can be considered a substrate, as it is changed by the reaction and then released from the enzyme.

- Substrate concentration affects the rate of an enzyme-catalyzed reaction. **Review Figure 8.13**

8.5 Enzyme Activities Can Be Regulated

- Metabolism is organized into pathways in which the product of one reaction is a reactant for the next reaction. Each reaction in the pathway is catalyzed by a different enzyme. **Review Activity 8.3**

- Enzyme activity is subject to regulation. Some inhibitors bind irreversibly to enzymes. Others bind reversibly. **Review Figures 8.15, 8.16, Animation 8.1**

- An allosteric effector binds to a site other than the active site and stabilizes the active or inactive form of an enzyme. **Review Figure 8.17, Animation 8.2**

- The end product of a metabolic pathway may inhibit an enzyme that catalyzes the commitment step of that pathway. **Review Figure 8.18**

- Reversible phosphorylation is an important mechanism for regulating enzyme activity.

- Enzymes are sensitive to their environments. Both pH and temperature affect enzyme activity. **Review Figures 8.19, 8.20**

> Go to **LearningCurve** (in **LaunchPad**) for dynamic quizzing that helps you solidify your understanding of this chapter. **LearningCurve** adapts to your responses, giving you the practice you need to master each key concept.

Apply What You've Learned

Review

8.2 ATP releases usable energy for a cell when hydrolyzed to ADP and P_i.

8.2 Through coupling reactions, ATP drives endergonic reactions using energy derived from exergonic reactions.

8.3 Catalysts speed up the rate of a reaction but do not allow the occurrence of a reaction that would not otherwise take place.

8.4 An enzyme may undergo a change in shape called induced fit as the result of binding its substrate.

8.5 Environmental factors such as pH and temperature affect enzyme activity.

Original Papers: McElroy, W. D. 1947. The energy source for bioluminescence in an isolated system. *Proceedings of the National Academy of Sciences USA* 33: 342–345.

McElroy, W. D. and B. L. Strehler. 1954. Bioluminescence. *Bacteriological Reviews* 18: 177–194.

Thompson, J. F. et al. 1997. Mutation of a protease-sensitive region in firefly luciferase alters light emission properties. *Journal of Biological Chemistry* 272: 18766–18771.

Many people enjoy watching and catching fireflies on warm summer nights, but the attraction isn't mutual; a firefly's primary goal is to find a mate. Male and female fireflies use specific patterns of light flashes to attract one another. The flashes are emitted from the firefly lantern, an organ located in the firefly abdomen.

Light is a form of energy, which means that fireflies must divert some of their energy to produce light flashes. Researchers used a biochemical approach to learn how this is done. When the lanterns of fireflies are ground up and extracted with water, the extract briefly produces an intense light, which then fades away. If ATP is added to the extract, light is again emitted, and again, it fades. The graph below shows the duration of light emission as a function of the amount of ATP added. When anaerobic conditions were used, no light was observed.

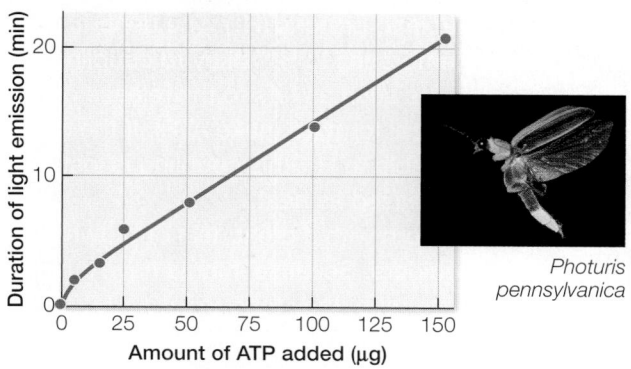

Photuris pennsylvanica

Through additional investigations, the researchers discovered that a compound called luciferin was responsible for the light emission as luciferin underwent a chemical reaction. The researchers purified an enzyme from extracts of firefly lanterns. The enzyme catalyzes the reaction in which luciferin emits light. In further studies, amounts of luciferin, ATP, and magnesium ion, Mg^{2+}, were found to give the maximum light intensity (100%) when mixed with a fixed amount of enzyme. The table below shows how the relative light intensity was affected when variables were changed.

Variable	Light intensity (%)
No enzyme	0
Enzyme heated before being added	0
No magnesium ion	4
1 mM magnesium ion	70
10 mM magnesium ion	100
pH 6.5	30
pH 7.6	100
pH 9.0	64

Questions

1. Based on the information given, what set of molecules is necessary for the production of light by firefly lanterns, and what role does each molecule serve?

2. Explain how energy taken in by the firefly from its environment is made available for the production of light in a firefly lantern.

3. Additional research has shown that an activated intermediate forms at the active site of the enzyme as luciferin reacts. This intermediate consists of luciferin covalently bonded to adenosine monophosphate (AMP). Still at the active site, this intermediate then reacts with oxygen to make an oxidized form of luciferin with the emission of light. The diagram below shows these steps. Redraw the diagram, adding steps to illustrate how coupling reactions could be involved to provide the energy needed for light emission.

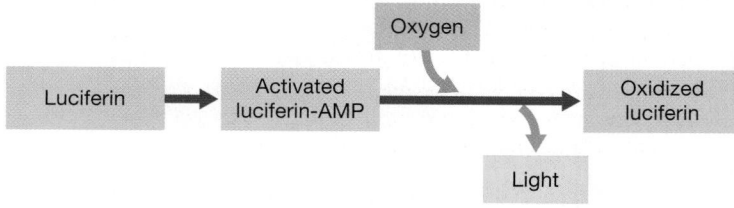

4. What evidence is there that the enzyme is sensitive to environmental factors? Suggest possible reasons why this enzyme responds as it does to these factors.

5. Some studies have shown that the enzyme undergoes a shape change upon binding ATP and luciferin. This change makes it impossible for a molecule of water to fit into the active site along with the substrates. What property of an enzyme does this represent, and how does this property help this particular enzyme carry out its role in catalysis?

Go to **LaunchPad** for the eBook, LearningCurve, animations, activities, flashcards, and additional resources and assignments.

9

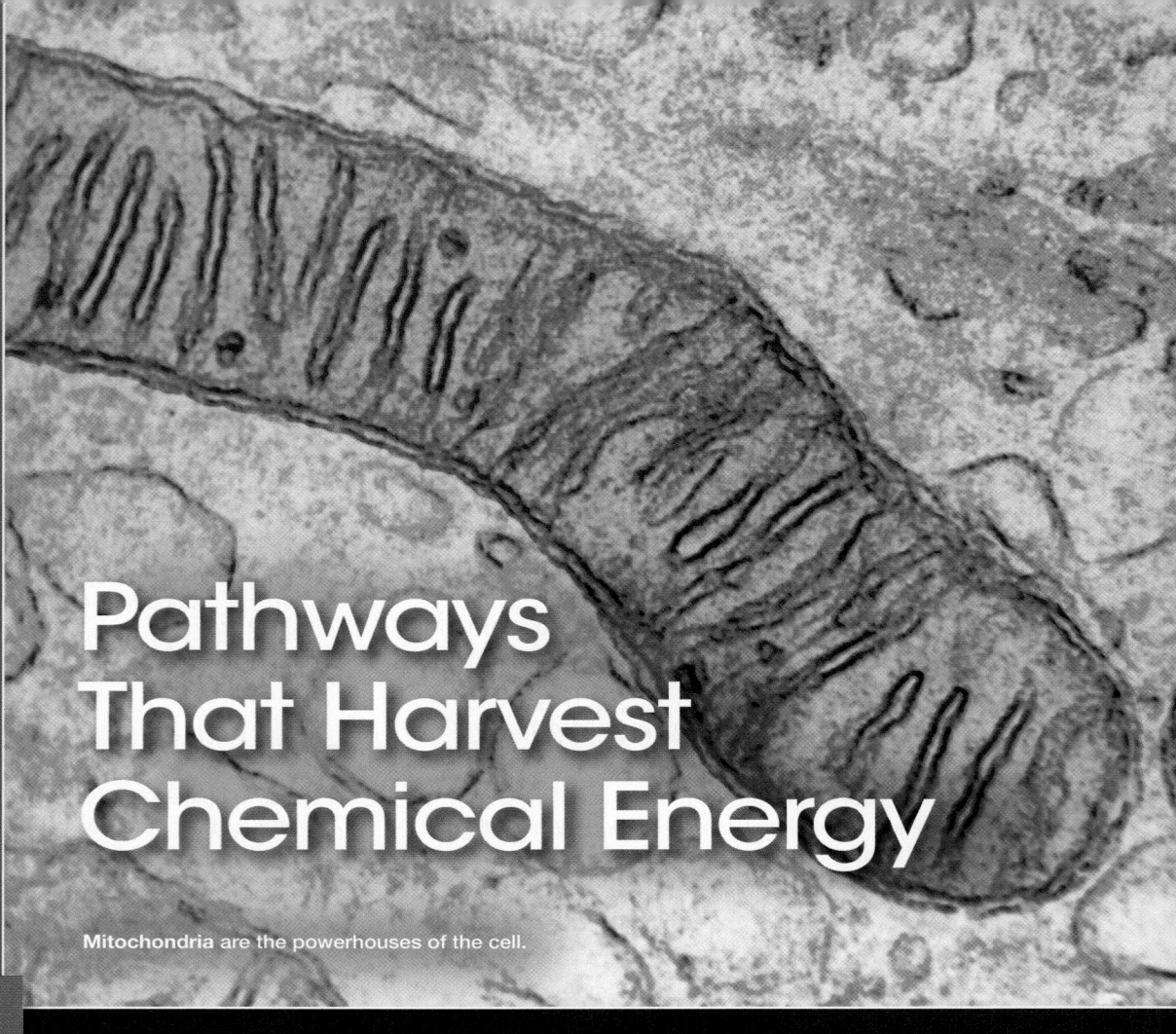

Pathways That Harvest Chemical Energy

Mitochondria are the powerhouses of the cell.

▶ investigatinglife

A Weighty Matter

With 17 percent of children and 35 percent of adults in the United States now described as obese, physicians have declared an "obesity epidemic." Dramatic increases in diseases associated with obesity—among them diabetes, heart disease, and cancer—have understandably alarmed the medical establishment.

While nutritional science has shown that not all foods affect all people in the same ways, in most cases the significant excess body weight of obese people can be prevented or reduced if they eat less and exercise more. It is a matter of energy: if we eat more energy-yielding molecules than we need to build our bodies and to fuel activities such as brain functions and physical activity, we will store the unneeded energy as fat. There is an evolutionary advantage in storing fat: the energy in the C—C and C—H bonds in fat can be used later when food is scarce. But excess fat has adverse consequences.

Not all fat (adipose) tissues are the same. White adipose tissue (sometimes referred to simply as "white fat") is used primarily to store energy.

Brown adipose tissue has high concentrations of mitochondria, which have iron-containing pigments. When energy-rich molecules in brown fat are catabolized, the stored energy is released not as chemical energy but as heat. The cells in brown fat make a protein called UCP1 (uncoupling protein 1) that inserts into the inner membranes of mitochondria, making them permeable to protons (H^+). The general impermeability of these membranes to H^+ is key to coupling the catabolism of molecules such as fats to the release of their stored energy in chemical form (to make ATP). If the membranes become permeable to H^+, this coupling is lost and the stored energy is released as heat.

Human infants are born with a great deal of brown fat in their back and shoulders. Because infants have a high surface area-to-volume ratio, they tend to lose a lot of heat. One way they keep warm is to produce heat in their brown fat tissues. As a child grows up, the brown fat content of the body is reduced. Adults have mostly white fat, which has less UCP1 and generates less heat when the fat is catabolized. But biologists recently found brown fat in adults, and its significance is gaining increasing attention.

Q/A How are mitochondria related to obesity?

9.1 Cells Harvest Chemical Energy from Glucose Oxidation

Energy is stored in the covalent bonds of fuels, and it can be released and transformed. You will recall from Chapter 8 that energy transformations can involve light, heat, chemical, mechanical, and electrical energy. For example, when you burn wood, it releases a great deal of its energy as heat and light. In cells, molecules used as fuels release chemical energy that is used to make ATP, which in turn drives endergonic reactions. Photosynthetic cells and organisms use energy from sunlight to synthesize their own fuels, as we will describe in Chapter 10. In nonphotosynthetic cells, the most common chemical fuel is the monosaccharide glucose. Other molecules, including other carbohydrates, fats, and proteins, can supply energy to the whole organism. However, to release their energy they must be converted into glucose or intermediate compounds that can enter into the various pathways of glucose metabolism.

focus your learning

- Basic principles of chemistry govern metabolic pathways in cells.
- Energy is transferred as electrons move from one molecule to another via an oxidation–reduction reaction.
- Electron carrier molecules function as coenzymes to pick up or donate electrons during biological redox reactions.

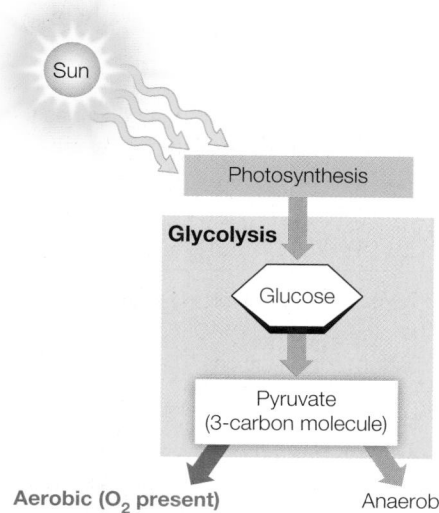

How do cells obtain energy from glucose?

Cells obtain *energy from glucose by the chemical process of oxidation, which is carried out through a series of metabolic pathways. As you read this chapter, keep in mind five principles that govern metabolic pathways:

1. A complex chemical transformation occurs in a series of separate reactions that form a metabolic pathway.
2. Each reaction is catalyzed by a specific enzyme.
3. Many metabolic pathways are similar in all organisms, from bacteria to humans.
4. In eukaryotes, many metabolic pathways are compartmentalized, with certain reactions occurring inside specific organelles, or even specific regions of an organelle.
5. Some key enzymes in each metabolic pathway can be inhibited or activated to alter the rate of the pathway.

*connect the concepts   The principles of energy transformations in living and nonliving systems are discussed in Key Concept 8.1.

As you saw in Key Concept 2.3, the familiar process of combustion (burning) is similar to the chemical processes that release energy in cells. If glucose is burned in a flame or is in a typical cell, it reacts with oxygen gas (O_2), forming carbon dioxide and water and releasing energy in the form of heat. The balanced equation for the complete reaction is

$$C_6H_{12}O_6 + 6\ O_2 \rightarrow 6\ CO_2 + 6\ H_2O + \text{free energy}$$
$$(\Delta G = -686\ \text{kcal/mol})$$

This is an oxidation–reduction reaction (more on this shortly), in which glucose loses electrons (becomes oxidized) and oxygen gains them (becomes reduced). The large standard free energy change (ΔG) is negative, indicating that the overall conversion is highly exergonic and can drive the endergonic formation of a lot of ATP from ADP and phosphate:

$$ADP + P_i + \text{free energy} \rightarrow ATP$$

The oxidation of glucose when you heat it in the lab happens all at once. But in cells, glucose catabolism happens in many steps in a pathway. Each step is catalyzed by an enzyme, and the process is compartmentalized. Unlike combustion, glucose catabolism is tightly regulated and occurs at temperatures compatible with life.

Three catabolic processes harvest the energy in the chemical bonds of glucose: glycolysis, cellular respiration, and fermentation (**Figure 9.1**). All three processes involve pathways made up of many distinct chemical reactions.

1. **Glycolysis** begins glucose catabolism. Through a series of chemical rearrangements, glucose is converted to two molecules of the three-carbon product **pyruvate**, and a small amount of energy is captured in usable forms. Glycolysis is an **anaerobic** process because it does not require O_2.

Figure 9.1 Energy for Life   Many prokaryotes and all eukaryotes obtain their energy from the food compounds produced by photosynthesis. They often convert these compounds into glucose, which they metabolize to trap energy in ATP.

2. **Cellular respiration** uses O_2 from the environment and thus is **aerobic**. Each pyruvate molecule is completely converted into three molecules of CO_2 through a set of catabolic pathways including pyruvate oxidation, the citric acid cycle, and an electron transport system (the respiratory chain). In the process, a great deal of the energy stored in the covalent bonds of pyruvate is captured to form ATP.

3. **Fermentation** does not involve O_2 (it is anaerobic). With the exception of many microorganisms, fermentation converts pyruvate into lactic acid or ethyl alcohol (ethanol), both of which are still relatively energy-rich molecules. Because the breakdown of glucose is incomplete, much less energy is released when glycolysis is coupled to fermentation than when it is coupled to cellular respiration.

Redox reactions transfer electrons and energy

As you saw in Key Concept 8.2, the addition of a phosphate group to ADP to make ATP is an endergonic reaction (see Figure 8.6). It is achieved by coupling an exergonic reaction to ATP production: the energy released in the exergonic reaction is used to drive ATP synthesis. Electrons are transferred in the exergonic reaction. A reaction in which one substance transfers one or more electrons to another substance is called an **oxidation–reduction**, or **redox**, **reaction**.

- **Reduction** is the gain of one or more electrons by an atom, ion, or molecule.
- **Oxidation** is the loss of one or more electrons.

Oxidation and reduction *always occur together*: as one chemical is oxidized, the electrons it loses are transferred to another chemical, reducing it. In a redox reaction, we call the reactant that becomes reduced an oxidizing agent and the one that becomes oxidized a reducing agent:

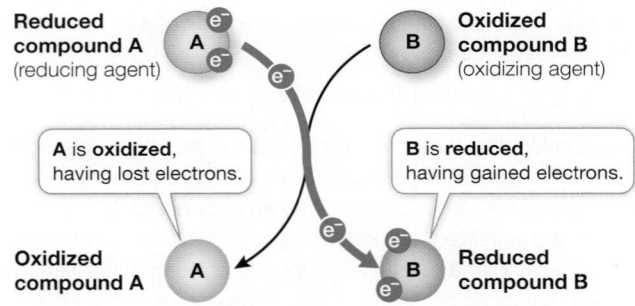

Reduced compound A (reducing agent)

Oxidized compound B (oxidizing agent)

A is **oxidized**, having lost electrons.

B is **reduced**, having gained electrons.

Oxidized compound A

Reduced compound B

In the metabolism of glucose:

- Glucose is the reducing agent (electron donor).
- O_2 is the oxidizing agent (electron acceptor).

Although oxidation and reduction are always defined in terms of electrons, it is often simpler to think in terms of the gain or loss of hydrogen atoms. The transfer of electrons is often associated with the transfer of hydrogen ions (a H atom contains H^+ + e^-). So when a molecule loses hydrogen atoms, it becomes oxidized.

The more reduced a molecule is, the more energy is stored in its covalent bonds (**Figure 9.2**). In a redox reaction, some energy is transferred from the reducing agent to the reduced product. The

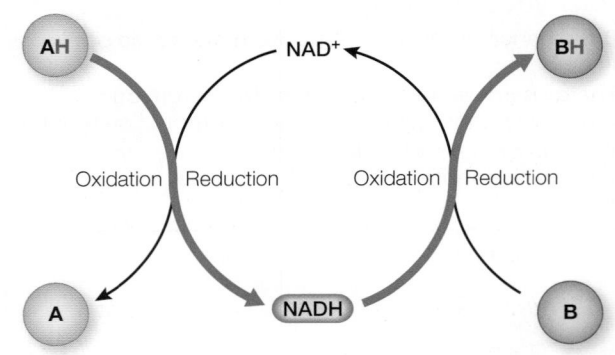

Methane (CH_4) Methanol (CH_3OH) Formaldehyde (CH_2O) Formic acid (HCOOH) Carbon dioxide (CO_2)

Most reduced state
Highest free energy

Most oxidized state
Lowest free energy

Figure 9.2 Oxidation, Reduction, and Energy The more oxidized a carbon atom in a molecule is, the less its stored free energy.

rest remains in the reducing agent or is lost to entropy. As you will see, some of the key reactions of glycolysis and cellular respiration are highly exergonic redox reactions.

The coenzyme NAD^+ is a key electron carrier in redox reactions

The *coenzyme nicotinamide adenine dinucleotide (NAD^+) acts as an electron carrier in redox reactions. Note the flow of electrons along the blue lines below:

AH NAD^+ BH

Oxidation Reduction Oxidation Reduction

A NADH B

---

*connect the concepts Key Concept 8.4 describes the role of coenzymes, small molecules that assist in enzyme-catalyzed reactions. ADP acts as a coenzyme when it picks up energy released in an exergonic reaction and packages it to form ATP.

As you can see, NAD^+ exists in two chemically distinct forms, one oxidized (NAD^+) and the other reduced (NADH) (**Focus: Key Figure 9.3**). Both forms participate in redox reactions. The reduction reaction

$$NAD+ + H+ + 2\ e^- \rightarrow NADH$$

is actually the transfer of a proton (the hydrogen ion, H^+) and two electrons, which are released by the accompanying oxidization reaction.

The electrons do not remain with the coenzyme. Oxygen is highly electronegative and readily accepts electrons from NADH. The oxidation of NADH by O_2 (which occurs in several steps)

$$NADH + H^+ + \tfrac{1}{2}\ O_2 \rightarrow NAD^+ + H_2O$$

is exergonic, with a standard free energy change at pH 7 ($\Delta G°$) of −52.4 kcal/mol (−219 kJ/mol). Note that the oxidizing agent appears here as "½ O_2" instead of "O." This notation emphasizes that it is molecular oxygen, O_2, that acts as the oxidizing agent.

focus: key figure

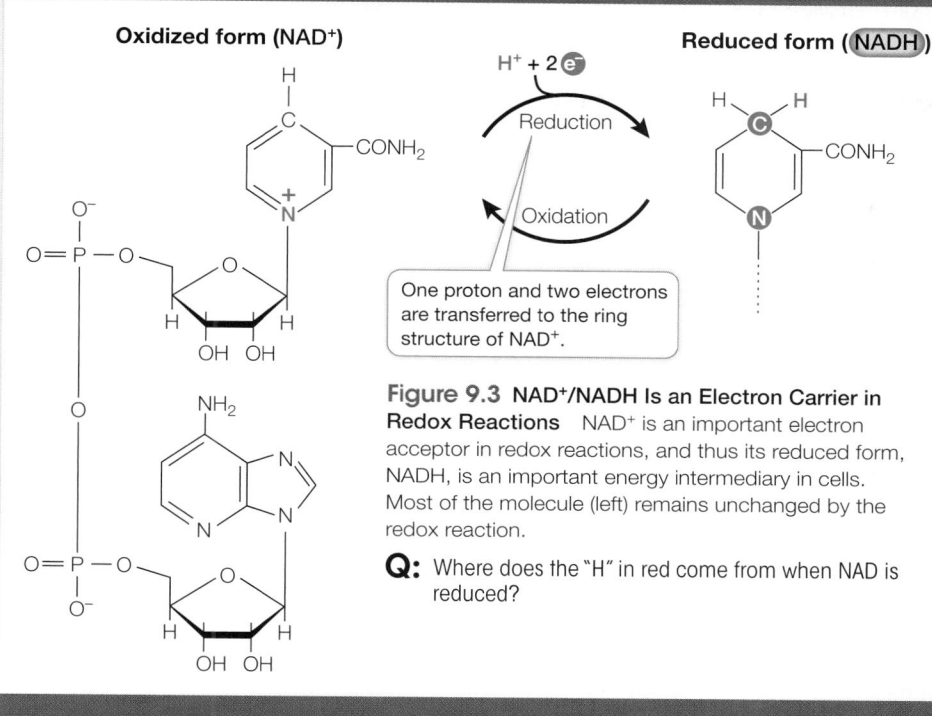

Oxidized form (NAD⁺)

$H^+ + 2\,e^-$

Reduction

Oxidation

Reduced form (NADH)

One proton and two electrons are transferred to the ring structure of NAD⁺.

Figure 9.3 NAD⁺/NADH Is an Electron Carrier in Redox Reactions NAD⁺ is an important electron acceptor in redox reactions, and thus its reduced form, NADH, is an important energy intermediary in cells. Most of the molecule (left) remains unchanged by the redox reaction.

Q: Where does the "H" in red come from when NAD is reduced?

Just as ATP can be thought of as a package of 7.3 kcal/mol (30.5 kJ/mol) of free energy, NADH can be thought of as a larger package of free energy (52.4 kcal/mol; see above). NAD⁺ is a common electron carrier in cells, but not the only one. Another carrier, flavin adenine dinucleotide (FAD), also transfers electrons during glucose metabolism.

An overview: Harvesting energy from glucose

Both eukaryotic and prokaryotic cells can harvest energy from glucose using different combinations of the following metabolic pathways:

- Under aerobic conditions, when O_2 is available as the final electron acceptor, four pathways operate (**Figure 9.4A**).

table **9.1** Cellular Locations for Major Energy Pathways in Eukaryotes and Prokaryotes	
Eukaryotes	**Prokaryotes**
In cytoplasm	In cytoplasm
Glycolysis	Glycolysis
Fermentation	Fermentation
	Citric acid cycle
Inside mitochondrion	On cell membrane
Matrix	Pyruvate oxidation
Citric acid cycle	Respiratory chain
Pyruvate oxidation	
Inner membrane	
Respiratory chain	

Glycolysis is followed by the three pathways of cellular respiration: pyruvate oxidation, the citric acid cycle (also called the Krebs cycle or the tricarboxylic acid cycle), and electron transport/ATP synthesis (also called oxidative phosphorylation).

- In eukaryotes and many prokaryotes, pyruvate oxidation, the citric acid cycle, and oxidative phosphorylation do not function under anaerobic conditions. The pyruvate produced by glycolysis is further metabolized by fermentation (**Figure 9.4B**). Some prokaryotes, however, are able to harvest energy in pathways involving oxidative phosphorylation even in the absence of oxygen (anaerobic respiration; see Key Concept 9.3).

The five pathways shown in Figure 9.4 occur in different locations in the cell (**Table 9.1**).

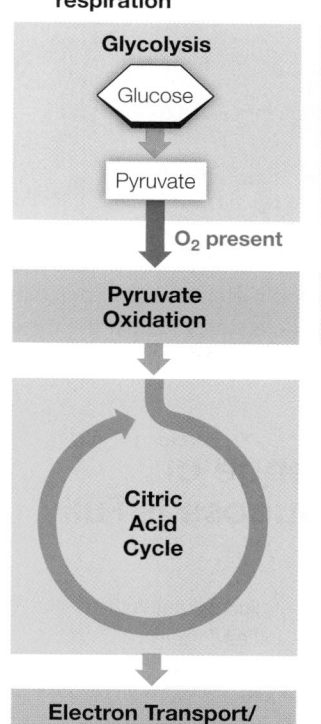

(A) Glycolysis and cellular respiration

Glycolysis

Glucose

Pyruvate

O_2 present

Pyruvate Oxidation

Citric Acid Cycle

Electron Transport/ATP Synthesis

CO_2 and H_2O

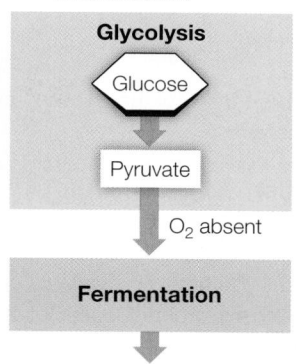

(B) Glycolysis and fermentation

Glycolysis

Glucose

Pyruvate

O_2 absent

Fermentation

Lactate or alcohol

Figure 9.4 Energy-Yielding Metabolic Pathways Energy-yielding reactions can be grouped into five metabolic pathways: glycolysis, pyruvate oxidation, the citric acid cycle, the respiratory chain/ATP synthesis, and fermentation. **(A)** The three lower pathways occur only in the presence of O_2 and are collectively referred to as cellular respiration. **(B)** When O_2 is unavailable, glycolysis is followed by fermentation.

▶ Activity 9.2 **Energy Pathways in Cells**
www.Life11e.com/ac9.2

▶ Activity 9.1 **Glycolysis and Fermentation**
www.Life11e.com/ac9.1

9.1 recap

The free energy released from the oxidation of glucose is trapped in the form of ATP. In many prokaryotes and all eukaryotes, five major pathways combine in different ways to produce ATP, which supplies the energy for myriad other reactions in living cells.

learning outcomes

You should be able to:

- Describe the principles of chemistry that govern metabolic pathways in cells.
- Identify the change in energy that occurs during an oxidation–reduction reaction.
- Describe the roles of electron acceptors and donors in biological redox reactions.
- Analyze biological redox reactions to identify the oxidizing and reducing agents involved.

1. What principles govern metabolic pathways in cells?
2. What are the roles of NAD^+ and O_2 with respect to electrons in a redox reaction?
3. The following reaction occurs in the citric acid cycle:

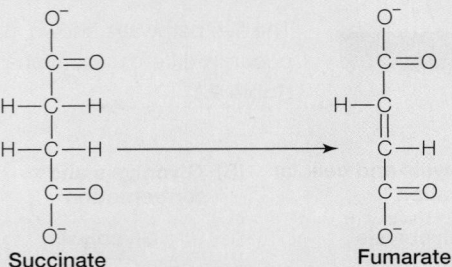

Succinate   Fumarate

Answer each of the following questions, and explain your answers:

a. Is this reaction an oxidation or a reduction?
b. Is the reaction exergonic or endergonic?
c. What kind of coenzyme does this reaction require?

Now that you have an overview of the metabolic pathways that harvest energy from glucose, let's take a closer look at the three pathways involved in aerobic glucose catabolism: glycolysis, pyruvate oxidation, and the citric acid cycle.

key concept

9.2 In the Presence of Oxygen, Glucose Is Fully Oxidized

The aerobic pathways of glucose catabolism oxidize glucose completely to CO_2 and H_2O. Initially, the glycolysis reactions convert the six-carbon glucose molecule to two three-carbon pyruvate molecules (**Figure 9.5**). Pyruvate is then converted to CO_2 in a second series of reactions beginning with pyruvate oxidation

Figure 9.5 Glycolysis Converts Glucose into Pyruvate   Glucose is converted to pyruvate in ten enzyme-catalyzed steps. Along the way, two ATP are used (Steps 1 and 3), two NAD^+ are reduced to two NADH (Step 6), and four ATP are produced (Steps 7 and 10).

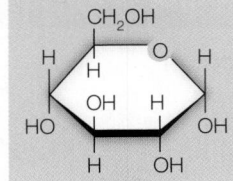

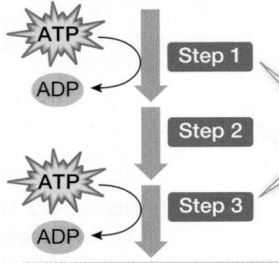

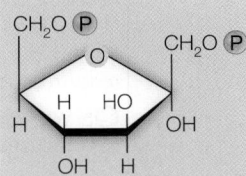

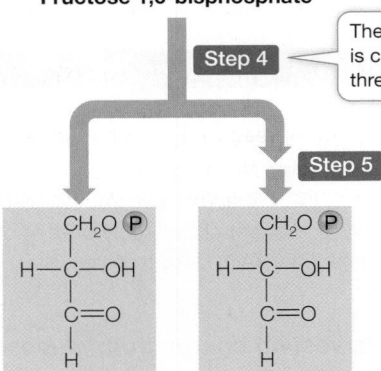

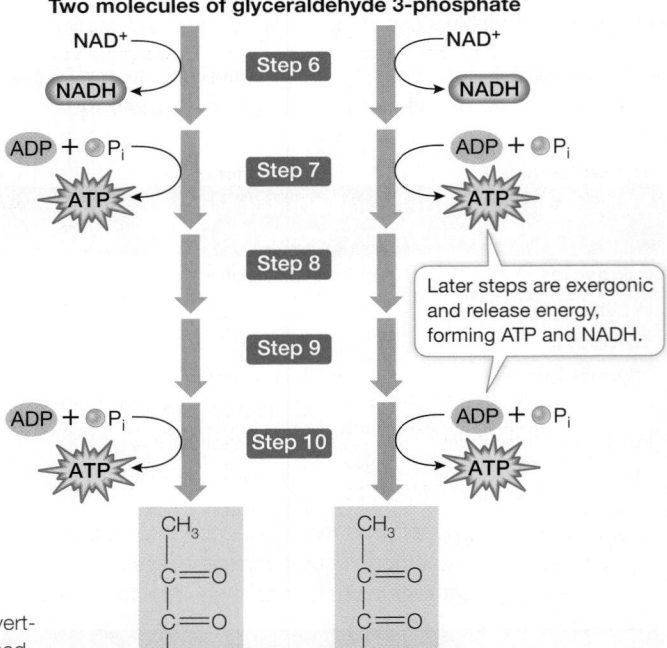

One molecule of glucose

Step 1

Two of the first three steps are endergonic and require energy from ATP hydrolysis.

Step 2

Step 3

Fructose 1,6-bisphosphate

The six-carbon sugar is cleaved into two three-carbon sugars.

Step 4

Step 5

Two molecules of glyceraldehyde 3-phosphate

Step 6

Step 7

Step 8

Later steps are exergonic and release energy, forming ATP and NADH.

Step 9

Step 10

Two molecules of pyruvate

and followed by the citric acid cycle. In addition to generating CO_2, the oxidation events are coupled with the reduction of electron carriers, mostly NAD^+.

focus your learning

- Glucose is partially oxidized to pyruvate during glycolysis.
- The generation of acetyl coenzyme A by pyruvate oxidation enables the entry of carbon from glucose into the citric acid cycle for further oxidation.
- A large amount of energy is captured in electron carriers and in GTP as acetyl CoA is fully oxidized during the citric acid cycle.

We will begin our consideration of the catabolism of glucose with a closer look at glycolysis.

In the glycolysis pathway, glucose is partially oxidized

Glycolysis takes place in the cytoplasm and involves ten enzyme-catalyzed reactions. During glycolysis, some of the covalent bonds between carbon and hydrogen atoms in the glucose molecule are oxidized, releasing some of the stored energy. The products are two molecules of pyruvate (pyruvic acid), two molecules of ATP, and two molecules of NADH. Glycolysis can be divided into two stages: the initial energy-investing reactions that consume ATP, and the energy-harvesting reactions that produce ATP and NADH (see Figure 9.5).

To help you understand the process without getting into extensive detail, we will focus on two consecutive reactions in this pathway (Steps 6 and 7 in Figure 9.5).

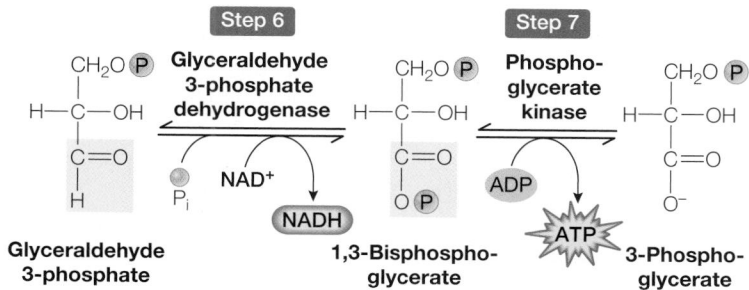

Steps 6 and 7 are examples of two types of reactions that occur repeatedly in glycolysis and in many other metabolic pathways:

1. *Oxidation–reduction*: The first reaction is exergonic—more than 50 kcal/mol of energy are released in the oxidation of glyceraldehyde 3-phosphate. (Look at the bottom carbon atom, where an H is replaced by an O.) The energy is trapped via the reduction of NAD^+ to NADH.

2. *Substrate-level phosphorylation*: The second reaction in this series is also exergonic, but in this case less energy is released, sufficient to transfer a phosphate directly from the substrate to ADP, forming ATP.

The end product of glycolysis, pyruvate, is somewhat more oxidized than glucose. In the presence of O_2, further oxidation can occur. In

prokaryotes these subsequent reactions take place in the cytoplasm, but in eukaryotes they take place in the mitochondrial matrix.

To summarize:

- The initial steps of glycolysis use the energy of hydrolysis of two ATP molecules per glucose molecule.
- The remaining steps produce four ATP molecules per glucose molecule, so the net production of ATP is two molecules.
- Glycolysis produces two molecules of NADH.

If O_2 is present, glycolysis is followed by the three stages of cellular respiration: pyruvate oxidation, the citric acid cycle, and the respiratory chain/ATP synthesis.

Pyruvate oxidation links glycolysis and the citric acid cycle

In eukaryotes, pyruvate is transported into the mitochondrial matrix (see Figure 5.11), where the next step in the aerobic catabolism of glucose occurs. This step involves the oxidation of pyruvate to a two-carbon acetate molecule and CO_2. The acetate is then bound to coenzyme A to form **acetyl coenzyme A** (**acetyl CoA**); CoA is used in various biochemical reactions as a carrier of the acetyl group ($H_3C-C=O$).

Pyruvate is the link between glycolysis and further oxidative reactions (see Figure 9.4).

The formation of acetyl CoA is a multistep reaction catalyzed by the pyruvate dehydrogenase complex, which contains 60 individual proteins and 5 different coenzymes. The overall reaction is exergonic, in which one molecule of NAD^+ is reduced to NADH. However, the main role of acetyl CoA is to donate its acetyl group to the four-carbon compound oxaloacetate, forming the six-carbon molecule citrate. This initiates the citric acid cycle, one of life's most important energy-harvesting pathways.

The citric acid cycle completes the oxidation of glucose to CO_2

Acetyl CoA is the starting point for the citric acid cycle. This pathway of eight reactions completely oxidizes the two-carbon acetyl group to two molecules of CO_2. The free energy released from these reactions is captured by GDP (guanosine diphosphate) and the electron carriers NAD^+ and FAD (**Figure 9.6**). (Remember from Key Concept 7.2 that GDP is a nucleoside diphosphate like ADP.) This is a cycle because the starting material, oxaloacetate, is regenerated in the last step and is ready to accept another acetate group from acetyl CoA. The citric acid cycle operates twice for each glucose molecule that enters glycolysis (once for each pyruvate that enters the mitochondrion).

Let's focus on the final reaction of the cycle (Step 8 in Figure 9.6), as an example of the kind of reaction that occurs:

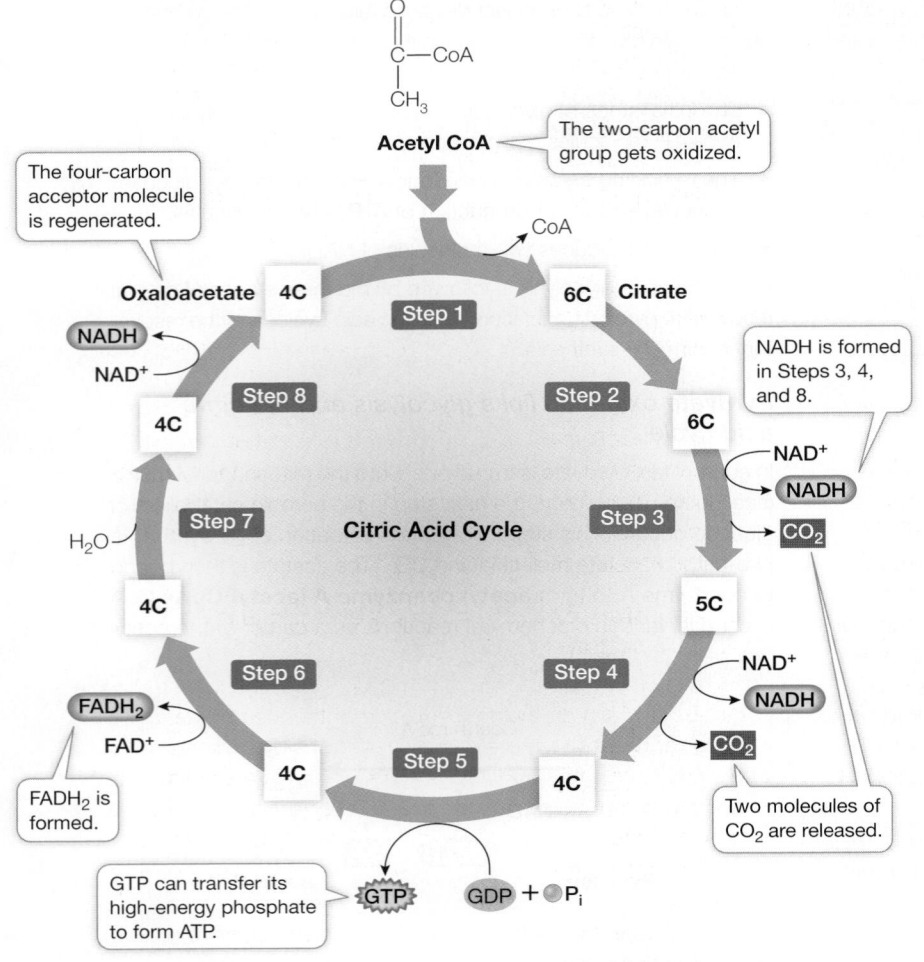

The four-carbon acceptor molecule is regenerated.

The two-carbon acetyl group gets oxidized.

NADH is formed in Steps 3, 4, and 8.

Two molecules of CO_2 are released.

$FADH_2$ is formed.

GTP can transfer its high-energy phosphate to form ATP.

Figure 9.6 The Citric Acid Cycle The citric acid cycle has eight steps; in the last step, the starting material acceptor, oxaloacetate, is regenerated. Energy is released and captured by reducing NAD⁺ or FAD, or by producing GTP. "6C," "5C," and so on indicate the number of carbon atoms in each intermediate in the cycle.

 Activity 9.3 The Citric Acid Cycle
www.Life11e.com/ac9.3

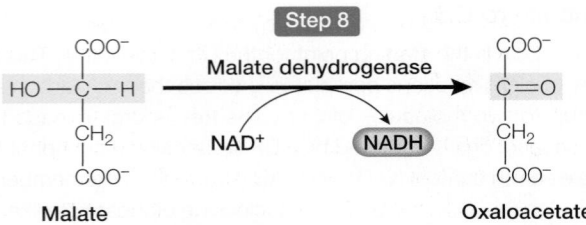

This oxidation reaction (see the carbon atom highlighted in blue) is exergonic, and the released energy is trapped by NAD⁺, forming NADH. With four such reactions (the $FADH_2$ produced in Step 6 is a reduced coenzyme similar to NADH), the citric acid cycle harvests a great deal of chemical energy from the oxidation of acetyl CoA.

To summarize:

- The inputs to the citric acid cycle are acetate (in the form of acetyl CoA), water, GDP, and the oxidized electron carriers NAD⁺ and FAD.

- The outputs are carbon dioxide, reduced electron carriers (NADH and $FADH_2$), and a small amount of GTP. The energy in the terminal phosphate of GTP is transferred to ATP:

$$GTP + ADP \rightarrow ATP + GDP$$

Thus the citric acid cycle releases two carbons as CO_2 and produces four reduced electron carrier molecules.

Overall, for each molecule of glucose that is oxidized, two molecules of pyruvate are produced during glycolysis, and after oxidation these feed two turns of the citric acid cycle. So the oxidation of one glucose molecule yields:

- Six CO_2

- Ten NADH (two in glycolysis, two in pyruvate oxidation, and six in the citric acid cycle)

- Two $FADH_2$

- Four ATP

Pyruvate oxidation and the citric acid cycle are regulated by the concentrations of starting materials

You have now seen that pyruvate, a three-carbon molecule, is completely oxidized to CO_2 by pyruvate dehydrogenase and the citric acid cycle. For the cycle to continue, the starting molecules—acetyl CoA and oxidized electron carriers—must all be replenished. The electron carriers are reduced during the citric acid cycle and in Step 6 of glycolysis (see Figure 9.5), and they must be reoxidized:

$$NADH \rightarrow NAD^+ + H^+ + 2\ e^-$$
$$FADH_2 \rightarrow FAD + 2\ H^+ + 2\ e^-$$

These oxidation reactions are coupled to reactions in which other molecules get reduced. When it is present, O_2 is the molecule that eventually accepts these electrons, and it is reduced to form H_2O.

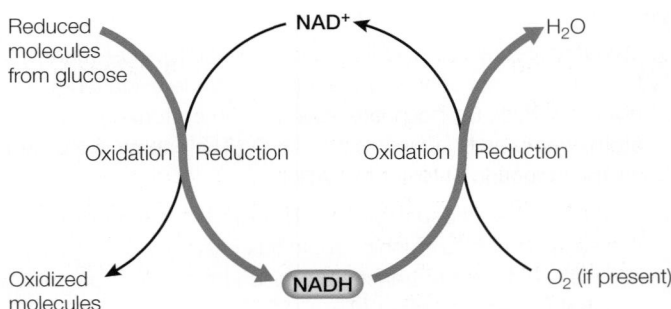

9.2 recap

The oxidation of glucose in the presence of O_2 involves glycolysis, pyruvate oxidation, and the citric acid cycle. In glycolysis, glucose is converted to pyruvate with some energy capture. Pyruvate is oxidized first to acetyl CoA by pyruvate dehydrogenase, then completely to CO_2 by the citric acid cycle, releasing energy that is captured in the form of reduced electron carriers.

learning outcomes

You should be able to:

- Describe the energy inputs and outputs of glycolysis.
- Explain how pyruvate oxidation connects the transformations of glycolysis and the citric acid cycle.
- Compare and contrast glycolysis and the citric acid cycle.

1. What is the net energy yield of glycolysis in terms of energy in ATP and reduced coenzymes invested and energy harvested?
2. What role does pyruvate oxidation play in relation to the citric acid cycle?
3. How is the citric acid cycle a "cycle" but glycolysis is not?

Pyruvate oxidation and the citric acid cycle cannot continue operating unless O_2 is available to receive electrons during the reoxidation of reduced electron carriers. However, these electrons are not passed directly to O_2, as you will learn next.

key concept 9.3 Oxidative Phosphorylation Forms ATP

In the presence of O_2, **oxidative phosphorylation** is a process of ATP synthesis resulting from the reoxidation of electron carriers in the presence of O_2. In this section we describe oxidative phosphorylation as it occurs in mitochondria, but the same process occurs in prokaryotes (see Table 9.1).

focus your learning

- Chemiosmosis converts the potential energy of a proton concentration gradient to chemical energy in ATP.
- Experimental results demonstrated the relationship between a proton gradient across a membrane and ATP synthesis.

What are the steps in oxidative phosphorylation?

Two components of the process can be distinguished:

1. *Electron transport*. The electrons from NADH and $FADH_2$ pass through the **respiratory chain**, a series of membrane-associated electron carriers. The flow of electrons along this pathway results in the active transport of protons out of the mitochondrial matrix and across the inner mitochondrial membrane, creating a proton concentration gradient.

2. *Chemiosmosis*. The protons diffuse back into the mitochondrial matrix through a channel protein, **ATP synthase**, which

couples this diffusion to the synthesis of ATP. As we mentioned in the chapter opening, the inner mitochondrial membrane is normally impermeable to protons, so the only way for them to follow their concentration gradient is through the channel.

Before we proceed with the details of these pathways, let's consider an important question: Why should the respiratory chain be such a complex process? Why don't cells use the following single step?

$$2\ NADH + 2\ H^+ + O_2 \rightarrow 2\ NAD^+ + 2\ H_2O$$

The answer is that this reaction would simply release too much energy to be efficiently trapped to make ATP. Oxidizing NADH to NAD^+ is extremely exergonic—doing it in one step would be like setting off a stick of dynamite in the cell. There is no biochemical way to harvest that burst of energy efficiently and put it to physiological use (that is, no single metabolic reaction is so endergonic as to consume a significant fraction of that energy in a single step). To control the release of energy during the oxidation of glucose, cells have evolved a lengthy respiratory chain: a series of reactions, each of which releases a small amount of energy, one step at a time.

The respiratory chain transfers electrons and protons, and releases energy

The respiratory chain is located in the inner mitochondrial membrane. Because of the extensive folding of the membrane, there is more room for the proteins involved in the chain than there would be in a membrane with less surface area. There are several interacting components, including large integral proteins, a small peripheral protein, and a small lipid molecule. **Figure 9.7** shows a plot of the free energy released as electrons are passed between the carriers.

- Four large protein complexes (I, II, III, and IV) contain electron carriers and associated enzymes. In eukaryotes they are integral proteins of the inner mitochondrial membrane (see Figure 5.11), and three are transmembrane proteins.

- Cytochrome *c* is a small peripheral protein that lies in the intermembrane space. It is loosely attached to the outer surface of the inner mitochondrial membrane.

- Ubiquinone (often referred to as coenzyme Q10; abbreviated Q) is a small, nonpolar, lipid molecule that moves freely within the hydrophobic interior of the phospholipid bilayer of the inner mitochondrial membrane.

As illustrated in Figure 9.7, NADH passes electrons to protein complex I (called NADH-Q reductase), which in turn passes the electrons to Q. This electron transfer is accompanied by a large drop in free energy. Complex II (succinate dehydrogenase) passes electrons to Q from $FADH_2$, which was generated in Step 6 of the citric acid cycle (see Figure 9.6). These electrons enter the chain later than those from NADH and will ultimately produce less ATP.

Complex III (cytochrome *c* reductase) receives electrons from Q and passes them to cytochrome *c*. Complex IV (cytochrome *c* oxidase) receives electrons from cytochrome *c* and passes them to oxygen. Finally the reduction of oxygen to H_2O occurs:

$$O_2 + 4\ H^+ + 4\ e^- \rightarrow 2\ H_2O$$

Notice that four protons (H^+) are also consumed in this reaction. This contributes to the proton concentration gradient across the inner

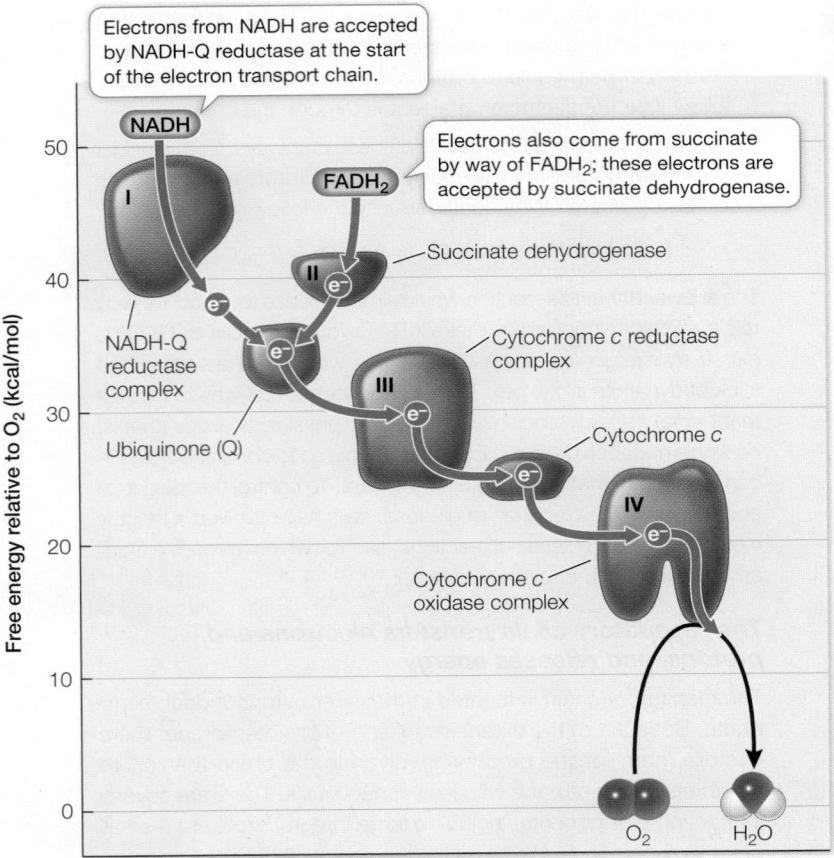

Electrons from NADH are accepted by NADH-Q reductase at the start of the electron transport chain.

Electrons also come from succinate by way of FADH₂; these electrons are accepted by succinate dehydrogenase.

Figure 9.7 The Oxidation of NADH and FADH₂ in the Respiratory Chain Electrons from NADH and FADH₂ are passed along the respiratory chain, a series of protein complexes in the inner mitochondrial membrane containing electron carriers and enzymes. The carriers gain free energy when they become reduced and release free energy when they are oxidized. This illustration shows the standard free energy changes along the respiratory chain.

Q: What is the ΔG for the transfer of electrons from cytochrome c to O_2?

 Activity 9.4 **Respiratory Chain**
www.Life11e.com/ac9.4

mitochondrial membrane by reducing the proton concentration in the mitochondrial matrix.

ATP is made through chemiosmosis

During electron transport, protons are also actively transported across the membrane: electron transport within each of the three transmembrane complexes (I, III, and IV) results in the transfer of protons from the matrix to the intermembrane space (**Figure 9.8**). Since the lipid bilayer does not allow charged H⁺ to diffuse across it, transfer of H⁺ across it through the electron transport chain sets up a gradient, with the concentration of H⁺ in the intermembrane space higher than in the matrix. In addition, because H⁺ carries a charge, there is more positive charge in the intermembrane space. These two gradients—of concentration and charge—set up a **proton-motive force** which is a key factor in energy metabolism in cells:

The gradient of H⁺ across the inner membrane is a source of potential energy.

How can this energy be tapped for use by the cell? The answer is that another protein, ATP synthase, allows the H⁺ to diffuse back into the matrix down its concentration gradient. In the process, potential energy is captured and used for the formation of ATP. The coupling of the proton-motive force and ATP synthesis is called the chemiosmotic mechanism—or **chemiosmosis**—and is found in all respiring cells.

To summarize, the energy originally contained in glucose and other fuel molecules is ultimately captured in the cellular energy currency, ATP. For each pair of electrons passed along the chain from NADH to oxygen, about 2.5 molecules of ATP are formed. FADH₂ oxidation produces about 1.5 ATP molecules because it enters the electron transport chain at a later step than NADH (see Figure 9.8).

ATP synthesis is a reversible reaction, and ATP synthase can also act as an ATPase, hydrolyzing ATP to ADP and P_i:

$$ATP \rightleftharpoons ADP + P_i + \text{free energy}$$

If the reaction goes to the right, free energy is released. In the mitochondrion, it is used to transfer H⁺ out of the mitochondrial matrix—not the usual mode of operation. If the reaction goes to the left, it uses the free energy from H⁺ diffusion into the matrix to make ATP. What makes it prefer ATP synthesis? There are two answers to this question:

1. ATP leaves the mitochondrial matrix for use elsewhere in the cell as soon as it is made, keeping the ATP concentration in the matrix low, and driving the reaction toward the left.

2. The H⁺ gradient is maintained by electron and proton transport.

Every day a person hydrolyzes about 10^{25} ATP molecules to ADP. This amounts to 9 kg, a significant fraction of the person's entire body weight! The vast majority of this ADP is "recycled"—converted back to ATP—using free energy from the oxidation of glucose.

Experiments demonstrate chemiosmosis

Because it is so important, chemiosmosis has been subjected to many lab experiments. Let's look at two lines of evidence; first, a direct demonstration in the lab that a H⁺ gradient can drive ATP synthesis, and second, natural disruption of the coupling of electron transport to ATP synthesis.

DIRECT DEMONSTRATION OF CHEMIOSOMOSIS The key experiment that demonstrated that a proton (H⁺) gradient across a membrane could drive ATP synthesis was first performed using chloroplasts, the organelles in plants that convert the energy in sunlight into chemical energy (photosynthesis; see Chapter 10) (**Figure 9.9**). Soon thereafter, the same mechanism was shown to work in mitochondria.

 Animation 9.2 **Two Experiments Demonstrate the Chemiosmotic Mechanism**
www.Life11e.com/a9.2

Glycolysis
Glucose
Pyruvate

Pyruvate Oxidation

Citric Acid Cycle

Electron Transport/ ATP Synthesis → ATP

CO_2 and H_2O

Mitochondrion

Activity 9.5 **Electron Transport Simulation**
www.Life11e.com/ac9.5

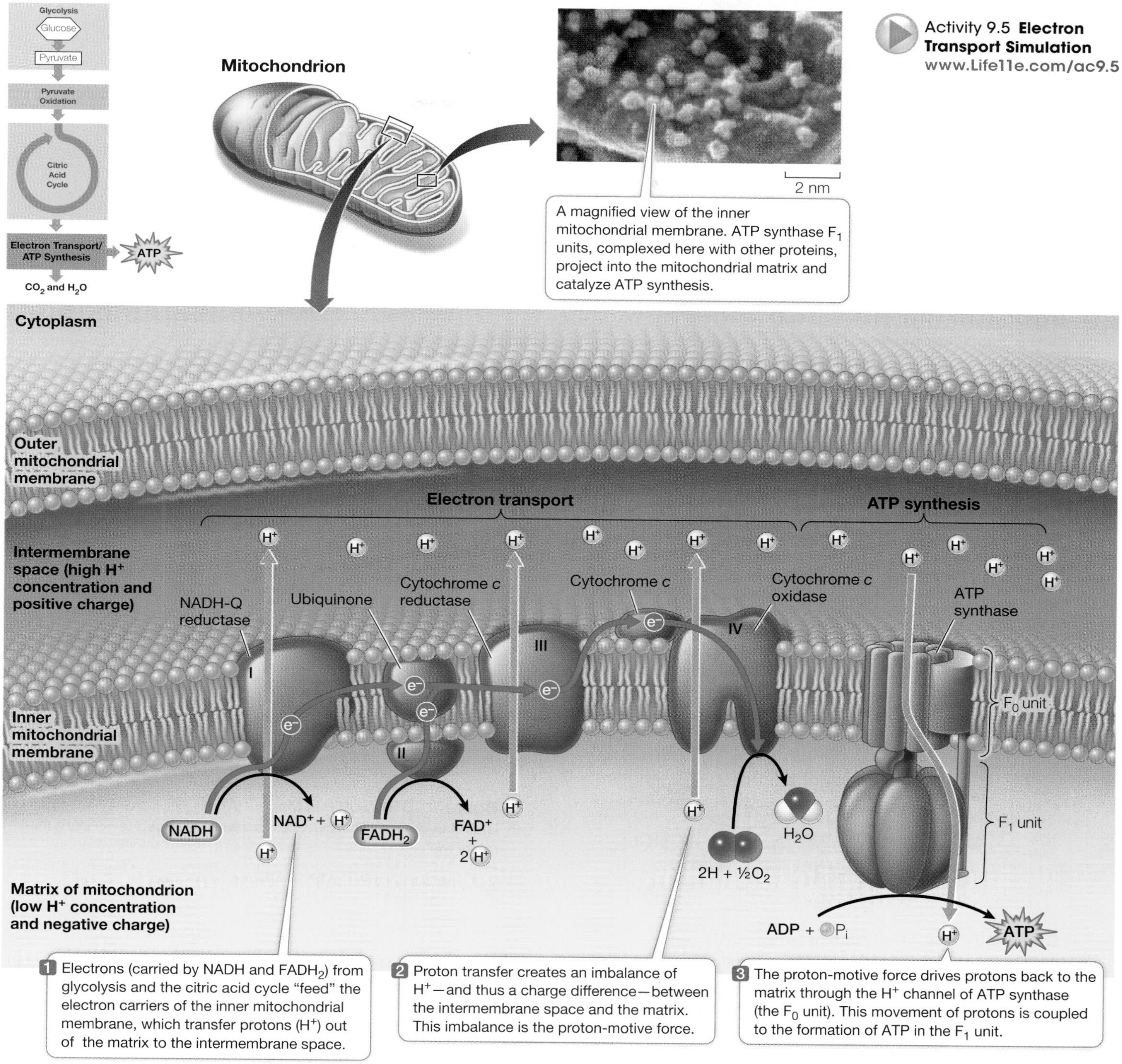

2 nm

A magnified view of the inner mitochondrial membrane. ATP synthase F_1 units, complexed here with other proteins, project into the mitochondrial matrix and catalyze ATP synthesis.

Cytoplasm

Outer mitochondrial membrane

Electron transport

ATP synthesis

Intermembrane space (high H^+ concentration and positive charge)

H^+ H^+ H^+ H^+ H^+ H^+ H^+ H^+ H^+ H^+ H^+ H^+

NADH-Q reductase
Ubiquinone
Cytochrome c reductase
Cytochrome c
Cytochrome c oxidase
ATP synthase

e^-
III
e^-
IV
e^-
I
e^-
e^-

F_0 unit

Inner mitochondrial membrane

II

F_1 unit

NADH
$NAD^+ + H^+$
H^+
FADH$_2$
$FAD^+ + 2 H^+$
H^+
H^+
H_2O
$2H + \frac{1}{2}O_2$

Matrix of mitochondrion (low H^+ concentration and negative charge)

$ADP + $ P_i
H^+
ATP

1 Electrons (carried by NADH and FADH$_2$) from glycolysis and the citric acid cycle "feed" the electron carriers of the inner mitochondrial membrane, which transfer protons (H^+) out of the matrix to the intermembrane space.

2 Proton transfer creates an imbalance of H^+—and thus a charge difference—between the intermembrane space and the matrix. This imbalance is the proton-motive force.

3 The proton-motive force drives protons back to the matrix through the H^+ channel of ATP synthase (the F_0 unit). This movement of protons is coupled to the formation of ATP in the F_1 unit.

Figure 9.8 The Respiratory Chain and ATP Synthase Produce ATP by a Chemiosmotic Mechanism As electrons pass through the transmembrane protein complexes in the respiratory chain, protons are transferred from the mitochondrial matrix into the intermembrane space. As the protons return to the matrix through ATP synthase, ATP is formed.

Animation 9.1 **Electron Transport and ATP Synthesis**
www.Life11e.com/a9.1

UNCOUPLING ELECTRON TRANSPORT FROM ATP PRODUCTION
As you have seen, the coupling of electron transport (which generates the proton gradient) with chemiosmosis is vital for the capture of free energy in the form of ATP. Uncoupling protein 1 (UCP1),

which is found in the mitochondria of brown fat cells (described in the chapter opening), demonstrates the importance of this coupling. By disrupting the gradient, UCP1 allows the energy released during electron transport to be in the form of heat, rather than chemical energy trapped in ATP.

The relationship between UCP1 in brown fat and body weight has been investigated experimentally in a study of a genetic strain

experiment

Figure 9.9 An Experiment Demonstrates the Chemiosmotic Mechanism

Original Paper: Jagendorf, A. T. and E. Uribe. 1966. ATP formation caused by acid-base transition of spinach chloroplasts. *Proceedings of the National Academy of Sciences* USA 55: 170–177.

The chemiosmosis hypothesis was a bold departure from the conventional scientific thinking of the time. It required an intact compartment enclosed by a membrane. Could a proton gradient drive the synthesis of ATP? The first experiments to answer this question used chloroplasts, plant organelles that use the same mechanism as mitochondria to synthesize ATP.

HYPOTHESIS▶ A H^+ gradient across a membrane that contains ATP synthase is sufficient to drive ATP synthesis.

METHOD

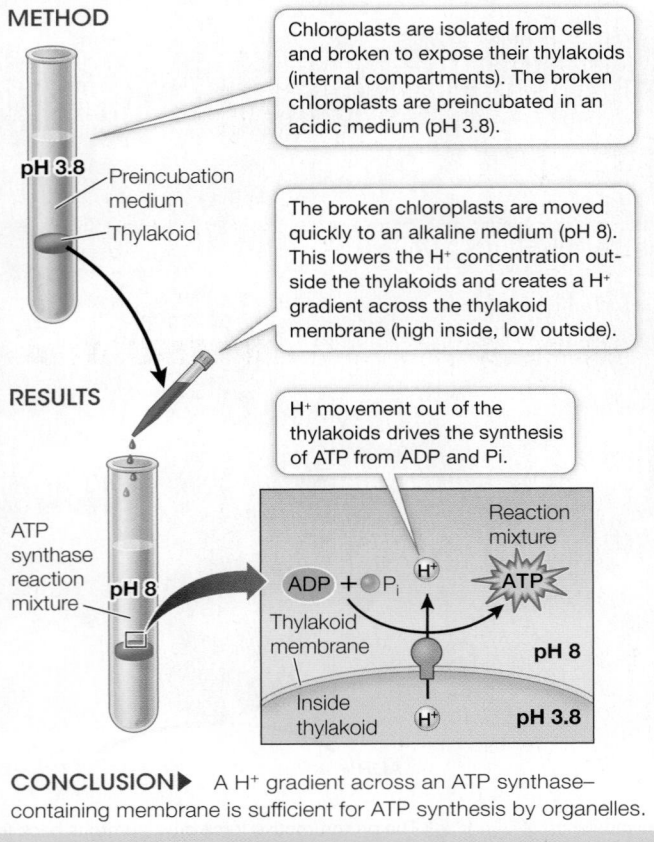

Chloroplasts are isolated from cells and broken to expose their thylakoids (internal compartments). The broken chloroplasts are preincubated in an acidic medium (pH 3.8).

pH 3.8
Preincubation medium
Thylakoid

The broken chloroplasts are moved quickly to an alkaline medium (pH 8). This lowers the H^+ concentration outside the thylakoids and creates a H^+ gradient across the thylakoid membrane (high inside, low outside).

RESULTS

H^+ movement out of the thylakoids drives the synthesis of ATP from ADP and Pi.

ATP synthase reaction mixture

pH 8

Reaction mixture

ADP + Pi → ATP

H^+

Thylakoid membrane

pH 8

Inside thylakoid

H^+

pH 3.8

CONCLUSION▶ A H^+ gradient across an ATP synthase–containing membrane is sufficient for ATP synthesis by organelles.

A **work with the data** exercise that accompanies this figure may be assigned in **LaunchPad**.

(A) The structure of ATP synthase

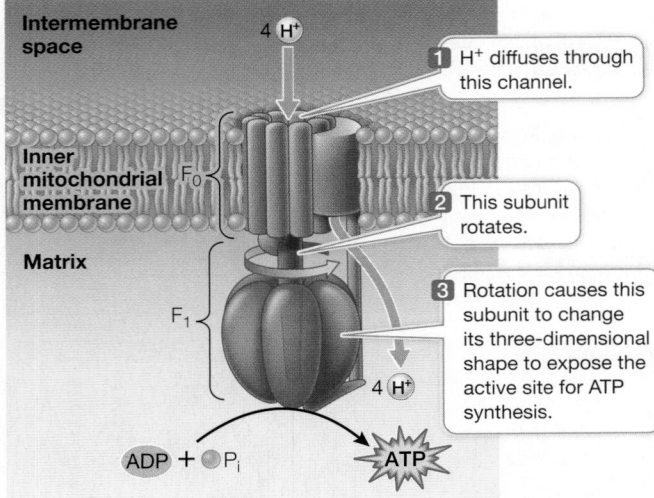

Intermembrane space

4 H^+

1 H^+ diffuses through this channel.

Inner mitochondrial membrane

F_0

Matrix

2 This subunit rotates.

F_1

3 Rotation causes this subunit to change its three-dimensional shape to expose the active site for ATP synthesis.

4 H^+

ADP + Pi → ATP

(B) Demonstration that ATP synthase has a rotary motor

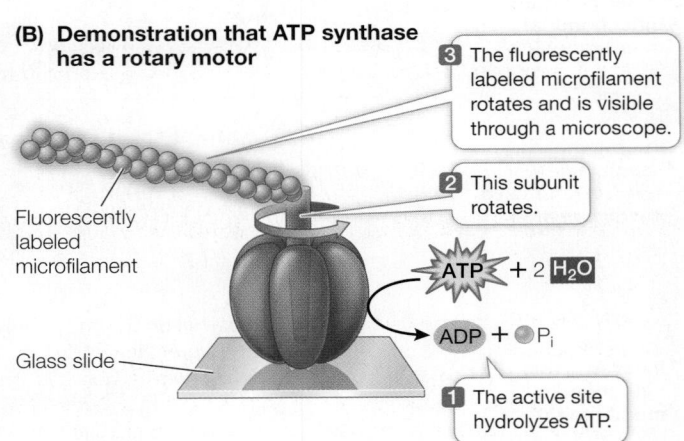

3 The fluorescently labeled microfilament rotates and is visible through a microscope.

2 This subunit rotates.

Fluorescently labeled microfilament

ATP + 2 H_2O

ADP + Pi

Glass slide

1 The active site hydrolyzes ATP.

Figure 9.10 How ATP Is Made **(A)** Mitochondrial ATP synthase is a rotary motor. **(B)** A clever experiment was used to visualize the rotary motor.

▶ Media Clip 9.1 **ATP Synthase in Motion**
www.Life11e.com/mc9.1

but since dinitrophenol acted on all mitochondria everywhere in the body, and not just in fat cells, severe side effects occurred as ATP production all over the body became critically low—in some cases too low for survival. So the idea of an uncoupler for weight loss was dropped.

of mice that—unlike normal mice—do not get fat as they age (people tend to put on weight as they age too). As **Investigating Life: Mitochondria, Genetics, and Obesity** reveals, it appears that the leaner mice make more UCP1 and burn off more fat than normal mice do. Might this observation point to ways to control weight gain in people? In a related study of obesity conducted in the 1930s, a chemical, dinitrophenol, was shown to uncouple oxidation and phosphorylation, just as UCP1 (which had not been discovered yet) does. Physicians got the idea that giving obese people the uncoupler would cause them to oxidize (burn off) their fat. It worked,

HOW ATP SYNTHASE WORKS: A MOLECULAR MOTOR Now that we have established that the H^+ gradient is needed for ATP synthesis, a question remains: How does the enzyme ATP synthase actually make ATP from ADP and P_i? This is certainly a fundamental question in biology, as it underlies energy harvesting in most cells. The structure and mechanism of ATP synthase, illustrated in **Figure 9.10A**, are shared by living organisms as diverse as bacteria and humans. ATP synthase is a molecular motor composed of two parts: the F_0 unit, a transmembrane region that is the H^+ channel; and the F_1 unit, which contains the active sites for ATP synthesis. F_1 consists

experiment

Original Paper: Ma, X., L. Lin, G. Qin, X. Lu, M. Fiorotto, V. Dixit and X. Sun. 2011. Ablations of ghrelin and ghrelin receptor exhibit differential metabolic phenotypes and thermogenic capacity during aging. *PLoS One* 6: e16391.

As people (and mice) get older, they tend to accumulate fat. In the course of investigating a strain of mice genetically unable to make the receptor for a hormone called ghrelin that is involved in controlling appetite, Yuxiang Sun and her team at Baylor College of Medicine compared these mice with normal mice as they aged. They were surprised to find that the genetically changed mice did not put on as much weight as their normal counterparts. The researchers discovered that the genetic mutation in these mice caused uncoupling of oxidative phosphorylation and burning of body fat. Their experiment investigated whether levels of mitochondrial uncoupling protein 1 (UCP1) were different in the genetically changed mice than in normal mice.

HYPOTHESIS▶   Mice that make more uncoupling protein (UCP1) burn more body fat.

CONCLUSION▶   Increased UCP1 correlates with less fat and a lower body weight.

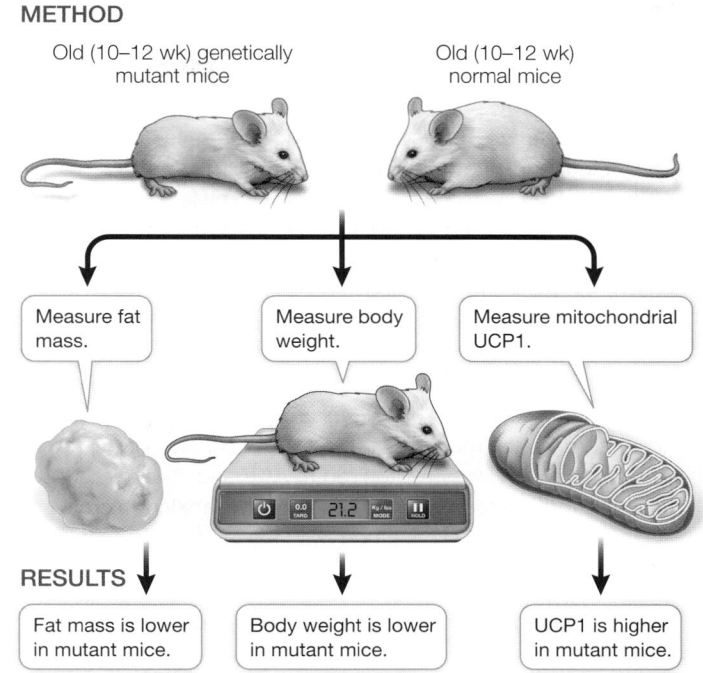

METHOD

Old (10–12 wk) genetically mutant mice

Old (10–12 wk) normal mice

Measure fat mass.

Measure body weight.

Measure mitochondrial UCP1.

RESULTS

Fat mass is lower in mutant mice.

Body weight is lower in mutant mice.

UCP1 is higher in mutant mice.

work with the data

In an effort to determine whether weight gain in normal mice might be due to inactivity or to eating too much (two of the well-known causes of weight gain), as opposed to a biochemical abnormality (such as unusual UCP1 levels), Sun and her team examined young (3–4 months old) and older (10–12 months old) mice of both strains. They measured weight and body composition in terms of fat and lean body tissues. The results are shown in **Figure A**.

QUESTIONS▶

1. Did the mutant mice and normal mice gain weight as they aged?

2. The researchers wondered whether the weight changes noted in Figure 1 were due to changes in eating patterns or exercising. So they measured how much food the mice ate per day and measured their movements over time in a special chamber. The results are shown in **Figure B**. What can you conclude about the role of food intake and exercise on the weight differences between the two strains of mice?

3. The uncoupling protein UCP1 found in the inner mitochondrial membrane of brown fat cells uncouples mitochondrial electron transport (oxidation) and ATP production (phosphorylation), so that instead of being trapped as chemical energy in the formation of ATP, the energy released by oxidation is released as heat. Sun and her team measured UCP1 levels in brown fat mitochondria of the two strains of mice. The results are shown in **Figure C**. What can you conclude about the role of UCP1 in the weight differences between the two strains of mice?

A similar **work with the data** exercise may be assigned in **LaunchPad**.

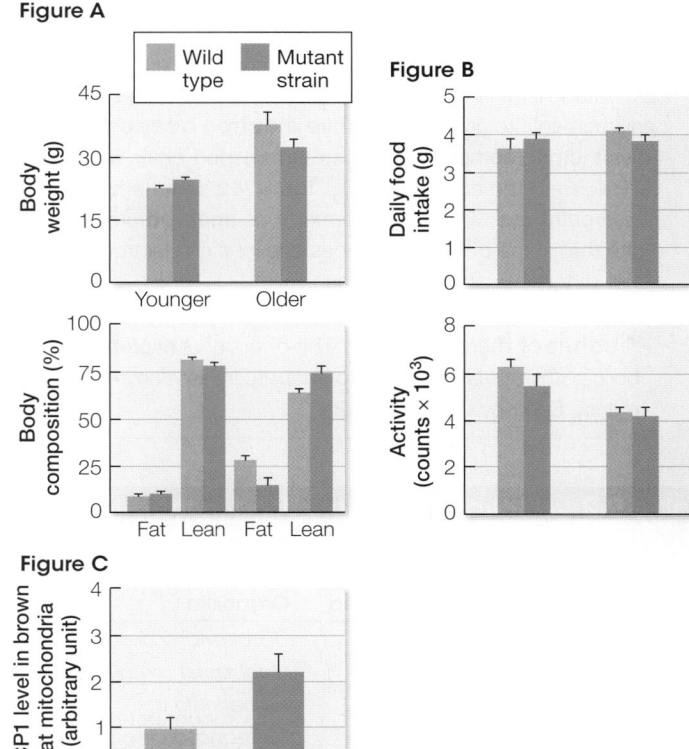

Figure A

Figure B

Figure C

of six subunits (three each of two polypeptide chains), arranged like the segments of an orange around a central shaftlike polypeptide that interacts with the membrane-embedded F_0. Electron transport sets up an H^+ gradient across the membrane. This gradient has potential energy, and when H^+ diffuses through the channel, the potential energy is converted to kinetic energy, causing the central polypeptide to rotate. The energy is transmitted to the catalytic subunits of F_1, resulting in ATP synthesis. These molecular motors make ATP at rates up to 100 molecules per second.

An ingenious experiment confirmed this rotary motor mechanism. Masasuke Yoshida and his colleagues at the Tokyo Institute of Technology isolated the F_1 portion of the ATP synthase and attached it to a glass slide. Fluorescently labeled microfilaments were attached to the central peptide, and the slide was incubated in a solution containing ATP. In this case there was no proton gradient to drive the molecular motor in the direction of ATP synthesis. Instead, ATP was hydrolyzed to ADP and P_i, and this energy caused the motor to spin. Rotation of the labeled microfilament was visible under a microscope, with the labeled filaments clearly rotating like propellers.

Some microorganisms use non-O_2 electron acceptors

A more general way to describe the last reaction in electron transport is:

$$X_{oxidized} + e^- \rightarrow X_{reduced}$$

Part of the amazing success of bacteria and archaea is that they have evolved biochemical pathways that allow them to exist in environments where O_2 is scarce or absent. As you will see in the next section, for most animals and plants, the anaerobic (no O_2) catabolism of glucose generally yields much less energy than aerobic catabolism. However, many bacteria and archaea exploit their environments to use *alternative electron acceptors. This allows them to complete the electron transport chain and produce ATP even in the absence of O_2. **Table 9.2** summarizes some of these pathways, which are referred to as **anaerobic respiration**. Note that some of these microbes use ions as electron acceptors while others use small molecules.

---

*connect the concepts   The diversity of microbial metabolic pathways allows them to adapt to life in many environments. See Key Concept 25.2.

---

table 9.2   Electron Acceptors Used in the Respiratory Chain of Anaerobic Microbes

Terminal electron acceptor	Product formed	Organism
SO_4^{-2}	H_2S	*Desulfovibrio desulfuricans*
Fe^{3+}	Fe^{2+}	*Geobacter metallireducens*
NO_3^-	NO_2^-	*Escherichia coli*
CO_2	CH_4	*Methanosarcina barkeri*
CO_2	CH_3COO^-	*Clostridium aceticum*
Fumarate	Succinate	*Wolinella succinogenes*

9.3 recap

The oxidation of reduced electron carriers in the respiratory chain drives the active transport of protons across the inner mitochondrial membrane, generating a proton-motive force. Diffusion of protons down their electrochemical gradient through ATP synthase is coupled to the synthesis of ATP. Electron transport can form toxic intermediates. Some bacteria and archaea can respire using alternative electron acceptors instead of O_2.

learning outcomes

- Describe how the proton motive force is established by the electron carriers and enzymes of the respiratory chain.
- Analyze experimental results that relate to electron transport and chemiosmosis.
- Predict results from experiments designed to explore aspects of electron transport and chemiosmosis.

1. How are protons transported from the mitochondrial matrix to the intermembrane space during electron transport?
2. How do the experiments described in Figures 9.9 and 9.10 demonstrate the chemiosmotic mechanism?
3. Trace the sequence of changes in redox reactions that occur in mammalian tissue when the oxygen supply is cut off. The first change is that all of the cytochrome *c* becomes reduced, because electrons can still flow from cytochrome *c*, but there is no oxygen to accept electrons from cytochrome *c* oxidase. What happens after this?
4. The drug antimycin A blocks electron transport in mitochondria and chloroplasts. Explain what would happen if the experiment in Figure 9.9 were repeated in the presence of this drug.

Oxidative phosphorylation captures a great deal of energy in ATP. But it does not occur if O_2 is absent. We will turn now to the metabolism of glucose in anaerobic conditions.

key concept 9.4   In the Absence of Oxygen, Some Energy Is Harvested from Glucose

In eukaryotes, in the absence of O_2 (anaerobic conditions), a small amount of ATP can be produced by glycolysis and fermentation. Like glycolysis, fermentation pathways occur in the cytoplasm. There are many different types of fermentation, but they all operate to regenerate NAD^+ so that the NAD^+-requiring reactions of glycolysis can continue. The two best-understood fermentation pathways are found in a wide variety of organisms, including eukaryotes:

- Lactic acid fermentation, the end product of which is lactic acid (lactate)
- Alcoholic fermentation, the end product of which is ethyl alcohol (ethanol)

focus your learning

- Lactic acid fermentation and alcoholic fermentation are pathways that allow glucose to be oxidized in the absence of oxygen by regenerating oxidized electron carriers.

In **lactic acid fermentation**, pyruvate serves as the electron acceptor and lactate is the product (**Figure 9.11A**). This process takes place in many microorganisms and complex organisms, including higher plants and vertebrates. A notable example of lactic acid fermentation occurs in vertebrate muscle tissue. Usually, vertebrates get their energy for muscle contraction aerobically, with the circulatory system supplying O_2 to muscles. In small vertebrates, this is almost always adequate: for example, birds can fly long distances without resting. But in larger vertebrates such as humans, the circulatory system is not up to the task of delivering enough O_2 when the need is great, such as during intense activity. At this point, the muscle cells break down glycogen (a stored polysaccharide; see Figure 3.18) and undergo lactic acid fermentation.

Lactate buildup becomes a problem after prolonged periods of intense exercise because it is associated with an increase in the H^+ concentration in the cell, lowering the pH. This affects cellular activities, which return to normal upon resting. Lactate dehydrogenase, the enzyme that catalyzes the fermentation reaction, works in both directions. That is, when O_2 is available it can catalyze the oxidation of lactate to form pyruvate, which is then catabolized to CO_2 with concomitant energy release to form ATP. When lactate levels are decreased, muscle activity can resume.

Alcoholic fermentation takes place in certain yeasts (eukaryotic microbes) and some plant cells under anaerobic conditions. This process requires two enzymes, pyruvate decarboxylase and alcohol dehydrogenase, which metabolize pyruvate to ethanol (**Figure 9.11B**). As with lactic acid fermentation, the reactions are essentially reversible. For thousands of years, humans have used anaerobic fermentation by yeast cells to produce alcoholic beverages. The cells use sugars from plant sources (glucose from grapes or maltose from barley) to produce the end product, ethanol, in wine and beer.

By recycling NAD^+, fermentation allows glycolysis to continue, thus producing small amounts of ATP through substrate-level phosphorylation. The net yield of two ATPs per glucose molecule is much lower than the energy yield from cellular respiration. For this reason, most organisms existing in anaerobic environments are small microbes that grow relatively slowly.

(A) Lactic acid fermentation

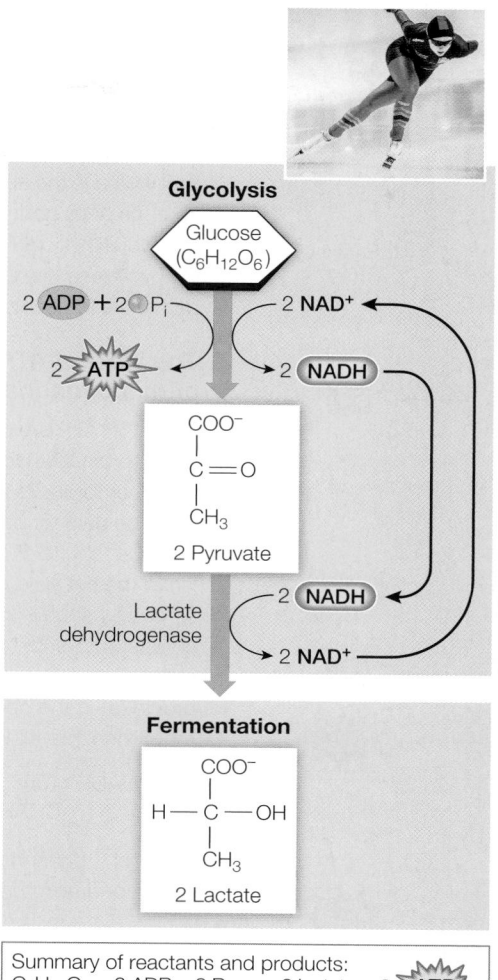

(B) Alcoholic fermentation

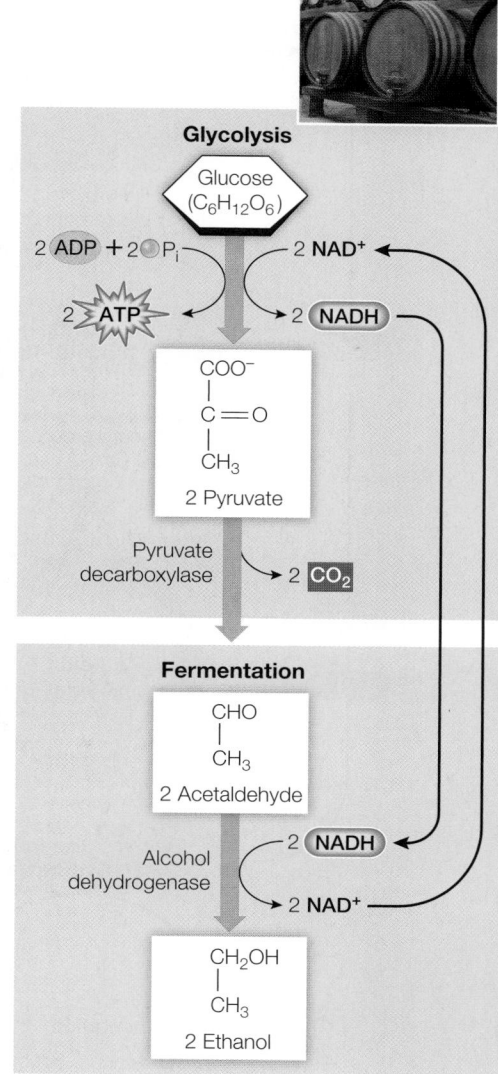

Figure 9.11 Fermentation   Glycolysis produces pyruvate, ATP, and NADH from glucose. **(A)** Lactic acid fermentation uses NADH as a reducing agent to reduce pyruvate to lactate, thus regenerating NAD^+ to keep glycolysis operating. **(B)** In alcoholic fermentation, pyruvate from glycolysis is converted into acetaldehyde, and CO_2 is released. NADH from glycolysis is used to reduce acetaldehyde to ethanol, thus regenerating NAD^+ to keep glycolysis operating.

Cellular respiration yields much more energy than fermentation

The total net energy yield from glycolysis plus fermentation is two molecules of ATP per molecule of glucose oxidized. The maximum yield of ATP that can be harvested from a molecule of glucose through glycolysis followed by cellular respiration is much greater—about 32 molecules of ATP (**Figure 9.12**). (Review Figures 9.5, 9.6, and 9.8, and p. 180 to see where all the ATP molecules come from.)

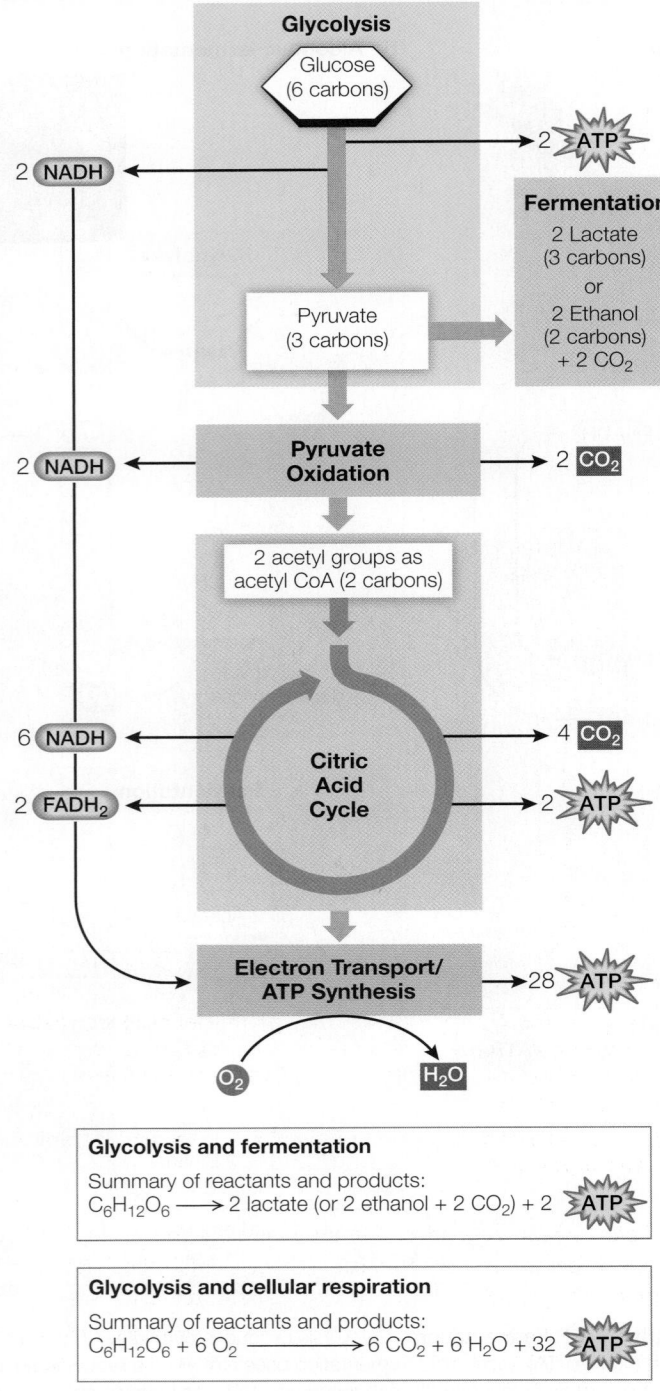

Glycolysis and fermentation

Summary of reactants and products:
$C_6H_{12}O_6 \longrightarrow$ 2 lactate (or 2 ethanol + 2 CO_2) + 2 **ATP**

Glycolysis and cellular respiration

Summary of reactants and products:
$C_6H_{12}O_6 + 6\ O_2 \longrightarrow 6\ CO_2 + 6\ H_2O + 32$ **ATP**

Figure 9.12 Cellular Respiration Yields More Energy Than Fermentation Electron carriers are reduced in pyruvate oxidation and the citric acid cycle, then oxidized by the respiratory chain. These reactions produce ATP via chemiosmosis.

 Activity 9.6 Energy Levels
www.Life11e.com/ac9.6

Why do the metabolic pathways that operate in aerobic environments produce so much more ATP? Glycolysis and fermentation only partially oxidize glucose. Much more energy remains in the end products of fermentation (lactic acid and ethanol) than in CO_2, the

end product of cellular respiration. In cellular respiration, carriers (mostly NAD$^+$) are reduced during pyruvate oxidation and the citric acid cycle. Then the reduced carriers are oxidized by the respiratory chain, with the accompanying production of ATP by chemiosmosis (about 2.5 ATP for each NADH and 1.5 ATP for each FADH$_2$). In an aerobic environment, a cell or organism capable of aerobic metabolism will have the advantage over one that is limited to fermentation, in terms of its ability to harvest chemical energy. Two key events in the evolution of multicellular organisms were the rise in atmospheric O_2 levels (see Key Concept 25.2) and the development of metabolic pathways to use that O_2.

The yield of ATP is reduced by the impermeability of mitochondria to NADH

About 32 ATP are produced from the oxidation of one molecule of glucose to CO_2. However, in many eukaryotes the inner mitochondrial membrane is impermeable to NADH, and a "toll" of one ATP must be paid for each NADH molecule produced in glycolysis that must be "shuttled" into the mitochondrial matrix. So in these organisms, the net yield of ATP is 30.

NADH shuttle systems transfer the electrons captured by glycolysis onto substrates that are capable of movement across the mitochondrial membranes. In muscle and liver tissues (and the brown fat in the opening story), an important shuttle involves glycerol 3-phosphate. In the cytoplasm,

NADH (from glycolysis) + dihydroxyacetone phosphate
(DHAP) → NAD$^+$ + glycerol 3-phosphate

Glycerol 3-phosphate is transferred to the outer surface of the inner mitochondrial membrane. At that surface,

FAD + glycerol 3-phosphate → FADH$_2$ + DHAP

The electrons flow from FADH$_2$ into the electron transport chain via ubiquinone (Q) (see Figure 9.8). DHAP is able to move back to the cytoplasm, where it is available to repeat the process. Note that the reducing electrons are transferred from NADH to FADH$_2$. As you know from Figure 9.8, the energy yield in terms of ATP from FADH$_2$ is lower than that from NADH. This lowers the overall energy yield.

9.4 recap

In the absence of O_2, fermentation pathways use NADH formed by glycolysis to reduce pyruvate and regenerate NAD$^+$. The energy yield of glycolysis coupled to fermentation is low because glucose is only partially oxidized. When O_2 is present, the electron carriers of cellular respiration allow for the full oxidation of glucose, so the energy yield from glucose is much higher.

learning outcomes

You should be able to:

• Describe the process of fermentation.

1. Why is replenishing NAD$^+$ required to continue glycolysis in the absence of O_2?
2. How does fermentation replenish NAD$^+$?

Now that you've seen how cells harvest energy, let's see how that energy moves through other metabolic pathways in the cell.

key concept 9.5 Metabolic Pathways Are Interrelated and Regulated

Glycolysis and the pathways of cellular respiration do not operate in isolation. Rather, there is an interchange of molecules into and out of these pathways, to and from the metabolic pathways for the synthesis and breakdown of amino acids, nucleotides, fatty acids, and other building blocks of life (see Figure 8.14). Carbon skeletons (i.e., the carbon backbones of organic molecules) can enter the catabolic pathways and be broken down to release their energy, or they can enter anabolic pathways to be used in the formation of the macromolecules that are the major constituents of the cell. These relationships are summarized in **Figure 9.13**. In this section we will explore how pathways are interrelated by the sharing of intermediate molecules, and we will see how pathways are regulated by the inhibitors of key enzymes.

focus your learning

• The synthesis and breakdown of macromolecules in a cell are linked through common metabolic pathways.
• Metabolic pathways are regulated to ensure efficiency and proper functioning of the cell.

Catabolism and anabolism are linked

A hamburger or veggie burger on a bun contains three major sources of carbon skeletons: carbohydrates, mostly in the form of starch (a polysaccharide); lipids, mostly as triglycerides (three fatty acids attached to glycerol); and proteins (polymers of amino acids). Look at Figure 9.13 to see how each of these three types of macromolecules can be hydrolyzed and used in catabolism or anabolism.

CATABOLIC INTERCONVERSIONS   Polysaccharides, lipids, and proteins can all be broken down to provide energy:

• *Polysaccharides* are hydrolyzed to glucose. Glucose then passes through glycolysis and cellular respiration, where its energy is captured in ATP.

• *Lipids* are broken down into their constituents, glycerol and fatty acids. Glycerol is converted into dihydroxyacetone phosphate (DHAP), an intermediate in glycolysis. Fatty acids are highly reduced molecules that are converted to acetyl CoA inside the mitochondrion by a series of oxidation enzymes, in a process known as β-oxidation. For example, the β-oxidation of a 16-carbon (C_{16}) fatty acid occurs in several steps:

$$C_{16} \text{ fatty acid} + \text{CoA} \rightarrow C_{16} \text{ fatty acyl CoA}$$
$$C_{16} \text{ fatty acyl CoA} + \text{CoA} \rightarrow C_{14} \text{ fatty acyl CoA} + \text{acetyl CoA}$$
$$\text{Repeat 6 times} \rightarrow 8 \text{ acetyl CoA}$$

The acetyl CoA can then enter the citric acid cycle and be catabolized to CO_2.

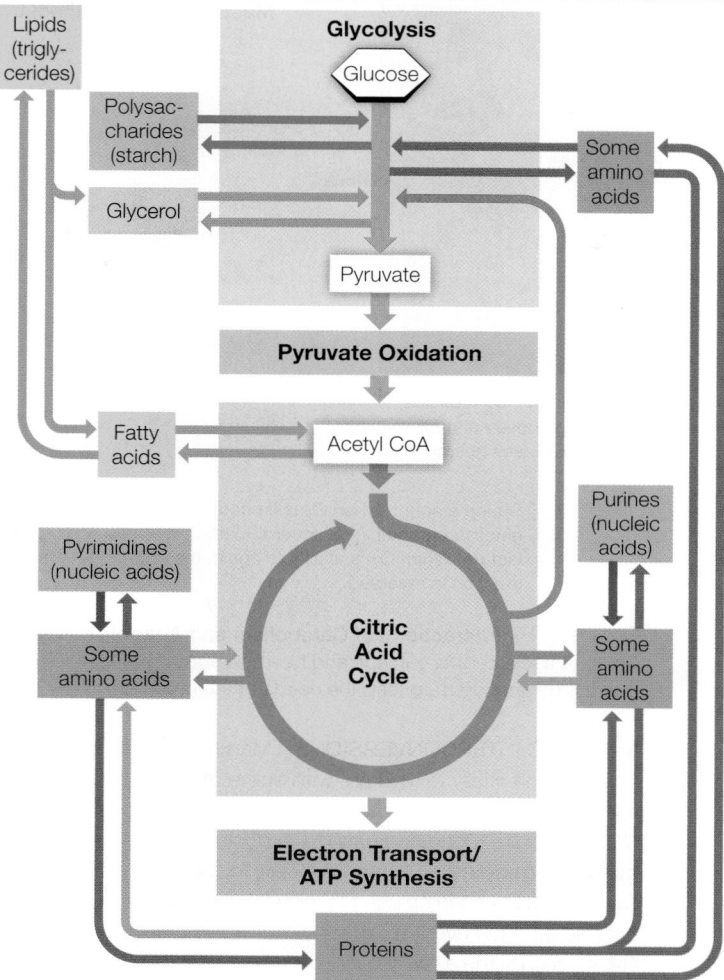

Figure 9.13 Relationships among the Major Metabolic Pathways of the Cell   Note the central positions of glycolysis and the citric acid cycle in this network of metabolic pathways. Also note that many of the pathways can operate essentially in reverse.

Q: Can DNA be an energy source? Explain. Why do you think it is not usually employed in this way?

• *Proteins* are hydrolyzed to their amino acid building blocks. The 20 different amino acids feed into glycolysis or the citric acid cycle at different points determined by their structures. For example, the amino acid glutamate is converted into α-ketoglutarate, an intermediate in the citric acid cycle (the five-carbon molecule in Figure 9.6).

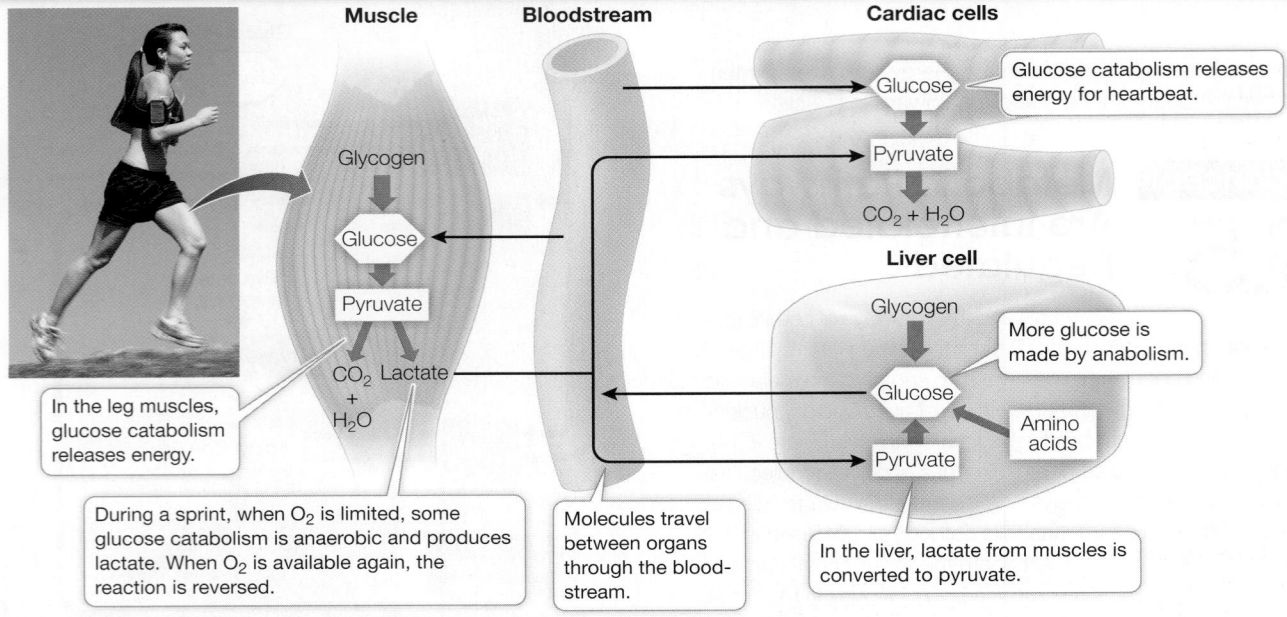

Figure 9.14 Interactions of Catabolism and Anabolism during Exercise   When we walk or run, muscle cells in our legs and heart catabolize glucose for energy, while liver cells produce glucose by anabolism to feed the needs of the muscles.

ANABOLIC INTERCONVERSIONS   Many catabolic pathways can operate essentially in reverse, with some modifications. Glycolytic and citric acid cycle intermediates, instead of being oxidized to form CO_2, can be reduced and used to form glucose in a process called **gluconeogenesis** (which means "new formation of glucose"). Likewise, acetyl CoA can be used to form fatty acids. The most common fatty acids have even numbers of carbons: 14, 16, or 18. These are formed by the addition of two-carbon acetyl CoA "units" one at a time until the appropriate chain length is reached. Acetyl CoA is also a building block for various pigments, plant growth substances, rubber, steroid hormones, and other molecules.

Some intermediates in the citric acid cycle are reactants in pathways that synthesize important components of nucleic acids. For example, α-ketoglutarate is a starting point for purines, and oxaloacetate for pyrimidines. In addition, α-ketoglutarate is a starting point for the synthesis of chlorophyll (used in photosynthesis; see Chapter 10) and the amino acid glutamate (used in protein synthesis).

Catabolism and anabolism are integrated

A carbon atom from a protein in your burger can end up in DNA, fat, or CO_2, among other fates. How does the organism "decide" which metabolic pathways to follow, in which cells? With all of the possible interconversions, you might expect that cellular concentrations of various biochemical molecules would vary widely. Remarkably, the levels of these substances in what is called the metabolic pool—the sum total of all the biochemical molecules in a cell—are quite constant. Organisms regulate the *enzymes in various cells in order to maintain a steady state between catabolism and anabolism. For example, let's look at how glucose levels are maintained in the body during exercise (**Figure 9.14**).

***connect the concepts**   Control of metabolic pathways often rests with regulation of enzymes already present. Review the mechanisms of enzyme regulation in Key Concept 8.4.

When you're walking or jogging, the muscles that most need energy are the leg muscles that power movement and the heart muscles that circulate blood. The energy comes from the catabolism of glucose by glycolysis, the citric acid cycle, and oxidative phosphorylation. These muscles therefore need a lot of glucose. Some glycogen is stored in the leg muscles and is hydrolyzed to glucose monomers. In addition, glycogen is hydrolyzed in the liver, which releases the glucose into the blood. As this glucose is used up by the working muscles, more glucose is made in the liver by anabolism from amino acids and pyruvate. Some of this pyruvate comes from lactate formed by fermentation in the leg muscles and that has been transported back to the liver in the blood.

The exquisite integration of catabolism and anabolism cannot be accomplished without control points in the biochemical pathways. For example, something must "tell" the liver to make glucose rather than catabolizing it or storing it. How are the systems integrated?

Metabolic pathways are regulated systems

The regulation of interconnecting biochemical pathways is a problem of systems biology, which seeks to understand how biochemical pathways interact (see Key Concept 8.5). It is a bit like trying to predict traffic patterns in a city: if an accident blocks traffic on a major road, drivers take alternate routes, where the traffic volume consequently changes.

Several mechanisms can be used to regulate the rate of each step in a biochemical pathway:

- *Changing the amount of active enzyme*: The cell can increase the expression of a gene encoding an enzyme.
- *Changing enzyme activity by covalent modifications*: Adding phosphate groups by protein kinase can alter the activity of an enzyme.
- *Feedback inhibition*: Allosteric changes in an enzyme due to binding of a product in a pathway can cause the entire pathway to shut down.

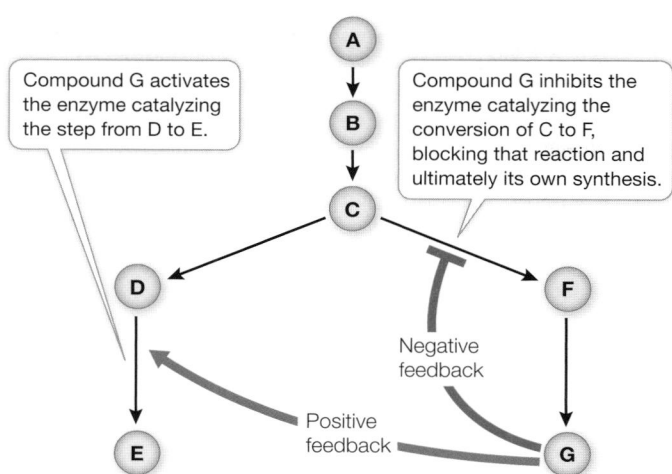

Compound G activates the enzyme catalyzing the step from D to E.

Compound G inhibits the enzyme catalyzing the conversion of C to F, blocking that reaction and ultimately its own synthesis.

Negative feedback

Positive feedback

Figure 9.15 Regulation by Negative and Positive Feedback Allosteric feedback regulation plays an important role in metabolic pathways. The accumulation of some products can shut down their synthesis, or can stimulate other pathways that require the same raw materials.

 Activity 9.7 **Regulation of Energy Pathways**
www.Life11e.com/ac9.7

- *Substrate availability*: If the substrate of a particular enzyme is used up by another pathway, the first enzyme can no longer function and the pathway shuts down.

Consider what happens to the starch in your burger bun. In the digestive system, starch is hydrolyzed to glucose, which enters the blood for distribution to the rest of the body. But before the glucose is distributed, a regulatory check must be made: If there is already enough glucose in the blood to supply the body's needs, the excess glucose is converted into glycogen and stored in the liver and muscles. If not enough glucose is supplied by food, glycogen is broken down, or other molecules are used to make glucose by gluconeogenesis. The end result is that the level of glucose in the blood is remarkably constant. How does the body accomplish this?

Glycolysis and the citric acid cycle are subject to allosteric regulation (see Key Concept 8.5) of key enzymes involved. In a metabolic pathway, a high concentration of the final product can inhibit the action of an enzyme that catalyzes an earlier reaction (see Figure 8.18). Furthermore, an excess of the product of one pathway can activate an enzyme in another pathway, speeding up its reactions and diverting raw materials away from synthesis of the first product (**Figure 9.15**). These negative and positive feedback mechanisms are used at many points in the energy-harvesting pathways and are summarized in **Figure 9.16**.

- The main control point in glycolysis is the glycolytic enzyme phosphofructokinase, which catalyzes Step 3 in Figure 9.5. This enzyme is allosterically inhibited by ATP or citrate, and activated by ADP or AMP. Under anaerobic conditions, fermentation yields a relatively small amount of ATP, and phosphofructokinase operates at a high rate. However, when conditions are aerobic, respiration makes 16 times more ATP than fermentation does, and the abundant ATP allosterically inhibits phosphofructokinase. So glycolysis slows down.

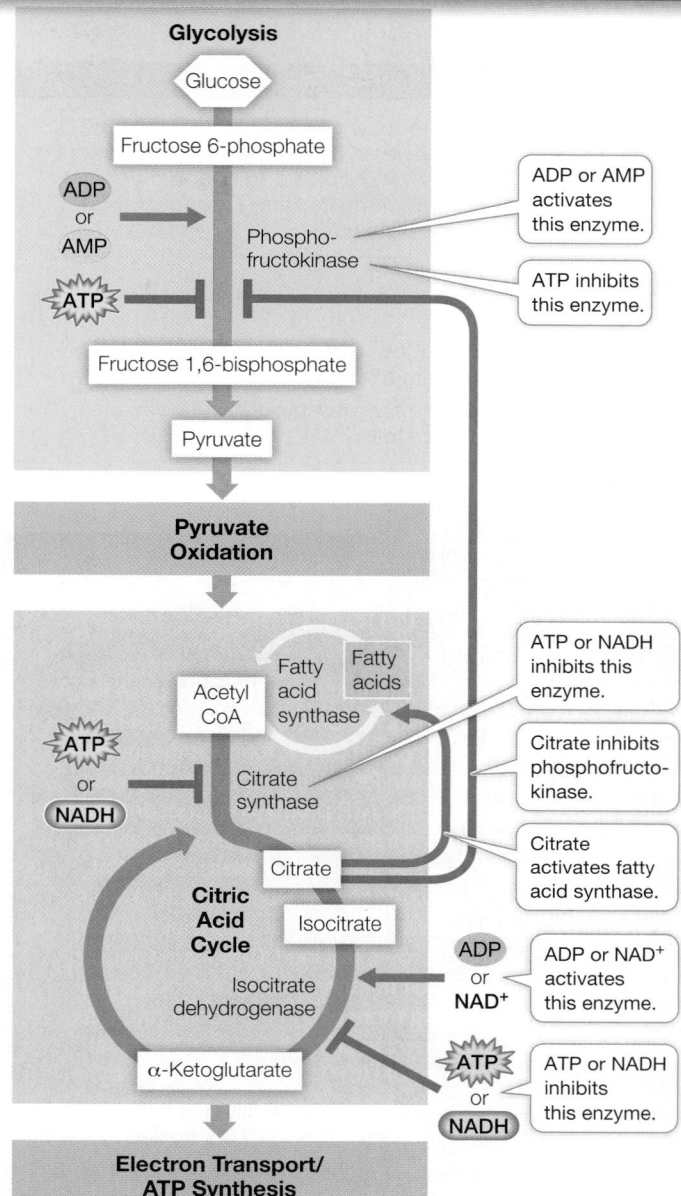

Figure 9.16 Allosteric Regulation of Glycolysis and the Citric Acid Cycle Allosteric regulation controls glycolysis and the citric acid cycle at crucial early steps, increasing their efficiency and preventing the excessive buildup of intermediates.

- The main control point in the citric acid cycle is the enzyme isocitrate dehydrogenase (which catalyzes Step 3 in Figure 9.6). This enzyme is activated by increases in substrate concentrations (ADP, NAD^+, and isocitrate) and is inhibited by products of the citric acid cycle: ATP and NADH. If too much ATP or NADH accumulates, the reaction is slowed, and the citric acid cycle shuts down.

- Another control point involves acetyl CoA. If the level of ATP is high and the citric acid cycle shuts down, the accumulation of citrate activates fatty acid synthase, diverting acetyl CoA to the synthesis of fatty acids for storage (contributing to a person's accumulation of fat). These fatty acids may be metabolized later to produce more acetyl CoA.

9.5 recap

Glucose can be made from intermediates in glycolysis and the citric acid cycle by gluconeogenesis. The metabolic pathways for the production and breakdown of lipids and amino acids are linked to those of glucose metabolism. Reaction products regulate key enzymes in the various pathways.

learning outcomes

You should be able to:

- Describe catabolic and anabolic pathways that link carbohydrate, lipid, and protein metabolism.
- Trace the path of an atom as it moves through various metabolic pathways within a cell.

- Explain how allosteric changes in an enzyme can be used to regulate flow through a metabolic pathway.

1. Give examples of the pathways relating to glycolysis and the citric acid cycle for catabolism of a lipid and the anabolism of a protein.
2. How does phosphofructokinase serve as a control point for glycolysis?
3. You eat a burger that contains polysaccharides, proteins, and lipids. How could carbon atoms in the proteins' amino acids and the glucose in the polysaccharides end up as fats?

investigatinglife

Q&A How are mitochondria related to obesity?

Recall from the opening of this chapter that brown fat is adipose tissue with a high concentration of mitochondria, which have iron-containing pigments. The uncoupling protein UCP1, found in the inner membranes of mitochondria in brown fat cells, uncouples mitochondrial electron transport (oxidation) and ATP production (phosphorylation), so that instead of being trapped as chemical energy in the formation of ATP, the energy released by oxidation is released as heat.

Hibernating animals take advantage of the heat emitted by the catabolism of energy-rich molecules in brown fat, especially when aroused. In hibernating animals, brown fat and UCP1 increase during the winter. It turns out that adult humans also experience a seasonal variation in brown fat, having more of it in the cooler months. Other factors such as diet and exercise being equal, increased fat catabolism occurs in the presence of UCP1 (as demonstrated by the experiment in Investigating Life: Mitochondria, Genetics, and Obesity).

Genetic studies confirm the role of UCP1 in fat breakdown as well. For example, some people have a genetically determined

variant of UCP1 that has lower uncoupling activity than normal UCP1. As people with the variant get older, they accumulate more body fat than their peers with normal UCP1 do.

Future directions

The recent finding that not just babies but also adults have mitochondria-laden brown fat has sparked interest in the possible activation of these cells to treat obesity. In addition to occurring in well-known locations such as the neck and shoulders, brown fat cells have been found mixed in with non–brown fat all over the body, and can be seen by PET (positron emission tomography) scans. At the U.S. National Institutes of Health, a research team led by Aaron Cypess has found unexpectedly that a drug used to treat people with an overactive bladder also activates fat breakdown in brown fat cells. The drug, mirabegron, targets a receptor on bladder cells that is also expressed on brown fat cells. When people without bladder disease were given this drug, brown fat cells were activated. There was increased glucose uptake and heat production due to mitochondrial uncoupling, and also an eightfold increase in fatty acid breakdown by the mitochondria. This could be a promising treatment for disorders that lead to obesity.

Chapter Summary 9

▶ **9.1 Cells Harvest Chemical Energy from Glucose Oxidation**

- As a material is **oxidized**, the electrons it loses are transferred to another material, which is thereby **reduced**. Such **oxidation-reduction**, or **redox**, **reactions** transfer large amounts of energy. Review Figure 9.2
- The coenzyme NAD^+ is a key electron carrier in biological redox reactions. It exists in two forms, one oxidized (NAD^+) and the other reduced (NADH). Review Focus: Key Figure 9.3
- **Glycolysis** does not use O_2. Under **aerobic** conditions, **cellular respiration** continues the process of breaking down glucose.

Under **anaerobic** conditions, **fermentation** occurs. Review Figures 9.1, 9.4, Activities 9.1, 9.2

- The pathways of cellular respiration after glycolysis are pyruvate oxidation, the citric acid cycle, and electron transport/ATP synthesis.

▶ **9.2 In the Presence of Oxygen, Glucose Is Fully Oxidized**

- Glycolysis consists of ten enzyme-catalyzed reactions that occur in the cell cytoplasm. Two **pyruvate** molecules are produced for each partially oxidized molecule of glucose, providing the starting material for both cellular respiration and fermentation. Review Figure 9.5

- Pyruvate oxidation follows glycolysis and links glycolysis to the citric acid cycle. This pathway converts pyruvate into **acetyl CoA**.
- Acetyl CoA is the starting point of the **citric acid cycle**. It reacts with oxaloacetate to produce citrate. A series of eight enzyme-catalyzed reactions oxidize citrate and regenerate oxaloacetate, continuing the cycle. Review Figure 9.6, Activity 9.3

▶9.3 Oxidative Phosphorylation Forms ATP
- Oxidation of electron carriers in the presence of O_2 releases energy that can be used to form ATP in a process called **oxidative phosphorylation**.
- The NADH and $FADH_2$ produced in glycolysis, pyruvate oxidation, and the citric acid cycle are oxidized by the **respiratory chain**, regenerating NAD^+ and FAD. Oxygen (O_2) is the final acceptor of electrons and protons, forming water (H_2O). Review Figure 9.7, Activity 9.4
- The respiratory chain not only transports electrons, but also transfers protons across the inner mitochondrial membrane, creating the **proton-motive force**.
- Protons driven by the proton-motive force can return to the mitochondrial matrix via **ATP synthase**, a molecular motor that couples this movement of protons to the synthesis of ATP. This process is called **chemiosmosis**. Review Figure 9.8, Activity 9.5
- There is considerable experimental evidence for chemiosmosis. Review Figures 9.9, 9.10, Animations 9.1, 9.2

▶9.4 In the Absence of Oxygen, Some Energy Is Harvested from Glucose
- In the absence of O_2 in most organisms, glycolysis is followed by fermentation. Together, these pathways partially oxidize pyruvate and generate end products such as lactic acid or ethanol. In the process, NAD^+ is regenerated from NADH so that glycolysis can continue, thus generating a small amount of ATP. Review Figure 9.11
- For each molecule of glucose used, glycolysis plus fermentation yields two molecules of ATP. In contrast, glycolysis operating with pyruvate oxidation, the citric acid cycle, and oxidative phosphorylation yields up to 32 molecules of ATP per molecule of glucose. Review Figure 9.12, Activity 9.6

▶9.5 Metabolic Pathways Are Interrelated and Regulated
- The catabolic pathways for the breakdown of carbohydrates, fats, and proteins feed into the energy-harvesting metabolic pathways. Review Figure 9.13
- Anabolic pathways use intermediate components of the energy-harvesting pathways to synthesize fats, amino acids, and other essential building blocks.
- The formation of glucose from intermediates of glycolysis and the citric acid cycle is called **gluconeogenesis**.
- The rates of glycolysis and the citric acid cycle are controlled by allosteric regulation and by the diversion of excess acetyl CoA into fatty acid synthesis. Key regulated enzymes include phosphofructokinase, citrate synthase, isocitrate dehydrogenase, and fatty acid synthase. See Figure 9.16, Activity 9.7

> Go to **LearningCurve** (in **LaunchPad**) for dynamic quizzing that helps you solidify your understanding of this chapter. **LearningCurve** adapts to your responses, giving you the practice you need to master each key concept.

▶ Apply What You've Learned

Review
9.5 The synthesis and breakdown of macromolecules in a cell are linked through common metabolic pathways.

9.5 Metabolic pathways are regulated to ensure efficiency and proper functioning of the cell.

Original Papers: Cahill, Jr., G. F. 2006. Fuel metabolism in starvation. *Annual Review of Nutrition* 26: 1–22.

Exton, J. H. and C. R. Park. 1967. Control of gluconeogenesis in liver: I. General features of gluconeogenesis in the perfused livers of rats. *Journal of Biological Chemistry* 242: 2622–2636.

People desperate to lose weight quickly sometimes resort to crash diets, which involve severely limiting foods rich in energy. But how effective is this strategy? What happens to the body when it is starved of food? Scientists have studied starvation in mammals to find out how their bodies compensate for a lack of food. Their findings can be used to better understand the effects of severe diets on the human body.

Studies have shown that mammalian bodies need a constant supply of glucose to maintain homeostasis. Food provides a regular supply of glucose, and when there is an excess, the body stores it as glycogen. **Figure A** shows how cells in the human body shift in their rate of glucose oxidation and the source of glucose oxidized after food intake has stopped.

Figure A

Time after food intake has stopped

(continued)

These data indicate that once glycogen reserves are used up, the body uses other molecules as sources of glucose. What are these molecules? To find out, researchers studied the biochemistry of the liver, since it is the major organ where glucose is made (gluconeogenesis). They surgically removed livers from healthy rats and kept the livers in an active physiological state using perfusion. Perfusion is a technique that pumps fluids through the blood vessels of an organ to mimic the normal blood flow taking place inside a living animal. Using perfused rat livers, the researchers tested the production of glucose inside liver cells when various metabolic compounds were pumped through the isolated livers. They did this by measuring glucose concentration (mM) in the perfusion fluid leaving the livers. The results of their studies are shown in **Figure B** and the table. (The data in Figure B were gathered from the livers of fasted rats.)

Figure B

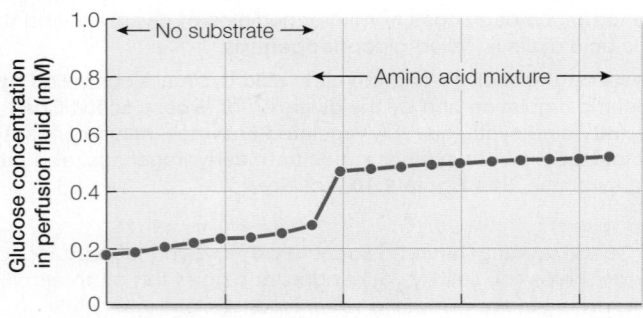

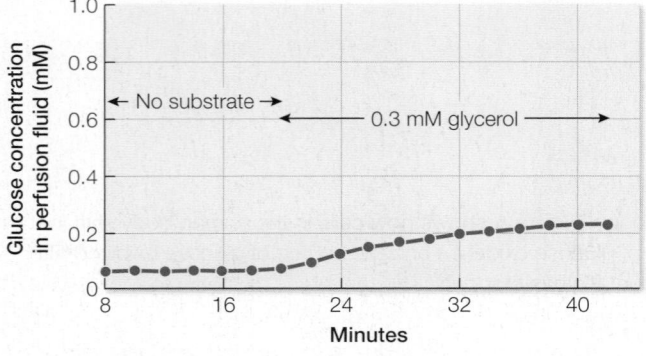

Experiment number	Condition	Additions	Rate of glucose production through gluconeogenesis (µmoles/g liver/hr)
1	Fasted rats	Protein	55.0
2	Fasted rats	Fatty acid + protein	58.9
3	Fed rats	No addition	105.4
4	Fed rats	Fatty acid	99.0

Questions

1. According to data in Figure A, what does the human body do to supply glucose to its organs once food intake stops?

2. Most people want to lose fat when they diet. Ideally, these people would like their fat to be used as fuel when they restrict their food intake. Recall that triglycerides are one form of fat stored in the body, and that a triglyceride is composed of a glycerol molecule bonded to three fatty acid molecules. Also recall that proteins are composed of amino acids. Assuming that all mammals respond as rats do, what do the data from the rat perfusion studies suggest about the identities of the molecules that mammalian bodies use to make glucose under starvation conditions?

3. Considering your answer to Question 2, and assuming that all mammals respond as rats do, what is likely to happen to the muscles of a person who goes on an extreme calorie-restriction diet as a way to lose weight? Is this advisable? Do you think people realize these consequences when they attempt losing weight via crash diets?

4. Alanine is an amino acid that has been shown to be an important intermediate in the gluconeogenesis pathway. Find and compare the structures of alanine and pyruvate. Show how an atom in an alanine molecule could be radiolabeled so that it could be traced through the formation of pyruvate as part of the gluconeogenesis pathway.

Go to **LaunchPad** for the eBook, LearningCurve, animations, activities, flashcards, and additional resources and assignments.

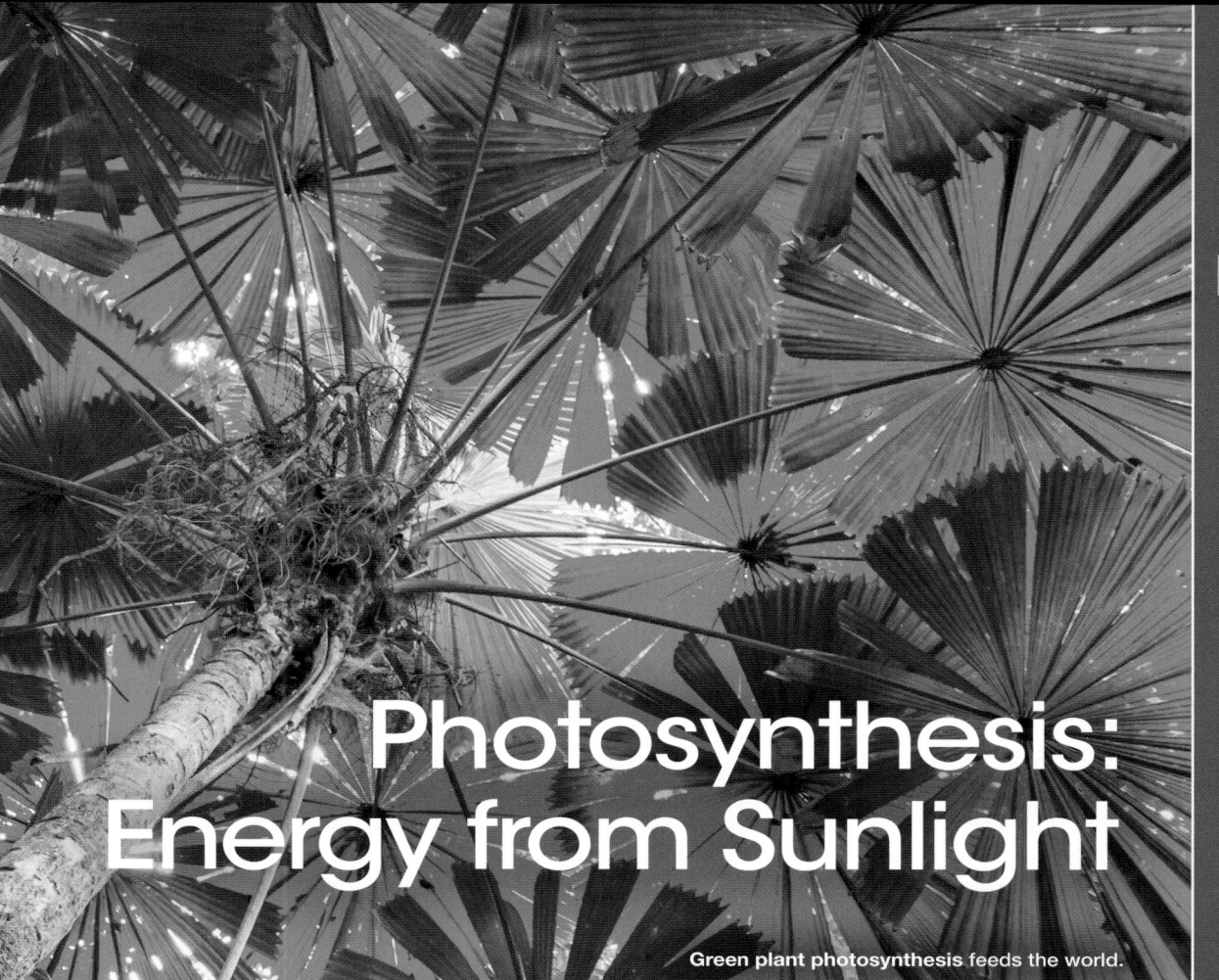

10

Photosynthesis: Energy from Sunlight

Green plant photosynthesis feeds the world.

investigating**life**

Saving FACE

In July 2014, a satellite was launched into space to measure photosynthesis over all of Earth. The Orbiting Carbon Observatory has instruments aboard that map not only the concentration of carbon dioxide, but also the presence of the green plant pigment chlorophyll. Why these measurements, and why now?

As you may know, green plants use the pigment chlorophyll in photosynthesis, which has the general equation

$$CO_2 + H_2O \xrightarrow{\text{Sunlight}} \text{carbohydrates} + O_2$$

The orbiting satellite is measuring chlorophyll, a reflection of the presence of green plants, and CO_2 over time because the global atmosphere and biological communities are changing. Over the past 200 years, the concentration of atmospheric CO_2 has increased—from 280 parts per million (ppm) in the year 1800 to 400 ppm in 2016—and the increase will probably continue for some time. Carbon dioxide is a "greenhouse gas" that traps heat in the atmosphere, and the rising CO_2 level is resulting in global climate change. Plant biologists are being asked two questions about the rise in CO_2: will it lead to an increased rate of photosynthesis, and if so, will it lead to increased plant growth?

To answer these questions, scientists developed a way to expose plants to high levels of CO_2 in the field. *Free-air concentration enrichment* (FACE) uses rings of pipes to release CO_2 into the air surrounding plants in fields or forests. Wind speed and direction are monitored by a computer, which constantly controls which pipes release CO_2. These experiments confirm that photosynthetic rates increase as the concentration of CO_2 rises, and indicate that as atmospheric CO_2 rises globally, there will be an increase in photosynthesis.

Will this increase in photosynthesis result in an increase in plant growth? Keep in mind that plants, like all organisms, use carbohydrates as an energy source. They perform cellular respiration with the general equation

$$\text{Carbohydrates} + O_2 \rightarrow CO_2 + H_2O + \text{energy}$$

The challenge facing plant biologists is to determine the balance between photosynthesis and respiration and how this affects the rate of plant growth. The FACE experiments indicate that plant growth and crop yields increase under higher CO_2 concentrations, suggesting that the overall increase in photosynthesis is greater than the increase in respiration.

Q What is the chemistry of photosynthesis, and how will it be affected by rising atmospheric CO_2?

10.1 Photosynthesis Uses Light to Make Carbohydrates

Recall that catabolism—the breakdown of complex molecules into simpler units—is the reverse of anabolism—the construction of complex molecules from simpler units. Chapter 9 described several energy-releasing catabolic pathways. The energy stored in chemical bonds in almost all organisms (with the exception of those living near deep-sea vents) ultimately comes from the sun. **Photosynthesis** (literally, "synthesis from light") is an anabolic process by which the energy of sunlight is captured and used to convert carbon dioxide (CO_2) into more complex carbon-containing compounds.

focus your learning

- Water molecules supply the protons and electrons needed to reduce carbon dioxide to carbohydrate in oxygenic photosynthesis.
- Photosynthesis occurs via two sequential reactions: light reactions followed by light-independent reactions.

Photosynthesis involves light and gas exchange

Plants, algae, and cyanobacteria live in aerobic environments and carry out oxygenic photosynthesis: the conversion of CO_2 and water (H_2O) into carbohydrates (which we will represent as a six-carbon sugar, $C_6H_{12}O_6$) and oxygen gas (O_2) (**Figure 10.1**).

$$6\ CO_2 + 6\ H_2O \rightarrow C_6H_{12}O_6 + 6\ O_2 \qquad (10.1)$$

Sugars, the organic products of **photosynthesis**, are transported throughout the plant body.

Sunlight

CO_2 H_2O O_2

CO_2 enters and O_2 and H_2O exit the leaves through pores on the leaf surface called stomata.

Sugars

Sugars

Leaf

Stem

H_2O

Root

Some kinds of bacteria live in anaerobic environments and carry out a type of photosynthesis in which energy from the sun is used to convert CO_2 to more complex molecules without the production of O_2. We will describe this process in more detail below, but first let's look at oxygenic photosynthesis.

Equation 10.1 describes an endergonic reaction. From experiments such as the ones described in the chapter-opening story using FACE, the role of CO_2 is well established. But while the equation is essentially correct, it is too general for a real understanding of the processes involved. Several questions arise: What are the precise chemical reactions of photosynthesis? What role does light play in these reactions? How do carbons become linked to form carbohydrates? What carbohydrates are formed? And does the oxygen gas come from CO_2 or H_2O?

Experiments with isotopes show that O_2 comes from H_2O in oxygenic photosynthesis

In 1941 Samuel Ruben and Martin Kamen at the University of California, Berkeley performed experiments using the isotopes ^{18}O and ^{16}O to identify the source of the O_2 produced during photosynthesis (**Investigating Life: What Is the Chemistry of Photosynthesis, and How Will Increasing CO_2 in the Atmosphere Affect It?**). Their results showed that all the oxygen gas produced during photosynthesis comes from water, as is reflected in the revised balanced equation:

$$6\ CO_2 + 12\ H_2O \rightarrow C_6H_{12}O_6 + 6\ O_2 + 6\ H_2O \qquad (10.2)$$

Water appears on both sides of the equation because it is both used as a reactant (the 12 molecules on the left) and released as a product (the 6 new molecules on the right). This revised equation accounts for all the water molecules needed for all the oxygen gas produced.

The realization that water was the source of photosynthetic O_2 led to an understanding of photosynthesis in terms of oxidation and reduction. As you learned in Chapter 9, oxidation–reduction (redox) reactions are coupled: when one molecule becomes oxidized in a reaction, another gets reduced. In this case, oxygen atoms in the reduced state in H_2O get oxidized to O_2:

$$12\ H_2O \rightarrow 24\ H^+ + 24\ e^- + 6\ O_2 \qquad (10.3)$$

while carbon atoms in the oxidized state in CO_2 get reduced to carbohydrate, with the simultaneous production of water:

$$6\ CO_2 + 24\ H^+ + 24\ e^- \rightarrow C_6H_{12}O_6 + 6\ H_2O \qquad (10.4)$$

Adding Equations 10.3 and 10.4 (chemistry students will recognize them as half-cell reactions) gives the overall Equation 10.2 shown above.

As you have just seen, water is the donor of protons and electrons in oxygenic photosynthesis. Earlier, we mentioned a type of photosynthesis that does not produce O_2. In these cases, other molecules are used

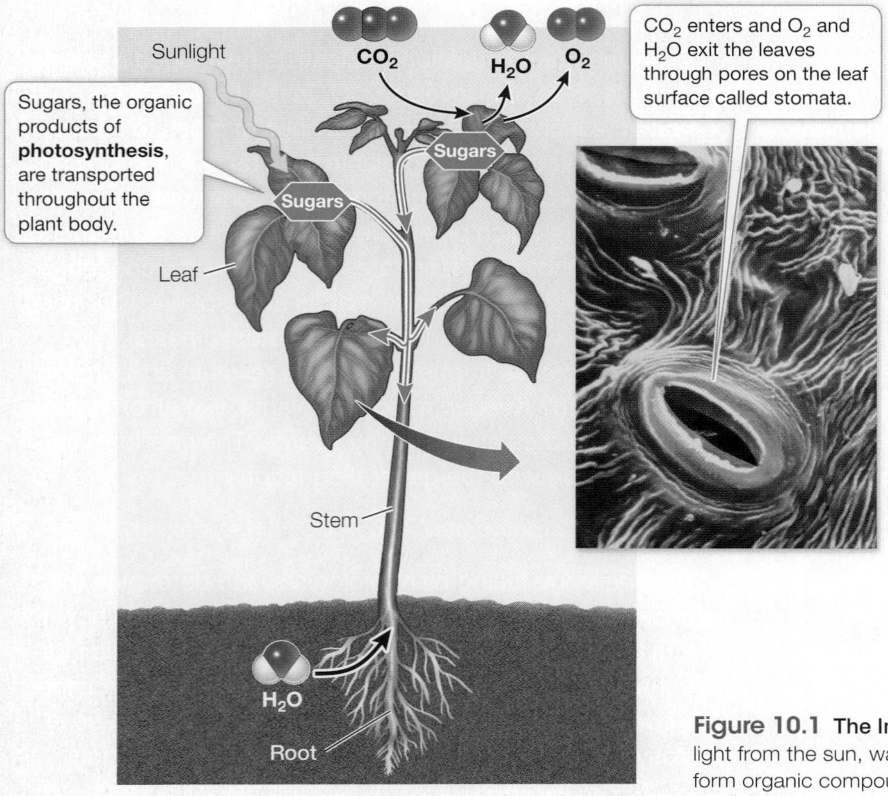

Figure 10.1 The Ingredients for Photosynthesis   A typical terrestrial plant uses light from the sun, water from the soil, and carbon dioxide from the atmosphere to form organic compounds by photosynthesis.

What is the Chemistry of Photosynthesis, and How Will Increasing CO_2 in the Atmosphere Affect It?

experiment

Original Paper: Ruben, S., M. Randall, M. D. Kamen and J. L. Hyde. 1941. Heavy oxygen (^{18}O) as a tracer in the study of photosynthesis. *Journal of the American Chemical Society* 63(3): 877–879.

Understanding the chemical reactions of photosynthesis is key to understanding the effects of increasing atmospheric CO_2. In particular, the source of O_2 was not known. Two possibilities were the reactants, CO_2 and H_2O. In two separate experiments, Samuel Ruben and Martin Kamen labeled the oxygen in these molecules with the isotope ^{18}O, then tested the O_2 produced by a green plant to find out which molecules contributed the oxygen.

HYPOTHESIS▶ The oxygen released by photosynthesis comes from water rather than CO_2.

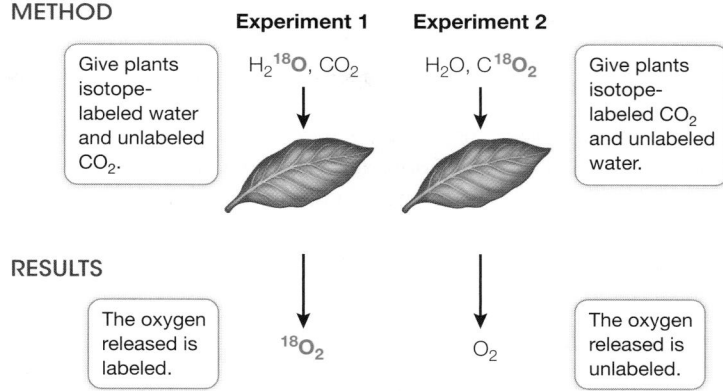

METHOD

Experiment 1   **Experiment 2**

Give plants isotope-labeled water and unlabeled CO_2.

$H_2{}^{18}O$, CO_2 H_2O, $C^{18}O_2$

Give plants isotope-labeled CO_2 and unlabeled water.

RESULTS

The oxygen released is labeled.

$^{18}O_2$ O_2

The oxygen released is unlabeled.

CONCLUSION▶ Water is the source of the oxygen atoms in the O_2 produced by photosynthesis.

work with the data

In the 1930s, Cornelius van Niel, then a graduate student at Stanford University, was the first to propose that the oxygen released during photosynthesis is not actually derived from carbon dioxide, but rather from the water molecules consumed in the reaction. This hypothesis was formed on the basis of the discovery that the anaerobic purple sulfur bacteria do not release oxygen during photosynthesis. Instead, these organisms convert hydrogen sulfide (H_2S) into elemental sulfur in their photosynthetic pathway (see Equation 10.5). This hypothesis was later confirmed by the experiment outlined above, which employed the "heavy" isotope of oxygen, ^{18}O, to trace the flow of oxygen in plants.

As part of the expanding research on radioisotopes during World War II, the U.S. government set up a radiation laboratory at the University of California, Berkeley. Out of this lab came key experiments that described the light-dependent and light-independent pathways of photosynthesis. In this set of experiments, *Chlorella* algal cells were exposed to water and CO_2, the latter coming from potassium carbonate (K_2CO_3) and potassium bicarbonate ($KHCO_3{}^-$), which dissolve in water to form CO_2. In Experiment 1, water had a higher isotopic ratio of $^{18}O/^{16}O$ than CO_2 and in Experiment 2, CO_2 had a higher isotopic ratio than water. A mass spectrometer was used to measure the isotopic contents of the reactants and the O_2 produced, and the data were presented as the isotopic ratio ($^{18}O/^{16}O$). These data are shown in the table.

QUESTIONS▶

1. In Experiment 1, was the isotopic ratio of O_2 similar to that of H_2O or to that of CO_2? What about in Experiment 2?
2. What can you conclude from these data?

	Time before start of O_2 collection (min)	Time at end of O_2 collection (min)	$^{18}O/^{16}O$ (proportion ^{18}O in compound)		
			H_2O	$HCO_3{}^- + CO_3{}^{2-}$ (CO_2 sources)	O_2
Experiment 1: 0.09 M $KHCO_3$ + 0.09 M K_2CO_3 (^{18}O in H_2O)	0		0.85	0.20	
	45	110	0.85	0.41	0.84
	110	223	0.85	0.55	0.85
	225	350	0.85	0.61	0.86
Experiment 2: 0.14 M $KHCO_3$ + 0.06 M K_2CO_3 (^{18}O in CO_2)	0		0.20		
	40	110	0.20	0.50	0.20
	110	185	0.20	0.40	0.20

Animation 10.1 The Source of the Oxygen Produced by Photosynthesis
www.Life11e.com/a10.1

A similar **work with the data** exercise may be assigned in **LaunchPad**.

as electron donors in the reduction of CO_2 to carbohydrates. For example, purple sulfur bacteria use hydrogen sulfide (H_2S) as the electron donor:

$$12\ H_2S + 6\ CO_2 + \text{light} \rightarrow C_6H_{12}O_6 + 6\ H_2O + 12\ S \qquad (10.5)$$

Green sulfur bacteria use sulfide ions, hydrogen, or ferrous iron as electron donors, whereas another group of bacteria use compounds derived from arsenic. The remainder of this chapter will focus on oxygenic photosynthesis, which produces the vast majority of the organic carbon used by life on Earth today and replenishes the O_2 in our atmosphere.

Photosynthesis involves two pathways

Equation 10.2 above summarizes the overall process of photosynthesis, but not the steps by which it happens. Water serves as the electron donor, but there is an intermediary carrier of the H^+ and electrons between the oxidation and reduction reactions. The carrier is the coenzyme nicotinamide adenine dinucleotide phosphate ($NADP^+$).

Like glycolysis and the other metabolic pathways that harvest energy in cells, photosynthesis has many steps. These reactions are commonly divided into two main pathways:

1. The **light reactions** convert light energy into chemical energy in the form of ATP and the reduced electron carrier NADPH. This molecule is similar to the coenzyme NADH (see Key Concept 9.1) but with an additional phosphate group attached to the sugar of its adenosine. In general, NADPH acts as a reducing agent in photosynthesis and other anabolic reactions.

2. The **light-independent reactions** (carbon-fixation reactions) do not use light directly, but instead use ATP, NADPH (made by the light reactions), and CO_2 to produce carbohydrate.

The light-independent reactions are sometimes called the dark reactions because they do not directly require light energy. They are also called the carbon-fixation reactions because by these reactions inorganic carbon is assimilated, or "fixed," into organic compounds. However, in most plants both the light reactions and the light-independent reactions *stop in the dark* because ATP synthesis and $NADP^+$ reduction require light. So to be accurate, we use the term light-independent reactions. The reactions of both pathways proceed within the chloroplast, but they are compartmentalized, occurring in different parts of that organelle (**Focus: Key Figure 10.2**).

As we describe these two series of reactions in more detail, you will see that they conform to the principles of biochemistry that we discussed in Chapters 8 and 9: energy transformations, oxidation–reduction, and the stepwise nature of biochemical pathways.

focus: key figure

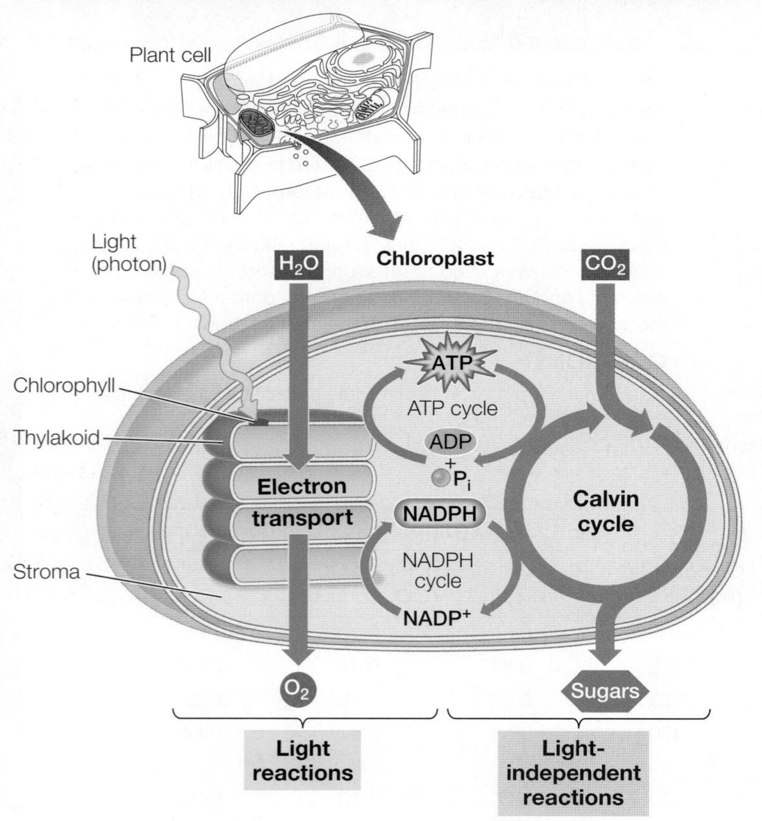

Figure 10.2 An Overview of Photosynthesis Photosynthesis consists of two pathways: the light reactions and the light-independent reactions. These reactions take place in the thylakoids and the stroma of chloroplasts, respectively. See Figure 5.12 for more on the structure of chloroplasts.

Q: In the cell, where does the reduction of CO_2 occur and what is the reducing agent?

10.1 recap

The light reactions of photosynthesis convert light energy into chemical energy. The light-independent reactions use that chemical energy to reduce CO_2 to carbohydrates. While most photosynthetic organisms use water as the electron donor for reduction of CO_2, some use other molecules, such as hydrogen sulfide (H_2S).

learning outcomes

You should be able to:

- Use experimental evidence to explain the source of the O_2 produced in oxygenic photosynthesis.
- Describe the relationship between the light reactions and the light-independent reactions of photosynthesis.

1. What is the relationship between the light reactions and the light-independent reactions of photosynthesis?
2. What is the experimental evidence that water is the source of the O_2 produced during photosynthesis?

We will describe the light reactions and the light-independent re-actions separately. We'll begin the next section by discussing the physical nature of light and the specific photosynthetic molecules that capture its energy.

key concept 10.2 Photosynthesis Converts Light Energy into Chemical Energy

Light is a form of energy, and it can be converted to other forms of energy such as heat or chemical energy. Our focus here will be on light as the source of energy to drive the formation of ATP (from ADP and P_i) and NADPH (from $NADP^+$ and H^+).

focus your learning

- Plants have pigment molecules that receive light energy and convert it to chemical energy in the form of excited electrons.
- Electron transport systems and two coordinated photosystems harvest the chemical energy of excited electrons to produce NADPH and ATP.

Light energy is absorbed by pigments in photosynthesis

It is helpful here to discuss light in terms of its photochemistry and photobiology.

PHOTOCHEMISTRY   Light is a form of **electromagnetic radiation**. Electromagnetic radiation is said to have a dual nature. Although it is propagated in waves, it also has particle-like behaviors. Particles of light can be described as packets of energy called **photons**, which have no mass. The amount of energy in electromagnetic radiation is inversely proportional to its **wavelength**—the shorter the wavelength, the greater the energy. The visible portion of the electromagnetic spectrum (**Figure 10.3**) encompasses a wide range of wavelengths and energy levels. In plants and other photosynthetic organisms, receptive molecules absorb photons and thereby harvest their energy for biological processes. These receptive molecules absorb only spe-cific wavelengths of light—photons with specific amounts of energy.

When a photon meets a molecule, one of three things can happen:

1. The photon may bounce off the mol-ecule—it may be scattered or reflected.

2. The photon may pass through the mol-ecule—it may be transmitted.

3. The photon may be absorbed by the molecule, adding energy to the molecule.

Neither of the first two outcomes causes any change in the mol-ecule. However, in the case of absorption, the photon disappears and its energy is absorbed by the molecule. The photon's energy cannot disappear, because according to the first law of thermody-namics, energy is neither created nor destroyed. When the molecule acquires the energy of the photon, it is raised from a ground state (with lower energy) to an excited state (with higher energy):

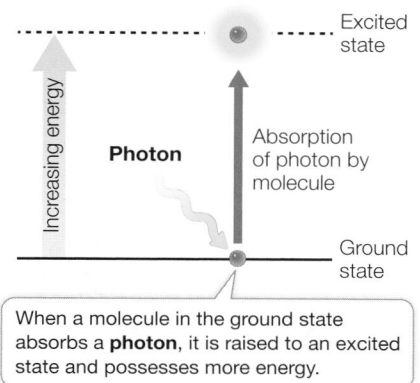

When a molecule in the ground state absorbs a **photon**, it is raised to an excited state and possesses more energy.

The difference in free energy between the molecule's excited state and its ground state is approximately equal to the free energy of the absorbed photon (a small amount of energy is lost to entropy, according to the second law of thermodynamics). The increase in energy boosts one of the electrons in the molecule into a *shell farther from its nucleus; this electron is now held less firmly, making the molecule unstable and more chemically reactive.

*connect the concepts   As discussed in Key Concept 2.1, the electrons orbiting an atom are distributed in a series of electron shells, or energy levels, around the nucleus. The far-ther an electron is from the nucleus, the higher its energy level.

PHOTOBIOLOGY   The specific wavelengths absorbed by a particu-lar molecule are characteristic of that type of molecule. Molecules that absorb wavelengths in the visible spectrum are called **pigments**.

When a beam of white light (containing all the wavelengths of visible light) falls on a pigment, certain wavelengths are absorbed. The remaining wavelengths are scattered or transmitted and make the pigment appear to us as colored. For example, the pigment

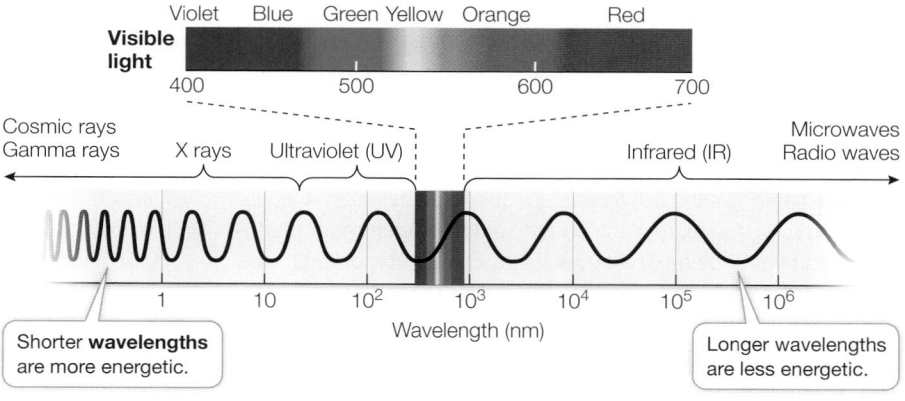

Figure 10.3 The Electromagnetic Spectrum The portion of the electromagnetic spectrum that is visible to humans as light is shown in detail at the top.

(A) The absorption spectra of *Anacharis*

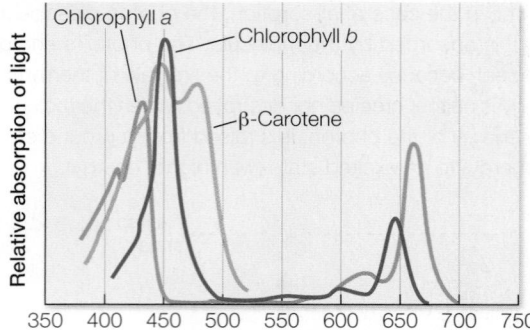

(B) The action spectrum for *Anacharis*

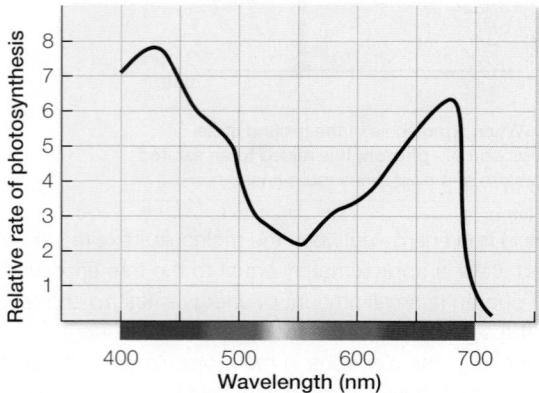

Figure 10.4 Absorption and Action Spectra **(A)** The absorption spectra of purified pigments from the common aquarium plant *Anacharis*. **(B)** The action spectrum for photosynthesis of that plant.

Q: Phycobilins are pigments that absorb yellow light. These pigments can transfer their absorbed energy to chlorophyll. How does this occur, thermodynamically?

chlorophyll absorbs both blue and red light, and we see the remaining light, which is primarily green. If we plot light absorbed by a purified pigment against wavelength, the result is an **absorption spectrum** for that pigment.

In contrast to the absorption spectrum, an **action spectrum** is a plot of the rate of photosynthesis carried out by an organism against the wavelengths of light to which it is exposed. An action spectrum for photosynthesis can be determined as follows:

1. Place the organism in a closed container.
2. Expose it to light of a certain wavelength for a period of time.
3. Measure the rate of photosynthesis by the amount of O_2 released.
4. Repeat with light of other wavelengths.

Figure 10.4 shows the absorption spectra of pigments in *Anacharis*, a common aquarium plant, and the action spectrum for photosynthesis by that plant. The two spectra can be compared to see which pigments in *Anacharis* contribute most to light harvesting for photosynthesis.

The major pigment used to drive the light reactions of oxygenic photosynthesis is chlorophyll *a*. (In Figure 10.4 you can see that the wavelengths at which photosynthesis is highest in *Anacharis* are the

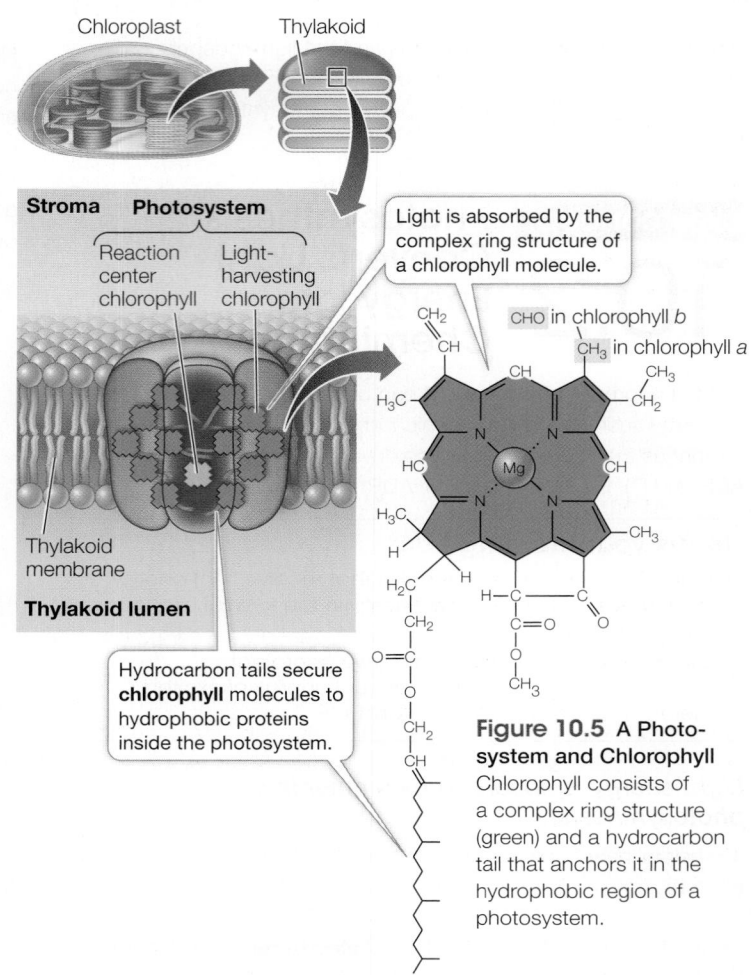

Figure 10.5 A Photosystem and Chlorophyll Chlorophyll consists of a complex ring structure (green) and a hydrocarbon tail that anchors it in the hydrophobic region of a photosystem.

same wavelengths at which chlorophyll *a* absorbs the most light.) Chlorophyll *a* has a complex ring structure (**Figure 10.5**), similar to that of the heme group of hemoglobin, with a magnesium ion at the center. A long hydrocarbon "tail" anchors the molecule to proteins within a large multi-protein complex called a **photosystem**, which spans the thylakoid membrane. Molecules of chlorophyll *a* and various accessory pigments (such as chlorophylls *b* and *c*, carotenoids, and phycobilins; see below) are arranged into **light-harvesting complexes**, also called antenna systems. Multiple light-harvesting complexes surround a single **reaction center** within the photosystem.

Light energy is captured by the light-harvesting complexes and transferred to the reaction center, where chlorophyll *a* molecules participate in redox reactions that convert the light energy to chemical energy.

Chlorophyll absorbs blue and red light, which are near the two ends of the visible spectrum (see Figure 10.3). The various accessory pigments absorb light in other parts of the spectrum and thus function to broaden the range of wavelengths that can be used for photosynthesis. You can see this in the action spectrum in Figure 10.4B: *Anacharis* is able to photosynthesize at wavelengths of light that chlorophyll *a* doesn't absorb but that other pigments absorb (e.g., 500 nanometers [nm]). Different photosynthetic organisms have different combinations of accessory pigments. Higher plants and green algae have chlorophyll *b* (with a structure and absorption spectrum very similar to that of chlorophyll *a*) and carotenoids such as

Figure 10.6 Energy Transfer and Electron Transport **(A)** The molecular structure of a single light-harvesting complex shows the polypeptide in brown with three helices that span the thylakoid membrane. Pigment molecules (carotenoids and chlorophylls *a* and *b*) are bound to the polypeptide. **(B)** This simplified illustration of the entire photosystem uses chlorophyll molecules to represent the light-harvesting complexes. Energy from a photon is transferred from one pigment molecule to another, until it reaches a chlorophyll *a* molecule in the reaction center. The chlorophyll *a* molecule can give up its excited electron to an electron acceptor.

β-carotene, which absorb photons in the blue and blue-green wavelengths. Phycobilins, which are found in red algae and cyanobacteria, absorb various yellow-green, yellow, and orange wavelengths.

Light absorption results in photochemical change

When a pigment molecule absorbs light, it enters an excited state. This is an unstable situation, and the molecule rapidly returns to its ground state, releasing most of the absorbed energy. So rapid is this process that it is measured in picoseconds (trillionths of a second). Within the antennae system of a photosystem (**Figure 10.6A**), the energy released by a pigment molecule (for example, chlorophyll *b*) is absorbed by other, adjacent pigment molecules. The energy (not as electrons but in the form of chemical energy called resonance) is passed from molecule to molecule until it reaches a chlorophyll *a* molecule at the reaction center of the photosystem (**Figure 10.6B**).

A ground-state chlorophyll *a* molecule at the reaction center (symbolized by Chl) absorbs the energy from the adjacent chlorophylls and becomes excited (Chl*), but to return to the ground state this chlorophyll does not pass the energy to another pigment molecule—something very different occurs. *The reaction center converts the absorbed light energy into chemical energy* (**Figure 10.7**). The chlorophyll molecule in the reaction center absorbs sufficient energy that it actually *gives up its excited electron to a chemical acceptor*:

$$\text{Chl*} + \text{acceptor} \rightarrow \text{Chl}^+ + \text{acceptor}^- \qquad (10.6)$$

This, then, is the first consequence of light absorption by chlorophyll: *the reaction center chlorophyll (Chl*) loses its excited electron in a redox reaction and becomes Chl$^+$.* As a result of this transfer of an electron, the chlorophyll gets oxidized, while the acceptor molecule is reduced.

Reduction leads to ATP and NADPH formation

The electron acceptor that is reduced by Chl* is the first in a chain of electron carriers in the thylakoid membrane. Electrons are passed from one carrier to another in an energetically "downhill" series of reductions and oxidations. The final electron acceptor is NADP$^+$, which gets reduced:

$$\text{NADP}^+ + \text{H}^+ + 2\,\text{e}^- \rightarrow \text{NADPH} \qquad (10.7)$$

As in mitochondria, ATP is produced chemiosmotically in the thylakoid membrane during the process of photophosphorylation, which we will illustrate shortly. The overall process involves two *electron transport processes, noncylic and cylic. The **noncyclic electron transport** reactions that use the energy from light to generate ATP and NADPH are illustrated in Figure 10.7.

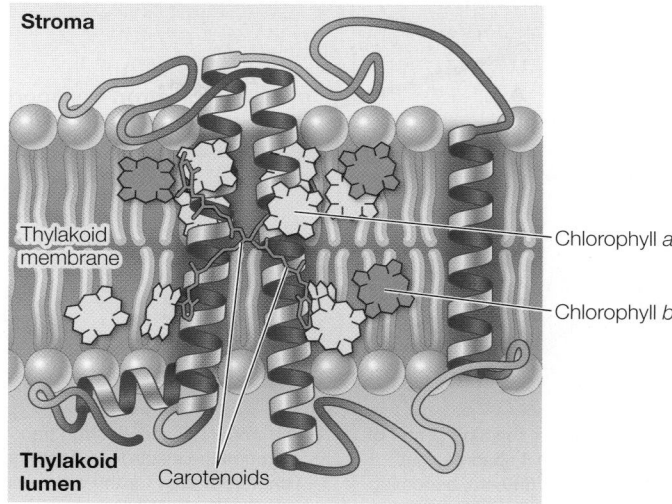

(A) A light-harvesting complex

Stroma

Thylakoid membrane

Thylakoid lumen

Chlorophyll *a*

Chlorophyll *b*

Carotenoids

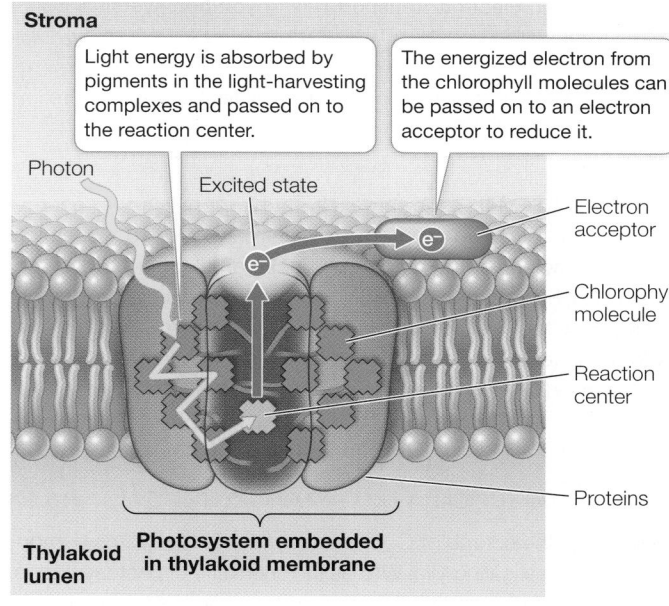

(B) A phototsystem

Stroma

Light energy is absorbed by pigments in the light-harvesting complexes and passed on to the reaction center.

The energized electron from the chlorophyll molecules can be passed on to an electron acceptor to reduce it.

Photon

Excited state

Electron acceptor

Chlorophyll molecule

Reaction center

Proteins

Thylakoid lumen

Photosystem embedded in thylakoid membrane

---

*connect the concepts The thylakoid membrane has an electron transport system similar to the respiratory chain of mitochondria (see Key Concept 9.3).

As seen in Figure 10.7, two coordinated photosystems, each with its own reaction center, collaborate to produce ATP and NADPH:

1. **Photosystem I** (containing the "P$_{700}$" chlorophylls at its reaction center) absorbs light energy best at 700 nm and passes its excited electrons (via intermediate molecules) to NADP$^+$, reducing it to NADPH.

2. **Photosystem II** (with "P$_{680}$" chlorophylls at its reaction center) absorbs light energy best at 680 nm, oxidizes water molecules, and passes its energized electrons through a series of carriers to produce ATP.

Let's look in more detail at these photosystems, beginning with photosystem II.

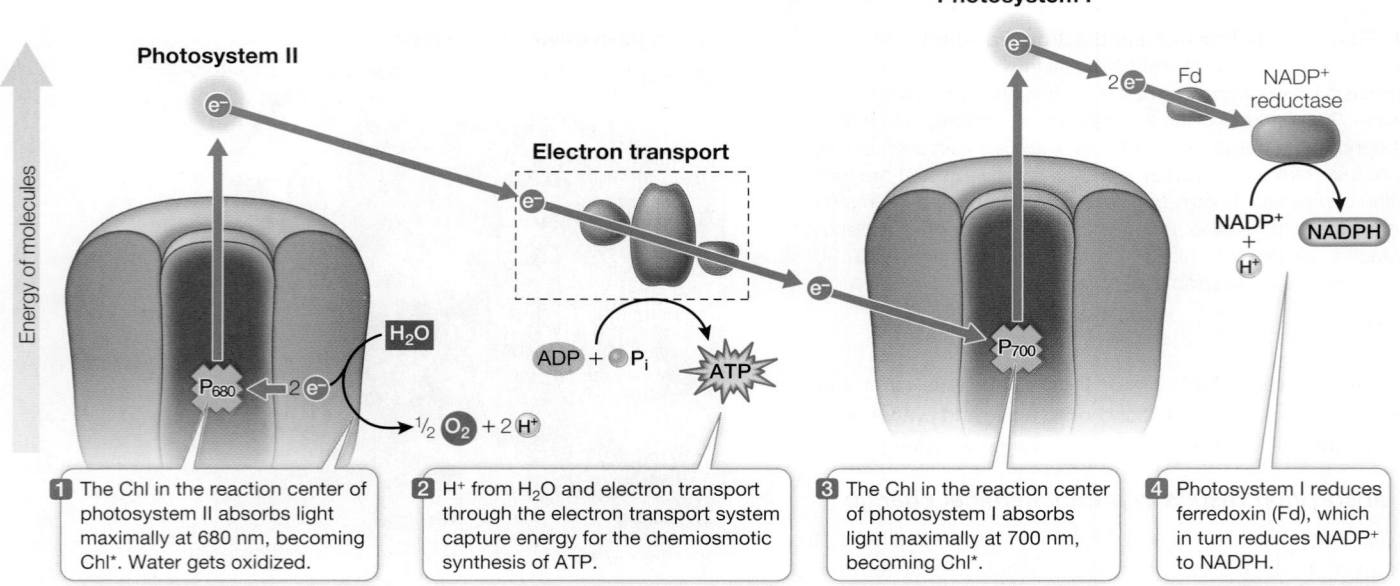

Figure 10.7 Noncyclic Electron Transport Uses Two Photosystems
Absorption of light energy by chlorophyll molecules in the reaction centers of photosystems I and II allows them to pass electrons into a series of redox reactions.

Q: An herbicide can act as an electron acceptor, becoming reduced by ferredoxin (Fd). What effect would this have on a plant?

PHOTOSYSTEM II After an excited chlorophyll in the reaction center (Chl*) gives up its energetic electron to reduce a chemical acceptor molecule, the chlorophyll lacks an electron and is very unstable. It has a strong tendency to obtain an electron from another molecule to replace the one it lost—in chemical terms, it is a strong oxidizing agent. The replenishing electrons come from water, splitting the H—O—H bonds:

$$H_2O \rightarrow \tfrac{1}{2} O_2 + 2 H^+ + 2 e^- \qquad (10.8)$$

$$2 e^- + 2 Chl^+ \rightarrow 2 Chl \qquad (10.9)$$

$$\text{Overall: } 2 Chl^+ + H_2O \rightarrow 2 Chl + 2 H^+ + \tfrac{1}{2} O_2 \qquad (10.10)$$

Notice that the source of O_2 in photosynthesis is H_2O (as demonstrated in Investigating Life: What is the chemistry of photosynthesis, and how will increasing CO_2 in the atmosphere affect it?).

Back to the electron acceptor in the electron transport system: the energetic electrons are passed through a series of membrane-bound carriers to a final acceptor at a lower energy level. As in the mitochondrion, a proton gradient is generated and is used by ATP synthase to make ATP (see below).

PHOTOSYSTEM I In photosystem I, an excited electron from the Chl* at the reaction center reduces an acceptor. The oxidized chlorophyll (Chl+) now obtains an electron, but in this case the electron comes from the last carrier in the electron transport system. This links the two photosystems chemically. They are also linked spatially, with the two photosystems adjacent to one another in the thylakoid membrane. The energetic electrons from photosystem I pass through several molecules and end up reducing NADP+ to NADPH.

Next in the process of harvesting light energy to produce carbohydrates is the series of carbon-fixation reactions. These reactions require more ATP than NADPH. If the pathway we just

described—the linear or noncyclic pathway—were the only set of light reactions operating, there might not be sufficient ATP for carbon fixation. **Cyclic electron transport** makes up for this imbalance. This pathway uses photosystem I and the electron transport system to produce ATP but not NADPH; it is cyclic because an electron is passed from an excited chlorophyll and recycles back to the same chlorophyll (**Figure 10.8**).

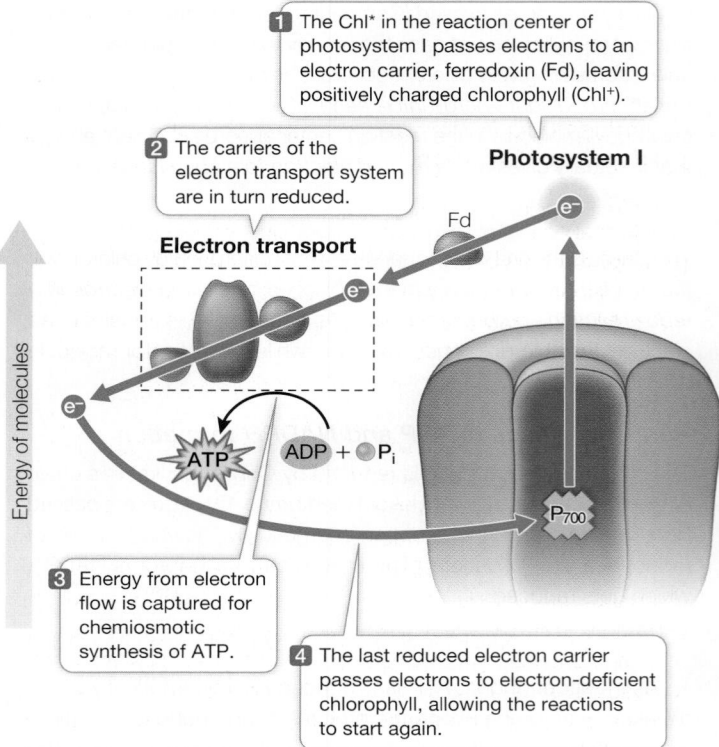

Figure 10.8 Cyclic Electron Transport Traps Light Energy as ATP

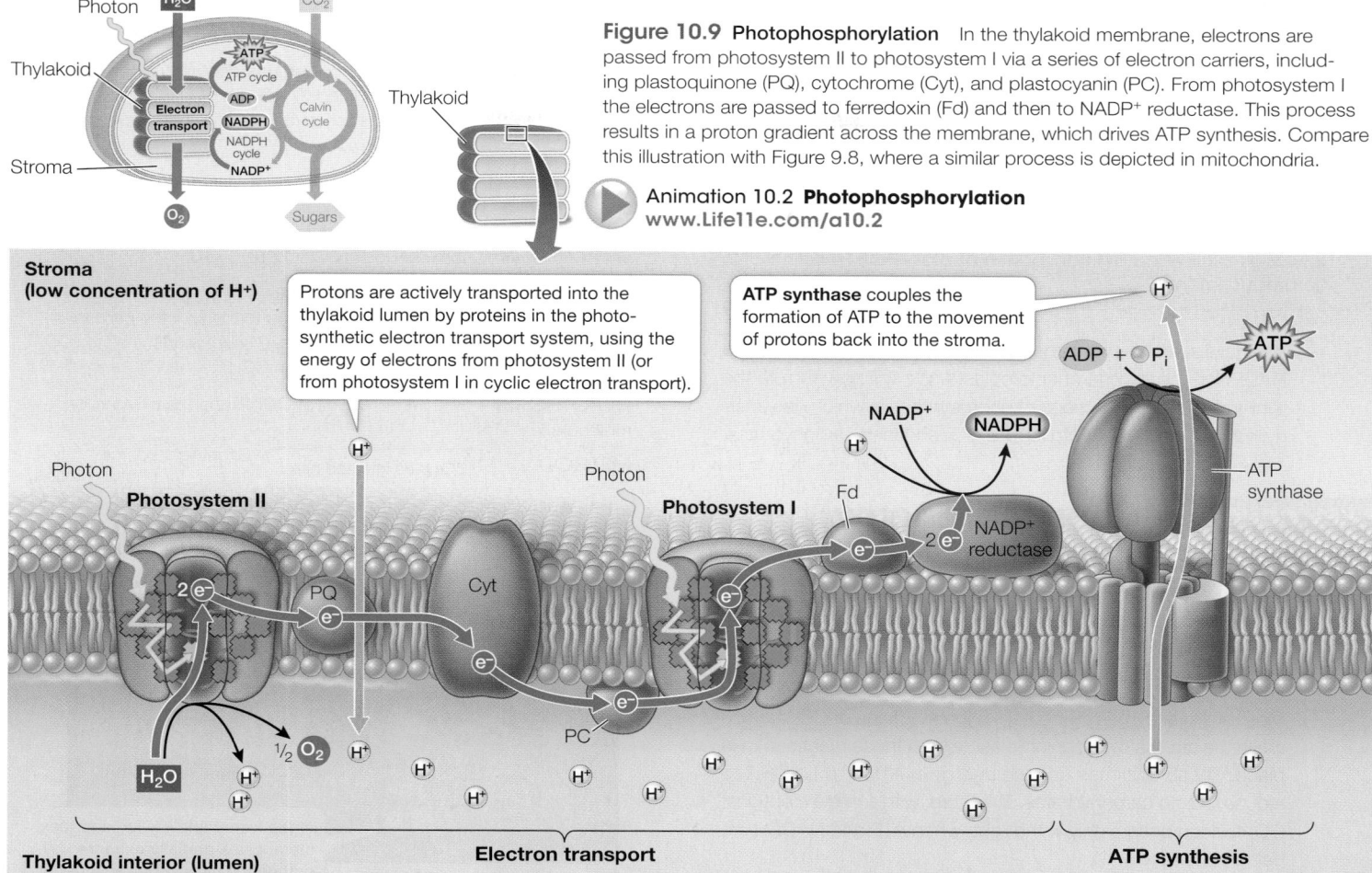

Figure 10.9 Photophosphorylation In the thylakoid membrane, electrons are passed from photosystem II to photosystem I via a series of electron carriers, including plastoquinone (PQ), cytochrome (Cyt), and plastocyanin (PC). From photosystem I the electrons are passed to ferredoxin (Fd) and then to $NADP^+$ reductase. This process results in a proton gradient across the membrane, which drives ATP synthesis. Compare this illustration with Figure 9.8, where a similar process is depicted in mitochondria.

Animation 10.2 **Photophosphorylation**
www.Life11e.com/a10.2

Chemiosmosis is the source of the ATP produced in photophosphorylation

In Chapter 9 you learned about the chemiosmotic mechanism for ATP formation in the mitochondrion. A similar mechanism, called **photophosphorylation**, operates in the chloroplast, where electron transport is coupled to the transport of protons (H^+) across the thylakoid membrane, resulting in a proton gradient across the membrane (**Figure 10.9**).

The electron carriers in the thylakoid membrane are oriented so that protons are transferred from the stroma into the lumen of the thylakoid. Thus the lumen becomes more acidic (higher concentration of protons) compared with the stroma, resulting in an electrochemical gradient across the thylakoid membrane, whose bilayer is not permeable to H^+. Water oxidation creates more H^+ in the thylakoid lumen, and $NADP^+$ reduction removes H^+ in the stroma. Both reactions contribute to the H^+ gradient. The high concentration of H^+ in the thylakoid space drives the movement of H^+ back into the stroma through protein channels in the membrane. These channels are also enzymes—ATP synthases—that couple the movement of protons to the formation of ATP, as they do in mitochondria (see Figure 9.8). Indeed, the amino acid sequence of chloroplast ATP synthase is about 60 percent identical to human mitochondrial ATP synthase—a remarkable similarity, given that plants and animals had

their most recent common ancestor more than a billion years ago. This is testimony to the evolutionary unity of life.

The mechanisms of the two enzymes are similar, but their orientations differ. In chloroplasts, protons flow through the ATP synthase out of the thylakoid lumen into the stroma (where the ATP is synthesized). In mitochondria, the protons flow out of the intermembrane space into the mitochondrial matrix.

▶ 10.2 recap

Conversion of light energy into chemical energy occurs when photons are absorbed by chlorophylls and accessory pigments in the light-harvesting complex within the chloroplast. Light energy is used to drive a series of protein-associated redox reactions in the thylakoid membranes of the chloroplast. In the process of photophosphorylation, two linked photosystems establish a proton gradient across the membrane that drives ATP synthesis.

learning outcomes

You should be able to:

• Describe the transfer of energy within the reaction center of a light-harvesting complex.

(continued)

10.2 recap (continued)

- Compare and contrast a pigment's absorption spectrum and its action spectrum.
- Explain how ATP and NADPH are produced in a chloroplast.

1. How can a pigment molecule lose the energy of an absorbed photon?
2. What is the difference between an absorption spectrum and an action spectrum?
3. How does cyclic electron transport in photosystem I result in the production of ATP?

You have seen how light energy drives the synthesis of ATP and NADPH in the stroma of chloroplasts. We will now turn to the light-independent reactions of photosynthesis, which use energy-rich ATP and NADPH to reduce CO_2 and form carbohydrates.

key concept 10.3 Chemical Energy Trapped in Photosynthesis Is Used to Synthesize Carbohydrates

Most of the enzymes that catalyze the reactions of CO_2 fixation are in the stroma of the chloroplast, where those reactions take place. These enzymes use the energy in ATP and NADPH to reduce CO_2 to carbohydrates. Therefore, with some exceptions, CO_2 fixation occurs only in the light, when ATP and NADPH are being generated.

focus your learning

- The Calvin cycle describes the chemical steps that produce carbohydrates from carbon dioxide in the stroma of the chloroplast.
- Light affects photosynthetic carbon fixation.

How were the steps in carbohydrate synthesis elucidated?

To identify the reactions by which the carbon from CO_2 ends up in carbohydrates, scientists found a way to label CO_2 so they could isolate and identify the compounds formed from it during photosynthesis. In the 1950s, Melvin Calvin, Andrew Benson, and their colleagues used radioactively labeled CO_2 in which some of the carbon atoms were the radioisotope ^{14}C rather than the normal ^{12}C. They were able to trace the chemical pathway of CO_2 fixation (**Figure 10.10**). The first molecule that appeared in the pathway was a three-carbon sugar phosphate called 3-phosphoglycerate (3PG) (the ^{14}C is shown in red):

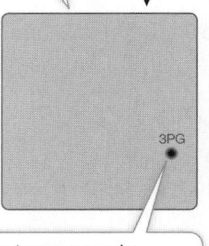

3-Phosphoglycerate (3PG)

Carboxyl group

---

experiment

Figure 10.10A   Tracing the Pathway of CO_2

Original Papers: Calvin and his colleagues described their experiments in a series of 26 papers titled "The Path of Carbon in Photosynthesis." Perhaps the most important was one that showed how labeled CO_2 could be used a tracer:

Benson, A. A., J. A. Bassham, M. Calvin, T. C. Goodale, V. A. Haas and W. Stepka. 1950. The path of carbon in photosynthesis. V. Paper chromatography and radioautography of the products. *Journal of the American Chemical Society* 72: 1710–1718.

How is CO_2 incorporated into carbohydrate during photosynthesis? What is the first stable covalent linkage that forms with the carbon of CO_2? Melvin Calvin and his colleagues used short exposures to $^{14}CO_2$ to identify the first compound formed from CO_2.

HYPOTHESIS▶ The first product of CO_2 fixation is a 3-carbon molecule.

METHOD

$^{14}CO_2$ was injected here.

Bright light source (energy for photosynthesis)

Algae were rapidly killed and their metabolites partially extracted by putting the cells in boiling ethanol.

Thin flask of green algae

The algal extract was applied here as a spot and run in two directions to separate compounds from one another.

First run

Second run

Paper chromatogram

After separation of the compounds, the chromatogram was overlaid with X-ray film. Radioactive compounds show up as dark spots.

RESULTS

GLUT
ALA
GLY SER
ASP CIT
SUC G3P
3PG
HEXOSE-P
3PG

A chromatogram made after 3 seconds of exposure to $^{14}CO_2$ shows ^{14}C only in 3PG (3-phosphoglycerate).

A chromatogram made after 30 seconds of exposure to $^{14}CO_2$ shows ^{14}C in many molecules.

CONCLUSION▶ The initial product of CO_2 fixation is 3PG. Later, the carbon from CO_2 ends up in many molecules.

Figure 10.10B **work with the data** follows on the next page.

work with the data

Figure 10.10B Tracing the Pathway of CO_2

Original Paper: Benson, A. A. et al. 1950.

To elucidate the sequence of reactions that allow carbon fixation, Melvin Calvin and colleagues exposed suspensions of the green alga *Chlorella* to $^{14}CO_2$ (see Figure 10.10A). After 3 sec of photosynthesis, the ^{14}C from $^{14}CO_2$ was found only in 3-phosphoglycerate (3PG), but after 30 sec many compounds were radioactive. Calvin and his colleagues then expanded on these results and were able to determine the exact sequence of reactions and reaction intermediates in the Calvin cycle by exposing the cells to $^{14}CO_2$ for various periods of time.

The first reaction in CO_2 fixation can occur in the dark. To show this, Calvin and his colleagues exposed *Chlorella* cells to $^{14}CO_2$ under bright lights for 20 min and harvested the cells. Then they repeated the experiment, but following the 20 min of light with varying periods of darkness (30 sec, 2 min, and 5 min). They harvested cells and made chromatograms to identify the labeled compounds, and used a radioactivity detector to quantify the amount of ^{14}C in each compound. The data are shown in the table.

Compound		Relative amount of radioactivity after:		
	20 min light	20 min light + 30 sec dark	20 min light + 2 min dark	20 min light + 5 min dark
3PG	5,500	10,100	10,000	5,200
RuBP	4,900	680	1,850	1,800
Sucrose	13,000	13,500	15,000	14,750

QUESTIONS▶

1. Plot radioactivity in 3PG versus time. What do the data show?
2. Why did the amount of radioactively labeled RuBP decline after 30 seconds in the dark?

 Animation 10.3 **Tracing the Pathway of CO_2**
www.Life11e.com/a10.3

A similar **work with the data** exercise may be assigned in **LaunchPad**.

Using successively longer exposures to $^{14}CO_2$, Calvin and his colleagues were able to trace the route of ^{14}C as it moved through a series of compounds, including monosaccharides and amino acids. It turned out the ^{14}C moved through a cyclical pathway. In this cycle, the CO_2 initially bonds covalently to a five-carbon acceptor molecule. The resulting six-carbon intermediate quickly breaks into two three-carbon molecules. As the cycle repeats, a carbohydrate is produced and the initial CO_2 acceptor is regenerated. This pathway was appropriately named the **Calvin cycle**.

The initial reaction in the Calvin cycle adds the one-carbon CO_2 to the five-carbon acceptor molecule ribulose 1,5-bisphosphate (RuBP). The product is an intermediate six-carbon compound, which quickly breaks down and forms two molecules of 3PG (**Figure 10.11**). The intermediate compound is broken down so rapidly that Calvin did not observe radioactive carbon appearing in it first. The enzyme that catalyzes its formation, **ribulose bisphosphate carboxylase/oxygenase** (**rubisco**), is the most abundant protein in the world! It constitutes up to 50 percent of all the protein in every plant leaf.

The Calvin cycle is made up of three processes

The Calvin cycle uses the ATP and NADPH made in the light to reduce CO_2 in the stroma to a carbohydrate. As in all biochemical pathways, each reaction is catalyzed by a specific enzyme. The cycle is composed of three distinct processes (**Figure 10.12**):

1. *Fixation* of CO_2. As we have seen, this reaction is catalyzed by rubisco, and its stable product is 3PG.
2. *Reduction* of 3PG to form glyceraldehyde 3-phosphate (G3P). This series of reactions involves a phosphorylation (using the ATP made in the light reactions) and a reduction (coupled to the oxidation of NADPH made in the light reactions).
3. *Regeneration* of the CO_2 acceptor, RuBP. Most of the G3P ends up as ribulose monophosphate (RuMP), and ATP is used to convert this compound into RuBP. So for every "turn" of the cycle, one CO_2 is fixed and one CO_2 acceptor is regenerated.

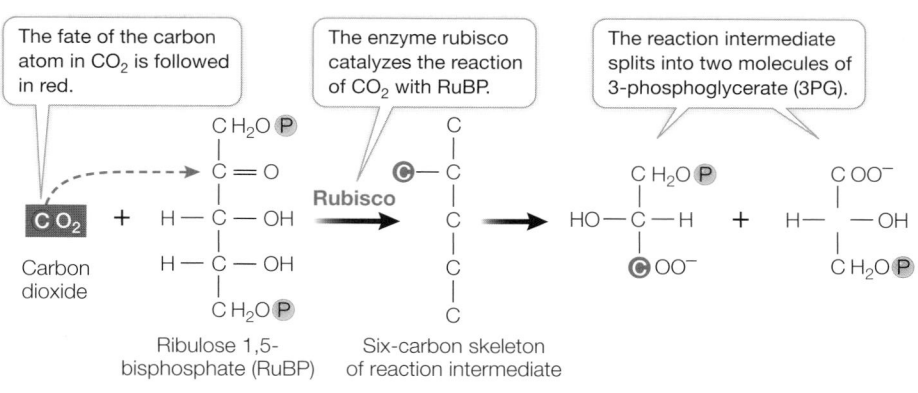

Figure 10.11 RuBP Is the Carbon Dioxide Acceptor CO_2 is added to a five-carbon compound, RuBP. The resulting six-carbon compound immediately splits into two molecules of 3PG.

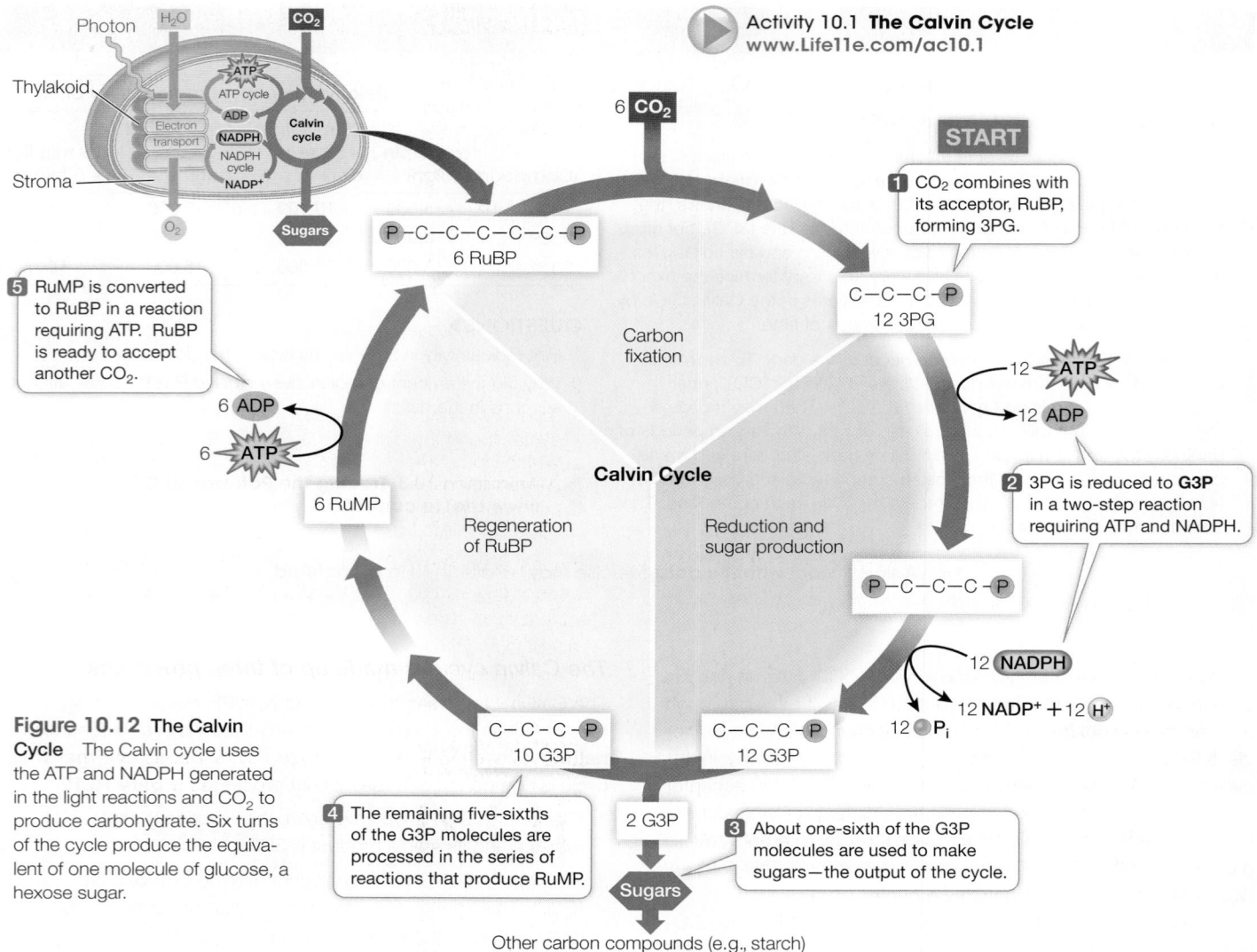

Activity 10.1 **The Calvin Cycle**
www.Life11e.com/ac10.1

Figure 10.12 The Calvin Cycle The Calvin cycle uses the ATP and NADPH generated in the light reactions and CO_2 to produce carbohydrate. Six turns of the cycle produce the equivalent of one molecule of glucose, a hexose sugar.

The product of this cycle is **glyceraldehyde 3-phosphate** (**G3P**), which is a three-carbon sugar phosphate, also called triose phosphate:

Glyceraldehyde 3-phosphate (G3P)

In a typical leaf, five-sixths of the G3P is recycled into RuBP. There are two fates for the remaining G3P, depending on the time of day and the needs of different parts of the plant:

1. Some of the G3P is exported out of the chloroplast to the cytoplasm, where it is converted to hexoses (glucose and fructose). These molecules may be used in glycolysis and mitochondrial respiration to power the activities of photosynthetic cells (see Chapter 9) or they may be converted into the disaccharide sucrose, which is transported out of the leaf to other organs in the plant. There the sucrose is hydrolyzed to its constituent monosaccharides, which can be used as sources of energy or as building blocks for other molecules.

2. Some of the G3P is used to synthesize glucose inside the chloroplast. As the day wears on, glucose molecules accumulate and are linked together to form the polysaccharide starch. This stored carbohydrate can then be drawn on during the night so that the photosynthetic tissues can continue to export sucrose to the rest of the plant, even when photosynthesis is not taking place. In addition, starch is abundant in nonphotosynthetic organs such as roots, underground stems, and seeds, where it provides glucose to fuel cellular activities, including plant growth.

The plant uses the carbohydrates produced in photosynthesis to make other molecules, including amino acids, lipids, and the building blocks of nucleic acids—in fact, all the organic molecules in the plant.

The products of the Calvin cycle are of crucial importance to Earth's entire biosphere. For the majority of living organisms on Earth, the

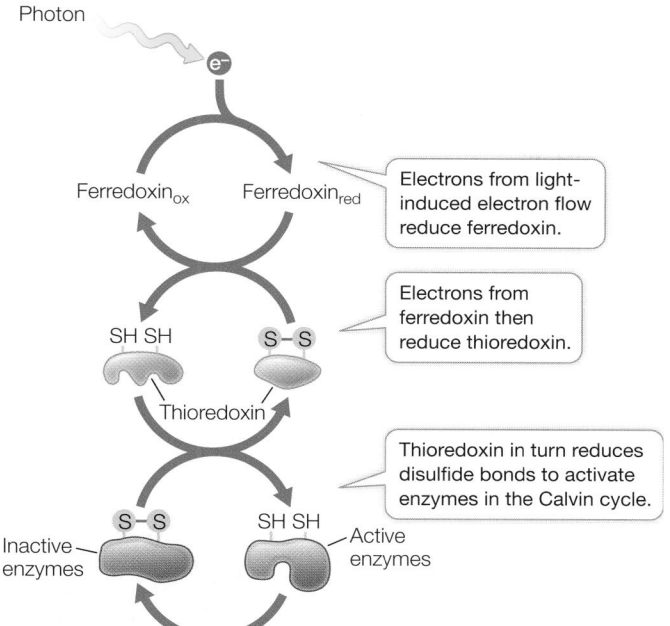

Figure 10.13 The Photochemical Reactions Stimulate the Calvin Cycle By reducing (breaking) disulfide bridges, electrons from the light reactions activate enzymes in CO_2 fixation.

Electrons from light-induced electron flow reduce ferredoxin.

Electrons from ferredoxin then reduce thioredoxin.

Thioredoxin in turn reduces disulfide bonds to activate enzymes in the Calvin cycle.

C—C and C—H covalent bonds generated by the cycle provide almost all of the energy for life. Photosynthetic organisms, which are also called **autotrophs** ("self-feeders"), release most of this energy by glycolysis and cellular respiration, and use it to support their own growth, development, and reproduction. But plants are also the source of energy for other organisms. Much plant matter ends up being consumed by **heterotrophs** ("other-feeders"), such as animals, including you, which cannot photosynthesize. Heterotrophs depend on autotrophs for both raw materials and energy. Free energy is released from food by glycolysis and cellular respiration in the cells of heterotrophs.

Light stimulates the Calvin cycle

As you have seen, the Calvin cycle uses NADPH and ATP, which are generated using energy from light. Two other processes connect the light reactions with this CO_2 fixation pathway. Both connections are indirect but significant:

1. *Light-induced pH changes in the stroma activate some Calvin cycle enzymes.* Proton transfer from the stroma into the thylakoid lumen causes an increase in the pH of the stroma from 7 to 8 (a tenfold decrease in H^+ concentration) that favors the activation of rubisco.

2. *Light-induced electron transport reduces disulfide bridges in four of the Calvin cycle enzymes, thereby activating them* (**Figure 10.13**). When ferredoxin is reduced in photosystem I (see Figure 10.7), it passes some electrons to a small, soluble protein called thioredoxin, and this protein passes electrons to four enzymes in the CO_2 fixation pathway. Reduction of the sulfurs in the disulfide bridges of these enzymes (see Figure 3.5) forms SH groups and breaks the bridges. The resulting changes in their three-dimensional shapes activate the enzymes and increase the rate at which the Calvin cycle operates.

10.3 recap

ATP and NADPH produced in the light reactions power the synthesis of carbohydrates by the Calvin cycle. This cycle fixes CO_2, reduces it, and regenerates the acceptor, RuBP, for further fixation.

learning outcomes

You should be able to:
- Describe the experiment that led to identification of the steps in the Calvin cycle.
- Explain how light reactions stimulate the Calvin cycle.

1. What experiments led to the identification of 3PG as the initial product of carbon fixation?
2. In what ways does light stimulate the Calvin cycle?
3. Both photosynthetic electron transport and the Calvin cycle stop in the dark. Which specific reaction stops first? Which stops next? Continue answering the question "Which stops next?" until you have explained why both pathways have stopped.

Although all green plants carry out the Calvin cycle, some plants have evolved variations on, or additional steps in, the light-independent reactions. These variations and additions have permitted plants to adapt to and thrive in certain environmental conditions. Let's look at these environmental limitations and the metabolic bypasses that have evolved to circumvent them.

key concept 10.4   Plants Have Adapted Photosynthesis to Environmental Conditions

In addition to fixing CO_2 during photosynthesis, rubisco can react with O_2. This reaction, which leads to a process called photorespiration, lowers the overall rate of CO_2 fixation in some plants. After examining this problem, we'll look at some biochemical pathways and features of plant anatomy that compensate for the limitations of rubisco.

focus your learning

- C_4 plants have evolved to avoid photorespiration under hot, dry conditions, whereas C_3 plants have not.
- CAM plants differ from C_4 plants by separating carbon fixation from the Calvin cycle according to different times of the day.

How do some plants overcome the limitations of CO_2 fixation?

As its full name indicates, rubisco (ribulose bisphosphate carboxylase/oxygenase) is an **oxygenase** as well as a **carboxylase**—it can add O_2 or CO_2, respectively, to the acceptor molecule RuBP:

- Rubisco is a carboxylase when it adds CO_2 to RuBP.
- Rubisco is an oxygenase when it adds O_2 to RuBP.

(A) The organelles of photorespiration

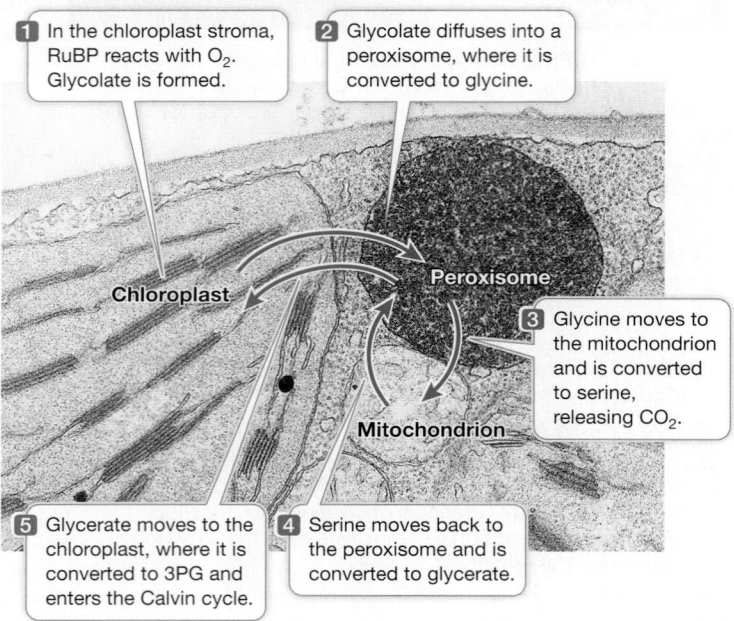

1 In the chloroplast stroma, RuBP reacts with O_2. Glycolate is formed.

2 Glycolate diffuses into a peroxisome, where it is converted to glycine.

Chloroplast

Peroxisome

3 Glycine moves to the mitochondrion and is converted to serine, releasing CO_2.

Mitochondrion

5 Glycerate moves to the chloroplast, where it is converted to 3PG and enters the Calvin cycle.

4 Serine moves back to the peroxisome and is converted to glycerate.

(B) Two catalytic activites of rubisco

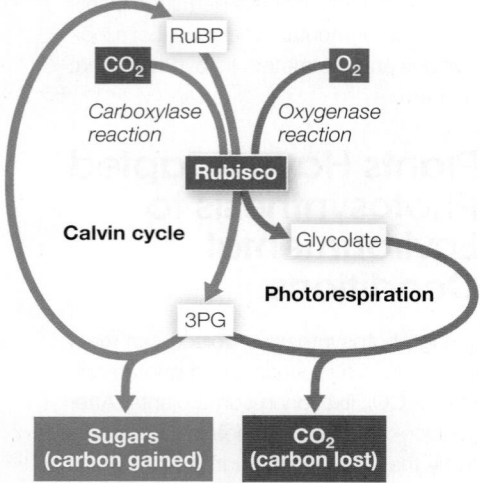

RuBP

CO_2 O_2

Carboxylase
reaction

Oxygenase
reaction

Rubisco

Calvin cycle   Glycolate

Photorespiration

3PG

Sugars
(carbon gained) CO_2
(carbon lost)

Figure 10.14 The Photorespiration Pathway   **(A)** The reactions of photorespiration take place in the chloroplasts, peroxisomes, and mitochondria. **(B)** Overall, photorespiration consumes O_2 and releases CO_2.

 Media Clip 10.1 Chloroplasts in Close-Up
www.Life11e.com/mc10.1

Think about the concentrations of O_2 and CO_2 in the typical air that you breathe. The concentration of CO_2 in Earth's atmosphere is about 400 ppm, or 0.04%. The concentration of O_2 in the air is about 20%. You would think that if there were an "even competition" between CO_2 and O_2 for rubisco that the latter, with its higher concentration, would be favored. But such is not the case. It turns out that the affinity (binding) of rubisco for CO_2 is much stronger than it is for O_2, so in normal air inside a leaf, CO_2 fixation is favored even though the concentration of CO_2 in the air is far lower than that of O_2. It is when there is an even higher concentration of

O_2 in the leaf relative to outside that O_2 competes with the CO_2, and rubisco combines RuBP with O_2 rather than with CO_2. This oxygenase activity reduces the overall amount of CO_2 that can be converted into carbohydrates, and may play a role in limiting plant growth.

- Relatively low O_2 and high CO_2 concentrations within the leaf favor carboxylase activity.
- Relatively high O_2 and low CO_2 concentrations within the leaf favor oxygenase activity.

Under what conditions inside the leaf does the air have higher O_2 and lower CO_2, favoring the oxygenase activity? On a hot, dry day, small pores in the leaf surface called **stomata** close to prevent water from evaporating from the leaf (see Figure 10.1). But stomata closure also prevents gases from entering and leaving the leaf. If stomata are closed, the CO_2 concentration in the leaf falls as CO_2 is used up in photosynthetic reactions, and the O_2 concentration rises because of these same reactions where water is used to form O_2 (see Investigating Life: What is the chemistry of photosynthesis, and how will increasing CO_2 in the atmosphere affect it?). As the ratio of CO_2 to O_2 falls, the oxygenase activity of rubisco is favored.

The consequences of oxygenase activity and lower carboxylase activity are significant. When O_2 is added to RuBP, one of the products is a two-carbon compound, phosphoglycolate:

$$\text{RuBP} + O_2 \rightarrow \text{phosphoglycolate} + \text{3-phosphoglycerate (3PG)}$$
(10.11)

The 3PG formed by rubisco's oxygenase activity enters the Calvin cycle, but the phosphoglycolate does not. Plants have evolved a metabolic pathway that can partially recover the carbon in phosphoglycolate. The phosphoglycolate is hydrolyzed to glycolate, which diffuses into peroxisomes (**Figure 10.14**). There, a series of reactions converts it into the amino acid glycine:

$$\text{Glycolate} + O_2 \rightarrow \text{glycine}$$
(10.12)

The glycine then diffuses into a mitochondrion, where two glycine molecules are converted in a series of reactions into the amino acid serine, releasing CO_2:

$$2 \text{ Glycine} \rightarrow \text{serine} + CO_2$$
(10.13)

The serine moves into the peroxisome, where it is converted to glycerate. The glycerate then moves into the chloroplast, where it is phosphorylated to make 3PG, which enters the Calvin cycle. Note that it takes two phosphoglycolate molecules from Equation 10.11 to produce the two glycines used in Equation 10.13. So overall:

$$2 \text{ Phosphoglycolate (4 carbons)} + O_2 \rightarrow \text{3PG (3 carbons)} + CO_2$$
(10.14)

This pathway thus reclaims 75 percent of the carbons from phosphoglycolate for the Calvin cycle. In other words, the reaction of RuBP with O_2 instead of CO_2 *reduces the net carbon fixed by the Calvin cycle by 25 percent*. The pathway is called **photorespiration** because it consumes O_2 and releases CO_2 and because it occurs only in the light (mediated by the same enzyme activation processes that we mentioned above with regard to the Calvin cycle).

(A) Arrangement of cells in a C$_3$ leaf

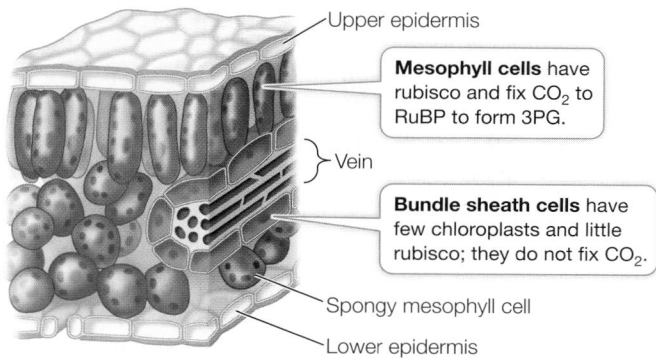

Upper epidermis

Mesophyll cells have rubisco and fix CO$_2$ to RuBP to form 3PG.

Vein

Bundle sheath cells have few chloroplasts and little rubisco; they do not fix CO$_2$.

Spongy mesophyll cell

Lower epidermis

(B) Arrangement of cells in a C$_4$ leaf

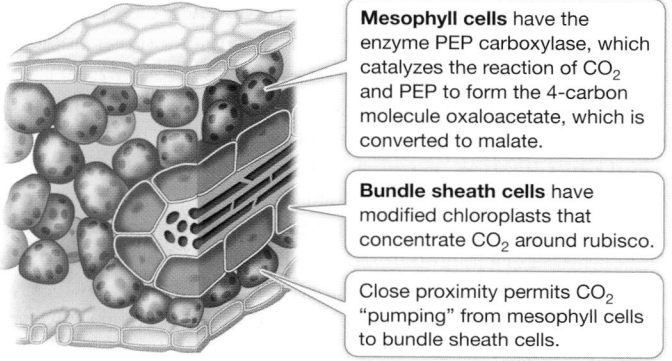

Mesophyll cells have the enzyme PEP carboxylase, which catalyzes the reaction of CO$_2$ and PEP to form the 4-carbon molecule oxaloacetate, which is converted to malate.

Bundle sheath cells have modified chloroplasts that concentrate CO$_2$ around rubisco.

Close proximity permits CO$_2$ "pumping" from mesophyll cells to bundle sheath cells.

Figure 10.15 Leaf Anatomy of C$_3$ and C$_4$ Plants Carbon dioxide fixation occurs in different organelles and cells of the leaves in **(A)** C$_3$ plants and **(B)** C$_4$ plants. Cells that are tinted blue have rubisco.

 Activity 10.2 **C$_3$ and C$_4$ Leaf Anatomy**
www.Life11e.com/ac10.2

C$_3$ plants undergo photorespiration but C$_4$ plants do not

Plants differ in how they fix CO$_2$, and can be distinguished as C$_3$ or C$_4$ plants, based on whether the first product of CO$_2$ fixation is a three- or four-carbon molecule. In **C$_3$ plants** such as roses, wheat, and rice, the first product is the three-carbon molecule 3PG—as we have just described for the Calvin cycle. In these plants the cells of the mesophyll, which makes up the main body of the leaf, are full of chloroplasts containing rubisco (**Figure 10.15A**). On a hot day, these leaves close their stomata to conserve water, and as a result, rubisco acts as an oxygenase as well as a carboxylase, and photorespiration occurs.

C$_4$ plants, which include corn, sugarcane, and tropical grasses, make the four-carbon molecule **oxaloacetate** as the first product of CO$_2$ fixation (**Figure 10.15B**). On a hot day, they partially close their stomata to conserve water, but their rate of photosynthesis does not fall. What do they do differently?

C$_4$ plants have evolved a mechanism that increases the concentration of CO$_2$ around the rubisco enzyme while at the same time isolating the rubisco from atmospheric O$_2$. Thus in these plants the carboxylase reaction is favored over the oxygenase reaction; the Calvin cycle operates, but photorespiration does not occur. This

(A)

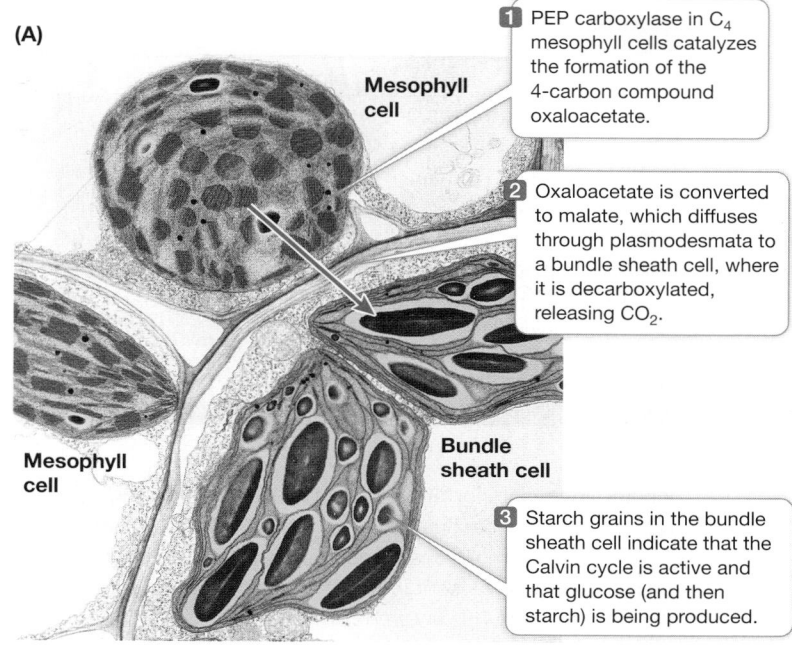

1 PEP carboxylase in C$_4$ mesophyll cells catalyzes the formation of the 4-carbon compound oxaloacetate.

Mesophyll cell

2 Oxaloacetate is converted to malate, which diffuses through plasmodesmata to a bundle sheath cell, where it is decarboxylated, releasing CO$_2$.

Mesophyll cell

Bundle sheath cell

3 Starch grains in the bundle sheath cell indicate that the Calvin cycle is active and that glucose (and then starch) is being produced.

(B)

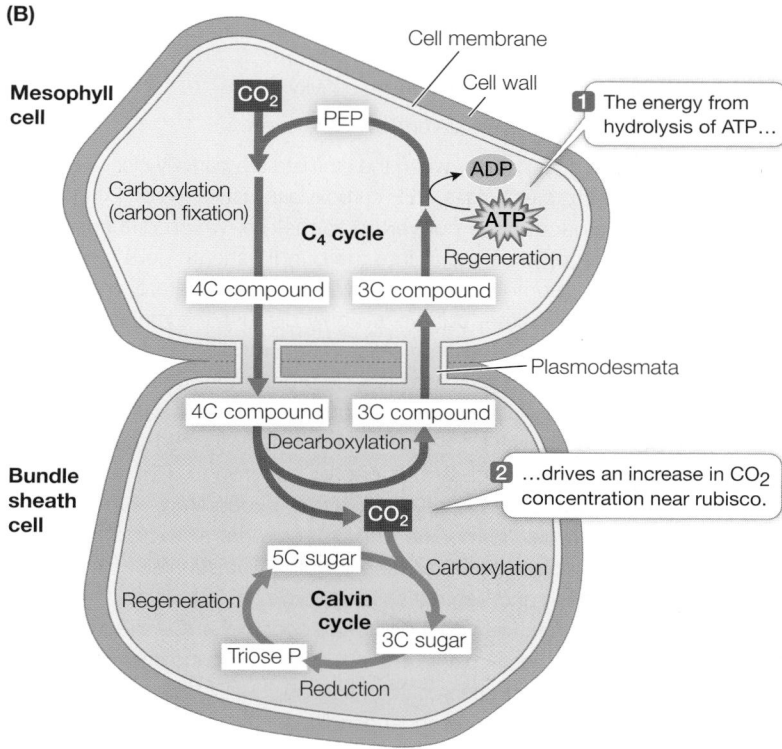

Mesophyll cell

Cell membrane

Cell wall

CO$_2$

PEP

ADP

1 The energy from hydrolysis of ATP…

Carboxylation (carbon fixation)

ATP

C$_4$ cycle

Regeneration

4C compound   3C compound

Plasmodesmata

4C compound   3C compound

Decarboxylation

2 …drives an increase in CO$_2$ concentration near rubisco.

Bundle sheath cell

CO$_2$

5C sugar

Carboxylation

Regeneration

Calvin cycle

Triose P   3C sugar

Reduction

Figure 10.16 The Anatomy and Biochemistry of C$_4$ Carbon Fixation
(A) Carbon dioxide is fixed initially in the mesophyll cells but enters the Calvin cycle in the bundle sheath cells. **(B)** The two cell types share an interconnected biochemical pathway for CO$_2$ assimilation.

mechanism involves the initial fixation of CO$_2$ in the mesophyll cells and then the transfer of the fixed carbon (as a four-carbon molecule) to the **bundle sheath cells**, where the fixed CO$_2$ is released for use in the Calvin cycle (**Figure 10.16**). The bundle sheath cells are

table 10.1 Comparison of Photosynthesis in C_3, C_4, and CAM Plants

	C_3 plants	C_4 plants	CAM plants
Calvin cycle used?	Yes	Yes	Yes
Primary CO_2 acceptor	RuBP	PEP	PEP
Primary CO_2-fixing enzyme	Rubisco	PEP carboxylase	PEP carboxylase
First product of CO_2 fixation	3PG (3-carbon)	Oxaloacetate (4-carbon)	Oxaloacetate (4-carbon)
Affinity of primary fixing enzyme for CO_2	Moderate	High	High
Photosynthetic cells of leaf	Mesophyll	Mesophyll and bundle sheath	Mesophyll with large vacuoles
Photorespiration	Extensive	Minimal	Minimal

located in the interior of the leaf where less atmospheric O_2 can reach them than reaches cells near the surface of the leaf.

The first enzyme in this C_4 carbon fixation process, called **PEP carboxylase**, is present in the cytosols of mesophyll cells near the leaf's surface. This enzyme combines CO_2 with a three-carbon acceptor compound, **phosphoenolpyruvate** (**PEP**), to produce the four-carbon fixation product, oxaloacetate. PEP carboxylase has two advantages over rubisco:

1. It does not have oxygenase activity.

2. It fixes CO_2 even at very low CO_2 levels.

So even on a hot day when the stomata are partially closed and the ratio of O_2 to CO_2 rises, PEP carboxylase just keeps on fixing CO_2.

Oxaloacetate is converted to malate, which diffuses out of the mesophyll cells and into the bundle sheath cells (see Figure 10.15B). (Some C_4 plants convert the oxaloacetate to aspartate instead of malate, but we will only discuss the malate pathway here.) The bundle sheath cells contain modified chloroplasts that are designed to concentrate CO_2 around the rubisco. There, the four-carbon malate loses one carbon (is decarboxylated), forming CO_2 and pyruvate. The latter moves back to the mesophyll cells where the three-carbon acceptor compound, PEP, is regenerated at the expense of ATP. So the "expenditure" of ATP in the mesophyll cell "pumps up" the CO_2 concentration around rubisco in the bundle sheath cell, ensuring that rubisco will function as a carboxylase and begin the Calvin cycle.

Under relatively cool or cloudy conditions, C_3 plants have an advantage over C_4 plants in that they don't expend energy to "pump up" the concentration of CO_2 near rubisco. But this advantage begins to be outweighed under conditions that favor photorespiration, such as warmer seasons and climates. Under these conditions C_4 plants have the advantage, especially if there is ample light to supply the extra ATP required for C_4 photosynthesis. For example, Kentucky bluegrass is a C_3 plant that thrives on lawns in April and May. But in the heat of summer it does not do as well, and Bermuda grass, a C_4 plant, takes over the lawn. The same is true on a global scale for crops: C_3 plants such as soybean, wheat, and barley have been adapted for human food production in temperate climates, whereas C_4 plants such as corn and sugarcane originated and are still grown in the tropics.

THE EVOLUTION OF CO_2 FIXATION PATHWAYS C_3 plants are more ancient than C_4 plants. Whereas C_3 photosynthesis appears to have begun about 2.5 billion years ago, C_4 plants appeared about 12 million years ago. A possible factor in the emergence of the C_4 pathway is the decline in atmospheric CO_2. When dinosaurs dominated Earth 100 million years ago, the concentration of CO_2 in the atmosphere was four times what it is now. As CO_2 levels declined thereafter, the C_4 plants would have gained an advantage over their C_3 counterparts in high-temperature, high-light environments.

As described in the opening of this chapter, atmospheric CO_2 levels have been increasing over the past 200 years. Currently, the level of CO_2 is not enough for maximal CO_2 fixation by rubisco, so photorespiration occurs, reducing the growth rates of C_3 plants. Under hot conditions, C_4 plants are favored. But if CO_2 levels in the atmosphere continue to rise, the reverse will occur and C_3 plants will have a comparative advantage. The overall growth rates of crops such as rice and wheat should increase. This may or may not translate into more food, given that other effects of the human-spurred CO_2 increase (such as global climate change) will also alter Earth's ecosystems.

CAM plants also use PEP carboxylase

Other plants besides the C_4 plants use PEP carboxylase to fix and accumulate CO_2. They include some water-storing plants (succulents) of the family Crassulaceae, many cacti, pineapples, and several other kinds of flowering plants. The CO_2 metabolism of these plants is called **crassulacean acid metabolism**, or **CAM**, after the family of succulents in which it was discovered. CAM is much like the metabolism of C_4 plants in that CO_2 is initially fixed into a four-carbon compound. But in CAM plants the initial CO_2 fixation and the Calvin cycle are separated in time rather than space.

- At night, when it is cooler and water loss is minimized, the stomata open. CO_2 is fixed in mesophyll cells to form the four-carbon compound oxaloacetate, which is converted into malate and stored in the vacuole.

- During the day, when the stomata close to reduce water loss, the accumulated malate is shipped from the vacuole to the chloroplasts, where its decarboxylation supplies the CO_2 for the Calvin cycle and the light reactions supply the necessary ATP and NADPH.

10.4 recap

Rubisco catalyzes the carboxylation of RuBP to form two 3PG, and the oxygenation of RuBP to form one 3PG and one phosphoglycolate. The diversion of rubisco to its oxygenase function decreases net CO_2 fixation. C_4 photosynthesis and CAM allow plants to fix CO_2 under warm, dry conditions when stomata are closed and CO_2 entry into the leaf is limited.

learning outcomes

You should be able to:

- Explain how and why C_4 plants maintain a high concentration of CO_2 around rubisco.
- Explain how CAM plants carry out carbon fixation separately from Calvin cycle reactions.

1. How do C_4 plants keep the concentration of CO_2 around rubisco high, and why?
2. Describe CAM plants and explain how they can temporally segregate CO_2 fixation from the Calvin cycle.

Table 10.1 compares photosynthesis in C_3, C_4, and CAM plants. Now that you understand how photosynthesis produces carbohydrates, let's see how the pathways of photosynthesis are connected to other metabolic pathways.

key concept 10.5 Photosynthesis Is an Integral Part of Plant Metabolism

Green plants are autotrophs and can synthesize all the molecules they need from simple starting materials: CO_2, H_2O, phosphate, sulfate, ammonium ions (NH_4^+), and small quantities of other mineral nutrients. Plants use the carbohydrates generated in photosynthesis to provide energy for processes such as active transport and anabolism. Both cellular respiration and fermentation can occur in plants, although the former is far more common. Unlike photosynthesis, plant cellular respiration occurs all the time in both the light and the dark.

focus your learning

- Photosynthesis is linked to cellular respiration, gluconeogenesis, and other metabolic pathways.
- Photosynthesis is responsible for providing the energy needed by most organisms on Earth.

Photosynthesis interacts with other metabolic pathways

Photosynthesis and respiration are closely linked through the Calvin cycle (**Figure 10.17**). The partitioning of G3P, the product of the Calvin cycle, is particularly important:

- Some G3P from the Calvin cycle enters glycolysis and is converted into pyruvate in the cytosol. This pyruvate can be used in cellular respiration for energy, or its carbon skeletons can be used in anabolic reactions to make lipids, proteins, and other carbohydrates (see Figure 9.13).

- Some G3P can enter a pathway that is the reverse of glycolysis (gluconeogenesis; see Key Concept 9.5). In this case, hexose phosphates (hexose P) and then sucrose are formed and transported to the nonphotosynthetic tissues of the plant (such as the root).

Energy flows from sunlight to reduced carbon in photosynthesis, then to ATP in respiration. Energy can also be stored in the bonds of macromolecules such as polysaccharides, lipids, and proteins. For

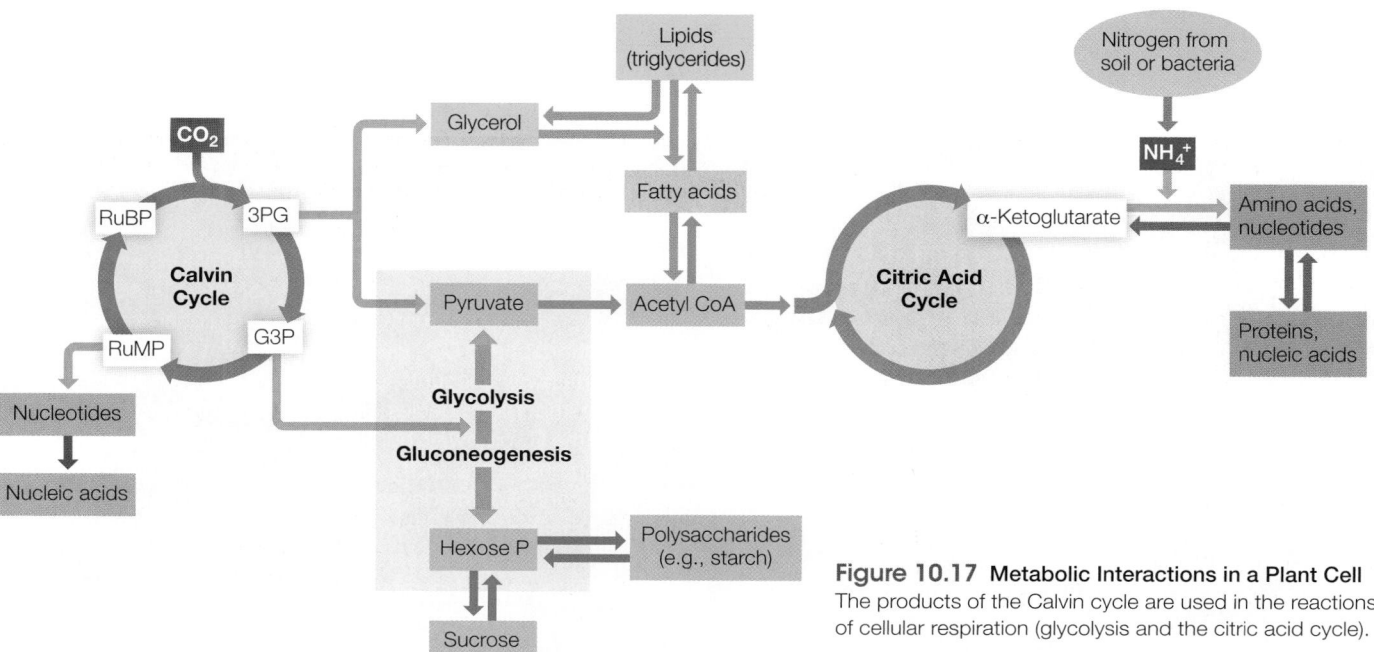

Figure 10.17 Metabolic Interactions in a Plant Cell
The products of the Calvin cycle are used in the reactions of cellular respiration (glycolysis and the citric acid cycle).

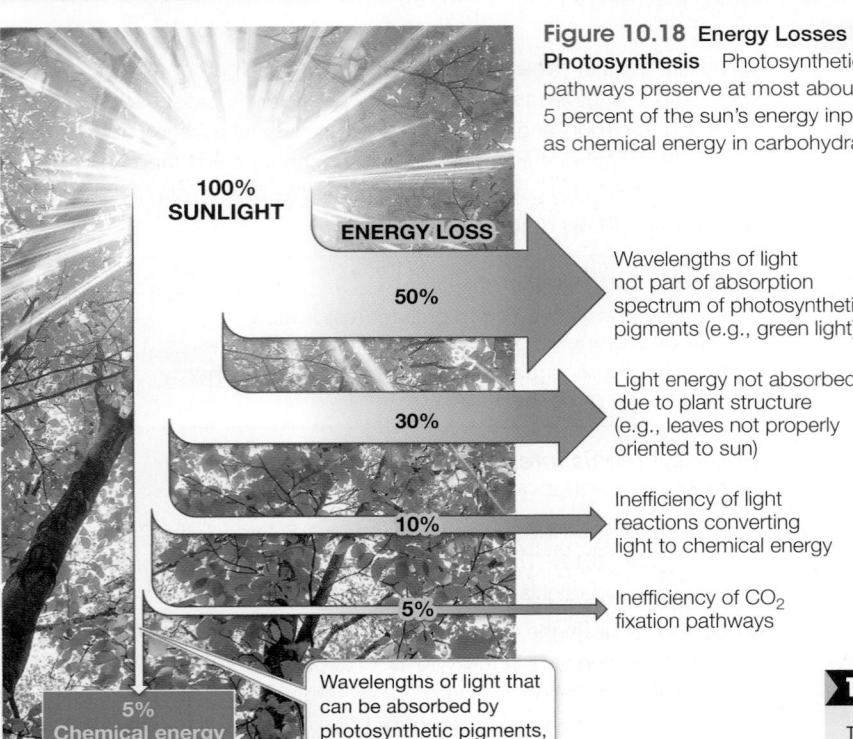

Figure 10.18 Energy Losses in Photosynthesis Photosynthetic pathways preserve at most about 5 percent of the sun's energy input as chemical energy in carbohydrates.

100% SUNLIGHT

ENERGY LOSS

50% — Wavelengths of light not part of absorption spectrum of photosynthetic pigments (e.g., green light)

30% — Light energy not absorbed due to plant structure (e.g., leaves not properly oriented to sun)

10% — Inefficiency of light reactions converting light to chemical energy

5% — Inefficiency of CO_2 fixation pathways

5% Chemical energy stored in carbohydrates

Wavelengths of light that can be absorbed by photosynthetic pigments, and chemical energy available for CO_2 fixation.

Photosynthesis provides most of the energy that we need for life. Given the uncertainties about the future of photosynthesis (because of changes in CO_2 levels and climate change), it would be wise to seek ways to improve photosynthetic efficiency. **Figure 10.18** shows the various ways in which solar energy is used by plants or lost. In essence, only about 5 percent of the sunlight that reaches Earth is converted into plant growth. The inefficiencies of photosynthesis involve basic chemistry and physics (some light energy is not absorbed by photosynthetic pigments) as well as biology (plant anatomy and leaf exposure, the oxygenase reaction of rubisco, and inefficiencies in metabolic pathways). While it is hard to change chemistry and physics, biologists might be able to use their knowledge of plants to improve on the basic biology of photosynthesis. This could result in a more efficient use of resources and better food production.

a plant to grow, energy storage (as body structures) must exceed energy release; that is, overall carbon fixation by photosynthesis must exceed respiration. This principle is the basis of the ecological *food chain, as you will see in later chapters.

> *connect the concepts   Chapter 57 describes the sequence of energy flow through a community, including food chains and food webs.

10.5 recap

The products of photosynthesis are used in glycolysis and the citric acid cycle, as well as in the synthesis of lipids, proteins, and other large molecules.

learning outcomes

You should be able to:

- Identify the metabolic intermediates that link photosynthesis to other metabolic pathways.
- Explain the efficiency of energy flow as it moves from the environment through living organisms on Earth.

1. How do common intermediates link the pathways of glycolysis, the citric acid cycle, and photosynthesis?
2. Why is at most 5 percent of the solar radiation that reaches Earth converted to plant growth?

investigatinglife

Q&A What is the chemistry of photosynthesis, and how will it be affected by rising atmospheric CO_2?

Crops could be affected by increased atmospheric CO_2 in multiple ways. Higher CO_2 levels generally lead to increased photosynthesis. This is especially true for C_3 plants, which are more sensitive than C_4 plants to CO_2 levels. Because increased photosynthesis leads to greater plant growth, C_3 crops such as wheat and rice will tend to grow more. However, it is unclear whether this growth will be in the vegetative parts of the plant (stems and leaves) or in the part we eat (grain). To further complicate matters, such increases in plant growth may be counteracted by the effects of increased CO_2 on climate. For example, increased temperatures would increase the rate of photosynthesis and extend the growing season, but might alter rainfall patterns. In some areas of the world there might be less rain, and this could limit plant growth.

Future directions

As the human population rises and land for farming is limited, we are challenged to improve the productivity of crops. A significant increase in overall crop productivity would have global consequences. Increased atmospheric CO_2 will not have a great effect on C_4 plants. But the C_4 pathway accounts for about 25 percent of all net carbon fixed. Some important species, such as corn, sorghum, and sugarcane, have the C_4 pathway, but the major grain crops such as rice do not. An intensive research program is under way to understand the genetic control of the development of the C_4 pathway, including not just the enzymes involved but also how the distinctive leaf anatomy develops. This research may lead to greater use of the C_4 pathway, and its related increase in productivity in more plant species in a higher-CO_2 world.

Chapter Summary 10

10.1 Photosynthesis Uses Light to Make Carbohydrates

- In the process of **photosynthesis**, the energy of sunlight is captured and used to convert CO_2 into more complex carbon-containing compounds. **See Animation 10.1**

- Isotope-labeling experiments demonstrated that water is the source of the oxygen atoms in the O_2 produced by photosynthesis. **Review Investigating Life: What Is the Chemistry of Photosynthesis, and How Will Increasing CO_2 in the Atmosphere Affect It?**

- Plants, algae, and cyanobacteria live in aerobic environments and carry out oxygenic photosynthesis: the conversion of CO_2 and H_2O into carbohydrates and release of O_2.

- Some bacteria that live in anaerobic environments carry out anoxygenic photosynthesis, in which energy from the sun is used to fix CO_2 without the use of H_2O and the production of O_2.

- The **light reactions** of photosynthesis convert light energy into chemical energy in the form of ATP and NADPH. **Review Focus: Key Figure 10.2**

- The **light-independent reactions** do not use light directly but instead use ATP and NADPH to reduce CO_2, forming carbohydrates.

10.2 Photosynthesis Converts Light Energy into Chemical Energy

- Light is a form of **electromagnetic radiation**. It is emitted in particle-like packets called **photons** but has wavelike properties.

- Molecules that absorb light in the visible spectrum are called **pigments**. Photosynthetic organisms have several pigments, most notably **chlorophylls**.

- Absorption of a photon puts an electron of a pigment molecule in an excited state that has more energy than its ground state.

- Each pigment has a characteristic **absorption spectrum**. An **action spectrum** reflects the rate of photosynthesis carried out by a photosynthetic organism at a given wavelength of light. **Review Figure 10.4**

- The pigments in photosynthetic organisms are arranged into **light-harvesting complexes** that absorb energy from light and funnel this energy to chlorophyll *a* molecules in the reaction center of the **photosystem**. Chlorophyll can act as a reducing agent, transferring excited electrons to other molecules. **Review Figure 10.6**

- **Noncyclic electron transport** uses photosystems I and II to produce ATP, NADPH, and O_2. **Cyclic electron transport** uses only photosystem I and produces only ATP. Both systems generate ATP via the electron transport system. **Review Figures 10.7, 10.8**

- Chemiosmosis is the mechanism of ATP production in **photophosphorylation**. **Review Figure 10.9, Animation 10.2**

10.3 Chemical Energy Trapped in Photosynthesis Is Used to Synthesize Carbohydrates

- The **Calvin cycle** makes carbohydrates from CO_2. The cycle consists of three processes: fixation of CO_2, reduction and carbohydrate production, and regeneration of RuBP. **See Animation 10.3**

- RuBP is the initial CO_2 acceptor, and 3PG is the first stable product of CO_2 fixation. The enzyme **ribulose bisphosphate carboxylase/oxygenase (rubisco)** catalyzes the reaction of CO_2 and RuBP to form 3PG. **Review Figures 10.10, 10.11**

- ATP and NADPH formed by the light reactions are used in the reduction of 3PG to form **glyceraldehyde 3-phosphate (G3P)**. **Review Figure 10.12, Activity 10.1**

- Light stimulates enzymes in the Calvin cycle, further integrating the light-dependent and light-independent pathways. **Review Figure 10.13**

10.4 Plants Have Adapted Photosynthesis to Environmental Conditions

- Rubisco can catalyze a reaction between O_2 and RuBP in addition to the reaction between CO_2 and RuBP. At high temperatures and low CO_2 concentrations, the oxygenase function of rubisco is favored over its carboxylase function.

- The oxygenase reaction catalyzed by rubisco significantly reduces the efficiency of photosynthesis. The subsequent reactions of **photorespiration** recover some of the fixed carbon that otherwise would be lost. **Review Figure 10.14**

- In **C4 plants**, CO_2 reacts with **phosphoenolpyruvate (PEP)** to form a four-carbon intermediate in mesophyll cells. The four-carbon product releases its CO_2 to rubisco in the **bundle sheath cells** in the interior of the leaf. **Review Figures 10.15, 10.16, Activity 10.2**

- **Crassulacean acid metabolism (CAM)** plants operate much like C_4 plants, but their initial CO_2 fixation by PEP carboxylase is temporally separated from the Calvin cycle, rather than spatially separated as in C_4 plants.

10.5 Photosynthesis Is an Integral Part of Plant Metabolism

- Photosynthesis and cellular respiration are linked through the Calvin cycle, the citric acid cycle, and glycolysis. **Review Figure 10.17**

- To survive, a plant must photosynthesize more than it respires.

- Photosynthesis uses only a small portion of the energy of sunlight. **Review Figure 10.18**

Apply What You've Learned

Review

10.3 Light affects photosynthetic carbon fixation.

Original Paper: Loach, K. 1967. Shade tolerance in tree seedlings: I. Leaf photosynthesis and respiration in plants raised under artificial shade. *New Phytologist* 66: 607–621.

Some plants are shade-tolerant, and others thrive in sunlight. This phenomenon raises questions, because all plants use the same basic photosynthetic process. What makes shade-tolerant plants different from shade-intolerant plants, allowing them to grow well in low light? Can plants adapt to cope with unfavorable lighting conditions?

    Researchers explored these questions by studying seedlings of two tree species—one shade-tolerant (beech) and one shade-intolerant (quaking aspen). The researchers planted newly germinated seedlings of both species under frames and covered the frames with shade cloth, limiting the amount of sunlight getting through to either 3% or 44% of normal daylight.

    After 5 weeks the researchers moved one leaf, still attached to each plant, out of the frame so it could be studied briefly. They exposed each leaf to varying light intensities for several minutes, while measuring the rate of photosynthesis in its cells. Then leaves were removed and analyzed for chlorophyll content by mass, density of chlorophyll in the leaf, and leaf surface area. The last measurement was expressed as surface area per gram of leaf tissue so that the two species, which have different leaf sizes, could be compared. The results are shown in the graphs at right and in the table below. Note on the x-axis that normal daylight is about 4.5 thousand foot-candles.

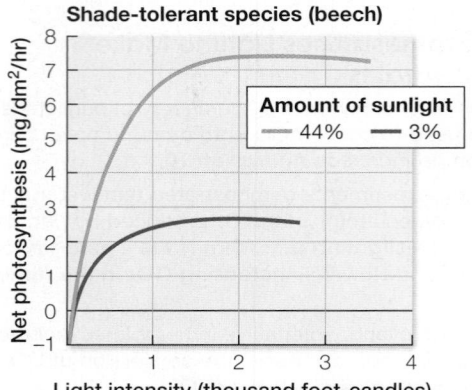

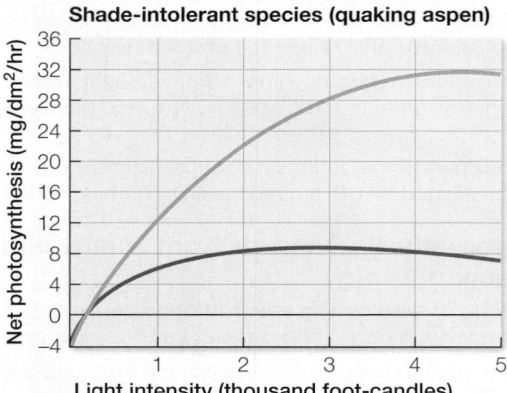

| Species | Shade tolerance | Cultivation light intensity (% sunlight) | Total chlorophyll | | Specific leaf surface area (dm²/g) |
			(mg/g dry weight leaf)	(mg/dm² leaf)	
Beech	Very tolerant	44	3.26	1.53	2.13
		3	7.02	2.82	2.49
Quaking aspen	Very intolerant	44	6.34	3.62	1.57
		3	8.23	4.38	1.88

Questions

1. In the graphs showing photosynthesis rates as a function of light intensity, every curve has the same basic shape. Why does the photosynthesis rate represented by each curve increase as the leaf is exposed to higher light intensities?

2. Look at the *y*-axis values on the two graphs, and compare the maximum photosynthesis rates for the two tree species. What does this suggest about a difference between shade-tolerant plants and shade-intolerant plants?

3. Analyze the chlorophyll data in the table. In general, how and why do plants adjust their levels of chlorophyll when raised in different light conditions?

4. Analyze the leaf surface area data in the table. How do shade-tolerant plants compare with shade-intolerant plants in specific leaf surface area, and why does this support adaptation to shade?

5. Predict which type of plant (shade-tolerant or shade-intolerant) will have the greatest fluctuation in photosynthesis rates over time in response to changes that occur in a single day as the skies go from overcast to clear and back to overcast.

11

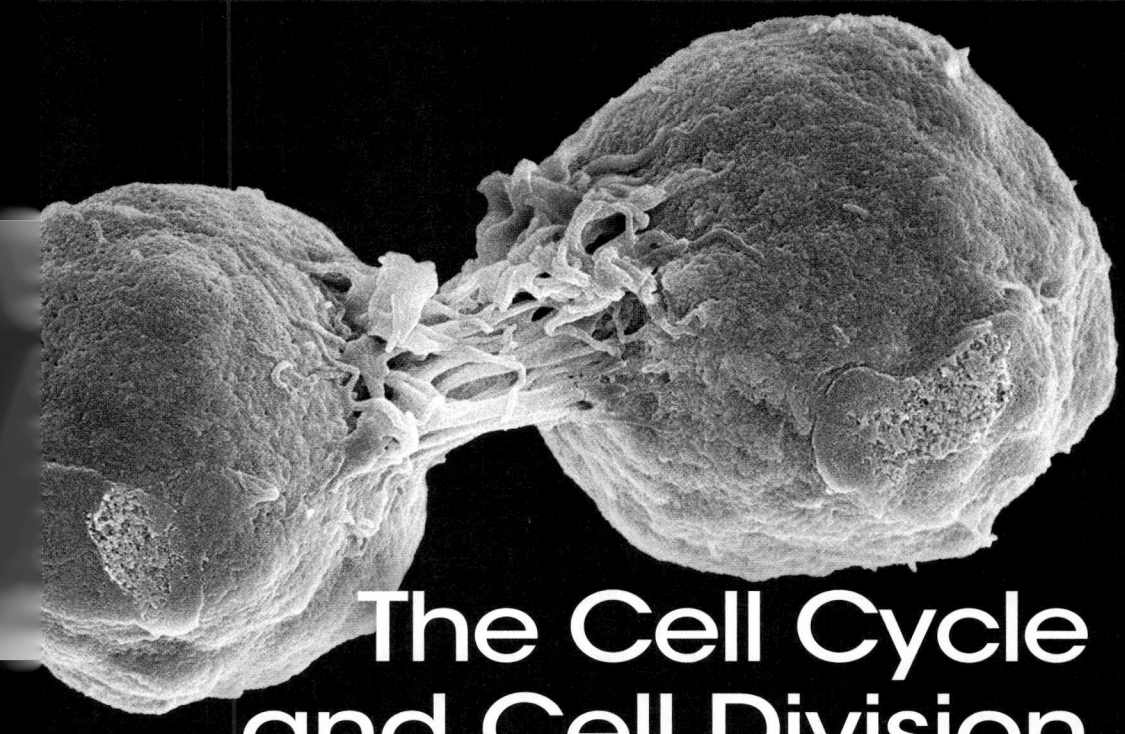

The Cell Cycle and Cell Division

Henrietta Lacks's tumor cells have far outlived her sad demise from cancer. They reproduce rapidly on the surface of a solid medium and have been grown in labs since her death in 1951.

investigatinglife

Immortal Cells

On January 29, 1951, 30-year-old Henrietta Lacks visited the nearby Johns Hopkins Hospital in Baltimore, Maryland, because she had been bleeding abnormally after the birth of her last child. The physician found the reason for the bleeding: a tumor the size of a quarter on her cervix. A piece of the tumor was sent to a pathologist in a clinical laboratory, who reported that the tumor was malignant.

A week later Ms. Lacks returned to the hospital, where physicians treated her tumor with radiation to try to kill it. But before the treatment began, they took a small sample of cells and sent them without her permission—a practice common then but not ethical today—to the research lab of George and Margaret Gey, two scientists at the hospital who had been trying for 20 years to coax human cells to live and multiply outside the body. If they could do so, they thought, they might find a cure for cancer. The Geys hit pay dirt with Lacks's cells; they grew and reproduced more vigorously than any cells they had ever seen and were given the name "HeLa cells." Unfortunately, they also grew fast in Lacks's body, and in a few months they had spread to almost all of her organs. Lacks died on October 4, 1951.

Because of their robust ability to reproduce, HeLa cells quickly became a staple of cell biology research. In controlled settings they could be infected with viruses, and they were instrumental in developing the supply of polioviruses that led to the first vaccine against that dread disease. HeLa cells have been used for important basic and applied research ever since, especially on the ways that human cells reproduce by cell division. Although Lacks had never been outside Virginia and Maryland, her cells have traveled all over the world and even into space on the space shuttle. Over the past 60 years, tens of thousands of research articles have been published using information obtained from Lacks's cells. You'll study one of them in this chapter.

Understanding the cell division cycle and its control is clearly an important subject for understanding cancer. But cell division is not just important in medicine. It underlies the growth, development, and reproduction of all organisms.

 What controls the reproduction of cancer cells?

11.1 All Cells Derive from Other Cells

The life cycle of an organism, from birth to death, is intimately linked to cell division. Cell division plays important roles in the development of an organism from a single cell, in the growth and repair of tissues in multicellular organisms, and in the reproduction of all organisms (**Figure 11.1**). Cell division in both prokaryotes and eukaryotes entails the following four events:

1. A **reproductive signal**. An intracellular or extracellular signal initiates cell division.

2. **Replication** of DNA. The cell's genetic material must be duplicated so that each of the two new cells will have a complete, identical set of genes.

3. **Segregation** of replicated DNA. The replicated DNA must be equally distributed to each of the two new cells.

4. **Cytokinesis**. Enzymes and organelles for the new cells must be synthesized, and new material must be added to the cell membrane (and the cell wall, in organisms that have one), in order to separate the two new cells.

As you will see, these four events happen differently in prokaryotes and eukaryotes.

---

focus your learning

- All cell division processes involve four main events: initiation, DNA replication, DNA segregation, and cytokinesis.
- Cell division in prokaryotes is rapid, occurs in response to environmental signals, and results in new individual cells, which are often the entire organism.
- Cell division in eukaryotes is complex, occurs in response to internal signals, and may reproduce the entire organism for single-celled eukaryotes, or result in more cells within a multicellular organism.

Prokaryotes divide by binary fission

In prokaryotes, cell division results in the reproduction of the entire single-celled organism. The cell grows in size, replicates its DNA, and then separates the cytoplasm and DNA into two new cells in a process called **binary fission**.

REPRODUCTIVE SIGNALS   External factors such as environmental conditions and nutrient concentrations are common signals for the initiation of cell division in prokaryotes. For example, the bacterium *Bacillus subtillis* divides every 120 minutes if abundant sources of carbohydrate and mineral nutrients are available. When nutrient levels are low, it slows down its growth and eventually stops dividing. When conditions improve, growth and division resume. The bacterium *Escherichia coli*, which is widely used in genetic studies, divides every 20 minutes if abundant sources of carbohydrates and mineral nutrients are available.

REPLICATION OF DNA   As you saw in Key Concept 5.3, a **chromosome** consists of a long, thin DNA molecule with proteins attached to it. When a cell divides, all of its chromosomes, which contain the genetic information for the organism, must be replicated, and one copy of each chromosome must find its way into each of the two new cells.

Most prokaryotes have just one main chromosome. If the *E. coli* chromosome were spread out into a full circle, the circle would be about 500 micrometers (μm) in diameter, more than 200 times larger than the cell! To fit into the cell, bacterial DNA must be compacted. The DNA folds in on itself, and positively charged (basic) proteins bound to the negatively charged (acidic) DNA contribute to this folding.

Two regions of the prokaryotic chromosome play functional roles in cell reproduction:

1. *ori*: the site where replication of the circular chromosome starts (the *ori*gin of replication)

2. *ter*: the site where replication ends (the *ter*minus of replication)

Chromosome replication takes place as the DNA is threaded through a replication complex of proteins near the center of the cell. Replication begins at the *ori* site and moves bidirectionally toward the *ter* site. While the DNA replicates, anabolic metabolism is active, resulting in cell growth. When replication is complete, the two daughter DNA molecules separate and segregate from one another at opposite ends of the cell. In rapidly dividing prokaryotes, DNA replication occupies the entire time between cell divisions.

SEGREGATION OF DNA MOLECULES   Replication begins near the center of the cell, and as it proceeds, the *ori* regions move toward opposite ends of the cell (**Figure 11.2A**). DNA sequences adjacent to the *ori* region bind proteins that are essential for this segregation. This is an energy-requiring process, since the binding proteins hydrolyze ATP.

(A) Reproduction

These eukaryotic yeast cells divide by budding.

(B) Growth

Cell division contributes to the growth of this root tissue.

(C) Regeneration

Cell division contributes to the regeneration of a lizard's tail.

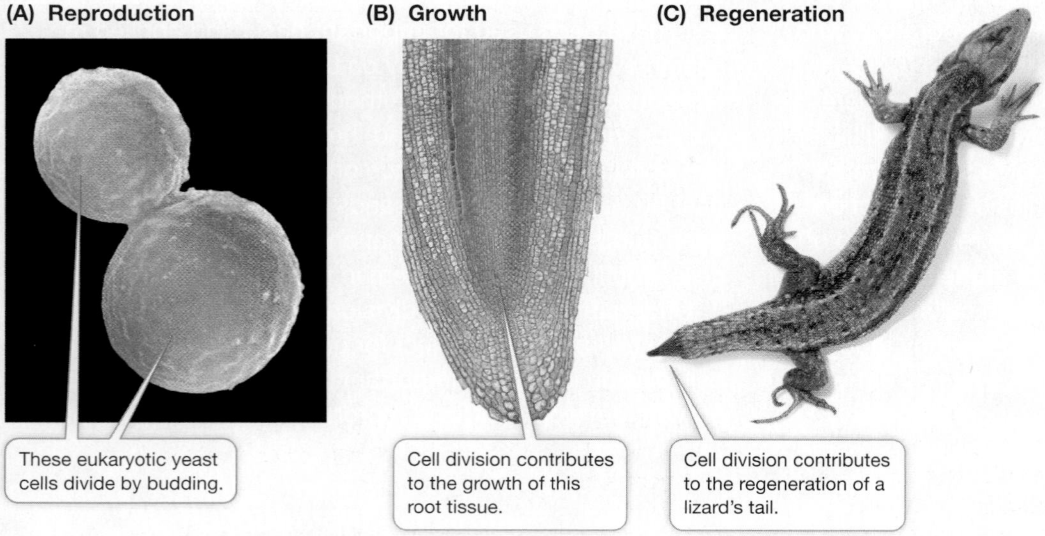

Figure 11.1 Important Consequences of Cell Division   Cell division is the basis for **(A)** reproduction, **(B)** growth, and **(C)** repair and regeneration of tissues.

(A) Binary fission in a bacterium

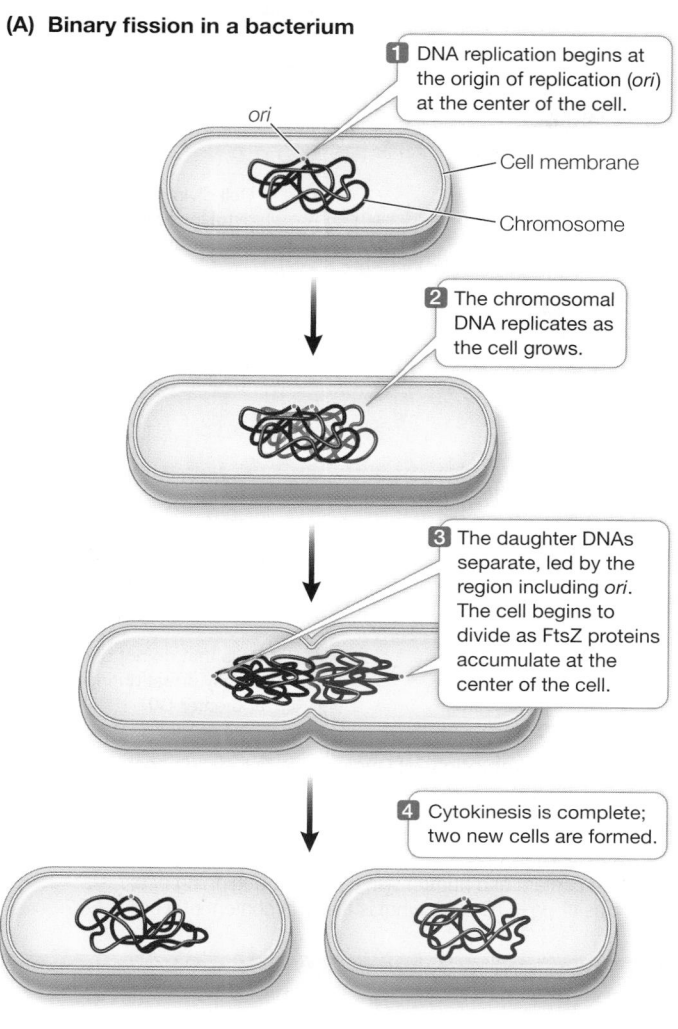

1 DNA replication begins at the origin of replication (*ori*) at the center of the cell.

ori

Cell membrane

Chromosome

2 The chromosomal DNA replicates as the cell grows.

3 The daughter DNAs separate, led by the region including *ori*. The cell begins to divide as FtsZ proteins accumulate at the center of the cell.

4 Cytokinesis is complete; two new cells are formed.

(B) Cytokinesis in a bacterium

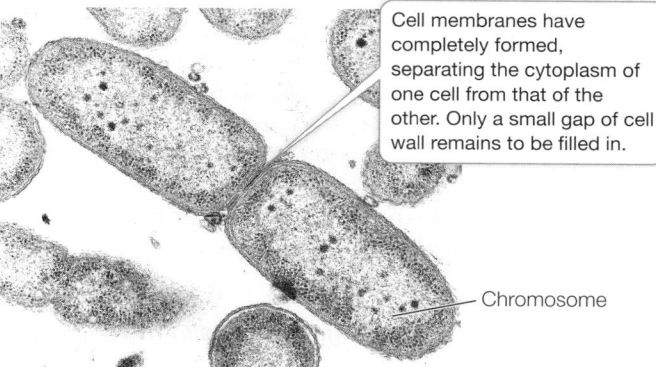

Cell membranes have completely formed, separating the cytoplasm of one cell from that of the other. Only a small gap of cell wall remains to be filled in.

Chromosome

Figure 11.2 Prokaryotic Cell Division

CYTOKINESIS Immediately after chromosome replication is finished, cytokinesis begins. At first, the cell membrane pinches in to form a ring of fibers similar to a purse string. The major component of these fibers is a protein called FtsZ that is related to eukaryotic tubulin (which makes up microtubules; see Figure 5.14). As the membrane pinches in, new cell wall materials are deposited, which finally separate the two cells (**Figure 11.2B**).

Eukaryotic cells divide by mitosis followed by cytokinesis

As in prokaryotes, cell reproduction in eukaryotes entails reproductive signals, DNA replication, segregation, and cytokinesis. The details, however, are quite different:

- *Reproductive signal*: Unlike prokaryotes, eukaryotic cells do not constantly divide whenever environmental conditions are adequate. In fact, most eukaryotic cells that are part of a multicellular organism and have become specialized seldom divide. In a eukaryotic organism, the signals for cell division are usually not related to the environment of a single cell, but to the function of the entire organism.

- *Replication*: Whereas most prokaryotes have a single main chromosome, eukaryotes usually have many (humans have 46). Consequently the processes of replication and segregation are more intricate in eukaryotes than in prokaryotes. In eukaryotes, DNA replication is usually limited to part of the period between cell divisions.

- *Segregation*: In eukaryotes, the newly replicated chromosomes are closely associated with each other (thus they are known as **sister chromatids**), and a mechanism called **mitosis** segregates them into two new nuclei.

- *Cytokinesis*: Cytokinesis proceeds differently in plant cells (which have a cell wall) than in animal cells (which do not).

The cells resulting from mitosis are identical to the parent cell in the amount and kind of DNA they contain.

There is another mechanism for nuclear division called **meiosis**, which is involved in gamete formation. We will return to meiosis and describe differences with mitosis in Key Concept 11.5.

11.1 recap

Four events are required for cell division: a reproductive signal, replication of the genetic material (DNA), segregation of replicated DNA, and separation of the two daughter cells (cytokinesis). Prokaryotes often have just one chromosome, and cell division can be rapid. Eukaryotes usually have multiple chromosomes, and the process of cell division is more intricate, involving either mitosis or meiosis.

learning outcomes

You should be able to:

- Analyze and explain the steps found in all cell division processes and their sequence.
- Describe the cell division process in prokaryotes.
- Explain how the four events of cell division occur in eukaryotic cells.
- Compare and contrast cell division in prokaryotes and eukaryotes.

1. What commonly limits the reproduction of a prokaryotic cell by binary fission?
2. Why must DNA be replicated and segregated before the cell can divide?
3. Why are DNA replication and cell division more complex in eukaryotes than in prokaryotes?

What determines whether a eukaryotic cell will divide? How does mitosis lead to identical cells? In the sections that follow, we will describe the details of mitosis.

key concept

11.2 The Eukaryotic Cell Division Cycle Is Regulated

As you will see throughout this book, different cells have different rates of cell division. Some cells, such as those in an early embryo, divide rapidly and continuously. Others, such as neurons in the brain, don't divide at all. This suggests that the signaling pathways for cells to divide are highly controlled.

focus your learning

- Dividing eukaryotic cells undergo an orderly sequence of events that together make up the cell cycle.
- Events of the eukaryotic cell cycle are internally regulated.
- External factors stimulate eukaryotic cells in the G0 state to begin dividing.

In eukaryotes, the period from one cell division to the next is known as the **cell cycle**. The cell cycle can be divided into mitosis/cytokinesis and interphase. During **interphase**, the cell nucleus is visible and typical cell functions occur, including DNA replication. Interphase begins when cytokinesis is completed and ends when mitosis (M) begins (**Focus: Key Figure 11.3**). In this section we will describe the events of interphase, especially those that trigger mitosis.

The duration of the cell cycle varies considerably in different cell types. In the early embryo the cell cycle may be as short as 30 minutes, whereas rapidly dividing cells in an adult human typically complete the cycle in about 24 hours. In general, cells spend most of their time in interphase. So if we take a snapshot through the microscope of a cell population, only a few cells will be in mitosis or cytokinesis at any given moment. Interphase has three subphases called G1, S, and G2. In a cell cycle of 24 hours, these subphases would typically last for 11 hours (G1), 8 hours (S), and 4 hours (G2), with the remaining 1 hour spent in mitosis.

- *G1 phase.* During **G1 phase**, each chromosome is a single, unreplicated DNA molecule with associated proteins. Variations in the duration of G1 account for most of the variability in the length of the cell cycle in different cell types. Some rapidly dividing embryonic cells dispense with it entirely, whereas other cells may remain in G1 for weeks or even years.

- *The G1-to-S transition.* At the **G1-to-S transition** the commitment is made to DNA replication and subsequent cell division.

- *S phase.* DNA replication occurs during **S phase** (see Key Concept 13.3 for a detailed description of DNA replication). Each chromosome is duplicated and thereafter consists of two sister chromatids (the products of DNA replication). The sister chromatids remain joined together until mitosis, when they segregate into two daughter cells.

- *G2 phase.* During **G2 phase**, the cell makes preparations for mitosis—for example, by synthesizing and assembling the

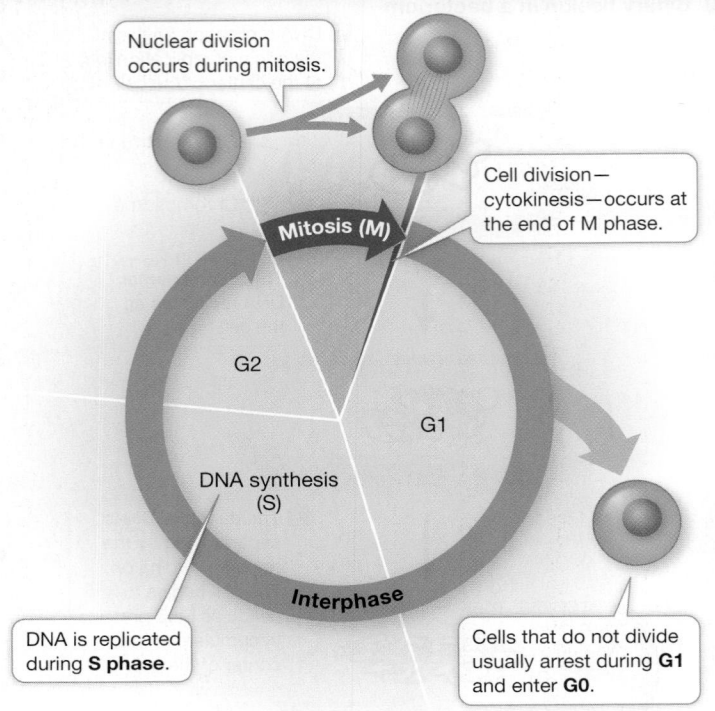

focus: key figure

Nuclear division occurs during mitosis.

Cell division—cytokinesis—occurs at the end of M phase.

Mitosis (M)

G2

G1

DNA synthesis (S)

Interphase

DNA is replicated during **S phase**.

Cells that do not divide usually arrest during **G1** and enter **G0**.

Figure 11.3 The Eukaryotic Cell Cycle   The cell cycle consists of a mitotic (M) phase, during which mitosis and cytokinesis take place, and a long period of growth known as interphase. Interphase has three subphases (G1, S, and G2) in cells that divide.

Q: If the DNA content of a human skin cell at the beginning of mitosis is 12 picograms (1 pg = 10^{-12} g), what is the DNA content of a cell in G1?

structures that move the chromatids to opposite ends of the dividing cell.

The initiation, termination, and operations of these phases are regulated by specific signals. Not all cells follow this "march through the phases." Some cells in G1 enter an inactive resting phase of the cell cycle called **G0**. These cells sometimes return to G1 and the rest of the cell cycle under certain environmental conditions, such as an extracellular signal. Otherwise, cells may stay in G0; examples in humans are cells of the heart (cardiac muscle) and brain (neurons).

Specific internal signals trigger events in the cell cycle

Cell fusion experiments were used to reveal the existence of internal signals that control the transitions between stages of the cell cycle. For example, an experiment involving the fusion of HeLa cells (the cells described in the opening story) at different phases of the cell cycle showed that a cell in S phase produces a substance that activates DNA replication (**Investigating Life: What Controls the Reproduction of Cancer Cells?**). Similar experiments pointed to the existence of signals controlling entry into M phase. Additional experiments showed that cell cycle progression signals are controlled by protein kinase activity.

experiment

Original Paper: Rao, P. N. and R. T. Johnson. 1970. Mammalian cell fusion: Studies on the regulation of DNA synthesis and mitosis. *Nature* 225: 159–164.

Nuclei of cells in G1 phase do not undergo DNA replication, but nuclei in S phase do. Rao and Johnson wondered whether substances present in cells in the S phase could be used to induce DNA replication in cells in the G1 phase.

HYPOTHESIS▶ A cell in S phase contains an activator of DNA replication.

METHOD

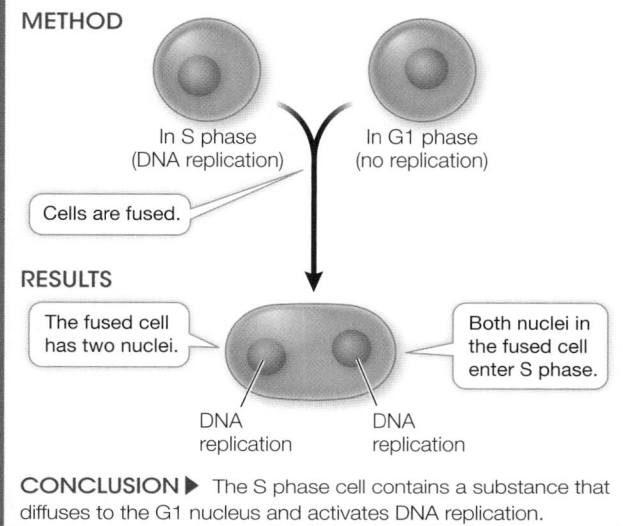

In S phase (DNA replication)   In G1 phase (no replication)

Cells are fused.

RESULTS

The fused cell has two nuclei.   Both nuclei in the fused cell enter S phase.

DNA replication   DNA replication

CONCLUSION▶ The S phase cell contains a substance that diffuses to the G1 nucleus and activates DNA replication.

Progress through the cell cycle depends on the activities of **cyclin-dependent kinases**, or **Cdk's**. Recall from Key Concept 7.2 that a protein kinase is an enzyme that catalyzes the transfer of a phosphate group from ATP to a target protein; this phosphate transfer is called phosphorylation.

$$\text{Protein + ATP} \xrightarrow{\text{Protein kinase}} \text{protein-P + ADP}$$
(Changed shape and activity)

By catalyzing the phosphorylation of certain target proteins, Cdk's play important roles at various points in the cell cycle. The discovery that Cdk's induce cell division is a beautiful example of how research on different organisms and different cell types can converge on a single idea. One group of scientists, led by James Maller at the University of Colorado, was studying immature frog eggs, trying to find out how they are stimulated to divide and eventually form mature eggs. They found that when they added a protein from maturing eggs to immature eggs cells, the latter were stimulated to divide. They named the protein, maturation promoting factor.

Meanwhile, Leland Hartwell at the University of Washington was studying the cell cycle in yeast (a single-celled eukaryote; see Figure 11.1A) and found a strain that was stalled at the G1–S boundary and determined that it lacked a Cdk. It turned out that this yeast Cdk and the sea urchin maturation promoting factor had similar properties, and further work confirmed that the sea urchin protein was indeed a Cdk!

work with the data

The fusion of cellular membranes is a natural process; it occurs during endocytosis and exocytosis, and in fertilization (the fusion of gametes). Membrane fusion also occurs when membrane-enclosed viruses infect their host cells. Occasionally these viruses also induce the fusion of adjacent host cells, creating a multinucleate cell. This observation led to the use of Sendai virus, a membrane-enclosed mouse respiratory virus, as a tool in the laboratory to fuse cells experimentally. Rao and Johnson used this strategy to study the regulation of the cell cycle.

In their experiment, Rao and Johnson used HeLa cells, which divide continuously (see the opening story of this chapter). First, they isolated cells in either G1 or S phase. Before fusion, the cells in S phase were exposed to a radioactively labeled component of DNA (thymidine). The radioactivity was incorporated into these cells' newly replicated DNA, labeling their nuclei. The S and G1 cells were then fused using Sendai virus (resulting in G1/S fusions) and again exposed to labeled thymidine. At various times after fusion, the scientists calculated the percent of previously unlabeled (G1) nuclei that had incorporated new label (i.e., had replicated their DNA) (**Figure A**). In a second series of experiments, S and G2 cells were fused in various combinations and then the numbers of cells in mitosis were counted and expressed as a percent of all cells in the population (**Figure B**).

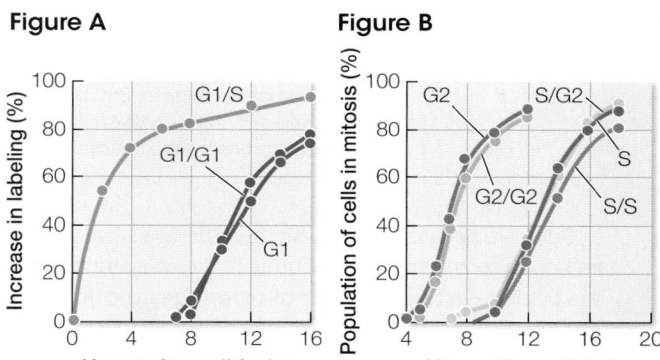

Figure A   **Figure B**

QUESTIONS▶

1. According to Figure A, how long did it take for all the G1 nuclei in the G1/S cells to become labeled?

2. Examine the data for fused G1/G1 cells and unfused G1 cells in Figure A. Explain why these were appropriate controls for the experiment. When did these nuclei become labeled? Compare these times with each other and with that for the G1/S nuclei and discuss.

3. Examine the data in Figure B. Why did it take longer for the S phase cells to begin mitosis than the G2 cells?

4. According to Figure B, did fusion with G2 cells alter the timing of mitosis in the S cell nuclei? Explain what this means in terms of regulation of the cell cycle.

A similar **work with the data** exercise may be assigned in **LaunchPad**.

Similar Cdk's were soon found to control the G1-to-S transition in many other organisms, including humans. This control point in the cell cycle is now called the **restriction (R) point**. Other Cdk's were found to control other parts of the cell cycle.

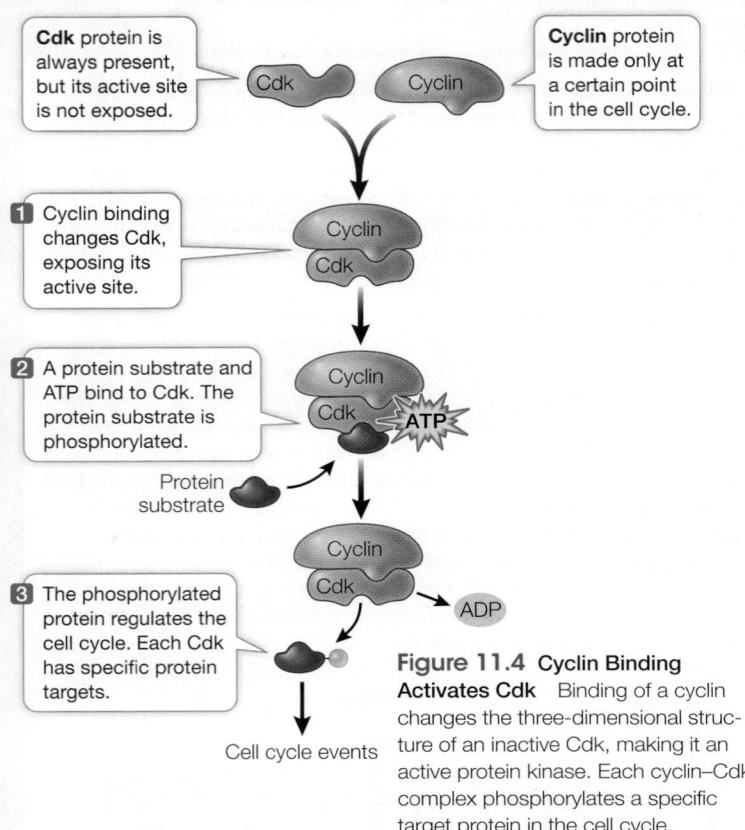

Cdk protein is always present, but its active site is not exposed.

Cyclin protein is made only at a certain point in the cell cycle.

1 Cyclin binding changes Cdk, exposing its active site.

2 A protein substrate and ATP bind to Cdk. The protein substrate is phosphorylated.

Protein substrate

ATP

3 The phosphorylated protein regulates the cell cycle. Each Cdk has specific protein targets.

ADP

Cell cycle events

Figure 11.4 Cyclin Binding Activates Cdk Binding of a cyclin changes the three-dimensional structure of an inactive Cdk, making it an active protein kinase. Each cyclin–Cdk complex phosphorylates a specific target protein in the cell cycle.

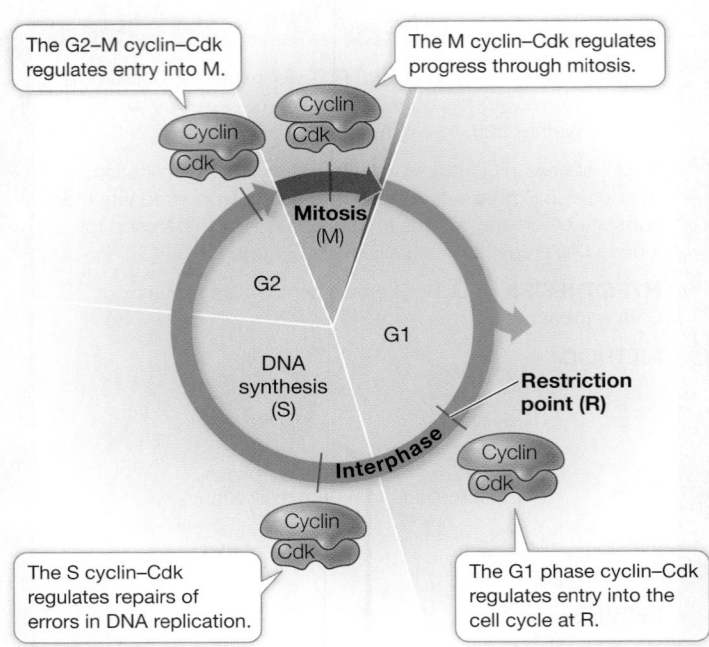

The G2–M cyclin–Cdk regulates entry into M.

The M cyclin–Cdk regulates progress through mitosis.

Mitosis (M)

G2

G1

DNA synthesis (S)

Restriction point (R)

Interphase

The S cyclin–Cdk regulates repairs of errors in DNA replication.

The G1 phase cyclin–Cdk regulates entry into the cell cycle at R.

Figure 11.5 Cyclin-Dependent Kinases Regulate Progress through the Cell Cycle By acting at checkpoints (red lines), different cyclin–Cdk complexes regulate the orderly sequence of events in the cell cycle.

Cdk's are not enzymatically active as protein kinases unless they are bound to another class of protein, the activator called **cyclin**. This binding—an example of *__allosteric regulation__—activates the Cdk by altering its shape and exposing its active site to substrates (**Figure 11.4**). The cyclin–Cdk that controls passage from G1 to S phase is not the only such complex involved in regulating the eukaryotic cell cycle. There are different cyclin–Cdk complexes, composed of various cyclins and Cdk's that act at different stages of the cycle (**Figure 11.5**). The details of how these complexes form and function vary among eukaryotic organisms, but we will focus here on the complexes found in mammalian cells. As an example, let's take a closer look at the cyclin–Cdk complex that controls the G1-to-S transition.

---

*connect the concepts As discussed in Key Concept 8.5, allosteric regulation occurs when another molecule induces a change in the three-dimensional structure of an enzyme. This changes the ability of the enzyme to bind its substrate, and so changes the rate of the reaction catalyzed by the enzyme.

Cyclin–Cdk catalyzes the phosphorylation of a protein called retinoblastoma protein (RB, named because of its role in cancer; see Key Concept 11.7). In many cells, RB or a protein like it acts as an inhibitor of the cell cycle at the R point. To begin S phase, a cell must get by the RB block. Here is where cyclin–Cdk comes in: it catalyzes the phosphorylation at multiple sites on the RB molecule. This causes a change in the three-dimensional structure of RB,

thereby inactivating it. With RB out of the way, the cell cycle can proceed. To summarize:

$$\text{RB} \xrightarrow{\text{Cyclin–Cdk}} \text{RB-P}$$
(Active—blocks cell cycle)   (Inactive—allows cell cycle)

Progress through the cell cycle is regulated by the activities of Cdk's, and so regulating *them* is a key to regulating cell division. An effective way to regulate Cdk's is to regulate the presence or absence of cyclins (**Figure 11.6**). Simply put, if a cyclin is not present, its partner Cdk is not active. As their name suggests, cyclins are present cyclically: they are made only at certain times in the cell cycle.

The different cyclin–Cdk's act at **cell cycle checkpoints**, signaling pathways that regulate the cell cycle's progress. For example, if a cell's DNA is substantially damaged by radiation or toxic chemicals, the cell may be prevented from successfully completing a cell cycle. During interphase, there are three checkpoints, with a fourth during mitosis (see Figure 11.5; **Table 11.1**). The table lists the triggers that will cause the cell cycle to pause at each point.

As an example, let's consider the G1 checkpoint (R). If DNA is damaged by radiation during G1, a signaling pathway results in the production of a protein called p21. (The *p* stands for "protein" and the *21* stands for its molecular weight—about 21,000.) The p21 protein binds to the G1–S cdk, preventing cyclin binding. This keeps the cdk in its inactive state and the cell cycle pauses while repairs are made to the DNA (you will learn more about DNA repair in Key Concept 13.4). When the DNA damage pathway is no longer operating, p21 breaks down, allowing the cyclin–Cdk's to function, and the cell cycle proceeds. If DNA damage is severe and it cannot be repaired, the cell will undergo programmed cell death (apoptosis,

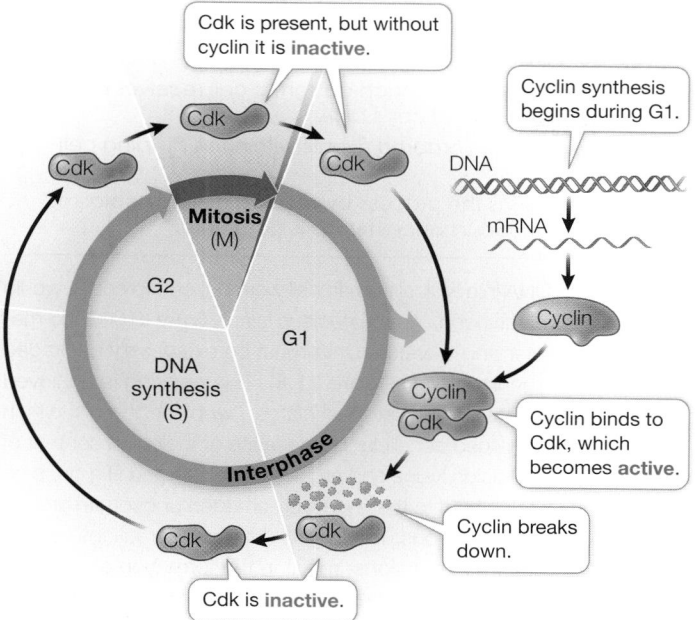

Figure 11.6 Cyclins Are Transient in the Cell Cycle Cyclins are made at a particular time and then break down. In this case, the cyclin is present during G1 and activates a Cdk at that time.

which we will discuss later in this chapter). Such controls prevent defective cells from proliferating and potentially harming an organism.

Growth factors can stimulate cells to divide

Cyclin–Cdk's provide cells with internal controls of their progress through the cell cycle. But the cell cycle is also influenced by external signals. Not all cells in an organism go through the cell cycle on a regular basis. Some cells either no longer go through the cell cycle and enter G0, or go through it slowly and divide infrequently. If such cells are to divide, they must be stimulated by external chemical signals called **growth factors**. Here are a few examples:

- If you cut yourself and bleed, specialized cell fragments called platelets gather at the wound to initiate blood clotting. The platelets produce and release a protein called platelet-derived growth factor which diffuses to the adjacent cells in the skin and stimulates them to divide and heal the wound.

- Red and white blood cells have limited lifetimes and must be replaced through the division of immature, unspecialized blood cell precursors in the bone marrow. Two types of growth factors, interleukins and erythropoietin, stimulate the division and specialization, respectively, of precursor cells of white blood cells and red blood cells.

table **11.1** Cell Cycle Checkpoints

Cell cycle phase	Checkpoint trigger
G1	DNA damage
S	Incomplete replication or DNA damage
G2	DNA damage
M	Chromosome unattached to spindle

Growth factors bind to specific receptors on target cells and activate signal transduction pathways (see Chapter 7) that end with cyclin synthesis, thereby activating Cdk's and the cell cycle.

11.2 recap

The eukaryotic cell cycle is under both external and internal control. Cdk's control the eukaryotic cell cycle, and their activities are regulated by binding of cyclins. External signals such as growth factors can initiate the cell cycle.

learning outcomes

You should be able to:

- Identify each phase of the cell cycle using diagrams or descriptions.
- Explain how the progress of a eukaryotic cell through the cell cycle is controlled.
- Describe the ways in which progress through the cell cycle is modulated.
- Predict the effects of external factors on the eukaryotic cell cycle.

1. Draw a cell cycle diagram and describe the three subphases of interphase.
2. How do cyclin–Cdk's control the progress of the cell cycle?
3. What are growth factors, and how do they act to control the cell cycle?
4. The protein p16 acts on the cell cycle in a fashion similar to that of p21. Recently it was found that there is more p16 protein in the cells of older people than in younger people. What are the implications of this for aging?

Once the cell cycle is initiated and interphase completed, the cell enters mitosis, where the replicated DNA is segregated into two daughter cells.

key concept
11.3
Eukaryotic Cells Divide by Mitosis

DNA is a very long polymer, up to several μm long. At every stage of the cell cycle, these threads must be packaged into compact structures. A eukaryotic chromosome consists of one or two linear, double-stranded DNA molecules bound with many proteins (the complex of DNA and proteins is referred to as **chromatin**). Before S phase, each chromosome contains only one double-stranded DNA molecule. After it replicates during S phase, however, there are two double-stranded DNA molecules: the sister chromatids (**Figure 11.7**). Throughout G2 the sister chromatids are held together along most of their length by a protein complex called cohesin. At mitosis most of the cohesin is removed, except in a region called the **centromere**, where the chromatids remain held together. At the end of G2 and the beginning of mitosis, a second group of proteins called condensins coats the DNA molecules and makes them more compact.

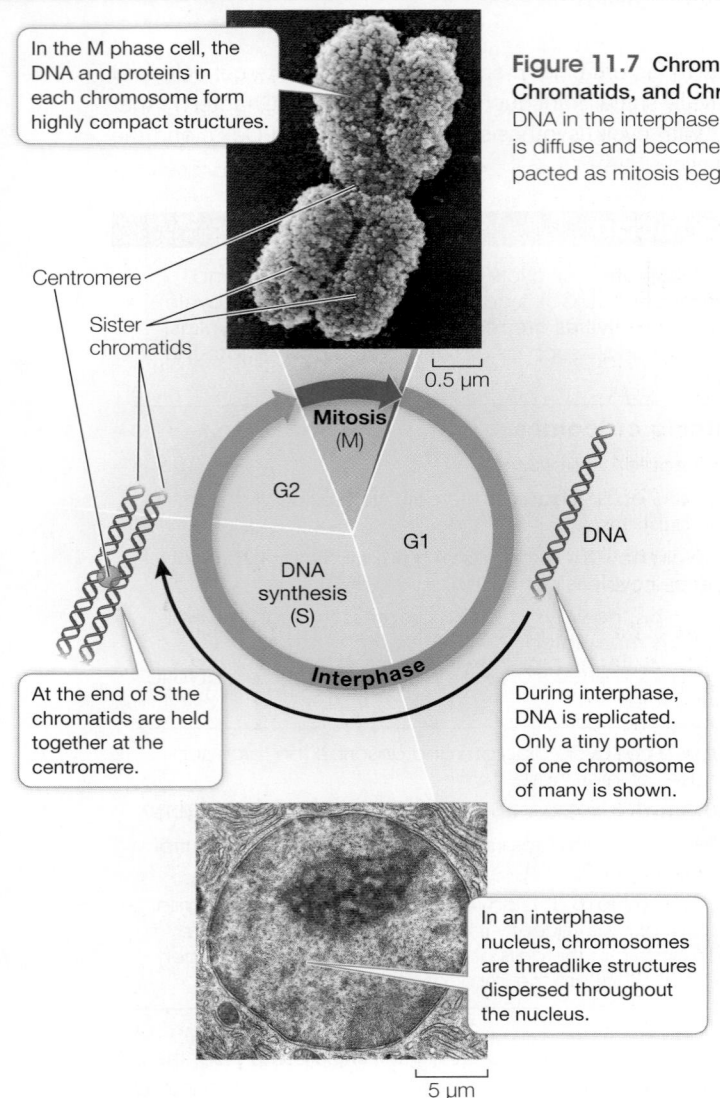

In the M phase cell, the DNA and proteins in each chromosome form highly compact structures.

Centromere

Sister chromatids

0.5 μm

Mitosis (M)

G2

G1

DNA synthesis (S)

Interphase

DNA

Figure 11.7 Chromosomes, Chromatids, and Chromatin DNA in the interphase nucleus is diffuse and becomes compacted as mitosis begins.

At the end of S the chromatids are held together at the centromere.

During interphase, DNA is replicated. Only a tiny portion of one chromosome of many is shown.

In an interphase nucleus, chromosomes are threadlike structures dispersed throughout the nucleus.

5 μm

focus your learning

- Mitosis ensures that each daughter cell receives one copy of the parent cell's DNA.
- Mitosis involves ordered events within the dividing cell nucleus.
- Cytokinesis is the process by which a cell's cytoplasm divides. It occurs once mitosis is complete.

If all of the DNA in a typical human cell were put end to end, it would be nearly 2 meters long. Yet the nucleus is only 5 μm (0.000005 meters) in diameter and therefore DNA must be extensively packaged in a highly organized way (**Figure 11.8**). This packing is achieved largely by proteins called histones (*histos*, "web" or "loom"), which are positively charged at cellular pH because of their high content of the basic amino acids lysine and arginine. The charged R groups on these amino acids bind to the negatively charged phosphate groups on DNA by ionic attractions. These DNA–histone interactions, as well as histone–histone interactions, result in the formation of beadlike units called **nucleosomes** (see Figure 11.8).

During interphase, the chromatin that makes up each chromosome consists of single DNA molecules running around vast numbers of nucleosomes that resemble beads on a string. During this phase of the cell cycle, the DNA is accessible to proteins involved in replication and transcription. Once a mitotic chromosome is formed, its compact nature makes it inaccessible to replication and transcription factors, and so these processes cannot occur. Further coiling of the chromatin continues up to the time at which the chromatids begin to move apart.

In mitosis, a single nucleus gives rise to two nuclei that are genetically identical to each other and to the parent nucleus. Mitosis

Figure 11.8 DNA Is Packed into a Mitotic Chromosome The nucleosome, formed by DNA and histones, is the essential building block in this highly compacted structure.

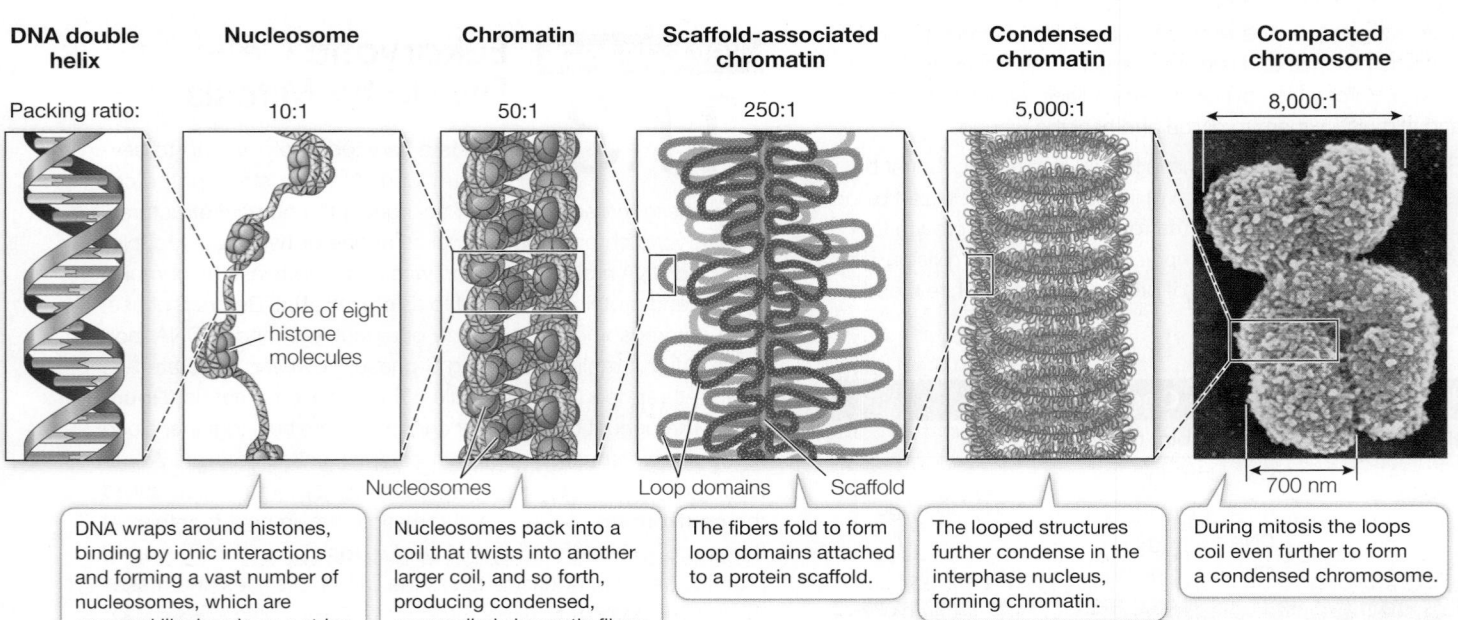

DNA double helix	Nucleosome	Chromatin	Scaffold-associated chromatin	Condensed chromatin	Compacted chromosome
Packing ratio:	10:1	50:1	250:1	5,000:1	8,000:1

Core of eight histone molecules

Nucleosomes

Loop domains   Scaffold

700 nm

DNA wraps around histones, binding by ionic interactions and forming a vast number of nucleosomes, which are arranged like beads on a string.

Nucleosomes pack into a coil that twists into another larger coil, and so forth, producing condensed, supercoiled chromatin fibers.

The fibers fold to form loop domains attached to a protein scaffold.

The looped structures further condense in the interphase nucleus, forming chromatin.

During mitosis the loops coil even further to form a condensed chromosome.

(the M phase of the cell cycle) ensures the accurate segregation of the eukaryotic cell's multiple chromosomes into the daughter nuclei. While mitosis is a continuous process in which each event flows smoothly into the next, it is convenient to subdivide it into a series of stages: prophase, prometaphase, metaphase, anaphase, and telophase (**Figure 11.9, Table 11.2**).

Let's take a closer look at two cellular structures that contribute to the orderly segregation of the chromosomes during mitosis—the centrosomes and the spindle.

 Animation 11.1 **Mitosis**
www.Life11e.com/a11.1

The centrosomes determine the plane of cell division

The **spindle apparatus** (also called the mitotic spindle or simply the spindle) is a dynamic microtubule structure that moves sister chromatids apart during mitosis. Before the spindle can form, its orientation is determined by the **centrosome** ("central body"), an organelle in the cytoplasm near the nucleus. In many organisms the centrosome consists of a pair of **centrioles**, each one a hollow tube formed by nine *****microtubule** triplets. During S phase the centrosome doubles, and at the beginning of prophase the two centrosomes separate from one another, moving to opposite ends of the nuclear envelope. These identify the "poles" toward which chromosomes move during anaphase. The cells of plants and fungi lack centrosomes, but distinct microtubule organizing centers at each end of the cell play the same role.

*****connect the concepts** As described in Key Concept 5.3, microtubules are made from dimers of tubulin monomers, which can be added or subtracted rapidly to adjust the length of the microtubule.

The positions of the centrosomes determine the plane at which the animal cell will divide; therefore they determine the spatial relationship between the two new cells. This relationship may be of little consequence to single free-living cells such as yeasts, but it is important for cells in a multicellular organism. For example, during development from a fertilized egg to an embryo, the daughter cells from some divisions must be positioned correctly to receive signals to form new tissues.

The spindle begins to form during prophase

During interphase, only the nuclear envelope, the nucleoli (see Key Concept 5.3), and a barely discernible tangle of chromatin are visible under the light microscope. The appearance of the nucleus changes as the cell enters prophase. At this stage, most of the cohesin that has held the two products of DNA replication together since S phase is removed, so the individual chromatids become visible. They are still held together by a small amount of cohesin at the centromere. Late in prophase, specialized structures called **kinetochores** develop in the centromere region, one on each chromatid. These structures will be important for chromosome movement.

Each of the two centrosomes, now on opposite sides of the nucleus, serves as a mitotic center, or pole, toward which the chromosomes will move (**Figure 11.10A**). During prophase and prometaphase, microtubules form between the poles and the

chromosomes to make up the spindle. The spindle serves as a structure to which the chromosomes attach and as a framework keeping the two poles apart. Each half of the spindle develops as tubulin dimers aggregate from around the centrioles and form long fibers that extend into the middle region of the cell. The microtubules are initially unstable, constantly forming and falling apart, until they contact kinetochores or microtubules from the other half-spindle and become more stable.

There are two groups of microtubules in the spindle:

1. Polar microtubules form the framework of the spindle and run from one pole to the other.

2. Kinetochore microtubules, which form later, attach to the kinetochores on the chromosomes. The two sister chromatids in each chromosome pair become attached to kinetochore microtubules in opposite halves of the spindle (**Figure 11.10B**). This ensures that the two chromatids will eventually move to opposite poles.

Movement of the chromatids achieves the central goal of mitosis. It accomplishes the segregation of the genetic material that must occur before the cell can divide and complete the cell cycle. Prophase prepares for this movement, and the actual segregation takes place in the next three phases of mitosis.

Chromosome separation and movement are highly organized

During the next three phases of mitosis—prometaphase, metaphase, and anaphase—dramatic changes take place in the cell and the chromosomes (see Figure 11.9). During prometaphase, the nuclear envelope breaks down and spindle formation is completed. During metaphase, the chromosomes line up at the equatorial position of the cell. Now let's consider two key processes of anaphase: separation of the chromatids, and the mechanism of their actual movement toward the poles.

table **11.2**	Summary of Cell Cycle Events
Phase	**Events**
Interphase:	
G1	Growth; restriction point at end
S	DNA replication
G2	Spindle synthesis begins; preparation for mitosis
Mitosis:	
Prophase	Condensation of chromosomes; spindle assembly
Prometaphase	Nuclear envelope breakdown; chromosome attachment to spindle
Metaphase	Alignment of chromosomes at equatorial plate
Anaphase	Separation of chromatids; migration to poles
Telophase	Chromosomes decondense; nuclear envelope re-forms
Cytokinesis	Cell separation; cell membrane and/or wall formation

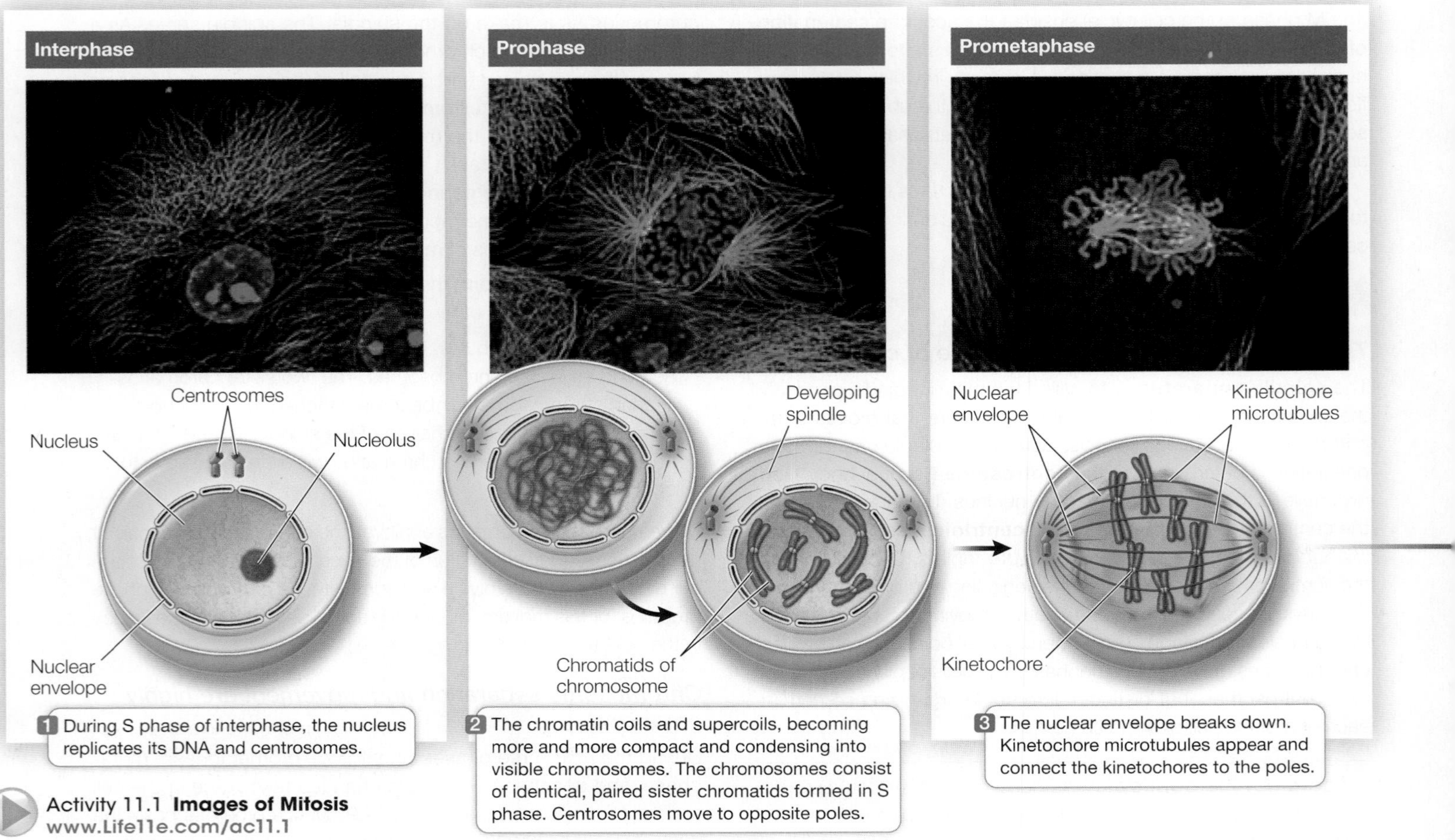

Interphase

Centrosomes

Nucleus

Nucleolus

Nuclear
envelope

1 During S phase of interphase, the nucleus
replicates its DNA and centrosomes.

Activity 11.1 **Images of Mitosis**
www.Life11e.com/ac11.1

Prophase

Developing
spindle

Chromatids of
chromosome

2 The chromatin coils and supercoils, becoming
more and more compact and condensing into
visible chromosomes. The chromosomes consist
of identical, paired sister chromatids formed in S
phase. Centrosomes move to opposite poles.

Prometaphase

Nuclear
envelope

Kinetochore
microtubules

Kinetochore

3 The nuclear envelope breaks down.
Kinetochore microtubules appear and
connect the kinetochores to the poles.

Figure 11.9 The Phases of Mitosis in an Animal Cell   Mitosis
results in two new genetically identical nuclei, each one destined for
one of the two daughter cells produced when a cell divides. In the
micrographs, the green dye stains microtubules (and thus the spindle);
the red dye stains the chromosomes. The chromosomes in the diagrams are stylized to emphasize the fates of the individual chromatids.

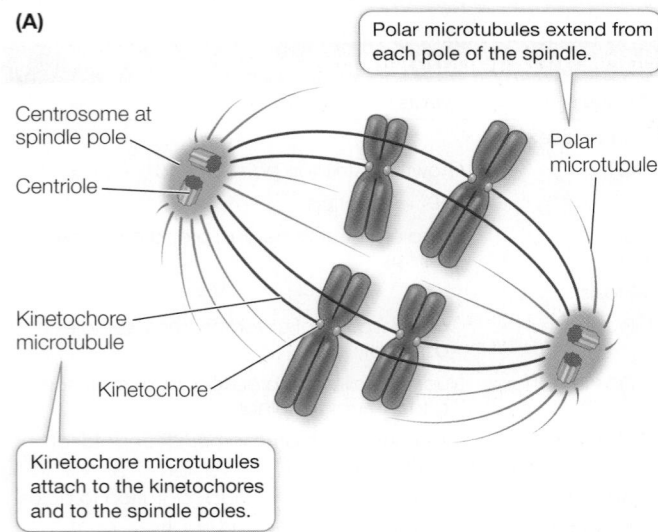

(A)

Centrosome at
spindle pole

Centriole

Kinetochore
microtubule

Kinetochore

Polar microtubules extend from
each pole of the spindle.

Polar
microtubule

Kinetochore microtubules
attach to the kinetochores
and to the spindle poles.

Activity 11.2 **The Mitotic Spindle**
www.Life11e.com/ac11.2

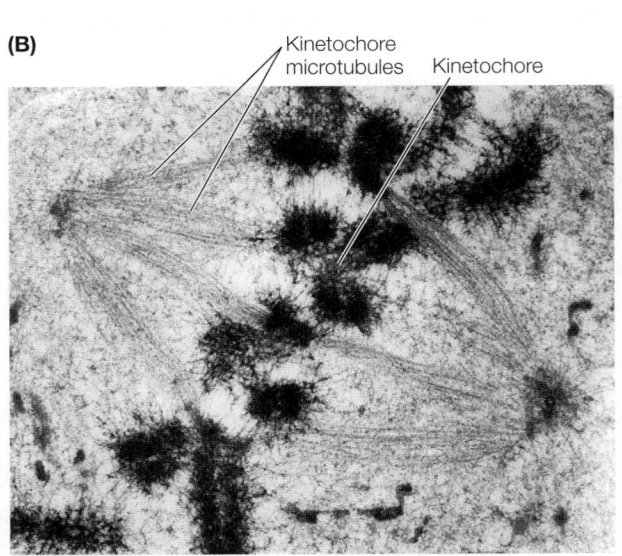

(B)

Kinetochore
microtubules   Kinetochore

Figure 11.10 The Mitotic Spindle Consists of Microtubules   **(A)** The
spindle apparatus in an animal cell at metaphase. In plant cells, centrioles
are not present. **(B)** An electron micrograph of metaphase, emphasizing the
kinetochore microtubules.

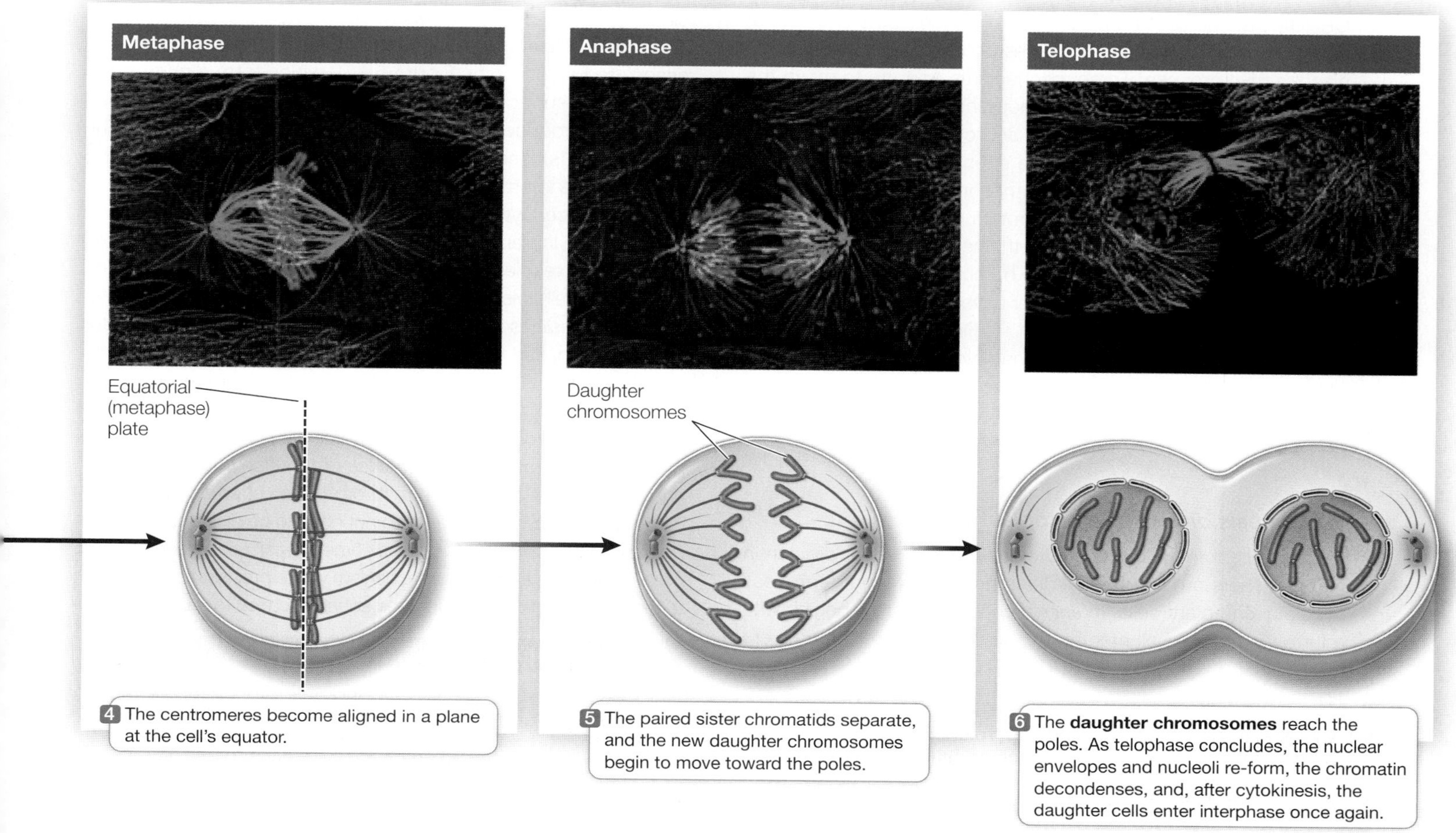

Metaphase

Equatorial (metaphase) plate

4 The centromeres become aligned in a plane at the cell's equator.

Anaphase

Daughter chromosomes

5 The paired sister chromatids separate, and the new daughter chromosomes begin to move toward the poles.

Telophase

6 The **daughter chromosomes** reach the poles. As telophase concludes, the nuclear envelopes and nucleoli re-form, the chromatin decondenses, and, after cytokinesis, the daughter cells enter interphase once again.

CHROMATID SEPARATION The separation of chromatids occurs at the beginning of anaphase. It is controlled by an M phase cyclin–Cdk (see Figure 11.5), which activates another protein complex called the anaphase-promoting complex (APC). Separation occurs because one subunit of the cohesin protein holding the sister chromatids together is hydrolyzed by a specific protease, appropriately called separase (**Figure 11.11**). A cell cycle checkpoint, often called the spindle assembly checkpoint, occurs at the end of metaphase to inhibit the APC if one of the chromosomes is not attached properly to the spindle. When all chromosomes are attached, the APC is activated and the chromatids separate. After separation the chromatids are called **daughter chromosomes**. Note the difference between chromatids and chromosomes:

- Chromatids share a centromere.
- Chromosomes have their own centromere.

Figure 11.11 Chromatid Attachment and Separation
The cohesin protein complex holds sister chromatids together at the centromere. The enzyme separase hydrolyzes cohesin at the end of metaphase, allowing the chromatids to separate into daughter chromosomes.

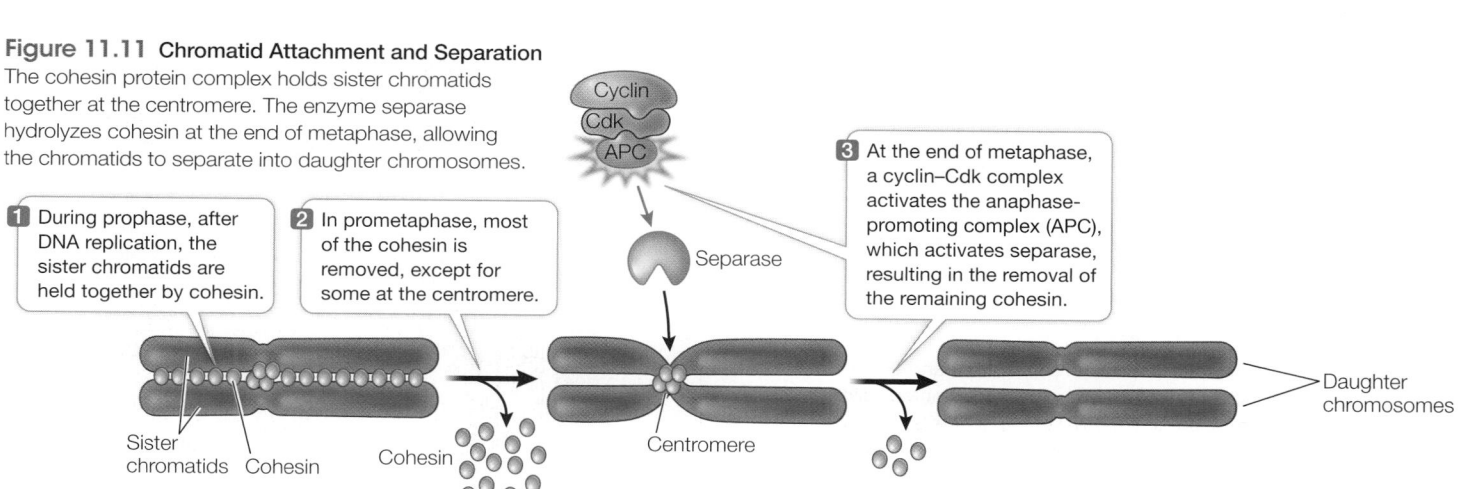

1 During prophase, after DNA replication, the sister chromatids are held together by cohesin.

2 In prometaphase, most of the cohesin is removed, except for some at the centromere.

Cyclin
Cdk
APC

Separase

3 At the end of metaphase, a cyclin–Cdk complex activates the anaphase-promoting complex (APC), which activates separase, resulting in the removal of the remaining cohesin.

Sister chromatids   Cohesin

Cohesin

Centromere

Daughter chromosomes

CHROMOSOME MOVEMENT The migration of the two sets of daughter chromosomes to the poles of the cell is a highly organized, active process. Three mechanisms operate to move the chromosomes along. First, the kinetochores contain molecular motor proteins, including kinesins and cytoplasmic dynein (see Figures 5.18 and 5.19), which use energy from ATP hydrolysis to do the work of moving the chromosomes along the microtubules. Second, the kinetochore microtubules shorten, drawing the chromosomes toward them. Third, the centrosomes move apart, aiding in separation. The last stage of mitosis is telophase, when the spindle disappears and a nuclear envelope forms around each set of chromosomes (see Figure 11.9). Finally, the cytoplasms of the two daughter cells separate during cytokinesis: the last stage of cell division.

 Media Clip 11.1 Mitosis: Live and Up Close
www.Life11e.com/mc11.1

Cytokinesis divides the cytoplasm

Cytokinesis divides the cell's cytoplasm after nuclear division in mitosis. There are substantial differences between the process in animal and plant cells. In animal cells, cytokinesis begins with a furrowing of the cell membrane, as if an invisible thread were cinching the cytoplasm between the two nuclei (**Figure 11.12A**). This contractile ring is composed of microfilaments of actin and associated myosin (see Figure 5.15), which form a ring on the cytoplasmic surface of the cell membrane. These two proteins interact to produce a contraction, pinching the cell in two. The microfilaments assemble rapidly from actin monomers that are present in the interphase cytoskeleton.

The plant cell cytoplasm divides differently because plants have rigid cell walls. In plant cells, as the spindle breaks down after mitosis, membranous vesicles derived from the Golgi apparatus appear along the plane of cell division, roughly midway between the two daughter nuclei. The vesicles are propelled along microtubules by the motor protein kinesin, and fuse to form a new cell membrane. At the same time they contribute their contents to a cell plate, which is the beginning of a new cell wall (**Figure 11.12B**).

Following cytokinesis, each daughter cell contains all the components of a complete cell. A precise distribution of chromosomes is ensured by mitosis. In contrast, organelles such as ribosomes, mitochondria, and chloroplasts do not need to be distributed equally between daughter cells, as long as some of each are present in each cell. Accordingly, there is no mechanism with a precision comparable to that of mitosis to provide for their equal allocation to daughter cells.

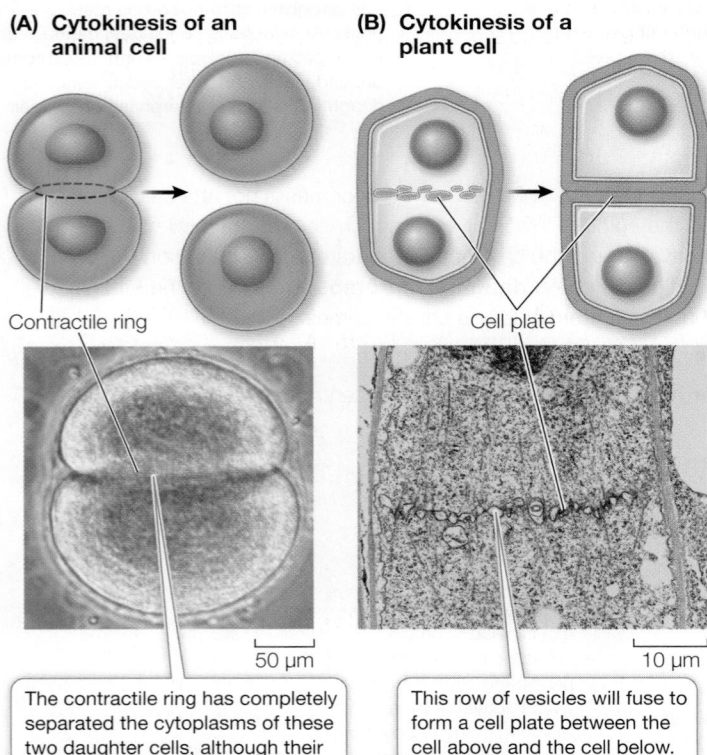

Contractile ring

Cell plate

50 μm

10 μm

The contractile ring has completely separated the cytoplasms of these two daughter cells, although their surfaces remain in contact.

This row of vesicles will fuse to form a cell plate between the cell above and the cell below.

Figure 11.12 Cytokinesis Differs in Animal and Plant Cells
(A) A sea urchin zygote (fertilized egg) that has just completed cytokinesis at the end of the first cell division of its development into an embryo. (B) A dividing plant cell in late telophase. Plant cells divide differently from animal cells because plant cells have cell walls.

11.3 recap

Mitosis is the ordered division of a eukaryotic cell nucleus into two nuclei with identical sets of chromosomes. The process of mitosis, while continuous, can be viewed as a series of events (prophase, prometaphase, metaphase, anaphase, and telophase). Once two nuclei have formed, the cell divides into two cells by cytokinesis.

learning outcomes

You should be able to:

- Show how packaging of DNA into chromosomes ensures that each daughter cell receives an entire set.
- Explain how cell structures are involved in mitosis.
- Describe cytokinesis.

1. Differentiate among a chromosome, a chromatid, and a daughter chromosome, and identify their roles in a dividing cell.
2. What are the various levels of "packing" by which the genetic information contained in linear DNA is condensed during prophase?
3. Taxol is an anticancer drug that binds to microtubules, preventing them from adding or removing subunits and causing them to form bundles. How do you think this drug works to stop cancer? Is Taxol specific to cancer cells?
4. How does cytokinesis differ in plant and animal cells?
5. How would the cell cycle be affected if there were a nonfunctional cohesin protein?

Mitosis results in two cells that are genetically identical. But there is another eukaryotic cell division process, meiosis, that results in genetic diversity. In the next section we will discuss the roles of mitosis and meiosis in sexual reproduction, and then turn to the details of meiosis in Key Concept 11.5.

key concept 11.4 Cell Division Plays Important Roles in the Sexual Life Cycle

The mitotic cell cycle repeats itself, and by this process a single cell can give rise to many cells with identical nuclear DNA. Meiosis, by contrast, produces just four daughter cells. Mitosis and meiosis are both involved in reproduction but in different ways: asexual reproduction involves only mitosis, whereas the sexual reproduction cycle involves both mitosis and meiosis.

*Asexual reproduction, sometimes called vegetative reproduction, is based on the mitotic division of the nucleus. An organism that reproduces asexually may be single-celled like yeast, reproducing itself with each cell cycle, or it may be multicellular like an aspen in a forest in the Wasatch Mountains of Utah (**Figure 11.13**). Aspen can reproduce sexually, with male and female plants, but in many aspen stands all the trees are the same sex, and DNA analyses have shown that they are **clones** of a single parent organism; the offspring are genetically identical to the parent. Any genetic variation among the trees is most likely due to small environmentally caused changes in the DNA. As you will see, this small amount of variation contrasts with the extensive variation possible in sexually reproducing organisms.

---

*connect the concepts Asexual reproduction is common in land plants, and you can learn about this in Key Concept 37.3.

Unlike asexual reproduction, **sexual reproduction** results in an organism that is not identical to its parents. Sexual reproduction requires **gametes** created by meiosis; two parents each contribute one gamete to each of their offspring. Meiosis can produce gametes—and thus offspring—that differ genetically from each other and from the parents. Because of this genetic variation, some offspring may be better adapted than others to survive and reproduce in a particular environment. While asexually reproducing organisms can have genetic diversity, meiosis generates much more diversity, which is available for natural selection and evolution.

focus your learning

- Asexual reproduction occurs via mitosis whereas sexual reproduction occurs via meiosis and mitosis.
- All sexual life cycles involve haploid and diploid phases.

Sexual life cycles produce haploid and diploid cells

In most multicellular organisms, the body cells that are *not* specialized for reproduction, called **somatic cells**, each contain two sets of chromosomes, which are found in pairs. One chromosome of each pair comes from each of the organism's two parents; in humans, for example, 23 of the 46 chromosomes come from the mother and 23 from the father. The members of such a **homologous pair** are similar in size and appearance. The two chromosomes in a homologous pair (called **homologs**) bear corresponding, though not identical, genetic information.

Figure 11.13 Asexual Reproduction on a Large Scale In this forest, the aspen trees arose from a single tree by asexual reproduction. They are virtually identical genetically.

For example, a homologous pair of chromosomes in a plant may carry different versions of a gene that controls seed shape. One homolog may carry the version for wrinkled seeds while the other may carry the version for smooth seeds. You'll see these genes in the next chapter.

There is no simple relationship between the size of an organism and the number of chromosomes in its genome. A housefly has 5 chromosome pairs and a horse has 32, but the smaller carp (a fish) has 52 pairs. Probably the highest number of chromosomes in any organism is in the fern *Ophioglossum reticulatum*, which has 1,260 (630 pairs)!

In contrast to somatic cells, gametes contain only a single set of chromosomes—that is, one homolog from each pair. The number of chromosomes in a gamete is denoted by n, and the cell is said to be **haploid**. During reproduction, two haploid gametes fuse to form a **zygote** in a process called **fertilization**. The zygote thus has two sets of chromosomes, just as the somatic cells do in a diploid organism. Its chromosome number is denoted by $2n$, and the zygote is said to be **diploid**. Depending on the organism, the zygote may divide by either meiosis or mitosis. Either way, a new mature organism develops that is capable of sexual reproduction.

All sexual life cycles involve meiosis to produce haploid cells. **Figure 11.14** presents three types. In some life cycles the products of meiosis undergo cell division, resulting in a mature organism with haploid cells. Specialized cells in these organisms become gametes. In other life cycles the gametes form directly from the products of meiosis. In all cases, the gametes fuse to produce a zygote, beginning the diploid stage of the life cycle. Since the origin of sexual reproduction, evolution has generated many different versions of the sexual life cycle.

The essence of sexual reproduction is the *random selection of half of the diploid chromosome set*—one member of each genetically similar chromosome pair—to make a haploid gamete, followed by fusion of two haploid gametes to produce a diploid cell. Both of these steps contribute to a shuffling of genetic information in the population, so that no two individuals have exactly

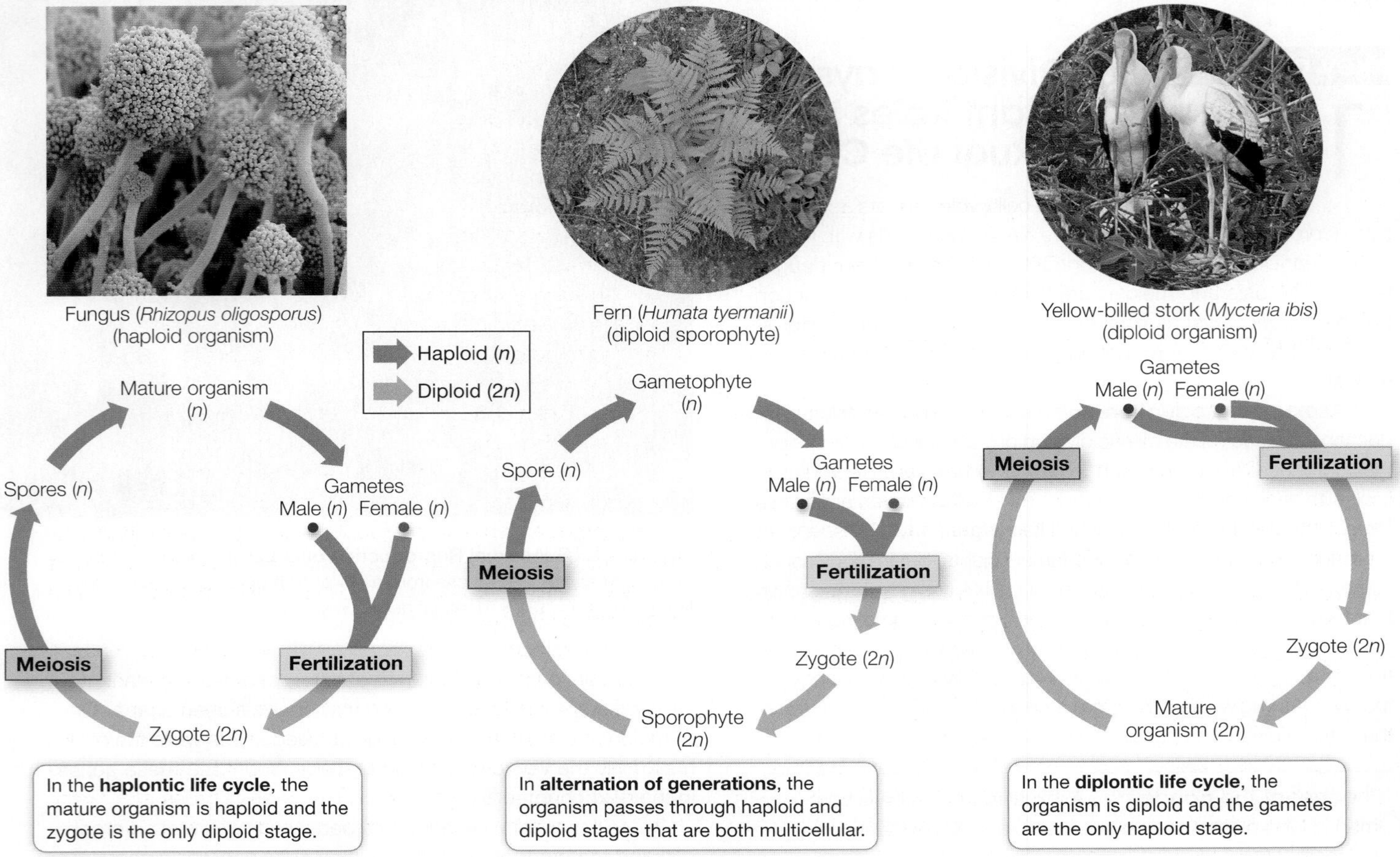

Fungus (*Rhizopus oligosporus*)
(haploid organism)

Fern (*Humata tyermanii*)
(diploid sporophyte)

Yellow-billed stork (*Mycteria ibis*)
(diploid organism)

→ Haploid (*n*)
→ Diploid (2*n*)

Fungus life cycle:
Mature organism (*n*)
Spores (*n*)
Gametes Male (*n*) Female (*n*)
Meiosis
Fertilization
Zygote (2*n*)

In the **haplontic life cycle**, the mature organism is haploid and the zygote is the only diploid stage.

Fern life cycle:
Gametophyte (*n*)
Spore (*n*)
Gametes Male (*n*) Female (*n*)
Meiosis
Fertilization
Zygote (2*n*)
Sporophyte (2*n*)

In **alternation of generations**, the organism passes through haploid and diploid stages that are both multicellular.

Stork life cycle:
Gametes Male (*n*) Female (*n*)
Meiosis
Fertilization
Zygote (2*n*)
Mature organism (2*n*)

In the **diplontic life cycle**, the organism is diploid and the gametes are the only haploid stage.

Figure 11.14 Fertilization and Meiosis Alternate in Sexual Reproduction In sexual reproduction, haploid (*n*) cells or organisms alternate with diploid (2*n*) cells or organisms.

Activity 11.3 **Sexual Life Cycle**
www.Life11e.com/ac11.3

the same genetic makeup (unless they are identical twins). The diversity provided by sexual reproduction opens up enormous opportunities for evolution.

▶ 11.4 recap

Meiosis is necessary for sexual reproduction, in which haploid gametes fuse to produce a diploid zygote. Sexual reproduction increases genetic diversity, the raw material of evolution by natural selection.

learning outcomes

You should be able to:

• Compare and contrast asexual and sexual reproduction.
• Differentiate between haploid and diploid cells.
• Compare and contrast sexual life cycles in different organisms.

1. In terms of chromosomes, how does fertilization produce a diploid organism? How would you know the cell was diploid?
2. What general features do all sexual life cycles have in common?
3. You have a sample of a small green plant. How would you tell if it is haploid or diploid?
4. Diagram haplontic and diplontic life cycles, clearly showing haploid and diploid generations.

Meiosis, unlike mitosis, results in daughter cells that differ genetically from the parent cell. We will now look at the details of meiosis to see how this genetic shuffling occurs.

key concept

11.5 Meiosis Leads to the Formation of Gametes

In the last section you learned about the role and importance of meiosis in sexual reproduction. Now you'll see how meiosis produces haploid cells. Unlike the products of mitosis, the products of meiosis are genetically different from one another and from the parent cell.

Meiosis has three overall goals:

1. It reduces the chromosome number from diploid (2*n*) to haploid (*n*).

2. It ensures that each of the haploid products has a complete set of chromosomes.

3. It generates genetic diversity among the products.

focus your learning

- Gametes, which are the products of meiosis, are genetically different from each other and from the parent cell.
- Recombination events during meiosis increase genetic variability in gametes.
- Random gene combinations in gametes result from events during meiosis.
- Meiotic errors and other events introduce additional genetic variation in gametes.

Meiosis reduces the chromosome number

Meiosis consists of *two* nuclear divisions that together reduce the number of chromosomes to the haploid number, in preparation for sexual reproduction. *Although the nucleus divides twice during meiosis, the DNA is replicated only once.* The events of meiosis are illustrated in **Figure 11.15**. You should refer to it often as you read about meiosis in detail. Two unique features characterize the first nuclear division, called **meiosis I**:

1. *Homologous chromosomes come together to pair* along their entire lengths. No such pairing occurs in mitosis.

2. *The homologous chromosome pairs separate*, but the individual chromosomes, each consisting of two sister chromatids, remain intact. (The chromatids will separate during meiosis II.)

Like mitosis, meiosis I is preceded by an interphase with an S phase, during which each chromosome is replicated. As a result, each chromosome consists of two sister chromatids, held together by cohesin proteins. At the end of meiosis I, two nuclei form, each with half of the original chromosomes (one member of each homologous pair). Since the centromeres did not separate, these chromosomes are still composed of two sister chromatids. The sister chromatids are separated during **meiosis II**, which is *not* preceded by DNA replication. As a result, the products of meiosis I and II are four cells, each containing the haploid number of chromosomes. But these four cells are not genetically identical.

 Animation 11.2 Meiosis
www.Life11e.com/a11.2

 Activity 11.4
Meiosis Simulation
www.Life11e.com/ac11.4

Chromatid exchanges during meiosis I generate genetic diversity

Meiosis I begins with a long prophase I (the first three panels of Figure 11.15), during which the chromosomes change markedly. The homologous chromosomes pair by adhering along their lengths in a process called **synapsis**. (This does not usually happen in mitosis.) This pairing process lasts from prophase I to the end of metaphase I. The four chromatids of each pair of homologous chromosomes form a **tetrad**, or bivalent. For example, in a human cell at the end of prophase I there are 23 tetrads, each consisting of four chromatids. The four chromatids come from the two partners in each homologous pair of chromosomes.

Throughout prophase I and metaphase I, the chromatin continues to coil and compact, so that the chromosomes appear ever thicker. At a certain point, the homologous chromosomes begin to be pulled apart by spindle microtubules, especially near the centromeres, but they remain held together by physical attachments mediated by cohesins. Later in prophase, regions having these attachments take on an X-shaped appearance (**Figure 11.16**) and are called **chiasmata** (singular *chiasma*, "cross").

A chiasma reflects an exchange of genetic material between non-sister chromatids on homologous chromosomes—what geneticists call **crossing over** (**Figure 11.17**). The chromosomes usually begin exchanging material shortly after synapsis begins, but chiasmata do not become visible until later, when the homologs are repelling each other. Crossing over results in **recombinant chromatids**, and it increases genetic variation among the products of meiosis by shuffling genetic information among the homologous pairs. In Chapter 12 we will explore further the genetic consequences of crossing over.

At this point, pause and think about what is going on at the molecular level: Each chromatid is a double-stranded DNA molecule, and the homologous chromatids therefore are homologous in DNA nucleotide sequence. Breakage and reunion of DNA molecules to form recombinant chromatids involves breakage of phosphodiester bonds that hold adjacent nucleotides on DNA together (see Figure 4.2) and formation of new bonds linking DNA molecules of two homologous chromatids.

Mitosis seldom takes more than an hour or two, but meiosis can take much longer. In human males, the cells in the testis that undergo meiosis take about a week for prophase I and about a month for the entire meiotic cycle. In females, prophase I begins long before a woman's birth, during her early fetal development, resumes and proceeds to meiosis II as much as decades later, during the monthly ovarian cycle, and is completed only when fertilization occurs (see Key Concept 42.4).

During meiosis homologous chromosomes separate by independent assortment

A diploid organism has two sets of chromosomes (*2n*): one set derived from its male parent, and the other from its female parent. As the organism grows and develops, its cells undergo mitotic divisions. In mitosis, each chromosome behaves independently of its homolog, and its two chromatids are sent to opposite poles during anaphase. Each daughter nucleus ends up with *2n* chromosomes. In meiosis, things are very different. **Figure 11.18** compares the two processes.

In meiosis I, chromosomes of maternal origin pair with their paternal homologs during synapsis. To clarify, you have two copies of chromosome 1 of the human genome in your cells. One copy came from your mother and the other from your father. In meiosis I, these two chromosome 1's pair up. And so on with the other 22 pairs of chromosomes. *Pairing of maternal and paternal homologs does not occur in mitosis.* Segregation of the homologs during meiotic anaphase I ensures that each newly formed cell receives one member of each homologous pair (see steps 4–6 of Figure 11.15). Returning to the example of chromosome 1, at the end of meiosis I in your cells, each daughter nucleus contains one copy of chromosome 1, whereas the diploid cell that began meiosis had two copies. Taking the other 22 pairs of chromosomes into account, at the end of meiosis I in humans, each of the two daughter cells has 23 of the original 46 chromosomes.

Recall that before meiosis begins each chromosome is duplicated in S phase of interphase. So the beginning cell of meiosis

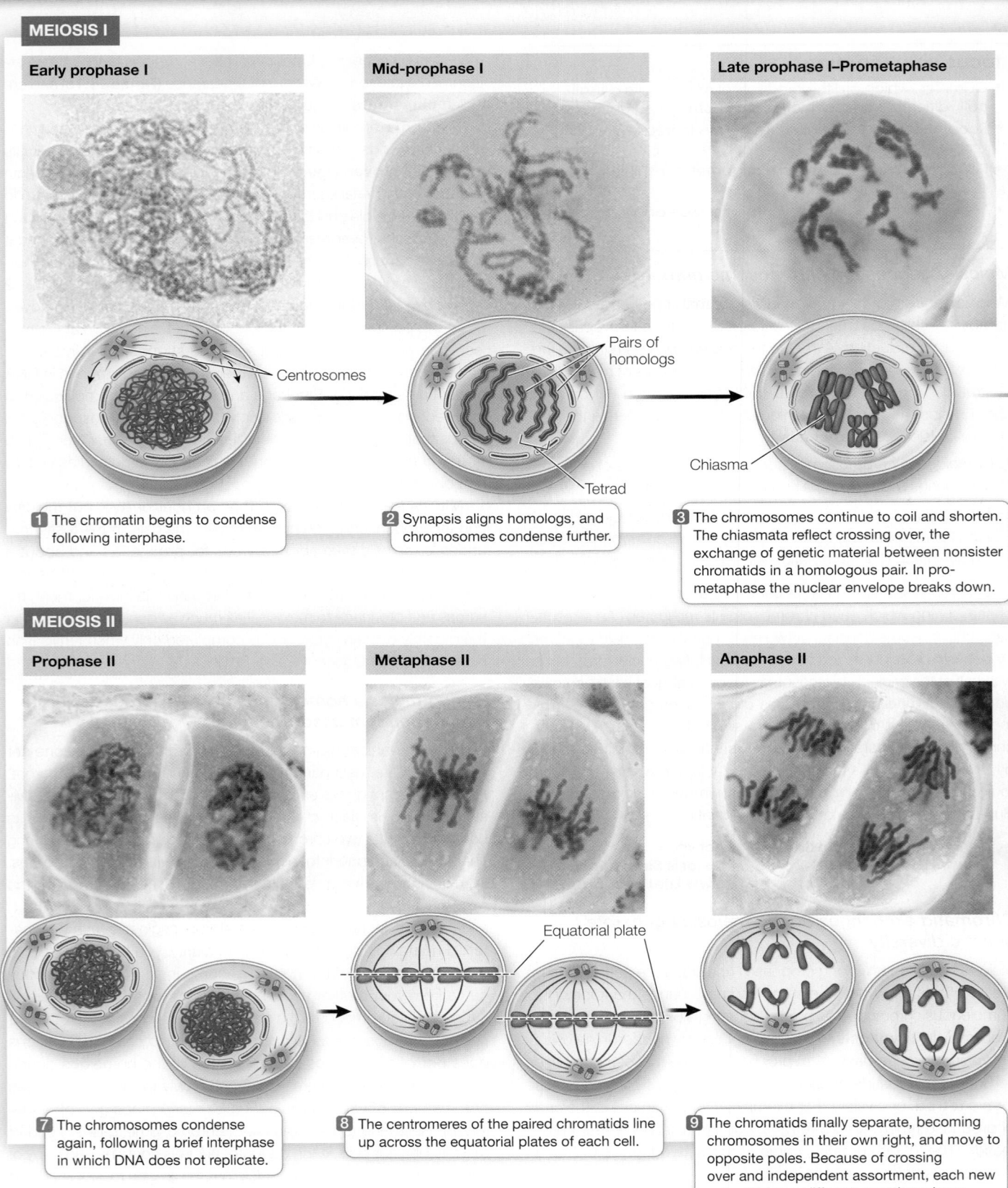

MEIOSIS I

Early prophase I

Centrosomes

1 The chromatin begins to condense following interphase.

Mid-prophase I

Pairs of homologs

Tetrad

2 Synapsis aligns homologs, and chromosomes condense further.

Late prophase I–Prometaphase

Chiasma

3 The chromosomes continue to coil and shorten. The chiasmata reflect crossing over, the exchange of genetic material between nonsister chromatids in a homologous pair. In prometaphase the nuclear envelope breaks down.

MEIOSIS II

Prophase II

7 The chromosomes condense again, following a brief interphase in which DNA does not replicate.

Metaphase II

Equatorial plate

8 The centromeres of the paired chromatids line up across the equatorial plates of each cell.

Anaphase II

9 The chromatids finally separate, becoming chromosomes in their own right, and move to opposite poles. Because of crossing over and independent assortment, each new cell will have a different genetic makeup.

had four copies of each chromosomal DNA molecule. After meiosis I each daughter cell therefore has two copies. In meiosis II, each product of meiosis I divides further to form two haploid gametes, with one copy of each chromosomal DNA molecule. So the end product of meiosis II is four haploid gametes.

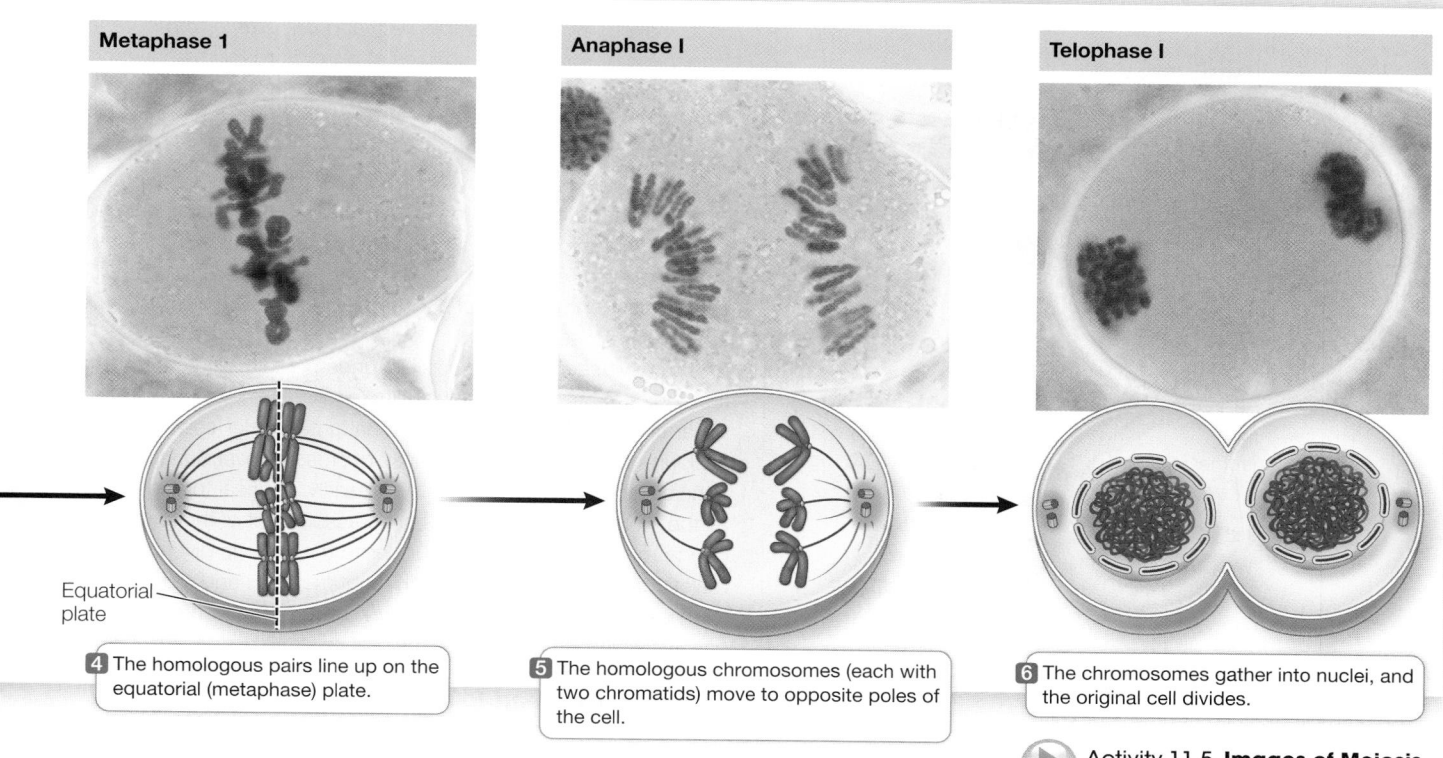

Metaphase 1

Equatorial plate

4 The homologous pairs line up on the equatorial (metaphase) plate.

Anaphase I

5 The homologous chromosomes (each with two chromatids) move to opposite poles of the cell.

Telophase I

6 The chromosomes gather into nuclei, and the original cell divides.

▶ Activity 11.5 **Images of Meiosis**
www.Life11e.com/ac11.5

Figure 11.15 Meiosis: Generating Haploid Cells   In meiosis, four daughter nuclei are produced, each of which has half as many chromosomes as the original cell. Four haploid cells are the result of two successive nuclear divisions. The micrographs show meiosis in the male reproductive organ of a lily; the diagrams show the corresponding phases in an animal cell. (For instructional purposes, the chromosomes from one parent are colored blue and those from the other parent are red.)

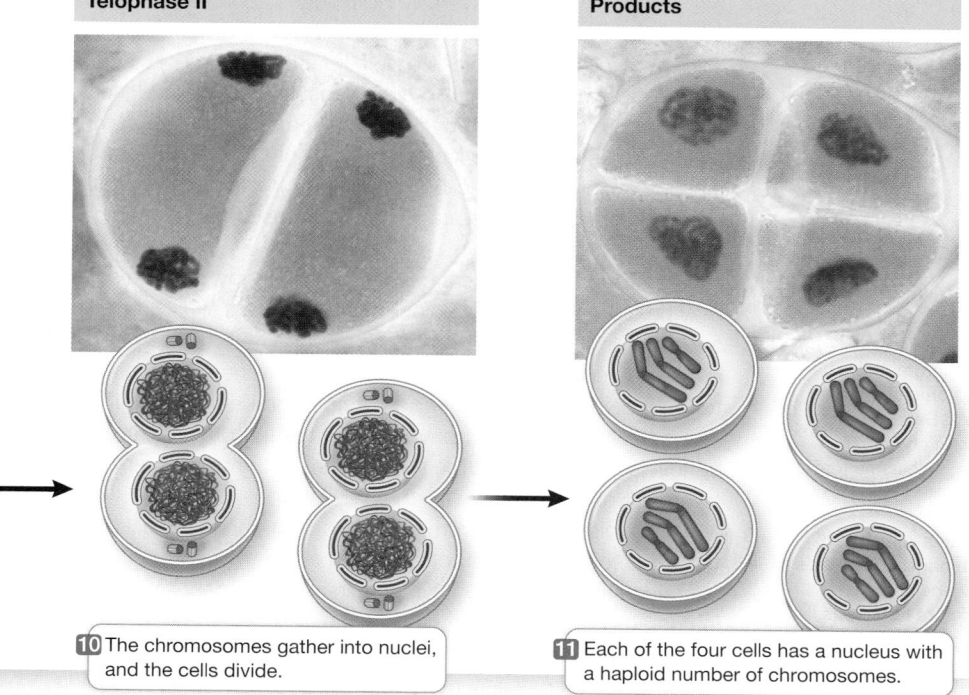

Telophase II

10 The chromosomes gather into nuclei, and the cells divide.

Products

11 Each of the four cells has a nucleus with a haploid number of chromosomes.

there are two homologous pairs of chromosomes in the diploid parent nucleus. A particular daughter nucleus could receive the paternal chromosome 1 and the maternal chromosome 2. Or it could get paternal 2 and maternal 1, or both maternal, or both paternal. It all depends on the way in which the homologous pairs line up at metaphase I.

***connect the concepts**   Mendel proposed the concept of independent assortment after he observed the results of genetic crosses with pea plants. See Key Concept 12.1.

Crossing over is one reason for the genetic diversity among the products of meiosis; another is ***independent assortment**, whereby each haploid cell receives an entire set of genes, but only one of each pair from each parent, from the diploid cell in a random fashion. It is a matter of chance which member of a homologous pair goes to which daughter cell at anaphase I. For example, imagine

Note that of the four possible chromosome combinations just described, only two produce daughter nuclei with full complements of either maternal or paternal chromosome sets (apart from the material exchanged by crossing over). *The greater the number of chromosomes, the less probable it is that the original parental*

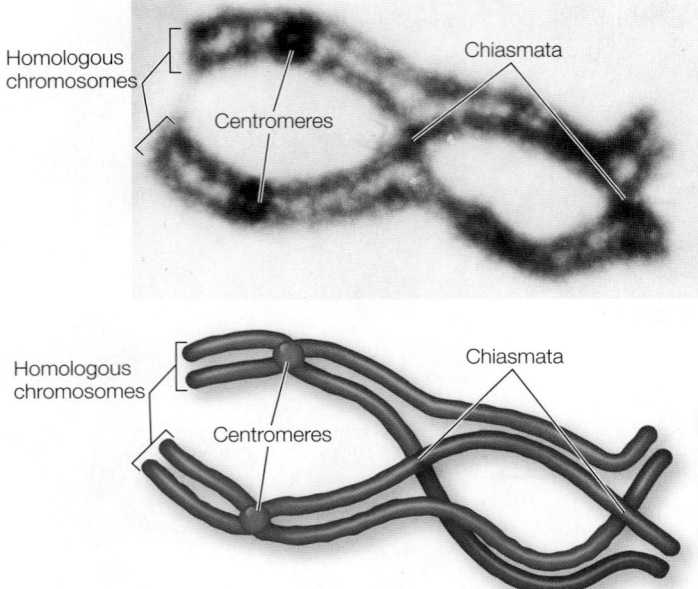

Figure 11.16 Chiasmata: Evidence of Genetic Exchange between Chromatids This micrograph shows a pair of homologous chromosomes, each with two chromatids, during prophase I of meiosis in a salamander. Two chiasmata are visible.

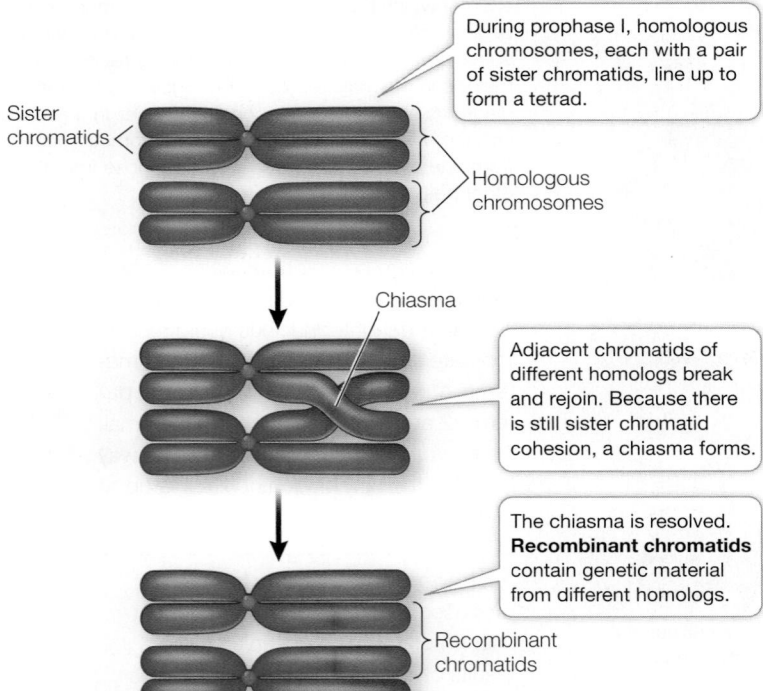

During prophase I, homologous chromosomes, each with a pair of sister chromatids, line up to form a tetrad.

Adjacent chromatids of different homologs break and rejoin. Because there is still sister chromatid cohesion, a chiasma forms.

The chiasma is resolved. **Recombinant chromatids** contain genetic material from different homologs.

Figure 11.17 Crossing Over Forms Genetically Diverse Chromosomes The exchange of genetic material by crossing over results in new combinations of genetic information on the recombinant chromosomes. The two different colors distinguish the chromosomes contributed by the male and female parents.

combinations will be reestablished, and the greater the potential for genetic diversity. Most species of diploid organisms have more than two pairs of chromosomes. In humans, with 23 chromosome pairs, 2^{23} (8,388,608) different combinations can be produced just by the mechanism of independent assortment. Taking the extra genetic shuffling afforded by crossing over into account, the number of possible combinations is very large indeed! Crossing over and independent assortment, along with the processes that result in mutations, provide the genetic diversity needed for the differential survival and reproduction of diverse individuals—the basis of evolution by natural selection.

You have seen how meiosis I is fundamentally different from mitosis. In contrast, meiosis II is similar to mitosis in that it involves the separation of sister chromatids into daughter nuclei (see steps 7–11 in Figure 11.15). However, because of crossing over during meiosis I, the sister chromatids are not necessarily identical to one another as they would be in mitosis. Chance assortment of the chromatids during meiosis II contributes further to the genetic diversity of the meiotic products. The final products of meiosis I and meiosis II are four haploid daughter cells, each with one set (*n*) of chromosomes, each of which is genetically non-identical.

Meiotic errors lead to abnormal chromosome structures and numbers

In the complex processes of mitosis and meiosis, things occasionally go wrong. In meiosis I, a pair of homologous chromosomes may fail to separate, and in mitosis or meiosis II, sister chromatids may fail to separate. Conversely, homologous chromosomes may fail to remain together during metaphase I of meiosis, and then both may migrate to the same pole in anaphase I. These are all examples of **nondisjunction**, which results in the production of aneuploid cells. **Aneuploidy** is a condition in which one or more chromosomes are either lacking or present in excess. If nondisjunction occurs during meiosis, it can lead to offspring with either one too many, or one too few, chromosomes in all of their cells (**Figure 11.19**).

There are many different causes of aneuploidy, but one cause may be a breakdown in the cohesins that keep sister chromatids and tetrads joined together during prophase. These and other proteins ensure that when the chromosomes line up at the equatorial plate at metaphase I, for example, one homolog will face one pole and the other homolog will face the other pole. If the cohesins break down at the wrong time, both homologs may go to one pole.

Aneuploidy resulting from nondisjunction during meiosis is often lethal for the affected offspring. In a few cases, the affected offspring survive but may have certain abnormalities. An example in humans is Down syndrome, which occurs when a gamete has two copies of chromosome 21. If, for example, an egg with two of these chromosomes is fertilized by a normal sperm, the resulting zygote will have three copies of the chromosome: it will be **trisomic** for chromosome 21. A child with Down syndrome has mild to moderately impaired intellectual ability; characteristic abnormalities of the hands, tongue, and eyelids; and an increased susceptibility to cardiac abnormalities. About 1 child in 800 is born with Down syndrome. If an egg that did not receive chromosome 21 is fertilized

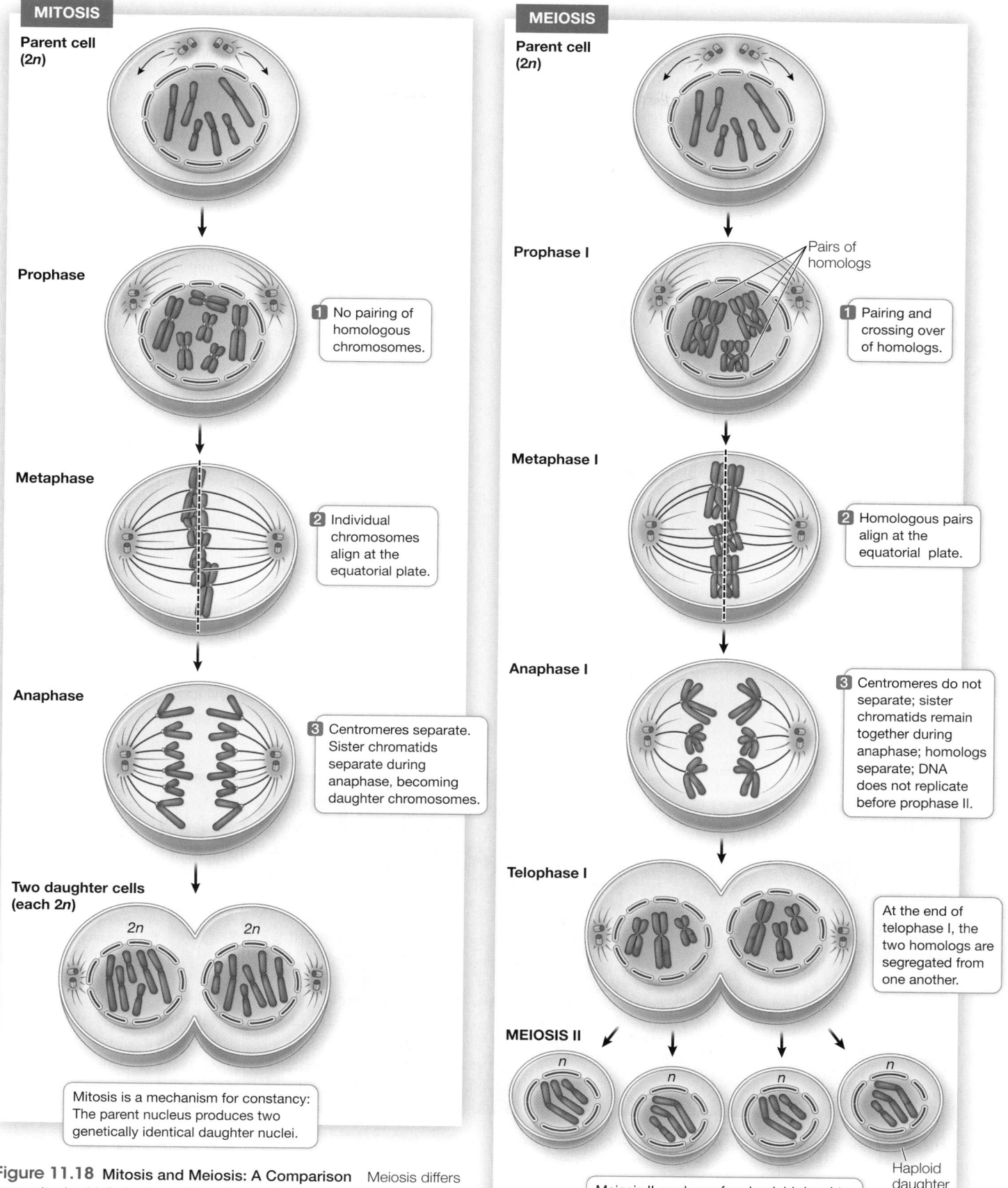

MITOSIS

Parent cell (2n)

Prophase

1 No pairing of homologous chromosomes.

Metaphase

2 Individual chromosomes align at the equatorial plate.

Anaphase

3 Centromeres separate. Sister chromatids separate during anaphase, becoming daughter chromosomes.

Two daughter cells (each 2n)

2n 2n

Mitosis is a mechanism for constancy: The parent nucleus produces two genetically identical daughter nuclei.

MEIOSIS

Parent cell (2n)

Prophase I

Pairs of homologs

1 Pairing and crossing over of homologs.

Metaphase I

2 Homologous pairs align at the equatorial plate.

Anaphase I

3 Centromeres do not separate; sister chromatids remain together during anaphase; homologs separate; DNA does not replicate before prophase II.

Telophase I

At the end of telophase I, the two homologs are segregated from one another.

MEIOSIS II

n n n n

Meiosis II produces four haploid daughter cells that are genetically distinct.

Haploid daughter cell (n)

Figure 11.18 Mitosis and Meiosis: A Comparison Meiosis differs from mitosis chiefly by the pairing of homologs and by continued association of sister chromatids at the end of metaphase I.

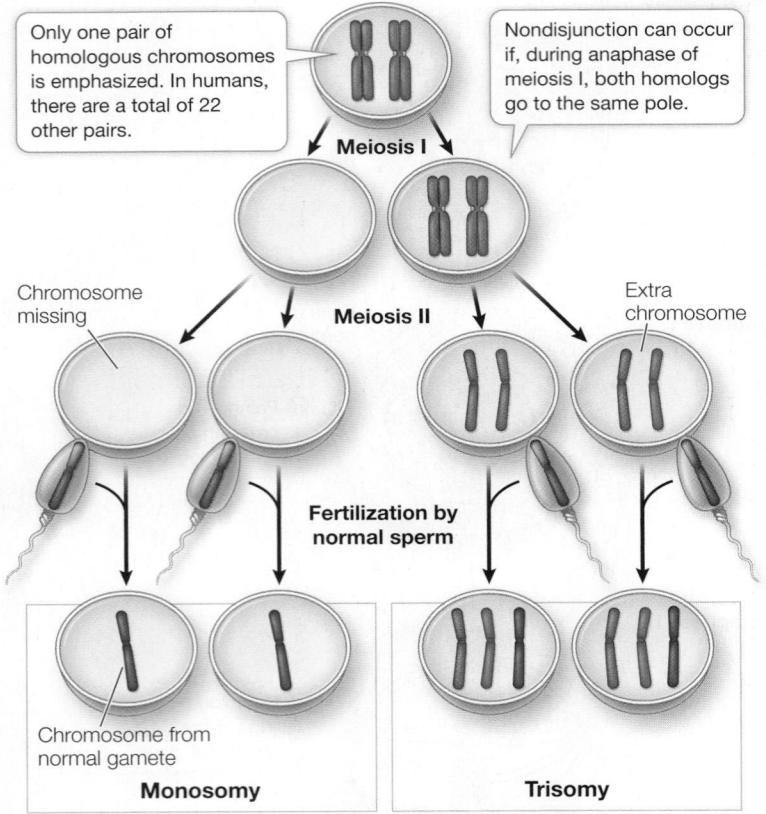

Only one pair of homologous chromosomes is emphasized. In humans, there are a total of 22 other pairs.

Nondisjunction can occur if, during anaphase of meiosis I, both homologs go to the same pole.

Meiosis I

Chromosome missing

Meiosis II

Extra chromosome

Fertilization by normal sperm

Chromosome from normal gamete

Monosomy

Trisomy

Figure 11.19 Nondisjunction Leads to Aneuploidy Nondisjunction occurs if homologous chromosomes fail to separate during meiosis I, as illustrated here, or if chromatids fail to separate during mitosis or meiosis II. The first case is shown here. The result is aneuploidy: one or more chromosomes are either lacking or present in excess. Generally, aneuploidy is lethal to the developing embryo.

by a normal sperm, the zygote will have only one copy: it will be **monosomic** for chromosome 21, and this is lethal.

Trisomies and the corresponding monosomies are surprisingly common in human zygotes, with 10–30 percent of all conceptions showing aneuploidy. But most of the embryos that develop from such zygotes do not survive to birth, and those that do often die before the age of 1 year (trisomies for chromosome 21 are the viable exception). At least one-fifth of all recognized pregnancies are spontaneously terminated (miscarried) during the first 2 months, largely because of trisomies and monosomies. The actual proportion of spontaneously terminated pregnancies is certainly higher, because the earliest ones often go unrecognized.

Other abnormal chromosomal events can also occur. In a process called **translocation**, a piece of a chromosome may break away and become attached to another chromosome. For example, a large part of one chromosome 21 may be translocated to another chromosome. Individuals who inherit this translocated piece along with two normal copies of chromosome 21 will also have Down syndrome.

The number, shapes, and sizes of the metaphase chromosomes constitute the karyotype

When cells are in metaphase of mitosis, it is often possible to count and characterize their individual chromosomes. If a photomicrograph of the entire set of chromosomes is made, the images of the individual chromosomes can be manipulated to pair them and place them in an orderly arrangement. Such a rearranged photomicrograph reveals the number, shapes, and sizes of the chromosomes in a cell, which together constitute its **karyotype** (Figure 11.20). In humans, karyotypes can aid in the diagnosis of chromosomal abnormalities such as trisomies or translocations, and this has led to an entire branch of medicine called cytogenetics. However, as you will see in Chapter 15, chromosome analysis with the microscope is replaced in some cases by direct analysis of DNA.

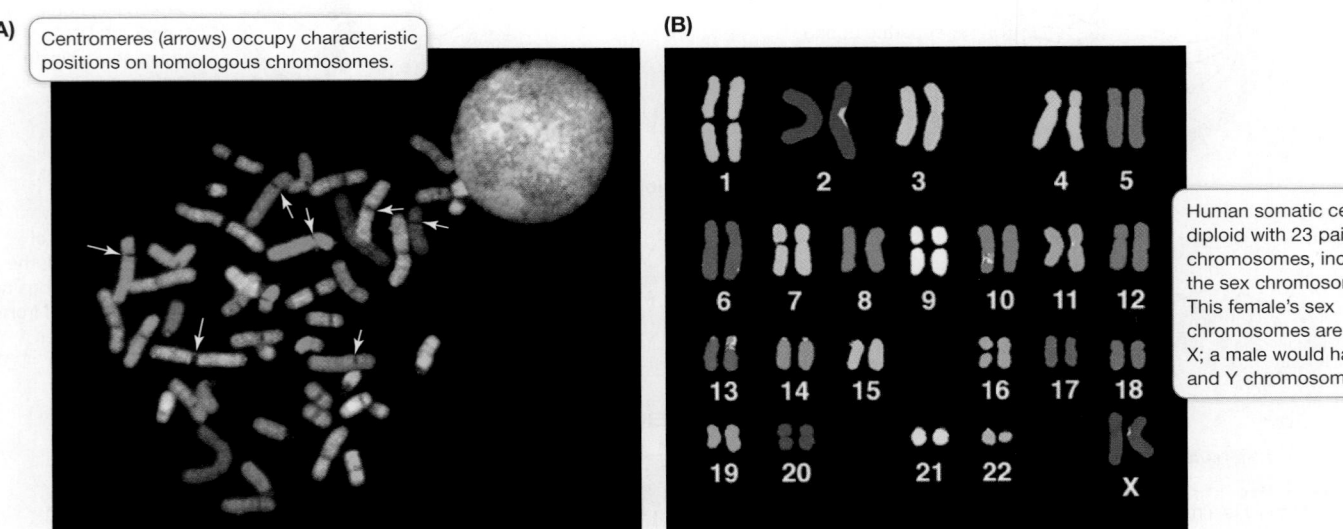

(A)

Centromeres (arrows) occupy characteristic positions on homologous chromosomes.

(B)

Human somatic cells are diploid with 23 pairs of chromosomes, including the sex chromosomes. This female's sex chromosomes are X and X; a male would have X and Y chromosomes.

Figure 11.20 The Human Karyotype **(A)** Chromosomes from a human cell in metaphase. The DNA of each chromosome pair has a specific nucleotide sequence that is stained by a particular colored dye, so that the chromosomes in a homologous pair share a distinctive color. Each chromosome at this stage is composed of two chromatids, but they cannot be distinguished. At the upper right is an interphase nucleus. **(B)** This karyogram, produced by computerized analysis of the image on the left, shows homologous pairs lined up together and numbered, clearly revealing the individual's karyotype.

Polyploids have more than two complete sets of chromosomes

As mentioned in Key Concept 11.4, mature organisms are often either diploid (for example, most animals) or haploid (for example, most fungi). Under some circumstances, triploid (3n), tetraploid (4n), or higher-order **polyploid** nuclei may form. Each of these ploidy levels represents an increase in the number of complete chromosome sets present. If there is nondisjunction of all of the chromosomes during meiosis I (see above), diploid gametes will form. This can lead to *auto*polyploidy after fertilization. Autotriploids and autotetraploids have been important in some cases in species formation. A diploid nucleus can undergo normal meiosis because there are two sets of chromosomes to make up homologous pairs, which separate during anaphase I. Similarly, a tetraploid nucleus has an even number of each kind of chromosome, so each chromosome can pair with its homolog. However, a triploid nucleus cannot undergo normal meiosis because one-third of the chromosomes would lack partners. Triploid individuals are usually sterile.

Because polyploid nuclei have more chromosome sets, their cells tend to be larger. This has led to the use of polyploid plants in agriculture. Diploid bananas (2n = 22) are smaller and produce inedible seeds; triploid bananas (3n = 33) are larger and seedless. A similar phenomenon is seen in triploid seedless watermelon. Perhaps the best known, and certainly the most important, polyploid crop plant is wheat. In this case, hybridization occurred between species, forming new *allo*polyploid conditions:

- Haploid gametes from two species (A and B) mated to form a diploid zygote (chromosomes AB).

- Nondisjunction of all chromosomes occurred during mitosis in the fertilized egg, resulting in a tetraploid (AABB), which grew up to be a fertile adult.

Modern bread wheat is the result of two such events, which occurred around 8,000 to 10,000 years ago, resulting in a hexaploid. Wheat's properties of grain formation and environmental adaptation thus come from three different ancestral species. Other allopolyploid crops include cotton, oats, and sugarcane.

11.5 recap

Meiosis produces four daughter cells in which the chromosome number is reduced from diploid to haploid. Because of the independent assortment of chromosomes and the crossing over of homologous chromatids, the four products of meiosis are not genetically identical. Meiotic errors, such as the failure of a homologous chromosome pair to separate, can lead to abnormal numbers of chromosomes. Several important crop plants, such as wheat, are polyploid.

learning outcomes

You should be able to:

- Compare and contrast the phases of mitosis with the phases of meiosis.
- Justify the claim that genetic diversity results from recombination events during meiosis.
- Use a model to illustrate how independent assortment of homologous chromosomes increases genetic diversity.

11.5 recap

- Explain how errors in meiosis can lead to unusual patterns in chromosomes.
- Describe polyploidy and explain how it develops.

1. How do crossing over and independent assortment result in unique daughter nuclei?
2. Compare prophase I of meiosis with prophase of mitosis. Compare anaphase I of meiosis with anaphase of mitosis.
3. Rarely, a person is born with an extra X chromosome (XXY). How can this arise by aneuploidy?
4. What is polyploidy, and how does it arise?

An essential role of cell division in complex eukaryotes is to replace dead cells. What causes cells to die?

key concept 11.6 Cell Death Is Important in Living Organisms

Cells within a living organism can die in one of two ways. The first type of cell death, **necrosis**, occurs when cells and tissues are damaged by mechanical means or toxins, or are starved of oxygen or nutrients. Necrotic cells often swell up and burst, releasing their contents into the extracellular environment and possibly causing inflammation (see Key Concept 41.2). **Apoptosis**, or programmed cell death, is a genetically determined series of events that occurs during normal developmental processes and in adult tissues as well.

focus your learning

- Cells die through one of two processes, necrosis or apoptosis.
- Apoptosis plays a crucial role in organism development and survival.
- Apoptosis occurs through a controlled series of cellular events.

Programmed cell death removes cells that do not benefit the organism

In animals, there are two possible reasons for apoptosis:

1. *The cell is no longer needed by the organism.* For example, before birth, a human fetus has weblike hands, with connective tissue between the fingers. As development proceeds, this unneeded tissue disappears as the cells undergo apoptosis in response to specific signals.

2. *The longer cells live, the more prone they are to genetic damage that could lead to cancer.* This is especially true of epithelial cells on the surface of an organism, which may be exposed to radiation or toxic substances. Such cells normally die after only days or weeks and are replaced by new cells.

The events of apoptosis are similar in many organisms. The cell becomes detached from its neighbors and its chromatin is digested by enzymes that cut the DNA (between the nucleosomes)

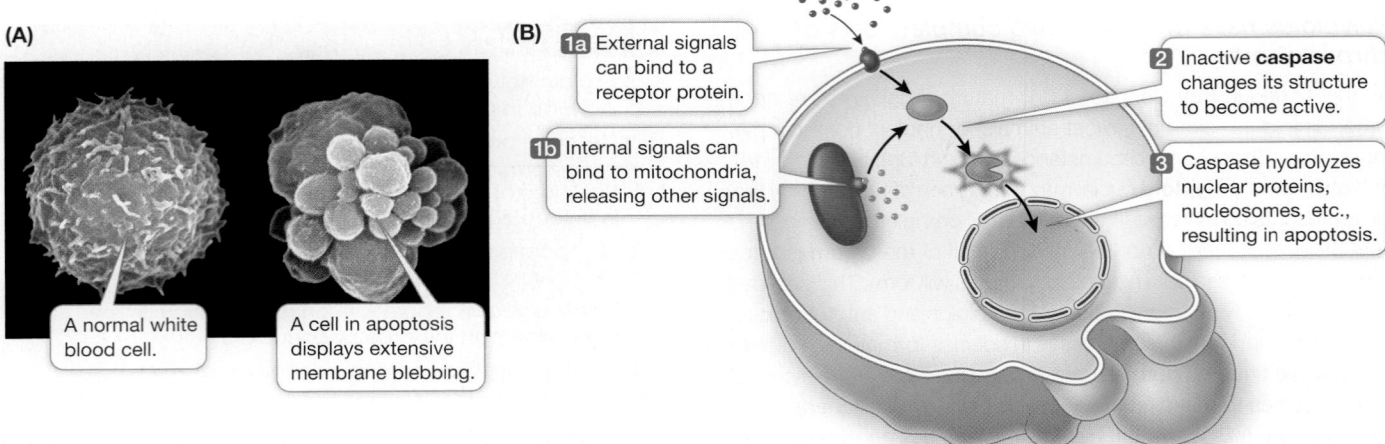

(A)

A normal white blood cell.

A cell in apoptosis displays extensive membrane blebbing.

(B)

1a External signals can bind to a receptor protein.

1b Internal signals can bind to mitochondria, releasing other signals.

2 Inactive **caspase** changes its structure to become active.

3 Caspase hydrolyzes nuclear proteins, nucleosomes, etc., resulting in apoptosis.

Figure 11.21 Apoptosis: Programmed Cell Death **(A)** Many cells are programmed to "self-destruct" when they are no longer needed, or when they have lived long enough to accumulate a burden of DNA damage that might harm the organism. **(B)** Both external and internal signals stimulate caspases, the enzymes that break down specific cell constituents, resulting in apoptosis.

Q: Apoptosis is present in most organisms, with a common molecular pathway. This indicates that apoptosis is important in terms of evolution. What might be a selective advantage for apoptosis?

into fragments of about 180 base pairs. The cell forms membranous lobes, or "blebs," that break up into cell fragments (**Figure 11.21A**). In a remarkable example of the economy of nature, the surrounding, living cells usually ingest the remains of the dead cell by phagocytosis. Neighboring cells digest the apoptotic cell contents in their lysosomes, and the digested components are recycled.

Apoptosis is also used by plant cells, in an important defense mechanism called the hypersensitive response. Plants can protect themselves from disease by undergoing apoptosis at the site of infection by a fungus or bacterium. With no living tissue to grow in, the invading organism is not able to spread to other parts of the plant. Because of their rigid cell walls, plant cells do not form blebs the way animal cells do. Instead, they digest their own cell contents in the vacuole and then release the digested components into the vascular system. You will learn more about this in Chapter 38.

A variety of signals, either external or from inside the cell, can lead to programmed cell death (**Figure 11.21B**). Such signals include hormones, growth factors, viral infection, certain toxins, or extensive DNA damage. These signals activate specific receptors, which in turn activate signal transduction pathways leading to apoptosis. Some apoptotic pathways target the mitochondria, for example, by increasing the permeability of mitochondrial membranes. The cell quickly dies if its mitochondria can't carry out cellular respiration. An important class of enzymes called **caspases** are activated during apoptosis. These enzymes are proteases that hydrolyze target molecules in a cascade of events. The cell dies as the caspases hydrolyze proteins of the nuclear envelope, nucleosomes, cytoskeleton, and cell membrane.

11.6 recap

Cell death can occur either by necrosis or by apoptosis. Apoptosis is governed by precise molecular controls.

learning outcomes

You should be able to:

• Compare and contrast necrosis and apoptosis.
• Justify the role of apoptosis in an organism's survival.
• Describe how apoptosis is regulated.

1. What types of events initiate necrosis, and how do these events differ from those that initiate apoptosis?
2. In what situations is apoptosis necessary?
3. How is apoptosis regulated?

Mitosis adds cells to organisms, and apoptosis removes them. Under normal circumstances these processes are balanced so as to benefit the organism as a whole. In the next section we will examine what happens when this balance is disturbed and cell production runs out of control.

key concept 11.7 Unregulated Cell Division Can Lead to Cancer

Perhaps no malady affecting people in the industrialized world instills more fear than cancer, and most people realize that it involves an inappropriate increase in cell numbers. One in three Americans will have some form of cancer in his or her lifetime, and at present, one in four will die of it. With 1.5 million new cases and half a million deaths in the United States annually, cancer ranks second only to heart disease as a killer.

focus your learning

• Improperly controlled cell division and the ability to migrate are two characteristics that distinguish cancer cells from normal cells.

- Cell cycle abnormalities play an important role in the development of cancer.
- Cancer therapies are targeted to events in the cell cycle.

Cancer cells differ from normal cells in important ways

Cancer cells differ from the normal cells from which they originate in two ways:

1. Cancer cells lose control over cell division.
2. Cancer cells can migrate to other locations in the body.

Most cells in the body divide only if they are exposed to extracellular signals such as growth factors. Cancer cells do not respond to these controls, and instead divide more or less continuously, ultimately forming **tumors** (large masses of cells). By the time a physician can feel a tumor or see one on an X-ray film or CT scan, it already contains millions of cells. Tumors can be benign or malignant:

- **Benign** tumors resemble the tissue they came from, grow slowly, and remain localized where they develop. For example, a lipoma is a benign tumor of fat cells that may arise in the armpit and remain there. Benign tumors are not cancers, but they must be removed if they impinge on an organ, obstructing its function.

- **Malignant** tumors do not look like their parent tissue at all. A flat, specialized epithelial cell in the lung wall may turn into a relatively featureless, round, malignant lung cancer cell (**Figure 11.22**). Malignant cells often have irregular structures, such as variable nucleus sizes and shapes. This characteristic was used to identify the cells in Henrietta Lacks's tumor as malignant (see the opening investigation of this chapter).

The second and most fearsome characteristic of cancer cells is their ability to invade surrounding tissues and spread to other parts of the body by traveling through the bloodstream or lymphatic ducts. When malignant cells become lodged in some distant part of the body, they go on dividing and growing, establishing a tumor at that new site. This spreading of cancer, called **metastasis**, results in organ failures and makes the cancer very hard to treat.

Cancer cells lose control over the cell cycle and apoptosis

Earlier in this chapter you learned about proteins that regulate the progress of a eukaryotic cell through the cell cycle:

- Positive regulators such as growth factors stimulate the cell cycle: they are like "gas pedals."
- Negative regulators such as retinoblastoma protein (RB) inhibit the cell cycle: they are like "brakes."

Just as driving a car requires stepping on the gas pedal *and* releasing the brakes, a cell will go through a division cycle only if the positive regulators are active and the negative regulators are inactive.

In most cells, the two regulatory systems ensure that the cells divide only when needed. In cancer cells, these two processes are abnormal.

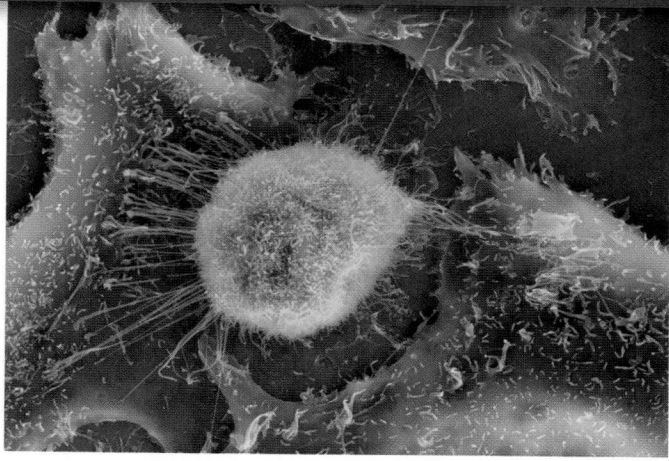

Figure 11.22 A Cancer Cell with its Normal Neighbors   This lung cancer cell (yellow-green) is quite different from the normal lung cells surrounding it. The cancer cell can divide more rapidly than its normal counterparts, and it can spread to other organs. This form of small-cell cancer is lethal, with a 5-year survival rate of only 10 percent. Most cases are caused by tobacco smoking.

- **Oncogene** proteins are positive regulators in cancer cells. They are derived from normal positive regulators that have become mutated to be overly active or that are present in excess, and they stimulate the cancer cells to divide more often. Oncogene products could be growth factors, their receptors, or other components in the signal transduction pathways (see Chapter 7) that stimulate cell division. An example of an oncogene protein is a growth factor receptor in a breast cancer cell (**Figure 11.23A**). Normal breast cells have relatively low numbers of the human epidermal growth factor receptor HER2. So breast cells are not usually stimulated to multiply in the presence of this growth factor. In about 25 percent of breast cancers, a DNA change results in the increased production of the HER2 receptor. This results in positive stimulation of the cell cycle, and a rapid proliferation of cells with the altered DNA.

- **Tumor suppressors** are negative regulators in both cancer and normal cells, but in cancer cells they are inactive. An example is the RB protein, which when active acts at R (the restriction point) in G1 to block the cell cycle (see Figure 11.5). In some cancer cells RB is inactive, allowing the cell cycle to proceed. Some viral proteins can inactivate tumor suppressors. For example, the human papillomavirus infects cells of the cervix and produces a protein called E7. E7 binds to the RB protein and prevents it from inhibiting the cell cycle (**Figure 11.23B**). Another important tumor suppressor is p53, a transcription factor that is involved in cell cycle checkpoint pathways and apoptosis. The importance of p53 as a tumor suppressor is illustrated by the fact that more than 50 percent of human tumors have mutations in the gene that encodes p53.

With oncogenes (the gas pedal) and tumor suppressor genes (the brakes), it takes more than one protein to allow the cancer cell cycle to proceed. There may be several oncogenes and tumor supressor genes involved in a single tumor. For example, two important oncogenes in mouse cells are *Myc*, whose expression

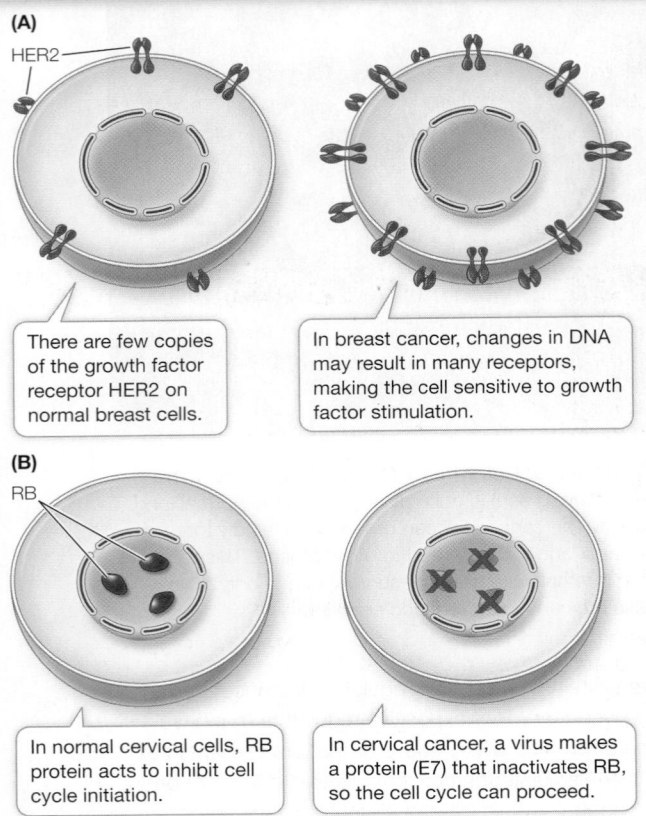

(A)

HER2

There are few copies of the growth factor receptor HER2 on normal breast cells.

In breast cancer, changes in DNA may result in many receptors, making the cell sensitive to growth factor stimulation.

(B)

RB

In normal cervical cells, RB protein acts to inhibit cell cycle initiation.

In cervical cancer, a virus makes a protein (E7) that inactivates RB, so the cell cycle can proceed.

Figure 11.23 Molecular Changes in Cancer Cells In cancer, oncogene proteins become active **(A)** and tumor suppressor proteins become inactive **(B)**.

stimulates the cell cycle and prevents apoptosis, and *Ras*, a signaling molecule you saw in Chapter 7 (see Figure 7.10). Experiments show that it takes the expression of both of these oncogenes to set the cell cycle in motion in mouse cells and convert them into tumor cells (**Figure 11.24**).

The discovery of apoptosis (see Key Concept 11.6) changed the way biologists think about cancer. In a population of cells, the net increase in cell numbers over time (the growth rate) is a function of cells added (the rate of cell division) and cells lost (the rate of apoptosis):

$$\text{Growth rate of cell population} = \text{rate of cell division} - \text{rate of apoptosis}$$

In normal nongrowing tissues, the rate of cell division equals the rate of apoptosis, so the cell population as a whole does not grow. Cancer cells are defective in their regulation of the cell cycle, resulting in increased rates of cell division. In addition, cancer cells can lose the ability to respond to positive regulators of apoptosis (see Figure 11.22), and this results in lowered rates of cell death. Both of these defects favor an increased growth rate of the cancer cell population.

Cancer treatments target the cell cycle

The most successful and widely used treatment for cancer is surgery. While physically removing a tumor is optimal, it is often difficult for a surgeon to get all of the tumor cells. (A tumor about 1 centimeter in diameter already has a billion cells!) Tumors are

experiment

Figure 11.24 Does It Take Multiple Events to Trigger the Cancer Cell Cycle?

Original Papers: Land, H., L. Parada and R. A. Weinberg. 1983. Tumorigenic conversion of primary embryo fibroblasts required at least two cooperating oncogenes. *Nature* 304: 596–602.

Sinn, E., W. Muller, P. Pattengale, I. Tepler, R. Wallace and P. Leder. 1987. Coexpression of MMTV/v-Ha-*ras* and MMTV/c-*myc* genes in transgenic mice: Synergistic action of oncogenes in vivo. *Cell* 49: 465–475.

Experiments using mouse cells in the laboratory showed that the expression of more than one oncogene is required to transform a normal cell into a cancer cell.

HYPOTHESIS ▶ The expression of both *Ras* and *Myc* oncogenes is needed to transform normal mouse cells into tumor cells.

METHOD

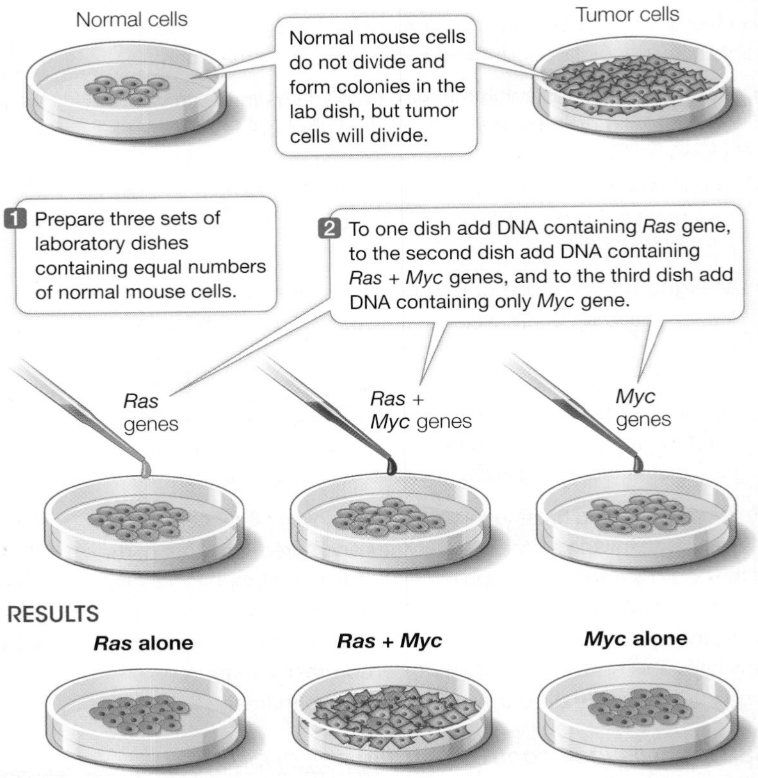

Normal cells

Normal mouse cells do not divide and form colonies in the lab dish, but tumor cells will divide.

Tumor cells

1 Prepare three sets of laboratory dishes containing equal numbers of normal mouse cells.

2 To one dish add DNA containing *Ras* gene, to the second dish add DNA containing *Ras* + *Myc* genes, and to the third dish add DNA containing only *Myc* gene.

Ras genes

Ras + *Myc* genes

Myc genes

RESULTS

Ras alone

Ras + *Myc*

Myc alone

CONCLUSION ▶ The expression of both *Myc* and *Ras* oncogenes is required for tumor cell formation.

A **work with the data** exercise that accompanies this figure may be assigned in **LaunchPad**.

generally embedded in normal tissues. Added to this is the probability that cells of the tumor may have broken off and spread to other organs. This makes it unlikely that localized surgery will be curative. So other approaches are taken to treat or cure cancer, and these generally target the cell cycle (**Figure 11.25**). The goal is to decrease the rate of cell division and/or increase the rate of apoptosis so that the cancer cell population decreases.

An example of a cancer drug that targets the cell cycle is 5-fluorouracil, which blocks the synthesis of thymine, one of the four bases in DNA. The drug paclitaxel prevents the functioning of microtubules

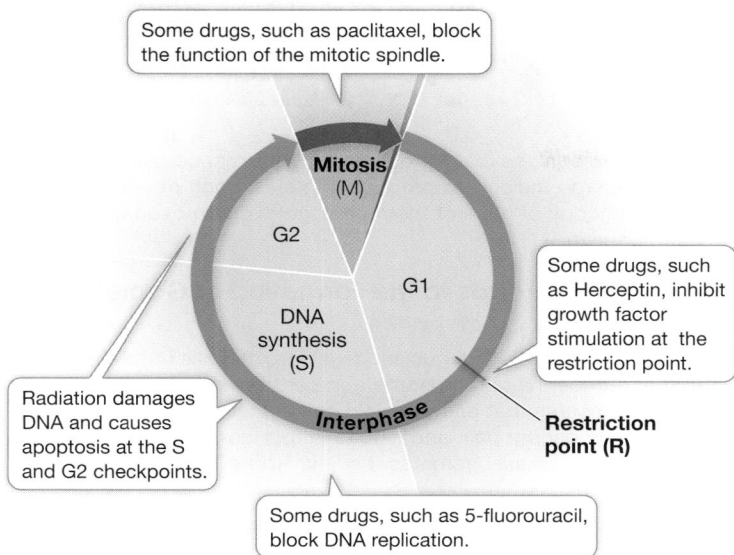

Some drugs, such as paclitaxel, block the function of the mitotic spindle.

Radiation damages DNA and causes apoptosis at the S and G2 checkpoints.

Some drugs, such as Herceptin, inhibit growth factor stimulation at the restriction point.

Some drugs, such as 5-fluorouracil, block DNA replication.

Figure 11.25 Cancer Treatment and the Cell Cycle   To prevent cancer cells from dividing, physicians use combinations of therapies that attack the cell cycle at different points.

Q: How do these treatments differ from treatments based on specific targets (see Figure 11.24)? Would the side effects of the two types of treatments be different?

in the mitotic spindle. Both drugs inhibit the cell cycle, and apoptosis causes tumor shrinkage. More dramatic is radiation treatment, in which a beam of high-energy radiation is focused on the tumor. DNA damage is extensive, and the cell cycle checkpoint for DNA repair is overwhelmed. As a result, the cell undergoes apoptosis. A major problem with these treatments is that they affect normal cells as well as the tumor cells. These treatments are toxic to tissues with large populations of normal dividing cells such as those in the intestine, skin, and bone marrow (where blood cells are produced).

A major effort in cancer research is to find treatments that target only cancer cells. A promising recent example is Herceptin, which targets the HER2 growth factor receptor that occurs at high levels on

the surfaces of some breast cancer cells (see Figure 11.24A). Herceptin binds specifically to the HER2 receptor but does not stimulate it. This prevents the natural growth factor from binding and stimulating cell division. Because cell division is blocked and the apoptosis rate remains the same, the tumor shrinks.

▶ 11.7 recap

Cancer cells differ from normal cells in terms of their rapid cell division and their ability to spread (metastasis). Many proteins regulate the cell cycle, either positively (oncogenes) or negatively (tumor suppressors). In cancer, one or more of these proteins is altered in some way, making its activity abnormal. Radiation and many cancer drugs target genes and proteins involved in the cell cycle.

learning outcomes

You should be able to:

• Compare and contrast cancer cells and normal cells.

• Explain how regulation of the cell cycle can become disrupted in cancer cells.

• Justify the rationale for targeting events in the cell cycle as a means of treating cancer.

1. Compare cells and genes in normal tissue, malignant tumors, and benign tumors.

2. How are oncogene proteins and tumor suppressor proteins involved in cell cycle control in normal and cancer cells?

3. Cancer-treating drugs are usually given in combination to target several stages of the cell cycle. Why might this be a better approach than using a single drug?

We have now looked at the cell cycle and at cell division by binary fission, mitosis, and meiosis. We have described the normal cell cycle and how its regulation is disrupted in cancer. We have seen how meiosis produces haploid cells in sexual life cycles. In the coming chapters we will examine heredity, genes, and DNA.

investigatinglife

◯A What controls the reproduction of cancer cells?

In normal tissues, the rate of cell division is offset by the rate of cell death. Unlike most normal cells, HeLa cells keep growing because they have a genetic imbalance that heavily favors cell reproduction over cell death. Henrietta Lacks was infected with the human papillomavirus, which stimulates cell division in cervical cells. In addition, an enzyme called telomerase, which keeps DNA intact and prevents cell death, is overexpressed in HeLa cells. This combination of traits—increased cell division plus decreased apoptosis—leads to the extraordinary growth rate of HeLa cells.

Because HeLa cells came from a tumor, you might expect that they are still cancer cells, and that they can form a tumor. But such is not the case. They are just cell division machines. What do you

Future directions

Knowledge of which genes are abnormal in tumors has led to the development of drugs targeted to the specific proteins that are made from these genes. In some instances, such as breast cancers expressing HER2, the results in the clinic have been extraordinary. Patients are living longer, and even being cured by these targeted drugs. Less attention has been given to the other part of the cell population equation—cell death. Many of the widely used drugs that affect the cell cycle in general (see Figure 11.26) work not just by stopping the cell cycle but by damaging cells, triggering apoptosis. But as you may know, these drugs have widespread side effects, stopping cell division and triggering apoptosis in normal cell as well as tumor cells. We know about the molecular events in apoptosis. Perhaps targeted drugs that stimulate this process only in tumor

Chapter Summary 11

▶ 11.1 All Cells Derive from Other Cells

- Cell division is necessary for the reproduction, growth, and repair of organisms.
- Cell division must be initiated by a **reproductive signal**. Before a cell can divide, the genetic material (DNA) must be **replicated** and **segregated** to separate portions of the cell. **Cytokinesis** then divides the cytoplasm into two cells.
- In prokaryotes, most cellular DNA is a single molecule, a **chromosome**. Prokaryotes reproduce by **binary fission**. Review Figure 11.2
- In eukaryotes, cells divide by either mitosis or meiosis. In contrast to prokaryotes, eukaryotic cells have a distinct nucleus whose chromosomes are replicated prior to separating into two daughter cells.

▶ 11.2 The Eukaryotic Cell Division Cycle Is Regulated

- The eukaryotic cell cycle has two main phases: **interphase**, during which cells are not dividing and the DNA is replicating, and mitosis or M phase, when the cells are dividing.
- During most of the eukaryotic cell cycle, the cell is in interphase, which is divided into three subphases: S, G1, and G2. DNA is replicated during **S phase**. Mitosis (M phase) and cytokinesis follow. Review Focus: Key Figure 11.3
- Cyclin–Cdk complexes regulate the passage of cells through checkpoints in the cell cycle. Retinoblastoma protein (RB) inhibits the cell cycle at the **restriction (R) point**. One cyclin–Cdk functions by inactivating RB and allows the cell cycle to progress. Review Figures 11.4, 11.5
- External controls such as **growth factors** can stimulate the cell to begin a division cycle.

▶ 11.3 Eukaryotic Cells Divide by Mitosis
See Animation 11.1

- In mitosis, a single nucleus gives rise to two nuclei that are genetically identical to each other and to the parent nucleus.
- DNA is wrapped around proteins called histones, forming bead-like units called **nucleosomes**. A eukaryotic chromosome contains strings of nucleosomes bound to proteins in a complex called **chromatin**. Review Figure 11.8
- At mitosis, the replicated chromosomes (**sister chromatids**) are held together at the **centromere**. Each chromatid consists of one double-stranded DNA molecule. During mitosis sister chromatids, attached by cohesin, line up at the equatorial plate and attach to the **spindle apparatus**. The chromatids separate (becoming **daughter chromosomes**) and migrate to opposite ends of the cell. Review Figures 11.9, 11.10, Activities 11.1, 11.2
- Mitosis can be divided into several phases called prophase, prometaphase, metaphase, anaphase, and telophase.
- Nuclear division is usually followed by cytokinesis. Animal cell cytoplasms divide via a contractile ring made up of actin microfilaments and myosin. In plant cells, cytokinesis is accomplished by vesicles that fuse to form a cell plate. Review Figure 11.12

▶ 11.4 Cell Division Plays Important Roles in the Sexual Life Cycle

- **Asexual reproduction** produces **clones**, new organisms that are genetically identical to the parent. Any genetic variation is the result of changes in genes.
- In **sexual reproduction**, two **haploid** gametes—one from each parent—unite in **fertilization** to form a genetically unique, **diploid** **zygote**. Sexual life cycles can be haplontic, diplontic, or involve alternation of generations. Review Figure 11.14, Activity 11.3

- In non-haplontic sexually producing organisms, certain cells in the adult undergo meiosis to produce gametes, each of which contains one copy of each of the homologous chromosome pairs of the organism.

▶ 11.5 Meiosis Leads to the Formation of Gametes
See Animation 11.2, Activity 11.4

- Meiosis consists of two nuclear divisions, **meiosis I** and **meiosis II**, that reduce the chromosome number from diploid to haploid. Meiosis ensures that each haploid cell contains one member of each chromosome pair, and results in four genetically diverse haploid cells, usually gametes. Review Figure 11.15, Activity 11.5
- In meiosis I, entire chromosomes, each with two chromatids, migrate to the poles. In meiosis II, the sister chromatids separate.
- During prophase I, homologous chromosomes undergo **synapsis** to form pairs in a **tetrad**. Chromatids can form junctions called **chiasmata**, and genetic material may be exchanged between the two homologs by **crossing over**. Review Figures 11.16, 11.17
- Both crossing over during prophase I and **independent assortment** of the homologs as they separate during anaphase I ensure that the gametes are genetically diverse.
- In **nondisjunction**, two members of a homologous pair of chromosomes go to the same pole during meiosis I, or two chromatids go to the same pole during meiosis II or mitosis. This leads to one gamete having an extra chromosome and another lacking that chromosome. Review Figure 11.19
- The union between a gamete with an abnormal chromosome number and a normal haploid gamete results in **aneuploidy**. Such genetic abnormalities can be harmful or lethal to the organism.
- The numbers, shapes, and sizes of the metaphase chromosomes constitute the **karyotype** of an organism.
- **Polyploids** have more than two sets of haploid chromosomes. Sometimes these sets come from different species.

▶ 11.6 Cell Death Is Important in Living Organisms

- A cell may die by **necrosis**, or it may self-destruct by **apoptosis**, a genetically programmed series of events that includes the fragmentation of the cell's nuclear DNA.
- Apoptosis is regulated by external and internal signals. These signals result in activation of a class of enzymes called **caspases** that hydrolyze proteins in the cell. Review Figure 11.21

▶ 11.7 Unregulated Cell Division Can Lead to Cancer

- Cancer cells divide more rapidly than normal cells and can be **metastatic**, spreading to distant organs in the body.
- Cancer can result from changes in either of two types of proteins that regulate the cell cycle. **Oncogene** proteins stimulate cell division and are activated in cancer. **Tumor suppressor** proteins normally inhibit the cell cycle, but in cancer they are inactive. Review Figure 11.23
- Multiple genetic events are necessary to form a malignant cancer cell. Review Figure 11.24
- Cancer treatment often targets the cell cycle in tumor cells. Review Figure 11.25

Go to **LearningCurve** (in **LaunchPad**) for dynamic quizzing that helps you solidify your understanding of this chapter. **LearningCurve** adapts to your responses, giving you the practice you need to master each key concept.

Apply What You've Learned

Review

11.7 Cell cycle abnormalities play an important role in the development of cancer.

11.7 Cancer therapies are targeted to events in the cell cycle.

The healing properties of plants have been known for thousands of years. When laboratory techniques became sophisticated enough to begin isolating such compounds from living things, scientists turned to plants as potential sources of anticancer drugs. Since then, countless candidates—from tiny flowering plants to giant pine trees—have been screened for anticancer properties.

Researchers use water or other solvents to make extracts from plants. They test the extracts first on cancer cells and then on animal models. Those extracts that have anticancer activities are studied to see how they work.

One way in which cells become cancerous is that they lose control of the cell cycle. There are four checkpoints involved in this control, each with cyclin and cyclin-dependent kinase (Cdk).

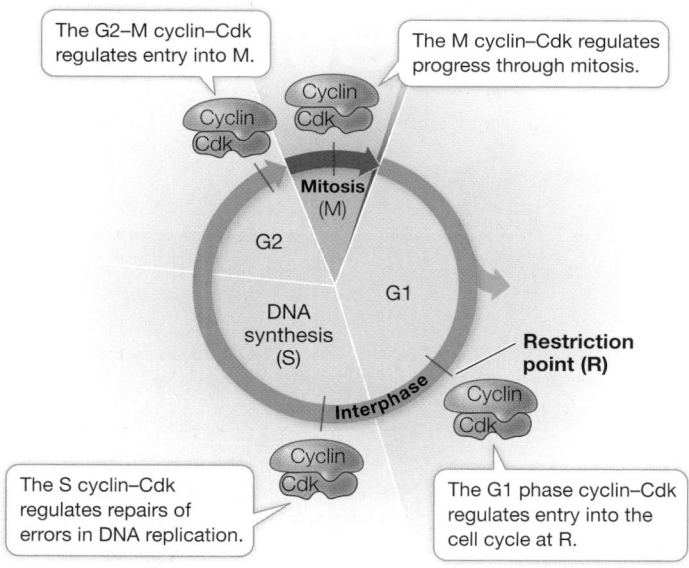

The G2–M cyclin–Cdk regulates entry into M.

The M cyclin–Cdk regulates progress through mitosis.

The S cyclin–Cdk regulates repairs of errors in DNA replication.

The G1 phase cyclin–Cdk regulates entry into the cell cycle at R.

The table below summarizes results from tests in which plant extracts were tested for their ability to inhibit cyclin–Cdk enzyme activity. Note that the naming of cyclins (A, B, etc.) and Cdks (1, 2, etc.) is arbitrary and does not indicate where they act in the cell cycle. Enzyme activity is expressed as +++ for full activity and as 0 for no activity.

Plant extract	Cyclin D–Cdk4 activity	Cyclin E–Cdk2 activity	Cyclin A–Cdk2 activity	Cyclin B–Cdk1 activity
1	+++	+++	0	+++
2	+++	+++	+++	0
3	0	+++	+++	+++
4	+++	0	+++	+++

Questions

1. Scientists can coax a group of cells to grow in synchrony so that all cells pass through the same cell cycle phases at the same time. What would you expect to observe if, in separate experiments, you added each of the four plant extracts to cancer cells in early G1?

2. Tumors have billions of cells. Very few tumors respond to a single drug that affects the cell cycle. Combination chemotherapy, targeting several checkpoints, is often used. Can you suggest a reason for this? (Hint: Not all cancer cells are the same.)

3. What would a researcher have to do to develop drugs from the plant extracts that could be administered to human cancer patients for halting tumor growth?

4. In some cancer cells, hyperactivation of cyclin–Cdk complexes results from greater-than-normal quantities of either cyclin or Cdk proteins. Suppose the following data were collected from various cells. Based on these data, which plant extract would be most useful for treating the tumor from which the cancer cells were taken?

	Ratio of protein expressed in cancer cells versus noncancer cells
Cyclin A	0.98
Cyclin B	0.99
Cyclin D	1.02
Cyclin E	1.01
Cdk1	4.25
Cdk2	0.95
Cdk4	0.98
Cdk6	1.04

5. Each Cdk shown above is regulated by another group of proteins. For each Cdk, there is a kinase that inhibits Cdk from binding to cyclin and a phosphatase that activates Cdk binding to cyclin. The diagram below shows kinase inactivation using a plus symbol and phosphatase activation using a minus symbol. Suppose researchers found a plant extract that halted kinase activity and another that halted phosphatase activity. Would the researchers be interested in one or both extracts as potential sources of an anticancer drug? Explain.

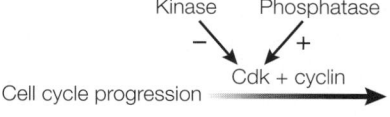

Go to **LaunchPad** for the eBook, LearningCurve, animations, activities, flashcards, and additional resources and assignments.

12

Inheritance, Genes, and Chromosomes

Red hair runs in families.

> investigating**life**

What Are the Rules of Inheritance?

Red hair is the rarest natural hair color in humans, with a frequency of about 1 percent. It is most common in northern Europe, with Scotland leading at a frequency of about 15 percent. Throughout history people have attempted to find a correlation between red-headedness and other traits. Red hair and hot temper, for example, are often associated. A recent study by the Hamburg Research Institute in Germany on the sexual activity of hundreds of women came to the conclusion that redheads were having more sex than any other hair color group. Attempts have been made to prove that redheads have a lower tolerance to pain than do people with other hair color.

Given its rarity, it is not surprising that red hair was an early subject of investigation in genetics, the science of heredity. As you will see in this chapter, the foundations of modern genetics were laid late in the nineteenth century through careful experimentation on pea plants by Gregor Mendel. By the early twentieth century, biologists examining human families noticed that in most cases, red-haired children were born from (1) two red-haired parents, (2) one parent with dark hair and one with red, or (3) two dark-haired parents. This fit the pattern for recessive inheritance, in which the gene that

determines red hair can be masked by a dominant gene for dark hair. Exceptions were noted, however, and how red hair is inherited was uncertain until the late 1950s, when Brenda Ellis took a genetics course from Professor Ralph Singleton at Wake Forest University in North Carolina.

After hearing her professor discuss theories about the inheritance of red hair, Ellis investigated her family tree dating back six generations, noting which relatives had red hair. She surmised that all the redheads in her family were the offspring of either two dark-haired parents or one with dark hair and one with red hair. Each of the dark-haired people had a red-haired ancestor—presumably a carrier of a gene that determines red hair. The genetic explanation that Ellis and her professor gave for the striking pattern of recessive inheritance of red hair in her family has since become widely accepted.

Sometimes, however, red hair pops up in families with no genetic history of red hair. In this chapter we'll examine various patterns of inheritance and the genetic mechanisms by which they arise.

Q How do inherited characteristics such as red hair skip generations?

12.1
Inheritance of Genes Follows Mendelian Laws

Genetics, the field of biology concerned with inheritance, has a long history. There is good evidence that people were deliberately breeding animals (horses) and plants (the date palm tree) for desirable characteristics as long as 5,000 years ago. The general idea was to examine the natural variation among the individuals of a species and "breed the best to the best and hope for the best." This was a hit-or-miss method—sometimes the resulting offspring had all the good characteristics of the parents, but often they did not.

focus your learning

- Two hypotheses have been proposed to explain how characteristics are passed from one generation to the next.
- Mendel's crosses of pea plants led to the formulation of laws of genetic inheritance.
- Mendel's laws of inheritance apply to all sexually reproducing organisms, including humans.

By the mid-nineteenth century, two theories had emerged to explain the results of breeding experiments:

1. The hypothesis of *blending inheritance* proposed that gametes (sex cells such as eggs or sperm) contained hereditary determinants (what we now call genes) that blended when the gametes fused during fertilization. Like inks of different colors, the two different determinants lost their individuality after blending and could never be separated. For example, if a plant that made smooth, round seeds was crossed (bred) with a plant that made wrinkled seeds, the offspring would be intermediate between the two and the determinants for the two parental characteristics would be lost.

2. The hypothesis of *particulate inheritance* proposed that each determinant had a physically distinct nature; when gametes fused in fertilization, the determinants remained intact. According to this hypothesis, if a plant that made round seeds was crossed with a plant that made wrinkled seeds, the offspring (no matter the shape of their seeds) would still contain the determinants for the two characteristics.

The story of how these competing hypotheses were tested provides a striking example of how the scientific method can be used to support one theory and reject another. In the following sections we will look in detail at experiments performed in the 1860s by an Austrian monk and scientist, Gregor Mendel, whose work clearly supported the particulate hypothesis.

Mendel's laws arose from controlled crosses of pea plants

After entering the priesthood at a monastery in Brno, in what is now the Czech Republic, Gregor Mendel was sent to the University of Vienna, where he studied biology, physics, and mathematics. He returned to the monastery in 1853 to teach. The abbot in charge had set up a small plot of land to do experiments with plants and encouraged Mendel to continue with the experiments. Over seven years, Mendel made crosses with many thousands of plants. Analysis of his meticulously gathered data suggested to him that inheritance was due to particulate factors.

Mendel presented his theories in two public lectures in 1865 and a detailed written publication in 1866, but his work was ignored by mainstream scientists until 1900. By that time, the discovery of chromosomes had suggested to biologists that genes might be carried on chromosomes. When they read Mendel's work on particulate inheritance, the biologists connected the dots between genes and chromosomes.

Mendel chose to study the common garden pea because of its ease of cultivation and the feasibility of making controlled crosses. Pea flowers have both male and female sex organs: stamens and pistils, which produce gametes that are contained within the pollen and ovules, respectively.

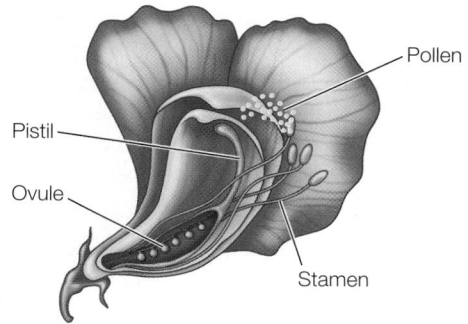

Pea flowers normally self-fertilize. That is, gametes of male and female are on the same flower and can combine. However, the male organs can be removed from a flower so that it can be manually fertilized with pollen from a different flower. That's what Mendel did.

There are many varieties of pea plants with easily recognizable characteristics. A **character** is an observable physical feature, such as seed shape. A **trait** is a particular form of a character, such as round or wrinkled seeds. Mendel worked with varieties with contrasting traits for seven characters such as seed shape, seed color, and flower color. These varieties were true-breeding: that is, when he crossed a plant that produced wrinkled seeds with another of the same variety, all of the offspring plants produced wrinkled seeds.

As you will see, Mendel proposed a set of hypotheses to explain the inheritance of particular pea traits, and then designed crossing experiments to test his hypotheses. He performed his crosses in the following manner:

- He removed the stamens (male parts) from flowers of one parental variety so that it couldn't self-fertilize. Then he collected pollen from another parental variety and placed it on the pistils of the emasculated flowers. The plants providing and receiving the pollen were the **parental generation**, designated **P**.

- In due course, seeds formed and were planted. The seeds and the resulting new plants constituted the **first filial generation**, or **F_1**. (The word "filial" refers to the relationship between offspring and parents, from the Latin *filius*, "son.")

Mendel examined each F$_1$ plant to see which traits it bore and then recorded the number of F$_1$ plants expressing each trait.

- In some experiments the F$_1$ plants were allowed to self-pollinate and produce a **second filial generation**, or **F$_2$**. Again, each F$_2$ plant was characterized and counted.

Mendel's first experiments involved monohybrid crosses

The term "hybrid" refers to the offspring of crosses between organisms differing in one or more characters. In Mendel's first experiments, he crossed parental (P) varieties with contrasting traits for a single character, producing monohybrids (from the Greek *monos*, "single") in the F$_1$ generation. He subsequently planted the F$_1$ seeds and allowed the resulting plants to self-pollinate to produce the F$_2$ generation. This technique is referred to as a **monohybrid cross**.

Mendel performed the same experiment for seven pea characters. You can follow his method in **Investigating Life: Mendel's Monohybrid Cross**, using seed shape as an example. When he crossed a variety that made round seeds with one that made wrinkled seeds, all of the F$_1$ seeds were round—it was as if the wrinkled seed trait had disappeared completely. However, when F$_1$ plants were allowed to self-pollinate to produce F$_2$ seeds, about one-fourth of the seeds were wrinkled. These two kinds of crosses were key to distinguishing the blending and particulate theories:

1. The F$_1$ offspring were not a blend of the two traits of the parents. Only one of the traits was present (in this case, round seeds).

2. Some F$_2$ offspring had wrinkled seeds. The trait had not disappeared because of blending.

These observations led to a rejection of the blending theory of inheritance and provided support for the particulate theory. We now know that hereditary determinants are not actually "particulate," but they *are* physically distinct entities: sequences of DNA carried on chromosomes, which we now call **genes**.

All seven crosses between varieties with contrasting traits gave the same kind of data (**Table 12.1**). In the F$_1$ generation only one of the two traits was seen, but the other trait reappeared in about one-fourth of the offspring in the F$_2$ generation. Mendel called the trait that appeared in the F$_1$ and was more abundant in the F$_2$ the **dominant** trait, and the other trait **recessive**. In the F$_2$ generation, the *ratio* of exhibiting dominant to plants with recessive traits was about 3:1. (To calculate the ratios shown in Table 12.1, divide the number of F$_2$ plants with the dominant trait by the number with the recessive trait.)

You can see in Table 12.1 that for each character, Mendel counted hundreds or even thousands of F$_2$ seeds or plants to see how many carried each trait. As we will discuss in more detail below, the probability of a plant inheriting a particular trait is independent of the probability of another plant inheriting the same trait. If Mendel had looked at only a few F$_2$ progeny from the "round × wrinkled" cross, he might, by chance, have found only round seeds. Or he might have found a higher proportion of wrinkled seeds than he did. In order to discover recurring patterns and to develop his laws of inheritance, Mendel used very large numbers of plants.

Mendel went on to expand on the particulate theory. He proposed that hereditary determinants—which we will refer to as genes, although Mendel did not use that term—occur in pairs and segregate (separate) from one another during the formation of gametes. He concluded that each pea plant has two copies of the gene for each character (such as seed shape), one inherited from each parent. We now use the term **diploid** (**2n**) to describe the state of having two copies of each gene; **haploids** (**n**) have just a single copy. You saw these terms in the last chapter when you learned about the cell cycle and meiosis.

Mendel concluded that while each gamete contains one copy of each gene, the resulting zygote contains two copies, because it is

table 12.1 Mendel's Results from Monohybrid Crosses

Parental generation phenotypes		F$_2$ generation phenotypes			
Dominant	Recessive	Dominant	Recessive	Total	Ratio
Round seeds × Wrinkled seeds		5,474	1,850	7,324	2.96:1
Yellow seeds × Green seeds		6,022	2,001	8,023	3.01:1
Purple flowers × White flowers		705	224	929	3.15:1
Inflated pods × Constricted pods		882	299	1,181	2.95:1
Green pods × Yellow pods		428	152	580	2.82:1
Axial flowers × Terminal flowers		651	207	858	3.14:1
Tall stems × Dwarf stems (1 m) (0.3 m)		787	277	1,064	2.84:1

experiment

Original Paper: The original German version of Mendel's paper, *Versuche uber Pflanzen-Hybriden*, with an English translation and extensive explanatory notes, is available online: www.mendelweb.org/Mendel.plain.html

Mendel performed crosses with pea plants and carefully analyzed the outcomes to show that genetic determinants are particulate.

HYPOTHESIS ▶ When two strains of peas with contrasting traits are bred, their characteristics are irreversibly blended in succeeding generations.

METHOD

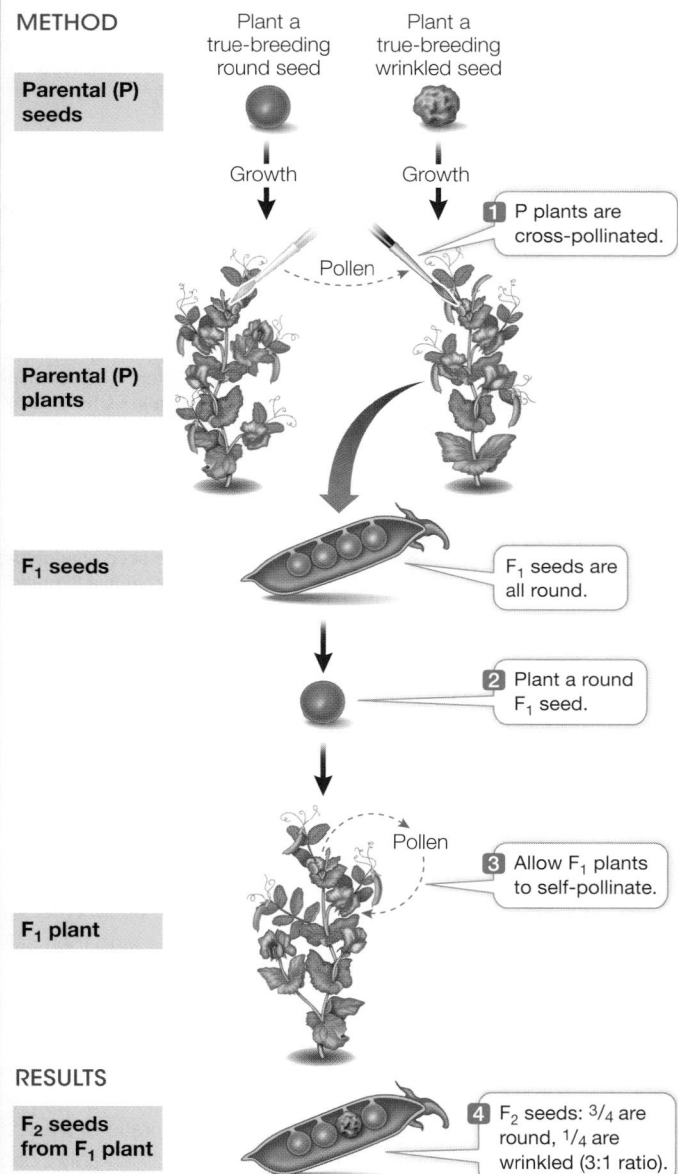

Plant a true-breeding round seed

Plant a true-breeding wrinkled seed

Parental (P) seeds

Growth

Growth

1 P plants are cross-pollinated.

Pollen

Parental (P) plants

F₁ seeds

F₁ seeds are all round.

2 Plant a round F₁ seed.

F₁ plant

Pollen

3 Allow F₁ plants to self-pollinate.

RESULTS

F₂ seeds from F₁ plant

4 F₂ seeds: ³/₄ are round, ¹/₄ are wrinkled (3:1 ratio).

CONCLUSION ▶ The hypothesis is rejected. There is no irreversible blending of characteristics, and a recessive trait can reappear in succeeding generations.

work with the data

Mendel's monohybrid crosses were key to his rejection of the theory of blending inheritance. While the experiment at left illustrates a monohybrid cross with round and wrinkled seeds, another of his monohybrid crosses was between true-breeding green-seeded and yellow-seeded pea plants. All of the pea plants in the F₁ generation of this cross had yellow seeds. Mendel allowed the F₁ plants to self-pollinate, and then analyzed the seed colors of the resulting F₂ generation. The table shows actual data from individual plants in the F₂ generation as reported in Mendel's paper. Mendel made mathematical calculations and presented the overall ratios for these two traits. He did not, however, perform a statistical analysis to determine whether the variations in the data reflected a general pattern of inheritance or were simply due to chance.

QUESTIONS ▶

1. Use the hypothesis that the ratio of yellow to green seeds in the F₂ generation would be 3:1 and perform a chi-square test to analyze the results for each plant in the table (refer to Appendix B for information about the chi-square test). What can you conclude about this hypothesis from the individual plants? How many crosses have P-values > 0.05?

2. Now total the data from all the plants and rerun the chi-square analysis. What can you conclude? What does your analysis indicate about the need for using a large number of organisms in studies of genetics?

	Seed color	
Plant	Yellow	Green
1	25	11
2	32	7
3	14	5
4	70	27
5	24	13
6	20	6
7	32	13
8	44	9
9	50	14
10	44	18

A similar **work with the data** exercise may be assigned in **LaunchPad**.

produced by the fusion of two gametes. Furthermore, he surmised that different traits arise from different forms of a gene (now called **alleles**) for a particular character. For example, Mendel studied two alleles for seed shape: one that resulted in round seeds and the other resulting in wrinkled seeds.

- An organism that is **homozygous** for a gene has two alleles that are the same (for example, two copies of the allele for round seeds).

- An organism that is **heterozygous** for a gene has two different alleles (for example, one allele for round seeds and one allele for wrinkled seeds).

In a heterozygote, one of the two alleles may be dominant (such as round, *R*) and the other recessive (wrinkled, *r*). By convention, dominant alleles are designated with uppercase letters and recessive alleles with lowercase letters. Note that the terms dominant and recessive refer only to which phenotype is expressed when the two allele are present together. They do not refer to which alleles are stronger, better, or more common.

The physical appearance of an organism is its **phenotype**. Mendel proposed that the phenotype is the result of the **genotype**, or genetic constitution, of the organism showing the phenotype. Round seeds and wrinkled seeds are two phenotypes resulting from three possible genotypes: the wrinkled seed phenotype is produced by the genotype *rr*, whereas the round seed phenotype is produced by either of the genotypes *RR* or *Rr* (because the *R* allele determines a dominant trait and the *r* allele determines a recessive trait).

Mendel's first law states that the two copies of a gene segregate

How do Mendel's theories explain the proportions of traits seen in the F₁ and F₂ generations of his monohybrid crosses? Mendel's first law—the **law of segregation**—states that *when any individual produces gametes, the two copies of a gene separate, so that each gamete receives only one copy*. Thus gametes from a parent with the *RR* genotype will all carry the *R allele*; gametes from an *rr* parent will all carry the *r allele*. Half of the gametes in an *Rr* individual will carry the *R* allele and the other half will carry *r*. What genotypes are produced when these gametes fuse to form the next generation? The allele combinations that will result from a cross can be predicted using a **Punnett square**. This device ensures that you consider all possible combinations of gametes when calculating expected genotype frequencies of the resulting offspring. A Punnett square looks like this:

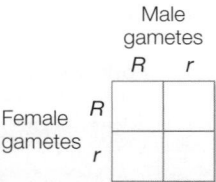

It is a simple grid with all possible male gamete (haploid sperm) genotypes shown along the top and all possible female gamete (haploid egg) genotypes along the left side. The grid is completed by filling in each square with the diploid genotype that can be generated from each combination of gametes.

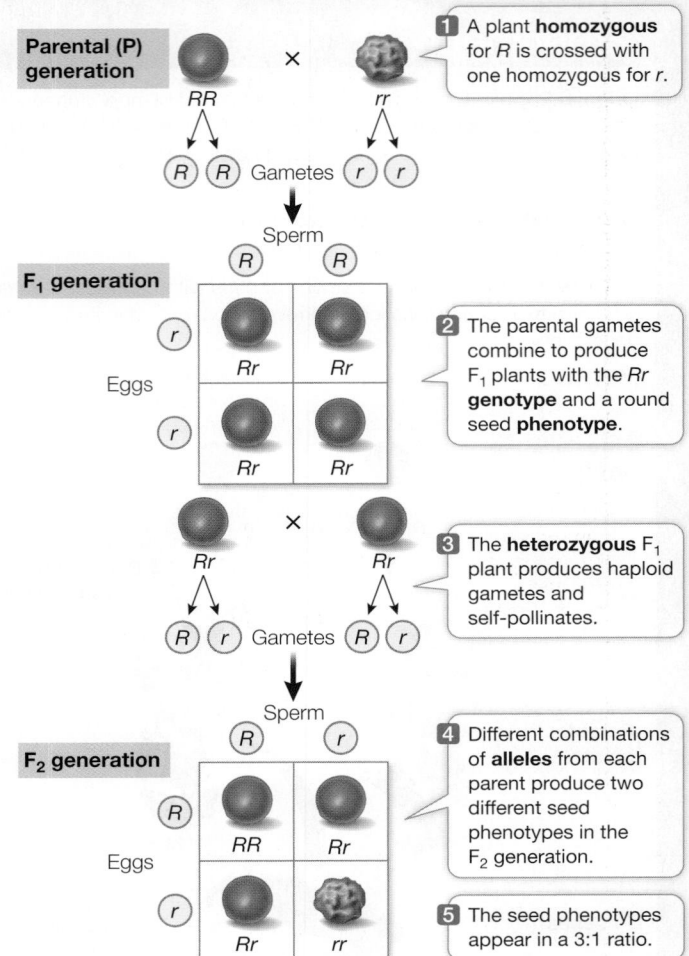

Figure 12.1 Mendel's Explanation of Inheritance Mendel concluded that inheritance depends on discrete factors from each parent that do not blend in the offspring.

Let's use a Punnett square to derive the progeny of an *RR* × *rr* cross (**Figure 12.1**). To determine the genotypes in the F₁ generation, into each square we put an *R* from the male gamete (a sperm cell from the pollen tube) and *r* from the female gamete (the egg cell). All the F₁ offspring of this cross (the F₁ generation) have the *Rr* genotype, which produces seeds with a round phenotype. What happens when the F₁ generation are crossed among themselves to produce the F₂ generation? Once the Punnett square is filled in, we readily see that there are four possible combinations of alleles in the F₂ generation: *RR*, *Rr*, *rR*, and *rr* (see Figure 12.1). Since *R* is dominant, there are three ways to get round seeds in the F₂ generation (genotype *RR*, *Rr*, or *rR*), but only one way to get wrinkled seeds (genotype *rr*). Therefore we predict a 3:1 ratio of the round and wrinkled phenotypes in the F₂ generation, remarkably close to the ratios Mendel found experimentally for all the traits he compared (see Table 12.1).

Mendel did not live to see his theories placed on a sound physical footing with the discoveries of chromosomes and DNA. Genes are now known to be sequences of DNA found on the much longer DNA molecules that make up chromosomes. With your knowledge of *meiosis, you can envision the different alleles of a gene segregating as chromosomes separate during meiosis I (**Focus: Key Figure 12.2**).

focus: key figure

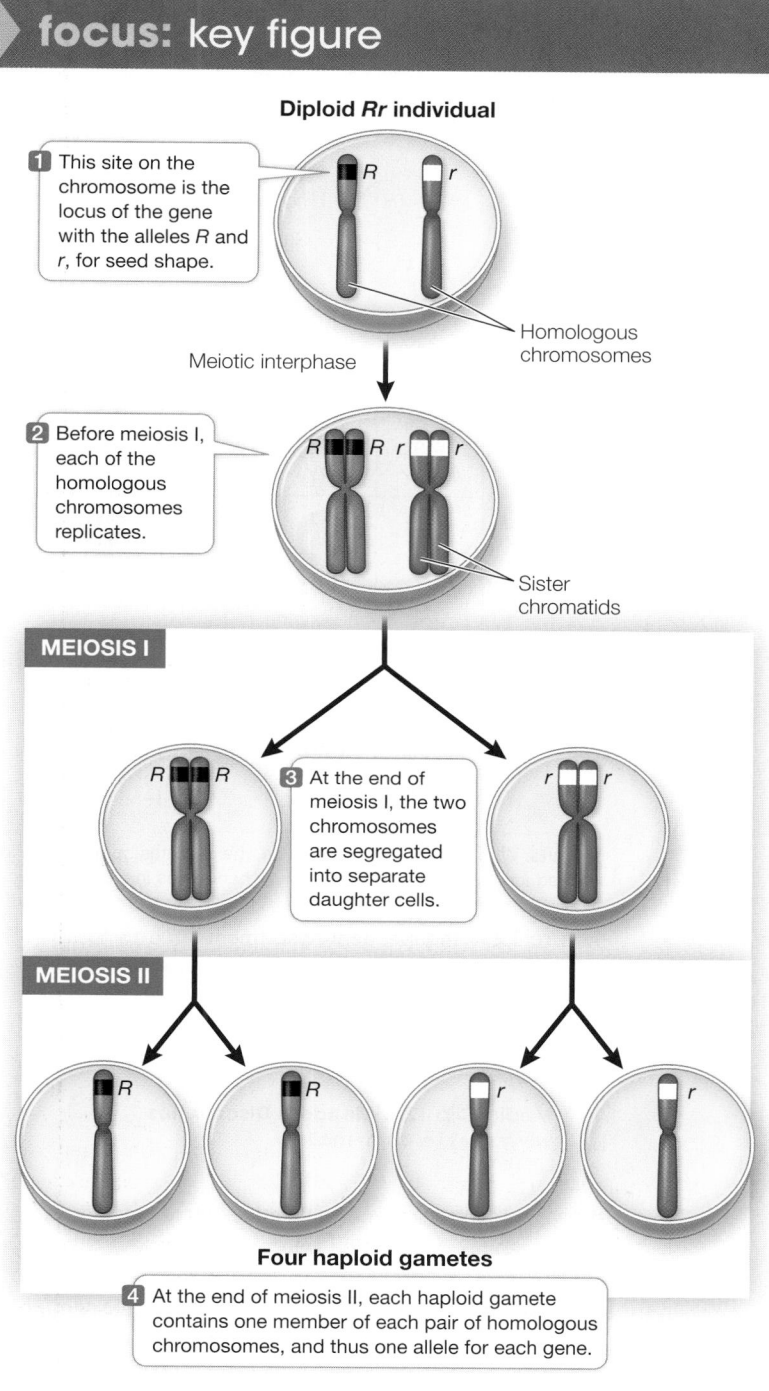

Diploid *Rr* individual

1 This site on the chromosome is the locus of the gene with the alleles *R* and *r*, for seed shape.

Homologous chromosomes

Meiotic interphase

2 Before meiosis I, each of the homologous chromosomes replicates.

Sister chromatids

MEIOSIS I

3 At the end of meiosis I, the two chromosomes are segregated into separate daughter cells.

MEIOSIS II

Four haploid gametes

4 At the end of meiosis II, each haploid gamete contains one member of each pair of homologous chromosomes, and thus one allele for each gene.

Figure 12.2 Meiosis Accounts for the Segregation of Alleles Although Mendel had no knowledge of chromosomes or meiosis, we now know that a pair of alleles resides on homologous chromosomes, and that those alleles segregate during meiosis.

Q: Cells in a pea leaf have 14 chromosomes. How many chromosomes are actually in each gamete?

*connect the concepts   As illustrated in Figure 11.15, homologous chromosomes separate in meiosis, which ultimately produces gametes with a haploid (*n*) number of chromosomes.

We know now that genes determine phenotypes mostly by producing proteins with particular functions, such as enzymes. In many cases a dominant gene is expressed (transcribed and translated) to produce a functional protein, while a recessive gene is mutated so that it is no longer expressed, or it encodes a mutant protein that is nonfunctional. For example, the wrinkled seed phenotype of *rr* peas is caused by the absence of an enzyme called starch branching enzyme 1 (SBE1), which is essential for starch synthesis. With less starch, the developing seed has more sucrose and this causes an inflow of water by osmosis. When the seed matures and dries out, the water is lost, leaving a shrunken seed. A single copy of the *R* allele produces enough functional SBE1 to prevent the wrinkled phenotype, which accounts for the dominance of *R* over *r*.

Mendel verified his hypotheses by performing test crosses

Mendel arrived at his laws of inheritance by developing a series of hypotheses and then designing experiments to test them. One hypothesis was that there are two possible allele combinations (*RR* or *Rr*) for seeds with the round phenotype. Mendel verified this hypothesis by performing test crosses with F_1 seeds derived from a variety of other crosses. A **test cross** is used to determine whether an individual showing a dominant trait is homozygous or heterozygous. The individual in question is crossed with an individual that is homozygous for the recessive allele—easy to identify, because all individuals with the recessive phenotype are homozygous for that trait.

The recessive homozygote for the seed shape gene has wrinkled seeds and the genotype *rr*. The individual being tested may be described initially as *R_* because we do not yet know the identity of the second allele. We can predict two possible results:

1. If the individual being tested is homozygous dominant (*RR*), all offspring of the test cross will be *Rr* and show the dominant trait (round seeds) (**Figure 12.3, left**).

2. If the individual being tested is heterozygous (*Rr*), then approximately half the offspring of the test cross will be heterozygous and show the dominant trait (*Rr*), and the other half will be homozygous for the recessive allele (*rr*) (**Figure 12.3, right**).

Mendel obtained results consistent with both of these predictions; thus his hypothesis accurately predicted the results of his test crosses.

experiment

Figure 12.3 Homozygous or Heterozygous?

Original Paper: The original German version of Mendel's paper, *Versuche uber Pflanzen-Hybriden*, with an English translation and extensive explanatory notes, is available online: www.mendelweb.org/Mendel.plain.html

An individual with a dominant phenotype may have either a homozygous or a heterozygous genotype. The test cross determines genotype.

HYPOTHESIS▶ The progeny of a test cross can reveal whether an organism is homozygous or heterozygous.

METHOD

1a Test round peas of undetermined genotype... → $R\_$ × rr ← **1b** ...by crossing them with wrinkled peas with a known genotype (homozygous recessive).

2a If the plant being tested is homozygous...

RR × rr

Gametes: R R r r

2b If the plant being tested is heterozygous...

Rr × rr

Gametes: R r r r

RESULTS

Sperm: r r

	r	r
R	Rr	Rr
R	Rr	Rr

Eggs

Sperm: r r

	r	r
R	Rr	Rr
r	rr	rr

Eggs

3a ...then all progeny will show the dominant phenotype (round).

3b ...then half the seeds from the cross will be wrinkled, and half will be round.

CONCLUSION▶ The plant being tested is likely homozygous.

CONCLUSION▶ The plant being tested is heterozygous.

▶ Activity 12.1 **Homozygous or Heterozygous?** www.Life11e.com/ac12.1

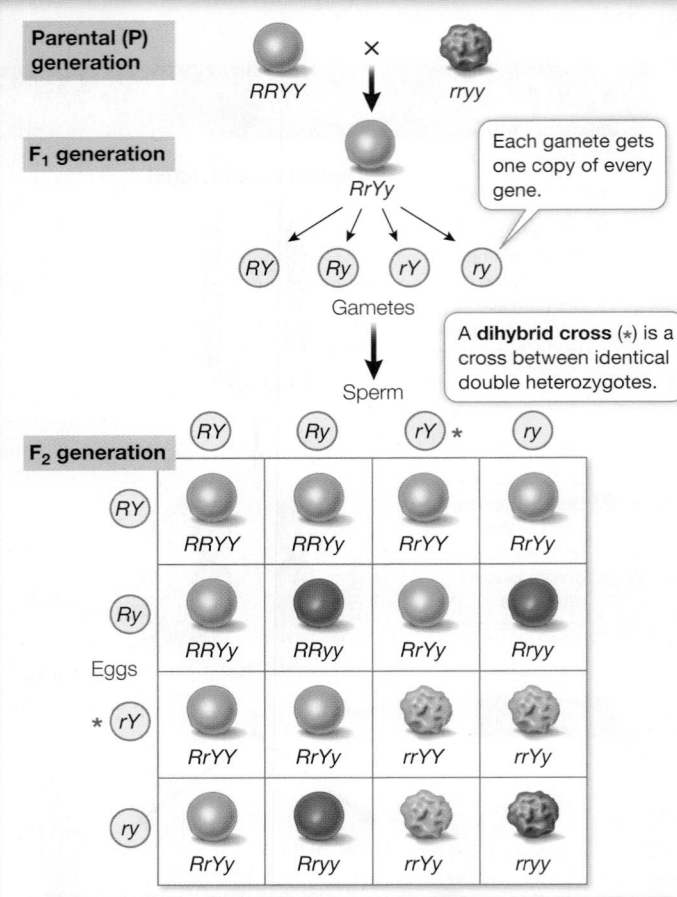

Parental (P) generation $RRYY$ × $rryy$

F₁ generation $RrYy$

Each gamete gets one copy of every gene.

Gametes: RY Ry rY ry

Sperm

A **dihybrid cross** (*) is a cross between identical double heterozygotes.

F₂ generation

	RY	Ry	rY *	ry
RY	$RRYY$	$RRYy$	$RrYY$	$RrYy$
Ry	$RRYy$	$RRyy$	$RrYy$	$Rryy$
*rY	$RrYY$	$RrYy$	$rrYY$	$rrYy$
ry	$RrYy$	$Rryy$	$rrYy$	$rryy$

Eggs

Results: When F₁ plants self-pollinate, the gametes combine randomly to produce an F₂ generation with four phenotypes in a 9:3:3:1 ratio.

Figure 12.4 Independent Assortment The 16 possible combinations of gametes in this dihybrid cross result in nine different genotypes. Because R and Y are dominant over r and y, respectively, the nine genotypes result in four phenotypes in a ratio of 9:3:3:1. These results show that the two genes segregate independently.

▶ Media Clip 12.1 **Mendel's Discoveries** www.Life11e.com/mc12.1

Mendel's second law states that copies of different genes assort independently

Now let's consider inheritance patterns involving two different genes. Consider an organism that is heterozygous for genes for seed color (yellow or green) and seed shape (round or wrinkled). In our example, the dominant R and Y alleles came from one parent, and the recessive r and y alleles came from the other parent. When this organism produces gametes, do the R and Y alleles always go together in one gamete, and the r and y alleles in another? Or can a single gamete receive one recessive and one dominant allele (R and y or r and Y)?

Mendel performed another series of experiments to answer these questions. He began with peas that differed in *two* characters: seed shape and seed color. One parental variety produced only round, yellow seeds ($RRYY$), and the other produced only wrinkled, green ones ($rryy$). A cross between these two varieties produced an F₁ generation in which all the plants were $RrYy$. Because the R and Y alleles were dominant, the F₁ seeds were all round and yellow.

Mendel continued this experiment into the F₂ generation by performing a **dihybrid cross**—a cross between individuals that are identical double heterozygotes. In this case, he simply allowed the F₁ plants, which were all double heterozygotes, to self-pollinate. Depending on whether the alleles of the two genes are inherited together or separately, there are two possible outcomes, as Mendel saw:

1. *The alleles could maintain the associations they had in the parental generation*. If this were the case, the F₁ plants would produce two types of gametes (RY and ry). The F₂ progeny resulting from self-pollination of these F₁ plants would consist of *two phenotypes*: round yellow and wrinkled green in the ratio of 3:1, just as in the monohybrid cross.

2. *The segregation of R from r could be independent of the segregation of Y from y—the two genes could be unlinked.* In this case, four kinds of gametes would be produced in equal numbers: *RY*, *Ry*, *rY*, and *ry*. When these gametes combine at random, they should produce an F$_2$ generation with *four phenotypes* (round yellow, round green, wrinkled yellow, wrinkled green). Putting these possibilities into a Punnett square, we can predict that these four phenotypes would occur in a ratio of 9:3:3:1.

The results supported the second prediction: four different phenotypes appeared in the F$_2$ generation in a ratio of about 9:3:3:1 (**Figure 12.4**). On the basis of such experiments, Mendel proposed his second law—the **law of independent assortment**: *alleles of different genes assort independently of one another during gamete formation*. In the example above, the segregation of the *R* and *r* alleles is independent of the segregation of the *Y* and *y* alleles. Mendel's second law is now understood in the context of meiosis (see Figure 11.15): *chromosomes segregate independently* during the formation of gametes, and so do any two genes located on separate chromosome pairs (**Focus: Key Figure 12.5**).

Probability can be used to predict inheritance

One key to Mendel's success was his use of large numbers of plants. By counting many offspring from each cross, he observed clear patterns that allowed him to formulate his theories. After his work became widely recognized, geneticists began using simple probability calculations to predict the ratios of genotypes and phenotypes in the progeny of a given cross or mating. They use statistics to determine whether the actual results match the prediction (as explored in the work with the data exercise on p. 243).

focus: key figure

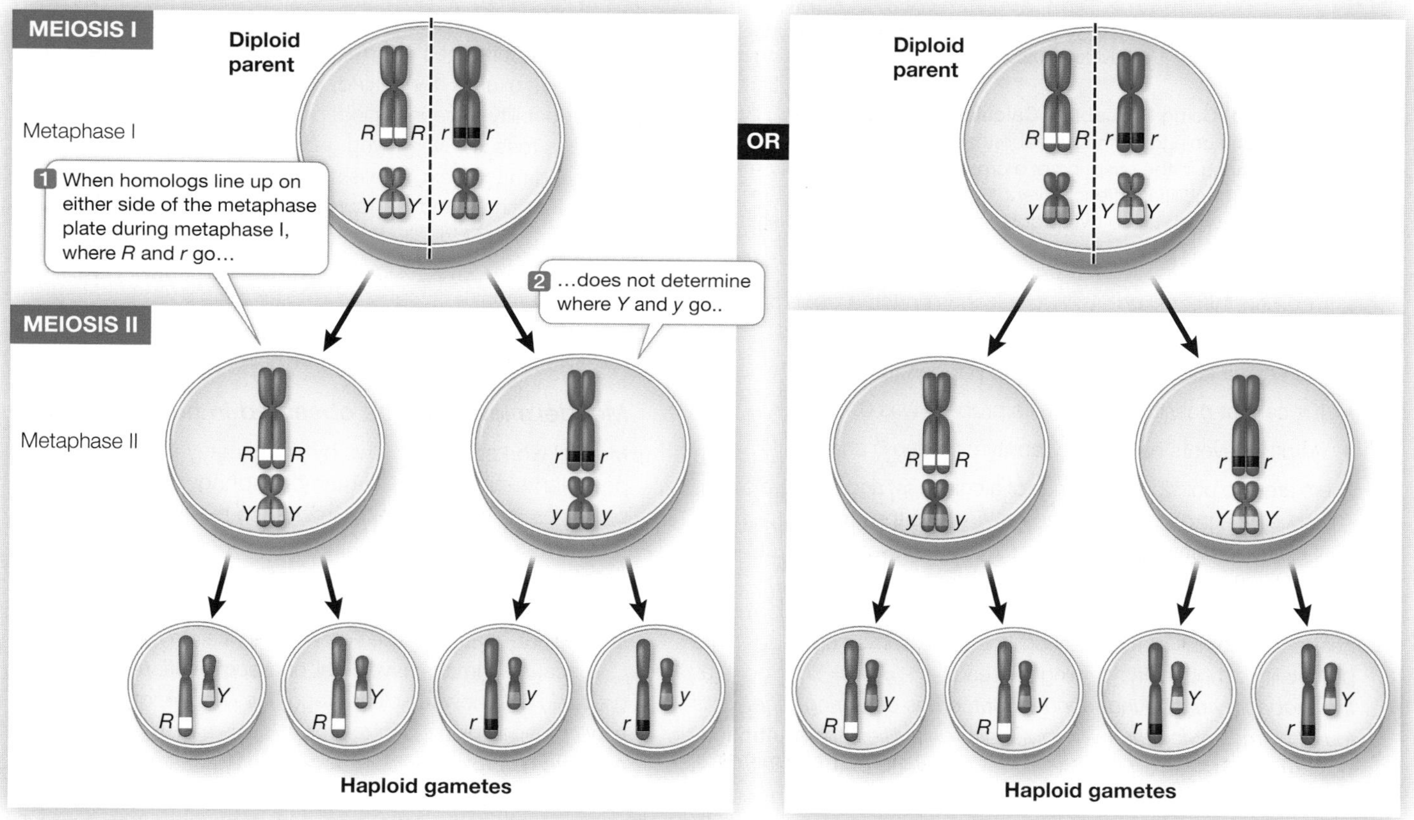

Figure 12.5 Meiosis Accounts for Independent Assortment of Alleles We now know that copies of genes on different chromosomes are segregated independently during metaphase I of meiosis. Thus a parent of genotype *RrYy* can form gametes with four different genotypes.

Animation 12.1 **Independent Assortment of Alleles**
www.Life11e.com/a12.1

Q: A diploid heterozygote with a single gene can produce haploid gametes of only two genetically distinct types. A heterozygote with two genes can produce four genetically distinct types of gametes, as illustrated here. How many genetically distinct gametes could be produced by a heterozygote with four genes assorting independently? (Answer this without drawing a diagram!)

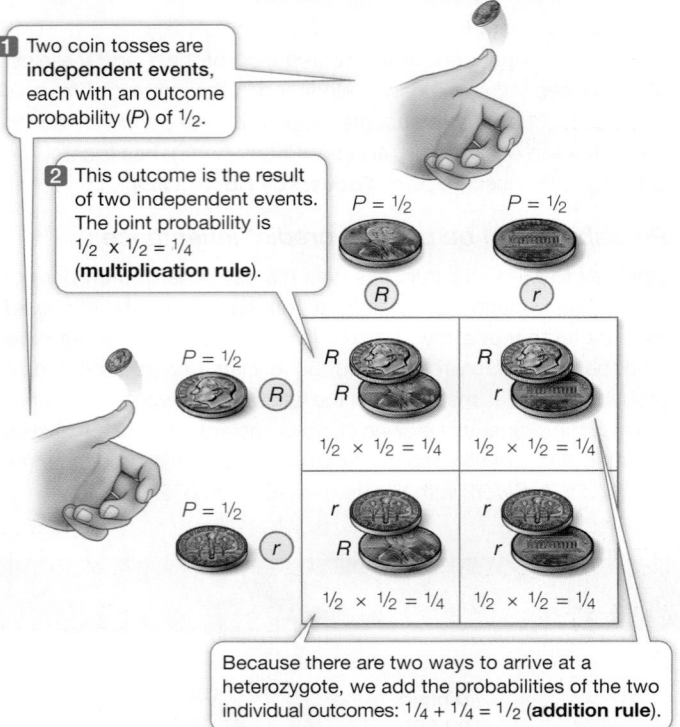

1 Two coin tosses are independent events, each with an outcome probability (*P*) of ½.

2 This outcome is the result of two independent events. The joint probability is ½ × ½ = ¼ (**multiplication rule**).

$P = ½$ $P = ½$

$P = ½$

$P = ½$

	R	r
R	R R	R r
	½ × ½ = ¼	½ × ½ = ¼
r	r R	r r
	½ × ½ = ¼	½ × ½ = ¼

Because there are two ways to arrive at a heterozygote, we add the probabilities of the two individual outcomes: ¼ + ¼ = ½ (**addition rule**).

Figure 12.6 Using Probability Calculations in Genetics Like the results of a coin toss, the probability of any given combination of alleles appearing in the offspring of a cross can be obtained by multiplying the probabilities of each event. Since a heterozygote can be formed in two ways, these two probabilities are added together.

You can think of probabilities by considering a coin toss. The basic conventions of probability are simple:

- If an event is absolutely certain to happen, its probability is 1.
- If it cannot possibly happen, its probability is 0.
- All other events have a probability between 0 and 1.

There are two possible outcomes of a coin toss, and both are equally likely, so the probability of heads is ½—as is the probability of tails.

If two coins (say a penny and a dime) are tossed, each acts independently of the other (**Figure 12.6**). What is the probability of both coins coming up heads? In half of the tosses, the penny comes up heads, and in half of that fraction, the dime comes up heads. The probability of both coins coming up heads is ½ × ½ = ¼. In general, *the probability of two independent outcomes occurring together is found by multiplying the two individual probabilities* (the **multiplication rule**). The multiplication rule can be seen in the results of a monohybrid cross (see Figure 12.1). After the self-pollination of an *Rr* F_1 plant, the probability that an F_2 plant will have the genotype *RR* is ½ × ½ = ¼, because the chance that the sperm will have the genotype *R* is ½, and the chance that the egg will have the genotype *R* is also ½. Similarly, the probability of obtaining an *rr* offspring is also ¼.

Probability can also be used to predict the proportions of phenotypes in a dihybrid cross. Let's see how this works for the experiment shown in Figure 12.4. Using the principles we just described, you can calculate the probability of an F_2 seed being round. This is found by adding the probability of obtaining an *Rr* heterozygote (½) to the probability of an *RR* homozygote (¼): a total of ¾ (the **addition**

rule). By the same reasoning, the probability that a seed will be yellow is also ¾. The two characters are determined by separate genes and are independent of each other, so:

- The joint probability for both round and yellow is ¾ × ¾ = ⁹⁄₁₆.

What is the probability of F_2 seeds being both wrinkled and yellow? The probability of being yellow is again ¾; the probability of being wrinkled is ½ × ½ = ¼, so:

- The joint probability for both wrinkled and yellow is ¼ × ¾ = ³⁄₁₆.

By the same reasoning:

- The joint probability for both round (¾) and green (¼) is also ³⁄₁₆.

Finally:

- The joint probability for both wrinkled and green is ¼ × ¼ = ¹⁄₁₆.

Looking at all four phenotypes, you can see that they are expected to occur in the ratio of 9:3:3:1.

A Punnett square or these simple probability calculations can be used to determine the *expected* proportions of offspring with particular phenotypes. In the dihybrid cross discussed above, about one-sixteenth of the F_2 seeds are expected to be wrinkled and green. But this does not mean that among 16 F_2 seeds there will always be exactly 1 wrinkled, green seed. For any toss of a coin, the probability of heads is independent of what happened in all the previous tosses. Even if you get three heads in a row, the chance of a head in the next toss is still ½, and it is quite possible to toss a coin four times and get four heads. But if you toss the coin many times, you are highly likely to get heads in about half of the tosses. If Mendel had examined only a few progeny in each of his crosses, it is unlikely that he would have observed the phenotypic ratios that he did observe. It was his large sample sizes that allowed him to identify the underlying patterns of inheritance.

Mendel's laws can be observed in human pedigrees

Mendel worked out his laws by doing many planned crosses with pea plants and counting many offspring. Neither of these approaches is possible with humans, so human geneticists rely on **pedigrees**: family trees that show the occurrence of phenotypes (and alleles) in several generations of related individuals. You saw an example of this in the opening story of this chapter. On the next page is the eight-generation pedigree for red hair in Brenda Ellis's family.

Because humans have small numbers of offspring, human pedigrees do not show the clear proportions of phenotypes that Mendel saw in his pea plants. For example, when a man and a woman who are both heterozygous for a recessive allele (say, *Aa*) have children together, each child has a ¼ probability of being a recessive homozygote (*aa*). If this couple were to have several dozen children, about one-fourth of them would be recessive homozygotes. But the offspring of a single couple are likely to be too few to show the exact one-fourth proportion. In a family with only two children, for example, both children could easily be *aa* (or *Aa*, or *AA*).

Figure 12.7A is a pedigree showing the pattern of inheritance of a rare dominant allele. The following are the key features to look for in such a pedigree:

- Every affected person has an affected parent.

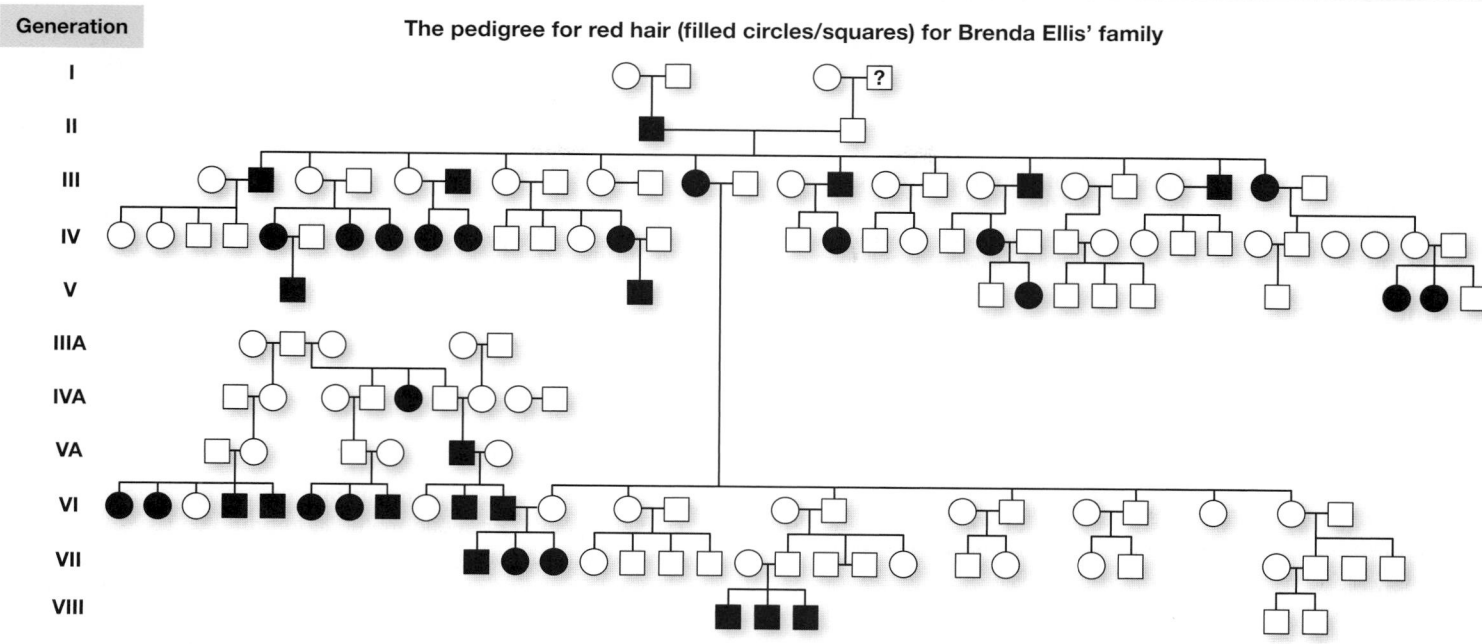

The pedigree for red hair (filled circles/squares) for Brenda Ellis' family

- About half of the offspring of an affected parent are also affected. (This is easiest to observe among the 12 cousins in generation III.)

Compare this pattern with the one shown in **Figure 12.7B**, which is typical for the inheritance of a rare recessive allele:

- Affected people can have two parents who are not affected.
- Only a small proportion of people are affected: about one-fourth of children whose parents are both heterozygotes.

In the families of individuals who have a rare recessive phenotype, it is not uncommon to find a marriage of two relatives. This observation is a result of the rarity of recessive alleles that give rise to abnormal phenotypes. For two phenotypically normal parents to have an affected child (*aa genotype*), the parents must both be heterozygous (*Aa*). If a particular recessive allele is rare in the general population, the chance of two people mating to produce offspring who are both carrying that allele is quite low. However, if that allele is present in a family, two cousins might share it (see Figure 12.7B). For this reason, studies on populations that are isolated either culturally (by religion, as with the Amish in the United States) or geographically (as on islands) have been extremely valuable to human geneticists. People in these groups are more likely to marry relatives who may carry the same rare recessive alleles.

(A) Dominant inheritance

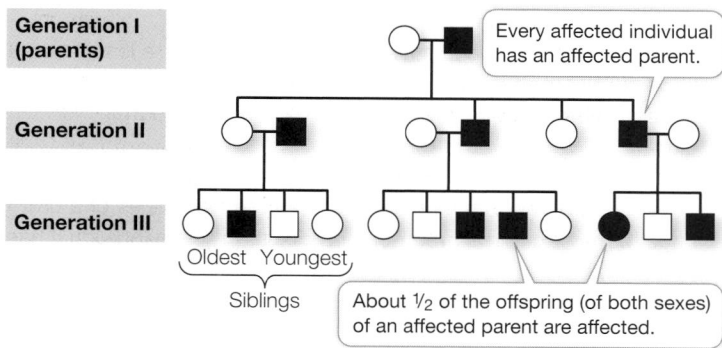

(B) Recessive inheritance

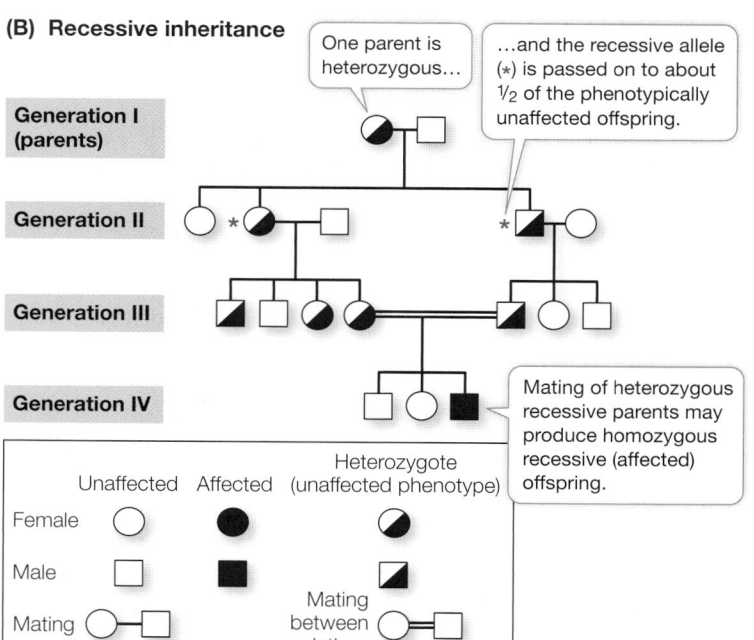

Figure 12.7 Pedigree Analysis and Inheritance **(A)** This pedigree represents a family affected by Huntington's disease, which results from a rare dominant allele. Everyone who inherits this allele is affected. **(B)** The family in this pedigree carries the allele for albinism, a recessive trait. Because the trait is recessive, heterozygotes do not have the albino phenotype, but they can pass the allele on to their offspring. In this family, in generation III the heterozygous parents are cousins; however, the same result could occur if the parents were unrelated but heterozygous.

 Activity 12.2 Pedigree Analysis Simulation
www.Life11e.com/a12.2

12.1 recap

Mendel showed that genetic determinants are particulate and do not "blend" when the two gametes combine. Mendel's first law of inheritance states that two copies of a gene segregate during gamete formation. His second law states that genes assort independently during gamete formation. The frequencies with which different allele combinations will be expressed in offspring can be calculated with a Punnett square or using probability theory.

learning outcomes

You should be able to:

- Compare and contrast hypotheses of inheritance.
- Give examples of dominance and recessiveness.
- Explain the law of segregation.
- Predict the results of a monohybrid cross using a Punnett square.
- Use a model to illustrate the law of independent assortment.
- Apply probability concepts to predict the results of crosses.
- Apply Mendel's laws to analyze aspects of human pedigrees.

1. What results seen in the F_1 and F_2 generations of Mendel's monohybrid cross experiments refuted the blending theory of inheritance?

2. Diagram the processes of meiosis that underlie the laws of segregation and independent assortment.

3. In guinea pigs, black body color (B) is completely dominant over albino (b). For the crosses below, give the genotypes of the parents:

Parental phenotypes	Black offspring	Albino offspring	Parental genotypes?
Black × albino	12	0	
Albino × albino	0	12	
Black × albino	5	7	
Black × black	9	3	

4. The pedigree below shows the inheritance of a rare mutant phenotype in humans, congenital cataracts (black symbols).

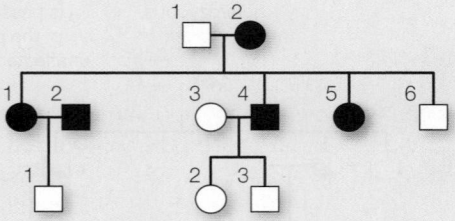

a. Are cataracts inherited as an autosomal dominant trait? Autosomal recessive?

b. Person #5 in the second generation mates with a man who does not have cataracts. Two of their four children, a boy and a girl, develop cataracts. What is the chance that their next child will be a girl with cataracts?

Mendel's laws of inheritance remain valid today; his discoveries laid the groundwork for all future studies of genetics. Inevitably, however, we have learned that things are more complicated than they seemed at first. Let's take a look at some of these complications, beginning with the interactions between alleles.

---

key concept 12.2 Alleles Can Produce Multiple Phenotypes

Over time genes accumulate changes, giving rise to new alleles. Thus there can be many alleles for a single character. In addition, alleles do not always show simple dominant–recessive relationships. Furthermore, a single allele may have multiple phenotypic effects.

focus your learning

- Alleles can interact to produce intermediate phenotypes.

New alleles arise by mutation

Genes are subject to **mutations**, which are stable, inherited changes in the genetic material. In other words, an allele can mutate to become a different allele. For example, imagine that at one time all pea plants were tall, and that this trait was attributable to a height allele, T. At some point, a mutation occurred in T that resulted in a new allele, t. If this *****mutation** were in a cell that underwent meiosis to form gametes, some of the resulting gametes would carry the t allele, and some offspring of this pea plant would carry the t allele. By creating variety, mutations provide the raw material for evolution.

> ***connect the concepts**   Key Concept 14.1 describes a mutation as a change in the nucleotide sequence of DNA, and Key Concept 15.1 describes how various types of mutations affect phenotype in different ways.

Geneticists usually define one particular allele of a gene as the **wild type**; this allele is the one that is present in most individuals in nature ("the wild") and gives rise to an expected trait or phenotype. Other alleles of that gene, often called mutant alleles, may produce a different phenotype. The wild-type and mutant alleles reside at the same genetic **locus**, which is their specific position on a chromosome. A genetic locus that carries a wild-type allele less than 99 percent of the time (the rest of the alleles being mutant) is said to be **polymorphic** (Greek *poly*, "many," + *morph*, "form").

Many genes have multiple alleles

Because of random mutations, more than two alleles of a given gene may exist in a group of individuals. Any one individual has only two alleles—one from its mother and one from its father. But among multiple individuals there may be several different alleles. In fact, there are many examples of such multiple alleles, and they often show a hierarchy of dominance. An example is coat color in rabbits, determined by four alleles of the C gene:

1. C determines dark gray.
2. c^{chd} determines chinchilla, a lighter gray.
3. c^h determines Himalayan, where pigment is restricted to the extremities (point restricted).
4. c determines albino.

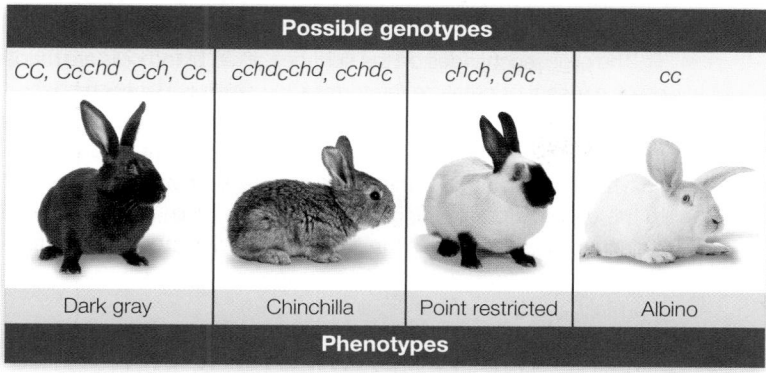

Figure 12.8 Multiple Alleles for Coat Color in Rabbits These photographs show the phenotypes conferred by four alleles of the *C* gene for coat color in rabbits. Different combinations of two alleles give different coat colors and pigment distributions.

The hierarchy of dominance for these alleles is $C > C^{chd}$, $C^h > c$. Any rabbit with the C allele (paired with itself or any other allele) is dark gray, and a cc rabbit is albino. Intermediate colors result from different allele combinations, as shown in **Figure 12.8**. As this example illustrates, multiple alleles can increase the number of possible phenotypes. At the biochemical level, these alleles determine various amounts of a pigment because of an enzyme with various levels of activity.

A dramatic and serious example of multiple alleles involves the gene product targeted by certain widely used drugs designed to undermine the effects of the parasite that causes malaria in humans. There are numerous alleles of this parasite gene, some of which make the gene product insensitive to the drugs (they cannot bind to the target protein). Since the parasite is haploid, when a parasite has one of these resistance alleles and is in a human who is being treated with the drugs, that parasite will survive and reproduce while the vast majority of parasites with drug-sensitive wild-type alleles will be killed. Drug-resistant malaria causes great human suffering. The World Health Organization estimates that there are about 200 million new cases of malaria every year, over half a million of them fatal.

Dominance is not always complete

In the pairs of alleles studied by Mendel, dominance is complete in heterozygous individuals. That is, an *Rr* individual always expresses the *R* phenotype. However, many genes have alleles that are not dominant or recessive to one another. Instead, the heterozygotes show an intermediate phenotype—at first glance, that appeared to support the old blending theory of inheritance. For example, if a true-breeding eggplant that produces the familiar purple fruit is crossed with a true-breeding white eggplant, all the F_1 plants produce violet fruit, an intermediate between the two parents. However, further crosses indicate that this apparent blending phenomenon can still be explained in terms of Mendelian genetics (**Figure 12.9**). The purple and white alleles have not disappeared, as those colors reappear when the F_1 plants are interbred.

When heterozygotes show a phenotype that is intermediate between those of the two homozygotes, the gene is said to be governed by **incomplete dominance**. In other words, neither of the two alleles is dominant. Incomplete dominance is common in nature; in fact, Mendel's study of pea-plant traits is unusual in that all the traits happened to be characterized by complete dominance.

In codominance, both alleles at a locus are expressed

Sometimes the two alleles at a locus produce two different phenotypes that *both* appear in heterozygotes, a phenomenon called **codominance**. Note that this is different from incomplete dominance, where the phenotype of a heterozygote is an intermediate between the phenotypes of the parents. A good example of codominance is seen in the ABO blood group system in humans (this is also an example of multiple alleles).

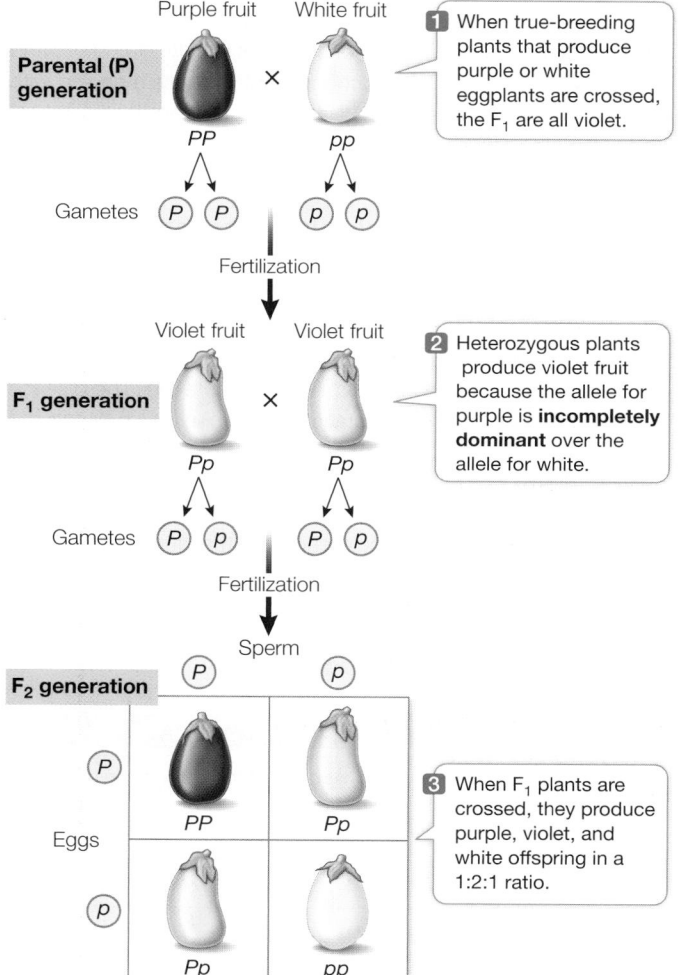

Figure 12.9 Incomplete Dominance Follows Mendel's Laws An intermediate phenotype can occur in heterozygotes when neither allele is dominant. The heterozygous phenotype (here, violet fruit) may give the appearance of a blended trait, but the traits of the parental generation reappear in their original forms in succeeding generations, as predicted by Mendel's laws of inheritance.

Blood type of cells	Geno-type	Blood cell types that body rejects	Blood type's reaction to added antibodies	
			Anti-A	Anti-B
A	$I^A I^A$ or $I^A I^O$	B		
B	$I^B I^B$ or $I^B I^O$	A		
AB	$I^A I^B$	Neither A nor B		
O	$I^O I^O$	A, B, and AB		

Red blood cells that do not react with antibody remain evenly dispersed.

Red blood cells that react with antibody clump together (speckled appearance).

Figure 12.10 ABO Blood Reactions Are Important in Transfusions This table shows the results of mixing red blood cells of types A, B, AB, and O with serum containing anti-A or anti-B antibodies. As you look down the columns, note that each of the types, when mixed separately with anti-A and with anti-B, gives a unique pair of results; this is the basic method by which blood is typed. People with type O blood are the universal blood donors. People with type AB blood are good recipients, since they make neither type of antibody. When blood transfusions are incompatible, the reaction (clumping of red blood cells) can have severely adverse consequences for the recipient.

Early attempts at blood transfusion frequently killed the patient. Around 1900, the Austrian scientist Karl Landsteiner mixed blood cells and serum (blood from which cells have been removed) from different individuals. He found that only certain combinations of blood and serum are compatible. In other combinations, the red blood cells from one individual form clumps in the presence of serum from the other individual, a discovery that led to our ability to administer compatible blood transfusions that do not kill the recipient.

Clumps in mixed blood form as a consequence of mechanisms of the immune system that protect the body from invasion by "nonself" molecules or organisms. Specific proteins in human serum, called antibodies, react with foreign molecules or particles. The specific part of a molecule that is recognized by an antibody is called an antigen. Oligosaccharides on the surfaces of red blood cells can function as antigens. For example, people with type A blood make the A antigen, and those in the B group make the B antigen; A and B antigens are specific oligosaccharides on the surfaces of their red blood cells. If a person with type A blood is given a transfusion of type B blood, that individual's immune system will recognize the B antigen as nonself and make antibodies against it (**Figure 12.10**). Likewise, a person with type B blood will make antibodies against the A antigen. However, someone with the codominant AB blood type makes *both* the A and the B antigens and will not make antibodies against either antigen. So a person with type AB blood can receive a transfusion from just about anyone. Individuals with type O blood make neither the A nor the B antigen (see Figure 12.10); they can receive blood only from a type O individual but can donate blood to individuals of any ABO blood group.

The oligosaccharides on the surfaces of red blood cells are made by enzymes that catalyze the formation of bonds between specific sugars. The ABO genetic locus encodes one such enzyme and has three alleles, I^A, I^B, and I^O, each producing a different version of the enzyme. The product of the I^A allele adds N-acetylgalactosamine to the end of a pre-existing oligosaccharide chain, resulting in the A antigen. The product of the I^B allele adds galactose to the same chain, making the B antigen. The I^O allele is recessive and encodes a protein that has no enzymatic activity:

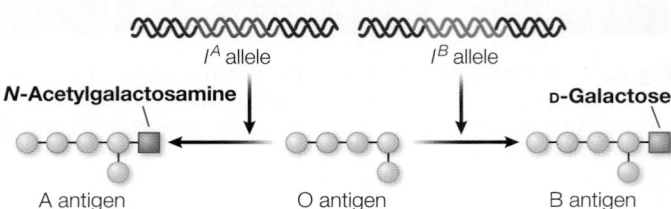

Since people inherit one allele from each parent, they may have at the ABO locus any combination of these alleles: $I^A I^B$, $I^A I^O$, $I^A I^A$, and so on. The I^A and I^B alleles are codominant because a person with both alleles makes both the A and the B antigens, and both kinds of oligosaccharide occur on their red blood cells. You'll learn more about the functions of antibodies and antigens in Chapter 42.

Some alleles have multiple phenotypic effects

Mendel's principles were further extended when it was discovered that a single allele can influence more than one phenotype. In such a case, we say that the allele is **pleiotropic**. An example is the heritable human disease phenylketonuria, which causes mental retardation and reduced hair and skin pigmentation. The disease occurs in people who have a mutation in the gene for a liver enzyme that converts the amino acid phenylalanine to tyrosine. Without a functional form of this enzyme, phenylalanine builds up in the body to toxic levels, and this affects development in a variety of ways. Given what we now know about genes and their functions, it is not surprising that a gene with such an important metabolic role should have pleiotropic effects. Other examples of pleiotropy include plant and animal genes whose products affect hormone levels, since many hormones play multiple roles in the body.

12.2 recap

Genes are subject to random mutations that give rise to new alleles; thus many genes have more than two alleles within a population. Dominance is not necessarily an all-or-nothing phenomenon. Some genes have multiple effects on phenotype.

learning outcomes

You should be able to:

• Analyze genetic crosses involving incomplete dominance.
• Explain the concept of codominance.
• Distinguish between incomplete dominance, codominance, and pleiotropic effects.

12.2 recap

1. What is the difference between codominance and incomplete dominance?

2. Explain how blood type AB results from codominance.

3. In a particular plant species, two alleles control flower color, which can be yellow, green, or blue. Crosses of these plants produce the following offspring:

Parental phenotypes	Offspring phenotypes (ratio)
Yellow × yellow	All yellow
Green × yellow	Green or yellow (1:1)
Green × blue	Blue or green (1:1)
Blue × blue	All blue

What will be the phenotype, and ratio, of the offspring of a cross of green × green?

Thus far we have discussed phenotypic characters that are affected by single genes. In many cases, however, several genes interact to determine a phenotype. To complicate things further, the physical environment may influence which genes of an individual are expressed, meaning that phenotype is sometimes determined by both genes and the environment.

key concept 12.3 Genes Can Interact to Produce a Phenotype

You have just seen how two alleles of the same gene can interact to produce a phenotype. Some phenotypes, such as human height, are influenced by the products of many genes, not just alleles of the same gene. We now turn to the genetics of such gene interactions.

focus your learning

- One gene may affect the expression of another gene.
- The environment can influence phenotype.

Epistasis ("to stand upon") occurs when the phenotypic expression of one gene is affected by another gene. For example, two genes (*B* and *E*) encode proteins that determine coat color in Labrador retrievers:

1. Allele *B* (black pigment) is dominant to *b* (brown).

2. Allele *E* (pigment deposition in hair) is dominant to *e* (no deposition, so hair is yellow).

An *EE* or *Ee* dog with *BB* or *Bb* is black, and one with *bb* is brown. An *ee* dog is yellow regardless of the *B* gene alleles present (**Figure 12.11**). Because the product of the *E* allele is needed for the expression of both the *B* and the *b* alleles, *E* is said to be epistatic to *B*.

Hybrid vigor results from new gene combinations and interactions

In 1876, Charles Darwin reported that when he crossed two different true-breeding, homozygous genetic strains of corn, the offspring were 25 percent taller than either of the parent strains. Darwin's observation was largely ignored for the next 30 years. In 1908, George Shull "rediscovered" this idea, reporting that not just plant height but the weight of the corn grain produced was dramatically higher in the offspring (**Figure 12.12**). Agricultural scientists took note, and Shull's paper had a lasting impact on the field of applied genetics. The cultivation of hybrid corn spread rapidly in the United States and all over the world, quadrupling grain production. The practice of hybridization has spread to many other crops and animals used in agriculture. For example, beef cattle that are crossbred are larger and live longer than cattle bred within their own genetic strain.

Farmers have known for centuries that matings among close relatives (known as **inbreeding**) can result in **inbreeding depression**: reduced biological fitness arising from mating close relatives that tend to have the same recessive, sometimes deleterious, alleles.

A dog with alleles *B* and *E* is black.

Black (*B_E_*)

A dog with alleles *bb* and *E* is brown.

Chocolate (*bbE_*)

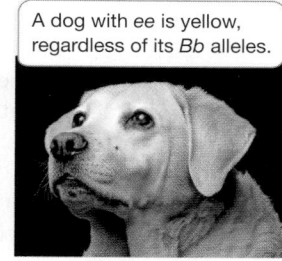
A dog with *ee* is yellow, regardless of its *Bb* alleles.

Yellow (*__ee*)

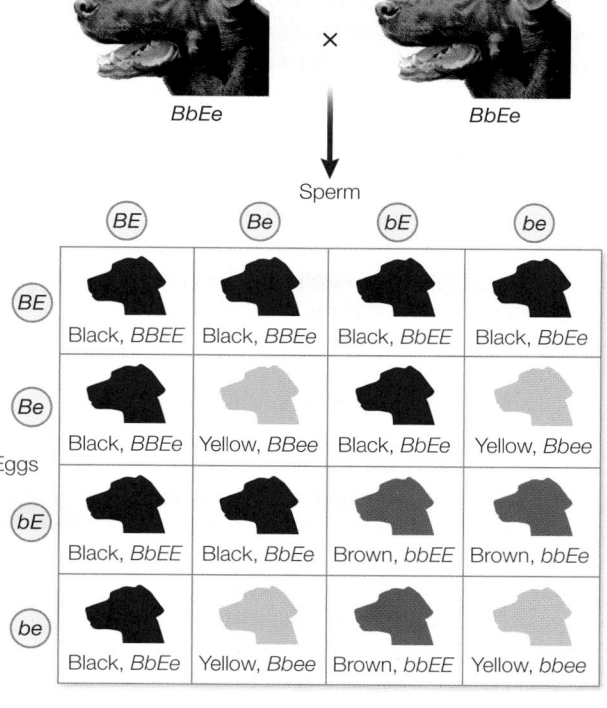

Figure 12.11 Genes May Interact Epistatically Epistasis occurs when one gene alters the phenotypic effect of another gene. In Labrador retrievers, the *E* gene determines the expression of the *B* gene.

Q: A mating of a yellow and brown retriever produced a black pup. What were the possible genotypes of the parents?

B73   Hybrid   Mo17

Figure 12.12 Hybrid Vigor in Corn Two homozygous parent lines of corn, B73 and Mo17, were crossed to produce the more vigorous hybrid line.

Inbreeding can produce offspring of smaller size and/or lower quality than matings between unrelated individuals. In contrast, superior qualities can arise from crossing inbred lines, a phenomenon called **hybrid vigor**, or **heterosis** (short for heterozygosis).

The mechanism by which heterosis arises is not known but is important to understand, given how agriculture relies on heterosis for improving food production. Two competing hypotheses have been proposed. The dominance hypothesis explains the extra growth in hybrids as an absence of inbreeding depression—by the assumption that hybrids are unlikely to be homozygous for deleterious recessive alleles. The overdominance hypothesis postulates that in hybrids, new combinations of alleles from the parental strains interact with one another, resulting in superior traits that cannot occur in the parental lines. Because many of the characters seen in hybrid species are in fact controlled by multiple genes, recent studies suggest that both dominance and overdominance can contribute to heterosis in specific characters.

The environment affects gene action

The phenotype of an individual does not result from its genotype alone. *Genotype and environment interact to determine the phenotype of an organism.* This is especially important to remember in the era of genome sequencing (covered in Chapter 17). When the sequence of the human genome was completed in 2003, it was hailed as the "book of life," and public expectations of the benefits gained from this knowledge were (and are) high. But this kind of "genetic determinism" is wrong. You already know that environmental variables such as light, temperature, and nutrition can affect the phenotypic expression of a genotype.

A familiar example of this phenomenon involves "point restriction" coat patterns found in Siamese cats and certain rabbit breeds (**Figure 12.13**). These animals carry a mutant allele of a gene that controls the growth of dark fur all over the body. As a result of this mutation, the enzyme encoded by the gene is inactive at temperatures above a certain point (usually around 35°C). The animals maintain a body temperature above this point, and so their fur is mostly light. However, the extremities—feet, ears, nose, and tail—are cooler, about 25°C, so the fur on these regions is dark. These animals are all white when they are born, because the extremities were kept warm in the mother's womb.

A simple experiment shows that the dark fur is temperature-dependent. If a patch of white fur on a point-restricted rabbit's back is removed and an ice pack is placed on the skin where the patch was, the fur that grows back will be dark. This indicates that although the gene for dark fur was expressed all along, the environment inhibited the activity of the mutant enzyme.

Two parameters describe the effects of genes and environment on the phenotype:

1. **Penetrance** is the proportion of individuals in a group with a given genotype that actually show the expected phenotype. For example, many people who inherit a mutant allele of the gene *BRCA1* develop breast cancer. But, some people with the mutation do not. So the *BRCA1* allele is said to be incompletely penetrant.

2. **Expressivity** is the degree to which a genotype is expressed in an individual. For example, a woman with the mutant *BRCA1* allele may develop both breast and ovarian cancer as part of the phenotype, but another woman with the same mutation may develop only breast cancer. So the mutant allele is said to have variable expressivity.

Most complex phenotypes are determined by multiple genes and the environment

Certain simple characters, such as those that Mendel studied in peas, differ in discrete, **qualitative** ways. Mendel used true-breeding parental pea plants that were either short or tall, had purple or

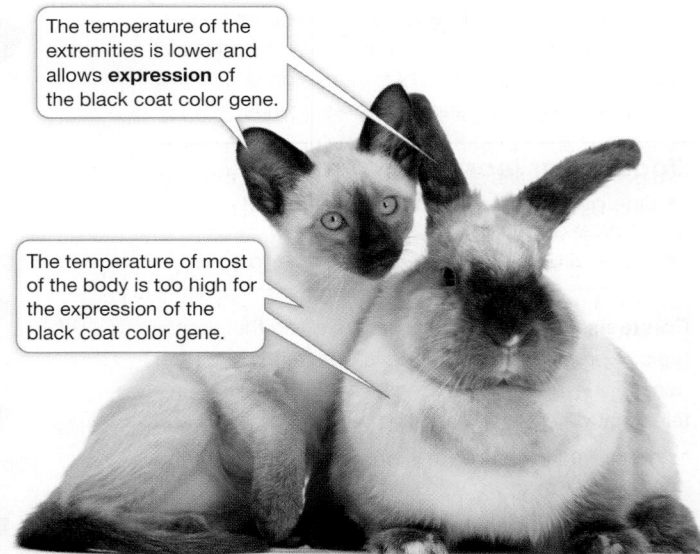

The temperature of the extremities is lower and allows **expression** of the black coat color gene.

The temperature of most of the body is too high for the expression of the black coat color gene.

Figure 12.13 The Environment Influences Gene Expression The rabbit and cat express a coat pattern called "point restriction." Their genotypes specify dark hair/fur, but the enzyme for dark color is inactive at normal body temperature, so only the extremities—the coolest regions of the body—express the phenotype.

Figure 12.14 Quantitative Variation   Quantitative variation is produced by the interaction of genes at multiple loci and the environment. These students (women [in white] are shorter; men [in blue] are taller) show continuous variation in height that is the result of interactions between many genes and the environment.

white flowers, or had round or wrinkled seeds. But for most complex characters, such as height in humans, the phenotype varies more or less continuously over a range. Some people are short, others are tall, and many are in between the two extremes. Such variation within a population is called **quantitative**, or continuous, variation (**Figure 12.14**).

Sometimes this variation results largely from the alleles that an individual possesses. For instance, much of human eye color is the result of a number of genes controlling the synthesis and distribution of dark melanin pigment. Dark eyes have a lot of it, brown eyes less, and green, gray, and blue eyes even less. In the latter cases, the distribution of other pigments in the eye is what determines light reflection and color.

In most cases, however, *quantitative variation is due to both genes and environment*. Height in humans certainly falls into this category. If you look at families, you often see that parents and their offspring all tend to be tall or short. However, nutrition also plays a role in height: American 18-year-olds today are about 6 percent taller than their great-grandparents were at the same age. Three generations are not enough time for mutations that would exert such a dramatic effect to spread throughout the general population, so the height difference must be due to environmental factors.

Geneticists call the genes that together determine such complex characters **quantitative trait loci**. Identifying these loci is a major challenge, and an important one. For example, the amount of grain that a variety of rice produces in a growing season is determined by many interacting genetic factors. Crop plant breeders have worked hard to decipher these factors in order to breed higher-yielding rice strains. In a similar way, human characteristics such as disease susceptibility and behavior are caused in part by quantitative trait loci. Recently, one of the many genes involved with human height was identified. The gene, *HMGA2*, has an allele that apparently has the potential to add 4 millimeters to human height.

12.3 recap

In epistasis, one gene affects the expression of another. Perhaps the most challenging problem for genetics is the explanation of complex phenotypes that are caused by many interacting genes and the environment.

learning outcomes

You should be able to:

• Analyze genetic data involving epistasis.
• Compare and contrast penetrance and expressivity.
• Compare and contrast quantitative and qualitative variation.

1. Explain the difference between penetrance and expressivity.

2. How is quantitative variation different from qualitative variation?

3. If the dominant allele *A* is necessary for hearing in humans, and another allele, *B*, located on a different chromosome, results in deafness no matter what other genes are present, what percentage of the offspring of a parental cross of *aaBb* × *Aabb* will be deaf?

In the next section we'll see how the discovery that genes occupy specific positions on chromosomes enabled Mendel's successors to provide a physical explanation for his model of inheritance, and to provide an explanation for those cases where Mendel's second law does not apply.

key concept

12.4

Genes Are Carried on Chromosomes

There are far more genes than chromosomes. Studies of different genes that are physically linked on the same chromosome reveal inheritance patterns that are not Mendelian. These patterns have been useful for identifying genes that are linked to one another, and for determining how far apart they are on the chromosome.

focus your learning

- Genes on the same chromosome are linked.
- Recombinant frequencies can be used to infer gene locations on a chromosome.
- The patterns of inheritance of phenotypes determined by genes on sex chromosomes differ from those of genes carried on autosomes.

Genetic linkage was first discovered in the fruit fly *Drosophila melanogaster*. Its small size, the ease with which it can be bred, its few chromosomes, and its short generation time make this animal an attractive experimental subject. Beginning in 1909, Thomas Hunt Morgan and his students at Columbia University pioneered the study of *Drosophila*, and today it remains a very important model organism for studies of genetics.

 Animation 12.2 **Alleles That Do Not Assort Independently**
www.Life11e.com/a12.2

Linked genes are inherited together

Some of the crosses Morgan performed with fruit flies yielded phenotypic ratios that were not in accordance with those predicted by Mendel's law of independent assortment. Morgan crossed *Drosophila* with known genotypes at two loci, *B* and *Vg*:

1. *B* (wild-type gray body) is dominant over *b* (black body).

2. *Vg* (wild-type wing) is dominant over *vg* (vestigial, a very small wing).

Morgan first made an F_1 generation by crossing homozygous dominant *BBVgVg* flies with homozygous recessives (*bbvgvg*). He then performed a test cross with the F_1 flies: *BbVgvg × bbvgvg*.[1] Morgan expected to see four phenotypes in a ratio of 1:1:1:1, but that is not what he observed. The body color gene and the wing size gene were not assorting independently; rather, they were usually inherited together (**Figure 12.15**).

Morgan knew about chromosomes and meiosis. To explain the data, he proposed that the two *Drosophila* loci are on the same chromosome—that is, that they might be linked. Suppose that the *B* and *Vg* loci are indeed located on the same chromosome. Why didn't all of Morgan's F_1 flies have the parental phenotypes—that is, why didn't his cross result in gray flies with normal wings and black flies with vestigial wings, in a 1:1 ratio? If linkage were absolute—that is, if chromosomes always remained intact and unchanged—we would expect to see just those two types of progeny. However, this does not always happen.

Genes can be exchanged between chromatids and mapped

If linkage were absolute, Mendel's law of independent assortment would apply only to loci on different chromosomes. Instead, genes

[1]Do you recognize this type of cross? It is a test cross for the two gene pairs; see Figure 12.3.

experiment

Figure 12.15A   Some Alleles Do Not Assort Independently

Original Paper: Morgan, T. H. 1912. Complete linkage in the second chromosome of *Drosophila. Science* 36: 719–720.

Morgan's studies showed that the genes for body color and wing size in *Drosophila* are linked, so that their alleles do not assort independently.

HYPOTHESIS▶   Alleles for different characteristics always assort independently.

METHOD

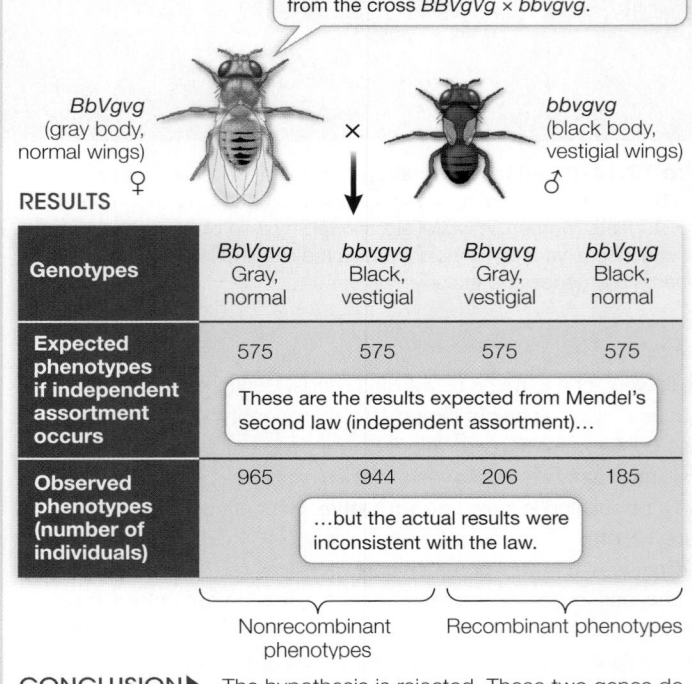

This heterozygous individual was produced from the cross *BBVgVg × bbvgvg*.

BbVgvg
(gray body, normal wings) ♀

×

bbvgvg
(black body, vestigial wings) ♂

RESULTS

Genotypes	*BbVgvg* Gray, normal	*bbvgvg* Black, vestigial	*Bbvgvg* Gray, vestigial	*bbVgvg* Black, normal
Expected phenotypes if independent assortment occurs	575	575	575	575
Observed phenotypes (number of individuals)	965	944	206	185

These are the results expected from Mendel's second law (independent assortment)…

…but the actual results were inconsistent with the law.

Nonrecombinant phenotypes | Recombinant phenotypes

CONCLUSION▶   The hypothesis is rejected. These two genes do not assort independently, but are linked (on the same chromosome).

Figure 12.15B **work with the data** follows on next page.

at different loci on the same chromosome *do* sometimes separate from one another during meiosis. Genes may recombine when two homologous chromosomes physically exchange corresponding segments during prophase I of meiosis—that is, by crossing over (**Figure 12.16**). As discussed in Key Concept 11.2, DNA is replicated during S phase, so that by prophase I, when homologous chromosome pairs come together to form tetrads, each chromosome consists of two chromatids.

Note that one exchange event involves *only two of the four chromatids* in a tetrad, one from each member of the homologous pair, and can occur at any point along the length of the chromosome. The chromosome segments involved are exchanged reciprocally, so both chromatids involved in crossing over become recombinant (that is, each chromatid ends up with genes from both of the organism's parents). Usually several exchange events occur along the length of each homologous pair.

work with the data

Figure 12.15B Some Alleles Do Not Assort Independently

Original Paper: Morgan, T. H. 1912.

Mendel's work was "rediscovered" 40 years after its publication. At that time, biologists began to find some exceptions to the rules of inheritance that Mendel had proposed. Thomas Hunt Morgan and his colleagues made dihybrid test crosses in fruit flies. They proposed that the clearest way to test for linkage was not to look at aberrations in the 9:3:3:1 phenotypic ratio expected from an $F_1 \times F_1$ cross, but to examine aberrations in the 1:1:1:1 ratio expected from an $F_1 \times$ homozygous recessive test cross (see Figure 12.3). Morgan's group then hypothesized that linkage had a physical basis, namely that genes are linked together on chromosomes and that rare crossing over during meiosis gives rise to the less frequent phenotypes. Examination of actual chromosomal events confirmed this.

QUESTIONS▶

1. Morgan first performed a dihybrid cross between black, normal-winged flies (*bbVgVg*) and gray, vestigial-winged flies (*BBvgvg*). The F_1 flies were interbred, yielding the F_2 phenotypes shown in the table (Experiment 1). Compare these data with the expected data in a 9:3:3:1 ratio by using the chi-square test (see Appendix B for information about the chi-square test). Are there differences, and are they significant?

2. To quantify linkage, Morgan crossed homozygous black, normal-winged females with homozygous gray, vestigial-winged males. He then crossed the F_1 females with black, vestigial-winged males. (You should note that this is not the same test cross as the one shown in the experiment in Figure 12.15A. In that case, the original parents were *BBVgVg* and *bbvgvg*.) The results of this test cross are shown in the table (Experiment 2). Are these genes linked? If they are linked, what is the map distance between the genes? Explain why these data are so different from the data shown in Figure 12.15A.

3. In a third experiment, Morgan crossed two genetic strains of flies that were homozygous for the body color and wing genes. The F_1 flies were all gray and normal-winged, and these were interbred. The results are shown in the table (Experiment 3). What were the genotypes and phenotypes of the original parents that produced the F_1?

	Number of progeny showing each phenotype			
Experiment	Gray, normal	Black, normal	Gray, vestigial	Black, vestigial
1	2,316	1,146	737	0
2	578	1,413	1,117	307
3	246	9	65	18

A similar **work with the data** exercise may be assigned in **LaunchPad**.

When crossing over takes place between two linked genes, not all the progeny of a cross have the parental phenotypes. Instead, recombinant offspring appear as well, as they did in Morgan's test cross (see Figure 12.15). They appear in proportions called

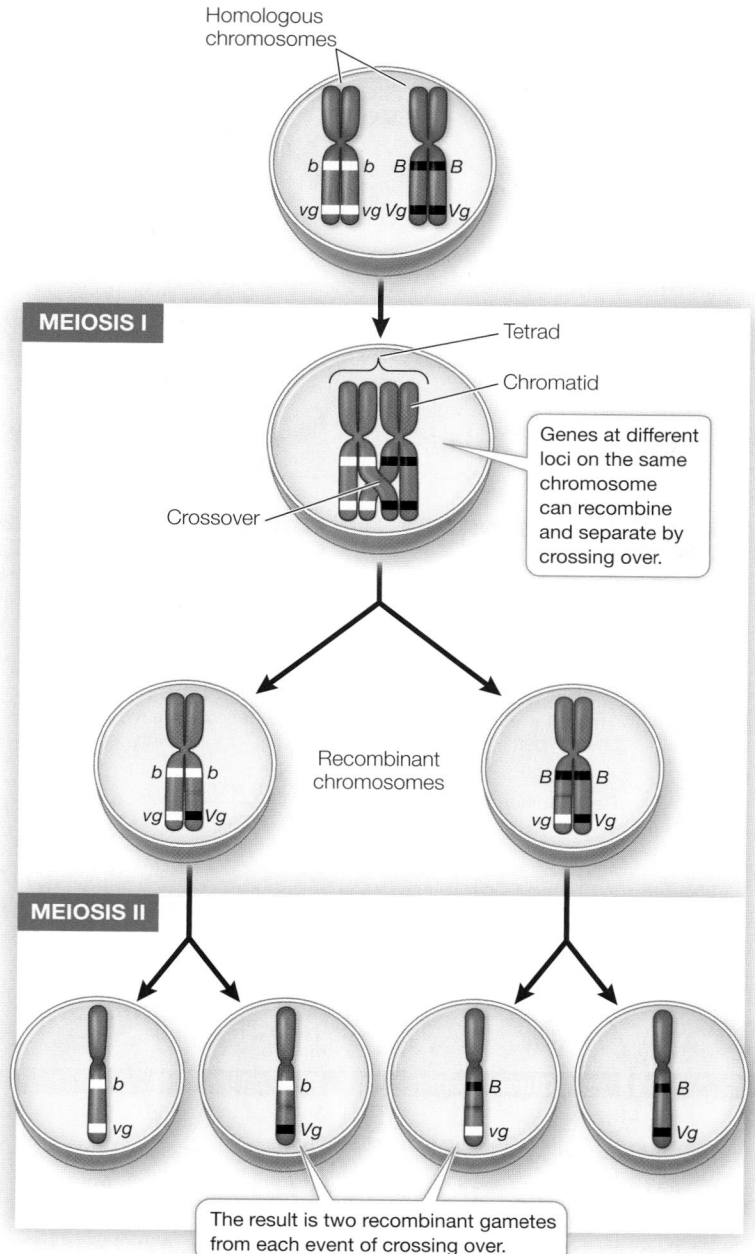

Figure 12.16 Crossing Over Results in Genetic Recombination
Recombination accounts for why linked alleles are not always inherited together. Alleles at different loci on the same chromosome can be recombined by crossing over, and separated from one another. Such recombination occurs during prophase I of meiosis.

recombinant frequencies, which are calculated by dividing the number of recombinant progeny by the total number of progeny (**Figure 12.17**). Recombinant frequencies will be *greater for loci that are farther apart* on the chromosome than for loci that are closer together because an exchange event is more likely to occur between genes that are far apart.

By calculating recombinant frequencies, geneticists can infer the locations of genes along a chromosome and generate a genetic map. Below is a map showing five genes on a fruit fly chromosome

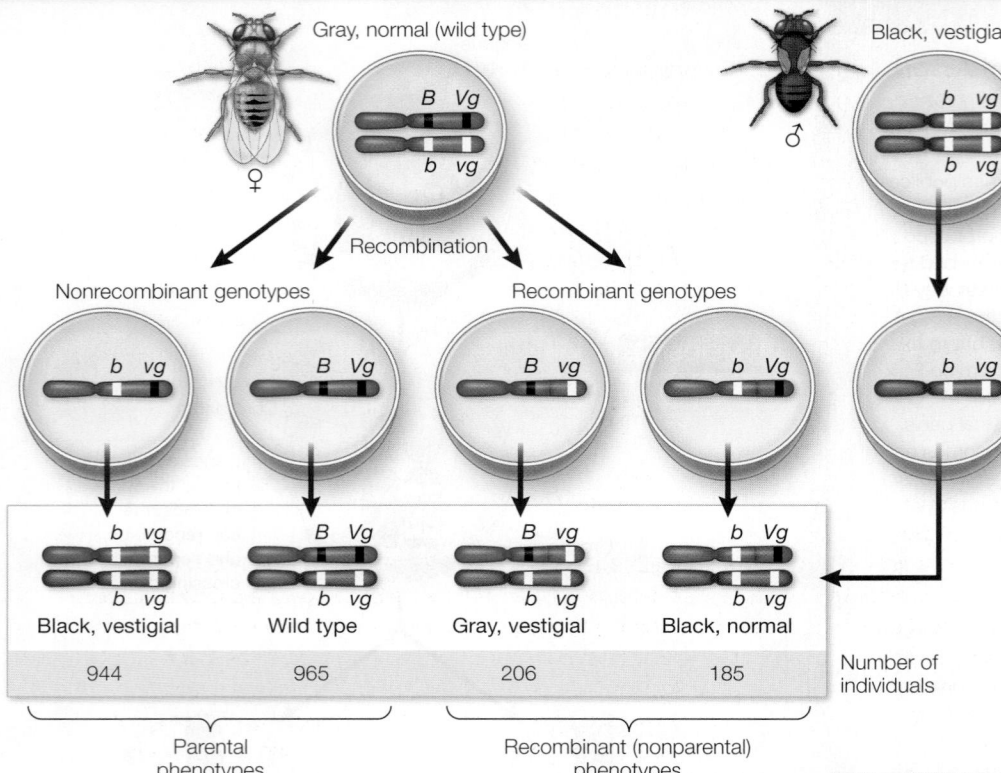

Gray, normal (wild type)

Black, vestigial

Recombination

Nonrecombinant genotypes

Recombinant genotypes

Black, vestigial   Wild type   Gray, vestigial   Black, normal

| 944 | 965 | 206 | 185 |

Number of individuals

Parental phenotypes

Recombinant (nonparental) phenotypes

$$\text{Recombinant frequency} = \frac{391 \text{ recombinants}}{2,300 \text{ total offspring}} = 0.17$$

Figure 12.17 Recombinant Frequencies The frequency of recombinant offspring (those with a phenotype different from either parent) can be calculated.

constructed using the recombination frequencies generated by test crosses involving various pairs of genes:

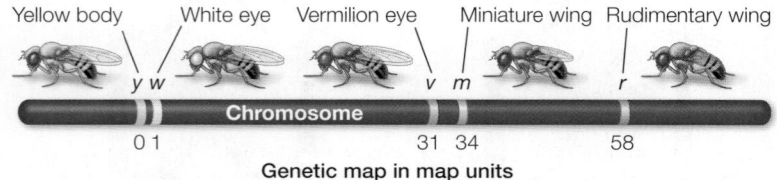

Yellow body   White eye   Vermilion eye   Miniature wing   Rudimentary wing

y w   v m   r

Chromosome

0 1   31 34   58

Genetic map in map units

In the chromosome shown above, the recombination frequency between *y* and *w* is low, so they are close together on the map. Recombination between *y* and *v* is more frequent, so they are farther apart. The recombination frequencies are converted to map units (also called centiMorgans, cM); one map unit is equivalent to an average recombination frequency of 0.01 (1%).

   *Gene sequencing of DNA has made chromosome mapping less important in some areas of genetics research. However, mapping is still a way to verify that a particular DNA sequence corresponds with a particular phenotype. Linkage has allowed biologists to isolate genes and to identify genetic markers linked to important genes. This is important in breeding new crops and animals for agriculture, and for identifying people carrying medically significant mutations.

---

   *connect the concepts   As described in Key Concept 17.1, gene sequencing techniques reveal the nucleotide sequences of genes and identify where genes begin and end on the chromosome.

Linkage is revealed by studies of the sex chromosomes

In Mendel's work, reciprocal crosses always gave similar results; it did not matter whether a dominant allele was contributed by the female parent or the male parent. But in some cases, the parental origin of a chromosome does matter. For example, human males inherit a bleeding disorder called hemophilia from their mother, not from their father. To understand the types of inheritance in which the parental origin of an allele is important, we must consider the ways in which sex is determined in different species.

SEX DETERMINATION BY CHROMOSOMES   In corn, every diploid adult has both male and female reproductive structures. The tissues in these two types of structure are genetically identical, just as roots and leaves are genetically identical. Organisms such as corn, in which the same individual produces both male and female gametes, are said to be **monoecious** (Greek, "one house"). Other organisms, such as date palms and most animals, are **dioecious** ("two houses"), meaning that some individuals can produce only male gametes and others can produce only female gametes. In other words, in dioecious organisms the different sexes are different individuals.

   In mammals and birds, sex is determined by differences in the chromosomes, but such determination operates in different ways in different groups of organisms. For example, in many animals, including mammals, sex is determined by a single pair of **sex chromosomes**, which differ from one another. The remaining chromosomes, called **autosomes**, occur in pairs in males and females. For example, in humans there are 22 pairs of autosomes in males and females, and 1 pair of sex chromosomes. The chromosomal bases for sex determination in various groups of animals are summarized in **Table 12.2**.

table **12.2** Sex Determination in Animals	
Animal group	**Mechanism**
Bees	Males are haploid, females are diploid
Fruit flies	Fly is female if ratio of X chromosomes to sets of autosomes is 1 or more
Birds	Males ZZ (homogametic), females WZ (heterogametic)
Mammals	Males XY (heterogametic), females XX (homogametic)

The sex chromosomes of female mammals consist of a pair of X chromosomes. Male mammals, by contrast, have one X chromosome and a sex chromosome that is not found in females, the Y chromosome. Females may be represented as XX (homogametic) and males as XY (heterogametic).

MALE MAMMALS PRODUCE TWO KINDS OF GAMETES   Each gamete produced by a male mammal has a complete set of autosomes, but just as each gamete carries one copy of each pair of autosomes, half carry a Y. When an X-bearing sperm fertilizes an egg, the resulting XX zygote is female; when a Y-bearing sperm fertilizes an egg, the resulting XY zygote is male.

SEX CHROMOSOME ABNORMALITIES REVEALED THE GENE THAT DETERMINES SEX   Can we determine which chromosome, X or Y, carries the sex-determining gene, and can the gene be identified? One way to determine cause (e.g., the presence of a gene on the Y chromosome) and effect (e.g., maleness) is to look at cases of biological error, in which the expected outcome does not happen.

We can learn something about the functions of X and Y chromosomes from abnormal sex chromosome arrangements resulting from nondisjunction during meiosis or mitosis (see Key Concept 11.5). As you will recall, nondisjunction occurs when a pair of homologous chromosomes (in meiosis I) or sister chromatids (in mitosis or meiosis II) fails to separate. As a result, a gamete may have one too few or one too many chromosomes. If this gamete fuses with another gamete that has the full haploid chromosome set, the resulting offspring will be aneuploid, with fewer or more chromosomes than normal.

In humans, XO individuals sometimes appear. The O indicates that a chromosome is missing—that is, individuals that are XO have only one sex chromosome (an X). Human XO individuals are females who are moderately abnormal physically but normal mentally; usually they are also sterile. The XO condition in humans is called Turner syndrome. It is the only known case in which a person can survive with only one member of a chromosome pair (here, the XY pair), although most XO conceptions are spontaneously terminated early in development. XXY individuals also occur; this condition, which affects males, is called Klinefelter syndrome and results in overlong limbs and sterility.

These observations suggested that the gene controlling maleness is located on the Y chromosome. Observations of people with other types of chromosomal abnormalities helped researchers pinpoint the location of that gene:

- Some women are genetically XY but lack a small portion of the Y chromosome.

- Some men are genetically XX but have a small piece of the Y chromosome attached to another chromosome.

The Y fragments that are respectively missing and present in these two cases are the same and contain the maleness-determining gene, which was named *SRY* (*sex-determining region on the Y* chromosome).

The *SRY* gene encodes a protein involved in **primary sex determination**—that is, the determination of the kinds of gametes that an individual will produce and the organs that will make them (the male and female gonads). In the presence of the functional SRY protein, an embryo develops sperm-producing testes. (Notice that *italic type*

is used for the name of a gene, but roman type is used for the name of a protein.) If the embryo has no Y chromosome, the *SRY* gene is absent, and thus the SRY protein is not made. In the absence of the SRY protein, the embryo develops egg-producing ovaries. In this case, a gene on the X chromosome called *DAX1* produces an anti-testis factor. So the role of SRY in a male is to inhibit the maleness inhibitor encoded by *DAX1*. The SRY protein does this in male cells, but since it is not present in females, DAX1 can act to inhibit maleness.

One function of the gonads is to produce hormones (such as testosterone and estrogen) that send signals to the rest of the body and control the development of **secondary sex characteristics**. These are outward manifestations of maleness and femaleness, such as differences in body type, breast development, body hair, and voice. Secondary sex characteristics distinguish males and females but are not directly part of the reproductive system.

SEX-LINKED INHERITANCE IN FRUIT FLIES   As noted in Table 12.2, sex determination in fruit flies is based on the *proportions* of sex chromosomes, because the numbers of these chromosomes can vary. But most commonly, the fruit fly genome has four pairs of chromosomes: three pairs of autosomes and (as in humans) a pair of sex chromosomes that differ in size. In this case, the female fly has two X chromosomes and the male has only one, the other being the Y chromosome—so the female is XX and the male is XY. As in other organisms, the X and Y chromosomes are not true homologs of one another: *many genes on the X chromosome are not present on the Y*. The X chromosome of *Drosophila* was one of the first to have specific genes assigned to it.

Thomas Morgan identified a **sex-linked** gene that controls eye color in *Drosophila*. The wild-type allele of the gene confers red eyes, whereas a recessive mutant allele confers white eyes. Morgan's experimental crosses demonstrated that this eye color locus is on the X chromosome. If we abbreviate the eye color alleles as *R* (red eyes) and *r* (white eyes), the presence of the alleles on the X chromosome is designated by X^R and X^r.

Morgan crossed a homozygous red-eyed female ($X^R X^R$) with a white-eyed male. The male is designated $X^r Y$ because the Y chromosome does not carry any allele for this gene. (Any gene that is present as a single copy in a diploid organism is called **hemizygous**.) All the sons and daughters from this cross had red eyes, because the red phenotype is dominant over white and all the progeny had inherited a wild-type X chromosome (X^R) from their mother (**Figure 12.18A**). This phenotypic outcome would have occurred even if the *R* gene had been present on an autosome rather than a sex chromosome. In that case, the male would have been homozygous recessive—*rr*.

When Morgan performed the reciprocal cross, in which a white-eyed female ($X^r X^r$) was mated with a red-eyed male ($X^R Y$), the results were unexpected: Instead of all offspring being heterozygous with red eyes, *all the sons were white-eyed and all the daughters were red-eyed* (**Figure 12.18B**). The sons from the reciprocal cross inherited their only X chromosome from their white-eyed mother and were therefore hemizygous for the white allele. The daughters, however, got an X chromosome bearing the *r* allele from their mother and an X chromosome bearing the *R* allele from their father; therefore they were red-eyed heterozygotes. When these heterozygous females were mated with red-eyed males, half their sons had white eyes but all their

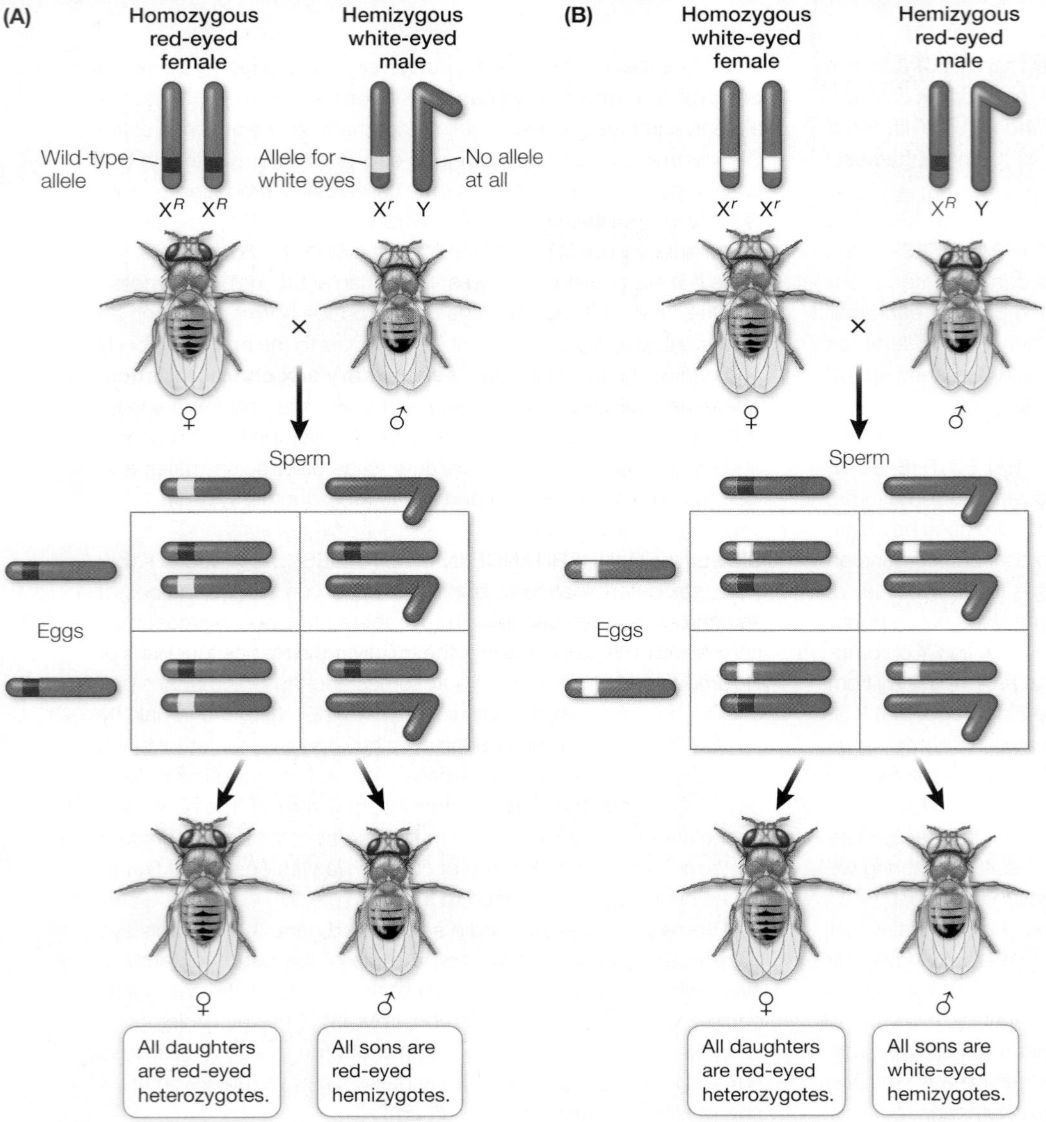

(A)

Homozygous red-eyed female — Wild-type allele — X^R X^R

Hemizygous white-eyed male — Allele for white eyes / No allele at all — X^r Y

♀ × ♂

Sperm

Eggs

All daughters are red-eyed heterozygotes.

All sons are red-eyed hemizygotes.

(B)

Homozygous white-eyed female — X^r X^r

Hemizygous red-eyed male — X^R Y

♀ × ♂

Sperm

Eggs

All daughters are red-eyed heterozygotes.

All sons are white-eyed hemizygotes.

Figure 12.18 Eye Color Is a Sex-Linked Trait in *Drosophila*   Morgan demonstrated that a mutant allele that causes white eyes in *Drosophila* is carried on the X chromosome. Note that in this case, the reciprocal crosses do not have the same results.

Q: If a red-eyed female was crossed with a white-eyed male and there were white-eyed males in the offspring, what was the genotype of the red-eyed female?

daughters had red eyes. Together, these results showed that eye color was carried on the X chromosome and not on the Y.

These and other experiments led to the term **sex-linked inheritance**: inheritance of a gene that is carried on a sex chromosome. (This term is somewhat misleading because "sex-linked" inheritance is not really linked to the sex of an organism—after all, both males and females carry X chromosomes.) In mammals, the X chromosome is larger and carries more genes than the Y. For this reason, most examples of sex-linked inheritance involve genes that are carried on the X chromosome.

Many sexually reproducing species, including humans, have sex chromosomes. As in most fruit flies, human males are XY, females are XX, and relatively few of the genes that are present on the X chromosome are present on the Y. Pedigree analyses of X-linked recessive phenotypes like the one in **Figure 12.19** reveal the following patterns (compare with the pedigrees of non–X-linked phenotypes in Figure 12.7):

- The phenotype appears much more often in males than in females, because only one copy of the rare allele is needed for its expression in males, whereas two copies must be present in females.

- A male with the mutation can pass it on only to his daughters; all his sons get his Y chromosome.

- Daughters who receive one X-linked mutation are heterozygous carriers. They are phenotypically normal, but they can pass the mutant allele to their sons or daughters. On average, half their children will inherit the mutant allele since half of their X chromosomes carry the normal allele.

- The mutant phenotype can skip a generation if the mutation passes from a male to his daughter (who will be phenotypically normal) and then to her son.

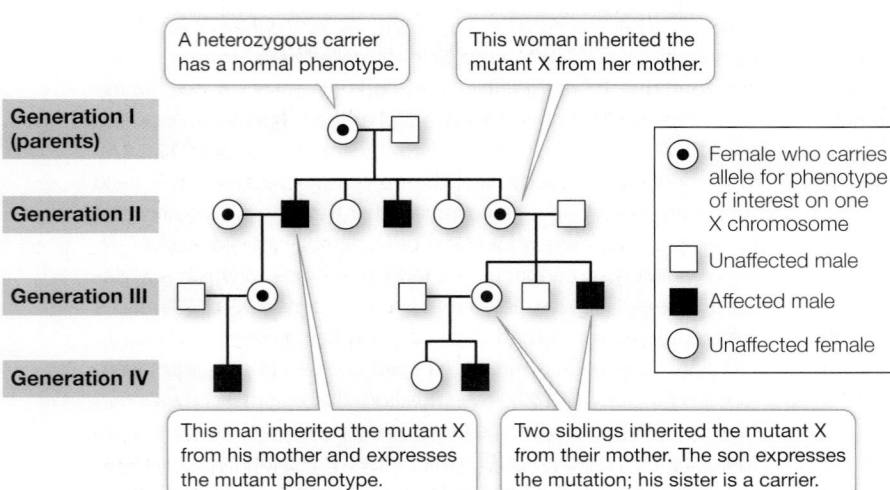

A heterozygous carrier has a normal phenotype.

This woman inherited the mutant X from her mother.

Generation I (parents)

Generation II

Generation III

Generation IV

Female who carries allele for phenotype of interest on one X chromosome

Unaffected male

Affected male

Unaffected female

This man inherited the mutant X from his mother and expresses the mutant phenotype.

Two siblings inherited the mutant X from their mother. The son expresses the mutation; his sister is a carrier.

Figure 12.19 Red-Green Color Blindness Is a Sex-Linked Trait in Humans   The mutant allele for red-green color blindness is expressed as an X-linked recessive trait, and therefore is always expressed in males when they carry that allele.

▶ **12.4 recap**

Simple Mendelian ratios are not observed when genes are linked on the same chromosome. Linkage results in atypical frequencies of phenotypes in the offspring from a test cross. Calculating recombination frequencies between genes can lead to a genetic map of the genes. Sex linkage in humans refers to genes on one sex chromosome (usually the X) that have no counterpart on the other sex chromosome.

learning outcomes

You should be able to:

- Describe evidence that genes on the same chromosome are linked.
- Use a model to illustrate how recombinant frequencies relate to gene locations on a chromosome.
- Analyze data involving gene linkage.
- Analyze data involving sex-linked genes.
- Construct a diagram to illustrate differences between sex-linked and autosomal genes.

1. Describe the concept of genetic linkage and its implications in genetic crosses.

2. Using a diagram of a chromosome, illustrate how the physical distance that separates genes on a chromosome affects their recombination frequency.

3. Use a Punnett square to show why red-green colorblindness is more common in males than in females.

4. How does a sex-linked gene behave differently in genetic crosses from a gene on an autosome?

5. Sometimes scientists get lucky. Consider Mendel's dihybrid cross shown in Figure 12.4. Peas have a haploid number of seven chromosomes, so many of their genes are linked. What would Mendel's results have been if the genes for seed color and seed shape were linked with a map distance of 10 units? Now, consider Morgan's fruit flies (see Figure 12.17). Suppose that the genes for body color and wing shape were not linked. What results would Morgan have obtained?

6. In *Drosophila*, three autosomal genes have alleles as follows:
Gray body color (*B*) is dominant over black (*b*).
Normal wings (*VG*) is dominant over vestigial (*vg*).
Red eye (*R*) is dominant over sepia (*r*).
Two crosses were performed, with the following results:

Cross I: Parents: heterozygous
red, normal × sepia, vestigial
Offspring: 131 red, normal
120 sepia, vestigial
122 red, vestigial
127 sepia, normal

Cross II: Parents: heterozygous
gray, normal × black, vestigial
Offspring: 236 gray, normal
253 black, vestigial
50 gray, vestigial
61 black, normal

Are any of the three genes linked on the same chromosome? If so, what is the distance between the linked genes (in map units)?

The genes we've discussed so far in this chapter are all in the cell nucleus. You might be surprised to learn that other organelles, including mitochondria and plastids, also carry genes. What are these genes, and how are they inherited?

▶ **key concept** Some Eukaryotic
12.5 Genes Are Outside the Nucleus

The nucleus is not the only organelle in a eukaryotic cell that carries genetic material. ***Mitochondria and plastids** contain small numbers of genes. For example, in humans there are about 21,000 genes coding for proteins in the nuclear genome and 37 in the mitochondrial genome. Plastid genomes are about five times larger than those of mitochondria.

focus your learning

- Mitochondria and plastids contain genes that follow inheritance patterns that differ from inheritance patterns of nuclear genes.

***connect the concepts**   Key Concept 5.5 describes the endosymbiotic theory, which proposes that some organelles—particularly mitochondria and plastids—may be descended from prokaryotes that were engulfed by other, larger cells.

The inheritance of organelle genes differs from that of nuclear genes for several reasons:

- In most organisms, mitochondria and/or plastids are inherited only from the female. As you will learn in Chapter 43, eggs contain abundant cytoplasm and organelles, but the only part of the sperm that survives to take part in the union of haploid gametes is the nucleus. So you have inherited your mother's mitochondria (with their genes), but not your father's.

- There may be hundreds of mitochondria and/or plastids in a cell. So a cell is not diploid for organelle genes.

- Organelle genes tend to mutate at much faster rates than nuclear genes, so organelle genes often have multiple alleles.

Although most genes for organelle proteins are in the nucleus, several of the genes carried by cytoplasmic organelles are important for organelle assembly and function, and mutations of these genes can have profound effects on the organism. The phenotypes resulting from such mutations reflect the organelles' roles. For example, in plants and some photosynthetic protists, certain plastid gene mutations affect the proteins that assemble chlorophyll molecules into photosystems. These mutations result in a phenotype that is essentially white instead of green. The inheritance of this phenotype follows a non-Mendelian, maternal pattern (**Figure 12.20**). Mitochondrial gene mutations that affect the respiratory chain result in less ATP production. In animals, these mutations have particularly noticeable effects in tissues with high energy requirements, such as the nervous system, muscles, and kidneys.

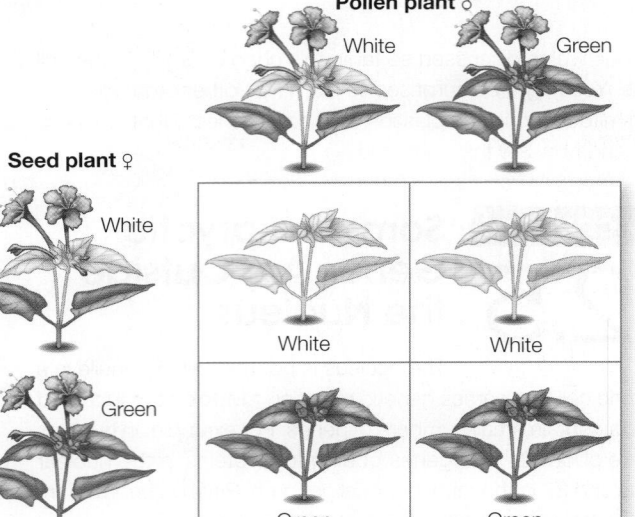

Figure 12.20 Cytoplasmic Inheritance In four o'clock plants, leaf color is inherited through the female plant only. The white leaf color is caused by a chloroplast mutation.

12.5 recap

Genes in the genomes of organelles, specifically plastids and mitochondria, do not behave in a Mendelian fashion.

learning outcome

You should be able to:
- Predict inheritance patterns of genes present in organelles.

1. A disease called Leber's hereditary optic neuropathy is caused by a mutation in a gene carried on mitochondrial DNA. What would be the phenotype of their first child if a man with this disease mated with a woman who did not have the disease? What would be the result if the wife had the disease and the husband did not?

Mendel and those who followed him focused on eukaryotes, with diploid organisms and haploid gametes. A half-century after the rediscovery of Mendel's work, a process that allows genetic recombination was discovered in prokaryotes as well. We will now turn to that process.

key concept 12.6 Prokaryotes Can Transmit Genes by Mating

As described in Chapter 5, prokaryotic cells lack nuclei; they contain their genetic material mostly as single chromosomes in central regions of their cells. Prokaryotes reproduce asexually by binary fission, a process that gives rise to progeny that are virtually identical genetically. That is, the offspring of cell reproduction in prokaryotes constitute a clone (see Chapter 11). You might expect, therefore, that there is no way for individuals of these organisms to exchange genes, as in sexual reproduction. But mutations do occur in prokaryotes just as they do in eukaryotes, and the resulting new alleles increase genetic diversity. In addition, it turns out that prokaryotes do have a sexual process for transferring genes between cells.

focus your learning

- Prokaryotes use conjugation to exchange genetic material on plasmids.

Bacteria exchange genes by conjugation

To illustrate the kind of experiment that led to the discovery of bacterial DNA transfer, consider two strains of the bacterium *E. coli* with different alleles for each of six genes. One stain carries the dominant (wild-type) alleles for three of the genes and the recessive (mutant) alleles for the other genes. This situation is reversed in the other strain. Simply put, the two strains have the following genotypes (remember that bacteria are haploid and thus have only one copy of each gene):

ABCdef   and   *abcDEF*

where capital letters indicate wild-type alleles and lowercase letters indicate mutant alleles.

When the two strains are grown together in the laboratory, most of the cells produce clones. That is, almost all of the cells that grow have the original genotypes. However, out of millions of bacteria, a few occur that have the genotype

ABCDEF

How could these completely wild-type bacteria arise? One possibility is mutation: in the *abcDEF* bacteria, the *a* allele could have mutated to *A*, the *b* allele to *B*, and the *c* allele to *C*. The problem with this explanation is that a mutation at any particular point in an organism's DNA sequence is a very rare event. The probability of all three events occurring in the same cell is extremely low—much lower than the actual rate of appearance of cells with the *ABCDEF* genotype. So the mutant cells must have acquired wild-type genes some other way—and this turns out to be the transfer of DNA between cells.

Electron microscopy shows that genetic transfers between bacteria can happen via physical contact between the cells (**Figure 12.21A**). Contact is initiated by a thin projection called a **sex pilus** (plural *pili*), which extends from one cell (the donor), attaches to another (the recipient), and draws the two cells together. Genetic material can then pass from the donor cell to the recipient through a thin cytoplasmic bridge called a conjugation tube. There is no reciprocal transfer of DNA from the recipient to the donor. This process is referred to as **bacterial conjugation**.

Once the donor DNA is inside the recipient cell, it can recombine with the recipient cell's genome. In much the same way that chromosomes pair up, gene for gene, in prophase I of meiosis, the donor DNA can line up beside its homologous genes in the recipient, and crossing over can occur. Some of the genes from the donor can become integrated into the genome of the recipient, thus changing the recipient's genetic constitution (**Figure 12.21B**). When the recipient cells proliferate, the integrated donor genes are passed on to all progeny cells.

Bacterial conjugation is controlled by plasmids

In addition to their main chromosome, many bacteria harbor additional smaller, circular DNA molecules called **plasmids**, which replicate independently of the main chromosome. *Plasmids typically contain at most a few dozen genes, which may fall into one of several categories:

(A)

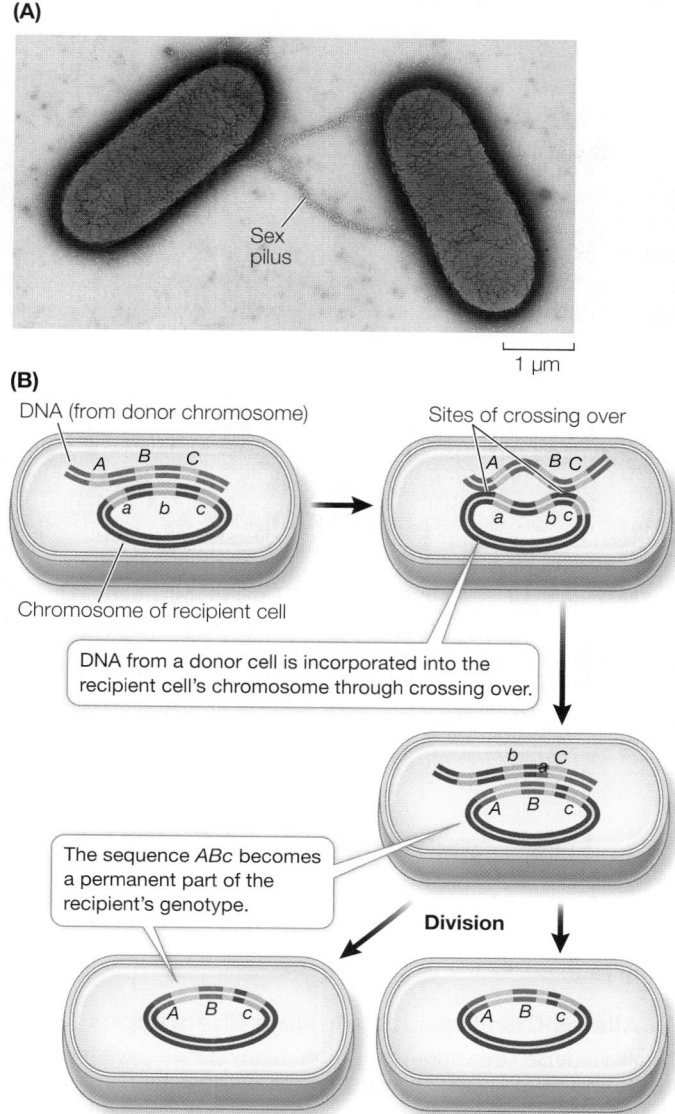

(B)

DNA (from donor chromosome)

Sites of crossing over

Chromosome of recipient cell

DNA from a donor cell is incorporated into the recipient cell's chromosome through crossing over.

The sequence *ABc* becomes a permanent part of the recipient's genotype.

Division

Figure 12.21 Bacterial Conjugation and Recombination **(A)** Sex pili draw two bacteria into close contact, so that a cytoplasmic conjugation tube can form. DNA is transferred from one cell to the other via the conjugation tube. **(B)** DNA from a donor cell can become incorporated into a recipient cell's chromosome through crossing over.

- *Genes for unusual metabolic capacities.* For example, bacteria carrying plasmids that confer the ability to break down hydrocarbons can be used to clean up oil spills.

- *Genes for antibiotic resistance.* Plasmids carrying genes for antibiotic resistance are called R factors, and since they can be transferred between bacteria via conjugation, they are a major threat to human health.

- *Genes that confer the ability to make a sex pilus.*

*connect the concepts As discussed in Key Concept 18.2, biologists can use plasmids as vectors to carry new DNA into other cells. This recombinant DNA technique may be used to produce many copies of a gene of interest, either for analysis or to produce the gene's protein product in quantity.

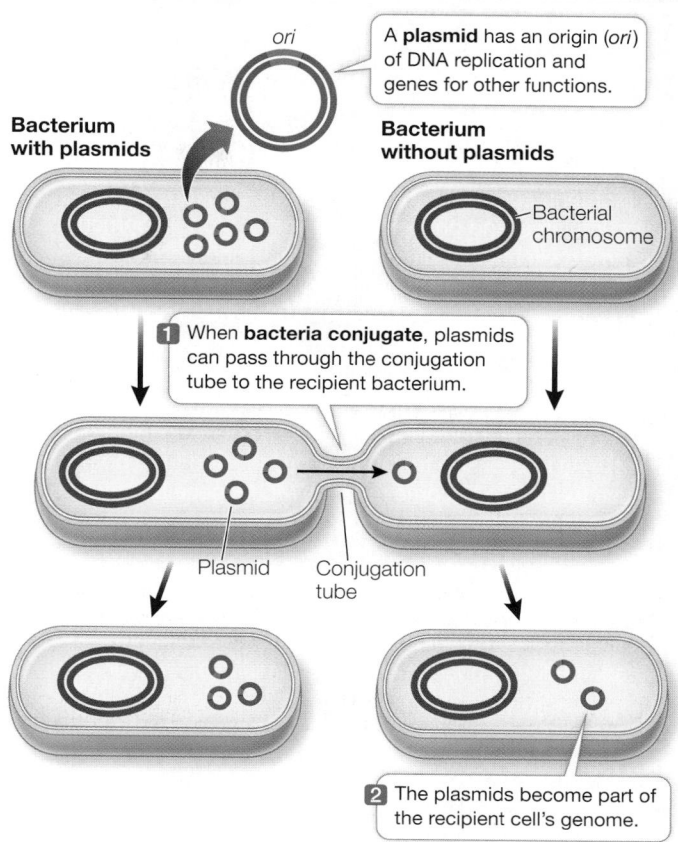

A **plasmid** has an origin (*ori*) of DNA replication and genes for other functions.

Bacterium with plasmids

Bacterium without plasmids

Bacterial chromosome

1 When **bacteria conjugate**, plasmids can pass through the conjugation tube to the recipient bacterium.

Plasmid

Conjugation tube

2 The plasmids become part of the recipient cell's genome.

Figure 12.22 Gene Transfer by Plasmids When plasmids enter a cell via conjugation, their genes can be expressed in the recipient cell.

During bacterial conjugation it is usually plasmids that are transferred from one bacterium to another (**Figure 12.22**). A single strand of the donor plasmid is transferred to the recipient, and then synthesis of a complementary DNA strand results in two complete copies of the plasmid, one in the donor and one in the recipient.

⟩12.6 recap

Although they are haploid and reproduce asexually, prokaryotes have the ability to transfer genes from one cell to another. These genes are usually carried on small, circular DNA molecules called plasmids, but chromosomal DNA is sometimes transferred as well.

learning outcomes

You should be able to:

- Outline the steps involved in the transfer of genetic material by bacterial conjugation.
- Explain the significance of mating and genetic recombination in prokaryotes.

1. Discuss the genetic significance of mating and recombination in bacteria.
2. What is the potential consequence of a plasmid being integrated into the main chromosome of a prokaryotic cell?
3. How does antibiotic resistance spread via plasmids?

investigatinglife

Q&A How do inherited characteristics such as red hair skip generations?

Hair color is determined by the pigment melanin, which is synthesized from the amino acid tyrosine (see Table 3.2). There are two forms of this pigment, eumelanin (black or brown, depending on how much accumulates) and phaeomelanin (red or blond). Pigment production is an excellent example of intercellular signaling at work (see Chapter 7). Skin cells called melanocytes have a receptor, MC1R (*melanocortin 1 receptor*), for the signal melanocortin. Receptor binding sets off a signal transduction pathway that activates the cells to make much more eumelanin than phaeomelanin.

As is common with recessive traits, red hair results from a mutation in a gene, in this case a mutation in the *MC1R* gene, located on chromosome 16. As a recessive trait it must be inherited from both parents to cause the hair to become red. The *MC1R* gene has multiple alleles: the wild type (*R*), which produces a normal receptor, and three recessive (*r*, mutated) alleles, each of which leads to red hair.

So what happens when a person inherits two of these recessive alleles? MC1R is not expressed, so it does not bind the signal that would activate the eumelanin pathway, and phaeomelanin is made instead. If enough phaeomelanin accumulates, red hair

results. The type of red hair that is inherited—fair, strawberry blonde, or auburn—depends on which recessive allele is inherited.

Future directions

As biologists learn more about individual genes and their alleles, they are beginning to make predictions about how gene products interact with one another and with factors in the environment. You probably know that smoking can cause lung cancer, yet there are some people who smoke a lot and don't get the disease. It turns out that these people often carry an allele that alleviates their encounter with the harmful substances in cigarette smoke. We all have genes whose products change environmental substances. In the case of cigarette smoke, the gene product of the typical allele turns the molecules in smoke into DNA-damaging, cancer-causing toxins. People who carry a mutant form of this gene are relatively resistant to getting lung cancer from smoking because the product of this mutant gene does not carry out the activation reactions. The genomes of all organisms carry such genes that are involved in environmental interaction. Describing these genes holds promise in understanding the relative contributions of environment and genetics to establishing a phenotype.

Chapter Summary 12

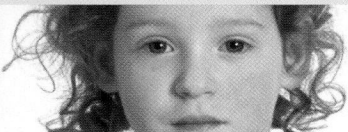

12.1 Inheritance of Genes Follows Mendelian Laws

- Physical features of organisms, or **characters**, can exist in different forms, or **traits**. A heritable trait is one that can be passed from parent to offspring. A **phenotype** is the physical appearance or other detectable characteristic of an organism; a **genotype** is the genetic constitution of the organism.

- The different forms of a **gene** are called **alleles**. Diploid organisms that have two identical alleles for a trait are called **homozygous**; organisms that have two different alleles for a trait are called **heterozygous**. A gene resides at a particular site on a chromosome called a **locus**.

- Mendel's experiments included reciprocal crosses and **monohybrid crosses** between true-breeding pea plants. Analysis of his meticulously tabulated data led Mendel to propose a particulate theory of inheritance stating that discrete units (now called genes) are responsible for the inheritance of specific traits, to which both parents contribute equally.

- Mendel's first law, the **law of segregation**, states that when any individual produces gametes, the two copies of a gene separate, so that each gamete receives only one member of the pair. Thus every individual in the F_1 inherits one copy from each parent. **Review Figure 12.1, Focus: Key Figure 12.2**

- Mendel used a **test cross** to find out whether an individual showing a dominant phenotype was homozygous or heterozygous. **Review Figure 12.3, Activity 12.1**

- Mendel's use of **dihybrid crosses** to study the inheritance of two characters led to his second law: the **law of independent assortment**. The independent assortment of chromosomes in meiosis leads to new combinations of phenotypes in the offspring of a dihybrid cross. **Review Figure 12.4, Focus: Key Figure 12.5, Animation 12.1**

- Probability calculations and **pedigrees** help geneticists trace Mendelian inheritance patterns. **Review Figures 12.6, 12.7, Activity 12.2**

12.2 Alleles Can Produce Multiple Phenotypes

- New alleles arise by random **mutation**. Many genes have multiple alleles. A wild-type allele gives rise to the predominant form of a trait. When the wild-type allele is present at a locus less than 99 percent of the time, the locus is said to be **polymorphic**. **Review Figure 12.8**

- In **incomplete dominance**, neither of two alleles is dominant. The heterozygous phenotype is intermediate between the homozygous phenotypes. **Review Figure 12.9**

- **Codominance** exists when two alleles at a locus produce two different phenotypes that both appear in heterozygotes. **Review Figure 12.10**

- An allele that affects more than one trait is said to be **pleiotropic**.

12.3 Genes Can Interact to Produce a Phenotype

- In **epistasis**, one gene affects the expression of another. **Review Figure 12.11**

- Environmental conditions can affect the expression of a genotype.

- **Penetrance** is the proportion of individuals in a group with a given genotype that show the expected phenotype. **Expressivity** is the degree to which a genotype is expressed in an individual.

- Variations in phenotypes can be **qualitative** (discrete) or **quantitative** (graduated, continuous). Most quantitative traits result from the effects of several genes and the environment. Genes that together determine quantitative characters are called **quantitative trait loci**.

12.4 Genes Are Carried on Chromosomes

- Each chromosome carries many genes.
- Genes on the same chromosome can recombine by crossing over. The resulting recombinant chromosomes have new combinations of alleles. Review Figures 12.16, 12.17
- **Sex chromosomes** are a pair, one of which often determines whether the organism will produce male or female gametes. All other chromosomes are called **autosomes**. The specific functions of sex chromosomes differ among different groups of organisms.
- **Primary sex determination** in mammals is usually a function of the presence or absence of the *SRY* gene. **Secondary sex characteristics** are the outward manifestations of maleness and femaleness.
- In fruit flies and mammals, the X chromosome carries many genes, but the Y chromosome has only a few. Males have only one allele (are **hemizygous**) for X-linked genes, so recessive **sex-linked** mutations are expressed phenotypically more often in males than in females. Females may be unaffected carriers of such alleles. Review Figure 12.18

12.5 Some Eukaryotic Genes Are Outside the Nucleus

- Cytoplasmic organelles such as plastids and mitochondria contain small numbers of genes. In many organisms, cytoplasmic genes are inherited only from the mother because the male gamete contributes only its nucleus (i.e., no cytoplasm) to the zygote at fertilization. Review Figure 12.20

12.6 Prokaryotes Can Transmit Genes by Mating

- Prokaryotes reproduce primarily asexually but can exchange genes in a sexual process called **bacterial conjugation**. Review Figure 12.21
- **Plasmids** are small, extra chromosomes in bacteria that carry genes involved in important metabolic processes and that can be transmitted from one cell to another. Review Figure 12.22

See Activity 12.3 for a concept review of this chapter.

Go to **LearningCurve** (in **LaunchPad**) for dynamic quizzing that helps you solidify your understanding of this chapter. **LearningCurve** adapts to your responses, giving you the practice you need to master each key concept.

▶ Apply What You've Learned

Review

12.4 Genes on the same chromosome are linked.

12.4 Recombinant frequencies can be used to infer gene locations on a chromosome.

Fruit flies (*Drosophila melanogaster*) seem to appear out of nowhere, hovering around bananas, peaches, and other ripe fruit. Biologists, having traced fruit fly origins to West Africa, have hypothesized that expansion of the range of *Drosophila* occurred with human assistance. *Drosophila* can now be found almost everywhere in the world. Though pesky in the kitchen, fruit flies are important laboratory organisms—particularly in genetics research—because of their short life span, ease of care, and many traits that can be observed by simple inspection.

Suppose you were studying fruit flies, looking at easily identified traits encoded by three genes on the same autosomal chromosome. The gene loci are summarized in the figure below. The lowercase letter in each represents the recessive allele; the capital letter represents the dominant allele. The numbers indicate the location in map units of each gene on the chromosome.

Questions

1. Give the genotypes and phenotypes of the F$_1$ and F$_2$ generations produced from the cross *DDbb* × *ddBB*, assuming no crossing over occurs. Also assume that the F$_2$ generation is produced by interbreeding the F$_1$ generation.

2. Give the genotypes and phenotypes of the F$_1$ and F$_2$ generations produced from the cross *BBcc* × *bbCC*, assuming no crossing over occurs. Also assume that the F$_2$ generation is produced by interbreeding the F$_1$ generation.

3. For each of the crosses above, what further phenotypes would appear in the offspring if crossing over occurred? In which cross would you find a higher frequency of recombinants resulting from crossing over? Explain.

4. Draw the chromosomes at the stage where crossing over occurs in meiosis in the *DDbb* × *ddBB* cross in Question 1 above. Show the chromosomes and chromatids before, during, and after crossing over. Clearly label the genes and alleles.

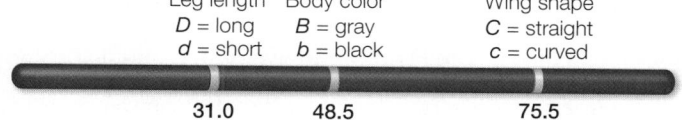

Leg length	Body color	Wing shape
D = long	*B* = gray	*C* = straight
d = short	*b* = black	*c* = curved

31.0   48.5   75.5

Go to **LaunchPad** for the eBook, LearningCurve, animations, activities, flashcards, and additional resources and assignments.

13

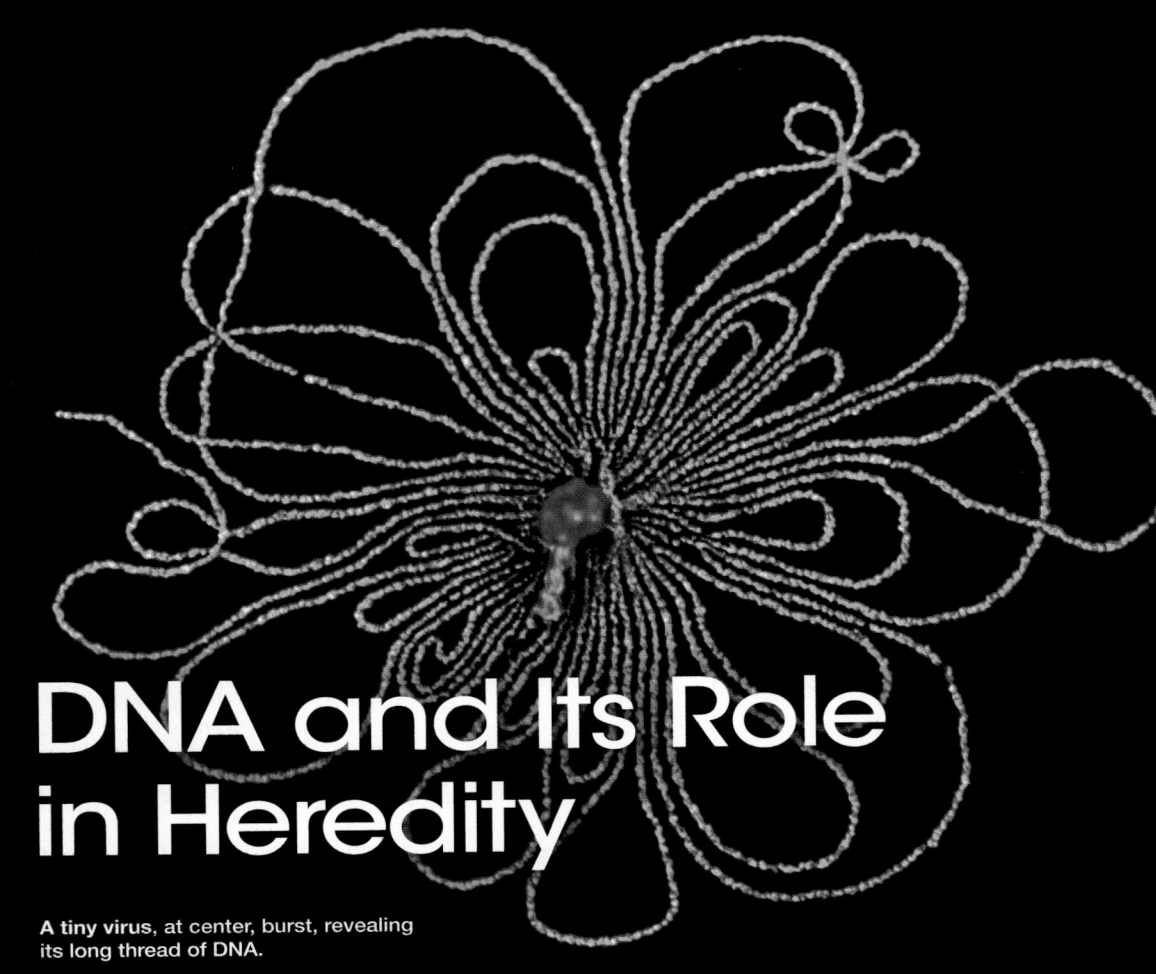

DNA and Its Role in Heredity

A tiny virus, at center, burst, revealing its long thread of DNA.

investigatinglife

Targeting DNA Replication in Cancer Therapy

Testicular cancer occurs with rapid cell divisions of germ cells and often spreads to other organs. It is the most common cancer in young men and is one of the few tumors of adults that is highly curable, due to a drug called cisplatin.

Dr. Barnett Rosenberg, a scientist at Michigan State University, was curious about how electric fields might affect cells. He put bacteria into a growth medium with platinum electrodes connected to a battery. The result was striking: the bacteria stopped dividing. To generalize the effect of electromagnetism on cells, Rosenberg tried the experiment again, this time using copper and zinc electrodes. (You are probably familiar with the Cu/Zn system if you've taken a chemistry course.) This time the bacteria kept dividing. Only platinum electrodes inhibited cell division.

In light of the data, Rosenberg revised his hypothesis to propose that something leaked out of the platinum electrodes into the medium, and that this "something" blocked cell division. He confirmed his hypothesis by treating bacteria with the medium in which the platinum electrodes had been inserted; the bacteria did not divide.

Realizing that cancer cells have uncontrolled cell division, he duplicated his experiments with tumor cells in a laboratory dish. His success at stopping tumor cell growth led to the isolation and development of cisplatin. The drug was so successful with testicular cancer that it has also been used with some success on other tumors. How does it work?

Cisplatin works by making the replication of DNA impossible. As you learned in the last chapter, an essential event for cell division is the complete and precise duplication of the genetic material, DNA. The two strands of DNA unwind and separate, each strand acting as a template for the building of a new strand. Strand separation is possible because the two strands are held together by weak forces, including hydrogen bonds. Cisplatin forms covalent bonds with nucleotides on opposite strands of the DNA, irreversibly cross-linking the two strands together. As a result, the DNA strands cannot separate for replication or expression. With such severe damage to its DNA, the cell then undergoes programmed cell death.

Q **A** What do we need to know about DNA replication to describe the mechanism of a drug that blocks it?

CIsplatin INJECTION 200 mg/200 mL (1 mg/mL) For Intravenous Use 200 mL   Rx only Multiple Dose Vial

key concept

13.1 Experiments Revealed the Function of DNA as Genetic Material

By the early twentieth century, geneticists had associated the presence of genes with chromosomes. Research began to focus on exactly which chemical component of chromosomes composed the genetic material.

focus your learning

- Cell staining techniques provided the first pieces of evidence that DNA was the genetic material.
- Experiments showed that one strain of bacteria could be genetically transformed to another strain and that DNA was the transforming agent.
- Experiments using radiolabeled protein and DNA identified DNA as the material injected by a virus during its infection of a host cell.

Circumstantial evidence indicates that the genetic material is DNA

A hundred years ago, scientists knew that chromosomes were made up of DNA and proteins. At this time a new dye was developed that could bind specifically to DNA and that stained cell nuclei red in direct proportion to the amount of DNA present in the cell. This technique provided circumstantial evidence that DNA was the genetic material:

- *DNA was in the right place*. DNA was confirmed to be an important component of the nucleus and the chromosomes, which were known to carry genes.
- *DNA was present in the right amounts*. The amount of DNA in somatic cells (body cells not specialized for reproduction) was twice that in reproductive cells (eggs or sperm)—as might be expected for diploid and haploid cells, respectively.
- *DNA varied among species*. When cells from different species were stained with the dye and their color intensity measured, each species appeared to have its own specific amount of nuclear DNA.

You should already realize that circumstantial evidence is not a scientific demonstration of cause and effect. After all, proteins are also present in cell nuclei. Science relies on experiments to test hypotheses. The convincing demonstration that DNA is the genetic material came from two sets of experiments, one with bacteria and the other with viruses.

DNA from one type of bacterium genetically transforms another type

In the 1920s, English physician Frederick Griffith was studying the bacterium *Streptococcus pneumoniae*, or pneumococcus, one of the agents that cause pneumonia in humans (**Figure 13.1**). He was

experiment

Figure 13.1   Genetic Transformation

Original Paper: Griffith, F. 1928. The significance of pneumococcal types. *Journal of Hygiene* 27: 113–159.

Griffith's experiments demonstrated that something in the virulent S strain of pneumococcus could transform nonvirulent R strain bacteria into a lethal form, even when the S strain bacteria had been killed by high temperatures.

HYPOTHESIS▶   Material in dead bacterial cells can genetically transform living bacterial cells.

METHOD

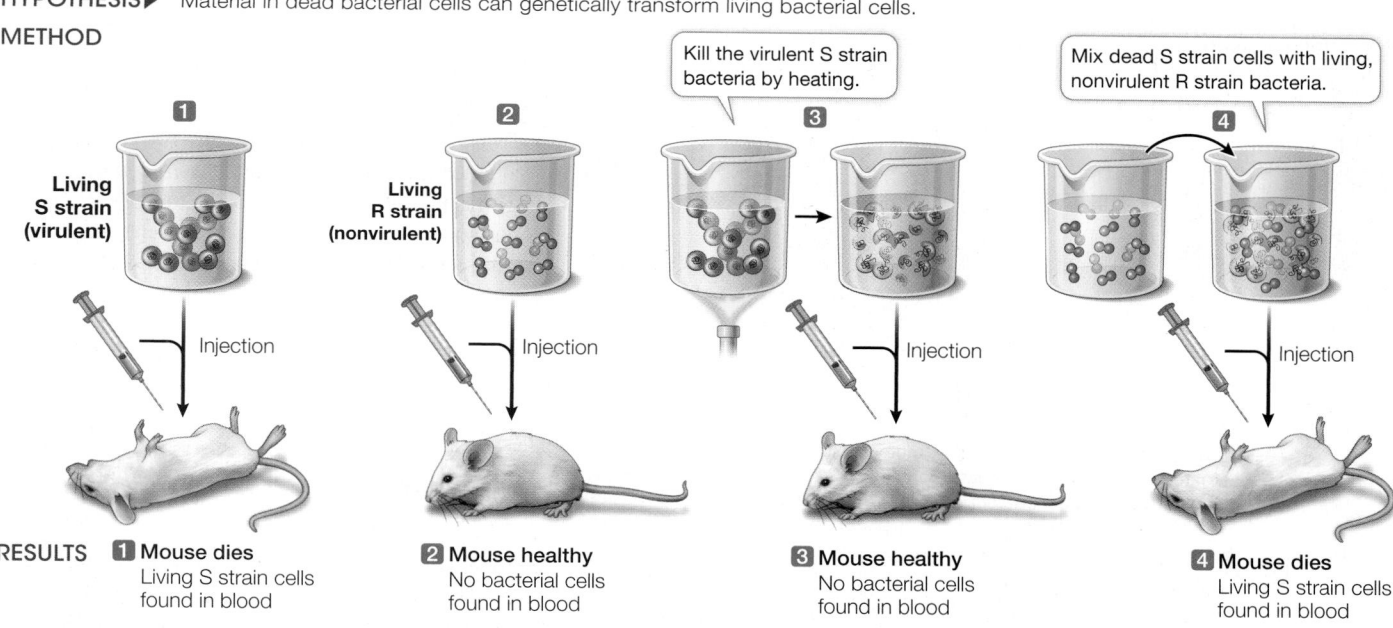

CONCLUSION▶   A chemical substance from one cell is capable of genetically transforming another cell.

experiment

Figure 13.2   Genetic Transformation by DNA

Original Paper: Avery, O. T., C. M. MacLeod and M. McCarty. 1944. Studies on the chemical nature of the substance inducing transformation of the pneumococcal types. *Journal of Experimental Medicine* 79: 137–158.

Experiments by Avery and his colleagues showed that DNA from the virulent S strain of pneumococcus was responsible for the transformation in Griffith's experiments (see Figure 13.1).

HYPOTHESIS▶   The chemical nature of the transforming substance from pneumococcus is DNA.

METHOD

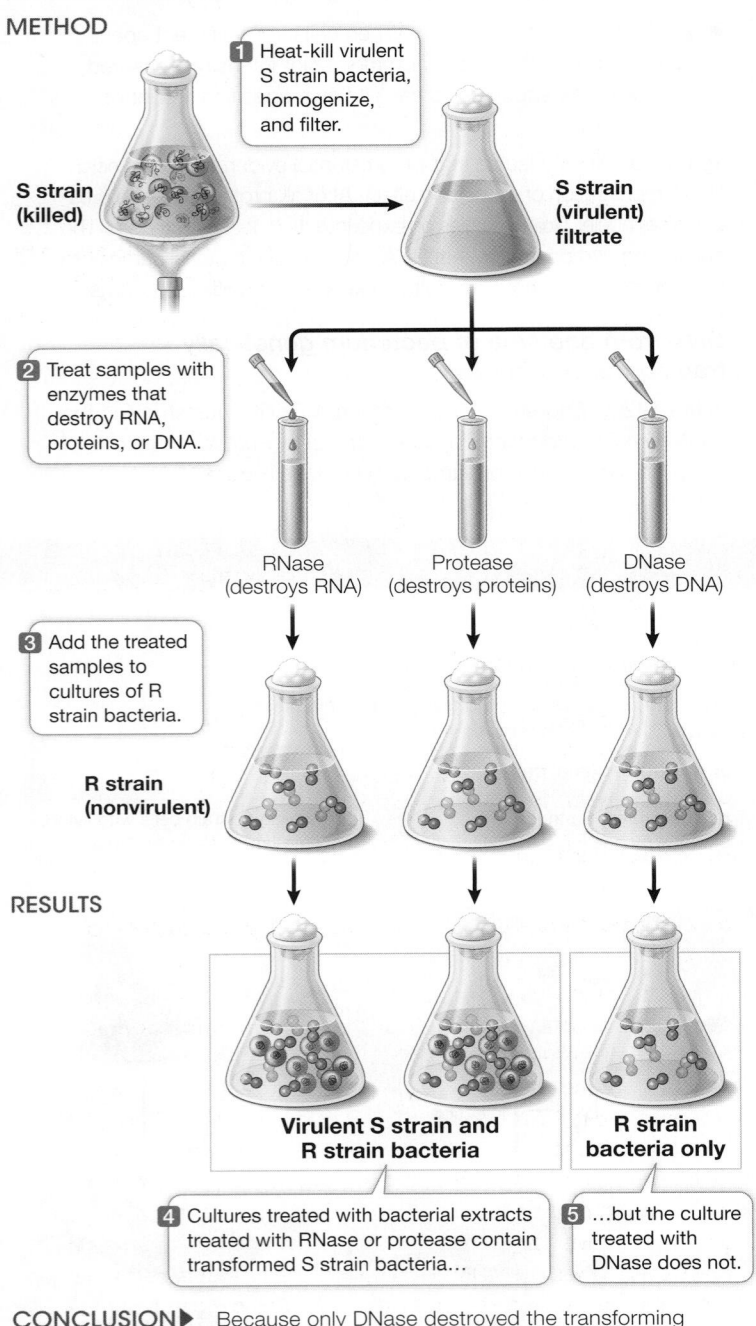

1 Heat-kill virulent S strain bacteria, homogenize, and filter.

S strain (killed)

S strain (virulent) filtrate

2 Treat samples with enzymes that destroy RNA, proteins, or DNA.

RNase (destroys RNA)   Protease (destroys proteins)   DNase (destroys DNA)

3 Add the treated samples to cultures of R strain bacteria.

R strain (nonvirulent)

RESULTS

Virulent S strain and R strain bacteria   **R strain bacteria only**

4 Cultures treated with bacterial extracts treated with RNase or protease contain transformed S strain bacteria…

5 …but the culture treated with DNase does not.

CONCLUSION▶   Because only DNase destroyed the transforming substance, the transforming substance must be DNA.

---

trying to develop a vaccine against this devastating illness (antibiotics had not yet been discovered). Griffith was working with two strains of pneumococcus:

1. Cells of the S strain produced colonies that looked smooth (S). Covered by a polysaccharide capsule, these cells were protected from attack by a host's immune system. When S cells were injected into mice, they reproduced and caused pneumonia (the strain was virulent).

2. Cells of the R strain produced colonies that looked rough (R), lacked the protective capsule, and were not virulent.

When Griffith injected mice with heat-killed S-type pneumococcus cells, the cells did not produce infection. However, when he injected other mice with a mixture of living R-type cells and heat-killed S-type cells, to his astonishment, the mice died of pneumonia. When he examined blood from these mice, he found it full of living bacteria—many of them with characteristics of the virulent S strain! Griffith concluded that in the presence of the dead S-type pneumococcus cells, some of the living R-type cells had been transformed into virulent S cells. These cells were able to grow and multiply in the mice, causing pneumonia. The fact that these S-type cells reproduced to make more S-type cells showed that the change from R-type to S-type was genetic.

Did this transformation of the bacteria need to occur in the mouse's body? No. The same transformation could be achieved in a test tube by mixing living R cells with heat-killed S cells, or even with a cell-free extract of the heat-killed S-type cells. (A cell-free extract contains all the contents of ruptured cells, but no intact cells.) These results demonstrated that some substance from the dead S pneumococcus cells could cause a heritable change in the living cells.

Oswald Avery and his colleagues at what is now The Rockefeller University identified the substance causing bacterial transformation in two ways:

1. *Eliminating other possibilities.* Cell-free extracts containing the transforming substance were treated with enzymes that destroyed candidates for the genetic material, such as proteins, RNA, and DNA. When the treated samples were tested, the ones treated with RNase and protease (which destroy RNA and proteins, respectively) were still able to transform R-type bacteria into the S-type. But the transforming activity was lost in the extract treated with DNase (which destroys DNA) (**Figure 13.2**).

2. *Positive experiment.* The researchers isolated virtually pure DNA from a cell-free extract containing the transforming substance. The DNA alone caused bacterial transformation.

We now know that the gene for the enzyme that catalyzes the synthesis of the polysaccharide capsule, which makes the bacterial colony look "smooth," was transferred into the R cells during transformation.

Viral infection experiments confirmed that DNA is the genetic material

Even with the bacterial transformation experiments, many biologists were still not convinced that DNA is the genetic material. One problem was that DNA, being made up of only four nucleotides (see Key Concept 4.1), seemed too uniform a substance to be able to confer instructions for all the functions and variety of life. The possibility remained that proteins, with all their chemical and structural diversity (20 amino acids), fulfilled that role. Experiments with a virus were designed to distinguish between these alternatives.

Alfred Hershey and Martha Chase of the Cold Spring Harbor Laboratory in New York studied bacteriophage T2 (phage T2), which infects the bacterium *Escherichia coli*. T2 phage consists of a DNA core packed inside a protein coat (**Figure 13.3**). When it attacks a bacterium, part (but not all) of the virus enters the bacterial cell. About 20 minutes later, the cell bursts, releasing dozens of particles that are virtually identical to the infecting virus particle. Clearly the virus is somehow able to hijack the host cell's molecular machinery and convert it into a viral replication machine. Hershey and Chase set out to determine what part of the virus—DNA or protein—enters the host cell to bring about this genetic change. To trace the two components of the virus over its life cycle, the scientists labeled each component with a specific radioisotope:

- *Proteins were labeled with radioactive sulfur*. Proteins contain some sulfur (in the amino acids cysteine and methionine), but DNA does not. Sulfur has a radioactive isotope, ^{35}S. Hershey and Chase grew bacteriophage T2 in a bacterial culture in the presence of ^{35}S, so the proteins of the resulting viruses were labeled with (contained) the radioisotope.

- *DNA was labeled with radioactive phosphorus*. DNA contains a lot of phosphorus (in the deoxyribose–phosphate backbone—see Figure 4.4), whereas proteins contain little or none. Phosphorus also has a radioisotope, ^{32}P. The researchers grew another batch of T2 in a bacterial culture in the presence of ^{32}P, thus labeling the viral DNA with ^{32}P.

Hershey and Chase used these radioactively labeled viruses in their experiments (**Figure 13.4**). In one experiment, they allowed ^{32}P-labeled bacteriophage to infect bacteria; in the other, the bacteria were infected with ^{35}S-labeled bacteriophage. After a few minutes they agitated each mixture vigorously in a kitchen blender, stripping away the parts of the viruses that had not penetrated the bacteria, without bursting the bacteria. Then they separated the bacteria from the rest of the material (the remains of the viruses) from each experiment individually in a centrifuge. The result was that the bacterial cells in the centrifuge pellet contained most of the ^{32}P (and thus the viral DNA), and the supernatant fluid with the viral remains contained most of the ^{35}S (and thus the viral protein). These results indicated that it was the DNA that had been transferred into the bacteria, and that DNA was the molecule responsible for redirecting the genetic program of the bacterial cell.

 Animation 13.1 The Hershey–Chase Experiment
www.Life11e.com/a13.1

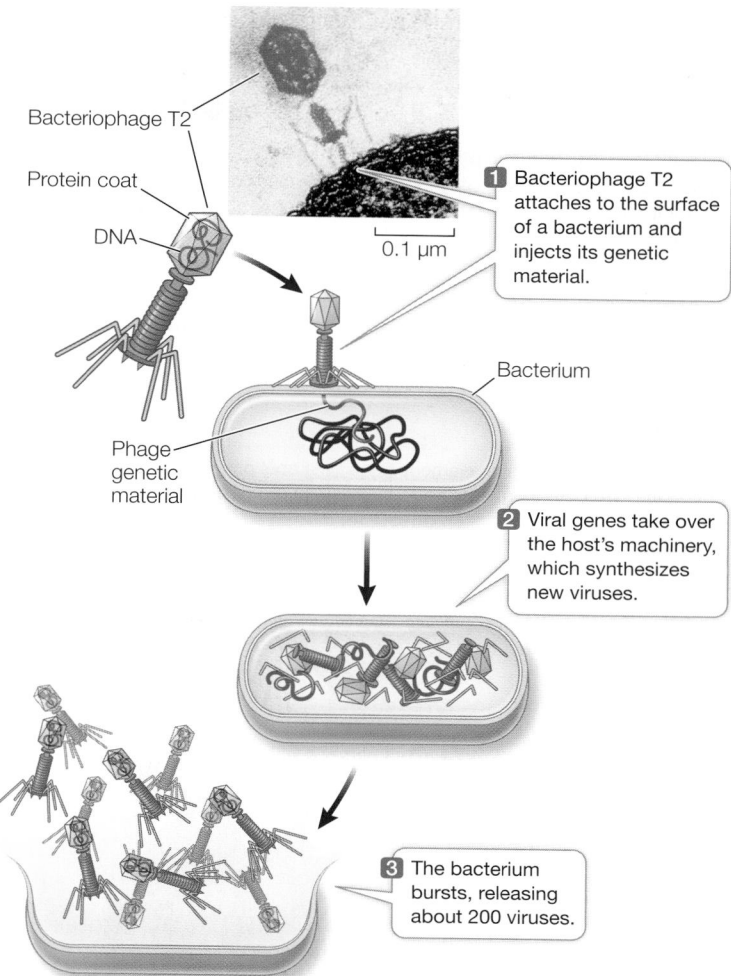

Figure 13.3 Bacteriophage T2: Reproduction Cycle Bacteriophage T2 is parasitic on *E. coli*, dependent on the bacterium to produce new viruses. The external structures of bacteriophage T2 consist entirely of protein, and the DNA is contained within the protein coat. When the virus infects an *E. coli* cell, its genetic material is injected into the host bacterium.

Bacteriophage T2

Protein coat

DNA

0.1 µm

1 Bacteriophage T2 attaches to the surface of a bacterium and injects its genetic material.

Bacterium

Phage genetic material

2 Viral genes take over the host's machinery, which synthesizes new viruses.

3 The bacterium bursts, releasing about 200 viruses.

Eukaryotic cells can also be genetically transformed by DNA

The transformation of eukaryotic cells by DNA is often called **transfection**. This can be demonstrated using a **genetic marker**, a gene whose presence in the recipient cells confers an observable phenotype. When transforming both prokaryotes and eukaryotes, researchers often use a genetically determined selection marker, for example, antibiotic resistance or a nutritional requirement, which permits the growth of transformed cells but not of nontransformed cells. A common marker in mammalian transfection experiments is a gene that confers resistance to the antibiotic neomycin. Transfection is achieved by various methods, including chemical treatments that allow the DNA to be taken up by the cells. Any cell can be transfected, even an egg cell. In the latter case, a whole new genetically transformed organism can result, known as a **transgenic organism**. Transformation in eukaryotes offers conclusive evidence that DNA is the genetic material.

experiment

Figure 13.4   The Hershey–Chase Experiment

Original Paper: Hershey, A. D. and M. Chase. 1952. Independent functions of viral protein and nucleic acid in growth of bacteriophage. *The Journal of General Physiology* 36: 39–56.

When Hershey and Chase infected bacterial cells with radioactively labeled T2 bacteriophage, only labeled DNA was found in the bacteria. The infected cells were agitated to remove the viral coats from the bacteria and were then centrifuged to pellet the bacteria. The labeled protein remained in the supernatant. This showed that DNA, not protein, is the genetic material.

HYPOTHESIS▶   Either component of a bacteriophage—DNA or protein—might be the hereditary material that enters a bacterial cell to direct the assembly of new viruses.

METHOD

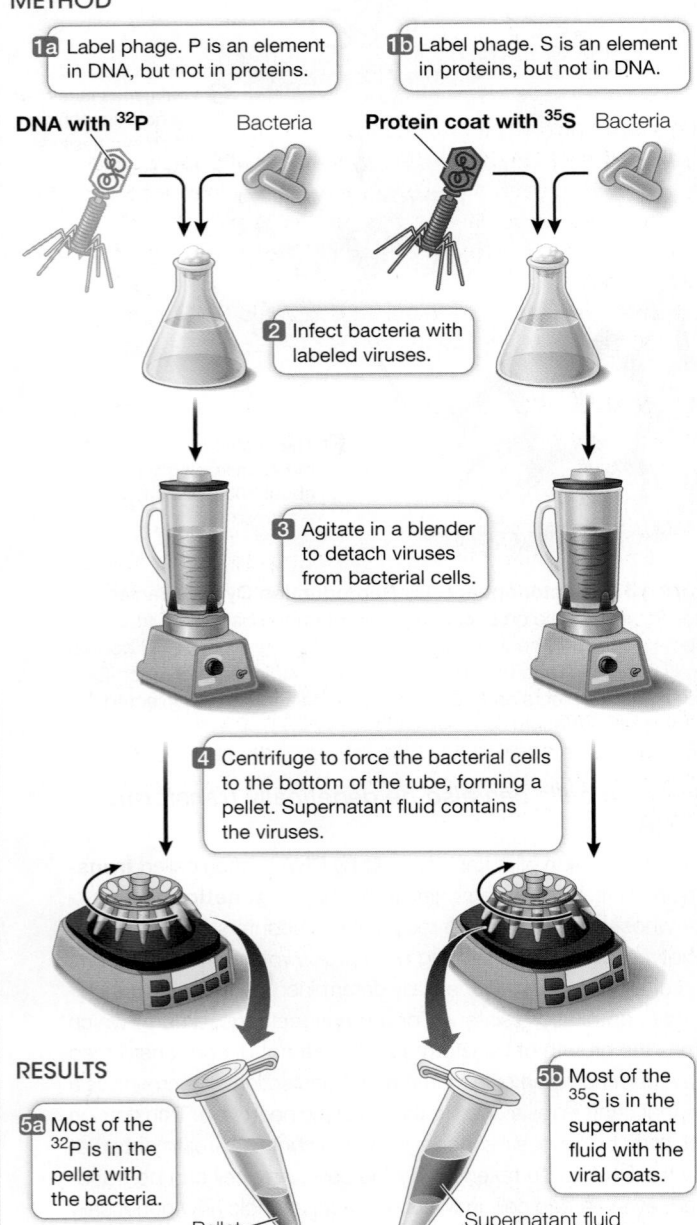

1a Label phage. P is an element in DNA, but not in proteins.

1b Label phage. S is an element in proteins, but not in DNA.

DNA with ^{32}P   Bacteria

Protein coat with ^{35}S   Bacteria

2 Infect bacteria with labeled viruses.

3 Agitate in a blender to detach viruses from bacterial cells.

4 Centrifuge to force the bacterial cells to the bottom of the tube, forming a pellet. Supernatant fluid contains the viruses.

RESULTS

5a Most of the ^{32}P is in the pellet with the bacteria.

Pellet

5b Most of the ^{35}S is in the supernatant fluid with the viral coats.

Supernatant fluid

CONCLUSION▶   DNA, not protein, enters bacterial cells and directs the assembly of new viruses.

A work with the data exercise that accompanies this figure may be assigned in LaunchPad.

13.1 recap

Experiments on bacteria and on viruses demonstrated that DNA is the genetic material. DNA from one genetic strain of bacteria was able to genetically transform another strain into the donor strain. Viral DNA was shown to be injected into a host cell and to genetically change that cell into a virus factory.

learning outcomes

You should be able to:

- Describe evidence from cell staining techniques that DNA was the genetic material.
- Explain how experiments conducted by Griffith and Avery provided evidence that DNA was the genetic material.
- Justify the experimental design used by Hershey and Chase.
- Design an experiment to test the hypothesis that DNA and not protein is the cellular material that carries genetic information.

1. At the time of Griffith's experiments in the 1920s, what circumstantial evidence suggested to scientists that DNA might be the genetic material?
2. How did the experiments of Avery and his colleagues rule out protein as the genetic material?
3. What attributes of bacteriophage T2 were key to the Hershey–Chase experiments demonstrating that DNA, rather than protein, is the genetic material?
4. Outline a series of experiments using radioactive isotopes (such as ^{32}P and ^{35}S) to show that it is DNA and not protein that moves from the donor cell to the recipient cell and is responsible for bacterial transformation.

The transformation and viral infection experiments convinced biologists that the genetic material is DNA. It had been known for several decades that chemically, DNA is a polymer of nucleotides. Next, scientists turned to DNA's precise three-dimensional structure.

▶key concept

13.2   DNA Has a Structure That Suits Its Function

In determining the structure of DNA, scientists hoped to answer two questions: (1) how is DNA replicated between cell divisions, and (2) how does it direct the synthesis of specific proteins? DNA's structure was deciphered only after many types of experimental evidence were considered together in a theoretical framework.

focus your learning

- Watson and Crick used X-ray diffraction data in conjunction with chemical evidence to build the double helix model for DNA.
- The genetic material performs four important functions and is well suited to its biological functions.

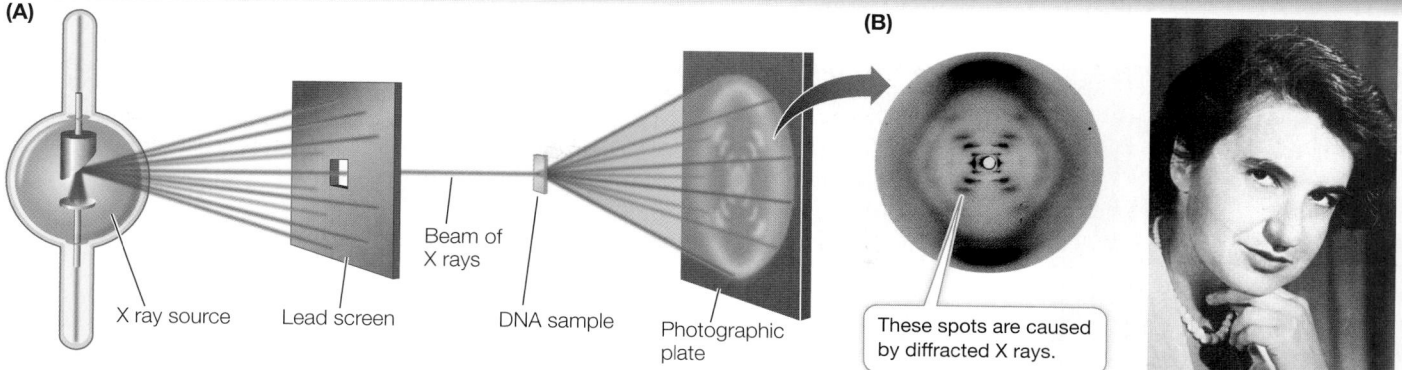

(A)

X ray source   Lead screen   Beam of X rays   DNA sample   Photographic plate

(B)

These spots are caused by diffracted X rays.

Figure 13.5 X-Ray Diffraction Helped Reveal the Structure of DNA (A) The positions of atoms in a crystallized chemical substance can be inferred by the pattern of diffraction of X rays passed through it. The pattern of DNA is both highly regular and repetitive. (B) Rosalind Franklin's diffraction and her "photograph 51" (shown) helped scientists visualize the helical structure of the DNA molecule.

How did Watson and Crick deduce the structure of DNA?

Once pure DNA fibers could be isolated, biophysicists and biochemists examined the DNA for hints about its structure. The evidence eventually used to solve DNA's structure included crucial data obtained using X-ray crystallography and a thorough characterization of the chemical composition of DNA.

PHYSICAL EVIDENCE FROM X-RAY DIFFRACTION   Some chemical substances, when they are isolated and purified, can be made to form crystals. The positions of atoms in a crystallized substance can be inferred from the diffraction pattern of X rays passing through the substance. In the early 1950s the New Zealand–born biophysicist Maurice Wilkins discovered a way to make highly ordered fibers of DNA that were suitable for X-ray diffraction studies. His samples were analyzed by Rosalind Franklin of Kings College, London (**Figure 13.5**). Franklin's data suggested that DNA was a double (two-stranded) helix with ten nucleotides in each full turn, and that each full turn was 3.4 nanometers (nm) in length. The molecule's diameter of 2 nm suggested that the sugar–phosphate backbone of each DNA strand must be on the outside of the helix.

CHEMICAL EVIDENCE FROM BASE COMPOSITION   Biochemists knew that DNA was a polymer of nucleotides. Each nucleotide consists of a molecule of the sugar deoxyribose, a phosphate group, and a nitrogen-containing base (see Figure 4.1). The only differences among the four nucleotides of DNA are their nitrogenous bases: the purines **adenine (A)** and **guanine (G)**, and the pyrimidines **cytosine (C)** and **thymine (T)**.

In the early 1950s, biochemist Erwin Chargaff and his colleagues at Columbia University reported that DNA from many different species—and from different sources within a single organism—exhibits certain regularities. This led to the following rule: In any DNA sample, the amount of adenine equals the amount of thymine (A = T), and the amount of guanine equals the amount of cytosine (G = C). As a result, the total abundance of purines (A + G) equals the total abundance of pyrimidines (T + C):

In DNA, A + G...   A = T   G = C   ...is always equal to T + C.

Purines   =   Pyrimidines

Chargaff's rule provided an important clue about the way the bases are arranged in a DNA double helix. While Chargaff and colleagues found that this rule held for every organism they examined, they noted that the relative abundances of A + T versus G + C vary among organisms. In human DNA, A and T each account for 30 percent of the nitrogenous bases present, while G and C each account for 20 percent. Put another way, in human DNA there is a ratio of 60 percent (A + T) to 40 percent (C + G).

WATSON AND CRICK'S MODEL   If you have taken chemistry courses, you may be familiar with model building, where balls (atoms) and sticks (bonds) are used to put together molecules based on known physical and chemical properties and bond angles. A physicist, Francis Crick, and a geneticist, James D. Watson (**Figure 13.6A**), who were then at the Cavendish Laboratory of Cambridge University, used model building to solve the structure of DNA. They used the physical and chemical evidence we just described:

- To be consistent with Franklin's X-ray diffraction images, Watson and Crick's model had the nucleotide bases on the interior of the two strands, with a sugar–phosphate "backbone" on the outside. In addition, the two DNA strands ran in opposite directions, that is, they were **antiparallel**. The two strands would not fit together otherwise:

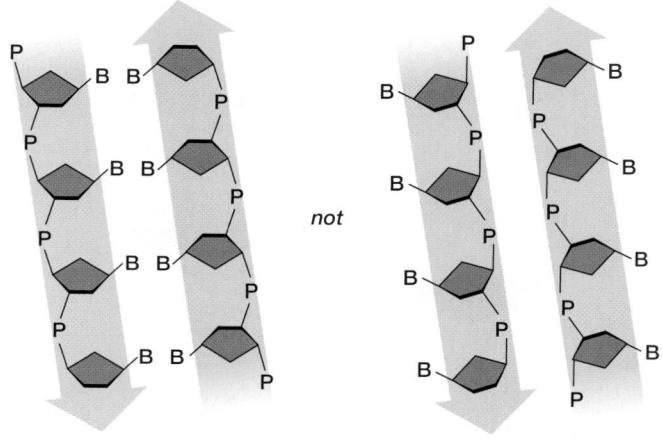

not

(A)

Figure 13.6 DNA Is a Double Helix **(A)** James Watson (left) and Francis Crick (right) proposed that the DNA molecule has a double-helical structure. **(B)** Biochemists can now pinpoint the position of every atom in a DNA molecule. To see that the essential features of the original Watson–Crick model have been verified, follow with your eyes the double-helical chains of sugar–phosphate groups and note the horizontal rungs of the bases.

Q: In DNA, where do the following chemical forces occur: hydrogen bonds, covalent bonds, and van der Waals forces?

(B)

Phosphorus

Carbon in sugar–phosphate backbone

Hydrogen

Oxygen

Bases

Major groove

Minor groove

The blue bands represent the two sugar–phosphate backbones, which run in opposite directions:

5′ 3′
3′ 5′

5′

3′

G C
A
G
A
G C
A
G
C G
G C
A
C
T A
A T

5′   3′

3.4 nm

 Media Clip 13.1 **Discovery of the Double Helix**
www.Life11e.com/mc13.1

- To satisfy Chargaff's rule (A = T and G = C), Watson and Crick's model always paired a purine on one strand with a pyrimidine on the opposite strand. These **base pairs** (A-T and G-C) have the same width down the double helix, a uniformity that was also confirmed by X-ray diffraction:

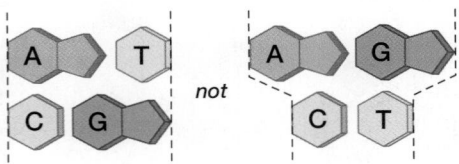

Crick and Watson built their tin model of DNA in late February 1953. This structure explained the known chemical properties of DNA, and it opened the door to understanding its biological functions.

Four key features define DNA structure

Four features summarize the molecular architecture of the DNA molecule (**Figure 13.6B**):

1. DNA is a *double-stranded helix*, with a sugar–phosphate backbone on the outside and base pairs lined up on the inside.

2. DNA is usually a *right-handed helix*. If you curl the fingers of your right hand and point your thumb upward, the curve of the helix follows the direction of your fingers, and it winds upward in the direction of your thumb. (See Figure 3.8, of right- versus left-handed helices.)

3. DNA is *antiparallel* (the two strands run in opposite directions).

4. DNA has *major and minor grooves* in which the outer edges of the nitrogenous bases are exposed.

THE HELIX The sugar–phosphate backbones of the polynucleotide chains form a coil around the outside of the helix, and the nitrogenous bases point toward the center. The chains are held together by two chemical forces:

1. *Hydrogen bonding* between specifically paired bases. Consistent with Chargaff's rule, adenine (A) pairs with thymine (T) by forming two hydrogen bonds, and guanine (G) pairs with cytosine (C) by forming three hydrogen bonds:

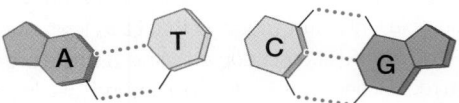

Every base pair consists of one purine (A or G) and one pyrimidine (T or C). This pattern is known as **complementary base pairing**.

2. *van der Waals* forces between adjacent bases on the same strand. When the base rings come near one another, they tend to stack like poker chips because of these weak attractions.

ANTIPARALLEL STRANDS The backbone of each DNA strand contains repeating units of the five-carbon monosaccharide deoxyribose:

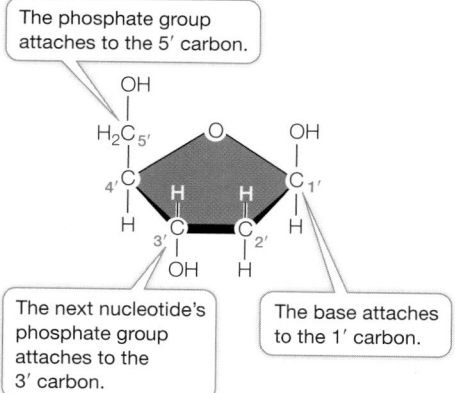

The phosphate group attaches to the 5′ carbon.

The next nucleotide's phosphate group attaches to the 3′ carbon.

The base attaches to the 1′ carbon.

The number followed by a prime (′) designates the position of a carbon atom in this sugar molecule. In the sugar–phosphate backbone of DNA, the phosphate groups are connected to the 3′ carbon of one deoxyribose molecule and the 5′ carbon of the next, linking successive sugars together.

Thus the two ends of a polynucleotide chain differ. At one end of a chain is a free (not connected to another nucleotide) 5′ phosphate group ($-OPO_3^-$); this is called the **5′ end**. At the other end is a free 3′ hydroxyl group ($-OH$); this is called the **3′ end**. In a DNA double helix, the 5′ end of one strand is paired with the 3′ end of the other strand, and vice versa. In other words, if you drew an arrow for each strand running from 5′ to 3′, the arrows would point in opposite directions (see also Figure 4.4A).

BASE EXPOSURE IN THE GROOVES   Look back at Figure 13.6B and note the major and minor grooves in the helix. These grooves exist because the backbones of the two strands are closer together on one side of the double helix (forming the minor groove) than on the other side (forming the major groove). **Figure 13.7** shows a "top-down" view of the four possible configurations of the flat, hydrogen-bonded base pairs in the major and minor grooves. The exposed outer edges of the base pairs are accessible for additional hydrogen bonding. Note that the arrangements of unpaired atoms and groups differ in A-T and G-C pairs. Thus, the *surfaces of the A-T and C-G base pairs are chemically distinct*, allowing other molecules such as proteins to recognize specific base pair sequences and bind to them, which is crucial for DNA function. *The binding of proteins to specific base pair sequences is the key to protein–DNA interactions*, which are necessary for the replication and expression of the genetic information in DNA.

The double-helical structure of DNA is essential to its function

The genetic material performs four important functions, and the DNA structure proposed by Watson and Crick was elegantly suited to three of them.

1. *The genetic material stores an organism's genetic information.* With its millions of nucleotides, the base sequence of a DNA molecule can encode and store an enormous amount of information. Variations in DNA sequences can account for species and individual differences. DNA fits this role nicely.

2. *The genetic material is susceptible to mutations* (permanent changes) *in the information it encodes.* For DNA, mutations might be simple changes in the linear sequence of base pairs.

3. *The genetic material is precisely replicated in the cell division cycle.* Replication could be accomplished by complementary

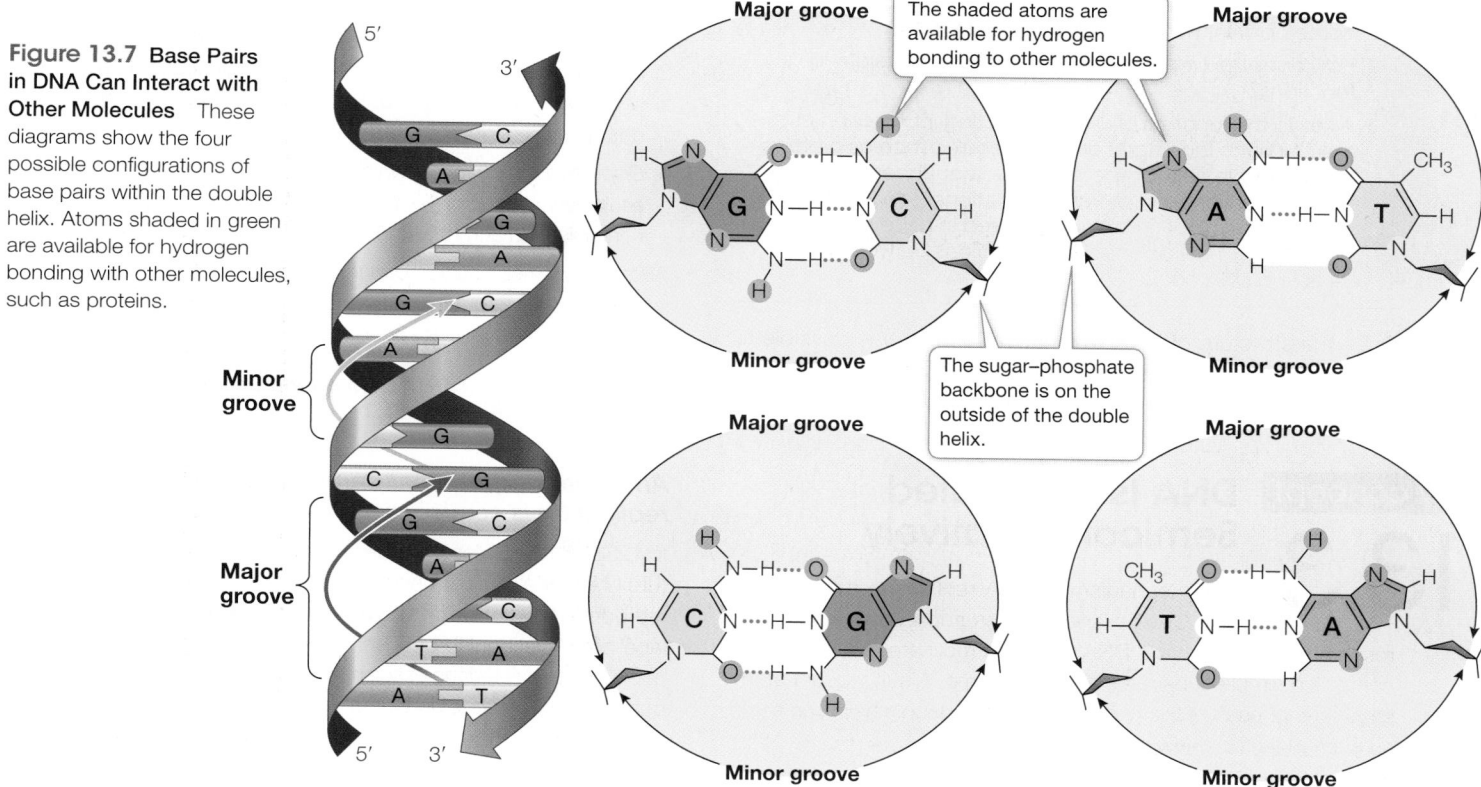

Figure 13.7 Base Pairs in DNA Can Interact with Other Molecules   These diagrams show the four possible configurations of base pairs within the double helix. Atoms shaded in green are available for hydrogen bonding with other molecules, such as proteins.

The shaded atoms are available for hydrogen bonding to other molecules.

The sugar–phosphate backbone is on the outside of the double helix.

base pairing, A with T and G with C. In the original publication of their findings in 1953, Watson and Crick coyly pointed out, "It has not escaped our notice that the specific pairing we have postulated immediately suggests a possible copying mechanism for the genetic material."

4. *The genetic material (the coded information in DNA) is expressed as the phenotype.* This function is not obvious in the structure of DNA. However, as we will see in the next chapter, the nucleotide sequence of DNA is copied into RNA, which uses the coded information to specify a linear sequence of amino acids—a protein. The folded forms of proteins determine many of the phenotypes of an organism.

▶ 13.2 recap

DNA is a double helix made up of two antiparallel polynucleotide chains. The two chains are joined by hydrogen bonds between the nucleotide bases, which pair specifically: A with T, and G with C. Chemical groups on the bases that are exposed in the grooves of the helix are available for hydrogen bonding with other molecules, such as proteins. Various molecules that interact with DNA can recognize specific sequences of nucleotide bases because the surfaces of the base pairs are chemically distinct.

learning outcomes

You should be able to:

- Explain how Watson and Crick developed their model of DNA.
- Relate structural aspects of the DNA molecule to its biological functions.

1. Describe the evidence that Watson and Crick used to come up with the double helix model for DNA.

2. How does the two-stranded structure of DNA relate to its functions?

3. A key to the function of DNA in directing information in the cell is the binding of other molecules such as proteins to the double helix at specific base sequences. Recalling the forces involved in protein interactions (see Figure 3.12) and the chemical groupings exposed in base pairs (see Figure 13.7), how do specific interactions between proteins and DNA occur?

Once the structure of DNA was understood, it was possible to investigate how DNA replicates itself. Next we will examine the experiments that taught us how this elegant process works.

▶ key concept

13.3 DNA Is Replicated Semiconservatively

The mechanism of DNA replication that Watson and Crick proposed was soon confirmed. First, researchers showed that DNA could be replicated in a test tube containing simple substrates and an enzyme. A subsequent study showed that each of the two strands of the double helix can serve as a template for a new strand of DNA.

focus your learning

- The Meselson-Stahl experiment provided evidence to support the semiconservative model of DNA replication.
- Semiconservative DNA replication requires deoxyribonucleoside triphosphates, DNA polymerase, and a DNA template.
- Telomerases provide an important function to maintain the integrity of DNA in an organism.
- DNA replication proceeds bidirectionally from an origin of replication (*ori*); *E. coli* has a single *ori*, and eukaryotic chromosomes have multiple ones.
- The first replication event at the *ori* uses DNA helicases and single-stranded binding proteins.

Test tube synthesis of DNA needed the following substances:

- The deoxyribonucleoside triphosphates dATP, dCTP, dGTP, and dTTP. These are the monomers from which the DNA polymers are formed.

- DNA molecules of a particular sequence that serve as **templates** to direct the sequence of nucleotides in the new molecules.

- A **DNA polymerase** enzyme to catalyze the polymerization reaction.

- Salts and a pH buffer, to create an appropriate chemical environment for the DNA polymerase.

The fact that DNA could be synthesized in a test tube confirmed that a DNA molecule contains the information needed for its own replication. The next challenge was to determine which of three possible replication patterns occurs during DNA replication:

1. *Semiconservative replication*, in which each parent strand serves as a template for a new strand, and the two new DNA molecules each have one old and one new strand (**Figure 13.8A**)

2. *Conservative replication*, in which the original double helix serves as a template for, but does not contribute to, a new double helix (**Figure 13.8B**)

3. *Dispersive replication*, in which fragments of the original DNA molecule serve as templates for assembling two new molecules, each containing old and new parts, perhaps at random (**Figure 13.8C**)

Watson and Crick's original paper suggested that DNA replication was semiconservative, but the test tube demonstration described above did not provide a basis for choosing among these three models.

 Animation 13.2 **DNA Replication and Polymerization**
www.Life11e.com/a13.2

An elegant experiment demonstrated that DNA replication is semiconservative

In 1958 Matthew Meselson and Franklin Stahl at the California Institute of Technology convinced the scientific community that DNA is reproduced by **semiconservative replication**. They used density labeling to distinguish between parent strands of DNA and newly copied ones. Historians of science have called this one of the "most beautiful experiments" in biology. For that reason alone, as well as its results, it's worth taking a close look at it.

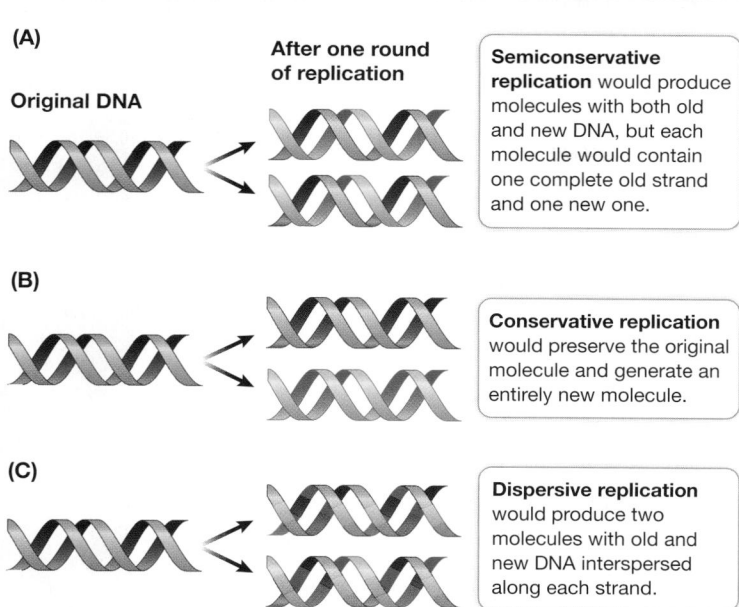

(A)

Original DNA

After one round of replication

Semiconservative replication would produce molecules with both old and new DNA, but each molecule would contain one complete old strand and one new one.

(B)

Conservative replication would preserve the original molecule and generate an entirely new molecule.

(C)

Dispersive replication would produce two molecules with old and new DNA interspersed along each strand.

Figure 13.8 Three Models for DNA Replication   In each model, the original DNA is shown in blue and the newly synthesized DNA is in red.

The key to the experiment was the use of a "heavy" isotope of nitrogen. Heavy nitrogen (^{15}N) is a rare, nonradioactive isotope that makes molecules containing it denser than chemically identical molecules containing the common isotope ^{14}N. Two cultures of the bacterium *E. coli* were grown for many generations, one in a medium containing ^{15}N and the other in a medium containing ^{14}N. When DNA extracts from the two cultures were combined and centrifuged in a solution of cesium chloride, which forms a density gradient under centrifugation, two separate bands of DNA formed in the centrifugation tube. The DNA from the ^{15}N culture was heavier than the DNA from the ^{14}N culture, so it formed a band at a different position in the density gradient. Figure A in the work with the data portion of **Investigating Life: The Meselson–Stahl Experiment** shows a photo of the two bands.

Next, Meselson and Stahl grew another *E. coli* culture in ^{15}N medium, then transferred the bacteria to normal ^{14}N medium and allowed them to continue growing. The cells replicated their DNA and divided every 20 minutes. Meselson and Stahl collected some of the bacteria at time intervals and extracted DNA from the samples. You can follow their results for the first two generations in Investigating Life: The Meselson–Stahl Experiment. The results can be explained only by the semiconservative model of DNA replication. The crucial observations demonstrating this model were that all the DNA at the end of the first generation was of intermediate density, while at the end of the second generation there were two discrete bands: one of intermediate and one of light DNA. If the conservative model had been true, there would have been no intermediate density DNA. If the dispersive model were correct, then the DNA would all have been intermediate for the first few generations, with the single intermediate band becoming progressively lighter.

Animation 13.3 **The Meselson-Stahl Experiment**
www.Life11e.com/a13.3

Not long after this experiment, other scientists investigated DNA replication in the presence of cisplatin (described in the opening story of this chapter). Repeating the Meselson–Stahl experiment with cisplatin added to the growth media, they found no change in the density of DNA after several generations—that is, only one DNA band formed in the tubes subjected to centrifugation. From this the researchers deduced that in the presence of cisplatin the two strands of DNA must not be separating during replication. We will discuss more specifically how cisplatin prevents DNA strand separation at the end of this chapter. First, let's consider the chemistry involved in DNA replication.

There are two steps in DNA replication

Semiconservative DNA replication in the cell involves several different enzymes and other proteins. It takes place in two general steps:

1. The DNA double helix is unwound to separate the two template strands and make them available for new base pairing.

2. As new nucleotides form complementary base pairs with template DNA, they are covalently linked together by phosphodiester bonds, forming a polymer whose base sequence is complementary to the bases in the template strand.

The nucleotides that make up DNA are deoxyribonucleoside monophosphates because they each contain deoxyribose and one phosphate group (see Figure 4.1). The four free monomers that are brought together to form DNA are the deoxyribonucleoside triphosphates dATP, dTTP, dCTP, and dGTP, collectively referred to as dNTPs. They are called triphosphates because each has three phosphate groups. The three phosphate groups are attached to the 5′ carbon on the deoxyribose sugar (as shown in the incoming nucleotide in **Figure 13.9**).

Note that during DNA replication *nucleotides are added to the growing new strand at the 3′ end*—the end at which the DNA strand has a free hydroxyl (—OH) group on the 3′ carbon of its terminal deoxyribose. In the formation of the phosphodiester linkage (a condensation reaction), two of the phosphate groups on an incoming dNTP are removed [as pyrophosphate (PP$_i$)], and the remaining phosphate is bonded to the 3′ carbon on the terminal deoxyribose (see Figure 4.2). Just as energy is released when ATP is hydrolyzed to AMP (with subsequent hydrolysis of PP$_i$ to two phosphates), energy is released by the hydrolysis of the dNTP, and this energy is used to drive the condensation reaction.

DNA polymerases add nucleotides to the growing chain

DNA replication begins with the binding of a large protein complex (the pre-replication complex) to a specific site on the DNA molecule. This complex contains several different proteins, including the enzyme DNA polymerase, which catalyzes the addition of nucleotides as the new DNA chain grows. All chromosomes have at least one region called the **origin of replication** (**ori**), to which

experiment

Original Paper: Meselson, M. and F. Stahl. 1958. The replication of DNA in *Escherichia coli*. *Proceedings of the National Academy of Sciences USA* 44: 671–682.

A centrifuge was used to separate DNA molecules labeled with isotopes of different densities. This experiment revealed a pattern that supports the semiconservative model of DNA replication.

HYPOTHESIS▶   DNA replicates semiconservatively.

METHOD

1 Grow bacteria in ^{15}N (heavy) medium.

2 Transfer some bacteria to ^{14}N (light) medium; bacterial growth continues.

3 Samples are taken after 0 minutes, 20 minutes (after one round of replication), and 40 minutes (two rounds of replication).

Sample at 0 minutes

Sample after 20 minutes

Sample after 40 minutes

4 Extract DNA from bacteria. Put DNA into a solution.

RESULTS

^{14}N/^{14}N (light) DNA

^{14}N/^{15}N (intermediate) DNA

^{15}N/^{15}N (heavy) DNA

Parent (all heavy)

First generation (all intermediate)

Second generation (half intermediate, half light)

INTERPRETATION

Before the bacteria reproduce for the first time in the light medium (at 0 minutes), all DNA parental) is heavy.

Parent strand ^{15}N   New strand ^{14}N

After two generations, half the DNA is intermediate and half is light; there was no heavy DNA.

CONCLUSION▶   This pattern could only have been observed if each DNA molecule contains a template strand from the parental DNA; thus DNA replication is semiconservative.

---

the pre-replication complex binds. Binding occurs when proteins in the complex recognize specific DNA sequences within the *ori*.

ORIGINS OF REPLICATION   The single circular chromosome of the bacterium *E. coli* has 4×10^6 base pairs (bp) of DNA. The 245 bp *ori* sequence is at a particular location on the chromosome. Once the pre-replication complex binds to it, the DNA is unwound and replication proceeds in both directions around the circle, forming two **replication forks** (**Figure 13.10A**). The replication rate in *E. coli* is approximately 1,000 bp per second, so it takes about 40 minutes to fully replicate the chromosome (with two replication forks). Rapidly

dividing *E. coli* cells divide every 20 minutes. In these cells, new rounds of replication begin at the *ori* of each new chromosome before the first chromosome has fully replicated. In this way the cells can divide more frequently than the time needed to finish replicating the original chromosome.

Eukaryotic chromosomes are typically much longer than those of prokaryotes—up to 1 billion bp—and are linear, not circular. If replication occurred from a single *ori* with two forks growing away from each other, it would take weeks to fully replicate a chromosome. So eukaryotic chromosomes have multiple origins of replication, scattered at intervals of 10,000 to 40,000 bp (**Figure 13.10B**).

work with the data

The Meselson–Stahl experiment has been called one of the "most beautiful experiments" in biology because of its essential simplicity. Meselson and Stahl used density gradients to examine how DNA molecules replicate, as illustrated in the experimental results shown in **Figure A**. The peaks on the drawing are proportional to the amount of DNA.

Figure B shows results of the experiment after each of four generations. Each sample contained the same number of bacteria, so the total amount of DNA in each panel was the same.

Figure A

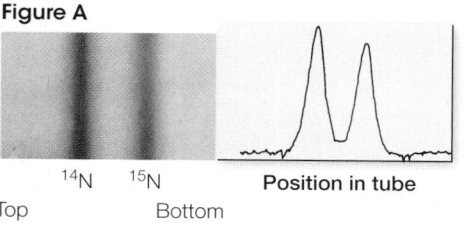

^{14}N ^{15}N

Position in tube

Top of tube Bottom of tube

Figure B

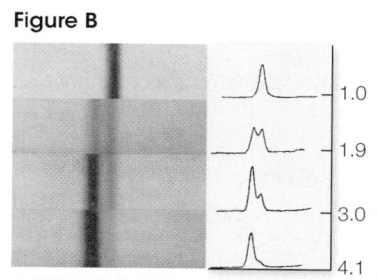

1.0
1.9
3.0
4.1

QUESTIONS▶

1. Use the heights of the peaks to estimate the percent of total DNA that was heavy, intermediate, and light at each generational stage. Create a table summarizing these calculations and discuss whether they support the authors' conclusions.

2. What would the data look like if the bacteria had been allowed to divide for three more generations?

3. If Meselson and Stahl had done their experiment starting with light DNA and then added ^{15}N for succeeding generations, what would the bands look like?

4. What would the data look like if conservative replication were the correct model? What would the data look like if dispersive replication were correct?

A similar **work with the data** exercise may be assigned in **LaunchPad**.

DNA REPLICATION BEGINS WITH A PRIMER A DNA polymerase elongates a polynucleotide strand by covalently linking new nucleotides to a preexisting strand. However, it cannot start this process without a short "starter" strand, called a **primer**. In most organisms this primer is a short single strand of RNA (**Figure 13.11**), but in some organisms it is DNA. The primer is complementary to the DNA template and is synthesized one nucleotide at a time by an enzyme called a **primase**. The DNA polymerase then adds nucleotides to the 3′ end of the primer and continues until the replication of that section of DNA has been completed. Then the RNA primer is degraded, DNA is added in its place, and the resulting DNA fragments are connected by the action of other enzymes. When DNA replication is complete, each new strand consists only of DNA.

DNA POLYMERASES ARE LARGE DNA polymerases are much larger than their substrates (the dNTPs) and the template DNA, which is very thin. Molecular models of the enzyme–substrate–template complex from bacteria show that the enzyme

is shaped like an open right hand with a palm, a thumb, and fingers (**Figure 13.12**). Within the "palm" is the active site of the enzyme, which brings together each dNTP substrate and the template. The "finger" regions have precise shapes that can recognize the different shapes of the four nucleotide bases. They bind to the bases

Figure 13.9 Each New DNA Strand Grows from Its 5′ End to Its 3′ End The DNA strand at the right (blue) is the template for the synthesis of the complementary strand that is growing at the left (pink). Here dCTP (circled) is being added.

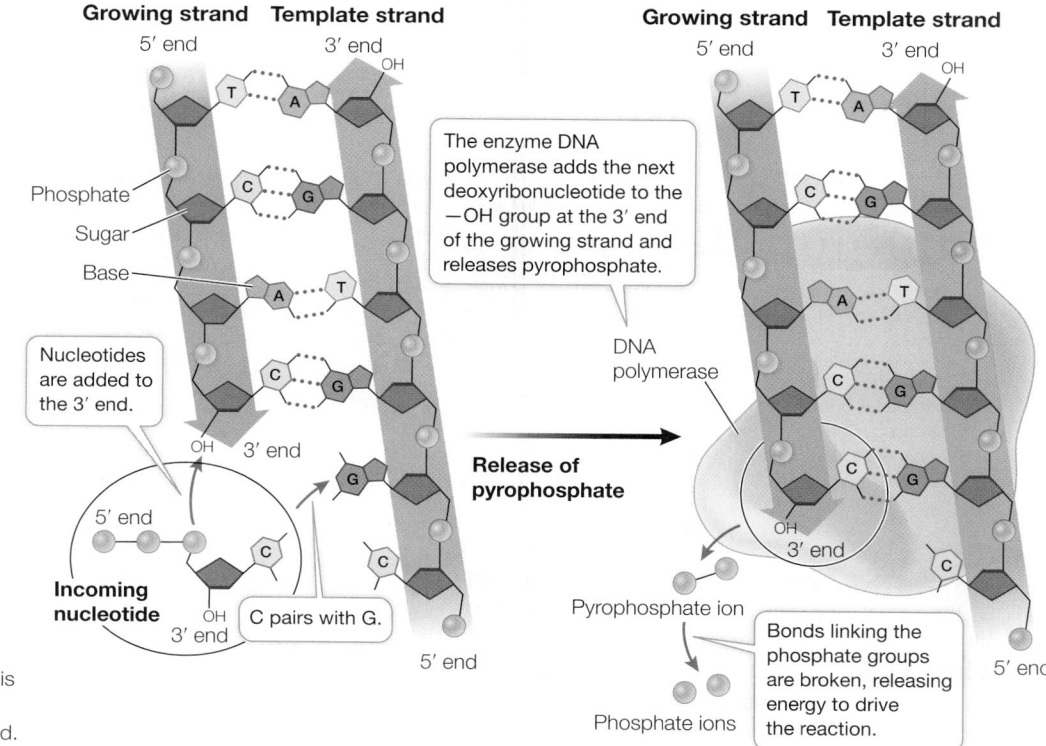

(A) A single *ori* in a prokaryotic chromosome

1 The *ori* sequence binds the pre-replication complex.

2 Two replication forks move away from one another.

ori

Pre-replication complex

ori

ori

ori

(B) Multiple *ori* in a eukaryotic chromosome

ori *ori* *ori*

There are multiple origins of replication.

Initiation of replication

Replication forks move away from each other.

Figure 13.10 The Origin of DNA Replication **(A)** Prokaryotic chromosomes usually have a single origin where DNA replication starts and proceeds in both directions. **(B)** The usually much larger eukaryotic chromosome typically has multiple replication origins.

by hydrogen bonding and rotate inward. Most cells contain more than one kind of DNA polymerase, but only one of them is responsible for chromosomal DNA replication. The others are involved in primer removal and DNA repair. Fifteen DNA polymerases have been identified in humans, whereas the bacterium *E. coli* has five DNA polymerases.

Many other proteins assist with DNA polymerization

Various other proteins play roles in other replication tasks; some of these are shown in **Focus: Key Figure 13.13**. The first event at the origin of replication is the localized unwinding and separation (denaturation) of the DNA strands. As we discussed in Key Concept 13.2, the two strands are held together by hydrogen bonds and van der Waals forces. An enzyme called **DNA helicase** uses energy from ATP hydrolysis to unwind and separate the strands, and **single-strand binding proteins** bind to the unwound strands to keep them from reassociating into a double helix. This process makes each of the two template strands available for complementary base pairing.

The two DNA strands grow differently at the replication fork

The DNA at the replication fork—the site where DNA unwinds to expose the bases so that they can act as templates—opens up like a zipper in one direction. Study **Figure 13.14** and try to imagine what is happening over a short period of time. Remember that the two DNA strands are antiparallel; that is, the 3′ end of one strand is paired with the 5′ end of the other.

- One newly replicating strand (the **leading strand**) is oriented so that it can grow continuously at its 3′ end as the fork opens up.

- The other new strand (the **lagging strand**) is oriented so that as the fork opens up, its exposed 3′ end gets farther and farther away from the fork, and an unreplicated gap is formed. This gap would get bigger and bigger if there were not a special mechanism to overcome this problem.

Synthesis of the lagging strand requires the synthesis of relatively small, discontinuous stretches of DNA (100–200 nucleotides in eukaryotes; 1,000–2,000 nucleotides in prokaryotes). These discontinuous stretches are synthesized just as the leading strand is, by the addition of new nucleotides one at a time to the 3′ end of the new strand, but the synthesis of this new strand moves in the

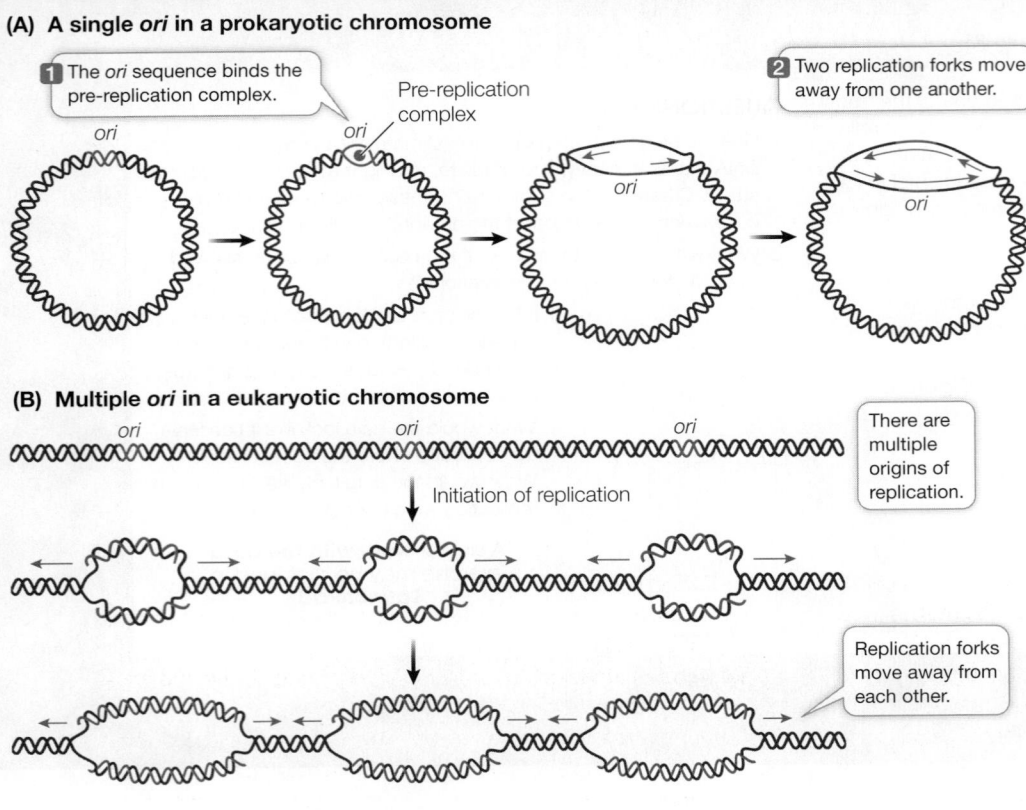

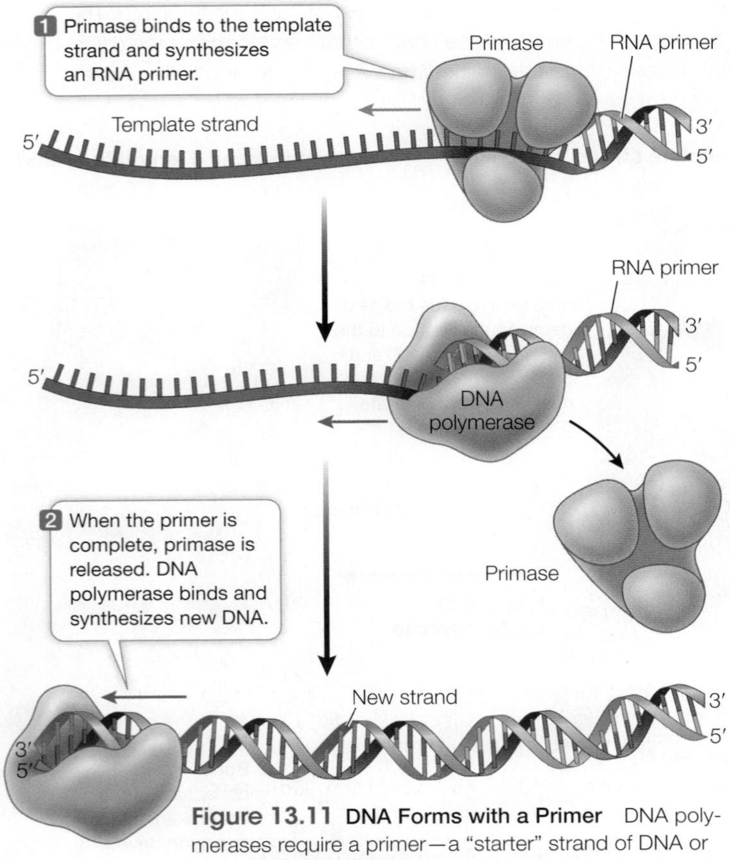

1 Primase binds to the template strand and synthesizes an RNA primer.

Primase

RNA primer

Template strand

5′

3′
5′

RNA primer

3′
5′

DNA polymerase

Primase

2 When the primer is complete, primase is released. DNA polymerase binds and synthesizes new DNA.

New strand

3′
5′

3′
5′

Figure 13.11 DNA Forms with a Primer DNA polymerases require a primer—a "starter" strand of DNA or RNA to which they can add new nucleotides.

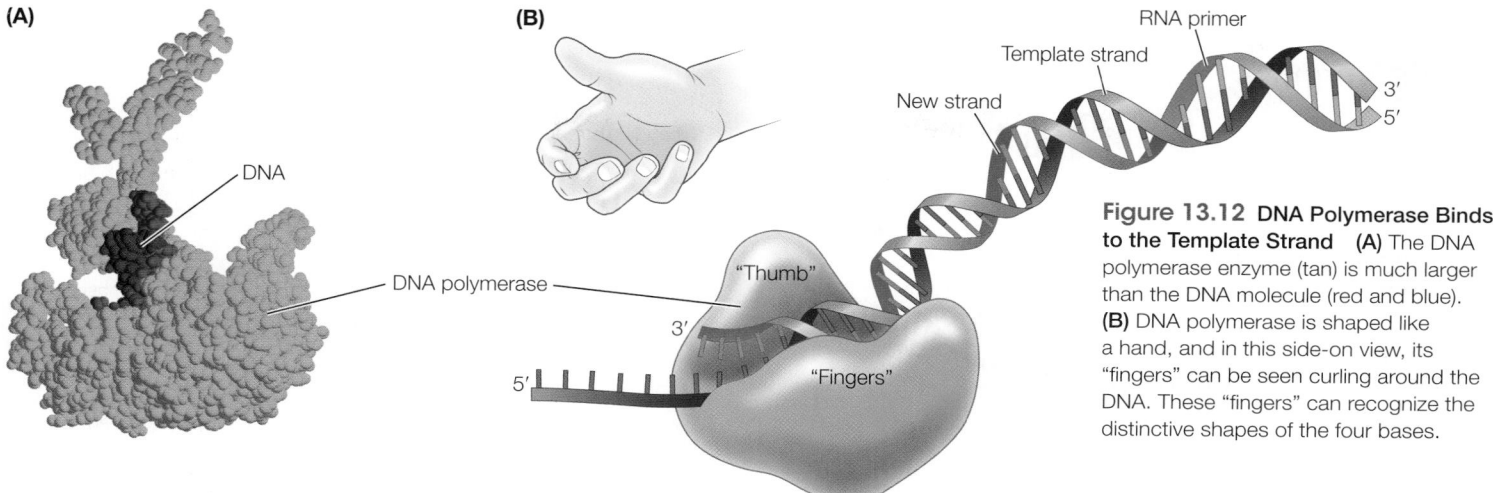

(A)

DNA

DNA polymerase

(B)

RNA primer

Template strand

New strand

3′
5′

"Thumb"

3′

"Fingers"

5′

Figure 13.12 DNA Polymerase Binds to the Template Strand **(A)** The DNA polymerase enzyme (tan) is much larger than the DNA molecule (red and blue). **(B)** DNA polymerase is shaped like a hand, and in this side-on view, its "fingers" can be seen curling around the DNA. These "fingers" can recognize the distinctive shapes of the four bases.

direction opposite to that in which the replication fork is moving. These stretches of new DNA are called **Okazaki fragments** (after their discoverer, the Japanese biochemist Reiji Okazaki). While the leading strand grows continuously "forward," the lagging strand grows in shorter, "backward" stretches with gaps between them.

A single primer is needed for synthesis of the leading strand, but each Okazaki fragment requires its own primer to be synthesized by the primase. In bacteria, DNA polymerase III then synthesizes an Okazaki fragment by adding nucleotides to one primer until it reaches the primer of the previous fragment (**Figure 13.15**). At this point, DNA polymerase I removes the old primer and replaces it with DNA. Left behind is a tiny nick—the final phosphodiester linkage between the adjacent Okazaki fragments is missing. The enzyme **DNA ligase** catalyzes the formation of that bond, linking the fragments and making the lagging strand whole.

DNA replication involves remarkable teamwork among various proteins that act on the DNA strands. Let's review the proteins involved in DNA replication in the order of their activity at the replication fork:

- *DNA helicase* unwinds the double helix and separates the two strands.

focus: key figure

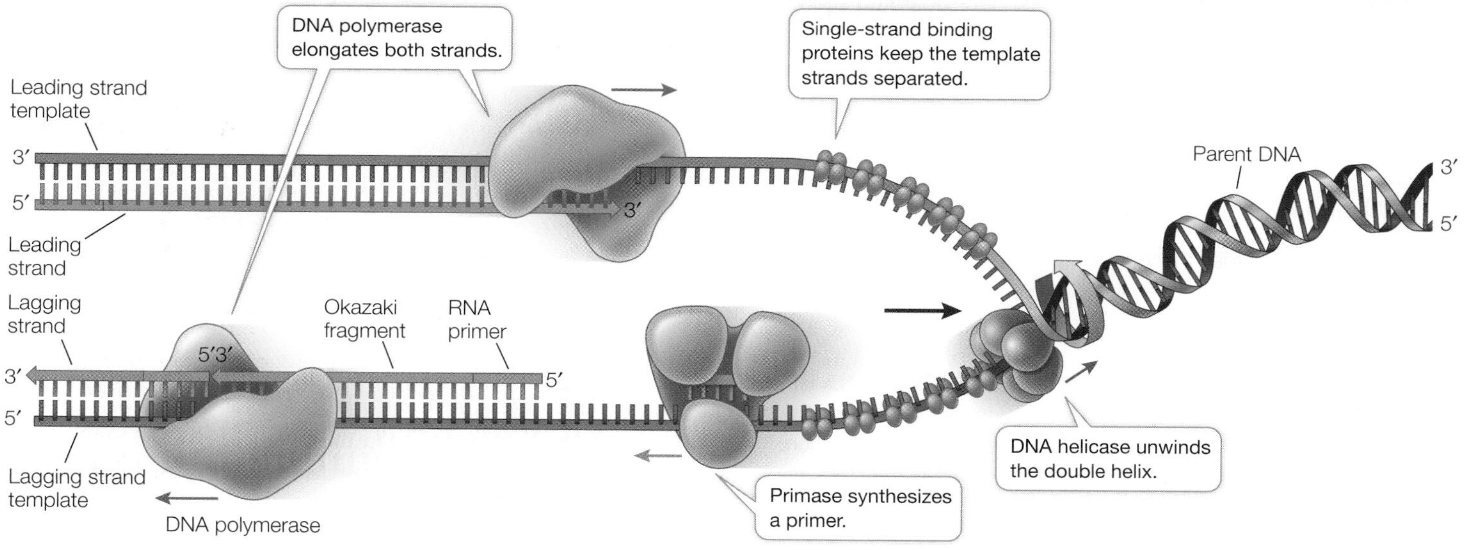

DNA polymerase elongates both strands.

Single-strand binding proteins keep the template strands separated.

Parent DNA

Leading strand template

3′
5′

3′

Leading strand

Lagging strand

Okazaki fragment

RNA primer

5′3′

3′
5′

5′

Lagging strand template

DNA polymerase

3′
5′

DNA helicase unwinds the double helix.

Primase synthesizes a primer.

Figure 13.13 Many Proteins Collaborate in the Replication Complex Several proteins in addition to DNA polymerase are involved in DNA replication. The two molecules of DNA polymerase shown here are actually part of the same complex.

Q: In bidirectional replication from a single origin point, with two replication forks moving in opposite directions, will the same DNA strands be leading and lagging?

Activity 13.1 **The Replication Complex**
www.Life11e.com/ac13.1

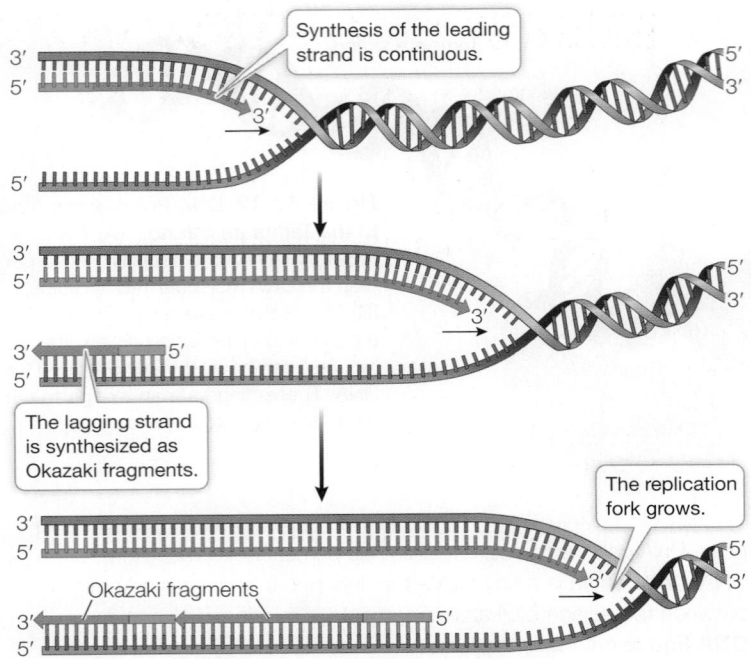

Figure 13.14 The Two New Strands Form in Different Ways As the parent DNA unwinds, both new strands are synthesized in the 5′-to-3′ direction, although their template strands are antiparallel. The leading strand grows continuously forward, but the lagging strand grows in short discontinuous stretches called Okazaki fragments. Eukaryotic Okazaki fragments are hundreds of nucleotides long, with gaps between them.

 Animation 13.4 Leading and Lagging Strand Synthesis
www.Life11e.com/a13.4

- *Single-strand binding proteins* bind to separated strands and prevent them from re-forming the double helix.
- *DNA primase* makes RNA primers.
- *DNA polymerase* links new nucleotides to form the new DNA strands and removes the primers.
- *DNA ligase* connects Okazaki fragments made by DNA polymerase to one another.

Working together, these proteins make new DNA at a rate in excess of 1,000 base pairs per second, committing errors in fewer than 1 base in a million.

A SLIDING CLAMP INCREASES THE RATE OF DNA REPLICATION How do DNA polymerases work so fast? We saw in Key Concept 8.3 that an enzyme catalyzes a chemical reaction:

Substrate binds to enzyme → *one* product is formed →
enzyme is released → cycle repeats

DNA replication would not proceed as rapidly as it does if it went through such a cycle for each nucleotide. Instead, DNA polymerases are **processive**—that is, they catalyze the formation of many phosphodiester linkages each time they bind to a DNA molecule:

Substrates bind to enzyme → *many* products are formed →
enzyme is released → cycle repeats

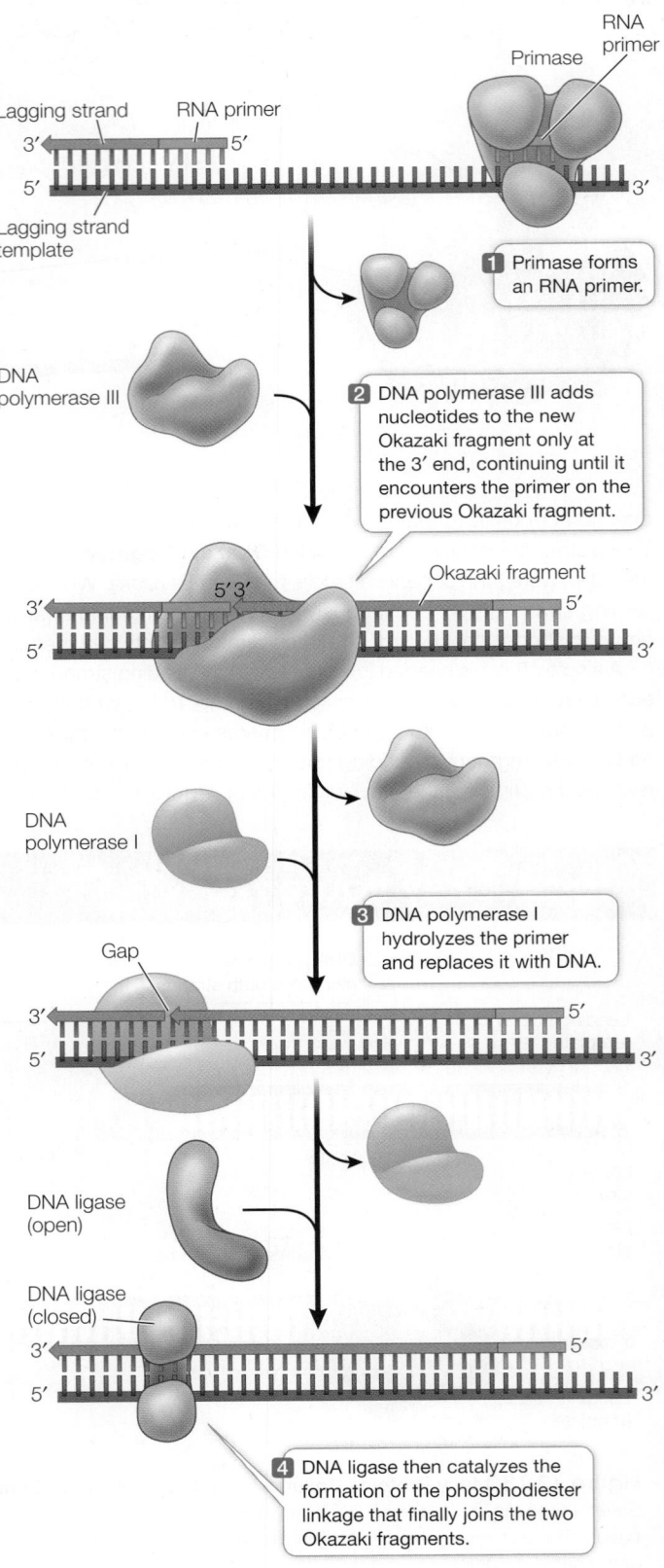

Figure 13.15 The Lagging Strand Story In bacteria, DNA polymerase I and DNA ligase cooperate with DNA polymerase III to complete the complex task of synthesizing the lagging strand.

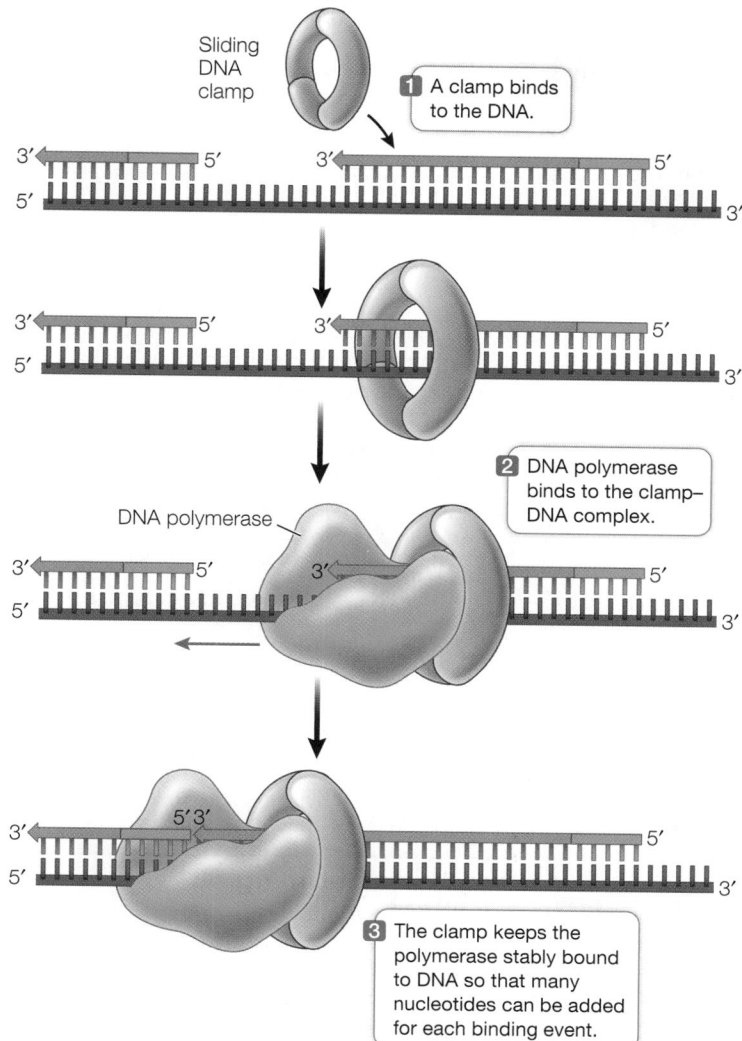

Figure 13.16 **A Sliding DNA Clamp Increases the Efficiency of DNA Polymerization**   The clamp increases the efficiency of polymerization by keeping the enzyme bound to the substrate, so the enzyme does not have to repeatedly bind to template and substrate.

The DNA polymerase–DNA complex is stabilized by a **sliding DNA clamp**, which has multiple identical subunits assembled into a doughnut shape (**Figure 13.16**). The doughnut's "hole" is just large enough to encircle the DNA double helix, along with a thin layer of water molecules for lubrication. The clamp binds to the DNA polymerase–DNA complex, keeping the enzyme and the DNA associated tightly with each other. If the clamp is absent, DNA polymerase dissociates from DNA after forming 20 to 100 phosphodiester linkages. With the clamp, it can polymerize up to 50,000 nucleotides before it detaches.

DNA IS THREADED THROUGH A REPLICATION COMPLEX

So far, you are probably envisioning DNA replication as a locomotive (the replication complex) moving along a railroad track (the DNA). But this is not so. Commonly in eukaryotes, the replication complexes seem to be stationary, attached at specific positions in the nucleus. It is the DNA that moves, essentially sliding into the replication complex as one double-stranded molecule and emerging as two double-stranded molecules.

Telomeres are not fully replicated and are prone to repair

As we will discuss in Key Concept 13.4, DNA may be damaged by radiation or chemicals. When this happens, DNA repair mechanisms are activated, and breaks exposing free 3′ and 5′ groups in DNA are rejoined via a combination of DNA synthesis and DNA ligase activity. So the ends of chromosomes are a potential problem: the DNA repair system might recognize the ends as breaks, and join two chromosomes together. This would create havoc with genomic integrity.

In many eukaryotes, there are repetitive sequences at the ends of chromosomes called **telomeres**. In humans, the telomere sequence is TTAGGG-3′, and it is repeated about 2,500 times at each chromosome end. These repeats bind special proteins that prevent the DNA repair system from recognizing the chromosome ends as breaks. In addition, the repeats may form loops that have a similar protective role.

But there is another problem with chromosome ends. As you have seen, replication of the lagging strand occurs by the addition of Okazaki fragments to RNA primers. When the terminal RNA primer is removed, no DNA can be synthesized to replace it because there is no 3′ end to extend. In most cells, the short piece of single-stranded DNA at each end of the chromosome is removed. Thus the chromosome becomes slightly shorter with each cell division (**Figure 13.17**).

Each human chromosome can lose 50 to 200 base pairs of telomeric DNA after each round of DNA replication and cell division. After many cell divisions, the genes near the ends of the chromosomes can be lost, and the cell dies. This phenomenon explains, in part, why many cell lineages do not last the entire lifetime of the organism: their telomeres are lost. Continuously dividing cells, such as bone marrow stem cells and gamete-producing cells, have a special mechanism for maintaining their telomeric DNA. An enzyme called **telomerase** catalyzes the addition of any lost telomeric sequences in these cells (see Figure 13.18). Telomerase contains an RNA sequence that acts as a template for the telomeric DNA repeat sequence.

Telomerase is expressed in more than 90 percent of human cancers, and may be an important factor in the ability of cancer cells to divide continuously. Since most normal cells do not have this ability, telomerase is an attractive target for drugs designed to attack tumors specifically. There is also interest in telomerase and aging. When cultured human cells are transformed with a telomerase gene that is expressed at high levels, their telomeres do not shorten. Instead of living 20 to 30 cell generations and then dying, the cells become immortal. It remains to be seen how this finding relates to the aging of a whole organism.

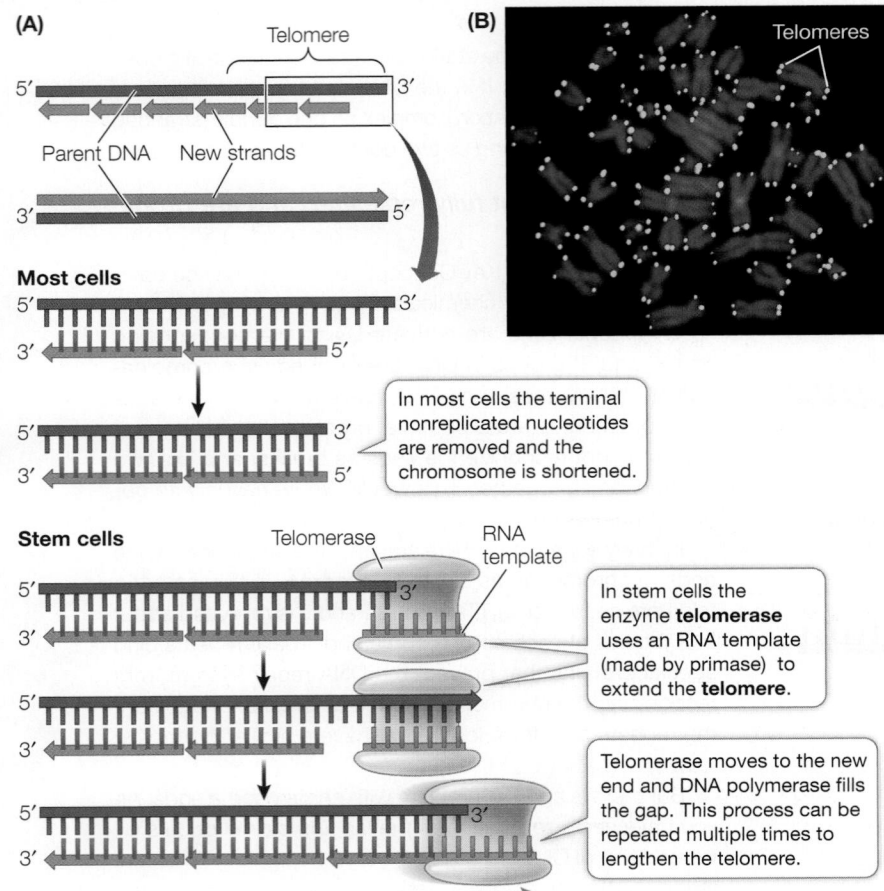

(A)

Telomere

5′ →

Parent DNA New strands

3′

Most cells

In most cells the terminal nonreplicated nucleotides are removed and the chromosome is shortened.

Stem cells Telomerase RNA template

In stem cells the enzyme **telomerase** uses an RNA template (made by primase) to extend the **telomere**.

Telomerase moves to the new end and DNA polymerase fills the gap. This process can be repeated multiple times to lengthen the telomere.

(B) Telomeres

Figure 13.17 Telomeres and Telomerase (A) In most cells, the chromosome shortens with each replication because the nonreplicated DNA at the 3′ end of the template DNA is removed. However, in stem cells telomerase uses an RNA template to extend the telomere and prevent chromosome shortening. **(B)** Bright fluorescent staining marks the telomeric regions on these blue-stained human chromosomes.

Q: Why do the cells that form gametes express telomerase?

The complex process of DNA replication is amazingly accurate, but it is not perfect. What happens when things go wrong?

> **key concept**
> # 13.4 Errors in DNA Can Be Repaired

DNA must be accurately replicated and maintained. This is essential for the proper functioning of every cell, whether a prokaryote or a cell in a complex, multicellular organism. Yet the replication of DNA is not perfectly accurate, and DNA is subject to change by chemicals and other environmental agents. In the face of these threats, how has life gone on for so long?

focus your learning

- Errors in DNA can occur during replication and as the result of chemical changes in bases.

Repair mechanisms preserve DNA

DNA polymerases make mistakes in assembling polynucleotide strands—typically about 1 wrong base is inserted per 100,000 replicated. While this may seem insignificant considering that a human cell has around 3 billion base pairs, the mistakes add up: if polymerase mistakes were not repaired, there would be about 60,000 incorrect bases in new strands every time a human cell divided. It gets worse: errors in base pairs can arise spontaneously as well. Because the bases themselves are chemically instable, outside agents like radiation can damage them, causing *mutations that prevent them from pairing properly.

---

*connect the concepts The molecular biology of DNA mutations is discussed in Key Concept 15.1 and the genetic consequences of mutations in Key Concept 12.2.

Fortunately, our cells correct DNA replication errors and repair damaged nucleotides. Cells have at least three DNA repair mechanisms at their disposal:

> **13.3 recap**

Meselson and Stahl showed that DNA replication is semiconservative: each parent DNA strand serves as a template for a new strand. A complex of proteins, most notably DNA polymerase, is involved in replication. New DNA is polymerized in one direction only, and since the two strands are antiparallel, one strand is made continuously and the other is synthesized in short Okazaki fragments that are eventually joined.

learning outcomes

You should be able to:

- Explain why the Meselson–Stahl experiment provided evidence in support of only one model of DNA replication.
- Identify the function of each component needed for DNA replication.
- Explain the roles of DNA helicases and single-stranded binding proteins.
- Explain the need for telomerases as a consequence of the semiconservative replication of DNA.

1. How did the Meselson–Stahl experiment differentiate between the three models for DNA replication?
2. Name five proteins needed for DNA replication. What are their roles?
3. Why are there special sequences at the ends of chromosomes, and how are they replicated?
4. The drug dideoxycytidine, used to treat certain viral infections, is a nucleotide made with 2′,3′-dideoxyribose. This sugar lacks —OH groups at both the 2′ and the 3′ positions. Explain why this drug stops the growth of a DNA chain when added to DNA.

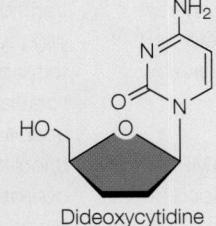

Dideoxycytidine

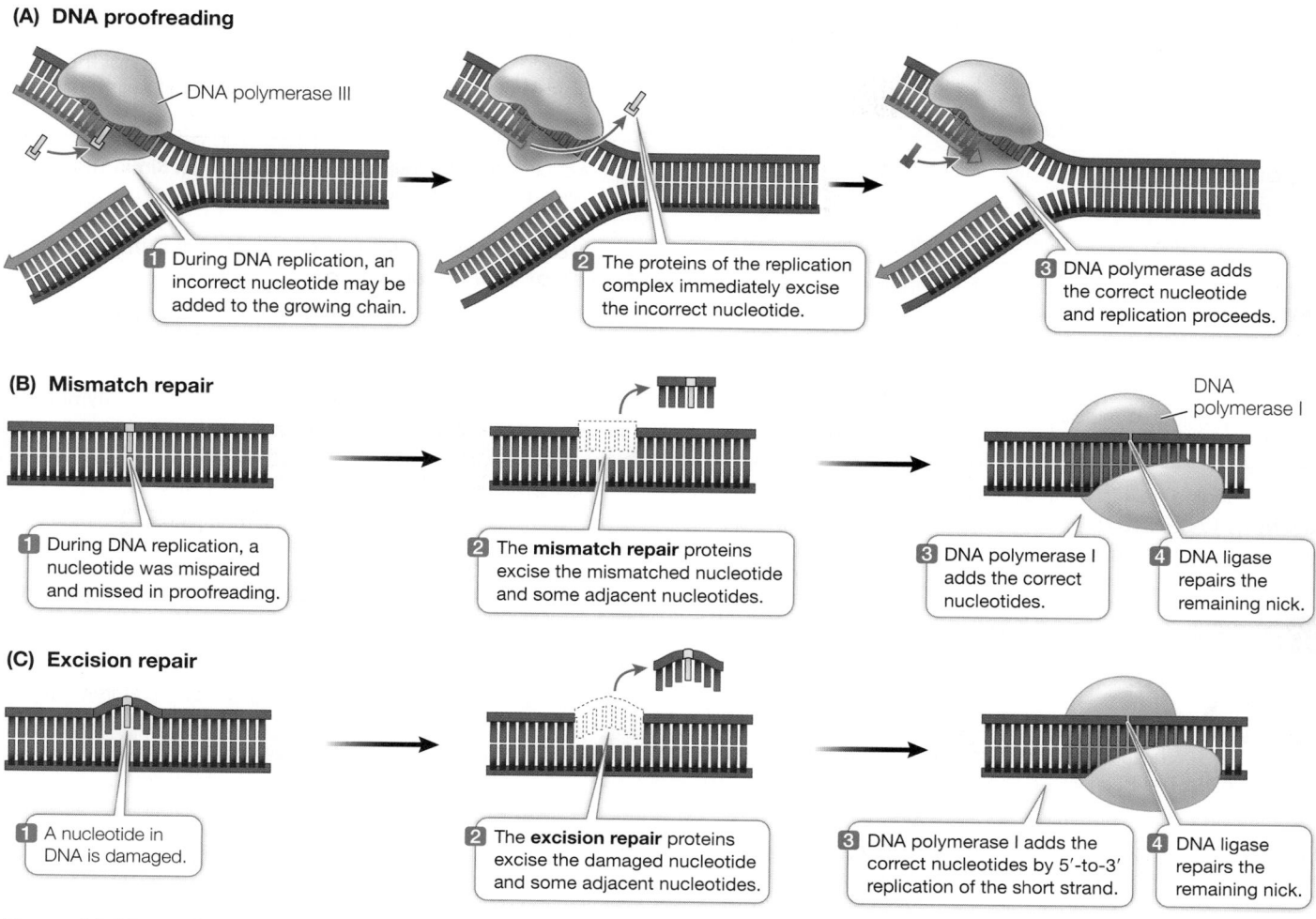

(A) DNA proofreading

DNA polymerase III

1 During DNA replication, an incorrect nucleotide may be added to the growing chain.

2 The proteins of the replication complex immediately excise the incorrect nucleotide.

3 DNA polymerase adds the correct nucleotide and replication proceeds.

(B) Mismatch repair

DNA polymerase I

1 During DNA replication, a nucleotide was mispaired and missed in proofreading.

2 The **mismatch repair** proteins excise the mismatched nucleotide and some adjacent nucleotides.

3 DNA polymerase I adds the correct nucleotides.

4 DNA ligase repairs the remaining nick.

(C) Excision repair

1 A nucleotide in DNA is damaged.

2 The **excision repair** proteins excise the damaged nucleotide and some adjacent nucleotides.

3 DNA polymerase I adds the correct nucleotides by 5'-to-3' replication of the short strand.

4 DNA ligase repairs the remaining nick.

Figure 13.18 DNA Repair Mechanisms The proteins of the replication complex function in DNA repair mechanisms, reducing the rate of errors in the replicated DNA. Another mechanism (excision repair) repairs damage to existing DNA molecules.

1. A **proofreading** mechanism corrects errors in replication as DNA polymerase makes them.

2. A **mismatch repair** mechanism scans DNA immediately after it has been replicated and corrects any base-pairing mismatches.

3. An **excision repair** mechanism removes abnormal bases that have formed because of chemical damage and replaces them with functional bases.

Most DNA polymerases perform a proofreading function each time they introduce a new nucleotide into a growing DNA strand (**Figure 13.18A**). When a DNA polymerase recognizes a mispairing of bases, it removes the improperly introduced nucleotide and tries again. (Other proteins in the replication complex also play roles in proofreading.) The error rate for this process is only about 1 in 10,000 repaired base pairs, and it lowers the overall error rate for replication to about one error in every 10^{10} bases replicated.

After the DNA has been replicated, a second set of proteins surveys the newly replicated molecule and looks for mismatched base pairs that were missed in proofreading (**Figure 13.18B**). For example, this mismatch repair mechanism might detect an A-C base pair instead of an A-T pair. The repair system recognizes which of the two bases in the A-C pair is the wrong one and makes the repair. If the correct pair should be A-T, the repair system replaces C with T, restoring an A-T pair. Alternatively, if it recognizes that the correct pair should be C-G, it replaces A with G. When mismatch repair fails, DNA sequences are altered. One form of colon cancer arises in part from a failure of mismatch repair.

DNA molecules can also be damaged during the life of a cell (e.g., when it is in G1). High-energy radiation, chemicals from the environment, and spontaneous chemical changes can damage DNA. For example, when adjacent thymines on the same DNA strand absorb ultraviolet light (at about 260 nm), they form a covalent bond between the bases, making a thymine dimer. These dimers interfere with base pairing during replication, leading to random bases being inserted. This is the primary cause of skin cancer in humans. Excision repair mechanisms deal with these kinds of damage (**Figure 13.18C**).

13.4 recap

DNA replication is not perfect. In addition, DNA may be altered or damaged by environmental factors. Repair mechanisms detect and repair mismatched or damaged DNA.

learning outcome

You should be able to:
• Predict the consequences of DNA errors if not corrected.

1. From what you know about the role of DNA in the cell, what would be the genetic consequences of an unrepaired DNA change in S phase before meiosis?

2. One form of colon cancer is caused by a mutation in a gene encoding a protein involved in mismatch repair. What would be the consequences of such a mutation in cancer formation? Recall cell cycle control genes involved in cancer formation (see Figure 11.24).

Understanding how DNA is replicated and repaired has allowed scientists to develop techniques for studying genes. We'll look at just one of those techniques next.

key concept

13.5 The Polymerase Chain Reaction Amplifies DNA

The principles underlying DNA replication in cells have been used to develop an important laboratory technique that has been vital in analyzing genes and genomes. This technique allows researchers to make multiple copies of short DNA sequences.

focus your learning

• The polymerase chain reaction was developed for the purpose of rapidly making multiple copies of segments of DNA.

 Activity 13.2 **Polymerase Chain Reaction Simulation**
www.Life11e.com/ac13.2

The polymerase chain reaction can make multiple copies of a DNA sequence

In order to study DNA and perform genetic manipulations in the lab, it is often necessary to make multiple copies of a DNA sequence. DNA amplification is needed because the amount of DNA isolated from a biological sample is often too small to work with. The **polymerase chain reaction (PCR)** technique essentially automates this replication process by copying a short region of DNA many times in a test tube. The PCR reaction mixture contains:

• A sample of double-stranded DNA from a biological sample, to act as the template

• Two short, artificially synthesized primers that are complementary to the ends of the sequence to be amplified

• The four dNTPs (dATP, dTTP, dCTP, and dGTP)

• A DNA polymerase that can tolerate high temperatures without becoming degraded

• Salts and a buffer to maintain a near-neutral pH

PCR amplification is a cyclic process in which a sequence of steps is repeated over and over (**Figure 13.19**):

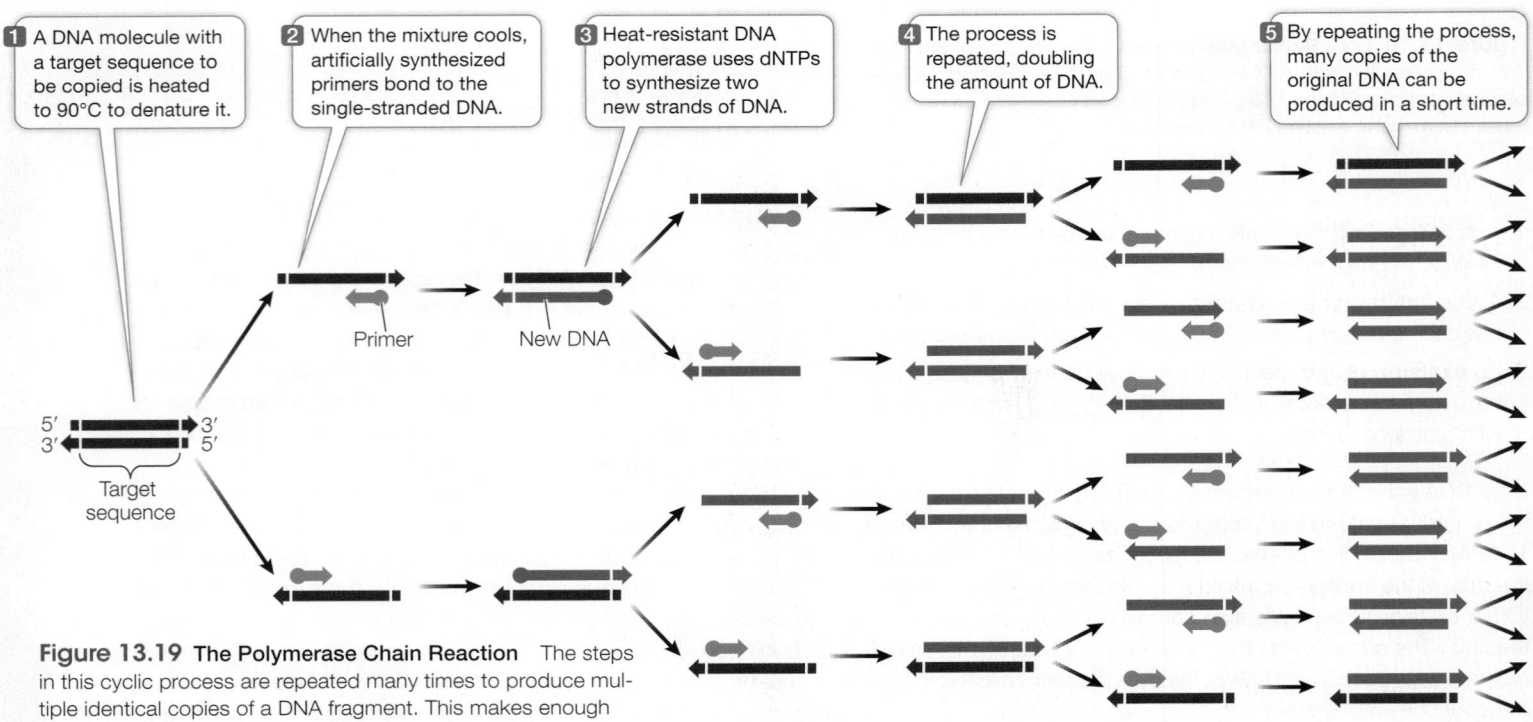

1 A DNA molecule with a target sequence to be copied is heated to 90°C to denature it.

2 When the mixture cools, artificially synthesized primers bond to the single-stranded DNA.

3 Heat-resistant DNA polymerase uses dNTPs to synthesize two new strands of DNA.

4 The process is repeated, doubling the amount of DNA.

5 By repeating the process, many copies of the original DNA can be produced in a short time.

Figure 13.19 The Polymerase Chain Reaction The steps in this cyclic process are repeated many times to produce multiple identical copies of a DNA fragment. This makes enough DNA for chemical analysis and genetic manipulations.

Q: How do PCR primers relate to DNA replication primers?

1. The first step involves heating the reaction mixture to near boiling point, to separate (denature) the two strands of the DNA template.

2. The reaction is then cooled to allow the primers to bind (or anneal) to the template strands.

3. Next, the reaction is warmed to an optimum temperature for the DNA polymerase to catalyze the production of the complementary new strands.

A single cycle takes a few minutes to produce two copies of the target DNA sequence, leaving the new DNA in the double-stranded state. Repeating the cycle many times leads to an exponential increase in the number of copies of the DNA sequence.

The PCR technique requires that the base sequences at the 3' end of each strand of the target DNA sequence be known, so that complementary primers, usually 15 to 30 bases long, can be made in the laboratory. Because of the uniqueness of DNA sequences, a pair of primers this length will usually bind to only a single region of DNA in an organism's genome. This specificity, despite the incredible diversity of DNA sequences, is a key to the power of PCR.

One initial problem with PCR was its temperature requirements. To denature DNA, it must be heated to more than 90°C—a temperature that destroys most DNA polymerases. When the PCR technique was first being developed, new enzyme had to be added after denaturation in each cycle, which made the technique impractical for widespread use. This problem was solved by nature: in the hot springs at Yellowstone National Park, as well as in other high-temperature locations, there lives a bacterium called, appropriately, *Thermus aquaticus* ("hot water"). The means by which this organism survives temperatures of up to 95°C was investigated by Thomas

Brock and his colleagues at the University of Wisconsin, Madison. They discovered that *T. aquaticus* has metabolic machinery that is heat-resistant, including a DNA polymerase that does not denature at these high temperatures.

Scientists pondering the problem of copying DNA by PCR read Brock's basic research articles and got a clever idea: why not use *T. aquaticus* DNA polymerase in the PCR technique? It could withstand the 90°C denaturation temperature and would not have to be added during each cycle. The idea worked, and it earned biochemist Kary Mullis a Nobel prize. PCR has had an enormous impact on genetic research. Some of its most striking applications will be described in Chapters 15–18. These applications range from amplifying DNA in order to identify an individual person or organism, to detection of diseases.

13.5 recap

Knowledge of the mechanisms of DNA replication led to the development of a technique for making multiple copies of DNA sequences.

---

learning outcomes

You should be able to:

- Explain the purpose for each step in the PCR process.
- Explain why PCR is useful in different research scenarios.

---

1. What is the role of primers in PCR?
2. One way to quickly identify an organism in the field is to use PCR: If there is amplification of a particular sequence, then the organism is identified. What is the basis for this test?

---

investigatinglife

What do we need to know about DNA replication to describe the mechanism of a drug that blocks it?

Recall that the opening story to this chapter described the discovery that platinum electrodes inhibit cell division, which lead researchers to wonder if they could invent a drug containing platinum that might inhibit the uncontrolled growth of cancer cells. The drug that Dr. Barnett Rosenberg developed, cisplatin, was remarkably efficient at halting certain kinds of tumor growth. Understanding how cisplatin works came with an understanding of the chemistry of DNA and the mechanism by which it replicates.

Investigating Life: The Meselson–Stahl Experiments discusses the experiments which revealed that DNA replicates semiconservatively—that separated strands of DNA serve as templates for new strands. We mentioned also the finding that semiconservative replication does not occur in the presence of cisplatin, suggesting that cisplatin interferes with strand separation, which is a prerequisite for replication. How does this come about, and what does it have to do with platinum?

Cisplatin contains a platinum atom bonded to two amino

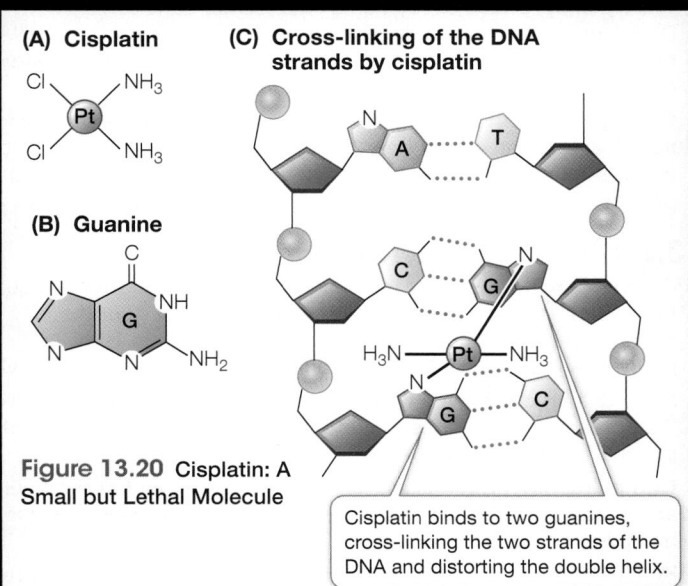

(A) Cisplatin

(B) Guanine

(C) Cross-linking of the DNA strands by cisplatin

Figure 13.20 Cisplatin: A Small but Lethal Molecule

Cisplatin binds to two guanines, cross-linking the two strands of the DNA and distorting the double helix.

experiments, this compound was formed when the platinum electrode reacted with salts in the surrounding solution. The bonds between the platinum atom and the chlorine atoms are weak, and the latter can be displaced by electron-rich substances (you may know from chemistry that these are called nucleophiles). In DNA, one of the nitrogen atoms of guanine (Figure 13.20B) displaces one of the chlorines, forming a strong covalent bond. If there is a nearby guanine on the opposite DNA strand, the other chlorine of the cisplatin molecule can be displaced as well, so that the cisplatin is effectively bonded to both of the DNA strands. In short, the DNA strands become cross-linked and cannot separate as needed for DNA replication (Figure 13.20C). Without replication, the cell cannot divide, and it undergoes apoptosis. The type of DNA lesion caused by cisplatin cannot be repaired by any of the cell's usual DNA repair mechanisms.

Future directions

Eukaryotic chromosomes have long strands of DNA, each with up to tens of millions of base pairs. During S phase of the cell cycle, each chromosome must be replicated only once and completely. Since replication is bidirectional from an origin, and since S phase typically is not long enough to replicate an entire chromosome, there must be many starting points for replication. Understanding how this happens is a major challenge for biologists. A recent discovery may give a clue as to how a DNA replication complex arrives at each replication origin. While most RNA molecules transcribed from DNA leave their site of synthesis, some do not. These short RNAs, complementary to their DNA template, turn back instead and bind to the DNA by base pairing. Indeed, this rapid binding displaces the non-template strand of DNA, so it forms a loop. At these spots on the long chromosome, there are loops with a region of DNA that is unpaired with its complementary strand, because that strand at that location is bound up with RNA. Emerging evidence suggests that the "R-loops," with their exposed DNA bases, act to recognize binding of the proteins of the DNA replication complex.

Chapter Summary 13

▶ 13.1 Experiments Revealed the Function of DNA as Genetic Material

- Griffith's experiments in the 1920s demonstrated that some substance in cells can cause heritable changes in other cells. **Review Figure 13.1**

- The location and quantity of DNA in the cell suggested that DNA might be the genetic material. Avery and his colleagues isolated the transforming principle from bacteria and identified it as DNA. **Review Figure 13.2**

- The Hershey–Chase experiments established conclusively that DNA (and not protein) is the genetic material, by tracing the DNA of radioactively labeled viruses, which were used to infect bacterial cells. **Review Figure 13.4, Animation 13.1**

- Genetic transformation of eukaryotic cells is often called **transfection**. Transformation and transfection can be studied with the aid of a **genetic marker** gene that confers a known and observable phenotype.

▶ 13.2 DNA Has a Structure That Suits Its Function

- Chargaff's rule states that the amount of **adenine** in DNA is equal to the amount of **thymine**, and that the amount of **guanine** is equal to the amount of **cytosine**; thus the total abundance of purines (A + G) equals the total abundance of pyrimidines (T + C).

- X-ray diffraction showed that the DNA molecule is a double helix. Watson and Crick proposed that the two strands in DNA are **antiparallel**. **Review Figure 13.6**

- **Complementary base pairing** between A and T and between G and C accounts for Chargaff's rule. The bases are held together by hydrogen bonding.

- Reactive groups are exposed in the paired bases, allowing for recognition by other molecules such as proteins. **Review Figure 13.7**

▶ 13.3 DNA Is Replicated Semiconservatively

See Animation 13.2

- Meselson and Stahl showed that DNA undergoes **semiconservative replication**. Each parent strand acts as a **template** for the synthesis of a new strand; thus the two replicated DNA molecules each contain one parent strand and one newly synthesized strand. **Review Investigating Life: The Meselson–Stahl Experiment, Animation 13.3**

- In DNA replication, the enzyme **DNA polymerase** catalyzes the addition of nucleotides to the 3′ end of each strand. Which nucleotides are added is determined by complementary base pairing with the template strand. **Review Figure 13.9**

- The pre-replication complex is a huge protein complex that attaches to the chromosome at the **origin of replication** (ori).

- Replication proceeds from the origin of replication on both strands in the 5′-to-3′ direction, forming a **replication fork**. **Review Figure 13.10**

- **Primase** catalyzes the synthesis of a short RNA **primer** to which nucleotides are added by DNA polymerase. **Review Figure 13.11**

- Many proteins assist in DNA replication. **DNA helicase** separates the strands, and **single-strand binding proteins** keep the strands from reassociating. **Review Figure 13.13, Activity 13.1**

- The **leading strand** is synthesized continuously and the **lagging strand** in pieces called **Okazaki fragments**. The fragments are joined together by **DNA ligase**. **Review Focus: Key Figure 13.13, Figure 13.14, Animation 13.4**

- The speed with which DNA polymerization proceeds is attributed to the **processive** nature of DNA polymerases, which can catalyze many polymerizations at a time. A **sliding DNA clamp** helps ensure the stability of this process. **Review Figure 13.16**

- At the ends of eukaryotic chromosomes are regions of repetitive DNA sequence called **telomeres**. Unless the enzyme **telomerase** is present, a short segment at the end of each telomere is lost each time the DNA is replicated. After multiple cell cycles, the telomeres shorten enough to cause chromosome instability and cell death. Review Figure 13.17

▶ 13.4 Errors in DNA Can Be Repaired

- DNA polymerases make about one error in 100,000 bases replicated. DNA is also subject to natural alterations and chemical damage. DNA can be repaired by at least three different mechanisms: **proofreading**, **mismatch repair**, and **excision repair**. Review Figure 13.18

▶ 13.5 The Polymerase Chain Reaction Amplifies DNA

- The **polymerase chain reaction** technique uses DNA polymerase to make multiple copies of DNA in the laboratory. Review Figure 13.19, Activity 13.2

Go to **LearningCurve** (in **LaunchPad**) for dynamic quizzing that helps you solidify your understanding of this chapter. **LearningCurve** adapts to your responses, giving you the practice you need to master each key concept.

▶ Apply What You've Learned

Review

13.3 DNA replication proceeds bidirectionally from an origin of replication (*ori*); *E. coli* has a single *ori*, and eukaryotic chromosomes have multiple ones.

Original Paper: Huberman, J. A. and Riggs A. D. 1968. On the mechanism of DNA replication in mammalian chromosomes. *Journal of Molecular Biology* 32: 327–341.

Biologists use a variety of methods to develop hypotheses about living systems. A mathematical approach, for example, can help you learn about DNA replication in human cells.

DNA replication is not an ongoing process in cells but is restricted to the S phase of the cell cycle. Typically, in dividing mammalian cells, S phase is 8 hours long. The rate at which DNA polymerase links nucleotides to form a new strand of DNA is about 50 base pairs per second. The shortest chromosomal DNA in a human cell contains approximately 50×10^6 base pairs; the longest chromosome contains approximately 250×10^6 base pairs.

Questions

1. Using the information above, calculate how long will it take a single DNA polymerase to replicate the entire DNA molecule in the shortest chromosome if it starts at one end of the DNA molecule and proceeds uninterrupted to the other end. Compare this to the 8 hours that the cell has to complete its DNA replication. Do the results suggest that DNA replication involves just one starting point, or do they point to an alternate hypothesis?

   Now consider the results from an investigation. As you know, thymine is one of the four bases in DNA. In the lab, scientists can replace the normal isotope of hydrogen (¹H) with the radioactive isotope (³H) in thymidine, the nucleotide containing the base thymine. If dividing cells in S phase are given all four nucleotides, but with ³H-thymidine for a short time (called a "pulse"), the newly synthesized DNA strand will be radioactive. The radioactive strand can be viewed under the microscope by a special process if the DNA is spread out on a microscope slide.

2. The diagram below shows radioactivity for a small segment of a single DNA strand extracted from human S phase cells after a pulse with ³H-thymidine during the last 4 hours of S phase.

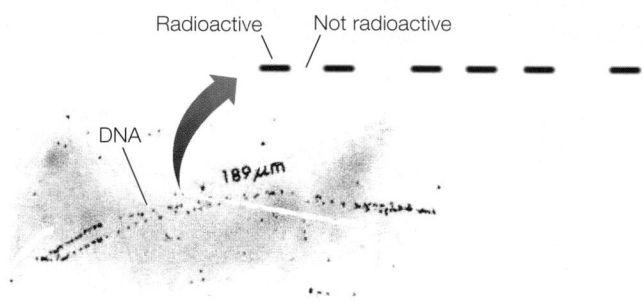

   What can you conclude about the number of starting points on a single DNA molecule? Is this consistent with your results in question 1?

   In a second experiment, the cells were given a pulse with ³H-thymidine just after they began S phase. After several hours the cells were given non-radioactive ¹H-thymidine (a "chase" to follow the "pulse"), so DNA made from that point on was not labeled. After some time, the DNA was extracted and the same chromosome was examined. The diagram below shows the results of this experiment on the same small segment of a single DNA strand as above.

3. Was DNA replication unidirectional or bidirectional from its starting point? Explain your answer.

4. Draw the double-stranded DNA that would result from the second experiment, showing where labeled and unlabeled nucleotides would be found.

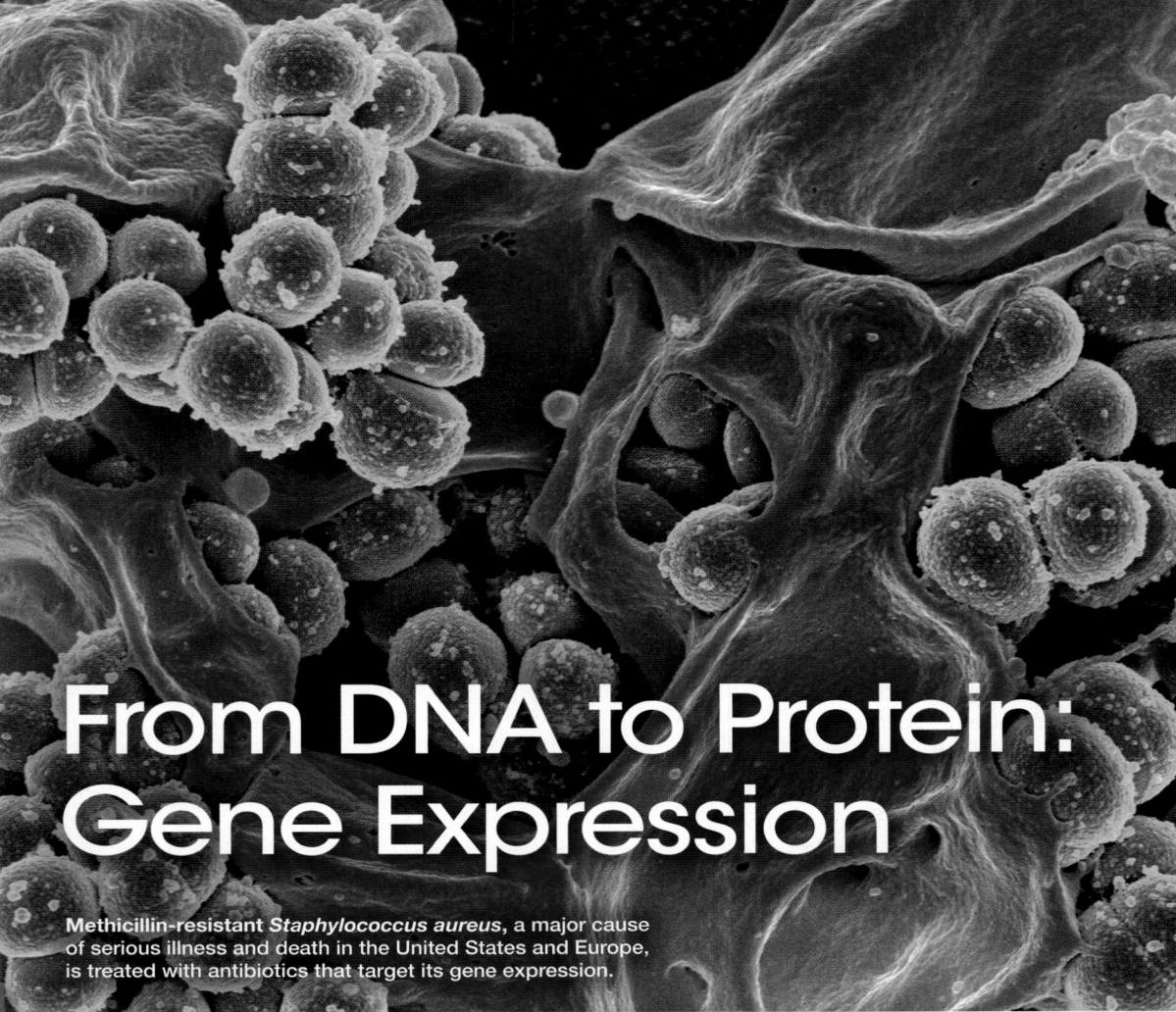

14

From DNA to Protein: Gene Expression

Methicillin-resistant *Staphylococcus aureus*, a major cause of serious illness and death in the United States and Europe, is treated with antibiotics that target its gene expression.

▶ investigating**life**

Employing the Genetic Code to Combat Superbugs

Staphylococcus aureus is among billions of bacteria that inhabit the human skin and nose, mostly with no ill effect. But sometimes, especially when the immune system has been weakened by age or disease, *S. aureus* can cause major skin infections and may enter the body through the nose or a wound site, causing serious and sometimes fatal infections of organs such as the heart and lungs.

Until recently, most *S. aureus* infections were successfully treated with penicillin and related drugs, including methicillin. These antibiotics bind and inactivate several related enzymes involved in the assembly of bacterial cell walls. New bacterial cells produced by cell division of antibiotic-treated bacteria do not survive.

Unfortunately, some *S. aureus* strains have acquired mutant versions of a penicillin-binding protein that are resistant to antibiotics—the mutant enzyme can catalyze the assembly of cell walls even in the presence of antibiotics. The mutant *mecA* gene that encodes this mutant protein can be passed from one bacterium to another by bacterial conjugation (see Key Concept 12.6).

What makes the mutant version of *mecA* different, and how does this difference lead to antibiotic resistance? Compared to the wild-type allele, the mutant *mecA* gene has a small change in its nucleotide sequence. This results in changes in the amino acid sequence of the expressed protein. This alteration in the protein's secondary structure affects its tertiary structure: the protein folds into a shape that doesn't bind the antibiotics.

The fundamental lesson conveyed by this example is that *a gene is expressed as a protein*, and more specifically, that *the sequence of nucleotides in DNA is expressed as a sequence of amino acids in a protein*. Understanding how the nucleotide sequence of a gene produces a specific protein came with the discovery of the genetic code, a landmark in biological research.

Mutant bacterial strains that are stubbornly resistant to antibiotics are known as "superbugs." One greatly feared superbug is methicillin-resistant *S. aureus*, or MRSA. About 1 person in 50 carries MRSA, so it represents a real threat to public health. MRSA incidence is declining, thanks to careful monitoring and treatment in hospitals and nursing homes, where people are especially vulnerable. The infection is treated with a new class of antibiotics that target bacterial protein synthesis.

QA How does knowledge of the genetic code help us understand the actions of some antibiotics?

▶key concept
14.1

Genes Code for Proteins

In Chapter 4 we introduced DNA and its role in gene expression. Then in Chapter 13 we presented evidence that DNA is the carrier of genetic information and described how DNA is replicated prior to cell division. Here we will focus on the evidence that proteins are the major products of gene expression, and we will describe how a gene is expressed as protein.

focus your learning

- Experimental evidence supports the claim that each gene encodes one protein.

Scientists had a molecular understanding of phenotypes before it was known that DNA was the genetic material; they had studied the chemical differences between individuals carrying wild-type and mutant alleles in organisms as diverse as humans and bread molds and knew that the major phenotypic differences resulted from differences in specific proteins. But just *how* different proteins arise in different individuals was not understood.

Observations in humans led to the proposal that genes determine enzymes

The identification of a gene product as a protein began with a mutation. In the early twentieth century, English physician Archibald Garrod saw several children with a rare disease. One symptom of the disease was that the urine turned dark brown when exposed to air, which was especially noticeable on the infants' diapers. The disease was given the descriptive name alkaptonuria ("black urine").

Garrod noticed that the disease was most common in children whose parents were first cousins. Mendelian genetics had just been "rediscovered," and Garrod realized that because first cousins share on average 1/8 of their alleles, the children of first cousins might inherit a rare mutant allele from both parents. He proposed that alkaptonuria was a recessive phenotype caused by a mutant allele that causes the disease phenotype.

Garrod took the analysis further by identifying the biochemical abnormality in the affected children. He isolated from them an unusual substance, homogentisic acid, which accumulated in blood, joints (where it crystallized and caused severe pain), and urine (where it turned black). The chemical structure of homogentisic acid is similar to that of the amino acid tyrosine:

Homogentisic acid   Tyrosine

The function of enzymes as biological catalysts had just been discovered. Garrod proposed that homogentisic acid was a breakdown product of tyrosine. He suggested that while homogentisic acid ordinarily would be converted to a harmless product, it was not

being converted in children with alkaptonuria. Garrod further hypothesized that the enzyme required for the breakdown of homogentisic acid was not being produced in these children. He suggested that a normal human allele (for the wild-type phenotype) was required for the synthesis of an enzyme that catalyzed this conversion:

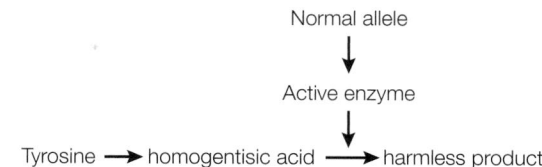

If the allele were mutated, the enzyme would be inactive and homogentisic acid would accumulate instead:

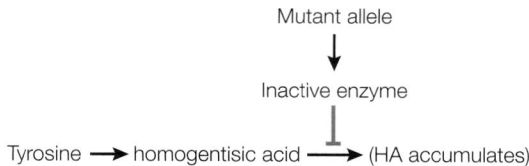

Garrod concluded that there must be *one gene to one enzyme* and coined the term "inborn error of metabolism" to describe this genetically determined biochemical disease. But his hypothesis was not confirmed until the actual enzyme and the specific gene mutation involved were identified. The enzyme, homogentisic acid oxidase, was identified as active in healthy people and inactive in alkaptonuria patients in 1958, and the specific DNA mutation was described in 1996.

To relate genes and enzymes more generally, biologists turned to simpler organisms that could be manipulated in the laboratory.

Experiments on bread mold established that genes determine enzymes

As they work to explain the principles that govern life, biologists often turn to organisms that are easy to manipulate experimentally. Such **model organisms** have certain characteristics that make them attractive experimental subjects. For example, model organisms:

- are easy to grow in the laboratory or greenhouse.
- have short generation times.
- are easy to manipulate genetically, by crossing or by other methods.
- often produce large numbers of progeny.

Biologists have used model organisms to develop principles of genetics that can then be applied more generally to other organisms. You have seen some of these organisms in previous chapters:

- Pea plants (*Pisum sativum*) were used by Mendel in his genetics experiments.
- Fruit flies (*Drosophila*) were used by Morgan in his genetics experiments.
- *Escherichia coli* was used by Meselson and Stahl to study DNA replication.

To this list we now add the bread mold *Neurospora*, an ascomycete fungus (see Chapter 30). This mold is haploid for most of its life cycle, so there are no dominant or recessive alleles: all alleles

experiment

Figure 14.1A   One Gene, One Enzyme

Original Paper: Srb, A. M. and N. H. Horowitz. 1944. The ornithine cycle in *Neospora* and its genetic control. *Journal of Biological Chemistry* 154: 129–139.

Srb and Horowitz developed several mutant strains of *Neurospora* that could not make arginine (arg). Several molecules including ornithine and citrulline are needed for arginine synthesis. By systematically adding each of these molecules to the growth media for the mutant strains, the researchers deduced that each mutant strain was deficient in one enzyme along a biochemical pathway.

HYPOTHESIS▶   Each gene determines an enzyme in a biochemical pathway.

METHOD   Place spores (single cells that divide to produce mold colonies) of each arg mutant strain on a minimal nutritional medium with and without supplements.

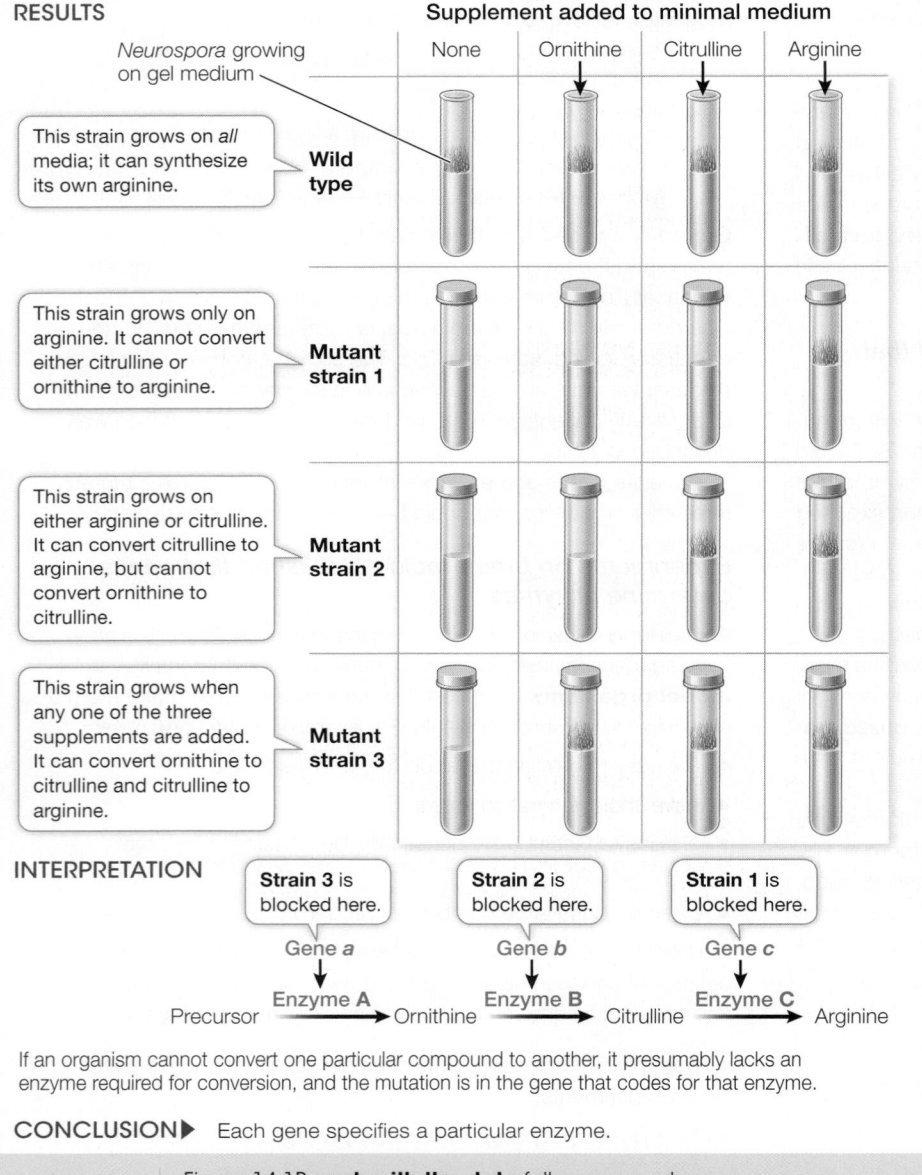

INTERPRETATION

If an organism cannot convert one particular compound to another, it presumably lacks an enzyme required for conversion, and the mutation is in the gene that codes for that enzyme.

CONCLUSION▶   Each gene specifies a particular enzyme.

Figure 14.1B **work with the data** follows on next page.

undertook studies to biochemically define the phenotypes in *Neurospora*.

Like Garrod, Beadle and Tatum hypothesized that the expression of a specific gene results in the activity of a specific enzyme. They set out to test this hypothesis directly. They grew *Neurospora* on a nutritional medium containing sucrose, minerals, and biotin, which is the only vitamin that wild-type *Neurospora* cannot synthesize itself. Using this minimal medium, the enzymes of wild-type *Neurospora* could catalyze all the metabolic reactions needed for growth.

The scientists then treated the wild-type *Neurospora* with X rays, which function as a mutagen. A **mutagen** is something that damages DNA, causing **mutations**: heritable alterations in the DNA sequence. After the X-ray treatment, some *Neurospora* strains could no longer grow on the minimal medium. These mutant strains grew only if they were supplied with specific additional nutrients, such as particular vitamins. Beadle and Tatum hypothesized that these genetic strains had mutations in the genes that code for production of enzymes needed to synthesize the additional nutrients. For each mutant strain, the scientists were able to find a single compound that, when added to the minimal medium, supported the growth of that strain. These results suggested that mutations have simple effects, and that each mutation causes a defect in only one enzyme in a metabolic pathway. These conclusions confirmed Garrod's **one-gene, one-enzyme hypothesis**.

Mutations provide a powerful way to determine cause and effect in biology. Nowhere has this been more evident than in the elucidation of biochemical pathways. Such pathways consist of sequential events (chemical reactions) in which each event is dependent on the occurrence of the preceding event. The general reasoning is as follows:

- *Observation.* A particular gene (*a*) is present and a particular reaction catalyzed by a particular enzyme (A) occurs; the two are correlated.
- *Hypothesis.* Gene *a* determines the synthesis of enzyme A.
- *Test of hypothesis.* Mutate gene *a*. Prediction: no functional enzyme is made, and the reaction does not occur.

are expressed phenotypically and not masked by a heterozygous condition. *Neurospora* is easy to grow in the laboratory. Biologists at Stanford University led by George Beadle and Edward Tatum

Two colleagues of Beadle and Tatum, Adrian Srb and Norman Horowitz, used this experimental approach to isolate *Neurospora*

work with the data

Figure 14.1B One Gene, One Enzyme

Original Paper: Srb, A. M. and N. H. Horowitz. 1944.

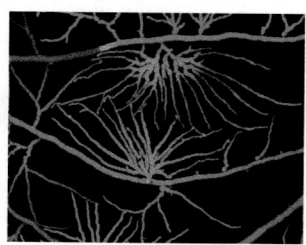

Neurospora (left) is haploid for most of its life cycle, except for the formation of a diploid cell when it undergoes mating; the cell then undergoes meiosis to form haploid spores. Beadle and Tatum used X rays to cause mutations in *Neurospora*. They isolated mutant strains that were unable to grow on minimal medium but were able to grow if the medium was supplemented with particular compounds. Their colleagues Adrian Srb and Norman Horowitz analyzed 15 mutant strains (the *arg* mutants) that could not synthesize arginine but could grow on medium supplemented with arginine. The scientists tested various compounds and found two, ornithine and citrulline, that could be used instead of arginine to support the growth of some of the mutant strains (as seen Figure 14.1A). The results for three of the strains are shown in the first three rows of the table, with growth expressed as dry weight of fungal material after growth for 5 days.

Strain	No addition	Ornithine added	Citrulline added	Arginine added
34105	1.1	25.5	30.0	33.2
33442	2.3	2.5	42.7	43.8
36703	0.0	0.0	0.0	20.4
Double mutant	0.0	0.0	0.0	22.0

QUESTIONS▶

1. Based on the biochemical pathway for arginine synthesis shown in the experiment (Figure 14.1A), which gene (*a*, *b*, or *c*) encoding proteins (A, B, or C) was mutated in each strain?

2. Why was there some growth in strains 34105 and 33442 even when there were no additions to the growth medium?

3. Nineteen other amino acids were tested as substitutes for arginine in the three strains. In all cases, there was no growth. Explain these results.

4. Sexual reproduction in *Neurospora* was used to create double mutants, which carried the mutations from both parental strains. A double mutant derived from strains 33442 and 36703 had the growth characteristics shown in the last row of the table. Explain these data in terms of the genes, mutations, and biochemical pathway.

A similar **work with the data** exercise may be assigned in **LaunchPad**.

---

mutants that could not survive without the amino acid arginine in their growth medium. By adding particular compounds to the medium, Srb and Horowitz were able to identify a series of steps in the biochemical pathway leading to the synthesis of arginine (**Figure 14.1**).

One gene determines one polypeptide

The one-gene, one-enzyme relationship has undergone several modifications in light of our current knowledge of molecular biology. Many proteins, including many enzymes, are composed of more than one *polypeptide chain, or subunit. So it is more correct to speak of a **one-gene, one-polypeptide relationship**.

> *connect the concepts Hemoglobin, illustrated in Figure 3.11 is an example of a protein composed of more than one polypeptide chain. Hemoglobin has four polypeptides—two α and two β subunits, and each subunit is encoded by a separate gene.

So far we have seen that in terms of protein synthesis, the *function of a gene is to prescribe the production of a single, specific polypeptide*. But not all genes code for polypeptides. As we will see below and in Chapter 16, many DNA sequences are transcribed to RNA molecules that are not translated into polypeptides, but instead have other functions.

▶14.1 recap

Studies of mutations in humans and bread molds led to our understanding of the one-gene, one-polypeptide relationship. In most cases, the function of a gene is to code for a specific polypeptide.

learning outcomes

You should be able to:

- Justify the selection of various model organisms used for genetic studies.
- Explain how experimental design has played an important role in enabling researchers to elucidate information about metabolic pathways.
- Analyze data to link genes to their products.

1. What is a model organism, and why is *Neurospora* a good model for studying biochemical genetics?

2. How were the experiments on mutant strains of *Neurospora* set up to determine the order of steps in a biochemical pathway?

3. Beadle and Tatum's experiments showed that a biochemical pathway could be deduced from mutant strains. In bacteria, the biosynthesis of the amino acid tryptophan (T) from the precursor chorismate (C) involves four intermediate chemical compounds, which we will call D, E, F, and G. Here are the phenotypes of various mutant strains. Each strain has a mutation in a gene for a different enzyme; + means growth with the indicated compound added to the medium, and 0 means no growth. Based on these data, order the compounds (C, D, E, F, G, and T) and enzymes (1, 2, 3, 4, and 5) in a biochemical pathway.

Mutant strain	Addition to medium					
	C	D	E	F	G	T
1	0	0	0	0	+	+
2	0	+	+	0	+	+
3	0	+	0	0	+	+
4	0	+	+	+	+	+
5	0	0	0	0	0	+

Now that you have seen the evidence for the one-gene, one-polypeptide relationship, how does it work? That is, how is the information encoded in DNA used to produce a particular polypeptide?

key concept

14.2 Information Flows from Genes to Proteins

As we discussed in Chapter 13 and Key Concept 14.1, DNA is the hereditary material and codes for proteins and RNAs. In the remainder of this chapter we will focus on the processes that occur when a protein-coding gene is expressed. We briefly outlined gene expression in Key Concept 4.1. To review, this process occurs in two major steps:

1. During **transcription**, the information in a DNA sequence (a gene) is copied into a complementary RNA sequence.
2. During **translation**, this RNA sequence is used to create the amino acid sequence of a polypeptide.

Francis Crick and James Watson, who deciphered the structure of DNA, first proposed this model for gene expression. They took the concept further by suggesting that gene expression can go in only one direction: DNA can be used to create a protein, but a protein can never be used to create DNA. At the time, Crick called this "the **central dogma** of molecular biology."

focus your learning

- Gene expression involves the processes of transcription and translation.
- Some virus genomes are composed of single-strand RNA, which is used to direct DNA synthesis by reverse transcription when the virus infects a host cell.

Three types of RNA have roles in the information flow from DNA to protein

There are numerous types of RNA. Three of them have vital roles in gene expression.

MESSENGER RNA AND TRANSCRIPTION   When a protein-coding gene is expressed, one of the two DNA strands in the gene is transcribed to produce a complementary RNA strand, which is then processed to produce **messenger RNA** (**mRNA**). In eukaryotic cells, the mRNA travels from the nucleus to the cytoplasm, where it is translated into a polypeptide (**Focus: Key Figure 14.2**). The nucleotide sequence of the mRNA determines the ordered sequence of amino acids in the polypeptide chain, which is built by a ribosome.

RIBOSOMAL RNA AND TRANSLATION   The **ribosome** is essentially a protein synthesis factory composed of multiple proteins and several **ribosomal RNAs** (**rRNAs**). One of the rRNAs catalyzes peptide bond formation between amino acids, to form a polypeptide.

TRANSFER RNA MEDIATES BETWEEN mRNA AND PROTEIN   Another type of RNA called **transfer RNA** (**tRNA**) can both bind a specific amino acid and recognize specific sequences of nucleotides in mRNA. It is the tRNA that recognizes which amino acid should be added next to a growing polypeptide chain.

 Media Clip 14.1 **Protein Synthesis: An Epic on a Cellular Level** www.Life11e.com/mc14.1

In some cases, RNA determines the sequence of DNA

While we have said that DNA is the genetic material, some viruses present exceptions to the general process of gene expression outlined above. As we saw in Key Concept 13.1, a virus is a non-cellular

focus: key figure

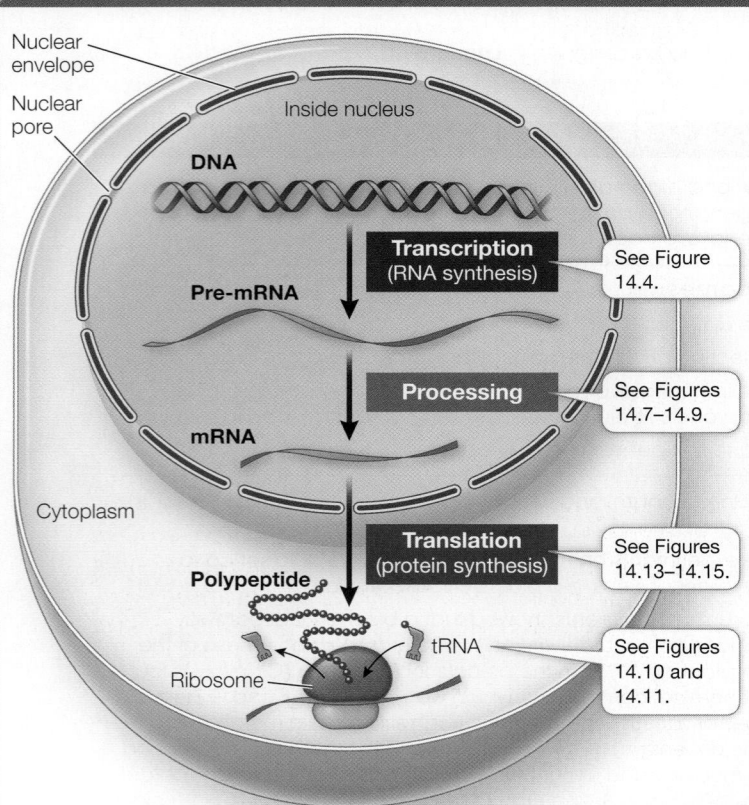

Figure 14.2 From Gene to Protein   This diagram summarizes the processes of gene expression in eukaryotes. Note that the nucleus is typically about ¼ the size shown here.

Q: In general, how would this diagram be different for a prokaryotic cell?

 Activity 14.1 **Eukaryotic Gene Expression** www.Life11e.com/ac14.1

infectious particle that reproduces inside cells. Many viruses, such as the ones that cause influenza and polio, have RNA rather than DNA as their genetic material. That is, the nucleotide sequence of viral RNA acts as an information carrier and can be expressed as a protein. Because RNA is usually single-stranded, the question arises: how do these viruses replicate? More specifically, how do they duplicate their genetic material? Most viruses replicate by transcribing from RNA to RNA, making an RNA strand that is complementary to their genomes. This "opposite" strand is then used to make multiple copies of the viral genome by transcription:

Not all viruses whose genomes consist of RNA replicate by transcribing from RNA to RNA. Some, such as human immunodeficiency viruses (HIVs) and certain rare tumor viruses, make a DNA copy of their genome after infecting a host cell. This DNA copy is then incorporated into the host's genome. Synthesis of DNA from RNA is called **reverse transcription**, and viruses that employ this kind of transcription are called **retroviruses**. Retroviruses rely on the host cell's transcription machinery to make more RNA. This RNA can either be translated to produce viral proteins, or incorporated as the viral genome into new viral particles.

14.2 recap

The expression of protein-encoding genes can be broken down into two fundamental steps: transcription and translation. In transcription the DNA code is used to produce mRNA, whose sequence determines the mature mRNA transcript. The mRNA sequence determines the order of amino acids in a polypeptide. Translation is the process by which this information in mRNA is converted into a polypeptide chain.

learning outcomes

You should be able to:
- Describe the central dogma of molecular biology.
- Evaluate gene expression systems to determine whether they adhere to the central dogma of molecular biology.

1. What is the central dogma of molecular biology?
2. Do retroviruses violate the central dogma? Explain.

We will revisit viral genetics in later chapters (in Key Concepts 16.3 and 25.4). In the rest of this chapter we will focus on gene expression in prokaryotes and eukaryotes. Understanding this process is essential for understanding how organisms function at the molecular level and is key to the application of biology to human welfare, in areas such as agriculture and medicine. We'll begin by describing how the information in DNA is transcribed to produce RNA.

key concept 14.3   DNA Is Transcribed to Produce RNA

The synthesis of RNA is directed by DNA. The base sequence of one strand of DNA is used as a template for RNA synthesis, so that the RNA made is complementary in sequence to the DNA strand, with the exception that in RNA there is uracil (U) instead of thymine (T) and in RNA the sugar is ribose instead of deoxyribose. It is important to realize that although the RNA made is a mirror image of its DNA template, the RNA has the same sequence as the other, non-template strand of DNA. So the information content of DNA is indeed preserved in RNA.

focus your learning
- All RNA polymerases have certain characteristics in common.
- The genetic code allows specific nucleotide sequences in RNA to be translated into specific amino acid sequences in polypeptides.

Transcription—the formation of a specific RNA sequence from a specific DNA sequence—requires several components:

- A DNA template for complementary base pairing; one of the two strands of DNA
- The four ribonucleoside triphosphates ATP, GTP, CTP, and UTP, to act as substrates
- An RNA polymerase enzyme
- Salts and a pH buffer to create an appropriate chemical environment for RNA polymerase (if transcription is performed in a test tube)

Several kinds of RNA are made from DNA templates. The most important from a genetic point of view is mRNA. But transcription also produces tRNA and rRNA, whose roles in protein synthesis will be described below. Like polypeptides, these last two RNAs are encoded by specific genes. Eukaryotes also make many kinds of small RNAs, including small nuclear RNA (snRNA), microRNA (miRNA), and small interfering RNA (siRNA), which are also transcribed. **Table 14.1** summarizes some of the RNAs found in eukaryotic cells. We will discuss the roles of miRNA and siRNA in Chapter 16.

RNA polymerases share common features

RNA polymerases from both prokaryotes and eukaryotes catalyze the synthesis of RNA from the DNA template. There is only one kind of RNA polymerase in bacteria, whereas there are several kinds in eukaryotes; however, they all share a common structure (**Figure 14.3**). Like DNA polymerases, RNA polymerases catalyze the addition of nucleotides in a 5'-to-3' direction and are processive; that is, a single enzyme–template binding event results in the polymerization of hundreds of RNA bases. But unlike DNA polymerases (see Figure 13.11), RNA polymerases *do not require a primer*.

table 14.1   Some RNAs in Eukaryotic Cells

RNA type	Location of activity	Role
Ribosomal RNA (rRNA)	Cytoplasm (ribosome)	Binding of mRNA and tRNA and protein synthesis
Messenger RNA (mRNA)	Cytoplasm	Carrier of gene sequence
Transfer RNA (tRNA)	Cytoplasm	Adaptor between mRNA and protein sequences
MicroRNA (miRNA)	Nucleus and cytoplasm	Regulates transcription and translation
Small interfering RNA (siRNA)	Nucleus and cytoplasm	Regulates other RNAs
Small nuclear RNA (snRNA)	Nucleus	Mediates mRNA processing

Transcription occurs in three steps

Transcription can be divided into three distinct processes: (1) **initiation**, (2) elongation, and (3) termination. Follow these processes in **Figure 14.4**.

INITIATION   Transcription begins when RNA polymerase binds to a special sequence of DNA called a **promoter** (see Figure 14.4A). Eukaryotic genes generally have one promoter each, whereas in prokaryotes and viruses, several genes often share one promoter. Promoters are important control sequences that "tell" the RNA polymerase two things:

1. Where to start transcription
2. Which strand of DNA to transcribe

A promoter reads in a particular direction, so it orients the RNA polymerase and thus "aims" it at the appropriate strand to use as a template. Part of each promoter is the **initiation site**, where transcription begins. Groups of nucleotides lying "upstream" from the initiation site (5′ on the non-template strand and 3′ on the template strand) help the RNA polymerase bind. Other proteins, which can bind to specific DNA sequences and to RNA polymerase, help direct the polymerase onto the promoter. These proteins, called **sigma factors** in prokaryotes and **transcription factors** in eukaryotes, help determine which specific genes are expressed at a particular time in the cell.

Although every gene has a promoter, not all promoters are identical. Some are more effective at transcription initiation than others. Furthermore, there are differences between transcription initiation in prokaryotes and in eukaryotes. We will discuss promoters and their roles in the regulation of gene expression in Chapter 16.

ELONGATION   After RNA polymerase has bound to the promoter, it begins the process of **elongation** (see Figure 14.4B). DNA unwinds about 10 base pairs at a time and RNA polymerase reads the template strand in the 3′-to-5′ direction. The first nucleotide in the new RNA forms its 5′ end, and subsequent nucleotides complementary to the DNA template are added to its 3′ end. Thus the RNA transcript is antiparallel to the DNA template strand.

You may recall from Key Concept 13.3 that DNA polymerase uses dNTPs (deoxyribonucleoside triphosphates) as substrates, and forms covalent bonds between each incoming dNTP and the 3′ end of the growing polynucleotide chain (see Figure 13.10). Energy released by the removal of two phosphate groups from the dNTP is used to drive the reaction. Similarly, RNA polymerase uses (ribo)nucleoside triphosphates (NTPs) as substrates, removing two phosphate groups from each substrate molecule and using the released energy to drive the polymerization reaction.

Because RNA polymerases do not proofread, transcription errors occur at a rate of one for every 10^4 to 10^5 bases. Because many copies of RNA are made, however, and because they often have only a relatively short life span, these errors are not as potentially harmful as mutations in DNA.

TERMINATION   Just as initiation sites in the DNA template strand specify the starting point for transcription, particular base sequences specify its **termination** (see Figure 14.4C). There are two mechanisms for ending transcription. For some genes, the newly formed transcript folds back on itself and forms internal hydrogen bonds between bases. A loop forms, and this structure causes the transcript to fall away from the DNA template and the RNA polymerase. In other cases, a protein binds to specific sequences on the transcript and causes the RNA to detach from the DNA template.

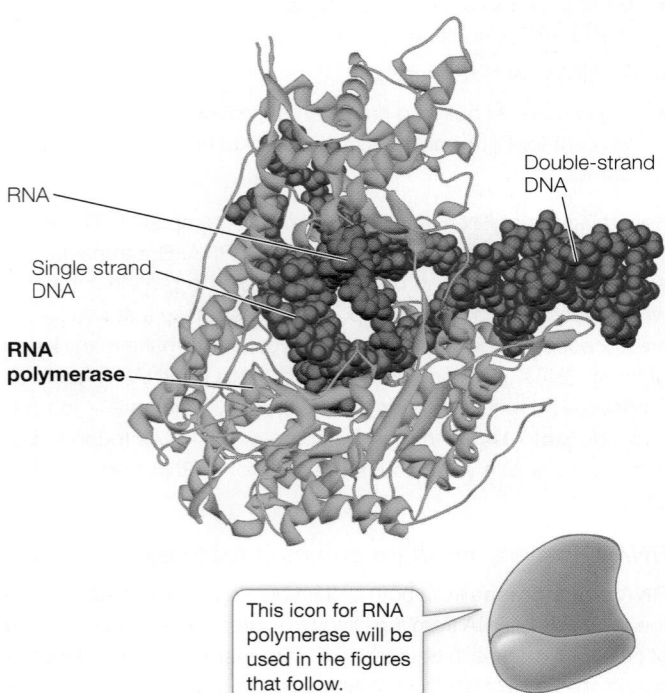

RNA

Single strand DNA

RNA polymerase

Double-strand DNA

This icon for RNA polymerase will be used in the figures that follow.

Figure 14.3 RNA Polymerase Interacting with DNA   The RNA polymerase enzyme depicted (in green) is from bacteriophage T7, but it is representative of most other RNA polymerases. Note the size relationship between enzyme and DNA. The inset indicates how this enzyme will be represented in the figures that follow.

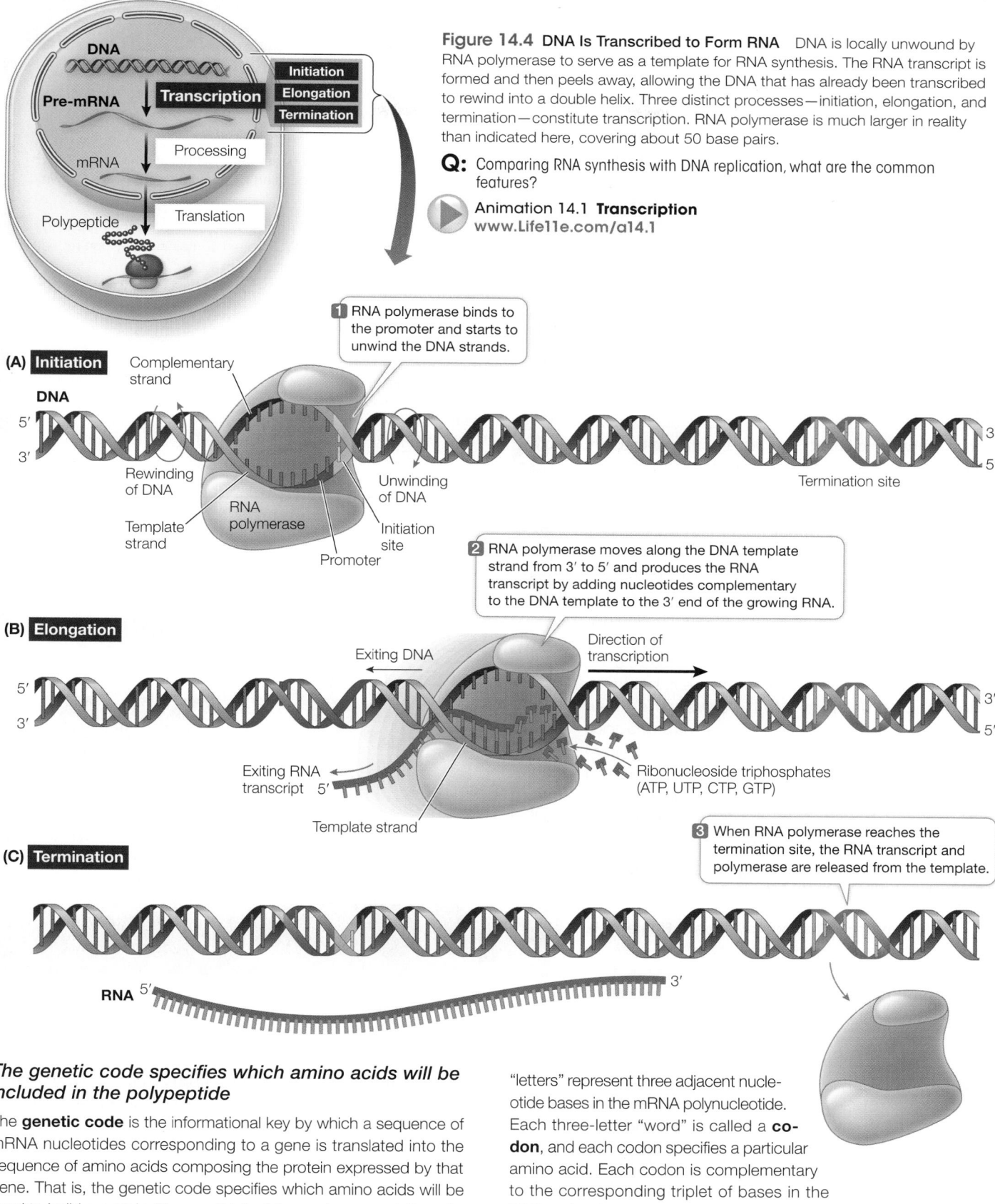

Figure 14.4 DNA Is Transcribed to Form RNA   DNA is locally unwound by RNA polymerase to serve as a template for RNA synthesis. The RNA transcript is formed and then peels away, allowing the DNA that has already been transcribed to rewind into a double helix. Three distinct processes—initiation, elongation, and termination—constitute transcription. RNA polymerase is much larger in reality than indicated here, covering about 50 base pairs.

Q: Comparing RNA synthesis with DNA replication, what are the common features?

Animation 14.1 **Transcription**
www.Life11e.com/a14.1

1 RNA polymerase binds to the promoter and starts to unwind the DNA strands.

(A) Initiation

2 RNA polymerase moves along the DNA template strand from 3′ to 5′ and produces the RNA transcript by adding nucleotides complementary to the DNA template to the 3′ end of the growing RNA.

(B) Elongation

3 When RNA polymerase reaches the termination site, the RNA transcript and polymerase are released from the template.

(C) Termination

The genetic code specifies which amino acids will be included in the polypeptide

The **genetic code** is the informational key by which a sequence of mRNA nucleotides corresponding to a gene is translated into the sequence of amino acids composing the protein expressed by that gene. That is, the genetic code specifies which amino acids will be used to build a protein. You can think of the code as consisting of a series of sequential, non-overlapping, three-letter "words." The three "letters" represent three adjacent nucleotide bases in the mRNA polynucleotide. Each three-letter "word" is called a **codon**, and each codon specifies a particular amino acid. Each codon is complementary to the corresponding triplet of bases in the DNA molecule from which it was transcribed. In short, the genetic code relates codons to their specifically encoded amino acids.

CHARACTERISTICS OF THE CODE   Molecular biologists "broke" the genetic code in the early 1960s. The problem they addressed was perplexing: how could more than 20 "code words" be written with an "alphabet" consisting of only four "letters"? In other words, how could four bases (A, U, G, and C) code for 20 different amino acids?

A triplet code, based on three-letter codons, was considered likely. Since there are only four letters (A, G, C, and U), a one-letter code clearly could not unambiguously encode 20 amino acids; it could encode only four of them. A two-letter code could have only 4 × 4 = 16 unambiguous codons—still not enough. But a triplet code could have 4 × 4 × 4 = 64 codons, more than enough to encode the 20 amino acids.

Marshall W. Nirenberg and J. H. Matthaei, at the U.S. National Institutes of Health, made the first decoding breakthrough in 1961 when they realized that the code would be easier to break if they were working with a very simple, known mRNA sequence rather than with a complex natural mRNA molecule. They set out to synthesize mRNA molecules consisting of just one type of nucleotide base; poly U mRNA, for example, consisted of just uracil nucleotides. Nirenberg and Matthaei's goal was to then identify, through a translation process conducted in a test tube, the polypeptide that the artificial messenger encoded. Their experiment, which is laid out in **Investigating Life: Deciphering the Genetic Code**, led to the identification of the first codons. Other scientists soon identified the rest of the code. This was a major achievement, linking the information in DNA (the gene) to its expression in a protein (the phenotype). The understanding that the amino acids for each protein are spelled out via codons not only led to our understanding of the fundamentals of genetics and mutation, but to investigations into the genetic underpinnings of disease, such as the development of resistance in MRSA described in the opening of this chapter.

 Animation 14.2 Deciphering the Genetic Code
www.Life11e.com/a14.2

The complete genetic code is shown in **Figure 14.5**. Notice that there are many more codons than there are different amino acids in proteins. Proteins are built from just 20 amino acids, but all possible combinations of the four available "letters" (the bases) give 64 (4^3) different three-letter codons. Why are there more codons than amino acids? One reason is that there is more than one codon for almost all amino acids. For example, leucine is represented by six different codons (see Figure 14.5). Only methionine and tryptophan are represented by just one codon each. Thus we say that the genetic code is redundant (or degenerate). Also, a few select codons serve functions other than coding for amino acids. AUG, for example, not only codes for methionine but is also the **start codon**, the initiation signal for translation. Three codons (UAA, UAG, UGA) are **stop codons**, or termination signals for translation. When the translation machinery reaches one of these codons, translation stops and the polypeptide is released from the translation complex.

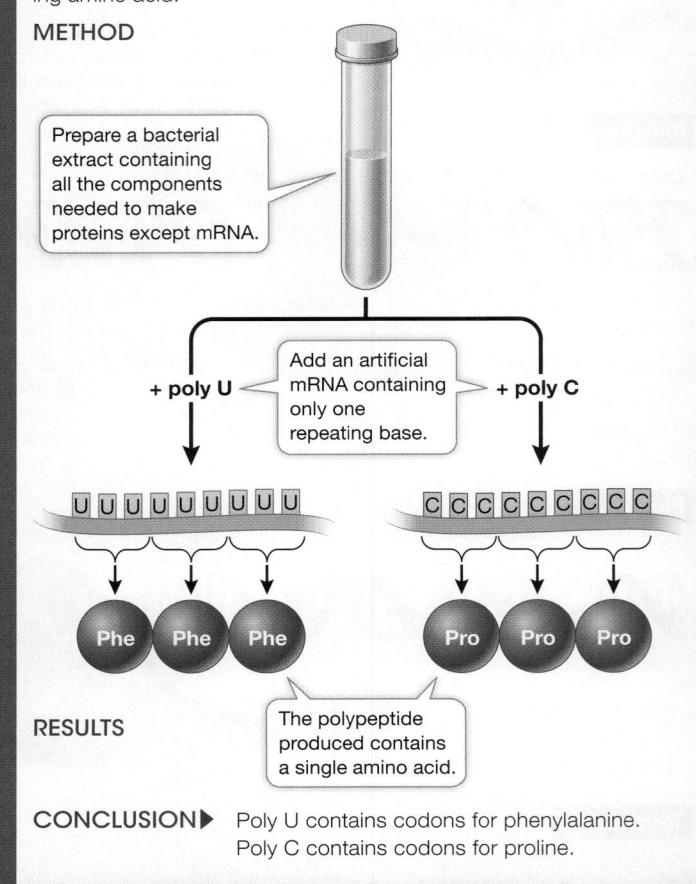

Original Paper: Nirenberg, M. and H. Matthaei. 1961. The dependence of cell-free protein synthesis in *E. coli* upon naturally occurring or synthetic polyribonucleotides. *Proceedings of the National Academy of Sciences USA* 47: 1588–1602.

investigating life   Deciphering the Genetic Code

experiment

Nirenberg and Matthaei used a test tube protein synthesis system to determine the amino acids specified by synthetic mRNAs of known compositions.

HYPOTHESIS▶   An artificial mRNA containing only one repeating base will direct the synthesis of a protein containing only one repeating amino acid.

METHOD

Prepare a bacterial extract containing all the components needed to make proteins except mRNA.

Add an artificial mRNA containing only one repeating base.

+ poly U

+ poly C

UUUUUUUUU

CCCCCCCCC

Phe Phe Phe

Pro Pro Pro

RESULTS

The polypeptide produced contains a single amino acid.

CONCLUSION▶   Poly U contains codons for phenylalanine. Poly C contains codons for proline.

Don't confuse a *redundant* code with an *ambiguous* code. If the code were ambiguous, a single codon could specify two (or more) different amino acids, and there would be doubt about which amino acid should be incorporated into a growing polypeptide chain. Redundancy in the code simply means there is more than one clear way to say "Put leucine here." The genetic code is not ambiguous: a given amino acid may be encoded by more than one codon, but a codon can code for only one amino acid.

THE GENETIC CODE IS (NEARLY) UNIVERSAL   The same basic genetic code is used by all the species on our planet. Thus the code must be an ancient one that has been maintained intact

After the relationship between DNA and proteins was established, genetic evidence pointed to triplets of nucleotides on RNA specifying each amino acid. The race was on to identify which triplet coded for which amino acid. Test tube systems were developed in which cell extracts were made and protein synthesis occurred. It was detected by supplying all 20 amino acids, with one of them being radioactive. Marshall Nirenberg, a scientist at the U.S. National Institutes of Health, and Heinrich Matthaei, a postdoctoral fellow from Germany, made a synthetic RNA consisting of the base uracil only (called poly U, codon UUU) and tested it in 20 tubes. Each test tube was supplied with all 20 amino acids, but in each one a different amino acid was tagged with a radioactive marker. In only one of them was a polypeptide made: a protein consisting of the amino acid phenylalanine bonded repeatedly to itself.

QUESTIONS▶

1. Poly U, an artificial mRNA, was added to a test tube with all the other components for protein synthesis ("complete system"). Other test tubes differed from the complete system as indicated in the table. Samples were tested for radioactive phenylalanine incorporation, with the results in **Table A**. Explain the results for each of the conditions.

Table A

Condition	Counts/minute (units of radioactivity)
Complete system	29,500
Minus poly U mRNA	70
Minus ribosomes	52
Minus ATP	83
Plus RNase (hydrolyzes RNA)	120
Plus DNase (hydrolyzes DNA)	27,600
Radioactive glycine instead of phenylalanine	33
Mixture of 19 radioactive amino acids minus phenylalanine	276

2. Poly U (red dots) was added to the test tubes at various intervals. Samples were tested for protein synthesis by radioactive amino acid incorporation after various times (and were compared with results of a control run, in which no RNA was added, indicated by blue dots). The results are shown in the figure below. What do these data show about the dependence of protein synthesis on added RNA?

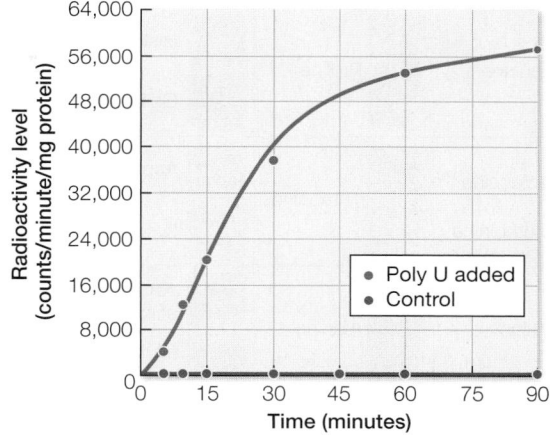

3. The experiment described in Question 2 was repeated with different amino acids; the results are in **Table B**. Explain these results in terms of the codon specificity of poly U.

Table B

Radioactive amino acid	Counts/minute/mg protein (radioactivity level)
Phenylalanine	38,300
Glycine, alanine, serine, aspartic acid, glutamic acid	33
Leucine, isoleucine, threonine, methionine, arginine, histidine, valine, lysine, tyrosine, proline, tryptophan	276
Cysteine	113

A similar **work with the data** exercise may be assigned in **LaunchPad**.

---

throughout the evolution of living organisms. Exceptions are known: for example, the code for ***mitochondrial DNA and chloroplast DNA** differs slightly from that used by prokaryotes and for the nuclear DNA of eukaryotic cells; and in one group of protists, UAA and UAG code for glutamine rather than for stop codons. The significance of these differences is not yet clear. What is clear is that the exceptions are few.

***connect the concepts**   As discussed in Key Concept 12.5, some cytoplasmic organelles, notably the mitochondria and chloroplasts, contain small numbers of genes that are remnants of the genomes of the prokaryotes that eventually gave rise to these organelles. The evolutionary process of endosymbiosis that is responsible for the assimilation of these organelles into eukaryotic cells is discussed in Key Concept 27.1.

A common genetic code provides a common language for evolution. Natural selection acts on phenotypic variations that result from genetic variation. The genetic code probably originated early in the evolution of life. As we saw in Chapter 4, simulation experiments indicate the plausibility of individual nucleotides and nucleotide polymers arising spontaneously on primeval Earth. The common code also has profound implications for genetic engineering, as we will see in Chapter 18, since it means that the code for a human gene is the same as that for a bacterial gene. It is therefore impressive, but not surprising, that a human gene can be expressed in *E. coli* via laboratory manipulations, since these cells speak the same "molecular language."

The codons illustrated in the Nirenberg and Matthaei experiment in Investigating Life: Deciphering the Genetic Code are mRNA codons. The base sequence of the DNA strand that is transcribed to produce the mRNA is complementary and antiparallel to these

First letter	Second letter: U	Second letter: C	Second letter: A	Second letter: G	Third letter
U	UUU Phenylalanine	UCU Serine	UAU Tyrosine	UGU Cysteine	U
U	UUC Phenylalanine	UCC Serine	UAC Tyrosine	UGC Cysteine	C
U	UUA Leucine	UCA Serine	**UAA** Stop codon	**UGA** Stop codon	A
U	UUG Leucine	UCG Serine	**UAG** Stop codon	UGG Tryptophan	G
C	CUU Leucine	CCU Proline	CAU Histidine	CGU Arginine	U
C	CUC Leucine	CCC Proline	CAC Histidine	CGC Arginine	C
C	CUA Leucine	CCA Proline	CAA Glutamine	CGA Arginine	A
C	CUG Leucine	CCG Proline	CAG Glutamine	CGG Arginine	G
A	AUU Isoleucine	ACU Threonine	AAU Asparagine	AGU Serine	U
A	AUC Isoleucine	ACC Threonine	AAC Asparagine	AGC Serine	C
A	AUA Isoleucine	ACA Threonine	AAA Lysine	AGA Arginine	A
A	**AUG** Methionine; start codon	ACG Threonine	AAG Lysine	AGG Arginine	G
G	GUU Valine	GCU Alanine	GAU Aspartic acid	GGU Glycine	U
G	GUC Valine	GCC Alanine	GAC Aspartic acid	GGC Glycine	C
G	GUA Valine	GCA Alanine	GAA Glutamic acid	GGA Glycine	A
G	GUG Valine	GCG Alanine	GAG Glutamic acid	GGG Glycine	G

Figure 14.5 The Genetic Code Genetic information is encoded in mRNA in three-letter units—codons—made up of nucleoside monophosphates with the bases uracil (U), cytosine (C), adenine (A), and guanine (G) and is read in a 5′-to-3′ direction on mRNA. To decode a codon, find its first letter in the left column, then read across the top to its second letter, then read down the right column to its third letter. The amino acid the codon specifies is given in the corresponding row. For example, AUG codes for methionine, and GUA codes for valine.

Activity 14.2 **The Genetic Code**
www.Life11e.com/ac14.2

codons. Thus, for example, 3′-AAA-5′ in the template DNA strand corresponds to phenylalanine (which is encoded by the mRNA codon 5′-UUU-3′).

- 3′-ACC-5′ in the template DNA corresponds to tryptophan (which is encoded by the mRNA codon 5′-UGG-3′).

The non-template strand of DNA has the same sequence as the mRNA (but with T's instead of U's), and is often referred to as the "coding strand." By convention, DNA sequences are usually shown beginning with the 5′ end of the coding sequence.

You might be thinking that in a long DNA molecule with many protein-coding regions, one strand is the coding strand for all the genes and the other strand is the template strand. In fact, some genes are transcribed from one strand and some genes are transcribed from the other. There is strand switching along a long DNA molecule in terms of the roles of the two strands. So while one strand may not be said to be "the" coding strand for all genes in a particular DNA molecule, for a given gene it is correct to say that one strand is coding and its complementary strand is template.

14.3 recap

Transcription, which is catalyzed by an RNA polymerase, proceeds in three steps: initiation, elongation, and termination. The genetic code relates the information in mRNA (as a linear sequence of codons) to protein (a linear sequence of amino acids).

14.3 recap

learning outcomes

You should be able to:

- Describe characteristics of RNA polymerases.
- Justify the need for a triplet code.
- Analyze the biological consequences of a triplet code.

1. Explain why the genetic code has triplets (e.g., AUA) of nucleotides, rather than singlets (e.g., A) or doublets (e.g., AU).
2. Describe the actions of RNA polymerase during transcription.
3. Errors in transcription occur about 100,000 times more often than errors in DNA replication. Why can this higher rate be tolerated in RNA but not in DNA synthesis?

The general features of transcription that we have described were first elucidated in model prokaryotes, such as *E. coli*. Biologists then used the same methods to analyze this process in eukaryotes, and although the basics are the same, there are some notable (and important) differences. We will now turn to a more detailed description of eukaryotic gene expression.

key concept 14.4 Eukaryotic Pre-mRNA Transcripts Are Processed prior to Translation

Since all organisms share the same genetic code, you might expect the process of gene expression to be the same in eukaryotes as it is in prokaryotes. And basically it is. In the last section, you learned about common features for transcription in all organisms. Our focus here is on the differences. **Table 14.2** lists the primary differences between prokaryotic and eukaryotic gene transcription.

focus your learning

- Eukaryotic mRNA transcripts undergo splicing and end modifications before translation.

In prokaryotes and eukaryotes, the sequence of an mRNA that reaches the ribosome is complementary to the sequence of a gene in the organism's DNA. One way to show this is by the technique of **nucleic acid hybridization**, shown in **Figure 14.6A**. This technique involves two steps:

1. A sample of chromosomal DNA containing the gene is denatured to break the hydrogen bonds between the base pairs and separate the two strands.
2. The single-strand mRNA (called a **probe**) is incubated with the denatured DNA. If the probe has a base sequence complementary to the target DNA, a probe–target double helix forms

(A) The process of nucleic acid hybridization

Target DNA

Denaturation

1 Upon being slowly heated or placed in a basic solution, the two strands of a DNA molecule denature (separate).

2 If a **probe** with a complementary base sequence is added to the denatured DNA...

Probe   3′

Hybridization

3 ...it binds the target DNA strand, forming a *double-stranded* hybrid molecule.

Figure 14.6 Nucleic Acid Hybridization and Introns   **(A)** Base pairing permits the detection of a sequence that is complementary to a probe. **(B)** Hybridization experiments show that there are introns in eukaryotic genes but generally not in prokaryotic genes.

(B) Hybridization experiments reveal noncoding introns in eukaryotic genes

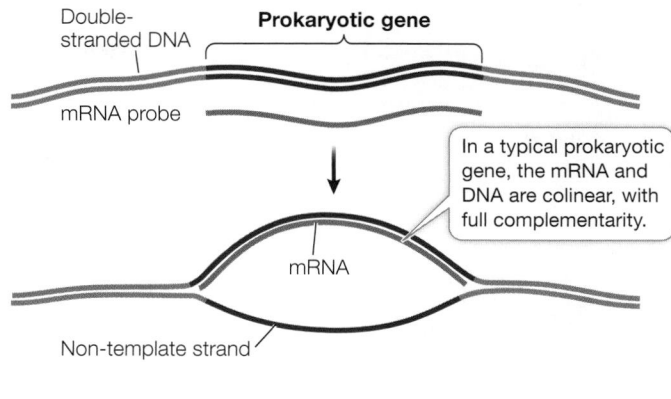

In a typical prokaryotic gene, the mRNA and DNA are colinear, with full complementarity.

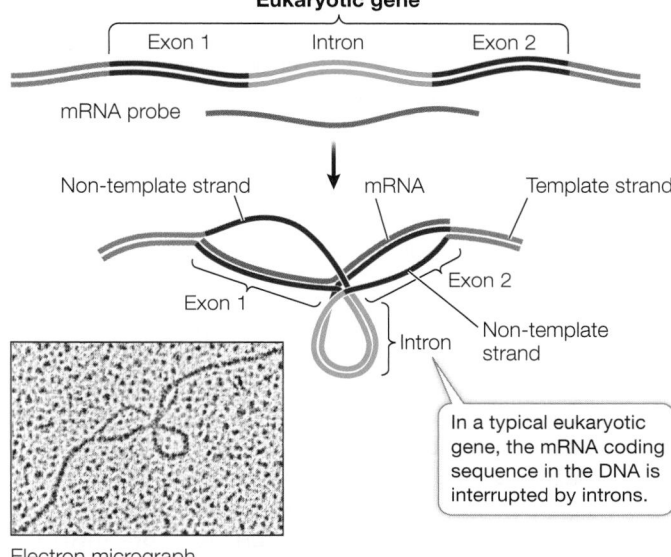

In a typical eukaryotic gene, the mRNA coding sequence in the DNA is interrupted by introns.

Electron micrograph of mRNA–DNA hybrid

by hydrogen bonding between the bases. Because the two strands are from different sources, the resulting double-strand region is called a hybrid.

Hybridization experiments can be performed with various combinations of DNA and RNA (RNA as target and DNA as probe; DNA as both target and probe, etc.). In many hybridization experiments, the probe is labeled in some way so that its binding to a specific target sequence can be detected. The double-strand hybrids can also be viewed by electron microscopy.

Noncoding sequences called introns often appear between genes in eukaryotic chromosomes

Differences between prokaryotic and eukaryotic transcription are revealed when mRNA probes from prokaryotes and eukaryotes are incubated with their respective chromosomal DNAs:

- In prokaryotes (**Figure 14.6B, top**), there is usually a 1:1 linear complementarity between the base sequence of the mRNA at the ribosome and that of the chromosomal DNA.

table **14.2** Differences between Prokaryotic and Eukaryotic Gene Expression		
Characteristic	Prokaryotes	Eukaryotes
Transcription and translation occurrence	At the same time in the cytoplasm	Transcription in the nucleus, then translation in the cytoplasm
Gene structure	DNA sequence is read in the same order as the amino acid sequence	Noncoding introns within coding sequence
Modification of mRNA after initial transcription but before translation	Usually none	Introns spliced out; 5′ end cap and 3′ poly A tail added

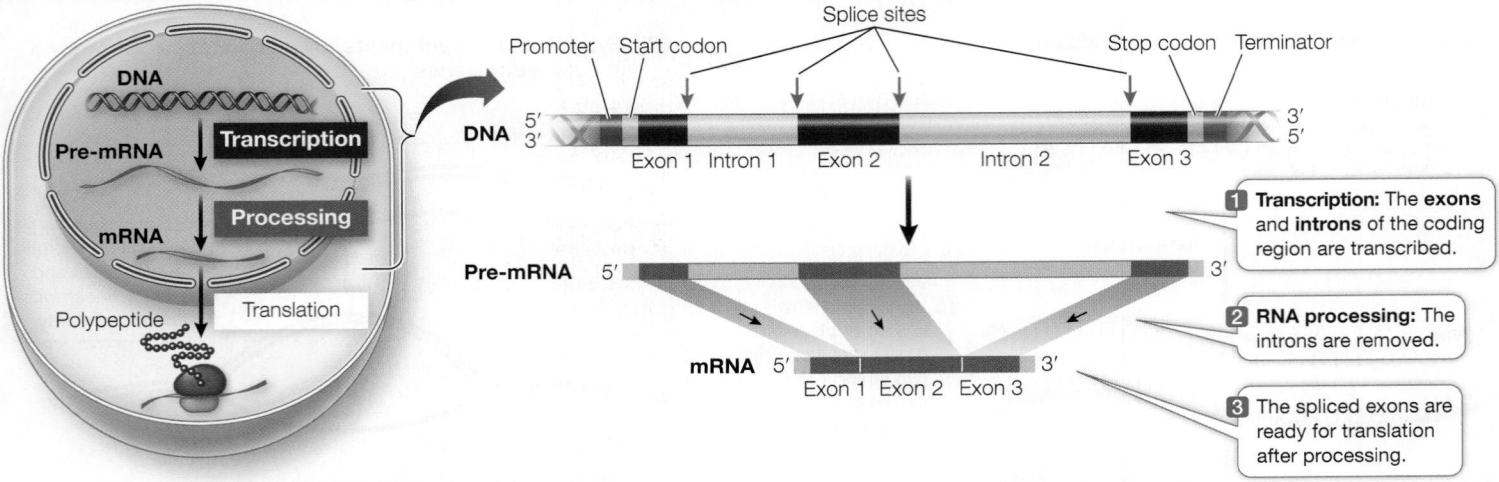

Figure 14.7 Transcription of a Eukaryotic Gene The β-globin gene diagrammed here is about 1,600 base pairs (bp) long. The three exons—the protein-coding sequences—contain codons for 146 amino acids plus a stop codon. The two introns—noncoding sequences of DNA containing almost 1,000 bp among them—are initially transcribed but are spliced out of the pre-mRNA transcript.

- In eukaryotes (**Figure 14.6B, bottom**), one or more non-hybridized DNA loops are often observed to extrude out of the mRNA–DNA hybrid, indicating that there are stretches of DNA sequence that do not have a complementary sequence in the mRNA that is translated at the ribosome.

The discovery that some stretches of eukaryotic DNA sequence are excluded from the mRNA that reaches the ribosome for translation initiated the question of whether this "extra" DNA actually gets transcribed. Does transcription "skip" these "extra" sequences, or are they transcribed and then somehow edited out of the mRNA transcript before it arrives at the ribosome? To find out, an experiment can be conducted in which the initial mRNA transcript in the cell nucleus—the **precursor mRNA**, or **pre-mRNA** (see Figure 14.2)—is hybridized with chromosomal DNA. If the pre-mRNA exhibits its full, linear, loopfree hybridization with the template DNA, we may conclude that the noncoding, intervening regions of DNA, known as **introns**, do get transcribed but are then spliced out of the pre-mRNA in the nucleus. The experiment demonstrates that only the coding sequences (**exons**) remain in the mRNA that reaches the ribosome. The step of editing out the introns is known as **RNA processing** and is shown in **Figure 14.7**.

Introns *interrupt, but do not scramble*, the DNA sequence of a gene. The base sequences of the exons in the template strand, if joined and taken in order, form a continuous sequence that is complementary to that of the mature mRNA. In some cases, the separated exons often encode different functional regions, or **domains**, of the protein. For example, the globin polypeptides that make up hemoglobin each have two domains: one for binding to a nonprotein pigment called heme, and another for binding to the other globin subunits. These two domains are encoded by different exons in the globin genes. Most (but not all) eukaryotic genes contain introns, and in rare cases, introns are also found in prokaryotes. The largest human gene encodes a muscle protein called titin; it has 363 exons, which together code for 38,138 amino acids.

Pre-mRNA processing prepares the mRNA transcript for translation

The transcript of a eukaryotic gene is modified in several ways before it leaves the nucleus: both ends of the pre-mRNA are modified, and the introns are removed.

MODIFICATION AT BOTH ENDS Two steps in the processing of pre-mRNA take place in the nucleus, one at each end of the molecule (**Figure 14.8**):

1. A **5′ cap** is added to the 5′ end of the pre-mRNA as it is transcribed. The 5′ cap, which is a chemically modified molecule of guanosine triphosphate (GTP), facilitates the binding of mRNA to the ribosome for translation, and it protects the mRNA from being digested by ribonucleases that break down RNAs.

2. A **poly A tail** is added to the 3′ end of the pre-mRNA at the end of transcription. Transcription ends downstream of the termination codon in DNA. In eukaryotes there is usually a "polyadenylation" sequence (AAUAAA) near the 3′ end of the pre-mRNA, after the last codon. This sequence acts as a signal for an enzyme to cut the pre-mRNA. Immediately after this cleavage, another

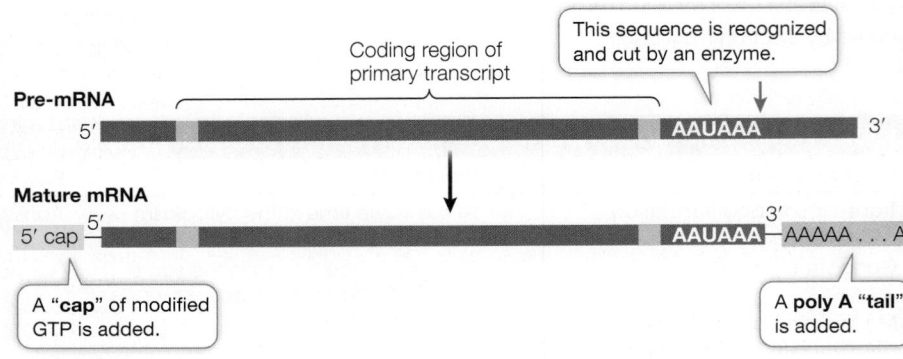

Figure 14.8 Processing the Ends of Eukaryotic Pre-mRNA Modifications at each end of the pre-mRNA transcript—the 5′ cap and the poly A tail—are important for mRNA function.

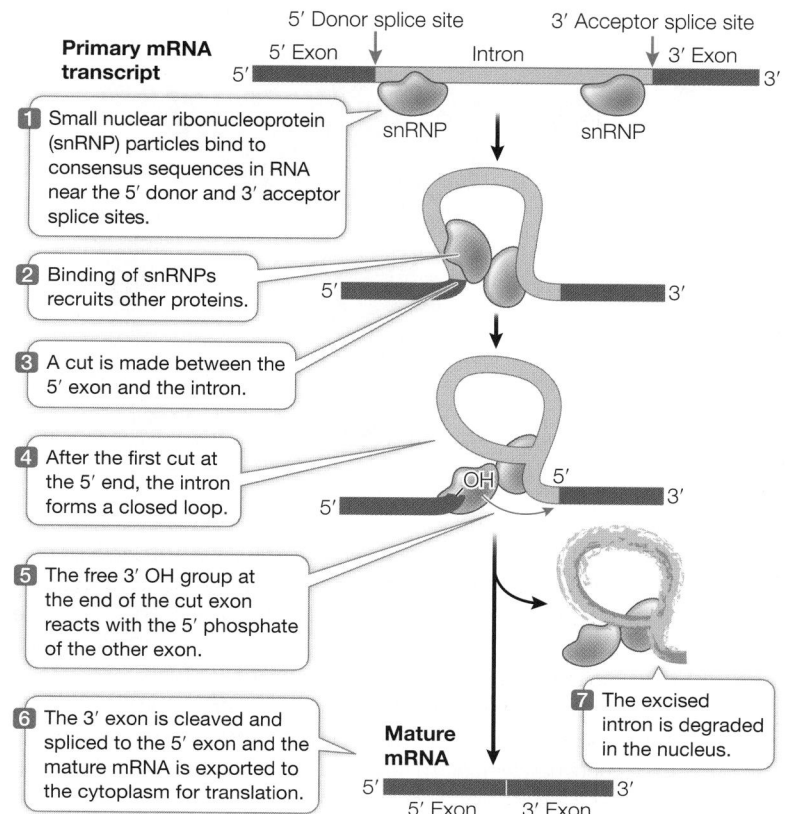

Primary mRNA transcript

5′ Donor splice site   3′ Acceptor splice site
5′ Exon   Intron   3′ Exon

1 Small nuclear ribonucleoprotein (snRNP) particles bind to consensus sequences in RNA near the 5′ donor and 3′ acceptor splice sites.

snRNP   snRNP

2 Binding of snRNPs recruits other proteins.

3 A cut is made between the 5′ exon and the intron.

4 After the first cut at the 5′ end, the intron forms a closed loop.

5 The free 3′ OH group at the end of the cut exon reacts with the 5′ phosphate of the other exon.

6 The 3′ exon is cleaved and spliced to the 5′ exon and the mature mRNA is exported to the cytoplasm for translation.

7 The excised intron is degraded in the nucleus.

Mature mRNA
5′ Exon   3′ Exon

Figure 14.9 RNA Splicing The binding of snRNPs to consensus sequences bordering the introns on the pre-mRNA results in a series of proteins binding and cutting the pre-mRNA with great precision.

 Animation 14.3 **RNA Splicing**
www.Life11e.com/a14.3

14.4 recap

Most eukaryotic genes contain noncoding sequences called introns, which are removed from the pre-mRNA transcript in the nucleus. The mature mRNA transcript is then exported through a nuclear pore to ribosomes in the cytoplasm, where translation takes place.

learning outcomes

You should be able to:
- Describe the mRNA processing steps in eukaryotes.
- Predict the outcome of a defect in mRNA processing in a eukaryotic cell.
- Compare the features of a eukaryotic gene, noting introns, exons, and signals, with its mature mRNA transcript.

1. How and why is the pre-mRNA transcript modified at the 5′ and 3′ ends during mRNA processing?
2. Outline the steps involved in RNA splicing. What are the consequences if it does not happen correctly?
3. A gene that is 1,440 base pairs long codes for a polypeptide that is 192 amino acids long. Discuss the discrepancy.

enzyme adds 100 to 300 adenine nucleotides (the poly A tail) to the 3′ end of the pre-mRNA. This tail helps in the export of mature mRNA from the nucleus and is important for mRNA stability.

SPLICING TO REMOVE INTRONS   Within the nucleus, introns are removed from eukaryotic pre-mRNA. If these RNA sequences were not removed, a very different amino acid sequence, and possibly a nonfunctional protein, would result. A process called **RNA splicing** removes the introns and splices the exons together. You can follow the steps of this important process in **Figure 14.9**.

Molecular studies of human genetic diseases have provided insights into RNA splicing. For example, people with the genetic disease beta thalassemia have a defect in the production of one of the hemoglobin subunits. These people suffer from severe anemia because they have an inadequate supply of red blood cells. In some cases, the genetic mutation that causes the disease occurs at an intron consensus sequence, where the splicing machinery binds to the RNA (see Figure 14.9, step 1) in the β-globin gene. Consequently, β-globin pre-mRNA cannot be spliced correctly, and β-globin mRNA that encodes a nonfunctional polypeptide is made. This finding offers another example of how biologists can use mutations to elucidate biological processes.

After processing is completed in the nucleus, the mature mRNA moves out into the cytoplasm through the nuclear pores. A protein bound to the 5′ nucleotide cap during processing is recognized by a receptor at the nuclear pore. Together, these proteins lead the mRNA through the pore. Unprocessed or incompletely processed pre-mRNAs remain in the nucleus.

Transcription and posttranscriptional events produce an mRNA that is ready to be translated into a sequence of amino acids in a polypeptide. We will turn now to the events of translation.

key concept 14.5 The Information in mRNA Is Translated into Proteins

When two languages are different (e.g., English and German) and you want to know what a word in one language means in the other, you may need to consult a foreign-language dictionary or an online translator. Likewise, a "translator" is required to convert the information in mRNA (a language composed of a sequence of nucleotides) into the language of protein (consisting of a sequence of amino acids). In biology, that translator is a special kind of RNA molecule called transfer RNA (tRNA). To ensure accurate translation—that is, to ensure that the protein made is the one specified by the mRNA—the tRNAs must (1) read each codon correctly, and (2) fetch the amino acids corresponding to each codon and deliver them to the ribosome.

focus your learning
- Ribosomes catalyze the translation of mRNA into polypeptide chains according to the following sequence of events: initiation, elongation, and termination.
- Polysomes allow simultaneous synthesis of multiple polypeptide chains from the same mRNA molecule.

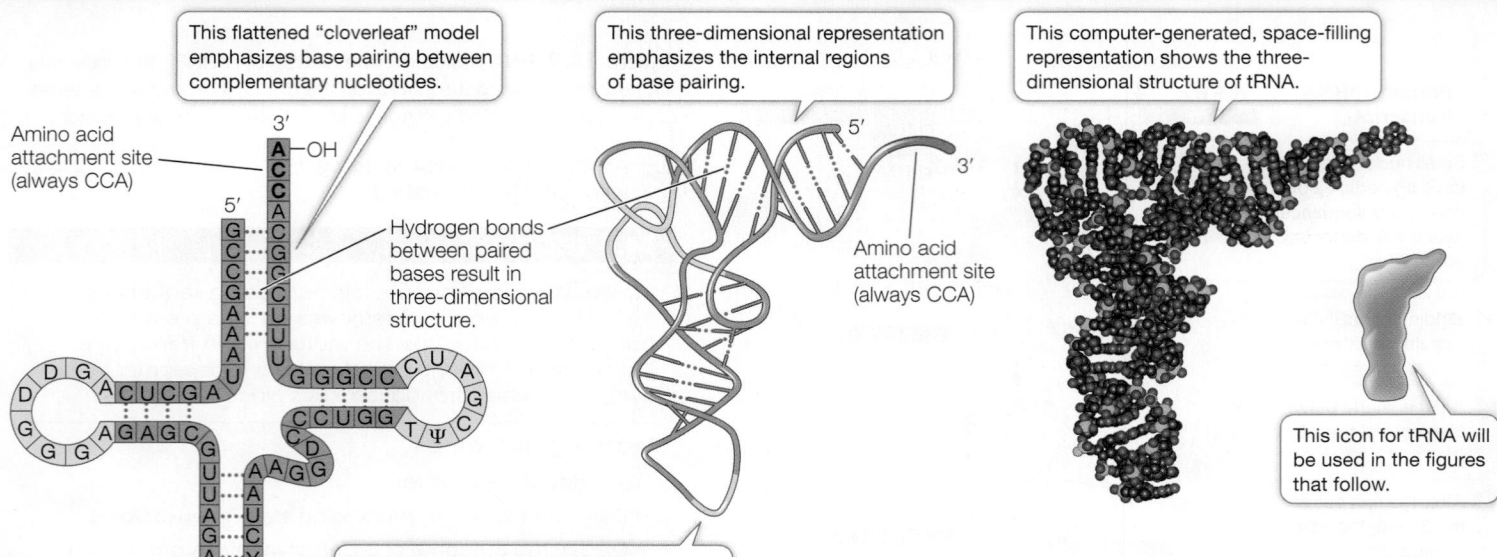

Figure 14.10 Transfer RNA The stem and loop structure of a tRNA molecule is well suited to its functions: binding to amino acids, associating with mRNA molecules, and interacting with ribosomes.

Once the tRNAs "decode" the mRNA and deliver the appropriate amino acids, components of the ribosome catalyze the formation of peptide bonds between amino acids. Let's now look at how the tRNAs read codons and get the appropriate amino acids into the ribosome.

 Animation 14.4 Translation
www.Life11e.com/a14.4

A transfer RNA carries a specific amino acid and binds to a specific mRNA codon

There is at least one specific tRNA molecule for each of the 20 amino acids. Each tRNA has three functions that are fulfilled by its structure and base sequence (**Figure 14.10**):

1. *tRNAs bind to particular amino acids.* Each tRNA binds to a specific enzyme that attaches it to only 1 of the 20 amino acids. The covalent attachment to an amino acid takes place at the 3' end of the tRNA. When it is carrying an amino acid, the tRNA is said to be "charged."

2. *tRNAs bind to mRNA.* At about the midpoint on the tRNA polynucleotide chain there is a triplet of bases called the **anticodon**, which is complementary to the mRNA codon for the particular amino acid that the tRNA carries. For example, the mRNA codon for arginine is 5'-CGG-3', and the complementary tRNA anticodon is 3'-GCC-5'. Like the two strands of DNA, the codon and anticodon bind together via noncovalent hydrogen bonds.

3. *tRNAs interact with ribosomes.* The ribosome has several sites on its surface that just fit the three-dimensional structure of a tRNA molecule. Interaction between the ribosome and the tRNA is noncovalent.

Recall that 61 different codons encode the 20 amino acids in proteins (see Figure 14.5). Does this mean that the cell must produce 61 different tRNA species, each with a different anticodon? No. The cell gets by with about two-thirds of that number of tRNA species because the specificity for the base at the 3' end of the codon (and the 5' end of the anticodon) is not always strictly observed. This phenomenon is called wobble, and it is possible because in some cases unusual or modified nucleotide bases occur in the 5' position of the anticodon. One such unusual base is inosine (I), which can pair with A, C, and U. For example, the presence of inosine in the tRNA with the anticodon 3'-CGI-5' allows it to recognize and bind to three of the alanine codons: GCA, GCC, and GCU. Wobble occurs in some matches but not in others; of most importance, it does not allow the genetic code to be ambiguous. That is, *each mRNA codon binds to just one tRNA species, carrying a specific amino acid.*

Each tRNA is specifically attached to an amino acid

The charging of each tRNA with its correct amino acid is achieved by a family of enzymes known as aminoacyl-tRNA synthetases. Each enzyme is specific for one amino acid and for its corresponding tRNA. The reaction uses ATP, forming a high-energy bond between the amino acid and the tRNAs (**Figure 14.11**). The energy in this bond is later used in the formation of peptide bonds between amino acids in a growing polypeptide chain.

The specificity between the tRNA and its corresponding amino acid is essential. These reactions, for example, are highly specific:

$$\text{Cysteine} + \text{tRNAcys} \xrightarrow{\text{Cys-tRNA synthetase}} \text{cys-tRNA}$$
(anticodon ACA)

$$\text{Alanine} + \text{tRNAala} \xrightarrow{\text{Ala-tRNA synthetase}} \text{ala-tRNA}$$
(anticodon CGA)

A clever experiment by Seymour Benzer and his colleagues at Purdue University demonstrated the importance of this specificity. They took the cys-tRNA molecule and chemically modified the cysteine, converting it into alanine. Which component—the amino acid or the tRNA—would be recognized when this hybrid charged tRNA

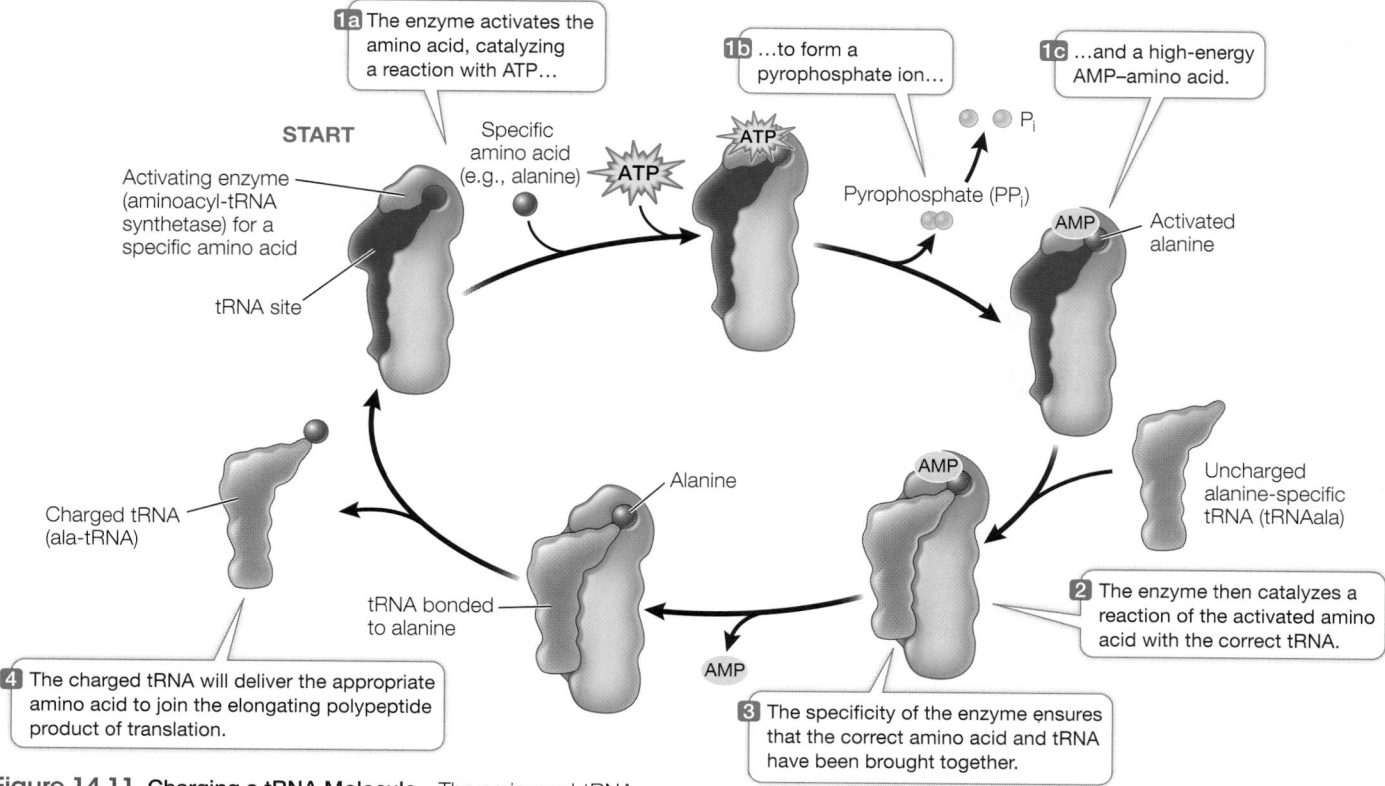

1a The enzyme activates the amino acid, catalyzing a reaction with ATP...

1b ...to form a pyrophosphate ion...

1c ...and a high-energy AMP–amino acid.

START

Activating enzyme (aminoacyl-tRNA synthetase) for a specific amino acid

Specific amino acid (e.g., alanine)

tRNA site

Pyrophosphate (PP$_i$)

Activated alanine

Charged tRNA (ala-tRNA)

Alanine

Uncharged alanine-specific tRNA (tRNAala)

tRNA bonded to alanine

2 The enzyme then catalyzes a reaction of the activated amino acid with the correct tRNA.

4 The charged tRNA will deliver the appropriate amino acid to join the elongating polypeptide product of translation.

3 The specificity of the enzyme ensures that the correct amino acid and tRNA have been brought together.

Figure 14.11 Charging a tRNA Molecule The aminoacyl-tRNA synthetase activates a specific amino acid and charges a specific tRNA with that amino acid.

was put into a protein synthesizing system? The answer was the tRNA. Everywhere in the synthesized protein where cysteine was supposed to be, alanine appeared instead. The cysteine-specific tRNA had delivered its cargo (alanine) to every mRNA codon for cysteine. This experiment showed that the protein synthesis machinery recognizes the anticodon of the charged tRNA, not the amino acid attached to it.

The ribosome is the workbench for translation

The ribosome is the molecular workbench where the task of translation is accomplished. Its structure enables it to hold mRNA and

charged tRNAs in the correct positions, thus allowing a polypeptide chain to be assembled efficiently. A given ribosome does not specifically produce just one kind of protein. A ribosome can use any mRNA and all species of charged tRNAs, and thus can be used to make many different polypeptide products. Ribosomes can be used over and over again, and there are thousands of them in a typical cell.

Although ribosomes are small relative to other cellular structures, their mass of several million daltons makes them large in comparison with charged tRNAs. Each ribosome consists of two subunits, a large one and a small one (**Figure 14.12**). The two subunits and several dozen other molecules interact noncovalently. In fact, when hydrophobic interactions between the proteins and RNAs

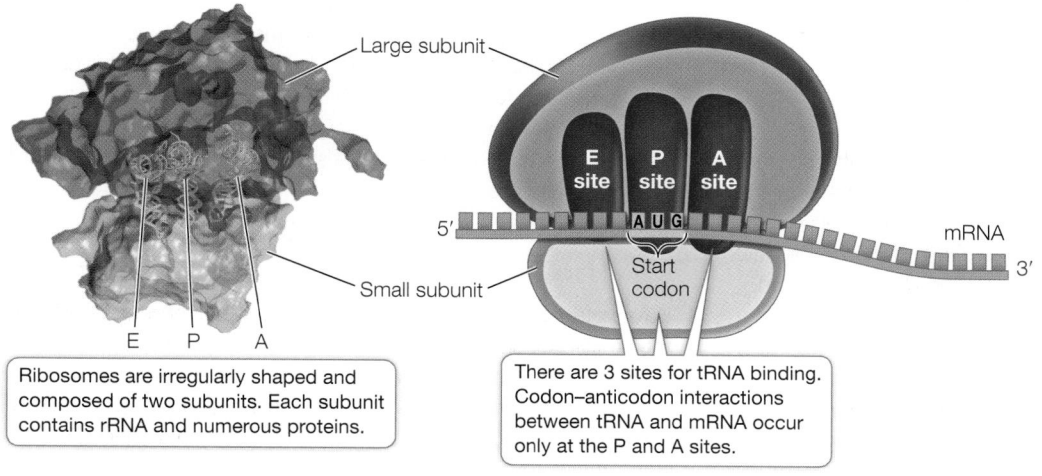

Large subunit

E site   P site   A site

5′   AUG   mRNA   3′

Start codon

Small subunit

E   P   A

Ribosomes are irregularly shaped and composed of two subunits. Each subunit contains rRNA and numerous proteins.

There are 3 sites for tRNA binding. Codon–anticodon interactions between tRNA and mRNA occur only at the P and A sites.

Figure 14.12 Ribosome Structure Each ribosome consists of a large and a small subunit. The subunits remain separate when they are not being used for protein synthesis.

Q: The ribosome consists of several dozen proteins and several RNA molecules, held together noncovalently. What are the chemical forces involved? How can these forces be disrupted and the molecules separated from one another?

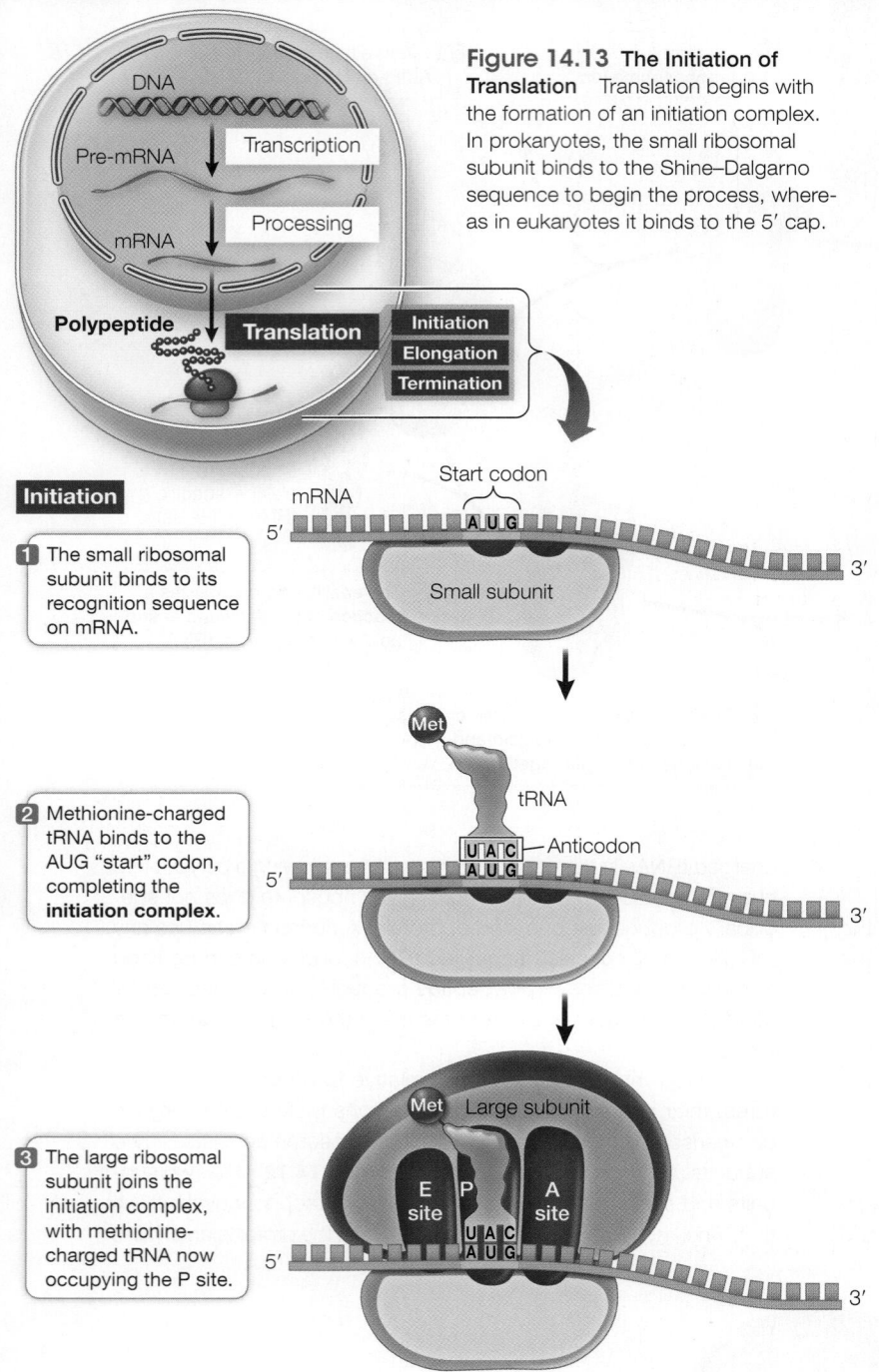

Figure 14.13 The Initiation of Translation Translation begins with the formation of an initiation complex. In prokaryotes, the small ribosomal subunit binds to the Shine–Dalgarno sequence to begin the process, whereas in eukaryotes it binds to the 5′ cap.

DNA

Transcription

Pre-mRNA

Processing

mRNA

Polypeptide

Translation — Initiation / Elongation / Termination

Initiation

1 The small ribosomal subunit binds to its recognition sequence on mRNA.

Start codon
mRNA
5′ AUG 3′
Small subunit

2 Methionine-charged tRNA binds to the AUG "start" codon, completing the **initiation complex**.

Met
tRNA
UAC — Anticodon
AUG
5′ 3′

3 The large ribosomal subunit joins the initiation complex, with methionine-charged tRNA now occupying the P site.

Met Large subunit
E site P A site
UAC
AUG
5′ 3′

and their ribosomal proteins and RNAs are different. Mitochondria and chloroplasts also contain ribosomes, some of which are similar to those of prokaryotes (see Chapter 5).

There are three sites to which a tRNA can bind on the large subunit of the ribosome, designated A, P, and E (see Figure 14.12). The mRNA and ribosome move in relation to one another, and as they do so, a charged tRNA traverses these three sites in order:

1. The *A (aminoacyl-tRNA) site* is where the charged tRNA anticodon binds to the mRNA codon, thus lining up the correct amino acid to be added to the growing polypeptide chain.

2. The *P (peptidyl-tRNA) site* is where the tRNA adds its amino acid to the polypeptide chain.

3. The *E (exit) site* is where the tRNA, having given up its amino acid, resides before being released from the ribosome and going back to the cytosol to pick up another amino acid and begin the process again.

The ribosome has a fidelity function that ensures that the mRNA–tRNA interactions are accurate; that is, that a charged tRNA with the correct anticodon (e.g., 3′-UAC-5′) binds to the appropriate codon in mRNA (e.g., 5′-AUG-3′). When proper binding occurs, hydrogen bonds form between the paired bases. The rRNA of the small ribosomal subunit plays a role in validating the three-base-pair match. If hydrogen bonds have not formed between all three base pairs, the tRNA must be the wrong one for that mRNA codon, and the incorrect tRNA is ejected from the ribosome.

Translation takes place in three steps

Translation is the process by which the information in mRNA (derived from DNA) is used to specify and link a specific sequence of amino acids, producing a polypeptide. Like transcription, translation occurs in three steps: initiation, elongation, and termination.

INITIATION The translation of mRNA begins with the formation of an **initiation complex**, which consists of a charged tRNA and a small ribosomal subunit, both bound to the mRNA (**Figure 14.13**).

In prokaryotes, the rRNA of the small ribosomal subunit first binds to a complementary ribosome binding site (AGGAGG; known as the Shine–Dalgarno sequence) on the mRNA. This sequence is less than 10 bases upstream of the actual start codon but lines up the start codon so that it is adjacent to the P site of the large subunit:

mRNA 5′………A G G A G G……(start codon)…..3′

rRNA 3′………. U C C U C C………(P site)……….5′

Eukaryotes load the mRNA onto the ribosome somewhat differently: the small ribosomal subunit binds to the 5′ cap on the mRNA and then moves along the mRNA until it reaches the start codon.

Recall that the mRNA start codon in the genetic code is AUG (see Figure 14.5). The anticodon (UAC) of a methionine-charged tRNA binds to this start codon by complementary base pairing to complete the initiation complex. Thus the first amino acid in a

are disrupted, the ribosome falls apart. The two subunits separate and all the RNAs and proteins separate from one another. If the disrupting agent is removed, the complex structure self-assembles perfectly! This is amazing: think of throwing the pieces of a jigsaw puzzle up in the air and having them fit together when they land. The ribosome reflects the high specificity of molecular machines in a cell composed of many molecules.

In eukaryotes, the large subunit consists of three different molecules of ribosomal RNA (rRNA) and 49 different protein molecules, arranged in a precise configuration. The small subunit consists of 1 rRNA molecule and 33 different protein molecules. The ribosomes of prokaryotes are somewhat smaller than those of eukaryotes,

polypeptide chain is always methionine. However, not all mature proteins have methionine as their N-terminal amino acid. In many cases, the initial methionine is removed by an enzyme after translation.

After the methionine-charged tRNA has bound to the mRNA, the large subunit of the ribosome joins the complex. The methionine-charged tRNA now lies in the P site of the ribosome, and the A site is aligned with the second mRNA codon. These ingredients—mRNA, two ribosomal subunits, and methionine-charged tRNA—are assembled by a group of proteins called initiation factors.

ELONGATION   A charged tRNA whose anticodon is complementary to the second codon of the mRNA now enters the open A site of the large ribosomal subunit (**Figure 14.14**). The large subunit then catalyzes two reactions:

1. It breaks the bond between the tRNA and its amino acid in the P site.

2. It catalyzes the formation of a peptide bond between the amino acid that has just been released from the P site and the one attached to the tRNA in the A site.

Because the large ribosomal subunit performs these two actions, it is said to have **peptidyl transferase** activity. In this way, methionine (the amino acid in the P site) becomes the N terminus of the new protein. The second amino acid is now bound to methionine but remains attached to its tRNA at the A site.

How does the large ribosomal subunit catalyze peptide bond formation? Harry Noller and his colleagues at the University of California at Santa Cruz did a series of experiments and found that:

- if they removed almost all of the proteins from the large subunit, it still catalyzed peptide bond formation.

- if the rRNA was extensively modified, peptidyl transferase activity was destroyed.

The experiment showed that *rRNA is the catalyst*. The purification and crystallization of ribosomes has allowed scientists to examine ribosome structure in detail, and the catalytic role of rRNA in peptidyl transferase activity has been confirmed. These findings support the hypothesis that RNA, and *catalytic RNA in particular, evolved before DNA.

After the first tRNA releases its methionine, it moves to the E site and is then dissociated from the ribosome, returning to the cytosol to become charged with another methionine. The second tRNA, now bearing a dipeptide (a two-amino acid chain), is shifted to the P site as the ribosome moves one codon along the mRNA in the 5′-to-3′ direction. The elongation process continues, and the polypeptide chain grows, as these steps are repeated. Follow the process in Figure 14.14. All these steps are assisted by ribosomal proteins called elongation factors.

---

*connect the concepts   As discussed in Key Concept 4.3, the folded, three-dimensional surface of an RNA molecule can be just as specific as that of a

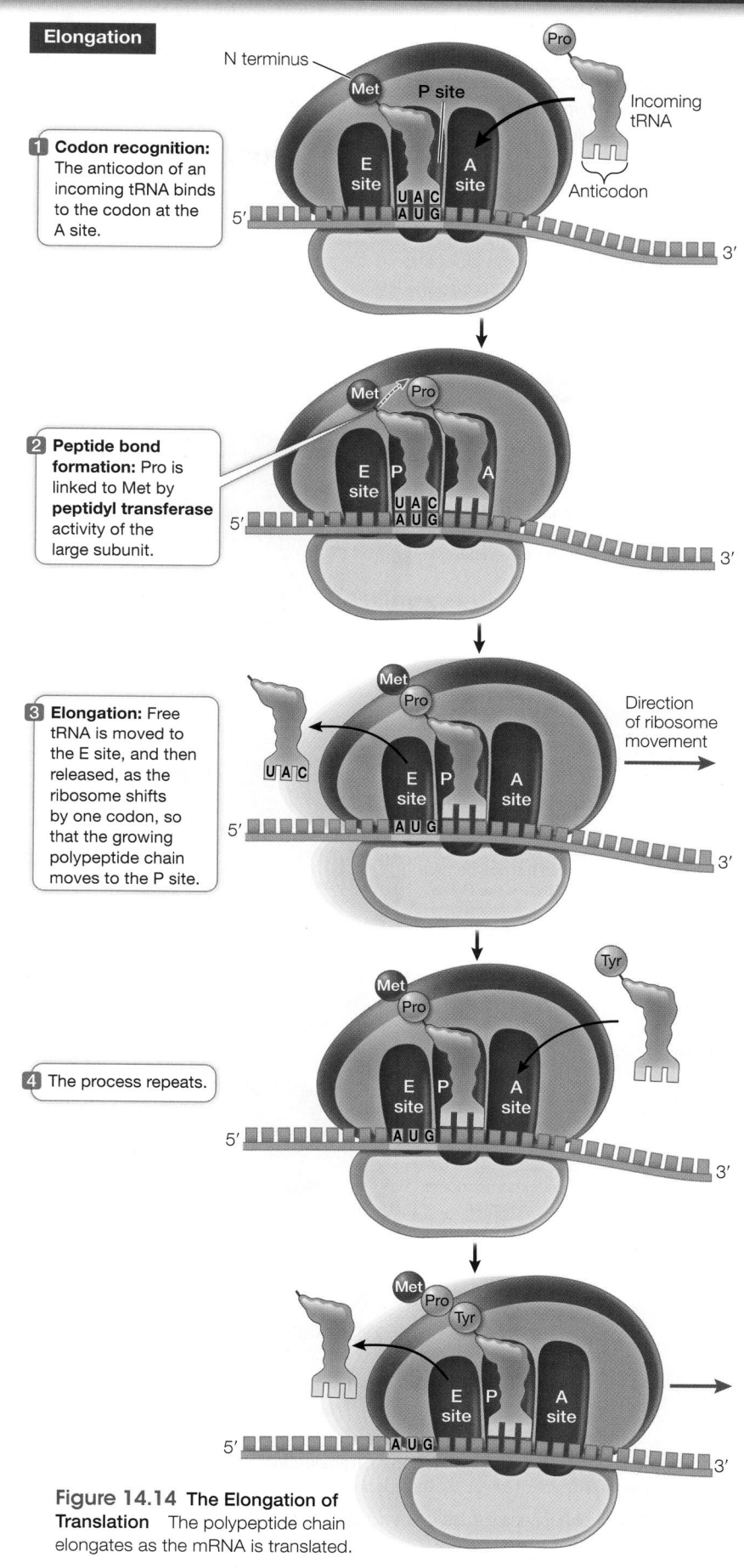

Elongation

1 **Codon recognition:** The anticodon of an incoming tRNA binds to the codon at the A site.

2 **Peptide bond formation:** Pro is linked to Met by **peptidyl transferase** activity of the large subunit.

3 **Elongation:** Free tRNA is moved to the E site, and then released, as the ribosome shifts by one codon, so that the growing polypeptide chain moves to the P site.

4 The process repeats.

Figure 14.14 The Elongation of Translation The polypeptide chain elongates as the mRNA is translated.

protein, and may thus take on a catalytic function. The "RNA world" hypothesis, which proposed that RNA serves as a catalyst for its own replication, was boosted by the discovery of ribozymes, catalytic RNAs that can speed up biological reactions, including those that involve their own nucleotides.

TERMINATION   The elongation cycle ends, and translation is terminated, when a stop codon—UAA, UAG, or UGA—enters the A site (**Figure 14.15**). Stop codons do not correspond with any amino

table **14.3**	Signals that Start and Stop Transcription and Translation	
	Transcription	Translation
Initiation	Promoter DNA	AUG start codon in the mRNA
Termination	Terminator DNA	UAA, UAG, or UGA in the mRNA

acids, nor do they bind any tRNAs. Rather, they bind a protein release factor, which allows hydrolysis of the bond between the polypeptide chain and the tRNA in the P site. The newly completed polypeptide thereupon separates from the ribosome. Its C terminus is the last amino acid to join the chain. Its N terminus, at least initially, is methionine, as a consequence of the AUG start codon. In its amino acid sequence, it contains information specifying its conformation, as well as its ultimate cellular destination.

Table 14.3 lists the nucleic acid signals for initiation and termination of transcription and translation.

Polysome formation increases the rate of protein synthesis

Several ribosomes can work simultaneously at translating a single mRNA molecule, producing multiple polypeptides at the same time. As soon as the first ribosome has moved far enough from the site of translation initiation, a second initiation complex can form, then a third, and so on. An assemblage consisting of a strand of mRNA with its beadlike ribosomes and their growing polypeptide chains is called a **polyribosome**, or **polysome** (**Figure 14.16**). Cells that are

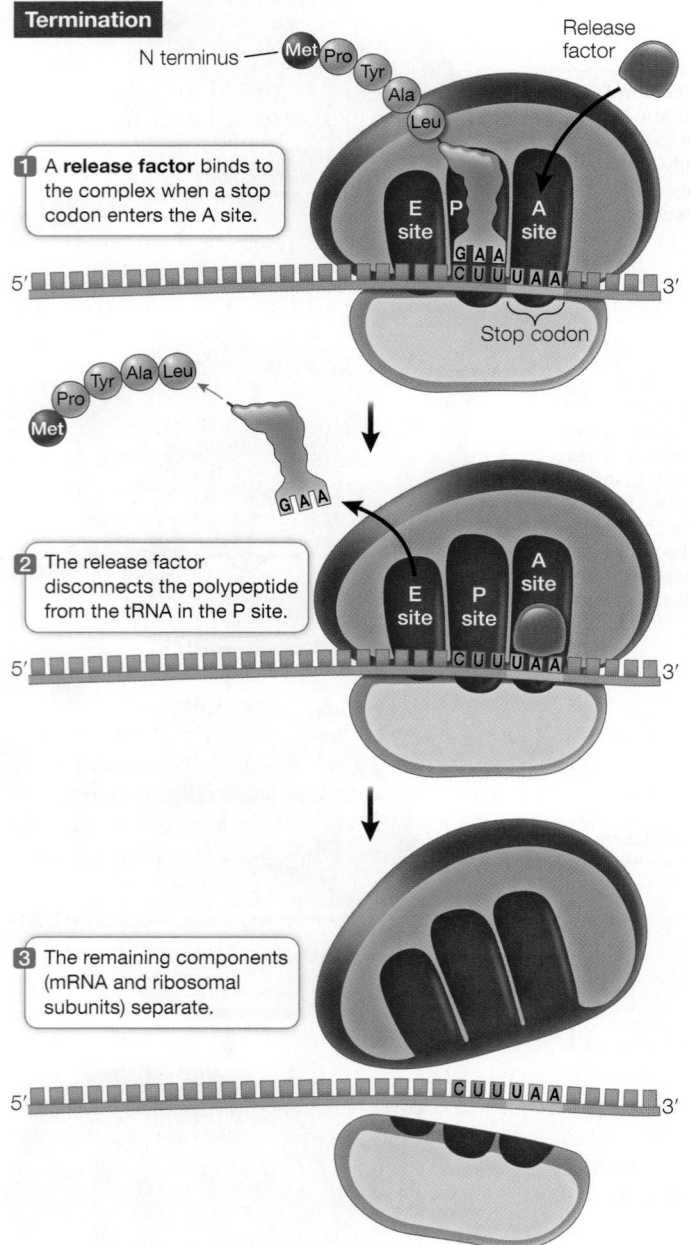

Termination

N terminus

1 A **release factor** binds to the complex when a stop codon enters the A site.

Release factor

Stop codon

2 The release factor disconnects the polypeptide from the tRNA in the P site.

3 The remaining components (mRNA and ribosomal subunits) separate.

Figure 14.15 The Termination of Translation Translation terminates when the A site of the ribosome encounters a stop codon on the mRNA.

Q: What happens if there is not a stop codon?

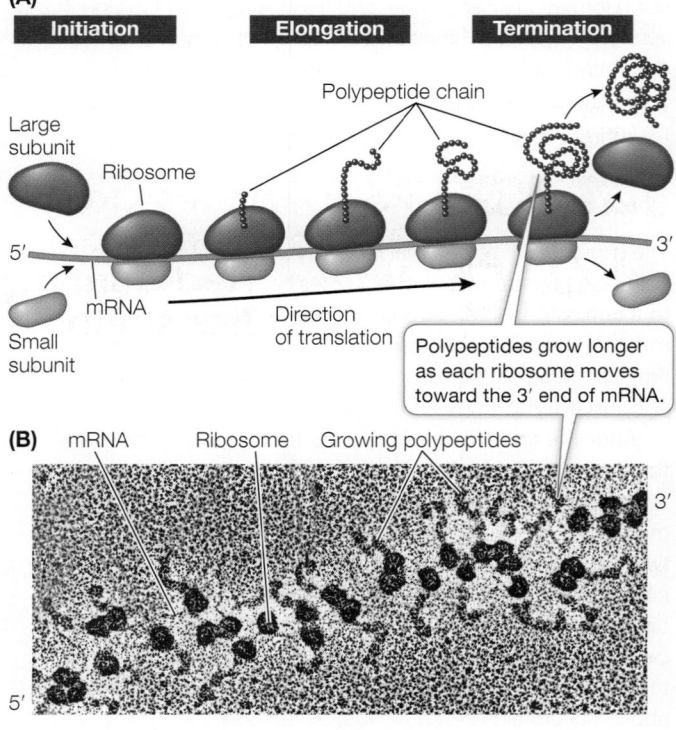

(A)

Initiation   Elongation   Termination

Polypeptide chain

Large subunit

Ribosome

Small subunit

mRNA

Direction of translation

Polypeptides grow longer as each ribosome moves toward the 3′ end of mRNA.

(B)   mRNA   Ribosome   Growing polypeptides

Figure 14.16 A Polysome **(A)** A polysome consists of multiple ribosomes and their growing polypeptide chains moving along an mRNA molecule. **(B)** An electron micrograph of a polysome.

actively synthesizing proteins contain large numbers of polysomes and few free ribosomes or ribosomal subunits.

14.5 recap

A key step in protein synthesis is the attachment of an amino acid to its proper tRNA, which is facilitated by a family of specific activating enzymes called aminoacyl-tRNA synthetases. Translation of the genetic information from mRNA into protein occurs at the ribosome in three stages: initiation, elongation, and termination. Multiple ribosomes may act on a single mRNA to make multiple copies of the protein that it encodes.

learning outcomes

You should be able to:

• Describe the process of translation.
• Translate a sequence of DNA into a peptide.
• Explain the biological significance of polysomes.

1. What are the roles of rRNA molecules in the ribosome?
2. Given the DNA sequence:

    5'-ATGCCCGGGTTAAGATATTTAAATGA-3',

    a. write out the sequence of the complementary DNA strand.
    b. indicate which strand is used as a template for transcription (and how you know this).
    c. write out the sequence of the transcribed mRNA, and provide the amino acid sequence of the translated peptide.
3. What are the structure and significance of a polysome?

The polypeptide chain that is released from the ribosome is not necessarily a functional protein. Let's look at some of the posttranslational changes that can affect the fate and function of a polypeptide.

key concept 14.6 Polypeptides Can Be Modified and Transported during or after Translation

The site of a polypeptide's function may be far away from its point of synthesis in the cytoplasm. This is especially true for eukaryotes. The polypeptide may be moved into an organelle, or even out of the cell. In addition, polypeptides are often modified by the addition of new chemical groups that have functional significance. In this section we examine these posttranslational aspects of protein synthesis.

focus your learning

• Direct transport of a protein within a cell or its export out of a cell can be specified by signal sequences attached to the protein.
• Posttranslational modifications of proteins can involve proteolytic cleavage, glycosylation, phosphorylation, and other chemical modifications.

How are proteins directed to their cellular destinations?

As a polypeptide chain emerges from the ribosome it may simply fold into its three-dimensional shape and perform its cellular role locally in the cytosol. However, if a newly formed polypeptide is meant to do its work elsewhere, a **signal sequence** (or **signal peptide**)—a short stretch of amino acids attached to the polypeptide—will tell the polypeptide where in the cell it belongs. Proteins destined for different locations have different signals.

Protein synthesis always begins on free ribosomes, and the "default" location for a protein is the cytosol. In the absence of a signal sequence, the protein will remain in the same cellular compartment in which it was synthesized. Some proteins contain signal sequences that "target" them to the nucleus, mitochondria, plastids, or peroxisomes (**Figure 14.17**). A signal sequence binds to a specific receptor protein at the surface of the organelle. Once it has bound, the targeted protein moves into the organelle. For example, here is a nuclear localization signal (NLS):

-Pro-Pro-Lys-Lys-Lys-Arg-Lys-Val-

How do we know this signal sequence directs the protein to the nucleus? The function of this NLS peptide was established using experiments like the one illustrated in **Figure 14.18**. Proteins were made in the laboratory with or without the peptide, and then tested by injecting them into cells. Only proteins with the NLS were found in the nucleus.

If a polypeptide carries a signal of about 20 hydrophobic amino acids at its N terminus, it will be directed to the rough endoplasmic reticulum (RER) for further processing (see Figure 14.17). Note that this is not a specific sequence of amino acids, just a generally hydrophobic sequence at the N terminus that is first translated. Translation will pause, and the ribosome will bind to a receptor at the RER membrane. Once the polypeptide–ribosome complex is bound, translation will resume, and as elongation continues, the protein will traverse the RER membrane. Such proteins may be retained in the lumen (the inside of the RER) or in membrane of the RER, or they may move elsewhere within the endomembrane system (Golgi apparatus, lysosomes, and cell membrane). If the proteins lack specific signals or modifications (see below) that specify destinations within the endomembrane system, they are usually secreted from the cell via vesicles that fuse with the cell membrane.

The importance of signals is shown by Inclusion-cell (I-cell) disease, an inherited disease that causes death in early childhood. People with this disease have a mutation in the gene encoding a Golgi enzyme that adds specific sugars to proteins destined for the lysosomes. These sugars act like signal sequences; without them, enzymes that are essential for the hydrolysis of various macromolecules cannot reach the lysosomes, where the enzymes are normally active. Without these enzymes, the macromolecules accumulate in the lysosomes, and this lack of cellular recycling has drastic effects, resulting in early death.

Mitochondria and chloroplasts make some of their own proteins and import others

As you saw in Chapter 5, mitochondria and chloroplasts are semiautonomous organelles, each having a small amount of DNA. Both

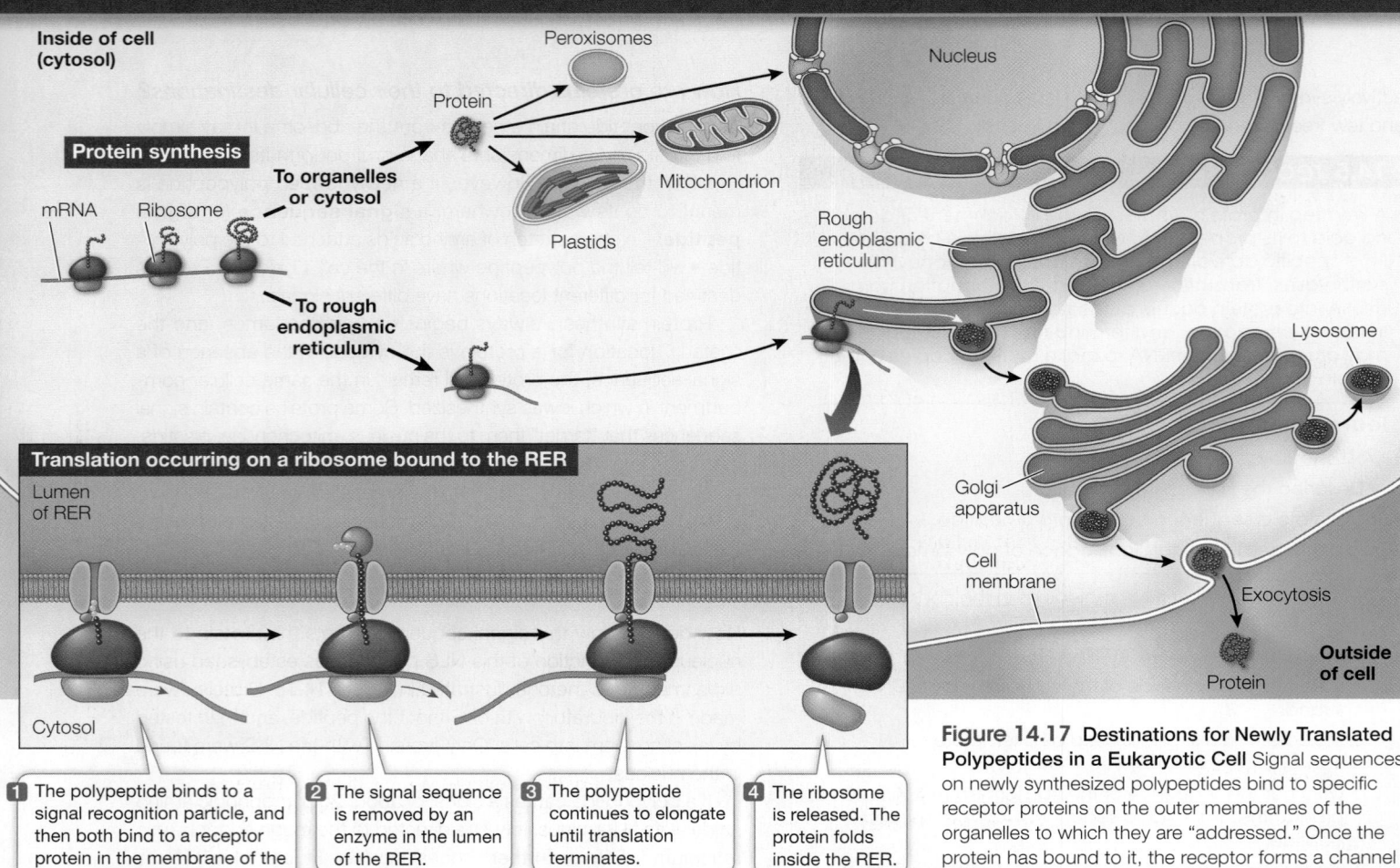

Figure 14.17 Destinations for Newly Translated Polypeptides in a Eukaryotic Cell Signal sequences on newly synthesized polypeptides bind to specific receptor proteins on the outer membranes of the organelles to which they are "addressed." Once the protein has bound to it, the receptor forms a channel in the membrane, and the protein enters the organelle.

Q: What happens to a protein that has no amino acid sequence "address"?

organelles contain the entire protein synthesis machinery, including a specific RNA polymerase, and their own specialized ribosomes that differ from the ones in the cytoplasm. The genetic coding capacity of the organelle DNA is small, however:

• Mitochondrial DNA encodes mitochondrial rRNAs, a few tRNAs, and 12 to 20 proteins, most of them involved in the electron transport chain.

• Chloroplast DNA encodes chloroplast rRNAs, some tRNAs, and about 40 proteins, many of which are involved in photosynthesis.

These organelles contain dozens more proteins, all of which are imported from the nuclear–cytoplasm protein synthesis system. In some cases (e.g., rubisco in the chloroplast) an organelle protein has several subunits, some of which are made locally inside the organelle and others of which are imported from the nucleus–cytoplasm system. As you can imagine, putting this together requires coordination between the two protein synthesis systems.

Many proteins are modified after translation

The amino acid sequences of most mature proteins are not identical to those in the polypeptide chains that are translated from mRNA on the ribosomes, because polypeptides are often modified in any of several ways after translation (**Figure 14.19**). These modifications are essential to the final functioning of the protein:

• **Proteolysis** is the cutting of a polypeptide chain, a reaction catalyzed by enzymes called proteases (also called peptidases or proteinases). Cleavage of the signal sequence from the growing polypeptide chain in the RER is an example of proteolysis (see Figure 14.17); the protein might move back out of the RER through the membrane channel if the signal sequence were not cut off. Some proteins are actually made from polyproteins (long polypeptides) that are cut into final products by proteases. These proteases are essential to some viruses, including human immunodeficiency viruses (HIVs), because the large viral polyprotein cannot fold properly unless it is cut. Certain drugs used to treat acquired immune deficiency syndrome (AIDS) work by inhibiting the HIV protease, thereby preventing the formation of proteins needed for viral reproduction.

• **Glycosylation** is the addition of sugars to proteins to form glycoproteins. In both the RER and the Golgi apparatus, resident enzymes catalyze the addition of various sugars or short sugar chains to certain amino acid R groups on proteins. One such type of "sugar coating" is essential for directing proteins to lysosomes, as mentioned above. Other types are important in the conformation of proteins and their recognition functions at the cell surface (see Key Concept 6.2). Other attached sugars help stabilize extracellular proteins, or proteins stored in vacuoles in plant seeds.

experiment

Figure 14.18 Testing the Signal

Original Paper: Dingwall, C. et al. 1988. The nucleoplasmin nuclear location sequence is larger and more complex than that of SV-40 large T antigen. *Journal of Cell Biology* 107: 841–849.

A. Richardson and his colleagues performed a series of experiments to test whether the nuclear localization signal (NLS) is all that is needed to direct a protein to the nucleus.

HYPOTHESIS▶ An NLS is necessary for import of a protein into the nucleus.

METHOD

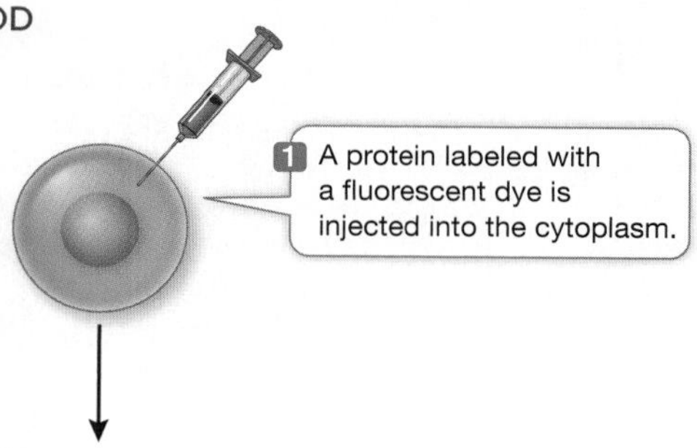

1 A protein labeled with a fluorescent dye is injected into the cytoplasm.

RESULTS

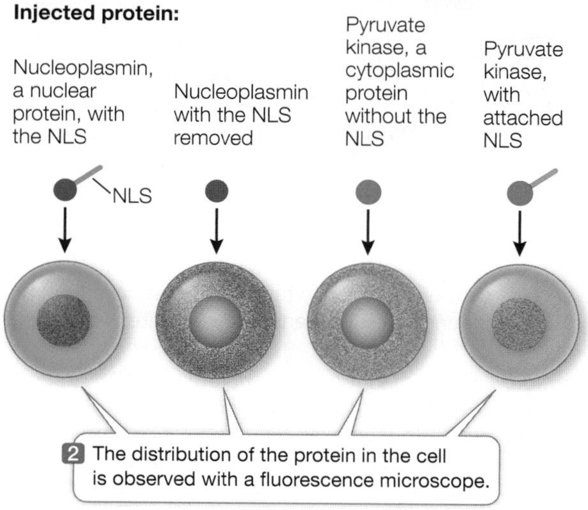

Injected protein:

| Nucleoplasmin, a nuclear protein, with the NLS | Nucleoplasmin with the NLS removed | Pyruvate kinase, a cytoplasmic protein without the NLS | Pyruvate kinase, with attached NLS |

NLS

2 The distribution of the protein in the cell is observed with a fluorescence microscope.

CONCLUSION▶ The NLS is essential for nuclear protein import and is sufficient to direct a normally cytoplasmic protein to the nucleus.

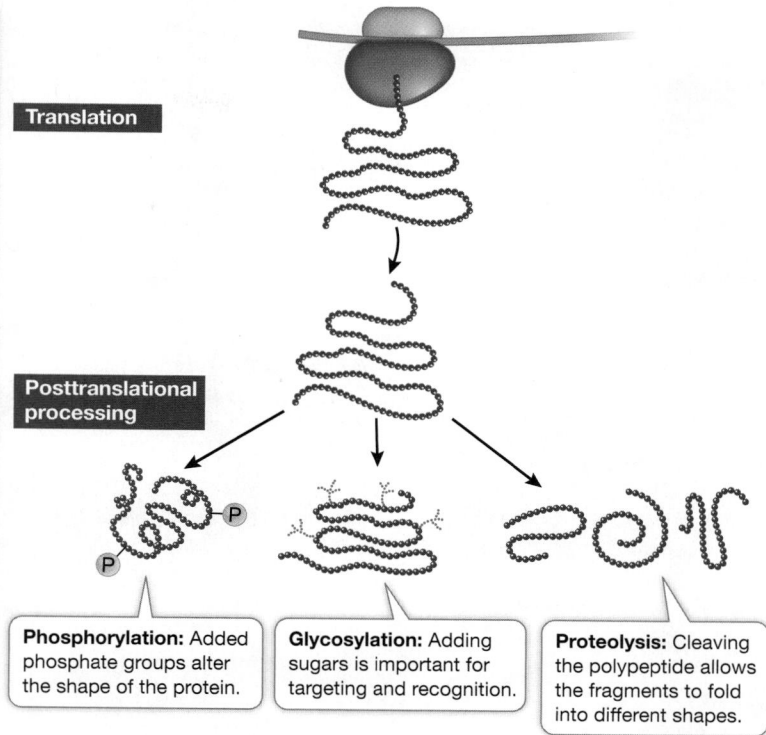

Translation

Posttranslational processing

Phosphorylation: Added phosphate groups alter the shape of the protein.

Glycosylation: Adding sugars is important for targeting and recognition.

Proteolysis: Cleaving the polypeptide allows the fragments to fold into different shapes.

Figure 14.19 Posttranslational Modifications of Proteins
Most polypeptides must be modified after translation in order to become functional proteins.

▶14.6 recap

Signal sequences in polypeptides direct them to their appropriate destinations inside or outside the cell. Many polypeptides are modified after translation.

learning outcomes

You should be able to:

- Outline the steps by which a newly synthesized protein is moved into the nucleus or exported outside of the cell.
- Justify the claim that posttranslational modification of a protein is ultimately directed by the DNA sequence of the gene encoding the protein.

1. How do signal sequences determine where a protein will go after it is made?
2. The genetic code determines the amino acid sequence of a protein. Is there a "code" that determines posttranslational modifications such as phosphorylation or glycosylation? How might this relate to the genetic code?

- **Phosphorylation** is the addition of phosphate groups to proteins, and is catalyzed by protein kinases. The charged phosphate groups change the conformation of a protein, often exposing the active site of an enzyme or the binding site for another protein. We have seen the important role of phosphorylation in cell signaling (see Chapter 7) and the cell cycle (see Chapter 11).

All of the processes we have just described result in a functional protein, but only if the amino acid sequence of that protein is correct. If the sequence is not correct, cellular dysfunction may result. Changes in the DNA—mutations—are a major source of errors in amino acid sequences. This is the subject of the next chapter.

Q&A How does knowledge of the genetic code help us understand the actions of some antibiotics?

MRSA can be treated if detected early. Sequencing the DNA of the bacteria can be used to detect mutations and predict the severity of the infection. Antibiotics such as tetracyclines, which target bacterial protein synthesis, are effective in some strains. Tetracyclines kill bacteria by interrupting translation, binding to the small subunit of bacterial ribosomes so that charged tRNA cannot bind to the A site on the ribosome. Tetracyclines do not kill eukaryotic cells, because eukaryotic ribosomes have different proteins and RNAs that do not offer a binding site for tetracyclines.

Whether tetracyclines will continue to offer a viable treatment for MRSA, however, is in doubt. Some strains of MRSA are already resistant to tetracyclines, having acquired genes that convey resistance through bacterial conjugation. The DNA of *S. aureus* has been altered through acquisition of genes from a plasmid so important that it has been given its own name, the resistome. As with methicillin resistance, the alleles of the resistome encode a ribosomal protein with an altered amino acid sequence that consequently does not bind the antibiotic.

Can you think of other ways to target protein synthesis in MRSA?

Future directions

While the expression of protein-coding genes has been well worked out, recent analyses of eukaryotic genomes show that a lot of transcription remains unaccounted for. International cooperation among laboratories in huge projects such as ENCODE (Encyclopedia of DNA Elements) has revealed, for example, that up to 70 percent of human DNA is transcribed at some point in some cells. With only 2 to 3 percent of DNA consisting of protein-coding genes, the question immediately arises as to what purpose the rest of the RNAs serve. A small fraction are tRNAs and rRNAs, and you will see in Chapter 16 that some of these noncoding RNAs are small (< 200 nucleotides) and have roles in regulating the expression of protein-coding genes. The roles of the longer RNAs (> 200 nucleotides) are much less clear. In humans, over 25,000 of the longer RNAs have been identified. They have much in common with mRNAs: they are transcribed using the same RNA polymerase, undergoing poly A addition, capping at the 5′ and 3′ ends, and splicing. However, the longer RNAs are not as long as mRNAs and are not translated at the ribosome. Some never leave the nucleus, where they may be involved in pre-mRNA events such as transcription and splicing. Others actually arrive at the ribosome, where they may regulate the translation of particular mRNAs. These longer untranslated RNAs are found in many organisms, indicating that they must have important roles that have been selected for in evolution.

Chapter Summary 14

▶ 14.1 Genes Code for Proteins

- Experiments on metabolic enzymes in the bread mold *Neurospora* led to the **one-gene, one-enzyme hypothesis**. We now know that there is a **one-gene, one-polypeptide relationship**. Review Figure 14.1

▶ 14.2 Information Flows from Genes to Proteins

- The **central dogma** of molecular biology states that DNA encodes RNA, and RNA encodes proteins. Proteins do not encode proteins, RNA, or DNA.
- The process by which the information in DNA is copied to RNA is called **transcription**. The process by which a protein is built from the information in RNA is called **translation**. Review Focus: Key Figure 14.2, Activity 14.1
- A product of transcription is **messenger RNA (mRNA)**. **Transfer RNAs (tRNAs)** translate the genetic information in the mRNA into a corresponding sequence of amino acids to produce a polypeptide.
- Certain RNA viruses present exceptions to the central dogma. For example, **retroviruses** synthesize DNA from RNA in a process called **reverse transcription**.

▶ 14.3 DNA Is Transcribed to Produce RNA

- In a given gene, only one of the two strands of DNA (the template strand) acts as a template for transcription. **RNA polymerase** is the catalyst for transcription.
- RNA transcription from DNA proceeds in three steps: **initiation**, **elongation**, and **termination**. Review Figure 14.4, Animation 14.1

- Initiation requires a **promoter**, to which RNA polymerase binds. Part of each promoter is the **initiation site**, where transcription begins.
- Elongation of the RNA molecule proceeds from the 5′ to 3′ end.
- Particular base sequences specify termination, at which point transcription ends and the RNA transcript separates from the DNA template.
- The **genetic code** is a "language" of triplets of mRNA nucleotide bases (**codons**) corresponding to 20 specific amino acids; there are **start** and **stop codons** as well. The code is redundant (an amino acid may be represented by more than one codon) but not ambiguous (no single codon represents more than one amino acid). Review Investigating Life: Deciphering the Genetic Code, Figure 14.5, Animation 14.2, Activity 14.2

▶ 14.4 Eukaryotic Pre-mRNA Transcripts Are Processed prior to Translation

- Unlike prokaryotes, where transcription and translation occur in the cytoplasm and are coupled, in eukaryotes transcription occurs in the nucleus and translation later in the cytoplasm. Review Table 14.2
- Eukaryotic genes contain **introns**, which are noncoding sequences within the transcribed regions of genes. Review Figures 14.6B, 14.7
- The initial transcript of a eukaryotic protein-coding gene is modified with a **5′ cap** and a **3′ poly A tail**. Review Figure 14.8
- Pre-mRNA introns are removed in the nucleus via **RNA splicing**. Then the mRNA passes through the nuclear pore into the cytoplasm, where it is translated through **ribosomes**. Review Figure 14.9, Animation 14.3

14.5 The Information in mRNA Is Translated into Proteins

See Animation 14.4

- During translation, amino acids are linked together in the order specified by the codons in the mRNA. This task is achieved by tRNAs, each of which binds to (is charged with) a specific amino acid and has an **anticodon** complementary to a specific mRNA codon. **Review Figures 14.10, 14.11**

- The ribosome is the molecular workbench where translation occurs. It has one large and one small subunit, each made of **ribosomal RNA** and proteins.

- Three sites on the large ribosomal subunit interact with tRNA. The A site is where the charged tRNA anticodon binds to the mRNA codon; the P site is where the tRNA adds its amino acid to the growing polypeptode chain; and the E site is where the tRNA is released. **Review Figure 14.12**

- Translation occurs in three steps: initiation, elongation, and termination. The **initiation complex** consists of tRNA bearing the first amino acid, the small ribosomal subunit, and mRNA. **Review Figure 14.13**

- The growing polypeptide chain is elongated by the formation of peptide bonds between amino acids, catalyzed by rRNA. **Review Figure 14.14**

- When a stop codon reaches the A site, it terminates translation by binding a release factor. **Review Figure 14.15**

- In a **polysome**, more than one ribosome moves along a strand of mRNA at one time. **Review Figure 14.16**

14.6 Polypeptides Can Be Modified and Transported during or after Translation

- **Signal sequences** of amino acids direct polypeptides to the cellular destinations, such as organelles. **Review Figures 14.17, 14.18**

- Proteins addressed to the RER bind to a receptor protein in the RER membrane. **Review Figure 14.17**

- Posttranslational modifications of polypeptides include **proteolysis**, whereby a polypeptide is cut to smaller fragments; **glycosylation**, whereby sugars are added; and **phosphorylation**, whereby phosphate groups are added. **Review Figure 14.19**

> Go to **LearningCurve** (in **LaunchPad**) for dynamic quizzing that helps you solidify your understanding of this chapter. **LearningCurve** adapts to your responses, giving you the practice you need to master each key concept.

Apply What You've Learned

Review

14.1 Experimental evidence supports the claim that each gene encodes one protein.

Original Paper: Gross, S. R. 1965. The regulation of synthesis of leucine biosynthetic enzymes in *Neurospora. Proceedings of the National Academy of Sciences* USA 54:1538–1546.

The fungus *Neurospora crassa* first drew researchers' attention as a common contaminant in French bakeries, and it quickly proved to be an exceptional model for research. Its nutritional requirements are simple, it grows quickly and easily in the lab, and it follows the rules of Mendelian genetics.

In one study, researchers experimented with *Neurospora* to identify the steps in the biosynthesis of the amino acid leucine. The figure below shows a partial pathway that was proposed early in this research.

Neurospora is haploid for most of its life cycle. During sexual reproduction, haploid cells of two different mating types fuse, producing a diploid cell that carries both sets of parental genes. *Neurospora* grows on minimal medium without added amino acids because it can synthesize all of the amino acids and other compounds it needs to carry out its metabolic functions.

With these properties in mind, the researchers isolated two mutant *Neurospora* strains and tested them for growth on various media. Fused cells produced from mating between the two *Neurospora* mutants were isolated and tested in the same way. The results are shown in the table, where "+" indicates growth and "–" indicates no growth.

Questions

1. Analyze the data to identify which mutation is associated with which step (A, B, C, D) in the biosynthetic pathway. Explain your reasoning.

2. Explain why the fusion of two mutant cells is able to restore the wild-type phenotype.

3. Predict the data that would be collected if the activities of the enzymes catalyzing steps A, B, C, and D were measured in each haploid mutant and in the fused diploid cells.

4. Suppose the enzyme catalyzing a single step in a metabolic pathway consists of four subunits, and that the quaternary structure for this enzyme has the structure $\alpha_2\beta_2$, where α and β each represent a different polypeptide chain. Experiments with haploid cells of two mutant strains of *Neurospora* show that they are deficient in this metabolic step. However, diploid cells resulting from fusion between these two strains results in restoration of the wild-type phenotype. Explain these observations.

Strain	Minimal medium	Minimal medium + leucine	Minimal medium + 2-isopropylmalate	Minimal medium + 3-isopropylmalate	Minimal medium + α-ketoisocaproate
Wild type (haploid)	+	+	+	+	+
Leu-1 (haploid)	–	+	–	–	+
Leu-2 (haploid)	–	+	–	+	+
Fused cells (diploid): Leu-1, Leu-2	+	+	+	+	+

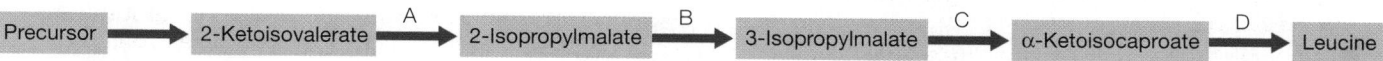

Precursor → 2-Ketoisovalerate →[A] 2-Isopropylmalate →[B] 3-Isopropylmalate →[C] α-Ketoisocaproate →[D] Leucine

15

Gene Mutation and Molecular Medicine

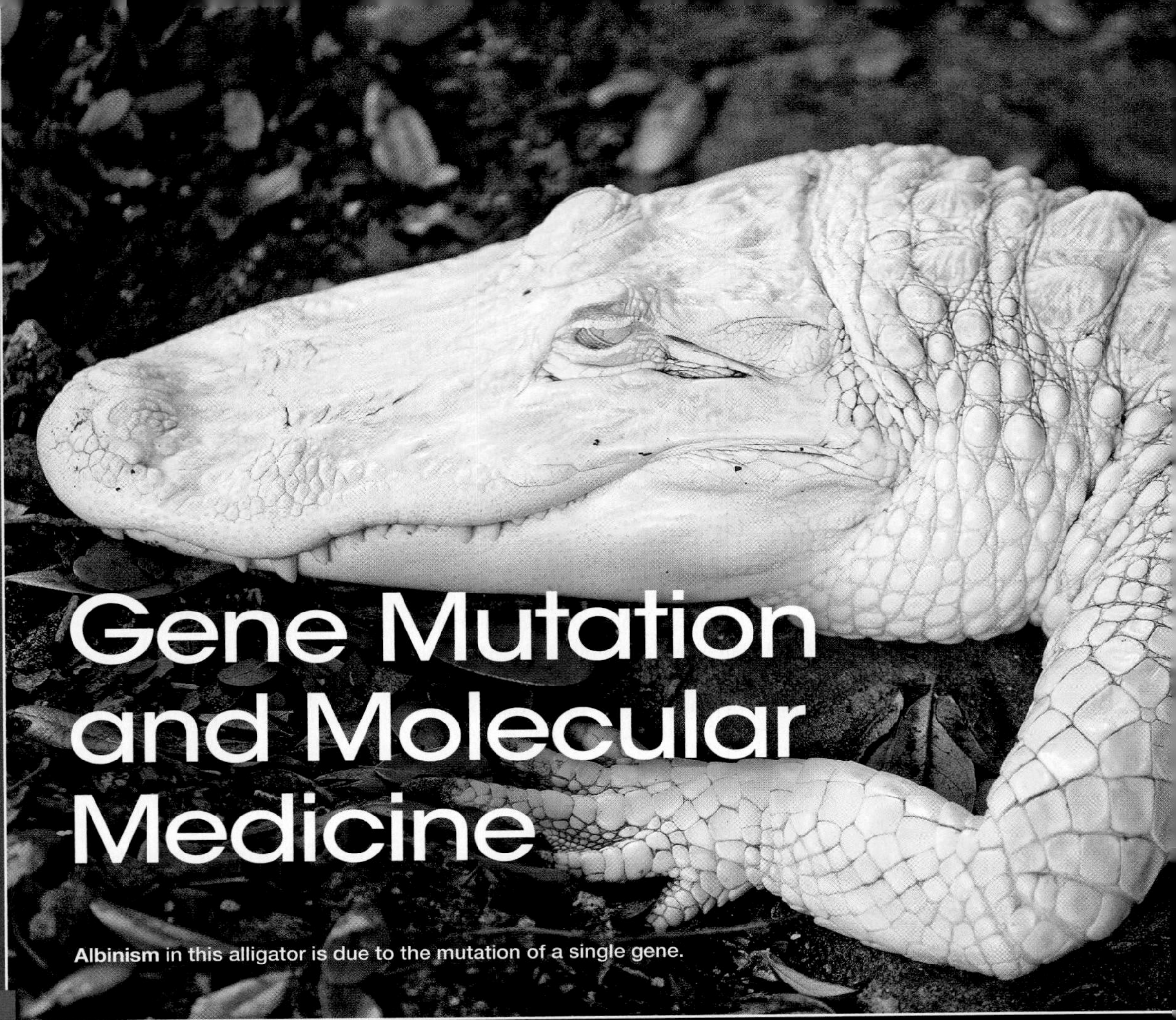

Albinism in this alligator is due to the mutation of a single gene.

> investigating**life**

The Angelina Jolie Effect

The actress Angelina Jolie surprised many people by announcing that she had had both breasts removed upon learning that she carries a mutation in the "breast cancer gene." Ms. Jolie had good reason to be concerned: even without genetic tests, her family history suggests a high risk for cancer. Her mother had breast cancer and died at age 56 of ovarian cancer. Her aunt died of breast cancer. Her grandmother died of ovarian cancer. Jolie learned that a mutation in a gene called *BRCA1* can be responsible for hereditary ovarian and breast cancers and that her DNA for this gene could be sequenced and analyzed for mutations before cancer symptoms arise. A woman with a harmful mutation in the *BRCA1* gene has an 80 percent chance of getting breast cancer by age 60, far in excess of the 8 percent likelihood for women without a mutation. Removal of breast tissue lowers this risk to 10 percent.

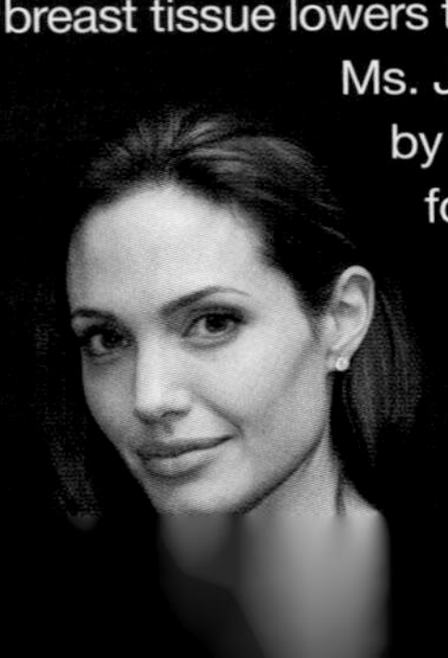

Ms. Jolie's announcement was widely praised by health-care professionals. Women at risk for a *BRCA1* mutation increasingly have elected to be tested, and more of those identified to be at an increased risk for breast cancer have chosen preventive surgery. Jolie's announcement has helped women who have made these decisions feel less emotionally isolated.

The *BRCA1* mutation test is the culmination of research in the molecular genetics of cancer. As we discussed in Chapter 11, cancer can arise through mutations of either oncogenes (resulting in overstimulation of the cell cycle) or tumor suppressor genes (which would otherwise function to slow down the cell cycle). *BRCA1* is a tumor suppressor gene involved in DNA repair. When the protein encoded by the gene is faulty because of gene mutation, unrepaired mutations accumulate in breast cell DNA, and some of these mutations improperly activate the cell cycle, resulting in uncontrolled cell division and other adverse effects of a growing tumor. Because *BRCA1* is also expressed in ovary tissue, women who are mutation carriers often have their ovaries removed as well; Ms. Jolie had this surgery two years after her breasts were removed.

The precise description of mutations in genetically caused diseases exemplifies the emerging field of molecular medicine. Our ability to identify mutations is leading to more precise testing, diagnoses, and treatments for the diseases they cause.

 What is the breast cancer gene, and what

Mutations Are Heritable Changes in DNA

15.1

In Chapter 12 we defined mutations as stable, inherited changes in genes, and we saw that different alleles may produce different phenotypes (short pea plants versus tall, for example). In the two chapters that followed we described the chemical nature of genes as DNA sequences, and how they are expressed as phenotypes (in particular, proteins). To reiterate, we can now define a **mutation** as a change in the nucleotide sequence of DNA that can be passed on from one cell, or organism, to another.

focus your learning

- Mutations that occur at the nucleotide sequence level can be classified by the specific changes they cause in DNA and phenotypic effects in proteins.
- Mutations that occur at the chromosomal level can be classified by the specific changes they cause in chromosomes.
- Induced mutations are caused by mutagens, which are environmental factors such as chemicals and radiation that alter DNA structure.
- Mutation hot spots occur where bases are more susceptible to mutation.

Mutations have different phenotypic effects

Mutations are classified generally by the type of cell in which they occur:

- **Somatic mutations** occur in somatic (nongamete) cells. These mutations are passed on to the daughter cells during mitosis, and to the daughters of those cells in turn, but are not passed on to sexually produced offspring. For example, a mutation in a single human skin cell could result in a patch of skin cells that all have the same mutation, but it would not be passed on to the person's children.

- **Germ line mutations** occur in the cells of the germ line—the specialized cells that give rise to gametes. A gamete with the mutation passes it on to a new organism at fertilization. The new organism will have the mutation in every cell of its body and will be able to pass the mutation on to its offspring. The *BRCA1* mutation that Angelina Jolie, her aunt, and her mother inherited from Angelina's grandmother is a germ line mutation.

Does every mutation have a phenotypic effect? Not necessarily. Some mutations have effects on proteins and their function, and some do not (**Focus: Key Figure 15.1**):

- A **silent mutation** does not usually affect protein function (see Figure 15.1B). It can be a mutation in a region of DNA that does not encode a protein, or it can be in the coding region of a gene but not affect the amino acid sequence. Because of the redundancy of the genetic code, a base change in a coding region will not always cause a change in the amino acid sequence when the altered mRNA is translated (see Figure 15.2). Silent mutations are common, and they usually result in genetic diversity that is not expressed as phenotypic differences.

- A **loss-of-function mutation** affects protein function (see Figure 15.1C). Such a mutation may cause a gene to not be expressed at all, or the gene may be expressed but produces a dysfunctional protein that can no longer play its cellular role, such as its catalytic function if it is an enzyme. Loss-of-function mutations almost always show recessive inheritance in diploid organisms, because the presence of one wild-type allele will usually result in sufficient functional protein for the cell. For example, recall from Key Concept 12.1 that the familiar wrinkled seed allele in pea plants, originally studied by Mendel, is due to a recessive loss-of-function mutation in the *SBE1* (starch branching enzyme) gene. Normally the protein made by this gene catalyzes the branching of starch as seeds develop. In the mutant, the SBE1 protein is not functional, and that leads to osmotic changes, causing the wrinkled appearance.

- A **gain-of-function mutation** leads to a protein with an altered function (see Figure 15.1D). A gain-of-function mutation usually shows dominant inheritance, because the presence of the wild-type allele does not prevent the mutant allele from functioning. This type of mutation is common in cancer. For example, there are mutations in oncogenes that result in proteins that constantly stimulate cell division.

Some mutations have effects on phenotype only under certain conditions. For example, a **conditional mutation** affects phenotype only under restrictive conditions and is not detectable under what are called permissive conditions. Many conditional mutations are temperature-sensitive, resulting in proteins with reduced stability at high temperatures. For example, the point restriction phenotype in rabbits and Siamese cats (see Figure 12.13) is due to a temperature-sensitive (conditional, loss-of-function) mutation in a coat color gene. At body temperature, the protein encoded by the gene is unstable and nonfunctional, so that the animal has dark fur only in its cooler extremities.

Most point mutations can be reversed; **reversion mutations** result when a gene is mutated a second time so that the DNA reverts to its original sequence or to a coding sequence that results in the non-mutant phenotype. When reversion mutations occur, the phenotype reverts to wild type.

Now let's look more closely at the DNA level to see how various kinds of mutations play out.

Point mutations are changes in single nucleotides

A **point mutation** is the addition or subtraction of a single nucleotide, or the substitution of one nucleotide base for another. There are two kinds of base substitution:

1. A **transition** is the substitution of one purine for the other purine, or one pyrimidine for the other:

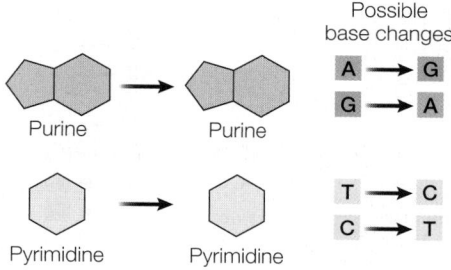

Possible base changes

A → G
G → A

T → C
C → T

Purine → Purine

Pyrimidine → Pyrimidine

focus: key figure

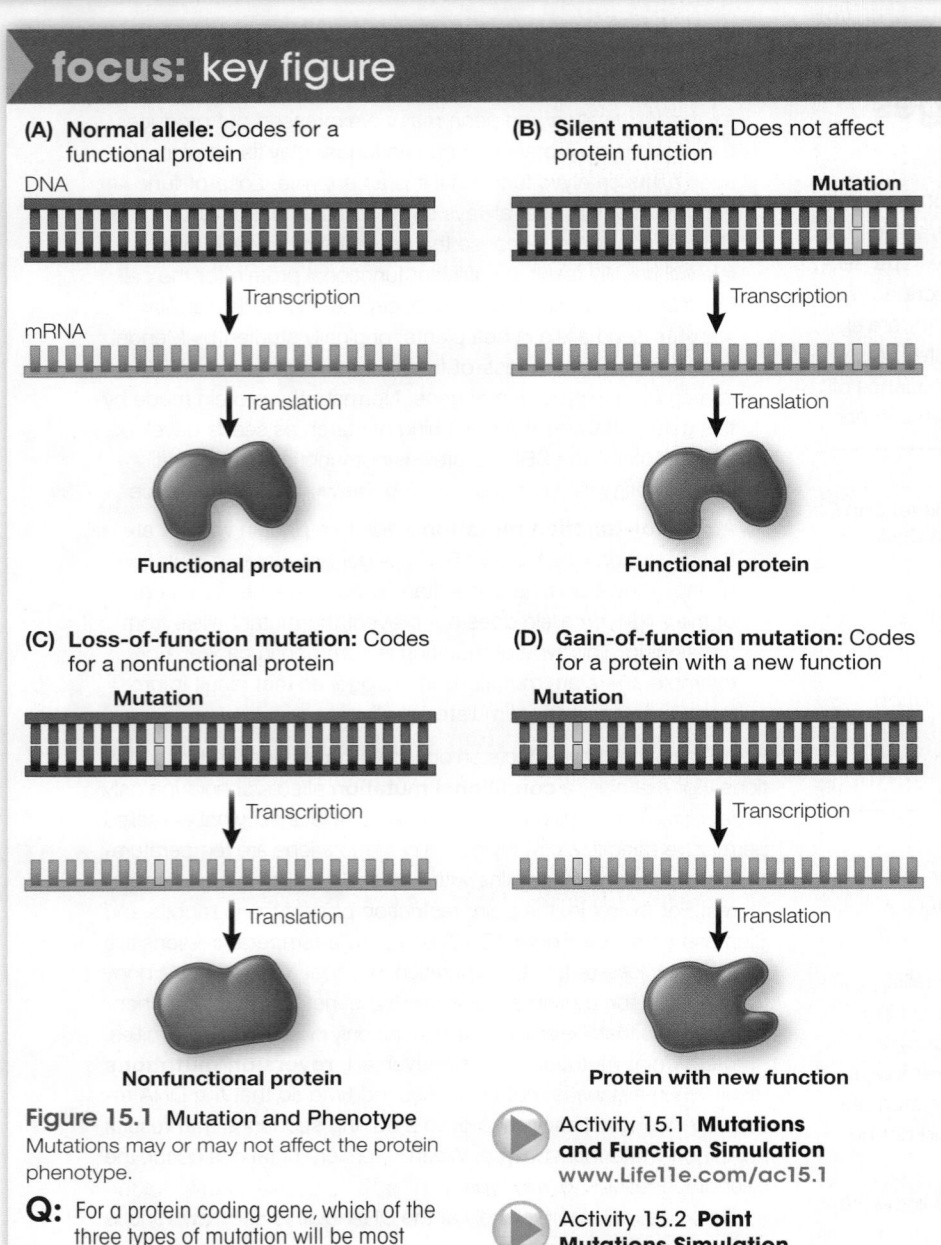

(A) Normal allele: Codes for a functional protein

DNA

↓ Transcription

mRNA

↓ Translation

Functional protein

(B) Silent mutation: Does not affect protein function

Mutation

↓ Transcription

↓ Translation

Functional protein

(C) Loss-of-function mutation: Codes for a nonfunctional protein

Mutation

↓ Transcription

↓ Translation

Nonfunctional protein

(D) Gain-of-function mutation: Codes for a protein with a new function

Mutation

↓ Transcription

↓ Translation

Protein with new function

Figure 15.1 Mutation and Phenotype
Mutations may or may not affect the protein phenotype.

Q: For a protein coding gene, which of the three types of mutation will be most common, and why?

▶ Activity 15.1 **Mutations and Function Simulation**
www.Life11e.com/ac15.1

▶ Activity 15.2 **Point Mutations Simulation**
www.Life11e.com/ac15.2

2. A **transversion** is the substitution of a purine for a pyrimidine, or vice versa:

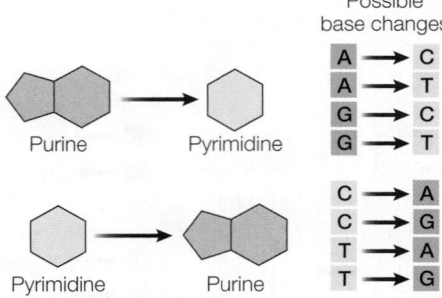

Purine → Pyrimidine

Pyrimidine → Purine

Possible base changes

A → C
A → T
G → C
G → T

C → A
C → G
T → A
T → G

A point mutation in the coding region of a gene will result in an alteration in the mRNA sequence. But remember: a change in the mRNA does not necessarily mean a change in the protein that gets translated from it at the ribosome. As we just mentioned, a silent mutation has no effect on the amino acid sequence of an encoded polypeptide. By contrast, missense, nonsense, and frame-shift mutations do result in changes in the protein, some of them drastic (**Figure 15.2**).

MISSENSE MUTATIONS Missense mutations are *base substitution changes that alter the genetic code such that one amino acid substitutes for another in a protein (see Figure 15.2C). A specific example is the mutation that causes sickle-cell disease, a serious heritable blood disorder. The disease occurs in people who carry two copies of the sickle allele of the gene for β-globin—a subunit of hemoglobin, the protein in human blood that carries oxygen. The sickle allele differs from the wild-type allele by one base pair, resulting in a polypeptide that differs by one amino acid from the wild-type protein. Individuals who are homozygous for this recessive allele have defective, sickle-shaped red blood cells:

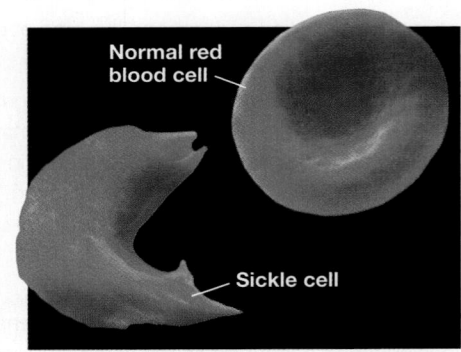

Normal red blood cell

Sickle cell

***connect the concepts** Many base substitution mutations result in changes in protein structure due to amino acid substitutions. Review the relationships between amino acid sequence and protein three-dimensional structure in Key Concept 3.2.

▶ Media Clip 15.1 **Sickle Cells: Deformed by a Mutation**
www.Life11e.com/mc15.1

A missense mutation may result in a defective protein, but often has no effect on the protein's function. For example, a hydrophilic amino acid may be substituted for another hydrophilic amino acid, so that the shape of the protein is unchanged. Or a missense mutation might reduce the functional efficiency of a protein rather than completely inactivating it. Therefore individuals homozygous for a

(A) Wild type (normal)

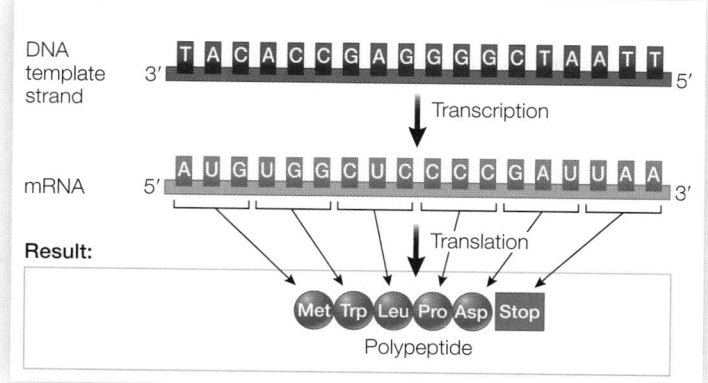

(B) Silent mutation

Mutation at position 12 in DNA: **A** instead of **C**

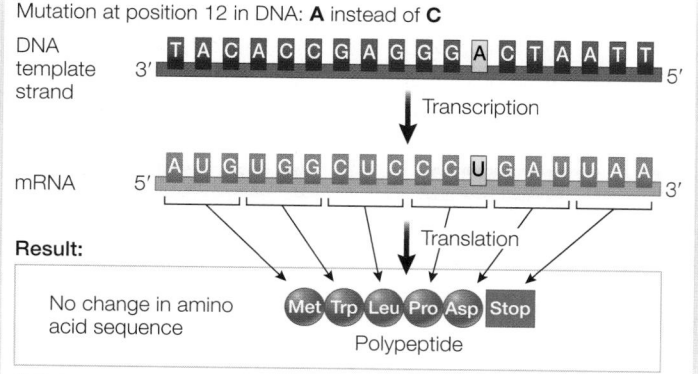

(C) Missense mutation

Mutation at position 14 in DNA: **A** instead of **T**

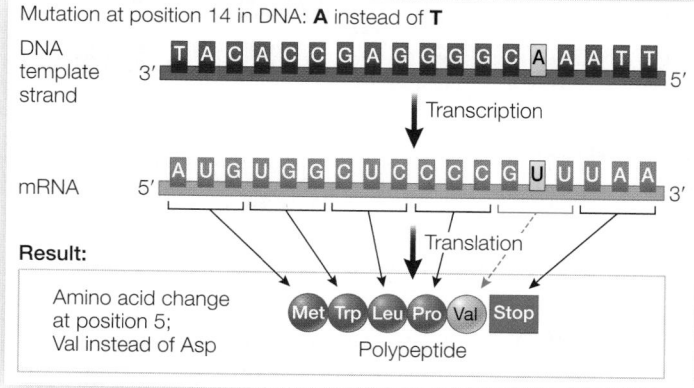

(D) Nonsense mutation

Mutation at position 5 in DNA: **T** instead of **C**

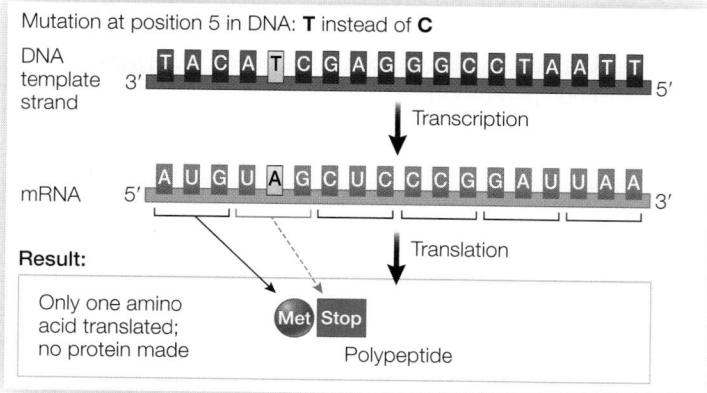

(E) Frame-shift mutation

Mutation by insertion of **T** between bases 6 and 7 in DNA

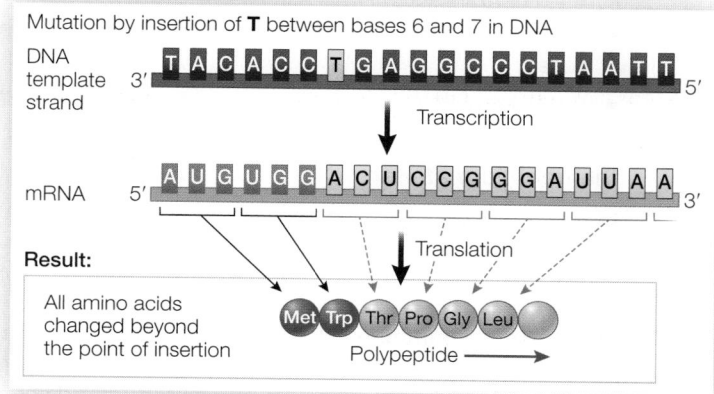

Figure 15.2 Point Mutations   When they occur in the coding regions of proteins, single base changes can cause silent, missense, nonsense, or frame-shift mutations.

missense mutation in a protein essential for life may survive if enough of the protein's function is retained.

In some cases, a gain-of-function missense mutation occurs. An example is a mutation in the human *P53* gene, which codes for a tumor suppressor—a protein that inhibits the cell cycle (see Key Concept 11.7). Certain mutations of the *P53* gene cause this protein to no longer inhibit cell division, but instead to promote it and prevent programmed cell death.

NONSENSE MUTATIONS   A **nonsense mutation** involves a base substitution that causes a stop codon (for translation) to form somewhere in the mRNA (see Figure 15.2D). A nonsense mutation results in a shortened protein, since translation does not proceed beyond the point where the mutation occurred. For example, a common mutation causing thalassemia (another blood disorder affecting hemoglobin) in Mediterranean populations is a nonsense mutation that drastically shortens the β-globin subunit. Shortened proteins are usually not functional; however, if the nonsense mutation occurs near the 3′ end of the gene, it may have no effect on function.

FRAME-SHIFT MUTATIONS   Not all point mutations are base substitutions. One or two nucleotides may be inserted into, or deleted from, a sequence of DNA. Such mutations in coding sequences are known as **frame-shift mutations** because they alter the reading frame (that is, consecutive triplets) in which the codons are read during translation (see Figure 15.2E). Think again of *codons as three-letter words, each corresponding to a particular amino acid. Translation proceeds codon by codon; if a nucleotide is added to the mRNA or subtracted from it, then the three-letter "words" are altered as translation proceeds beyond that point, and the result is a completely different amino acid sequence. Frame-shift mutations almost always lead to the production of nonfunctional proteins.

*connect the concepts   Even in a protein-encoding region, some mutations do not affect amino acid sequence because of redundancies in the genetic code. The genetic code provides more than one codon for most of the amino acids. See Key Concept 14.3.

MUTATIONS OUTSIDE THE CODING REGION   As discussed in Chapter 17, in genomes like ours, much of the DNA does not contain protein-coding genes. If there is a mutation in a region that is not involved with a protein-coding gene, it often does not have a phenotypic effect, although it is still a mutation—an inherited change in DNA. Recall from Chapter 14 that protein-coding genes have stretches of DNA that do not encode amino acids, such as promoters and introns. Mutations in these regions may have significant effects. For example:

- promoter mutations may alter the rate of transcription of the gene.

- mutations at RNA splicing sites may lead to abnormal mRNA.

Chromosomal mutations are extensive changes in the genetic material

Changes in single nucleotides are not the most dramatic changes that can occur in the genetic material. Whole DNA molecules (that is, whole chromosomes) can break and rejoin, grossly disrupting the sequence of genetic information. There are four types of such **chromosomal mutations**: deletions, duplications, inversions, and translocations. These mutations can be caused by severe damage to chromosomes resulting from mutagens or by drastic errors in chromosome replication.

- A **deletion** occurs by the removal of part of the genetic material and can happen if a chromosome breaks at two points and then rejoins, leaving out the DNA between the breaks (**Figure 15.3A**).

- A **duplication** can be produced at the same time as a deletion and can occur if homologous chromosomes break at different positions and then reconnect to the wrong partners (**Figure 15.3B**). One of the two chromosomes ends up with a deleted segment, and the other has two copies (a duplication) of the same segment.

- An **inversion** can also result from the breaking and rejoining of a chromosome, and can occur if a segment of DNA becomes "flipped," so that it runs in the opposite direction from its original orientation (**Figure 15.3C**).

- A **translocation** results when a segment of a chromosome breaks off and becomes attached to a different chromosome. As we mentioned in Key Concept 11.5, a translocation of a large segment of chromosome 21 is one cause of Down syndrome. Translocations may involve reciprocal exchanges of chromosome segments, as in **Figure 15.3D**.

Retroviruses and transposons can cause loss-of-function mutations or duplications

Key Concept 14.2 described how certain viruses called retroviruses can insert their genetic material into the host cell's genome. Such insertions happen at random, and if one occurs within a gene, it can cause a loss-of-function mutation in that gene. In many cases, the viral DNA remains in the host genome and is passed on from one generation to the next. When this happens the virus is called an endogenous retrovirus. Endogenous retroviruses are common—in fact, they make up 5 to 8 percent of the human genome.

Another form of DNA, called a transposon or transposable element, can also insert itself into genes and cause mutations. As you will see in Chapter 17, transposons are widespread in both prokaryotic and eukaryotic genomes. A transposon is a DNA sequence of a few hundred to a few thousand base pairs that can move from one position in the genome to another. It usually carries genes that encode the enzymes needed for this movement. Some transposons remove themselves from their positions in the genome and then insert themselves into other sites (the "cut and paste" mode of transposition). These transposons do not always excise cleanly, but leave behind short sequences of a few base pairs that become permanent mutations in the affected genes. Other transposons first

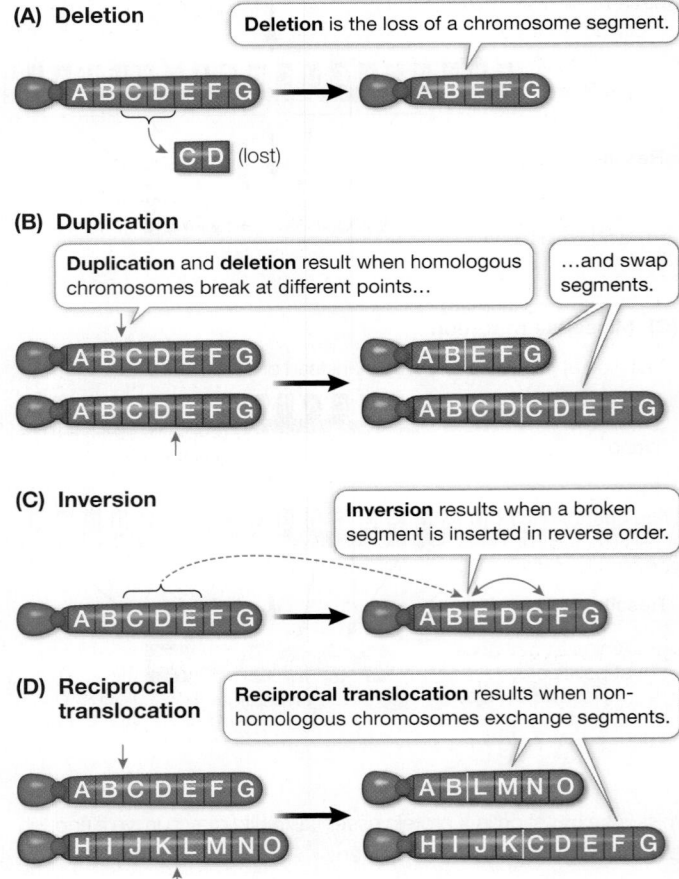

(A) Deletion

> **Deletion** is the loss of a chromosome segment.

A B C D E F G → A B E F G

C D (lost)

(B) Duplication

> **Duplication** and **deletion** result when homologous chromosomes break at different points... ...and swap segments.

A B C D E F G → A B E F G

A B C D E F G → A B C D C D E F G

(C) Inversion

> **Inversion** results when a broken segment is inserted in reverse order.

A B C D E F G → A B E D C F G

(D) Reciprocal translocation

> **Reciprocal translocation** results when nonhomologous chromosomes exchange segments.

A B C D E F G → A B L M N O

H I J K L M N O → H I J K C D E F G

Figure 15.3 Chromosomal Mutations   Chromosomes may break during replication, and parts of chromosomes may then rejoin incorrectly. This can result in deletions (**A, B**), duplications (**B**), inversions (**C**), or reciprocal translocations (**D**). Note that the letters on these illustrations represent large segments of the chromosomes. Because chromosomes contain regions of noncoding DNA, each segment may include anywhere from zero to hundreds or thousands of genes.

Q: How are chromosomal mutations detected? Hint: See Figure 11.20.

replicate themselves, and then the new copies are inserted into new sites in the genome (the "copy and paste" mode). A sequence of genomic DNA is sometimes carried along with the transposon DNA when it moves, and gene duplication occurs. As you'll see, gene duplication plays an important role in evolution.

Mutations can be spontaneous or induced

It is useful to distinguish between mutations that are spontaneous or induced, based on their causes. **Spontaneous mutations** are permanent changes in the genetic material that occur without any outside influence. The movement of transposons is an example of spontaneous mutation. Spontaneous mutations can also occur because cellular processes are imperfect, and may occur by several mechanisms:

- *A transient rearrangement in the structure of a nucleotide base can result in mistakes during replication.* Each base can exist in two different forms (called tautomers), one of which is common and one rare. When a base temporarily forms its rare tautomer, it can pair with the wrong base. For example, C normally pairs with G, but if C takes on the form of its rare tautomer at the time of DNA replication, it pairs with (and DNA polymerase will insert) an A. If this is passed on to a daughter cell after cell division, the result is a point mutation: G → A (**Figure 15.4A and C**).

- *A chemical reaction may alter the structure of a DNA base.* For example, a deamination reaction can result in loss of the amino group (NH$_2$) attached to carbon 4 in cytosine. If this occurs in a

DNA molecule, the error will usually be repaired. However, since the repair mechanism is not perfect, the altered nucleotide will sometimes remain during replication. In these cases, DNA polymerase will add an A (which base-pairs with U on the template DNA) instead of G (which normally pairs with C).

- *DNA polymerase can make errors in replication* (see Key Concept 13.4)—for example, by inserting a T opposite a G. Most of these errors are repaired by the proofreading function of the replication complex, but some errors escape detection and become permanent.

- *Meiosis is not perfect.* Nondisjunction—the failure of homologous chromosomes to separate during meiosis—can occur, leading to one too many chromosomes or one too few (see Figure 11.19). Random chromosome breakage and rejoining can produce deletions, duplications, inversions, or translocations.

Induced mutations occur when some agent from outside the cell—a **mutagen**—causes a permanent change in DNA. As we mentioned above, retroviruses can function as mutagens. In addition, certain chemicals and radiation can cause mutations:

- *Some chemicals can alter nucleotide bases.* For example, nitrous acid (HNO$_2$) and similar molecules can react with cytosine and convert it to uracil by deamination. More specifically, they convert an amino group on the cytosine (—NH$_2$) into a keto group (—C=O) (**Figure 15.4B**). This alteration has

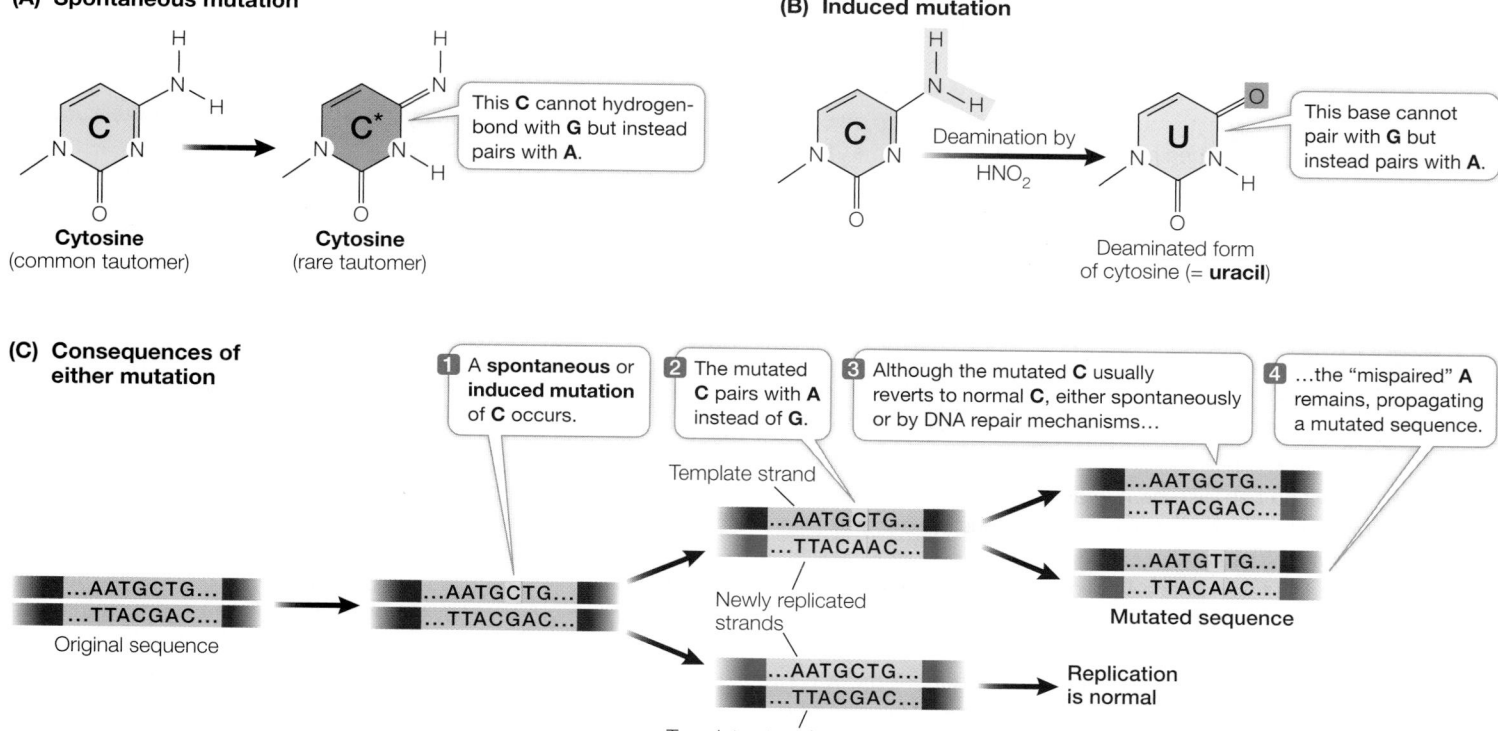

Figure 15.4 Spontaneous and Induced Mutations **(A)** All four nitrogenous bases in DNA exist in both a prevalent (common) form and a rare form. When a base spontaneously forms its rare tautomer, it can pair with a different base. **(B)** Mutagens such as nitrous acid (HNO$_2$) can induce changes in the bases. **(C)** The results of both spontaneous and induced mutations are permanent changes in the DNA sequence following replication.

the same result as spontaneous deamination: instead of a G, DNA polymerase inserts an A (see Figure 15.4C).

- *Some chemicals add groups to the bases.* For instance, benzopyrene, a component of cigarette smoke, adds a large chemical group to guanine, making it unavailable for base pairing. When DNA polymerase reaches such a modified guanine, it inserts any one of the four bases at random. Three-fourths of the time the inserted base is not cytosine, and a mutation results.

- *Radiation damages the genetic material.* Radiation can damage DNA in two ways. First, ionizing radiation (including X rays, gamma rays, and radiation from unstable isotopes) produces highly reactive chemicals called free radicals. Free radicals can change bases in DNA to forms that are not recognized by DNA polymerase. Ionizing radiation can also break the sugar–phosphate backbone of DNA, causing chromosomal abnormalities. Second, ultraviolet radiation (from the sun or a tanning lamp) is absorbed by thymine, causing it to form covalent bonds with adjacent bases. This, too, plays havoc with DNA replication by distorting the double helix.

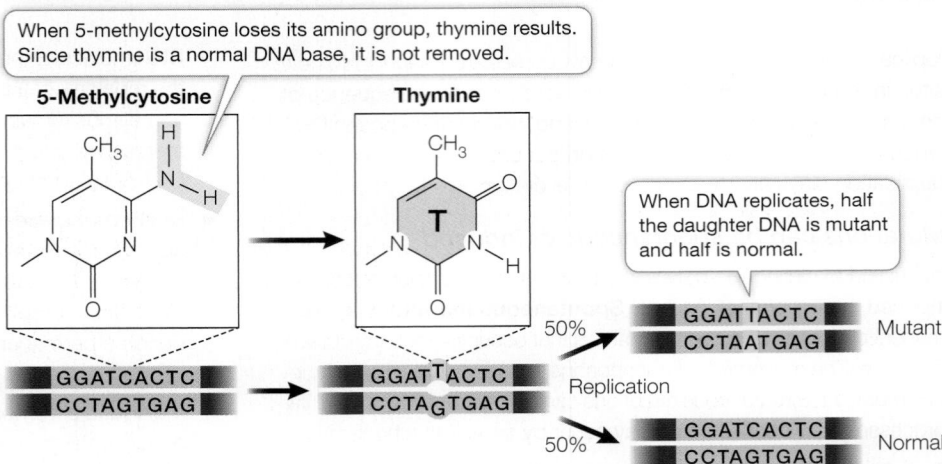

Figure 15.5 **5-Methylcytosine in DNA Is a "Hot Spot" for Mutations** If cytosine has been methylated to 5-methylcytosine, the mutation is unlikely to be repaired and a C-G base pair is replaced with a T-A pair.

Mutagens can be natural or artificial

Many people associate mutagens with materials made by humans, but many mutagenic substances are found in nature. An example of a naturally occurring mutagen is aflatoxin, which is made by the mold *Aspergillus*. When mammals ingest the mold, the aflatoxin is converted by the smooth endoplasmic reticulum in liver cells into a product that, like benzopyrene from cigarette smoke, binds to guanine; this also causes mutations. Plants (and to a lesser extent animals) make thousands of small molecules with a variety of functions, some of which are mutagenic and potentially carcinogenic. Examples of human-made mutagens are nitrites, which are used to preserve meats. Once in mammals, nitrites get converted by the smooth endoplasmic reticulum (SER) to nitrosamines, which are strongly mutagenic because they cause deamination of cytosine (see Figure 15.4).

Radiation can also be human-made or natural. Some of the isotopes made in nuclear reactors and nuclear bomb explosions are certainly harmful. For example, extensive studies have shown increased mutations in the survivors of the atom bombs dropped on Japan in 1945. As previously mentioned, natural ultraviolet radiation in sunlight also causes mutations.

By now, you may be getting worried about all the ways your DNA can be mutated. But for us, and most organisms, DNA repair mechanisms have evolved so that most changes in DNA do not get passed on to either the daughter cells or the next generation (see Figure 13.18). Biochemists have estimated how much DNA damage occurs in the human genome under normal circumstances: among the genome's 3.2 billion base pairs, there are about 16,000 DNA-damaging events per cell per day, of which 80 percent are repaired.

Some base pairs are more vulnerable than others to mutation

In certain regions of DNA, many of the cytosine residues have methyl groups added at their 5 positions, forming 5–methylcytosine. Methylation plays an important role in gene regulation (discussed in Key Concept 16.4). DNA sequencing has revealed that mutation "hot spots" are often located where cytosines have been methylated. Figure 15.4 shows unmethylated cytosine, which can lose its amino group, either spontaneously (see Figure 15.4A) or because of a chemical mutagen, to form uracil (see Figure 15.4B). This type of error is usually detected by the cell and repaired, because uracil is recognized as inappropriate for DNA. However, when 5–methylcytosine (methylated cytosine) loses its amino group, the product is thymine, a natural base for DNA (**Figure 15.5**). The DNA repair mechanism ignores this thymine. During replication, however, the mismatch repair mechanism recognizes that G-T is a mismatched pair, although it cannot tell which base is incorrect. Half of the time it matches a new C to the G, but the other half of the time it matches a new A to the T, resulting in a mutation.

Mutations have both benefits and costs

As we will see in Part Seven of this book, mutations are the raw material of evolution: they provide the genetic diversity that makes natural selection possible. This diversity can be beneficial in two ways. First, a mutation in a somatic cell may benefit the organism immediately. Second, a mutation in a germ line cell may have no immediate selective advantage to the organism, but it may cause a phenotypic change in the organism's offspring. If the environment changes in a later generation, this mutation may be advantageous, and thus selected for, under the new conditions.

We noted that gene duplication can arise through either chromosomal rearrangements or through the movements of transposons. Gene duplication is not always harmful and is an important source of genetic variation. In a duplicated pair of genes, one gene may continue to play its original role in the cell, while the other may acquire a gain-of-function mutation that produces a new phenotype. As with any other mutation, this may be of immediate benefit for the organism, or it may provide a later generation with a selective advantage.

By contrast, mutations in genes whose products are needed for normal cellular processes are often harmful, especially if they occur in germ line cells that produce eggs or sperm. In such cases, some offspring can inherit harmful recessive alleles in the homozygous condition. In their extreme form, such mutations produce phenotypes that are lethal, killing the organism during early development. Likewise, mutations in somatic cells can be harmful, as in the case of cancer. Think about what would happen if an oncogene that stimulates cell division in a colon cell had a gain-of-function mutation in an adult. The result would be the same as inheriting the mutant allele: colon cancer.

What can we do about mutation? While spontaneous mutagenesis is not in our control, we can certainly try to avoid mutagenic substances and radiation. Not surprisingly, many things that cause cancer (carcinogens) are also mutagens. A good example is benzopyrene (discussed above), which is found in coal tar, car exhaust fumes, and charbroiled foods, as well as in cigarette smoke. A major public-policy goal is to reduce the effects of both human-made and natural mutagens on human health. An example is the Montreal Protocol, an international environmental agreement signed and adhered to by all members of the United Nations. It bans chlorofluorocarbons and other substances that cause depletion of the ozone layer in the upper atmosphere of Earth. The thinning of the ozone layer can result in increased ultraviolet radiation reaching Earth's surface, which is implicated in somatic mutations that lead to skin cancer.

15.1 recap

Mutations are uncorrected alterations in the nucleotide sequence of DNA. Mutations can involve changes in single nucleotides or extensive rearrangements of chromosomes. If they occur in somatic cells, they will be passed on to daughter cells; if they occur in germ line cells, they may be passed on to offspring.

learning outcomes

You should be able to:

- Use a defined DNA sequence to illustrate how the following occur, and describe their effect on phenotype: transitions, transversions, missense mutations, nonsense mutations, frame-shift mutations, and mutations outside the coding region of a gene.
- Compare the following chromosomal mutations: deletion, duplication, inversion, and translocation.
- Compare and contrast spontaneous and induced mutations and give examples of each.
- Explain why some regions of DNA are considered mutation hot spots.

1. What are the differences between point mutations that cause phenotypic changes and those that don't?
2. What distinguishes the various kinds of chromosomal mutations: deletions, duplications, inversions, and translocations?
3. What are the differences between spontaneous and induced mutagenesis? Give an example of each.
4. Why do many mutations involve G-C base pairs?

You have seen that there are many different ways in which DNA can be altered, in terms of both the types of changes and the mechanisms by which they occur. We will turn now to the ways in which mutations can cause disease.

key concept 15.2 Mutations in Humans Can Lead to Diseases

The biochemistry that relates genotype (DNA) and phenotype (proteins) has been most completely described for model organisms, such as the prokaryote *E. coli* and the eukaryotes yeast and *Drosophila*. While the details vary, there is great similarity in the fundamental processes among these forms of life. These similarities have permitted the application of knowledge and methods discovered using these model organisms to the study of human biochemical genetics. Our focus in this chapter is mutations that affect human phenotypes, leading to diseases.

focus your learning

- Genetic diseases often result from mutations that render proteins nonfunctional.
- Mutations that cause disease include the full range of affected DNA from point mutations to entire chromosomes.
- Most diseases result from a combination of factors that include both genetics and the environment.

Disease-causing mutations may make proteins dysfunctional

Genetic mutations are often expressed phenotypically as proteins that differ from normal (wild-type) proteins. Abnormalities in enzymes, receptor proteins, transport proteins, structural proteins, and most of the other functional classes of proteins have all been implicated in genetic diseases.

LOSS OF ENZYME FUNCTION In 1934, the urine of two intellectually disabled young siblings was found to contain phenylpyruvic acid, an unusual by-product of the metabolism of the amino acid phenylalanine. It took two decades for scientists to trace the complex clinical phenotype of the disease that afflicted these children, called phenylketonuria (PKU), back to its molecular cause. The disease results from an abnormality in a single enzyme, phenylalanine hydroxylase (PAH), which catalyzes the conversion of dietary phenylalanine to tyrosine (**Figure 15.6**). This enzyme is not active in the livers of PKU patients, leading to excesses of phenylalanine and phenylpyruvic acid in the blood. Since then, the nucleotide sequence of the *PAH* gene has been compared between healthy people and those with the PKU disease, and more than 400 different disease-causing mutations have been found. The most common one is a missense mutation that results in tryptophan instead of arginine at position 408 in the polypeptide chain (**Table 15.1**). As is often the case with loss-of-function mutations, the mutant alleles are recessive, because one functional allele is all that is needed to produce enough functional PAH to prevent the disease.

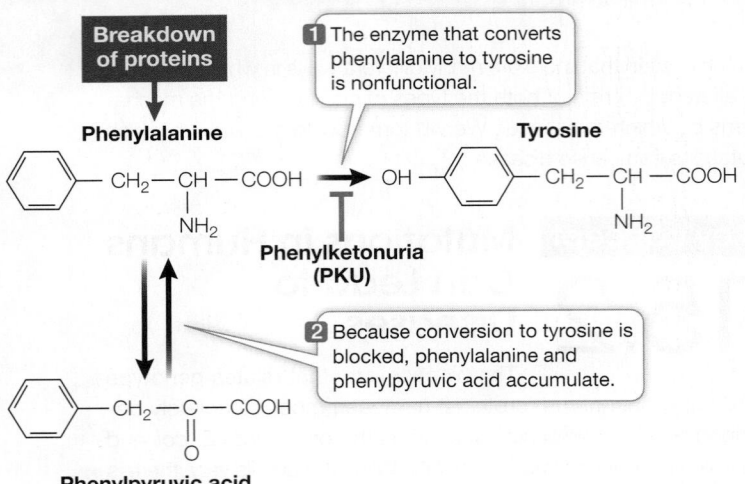

Figure 15.6 One Gene, One Enzyme Phenylketonuria is caused by an abnormality in a specific enzyme that metabolizes the amino acid phenylalanine. Knowing the molecular causes of such single-gene, single-enzyme metabolic diseases can aid researchers in developing screening tests as well as treatments.

	Amino acid position (of 146)								
	2	6	7	16	24	26	56	63	95
A (wild type)	His	Glu	Glu	Gly	Gly	Glu	Gly	His	Lys
Tokuchi	Tyr								
S		Val							
C		Lys							
G			Gly						
J Baltimore				Asp					
Savannah					Val				
E						Lys			
Bangkok							Asp		
Zürich								Arg	
M Saskatoon								Tyr	
N Baltimore									Glu

Only three hemoglobin variants (S, C, and E) lead to clinical problems.

Variants of β-globin

Figure 15.7 Hemoglobin Polymorphism Each of these mutant alleles codes for a protein with a single amino acid change in the 146-amino acid chain of β-globin. Only three of the hundreds of known variants of β-globin, shown on the left, are known to lead to clinical abnormalities. "S" is the sickle-cell anemia allele.

Hundreds of human genetic diseases that result from enzyme abnormalities have been discovered, some of which lead to intellectual disability and premature death. Most of these diseases are rare; PKU, for example, shows up in 1 out of every 12,000 newborns. As you by now realize, for any given gene there may be numerous alleles. Some alleles encode proteins that function normally, whereas others produce proteins that cause disease. Let's take a look at a gene whose alleles illustrate both of these circumstances—the gene for one of the polypeptide chains in hemoglobin.

ABNORMAL HEMOGLOBIN As mentioned in Key Concept 15.1, sickle-cell disease is caused by a recessive, missense mutation. This blood disorder most often afflicts people whose ancestors came from the tropics or from the Mediterranean region.

Recall that human hemoglobin is composed of four globin subunits—two α-chains and two β-chains—as well as the pigment heme (see Figure 3.11). In sickle-cell disease, one of the 146 amino acids in the β-globin polypeptide chain is abnormal: at position 6, glutamic acid is replaced by valine. This replacement changes the charge of the protein (glutamic acid is hydrophilic and valine is hydrophobic), causing it to form long, needlelike aggregates in the red blood cells. The phenotypic result is sickle-shaped red blood cells and an impaired ability of the blood to carry oxygen. The sickled cells tend to block narrow blood capillaries, resulting in tissue damage and eventually death by organ failure.

Because hemoglobin is easy to isolate and study, its variations in the human population have been extensively documented (**Figure 15.7**). Hundreds of single amino acid alterations in β-globin, all due to mutations forming alleles, have been reported. For example, at the same position that is mutated in sickle-cell disease (resulting in hemoglobin S), glutamic acid may be replaced by lysine, causing hemoglobin C disease. In this case, the resulting anemia is usually not severe. Many alleles that result in alterations of the amino acid sequence of hemoglobin do not affect its function. In fact, about 5 percent of us carry at least one missense point mutation in a β-globin allele.

Some of the more common examples of inherited diseases caused by specific protein defects are listed in **Table 15.2**. These mutations can be dominant, codominant, or recessive, and some are sex-linked.

Disease-causing mutations may involve any number of base pairs

Disease-causing mutations may involve a single base pair, a long stretch of DNA, multiple segments of DNA, or even entire chromosomes (as we saw for Down syndrome in Key Concept 11.5).

table 15.1 Two Common Mutations That Cause Phenylketonuria					
	Codon 408 (20% of PKU cases)			Codon 280 (2% of PKU cases)	
	Normal	Mutant		Normal	Mutant
Length of PAH protein	452 amino acids	452 amino acids		452 amino acids	452 amino acids
DNA at codon	...CGG...	...TGG...		...GAA...	...AAA...
	...GCC...	...ACC...		...CTT...	...TTT...
mRNA at codon	...CGG...	...UGG...		...GAA...	...AAA...
Amino acid at codon	Arginine	Tryptophan		Glutamic acid	Lysine
Active PAH enzyme?	Yes	No		Yes	No

table 15.2 Some Human Genetic Diseases

Disease name	Inheritance pattern; births frequency	Gene mutated; protein product	Clinical phenotype
Familial hypercholesterolemia	Autosomal codominant; 1 in 500 heterozygous	*LDLR*; low-density lipoprotein receptor	High blood cholesterol, heart disease
Cystic fibrosis	Autosomal recessive; 1 in 4,000	*CFTR*; chloride ion channel in membrane	Immune, digestive, and respiratory illness
Duchenne muscular dystrophy	Sex-linked recessive; 1 in 3,500 males	*DMD*; the muscle membrane protein dystrophin	Muscle weakness
Hemophilia A	Sex-linked recessive; 1 in 5,000 males	*HEMA*; factor VIII blood clotting protein	Inability to clot blood after injury, hemorrhage

POINT MUTATIONS Sickle-cell anemia is just one of many diseases caused by a point mutation. In some cases (sickle-cell anemia, for example), everyone with the disease has the same genetic mutation. In other cases, different loss-of-function point mutations in one gene can lead to the same disease (as we saw above for PKU). Think about it: the three-dimensional structure of an enzyme protein depends on its secondary structure, so any change in the amino acid sequence of a protein has the potential to affect its structure, and consequently its function.

LARGE DELETIONS Larger mutations may involve many base pairs of DNA. For example, deletions in the X chromosome that include the gene for the protein dystrophin result in Duchenne muscular dystrophy. Dystrophin is important in organizing the structure of muscles, and people who have only the abnormal form have severe muscle weakness. Sometimes only part of the dystrophin gene is missing, leading to an incomplete but partly functional protein and a mild form of the disease. In other cases, deletions span the entire sequence of the gene, so that the protein is missing entirely, resulting in a severe form of the disease. In yet other cases, deletions involve millions of base pairs and cover not only the dystrophin gene but adjacent genes as well; the result may be several diseases in the same person.

CHROMOSOMAL ABNORMALITIES Chromosomal abnormalities also cause human diseases. Such abnormalities result from the gain or loss of complete chromosomes (aneuploidy) (see Figure 11.19), or from the gain or loss of chromosomal segments (see Figure 15.3). About 1 newborn in 200 has a chromosomal abnormality. This may be inherited from a parent who also has the abnormality, or it may result from an error in meiosis during the formation of gametes in one of the parents. One example is fragile-X syndrome, which is a constriction in the tip of the X chromosome that can result in intellectual disability (**Figure 15.8**). About 1 male in 3,000 and 1 female in 7,000 are affected. Although the basic pattern of inheritance is that of an X-linked recessive trait, there are departures from this pattern. Not all people with the fragile-X chromosomal abnormality are intellectually disabled, as we will see.

Expanding triplet repeats demonstrate the fragility of some human genes

About one-fifth of all males who have the fragile-X chromosomal abnormality are phenotypically normal, as are most of their daughters. But many of those daughters' sons are intellectually disabled. In a

family in which fragile-X syndrome appears, later generations tend to show earlier onset and more severe symptoms of the disease. It is almost as if the abnormal allele itself is changing—and getting worse. And that's exactly what is happening.

The gene associated with fragile-X syndrome (*FMR1*) contains a repeated triplet, CGG, at a certain point in the promoter region (**Figure 15.9**). In normal people, this triplet is repeated 6 to 54 times (the average is 29). In intellectually disabled people with fragile-X syndrome, the CGG sequence is repeated 200 to 2,000 times.

Males carrying a moderate number of repeats (55–199) show no symptoms and are called premutated. These repeats become more numerous as the daughters of these men pass the chromosome on to their children. With 200 or more repeats, increased methylation of the cytosines in the CGG triplets is likely, which inhibits transcription of the *FMR1* gene. The normal role of the protein product of this gene is to bind to mRNAs involved in neuron function and to regulate their translation at the ribosome. When the FMR1 protein is not made in adequate amounts, these mRNAs are not properly translated, and nerve cells die. Their loss often results in intellectual disability. The methylated cytosines bind proteins that cause the chromosome constriction and fragile appearance.

This phenomenon of **expanding triplet repeats** has been found in more than a dozen other diseases, such as myotonic dystrophy (involving repeated CTG triplets) and Huntington's disease (in which CAG is repeated). Such repeats, which may be found within a protein-coding region or outside it, appear to be present in many other genes without causing harm. How the repeats expand is not known; one hypothesis is that DNA polymerase may slip after copying a repeat and then fall back to copy it again.

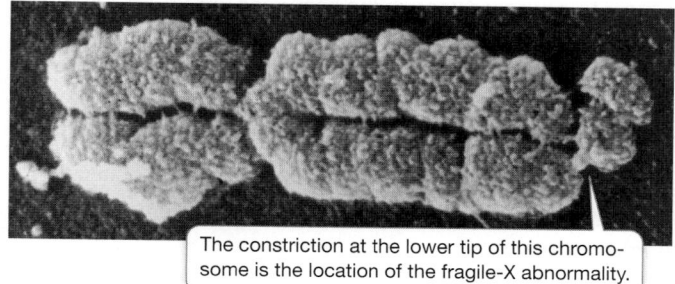

The constriction at the lower tip of this chromosome is the location of the fragile-X abnormality.

Figure 15.8 A Fragile-X Chromosome at Metaphase The chromosomal abnormality associated with fragile-X syndrome shows up under the microscope as a constriction in the chromosome. This occurs during preparation of the chromosome for microscopy.

Normal allele

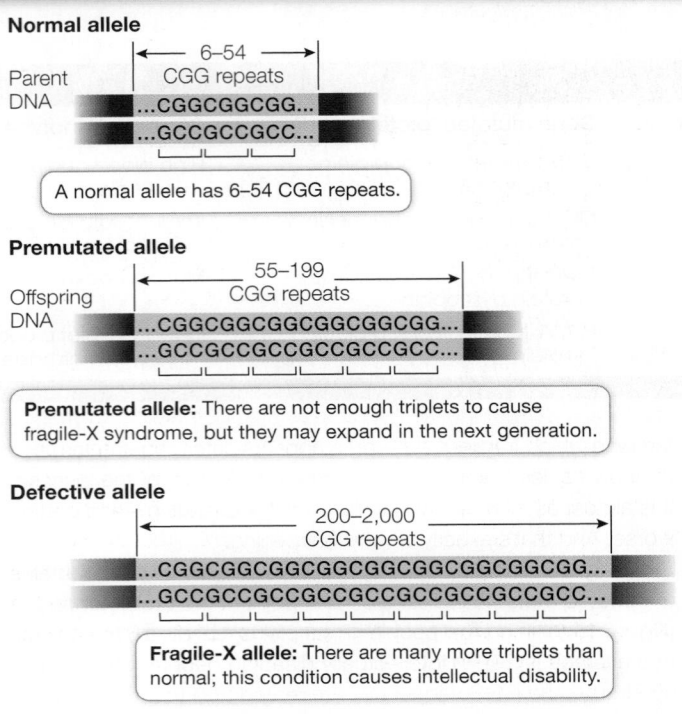

Parent
DNA

6–54
CGG repeats

...CGGCGGCGG...
...GCCGCCGCC...

A normal allele has 6–54 CGG repeats.

Premutated allele

Offspring
DNA

55–199
CGG repeats

...CGGCGGCGGCGGCGGCGG...
...GCCGCCGCCGCCGCCGCC...

Premutated allele: There are not enough triplets to cause fragile-X syndrome, but they may expand in the next generation.

Defective allele

200–2,000
CGG repeats

...CGGCGGCGGCGGCGGCGGCGGCGGCGG...
...GCCGCCGCCGCCGCCGCCGCCGCCGCC...

Fragile-X allele: There are many more triplets than normal; this condition causes intellectual disability.

Figure 15.9 The CGG Repeats in the FMR1 Gene Expand with Each Generation   The genetic defect in fragile-X syndrome is caused by 200 or more repeats of the CGG triplet.

Cancer often involves somatic mutations

Many chromosomal and point mutations have been described in cancer cells. Such mutations affect *oncogenes, whose products stimulate cell division, or tumor suppressor genes, whose products inhibit cell division. More than two gene mutations are usually needed for full-blown cancer.

---

*connect the concepts   You can learn more about the roles of oncogene and tumor suppressor gene mutations in the abnormal cell cycle of cancer cells in Key Concept 11.7.

Because colon cancer progresses to full malignancy slowly, it has been possible to identify the gene mutations that lead to each stage. **Figure 15.10** outlines the "molecular biography" of this form of cancer. At least three tumor suppressor genes and one oncogene must be mutated in sequence in a cell in the colon lining for cancer to develop. Although the occurrence of all of these events in a single cell might seem unlikely, remember that the colon lining has millions of cells, that these cells arise from stem cells that are constantly dividing, and that these changes take place over many years of exposure to natural and synthetic substances in foods, which may act as mutagens.

Most diseases are caused by multiple genes and environment

The example of cancer illustrates how many common phenotypes, including ones that cause disease, are **multifactorial**; that is, they are caused by the interactions of many genes and proteins with one or more factors in the environment. When studying genetics, we tend to call individuals either normal (wild type) or abnormal

(A) Colon cancer development

Section through colon (large intestine)

Normal cells

Loss of normal tumor suppressor gene *APC*

1 A polyp (small growth) forms on the colon wall.

2 A benign, precancerous tumor grows.

Activation of oncogene *RAS*

3 A class II adenoma (benign) grows.

Loss of tumor suppressor gene *DCC*

4 A class III adenoma (benign) grows.

Loss of tumor suppressor gene *p53*

5 A carcinoma (malignant tumor) develops.

Other changes; loss of anti-metastasis gene

6 The cancer metastasizes (spreads to other tissues).

(B) Normal versus cancerous colon

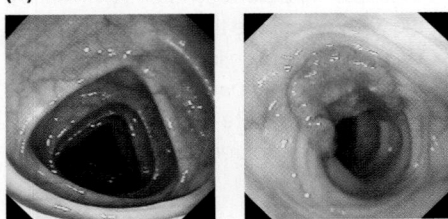

Figure 15.10 Multiple Somatic Mutations Transform a Normal Colon Epithelial Cell into a Cancer Cell   **(A)** At least five genes must be mutated in a single cell to produce metastatic colon cancer. **(B)** These images from a screening test reveal a normal colon (left) and colon cancer (right).

(mutant); however, in reality every individual contains thousands or millions of genetic variations that arose through mutations. Our susceptibility to disease is often determined by complex interactions between these genotypes and factors in the environment, such as the foods we eat or the pathogens we encounter. For example, a complex set of genotypes determines who among us can eat a high-fat diet and not experience a heart attack, or who will succumb to disease when exposed to infectious bacteria. Estimates suggest that up to 60 percent of all people are affected by diseases that are genetically influenced. Identifying these genetic influences is another major task of molecular medicine and human genome sequencing.

15.2 recap

Many genetic mutations are expressed as nonfunctional enzymes, structural proteins, or membrane proteins. Human genetic diseases may be inherited in dominant, codominant, or recessive patterns, and they may be sex-linked.

learning outcomes

You should be able to:

- Describe an abnormal protein in humans that results from a genetic mutation and causes a disease.
- Show how the following mutations can lead to phenotypic change: point mutations, large deletions, and chromosomal abnormality.
- Justify the claim that most diseases arise from complex interactions between genetic and environmental factors.

1. Alkaptonuria is an inborn error of metabolism, caused by defects in an enzyme in the pathway that breaks down tyrosine (see Key Concept 14.1). Humans who are homozygous for one of these mutations make nonfunctional enzyme and accumulate the enzyme's substrate, homogentisic acid, which causes their disease symptoms. In 1996, researchers in Spain cloned and sequenced the gene for the enzyme, and characterized several mutant alleles. Here is the wild-type coding strand sequence for part of the gene, with the amino acid sequence of the protein below:

    ... TTG   ATA   CCC   ATT   GCC ...
    ... Leu   Ile   Pro   Ile   Ala ...

    Here is the sequence for one of the mutant alleles:

    ... TTG   ATA   TCC   ATT   GCC ...

    a. What is the amino acid sequence produced by the mutant allele? What type of mutation is this: silent, nonsense, missense, or frame-shift?
    b. Why is this mutation likely to affect the function of the enzyme? (Hint: see Key Concept 3.2, especially Table 3.2.)

2. Not all women with a mutation in the *BRCA1* gene get early-onset breast cancer. Explain.

3. Why do some members of a family show symptoms of fragile-X syndrome and others do not?

In the previous section we described the ways in which mutations can lead to human disease. We will turn now to the ways that biologists detect mutations in DNA.

key concept 15.3 Mutations Can Be Detected and Analyzed

A challenge for biologists studying mutations is to precisely describe the DNA changes that lead to specific protein changes—an area of research called molecular genetics. Of course, the most direct and comprehensive way to analyze DNA is to determine its sequence of bases. DNA sequencing technologies are continually improving, and the entire genomes of many organisms have now been sequenced completely. Furthermore, the genomes of closely related organisms have been compared in order to identify mutations. We will discuss sequencing technology in Chapter 17. We are approaching the era where DNA sequencing will be the method of choice in searching for mutations. DNA sequencing is not yet, however, in routine use in medicine. In this section we will look at some of the techniques that are used in combination with DNA sequencing to study DNA, and to identify mutations that cause disease.

focus your learning

- DNA sequencing techniques provide great promise for identifying mutations as part of medical diagnosis and treatment.
- Mutations in DNA can be used to look for disease.

Cleavage of DNA by restriction enzymes can be used to rapidly detect mutations

All organisms, including bacteria, must have ways of dealing with their enemies. As we saw in Key Concept 13.1, bacteria can be attacked by viruses called bacteriophages. These viruses inject their genetic material into the host cell and turn it into a virus-producing factory, eventually killing the cell. Some bacteria defend themselves against such invasions by producing **restriction enzymes** (also known as restriction endonucleases), which cut double-stranded DNA molecules—such as those injected by bacteriophages—into smaller, noninfectious fragments (**Figure 15.11**). These enzymes break the bonds of the DNA backbone between the 3' hydroxyl group of one nucleotide and the 5' phosphate group of the next nucleotide. This cutting process is called **restriction digestion**.

There are many such restriction enzymes, each of which cleaves DNA at a specific sequence of bases called a **recognition sequence** or a **restriction site**. Most recognition sequences are four to six base pairs long. Because each sequence of bases has a unique structure (see Key Concept 13.2), it can be specifically recognized by a particular restriction enzyme. Cells protect themselves from being digested by their own enzymes by modifying their DNA, often with methyl groups, to prevent binding by the restriction enzymes.

Restriction enzymes can be isolated from the cells that make them and used as biochemical reagents in the laboratory to give information about the nucleotide sequences of DNA molecules from other organisms. If DNA from any organism is incubated in a test tube with a restriction enzyme (along with buffers and salts that help the enzyme function), that DNA will be cut wherever the restriction

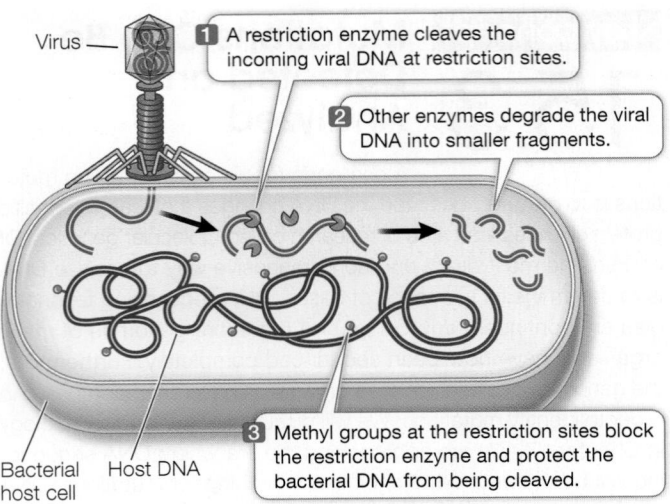

Virus

1 A restriction enzyme cleaves the incoming viral DNA at restriction sites.

2 Other enzymes degrade the viral DNA into smaller fragments.

3 Methyl groups at the restriction sites block the restriction enzyme and protect the bacterial DNA from being cleaved.

Bacterial host cell   Host DNA

Figure 15.11 Bacteria Fight Invading Viruses by Making Restriction Enzymes

site occurs. A specific sequence of bases defines each restriction site. For example, the enzyme *Eco*RI (named after its source strain of the bacterium *E. coli*) cuts DNA only where it encounters the following paired sequence in the DNA double helix:

$$5' \ldots \text{GAATTC} \ldots 3'$$
$$3' \ldots \text{CTTAAG} \ldots 5'$$

Note that this sequence is palindromic, like the word "mom." This means that both strands have the same sequence when they are read from their 5' (or their 3') ends. The *Eco*RI enzyme has two identical active sites on its two subunits, which cleave the two strands simultaneously between the G and the A of each strand:

$$5' \ldots \text{GAATTC} \ldots 3' \longrightarrow 5' \ldots \text{G} \qquad \text{AATTC} \ldots 3'$$
$$3' \ldots \text{CTTAAG} \ldots 5' \qquad \qquad 3' \ldots \text{CTTAA} \qquad \text{G} \ldots 5'$$

The *Eco*RI recognition sequence occurs, on average, about once in every 4,000 base pairs in a typical prokaryotic genome, or about once per four prokaryotic genes. So *Eco*RI can chop a large piece of DNA into smaller pieces containing, on average, just a few genes. Using *Eco*RI in the laboratory to cut small genomes, such as those of viruses that have tens of thousands of base pairs, may result in just a few fragments. For a huge eukaryotic chromosome with tens of millions of base pairs, a very large number of fragments will be created.

Of course, "on average" does not mean that the enzyme cuts all stretches of DNA at regular intervals. For example, the *Eco*RI recognition sequence does not occur even once in the 40,000 base pairs of the T7 bacteriophage genome—a fact that is crucial to the survival of this virus, since its host is *E. coli*. Fortunately for *E. coli*, the *Eco*RI recognition sequence does appear in the DNA of other bacteriophages.

Gel electrophoresis separates DNA fragments

Restriction enzyme digestion is used to manipulate DNA in the laboratory so that mutations can be identified and analyzed. After a laboratory sample of DNA has been cut with one or more restriction enzymes, the DNA is in fragments, which must be separated to identify (map) where the cuts were made. Because the recognition sequence does not occur at regular intervals, the fragments are not all the same size, and these size differences can be used to separate the fragments from one another. Separating the fragments is necessary to determine the number and molecular sizes (in base pairs) of the fragments produced, or to identify and purify an individual fragment for further analysis or for use in an experiment.

Gel electrophoresis is a common and convenient technique for separating or purifying DNA fragments. Samples containing the fragments are placed in wells at one end of a semisolid gel (usually made of agarose or polyacrylamide), and an electric field is applied to the gel (**Figure 15.12**). Because of its phosphate groups, DNA is negatively charged at neutral pH; therefore, because opposite charges attract, the DNA fragments move through the gel toward the positive end of the field. Because the spaces between the polymers of the gel are small, small DNA molecules can move through the gel faster than larger ones. Thus DNA fragments of different sizes separate from one another, forming bands that can be detected with a dye. This provides three types of information:

1. *The number of fragments*. The number of fragments produced by digestion of a DNA sample with a given restriction enzyme depends on how many times that enzyme's recognition sequence occurs in the sample. Thus gel electrophoresis can provide some information about the presence of specific DNA sequences (the restriction sites) in the DNA sample.

2. *The sizes of the fragments*. DNA fragments of known size (size markers) are often placed in one well of the gel to provide a standard for comparison. The size markers are used to determine the sizes of the DNA fragments in samples in the other wells. By comparing the fragment sizes obtained with two or more restriction enzymes, the locations of their recognition sites relative to one another can be worked out (mapped).

3. *The relative abundance of a fragment*. In many experiments, the investigator is interested in how much DNA is present. The relative intensity of a band produced by a specific fragment can indicate the amount of that fragment.

DNA fingerprinting combines PCR with restriction analysis and electrophoresis

The methods we have just described are used in **DNA fingerprinting**, which identifies individuals based on differences in their DNA sequences. DNA fingerprinting works best with sequences that are highly polymorphic—that is, sequences that have multiple alleles (because of many point mutations during the evolution of the organism) and are therefore likely to be different in different individuals. Two types of polymorphisms are especially informative:

1. **Single nucleotide polymorphisms (SNPs**; pronounced "snips") are inherited variations involving a single nucleotide base—they are point mutations. These polymorphisms have been mapped for many organisms. If one parent is homozygous for the base A at a certain point in the genome, and the other parent is homozygous for a G at that point, the offspring will be heterozygous: one chromosome will have A at that point and the other will have G. If a SNP occurs in a restriction enzyme recognition site, such that one variant is recognized by the enzyme and the other isn't, then individuals can be distinguished from one another very easily using the *polymerase chain reaction (PCR). A fragment containing the polymorphic sequence is amplified by PCR from samples of total DNA isolated from each individual. The fragments are then cut with the restriction enzyme and analyzed by gel electrophoresis.

 Activity 15.3 **Allele-Specific Cleavage**
www.Life11e.com/ac15.3

2. **Short tandem repeats (STRs)** are short, repetitive DNA sequences that occur side by side on the chromosomes, usually in the noncoding regions. These repeat patterns, which contain one to five base pairs, are also inherited. For example, at a particular locus on chromosome 15 there may be an STR of "AGG." An individual may inherit an allele with six copies of the repeat (AGGAGGAG-GAGGAGGAGG) from her mother and an allele with two copies (AGGAGG) from her father. Again, PCR is used to amplify DNA fragments containing these repeated sequences, and then the amplified fragments, which have different sizes because of the different lengths of the repeats, are distinguished by gel electrophoresis (**Figure 15.13A**).

***connect the concepts** As described in Key Concept 13.5, PCR is a technique used to make multiple copies of a DNA sequence—that is, to amplify it. The process essentially automates DNA replication; over many cycles the amount of DNA can be increased exponentially.

The method of DNA fingerprinting used most commonly today hydrolyzes STR analysis. The Federal Bureau of Investigation in the United States uses 13 STR loci in its Combined DNA Index System (CODIS) database (**Table 15.3**). An analysis of these loci in your DNA would reveal your particular DNA fingerprint. Looking at Table 15.3, you might inherit:

- From your mother: allele 72 from chromosome 4; allele 23 from chromosome 7; allele 14 from chromosome 11; and allele 12 from chromosome 18

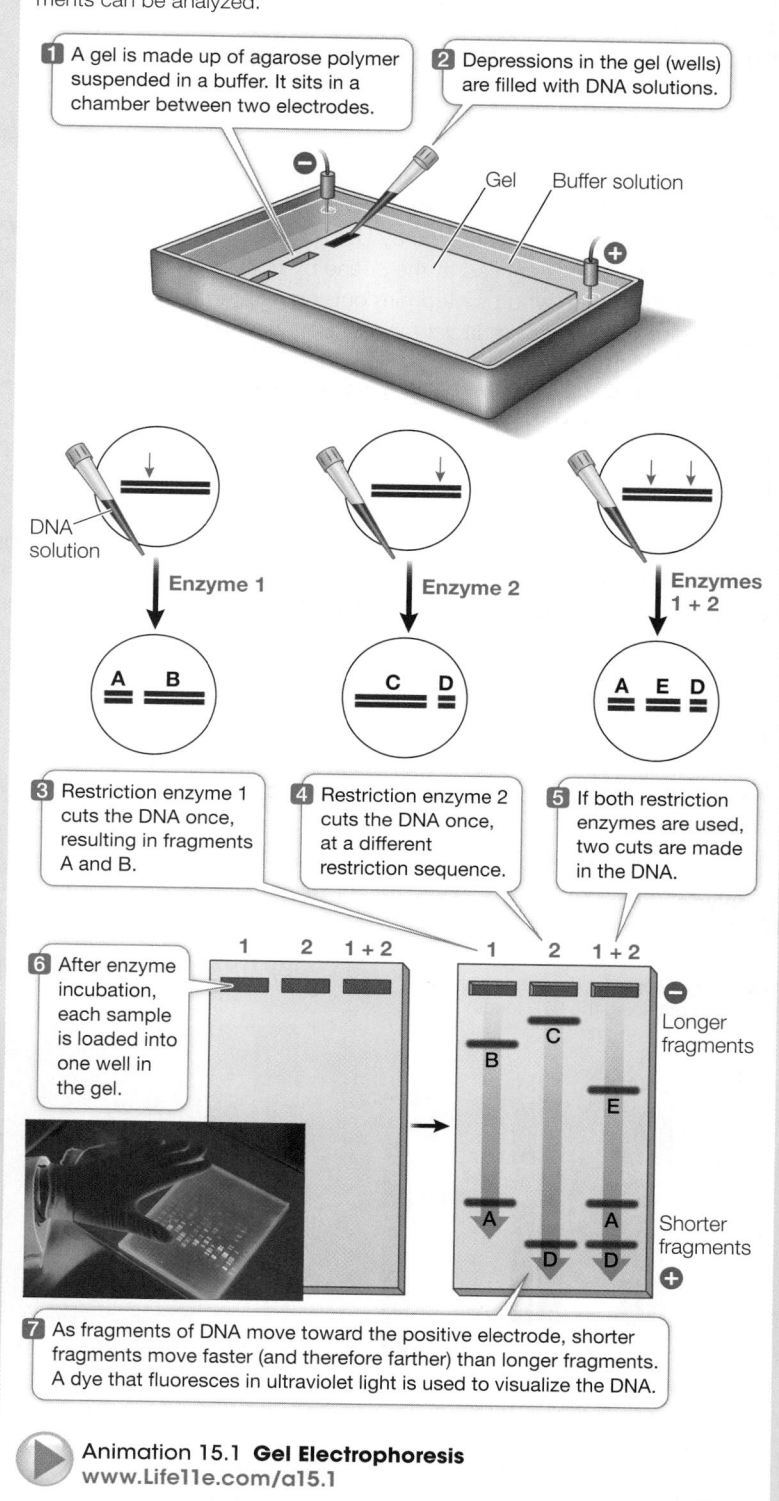

research tools

Figure 15.12 Separating Fragments of DNA by Gel Electrophoresis
A mixture of DNA fragments is placed in a gel, and an electric field is applied across the gel. The negatively charged DNA moves toward the positive end of the field, with smaller molecules moving faster than larger ones. After minutes to hours for separation, the electric power is shut off and the separated fragments can be analyzed.

1 A gel is made up of agarose polymer suspended in a buffer. It sits in a chamber between two electrodes.

2 Depressions in the gel (wells) are filled with DNA solutions.

Gel   Buffer solution

DNA solution

Enzyme 1   Enzyme 2   Enzymes 1 + 2

A B   C D   A E D

3 Restriction enzyme 1 cuts the DNA once, resulting in fragments A and B.

4 Restriction enzyme 2 cuts the DNA once, at a different restriction sequence.

5 If both restriction enzymes are used, two cuts are made in the DNA.

6 After enzyme incubation, each sample is loaded into one well in the gel.

1   2   1 + 2   1   2   1 + 2

Longer fragments

Shorter fragments

7 As fragments of DNA move toward the positive electrode, shorter fragments move faster (and therefore farther) than longer fragments. A dye that fluoresces in ultraviolet light is used to visualize the DNA.

Animation 15.1 **Gel Electrophoresis**
www.Life11e.com/a15.1

- From your father: allele 56 from chromosome 4; allele 22 from chromosome 7; allele 16 from chromosome 11; and allele 12 from chromosome 18

Note that in this case you are heterozygous for the alleles on three of the chromosomes and homozygous for the allele on chromosome 18. With all the alleles and 13 loci, the probability of two people sharing the same alleles is very small. So a DNA sample from a crime scene can be used to determine whether a particular suspect left that sample at the scene.

As we have just shown, DNA fingerprinting can be used to help prove the innocence or guilt of a suspect, but it can also be used to identify individuals who are related to one another. On May 2, 2011, Osama bin Laden was killed by U.S. soldiers at his home in Pakistan. He was identified at the scene by comparison with photographs, a wife who pointed him out, and instant analysis using a digital camera with facial recognition software. In addition, DNA fingerprinting was used. Bin Laden's son Khalid was also killed in the raid, and a sister had previously died in the U.S. Analyses of their DNA along with that of Osama indicated that the three shared many polymorphisms and were highly likely to be closely related. The same methods were also used to identify Saddam Hussein, who was captured in 2003 and later executed in Iraq (**Figure 15.13B**).

DNA analysis with genetic markers such as SNPs and STRs has applications throughout all areas of biological research. For example, these markers are used to analyze the organization of genomes, to identify species or individuals within species, to compare species or organisms to see how closely related they are, and to analyze particular genes and the phenotypes associated with them. In the remainder of this chapter we will focus on the use of these markers and other technologies that are used to study and treat genetic diseases.

table **15.3**	Four of the Genetic Loci Used for Identification in the CODIS Database		
Human chromosome	Locus name	Repeated sequence	Number of alleles
4	FGA	CTTT	80
7	D7S820	GATA	30
11	TH01	TCAT	20
18	S18S51	AGAA	51

DNA analysis can be used to identify mutations that lead to disease

With diseases such as PKU and sickle-cell anemia, you have seen that the clinical phenotypes of inherited diseases could be traced to individual proteins, and that the genes could then be identified. For example, in sickle-cell anemia, the protein abnormality in hemoglobin was described first (a single amino acid change), and then the gene for β-globin was isolated and the DNA mutation was pinpointed.

Clinical phenotype → protein phenotype → gene

With the advent of new ways to identify DNA variations, a new pattern of human genetic analysis emerged. In these cases, the clinical phenotype is first related to a DNA variation, and then the gene and protein involved are identified. For cystic fibrosis (see Table 15.2), a mutant version of the gene *CFTR* was isolated first, and then the protein was characterized:

Clinical phenotype → gene → protein phenotype

Whichever approach is used, final identification of the protein(s) involved in a disease is important in designing specific therapies.

(A) STR analysis compares sequence for specific loci from two or more DNA samples.

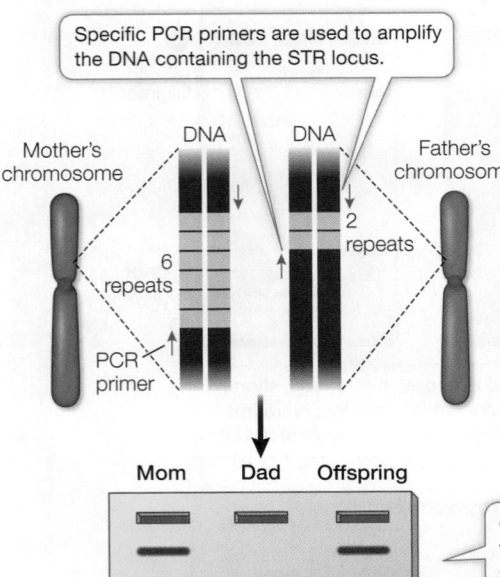

(B) DNA fingerprinting confirms individual's identities

Figure 15.13 DNA Fingerprinting with Short Tandem Repeats **(A)** A particular STR locus can be analyzed to determine the number of repeat sequences that were inherited by an individual from each parent. The two alleles can be identified in an electrophoresis gel on the basis of their sizes. When several STR loci are analyzed, the pattern can constitute a definitive identification of an individual. **(B)** When the dictator Saddam Hussein was captured in Iraq, a sample of his cheek epithelial cells was taken for DNA fingerprinting. A comparison with DNA fingerprints of relatives provided military scientists with evidence that the man in question was indeed Saddam Hussein.

Q: Should everyone have their DNA fingerprinted at birth to create a "genetic ID" for use in screening for genetic diseases? What would be the advantages and disadvantages of this?

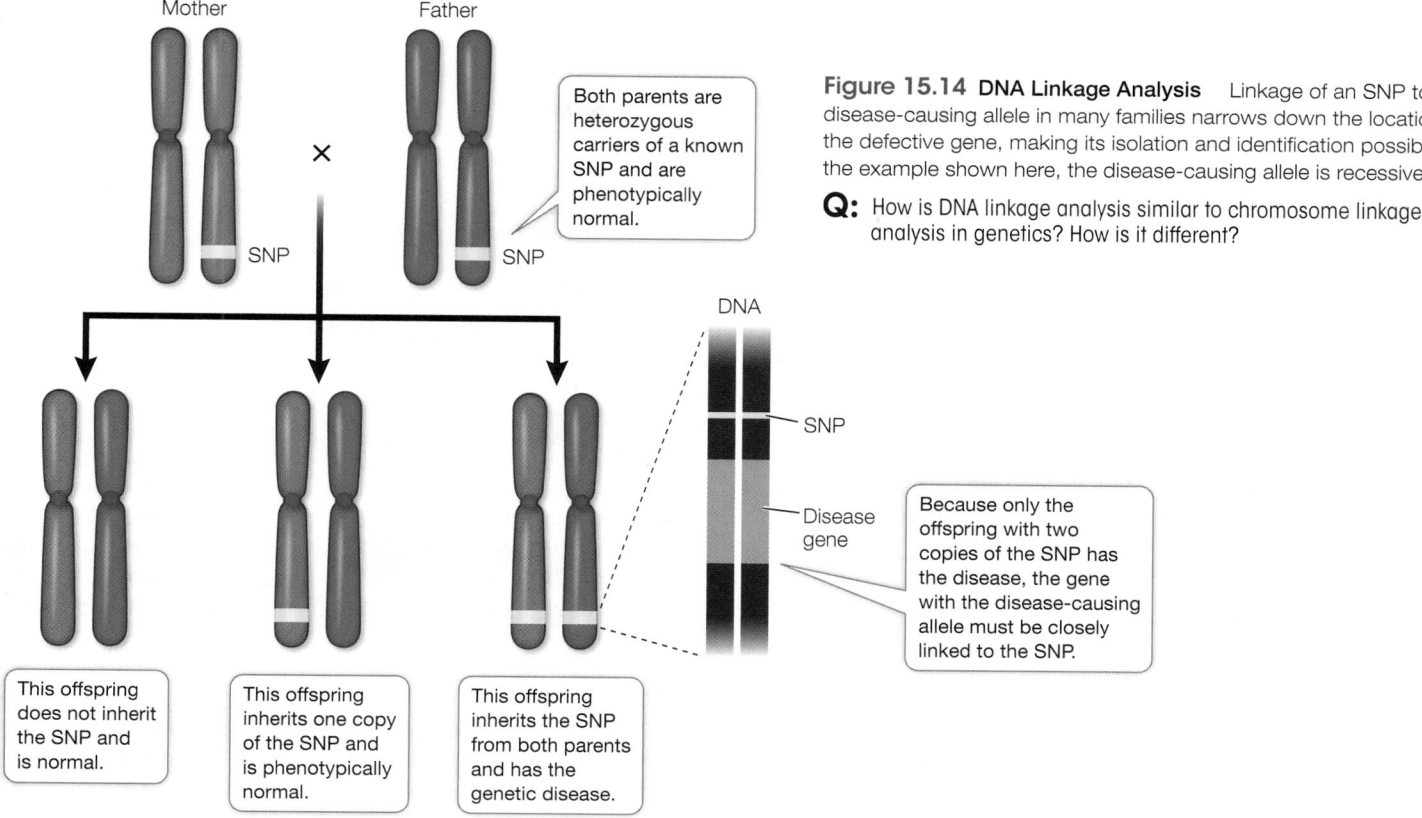

Figure 15.14 DNA Linkage Analysis   Linkage of an SNP to a disease-causing allele in many families narrows down the location of the defective gene, making its isolation and identification possible. In the example shown here, the disease-causing allele is recessive.

Q: How is DNA linkage analysis similar to chromosome linkage analysis in genetics? How is it different?

Both parents are heterozygous carriers of a known SNP and are phenotypically normal.

This offspring does not inherit the SNP and is normal.

This offspring inherits one copy of the SNP and is phenotypically normal.

This offspring inherits the SNP from both parents and has the genetic disease.

Because only the offspring with two copies of the SNP has the disease, the gene with the disease-causing allele must be closely linked to the SNP.

Genetic markers can be used to find disease-causing genes

Identifying a mutant gene requires finding a marker that is closely linked to the gene of interest, a process called **linkage analysis**. The reference points for gene isolation are **genetic markers**. Genetic markers such as STRs and SNPs can be used as landmarks to find a gene of interest, if the gene also has multiple alleles (for example, normal and disease-causing alleles). The key to this method is the well-established observation that if two genes are located near each other on the same chromosome, they are usually passed on together from parent to offspring (see Key Concept 12.4). The same holds true for any pair of DNA genetic markers. In the case of linkage analysis, the idea is to find markers that are progressively closer to the gene of interest.

As you have seen, SNPs and STRs are widespread in eukaryotic genomes. There is roughly one SNP for every 1,330 base pairs in the human genome, and any regions of the genome also contain repetitive DNA sequences such as those found in STRs. SNPs and STRs can be analyzed using the PCR techniques mentioned above. SNPs can also be detected using sophisticated chemical methods such as mass spectrometry.

To narrow down the location of a gene, a scientist must find a genetic marker that is *always inherited with the gene*. To do this, family medical histories are taken and pedigrees are constructed. If a genetic marker and a genetic disease are inherited together in many families, then they must be near each other on the same chromosome (**Figure 15.14**). This situation recalls the conclusion reached in the classic studies of inheritance undertaken

by Thomas Hunt Morgan and discussed in Key Concept 12.4: two genes do not always assort independently. Genes that are "linked" on the same chromosome will sometimes be inherited together—especially when the two loci are close to one another on the chromosome.

Linkage narrows down the location of the gene to a few hundred thousand base pairs. Once a linked DNA region is identified, many methods are available to identify the actual gene responsible for a genetic disease. The complete sequence of the region can be searched for candidate genes, using information available from databases of genome sequences. With luck a scientist can make an educated guess, based on biochemical or physiological information about the disease, along with information about the functions of candidate genes, as to which gene is responsible for the disease. The identification of DNA polymorphisms within candidate genes that correlate with the presence or absence of disease can also help narrow down the search. A variety of techniques, such as analyzing mRNA levels of candidate genes in diseased and healthy individuals, are used to confirm that the correct gene has been identified.

The isolation of the *BRCA1* gene that is involved in breast cancer (described in the chapter opening) offers a good illustration of molecular techniques used to identify genes associated with disease, and is described in **Investigating Life: How Was the *BRCA1* Gene Identified?** Analysis of the *BRCA1* mutations came after Marie-Claire King of the University of California at Berkeley worked with breast cancer patients and their families to identify SNPs closely linked to the breast cancer phenotype.

experiment

Original Paper: Miki, Y. et al. 1994. A strong candidate for the breast and ovarian cancer susceptibility gene *BRCA1*. *Science* 266: 66–71.

DNA linkage analysis of families led to the isolation of *BRCA1*, a gene that can lead to breast cancer.

HYPOTHESIS▶ A gene mutated in hereditary breast cancer can be isolated.

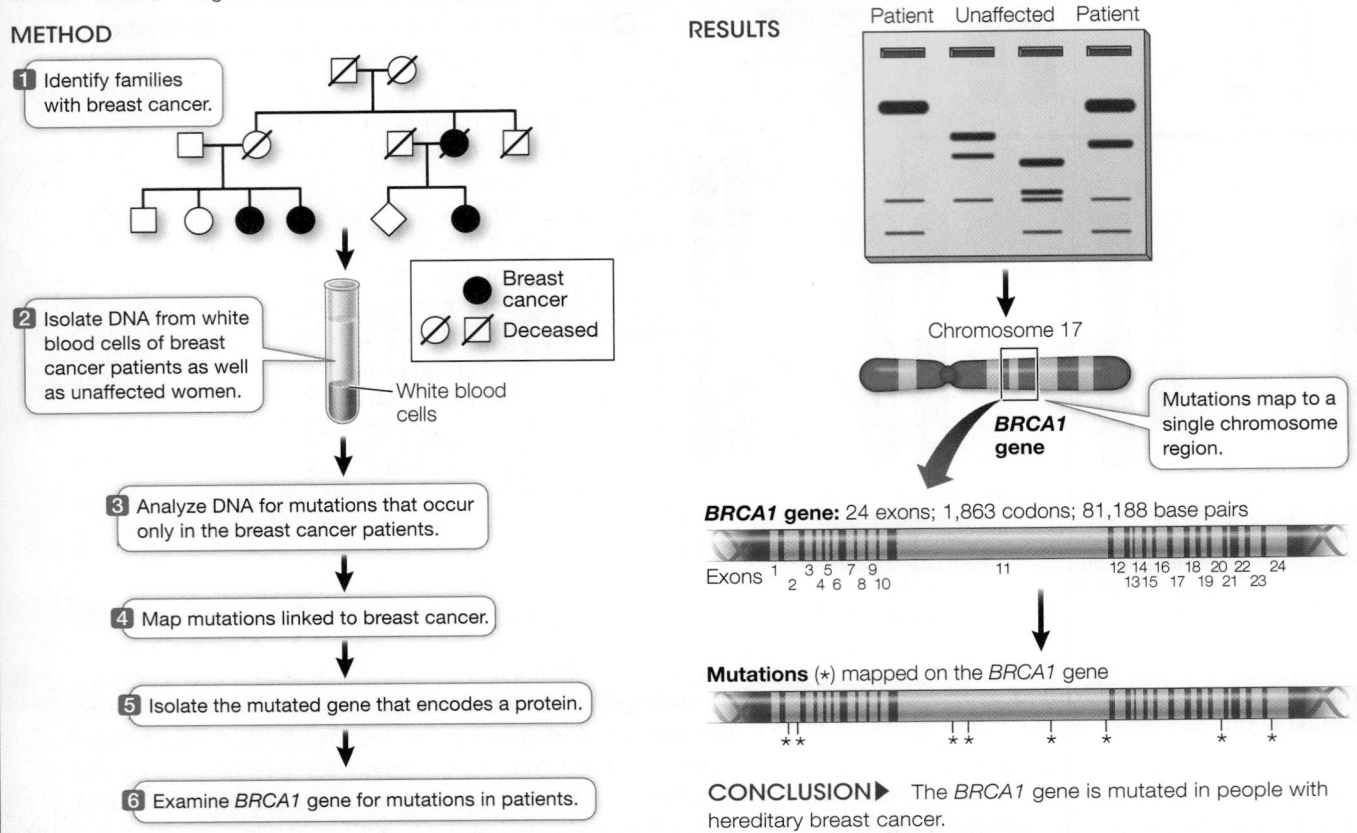

METHOD

1 Identify families with breast cancer.

2 Isolate DNA from white blood cells of breast cancer patients as well as unaffected women.

— White blood cells

3 Analyze DNA for mutations that occur only in the breast cancer patients.

4 Map mutations linked to breast cancer.

5 Isolate the mutated gene that encodes a protein.

6 Examine *BRCA1* gene for mutations in patients.

● Breast cancer
⊘ ⊘ Deceased

RESULTS

Patient   Unaffected   Patient

Chromosome 17

BRCA1 gene

Mutations map to a single chromosome region.

***BRCA1* gene:** 24 exons; 1,863 codons; 81,188 base pairs

Exons   1 3 5 7 9   11   12 14 16 18 20 22 24
          2 4 6 8 10   13 15 17 19 21 23

Mutations (*) mapped on the *BRCA1* gene

CONCLUSION▶ The *BRCA1* gene is mutated in people with hereditary breast cancer.

15.3 recap

Large DNA molecules can be cut into smaller pieces by restriction digestion and then sorted by gel electrophoresis. PCR is used to amplify sequences of interest from complex samples. These techniques are used in DNA fingerprinting to analyze DNA polymorphisms for the purpose of identifying individuals. Genes involved in disease can be identified by first detecting the abnormal DNA sequence and then the protein that the wild-type allele encodes. Scientists hope to be able to identify all species using DNA analysis.

learning outcomes

You should be able to:

- Explain how DNA sequencing can be used to identify mutations.
- Explain how a mutation can be identified as a cause for disease.

15.3 recap

1. Cystic fibrosis is an autosomal recessive disease in which thick mucus is produced in the lungs and airways. The gene responsible for this disease encodes a protein composed of 1,480 amino acids. In most patients with cystic fibrosis, the protein has 1,479 amino acids: a phenylalanine is missing at position 508. A baby is born with cystic fibrosis. He has an older brother who is not affected. How would you test the DNA of the older brother to determine whether he is a carrier for cystic fibrosis?

2. How can a gene mutation that causes a disease be mapped and detected before its protein product is known?

The determination of the precise molecular phenotypes and genotypes of various human genetic diseases has made it possible to diagnose these diseases even before symptoms appear. Let's take a detailed look at some of these genetic screening techniques.

About one in ten cases of breast cancer is hereditary. That is, the cancer arises from a germ line mutation (meaning it occurs in all cells of the body) that is expressed in breast cells as cancer. The other nine in ten cases come about from somatic mutations in breast cells. The identification of a gene involved in hereditary breast cancer was made by DNA analysis. First, the location of the gene was narrowed down by analysis of SNPs in families where several women had breast cancer (see the experiment). Then a team led by Mark Skolnick at the University of Utah and the company Myriad Genetics sequenced the DNA in the candidate region and looked for a sequence that had a promoter, transcription terminator, start and stop codons, and sequences at the ends of introns—in short, the sequences that mark a protein-coding gene. Once they found the candidate gene, they subjected it to a series of tests to see if it was the one associated with breast cancer.

QUESTIONS▶

1. Hereditary (as opposed to somatic) breast cancer is often suspected when the cancer develops relatively early in life. Hereditary cancers tend to arise earlier because they involve a germ line mutation present in all cells at birth, whereas a cancer due to a spontaneous somatic mutation in the DNA of a breast cell can happen any time (statistically, later). Skolnick's team examined the timing of breast cancer cases in patients who had different mutations in *BRCA1*. The results are shown in **Table A**. What conclusions can you draw?

Table A

Family	Total cases	Cases before age 50
A	31	20
B	22	14
C	10	7

2. DNA sequencing evaluated *BRCA1* in the cancer patients and their families, as well as in controls who did not have breast cancer themselves or in their family. The results are shown in **Table B**. Were the mutations in *BRCA1* present only in the breast cancer patients? What do you think the effects of the mutations were in each of the three families?

Table B

Family	Codon number in *BRCA1*	Change in patients	Change in controls?
A	1313	C → T	None (0/170)
B	1775	T → G	None (0/120)
C	24	deletion 11 bp	None (0/180)

3. Human tissues from people without breast cancer were examined by nucleic acid hybridization (see Figure 14.6) for the level of mRNA for *BRCA1*. The results are shown in **Table C**. What can you conclude about the kinds of cancer that might be caused by mutations in *BRCA1*?

Table C

Tissue	mRNA detected?
Breast	Yes
Colon	No
Intestine	No
Ovary	Yes
Prostate	No
Spleen	No
Testis	No
Thymus	Yes

A similar **work with the data** exercise may be assigned in **LaunchPad**.

▶key concept

15.4 Genetic Screening Is Used to Detect Diseases

Genetic screening is the use of a test to identify people who have, are predisposed to, or are carriers of a genetic disease. It can be done at many times of life and used for many purposes:

- *Prenatal screening* can be used to identify an embryo or fetus with a disease so that medical intervention can be applied or decisions can be made about whether or not to continue the pregnancy.

- *Newborn babies* can be screened so that proper medical intervention can be initiated quickly for those babies who need it.

- *Asymptomatic people* who have relatives with a genetic disease can be screened to determine whether they are carriers of the disease-associated allele or are likely to develop the disease themselves.

Genetic screening can be done at the level of either the phenotype or the genotype.

focus your learning

- Genetic screening can involve phenotype or genotype analysis.

Genetic screening can be done by examining the phenotype

Genetic screening can involve examining a protein or other chemical that is relevant to a phenotype associated with a particular disease. Perhaps the best example is the test for phenylketonuria (PKU), which has made it possible to identify the disease in newborns, so that treatment can be started immediately. It is very likely that you were screened as a newborn for PKU.

Initially, babies born with PKU have a normal phenotype because excess phenylalanine in their blood before birth diffuses across the placenta to the mother's circulatory system. Since the mother is almost always heterozygous, and therefore has adequate phenylalanine hydroxylase activity, her body metabolizes the excess phenylalanine from the fetus. After birth, however, the baby begins to consume protein-rich food (milk) and to break down some of his or her own proteins. Phenylalanine begins to accumulate in the blood.

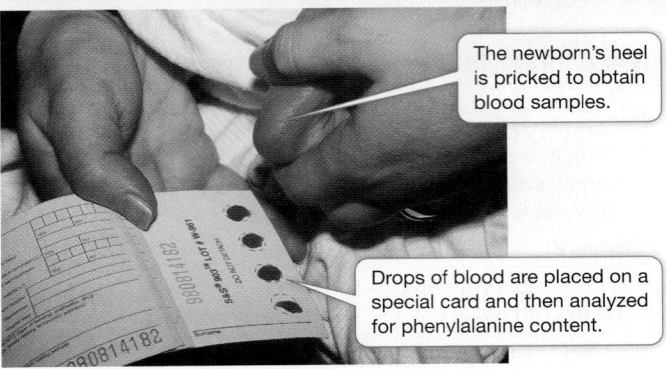

The newborn's heel is pricked to obtain blood samples.

Drops of blood are placed on a special card and then analyzed for phenylalanine content.

Figure 15.15 Genetic Screening of Newborns for Phenylketonuria
A blood test is used to screen newborns for phenylketonuria. Small samples of blood are taken from a newborn's heel. The samples are placed in a machine that measures the phenylalanine concentration in the blood. Early detection means that the symptoms of the condition can be prevented by putting the baby on a therapeutic diet.

After a few days, the phenylalanine level in the baby's blood may be ten times higher than normal. Within days, the developing brain is damaged, and untreated children with PKU become severely intellectually disabled. But if detected early, PKU can be treated with a special diet low in phenylalanine to avoid the brain damage that would otherwise result. Thus early detection is imperative.

Routine newborn screening for PKU and other diseases began in 1963 with the development of a simple, rapid test for the presence of excess phenylalanine in blood serum (**Figure 15.15**). This method uses dried blood spots from newborn babies and can be automated so that a screening laboratory can process many samples in a day. Newborn babies' blood is now screened for up to 35 genetic diseases. Some are common, such as congenital hypothyroidism, which occurs about once in 4,000 births and causes reduced growth and intellectual disability because of low levels of thyroid hormone. With early intervention, many of these infants can be successfully treated. So it is not surprising that newborn screening is legally mandatory in many countries, including the United States and Canada. You were probably screened for several genetic diseases in the first days of your life. Ask your parents which ones!

DNA testing is the most accurate way to detect abnormal genes

The level of phenylalanine in the blood is an indirect measure of phenylalanine hydroxylase activity in the liver. But how can we screen for *genetic diseases that are not detectable by blood tests? What if blood is difficult to obtain, as it is in a fetus? How are genetic abnormalities in heterozygotes, who express the normal protein at some level, identified?

---

*connect the concepts Genetic testing is a rapidly changing field, with tests being done at different stages of prenatal development. See Key Concept 43.2 for an overview of the stages of human development, to place the timing of genetic tests in perspective.

DNA testing is the direct analysis of DNA for a mutation, and it offers the most direct and accurate way of detecting an abnormal allele. Now that the mutations responsible for many human diseases have been identified, any cell in the body can be examined at any time of life for mutations. The amplification power of PCR means that only one or a few cells are needed for testing. These methods work best for diseases caused by only one or a few different mutations.

research tools

Figure 15.16 DNA Testing by Allele-Specific Oligonucleotide Hybridization   Testing of this family reveals that three of them are heterozygous carriers of the sickle allele. The first child, however, has inherited two normal alleles and is neither affected by the disease nor a carrier.

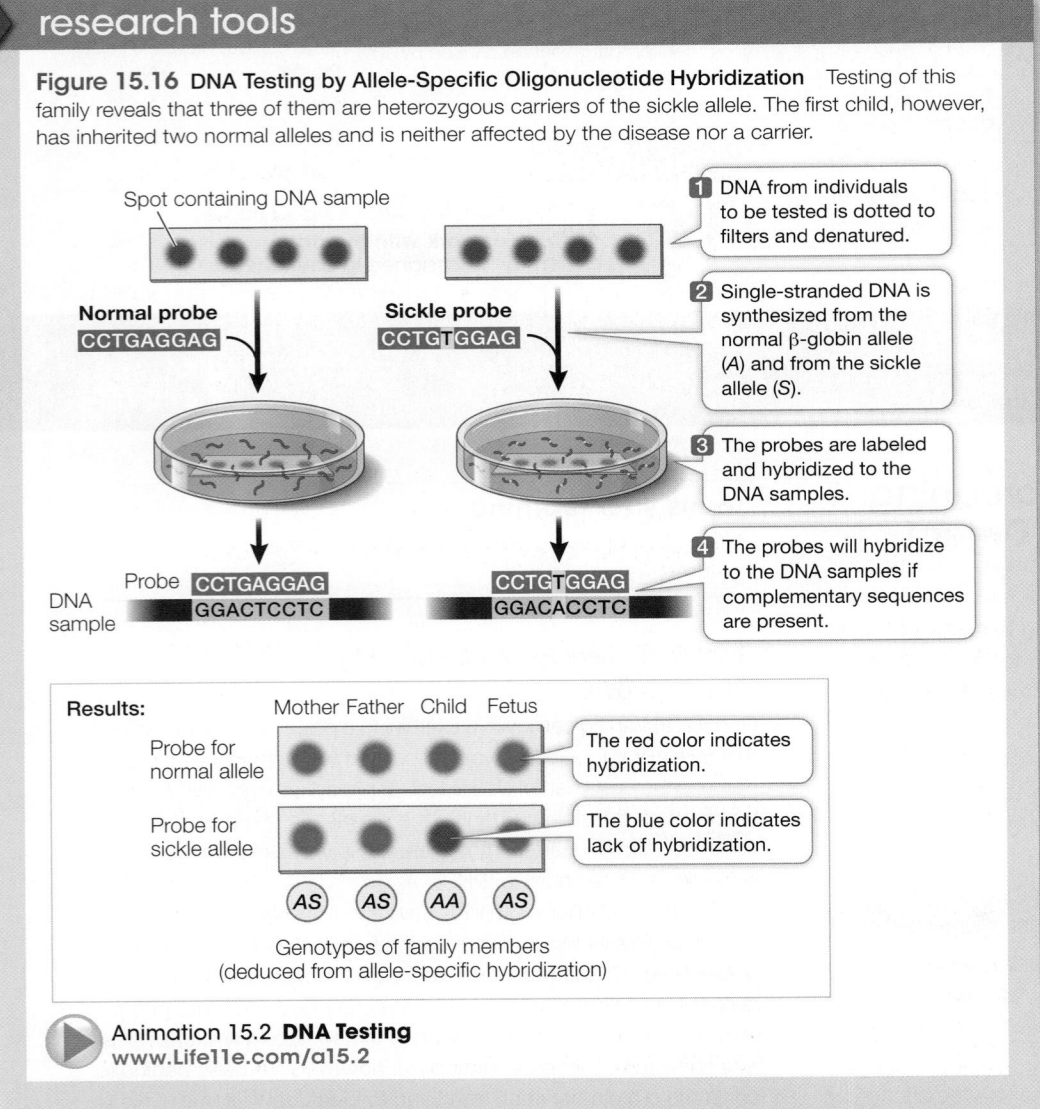

Spot containing DNA sample

1 DNA from individuals to be tested is dotted to filters and denatured.

Normal probe
CCTGAGGAG

Sickle probe
CCTGTGGAG

2 Single-stranded DNA is synthesized from the normal β-globin allele (A) and from the sickle allele (S).

3 The probes are labeled and hybridized to the DNA samples.

Probe CCTGAGGAG   CCTGTGGAG
DNA   GGACTCCTC   GGACACCTC
sample

4 The probes will hybridize to the DNA samples if complementary sequences are present.

Results: Mother Father Child Fetus

Probe for normal allele

The red color indicates hybridization.

Probe for sickle allele

The blue color indicates lack of hybridization.

(AS) (AS) (AA) (AS)

Genotypes of family members (deduced from allele-specific hybridization)

▶ Animation 15.2 **DNA Testing**
www.Life11e.com/a15.2

Consider, for example, two parents who are both heterozygous for the cystic fibrosis allele but who want to have a healthy child. If treated with the appropriate hormones, the mother can be induced to "superovulate," releasing several eggs. Eggs can be harvested from the mother and artificially inseminated. The fertilized egg can be allowed to divide to the eight-cell stage. If one of these embryonic cells is removed, it can be tested for the presence of the cystic fibrosis allele. If the test is negative, the remaining seven-cell embryo can be implanted in the mother's womb, where with luck it will develop normally.

Such preimplantation screening is performed only rarely. More typical are analyses of fetal cells after normal fertilization and implantation in the womb. Fetal cells or even DNA that has leaked out of dead cells can be analyzed at about the tenth week of pregnancy by chorionic villus sampling, or during the thirteenth to seventeenth weeks by amniocentesis. In either case, only a few fetal cells are necessary to perform DNA testing. Recently, very sensitive methods were developed so that DNA testing can be done with the few fetal cells that are released into the mother's blood. A 10-milliliter blood sample from a pregnant woman has enough fetal cells for the analysis of many disorders, including Down syndrome and cystic fibrosis. This relatively noninvasive procedure could replace amniocentesis and chorionic villus sampling—which both carry a slight risk of causing a miscarriage—in the near future.

DNA testing can also be performed with newborns. The blood samples used for screening for PKU and other disorders contain enough of the baby's blood cells to permit DNA analysis using PCR-based techniques. DNA analysis is now being used to screen for sickle-cell disease and cystic fibrosis; similar tests for other diseases will surely follow. Of the numerous methods of DNA testing available, we will describe DNA hybridization, using sickle-cell anemia as an example.

Allele-specific oligonucleotide hybridization can detect mutations

Nucleic acid hybridization (see Figure 14.6) can be used to detect the presence of a specific DNA sequence, such as a sequence containing a particular mutation. Samples of DNA are collected from people who may or may not carry the mutation, and PCR is used to amplify the region of DNA where the mutation may occur. Short synthetic DNA strands called oligonucleotide probes are hybridized with the denatured PCR products. The probe is labeled in some way (e.g., with radioactivity or a fluorescent dye) so that hybridization can be readily detected (**Figure 15.16**).

Detection of a mutation by DNA screening can be used for diagnosis of a genetic disease, so that appropriate treatment can begin. In addition, DNA screening provides a person with important information about his or her genome.

15.4 recap

Genetic screening can be used to identify people who have, are predisposed to, or are carriers of genetic diseases. Screening can be done at the phenotype level by identifying an abnormal protein such as an enzyme with altered activity. It can also be done at the genotype level by direct testing of DNA.

15.4 recap

learning outcomes

You should be able to:

- Compare and contrast phenotypic and genotypic genetic screening.
- Explain how DNA testing is used for genetic screening.

1. What is the advantage of screening for genetic mutations by allele-specific oligonucleotide hybridization relative to screening phenotype differences in enzyme activity?
2. Explain how allele-specific oligonucleotide hybridization could be used to do genetic screening for sickle-cell anemia.

Ongoing research has resulted in the development of increasingly accurate diagnostic tests and a better understanding of various genetic diseases at the molecular level. This knowledge is now being applied to the development of new treatments for genetic diseases. In the next section we will survey various approaches to treatment, including modifications of the mutant phenotype and gene therapy, in which the normal version of a mutant gene is supplied.

> key concept
> # 15.5

Genetic Diseases Can Be Treated

Most treatments for genetic diseases simply try to alleviate the patient's symptoms. But to effectively treat these diseases—whether they affect all cells, as in inherited disorders such as PKU, or only somatic cells, as in cancer—physicians must be able to diagnose the disease accurately, understand how the disease works at the molecular level, and intervene early, before the disease ravages or kills the individual. There are two main approaches to treating genetic diseases:

1. Modifying the disease phenotype
2. Replacing the defective gene (modifying the genotype)

focus your learning

- Some genetic diseases are treated at the phenotype level.
- Other genetic diseases can be treated at the level of the gene using gene therapy.

Genetic diseases can be treated by modifying the phenotype

Altering the phenotype of a genetic disease so that it no longer harms an individual is commonly done in one of three ways: by restricting the substrate of a deficient enzyme, by inhibiting a harmful metabolic reaction, or by supplying a missing protein product (**Figure 15.17**).

RESTRICTING THE SUBSTRATE   Restricting the substrate of a deficient enzyme is the approach taken when a newborn is diagnosed with PKU. In this case, the deficient enzyme is phenylalanine

Cause of disease

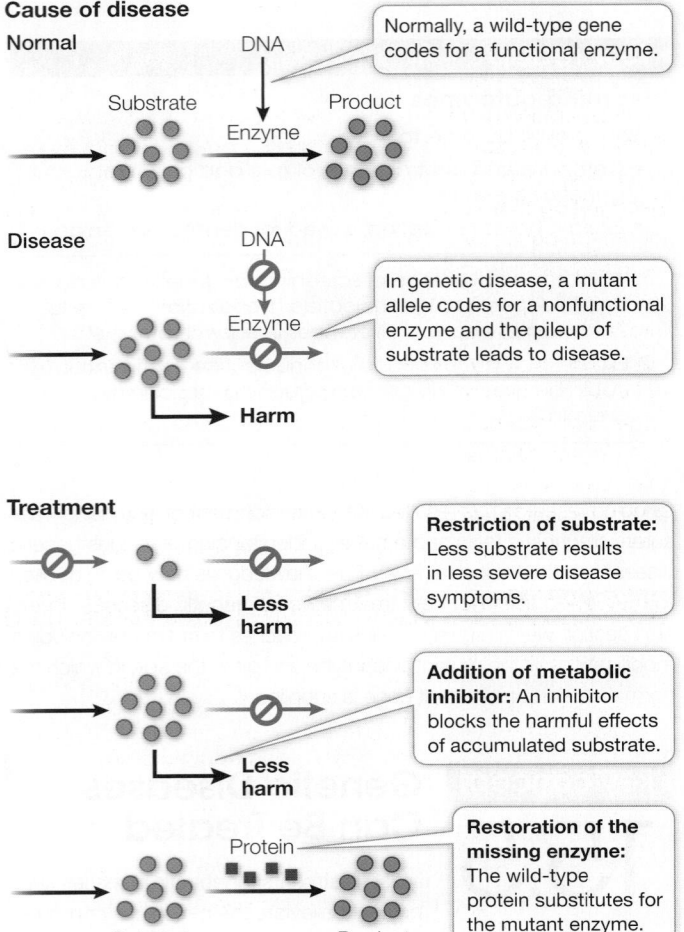

Figure 15.17 Strategies for Treating Genetic Diseases

hydroxylase, and the substrate is phenylalanine (see Figure 15.6). The infant's inability to break down phenylalanine in food leads to a buildup of the substrate, which causes the clinical symptoms. So the infant is immediately put on a special diet that contains only enough phenylalanine for immediate use. Lofenelac, a milk-based product that is low in phenylalanine, is fed to these infants just like formula. Later, certain fruits, vegetables, cereals, and noodles low in phenylalanine can be added to the diet. Meat, fish, eggs, dairy products, and bread, which contain high amounts of phenylalanine, must be avoided, especially during childhood, when brain development is most rapid. The artificial sweetener aspartame must also be avoided because it is made of two amino acids, one of which is phenylalanine.

People with PKU are generally advised to stay on a low-phenylalanine diet for life. Although maintaining these dietary restrictions may be difficult, it is effective. Numerous follow-up studies since newborn screening was initiated have shown that people with PKU who stay on the diet are no different from the rest of the population in terms of mental ability. This is an impressive achievement in public health, given the severity of intellectual disability in untreated patients.

METABOLIC INHIBITORS In Key Concept 11.7 we described how drugs that are inhibitors of various cell cycle processes are used to treat cancer. Drugs are also used to treat the symptoms of many

genetic diseases. As biologists have gained insight into the molecular characteristics of these diseases and the specific proteins involved, a more specific approach to treatment is taking shape. Targeted therapies are being developed, especially for cancer, and some have resulted in life-saving interventions. This is a major area of applied research.

SUPPLYING THE MISSING PROTEIN An obvious way to treat a disease caused by the lack of a functional protein is to supply that protein. This approach is used to treat hemophilia A, a disease in which blood factor VIII is missing and blood clotting is impaired (see Table 15.2). In the past, the missing protein was obtained from blood. Sometimes, however, blood carries contaminants, such as viruses (e.g., HIV) or other pathogens that could harm the recipient. Now human clotting proteins are produced by recombinant DNA technology (see Chapter 18), making it possible to provide the protein in a much purer form.

Unfortunately, the phenotypes of many diseases caused by genetic mutations are very complex. In these cases, simple interventions like those we have just described do not work. Indeed, a recent survey of 351 diseases caused by single-gene mutations showed that current therapies increased patients' life spans by an average of only 15 percent.

Gene therapy offers the hope of specific treatments

If a cell lacks an allele that encodes a functional product, an optimal treatment would be to provide a functional allele. The objective of **gene therapy** is to add a new gene that will be expressed in appropriate cells in a patient. What cells should be targeted? There are two approaches:

1. **Germ line gene therapy**: The new gene is inserted into a gamete (usually an egg) or the fertilized egg. In this case, all cells of the adult will carry the new gene. Ethical considerations preclude the use of this method in humans.

2. **Somatic cell gene therapy**: The new gene is inserted into somatic cells involved in the disease. This method is being tried for numerous diseases, ranging from inherited genetic disorders to cancer.

There are two approaches to somatic cell gene therapy:

1. **Ex vivo gene therapy**: Target cells are removed from the patient, given the new gene, and then reinserted into the patient. This approach is being used, for example, for diseases caused by defects in genes that are expressed in white blood cells.

2. **In vivo gene therapy**: The gene is actually inserted directly into a patient, targeted to the appropriate cells. An example is a treatment for lung cancer in which a solution with a therapeutic gene is squirted onto a tumor.

Armed with knowledge of how genes are expressed (see Chapter 14) and regulated (see Chapter 16), physicians can design a therapeutic gene that contains not only a normal protein-coding sequence but also other sequences—such as an appropriate promoter—required for the gene's expression in targeted cells.

A major challenge has been getting the therapeutic gene into cells. Uptake of DNA into eukaryotic cells is a rare event, and once the DNA is inside a cell, its entry into the nucleus and expression are rarer still. One solution to these problems is to insert the gene into a carrier virus (a

experiment

Figure 15.18 Gene Therapy

Original Paper: LeWitt, P. A. et al. 2011. AAV2-GAD gene therapy for advanced Parkinson's disease: A double-blind, sham-surgery controlled randomized trial. *Lancet Neurology* 10: 309–319.

Andrew Feigin and his colleagues showed that a virus can be used to insert a therapeutic gene into the brains of patients with Parkinson's disease.

HYPOTHESIS▶ Adding a gene for glutamate decarboxylase to brain cells can alleviate symptoms caused by GABA deficiency.

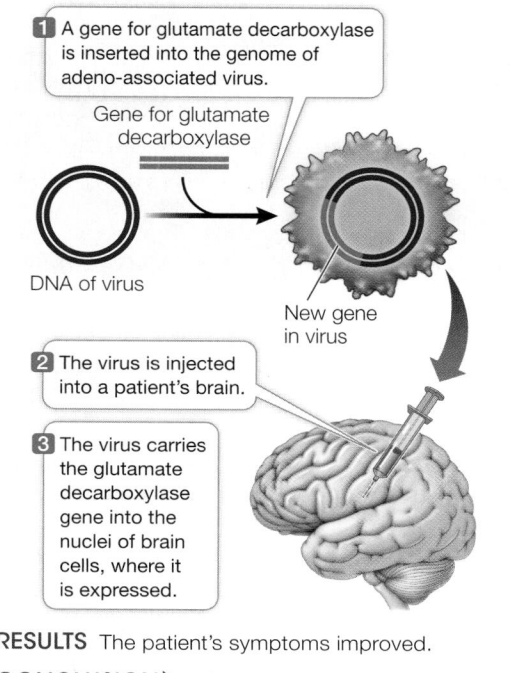

1 A gene for glutamate decarboxylase is inserted into the genome of adeno-associated virus.

Gene for glutamate decarboxylase

DNA of virus

New gene in virus

2 The virus is injected into a patient's brain.

3 The virus carries the glutamate decarboxylase gene into the nuclei of brain cells, where it is expressed.

RESULTS The patient's symptoms improved.

CONCLUSION▶ Gene therapy can relieve symptoms in a patient with a neurotransmitter deficiency.

A **work with the data** exercise that accompanies this figure may be assigned in **LaunchPad**.

viral vector) that can infect human cells but has been altered genetically to prevent viral replication. An example is the DNA virus called **adeno-associated virus**, which has been widely used in human gene therapy clinical trials. This virus has a small genome that can be spliced into a human gene; infects most human cells, including nondividing cells such as neurons; is harmless to humans; does not provoke rejection by the immune system; and enters the cell nucleus, where its DNA with the new gene can be expressed.

Adeno-associated virus has been used successfully in human gene therapy targeting Parkinson's disease (**Figure 15.18**). Parkinson's disease is a neurological condition that affects about 1 in 200 persons in their 60s, and 1 in 25 persons in their 80s. Its symptoms include muscle stiffness and shaking and—in about half of the patients—dementia. Patients with Parkinson's

disease produce inadequate amounts of the neurotransmitter γ-aminobutyric acid (GABA) in a particular part of the brain due to degeneration of the neurons that normally produce GABA. Low GABA levels result in poor coordination of movement. Raising GABA levels would thus be a reasonable approach to treatment. Recognizing that synthesis of GABA is catalyzed by the enzyme glutamate decarboxylase, a team led by Dr. Andrew Feigin at the Feinstein Institute in New York packaged a glutamate decarboxylase gene into adeno-associated virus that was then injected into the affected part of the brain. After six months, the patients who received the virus-encapsulated gene had increased levels of GABA and significant improvement in their disease symptoms. Other diseases being treated with gene therapy using this virus include cystic fibrosis, hemophilia, and muscular dystrophy. While this looks exciting on paper, bear in mind that this type of therapy is very rare, at the frontier of medicine. We aren't close to the day when we can go see our local gene therapist for a cure.

▶15.5 recap

Treatment of a human genetic disease may involve an attempt to modify the abnormal phenotype by restricting the substrate of a deficient enzyme, inhibiting a harmful metabolic reaction, or supplying a missing protein. By contrast, gene therapy aims to address a genetic defect by inserting a normal allele into a patient's cells.

learning outcomes

You should be able to:

- Explain how a genetic disease can be treated by modifying the disease phenotype.
- Compare ex vivo and in vivo approaches to gene therapy.

1. How do metabolic inhibitors function in treating genetic diseases such as cancer?

2. How does in vivo gene therapy work? Can you give an example?

3. In the past, it was common for people with phenylketonuria (PKU) who were placed on a low-phenylalanine diet after birth to be allowed to return to a normal diet during their teenage years. Although the levels of phenylalanine in their blood were high, their brains were thought to be beyond the stage when they could be harmed. If a woman with PKU becomes pregnant, however, a problem arises. Typically, the fetus is heterozygous but is unable, at early stages of development, to metabolize the high levels of phenylalanine that arrive from the mother's blood.

   a. Why is the fetus likely to be heterozygous?

   b. What do you think would happen to the fetus during this "maternal PKU" situation?

   c. What would be your advice to a woman with PKU who wants to have a child?

In this chapter you have learned about mutations, focusing on DNA changes that affect phenotypes through specific protein products. But there is much more to molecular genetics than the sequences of genes and proteins. Determining which genes will be expressed when and where is a major function of the genome. In Chapter 16 we will turn to gene regulation.

Q&A What is the breast cancer gene, and what are the issues with DNA testing for cancer?

Angelina Jolie's public announcement that she had elected to have her breasts removed in light of her cancer gene mutations attracted worldwide interest. In fact, the phrase "mutation" had never been so popular on internet search engines! Requests for *BRCA1* testing from worried women have increased dramatically. Two issues have arisen: who should be tested, and who should perform the test? Because breast cancer is relatively common (women have about a 10 percent lifetime chance of having it), having several relatives who have had the disease is not unusual. Medical geneticists have come up with several criteria for defining when *BRCA1* DNA analysis is warranted. These include having several close relatives who had or have the disease, with early onset (typically before age 50) and the presence of multiple tumors.

Even when a test shows no *BRCA1* mutations in a woman with a family history of breast or ovarian cancer, we cannot say that breast or ovarian cancer may not arise. It turns out that *BRCA1* mutations only account for about one in five hereditary breast cancer cases. Other genes are involved; we just don't know what they are,

so we can't analyze their DNA. Remember: hereditary breast cancer (a germ line mutation) accounts for only one case in ten. The other 90 percent arise from somatic mutation of breast cells.

As to who does the test, the *BRCA1* gene was isolated by a private company that patented the DNA sequence. Until recently, that company was the only lab that could do *BRCA1* analysis, and not surprisingly, the price was high. In 2013, the U.S. Supreme Court invalidated the patent, and now more labs can do it, and more cheaply.

Future directions

Are there genetic tests for other hereditary cancers? About 10 percent of all cancers are caused by inheritance of mutated oncogenes or tumor suppressor genes (as opposed to mutations that occur after birth). As with the situation in breast cancer, a person who inherits a cancer-causing mutation has a much higher chance of developing cancer than one who does not have that mutation. As whole genome DNA sequencing is developed (see Chapter 18), mutations involved with other types of hereditary cancer are being identified and screening of relatives with a propensity for cancer is on the horizon.

Chapter Summary 15

▶ 15.1 Mutations Are Heritable Changes in DNA

- **Mutations** are stable, heritable changes in DNA. **Somatic mutations** are passed on to daughter cells, but only **germ line mutations** are passed on to sexually produced offspring. Review Focus: Key Figure 15.1, Activity 15.1, Activity 15.2

- **Point mutations** result from alterations in single base pairs of DNA. **Silent mutations** can occur in noncoding DNA or in coding regions of genes and do not affect the amino acid sequences of proteins. **Missense**, **nonsense**, and **frame-shift mutations** all cause changes in protein sequences. Review Figure 15.2

- Chromosomal mutations (**deletions**, **duplications**, **inversions**, and **translocations**) involve large regions of chromosomes. Review Figure 15.3

- **Spontaneous mutations** occur because of instabilities in DNA or chromosomes. **Induced mutations** occur when a mutagen damages DNA. Review Figure 15.4

- Mutations can occur in "hot spots" where cytosine has been methylated to 5-methylcytosine. Review Figure 15.5

- Mutations, although often detrimental to an individual organism, are the raw material of evolution.

▶ 15.2 Mutations in Humans Can Lead to Diseases

- Abnormalities in proteins have been implicated in genetic diseases.
- While a single amino acid difference can be the cause of disease, amino acid variations have been detected in many functional proteins. Review Figures 15.6, 15.7
- Point mutations, deletions, and chromosome abnormalities are associated with genetic diseases. Review Figure 15.8

- The effects of fragile-X syndrome worsen with each generation. This pattern is the result of an **expanding triplet repeat**. Review Figure 15.9

- A series of genetic mutations can lead to colon cancer. Review Figure 15.10

- **Multifactorial** diseases are caused by the interactions of many genes and proteins with the environment. They are much more common than diseases caused by mutations in a single gene.

▶ 15.3 Mutations Can Be Detected and Analyzed

- **Restriction enzymes**, which are made by microorganisms as a defense against viruses, bind to and cut DNA at specific **recognition sequences** (also called **restriction sites**), producing smaller fragments of DNA. This cutting process is known as **restriction digestion**. Restriction enzymes can be used in the laboratory to produce small fragments of DNA for study. Review Figure 15.11

- DNA fragments can be separated by size using **gel electrophoresis**. Review Figure 15.12, Animation 15.1

- **DNA fingerprinting** is used to distinguish among specific individuals or to reveal which individuals are most closely related to one another. It involves the detection of DNA polymorphisms, including **single nucleotide polymorphisms (SNPs)** and **short tandem repeats (STRs)**. Review Figure 15.13, Activity 15.3

- It is possible to isolate both the mutant genes and the abnormal proteins responsible for human diseases. Review Figure 15.14, Investigating Life: How Was the *BRCA1* Gene Identified?

15.4 Genetic Screening Is Used to Detect Diseases

- **Genetic screening** is used to detect human genetic diseases, alleles predisposing people to those diseases, or carriers of those disease alleles.
- Genetic screening can be done by looking for abnormal protein expression. **Review Figure 15.15**
- **DNA testing** is the direct identification of mutant alleles. Any cell can be tested at any time in the life cycle. **Review Figure 15.16, Animation 15.2**

15.5 Genetic Diseases Can Be Treated

- There are three ways to modify the phenotype of a genetic disease: restrict the substrate of a deficient enzyme, inhibit a harmful metabolic reaction, or supply a missing protein. **Review Figure 15.17**
- Cancer sometimes can be treated with metabolic inhibitors.
- In gene therapy, a mutant gene is replaced with a normal gene. Both ex vivo and in vivo therapies are being developed. **Review Figure 15.18**

> Go to **LearningCurve** (in **LaunchPad**) for dynamic quizzing that helps you solidify your understanding of this chapter. **LearningCurve** adapts to your responses, giving you the practice you need to master each key concept.

▶ Apply What You've Learned

Review

15.1 Use a defined DNA sequence to illustrate how the following occur and describe their effect upon phenotype: transitions, transversions, missense mutations, nonsense mutations, frame-shift mutations, and mutations outside the coding region of a gene.

15.2 Genetic diseases often result from mutations that render proteins nonfunctional.

15.2 Mutations that cause disease include the full range of affected DNA from point mutations to entire chromosomes.

Hemoglobin, composed of four polypeptide subunits and an oxygen-binding heme group, carries O_2 in human blood, delivering it to tissues. Because of the importance of O_2 for cellular respiration, a genetic mutation that alters the structure and function of hemoglobin can have severe consequences. In an adult, hemoglobin has two types of polypeptide subunits: alpha-globin (α-globin), consisting of 141 amino acids, and beta-globin (β-globin), consisting of 146 amino acids (see figure below).

People with thalassemia typically make little to no α- or β-globin and so have anemia—inadequate quantities of normal red blood cells. They need blood transfusions to survive. Most common in Mediterranean countries, thalassemia is typically inherited as an autosomal recessive. While homozygotes with the mutant alleles have severe symptoms,

people who are heterozygous carriers for thalassemia often have mild anemia, as they have only one of two functional alleles for the globin gene in question.

Consider the following scenario: Two people have a mild, persistent anemia. Because each has relatives who suffered from thalassemia, they are concerned that they may be carriers for a thalassemia mutation. Tissue samples of both people are taken and DNA for globin genes sequenced. Here is a region for the normal β-globin gene in the coding strand of DNA:

Codon #	36	37	38	39	40	41
Normal:	CCT	TGG	ACC	CAG	AGG	TTC

Below are the DNA sequences of the coding strands for the same region of the β-globin gene in the two people with mild anemia. Person 1 has two different sequences in this region:

> Sequence 1: CCT TGG ACC CAG AGG TTC
> Sequence 2: CCT TGG ACC TAG AGG TTC

Person 2 also has two different sequences in this region:

> Sequence 1: CCT TGG ACC CAG AGG TTC
> Sequence 2: CCT GGA CCC AGA GGT TCT

Questions

1. Analyze the DNA sequences to identify the type of mutation (silent, nonsense, missense, or frame-shift) in these two patients. Then evaluate each mutation's effect on the protein chain.

2. Describe a mutation that would have an effect similar to that of the two mutations described but that would involve a change outside the coding region for the β-globin gene.

3. If person 1 were to have a child with person 2, what are the risks involved for this couple's children? Draw a pedigree chart to illustrate the possible outcomes.

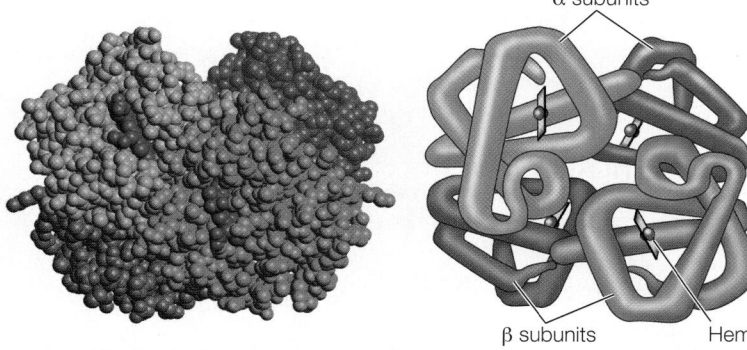

α subunits

β subunits   Heme

Go to **LaunchPad** for the eBook, LearningCurve, animations, activities, flashcards, and additional resources and assignments.

16

Regulation of Gene Expression

Worker bees care for a queen. Gene expression differences underlie the differences in behavior of worker and queen bees.

investigatinglife

Gene Expression and Behavior

Any teenager or parent of a teenager will tell you that those years can be rough. But some young people have a rougher time than others. Teens whose mothers were extremely stressed during pregnancy often have more behavioral problems than those whose mothers had calmer pregnancies. A recent study led by psychologist Thomas Elbert and evolutionary biologist Axel Meyer at the University of Konstanz in Germany suggests an underlying genetic cause for these behavioral differences. The study examined a gene for the glucocorticoid receptor, which is involved in regulating hormonal responses to stress. The researchers found that teenagers whose mothers had suffered physical abuse during pregnancy had higher rates of cytosine methylation in the promoter of this gene than did teenagers whose mothers had not suffered such abuse.

A major control point for gene expression is the promoter, a sequence of DNA adjacent to the coding region of a gene where proteins bind and control the rate of transcription. The ability of these proteins to bind to the promoter is affected by the level of DNA methylation in the promoter. As we mentioned in Key Concept 15.1,

in certain regions of DNA many of the cytosine residues have methyl groups added at their 5 positions, forming 5-methyl-cytosine. If a gene promoter has a high degree of methylation, some proteins that promote transcription can't bind to it. Instead, other proteins bind to the methylated DNA and prevent expression of the gene. DNA methylation plays an important role in gene regulation and is a normal part of development. But the level of methylation can change over time and can vary among individuals, as was the case for the teenagers in Elbert and Meyer's study. Their finding is interesting because it shows a correlation between maternal stress in humans and DNA methylation in their offspring.

Such studies have spawned the new field of behavioral epigenetics. Epigenetics is the study of heritable changes in gene expression that do not involve changes in the DNA sequence. These changes may underlie behaviors in other animals as well. A fascinating example is the well-known behavior of a queen bee, who "lords it over" her genetically identical worker sisters. Genes in the brains of these two types of bees have very different patterns of expression, and these in turn are caused by epigenetic changes.

key concept 16.1 Prokaryotic Gene Expression Is Regulated in Operons

Prokaryotes conserve energy and resources by making certain proteins only when they are needed. The protein content of a bacterium can change rapidly when conditions warrant. Based on what you learned about gene expression in Chapter 14, you might suggest several ways in which a prokaryotic cell could shut off the supply of an unneeded protein. The cell could:

- decrease the rate of transcription of mRNA for that protein;
- hydrolyze the mRNA after it is made, preventing translation;
- prevent translation of the mRNA at the ribosome;
- hydrolyze the protein after it is made; or
- inhibit the function of the protein.

Whichever mechanism is used, it must be both responsive to environmental signals and efficient. The earlier the cell intervenes in the process of protein synthesis, the less energy it wastes. Selective blocking of transcription is far more efficient than transcribing the gene, translating the message, and then degrading or inhibiting the protein. While all five mechanisms for regulating protein levels are found in nature, prokaryotes generally use the most efficient one: transcriptional regulation.

focus your learning

- The *lac* operon is an example of an inducible operon regulated by a repressor protein.
- Sigma factors provide another type of transcriptional regulation by directing RNA polymerases to specific promoters.

Gene expression begins at the *promoter, where RNA polymerase binds to initiate transcription. Note, however, that in a given cell at a given point in time, not all promoters are active. This observation suggests that gene transcription must be selective. The "decision" regarding which genes to activate involves two types of regulatory proteins that bind to DNA: repressor proteins and activator proteins. In both cases, these proteins bind to the promoter to regulate the gene (Focus: Key Figure 16.1):

- In **negative regulation**, binding of a repressor protein prevents transcription.
- In **positive regulation**, an activator protein binds DNA to stimulate transcription.

*connect the concepts   A key DNA sequence involved in regulation is the promoter, where transcription begins and various regulatory proteins bind. Learn details about proteins at the promoter in Key Concept 14.3.

You will see examples of these mechanisms, or combinations of them, as we examine regulation in prokaryotes, eukaryotes, and viruses. We'll focus first on a regulatory system for the use of the sugar lactose.

focus: key figure

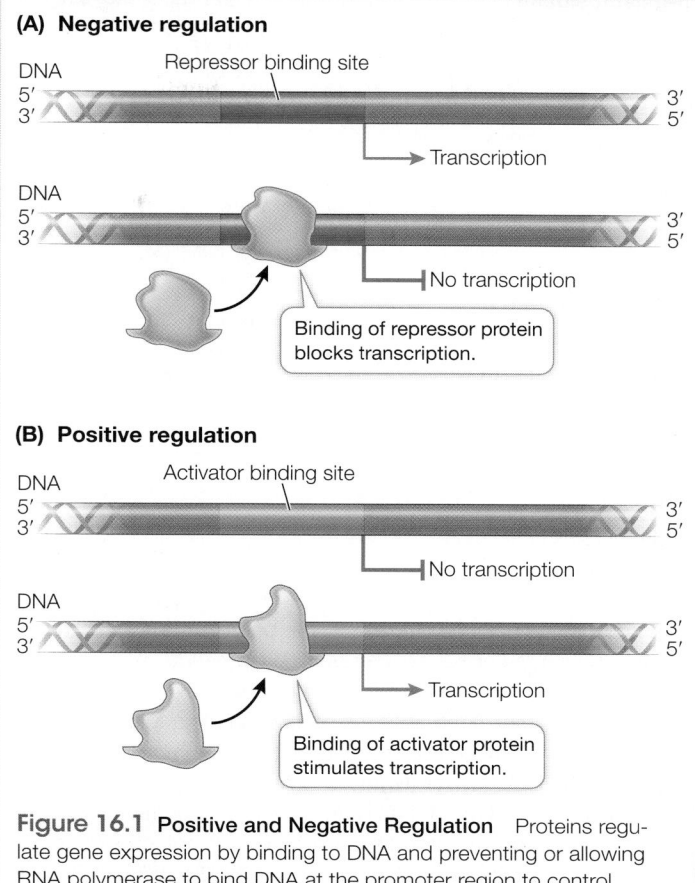

Figure 16.1 **Positive and Negative Regulation** Proteins regulate gene expression by binding to DNA and preventing or allowing RNA polymerase to bind DNA at the promoter region to control transcription of the gene.

Q: Could a gene be under both positive and negative regulation?

Regulating gene transcription conserves energy

As a normal inhabitant of the human intestine, *E. coli* must be able to adjust to sudden changes in its chemical environment. Its host may present it with one foodstuff one hour (e.g., glucose in fruit) and another the next (e.g., lactose in milk). Such changes in nutrients present the bacterium with a metabolic challenge. Glucose is its preferred energy source, and is the easiest sugar to metabolize. Lactose is a β-galactoside—a disaccharide containing galactose β-linked to glucose (see Key Concept 3.3). Three proteins are involved in the initial uptake and metabolism of lactose by *E. coli*:

1. β-Galactoside permease is a carrier protein in the bacterial cell membrane that moves the sugar into the cell.
2. β-Galactosidase is an enzyme that hydrolyses lactose to glucose and galactose.
3. β-Galactoside transacetylase transfers acetyl groups from acetyl CoA to certain β-galactosides. Its role in the metabolism of lactose is not clear.

When *E. coli* grows and reproduces in a lab medium that contains glucose but no lactose or other β-galactosides, the levels of these three proteins are extremely low—the cell does not waste energy

(A) Lactose induces the synthesis of β-galactosidase

When lactose is added to the growth medium, an enzyme essential to the metabolism of lactose is made after a lag period.

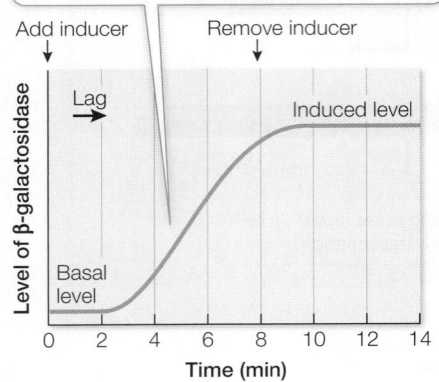

(B) Effect of inducer on mRNA level

During the lag period, the mRNA for the enzyme is made.

The amount of mRNA decreases shortly after the **inducer** is removed.

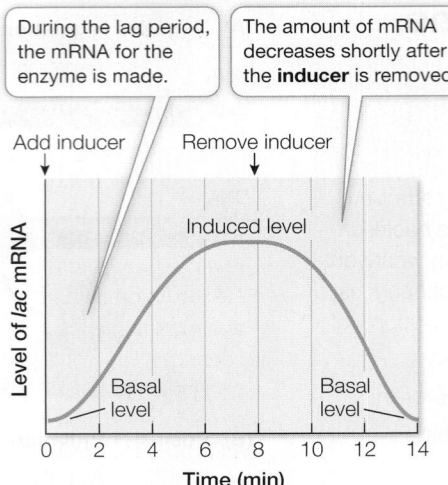

Figure 16.2 An Inducer Stimulates the Expression of a Gene for an Enzyme **(A)** When lactose is added to the growth medium for the bacterium *E. coli*, the synthesis of β–galactosidase begins only after an initial lag period. **(B)** There is a lag period because the mRNA for β-galactosidase has to be made before the protein can be made. The amount of mRNA decreases rapidly after the lactose is removed, indicating that transcription is no longer occurring. These changes in mRNA levels indicate that the mechanism of induction by lactose is transcriptional regulation.

and materials making the unneeded enzymes. But if the environment changes such that lactose is the predominant sugar available and very little glucose is present, the bacterium promptly begins making all three enzymes after a short lag period. While few molecules of β-galactosidase are present in an *E. coli* cell in the presence of glucose, in the absence of glucose the addition of lactose can induce the synthesis of about 1,500 times more molecules of β-galactosidase per cell (**Figure 16.2A**)!

What's behind this dramatic increase? An important clue comes from measuring the amount of mRNA for β-galactosidase. The mRNA level increases during the lag period after lactose is added to the medium, and this mRNA is translated into protein (**Figure 16.2B**). Moreover, the high mRNA level depends on the presence of lactose, because if the lactose is removed, the mRNA level goes down. *The response of the bacterial cell to lactose is clearly at the level of transcription.*

Compounds such as lactose that stimulate the synthesis of a protein are called **inducers**. The proteins that are produced are called **inducible proteins**, whereas proteins that are made all the time at a constant rate are called **constitutive proteins**. (Think of the constitution of a country, a document that does not change under normal circumstances.)

We have now seen two basic ways of regulating the rate of a metabolic pathway. In Key Concept 8.5 we described the allosteric regulation of enzyme activity, which allows the rapid fine-tuning of metabolism. Regulation of protein synthesis—that is, regulation of the concentration of enzymes—is slower but results in greater savings of energy and resources. Protein synthesis is a highly

endergonic process, since assembling mRNA, charging tRNA, and moving the ribosomes along mRNA all require the hydrolysis of nucleoside triphosphates such as ATP. **Figure 16.3** compares these two modes of regulation.

Operons are units of transcriptional regulation in prokaryotes

The genes in *E. coli* that encode the three enzymes for using lactose are structural genes; **structural genes** specify the primary structures (the amino acid sequences) of protein molecules that act as enzymes or cytoskeletal proteins. The three genes are adjacent to one another on the *E. coli* chromosome. This arrangement is no coincidence: the genes share a single promoter, and their DNA is transcribed into a single, continuous molecule of mRNA. Because this particular mRNA governs the synthesis of all three lactose-metabolizing enzymes, either all or none of these enzymes are made, depending on whether their common message—their mRNA—is present in the cell.

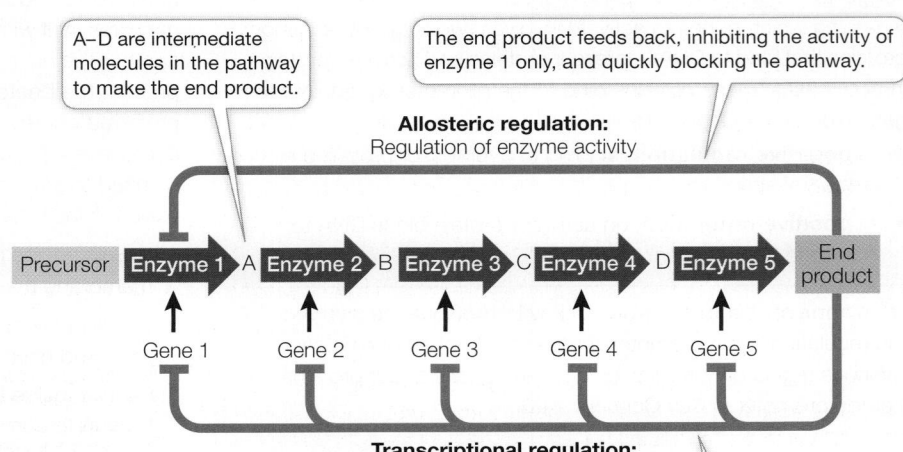

A–D are intermediate molecules in the pathway to make the end product.

The end product feeds back, inhibiting the activity of enzyme 1 only, and quickly blocking the pathway.

Allosteric regulation:
Regulation of enzyme activity

Transcriptional regulation:
Regulation of enzyme concentration

The end product blocks the transcription of all five genes. No enzymes are produced.

Figure 16.3 Two Ways to Regulate a Metabolic Pathway Feedback from the end product of a metabolic pathway can block enzyme activity (allosteric regulation), or it can stop the transcription of genes that code for the enzymes in the pathway (transcriptional regulation).

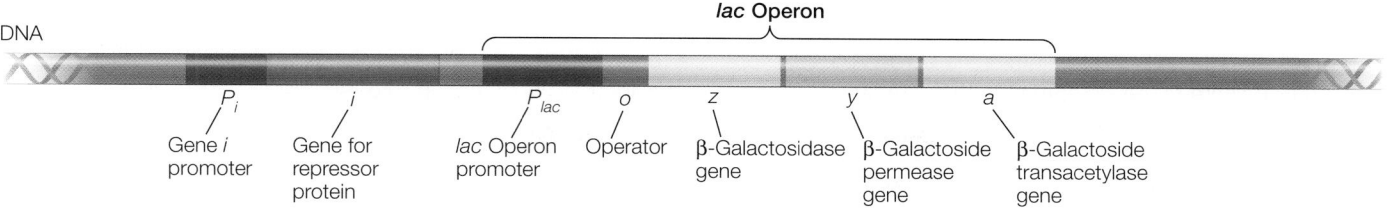

DNA

P_i — Gene *i* promoter
i — Gene for repressor protein
P_{lac} — *lac* Operon promoter
o — Operator
z — β-Galactosidase gene
y — β-Galactoside permease gene
a — β-Galactoside transacetylase gene

lac Operon

Figure 16.4 The *lac* Operon of *E. coli*   The *lac* operon of *E. coli* is a segment of DNA that includes a promoter, an operator, and the three structural genes that code for lactose-metabolizing enzymes.

A cluster of genes with a single promoter is called an **operon**, and the operon that encodes the three lactose-metabolizing enzymes in *E. coli* is called the *lac* operon. The *lac* operon promoter can be very efficient (the maximum rate of mRNA synthesis can be high), but mRNA synthesis can be shut down when the enzymes are not needed. In addition to the promoter, an operon has other **regulatory sequences** that are not transcribed. A typical operon consists of a promoter, an operator, and two or more structural genes (**Figure 16.4**). The **operator** is a short stretch of DNA that lies between the promoter and the structural genes. It can bind very tightly with regulatory proteins that either activate or repress transcription.

There are numerous mechanisms to control the transcription of operons; we will describe three examples:

1. An inducible operon regulated by a repressor protein
2. A repressible operon regulated by a repressor protein
3. An operon regulated by an activator protein

Operator–repressor interactions control transcription in the lac *and* trp *operons*

The *lac* operon contains a promoter, to which RNA polymerase binds to initiate transcription, and an operator, to which a **repressor** protein can bind. The gene that encodes this repressor is located near the *lac* operon on the *E. coli* chromosome. When the repressor is bound, transcription of the operon is blocked. This example of negative regulation was elegantly worked out by Nobel Prize winners François Jacob and Jacques Monod.

The repressor protein has two binding sites: one for the operator and the other for the inducer. The environmental signal that induces the *lac* operon (for example, in the human digestive tract) is lactose, but the actual inducer is allolactose, a molecule that forms from lactose once it enters the cell. In the absence of the inducer, the repressor protein fits into the major groove of the operator DNA and recognizes and binds to a specific nucleotide base sequence. This prevents the binding of RNA polymerase to the promoter, and the operon is not transcribed (**Figure 16.5A**). When the inducer is present, it binds to the repressor and changes the shape of the

repressor. This change in three-dimensional structure (conformation) prevents the repressor from binding to the operator. As a result, RNA polymerase can bind to the promoter and start transcribing the structural genes of the *lac* operon (**Figure 16.5B**).

You can see from this example that a key to transcriptional control of gene expression is the presence of regulatory sequences that do not code for proteins, but are binding sites for regulatory proteins and other proteins involved in transcription.

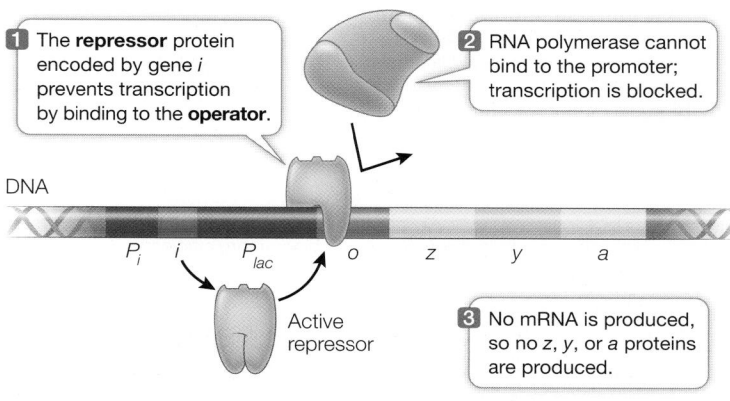

(A) Lactose absent

1 The **repressor** protein encoded by gene *i* prevents transcription by binding to the **operator**.

2 RNA polymerase cannot bind to the promoter; transcription is blocked.

DNA

P_i *i* P_{lac} *o*   *z*   *y*   *a*

Active repressor

3 No mRNA is produced, so no *z*, *y*, or *a* proteins are produced.

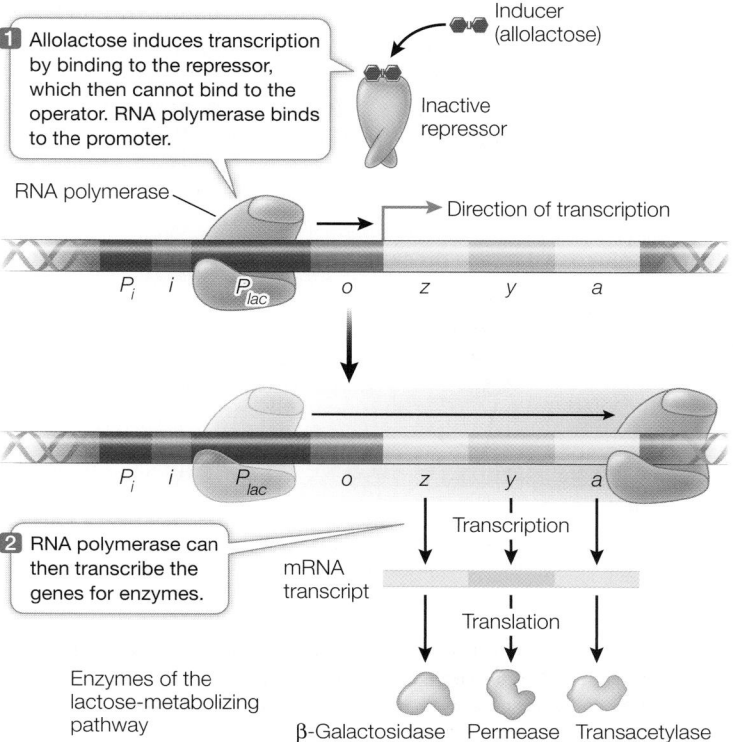

(B) Lactose present

1 Allolactose induces transcription by binding to the repressor, which then cannot bind to the operator. RNA polymerase binds to the promoter.

Inducer (allolactose)

Inactive repressor

RNA polymerase

Direction of transcription

P_i *i* P_{lac} *o*   *z*   *y*   *a*

P_i *i* P_{lac} *o*   *z*   *y*   *a*

2 RNA polymerase can then transcribe the genes for enzymes.

mRNA transcript

Transcription

Translation

Enzymes of the lactose-metabolizing pathway

β-Galactosidase   Permease   Transacetylase

Figure 16.5 The *lac* Operon: An Inducible System   **(A)** When lactose is absent, the synthesis of enzymes for its metabolism is inhibited. **(B)** Lactose (the inducer) leads to synthesis of the enzymes in the lactose-metabolizing pathway by binding to the repressor protein and preventing its binding to the operator.

▶ Animation 16.1 **The *lac* Operon**
www.Life11e.com/a16.1

In contrast to the inducible system of the *lac* operon, other operons in *E. coli* are repressible; that is, they are repressed only under specific conditions. In such a system, the repressor is not normally bound to the operator. But if another molecule called a **co-repressor** binds to the repressor, the repressor changes shape and binds to the operator, thereby inhibiting transcription. An example is the *trp* operon, whose structural genes catalyze the synthesis of the amino acid tryptophan:

5 Enzyme-catalyzed reactions
Precursor molecules → → → → → tryptophan

When tryptophan is present in the cell in adequate concentrations, it is advantageous to stop making the enzymes for tryptophan synthesis. To do this, the cell uses a repressor that binds to an operator in the *trp* operon. But the repressor of the *trp* operon is not normally

bound to the operator; it only binds when its shape is changed by binding to tryptophan, the co-repressor.

Let's pause to summarize the differences between these two types of operons:

● In *inducible* systems, the substrate of a metabolic pathway (the inducer) interacts with a regulatory protein (the repressor), rendering the repressor incapable of binding to the operator and thus allowing transcription.

● In *repressible* systems, the product of a metabolic pathway (the co-repressor) binds to a regulatory protein, which is then able to bind to the operator and block transcription.

Usually, inducible systems control *catabolic pathways (which are turned on only when the substrate is available), whereas repressible systems control anabolic pathways (which are turned on until the concentration of the product becomes excessive). In both systems, the regulatory protein is a repressor that functions by binding to the operator. Next we will consider an example of positive control involving an activator.

*connect the concepts** As discussed in Key Concept 8.1, there are two kinds of metabolism: Catabolic pathways break down complex molecules into simpler ones, releasing energy formerly stored in the chemical bonds. Anabolic pathways link simple molecules to form more complex molecules, a process that requires an input of energy.

Animation 16.2 The *trp* Operon
www.Life11e.com/a16.2

Protein synthesis can be controlled by increasing promoter efficiency

In negative control, transcription is *decreased* in the presence of a repressor protein. *E. coli* can also use positive control to *increase* transcription through the presence of an **activator** protein. For an example we return to the *lac* operon, where the relative levels of glucose and lactose determine the amount of transcription. We have seen that in the presence of lactose the *lac* repressor is unable to bind to the *lac* operator to repress transcription (see Figure 16.5B). But glucose is the preferred source of energy for the cell, so if glucose and lactose levels are both high, the *lac* operon is still not transcribed efficiently. This is because efficient transcription of the *lac* operon requires binding of an activator protein to its promoter.

Low levels of glucose in the cell set off a signaling pathway that leads to increased levels of the second messenger cyclic AMP (cAMP) (see Key Concept 7.3). Cyclic AMP binds to an activator protein called *cAMP receptor protein* (CRP), producing a conformational change in CRP that allows it to bind to the *lac* promoter. CRP is an activator of transcription, because its binding results in more efficient binding of RNA polymerase to the promoter, and thus increased transcription of the structural genes (**Figure 16.6**). In the presence of abundant glucose, cAMP levels are low, CRP does not bind to the promoter, and the efficiency of transcription of the *lac* operon is reduced. This is an example of **catabolite repression**, a system of gene regulation in which the presence of the preferred energy source represses other catabolic pathways. The

(A) Low glucose, lactose present

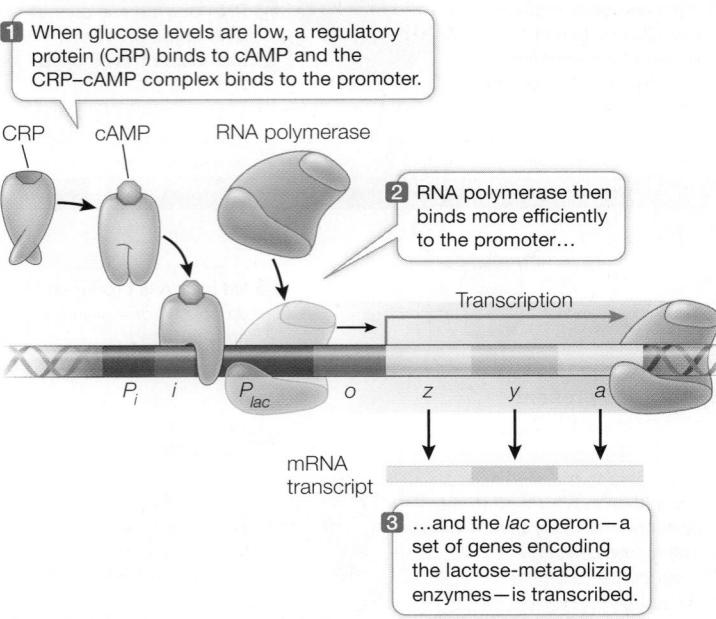

1 When glucose levels are low, a regulatory protein (CRP) binds to cAMP and the CRP–cAMP complex binds to the promoter.

2 RNA polymerase then binds more efficiently to the promoter...

3 ...and the *lac* operon—a set of genes encoding the lactose-metabolizing enzymes—is transcribed.

(B) High glucose, lactose present

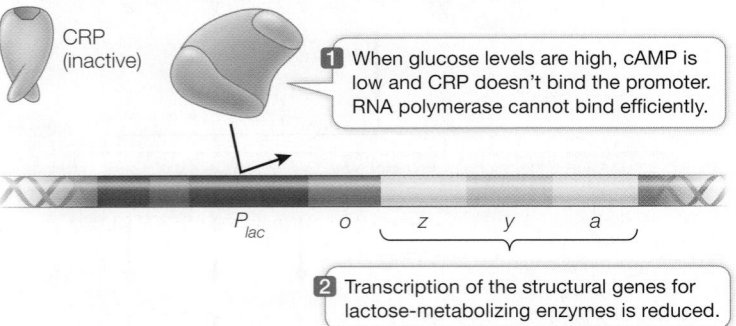

1 When glucose levels are high, cAMP is low and CRP doesn't bind the promoter. RNA polymerase cannot bind efficiently.

2 Transcription of the structural genes for lactose-metabolizing enzymes is reduced.

Figure 16.6 Catabolite Repression Regulates the *lac* Operon (A) The promoter for the *lac* operon does not function efficiently in the absence of cAMP, as occurs when glucose levels are high. **(B)** High glucose levels thus repress the enzymes that metabolize lactose.

table 16.1 Positive and Negative Regulation in the *lac* Operon

Glucose	cAMP levels	RNA polymerase binding to promoter	Lactose	*lac* Repressor	Transcription of *lac* genes?	Lactose used by cells?
Present	Low	Absent	Absent	Active and bound to operator	No	No
Present	Low	Present, not efficient	Present	Inactive and not bound to operator	Low level	No
Absent	High	Present, very efficient	Present	Inactive and not bound to operator	High level	Yes
Absent	High	Absent	Absent	Active and bound to operator	No	No

mechanisms controlling positive and negative regulation of the *lac* operon are summarized in **Table 16.1**.

RNA polymerases can be directed to particular classes of promoters

Thus far we have described a promoter as a specific DNA sequence located upstream of a transcription initiation site. The promoter binds RNA polymerase so that the enzyme can then catalyze the synthesis of RNA from a gene-encoding region of DNA. The promoter also orients the polymerase so that it transcribes the correct one of the two DNA strands. Not all promoters are identical, but they all have similar sequences by which they are recognized by the RNA polymerase and other proteins. Prokaryotic promoters generally have two sites for these recognition sequences, which begin 10 and 35 base pairs upstream of the transcription start site (the −10 element and the −35 element). Different classes of promoters have different recognition sequences at these two sites. The largest class consists of promoters for "housekeeping genes," which are all the genes that are normally expressed in actively growing cells. In these genes, the −10 element is 5′-TATAAT-3′, and the −35 element is 5′-TTGACAT-3′ (N stands for any nucleotide):

```
5′-NNNTTGACATNNNNNNNNNNNNNNNNNNNTATAATNNNN*NNNNNNNNNNNNN-3′
3′-NNNAACTGTANNNNNNNNNNNNNNNNNNNATATTANNNN*NNNNNNNNNNNNN-5′
```
     −35 Element   −10 Element Transcription start site

Other classes of genes have different recognition sequences at their −10 and −35 sites. Why do DNA recognition sequences differ in different classes of promoters? After all, aren't they all binding the same protein, RNA polymerase? The answer lies in the fact that these DNA sequences bind not just RNA polymerase but other proteins as well. And it is those other proteins that enhance RNA polymerase binding at certain promoters, thereby giving the system a degree of specificity.

Sigma factors are the proteins in prokaryotic cells that bind to RNA polymerase and direct it to specific classes of promoters. The RNA polymerase must be bound to a sigma factor before it can recognize a promoter and begin transcription. For example, the sigma-70 factor is active most of the time and binds to the recognition sequences of housekeeping genes; other sigma factors are activated only under specific conditions. When *E. coli* cells experience conditions such as DNA damage or osmotic stress, the sigma-38 factor is activated, and it directs RNA polymerase to the promoters of various genes that are expressed under stress conditions. *E. coli* has seven sigma factors; this number varies in other prokaryotes.

Regulation of proteins directing RNA polymerase to certain promoters is not unusual. In fact, you'll see in the next section that it is also common in eukaryotes.

▶ 16.1 recap

Gene expression in prokaryotes is most commonly regulated through control of transcription. An operon consists of a set of closely linked structural genes and a set of DNA sequences (promoter and operator) that control their transcription. Operons can be regulated by both negative and positive controls. Sigma factors control the expression of specific classes of prokaryotic genes that share recognition sequences in their promoters.

learning outcomes

You should be able to:

- Analyze conditions that affect expression of genes in the *lac* operon.
- Describe the roles of various DNA sequences in the *lac* operon.
- Explain how sigma factors and genetic sequences are involved in gene regulation in prokaryotes.

1. A prokaryotic cell can metabolize sugar "X," using the enzyme "Xase." When there is a low concentration of X in the environment, there is low activity of Xase in the cells; but when the X concentration is high, Xase is also high. What five mechanisms could the cell use to reduce the activity of Xase in the absence of X?

2. Compare the molecular conditions at the *lac* operon promoter in the presence versus absence of lactose.

3. How do sigma factors and recognition sequences act to affect the expression of classes of genes?

4. The repressor protein that acts on the *lac* operon of *E. coli* is encoded by a regulatory gene. The repressor is made in small quantities and at a constant rate. Would you surmise that the promoter for this repressor protein is efficient or inefficient? Is synthesis of the repressor constitutive, or is it inducible and under environmental control?

Studies of bacteria have provided a basic understanding of mechanisms that regulate gene expression and of the roles of regulatory proteins in both positive and negative regulation. You'll see these same types of mechanisms again as we now turn to the transcriptional control of gene expression in eukaryotes.

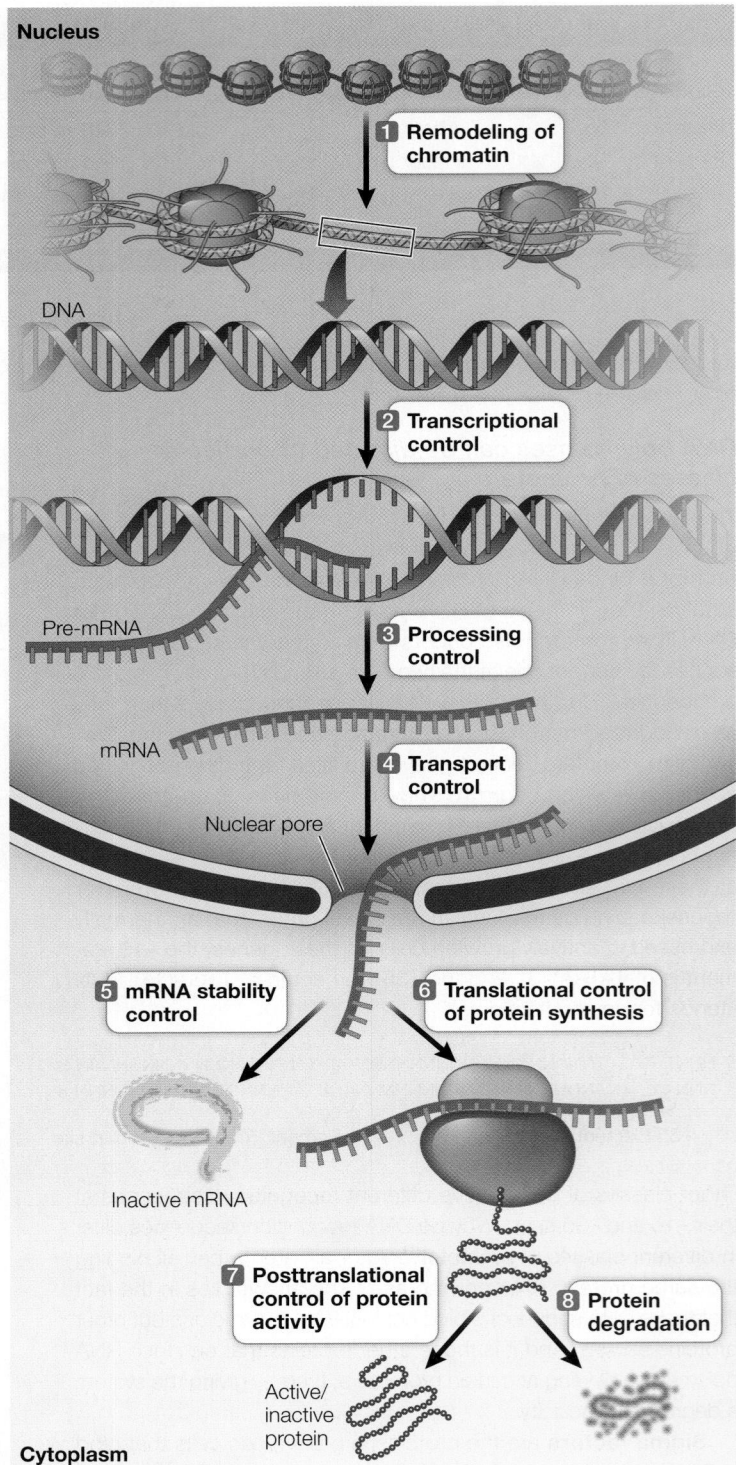

key concept 16.2 Eukaryotic Gene Expression Is Regulated by Transcription Factors

For cell function in single-celled eukaryotes as well as the normal development of a multicellular organism from fertilized egg to adult, certain proteins must be made at just the right times and in just the right cells; these proteins must not be made at other times in other cells. Here are two examples from humans:

1. In human pancreatic exocrine cells, the digestive enzyme procarboxypeptidase A makes up 7.6 percent of all the protein in the cell; in other cell types it is usually undetectable.

2. In human breast duct cells, alpha-lactalbumin, a protein in breast milk, is made only late in pregnancy and during lactation. Alpha-lactalbumin is not made in any other cell types.

Clearly the expression of eukaryotic genes is regulated.

focus your learning

- The rate of transcription of a eukaryotic gene depends on the combination of transcription factors and other proteins binding to regulatory sequences associated with the gene.
- DNA binding proteins have certain structural motifs in common that are important in their binding function.
- Eukaryotic genes whose expression is coordinated share the same transcription factors.

As in prokaryotes, gene expression in eukaryotes can be regulated at several different points in the process of transcribing and translating the gene into a protein (**Figure 16.7**). In this section we will describe the mechanisms that result in the selective transcription of specific genes. The mechanisms for regulating gene expression in eukaryotes have similar themes as in those of prokaryotes. Both types of cells use DNA–protein interactions and negative and positive control. However, there are many differences, some of them dictated by the presence of a nucleus, which physically separates transcription and translation (**Table 16.2**).

General transcription factors act at eukaryotic promoters

As in prokaryotes, a promoter in a eukaryotic gene is a sequence of DNA near the 5′ end of the coding region, where RNA polymerase binds and initiates transcription. Although eukaryotic promoters are more diverse than those of prokaryotes, many contain a nucleotide sequence similar to the –10 element in prokaryotic promoters. This element is usually located close to the transcription start site and is called the **TATA box** because it is rich in AT base pairs. The TATA box is the site where DNA begins to denature so that the template strand can be exposed. In addition to the TATA box, eukaryotic promoters typically include multiple regulatory sequences that are recognized and bound by **transcription factors**: regulatory proteins that help control transcription.

Figure 16.7 Potential Points for the Regulation of Gene Expression Gene expression can be regulated before transcription (1), during transcription (2, 3), after transcription but before translation (4, 5), at translation (6), or after translation (7).

Q: In prokaryotes, transcription and translation are often coupled in time and space. But in eukaryotes they are separated. What are the advantages of the nucleus as a compartment?

 Activity 16.1 Eukaryotic Gene Expression Control Points
www.Life11e.com/ac16.1

Like the prokaryotic RNA polymerase, eukaryotic RNA polymerase II cannot simply bind to the promoter and initiate transcription. Rather, it does so only after various **general transcription factors** have assembled on the chromosome (**Figure 16.8**). General transcription factors bind to most promoters and are distinct from transcription factors that act only at certain promoters or classes of promoters. First, the protein complex called TFIID ("TF" stands for transcription factor) binds to the TATA box. Binding of TFIID changes both its own shape and that of the DNA, presenting a new surface that attracts the binding of other general transcription factors to form a transcription initiation complex. RNA polymerase II binds only after several other proteins have bound to this complex.

Each general transcription factor has a role in gene expression:

- TFIIB binds both RNA polymerase and TFIID, and helps identify the transcription initiation site.

- TFIIF prevents nonspecific binding of the complex to DNA and helps recruit RNA polymerase to the complex; it is similar in function to a bacterial sigma factor.

- TFIIE binds to the promoter and stabilizes the denaturation of the DNA.

- TFIIH opens up the DNA for transcription.

Some regulatory DNA sequences, such as the TATA box, are common to the promoters of many eukaryotic genes and are recognized by general transcription factors that are found in all the cells of an organism. Other regulatory sequences are present in only a few genes and are recognized by specific transcription factors. These factors may be found only in certain types of cells or at certain stages of the cell cycle, or they may be activated by signaling pathways in response to cellular or environmental signals (see Chapter 7).

Specific proteins can recognize and bind to DNA sequences and regulate transcription

Some regulatory DNA sequences are positive elements termed **enhancers**: they bind transcription factors that either activate transcription or increase the rate of transcription. Other regulatory elements are **silencers**: they bind factors that repress transcription. Most of the regulatory elements needed for correct expression of a gene are found within a few hundred base pairs of the transcription start site. For example, the mouse albumin gene promoter contains all the information needed for liver cell–specific expression within 170 base pairs upstream of the transcription start site. But some regulatory elements may be located thousands of base pairs away and they may affect the expression of several nearby genes. When transcription factors bind to these elements, they interact with the RNA polymerase complex, causing the DNA to bend (**Figure 16.9**).

The combination of transcription factors binding to a gene determines the rate of transcription. For example, the immature red blood cells in bone marrow make large amounts of β-globin. At least

Figure 16.8 The Initiation of Transcription in Eukaryotes Apart from TFIID, which binds to the TATA box, each general transcription factor in this transcription complex has binding sites only for the other proteins in the complex, and does not bind directly to DNA. B, E, F, and H are general transcription factors.

 Animation 16.3 **Initiation of Transcription**
www.Life11e.com/a16.3

table **16.2** Transcription in Prokaryotes and Eukaryotes		
	Prokaryotes	Eukaryotes
Locations of functionally related genes	Often clustered in operons	Often distant from one another with separate promoters
RNA polymerases	One	Three: I transcribes rRNA II transcribes mRNA III transcribes tRNA and small RNAs
Promoters and other regulatory sequences	Few	Many
Initiation of transcription	Binding of RNA polymerase to promoter	Binding of many proteins, including RNA polymerase

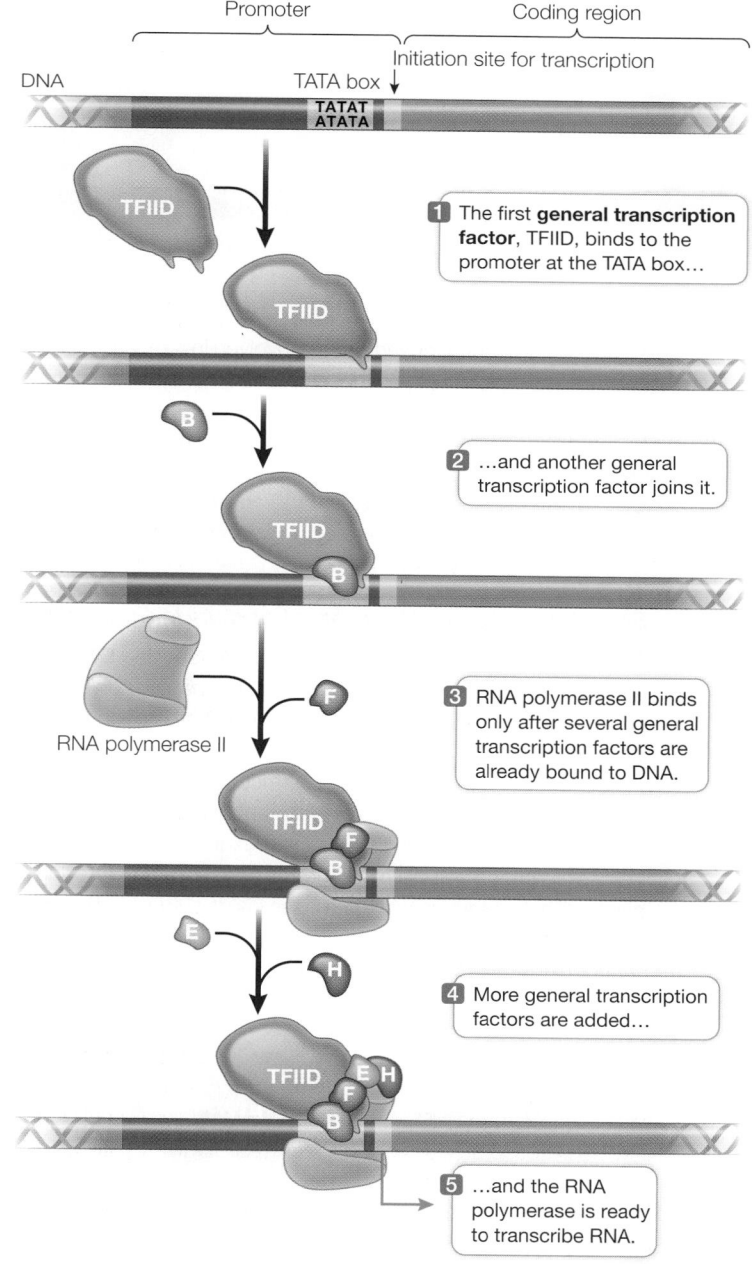

1 The first **general transcription factor**, TFIID, binds to the promoter at the TATA box...

2 ...and another general transcription factor joins it.

3 RNA polymerase II binds only after several general transcription factors are already bound to DNA.

4 More general transcription factors are added...

5 ...and the RNA polymerase is ready to transcribe RNA.

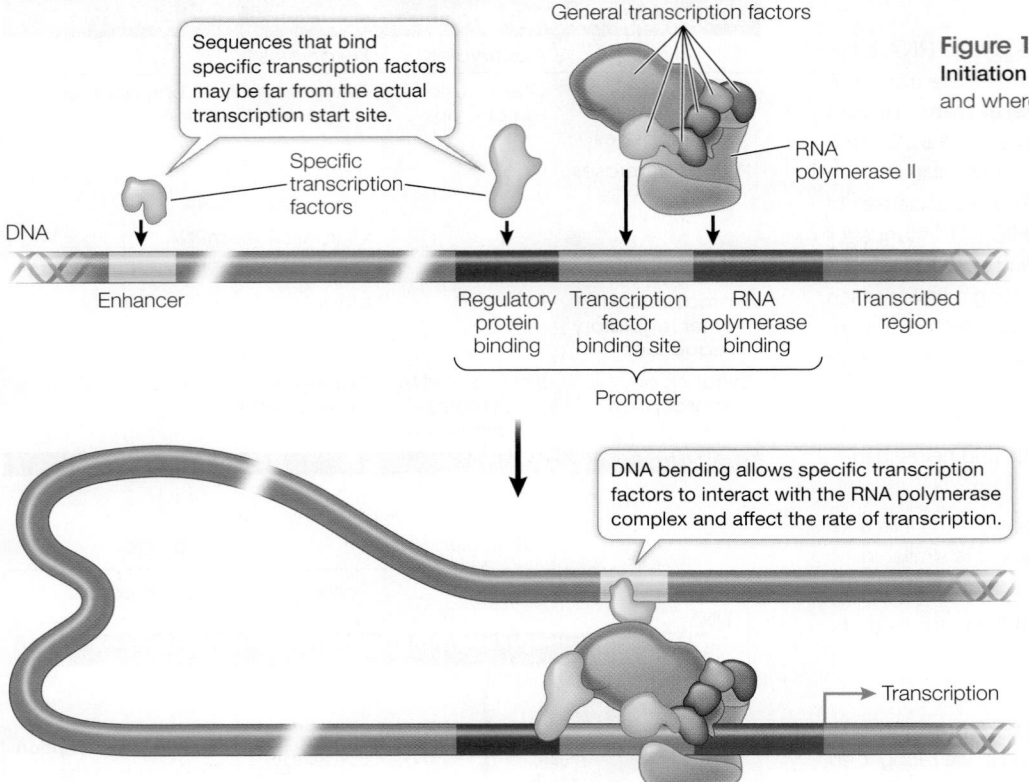

Sequences that bind specific transcription factors may be far from the actual transcription start site.

General transcription factors

RNA polymerase II

Specific transcription factors

DNA

Enhancer

Regulatory protein binding

Transcription factor binding site

RNA polymerase binding

Transcribed region

Promoter

DNA bending allows specific transcription factors to interact with the RNA polymerase complex and affect the rate of transcription.

Transcription

Figure 16.9 Transcription Factors and Transcription Initiation The actions of many proteins determine whether and where RNA polymerase II will transcribe DNA.

connect the concepts The structure and chemistry of DNA are key to its recognition by proteins. How the shapes and chemical structures of proteins allow them to bind noncovalently to other molecules is covered in Key Concept 3.2.

How does a protein recognize a sequence in DNA? As you learned in Key Concept 3.2, the complementary bases in DNA not only form hydrogen bonds with each other, but also can form additional hydrogen bonds with proteins, particularly at points exposed in the major and minor grooves. In this way, an intact DNA double helix can be recognized by a protein motif whose structure:

- fits into the major or minor groove;
- has amino acids that can project into the interior of the double helix; and
- has amino acids that can form hydrogen bonds with the interior bases.

Transcription factors underlie cell differentiation

During the development of a complex organism from fertilized egg to adult, cells become more and more differentiated (specialized). Differentiation is mediated in many cases by changes in gene expression, resulting from the activation (and inactivation) of various transcription factors. We will discuss this topic in more detail in Chapter 19. For now, remember that all differentiated cells contain the entire genome, and that their specific characteristics arise from differential gene expression.

The expression of sets of genes can be coordinately regulated by transcription factors

How do eukaryotic cells coordinate the regulation of several genes whose transcription must be turned on at the same time? Prokaryotes solve this problem by arranging multiple genes in an operon that is controlled by a single promoter, and by using sigma factors to recognize particular classes of promoters. Most eukaryotic genes have their own separate promoters, and genes that are coordinately regulated may be far apart. In these cases, the expression of genes can be coordinated if they share regulatory sequences that bind the same transcription factors.

Shared regulatory sequences enable organisms to respond to stress—plants, for example, use shared regulatory sequences to respond to drought. Under conditions of drought stress, a plant must simultaneously synthesize several proteins whose genes are

13 different transcription factors are involved in regulating transcription of the β-globin gene in these cells. Not all of these factors are present or active in other cells, such as the immature white blood cells produced by the same bone marrow. As a result, the β-globin gene is not transcribed in those cells. So although the same genes are present in all cells, the fate of the cell is determined by which of its genes are expressed. How do transcription factors recognize specific DNA sequences?

Specific protein–DNA interactions underlie binding

As we have seen, transcription factors with specific DNA-binding domains are involved in the activation and inactivation of specific genes. There are several common structural themes in the *protein domains* that bind to DNA. These themes, or **structural motifs**, consist of different combinations of structural elements (protein conformations) and may include special components such as zinc. One of the common structural motifs is the helix-turn-helix, in which two α helices are connected via a non-helical turn. The interior-facing "recognition" helix interacts with the bases inside the DNA. The exterior-facing helix sits on the sugar–phosphate backbone, ensuring that the interior helix is presented to the bases in the correct configuration:

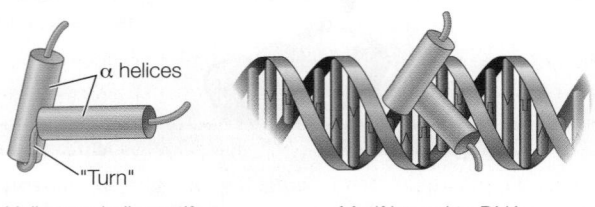

α helices

"Turn"

Helix-turn-helix motif   Motif bound to DNA

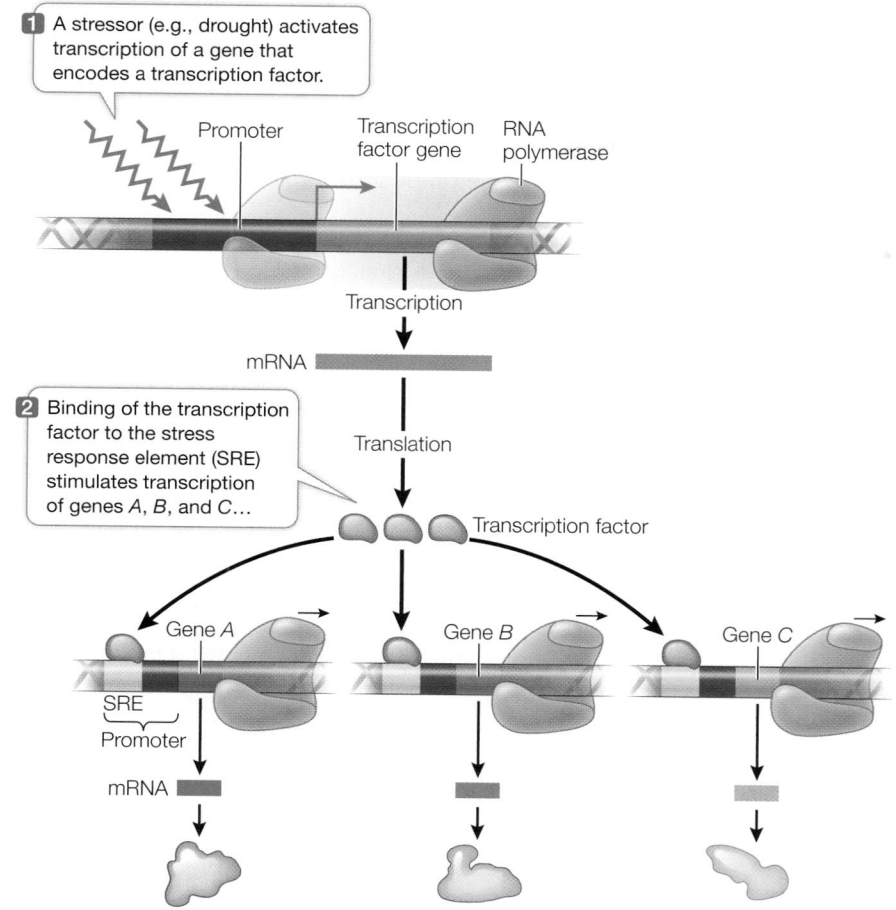

1 A stressor (e.g., drought) activates transcription of a gene that encodes a transcription factor.

Promoter   Transcription factor gene   RNA polymerase

Transcription

mRNA

2 Binding of the transcription factor to the stress response element (SRE) stimulates transcription of genes *A*, *B*, and *C*…

Translation

Transcription factor

Gene *A*

SRE
Promoter

mRNA

Gene *B*

Gene *C*

3 … which produce different proteins participating in the stress response.

Figure 16.10 Coordinating Gene Expression A single environmental signal, such as drought stress, causes the synthesis of a transcription factor that acts on many genes.

scattered throughout the genome. To coordinate expression of the stress response, each of the associated genes has a specific regulatory sequence near its promoter called the stress response element (SRE). A transcription factor binds to this element and stimulates mRNA synthesis (**Figure 16.10**). The stress response proteins not only help the plant conserve water, but also protect the plant against excess salt in the soil and freezing. This finding has considerable importance for agriculture because crops are often grown under less than optimal conditions or are affected by weather.

▶ 16.2 recap

Eukaryotes can increase or decrease transcription in various ways to help regulate gene expression. A number of general transcription factors must bind to a eukaryotic promoter before RNA polymerase will bind to it and begin transcription. Other, specific transcription factors bind to regulatory DNA sequences and interact with the RNA polymerase complex to control differential gene expression.

▶ 16.2 recap

learning outcomes

You should be able to:

- Explain the various ways that transcription factors regulate gene transcription.
- Describe the elements present in transcription factors that allow them to recognize and bind specific regions of DNA.
- Give an example of how eukaryotic genes widely separated by location are expressed in a coordinated fashion.

1. What are the ways whereby transcription factors regulate the rate of gene transcription?

2. How do transcription factors recognize specific DNA sequences?

3. How can more than one gene be regulated at the same time in a eukaryote?

We have seen how prokaryotes and eukaryotes regulate the transcription of their genes and operons. In the next section we will see how viruses can hijack prokaryotic and eukaryotic transcription mechanisms in order to complete their life cycles.

▶ key concept

16.3 Viruses Regulate Their Gene Expression during the Reproductive Cycle

"A virus is a piece of bad news wrapped in protein." This quote from immunologist Sir Peter Medawar is certainly true for the cells that viruses infect. As we described in Chapter 13, a bacterial virus (**bacteriophage**) injects its genetic material into a host bacterium and turns that cell into a virus factory (see Figure 13.3). Other viruses enter cells intact and then shed their coats and take over the cell's replication machinery. Viral life cycles can be very efficient. An example is the poliovirus: a single poliovirus infecting a mammalian cell can produce more than 100,000 new virus particles!

focus your learning

- A virus life cycle may be lytic or lysogenic.
- Understanding the infection cycles of a virus makes it possible to design therapeutic agents to fight infections.

Viruses are small infectious agents that infect cellular organisms and that cannot reproduce outside their host cells. Most virus particles, called **virions**, consist of only two or three components: the genetic material made up of DNA or RNA, a protein coat that protects the genetic material, and in some cases, an envelope of lipids that surrounds the protein coat. As we will see in this section, viral genomes include sequences that encode regulatory proteins. These proteins "hijack" the host cells' transcriptional machinery, allowing the viruses to complete their reproductive cycles.

Viruses undertake two kinds of reproductive cycles

After a viral genome enters a cell, typically the invader takes over the cells' molecular genetic machinery. But in some cases, there is an alternate series of events, in which the viral genome becomes integrated into the host genome.

LYTIC CYCLE   The Hershey–Chase experiment (see Figure 13.4) involved a typical **lytic** viral reproductive cycle, so named because soon after infection, the host cell bursts (lyses), releasing progeny viruses.

In this cycle, the viral genetic material takes over the host's synthetic machinery for its own reproduction immediately after infection. In the case of some bacteriophages, the process is extremely rapid—within 15 minutes, new phage particles appear in the bacterial cell. Ten minutes later, the "game is over," and these particles are released from the lysed cell. What happened?

At the molecular level, the reproductive cycle of a typical lytic virus has two stages: early and late, as illustrated in **Figure 16.11**. Follow along in the text and Figure 16.11 and you'll see examples of both positive and negative regulation, which stimulate and inhibit, respectively, gene expression:

- The viral genome contains a promoter that binds host RNA polymerase. In the early stage (1–2 min after phage DNA entry), viral genes that lie adjacent to this promoter are transcribed (positive regulation).

- These early genes often encode proteins that shut down host transcription (negative regulation) and stimulate viral genome replication and transcription of viral late genes (positive regulation). Three minutes after DNA entry, viral nuclease enzymes digest the host's chromosome, providing nucleotides for the synthesis of viral genomes.

- In the late stage, viral late genes are transcribed (positive regulation); they encode the proteins that make up the **capsid** (the outer shell of the virus) and other protein components of the virus and enzymes that lyse the host cell to release the new virions. This begins 9 min after DNA entry and 6 min before the first new phage particles appear.

The entire process—from binding and infection to release of new phage—takes about 30 min. During this period, the sequence of transcriptional events is carefully controlled to produce complete, infective virions.

LYSOGENIC CYCLE   Like all nucleic acid genomes, those of viruses can mutate and evolve by natural selection. Some viruses have evolved an advantageous process called **lysogeny** that postpones the lytic cycle. In lysogeny, the viral DNA becomes integrated into the host DNA and becomes a **prophage** (Figure 16.12). As the host cell divides, the viral DNA gets replicated along with that of the host. The prophage

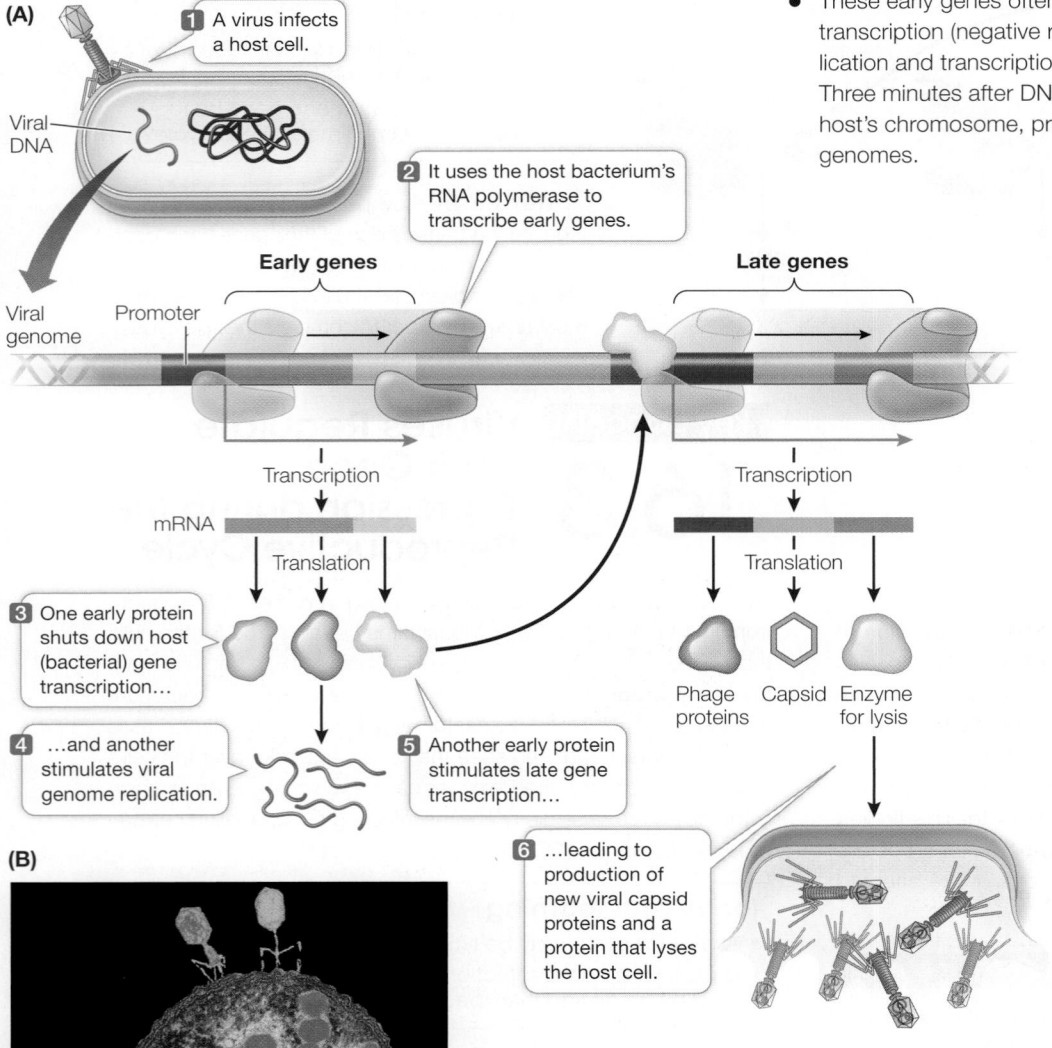

Figure 16.11 The Lytic Cycle: A Strategy for Viral Reproduction   **(A)** In a host cell infected with a virus, the viral genome uses its early genes to shut down host transcription while it replicates itself. Once the viral genome is replicated, its late genes produce capsid proteins that package the genome and other proteins that lyse the host cell. **(B)** Bacteriophages have attached to this *E. coli* cell, and the reproductive cycle is underway, producing new phage particles. The cell is viewed in transverse section.

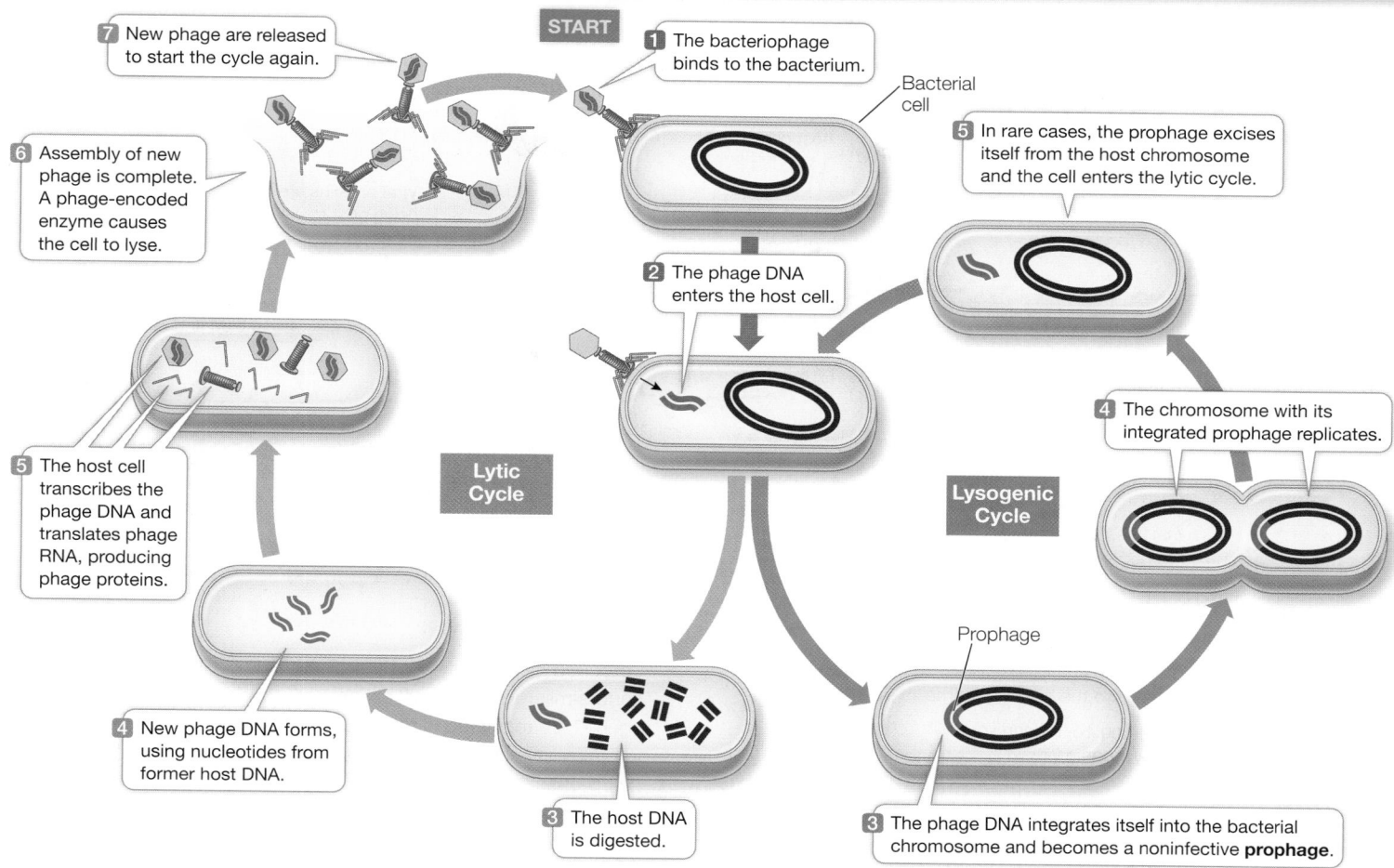

7 New phage are released to start the cycle again.

START

1 The bacteriophage binds to the bacterium.

Bacterial cell

5 In rare cases, the prophage excises itself from the host chromosome and the cell enters the lytic cycle.

6 Assembly of new phage is complete. A phage-encoded enzyme causes the cell to lyse.

2 The phage DNA enters the host cell.

4 The chromosome with its integrated prophage replicates.

Lytic Cycle

Lysogenic Cycle

5 The host cell transcribes the phage DNA and translates phage RNA, producing phage proteins.

Prophage

4 New phage DNA forms, using nucleotides from former host DNA.

3 The host DNA is digested.

3 The phage DNA integrates itself into the bacterial chromosome and becomes a noninfective **prophage**.

Figure 16.12 The Lytic and Lysogenic Cycles of Bacteriophages
In the lytic cycle, infection of a bacterium by viral DNA leads directly to multiplication of the virus and lysis of the host cell. In the lysogenic cycle, an inactive prophage is integrated into the host DNA where it is replicated during the bacterial life cycle.

can remain inactive within the bacterial genome for thousands of generations, producing many copies of the original viral DNA.

However, if the host cell is not growing well, the virus "cuts its losses." It switches to a lytic cycle, in which the prophage excises itself from the host chromosome and reproduces. In other words, the virus is able to enhance its chances of multiplication and survival by inserting its DNA into the host chromosome, where it sits as a silent passenger until conditions are right for lysis.

Eukaryotic viruses can have complex life cycles

Eukaryotes are susceptible to infection by various kinds of viruses whose genomes may consist of RNA or DNA. A subgroup of RNA viruses are called retroviruses.

- *DNA viruses*. Many viral particles contain double-stranded DNA. However, some contain single-stranded DNA, and a complementary strand is made after the viral genome enters the host cell. Like some bacteriophages, DNA viruses that infect eukaryotes are capable of undergoing both lytic and lysogenic life cycles. Examples include the herpes viruses and papillomaviruses (which cause warts).

- *RNA viruses*. Some viral genomes are made up of RNA that is usually, but not always, single-stranded. The RNA is translated by the host's machinery to produce viral proteins, some of which are involved in replication of the RNA genome. The influenza virus has an RNA genome.

- *Retroviruses*. As we described in Key Concept 14.2, a **retrovirus** is an RNA virus that carries a gene for **reverse transcriptase**, a protein that synthesizes DNA from an RNA template. The **retrovirus* uses this protein to make a DNA copy of its genome, which then becomes integrated into the host genome. The integrated DNA acts as a template for both mRNA and new viral genomes. HIV is a retrovirus that infects cells of the immune system and causes acquired immune deficiency syndrome (AIDS).

---

*connect the concepts   Viral diversity is discussed in Key Concept 25.4, which explains why the genomes of some viruses consist of single-stranded RNA and how RNA retroviruses reproduce themselves by reverse transcription.

HIV gene regulation occurs at the level of transcription elongation

As we have discussed so far, many instances of gene regulation occur at the level of transcription *initiation*, involving both activator and repressor proteins that bind to the promoters of genes. However,

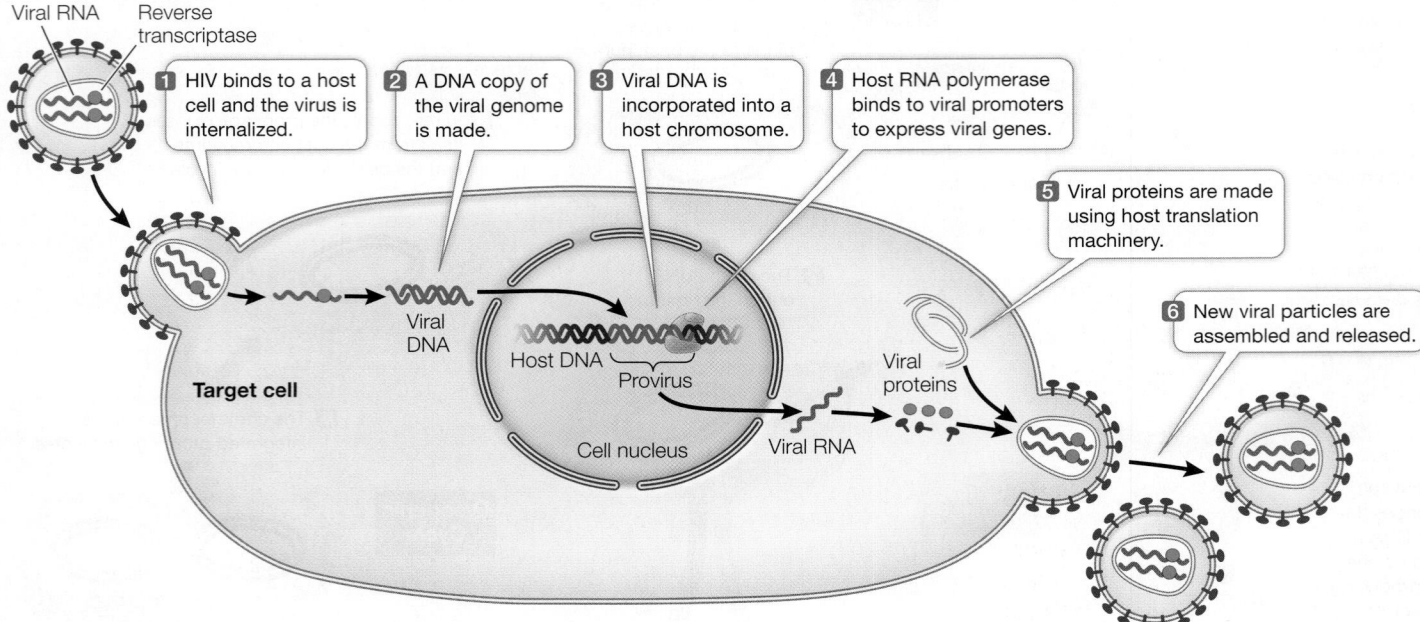

Figure 16.13 The Reproductive Cycle of HIV This retrovirus enters a host cell via fusion of its envelope with the host's cell membrane. Reverse transcription of retroviral RNA then produces a DNA provirus—a molecule of complementary DNA that inserts itself into the host's genome.

Q: Looking at Figures 16.12 and 16.13, can you classify HIV as a lytic or lysogenic virus?

studies of *HIV and other viruses have revealed that transcription can also be controlled at the *elongation* stage.

***connect the concepts** Learn about HIV and its effects on the immune system in Key Concept 41.6.

HIV is an **enveloped virus**; it is enclosed within a phospholipid membrane derived from its host cells (a specific type of immune system cell) (**Figure 16.13**). During infection, proteins in this membrane interact with proteins on the host cell surface, and the viral envelope fuses with the host cell membrane. After the virus enters the cell, its capsid is broken down. The viral reverse transcriptase then uses the virus's RNA template to produce a complementary DNA (cDNA) strand, while at the same time degrading the viral RNA. The enzyme then makes a complementary copy of the cDNA, and the resulting double-stranded DNA is inserted into the host's chromosome by an enzyme appropriately named integrase. The integrated DNA is referred to as the **provirus**. Both the reverse transcriptase and the integrase are carried inside the HIV virion, along with other proteins needed at the very early stages of infection.

The HIV provirus resides permanently in the host chromosome, and can remain in a latent (inactive) state for many years. During this time transcription of the viral DNA is initiated, but host cell proteins prevent the RNA from elongating, and transcription is terminated prematurely. Under some circumstances, such as when the host immune cell is activated, the level of transcription initiation increases and some viral RNA is made. One of these viral genes encodes a protein called tat (*trans*activator of *t*ranscription), which binds to a stem-and-loop structure at the 5' end of the viral RNA. As a result of tat binding, the production of full-length viral RNA is dramatically increased, and the rest of the viral reproductive cycle is able to

proceed. It was only after the discovery of this mechanism in HIV and similar viruses that researchers found that many eukaryotic genes are regulated at the level of transcription elongation.

Almost every step in the reproductive cycle of HIV is, in principle, a potential target for drugs to treat AIDS. The classes of anti-HIV drugs currently in use include:

- reverse transcriptase inhibitors that block viral DNA synthesis from RNA (at step 2 in Figure 16.13);
- integrase inhibitors that block the incorporation of viral DNA into the host chromosome (at step 3); and
- protease inhibitors that block the posttranslational processing of viral proteins (at step 5).

Combinations of drugs from these classes have been spectacularly successful in treating HIV infection.

▶ 16.3 recap

A virus consists of nucleic acids, a few proteins, and in some cases, a lipid envelope. Viruses require host cells to reproduce. Viral life cycles can include lytic and lysogenic stages. Bacteriophages λ use both positive and negative regulators of transcription initiation. Studies of HIV revealed a new mechanism for gene regulation: the regulation of transcription elongation.

learning outcomes

You should be able to:

- Identify instances of positive and negative regulation of viral gene expression in prokaryotes.
- Predict how an anti-HIV drug would work at the molecular level.

---

1. Describe positive and negative regulation of gene expression in the bacteriophage λ and HIV life cycles.
2. If the function of the protease that cuts HIV protein was specifically blocked by a drug, what would be the effect on the HIV reproductive cycle?

So far we have discussed mechanisms that cells and viruses use to control gene transcription. These mechanisms usually involve the interaction of regulatory proteins with specific DNA sequences. However, there are other mechanisms for controlling gene expression that do not depend on specific DNA sequences. We will discuss these mechanisms in the next section.

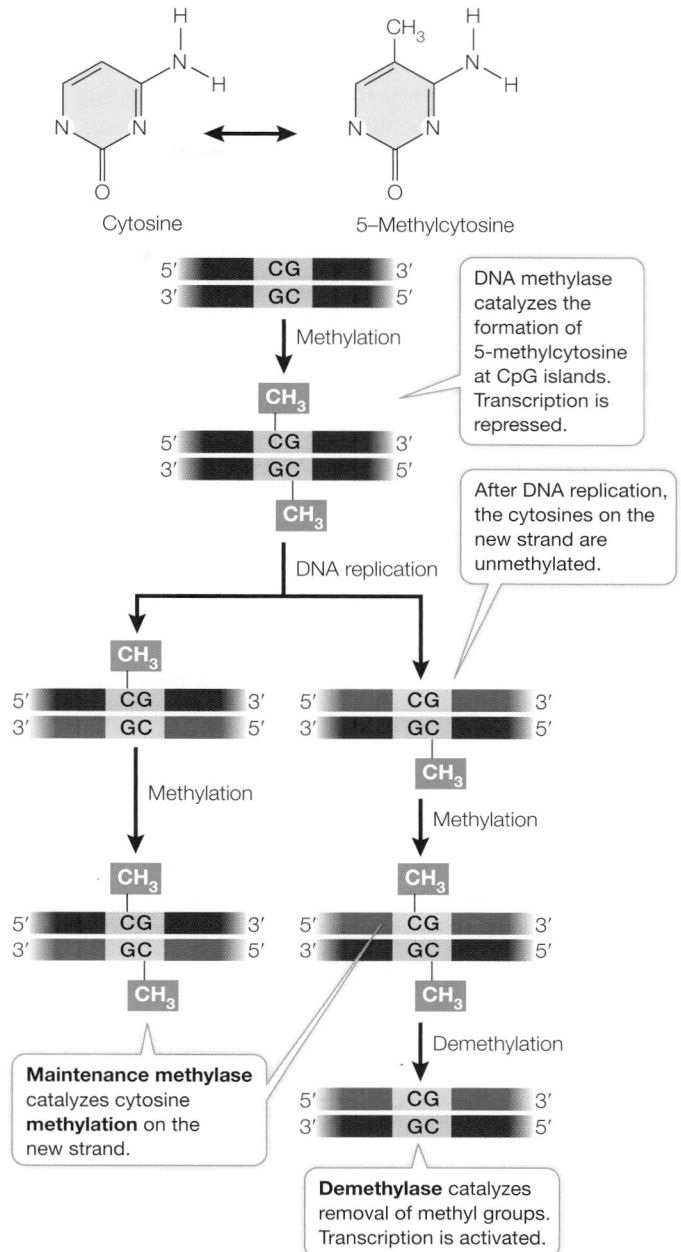

Figure 16.14 **DNA Methylation: An Epigenetic Change** The reversible formation of 5–methylcytosine in DNA can alter the rate of transcription.

Q: 5–methylcytosine is a mutational "hot spot" (see Figure 15.5). How might this relate to the importance of epigenetics in gene regulation?

>key concept 16.4 Epigenetic Changes Regulate Gene Expression

In the mid-twentieth century, the developmental biologist Conrad Hal Waddington coined the term "epigenetics" and defined it as "that branch of biology which studies the causal interactions between genes and their products which bring the phenotype into being." Today **epigenetics** is defined more specifically, referring to the study of changes in gene expression that are prompted without changes in the DNA sequence.

focus your learning

- Methylation of cytosine bases in DNA can enhance the binding of repressor proteins to promoter regions, resulting in silencing of gene expression.
- Acetylation and deacetylation of histone proteins alter the affinity of the histones for DNA, changing the accessibility of regions of the DNA to RNA polymerase.
- Environmental factors can cause epigenetic changes.
- Some heterochromatin such as the inactive X chromosome in female mammals results from extensive DNA methylation.

Epigenetic changes are reversible but sometimes are stable and heritable. You saw an example of this phenomenon in the opening story of this chapter, which described how genes inherited from mothers who had particularly stressful pregnancies were expressed at different levels than they were in children whose mothers had not suffered during pregnancy. In that example, stress—an environmental change—prompted a higher degree of DNA **methylation** in the promoter, reducing expression of a gene associated with behavior in teenagers. Here we examine more closely how epigenetic effects are caused either by DNA methylation or by alterations in chromosomal proteins.

DNA methylation occurs at the promoter and silences transcription

Depending on the organism, from 1 to 5 percent of cytosine residues in the organism's DNA are chemically modified by the addition of a methyl group ($-CH_3$) to the 5–carbon, to form 5–methylcytosine (**Figure 16.14**). This covalent addition is catalyzed by the enzyme **DNA methyltransferase** and, in mammals, usually occurs in C residues that are adjacent to G residues. DNA regions rich in these doublets are called **CpG islands**, and are especially abundant in promoters.

This covalent change in DNA is heritable: when DNA is replicated, a **maintenance methylase** catalyzes the formation of 5–methyl-cytosine in the new DNA strand. However, the pattern of cytosine methylation can also be altered, because methylation is reversible:

a third enzyme, appropriately called **demethylase**, catalyzes the removal of the methyl group from cytosine (see Figure 16.14).

What is the effect of DNA methylation? During replication and transcription, 5–methylcytosine behaves just like plain cytosine: it base-pairs with guanine. But extra methyl groups in a promoter attract proteins that bind methylated DNA. These proteins are generally involved in the repression of gene transcription; thus heavily methylated genes tend to be inactive. *This form of genetic regulation is epigenetic because it affects gene expression patterns without altering the DNA sequence.*

DNA methylation is important in development from egg to embryo. For example, when a mammalian sperm enters an egg, many genes in first the male and then the female genome become demethylated. Thus many genes that are usually inactive are expressed during early development. As the embryo develops and its cells become more specialized, genes whose products are not needed in particular cell types become methylated. These methylated genes are "silenced"; their transcription is repressed. However, unusual or abnormal events can sometimes turn silent genes back on.

For example, DNA methylation may play roles in the genesis of some cancers. In cancer cells, oncogenes get activated and promote cell division, and tumor suppressor genes (which normally inhibit cell division) are turned off (see Chapter 11). This misregulation can occur when the promoters of oncogenes become demethylated while those of tumor suppressor genes become methylated, as is the case in colorectal cancer (see Figure 15.10).

Histone protein modifications affect transcription

Another mechanism for epigenetic gene regulation is the alteration of chromatin structure, or chromatin remodeling. As we saw in Chapter 11, DNA is packaged with histone proteins into nucleosomes (see Figure 11.8), which can make DNA physically inaccessible to RNA polymerase and the rest of the transcription apparatus. Each histone protein has a "tail" of approximately 20 amino acids at its N terminus that sticks out of the compact structure and contains certain positively charged amino acids (notably lysine). Ordinarily there is strong ionic attraction between the positively charged histone proteins and DNA, which is negatively charged because of its phosphate groups. However, enzymes called histone acetyltransferases can add acetyl groups to these positively charged amino acids, thus changing their charges:

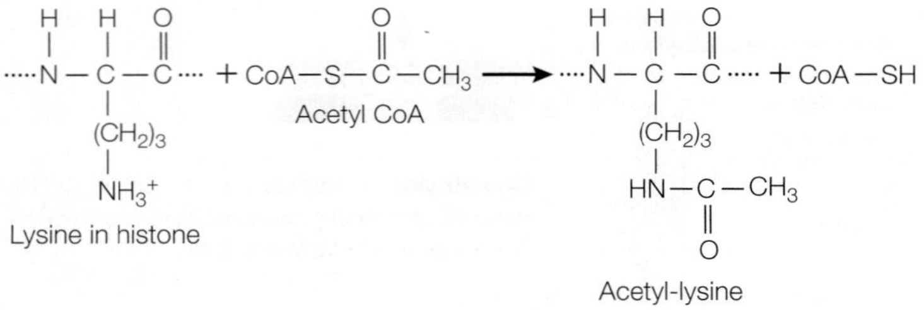

Reducing the positive charges of the histone tails reduces the affinity of the histones for DNA, opening up the compact nucleosome (**Figure 16.15**). Additional chromatin remodeling proteins can bind to the loosened nucleosome–DNA complex, opening up the DNA for gene expression. Histone acetyltransferases can thus activate transcription.

Another kind of chromatin remodeling protein, histone deacetylase, can remove the acetyl groups from histones and thereby repress transcription. Histone deacetylases are targets for drug development to treat some forms of cancer. As noted above, certain genes block cell division in normal specialized tissues. In some cancers these genes are less active than in normal cells, and the histones near them show excessive levels of deacetylation. Theoretically, a drug acting as a histone deacetylase inhibitor could tip the balance toward acetylation, and this might activate genes that normally inhibit cell division.

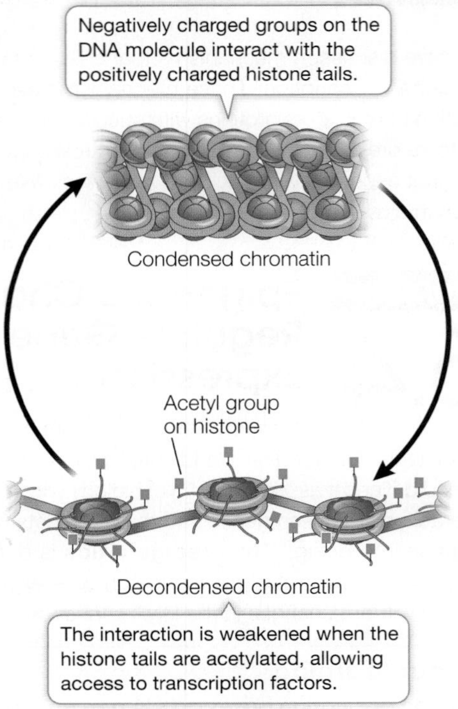

Figure 16.15 Epigenetic Remodeling of Chromatin for Transcription   Initiation of transcription requires that nucleosomes change their structure, becoming less compact. This chromatin remodeling makes DNA accessible to the transcription initiation complex (see Figure 16.8).

Acetylation isn't the only type of histone modification that can affect gene activation and repression. For example, histone methylation (not to be confused with DNA methylation) is associated with gene inactivation, and histone phosphorylation also affects gene expression, the specific effects depending on which amino acids are modified. All of these effects are reversible, and so the activity of a eukaryotic gene may be determined by very complex patterns of histone modification.

Epigenetic changes can be induced by the environment

Female honey bees all have the same genetic makeup. When they are in the immature stage called a larva, however, one female in the hive eats a protein-rich substance called royal jelly that dramatically alters the expression of many genes. The queen grows much larger than her peers, stays in the hive, and is tended to by the other bees. Above all, the queen lives up to several years and is fertile, laying the eggs that will produce more bees. The multitudes of her female compatriot larvae express a different set of genes, becoming workers who build the honeycomb, forage for food, and tend to the queen for their short lives of a few weeks. All of these differences ultimately come from the environment, specifically the royal jelly diet. Recent investigations have shown that the differences in gene expression between these very different yet genetically identical bees are due to differences in DNA methylation (**Investigating Life: Gene Expression and Behavior**).

Although they are reversible, many epigenetic changes such as DNA methylation and histone modification can permanently alter

experiment

Original Paper: Kucharski, R., J. Maleszka, S. Foret and R. Maleszka. 2008. Nutritional control of reproductive status in honeybees via DNA methylation. *Science* 319: 1817–1830.

Most female honey bee larvae grow up to be workers; just one eats royal jelly and becomes the queen. To understand differences in gene expression in queen bees versus their sisters with identical genomes, Ryszard Maleszka and colleagues at Australian National University developed an experiment to test for possible epigenetic effects.

The experiment involved injecting female honey bee larvae with a substance that inhibits expression of the enzyme DNA cytosine-5-methyltransferase (DNMT)—the enzyme that catalyzes the formation of 5–methylcytosine from cytosine in DNA. In a technique called RNAi, the scientists injected female larvae with a small RNA complementary to the mRNA for DNMT to specifically block expression of the DNMT gene.

HYPOTHESIS▶ DNA methylation underlies the developmental and behavioral differences between queen and worker honey bees.

METHOD

RESULTS

CONCLUSION▶ Decreasing DNA methylation mimics the effects of royal jelly, causing larvae destined to become worker bees to instead become queens.

work with the data

QUESTIONS▶

1. After injection with either an inhibitor of DNMT or a control, the level of DNMT mRNA was measured in the heads of the larvae and compared with the level of a control mRNA that is always expressed at a high level. The results are shown in **Table A**.

Table A

Time (h)	DNMT mRNA level compared to control (%)
23	105
48	41

  a. Why was mRNA measured in the heads of the larvae?

  b. What can you conclude about the effectiveness of the inhibition of DNMT expression?

2. The DNA of a gene that is normally expressed in the brain (head) of the larvae was sequenced and the percentage of 5–methylcytosine measured to determine the extent of cytosine methylation in CpG regions. **Table B** shows the results. What can you conclude about the effect of DNMT inhibition?

Table B

Condition	Percent 5–methylcytosine
Control	79
DNMT expression inhibited	63

3. Larvae that had been injected with the inhibitor for DNMT gene expression or a control were allowed to develop into adults. The phenotypes of the adults were evaluated, and the data are shown in **Table C**. What do the data show about the effect of the inhibition of DNA methylation? Does the extent of DNA methylation strictly correlate with the extent of phenotypic change?

Table C

Condition	Number of workers	Number of queens
Control	238	73
Inhibited	74	188

A similar **work with the data** exercise may be assigned in **LaunchPad**.

gene expression patterns in a cell. In a germ line cell that forms gametes, epigenetic changes can be passed on to the next generation. But what determines these epigenetic changes? A clue comes from a recent study of monozygotic (identical) twins. Monozygotic twins come from a single fertilized egg that divides to produce two separate cells; each of these goes on to develop a separate individual.

(A)

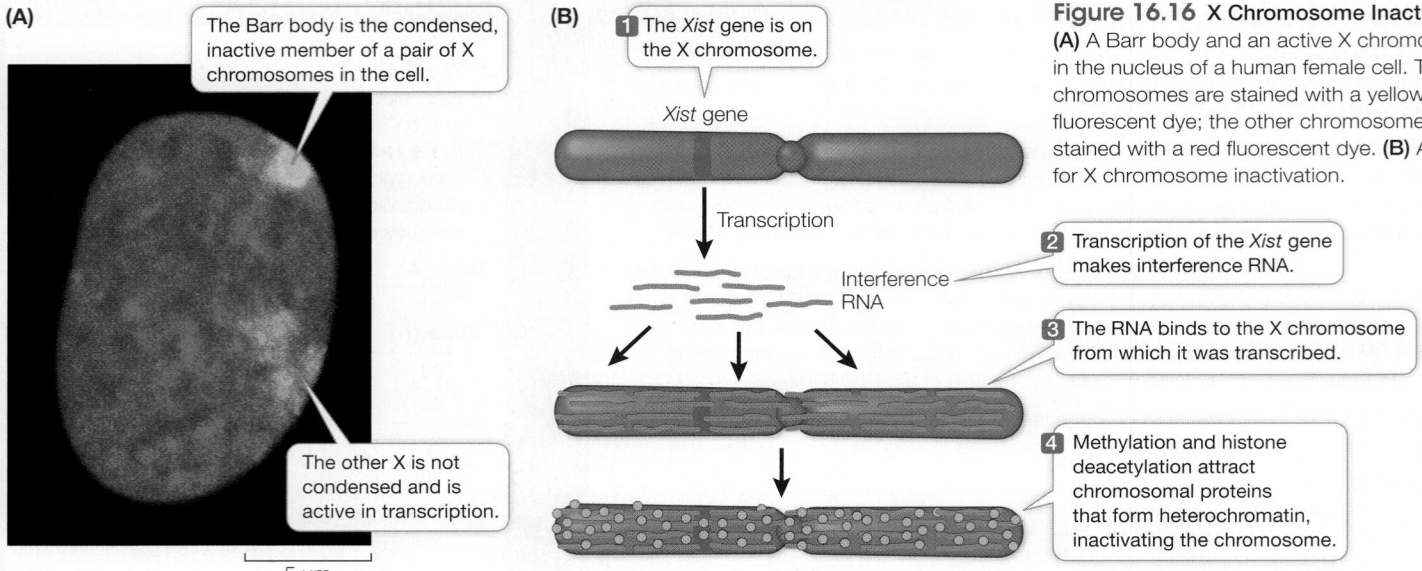

The Barr body is the condensed, inactive member of a pair of X chromosomes in the cell.

The other X is not condensed and is active in transcription.

5 μm

(B)

1 The *Xist* gene is on the X chromosome.

Xist gene

Transcription

Interference RNA

2 Transcription of the *Xist* gene makes interference RNA.

3 The RNA binds to the X chromosome from which it was transcribed.

4 Methylation and histone deacetylation attract chromosomal proteins that form heterochromatin, inactivating the chromosome.

Figure 16.16 X Chromosome Inactivation **(A)** A Barr body and an active X chromosome in the nucleus of a human female cell. The X chromosomes are stained with a yellow-green fluorescent dye; the other chromosomes are stained with a red fluorescent dye. **(B)** A model for X chromosome inactivation.

Monozygotic twins thus have identical genomes. But are they identical in their epigenomes? A comparison of DNA in hundreds of such twin pairs shows that in tissues of 3-year-olds, the DNA methylation patterns are virtually the same. But by age 50, by which time the twins have usually been living apart in different environments for decades, the patterns are quite different. This indicates that the *environment plays an important role in epigenetic modifications*, and thus in the regulation of genes that these modifications affect.

 Media Clip 16.1 The Surprising Epigenetics of Identical Twins
www.Life11e.com/mc16.1

Global chromosome changes involve DNA methylation

Like single genes, large regions of chromosomes or even entire chromosomes can have distinct patterns of DNA methylation. Under a microscope, two kinds of chromatin can be distinguished in the stained interphase nucleus: **euchromatin** and **heterochromatin**. The euchromatin appears diffuse and stains lightly; it contains the DNA that is transcribed into mRNA. Heterochromatin is condensed and stains darkly; any genes it contains are generally not transcribed.

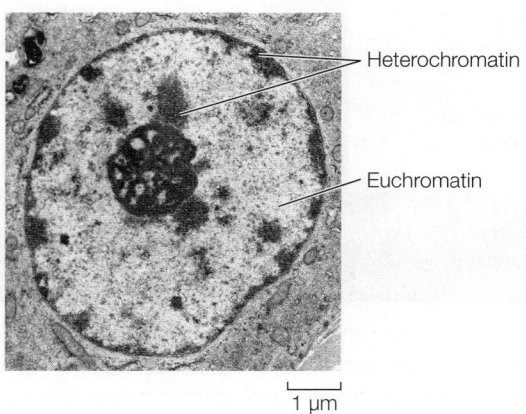

Heterochromatin

Euchromatin

1 μm

Perhaps the most dramatic example of heterochromatin is the inactive X chromosome of mammals. A normal female mammal has two X chromosomes; a normal male has an X and a Y (see Key Concept 12.4). The X and Y chromosomes probably arose from a pair of autosomes (non–sex chromosomes) about 300 million years ago. Over time, mutations in the Y chromosome resulted in maleness-determining genes, and the Y chromosome gradually lost most of the genes it once shared with its X homolog. As a result, females and males differ greatly in the "dosage" of X-linked genes. Each female cell has two copies of each gene on the X chromosome and therefore has the potential to produce twice as much of each protein product. Nevertheless, for 75 percent of the genes on the X chromosome, transcription is generally the same in males and in females. How does this happen?

During early embryonic development, one of the X chromosomes in each cell of a female is largely inactivated with regard to transcription. The same X chromosome remains inactive in all of that cell's descendants. In a given embryonic cell, the "choice" of which X in the pair to inactivate is random. Recall that one X in a female comes from her father and one from her mother. Thus in one embryonic cell the paternal X might be the one remaining transcriptionally active, but in a neighboring cell the maternal X might be active.

The inactivated X chromosome is identifiable within the nucleus because it is very compact, even during interphase. Typically, a nuclear structure called a Barr body (after its discoverer, Murray Barr) can be seen in human female cells under the light microscope (**Figure 16.16A**). This clump of heterochromatin, which is not present in normal males, is the inactivated X chromosome, and it consists of heavily methylated DNA. A female with the normal two X chromosomes will have one Barr body, whereas a rare female with three Xs will have two, and an XXXX female will have three. Males that are XXY will have one. These observations suggest that the interphase cells of each person, male or female, have a single active X chromosome, and thus a constant dosage of expressed X chromosome genes.

Condensation of the inactive X chromosome makes its DNA sequences physically unavailable to the transcriptional machinery. Most of the genes of the inactive X are heavily methylated. However, one gene, *Xist* (for *X i*nactivation-specific *t*ranscript), is only lightly methylated and is transcriptionally active. On the active X chromosome, *Xist* is heavily methylated and not transcribed. The RNA transcribed from *Xist* binds to the X chromosome from which it is transcribed, and this binding leads to a spreading of inactivation along the chromosome. The *Xist* RNA transcript is an example of **interference RNA** (Figure 16.16B).

16.4 recap

Epigenetics describes stable changes in gene expression that do not involve changes in DNA sequences. These changes involve modifications of DNA (cytosine methylation) or of histone proteins bound to DNA. Epigenetic changes can be affected by the environment. Large stretches of DNA can be epigenetically modified, leading to inactivtion of many genes.

learning outcomes

You should be able to:

- Describe the role of methylation in gene expression.
- Explain how acetylation of proteins changes the structure of chromatin and affects the rate of transcription.
- Cite evidence supporting the claim that environmental effects have a strong role in epigenetic changes.
- Describe how and where X chromosome inactivation occurs, and explain why it is believed to occur.

1. How do histone modifications affect transcription?
2. What is the evidence that epigenetic modifications affect behavior?
3. How does X chromosome inactivation occur, and why is it believed to occur?
4. In colorectal cancer, certain tumor suppressor genes are not active, and this results in uncontrolled cell division. Two possible explanations for the inactive genes are (a) mutations in the coding regions, resulting in inactive proteins, and (b) epigenetic silencing at the promoters of the genes, resulting in reduced transcription. How would you investigate these two possibilities?

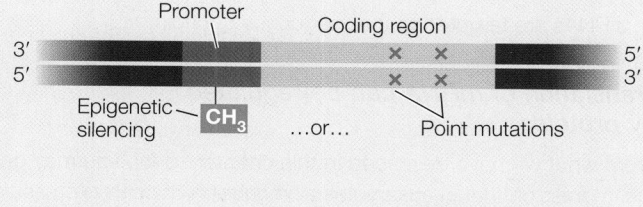

Gene expression involves transcription and then translation. So far we have described how gene expression is regulated at the transcriptional level. But as Figure 16.7 shows, there are many points at which regulation can occur after the initial gene transcript is made.

Eukaryotic Gene Expression Can Be Regulated after Transcription

Eukaryotic gene expression can be regulated both in the nucleus prior to mRNA export, and after the mRNA leaves the nucleus. This can happen in several ways.

focus your learning

- Alternative splicing can give rise to multiple mRNA molecules from a single primary mRNA transcript.
- MicroRNAs (miRNAs) are short single-stranded RNA molecules complementary to mRNAs that bind to and inhibit mRNAs from being translated.
- Small interfering RNAs (siRNAs) perform a similar role in the cell as miRNAs but have a viral origin.
- Single-stranded RNA can form secondary structures that inhibit its translation.
- The cell tags proteins to be degraded with ubiquitin, which leads to its binding to a proteasome, where degradation occurs.

RNA splicing can result in different mRNAs being made from the same gene

Most primary mRNA transcripts contain several introns (see Figure 14.7). Before the RNA is exported from the nucleus, a splicing mechanism recognizes the boundaries between exons and introns and converts pre-mRNA, which has the introns, into mature mRNA, which does not:

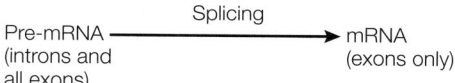

For many genes, **alternative splicing** can occur, whereby some exons are spliced out along with the introns (**Figure 16.17**). This mechanism generates a family of different proteins, with different functions, from a single gene. Recent surveys show that about half of all human genes are alternatively spliced. Alternative splicing may be a key to the differences in levels of complexity among organisms. For example, although humans and chimpanzees have similar-sized genomes, there is more alternative splicing in the human brain than in the chimpanzee brain.

Alternative RNA splicing is controlled both by regulatory elements in the RNA sequence that bind specific proteins (similar to regulatory sequences in DNA) and by secondary RNA structures that form by hybridization between nucleotides in the single-stranded RNA molecule.

Small RNAs are important regulators of gene expression

As you'll see in Chapter 17, less than 5 percent of the genome in most plants and animals codes for proteins. Some of the genome encodes ribosomal RNA and transfer RNAs, but until recently biologists thought that the rest of the genome was not transcribed;

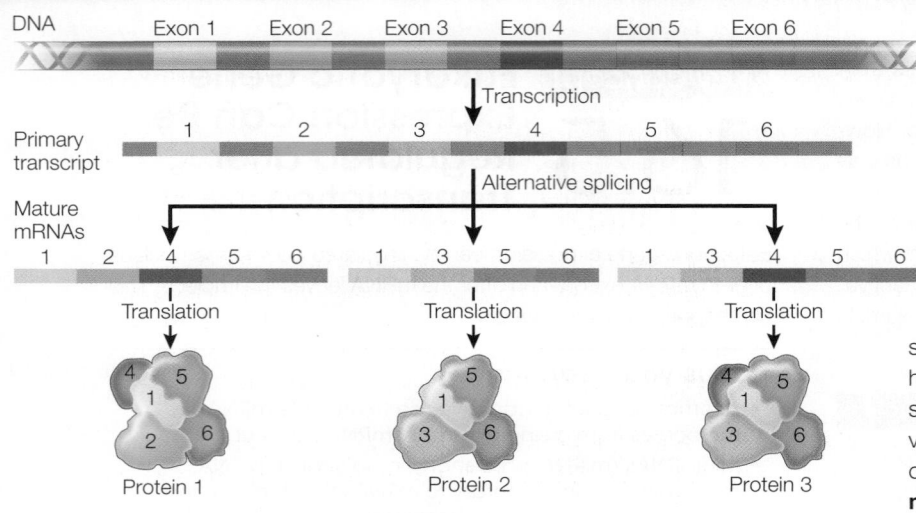

Figure 16.17 Alternative Splicing Results in Different Mature mRNAs and Proteins Pre-mRNA can be spliced differently in different tissues, resulting in different proteins.

Q: A protein-encoding gene in a eukaryote has four introns. How many different proteins could be made by alternative splicing of the pre-mRNA from this gene?

some even called it "junk." Recent investigations, however, have shown that some of these noncoding regions are transcribed. The RNAs produced from these regions are often very small and therefore difficult to detect. In both prokaryotes and eukaryotes, these tiny RNA molecules are called **microRNA (miRNA)**.

There are as many as 5,000 miRNAs, with about 1,000 of them encoded in the human genome. Each miRNA is about 22 bases long and usually has dozens of mRNA targets because the base pairing between the miRNA and the target mRNA doesn't have to be perfect. MicroRNAs are transcribed as longer precursors that fold into double-stranded RNA molecules and are then processed through a series of steps into single-stranded miRNAs. A protein complex guides the miRNA to its target mRNA, where translation is inhibited (**Figure 16.18A**). The remarkable conservation of the miRNA gene-silencing mechanism indicates that it is evolutionarily ancient and biologically important.

In addition to miRNAs, there is a similarly acting class of molecules called **small interfering RNAs (siRNAs)**. These often arise from viral infections, when two complementary strands of a viral genome are transcribed. Large double-stranded RNAs are formed, and as with miRNAs, these are converted into shorter single-stranded sequences; these bind to the target RNA and cause its degradation (**Figure 16.18B**). Small interfering RNAs are also derived from transposon sequences, which are widespread in eukaryotic genomes (see Key Concept 15.1). Therefore it is likely that gene silencing involving siRNAs evolved as a defense mechanism to prevent the translation of viral and transposon sequences. MicroRNAs and siRNAs are similar molecules that are processed by the same cellular enzymes. A major difference between them is that:

- miRNAs are synthesized from DNA sequences separate from their target, whereas
- siRNAs are targeted to their sequence of origin.

Translation of mRNA can be regulated by proteins

From what we have described in this chapter so far, you may get the impression that in eukaryotes all regulation of gene expression is at the level of transcription. But is the amount of a protein in a cell really determined only by the amount of its mRNA? The answer is no. For example, a survey of genes and their expression in yeast cells showed that for about one-third of the genes, there was a clear correlation between mRNA and protein: more of one led to more of the other. But for two-thirds of the genes, there was no apparent

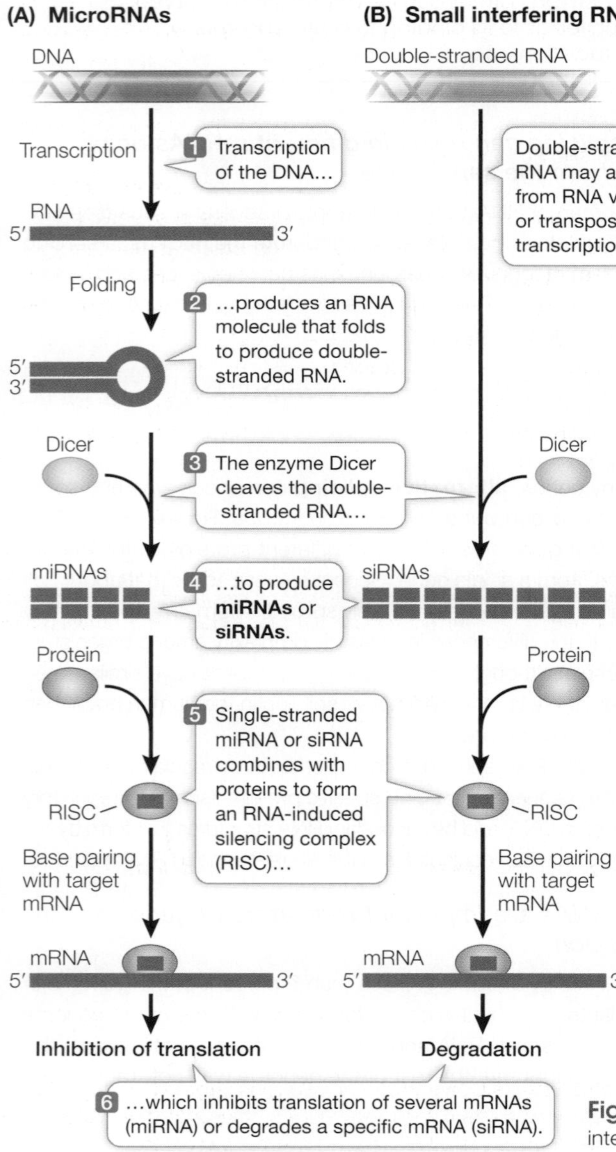

(A) MicroRNAs

DNA

Transcription

1 Transcription of the DNA...

RNA

5' ———— 3'

Folding

2 ...produces an RNA molecule that folds to produce double-stranded RNA.

5'
3'

Dicer

3 The enzyme Dicer cleaves the double-stranded RNA...

miRNAs

4 ...to produce **miRNAs** or **siRNAs**.

Protein

5 Single-stranded miRNA or siRNA combines with proteins to form an RNA-induced silencing complex (RISC)...

RISC

Base pairing with target mRNA

mRNA

5' ———— 3'

Inhibition of translation

6 ...which inhibits translation of several mRNAs (miRNA) or degrades a specific mRNA (siRNA).

(B) Small interfering RNAs

Double-stranded RNA

Double-stranded RNA may arise from RNA viruses or transposon transcription.

Dicer

siRNAs

Protein

RISC

Base pairing with target mRNA

mRNA

5' ———— 3'

Degradation

Figure 16.18 mRNA Inhibition by RNAs MicroRNAs and small interfering RNAs can inhibit translation by binding to target mRNAs.

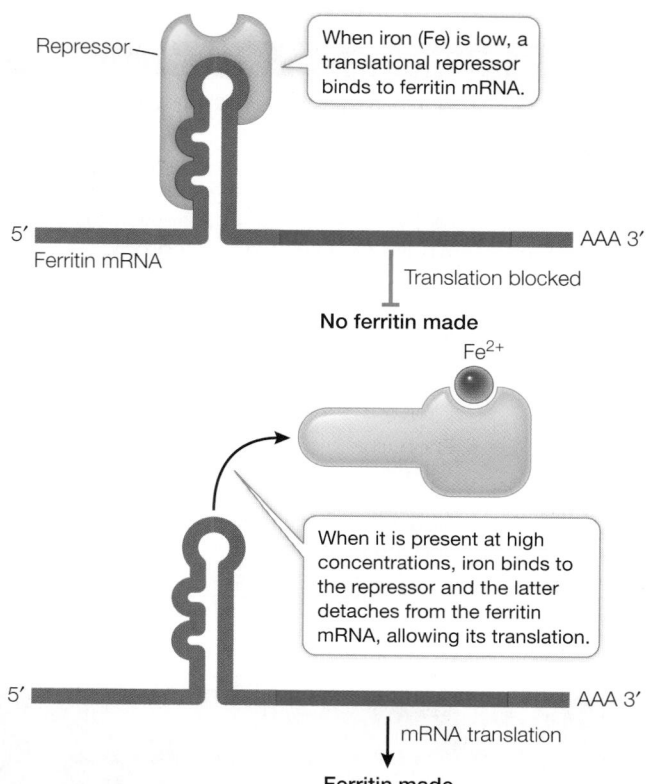

Repressor —

When iron (Fe) is low, a translational repressor binds to ferritin mRNA.

5′   Ferritin mRNA   AAA 3′

Translation blocked

No ferritin made

Fe^{2+}

When it is present at high concentrations, iron binds to the repressor and the latter detaches from the ferritin mRNA, allowing its translation.

5′   AAA 3′

mRNA translation

Ferritin made

Figure 16.19 A Translational Repressor Can Repress Translation Binding of a protein to a target mRNA can inhibit its translation.

relationship between the two: sometimes there was lots of mRNA and little or no protein, or lots of protein and little mRNA. The concentrations of these proteins must therefore have been determined by factors acting after the mRNA was made. Cells have two major ways to control the amount of a protein after transcription:

1. They can regulate translation of the protein's mRNA.

2. They can regulate how long a newly synthesized protein persists in the cell (protein longevity).

REGULATION OF TRANSLATION   There are a variety of ways in which the translation of mRNA can be regulated. One way, as we saw in the previous section, is to inhibit translation with siRNAs and miRNAs. A second way involves modification of the guanosine triphosphate cap on the 5′ end of the mRNA (see Key Concept 14.4). An mRNA that is capped with an unmodified GTP molecule is not translated. For example, stored mRNAs in the egg cells of the tobacco hornworm moth are capped with unmodified GTP molecules and are not translated. After the egg is fertilized, however, the caps are modified, allowing the mRNA to be translated to produce the proteins needed for early embryonic development.

In another system, repressor proteins directly block translation. For example, in mammalian cells the protein ferritin binds free iron ions (Fe^{2+}). When iron is present in excess, ferritin synthesis rises

dramatically, but the amount of ferritin mRNA remains constant, indicating that the increase in ferritin synthesis is due to an increased rate of mRNA translation. Indeed, when the iron level in the cell is low, a translational repressor protein binds to the 5′ noncoding region of ferritin mRNA and prevents its translation by blocking its attachment to a ribosome. When the iron level rises, some of the excess Fe^{2+} ions bind to the repressor and alter its three-dimensional structure, causing the repressor to detach from the mRNA and allowing translation to proceed (**Figure 16.19**). The binding site for the translational repressor on mRNA is a stem-and-loop region with sufficient three-dimensional structure for recognition by a protein or small molecule.

REGULATION OF PROTEIN LONGEVITY   The protein content of a cell at any given time is a function of both protein synthesis and protein degradation. Certain proteins can be targeted for destruction in a chain of events that begins when an enzyme attaches a 76-amino acid protein called **ubiquitin** (so named because it is ubiquitous, or widespread) to a lysine residue of the protein to be destroyed. Other ubiquitins then attach to the primary one, forming a polyubiquitin chain. The protein–polyubiquitin complex then binds to a huge protein complex called a **proteasome** (from *protease* + *soma*, "body") (**Figure 16.20**). Upon entering the proteasome, the polyubiquitin is removed and ATP energy is used to unfold the target protein. Three different proteases then digest the protein into small peptides and amino acids.

You may recall from Key Concept 11.2 that cyclins are proteins that regulate the activities of key enzymes at specific points in the cell cycle. Cyclins must be broken down at just the right time, and this is done by attaching ubiquitin to them and degrading them in the proteasomes. Viruses can hijack this system. For example, some strains of the human papillomavirus (HPV) add ubiquitin to the p53 and retinoblastoma proteins, targeting them for proteasomal degradation. These proteins normally inhibit the cell cycle, so the result of this HPV activity is unregulated cell division (cancer).

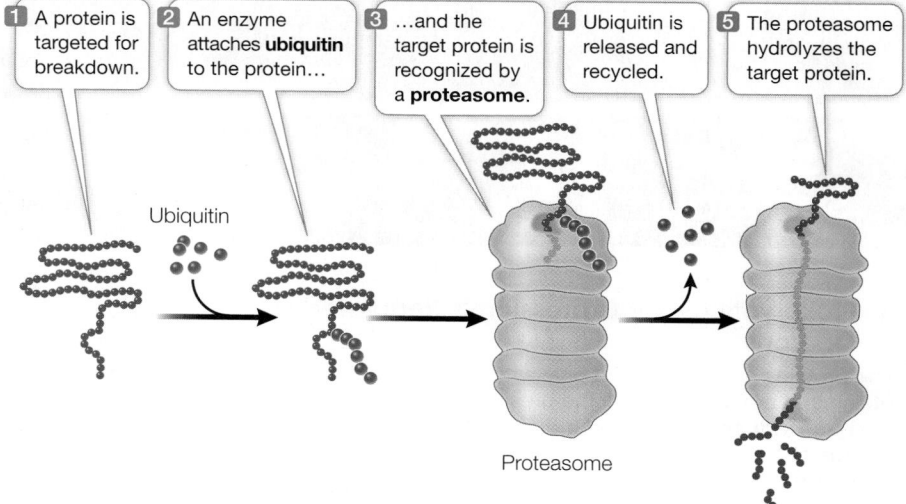

1 A protein is targeted for breakdown.

2 An enzyme attaches **ubiquitin** to the protein…

3 …and the target protein is recognized by a **proteasome**.

4 Ubiquitin is released and recycled.

5 The proteasome hydrolyzes the target protein.

Ubiquitin

Proteasome

Figure 16.20 A Proteasome Breaks Down Proteins   Proteins targeted for degradation are bound to ubiquitin, which then binds the targeted protein to a proteasome. The proteasome is a complex structure where proteins are digested by several powerful proteases.

16.5 recap

One of the most important means of posttranscriptional regulation is alternative RNA splicing, which allows more than one protein to be made from a single gene. The stability of mRNA in the cytoplasm can also be regulated. MicroRNAs, siRNAs, mRNA modifications, and translational repressors can prevent mRNA translation. Proteins in the cell can be targeted for breakdown by ubiquitin and then hydrolyzed in proteasomes.

learning outcomes

You should be able to:

- Explain the differences in the number of protein-coding genes and the number of proteins in a cell.
- Explain how miRNAs regulate translation of mRNA.
- Compare and contrast miRNAs and siRNAs.
- Suggest a way that the structure of an mRNA molecule could affect its translation.
- Describe the function of the proteasome in a cell.

1. How do miRNAs and siRNAs regulate gene expression?

2. How does the three-dimensional structure of mRNA contribute to the regulation of its expression?

3. What is the role of the proteasome?

4. Before the human genome was sequenced, most scientists thought it contained 80,000 to 150,000 protein-coding genes. The actual sequence revealed only about 21,000. How can mRNA explain this discrepancy?

5. Pharmaceutical companies are interested in developing miRNA drugs. How might they work in cancer? (Hint: See Figure 11.23, about oncogene and tumor suppressor proteins.)

investigatinglife

Q&A Can epigenetic changes be manipulated?

In the opening story, you saw how epigenetic changes in cytosine methylation of a gene encoding a receptor resulted in behavioral changes in people's response to stress. In Investigating Life: Gene Expression and Behavior, you learned that epigenetic changes underlie the differences between worker and queen bees. Both examples involve changes in DNA that were triggered by the environment and resulted in altered gene expression.

As in bees, nutrients in the diets of mammals (including humans), such as folic acid and SAM-e (S-adenosyl methinione), contain methyl groups that can participate in reactions that modify DNA. Experiments with mice have shown that feeding young animals a diet enriched with these nutrients causes changes in epigenetic patterns and gene expression that remain throughout life. A new field with the name nutriepigenomics (could you have guessed the name?) has developed to investigate the possibility of altering gene expression in mammals by diet.

Future directions

In addition to the altering of DNA methylation by diet, specific drugs have been developed that target the enzymes involved in this phenomenon. In cancer, gene expression can be altered in many ways, such as by mutation (as you saw in Chapter 11). Cancer cells can alter gene expression though epigenetic mechanisms as well. For example, in some colon cancers, the expression of an important tumor suppressor gene involved in DNA repair is silenced by extensive DNA methylation at the promoter, resulting in uncontrolled cell division. The nucleotide analog 5′-azacytidine blocks DNA methyltransferase and can be employed to combat some cancers, by reducing DNA methylation and halting uncontrolled cell division. Other drugs are being used to block histone modifications that in the tumor cell result in gene expression changes.

Chapter Summary 16

▶ 16.1 Prokaryotic Gene Expression Is Regulated in Operons

- Some proteins are synthesized only when they are needed. Proteins that are made only in the presence of a particular compound—an **inducer**—are called **inducible proteins**. Proteins that are made at a constant rate regardless of conditions are called **constitutive proteins**. Review Focus: Key Figure 16.1, Figure 16.2

- An **operon** consists of a promoter, an **operator**, and two or more **structural genes**. Promoters and operators do not code for proteins, but serve as binding sites for regulatory proteins. Review Figure 16.4

- Regulatory genes code for regulatory proteins, such as **repressors**. When a repressor binds to an operator, transcription of the structural gene is inhibited. Review Figure 16.5, Animations 16.1, 16.2

- The *lac* operon is an example of an inducible system, in which the presence of an inducer (lactose) keeps the repressor from binding the operator, allowing the transcription of structural genes for lactose metabolism.
- Transcription can be enhanced by the binding of an **activator** protein to the promoter. Review Figure 16.6
- **Catabolite repression** is the inhibition of a catabolic pathway for one energy source by a different, preferred energy source.

16.2 Eukaryotic Gene Expression Is Regulated by Transcription Factors

- Eukaryotic gene expression can be regulated before, during, and after transcription. Review Figure 16.7, Activity 16.1
- **Transcription factors** and other proteins bind to DNA and affect the rate of initiation of transcription at the promoter. Review Figures 16.8, 16.9, Animation 16.3
- The interactions of these proteins with DNA are highly specific and depend on protein domains and DNA sequences.
- Genes at distant locations from one another can be coordinately regulated by transcription factors and promoter elements. Review Figure 16.10

16.3 Viruses Regulate Their Gene Expression during the Reproductive Cycle

- **Viruses** are not cells, and rely on host cells to reproduce.
- The basic unit of a virus is a **virion**, which consists of a nucleic acid genome (DNA or RNA); a protein coat, called a **capsid**; and in some cases, an envelope of lipids that surrounds the capsid.
- **Bacteriophages** are viruses that infect bacteria.
- Viruses undergo a **lytic** cycle, which causes the host cell to burst, releasing new virions.
- Some viruses have promoters that bind host RNA polymerase, which they use to transcribe their own genes and proteins. Review Figure 16.11
- Some viruses can also undergo **lysogeny**, in which a molecule of their DNA, called a **prophage**, is inserted into the host

chromosome, where it replicates for generations. Review Figure 16.12

- A **retrovirus** uses **reverse transcriptase** to generate a cDNA **provirus** from its RNA genome. The provirus is incorporated into the host's DNA and can be activated to produce new virions. Review Figure 16.13

16.4 Epigenetic Changes Regulate Gene Expression

- **Epigenetics** refers to changes in gene expression that do not involve changes in DNA sequences.
- **Methylation** of cytosine residues generally inhibits transcription. Review Figure 16.14
- Modifications of histone proteins in nucleosomes make transcription either easier or more difficult. Review Figure 16.15
- Changes in the environment, such as drought stress, can prompt epigenetic changes to occur. Review Investigating Life: Gene Expression and Behavior

16.5 Eukaryotic Gene Expression Can Be Regulated after Transcription

- **Alternative splicing** of pre-mRNA can produce different proteins. Review Figure 16.17
- Small RNAs (**microRNAs** and **small interfering RNAs**) do not code for proteins but regulate the translation and longevity of mRNA. Review Figure 16.18
- The translation of mRNA to proteins can be regulated by translational repressors.
- A **proteasome** can break down proteins, thus affecting protein longevity. Review Figure 16.20

See Activity 16.2 for a concept review of this chapter.

Go to **LearningCurve** (in **LaunchPad**) for dynamic quizzing that helps you solidify your understanding of this chapter. **LearningCurve** adapts to your responses, giving you the practice you need to master each key concept.

Apply What You've Learned

Review

16.4 Methylation of cytosine bases in DNA can enhance the binding of repressor proteins to promoter regions, resulting in silencing of gene expression.

16.4 Acetylation and deacetylation of histone proteins alter the affinity of the histones for DNA, changing the accessibility of regions of the DNA to RNA polymerase.

Original Paper: Bovenzi, V. and R. L. Momparler, 2001. Antineoplastic action of 5-aza-2'deoxycytidine and histone deacetylase inhibitor and their effect on the expression of the retinoic acid receptor β and estrogen receptor α genes in breast carcinoma cells. *Cancer Chemotherapy and Pharmacology* 48: 71–76.

Cancer cells do not stop dividing. In some types of cancer, this happens because oncogenes that signal the go-ahead at cell cycle checkpoints are continually expressed, while tumor suppressor genes that inhibit progression through checkpoints are silenced. So when a group of

researchers observed a pattern of gene silencing in breast cancer cells, they wondered if they could intervene. If the genes had been silenced as the result of DNA methylation or histone deacetylation, perhaps the researchers could induce expression of the genes by reversing DNA methylation and/or histone deacetylation.

Recall that DNA methyltransferase adds methyl groups onto cytosines in DNA, while a demethylase catalyzes the reverse reaction. Methylation of promoter regions in DNA inhibits gene expression; demethylation activates gene expression.

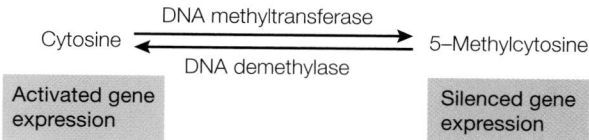

Recall, too, that chromatin can undergo remodeling as histones undergo reversible acetylation and deacetylation of lysine residues. Acetylation of histones activates gene expression; deacetylation inhibits gene expression.

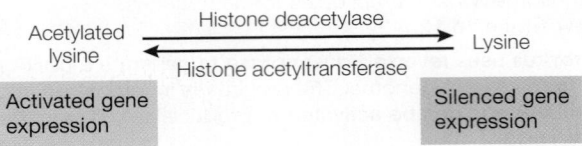

With this in mind, the researchers decided to treat the breast cancer cells with two different inhibitors: AZA, a compound that inhibits DNA methyltransferase; and TSA, a compound that inhibits histone deacetylase. The researchers looked at the effects of both inhibitors, first separately and then together (see graph at right), on the growth of the cancer cells.

In another experiment, the researchers measured the expression of two genes that have been shown to be silenced in these cancer cells. At the beginning of the experiment, cancer cells were either incubated with each inhibitor separately or with the two inhibitors combined. The researchers then isolated RNA from the cells and used the polymerase chain reaction (PCR) to amplify mRNA that had been expressed from the genes of interest. They compared the quantities of the amplified products, which were proportional to the quantities of the original mRNA transcripts that had been expressed in the cells during their treatment. The results are shown in the table below.

Untreated control of mRNA transcript of:	Treatment		
	AZA	TSA	AZA + TSA
Gene 1	7	23	25
Gene 2	<1	<1	11

Questions

1. Analyze the results shown for the single drugs. What do the data suggest about the roles of DNA methylation and histone deacetylation in gene expression in this cell line?

2. Analyze and explain the results shown for both drugs together as they compare with the results for the single drugs. What do the results suggest about the combined effect of DNA methylation and histone acetylation on gene expression in these cells?

3. Analyze the gene expression data in the table. What do these data suggest about the mechanism of silencing of each gene?

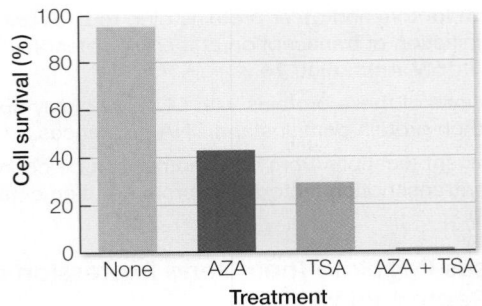

Go to **LaunchPad** for the eBook, LearningCurve, animations, activities, flashcards, and additional resources and assignments.

17

Genomes

Genome sequencing has revealed insights into how size is controlled by genes.

investigating life

The Dog Genome Project

Canis lupus familiaris, the dog, was domesticated by humans thousands of years ago. While there are many kinds of wolves, they all look about the same. Not so with "man's best friend." The American Kennel Club recognizes about 155 different breeds, which not only look different but also vary greatly in size. For example, an adult Chihuahua weighs just 1.5 kg, whereas a Scottish deerhound weighs 70 kg. No other mammal shows such large phenotypic variation. Also, there are hundreds of genetic diseases in dogs, many of which have counterparts in humans. To study the genes behind the phenotypic variation, and to elucidate the relationships between genes and diseases, the Dog Genome Project was started in the late 1990s.

Two dogs—a boxer and a poodle—were the first to have their entire genomes sequenced. The dog genome contains 2.8 billion base pairs of DNA in 39 pairs of chromosomes. There are 19,000 protein-coding genes, most with close counterparts in other mammals, including humans. Given the whole genome sequence, researchers began to map genetic markers—specific nucleotides or short sequences of DNA—at particular locations on the genome that differ among individual dogs or breeds.

Genetic markers can map the locations of (and thus identify) genes that control particular traits. For example, Dr. Elaine Ostrander and her colleagues at the National Institutes of Health studied Portuguese water dogs to identify genes that control size. Samples of cells for DNA isolation were obtained by sweeping a cotton swab over the inside of the cheek. It turned out that the gene for *i*nsulin-like *g*rowth *f*actor 1 (IGF-1) is important in determining size: large breeds have an allele that codes for an active IGF-1, and small breeds have a different allele that codes for a less active IGF-1.

Inevitably, some scientists set up companies to test dogs for genetic variations, using DNA supplied by people anxious to validate the "purebred" status of their dogs. Not to be outdone, cat fanciers have sequenced the genomes of domestic and wild cats. Comparisons of these animal genomes help establish the evolutionary history of mammalian lineages as well as identify the genes responsible for diseases and traits seen in various mammalian species.

 What have we learned from sequencing animal genomes?

key concept

17.1 Genomes Can Be Sequenced Rapidly

Genome sequencing involves determining the nucleotide base sequence of the entire genome of an organism. For a prokaryotic organism with a single chromosome, the genome sequence is one continuous series of base pairs (bp). In the case of a diploid, sexually reproducing species with multiple autosomes and a pair of sex chromosomes (see Key Concept 12.4), the "sequenced genome" usually means the sequence of all the bases in one set of autosomes and in each of the two sex chromosomes.

focus your learning

- Genomes are sequenced in short fragments that are ordered through overlaps.
- Functional genomics uses sequence information to identify the functions of various parts of the genome.
- Comparative genomics compares sequences in genomes from different organisms.

With advances in the technology for DNA sequencing, there has been an explosion of genetic information that scientists can use in a variety of ways:

- The genomes of different species can be compared to find out how they differ at the DNA level, and this information can be used to trace evolutionary relationships.
- The sequences of individuals within a species can be compared to identify mutations that affect particular phenotypes.
- Sequence information can be used to identify genes for particular traits, such as genes associated with diseases.

The base sequence of a short DNA fragment can be determined quickly

The notion of sequencing the entire genome of a complex organism was not contemplated until 1986. The Nobel laureate Renato Dulbecco and others proposed at that time that the world scientific community be mobilized to undertake the sequencing of the entire human genome. One motive was to detect DNA damage in people who had survived the atomic bomb attacks and been exposed to radiation in Japan during World War II. But in order to detect changes in the human genome, scientists first needed to know its sequence.

The result was the publicly funded **Human Genome Project**, an enormous undertaking that was successfully completed in 2003. This effort was aided and complemented by privately funded groups. The genome project benefited from the development of many new methods that were first developed to sequence smaller genomes—those of prokaryotes and simple eukaryotes, the model organisms you have encountered in earlier chapters of this book. Many of these methods are still applied widely, and powerful new methods for sequencing genomes have emerged. These methods are complemented by new ways to examine phenotypic diversity in a cell's proteins and in the metabolic products of the cell's enzymes.

Many prokaryotes have a single chromosome, whereas eukaryotes have many. Because of their differing sizes, chromosomes are easy to separate from one another. You might think that the straightforward way to sequence a chromosome would be to start at one end and simply sequence the DNA molecule one nucleotide at a time. The task is somewhat simplified because only one of the two strands needs to be sequenced, the other being complementary. So if you have the sequence

5′ AAGCTCA.....3′

you know that the other strand must be

3′ TTCGAGT.....5′.

But sequencing a DNA molecule millions of bp long from one end to the other is just not possible using current methods. At most, about a few thousand bp can be sequenced at a time using current methods. The key to determining genome sequences is to break up long chromosomes into smaller DNA fragments and then thousands of fragments are sequenced individually at the same time.

In the 1970s Frederick Sanger and his colleagues invented a way to sequence DNA by using chemically modified nucleotides that were originally developed to stop cell division in cancer. This method, or a variation of it, was used to obtain the first human genome sequence as well as those of several model organisms. However, it was relatively slow, expensive, and labor-intensive. The first decade of the new millennium saw the development of faster and less expensive methods, often referred to under the general term **high-throughput sequencing**. These methods use miniaturization techniques first developed for the electronics industry, as well as the principles of DNA replication, often in combination with the ***polymerase chain reaction (PCR)**.

---

*connect the concepts   PCR can be automated and is a key to sequencing small amounts of DNA. Learn about PCR in Key Concept 13.5.

High-throughput sequencing methods are rapidly improving. Just one of the many approaches is outlined here and illustrated in **Figure 17.1**. First the DNA is prepared for sequencing by attaching it to a solid surface and amplifying the DNA by PCR (see Figure 17.1A):

1. A large molecule of DNA is cut into small fragments of about 100 bp each. This can be done physically, using mechanical forces to shear (break up) the DNA, or by using enzymes that hydrolyze the phosphodiester bonds between nucleotides at intervals in the DNA backbone.

2. The DNA is denatured by heat, breaking the hydrogen bonds that hold the two strands together. Each single strand acts as a template for the synthesis of new, complementary DNA.

3. Short, synthetic oligonucleotides are attached to each end of each fragment, and these are attached to a solid support.

4. The DNA is amplified by PCR using primers complementary to the synthetic oligonucleotides attached to the ends of each DNA. The multiple (approximately 1,000) copies of the DNA at a single location allow for easy detection of added nucleotides during the sequencing steps.

(A) Amplification of DNA fragments by PCR

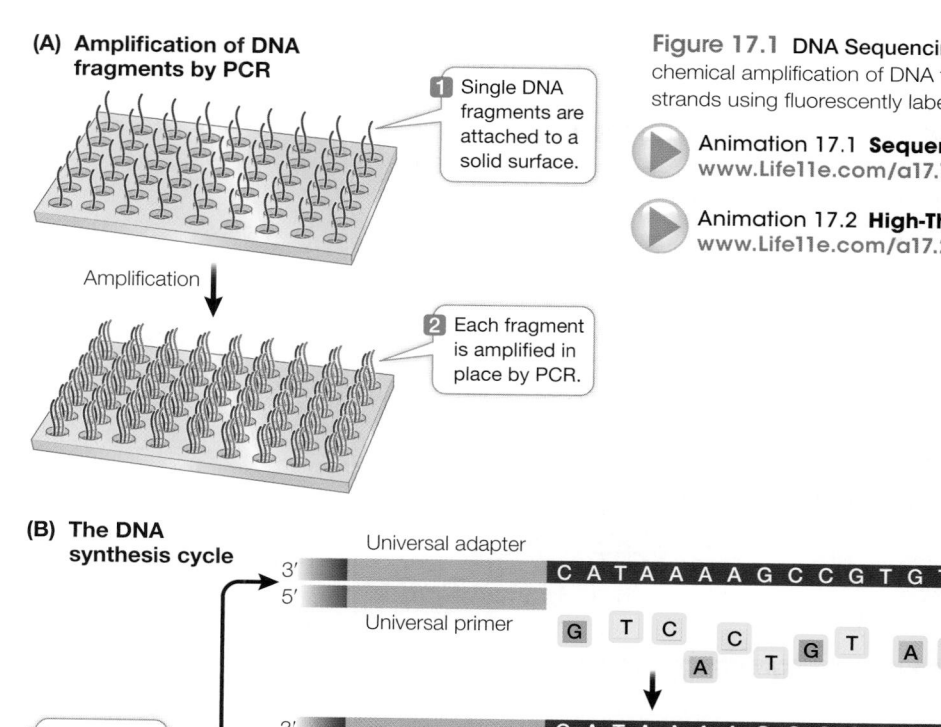

1 Single DNA fragments are attached to a solid surface.

Amplification

2 Each fragment is amplified in place by PCR.

Figure 17.1 DNA Sequencing High-throughput sequencing involves **(A)** the chemical amplification of DNA fragments and **(B)** the synthesis of complementary strands using fluorescently labeled nucleotides.

▶ Animation 17.1 **Sequencing the Genome**
www.Life11e.com/a17.1

▶ Animation 17.2 **High-Throughput Sequencing**
www.Life11e.com/a17.2

(B) The DNA synthesis cycle

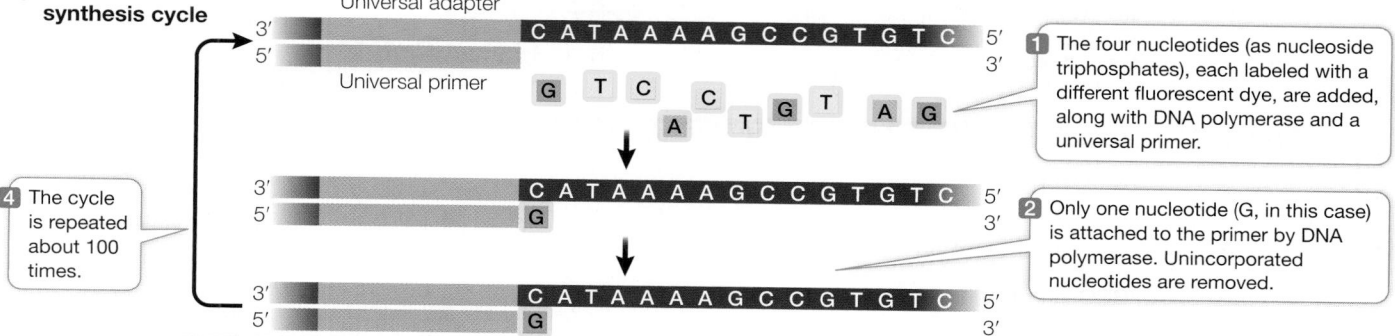

4 The cycle is repeated about 100 times.

1 The four nucleotides (as nucleoside triphosphates), each labeled with a different fluorescent dye, are added, along with DNA polymerase and a universal primer.

2 Only one nucleotide (G, in this case) is attached to the primer by DNA polymerase. Unincorporated nucleotides are removed.

3 The newly added nucleotide is detected by a camera.

Once the DNA has been attached to a solid substrate and amplified, it is ready for sequencing (see Figure 17.1B):

1. At the beginning of each sequencing cycle, the fragments are heated to denature them. A solution containing a universal primer (complementary to one of the same synthetic oligonucleotides used for the PCR amplification step), DNA polymerase, and the four deoxyribonucleoside triphosphates (dNTPs: dATP, dGTP, dCTP, and dTTP) is then added to the DNA. Recall that dNTPs are the substrates that the DNA polymerase uses in DNA synthesis (see Key Concept 13.3). Each of the four kinds of dNTP is tagged with a different colored fluorescent dye.

2. The DNA synthesis reaction is set up so that only one nucleotide is added to the new DNA strand in each sequencing cycle. After each addition, the unincorporated dNTPs are removed.

3. The fluorescence of the new nucleotide at each location is detected with a camera. The color of the fluorescence indicates which of the four nucleotides was added.

4. The fluorescent tag is removed from the nucleotide that is already attached, and then the DNA synthesis cycle is repeated. Images are captured after each nucleotide is added. The series of colors at each location indicate the sequence of nucleotides in the growing DNA strand at that location.

The power of this method derives from the fact that:

● It is fully automated and miniaturized.

● Millions of different fragments are sequenced at the same time.

● It is an inexpensive way to sequence large genomes. For example, at the time of this writing, a complete human genome could be sequenced in less than a day for $1,000. This is in contrast to the Human Genome Project, which took 13 years and $2.7 billion to sequence one genome!

Sequencing millions of short DNA fragments is only part of the process of constructing the genome. Once these sequences have been determined, they need to be put together in the correct order. That is, they need to be arranged to reflect the sequence of the chromosomes from which they came. Imagine if you cut out every word in this book (there are more than half a million of them), put them on a table, and tried to arrange them in their original order! The enormous task of determining DNA sequences is possible because the original DNA fragments are overlapping.

Let's illustrate the process using a single 10-bp DNA molecule. (This is a double-stranded molecule, but for convenience we show only the sequence of one strand.) The molecule is cut three ways (for example, using three different restriction enzymes). Cutting with the first enzyme generates the fragments:

TG, ATG, and CCTAC

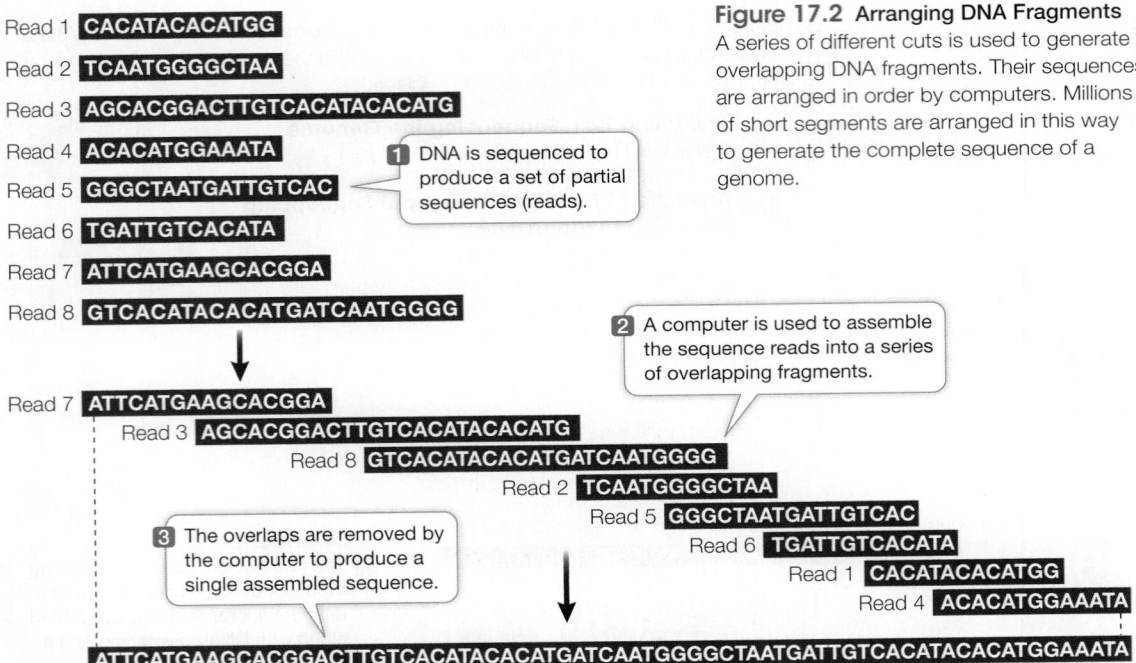

Figure 17.2 Arranging DNA Fragments
A series of different cuts is used to generate overlapping DNA fragments. Their sequences are arranged in order by computers. Millions of short segments are arranged in this way to generate the complete sequence of a genome.

Cutting the same molecule with the second enzyme generates the fragments:

AT, GCC, and TACTG

Cutting with the third enzyme results in:

CTG, CTA, and ATGC

Can you put the fragments in the correct order? (The answer is ATGCCTACTG.) For genome sequencing, the sequence fragments are called "reads" (**Figure 17.2**). Of course, the problem of ordering 2.5 million fragments from human chromosome 1 (246 million bp) is more challenging than our 10-bp example above! The field of **bioinformatics** was developed to analyze DNA sequences using sophisticated mathematics and computer programs to handle the large amounts of data generated in genome sequencing.

Genome sequences yield several kinds of information

New genome sequences are being published at an accelerating pace, creating a torrent of biological information. This information is used in two related fields of research, both focused on studying genomes. In **functional genomics**, biologists use sequence information to identify the functions of various parts of genomes (the meaningful sequences in genomes such as those encoding mRNA, tRNA, and regulatory sequences; see Key Concept 14.4). These parts include:

- *Open reading frames*, which are the coding regions of genes. For protein-coding genes, these regions can be recognized by the start and stop codons for translation, and by recognition sequences that indicate the locations of introns. A major goal of functional genomics is to understand the function of every open reading frame in each genome.
- *Amino acid sequences* of proteins, which can be deduced by applying the genetic code to the DNA sequences of open reading frames.

- *Regulatory sequences*, such as promoters, enhancers, and terminators for transcription. These are identified by their proximity to open reading frames and because they contain recognition sequences for the binding of specific transcription factors.
- *RNA genes*, including rRNA, tRNA, small nuclear RNA, and microRNA genes.
- *Other noncoding sequences* that can be classified into various categories, including centromeric and telomeric regions, transposons, and other repetitive sequences.

Sequence information is also used in **comparative genomics**: the comparison of a newly sequenced genome (or parts thereof) with sequences from other organisms. The Dog Genome Project described at the opening of this chapter, for example, has yielded information not only about dogs, but on how the dog genome compares with other animal genomes. Genome comparisons can provide further information about the functions of sequences and can be used to trace evolutionary relationships among different species. Each animal genome sequenced offers new insights. **Investigating Life: Comparative Analysis of the Tiger Genome** describes the recent sequencing of the tiger genome and how it relates to the genomes of other cats and other mammals.

▶ 17.1 recap

The sequencing of genomes involves cutting chromosomes into fragments, sequencing the fragments, and then assembling the fragment sequences into continuous sequences for entire chromosomes. Current sequencing methods use automation and powerful computers. These methods use labeled nucleotides that are detected at the ends of growing polynucleotide chains.

(continued)

experiment

Original Paper: Cho, Y. S. et al. 2013. The tiger genome and comparative analysis with lion and snow leopard genomes. *Nature Communications* 4: 1–7.

Panthera tigris, the tiger, is possibly the most famous endangered species. Fewer than 4,000 individuals remain in the wild. While a century ago there were nine recognized genetically distinct subspecies, four have become extinct. The five extant subspecies include the Bengal tiger seen at zoos and the white Amur tiger that lives in snowy regions of Russia, China, and North Korea. While the genomes of other cats, such as the lion, snow leopard, and domestic cat, had already been sequenced, that of the tiger had not until this study.

HYPOTHESIS▶ Sequencing the genomes of big cats will reveal that the adaptive phenotypic variation seen between cats is due to genetic variation.

METHOD

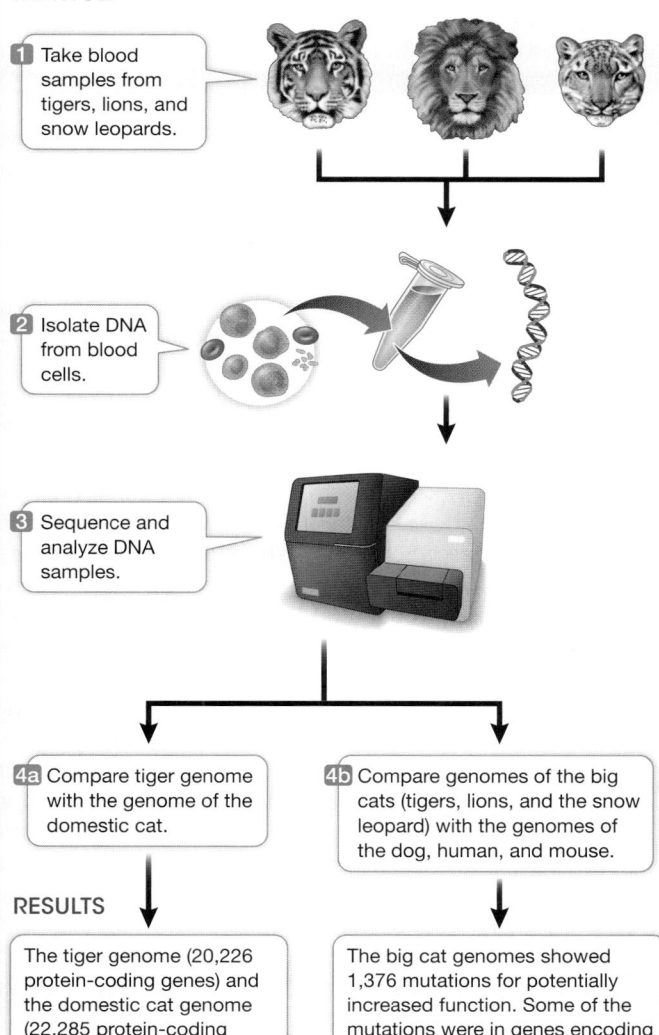

1 Take blood samples from tigers, lions, and snow leopards.

2 Isolate DNA from blood cells.

3 Sequence and analyze DNA samples.

4a Compare tiger genome with the genome of the domestic cat.

4b Compare genomes of the big cats (tigers, lions, and the snow leopard) with the genomes of the dog, human, and mouse.

RESULTS

The tiger genome (20,226 protein-coding genes) and the domestic cat genome (22,285 protein-coding genes) showed a 95.6 percent sequence similarity.

The big cat genomes showed 1,376 mutations for potentially increased function. Some of the mutations were in genes encoding muscle protein development.

CONCLUSION▶ What would you conclude about the phenotype of big cats? Explore the possibilities in the following work with the data.

work with the data

An international team led by Jong Bhak at the Genome Research Foundation in Suwon, South Korea, determined the sequence of tiger DNA and compared it with lion and snow leopard DNA, as well as that of the domestic cat.

QUESTIONS▶

1. The overall sequence of tiger DNA was compared with that of the domestic cat. For reference, a comparison between the human and gorilla genomes was included. Evolutionary distances (last common ancestor) were also estimated. The results are shown in the table. What can you conclude about the rate of evolutionary change between the two cat genomes and between the human and gorilla genomes?

Comparison groups	Last common ancestor (millions of years ago)	Genome sequence similarity (%)
Domestic cat and tiger	10.8	95.6
Human and gorilla	8.8	94.8

2. While there are 20,226 protein-coding genes in the tiger genome, some of them are in gene families. That is, some genes are highly similar to one another with the same function. There are 14,528 protein-coding genes in the tiger genome. The research team looked up the genome sequences of other mammals and asked whether the gene families identified in the tiger genome are also present in those other species. The genomics comparisons are shown in the Venn diagram. The number in an overlapping region represents the number of gene families shared with an overlapping genome. For example, the tiger and the domestic cat genomes share 99 gene families with the opossum genome.

   a. Looking at the Venn diagram, how many gene families are shared by all the mammalian genomes examined?

   b. How many gene families are unique to the tiger and domestic cat genomes? How many are unique to the human and mouse genomes?

   c. What do these data indicate about the basic mammalian genome?

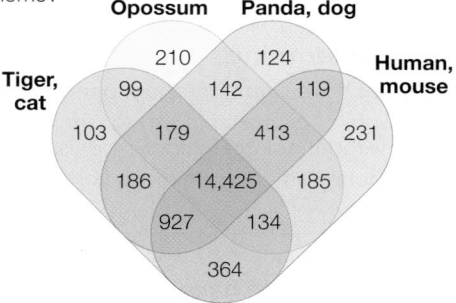

3. The tiger genome was analyzed for sequences of protein-coding genes of known function. The following types of genes were found in relatively higher numbers compared to the domestic cat:

   Olfactory (smell) receptors: 289 genes

   G protein-coupled signaling complexes: 302 genes

   Signal transduction: 295 genes

   Protein metabolism: 220 genes

   What do these genotype data indicate about the phenotype of the tiger?

   A similar **work with the data** exercise may be assigned in **LaunchPad**.

table 17.1 Gene Functions in Three Bacteria

Category	Number of genes in: E. coli	H. influenzae	M. genitalium
Total protein-coding genes	4,288	1,727	482
Biosynthesis of amino acids	131	68	1
Biosynthesis of cofactors	103	54	5
Biosynthesis of nucleotides	58	53	19
Cell envelope proteins	237	84	17
Energy metabolism	243	112	31
Intermediary metabolism	188	30	6
Lipid metabolism	48	25	6
DNA replication, recombination, and repair	115	87	32
Protein folding	9	6	7
Regulatory proteins	178	64	7
Transcription	55	27	12
Translation	182	141	101
Uptake of molecules from the environment	427	123	34

17.1 recap (continued)

learning outcomes

You should be able to:

- Explain why overlapping sequence fragments are essential in sequencing genomes and how they are arranged to give the final sequence.
- Identify the characteristics of open reading frames.
- Describe exmples of comparative genomics.

1. In sequencing genomes, why are overlapping sequences obtained, and how are they arranged to give the final sequence?
2. How are open reading frames recognized in a genomic sequence? What kind of information can be derived from an open reading frame?
3. What is comparative genomics? Give an example of how it is used.

The first genomes to be fully sequenced were those of viruses and prokaryotes. Next we will discuss the information provided by the relatively simple prokaryotic genomes.

key concept 17.2 Prokaryotic Genomes Are Compact

In 1968 Robert W. Holley, was awarded the Nobel Prize for the first sequencing of a nucleic acid—a tRNA with just 80 bases. It took Holley and his team 5 years to discover the sequence. A few years later the first genome was sequenced, that of a 5,386-bp bacteriophage. As methods gradually improved, the genome of a prokaryote was sequenced in the 1990s. With the new automated sequencing techniques we just described, we now have genome sequences for many prokaryotes, to the great benefit of microbiology and medicine.

focus your learning

- Prokaryotes have small, compact genomes.
- Metagenomics allows biologists to sample genomic diversity without isolating intact organisms.
- Transposable elements can move from place to place in the genome.

Prokaryotic genomes have distinctive features

Bacterial and archaeal genomes have several notable features:

- *They are relatively small.* Prokaryotic genomes range from about 160,000 to 12 million bp and are usually organized into a single chromosome.
- *They are compact.* Typically, more than 85 percent of the DNA is in protein-coding sequences or RNA genes, with only short sequences between genes.
- *Prokaryotic gene sequences usually are not interrupted by introns.* An exception is the rRNA and tRNA genes of archaea, which frequently contain introns.

- *In addition to the main chromosome, prokaryotes often have smaller, circular molecules of DNA called plasmids*, which may be transferred between cells (see Key Concept 12.6).

Beyond these similarities, there is great diversity among these single-celled organisms, reflecting the huge variety of environments in which they are found (described in Chapter 25). Let's examine prokaryotic genomes in terms of functional and comparative genomics.

FUNCTIONAL GENOMICS　As we described in Key Concept 17.1, functional genomics is the biological discipline that assigns functions to the products of gene sequences. You can see the various functions encoded by the genomes of three prokaryotes in **Table 17.1**. As an example, look at the first bacterial genome sequenced, that of *Haemophilus influenzae*, which lives in the human respiratory tract and can cause ear infections or, more seriously, meningitis in children. Its single circular chromosome has 1,830,138 bp. In addition to its origin of replication and the genes coding for rRNAs and tRNAs, this bacterial chromosome has 1,727 open reading frames with promoters nearby.

When the *H. influenzae* sequence was first announced 20 years ago, only 1,007 (58 percent) of the open reading frames coded for proteins with known functions. Since then, scientists have identified the functions of the rest of the encoded proteins. All of the major biochemical pathways and molecular functions are represented. For example, the genes that encode enzymes involved in glycolysis, fermentation, and electron transport have been identified. Gene sequences that code for membrane proteins have been identified as well, including those involved in active transport. An important finding was that highly infective strains of *H. influenzae*—but not noninfective strains—have genes for surface proteins that attach the bacterium to the human respiratory tract. These surface proteins are now a focus of research on possible treatments for *H. influenzae* infections.

COMPARATIVE GENOMICS　Soon after the sequence of *H. influenzae* was announced, smaller (*Mycoplasma genitalium*: 580,073 bp) and larger (*E. coli*: 4,639,221 bp) prokaryotic sequences were completed. Thus began the era of comparative genomics. Scientists can identify genes that are present in one bacterium and missing in another, allowing them to relate these genes to bacterial function.

M. genitalium, for example, lacks enzymes needed to synthesize amino acids, whereas *E. coli* and *H. influenzae* both possess such enzymes. This finding reveals that *M. genitalium* must obtain all its amino acids from its environment (usually the human urogenital tract). Furthermore, *E. coli* has 55 genes that encode transcriptional activators, whereas *M. genitalium* has only 12. This relative lack of control over gene expression suggests that the biochemical flexibility of *M. genitalium* must be limited compared with that of *E. coli*.

The sequencing of prokaryotic and viral genomes has many potential benefits

Prokaryotic genome sequencing is providing insights into microorganisms that are important for agriculture and medicine. Scientists who analyze the sequences have discovered previously unknown genes and proteins that can be targeted for isolation and functional study. They have also discovered surprising relationships between some organisms, suggesting that genes may be transferred between different groups.

- *Rhizobium* species are bacteria that form symbiotic associations with plants, living inside the roots of legumes such as beans, peas, and clover. The bacteria fix atmospheric nitrogen from the air and convert it into forms usable by the plants, reducing the need for nitrogen-containing fertilizers. Genome sequences from several *Rhizobium* species have been used to identify the genes involved in successful symbiosis, and this knowledge is being used both to improve the efficiency of this process and to broaden the range of plants that can form these beneficial associations.

- *E. coli* strain O157:H7 causes illness (sometimes severe) in at least 70,000 people a year in the United States. Its genome has 5,416 genes, of which 1,387 are different from those in the familiar (and harmless) laboratory strains of this bacterium. Many of these unique genes are also present in other pathogenic bacteria, such as *Salmonella* and *Shigella*. This finding suggests that there is extensive genetic exchange among these species, and indeed, "superbugs" that have acquired multiple genes for antibiotic resistance are already a major problem. Sequencing of a particular strain of *E. coli* causing food-borne illness, and comparisons of that sequence with the bacterium in different lots of meat, for example, can allow scientists to pinpoint the meat causing the infection.

- Severe acute respiratory syndrome (SARS) was first detected in southern China in 2002 and rapidly spread in 2003. There is no effective treatment, and 10 percent of infected people die. Isolation of the causative agent, a virus, and the rapid sequencing of its genome revealed several novel proteins that are possible targets for antiviral drugs or vaccines. Research is under way on both fronts, since further outbreaks of this and other newly revealed pathogens are anticipated.

- Some archaean prokaryotes, such as *Methanobrevibacter*, produce methane gas (CH_4) in the stomachs of cows and other ruminants. By contrast, certain bacterial species, such as *Methylococcus*, remove methane from the air and use it as an energy source. The genomes of both of these prokaryotes have been sequenced. Understanding the genes involved in methane production and consumption is stimulating thinking about ways to slow the progress of global warming, given that methane is the second most prevalent greenhouse gas emitted in the United States and a significant contributor to climate change (see Key Concept 53.1).

Metagenomics allows us to describe new organisms and ecosystems

If you take a microbiology laboratory course you'll learn how to identify various prokaryotes on the basis of their growth on particular artificial media. For example, staphylococci are a group of bacteria that infect skin and nasal passages. When grown on a medium called blood agar, they form round, raised colonies. Microorganisms can also be identified by their nutritional requirements or the conditions under which they will grow (for example, aerobic versus anaerobic). Such culture methods have been the mainstay of microbial identification for more than a century and are still useful and important. However, scientists can now use PCR and DNA sequencing to identify microbes without culturing them in the laboratory.

In 1985 Norman Pace, then at Indiana University, came up with the idea of isolating DNA directly from environmental samples. He used PCR to amplify specific rRNA-encoding sequences from the samples to determine whether particular microbes were present. The PCR products were sequenced to explore their diversity. The term **metagenomics** was coined to describe this approach of analyzing genes without isolating the intact organism. It is now possible to sequence DNA samples from almost any environment. The sequences can be used to detect the presence of both known microbes and heretofore unidentified organisms (**Figure 17.3**). For example:

- Sequencing of DNA from 200 liters of seawater indicated that it contained 5,000 different viruses and 2,000 different bacteria, many of which had not been described previously.

- One kilogram of marine sediment contained 1 million different viruses, most of them new.

- Water runoff from a mine contained many previously unknown species of prokaryotes thriving in what was previously thought of as an inhospitable environment. Some of these organisms exhibited metabolic pathways that were previously unknown to biologists. These organisms and their capabilities may be useful in cleaning up pollutants from the water.

- Gut samples from 124 Europeans revealed that each person harbored at least 160 species of bacteria (making up their gut microflora, or *****microbiome**). Many of these species were found in all individuals, but the presence of other bacteria varied from person to person. Such variations may be associated with obesity or bowel diseases.

*****connect the concepts**   Learn more about how disruptions of the human microbiome are associated with various autoimmune diseases in Key Concept 25.3.

These and other discoveries are truly extraordinary and potentially very important. It is estimated that 90 percent of the microbial

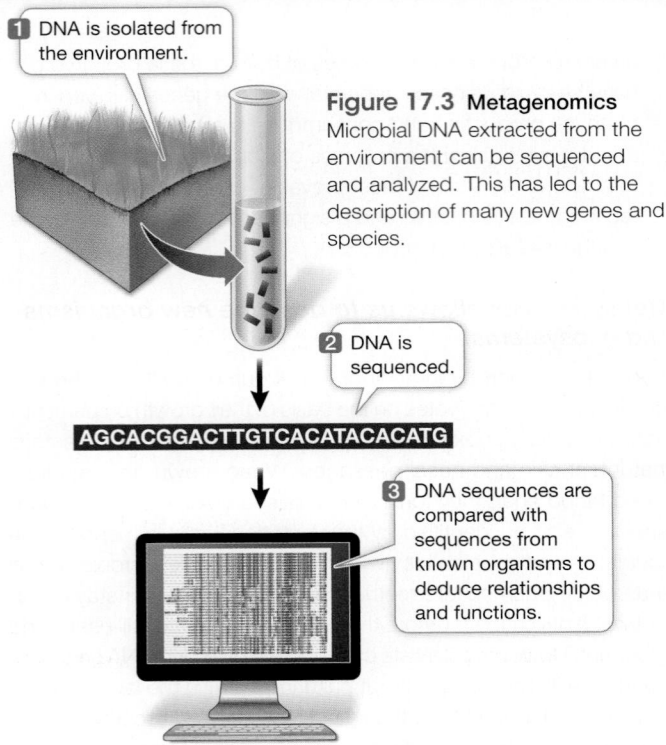

1 DNA is isolated from the environment.

Figure 17.3 **Metagenomics** Microbial DNA extracted from the environment can be sequenced and analyzed. This has led to the description of many new genes and species.

2 DNA is sequenced.

AGCACGGACTTGTCACATACACATG

3 DNA sequences are compared with sequences from known organisms to deduce relationships and functions.

world has been invisible to biologists and is only now being revealed by metagenomics. Entirely new ecosystems of bacteria and viruses are being discovered in which, for example, one species produces a molecule that another metabolizes. It is hard to overemphasize the importance of such an increase in our knowledge of the hidden world of microbes. This knowledge will help us understand natural

ecological processes, and has the potential to help us find better ways to manage environmental catastrophes such as oil spills, or to remove toxic heavy metals from soil.

Some sequences of DNA can move about the genome

As we mentioned in Key Concept 15.1, **transposable elements** (or **transposons**) are segments of DNA that can move from place to place in the genome. Genome sequencing has allowed scientists to study these elements more broadly, and they are now known to be widespread in both prokaryotes and eukaryotes. Prokaryotic transposable elements are often short sequences of 1,000 to 2,000 bp, and they can be found both in chromosomes and in plasmids. A transposable element might be at one location in the genome of one *E. coli* strain, and at a different location in another strain. The insertion of this movable DNA sequence from elsewhere in the genome into the middle of a protein-coding gene can disrupt that gene (**Figure 17.4A**). In a prokaryote, the mRNA transcribed from the disrupted gene will contain the extra sequence, and the protein it expresses will be altered and almost certainly nonfunctional. So in prokaryotes transposable elements can produce significant phenotypic effects. Sometimes, two transposable elements located near one another (within a few thousand bp) will transpose together and carry the intervening DNA sequence with them, creating what is called a **composite transposon** (**Figure 17.4B**). Genes for antibiotic resistance can be multiplied and transferred between bacteria in this way: composite transposons carrying genes for antibiotic resistance can insert into a plasmid that then moves between bacteria by conjugation (see Figure 12.22).

As you saw in Chapter 15, there are several ways for transposable elements to move around the genome. For example, a transposable element may be replicated, and then the copy can be inserted into another site in the genome (the "copy and paste" mode). Or the element might splice out of one location and move to another location ("cut and paste"). Transposable elements usually carry genes for enzymes such as transposases, which catalyze the reactions needed for transposition. Often the elements are flanked by inverted repeat DNA sequences that are recognized by these enzymes.

Will defining the genes required for cellular life lead to artificial life?

When we compare the genomes of prokaryotes and eukaryotes, a striking conclusion arises: certain genes are present in all organisms (universal genes). Not surprisingly, these include genes whose products are involved in DNA replication, transcription, and RNA translation to form proteins. There are also some (nearly) universal gene segments that are present in many genes in many organisms; for

(A) A simple transposon disrupts a coding sequence

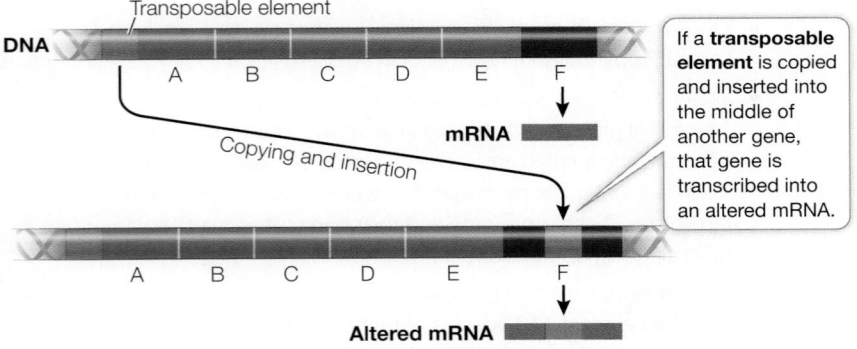

Transposable element

DNA

A   B   C   D   E   F

mRNA

Copying and insertion

A   B   C   D   E   F

Altered mRNA

If a **transposable element** is copied and inserted into the middle of another gene, that gene is transcribed into an altered mRNA.

(B) A composite transposon inserts an additional coding sequence

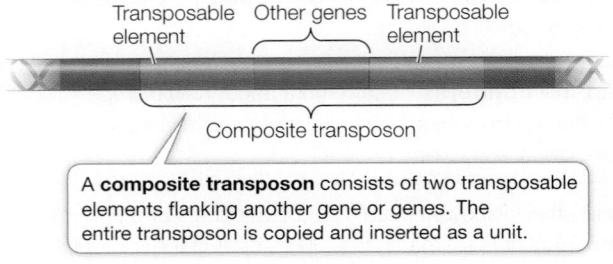

Transposable element   Other genes   Transposable element

Composite transposon

A **composite transposon** consists of two transposable elements flanking another gene or genes. The entire transposon is copied and inserted as a unit.

Figure 17.4 **DNA Sequences That Move**   Transposable elements are DNA sequences that move from one location to another. **(A)** In one method of transposition ("copy and paste"), the DNA sequence is replicated and the copy inserts elsewhere in the genome. **(B)** Composite transposons contain additional genes flanked by two transposable elements.

Q: One hypothesis for the origin of some transposons is through a retroviral infection. Based on what you know of the life cycle of a retrovirus (see Key Concept 16.3), how might this occur?

example, the sequence that codes for an ATP binding site in a protein. These findings suggest that there is some ancient, minimal set of DNA sequences that is common to all cells. One way to identify these sequences is to look for them in computer analyses of sequenced genomes.

Another way to define the minimal genome is to take an organism with a simple genome and deliberately mutate one gene at a time to see what happens. *M. genitalium* has one of the smallest known genomes—only 482 protein-coding genes. Even so, some of its genes are dispensable under some circumstances. For example, *M. genitalium* has genes for metabolizing both glucose and fructose, but it can survive in the laboratory on a medium containing only one of these sugars. Under such conditions it doesn't need the genes for metabolizing the other sugar.

What about other genes? A team led by Craig Venter has addressed this question with experiments involving the use of transposons as mutagens. When transposons in the bacterium are activated, they insert themselves into genes at random, mutating and inactivating them (**Figure 17.5**). The mutated bacteria are tested for growth and survival, and DNA from interesting mutants is sequenced to find out which genes contain transposons. The astonishing result of these studies is that *M. genitalium* can survive in the laboratory with a minimal genome of only 382 protein-coding genes! In yeast, the minimal genome has been found to consist of only about 10 percent of the 5,000 protein-coding genes; in the nematode *Caenorhabditis elegans*, it is a similar proportion.

One goal of this research is to design new life forms with specific purposes, such as bacteria that will clean up oil spills. You'll learn about this approach, called synthetic genetics, in the next chapter.

17.2 recap

DNA sequencing is used to study the genomes of prokaryotes that are important to humans and to ecosystems. Functional genomics uses gene sequences to determine the functions of the gene products. Comparative genomics compares gene sequences from different organisms to help identify their functions and evolutionary relationships. Transposable elements, including composite transposons, move from one place to another in the genome. Studies of the minimal genome may lead to the creation of artificial species.

learning outcomes

You should be able to:

- Describe the characteristic features of prokaryotic genomes
- Develop experimental approaches to address metagenomic questions.
- Summarize the ways in which transposable elements can move about the genome.

1. Suggest evolutionary reasons to explain why prokaryotic genomes are small, sometimes contain plasmids, and have genes that move about the genome.
2. How do transposons move about the genome?
3. There are many bacterial species on the skin of the human belly button. How would you use metagenomics to develop a microbial "fingerprint"?

experiment

Figure 17.5   Using Transposon Mutagenesis to Determine the Minimal Genome

Original Papers: Hutchison, C. et al. 1999. Global transposon mutagenesis and a minimal *Mycoplasma* genome. *Science* 286: 2165–2169.

Glass, J. I., N. Assad-Garcia, N. Alperovich, S. Yooseph, M. R. Lewis, M. Maruf, C. A. Hutchison III, H. O. Smith and J. C. Venter. 2006. Essential genes of a minimal bacterium. *Proceedings of the National Academy of Sciences USA* 103: 425–430.

Mycoplasma genitalium has one of the smallest known genomes of any prokaryote. But are all of its genes essential to life? By inactivating the genes one by one, scientists determined which of them are essential for the cell's survival.

HYPOTHESIS▶   Only some of the genes in a bacterial genome are essential for cell survival.

METHOD

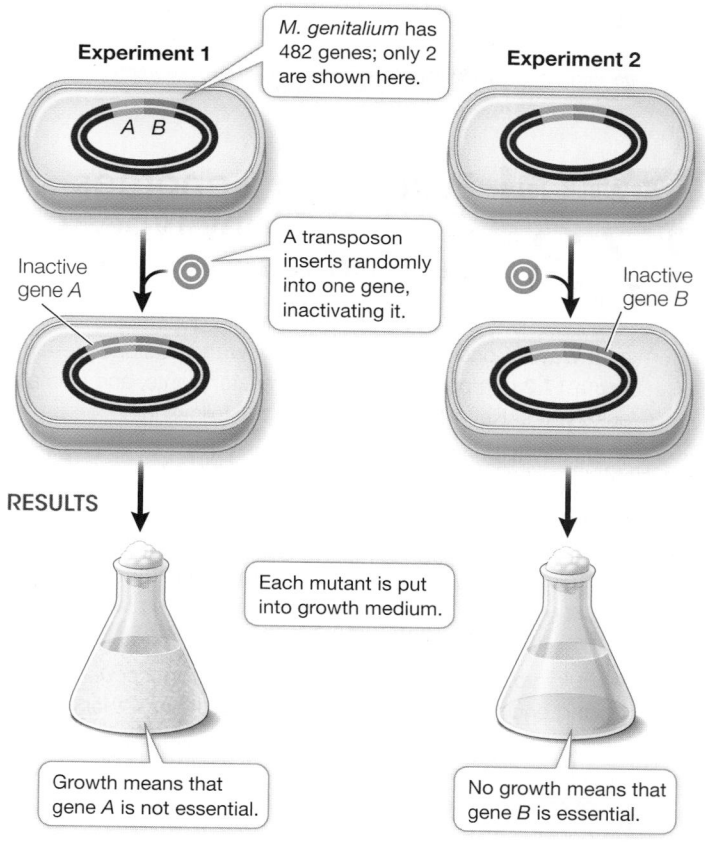

CONCLUSION▶   If each gene is inactivated in turn, a "minimal essential genome" can be determined.

Advances in DNA sequencing and sequence analysis have led to the rapid sequencing of eukaryotic genomes. We will now look at some of the new insights that have come from these studies.

table 17.2 Representative Sequenced Genomes

Organism	Haploid genome size (Mb)[a]	Number of protein-coding genes	Percent of genome that codes for proteins	Notable attributes
Bacteria				
Mycoplasma genitalium	0.58	482	88	Minimal genome
Haemophilus influenzae	1.83	1,727	89	
Escherichia coli	4.6	4,288	88	Well-studied enteric bacterium
Yeasts				Targeting; cell organelles
Saccharomyces cerevisiae	12.2	6,275	70	
Schizosaccharomyces pombe	13.8	4,824	60	
Plants				Photosynthesis; cell walls
Arabidopsis thaliana	125	27,416	25	Small plant genome
Oryza sativa (rice)	420	40,838	12	Water tolerance for roots
Glycine max (soybean)	973	46,430	7	Lipid synthesis, storage
Animals				
Caenorhabditis elegans (nematode)	100	20,470	25	Tissue formation
Drosophila melanogaster (fruit fly)	140	13,918	13	Embryonic development
Homo sapiens (human)	3,200	~21,000	1.2	Language

[a] = Millions of base pairs

key concept 17.3 Eukaryotic Genomes Contain Many Types of Sequences

Eukaryotic and prokaryotic genomes differ in significant ways. Consider, for example, how the bacterial genomes in **Table 17.2** compare with those of yeasts, plants, and animals—all eukaryotes. Key differences include the following:

- *Eukaryotic genomes are larger than those of prokaryotes*, and they have more protein-coding genes. This difference is not surprising given that multicellular organisms have many cell types with specialized functions. As we saw above, the simple prokaryote *Mycoplasma* has several hundred protein-coding genes in a genome of 0.58 million bp. A rice plant, in contrast, has almost 55,986 genes!

- *Eukaryotic genomes have more regulatory sequences*— and encode many more regulatory proteins—than prokaryotic genomes. The greater complexity of eukaryotes requires much more regulation, which is evident in the many points of control associated with the expression of eukaryotic genes (see Figure 16.7).

- *Much of eukaryotic DNA is noncoding*. Distributed throughout many eukaryotic genomes are various kinds of DNA sequences that are not transcribed into mRNA, most notably introns and gene control sequences. As we discussed in Chapter 16, some noncoding sequences are transcribed into microRNAs. In addition, eukaryotic genomes contain various kinds of repeated sequences. Prokaryotes, in contrast, seldom contain long noncoding and repeated sequences.

- *Eukaryotes have multiple chromosomes*, whereas prokaryotes usually have a single, circular chromosome. As we

have described in previous chapters, eukaryotic chromosomes have multiple origins of replication, a centromere region that holds the replicated chromosomes together before mitosis, and a telomeric sequence at each end of the chromosome that maintains chromosome integrity (**Focus: Key Figure 17.6**).

focus your learning

- Nematodes are useful model organisms for studying eukaryotic development.
- *Arabidopsis* is an important plant model.
- Gene families result from gene duplications and evolutionary divergence.
- Some moderately repetitive sequences are transcribed but others are not.

Genome sequences of model organisms provide important information

Most of our information about eukaryotic genomes has come from model organisms that have been studied extensively: the yeast *Saccharomyces cerevisiae*, the nematode (roundworm) *Caenorhabditis elegans*, the fruit fly *Drosophila melanogaster*, and the thale cress plant (*Arabidopsis thaliana*). Model organisms have been chosen because they are relatively easy to grow and study in a laboratory, their genetics are well studied, and they exhibit characteristics that represent a larger group of organisms.

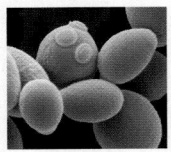 **YEAST: THE BASIC EUKARYOTIC MODEL** Yeasts are single-celled eukaryotes. Like most eukaryotes, they have membrane-enclosed organelles, such as the nucleus and endoplasmic reticulum, and a haplontic life cycle that alternates between

focus: key figure

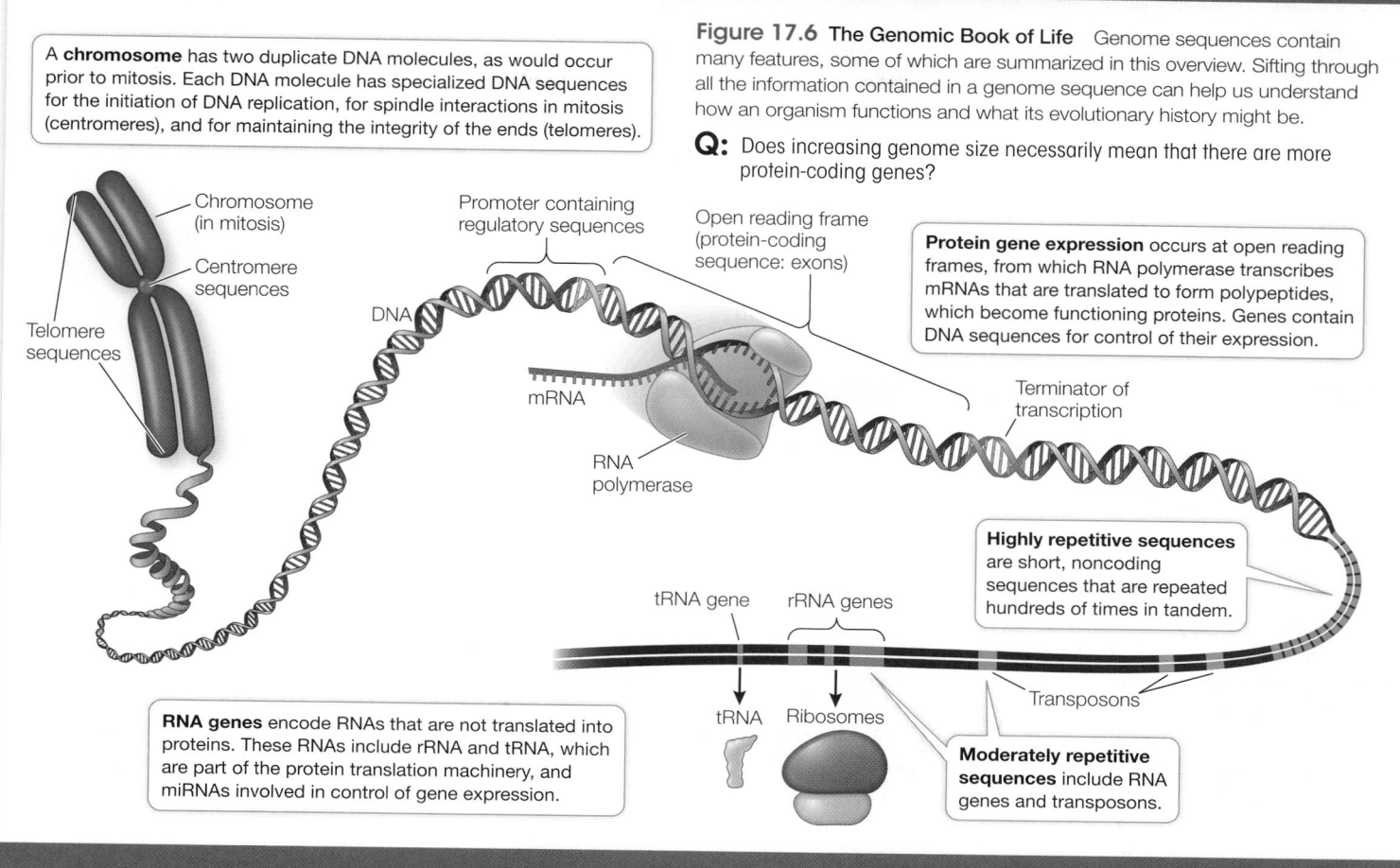

A **chromosome** has two duplicate DNA molecules, as would occur prior to mitosis. Each DNA molecule has specialized DNA sequences for the initiation of DNA replication, for spindle interactions in mitosis (centromeres), and for maintaining the integrity of the ends (telomeres).

Figure 17.6 The Genomic Book of Life Genome sequences contain many features, some of which are summarized in this overview. Sifting through all the information contained in a genome sequence can help us understand how an organism functions and what its evolutionary history might be.

Q: Does increasing genome size necessarily mean that there are more protein-coding genes?

Chromosome (in mitosis)

Centromere sequences

Telomere sequences

Promoter containing regulatory sequences

DNA

Open reading frame (protein-coding sequence: exons)

mRNA

RNA polymerase

Protein gene expression occurs at open reading frames, from which RNA polymerase transcribes mRNAs that are translated to form polypeptides, which become functioning proteins. Genes contain DNA sequences for control of their expression.

Terminator of transcription

Highly repetitive sequences are short, noncoding sequences that are repeated hundreds of times in tandem.

tRNA gene   rRNA genes

tRNA   Ribosomes

Transposons

Moderately repetitive sequences include RNA genes and transposons.

RNA genes encode RNAs that are not translated into proteins. These RNAs include rRNA and tRNA, which are part of the protein translation machinery, and miRNAs involved in control of gene expression.

---

haploid and diploid generations (see Figure 11.14). So it is not surprising that single-celled yeast has a larger genome with more protein-coding genes than a single-celled bacterium (see Table 17.2). Gene inactivation studies similar to those carried out for *M. genitalium* (see Figure 17.6) indicate that fewer than 10 percent of the yeast's genes are essential to survival. The most striking difference between the yeast genome and that of *E. coli* is the number of genes for targeting proteins to organelles (**Table 17.3**). Both of these single-celled organisms appear to use about the same number of genes to perform the basic functions of cell survival. It is the compartmentalization of the eukaryotic yeast cell into organelles that requires it to have many more genes. This finding is direct, quantitative confirmation of something we have known for a century: the eukaryotic cell is structurally more complex than the prokaryotic cell.

THE NEMATODE: UNDERSTANDING EUKARYOTIC DEVELOPMENT A simple organism in which to study multicellularity is *Caenorhabditis elegans*, a 1-mm-long nematode (roundworm) that normally lives in the soil. It can also live in the laboratory, where it has become a favorite model organism of developmental biologists (see Key Concept 19.2). The

nematode has a transparent body that develops over 3 days from a fertilized egg to an adult worm made up of nearly 1,000 cells. In spite of its small number of cells, the nematode has a nervous system, digests food, reproduces sexually, and ages. So it is not surprising that an intense effort was made to sequence the genome of this model organism.

table 17.3 Comparison of the Genomes of *E. coli* and *S. cerevisiae*

	E. coli	Yeast
Genome length (base pairs)	4,640,000	12,157,000
Number of protein-coding genes	4,288	6,275
Proteins with roles in:		
Metabolism	650	650
Energy production/storage	240	175
Membrane transport	280	250
DNA replication/repair/recombination	115	175
Transcription	55	400
Translation	182	350
Protein targeting/secretion	35	430
Cell structure	180	250

table 17.4 *C. elegans* Genes Essential to Multicellularity

Function	Protein/domain	Number of genes
Transcription control	Zinc finger; homeobox	540
RNA processing	RNA binding domains	100
Nerve impulse transmission	Gated ion channels	80
Tissue formation	Collagens	170
Cell interactions	Extracellular domains; glycotransferases	330
Cell–cell signaling	G protein-coupled receptors; protein kinases; protein phosphatases	1,290

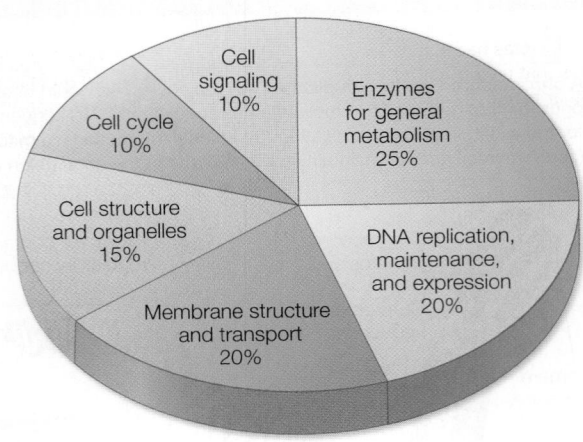

Figure 17.7 Functions of the Eukaryotic Genome   The distribution of gene functions in *Drosophila melanogaster* shows a pattern that is typical of many complex organisms.

The *C. elegans* genome (100 million bp) is 8 times larger than that of the yeast *Saccharomyces cerevisiae* and has 3.3 times as many protein-coding genes (see Table 17.2). Gene inactivation studies have shown that the worm can survive in laboratory cultures with only 10 percent of these genes. So the minimal genome of the worm is about twice the size of that of the yeast (about 5,000 genes), which in turn is about twelve times the size of the minimal genome for *Mycoplasma*. What do these extra genes do? All cells must have genes for survival, growth, and division. In addition, the cells of multicellular organisms must have genes for holding cells together to form tissues, for cell differentiation, and for intercellular communication. Looking at **Table 17.4**, you will recognize functions that we discussed in earlier chapters, including gene regulation (see Chapter 16) and cell communication (see Chapter 7).

DROSOPHILA MELANOGASTER: RELATING GENETICS TO GENOMICS   The fruit fly *Drosophila melanogaster* is a famous model organism. Studies of fruit flies resulted in the formulation of many basic principles of genetics (see Key Concept 12.4). More than 2,500 mutations of *D. melanogaster* have been described, and this fact alone was a good reason for sequencing the fruit fly's DNA. The fruit fly is a much larger organism than *C. elegans*, both in size (it has ten times more cells) and complexity, and it undergoes complicated developmental transformations from egg to larva to pupa to adult. **Figure 17.7** summarizes the functions of the *Drosophila* genes that have been characterized so far; this distribution is typical of complex eukaryotes.

ARABIDOPSIS: STUDYING THE GENOMES OF PLANTS   About 250,000 species of flowering plants dominate the land and fresh water. But in the context of the history of life, the flowering plants are fairly young, having evolved only about 200 million years ago. The genomes of some plants are huge—for example, the genome of corn is about 3 billion bp, and that of wheat is 17 billion bp. So although we are naturally most interested in the genomes of plants we use as food and fiber, it is not surprising that scientists first chose to sequence a simpler flowering plant.

Arabidopsis thaliana, thale cress, is a member of the mustard family and has long been a favorite model organism of plant biologists. It is small (hundreds could grow and reproduce in the space occupied by this page), it is easy to manipulate, and it has a relatively small genome of 125 million bp. The *Arabidopsis* genome has more than 27,000 protein-coding genes (see Table 17.2), but remarkably, many of these genes are duplicates and probably originated by chromosomal rearrangements. When these duplicate genes are subtracted from the total, about 15,000 unique genes are left—similar to the number of genes found in fruit flies. Indeed, many of the genes found in fruit flies have homologs (related genes) in *Arabidopsis* and other plants, suggesting that plants and animals have a common ancestor.

Arabidopsis does, of course, have some genes that distinguish it as a plant (**Table 17.5**). From what you know about plants, you can guess what these are: genes involved in photosynthesis, in the transport of water into the root and throughout the plant, in the assembly of the cell wall, in the uptake and metabolism of inorganic substances from the environment, and in the synthesis of specific molecules used for defense against microbes and herbivores (organisms that eat plants). The plant-specific genes in *Arabidopsis* are also found in the genomes of other plants, including rice (*Oryza sativa*), the first major crop plant whose sequence was determined. Rice is the world's most important crop; it is a staple in the diet of 3 billion people. Despite its larger genome, rice has a set of genes remarkably similar to those of *Arabidopsis*. More recently the genome of the poplar tree *Populus trichocarpa* was sequenced. This rapidly growing tree is widely used for manufacturing paper and is a potential source of fixed carbon for making fuel. A comparison

table 17.5 *Arabidopsis* Genes Unique to Plants

Function	Number of genes
Cell wall and growth	42
Water channels	300
Photosynthesis	139
Defense and metabolism	94

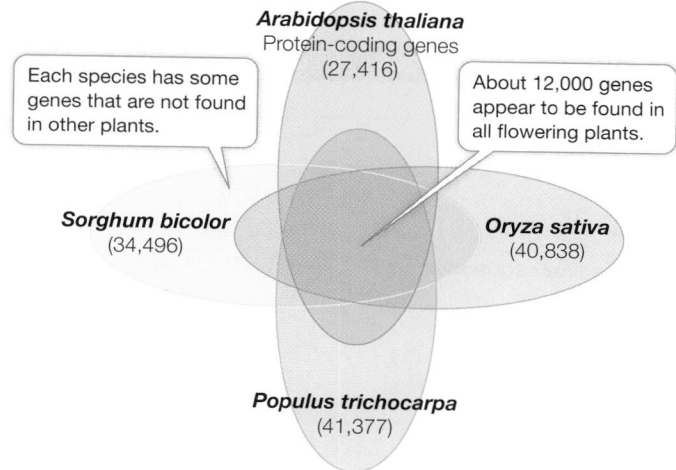

Figure 17.8 **Plant Genomes** Four plant genomes share a common set of approximately 12,000 genes that appear to comprise the minimal plant genome.

of four plant genomes shows many genes in common, comprising the basic minimal plant genome (**Figure 17.8**).

Eukaryotes have gene families

For about half of all eukaryotic protein-coding genes, only a single copy will be found in the haploid genome (two copies in somatic cells). But there are multiple copies of most other protein-coding genes, a phenomenon resulting from *****gene duplications** occurring over generations. In some cases, over evolutionary time, different copies of genes have undergone separate mutations, giving rise to groups of closely related genes called **gene families**. Some gene families, such as those encoding the globin proteins that make up hemoglobin, contain only a few members. Other families, such as the genes encoding the immunoglobulins that make up antibodies, have hundreds of members. In the human genome there are about 21,000 protein-coding genes, but 16,000 of these are members of gene families. So only about one-third of the human genes are unique.

*****connect the concepts**   Several specific examples of gene duplication resulting in gene homologs are discussed in Chapter 19, such as the evolution of remarkable homology across animal species of the Hox genes, which regulate early development.

The DNA sequences in a gene family are usually different from one another. As long as at least one member encodes a functional protein, the other members may mutate in ways that change the functions of the proteins they encode. During evolution, the availability of multiple copies of a gene allows for selection of mutations that provide advantages under certain circumstances. If a mutated gene is useful, it may be selected for in succeeding generations. If the mutated gene is a total loss, the functional copy is still there to carry out its role.

Let's take a look at one gene family, the globin genes in vertebrates. The globins are components of hemoglobin and myoglobin (an oxygen-binding protein present in muscle). The globin genes all arose long ago from a single common ancestral gene. In humans there are three functional members of the α-globin cluster and five in

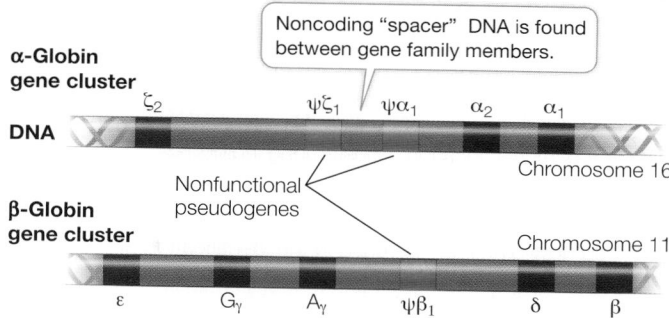

Figure 17.9 **The Globin Gene Family**   The α-globin and β-globin clusters of the human globin gene family are located on different chromosomes. The genes of each cluster are separated by noncoding "spacer" DNA. The nonfunctional pseudogenes are indicated by the Greek letter psi (ψ). The γ gene has two variants, A_γ and G_γ.

the β-globin cluster (**Figure 17.9**). In adults, each hemoglobin molecule is a tetramer containing two identical α-globin subunits, two identical β-globin subunits, and four heme pigments (see Figure 3.11).

During human development, different members of the globin gene cluster are expressed at different times and in different tissues. This differential gene expression has great physiological significance. For example, hemoglobin that contains γ-globin, a subunit found in the hemoglobin of the human fetus, binds O_2 more tightly than adult hemoglobin does. This specialized form of hemoglobin ensures that in the placenta, O_2 will be transferred from the mother's blood to the developing fetus's blood. Just before birth the liver stops synthesizing fetal hemoglobin and the bone marrow cells take over, making the adult form (2 α and 2 β). Thus hemoglobins with different binding affinities for O_2 are provided at different stages of human development.

In addition to genes that encode proteins, many gene families include nonfunctional **pseudogenes**, which are designated with the Greek letter psi (ψ) (see Figure 17.9). Pseudogenes result from mutations that cause a loss of function rather than an enhanced or new function. The DNA sequence of a pseudogene may not differ greatly from that of other family members. It may simply lack a promoter, for example, and thus fail to be transcribed. Or it may lack a recognition site needed for the removal of an intron, so that the transcript it makes is not correctly processed into a useful mature mRNA. In some gene families pseudogenes outnumber functional genes. In such cases, there appears to be no evolutionary advantage for the deletion of the pseudogenes, even though they have no apparent function.

Eukaryotic genomes contain repetitive sequences

Eukaryotic genomes contain numerous repetitive DNA sequences that do not code for polypeptides. These are typically not in protein-coding genes, which take up only a small percentage of the DNA. These include highly repetitive sequences, moderately repetitive sequences, and transposons.

Highly repetitive sequences are short (less than 100 bp) sequences that are repeated thousands of times in tandem (side-by-side) arrangements in the genome. They are not transcribed. Their proportion in eukaryotic genomes varies, from 10 percent in humans to about half the genome in some species of fruit flies. Often they are associated with heterochromatin, the densely packed, transcriptionally inactive part of the genome. Other highly repetitive sequences are scattered around the genome. For example, *short tandem repeats (STRs) of 1 to 5 bp can be repeated up to 100 times at a particular chromosomal location. The copy number of an STR at a particular location varies among individuals.

---

connect the concepts Key Concept 15.3 described how STRs can be used in the identification of individuals (DNA fingerprinting).

Moderately repetitive sequences are repeated 10 to 1,000 times in the eukaryotic genomes. These sequences include the genes that are transcribed to produce tRNAs and rRNAs, which are used in protein synthesis. The cell makes tRNAs and rRNAs constantly, but even at the maximum rate of transcription, single copies of the tRNA and rRNA genes would be inadequate to supply the large amounts of these molecules needed by most cells. Thus the genome has multiple copies of these genes.

In mammals, four different rRNA molecules make up the ribosome: the 18S, 5.8S, 28S, and 5S rRNAs. (The S stands for Svedberg unit, which is a measure of size.) The 18S, 5.8S, and 28S rRNAs are transcribed together as a single precursor RNA molecule (**Figure 17.10**). As a result of several posttranscriptional steps, the precursor is cut into the final three rRNA products, and the noncoding "spacer" RNA is discarded. The sequence encoding these

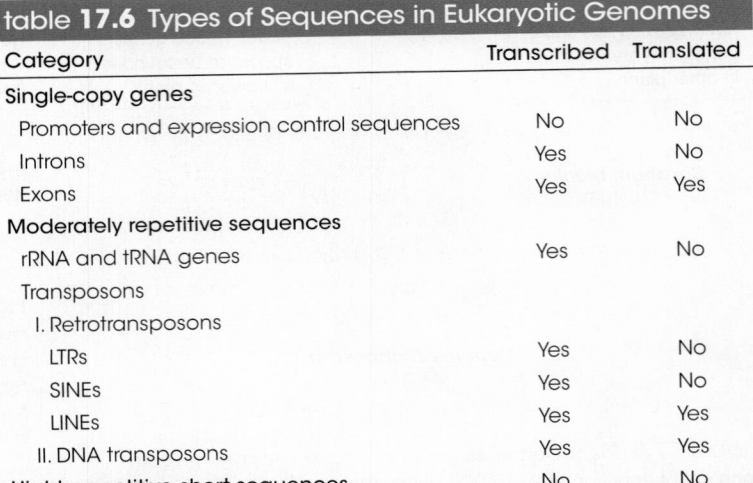

table **17.6** Types of Sequences in Eukaryotic Genomes		
Category	Transcribed	Translated
Single-copy genes		
Promoters and expression control sequences	No	No
Introns	Yes	No
Exons	Yes	Yes
Moderately repetitive sequences		
rRNA and tRNA genes	Yes	No
Transposons		
I. Retrotransposons		
LTRs	Yes	No
SINEs	Yes	No
LINEs	Yes	Yes
II. DNA transposons	Yes	Yes
Highly repetitive short sequences	No	No

RNAs is moderately repetitive in humans: a total of 280 copies of the sequence are located in clusters on five different chromosomes.

Apart from the RNA genes, most moderately repetitive sequences are transposons, which, like the prokaryotic transposons we discussed earlier, can move about in the genome. Transposons make up more than 40 percent of the human genome and about 50 percent of the corn genome, but in many other eukaryotes the percentage is smaller (3–10 percent).

Table 17.6 summarizes the major types of transposons in eukaryotes. Retrotransposns fall into three groups, based on the types of repetitive sequences they contain: long terminal repeats (LTRs), long interspersed elements (LINEs), and short interspersed elements (SINEs). Retrotransposons move about the genome in a distinctive way: they are transcribed into RNA, which then acts as a template for new DNA. The new DNA becomes inserted at a new location in the genome. This "copy and paste" mechanism results in two copies of the transposon: one at the original location and the other at a new location. A single type of SINE retrotransposon, the 300-bp Alu element, accounts for 11 percent of the human genome; it is present in a million copies!

Transposons of the fourth type, the DNA transposons, do not use RNA intermediates. Like some prokaryotic transposable elements, they are excised from the original location and

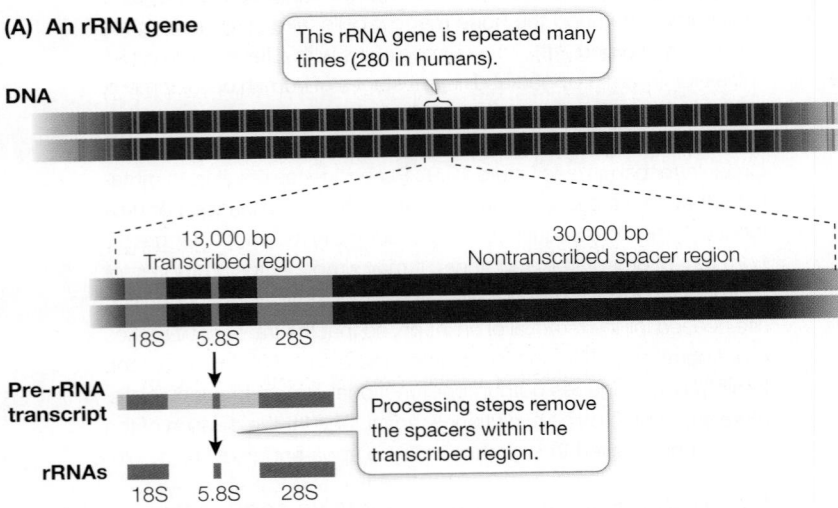

(A) An rRNA gene

This rRNA gene is repeated many times (280 in humans).

DNA

13,000 bp
Transcribed region

30,000 bp
Nontranscribed spacer region

18S 5.8S 28S

Pre-rRNA transcript

Processing steps remove the spacers within the transcribed region.

rRNAs
18S 5.8S 28S

(B) Transcription of multiple rRNA genes

Strands of rRNA

Transcription begins here...

...the RNA elongates...

...and elongates until it is released here.

DNA

Many rRNA precursors are being transcribed from multiple rRNA genes.

Figure 17.10 A Moderately Repetitive Sequence Codes for rRNA **(A)** This rRNA gene, along with its nontranscribed spacer region, is repeated 280 times in the human genome, with clusters on five chromosomes. **(B)** This electron micrograph shows transcription of multiple rRNA genes.

Q: Are there similarities between the process depicted in the photo here and the process of translation via polysome shown in Figure 14.16?

become inserted at a new location without being replicated (a "cut and paste" mechanism).

What role do these moving sequences play in the cell? The best answer so far seems to be that transposons are simply molecular parasites that can be replicated. The insertion of a transposon at a new location can have important consequences. For example, the insertion of a transposon into the coding region of a gene results in a mutation (see Figure 17.5). This phenomenon accounts for rare forms of several genetic diseases in humans, including hemophilia and muscular dystrophy. If the insertion of a transposon takes place in the germ line, a gamete with a new mutation results. If the insertion takes place in a somatic cell, cancer may result.

Sometimes an adjacent gene can be replicated along with a transposon, resulting in a gene duplication. A transposon can carry a gene, or a part of it, to a new location in the genome, shuffling the genetic material and creating new genes. Clearly, transposition stirs the genetic pot in the eukaryotic genomes and thus contributes to genetic variation.

17.3 recap

The genomes of eukaryotes contain more genes than those of prokaryotes. Some of these "additional" genes encode functions associated with the compartmentalization of eukaryotic cells; others are needed to support multicellularity. The genome sequences of model organisms have been compared to identify common features of the eukaryotic genomes, including the presence of abundant regulatory sequences, repetitive sequences, and noncoding DNA. Some eukaryotic genes comprise gene families, which may include members that are mutated and nonfunctional.

learning outcomes

You should be able to:

- Distinguish between the features of genomes or model organisms.
- Predict why different plant species will differ in the features of their genomes.
- Provide an evolutionary explanation for how gene families promote further evolution in vertebrates.
- Summarize why the rRNA and tRNA genes involved in protein synthesis are present in multiple copies.

1. Describe the roles of a class of proteins found in *Caenorhabditis elegans* that has few counterparts in yeasts.
2. What is the importance of gene families for evolution?
3. Why are there multiple copies of sequences coding for rRNA in the mammalian genome?
4. The genomes of rice, wheat, and corn are similar to one another and to that of *Arabidopsis*. What processes might explain why these four plants are nevertheless distinguished by unique proteins?

The analysis of eukaryotic genomes has resulted in an enormous amount of useful information, as we have seen. In the next section we will look more closely at the human genome.

key concept 17.4   Human Biology Is Revealed through the Genome

Since the first human genome sequences were completed early in the first decade of this millennium, the genomes of many other individuals have been sequenced and published. With the rapid development of sequencing technologies, a human genome can be sequenced for less than $1,000.

focus your learning

- The human genome and its genes exemplify the characteristics of complex eukaryote genomes.
- Comparisons of haplotypes of individuals with and without a particular genetic disease can identify loci associated with the disease.
- Pharmacogenomics is the study of how an individual's genome affects responses to drugs or other outside agents.

The following are just some of the interesting facts that sequencing of the human genome has brought to light:

- Of the 3.2 billion bp in the haploid human genome, only an estimated 1.2 percent (about 21,000 genes) make up protein-coding regions. This was a surprise. Before sequencing began, the diversity of human proteins suggested there might be 80,000 to 150,000 genes in the human genome. The actual number of genes—barely more than in the nematode *C. elegans*—means that posttranscriptional mechanisms (such as alternative splicing) must account for the observed number of proteins in humans. That is, the average human gene must code for several different proteins.

- The average human gene has 27,000 bp. Gene sizes vary greatly, from about 1,000 bp to 2.4 million bp. Variation in gene size was expected given that human proteins (and RNAs) vary in size, from 100 to about 5,000 amino acids per polypeptide chain.

- Most human genes have many introns. The "average" human gene has eight of them, with an average length of 3,300 bp.

- About half of the genome is made up of transposons (see Table 17.6) and other highly repetitive sequences.

- When the genomes of two unrelated individuals of the same species are compared, most of the sequence—about 99.9 percent—is identical. Before you jump to the conclusion that we are all the same, remember that we have 3.2 billion bp, so that 0.1 percent difference is 3 million bp! The main differences are in copy number variations of repeated sequences and single nucleotide polymorphisms (SNPs). And don't forget epigenetic differences as well (see Key Concept 16.4).

- Genes are not evenly distributed over the genome. Chromosome 19 is packed densely with genes, whereas chromosome 8 has long stretches without coding regions. The Y chromosome has the fewest genes (about 230), and chromosome 1 has the most (about 3,000).

Media Clip 17.1 **A Big Surprise from Genomics**
www.Life11e.com/mc17.1

Comparative genomics reveals the evolution of the human genome

Comparisons among sequenced genomes from prokaryotes and eukaryotes have revealed some of the evolutionary relationships among genes. Some genes are present in both prokaryotes and eukaryotes, others are only in eukaryotes, and still others are only in animals or only in vertebrates (**Figure 17.11**).

The genomes of various other primates, including all of the great apes, have now been sequenced. The search is on for a set of human genes that differ from those found in other primates and that make us unique. Chimpanzees are our closest living relatives, sharing nearly 99 percent of our DNA sequence. About 500 protein-coding genes have undergone accelerated *evolution in humans and chimpanzees, including genes involved in hearing and brain development. Further analyses of these sequences may reveal genes that distinguish us from other apes, and that "make humans human."

---

connect the concepts How comparisons of primate and human genomes have refined our understanding of human evolution is discussed in Key Concept 32.5.

Other clues about "human" genes have come from sequencing the genomes of ancient human relatives. An international team of scientists led by Svante Pääbo at the Max Planck Institute for Evolutionary Anthropology in Germany has extracted and sequenced DNA from the bones of Neanderthals, who lived in Europe up to 50,000 years ago. The entire Neanderthal genome has been sequenced. It

is more than 99 percent identical to our human DNA, justifying the classification of Neanderthals as part of the same genus, *Homo*.

Comparisons of the human and Neanderthal genomes are ongoing and have already revealed several interesting facts:

- The gene *MC1R* is involved in skin and hair pigmentation. A point mutation found in Neanderthals but not humans caused lower activity of the MC1R protein when it was introduced into cell cultures. Such lower activity of MC1R is known to result in fair skin and red hair in humans. So it appears that at least some Neanderthals may have had pale skin and red hair.

- The gene *FOXP2* is involved in vocalization in many organisms, including birds and mammals. In humans, mutations in this gene result in severe speech impairment. The Neanderthal *FOXP2* gene is identical to that of humans, whereas that of chimpanzees is slightly different. This has led to speculation that Neanderthals were capable of speech.

- While the human and Neanderthal genome sequences are very similar, there are differences in many point mutations and larger chromosomal arrangements. There are distinctive "human" DNA sequences and also distinctive "Neanderthal" sequences. There is some mixture of the two, indicating that humans and Neanderthals interbred, with transfer of DNA between the two.

Human genomics has potential benefits in medicine

Most complex phenotypes are determined not by single genes but by multiple genes interacting with the environment. The single-allele explanations of phenylketonuria and sickle-cell anemia (see Key Concept 15.2) do not apply to such common disorders as diabetes, heart disease, and Alzheimer's disease. To understand the genetic bases of these diseases, biologists are now using rapid genotyping technologies to create "haplotype maps," which are used to identify SNPs that are linked to genes involved in disease.

HAPLOTYPE MAPPING   The SNPs that differ among individuals are not inherited as independent alleles. Rather, a set of SNPs that are present on a segment of chromosome are usually inherited as a unit. This linked piece of a chromosome is called a **haplotype**. You can think of the chromosome segment as a sentence, the haplotype as a word, and the SNP as a letter in the word. Analyzing SNPs is faster and less expensive than sequencing whole genomes, so haplotype mapping provides a shortcut for identifying the locations of genes and mutations involved in particular diseases (see Key Concept 15.3). By comparing the haplotypes of individuals with and without a particular genetic disease, the genetic loci associated with the disease can be identified (**Figure 17.12**).

A microarray of 500,000 SNPs has been used to analyze thousands of people to find out which SNPs are associated with specific diseases. The amount of data is prodigious: 500,000 SNPs, thousands of people, thousands of medical records. With so much natural variation, statistical measures of association between a haplotype and a disease need to be very rigorous.

GENOTYPING TECHNOLOGY AND PERSONAL GENOMICS Association tests like the one described in Figure 17.12 have revealed

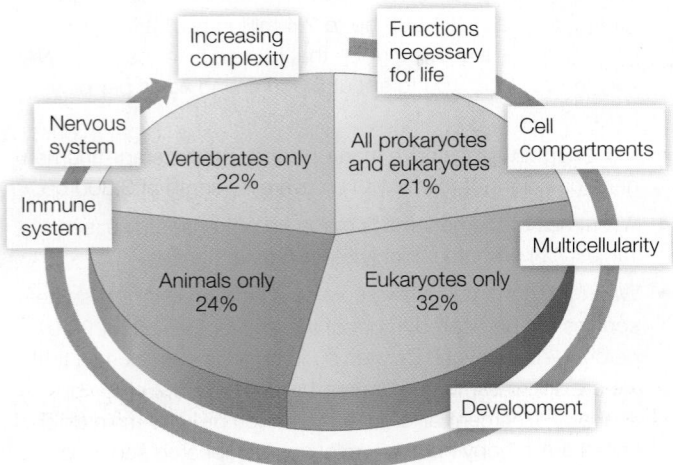

Figure 17.11 Evolution of the Genome A comparison of the human and other genomes has revealed how genes with new functions have been added over the course of evolution. Each percentage refers to genes in the human genome. Thus 21 percent of human genes have homologs in prokaryotes and other eukaryotes, 32 percent of human genes occur only in other eukaryotes, and so on.

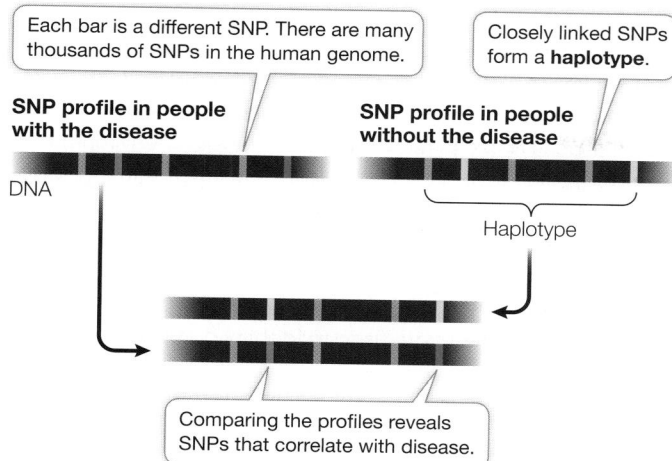

Each bar is a different SNP. There are many thousands of SNPs in the human genome.

Closely linked SNPs form a **haplotype**.

SNP profile in people with the disease

SNP profile in people without the disease

DNA

Haplotype

Comparing the profiles reveals SNPs that correlate with disease.

Figure 17.12 SNP Genotyping and Disease   Scanning the genomes of people with and without particular diseases reveals correlations between SNPs and complex diseases.

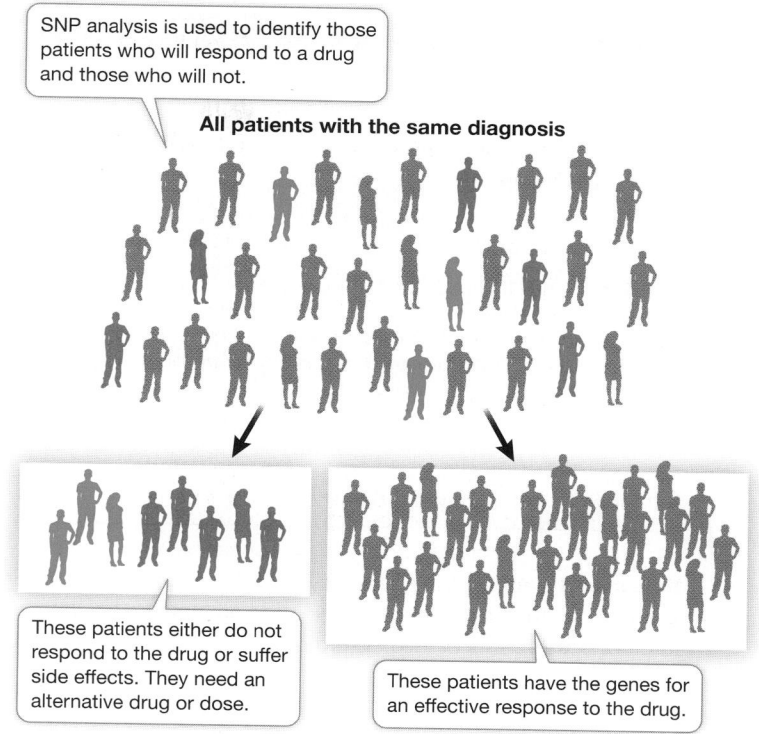

SNP analysis is used to identify those patients who will respond to a drug and those who will not.

All patients with the same diagnosis

These patients either do not respond to the drug or suffer side effects. They need an alternative drug or dose.

These patients have the genes for an effective response to the drug.

Figure 17.13 Pharmacogenomics   Correlations between genotypes and responses to drugs will help physicians develop personalized medical care. The different colors indicate individuals with different SNPs.

particular haplotypes or alleles that are associated with modestly increased risks for diseases. For example, a particular SNP on chromosome 10 is associated with a 65% increased risk for adult-onset diabetes when heterozygous, and almost three-fold when homozygous. This information can have predictive value in counseling and treating patients. Private companies can now scan a human genome for these variants—and the price for this service keeps getting lower. However, at this point it is unclear what a person without symptoms should do with the information, since multiple genes, environmental influences, and epigenetic effects all contribute to the development of these diseases.

Of course, the most comprehensive way to analyze a person's genome is by actually sequencing it. As the cost of genome sequencing and analysis goes down, SNP testing will be superseded. In 2015 the U.S. government announced a plan to sequence the genomes of 1 million Americans and link the sequences to medical records. This approach is not new, but the large scale of it is. The goal is to relate genome changes to diseases such as cancer and diabetes, and to use these changes for diagnosis and, if they lead to changes in encoded proteins, to therapies targeted at the altered proteins.

PHARMACOGENOMICS   Genetic variation can affect how an individual responds to a particular drug. For example, a drug may be chemically modified in the liver to make it more or less active. Consider an enzyme that catalyzes the following reaction:

Active drug → less active drug

A mutation in the gene that encodes this enzyme may make the enzyme less active. For a given dose of the drug, a person with the mutation would have more active drug in the bloodstream than a person without the mutation. So the effective dose of the drug would be lower in the person with the mutation.

Now consider a different case, in which the liver enzyme is needed to make the drug active:

Inactive drug → active drug

A person carrying a mutation in the gene encoding this liver enzyme would not be affected by the drug, since the activating enzyme is not present.

The study of how an individual's genome affects his or her response to drugs or other outside agents is called **pharmacogenomics**. Just as it is possible to identify haplotypes or SNPs that are associated with particular disease susceptibilities, it is also possible to identify SNPs that are associated with specific drug responses. This type of analysis makes it possible to predict whether a drug will be effective. The objective is to personalize drug treatment so that a physician can know in advance whether an individual will benefit from a particular drug (**Figure 17.13**). This approach might also be used to reduce the incidence of adverse drug reactions by identifying individuals who will metabolize a drug slowly, which can lead to a dangerously high level of the drug in the body.

17.4 recap

The haploid human genome has 3.2 billion bp, but only about 1.2 percent of the genome codes for proteins. Most human genes are subject to alternative splicing, which may account for why there are more proteins than genes. Haplotype mapping to find correlations of specific SNPs with disease and drug susceptibility holds promise for personalized, precision medicine.

(continued)

17.4 recap (continued)

learning outcomes

You should be able to:

- Describe the characteristics of the human genome and its genes.

- Use haplotype information to determine whether a gene is associated with a phenotypic characteristic.

- Summarize the use of pharmacogenomics.

1. How do human cells express a larger number of proteins than is accounted for by the number of protein-coding genes in the human genome?

2. A patient is prescribed a drug whose metabolism in the body is under the control of known genes and alleles. How could haplotype mapping be used to get information on whether this drug would be effective in this patient?

Genome sequencing has advanced our understanding of biology enormously. High-throughput technologies are now being applied to other components of the cell: proteins and metabolites. We will now turn to the results of these studies.

key concept 17.5 Proteomics and Metabolomics Can Provide Insights beyond the Genome

Statements like "The human genome is the book of life" were commonly made at the time the human genome sequence was first revealed, reflecting a belief in "genetic determinism"—that a person's phenotype is determined by his or her genotype. But as we have repeatedly stressed, an organism is not just a product of gene expression. The proteins and small molecules present in any cell at a given point in time reflect not just gene expression but also the influences of the intracellular and extracellular environments. Complementing genomics, the fields of proteomics and metabolomics have emerged to provide a more complete understanding of the relationship between genome and organism.

focus your learning

- Chemical methods are used to identify components of the proteome.

- The metabolome is the complete set of small molecules, called metabolites, that are present in the cell.

The proteome is the complete set of proteins in a cell, tissue, or organism at a given time

As mentioned above, many genes encode more than a single protein (**Figure 17.14**). *__Alternative splicing__ of a given gene can allow different combinations of exons to be transcribed into the mature mRNAs. Posttranslational modifications also increase the number of protein variants that can be derived from one gene. It should also be noted that in a multicellular organism many proteins are produced only by certain cells and only under specific conditions. Even single-celled organisms express only a subset of their genes at any particular time. The **proteome** is the sum total of the proteins produced by a cell, tissue, or organism at a given time, under defined conditions.

---

*__connect the concepts__   Recall from Key Concept 16.5 that different mRNAs can be made from the same gene by alternative splicing, resulting in a family of different proteins, with different functions, from a single gene. Recall also that proteins can be modified in posttranslational processes such as proteolysis, glycosylation, and phosphorylation (see Figure 14.18).

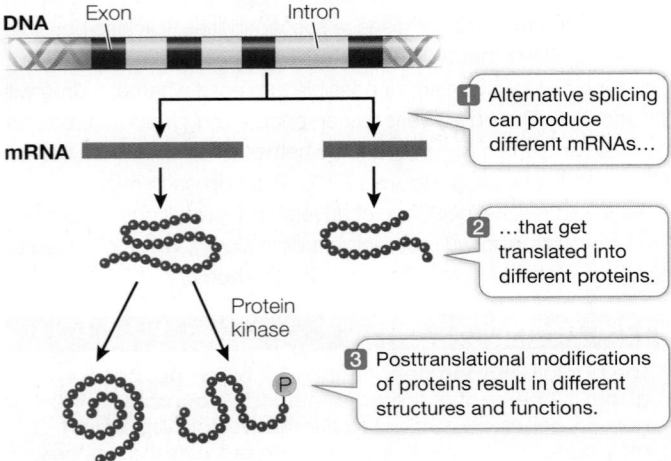

Figure 17.14 **Proteomics**   A single gene can code for multiple proteins.

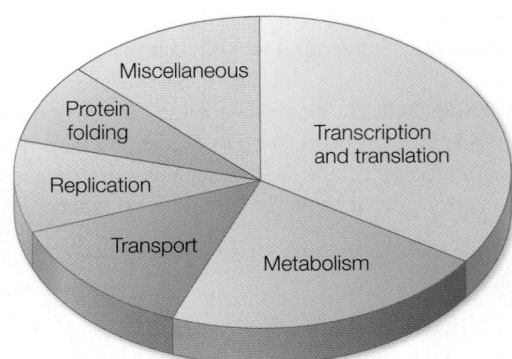

Figure 17.15 Proteins of the Eukaryotic Proteome   About 1,300 proteins are common to all eukaryotes and fall into these categories. Although their amino acid sequences may differ to a limited extent, they perform the same essential functions in all eukaryotes.

The proteome is analyzed by mass spectrometry, a technique that uses electromagnets to identify proteins by the mass. The ultimate aim of proteomics is just as ambitious as that of genomics. Whereas genomics seeks to describe the genome and its expression, proteomics seeks to identify and characterize all of the expressed proteins.

Comparisons of the proteomes of humans and other eukaryotic organisms have revealed a common set of proteins that can be categorized into groups with similar amino acid sequences and similar functions. When considered on a whole-organism basis, 46 percent of the yeast proteome, 43 percent of the nematode proteome, and 61 percent of the fruit fly proteome are shared by the human proteome. Functional analyses indicate that this set of 1,300 proteins provides the basic metabolic functions of a eukaryotic cell, such as glycolysis, the citric acid cycle, membrane transport, protein synthesis, DNA replication, and so on (**Figure 17.15**).

There are, of course, many uniquely human proteins in addition to those we share with other eukaryotic organisms. As we have mentioned before, proteins have different functional regions called domains (for example, a domain for binding a substrate, or a domain for spanning a membrane). While a particular organism may have many unique proteins, those proteins are often just unique combinations of domains that exist in other organisms. *This reshuffling of the genetic deck is a key to evolution*.

Proteins seldom exist in isolation. Most interact with other molecules, such as nucleic acids (e.g., transcription factors and DNA), other proteins (e.g., the respiratory chain complexes in mitochondria), and lipids (receptors in the cell membrane). A major task of proteomics is analyzing these interactions.

Metabolomics is the study of chemical phenotype

Studying genes and proteins gives a limited picture of what is going on in a cell. As we have seen, both gene function and protein function are affected by the internal and external environments of the cell. Many proteins are enzymes, and their activities affect the concentrations of their substrates and products. So as the proteome changes, so will the abundances of small molecules called metabolites. The **metabolome** is the complete set of small molecules present in a cell, tissue, or organism under defined conditions. These include:

- *Primary metabolites* involved in normal processes, such as intermediates in pathways such as glycolysis. This category also includes hormones and other signaling molecules.

- *Secondary metabolites*, which are often unique to particular organisms or groups of organisms. They are often involved in special responses to the environment. Examples are antibiotics made by microbes, and the many chemicals made by plants that are used in defense against pathogens and herbivores.

Not surprisingly, measuring metabolites involves sophisticated analytical instruments. If you have studied organic or analytical chemistry, you may be familiar with gas chromatography and high-performance liquid chromatography, which separate molecules, and with mass spectrometry and nuclear magnetic resonance spectroscopy, which are used to identify them. These measurements result in "chemical snapshots" of cells or organisms, which can be related to physiological states.

There has been some progress in defining the human metabolome. A database created by David Wishart and colleagues at the University of Alberta contains more than 6,500 metabolite entries. The challenge now is to relate levels of these substances to physiology. For example, you probably know that high levels of glucose in the blood are associated with diabetes. But what about the early stages of heart disease? There may be a pattern of metabolites that is diagnostic of this disease. This could aid in early diagnosis and treatment.

Plant biologists are ahead of medical researchers in the field of metabolomics. Over the years, tens of thousands of secondary metabolites have been identified in plants, many of them made in response to environmental challenges. Some of these will be discussed in Chapter 38. The metabolome of the model organism *Arabidopsis thaliana* is being described, and will give insight into how a plant copes with stresses such as drought or pathogen attack. This knowledge could be helpful in optimizing plant growth for agriculture.

▶ 17.5 recap

The proteome is the set of all proteins produced by a cell, tissue, or organism under specific conditions. The metabolome is the total content of small molecules, such as intermediates in primary metabolism, hormones, and secondary metabolites. The proteome and the metabolome can be analyzed using chemical methods that separate and identify molecules.

learning outcomes

You should be able to:

- Summarize how chemical methods are used in proteomics.

- Provide examples of applications of metabolomic data for human health.

1. How is the proteome analyzed?

2. Why would it be useful to have a database of primary and secondary metabolites involved in human metabolism?

What have we learned from sequencing animal genomes?

In certain dog breeds, such as whippets, the muscles of some individuals can be much bulkier than is normal. Genome analyses have shown that the gene for myostatin, a protein that inhibits muscle growth, is mutated in whippets that have bulky muscles. When the myostatin gene is mutated, the protein that inhibits muscle growth is dysfunctional (Figure 17.16). Comparative genomics reveals that the myostatin gene is also mutated in certain breeds of cattle and sheep that are noted for overdeveloped muscles.

With the knowledge that myostatin affects muscle development, some consideration has been given to the possibility of manipulating myostatin in humans to treat muscle-wasting diseases such as muscular dystrophy. As you might expect, athletes anxious to have bulkier muscles are also very interested in this gene and its protein product.

Future directions

Few scientific endeavors have received as much hype and hope as genome sequencing. A major effort is under way to sequence the genomes of tumors in as many people as possible, to screen for mutations. When the *BRCA1* gene was identified as mutated in breast cancer (see Chapter 15), some medical scientists hoped that cancer would be like other genetic diseases such as sickle-cell anemia—caused by a single mutation in a single gene. So far, the lessons learned from cancer genomes include that there are many mutations in a tumor, and that some of them drive the tumor formation

Figure 17.16 **Muscular Gene** These dogs are both whippets, but the muscle-bound dog (right) has a mutation in the myostatin gene.

while others do not; that the mutations in one person's tumor of a particular type may differ from those in the same tumor in someone else; and that tumor genomes evolve over time as the tumor grows and spreads. Understanding tumor genomes is not straightforward. Nevertheless, the goal is to study genomes to determine how gene products and the environment interact to form the tumor phenotype, and to design therapy for that tumor in individual people. This is called precision medicine, and it is in the future. It is a model for a goal of genomics: to determine the relationship between genotype, the environment, and phenotype.

Chapter Summary 17

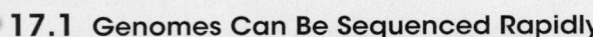

17.1 Genomes Can Be Sequenced Rapidly
- Methods of DNA sequencing involve miniaturization and computerized analysis. Review Figure 17.1, Animations 17.1, 17.2
- Genomes are sequenced in overlapping fragments, and then the fragments are lined up to give the final sequence. Review Figure 17.2
- The analysis of genome sequences gives information about protein-coding and noncoding regions.

17.2 Prokaryotic Genomes Are Compact
- DNA sequencing is used to study the genomes of prokaryotes that are important to humans and ecosystems.
- **Metagenomics** is the identification of DNA sequences from multiple organisms from a bulk environmental sample (such as soil or seawater). Review Figure 17.3
- **Transposable elements** (or **transposons**) and **composite transposons** can move about the genome. Review Figure 17.4
- Transposon mutagenesis can be used to inactivate genes one by one. The mutated organism can be tested for survival and a minimal set of essential genes described. Review Figure 17.5

17.3 Eukaryotic Genomes Contain Many Types of Sequences
Review Focus: Key Figure 17.6
- Comparisons of genome sequences from model organisms have revealed some common features of the eukaryotic genomes as well as specialized genes for cellular compartmentalization, development, and features unique to plants. Review Tables 17.2–17.5, Figures 17.7, 17.8
- Some eukaryotic genes exist as members of **gene families**. Proteins may be made from these closely related genes at different times and in different tissues. Some members of gene families may be nonfunctional **pseudogenes**. Review Figure 17.9
- Repeated sequences are present in the eukaryotic genomes. Review Table 17.6
- **Moderately repetitive sequences** include those coding for rRNA and transposons. Review Figure 17.10

17.4 Human Biology Is Revealed through the Genome
- The haploid human genome has 3.2 billion bp.
- Only about 1.2 percent of the genome codes for proteins; the rest consists of repeated sequences and noncoding DNA.

- Most human genes have introns, and alternative splicing leads to the production of more than one protein per gene.
- SNP genotyping (haplotype mapping) correlates variations in the genome with diseases or drug sensitivity. It may lead to precision medicine. Review Figure 17.12
- **Pharmacogenomics** uses DNA and amino acid sequence data to inform drug development and testing. Review Figure 17.13

▶ **17.5 Proteomics and Metabolomics Can Provide Insights beyond the Genome**

- The **proteome** is the complete set of proteins in a cell, tissue, or organism at a given time.

- There are more proteins than there are protein-coding genes in the genome.
- The proteome can be analyzed using chemical methods that separate and identify proteins.
- The **metabolome** is the complete set of small molecules present in a cell, tissue, or organism under defined conditions.

See Activity 17.1 for a concept review of this chapter.

Go to **LearningCurve** (in **LaunchPad**) for dynamic quizzing that helps you solidify your understanding of this chapter. **LearningCurve** adapts to your responses, giving you the practice you need to master each key concept.

▷ Apply What You've Learned

Review

17.4 Comparisons of haplotypes of individuals with and without a particular genetic disease can identify loci associated with the disease.

17.4 Pharmacogenomics is the study of how an individual's genome affects responses to drugs or other outside agents.

17.5 Chemical methods are used to identify components of the proteome.

Different people metabolize the (hypothetical) antianxiety drug Calm at different rates. This variation is potentially a problem because too-high concentrations of the drug can have harmful side effects, while too-low concentrations may not give the desired therapeutic effect. Knowledge about how genotypes affect the metabolism of Calm would be very useful.

In a study, people who had been genotyped at various single nucleotide polymorphisms (SNPs) across the genome were given Calm. Twelve hours later, they had their blood drawn to assess the concentration of the drug. Individual concentrations were normalized by dividing by the global mean concentration and multiplying by 100. The results for the three SNPs with the greatest effects are shown in the figure below. Note that the double letters correspond to the nucleotides on homologous chromosomes (e. g., "AC" means there is A on one chromosome and C on its homolog).

Questions

1. Describe the relationship between genotype and Calm concentration and between genotype and Calm metabolism for each of the three SNPs.

2. Based on the data, would you recommend a higher dose or a lower dose of the drug for individuals with the *AA* genotype than for those with the *GG* genotype at SNP 5689? Explain your answer.

3. Suppose SNP 5689 were at a gene that encoded an enzyme that broke down the drug. Would you expect the enzyme encoded by the *A* allele of the gene to be more or less active than the enzyme encoded by the *G* allele? Explain your answer.

4. Suppose SNP 8835 were at a gene that encoded an inhibitor for the enzyme in Question 3. Would you expect the enzyme encoded by the *A* allele of the gene to be more or less active than the enzyme encoded by the *C* allele? Explain your answer.

5. The drug Calm appears to affect protein synthesis at several genes in liver tissue. Briefly describe how a medical researcher could examine which proteins have altered expression in a mouse model, highlighting the technique that would be used.

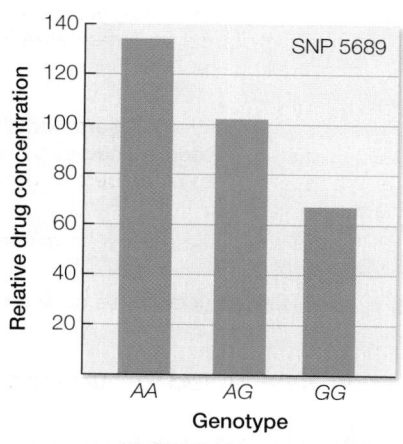
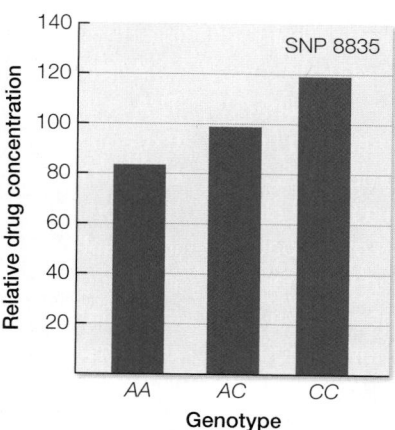
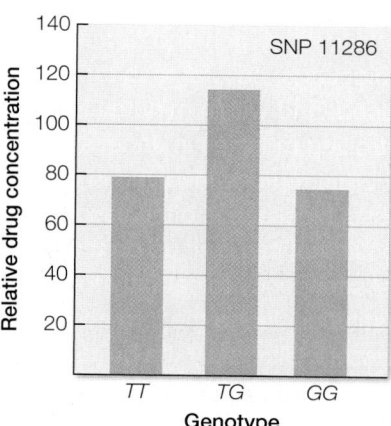

18

Nucleic acids and proteins can be analyzed and changed in small volumes, in this case far less than 1 mL.

key concepts

18.1 DNA from Different Sources Forms Recombinant DNA

18.2 There Are Several Ways to Insert DNA into Cells

18.3 Any Sequence of DNA Can Be Used for Cloning

18.4 Several Tools Are Used to Modify DNA and Study Its Function

18.5 DNA Can Be Manipulated for Human Benefit

Recombinant DNA and Biotechnology

DNA Technology Meets Medicine

Janet was arriving at her office when she first felt numb on the left side of her face. She became confused and had difficulty finding her office. After a passerby called 911, an emergency response team transported Janet to a nearby hospital. She had had a stroke, but amazingly, she was back at work and fully functional two days later.

A stroke can occur when a large blood vessel that supplies the brain with oxygen and nutrients is blocked. Within minutes of blockage, cellular respiration in affected brain cells is reduced, and within minutes to hours, the cells die. Nerves controlling muscles in Janet's face were affected first, followed by brain regions involved in cognitive processes.

Blood vessel blockage is often associated with a blood clot on a fat deposit that has built up on the inner walls of a blood vessel. Janet's tendency to eat fatty foods, coupled with her genetic constitution, led to the formation of fatty deposits in her blood vessels. Ultimately the buildup of these fatty plaques can narrow blood vessels, causing blood clots to form that can block blood flow. Blockage of a vessel that feeds the brain results in a stroke.

Janet was treated with tissue plasminogen activator, or TPA, a protein involved in dissolving blood clots. Your body produces TPA when you have a cut; its action is seen when the clot "goes away" a few days later. Janet's cells would have gradually made TPA to slowly dissolve the clot blocking the vessel to her brain, but in the meantime her brain cells would have died. To speed up the process, Janet was injected with TPA right around the clot.

Where did the TPA used in Janet's treatment come from? TPA is a protein made in tiny amounts by blood vessel cells. Extracting enough TPA from tissues is impractical; recombinant DNA technology is required to get enough of it for therapeutic injection into a bloodstream. As you will learn in Investigating Life: Producing TPA, the process involves inserting the gene that encodes TPA into living bacterial cells in the lab. These cells are then biochemically coaxed into expressing TPA protein in quantity. The production of TPA is but one example of how knowledge of molecular biology and advances in biotechnology have transformed biological research and launched new industries that employ microbes to produce proteins and other useful chemicals.

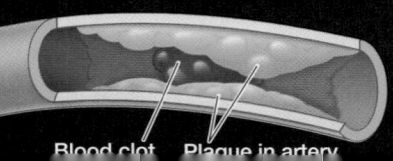

Blood clot   Plaque in artery

 How is biotechnology changing medicine?

key concept

18.1 DNA from Different Sources Forms Recombinant DNA

Recombinant DNA is a DNA molecule that has been made in the laboratory using at least two different sources of DNA. The DNAs used for recombination can be from the same organism (e.g., from different chromosomes), from different members of the same species, or from entirely different organisms.

focus your learning

- Recombinant DNA technology uses restriction enzymes and DNA ligase to respectively cut and paste DNA.

Restriction enzymes (*restriction endonucleases**) can cleave double-stranded DNA. In the late 1960s scientists discovered **DNA ligase**, which catalyzes the joining of DNA fragments; one of its functions is to join Okazaki fragments during DNA replication (see Figure 13.15). With restriction enzymes in hand, scientists saw the potential to cut DNA molecules and then splice the fragments together in new combinations ("recombine" the DNA fragments). In 1973 Stanley Cohen at Stanford University, Herbert Boyer at the University of California, and their colleagues did just that. As illustrated in **Figure 18.1**, they isolated two different plasmids (small circular DNA molecules that replicate independently of the main chromosome in bacterial cells; see Figure 12.22) from

Escherichia coli. Each of the two plasmids contained different antibiotic resistance genes. The scientists cut the two plasmids with restriction enzymes, mixed the fragments together, and then used DNA ligase to rejoin them. The recombinant plasmids of this ligation reaction were then inserted into new *E. coli* cells, and the cells were grown in the presence of both antibiotics. A few of the bacteria had been transformed with a recombinant plasmid containing both of the antibiotic resistance genes and could grow on the medium, whereas no *E. coli* lacking a recombinant plasmid were resistant to both antibiotics. With this experiment, recombinant DNA technology was born.

***connect the concepts** As discussed in Key Concept 15.3, restriction endonucleases are a class of enzymes that cut DNA molecules. Each enzyme recognizes a unique sequence of nucleotides in the DNA strand, usually about 4–6 base pairs long.

Let's look more closely at what happens when DNA is cut with a restriction enzyme and then rejoined with DNA ligase. Many restriction enzymes recognize palindromic DNA sequences—sequences that read the same way in both directions. For example, you can read the DNA recognition sequence for the restriction enzyme *Eco*RI from 5′ to 3′ as GAATTC on both strands:

$$5'.......GAATTC......3'$$
$$3'.......CTTAAG......5'$$

---

experiment

Figure 18.1 Can DNA from Two Different Sources Be Recombined into a Single Functional DNA Molecule?

Original Paper: Cohen, S. N., A. C. Y. Chang, H. W. Boyer and R. B. Helling. 1973. Construction of biologically functional bacterial plasmids in vitro. *Proceedings of the National Academy of Sciences USA* 70: 3240–3244.

With the discovery of restriction enzymes and DNA ligase, it became possible to combine DNA fragments from different sources in the laboratory. But would such "recombinant DNA" be functional when inserted into a living cell? The results of this experiment by Stanley Cohen and Herbert Boyer completely changed the scope of genetic research, increasing our knowledge of gene structure and function, and ushered in the new field of biotechnology.

HYPOTHESIS▶ Biologically functional recombinant plasmids can be made in the laboratory.

 Media Clip 18.1 **Striking Views of Recombinant DNA Being Made** www.Life11e.com/mc18.1

METHOD
E. coli plasmids carrying a gene for resistance to either the antibiotic kanamycin (K) or tetracycline (T) are cut with a restriction enzyme.

Experimental group
Plasmids are cut.

Kʳ   Tʳ

E. coli plasmid

The cut plasmids are mixed with DNA ligase to form recombinant DNA.

Kʳ   Tʳ

The plasmids are inserted into *E. coli*.

RESULTS

Some *E. coli* are resistant to both antibiotics.

Control group
Plasmids are not cut.

Kʳ   Tʳ

No *E. coli* are resistant to both antibiotics.

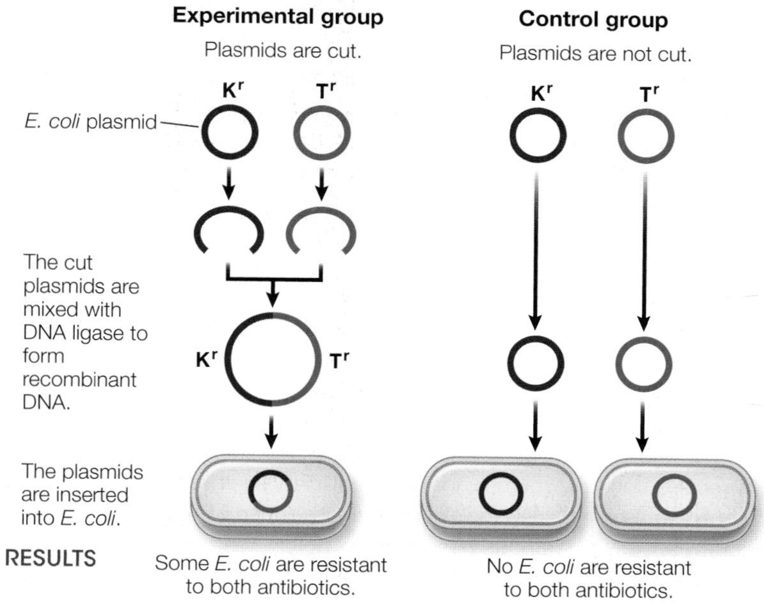

CONCLUSION▶ Two DNA fragments with different genes can be joined to make a functional recombinant DNA molecule.

A **work with the data** exercise that accompanies this figure may be assigned in **LaunchPad**.

focus: key figure

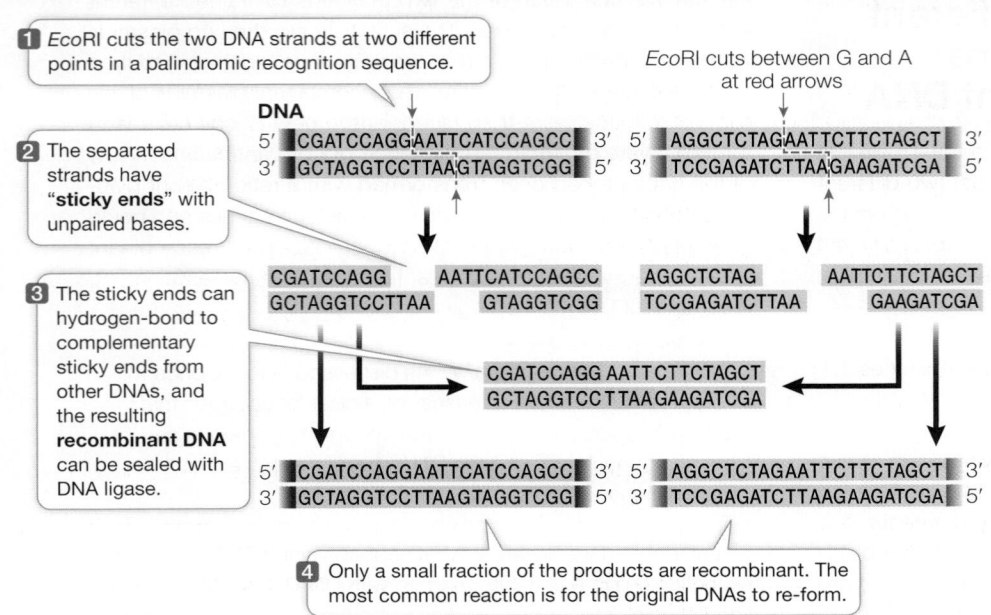

1 *Eco*RI cuts the two DNA strands at two different points in a palindromic recognition sequence.

2 The separated strands have "**sticky ends**" with unpaired bases.

3 The sticky ends can hydrogen-bond to complementary sticky ends from other DNAs, and the resulting **recombinant DNA** can be sealed with DNA ligase.

*Eco*RI cuts between G and A at red arrows

DNA
5′ CGATCCAGGAATTCATCCAGCC 3′   5′ AGGCTCTAGAATTCTTCTAGCT 3′
3′ GCTAGGTCCTTAAGTAGGTCGG 5′   3′ TCCGAGATCTTAAGAAGATCGA 5′

CGATCCAGG   AATTCATCCAGCC   AGGCTCTAG   AATTCTTCTAGCT
GCTAGGTCCTTAA   GTAGGTCGG   TCCGAGATCTTAA   GAAGATCGA

CGATCCAGG AATTCTTCTAGCT
GCTAGGTCC TTAAGAAGATCGA

5′ CGATCCAGGAATTCATCCAGCC 3′   5′ AGGCTCTAGAATTCTTCTAGCT 3′
3′ GCTAGGTCCTTAAGTAGGTCGG 5′   3′ TCCGAGATCTTAAGAAGATCGA 5′

4 Only a small fraction of the products are recombinant. The most common reaction is for the original DNAs to re-form.

Figure 18.2 Cutting, Splicing, and Joining DNA   Some restriction enzymes (*Eco*RI is shown here) make staggered cuts in DNA. *Eco*RI can be used to cut two different DNA molecules (blue and orange). The exposed bases can hydrogen-bond with complementary exposed bases on other DNA fragments, forming recombinant DNA. DNA ligase stabilizes the recombinant molecule by forming covalent bonds in the DNA backbone.

Q: If restriction enzymes make blunt (non-sticky) ends in DNA, how can two DNA molecules be spliced together?

Some restriction enzymes cut the DNA straight through the middle of the palindrome, generating "blunt-ended" fragments. Others, such as *Eco*RI, make staggered cuts—they cut the phosphodiester bond of one strand of the double helix several bases away from where they cut the other (**Focus: Key Figure 18.2**). After *Eco*RI makes its two cuts in the complementary strands, the ends of the strands are held together only by the hydrogen bonds between four base pairs. At warm temperatures (above room temperature) these hydrogen bonds are too weak to hold the two strands together, so the DNA separates into fragments when it is warmed. Each fragment carries a single-stranded "overhang" at the location of each cut. These overhangs are called **sticky ends** because they have specific base sequences that can bind by base pairing with complementary sticky ends.

Any two sticky ends that are complementary can form hydrogen bonds with one another. The sticky ends of the original molecule may rejoin, or two different fragments may join (see Figure 18.2). Indeed, a fragment from one source, such as a human, can be joined to a fragment from another source, such as a bacterium. Initially the fragments are held together only by weak hydrogen bonds, but the enzyme DNA ligase catalyzes the formation of covalent bonds between adjacent nucleotides at the ends of the fragments, joining them to form a single, larger molecule.

Besides *Eco*RI, there are hundreds of other restriction enzymes, each with its own recognition sequence. With these tools—restriction enzymes and DNA ligase—scientists can cut and rejoin different DNA molecules from any sources, including artificially synthesized DNA sequences. Recently, new tools have been developed for making recombinant DNA. Methods based on the polymerase chain reaction (PCR) allow the joining of any two DNA molecules without the need for conveniently placed restriction enzyme sites. Despite these advances, restriction enzymes and DNA ligase are still used routinely for recombinant DNA construction in biology labs.

▶ 18.1 recap

DNA fragments from different sources can be linked together to make recombinant DNA. DNA can be cut with a restriction enzyme and then rejoined with DNA ligase. There are many different restriction enzymes, each with its own recognition sequence.

learning outcomes

You should be able to:

- Show how staggered cuts in DNA made by restriction enzymes lead to the creation of sticky ends.
- Summarize the Cohen and Boyer experiments on creating recombinant DNA.

1. How does a staggered cut in DNA create a sticky end?
2. How did Cohen and Boyer make the first recombinant DNA?

Recombinant DNA has no biological significance until it is inserted inside a living cell, which can replicate and transcribe the transplanted genetic information. How can recombinant DNA made in the laboratory be inserted and expressed in living cells?

▶ key concept 18.2 There Are Several Ways to Insert DNA into Cells

One reason for making recombinant DNA is to **clone**—that is, to produce many identical copies of—a particular gene or other DNA sequence. We have seen the term "clone" used in the context of whole cells or organisms (see Chapters 11 and 12) that are genetically identical to one another. A gene can be cloned by inserting it into a bacterial cell such as *E. coli*. The bacterium is allowed to reproduce and multiply into millions of identical cells, all carrying copies of the gene. Cloning might be done to get enough DNA for sequencing and subsequent analysis, to produce a protein product in quantity, or as a step toward creating an organism with a new phenotype.

focus your learning

- Reporter genes carried by a vector can be used to detect the presence of recombinant DNA in host cells.

Recombinant DNA is cloned by inserting it into host cells in a process known as **transformation**—or **transfection** if the host cells are derived from an animal. (See Key Concept 13.1 for another example of transformation in bacteria.) A host cell or organism that contains recombinant DNA is referred to as a **transgenic** cell or organism, and the non-native DNA is called a transgene. Later in this chapter we will encounter many examples of transgenic organisms, including yeast, rice plants, and even cattle.

Selectable genetic markers are used to identify host cells containing recombinant DNA

Various methods are used to create transgenic cells. Generally these methods are inefficient in that only a few of the cells that are exposed to the recombinant DNA actually become transformed with it. In order to isolate and reproduce only transgenic cells, **selectable marker** genes, such as genes that confer resistance to antibiotics, are often included as part of the recombinant DNA molecule. When an antibiotic resistance gene is used as the selectable marker, the cells from the transformation experiment are grown in the presence of the antibiotic; the antibiotic kills all nontransgenic cells, leaving only the transgenic cells. Antibiotic resistance genes were the markers used in Cohen and Boyer's experiment (see Figure 18.1).

Genes can be inserted into prokaryotic or eukaryotic cells

In theory, any cell or organism can act as a host for the introduction of recombinant DNA. Most research has been done using model organisms:

- *Bacteria* are easily grown and manipulated in the laboratory. Much of their molecular biology is known, especially for well-studied bacteria such as *E. coli*. Furthermore, bacteria contain plasmids, which are easily manipulated to carry recombinant DNA into the cell. But since the processes of transcription, translation, and posttranslational modification proceed differently in prokaryotes than they do in eukaryotes, bacteria are not always well suited as hosts to express eukaryotic genes.

- *Yeasts* such as *Saccharomyces cerevisiae* are commonly used as eukaryotic hosts for recombinant DNA studies. The advantages of using yeasts include rapid cell division (a life cycle completed in 2–4 hours), ease of growth in the laboratory, and a relatively small genome size (see Table 17.2). In addition, yeast cells have most of the characteristics of other eukaryotic cells, with a notable exception being those associated with multicellularity.

- *Plant cells* are good hosts because of their ability to make totipotent *stem cells from mature plant tissues. The unspecialized cells can be transformed with recombinant DNA and then studied in culture, or grown into new plants. There are also methods for making whole transgenic plants without going through the cell culture step. These methods result in plants that carry the recombinant DNA in all their cells, including the germ line cells.

- *Animal cells grown in lab culture* can be used to study expression of human or animal genes, for example for medical purposes. Whole transgenic animals can also be made by inserting new DNA into egg cells.

---

*connect the concepts   As explained in Key Concept 19.5, stem cells occur in both plants and animals and are cells that are continuously dividing, which can specialize if given the right signals.

Inserted DNA is usually integrated into the host chromosome

Cells can be chemically treated to make their outer membranes more permeable, and then mixed with the DNA so that it can diffuse into the cells. Another approach is called electroporation: a short electric shock is used to create temporary pores in the membranes through which the DNA can enter. Viruses can be altered so that they carry recombinant DNA into cells. A common method for transforming plants involves a specific bacterium that inserts DNA into plant cells. Transgenic animals can be produced by injecting recombinant DNA into the nuclei of fertilized eggs. There are even "gene guns" that "shoot" the host cells with tiny metal particles coated with the DNA.

The challenge of inserting new DNA into a cell lies not only in getting it into the host cell, but also in getting it to replicate as the host cell divides. As you saw in Key Concept 13.3, DNA polymerase does not bind to and copy just any sequence. If the new DNA is to be replicated, it must become part of a segment of DNA that contains an origin of replication. Such a DNA molecule is called a **replicon**, or replication unit.

There are two general ways in which the newly introduced DNA can become part of a replicon within the host cell:

1. It can be spliced into a host chromosome directly.

2. It can enter the host cell as part of a carrier DNA sequence, called a **vector**, and can then either integrate into a host chromosome or be replicated from its own origin of replication.

Several types of vectors are used to get DNA into cells, some of which we will now describe in more detail.

PLASMIDS AS VECTORS   As we described in Chapter 12, plasmids are small, circular DNA molecules that replicate autonomously in many prokaryotic cells. Several characteristics make plasmids useful as transformation vectors:

- Plasmids are relatively small (an *E. coli* plasmid usually has 2,000–6,000 base pairs) and are therefore easy to manipulate in the laboratory.

- A plasmid often has one or more restriction enzyme recognition sequences that each occur only once in the plasmid sequence. This makes it easy to insert new DNA into the prokaryotic plasmid before it is used to transform host cells.

- Many plasmids contain genes that confer resistance to antibiotics, which can serve as selectable markers.

- Plasmids have a bacterial origin of replication (*ori*) and can replicate independently of the host chromosome. It is not uncommon for a bacterial cell to contain hundreds of copies of a recombinant plasmid. For this reason, the power of bacterial transformation to amplify a gene is extraordinary. A 1-liter culture of bacteria harboring the human β-globin gene in a typical plasmid has as many copies of that gene as there are cells in a typical adult human (10^{14}).

A plasmid used as a vector in the laboratory has been extensively altered to include convenient features:

- Multiple cloning sites, often with 20 or more unique restriction enzyme sites for cloning purposes (see *Hind*III, *Bam*HI, *Sal*I, and *Pst*I in the diagram below)

- An origin of replication that will function in a particular type of host cell (see *ori* in the diagram below)

- One or more reporter genes, such as selectable marker genes (see ampicillin resistance and tetracycline resistance in the diagram below)

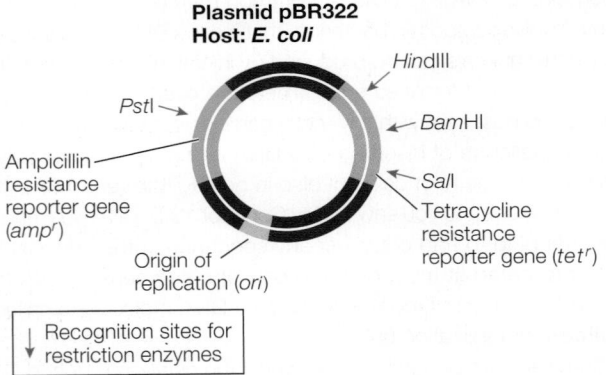

Plasmid pBR322
Host: *E. coli*

*Hind*III
*Pst*I
*Bam*HI
Ampicillin resistance reporter gene (*amp*r)
*Sal*I
Tetracycline resistance reporter gene (*tet*r)
Origin of replication (*ori*)

↓ Recognition sites for restriction enzymes

PLASMID VECTORS FOR PLANTS   An important vector for carrying new DNA into many types of plants is a plasmid found in the bacterium *Agrobacterium tumefaciens*. This bacterium lives in the soil, infects plants, and causes a disease called crown gall, which is characterized by the presence of growths (or tumors) on the plant. *Agrobacterium* contains a plasmid called Ti (for *tumor-inducing*). When the bacterium infects a plant cell, a region of the Ti plasmid called the T DNA is inserted into the cell, where it becomes incorporated into one of the plant's chromosomes. The Ti plasmid carries the genes needed for this transfer and incorporation of the T DNA. The T DNA carries genes that are expressed by the host cell, causing the growth of tumors and the production of specific sugars that the bacterium uses as sources of energy. Scientists have exploited this remarkable natural "genetic engineer" to insert foreign DNA into the genomes of plants.

When used as a vector for plant transformation, the tumor-inducing and sugar-producing genes on the T DNA are removed and replaced with foreign DNA. The recombinant Ti plasmids are first used to transform *Agrobacterium* cells from which the original Ti plasmids have been removed. Then the *Agrobacterium* cells are used to infect plant cells.

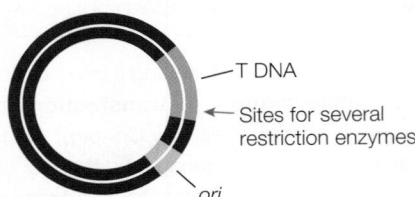

Ti plasmid
Hosts: *Agrobacterium tumefaciens* (plasmid) and infected plants

T DNA
Sites for several restriction enzymes
ori

VIRUSES AS VECTORS   Constraints on plasmid replication limit the size of the new DNA that can be inserted into a plasmid to about 10,000 base pairs. Although many prokaryotic genes may be smaller than this, most eukaryotic genes—with their introns and extensive flanking sequences—are bigger. A vector that accommodates larger DNA inserts is needed for these genes.

Both prokaryotic and eukaryotic viruses are used as vectors for eukaryotic DNA. Bacteriophage λ, which infects *E. coli*, has a DNA genome of about 45,000 base pairs. About 20,000 base pairs are not necessary for the bacteriophage to complete its life cycle (see Figure 16.12). These 20,000 base pairs can be deleted and replaced with DNA from another organism, which then gets replicated along with the virus DNA. Because viruses infect cells naturally, they offer a great advantage over plasmids, which often require artificial means to coax them to enter host cells.

Reporter genes help select or identify host cells containing recombinant DNA

Even when a population of host cells is exposed to an appropriate vector, only a small proportion of the cells actually take up the vector. Furthermore, the process of making recombinant DNA is far from perfect. During a ligation reaction, DNA molecules can combine in various ways, many of which do not produce the desired recombinant molecule (see Figures 18.1 and 18.2). Methods have been developed to improve the chance that a desired combination will occur. A simple approach is to cut the vector with two different restriction enzymes that have sites near each other. This produces a molecule with incompatible sticky ends, which is much less likely to simply recircularize during the ligation reaction. The desired insert molecule is cut with the same two enzymes, so that theoretically a functional circular plasmid can be produced only if the vector and insert ligate with one another. Even so, it is often the case that only a small proportion of the ligation products have the desired recombinant sequence.

How can we identify or select the host cells that contain the desired sequence? One way is to use selectable markers, such as those for antibiotic resistance used in early experiments with recombinant DNA (see Figure 18.1). Selectable markers are one type of **reporter gene**, which is any gene whose expression is easily observed. Here are several types of reporter genes:

- As described earlier, an antibiotic resistance gene in a plasmid or other vector allows the detection of transformed host cells grown in the presence of the selected antibiotic. If the host cells are normally sensitive to the antibiotic, they will grow on medium containing the antibiotic only if they have been transformed by the vector. This approach is used in the selection of transgenic prokaryotic and eukaryotic cells, including those of plants and animals.

- The β-galactosidase (*lacZ*) gene in the *E. coli lac* operon (see Figure 16.4) codes for an enzyme that can convert the artificial substrate X-Gal into a bright blue product. Many plasmids contain the *lacZ* gene with a multiple cloning site (i.e., multiple unique restriction enzyme sites where target DNA sequences can be inserted) within its sequence. These plasmids also carry genes for antibiotic resistance. Bacterial colonies containing the plasmid are selected on a solid medium containing the antibiotic. X-Gal is also included in the medium. If a bacterial colony contains a recombinant plasmid that carries the foreign DNA inserted into the *lacZ* gene, it will not make β-galactosidase and the colony will be white. Clones that contain the original plasmid with no insert express the *lacZ* gene and make blue colonies (**Figure 18.3**).

- Green fluorescent protein (GFP), which normally occurs in the jellyfish *Aequorea victoria*, emits visible green light when exposed to ultraviolet light. The gene for this protein has been isolated and incorporated into vectors. GFP is now widely used as a reporter gene (**Figure 18.4**). GFP has also been modified to emit other colors when exposed to ultraviolet light, and these new variants are widely used by molecular biologists.

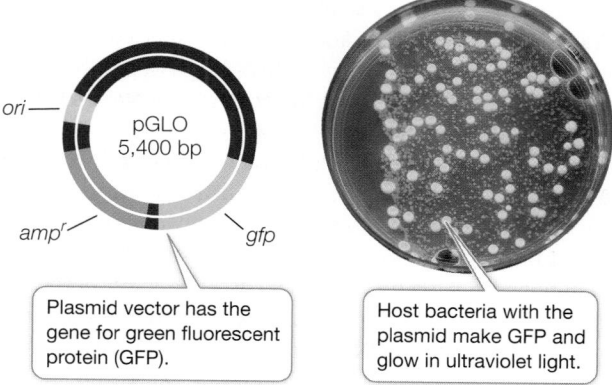

Figure 18.4 Green Fluorescent Protein as a Reporter   The presence of a plasmid with the gene for green fluorescent protein (GFP) is readily apparent in transgenic cells because the GFP that the cells produce glows under ultraviolet light. This allows the identification of cells carrying a plasmid without the use of selection on antibiotics.

Q: Are cells killed during GFP selection? Why is this important?

Figure 18.3 Selection for Recombinant DNA   Selectable marker (reporter) genes are used by biologists to select for bacteria that have taken up a plasmid. In a typical experiment, most of the bacteria will not take up any DNA. Of those that do, only a small fraction will take up recombinant DNA.

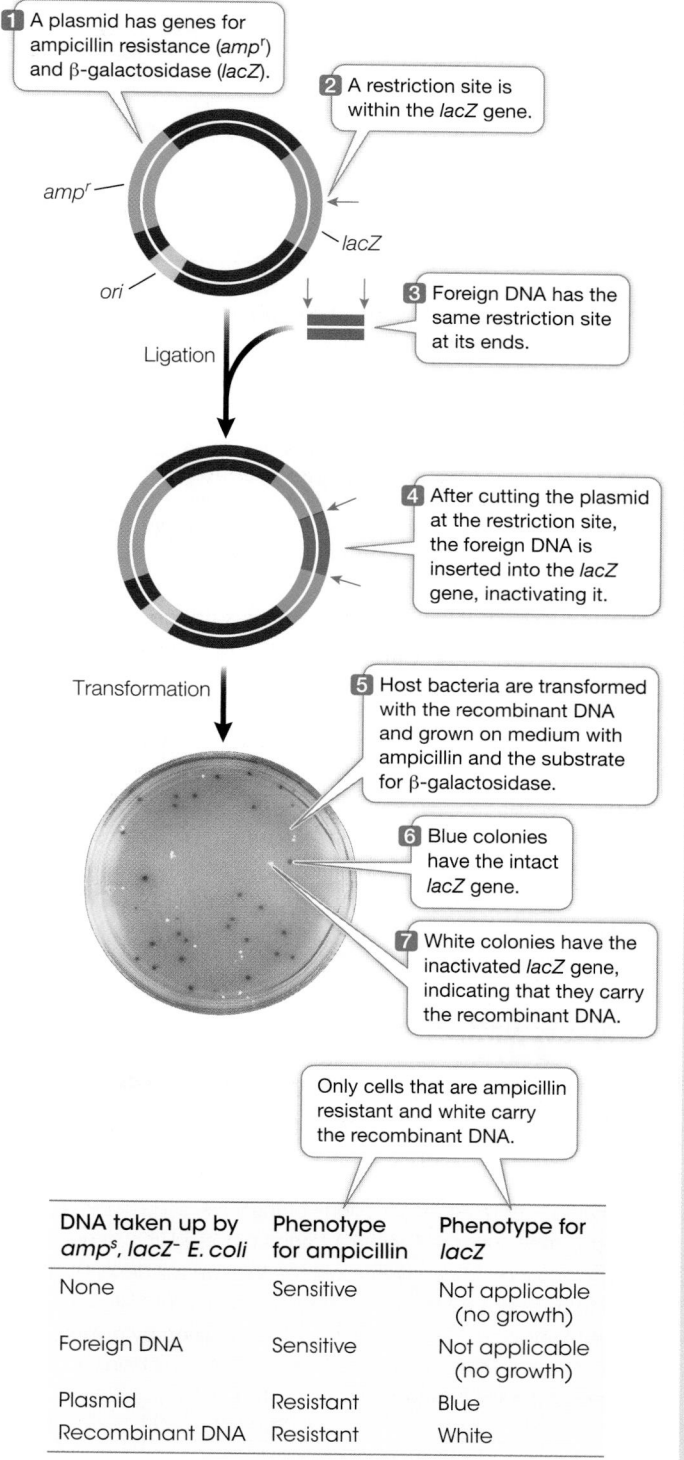

DNA taken up by amp^s^, lacZ⁻ E. coli	Phenotype for ampicillin	Phenotype for lacZ
None	Sensitive	Not applicable (no growth)
Foreign DNA	Sensitive	Not applicable (no growth)
Plasmid	Resistant	Blue
Recombinant DNA	Resistant	White

18.2 recap

Recombinant DNA can be cloned by using a vector to insert it into a suitable host cell. The vector often has a selectable marker or other reporter gene that gives the host cell a phenotype by which transgenic cells can be identified.

learning outcomes

You should be able to:

- Compare and contrast different types of vectors and reporter genes.
- Outline the steps involved in creating an organism or cell with recombinant DNA, including detection of the cells containing that DNA.

1. Compare the use of a GFP reporter gene with that of an antibiotic resistance gene in detecting cells harboring a foreign DNA sequence.

2. Assume you are using a plasmid vector that contains genes encoding ampicillin resistance and the green fluorescent protein, with a restriction site in the latter gene only. Arrange a–e below to reflect the sequence of steps for inserting a piece of foreign DNA into this plasmid, introducing the recombinant plasmid into bacteria, and verifying that the plasmid and the foreign gene are both present in the bacteria.

    a. Transform host cells.

    b. Select colonies for antibiotic resistance.

    c. Select colonies that do not glow under ultraviolet light.

    d. Digest vector and foreign DNA with a restriction enzyme.

    e. Ligate the digested plasmid together with the foreign DNA.

We have described how DNA can be cut with restriction enzymes, inserted into a vector, and introduced into host cells. We have also seen how host cells carrying recombinant DNA can be identified. Now let's consider where the genes or DNA fragments used in these procedures come from.

key concept 18.3 Any Sequence of DNA Can Be Used for Cloning

A major goal of molecular cloning experiments is to learn the functions of the DNA sequences and the proteins they encode. The DNA fragments used in cloning procedures are obtained from several sources that we will describe in this section: random fragments of chromosomes that are maintained as gene libraries, complementary DNA obtained by reverse transcription from mRNA, products of the polymerase chain reaction (PCR), and artificially synthesized or mutated DNA.

focus your learning

- A cDNA library represents the RNAs made by a particular cell or tissue type at a particular point in time.

- Reverse transcriptase can be used in PCR procedures to create and amplify a specific cDNA sequence such that construction of a library is not necessary.
- Artificial DNA can be synthesized using organic chemistry.

DNA for cloning can come from a library

A **genomic library** is a collection of DNA fragments that make up the genome of an organism. Restriction enzyme digestion or other means, such as mechanical shearing, can be used to break chromosomes into smaller pieces. These smaller DNA fragments collectively still constitute a genome (**Figure 18.5A**), but the information is now in many smaller "volumes." Any desired fragment can be inserted into a vector, which is then taken up by a host cell. Proliferation of a single such transformed cell produces a colony of cells, each of which harbors many copies of the same fragment of DNA.

If a plasmid vector is used, about 700,000 separate fragments are required to make a library of the human genome. By using bacteriophage λ, which can carry about four times as much DNA as a typical plasmid, the number of "volumes" in the library can be reduced to about 160,000. Although this still seems like a large number, a single petri plate can hold thousands of phage colonies, or plaques, which are easily screened for the presence of a particular DNA sequence by *hybridization to an appropriate nucleic acid probe.

---

*connect the concepts The technique of nucleic acid hybridization, illustrated in Figure 14.6, involves using a single-stranded nucleic acid probe with a complementary sequence to the target DNA. If the probe forms a double-stranded hybrid molecule, the target DNA fragment has been identified.

---

cDNA is made from mRNA transcripts

A much smaller DNA library—one that includes only the genes transcribed in a particular tissue at a particular point in time—can be made from **complementary DNA**, or **cDNA** (**Figure 18.5B**). Making cDNA involves isolating mRNA from cells, then making cDNA copies of that mRNA by complementary base pairing in a process called **reverse transcription**. Reverse transcription is catalyzed by the enzyme **reverse transcriptase.**

A collection of cDNAs from a particular tissue at a particular time in the life cycle of an organism is called a **cDNA library**. Because mRNAs do not last long in the cytoplasm, the types and amounts of mRNAs present in a cell are a good representation of the transcription rates of all the genes in that cell. So a cDNA library is a "snapshot" that captures the transcription pattern of a set of cells at a given point in time. cDNA libraries have been invaluable for comparing gene expression in different tissues at different stages of development. For example, researchers have found that up to one-third of all the genes of an animal are expressed only during development. cDNA is also a good starting point for cloning eukaryotic genes, because the clones contain only the coding sequences of the genes (the introns having been spliced out; see Figure 14.7). Also, if a eukaryotic gene is highly expressed in a particular tissue, a cDNA library made from that tissue will be enriched for the gene, making it easier to identify and clone the gene.

Reverse transcriptase can be used in *PCR procedures to create and amplify a specific cDNA sequence so that it is not necessary to construct a library. In this case, RNA is isolated from an organism or tissue, and reverse transcriptase is used to make cDNA from the RNA. Then PCR is used to amplify a specific sequence directly from the sample of cDNA. This procedure, called **RT-PCR**, has become an invaluable tool for studying the expression of particular genes in cells and organisms.

---

*connect the concepts As illustrated in Figure 13.19, the polymerase chain reaction (PCR) automates replication of a DNA fragment in a test tube. Also called DNA amplification, the process involves repeatedly synthesizing DNA complementary to a target sequence to make enough DNA for chemical analysis and genetic manipulations.

Synthetic DNA can be made by PCR or by organic chemistry

PCR can begin with as little as 10^{-12} g of DNA (a picogram). Any fragment of DNA can be amplified by PCR as long as appropriate primers are available. DNA replication (by PCR or in a cell) requires not just a template onto which DNA polymerase adds complementary nucleotides, but also a short oligonucleotide primer where replication begins (see Figure 13.11). If the appropriate primers (two are needed—one for each strand of DNA) are added to template DNA in a PCR reaction, millions of copies of the DNA region between the primers can be produced in just a few hours. This amplified DNA can then be inserted into a vector to create recombinant DNA and be cloned in host cells.

Artificial DNA can be synthesized using organic chemistry to link nucleotides together in a specified sequence. This process is now fully automated, and a laboratory can make large numbers of short- to medium-length sequences overnight. PCR primers, for example, are made in this way. The synthesis of *artificial DNA does not require a template, so DNA with any sequence can be made. This flexibility is useful for creating DNA fragments with desirable characteristics, such as convenient restriction sites or specific mutations. Longer synthetic sequences can be pieced together to construct completely artificial genes that have been designed for specific purposes. For example, a gene might be designed to be highly expressed in a particular cell type, or to encode a highly active enzyme.

---

*connect the concepts Plasmids containing synthetic transposon DNA have been used to transfect host cells to inactivate genes one by one and give insight into the minimal genome. See Key Concept 17.2.

research tools

Figure 18.5 Constructing Libraries Intact genomic DNA is too large to be introduced into host cells. **(A)** A genomic library can be made by breaking the DNA into small fragments, incorporating the fragments into a vector, and then transforming host cells with the recombinant vectors. Each colony of cells contains many copies of a small part of the genome. **(B)** The many mRNAs in a cell can be copied into cDNAs and a library made from them. The DNA in these colonies can then be isolated for analysis.

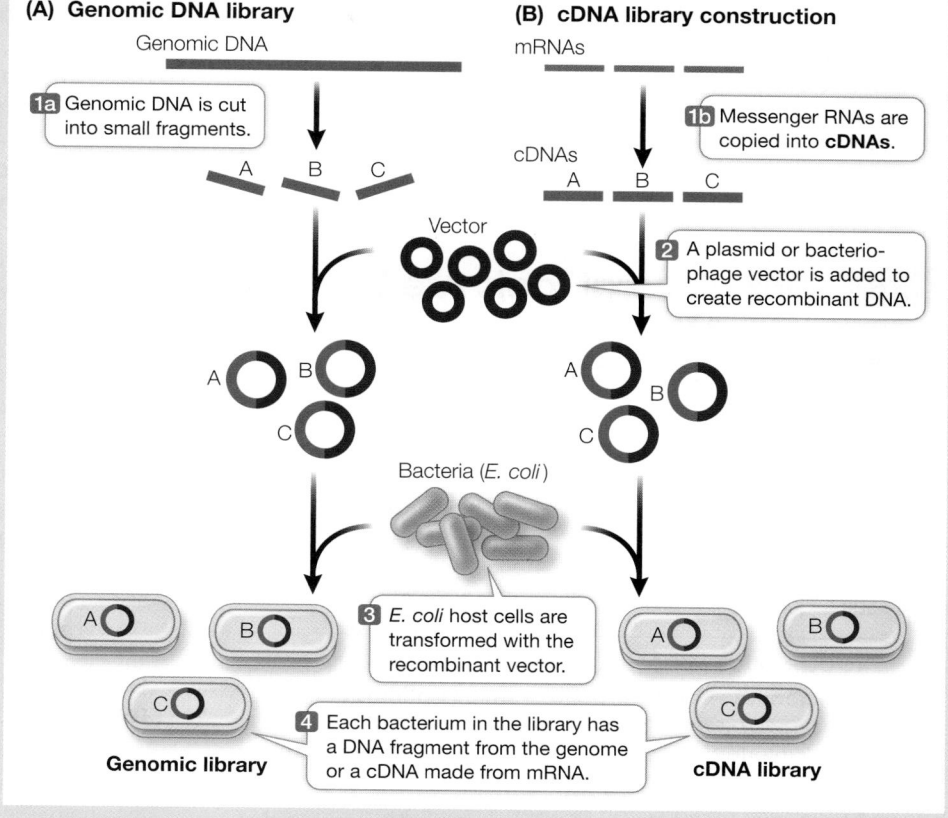

(A) Genomic DNA library

Genomic DNA

1a Genomic DNA is cut into small fragments.

A B C

Vector

A B
C

(B) cDNA library construction

mRNAs

1b Messenger RNAs are copied into **cDNAs**.

cDNAs
A B C

2 A plasmid or bacteriophage vector is added to create recombinant DNA.

A B
C

Bacteria (E. coli)

3 E. coli host cells are transformed with the recombinant vector.

A B
C

Genomic library

4 Each bacterium in the library has a DNA fragment from the genome or a cDNA made from mRNA.

A B
C

cDNA library

18.3 recap

DNA for cloning can be obtained from genomic libraries, cDNA, or artificially synthesized DNA fragments.

learning outcomes

You should be able to:

- Determine which type of library would be most appropriate to construct for a given scenario.
- Predict the expected results in terms of quantities and sequences of cDNA libraries taken from different tissues.
- Show how RT-PCR is used to amplify a specific gene sequence.
- Summarize uses of synthetic DNA.

1. What type of gene library would you use for analyzing each of the following, and why?
   a. The expression of a gene encoding amylase, a starch hydrolyzing enzyme, in seeds as they germinate
   b. The sequence of the amylase gene in two different genetic strains of corn

(continued)

We've explored the various sources of DNA that can be used to make recombinant DNA molecules, and how organisms are transformed with recombinant DNA. We will now turn to some of the ways that recombinant DNA and transformation methods can be used to study the functions of genes and proteins.

key concept 18.4 Several Tools Are Used to Modify DNA and Study Its Function

So far in this chapter we have seen how recombinant DNA is made and how organisms are transformed with recombinant DNA. In this section we will examine several additional techniques for studying DNA. These approaches include expression of genes in different biological systems, mutagenesis, methods to block gene expression, and DNA microarrays to analyze large numbers of nucleotide sequences.

focus your learning

- Microarrays can be used to examine patterns of gene expression simultaneously across the genome.

- RNA interference can be used to block specific gene expression, usually at the translational level.

- CRISPR technology can be used to alter and inactivate specific genes.

Gene expression can be modulated by DNA technology

DNA can be manipulated to be expressed in a different location or at a different rate. Researchers have learned a lot about gene expression and function by exploring questions such as these:

- *What happens when a gene is moved to a new location?* A scientist may want to study a gene's function and how it interacts with other genes by expressing the gene in a different organism—for example, by inserting a human gene into a bacterium, or a bacterial gene into a plant. Researchers have learned that expression of a transplanted gene requires a promoter and other regulatory sequences from the host organism: a bacterial promoter will not function in a plant cell, for example. The coding region of the gene of interest needs to be inserted between a promoter and a transcription termination sequence derived from the host organism, or from an organism that uses similar mechanisms for gene regulation.

- *What happens when the rate of expression of a gene is changed?* Researchers have learned that a gene can be overexpressed in cells where it is normally expressed at lower levels, so that much more of its protein product is made. Overexpression is achieved by inserting a gene adjacent to a promoter that is very active in a cell, and then cloning the recombinant promoter–gene combination in a host cell.

Many thousands of experiments have shed light on the functions of genes and their protein products. One example involves a genetic system that probably evolved to prevent inbreeding in plants. Most plants produce flowers with both male and female parts, but many plant species are self-incompatible; they cannot self-pollinate (see Key Concepts 28.3 and 37.1). Their flowers produce a protein that recognizes "self" pollen, and prevents the pollen from fertilizing egg cells in the same flower (see Figure 37.4). Genetic crosses suggested that a particular multi-allelic gene, called the *S* gene, was responsible for self-incompatibility. Definitive proof was obtained when plants were transformed with recombinant DNA containing an *S* allele different from their own *S* alleles. The transgenic plants rejected not only their own pollen but also pollen from flowers that naturally carried the foreign *S* allele.

DNA mutations can be created in the laboratory

Mutations that occur in nature have been important in demonstrating cause-and-effect relationships for a specific gene. However, mutations in nature are rare events. DNA technology allows us to explore the consequences of various mutations by creating them artificially. Synthetic DNA can be made with any desired sequence, and its effect on the organism can be observed when the mutant DNA is expressed in host cells.

Genes can be inactivated and changed by CRISPR technology

As we have just discussed, a gene can be studied by expressing it in cells where it is not normally expressed or by expressing it at higher than normal levels. Another way to study a gene is to inactivate it, so that it is not transcribed and translated into a functional protein, or to change its DNA sequence so that an altered gene product is made. CRISPR technology offers a powerful new approach for "gene editing" (adding, disrupting, or changing the sequence of specific genes) and gene regulation (**Figure 18.6**). As with many other tools of biotechnology, the method emulates a mechanism found in nature, one that some bacteria and many archaea use to defend themselves against viral infections. The genomes of these species have many short (24–48 bp) palindromic sequences of DNA (see Key Concept 18.1), with unique 24-bp spacer DNA in between, which are known by the acronym **CRISPR** (clustered regularly interspaced short palindromic repeat sequences). While the CRISPR sequence is the same for each repeating unit, the spacers are different. Each spacer is a fragment of DNA from a virus that previously infected the cell but did not end up destroying its host. The spacer sequences thus provide "genetic mug shots" that inform the cell carrying them what viruses to look out for. When a virus carrying a sequence similar to the spacer DNA injects its DNA into the cell, the unit—CRISPR and spacer—is transcribed into RNA, to which a second short RNA attaches. This RNA complex then performs the following two functions:

1. Because the spacer sequence is complementary to part of the viral genome, it attaches to it by base pairing.

2. The complex binds a protein called CAS9, which is a nuclease that cuts the invading viral DNA, destroying its ability to produce new viruses and thereby protecting the host cell.

Therefore any cell that was once infected by a virus has the "memory" of that infection encoded in its genome, and those of all cells that are produced from it.

Once this natural immune system was described, it occurred to Emmanuelle Charpentier at Umeå University in Sweden and Jennifer Doudna at the University of California, Berkeley that it might be adapted to cut any DNA at a specific location: all that was needed was the RNA complementary to the target (a "guide" RNA) and the CAS9 enzyme. Tens of thousands of unique guide RNAs are now available commercially, each targeted to a specific gene in a specific organism, prokaryotic or eukaryotic, making possible a diversity of studies of inactivated genes and exploration of a multitude of "What if?" questions.

The simplicity of CRISPR technology to inactivate genes holds great promise. To give just one early example, a fungal disease called powdery mildew can devastate a wheat crop. Many wheat varieties have a gene whose expression inhibits wheat's natural defenses against this disease, so the only way to control the disease is through heavy application of fungicides. Scientists in China used CRISPR technology to inactivate this gene, so that natural resistance is favored.

CRISPR technology can be used to create specific mutations. When CAS9 cuts target DNA in a living cell, the cell attempts to repair the damage (see Figure 13.18). Often this repair is imperfect, so that the rejoined DNA has a mutation. Biologists can specify the mutation by synthesizing a short DNA sequence complementary to the region at the break so that the new sequence will have a specific alteration.

Perhaps more dramatic than the ability to induce mutations is the possibility of correcting mutations, particularly those that cause disease. For example, the disease cystic fibrosis is caused by mutation of a membrane transport protein. Scientists in the Netherlands have used CRISPR to correct the mutation in cells from patients with this disease. While this is not a cure (the corrected gene would have to function in many tissues), the possibility arises that the mutated gene could be corrected in germ line cells, such as those of an early embryo. However, the ease with which the human germ line genome can be altered using CRISPR technology has elicited ethical and legal concerns. In some countries altering the germ line in a human is illegal.

Complementary RNA can prevent the expression of specific genes

Another way to study the expression of a specific gene is to knock it out by blocking the translation of its mRNA. The process developed for knocking out genes is yet another example of scientists imitating nature. As described in Key Concept 16.5, gene expression is sometimes regulated by the production of double-stranded RNA molecules that are processed to produce short, single-stranded microRNA (miRNA). These miRNAs are complementary to specific mRNA sequences (see Figure 16.18A), and when they bind to their target mRNAs, they inhibit translation. The hybrid mRNA molecules consisting of miRNA and the target mRNA tend to break down rapidly in the cytoplasm. Although the target gene continues to be transcribed, translation does not take place. Scientists have applied the idea of hybrid RNAs in developing methods for blocking the expression of particular genes (**Figure 18.7**). An organism can be transformed to produce mRNAs that are

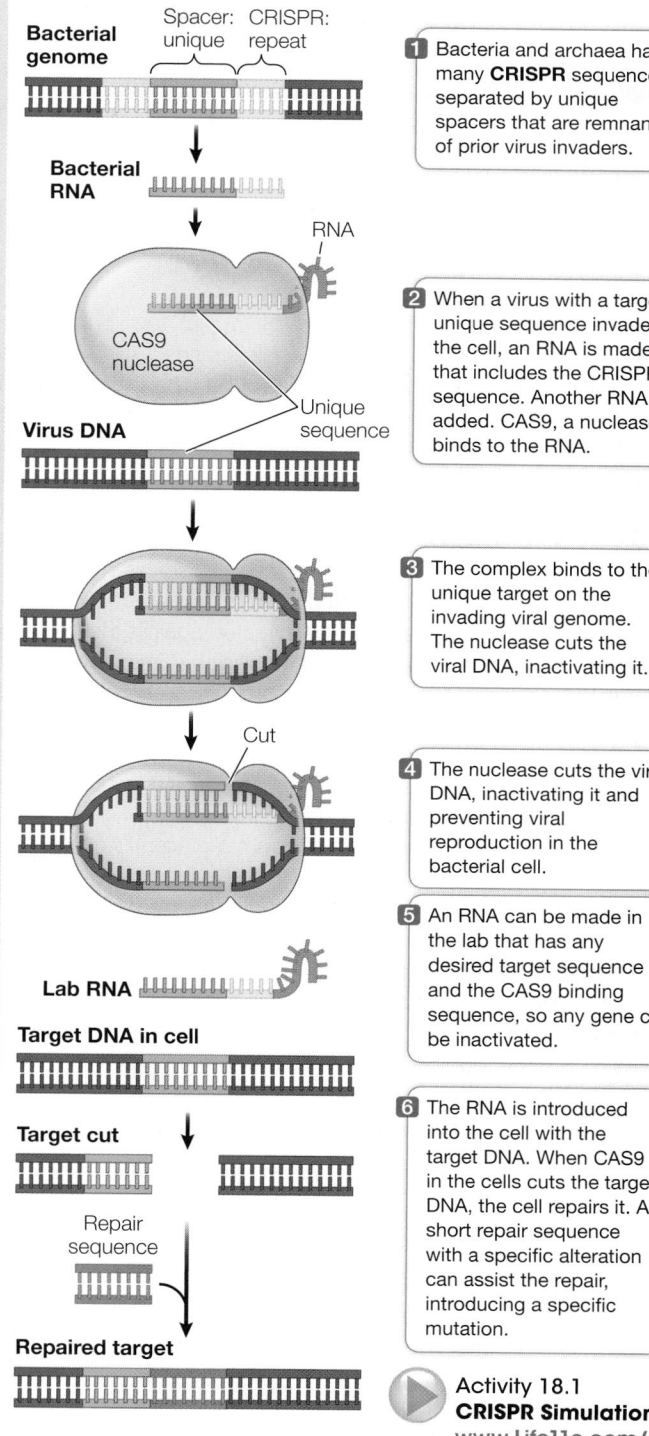

Figure 18.6 Inactivating or Mutating a Gene by CRISPR
A molecular system used by bacteria to combat invading viruses has been adapted to inactivate or mutate any gene.

1 Bacteria and archaea have many **CRISPR** sequences separated by unique spacers that are remnants of prior virus invaders.

2 When a virus with a target unique sequence invades the cell, an RNA is made that includes the CRISPR sequence. Another RNA is added. CAS9, a nuclease, binds to the RNA.

3 The complex binds to the unique target on the invading viral genome. The nuclease cuts the viral DNA, inactivating it.

4 The nuclease cuts the viral DNA, inactivating it and preventing viral reproduction in the bacterial cell.

5 An RNA can be made in the lab that has any desired target sequence and the CAS9 binding sequence, so any gene can be inactivated.

6 The RNA is introduced into the cell with the target DNA. When CAS9 in the cells cuts the target DNA, the cell repairs it. A short repair sequence with a specific alteration can assist the repair, introducing a specific mutation.

Activity 18.1
CRISPR Simulation
www.Life11e.com/ac18.1

complementary to and thus bind to specific endogenous mRNAs, which prevents expression of the endogenous mRNA. Alternatively, cells can be injected with synthetic complementary sequences (see

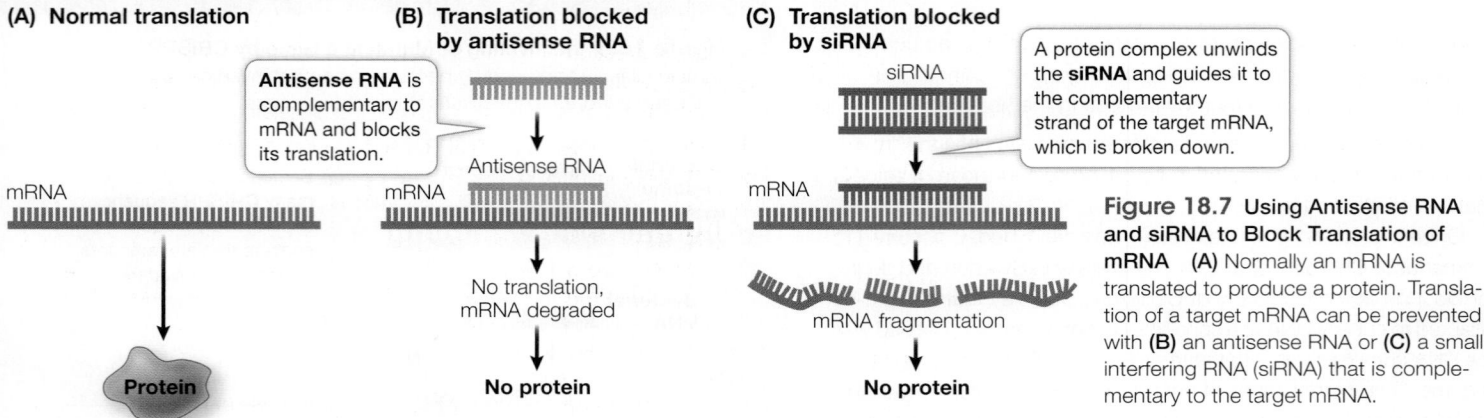

Figure 18.7 Using Antisense RNA and siRNA to Block Translation of mRNA (A) Normally an mRNA is translated to produce a protein. Translation of a target mRNA can be prevented with (B) an antisense RNA or (C) a small interfering RNA (siRNA) that is complementary to the target mRNA.

Figure 18.7B). The complementary single-stranded RNA sequence is called **antisense RNA** because it binds by base pairing to the "sense" bases on the mRNA.

It may be possible to make antisense drugs to treat diseases that involve gene expression. For example, people with the genetic disease familial hypercholesterolemia are unable to break down cholesterol in the blood and accumulate fat deposits in blood vessels (as was the case with Janet whose story opened this chapter). Their excess cholesterol is transported in blood as part of a lipoprotein. Mipomersen is an antisense RNA directed against the translation of the mRNA for the "protein" part of this lipoprotein. When people with hypercholesterolemia take this drug, the level of the lipoprotein is reduced and so is the level of cholesterol in the bloodstream. Fat accumulation in blood vessels is lowered.

A technique related to antisense RNA takes advantage of **RNA interference** (**RNAi**), a natural mechanism for inhibiting mRNA translation that involves small interfering RNA (siRNA; see Figure 16.18B). siRNAs bind to complementary regions on their target mRNAs, which are then degraded. Since the discovery of RNAi in the late 1990s, scientists have synthesized artificial double-stranded siRNAs to inhibit the expression of known genes (see Figure 18.7C). Because these double-stranded siRNAs are more stable than antisense RNAs, the use of siRNAs is the preferred approach for blocking translation. An RNAi-based therapy has been developed to treat macular degeneration, an eye disease that results in near-blindness when blood vessels proliferate in the eye. The signaling molecule that stimulates vessel proliferation is a growth factor. An siRNA that targets this growth factor's mRNA shows promise in stopping and even reversing the progress of the disease. Although medical applications for RNAi are mostly at the experimental stage, antisense RNA and RNAi have been widely used to test cause-and-effect relationships in biological research.

DNA microarrays reveal RNA expression patterns

The science of genomics faces two major quantitative challenges. First, there are very large numbers of genes in eukaryotic genomes. Second, there are myriad distinct patterns of gene expression in different tissues at different times. For example, the cells of a skin cancer at its early stage may have a unique mRNA "fingerprint" that differs from those of normal skin cells and cells from a more advanced skin cancer. In such a case, the pattern of gene expression could provide invaluable information to a clinician trying to characterize a patient's tumor (see below).

To identify distinct patterns of gene expression, scientists could isolate mRNA from cells and measure the amount of each gene's mRNA one gene at a time by hybridization or RT-PCR. But that would involve many steps and take a very long time. It is far simpler to do these hybridizations all in one step, and that is made possible with **DNA microarray** technology.

DNA microarrays ("gene chips") contain a series of DNA sequences attached to a glass slide in a precise order. The slide is

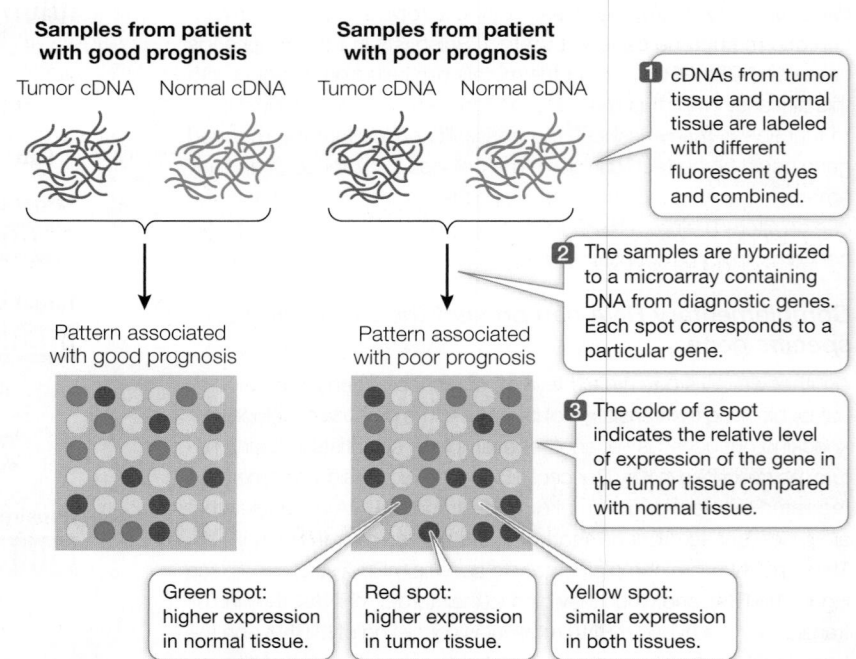

Figure 18.8 Using DNA Microarrays for Clinical Decision-Making The pattern of expression of many genes in tumor tissues (the pattern of colored spots) indicates whether breast cancer is likely to recur. Actual arrays have more spots than shown here.

 **Animation 18.1 DNA Microarray Technology**
www.Life11e.com/a18.1

divided into a grid of microscopic spots, or "wells." Each spot contains thousands of copies of a particular oligonucleotide of 20 or more bases. A computer controls the addition of these oligonucleotide sequences in a predetermined pattern. Each oligonucleotide can hybridize with only one DNA or RNA sequence and thus is a unique identifier of a gene. Many thousands of different oligonucleotides can be placed on a single microarray to detect genetic variation. For example, DNA with a mutant allele may not hybridize to a target wild-type gene fragment on an array. Since arrays can contain thousands of gene fragments, mutations in many genes can be screened for at the same time. With different patterns of gene expression, differences in mRNAs in tissues such a breast cancers can be detected by microarrays (**Figure 18.8**).

> ## 18.4 recap
>
> Researchers can study the function of a gene by expressing it in cells where it is not normally expressed, by overexpressing it, or by inactivating the gene in a living organism. CRISPR technology allows selective inactivation or alteration of gene sequences. Antisense RNAs and siRNAs prevent gene expression by selectively blocking mRNA translation. DNA microarrays allow the simultaneous analysis of many different mRNA transcripts.
>
> ### learning outcomes
>
> You should be able to:
>
> - Outline experiments that use RNA interference to develop new drugs for diseases.
> - Describe how microarrays are used to examine patterns of gene expression
>
> 1. In some bladder cancers, the gene for the cell signaling molecules, *Ras*, is mutated, resulting in an altered protein that stimulates cell division. How might siRNA technology be used to develop a drug to treat this disease? Would the drug be specific for the cancer?
> 2. As pea seeds germinate, the embryo changes which genes are expressed. How would you use microarrays to analyze this?
> 3. Both antisense RNA and CRISPR technology can be used to block the expression of specific genes. What is the additional advantage of CRISPR?

Now that you've seen how DNA can be fragmented, recombined, manipulated, and put back into living organisms, let's see some examples of how these techniques are used to make useful products.

key concept 18.5   DNA Can Be Manipulated for Human Benefit

Biotechnology is the use of cells or whole living organisms to produce materials useful to people, such as foods, medicines, and chemicals. People have been using various forms of biotechnology for a long time. For example, the use of yeasts to brew beer and wine dates back at least 8,000 years, and the use of bacterial cultures to make cheese and yogurt is a technique many centuries old. For a long time people exploited these biochemical transformations without being aware of the organisms and genes involved. Today with recombinant DNA techniques, almost any gene can be inserted into bacteria, yeasts, mammalian cells, plant cells, or even whole organisms, inducing them to make a gene product in large amounts.

focus your learning

- Expression vectors are constructed to ensure the expression of foreign genes in host cells.
- Modern recombinant DNA technology has significant advantages over traditional breeding methods used in agriculture.
- The use of biotechnology has ethical and environmental ramifications.

Cells can be turned into factories for a desired protein

If a eukaryotic gene is inserted into a typical bacterial plasmid and used to transform *E. coli*, little if any of the gene product will be made unless other key prokaryotic DNA sequences are included with the gene. A bacterial promoter, a signal for transcription termination, and a special sequence that is necessary for binding of bacterial ribosomes to the mRNA must all be included in the transformation vector if the gene is to be expressed in a bacterial cell.

To ensure expression of foreign genes by host cells, scientists make **expression vectors** having not only the characteristics of cloning vectors, but the extra sequences needed for the foreign gene (also called a transgene) to be expressed in the host cell. The insertion of an expression vector for a bacterial host is illustrated in **Figure 18.9**. An expression vector for a eukaryotic host would require the poly A–addition sequence for mRNA stability and a promoter that contains all the elements needed for expression in a eukaryotic cell.

An expression vector can be designed to deliver transgenes to any class of prokaryotic or eukaryotic host and may include additional features:

- An *inducible promoter*, which responds to a specific signal, can be included. For example, a promoter that responds to hormonal stimulation can be used so that the transgene will be expressed at high levels only when the hormone is added.

- A *tissue-specific promoter*, which is expressed only in a certain tissue at a certain time, can be used if localized expression is desired. For example, many seed proteins are expressed only in the plant embryo. Coupling a transgene to a seed-specific promoter will allow the gene to be expressed only in seeds.

- *Signal sequences* can be added so that the gene product is directed to an appropriate destination. For example, when a protein is made by yeast or bacterial cells in a liquid medium, it is economical to include a signal directing the protein to be secreted into the extracellular medium for easier recovery.

Medically useful proteins can be made using biotechnology

Biotechnology produces many medically useful products (**Table 18.1**), and more are in various stages of development. A good illustration of a medical application of biotechnology is the manufacture

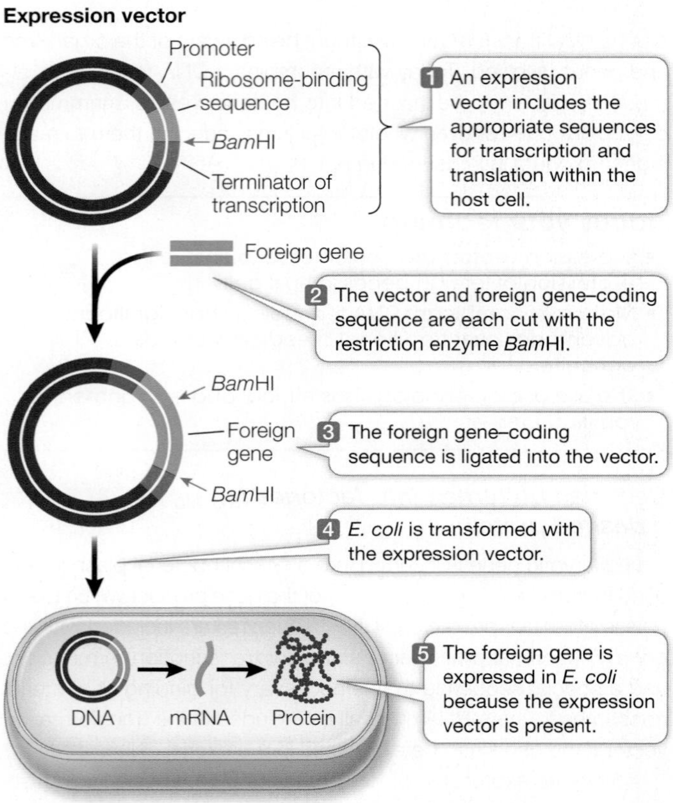

Expression vector

1. An expression vector includes the appropriate sequences for transcription and translation within the host cell.

2. The vector and foreign gene–coding sequence are each cut with the restriction enzyme *Bam*HI.

3. The foreign gene–coding sequence is ligated into the vector.

4. *E. coli* is transformed with the expression vector.

5. The foreign gene is expressed in *E. coli* because the expression vector is present.

Figure 18.9 Expression of a Transgene in a Host Cell Produces Large Amounts of Its Protein Product To be expressed in *E. coli*, a gene derived from a eukaryote requires bacterial sequences for transcription initiation (promoter), transcription termination, and ribosome binding. Expression vectors contain these additional sequences, enabling the eukaryotic protein to be synthesized in the prokaryotic cell.

▶ Activity 18.2 **Expression Vectors**
www.Life11e.com/ac18.2

table 18.1 Some Medically Useful Products of Biotechnology

Product	Use
Colony-stimulating factor	Stimulates production of white blood cells in patients with cancer and AIDS
Erythropoietin	Prevents anemia in patients undergoing kidney dialysis and cancer therapy
Factor VIII	Replaces clotting factor missing in patients with hemophilia A
Growth hormone	Replaces missing hormone in people of short stature
Insulin	Stimulates glucose uptake from blood in people with insulin-dependent (type I) diabetes
Platelet-derived growth factor	Stimulates wound healing
Tissue plasminogen activator	Dissolves blood clots after heart attacks and strokes
Vaccine proteins: Hepatitis B, herpes, influenza, Lyme disease, meningitis, pertussis, etc.	Prevent and treat infectious diseases

of tissue plasminogen activator (TPA) discussed in the opening of this chapter. As noted there, TPA is a human protein involved in hydrolysis of blood clots, and can be used to remove clots that block blood flow in a stroke.

To make TPA in amounts needed for clinical use, scientists first isolated the TPA gene by taking the mRNA for TPA and making a cDNA copy in the lab. The cDNA was then inserted into an expression vector and used to transform mammalian cells (**Investigating Life: Producing TPA**). The transgenic cells made the protein in quantity, and it soon became available commercially. As you saw at the start of the chapter, the drug has been a success in treating strokes.

Another way of making medically useful products in large amounts is **pharming**: the production of pharmaceuticals in farm animals or plants. For example, a gene encoding a useful protein might be placed next to the promoter of the gene that encodes lactoglobulin, an abundant milk protein. Transgenic animals carrying this recombinant DNA will secrete large amounts of the foreign protein into their milk. These natural "bioreactors" can produce abundant supplies of the protein, which can be separated easily from the other components of the milk (**Figure 18.10**).

DNA manipulation is changing agriculture

The cultivation of plants and the husbandry of animals provide the world's oldest examples of biotechnology, dating back more than 10,000 years. Over the centuries, people have adapted crops and farm animals to their needs, producing organisms with desirable characteristics such as large seeds, high fat content in milk, or resistance to disease.

Until recently, the most common way to improve crop plants and farm animals was to identify individuals with desirable phenotypes that existed as a result of natural variation. Through many deliberate *crosses—a process called selective breeding—the genes responsible for the desirable trait could be introduced into a widely used variety or breed of that organism.

---

*connect the concepts In genetics experiments, typically only a few genes and alleles are involved. But crop plants have phenotypes determined by many genes and alleles. This makes selecting genetically stable, desirable offspring from crosses between two varieties challenging. See Key Concepts 12.2 and 12.3.

Despite some spectacular successes—among them the breeding of high-yielding varieties of wheat, rice, and hybrid corn—such deliberate crossing can be a hit-or-miss affair. Many desirable traits are controlled by multiple genes, and it is hard to predict the results of a cross or to maintain a prized combination as a true-breeding variety year after year, especially because in sexual reproduction combinations of desirable genes are quickly separated by meiosis. Furthermore, traditional breeding takes a long time: many plants and animals take years to reach maturity and then can reproduce only once or twice a year—a far cry from the rapid reproduction of bacteria.

Here are just some of the advantages that modern recombinant DNA technology offers over traditional methods of breeding:

● *The ability to identify specific genes.* The development of genetic markers allows breeders to select for specific desirable genes, making the breeding process more precise and rapid.

experiment

For a protein such as TPA to become useful in the clinic, the expression of recombinant DNA (in this case, the TPA gene coupled to an active promoter) in cells must result in a large amount of the protein. The protein is then purified and used to treat patients with a stroke, by catalyzing the dissolution of a blood clot in an artery leading to the brain.

Original Paper: Collen, D. et al. 1984. Biological properties of human tissue-type plasminogen activator obtained by expression of recombinant DNA in mammalian cells. *Journal of Pharmacology and Experimental Therapeutics* 231: 146–152.

HYPOTHESIS▶ A clinically useful TPA can be made by recombinant DNA technology.

METHOD

mRNA for TPA

1 Isolate mRNA for the TPA gene.

Reverse transcription

cDNA for TPA

2 Using reverse transcription, copy mRNA for the TPA gene into cDNA.

Expression vector for mammalian cell

3 Create an expression vector suitable for a mammalian host.

← *Bam*HI

4 Insert the TPA gene into the expression vector.

TPA gene

RESULTS

Mammalian cell

DNA

5 The transformed cell produces TPA in large amounts.

TPA

6 TPA is harvested, purified, and injected into a stroke patient to dissolve the blood clot.

CONCLUSION▶ TPA made by biotechnology can be used to treat patients.

work with the data

It is essential that a purified molecule made by biotechnology have the desired biological properties. In the case of TPA, the goal was to mimic the properties of the natural clot-dissolving molecule. Désiré Collen, a physician-scientist in Belgium, led an international team that compared laboratory-made TPA with the natural molecule made by human cells. They performed two types of experiments, one on clots suspended in human blood in the lab, and the other on clots that had developed in rabbits. The team also investigated the time course of clot dissolution and the dose of TPA needed for the effect.

QUESTIONS▶

1. Human blood clots were prepared in the lab using a radioactively labeled clotting molecule. Dissolution of the clot resulted in the radioactive molecules becoming hydrolysis products (monomers), so comparing soluble to insoluble radioactivity was a measure of clot dissolution. **Table A** shows the results of clot dissolution over time after injection with natural TPA, TPA made by biotechnology, and a control injection without TPA. Plot the data in a graph of percent clot dissolution over time. What can you conclude about the experiment?

2. To determine whether lab-made TPA could dissolve clots in the mammalian blood system, clots were induced in a large vein in rabbits. After 20 h, either natural or lab-made TPA was injected at the clot region and the dissolution of the clot measured 4 h later. The results are shown in **Table B**. What can you conclude about this experiment? The results show the SEM (standard error of the mean) for the data. How would you use a statistical test to examine the significance, if any, between the two treatment groups and compare them with the control?

Table A

	Percent clot dissolution		
Time (h)	No TPA (Control)	Natural TPA	Lab-made TPA
1	0	4	5
2	0	10	20
3	1	20	35
4	1	35	55
5	2	50	65

Table B

	Percent clot dissolution (SEM)	
Dose of TPA (units)	Natural TPA	Lab-made TPA
0 (control)	14.3 (1.4)	14.3 (1.4)
12,000	19.8 (5.4)	22.3 (6.0)
24,000	24.5 (7.9)	30.6 (0.8)
48,000	38.9 (4.8)	49.3 (9.7)
96,000	66.0 (6.3)	75.4 (3.9)

A similar **work with the data** exercise may be assigned in **LaunchPad**.

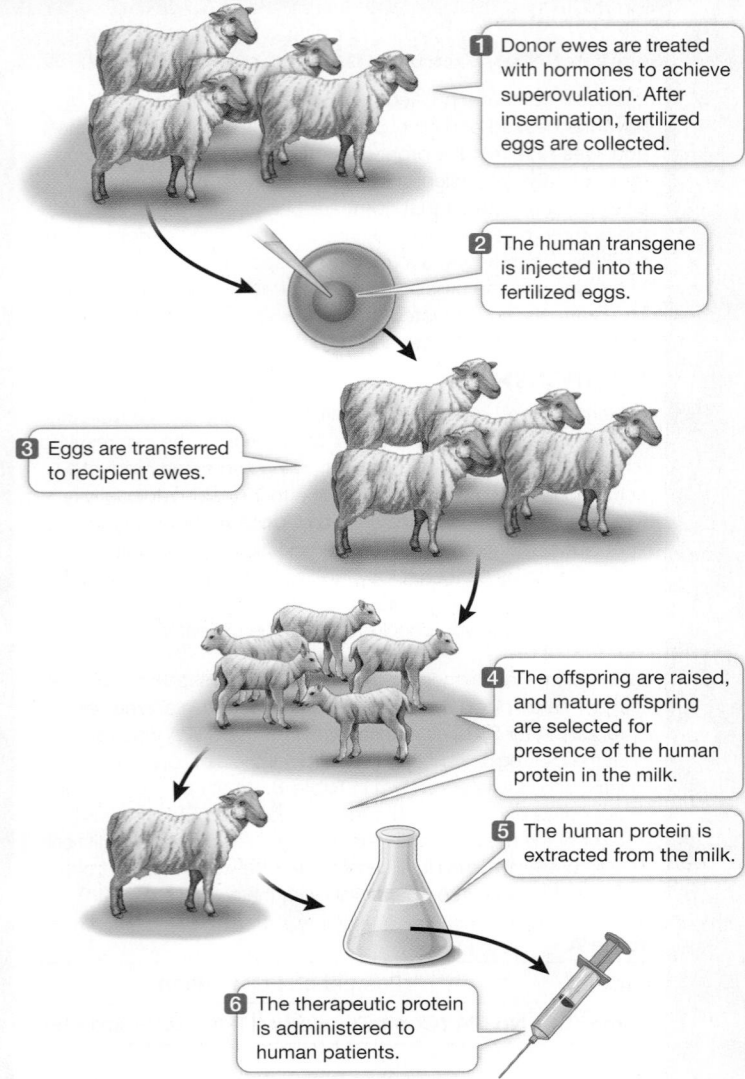

1 Donor ewes are treated with hormones to achieve superovulation. After insemination, fertilized eggs are collected.

2 The human transgene is injected into the fertilized eggs.

3 Eggs are transferred to recipient ewes.

4 The offspring are raised, and mature offspring are selected for presence of the human protein in the milk.

5 The human protein is extracted from the milk.

6 The therapeutic protein is administered to human patients.

Figure 18.10 Pharming An expression vector carrying a desired gene can be put into an animal egg, which is implanted into a surrogate mother. The transgenic offspring produces the new protein in its milk. The milk is easily harvested and the protein isolated, purified, and made clinically available to patients.

Q: A drug that is an enzyme from bacteria has been developed to treat the genetic disorder Gaucher's disease, using carrot cells in tissue culture. The targeted enzyme functions inside the lysosome. In addition to inserting the gene for the enzyme into carrot cells, what other DNA sequences need to be inserted into the plant cells so that the enzyme will be functional in humans?

table **18.2** Agricultural Applications of Biotechnology under Development	
Goal	**Technology/genes**
Improving the environmental adaptations of plants	Genes for drought tolerance, salt tolerance
Improving nutritional traits	High-lysine seeds; β-carotene in rice
Improving crops after harvest	Delay of fruit ripening; sweeter vegetables
Using plants as bioreactors	Plastics, oils, and drugs produced in plants

- *The ability to introduce any gene from any organism into a plant or animal species*. This ability, combined with mutagenesis techniques, vastly expands the range of possible new traits.

- *The ability to generate new organisms quickly*. Manipulating cells in the laboratory and regenerating a whole plant or animal by cloning is much faster than traditional breeding.

Consequently, recombinant DNA technology has found many applications in agriculture (**Table 18.2**). We will describe a few examples to demonstrate the approaches that plant scientists have used to improve crop plants.

PLANTS THAT MAKE THEIR OWN INSECTICIDES Plants are subject to infections by viruses, bacteria, and fungi, but probably the most important crop pests are herbivorous insects. From the locusts of biblical (and modern) times to the cotton boll weevil, insects have continually eaten the crops people grow.

The development of insecticides has improved the situation, but insecticides have their own problems. Many, including the organophosphates, are relatively nonspecific and kill beneficial insects in the broader ecosystem as well as crop pests. Some pesticides even have toxic effects on other groups of organisms, including people. What's more, many insecticides persist in the environment for a long time.

Some bacteria protect themselves by producing proteins that can kill insects. For example, the bacterium *Bacillus thuringiensis* produces a protein that is toxic to the insect larvae that prey on it.

Bacillus thuringiensis

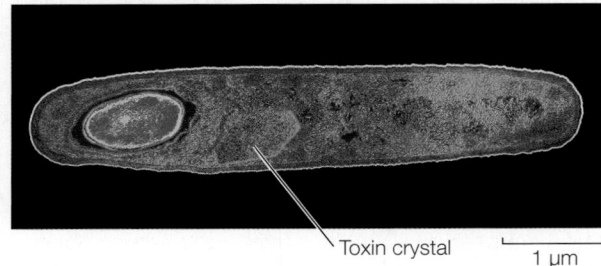

Toxin crystal   1 µm

The toxicity of this protein to insects is 80,000 times greater than that of a typical commercial insecticide. When a hapless larva eats the bacteria, the toxin becomes activated and binds specifically to the insect's gut, producing holes and killing the insect. Dried preparations of *B. thuringiensis* have been sold for decades as safe insecticides that break down rapidly in the environment. But the biodegradation of these preparations is their limitation, because it means that the dried bacteria must be applied repeatedly during the growing season.

A longer-acting approach is to have the crop plants themselves make the toxin, and this is exactly what plant scientists have done. The toxin gene from *B. thuringiensis* has been isolated, cloned, and extensively modified by the addition of a plant promoter and other regulatory sequences. Transgenic corn, cotton, soybeans, tomatoes, and other crops are now being grown successfully with this added gene. Farmers growing these transgenic crops use less of other pesticides.

CROPS THAT ARE RESISTANT TO HERBICIDES Herbivorous insects are not the only threat to agriculture. Weeds may grow in

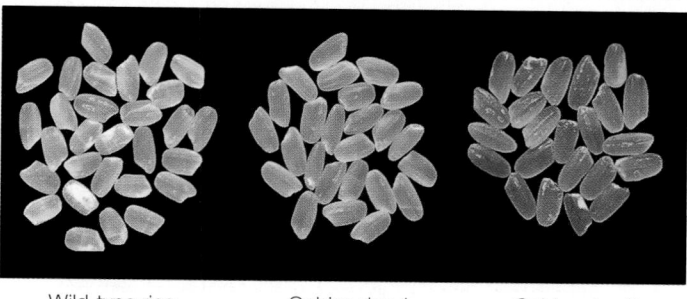

Wild-type rice   Golden rice 1   Golden rice 2

Figure 18.11 **Transgenic Rice Rich in β-Carotene** Middle and right: The grains from these transgenic rice strains are colored because they make the pigment β-carotene, which is converted to vitamin A in the human body. Left: Wild-type rice grains do not contain β-carotene.

fields and compete with crop plants for water and soil nutrients. Glyphosate is a widely used and effective herbicide, or weed killer, that works only on plants. It inhibits an enzyme system in the chloroplast that is involved in the synthesis of amino acids. Glyphosate is a broad-spectrum herbicide that kills most weeds, but unfortunately it also kills crop plants. One solution to this problem is to use it to rid a field of weeds before the crop plants start to grow. But as any gardener knows, when the crop begins to grow, the weeds reappear. If the crop were not affected by the herbicide, the herbicide could be applied to the field at any time without harming the crop.

Scientists have used expression vectors to make plants that synthesize a different form of the target enzyme for glyphosate that is unaffected by the herbicide. The gene for this enzyme has been inserted into corn, cotton, and soybean plants, making them resistant to glyphosate. This technology has expanded rapidly, and a large proportion of cotton and soybean plants now carry this gene.

GRAINS WITH IMPROVED NUTRITIONAL CHARACTERISTICS To remain healthy, humans must consume adequate amounts of β-carotene, which the body converts into vitamin A. About 400 million people worldwide suffer from vitamin A deficiency, which makes them susceptible to infections and blindness. One reason is that rice grains, which do not contain β-carotene, make up a large part of their diets. Rice grains lack the two-enzyme biochemical pathway that synthesizes β-carotene.

Plant biologists Ingo Potrykus and Peter Beyer isolated one of the genes for the β-carotene pathway from the bacterium *Erwinia uredovora* and the other from daffodil plants. They added a promoter and other signals for expression in the developing rice grain and then transformed rice plants with the two genes. The resulting rice plants produce grains that look yellow because of their high β-carotene content. A newer variety with a corn gene replacing the one from daffodils makes even more β-carotene and is golden in color (**Figure 18.11**). A daily intake of about 150 g of this cooked rice can supply all the β-carotene a person needs. This new transgenic strain has been crossed with strains adapted for various local environments, in the hope of improving the diets of millions of people.

CROPS THAT ADAPT TO THE ENVIRONMENT Agriculture depends on ecological management—tailoring the environment to the needs of crop plants and animals. A farm field is an unnatural, human-designed system that must be carefully managed to maintain optimal conditions for crop growth. For example, excessive irrigation can cause increases in *soil salinity. The Fertile Crescent, the region between the Tigris and Euphrates rivers in the Middle East where agriculture probably originated 10,000 years ago, is no longer fertile. It is now a desert, largely because the soil has a high salt concentration. Few plants can grow on salty soils, partly because of osmotic effects that result in wilting, and partly because excess salt ions are toxic to plant cells.

---

*connect the concepts Learn about the effects of salt and heavy metals on plants in Key Concept 38.3.

Some plants can tolerate salty soils because they have a protein that transports Na^+ ions out of the cytoplasm and into the vacuole, where the ions can accumulate without harming plant growth (see Key Concept 5.3 for a description of the plant vacuole). Scientists developed a highly active version of the gene for this transporter protein and used it to transform crop plants that are less tolerant to salt, including rapeseed, wheat, and tomatoes. When this gene was added to tomato plants, they grew in water that was four times as salty as the typical lethal level (**Figure 18.12**). This finding raises the prospect of growing crops on what were previously unproductive soils.

Perhaps the most spectacular transgenic crop developed so far is a strain of rice that is resistant to both salt stress (using the gene above) and drought stress (using a gene from a bacterium) and has the ability to grow on nitrogen-deficient soils (using a gene from barley). This triply-tolerant rice is undergoing field tests.

These examples illustrate what could become a fundamental shift in the relationship between crop plants and the environment. *Instead of manipulating the environment to suit the plant, biotechnology may allow us to adapt the plant to the environment.* As a result, some of the negative effects of agriculture, such as water pollution, could be lessened.

(A)

(B)

Figure 18.12 **Salt-Tolerant Tomato Plants** Transgenic plants containing a gene for salt tolerance thrive in salty water **(A)**, whereas plants without the transgene die **(B)**. This technology may allow crops to be grown on salty soils.

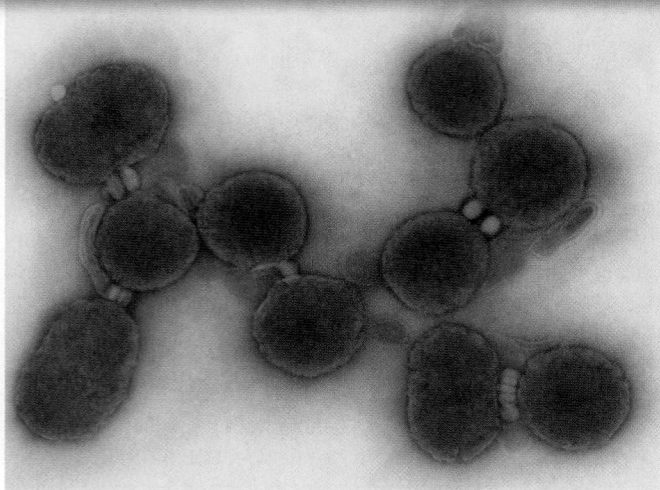

Figure 18.13 Synthetic Cells These cells of *Mycoplasma mycoides* contain a synthetic genome.

Q: What are some potential uses of synthetic cells?

Synthetic biology can create living factories for new products

In Key Concept 17.2, you learned how inactivating genes one at a time can help define which genes are necessary for life in a particular organism (the minimal genome) and which ones confer special properties to the organism. One goal of this research is to design new life forms with specific purposes. That is, new genes could be added to the minimal genome. An important step toward this goal has been the laboratory synthesis of an artificial genome and its insertion into bacterial cells. This was first done using the bacterium *Mycoplasma mycoides*. The artificial genome was then inserted into empty cells of a related species, *Mycoplasma capricolum*, whose own DNA had been hydrolyzed. The new DNA directed the cell to perform the biochemical functions of life, including reproduction (**Figure 18.13**).

The aims of synthetic biology are bold: to create bacteria or other organisms with new functions, such as producing hydrocarbons for fuel, new drugs for medicine, bacteria that make plastics and other polymers, and so on. But before these functions can be added to the genome, the genes involved must be clearly described, including regulatory genes and sequences such as promoters that function in host cells. Thousands of such genes and chemicals that regulate their expression have now been associated with many known gene functions.

For example, a single function (bacteria making ammonium sulfate that could be used for fertilizer in farming) requires more than 70 genes.

There is public concern about biotechnology

Concerns have been raised about the wisdom and safety of genetically modifying crops and other organisms. These concerns are centered on three claims:

1. Genetic manipulation is an unnatural interference with nature.
2. Genetically altered foods are unsafe to eat.
3. Genetically altered crop plants are dangerous to the environment.

Advocates of biotechnology tend to agree with the first claim. However, they point out that all crops are unnatural in the sense that they come from artificially bred plants growing in a manipulated environment (a farmer's field). Recombinant DNA technology just adds another level of sophistication to these technologies. Synthetic biology, however, takes genetic manipulation dramatically further.

To counter concern about whether genetically engineered crops are safe for human consumption, biotechnology advocates point out that only single genes are added and that these genes are specific for plant function. For example, the *B. thuringiensis* toxin produced by transgenic plants has no effect on people. However, as plant biotechnology moves from adding genes that improve plant growth to adding genes that affect human biology, such concerns will become more pressing.

Various negative environmental impacts have been envisaged. The concern has been raised, for example, about the possible "escape" of transgenes from crops to other species. If the gene for herbicide resistance, for example, were inadvertently transferred from a crop plant to a closely related weed, that weed could thrive in herbicide-treated areas. Increased use of herbicides has already resulted in the selection of weeds with naturally occurring mutations that make them resistant to that herbicide. Widespread use of glyphosate on fields of glyphosate-resistant crops, for example, has resulted in the selection of rare mutations in weeds that make them resistant to glyphosate. Field-evolved resistance now occurs in half the major weed species examined.

Because of the potential benefits of biotechnology, scientists believe that it is wise to proceed, albeit with caution.

18.5 recap

Expression vectors maximize the expression of transgenes inserted into host cells. Biotechnology has been used to produce medicines and to develop transgenic plants with improved agricultural and nutritional characteristics.

learning outcomes

You should be able to:

- Describe the essential features of expression vectors.
- Summarize the advantages of modern recombinant DNA technology over selective breeding in agriculture.
- Evaluate the ethical and environmental concerns posed by specific uses of biotechnology, taking into account the risks and rewards that would likely come from their use.

1. What are the essential features of expression vectors?
2. Compared with traditional methods, what are the advantages of using biotechnology for plant breeding?
3. After Cohen and Boyer did their experiment (see Key Concept 18.1), a group of scientists, public health officials, and bioethicists held a conference and agreed to a temporary moratorium on most recombinant DNA research. What do you think were the concerns that led to the moratorium? How was safety assured so that the moratorium was lifted?

investigatinglife

How is biotechnology changing medicine?

What are the limitations of treating patients with the clot-busting drug TPA, made by biotechnology? We opened this chapter with the story of Janet's stroke, pointing out how important it is to treat stroke patients quickly. An analysis published in 2014 of many studies done in the 20 years since TPA was first clinically used showed that the probability of having no significant disability is 75 percent higher in people who get TPA within 3 hours of stoke onset compared with people who don't get TPA. This benefit is reduced if a patient gets the drug later, and is almost nonexistent at 6 hours poststroke.

One of the problems that arose when TPA treatments were initially administered was that the drug tended to break down rapidly in the bloodstream after injection. Biotechnologists solved this problem by altering the TPA gene slightly so that the protein gets glycosylated (adding sugars), which leads to a longer lifetime in the bloodstream. As other drugs are developed by biotechnology, the TPA story is being repeated: initial success, problems with the chemistry, and then modification.

While the commercial production of drugs by DNA biotechnology has had some successes, the examples are relatively limited and certainly not as notable as those in traditional biotechnology, where bacteria and other microbes are manipulated by their environment to make large amounts of molecules, such as antibiotics. However, as we described, DNA manipulation technologies such as mircoarrays and CRISPR hold great promise to profoundly change medical diagnosis and treatment.

Future directions

A striking example of "old" biotechnology (in which a microorganism is coaxed into making a valuable product) merging with "new" biotechnology (which uses recombinant DNA) is the emerging field of cyanobacteria-produced plant secondary metabolites. Plants make these small-molecule metabolites to protect themselves against infections, injuries, and environmental factors. Many of the metabolites are strong antioxidants that react with and eliminate harmful oxidants such as oxygen atoms with unpaired electrons (superoxides) in cells under stress. Because human cells in disease states produce such oxidants that harm tissues, there is intense interest in using certain secondary metabolites that plants produce as drugs. Unfortunately, it takes a huge quantity of plants to obtain enough of a given antioxidant for clinical use.

This is where cyanobacteria come in. Cyanobacteria (sometimes incorrectly called blue-green algae) are photosynthetic single-celled organisms that grow well under light conditions in ponds or vats in the lab. Although they are not plants, cyanobacteria have genes encoding some of the enzymes involved with secondary metabolite production. Using recombinant DNA technology, genes from plants that encode other needed compounds in the pathway of certain antioxidants (specifically, phenylpropanoids) have been inserted into cyanobacteria, turning these organisms into factories that produce plant metabolites for clinical use.

Chapter Summary 18

18.1 DNA from Different Sources Forms Recombinant DNA

- **Recombinant DNA** is formed in the laboratory by combining at least two DNA sequences from different sources. Review Figure 18.1
- Many restriction enzymes make staggered cuts in the two strands of DNA, creating fragments that have **sticky ends** with unpaired bases.
- DNA fragments cut with restriction enzymes can be used to create recombinant DNA. DNA molecules from different sources can be cut with the same restriction enzyme and spliced together using **DNA ligase**. Review Focus: Key Figure 18.2

18.2 There Are Several Ways to Insert DNA into Cells

- One reason to make recombinant DNA is to **clone** a particular gene or other DNA sequence, either for sequencing and subsequent analysis or to produce a protein product in quantity.
- Bacteria, yeasts, and cultured plant and animal cells are commonly used as hosts for recombinant DNA. The insertion of foreign DNA into host cells is called **transformation** or (for animal cells) **transfection**. Transformed or transfected cells are called **transgenic** cells.
- Various methods are used to get recombinant DNA into cells. These include chemical or electrical treatment of the cells, the use of viral vectors, and injection. *Agrobacterium tumefaciens* is often used to insert DNA into plant cells.
- To identify host cells that have taken up a foreign gene, the inserted sequence can be tagged with one or more **reporter genes**,

which are genetic markers with easily identifiable phenotypes. **Selectable markers** allow for the selective growth of transgenic cells. Review Figures 18.3, 18.4

- Replication of the foreign gene in the host cell requires that it become part of a segment of DNA that contains a **replicon** (origin and terminus of replication).
- **Vectors** are DNA sequences that can carry new DNA into host cells. Plasmids and viruses are commonly used as vectors.

18.3 Any Sequence of DNA Can Be Used for Cloning

- DNA fragments from a genome can be inserted into host cells to create a **genomic library**. Review Figure 18.5A
- The mRNAs produced in a certain tissue at a certain time can be extracted and used to create **complementary DNA (cDNA)** by **reverse transcription**. Review Figure 18.5B
- PCR products can be used for cloning. Synthetic DNA containing any desired sequence can be made in the laboratory.

18.4 Several Tools Are Used to Modify DNA and Study Its Function

- **CRISPR** technology can be used to inactivate or mutated any gene. Review Figure 18.6, Activity 18.1
- Gene silencing techniques can be used to inactivate the mRNA transcript of a gene, which may provide clues to the gene's function.

(continued)

Chapter Summary 18 (continued)

Artificially created **antisense RNA** or siRNA can be added to a cell to prevent translation of a specific mRNA. Review Figure 18.7

- **DNA microarray** technology permits the screening of thousands of cDNA sequences at the same time. Review Figure 18.8, Animation 18.2

▶ 18.5 DNA Can Be Manipulated for Human Benefit

- **Biotechnology** is the use of living cells to produce materials useful to people.
- **Expression vectors** allow a transgene to be expressed in a host cell. Review Figure 18.9, Activity 18.1
- Recombinant DNA techniques have been used to make medically useful proteins. Review Investigating Life: Producing TPA
- **Pharming** is the use of transgenic plants or animals to produce pharmaceuticals. Review Figure 18.10

- Because recombinant DNA technology has several advantages over traditional agricultural biotechnology, it is being extensively applied to agriculture.
- Biotechnology may allow us to adapt the plant to the environment, rather than vice versa.
- Synthetic biology involves creating new cells with artificial DNA made in the laboratory.
- There is public concern about the application of recombinant DNA technology to food production.

Go to **LearningCurve** (in **LaunchPad**) for dynamic quizzing that helps you solidify your understanding of this chapter. **LearningCurve** adapts to your responses, giving you the practice you need to master each key concept.

▶ Apply What You've Learned

Review

18.4 CRISPR technology can be used to alter and inactivate specific genes.

18.5 The use of biotechnology has ethical and environmental ramifications.

Original Paper: Gantz, V. M. and E. Bier. 2015. The mutagenic chain reaction: A method for converting heterozygous to homozygous mutations. *Science* 348: 442–444.

Mosquitoes in the genus *Anopheles* transmit the parasite *Plasmodium* to humans, resulting in more than half a million deaths from malaria every year. Some genetic strains of *Anopheles* carry a recessive gene that, when homozygous, makes them inhospitable to the parasite. Gene knockout methods have been used to produce large numbers of mosquitoes carrying the resistance mutation, but since only one of the two alleles is mutated by this method, release of these insects into the wild will not result in a rapid spread of malaria resistance. A method to mutate both alleles, producing mosquitoes homozygous for resistance, would result in a more rapid spread of resistance.

Gantz and colleagues developed a method called the mutagenic chain reaction (MCR) to address this. They used CRISPR technology to convert heterozygous mutations to homozygous ones in the model organism *Drosophila melanogaster*. The CAS9 enzyme was used to cut the body-color gene carried on the X chromosome. This produced the first mutation: to yellow body (**Figure A**). This mutated gene was replicated by the fly's own DNA repair system resulting in a homozygous yellow fly (**Figure B**).

Figure C shows how MCR inheritance works. CRISPR technology created a male fly with the mutated yellow gene (marked in yellow on Figure B) on the X chromosome. When this fly was crossed with a homozygous wild-type female, the female offspring inherited one X chromosome with the yellow allele from the male parent, and one X chromosome from the female parent that had the wild-type allele that was immediately converted to a mutated (yellow) allele. So these females were yellow.

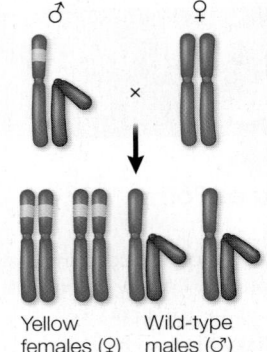

Figure A

Figure B

Black body allele

⬇ RNA and CRISPR

Yellow body allele

⬇ Mating

Heterozygote

MCR ⬇ Cut and repair

Homozygous yellow

Figure C

MCR inheritance

♂   ♀

×

⬇

Yellow   Wild-type
females (♀)   males (♂)

Questions

1. How do the genotypes in the CRISPR cross compare with those produced by sex-linked Mendelian inheritance?

2. Why do none of the male offspring have the mutated allele?

3. Explain why this method would result in a rapid increase in the frequency of the mutated alleles in the population if the flies were released to the wild.

4. The researchers plan to use similar technology to fight malaria. What else would be needed to enable this technology to combat malaria?

5. What are some possible risks of this technology? How can these risks be minimized?

Genes, Development, and Evolution

On the point of a pin. Even at this early point, an embryo is undergoing molecular changes to set the stage for later development.

▶ investigatinglife

Stem Cell Therapy

Ruby was getting old, and like many in her age group, she developed arthritis in her hips. "Arthritis" means joint inflammation. A breakdown of cartilage between the bones was causing inflammation and pain. Her loved ones could hear the bones in her hips grinding against one another as she tried to walk. The doctor recommended a hip replacement, whereby an artificial joint is inserted. But Karen and Dave did something different for their beloved collie.

A veterinary surgeon removed fat cells from Ruby's body and isolated mesenchymal stem cells from these tissues. Stem cells are actively dividing, unspecialized cells that have the potential to produce different cell types depending on the signals they receive from the body. Mesenchymal stem cells are able to differentiate into various kinds of connective tissue, including bone, cartilage, blood vessels, tendons, and muscle. Ruby's stem cells were injected into her hip, and months later she was much better. She was back to her old frisky self, enjoying walks and car rides and even walking up stairs without assistance. Her owners were ecstatic.

Stem cell therapy in humans has been much in the news of late. The basic idea is to inject stem cells into damaged tissues, where they will differentiate and form new, healthy tissues. Many procedures are being tried experimentally, and widespread clinical use is on the horizon. Stem cell treatments have, however, been given to animals for more than a decade. Dogs like Ruby have been returned to their owners with a new lease on life, horses have been returned to racing or to the show ring, and even zoo animals have been treated successfully.

The processes by which an unspecialized stem cell proliferates and forms specialized cells and tissues with distinctive appearances and functions are similar to the developmental processes that occur in the embryo. Much of our knowledge of developmental biology has come from studies on model organisms such as fruit flies, nematodes, frogs, sea urchins, mice, and the small flowering plant *Arabidopsis thaliana*. Eukaryotes share many similar genes, and the cellular and molecular principles underlying their development also turn out to be similar. Thus discoveries from one organism aid us in understanding other organisms, including ourselves.

▶key concept 19.1 The Four Major Processes of Development Are Determination, Differentiation, Morphogenesis, and Growth

Development is the process by which a multicellular organism, beginning with a single cell, goes through a series of changes, taking on the successive forms that characterize its life cycle (**Figure 19.1**). After the egg is fertilized it is called a zygote, and in the earliest stages of development a plant or animal is called an **embryo**. Sometimes the embryo is contained within a protective structure such as a seed coat, an eggshell, or a uterus. Embryogenesis results in a new organism with a body plan characteristic of its species. Many organisms continue to develop throughout their life cycles, with development ceasing only at death.

focus your learning

- In the embryo, determination of the fate of a cell occurs before actual cell differentiation.
- Experimental manipulations indicate that a differentiated cell in a plant or animal has the whole genome of the organism and the potential to form all other cells of that organism.
- Cloning of plants and animals using differentiated cells or cell nuclei has several potential uses.
- Pluripotent stem cells can be obtained from animal embryos or formed from differentiated cells in the laboratory.

Development involves four distinct but overlapping processes

As an organism progresses from a fertilized egg to a mature adult, four processes occur:

1. **Determination** sets the developmental *fate* of a cell—what type of cell it will become—even before any characteristics of that cell type are observable. For example, the mesenchymal stem cells described in the story opening this chapter look unspecialized, but their fate to become connective tissue cells has already been determined.

2. **Differentiation** is the process by which different types of cells arise, leading to cells with specific structures and functions. For example, mesenchymal stem cells differentiate to become muscle, fat, tendon, or other connective tissue cells.

3. **Morphogenesis** (Greek for "origin of form") is the organization and spatial distribution of differentiated cells into the multicellular body and its organs.

4. **Growth** is the increase in size of the body and its organs by cell division and cell enlargement.

Determination and differentiation occur largely because of regulation of gene expression, a topic explored in Chapter 16. You'll see many

Figure 19.1 From Fertilized Egg to Adult   The stages of development from zygote to maturity are shown for an animal and for a plant. The blastula is a hollow sphere of cells; the gastrula has three cell layers.

▶ Activity 19.1 **Stages of Development**
www.Life11e.com/ac19.1

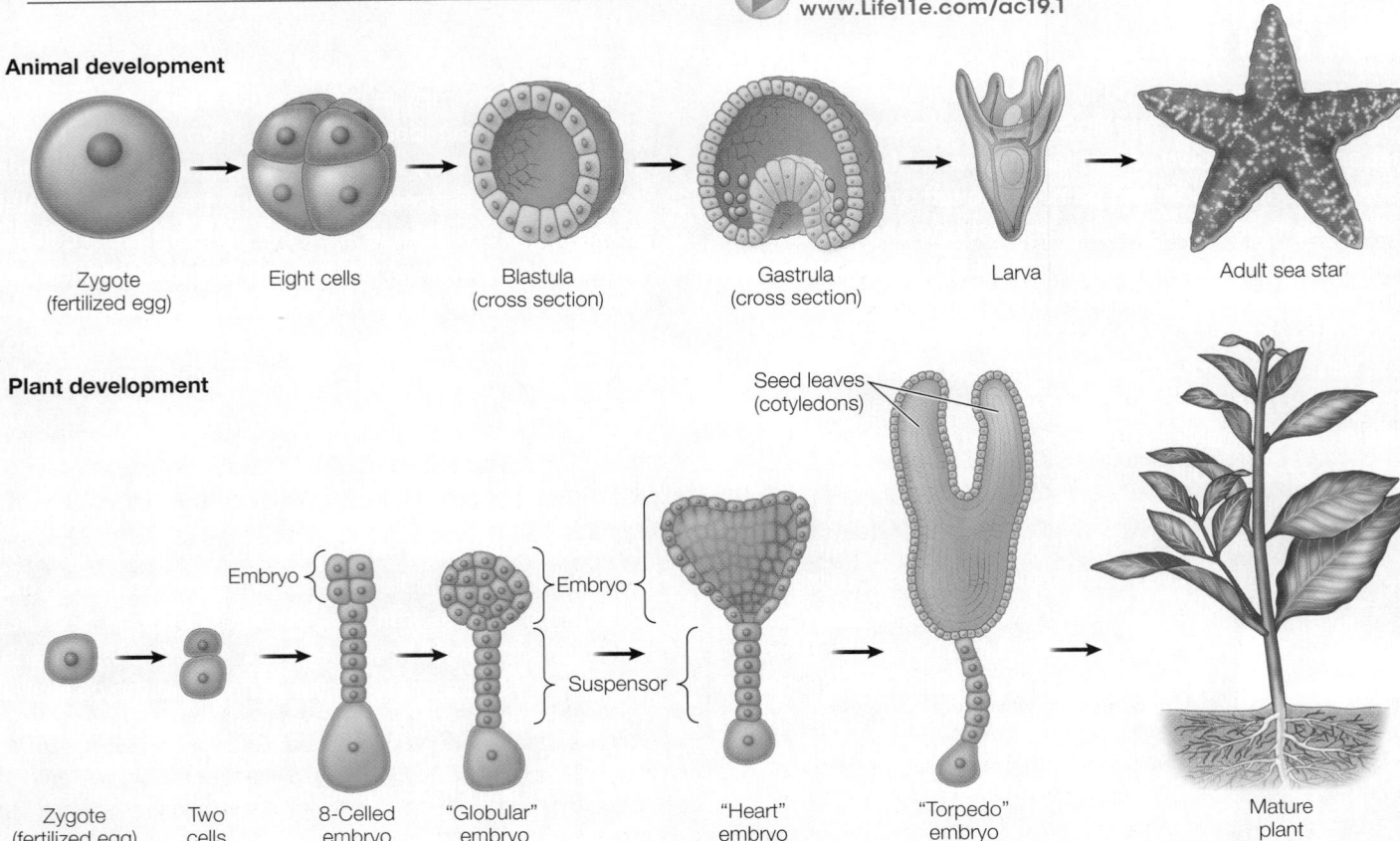

Animal development

Zygote (fertilized egg) → Eight cells → Blastula (cross section) → Gastrula (cross section) → Larva → Adult sea star

Plant development

Embryo { ... }   Embryo }   Seed leaves (cotyledons)

Suspensor {

Zygote (fertilized egg) → Two cells → 8-Celled embryo → "Globular" embryo → "Heart" embryo → "Torpedo" embryo → Mature plant of the mechanisms described in that chapter here. The cells that arise from repeated mitoses in the early embryo may look the same superficially, but they soon begin to differ in terms of which of the thousands of genes in the genome are expressed.

Morphogenesis involves not just differential gene expression but also intercellular signaling (see Chapter 7). Morphogenesis proceeds in several ways, influenced by:

- Cell division—a key event in both plants and animals

- Cell expansion—especially important in plant development, in which a cell's position and shape are constrained by the cell wall

- Cell movements—very important in animal morphogenesis

- Apoptosis (programmed cell death)—essential in organ development

Growth occurs by cell enlargement. In some cases, cell enlargement is coupled to cell division, so the average cell size remains the same as the tissue grows; in other cases (especially in plant tissues), cells enlarge without dividing, so the average cell size increases. Growth continues throughout the individual's life in some organisms, but reaches a more or less stable end point in others.

As development proceeds, cell fates become restricted

During development, an undifferentiated cell will become part of a particular type of tissue—this is referred to as **cell fate**. Cell fate determination occurs as the embryo develops. The timing of this determination varies with the organism, but it is typically quite early. One way the timing can be revealed is to transplant cells from one embryo to a different region of a recipient embryo (**Figure 19.2**). Do the transplanted cells adopt the differentiation pattern of their new surroundings, or do they continue on their own path, with their fate already sealed?

Experiments on amphibian embryos indicate that determination happens early in development. If the donor tissue is from an early-stage embryo (blastula), it adopts the fate of the new surroundings. In this case, cell fate has not been determined and is influenced by the extracellular environment. But if the donor tissue is from an older embryo (gastrula), it continues on its original developmental path. In this case, cell fate has already been determined and is no longer influenced by the extracellular environment.

Cell fate determination is influenced by changes in gene expression as well as the extracellular environment. You can't see cell fate determination by looking at an embryo under the microscope—cells do not change their appearance when they become determined. Rather, the changes that mark determination are within the cell's molecular makeup. Determination is followed by differentiation—the actual changes in structure and function that result in different cell types. *Determination is a commitment; the final realization of that commitment is differentiation*.

During animal development, cell fate becomes progressively more restricted. This can be thought of in terms of **cell potency**, which is a cell's potential to differentiate into other cell types:

- The cells of an early embryo are **totipotent** (*toti*, "all," + *potent*, "capable"); they have the potential to differentiate into any cell type, including more embryonic cells.

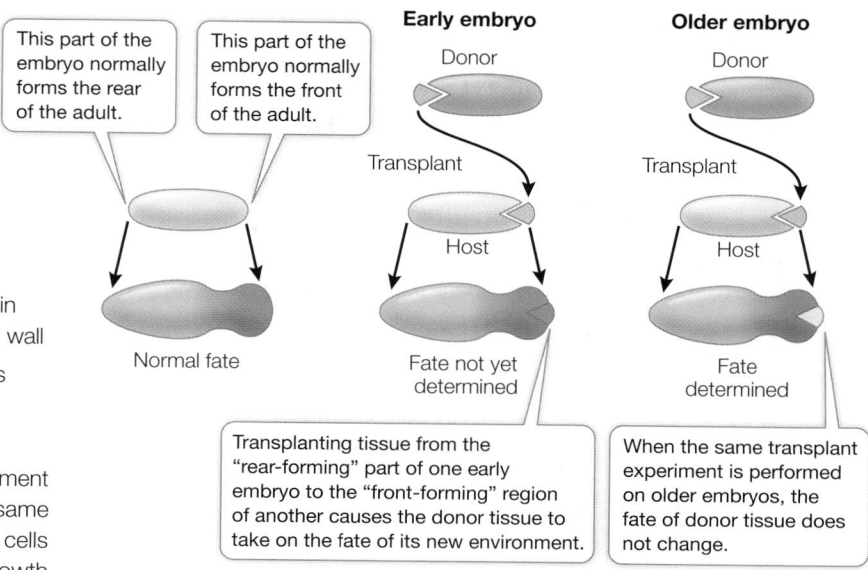

Figure 19.2 A Cell's Fate Is Determined in the Embryo Transplantation experiments using amphibian embryos show that the fate of cells is determined as the early embryo develops.

 Activity 19.2 **Cell Fates Simulation** www.Life11e.com.ac19.2

- In later stages of the embryo, many cells are **pluripotent** (*pluri*, "many"); they have the potential to develop into most other cell types, but they cannot form new embryos.

- Through later developmental stages, including adulthood, certain stem cells are **multipotent**; they can differentiate into several different, related cell types. Mesenchymal stem cells (see the opening story of this chapter) are one kind of multipotent stem cell.

- Many cells in the mature organism are **unipotent**; they can produce only one cell type—their own.

Cell differentiation is sometimes reversible

Once a cell's fate is determined, the cell differentiates. However, under the right experimental conditions, a determined or differentiated cell can become undetermined again. In some cases the cell can become totipotent, able to form an entire organism with all of its differentiated cells.

PLANT CELL TOTIPOTENCY   A carrot root cell normally faces a dark future. It cannot photosynthesize and generally does not give rise to new carrot plants. However, in 1958 Frederick Steward at Cornell University showed that if he isolated cells from a carrot root and maintained them in a suitable nutrient medium, he could induce them to dedifferentiate—to lose their differentiated characteristics. The cells could divide and give rise to masses of undifferentiated cells called calli (singular "callus"), which could be maintained in culture indefinitely. Furthermore, if they were provided with the right chemical cues, the cells could develop into embryos and eventually into complete new plants (**Figure 19.3**). Since the new plants were genetically identical to the cells from which they came, they were clones of the original carrot plant.

experiment

Figure 19.3   Cloning a Plant

Original Paper: Steward, F. C., M. O. Mapes and K. Mears. 1958. Growth and organized development of cultured cells. II. Organization in cultures grown from freely suspended cells. *American Journal of Botany* 45: 705–708.

When cells were removed from a plant and put into a medium with nutrients and hormones, they lost many of their specialized features—in other words, they dedifferentiated. Did these cells retain the ability to differentiate again? Frederick Steward found that a cultured carrot cell did indeed retain the ability to develop into an embryo and a new plant.

HYPOTHESIS▶   Differentiated carrot cells can be induced to generate all types of the plant's cells.

METHOD

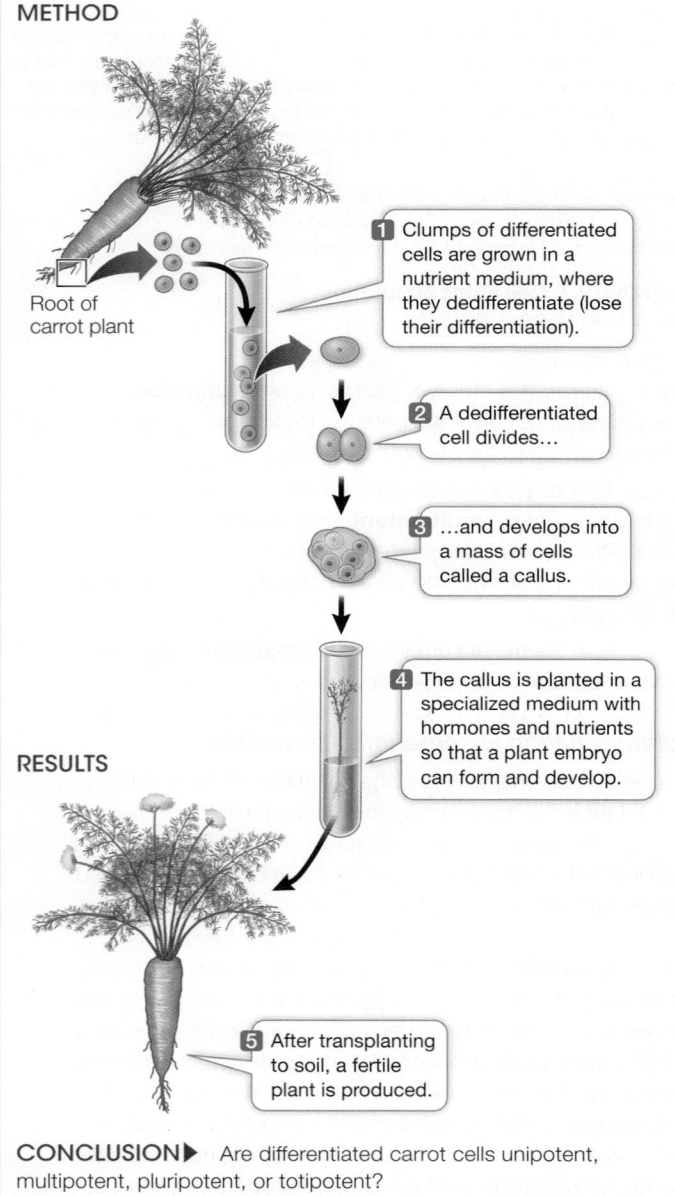

1 Clumps of differentiated cells are grown in a nutrient medium, where they dedifferentiate (lose their differentiation).

Root of carrot plant

2 A dedifferentiated cell divides…

3 …and develops into a mass of cells called a callus.

4 The callus is planted in a specialized medium with hormones and nutrients so that a plant embryo can form and develop.

RESULTS

5 After transplanting to soil, a fertile plant is produced.

CONCLUSION▶   Are differentiated carrot cells unipotent, multipotent, pluripotent, or totipotent?

Cloning an entire carrot plant from a ***differentiated root cell** showed that the root cell contained a complete carrot genome, and that under the right conditions, the cell and its descendants could express the appropriate genes in the right sequence to form a new plant. Many types of cells from other plant species show similar behavior in the laboratory, and in some cases in nature (asexual reproduction). Being able to generate a whole plant from a single cell in the lab is invaluable in agriculture and forestry. For example, trees from planted forests are used in making paper, lumber, and other products. To replace the trees reliably, forestry companies regenerate new trees from the leaves of selected trees with desirable traits. The characteristics of these clones are more uniform and predictable than those of trees grown from seeds.

*connect the concepts   Plant cell differentiation from stem cells is discussed in Key Concept 33.3.

NUCLEAR TOTIPOTENCY IN ANIMALS   Animal somatic cells cannot be manipulated as easily as plant cells can. However, as you saw with the transplantation experiments on sea urchins, the cells of very early animal embryos are totipotent. In humans this totipotency permits genetic screening (see Key Concept 15.4) of an embryo. One or two cells can be removed from the embryo and examined to determine whether a certain genetic condition is present. Because of their totipotency, the remaining cells can develop into a complete embryo, which can be implanted into the mother's uterus, where it develops into a normal fetus and infant. A later animal embryo cell won't develop into a complete organism, but the *nucleus* of such a cell has the genetic potential to do so.

Nuclear transfer experiments demonstrate that the genetic information from a single animal cell can be used to create cloned animals. Robert Briggs and Thomas King performed the first such experiments in the 1950s using frog embryos. First they removed the nuclei from unfertilized eggs, forming enucleated eggs. Then, with very fine glass needles, they punctured cells from early embryos and drew up parts of their contents, including the nuclei. Each nucleus was injected into an enucleated egg. They stimulated the eggs to divide, and many went on to form embryos, and eventually frogs, which were clones from the original implanted nuclei. These experiments led to two important conclusions:

1. No information is lost from the nuclei of cells as they pass through the early stages of embryonic development. This fundamental principle of developmental biology is known as **genomic equivalence**.

2. The cytoplasmic environment around a cell nucleus can modify its fate.

More recent studies have demonstrated that a cell from a fully developed animal can be induced to dedifferentiate and give rise to an entire new individual. In 1996 Ian Wilmut and his colleagues at the Roslin Institute in Edinburgh cloned the first mammal by somatic cell nuclear transfer. This method involves the fusion of a somatic (nonreproductive) cell from an adult animal, containing the donor nucleus, with an enucleated egg. The fully differentiated donor cells were isolated from a Finn Dorset ewe's udder and starved of

research tools

Figure 19.4 Cloning a Mammal   The experimental procedure described here produced the first cloned mammal, a Finn Dorset sheep named Dolly (shown on the left in the photo). As an adult, Dolly mated and subsequently gave birth to a normal offspring, thus proving the genetic viability of cloned mammals.

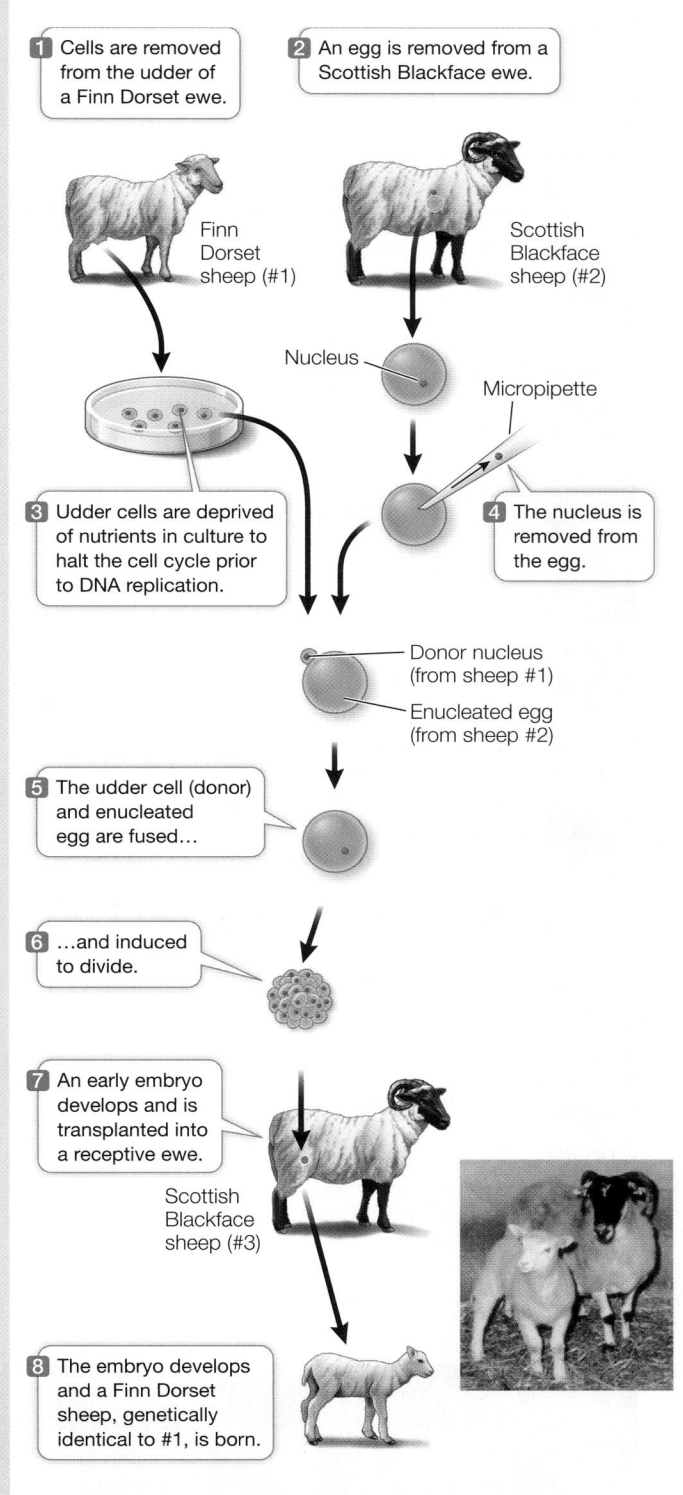

1 Cells are removed from the udder of a Finn Dorset ewe.

2 An egg is removed from a Scottish Blackface ewe.

Finn Dorset sheep (#1)

Scottish Blackface sheep (#2)

Nucleus

Micropipette

3 Udder cells are deprived of nutrients in culture to halt the cell cycle prior to DNA replication.

4 The nucleus is removed from the egg.

Donor nucleus (from sheep #1)

Enucleated egg (from sheep #2)

5 The udder cell (donor) and enucleated egg are fused...

6 ...and induced to divide.

7 An early embryo develops and is transplanted into a receptive ewe.

Scottish Blackface sheep (#3)

8 The embryo develops and a Finn Dorset sheep, genetically identical to #1, is born.

A **work with the data** exercise that accompanies this figure may be assigned in **LaunchPad**.

nutrients for a week, halting the cells in the G1 phase of the cell cycle. One of these cells was fused with an enucleated egg from a Scottish Blackface ewe, and this fused cell began to divide. After several cell divisions, the resulting early embryo was transplanted into the womb of a surrogate mother. Eventually a lamb named Dolly was born (**Figure 19.4**). Dolly showed all the characteristics of a Finn Dorset sheep: she carried the same genetic material as the nuclear donor, and thus was a clone of that donor.

The production of Dolly demonstrated that a fully differentiated cell from a mature organism can revert to a totipotent state, and that this cell can be used to create a new animal. More than 20 animal species, including cats, deer, dogs, horses, pigs, rabbits, and mice, have since been cloned by nuclear transfer. There are several reasons to clone animals:

- *Expansion of the numbers of valuable animals*: One goal of Wilmut's experiments was to develop a method for cloning transgenic animals with useful phenotypes. For example, a cow was genetically engineered to make human growth hormone in her milk. This animal was then cloned to produce additional cows that do the same thing. Only 15 such cows are needed to supply the world's demand for this medication, which is used to treat short stature caused by growth hormone deficiency.

- *Preservation of endangered species*: The mouflon, a small sheep native to the European islands of Sardinia and Corsica, was the first endangered animal to be successfully cloned. Cloning may be the only way to save some endangered species with low rates of natural reproduction, such as the giant panda.

- *Perpetuation of pets*: Many people get great personal benefit from pets, and the death of a pet can be devastating. Companies have been set up to clone cats and dogs from cells provided by their owners. Owners have paid more $100,000 each to have their pets cloned. Of course, the behavioral characteristics of the beloved pet, which are certainly derived in part from the environment, may not be the same in the cloned pet as in its genetic parent.

Multipotent stem cells differentiate in response to environmental signals

As you learned in the opening story, **stem cells** are rapidly dividing, undifferentiated cells that can differentiate into diverse cell types. In plants, stem cells occur in the meristems (see Key Concept 33.1). In mammals, stem cells are found in adult tissues that need frequent cell replacement, such as the skin, the inner lining of the intestine, and the bone marrow, where blood cells and other types of cells are formed.

The stem cells found in adult animals are not totipotent, because their ability to differentiate is limited to a relatively few cell types. In other words, they are multipotent. For example, there are two types of multipotent stem cells in the bone marrow. Hematopoietic stem cells produce red and white blood cells, whereas mesenchymal stem cells (like those used to

experiment

Original Paper: Black, L. L., J. Gaynor, D. Gahring and C. Adams. 2007. Effect of adipose-derived mesenchymal stem and regenerative cells on lameness in dogs with chronic osteoarthritis of the coxofemoral joints: A randomized, double-blinded multicenter controlled trial. *Veterinary Therapeutics* 8: 272–284.

Arthritis causes significant disability in dogs, especially older ones. In the United States, more than 10 million dogs suffer from arthritis at some point in their lives. A major reason for arthritis at joints is a breakdown of cartilage tissue that acts as a lubricant between the bones. Without it, there is pain and lameness. While working at a biopharmaceutical company serving veterinarians, Linda Black coordinated a study to determine whether multipotent stem cells contained within fat tissue could, in the right environment, make cartilage that would relieve pain and mobility problems in arthritic dogs.

HYPOTHESIS ▶ Stem cells derived from fat can clinically benefit dogs with arthritis.

METHOD

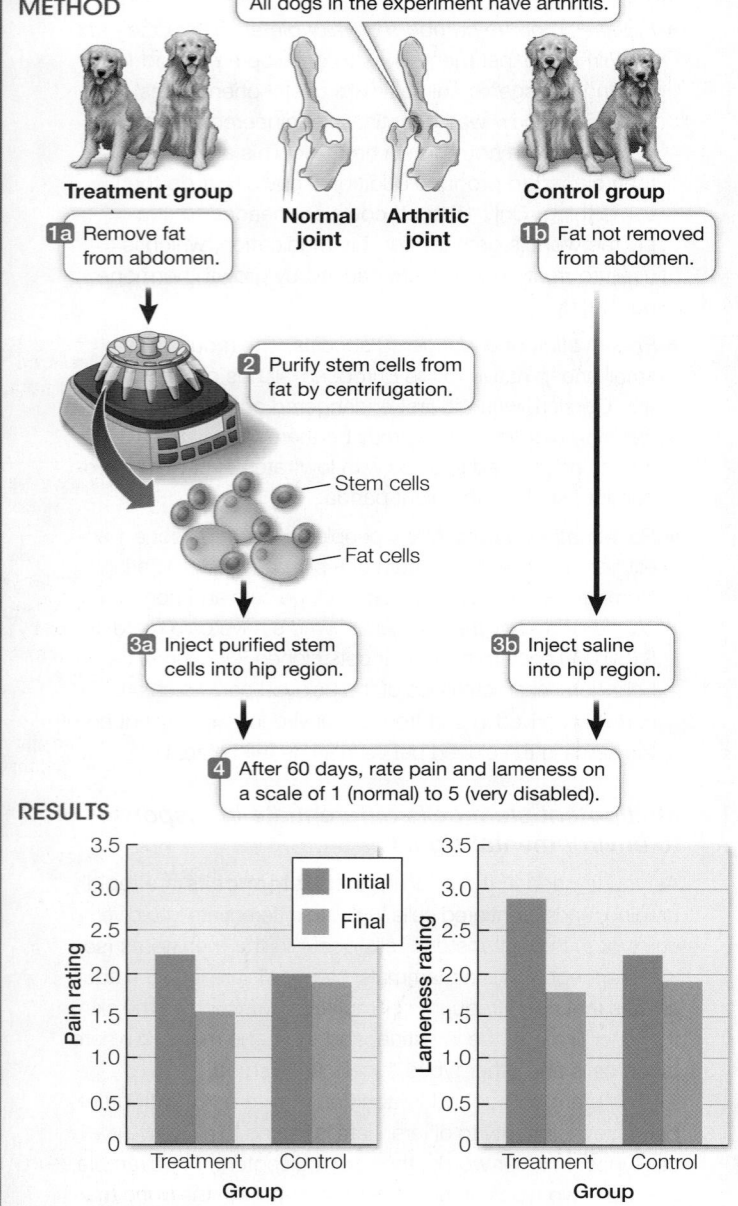

CONCLUSION ▶ Stem cell treatment improves hip function in dogs with arthritis.

work with the data

The experiment was conducted as a randomized, double-blinded, controlled clinical trial (RCT). "Randomized" means that the dogs all had similar clinical presentation and then were assigned to treated or untreated groups at random, with no selection. "Double-blinded" means that neither the scientists giving the injections nor the scientists evaluating the dogs before and after treatment knew which group the dogs were in. "Controlled" means that the dogs treated with stem cells or not were otherwise treated the same (e.g., food, living conditions). **Table A** shows the results of the experiment. Data are presented as mean score (± SEM).

Table A

Condition	Treatment	Initial	60 days later
Lameness	Stem cell	2.89 (0.20)	1.78 (0.32)
	Control	2.22 (0.15)	1.89 (0.31)
Pain	Stem cell	2.22 (0.15)	1.56 (0.18)
	Control	2.00 (0.17)	1.89 (0.20)
Motion	Stem cell	2.89 (0.20)	1.89 (0.26)
	Control	2.33 (0.17)	2.11 (0.20)

QUESTIONS ▶

1. Why is the RCT considered the "gold standard" for clinical research?

2. How did researchers reach the conclusion that the stem cell treatment given to the dogs in the experimental group was effective? What statistical test would you use to evaluate the significance of any differences between the initial and the final observations in Table A (see Appendix B)?

3. To determine the time course of any changes in clinical status, dogs were examined at various time periods after injection. The scores for range of motion are reported in **Table B**. Plot these data as percent change in clinical score versus time. What can you conclude about the time course of the effect of stem cells?

Table B

Time (days)	Stem cell group	Control group
Initial	2.89	2.33
30	1.73	2.05
60	1.89	2.11
90	1.85	2.30

A similar **work with the data** exercise may be assigned in **LaunchPad**.

treat the dog Ruby in the opening story) produce bone and connective tissues, including muscle.

Multipotent stem cells proliferate and differentiate "on demand" in response to environmental signals. For example, hematopoietic stem cells proliferate in the bone marrow in response to growth factors, and the extra stem cells are released into the blood. An important therapy called hematopoietic stem cell transplantation helps replace cells damaged in some kinds of cancer treatment. Stem cells can be harvested from the bone marrow or blood of the patient (prior to cancer treatment) or of a donor, and then the cells are injected back into the patient after cancer treatment.

Signals from adjacent cells can stimulate stem cell differentiation. Many controlled experiments have shown that damaged animal tissues (for example, hearts and tendons) that are injected with stem cells can heal more effectively than tissues that don't receive this treatment. As mentioned in the opening story of this chapter, therapy for degeneration of joints due to age or injury is now a well-established practice in veterinary medicine (**Investigating Life: Stem Cell Therapy**).

How stem cells work after injection is not clear. They may actually insert themselves into the damaged tissue and differentiate into new cells of that tissue. Alternatively, injected cells may secrete growth factors and other molecules that induce the cells in the surrounding tissue to regenerate into healthy tissue. Whatever the mechanisms by which multipotent stem cells contribute to the healing of damaged tissues, their use in treating diseases is very promising.

Pluripotent stem cells can be obtained in two ways

As stated earlier, totipotent stem cells that can differentiate into any cell type are found only in very early embryos. In both mice and humans, a slightly later embryonic stage is a hollow sphere of cells called a **blastocyst** (see Figure 43.4). A group of cells within the blastocyst is pluripotent: they can differentiate into most cell types but cannot give rise to a complete organism. In mice, these **embryonic stem cells** (**ESCs**) can be removed from the blastocyst and grown in laboratory culture almost indefinitely, given the right conditions. When cultured mouse ESCs are injected back into another mouse blastocyst, the stem cells mix with the resident cells and differentiate to form all the cell types in the mouse, which indicates that ESCs retain their developmental potential while growing in the laboratory.

ESCs growing in the laboratory can also be induced to differentiate in a particular way if the right signal is provided (**Figure 19.5A**). For example, treatment of

mouse ESCs with a derivative of vitamin A causes them to form neurons, whereas other growth factors induce them to form blood cells. Such experiments demonstrate both the cells' developmental potential and the roles of environmental signals. This finding raises the possibility of using ESC cultures as sources of differentiated cells to repair specific tissues, such as a damaged pancreas in diabetes, or a brain that malfunctions in Parkinson's disease.

ESCs can be harvested from human embryos conceived by in vitro ("under glass"—in the laboratory) fertilization, with the consent of the parents that conceived them. Since more than one embryo is usually conceived in this procedure, there is great interest in clinical studies of ESCs from embryos not used for reproduction. While there is substantial promise in the possibility of using ESCs from human embryos as sources of tissues for transplantation into patients with tissue damage, serious concerns must first be addressed, including the ethical issue of whether human embryos should be destroyed for this (or any) purpose, and the possibility that stem cells, and tissues derived from them, might provoke an immune response in a recipient (see Chapter 41).

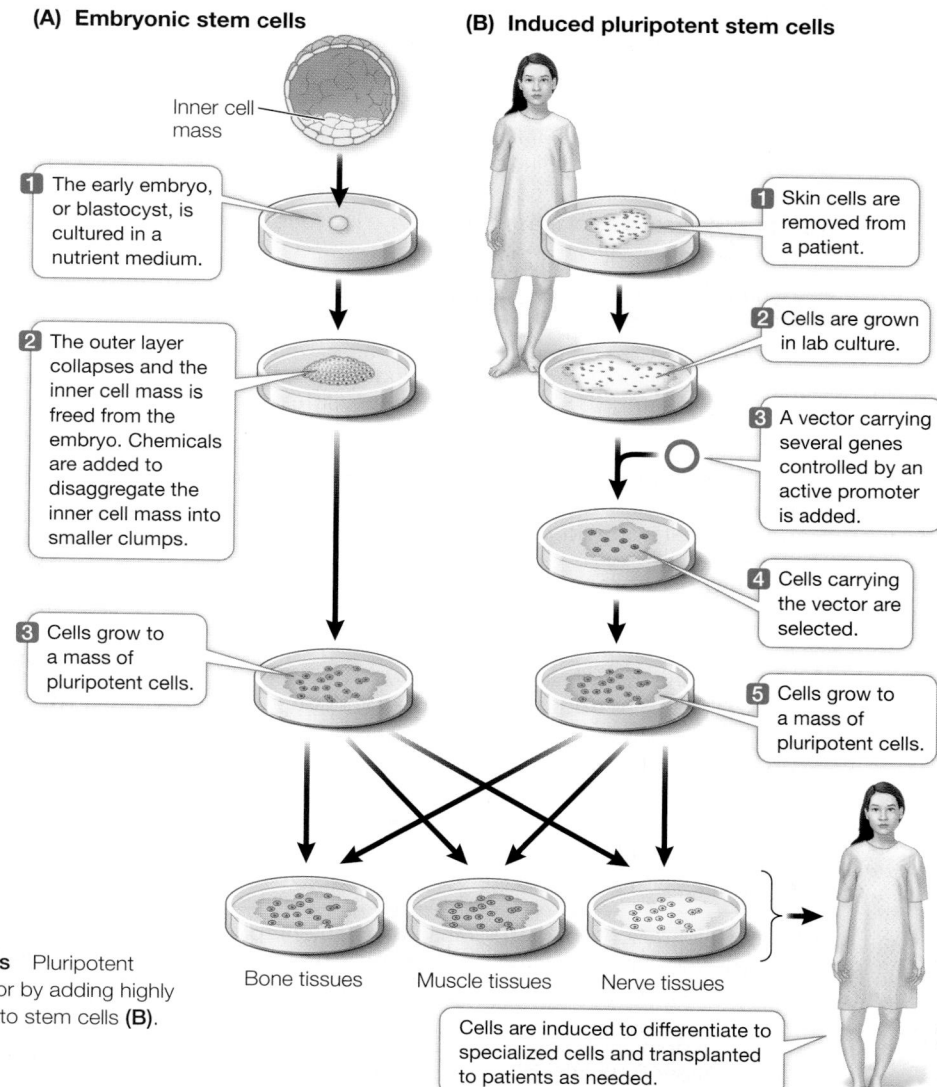

(A) Embryonic stem cells

Inner cell mass

1 The early embryo, or blastocyst, is cultured in a nutrient medium.

2 The outer layer collapses and the inner cell mass is freed from the embryo. Chemicals are added to disaggregate the inner cell mass into smaller clumps.

3 Cells grow to a mass of pluripotent cells.

(B) Induced pluripotent stem cells

1 Skin cells are removed from a patient.

2 Cells are grown in lab culture.

3 A vector carrying several genes controlled by an active promoter is added.

4 Cells carrying the vector are selected.

5 Cells grow to a mass of pluripotent cells.

Bone tissues   Muscle tissues   Nerve tissues

Cells are induced to differentiate to specialized cells and transplanted to patients as needed.

Figure 19.5 Two Ways to Obtain Pluripotent Stem Cells   Pluripotent stem cells can be obtained either from human embryos **(A)** or by adding highly expressed stem cell genes to skin cells to transform them into stem cells **(B)**.

 Animation 19.1 Embryonic Stem Cells
www.Life11e.com/a19.1

Shinya Yamanaka and coworkers at Kyoto University in Japan have developed a way to produce pluripotent stem cells that gets around these two problems (**Figure 19.5B**). Instead of extracting ESCs from blastocysts, they make pluripotent stem cells from skin cells. They developed this method systematically:

1. First they used microarrays (see Figure 18.8) to compare the genes expressed in ESCs with non-stem cells. They found several genes that were uniquely expressed at high levels in ESCs. These genes were believed to be essential to the undifferentiated state and function of stem cells.

2. Next they isolated the genes, coupled them to highly expressing promoters, and inserted them into skin cells (see Key Concept 18.5). They found that the skin cells now expressed the newly added genes at high levels.

3. Finally, they showed that the altered skin cells were pluripotent and could be induced to differentiate into many tissues. They called these cells **induced pluripotent stem cells** (**iPS cells**).

Because the iPS cells can be made from skin cells of the individual who is to be treated, an immune response may be avoided. Such cells have already been used for cell therapy in animals for diseases similar to human Parkinson's disease (a brain disorder), diabetes, and sickle-cell anemia. If it can be shown conclusively that iPS cells have the same properties as ESCs, human uses are sure to follow. At this writing, there are numerous clinical trials under way using iPS cells. For example, in Japan they are being used to regenerate cells of the retina in the eye in efforts to reverse macular degeneration.

▶ 19.1 recap

Development takes place via the processes of determination, differentiation, morphogenesis, and growth. Cells in the very early embryo have not yet had their fates determined; as development proceeds, their potential fates become more and more restricted. Differentiated cells retain their ability to differentiate into other cell types, given appropriate chemical signals—a fact that has made cloning and stem cell technologies possible.

learning outcomes

You should be able to:

- Interpret expected results from cell transplantation experiments in amphibian embryos with respect to what they tell us about cell fate determination.
- Outline experiments to compare gene expression before and after the reprogramming of gene expression that occurs during cloning.
- Describe the potential uses of reproductive cloning in humans and other animals.
- Compare human embryonic and adult stem cells.

1. What information did transplantation experiments provide about when cell fates are determined and the extent to which the environment can affect determination in early and late embryos?

2. How are stem cells in adult body tissues different from early embryonic stem cells?

▶ 19.1 recap

3. Cloning involves considerable reprogramming of gene expression in a differentiated cell nucleus so that it acts as it does in a totipotent cell. How would you investigate this reprogramming?

4. There are two considerations for human cloning. In therapeutic cloning, cells from an individual—either genetically altered or not—would be grown into organs for transplantation. In reproductive cloning, nuclear transfer would be used to clone a person. Cloning to produce a new organism has been successfully achieved in animals but not in humans. What are some of the uses for cloning in animals and some potential uses for cloning in humans?

You have learned about the basic processes that occur during development, and seen that cell fate determination occurs before cells differentiate and become specialized. We will now turn to the mechanisms of cell fate determination.

▶ key concept 19.2 — Gene Expression Differences Determine Cell Fate and Cell Differentiation

The fertilized egg undergoes many cell divisions to produce the many differentiated cells in the body (such as liver, muscle, and nerve cells). How can one cell produce so many different cell types? Determination occurs in two ways:

1. **Cytoplasmic segregation** (unequal cytokinesis). Factors that play important roles in setting up the signaling cascades that orchestrate the major events of development may be unequally distributed in the cytoplasm of the egg, zygote, or precursor cell. After cell division, one or more of these factors, generally called **cytoplasmic determinants**, end up in some daughter cells or regions of cells, but not others.

2. **Induction** (cell-to-cell communication). A factor is actively produced and secreted by certain cells to induce other cells to become determined.

focus your learning

- Polarity may take place early in development, even within the fertilized egg.
- Induction, whereby one part of the embryo signals another part to develop in a specific way, is exemplified by the development of the vulva in a nematode.
- Cell differentiation during development involves changes in gene expression, as shown by the activation of the transcription factor MyoD in muscle differentiation.

Cytoplasmic segregation can determine cell fate

Some differences in gene expression patterns and determination of a cell's fate happen as a result of cytoplasmic differences among cells. One such cytoplasmic difference is the emergence of distinct "top" and "bottom" ends of an organism or structure; a difference that establishes **polarity**. Many examples of polarity are observed as development proceeds. An axis of polarity is established by our heads and "tails," as well as between the distal (far) ends of our arms and legs (wrists, ankles, fingers, toes) and their proximal (near) ends (shoulders and hips).

Polarity may develop early; even within the fertilized egg, the yolk and other factors are often distributed asymmetrically. During *early development in animals*, polarity is specified by what is called the **animal pole** at the top of the zygote and the **vegetal pole** at the bottom. This polarity can lead to determination of cell fates at a very early stage of development. For example, sea urchin embryos can be bisected at the eight-cell stage in two ways:

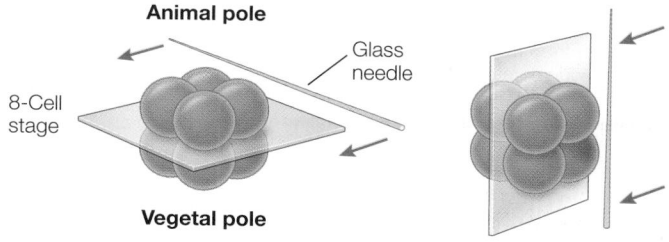

connect the concepts   The processes of early development in animals, including the influences of cytoplasmic determinants on cell fates and the ways in which the yolk and other factors are distributed in early sequences of cell division (i.e., in patterns of cleavage), are further considered in Key Concepts 43.1 and 43.2.

If the two halves of these embryos (each with four cells) are allowed to develop, the results are dramatically different for the two different cuts:

1. If the embryo is cut into a top half and a bottom half (left, above), the bottom half develops into a small sea urchin and the top half does not develop at all but instead gives rise to a hollow ball of undifferentiated cells.

2. If the embryo is cut into two side halves (right, above), both halves develop into small sea urchins.

These results indicate that the top and bottom halves of an eight-celled sea urchin embryo have *already developed distinct fates*. Such observations led to the model of cytoplasmic segregation shown in **Figure 19.6**. The model states that the cytoplasmic determinants are distributed unequally in the egg cytoplasm. Cytoplasmic determinants include specific proteins, small regulatory RNAs, and mRNAs, and they play roles in directing the embryonic development of many organisms. What accounts for the unequal distribution of these determinants?

 Animation 19.2 **Early Asymmetry in the Embryo**
www.Life11e.com/a19.2

Inducers passing from one cell to another can determine cell fate

The term "induction" has different meanings in different contexts. In biology it can be used broadly to refer to the initiation of, or cause of,

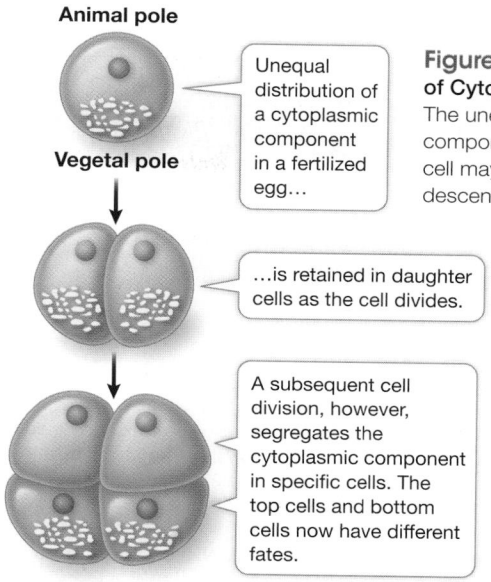

Figure 19.6 The Principle of Cytoplasmic Segregation The unequal distribution of some component in the cytoplasm of a cell may determine the fates of its descendants.

Unequal distribution of a cytoplasmic component in a fertilized egg…

…is retained in daughter cells as the cell divides.

A subsequent cell division, however, segregates the cytoplasmic component in specific cells. The top cells and bottom cells now have different fates.

a change or process. But in the context of cellular differentiation, it refers to the signaling events by which cells in a developing organism communicate and influence one another's developmental fate. Induction involves chemical signals and signal transduction mechanisms. An example of embryonic induction occurs during the development of a reproductive structure of the nematode *Caenorhabditis elegans*.

Caenorhabditis elegans develops from fertilized egg to larva in only about 12 hours, and the worm reaches the adult stage in just 3.5 days. The process is easily observed using a low-magnification dissecting microscope because the body covering is transparent (**Figure 19.7A**). Most adult nematodes are hermaphroditic, containing both male and female reproductive organs. They lay eggs through a pore called the vulva on the ventral (lower) surface. During development, a single cell in the overlying gonad, called the anchor cell, induces the vulva to form from six cells on a worm's ventral surface (**Figure 19.7B**). In this case there are two molecular signals: the primary (1°) inducer and the secondary (2°) inducer. Each of the six ventral cells has three possible fates: it may become a primary vulval precursor cell, a secondary vulval precursor cell, or simply become part of the worm's skin—an epidermal cell. You can follow the sequence of events in Figure 19.7B. The concentration gradient of the primary inducer, LIN-3, is key. (LIN stands for abnormal cell *line*age.) The anchor cell produces the LIN-3 protein, which diffuses out of the cell and forms a concentration gradient with respect to adjacent cells. Three cells receive more LIN-3 than the others and become vulval precursor cells; cells farther from the anchor cell receive less LIN-3 and become epidermal cells. The cell closest to the anchor cell receives the most LIN-3—enough LIN-3 to trigger a signal transduction response in the closest neighboring cell that results in the expression of the secondary inducer. The secondary inducer then acts on the two adjacent cells. This second induction event results in the two classes of vulval precursor cells: primary and secondary.

Induction involves the activation or inactivation of specific sets of genes through signal transduction cascades in the responding cells. When an inducer binds to its specific receptor on the surface of a cell, a signal transduction pathway leads to the activation of one or

(A) Basic anatomy of *C. elegans*

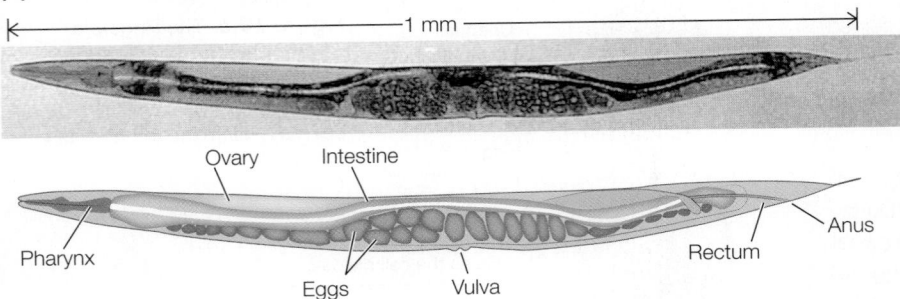

Ovary   Intestine

Pharynx

Eggs   Vulva

Rectum   Anus

Figure 19.7 Induction during Vulval Development in *Caenorhabditis elegans* **(A)** In the nematode *C. elegans* (shown in false color here), it is possible to follow all of the cell divisions from the fertilized egg to the 959 somatic cells found in the adult. **(B)** During vulval development, a molecule secreted by the anchor cell (the LIN-3 protein) acts as the primary (1°) inducer. The primary precursor cell (the one that receives the highest concentration of LIN-3) then secretes a secondary (2°) inducer that acts on its neighbors. The gene expression patterns triggered by these molecular switches determine cell fates.

(B) The influence of LIN-3 on the primary and secondary inducers involved in cell differentiation in *C. elegans*

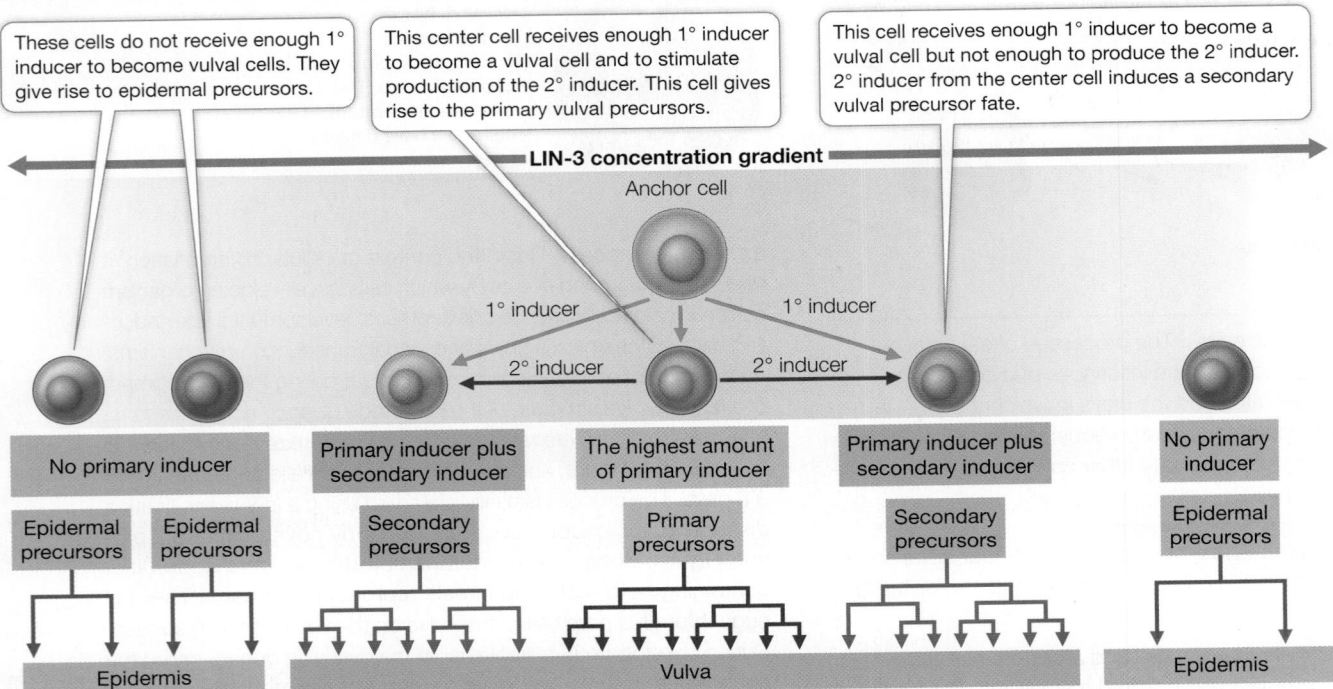

These cells do not receive enough 1° inducer to become vulval cells. They give rise to epidermal precursors.

This center cell receives enough 1° inducer to become a vulval cell and to stimulate production of the 2° inducer. This cell gives rise to the primary vulval precursors.

This cell receives enough 1° inducer to become a vulval cell but not enough to produce the 2° inducer. 2° inducer from the center cell induces a secondary vulval precursor fate.

LIN-3 concentration gradient

Anchor cell

1° inducer   1° inducer

2° inducer   2° inducer

| No primary inducer | Primary inducer plus secondary inducer | The highest amount of primary inducer | Primary inducer plus secondary inducer | No primary inducer |

| Epidermal precursors | Epidermal precursors | Secondary precursors | Primary precursors | Secondary precursors | Epidermal precursors |

Epidermis   Vulva   Epidermis

---

more transcription factors. Recall from Key Concept 16.2 that these DNA-binding proteins regulate the expression of specific genes. You can follow this process in **Focus: Key Figure 19.8**, where the cell on the left is exposed to a high concentration of inducer. The inducer activates a transcription factor in the cytoplasm, which enters the nucleus to switch on the expression of a specific gene. The cell on the right is exposed to a lower concentration of the inducer, and as a result, gene expression is not activated.

Differential gene transcription is a hallmark of cell differentiation

One well-studied example of cell differentiation is the conversion of un-differentiated muscle precursor cells into cells destined to form muscle (**Figure 19.9**). In the vertebrate embryo these cells come from a tissue layer called the *****mesoderm**. A key event in the commitment of these cells to become muscle is that they stop dividing. Indeed, in many parts of the embryo, *cell division and cell differentiation are mutually exclusive.* Cell signaling activates the gene for a transcription factor called **MyoD** (for *myo*blast-*d*etermining gene); this in turn activates

the gene for p21, which is an inhibitor of the cyclin-dependent kinases (Cdk's) that normally stimulate the cell cycle at G1 (see Figure 11.5). Expression of the *p21* gene causes the cell cycle to stop, and other transcription factors then enter the picture so that differentiation can proceed. Interestingly, MyoD is also activated in the stem cells that are present in adult muscle, indicating a role of this transcription factor in the repair of muscle tissue as it gets damaged and worn out.

*****connect the concepts**   The mesoderm is one of three germ layers (primitive tissue layers) formed in animal embryos in a developmental phase called gastrulation (see Key Concept 43.3).

Genes such as *myoD* that direct the most fundamental decisions in development (often by regulating other genes on other chromosomes) usually encode transcription factors. In some cases a single transcription factor can cause a cell to differentiate in a certain way. In others, complex interactions between genes and proteins determine a sequence of transcriptional events that leads to differentiation.

focus: key figure

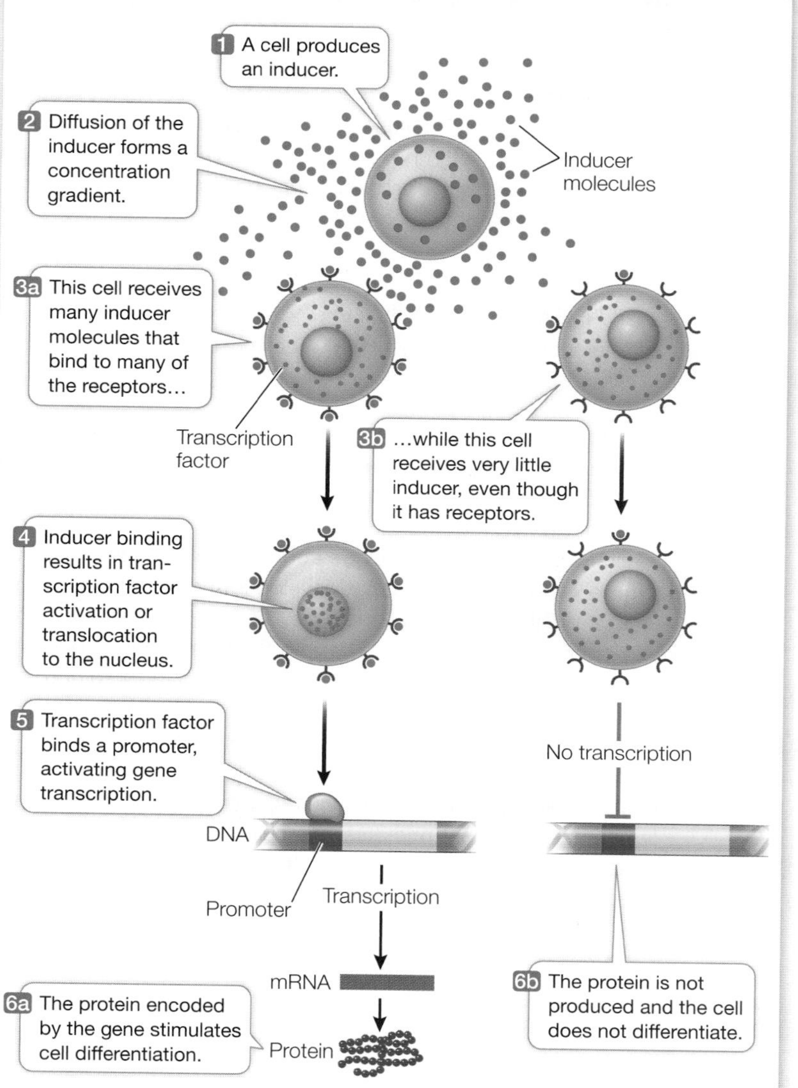

1 A cell produces an inducer.

2 Diffusion of the inducer forms a concentration gradient.

Inducer molecules

3a This cell receives many inducer molecules that bind to many of the receptors…

Transcription factor

3b …while this cell receives very little inducer, even though it has receptors.

4 Inducer binding results in transcription factor activation or translocation to the nucleus.

5 Transcription factor binds a promoter, activating gene transcription.

No transcription

DNA

Promoter   Transcription

mRNA

6a The protein encoded by the gene stimulates cell differentiation.

Protein

6b The protein is not produced and the cell does not differentiate.

Figure 19.8 Induction   The concentration of an inducer directly affects the degree to which a transcription factor is activated. The inducer acts by binding to a receptor on the target cell. This binding is followed by signal transduction involving transcription factor activation or translocation from the cytoplasm to the nucleus. In the nucleus it acts to stimulate the expression of genes involved in cell differentiation.

Q: Could differential expression of receptor genes result in differential gene expression through induction?

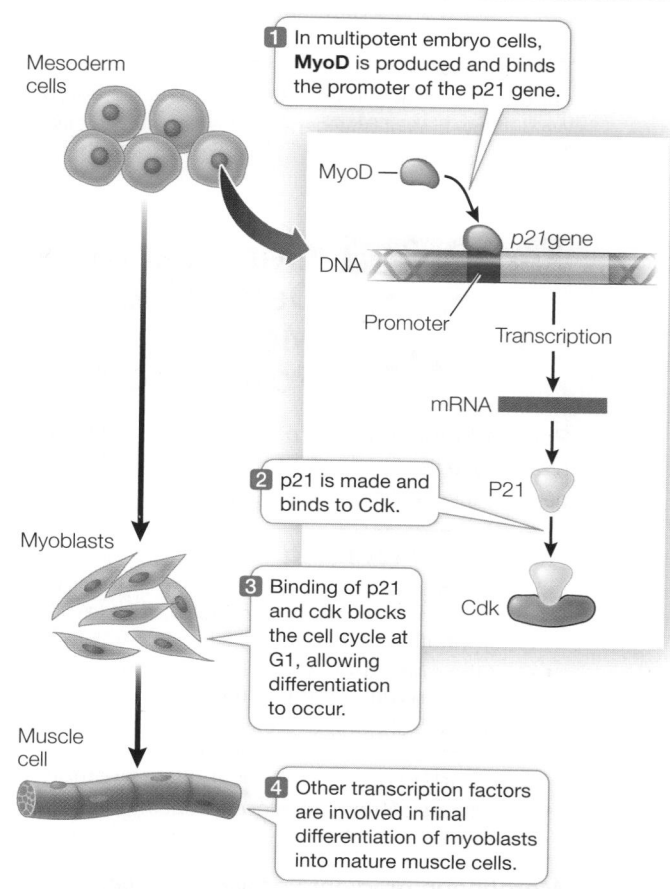

Mesoderm cells

1 In multipotent embryo cells, **MyoD** is produced and binds the promoter of the p21 gene.

MyoD

p21 gene

DNA

Promoter   Transcription

mRNA

2 p21 is made and binds to Cdk.

P21

Myoblasts

3 Binding of p21 and cdk blocks the cell cycle at G1, allowing differentiation to occur.

Cdk

Muscle cell

4 Other transcription factors are involved in final differentiation of myoblasts into mature muscle cells.

Figure 19.9 Transcription and Differentiation in the Formation of Muscle Cells   Production of the transcription factor MyoD is important in muscle cell differentiation.

19.2 recap

Cell fate determination involves cytoplasmic segregation and induction. Cytoplasmic segregation is the unequal distribution of gene products in the egg, zygote, or early embryo. Induction occurs when one cell or tissue sends a chemical signal to another. Cell fate determination involves the activation of signal transduction pathways that lead to differential gene expression. Differentiation involves selective gene expression, controlled at the level of transcription by transcription factors.

learning outcomes

You should be able to:

- Summarize how cytoplasmic segregation results in polarity in a fertilized egg and how this polarity affects cell fate determination.
- Predict the consequences of manipulations of inducer-producing cells on development and differentiation.
- Predict cellular, gene-expression, and phenotypic consequences of experimental manipulations of a transcription factor, such as MyoD concentrations on the development of muscle tissue.

1. How does cytoplasmic segregation result in polarity in a fertilized egg, and how does polarity affect cell fate determination?

2. What would happen if an artificial, impermeable membrane were placed between the anchor cell and the underlying cells of the developing nematode? Explain your answer.

3. Molecular biologists can attach genes to active promoters and insert them into cells. What would happen if *myoD* in undifferentiated myoblasts were inserted and overexpressed? Explain your answer.

You have seen how cell fate is determined and examined the roles of gene expression in cell fate determination and differentiation. We will now look at how gene expression affects differentiation and morphogenesis.

key concept

19.3 Gene Expression Determines Morphogenesis and Pattern Formation

Pattern formation is the process that results in the spatial organization of a tissue or organism. It is inextricably linked to morphogenesis, the creation of body form.

---

focus your learning

- In development of flowering plants, organ identity genes encode transcription factors that result in the differentiation of floral organs.
- Genetic and experimental lines of evidence support the floral organ determination model.
- A cascade of gene expression events early in development establishes the differentiation of growth axes and organ formation in the fruit fly embryo.
- Alterations of Hox genes can result in homeotic mutations.

Morphogen gradients provide positional information

During development, the key cellular question "What will I be?" is often answered in part by "Where am I?" Think of the cells in the developing nematode, which can develop into different parts of the vulva depending on their positions relative to the anchor cell (see Figure 19.7). This spatial "sense" is called **positional information**. Positional information often comes in the form of an inducer called a **morphogen**, which diffuses from one cell or group of cells to surrounding cells, setting up a concentration gradient (as you saw for LIN-3 in *C. elegans* vulval induction). There are two requirements for a signal to be considered a morphogen:

1. It must directly affect target cells, rather than triggering a secondary signal that affects target cells.
2. Different concentrations of the signal must cause different effects.

Developmental biologist Lewis Wolpert uses the "French flag model" to explain morphogens (**Figure 19.10A**). This model can be applied to the differentiation of the vulva in *C. elegans* and to the development of a vertebrate limb.

The vertebrate limb develops from a paddle-shaped limb bud (**Figure 19.10B**). The cells that develop into different digits must receive positional information; if they do not, the limb will be totally disorganized—imagine a hand with only thumbs or only little fingers. A group of cells at the posterior base of the limb bud, just where it joins the body wall, is called the zone of polarizing activity (ZPA). The cells of the ZPA secrete a protein morphogen called *Sonic hedgehog* (Shh). Shh forms a gradient that determines the posterior–anterior (little finger to thumb) axis of the developing limb. In humans and other primates, the cells exposed to the highest dose of Shh form the little finger; those that receive the lowest dose develop into the thumb.

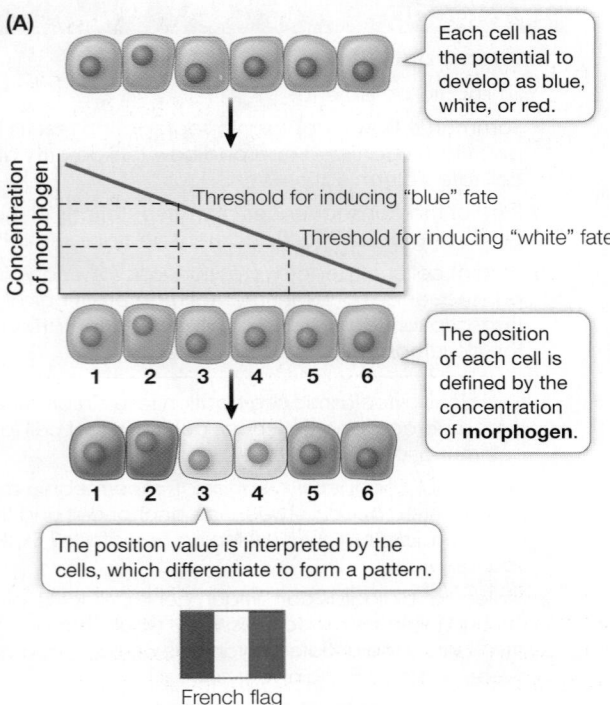

(A)

Each cell has the potential to develop as blue, white, or red.

Threshold for inducing "blue" fate
Threshold for inducing "white" fate

The position of each cell is defined by the concentration of **morphogen**.

The position value is interpreted by the cells, which differentiate to form a pattern.

French flag

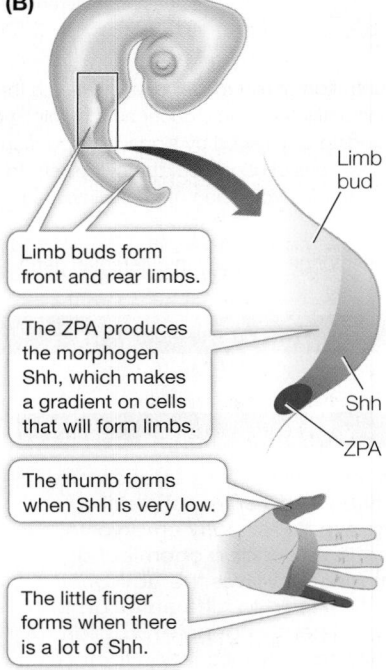

(B)

Limb bud

Limb buds form front and rear limbs.

The ZPA produces the morphogen Shh, which makes a gradient on cells that will form limbs.

Shh

ZPA

The thumb forms when Shh is very low.

The little finger forms when there is a lot of Shh.

Figure 19.10 The French Flag Model **(A)** In the "French flag" model, a concentration gradient of a diffusible morphogen signals each cell to specify its position. **(B)** The zone of polarizing activity (ZPA) in the limb bud of the embryo secretes the morphogen *Sonic hedgehog* (Shh). Cells in the bud form different digits depending on the concentration of Shh.

(A)

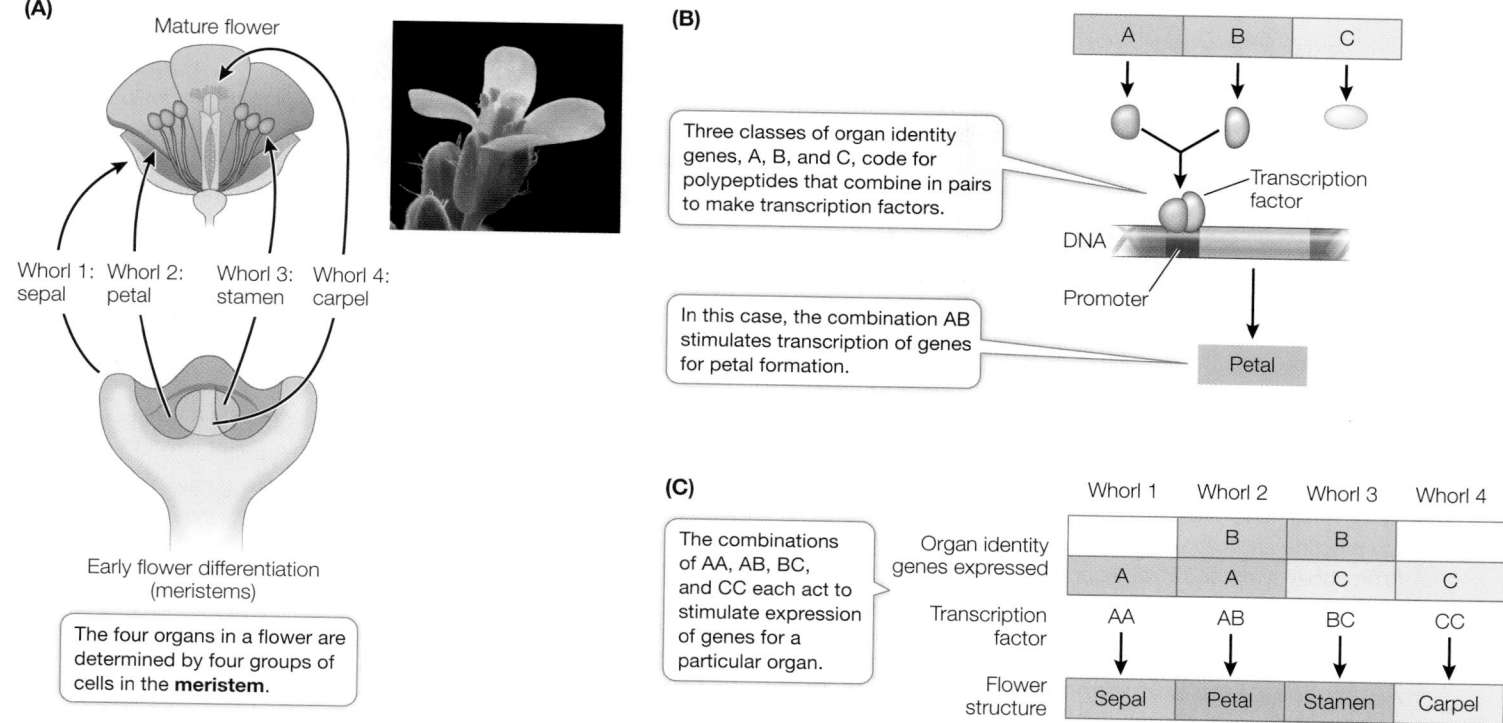

(B)

Three classes of organ identity genes, A, B, and C, code for polypeptides that combine in pairs to make transcription factors.

In this case, the combination AB stimulates transcription of genes for petal formation.

(C)

The combinations of AA, AB, BC, and CC each act to stimulate expression of genes for a particular organ.

	Whorl 1	Whorl 2	Whorl 3	Whorl 4
Organ identity genes expressed		B	B	
	A	A	C	C
Transcription factor	AA	AB	BC	CC
Flower structure	Sepal	Petal	Stamen	Carpel

Figure 19.11 Organ Identity Genes in *Arabidopsis* Flowers   (A) The four organs of a flower—carpels (yellow), stamens (green), petals (purple), and sepals (pink)—grow in whorls that develop from the floral meristem. **(B)** Floral organs are determined by three classes of organ identity genes whose polypeptide products combine in pairs to form transcription factors.

(C) Combinations of polypeptide subunits in transcription factors activate gene expression for specific organs.

 Activity 19.3 Genes and Development Simulation
www.Life11e.com/ac19.3

Expression of transcription factor genes determines organ differentiation in plants

Like animals, plants have organs—for example, leaves and roots. Many plants form flowers, and many flowers are composed of four types of organs: sepals, petals, stamens (male reproductive organs), and carpels (female reproductive organs). These floral organs occur in concentric whorls, with groups of each organ type encircling a central axis. The sepals are on the outside and the carpels are on the inside (**Figure 19.11A**).

In *Arabidopsis thaliana* (thale cress), flowers develop in a radial pattern around the shoot (stem and leaf) apex as it develops and elongates. At the shoot apex and in other parts of the plant where growth and differentiation occur (such as the root tip), there are groups of undifferentiated, rapidly dividing cells called **meristems**. Each flower begins as a floral meristem of about 700 undifferentiated cells arranged in a dome, and the four whorls develop from this meristem. How is the identity of a particular whorl determined? Three classes of genes called **organ identity genes** encode proteins that act in combination to produce specific whorl features (**Figure 19.11B and C**):

1. Genes in class A are expressed in whorls 1 and 2 (which form sepals and petals, respectively).

2. Genes in class B are expressed in whorls 2 and 3 (which form petals and stamens).

3. Genes in class C are expressed in whorls 3 and 4 (which form stamens and carpels).

These genes encode transcription factors that are active as dimers, that is, proteins with two polypeptide subunits. The composition of the dimer determines which genes the transcription factor activates. For example, a dimer made up of two class A monomers activates transcription of the genes that make sepals; a dimer made up of A and B monomers results in petals, and so forth. A common feature of the A, B, and C proteins, as well as many other plant transcription factors, is a DNA-binding domain called the **MADS box**. The name "MADS" comes from the initials of four genes encoding proteins with this domain.

Two lines of experimental evidence support this model for floral organ determination:

1. *Loss-of-function mutations*: for example, a mutation in a class A gene results in no sepals or petals.

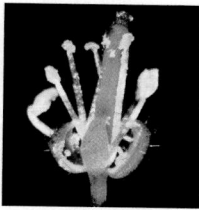

2. *Gain-of-function mutations*: for example, a promoter for a class C gene can be artificially coupled to a class A gene. In this case, the class A gene is expressed in all four whorls, resulting in only sepals and petals. In any organism, the replacement of

one organ for another is called homeosis, and this type of mutation is a **homeotic mutation**.

Transcription of the floral organ identity genes is controlled by other gene products, including the LEAFY protein. The wild-type LEAFY protein is a transcription factor that stimulates expression of the class A, B, and C genes so that they produce flowers. Plants with loss-of-function mutations in the *LEAFY* gene make stems instead of flowers, with increased numbers of modified leaves called bracts. This finding has practical applications. It usually takes 6–20 years for a citrus tree to produce flowers and fruits. Scientists have made transgenic orange trees expressing the *LEAFY* gene coupled to a strongly expressed promoter. These trees flower and fruit years earlier than normal trees.

A cascade of transcription factors establishes body segmentation in the fruit fly

Perhaps the best-studied example of how morphogens determine cell fate is body segmentation in the fruit fly *Drosophila melanogaster*. The body segments of a fruit fly are different from one another. The adult fly has an anterior head (composed of several fused segments), three different thoracic segments, and eight abdominal segments at the posterior end. Each segment develops into different body parts: for example, antennae and eyes develop from head segments, wings from the thorax, and so on.

The life cycle of *Drosophila* from fertilized egg to adult takes about 2 weeks at room temperature. The egg hatches into a larva, which then forms a pupa, which finally is transformed into the adult fly. By the time a larva appears—about 24 hours after fertilization—there are recognizable segments. The thoracic and abdominal segments all look similar, *but the fates of their cells to become different adult segments have already been determined*.

As in other organisms, fertilization in *Drosophila* leads to a rapid series of mitoses. However, the first 12 cycles of nuclear division are not accompanied by cytokinesis. So a multinucleate embryo forms instead of a multicellular embryo (the nuclei are brightly stained in the micrographs below):

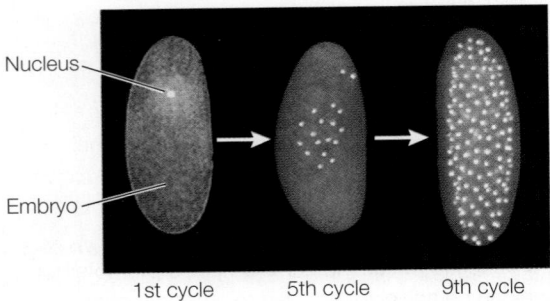

Nucleus

Embryo

1st cycle   5th cycle   9th cycle

With no cell membranes to cross, morphogens can diffuse easily within the embryo. We focus here on the determination events that occur in the first 24 hours after fertilization. We'll present what may appear to be a lot of detail but is really just an outline of the molecular events. In describing them, biologists are getting closer to answering a key question: How does a complex organism develop from a single cell?

Biologists used experimental genetics to describe the events leading to cell fate determination:

- First, developmental mutations were identified. For example, a mutant strain might produce larvae with two heads or no segments.

- Then the mutant flies were compared with wild-type flies, and the gene responsible for the developmental mistake, and the gene's protein product (if appropriate), were isolated.

- Finally, experiments with the gene (making transgenic flies) and protein (injecting the protein into an egg or into an embryo) were done to confirm their roles in the proposed developmental pathway.

These approaches revealed an amazing cascade of gene expression events that result in the determination of each segment within 24 hours after fertilization. Several classes of genes are involved:

- **Maternal effect genes** set up the major axes (anterior–posterior and dorsal–ventral) of the egg.

- **Segmentation genes** determine the boundaries and polarity of each segment.

- **Hox genes** determine which organ will be made at a given location.

MATERNAL EFFECT GENES   Like the eggs and early embryos of sea urchins, *Drosophila* eggs and larvae are characterized by unevenly distributed cytoplasmic determinants (see Figure 19.6). These molecular determinants are the products of specific maternal effect genes that are transcribed in the cells of the mother's ovary. Two maternal effect genes called *bicoid* and *nanos* help determine the anterior–posterior axis of the egg. (The dorsal–ventral axis is determined by other maternal effect genes that will not be described here.)

The mRNAs for *bicoid* and *nanos* diffuse from the mother's cells and are passed to the egg by cytoplasmic bridges into what will be the anterior end of the egg. The *bicoid* mRNA is translated into Bicoid protein, a transcription factor that diffuses away from the anterior end, establishing a gradient in the egg cytoplasm (**Figure 19.12A**). Meanwhile, the egg's cytoskeleton transports the *nanos* mRNA from the anterior end of the egg, where it was deposited, to the posterior end, where it is translated (**Figure 19.12B**).

The actions of Bicoid and Nanos establish a gradient of yet another protein, called Hunchback, which determines the anterior and posterior ends of the embryo. Initially, the *hunchback* mRNA is evenly distributed in the embryo, but Nanos inhibits its translation, thus preventing Hunchback protein accumulation at the posterior end of the embryo (**Figure 19.12C**). Meanwhile, at the anterior end of the embryo, Bicoid stimulates increased transcription of the *hunchback* gene, thus increasing the amount of *hunchback* mRNA (and thus Hunchback protein) and further strengthening the Hunchback gradient.

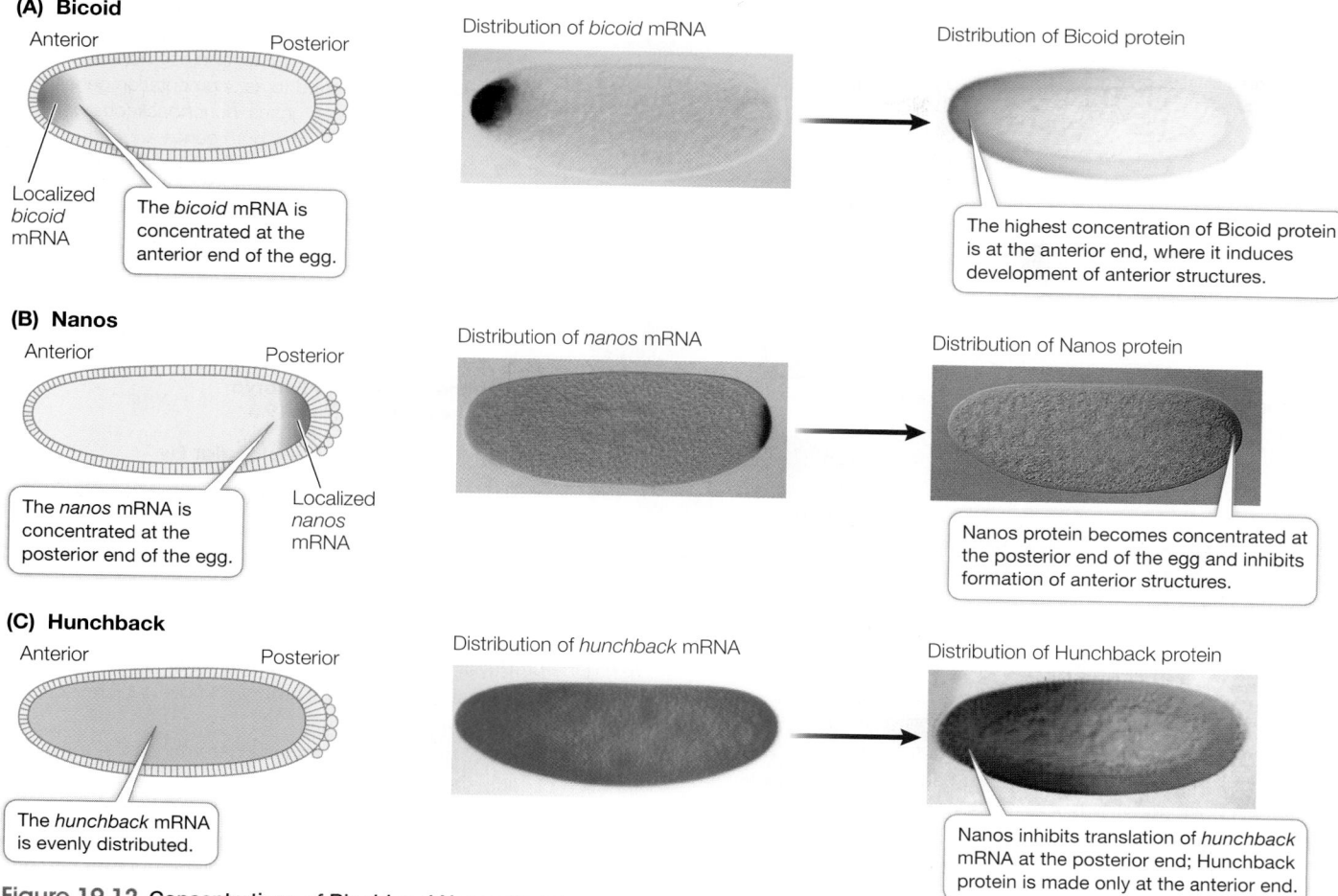

(A) Bicoid

Anterior   Posterior

Localized *bicoid* mRNA

The *bicoid* mRNA is concentrated at the anterior end of the egg.

Distribution of *bicoid* mRNA

Distribution of Bicoid protein

The highest concentration of Bicoid protein is at the anterior end, where it induces development of anterior structures.

(B) Nanos

Anterior   Posterior

The *nanos* mRNA is concentrated at the posterior end of the egg.

Localized *nanos* mRNA

Distribution of *nanos* mRNA

Distribution of Nanos protein

Nanos protein becomes concentrated at the posterior end of the egg and inhibits formation of anterior structures.

(C) Hunchback

Anterior   Posterior

The *hunchback* mRNA is evenly distributed.

Distribution of *hunchback* mRNA

Distribution of Hunchback protein

Nanos inhibits translation of *hunchback* mRNA at the posterior end; Hunchback protein is made only at the anterior end.

Figure 19.12 Concentrations of Bicoid and Nanos Proteins Determine the Anterior–Posterior Axis   The anterior–posterior axis of *Drosophila* arises from gradients of the morphogens encoded by **(A)** *bicoid* and **(B)** *nanos*. Together, Bicoid and Nanos establish a concentration gradient of Hunchback **(C)**.

How did biologists elucidate these pathways? Let's look at the experimental approaches used in this case.

- Females that are homozygous for a particular *bicoid* mutation produce larvae with no head and no thorax; thus the Bicoid protein must be needed for the anterior structures to develop.

- If the eggs of these *bicoid* mutant flies are injected at the anterior end with cytoplasm from the anterior region of a wild-type egg, the injected eggs develop into normal larvae. This experiment also shows that the Bicoid protein is involved in the development of anterior structures.

- If cytoplasm from the anterior region of a wild-type egg is injected into the posterior region of another egg, anterior structures develop there. The degree of induction depends on how much cytoplasm is injected.

- Eggs from homozygous *nanos* mutant females develop into larvae with missing abdominal segments.

- If cytoplasm from the posterior region of a wild-type egg is injected into the posterior region of a *nanos* mutant egg, it will develop normally.

The events involving *bicoid*, *nanos*, and *hunchback* begin before fertilization and continue after it, during the multinucleate stage, which lasts a few hours. At this stage the embryo looks like a bunch of indistinguishable nuclei under the light microscope. But a lot is going on at the molecular level, as cell fates have already begun to be determined. After the anterior and posterior ends have been established, the next step in pattern formation is the determination of segment number and locations.

SEGMENTATION GENES   These genes determine the number and polarity of the *Drosophila* larval segments and are expressed when there are about 6,000 nuclei in the embryo (about 3 hours after fertilization). Three classes of segmentation genes act one after the other to regulate finer and finer details of the segmentation pattern:

1. **Gap genes** organize broad areas along the anterior–posterior axis. Mutations in gap genes result in gaps in the body plan—the omission of several consecutive larval segments.

2. **Pair rule genes** divide the embryo into units of two segments each. Mutations in pair rule genes result in embryos missing every other segment.

3. **Segment polarity genes** determine the boundaries and anterior–posterior organization of the individual segments.

(A)

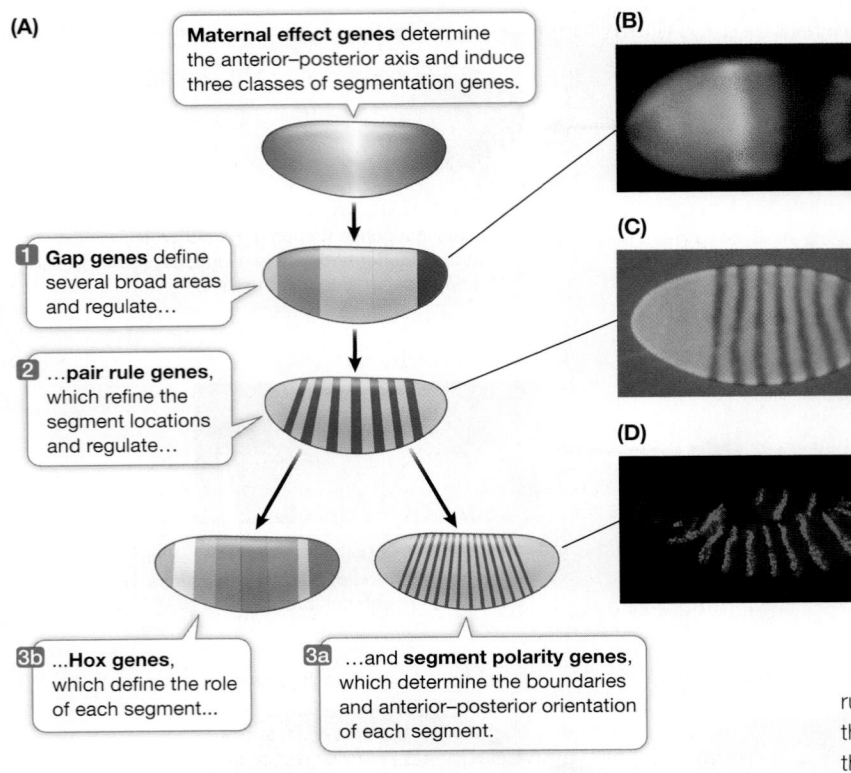

Maternal effect genes determine the anterior–posterior axis and induce three classes of segmentation genes.

1 Gap genes define several broad areas and regulate...

2 ...pair rule genes, which refine the segment locations and regulate...

3b ...Hox genes, which define the role of each segment...

3a ...and segment polarity genes, which determine the boundaries and anterior–posterior orientation of each segment.

(B)

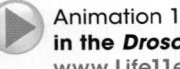

(C)

(D)

Figure 19.13 A Gene Cascade Controls Pattern Formation in the *Drosophila* Embryo (A) Maternal effect genes induce gap, pair rule, and segment polarity genes—collectively referred to as segmentation genes. **(B)** Expression of two gap genes, *hunchback* (orange) and *Krüppel* (green), overlaps; both genes are transcribed in the yellow area. **(C)** The pair rule gene *fushi tarazu* is transcribed in the dark blue areas. **(D)** The segment polarity gene *engrailed* (bright green) is seen here at a slightly more advanced stage than is depicted in (A). By the end of this cascade, a group of nuclei at the anterior of the embryo, for example, is determined to become the first head segment in the adult fly.

▶ Animation 19.3 **Pattern Formation in the *Drosophila* Embryo** www.Life11e.com/a19.3

▶ Media Clip 19.1 **Spectacular Fly Development in 3D** www.Life11e.com/mc19.1

Mutations in segment polarity genes can result in segments in which posterior structures are replaced by reversed (mirror-image) anterior structures.

The expression of these genes is sequential (**Figure 19.13**). The products of the gap genes activate pair rule genes, and the pair rule gene products activate segment polarity genes. By the end of this cascade, nuclei throughout the embryo "know" which segment they will be part of in the adult fly.

The next set of genes in the cascade determines the form and function of each segment.

HOX GENES Hox (for "Homeobox") genes encode a family of transcription factors that are expressed in different combinations along the length of the embryo, and help determine cell fate within each segment. Hox gene expression tells the cells of a segment in the head to make eyes, those of a segment in the thorax to make wings, and so on. The *Drosophila* Hox genes occur in two clusters on chromosome 3, in the same order as the segments whose function they determine (**Figure 19.14**). By the time the fruit fly larva hatches, its segments are completely determined. Hox genes are shared by all animals and are homeotic genes—that is, a mutation in a Hox gene can result in one organ being replaced by another.

In *Drosophila*, the maternal effect genes, segmentation genes, and Hox genes interact to "build" a larva step by step, beginning with the unfertilized egg. How do we know that the Hox genes determine segment identity? A clue comes from homeotic mutations. A mutation in the Hox gene *Antennapedia* causes legs to grow on the head in place of antennae.

In the adult fly, this segment...

...is determined by this gene.

This diagram approximates the positions of gene expression in the embryo.

Head | Thorax (T1–T3) | Abdomen (A1–A8)

Drosophila embryo (10 hours)

lab *pb* *Dfd* *Scr* *Antp* *Ubx* *AbdA* *AbdB*

Antennapedia cluster | *Bithorax* cluster

Figure 19.14 Hox Genes in *Drosophila* Determine Segment Identity Two clusters of Hox genes on chromosome 3 (center) determine segment function in the adult fly (top). These genes are expressed in the embryo (bottom) long before the structures of the segments actually appear.

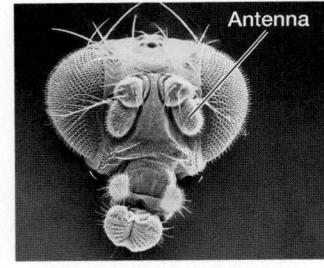
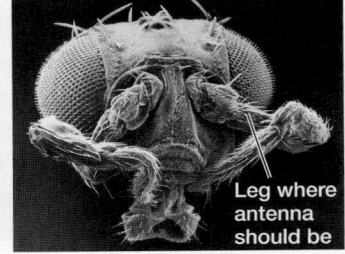

Antenna

Leg where antenna should be

When another Hox gene—*Ultrabithorax*—is mutated, an extra pair of wings grows in a thoracic segment where wings do not normally occur. So the normal (wild-type) functions of Hox genes must be to "tell" a segment what organ to form.

The *Antennapedia* and *Ultrabithorax* genes both encode transcription factors and have a common 180 base-pair sequence called the **homeobox**. The homeobox encodes a 60 amino acid sequence called the **homeodomain**. The homeodomain recognizes and binds to a specific DNA sequence in the promoters of its target genes. This protein domain is found in transcription factors that regulate development in many other animals with an anterior–posterior axis. We will discuss the evolutionary significance of these common pathways for development in the next section.

▶ 19.3 recap

Cascades of transcription factors govern pattern formation and the subsequent development of animal and plant organs. Often these transcription factors create or respond to morphogen gradients. In plants, cell fate is often determined by MADS box genes; in animal embryos, cell fate is determined in part by Hox genes.

learning outcomes

You should be able to:

- Summarize the function of floral organ identity genes.
- Use an example to explain how organ identity genes work in combination to produce flower organs.
- Determine the appropriate homeotic mutation to use to generate flowers with particular alterations in their organs.
- Summarize how mutations were used to identify organ identity genes.
- Predict the consequences of altering *nanos* expression for the developing fly embryo.

1. What are organ identity genes, and how are they identified by mutations? How do they act in combination to produce different organs in a developing plant?

2. If you wanted a rose plant to make flowers with only petals, what kind of homeotic mutation would you seek in the rose genome?

3. Biotechnology enables the insertion of foreign genes into host cells along with promoters that maximize its expression (see the discussion of expression vectors in Key Concept 18.5). What would happen if *nanos* were inserted and over-expressed at the anterior end of the *Drosophila* embryo?

We have described some of the details of the role of gene expression in the control of development in two model organisms—*Arabidopsis* and *Drosophila*. Both involve genes whose expression as transcription factors control the expressions of other genes that result in differentiation and organ formation. Are these mechanisms basic to developmental biology, and have they been conserved through evolution?

▶ key concept 19.4   Changes in Gene Expression Underlie the Evolution of Development

The discovery of the genes that control the development of *Drosophila* provided biologists with tools to investigate the development of other organisms. For example, when scientists used homeobox DNA as a hybridization probe (see Key Concept 14.4) to search for similar genes elsewhere, they found the homeobox sequence in many genes in other organisms. This and other discoveries that followed revealed similarities in the molecular events underlying morphogenesis in organisms ranging from flies to fish to mammals. These findings suggested that just as the forms of organisms evolved through descent with modification from a common ancestor, so did the molecular mechanisms that produce those forms. Biologists started to ask new questions about the interplay between evolutionary and developmental processes, a field of study called **evolutionary developmental biology**, or **evo-devo**.

focus your learning

- Common developmental pathways are involved in embryonic development of many animals, such as the formation of eyes in both insects and vertebrates.
- DNA sequences and regulatory proteins common in developmental pathways comprise a "toolkit." Genetic switches control how the toolkit is expressed.
- Heterochrony has significant importance, as illustrated by the extreme length of a giraffe's neck.
- As organisms have evolved, there are changes in where (heterotopy), when (heterochrony), and to what extent (heterometry) developmental genes are expressed.
- Spatial differences in the expression of a developmental gene are known as heterotopy.

What is evo-devo?

The basic principles of evo-devo are:

- Organisms share similar molecular mechanisms for development that include a "toolkit" of regulatory molecules that control the expression of genes.

- Toolkit regulatory molecules are able to act independently in different tissues and regions of the body, enabling modular evolutionary change.

- Developmental differences can arise from changes in the timing of regulatory molecule action, the location of its action, or the quantity of its action.

- Differences among species can arise from alterations in the expression of developmental genes.

- Developmental changes can arise from environmental influences on developmental processes.

As you saw earlier in the chapter, the development of a multicellular organism from a fertilized egg involves sequential gene expression. The complexities for each organism led biologists to predict that the pathways that make a mouse, for example, would be very different from those that make a fruit fly. But they were in for a surprise. It turns out that the regulatory genes that control the formation of organs are similar in very different organisms.

Developmental genes in distantly related organisms are similar

About a dozen major *kinds of eyes are found among the different animals, including the camera-like eyes of humans and the compound eyes of insects. Although the eyes of insects and vertebrates evolved independently, a remarkable discovery showed that common developmental pathways are involved in the embryonic formation of eyes in both groups.

---

*connect the concepts   The structures of vertebrate and invertebrate eyes are discussed and compared in Key Concept 45.4.

More than a century ago, a mutation was described in *Drosophila* called *eyeless*, with a phenotype of—you guessed it—no eyes.

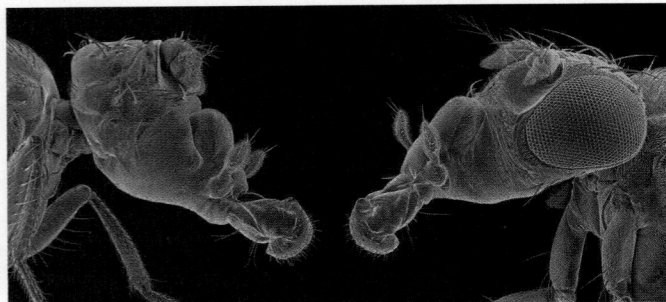

This mutation remained a laboratory curiosity for many decades until its molecular nature was elucidated in the 1990s. Swiss developmental biologists Rebecca Quiring and Walter Gehring isolated the protein product of the wild-type version of the *eyeless* gene, which they determined was a transcription factor that controls the expression of genes responsible for eye development. How did they know this? By making recombinant DNA constructs that allowed the wild-type *eyeless* gene to be expressed in different embryonic tissues of transgenic flies, they were able to produce flies with extra eyes on various body parts such as the legs, the antennae, and under the wings.

The big surprise came when the biologists sequenced the *eyeless* gene and used computers to query a database of genes with known sequences. The *eyeless* gene sequence was similar to that of the *Pax6* gene in mice; *Pax6*, when mutated, leads to the development of abnormally small eyes. Could the extremely different eyes of flies and mice be variations on a common developmental theme? To test for functional similarity between the insect and mammalian genes, the biologists repeated their experiments on flies, but using the mouse *Pax6* gene instead of the fly *eyeless* gene. Once again, eyes developed at various sites on the transgenic flies. Thus a gene whose expression normally leads to the development of a mammalian eye now led to the development of the very different insect eye.

We have to look very far back in evolutionary time for a common ancestor of fruit flies and mice. Yet the *eyeless* and *Pax6* genes

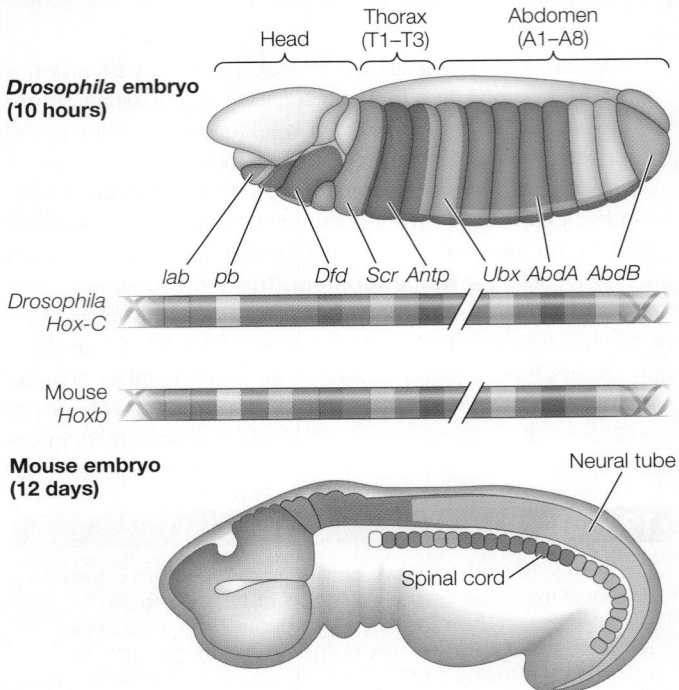

Figure 19.15 Regulatory Genes Show Similar Expression Patterns   Homologous genes encoding similar transcription factors are expressed in similar patterns along the anterior–posterior axes of both insects and vertebrates. The mouse (and human) Hox genes are present in multiple copies; this prevents a single mutation from having drastic effects.

contain sequences that are highly conserved, not only in these two species but in others as well. Biologists call such genes **homologous**, meaning that they evolved from a gene present in a common ancestor.

In recent years a large number of homologous genes (the regulatory "toolkit") have been shown to control development in distantly related species. For example, the fruit fly homeotic genes such as *Antennapedia* and *bithorax* are similar to mouse (and human) genes that play similar developmental roles. This discovery indicates that the positional information controlled by these genes has been conserved, even as the structures formed at each position have changed. Remarkably, these genes are arranged along a chromosome in both fruit fly and mouse in the same order as they are expressed along the anterior–posterior axis of the embryos (**Figure 19.15**). These and other examples have led biologists to the idea that certain developmental mechanisms are controlled by specific sequences comprising a **genetic toolkit**, the contents of which have been changed and reshuffled over the course of evolutionary time to produce the amazing diversity of plants, animals, and other organisms that we see today.

One way that the toolkit can change is through *gene duplication. As noted in Figure 19.14, *Antennapedia* and *bithorax* are not just single genes, but gene clusters, with each gene slightly different from the others. When we described gene families in Key Concept 17.3, we observed that each member of the family evolved from a single, ancestral gene. So it is with these developmental genes. Through duplication and subsequent mutation, the Hox genes diverged to encode different structures in different segments:

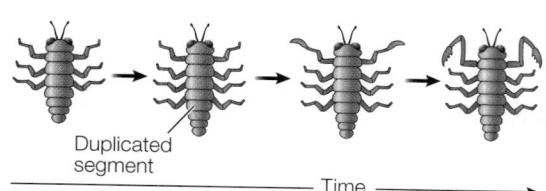

Duplicated segment

Time

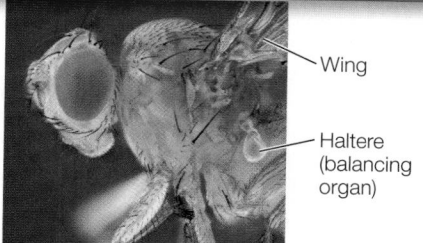

Wing

Haltere (balancing organ)

---

*connect the concepts   Key Concept 23.3 explains that most new functions arise as a result of gene duplication.

Genetic switches govern how the genetic toolkit is used

Developmental modules based on a common set of genetic instructions can evolve separately within a species because **genetic switches** control how the toolkit is used. These switches include:

- gene promoters and the *transcription factors that bind to promoters.

- enhancers and repressors that can modulate the interactions of transcription factors and promoters.

- signaling molecules that activate pathways or gene expression.

- signal transduction components that mediate signaling and its effects.

---

*connect the concepts   How transcription factors act as inducers or repressors of gene expression is explained in Key Concept 16.2.

Multiple switches control each gene, creating different expression patterns in different locations. In this way, elements of the genetic toolkit can be involved in multiple developmental processes and still allow individual modules to develop and evolve independently.

 Animation 19.4 **Modularity**
www.Life11e.com/a19.4

During evolution, changes in the functions of genetic switches have led to changes in the forms and functions of organisms. To illustrate this, let's look at the development of wings in *Drosophila* and other insects. *Drosophila* species are members of the insect group Diptera, which means "two wings"—that is, they have a single pair of wings, whereas most insects have two pairs of wings (i.e., four wings). The single pair of wings of dipterans develops on the second thoracic segment, where the Hox gene *Antennapedia* (*Antp*) is expressed. *Antp* is also expressed in the third thoracic segment, but in that segment a pair of balancing organs called halteres develops in dipterans. A critical difference between thoracic segments 2 and 3 is that another Hox gene, *Ultrabithorax* (*Ubx*), is expressed along with *Antp* in segment 3 (**Figure 19.16**). *Ubx* represses *Antp* function in dipterans. If *Ubx* is inactivated by mutation, a second pair of wings forms in thoracic segment 3, as is typical of many other insect groups. Thus some major morphological differences among groups of animals can result from relatively small changes in gene expression.

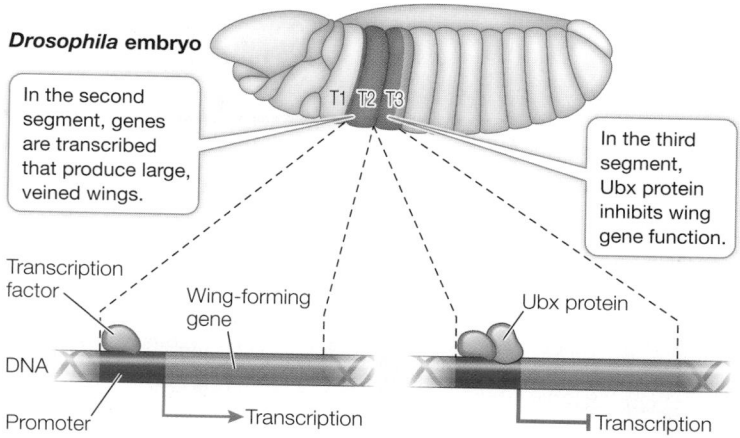

Drosophila embryo

In the second segment, genes are transcribed that produce large, veined wings.

T1 T2 T3

In the third segment, Ubx protein inhibits wing gene function.

Transcription factor

Wing-forming gene

Ubx protein

DNA

Promoter

Transcription

Transcription

Figure 19.16 Segments Differentiate under Control of Genetic Switches The binding of a single protein, Ultrabithorax (Ubx), determines whether a thoracic segment produces full wings or halteres.

Q: What would happen if *Ubx* was genetically engineered to be expressed in all thoracic segments in a butterfly embryo?

Modularity allows for differences in the patterns of gene expression

The modularity of development means that the molecular pathways for developmental processes such as organ formation operate independently from one another. For instance, the *Antennapedia* mutant fly shown on p. 414 grew a leg where an antenna should be, but the rest of the segments formed structures normally. Modularity allows the timing and position of a developmental process to change without changing the whole organism. Several patterns have been noted.

HETEROMETRY   A striking example of **heterometry** ("different measure") is beak development in Galápagos finches, in which the beaks of some species are large and robust for cracking seeds and in other species are thin and long for probing for food. When he visited these islands, Charles Darwin wrote that "seeing this gradation and diversity of structure in one small intimately related group of birds, one might really fancy that from an original paucity of birds in this archipelago, one species had been taken and modified for different ends."

Darwin had no idea of the genetic basis for such modification. Now we do. Beak shapes are evident when birds hatch from their eggs, so they must be determined during development. The beak develops from tissues at the anterior of the embryo that will form the facial bones. Cell divisions in this embryonic tissue are controlled by signaling proteins, one of which is called bone morphogenetic protein 4 (BMP4); another is the protein calmodulin. If BMP4 is present early and in large amounts, the beak becomes broad and deep. If calmodulin is present early and in large amounts, the beak grows

(A)

Length

Depth

Width

(B)

Mixed diet of
seeds and insects

Ancestral sharp-beaked finch

Low BMP4: low beak depth/width
Low CaM: short beak

Probing cactus flowers and fruit	Probing cactus flowers and fruit	Crushing seeds	Crushing hard/large seeds
Low BMP4: low beak depth/width	Low-moderate BMP4: moderate beak depth/width	Moderate BMP4: moderate beak depth/width	Early/high BMP4: high beak depth/width
High CaM: elongated beak	High CaM: elongated beak	Low CaM: short beak	Low CaM: short beak
Cactus finch (*G. scandens*)	Large cactus finch (*G. conirostris*)	Medium ground finch (*G. fortis*)	Large ground finch (*G. magnirostris*)

Figure 19.17 Heterometry and the Beaks of Finches **(A)** Bird beaks can be measured in three dimensions (length, width, and depth) for the purpose of comparing different species. **(B)** Among Galápagos finches, mutations resulting in different patterns of BMP4 and calmodulin (CaM) expression may be responsible for the different species' beak sizes and shapes, which are suited for exploiting different types of food.

longer and thinner (**Figure 19.17**). Thus beak structure is affected by changes in protein production.

HETEROCHRONY The evolution of the giraffe's neck provides an example of **heterochrony** ("different time"). Giraffes, like all mammals except manatees and sloths, have seven cervical (neck) vertebrae. So giraffes did not get longer necks by adding more vertebrae. However, the cervical vertebrae of giraffes are much longer than those of other mammals (**Figure 19.18**).

Bone growth in mammals is the result of the proliferation of cartilage-producing cells called chondrocytes. Bone growth is stopped by a genetic signal that results in apoptosis, or cell death, of chondrocytes and calcification of the bone matrix (see Key Concept 47.3). In giraffes this signaling process is delayed in the cervical vertebrae, so that these vertebrae grow longer. Thus the evolution of longer necks resulted from *changes in the timing* of expression of the genes that control bone formation.

HETEROTOPY Spatial differences in the expression of a developmental gene are known as **heterotopy** ("different place"), exemplified by the different development of feet in ducks and chickens. The feet of all bird embryos have webs of skin that connect their toes. This

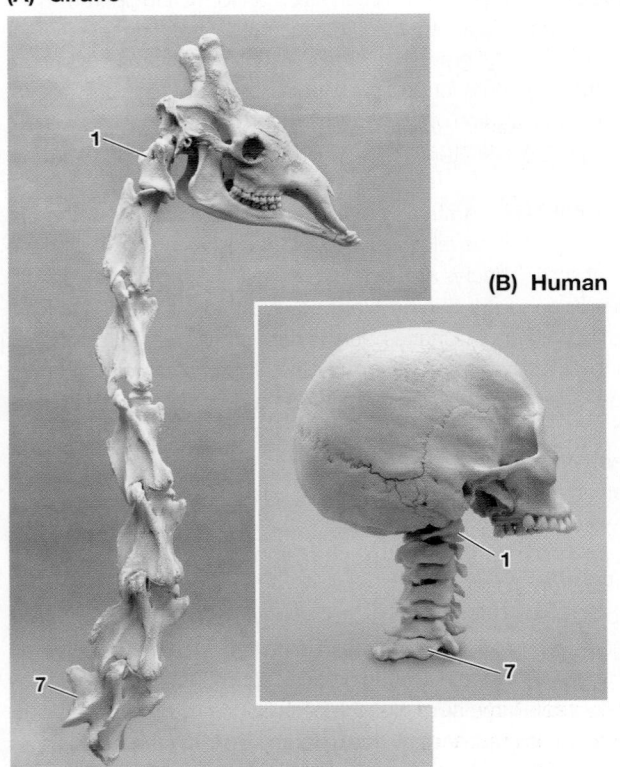

(A) Giraffe

(B) Human

Figure 19.18 Heterochrony in the Development of a Longer Neck
There are seven vertebrae in the neck of the giraffe **(A)** and human (**B**; not to scale). But the vertebrae of the giraffe are much longer (25 cm compared with 1.5 cm) because during development, growth continues for a longer period of time. This timing difference is called heterochrony.

Chick hindlimb   Duck hindlimb

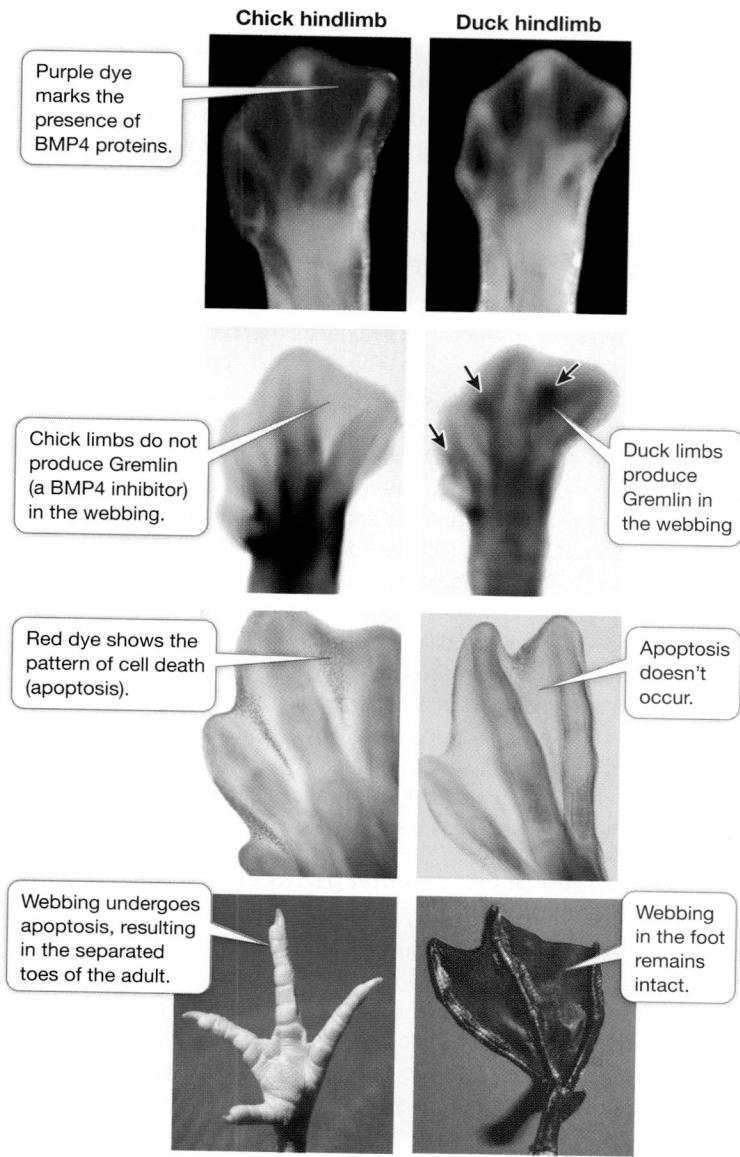

Purple dye marks the presence of BMP4 proteins.

Chick limbs do not produce Gremlin (a BMP4 inhibitor) in the webbing.

Duck limbs produce Gremlin in the webbing

Red dye shows the pattern of cell death (apoptosis).

Apoptosis doesn't occur.

Webbing undergoes apoptosis, resulting in the separated toes of the adult.

Webbing in the foot remains intact.

Figure 19.19 Heterotopy in Gremlin Expression Correlates with Changes in Hindlimb Structure The left column of photos shows foot development in a chicken; the right column shows foot development in a duck. Gremlin protein in the webbing of the duck foot inhibits BMP4 signaling, thus preventing the embryonic webbing from undergoing apoptosis.

Q: What do you predict would happen if you blocked the expression of Gremlin in the feet of duck embryos?

webbing is retained in adult ducks (and other aquatic birds) but not in adult chickens (and other nonaquatic birds). The loss of webbing is controlled by the BMP4 signaling protein, which as we have noted is also involved in beak development (another example of use of a common genetic toolkit to produce different kinds of change).

BMP4 protein induces the cells that produce webbing to undergo apoptosis and thus eliminate the webbing between the toes. The hindlimbs of both duck and chicken embryos express the *BMP4* gene in the webbing between the toes; however, they differ in expression of the *Gremlin* gene, which encodes a protein that inhibits *BMP4* expression (**Figure 19.19**). In ducks, but not in chickens, *Gremlin* is expressed in the webbing cells and Gremlin protein

inhibits *BMP4* expression. With no BMP4 protein to stimulate apoptosis, a webbed foot develops. If chick hindlimbs are experimentally exposed to Gremlin during development, the adult chicken will have ducklike webbed feet.

19.4 recap

A genetic toolkit consisting of highly conserved regulatory genes governs pattern formation in multicellular organisms. Genetic toolkit genes act independently among modules of a developing embryo. The expression of these developmental genes can differ among species in various ways, often resulting in major morphological differences. They can differ in amount of expression (heterometry), in the timing of expression (heterochrony), or in the location of expression (heterotopy).

learning outcomes

You should be able to:

- Use a specific example to defend the claim "Genes controlling development are highly conserved."
- Provide plausible evolutionary explanations for cases where different populations or closely related species differ in genetic switches.
- Interpret experimental results to assess the role of BMP4 in beak development.

1. How does the story of the eye-determining genes *Pax6* and *eyeless* support the claim that genes controlling development are highly conserved?

2. In a series of experiments on chick embryos, researchers applied different concentrations of BMP4 to the embryos' beak growth region. They measured the size of the beak cartilage at a later stage of development. Based on the data in the table below, what could you conclude about the role of BMP4 in beak growth?

Amount of BMP4	Cartilage diameter (mm)
None (control)	0.5
0.1 unit	0.7
0.3 unit	1.0
1.0 unit	1.8

3. *Plasmodium vivax* is a protist that causes a form of malaria. When *P. vivax* enters the blood, it attaches to a glycoprotein on the red blood cells. Some human populations in Africa are immune to *P. vivax* because they lack this particular glycoprotein on their red blood cells. In a variety of tissues the transcription of this glycoprotein is under the influence of enhancers, and the enhancer that is normally expressed in red blood cell precursors is mutated in individuals immune to this form of malaria. What does this case illustrate in evolutionary developmental terms?

You have seen how the genetic toolkit guides morphogenesis in individual organisms, and how differences in genetic switches contribute to differences among species. Now we discuss the roles that some of these same tools play in the evolution of new forms and new species.

key concept 19.5 Developmental Gene Changes Can Shape Evolution

The genetic switches that allow different structures to develop in different regions of the embryo can also give rise to major morphological differences among species. Changes in timing and position of a genetic switch (e.g., *Gremlin* expression) can generate morphological changes (e.g., webbed or unwebbed feet) that can be acted on by natural selection.

focus your learning

- A major developmental change can be caused by an alteration in a regulatory molecule.
- Similar changes in gene expression in isolated populations of organisms can lead to the same evolutionary result, a process called parallel evolution.

French geneticist François Jacob suggested that evolution works like a tinker, assembling new structures by combining and modifying the available materials, and not like an engineer, who is free to develop dramatically different designs (a jet engine to replace a propeller-driven engine, for example). You have seen that morphological evolution is not usually governed by the acquisition of radically new genes, but proceeds primarily by "tinkering" with the expression patterns of existing genes. Thus developmental genes and their expression constrain evolution in two major ways:

1. Nearly all evolutionary innovations are modifications of previously existing structures.
2. The basic set of regulatory genes that control development is broadly conserved, changing only slowly over the course of evolution.

Mutations in developmental genes can cause major morphological changes

Sometimes a major developmental change is caused by an alteration in the regulatory molecule itself rather than a change in where, when, or how much it is expressed. The gene that controls the number of legs in arthropods offers an excellent example. Arthropods all have head, thoracic, and abdominal regions with variable numbers of segments. Insects, such as *Drosophila*, have three pairs of legs on their three thoracic segments, whereas centipedes have many legs on both thoracic and abdominal segments. All arthropods express a gene called *Distal-less* (*Dll*) that controls segmental leg development. In insects, *Dll* expression is repressed in abdominal segments by the Hox gene *Ubx*. *Ubx* is expressed in the abdominal segments of all arthropods, but it has different effects in different species. In centipedes, *Ubx* is co-expressed with *Dll* to promote the formation of legs. During the evolution of insects, a mutation in the *Ubx* gene sequence resulted in a modified Ubx protein that *represses Dll* expression in abdominal segments. A phylogenetic tree of arthropods shows that this change in *Ubx* occurred in the ancestor of insects, at the same time that abdominal legs were lost (**Figure 19.20**).

Conserved developmental genes can lead to parallel evolution

The existence of highly conserved developmental genes makes it likely that similar traits will evolve repeatedly, especially among closely related species. This process is known as **parallel evolution**, and a good example is provided by a small fish, the three-spined stickleback (*Gasterosteus aculeatus*). Sticklebacks are widely distributed throughout the Atlantic and Pacific oceans; they are

Figure 19.20 A Mutation in a Hox Gene Changed the Number of Legs in Insects In the insect lineage (blue box) of the arthropods, a change to the *Ubx* gene resulted in a protein that inhibits the *Dll* gene, which is required for legs to form. Because insects express this modified *Ubx* gene in their abdominal segments, no legs grow from these segments. Other arthropods, such as centipedes, produce an unmodified Ubx protein and do grow legs from their abdominal segments.

Q: What do you predict the phenotype would be if you genetically engineered *Drosophila* to express the centipede version of *Ubx*?

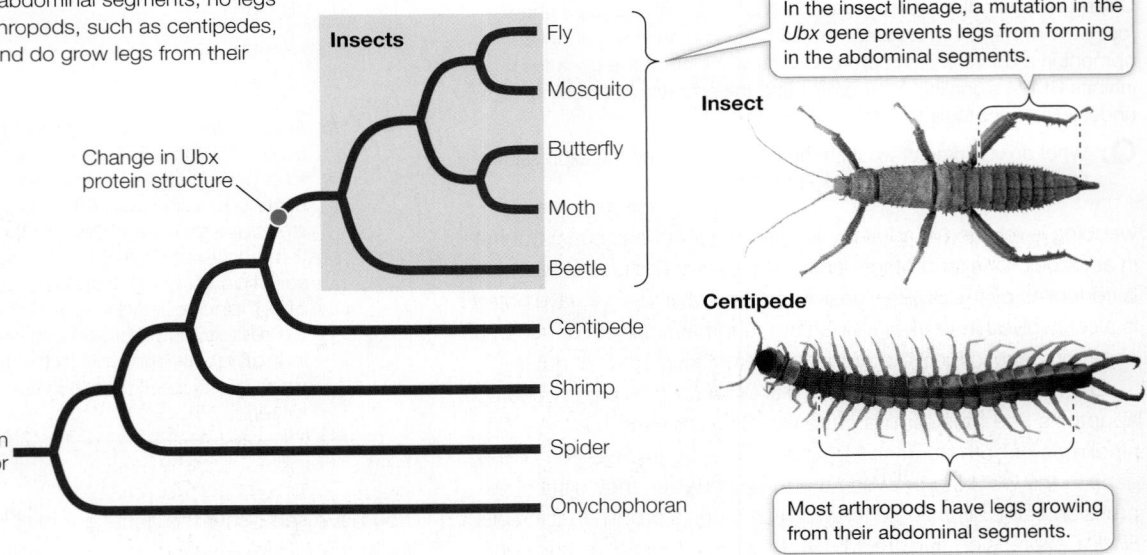

In the insect lineage, a mutation in the *Ubx* gene prevents legs from forming in the abdominal segments.

Most arthropods have legs growing from their abdominal segments.

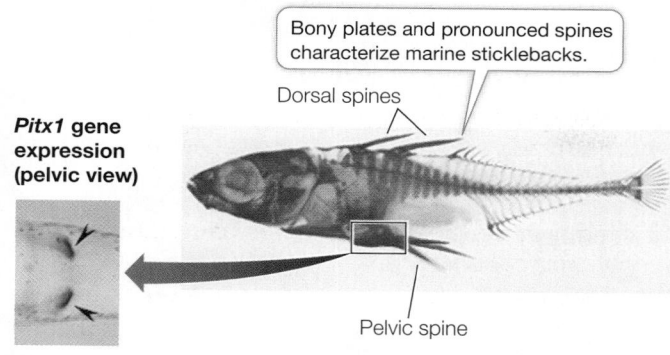

Pitx1 gene expression (pelvic view)

Bony plates and pronounced spines characterize marine sticklebacks.

Dorsal spines

Pelvic spine

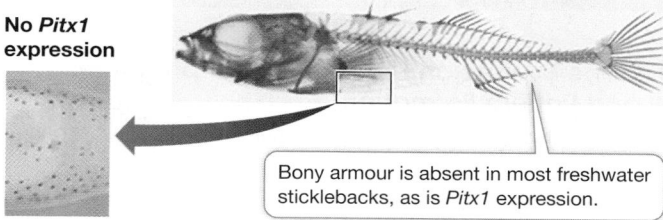

No *Pitx1* expression

Bony armour is absent in most freshwater sticklebacks, as is *Pitx1* expression.

Figure 19.21 Parallel Phenotypic Evolution in Sticklebacks
A developmental gene, *Pitx1*, encodes a transcription factor that stimulates the production of plates and spines. This gene is active in marine sticklebacks, but mutated and inactive in various freshwater populations of the fish. The fact that this mutation is found in geographically distant and isolated freshwater populations is evidence for parallel evolution.

also found in many freshwater lakes and rivers. Marine sticklebacks spend most of their lives at sea but return to fresh water to breed. However, freshwater populations that are isolated in lakes never encounter salt water.

Genetic evidence shows that freshwater populations have arisen from marine populations many times, and independently. Marine sticklebacks have structures that protect them from predatory marine fish; these are bony plates and well-developed pelvic bones with pelvic spines that lacerate the mouths of predators. Freshwater sticklebacks do not face such predatory dangers; their body armor is greatly reduced, and their dorsal and pelvic spines are shorter or even lacking (**Figure 19.21**).

The differences between marine and freshwater sticklebacks are not induced by environmental conditions. Marine species reared in

fresh water still grow armor and spines. The differences are due to the expression of a developmental regulatory gene, *Pitx1*. *Pitx1* codes for a transcription factor normally expressed in regions of the developing embryo that in marine sticklebacks form the head, trunk, tail, and pelvis. However, in separate and long-isolated populations of freshwater sticklebacks from Canada, the United Kingdom, the United States, and Iceland, the *Pitx1* gene is no longer expressed in the pelvis, and spines do not develop. This same change in regulatory gene expression has evolved to produce similar phenotypic changes in several independent populations, and is thus a good example of parallel evolution.

19.5 recap

A mutation in a gene controlling a developmental pathway can result in major changes in morphology, which provide the raw material for evolution by natural selection. The conservation of many developmental regulatory genes makes it likely that similar traits will evolve repeatedly.

learning outcomes

You should be able to:

- Summarize how mutations in developmental genes have led to diversity in arthropod body plans.
- Show how differences between marine and freshwater sticklebacks exemplify parallel evolution via changes in gene regulation.

1. How have diverse body forms in arthropods evolved by means of mutations in developmental genes?
2. How do the differences between marine and freshwater sticklebacks exemplify parallel evolution via changes in gene regulation?

Many novel traits have arisen during the course of evolution, but most of them failed to persist beyond even a single generation. Part Six of this book will examine the processes of evolution—the powerful forces that influence the survival and reproductive success of various life forms. You will see how different adaptations become prevalent in different environments, resulting in the extraordinary diversity of life on Earth today, which we will describe in further detail in Part Seven.

investigatinglife

What are the potential uses of stem cells?

A procedure has been developed to isolate stem cells from human patients in the operating room. Large quantities of stem cells from fat can be derived from the cosmetic procedure known as liposuction. These stem cells are stored at very low temperatures and become viable when they are warmed up. Stem cells retrieved from liposuction procedures have been used to repair tissues. For example, they have been used in stem cell transplantation after surgery

for breast cancer. In other cases, they have been used to help heal skull fractures. Mesenchymal cells from fat as well as bone marrow have been used to help the healing of connective tissues such as muscles and tendons in athletic injuries as well.

Future directions

Millions of people have type 1 diabetes, whereby β (beta) cells in the pancreas do not make the hormone insulin. These people must take insulin as a medication, made from recombinant DNA technology (see Key Concept 18.5). However, the timing and regulation of the

(continued)

amount of insulin taken can be problematic, and symptoms often persist. Supplying a patient with functional β cells could potentially cure type I diabetes. Both embryonic and induced human pluripotent stem cells have been shown in lab experiments to make insulin. Moreover, they also regulate insulin release in response to environmental conditions and restore normal function when transplanted into the pancreas of diabetic mice. Clinical trials in diabetic humans are planned.

In the United States, the only clinical use of stem cells for humans that has been rigorously tested and approved by the government is hematopoietic (bone marrow) transplantation. Nevertheless, clinics offering stem cell treatments for many diseases are springing up in South America, Asia, and the United States (without government approval), treating growing numbers of patients each year. While claims of cures abound, more research is required to establish the efficacy and safety of stem cell therapy.

Chapter Summary 19

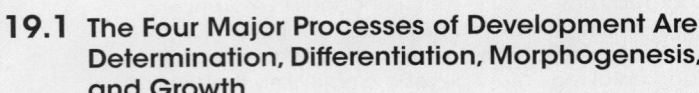

▶ 19.1 The Four Major Processes of Development Are Determination, Differentiation, Morphogenesis, and Growth

- A multicellular organism begins its development as an **embryo**. Embryogenesis results in a new organism with a body plan characteristic of its species. Review Figure 19.1, Activity 19.1

- The processes of development are **determination**, **differentiation**, **morphogenesis**, and **growth**.

- Differential gene expression is responsible for the differences among cell types. **Cell fate** is determined by environmental factors, such as the cell's position in the embryo, as well as by intracellular influences. Review Figure 19.2, Activity 19.2

- Over the course of development, embryo cells decrease in **cell potency**. **Totipotent** cells (such as a zygote) are capable of forming every cell type in the adult body. **Pluripotent** cells can give rise to most cell types, **multipotent** cells to several cell types, and **unipotent** cells to only one cell type.

- The ability to create clones from differentiated cells demonstrates the principle of **genomic equivalence**. Review Figures 19.3, 19.4

- **Stem cells** produce daughter cells that differentiate when provided with appropriate intercellular signals. Some multipotent stem cells in the adult body can differentiate into a limited number of cell types to replace dead cells and maintain tissues. Review Investigating Life: Stem Cell Therapy

- **Embryonic stem cells (ESCs)** are pluripotent and can be cultured in the laboratory. Under suitable environmental conditions, these cells can differentiate into almost any tissue type. **Induced pluripotent stem cells (iPS cells)** have similar characteristics, making possible technologies designed to replace cells or tissues damaged by injury or disease. Review Figure 19.5, Animation 19.1

▶ 19.2 Gene Expression Differences Determine Cell Fate and Cell Differentiation

- **Cytoplasmic segregation**—the unequal distribution of **cytoplasmic determinants** in the egg, zygote, or early embryo—can establish **polarity** and lead to cell fate determination. Review Figure 19.6, Animation 19.2

- **Induction** is a process by which embryonic animal tissues direct the development of neighboring cells and tissues by secreting chemical signals, called inducers. Review Figure 19.7

- Inducers act through signaling pathways to determine cell fate. Review Focus: Key Figure 19.8

- Differential gene expression results in cell differentiation. Transcription factors are especially important in regulating gene expression during differentiation. Review Figure 19.9

▶ 19.3 Gene Expression Determines Morphogenesis and Pattern Formation

- **Pattern formation** is the process that results in the spatial organization of a tissue or organism.

- Both plants and animals use **positional information** as a basis for pattern formation. Positional information usually comes in the form of a signal called a **morphogen**. Different concentrations of the morphogen cause different effects. Review Figure 19.10

- Sepals, petals, stamens, and carpels form in plants as a result of combinatorial interactions between transcription factors encoded by **organ identity genes**. Review Figure 19.11, Activity 19.3

- In the fruit fly *Drosophila melanogaster*, a cascade of transcriptional activation sets up the axes of the embryo, the development of the segments, and finally the determination of cell fate in each segment. The cascade involves the sequential expression of **maternal effect genes**, **gap genes**, **pair rule genes**, **segment polarity genes**, and **Hox genes**. Review Figures 19.12, 19.13, Animation 19.3

- Hox genes help determine cell fate in the embryos of all animals. The **homeobox** is a DNA sequence found in Hox genes and other genes that code for transcription factors. The sequence of amino acids encoded by the homeobox is called the **homeodomain**. Review Figure 19.14

▶ 19.4 Changes in Gene Expression Underlie the Evolution of Development

- **Evolutionary developmental biology**, or **evo-devo**, is the study of the evolutionary aspects of development. This field focuses on the molecular mechanisms that underlie the development of phenotypic diversity.

- Similarities in the basic mechanisms of development between widely divergent organisms reflect common ancestry.

- Genes encoding transcription factors and other regulatory proteins that govern pattern formation in the developing bodies of multicellular organisms comprise what is called a **genetic toolkit**. These regulatory genes have been highly conserved throughout evolution. Review Figure 19.15

- The bodies of developing and mature organisms are organized into self-contained units, or modules, that can be modified independently. See Animation 19.4

- The genetic toolkit involves **genetic switches**—promoters, enhancers and repressors, signaling molecules, and signal transduction components—that can alter the expression of developmental genes in different modules independently of one another. Review Figure 19.16

- Developmental genes can be expressed in a modular fashion in different amounts (**heterometry**), at different times (**heterochrony**), or in different locations (**heterotopy**). Review Figures 19.17–19.19

▶ **19.5 Developmental Gene Changes Can Shape Evolution**

- Morphological differences among species can result from mutations in the genes that regulate the development of modules such as body segments or wings. Review Figure 19.20

- Because many genes that govern development have been highly conserved, similar traits are likely to evolve repeatedly, especially among closely related species. This process is called **parallel evolution**. Review Figure 19.21

> Go to **LearningCurve** (in **LaunchPad**) for dynamic quizzing that helps you solidify your understanding of this chapter. **LearningCurve** adapts to your responses, giving you the practice you need to master each key concept.

▶ Apply What You've Learned

Review

19.3 Alterations of Hox genes can result in homeotic mutations.

19.4 Heterochrony has significant importance, as illustrated by the extreme length of a giraffe's neck.

19.4 Spatial differences in the expression of a developmental gene are known as heterotopy.

19.5 A major developmental change can be caused by an alteration in a regulatory molecule.

Differences in bristle patterns in flies often have implications for how the flies sense the world—and can even affect their courtship behavior. Bristles can be important in the evolution of these flies.

Two hypothetical, closely related fly species differ primarily in the length of bristles on the third segment of the thorax. The transcription factor Hairball is thought to be involved in determining the differences between these structures. The figure below shows the gene expression pattern for both species over the course of time.

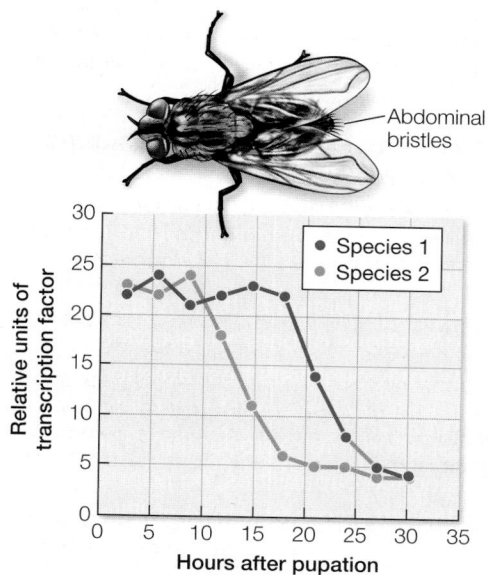

Questions

1. Describe the results and formulate a hypothesis for the role of the transcription factor in the two species. Assess whether this is an example of heterotopy, heterometry, or heterochrony.

2. Outline an experimental approach that would test the hypothesis generated in Question 1, and provide the expected results.

3. Bristles ordinarily do not grow on the other two thoracic segments of either of these species, and *hairball* is not expressed at high levels in these segments. A third species, Species 3, has hairs on the other thoracic segments and expresses *hairball* during development in these segments. Predict what would happen if *hairball* were expressed in the other thoracic segments in Species 1 and 2. Assess whether this is an example of heterotopy, heterometry, or heterochrony.

4. Suppose a *Hox* gene regulates *hairball* expression, and Species 3 had a variant of the gene that differed by one nucleotide from the variant in Species 1 and 2. How would you test whether this *Hox* gene was responsible for the differences between Species 3 and Species 1 and 2?

Go to **LaunchPad** for the eBook, LearningCurve, animations, activities, flashcards, and additional resources and assignments.

20

▶ investigating**life**

Processes of Evolution

Mexican free-tailed bats taking flight from a bridge in Austin, Texas.

An Evolutionary Arms Race Between Bats and Moths

Many species of bats roost by day and fly out as night approaches. The mass exodus of thousands or even millions of bats from a roost is an impressive sight, often attracting crowds of human bat-watchers. In urban areas, the cavities under bridges can attract enormous colonies of bats. People once feared these urban bat colonies, but we now know that many species of bats prey on night-flying insects, including mosquitoes and agricultural pests. Each bat can consume a third of its weight in insects per night, so just 1,000 bats consume about 4 tons of insects every year.

Most bats catch insects in flight at night. About 50 million years ago, bats evolved the ability to echolocate, which allows them to detect flying insects (such as moths) in darkness. The bats produce ultrasonic sounds, and compare these outgoing sound pulses with the returning echoes of the pulses that bounce off the flying moths. Because bats are among the greatest threats to night-flying insects, any insect that can escape detection by bats is much more likely to survive and reproduce. Inherited traits that lead to greater survival increase in the population over time as surviving moths pass these traits on to their offspring.

Many groups of night-flying insects, including several groups of moths, have evolved the ability to detect the ultrasonic sounds of bats. When such a moth hears a bat, it flies away from the source of the sound. If the bat is too close, the moth may dive rapidly to avoid capture. The moths in each generation that are better able to detect and avoid bats leave more offspring, so moths gradually evolve more effective escape mechanisms.

As moths get better at avoiding bats, why don't the bats starve? Because populations of bats are evolving too. In each generation, the individual bats best able to detect and capture flying insects get the most food, which allows them to put the most effort into reproduction. Their offspring inherit their prey-detection capabilities, so populations of bats evolve more efficient insect-catching abilities, even as the moth populations evolve better escape strategies. This gradual change in both moth and bat populations is known as evolution by natural selection—a process that is constantly shaping and changing all species of life on our planet.

Q&A How do complex traits like echolocation, or the ability to avoid detection by echolocation, evolve in the first place?

key concept 20.1 Evolution Is Both Factual and the Basis of Broader Theory

All biological populations change in their genetic makeup over time. Change in the genetic composition of populations over time is called **evolution**. We can, and do, observe evolutionary change on a regular basis, both in laboratory experiments and in natural populations. We measure the rate at which new mutations arise, observe the spread of new genetic variants through a population, and see the effects of genetic change on the form and function of organisms. In the fossil record, we observe the long-term morphological changes (which are the result of underlying genetic changes) that have occurred among living organisms. These underlying changes in the genetic makeup of populations drive the origin and extinction of species and fuel the diversification of life.

focus your learning

- Evolution is directly observable and is a universal principle of life.
- A scientific theory is a well-supported, evidence-based explanation of natural phenomena.
- Natural selection favors traits that convey survival and reproductive advantages.
- Evolution is responsible for the vast diversity of life on Earth.

In addition to observing and recording physical changes over evolutionary time, biologists have accumulated a large body of evidence about *how* these changes occur, and about *what* evolutionary changes have occurred in the past. The resulting understanding of the processes of evolutionary change is known as **evolutionary theory**.

Evolutionary theory has many useful applications. We constantly apply it to the study and treatment of diseases. Evolutionary theory is critical to the development of better agricultural crops and practices, and to the development of industrial processes that produce new molecules with useful properties. At a more basic level, knowledge of evolutionary theory allows biologists to understand how life diversified. It also helps us make predictions about the biological world.

In everyday speech, people tend to use the word "theory" to mean an untested hypothesis, or even a guess. But evolutionary theory does not refer to any single hypothesis, and it certainly is not guesswork. A vast and rich array of geological, morphological, behavioral, and molecular data all support the factual basis of evolution. Observations of fossils and natural populations are supported by experiments that demonstrate the basic operation of evolutionary processes.

When we refer to evolutionary theory, we are referring to our understanding of the processes that result in genetic changes in populations over time. We then apply that understanding to interpret the changes we observe in natural populations. We can directly observe the evolution of many living organisms. We can observe the yearly changes in the viruses that give us the flu, but it is evolutionary theory that allows us to apply our observations to the task of developing more effective vaccines against these viruses. Several processes of evolutionary change are recognized, and the scientific community is continually using evolutionary theory to expand its understanding of how and when these processes apply to particular biological problems.

 Media Clip 20.1 **Watching Evolution in Real Time**
www.Life11e.com/mc20.1

Darwin and Wallace introduced the idea of evolution by natural selection

In the early 1800s, it was not yet evident to many people that populations of living organisms evolve. But several biologists had suggested that the species living on Earth had changed over time—that is, that evolution had taken place. Jean-Baptiste Lamarck, for one, presented strong evidence for the fact of evolution in 1809, but his ideas about *how* it occurred were not convincing. At that time, no one had yet envisioned a viable process for evolution.

In the 1820s, a young Charles Darwin became passionately interested in the subjects of geology (with its new sense of Earth's great age) and natural history (the scientific study of how different organisms function and carry out their lives in nature). Despite these interests, he planned, at his father's behest, to become a doctor. But surgery conducted without anesthesia nauseated Darwin, and he gave up medicine to study at Cambridge University for a career as a clergyman in the Church of England. Always more interested in science than in theology, he gravitated toward scientists on the faculty, especially the botanist John Henslow. In 1831, Henslow recommended Darwin for a position on HMS *Beagle*, a Royal Navy vessel that was preparing for a survey voyage around the world (**Figure 20.1**).

Whenever possible during the five-year voyage, Darwin went ashore to study rocks and to observe and collect plants and animals. He noticed striking differences between the species he saw in South America and those of Europe. He observed that the species of the temperate regions of South America (Argentina and Chile) were more similar to those of tropical South America (Brazil) than they were to temperate European species. When he explored the islands of the Galápagos archipelago, west of Ecuador, he noted that most of the animals were endemic to the islands (meaning they were unique and found nowhere else), although they were similar to animals found on the mainland of South America. Darwin also observed that the fauna of the Galápagos differed from island to island. He postulated that some animals had come to the archipelago from mainland South America and had subsequently undergone different changes on each of the islands. He wondered what might account for these changes.

When he returned to England in 1836, Darwin continued to ponder his observations. His thoughts were strongly influenced by the geologist Charles Lyell, who had recently popularized the idea that Earth had been shaped by slow-acting forces that are still at work today. Darwin reasoned that similar thinking could be applied to the living world. Within a decade, he had developed the framework of an explanatory theory for evolutionary change based on three major propositions:

1. Species are not immutable; they change over time.
2. Divergent species share a common ancestor and have diverged from one another gradually over time (a concept Darwin termed **descent with modification**).

Charles Robert Darwin

Figure 20.1 Darwin and the Voyage of the Beagle The mission of HMS *Beagle* was to chart the oceans and collect oceanographic and biological information from around the world. The world map indicates the ship's path; the inset map shows the Galápagos Islands, whose organisms were an important source of Darwin's ideas on natural selection. The portrait is of Charles Darwin at age 27, shortly after the *Beagle* returned to England.

▶ Activity 20.1 **Darwin's Voyage**
www.Life11e.com/ac20.1

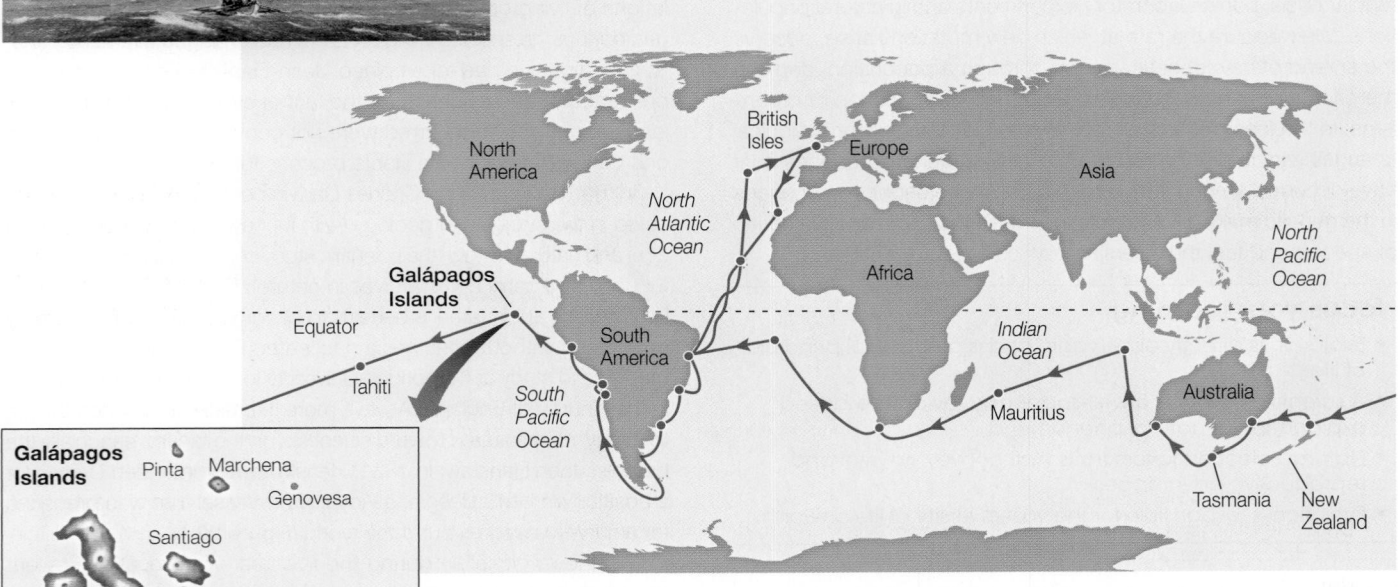

3. Changes in species over time can be explained by **natural selection**: the increased survival and reproduction of some individuals compared with others, based on differences in their traits.

The first of these propositions was not unique to Darwin; several earlier authors had argued for the fact of evolution. A more revolutionary idea was his second proposition, that divergent species are related to one another through common descent. But Darwin is probably best known for his third proposition, that of natural selection.

Darwin realized that many more individuals of most species are born than survive to reproduce. He also knew that, although offspring usually resemble their parents, offspring are not identical to one another or to either parent. Finally, he was well aware that human breeders of plants and animals often selected their breeding stock based on the occurrence of particular traits. Over time, this selection resulted in dramatic changes in the appearance of the descendants of those plants or animals. In natural populations, wouldn't the individuals with the best chances of survival and reproduction be similarly "selected," and thus pass their traits on to the next generation?

Darwin's simple but powerful idea was that nature did the selecting in natural populations on the basis of traits that resulted in greater survival and, eventually, greater likelihood of reproduction.

In 1844, Darwin wrote a long essay describing the role of natural selection as a process of evolution. But he was reluctant to publish it, preferring to assemble more evidence first. Darwin's hand was forced in 1858, when he received a letter and manuscript from another traveling English naturalist, Alfred Russel Wallace, who was studying the plants and animals of the Malay Archipelago. Wallace asked Darwin to evaluate his manuscript, which included an explanation of natural selection almost identical to Darwin's. Darwin was at first dismayed, believing Wallace to have preempted his idea. Parts of Darwin's 1844 essay, together with Wallace's manuscript, were presented to the Linnaean Society of London on July 1, 1858, thereby crediting both men for the idea of natural selection. Darwin then worked quickly to finish his full-length book, *On the Origin of Species*, which was published the following year.

▶ Animation 20.1 **Natural Selection**
www.Life11e.com/a20.1

Although Darwin and Wallace independently articulated the concept of natural selection, Darwin developed his ideas first. Furthermore, *On the Origin of Species* proved to be a stunning work of scholarship that provided exhaustive evidence from many fields supporting both the premise of evolution itself and the understanding of natural selection as a process of evolution. Thus both concepts are more closely associated with Darwin than with Wallace.

The publication of *On the Origin of Species* in 1859 stirred considerable interest (and controversy) among scientists and the public alike. Scientists spent much of the rest of the nineteenth century amassing biological and paleontological data to test evolutionary ideas and document the history of life on Earth. By 1900, the fact of biological evolution (defined at that time as change in the physical characteristics of populations over time) was established beyond any reasonable doubt. As biologists discovered the details of genetic inheritance in the twentieth century, the genetic mechanisms of evolution became clear. The development of methods for sequencing DNA in the late 1970s allowed biologists to document evolutionary changes within and between species with great precision. This technology led to explosive growth in the field of evolutionary biology. In the past three decades, well over a quarter of a million scientific papers on evolutionary observations, experiments, and theory have been published.

▶ 20.1 recap

Evolution, or changes in gene frequencies in biological populations over time, is directly observable. Natural selection occurs when specific alleles increase the rates of survival or reproduction of individuals in a population relative to individuals that do not possess those alleles. Under such conditions, the frequency of the favored allele increases in the population from one generation to the next.

learning outcomes

You should be able to:

- Describe an example in which evolution by natural selection can be observed directly within a human lifetime.
- Differentiate between the everyday use of the word "theory" and a scientific theory.
- Apply the principles of selection to predict and explain evolutionary outcomes.
- Explain how evolution produces diversity and provides evidence of common ancestry.

1. Why do biologists speak of "evolutionary theory" if the facts of evolution are not in doubt?

2. Antibiotics are drugs that kill most bacteria, but genetic mutations can allow some individual bacteria to survive short-term exposure to these drugs. How would you expect the frequency of resistant bacteria to change over time in populations of bacteria that were exposed to an antibiotic drug on a regular basis? Why do you think antibiotics come with a warning to take the full course of the treatment, rather than stopping after you begin to feel better?

3. In what ways does selection by humans in developing agricultural crops differ from natural selection? Can you give an example of a trait that might be favored by artificial selection in agriculture, but selected against by natural selection in a wild population?

4. Natural selection cannot adapt populations to conditions they have not experienced. Yet many organisms appear to respond to natural events before they happen. For example, many mammals go into hibernation while it is still quite warm. Similarly, many birds leave the temperate zone for their southern wintering grounds long before winter has arrived. How do you think such "anticipatory" behaviors evolve?

▶ 20.1 recap

5. As more humans live longer, many people face degenerative conditions such as Alzheimer's disease that (in most cases) are linked to advancing age. Assuming that some individuals may be genetically predisposed to successfully combat these conditions, is it likely that natural selection alone would act to favor such a predisposition in human populations? Why or why not?

Although the importance of natural selection to evolution has been confirmed in many thousands of scientific studies, it is not the only process that drives evolution. In the next section we'll consider a more complete view of evolutionary processes and how they operate.

▶ key concept

20.2 Mutation, Selection, Gene Flow, Genetic Drift, and Nonrandom Mating Result in Evolution

The word "evolution" is often used in a general sense to mean simply "change," but in a biological context "evolution" refers specifically to change in the genetic makeup of populations over time. Developmental changes that occur in a single individual over the course of the life cycle are not the result of evolutionary change. Evolution is genetic change across generations of a **population**—a group of individuals of a single species that live and interbreed in a particular geographic area at the same time. It is important to remember that *individuals do not evolve; populations do*.

focus your learning

- The terms "adapt," "evolve," and "population" have specific scientific meanings.
- Evolution is the result of five major processes: mutation, natural selection, gene flow, genetic drift, and nonrandom mating.

The premise of natural selection was one of Darwin's principal insights and has been demonstrated to be an important process of evolution, but natural selection does not act alone. Four additional processes—mutation, gene flow, genetic drift, and nonrandom mating—affect the genetic makeup of populations over time. Before we consider how the other processes can change the frequencies of gene variants in a population, we need to understand how mutation brings such variants into existence.

Mutation generates genetic variation

The origin of genetic variation is mutation. As described in Key Concept 15.1, a mutation is any change in the nucleotide sequences of an organism's DNA. The process of DNA replication is not perfect, and some changes appear almost every time a genome is replicated. Mutations occur randomly with respect to an organism's needs; it is natural selection acting on this random variation that results in adaptation. Most mutations are either harmful to their bearers (deleterious mutations) or have no effect (neutral mutations). But a few mutations are beneficial, and even previously deleterious or neutral alleles may become advantageous if environmental conditions change. In addition, mutation can restore genetic variation that other evolutionary processes have removed. Thus mutation both creates and helps maintain genetic variation in populations.

Mutation rates can be high, especially in viruses and bacteria, which explains the rapid evolution of many pathogens. But in many protein-coding genes of multicellular organisms, the mutation rate is very low (on the order of 10^{-8} to 10^{-9} changes per base pair of DNA per generation). Even low overall mutation rates, however, create considerable genetic variation, because each of a large number of genes may change, and populations often contain large numbers of individuals. For example, if the probability of a point mutation (an addition, deletion, or substitution of a single base) were 10^{-9} per base pair per generation, then each human gamete—the DNA of which contains 3×10^9 base pairs—would average three new point mutations ($3 \times 10^9 \times 10^{-9} = 3$), and each zygote would carry an average of six new mutations. The current human population of about 7 billion people would thus be expected to carry about 42 billion new mutations (i.e., changes in the nucleotide sequences of their DNA that were not present one generation earlier). So even though the mutation rate in humans is low, human populations still contain enormous genetic variation on which other evolutionary processes can act.

As a result of mutation, different forms of a gene, known as **alleles**, may exist at a particular chromosomal locus (see Key Concept 12.1). At any particular locus, a single diploid individual has no more than two of the alleles found in the population to which it belongs. The sum of all copies of all alleles at all loci found in a population constitutes that population's **gene pool** (Figure 20.2). (We can also refer to the gene pool for a particular chromosomal locus or loci.) The gene pool is the sum of the genetic variation in the population. The proportion of each allele in the gene pool is the *allele frequency*. Likewise, the proportion of each genotype among individuals in the population is the genotype frequency.

A simple experiment demonstrates how mutations accumulate in populations in a continuous, almost constant fashion over time (**Figure 20.3**). Lines of the bacterium *E. coli* were grown in the laboratory for 20,000 generations, and the genomes were sequenced from individuals in the experimental lines at least once every 5,000 generations. Over the experiment, the lines accumulated about 45 changes to their genomes, and these changes appeared at a fairly constant rate over time. All living organisms experience a similar accumulation of mutations over time (although the rate of change differs among species), and these changes provide the raw material for evolution.

---

*connect the concepts   As discussed in Key Concept 12.1, alleles of different genes assort independently in meiosis. The frequencies with which different allele combinations will be expressed in offspring can be calculated using a Punnett square or probability theory.

Selection acting on genetic variation leads to new phenotypes

As a result of mutation, the gene pools of nearly all populations contain variation for many traits. Selection that favors different traits can lead to distinctly different lineages that descend from the same

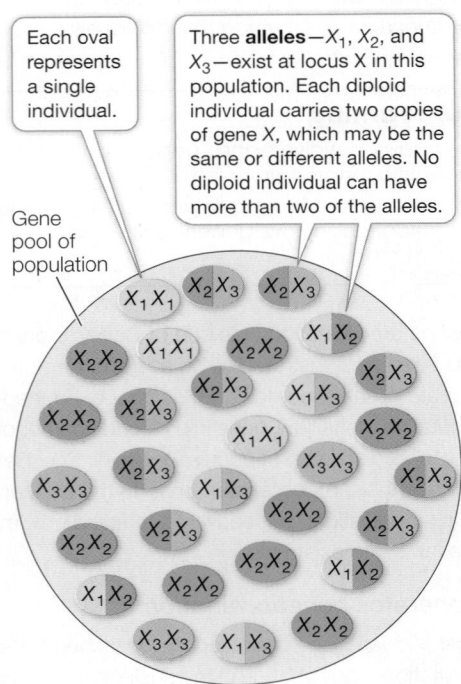

Figure 20.2 A Gene Pool   A gene pool is the sum of all the alleles found in a population or at a particular locus in that population. This figure shows the gene pool for one locus, *X*, in a population of diploid organisms. The allele frequencies in this case are 0.20 for X_1, 0.50 for X_2, and 0.30 for X_3 (see Figure 20.10).

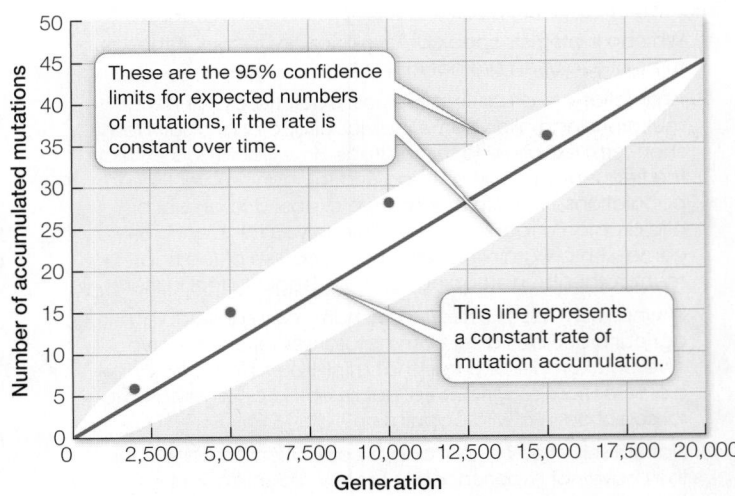

Figure 20.3 Mutations Accumulate Continuously   An experimental lineage of the bacterium *Escherichia coli* was propagated in the laboratory for 20,000 generations. Genomes were sequenced from individuals sampled at various points during the experiment and were compared with the genome of the ancestral clone. Note that mutations accumulated at a relatively constant rate throughout the experiment.

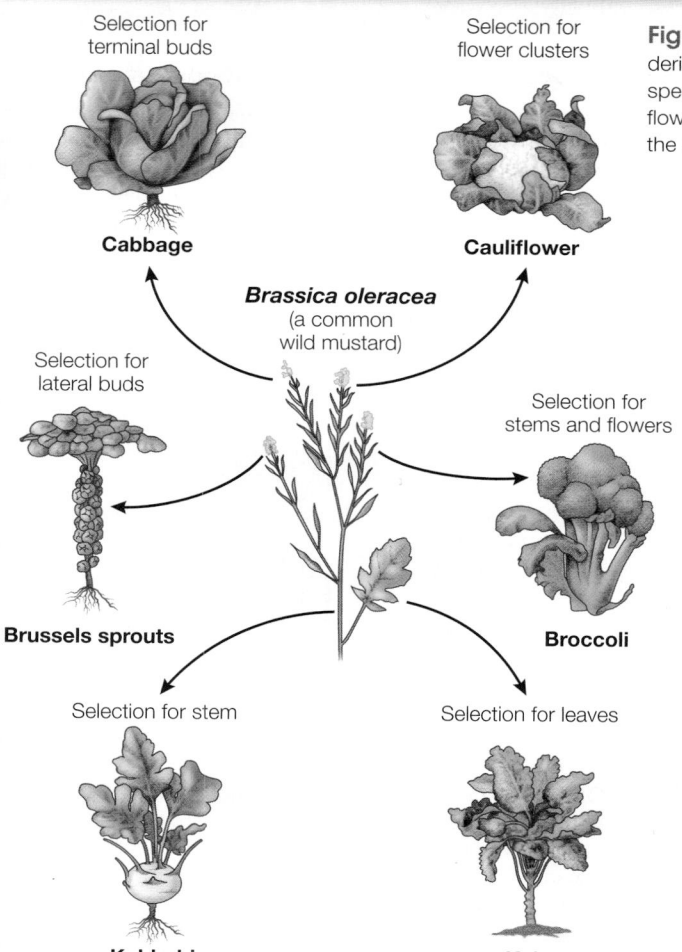

Selection for terminal buds

Cabbage

Selection for flower clusters

Cauliflower

Brassica oleracea
(a common wild mustard)

Selection for lateral buds

Selection for stems and flowers

Brussels sprouts

Broccoli

Selection for stem

Selection for leaves

Kohlrabi

Kale

Figure 20.4 Many Vegetables from One Species   All the crop plants shown here derive from a single wild mustard species. European agriculturalists produced these crop species by selecting and breeding plants with unusually large buds, stems, leaves, or flowers. The results substantiate the vast amount of variation present in the gene pool of the ancestral species.

nature of variation and selection came from domesticated plants and animals. Darwin bred pigeons and thus knew firsthand the astonishing diversity in color, size, form, and behavior that breeders could achieve (**Figure 20.5**). He recognized close parallels between selection by breeders and selection in nature. Whereas artificial selection resulted in traits that were preferred by the human breeders, natural selection resulted in traits that helped organisms survive and reproduce more effectively. In both cases, selection simply increased the frequency of the favored trait from one generation to the next.

Laboratory experiments also demonstrate the existence of considerable genetic variation in populations, and show how this variation can lead to evolution through selection. In one such experiment, investigators bred populations of the fruit fly *Drosophila melanogaster* with high or low numbers of bristles on their abdomens from an initial population with intermediate numbers of bristles. After 35 generations, all flies in both the high- and low-bristle lineages had bristle numbers that fell well outside the range found in the original population (**Figure 20.6**). Selection for high and low bristle numbers resulted in new combinations of the many different genes that were present in the original population, so that the phenotypic variation seen in subsequent generations fell outside the phenotypic variation seen in the original population.

Natural selection increases the frequency of beneficial mutations in populations

Darwin knew that far more individuals of most species are born than survive to reproduce. He also knew that, although offspring tend to resemble their parents, the offspring of most organisms are not identical either to their parents or to one another. He suggested that slight differences among individuals affect the chance that a given individual will survive and reproduce, which increases the frequency of the favored trait in the next generation. A favored trait that spreads through

ancestor. For example, artificial selection for different traits in a single European species of wild mustard produced many important crop plants (**Figure 20.4**). Agriculturalists were able to achieve these results because the original mustard population had genetic variation for the characters of interest (such as stem thickness or number of leaves).

Darwin compared artificial selection, which was commonly practiced by animal and plant breeders, with natural selection that occurred in natural populations. Many of Darwin's observations on the

Figure 20.5 Artificial Selection   Charles Darwin, who raised pigeons as a hobby, noted similar forces at work in artificial and natural selection. The "fancy" pigeons shown here represent 3 of the more than 300 varieties derived from the wild rock pigeon (*Columba livia*; left) by artificial selection on characters such as color and feather distribution.

experiment

Original Paper: Barber, J. R., B. C. Leavell, A. L. Keener, J. W. Brein-holt, B. A. Chadwell, C. J. W. McClure, G. M. Hill and A. Y. Kawahara. 2015. Moth tails divert bat attack: Evolution of acoustic deflection. *Proceedings of the National Academy of Sciences USA* 112: 2812–2816.

Many groups of moths have evolved the ability to hear the ultrasounds that bats produce for echolocation. These moths can then use evasive behavior to avoid predation by the bats. But almost half of all moths lack ears to hear the sounds produced by bats. Jesse Barber and his colleagues noticed that many non-hearing moths have long, posterior wing extensions ("wing tails"), and wondered if these wing tails might serve to deflect attacks by echolocating bats.

HYPOTHESIS▶ As the long tails on moth wings flutter in flight, they serve to deflect the attacks of echolocating bats.

METHOD

1. Remove the wing tails from an experimental group of luna moths (*Actias luna*).

2. In an experimental setting, compare the capture rate by big brown bats (*Eptesicus fuscus*) of moths with and without wing tails.

3. Control for the effects of wing tail size by conducting comparable experiments on moths with various wing tail sizes, in species that naturally lack wing tails.

4. Compare the flying efficiency of luna moths with and without wing tails, to test the possibility that wing tails improve flying performance.

RESULTS

The successful capture rate by bats on luna moths was much higher in moths that had their wing tails removed. Large wing tail size does seem to have an effect on reducing successful predation as well, but the long wing tails of luna moths provide a much greater benefit to predation avoidance than predicted by the increase in wing tail size alone. The removal of wing tails from luna moths had no significant effect on the flight performance of the moths. The graph shows successful capture rate by bats on luna moths, with and without wing tails. Brackets show 95% confidence intervals.

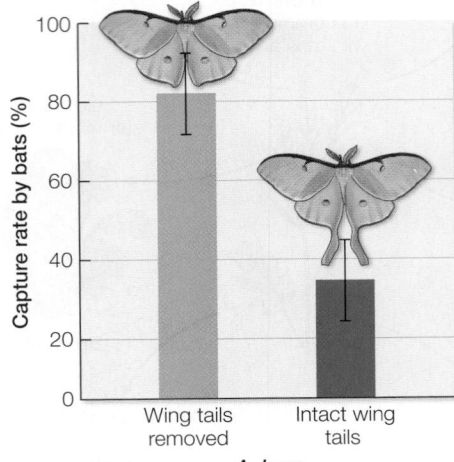

A. luna

CONCLUSION▶ The tail extensions on the wings of luna moths reduce the rate of successful bat attacks on the moths. Bats attack the fluttering wing tails of luna moths rather than their bodies, allowing the moths to escape. The wing tails do not provide improvements in flight performance, but rather in predation avoidance.

---

a population by natural selection is known as an **adaptation**; this word is used to describe both the trait itself and the process that produces the trait.

Biologists regard an organism as being adapted to a particular environment when they can demonstrate that a slightly different organism reproduces and survives less well in that environment. To understand adaptation, biologists compare the performances of individuals that differ in their traits.

In the story that opens this chapter, we learned that many groups of moths have evolved the ability to detect ultrasounds of bats, which allows the moths to fly away from the bats or use evasive flight behavior to avoid predation. But many species of moths are unable to hear bats, so they would seem to be at a huge disadvantage. As described in **Investigating Life: Do Long Wing Tails Help Moths Escape Bat Predation?**, Jesse Barber and his colleagues noticed that several groups of non-hearing moths have long extensions of their

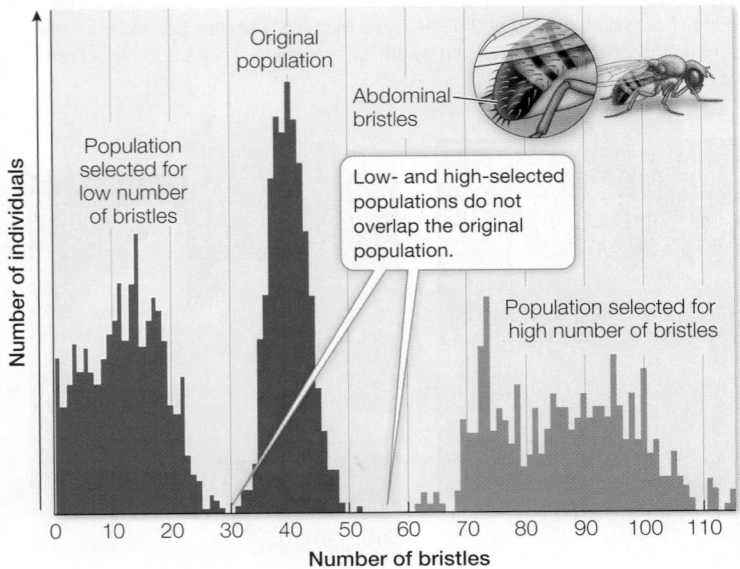

Figure 20.6 Artificial Selection Reveals Genetic Variation When investigators subjected *Drosophila melanogaster* to artificial selection for abdominal bristle number, that character evolved rapidly. The graph shows the number of flies with different numbers of bristles in the original population and after 35 generations of artificial selection for low and for high bristle numbers.

Q: How did flies in the selected populations end up with fewer or more bristles than were observed in any of the flies in the original population?

Tail/body size ratio

0   3.3

Length = 0.232

Saturniidae

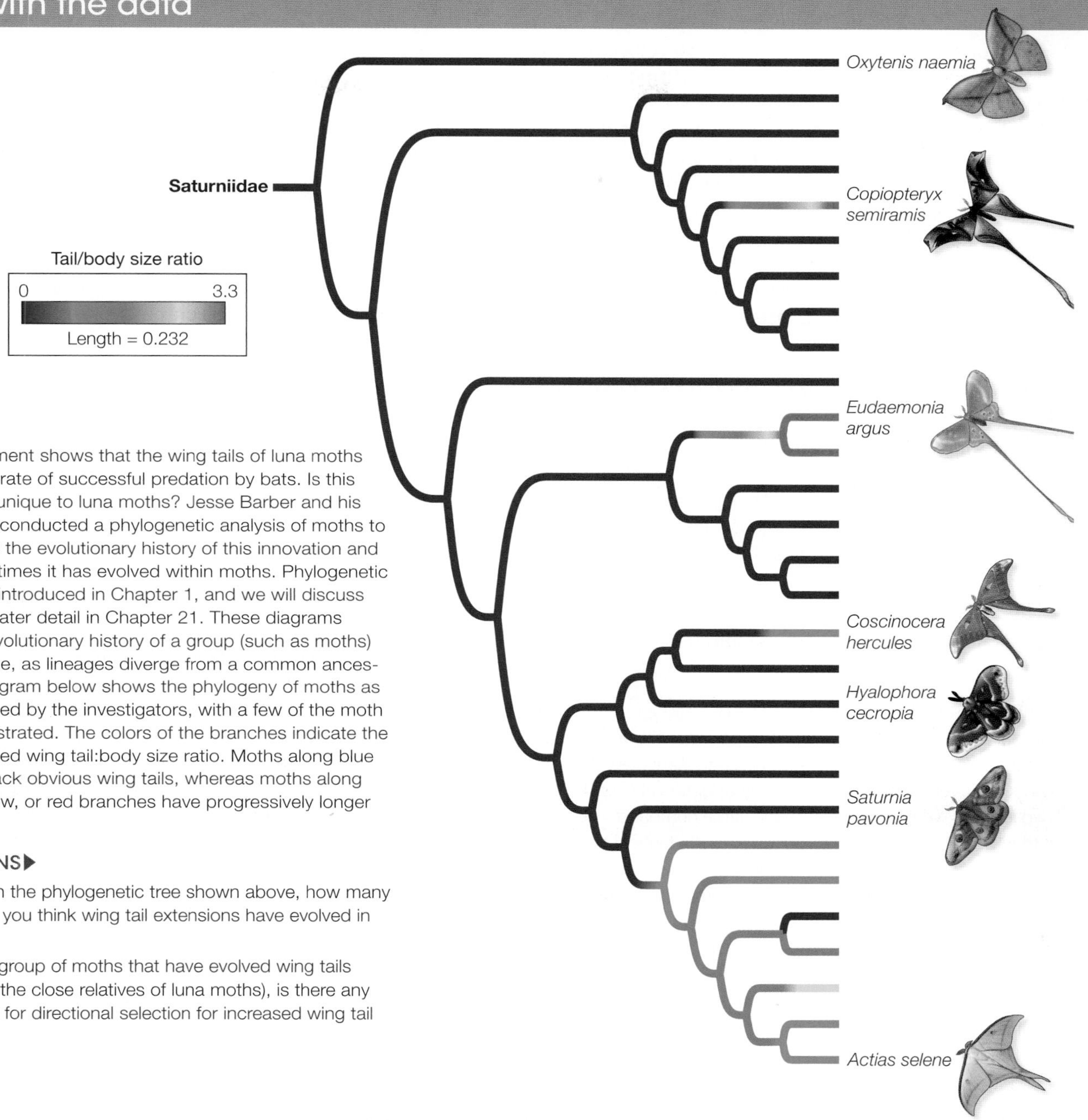

The experiment shows that the wing tails of luna moths reduce the rate of successful predation by bats. Is this innovation unique to luna moths? Jesse Barber and his colleagues conducted a phylogenetic analysis of moths to understand the evolutionary history of this innovation and how many times it has evolved within moths. Phylogenetic trees were introduced in Chapter 1, and we will discuss them in greater detail in Chapter 21. These diagrams trace the evolutionary history of a group (such as moths) through time, as lineages diverge from a common ancestor. The diagram below shows the phylogeny of moths as reconstructed by the investigators, with a few of the moth species illustrated. The colors of the branches indicate the reconstructed wing tail:body size ratio. Moths along blue branches lack obvious wing tails, whereas moths along green, yellow, or red branches have progressively longer wing tails.

QUESTIONS▶

1. Based on the phylogenetic tree shown above, how many times do you think wing tail extensions have evolved in moths?

2. Within a group of moths that have evolved wing tails (such as the close relatives of luna moths), is there any evidence for directional selection for increased wing tail length?

Oxytenis naemia

Copiopteryx semiramis

Eudaemonia argus

Coscinocera hercules

Hyalophora cecropia

Saturnia pavonia

Actias selene

wings ("wing tails"), and they wondered if these structures might serve to distract echolating bats and allow the moths to escape. The researchers cut the wing tails off some individual moths and compared the rates of bat predation, and flight performance, in the moths with and without wing tails. They found that the wing tails had very little effect on the moth's flying efficiency, but conveyed a huge advantage in reducing bat predation. The wing tails flutter as the moths fly, and the echolocation system of the bats is more likely to locate the fluttering nonessential wing tail than the body of the moth,

thereby allowing a moth to survive a bat attack. Thus, the wing tails appear to be an adaptation that helps moths avoid bat predation.

 Media Clip 20.2 *A. luna* in Flight
www.Life11e.com/mc20.2

 Media Clip 20.3 Big Brown Bats (*E. fuscus*) versus *A. luna*
www.Life11e.com/mc20.3

 Activity 20.2 Natural Selection Simulation
www.Life11e.com/ac20.2

Gene flow may change allele frequencies

Few populations are completely isolated from other populations of the same species. Migration of individuals and movements of gametes (in pollen, for example) between populations—a phenomenon called *gene flow—can change allele frequencies in a population. If the arriving individuals survive and reproduce in their new location, they may add new alleles to the population's gene pool, or they may change the frequencies of alleles present in the original population.

---

*connect the concepts   If gene flow between two populations stops, those populations may diverge and become different species; see Key Concept 22.2.

Genetic drift may cause large changes in small populations

In small populations, **genetic drift**—random changes in allele frequencies from one generation to the next—may produce large changes in allele frequencies over time. Harmful alleles may increase in frequency, and rare advantageous alleles may be lost. Even in large populations, genetic drift can influence the frequencies of neutral alleles (which do not affect the survival and reproductive rates of their bearers).

 Activity 20.3 Genetic Drift Simulation
www.Life11e.com/ac20.3

To illustrate the effects of genetic drift, suppose there are only two females in a small population of normally brown mice, and one of these females carries a newly arisen dominant allele that produces black fur. Even in the absence of any selection, it is unlikely that the two females will produce exactly the same number of offspring. Even if they do produce identical litter sizes and identical numbers of litters, chance events that have nothing to do with genetic characteristics are likely to result in differential mortality among their offspring. If each female produces one litter, but a flood envelops the black female's nest and kills all of her offspring, the novel allele could be lost from the population in just one generation. In contrast, if the brown female's litter is lost, then the frequency of the newly arisen allele (and phenotype) for black fur will rise dramatically in just one generation.

Genetic drift is especially potent when a population is reduced dramatically in size. Even populations that are normally large may occasionally pass through environmental events that only a small number of individuals survive, a situation known as a **population bottleneck**. The effect of genetic drift in such a situation is illustrated in **Figure 20.7**, in which red and yellow beans represent two alleles of a gene. Most of the beans in the small sample of the "population" that "survives" the bottleneck event are, just by chance, red, so the new population has a much higher frequency of red beans than

the previous generation had. In a real population, the red and yellow allele frequencies would be described as having "drifted."

A population forced through a bottleneck is likely to lose much of its genetic variation. For example, when Europeans first arrived in North America, millions of greater prairie-chickens (*Tympanuchus cupido*) inhabited the midwestern prairies. As a result of hunting and habitat destruction by the new settlers, the Illinois population of this species plummeted from about 100 million birds in 1900 to fewer than 50 individuals in the 1990s. A comparison of DNA from birds collected in Illinois during the middle of the twentieth century with DNA from the surviving population in the 1990s showed that Illinois prairie-chickens have lost most of their genetic diversity. Loss of genetic variation in small populations is one of the problems facing biologists who attempt to protect endangered species.

Genetic drift can have similar effects when a few pioneering individuals colonize a new region. Because of its small size, the colonizing population is unlikely to possess all of the alleles found in the gene pool of its source population. The resulting change in genetic variation, called a **founder effect**, is equivalent to that in a large population reduced by a bottleneck.

Nonrandom mating can change genotype or allele frequencies

Mating patterns often alter genotype frequencies because the individuals in a population do not choose mates at random. For example, self-fertilization is common in many groups of organisms, especially plants. Any time individuals mate preferentially with other individuals of the same genotype (including themselves), homozygous genotypes will increase in frequency and heterozygous genotypes will decrease in frequency over time. The opposite effect (more heterozygotes, fewer homozygotes) is expected when individuals mate primarily or exclusively with individuals of different genotypes.

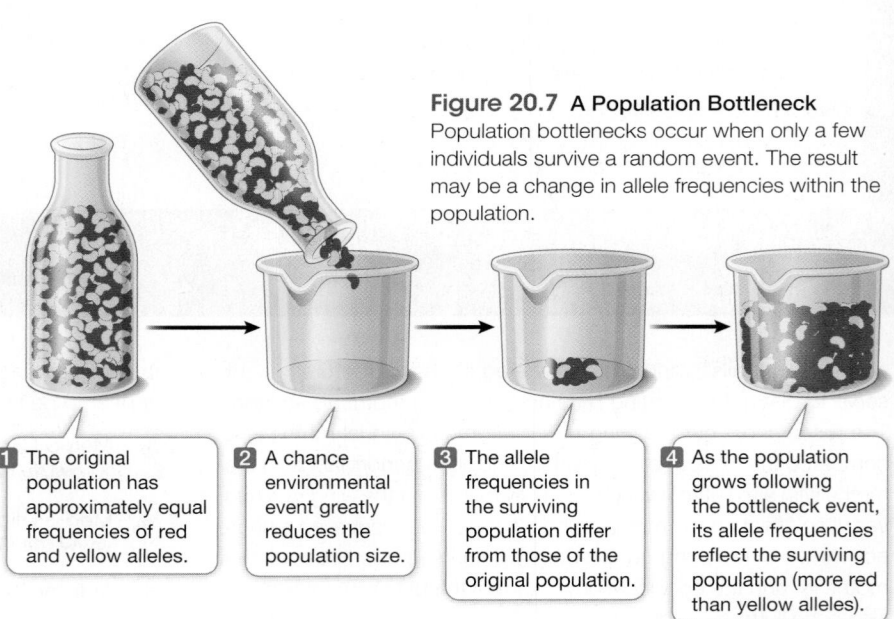

Figure 20.7 A Population Bottleneck
Population bottlenecks occur when only a few individuals survive a random event. The result may be a change in allele frequencies within the population.

1 The original population has approximately equal frequencies of red and yellow alleles.

2 A chance environmental event greatly reduces the population size.

3 The allele frequencies in the surviving population differ from those of the original population.

4 As the population grows following the bottleneck event, its allele frequencies reflect the surviving population (more red than yellow alleles).

Euplectes progne

Figure 20.8 What Is the Advantage?   The extensive tail of the male African long-tailed widowbird inhibits its ability to fly. Darwin attributed the evolution of this seemingly nonadaptive trait to sexual selection.

Many nonrandom mating systems produce changes in genotype frequencies but not in allele frequencies, and thus do not, by themselves, result in evolutionary change in a population. However, nonrandom mating systems that result in different reproductive success among individuals do produce allele frequency changes from one generation to the next. One such process, **sexual selection** occurs when individuals of one sex mate preferentially with particular individuals of the opposite sex rather than at random.

Sexual selection was first suggested by Charles Darwin, who developed the idea to explain the evolution of conspicuous traits that would appear to inhibit survival, such as bright colors and elaborate courtship displays in males of many species. He hypothesized that these features either improved the ability of their bearers to compete for access to mates (intrasexual selection) or made their bearers more attractive to members of the opposite sex (intersexual selection).

Darwin argued that while natural selection typically favors traits that enhance the survival of their bearers or their bearers' descendants, sexual selection is primarily about successful reproduction. An animal that survives but fails to reproduce makes no contribution to the next generation. Thus sexual selection may favor traits that enhance an individual's chances of reproduction even when these traits reduce its chances of survival. For example, females may be more likely to see or hear males with a given trait (and thus be more likely to mate with those males), even though the favored trait also increases the chances that the male will be seen or heard by a predator.

One example of a trait that Darwin attributed to sexual selection is the remarkable tail of the male African long-tailed widowbird (*Euplectes progne*), which is longer than the bird's head and body combined (**Figure 20.8**). Male widowbirds normally select, and defend from other males, a territory where they perform courtship displays to attract females. To investigate whether sexual selection drove the evolution of widowbird tails, a biologist clipped the tails of some captured male widowbirds and lengthened the tails of others by gluing on additional feathers. He then cut and reglued the tail feathers of still other males, which served as controls. Both short- and long-tailed males successfully defended their display territories, indicating that a long tail does not confer an advantage in male–male competition. However, males with artificially elongated tails attracted about four times more females than did males with shortened tails (**Figure 20.9**). Thus males with long tails pass on their genes to more offspring than do males with short tails, which leads to the evolution of this unusual trait.

experiment

Figure 20.9   Sexual Selection in Action

Original Paper: Andersson, M. 1982. Female choice selects for extreme tail length in a widowbird. *Nature* 299: 818–820.

Behavioral ecologist Malte Andersson tested Darwin's hypothesis that excessively long tails evolved in male widowbirds because female preference for longer-tailed males increased their mating and reproductive success.

HYPOTHESIS▶   Female widowbirds prefer to mate with the male that displays the longest tail; longer-tailed males thus are favored by sexual selection because they will father more offspring.

METHOD

1. Capture males and artificially lengthen or shorten tails by cutting or gluing on feathers. In a control group, cut and replace tails to their normal length (to control for the effects of tail-cutting).
2. Release the males to establish their territories and mate.
3. Count the nests with eggs or young on each male's territory.

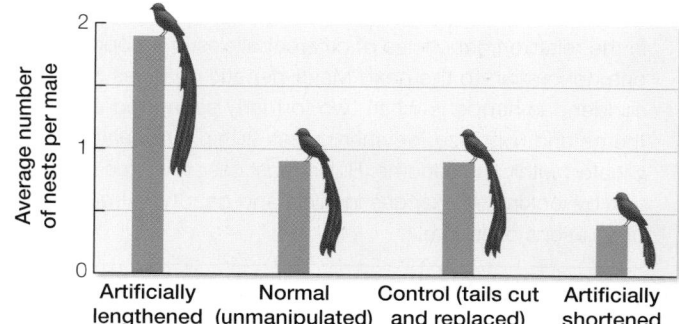

Average number of nests per male

| Artificially lengthened | Normal (unmanipulated) | Control (tails cut and replaced) | Artificially shortened |

RESULTS

Male widowbirds with artificially shortened tails established and defended display sites sucessfully but fathered fewer offspring than did control or unmanipulated males. Males with artificially lengthened tails fathered the most offspring.

CONCLUSION▶   Sexual selection in widowbirds has favored the evolution of long tails in the males.

20.2 recap

Evolutionary processes change the genetic structure of a population from one generation to the next. These processes include mutation, natural selection, gene flow, genetic drift, and nonrandom mating.

learning outcomes

You should be able to:

- Distinguish between the everyday and scientific uses of the terms "adapt" and "evolve."
- Describe how each of the five processes of evolution can result in changes in gene frequencies across generations of a population.
- Predict the effects of each of the five processes of evolution on a population.

1. How is mutation a necessary component of the evolutionary processes?
2. What kind of traits would be more likely to evolve by genetic drift than by natural selection?
3. Describe conditions where nonrandom mating does and does not lead to evolution.
4. Can you think of reasons why gene flow might slow the evolution of a population?

The processes of mutation, natural selection, gene flow, genetic drift, and nonrandom mating can all result in evolutionary change. We will consider next how evolutionary change that results from these processes is measured.

▶ key concept 20.3 Evolution Can Be Measured by Changes in Allele Frequencies

Much of evolution occurs through gradual changes in the relative frequencies of different alleles in a population from one generation to the next. Major genetic changes can also be sudden, as happens when two formerly separated populations merge and hybridize, or when genes within a population are duplicated within the genome. But in most cases, we measure evolution by looking at changes in allele and genotype frequencies in populations over time.

focus your learning

- Biologists detect and measure evolutionary changes in populations by calculating changes in allele frequencies over time.
- The evolutionary processes acting on a population can be identified using Hardy–Weinberg equilibrium as a basis for calculations.

To measure allele frequencies in a population precisely, we would need to count every allele at every locus in every individual in the population. Fortunately, we do not need to make such complete measurements because we can reliably estimate allele frequencies for a given locus by counting alleles in a sample of individuals from the population. The sum of all allele frequencies at a locus is equal to 1, so measures of allele frequency range from 0 to 1.

An allele's frequency is calculated using the following formula:

$$p = \frac{\text{number of copies of the allele in the population}}{\text{total number of copies of all alleles in the population}}$$

If only two alleles (we'll call them A and a) for a given locus are found among the members of a diploid population, those alleles can combine to form three different genotypes: AA, Aa, and aa. A population with more than one allele at a locus is said to be polymorphic ("many forms") at that locus. Applying the formula above, as shown in **Figure 20.10**, we can calculate the relative frequencies of alleles A and a in a population of N individuals as follows:

- Let N_{AA} be the number that are homozygous for the A allele (AA).
- Let N_{Aa} be the number that are heterozygous (Aa).
- Let N_{aa} be the number that are homozygous for the a allele (aa).

Note that $N_{AA} + N_{Aa} + N_{aa} = N$, the total number of individuals in the population, and that the total number of copies of both alleles present in the population is $2N$, because each individual is diploid. Each AA individual has two copies of the A allele, and each Aa individual has one copy of the A allele. Therefore the total number of A alleles in the population is $2N_{AA} + N_{Aa}$. Similarly, the total number of a alleles in the population is $2N_{aa} + N_{Aa}$. If p represents the frequency of A, and q represents the frequency of a, then

$$q = \frac{2N_{aa} + N_{Aa}}{2N}$$

and

$$p = \frac{2N_{AA} + N_{Aa}}{2N}$$

The calculations in Figure 20.10 demonstrate two important points. First, notice that for each population, $p + q = 1$, which means that $q = 1 - p$. So when there are only two alleles at a given locus in a population, we can calculate the frequency of one allele and obtain the second allele's frequency by subtraction. If there is only one allele at a given locus in a population, its frequency is 1: the population is then monomorphic at that locus, and the allele is said to be **fixed**.

Second, notice that population 1 (consisting mostly of homozygotes) and population 2 (consisting mostly of heterozygotes) have the same allele frequencies for A and a. Thus they have the same gene pool for this locus. Because the alleles in the gene pool are distributed differently among individuals, however, the genotype frequencies of the two populations differ.

The frequencies of the different alleles at each locus and the frequencies of the different genotypes in a population describe that population's **genetic structure**. Allele frequencies measure the amount of genetic variation in a population, whereas genotype frequencies show how a population's genetic variation is distributed among its members. Other measures, such as the proportion of loci that are polymorphic, are also used to measure variation in populations. With these measurements, it becomes possible to consider how the genetic structure of a population changes or remains the same over generations—that is, to measure evolutionary change.

research tools

Figure 20.10 Calculating Allele and Genotype Frequencies

Allele and genotype frequencies for a gene locus with two alleles in the population can be calculated using the equations in panel 1. When the equations are applied to two populations (panel 2), we find that the *frequencies of alleles A and a in the two populations* are the same, but the alleles are distributed differently between heterozygous and homozygous genotypes.

1 In any population, where N is the total number of individuals in the population:

$$\text{Frequency of allele A} = p = \frac{2N_{AA} + N_{Aa}}{2N} \qquad \text{Frequency of allele a} = q = \frac{2N_{aa} + N_{Aa}}{2N}$$

Frequency of genotype $AA = N_{AA}/N$
Frequency of genotype $Aa = N_{Aa}/N$
Frequency of genotype $aa = N_{aa}/N$

2 Compute the allele and genotype frequencies for two separate populations of $N = 200$:

Population 1 (mostly homozygotes)	Population 2 (mostly heterozygotes)
$N_{AA} = 90$, $N_{Aa} = 40$, and $N_{aa} = 70$	$N_{AA} = 45$, $N_{Aa} = 130$, and $N_{aa} = 25$
$p = \dfrac{180 + 40}{400} = 0.55$	$p = \dfrac{90 + 130}{400} = 0.55$
$q = \dfrac{140 + 40}{400} = 0.45$	$q = \dfrac{50 + 130}{400} = 0.45$
Freq. $AA = 90/200 = 0.45$	Freq. $AA = 45/200 = 0.225$
Freq. $Aa = 40/200 = 0.20$	Freq. $Aa = 130/200 = 0.65$
Freq. $aa = 70/200 = 0.35$	Freq. $aa = 25/200 = 0.125$

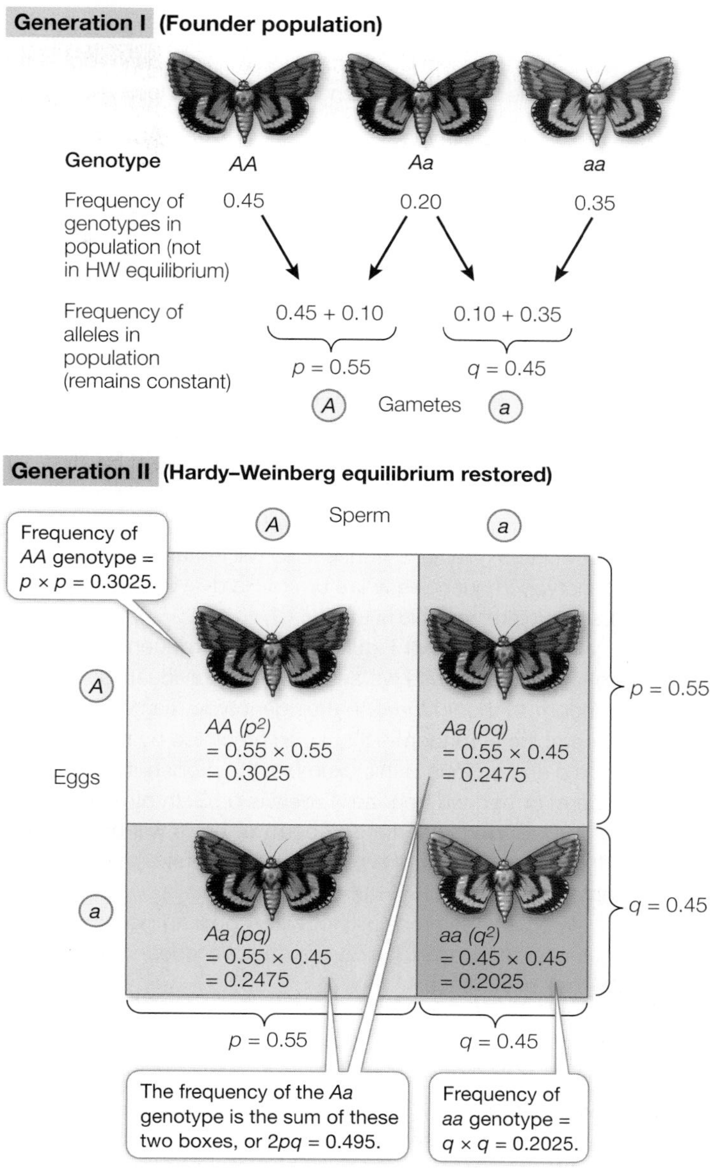

Evolution will occur unless certain restrictive conditions exist

In 1908, the British mathematician Godfrey Hardy and the German physician Wilhelm Weinberg independently deduced the conditions that must prevail if the genetic structure of a population is to remain the same over time. If the conditions they identified do not exist, then evolution will occur. The resulting principle is known as **Hardy–Weinberg equilibrium**. Hardy–Weinberg equilibrium describes a model in which allele frequencies do not change across generations and genotype frequencies can be predicted from allele frequencies (**Figure 20.11**). The principles of Hardy–Weinberg equilibrium apply only to sexually reproducing organisms. Several conditions must be met for a population to be at Hardy–Weinberg equilibrium. Note that the following conditions correspond inversely to the five principal processes of evolution (discussed in Key Concept 20.2):

- *There is no mutation*. The alleles present in the population do not change, and no new alleles are added to the gene pool.
- *There is no selection among genotypes*. Individuals with different genotypes have equal probabilities of survival and equal rates of reproduction.

Generation I (Founder population)

Genotype	AA	Aa	aa
Frequency of genotypes in population (not in HW equilibrium)	0.45	0.20	0.35

Frequency of alleles in population (remains constant)

$0.45 + 0.10 \qquad 0.10 + 0.35$
$p = 0.55 \qquad q = 0.45$

(A) Gametes (a)

Generation II (Hardy–Weinberg equilibrium restored)

Frequency of AA genotype = $p \times p = 0.3025$.

Sperm (A) (a)

Eggs

(A)
AA (p^2) = 0.55 × 0.55 = 0.3025
Aa (pq) = 0.55 × 0.45 = 0.2475
$p = 0.55$

(a)
Aa (pq) = 0.55 × 0.45 = 0.2475
aa (q^2) = 0.45 × 0.45 = 0.2025
$q = 0.45$

$p = 0.55 \qquad q = 0.45$

The frequency of the Aa genotype is the sum of these two boxes, or $2pq = 0.495$.

Frequency of aa genotype = $q \times q = 0.2025$.

Figure 20.11 One Generation of Random Mating Restores Hardy–Weinberg Equilibrium

Generation I of this population is made up of migrants from several source populations and so is not initially in Hardy–Weinberg equilibrium. After one generation of random mating, the allele frequencies are unchanged, and the genotype frequencies return to Hardy–Weinberg expectations. The lengths of the sides of each rectangle are proportional to the allele frequencies in the population; the areas of the rectangles are proportional to the genotype frequencies.

- *There is no gene flow*. There is no movement of individuals into or out of the population or reproductive contact with other populations.
- *Population size is infinite*. The larger a population, the smaller will be the effect of genetic drift.
- *Mating is random*. Individuals do not preferentially choose mates with certain genotypes.

If these idealized conditions hold, two major consequences follow. First, the frequencies of alleles at a locus remain constant from generation to generation. Second, following one generation of random mating, the genotype frequencies occur in the following proportions:

Genotype	AA	Aa	aa
Frequency	p^2	$2pq$	q^2

To understand why these consequences are important, start by considering a population that is *not* in Hardy–Weinberg equilibrium, such as generation I in Figure 20.11. This could occur, for example, if the initial population is founded by migrants from several other populations, thus violating the Hardy–Weinberg assumption of no gene flow. In this example, generation I has more homozygous individuals and fewer heterozygous individuals than would be expected under Hardy–Weinberg equilibrium (a condition known as heterozygote deficiency).

Even with a starting population that is not in Hardy–Weinberg equilibrium, we can predict that after a single generation of random mating, and if the other Hardy–Weinberg assumptions are not violated, the allele frequencies will remain unchanged, but the genotype frequencies will return to Hardy–Weinberg expectations. Let's explore why this is true.

In generation I of Figure 20.11, the frequency of the *A* allele (*p*) is 0.55. Because we assume that individuals select mates at random, without regard to their genotype, gametes carrying *A* or *a* combine at random—that is, as predicted by the allele frequencies *p* and *q*. Thus in this example, the probability that a particular sperm or egg will bear an *A* allele is 0.55. In other words, 55 out of 100 randomly sampled sperm or eggs will bear an *A* allele. Because *q* = 1 – *p*, the probability that a sperm or egg will bear an *a* allele is 1 – 0.55 = 0.45.

To obtain the probability of two *A*-bearing gametes coming together at fertilization, we multiply the two independent probabilities of their occurrence:

$$p \times p = p^2 = (0.55)^2 = 0.3025$$

Therefore 0.3025, or 30.25 percent, of the offspring in generation II will have homozygous genotype *AA*. Similarly, the probability of two *a*-bearing gametes coming together is

$$q \times q = q^2 = (0.45)^2 = 0.2025$$

which means that 20.25 percent of generation II will have the *aa* genotype.

There are two ways of producing a heterozygote: an *A* sperm may combine with an *a* egg, the probability of which is *p* × *q*; or an *a* sperm may combine with an *A* egg, the probability of which is *q* × *p*. Consequently, the overall probability of obtaining a heterozygote is 2*pq*, or 0.495. The frequencies of the *AA*, *Aa*, and *aa* genotypes in generation II of Figure 20.11 now meet Hardy–Weinberg expectations, and the frequencies of the two alleles (*p* and *q*) have not changed from generation I.

Under the assumptions of Hardy–Weinberg equilibrium, allele frequencies *p* and *q* remain constant from generation to generation. If Hardy–Weinberg assumptions are violated and the genotype frequencies in the parental generation are altered (say, by the loss of a large number of *AA* individuals from the population), then the allele frequencies in the next generation will be altered. However, based on the new allele frequencies, another generation of random mating will be sufficient to restore the genotype frequencies to Hardy–Weinberg equilibrium.

 Activity 20.4 **Hardy–Weinberg Equilibrium**
www.Life11e.com/ac20.4

Deviations from Hardy–Weinberg equilibrium show that evolution is occurring

You probably have realized that populations in nature never meet the stringent conditions necessary to be at Hardy–Weinberg equilibrium—which explains why all biological populations evolve. Why, then, is this model considered so important for the study of evolution? There are two reasons. First, the equation is useful for predicting the approximate genotype frequencies of a population from its allele frequencies. Second—and crucially—the model allows biologists to evaluate which processes are acting on the evolution of a particular population. The specific patterns of deviation from Hardy–Weinberg equilibrium can help us identify the various processes of evolutionary change.

20.3 recap

Hardy–Weinberg equilibrium describes the theoretical conditions required for evolution *not* to occur. Deviations from Hardy–Weinberg expectations provide information about how evolution *is* occurring in a given population.

learning outcomes
You should be able to:
- Calculate the allele frequencies for two alleles at a given locus in a diploid population.
- Calculate observed and expected genotype frequencies from allele frequencies based on the principles of Hardy–Weinberg equilibrium.
- Analyze data and formulate hypotheses to explain deviations from Hardy–Weinberg expectations in observed genotype frequencies.

1. The following sample lists the genotype at locus *A* for 10 individuals in a diploid population: *AA, AA, Aa, Aa, Aa, Aa, aa, aa, aa, aa*.
   a. Based on this sample, what is the observed frequency of allele *a*? The observed frequency of allele *A*?
   b. What are the observed frequencies of genotypes *aa, Aa*, and *AA*?
   c. After one generation of random mating, what would be the Hardy–Weinberg expectations for the frequencies of genotypes *aa, Aa*, and *AA*?

2. In a population of toads, allele *A* is present at a frequency of 0.2, and allele *a* is present at a frequency of 0.8. Based on Hardy–Weinberg expectations, what is the expected frequency of the genotype *Aa* in the population?

3. Assume that the observed frequency of the genotype *Aa* in the toad population in Question 2 is 0.15, rather than the expected frequency you calculated. What are some possible explanations?

Our discussion so far has focused on changes in allele frequencies at a single gene locus. Genes do not exist in isolation, however, but interact with one another (and with the environment) to produce an organism's phenotype. What effects can these interactions have on selection?

key concept
20.4

key concept 20.4 Selection Can Be Stabilizing, Directional, or Disruptive

Although evolution is defined as changes in the genetic makeup of a population from one generation to the next, natural selection acts directly on the phenotype—that is, on the physical features expressed by an organism with a given genotype—and therefore acts only indirectly on the genotype. The reproductive contribution of a phenotype to subsequent generations relative to the contributions of other phenotypes is called its **fitness**.

focus your learning

- Natural selection can alter the distribution of a quantitative trait in a population over time.

Changes in reproductive rate do not necessarily change the genetic structure of a population. For example, if all individuals in a population experience the same increase in reproductive rate (during an environmentally favorable year, for instance), the genetic structure of the population will not change. Changes in numbers of offspring are responsible for increases and decreases in the *size* of a population, but only changes in the *relative* success of different phenotypes in a population will lead to changes in allele frequencies from one generation to the next. The fitness of individuals of a particular phenotype is a function of the probability of those individuals surviving multiplied by the average number of offspring they produce over their lifetimes. In other words, the *fitness of a phenotype is determined by the relative rates of survival and reproduction of individuals with that phenotype.*

Until now, we have only discussed traits influenced by alleles at a single locus. Such traits are often distinguished by discrete qualities (black versus white, or smooth versus wrinkled) and so are called **qualitative traits**. Many traits, however, are influenced by alleles at more than one locus. Such traits are likely to show *continuous* quantitative variation rather than *discrete* qualitative variation, and so are known as **quantitative traits**. For example, body size is a continuous variable; the distribution of body sizes of individuals in a population, a trait that is influenced by genes at many loci as well as by the environment, is likely to resemble a continuous bell-shaped curve.

Natural selection can act on characters with quantitative variation in any one of several different ways, producing quite different results (**Focus: Key Figure 20.12**):

- **Stabilizing selection** preserves the average characteristics of a population by favoring average individuals.

- **Directional selection** changes the characteristics of a population by favoring individuals that vary in one direction from the mean of the population.

- **Disruptive selection** changes the characteristics of a population by favoring individuals that vary in both directions from the mean of the population.

focus: key figure

(A) Stabilizing selection

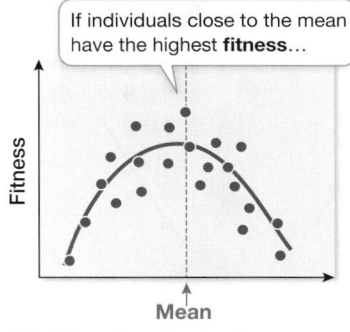

If individuals close to the mean have the highest **fitness**...

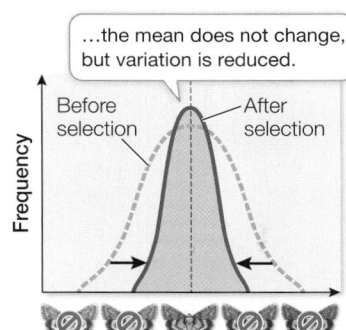

...the mean does not change, but variation is reduced.

(B) Directional selection

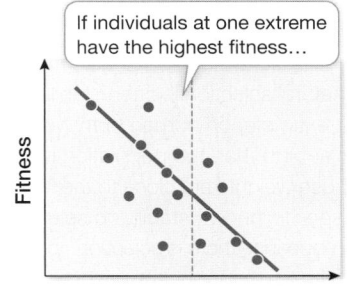

If individuals at one extreme have the highest fitness...

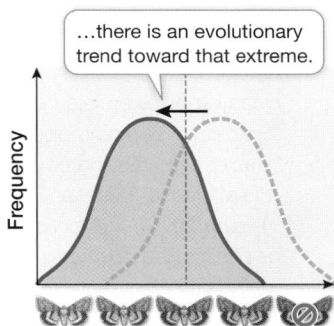

...there is an evolutionary trend toward that extreme.

(C) Disruptive selection

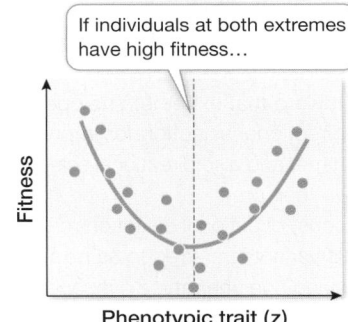

If individuals at both extremes have high fitness...

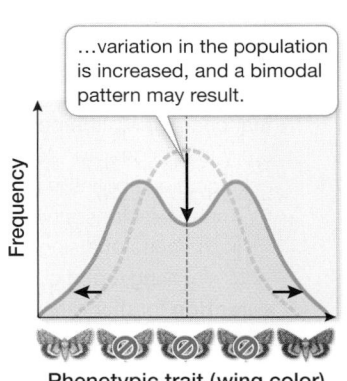

...variation in the population is increased, and a bimodal pattern may result.

Phenotypic trait (wing color)

Figure 20.12 Natural Selection Can Operate in Several Ways The graphs in the left-hand column show the fitness of individuals with different phenotypes for the same character. The right-hand graphs show the distribution of the phenotypes in the population before (light green) and after (dark green) the influence of selection.

Q: Why might selection for a trait in a population shift from directional to stabilizing selection over time, even if the optimal value for the trait stays constant?

Stabilizing selection reduces variation in populations

If the smallest and largest individuals in a population contribute fewer offspring to the next generation than do individuals closer to the average size, then stabilizing selection is operating on body size (see Figure 20.12A). Stabilizing selection reduces variation in populations, but it does not change the mean. Natural selection frequently acts in this way, countering increases in variation brought about by sexual recombination, mutation, or gene flow. Rates of phenotypic change in

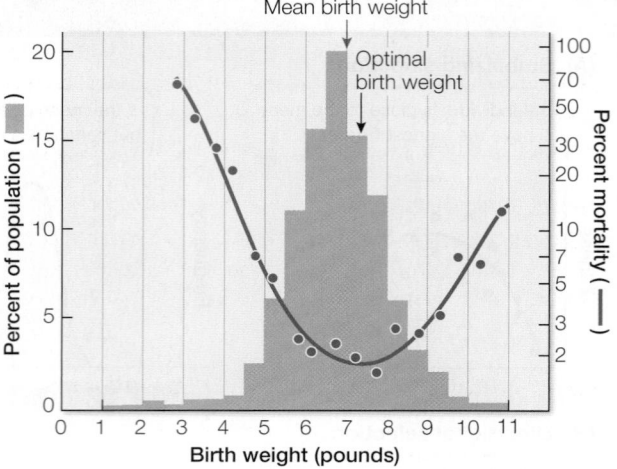

Mean birth weight

Optimal birth weight

Figure 20.13 **Human Birth Weight Is Influenced by Stabilizing Selection** Babies that weigh more or less than average are more likely to die soon after birth than are babies with weights closer to the population mean.

many species are slow because natural selection is often stabilizing. Stabilizing selection operates, for example, on human birth weight. Babies who are lighter or heavier at birth than the population mean die at higher rates than babies whose weights are close to the mean (**Figure 20.13**). In discussions of specific genes, stabilizing selection is often called **purifying selection** because there is selection against any deleterious mutations to the usual gene sequence.

Directional selection favors one extreme

Directional selection is operating when individuals at one extreme of a character distribution contribute more offspring to the next generation than other individuals do, shifting the average value of that character in the population toward that extreme. The opening story in this chapter describes directional selection for improved echolocation abilities in bats over time, and also directional selection for improved ultrasound hearing in their prey.

In the case of a single gene locus, directional selection may result in favoring a particular genetic variant—referred to as **positive selection** for that variant. By favoring one phenotype over another, directional selection results in an increase of the frequencies of alleles that produce the favored phenotype.

If directional selection operates over many generations, an evolutionary trend is seen in the population (see Figure 20.12B). Evolutionary trends often continue for many generations, but they can be reversed if the environment changes and different phenotypes are favored, or halted when an optimal phenotype is reached or trade-offs

between different adaptational advantages oppose further change. The character then undergoes stabilizing selection.

The long horns of Texas Longhorn cattle (**Figure 20.14**) are an example of a trait that has evolved through directional selection. Texas Longhorns are descendants of cattle brought to the New World by Christopher Columbus, who picked up a few cattle in the Canary Islands and brought them to the island of Hispaniola in 1493. The cattle multiplied, and their descendants were taken to the mainland of Mexico. Spaniards exploring what would become Texas and the southwestern United States brought these cattle with them, some of which escaped and formed feral herds. Populations of feral cattle increased greatly over the next few hundred years, but there was heavy predation from bears, mountain lions, and wolves, especially on the young calves. Cows with longer horns were more successful in protecting their calves against attacks, and over a few hundred years the average horn length in the feral herds increased considerably. In addition, the cattle evolved resistance to endemic diseases of the Southwest, as well as higher fecundity and longevity. Texas Longhorns often live and produce calves well into their twenties—about twice as long as many breeds of cattle that have been artificially selected by humans for traits such as high fat content or high milk production (which are examples of artificial directional selection).

Disruptive selection favors extremes over the mean

When disruptive selection operates, individuals at opposite extremes of a character distribution contribute more offspring to the next generation than do individuals close to the mean, which increases variation in the population (see Figure 20.12C).

The strikingly bimodal (two-peaked) distribution of bill sizes in the black-bellied seedcracker (*Pyrenestes ostrinus*), a West African finch (**Figure 20.15**), illustrates how disruptive selection can influence populations in nature. The seeds of two types of sedges (marsh plants) are the most abundant food source for these finches during part of the year. Birds with large bills can readily crack the hard seeds of the sedge *Scleria verrucosa*. Birds with small bills can crack *S. verrucosa* seeds only with difficulty; however, they feed more efficiently on the soft seeds of *S. goossensii* than do birds with larger bills. Young finches whose bills deviate markedly from the two predominant bill sizes do not survive as well as finches whose bills are close to one of the two sizes represented by the distribution peaks. Because there are few abundant food sources in the finches' environment, and because the seeds of the two sedges do not overlap in hardness, birds with intermediate-sized bills are less efficient in using either one of the species' principal food sources. Disruptive selection therefore maintains a bimodal bill size distribution.

Figure 20.14 **A Result of Directional Selection** In the American Southwest, long horns were advantageous for defending calves from attacks by predators, so cows with longer horns were more likely to raise calves successfully. As a result, horn length in feral herds of cattle increased between the early 1500s and the 1860s, leading to the Texas Longhorn breed. This evolutionary trend has been maintained by modern ranchers practicing artificial selection.

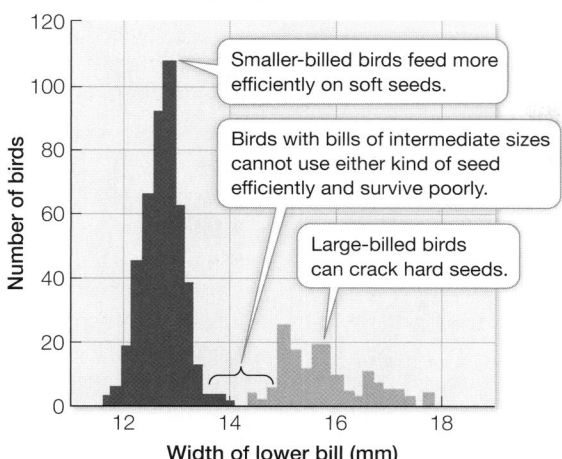

Figure 20.15 Disruptive Selection Results in a Bimodal Character Distribution   The bimodal distribution of bill sizes in the black-bellied seedcracker (*Pyrenestes ostrinus*) of West Africa is a result of disruptive selection, which favors individuals with larger and smaller bill sizes over individuals with intermediate-sized bills.

20.4 recap

Natural selection can both change and stabilize phenotypes within populations. Directional selection can result in continuous change to phenotypes; stabilizing selection slows change from a favored phenotype; and disruptive selection can result in one population that exhibits multiple distinct phenotypes.

learning outcomes

You should be able to:

- Describe the effect of each type of selection (stabilizing, directional, and disruptive) on phenotype distribution.
- Distinguish conditions in which selection is likely to be directional, stabilizing, or disruptive.

1. Describe the different expected outcomes of stabilizing, directional, and disruptive selection.
2. Why would you expect selection on human birth weight to be stabilizing rather than directional?
3. Can you think of examples of extreme phenotypes in animal or plant populations that could be explained by directional selection?

Genetic drift, stabilizing selection, and directional selection all tend to reduce genetic variation within populations. Nevertheless, as we have seen, most populations harbor considerable genetic variation. What processes produce and maintain genetic variation within populations?

key concept 20.5 Multiple Factors Account for the Maintenance of Variation in Populations

Genetic variation is the raw material on which mechanisms of evolution act. But if favored alleles increase in frequency in populations, then why aren't other alleles always lost from the population? In this section we will discuss several factors—neutral mutations, sexual recombination, frequency-dependent selection, and heterozygote advantage—that affect how genetic variation is established, how it is distributed among individuals, and how it is maintained within populations.

focus your learning

- Genetic variation is critical to the survival of a species.
- Several mechanisms maintain genetic variation in populations despite selection for favored alleles.

Neutral mutations accumulate in populations

An allele that does not affect the fitness of an organism—that is, an allele that is no better or worse than alternative alleles at the same locus—is called a **neutral allele**. Neutral alleles are added to a population over time through mutation, providing the population with considerable genetic variation. The frequencies of neutral alleles are not affected directly by natural selection. Even in large populations, neutral alleles may be lost, or may increase in frequency, purely by random genetic drift.

Much of the phenotypic variation we are able to observe is not neutral. However, modern techniques enable us to measure *neutral variation at the molecular level and provide the means to distinguish it from adaptive variation.

> *connect the concepts   Key Concept 23.2 describes how variation in neutral molecular traits can be used to study divergence among genes, populations, and species.

Sexual recombination amplifies the number of possible genotypes

In asexually reproducing organisms, each new individual is genetically identical to its parent unless there has been a mutation. When organisms reproduce sexually, however, offspring differ from their parents not only because they result from the combination of genetic material from two different gametes, but also through crossing over and independent assortment of chromosomes during meiosis, as described in Key Concept 11.5. Sexual recombination generates an endless variety of genotypic combinations that increase the evolutionary potential of populations—a long-term advantage of sex. Although many species reproduce asexually most of the time, few are strictly asexual over long periods of evolutionary time. Almost all have some means of achieving genetic recombination.

The evolution of the mechanisms of meiosis and the evolution of sexual recombination were crucial events in the history of life. Exactly how these attributes arose is puzzling, however, because sex has at least three striking disadvantages in the short term:

1. Recombination breaks up adaptive combinations of genes.
2. Sex reduces the rate at which females pass genes on to their offspring.
3. Dividing offspring into separate sexes greatly reduces the overall reproductive rate.

To see why this last disadvantage exists, consider an asexual female that produces the same number of offspring as a sexual female. Let's assume that both females produce two offspring, but that

50 percent of the sexual female's offspring will be males (and thus contribute only sperm). In this next (F₁) generation, both asexual females will produce two more offspring each, but there is only one sexual F_1 female to produce offspring. Thus the effective reproductive rate of the asexual lineage is twice that of the sexual lineage.

The evolutionary problem is to identify the advantages of sex that can overcome such short-term disadvantages. Several hypotheses have been proposed to explain the existence of sex, none of which are mutually exclusive. One is that sexual recombination facilitates repair of damaged DNA, because breaks and other errors in DNA on one chromosome can be repaired by copying the intact sequence from the homologous chromosome.

Another advantage is that sexual reproduction permits the elimination of deleterious mutations. As explained in Key Concept 13.4, DNA replication is not perfect. Errors are introduced in every generation, and most of these errors result in lower fitness. Asexual organisms have no mechanism to eliminate deleterious mutations. Hermann J. Muller noted that the accumulation of deleterious mutations in a nonrecombining genome is like a genetic ratchet. The mutations accumulate—"ratchet up"—at each replication. A mutation occurs and is passed on when the genome replicates, then two new mutations occur in the next replication, so three mutations are passed on, and so on. Over time, the least-mutated class of individuals is lost from the population as new mutations occur. Deleterious mutations cannot be eliminated except by the death of the lineage or a rare back mutation. This accumulation of deleterious mutations in lineages that lack genetic recombination is known as **Muller's ratchet**. In sexual species, by contrast, genetic recombination produces some individuals with more of these deleterious mutations and some with fewer. The individuals with fewer deleterious mutations are more likely to survive. Thus sexual reproduction allows natural selection to eliminate particular deleterious mutations from the population over time.

Another advantage of sex is the great variety of genetic combinations it creates in each generation. Sexual recombination does not directly influence the frequencies of alleles; rather, *it generates new combinations of alleles on which natural selection can act*. It expands variation in a character influenced by alleles at many loci by creating new genotypes. For example, genetic variation can be a defense against pathogens and parasites. Most pathogens and parasites have much shorter life cycles than their hosts and can rapidly evolve counteradaptations to host defenses. Sexual recombination can give the host's defenses a chance to keep up.

Frequency-dependent selection maintains genetic variation within populations

Natural selection often preserves variation as a polymorphism (the presence of two or more variants of a character in the same population). When the fitness of a given phenotype depends on its frequency in a population, a polymorphism may be maintained by a process known as **frequency-dependent selection**. *Perissodus microlepis*, a small fish that lives in Lake Tanganyika in East Africa, provides an example of frequency-dependent selection.

P. microlepis feeds on the scales of other fish, approaching its prey from behind and dashing in to bite off several scales from the prey's flank. Because of an asymmetrical jaw joint, the mouth of this scale-eating species opens either to the right or to the left; the

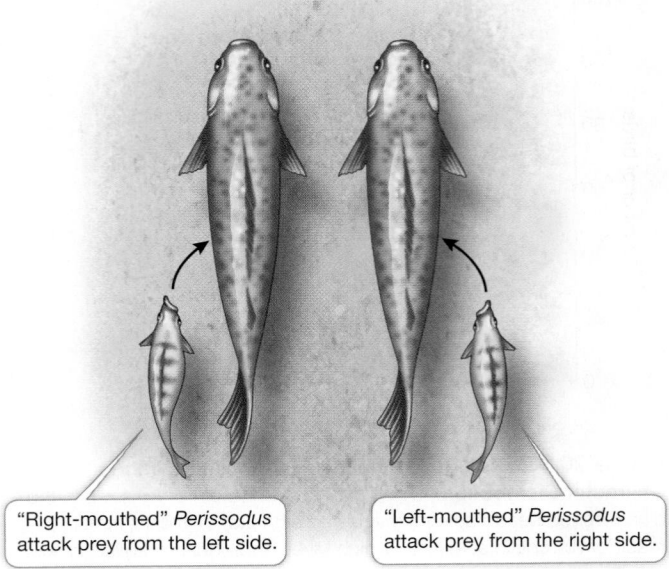

"Right-mouthed" *Perissodus* attack prey from the left side.

"Left-mouthed" *Perissodus* attack prey from the right side.

Figure 20.16 A Stable Polymorphism Frequency-dependent selection maintains equal proportions of left- and right-mouthed individuals of the scale-eating fish *Perissodus microlepis*.

direction is genetically determined (**Figure 20.16**). "Right-mouthed" individuals always attack from the victim's left, and "left-mouthed" individuals always attack from the victim's right. The distorted mouth enlarges the area of teeth in contact with the prey's flank, but only if the scale-eater attacks from the appropriate side.

Prey fish are alert to approaching scale-eaters, so attacks are more likely to be successful if the prey must watch both flanks. Vigilance by prey thus favors equal numbers of right- and left-mouthed scale-eaters in a population, because if attacks from one side were more common than the other, prey fish would pay more attention to potential attacks from that side. Over an 11-year study of *P. microlepis* in Lake Tanganyika, the genetic polymorphism was found to be stable, and the two phenotypes of the scale-eaters remained at about equal frequencies.

Heterozygote advantage maintains polymorphic loci

In many cases, different alleles of a particular gene are advantageous under different environmental conditions. Most organisms experience a wide variety of environmental conditions over time. A night is dramatically different from the preceding day. A cold, cloudy day differs from a clear, hot one. Day length and temperature change seasonally. For many genes, a single allele is unlikely to perform well under all these conditions. In such cases, heterozygous individuals (with two different alleles) are likely to outperform individuals that are homozygous (with only one of those two alleles).

Colias butterflies of the Rocky Mountains live in environments where dawn temperatures are often too cold, and afternoon temperatures too hot, for the butterflies to fly. Populations of these butterflies are polymorphic for a gene that encodes the enzyme phosphoglucose isomerase (PGI), which influences how well a butterfly flies at different temperatures. Butterflies with certain PGI genotypes can fly better during the cold hours of early morning; others perform better during midday heat (**Figure 20.17**). The optimal body temperature for flight is 35°C–39°C, but some butterflies can fly with body temperatures as low as 29°C or as high as 40°C. During spells of unusually hot

experiment

Figure 20.17A   A Heterozygote Mating Advantage

Original Paper: Watt, W. B., P. A. Carter and S. M. Blower. 1985. Adaptation at specific loci. IV. Differential mating success among glycolytic allozyme genotypes of *Colias* butterflies. *Genetics* 109: 157–175.

Among butterflies of the genus *Colias*, males that are heterozygous for two alleles of the PGI enzyme can fly farther under a broader range of temperatures than males that are homozygous for either allele. Does this ability give heterozygous males a mating advantage?

HYPOTHESIS▶ Heterozygous male *Colias* will have proportionally greater mating success than homozygous males.

METHOD

1. For each of two *Colias* species, capture mated female butterflies in the field. In the laboratory, allow them to lay eggs.
2. Determine the genotypes of the females and their offspring, and thus the genotypes of the fathers.
3. Compare the frequency of heterozygotes among successfully mating males with the frequency of heterozygotes among all viable males (i.e., males captured flying with females).

RESULTS

For both species, the proportion of heterozygotes among the males that mated successfully was higher than the proportion of heterozygotes among all viable males.

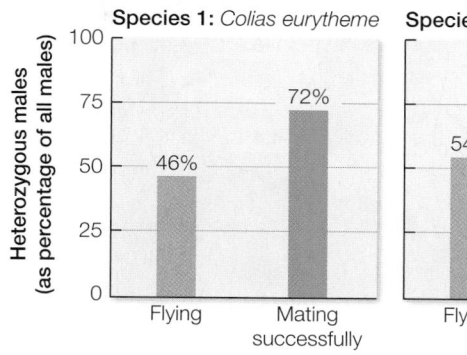 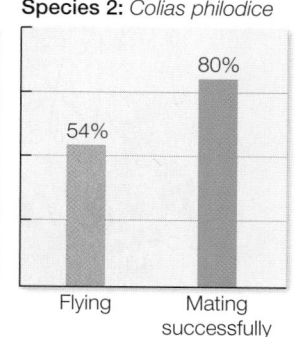

CONCLUSION▶ Heterozygous *Colias* males have a mating advantage over homozygous males.

work with the data

Figure 20.17B   A Heterozygote Mating Advantage

Original Paper: Watt, W. B., P. A. Carter and S. M. Blower. 1985.

Ward Watt and his colleagues tested the hypothesis that males with two different alleles for the PGI enzyme (heterozygotes) were more likely to mate successfully with females than were homozygous males. They reasoned that the heterozygous males could fly farther under a broader range of temperatures than could homozygous males, and that this ability would give heterozygous males greater access to receptive females. To test this hypothesis, they needed to know the frequency of heterozygotes among successfully mating males, and they needed to compare that frequency with the frequency of heterozygotes among males in the general population (i.e., all the potential mates available to females). To estimate the frequency of heterozygotes among mating males, Watt and his colleagues collected mated female butterflies in the field and allowed them to lay eggs in the laboratory. They hatched the eggs and determined the genotypes of the offspring, as well as the genotypes of the females. Using this information, they could determine the genotypes of the males that fathered the larvae. They then compared the estimated frequency of heterozygotes among the successful fathers with the frequency of heterozygotes among all viable males in the population. Samples of their data are given in the table.

Species	All viable males[a]		Mating males	
	Heterozygous/ total	Percent heterozygous	Heterozygous/ total	Percent heterozygous
C. philodice	32/74	43.2	31/50	62
C. eurytheme	44/92	47.8	45/59	76.3

[a] "Viable males" are males captured flying with females (hence the potential to mate).

QUESTIONS▶

1. If we assume that the proportions of each genotype among mating males should be the same as the proportions seen among all viable males, what is the number of *mating males* expected to be heterozygous and homozygous in each sample?

2. Use a chi-square test (see Appendix B) to evaluate the significance of the difference in the observed and expected numbers of heterozygous and homozygous individuals among the mating males. The critical value ($P = 0.05$) of the chi-square distribution with one degree of freedom is 3.841. Are the observed numbers of genotypes among mating males significantly different ($P < 0.05$) from the expected numbers in these samples?

3. The investigators determined the genotypes of enough larvae from each batch of eggs to judge the genotype of the father with 99% certainty. How many larvae did they need to measure to achieve that level of certainty?

   Hint: If the female is homozygous—say, genotype *ii*—the number needed is small. However, if a female is heterozygous—say, genotype *ij*—and only *ii* and *ij* progeny are found among her offspring, more larvae need to be genotyped. In this particular case, the father can be only *ii* or *ij*. If he were *ij*, the probability that any one offspring is not *jj* = 0.75, so the chance of getting only *ii* and *ij* among *n* offspring is 0.75^n. What value of *n* is required to reduce the probability of error in determining the father's genotype to 0.01?

A similar **work with the data** exercise may be assigned in **LaunchPad**.

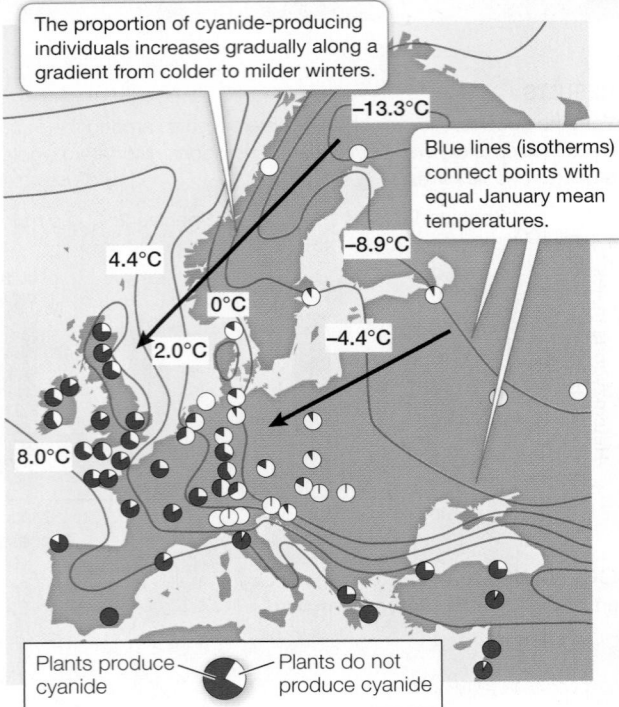

The proportion of cyanide-producing individuals increases gradually along a gradient from colder to milder winters.

−13.3°C

Blue lines (isotherms) connect points with equal January mean temperatures.

4.4°C

−8.9°C

0°C

−4.4°C

2.0°C

8.0°C

Plants produce cyanide — Plants do not produce cyanide

Figure 20.18 Geographic Variation in a Defensive Chemical The proportion of cyanide-producing individuals in European populations of white clover (*Trifolium repens*) depends on winter temperatures.

weather, heat-tolerant genotypes are favored; during spells of unusually cool weather, cold-tolerant genotypes are favored.

Heterozygous *Colias* butterflies can fly over a greater range of temperatures than homozygous individuals, which should give them an advantage in foraging and finding mates. A test of this prediction found that heterozygous males did indeed have a mating advantage, and further, that this advantage maintains the polymorphism in the population. Of course, the heterozygous genotype can never become fixed in the population, because the offspring of two heterozygotes will always include both classes of homozygotes in addition to heterozygotes.

Genetic variation within species is maintained in geographically distinct populations

Much of the genetic variation within species is preserved as differences among members living in different places (populations). Populations often vary genetically because they are subjected to different selective pressures in different environments. Some individuals of the white clover (*Trifolium repens*) produce the poisonous chemical cyanide. Poisonous individuals are less appealing to herbivores—particularly mice and slugs—than are nonpoisonous individuals. However, clover plants that produce cyanide are more likely to be killed by frost, because freezing damages cell membranes and releases cyanide into the plant's own tissues.

In European populations of *Trifolium repens*, the frequency of cyanide-producing individuals increases gradually from north to south and from east to west (**Figure 20.18**). A pattern of gradual change in phenotype across a geographic gradient is known as **clinal variation**. In the white clover cline, poisonous plants make

up a large proportion of populations only in areas where winters are mild. Cyanide-producing individuals are rare where winters are cold, even though herbivores graze clovers heavily in those areas.

20.5 recap

Neutral mutations, sexual recombination, frequency-dependent selection, and heterozygote advantage all act to maintain considerable genetic variation in most populations. Variation within species is also maintained among geographically distinct populations.

learning outcomes
You should be able to:
- Explain how genetic variation is useful for a species in the face of environmental variability.
- Propose a process for the maintenance of deleterious alleles in a population and a method for testing the hypothesis.

1. Human populations in central Africa have a high frequency of sickle-cell anemia, a genetic disorder that occurs in individuals who are homozygous for a particular allele of the hemoglobin gene. Despite strong selection against the homozygous sickle-cell condition, the sickle allele is maintained in relatively high frequencies in human populations that live in areas where malaria was historically common (compare the maps below). Formulate a hypothesis that explains how this deleterious allele could be maintained in the population, even though individuals who are homozygous for the sickle allele are strongly disadvantaged.

Historic range of malaria in human populations

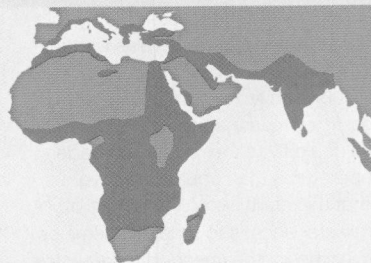

Distribution of sickle cell allele

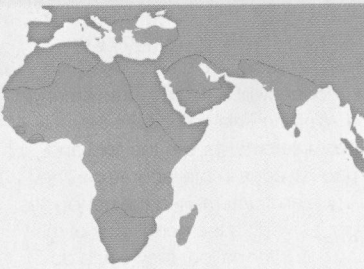

2. How could you test the hypothesis that you formulated in Question 1?

3. How could the presence of a great deal of genetic variation within a population increase the chances that some members of the population would survive an unprecedented environmental change? Why is there no guarantee that this would be the case?

The mechanisms of evolution have produced a remarkable variety of organisms. There are organisms that have adapted to nearly every environment on Earth. This natural variation, along with the success of breeders attempting to produce desired traits in domesticated plants and animals, suggests that evolution can produce a wide variety of adaptive traits. But are there limits to the adaptations evolution can produce?

key concept 20.6 Evolution Is Constrained by History and Trade-Offs

We would be mistaken to assume that evolutionary mechanisms can produce any trait we might imagine. Evolution is constrained in many ways. Lack of appropriate genetic variation, for example, prevents the development of many potentially favorable traits. If the allele for a given trait does not exist in a population, that trait cannot evolve, even if it would be highly favored by natural selection. Most possible combinations of genes and genotypes have never existed in any population and so have never been tested under natural selection.

focus your learning

- Evolutionary outcomes are limited by preexisting traits, cost–benefit trade-offs, and constraints of the natural world.
- Macroevolutionary (large-scale, long-term) patterns across species sometimes require additional explanations beyond microevolutionary (small-scale, short-term) processes within populations.

In addition, constraints are imposed on organisms by the dictates of physics and chemistry. The size of cells, for example, is constrained by the stringencies of surface area-to-volume ratios (see Figure 5.2). The ways in which proteins can fold are limited by the bonding capacities of their constituent molecules (see Key Concept 3.2). And the energy transfers that fuel life must operate within the laws of thermodynamics (see Key Concept 8.1). Keep in mind that evolution works within the boundaries of these universal constraints as well as the constraints described in this section.

Developmental processes constrain evolution

As Chapter 19 explained, developmental constraints on evolution are paramount *because all evolutionary innovations are modifications of previously existing structures*. Human engineers seeking to power an airplane can start "from scratch" to design a completely new type of engine (powered by jet propulsion) to replace the previous type (powered by propellers). Evolutionary changes, however, cannot happen in this way. Current phenotypes of organisms are constrained by historical conditions and past selective pressures.

A striking example of such developmental constraints is provided by the evolution of fish that spend most of their time on the sea bottom, where a ventrally flattened body is advantageous. One such lineage, the bottom-dwelling skates and rays, shares a common ancestor with sharks, whose bodies are already somewhat ventrally flattened and whose skeletal frame is made of flexible cartilage.

(A) *Taeniura lymma*

(B) *Bothus lunatus*

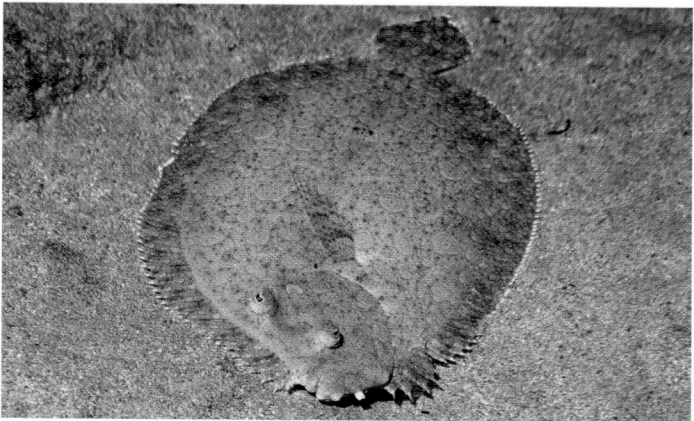

Figure 20.19 Two Solutions to a Single Problem (A) This stingray, whose ancestors were dorsoventrally flattened, lies on its belly. Stingrays' bodies are symmetrical around the dorsal backbone. **(B)** The flounder, whose ancestors were laterally flattened, lies on its side. (The backbone of this individual is at the left.) Flounders' eyes migrate during development so that both are on the same side of the body.

Skates and rays evolved a body type that further flattened their bellies, allowing them to swim along the ocean floor (**Figure 20.19A**).

By contrast, plaice, sole, and flounder are bottom-dwelling descendants of deep-bellied, laterally flattened ancestors with bony skeletons. The only way these fish can lie flat is to flop over on their sides. Their ability to swim is thus curtailed, but their bodies can lie still and are well camouflaged. During development, one eye of these flatfishes moves so that both eyes are positioned on the same side of the body (**Figure 20.19B**). Such shifts in eye position have evolved several times, and shifts have happened in both directions (that is, both left- and right-eyed flatfish have evolved independently). Small shifts in the position of one eye probably helped ancestral flatfish see better, resulting in the body forms found today. This path to producing a flattened body may not be optimal, but the fishes' developmental capabilities constrain the pathways that evolution can take.

Trade-offs constrain evolution

Adaptations frequently impose both costs and benefits. For an adaptation to be favored, the fitness benefits it confers must exceed the fitness costs it imposes—in other words, the **trade-off** must

Figure 20.20 Resistance to a Toxin Comes at a Cost **(A)** Garter snakes (*Thamnophis sirtalis*) prey on rough-skinned newts (*Taricha granulosa*). Rough-skinned newts defend themselves by sequestering a neurotoxin, TTX, in their skin. In turn, TTX-resistant sodium channels have evolved in some snake populations, allowing the snakes to eat toxic newts but resulting in slower movement by the snakes. **(B)** High TTX resistance in garter snakes is found only in regions where snake and newt populations overlap (tan area).

Q: Why doesn't TTX resistance spread throughout the range of the garter snake?

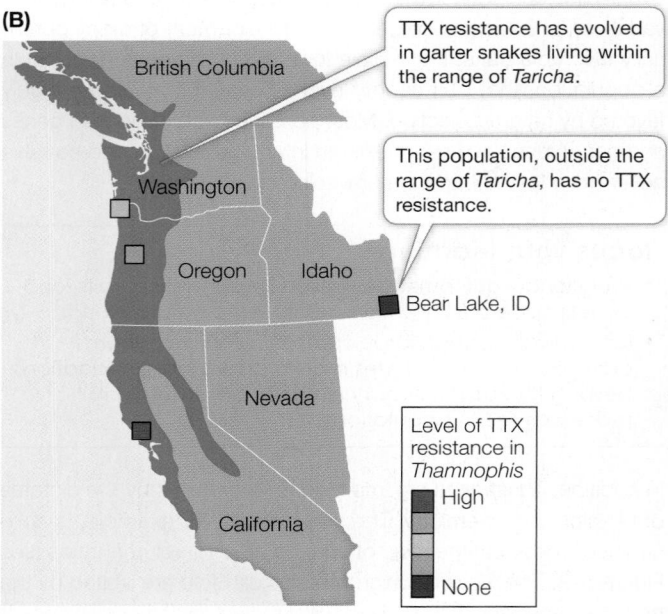

TTX resistance has evolved in garter snakes living within the range of *Taricha*.

This population, outside the range of *Taricha*, has no TTX resistance.

British Columbia

Washington

Oregon   Idaho

Bear Lake, ID

Nevada

California

Level of TTX resistance in *Thamnophis*

High

None

be worthwhile. For example, there are metabolic costs associated with developing and maintaining certain conspicuous features (such as antlers or horns) that males use to compete with other males for access to females. The fact that these features are common in many species suggests that the benefits derived from possessing them must outweigh the costs.

As a result of trade-offs, many traits that are adaptive in one context may be maladaptive in another. Consider the rough-skinned newt and one of its predators, the common garter snake (**Figure 20.20A**). The newt sequesters a potent neurotoxin called tetrodotoxin (TTX) in its skin. TTX paralyzes nerves and muscles by blocking sodium channels (see Key Concept 44.3). Most vertebrates—including many garter snakes—will die if they eat a rough-skinned newt. But some garter snakes can eat rough-skinned newts and survive: TTX-resistant sodium channels have evolved in the nerves and muscles of such individuals. However, for several hours after eating a newt, TTX-resistant snakes can move only slowly, and they never move as fast as nonresistant snakes. Resistant snakes are thus more vulnerable to their own predators than are TTX-sensitive snakes. This vulnerability leads to selection *against* TTX-resistant sodium channels in garter snake populations that occur outside the range of rough-skinned newts, even though there is selection *for* TTX resistance in many areas where newts are present (**Figure 20.20B**).

 Animation 20.2 Assessing the Costs of Adaptation www.Life11e.com/a20.2

Short-term and long-term evolutionary outcomes sometimes differ

The short-term changes in allele frequencies within populations that we have emphasized in this chapter, often termed **microevolutionary** changes, are an important focus of study for evolutionary biologists. These changes can be observed directly, they can be manipulated experimentally, and they demonstrate the actual processes by which evolution occurs.

Long-term patterns of **macroevolutionary** change can also be strongly influenced by events that occur so infrequently (such as a meteorite impact) or so slowly (such as continental drift) that they are unlikely to be observed during short-term studies. The evolutionary mechanisms at work may change over time with changing environmental conditions. Even among the descendants of a single ancestral species, different lineages may evolve in different directions. Additional types of evidence—evidence demonstrating the effects of rare and unusual events on trends in the fossil record—must be gathered if we

wish to understand the course of evolution over billions of years. We will consider these long-term aspects of evolution in the remaining chapters of this section.

20.6 recap

Developmental processes constrain evolution because all evolutionary innovations are modifications of previously existing structures. An adaptation can evolve only if the fitness benefits it confers exceed the fitness costs it imposes.

learning outcomes

You should be able to:

- Formulate a hypothesis to explain why potentially highly advantageous phenotypes might not evolve.
- Describe an example illustrating a macroevolutionary pattern that is not easily explained by microevolutionary processes.

20.6 recap

1. The largest living adult insects weigh less than 75 g (less than the weight of a relatively small songbird, such as a starling). Many flying birds are much larger than any insect, so clearly flight alone does not constrain the current size of insects. Formulate a hypothesis for why there are no giant insects presently roaming Earth.

2. Natural selection results in gradual changes *within species* over long periods of time. Describe an event that could result in significant and sudden changes across many species simultaneously.

investigating life

 **How do complex traits like echolocation, or the ability to avoid detection from echolocation, evolve in the first place?**

In the opening story of this chapter, we discussed the evolutionary arms race between echolocating bats and the moths that evolved the ability to detect bat ultrasounds, and then avoid the predators. Moths are not the only potential prey of bats that have evolved the ability to hear ultrasounds. Many other groups of insects can hear the sounds produced by bats, and act to avoid these predators. In many cases, the insects can already hear, as they have evolved complex communication systems to attract mates. This produces the natural variation in the trait (hearing) that is needed for natural selection to act.

As we learned in the experiment discussed in Investigating Life: Do Long Wing Tails Help Moths Escape Bat Predation?, many non-hearing insects have evolved other strategies for confusing or avoiding echolocating bats. Several species of moths have evolved long extensions of their wings, called wing tails. The wing tails flutter as the moths fly, and the fluttering wing tails serve to distract the echolocating bats. A bat that attacks the long wing tail of a moth may get nothing but a tiny piece of moth wing, while the moth survives the attack.

Some bats locate their prey using the sounds produced by the prey themselves. There are species of bats that locate prey by the sound of calling frogs, insects, or even the ripples made by fish on the surface of the water. These prey species often must limit or modify the sounds they produce to balance the benefits of producing a sound with the costs associated with bat predation.

Species rarely evolve to a state of perfection and then stop evolving. Prey species are constantly evolving new ways to avoid predation, so predator species must evolve improved means of capturing prey. Although there are often historical, physiological, or mechanical limits to what can evolve, species are constantly changing in response to changes in their physical environment and species interactions. This constant selection for change, over millions or even billions of years, has produced the enormous diversity of life we see on Earth.

Future directions

Bats are not the only animals that have evolved the ability to echolocate. Toothed whales and dolphins, as well as some burrowing shrews and cave-dwelling birds, also use echolocation to "see" their environment with sound. So echolocation has evolved repeatedly in species that live in environments in which sight is limited. But how does a trait like echolocation first begin to evolve in a species? Some rudimentary form of the trait has to be present before selection can act to refine this sensory mode. Often, species co-opt a trait that evolved for other purposes, and then selection refines it over time for a new purpose.

Most humans do not use echolocation. We rely heavily on sight, and most of us feel awkward moving around unfamiliar spaces in the dark. But what would happen if humans were forced to live in an environment where we could not see? We already hear, of course, so we have a system that could be co-opted for echolocation. We are also capable of making sounds. Echolocation simply requires the production of sounds, the ability to detect the reflected sounds, and the ability to process the information appropriately. Indeed, some blind humans have trained themselves to use echolocation to sense their environment (see Media Clip 20.4). An individual learning a behavior is not the same as evolution of a behavior within a species, however. Before natural selection can act, there has to be genetic variation in the population for the trait in question. The fact that a few humans are able to use simple forms of echolocation shows that variation for this trait is already present in our species. The same is true for many species, which is why it is not surprising that echolocation has evolved multiple times in species that live or feed in the dark.

 Media Clip 20.4 Human Echolocation
www.Life11e.com/mc20.4

Chapter Summary 20

20.1 Evolution Is Both Factual and the Basis of Broader Theory

- **Evolution** is genetic change in populations over time. Evolution can be observed directly in living populations as well as in the fossil record of life.

- **Evolutionary theory** refers to our understanding of the mechanisms of evolutionary change.

- Charles Darwin in best known for his ideas on the common ancestry of divergent species and on **natural selection** (the differential survival and reproduction of individuals based on variation in their traits) as a mechanism of evolution. See Animation 20.1, Activity 20.1

- Since Darwin's time, many biologists have contributed to the development of evolutionary theory, and rapid progress in our understanding continues today.

20.2 Mutation, Selection, Gene Flow, Genetic Drift, and Nonrandom Mating Result in Evolution

- Mutation is the source of the genetic variation on which mechanisms of evolution act. Review Figure 20.3

- The term **adaptation** refers both to a trait that evolves through natural selection and to the process that produces such traits.

- Within populations, selection acts to increase the frequency of beneficial alleles and to decrease the frequency of deleterious alleles. Review Figure 20.6, Investigating Life: Do Long Wing Tails Help Moths Escape Bat Predation?, Activity 20.2

- Movement of individuals or gametes between populations results in **gene flow**.

- In small populations, **genetic drift**—the random loss of individuals and the alleles they possess from one generation to the next—may produce large changes in allele frequencies over time and greatly reduce genetic variation. See Activity 20.3

- **Population bottlenecks** occur when only a few individuals survive a random event, resulting in a drastic shift in allele frequencies within the population and the loss of genetic variation. Similarly, a population established by a small number of individuals colonizing a new region may lose genetic variation via a **founder effect**. Review Figure 20.7

- Nonrandom mating may result in changes in genotype and allele frequencies in a population.

- **Sexual selection** results from differential reproductive success based on individuals' phenotypes. Review Figure 20.9

20.3 Evolution Can Be Measured by Changes in Allele Frequencies

- Allele frequencies measure the amount of genetic variation in a population. Genotype frequencies show how a population's genetic variation is distributed among its members. Together, allele and genotype frequencies describe a population's **genetic structure**. Review Figure 20.10

- **Hardy–Weinberg equilibrium** predicts genotype frequencies from allele frequencies in the absence of evolution. Deviation from these frequencies indicates that evolutionary mechanisms are at work. Review Figure 20.11, Activity 20.4

20.4 Selection Can Be Stabilizing, Directional, or Disruptive

- Natural selection can act on characters with quantitative variation in several different ways. Review Focus: Key Figure 20.12

- **Stabilizing selection** acts to reduce variation without changing the mean value of a trait. Review Figure 20.13

- **Directional selection** acts to shift the mean value of a trait toward one extreme. Review Figure 20.14

- **Disruptive selection** favors both extremes of a trait value, resulting in a bimodal character distribution. Review Figure 20.15

20.5 Multiple Factors Account for the Maintenance of Variation in Populations

- Neutral mutations, sexual recombination, frequency-dependent selection, and heterozygote advantage can all maintain genetic variation within populations.

- **Neutral alleles** do not affect the fitness of an organism, are not affected by natural selection, and may accumulate or be lost by genetic drift.

- Despite its short-term disadvantages, sexual reproduction generates countless genotypic combinations that increase the evolutionary potential and survivorship of populations.

- A polymorphism may be maintained by **frequency-dependent selection** when the fitness of a genotype depends on its frequency in a population. Review Figure 20.16

- A polymorphism may also be maintained by heterozygote advantage when the fitness of the heterozygote exceeds the fitness of either homozygote. Review Figure 20.17

- Genetic variation within species may be maintained by the existence of genetically distinct populations over geographic space. A gradual change in phenotype across a geographic gradient is known as **clinal variation**. Review Figure 20.18

20.6 Evolution Is Constrained by History and Trade-Offs

- Developmental processes constrain evolution because all evolutionary innovations are modifications of previously existing structures. Review Figure 20.19

- Most adaptations impose costs as well as benefits. An adaptation can evolve only if the benefits it confers exceed the costs it imposes. Review Figure 20.20, Animation 20.2

Go to **LearningCurve** (in **LaunchPad**) for dynamic quizzing that helps you solidify your understanding of this chapter. **LearningCurve** adapts to your responses, giving you the practice you need to master each key concept.

Apply What You've Learned

Review

20.1 Evolution is directly observable and is a universal principle of life.

20.2 Evolution is the result of five major processes: mutation, natural selection, gene flow, genetic drift, and non-random mating.

Original Paper: Stuart, Y. E., T. S. Campbell, P. A. Hohenlohe, R. G. Reynolds, L. J. Revell, and J. B. Losos. 2014. Rapid evolution of a native species following invasion by a congener. *Science* 346: 463–466.

Anoles, an abundant and diverse group of lizards in the American tropics and subtropics, are common subjects of ecological and evolutionary studies. The Carolina (or green) anole, *Anolis carolinensis* (**Figure A**), is the only anole native to the southeastern United States. The Cuban brown anole, *Anolis sagrei* (**Figure B**), was introduced into south Florida in the late 1800s and has been expanding its range northward since then. When living alone, *A. carolinensis* occupies habitats from ground level to treetops, whereas *A. sagrei* lives on the ground and lower perches. The two species compete, and the more aggressive *A. sagrei* displaces *A. carolinensis* from its preferred perches. Where the two species live together, *A. carolinensis* is restricted to high treetops.

Figure A

Figure B

Across species of *Anolis*, there is a strong association between perch height and the structure of the toe pads that the lizards use to cling to branches (**Figure C**). Researchers wondered if the shift in perching habitat would lead to rapid evolutionary change in the toe pads of displaced *A. carolinensis* populations. In 1995, researchers deliberately introduced populations of *A. sagrei* onto islands that previously supported only populations of *A. carolinensis*, then followed the evolutionary changes in the lizard populations over the next 15 years. Researchers examined the size of the toe pads, and the number of expanded scales with adhesive bristles (lamellae) on each pad. Other islands continued to support only *A. carolinensis*.

Figure C

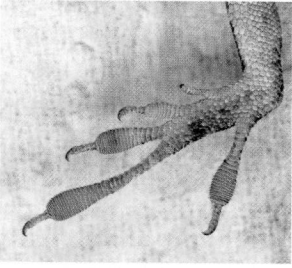

In 2010, the researchers measured toe-pad size and the number of lamellae on *A. carolinensis* lizards collected from both sets of islands (with and without *A. sagrei*) (**Figures D, E**). The researchers also raised offspring of lizards from these islands in a "common garden experiment,"
which uses a common, controlled environment to confirm that any differences between the populations have a genetic basis.

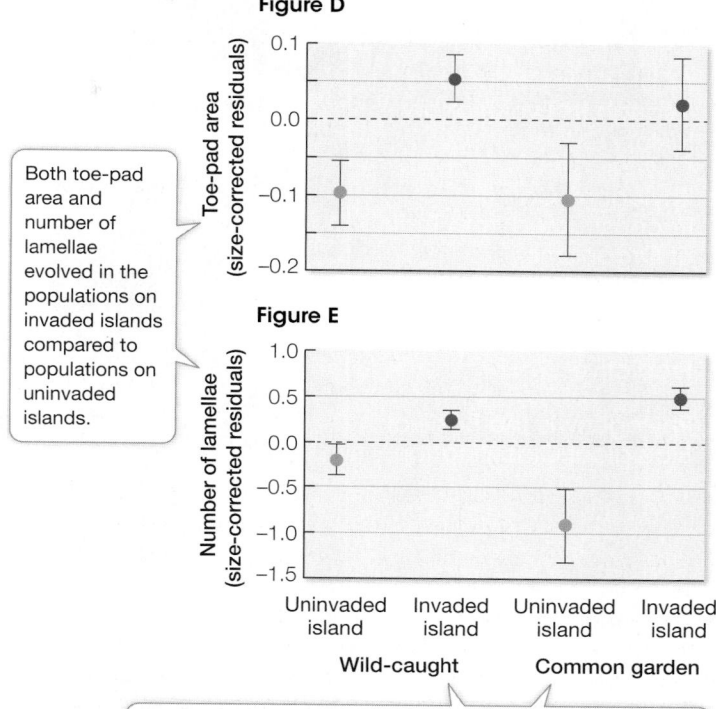

Figure D

Toe-pad area (size-corrected residuals)

Both toe-pad area and number of lamellae evolved in the populations on invaded islands compared to populations on uninvaded islands.

Figure E

Number of lamellae (size-corrected residuals)

Uninvaded island | Invaded island | Uninvaded island | Invaded island

Wild-caught   Common garden

Differences between wild-caught populations were retained in the common garden experiment. This shows that the observed changes were genetic rather than environmental, confirming that evolution (genetic changes in the population) occurred.

Questions

1. What is the evidence for toe-pad evolution in the populations of *A. carolinensis* on the invaded islands? Why have toe pads in these populations of *A. carolinensis* changed so rapidly?

2. Why was the common garden experiment needed? How would the conclusions of the study have been different if the observed differences in the wild populations were not also evident in the lizards raised in a common, controlled environment?

3. Which evolutionary process or processes are most likely to explain the observed evolutionary changes? Which processes do you think are less important, and why?

4. Describe the ways that these populations of *A. carolinensis* deviate from assumptions of Hardy–Weinberg equilibrium (the conditions that would be necessary for the populations to *not* evolve).

5. Instead of *A. sagrei*, suppose another species of *Anolis* had been introduced to the Florida islands. Suppose this lizard were an aggressive treetop specialist that excluded *A. carolinensis* from the highest perches and restricted it to the ground and low perches. In this case, how would you expect the displaced populations of *A. carolinensis* to have evolved?

Go to **LaunchPad** for the eBook, LearningCurve, animations, activities, flashcards, and additional resources and assignments.

21

Reconstructing and Using Phylogenies

Phylogenies allow us to understand the evolution of diversity, as within this single genus of South African plants (*Protea*).

> investigating**life**

Using Phylogeny to Improve a Genetic Tool

Green fluorescent protein (GFP) was discovered in 1962 when Osamu Shimomura led a team that was able to purify the protein from the tissues of the bioluminescent jellyfish *Aequorea victoria*. Some 30 years after GFP's initial discovery, Martin Chalfie had the idea (and the technology) to link the gene for GFP to other protein-coding genes so that the expression of specific genes of interest could be visualized in glowing green within cells and tissues of living organisms (see Figure 18.4). This work was extended by Roger Tsien, who changed some of the amino acids within GFP to create fluorescent proteins of several distinct colors. Different-colored fluorescent proteins meant that the expression of several different proteins could be visualized and studied in the same organism at the same time. These three scientists were awarded the 2008 Nobel Prize in Chemistry for the isolation and development of GFP for visualizing gene expression.

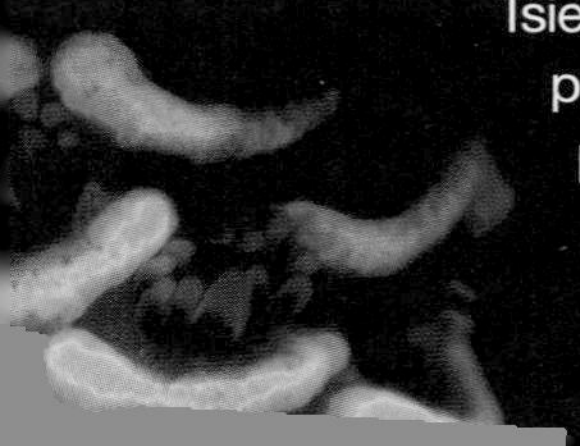

Tsien was able to produce different-colored proteins, but he could not produce a *red* protein. This was frustrating; a red fluorescent protein would be particularly useful to biologists because red light penetrates tissues more easily than do other colors. Tsien's work stimulated Mikhail Matz to

look for new fluorescent proteins in corals. Among the different coral species he studied, Matz found coral proteins that fluoresced in various shades of green, cyan (blue-green)—and red.

How had fluorescent red pigments evolved among the corals, given that the necessary molecular changes had eluded Tsien? To answer this question, Matz sequenced the genes of the fluorescent proteins and used these sequences to reconstruct the evolutionary history of the amino acid changes that produced different colors in different species of corals.

Matz's work showed that the ancestral fluorescent protein in corals was green, and that red fluorescent proteins evolved in a series of gradual steps. His analysis of evolutionary relationships allowed him to retrace these steps. Such an evolutionary history, as depicted in a tree of relationships among lineages, is called a phylogeny.

The evolution of many aspects of an organism's biology can be studied using phylogenetic methods. This information is used in all fields of biology to understand the structure, function, and behavior of organisms.

 Q **A** How are phylogenetic methods used to resurrect protein sequences from extinct organisms?

key concept 21.1 All of Life Is Connected through Its Evolutionary History

The sequencing of complete genomes from many diverse species has confirmed what biologists have long suspected: all of life is related through a common ancestor. The common ancestry of life explains why the general principles of biology apply to all organisms. Thus we can learn much about how the human genome works by studying the biology of model organisms because we share a common evolutionary history with those organisms. The evolutionary history of these relationships is known as **phylogeny**, and a **phylogenetic tree** is a diagrammatic reconstruction of that history.

focus your learning

- Phylogenetic trees represent evolutionary relationships.
- Phylogenies enable biologists to compare organisms and make predictions and inferences based on similarities and differences in traits.
- Only homologous traits are used in reconstructing phylogenetic trees.

Phylogenetic trees are commonly used to depict the evolutionary history of species, populations, and genes. For many years such trees have been constructed based on physical structures, behaviors, and biochemical attributes. Now, as genomes are sequenced for more and more organisms, biologists are able to reconstruct the history of life in ever greater detail.

In Chapter 20 we discussed why we expect populations of organisms to evolve over time. We call a series of ancestor and descendant populations a **lineage**, which we can depict as a line drawn on a time axis, as shown in **Figure 21.1**. What happens when a single lineage divides into two? For example, a geographic barrier may divide an ancestral population into two descendant populations that no longer interbreed with one another. We depict such an event as a split, or **node**, in a phylogenetic tree (see Figure 21.1). Each of the descendant populations gives rise to a new lineage, and as these independent lineages evolve, new traits arise in each. As the lineages continue to split over time, this history can be represented in the form of a branching tree that can be used to trace the evolutionary relationships from the ancient common ancestor of a group of species, through the various lineage splits, up to the present populations of the organisms.

A phylogenetic tree may portray the evolutionary history of all life forms. Phylogenetic trees can also depict the history of a major evolutionary group (such as the insects) or of a much smaller group of closely related species. In some cases, phylogenetic trees are used to show the history of individuals, populations, or genes within a species. The common ancestor of all the organisms in the tree forms the **root** of the tree.

The depictions of phylogenetic trees in this book are rooted at the left, with time flowing from left (earliest) to right (most recent) (**Focus: Key Figure 21.2A**). The timing of splitting events in lineages is shown by the position of nodes on a time axis. These splits represent events

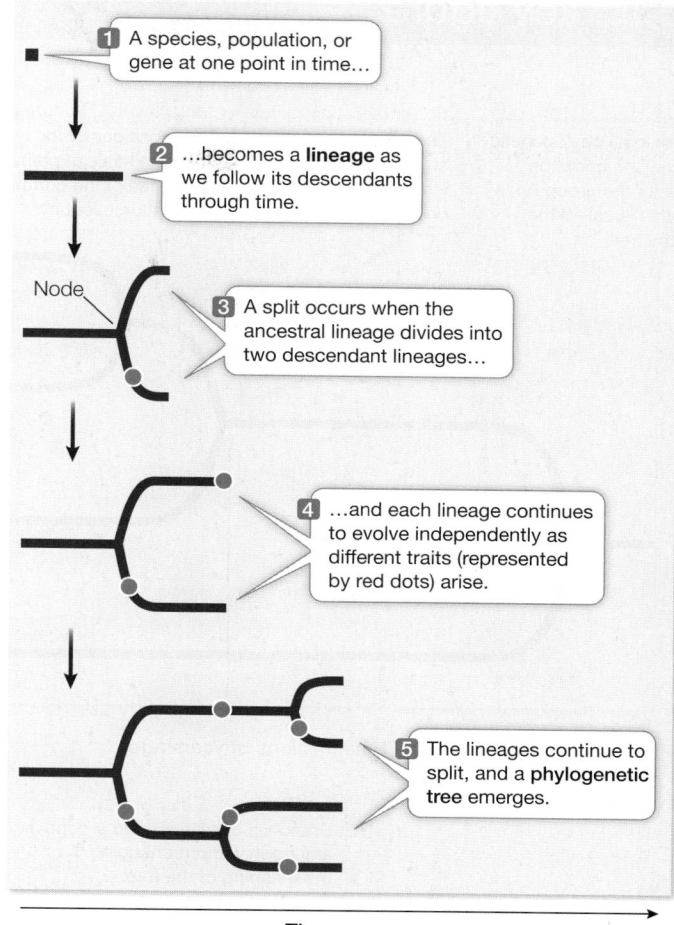

Figure 21.1 The Components of a Phylogenetic Tree
Evolutionary relationships among organisms can be represented in a treelike diagram.

where one lineage diverged into two, such as a speciation event (for a tree of species), a gene duplication event (for a tree of genes), or a transmission event (for a tree of viral lineages transmitted through a host population). The time axis may have an explicit scale, or it may simply show the relative timing of divergence events.

In this book's illustrations, the order in which nodes are placed along the horizontal (time) axis has meaning, but the vertical distance between the branches does not. Vertical distances have been adjusted for legibility and clarity of presentation; they do not correlate with the degree of similarity or difference among groups. Note too that lineages can be rotated around nodes in the tree, so the vertical order of lineages is also largely arbitrary (**Focus: Key Figure 21.2B**). The important information in the tree is the branching order along the time axis, as this indicates when the various lineages last shared a common ancestor.

Any group of species that we designate with a name is a **taxon** (plural *taxa*). Examples of familiar taxa include humans, primates, mammals, and vertebrates; in this series, each taxon is also a member of the next, more inclusive taxon. Any taxon that consists of all the evolutionary descendants of a common ancestor is called a **clade**. Clades can be identified by picking any point on a phylogenetic tree and from that point tracing all the descendant lineages to the tips of

focus: key figure

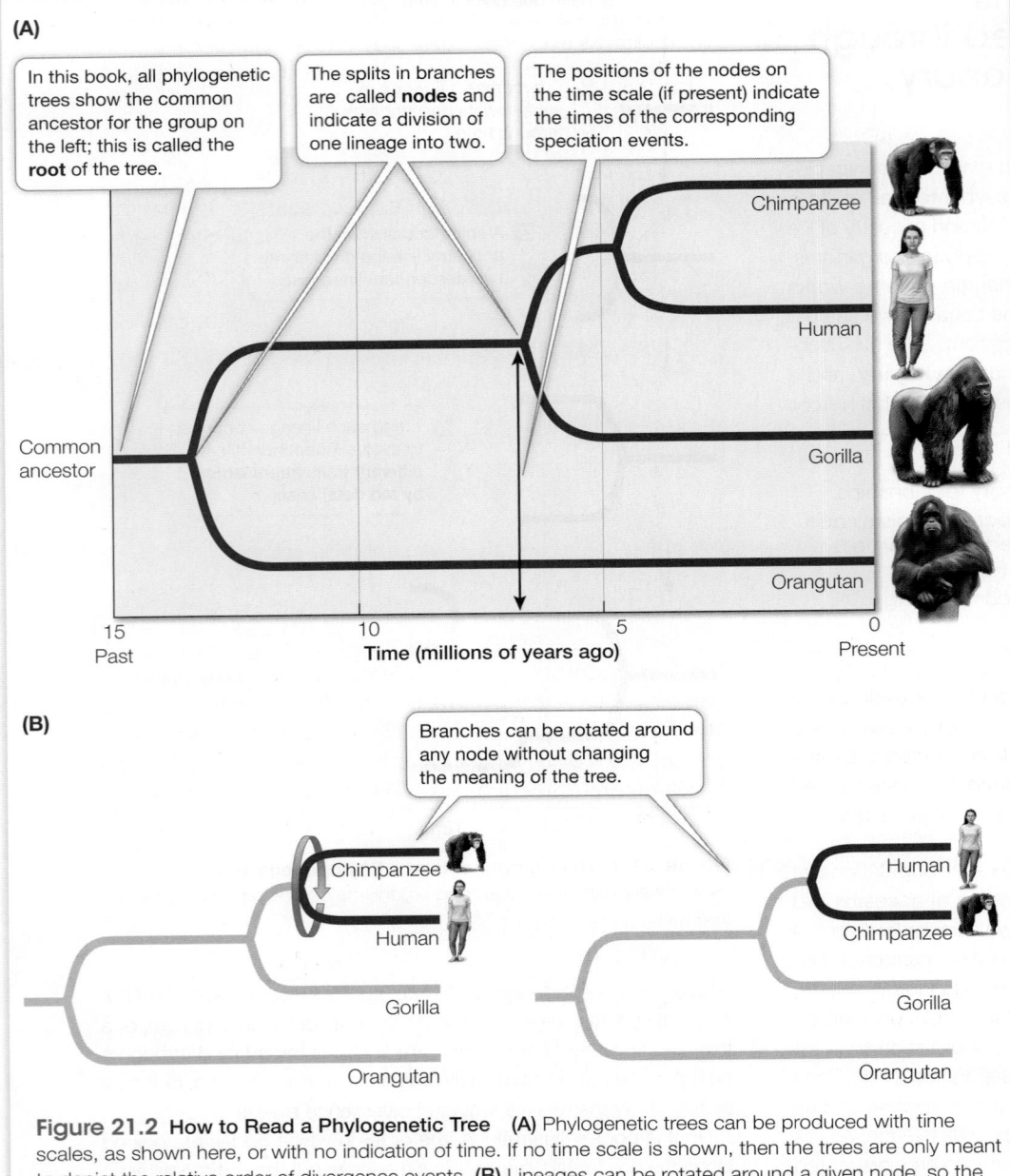

(A)

In this book, all phylogenetic trees show the common ancestor for the group on the left; this is called the **root** of the tree.

The splits in branches are called **nodes** and indicate a division of one lineage into two.

The positions of the nodes on the time scale (if present) indicate the times of the corresponding speciation events.

Common ancestor

Chimpanzee

Human

Gorilla

Orangutan

15 — Past

10

5

0 — Present

Time (millions of years ago)

(B)

Branches can be rotated around any node without changing the meaning of the tree.

Chimpanzee

Human

Gorilla

Orangutan

Human

Chimpanzee

Gorilla

Orangutan

Figure 21.2 How to Read a Phylogenetic Tree **(A)** Phylogenetic trees can be produced with time scales, as shown here, or with no indication of time. If no time scale is shown, then the trees are only meant to depict the relative order of divergence events. **(B)** Lineages can be rotated around a given node, so the vertical order of taxa is largely arbitrary.

Phylogenetic trees are the basis of comparative biology

In biology we study life at all levels of organization—from genes, cells, organisms, populations, and species to the major divisions of life. In most cases, however, no individual gene or organism (or other unit of study) is exactly like any other gene or organism that we investigate.

Consider the individuals in your biology class. We recognize each person as an individual human, but we know that no two are exactly alike. If we knew everyone's family tree in detail, the genetic similarity of any pair of students would be more predictable. We would find that more closely related students have many more traits in common (from the color of their hair to their susceptibility or resistance to diseases). Likewise, biologists use phylogenies to make comparisons and predictions about shared traits across genes, populations, and species.

The evolutionary relationships among species, as represented in the **tree of life**, form the basis for biological classification. Biologists estimate that there are tens of millions of species on Earth. So far, however, only about 1.8 million species have been classified—that is, formally described and named. New species are being discovered all the time and phylogenetic analyses are constantly reviewed and revised, so our knowledge of the tree of life is far from complete. Yet knowledge of evolutionary relationships is essential for making comparisons in biology, so biologists build phylogenies for groups of interest as the need arises. The tree of life's evolutionary framework allows us to make many predictions about the behavior, ecology, physiology, genetics, and morphology of species that have not yet been studied in detail.

the terminal branches (**Figure 21.3**). Two species that are each other's closest relatives are called **sister species**. Similarly, any two clades that are each other's closest relatives are **sister clades**.

Before the 1980s, phylogenetic trees tended to be seen only in the literature on evolutionary biology, especially in the area of **systematics**—the study and classification of biodiversity. But almost every journal in the life sciences published during the last few years contains phylogenetic trees. Trees are widely used in molecular biology, biomedicine, physiology, behavior, ecology, and virtually all other fields of biology. Why have phylogenetic studies become so widespread?

When biologists compare species, they observe traits that differ within the group of interest and try to understand when these traits evolved. In many cases, investigators are interested in how the evolution of a trait relates to environmental conditions or selective pressures. For instance, scientists have used phylogenetic analyses to discover changes in the genome of human immunodeficiency viruses (HIVs) that result in resistance to particular drug treatments. The association of a particular genetic change in HIV with a particular treatment provides a hypothesis about the evolution of resistance that can be tested experimentally.

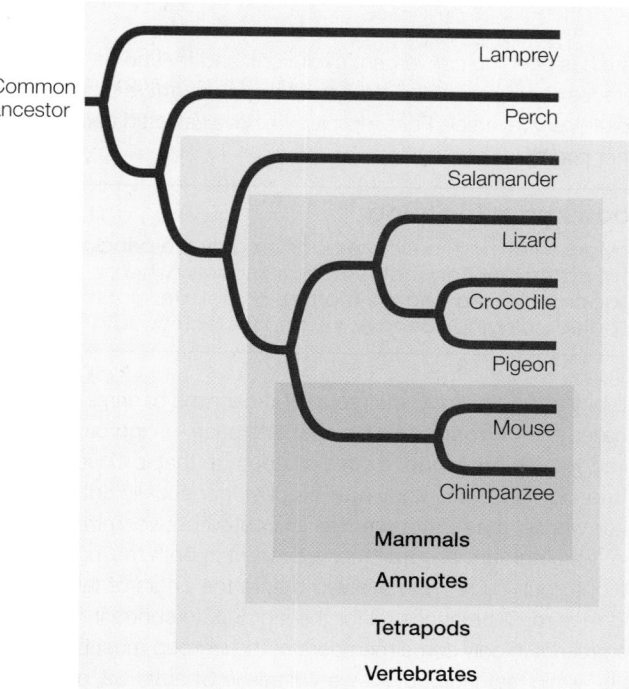

Any features shared by two or more species that have been inherited from a common ancestor are said to be **homologous**. Homologous features may be any heritable traits, including DNA sequences, protein structures, anatomical structures, and even some behavior patterns. For example, all living vertebrates have a vertebral column, as did the ancestral vertebrate. Therefore the vertebral column is judged to be homologous in all vertebrates.

Derived traits provide evidence of evolutionary relationships

In tracing the evolution of a character, biologists distinguish between ancestral and derived traits. Each character of an organism evolves from one condition (the **ancestral trait**) to another condition (the **derived trait**). Derived traits that are shared among a group of organisms and are also viewed as evidence of the common ancestry of the group are called **synapomorphies** (*syn*, "shared"; *apo*, "derived"; *morph*, "form," referring to the "form" of a trait). Thus the vertebral column is considered a synapomorphy—a shared, derived trait—of the vertebrates. (The ancestral trait was an undivided supporting rod.)

Not all similar traits are evidence of relatedness. Similar traits in unrelated groups of organisms can develop for either of the following reasons:

- Superficially similar traits may evolve independently in different lineages, a phenomenon called **convergent evolution**. For example, although the *wing bones* of bats and birds are homologous, having been inherited from a common tetrapod ancestor, the *wings* of bats and birds are not homologous because they evolved independently from the forelimbs of different nonflying ancestors (**Figure 21.4**). Functionally similar structures that have independent evolutionary origins are called **analogous** characters.

- A character may revert from a derived state back to an ancestral state in an event called an **evolutionary reversal**.

Figure 21.3 Clades Represent an Ancestor and All of Its Evolutionary Descendants   All clades are subsets of larger clades, with all of life as the most inclusive taxon. In this example, the groups called mammals, amniotes, tetrapods, and vertebrates represent successively larger clades. Only a few species within each clade are represented on the tree.

For example, the derived limbs of terrestrial tetrapods evolved from the ancestral fins of their aquatic ancestors. Then, within the mammals, the ancestors of modern cetaceans (whales and dolphins) returned to the ocean, and cetacean limbs evolved to once again resemble their ancestral state—fins. The superficial similarity of cetacean and fish fins does not suggest a close relationship between these groups. Instead, the similarity arises from evolutionary reversal.

Similar traits generated by convergent evolution and evolutionary reversals are called homoplastic traits or **homoplasies**.

▶ Media Clip 21.1 **Morphing Arachnids**
www.Life11e.com/mc21.1

A particular trait may be ancestral or derived, depending on our point of reference. For example, all birds have feathers. We infer from this that feathers (which are highly modified scales) were present in the common ancestor of modern birds. Therefore we consider the presence of feathers to be an ancestral trait for any particular group of modern birds, such as the songbirds. However, feathers are not present in any other living animals. In reconstructing a phylogeny of all living vertebrates, the presence of feathers is a derived trait found only among birds, and thus is a synapomorphy of the birds.

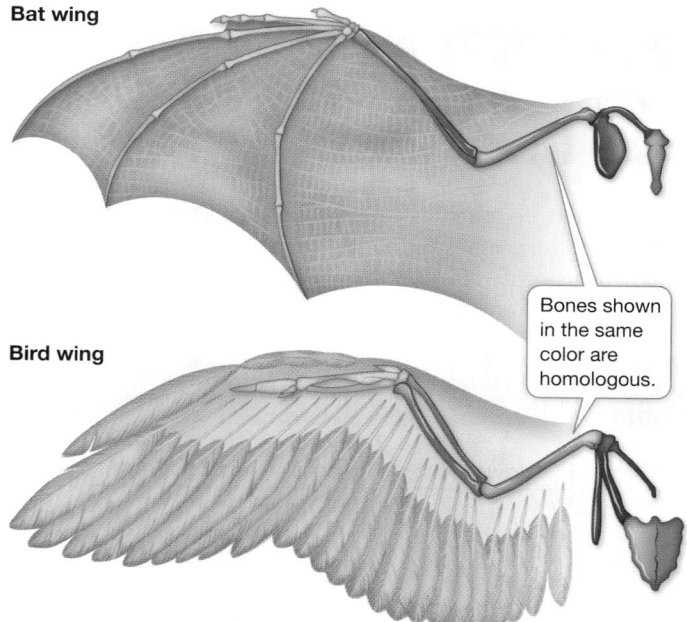

Figure 21.4 The Bones Are Homologous, the Wings Are Not
The supporting bone structures of both bat wings and bird wings are derived from a common four-limbed ancestor and are thus homologous. However, the wings themselves—an adaptation for flight—evolved independently in the two groups.

21.1 recap

A phylogenetic tree is a description of evolutionary relationships among organisms or their genes. All living organisms share a common ancestor and are related through the phylogenetic tree of life.

learning outcomes

You should be able to:

- Draw and label the parts of a phylogenetic tree and explain the biological interpretation of each part.
- Make inferences and predictions about evolutionary groups based on a phylogenetic tree.
- Explain how homoplasies (convergences and reversals of character states) are accounted for when reconstructing phylogenetic relationships.

1. What biological processes are represented in a phylogenetic tree?
2. Why is it important to consider only homologous characters in constructing phylogenetic trees?
3. What are some reasons that similar traits might arise independently in species that are only distantly related? Can you think of examples among familiar organisms? How do biologists account for these homoplasies in reconstructing phylogenies?

Phylogenetic analyses have become increasingly important to many types of biological research in recent years, and they are the basis for the comparative nature of biology. For the most part, however, evolutionary history cannot be observed directly. How, then, do biologists reconstruct the past? One way is by using phylogenetic analyses to construct a tree.

key concept

21.2 Phylogeny Can Be Reconstructed from Traits of Organisms

To illustrate how a phylogenetic tree is constructed, consider the eight vertebrate animals listed in **Table 21.1**: lamprey, perch, salamander, lizard, crocodile, pigeon, mouse, and chimpanzee. We will initially assume that any given derived trait arose only once during the evolution of these animals (that is,

there has been no convergent evolution), and that no derived traits were lost from any of the descendant groups (there has been no evolutionary reversal). For simplicity, we have selected traits that are either present (+) or absent (–).

focus your learning

- Modern phylogenetic methods employ the principle of parsimony and mathematical models (when appropriate) to analyze morphological, developmental, paleontological, behavioral, and molecular data.

In a phylogenetic study, the group of organisms of primary interest is called the **ingroup**. As a point of reference, an ingroup is compared with an **outgroup**: a species or group that is closely related to the ingroup but is known to be phylogenetically outside it. In other words, the root of the tree is located between the ingroup and the outgroup. Any trait that is present in both the ingroup and the outgroup must have evolved before the origin of the ingroup and thus must be ancestral for the ingroup. In contrast, traits that are present in only some members of the ingroup must be derived traits within that ingroup. As we will see in Chapter 32, a group of jawless fish called the lampreys is thought to have separated from the lineage leading to the other vertebrates before the jaw arose. Therefore we have included the lamprey as the outgroup for our analysis. Because derived traits are traits acquired by other members of the vertebrate lineage *after* they diverged from the outgroup, any trait that is present in both the lamprey and the other vertebrates is judged to be ancestral.

We begin by noting that the chimpanzee and mouse share two traits—mammary glands and fur—that are absent in both the outgroup and in the other species of the ingroup. Therefore we infer that mammary glands and fur are derived traits that evolved in a common ancestor of chimpanzees and mice after that lineage separated from the lineages leading to the other vertebrates. These characters are synapomorphies that unite chimpanzees and mice (as well as all other mammals, although we have not included other mammalian species in this example). By the same reasoning, we can infer that the other shared derived traits are synapomorphies for the various groups in which they are expressed. For instance, keratinous scales are a synapomorphy of the lizard, crocodile, and pigeon.

Table 21.1 also tells us that, among the animals in our ingroup, the pigeon has a unique trait: feathers. Feathers are a synapomorphy of birds and their extinct relatives. However, because we only have one bird in this example, the presence of feathers provides no clues concerning relationships among these eight species of vertebrates. However, gizzards are found in both birds and crocodiles, so this trait is evidence of a close relationship between birds and crocodilians.

By combining information about the various synapomorphies, we can construct a phylogenetic tree. We infer from our information that mice and

table 21.1 Eight Vertebrates and the Presence or Absence of Some Shared Derived Traits

Taxon	Jaws	Lungs	Claws or nails	Gizzard	Feathers	Fur	Mammary glands	Keratinous scales
Lamprey (outgroup)	–	–	–	–	–	–	–	–
Perch	+	–	–	–	–	–	–	–
Salamander	+	+	–	–	–	–	–	–
Lizard	+	+	+	–	–	–	–	+
Crocodile	+	+	+	+	–	–	–	+
Pigeon	+	+	+	+	+	–	–	+
Mouse	+	+	+	–	–	+	+	–
Chimpanzee	+	+	+	–	–	+	+	–

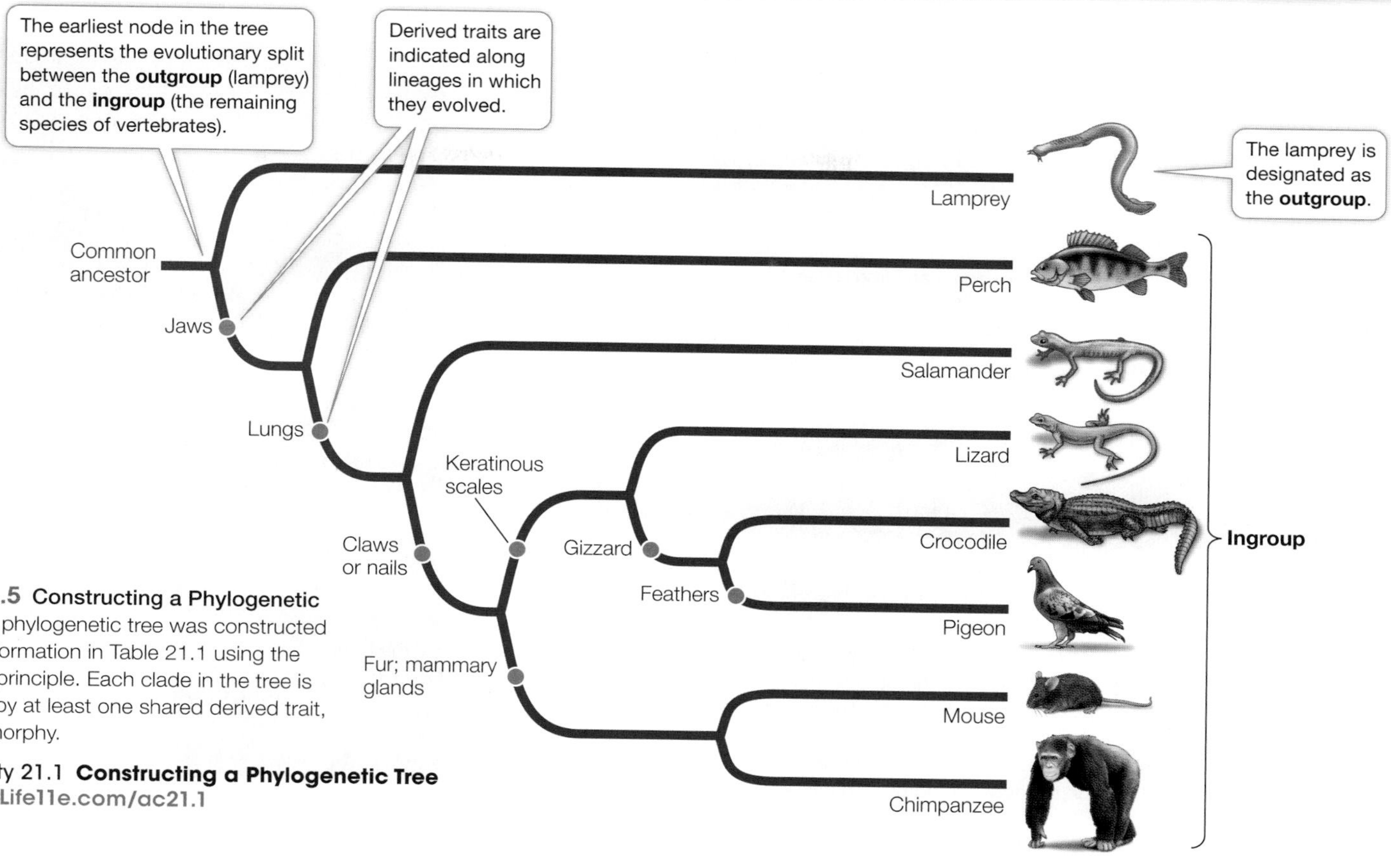

The earliest node in the tree represents the evolutionary split between the **outgroup** (lamprey) and the **ingroup** (the remaining species of vertebrates).

Derived traits are indicated along lineages in which they evolved.

The lamprey is designated as the **outgroup**.

Lamprey

Perch

Salamander

Lizard

Crocodile

Pigeon

Mouse

Chimpanzee

Common ancestor

Jaws

Lungs

Keratinous scales

Claws or nails

Gizzard

Feathers

Fur; mammary glands

Ingroup

Figure 21.5 Constructing a Phylogenetic Tree This phylogenetic tree was constructed from the information in Table 21.1 using the parsimony principle. Each clade in the tree is supported by at least one shared derived trait, or synapomorphy.

 Activity 21.1 Constructing a Phylogenetic Tree www.Life11e.com/ac21.1

chimpanzees—the only two animals that share fur and mammary glands—share a more recent common ancestor with each other than they do with pigeons and crocodiles. Otherwise we would need to assume that the ancestors of pigeons and crocodiles also had fur and mammary glands but subsequently lost them. There is no need to make these additional assumptions.

Figure 21.5 shows a phylogenetic tree for the vertebrates in Table 21.1, based on the shared derived traits we examined. This particular tree was easy to construct because it is based on a very small sample of traits, and the derived traits we examined evolved only once and were never lost after they appeared. Had we included a snake in the group, our analysis would not have been as straightforward. We would have needed to examine additional characters to determine that snakes evolved from a group of lizards that had limbs. In fact, the analysis of many characters shows that snakes evolved from burrowing lizards that became adapted to a subterranean existence.

Parsimony provides the simplest explanation for phylogenetic data

The phylogenetic tree shown in Figure 21.5 is based on only a very small sample of traits. Typically, biologists construct phylogenetic trees using hundreds or thousands of traits. With larger data sets, we would expect to observe traits that have changed more than once, and thus would expect to see convergence and evolutionary reversal. How do we determine which traits are synapomorphies and which are homoplasies? One way is to invoke the principle of parsimony.

 Activity 21.2 Phylogeny and Molecular Evolution Simulation www.Life11e.com/ac21.2

In its most general form, the **parsimony principle** states that the preferred explanation of observed data is the simplest explanation. Applying the principle of parsimony to the reconstruction of phylogenies entails minimizing the number of evolutionary changes that need to be assumed over all characters in all groups in the tree. In other words, the best hypothesis under the parsimony principle is one that requires the fewest homoplasies. This application of parsimony is a specific case of a general principle of reasoning called Occam's razor: the best explanation is the one that best fits the data while making the fewest assumptions. More complicated explanations are accepted only when the evidence requires them. Phylogenetic trees represent our best estimates about evolutionary relationships, given our current knowledge. They are continually modified as additional evidence becomes available.

Phylogenies are reconstructed from many sources of data

Naturalists have constructed various forms of phylogenetic trees for more than 150 years. In fact, the only figure in the first edition of *On the Origin of Species* was a phylogenetic tree. Tree construction has been revolutionized, however, by the advent of computer software that allows us to consider far more data and analyze far more traits than could ever before be processed. Combining these advances in methodology with the massive comparative data sets being generated through studies of genomes, biologists are learning details about the tree of life at a remarkable pace (see Appendix A: The Tree of Life).

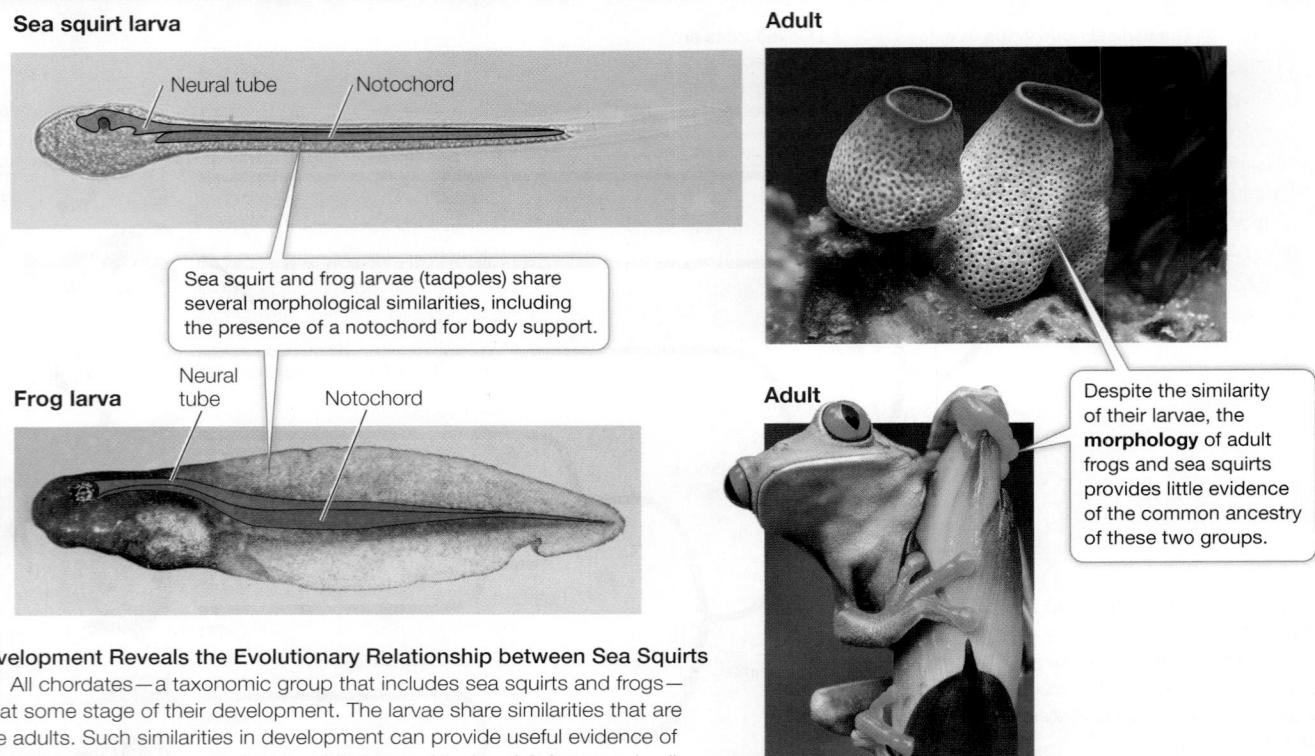

Sea squirt larva

Neural tube   Notochord

Sea squirt and frog larvae (tadpoles) share several morphological similarities, including the presence of a notochord for body support.

Frog larva

Neural tube   Notochord

Adult

Adult

Despite the similarity of their larvae, the **morphology** of adult frogs and sea squirts provides little evidence of the common ancestry of these two groups.

Figure 21.6 Development Reveals the Evolutionary Relationship between Sea Squirts and Vertebrates All chordates—a taxonomic group that includes sea squirts and frogs—have a notochord at some stage of their development. The larvae share similarities that are not apparent in the adults. Such similarities in development can provide useful evidence of evolutionary relationships. The notochord is lost in adult sea squirts. In adult frogs, as in all vertebrates, the vertebral column replaces the notochord as the support structure.

Any trait that is genetically determined, and therefore heritable, can be used in a phylogenetic analysis. Evolutionary relationships can be revealed through studies of morphology, development, the fossil record, behavioral traits, and molecular traits such as DNA and protein sequences. Let's take a closer look at the types of data used in modern phylogenetic analyses.

MORPHOLOGY   An important source of phylogenetic information is **morphology**: the presence, size, shape, and other attributes of body parts. Since living organisms have been observed, depicted, collected, and studied for millennia, we have a wealth of recorded morphological data as well as extensive museum and herbarium collections of organisms whose traits can be measured. New technological tools, such as the electron microscope and computed tomography (CT) scans, enable systematists to examine and analyze the structures of organisms at much finer scales than was formerly possible.

Most species are described and known primarily by their morphology, and morphology still provides the most comprehensive data set available for many taxa. The morphological features that are important for phylogenetic analysis are often specific to a particular group. For example, the presence, development, shape, and size of various features of the skeletal system are important in vertebrate phylogeny, whereas floral structures are important for studying the relationships among flowering plants.

Morphological approaches to phylogenetic analysis have some limitations, however. Some taxa exhibit little morphological diversity, despite great species diversity. For example, the phylogeny of the leopard frogs of North and Central America would be difficult to infer from morphological differences alone, because the many species look very similar, despite important differences in their behavior and physiology. At the other extreme, few morphological traits can be compared across distantly related species (earthworms and mammals, for example). Furthermore, some morphological variation has an environmental (rather than a genetic) basis and so must be excluded from phylogenetic analyses. An accurate phylogenetic analysis often requires information beyond that supplied by morphology.

DEVELOPMENT   Similarities in *developmental patterns* may reveal evolutionary relationships. Some organisms exhibit similarities only in early developmental stages. The larvae of marine creatures called sea squirts, for example, have a flexible gelatinous rod in the back—the notochord—that disappears as the larvae develop into adults. All vertebrate animals also have a notochord at some time during their development (**Figure 21.6**). This shared structure is one of the reasons for inferring that sea squirts are more closely related to vertebrates than would be suspected if only adult sea squirts were examined.

---

*connect the concepts   Evolutionary developmental biology, the subject of Chapter 19, compares the developmental processes of different organisms to determine the ancestral relationship between them, and to discover how developmental processes evolved. Genetic toolkit genes are expressed in different ways in different species, resulting in major morphological differences among species (see Key Concepts 19.4 and 19.5). The existence of highly conserved development genes makes it likely that similar traits will evolve repeatedly.

PALEONTOLOGY   The fossil record is another important source of information on evolutionary history. Fossils show us where and when organisms lived in the past and give us an idea of what they looked like. Fossils provide important evidence that helps us distinguish ancestral from derived traits. The fossil record can also reveal when lineages diverged and began their independent evolutionary histories. Furthermore, in groups with few species that have survived to the present, information on extinct species is often critical to an understanding of the large divergences among the surviving species. The fossil record has limitations, however. Few or no fossils have been found for some groups, and the fossil record for many groups is fragmentary.

BEHAVIOR   Some behavioral traits are culturally transmitted and others are genetically inherited. If a particular behavior is culturally transmitted, it may not accurately reflect evolutionary relationships (but may nonetheless reflect cultural connections). Many bird songs, for instance, are learned and may be inappropriate traits for phylogenetic analysis. Frog calls, however, are genetically determined and appear to be acceptable sources of information for reconstructing phylogenies.

MOLECULAR DATA   All heritable variation is encoded in DNA, and so the complete genome of an organism contains an enormous set of traits (the individual nucleotide bases of DNA) that can be used in phylogenetic analyses. In recent years, DNA sequences have become among the most widely used sources of data for constructing phylogenetic trees. Comparisons of nucleotide sequences are not limited to the DNA in the cell nucleus. Eukaryotes have genes in their mitochondria as well as in their nuclei. Plant cells also have genes in their chloroplasts. The chloroplast genome (cpDNA), which is used extensively in phylogenetic studies of plants, has changed slowly over evolutionary time, so it is often used to study relatively ancient phylogenetic relationships. Most animal mitochondrial DNA (mtDNA) has changed more rapidly, so mitochondrial genes are used to study evolutionary relationships among closely related animal species (the mitochondrial genes of plants evolve more slowly). Many nuclear gene sequences are also commonly analyzed, and now that entire genomes have been sequenced from many species, they too are used to construct phylogenetic trees. Information on gene products (such as the amino acid sequences of proteins) is also widely used for phylogenetic analyses.

Mathematical models expand the power of phylogenetic reconstruction

As biologists began to use DNA sequences to infer phylogenies in the 1970s and 1980s, they developed explicit mathematical models describing how DNA sequences change over time. These models account for multiple changes at a given position in a DNA sequence. They also take into account different rates of change at different positions in a gene, at different positions in a codon, and among different nucleotides. For example, transitions (changes between two purines or between two pyrimidines) are usually more likely than are transversions (changes between a purine and pyrimidine).

Mathematical models can be used to compute how a tree might evolve given the observed data. A **maximum likelihood** method will identify the tree that most likely produced the observed data, given the assumed model of evolutionary change. Maximum likelihood

methods can be used for any kind of characters, but they are most often used with molecular data, for which explicit mathematical models of evolutionary change are easier to develop. The principal advantages of maximum likelihood analyses are that they incorporate more information about evolutionary change than do parsimony methods, and they are easier to treat in a statistical framework. The principal disadvantages are that they are computationally intensive and require explicit models of evolutionary change (which may not be available for some kinds of character change).

The accuracy of phylogenetic methods can be tested

If phylogenetic trees represent reconstructions of past events, and if many of these events occurred before any humans were around to witness them, how can we test the accuracy of phylogenetic methods? Biologists have conducted experiments both in living organisms and with computer simulations that have demonstrated the effectiveness and accuracy of phylogenetic methods.

In one experiment designed to test the accuracy of phylogenetic analysis, a single viral culture of bacteriophage T7 was used as a starting point, and lineages were allowed to evolve from this ancestral virus in the laboratory (**Investigating Life: Testing the Accuracy of Phylogenetic Analysis**). The initial culture was split into two separate lineages, one of which became the ingroup for analysis and the other of which became the outgroup for rooting the tree. The lineages in the ingroup were split in two after every 400 generations, and samples of the virus were saved for analysis at each branching point. The lineages were allowed to evolve until there were eight lineages in the ingroup. Mutagens were added to the viral cultures to increase the mutation rate so that the amount of change and the degree of homoplasy would be typical of the organisms analyzed in average phylogenetic analyses. The investigators then sequenced samples from the end points of the eight ingroups and one outgroup lineages, as well as from the ancestors at the branching points. They then gave the sequences from the end points of the lineages to other investigators to analyze, without revealing the known history of the lineages or the sequences of the ancestral viruses.

After the phylogenetic analysis was completed, the investigators asked two questions. Did phylogenetic methods reconstruct the known history correctly? And were the sequences of the ancestral viruses reconstructed accurately? The answer in both cases was yes. The branching order of the lineages was reconstructed exactly as it had occurred, more than 98 percent of the nucleotide positions of the ancestral viruses were reconstructed correctly, and 100 percent of the amino acid changes in the viral proteins were reconstructed correctly.

The experiment shown in Investigating Life: Testing the Accuracy of Phylogenetic Analysis demonstrated that phylogenetic analysis was accurate under the conditions tested, but it did not examine all possible conditions. Other experimental studies have taken other factors into account, such as the sensitivity of phylogenetic analysis to convergent environments and highly variable rates of evolutionary change. In addition, computer simulations based on evolutionary models have been used extensively to study the effectiveness of phylogenetic analysis. These studies have also confirmed the accuracy of phylogenetic methods and have been used to refine those methods and extend them to new applications.

experiment

Original Paper: Hillis, D. M., J. J. Bull, M. E. White, M. R. Badgett, and I. J. Molineux. 1992. Experimental phylogenetics: Generation of a known phylogeny. *Science* 255: 589–592.

To test whether analysis of gene sequences can accurately reconstruct evolutionary phylogeny, we must have an unambiguously known phylogeny to compare against the reconstruction. Will the observed phylogeny match the reconstruction?

HYPOTHESIS▶ A phylogeny reconstructed by analyzing the DNA sequences of living organisms can accurately match the known evolutionary history of the organisms.

METHOD In the laboratory, researchers produced an unambiguous phylogeny of nine viral lineages, enhancing the mutation rate to increase variation among the lineages.

Viral sequences from the end points of each lineage (blue dots) were subjected to phylogenetic analysis by investigators who were unaware of the history of the lineages or the gene sequences of the ancestral viruses. These investigators reconstructed the phylogeny and ancestral DNA sequences based solely on their analyses of the descendants' genomes.

RESULTS The true phylogeny and ancestral DNA sequences were accurately reconstructed solely from the DNA sequences of the viruses at the tips of the tree branches.

CONCLUSION▶ Phylogenetic analysis of DNA sequences can accurately reconstruct evolutionary history.

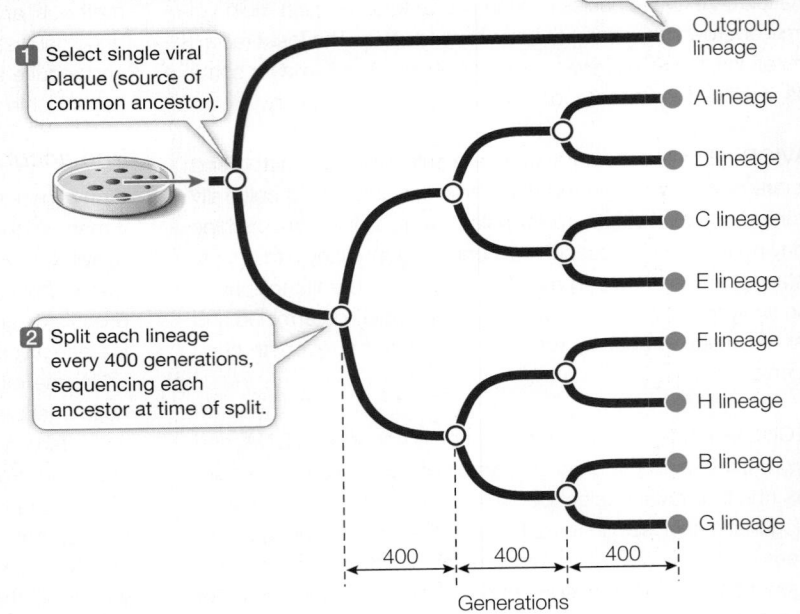

[1] Select single viral plaque (source of common ancestor).

[2] Split each lineage every 400 generations, sequencing each ancestor at time of split.

[3] Sequence genes from the end points of the lineages (blue dots).

Outgroup lineage
A lineage
D lineage
C lineage
E lineage
F lineage
H lineage
B lineage
G lineage

400 400 400

Generations

▶ Animation 21.1 **Using Phylogenetic Analysis to Reconstruct Evolutionary History**
www.Life11e.com/a21.1

▶ 21.2 recap

Phylogenetic trees can be constructed by using the parsimony principle to find the simplest explanation for phylogenetic data. Maximum likelihood methods incorporate more explicit mathematical models of evolutionary change to reconstruct evolutionary history.

learning outcomes

You should be able to:

• Analyze a phylogenetic tree to identify synapomorphies, homoplasies, and relationships among taxa.

• Reconstruct a phylogenetic tree from a data matrix of characters.

The matrix at right supplies data for seven land plants and an outgroup: an aquatic plant known as a stonewort. Each trait is scored as either present (+) or absent (–) in each of the plants. Use this data matrix to reconstruct the phylogeny of land plants and answer the questions that follow.

Taxon	Protected embryos	True roots	Persistently green sporophyte	Vascular cells	Stomata	Megaphylls (true leaves)	Seeds
Stonewort (outgroup)	–	–	–	–	–	–	–
Liverwort	+	–	–	–	–	–	–
Pine tree	+	+	+	+	+	+	+
Bracken fern	+	+	+	+	+	+	–
Club moss	+	+	+	+	+	–	–
Sphagnum moss	+	–	–	–	+	–	–
Hornwort	+	–	+	–	+	–	–
Sunflower	+	+	+	+	+	+	+

1. Which two of these taxa are most closely related?

2. Plants that produce seeds are known as seed plants. What is the sister group to the seed plants among these taxa?

3. Which two traits evolved along the same branch of your reconstructed phylogeny?

4. Are there any homoplasies in your reconstructed phylogeny?

Original Paper: Bull, J. J., C. W. Cunningham, I. J. Molineux, M. R. Badgett, and D. M. Hillis. 1993. Experimental molecular evolution of bacteriophage T7. *Evolution* 47: 993–1007.

The full DNA sequences for the viral lineages produced in this experiment are thousands of nucleotides long. However, 23 of the nucleotide positions are shown in the table below, and you can use these data to repeat the researchers' analysis. Each nucleotide position represents a separate character.

QUESTIONS▶

1. Construct a phylogenetic tree from the nucleotide positions using the parsimony principle (see Key Concept 21.2 and the examples in Table 21.1 and Figure 21.5). Use the outgroup to root your tree. Assume that all changes among nucleotides are equally likely.

2. Using your tree from Question 1, reconstruct the DNA sequences of the ancestral lineages.

3. Transitions are mutations that change one purine to the other (G ↔ A) or one pyrimidine to the other (C ↔ T), whereas transversions exchange a purine for a pyrimidine or vice versa (e.g., A → C or T; C → A or G). Which kind of mutation predominates in this phylogeny? Why might this be the case?

A similar **work with the data** exercise may be assigned in **LaunchPad**.

	Character at position																						
Lineage	1	2	3	4	5	6	7	8	9	10	11	12	13	14	15	16	17	18	19	20	21	22	23
A	T	C	G	G	G	C	C	C	C	C	C	C	C	A	A	C	C	G	A	T	A	C	A	A
B	C	C	G	G	G	T	C	C	C	T	C	C	G	A	T	T	A	G	C	G	T	G	G
C	C	C	G	G	G	C	C	C	T	C	C	T	A	A	C	C	G	G	T	A	C	A	A
D	T	C	A	G	G	C	C	C	C	C	C	C	A	A	C	C	G	A	T	A	C	A	A
E	C	T	G	G	G	C	C	C	C	C	C	T	A	A	C	C	G	G	T	A	C	A	A
F	C	T	G	A	A	C	C	C	C	C	C	C	G	A	C	T	G	G	C	G	C	G	G
G	C	C	G	G	G	T	T	C	C	T	C	C	G	A	T	T	A	G	C	G	C	G	G
H	C	C	G	G	A	C	C	C	C	C	C	C	G	C	C	T	G	G	C	G	C	G	G
Outgroup	C	C	G	G	G	C	C	T	C	C	T	C	G	A	C	C	G	G	C	A	C	G	G

Why do biologists expend the time and effort necessary to reconstruct phylogenies? Information about the evolutionary relationships among organisms is a useful source of data for scientists investigating a wide variety of biological questions. Next we will describe how phylogenetic trees are used to answer questions about the past, and to predict and compare traits of organisms in the present.

▶key concept 21.3 Phylogeny Makes Biology Comparative and Predictive

Once a phylogeny is reconstructed, what do we do with it? What beyond an understanding of evolutionary history does phylogeny offer us?

focus your learning

- Biologists use phylogenetic trees to investigate living organisms, explore instances of convergent evolution, and reconstruct ancestral states.
- The timing of an evolutionary event can be estimated using the average rate of change for a given gene or protein and known calibration dates.

Phylogenetic trees can be used to reconstruct past events

Reconstructing past events is important for understanding many biological processes. In the case of zoonotic diseases (diseases caused by infectious organisms transmitted to humans from another animal host), it is important to understand when, where, and how the disease first entered a human population. Human immunodeficiency virus (HIV) is the cause of such a zoonotic disease, acquired immunodeficiency syndrome, or AIDS. Phylogenetic analyses have become important for studying the transmission of viruses such as HIV. Phylogenies are also important for understanding the present global diversity of HIV and for determining the virus's origins in human populations. A broader phylogenetic analysis of immunodeficiency viruses shows that humans acquired these viruses from two different hosts: HIV-1 from chimpanzees, and HIV-2 from sooty mangabeys (**Figure 21.7**).

HIV-1 is the common form of the virus in human populations in central Africa, where chimpanzees are hunted for food, and HIV-2 is the common form in human populations in western Africa, where sooty mangabeys are hunted for food. Thus it seems likely that these viruses entered human populations through hunters who cut themselves while skinning chimpanzees and sooty mangabeys. The global pandemic of AIDS occurred when these infections in local African populations rapidly spread through human populations around the world.

Figure 21.7 Phylogenetic Tree of Immunodeficiency Viruses
The evolutionary relationships of immunodeficiency viruses show that these viruses have been transmitted to humans from two different simian hosts: HIV-1 from chimpanzees and HIV-2 from sooty mangabeys. (SIV stands for simian immunodeficiency virus.)

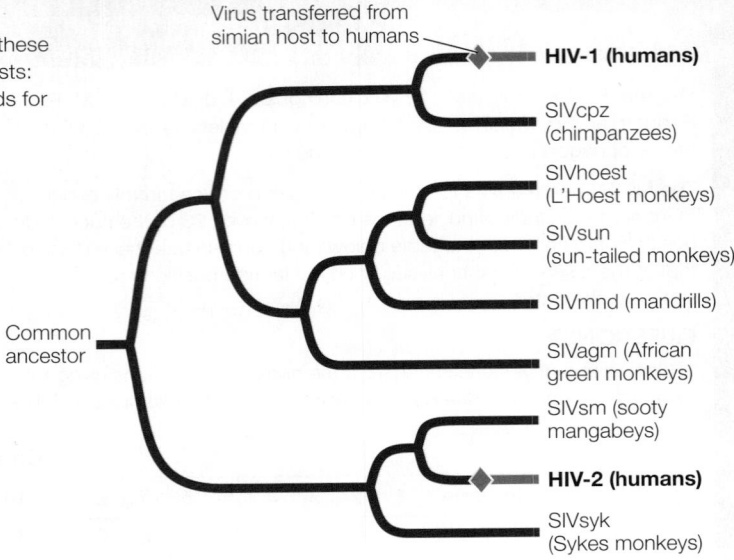

In recent years, phylogenetic analysis has become important in forensic investigations that involve viral transmission events. For example, phylogenetic analysis was critical for a criminal investigation of a physician who was accused of purposefully injecting blood from one of his HIV-positive patients into his former girlfriend in an attempt to kill her. The phylogenetic analysis revealed that the HIV strains present in the girlfriend were a subset of those present in the physician's patient (**Figure 21.8**). Other evidence was needed, of course, to connect the physician to this purposeful transmission event, but the phylogenetic

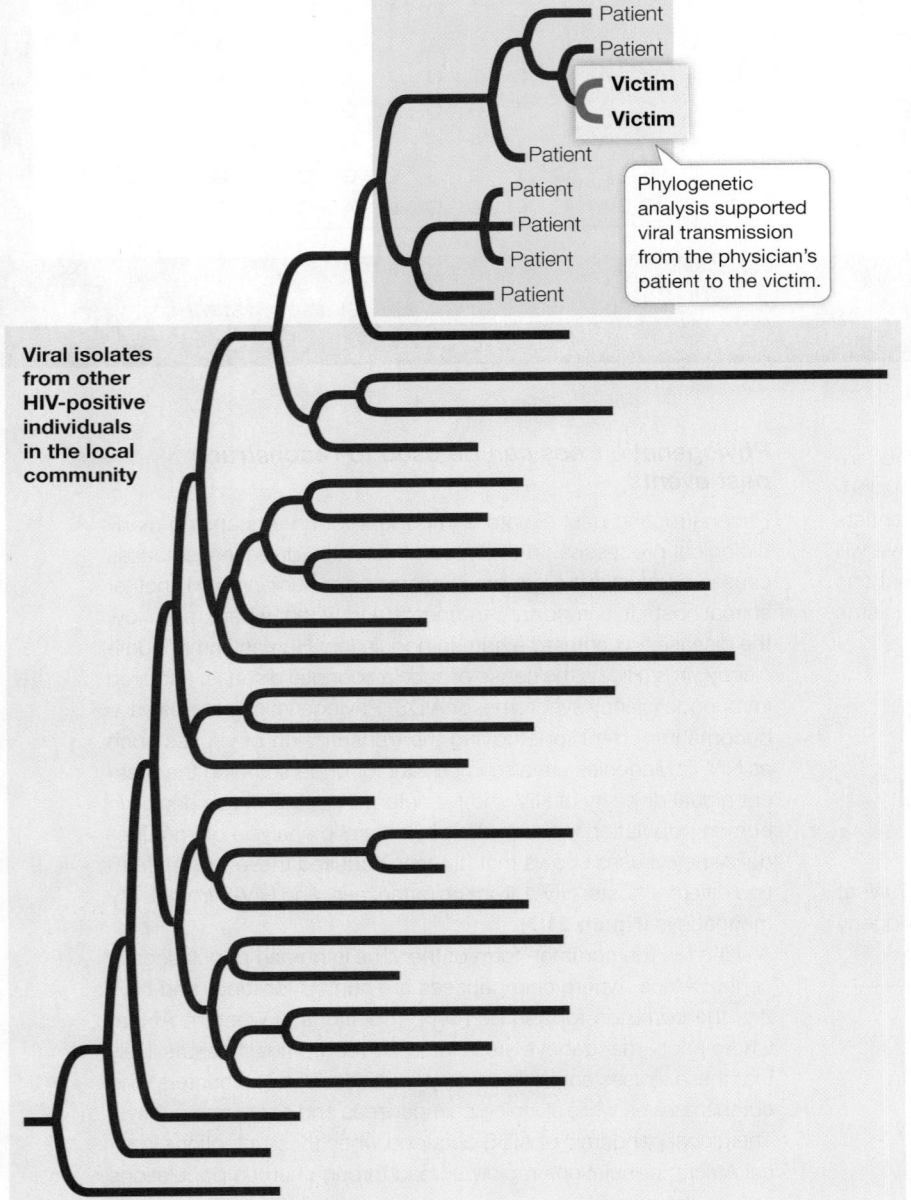

analysis was important to support the viral transmission event from the patient to the victim.

Phylogenies allow us to compare and contrast living organisms

Male swordtails—a group of fish in the genus *Xiphophorus*—have a long, colorful tail extension, and their reproductive success is closely associated with this appendage. Males with a long sword are more likely to mate successfully than are males with a short sword (an example of sexual selection; see Key Concept 20.2). Several explanations have been advanced for the evolution of this structure, including the hypothesis that the sword simply exploits a preexisting bias in the sensory system of the females. This sensory exploitation hypothesis suggests that female swordtails had a preference for males with long tails even before the tails evolved (perhaps because females assess the size of males by their total body length—including the tail—and prefer larger males).

To test the sensory exploitation hypothesis, phylogenetic analysis was used to identify the relatives of swordtails that had split most recently from their lineage before the evolution of swords. These closest relatives turned out to be fish in the genus *Priapella*. Even though male *Priapella* do not normally have swords, when researchers attached artificial swordlike structures to the tails of male *Priapella*, female *Priapella*

Figure 21.8 A Forensic Application of Phylogenetic Analysis   This phylogenetic analysis demonstrated that strains of HIV present in a victim (shown in red) were a phylogenetic subset of viruses isolated from a physician's patient (shown in blue). This analysis was part of the evidence used to show that the physician drew blood from his HIV-positive patient and injected it into the victim in an attempt to kill her. A jury found the physician guilty of attempted murder.

Q: What hypothetical phylogenetic results could have exonerated the physician?

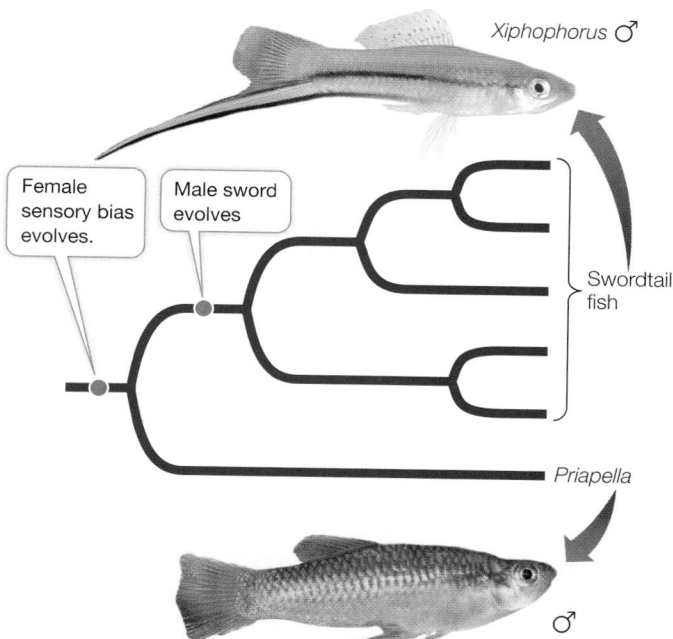

Figure 21.9 The Origin of a Sexually Selected Trait The tail extension of male swordtails (genus *Xiphophorus*) apparently evolved through sexual selection, as females mated preferentially with males that had long "swords." Phylogenetic analysis reveals that *Priapella* split from the swordtails before the evolution of the sword. The independent finding that female *Priapella* prefer male *Priapella* with an artificial sword further supports the idea that this appendage evolved as a result of a preexisting preference in females.

preferred those males. This result provided support for the hypothesis that female *Xiphophorus* had a preexisting sensory bias favoring tail extensions even before the trait evolved (**Figure 21.9**). Thus a long tail became a sexually selected trait because of the

preexisting preference of the females. The phylogeny allows us to understand when the trait evolved relative to the change in female preference.

Phylogenies can reveal convergent evolution

Like most animals, flowering plants (angiosperms) often reproduce by mating with another individual of the same species. But in many angiosperm species, the same individual produces both male and female gametes (contained within pollen and ovules, respectively). *Self-incompatible species have mechanisms to prevent fertilization of the ovule by the individual's own pollen, and so must reproduce by outcrossing with another individual. Individuals of some species, however, regularly fertilize their ovules using their own pollen; they are self-fertilizing or selfing species, and their gametes are self-compatible.

---

*connect the concepts Some mechanisms of self-incompatibility are discussed in Key Concept 37.1.

The evolution of angiosperm fertilization mechanisms was examined in *Leptosiphon*, a genus in the phlox family that exhibits a diversity of mating systems and pollination mechanisms. The self-incompatible (outcrossing) species of *Leptosiphon* have long petals and are pollinated by long-tongued flies. In contrast, self-pollinating species have short petals and do not require insect pollinators to reproduce successfully. Using ribosomal DNA sequences, investigators reconstructed a phylogeny of this genus (**Figure 21.10**). They then determined whether each species was self-compatible by artificially pollinating flowers with the plant's own pollen or with pollen from other individuals and observing whether viable seeds formed.

The reconstructed phylogeny suggests that self-incompatibility is the ancestral state and that self-compatibility evolved three times within this group of *Leptosiphon*. The change to self-compatibility

Figure 21.10 Phylogeny Reveals Convergent Evolution Self-compatibility apparently evolved independently three times among these species of the plant genus *Leptosiphon*. Because the appearance and structure of the flowers converged in the three selfing lineages, taxonomists mistakenly thought they were varieties of the same species.

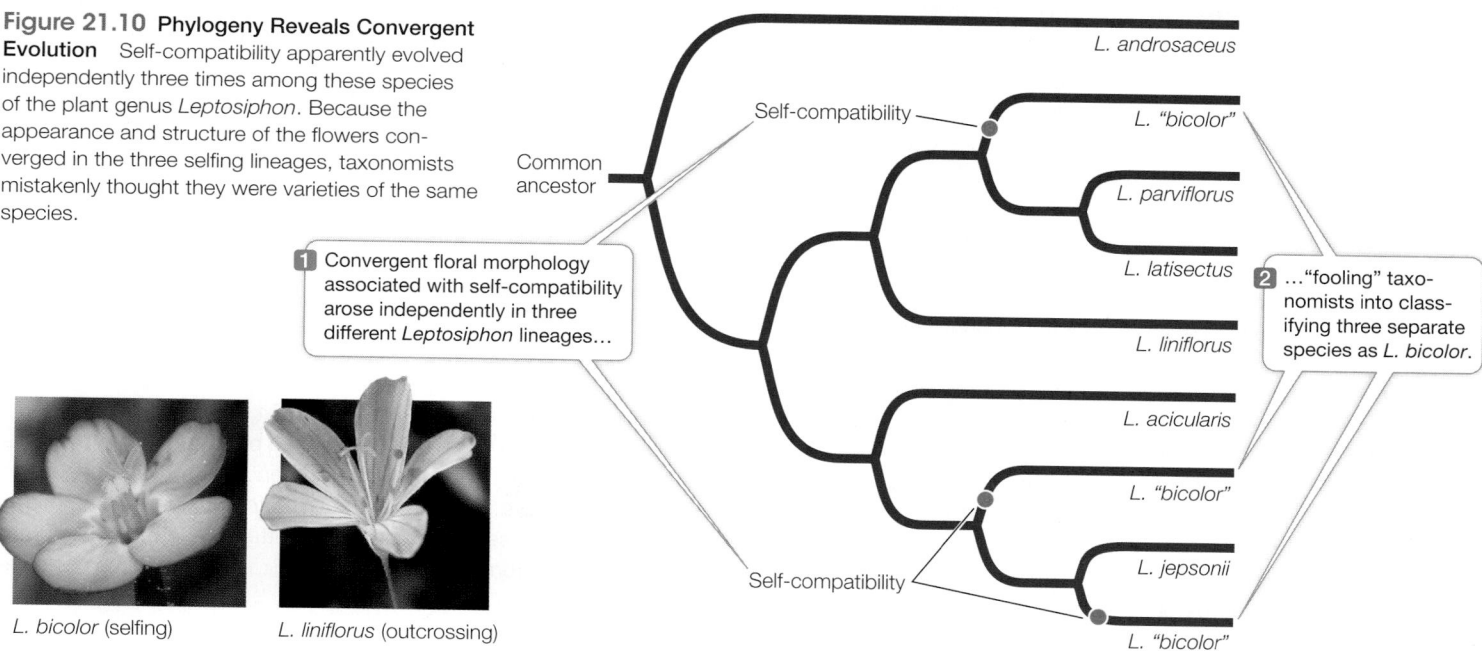

1 Convergent floral morphology associated with self-compatibility arose independently in three different *Leptosiphon* lineages...

2 ..."fooling" taxonomists into classifying three separate species as *L. bicolor*.

L. bicolor (selfing)   *L. liniflorus* (outcrossing)

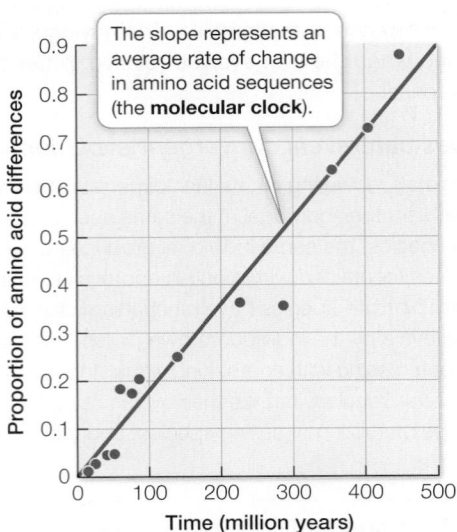

Figure 21.11 A Molecular Clock for the Protein Hemoglobin
Amino acid replacements in hemoglobin have occurred at a relatively constant rate over nearly 500 million years of evolution. The graph shows the relationship between the time of divergence and the proportion of amino acids that have changed for 13 pairs of vertebrate hemoglobin proteins. The average rate of change represents the molecular clock for hemoglobin in vertebrates.

eliminated the plants' dependence on an outside pollinator and has been accompanied by the evolution of reduced petal size. Indeed, the striking morphological similarity of the flowers in the self-compatible groups once led to their being classified as members of a single species (*L. bicolor*). Phylogenetic analysis, however, shows them to be members of three distinct lineages. From this information we can infer that self-compatibility and its associated floral structure are convergent in the three independent lineages that had been called *L. bicolor*.

Ancestral states can be reconstructed

In addition to using phylogenetic methods to infer evolutionary relationships, biologists can use these techniques to reconstruct the morphology, behavior, or nucleotide and amino acid sequences of ancestral species (as was demonstrated for the ancestral sequence of bacteriophage T7 in Investigating Life: Testing the Accuracy of Phylogenetic Analysis). In the opening of this chapter, we described how Mikhail Matz used phylogenetic analysis to reconstruct the sequence of changes in fluorescent proteins of corals to understand how red fluorescent proteins could be produced.

Reconstruction of ancient DNA sequences can also provide information about the biology of long-extinct organisms. For example, phylogenetic analysis was used to reconstruct an opsin protein in the ancestral archosaur (the most recent common ancestor of birds, dinosaurs, and crocodiles). Opsins are pigment proteins involved in vision; different opsins (with different amino acid sequences) are excited by different wavelengths of light. Knowledge of the opsin sequence in the ancestral archosaur would provide clues about the animal's visual capabilities and therefore about some of its probable behaviors. Investigators used phylogenetic analysis of opsin from living vertebrates to estimate the amino acid sequence of the pigment that existed in the ancestral archosaur. A protein with this same sequence was then constructed in the laboratory. The investigators tested the reconstructed opsin and found a significant shift toward the red end of the spectrum in the light sensitivity of this protein compared with that of most modern opsins. Modern species that exhibit similar sensitivity are adapted for nocturnal vision, so the investigators inferred that the ancestral archosaur might have been active at night. Thus, reminiscent of the movies *Jurassic Park* and *Jurassic World*, phylogenetic analyses are being used to reconstruct extinct species, one protein at a time.

Molecular clocks help date evolutionary events

For many applications, biologists want to know not only the order in which evolutionary lineages split but also the timing of those splits. In 1965, Emile Zuckerkandl and Linus Pauling hypothesized that rates of molecular change were constant enough that they could be used to predict evolutionary divergence times—an idea that has become known as the **molecular clock** hypothesis.

Of course, different genes evolve at different rates, and there are also differences in evolutionary rates among species related to differing generation times, environments, efficiencies of DNA repair systems, and other biological factors. Nonetheless, among closely related species, a given gene usually evolves at a reasonably constant rate. Therefore the protein encoded by the gene accumulates amino acid replacements at a relatively constant rate (**Figure 21.11**). A molecular clock uses the average rate at which a given gene or protein accumulates changes to gauge the time of divergence for a particular split in the phylogeny. Molecular clocks must be calibrated using independent data, such as the fossil record, known times of divergence, or biogeographic dates (e.g., the time of separations of continents). Using such calibrations, times of divergence have been estimated for many groups of species that have diverged over millions of years.

Molecular clocks are not only used to date ancient events; they are also used to study the timing of comparatively recent events. Most samples of HIV-1 have been collected from humans only since the early 1980s, although a few isolates from medical biopsies are available from as early as the 1950s. Biologists can use the observed changes in HIV-1 over the past several decades to project back to the common ancestor of all HIV-1 isolates, and estimate when HIV-1 first entered human populations from chimpanzees (**Figure 21.12**). This molecular clock was calibrated using the samples from the 1980s and 1990s, and then tested using the samples from the 1950s. As shown in Figure 21.12C, a sample from a 1959 biopsy is dated by molecular clock analysis at 1957 ± 10 years. Extrapolation back to the common ancestor of the samples suggested a date of origin for this group of viruses of about 1930. Although AIDS was unknown to Western medicine until the 1980s, this analysis shows that HIV-1 was present (probably at a very low frequency) in human populations in Africa for at least a half-century before its emergence as a global pandemic. Biologists have used similar analyses to conclude that immunodeficiency viruses have been transmitted repeatedly into human populations from multiple primates for more than a century (see also Figure 21.7).

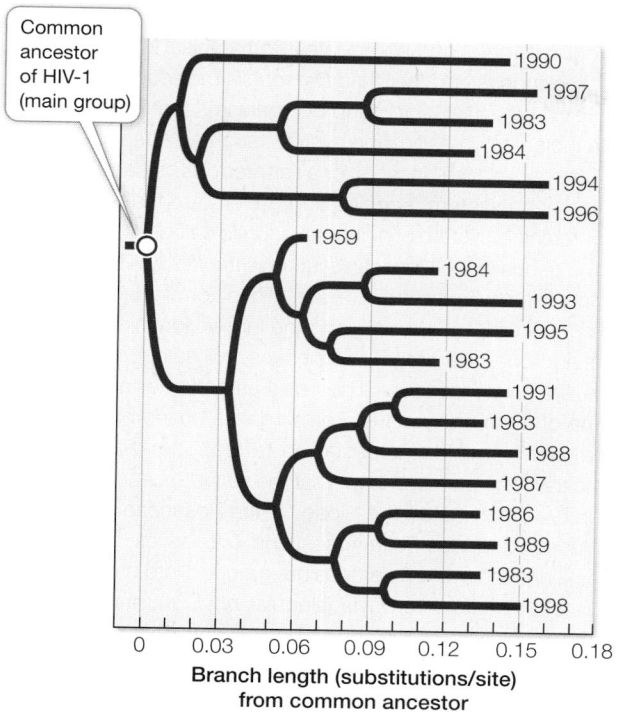

(A) Tree of HIV-1 samples plotted on a divergence axis

Common ancestor of HIV-1 (main group)

1990
1997
1983
1984
1994
1996
1959
1984
1993
1995
1983
1991
1983
1988
1987
1986
1989
1983
1998

0   0.03   0.06   0.09   0.12   0.15   0.18
Branch length (substitutions/site) from common ancestor

(B) Plot of average rate of divergence (the molecular clock)

Branch length from common ancestor

0.18
0.17
0.16
0.15
0.14
0.13
0.12
0.11
0.10

1980   1985   1990   1995   2000
Year

(C) Extrapolation: Origin of 1959 sample

0.18
0.15
0.12
0.09
0.06
0.03
0

Confidence limits

1959 sample

Predicted sampling date 1957±10 years

1900 1920 1940 1960 1980 2000
Year

Estimated date for origin of HIV-1 main group

Figure 21.12 Dating the Origin of HIV-1 in Human Populations
(A) A phylogenetic tree for samples of the main group of HIV-1 virus. The dates indicate the years in which the samples were taken. (For clarity, only a small fraction of the samples that were examined in the original study are shown.) **(B)** A plot of sample year versus genetic divergence from the common ancestor provided an average rate of divergence, or a molecular clock. **(C)** The molecular clock was used to date a sample taken in 1959 (as a test of the clock) and to estimate the date of origin of the HIV-1 main group (about 1930). Branch length from a common ancestor represents the average number of substitutions per nucleotide.

Q: What would be the expected branch length from the common ancestor of an HIV isolate from 1970?

21.3 recap

Phylogenetic trees are used to reconstruct the evolutionary history of lineages, to determine when and where traits arose, and to make biological comparisons among genes, populations, and species. They can also be used to reconstruct ancestral traits and to estimate the timing of evolutionary events.

learning outcomes

You should be able to:

- Use a phylogenetic tree to formulate a hypothesis about the origins of an epidemic.
- Calculate the rate of a molecular clock from a graph that shows change over time.

1. West Nile virus kills birds of many species and can cause fatal encephalitis (inflammation of the brain) in humans and horses. In the 1930s the virus was first isolated in Africa, where it is thought to be endemic. By the 1990s it had been found throughout much of Eurasia. West Nile virus was not found in North America until 1999, when it was first detected in New York, but since that time it has spread rapidly across most of the United States. Use the phylogenetic tree of West Nile virus isolates shown at right to construct a hypothesis about the origin of the virus lineage that was introduced into the United States. The isolates are identified by their place and date of isolation.

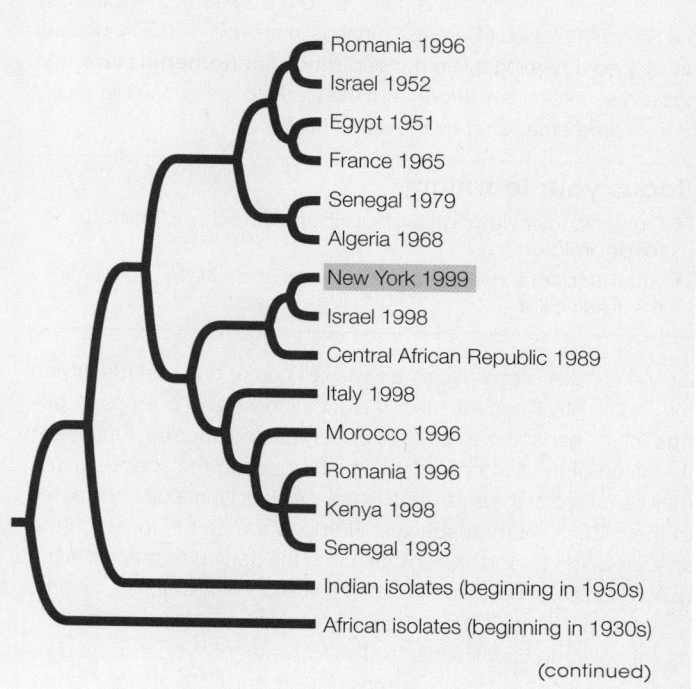

Romania 1996
Israel 1952
Egypt 1951
France 1965
Senegal 1979
Algeria 1968
New York 1999
Israel 1998
Central African Republic 1989
Italy 1998
Morocco 1996
Romania 1996
Kenya 1998
Senegal 1993
Indian isolates (beginning in 1950s)
African isolates (beginning in 1930s)

(continued)

2. Researchers examined the molecular divergence of a mitochondrial DNA gene (*cyt b*) among sister species of birds called honeycreepers on different Hawaiian Islands. They then plotted the measured molecular divergence against the estimated dates for separation of the islands (see below). Calculate the average rate of change in *cyt b* (the molecular clock) from the graph.

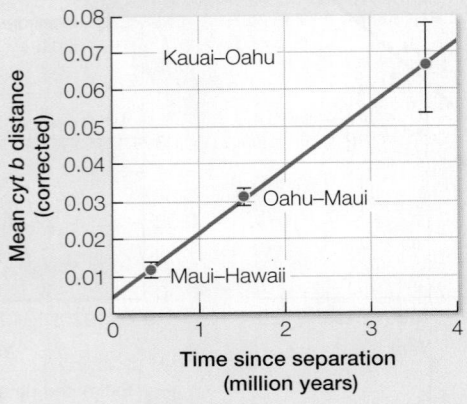

All of life is connected through evolutionary history, and the relationships among organisms provide a natural basis for making biological comparisons. For these reasons, biologists use phylogenetic relationships as the basis for organizing life into a coherent classification system.

21.4 Phylogeny Is the Basis of Biological Classification

The biological classification system in widespread use today is derived from a system developed by the Swedish biologist Carolus Linnaeus in the mid-1700s. Linnaeus developed a naming system called **binomial nomenclature** that has allowed scientists throughout the world to refer unambiguously to the same organisms by the same names.

focus your learning

- Only monophyletic groups are considered appropriate taxonomic units.
- Classifications are used to organize and name groups on the tree of life.

Linnaeus gave each species a two-part name, one part identifying the species itself and the other the genus to which it belongs. A **genus** (plural *genera*) is a group of closely related species. Optionally, the name of the taxonomist who first proposed the species name may be added at the end. Thus *Homo sapiens* Linnaeus is the name of the modern human species. *Homo* is the genus to which the species belongs, and *sapiens* identifies the particular species in the genus *Homo*; Linnaeus proposed the species name *Homo sapiens*.

You can think of *Homo* as equivalent to your surname and *sapiens* as equivalent to your first name. The first letter of the genus name is capitalized, and the specific name is lowercase. Both of these formal designations are italicized. Rather than repeating the name of a genus when it is used several times in the same discussion, biologists often spell it out only once and abbreviate it to the initial letter thereafter (e.g., *D. melanogaster* rather than *Drosophila melanogaster*).

As we noted earlier, any group of organisms that is treated as a unit in a biological classification system, such as all species in the genus *Drosophila*, or all insects, or all arthropods, is called a taxon. In the Linnaean system, species and genera are further grouped into a hierarchical system of higher taxonomic categories. The taxon above the genus in the Linnaean system is the family. The names of animal families end in the suffix "-idae." Thus Formicidae is the family that contains all ant species, and the family Hominidae contains humans and our recent fossil relatives, as well as our closest living relatives, the chimpanzees and gorillas. Family names are based on the name of a member genus; Formicidae is based on the genus *Formica*, and Hominidae is based on *Homo*. The same rules are used in classifying plants, except that the suffix "-aceae" is used for plant family names instead of "-idae." Thus Rosaceae is the family that includes the genus *Rosa* (roses) and its relatives.

In the Linnaean system, families are grouped into orders, orders into classes, classes into phyla (singular *phylum*), and phyla into kingdoms. However, the ranking of taxa within Linnaean classification is subjective. Whether a particular taxon is considered, say, an order or a class is informative only with respect to the *relative* ranking of other related taxa. Although families are always grouped within orders, orders within classes, and so forth, there is nothing that makes a "family" in one group equivalent (in number of genera or in evolutionary age, for instance) to a "family" in another group.

Linnaeus recognized the overarching hierarchy of life, but he developed his system before evolutionary thought had become widespread. Biologists today recognize the tree of life as the basis for biological classification and often name taxa without placing them into the various Linnaean ranks.

Evolutionary history is the basis for modern biological classification

Today's biological classifications express the evolutionary relationships of organisms. Taxa are expected to be **monophyletic**, meaning that the taxon contains an ancestor and all descendants of that ancestor, and no other organisms. In other words, a monophyletic taxon is a historical group of related species, or a complete branch on the tree of life. As noted earlier, this is also the definition of a clade. A true monophyletic group can be removed from a phylogenetic tree by a single "cut" in the tree, as shown in **Figure 21.13**.

Note that there are many monophyletic groups on any phylogenetic tree, and that these groups are successively smaller subsets of larger monophyletic groups. This hierarchy of biological taxa, with all of life as the most inclusive taxon and many smaller taxa within larger taxa, down to the individual species, is the modern basis for biological classification.

Although biologists seek to describe and name only monophyletic taxa, the detailed phylogenetic information needed to do so is not always available. A group that does not include its common ancestor is **polyphyletic**. A group that does not include all the descendants of a common ancestor is referred to as **paraphyletic** (see Figure 21.13). Virtually all taxonomists now agree that polyphyletic and paraphyletic groups are inappropriate as taxonomic units because they do not correctly reflect evolutionary history. Some

Figure 21.13 Monophyletic, Polyphyletic, and Paraphyletic Groups Monophyletic groups are the basis of taxa in modern biological classifications. Polyphyletic and paraphyletic groups are not appropriate for use in classifications because they do not accurately reflect evolutionary history.

Activity 21.3 Types of Taxa
www.Life11e.com/ac21.3

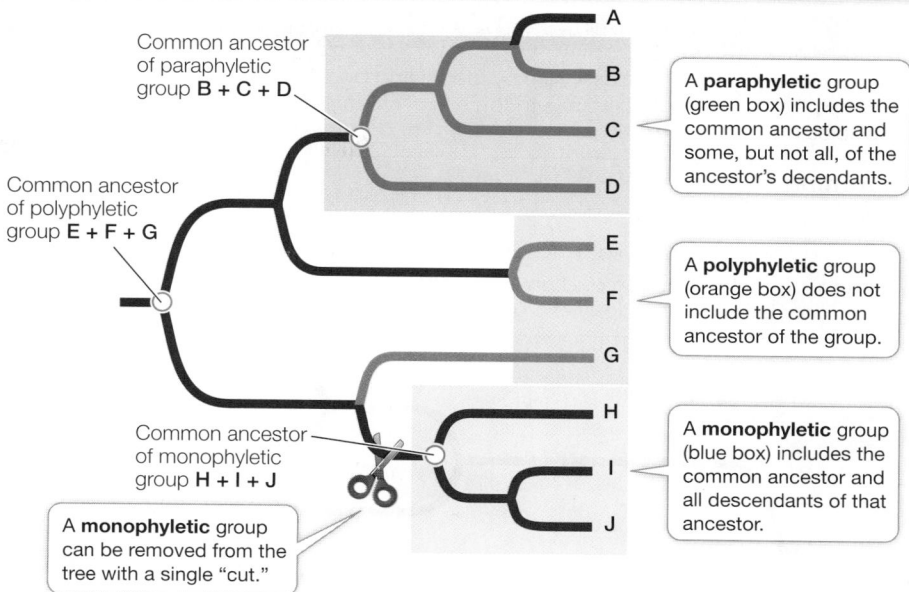

Common ancestor of paraphyletic group **B + C + D**

Common ancestor of polyphyletic group **E + F + G**

Common ancestor of monophyletic group **H + I + J**

A **monophyletic** group can be removed from the tree with a single "cut."

A **paraphyletic** group (green box) includes the common ancestor and some, but not all, of the ancestor's decendants.

A **polyphyletic** group (orange box) does not include the common ancestor of the group.

A **monophyletic** group (blue box) includes the common ancestor and all descendants of that ancestor.

classifications still contain such groups because some organisms have not been evaluated phylogenetically. As mistakes in prior classifications are detected, taxonomic names are revised and polyphyletic and paraphyletic groups are eliminated from the classifications.

Several codes of biological nomenclature govern the use of scientific names

Several sets of explicit rules govern the use of scientific names. Biologists around the world follow these rules voluntarily to facilitate communication and dialogue. There may be dozens of common names for an organism in many different languages, and the same common name may refer to more than one species (**Figure 21.14**). The rules of biological nomenclature are designed so that there is only one correct scientific name for any single recognized taxon, and (ideally) a given scientific name applies only to a single taxon (that is, each scientific name is unique). Sometimes the same species is named more than once (when more than one taxonomist has taken up the task). In these cases, the rules specify that the valid name is the first name that was proposed. If the same name is inadvertently given to two different species, then the species that was named second must be given a new name.

Because of the historical separation of the fields of zoology, botany (which originally included mycology, the study of fungi), and microbiology, different sets of taxonomic rules were developed for each of these groups. Yet another set of rules emerged later for classifying viruses. This separation of fields resulted in duplicated taxon names in groups governed by the different sets of rules. *Drosophila*, for example, is both a genus of fruit flies and a genus of fungi, and some species in both groups have identical names. Until recently these duplicated names caused little confusion, since traditionally biologists who studied fruit flies were unlikely to read the literature on fungi (and vice versa). Today, given the prevalence of large, universal biological databases (such as GenBank, which includes DNA sequences from across all life), it is increasingly important that each taxon have a unique and unambiguous name.

Biologists are working on a universal code of nomenclature that can be applied to all organisms, so that every species will have a unique identifying name or registration number. This will assist efforts to build an online *Encyclopedia of Life* that links all the information for all the world's species.

(A) *Asclepias tuberosa*

Figure 21.14 Same Common Name, Not the Same Species
All three of these distinct plant species are called "Indian paintbrush." Binomial nomenclature allows us to avoid the ambiguity of such common names and communicate exactly what is being described. **(A)** *Asclepias tuberosa* is a perennial milkweed native to eastern North America. **(B)** *Castilleja coccinea* is also native to eastern North America, but is a member of a very different group of plants called scrophs. **(C)** *Hieracium aurantiacum* is a European species of aster that has been widely introduced into North America.

(B) *Castilleja coccinea*

(C) *Hieracium aurantiacum*

21.4 recap

Biologists organize and classify life by identifying and naming monophyletic groups. Several sets of rules govern the use of scientific names so that each species and higher taxon can be identified and named unambiguously.

learning outcomes

You should be able to:
- Use a phylogeny to build a classification for a group of organisms.
- Analyze a classification and phylogenetic tree to identify monophyletic, polyphyletic, and paraphyletic groups.

Consider the phylogeny and three possible classifications shown below.

1. Which of these classifications contain a paraphyletic group?
2. Which of these classifications contains a polyphyletic group?
3. Which of these classifications is consistent with the goal of including only monophyletic groups in a biological classification?

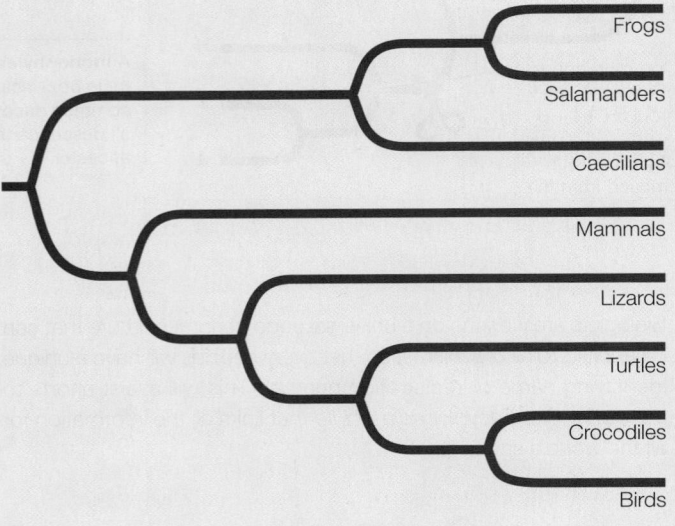

Classification one:

Named group	Included taxa
Amphibia	Frogs, salamanders, and caecilians
Mammalia	Mammals
Reptilia	Lizards, turtles, and crocodiles
Aves	Birds

Classification two:

Named group	Included taxa
Amphibia	Frogs, salamanders, and caecilians
Mammalia	Mammals
Reptilia	Lizards, turtles, crocodiles, and birds

Classification three:

Named group	Included taxa
Amphibia	Frogs, salamanders, and caecilians
Homothermia	Mammals and birds
Reptilia	Lizards, turtles, and crocodiles

Now that we have seen how evolution occurs and how phylogenies can be used to study evolutionary relationships, we are ready to consider the process of speciation. Speciation is what leads to the branching events on the tree of life, and it is the process that results in the millions of species that constitute biodiversity.

 **investigatinglife**

How are phylogenetic methods used to resurrect protein sequences from extinct organisms?

Most genes and proteins of organisms that lived millions of years ago have decomposed in the fossil remains of these species. Nonetheless, the sequences of many ancient genes and proteins can be reconstructed by the methods described in this chapter. As we saw in Investigating Life: Testing the Accuracy of Phylogenetic Analysis, we can reconstruct ancestral DNA sequences—if we have enough information about the genomes of their descendants. Biologists have reconstructed gene sequences from species that have been extinct for millions of years. Using this information, a laboratory can reconstruct real proteins that correspond to those gene sequences. This is how Mikhail Matz and his colleagues were able to resurrect fluorescent proteins from the extinct ancestors of modern corals, then visualize the colors produced by these proteins in the laboratory (Figure 21.15).

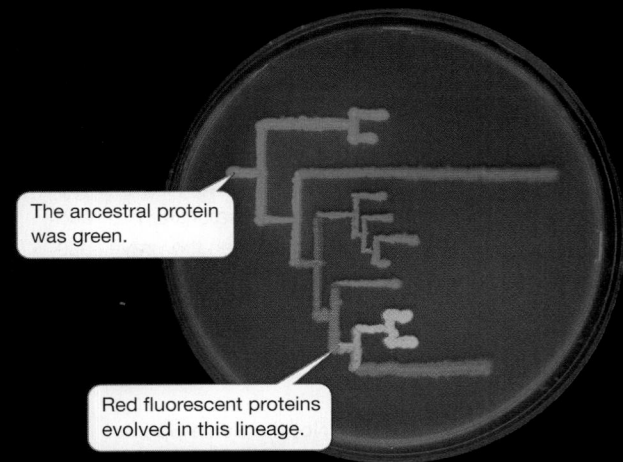

The ancestral protein was green.

Red fluorescent proteins evolved in this lineage.

Figure 21.15 Evolution of Fluorescent Proteins of Corals Mikhail Matz and his colleagues used phylogenetic analysis to reconstruct the sequences of fluorescent proteins that were present in the extinct ancestors of modern corals. They then expressed these proteins in bacteria and plated the bacteria in the form of a phylogenetic tree to

Future directions

Biologists are now using phylogenetic analysis to reconstruct many ancient protein sequences. These reconstructed protein sequences are then made into actual proteins in the laboratory. For example, when biologists measured the temperature optima for resurrected proteins that were present in the common ancestor of all life, they found that the proteins functioned best in the range of 55°C–65°C. This result is consistent with hypotheses that life evolved in a high-temperature environment.

To reconstruct protein sequences from species that have been extinct for millions or even billions of years, biologists use detailed mathematical models that take into account much of what we have learned about molecular evolution, as described in Key Concept 21.2. These models incorporate information on rates of replacement among different amino acids in proteins, information on different substitution rates among nucleotides, and changes in the rate of molecular evolution among the major lineages of life. These studies are opening up opportunities to see how proteins have evolved through time and how extinct species once functioned.

Chapter Summary 21

▶ 21.1 All of Life Is Connected through Its Evolutionary History

- **Phylogeny** is the history of evolutionary relationships among organisms or their genes. Groups of evolutionarily related species are represented as branches in a **phylogenetic tree**. Review Figures 21.1, Focus: Key Figure 21.2

- Named species and groups of species are called **taxa**. A taxon that consists of an ancestor and all of its evolutionary descendants is called a **clade**. Review Figure 21.3

- **Homologies** are similar traits that have been inherited from a common ancestor. Review Figure 21.4

- A derived trait that is shared by two or more taxa and is inherited from their common ancestor is called a **synapomorphy**.

- Distantly related species may show similar traits that do not result from common ancestry. **Convergent evolution** and **evolutionary reversals** can give rise to such traits, which are called **homoplasies**.

▶ 21.2 Phylogeny Can Be Reconstructed from Traits of Organisms

- Phylogenetic trees can be constructed from synapomorphies using the logic of **parsimony**. Review Figure 21.5, Activities 21.1, 21.2

- Sources of phylogenetic information include **morphology**, patterns of development, the fossil record, behavioral traits, and molecular traits such as DNA and protein sequences.

- Phylogenetic trees can also be constructed with **maximum likelihood** methods, which find the tree most likely to have generated the observed data under a specific model of evolution.

- Phylogenetic methods have been tested in both experimental and simulation studies, and have been shown to be accurate under a wide variety of conditions. See Animation 21.1

▶ 21.3 Phylogeny Makes Biology Comparative and Predictive

- Phylogenetic trees are used to make comparisons among living organisms. Review Figure 21.9

- Phylogenetic trees are used to reconstruct the past and to understand the origin of traits. Review Figure 21.10

- Biologists can use phylogenetic trees to reconstruct ancestral states.

- Phylogenetic trees may include estimates of divergence times of lineages determined by **molecular clock** analysis. Review Figure 21.12

▶ 21.4 Phylogeny Is the Basis of Biological Classification

- Biologists use phylogenetic relationships to organize life into a coherent classification system.

- Taxa in modern classifications are expected to be **monophyletic** groups. **Paraphyletic** and **polyphyletic** groups are not considered appropriate taxonomic units. Review Figure 21.13, Activity 21.3

- Several sets of rules govern the use of scientific names, with the goal of providing unique and universal names for taxa.

Go to **LearningCurve** (in **LaunchPad**) for dynamic quizzing that helps you solidify your understanding of this chapter. **LearningCurve** adapts to your responses, giving you the practice you need to master each key concept.

▷ Apply What You've Learned

Review

21.1 Phylogenetic trees represent evolutionary relationships.

21.1 Phylogenies enable biologists to compare organisms and make predictions and inferences based on similarities and differences in traits.

21.3 Biologists use phylogenetic trees to investigate living organisms, explore instances of convergent evolution, and reconstruct ancestral states.

Original Paper: Scaduto, D. I., J. M. Brown, W. C. Haaland, D. J. Zwickl, D. M. Hillis, and M. L. Metzker. 2010. Source identification in two criminal cases using phylogenetic analysis of HIV-1 DNA sequences. *Proceedings of the National Academy of Sciences USA* 107:21242–21247.

Phylogenetic trees are used throughout biology, but only in recent years have they become important for forensic investigations. In these cases, samples are "blinded" to the investigators by assigning numbers to each sample, rather than using people's names. Only after the conclusions are finalized do other investigators decode the numbers to reveal the results.

A recent criminal case in Texas charged a defendant with knowingly and intentionally infecting a series of women with HIV. A phylogenetic analysis was used to demonstrate that the defendant transmitted HIV to his victims. (Other evidence was needed to prove knowledge and intent.) In this case, sequences of HIV isolated from the victims and the defendant, together with the closest sequences from an HIV database (the outgroup), were compared and used to construct a phylogenetic tree of the viruses. Viruses from each individual in the case are colored alike on the tree to the right. The labels are the codes for the individuals in the case. All of the individuals labeled CC01–CC08 are known to have engaged in sex; they represent an epidemiological cluster.

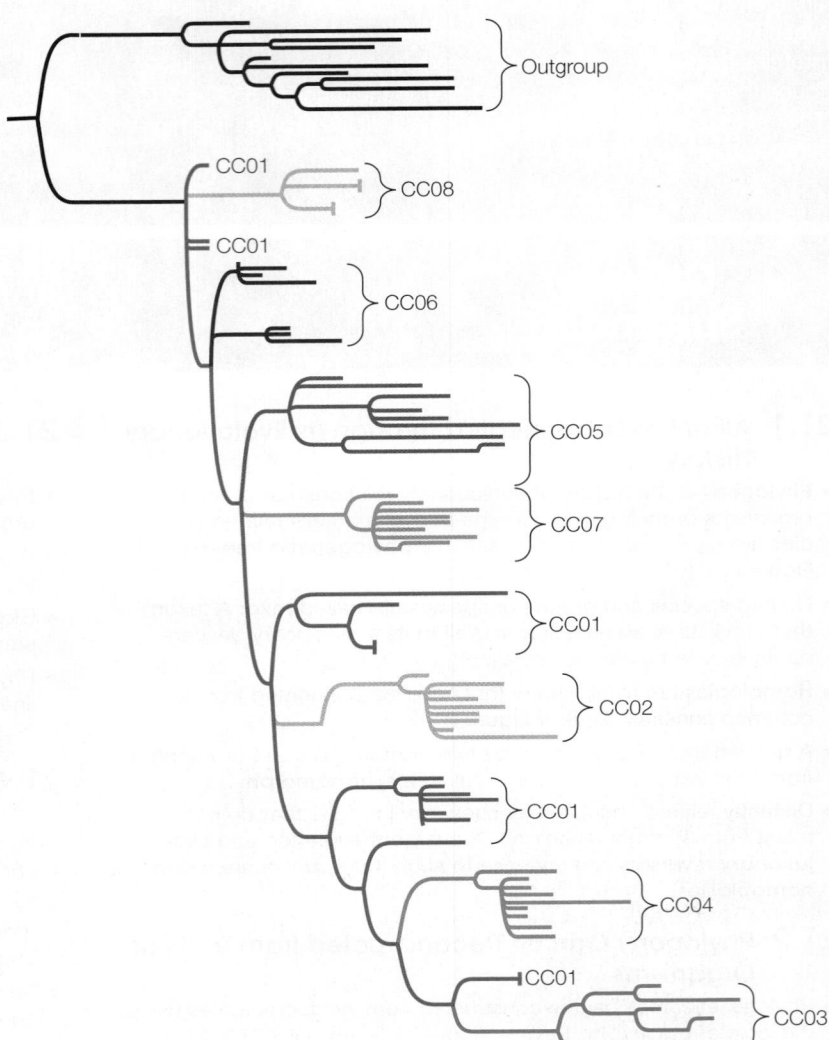

Questions

1. Which of the individuals labeled in the tree is consistent with being the source of this infection cluster? Why?

2. Why is the tree inconsistent with any of the other individuals being the source of infection within this cluster?

3. What was the purpose of including an outgroup made up of individuals that were outside the epidemiological cluster?

Speciation

This composite photograph shows a few of the nearly 1,000 species of haplochromine cichlids that are endemic to Lake Malawi, all of which are from a single founder species.

investigatinglife

Rapid Speciation in African Lake Cichlids

Not quite 2 million years ago, a tectonic split in the Great Rift Valley of East Africa led to the formation of Lake Malawi, which lies between the modern countries of Malawi, Tanzania, and Mozambique. A few fish species entered the new lake, including a type known as a haplochromine cichlid. Today the descendants of this early invader include nearly 1,000 species of haplochromine cichlids. All of them are endemic to Lake Malawi—they are found nowhere else. This vast array of cichlid species makes this the most diverse lake in the world in terms of its fish fauna. How did so many different species arise from a single ancestral species in less than 2 million years?

By studying the history and timing of speciation events in Lake Malawi, biologists have pieced together some of the processes that led to so many cichlid species. The earliest haplochromine cichlids to enter the new lake encountered diverse habitats in Lake Malawi, as some shores were rocky and others were sandy. Cichlid populations quickly adapted to these distinct habitat types. Fish in rocky habitats adapted to breeding and living in rocky conditions, and those in sandy habitats evolved specializations for life over sand. These changes resulted in an early speciation event.

Within each of these major habitat types, there were numerous opportunities for diet specialization. Various populations of cichlids became rock scrapers, bottom feeders, fish predators, scale biters, pelagic zooplankton eaters, or plant specialists. Each of these feeding specializations requires a different mouth morphology. The offspring of fish that bred with fish of similar morphology were more likely to survive than were fish with two very different parents. These differences in fitness led to the formation of many more new species, each adapted to a different feeding mode.

The Lake Malawi cichlids continued to diverge and form new species. Male cichlids compete for the attention of females through their bright body colors. Diversification of the body colors of males, and of the preferences of females for different body colors, led to many more new species of cichlids, each isolated from the other by its sexual preferences. Now biologists are studying the genomes of these Lake Malawi cichlids to understand the details of the genetic changes that have given rise to so many species over so little time.

Q|A If speciation typically takes thousands to millions of years, how do biologists conduct

22.1 Species Are Reproductively Isolated Lineages on the Tree of Life

Biological diversity does not vary in a smooth, continuous way. People have long recognized groups of similar organisms that mate with one another, and they have noticed that there are usually distinct morphological breaks between these groups. Groups of organisms that mate with one another are commonly called **species** (note that this is both the plural and singular form of the word). Species are the result of the process of **speciation**: the divergence of biological lineages and the emergence of reproductive isolation between lineages.

focus your learning

- Biologists approach the concept of species differently, depending on the focus of their research.
- The various species concepts are connected by the importance of reproductive isolation in understanding the origin of species.

Although "species" is a useful and common term, its usage varies among biologists who are interested in different aspects of speciation. Different biologists think about species differently because they ask different questions: How can we recognize and identify species? How do new species arise? How do different species remain separate? Why do rates of speciation differ among groups of organisms? In answering these questions, biologists focus on different attributes of species, leading to several different ways of thinking about what species are and how they form. Most of the various **species concepts** proposed by biologists are simply different ways of approaching the question "What are species?" Let's compare three major classes of species concepts to contrast the way that biologists think about species.

We can recognize many species by their appearance

Someone who is knowledgeable about a group of organisms, such as birds or flowering plants, can usually distinguish the different species found in a particular area simply by looking at them. Standard field guides to birds, mammals, insects, and wildflowers are possible only because many species change little in appearance over large geographic distances (**Figure 22.1A**).

More than 250 years ago, Carolus Linnaeus developed the system of binomial nomenclature by which species are named today (see Key Concept 21.4). Linnaeus described and named thousands of species, but because he knew nothing about the genetics or the mating behavior of the organisms he was naming, he classified them on the basis of their appearance alone. In other words, Linnaeus used a **morphological species concept**, a construct that assumes that a species comprises individuals that "look alike" and that individuals that do not look alike belong to different species. Although Linnaeus could not have known it, the members of most of the groups he classified as species look alike because they share many alleles of the genes that code for their morphological features.

Using morphology to define species has limitations. Members of the same species do not always look alike. For example, males, females, and young individuals do not always resemble one another closely (**Figure 22.1B**). Furthermore, morphology is of little use in the case of cryptic species—instances in which two or more species are morphologically indistinguishable but do not interbreed (**Figure 22.2**). Biologists therefore cannot rely on appearance alone in determining whether individual organisms are members of the same or different species. Today, biologists use several additional types of information—especially behavioral and genetic data—to differentiate species.

Reproductive isolation is key

The most important factor in the divergence of sexually reproducing lineages from one another is the evolution of **reproductive isolation**, a state in which two groups of organisms can no longer exchange genes. If individuals of group A mate and reproduce only with one another, then group A constitutes a distinct species within which genes recombine. In other words, group A is an independent evolutionary lineage—a separate branch on the tree of life.

Evolutionary biologist Ernst Mayr recognized the importance of reproductive isolation in maintaining species, and so he proposed the **biological species concept**: *"Species are groups of actually or potentially interbreeding natural populations which are reproductively*

(A)

Lophodytes cucullatus
Male, British Columbia

Lophodytes cucullatus
Male, New Mexico

(B)

Lophodytes cucullatus
Female

Figure 22.1 Not All Members of the Same Species Look Alike **(A)** It is easy to identify these two male hooded mergansers as members of the same species, even though they were photographed thousands of miles apart in British Columbia and New Mexico,

respectively. Despite their geographic separation, the two individuals are morphologically very similar. **(B)** Hooded mergansers are sexually dimorphic, which means the female's appearance is quite different from that of the male.

(A) *Hyla versicolor*

(B) *Hyla chrysoscelis*

Figure 22.2 Cryptic Species Look Alike but Do Not Interbreed
These two species of gray tree frogs cannot be distinguished by their external morphology, but they do not interbreed even when they occupy the same geographic range. **(A)** *Hyla versicolor* is a tetraploid species (four sets of chromosomes), whereas **(B)** *H. chrysoscelis* is diploid (two sets of chromosomes). And although they look alike, the males have distinctive mating calls; female frogs recognize and mate with males of their own species based on these calls.

isolated from other such groups." The phrase "actually or potentially" is an important element of this definition. "Actually" says that the individuals live in the same area and interbreed with one another. "Potentially" says that even though the individuals do not live in the same area, and therefore do not interbreed, other information suggests that they *would* do so if they were able to get together. This widely used species concept does not apply to organisms that reproduce asexually, and it is limited to a single point in evolutionary time.

The lineage approach takes a long-term view

Evolutionary biologists often think of species as branches on the tree of life. This idea can be termed a **lineage species concept**. In this framework for thinking about species, one species splits into two descendant species, which thereafter evolve as distinct lineages. A lineage species concept allows biologists to consider species over evolutionary time.

A **lineage** is an ancestor–descendant series of populations followed over time. Each species has a history that starts with a speciation event by which one lineage on the tree is split into two, and ends either at extinction or at another speciation event, at which time the species produces two daughter species. The process of lineage splitting may be gradual, taking thousands of generations to complete. At the other extreme, an ancestral lineage may be split in two within a few generations (as happens with polyploidy, which we'll discuss in Key Concept 22.3). The gradual nature of some splitting events means that at a single point in time, the final outcome of the process may not be clear. In these cases, it may be difficult to predict whether the incipient species will continue to diverge and become fully isolated from their sibling species, or if they will merge again in the future.

The different species concepts are not mutually exclusive

Many named variants of these three major classes of species concepts exist. These various concepts are not incompatible;

they simply emphasize different aspects of species or speciation. The morphological species concept emphasizes the practical aspects of recognizing species, although it sometimes results in underestimation or overestimation of the actual number of species. Mayr's biological species concept emphasizes that reproductive isolation is what allows sexual species to evolve independently of one another. The lineage species concept embraces the idea that sexual species are maintained by reproductive isolation, but extends the concept of a species as a lineage over evolutionary time. The species-as-lineage concept also accommodates species that reproduce asexually.

Virtually all species exhibit some degree of genetic recombination among individuals, even if recombination events are relatively rare. Significant reproductive isolation between species is therefore necessary for lineages to remain distinct over evolutionary time. Furthermore, reproductive isolation is responsible for the morphological distinctiveness of most species, because mutations that result in morphological changes cannot spread between reproductively isolated species. Therefore no matter which species concept we emphasize, the evolution of reproductive isolation is important for understanding the origin of species.

22.1 recap

Species are distinct lineages on the tree of life. Speciation is usually a gradual process as one lineage divides into two. Over time, lineages of sexual species remain distinct from one another because they have become reproductively isolated.

learning outcomes

You should be able to:

• Explain how different species concepts may be useful for different purposes.

• Describe the importance of reproductive isolation to understanding speciation.

1. Why do different biologists emphasize different attributes of species in formulating species concepts?

2. What makes reproductive isolation such an important component of each of the species concepts discussed here?

3. Why is the biological species concept not applicable to asexually reproducing organisms? Do you think this limits its applicability?

Although Charles Darwin titled his groundbreaking book *On the Origin of Species*, in fact it included very little about speciation as we understand it today. Darwin devoted most of his attention to demonstrating that individual species are altered over time by natural selection. The remaining sections of this chapter will discuss the many aspects of speciation that biologists have learned about since Darwin's time.

key concept 22.2 Speciation Is a Natural Consequence of Population Subdivision

Not all evolutionary changes result in new species. A single lineage may change over time without giving rise to a new species. Speciation requires the interruption of gene flow within a species whose members formerly exchanged genes. But if a genetic change prevents reproduction between individuals of a species, how can such a change spread through a species in the first place?

focus your learning

- The Dobzhansky–Muller model explains how reproductive isolation results from the accumulation of genetic incompatibilities following a splitting of the ancestral population (interruption of gene flow).
- Greater genetic differences between species are correlated with increased reproductive isolation.

Incompatibilities between genes can produce reproductive isolation

If a new allele that causes reproductive incompatibility arises in a population, it cannot spread through the population because no other individuals will be reproductively compatible with the individual that carries the new allele. So how can one reproductively cohesive lineage ever split into two reproductively isolated species? Several early geneticists, including Theodosius Dobzhansky and Hermann Joseph Muller, developed a genetic model to explain this apparent conundrum (**Figure 22.3**).

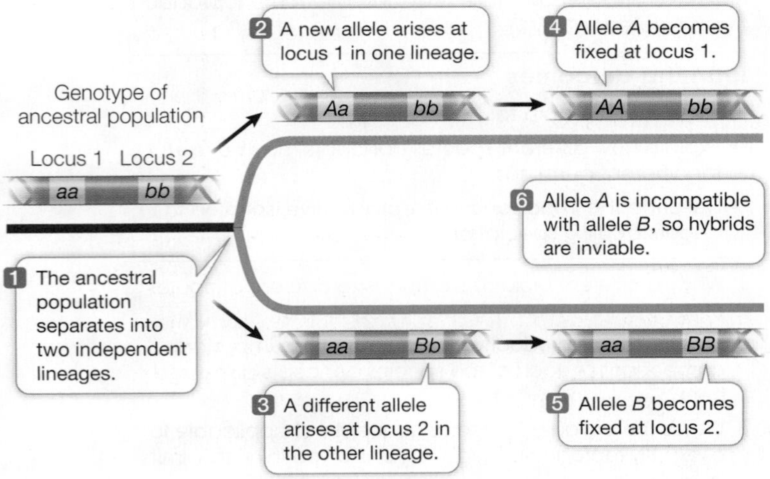

The Dobzhansky–Muller model is quite simple. First, assume that a single ancestral population is subdivided into two daughter populations (by the formation of a new mountain range, for instance), which then evolve as independent lineages. In one of the descendant lineages, a new allele (*A*) arises and becomes fixed (see Figure 22.3). In the other lineage, another new allele (*B*) becomes fixed *at a different gene locus*. Neither new allele at either locus results in any loss of reproductive compatibility. However, the two new forms of these two different genes have never occurred together in the same individual or population. Recall that the products of many genes must work together in an organism. It is possible that the new protein forms encoded by the two new alleles will not be compatible with each other. If individuals from the two lineages come back together after these genetic changes, they may still be able to interbreed. However, the hybrid offspring may have a new combination of genes that is functionally inferior, or even lethal. This will not happen with all new combinations of genes, but over time, isolated lineages will accumulate many allele differences at many gene loci. Some combinations of these differentiated genes will not function well together in hybrids. Thus genetic incompatibility between the two isolated lineages will develop over time.

Many empirical and experimental examples support the Dobzhansky–Muller model. This model works not only for pairs of individual genes but also for some kinds of chromosomal rearrangements. Bats of the genus *Rhogeessa*, for example, exhibit considerable variation in centric fusions of their chromosomes. The chromosomes of the various species contain the same basic chromosomal arms, but in some species two acrocentric (one-armed) chromosomes have fused at the centromere to form larger, metacentric (two-armed) chromosomes. A polymorphism in centric fusion causes few, if any, problems in meiosis because the respective chromosomes can still align and assort normally. Therefore a given centric fusion can become fixed in a lineage. However, if a *different* centric fusion becomes fixed in a second lineage, then hybrids between individuals of each lineage will not be able to produce normal gametes in meiosis (**Figure 22.4**). Most of the closely related species of *Rhogeessa* display different combinations of these centric fusions and are thereby reproductively isolated from one another.

Reproductive isolation develops with increasing genetic divergence

As pairs of species diverge genetically, they become increasingly reproductively isolated (**Figure 22.5**). Both the rate at which

Figure 22.3 The Dobzhansky–Muller Model In this simple two-locus version of the model, two lineages from the same ancestral population become physically separated from each other and evolve independently. A new allele becomes fixed in each descendant lineage, but at a different locus. Neither of the new alleles is incompatible with the ancestral alleles, but the two new alleles in the two different genes are incompatible with each other. Thus the two descendant lineages are reproductively incompatible.

Activity 22.1 Speciation Simulation
www.Life11e.com/ac22.1

Figure 22.3 content:

Genotype of ancestral population

Locus 1   Locus 2

aa   bb

1 The ancestral population separates into two independent lineages.

2 A new allele arises at locus 1 in one lineage.

Aa   bb

4 Allele *A* becomes fixed at locus 1.

AA   bb

3 A different allele arises at locus 2 in the other lineage.

aa   Bb

5 Allele *B* becomes fixed at locus 2.

aa   BB

6 Allele *A* is incompatible with allele *B*, so hybrids are inviable.

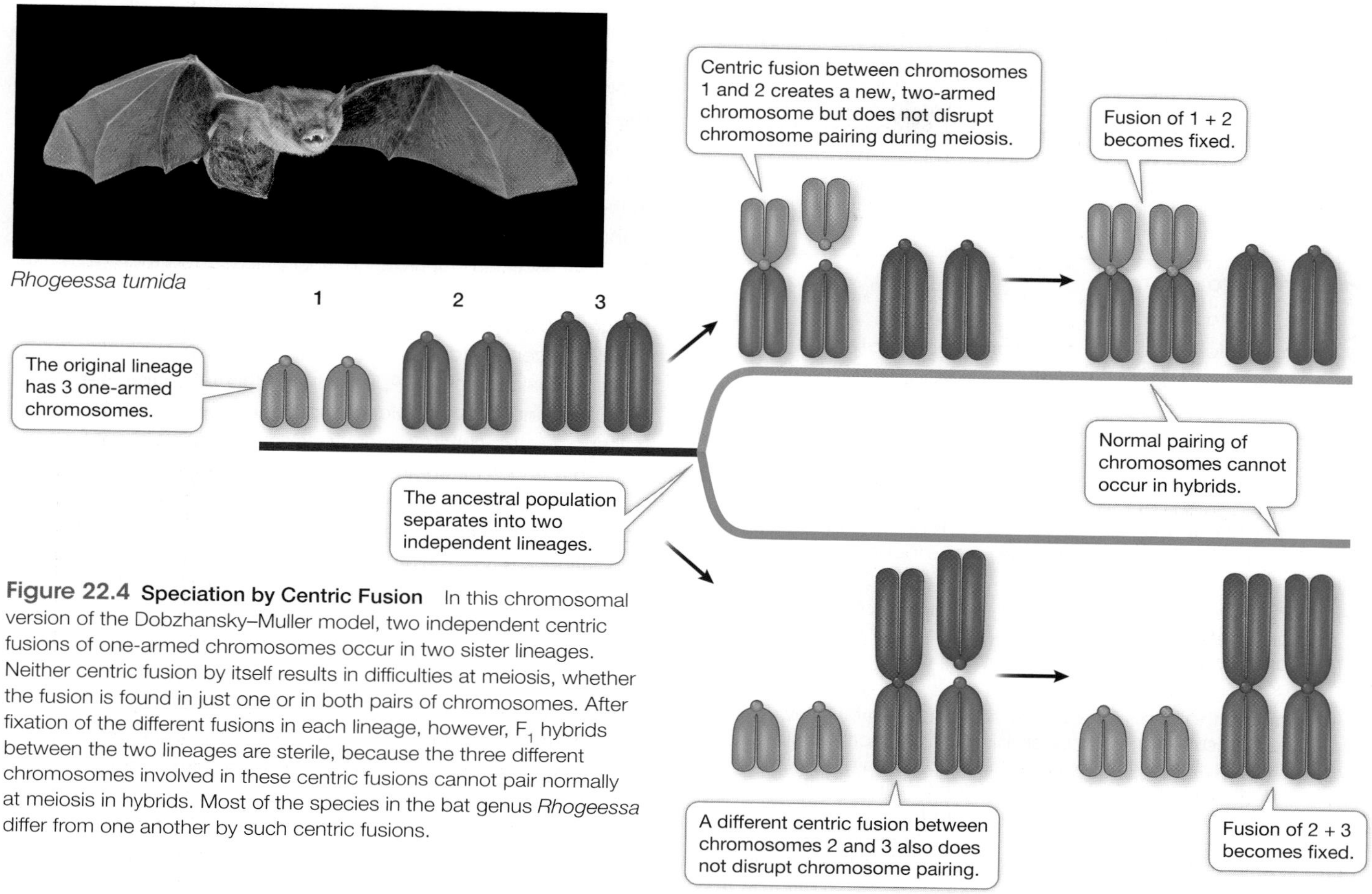

Rhogeessa tumida

The original lineage has 3 one-armed chromosomes.

1   2   3

The ancestral population separates into two independent lineages.

Centric fusion between chromosomes 1 and 2 creates a new, two-armed chromosome but does not disrupt chromosome pairing during meiosis.

Fusion of 1 + 2 becomes fixed.

Normal pairing of chromosomes cannot occur in hybrids.

A different centric fusion between chromosomes 2 and 3 also does not disrupt chromosome pairing.

Fusion of 2 + 3 becomes fixed.

Figure 22.4 Speciation by Centric Fusion   In this chromosomal version of the Dobzhansky–Muller model, two independent centric fusions of one-armed chromosomes occur in two sister lineages. Neither centric fusion by itself results in difficulties at meiosis, whether the fusion is found in just one or in both pairs of chromosomes. After fixation of the different fusions in each lineage, however, F_1 hybrids between the two lineages are sterile, because the three different chromosomes involved in these centric fusions cannot pair normally at meiosis in hybrids. Most of the species in the bat genus *Rhogeessa* differ from one another by such centric fusions.

reproductive isolation develops and the mechanisms that produce it vary from group to group. Reproductive incompatibility has been shown to develop gradually in many groups of plants, animals, and fungi, reflecting the slow pace at which incompatible genes accumulate in each lineage. In some cases, complete reproductive isolation may take millions of years. In other cases (as with the chromosomal fusions of *Rhogeessa* described above), reproductive isolation can develop over just a few generations.

Partial reproductive isolation has evolved in many strains of plants that have been artificially isolated by humans. In 1835, Thomas Drummond collected seeds of a newly discovered species of phlox in Texas and distributed them to nurseries in Europe. The European nurseries established more than 200 true-breeding strains of Drummond's phlox that differed in flower size, flower color, and plant growth form. The breeders did not select directly for reproductive incompatibility between strains, but in subsequent experiments in which strains were crossed and seed production was measured and compared, biologists found that reproductive compatibility between strains had been reduced by 14 to 50 percent, depending on the cross—even though the strains had been isolated from one another for less than two centuries.

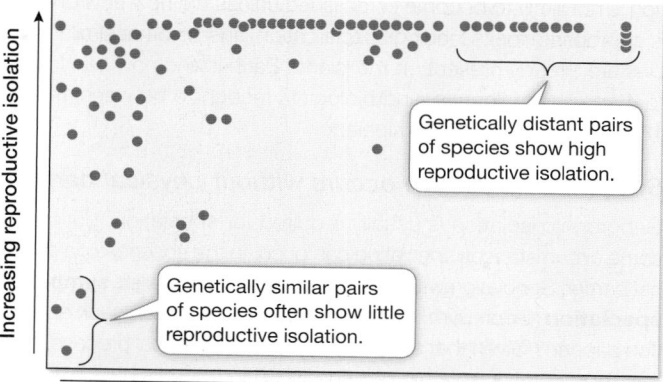

Genetically distant pairs of species show high reproductive isolation.

Genetically similar pairs of species often show little reproductive isolation.

Increasing reproductive isolation

Increasing genetic divergence

Figure 22.5 Reproductive Isolation Increases with Genetic Divergence   Among pairs of *Drosophila* species, the more the species differ genetically (x axis), the greater their reproductive isolation from each other (y axis). Each dot represents a comparison of one species pair. Such positive relationships between genetic distance and reproductive isolation have been observed in many groups of plants, animals, and fungi.

Q: Why might some genetically similar species pairs exhibit high levels of reproductive isolation?

22.2 recap

When two parts of a population become isolated from each other by some barrier to gene flow, they begin to diverge genetically. The Dobzhansky–Muller model describes how new alleles or chromosomal arrangements that arise in the two descendent lineages can lead to genetic incompatibility, and hence reproductive isolation, of the two lineages.

learning outcomes

You should be able to:

- Make inferences and draw conclusions based on the Dobzhansky–Muller model.
- Evaluate data from two populations to determine if they are reproductively isolated.

1. The Dobzhansky–Muller model suggests that divergence among alleles at *different* gene loci leads to genetic incompatibility between species. Why is genetic incompatibility between two alleles at the *same* locus considered less likely?

2. Why do some combinations of chromosomal centric fusions cause problems in meiosis? Can you diagram what would happen at meiosis in a hybrid of the divergent lineages shown in Figure 22.4?

3. Assume that the reproductive isolation seen in *Phlox* strains results from lethal combinations of incompatible alleles at several loci among the various strains. Given this assumption, why might the reproductive isolation seen among these strains be partial rather than complete?

We have now seen how the splitting of an ancestral population leads to genetic divergence and reproductive incompatibility in the two descendant lineages. Next we will consider ways in which the descendant lineages could have become separated in the first place.

key concept 22.3 Speciation May Occur through Geographic Isolation or in Sympatry

Many biologists who study speciation have concentrated on geographic processes that can result in the division of an ancestral species. Splitting of the geographic range of a species is one obvious way of achieving such a division, but it is not the only way.

focus your learning

- Speciation can be allopatric (resulting from the physical division of a population) or sympatric (due to disruptive selection or polyploidy).

Physical barriers give rise to allopatric speciation

Speciation that results when a population is divided by a physical barrier is known as **allopatric speciation** (Greek *allos*, "other," + *patria*, "homeland"). Allopatric speciation is thought to be the dominant mode of speciation in most groups of organisms. The physical barrier that divides the range of a species may be a body of water or a mountain range for terrestrial organisms, or dry land for aquatic

organisms—in other words, any type of habitat that is inhospitable to the species. Such barriers can form when continents drift, sea levels rise or fall, glaciers advance or retreat, or climates change. The populations separated by such barriers are often, but not always, initially large. The lineages that descend from these founding populations evolve differences for a variety of reasons, including mutation, genetic drift, and adaptation to different environments in the two areas. As a result, many pairs of closely related **sister species**—species that are each other's closest relatives—may exist on either side of the geographic barrier. An example of a physical geographic barrier that produced many pairs of sister species was the Pleistocene glaciation that isolated freshwater streams in the eastern highlands of the Appalachian Mountains from streams in the Ozark and Ouachita Mountains (**Focus: Key Figure 22.6**). This splitting event resulted in many parallel speciation events among isolated lineages of stream-dwelling organisms.

Allopatric speciation may also result when some members of a population cross an existing barrier and establish a new, isolated population. Many of the more than 800 species of *Drosophila* found in the Hawaiian Islands are restricted to a single island. We know that these species are the descendants of new populations founded by individuals dispersing among the islands when we find that the closest relative of a species on one island is a species on a neighboring island rather than a species on the same island. Biologists who have studied the chromosomes of these fruit flies estimate that speciation in this group of *Drosophila* has resulted from at least 45 such founder events (**Figure 22.7**).

The species of finches found in the islands of the Galápagos archipelago, some 1,000 km off the coast of Ecuador, are one of the most famous examples of allopatric speciation. Darwin's finches (as they are usually called, because Darwin was the first scientist to study them) arose in the Galápagos from a single South American finch species that colonized the islands. Today the Galápagos species differ strikingly not only from their closest mainland relative, but also from one another (**Figure 22.8**). The islands are sufficiently far apart that the birds move among them only infrequently. In addition, environmental conditions differ widely from island to island. Some islands are relatively flat and arid; others have forested mountain slopes. Sister lineages on different islands have diverged over hundreds of thousands of years, and several feeding specializations have arisen on different islands with different environments. Although finches occasionally fly between islands, an immigrant finch population is not likely to become established unless the new environment is appropriate for its feeding specialization, and no other similar species are already present on the island. Each island now has from 1 to 4 species of finches, and biologists recognize between 14 and 18 species across the archipelago.

Sympatric speciation occurs without physical barriers

Geographic isolation is usually required for speciation, but under some circumstances speciation can occur in the absence of a physical barrier. Speciation without physical isolation is called **sympatric speciation** (Greek *sym*, "together with"). But how can such speciation happen? Given that speciation is usually a gradual process, how can reproductive isolation develop when individuals have frequent opportunities to mate with one another?

focus: key figure

(A) Distribution of freshwater stream fish in the Pliocene

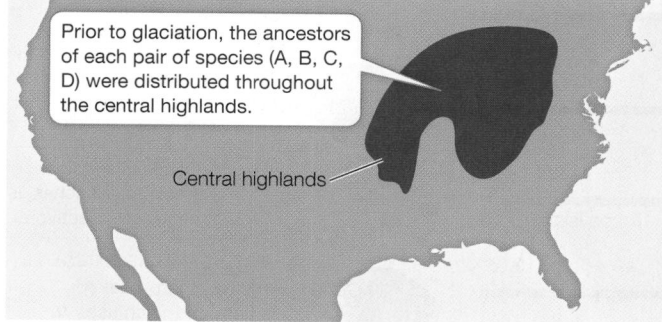

Prior to glaciation, the ancestors of each pair of species (A, B, C, D) were distributed throughout the central highlands.

Central highlands

(B) Distribution of freshwater stream fish in the Pleistocene

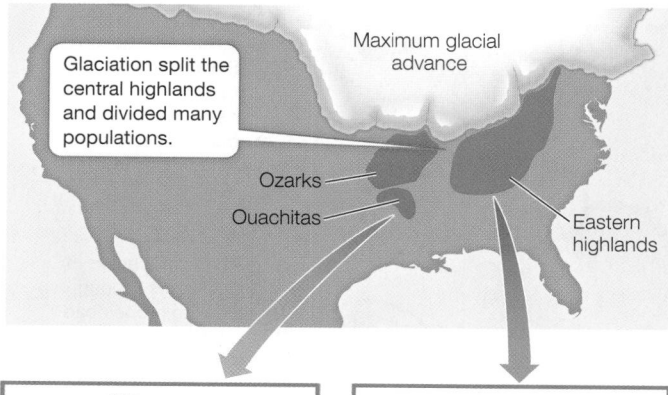

Maximum glacial advance

Glaciation split the central highlands and divided many populations.

Ozarks
Ouachitas
Eastern highlands

A₁ Missouri saddled darter
Etheostoma tetrazonum

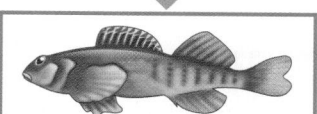

A₂ Variegated darter
E. variatum

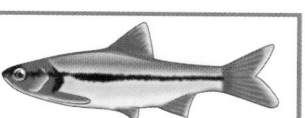

B₁ Bleeding shiner
Luxilus zonatus

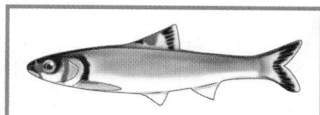

B₂ Warpaint shiner
L. coccogenis

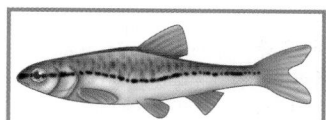

C₁ Ozark minnow
Notropis nubilus

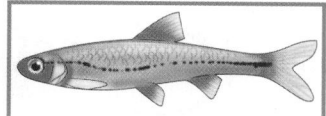

C₂ Tennessee shiner
N. leuciodus

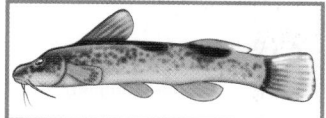

D₁ Ozark madtom
Noturus albater

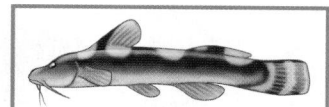

D₂ Elegant madtom
N. elegans

Figure 22.6 Allopatric Speciation Allopatric speciation may result when an ancestral population is divided into two separate populations by a physical barrier and those populations then diverge. **(A)** Many species of freshwater stream fishes were distributed throughout the central highlands of North America in the Pliocene epoch (about 5.3–2.6 million years ago). **(B)** During the Pleistocene (about 2.6 million years ago–10,000 years ago), glaciers advanced and isolated fish populations in the Ozark and Ouachita Mountains to the west from fish populations in the highlands of the Appalachian Mountains to the east. Numerous species diverged as a result of this separation, including the ancestors of the four pairs of sister species shown here.

Q: After the retreat of the glaciers, why did the fish species in the Ozarks and Ouachitas remain reproductively isolated from those in the Appalachians to the east?

 Animation 22.1 **Speciation Mechanisms**
www.Life11e.com/a22.1

DISRUPTIVE SELECTION Sympatric speciation may occur with some forms of disruptive selection (see Key Concept 20.4) in which individuals with certain genotypes have a preference for distinct microhabitats where mating takes place. For example, sympatric speciation via disruptive selection appears to be taking place in the apple maggot fly (*Rhagoletis pomonella*) of eastern North America. Until the mid-1800s, *Rhagoletis* flies courted, mated, and deposited their eggs only on hawthorn fruits. About 150 years ago, some flies began to lay their eggs on apples, which European immigrants had introduced into eastern North

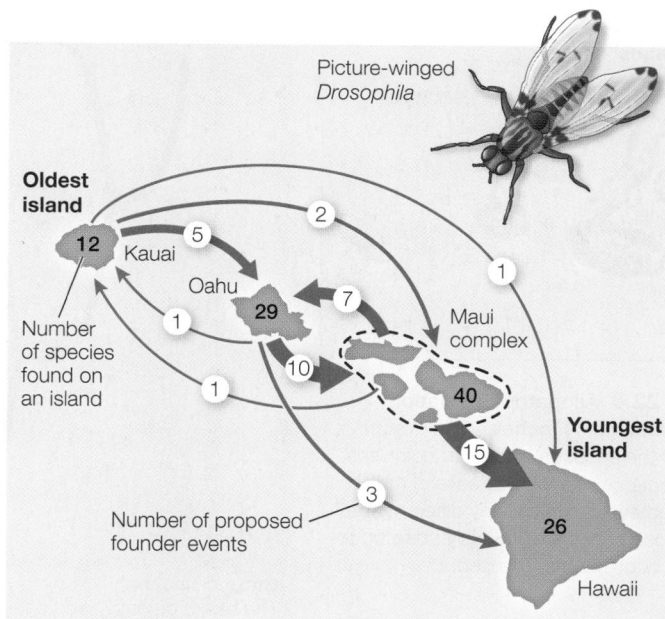

Figure 22.7 Founder Events Lead to Allopatric Speciation The large number of species of picture-winged *Drosophila* in the Hawaiian Islands is the result of founder events: the founding of new populations by individuals dispersing among the islands. The islands, which were formed in sequence as Earth's crust moved over a volcanic "hot spot," vary in age.

 Animation 22.2 **Founder Events and Allopatric Speciation** www.Life11e.com/a22.2

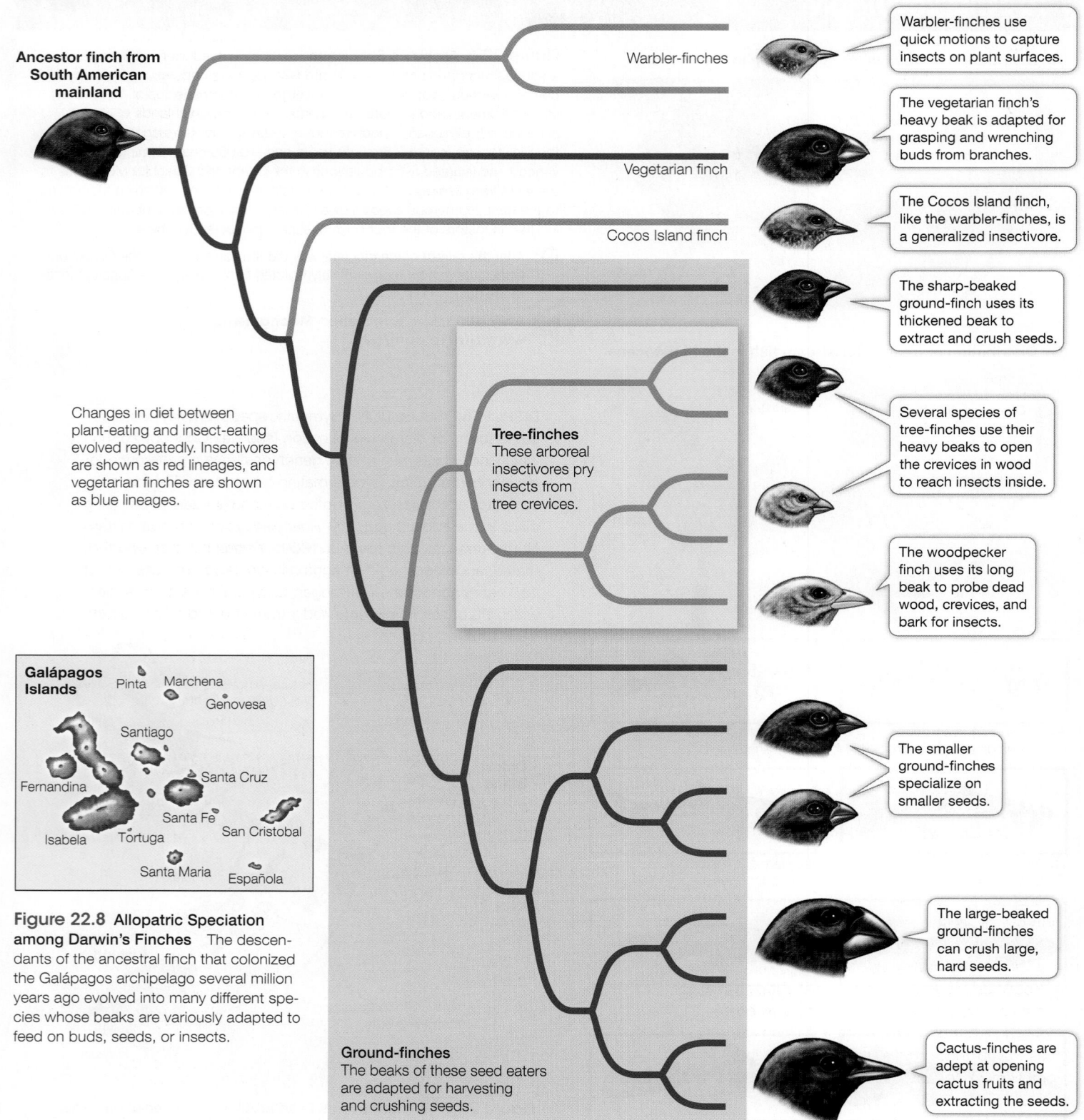

Ancestor finch from South American mainland

Warbler-finches

> Warbler-finches use quick motions to capture insects on plant surfaces.

Vegetarian finch

> The vegetarian finch's heavy beak is adapted for grasping and wrenching buds from branches.

Cocos Island finch

> The Cocos Island finch, like the warbler-finches, is a generalized insectivore.

> The sharp-beaked ground-finch uses its thickened beak to extract and crush seeds.

Changes in diet between plant-eating and insect-eating evolved repeatedly. Insectivores are shown as red lineages, and vegetarian finches are shown as blue lineages.

Tree-finches These arboreal insectivores pry insects from tree crevices.

> Several species of tree-finches use their heavy beaks to open the crevices in wood to reach insects inside.

> The woodpecker finch uses its long beak to probe dead wood, crevices, and bark for insects.

Galápagos Islands
Pinta
Marchena
Genovesa
Santiago
Fernandina
Santa Cruz
Santa Fe
Isabela
Tortuga
San Cristobal
Santa Maria
Española

> The smaller ground-finches specialize on smaller seeds.

Figure 22.8 Allopatric Speciation among Darwin's Finches The descendants of the ancestral finch that colonized the Galápagos archipelago several million years ago evolved into many different species whose beaks are variously adapted to feed on buds, seeds, or insects.

> The large-beaked ground-finches can crush large, hard seeds.

Ground-finches The beaks of these seed eaters are adapted for harvesting and crushing seeds.

> Cactus-finches are adept at opening cactus fruits and extracting the seeds.

America. Apple trees are closely related to hawthorns, but the smell of the fruits differs, and the apple fruits appear earlier than those of hawthorns. Some early-emerging female *Rhagoletis* laid their eggs on apples, and over time, a genetic preference for the smell of apples evolved among early-emerging insects. When the offspring of these flies sought out apple trees for mating and egg deposition,

they mated with other flies reared on apples, which shared the same preferences.

Today the two groups of *Rhagoletis pomonella* in the eastern United States appear to be on the way to becoming distinct species. One group mates and lays eggs primarily on hawthorn fruits, the other on apples. The incipient species are partially reproductively

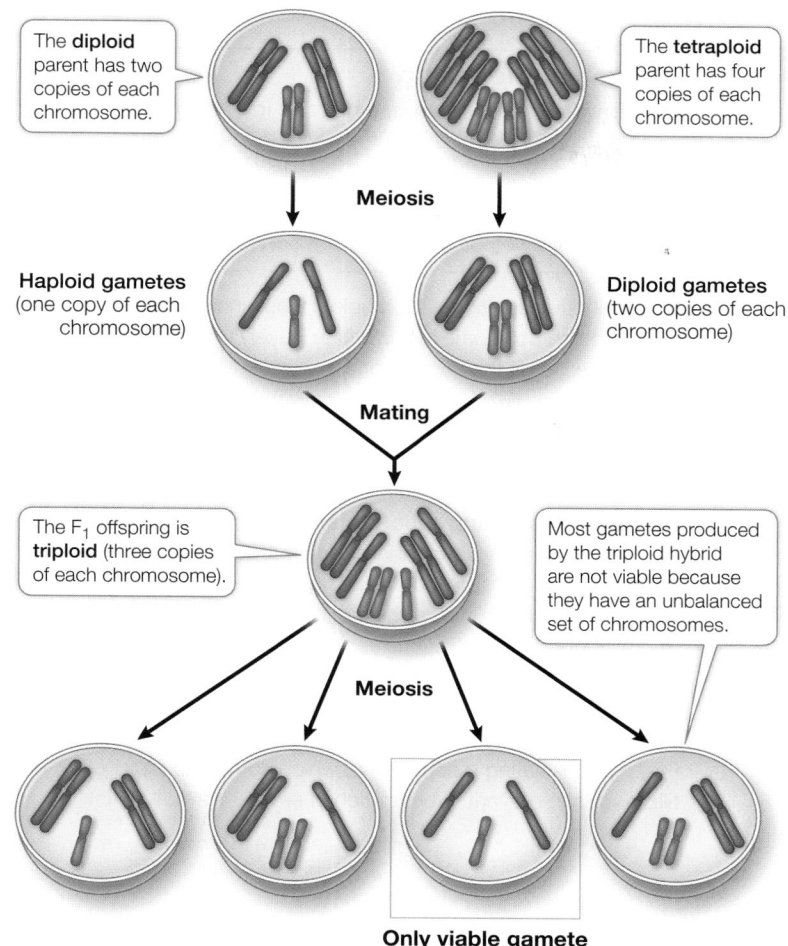

The **diploid** parent has two copies of each chromosome.

The **tetraploid** parent has four copies of each chromosome.

Meiosis

Haploid gametes (one copy of each chromosome)

Diploid gametes (two copies of each chromosome)

Mating

The F₁ offspring is **triploid** (three copies of each chromosome).

Most gametes produced by the triploid hybrid are not viable because they have an unbalanced set of chromosomes.

Meiosis

Only viable gamete

Figure 22.9 Tetraploids Are Reproductively Isolated from Their Diploid Ancestors   Even if the triploid offspring of a diploid and a tetraploid parent survives and reaches sexual maturity, most of the gametes it produces have aneuploid (unbalanced) numbers of chromosomes. Such triploid individuals are effectively sterile. (For simplicity, the diagram shows only three homologous chromosomes. Most species have many more chromosomes, so viable gametes are extremely rare.)

Allopolyploids may be produced when individuals of two different (but closely related) species interbreed. Such hybridization often disrupts normal meiosis, which can result in chromosomal doubling. Allopolyploids are often fertile because each of the chromosomes has a nearly identical partner with which to pair during meiosis.

Speciation by polyploidy has been particularly important in the evolution of plants, although it has contributed to speciation in animals as well (such as the tree frogs in Figure 22.2). Botanists estimate that about 70 percent of flowering plant species and 95 percent of fern species are the result of recent polyploidization. Some of these species arose from hybridization between two species followed by chromosomal duplication and self-fertilization. Other species diverged from polyploid ancestors, so that the new species shared their ancestors' duplicated sets of chromosomes. New species arise by polyploidy more easily among plants than among animals because plants of many species can reproduce by self-fertilization. In addition, if polyploidy arises in several offspring of a single parent, the siblings can fertilize one another.

isolated because they mate primarily with individuals raised on the same fruit and because they emerge from their pupae at different times of the year. In addition, the apple-feeding flies now grow more rapidly on apples than they originally did. Sympatric speciation that arises from such host-plant specificity may be widespread among insects, many of which feed only on a single plant species.

POLYPLOIDY   The most common means of sympatric speciation is **polyploidy**, or the duplication of sets of chromosomes within individuals. Polyploidy can arise either from chromosome duplication in a single species (**autopolyploidy**) or from the combining of the chromosomes of two different species (**allopolyploidy**).

An autopolyploid individual originates when, for example, two accidentally unreduced diploid gametes (with two sets of chromosomes) combine to form a tetraploid individual (with four sets of chromosomes). Tetraploid and diploid individuals of the same species are reproductively isolated because their hybrid offspring are triploid. Even if these offspring survive, they are usually sterile; they cannot produce normal gametes because their chromosomes do not segregate evenly during meiosis (**Figure 22.9**). So a tetraploid individual usually cannot produce viable offspring by mating with a diploid individual—but it *can* do so if it self-fertilizes or mates with another tetraploid. Thus polyploidy can result in complete reproductive isolation in two generations—an important exception to the general rule that speciation is a gradual process.

22.3 recap

Allopatric speciation results from the separation of populations by geographic barriers; it is the dominant mode of speciation among most groups of organisms. Sympatric speciation may result from disruptive selection that results in ecological isolation, but polyploidy is the most common cause of sympatric speciation among plants.

learning outcomes

You should be able to:

- Contrast the steps involved in different modes of speciation.
- Make inferences based on an understanding of the process of allotropic speciation.
- Diagram the relationship between geographic isolation and speciation and interpret the data to make and support predictions.

1. Explain how speciation via polyploidy can happen in only two generations.

2. If allopatric speciation is the most prevalent mode of speciation, what do you predict about the geographic distributions of many closely related species? Does your answer differ for species that are sedentary versus highly mobile?

(continued)

22.3 recap (continued)

3. The species of Darwin's finches shown in the phylogeny in Figure 22.8 have all evolved on islands of the Galápagos archipelago within the past 3 million years. Molecular clock analysis (see Key Concept 21.3) has been used to determine the dates of the various speciation events in that phylogeny. Geological techniques for dating rock samples (see Key Concept 24.1) have been used to determine the ages of the various Galápagos islands. The table shows the number of species of Darwin's finches and the number of islands that have existed in the archipelago at several times during the past 4 million years.

Time (mya)	Number of islands	Number of finch species
0.25	18	14
0.50	18	9
0.75	9	7
1.00	6	5
2.00	4	3
3.00	4	1
4.00	3	0

a. Plot the number of species of Darwin's finches and the number of islands in the Galápagos archipelago (dependent variables) against time (independent variable).

b. Are the data consistent with the hypothesis that isolation of populations on newly formed islands is related to speciation in this group of birds? Why or why not?

c. If no more islands form in the Galápagos archipelago, do you think that speciation by geographic isolation will continue to occur among Darwin's finches? Why or why not? What additional data could you collect to test your hypothesis (without waiting to see if speciation occurs)?

Most populations separated by a physical barrier become reproductively isolated only slowly and gradually. If two incipient species once again come into contact with each other, what keeps them from merging back into a single species?

key concept
22.4 Reproductive Isolation Is Reinforced When Diverging Species Come into Contact

As discussed in Key Concept 22.2, once a barrier to gene flow is established, reproductive isolation will begin to develop through genetic divergence. Over many generations, differences accumulate in the isolated lineages, reducing the probability that individuals from each lineage will mate successfully with individuals in the other when they come back into contact. In this way, reproductive isolation can evolve as a by-product of the genetic changes in the two diverging lineages.

focus your learning

- Reproductive isolation is reinforced in sympatry through selection for prezygotic mechanisms that prevent hybridization, which results from postzygotic mechanisms that reduce the fitness of hybrids.
- Hybrid zones may contain offspring of the two hybridizing species as well as offspring of hybrid individuals.

If reproductive isolation is incomplete when incipient species come back into contact, some hybridization is likely to occur. If hybrid individuals are less fit than non-hybrids, selection will favor parents that do not produce hybrid offspring. Under these conditions, selection will result in the strengthening, or **reinforcement**, of mechanisms that prevent hybridization.

Mechanisms that prevent hybridization from occurring are called **prezygotic isolating mechanisms**. Mechanisms that reduce the fitness of hybrid offspring are called **postzygotic isolating mechanisms**. Postzygotic isolating mechanisms result in selection against hybridization, which in turn leads to the reinforcement of prezygotic isolating mechanisms.

Prezygotic isolating mechanisms prevent hybridization

Prezygotic isolating mechanisms, which come into play before fertilization, can prevent hybridization in several ways.

Ophrys apifera

Bombus lucorum

Figure 22.10 Mechanical Isolation through Mimicry Many orchid species maintain reproductive isolation by means of flowers that look and smell like females of one—and only one—bee or wasp species. A male insect of the correct species must land on the flower and attempt to mate with it; only males of this particular species are physically configured to collect and transfer the orchid's pollen. The constraints of this method of pollen transfer reproductively isolate the plant from related orchid species that attract different insect pollinators. The species shown here are the two players in one such interspecific relationship; see Figure 55.13 for another example.

MECHANICAL ISOLATION   Differences in the sizes and shapes of reproductive organs may prevent the union of gametes from different species. With animals, there may be a match between the shapes of the reproductive organs of males and females of the same species, so that reproduction between individuals with mismatched reproductive structures is not physically possible. In plants, mechanical isolation may involve a pollinator. For example, some orchid species produce flowers that look and smell like the females of particular species of bee or wasp (**Figure 22.10**). When a male wasp visits and attempts to mate with the flower (thinking it is a female wasp of his species), his mating behavior results in the transfer of pollen to and from his body by appropriately configured anthers and stigmas on the flower. Insects that visit the flower but do not attempt to mate with it do not trigger the transfer of pollen between the insect and the flower.

TEMPORAL ISOLATION   Many organisms have distinct mating seasons. If two closely related species breed at different times of the year (or different times of day), they may never have an opportunity to hybridize. For example, in sympatric populations of three closely related leopard frog species, each species breeds at a different time of year (**Figure 22.11**). Although there is some overlap in the breeding seasons, the opportunities for hybridization are minimized.

BEHAVIORAL ISOLATION   Individuals may reject, or fail to recognize, individuals of other species as potential mating partners. For example, the mating calls of male frogs of related species diverge quickly (**Figure 22.12**). Female frogs respond to mating calls from males of their own species but ignore the calls of other species, even closely related ones. The evolution of female preferences for certain male coloration patterns among the cichlids of Lake Malawi, described at the opening story of this chapter, is another example of behavioral isolation.

Sometimes the mate choice of one species is mediated by the behavior of individuals of other species. For example, whether or not two plant species hybridize may depend on the food preferences of their *pollinators. The floral traits of plants, including their color and shape, can enhance reproductive isolation either by influencing which pollinators are attracted to the flowers or by altering where pollen is deposited on the bodies of pollinators. A plant whose flowers are pendant will be pollinated by an animal with different physical characteristics than will a plant in which the flowers grow upright (**Figure 22.13A and B**). Because each pollinator prefers (and is adapted to) a different type of flower, pollinators will rarely transfer pollen from one plant species to the other. Such isolation by pollinator behavior is seen in the mountains of California in two sympatric species of columbines (*Aquilegia*) that have diverged in flower color, structure, and orientation. *A. formosa* (**Figure 22.13C**) has pendant flowers with short spurs (spikelike, nectar-containing structures) and is pollinated by hummingbirds. *A. pubescens* (**Figure 22.13D**) has upright, lighter-colored flowers with long spurs and is pollinated by hawkmoths. The difference in pollinators means that these two species are effectively reproductively isolated even though they populate the same geographic range.

---

*connect the concepts   Some plants and their pollinators become so tightly adapted to each other that they develop mutually dependent relationships, as described in Key Concept 29.3.

HABITAT ISOLATION   When two closely related species evolve preferences for living or mating in different habitats, they may never come into contact during their respective mating periods. The *Rhagoletis* flies discussed in Key Concept 22.3 experienced such habitat isolation, as did the cichlid fish that first adapted to rocky or sandy habitats upon entering Lake Malawi, as described at the opening of this chapter.

GAMETIC ISOLATION   The sperm of one species may not attach to the eggs of another species because the eggs do not release the appropriate attractive chemicals, or the sperm may be unable to penetrate the egg because the two gametes are chemically incompatible. Thus even though the gametes of two species may come into contact, the gametes never fuse into a zygote.

Gametic isolation is extremely important for many aquatic species that spawn (release their gametes directly into the environment).

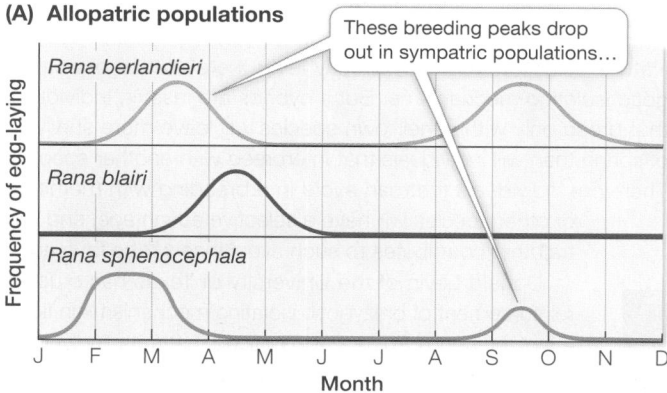

(A) Allopatric populations

These breeding peaks drop out in sympatric populations…

Rana berlandieri

Rana blairi

Rana sphenocephala

Frequency of egg-laying

J F M A M J J A S O N D
Month

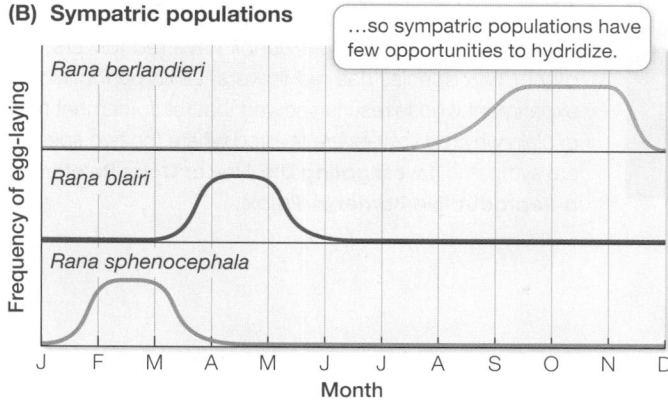

(B) Sympatric populations

…so sympatric populations have few opportunities to hybridize.

Rana berlandieri

Rana blairi

Rana sphenocephala

Frequency of egg-laying

J F M A M J J A S O N D
Month

Figure 22.11 Temporal Isolation of Breeding Seasons   (A) The peak breeding seasons of three species of leopard frogs (*Rana*) overlap when the species are physically separated (allopatry). (B) Where two or more species of *Rana* live together (sympatry), overlap between their peak breeding seasons is greatly reduced or eliminated.

Q: What means of selection helps reinforce this prezygotic isolating mechanism in areas of sympatry?

Gastrophryne olivacea ▮

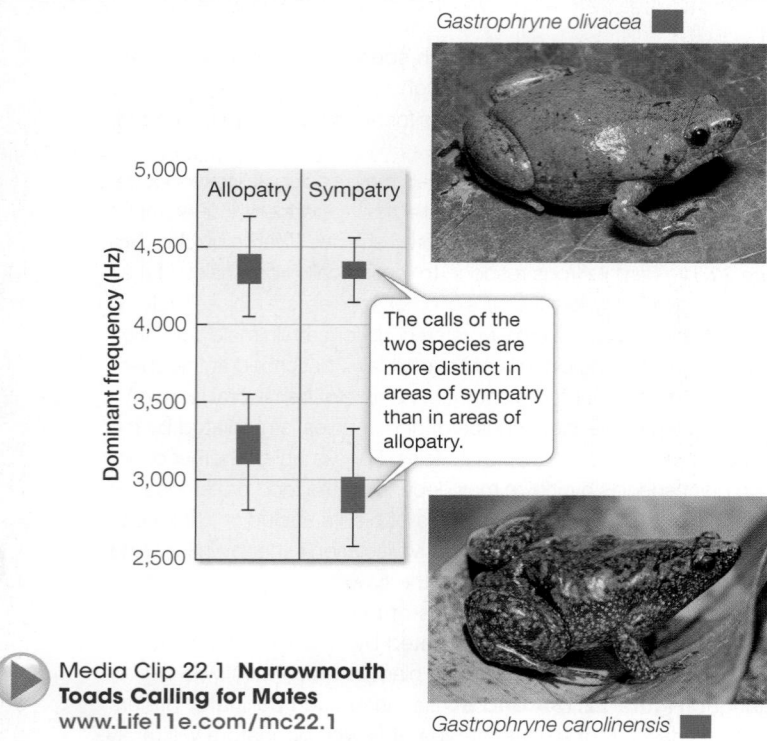

The calls of the two species are more distinct in areas of sympatry than in areas of allopatry.

▶ Media Clip 22.1 **Narrowmouth Toads Calling for Mates** www.Life11e.com/mc22.1

Gastrophryne carolinensis ▮

Figure 22.12 Behavioral Isolation in Mating Calls The males of most frog species produce species-specific calls. The calls of the two closely related frog species shown here differ in their dominant frequency (a high-frequency sound wave results in a high-pitched sound; a low frequency results in a low-pitched sound). Female frogs are attracted to the calls of males of their own species.

For example, gametic isolation has been extensively studied in spawning sea urchins. A protein known as bindin is found in sea urchin sperm and functions in attaching ("binding") the sperm to eggs. All sea urchin species studied produce this egg-recognition protein, but the bindin gene sequence diverges so rapidly that it becomes species-specific. Since sperm can attach only to eggs of the same species, no interspecific hybridization occurs.

Postzygotic isolating mechanisms result in selection against hybridization

Genetic differences that accumulate between two diverging lineages may reduce the survival and reproductive rates of hybrid offspring in any of several ways:

- *Low hybrid zygote viability.* Hybrid zygotes may fail to mature normally, either dying during development or developing phenotypic abnormalities that prevent them from becoming reproductively capable adults.

- *Low hybrid adult viability.* Hybrid offspring may have lower survivorship than non-hybrid offspring.

- *Hybrid infertility.* Hybrids may mature into infertile adults. For example, the offspring of matings between horses and donkeys—mules—are sterile. Although otherwise healthy, mules produce no descendants.

Natural selection does not directly favor the evolution of postzygotic isolating mechanisms. But if hybrids are less fit, individuals that breed only within their own species will leave more surviving offspring than will individuals that interbreed with another species. Therefore individuals that can avoid interbreeding with members of other species will have a selective advantage, and any trait that contributes to such avoidance will be favored.

Donald Levin of the University of Texas has studied reinforcement of prezygotic isolating mechanisms in flowers of the genus *Phlox*. Levin noticed that most individuals of *P. drummondii* in most of the range of the species in Texas have pink flowers. However, where *P. drummondii* is sympatric with its close relative, the pink-flowered *P. cuspidata*, most *P. drummondii* have red flowers. No other *Phlox* species has red flowers. Levin performed an experiment whose results showed that reinforcement may explain why red flowers are favored where the two species are sympatric (**Investigating Life: Flower Color Reinforces a Reproductive Barrier in *Phlox***).

(A) **(B)**

(C) *Aquilegia formosa* **(D)** *A. pubescens*

Figure 22.13 Reproductive Isolating Mechanisms May Be Mediated by Species Interactions **(A)** This hummingbird's morphology and behavior are adapted for feeding on nectar from pendant flowers. **(B)** The nectar-extracting proboscis of this hawkmoth is adapted to flowers that grow upright. **(C)** *Aquilegia formosa* flowers are normally pendant and are pollinated by hummingbirds. **(D)** Flowers of *A. pubescens* are normally upright and are pollinated by hawkmoths. In addition, their long floral spurs appear to restrict access by some other potential pollinators.

experiment

Original Paper: Levin, D. A. 1985. Reproductive character displacement in *Phlox*. *Evolution* 39: 1275–1281.

Most *Phlox drummondii* flowers are pink, but in regions where they are sympatric with *P. cuspidata*—which is always pink—most *P. drummondii* individuals are red. Most pollinators preferentially visit flowers of one color or the other. In this experiment, Donald Levin explored whether flower color reinforces a prezygotic reproductive barrier, lessening the chances of interspecific hybridization.

HYPOTHESIS▶ Red-flowered *P. drummondii* are less likely to hybridize with *P. cuspidata* than are pink-flowered *P. drummondii*.

METHOD

1. Introduce equal numbers of red- and pink-flowered *P. drummondii* individuals into an area with many pink-flowered *P. cuspidata*.

P. cuspidata

P. drummondii

2. After the flowering season ends, measure hybridization by assessing the genetic composition of the seeds produced by *P. drummondii* plants of both colors.

RESULTS Of the seeds produced by pink-flowered *P. drummondii*, 38% were hybrids with *P. cuspidata*. Only 13% of the seeds produced by red-flowered individuals were genetic hybrids.

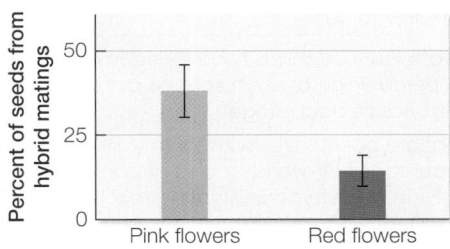

CONCLUSION▶ *P. drummondii* and *P. cuspidata* are less likely to hybridize if the flowers of the two species differ in color.

work with the data

Donald Levin of the University of Texas proposed that *Phlox drummondii* has red flowers only in locations where it is sympatric with pink-flowered *P. cuspidata* because having red flowers decreases interspecific hybridization. To test this hypothesis, Levin introduced equal numbers of red- and pink-flowered *P. drummondii* individuals into an area with many pink-flowered *P. cuspidata*. At the end of the flowering season, he assessed the genetic composition of the seeds produced by *P. drummondii*. The results are shown in the table below.

Morph (flower color)	Number of seeds (progeny)		
	P. drummondii	Hybrid	Total
Red	181 (87%)	27 (13%)	208
Pink	86 (62%)	53 (38%)	139

QUESTIONS▶

1. Check the 95% confidence intervals for the proportion of hybrid seeds in red- and pink-flowered *P. drummondii* in the graph in the Results of the experiment. There are many websites available for calculating confidence intervals; a good one is the Vassar College statistical computation site, VassarStats.net. You can go to this site and select "Proportions" from the left-hand menu, then select "The Confidence Interval of a Proportion." What are the numerical values of the 95% confidence intervals?

2. Notice that the proportions of hybrids among the seeds of red- versus pink-flowered samples are significantly different, because the 95% confidence intervals do not overlap. To quantify the significance of this difference, use the website suggested in Question 1, but select "Significance of the Difference between Two Independent Proportions" from the "Proportions" menu. What null hypothesis are you testing in this case? (See Appendix B if you need help.) What is the *P*-value of getting results at least as different as these two samples if your null hypothesis is true?

3. How would you extend or improve the experimental design of this study? What kinds of additional test sites or conditions would you want to examine? How might replicate or control sites make the study more convincing?

A similar **work with the data** exercise may be assigned in **LaunchPad**.

Likely cases of reinforcement are often detected by comparing sympatric and allopatric populations of potentially hybridizing species, as in the case of *Phlox*. If reinforcement is occurring, then sympatric populations of closely related species are expected to evolve more effective prezygotic reproductive barriers than do allopatric populations of the same species. As Figure 22.11 shows, the breeding seasons of sympatric populations of different leopard frog species overlap much less than do those of allopatric populations. Similarly, the frequencies of the frog mating calls illustrated in Figure 22.12 are more divergent in sympatric populations than in allopatric populations. In both cases, there appears to have been natural selection against hybridization in areas of sympatry.

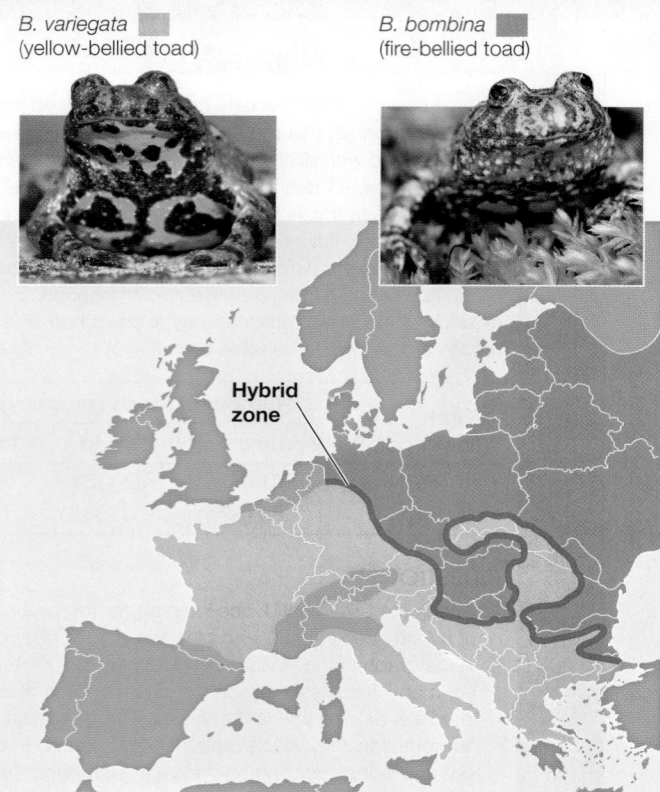

B. variegata
(yellow-bellied toad)

B. bombina
(fire-bellied toad)

Hybrid zone

Figure 22.14 A Hybrid Zone The narrow zone (shown in red) in which fire-bellied toads meet and hybridize with yellow-bellied toads has been stable for hundreds of years.

Hybrid zones may form if reproductive isolation is incomplete

Unless reproductive isolation is complete, closely related species may hybridize in areas where their ranges overlap, resulting in the formation of a **hybrid zone**. When a hybrid zone first forms, most hybrids are offspring of crosses between individuals of the two hybridizing species. However, subsequent generations include a variety of individuals with varying proportions of their genes derived from the original two species, so hybrid zones often contain recombinant individuals resulting from many generations of hybridization.

Detailed genetic studies can tell us why narrow hybrid zones may persist for long periods between the ranges of two species. The fire-bellied toad (*B. bombina*) lives in eastern Europe; the closely related yellow-bellied toad (*B. variegata*) lives in western and southern Europe. The ranges of the two species overlap in a long, but narrow zone stretching 4,800 km from eastern Germany to the Black Sea (**Figure 22.14**). Hybrids between the two species suffer from a range of defects, many of which are lethal. Those hybrids that survive often have skeletal abnormalities, such as a misshapen mouth, ribs that are fused to vertebrae, and a reduced number of vertebrae. By following the fates of thousands of toads from the hybrid zone, investigators found that a hybrid toad, on average, is only half as fit as an individual of either species. The hybrid zone remains narrow because there is strong selection against hybrids and because adult toads do not move over long distances. The zone has persisted for hundreds of years, however, because individuals of both species continue to move short distances into it, continually replenishing the hybrid population.

22.4 recap

Reproductive isolation may result from prezygotic or postzygotic isolating mechanisms. Lower fitness of hybrids can lead to the reinforcement of prezygotic isolating mechanisms.

learning outcomes

You should be able to:

- Create graphical representations of hybridization frequencies and analyze data to develop hypotheses about postzygotic isolation.
- Design an experiment to investigate a given isolating mechanism.
- Relate the width and/or persistence of a hybrid zone to the fitness of hybrid offspring.

1. As shown in Figure 22.11, the leopard frogs *Rana berlandieri* and *R. sphenocephala* usually have non-overlapping breeding seasons in areas of sympatry, but where they are allopatric, both species breed in both spring and fall. When new ponds are created where the ranges of the two species come close together, frogs from previously allopatric populations may colonize the new ponds and hybridize during their overlapping breeding seasons.

   Imagine you have collected and tabulated data on hybridization between these two frog species. You have sampled various life stages of frogs and their tadpoles for 2 years after an initial spring breeding season at a newly established pond. Use the data below to answer the following questions.

Life stage	*R. berlandieri*	*R. sphenocephala*	F_1 hybrids
Recently hatched tadpoles (spring, year 1)	155	125	238
Late-stage tadpoles (summer, year 1)	45	55	64
Newly metamorphosed froglets (fall, year 1)	32	42	15
Adult frogs (year 2)	10	15	1

   a. Create four pie charts (one for each life stage) showing the percentage of each species and the percentage of hybrids at each stage.

   b. What are possible reasons for the differences in the percentages of hybrids found at each life stage? Suggest some postzygotic isolating mechanisms that are consistent with your data.

   c. Over time, what changes might you expect in the breeding seasons of the two species at this particular pond, and why? How would future pie charts differ from those you created above, if your predictions about breeding seasons are correct?

2. In each of the columbine (*Aquilegia*) species shown in Figure 22.13, the orientation of the flowers and the length of flower spurs are associated with a particular type of pollinator (hummingbirds or hawkmoths). Columbine flowers vary in other ways as well—for example, in color and probably in fragrance. What experiments could you design to determine the traits that various pollinators use to distinguish among the flowers of different columbine species?

3. Why don't most narrow hybrid zones, such as the one between *Bombina bombina* and *B. variegata* shown in Figure 22.14, get wider over time?

Some groups of organisms have many species, others only a few. Hundreds of species of *Drosophila* evolved in the small area of the Hawaiian Islands over about 20 million years. In contrast, there are only a few species of horseshoe crabs in the world, and only one species of ginkgo tree, even though these latter groups have persisted for hundreds of millions of years. Why do different groups of organisms have such different rates of speciation?

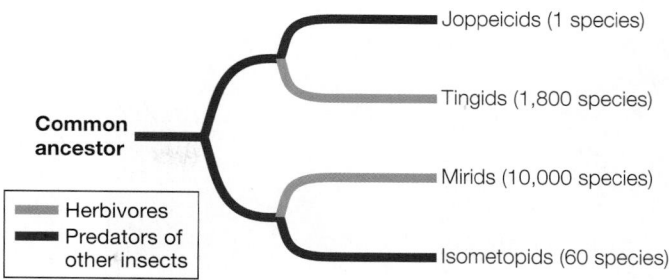

Figure 22.15 **Dietary Shifts Can Promote Speciation** Herbivorous groups of hemipteran insects have speciated several times faster than closely related predatory groups.

> ## key concept
> # 22.5
Speciation Rates Are Highly Variable across Life

Many factors influence the likelihood that a lineage will split to form two or more species. Therefore rates of speciation (the proportion of existing species that split to form new species over a given period) vary greatly among groups of organisms. What are some of the factors that influence the probability of a given lineage splitting into two?

restrict the number of pollinator species that visit the flowers, thus increasing opportunities for reproductive isolation (see Figure 22.13).

SEXUAL SELECTION   It appears that the mechanisms of sexual selection (see Key Concept 20.2) result in high rates of speciation, as we saw in the case of the cichlids of Lake Malawi. Some of the most striking examples of sexual selection are found in birds with polygynous mating systems. Birdwatchers travel thousands of miles to Papua New Guinea to witness the mating displays of male birds of paradise, which have brightly colored plumage (**Figure 22.16A**) and look distinctly different from females of their species—a phenomenon called sexual dimorphism. Males assemble at display grounds called leks, and females come there to choose a mate. After

focus your learning
- Speciation rates vary for different groups of organisms, due to several ecological and behavioral factors.
- The availability of unoccupied ecological niches increases the likelihood of adaptive radiation following colonization by a founder species.

Several ecological and behavioral factors influence speciation rates

Many factors can influence rates of speciation across groups, including the diet, behavioral complexity, and dispersal abilities of the respective species.

DIET SPECIALIZATION   Populations of species that have specialized diets may be more likely to diverge than those with more generalized diets. To investigate the effects of diet specialization on rates of speciation, Charles Mitter and colleagues compared species richness in some closely related groups of true bugs (hemipterans). The common ancestor of these groups was a predator that fed on other insects, but a dietary shift to herbivory (eating plants) evolved at least twice in the groups under study. Herbivorous bugs typically specialize on one or a few closely related species of plants, whereas predatory bugs tend to feed on many different species of insects. A high diversity of host-plant species can lead to a correspondingly high species diversity among herbivorous specialists. The Mitter et al. study showed that among these insects, the herbivorous groups do indeed contain many more species than do the related predatory groups (**Figure 22.15**).

POLLINATION   Speciation rates are faster in animal-pollinated than in wind-pollinated plants. Animal-pollinated groups have, on average, 2.4 times as many species as related groups pollinated by wind. Among animal-pollinated plants, speciation rates are correlated with pollinator specialization. In columbines (*Aquilegia*), the rate of evolution of new species has been about three times faster in lineages that have long nectar spurs than in lineages that lack spurs. Why do nectar spurs increase the speciation rate? Apparently it is because spurs

(A) *Paradisaea minor*

(B) *Manucodia comrii*

Figure 22.16 **Sexual Selection Can Lead to Higher Speciation Rates** **(A)** Birds of paradise (the brightly colored male is above the female on the branch) and **(B)** manucodes are closely related bird groups of the South Pacific. Speciation rates are much higher among the sexually dimorphic, polygynous birds of paradise (33 species) than among the sexually monomorphic, monogamous manucodes (5 species).

Madia sativa (tarweed)

Argyroxiphium sandwicense

Wilkesia gymnoxiphium

Dubautia menziesii

Figure 22.17 Rapid Evolution among Hawaiian Silverswords The Hawaiian silverswords, three closely related genera of the sunflower family, are believed to have descended from a single common ancestor (a plant similar to the tarweed *Madia sativa*) that colonized Hawaii from the Pacific coast of North America. The four plants shown here are more closely related than they appear to be based on their morphology.

mating, the females leave the display grounds, build their nests, lay their eggs, and feed their offspring with no help from the males. The males remain at the lek to court more females.

The closest relatives of the birds of paradise are the manucodes (**Figure 22.16B**). Male and female manucodes differ only slightly in size and plumage (they are sexually monomorphic). They form monogamous pair bonds, and both sexes contribute to raising the young. There are only 5 species of manucodes, compared with 33 species of birds of paradise. By itself, this one comparison would not be convincing evidence that sexually dimorphic clades have higher rates of speciation than do monomorphic clades. However, when biologists examined all the examples of birds in which one clade is sexually dimorphic and the most closely related clade is sexually monomorphic, the sexually dimorphic clades were significantly more likely to contain more species. But why would sexual dimorphism be associated with a higher rate of speciation?

Animals with complex sexually selected behaviors are likely to form new species at a high rate because they make sophisticated discriminations among potential mating partners. They distinguish members of their own species from members of other species, and they make subtle discriminations among members of their own species on the basis of size, shape, appearance, and behavior. Such discriminations can greatly influence which individuals are most successful in mating and producing offspring, so they may lead to rapid evolution of behavioral isolating mechanisms among populations.

DISPERSAL ABILITY Speciation rates are usually higher in groups with poor dispersal abilities than in groups with good dispersal abilities because even narrow barriers can be effective in dividing a species whose members are highly sedentary. Until recently, the Hawaiian Islands had about 1,000 species of land snails, many of which were restricted to a single valley. Because snails move only short distances, the high ridges that separate the valleys were effective barriers to their dispersal. Unfortunately, introductions of other species and changes in habitat have resulted in the recent extinction of most of these unique Hawaiian land snails.

Rapid speciation can lead to adaptive radiation

The rapid proliferation of a large number of descendant species from a single ancestor species is called an **evolutionary radiation**.

Evolutionary radiations often occur when a species colonizes a new area, such as an island archipelago that contains no other closely related species, because of the large number of open ecological niches. If such a rapid proliferation of species results in an array of species that live in a variety of environments and differ in the characteristics they use to exploit those environments, it is referred to as an **adaptive radiation**.

Several remarkable adaptive radiations have occurred in the Hawaiian Islands. In addition to its 1,000 species of land snails, the native Hawaiian biota includes 1,000 species of flowering plants, 10,000 species of insects, and more than 100 bird species. However, there were no amphibians, no terrestrial reptiles, and only one native terrestrial mammal (a bat) on the islands until humans introduced additional species. The 10,000 known native species of insects on Hawaii are believed to have evolved from about 400 immigrant species; only 7 immigrant species are believed to account for all the native Hawaiian land birds. Similarly, as we saw earlier in this chapter, an adaptive radiation in the Galápagos archipelago resulted in the many species of Darwin's finches, which differ strikingly in the size and shape of their bills and, accordingly, in the food resources they use (see Figure 22.8).

The 28 species of Hawaiian sunflowers called silverswords are an impressive example of an adaptive radiation in plants (**Figure 22.17**). DNA sequences show that these species share a relatively recent common ancestor with a species of tarweed from the Pacific coast of North America. Whereas all mainland tarweeds are small, upright herbs (nonwoody plants such as *Madia sativa*; see Figure 22.17), the silverswords include shrubs, trees, and vines as well as both upright and ground-hugging herbs. Silversword species occupy nearly all the habitats of the Hawaiian Islands, from sea level to above the timberline in the mountains. Despite their extraordinary morphological diversification, all silverswords are genetically very similar.

The Hawaiian silverswords are more diverse in size and shape than the mainland tarweeds because their tarweed ancestors first arrived on islands that harbored very few plant species. In particular, there were few trees and shrubs because such large-seeded plants rarely disperse to oceanic islands. Trees and shrubs have evolved from non-woody ancestors on many oceanic islands. On the mainland, however, tarweeds live in ecological communities that contain many tree and shrub species in lineages with long evolutionary histories. In those environments, opportunities to exploit the "tree" way of life have already been preempted.

KEY CONCEPT 22.5   483

▶ 22.5 recap

Dietary specialization, pollinator specialization, sexual selection, and poor dispersal abilities are correlated with high rates of speciation. Open ecological niches present opportunities for adaptive radiations.

learning outcomes

You should be able to:

- Explain the factors that commonly affect speciation rates.
- Describe the conditions that favor adaptive radiation.

1. How can pollinator specialization in plants and sexual selection in animals increase rates of speciation?

2. Why do adaptive radiations often occur when a founder species invades an isolated geographic area?

The result of 3.8 billion of years of evolution has been many millions of species, each adapted to live in a particular environment and to use environmental resources in a particular way. In the next chapter we will look at the molecular processes that gave rise to that diversity.

▶ investigating life

If speciation typically takes thousands to millions of years, how do biologists conduct experiments to study this process?

Although speciation usually takes thousands or millions of years, and although it is typically studied in natural settings such as Lake Malawi or in the field experiments described in Investigating Life: Flower Color Reinforces a Reproductive Barrier in *Phlox*, some aspects of speciation can be studied and observed in controlled laboratory experiments. Most such experiments use organisms with short generation times, in which evolution is expected to be relatively rapid.

William Rice and George Salt conducted an experiment in which fruit flies (*Drosophila melanogaster*) were allowed to choose food sources in different habitats. The habitats—where mating also took place—were vials in different parts of an experimental cage (Figure 22.18). The vials differed in three environmental factors: (1) light; (2) the direction (up or down) in which the fruit flies had to move to reach food; and (3) the concentrations of two aromatic chemicals, ethanol and acetaldehyde. In just 35 generations, the two groups of flies that chose the most divergent habitats had become reproductively isolated from each other, having evolved distinct preferences for the different habitats.

The experiment by Rice and Salt (see *American Naturalist* 131: 911–917, 1988) demonstrated an example of habitat isolation as a prezygotic isolating mechanism. Even though the different habitats were in the same cage, and individual fruit flies were capable of flying from one habitat to the other, habitat preferences were inherited by offspring from their parents, and populations from the two divergent habitats did not interbreed. Similar habitat isolation is thought to have resulted in the early split between cichlids that preferred the rocky versus the sandy shores of Lake Malawi. In controlled experiments like this one, biologists can observe many aspects of the process of speciation directly.

Future directions

Biologists can now sequence entire genomes of organisms, which allows them to compare divergent genes in pairs of recently divergent species, and to identify the genes responsible for prezygotic and postzygotic isolation. These same techniques also allow researchers to detect any level of hybridization between species. Thus we are now able to determine the genetic basis of reproductive isolation, and measure the rate at which diverging species become isolated from one another. With these advances, biologists can now appreciate the details and genetic basis for the origin of species.

Figure 22.18 **Evolution in the Laboratory** For their experiments on the evolution of prezygotic isolating mechanisms in *Drosophila melanogaster*, Rice and Salt built an elaborate system of varying habitats contained within vials inside a large fly enclosure. Some groups of flies developed preferences for widely divergent habitats and became reproductively isolated within 35 generations.

Chapter Summary 22

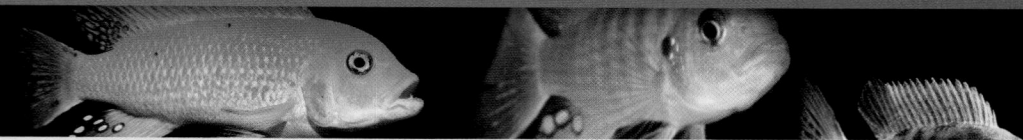

22.1 Species Are Reproductively Isolated Lineages on the Tree of Life

- **Speciation** is the process by which one species splits into two or more daughter species, which thereafter evolve as distinct lineages.

- The **morphological species concept** distinguishes species on the basis of physical similarities; it often underestimates or overestimates the actual number of reproductively isolated species.

- The **biological species concept** distinguishes species on the basis of **reproductive isolation**.

- The **lineage species concept**, which recognizes independent evolutionary lineages as species, allows biologists to consider species over evolutionary time.

22.2 Speciation Is a Natural Consequence of Population Subdivision

- Genetic divergence results from the interruption of gene flow within a population.

- The Dobzhansky–Muller model describes how reproductive isolation between two physically isolated populations can develop through the accumulation of incompatible genes or chromosomal arrangements. **Review Figures 22.3, 22.4, Activity 22.1**

- Reproductive isolation increases with increasing genetic divergence between populations. **Review Figure 22.5**

22.3 Speciation May Occur through Geographic Isolation or in Sympatry

- **Allopatric speciation**, which results when populations are separated by a physical barrier, is the dominant mode of speciation. This type of speciation may follow founder events, in which some members of a population cross a barrier and found a new, isolated population. **Review Focus: Key Figure 22.6, Figures 22.7, 22.8, Animations 22.1, 22.2**

- **Sympatric speciation** results when the genomes of two groups diverge in the absence of physical isolation. It can result from disruptive selection in two or more distinct microhabitats.

- Sympatric speciation can occur within two generations via **polyploidy**, an increase in the number of chromosomes sets. Polyploidy may arise from chromosome duplications within a species (**autopolyploidy**) or from hybridization that results in combining the chromosomes of two species (**allopolyploidy**). **Review Figure 22.9**

22.4 Reproductive Isolation Is Reinforced When Diverging Species Come into Contact

- **Prezygotic isolating mechanisms** prevent hybridization; **postzygotic isolating mechanisms** reduce the fitness of hybrids.

- Postzygotic isolating mechanisms lead to **reinforcement** of prezygotic isolating mechanisms by natural selection. **Review Figures 22.11, 22.12, Investigating Life: Flower Color Reinforces a Reproductive Barrier in *Phlox***

- **Hybrid zones** may form and persist if reproductive isolation between species is incomplete. **Review Figure 22.14**

22.5 Speciation Rates Are Highly Variable across Life

- Dietary specialization, pollinator specialization, sexual selection, and dispersal ability all influence speciation rates. **Review Figures 22.15, 22.16**

- **Evolutionary radiation** refers to the rapid proliferation of descendant species from a single ancestor species. This often occurs when a species colonizes a new area with unoccupied ecological niches. If evolutionary radiation results in an array of species that live in a variety of environments and differ in the characteristics they use to exploit those environments, it is referred to as **adaptive radiation**.

See Activity 22.2 for a concept review of this chapter.

Go to **LearningCurve** (in **LaunchPad**) for dynamic quizzing that helps you solidify your understanding of this chapter. **LearningCurve** adapts to your responses, giving you the practice you need to master each key concept.

Apply What You've Learned

Review

22.4 Reproductive isolation is reinforced in sympatry through selection for prezygotic mechanisms that prevent hybridization, which results from postzygotic mechanisms that reduce the fitness of hybrids.

Evolutionary biologists are particularly interested in reinforcement's role in speciation because this process directly connects natural selection to the origin of species.

Different *Drosophila* species vary in the concentrations of different hydrocarbons on their cuticles. These hydrocarbons often serve as reproductive signals, with females preferring to mate with males that have the suite of hydrocarbons from the females' own species over those of other species. As they often are reproductive signals, the hydrocarbons may be the target of the selection involved in reinforcement.

Two species of *Drosophila*, *D. serrata* and *D. birchii*, are found on Australia's east coast and overlap in range. *D. serrata* populations from areas where the species encounters *D. birchii* (sympatric populations) differ from *D. serrata* populations that do not encounter the other species (allopatric populations), and these differences are likely due to reinforcement. The species do produce some hybrids, but the fitness of the hybrids is low. The graph shows concentrations of three different hypothetical hydrocarbons taken from allopatric and sympatric populations of *D. serrata* and *D. birchii*. (Numbers do not add to 100 because there are other hydrocarbons that are not shown.)

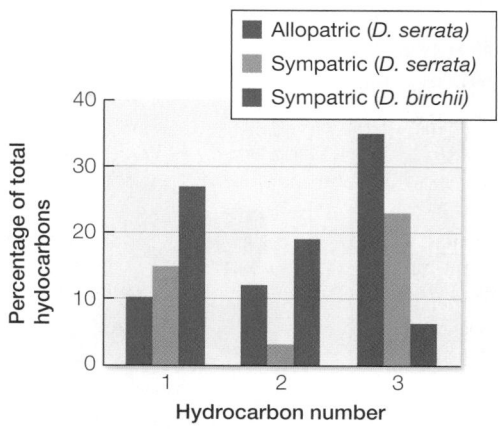

Questions

1. Based on the data, which hydrocarbon(s) show a pattern that is consistent with reinforcement? Explain your answer.

2. Suppose the allopatric populations of *D. serrata* were taken into the lab and exposed to *D. birchii* for many generations. If the artificial selection in this experiment acted like the natural selection in reinforcement, what would you expect to happen to the concentrations of each of the hydrocarbons in the allopatric populations?

3. If the artificial selection did act like the natural selection involved in reinforcement, would you expect the frequency of hybridization between allopatric populations and *D. birchii* to increase or decrease? Why?

4. Would you expect the extent of postzygotic reproductive isolation to change as a result of the artificial selection in the experimental evolution study? Explain.

Go to **LaunchPad** for the eBook, LearningCurve, animations, activities, flashcards, and additional resources and assignments.

23

Evolution of Genes and Genomes

An artist's view of an influenza virus. The spikes on the surface represent proteins that elicit an immune response in humans.

investigating life

Evolutionary Theory Helps Us Make Better Flu Vaccines

Combat in World War I ended in November 1918. But the death toll from four years of war was soon surpassed by the casualties of a massive influenza epidemic that killed more than 50 million people worldwide—more than twice the number of World War I combat deaths.

The 1918–1919 pandemic was noteworthy because the death rate among young adults—who are usually less likely to die from influenza than are the elderly or the very young—was 20 times higher than in flu epidemics before or since. Why was that particular virus so deadly, especially to typically hardy individuals? The 1918 flu strain triggered an especially intense reaction in the human immune system. This overreaction meant that people with strong immune systems were likely to be more severely affected.

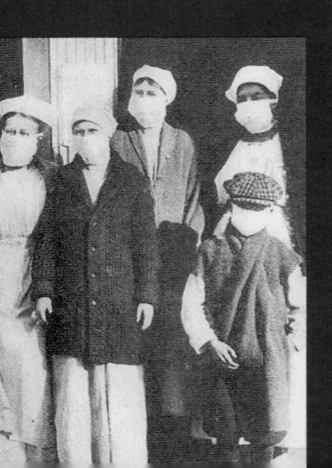

Usually we can count on our immune system to fight viruses, and the immune response is the basis of vaccination. Since 1945, programs to administer flu vaccines have helped keep the number and severity of influenza outbreaks in check. Last year's vaccine, however, will probably not be effective against this year's virus. New strains of flu virus are evolving continuously, producing genetic variation in the population. If these viruses did not evolve, we would become resistant to them and annual vaccination would become unnecessary. But because the viruses do evolve, biologists must develop a new and different flu vaccine each year.

The vertebrate immune response is initiated when the immune system recognizes proteins on the viral surface. Changes in viral surface proteins can allow the virus to escape immune detection. Virus strains with the greatest number of changes to their surface proteins are most likely to avoid detection and infect their hosts, and thus have an advantage over other strains. Biologists observe evolution in action by following changes in influenza virus proteins from year to year, and with this knowledge can produce more effective vaccines.

We learn a great deal about the molecular basis of evolution by examining rapidly evolving organisms such as viruses. Molecular studies of evolution, in turn, are put to practical uses, such as the development of better strategies for combating deadly diseases.

QA Why was the 1918–1919 flu pandemic worse than any before or since?

key concept
23.1
DNA Sequences Record the History of Gene Evolution

An organism's **genome** is the full set of genes it contains, as well as any noncoding regions of the DNA (or in the case of some viruses, RNA). Most of the genes of eukaryotic organisms are found on chromosomes in the nucleus, but genes are also present in chloroplasts and mitochondria. In organisms that reproduce sexually, both males and females contribute nuclear genes, but mitochondrial and chloroplast genes are usually transmitted only via the cytoplasm of one of the two gametes (usually from the female parent).

focus your learning

- Alignment of sequences allows biologists to compare nucleotide substitutions or amino acid replacements that have occurred between individuals or species.
- Simple counts of nucleotide substitutions or amino acid replacements between sequences often underestimate the underlying changes.

Genomes must be replicated to be transmitted from parents to offspring. DNA replication does not occur without error, however. Mistakes in DNA replication—mutations—provide much of the raw material for evolutionary change. Mutations are essential for the long-term survival of life because they are the initial source of the genetic variation that permits species to evolve in response to changes in their environment.

A particular allele of a gene will not be passed on to successive generations unless an individual carrying that allele survives and reproduces. The allele must function in combination with many other genes in the genome or it will quickly be selected against. Moreover, the degree and timing of a gene's expression are affected by its location in the genome. For these reasons, the genes of an individual organism can be viewed as interacting members of a group, among which there are divisions of labor but also strong interdependencies.

A genome, then, is not simply a random collection of genes in a random order along chromosomes. Rather, it is a complex set of integrated genes, regulatory sequences, and structural elements, interspersed with vast stretches of noncoding DNA that may have little direct function. Both the positions of genes and their sequences are subject to evolutionary change, as are the extent and location of noncoding DNA. All of these changes can affect the phenotype of an organism.

Biologists have now sequenced the complete genomes of a large number of organisms, including humans. The information in these sequences is helping us understand how and why organisms differ, how they function, and how they have evolved.

Evolution of genomes results in biological diversity

The field of **molecular evolution** investigates the mechanisms and consequences of the evolution of macromolecules—particularly nucleic acids (DNA and RNA) and proteins. Molecular evolutionists study relationships between the structures of genes and proteins and the functions of organisms. They also examine molecular variation to reconstruct evolutionary history and to study the mechanisms and consequences of evolution. Students of this field ask questions such as: What does molecular variation tell us about a gene's function? Why do the genomes of different organisms vary in size? What evolutionary forces shape patterns of variation among genomes? And a crucial question from an evolutionary perspective: How do genomes acquire new functions? Investigations into the evolution of particular nucleic acids and proteins are instrumental in reconstructing the evolutionary histories of genes. Ultimately, molecular evolutionary biologists hope to explain the molecular basis of biological diversity.

The evolution of nucleic acids and proteins depends on genetic variation introduced by mutations. One of several ways in which genes evolve is by means of nucleotide substitutions (the incorporation of point mutations in populations). In genes that encode proteins, nucleotide substitutions sometimes result in amino acid replacements that can change the charge, the structure, and other chemical and physical properties of the encoded protein. These changes in a protein molecule often affect the way that protein functions in the organism.

Evolutionary changes in genes and proteins can be identified by comparing nucleotide or amino acid sequences from different organisms. The longer two sequences have been evolving separately, the more differences they accumulate (bearing in mind that different genes in the same species evolve at different rates). Determining how long ago changes in nucleotide or amino acid sequences occurred is a useful step toward inferring their causes. Knowledge of the pattern and rate of evolutionary change in a given macromolecule is useful in reconstructing the evolutionary history of groups of organisms.

To compare genes or proteins from different organisms, biologists need a way to identify homologous parts of macromolecules. (Recall from Key Concept 21.1 that homologous features are those shared by two or more species that have been inherited from a common ancestor.) Homologous parts of a protein can be identified by their homologous amino acid sequences. And since nucleotide sequences encode amino acid sequences, the concept of homology extends down to the level of individual nucleotide positions. Therefore one of the first steps in studying the evolution of genes or proteins is to align homologous positions in the amino acid or nucleotide sequence of interest.

Genes and proteins are compared through sequence alignment

Once the nucleotide or amino acid sequences of molecules from different organisms have been determined, they can be compared. Homologous positions can be identified only if we first pinpoint the locations of deletions and insertions that have occurred in the molecules of interest in the time since the organisms diverged from a common ancestor. A simple hypothetical example illustrates this **sequence alignment** technique. In **Figure 23.1** we compare two amino acid sequences from homologous proteins in different organisms. The two sequences at first appear to differ in both the number and identity of their amino acids. If we insert a gap after the first amino acid in sequence 2 (after leucine), however, the similarities in the two sequences become obvious. This gap represents the occurrence of one of two evolutionary events: an insertion of an

research tools

Figure 23.1 Amino Acid Sequence Alignment   Amino acid sequence alignment is a way of arranging protein sequences to identify regions of homology between the sequences. Gaps are inserted between the amino acid residues to align similar residues in columns. Differences (number of amino acid differences plus insertion or deletion events) and similarities (number of identical amino acids) between each pair of aligned sequences are then summarized in a similarity matrix. Homologous DNA sequences can be aligned in a similar manner.

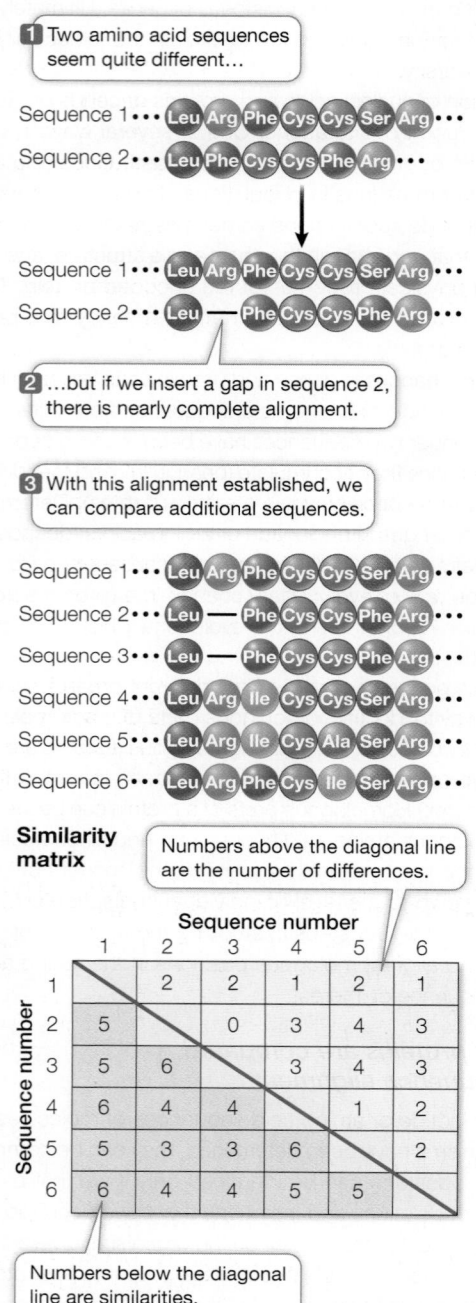

amino acid in the longer protein or a deletion of an amino acid in the shorter protein. Having adjusted for this insertion or deletion event, we can see that the two sequences differ by only one amino acid at position 6 (serine or phenylalanine).

 Activity 23.1 Amino Acid Sequence Alignment
www.Life11e.com/ac23.1

Adding a single gap—that is, identifying a deletion or an insertion—aligns the two sequences in Figure 23.1. Additional sequences can now be added to the alignment in a similar manner. Longer amino acid sequences, and those that have diverged more extensively, require more elaborate adjustments. Explicit models (incorporated into computer algorithms) have been developed to account for the relative probabilities of deletions, insertions, and particular amino acid replacements.

Having aligned the sequences, we can compare them by counting the number of nucleotides or amino acids that differ between them. Summing the numbers of the same and different amino acids in each pair of sequences allows us to construct a **similarity matrix**, which gives us a measure of the minimum number of changes that have occurred since the divergence of each pair of organisms (see Figure 23.1).

 Activity 23.2 Similarity Matrix Construction
www.Life11e.com/ac23.2

Models of sequence evolution are used to calculate evolutionary divergence

The sequence comparison procedure illustrated in Figure 23.1 gives a simple count of the number of similarities and differences between the proteins of two species. In the context of two aligned DNA sequences, we can count the number of differences at homologous nucleotide positions, and this count indicates the minimum number of nucleotide changes that must have occurred since the two sequences diverged from a common ancestral sequence.

Although it is useful in determining a *minimum* number of changes between two DNA sequences, the count provided by sequence alignment almost certainly underestimates the *actual* number of changes that have occurred since the sequences diverged. Any given change counted in a similarity matrix of DNA sequences may result from multiple substitution events that occurred at a given nucleotide position over time. As illustrated in **Figure 23.2**, any of the following events may have occurred at a given nucleotide position that would not be revealed by a simple count of similarities and differences between two DNA sequences:

● *Multiple substitutions*. More than one change has occurred at a given position between the ancestral sequence and at least one of the observed sequences.

● *Coincident substitutions*. At a given position, different substitutions have occurred between the ancestral sequence and each observed sequence.

● *Parallel substitutions*. The same substitution has occurred independently between the ancestral sequence and each observed sequence.

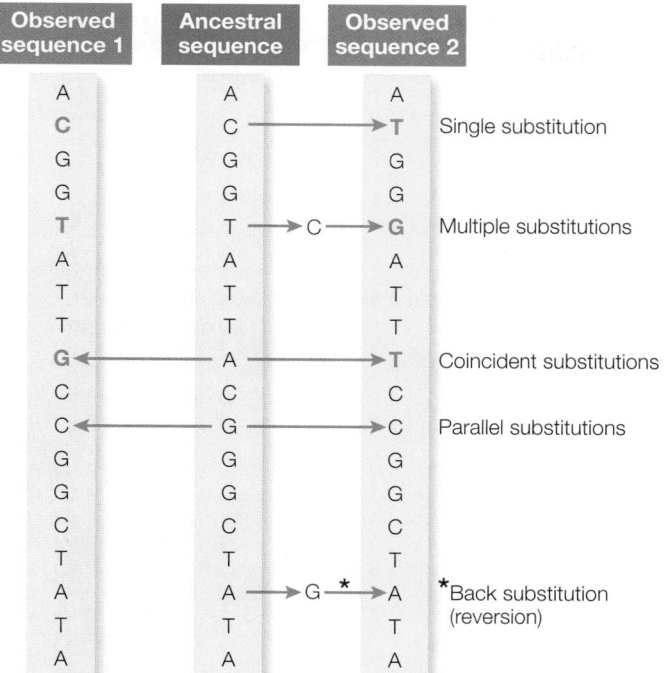

Figure 23.2 Multiple Substitutions Are Not Reflected in Pairwise Sequence Comparisons Two observed sequences descended from a common ancestral sequence (center) have undergone a series of substitutions. Although the two observed sequences differ by only three nucleotides (colored letters), these three differences result from a total of nine substitutions (arrows).

- *Back substitutions* (also called reversions). In a variation on multiple substitutions, after a change at a given position, a subsequent substitution has changed the position back to the ancestral state.

To correct for undercounting of substitutions, molecular evolutionists have developed mathematical models that describe how DNA (and protein) sequences evolve. These models take into account the relative rates of change from one nucleotide to another. For example, *transitions* (changes between the two purines, A ↔ G, or between the pyrimidines, C ↔ T) are typically more frequent than *transversions* (a purine is replaced by a pyrimidine, or vice versa). These models also include parameters such as the different rates of substitution across different parts of a gene and the proportions of each nucleotide present in a given sequence. Once such parameters have been estimated, the model is used to correct for multiple substitutions, coincident substitutions, parallel substitutions, and back substitutions. The revised estimate accounts for the *total* number of substitutions likely to have occurred between two sequences, which is almost always greater than the observed number of differences.

As sequence information becomes available for more and more genes in an ever-expanding database, sequence alignments can be extended across multiple homologous sequences, and the minimum number of insertions, deletions, and substitutions can be summed across homologous genes of an entire group of organisms. Similar databases have been constructed for homologous proteins. **Figure 23.3** shows aligned data for cytochrome *c* protein sequences in 33 species of animals, plants, and fungi. Such information is used extensively in determining evolutionary relationships among species.

Experimental studies examine molecular evolution directly

Although molecular evolutionists are often interested in naturally evolved genes and proteins, molecular evolution can also be observed directly in the laboratory. Increasingly, evolutionary biologists are studying evolution experimentally. Because substitution rates are related to generation time rather than to absolute time, most of these experiments use unicellular organisms or viruses with short generations. Viruses, bacteria, and unicellular eukaryotes (such as the yeasts) can be cultured in large populations in the laboratory, and many of these organisms can evolve rapidly. In the case of some RNA viruses, the natural substitution rate may be as high as 1 substitution per 1,000 nucleotides per generation. Therefore in a virus of a few thousand nucleotides, one or more substitutions are expected (on average) every generation, and these changes can easily be determined by sequencing the entire viral genome (because of its small size). Generation time may be only tens of minutes (rather than years or decades, as in many animals), so biologists can directly observe substantial molecular evolution in a controlled population over the course of days, weeks, or months.

Experimental molecular evolutionary studies are used for a wide variety of purposes and have greatly expanded the ability of evolutionary biologists to test evolutionary concepts and principles. Biologists now routinely study evolution in the laboratory, and as we will see later in this chapter, they can use in vitro evolutionary techniques to produce novel molecules that perform new functions with industrial and pharmaceutical uses.

▶ 23.1 recap

The genomes of all organisms evolve over time. Evolutionary changes can be detected by comparing the nucleic acid and protein sequences of different species. Experimental studies of molecular evolution under controlled conditions allow biologists to study many processes of evolution directly.

learning outcomes

You should be able to:

- Align a set of sequences and create a matrix that compares the similarities and differences in the sequences.
- Demonstrate why counts of nucleotide or amino acid differences between two sequences often underestimate the actual number of changes that have occurred between the sequences.

(continued)

Tuna

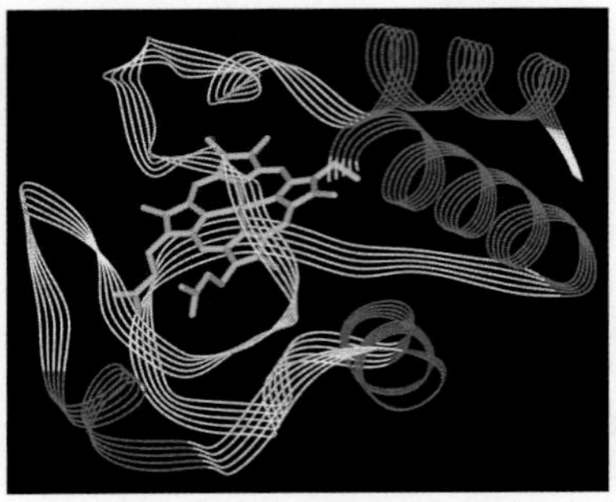

Rice

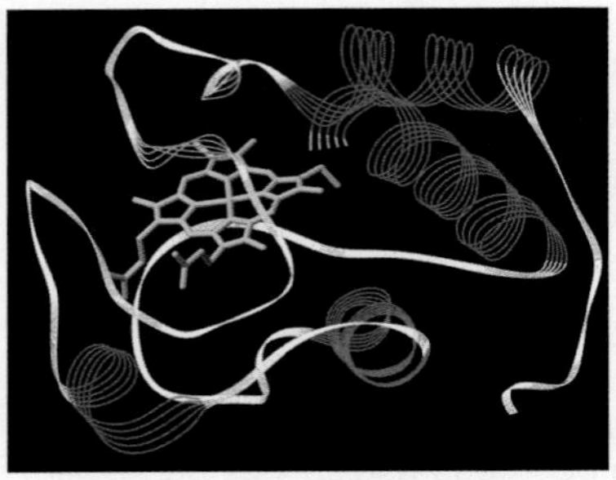

Acidic side chains
D Aspartic acid
E Glutamic acid

Basic side chains
H Histidine
K Lysine
R Arginine

Hydrophobic side chains
F Phenylalanine
I Isoleucine
L Leucine
M Methionine

V Valine
Y Tyrosine
W Tryptophan
A Alanine

Other
C Cysteine
P Proline
Q Glutamine
N Asparagine
S Serine
T Threonine
G Glycine

The number 1 indicates an invariant position in the cytochrome *c* molecule (i.e., all the organisms have the same amino acid in this position). Such a position is probably under strong purifying selection.

Amino acids at positions marked by red arrowheads have side chains that interact with the heme group.

Position in sequence   1   5   10   15   20   25   30

Number of amino acids at the position: 1 3 5 5 5 1 3 3 4 1 4 3 2 1 3 3 1 1 2 4 3 4 2 3 4 2 1 4 1 1 2 1 5 1

```
Human, chimpanzee          GDVEKGKKIFIMKCSQCHTVEKGGKHKTGPNLHG
Rhesus monkey              GDVEKGKKIFIMKCSQCHTVEKGGKHKTGPNLHG
Horse                      GDVEKGKKIFVQKCAQCHTVEKGGKHKTGPNLHG
Donkey                     GDVEKGKKIFVQKCAQCHTVEKGGKHKTGPNLHG
Cow, pig, sheep            GDVEKGKKIFVQKCAQCHTVEKGGKHKTGPNLHG
Dog                        GDVEKGKKIFVQKCAQCHTVEKGGKHKTGPNLHG
Rabbit                     GDVEKGKKIFVQKCAQCHTVEKGGKHKTGPNLHG
Gray whale                 GDVEKGKKIFVQKCAQCHTVEKGGKHKTGPNLHG
Gray kangaroo              GDVEKGKKIFVQKCAQCHTVEKGGKHKTGPNLNG

Chicken, turkey            GDIEKGKKIFVQKCSQCHTVEKGGKHKTGPNLHG
Pigeon                     GDIEKGKKIFVQKCSQCHTVEKGGKHKTGPNLHG
Pekin duck                 GDVEKGKKIFVQKCSQCHTVEKGGKHKTGPNLHG
Snapping turtle            GDVEKGKKIFVQKCAQCHTVEKGGKHKTGPNLHG
Rattlesnake                GDVEKGKKIFTMKCSQCHTVEKGGKHKVGPNLHG
Bullfrog                   GDVEKGKKIFVQKCAQCHTCEKGGKHKVGPNLYG
Tuna                       GDVAKGKKTFVQKCAQCHTVENGGKHKVGPNLWG
Dogfish                    GDVEKGKKVFVQKCAQCHTVENGGKHKTGPNLSG

Samia cynthia (moth)       GNAENGKKIFVQRCAQCHTVEAGGKHKVGPNLHG
Tobacco hornworm moth      GNADNGKKIFVQRCAQCHTVEAGGKHKVGPNLHG
Screwworm fly              GDVEKGKKIFVQRCAQCHTVEAGGKHKVGPNLHG
Drosophila (fruit fly)     GDVEKGKKLFVQRCAQCHTVEAGGKHKVGPNLHG
Baker's yeast              GSAKKGATLFKTRCELCHTVEKGGPHKVGPNLHG
Candida krusei (yeast)     GSAKKGATLFKTRCAECHTIEAGGPHKVGPNLHG
Neurospora crassa (mold)   GDSKKGANLFKTRCAECH--E-NLTQKIGPALHG
Wheat                      GNPDAGAKIFKTKCAQCHTVDAGA-HKQGPNLHG
Sunflower                  GDPTTGAKIFKTKCAQCHTVEKGA-HKQGPNLNG
Mung bean                  GDSKSGEKIFKTKCAQCHTVDKGA-HKQGPNLNG
Rice                       GNPKAGEKIFKTKCAQCHTVDKGA-HKQGPNLNG
Sesame                     GDVKSGEKIFKTKCAQCHTVDKGA-HKQGPNLNG
```

Gaps indicate insertion and/or deletion events.

Figure 23.3 Amino Acid Sequences of Cytochrome *c* The amino acid sequences shown in the table were obtained from analyses of the enzyme cytochrome *c* from 29 species of plants, fungi, and animals. Note the lack of variation across the sequences at positions 70–80, suggesting that this region is under strong purifying selection and that changing its amino acid sequence would impair the protein's function. The molecular models at the upper left are created from these sequences and show the three-dimensional structures of tuna and rice cytochrome *c*. Alpha helices are in red, and the molecule's heme group is shown in yellow.

Media Clip 23.1 **The Ubiquitous Protein**
www.Life11e.com/mc23.1

> **23.1 recap** (continued)

1. Align the following sequences. Then create a matrix that compares both the number of identical nucleotides and the number of differences (including insertion and deletion events).

   Sequence *a*   AATGCAGGGTATACG
   Sequence *b*   ATTCAGGGTATACC
   Sequence *c*   ATTGCAGCGTATAACC
   Sequence *d*   ATTGCAGGGTATACG

2. Explain why a simple count of nucleotide differences between two sequences often underestimates the actual number of nucleotide substitutions since the sequences diverged. Use an example based on the comparison of your alignment of Sequences *a* and *b* from Question 1.

We have seen that molecular evolutionists can directly observe the evolution of genomes over time, compare the genomes of different organisms, and reconstruct the changes that have occurred during their evolution. Let's turn now to the question of how genomes change, and examine some of the consequences of those changes.

> **key concept**
> ## 23.2 Genomes Reveal Both Neutral and Selective Processes of Evolution

A mutation, as we saw in Key Concept 12.2, is any change in the genetic material. A nucleotide substitution is the product of one type of mutation, incorporated into a population. Many nucleotide substitutions have no effect on phenotype, even if the change occurs in a gene that encodes a protein, because most *amino acids are specified by more than one codon (see Figure 14.4). A

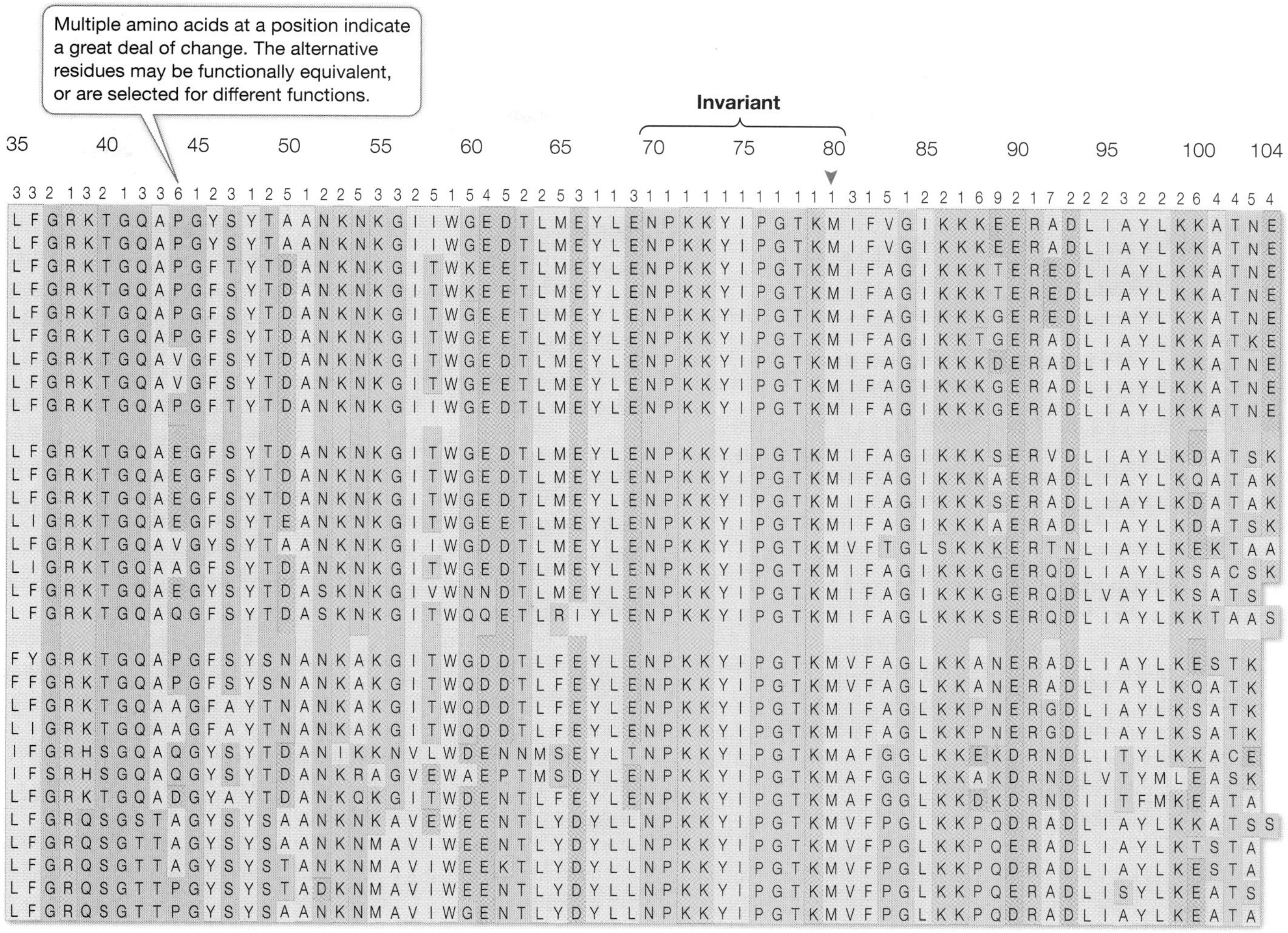

> Multiple amino acids at a position indicate a great deal of change. The alternative residues may be functionally equivalent, or are selected for different functions.

Invariant

```
 35    40    45    50    55    60    65      70    75    80     85    90    95   100  104

3 3 2 1 3 2 1 3 3 6 1 2 3 1 2 5 1 2 2 5 3 3 2 5 1 5 4 5 2 2 5 3 1 1 3 1 1 1 1 1 1 1 1 1 1 1 3 1 5 1 5 2 2 1 6 9 2 1 7 2 2 2 3 2 2 2 6 4 4 5 4

L F G R K T G Q A P G Y S Y T A A N K N K G I I W G E D T L M E Y L E N P K K Y I P G T K M I F V G I K K K E E R A D L I A Y L K K A T N E
L F G R K T G Q A P G Y S Y T A A N K N K G I I W G E D T L M E Y L E N P K K Y I P G T K M I F V G I K K K E E R A D L I A Y L K K A T N E
L F G R K T G Q A P G F T Y T D A N K N K G I T W K E E T L M E Y L E N P K K Y I P G T K M I F A G I K K K T E R E D L I A Y L K K A T N E
L F G R K T G Q A P G F S Y T D A N K N K G I T W K E E T L M E Y L E N P K K Y I P G T K M I F A G I K K K T E R E D L I A Y L K K A T N E
L F G R K T G Q A P G F S Y T D A N K N K G I T W G E E T L M E Y L E N P K K Y I P G T K M I F A G I K K K G E R E D L I A Y L K K A T N E
L F G R K T G Q A V G F S Y T D A N K N K G I T W G E D T L M E Y L E N P K K Y I P G T K M I F A G I K K K D E R A D L I A Y L K K A T N E
L F G R K T G Q A V G F S Y T D A N K N K G I T W G E E T L M E Y L E N P K K Y I P G T K M I F A G I K K K G E R A D L I A Y L K K A T K E
L F G R K T G Q A P G F T Y T D A N K N K G I I W G E D T L M E Y L E N P K K Y I P G T K M I F A G I K K K G E R A D L I A Y L K K A T N E

L F G R K T G Q A E G F S Y T D A N K N K G I T W G E D T L M E Y L E N P K K Y I P G T K M I F A G I K K K S E R V D L I A Y L K D A T S K
L F G R K T G Q A E G F S Y T D A N K N K G I T W G E D T L M E Y L E N P K K Y I P G T K M I F A G I K K K A E R A D L I A Y L K Q A T A K
L F G R K T G Q A E G F S Y T D A N K N K G I T W G E D T L M E Y L E N P K K Y I P G T K M I F A G I K K K S E R A D L I A Y L K D A T A K
L I G R K T G Q A E G F S Y T E A N K N K G I T W G E E T L M E Y L E N P K K Y I P G T K M I F A G I K K K A E R A D L I A Y L K D A T S K
L F G R K T G Q A V G Y S Y T A A N K N K G I I W G D D T L M E Y L E N P K K Y I P G T K M V F T G L S K K K E R T N L I A Y L K E K T A A
L I G R K T G Q A A G F S Y T D A N K N K G I I W G E D T L M E Y L E N P K K Y I P G T K M I F A G I K K K G E R Q D L I A Y L K S A C S K
L F G R K T G Q A E G Y S Y T D A S K N K G I V W N N D T L M E Y L E N P K K Y I P G T K M I F A G I K K K G E R Q D L V A Y L K S A T S
L F G R K T G Q A Q G F S Y T D A S K N K G I T W Q Q E T L R I Y L E N P K K Y I P G T K M I F A G L K K K S E R Q D L I A Y L K K T A A S

F Y G R K T G Q A P G F S Y S N A N K A K G I T W G D D T L F E Y L E N P K K Y I P G T K M V F A G L K K A N E R A D L I A Y L K E S T K
F F G R K T G Q A P G F S Y S N A N K A K G I T W Q D D T L F E Y L E N P K K Y I P G T K M V F A G L K K A N E R A D L I A Y L K Q A T K
L F G R K T G Q A A G F A Y T N A N K A K G I T W Q D D T L F E Y L E N P K K Y I P G T K M I F A G L K K P N E R G D L I A Y L K S A T K
L I G R K T G Q A A G F A Y T N A N K A K G I T W Q D D T L F E Y L E N P K K Y I P G T K M I F A G L K K P N E R G D L I A Y L K S A T K
I F G R H S G Q A Q G Y S Y T D A N I K K N V L W D E N N M S E Y L T N P K K Y I P G T K M A F G G L K K E K D R N D L I T Y L K K A C E
I F S R H S G Q A Q G Y S Y T D A N K R A G V E W A E P T M S D Y L E N P K K Y I P G T K M A F G G L K K A K D R N D L V T Y M L E A S K
L F G R K T G Q A D G Y A Y T D A N K Q K G I T W D E N T L F E Y L E N P K K Y I P G T K M A F G G L K K D K D R N D I I T F M K E A T A
L F G R Q S G S T A G Y S Y S A A N K N K A V E W E E N T L Y D Y L L N P K K Y I P G T K M V F P G L K K P Q D R A D L I A Y L K K A T S S
L F G R Q S G T T A G Y S Y S A A N K N M A V I W E E N T L Y D Y L L N P K K Y I P G T K M V F P G L K K P Q E R A D L I A Y L K T S T A
L F G R Q S G T T A G Y S Y S T A N K N M A V I W E E K T L Y D Y L L N P K K Y I P G T K M V F P G L K K P Q D R A D L I A Y L K E S T A
L F G R Q S G T T P G Y S Y S T A D K N M A V I W E E N T L Y D Y L L N P K K Y I P G T K M V F P G L K K P Q E R A D L I S Y L K E A T S
L F G R Q S G T T P G Y S Y S A A N K N M A V I W G E N T L Y D Y L L N P K K Y I P G T K M V F P G L K K P Q D R A D L I A Y L K E A T A
```

substitution that does not change the encoded amino acid is known as a **synonymous substitution** or **silent substitution** (Figure 23.4A). Synonymous substitutions do not affect the functioning of a protein (although they may have other effects, as described in Key Concept 15.1) and are therefore less likely than other types of substitutions to be subject to natural selection.

focus your learning

- Neutral evolution is distinguished from purifying and positive selection by its lack of effect on the survival and reproduction of the organism.
- The rate of fixation of neutral nucleotide changes within populations is independent of population size.
- Comparing rates of synonymous and nonsynonymous substitutions can be used to identify positive and purifying selection in protein genes.
- Genome sizes vary widely among organisms, even though the number of protein-coding genes shows much less variation.

***connect the concepts** The genetic code determines the amino acid that is encoded by each codon; see Figure 14.5.

A nucleotide substitution that *does* change the amino acid sequence encoded by a gene is known as a **nonsynonymous substitution** (Figure 23.4B). Nonsynonomous substitutions include

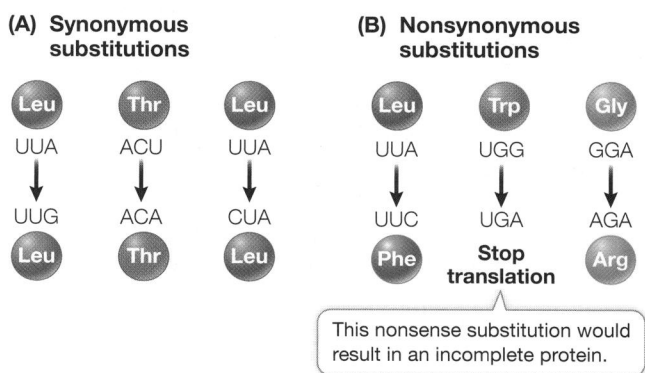

(A) Synonymous substitutions

Leu	Thr	Leu
UUA	ACU	UUA
↓	↓	↓
UUG	ACA	CUA
Leu	Thr	Leu

(B) Nonsynonymous substitutions

Leu	Trp	Gly
UUA	UGG	GGA
↓	↓	↓
UUC	UGA	AGA
Phe	Stop translation	Arg

> This nonsense substitution would result in an incomplete protein.

Figure 23.4 When One Nucleotide Does or Doesn't Make a Difference (A) Synonymous (silent) substitutions do not change the amino acid specified and do not affect protein function; such substitutions are unlikely to be subject to natural selection. (B) Nonsynonymous (missense and nonsense) substitutions do change the amino acid sequence and are likely to have an effect (often deleterious) on protein function; such substitutions are targets for natural selection.

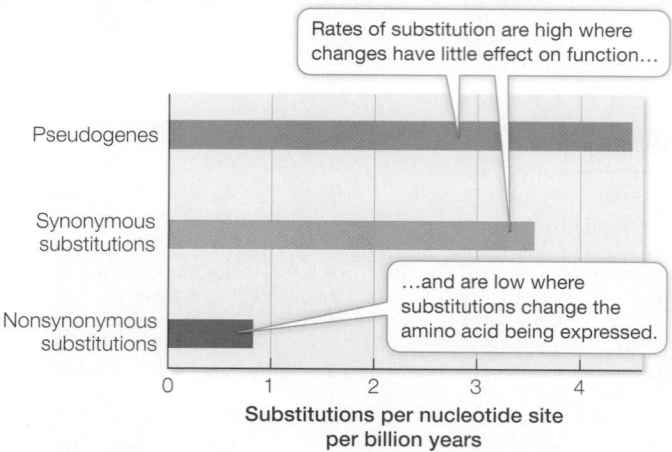

Figure 23.5 Rates of Substitution Differ Rates of nonsynonymous substitution typically are much lower than rates of synonymous substitution and the substitution rate in pseudogenes. This pattern reflects differing levels of functional constraints.

missense substitutions that change the specified amino acid, and **nonsense substitutions** that produce a stop codon and terminate the protein. In general, nonsynonymous substitutions are likely to be deleterious to the organism. But not every amino acid replacement alters a protein's shape and charge (and hence its functional properties), so some nonsynonymous substitutions may also be selectively neutral (or nearly so). Conversely, an amino acid replacement that confers an advantage to the organism would result in positive selection for the corresponding nonsynonymous substitution.

Investigators have measured the average rate of nonsynonymous nucleotide substitutions in some highly conserved protein-coding genes at about 0.9 substitutions per position per billion years. Synonymous substitutions in these genes have occurred about five times more frequently than nonsynonymous substitutions. In other words, *substitution rates are highest at nucleotide positions that do not change the amino acid being expressed* (**Figure 23.5**). Substitution rates are even higher in **pseudogenes**, which are duplicate, nonfunctional copies of genes.

Most natural populations harbor far more genetic variation than we would expect to find if genetic variation were influenced by natural selection alone. This discovery, combined with the knowledge that many mutations do not change molecular function, stimulated the development of the neutral theory of molecular evolution.

Much of evolution is neutral

In 1968, Motoo Kimura proposed the **neutral theory** of molecular evolution. Kimura suggested that, at the molecular level, the majority of the variants we observe in most populations are selectively neutral; that is, they confer neither an advantage nor a disadvantage on their bearers. These neutral variants accumulate through genetic drift rather than through positive selection.

The rate of fixation of neutral mutations by genetic drift is independent of population size. To see why this is so, consider a diploid population of size N and a neutral mutation rate μ (mu) per gamete per generation at a locus. The number of new mutations would be, on average, $\mu \times 2N$, because $2N$ gene copies are available to mutate. The probability that a given mutation will be fixed by drift alone is its frequency, which equals $1/(2N)$ for a newly arisen mutation. We can multiply these two terms to get the rate of fixation of neutral mutations (m) in a given population of N individuals:

$$m = 2N\mu \frac{1}{2N}$$

Therefore the rate of fixation m of neutral mutations depends only on the neutral mutation rate μ and is independent of population size. A given mutation is more likely to appear in a large population than in a small one, but any mutation that does appear is more likely to become fixed in a small population. These two influences of population size cancel each other out, so the rate of fixation of neutral mutations is equal to the mutation rate (i.e., $m = \mu$).

As long as the underlying mutation rate is constant, macromolecules evolving in separate populations should diverge from one another in neutral changes at a constant rate. Investigators have confirmed that the rate of evolution of particular genes and proteins is often relatively constant over time and can therefore can be used as a "molecular clock." As we described in Key Concept 21.3, molecular clocks can be used to calculate evolutionary divergence times between species.

Although much of the genetic variation we observe in populations is the result of neutral evolution, the neutral theory does not imply that most mutations have no effect on the organism. Many mutations are never observed in populations because they are lethal or strongly detrimental to the organism and are thus quickly removed from the population through natural selection. Similarly, mutations that confer a selective advantage tend to be quickly fixed in populations, so they do not result in variation at the population level either. Nonetheless, in any population, some amino acid positions will remain constant under purifying selection, others will vary through neutral genetic drift, and still others will differ between species as a result of positive selection for change. How can these evolutionary processes be distinguished?

Positive and purifying selection can be detected in the genome

As we have just seen, substitutions in a protein-coding gene can be either synonymous or nonsynonymous, depending on whether they change the resulting amino acid sequence of the protein. The relative rates of synonymous and nonsynonymous substitutions are expected to differ in regions of genes that are evolving neutrally, under positive selection for change, or staying unchanged under purifying selection:

- If a given amino acid in a protein can be one of many alternatives (without changing the protein's function), then an amino acid replacement is *neutral* with respect to the fitness of an organism. In this case, the rates of synonymous and nonsynonymous substitutions in the corresponding DNA sequences are expected to be very similar, so the ratio of the two rates should be close to 1.

- If a given amino acid position is under *positive* selection for change, the observed rate of nonsynonymous substitutions is expected to exceed the rate of synonymous substitutions in the corresponding DNA sequences.

- If a given amino acid position is under *purifying* selection, then the observed rate of synonymous substitutions is expected to be much higher than the rate of nonsynonymous substitutions in the corresponding DNA sequences.

By comparing the gene sequences that encode homologous proteins from many species, scientists can determine the history and timing of synonymous and nonsynonymous substitutions. This information can be mapped on a phylogenetic tree, as we saw in Chapter 21. Regions of genes that are evolving under neutral, purifying, or positive selection can be identified by comparing the nature and rates of substitutions across the phylogenetic tree.

By examining genome sequences for synonymous and nonsynonymous substitutions, biologists have determined which amino acid positions of the surface proteins of influenza viruses are evolving under positive selection (to escape immune detection in their hosts). This information allows researchers to select viruses from current influenza epidemics that are the most likely to escape human immune detection. These viruses are the ones that are most likely to give rise to the next flu epidemic, so they are the best candidates to use in vaccine production. As we discussed in the opening story, production of successful flu vaccines has greatly lowered the impact of flu epidemics in recent decades.

A study of the evolution of lysozyme illustrates how and why particular amino acid positions might be under different modes of selection (**Figure 23.6**). The enzyme lysozyme (see Figure 3.9) is found in almost all animals. It is produced in the tears, saliva, and milk of mammals and in the albumen (whites) of bird eggs. Lysozyme digests the cell walls of bacteria, rupturing and killing them. As a result, it plays an important role as a first line of defense against invading bacteria. Most animals defend themselves against bacteria by digesting them, which is probably why most animals have lysozyme. Some animals also use lysozyme in the digestion of food.

Among mammals, a mode of digestion called foregut fermentation has evolved twice. In mammals with this mode of digestion, the foregut—the posterior esophagus or the stomach—has been converted into a chamber in which bacteria break down ingested plant matter by fermentation. Foregut fermenters can extract nutrients from the otherwise indigestible cellulose that makes up a large proportion of plant tissue. Foregut fermentation evolved independently in

experiment

Figure 23.6A Convergent Molecular Evolution

Original Paper: Stewart, C.-B et al. 1987. Adaptive evolution in the stomach lysozymes of foregut fermenters. *Nature* 330: 401–404.

Langurs (a group of monkeys) and cattle are only distantly related but both have evolved foregut fermentation. They uniquely express the enzyme lysozyme in their stomachs (foreguts) to aid in breaking down bacteria that are involved in fermentation. Stewart and colleagues compared the gene sequences of lysozyme in mammals with and without foregut fermentation to see if there is convergence in the independently evolved amino acid sequences of lysozyme in langurs and cattle.

HYPOTHESIS▶ Similar selective conditions in distantly related mammals have resulted in convergence of adaptations for foregut fermentation in the amino acid sequences of lysozyme.

METHOD

1. Isolate and sequence lysozyme from two distantly related mammal species with foregut fermentation (langur and cattle) as well as other mammals that are more closely related to either langurs or to cattle but lack foregut fermentation.

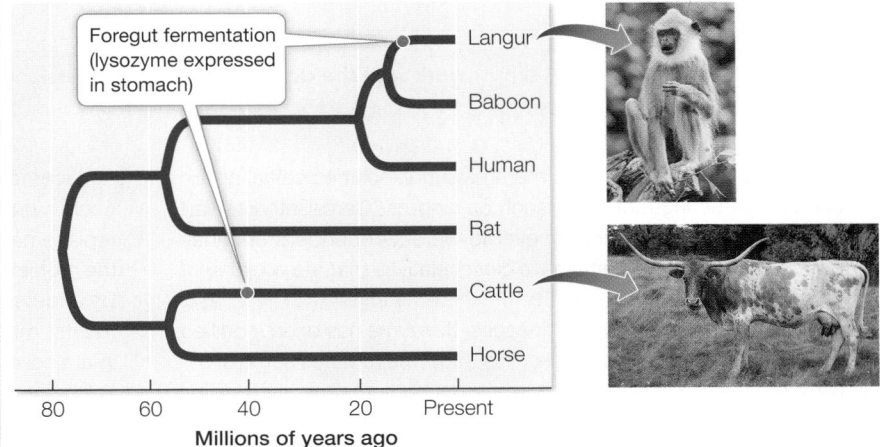

	Langur	Baboon	Human	Rat	Cattle	Horse
Langur		14	18	38	32	65
Baboon	0		14	33	39	65
Human	0	1		37	41	64
Rat	0	1	0		55	64
Cattle	5	0	0	0		71
Horse	0	0	0	0	1	

The lysozymes of langurs and cattle are convergent for 5 amino acids.

2. Tabulate the pairwise differences in the amino acid sequences. Plot the amino acid changes on the phylogenetic tree and count the number of convergent similarities between each pair of species. The results can then be plotted as a matrix.

RESULTS The matrix shows the number of all pairwise amino acid differences above the diagonal and the number of convergent similarities below the diagonal.

CONCLUSION▶ The lysozyme sequences of the two species with foregut fermentation account for the majority of the convergent amino acid replacements observed among these species, demonstrating molecular convergence associated with the independent evolution of foregut evolution.

Figure 23.6B **work with the data** follows on next page.

work with the data

Figure 23.6B Convergent Molecular Evolution

Original Paper: Stewart, C.-B., J. W. Schilling, and A. C. Wilson. 1987.

Caro-Beth Stewart and her colleagues collected lysozyme sequences from six species of mammals. A small sample of their data is shown in the table. The phylogeny of these six species is well supported from analysis of many genes and much morphological data.

Species	Amino acid position										
	2	14	17	21	50	63	75	87	117	118	130
Langur	I	K	L	K	E	Y	D	N	Q	N	V
Baboon	I	R	L	R	Q	Y	N	D	Q	N	V
Human	V	R	M	R	R	Y	N	D	Q	N	V
Rat	T	R	M	Y	Q	Y	N	D	K	N	V
Cattle	V	K	L	K	E	W	D	N	R	D	L
Horse	V	A	M	G	G	W	N	E	K	D	L
Ancestral state	V	R	M	R	Q	W	N	D	K	N	V

QUESTIONS ▶

1. Using the phylogenetic tree (see Figure 23.6A), plot the amino acid changes across the phylogeny of the six mammals. Assume that the ancestral state is the amino acid present at the base of the tree.

2. Which amino acid positions show unique convergence between the langur and cattle lineages (i.e., the derived state is found *only* in cattle and langurs)?

3. Which additional position is convergent between cattle and the ancestor of langurs and baboons?

4. Did you detect any other convergent amino acid changes between any other pair of lineages? What does this suggest about the convergent changes you observed between cattle and langurs?

A similar **work with the data** exercise may be assigned in **LaunchPad**.

---

ruminants (a group of hoofed mammals that includes cattle) and in certain leaf-eating monkeys, such as langurs. Caro-Beth Stewart knew that these evolutionary events were independent because both langurs and ruminants have close relatives that are not foregut fermenters.

In both foregut-fermenting lineages, lysozyme has been modified to play a new, nondefensive role. This lysozyme ruptures some of the bacteria that live in the foregut, releasing nutrients metabolized by the bacteria, which the mammal then absorbs. How many changes in the lysozyme molecule were needed to allow it to perform this function amid the digestive enzymes and acidic conditions of the mammalian foregut? To answer this question, Stewart and her colleagues compared the lysozyme-coding sequences in foregut fermenters with those in several of their nonfermenting relatives. They determined which amino acids differed and which were shared among the species, as well as the rates of synonymous and nonsynonymous substitutions in lysozyme genes across the evolutionary history of the sampled species.

For many of the amino acid positions of lysozyme, the rate of synonymous substitutions in the corresponding gene sequence was much higher than the rate of nonsynonymous substitutions. This observation indicates that many of the amino acids that make up lysozyme are evolving under purifying selection. In other words, there is selection against change in the protein at these positions, and the observed amino acids must therefore be critical for lysozyme function. At other positions, several different amino acids function equally well, and the corresponding gene sequences had similar rates of synonymous and nonsynonymous substitutions. The most striking finding was that amino acid replacements in lysozyme happened at a much higher rate in the lineage leading to langurs than in any other primate lineage. The high rate of nonsynonymous substitutions in the langur lysozyme gene shows that lysozyme went through a period of rapid change in adapting to the foregut of langurs. Moreover, the lysozymes of langurs and cattle share five unique amino acid

replacements, all of which lie on the surface of the lysozyme molecule, well away from the enzyme's active site. Two of these shared replacements involve changes from arginine to lysine, which makes the proteins more resistant to attack by the stomach enzyme pepsin. By understanding the functional significance of amino acid replacements, molecular evolutionists can explain the observed changes in amino acid sequences in terms of changes in the functioning of the protein.

A large body of fossil, morphological, and molecular evidence shows that langurs and ruminants do not share a recent common ancestor. Yet langur and ruminant lysozymes share several amino acids that neither mammal shares with the lysozymes of its own closer relatives. The lysozymes of these two mammals have undergone convergent evolution at some amino acid positions despite their very different ancestry. The amino acids they share give these lysozymes the ability to lyse the bacteria that ferment plant material in the foregut.

The hoatzin, an unusual leaf-eating South American bird and the only known avian foregut fermenter, offers another remarkable example of the convergent evolution of lysozyme. Many birds have an enlarged esophageal chamber called a crop. The crop of the hoatzin contains lysozyme and bacteria and acts as a fermenting chamber. Many of the amino acid replacements that occurred in the adaptation of hoatzin crop lysozyme are identical to those that evolved in ruminants and langurs. Thus even though the hoatzin and foregut-fermenting mammals have not shared a common ancestor in hundreds of millions of years, they have all evolved similar adaptations in their lysozymes that enable them to recover nutrients from their fermenting bacteria.

Genome size also evolves

We know that genome size varies tremendously among organisms. Across broad taxonomic categories, there is some correlation between genome size and organismal complexity. The genome of

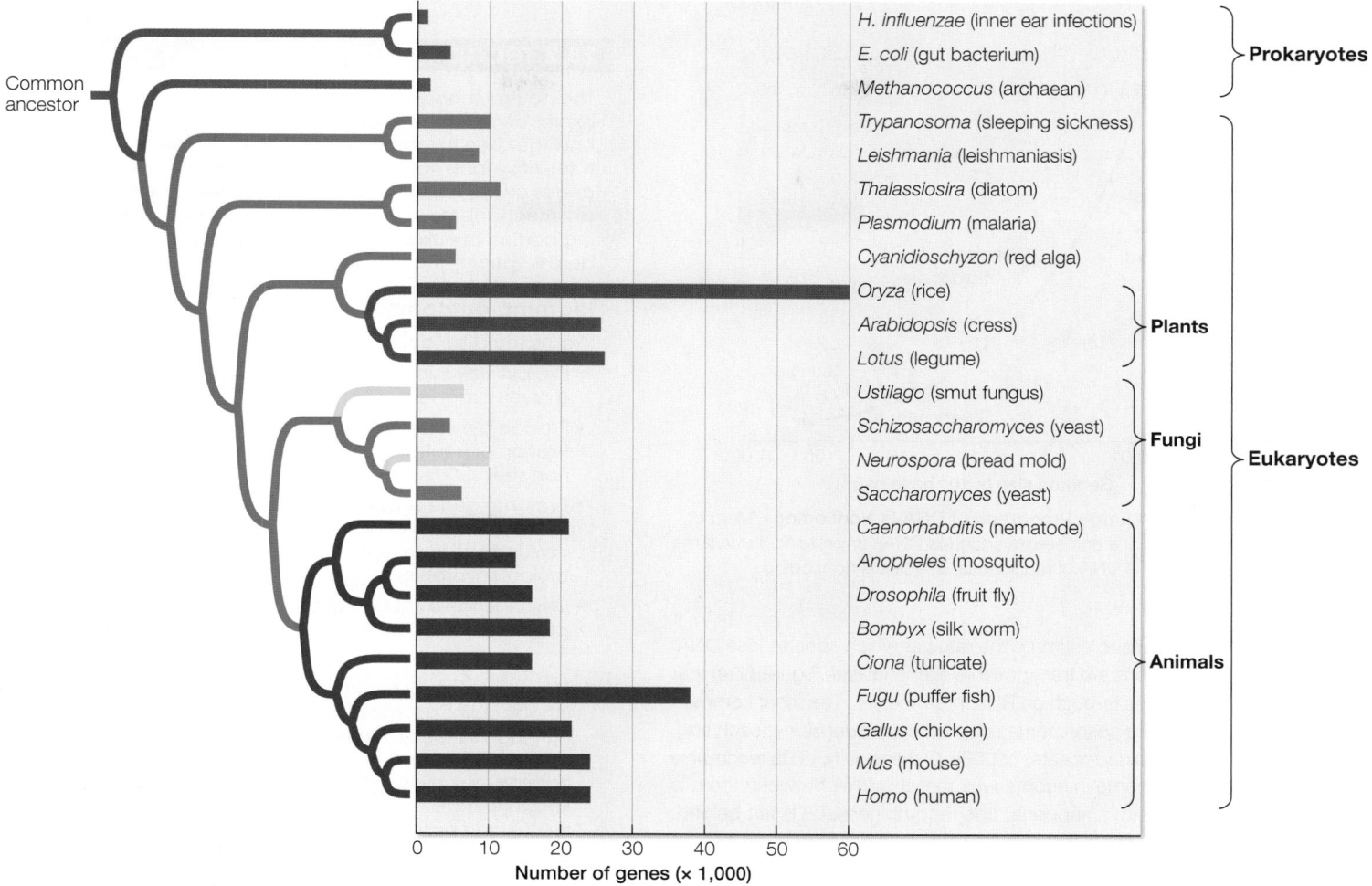

Common ancestor

H. influenzae (inner ear infections)	Prokaryotes
E. coli (gut bacterium)	
Methanococcus (archaean)	
Trypanosoma (sleeping sickness)	
Leishmania (leishmaniasis)	
Thalassiosira (diatom)	
Plasmodium (malaria)	
Cyanidioschyzon (red alga)	
Oryza (rice)	Plants
Arabidopsis (cress)	
Lotus (legume)	
Ustilago (smut fungus)	Fungi
Schizosaccharomyces (yeast)	
Neurospora (bread mold)	
Saccharomyces (yeast)	
Caenorhabditis (nematode)	Animals
Anopheles (mosquito)	
Drosophila (fruit fly)	
Bombyx (silk worm)	
Ciona (tunicate)	
Fugu (puffer fish)	
Gallus (chicken)	
Mus (mouse)	
Homo (human)	

Eukaryotes

Number of genes (× 1,000)
0 10 20 30 40 50 60

Figure 23.7 Genome Size Varies Widely   This tree shows the numbers of genes from a sample of organisms whose genomes have been fully sequenced, arranged by their evolutionary relationships. Bacteria and archaea typically have fewer genes than most eukaryotes. Among eukaryotes, multicellular organisms with tissue organization (plants and animals; dark green and blue branches) have more genes than single-celled organisms (aqua branches) or multicellular organisms that lack pronounced tissue organization (yellow branches).

the tiny bacterium *Mycoplasma genitalium* has only 470 genes. *Rickettsia prowazekii*, the bacterium that causes typhus, has 634 genes. *Homo sapiens*, by contrast, has about 21,000 protein-coding genes. **Figure 23.7** shows the numbers of genes in a sample of organisms whose genomes have been fully sequenced, arranged by their evolutionary relationships. As this figure reveals, a larger genome does not always indicate greater complexity (compare rice with the other plants, for example). It is not surprising that more complex genetic instructions are needed for building and maintaining a large multicellular organism than a small single-celled bacterium. What *is* surprising is that some multicellular organisms, such as lungfish, some salamanders, and lilies, have about 40 times as much DNA as humans do. Structurally, a lungfish or a lily is not 40 times more complex than a human. So why does genome size vary so much?

Differences in genome size are not so great if we take into account only the portion of the DNA that actually encodes RNAs or proteins. Although the organisms with the largest total amounts of nuclear DNA (some ferns and flowering plants) have 80,000 times as much DNA as do the bacteria with the smallest genomes, no species has more than about 100 times as many protein-coding genes as a bacterium. Therefore much of the variation in genome size lies not in the number of functional genes but in the amount of noncoding DNA (**Figure 23.8**).

Why do the cells of most eukaryotic organisms have so much non-coding DNA? As we noted earlier, some of the noncoding DNA has a regulatory function that controls the degree or timing of expression coding genes. But the genomes of many species have far more noncoding DNA that is used for gene regulation. Does this extra noncoding DNA have a function, or is it "junk"? Many regions of noncoding DNA consist of pseudogenes (nonfunctional copies of former genes) that are carried in the genome simply because the cost of doing so is very small. These pseudogenes may become the raw material for the evolution of new genes with novel functions. Some noncoding DNA functions solely in maintaining chromosomal structure. Still other sequences consist of "selfish" transposable elements that proliferate because they reproduce faster than the host genome.

DNA does not just accumulate in genomes over time; non-critical nucleotide sequences are also lost from genomes. Some species differ so much in genome size because they lose non-critical sequences at very different rates. Investigators can use

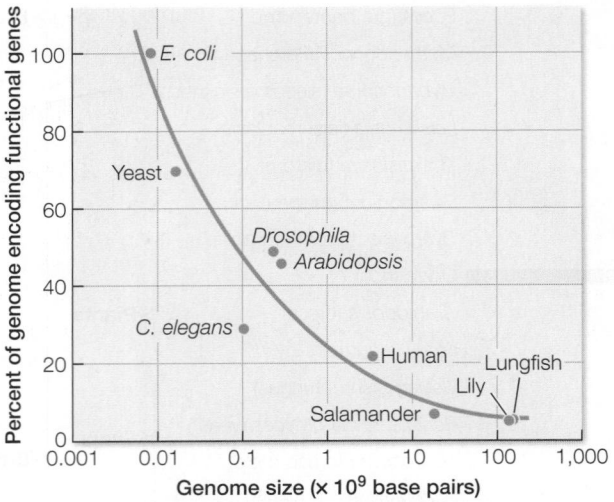

Figure 23.8 A Large Proportion of DNA Is Noncoding Most of the DNA of bacteria and yeasts encodes RNAs or proteins, but a large percentage of the DNA of multicellular species is noncoding.

retrotransposons to estimate the rates at which species lose DNA. Retrotransposons are transposable elements (see Figure 17.4) that copy themselves through an RNA intermediate. The most common type of retrotransposon carries duplicated sequences at each end, called long terminal repeats, or LTRs. Occasionally, LTRs recombine in the host genome in such a way that the DNA between them is excised. When this happens, one recombined LTR is left behind. The number of such "orphaned" LTRs in a genome is a measure of how many retrotransposons have been lost. By comparing the number of LTRs in the genomes of Hawaiian crickets (*Laupala*) and fruit flies (*Drosophila*), investigators found that *Laupala* loses DNA more than 40 times more slowly than does *Drosophila*. Therefore it is not surprising that the genome of *Laupala* is much larger than that of *Drosophila*.

Why do species differ so greatly in the rate at which they gain or lose apparently functionless DNA? One hypothesis is that genome size is related to the rate at which the organism develops, which may be under selection pressure. Large genomes can slow down the rate of development and thus alter the relative timing of expression of particular genes. As discussed in Key Concept 19.4, changes in the timing of gene expression—heterochrony—can produce major changes in phenotype. Thus although some noncoding DNA sequences may have no direct function, they may still affect the development of the organism.

Another hypothesis is that the proportion of noncoding DNA is related primarily to population size. Noncoding sequences that are only slightly deleterious to the organism are likely to be purged by selection most efficiently in species with large population sizes. In species with small populations, the effects of genetic drift can overwhelm selection against noncoding sequences that have small deleterious consequences. Therefore selection against the accumulation of noncoding sequences is most effective in species with large populations, and such species (such as bacteria and yeasts) have relatively little noncoding DNA compared with species with small populations (see Figure 23.9).

23.2 recap

The neutral theory of molecular evolution provides an explanation for the relatively constant rate of molecular change seen in many species. By examining the relative rates of synonymous and nonsynonymous substitutions in genes over time, biologists can distinguish the evolutionary mechanisms acting on individual genes. The noncoding portion of eukaryote genomes is more variable in size across species than is the coding portion.

learning outcomes

You should be able to:

- Explain why substitutions may be neutral or selective in a given organism.
- Provide a mathematical explanation for why the rate of fixation of neutral mutations is independent of population size.
- Evaluate gene sequences of protein-coding genes to identify codons that are evolving under purifying versus positive selection, and apply this knowledge to a biological problem.
- Describe the evolution of genomes in a way that incorporates consideration of noncoding DNA.

1. Over evolutionary history, many groups of organisms that inhabit caves have lost the organs of sight. For example, although surface-dwelling crayfish have functional eyes, several crayfish species that are restricted to underground habitats lack eyes. Opsins are light-sensitive proteins known to have an important function in vision (see Chapter 45), and opsin genes are expressed in eye tissues. Opsin genes are present in the genomes of eyeless, cave-dwelling crayfish. Two alternative hypotheses that might explain the presence of opsin genes in an eyeless organism are (1) the genes were inherited from an eyed ancestor, but are no longer functional; or (2) the genes have a new function other than vision. How would you investigate these two hypotheses using the sequences of the opsin genes in various species of crayfish?

2. Why is the rate of fixation of neutral mutations independent of population size?

3. An investigator compared many gene sequences encoding surface proteins from influenza viruses sampled over time and collected the data shown in the table below.

Codon position	Synonymous substitutions	Nonsynonymous substitutions
12	0	7
15	1	9
61	0	12
80	7	0
137	12	1
156	24	2
165	3	4
226	38	3

a. Which positions encode amino acids that are likely to have changed as a result of positive selection? Explain your answer.

23.2 recap (continued)

b. Which positions encode amino acids that are likely to have changed as a result of purifying selection? Explain your answer.

(Hint: To calculate rates of each substitution type, consider the number of synonymous and nonsynonymous substitutions present *relative to the number of possible substitutions of each type.* There are approximately three times as many possible nonsynonymous substitutions as there are synonymous substitutions.)

4. Suggest and contrast two hypotheses for the wide diversity of genome sizes among different organisms.)

We have examined some of the ways in which organisms can lose DNA without losing gene functions. But how do organisms gain new functions through time?

key concept
23.3 Lateral Gene Transfer and Gene Duplication Can Produce Major Changes

As we noted in the previous section, most multicellular organisms have many more genes than do most unicellular species. But multicellular organisms evolved from unicellular ancestors. How did the numbers of genes within the genomes of multicellular organisms increase over evolutionary time? There are two primary mechanisms that can result in such increases: genes can be transferred from other species, or genes can be duplicated within species.

focus your learning

- Genes can sometimes move among distantly related lineages on the tree of life.
- The duplication of a gene provides opportunities for the evolution of new functions.
- Some genes are present in multiple copies in the genome, and these copies often evolve together through time.

Lateral gene transfer can result in the gain of new functions

Chapter 22 described how, through the process of speciation, ancestral lineages divide into descendant lineages, and it is those speciation events that are captured by the branches in the tree of life. However, there are also processes of **lateral gene transfer**, which allow individual genes, organelles, or fragments of genomes to move horizontally from one lineage to another. Some species may pick up fragments of DNA directly from the environment. Other genes may be picked up in a viral genome and transferred to a new host when the virus becomes integrated into the new host's genome. Hybridization between species also results in the transfer of large numbers of genes.

Lateral gene transfer can be highly advantageous to a species that incorporates novel genes from a distant relative. Genes that confer antibiotic resistance, for example, are commonly transferred among different species of bacteria. Lateral gene transfer is another way, in addition to mutation and recombination, in which species can increase their genetic variation. That genetic variation then provides the raw material on which selection acts, resulting in evolution.

A phylogenetic tree constructed from a single laterally transferred genome fragment is likely to reflect only the evolutionary history of that fragment, rather than the overall organismal phylogeny (see Key Concept 25.1). Most biologists prefer to build trees from large samples of genes or their products, so that the underlying species tree (as well as any lateral gene transfer events) can be reconstructed. Depictions of lateral gene transfer events on the underlying species tree are known as **reticulations**.

The degree to which lateral gene transfer events occur in various parts of the tree of life is a matter of considerable current investigation and debate. Lateral gene transfer appears to be relatively uncommon among most eukaryote lineages, although the two major endosymbioses that gave rise to mitochondria and chloroplasts can be viewed as lateral transfers of entire bacterial genomes to the eukaryote lineage. Some groups of eukaryotes, most notably some plants, are subject to relatively high levels of hybridization among closely related species. Hybridization leads to the exchange of many genes among recently separated lineages of plants. The greatest degree of lateral transfer, however, appears to occur among bacteria. Many bacterial genes have been transferred repeatedly among lineages of bacteria, to the point that relationships among bacterial species are often hard to decipher. Nonetheless, the broad relationships of the major groups of bacteria can still be determined (as we will discuss in Part Seven of this book). Lateral transfer of genes also makes it difficult to identify the boundaries of bacterial species, which is one reason why fewer bacterial species have been named than are known to exist.

Most new functions arise following gene duplication

Gene duplication is yet another way in which genomes can acquire new functions. When a gene is duplicated, one copy of that gene is potentially freed from having to perform its original function. The initially identical copies of a duplicated gene can have any one of four subsequent fates:

1. Both copies of the gene may retain their original function (which can result in a change in the amount of gene product that is produced by the organism).

2. Both copies of the gene may retain the ability to produce the original gene product, but the expression of the genes may diverge in different tissues or at different times in development.

3. One copy of the gene may be incapacitated by the accumulation of deleterious substitutions and become a nonfunctional pseudogene, or may be eliminated from the genome altogether.

4. One copy of the gene may retain its original function while the second copy accumulates enough substitutions that it can perform a different function.

How often do *gene duplications arise, and which of these four outcomes is most likely? Investigators have found that rates of gene duplication are fast enough for a yeast or *Drosophila* population to acquire several hundred duplicate genes over the course of a million years. They have also found that most of the duplicated genes in these organisms are very young. Many extra genes are lost from a genome within 10 million years (which is rapid on an evolutionary time scale).

---

*connect the concepts   Key Concept 19.1 describes shared developmental mechanisms controlled by specific DNA sequences that have been modified and reshuffled to produce the remarkable diversity of plants, animals, and other organisms we know today. Similarity in the homeobox sequence common to the Hox genes suggests that the Hox genes arose through duplication of an ancestral gene, which then diverged to take on new functions.

Some genes may be duplicated many times, resulting in large numbers of related pseudogenes scattered throughout the genome. In the human genome, the functional copy of the ribosomal protein gene *RPL21* is located on chromosome pair 13, but pseudogenes derived from it are found on most of the other chromosome pairs (**Figure 23.9**). Although not all genes are represented by pseudogenes, there are nearly as many known pseudogenes in the human genome as there are functional protein-coding genes.

Although many extra genes disappear rapidly, some duplication events lead to the evolution of genes with new functions. Several successive rounds of duplication and mutation may result in a **gene family**: a group of homologous genes with related functions, often arrayed in tandem along a chromosome. An example of this process is provided by the globin gene family (see Figure 17.9). The globins were among the first proteins to be sequenced and compared. Comparisons of their amino acid sequences strongly suggest that the different globins arose via gene duplications. These comparisons also allow us to estimate how long the globins have been evolving separately, because differences among these proteins have accumulated with time.

Hemoglobin, a tetramer (four-subunit molecule) consisting of two α-globin and two β-globin polypeptide chains, carries oxygen in blood. Myoglobin, a monomer, is the primary O_2 storage protein in muscle. Myoglobin's affinity for O_2 is much higher than that of hemoglobin, but hemoglobin has evolved to be more diversified in its roles. Hemoglobin binds O_2 in the lungs or gills, where the O_2 concentration is relatively high, transports it to deep body tissues, where the O_2 concentration is low, and releases it in those tissues. With its more complex tetrameric structure, hemoglobin is able to carry four molecules of O_2, as well as hydrogen ions and carbon dioxide, in the blood.

To estimate the time of the globin gene duplication that gave rise to the α- and β-globin gene clusters, we can create a **gene tree**—a phylogenetic tree that describes the evolutionary history of particular genes or gene families, in this case the gene sequences that encode the various globins (**Figure 23.10**). The rate of molecular evolution of globin genes has been estimated from other studies, using the divergence times of groups of vertebrates that are well documented in the fossil record. These studies indicate an average rate of divergence for globin genes of about 1 nucleotide substitution every 2 million years. By applying this rate to the globin gene tree, we can estimate the divergence time of the two globin gene clusters at about 450 million years ago.

Many gene duplications affect only one or a few genes at a time, but entire genomes are duplicated in polyploid organisms (which include many plants). When all the genes are duplicated, there are massive opportunities for new functions to evolve. That is exactly what appears to have happened in the evolution of vertebrates. The genomes of the jawed vertebrates appear to have four diploid sets of many major genes, which led biologists to conclude that two genome-wide duplication events occurred in the ancestor of these species. These duplications have allowed considerable specialization of individual vertebrate genes, many of which are now highly tissue-specific in their expression.

Some gene families evolve through concerted evolution

Although the members of the globin gene family have diversified in form and function, the members of many other gene families do not evolve independently of one another. For instance, almost all

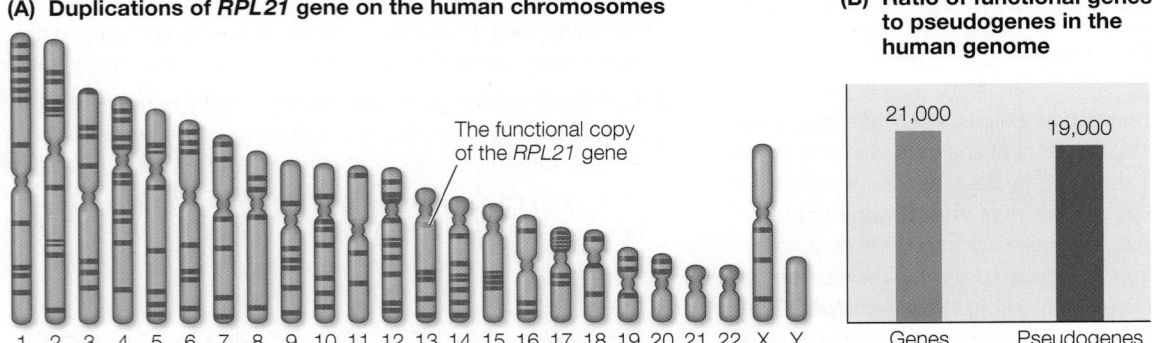

(A) Duplications of *RPL21* gene on the human chromosomes

The functional copy of the *RPL21* gene

1 2 3 4 5 6 7 8 9 10 11 12 13 14 15 16 17 18 19 20 21 22 X Y

(B) Ratio of functional genes to pseudogenes in the human genome

21,000   19,000

Genes   Pseudogenes

Figure 23.9 Some Functional Genes Are Duplicated Many Times as Nonfunctional Pseudogenes   (A) The functional gene that encodes ribosomal protein RPL21 is located on human chromosome 13 (indicated in orange). In addition, there are many nonfunctional pseudogenes of *RPL21* in the human genome, produced through repeated duplication events (indicated in blue). (B) Although *RPL21* represents a relatively extreme example of pseudogene duplication, there are almost as many known pseudogenes in the human genome as there are functional genes.

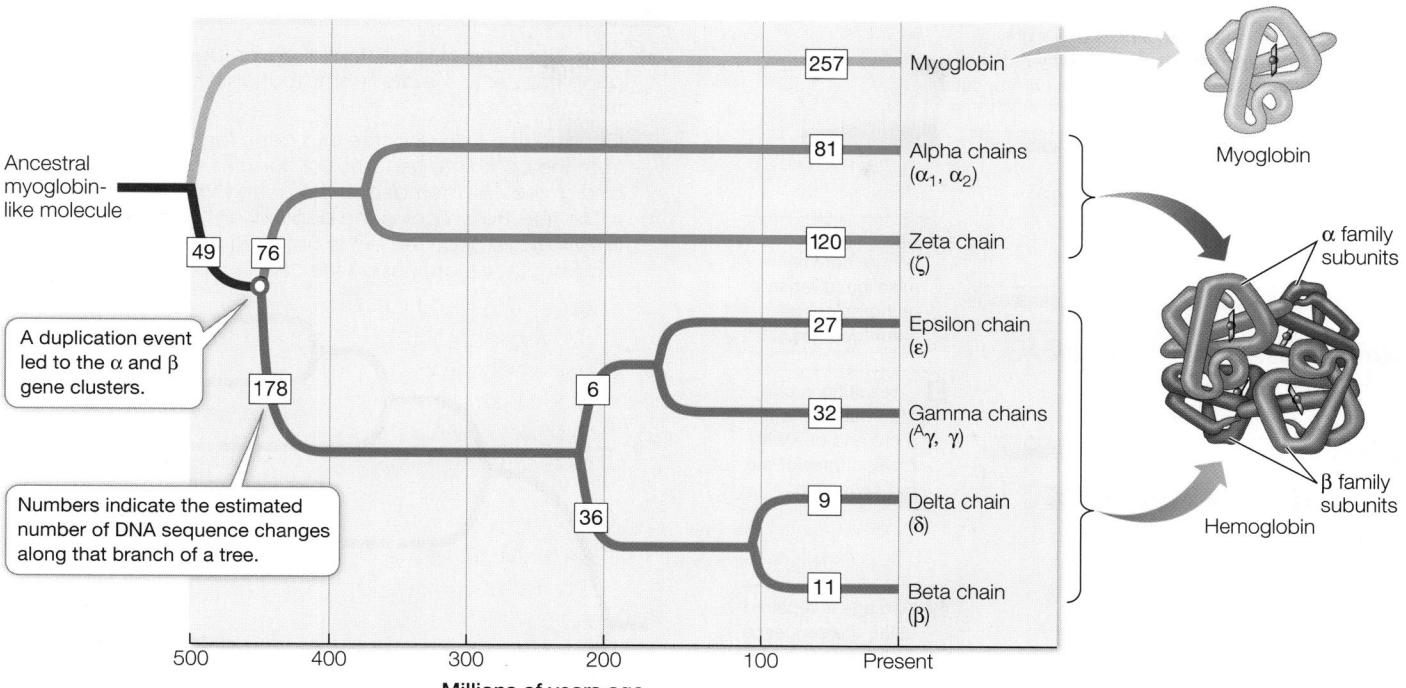

Ancestral myoglobin-like molecule

A duplication event led to the α and β gene clusters.

Numbers indicate the estimated number of DNA sequence changes along that branch of a tree.

257 Myoglobin

81 Alpha chains (α₁, α₂)

120 Zeta chain (ζ)

27 Epsilon chain (ε)

32 Gamma chains (ᴬγ, γ)

9 Delta chain (δ)

11 Beta chain (β)

49 76 178 6 36

500 400 300 200 100 Present
Millions of years ago

Myoglobin

α family subunits

β family subunits

Hemoglobin

Figure 23.10 A Globin Family Gene Tree A molecular clock analysis suggests that the α-globin (blue) and β-globin (green) gene clusters diverged about 450 million years ago, soon after the origin of the vertebrates.

 Animation 23.1 Concerted Evolution
www.Life11e.com/a23.1

Q: When did the gene duplication event that gave rise to the delta and beta chains occur?

 Activity 23.3 Gene Tree Construction
www.Life11e.com/ac23.3

organisms have many copies (up to thousands) of the ribosomal RNA genes. Ribosomal RNA (rRNA) is the principal structural element of ribosomes and, as such, has a primary role in protein synthesis. Every living species needs to synthesize proteins, often in large amounts (especially during early development). Having many copies of the rRNA genes ensures that organisms can rapidly produce many ribosomes and thereby maintain a high rate of protein synthesis.

Like all portions of the genome, ribosomal RNA genes evolve, and differences accumulate in the rRNA genes of different species. But within any one species, the multiple copies of the rRNA genes are very similar, both structurally and functionally. This similarity makes sense because, ideally, every ribosome in a species should synthesize proteins in the same way. In other words, within a given species, the multiple copies of these rRNA genes are evolving in concert with one another, a phenomenon called **concerted evolution**.

How does concerted evolution occur? Two different mechanisms appear to be responsible. The first of these is **unequal crossing over**. When DNA is replicated during meiosis in a diploid species, the homologous chromosome pairs align and recombine by crossing over (see Key Concept 11.4). In the case of highly repeated genes, however, it is easy for genes to become displaced in alignment, since so many copies of the same genes are present on the chromosomes (**Figure 23.11A**). The end result is that one chromosome

may gain extra copies of the repeat and the other chromosome may have fewer copies of the repeat. If a new substitution arises in one copy of the repeat, it can spread to new copies (or be eliminated) through unequal crossing over. Thus, over time, a novel substitution will either become fixed or it will be lost entirely. In either case, all copies of the repeat will remain very similar to one another.

The second mechanism that produces concerted evolution is **biased gene conversion**. This mechanism can be much faster than unequal crossing over and has been shown to be the primary mechanism for concerted evolution of rRNA genes. DNA strands are frequently broken and repaired (see Key Concept 13.4). At many times during the cell cycle, the genes for ribosomal RNA are clustered close together. If damage occurs to one of the genes, a copy of the rRNA gene on another chromosome may be used to repair the damaged copy, and the sequence that is used as a template can thereby replace the original sequence (**Figure 23.11B**). In many cases, this repair system appears to be biased in favor of using particular sequences as templates for repair, and thus the favored sequence rapidly spreads across all copies of the gene. In this way, changes may appear in a single copy and then rapidly spread to all the other copies.

Regardless of the mechanism responsible, the net result of concerted evolution is that the copies of a highly repeated gene do not evolve independently of one another. Mutations still occur, but once they arise in one copy, they either spread rapidly across all the copies or are lost from the genome completely. This process allows the products of each copy to remain similar over time in both sequence and function.

(A) Unequal crossing over

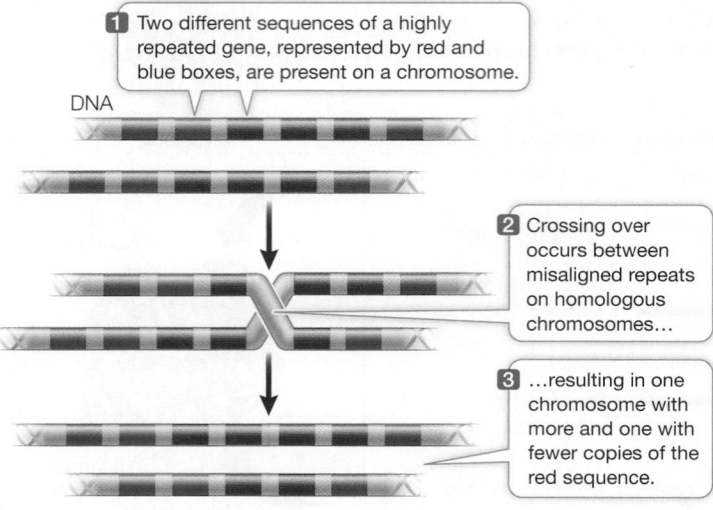

1 Two different sequences of a highly repeated gene, represented by red and blue boxes, are present on a chromosome.

DNA

2 Crossing over occurs between misaligned repeats on homologous chromosomes...

3 ...resulting in one chromosome with more and one with fewer copies of the red sequence.

(B) Biased gene conversion

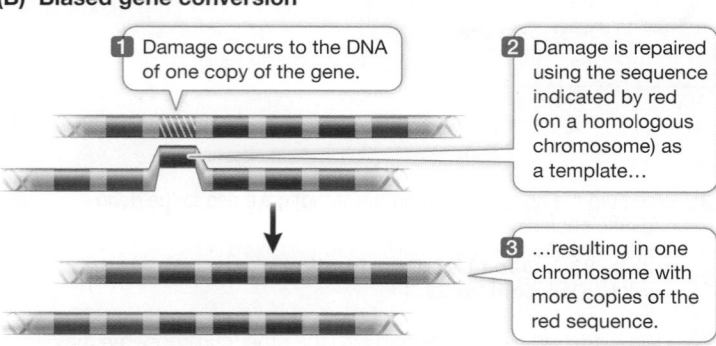

1 Damage occurs to the DNA of one copy of the gene.

2 Damage is repaired using the sequence indicated by red (on a homologous chromosome) as a template...

3 ...resulting in one chromosome with more copies of the red sequence.

Figure 23.11 Concerted Evolution Two mechanisms can produce concerted evolution of highly repeated genes. **(A)** Unequal crossing over results in deletions and duplications of a repeated gene. **(B)** Biased gene conversion can rapidly spread a new variant across multiple copies of a repeated gene.

We have seen how the principles and methods of molecular evolution have opened new vistas in evolutionary biology. Next we will consider some of the practical applications of this field.

23.3 recap

Lateral gene transfer can result in the transfer of genetic functions between distantly related species. Gene duplication can lead to the evolution of new functions. Some highly repeated genes undergo concerted evolution, which maintains uniform functionality.

learning outcomes

You should be able to:

- Describe how genes can move among different lineages, especially among bacteria, via lateral gene transfer.
- Use a phylogenetic tree of a gene family to infer gene duplications across the history of a group of species.
- Describe how a duplicated gene provides opportunities for the evolution of new functions.
- Diagram and distinguish the two primary processes that produce concerted evolution.

23.3 recap

1. What are some of the potential advantages of lateral gene transfer to the organisms that gain new genes by this mechanism?

2. Consider the following tree of a gene family sampled from the complete genomes of humans, chimpanzees, and gorillas. All members of the gene family are included in the tree. How many gene duplications and losses are likely to have occurred in the history of this gene family? Diagram their locations on the tree.

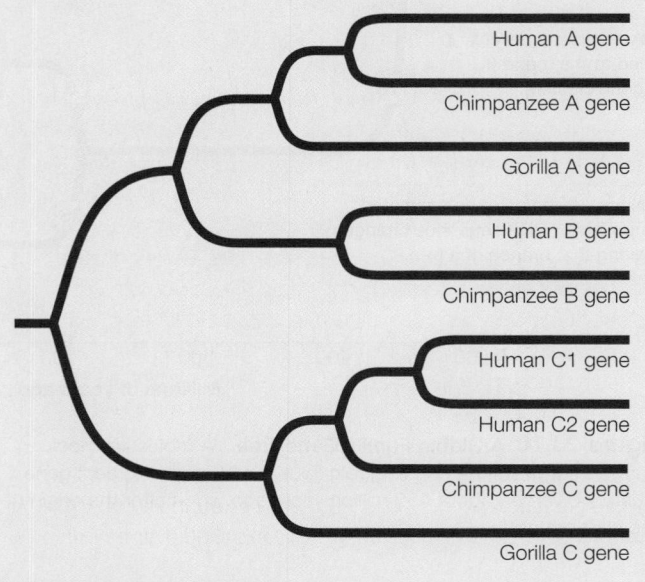

- Human A gene
- Chimpanzee A gene
- Gorilla A gene
- Human B gene
- Chimpanzee B gene
- Human C1 gene
- Human C2 gene
- Chimpanzee C gene
- Gorilla C gene

3. Why is gene duplication considered important for long-term evolutionary change?

4. Describe the pattern of concerted evolution among highly repeated genes, and diagram the processes that lead to concerted evolution.

> **key concept**
>
> # 23.4
> ## Molecular Evolution Has Many Practical Applications

Studies of molecular evolution have practical applications throughout biology, from understanding basic aspects of biological function to studies of human health.

focus your learning

- The evolutionary history of genes provides information about protein function.
- The principles of in vitro evolution can be applied to produce new molecules with novel and useful functions.
- Studies of molecular evolution can provide knowledge of vectors and disease transmission that is important for understanding and controlling epidemics.

Molecular sequence data are used to determine the evolutionary history of genes

A gene tree can show the evolutionary relationships of a single gene in different species, or it can trace the evolution of members of a gene family (as in Figure 23.11). The methods for constructing a gene tree are the same as those we described in Key Concept 21.2 for building phylogenetic trees of species. The process involves identifying differences between genes and using those differences to reconstruct the evolutionary history of the genes. Gene trees are often used to construct phylogenetic trees of species, but the two types of trees are not necessarily equivalent. Processes such as gene duplication can give rise to differences between the phylogenetic trees of genes and species. From a gene tree, biologists can reconstruct the history and timing of gene duplication events and learn how gene diversification has resulted in the evolution of new protein functions.

All the genes of a particular gene family have similar sequences because they have a common ancestry. As we discussed in Key Concept 21.1, features that are similar as a result of common ancestry are said to be homologous. When discussing gene trees, however, we usually need to distinguish between two forms of homology. Homologous genes that are found in different species and whose divergence we can trace to the speciation events that gave rise to those species are called **orthologs**. Homologous genes in the same or different species that are related through gene duplication events are called **paralogs**. When we examine a gene tree, the questions we wish to address determine whether we should compare orthologous or paralogous genes. If we wish to reconstruct the evolutionary history of the species that contain the genes, then our comparison should be restricted to orthologs (because they will reflect the history of speciation events). If we are interested in the changes in function that have resulted from gene duplication events, however,

then the appropriate comparison is among paralogs (because they will reflect the history of gene duplication events). If our focus is on the diversification of a gene family through both processes, then we will want to include both paralogs and orthologs in our analysis.

Figure 23.12 depicts a gene tree for the members of a gene family called *engrailed* (its members encode transcription factors that regulate development). At least three gene duplications have occurred in this family, resulting in up to four different *engrailed* genes (*En*) in some vertebrate species (such as the zebrafish). All of the *engrailed* genes are homologs because they have a common ancestor. Gene duplication events have generated paralogous *engrailed* genes (*En1* and *En2*) in some lineages of vertebrates. We could compare the orthologous sequences of the *En1* group of genes to reconstruct the history of the bony vertebrates (i.e., all the vertebrate species in Figure 23.12 except the lamprey), or we could use the orthologous sequences of the *En2* group of genes and expect the same answer (because there is only one history of the underlying speciation events). All bony vertebrates have both groups of *engrailed* genes because the two groups arose from a gene duplication event in the common ancestor of bony vertebrates. If we wanted to focus on the diversification that occurred as a result of this duplication, then the appropriate comparison would be between the paralogous genes of the *En1* versus *En2* groups.

Gene evolution is used to study protein function

Earlier in this chapter we discussed the ways in which biologists can detect regions of genes that are under positive selection for change. What are the practical uses of this information? Consider the evolution of the family of genes encoding *voltage-gated sodium channels. Sodium channels have many functions, including the control of nerve impulses in the nervous system. Sodium channels can be blocked by various toxins, such as tetrodotoxin (TTX), a neurotoxin present in the tissues of some puffer fish and several other animals. A human who eats those tissues of a puffer fish that contain TTX can become paralyzed and die because the toxin blocks sodium channels and prevents nerves and muscles from functioning.

---

*connect the concepts   The functions of voltage-gated sodium channels are discussed in detail in Key Concept 44.2.

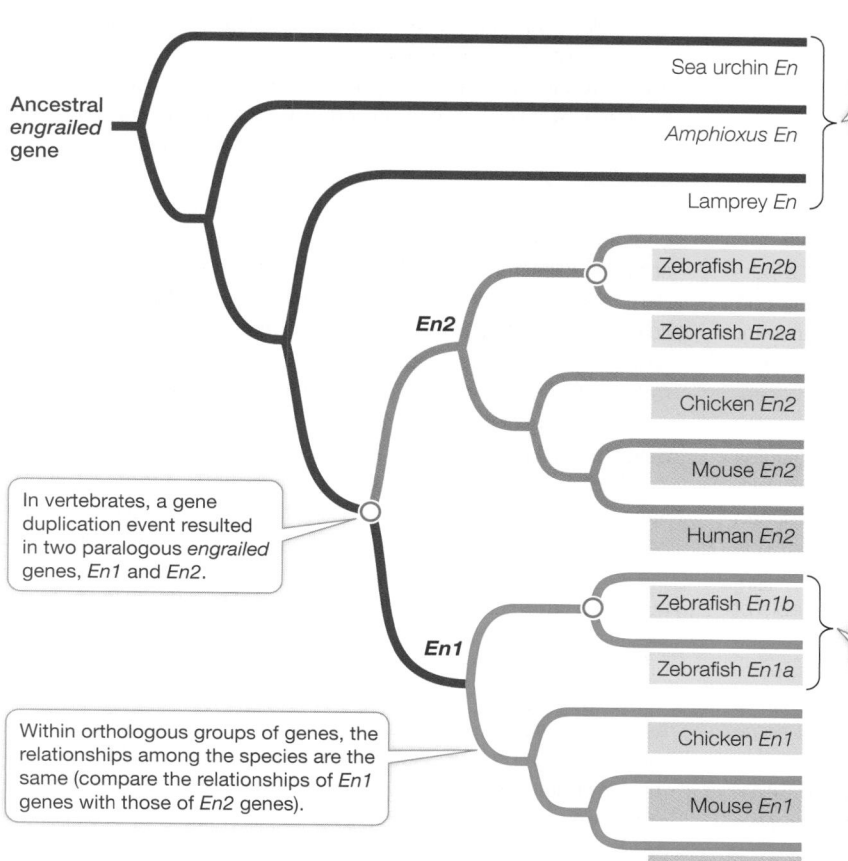

Ancestral *engrailed* gene

Sea urchin *En*

These species have a single *engrailed* gene.

Amphioxus *En*

Lamprey *En*

Zebrafish *En2b*

En2

Zebrafish *En2a*

Chicken *En2*

Mouse *En2*

Human *En2*

In vertebrates, a gene duplication event resulted in two paralogous *engrailed* genes, *En1* and *En2*.

Zebrafish *En1b*

Additional gene duplications occurred in the zebrafish lineage.

En1

Zebrafish *En1a*

Within orthologous groups of genes, the relationships among the species are the same (compare the relationships of *En1* genes with those of *En2* genes).

Chicken *En1*

Mouse *En1*

Human *En1*

Figure 23.12 Phylogeny of the *engrailed* Genes   The *engrailed* genes are homologous because they share a common ancestor. Speciation events have generated orthologous *engrailed* genes, and gene duplication events (open circles) have generated paralogous *engrailed* genes among bony vertebrates.

Q: How many gene duplication events occurred just within the zebrafish lineage?

But puffer fish have sodium channels too, so why doesn't the TTX cause paralysis in the puffer fish themselves? The sodium channels of puffer fish (and of other animals that sequester TTX, such as the rough-skinned newt shown in Figure 20.20) have evolved to become resistant to the toxin. Nucleotide substitutions in the puffer fish genome have resulted in changes in the proteins that make up the sodium channels, and those changes prevent TTX from binding to the sodium channel pore.

Several different substitutions that result in TTX resistance have evolved in the various duplicated sodium channel genes of the many species of puffer fish. Many other changes that have nothing to do with the evolution of TTX resistance have occurred in these genes as well. Biologists who study the function of sodium channels can learn a great deal about how the channels work (and about neurological diseases that are caused by mutations in the sodium channel genes) by understanding which changes have been selected for TTX resistance. They do this by comparing the rates of synonymous and nonsynonymous substitutions across the genes in various lineages that have evolved TTX resistance. In a similar manner, molecular evolutionary principles are used to understand function and diversification of function in many other proteins.

In vitro evolution is used to produce new molecules

As biologists studied the relationships among selection, evolution, and function in macromolecules, they realized that molecular evolution could be used in a controlled laboratory environment to produce new molecules with novel and useful functions. Thus were born the applications of **in vitro evolution**.

Living organisms produce thousands of compounds that humans have found useful. The search for such naturally occurring compounds, which can be used for pharmaceutical, agricultural, or industrial purposes, has been termed "bioprospecting." These compounds are the result of millions of years of molecular evolution across millions of species of living organisms. Yet biologists can also imagine molecules that *could* have evolved but, lacking the right combination of selection pressures and opportunities, have not. For instance, we might want to have a molecule that binds a particular environmental contaminant so that the contaminant can be isolated and extracted from the environment. But if the contaminant is synthetic (i.e., not produced naturally), it is unlikely that any living organism will have evolved a molecule with the function we desire. This problem was the inspiration for the field of in vitro evolution.

The principles of in vitro evolution are based on the principles of molecular evolution that we have learned from the natural world. Consider the evolution of a new RNA molecule that was produced in the laboratory using the principles of mutation and selection. This molecule's intended function was to join two other RNA molecules. That is, the lab created a ribozyme—an RNA molecule capable of catalyzing a biochemical reaction. The process started with a large pool of random RNA sequences (10^{15} different sequences, each about 300 nucleotides long), which were then selected for any ligase activity (**Figure 23.13**). None were very effective ligases, but some were slightly better than others. The best of the ribozymes were selected and reverse-transcribed into cDNA (using the enzyme reverse transcriptase). The cDNA molecules were then amplified using the polymerase chain reaction (PCR; see Figure 13.19).

PCR amplification is not perfect, and it introduced many new mutations into the pool of sequences. These sequences were then transcribed back into RNA molecules using RNA polymerase, and the process was repeated. The ligase activity of the RNAs evolved quickly; after 10 rounds of in vitro evolution, it had increased by about 7 million times. Similar techniques have since been used to create a wide variety of molecules with novel enzymatic and binding functions.

Molecular evolution is used to study and combat diseases

Many of the most problematic human diseases are caused by living, evolving organisms that present a moving target for modern medicine. Recall the example of influenza described in the opening story of this chapter and that of HIV in Chapter 21. The control of these and many other human diseases depends on techniques that can track the evolution of pathogenic organisms over time.

The transportation advances of the past century have allowed humans to move around the world with unprecedented speed and frequency. Unfortunately, this mobility has allowed pathogens to be transmitted among human populations, and between humans and other animals, at increasing rates. Cross-species transmission of viruses has led to the global emergence of many "new" diseases.

In the opening of this chapter, we described the devastating influenza pandemic of 1918–1919. Michael Worobey and his colleagues used evolutionary analyses of influenza genomes to explore the reasons why this particular flu epidemic was so much more severe than any before or since (**Investigating Life: Why Was the 1918–1919 Influenza Pandemic So Severe?**). They reconstructed the evolutionary history of influenza viruses in birds, pigs, horses, and humans and found that influenza viruses have repeatedly been transferred from birds to the various mammals. Most cross-species transfers have been accompanied by a major epidemic in the new host. The 1918–1919 epidemic occurred immediately after such a transfer between hosts. Human immune systems had no experience with the newly acquired virus, and the strong resulting reaction caused many of the strongest symptoms and deaths. Now that cross-species transmission has been associated with major flu epidemics, health workers around the world are constantly monitoring flu cases for indications of new cross-species transmissions. When emerging viruses are identified early on, efforts to control and irradicate them have a good chance of being successful. So far, our understanding of molecular evolution has prevented a repeat of the devastating flu pandemic of 1918–1919.

Studies of the genomes of many infectious agents have advanced our understanding and treatment of the diseases they cause. For example, rodent-borne hantaviruses were identified as the source of widespread respiratory illnesses, and the virus (and its host) that causes Sudden Acute Respiratory Syndrome (SARS) was identified using evolutionary comparisons of genes. Studies of the origins, the timing of emergence, and the global diversity of many human pathogens depend on the principles of molecular evolution, as do the efforts to develop and use effective vaccines against these pathogens. For example, the techniques used to develop modern polio vaccines, as well as the methods used to track their effectiveness in human populations, rely on molecular evolutionary approaches.

experiment

Original Papers: Tumpey, T. M. et al. 2005. Characterization of the reconstructed 1918 Spanish influenza pandemic virus. *Science* 310: 77–80.

Worobey, M., G.-Z. Han, and A. Rambaut. 2014. A synchronized global sweep of the internal genes of modern avian influenza virus. *Nature* 508: 254–257.

As conveyed in the story that opens this chapter, the 1918–1919 influenza pandemic killed more than 50 million people worldwide—many more than the number of World War I–related combat deaths. Unlike flu epidemics before or since, the flu viruses that circulated in the 1918–1919 epidemic triggered an especially intense reaction in the human immune system. What was different about the flu viruses circulating at that time?

Terrence Tumpey and his colleagues at the Centers for Disease Control and Prevention isolated and sequenced a complete flu virus from biopsies of 1918 victims of the pandemic. Michael Worobey of the University of Arizona and his colleagues then analyzed and compared the genomes of flu viruses that have been collected from 1918 to the present, from birds, pigs, horses, and humans. The viruses are largely transmitted within each of these host groups, but rarely jump between host groups. Worobey and his colleagues reconstructed the phylogeny of the viruses to determine if the 1918 virus had recently "jumped" into human populations from another species.

HYPOTHESIS▶ The 1918–1919 influenza epidemic was so severe because the 1918 virus had recently moved into human populations from another species.

METHOD

1. Sequence the genomes of influenza viruses isolated from samples of birds, pigs, horses, and humans throughout the past century.

2. Reconstruct a phylogenetic tree based on sequences of the polymerase protein PA from the influenza virus genomes. Repeat this analysis for each of the genes in the genome. (Note that only the results based on PA are shown below.)

3. Determine the rate of evolution of the influenza viruses in each species, and use this information to conduct a molecular clock analysis to date the origins of each cross-species transmission.

RESULTS

Species-specific rates of the polymerase protein PA from the genome of influenza viruses in each host species:

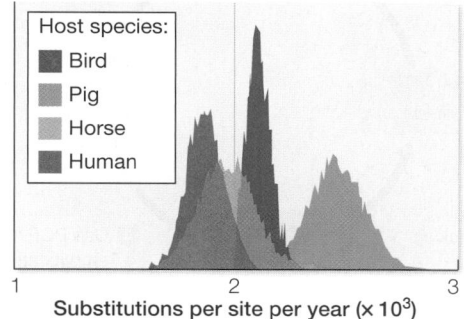

Time-calibrated phylogeny of the *PA* gene from influenza:

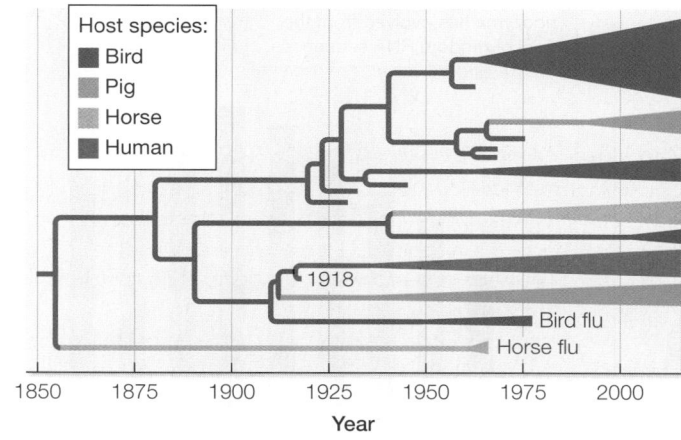

Viruses represented by blue lineages were isolated from birds; green lineages from pigs; orange lineages from horses; and red lineages from humans. The triangles represent many closely related sequences sampled from the respective species in the time period indicated; the width of the triangle is proportional to the number of sequences analyzed. The 1918 human influenza virus is labeled "1918."

CONCLUSION▶ What would you conclude about the 1918–1919 influenza pandemic from these results? Explore the possibilities in the following work with the data.

work with the data

Original Paper: Worobey, M., G.-Z. Han and A. Rambaut. 2014.

Michael Worobey and his colleagues were able to reconstruct many of the cross-species transmissions of influenza virus that have occurred over the past 150 years, and to determine a likely reason that the 1918–1919 pandemic in humans was so severe. Can you replicate their conclusions?

QUESTIONS▶

1. All the influenza viruses shown in the tree in the experiment are thought to have arisen in one group of host animals. Which group of animals do you think is the original source of the virus, and why?

2. How many cross-species transmissions are shown on the tree? What hosts were involved (as source and recipient) of each transmission event?

3. What do you think is a likely explanation for the severity of the 1918–1919 influenza pandemic in humans?

4. Influenza was known in humans well before the 1918–1919 pandemic. Why do you think that no human influenza viruses are indicated on the tree before about 1918?

A similar **work with the data** exercise may be assigned in **LaunchPad**.

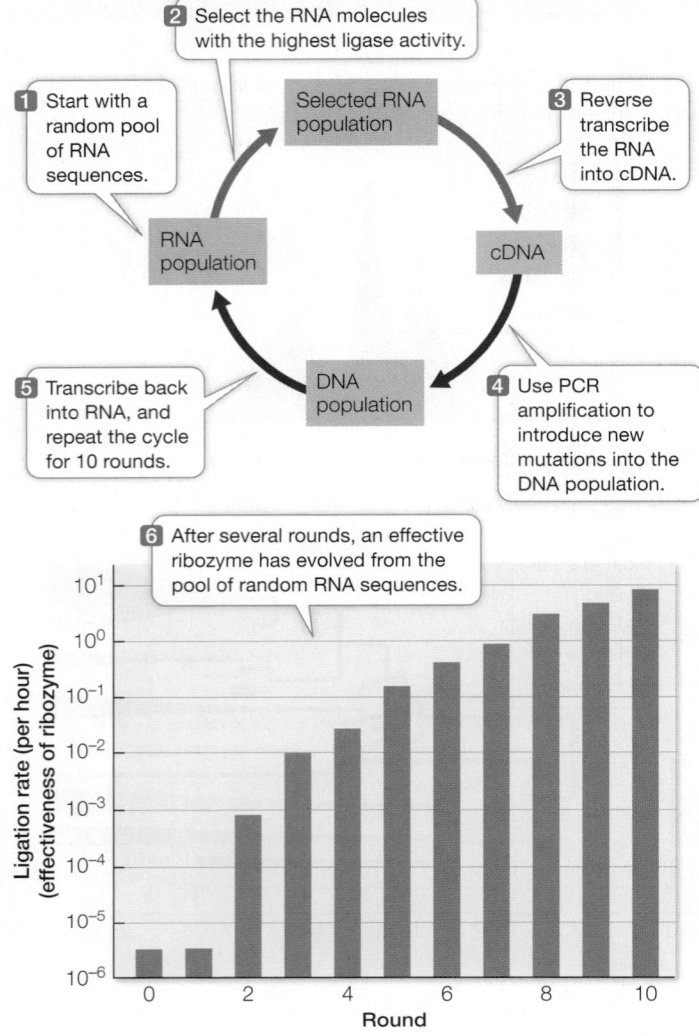

1 Start with a random pool of RNA sequences.

2 Select the RNA molecules with the highest ligase activity.

Selected RNA population

3 Reverse transcribe the RNA into cDNA.

RNA population

cDNA

5 Transcribe back into RNA, and repeat the cycle for 10 rounds.

DNA population

4 Use PCR amplification to introduce new mutations into the DNA population.

6 After several rounds, an effective ribozyme has evolved from the pool of random RNA sequences.

Figure 23.13 In Vitro Evolution Starting with a large pool of random RNA sequences, investigators produced a new ribozyme through rounds of mutation and selection for the ability to ligate RNA sequences.

In the future, molecular evolution will become even more critical to the identification of human (and other) diseases. Once biologists have collected data on the genomes of enough organisms, it will be possible to identify an infection by sequencing a portion of the infecting organism's genome and comparing this sequence with other sequences on an evolutionary tree. At present it is difficult to identify many common viral infections (those that cause "colds," for instance). As genomic databases and evolutionary trees increase, however, automated methods of sequencing and rapid phylogenetic comparison of the sequences will allow us to identify and treat a much wider array of human illnesses.

23.4 recap

Studies of molecular evolution have provided biologists with new tools to understand the functions of macromolecules and how those functions can change over time. These tools can be used to develop synthetic molecules and to identify and combat human diseases.

learning outcomes

You should be able to:

- Describe how studies of molecular evolution are used to understand protein function.
- Design in vitro evolution protocols that will generate sequences with desired functional characteristics.
- Evaluate evolutionary relationships among viral sequences to understand the evolution and transmission of a viral disease.

1. How can gene evolution be used to study protein function? Describe a specific example.
2. What are the key elements of in vitro evolution, and how do these elements correspond to natural evolutionary processes?
3. How are principles of evolutionary biology used to identify emerging diseases?

Now that we have discussed how organisms and biological molecules evolve, we are ready to consider the broader evolutionary history of life on Earth. Chapter 24 will describe the long-term evolutionary changes that have given rise to all of life's diversity.

investigatinglife

Why was the 1918–1919 flu pandemic worse than any before or since?

Many different strains of influenza virus circulate among human populations and other vertebrate hosts each year, but only a few of those strains survive and produce descendants. One of the ways in which influenza strains differ is in the configuration of proteins on their surface. These surface proteins are the targets of recognition by the host immune system. When changes occur in the surface

proteins of an influenza virus, the host immune system may no longer recognize the invading virus, so that virus is more likely to replicate successfully. Those virus strains with the greatest number of changes to their surface proteins are most likely to escape detection by the host immune system, and are therefore most likely to spread among the host population and result in future flu epidemics. In other words, there is positive selection for change in the surface proteins of influenza viruses.

As demonstrated by data presented in Investigating Life: Why Was the 1918–1919 Influenza Pandemic So Severe?, the 1918–1919

Chapter Summary 505

flu epidemic occurred shortly after the cross-species transfer of an avian flu virus into human populations. For humans, this was a new virus, different than any influenza viruses that had circulated before in human populations. As human immune systems had no previous exposure to this virus, they had not yet developed effective immune defenses. The virus overstimulated human immune systems, which led to secondary complications and death. Today, investigators carefully monitor influenza viruses circulating in human populations around the world. Early detection of new cross-species transmissions from birds or pigs into human populations is the key to their elimination before they can result in new global human influenza pandemics.

Future directions

By comparing the survival and proliferation rates of influenza virus strains that have different gene sequences coding for surface proteins, biologists can study the adaptation of the viruses over time.

If biologists can predict which of the currently circulating flu virus strains are most likely to escape host detection, they can then identify the strains that are most likely to be involved in upcoming influenza epidemics and can target those strains for vaccine production.

How can biologists make such predictions? By determining the ratio of synonymous to nonsynonymous substitutions in genes that encode viral surface proteins, biologists can detect which codon changes (i.e., mutations) are under positive selection. They can then assess which of the currently circulating flu strains show the greatest number of changes in these positively selected codons. It is these flu strains that are most likely to survive, proliferate, and lead to the flu epidemics of the future, so they are the logical targets for new vaccines. This practical application of evolutionary theory leads to more effective flu vaccines—and thus fewer illnesses and influenza-related deaths each year.

Chapter Summary 23

23.1 DNA Sequences Record the History of Gene Evolution

- A **genome** is an organism's full set of genes, regulatory sequences, and structural elements as well as noncoding DNA.
- The field of **molecular evolution** concerns relationships between the structures of genes and proteins and the functions of organisms.
- **Sequence alignments** of proteins or nucleic acids from different organisms allow us to compare the sequences and identify homologous positions. Review Figure 23.1, Activity 23.1
- The minimum number of changes between sequences can be calculated from a **similarity matrix**. Models of sequence evolution can be used to account for changes that cannot be observed directly. Review Figure 23.2, Activity 23.2

23.2 Genomes Reveal Both Neutral and Selective Processes of Evolution

- **Nonsynonymous substitutions** of nucleotides result in changes to the amino acid sequences of proteins, but **synonymous substitutions** do not. Review Figure 23.4
- The **neutral theory** of molecular evolution states that much of the molecular change in nucleotide sequences does not change genome function. The rate of fixation of neutral mutations is independent of population size and is equal to the mutation rate.
- Positive selection for change in a protein-coding gene may be detected by a higher rate of nonsynonymous than synonymous substitutions. The reverse is true of purifying selection.
- Common selective constraints can lead to convergent evolution of amino acid sequences in distantly related species. Review Figure 23.6
- The total size of genomes varies much more widely across multicellular species than does the number of functional genes. Review Figures 23.7, 23.8

23.3 Lateral Gene Transfer and Gene Duplication Can Produce Major Changes

- **Lateral gene transfer** can result in the rapid acquisition of new functions from distantly related species.
- **Gene duplications** can result in increased production of a gene's product, in nonfunctional **pseudogenes**, or in new gene functions. Several rounds of gene duplication can give rise to multiple genes with related functions, collectively known as a **gene family**. Review Figures 23.9, 23.10
- **Gene trees** describe the evolutionary history of particular genes or gene families. See Activity 23.3
- Some highly repeated genes undergo **concerted evolution**, in which the multiple copies within the genome maintain their similarity, even as the genes diverge among species. Review Figure 23.11, Animation 23.1

23.4 Molecular Evolution Has Many Practical Applications

- **Orthologs** are genes that are related through speciation events, whereas **paralogs** are genes that are related through gene duplication events. Review Figure 23.12
- Protein function can be studied by examining gene evolution. Detection of positive selection can be used to identify molecular changes that have resulted in functional changes.
- **In vitro evolution** is used to produce synthetic molecules with particular desired functions. Review Figure 23.13
- Many diseases are identified, studied, and combated through molecular evolutionary investigations. Review Investigating Life: Why Was the 1918–1919 Influenza Pandemic So Severe?

Go to **LearningCurve** (in **LaunchPad**) for dynamic quizzing that helps you solidify your understanding of this chapter. **LearningCurve** adapts to your responses, giving you the practice you need to master each key concept.

Apply What You've Learned

Review

23.1 Simple counts of nucleotide substitutions or amino acid replacements between sequences often underestimate the underlying changes.

23.2 The rate of fixation of neutral nucleotide changes within populations is independent of population size.

23.3 Comparing rates of synonymous and nonsynonymous substitutions can be used to identify positive and purifying selection in protein genes.

By studying molecular evolution patterns of genes, biologists are able to make inferences about various types of natural selection operating on them. These inferences are important to our understanding of how genes function and how they evolve through time in response to selection for new functions or new conditions.

A hypothetical gene encodes a transcription factor in various *Drosophila* species. The phylogeny of these species is shown below, with divergence times shown in millions of years.

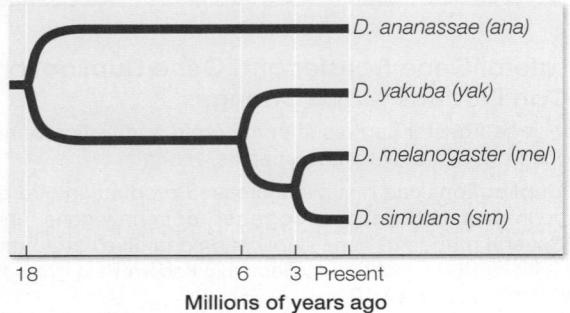

Millions of years ago

One exon of this gene has been sequenced in the four species. Below is a matrix representing the number of synonymous (above the diagonal) and nonsynonymous (below the diagonal) substitutions between pairs of species. For instance, there are 21 synonymous substitutions between *D. melanogaster* and *D. yakuba*. There are 2000 nonsynonymous and 600 synonymous sites in this exon, and there are five generations per year.

	D. ananassae	D. yakuba	D. melanogaster	D. simulans
D. ananassae		54	56	52
D. yakuba	20		21	16
D. melanogaster	17	7		10
D. simulans	17	5	3	

Answers to the questions below will provide information about the types of selection operating on this transcription factor and, potentially, its function.

Questions

1. Graph the relationship between the number of synonymous substitutions and the years of divergence between each pair of species. Label both axes. Use this graph to estimate the rates of synonymous substitutions per million years and per generation for the whole exon and on a per-site basis. Remember that evolution is occurring on both lineages.

2. Assume that synonymous substitutions are neutral. What is the expected per-site mutation rate based on the answer in Question 1?

3. Houseflies (*Musca domestica*) diverged from *D. melanogaster* about 90 million years ago. Suppose there are 174 synonymous substitutions between these two taxa. Assume that the number of generations remains constant at five per year across these species. Calculate the *expected* number of synonymous substitutions between these two species given the calculated synonymous substitution rate from Question 1, the time of divergence, and the number of generations per year. Provide an explanation for why there is a discrepancy between the observed and expected values, continuing to assume that the mutations are completely neutral. (Hint: Consider what you know about substitution patterns.)

4. Calculate the average per-site, per-generation, nonsynonymous substitution rate between *D. ananassae* and the other three species. Based on the differences between this rate and the synonymous substitution rate calculated in Question 1, what can you infer about the type of selection operating on the nonsynonymous sites?

5. Suppose a biologist asserts that this transcription factor is actually a pseudogene in flies. Based on the information you have obtained, do you agree with this biologist? Why or why not?

Go to **LaunchPad** for the eBook, LearningCurve, animations, activities, flashcards, and additional resources and assignments.

Layers of rocks in the Grand Canyon reveal almost 2 billion years of Earth's history.

The History of Life on Earth

investigatinglife

When Giant Insects Ruled the Skies

The largest dragonflies alive today have wingspans that can be covered by a human hand. Three hundred million years ago, however, dragonflies such as *Meganeuropsis permiana* had wingspans of more than 70 centimeters—well over 2 feet, matching or exceeding the wingspans of many modern birds of prey. These dragonflies were the largest flying predators of their time. Below, an artist's reconstruction of *M. permiana* is shown at the same scale as a modern dragonfly.

No flying insects alive today are anywhere near this size. But during the Carboniferous and Permian geological periods, 350 to 250 million years ago, many groups of flying insects contained gigantic species. *Meganeuropsis* probably ate huge mayflies and other giant flying insects that shared its home in the Permian swamps. These enormous insects were themselves eaten by giant amphibians. None of these insects or amphibians would be able to survive on Earth today. The oxygen concentrations in Earth's atmosphere were about 50 percent higher then than they are now, and those high oxygen concentrations are thought to

have been necessary to support giant insects and their huge amphibian predators.

Paleontologists have uncovered fossils of *Meganeuropsis permiana* in the rocks of Kansas. How do we know the age of these fossils, and how can we know how much oxygen that long-vanished atmosphere contained? The layering of rocks allows us to tell their ages relative to one another, but it does not by itself indicate a given layer's absolute age.

Scientists have developed sophisticated techniques that use the decay rates of radioisotopes, the ratios of certain molecules in rocks, and changes in Earth's magnetic field to infer conditions and events in the remote past and to date them accurately. It is those methods that allow us to age the fossils of *Meganeuropsis* and to calculate the concentration of oxygen in Earth's atmosphere at the time.

Earth is about 4.5 billion years old, and life has existed on it for about 3.8 billion of those years. That means human civilizations have occupied Earth for less than 0.0003 percent of the history of life. Discovering what happened before humans were around is an ongoing and exciting area of science.

Q|A How can modern experiments test hypotheses about the evolutionary impact

table 24.1 Earth's Geological History

Eon	Era	Period	Onset	Major physical changes on Earth
Phanerozoic (~0.5 billion years long)	Cenozoic	Quaternary (Q)	2.6 mya	Cold/dry climate; repeated glaciations
		Tertiary (T)	65.5 mya	Continents near current positions; climate cools
	Mesozoic	Cretaceous (K)	145.5 mya	Laurasian continents attached to one another; Gondwana begins to drift apart; meteorite strikes near current Yucatán Peninsula at end of period
		Jurassic (J)	201.6 mya	Two large continents form: Laurasia (north) and Gondwana (south); climate warm
		Triassic (Tr)	251.0 mya	Pangaea begins to drift apart; hot/humid climate
	Paleozoic	Permian (P)	299 mya	Extensive lowland swamps; O_2 levels 50% higher than present; by end of period continents aggregate to form Pangaea, and O_2 levels drop rapidly
		Carboniferous (C)	359 mya	Climate cools; marked latitudinal climate gradients
		Devonian (D)	416 mya	Continents collide at end of period; giant meteorite probably strikes Earth
		Silurian (S)	444 mya	Sea levels rise; two large land masses emerge; hot/humid climate
		Ordovician (O)	488 mya	Massive glaciation; sea level drops 50 meters
		Cambrian (C)	542 mya	Atmospheric O_2 levels approach current levels
Proterozoic	Collectively called the Precambrian (~4 billion years long)		2.5 bya	Atmospheric O_2 levels increase from negligible to about 18%; "snowball Earth" from about 750 to 580 mya
Archean			3.8 bya	Earth accumulates more atmosphere (still almost no O_2); meteorite impacts greatly reduced
Hadean			4.5–4.6 bya	Formation of Earth; cooling of Earth's surface; atmosphere contains almost no free O_2; oceans form; Earth under almost continuous bombardment from meteorites

Note: mya, million years ago; bya, billion years ago.

key concept 24.1 Events in Earth's History Can Be Dated

Some evolutionary changes happen rapidly enough to be studied directly and manipulated experimentally. Plant and animal breeding by agriculturalists and the evolution of surface proteins in influenza viruses are examples of rapid, short-term evolution that we have discussed in previous chapters. Such changes can take place in months, years, decades, or centuries. Other evolutionary changes, such as the appearance of new species and major evolutionary lineages, usually take place over a **geological time scale (Table 24.1)**.

focus your learning

- Geologists use several methods to date ancient events.
- Scientists have developed a geological time scale.
- Geologists construct geological maps based on the age of rocks.

To understand long-term patterns of evolutionary change, we must not only think in time scales spanning many millions of years, but also consider events and conditions very different from those we observe today. Earth of the distant past was so unlike our present Earth that it would seem like a foreign planet inhabited by strange organisms. The continents were not where they are now, and climates were sometimes dramatically different from those of today. We know this because much of Earth's history is recorded in its rocks.

We cannot tell the ages of rocks just by looking at them, but we can visually determine the ages of rocks *relative to one another*. The first person to formally recognize this fact was the seventeenth-century Danish physician Nicolaus Steno. Steno realized that in undisturbed **sedimentary rocks** (rocks formed by the accumulation of sediments), the oldest layers of rock, or **strata** (singular *stratum*), lie at the bottom, and successively higher strata are progressively younger.

Geologists subsequently combined Steno's insight with their observations of fossils contained in sedimentary rocks. They developed the following principles of **stratigraphy**:

- Fossils of similar organisms are found in widely separated places on Earth.
- Certain fossils are always found in younger strata, and certain other fossils are always found in older strata.
- Organisms found in younger strata are more similar to modern organisms than are those found in older strata.

These patterns revealed much about the relative ages of sedimentary rocks and the fossils they contained, as well as patterns in the evolution of life. But the geologists still could not determine the age of particular rocks. A method for absolute dating of rocks—that is, determining their actual age rather than just their age relative

Major events in the history of life
Humans evolve; many large mammals become extinct
Diversification of birds, mammals, flowering plants, and insects
Dinosaurs continue to diversify; mass extinction at end of period (~76% of species lost)
Diverse dinosaurs; radiation of ray-finned fishes; first fossils of flowering plants
Early dinosaurs; first mammals; marine invertebrates diversify; mass extinction at end of period (~65% of species lost)
Reptiles diversify; giant amphibians and flying insects present; mass extinction at end of period (~96% of species lost)
Extensive fern/horsetail/giant club moss forests; first reptiles; insects diversify
Jawed fishes diversify; first insects and amphibians; mass extinction at end of period (~75% of marine species lost)
Jawless fishes diversify; first ray-finned fishes; plants and animals colonize land
Mass extinction at end of period (~75% of species lost)
Rapid diversification of multicellular animals; diverse photosynthetic protists
Origin of photosynthesis, multicellular organisms, and eukaryotes
Origin of life; prokaryotes flourish
Life not yet present

 Media Clip 24.1 **The Age of the Earth**
www.Life11e.com/mc24.1

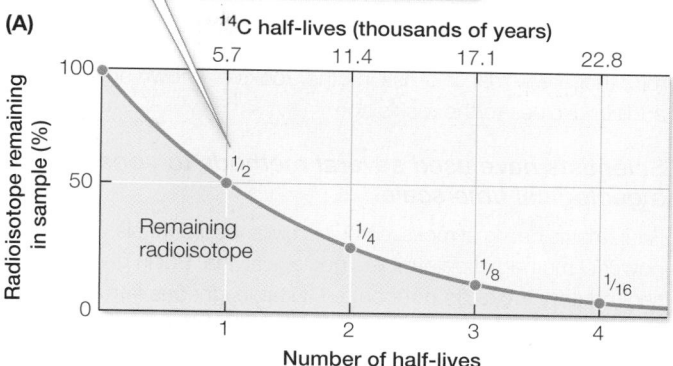

Fractions show the proportion of radioisotope remaining in the sample at the end of each half-life.

(A)

Figure 24.1 Radioactive Isotopes Allow Us to Date Ancient Rocks   The decay of radioactive isotopes into stable isotopes happens at a steady rate. A half-life is the time it takes for half of the remaining atoms to decay in this way. **(A)** The graph demonstrates the principle of half-life using carbon-14 (^{14}C) as an example. The half-life of ^{14}C is 5,700 years. **(B)** Different radioisotopes have different characteristic half-lives that allow us to estimate the ages of many rocks.

(B)

Radioisotope	Decay product	Half-life (years)	Useful dating range (years)
Carbon-14 (^{14}C)	Nitrogen-14 (^{14}N)	5,700	100–60,000
Uranium-234 (^{234}U)	Thorium-230 (^{230}Th)	80,000	10,000–500,000
Uranium-235 (^{235}U)	Lead-207 (^{207}Pb)	704 million	200,000–4.5 billion
Potassium-40 (^{40}K)	Argon-40 (^{40}Ar)	1.3 billion	10 million–4.5 billion

to one another—did not become available until after radioactivity was discovered at the beginning of the twentieth century.

Radioisotopes provide a way to date fossils and rocks

Radioactive isotopes of atoms—*radioisotopes—decay in a predictable pattern over long periods. Over a specific time interval, known as a **half-life**, half of the atoms in a radioisotope decay to become a different isotope (**Figure 24.1A**). The use of this knowledge to date fossils and rocks is known as **radiometric dating**.

---

*connect the concepts   As noted in Key Concept 2.1, stable isotopes of the same element have the same number of protons but different numbers of neutrons. Radioisotopes, however, are not stable; they spontaneously give off energy in the form of alpha, beta, or gamma radiation, which transforms the original atom.

To use a radioisotope to date a past event, we must know or estimate the concentration of that isotope at the time of that event, and we must know the radioisotope's half-life. In the case of carbon-14, a radioisotope of carbon, the production of new carbon-14 (^{14}C) in the upper atmosphere—by the reaction of neutrons with nitrogen-14 (^{14}N, a stable isotope of nitrogen)—just balances the natural radioactive

decay of ^{14}C into ^{14}N. Therefore the ratio of ^{14}C to the more common stable isotope of carbon, carbon-12 (^{12}C), is relatively constant in living organisms and in their environment. As soon as an organism dies, however, it ceases to exchange carbon compounds with its environment. Its decaying ^{14}C is no longer replenished, and the ratio of ^{14}C to ^{12}C in its remains decreases over time. Paleontologists can use the ratio of ^{14}C to ^{12}C in fossil material to date fossils that are less than 60,000 years old (and thus the sedimentary rocks that contain those fossils). If fossils are older than that, so little ^{14}C remains that the limits of detection using this particular isotope are reached, and other radioisotopes must be used instead.

Radiometric dating methods have been expanded and refined

Sedimentary rocks are formed from materials that existed for varying lengths of time before being weathered, fragmented, and transported, sometimes over long distances, to the site of their deposition. Therefore the radioisotopes in sedimentary rocks do not contain reliable information about the date of their formation. Radiometric dating of rocks older than 60,000 years requires estimating radioisotope concentrations in **igneous rocks**, which are formed when molten material cools. To date sedimentary strata, geologists search for places where volcanic ash or lava flows have intruded into the sedimentary rock.

A preliminary estimate of the age of an igneous rock determines which radioisotopes can be used to date it (**Figure 24.1B**). The decay of potassium-40 (which has a half-life of 1.3 billion years) to argon-40, for example, has been used to date many of the ancient events in the evolution of life. Fossils in the adjacent sedimentary rock that are similar to those in other rocks of known ages provide additional clues to the rock's age.

Scientists have used several methods to construct a geological time scale

Radiometric dating of rocks, combined with fossil analysis, is the most powerful method of determining geological age. But in places where sedimentary rocks do not contain suitable igneous intrusions and few fossils are present, paleontologists turn to other dating methods.

One method, known as paleomagnetic dating, relates the ages of rocks to patterns in Earth's magnetism, which change over time. Earth's magnetic field is not constant. Changes in both the strength and the polarity of Earth's magnetic field occur over time. Reversals of the poles occur periodically. Both sedimentary and igneous rocks preserve a record of Earth's magnetic field at the time they were formed, and that record can be used to determine the ages of those rocks. Other dating methods use information about continental drift, information about sea level changes, and *molecular clocks.

---

*connect the concepts   As Key Concept 21.3 explains, a molecular clock measures the average rate at which a given gene or protein product accumulates changes, and this rate of change can be used to deduce the geological time when two biological lineages diverged from one another.

Using all of these methods, geologists developed the geological time scale (see Table 24.1). They divided the broad history of life into four eons. The Hadean eon refers to the time on Earth before life evolved. The early history of life occurred in the Archean eon, which ended about the time that photosynthetic organisms first appeared on Earth. Prokaryotic life diversified rapidly in the Proterozoic eon, and the first eukaryotes in the fossil record date from this time. These three eons are sometimes referred to collectively as Precambrian time, or simply the **Precambrian**. The Precambrian lasted for approximately 4 billion years and thus accounts for the vast majority of geological time. It was in the Phanerozoic eon, however—a mere 542-million-year time span—that multicellular eukaryotes rapidly diversified. To emphasize the events of the Phanerozoic, Table 24.1 shows the subdivision of this eon into eras and periods. The boundaries between these divisions of time are based largely on the striking differences geologists observe in the assemblages of fossil organisms contained in successive strata. This geological record of life reveals a remarkable story of a world in which the continents and biological communities are constantly changing.

24.1 recap

Fossils in sedimentary rock strata enabled geologists to determine the relative ages of organisms, but absolute dating was not possible until the discovery of radioactivity. Geologists divide the history of life into eons, eras, and periods, based on assemblages of fossil organisms found in successive layers of rocks.

24.1 recap

learning outcomes

You should be able to:

- Design and interpret a geological map indicating the ages of exposed rocks.
- Suggest appropriate methods for dating fossils and rocks.

Imagine you have been asked to produce a geological map of volcanic rocks formed between 400 and 600 million years ago. You collect samples from ten sites, as indicated below, and determine the ratio of ^{206}Pb to ^{238}U for each site. Use the data below to answer the following questions.

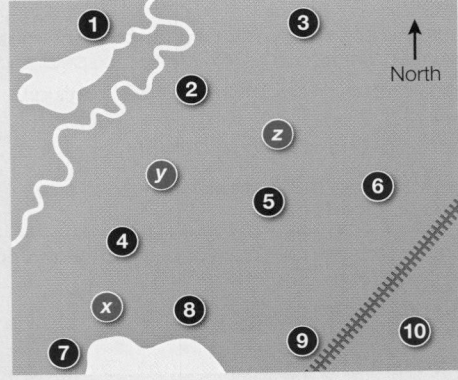

Site	^{206}Pb/^{238}U ratio	Estimated age (mya)
1	0.076	474
2	0.077	479
3	0.069	431
4	0.081	505
5	0.076	474
6	0.070	435
7	0.089	550
8	0.080	500
9	0.079	495
10	0.077	479

1. Assign each site to a geological period based on the information in this table and in Table 24.1.

2. Mark rough boundaries between the sites using the estimated ages and geological periods of the samples.

3. If you wanted to refine the boundary between the Ordovician and Silurian periods on your map, which sampling site would you add to your analysis: x, y, or z?

As geologists began to develop accurate ways to measure the age of Earth, they began to understand that Earth is far older than anyone had previously understood. During its 4.5-billion-year history, Earth has undergone massive physical changes. These changes have influenced the evolution of life, and life, in its turn, has influenced Earth's physical environment.

key concept 24.2 Changes in Earth's Physical Environment Have Affected the Evolution of Life

As we saw in the previous section, the Phanerozoic eon has been notable for the rapid diversification of multicellular eukaryotes. But the diversity of multicellular organisms has not simply increased steadily through time. New species have arisen, and species have gone extinct, throughout the history of life. In the absence of major, rapid environmental changes, the average rate of extinction is known as the background extinction rate. But there have been times during which extinction rates have increased dramatically over the background levels (Figure 24.2). These **mass extinction events** are the cause of some of the striking differences in fossil assemblages that geologists use to divide the Phanerozoic eon into eras and periods. After each mass extinction, the diversity of life rebounded, although recovery took millions of years. In this section we will discuss some of the physical changes on Earth that have resulted in dramatic changes in life's diversity.

focus your learning

- The evolution of life has changed the physical nature of Earth.
- The evolution of life and the physical nature of Earth are mutually related.
- Research on living organisms can help us understand the evolution of ancient organisms.

Earth's continents and climates have changed over time

The globes and maps that adorn our walls, shelves, and books give an impression of a static Earth. It would be easy for us to assume that the continents have always been where they are. But we would be wrong. The idea that Earth's land masses have changed their positions over the millennia, and that they continue to do so, was first put forth in 1912 by the German meteorologist and geophysicist Alfred Wegener. His idea, known as **continental drift**, was initially met with skepticism and resistance. By the 1960s, however, physical evidence and increased understanding of **plate tectonics**—the geophysics of the movement of major land masses—had convinced virtually all geologists of the reality of Wegener's vision. Plate tectonics provided the geological mechanism that explained Wegener's hypothesis of continental drift.

Earth's crust consists of several solid plates, which collectively make up the solid **lithosphere** ("stone sphere"). Thick continental and thinner oceanic lithospheric plates overlie a viscous, malleable layer of Earth's mantle, known as the **asthenosphere** ("weak sphere"). Heat produced by radioactive decay deep in Earth's core sets up large-scale convection currents in the mantle. New crust is formed as mantle material rises between diverging plates, pushing them apart.

Where oceanic plates and continental plates converge, the thinner oceanic plate is forced underneath the thicker continental plate, a process known as **subduction**. Subduction results in volcanism and mountain building on the continental boundary (**Focus: Key Figure 24.3A**). For example, in the Pacific Northwest of North America,

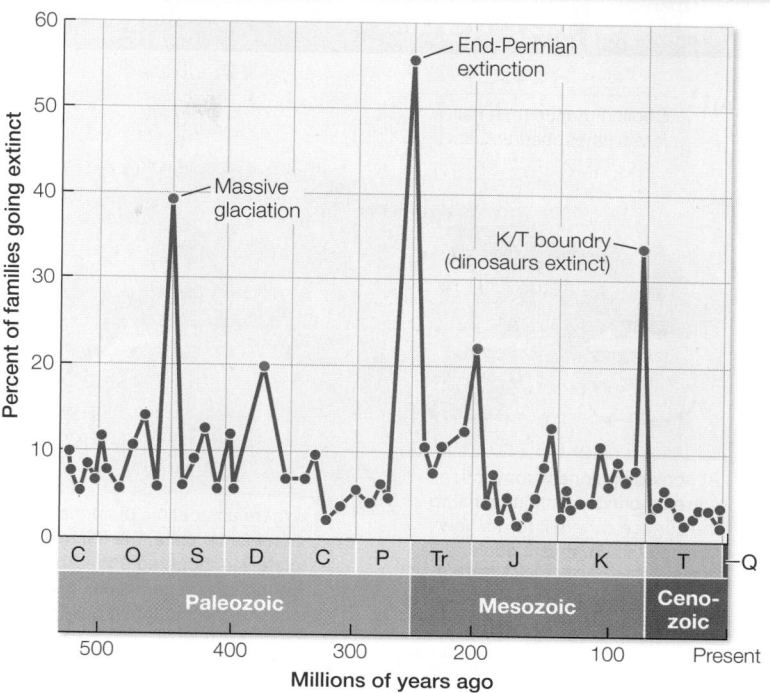

Figure 24.2 Periodic Mass Extinctions Mark Many Geologic Boundaries   Five sharp rises (marked by red dots) above the background extinction rate have occurred throughout the Phanerozoic. The most sweeping of these events, the end-Permian extinction, was associated with dramatic drops in sea level (see Figure 24.4), global temperature, and atmospheric oxygen level (see Figure 24.8).

a series of volcanoes formed the Cascade mountain range as the Juan de Fuca oceanic plate has been subducted beneath a portion of the continental North American Plate (**Focus: Key Figure 24.3B**). When two oceanic plates collide, one is subducted below the other, producing a deep oceanic trench and associated volcanic activity. The deepest part of the world's oceans—the Mariana Trench in the western Pacific—formed where two oceanic plates collided. Volcanic activity associated with the subduction at the Mariana Trench produced the nearby Mariana Islands.

When two thick continental plates collide, neither plate is subducted. Instead, the plates push up against one another, forming high mountain chains. The highest mountain chain in the world, the Himalayas, was formed this way when the Indian Plate collided with the Eurasian Plate. When continental plates diverge, new crust forms in the intervening spaces, resulting in deep clefts called rift valleys in which large freshwater lakes typically form. The Great Rift Valley lakes of eastern Africa, including Lake Malawi (discussed at the opening of Chapter 22), were formed in this way. Two plates can also slide past one another, forming a transform fault boundary (such as the San Andreas Fault that produces violent seismic activity in parts of California).

Many physical conditions on Earth have oscillated in response to plate tectonic processes. We now know that the movement of the plates has sometimes brought continents together and at other times has pushed them apart, as seen in the maps across the top of Figure 24.14. The positions and sizes of the continents influence oceanic circulation patterns, global climates, and sea levels. Sea level is influenced directly by plate tectonic processes (which

focus: key figure

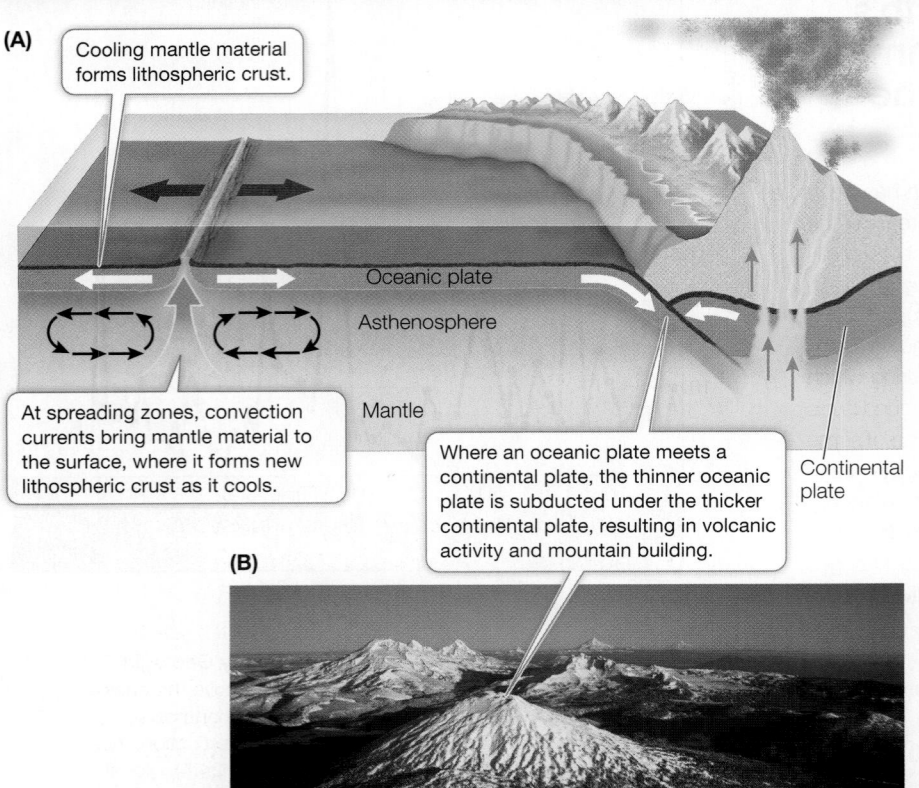

(A)

Cooling mantle material forms lithospheric crust.

At spreading zones, convection currents bring mantle material to the surface, where it forms new lithospheric crust as it cools.

Oceanic plate

Asthenosphere

Mantle

Where an oceanic plate meets a continental plate, the thinner oceanic plate is subducted under the thicker continental plate, resulting in volcanic activity and mountain building.

Continental plate

(B)

Figure 24.3 Plate Tectonics and Continental Drift **(A)** The heat of Earth's core generates convection currents in the viscous mantle material underlying the oceanic and continental plates. Those currents push the continental plates, along with the land masses they carry, together or apart. Where plates collide, one may slide under the other, creating mountain ranges and often volcanoes. **(B)** The Cascade Range of the Pacific Northwest of North America is an example of a mountain chain produced by subduction of an oceanic plate under a continental plate.

Q: Why are both shores of the Pacific Ocean ringed with volcanic mountain ranges?

 Media Clip 24.2 **Lava Flows and Magma Explosions**
www.Life11e.com/mc24.2

Earth's climate has shifted between hot and cold conditions

Through much of Earth's history, the climate was considerably warmer than it is today, and temperatures decreased more gradually toward the poles. At other times, Earth was colder than it is today. Periodic rapid drops in sea level throughout the history of Earth have resulted mainly from increased global glaciation (**Figure 24.4**). Many of these drops in sea level were accompanied by mass extinctions—particularly of marine organisms, which could not survive the disappearance of the shallow seas that covered vast areas of the continental shelves.

Earth's cold periods were separated by long periods of milder climates. Because we are living in one of the colder periods, it is difficult for us to imagine the mild climates that were found at high latitudes during much of the history of life. The Quaternary period has been marked by a series of glacial advances, interspersed with warmer interglacial intervals during which the glaciers retreated.

"Weather" refers to the daily events at a given location, such as individual storms and the high and low temperatures on a given day. "Climate" refers to long-term average expectations over the various seasons at a given location. Weather often changes rapidly, whereas climates typically change slowly. However, major climatic shifts have taken place over periods as short as 5,000 to 10,000 years, primarily as a result of changes in Earth's orbit around the sun. A few climatic shifts have been even more rapid. For example, during one Quaternary interglacial period, the ice-locked Antarctic Ocean became nearly ice-free in less than 100 years. Some climate changes have been so rapid that the extinctions caused by them appear to be nearly instantaneous in the fossil record. Such rapid changes are usually caused by sudden shifts in ocean currents.

We are currently living in a time of ***rapid climate change** caused by a buildup of atmospheric CO_2, primarily from the burning of fossil fuels by human populations. We are reversing the energy transformations accrued in the burial and decomposition of organic material that occurred (especially) in the Carboniferous, Permian, and Triassic, which gave rise to the fossil fuels we are using today. But we are burning these fuels over a few hundred years, rather than the many millions of years over which those deposits accumulated. The current rate of increase in atmospheric CO_2 is unprecedented in Earth's history. A doubling of the atmospheric CO_2 concentration—which may happen during the current century—is expected to increase the average temperature of Earth, change rainfall patterns, melt glaciers and ice caps, and raise sea levels.

can influence the depth of ocean basins) and indirectly by oceanic circulation patterns, which affect patterns of glaciation. As climates cool, glaciers form and tie up water over land masses; as climates warm, glaciers melt and release water.

 Animation 24.1 **Movement of the Continents**
www.Life11e.com/a24.1

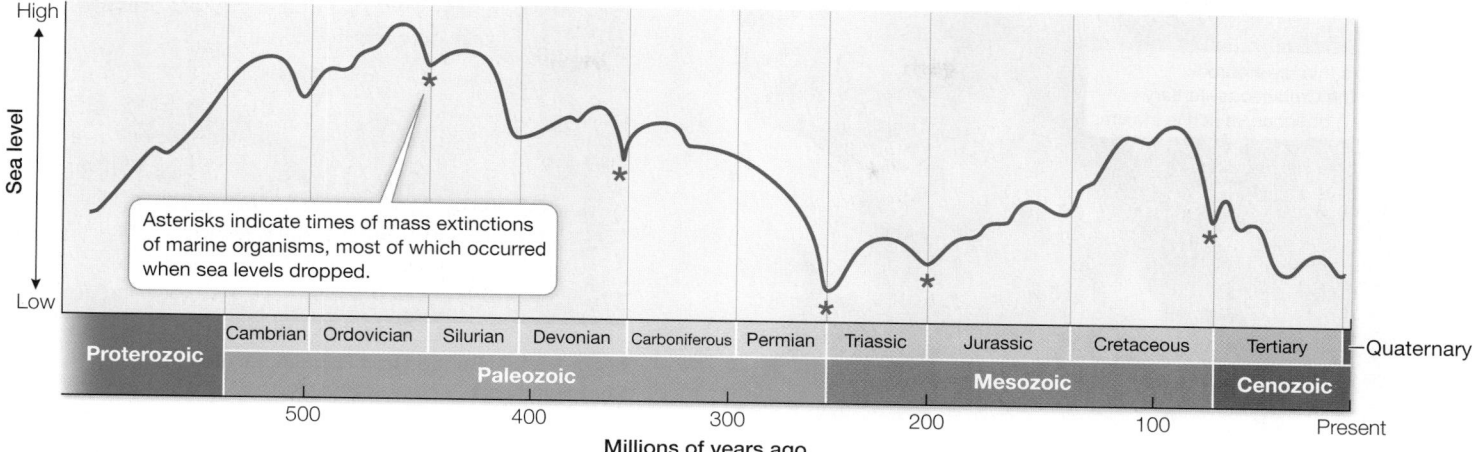

High

Sea level

Low

Asterisks indicate times of mass extinctions of marine organisms, most of which occurred when sea levels dropped.

| Proterozoic | Cambrian | Ordovician | Silurian | Devonian | Carboniferous | Permian | Triassic | Jurassic | Cretaceous | Tertiary | Quaternary |

Paleozoic   Mesozoic   Cenozoic

500   400   300   200   100   Present

Millions of years ago

Figure 24.4 Sea Levels Have Changed Repeatedly   Rapid drops in sea level are associated with periods of globally cooler temperatures and increased glaciation. Most mass extinctions of marine organisms have coincided with low sea levels.

Q: Why do cooler global temperatures result in lower sea levels?

---

*connect the concepts   Today's rapid climate changes are having an enormous effect on populations of many living organisms, as discussed in Key Concept 58.2.

Volcanoes have occasionally changed the history of life

Most volcanic eruptions produce only local or short-lived effects, but a few large volcanic eruptions have had major consequences for life. When Krakatau (a volcanic island in the Sunda Strait off Indonesia) erupted in 1883, it ejected more than 25 cubic kilometers of ash and rock as well as large quantities of sulfur dioxide gas (SO_2). The SO_2 was ejected into the stratosphere and carried by high-altitude winds around the planet. Its presence led to high concentrations of sulfurous acid (H_2SO_3) in high-altitude clouds, creating a "parasol effect" that reduced the amount of sunlight reaching Earth's surface. Global temperatures dropped by 1.2°C in the year following the eruption, and global weather patterns showed strong effects for another 5 years. More recently, the eruption of Mount Pinatubo in the Philippines in 1991 (**Figure 24.5**) temporarily reduced global temperatures by about 0.5°C.

Although these individual volcanoes had only relatively short-term effects on global temperatures, they suggest that the simultaneous eruption of many volcanoes could have a much stronger effect on Earth's climate. What would cause many volcanoes to erupt at the same time? The collision of continents during the Permian period, about 275 million years ago (mya), formed a single, gigantic land mass and caused a multitude of massive volcanic eruptions as the continental plates overrode one another (see Figure 24.3). Emissions from these eruptions blocked considerable sunlight, contributing to the advance of glaciers and a consequent drop in sea level (see Figure 24.4). Thus volcanic eruptions were probably responsible, at least in part, for the greatest mass extinction in Earth's history.

Extraterrestrial events have triggered changes on Earth

At least 30 meteorites of sizes between tennis and soccer balls strike Earth each year. Collisions with larger meteorites or comets are rare, but such collisions have probably been responsible for several mass extinctions. Several types of evidence tell us about these collisions. Their craters, and the dramatically disfigured rocks that result from their impact, are found in many places. Geologists have discovered compounds in these rocks that contain helium and argon with isotope ratios characteristic of meteorites, which are very different from the ratios found elsewhere on Earth.

A meteorite caused or contributed to a mass extinction at the end of the Cretaceous period (about 65.5 mya). The first clue that a meteorite was responsible came from the abnormally high concentrations

Figure 24.5 Volcanic Eruptions Can Cool Global Temperatures   When Mount Pinatubo erupted in 1991, it increased the concentrations of sulfurous acid in high-altitude clouds, which temporarily lowered global temperatures by about 0.5°C.

This thin layer of rock at the Cretaceous–Tertiary (K/T) boundary is rich in iridium.

Figure 24.6 Evidence of a Meteorite Impact The white layers of rock are Cretaceous in age, whereas the layers at the upper left were deposited in the Tertiary. Between the two is a thin, dark layer of clay that contains large amounts of iridium, a metal common in some meteorites but rare on Earth. Its high concentration in this sediment layer, deposited about 65.5 million years ago, suggests the impact of a large meteorite at that time.

5 cm

Figure 24.7 Banded Iron Formations Indicate Early Photosynthesis The alternating red and dark layers in this 2.25-billion-year-old sedimentary rock formation from Lake Superior resulted from a reaction between dissolved iron and the atmospheric oxygen produced by Earth's first photosynthetic organisms. The chemical reaction produced nearly pure iron oxide, or hematite, which forms the gray, metallic layers in this sample. The red bands are jasper tinged with much smaller amounts of iron oxide.

of the element iridium found in a thin layer separating rocks deposited during the Cretaceous from rocks deposited during the Tertiary (**Figure 24.6**). Iridium is abundant in some meteorites, but it is exceedingly rare on Earth's surface. When scientists discovered a circular crater 180 km in diameter buried beneath the northern coast of the Yucatán Peninsula of Mexico, they constructed the following scenario. When it collided with Earth, the meteorite released energy equivalent to that of 100 million megatons of high explosives, creating great tsunamis. A massive plume of debris rose into the atmosphere, spread around Earth, and descended. The descending debris heated the atmosphere to several hundred degrees and ignited massive fires. It also blocked the sun, preventing plants from photosynthesizing. The settling debris formed the iridium-rich layer. About a billion tons of soot with a composition matching that of smoke from forest fires were also deposited. These events had devastating effects on biodiversity. Many fossil species (including non-avian dinosaurs) that are found in Cretaceous rocks are not found in the overlaying Tertiary rocks.

Oxygen concentrations in Earth's atmosphere have changed over time

As the continents have moved over Earth's surface, the world has experienced other physical changes, including large increases and decreases in atmospheric oxygen concentrations. The atmosphere of early Earth probably contained little or no free oxygen gas (O_2). The increase in atmospheric O_2 came in two big steps more than a billion years apart.

The first step occurred at least 2.4 billion years ago (bya), when certain bacteria evolved the ability to use water as the source of hydrogen ions for photosynthesis. By chemically splitting H_2O, these bacteria generated O_2 as a waste product. They also made electrons available for reducing CO_2 to form the carbohydrate end-products of photosynthesis (see Key Concept 10.3). The O_2 they produced dissolved in water and reacted with dissolved iron. The reaction product then precipitated as iron oxide, which accumulated in alternating layers of red and dark rock known as banded iron formations (**Figure 24.7**). These formations provide evidence for the earliest photosynthetic organisms. As photosynthetic organisms continued to release O_2, oxygen gas began to accumulate in the atmosphere.

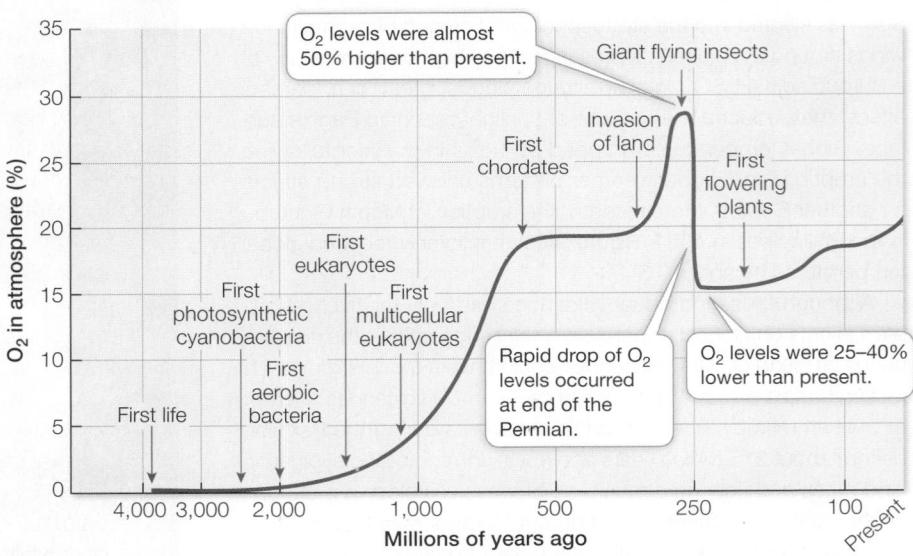

Figure 24.8 Atmospheric Oxygen Concentrations Have Changed over Time Changes in atmospheric oxygen concentrations have strongly influenced, and have been influenced by, the evolution of life. (Note that the horizontal axis of the graph is on a logarithmic scale.)

Q: Why did oxygen levels decrease at the end of the Permian?

(A)

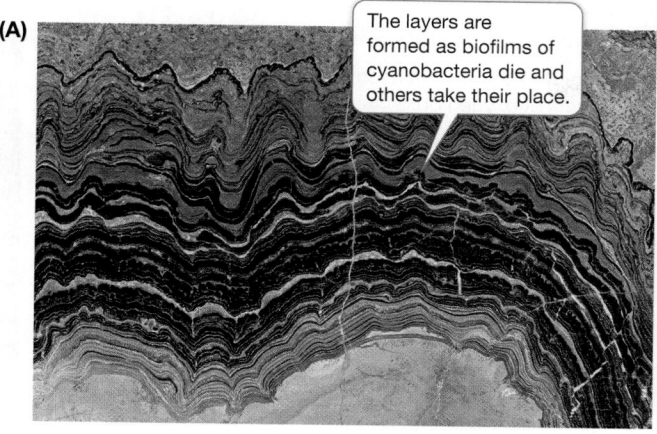

The layers are formed as biofilms of cyanobacteria die and others take their place.

5 cm

(B)

Living cyanobacteria are found in the upper parts of these stromatolites.

Figure 24.9 Stromatolites **(A)** A vertical section through a fossil stromatolite. **(B)** These rocklike structures are living stromatolites that thrive in the very salty waters of Shark Bay in Western Australia.

The second step occurred about a billion years later, when some of these photosynthetic bacteria became symbiotic within eukaryote cells, leading to the evolution of chloroplasts in photosynthetic plants and other eukaryotes. This change resulted in continued accumulation of O_2 in Earth's atmosphere (**Figure 24.8**).

One group of O_2-generating bacteria, the cyanobacteria, were a major component of complex communities that formed rocklike structures called stromatolites, which are abundantly preserved in the fossil record. To this day, stromatolites are still formed in a few very salty places (**Figure 24.9**). Cyanobacteria liberated enough O_2 to open the way for the evolution of oxidation reactions as the energy source for the synthesis of ATP.

Thus the evolution of life irrevocably changed the physical nature of Earth. Those physical changes, in turn, influenced the evolution of life. When it first appeared in the atmosphere, O_2 was toxic to most of the anaerobic prokaryotes that inhabited Earth at the time. Over

millennia, however, prokaryotes that evolved the ability to tolerate and use O_2 not only survived but gained the advantage. Aerobic metabolism proceeds more rapidly, and harvests energy more efficiently, than anaerobic metabolism. Organisms with aerobic metabolism replaced anaerobes in most highly oxygenated environments.

An atmosphere rich in O_2 also made possible larger and more complex organisms. Small single-celled aquatic organisms can obtain enough oxygen by simple diffusion even when dissolved oxygen concentrations in the water are very low. Larger single-celled organisms, however, have lower surface area-to-volume ratios. To obtain enough oxygen by simple diffusion, larger organisms must live in an environment with a relatively high oxygen concentration. Bacteria can thrive at 1 percent of the current oxygen concentration, but eukaryotic cells require levels that are at least 2–3 percent of the current concentration. For concentrations of dissolved oxygen in the oceans to have reached these levels, much higher atmospheric concentrations were needed.

Probably because it took many millions of years for Earth to develop an oxygenated atmosphere, only single-celled prokaryotes lived on Earth for more than 2 billion years. About 1.5 bya, atmospheric O_2 concentrations became high enough for larger eukaryotic cells to flourish. Further increases in atmospheric O_2 concentrations in the late Precambrian enabled several groups of multicellular organisms to evolve (see Figure 24.8).

Oxygen concentrations increased again during the Carboniferous and Permian periods because of the evolution of large vascular plants. These plants lived in the expansive lowland swamps that existed at the time (see Table 24.1). Massive amounts of organic material were buried in these swamps as the plants died, leading to the formation of Earth's vast coal deposits. Because the buried organic material was not subject to oxidation as it decomposed, and because the living plants were producing large quantities of O_2, atmospheric O_2 increased to concentrations that have not been reached again in Earth's history (see Figure 24.8). As mentioned at the opening of this chapter, these high concentrations of atmospheric O_2 allowed the evolution of giant flying insects and large amphibians that could not survive in today's atmosphere.

The drying of the lowland swamps at the end of the Permian reduced burial of organic matter as well as the production of O_2, so atmospheric O_2 concentrations dropped rapidly. Over the past 200 million years, with the diversification of flowering plants, O_2 concentrations have again increased, but not to the levels that characterized the Carboniferous and Permian periods.

Biologists have conducted experiments revealing the changing selection pressures that can accompany changes in atmospheric O_2 concentrations. When fruit flies (*Drosophila*) are raised in hyperoxic conditions (i.e., with artificially increased atmospheric concentrations of O_2), they evolve larger body sizes in just a few generations (**Investigating Life: The Relationship between Atmospheric Oxygen Concentration and Body Size in Insects**). The present atmospheric O_2 concentrations appear to constrain body size in these flying insects, whereas increases in O_2 appear to relax those constraints. This experiment demonstrates that the ***stabilizing selection** on body size at current O_2 concentrations can quickly switch to **directional selection** for a change in body size in response to a change in O_2 concentrations.

experiment

Original Paper: Klok, C. J., A. J. Hubb and J. F. Harrison. 2009. Single and multigenerational responses of body mass to atmospheric oxygen concentration in *Drosophila melanogaster*: Evidence for roles of plasticity and evolution. *Journal of Evolutionary Biology* 22: 2496–2504.

C. Jaco Klok and his colleagues asked whether insects raised in hyperoxic conditions (i.e., in an atmosphere having a higher partial pressure of oxygen than is normal today) would evolve to be larger than their counterparts raised under today's atmospheric conditions. They raised strains of fruit flies (*Drosophila melanogaster*) under both conditions to test the effects of increased O_2 concentrations on the evolution of body size.

HYPOTHESIS▶ In hyperoxic conditions, increased partial pressure of oxygen results in evolution of increased body size in flying insects.

METHOD

1. Separate a population of fruit flies into multiple lineages.
2. Raise half the lineages in current atmospheric (control) conditions; raise the other lineages in hyperoxic (experimental) conditions. Continue all lineages for seven generations.
3. Raise the F_8 individuals of all lineages under identical (current) atmospheric conditions.
4. Weigh 50 flies from each of the replicate lines and test for statistical differences in body weight.

RESULTS The average body mass of F_8 individuals of both sexes raised under hyperoxic conditions was significantly ($P < 0.001$) greater than that of individuals in the control lineages.

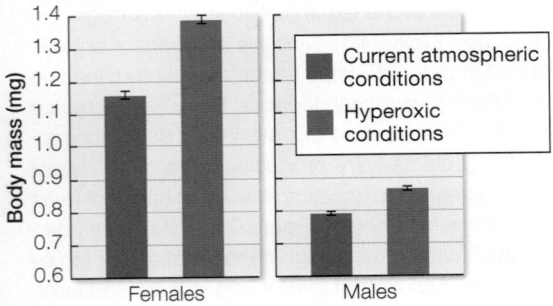

CONCLUSION▶ Increased O_2 concentrations led to evolution of larger body size in fruit flies, consistent with the trends seen among other flying insects in the fossil record.

work with the data

Original Papers: Harrison, J. F. and G. G. Haddad. 2011. Effects of oxygen on growth and size: Synthesis of molecular, organismal and evolutionary studies with *Drosophila melanogaster*. *Annual Review of Physiology* 73: 13.1–13.9.

Harrison, J. F., A. Kaiser and J. M. VandenBrooks. 2010. Atmospheric oxygen level and the evolution of insect body size. *Proceedings of the Royal Society of London B* 277: 1937–1946.

The data in the bar graphs show that the body mass of individuals in the experimental populations of *Drosophila* increased, on average, about 2 percent per generation under hyperoxic conditions (although the rate of increase was not constant over the experiment). Here you will extrapolate from Harrison et al.'s study to determine whether the observed rate of increase in body mass per generation that these researchers observed in *Drosophila* is sufficient to account for the giant dragonflies of the Permian period.

QUESTIONS▶

1. Suppose that the average rate of increase in dragonfly size during the Permian was much slower than the rate observed in the experiment. For this exercise, assume that the actual rate of increase for dragonflies was just 0.01 percent per generation, rather than the 2 percent observed over a few generations for *Drosophila*. Also assume that dragonflies complete only 1 generation per year (as opposed to 40 or more generations for *Drosophila*). Starting with an average body mass of 1 gram, calculate the projected increase in dragonfly body mass over 50,000 years. This calculation is similar to computing compound interest for a savings account. Use the formula $W = S(1 + R)^N$ where W = the final mass, S = the starting mass, R = the rate of increase per generation (0.0001 in this case), and N = the number of generations.

2. What percent of the Permian period does 50,000 years represent? Use Table 24.1 for your calculation.

3. Given your calculations, do you think that increased O_2 concentrations during the Permian were sufficient to account for the evolution of giant dragonflies? Why or why not?

A similar **work with the data** exercise may be assigned in **LaunchPad**.

---

*connect the concepts Stabilizing selection resists change from a current state, whereas directional selection results in change away from the current state (see Key Concept 20.4).

Extinction happens continuously, but mass extinctions result from sudden environmental changes

The physical and biological components of Earth's environment are constantly changing. As we saw in Chapter 23, new species are constantly being generated. At the same time, some species decline to the point that they can no longer maintain viable populations and go extinct. Both speciation and extinction have been occurring throughout the history of life. Why do species go extinct? Short-term or long-term environmental fluctuations can make growth or reproduction difficult. As new species arise, they may outcompete existing species for food, space, or other resources. If a prey species goes extinct, predators that depend on the prey may decline or disappear as well.

As population sizes decrease, the probability of extinction accelerates. At any population size, there is variation in reproductive success, and a particular individual may fail to reproduce successfully. When population sizes are very small, such variation has a bigger influence. Low reproduction by a few individuals in a small population is likely

experiment

Figure 24.10A   What Factors Influence Extinction in Small Populations?

Original Paper: Wootton, J. T. and C. A. Pfister. 2013. Experimental separation of genetic and demographic factors on extinction risk in wild populations. *Ecology* 94: 2117–2123.

J. Timothy Wootton and Catherine Pfister established new, isolated populations of sea palms (a species of kelp, *Postelsia palmaeformis*) on exposed rocky shores of the northwestern United States. They then studied the probability of population extinct as a function of genetic structure and population size.

HYPOTHESIS▶   Random variation in demographic events (individual births and deaths) will overwhelm genetic factors as contributions to short-term extinction probabilities in small populations. Extinction in larger populations will occur primarily from environmental fluctuation.

METHOD

1. Establish new isolated populations of *Postelsia palmaeformis* on rocky shorelines.

   Establish three types of populations:

   a. Inbred populations (populations derived from a single individual)

   b. Control populations (populations established from multiple individuals from the same source population)

   c. Outbred populations (populations established from multiple individuals from different source populations)

2. After populations are established, thin populations to small ($N = 20$) or large ($N = 50$) size. Follow the populations through time, recording population size each spring.

RESULTS

In the smallest populations, random variation in births and deaths was by far the largest contributing factor to population extinction. Genetic effects had very little contribution to short-term extinction rates in this species. Allee effects were most evident in populations of 1–10 individuals. Above $N = 10$, almost all population extinction resulted from environmental fluctuations.

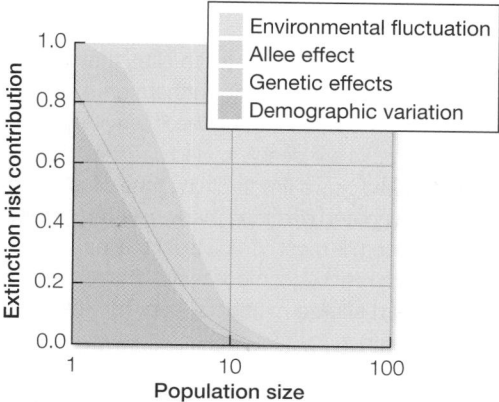

CONCLUSION▶   Chance variation in births and deaths has the highest effect on extinction at the smallest population sizes. At larger population sizes, most local extinction results from fluctuation in the environment.

work with the data

Figure 24.10B   What Factors Influence Extinction in Small Populations?

After establishing the experimental populations, the researchers counted the number of individuals present in each population each spring. They then recorded which populations were extant (still existing) or extinct in the following year. The table shows data for a few of the population sizes they examined:

Population size in previous year	Number of populations extant in following year	Number of populations extinct in following year
1–5	2	13
20	13	13
50	19	5
100–1,000	24	0

QUESTIONS▶

1. Draw a graph showing probability of extinction (*y* axis) as a function of population size (*x* axis) based on the data presented above. Use a log scale for the *x* axis. Draw a curve to approximate the relationship between probability of extinction and population size.

2. The graph you produced in Question 1 shows the extinction probability of a population after 1 year. If you started a population with 20 individuals, how likely do you think it would be to survive for 2 years? What additional data could you collect after 1 year to help answer this question?

A similar **work with the data** exercise may be assigned in **LaunchPad**.

to lead to loss of the population. The smaller the population, the larger the role that chance plays. Variation in reproductive rates is less likely to lead to extinction in large populations, where chance effects play a much smaller role.

Smaller populations can also have density-dependent disadvantages compared with larger populations—a phenomenon known as the **Allee effect**. Allee effects can occur, for example, from the inability of a small group to defend itself against predators, or from a decline in the ability of individuals to find mates. Small populations can also decline from the negative effects of inbreeding depression and genetic drift (see Chapter 21). These processes can result in loss of genetic variation, uncovering of deleterious alleles, and loss of beneficial alleles.

An experiment on sea palms (a species of kelp) illustrates the relative influence of various factors on extinction probability as it relates to population size (**Figure 24.10**). Biologists experimentally established isolated populations of sea palms on intertidal rocky shores and examined their survivorship through time. At small population sizes, random demographic factors were most likely to lead to local extinction, whereas environmental fluctuations were much more likely to cause the extinction of larger populations.

The five mass extinction events shown in Figure 24.2 are notable because they suddenly increased the extinction rate and affected a large percentage of living organisms at the same time. In each case, rapid and extreme environmental changes caused many species to go extinct simultaneously.

24.2 recap

Physical conditions on Earth have changed dramatically over time. Changes in Earth's climate and sea levels have had major effects on biological evolution. Continental drift, volcanic eruptions, and large meteorite strikes have contributed to major climate changes during Earth's history, and many of these climate shifts have resulted in mass extinction events. Changes in atmospheric concentrations of O_2 have also influenced the evolution of life, and the evolution of life has, in turn, affected the physical nature of the planet.

learning outcomes

You should be able to:

- Place important events in biological history onto a time line of Earth's history.
- Characterize the relative impact of weather events and climate changes on Earth's physical conditions.
- Explain how the production of atmospheric oxygen has affected Earth and its living organisms.
- Analyze the connections between a specific geological event or environmental change (e.g., fluctuations in ocean currents or sea levels, volcanic eruptions, meteorite strikes) and an evolutionary trend among living organisms.

1. Explain why an occasional major winter blizzard is irrelevant to discussions of global climate warming.
2. Describe changes in the environment that may have favored the evolution of groups of multicellular organisms near the end of the Precambrian.

The many dramatic physical events in Earth's history have influenced the nature and timing of evolutionary changes among Earth's living organisms. We now will look more closely at some of the major events that characterize the history of life on Earth.

key concept
24.3 Major Events in the Evolution of Life Can Be Read in the Fossil Record

How do we know how the physical changes described in the previous section affected the evolution of life? To reconstruct life's history, scientists rely heavily on the fossil record. As we have seen, geologists divided Earth's history into eons, eras, and periods based on distinct fossil assemblages (see Table 24.1). Biologists refer to the assemblage of all organisms of all kinds living at a particular time or place as a **biota**. All of the plants living at a particular time or place are its **flora**; all of the animals are its **fauna**.

focus your learning

- Earth's history is divided into geological periods that are associated with major events in biological evolution.
- Life's diversity has expanded and contracted many times over Earth's history.

Figure 24.11 Fossils in Amber   Chunks of amber—fossilized tree resin—often contain detailed fossils such as this spider, which was preserved when it became trapped in the sticky resin.

About 300,000 species of fossil organisms have been described and named, and the number steadily grows. The number of named species, however, is only a tiny fraction of the species that have ever lived. We do not know how many species lived in the past, but we have ways of making reasonable estimates. Of the present-day biota, about 1.8 million species have been named. The actual number of living species is estimated to be at least 10 million, and possibly much higher, because many species have not yet been discovered and described by biologists. So the number of described fossil species is only about 3 percent of the estimated minimum number of living species. Life has existed on Earth for about 3.8 billion years. Many species last only a few million years before undergoing speciation or going extinct. From this we know that Earth's biota must have turned over many times during geological history. So the total number of species that have lived over evolutionary time must vastly exceed the number living today. Why have only about 300,000 of these tens of millions of species been described from fossils to date?

Several processes contribute to the paucity of fossils

Only a tiny fraction of organisms ever become fossils, and only a tiny fraction of fossils are ever discovered by paleontologists. Most organisms live and die in oxygen-rich environments, in which they quickly decompose. Organisms are not likely to become fossils unless they are transported by wind or water to sites that lack oxygen, where decomposition proceeds slowly. Furthermore, geological processes transform many rocks, destroying the fossils they contain, and many fossil-bearing rocks are deeply buried and inaccessible. Paleontologists have studied only a tiny fraction of the sites that contain fossils, although they find and describe many new ones every year.

The fossil record is most complete for marine animals that had hard skeletons (which resist decomposition). Among the nine major animal groups with hard-shelled members, approximately 200,000 species have been described from fossils—roughly twice the number of living marine species in these same groups. Paleontologists lean heavily on these groups in their interpretations of the evolution of life. Arthropods are also relatively well represented in the fossil record because they are numerically abundant and have hard exoskeletons (**Figure 24.11**). The fossil record, though incomplete, is good enough to document clearly the factual history of the evolution of life.

By combining information about physical changes during Earth's history with evidence from the fossil record, scientists have composed portraits of what Earth and its inhabitants may have looked

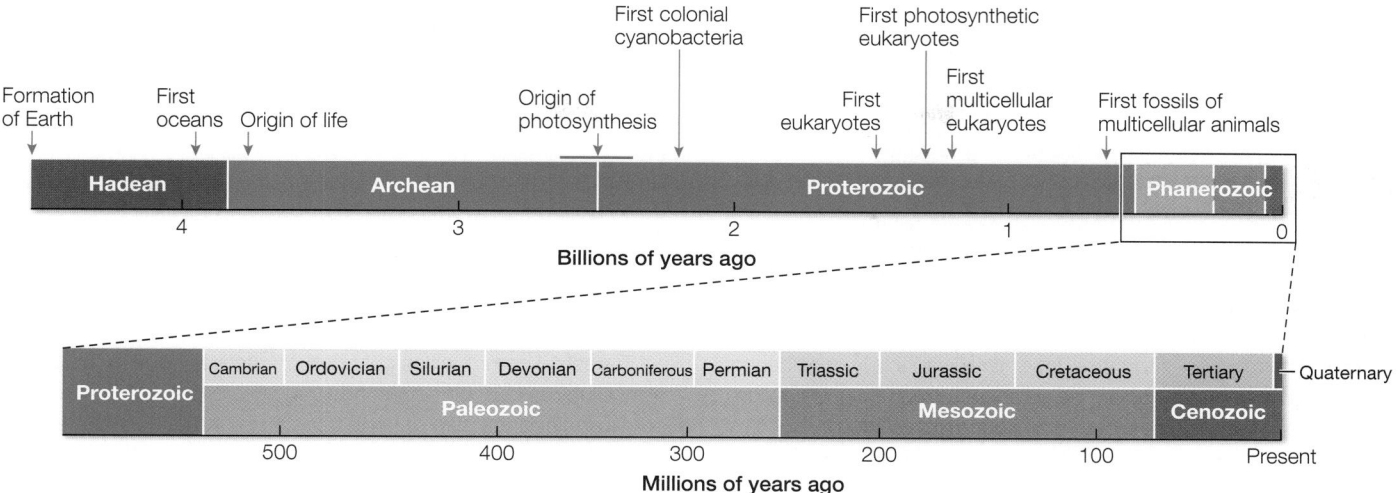

Figure 24.12 A Sense of Life's Time   The top timeline shows the 4.5-billion-year history of Earth. Most of this history is accounted for by the Precambrian, a 3.4-billion-year time span that saw the origin of life and the evolution of cells, photosynthesis, and multicellularity. The final 600 million years are expanded in the bottom timeline and detailed in Figure 24.14.

like at different times. We know in general where the continents were and how life changed over time, but many of the details are poorly known, especially for events in the more remote past.

Precambrian life was small and aquatic

Life first appeared on Earth about 3.8 bya (**Figure 24.12**). The fossil record of organisms that lived prior to the Phanerozoic is fragmentary, but it is good enough to establish that the total number of species and individuals increased dramatically in the late Precambrian.

For most of its history, life was confined to the oceans, and all organisms were small. For more than 3 billion years, all organisms lived in shallow seas. These seas slowly began to teem with microscopic prokaryotes. After the first eukaryotes appeared about 1.5 billion years ago, during the Proterozoic, unicellular eukaryotes and small multicellular animals fed on the microorganisms. Small floating organisms, known collectively as **plankton**, were strained from the water and eaten by slightly larger filter-feeding animals. Other animals ingested sediments on the seafloor and digested the remains of organisms within them. But it still took nearly a billion years before eukaryotes began to diversify rapidly into the many different morphological forms that we know today.

What limited the diversity of multicellular eukaryotes (in terms of their size and shape) for much of their early existence? It is likely that a combination of factors was responsible. We have already noted

that O_2 levels increased throughout the Proterozoic, and it is likely that high atmospheric and dissolved O_2 concentrations were needed to support large multicellular organisms. In addition, geologic evidence points to a series of intensely cold periods during the late Proterozoic, which would have resulted in seas that were largely covered by ice and continents that were covered by glaciers. The "snowball Earth" hypothesis suggests that cold conditions confined life to warm places such as hot springs, deep thermal vents, and perhaps a few equatorial oceans that avoided ice cover. The last of these Proterozoic glaciations ended about 580 million years ago, just before several major radiations of multicellular eukaryotes appear in the fossil record (**Figure 24.13**). Many of the multicellular organisms known from the late Proterozoic and early Phanerozoic were very different from any animals living today and may be members of groups that left no living descendants.

Figure 24.13 Diversification of Multicellular Organisms: The Cambrian "Explosion"   Shortly after the end of Proterozoic glaciations (about 580 mya), several major radiations of multicellular organisms appear in the fossil record. (A) These microscopic fossils from the Doushantuo rock formation of China are the remains of tiny one-, two-, four-, and eight-celled stages of multicellular organisms. (B) Unusual soft-bodied marine invertebrates, unlike any animals alive at present, characterize the fossilized fauna preserved at Ediacara in southern Australia. (C) By the early Phanerozoic, fossilized faunas such as those preserved in Canada's Burgess Shale include extinct representatives of some of the major animal groups alive today.

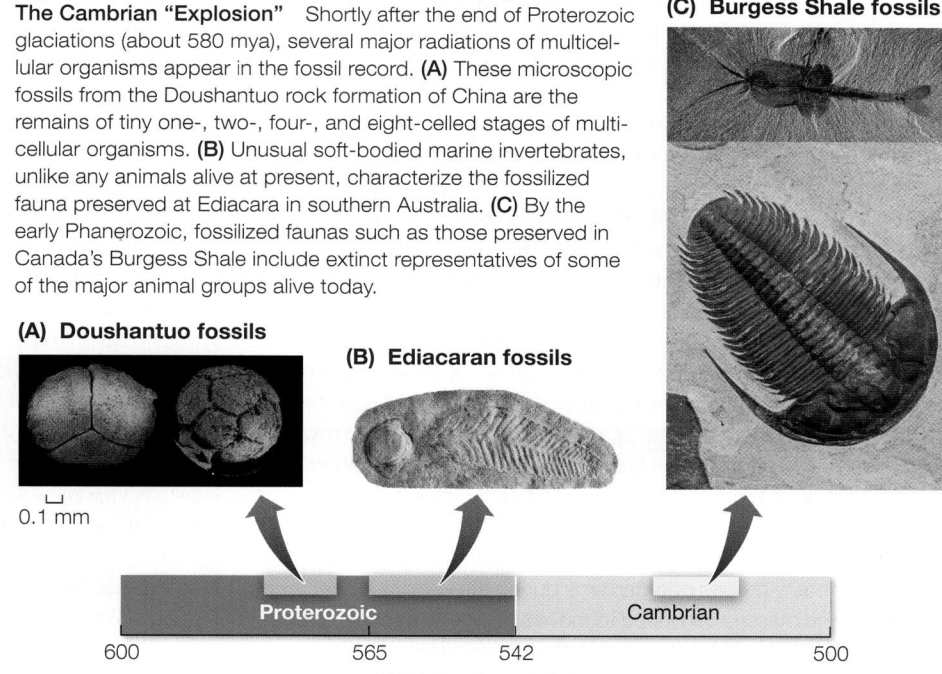

Life expanded rapidly during the Cambrian period

The Cambrian period (542–488 mya) marks the beginning of the Paleozoic, the first era of the Phanerozoic. The O$_2$ concentration in the Cambrian atmosphere was approaching its current level, and the glaciations of the late Proterozoic had ended nearly 40 million years earlier. A geologically rapid diversification of life took place that is often called the **Cambrian explosion**. This name is somewhat misleading, as the series of radiations it refers to actually began before the start of the Cambrian and continued for about 60 million years into the early Cambrian (see Figure 24.12). Nonetheless, 60 million years represents a relatively short amount of time, especially considering that the first eukaryotes had appeared about a billion (= 1,000 million) years earlier. Many of the major animal groups represented by species alive today first appeared during these evolutionary radiations. **Figure 24.14** provides an overview of the numerous continental and biotic innovations that have characterized the Phanerozoic.

For the most part, fossils tell us only about the hard parts of organisms, but in some well-studied Cambrian fossil beds the soft parts of many animals were preserved. Multicellular life was largely or completely aquatic during the Cambrian. If there was life on land at this time, it was probably restricted to microorganisms.

Many groups of organisms that arose during the Cambrian later diversified

Geologists divide the remainder of the Paleozoic era into the Ordovician, Silurian, Devonian, Carboniferous, and Permian periods. Each period is characterized by the diversification of specific groups of organisms. Mass extinctions marked the ends of the Ordovician, Devonian, and Permian.

THE ORDOVICIAN (488–444 MYA)　During the Ordovician period, the continents, which were located primarily in the Southern Hemisphere, still lacked multicellular life. Evolutionary radiation of marine organisms was spectacular during the early Ordovician, especially among animals, such as brachiopods and mollusks, that lived on the seafloor and filtered small prey from the water. At the end of the Ordovician, as massive glaciers formed over the southern continents, sea levels dropped about 50 meters, and ocean temperatures dropped. About 75 percent of all animal species became extinct, probably because of these major environmental changes.

THE SILURIAN (444–416 MYA)　During the Silurian period, the continents began to merge together. Marine life rebounded from the mass extinction at the end of the Ordovician. Animals able to swim in open water and feed above the ocean floor appeared for the first time. Jawless fishes diversified, and the first fishes with supporting rays in their fins appeared. The tropical sea was uninterrupted by land barriers, and most marine organisms were widely distributed. On land, the first vascular plants evolved late in the Silurian (about 420 mya). The first terrestrial arthropods—scorpions and millipedes—evolved at about the same time.

THE DEVONIAN (416–359 MYA)　Rates of evolutionary change accelerated in many groups of organisms during the Devonian period. The major land masses continued to move slowly toward each other. In the oceans there were great evolutionary radiations of corals and of shelled, squidlike cephalopod mollusks. Fishes diversified as

Figure 24.14 A Brief History of Multicellular Life on Earth ▶
The geologically rapid "explosion" of life shortly before and during the Cambrian saw the rise of many major animal groups that have representatives surviving today. The following three pages depict life's history through the Phanerozoic. The movements of the major continents during the past half-billion years are shown in the maps of Earth, and associated biotas for each time period are depicted. The artists' reconstructions are based on fossils such as those shown in the photographs.

jawed forms replaced jawless ones and as bony armor gave way to the less rigid scales of modern fishes.

Terrestrial communities changed dramatically during the Devonian. Club mosses, horsetails, and tree ferns became common, and some attained the size of large trees. Their roots accelerated the weathering of rocks, resulting in the development of the first forest soils. The first plants to produce seeds appeared in the Devonian. The earliest fossil centipedes, spiders, mites, and insects date to this period, as do the earliest terrestrial vertebrates.

A massive extinction of about 75 percent of all marine species marked the end of the Devonian. Paleontologists are uncertain about its cause, but two large meteorites that collided with Earth at about that time (one in present-day Nevada, the other in Western Australia) may have been responsible, or at least a contributing factor. The continued merging of the continents, with the corresponding reduction in the area of continental shelves, may have also contributed to this mass extinction.

THE CARBONIFEROUS (359–299 MYA)　Large glaciers formed over high-latitude portions of the southern land masses during the Carboniferous period, but extensive swamp forests grew on the tropical continents. These forests were dominated by giant tree ferns and horsetails with small leaves. Their fossilized remains formed the coal we now mine for energy. In the seas, crinoids (a group of echinoderms, related to sea stars and sea urchins) reached their greatest diversity, forming "meadows" on the seafloor.

The diversity of terrestrial animals increased greatly during the Carboniferous. Snails, scorpions, centipedes, and insects were abundant and diverse. Insects evolved wings, becoming the first animals to fly. Flight gave herbivorous insects easy access to tall plants, and plant fossils from this period show evidence of chewing by insects (**Figure 24.15**). The terrestrial vertebrates split into two lineages. The amphibians became larger and better adapted to terrestrial existence, while the sister lineage led to the amniotes: vertebrates with well-protected eggs that can be laid in dry places.

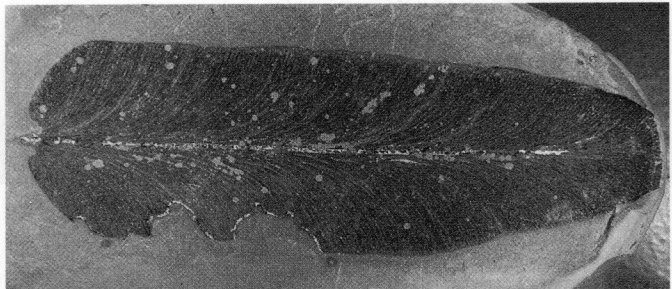

Figure 24.15 Evidence of Insect Diversification　The margins of this fossil fern leaf from the Carboniferous have been chewed by insects.

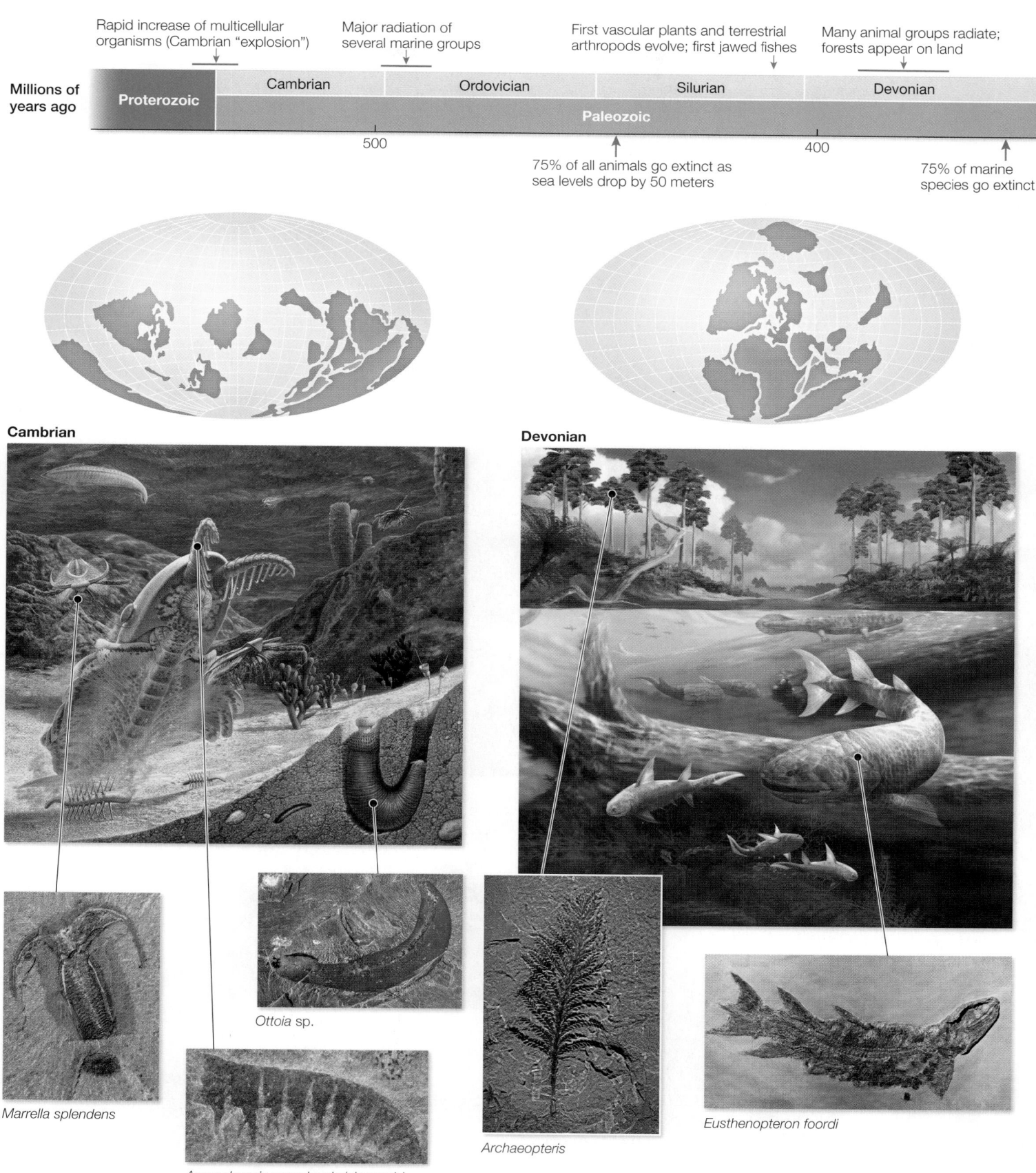

Rapid increase of multicellular organisms (Cambrian "explosion")

Major radiation of several marine groups

First vascular plants and terrestrial arthropods evolve; first jawed fishes

Many animal groups radiate; forests appear on land

Millions of years ago

| Proterozoic | Cambrian | Ordovician | Silurian | Devonian |

Paleozoic

500

400

75% of all animals go extinct as sea levels drop by 50 meters

75% of marine species go extinct

Cambrian

Devonian

Marrella splendens

Ottoia sp.

Anomalocaris canadensis (claw only)

Archaeopteris

Eusthenopteron foordi

Extensive swamp forests produce coal; origin of amniotes; great increase in terrestrial animal diversity

Giant amphibians and flying insects; ray-finned fish abundant in fresh water

On land, conifers become dominant plants; frogs and reptiles begin to diversify

First mammals appear

Dinosaurs, pterosaurs, ray-finned fish diversify

First known flowering plant fossils

Carboniferous	Permian	Triassic	Jurassic
Paleozoic		Mesozoic	

300

Extinction of 96% of Earth's species; oxygen levels drop rapidly

200

Mass extinction event, including about 65% of all species

Pangaea

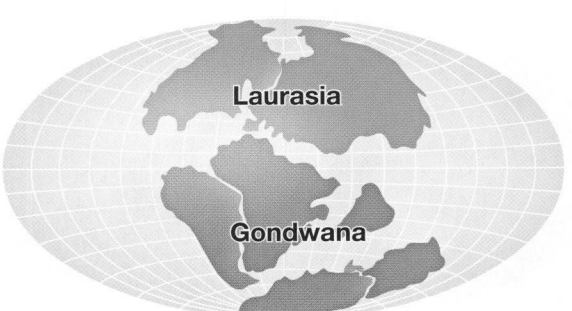

Laurasia

Gondwana

Permian

Jurassic

Europasaurus holgeri

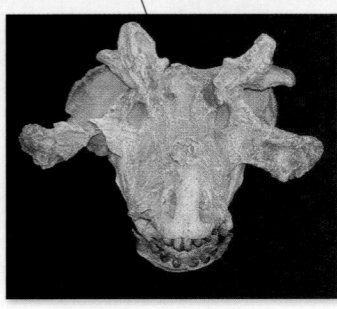

Estemmenosuchus sp.

Equisetum sp.

Ginkgo sp.

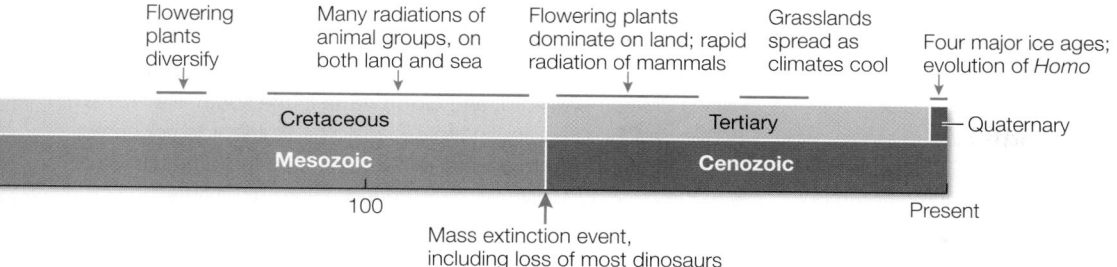

Flowering plants diversify

Many radiations of animal groups, on both land and sea

Flowering plants dominate on land; rapid radiation of mammals

Grasslands spread as climates cool

Four major ice ages; evolution of *Homo*

Cretaceous | Tertiary | Quaternary

Mesozoic | Cenozoic

100

Present

Mass extinction event, including loss of most dinosaurs

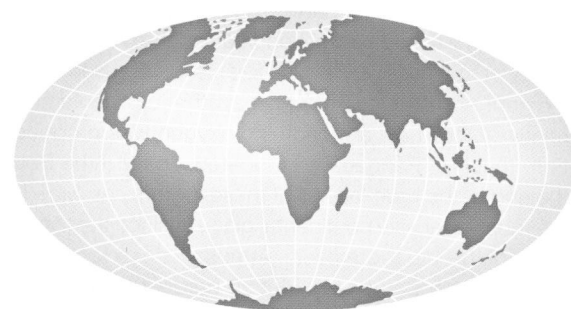

Cretaceous

Tertiary

Sapindopsis belviderensis (leaves)

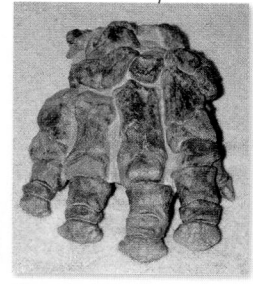

Coryphodon

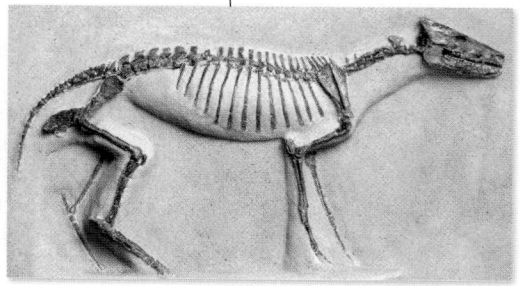

Hyracotherium leporinum

Chasmosaurus belli

THE PERMIAN (299–251 MYA)   During the Permian period, the continents merged into a single supercontinent called Pangaea. Permian rocks contain representatives of many of the major groups of insects we know today. By the end of the period the amniotes had split into two lineages: the reptiles, and a second lineage that would lead to the mammals. Ray-finned fishes became common in the freshwaters of Pangaea.

Toward the end of the Permian, conditions for life deteriorated. Massive volcanic eruptions resulted in outpourings of lava that covered large areas of Earth. The ash and gases produced by the volcanoes blocked sunlight and cooled the climate. The death and decay of the massive Permian forests rapidly used up atmospheric oxygen, and the loss of photosynthetic organisms meant that relatively little new atmospheric oxygen was produced. In addition, much of Pangaea was located close to the South Pole by the end of the Permian. All of these factors combined to produce the most extensive continental glaciers since the "snowball Earth" times of the late Proterozoic. Atmospheric oxygen concentrations gradually dropped from about 30 percent to 15 percent. At such low concentrations, most animals would have been unable to survive at elevations above 500 meters, so about half of the land area would have been uninhabitable at the end of the Permian. The combination of these changes resulted in the most drastic mass extinction in Earth's history. Scientists estimate that about 96 percent of all multicellular species became extinct at the end of the Permian.

Geographic differentiation increased during the Mesozoic era

The few organisms that survived the Permian mass extinction found themselves in a relatively empty world at the start of the Mesozoic era (251 mya). As Pangaea slowly began to break apart in the Mesozoic, the biotas of the newly separated continents began to diverge. The oceans rose and once again flooded the continental shelves, forming huge, shallow inland seas. Atmospheric oxygen concentrations gradually rose. Life once again proliferated and diversified, but different groups of organisms came to the fore. The three groups of phytoplankton (floating photosynthetic organisms) that dominate today's oceans—dinoflagellates, coccolithophores, and diatoms—became ecologically important at this time, and their remains are the primary origin of the world's oil deposits. Seed-bearing plants replaced the trees that had ruled the Permian forests.

The Mesozoic era is divided into three periods: the Triassic, Jurassic, and Cretaceous. The Triassic and Cretaceous were terminated by mass extinctions, probably caused by meteorite impacts.

THE TRIASSIC (251–201.6 MYA)   Pangaea remained largely intact through the Triassic. Many invertebrate groups diversified, and many burrowing animals evolved from groups living on the surfaces of seafloor sediments. On land, conifers and seed ferns were the dominant trees. The first frogs and turtles appeared. A great radiation of reptiles began, which eventually gave rise to crocodilians and dinosaurs (including birds). The end of the Triassic was marked by a mass extinction that eliminated about 65 percent of the species on Earth.

THE JURASSIC (201.6–145.5 MYA)   Late in the Jurassic period, Pangaea became fully divided into two large continents: **Laurasia**, which drifted northward, and **Gondwana** in the south. Ray-finned fishes rapidly diversified in the oceans. The first lizards appeared, and flying reptiles (pterosaurs) evolved. Most of the large terrestrial predators and herbivores of the period were dinosaurs. Several groups of mammals made their first appearance, and the earliest known fossils of flowering plants are from late in this period.

THE CRETACEOUS (145.5–65.5 MYA)   By the mid-Cretaceous, Laurasia and Gondwana had largely broken apart into the continents we know today (although the Indian subcontinent was still separated from Asia). A continuous sea encircled the tropics. Sea levels were high, and Earth was warm and humid. Life proliferated both on land and in the oceans. Marine invertebrates increased in diversity. On land, the reptile radiation continued as dinosaurs diversified further and the first snakes appeared. Early in the Cretaceous, flowering plants began the radiation that led to their current dominance of the land. By the end of the period, many groups of mammals had appeared.

As described in Key Concept 24.2, another meteorite-caused mass extinction took place at the end of the Cretaceous. In the seas, many planktonic organisms and bottom-dwelling invertebrates became extinct. On land, almost all animals larger than about 25 kg in body weight became extinct. Many species of insects died out, perhaps because the growth of their food plants was greatly reduced following the impact. Some species in northern North America and Eurasia survived in areas that were not subjected to the devastating fires that engulfed most low-latitude regions.

Modern biotas evolved during the Cenozoic era

By the early Cenozoic era (65.5 mya), the positions of the continents were getting closer to their present positions, but the Indian subcontinent was still separated from Asia, and the Atlantic Ocean was much narrower. The Cenozoic was characterized by an extensive radiation of mammals, but other groups were also undergoing important changes.

Flowering plants diversified extensively and came to dominate world forests except in the coolest regions, where the forests were composed primarily of gymnosperms. Mutations of two genes in one group of plants (the legumes) allowed them to use atmospheric nitrogen directly by forming symbioses with a few species of nitrogen-fixing bacteria. The evolution of this symbiosis was the first "green revolution" and dramatically increased the amount of nitrogen available for terrestrial plant growth. This symbiosis remains fundamental to the ecological base of life as we know it today.

The Cenozoic era is divided into the Tertiary and the Quaternary periods, which are commonly subdivided into **epochs (Table 24.2)**.

THE TERTIARY (65.5–2.6 MYA)   During the Tertiary period, the Indian subcontinent continued its northward drift. By about 55 mya it made initial contact with parts of southeastern Asia. By about 35 mya, the Indian Plate ran fully into the Eurasian Plate, and the Himalaya Mountains began to be pushed up as a result.

table 24.2 Subdivisions of the Cenozoic Era

Period	Epoch	Onset (mya)
Quaternary	Holocene (Recent)	0.01 (~10,000 years ago)
	Pleistocene	2.6
Tertiary	Pliocene	5.3
	Miocene	23
	Oligocene	34
	Eocene	55.8
	Paleocene	65.5

The early Tertiary was a hot and humid time, and the ranges of many plants shifted latitudinally. The tropics were probably too hot to support rainforest vegetation and instead were clothed in low-lying vegetation. In the middle of the Tertiary, however, Earth's climate became considerably cooler and drier. Many lineages of flowering plants evolved herbaceous (nonwoody) forms, and grasslands spread over much of Earth.

By the start of the Cenozoic era, invertebrate faunas had already come to resemble those of today. It is among the terrestrial vertebrates that evolutionary changes during the Tertiary were most rapid. Frogs, snakes, lizards, birds, and mammals all underwent extensive radiations during the Tertiary. Three waves of mammals dispersed from Asia to North America across one of the several land bridges that have intermittently connected the two continents during the past 55 million years. Rodents, marsupials, primates, and hoofed mammals appeared in North America for the first time.

THE QUATERNARY (2.6 MYA TO PRESENT)   We are living in the Quaternary period. It is subdivided into two epochs, the Pleistocene and the Holocene (the Holocene is also known as the Recent).

The Pleistocene was a time of drastic cooling and climate fluctuations. During 4 major and about 20 minor "ice ages," massive glaciers spread across the continents, and the ranges of animal and plant populations shifted toward the equator. The last of these glaciers retreated from temperate latitudes less than 15,000 years ago. Organisms are still adjusting to this change. Many high-latitude ecological communities have occupied their current locations for no more than a few thousand years.

It was during the Pleistocene that divergence within one group of mammals, the primates, resulted in the evolution of the hominoid lineage. Subsequent hominoid radiation eventually led to the species *Homo sapiens*—modern humans. Many large bird and mammal species became extinct in Australia and in the Americas when *H. sapiens* arrived on those continents about 45,000 and 15,000 years ago, respectively. Many paleontologists believe at least some of these extinctions were the result of hunting and other influences of *Homo sapiens*.

The tree of life is used to reconstruct evolutionary events

The fossil record reveals broad patterns in life's evolution. To reconstruct major events in the history of life, biologists also rely on the phylogenetic information in the tree of life (see Appendix A). We can use phylogeny, in combination with the fossil record, to reconstruct the timing of such major events as the acquisition of mitochondria in the ancestral eukaryotic cell, the several independent origins of multicellular organisms, and the movement of life onto dry land. We can also follow major changes in the genomes of organisms, and we can even reconstruct many gene sequences of species that are long extinct, as described in Key Concept 21.3.

Changes in Earth's physical environment have clearly influenced the diversity of organisms we see on the planet today. To study the evolution of that diversity, biologists examine the evolutionary relationships among species. Deciphering phylogenetic relationships is an important step in understanding how life has diversified on Earth. The next part of this book will explore the major groups of life and the different solutions these groups have evolved to meet major challenges such as reproduction, energy acquisition, dispersal, and escape from predation.

24.3 recap

Life evolved in the oceans about 3.8 billion years ago. It diversified as atmospheric oxygen approached its current level. Numerous climate changes and rearrangements of the continents, as well as meteorite impacts, contributed to five major mass extinctions.

learning outcomes

You should be able to:

- Explain how increases and decreases in biodiversity through time relate to major changes in Earth's environmental conditions.
- Characterize environmental conditions during successive geological periods and connect these periods with major events in biological evolution.
- Illustrate the relationship between a phylogenetic tree and geological time.

1. Give a likely reason for the "green revolution" of the Cenozoic era, and explain why it enabled flowering plants to diversify.

2. Compare the tree of life in Appendix A with the chronology of major events in Table 24.1. To represent the diversity of life in one figure, the tree in the appendix is not drawn proportionally in relation to time. Which geological eon, era, or period seems most distorted, and why? What major changes to the tree would be needed to show divergences in groups that are proportional?

Q&A How can modern experiments test hypotheses about the evolutionary impact of ancient environmental changes?

Several experiments have been conducted to test the link between O_2 concentrations and evolution of body size in flying insects (one of these is discussed in Investigating Life: The Relationship between Atmospheric Oxygen Concentration and Body Size in Insects). Results of these experiments are consistent with the evolution of larger body size in flying insects in hyperoxic (high-oxygen) environments. Experiments have also been conducted under hypoxic (low-oxygen) conditions, such as existed at the end of the Permian. Results of these experiments suggest that the evolution of body size is constrained under hypoxic conditions, even under strong artificial selection for larger body size. These latter results are consistent with the extinction of many of the large flying insects at the end of the Permian, the result of rapidly decreasing O_2 concentrations. Giant flying insects simply could not have survived the lower O_2 concentrations that existed at that time. The mass extinction at the end of the Permian is the only known mass extinction that involved considerable loss of insect diversity.

Future directions

Experiments with living organisms are providing information about how quickly various groups of organisms evolve. They also inform us about the factors involved in extinction of species (as in the example in Figure 24.10). As human activities affect the climate of Earth, experimental approaches will be important to help us predict how our actions may affect future directions of evolution and extinction.

Chapter Summary 24

▶ 24.1 Events in Earth's History Can Be Dated

- The dates of evolutionary events can be determined by the dating of fossils and the **strata** of **sedimentary rocks** in which the fossils are found.

- **Radiometric dating** techniques use a variety of radioisotopes with different **half-lives** to date events at different times in the remote past. Review Figure 24.1

- Geologists divide the history of life into eons, eras, and periods. These divisions are based largely on major differences in the fossil assemblages found in successive strata. Review Table 24.1

▶ 24.2 Changes in Earth's Physical Environment Have Affected the Evolution of Life

- Earth's crust consists of solid lithospheric plates that float on the viscous mantle. **Continental drift** is caused by convection currents in the mantle, which move the plates and the continents that lie on top of them. Review Focus: Key Figure 24.3, Animation 24.1

- Five episodes of **mass extinction events** punctuated the history of life in the Paleozoic and Mesozoic eras. Review Figure 24.2

- Major physical events on Earth, such as continental collisions and volcanic eruptions, have affected Earth's surface, climate, atmosphere, and sea levels. In addition, extraterrestrial events such as meteorite strikes have created sudden and dramatic environmental shifts. All of these changes affected the history of life. Review Figure 24.4

- Oxygen-generating cyanobacteria liberated enough O_2 to open the door to oxidation reactions in metabolic pathways. Aerobic prokaryotes were able to harvest more energy than anaerobic organisms and began to proliferate. Increases in atmospheric O_2 concentrations supported the evolution of large eukaryotic cells and, eventually, multicellular organisms. Review Figure 24.8, Investigating Life: The Relationship between Atmospheric Oxygen Concentration and Body Size in Insects

- Extinctions of species happen throughout evolutionary history, but major, rapid environmental changes can produce mass extinctions of many species simultaneously. Review Figure 24.10

▶ 24.3 Major Events in the Evolution of Life Can Be Read in the Fossil Record

- Paleontologists use fossils and evidence of geological changes to determine what Earth and its **biota** may have looked like at different times. Review Figure 24.12

- Before the Phanerozoic, life was almost completely confined to the oceans. Multicellular life diversified extensively during the **Cambrian explosion**, an example of an evolutionary radiation. Review Figure 24.13

- The periods of the Paleozoic era were each characterized by the diversification of specific groups of organisms. The Paleozoic ended with the most drastic mass extinction in Earth's history. During the Mesozoic era, distinct terrestrial biotas evolved on each continent. Review Figure 24.14

- Earth's **flora** has been dominated by flowering plants since the Cenozoic era began.

- The tree of life can be used to reconstruct the timing of evolutionary events.

See Activity 24.1 for a concept review of this chapter.

Go to **LearningCurve** (in **LaunchPad**) for dynamic quizzing that helps you solidify your understanding of this chapter. **LearningCurve** adapts to your responses, giving you the practice you need to master each key concept.

Apply What You've Learned

Review

24.1 Geologists use several methods to date ancient events.

24.3 Earth's history is divided into geological periods that are associated with major events in biological evolution.

The diagram below summarizes evidence gathered through the study of fossils and rocks from locations around the world. Assemblages of two groups of animal fossils are indicated as the Ediacaran biota and the Cambrian-type shelly biota.

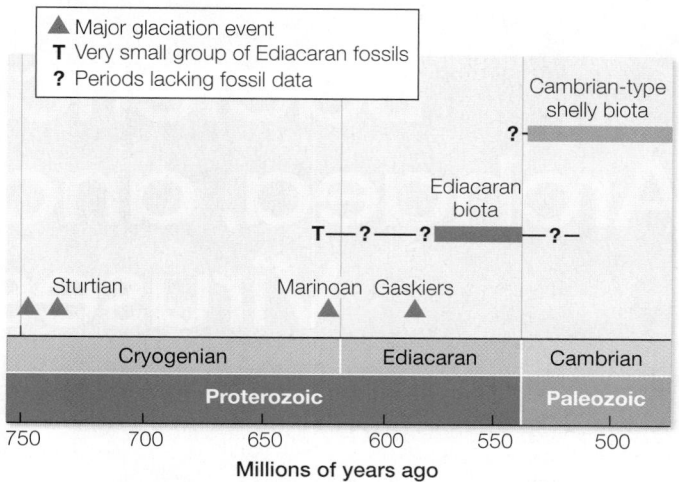

The Ediacaran biota represents the remains of the oldest large multicellular animals in Earth's history. These were soft-bodied marine animals that lacked skeletons and attached to the sea floor by a holdfast or foot. With body shapes resembling modern ferns, they grew up to 1 meter in length.

Ediacaran animals appear to have flourished for at least 23 million years (as indicated in the diagram) before going extinct. They were followed by the Cambrian-type shelly biota, which had mineralized exoskeletons and were also complex, multicellular animals.

Questions

1. How would the geological strata in which Ediacaran fossils were discovered compare with those in which Cambrian-type shelly biota fossils were discovered? Justify your answer with evidence from the diagram.

2. What factors were likely to have favored the success of the Ediacaran biota at that time in Earth's history, but not earlier?

3. What factors could have resulted in the extinction of the Ediacaran biota and the rise of the Cambrian-type shelly biota?

4. It has been stated that "absence of evidence is not evidence of absence" in reference to the interpretation of the fossil record. Evaluate this statement as it relates to the study of the Ediacaran biota.

5. An index fossil is one whose presence in a geological formation can indicate the age of the formation. Propose a scenario to illustrate how Ediacaran fossils might be used as index fossils. Explain your reasoning.

Go to **LaunchPad** for the eBook, LearningCurve, animations, activities, flashcards, and additional resources and assignments.

25

key concepts

25.1 Bacteria and Archaea Are the Two Primary Divisions of Life

25.2 Prokaryote Diversity Reflects the Ancient Origins of Life

25.3 Ecological Communities Depend on Prokaryotes

25.4 Viruses Have Evolved Many Times

Bacteria, Archaea, and Viruses

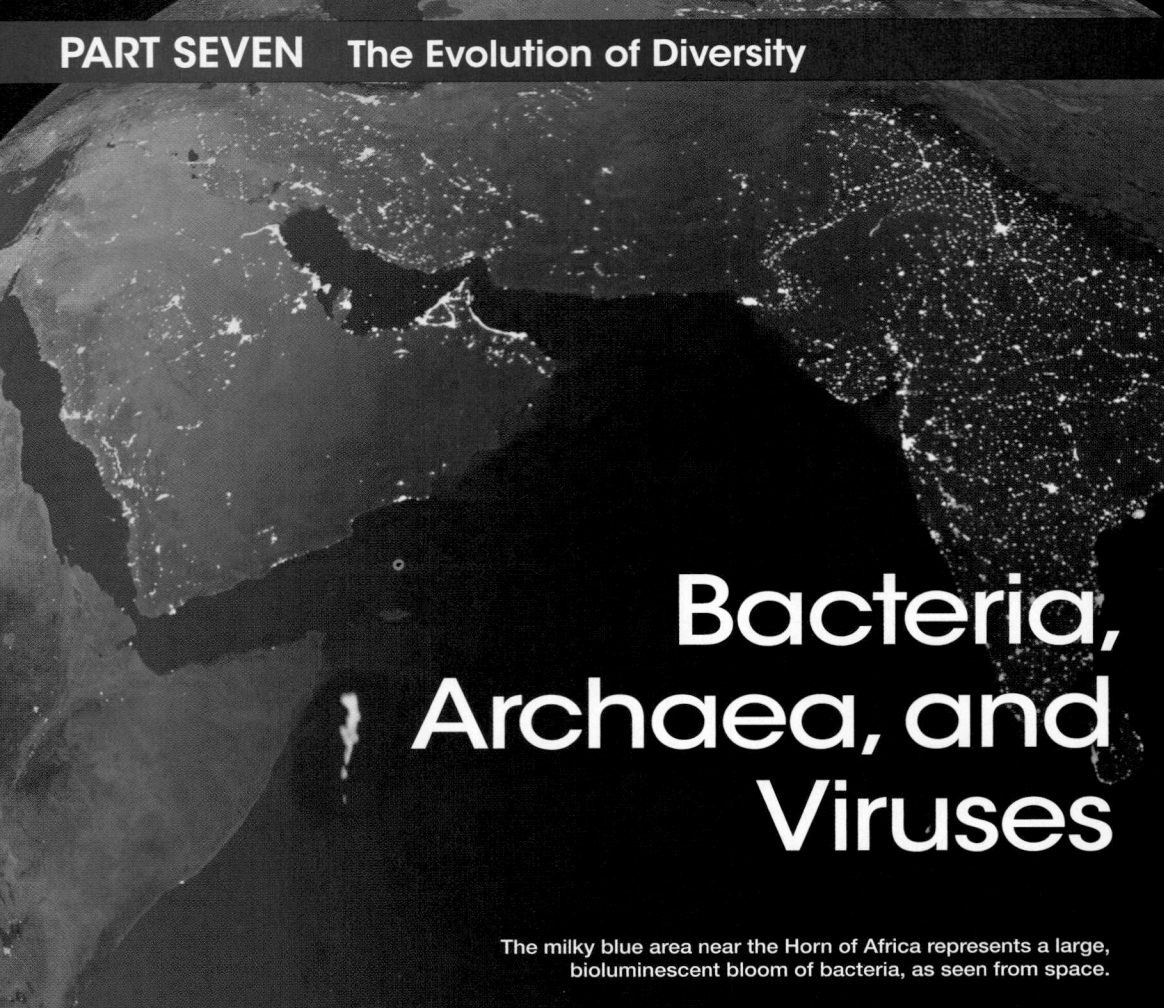

The milky blue area near the Horn of Africa represents a large, bioluminescent bloom of bacteria, as seen from space.

investigating life

Bacteria Light Up the Sea

One night in January of 1995, the British merchant vessel *Lima* was off the coast of Somalia, near the Horn of Africa. This area is infamous for bands of pirates, so the crew was keeping a watchful eye on the seas when they spotted an eerie, whitish glow on the horizon. It was directly in their path, and there was no way to avoid it. Was this strange sight the result of some strange trick of piracy?

Within 15 minutes of first sighting the glow, the *Lima* was surrounded by shining waters for as far as her crew could see. The ship's log recorded that "it appeared as though the ship was sailing over a field of snow or gliding over the clouds." Fortunately for the crew, the glow had nothing to do with pirates.

For centuries, mariners in this part of the world had reported occasional "milky seas" in which the sea surface produced a strange glow at night, extending from horizon to horizon. Scientists up to that point had never been able to confirm the reality or the cause of such phenomena. It was well established, however, that many organisms emit light by bioluminescence, a complex, enzyme-catalyzed biochemical reaction that emits light but not heat.

What kind of organism could cause the vast expanse of bioluminescence observed by the *Lima*? Some marine organisms emit flashes of light when they are disturbed, but they could not have produced the sustained and uniform glow seen in milky seas. The only organisms known to produce the level of sustained bioluminescence consistent with milky seas are certain prokaryotes, such as bacteria of the genus *Vibrio*. Using information supplied by the *Lima*, biologists scanned satellite images of the Indian Ocean for the specific light wavelengths emitted by *Vibrio*. The satellite images clearly identified thousands of square kilometers of *Vibrio*-produced milky seas.

Vibrio's bioluminescence requires a critical concentration of a specific chemical signal produced by the bacteria, so at low densities, free-living *Vibrio* populations do not glow. But as a colony establishes itself on phytoplankton, the bacteria's population density increases and concentrations of the luminescence signal build up. Eventually bacterial density becomes high enough for the huge colony to produce visible light. Such chemical-induced action among bacterial cells is referred to as quorum sensing.

key concept 25.1 Bacteria and Archaea Are the Two Primary Divisions of Life

You may think that you have little in common with a bacterium. But all multicellular eukaryotes—including you—share many attributes with organisms that lack a nucleus, which are called **prokaryotes**. For example, all organisms, whether eukaryotes or prokaryotes,

- have cell membranes and ribosomes (see Chapters 5 and 6).
- have a common set of metabolic pathways, such as glycolysis (see Chapters 8 and 9).
- replicate DNA semiconservatively (see Chapter 13).
- use DNA as the genetic material to encode proteins, and use similar genetic codes to produce those proteins by transcription and translation (see Chapter 14).

These shared features support the conclusion that all living organisms share a common ancestor. If life had multiple origins, there would be little reason to expect all organisms to use overwhelmingly similar genetic codes or to share structures as distinctive as ribosomes. Furthermore, similarities in the DNA sequences of genes that are shared by all organisms confirm the monophyly of life.

focus your learning

- Bacteria and prokaryotic archaea share some features and differ in others.
- Some groups of bacteria can be distinguished based on the Gram stain.
- Lateral gene transfer can lead to discordant gene trees.

Although all living things share many features, major differences have evolved across the diversity of life. Biologists have now sequenced the genomes of many living organisms, and these genomes allow us to reconstruct the details of evolutionary history. These studies clarify that there are two fundamental divisions of life: Bacteria and Archaea (**Figure 25.1**). Another major group evolved from within the Archaea: the Eukarya, which includes all plants, animals, and fungi.

When biologists speak of three major **domains** of life, they usually mean Bacteria, prokaryotic Archaea, and Eukarya. But if you examine Figure 25.1, you will see that eukaryotes evolved from within the prokaryotic Archaea. So in this sense, eukaryotes are a specialized group of Archaea that developed some important new features (including a cell nucleus) that allowed them to thrive. We call all the organisms that lack these specializations prokaryotes (Greek, "before the kernel," or before the evolution of a cellular nucleus). Prokaryotes are clearly not all most closely related to one another, but the name "prokaryotes" is nonetheless a useful way of talking about all organisms that are not eukaryotes. In a similar manner, when most biologists talk about "archaea," typically they are referring to just the prokaryotic Archaea.

All prokaryotic organisms are unicellular, although they may form large, coordinated colonies or communities consisting of many individuals. Eukaryotes, by contrast, include both unicellular as well as many multicellular life forms. As we saw in Chapter 5, prokaryotic cells differ from eukaryotic cells in some important ways:

- *Prokaryotic cells do not divide by mitosis.* Instead, after replicating their DNA, prokaryotic cells divide by their own method, binary fission (see Key Concept 11.1).
- *The organization of the genetic material differs.* The DNA of the prokaryotic cell is not organized within a membrane-enclosed nucleus. DNA molecules in prokaryotes are often circular. Many (but not all) prokaryotes have only one main chromosome and are effectively haploid, although many have additional smaller DNA molecules, called plasmids (see Key Concept 12.6).
- *Prokaryotes have none of the membrane-enclosed cytoplasmic organelles that are found in most eukaryotes.* However, the cytoplasm of a prokaryotic cell may contain a variety of infoldings of the cell membrane and photosynthetic membrane systems that are not found in eukaryotes.

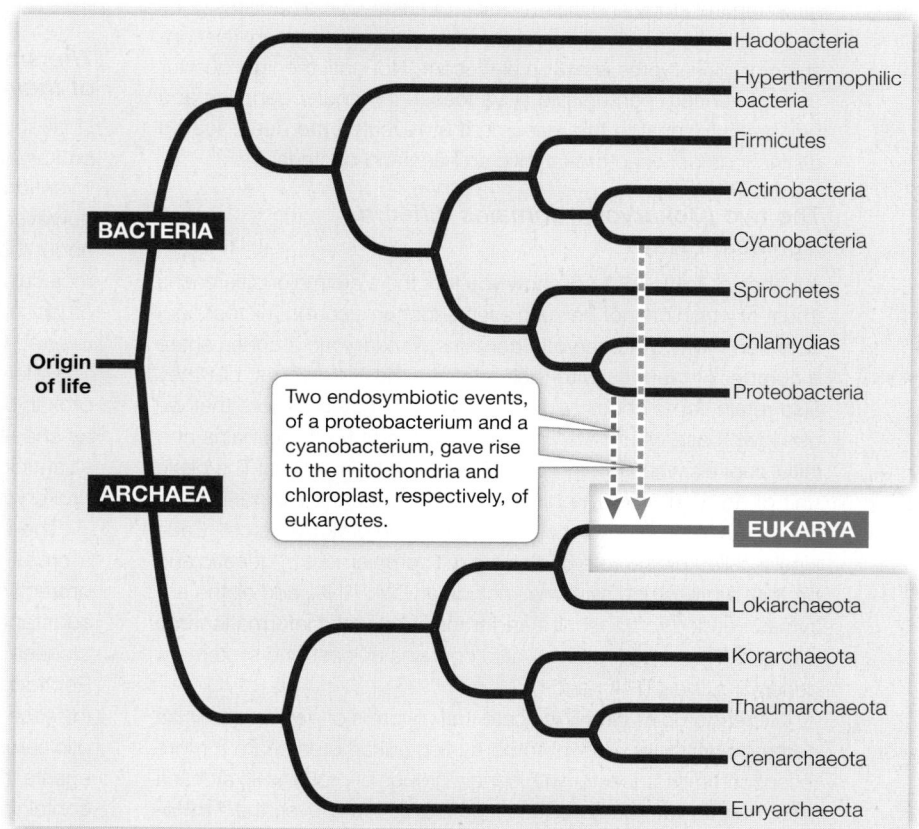

Two endosymbiotic events, of a proteobacterium and a cyanobacterium, gave rise to the mitochondria and chloroplast, respectively, of eukaryotes.

Figure 25.1 The Major Groups of the Living World This phylogenetic tree of Bacteria and Archaea shows their relationships to each other and to Eukarya. The relationships among the many clades of bacteria, not all of which are listed here, are incompletely resolved at this time.

 Animation 25.1 **The Evolution of the Three Domains**
www.Life11e.com/a25.1

table 25.1 The Three "Domains" of Life

Characteristic	Bacteria	Prokaryotic Archaea	Eukarya
Membrane-enclosed nucleus	Absent	Absent	**Present**
Membrane-enclosed organelles	Few	Absent	**Many**
Peptidoglycan in cell wall	**Present**	Absent	Absent
Membrane lipids	Ester-linked	**Ether-linked**	Ester-linked
	Unbranched	**Branched**	Unbranched
Ribosomes[a]	70S	70S	**80S**
Initiator tRNA	**Formylmethionine**	Methionine	Methionine
Operons	Yes	Yes	**Rare**
Plasmids	Yes	Yes	**Rare**
Number of RNA polymerases[b]	One	One	**Three**
Ribosomes sensitive to chloramphenicol and streptomycin	**Yes**	No	No
Ribosomes sensitive to diphtheria toxin	**No**	Yes	Yes

[a]70S ribosomes are smaller than 80S ribosomes.
[b]The structure of prokaryotic archaeal RNA polymerase is similar to that of eukaryotic polymerases.

Although the study and classification of eukaryotic organisms goes back centuries, much of our knowledge of the evolutionary relationships of prokaryotes is extremely recent. Not until the final quarter of the twentieth century did advances in molecular genetics and biochemistry enable the research that revealed the deep-seated distinctions between prokaryotic archaea and bacteria.

The two prokaryotic domains differ in significant ways

A glance at **Table 25.1** will show you that there are major differences (most of which cannot be seen even under an electron microscope) between the two prokaryotic domains. Prokaryotic archaea share a number of features with their relatives the eukaryotes, but they also retain some ancestral similarities with bacteria. (Note that we use lowercase when referring to members of these domains and initial capitals when referring to the domains themselves.) The basic unit of an archaeon (the term for a single archaeal organism) or bacterium (a single bacterial organism) is the prokaryotic cell. Each single-celled prokaryote contains a full complement of genetic and protein-synthesizing systems, including DNA, RNA, and all the enzymes needed to transcribe and translate genetic information into proteins. The prokaryotic cell also contains at least one system for generating the ATP it needs.

Genetic studies clearly indicate that all forms of life share a single common ancestor. As we noted earlier, eukaryotes share a more recent common ancestor with certain groups of prokaryotic archaea than they do with bacteria (see Figure 25.1). However, the *mito-chondria of eukaryotes and the **chloroplasts** of photosynthetic eukaryotes (such as plants) originated through endosymbiosis with bacteria. Some biologists prefer to view the origin of eukaryotes as a fusion of two equal partners (one ancestor that was related to modern prokaryotic archaea and another that was more closely related to modern bacteria). Others view the divergence of the early

eukaryotes from specific groups of prokaryotic archaea as an event separate from and earlier than the later endosymbioses. In either case, most eukaryote genes are more closely related to those of specific groups of prokaryotic archaea, whereas other genes (especially genes related to mitochondria and chloroplasts) are most closely related to those of bacteria. The tree of life therefore contains some merging of lineages as well as the predominant divergence of lineages.

---

*connect the concepts The origin of mitochondria and chloroplasts by endosymbiosis is described in Key Concepts 5.5 and 27.1.

Biologists estimate that the last common ancestor of the three domains lived about 3 billion years ago. We can deduce that it had DNA as its genetic material, and that its machinery for transcription and translation produced RNAs and proteins, respectively. This ancestor likely had a circular chromosome. All living organisms are the products of billions of years of mutation, natural selection, and genetic drift, and they are all well adapted to present-day environments. The earliest prokaryote fossils, which date back at least 3.5 billion years, indicate that there was considerable diversity among the prokaryotes even during those earliest days of life.

The small size of prokaryotes has hindered our study of their evolutionary relationships

Until about 300 years ago, nobody had even *seen* an individual prokaryote. Most prokaryotes remained invisible to humans until the invention of the first simple microscope. Prokaryotes are so small, however, that even the best light microscopes don't reveal much about them. It took advanced microscopic equipment and modern molecular techniques to open up the microbial world. (Microscopic organisms—both prokaryotes and eukaryotes—are often collectively referred to as "microbes.")

Before DNA sequencing became practical, taxonomists based prokaryote classification on observable phenotypic characters such as shape, color, motility, nutritional requirements, and sensitivity to antibiotics. One of the characters most widely used to classify prokaryotes is the structure of their cell walls.

The cell walls of almost all bacteria contain **peptidoglycan**, a cross-linked polymer of amino sugars that produces a firm, protective, meshlike structure around the cell. Peptidoglycan is a substance unique to bacteria; its absence from the cell walls of archaea is a key difference between the two prokaryotic domains. Peptidoglycan is also an excellent target for combating pathogenic (disease-causing) bacteria because it has no counterpart in eukaryotic cells. Antibiotics such as penicillin and ampicillin, as well as other agents that specifically interfere with the synthesis of peptidoglycan-containing cell walls, tend to have little, if any, effect on the cells of humans and other eukaryotes.

The **Gram stain** is a technique that can be used to separate most types of bacteria into two distinct groups. A smear of bacterial cells on a microscope slide is soaked in a violet dye and treated with iodine; it is then washed with alcohol and counterstained with a red dye called safranin. **Gram-positive bacteria** retain the violet dye and

Figure 25.2 The Gram Stain and the Bacterial Cell Wall
When treated with Gram-staining reagents, the cell walls of
bacteria react in one of two ways. **(A)** Gram-positive bacteria
have a thick peptidoglycan cell wall that retains the violet
dye and appears deep blue or purple. **(B)** Gram-negative
bacteria have a thin peptidoglycan layer that does not retain
the violet dye, but picks up the counterstain and appears
pink to red.

 Activity 25.1 Gram Stain and Bacteria
www.Life11e.com/ac25.1

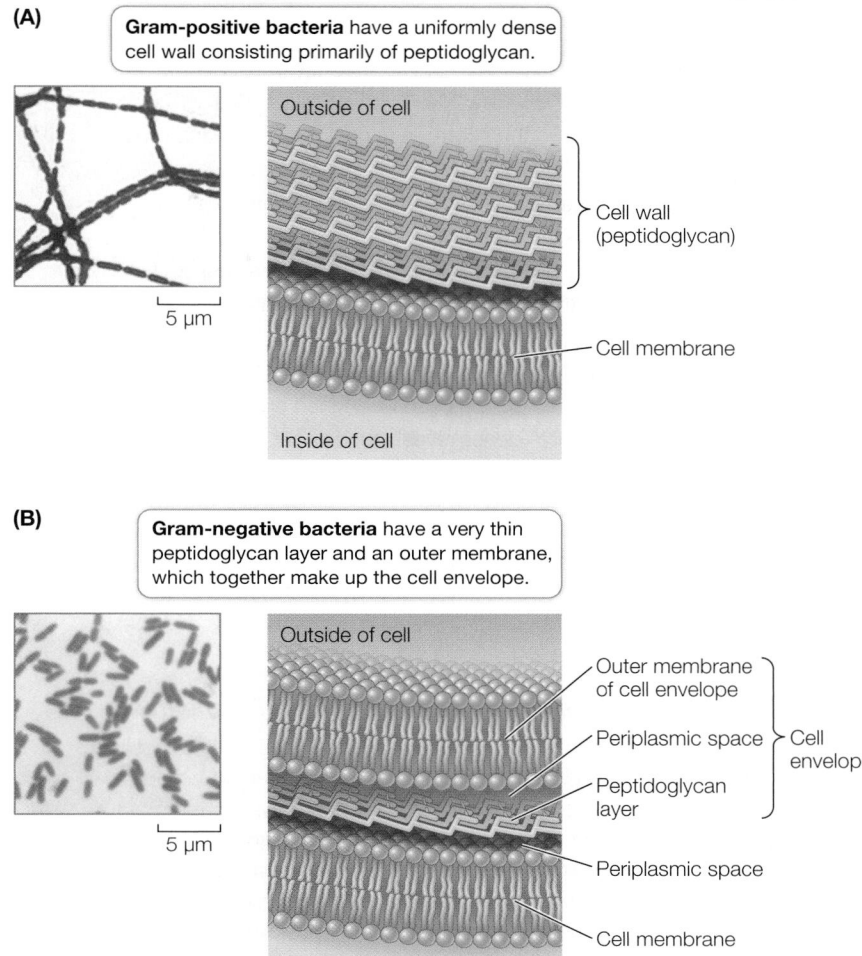

(A)

Gram-positive bacteria have a uniformly dense
cell wall consisting primarily of peptidoglycan.

Outside of cell

5 µm

Cell wall
(peptidoglycan)

Cell membrane

Inside of cell

(B)

Gram-negative bacteria have a very thin
peptidoglycan layer and an outer membrane,
which together make up the cell envelope.

Outside of cell

5 µm

Outer membrane
of cell envelope

Periplasmic space — Cell
envelope

Peptidoglycan
layer

Periplasmic space

Cell membrane

Inside of cell

appear blue to purple (**Figure 25.2A**). The alcohol washes
the violet stain out of **Gram-negative bacteria**, which then
pick up the safranin counterstain and appear pink to red
(**Figure 25.2B**). For most bacteria, the effect of the Gram
stain is determined by the chemical structure of the cell wall:

- A *Gram-negative cell wall* usually has a thin peptidoglycan layer, which is surrounded by a second, outer
  membrane quite distinct in chemical makeup from
  the cell membrane (see Figure 25.2B). Together the
  cell wall and the outer membrane are called the cell
  envelope. The space between the cell membrane
  and the outer membrane (known as the periplasmic
  space) contains proteins that are important in digesting some materials, transporting others, and detecting chemical gradients in the environment.

- A *Gram-positive cell wall* usually has about five times
  as much peptidoglycan as a Gram-negative cell wall.
  Its thick peptidoglycan layer is a meshwork that may
  serve some of the same purposes as the periplasmic
  space of the Gram-negative cell envelope.

Shape is another phenotypic characteristic that is useful for the basic identification of bacteria. The three most common
shapes are spheres, rods, and spiral forms (**Figure 25.3**). Many
bacterial names are based on these shapes. A spherical bacterium
is called a **coccus** (plural *cocci*). Cocci may live singly or may associate in two- or three-dimensional arrays such as chains, plates,
blocks, or clusters of cells. A rod-shaped bacterium is called a
bacillus (plural *bacilli*). A spiral bacterium (shaped like a corkscrew)
is called a **spirillum** (plural *spirilla*). Bacilli and spirilla may be single,
form chains, or gather in regular clusters. Among the other bacterial
shapes are long filaments and branched filaments.

Less is known about the shapes of prokaryotic archaea because many of these organisms have never been seen. Many
prokaryotic archaea are known only from samples of DNA from
the environment. However, the species whose morphologies are
known include cocci, bacilli, and even triangular and square species. Some flattened species grow on surfaces, arranged like
sheets of postage stamps.

The nucleotide sequences of prokaryotes reveal their evolutionary relationships

Analyses of the nucleotide sequences of ribosomal RNA (rRNA)
genes provided the first comprehensive evidence of evolutionary
relationships among prokaryotes. Comparisons of rRNA genes are

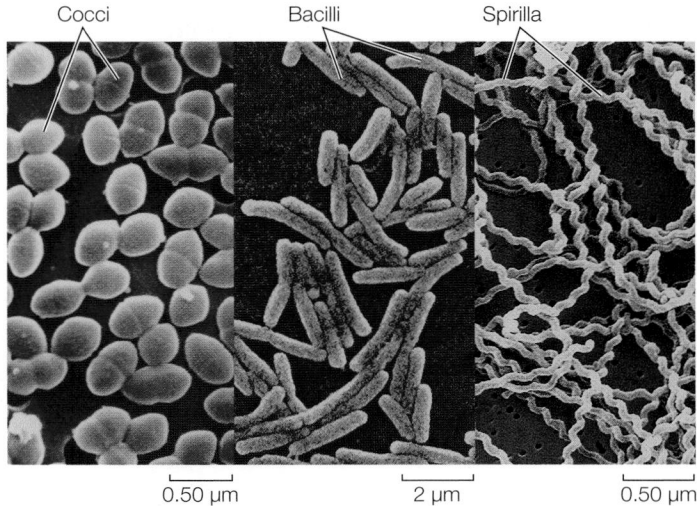

Cocci   Bacilli   Spirilla

0.50 µm   2 µm   0.50 µm

Figure 25.3 Bacterial Cell Shapes   This composite, colorized
micrograph shows the three most common bacterial shapes. Spherical
cells are called cocci; those pictured are a species of *Enterococcus* from
the mammalian gut. Rod-shaped cells are called bacilli; these *Escherichia
coli* also reside in the gut. The helix-shaped spirilla are *Leptospira inter-
rogans*, a human pathogen.

often used to identify microbes. For several reasons, rRNA is particularly useful for phylogenetic studies and identification purposes:

- rRNA was present in the common ancestor of all life and is therefore evolutionarily ancient.

- No free-living organism lacks rRNA, so rRNA genes can be compared across the tree of life.

- rRNA plays a critical role in translation in all organisms, so lateral transfer of rRNA genes among distantly related species is unlikely.

- rRNA has evolved slowly enough that gene sequences from even distantly related species can be aligned and analyzed.

Although studies of rRNA genes reveal much about the evolutionary relationships of prokaryotes, they don't always reveal the entire evolutionary history of these organisms. In some groups of prokaryotes, analyses of multiple gene sequences have suggested several different phylogenetic patterns. How could such differences among different gene sequences arise? Studies of whole prokaryotic genomes have revealed that even distantly related prokaryotes sometimes exchange genetic material.

Lateral gene transfer can lead to discordant gene trees

As noted earlier, prokaryotes reproduce by binary fission. If we could follow these divisions back through evolutionary time, we would be tracing the complete tree of life. At a much broader scale, these divisions of organisms lead to splits among the major evolutionary lineages, or species of life (represented in highly abbreviated form in Appendix A). Because binary fission is an asexual process that replicates whole genomes, we would expect phylogenetic trees of prokaryotes constructed from most gene sequences (see Chapter 22) to reflect these same relationships.

Even though binary fission is an asexual process, there are other processes—including ***transformation, conjugation, and transduction**—that allow the transfer of genetic information between some prokaryotes without reproduction. Thus prokaryotes can transfer and recombine their DNA with that of other individuals (this is sex in the genetic sense of the word), but this genetic exchange is not directly linked to reproduction, as it is in most eukaryotes.

---

*connect the concepts Prokaryote exchange of genetic material by conjugation and transformation is described in Key Concepts 12.6 and 13.1, respectively.

From early in evolution to the present day, some genes have been moving "sideways" from one prokaryote species to another, a phenomenon known as **lateral gene transfer** (sometimes called horizontal gene transfer). Lateral gene transfers are well documented among closely related species, and some have been documented even across the domains of life.

Consider, for example, the genome of *Thermotoga maritima*, a bacterium that can survive extremely high temperatures. By comparing the 1,869 gene sequences of *T. maritima* with sequences encoding the same proteins in other species, investigators found that some of this bacterium's genes have their closest relationships not with the genes of other bacterial species, but with the genes of archaea that live in similar extreme environments.

When genes involved in lateral transfer events are sequenced and analyzed, the resulting **gene trees** will not match the organismal tree in every respect (**Focus: Key Figure 25.4**). The individual gene trees will vary because the history of lateral transfer events is different for each gene. Biologists can reconstruct the underlying organismal phylogeny by comparing multiple genes (to produce a consensus tree) or by concentrating on genes that are unlikely to be involved in lateral gene transfer events. For example, genes that are involved in fundamental cellular processes (such as the rRNA genes discussed above) are unlikely to be replaced by the same genes from other species because functional, locally adapted copies of these genes are already present.

What kinds of genes are most likely to be involved in lateral gene transfer? Genes that result in a new adaptation that confers higher fitness on a recipient species are most likely to be transferred repeatedly among species. For example, genes that produce antibiotic resistance are often transferred among bacterial species on plasmids, especially under the strong selection pressure such as that imposed by modern antibiotic medications. Improper or overly frequent use of antibiotics can select for resistant strains of bacteria that are much harder to treat. This selection for antibiotic resistance explains why informed physicians have become more careful in prescribing antibiotics.

It is debatable whether lateral gene transfer has seriously complicated our attempts to resolve the tree of prokaryotic life. Recent work suggests that it has not. Lateral gene transfer rarely creates problems at higher taxonomic levels, even though it may complicate our understanding of the relationships among individual species. Some species clearly obtain some of their genes from otherwise distantly related species, so evolutionary histories of individual genes may differ within a single organism. But it is now possible to make nucleotide sequence comparisons involving entire genomes, and these studies are revealing a stable core of crucial genes that are uncomplicated by lateral gene transfer. Gene trees based on this stable core more accurately reveal the organismal phylogeny (see Figure 25.4). The problem remains, however, that only a very small proportion of the prokaryotic world has been described and studied.

The great majority of prokaryote species have never been studied

Most prokaryotes have defied all attempts to grow them in pure culture, causing biologists to wonder how many species, and possibly even major clades, we might be missing. A window onto this problem was opened with the introduction of a new way of examining nucleic acid sequences. When biologists are unable to work with the whole genome of a single prokaryote species, they can instead examine genomes collected from an environmental sample (such as a scoop of sediment from the sea floor). This technique is known as **environmental genomics**.

Biologists now routinely isolate gene sequences, or even whole genomes, from environmental samples such as soil and seawater. Comparing such sequences with previously known ones has revealed that an extraordinary number of the sequences represent new, previously unrecognized species. Biologists have described only about 10,000 species of bacteria and only a few hundred species of prokaryotic archaea (see Figure 1.9). The results of some environmental genomic studies suggest that there may be

focus: key figure

(A) Organismal tree

Species A

Species B

Species C

Species D

Gene x is transferred laterally between species C and D.

(B) Gene x tree

A

B

C

D

The apparent close relationship of C and D inferred from sequences of gene x reflects the lateral transfer of this gene rather than the phylogeny of the organisms.

(C) Consensus tree

A

B

C

D

A consensus tree based on multiple genes more accurately reflects the organismal phylogeny.

Figure 25.4 Lateral Gene Transfer Complicates Phylogenetic Relationships **(A)** The phylogeny of four hypothetical prokaryote species, two of which have been involved in a lateral transfer of gene x. **(B)** A tree based only on gene x shows the phylogeny of the laterally transferred gene, rather than the organismal phylogeny. **(C)** A consensus tree based on multiple genes is more likely to reflect the true organismal phylogeny, especially if those genes come from a stable core of genes involved in fundamental processes.

Q: Why are multiple lateral gene transfers between the same two branches on a phylogeny expected to be rare, at least compared with similarities inherited through the stable core?

millions—perhaps hundreds of millions—of prokaryote species. Other biologists put the estimate much lower, arguing that the high dispersal ability of many bacterial species greatly reduces endemism (i.e., the number of species restricted to a small geographic area). Only the magnitude of these estimates differs, however; all sides agree that we have just begun to uncover Earth's prokaryotic diversity.

25.1 recap

Bacteria and Archaea are the two major divisions of the tree of life. Eukaryotes evolved from a group of prokaryotic archaea that captured and incorporated at least two different bacteria through endosymbiosis. Environmental genomic studies have revealed a much higher diversity of prokaryotes than was previously known.

learning outcomes

You should be able to:

- Compare and contrast the features of bacteria and prokaryotic archaea in a phylogenetic context.
- Distinguish between two scenarios for the contributions of bacteria and archaea to eukaryotes.
- Draw phylogenetic trees demonstrating the effects of lateral gene transfer.

1. What findings led to the establishment of Bacteria and Archaea as separate domains?

2. The figure at right shows an organismal tree in which gene x has undergone a lateral transfer event. Draw the phylogenetic tree you would expect based on gene x, as well as the phylogenetic tree you would expect based on a consensus of non-transferred genes.

25.1 recap

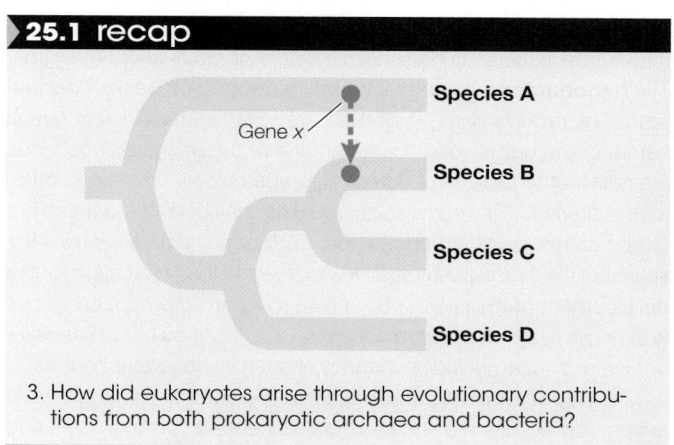

Gene x

Species A

Species B

Species C

Species D

3. How did eukaryotes arise through evolutionary contributions from both prokaryotic archaea and bacteria?

Despite the challenges of reconstructing prokaryote phylogeny, taxonomists are starting to establish evolutionary classification systems for these organisms. With a full understanding that new information requires periodic revisions in these classifications, we will next apply a current system of classification to organize our survey of prokaryote diversity.

key concept

25.2 Prokaryote Diversity Reflects the Ancient Origins of Life

The prokaryotes were alone on Earth for a very long time, adapting to new environments and to changes in existing environments. They have survived to this day, in massive numbers and incredible diversity, and they are found everywhere. In numbers of individuals, prokaryotes are far more abundant than eukaryotes. Individual prokaryotes in the oceans number more than 3×10^{28}—more than the number of stars in the universe. Closer to home, the individual bacteria living in your intestinal tract outnumber all the humans who have ever lived.

focus your learning
- Prokaryotes are classified based on the sequences of their genomes.

Given our still-fragmentary knowledge of prokaryote diversity, it is not surprising that there are many different hypotheses about the relationships of the major groups of prokaryotes. In this book we use a classification system that has considerable support from nucleotide sequence data. We will discuss eight major bacterial groups that have the broadest phylogenetic support and have received the most study: hadobacteria, hyperthermophilic bacteria, firmicutes, actinobacteria, cyanobacteria, spirochetes, chlamydias, and proteobacteria (see Figure 25.1). Many other major groups of bacteria are known but are less thoroughly studied. We will then describe the prokaryotic archaea, whose diversity is even less well studied than that of the bacteria.

Two early-branching lineages of bacteria live at very high temperatures

Several lineages of bacteria and archaea are **extremophiles**: they thrive under extreme conditions that would kill most other organisms. The **hadobacteria**, for example, are thermophiles (Greek, "heat lovers"). The group's name is derived from Hades, the ancient Greek name for the underworld. Hadobacteria of the genus *Deinococcus* are resistant to radiation and can degrade nuclear waste and other toxic materials. They can also survive extremes of cold as well as hot temperatures. Another hadobacterium, *Thermus aquaticus*, was the source of the thermally stable DNA polymerase that was critical for the development of the polymerase chain reaction. *Thermus aquaticus* was originally isolated from a hot spring, but it can be found wherever hot water occurs (including in many residential hot-water heaters).

The **hyperthermophilic bacteria** are another major group of extremophiles. Genera such as *Aquifex* live near volcanic vents and in hot springs, sometimes at temperatures near the boiling point of water. Some species of *Aquifex* need only hydrogen, oxygen, carbon dioxide, and mineral salts to live and grow. Species of the genus *Thermotoga* live deep underground in oil reservoirs as well as in other high-temperature environments.

Biologists have hypothesized that high temperatures characterized the ancestral conditions for life, given that most environments on early Earth were much hotter than those of today. Reconstructions of ancestral bacterial genes have supported this hypothesis by showing that the ancestral sequences functioned best at elevated temperatures. The presence of multiple lineages of extremophiles at the base of the bacterial tree (see Figure 25.1) also provides support for the origin of life in a high-temperature environment.

Firmicutes include some of the smallest cellular organisms

The **firmicutes** are also known as low-GC Gram-positive bacteria. The first part of this description (low-GC) derives from the relatively

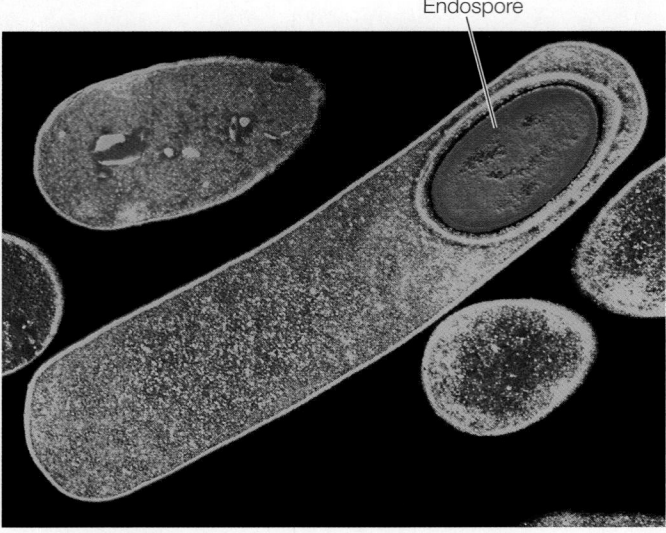

Endospore

Clostridium difficile   0.3 µm

Figure 25.5 A Structure for Waiting Out Bad Times   Under harsh conditions, some firmicutes can replicate their DNA and encase it in an endospore. The parent cell then breaks down, and the endospore survives in a dormant state until conditions improve.

low ratio of G-C to A-T nucleotide base pairs in their DNA. The second part (Gram-positive) is less accurate: some of the firmicutes are in fact Gram-negative, and some have no cell wall at all. Despite these differences, phylogenetic analyses of DNA sequences support the monophyly of this bacterial group.

One group of firmicutes can produce resting structures called **endospores** (**Figure 25.5**). When a key nutrient such as nitrogen or carbon becomes scarce, the bacterium replicates its DNA and encapsulates one copy, along with some of its cytoplasm, in a tough endospore wall heavily thickened with peptidoglycan and surrounded by a spore coat. The parent cell then breaks down, releasing the endospore. Endospore production is not a reproductive process, as the endospore merely replaces the parent cell. The endospore, however, can survive harsh environmental conditions that would kill the parent cell, such as high or low temperatures or drought, because it is dormant—its normal metabolic activity is suspended. Later, if it encounters favorable conditions, the endospore becomes metabolically active and divides, forming new cells that are like the parent cell. Members of this endospore-forming group include the many species of *Clostridium* and *Bacillus*. Some of their endospores can be reactivated after more than 1,000 years of dormancy. There are even credible claims of reactivation of *Bacillus* endospores that are millions of years old.

Endospores of *Bacillus anthracis* are the cause of anthrax. Anthrax is primarily a disease of cattle and sheep, but it can be fatal in humans. When the endospores sense macrophages in mammalian blood, they reactivate and release toxins into the bloodstream. *Bacillus anthracis* has been used as a bioterrorism agent because it is relatively easy to transport large quantities of its endospores and release them among human populations, where they may be inhaled or ingested.

Members of the genus *Staphylococcus*—the **staphylococci** (**Figure 25.6**)—are abundant on the human body surface; they are responsible for boils and many other skin problems. *Staphylococcus*

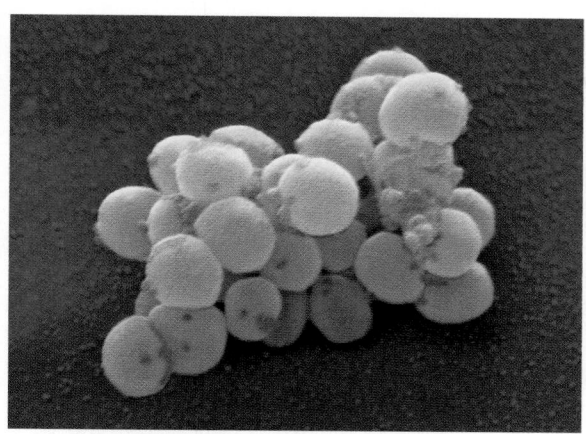

Staphylococcus aureus

⊢———⊣
1 µm

Figure 25.6 Staphylococci "Grape clusters" are the usual arrangement of these firmicutes, which are often the cause of skin or wound infections.

aureus is the best-known human pathogen in this genus; it is present in 20 to 40 percent of normal adults (and in 50–70% of hospitalized adults). In addition to skin diseases, *S. aureus* can cause respiratory, intestinal, and wound infections.

Another interesting group of firmicutes, the **mycoplasmas**, lack cell walls, although some have a stiffening material outside the cell membrane. The mycoplasmas are among the smallest cellular organisms known (**Figure 25.7**). The smallest mycoplasmas have a diameter of about 0.2 µm. They are small in another crucial sense as well: they have less than half as much DNA as most other prokaryotes. It has been speculated that the DNA in a mycoplasma, which codes for fewer than 500 proteins, may be close to the minimum amount required to encode the essential properties of a living cell.

Actinobacteria include major pathogens as well as valuable sources of antibiotics

Actinobacteria, also known as high-GC Gram-positive bacteria, have a higher ratio of G-C to A-T nucleotide base pairs than do the firmicutes (low-GC Gram-positives). These bacteria develop

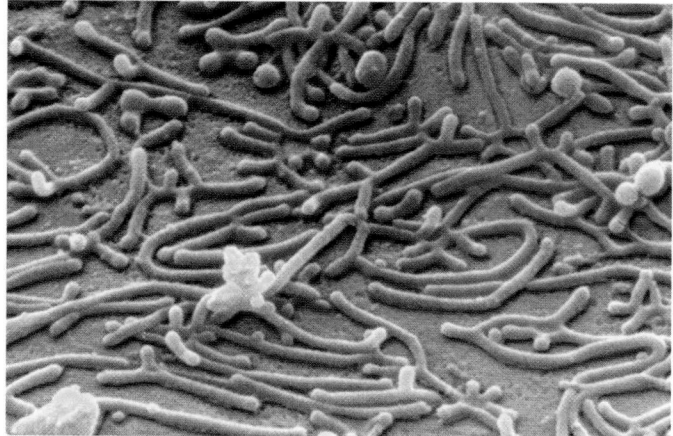

Mycoplasma sp.

⊢———⊣
0.7 µm

Figure 25.7 Tiny Cells With about one-fifth as much DNA as *E. coli*, mycoplasmas are among the smallest known bacteria.

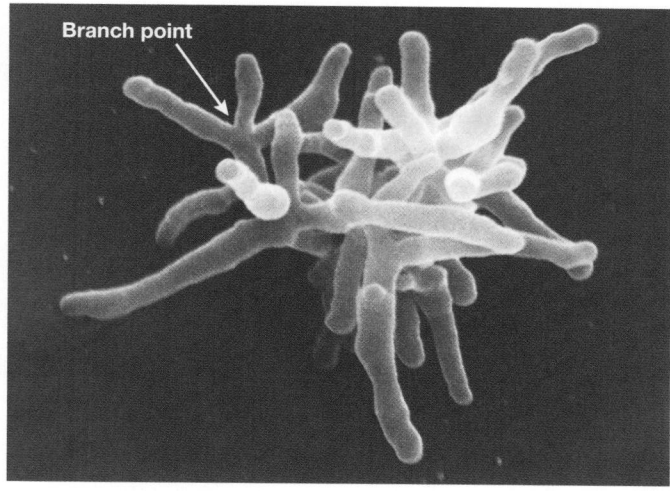

Branch point

Actinomyces sp.

⊢———⊣
2 µm

Figure 25.8 Actinobacteria Often Produce Branching Filaments The tangled, branching filaments seen in this scanning electron micrograph are typical of this medically important bacterial group.

an elaborately branched system of filaments (**Figure 25.8**) that resembles the filamentous growth habit of fungi, albeit at a smaller scale. Some actinobacteria reproduce by forming chains of spores at the tips of the filaments. In species that do not form spores, the branched, filamentous growth ceases and the structure breaks up into typical cocci or bacilli, which then reproduce by binary fission.

The actinobacteria include several medically important bacteria. *Mycobacterium tuberculosis* causes tuberculosis, which kills 3 million people each year. Genetic data suggest that this bacterium may have been infecting our ancestors for almost 3 million years, making it the oldest known human bacterial pathogen. The genus *Streptomyces* produces streptomycin as well as hundreds of other antibiotics. We derive most of our antibiotics from actinobacteria.

Cyanobacteria were the first photosynthesizers

Cyanobacteria, sometimes called blue-green bacteria because of their pigmentation, are photosynthetic. They use chlorophyll *a* for photosynthesis and release oxygen gas (O_2); many species also fix nitrogen (which we will discuss in Key Concept 25.3). The production of oxygen by these bacteria transformed the atmosphere of early Earth (see Key Concept 24.2).

Cyanobacteria carry out the same type of photosynthesis that is characteristic of eukaryotic photosynthesizers. They contain elaborate and highly organized internal membrane systems called **photosynthetic lamellae**. As mentioned in Key Concept 25.1, the chloroplasts of photosynthetic eukaryotes are derived from an endosymbiotic cyanobacterium.

Cyanobacteria may live free as single cells or associate in multicellular colonies. Depending on the species and on growth conditions, these colonies may range from flat sheets one cell thick to filaments to spherical balls of cells. Some filamentous colonies of cyanobacteria differentiate into three specialized cell types: vegetative cells, spores, and heterocysts (**Figure 25.9**). **Vegetative cells** photosynthesize, **spores** are resting stages that can survive harsh environmental conditions and eventually develop into new filaments, and **heterocysts** are cells specialized for nitrogen fixation. All of

(A) *Anabaena* sp.

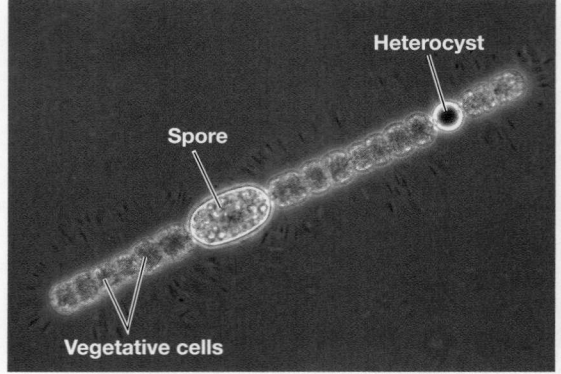

(B) *Nostoc punctiforme*

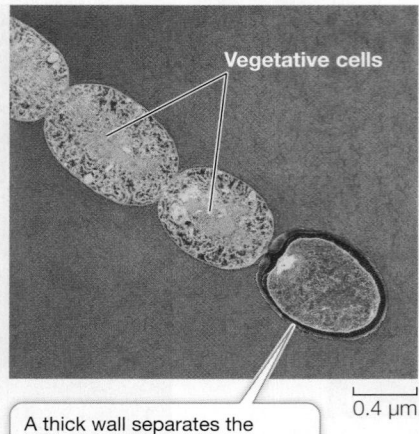

A thick wall separates the cytoplasm of the nitrogen-fixing heterocyst from the surrounding environment.

Figure 25.9 Cyanobacteria **(A)** Some cyanobacteria form filamentous colonies containing three cell types. **(B)** Heterocysts are specialized for nitrogen fixation and may serve as a breaking point when filaments reproduce. **(C)** This pond in Canada has experienced eutrophication: phosphorus and other nutrients generated by human activity have accumulated, feeding an immense green mat (commonly referred to as "pond scum") that is made up of several species of free-living cyanobacteria.

the known cyanobacteria with heterocysts fix nitrogen. Heterocysts also have a role in reproduction: when filaments break apart to reproduce, the heterocyst may serve as a breaking point.

 Media Clip 25.1 Cyanobacteria
www.Life11e.com/mc25.1

Spirochetes move by means of axial filaments

Spirochetes are Gram-negative, motile bacteria characterized by a unique structure known as an axial filament, which is composed of multiple internal flagella running through the periplasmic space (**Figure 25.10A**). The cell body is a long cylinder coiled into a helix

(**Figure 25.10B**). The internal flagella begin at either end of the cell and overlap in the middle. Motor proteins connect the axial filament to the cell wall, enabling the corkscrew-like movement of the bacterium. Many spirochetes are parasites of humans; a few are pathogens, including those that cause syphilis and Lyme disease. Others live free in mud or water.

Chlamydias are extremely small parasites

Chlamydias are among the smaller bacteria (0.2–1.5 μm in diameter). They can live only as parasites in the cells of other organisms. It was once believed that their obligate parasitism resulted from an inability to produce ATP—that chlamydias were "energy parasites." However, genome sequencing indicates that chlamydias have the genetic capacity to produce at least some ATP. They can augment this capacity by using an enzyme called a translocase, which allows them to take up ATP from the cytoplasm of their host in exchange for ADP from their own cells.

(A)

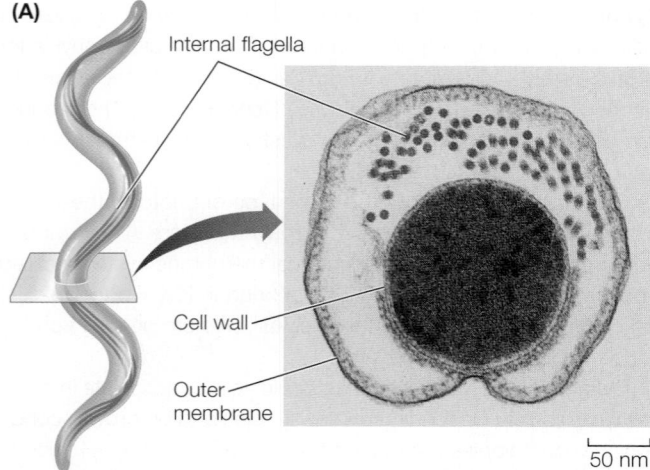

Figure 25.10 Spirochetes Get Their Shape from Axial Filaments
(A) A spirochete from the gut of a termite, seen in cross section, shows the internal flagella that compose the axial filament, which these helical prokaryotes use to produce a corkscrew-like movement. **(B)** This spirochete species causes syphilis in humans.

(B)

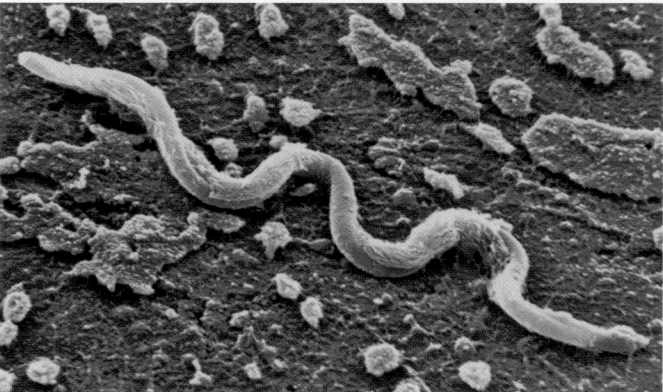

Treponema pallidum

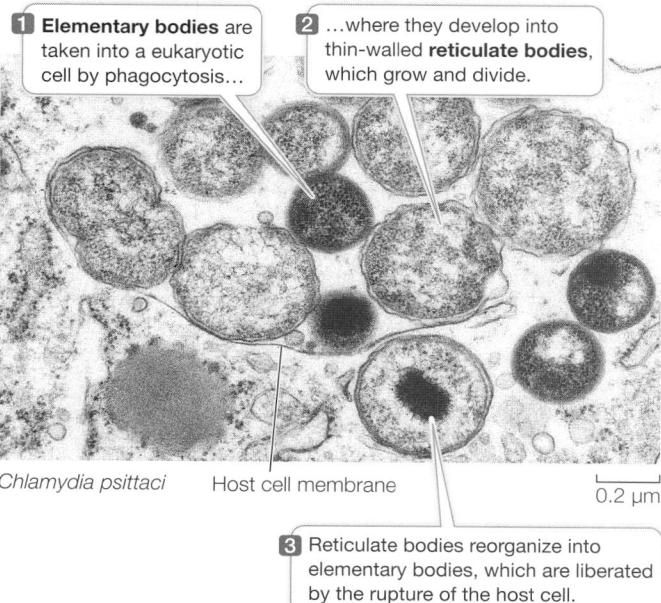

1 **Elementary bodies** are taken into a eukaryotic cell by phagocytosis...

2 ...where they develop into thin-walled **reticulate bodies**, which grow and divide.

Chlamydia psittaci   Host cell membrane   0.2 μm

3 Reticulate bodies reorganize into elementary bodies, which are liberated by the rupture of the host cell.

Figure 25.11 Chlamydias Change Form Elementary bodies and reticulate bodies are the two cell forms of the chlamydia life cycle.

These tiny, Gram-negative cocci are unique among prokaryotes because of a complex life cycle that involves two different forms of cells, elementary bodies and reticulate bodies (**Figure 25.11**). Various strains of chlamydias cause eye infections (especially trachoma), sexually transmitted diseases, and some forms of pneumonia in humans.

The proteobacteria are a large and diverse group

By far the largest bacterial group, in terms of numbers of described species, is the **proteobacteria**. The proteobacteria include many species of Gram-negative photoautotrophs (see Key Concept 25.3) that use light-driven reactions to metabolize sulfur, as well as dramatically diverse bacteria that bear no phenotypic resemblance to the photoautotrophic species. Genetic and morphological evidence indicates that the mitochondria of eukaryotes were derived from a proteobacterium by endosymbiosis.

Among the proteobacteria are some nitrogen-fixing genera, such as *Rhizobium*, and other bacteria that contribute to the global nitrogen and sulfur cycles. *Escherichia coli*, one of the most studied organisms on Earth, is a proteobacterium. So, too, are many of the most famous human pathogens, such as *Yersinia pestis* (which causes bubonic plague), *Vibrio cholerae* (cholera), and *Salmonella typhimurium* (gastrointestinal disease; **Figure 25.12**).

 Media Clip 25.2 **A Swarm of *Salmonella***
www.Life11e.com/mc25.2

The bioluminescent *Vibrio* we discussed at the opening of this chapter are also members of this group. There are many potential applications of the genes that encode bioluminescent proteins in bacteria. Already, these genes are being inserted into the genomes of other species in which the resulting bioluminescence is used as a marker of gene expression. Futuristic proposals for making use of bioluminescence in bioengineered organisms include crop

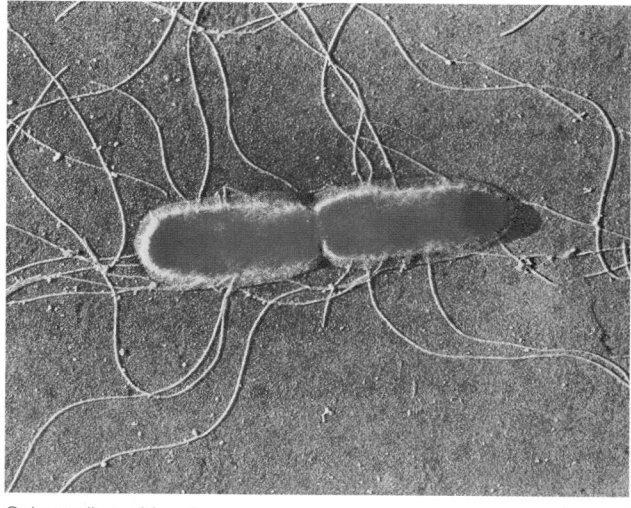

Salmonella typhimurium   0.5 μm

Figure 25.12 Proteobacteria Include Many Familiar Bacteria These conjugating cells of *Salmonella typhimurium* are exchanging genetic material. This pathogen causes a wide range of gastrointestinal illnesses in humans.

plants that glow when they become water-stressed and need to be irrigated, and glowing trees that could light highways at night in place of electric lights.

Although fungi cause most plant diseases, and viruses cause others, about 200 known plant diseases are of bacterial origin. Crown gall, with its characteristic tumors (**Figure 25.13**), is one of the most striking. The causal agent of crown gall is *Agrobacterium tumefaciens*, a proteobacterium that harbors a plasmid used in recombinant DNA studies as a vehicle for inserting genes into new plant hosts.

Euonymus sp.

Figure 25.13 Crown Gall Crown gall, a type of tumor shown here growing on the stem of a *Euonymus* shrub, is caused by the proteobacterium *Agrobacterium tumefaciens*.

experiment

Figure 25.14A   What Is the Highest Temperature Compatible with Life?

Original Paper: Kashefi, K. and D. R. Lovley. 2003. Extending the upper temperature limit for life. *Science* 301: 934.

Can any organism thrive at temperatures above 120°C? This is the temperature used for sterilization, known to destroy all previously described organisms. Kazem Kashefi and Derek Lovley isolated an unidentified prokaryote from water samples taken near a hydrothermal vent and found it survived and even multiplied at 121°C. The organism was dubbed "Strain 121," and its gene sequencing results indicate that it is a prokaryotic archaeal species.

HYPOTHESIS▶   Some prokaryotes can survive at temperatures above 120°C.

METHOD

1. Seal samples of unidentified, iron-reducing, thermal vent prokaryotes in tubes with a medium containing Fe^{3+} as an electron acceptor. Control tubes contain Fe^{3+} but no organisms.
2. Hold both tubes in a sterilizer at 121°C for 10 hours. If the iron-reducing organisms are metabolically active, they will reduce the Fe^{3+} to Fe^{2+} (as magnetite, which can be detected with a magnet).

RESULTS

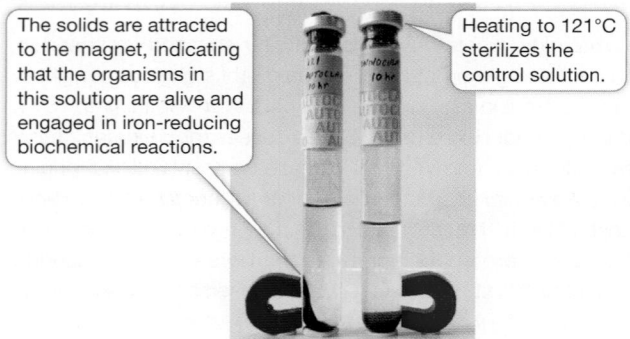

The solids are attracted to the magnet, indicating that the organisms in this solution are alive and engaged in iron-reducing biochemical reactions.

Heating to 121°C sterilizes the control solution.

CONCLUSION▶   Prokaryotic archaea of "Strain 121" can survive at temperatures above the previously defined sterilization limit.

Gene sequencing enabled biologists to differentiate Archaea from Bacteria

The original identification of Archaea as a group distinct from Bacteria was based on phylogenetic relationships determined from rRNA gene sequences. This separation was supported when biologists sequenced the first complete archaeal genome, which consisted of 1,738 genes—more than half of which were unlike any genes ever found in Bacteria.

Prokaryotic archaea are known for living in extreme habitats such as those with high salinity (salt content), low oxygen concentrations, high temperatures, or high or low pH (**Figure 25.14**). Many prokaryotic archaea are not extremophiles, however—they are common in soil, for example. Perhaps the largest numbers of prokaryotic archaea live in the ocean depths. None are known to be pathogens of any eukaryotes.

There are five major groups of prokaryotic archaea (see Figure 25.1): **Euryarchaeota**, **Crenarchaeota**, **Thaumarchaeota**,

work with the data

Figure 25.14B   What is the Highest Temperature Compatible with Life?

Original Paper: Kashefi, K. and D. R. Lovley. 2003.

After Kashefi and Lovley isolated Strain 121, they examined its growth at various temperatures. The table below shows generation time (time between cell divisions) at nine temperatures.

Temperature °C	Generation time (hr)
85	10
90	4
95	3
100	2.5
105	2
110	4
115	6
120	20
130	No growth, but cells not killed

QUESTIONS▶

1. Make a graph from these data showing time as a function of temperature.
2. Which temperature appears to be closest to the optimum for the growth of Strain 121?
3. Note that no growth occurred at 130°C, but that the cells were not killed. How would you demonstrate that these cells were still alive?

A similar **work with the data** exercise may be assigned in **LaunchPad**.

Korarchaeota, and **Lokiarchaeota**. The Lokiarchaeota are of particular interest because recent studies of their genomes show that they are the closest known living relatives of eukaryotes.

Two characteristics shared by all prokaryotic archaea are the absence of peptidoglycan in their cell walls and the presence of lipids of distinctive composition in their cell membranes (see Table 25.1). The unusual lipids in the membranes of prokaryotic archaea are found in no bacteria or eukaryotes. Most lipids in bacterial and eukaryotic membranes contain unbranched long-chain fatty acids connected to glycerol molecules by **ester linkages**:

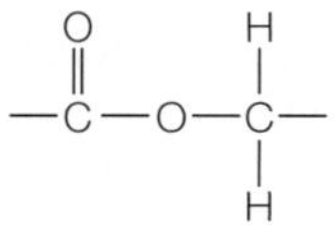

In contrast, some lipids in membranes of prokaryotic archaea contain long-chain hydrocarbons connected to glycerol molecules by **ether linkages**:

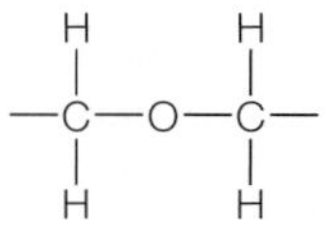

In addition, the long hydrocarbon chains in the lipids of prokaryotic archaea are branched. One class of archaeal lipids contains glycerol

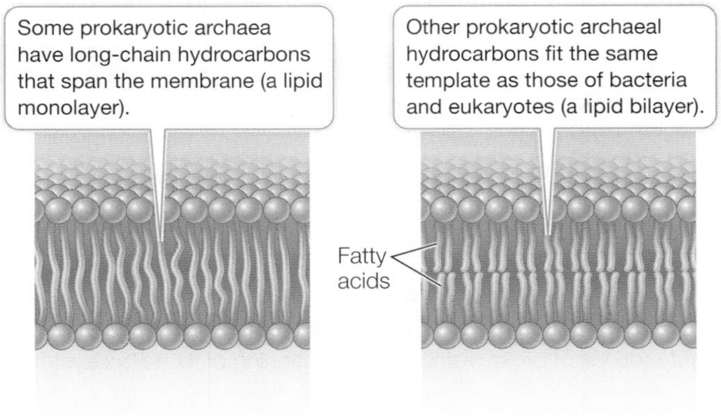

Some prokaryotic archaea have long-chain hydrocarbons that span the membrane (a lipid monolayer).

Other prokaryotic archaeal hydrocarbons fit the same template as those of bacteria and eukaryotes (a lipid bilayer).

Fatty acids

Figure 25.15 Membrane Architecture in Prokaryotic Archaea
The long-chain hydrocarbons of many prokaryotic archaeal lipids have glycerol molecules at both ends, so that the membranes they form consist of a lipid monolayer. In contrast, the membranes of other archaea, bacteria, and eukaryotes consist of a lipid bilayer.

at *both* ends of the hydrocarbons (**Figure 25.15**). These lipids form a lipid monolayer structure that is unique to prokaryotic archaea. They still fit into a biological membrane because they are twice as long as the typical lipids in the bilayers of other membranes. Lipid monolayers and bilayers are both found among the prokaryotic archaea. The effects, if any, of these structural features on membrane performance are unknown.

Prokaryotic archaea live in extremely diverse environments

Most known crenarchaeotes are either thermophilic, acidophilic (acid loving), or both. Members of the genus *Sulfolobus* live in hot sulfur springs at temperatures of 70°C to 75°C. They become metabolically inactive at 55°C (131°F). Hot sulfur springs are also extremely acidic. *Sulfolobus* grows best in the range from pH 2 to pH 3, but some members of this genus readily tolerate pH values as low as

0.9. Most acidophilic thermophiles maintain an internal pH of 5.5 to 7 (close to neutral) in spite of their acidic environment. These and other crenarchaeotes thrive where very few other organisms can even survive (**Figure 25.16**).

Some species of euryarchaeotes are **methanogens**: they produce methane (CH_4) by reducing carbon dioxide as the key step in their energy metabolism. All of the methanogens are obligate anaerobes (see Key Concept 25.3). Comparison of their rRNA gene sequences has revealed a close evolutionary relationship among these methanogenic species, which were previously assigned to several different groups of bacteria.

Methanogenic euryarchaeotes release approximately 2 billion tons of methane gas into Earth's atmosphere each year, accounting for 80 to 90 percent of the methane that enters the atmosphere, including that produced in some mammalian digestive systems. Approximately a third of this methane comes from methanogens living in the guts of ruminants such as cattle, sheep, and deer, and another large fraction comes from methanogens living in the guts of termites and cockroaches. Methane is increasing in Earth's atmosphere by about 1 percent per year and contributes to the greenhouse effect. Part of that increase is due to increases in cattle and rice production and the methanogens associated with both.

Another group of euryarchaeotes, the **extreme halophiles** (salt lovers), live exclusively in very salty environments. Because they contain pink carotenoid pigments, these archaea are sometimes easy to see (**Figure 25.17**). Extreme halophiles grow in the Dead Sea and in brines of all types. The reddish pink spots that can occur on pickled fish are colonies of halophilic archaea. Few other organisms can live in the saltiest homes that the extreme halophiles occupy— most would "dry" to death, losing too much water to the hypertonic environment. Extreme halophiles have been found in lakes with pH values as high as 11.5. These are the most alkaline environments inhabited by living organisms, and almost as alkaline as household ammonia.

Some of the extreme halophiles have a unique system for trapping light energy and using it to form ATP—without using any form of

Figure 25.16 Some Crenarchaeotes Like It Hot   Thermophilic crenarchaeotes can thrive in the intense heat of volcanic hot sulfur springs such as these in Yellowstone National Park.

Figure 25.17 Extreme Halophiles   Highly saline environments such as these commercial seawater evaporating ponds in San Francisco Bay are home to extreme halophiles. The prokaryotic archaea are easily visible here because of the rich red coloration of their carotenoid pigments.

chlorophyll—when oxygen is in short supply. They use the pigment retinal (also found in the vertebrate eye) combined with a protein to form a light-absorbing molecule called microbial rhodopsin.

Another member of the Euryarchaeota, *Thermoplasma*, has no cell wall. It is thermophilic and acidophilic, its metabolism is aerobic, and it lives in coal deposits. Its genome of 1,100,000 base pairs is among the smallest (along with that of the mycoplasmas) found in any free-living organism, although some parasitic organisms have even smaller genomes.

Most known prokaryotic archaea are crenarchaeotes or euryarchaeotes, but studies of extreme environments have identified several small lineages that are not closely related to either of these major groups. For example, the korarchaeotes and thaumarchaeotes are known only from DNA isolated directly from hot environments. Neither group has been grown successfully in pure culture. The thaumarchaeotes oxidize ammonia and may play an important role in the nitrogen cycle.

The lokiarchaeotes were discovered in 2015 by sequencing environmental samples from near a hydrothermal vent called Loki's Castle deep in the Arctic Ocean. An organism's genome was detected and sequenced from this deep-sea sample, although the organism has not yet been cultured. Sequence analysis revealed that the organism was a distinct lineage of archaea, which was named *Lokiarchaeum*. Of particular interest was the finding that the *Lokiarchaeum* genome contains a large number of genes with cell membrane–related functions—genes and functions that had previously been known to occur only in eukaryotes. Phylogenetic analysis of eukaryote, lokiarchaeote, and other archaeal genomes shows that *Lokiarchaeum* is more closely related to eukaryotes than it is to any other prokaryotes (see Figure 25.1). Some of the genes found in *Lokiarchaeum* are similar to genes that function in eukaryotes in controlling cell shape and cytoskeleton formation. These findings suggest that *Lokiarchaeum* is the closest known prokaryotic relative of eukaryotes, and this fascinating organism may share some characteristics that have previously been thought to be found only in eukaryotes.

25.2 recap

Bacteria and Archaea are highly diverse groups that survive in almost every imaginable habitat on Earth. Many prokaryotes can survive and even thrive in habitats where no eukaryotes can live, including extremely hot, acidic, or saline conditions. Eukaryotes are most closely related to some lineages of prokaryotic archaea, although endosymbioses of bacteria within eukaryotic cells contributed to the evolution of eukaryotic organelles.

learning outcomes

You should be able to:

- Interpret a phylogenetic tree of major lineages of Bacteria and Archaea and use it to explain (a) the difference between a prokaryote and a eukaryote, and (b) how prokaryotic archaea and bacteria both contributed to the evolution of eukaryotes.

1. Consider the difference between prokaryotes and eukaryotes shown in Table 25.1. Why might eukaryotes be more like prokaryotic archaea in some features and more like bacteria in others?

25.2 recap

2. Given that all species of life have evolved for the same amount of time since their common origin, how would you respond to someone who characterizes prokaryotes as "primitive"? Include at least two examples of major groups of prokaryotes to support your answer.

Prokaryotes are found almost everywhere on Earth and live in a wide variety of ecosystems. In the next section we will examine the contributions of prokaryotes to the functioning of those ecosystems.

key concept

25.3 Ecological Communities Depend on Prokaryotes

Prokaryotic cells do not usually live in isolation. Rather, they live in communities of many different species, often including microscopic eukaryotes. Whereas some microbial communities are harmful to humans, others provide important services. They help us digest our food, break down municipal waste, and recycle organic matter and chemical elements in the environment.

focus your learning

- Microbiomes are critical to the health of many eukaryotes.
- Prokaryotes differ with respect to oxygen metabolism.
- Prokaryotes play important roles in nitrogen cycling.

Many prokaryotes form complex communities

Some microbial communities form layers in sediments, and others form clumps a meter or more in diameter. Many microbial communities tend to form dense **biofilms**. Upon contacting a solid surface, the cells bind to that surface and secrete a sticky, gel-like polysaccharide matrix that traps other cells (**Figure 25.18**). Once a biofilm forms, the cells become more difficult to kill.

Biofilms are found in many places, and in some of those places they cause problems for humans. The material on our teeth that we call dental plaque is a biofilm. Pathogenic bacteria are difficult for the immune system—and modern medicine—to combat once they form a biofilm, which may be impermeable to antibiotics. Worse, some drugs stimulate the bacteria in a biofilm to lay down more matrix, making the film even more impermeable. Biofilms may form on just about any available surface, including contact lenses and artificial joint replacements. They foul metal pipes and cause corrosion, a major problem in steam-driven electricity generation plants. Fossil stromatolites—large, rocky structures made up of alternating layers of fossilized biofilm and calcium carbonate—are among the oldest remnants of life on Earth (see Figure 24.9B).

Some biologists are studying the chemical signals that prokaryotes use to communicate with one another and that trigger density-linked activities such as biofilm formation. We saw one example of this type of communication—called **quorum sensing**—in the chapter-opening discussion of bioluminescent *Vibrio*. How does

(A) Recruitment of organisms for biofilm community

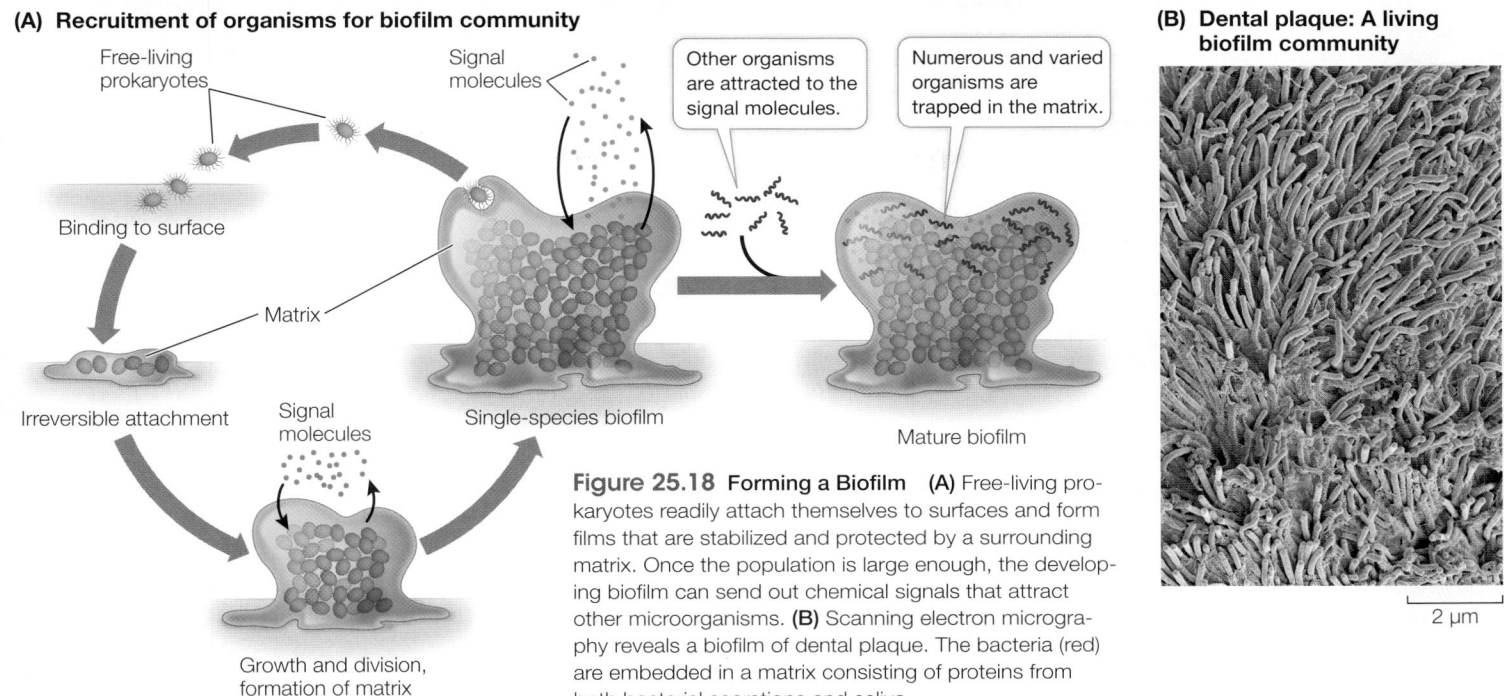

Free-living prokaryotes

Signal molecules

Other organisms are attracted to the signal molecules.

Numerous and varied organisms are trapped in the matrix.

Binding to surface

Matrix

Irreversible attachment

Signal molecules

Single-species biofilm

Mature biofilm

Growth and division, formation of matrix

(B) Dental plaque: A living biofilm community

2 µm

Figure 25.18 Forming a Biofilm **(A)** Free-living prokaryotes readily attach themselves to surfaces and form films that are stabilized and protected by a surrounding matrix. Once the population is large enough, the developing biofilm can send out chemical signals that attract other microorganisms. **(B)** Scanning electron micrography reveals a biofilm of dental plaque. The bacteria (red) are embedded in a matrix consisting of proteins from both bacterial secretions and saliva.

quorum sensing work? As demonstrated in **Investigating Life: How Do Bacteria Communicate with One Another?**, individual *Vibrio* bacteria can excrete a signal that is detected by other individuals, and this signal then functions to turn on the genes that produce luciferase—an enzyme that produces bioluminescence when it is active.

Microbiomes are critical to the health of many eukaryotes

Although only a few bacterial species are pathogens, popular notions of bacteria as "germs" and fear of the consequences of infection cause many people to assume that most bacteria are harmful. Increasingly, however, biologists are discovering that the health of humans (as well as that of most other eukaryotes) depends in large part on the health of our **microbiomes**: the communities of bacteria and prokaryotic archaea that live in and on our bodies. Other communities of microbes live in close association with other multicellular organisms.

Every surface of your body is covered with diverse communities of bacteria (**Figure 25.19**). A recent study identified more than 1,000 species of bacteria that live on human skin. Inside your body, your

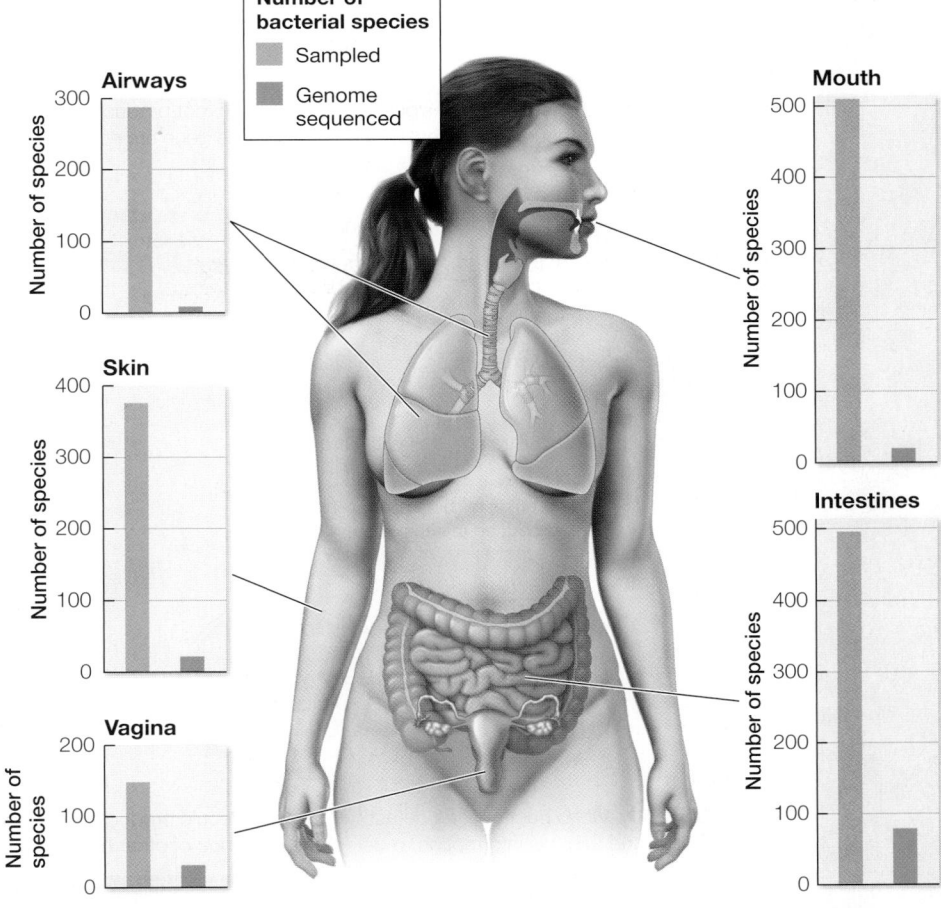

Number of bacterial species
- Sampled
- Genome sequenced

Airways

Skin

Vagina

Mouth

Intestines

Number of species

Figure 25.19 The Body's Microbiome Is Critical to the Maintenance of Health Surveys of the human microbiome have shown that this community includes thousands of diverse bacterial species that are adapted to grow in or on various parts of the body. Although we now know that the composition of this microbiome is closely associated with many aspects of human health, most of the component species are poorly characterized and remain largely unstudied by biologists. What has become clear is that, although the "subcommunities" in different parts of the body share similarities, each is a site-specific assemblage of many distinctive species.

experiment

Original Paper: Miller, M. B. and B. L. Bassler. 2001. Quorum sensing in bacteria. *Annual Review of Microbiology* 55:165–199.

Bonnie Bassler and her colleagues at Princeton University investigated how *Vibrio fischeri* bacteria communicate with one another. These bacteria produce bioluminescence when they are present in sufficiently high densities. In a normal *V. fischeri* bacterium, the following pathway produces bioluminescence when a bacterial colony becomes dense enough to produce sufficient signal:

Wild-type

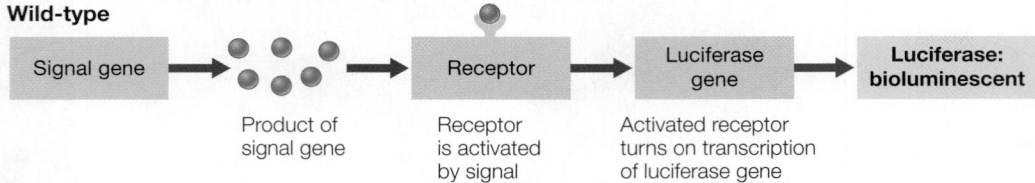

The fact that the bacteria emit light only when they are present in high densities suggests that the signal is used to communicate among nearby bacteria, alerting one another to their presence. But how can we tell that the signal produced by one bacterium is being received by another?

HYPOTHESIS▶ *V. fischeri* can transmit the signal molecule from one individual to another, thus communicating about the presence of other nearby bacteria.

METHOD

1. Select two mutant strains of *V. fischeri* incapable of emitting light on their own. In strain A, a mutation for the signal receptor renders the bacteria unable to detect the signal molecule, so they do not bioluminesce:

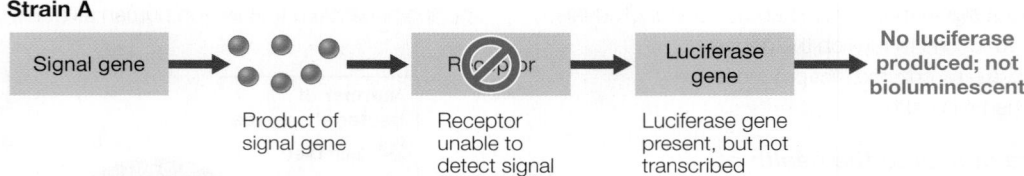

Strain B bacteria have a mutated signal gene, so they do not produce a signal molecule, although the receptor and the luciferase gene are normal:

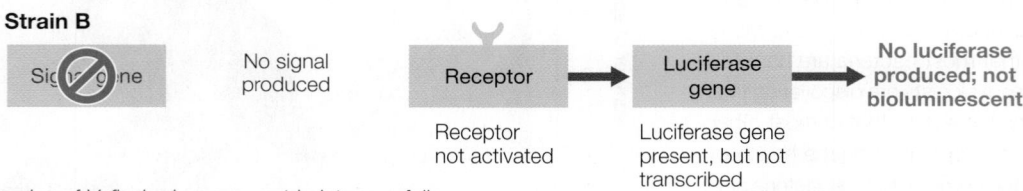

2. Incubate samples of *V. fischeri* on agar petri plates, as follows:
   Plate 1: Strain A only
   Plate 2: Strain B only
   Plate 3: Half of plate with Strain A, half with Strain B

RESULTS

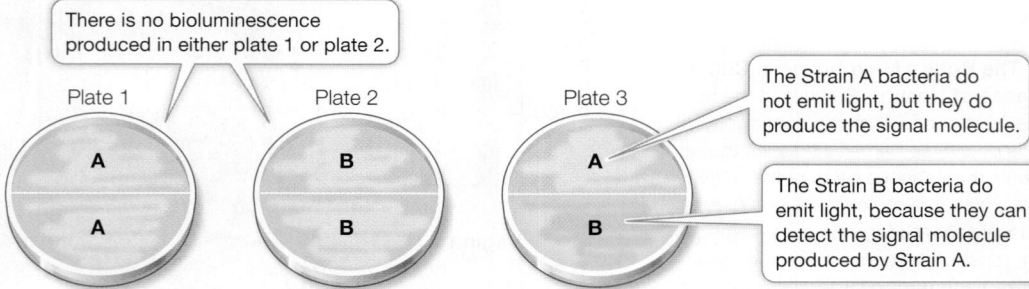

There is no bioluminescence produced in either plate 1 or plate 2.

The Strain A bacteria do not emit light, but they do produce the signal molecule.

The Strain B bacteria do emit light, because they can detect the signal molecule produced by Strain A.

CONCLUSION▶ *Vibrio* bacteria can detect the signal molecule produced by other nearby bacteria and use this information to detect the presence of other bacteria.

digestive system teems with bacteria. When these communities are disrupted, they must be restored before the body can function normally.

Biologists are discovering that many complex health problems are linked to the disruption of our microbiomes. These diverse microbial communities affect the expression of our genes and play a critical role in the development and maintenance of a healthy immune system. When our microbiomes contain an appropriate community of beneficial species, our bodies function normally. But these communities are strongly affected by our life experiences, by the food we eat, by the medicines we take, and by our exposure to various environmental toxins. The recent rapid increase in the rate of autoimmune diseases in humans—diseases in which the immune system begins to attack the body—has been linked to the changing diversity and composition of our microbiomes.

The early acquisition of an appropriate microbiome is critical for lifelong health. Normally, a human infant acquires much of its microbiome at birth, from the microbiome in its mother's vagina. Other components of the microbiome are also acquired from the mother, especially through breast feeding. Recent studies have shown that babies born by cesarean section, as well as babies that are bottle-fed on artificial milk formula, typically acquire microbes from a wider variety of sources. Many of the bacteria acquired in this way are not well suited for human health. Biologists have discovered that the incidence of many autoimmune diseases is much higher in people who were born by cesarean section and in those who were fed on formula as infants, compared with individuals who were born vaginally and breast-fed as infants. The difference appears to be related to the composition of the individual's original microbiome.

Humans use some of the metabolic products—especially vitamins B_{12} and K—produced by the microbiome living in the large intestine. Communities of bacteria line our intestines with a dense biofilm that is in intimate contact with the mucosal lining of the gut. This biofilm facilitates nutrient transfer from the intestine into the body, functioning like a specialized "tissue" that is essential to our health. This biofilm has a complex ecology that scientists have just begun to explore in detail—including the possibility that the species composition of an individual's gut microbiome may contribute to obesity.

Animals harbor a variety of microbes in their digestive tracts, many of which play important roles in digestion. Cattle depend on prokaryotes to break down plant material. Like most animals, cattle cannot produce cellulase, the enzyme needed to start the digestion of the cellulose that makes up the bulk of their plant food. However, bacteria living in a special section of the gut, called the rumen, produce enough cellulase to process the daily diet for the cattle.

A small minority of bacteria are pathogens

The late nineteenth century was a productive era in the history of medicine—a time when bacteriologists, chemists, and physicians proved that many diseases are caused by microbial agents. During this time, the German physician Robert Koch laid down a set of four rules for establishing that a particular microorganism causes a particular disease:

1. The microorganism is always found in individuals with the disease.

2. The microorganism can be taken from the host and grown in pure culture.

3. A sample of the culture produces the same disease when injected into a new, healthy host.

4. The newly infected host yields a new, pure culture of microorganisms identical to those obtained in the second step.

These rules, called **Koch's postulates**, were important tools in a time when it was not widely understood that microorganisms cause disease. Although modern medical science has more powerful diagnostic tools, Koch's postulates remain useful. For example, physicians were taken aback in the 1980s when stomach ulcers—long accepted and treated as the result of excess stomach acid—were shown by Koch's postulates to be caused by the bacterium *Helicobacter pylori* (**Figure 25.20**).

For an organism to be a successful pathogen, it must:

- arrive at the body surface of a potential host;
- enter the host's body;
- evade the host's defenses;
- reproduce inside the host; and
- infect a new host.

Failure to complete any of these steps ends the disease cycle of a pathogenic organism. Yet in spite of the many defenses available to potential hosts, some bacteria are very successful pathogens. Pathogenic bacteria are often surprisingly difficult to combat, even with today's arsenal of antibiotics. One source of this difficulty is their ability to form biofilms.

For the host, the consequences of a bacterial infection depend on several factors. One is the invasiveness of the pathogen: its ability to multiply in the host's body. Another is its toxigenicity: its ability to produce toxins (chemical substances that are harmful to the host's tissues). *Corynebacterium diphtheriae*, the agent that causes diphtheria, has low invasiveness and multiplies only in the throat, but its toxigenicity is so great that the entire body is affected. In contrast, *Bacillus anthracis*, which causes anthrax, has low toxigenicity but is so invasive that the entire bloodstream ultimately teems with the bacteria.

There are two general types of bacterial toxins: exotoxins and endotoxins. **Endotoxins** are released when certain bacteria grow or lyse (burst). Endotoxins are lipopolysaccharides (complexes consisting of a polysaccharide and a lipid component) that form part of the outer bacterial membrane. Endotoxins are rarely fatal to the host; they normally cause fever, vomiting, and diarrhea. Among the endotoxin producers are some strains of the proteobacteria *Salmonella* and *Escherichia*.

Exotoxins are soluble proteins released by living, multiplying bacteria. They are highly toxic—often fatal—to the host. Human diseases induced by bacterial exotoxins include tetanus (*Clostridium tetani*), cholera (*Vibrio cholerae*), and bubonic plague (*Yersinia pestis*). Anthrax is caused by three exotoxins produced by *Bacillus anthracis*. Botulism is caused by exotoxins produced by *Clostridium botulinum*; these exotoxins are among the most poisonous ever discovered. The lethal dose for humans of one exotoxin of *C. botulinum* is about one-millionth of a gram. Nonetheless, much smaller doses

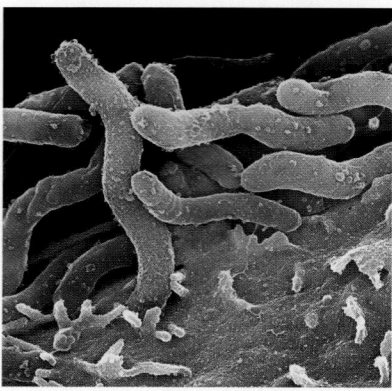

Figure 25.20 Satisfying Koch's Postulates   Robin Warren and Barry Marshall of the University of Western Australia won the 2005 Nobel Prize in Medicine for showing that ulcers are caused not by the action of stomach acid but by infection with the bacterium *Helicobacter pylori*.

Helicobacter pylori   ⊢── 2 μm ──⊣

Marshall and Warren set out to satisfy Koch's postulates:

Test 1

The microorganism must be present in every case of the disease.

Results: Biopsies from the stomachs of many patients revealed that the bacterium was always present if the stomach was inflamed or ulcerated.

Test 2

The microorganism must be cultured from a sick host.

Results: The bacterium was isolated from biopsy material and eventually grown in culture media in the laboratory.

Test 3

The isolated and cultured bacteria must be able to induce the disease.

Results: Marshall was examined and found to be free of bacteria and inflammation in his stomach. After drinking a pure culture of the bacterium, he developed stomach inflammation (gastritis).

Test 4

The bacteria must be recoverable from newly infected individuals.

Results: Biopsy of Marshall's stomach 2 weeks after he ingested the bacteria revealed the presence of the bacterium, now christened *Helicobacter pylori*, in the inflamed tissue.

Conclusion

Antibiotic treatment eliminated the bacteria and the inflammation in Marshall's stomach. The experiment was repeated on healthy volunteers, and many patients with gastric ulcers were cured with antibiotics. Thus Marshall and Warren demonstrated that the stomach inflammation leading to ulcers is caused by *H. pylori* infections in the stomach.

---

of this exotoxin, marketed under various trade names (e.g., Botox), are used to treat muscle spasms and for cosmetic purposes (temporary wrinkle reduction in the skin).

Prokaryotes have amazingly diverse metabolic pathways

Prokaryotic archaea and bacteria outdo the eukaryotes in terms of metabolic diversity. Although they are much more diverse in size and shape, eukaryotes draw on fewer metabolic mechanisms for their energy needs. In fact, much of the eukaryotes' energy metabolism is carried out in organelles—mitochondria and chloroplasts—that are endosymbiotic descendants of bacteria. The long evolutionary history of prokaryotes, during which they have had time to explore a wide variety of habitats, has led to the extraordinary diversity of their metabolic "lifestyles"—their use or nonuse of oxygen, their energy sources, their sources of carbon atoms, and the materials they release as waste products.

ANAEROBIC VERSUS AEROBIC METABOLISM   Some prokaryotes can live only by anaerobic metabolism because oxygen is poisonous to them. These oxygen-sensitive organisms are called **obligate anaerobes**. Other prokaryotes can shift their metabolism between anaerobic and aerobic modes and are thus called **facultative anaerobes**. Many facultative anaerobes alternate between anaerobic metabolism (such as fermentation) and cellular respiration as conditions dictate. **Aerotolerant anaerobes** cannot conduct cellular respiration, but they are not damaged by oxygen when it is present. By definition, an anaerobe does not use oxygen as an electron acceptor for its respiration.

At the other extreme from the obligate anaerobes, some prokaryotes are **obligate aerobes**, unable to survive for extended periods in the *absence* of oxygen. They require oxygen for cellular respiration.

NUTRITIONAL CATEGORIES   All living organisms face the same nutritional challenges: they must synthesize energy-rich compounds such as ATP to power their life-sustaining metabolic reactions, and they must obtain carbon atoms to build their own organic molecules. Biologists recognize four broad nutritional categories of organisms: photoautotrophs, photoheterotrophs, chemoautotrophs,

and chemoheterotrophs. Prokaryotes are represented in all four groups (**Table 25.2**).

Photoautotrophs perform photosynthesis. They use light as their energy source and carbon dioxide (CO_2) as their carbon source. The cyanobacteria, like green plants and other photosynthetic eukaryotes, use chlorophyll *a* as their key photosynthetic pigment and produce oxygen gas (O_2) as a by-product of noncyclic electron transport.

There are other photoautotrophs among the bacteria, but these organisms use bacteriochlorophyll as their key photosynthetic pigment, and they do not produce O_2. Instead, some of these photosynthesizers produce particles of pure sulfur, because hydrogen sulfide (H_2S), rather than H_2O, is their electron donor for photophosphorylation. Many proteobacteria fit into this category. Bacteriochlorophyll molecules absorb light of longer wavelengths than the chlorophyll molecules used by other photosynthesizing organisms. As a result, bacteria using this pigment can grow in water under fairly dense layers of algae, using light of wavelengths that are not absorbed by the algae (**Figure 25.21**).

Photoheterotrophs use light as their energy source but must obtain their carbon atoms from organic compounds made by other organisms. Their "food" consists of organic compounds such as carbohydrates, fatty acids, and alcohols. For example, compounds released from plant roots (as in rice paddies) or from decomposing photosynthetic bacteria in hot springs are taken up by photoheterotrophs and metabolized to form building blocks for other compounds. Sunlight provides the ATP necessary for metabolism through photophosphorylation.

table 25.2   How Organisms Obtain Their Energy and Carbon

Nutritional category	Energy source	Carbon source
Photoautotrophs (some bacteria, some eukaryotes)	Light	Carbon dioxide
Photoheterotrophs (some bacteria)	Light	Organic compounds
Chemoautotrophs (some bacteria, many prokaryotic archaea)	Inorganic substances	Carbon dioxide
Chemoheterotrophs (found in all three domains)	Usually organic compounds; sometimes inorganic substances	Organic compounds

Chemoautotrophs obtain their energy by oxidizing inorganic substances, and they use some of that energy to fix carbon. Some chemoautotrophs use reactions identical to those of the typical photosynthetic cycle, but others use alternative pathways for carbon fixation. Some bacteria oxidize ammonia or nitrite ions to form nitrate ions. Others oxidize hydrogen gas, hydrogen sulfide, sulfur, and other materials. Many prokaryotic archaea are chemoautotrophs.

Finally, **chemoheterotrophs** obtain both energy and carbon atoms from one or more complex organic compounds that have been synthesized by other organisms. Most known bacteria and prokaryotic archaea are chemoheterotrophs—as are all animals and fungi and many protists.

Although most chemoheterotrophs rely on the breakdown of organic compounds for energy, some chemoheterotrophic prokaryotes obtain their energy by breaking down inorganic substances. Organisms that obtain energy from oxidizing inorganic substances (both chemoautotrophs as well as some chemoheterotrophs) are also known as lithotrophs (Greek, "rock consumers").

Prokaryotes play important roles in element cycling

The metabolic diversity of the prokaryotes makes them key players in the cycles that keep elements moving through ecosystems. Many prokaryotes are decomposers: organisms that metabolize organic compounds in dead organic material and return the products to the environment as inorganic substances. Prokaryotes,

along with fungi, return tremendous quantities of carbon to the atmosphere as carbon dioxide, thus carrying out a key step in the carbon cycle.

The key metabolic reactions of many prokaryotes involve nitrogen or sulfur. For example, some bacteria carry out respiratory electron transport without using oxygen as an electron acceptor. These organisms use oxidized inorganic ions such as nitrate, nitrite, or sulfate as electron acceptors. Examples include the **denitrifiers**, which release nitrogen to the atmosphere as nitrogen gas (N_2). These normally aerobic bacteria, mostly species of the genera *Bacillus* and *Pseudomonas*, use nitrate (NO_3^-) as an electron acceptor in place of oxygen if they are kept under anaerobic conditions:

$$2 \ NO_3^- + 10 \ e^- + 12 \ H^+ \rightarrow N_2 + 6 \ H_2O$$

Denitrifiers play a key role in the cycling of nitrogen through ecosystems. Without denitrifiers, which convert nitrate ions back into nitrogen gas, all forms of nitrogen would leach from the soil and end up in lakes and oceans, making life on land much more difficult.

Nitrogen fixers convert atmospheric nitrogen gas into a chemical form (ammonia) that is usable by the nitrogen fixers themselves as well as by other organisms:

$$N_2 + 6 \ H \rightarrow 2 \ NH_3$$

All organisms require nitrogen in order to build proteins, nucleic acids, and other important compounds. ***Nitrogen fixation** is thus vital to life as we know it. This all-important biochemical process is carried out by a wide variety of prokaryotic archaea and bacteria (including cyanobacteria) but by no eukaryotes, so we depend on these prokaryotes for our very existence.

***connect the concepts**   For descriptions of the role of nitrogen in plant nutrition and in the global nitrogen cycle, see Key Concepts 35.4 and 57.4.

Ammonia is oxidized to nitrate in soil and in seawater by chemoautotrophic bacteria called **nitrifiers**. Bacteria of two genera, *Nitrosomonas* and *Nitrosococcus*, convert ammonia (NH_3) to nitrite ions (NO_2^-), and *Nitrobacter* oxidize nitrite to nitrate (NO_3^-), the form of nitrogen most easily used by many plants. What do the nitrifiers get out of these reactions? Their metabolism is powered by

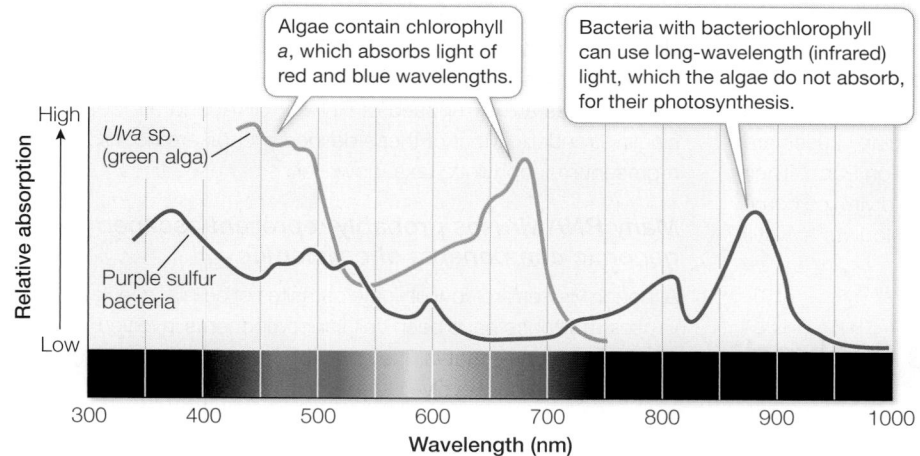

Figure 25.21 Bacteriochlorophyll Absorbs Long-Wavelength Light   The green alga *Ulva* contains chlorophyll, which absorbs no light of wavelengths longer than 750 nm. Purple sulfur bacteria, which contain bacteriochlorophyll, can conduct photosynthesis using longer infrared wavelengths. As a result, these bacteria can grow under layers of algae.

the energy released by the oxidation of ammonia or nitrite. For example, by passing the electrons from nitrite through an electron transport system, *Nitrobacter* can make ATP and, using some of this ATP, can also make NADH. With this ATP and NADH, the bacterium can convert CO_2 and H_2O into glucose.

We have already seen the importance of the cyanobacteria in the cycling of oxygen: in ancient times, the oxygen generated by their photosynthesis converted Earth's atmosphere from an anaerobic to an aerobic environment. Other prokaryotes—both bacteria and archaea—contribute to the cycling of sulfur. Deep-sea hydrothermal vent ecosystems depend on chemoautotrophic prokaryotes that are incorporated into large communities of crabs, mollusks, and giant worms, all living at a depth of 2,500 m—below any hint of sunlight. These bacteria obtain energy by oxidizing hydrogen sulfide and other substances released in the near-boiling water flowing from volcanic vents in the ocean floor.

25.3 recap

Many prokaryotes are beneficial and even necessary to other forms of life. Most animals, including humans, depend on a complex community of prokaryotes—a microbiome—to maintain health, especially of the immune and digestive systems. Pathogenic bacteria are the direct causes of diseases.

learning outcomes

You should be able to:

- Summarize the applications of biofilms.
- Devise experiments to test effects of microbiomes on human health.
- Compare and contrast nitrogen fixers, denitrifiers, and nitrifiers.

1. How do biofilms form, and why are they of special interest to researchers?
2. Why would elimination of all bacteria from a human gut be problematic from a health standpoint?
3. Why is nitrogen metabolism in prokaryotes vital to other organisms?

Before moving on to discuss the diversity of eukaryotic life, it is appropriate to consider another category of life that includes some pathogens: the viruses. Although they are not cellular, viruses are numerically among the most abundant forms of life on Earth. Their effects on other organisms are enormous. Where did viruses come from, and how do they fit into the tree of life? Biologists are still working to answer these questions.

key concept 25.4 Viruses Have Evolved Many Times

Some biologists do not think of viruses as living organisms, primarily because they are not cellular and must depend on cellular organisms for basic life functions such as replication and metabolism. But viruses are derived from the cells of living organisms. They use the same essential forms of genetic information storage and transmission as do cellular organisms. Viruses infect all cellular forms of life—both prokaryotes and eukaryotes. They replicate, mutate, evolve, and interact with other organisms, often causing serious diseases in their hosts. Finally, viruses clearly evolve independently of other organisms, so it is almost impossible not to treat them as a part of life.

focus your learning

- Several factors limit phylogenetic analysis of viruses.
- Some DNA viruses may have evolved from reduced cellular organisms.
- Viruses can be used to fight bacterial infections.

Viruses are abundant in many environments. In some freshwater and marine ecosystems, they can occur at densities of up to 10 million viruses per milliliter of water. Biologists estimate that there are approximately 10^{31} individual virus particles on Earth—about 1,000 times the number of cellular organisms on the planet. Viruses have an enormous effect on the ecology of the oceans. Every day, about one-half of the bacteria in the oceans are killed by viruses. Huge marine blooms of bacteria, such as the *Vibrio* bloom that produced the milky seas described at the opening of this chapter, do not last for long because viral blooms soon follow the initial bacterial bloom. As the viruses increase, they begin to kill bacteria faster than the bacteria can reproduce.

Although viruses are everywhere and play an important role in many ecosystems, many aspects of their ecology and evolution are still poorly known. For example, several factors make virus phylogeny difficult to resolve. The tiny size of many virus genomes restricts the phylogenetic analyses that can be conducted to relate viruses to cellular organisms. Their rapid mutation rate, which results in rapid evolution of virus genomes, tends to cloud evolutionary relationships over long periods. There are no known fossil viruses (viruses are too small and delicate to fossilize), so the paleontological record offers no clues to virus origins. Finally, viruses are highly diverse (**Figure 25.22**). Several lines of evidence support the hypothesis that viruses have evolved repeatedly within each of the major groups of life. The difficulty in resolving deep evolutionary relationships of viruses makes a phylogeny-based classification difficult. Instead, viruses are placed in one of several functionally similar groups on the basis of the structure of their genomes (for example, whether the genomes are composed of RNA or DNA, and are double- or single-stranded). Most of these defined groups are not thought to represent monophyletic taxa, however.

Many RNA viruses probably represent escaped genomic components of cellular life

Although viruses are now obligate parasites of cellular species, many viruses may once have been cellular components involved in basic cellular functions—that is, they may be "escaped" components of cellular life that now evolve independently of their hosts.

NEGATIVE-SENSE SINGLE-STRANDED RNA VIRUSES   An example of a likely "escaped" biochemical function comes from viruses with genomes composed of single-stranded **negative-sense RNA**: RNA that is the complement of the mRNA needed for protein translation. Many of these negative-sense single-stranded RNA viruses

(A)

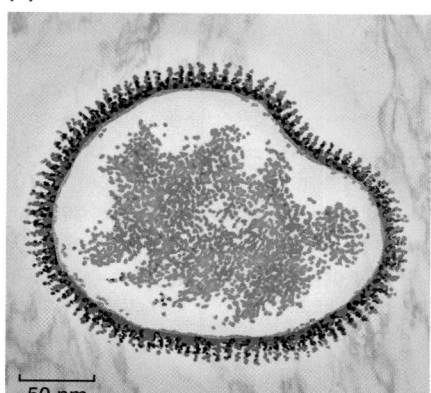

50 nm

A negative-sense single-stranded RNA virus: The influenza A virus. This virus is responsible for seasonal influenza epidemics in humans. Surface view.

(B)

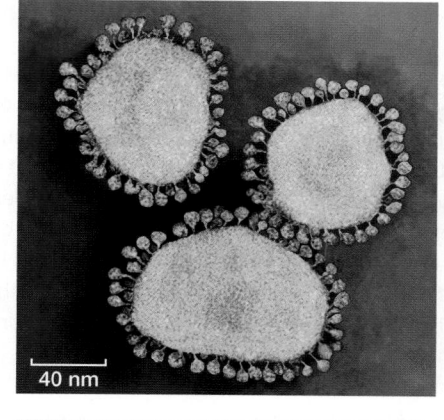

40 nm

A positive-sense single-stranded RNA virus: Coronavirus of a type thought to be responsible for severe acute respiratory syndrome (SARS). Surface view.

(C)

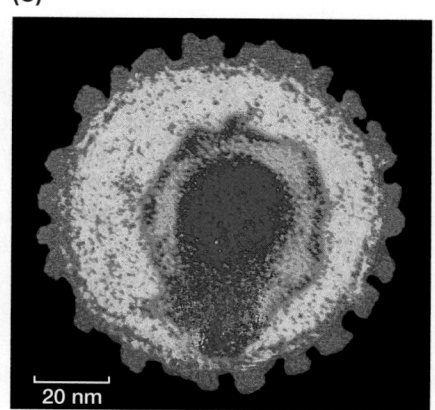

20 nm

An RNA retrovirus: One of the human immunodeficiency viruses (HIV) that causes AIDS. Cutaway view.

(D)

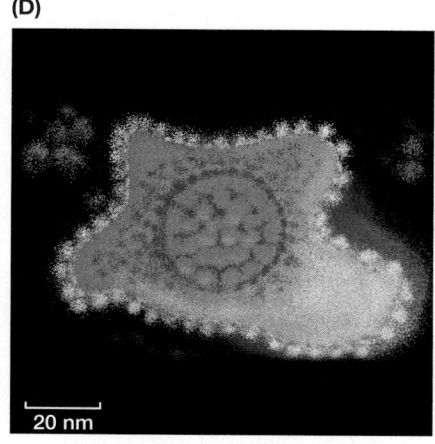

20 nm

A double-stranded DNA virus: One of the many herpes viruses (Herpesviridae). Different herpes viruses are responsible for many human infections, including chicken pox, shingles, cold sores, and genital herpes (HSV1/2). Surface view.

(E)

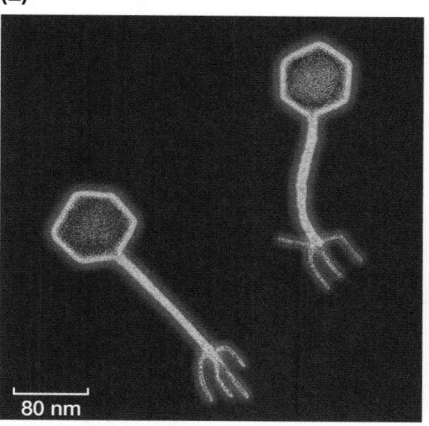

80 nm

A double-stranded DNA virus: Bacteriophage T4. Viruses that infect bacteria are referred to as bacteriophage (or simply phage). T4 attaches leglike fibers to the outside of its host cell and injects its DNA into the cytoplasm through its "tail" (pink structure in this rendition).

(F)

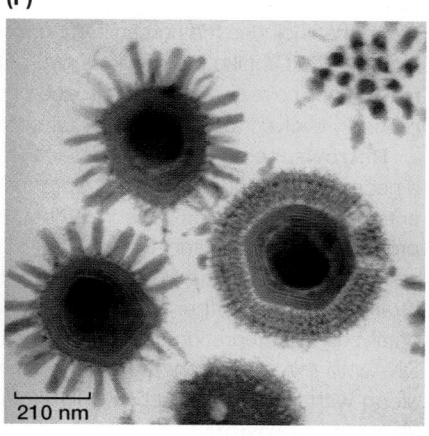

210 nm

A double-stranded DNA mimivirus: This Acanthamoeba polyphaga mimivirus (APMV) has a genome larger than some prokaryote genomes. Cutaway view.

Figure 25.22 Viruses Are Diverse Relatively small genomes and rapid evolutionary rates make it difficult to reconstruct phylogenetic relationships among viruses. Instead, viruses are classified largely by general characteristics of their genomes. The images here are computer artists' reconstructions based on cryoelectron micrographs.

have only a few genes, including one for an RNA-dependent RNA polymerase that allows them to make mRNA from their negative-sense RNA genome. Modern cellular organisms cannot generate mRNA in this manner (at least in the absence of viral infections), but scientists speculate that single-stranded RNA genomes may have been common in the distant past, before DNA became the primary molecule for genetic information storage.

A self-replicating RNA polymerase gene that began to replicate independently of a cellular genome could conceivably acquire a few additional protein-coding genes through recombination with its host's DNA. If one or more of these genes were to foster the development of a protein coat, the virus might then survive outside the host and infect new hosts. It is believed that this scenario has been repeated many times independently across the tree of life, given that many of the negative-sense single-stranded RNA viruses that infect organisms from bacteria to humans are not closely related to one another. In other words, negative-sense single-stranded RNA viruses do not represent a distinct taxonomic group, but rather exemplify a particular process of cellular escape that probably happened many different times.

Familiar examples of negative-sense single-stranded RNA viruses include the viruses that cause measles, mumps, rabies, and influenza (see Figure 25.22A).

POSITIVE-SENSE SINGLE-STRANDED RNA VIRUSES   The genome of another type of single-stranded RNA virus is composed of positive-sense RNA. Positive-sense genomes are already set for translation; no replication of the genome to form a complement strand is needed before protein translation can take place. Positive-sense single-stranded RNA viruses (see Figure 25.22B) are the most abundant and diverse class of viruses. Most of the viruses that cause diseases in crop plants are members of this group. These viruses kill patches of cells in the leaves or stems of plants, leaving live cells amid a patchwork of discolored dead tissue (giving them the name of mosaic or mottle viruses; **Figure 25.23**). Other viruses in this group infect bacteria, fungi, and animals. Human diseases caused by positive-sense single-stranded RNA viruses include polio, hepatitis C, and the common cold. As is true of the other functionally defined groups of viruses, these viruses appear to have evolved multiple times across the tree of life from different groups of cellular ancestors.

RNA RETROVIRUSES   The RNA retroviruses are best known as the group that includes the human immunodeficiency viruses (HIV; see Figure 25.22C). Like the previous two categories of viruses, RNA retroviruses have genomes composed of single-stranded RNA and probably evolved as escaped cellular components.

Retroviruses are so named because reverse transcription is a necessary component of their reproduction. When the retrovirus enters the nucleus of its vertebrate host, viral reverse transcriptase produces complementary DNA (cDNA) from the viral RNA genome and then replicates the single-stranded cDNA to produce double-stranded DNA. Another virally encoded enzyme called integrase catalyzes the integration of the new piece of double-stranded viral DNA into the host's genome. The viral genome is then replicated along with the host cell's DNA. The integrated retroviral DNA is known as a **provirus**.

Retroviruses are only known to infect vertebrates, although genomic elements that resemble portions of these viruses are a component of the genomes of a wide variety of organisms, including bacteria, plants, and many animals. Several retroviruses are associated with the development of various forms of cancer, as

cells infected with these viruses are likely to undergo uncontrolled replication.

As retroviruses become incorporated into the genomes of their hosts, many become nonfunctional copies that are no longer expressed as functional viruses. These sequences may provide a record of ancient viral infections that plagued our ancestors. Humans, for example, carry about 100,000 fragments of **endogenous retroviruses** in our genome. These fragments make up about 8 percent of our DNA—a considerably larger fraction of our genome than the fraction that comprises all our protein-coding genes (about 1.2 percent of our genome).

DOUBLE-STRANDED RNA VIRUSES   Double-stranded RNA viruses may have evolved repeatedly from single-stranded RNA ancestors—or perhaps vice versa. These viruses, which are not closely related to one another, infect organisms from throughout the tree of life. Many plant diseases are caused by double-stranded RNA viruses. Other viruses of this type cause many cases of infant diarrhea in humans.

Some DNA viruses may have evolved from reduced cellular organisms

Another class of viruses is composed of viruses that have a double-stranded DNA genome (see Figure 25.22D–F). This group is also almost certainly polyphyletic (with many independent origins). Many of the common phage that infect bacteria are double-stranded DNA viruses, as are the viruses that cause smallpox and herpes in humans.

Some biologists think that at least some of the DNA viruses may represent highly reduced parasitic organisms that have lost their cellular structure as well as their ability to survive as free-living species. For example, the mimiviruses, which are some of the largest DNA viruses (see Figure 25.22F), have a genome in excess of a million base pairs of DNA that encode more than 900 proteins. This genome is similar in size to the genomes of many parasitic bacteria and about twice as large as the genome of the smallest bacteria (**Figure 25.24**). Phylogenetic analyses of these DNA viruses suggest that they have evolved repeatedly from cellular organisms. Furthermore, recombination among different viruses may have allowed the exchange of various genetic modules, further complicating the history and origins of these viruses.

Viruses can be used to fight bacterial infections

Although some viruses cause devastating diseases, other viruses have been used to fight disease. Most bacterial diseases are treated today with antibiotics. Antibiotics were first discovered in the 1930s, but they were not widely used to treat bacterial diseases until the 1940s. So antibiotics were not yet available during World War I, when bacterial infections plagued the battlefields. Battlefield wounds were often infected by bacteria, and in the absence of antibiotics, these infections often led to the loss of limbs and lives. While trying to find a way to combat this problem, a physician named Felix d'Herelle discovered the first evidence of viruses that attack bacteria. He named these viruses bacteriophage, or "eaters of bacteria." Herelle extracted bacteriophage from the stool of infected patients. He then used these extracts to treat patients with deadly bacterial

Yellow areas are dead leaf cells, killed by the mosaic virus.

Figure 25.23 Mosaic Viruses Are a Problem for Agriculture Mosaic, or "mottle," viruses are the most diverse class of viruses. This leaf is from an apple tree infected with a mosaic virus.

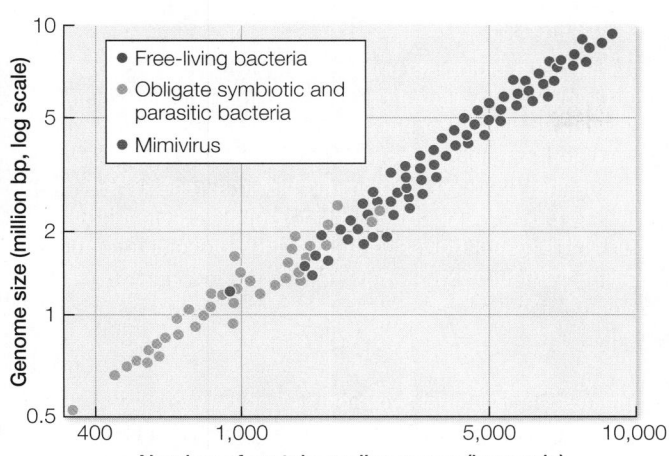

Figure 25.24 Mimiviruses Have Genomes Similar in Size to Those of Many Parasitic Bacteria The genome of *Acanthamoeba polyphaga* mimivirus contains 1,181,404 base pairs and encompasses 911 protein-encoding genes. This observation is consistent with the hypothesis that this virus evolved from a parasitic bacterium.

Q: Why do parasitic bacteria generally have smaller genomes and fewer protein-coding genes than free-living bacteria do?

infections, including dysentery, cholera, and bubonic plague. This practice became known as **phage therapy**. After the war, phage therapy was widely used among the general public to treat bacterial infections of the skin and intestines.

 Media Clip 25.3 **Bacteriophages Attack *E. coli***
www.Life11e.com/mc25.3

Phage therapy was mostly replaced by the use of antibiotics in the 1930s and 1940s as physicians grew concerned about treating patients with live viruses. Phage therapy continued to be used in the Soviet Union but largely disappeared from western medical practice. Today, however, many antibiotics are losing their effectiveness as bacterial pathogens evolve resistance to these drugs. Phage therapy is once again an active area of research, and it is likely that bacteriophage will become increasingly important as weapons against

bacterial diseases. One advantage that bacteriophage may have over antibiotics is that, like bacteria, bacteriophage can evolve. As bacteria evolve resistance to a strain of bacteriophage, biologists can select for new strains of bacteriophage that retain their effectiveness against the pathogens. In this way, biologists are using their understanding of evolution to combat the problem of antibiotic-resistant bacteria.

25.4 recap

Viruses are highly diverse and appear to have evolved independently from many different cellular organisms within each of the major groups of life. Some viruses appear to have evolved from escaped components of cellular organisms, whereas other viruses may have evolved from parasitic cellular ancestors.

learning outcomes

You should be able to:

- Explain why viruses are difficult to place within the tree of life.
- Contrast the "escaped component" and "reduced parasite" hypotheses for the origin of viruses, and name major groups of viruses that best support each hypothesis.
- Explain why phage therapy is once again an active area of research.

1. How does the size of viral genomes limit phylogenetic analyses of viruses?
2. What are the two main hypotheses of viral origins?
3. How can viruses be used to treat some human diseases?

It appears that the enormous diversity of viruses is, at least in part, a result of their multiple origins from many different cellular organisms. It may be best to view viruses as spin-offs from the various branches on the tree of life—sometimes evolving independently of cellular genomes, sometimes recombining with them. One way to think of viruses is as the "bark" on the tree of life: certainly an important component all across the tree, but not quite like the main branches.

 investigatinglife

How and why do bacteria communicate with one another?

As we explored in Investigating Life: How Do Bacteria Communicate with One Another?, *Vibrio* bacteria send a chemical signal that is received by other bacteria. This signal is needed to produce bioluminescence. As *Vibrio* populations increase in density, the bioluminescence becomes strong enough to be visible to humans and other animals. But how is bioluminescence an advantage to the bacteria? What is its function?

Although marine *Vibrio* are able to live independently, they truly thrive inside the guts of fish and other marine animals. Inside a fish, *Vibrio* cells attach themselves to food particles, including

phytoplankton, and are often expelled into the ocean as waste. How can they get back into their preferred environment? The bioluminescent glow produced by a dense colony of free-living *Vibrio* growing on phytoplankton attracts fish, which consume the phytoplankton and thus ingest the bacteria—which gets the bacteria into a new host fish.

The bioluminescent *Vibrio* described in Investigating Life: How Do Bacteria Communicate with One Another?, *V. fischeri*, have evolved a symbiotic relationship with the Hawaiian bobtail squid (*Euprymna scolopes*). Chemicals produced by a developing squid embryo specifically "recruit" *V. fischeri* from the surrounding seawater. The bacteria then preferentially migrate to specific tissues that develop into a bioluminescent "light organ" in the belly of the adult

(continued)

squid (Figure 25.25). The tiny (about 3 cm long) adult squid feed while floating near the sea surface at night. The soft glow produced by the bioluminescent bacteria mimics the moonlight above, so the squid are less visible to potential predators coming at them from below.

Vibrio fischeri live symbiotically inside the squid's light organ.

Future directions

Research that helps biologists understand how bacteria communicate, and allows us to disrupt or change communication signals among bacteria, is critical for developing treatment strategies against persistent bacterial infections. In the case of health-threatening bacteria that produce problematic biofilms, researchers hope to find ways to block the quorum-sensing signals that lead to the production of the matrix polysaccharides, thus preventing pathogenic biofilms from forming.

Figure 25.25 Bioluminescent Bacterial Symbionts *Vibrio* bacteria within the light organ emit bioluminescence downward as the Hawaiian bobtail squid floats near the ocean surface to feed. At night, this allows the squid to blend in with moonlight or starlight rather than becoming a target for a predator from below.

Chapter Summary 25

▶ 25.1 Bacteria and Archaea Are the Two Primary Divisions of Life

- The earliest split in the tree of life is between Bacteria and Archaea. One lineage within the Archaea branch, the Eukarya, evolved specializations (such as a nucleus and membrane-enclosed organelles) that allowed them flourish. All organisms that are not eukaryotes are called prokaryotes. **Review Figure 25.1, Table 25.1, Animation 25.1**

- The cell walls of almost all bacteria contain **peptidoglycan**, whereas the cell walls of prokaryotic archaea lack peptidoglycan.

- Some groups of bacteria can be differentiated using the **Gram stain**. **Gram-negative bacteria** have a periplasmic space between the cell membrane and a distinct outer membrane. **Gram-positive bacteria** have a thick cell wall containing about five times as much peptidoglycan as a Gram-negative cell wall. **Review Figure 25.2, Activity 25.1**

- The three most common bacterial shapes are **cocci** (spheres), **bacilli** (rods), and **spirilla** (helices). **Review Figure 25.3**

- Phylogenetic classification of prokaryotes is now based principally on nucleotide sequences of their genomes, especially of the rRNA genes.

- Prokaryotes reproduce asexually by binary fission but may exchange genetic material. Reproduction and genetic exchange are not directly linked in prokaryotes.

- Although **lateral gene transfer** has occurred throughout prokaryotic evolutionary history, elucidation of many aspects of prokaryote phylogeny is still possible. **Review Focus: Key Figure 25.4**

▶ 25.2 Prokaryote Diversity Reflects the Ancient Origins of Life

- Prokaryotes are the most numerous organisms on Earth, but only a small fraction of prokaryote diversity has been characterized to date.

- Several early-diverging groups of bacteria, such as **hadobacteria** and **hyperthermophilic bacteria**, live mostly in hot environments.

- The **firmicutes** include the **mycoplasmas**, which are among the smallest cellular organisms known.

- Some **actinobacteria** produce important antibiotics.

- The photosynthetic **cyanobacteria** release oxygen into the atmosphere. Cyanobacteria may live free as single cells or associate in multicellular colonies. **Review Figure 25.9**

- **Spirochetes** have unique structures called axial filaments that allow them to move in a corkscrew-like manner. **Review Figure 25.10**

- **Chlamydias** are tiny parasitic bacteria (0.2–1.5 μm in diameter) that live within the cells of other organisms. **Review Figure 25.11**

- The **proteobacteria** embrace the largest number of known species of bacteria.

- Prokaryotic archaea live in highly diverse environments and include many **extremophiles**. Most known crenarchaeotes are thermophilic, acidophilic, or both. Some euryarchaeotes are **methanogens**; **extreme halophiles** are also found among the euryarchaeotes. **Review Figure 25.14**

- **Ether linkages** in the branched long hydrocarbon chains of the lipids that make up the cell membranes are characteristic of prokaryotic archaea. **Review Figure 25.15**

▶ 25.3 Ecological Communities Depend on Prokaryotes

- Prokaryotes form complex communities, of which **biofilms** are one example. **Review Figure 25.18**

- **Microbiomes** are the communities of prokaryotes that live in and on the bodies of multicellular organisms. These communities are often important to the health of the hosts, and changes to the microbiome may lead to serious health consequences. **Review Figure 25.19**

- Communities of bacteria can communicate information about their density using chemical signals in a process known as **quorum sensing**. **Review Investigating Life: How Do Bacteria Communicate with One Another?**

- Prokaryotes inhabit the guts of many animals (including humans) and help them digest food.

- **Koch's postulates** establish the criteria by which an organism may be classified as a pathogen. Relatively few bacteria—and no prokaryotic archaea—are known to be pathogens. **Review Figure 25.20**
- Prokaryote metabolism is diverse. Some prokaryotes are anaerobic, others are aerobic, and still others can shift between these modes.
- Prokaryotes fall into four broad nutritional categories: **photoautotrophs, photoheterotrophs, chemoautotrophs,** and **chemoheterotrophs. Review Table 25.2**
- Prokaryotes play key roles in the cycling of elements such as nitrogen, oxygen, sulfur, and carbon.
- Some prokaryotes metabolize sulfur or nitrogen. **Nitrogen fixers** convert nitrogen gas into a form that organisms can metabolize. **Nitrifiers** convert that nitrogen into forms that can be used by plants, and **denitrifiers** return nitrogen to the atmosphere.

▶ 25.4 Viruses Have Evolved Many Times
- Viruses have evolved many times from many different groups of cellular organisms. They are placed in groups according to the structure of their genomes, but these groups are not thought to represent monophyletic taxa. **Review Figure 25.22**
- Some viruses are probably derived from escaped components of cellular organisms; others are thought to have evolved as highly reduced parasites. **Review Figure 25.24**
- **Endogenous retroviruses** can represent a substantial component of vertebrate genomes (8% in the case of the human genome).
- Viruses can be used to fight bacterial infections in a process known as **phage therapy**.

> Go to **LearningCurve** (in **LaunchPad**) for dynamic quizzing that helps you solidify your understanding of this chapter. **LearningCurve** adapts to your responses, giving you the practice you need to master each key concept.

▷ Apply What You've Learned

Review

25.1 Some groups of bacteria can be distinguished based on the Gram stain.

25.3 Prokaryotes differ with respect to oxygen metabolism.

25.3 Prokaryotes play important roles in nitrogen cycling.

Bacteria are increasingly used for a variety of purposes that range from treating medical conditions to mitigating toxic waste spills. Determining which bacterial species would be appropriate for a particular purpose requires the application of knowledge of the diverse phenotypic characteristics of these bacteria.

The table below lists the characteristics of four bacteria, *Clostridium novyi*, *Thermus aquaticus*, *Paracoccus denitrificans*, and *Trichodesmium thiebautii*. Use this information to address the questions that follow.

	C. novyi	T. aquaticus	P. denitrificans	T. thiebautii
Metabolism	Obligate anaerobic	Obligate aerobic	Facultative anaerobic	Facultative anaerobic
Gram status	Positive	Negative	Negative	Negative
Optimal temperature (°C)	10°–40°	50°–80°	5°–30°	10°–30°
Typical habitat	Terrestrial	Aquatic	Aquatic	Aquatic
Miscellaneous		Chemotroph	Denitrifier	Nitrogen fixer

Questions

1. The concentration of nitrates in a sample of municipal sewage water is well above acceptable levels. Which of these bacteria would be expected to be most appropriate for reducing nitrate concentrations? Explain your answer.

2. Because tumors often grow much faster than they can acquire blood supply, cancerous tumors can have large regions where oxygen concentrations are very low. These hypoxic conditions are generally not found elsewhere in the body. Biomedical researchers can take advantage of this unusual property to inject bacteria that would specifically target the tumor cells and have as little effect on the rest of the body as possible. Based on the information in the table, which bacterium would be expected to be most appropriate for this application? Explain your answer.

3. Adding small amounts of ammonium increased the productivity of an aquatic ecosystem, but the effect was short-lived. Based on that finding, which bacterium would most likely improve the productivity of the ecosystem?

4. The antibiotic vancomycin inhibits peptidoglycan synthesis in bacteria that have a thick cell wall with much peptidoglycan. Which bacterium would most likely be sensitive to this antibiotic? Explain your answer.

> Go to **LaunchPad** for the eBook, LearningCurve, animations, activities, flashcards, and additional resources and assignments.

26

key concepts

26.1 Eukaryotes Acquired Features from Both Archaea and Bacteria

26.2 Major Lineages of Eukaryotes Diversified in the Precambrian

26.3 Protists Reproduce Sexually and Asexually

26.4 Protists Are Critical Components of Many Ecosystems

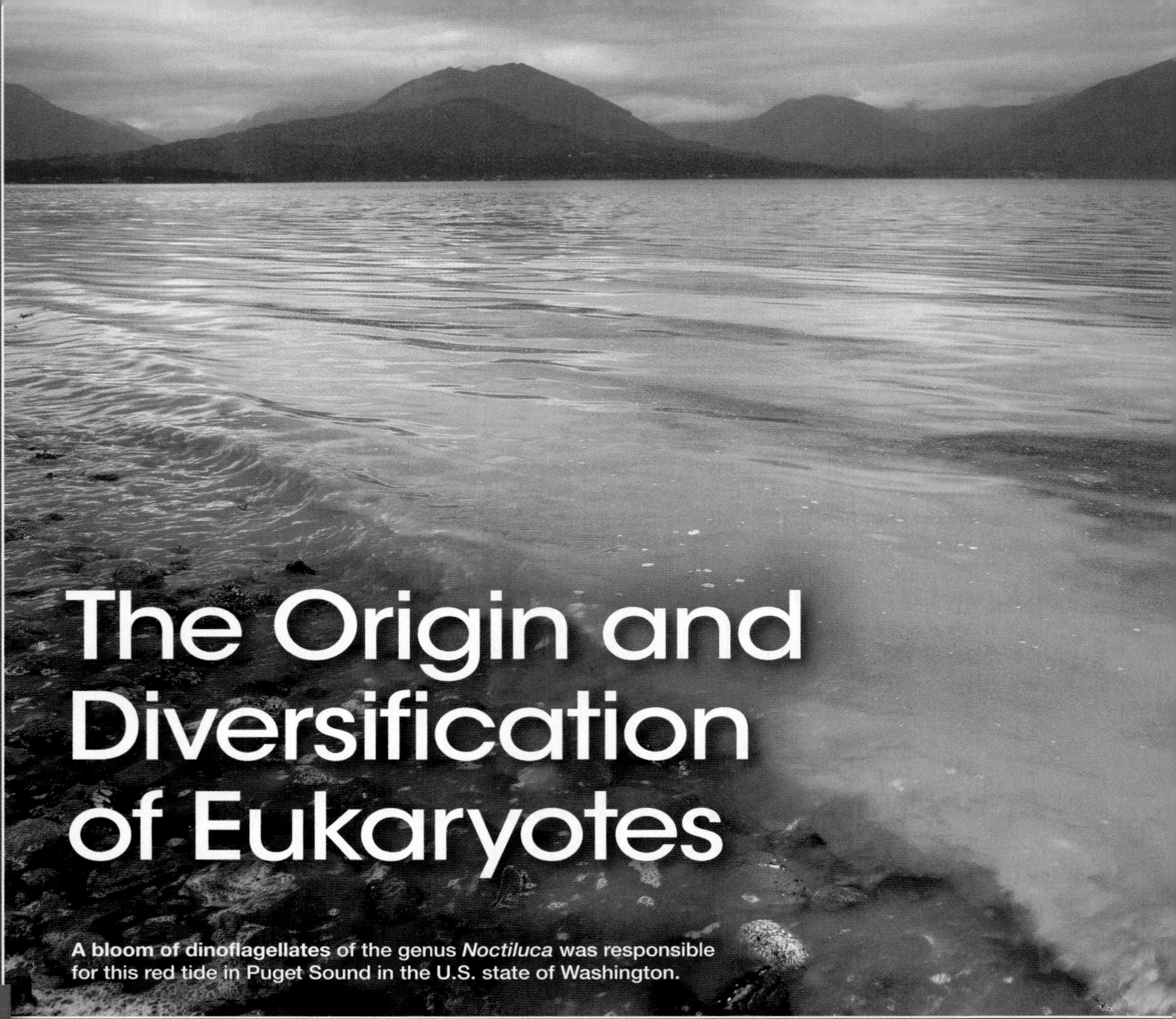

The Origin and Diversification of Eukaryotes

A bloom of dinoflagellates of the genus *Noctiluca* was responsible for this red tide in Puget Sound in the U.S. state of Washington.

investigatinglife

Predicting Toxic Red Tides

In summer 2005, a devastating red tide crippled the shellfish industry along the Atlantic coast of North America from Canada to Massachusetts. This red tide was produced by a bloom of dinoflagellates of the genus *Alexandrium*. These protists produce a powerful toxin that accumulates in clams, mussels, and oysters. A person who eats a mollusk contaminated with the toxin can experience a syndrome known as paralytic shellfish poisoning. Many people were sickened by eating mollusks that were harvested before the problem was diagnosed, and losses to the shellfish industry in 2005 were estimated at $50 million.

Several species of dinoflagellates produce toxic red tides in many parts of the world. Along the Gulf of Mexico, red tides caused by dinoflagellates of the genus *Karenia* produce a neurotoxin that affects the central nervous systems of fish, which become paralyzed and cannot respire effectively. Huge numbers of dead fish wash up on Gulf Coast beaches during a *Karenia* red tide. In addition, wave action can produce aerosols of the *Karenia* toxin, and these aerosols often cause asthma-like symptoms in humans on shore.

After the losses that resulted from the 2005 red tide, biologists at the Woods Hole Oceanographic Institution (WHOI) on Cape Cod began to monitor and model dinoflagellate populations off the New England coast. If biologists could accurately forecast future blooms, people in the area could be made aware of the problem in advance and adjust the shellfish harvest (and their eating habits) accordingly.

Biologists from WHOI monitored counts of dinoflagellates in the water and in seafloor sediments. They also monitored river runoff, water currents, water temperature and salinity, winds, and tides. An additional environmental factor was the "nor'easter" storms common along the New England coast. By correlating their measurements of these environmental factors with dinoflagellate counts, biologists produced a model that predicted growth of dinoflagellate populations.

In spring 2008, the WHOI team determined that all the factors were in place to produce another red tide like the one of 2005—if a nor'easter occurred to blow the dinoflagellates toward the coast. A nor'easter did occur at just the wrong time, and another red tide materialized in summer 2008, just as predicted. But this time, people were warned. Shellfish harvesters adjusted their harvest, and many fewer people were harmed by eating toxic mollusks.

Q Red tides are harmful, but how are

26.1 Eukaryotes Acquired Features from Both Archaea and Bacteria

We easily recognize trees, mushrooms, and insects as plants, fungi, and animals, respectively. But there is a dazzling assortment of other eukaryotic organisms—mostly microscopic—that do not fit into these three groups. Eukaryotes that are not plants, animals, or fungi have traditionally been called **protists**. But phylogenetic analyses reveal that many of the groups we commonly refer to as protists are not, in fact, closely related. Thus the term "protist" does not describe a monophyletic group, but is a convenience term for "all the eukaryotes that are not plants, animals, or fungi."

focus your learning

- "Protists" is a convenient term for all eukaryotes that are not plants, animals, or fungi, but protists are not a monophyletic group.
- An important step toward the modern eukaryotic cell was the loss of the cell wall.
- Endosymbiosis was a key step in the evolution of modern eukaryotes.
- All chloroplasts can be traced back to the same primary endosymbiosis event.

The unique characteristics of the eukaryotic cell lead scientists to conclude that the eukaryotes are monophyletic, and that a single eukaryotic ancestor diversified into the many different protist lineages as well as giving rise to the plants, fungi, and animals. As we saw in Key Concept 25.1, eukaryotes are a specialized group of archaea that acquired a cell nucleus. The mitochondria and chloroplasts of eukaryotes, however, are clearly derived from bacterial lineages (see Figure 25.1).

Biologists have hypothesized that the origin of eukaryotes from a specialized archaean ancestor was followed by the endosymbioses with bacterial lineages that led to the origin of mitochondria and chloroplasts. Some biologists prefer to view the origin of eukaryotes as the fusion of lineages from the two prokaryote groups. This difference is largely a semantic one that hinges on the subjective point at which we deem the eukaryote lineage to have become definitively "eukaryotic." In either case, we can make some reasonable inferences about the events that led to the evolution of a new cell type, bearing in mind that the environment underwent an enormous change—from low to high availability of free atmospheric oxygen—during the course of these events.

The modern eukaryotic cell arose in several steps

Several events were important in the origin of the modern eukaryotic cell (**Focus: Key Figure 26.1**):

- The origin of a flexible cell surface
- The origin of a cytoskeleton
- The origin of a nuclear envelope, which enclosed a genome organized into chromosomes
- The appearance of digestive vacuoles

- The acquisition of mitochondria and chloroplasts via endosymbiosis

FLEXIBLE CELL SURFACE   We presume that ancient prokaryotic organisms, like most present-day prokaryotic cells, had firm cell walls. The first step toward the eukaryotic condition was the loss of the cell wall by a prokaryotic archaean. This wall-less condition occurs in some present-day prokaryotes.

Consider the possibilities open to a flexible cell without a firm wall, starting with cell size. As a cell grows larger, its surface area-to-volume ratio decreases (see Figure 5.2). Unless the surface area can be increased, the cell volume will reach an upper limit. If the cell's surface is flexible, however, it can fold inward and become more elaborate, creating more surface area for gas and nutrient exchange. With a surface flexible enough to allow infolding, the cell can exchange materials with its environment rapidly enough to sustain a larger volume and more rapid metabolism (see Figure 26.1, steps 1–2). Furthermore, a flexible surface can pinch off bits of the environment, bringing them into the cell by endocytosis. These infoldings of the cell surface, which also exist in some modern prokaryotes, were important for the evolution of large eukaryotic cells.

CHANGES IN CELL STRUCTURE AND FUNCTION   Other early steps that were important for the evolution of the eukaryotic cell involved increased compartmentalization and complexity of the cell (see Figure 26.1, steps 3–7):

- The development of a more complex cytoskeleton
- The formation of ribosome-studded internal membranes, some of which surrounded the DNA
- The enclosure of the cell's DNA in a nucleus
- The formation of a flagellum from microtubules of the cytoskeleton
- The evolution of digestive vacuoles

Until a few years ago, biologists thought that cytoskeletons were restricted to eukaryotes. Improved imaging technology and molecular analyses have now revealed homologs of many cytoskeletal proteins in prokaryotes, so simple cytoskeletons evolved before the origin of eukaryotes. The cytoskeleton of a eukaryote, however, is much more developed and complex than that of a prokaryote. This greater development of microfilaments and microtubules supports the eukaryotic cell and allows it to manage changes in shape, to distribute daughter chromosomes, and to move materials from one part of its larger cell to other parts. In addition, the presence of microtubules in the cytoskeleton allowed some cells to develop the characteristic eukaryotic flagellum.

The DNA of a prokaryotic cell is attached to a site on its cell membrane. If that region of the cell membrane were to fold into the cell, the first step would be taken toward the evolution of a nucleus, a primary feature of the eukaryotic cell. The nuclear envelope appeared early in the eukaryote lineage. The next step was probably phagocytosis—the ability to engulf and digest other cells.

ENDOSYMBIOSIS   At the same time the processes outlined above were taking place, cyanobacteria were generating O_2 as a product

focus: key figure

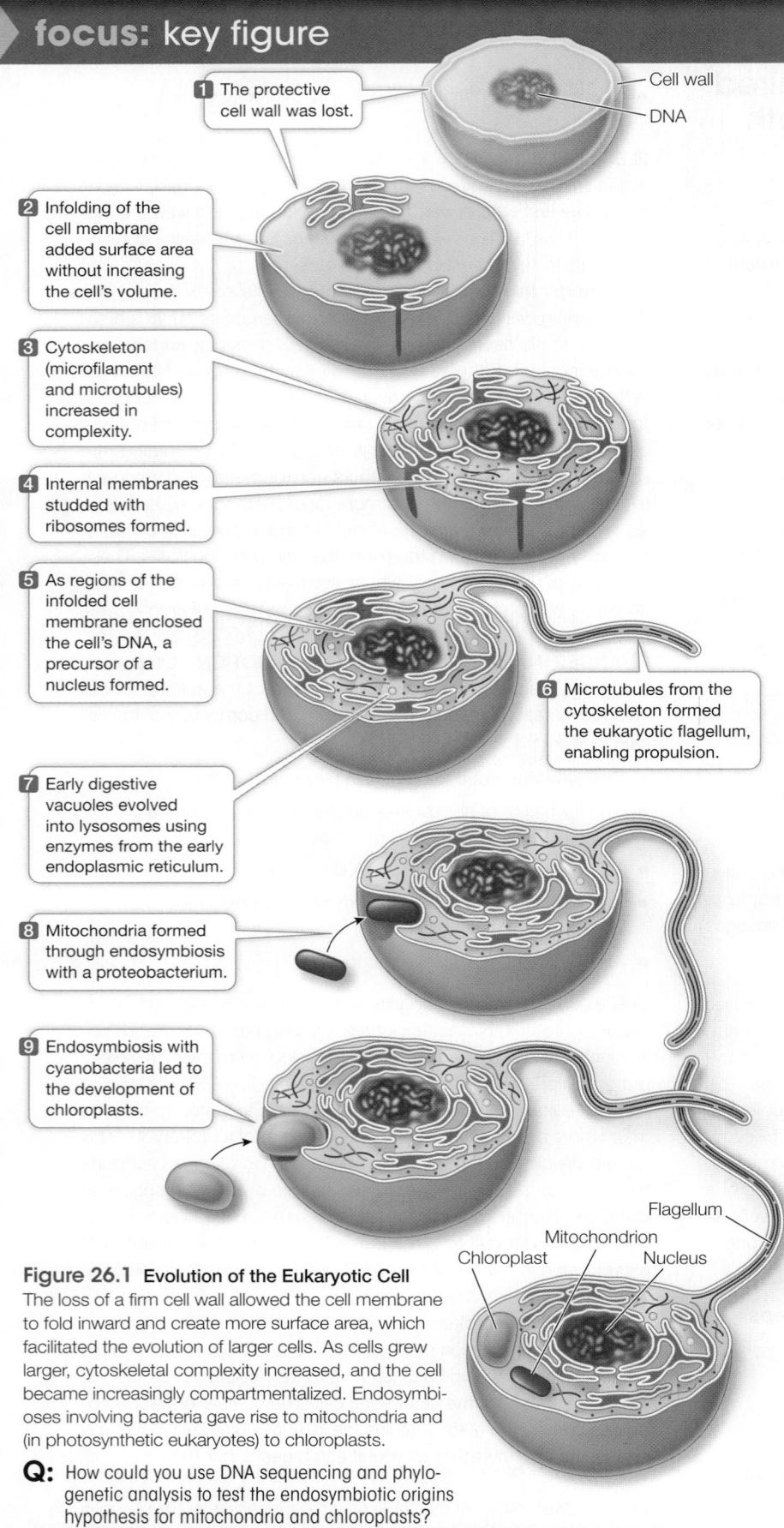

1 The protective cell wall was lost.

Cell wall
DNA

2 Infolding of the cell membrane added surface area without increasing the cell's volume.

3 Cytoskeleton (microfilament and microtubules) increased in complexity.

4 Internal membranes studded with ribosomes formed.

5 As regions of the infolded cell membrane enclosed the cell's DNA, a precursor of a nucleus formed.

6 Microtubules from the cytoskeleton formed the eukaryotic flagellum, enabling propulsion.

7 Early digestive vacuoles evolved into lysosomes using enzymes from the early endoplasmic reticulum.

8 Mitochondria formed through endosymbiosis with a proteobacterium.

9 Endosymbiosis with cyanobacteria led to the development of chloroplasts.

Flagellum
Mitochondrion
Chloroplast
Nucleus

Figure 26.1 Evolution of the Eukaryotic Cell
The loss of a firm cell wall allowed the cell membrane to fold inward and create more surface area, which facilitated the evolution of larger cells. As cells grew larger, cytoskeletal complexity increased, and the cell became increasingly compartmentalized. Endosymbioses involving bacteria gave rise to mitochondria and (in photosynthetic eukaryotes) to chloroplasts.

Q: How could you use DNA sequencing and phylogenetic analysis to test the endosymbiotic origins hypothesis for mitochondria and chloroplasts?

of photosynthesis. The increasing concentrations of O_2 in the oceans, and eventually in the atmosphere, had disastrous consequences for most organisms of the time, which were unable to tolerate the newly oxidizing environment. But some prokaryotes evolved strategies to use the increasing O_2, and—fortunately for us—so did some of the new phagocytic eukaryotes.

At about this time, endosymbioses began to play a role in eukaryote evolution (see Figure 26.1, steps 8–9). The theory of endosymbiosis proposes that certain organelles are the descendants of prokaryotes engulfed, but not digested, by ancient eukaryotic cells. One crucial event in the history of eukaryotes was the incorporation of a proteobacterium that evolved into the mitochondrion. Initially the new organelle's primary function was probably to detoxify O_2 by reducing it to water. Later this reduction became coupled with the formation of ATP in *__cellular respiration__*. After this step, the essential eukaryotic cell was complete.

__connect the concepts__ You may wish to review the reactions of cellular respiration in Key Concept 5.3.

Photosynthetic eukaryotes are the result of yet another endosymbiotic step: the incorporation of a prokaryote related to today's cyanobacteria, which became the chloroplast.

Chloroplasts have been transferred among eukaryotes several times

Eukaryotes in several different groups possess chloroplasts, and groups with chloroplasts appear in several distantly related eukaryote clades. Some of these groups differ in the photosynthetic pigments their chloroplasts contain. And not all chloroplasts are limited to a pair of surrounding membranes—in some microbial eukaryotes, chloroplasts are surrounded by three or more membranes. We now view these observations as evidence of a remarkable series of endosymbioses. This conclusion is supported by extensive evidence from electron microscopy and nucleic acid sequence comparisons.

All chloroplasts trace their ancestry back to the engulfment of one cyanobacterium by a larger eukaryotic cell. This event, the step that first gave rise to the photosynthetic eukaryotes, is known as **primary endosymbiosis** (**Figure 26.2A**). The cyanobacterium, a Gram-negative bacterium, had both an inner and an outer membrane (see Figure 25.2B). Thus the original chloroplasts had two surrounding membranes: the inner and outer

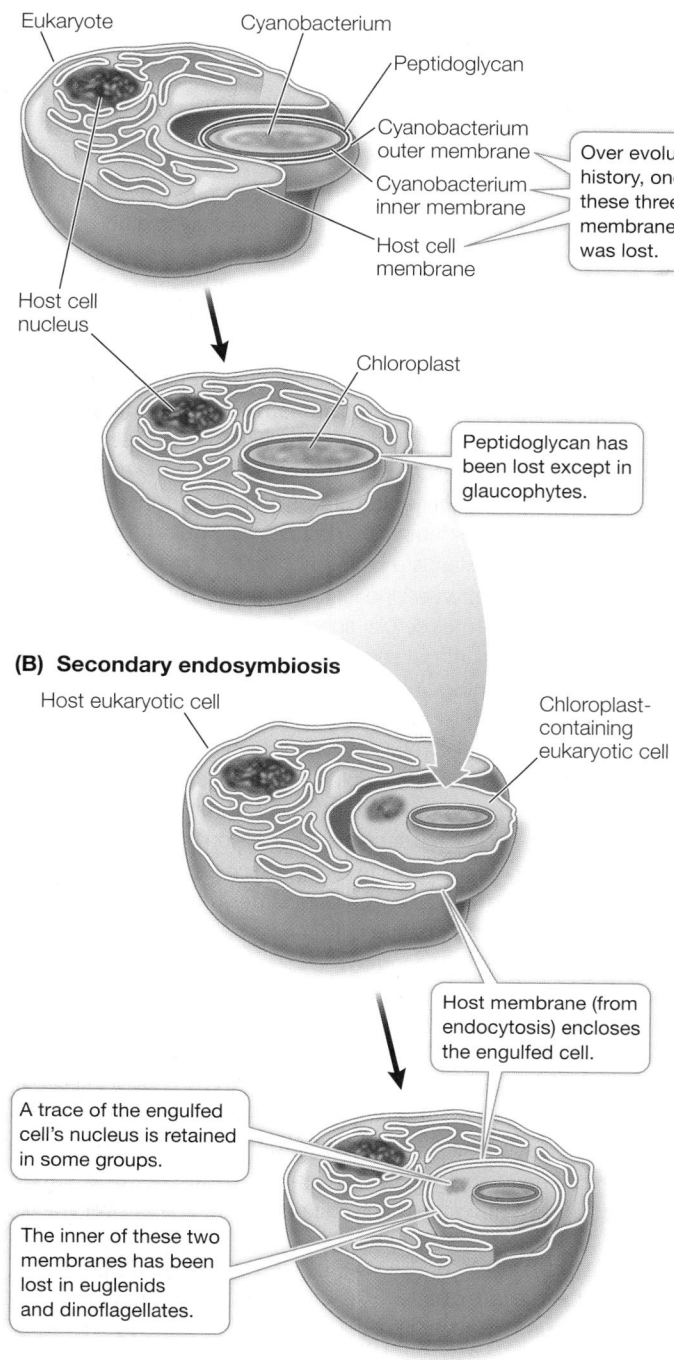

(A) Primary endosymbiosis

Eukaryote

Cyanobacterium

Peptidoglycan

Cyanobacterium outer membrane

Cyanobacterium inner membrane

Host cell membrane

Over evolutionary history, one of these three membranes was lost.

Host cell nucleus

Chloroplast

Peptidoglycan has been lost except in glaucophytes.

(B) Secondary endosymbiosis

Host eukaryotic cell

Chloroplast-containing eukaryotic cell

Host membrane (from endocytosis) encloses the engulfed cell.

A trace of the engulfed cell's nucleus is retained in some groups.

The inner of these two membranes has been lost in euglenids and dinoflagellates.

Figure 26.2 Endosymbiotic Events in the Evolution of Chloroplasts (A) A single instance of primary endosymbiosis ultimately gave rise to all of today's chloroplasts. (B) Secondary endosymbiosis—the uptake and retention of a chloroplast-containing cell by another eukaryotic cell—took place several times, independently.

 Animation 26.1 **Family Tree of Chloroplasts** www.Life11e.com/a26.1

membranes of the cyanobacterium. Remnants of the peptidoglycan-containing cell wall of the bacterium are present in the form of a bit of peptidoglycan between the chloroplast membranes of

glaucophytes, the first eukaryote group to branch off following primary endosymbiosis (as we will see in Chapter 28). Primary endosymbiosis also gave rise to the chloroplasts of the red algae, green algae, and land plants. The red algal chloroplast retains certain pigments of the original cyanobacterial endosymbiont that are absent in green algal chloroplasts.

Almost all remaining photosynthetic eukaryotes are the result of additional rounds of endosymbiosis. For example, the photosynthetic euglenids derived their chloroplasts from **secondary endosymbiosis (Figure 26.2B)**. Their ancestor took up a unicellular green alga, retaining its chloroplast and eventually losing the rest of the constituents of the alga. This history explains why the photosynthetic euglenids have the same photosynthetic pigments as the green algae and land plants. It also accounts for the third membrane of the euglenid chloroplast, which is derived from the euglenid's cell membrane (as a result of endocytosis). An additional round—**tertiary endosymbiosis**—occurred when a dinoflagellate apparently lost its chloroplast and took up another protist that had acquired its chloroplast through secondary endosymbiosis.

26.1 recap

The modern eukaryotic cell probably arose from an ancestral prokaryotic archaean in several steps, including the origin of a flexible cell surface and the enclosure of the genetic material in a nucleus. Later, endosymbioses of proteobacteria and cyanobacteria led to the origins of mitochondria and chloroplasts.

learning outcomes

You should be able to:

- Explain the use of the term "protist," and describe the relationships of major protist groups to other eukaryotes.
- Summarize the major changes that were important in the evolution of the eukaryotic cell.
- Explain how major changes in Earth's environment affected the evolution of the eukaryotic cell.
- Infer the history of genes based on whether their origins are nuclear, mitochondrial, or chloroplastic.

1. Why was the development of a flexible cell surface a key event for eukaryote evolution?

2. Explain how increased availability of atmospheric oxygen (O_2) could have influenced the evolution of the eukaryotic cell.

Background for Questions 3–4: Ribosomal RNA (rRNA) genes are present in the nuclear genome of eukaryotes. There are also rRNA genes in the genomes of mitochondria and chloroplasts. Therefore photosynthetic eukaryotes have three different sets of rRNA genes, which encode the structural RNA of three separate sets of ribosomes. The gene tree on the following page shows the evolutionary relationships among rRNA gene sequences isolated from the nuclear genomes of humans, yeast, and corn; from a prokaryotic archaean (*Halobacterium*), a proteobacterium (*E. coli*), and a cyanobacterium (*Chlorobium*); and from the mitochondrial and chloroplast genomes of corn.

(continued)

26.1 recap (continued)

Halobacterium rRNA

Human nuclear rRNA

Yeast nuclear rRNA

Corn nuclear rRNA

E. coli rRNA

Corn mitochondrial rRNA

Chlorobium rRNA

Corn chloroplast rRNA

3. Why are the three rRNA genes of corn *not* one another's closest relatives?

4. Why is the relationship of the mitochondrial rRNA gene of corn to the rRNA gene of *E. coli* closer than it is to the nuclear rRNA genes of other eukaryotes? Can you explain the relationship of the rRNA gene from the chloroplast of corn to the rRNA gene of the cyanobacterium?

5. Explain why the term "protists" does not refer to a formal taxonomic group.

The features that eukaryotes gained from prokaryotic archaea and bacteria have allowed them to exploit many different environments. This led to the evolution of great diversity among eukaryotes, beginning with a radiation that started in the Precambrian.

key concept

26.2 Major Lineages of Eukaryotes Diversified in the Precambrian

Most eukaryotes can be placed in one of eight major clades that began to diversify about 1.5 billion years ago: alveolates, stramenopiles, rhizarians, excavates, plants, amoebozoans, fungi, and animals (**Figure 26.3**). Plants, fungi, and animals each have close protist relatives (such as the choanoflagellate relatives of animals), which we will discuss along with those major multicellular eukaryote groups in Chapters 27–32.

focus your learning

- Multicellularity evolved dozens of times in the evolutionary history of eukaryotes.
- Dinoflagellates are important primary producers of aquatic organic matter.
- Apicomplexans possess an apical complex.
- Ciliates are named for their numerous hair-like cilia.
- Stramenopiles include the diatoms, the brown algae, and the oomycetes.
- Brown algae are multicellular stramenopiles that have an abundance of fucoxanthin in their chloroplasts.

- Rhizarians include the cercozoans, foraminiferans, and radiolarians.
- A single amoeboid cell is the vegetative unit of the cellular slime mold.
- Many protists are medically or economically important.

Each of the five major groups of protist eukaryotes covered in this chapter consists of organisms with enormously diverse body forms and nutritional lifestyles. Some protists are motile, whereas others do not move. Some protists are photosynthetic, whereas others are heterotrophic. Most protists are unicellular, but some are multicellular. Most protists are microscopic, but a few are huge (giant kelps, for example, can grow to half the length of a football field). We refer to the unicellular species of protists as **microbial eukaryotes**, but keep in mind that there are large, multicellular protists as well.

Multicellularity has arisen dozens of times across the evolutionary history of eukaryotes. Four of the origins of multicellularity resulted in large organisms that are familiar to most people: plants, animals, fungi, and brown algae (the last are a group of stramenopiles). In addition, there are dozens of smaller and less familiar groups among the eukaryotes that include multicellular species. Recent experimental studies have shown that artificial selection for multicellularity can produce repeated, convergent evolution of multicellular forms over just a few months in some normally unicellular eukaryotic species. In addition, many unicellular species retain individual identities but nonetheless associate in large multicellular colonies. There is a near-continuum between fully integrated, multicellular organisms on the one hand and loosely integrated multicellular colonies of cells on the other. Biologists do not always agree on where to draw the line between the two.

Biologists used to classify protists largely on the basis of their life histories and reproductive features. In recent years, however, electron microscopy and gene sequencing have revealed many new patterns of evolutionary relatedness among these groups. Analyses of slowly evolving gene sequences are making it possible to explore evolutionary relationships among eukaryotes in ever greater detail and with greater confidence. Nonetheless, some substantial areas of uncertainty remain, and lateral gene transfer may complicate efforts to reconstruct the evolutionary history of protists (as was also true for prokaryotes; see Key Concept 25.1). Today we recognize great diversity among the many distantly related protist clades.

Alveolates have sacs under their cell membranes

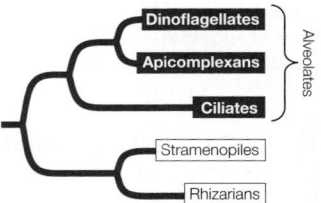

Alveolates are so named because they possess sacs, called alveoli, just beneath their cell membranes, which may play a role in supporting the cell surface. All alveolates are unicellular, and most are photosynthetic, but they are diverse in body form. The alveolate groups we will consider in detail here are the dinoflagellates, apicomplexans, and ciliates.

DINOFLAGELLATES   Most **dinoflagellates** are marine and photosynthetic; they are important primary producers of organic matter in the oceans. Although fewer *species* of dinoflagellates live in fresh

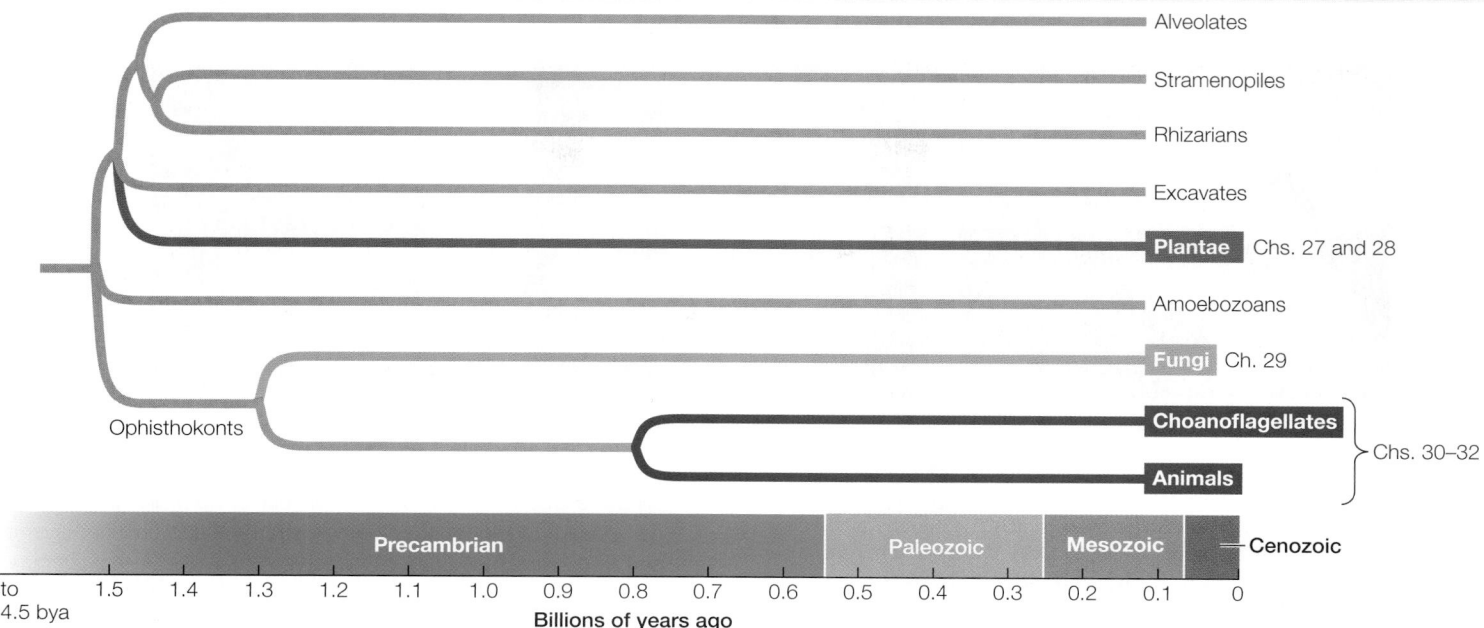

Figure 26.3 Precambrian Divergence of Major Eukaryote Groups A phylogenetic tree shows a current hypothesis and estimated time line for the origin of the major groups of eukaryotes. The rapid divergence of major lineages between 1.5 and 1.4 billion years ago makes reconstruction of their precise relationships difficult. The major multicellular groups (tinted boxes) will be covered in subsequent chapters.

water, individuals can be abundant in freshwater environments. The dinoflagellates are of great ecological, evolutionary, and morphological interest. A distinctive mixture of photosynthetic and accessory pigments gives their chloroplasts a golden brown color. Some dinoflagellate species cause red tides, as discussed at the start of this chapter. Other species are photosynthetic endosymbionts that live within the cells of other organisms, including invertebrate animals (such as corals; see Investigating Life: Can Corals Reacquire Dinoflagellate Endosymbionts Lost to Bleaching?) and other marine protists (see Figure 26.12A). Still others are nonphotosynthetic and live as parasites within other marine organisms.

Dinoflagellates have a distinctive appearance. They generally have two flagella, one in an equatorial groove around the cell, the other starting near the same point as the first and passing down a longitudinal groove before extending into the surrounding medium (**Figure 26.4**). Some dinoflagellates can take on different forms, including amoeboid ones, depending on environmental conditions. It has been claimed that the dinoflagellate *Pfiesteria piscicida* can occur in at least two dozen distinct forms, although this claim is highly controversial. In any case, this remarkable dinoflagellate, when present in large enough numbers, is harmful to fish and can both stun and feed on them.

APICOMPLEXANS The exclusively parasitic **apicomplexans** derive their name from the apical complex, a mass of organelles contained in the apical end (the tip) of the cell. These organelles help the apicomplexan invade its host's tissues. For example, the apical complex enables *Plasmodium*, the causative agent of malaria, to enter its target cells in the human body after transmission by a mosquito.

Like many obligate parasites, apicomplexans have elaborate life cycles featuring asexual and sexual reproduction through a series of very dissimilar life stages (see Figure 26.20). In many species, these life stages are associated with two different types of host organisms, as is the case with *Plasmodium*. Another apicomplexan, *Toxoplasma*, alternates between cats and rats to complete its life cycle. A rat infected with *Toxoplasma* loses its fear of cats, which makes it more likely to be eaten by, and thus transfer the parasite to, a cat.

Amphidiniopsis kofoidii

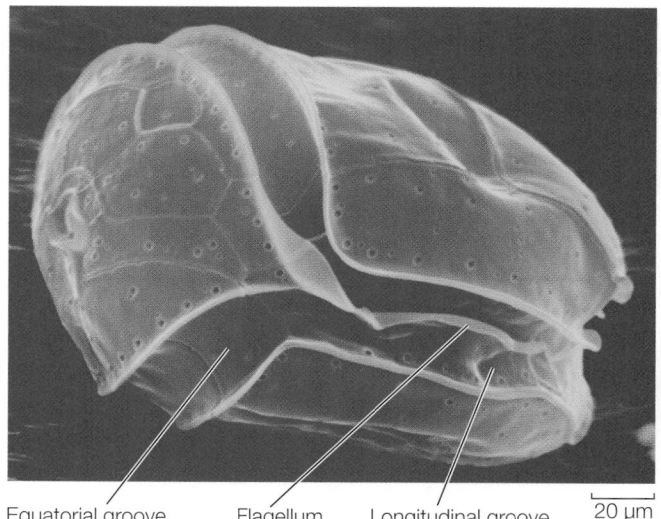

Equatorial groove   Flagellum   Longitudinal groove   20 µm

Figure 26.4 A Dinoflagellate The presence of two flagella is characteristic of many dinoflagellates, although these appendages are contained within deep grooves and thus are seldom visible. One flagellum lies within the equatorial groove and provides forward thrust and spin to the organism. The second flagellum originates in the longitudinal groove and acts like the rudder of a boat.

 Media Clip 26.1 **A Dinoflagellate Shows Off Its Flagellum** www.Life11e.com/mc26.1

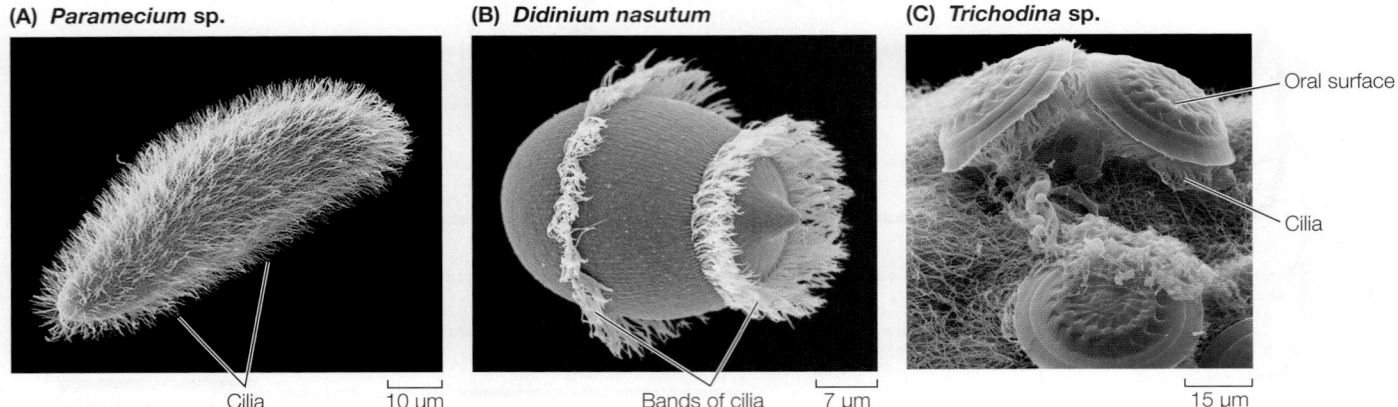

(A) *Paramecium* sp.

(B) *Didinium nasutum*

(C) *Trichodina* sp.

Cilia   10 µm

Bands of cilia   7 µm

Oral surface

Cilia

15 µm

Figure 26.5 Diversity among the Ciliates **(A)** A free-swimming organism, this *Paramecium* belongs to a ciliate group whose members have many cilia of uniform length. **(B)** The barrel-shaped *Didinium nasutum* feeds on other ciliates, including *Paramecium*. Its cilia occur in two separate bands. **(C)** The parasite *Trichodina* creates suction with its ring of cilia that help attach it to the surface of a fish host.

CILIATES   The **ciliates** are named for their numerous hairlike cilia, which are shorter than, but otherwise identical to, eukaryotic flagella. The ciliates are much more complex in body form than are most other unicellular eukaryotes (**Figure 26.5**). Their definitive characteristic is the possession of two types of nuclei (whose roles we will describe in Key Concept 26.3 when we discuss protist reproduction). Almost all ciliates are heterotrophic, although a few contain photosynthetic endosymbionts.

Paramecium, a frequently studied ciliate genus, exemplifies the complex structure and behavior of ciliates (**Figure 26.6**). The slipper-shaped cell is covered by an elaborate pellicle, a structure composed principally of an outer membrane and an inner layer of closely packed, membrane-enclosed sacs (the alveoli) that surround the bases of the cilia. Defensive organelles called trichocysts are also present in the pellicle. In response to a threat, a microscopic explosion expels the trichocysts in a few milliseconds, and they emerge as sharp darts, driven forward at the tip of a long, expanding filament.

The cilia provide *Paramecium* with a form of locomotion that is generally more precise than locomotion by flagella. A *Paramecium* can coordinate the beating of its cilia to propel itself either forward or backward in a spiraling manner. It can also back off swiftly when it encounters a barrier or a negative stimulus. The coordination of ciliary beating is probably the result of a differential distribution of ion channels in the cell membrane near the two ends of the cell.

Organisms living in fresh water are hypertonic to their environment. Many freshwater protists, including *Paramecium*, address this problem by means of specialized **contractile vacuoles** that excrete the excess water the organisms constantly take in by osmosis. The excess water collects in the contractile vacuoles, which then contract and expel the water from the cell.

Paramecium and many other protists engulf solid food by endocytosis, forming a **digestive vacuole** within which the food is digested (**Figure 26.7**). Smaller vesicles containing digested food pinch away from the digestive vacuole and enter the cytoplasm. These tiny vesicles provide a large surface area across which the products of digestion can be absorbed by the rest of the cell.

 Animation 26.2 Digestive Vacuoles
www.Life11e.com/a26.2

Micronuclei function in genetic recombination.

The macronucleus controls the cell's activities.

Contractile vacuole

Alveoli

Cilia

Digestive vacuole

Oral groove

Anal pore

Pellicle

Trichocyst

Fibrils

Alveolus

Cilium

Figure 26.6 Anatomy of *Paramecium* *Paramecium*, with its many specialized organelles, exemplifies the complex body form of ciliates.

 **Activity 26.1 Anatomy of *Paramecium***
www.Life11e.com/ac26.1

Stramenopiles typically have two unequal flagella, one with hairs

A morphological synapomorphy of most **stramenopiles** is the possession of rows of tubular hairs on the longer of their two flagella. Some stramenopiles lack flagella, but they are descended from ancestors that possessed flagella. The stramenopiles include the diatoms and the brown algae, which are photosynthetic, and the oomycetes, which are not.

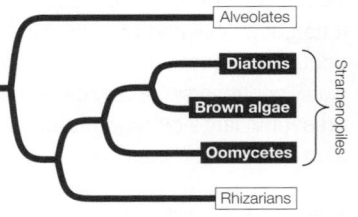

Alveolates

Diatoms

Brown algae

Oomycetes

Rhizarians

Stramenopiles

experiment

Figure 26.7 The Role of Vacuoles in Ciliate Digestion

Original Paper: Mast, S. O. 1947. The food-vacuole in *Paramecium*. *Biological Bulletin* 92: 31–72.

An acidic environment is known to aid digestion in many multicellular organisms. Do ciliates also use acid to obtain nutrients?

HYPOTHESIS▶ The digestive vacuoles of *Paramecium* produce an acidic environment that allows the organism to digest food particles.

METHOD

1. Feed *Paramecium* yeast cells stained with Congo red, a dye that is red at neutral or basic pH but turns green at acidic pH.

2. Under a light microscope, observe the formation and degradation of digestive vacuoles within the *Paramecium*. Note time and sequence of color (i.e., acid level) changes.

RESULTS

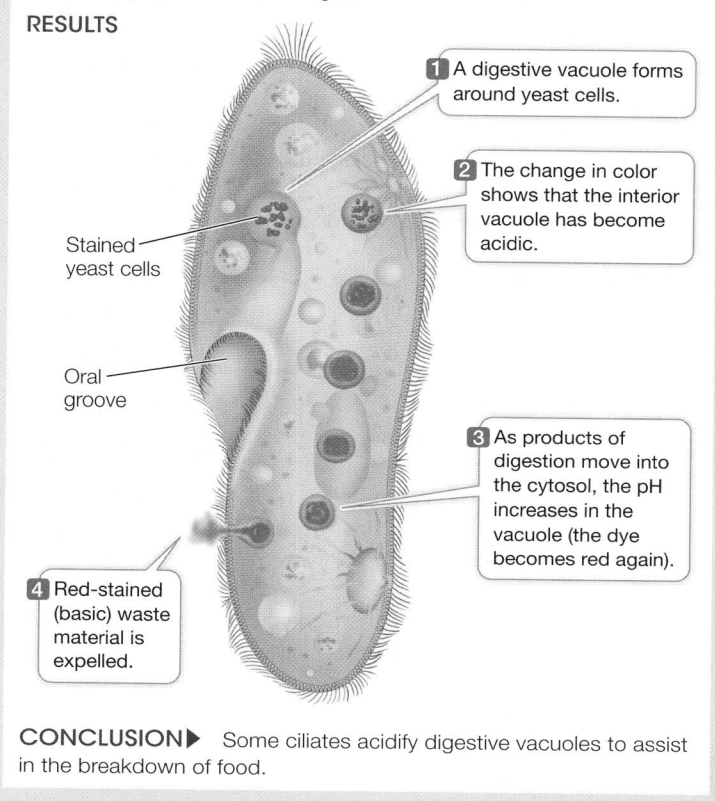

1 A digestive vacuole forms around yeast cells.

2 The change in color shows that the interior vacuole has become acidic.

Stained yeast cells

Oral groove

3 As products of digestion move into the cytosol, the pH increases in the vacuole (the dye becomes red again).

4 Red-stained (basic) waste material is expelled.

CONCLUSION▶ Some ciliates acidify digestive vacuoles to assist in the breakdown of food.

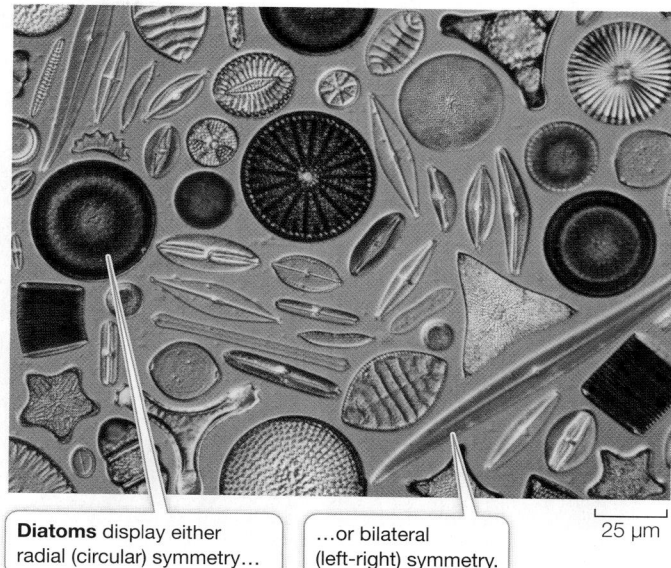

Diatoms display either radial (circular) symmetry…

…or bilateral (left-right) symmetry.

25 μm

Figure 26.8 Diatom Diversity This bright-field micrograph illustrates the variety of species-specific forms found among the diatoms.

▶ **Media Clip 26.2 Diatoms in Action**
www.Life11e.com/mc26.2

Diatoms reproduce both sexually and asexually. Asexual reproduction by binary fission is somewhat constrained by the stiff cell wall. Both the top and bottom of the "petri dish" become tops of new "dishes" without changing appreciably in size. As a result, the new cell made from the former bottom is smaller than the parent cell. If this process continued indefinitely, one cell line would simply vanish, but sexual reproduction largely solves this potential problem. Gametes are formed, shed their cell walls, and fuse. The resulting zygote then grows substantially in size before a new cell wall is laid down.

Diatoms are found in all the oceans and are frequently present in great numbers. They are major photosynthetic producers in coastal waters (see Key Concept 26.4) and are among the dominant organisms in the dense "blooms" of phytoplankton that occasionally appear in the open ocean. Diatoms are also common in fresh water and even occur on the wet surfaces of terrestrial mosses.

DIATOMS All of the **diatoms** are unicellular, although some species associate in filaments. Many have sufficient carotenoids in their chloroplasts to give them a yellow or brownish color. All of them synthesize carbohydrates and oils as photosynthetic storage products. Diatoms lack flagella except in male gametes.

Architectural magnificence on a microscopic scale is the hallmark of the diatoms. Almost all diatoms deposit silica (hydrated silicon dioxide) in their cell walls. The cell wall of a diatom is constructed in two pieces, with the top overlapping the bottom like the top of a petri dish. The silica-impregnated walls have intricate patterns unique to each species (**Figure 26.8**). Despite their remarkable morphological diversity, all diatoms are symmetrical—either bilaterally (with "right" and "left" halves) or radially (with the type of symmetry possessed by a circle).

BROWN ALGAE The **brown algae** obtain their namesake color from the carotenoid fucoxanthin, which is abundant in their chloroplasts. The combination of this yellow-orange pigment with the green of chlorophylls *a* and *c* yields a brownish tinge. All brown algae are multicellular, and some are extremely large. Giant kelps such as those of the genus *Macrocystis* may be up to 60 m long.

▶ **Media Clip 26.3 A Kelp Forest**
www.Life11e.com/mc26.3

The brown algae are almost exclusively marine. They are composed either of branched filaments (**Figure 26.9A**) or of leaflike growths (**Figure 26.9B**). Some float in the open ocean; the most famous example is the genus *Sargassum*, which forms dense

(A) Filamentous growth pattern   **(B) Leaf-like growth pattern**

Himanthalia elongata   *Postelsia paliformis*   Holdfasts

Figure 26.9 Brown Algae **(A)** This seaweed illustrates the filamentous growth form of the brown algae. **(B)** Sea palms exemplify the leaflike growth form of brown algae. Sea palms and many other brown algal species are "glued" to the rocks by tough, branched structures called holdfasts that can withstand the pounding of the surf.

mats in the Sargasso Sea in the mid-Atlantic. Most brown algae, however, attach themselves to rocks near the shore. A few thrive only where they are regularly exposed to heavy surf. All of the attached forms develop a specialized structure, called a holdfast, that literally glues them to the rocks. The "glue" of the holdfast is alginic acid, a gummy polymer found in the walls of many brown algal cells. In addition to its function in holdfasts, alginic acid

Saprolegnia sp.

3 mm

Figure 26.10 An Oomycete The filaments of a water mold radiate from the carcass of a beetle.

cements algal cells and filaments together. It is harvested and used by humans as an emulsifier in ice cream, cosmetics, and other products.

OOMYCETES The **oomycetes** are the water molds and their terrestrial relatives. Water molds are filamentous and stationary. They are **absorptive heterotrophs**—that is, they secrete enzymes that digest large food molecules into smaller molecules that they can absorb. They are all aquatic and **saprobic**—meaning they feed on dead organic matter. If you have seen a whitish, cottony mold growing on dead fish or dead insects in water, it was probably a water mold of the common genus *Saprolegnia* (**Figure 26.10**).

Some other oomycetes, such as the downy mildews, are terrestrial. Although most of the terrestrial oomycetes are harmless or helpful decomposers of dead matter, a few are plant parasites that attack crops such as avocados, grapes, and potatoes.

Oomycetes were once classified as fungi. However, we now know that their similarity to fungi is only superficial, and that the oomycetes are more distantly related to the fungi than are many other eukaryote groups, including humans (see Figure 26.3). For example, the cell walls of oomycetes are typically made of cellulose, whereas those of fungi are made of chitin.

Rhizarians typically have long, thin pseudopods

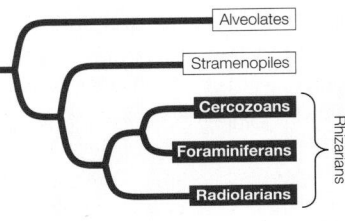

The three primary groups of **rhizarians**—cercozoans, foraminiferans, and radiolarians—are unicellular and mostly aquatic. The rhizarians have contributed their shells to ocean sediments, some of which have become terrestrial features over the course of geological history.

CERCOZOANS The **cercozoans** are a diverse group with many forms and habitats. Some are aquatic; others live in soil. One group of cercozoans possesses chloroplasts derived from a green alga by secondary endosymbiosis, and those chloroplasts contain a trace of the alga's nucleus.

FORAMINIFERANS Some **foraminiferans** secrete external shells of calcium carbonate (**Figure 26.11**). These shells have accumulated over time to produce much of the world's limestone. Some foraminiferans live as plankton; others live on the seafloor. Living foraminiferans have been found 10,896 m down in the western Pacific's Challenger Deep—the deepest point in the world's oceans. At that depth, however, they cannot secrete normal shells because the surrounding water is too poor in calcium carbonate.

In living planktonic foraminiferans, long, threadlike, branched pseudopods extend through numerous microscopic apertures in the shell and interconnect to create a sticky, reticulated net, which the foraminiferans use to catch smaller plankton. In some foraminiferan species, the pseudopods provide locomotion.

RADIOLARIANS **Radiolarians** are recognizable by their thin, stiff pseudopods, which are reinforced by microtubules (**Figure 26.12A**).

Figure 26.11 **Building Blocks of Limestone** Some foraminiferans secrete calcium carbonate to form shells. The shells of different species have distinctive shapes. Over millions of years, the shells of foraminiferans have accumulated to form limestone deposits.

(A) *Amphilonche heteracantha*

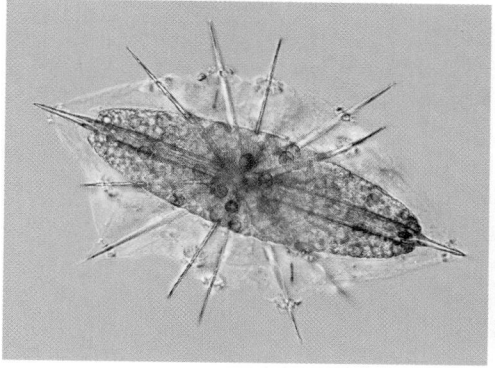

(B) *Thyrsocyrtis* sp.

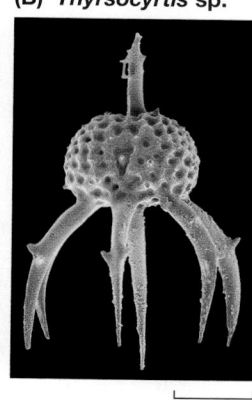

50 μm

100 μm

Figure 26.12 **Radiolarians Exhibit Distinctive Pseudopods and Radial Symmetry** **(A)** The radiolarians are distinguished by their thin, stiff pseudopods and by their radial symmetry. The pigmentation seen at the center of this radiolarian's glassy endoskeleton is imparted by endosymbiotic dinoflagellates. **(B)** The endoskeleton secreted by a radiolarian.

These pseudopods greatly increase the surface area of the cell, and they help the cell stay afloat in its marine environment.

Radiolarians also are immediately recognizable by their distinctive radial symmetry. Almost all radiolarian species secrete glassy endoskeletons (internal skeletons). The skeletons of the different species are as varied as snowflakes, and many have elaborate geometric designs (**Figure 26.12B**). A few radiolarians are among the largest of the unicellular eukaryotes, measuring several millimeters across.

Excavates began to diversify about 1.5 billion years ago

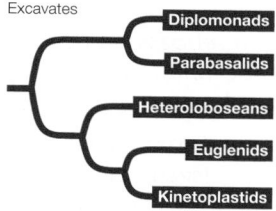

The **excavates** include a number of diverse groups that began to split from one another soon after the origin of eukaryotes. Several groups of excavates lack mitochondria, an absence that once led to the view that these groups might represent early-diverging eukaryotes that diversified before the evolution of mitochondria. However, the discovery of genes in the nucleus that are normally associated with mitochondria suggests that the absence of mitochondria is a derived condition in these organisms. In other words, ancestors of these excavate groups probably possessed mitochondria that were lost or reduced over the course of evolution. The existence of these organisms today shows that eukaryotic life is possible without mitochondria, at least among parasitic species.

DIPLOMONADS AND PARABASALIDS The **diplomonads** and the **parabasalids** are unicellular and lack mitochondria (although they have reduced organelles that are derived from mitochondria). The parasitic *Giardia lamblia*, a diplomonad, causes the intestinal disease giardiasis. *Giardia* infections may result from contact with contaminated water. In the United States, such infections are common among hikers and campers using spring or stream water in recreational areas, as well as among children kept in close

quarters (as in day-care centers). This tiny organism contains two nuclei bounded by nuclear envelopes, and it has a cytoskeleton and multiple flagella (**Figure 26.13A**).

(A) *Giardia muris*

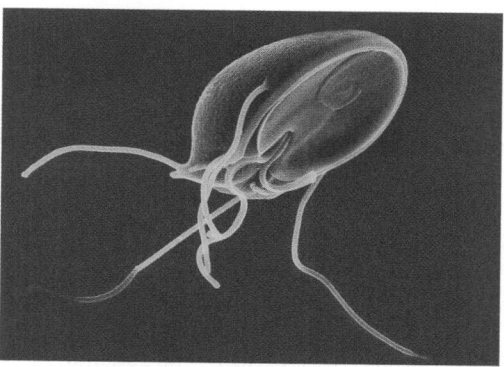

2 μm

(B) *Trichomonas vaginalis*

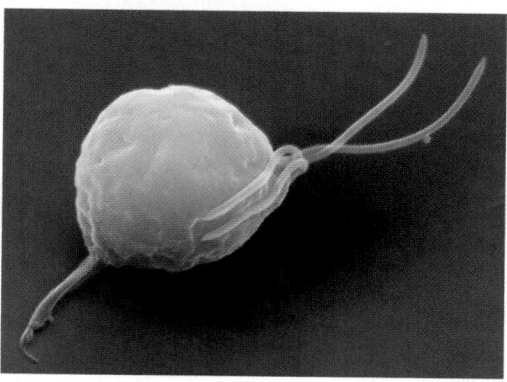

2 μm

Figure 26.13 **Some Excavate Groups Lack Mitochondria**
(A) *Giardia*, a diplomonad, has flagella and two nuclei. **(B)** *Trichomonas*, a parabasalid, has flagella and undulating membranes. Neither of these organisms possesses mitochondria.

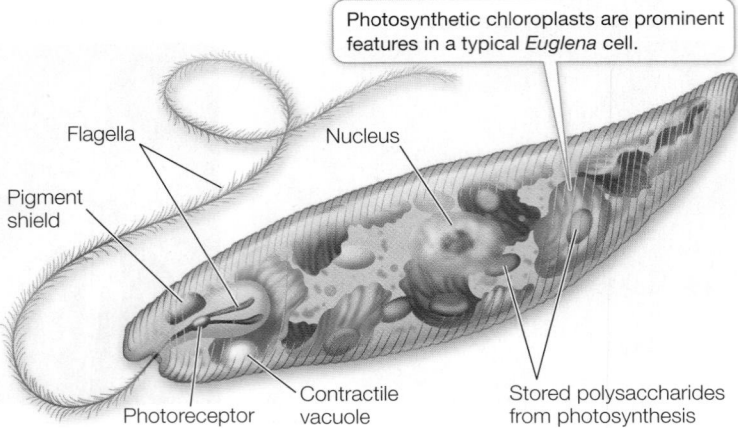

Photosynthetic chloroplasts are prominent features in a typical *Euglena* cell.

Flagella
Nucleus
Pigment shield
Photoreceptor
Contractile vacuole
Stored polysaccharides from photosynthesis

Figure 26.14 A Photosynthetic Euglenid In the *Euglena* species illustrated in this drawing, the second flagellum is rudimentary. Note that the primary flagellum originates at the anterior of the organism and trails toward its posterior.

In addition to flagella and a cytoskeleton, the parabasalids have undulating membranes that also contribute to the cell's locomotion. *Trichomonas vaginalis* (**Figure 26.13B**) is a parabasalid responsible for a sexually transmitted disease in humans. Infection of the male urethra, where it may occur without symptoms, is less common than infection of the vagina.

HETEROLOBOSEANS The amoeboid body form appears in several protist groups that are only distantly related to one another. The body forms of **heteroloboseans**, for example, resemble those of loboseans, an amoebozoan group that is not at all closely related to heteroloboseans (see the next section). Amoebas of the free-living heterolobosean genus *Naegleria*, some of which can enter the human body and cause a fatal disease of the nervous system, usually have a two-stage life cycle, in which one stage has amoeboid cells and the other flagellated cells.

EUGLENIDS AND KINETOPLASTIDS The **euglenids** and **kineto-plastids** together constitute a clade of unicellular excavates with flagella. Their mitochondria contain distinctive disc-shaped cristae, and their flagella contain a crystalline rod not found in other organisms. They reproduce primarily asexually by binary fission.

The flagella of euglenids arise from a pocket at the anterior end of the cell. Spiraling strips of proteins under the cell membrane

control the cell's shape. Some euglenids are photosynthetic. **Figure 26.14** depicts a typical cell of the genus *Euglena*. Like most other euglenids, this common freshwater organism has a complex cell structure. It propels itself through the water with the longer of its two flagella, which may also serve as an anchor to hold the organism in place. The second flagellum is often rudimentary.

▶ Media Clip 26.4 **Euglenids**
www.Life11e.com/mc26.4

The euglenids have diverse nutritional requirements. Many species are always heterotrophic. Other species, including species of *Euglena*, are fully autotrophic in sunlight, using chloroplasts to synthesize organic compounds through photosynthesis. When kept in the dark, these euglenids lose their photosynthetic pigment and begin to feed exclusively on dissolved organic material in the water around them. A "bleached" *Euglena* resynthesizes its photosynthetic pigment when it is returned to the light and becomes autotrophic again. But *Euglena* cells treated with certain antibiotics or mutagens lose their photosynthetic pigment completely; neither they nor their descendants are ever autotrophs again. However, those descendants function well as heterotrophs.

The kinetoplastids are unicellular parasites with two flagella and a single, large mitochondrion. The mitochondrion contains a kinetoplast, a unique structure housing multiple circular DNA molecules and associated proteins. Some of these DNA molecules encode "guide proteins" that edit mRNA within the mitochondrion.

The kinetoplastids include several medically important species of pathogenic trypanosomes (**Table 26.1**). Some of these organisms are able to change their cell surface recognition molecules frequently, allowing them to evade our best attempts to kill them and thus eradicate the diseases they cause.

Amoebozoans use lobe-shaped pseudopods for locomotion

Amoebozoans appear to have diverged from other eukaryotes about 1.5 billion years ago (see Figure 26.3). It is not yet clear whether they are more closely related to opisthokonts (which include fungi and animals) or to other major groups of eukaryotes.

The lobe-shaped pseudopods of amoebozoans (**Figure 26.15**) are a hallmark of the amoeboid body form. Amoebozoan pseudopods differ in form and function from the slender pseudopods of

table **26.1** Three Pathogenic Trypanosomes			
	Trypanosoma brucei	*Trypanosoma cruzi*	*Leishmania major*
Human disease	Sleeping sickness	Chagas disease	Leishmaniasis
Insect vector	Tsetse fly	Assassin bugs (many species)	Sand fly
Vaccine or effective cure	None	None	None
Strategy for survival	Changes surface recognition molecules frequently	Causes changes in surface recognition molecules on host cell	Reduces effectiveness of macrophage hosts
Site in human body	Bloodstream; in final stages, attacks nerve tissue	Enters cells, especially muscle cells	Enters cells, primarily macrophages
Approximate number of deaths per year	7,000	11,000	63,000

Amoeba proteus

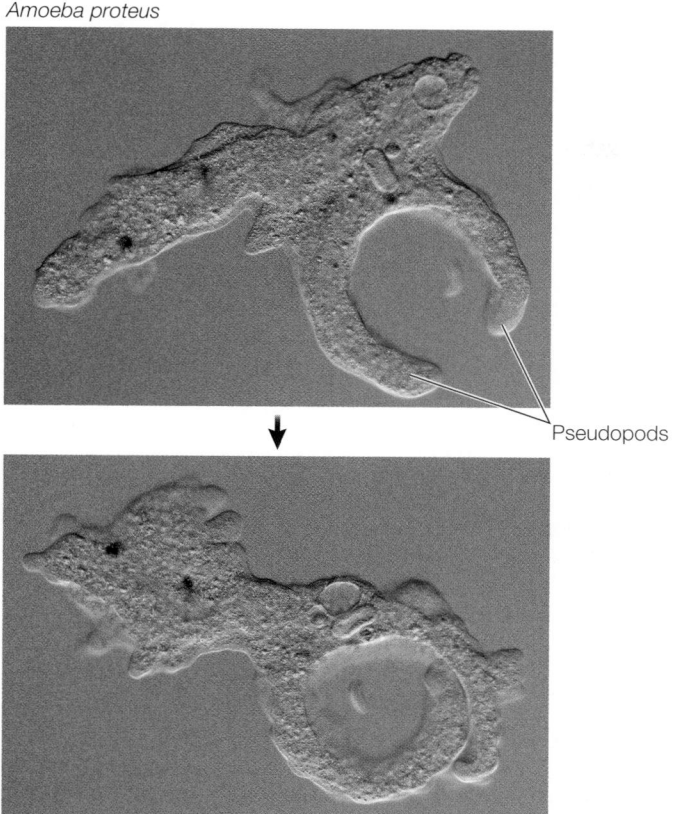

Pseudopods

120 µm

Figure 26.15 An Amoeba in Motion   The flowing pseudopods of this "chaos amoeba" (a lobosean) are constantly changing shape as it moves and feeds.

 Media Clip 26.5 Amoeboid Movement
www.Life11e.com/mc26.5

rhizarians. We consider three amoebozoan groups here: the loboseans and two groups known as slime molds.

LOBOSEANS   Loboseans are small amoebozoans that feed on other small organisms and particles of organic matter by phagocytosis, engulfing them with pseudopods. Many loboseans are adapted for life on the bottoms of lakes, ponds, and other bodies of water. Their creeping locomotion and their manner of engulfing food particles fit them for life close to a relatively rich supply of sedentary organisms or organic particles. Most loboseans exist as predators, parasites, or scavengers. Members of one group of loboseans, the testate amoebas, live inside shells. Some of these amoebas produce casings by gluing sand grains together (**Figure 26.16**). Other testate amoebas have shells secreted by the organism itself.

PLASMODIAL SLIME MOLDS   If the nucleus of an amoeba began rapid mitotic division, accompanied by a tremendous increase in cytoplasm and organelles, but no cytokinesis, the resulting organism would resemble the multinucleate mass of a **plasmodial slime mold**. During its vegetative (feeding, nonreproductive) stage, a plasmodial slime mold is a wall-less mass of cytoplasm with numerous diploid nuclei. This mass streams very slowly over its substrate in a remarkable network of strands called a plasmodium (**Figure 26.17A**). The plasmodium is an example of a **coenocyte**: many

Nebela collaris

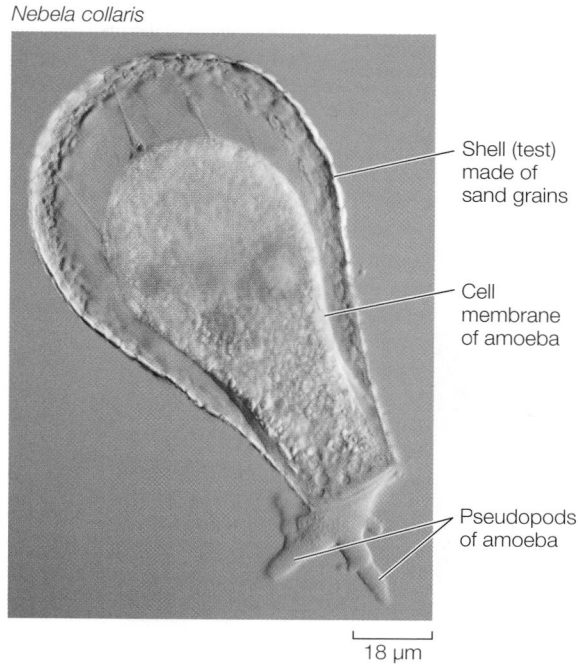

Shell (test) made of sand grains

Cell membrane of amoeba

Pseudopods of amoeba

18 µm

Figure 26.16 Life in a Glass House   This testate amoeba has built a lightbulb-shaped shell, or test, by gluing sand grains together. Its pseudopods extend through the single aperture in the test.

(A) A plasmodium

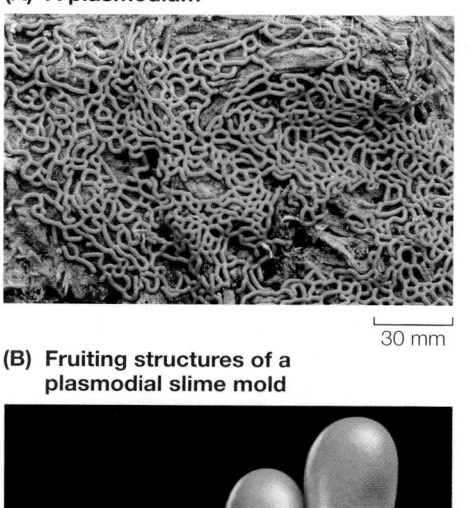

30 mm

(B) Fruiting structures of a plasmodial slime mold

1.5 mm

Figure 26.17 A Plasmodial Slime Mold   **(A)** The plasmodial form of the slime mold *Hemitrichia serpula* covers rocks, decaying logs, and other objects as it engulfs bacteria and other food items; it is also responsible for the organism's common name of "pretzel mold." **(B)** Fruiting structures of *Hemitrichia*.

 Media Clip 26.6 Plasmodial Slime Mold Growth
www.Life11e.com/mc26.6

Dictyostelium discoideum

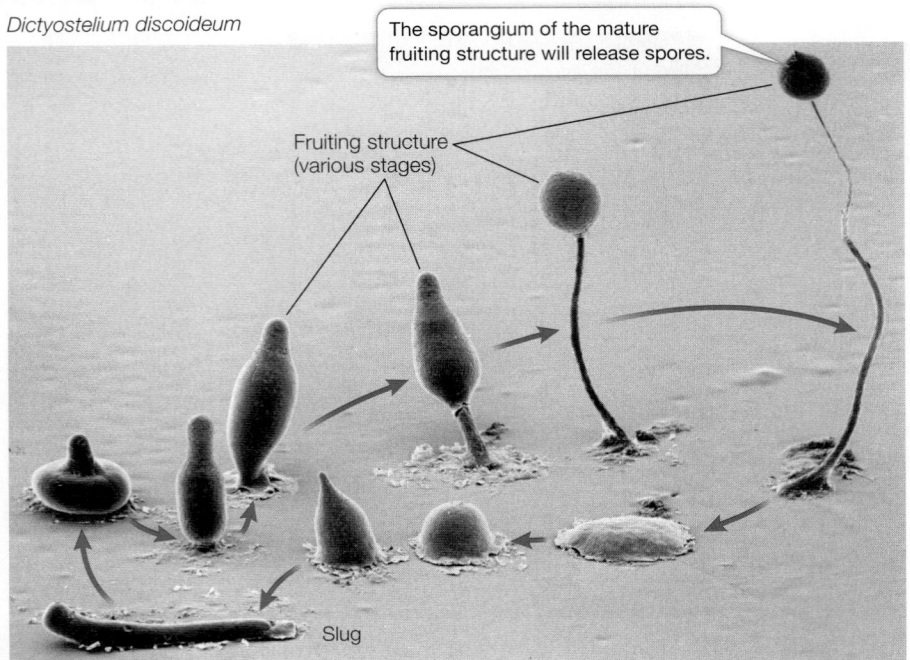

The sporangium of the mature fruiting structure will release spores.

Fruiting structure (various stages)

Slug

0.25 mm

Figure 26.18 A Cellular Slime Mold   This composite micrograph shows the life cycle of the slime mold *Dictyostelium*.

 Media Clip 26.7 Cellular Slime Mold Aggregation
www.Life11e.com/mc26.7

nuclei enclosed in a single cell membrane. The outer cytoplasm of the plasmodium (closest to the environment) is normally less fluid than the interior cytoplasm and thus provides some structural rigidity.

Plasmodial slime molds provide a dramatic example of movement by **cytoplasmic streaming**. The outer cytoplasm of the plasmodium becomes more fluid in places, and cytoplasm rushes into those areas, stretching the plasmodium. This streaming reverses its direction every few minutes as cytoplasm rushes into a new area and drains away from a previous one, moving the plasmodium over its substrate. Sometimes an entire wave of plasmodium moves across a surface, leaving strands behind. Microfilaments and a contractile protein called myxomyosin interact to produce the streaming movement. As it moves, the plasmodium engulfs food particles by endocytosis—predominantly bacteria, yeasts, spores of fungi, and other small organisms as well as decaying animal and plant remains.

A plasmodial slime mold can grow almost indefinitely in its plasmodial stage as long as the food supply is adequate and other conditions, such as moisture and pH, are favorable. If conditions become unfavorable, however, one of two things can happen. First, the plasmodium can form an irregular mass of hardened cell-like components. This resting structure rapidly becomes a plasmodium again when favorable conditions are restored.

Alternatively, the plasmodium can transform itself into spore-bearing fruiting structures (**Figure 26.17B**). These stalked or branched structures rise from heaped masses of plasmodium. They derive their rigidity from walls that form and thicken between their nuclei. The diploid nuclei of the plasmodium divide by meiosis as the

fruiting structure develops. One or more knobs, called sporangia, develop on the end of the stalk. Within a sporangium, haploid nuclei become surrounded by walls to form spores. Eventually, as the fruiting structure dries, it sheds its spores.

The spores germinate into wall-less, haploid cells called swarm cells, which can either divide mitotically to produce more haploid swarm cells or function as gametes. Swarm cells can live as separate individual cells that move by means of flagella or pseudopods, or they can become walled and resistant resting cysts when conditions are unfavorable; when conditions improve again, the cysts release swarm cells. Two swarm cells can also fuse to form a diploid zygote, which divides by mitosis (but without a wall forming between the nuclei) and thus forms a new coenocytic plasmodium.

CELLULAR SLIME MOLDS   Whereas the plasmodium is the basic vegetative unit of the plasmodial slime molds, a single amoeboid cell is the vegetative unit of the **cellular slime molds** (**Figure 26.18**). Cells called **myxamoebas**, which have single haploid nuclei, swarm together as they engulf bacteria and other food particles by endocytosis and reproduce by mitosis and fission. This simple life cycle stage, consisting of swarms of independent, isolated cells, can persist indefinitely as long as food and moisture are available.

When conditions become unfavorable, the cellular slime molds form fruiting structures, as do their plasmodial counterparts. The individual myxamoebas aggregate into a mass called a slug or **pseudoplasmodium**. Unlike the true plasmodium of the plasmodial slime molds, this structure is not simply a giant sheet of cytoplasm with many nuclei; the individual myxamoebas within the slug retain their cell membranes and therefore their identity.

A slug may migrate over a substrate for several hours before becoming motionless and reorganizing to construct a delicate, stalked fruiting structure. Cells at the top of the fruiting structure develop into thick-walled spores, which are eventually released. Later, under favorable conditions, the spores germinate, releasing myxamoebas.

The cycle from myxamoebas through slug and spores to new myxamoebas is asexual. Cellular slime molds also have a sexual cycle, in which two myxamoebas fuse. The product of this fusion develops into a spherical structure that ultimately germinates, releasing new haploid myxamoebas.

26.2 recap

The major lineages of eukaryotes began to diverge about 1.5 billion years ago. Major groups of eukaryotes are highly diverse in their habitat, nutrition, locomotion, and body form. Many protists are photosynthetic autotrophs, but heterotrophic lineages have evolved repeatedly. Although most protists are unicellular, multicellularity has arisen independently many times.

learning outcomes

You should be able to:

- Identify dinoflagellates on the basis of phenotypic characteristics.
- Identify stramenophiles on the basis of phenotypic characteristics.
- Distinguish among the major groups of eukaryotes.
- Justify the position that multicellularity is relatively easy to evolve.
- Explain why protists are important to our medical and economic concerns.

1. For each pair of groups below, describe how you could recognize members of the two groups and differentiate them from one another. Then describe features that the two groups in each pair share.
   a. Foraminiferans and radiolarians
   b. Ciliates and dinoflagellates
   c. Diatoms and brown algae
   d. Plasmodial slime molds and cellular slime molds
2. The fossil record of eukaryotes from the Precambrian is poor compared with that from the Cambrian and later geological periods, even though eukaryotes were diversifying for the last billion years of the Precambrian. Can you think of some reasons why the eukaryotic fossil record became more extensive in the Cambrian?
3. Give examples of alveolates, stramenopiles, and excavates that are important for medical or culinary reasons.

The ancient origins of the major eukaryote lineages and the adaptation of these lineages to a wide variety of lifestyles and environments resulted in enormous protist diversity. It is not surprising, then, that reproductive modes among protists are also highly diverse.

key concept 26.3 Protists Reproduce Sexually and Asexually

Although most protists engage in both asexual and sexual reproduction, sexual reproduction has yet to be observed in some groups. In some protists, as in all prokaryotes, the acts of sex and reproduction are not directly linked.

focus your learning

- Ciliate conjugation is a sexual process but is not a reproductive process.
- Organisms that engage in alternation of generations have distinct multicellular haploid and diploid phases.

Several asexual reproductive processes have been observed among the protists:

- The equal splitting of one cell into two by mitosis followed by cytokinesis
- The splitting of one cell into multiple (i.e., more than two) cells

- The outgrowth of a new cell from the surface of an old one (known as **budding**)
- The formation of specialized cells (spores) that are capable of developing into new individuals (know as **sporulation**)

Asexual reproduction results in offspring that are genetically nearly identical to their parents (they differ only by new mutations that may arise during DNA replication). Such asexually reproduced groups of nearly identical organisms are known as **clonal lineages**.

Sexual reproduction among the protists takes various forms. In some protists, as in animals, the gametes are the only haploid cells. In others, the zygote is the only diploid cell. In still others, both diploid and haploid cells undergo mitosis, giving rise to alternating multicellular diploid and haploid life stages.

Some protists reproduce without sex and have sex without reproduction

As noted in Key Concept 26.2, members of the genus *Paramecium* are ciliates, which commonly have two types of nuclei in a single cell (one macronucleus and from one to several micronuclei; see Figure 26.6). The micronuclei are typical eukaryotic nuclei and are essential for genetic recombination. Each macronucleus contains many copies of the genetic information, packaged in units containing only a few genes each. The macronuclear DNA is transcribed and translated to regulate the life of the cell.

When paramecia reproduce asexually, all of the nuclei are copied before the cell divides. Paramecia (and many other protists) also have an elaborate sexual behavior called **conjugation**, in which two individuals line up tightly against each other and fuse in the oral groove region of the body. Nuclear material is extensively reorganized and exchanged over the next several hours (**Figure 26.19**). Each cell ends up with two haploid micronuclei, one of its own and one from the other cell, which fuse to form a new diploid micronucleus. A new macronucleus develops from that micronucleus through a series of dramatic chromosomal rearrangements. The exchange of nuclei is fully reciprocal: each of the two paramecia gives and receives an equal amount of DNA. The two organisms then separate and go their own ways, each equipped with new combinations of alleles.

Conjugation in *Paramecium* is a sexual process of genetic recombination, but it is not a reproductive process. Two cells begin conjugation and two cells are there at the end, so no new cells are created. As a rule, each asexual clone of paramecia must conjugate periodically. Experiments have shown that if some species are not permitted to conjugate, the clones can live through only about 350 cell divisions before dying out.

Some protist life cycles feature alternation of generations

Alternation of generations is a type of life cycle found in many multicellular protists, all land plants, and some fungi (see Figure 27.6). A multicellular, diploid, spore-producing organism gives rise to a multicellular, haploid, gamete-producing organism. When two haploid gametes fuse, a diploid organism is produced. The haploid organism, the diploid organism, or both may also reproduce asexually. Note that alternation of generations is distinct from the familiar

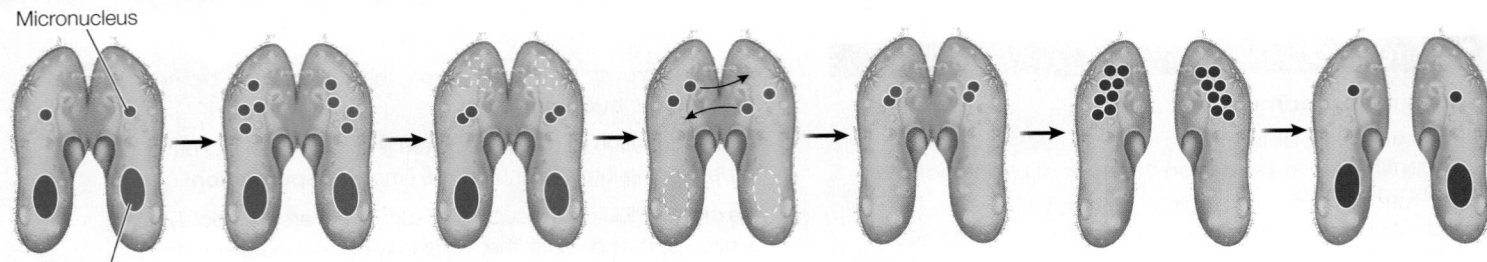

Micronucleus

Macronucleus

1. Two paramecia conjugate; all but one micronucleus in each cell disintegrate. The remaining micronucleus undergoes meiosis.

2. Three of the four haploid micronuclei disintegrate; the remaining micronucleus undergoes mitosis.

3. The paramecia donate micronuclei to each other. The macronuclei disintegrate.

4. The two micronuclei in each cell—each genetically different—fuse.

5. The new diploid micronuclei divide mitotically, eventually giving rise to a macronucleus and the appropriate number of micronuclei.

Figure 26.19 Conjugation in Paramecia The exchange of micronuclei by two conjugating *Paramecium* individuals results in genetic recombination. After conjugation, the cells separate and continue their lives as two individuals.

Q: Why is conjugation considered sex without reproduction?

reproductive system of animals, in which the only haploid stages are unicellular gametes produced by multicellular, diploid adults.

The two alternating (spore-producing and gamete-producing) generations differ genetically (one has diploid cells, the other haploid cells), but they may or may not differ morphologically. In **heteromorphic** alternation of generations, the two generations differ morphologically; in **isomorphic** alternation of generations, they do not. Examples of both heteromorphic and isomorphic alternation of generations are found among the brown algae.

The gamete-producing generation does not produce gametes by meiosis because the gamete-producing organism is already haploid. Instead, specialized cells of the diploid spore-producing organism, called **sporocytes**, divide meiotically to produce four haploid spores. The spores may eventually germinate and divide mitotically to produce the multicellular haploid generation, which then produces gametes by mitosis and cytokinesis.

Gametes, unlike spores, can produce new organisms only by fusing with other gametes. The fusion of two gametes produces a diploid zygote, which then undergoes mitotic divisions to produce a diploid organism. The diploid organism's sporocytes then undergo meiosis and produce haploid spores, starting the cycle anew.

▶ 26.3 recap

Protists reproduce both asexually and sexually, although sex occurs independently of reproduction in some species. Some multicellular protists exhibit alternation of generations, alternating between multicellular haploid and diploid life stages.

learning outcomes

You should be able to:

- Distinguish between sex (in the sense of exchange of genes leading to genetic recombination) and reproduction.
- Predict the evolutionary consequences of reproduction without sex for long periods of time.

▶ 26.3 recap

- Apply the definition of "alternation of generations" to determine whether a species is undergoing the process.

---

1. Why is conjugation between paramecia considered a sexual process but not a reproductive process?

2. Why do you think paramecia that are not allowed to conjugate begin to die out after about 350 rounds of asexual reproduction?

3. Although most diploid animals have haploid stages (e.g., eggs and sperm), their life cycles are not considered an example of alternation of generations. Why not?

Given the diversity of protists and of the environments in which they live, it is not surprising that they influence their environments in numerous ways.

▶ key concept 26.4 Protists Are Critical Components of Many Ecosystems

Some protists are food for marine animals, while others poison those animals. Some are packaged as nutritional supplements for humans, and some are human pathogens. The remains of some form the sands of many modern beaches, and others are a major source of the oil that sometimes fouls those beaches.

focus your learning

- *Plasmodium* is a specialized group of malaria-causing apicomplexans that are transmitted by mosquitoes.
- The overproduction of some diatoms and dinoflagellates can result in deadly red tides.
- Corals rely on dinoflagellate endosymbionts to obtain food.
- Diatomaceous earth comes from the decomposition of diatoms.

Phytoplankton are primary producers

A single protist clade, the diatoms, performs about one-fifth of all photosynthetic carbon fixation on Earth—about the same amount as all of Earth's rainforests. These spectacular unicellular organisms (see Figure 26.8) are the predominant component of the oceanic phytoplankton, but the phytoplankton include many other protists that also contribute heavily to global photosynthesis. Like green plants on land, these "floating photosynthesizers" are the gateway for energy from the sun into the rest of the living world; in other words, they are **primary producers**. These autotrophs are eaten by heterotrophs, including animals and many other protists. Those consumers are, in turn, eaten by other consumers. Most aquatic heterotrophs (with the exception of some species in the deep sea) depend on photosynthesis performed by phytoplankton.

Some microbial eukaryotes are deadly

Some microbial eukaryotes are pathogens that cause serious diseases in humans and other vertebrates. The best-known pathogenic protists are members of the genus *Plasmodium*, a highly specialized group of apicomplexans that spend part of their life cycle as parasites in human red blood cells, where they are the cause of malaria. In terms of the number of people affected, malaria is one of the world's three most serious infectious diseases: it infects more than 350 million people, and kills nearly a million people, each year. On average, about two people die from malaria every minute of every day—most of them in sub-Saharan Africa, although malaria occurs in more than 100 countries.

Mosquitoes of the genus *Anopheles* transmit *Plasmodium* to humans. The parasites enter the human circulatory system when an infected female *Anopheles* mosquito penetrates the skin in search of blood. The parasites find their way to cells in the liver and the lymphatic system, change their form, multiply, and reenter the bloodstream, where they invade red blood cells.

The parasites multiply inside the red blood cells, which then lyse (burst), releasing new swarms of parasites. These episodes of bursting red blood cells coincide with the primary symptoms of malaria, which include fever, shivering, vomiting, joint pains, and convulsions. If another *Anopheles* bites the victim, the mosquito takes in *Plasmodium* cells along with blood. Some of the ingested cells develop into gametes that unite in the mosquito, forming zygotes. The zygotes lodge in the mosquito's gut, divide several times, and move into its salivary glands, from which they can be passed on to another human host. Thus *Plasmodium* is an extracellular parasite in the mosquito vector and an intracellular parasite in the human host (**Figure 26.20**). Such an organism—that is, a parasite that requires more than one host—is said to have a **complex life cycle**.

Plasmodium has proved to be a singularly difficult pathogen to attack. The complex *Plasmodium* life cycle is best broken by the removal of stagnant water, in which mosquitoes breed. Using insecticides to reduce the *Anopheles* population can also be effective, but the benefits must be weighed against the ecological, economic, and health risks posed by the insecticides themselves.

Even some of the phytoplankton that are such important primary producers can be deadly, as described in this chapter's opening story. Some diatoms and dinoflagellates reproduce in enormous numbers when environmental conditions are favorable for their growth. In the resulting red tides, the concentration of dinoflagellates may reach 60 million per liter of ocean water and produce potent nerve toxins that harm or kill many vertebrates, especially fish.

Some microbial eukaryotes are endosymbionts

Endosymbiosis is common among the microbial eukaryotes, many of which live within the cells of animals. Many radiolarians harbor photosynthetic endosymbionts (see Figure 26.12A). As a result, these radiolarians, which are not photosynthetic themselves, appear greenish or golden, depending on the type of endosymbiont they contain. This arrangement is often mutually beneficial: the radiolarian can make use of the carbon compounds produced by its photosynthetic guest, and the guest may in turn make use of metabolites made by the host or receive physical protection. In some cases, the guest is exploited for its photosynthetic products while receiving little or no benefit itself.

Dinoflagellates are also common endosymbionts and can be found in both animals and other protists. Most, but not all, dinoflagellate endosymbionts are photosynthetic. Some dinoflagellates live endosymbiotically in the cells of corals, contributing the products of their photosynthesis to the partnership. Their importance to the corals is demonstrated when the dinoflagellates die or are expelled by the corals as a result of changing environmental conditions such as rising water temperatures or increased water turbidity. This phenomenon is known as **coral bleaching**. Unless the corals can acquire new endosymbionts, they are ultimately damaged or destroyed as a result of their reduced food supply (**Investigating Life: Can Corals Reacquire Dinoflagellate Endosymbionts Lost to Bleaching?**).

We rely on the remains of ancient marine protists

Diatoms are lovely to look at, but their importance to us goes far beyond aesthetics, and even beyond their role as primary producers. Diatoms store oil as an energy reserve and to keep themselves afloat at the correct depth in the ocean. Over millions of years, diatoms have died and sunk to the ocean floor, where they have undergone chemical changes. In this way, they have become a major source of petroleum and natural gas, two of our most important energy supplies and political concerns.

Because the silica-containing cell walls of dead diatoms resist decomposition, some sedimentary rocks are composed almost entirely of diatom skeletons that sank to the seafloor over time. Diatomaceous earth, which is obtained from such rocks, has many industrial uses, such as insulation, filtration, and metal polishing. It has also been used as an "Earth-friendly" insecticide that clogs the tracheae (breathing structures) of insects.

Other ancient marine protists have also contributed to today's world. Some foraminiferans, as we have seen, secrete shells of calcium carbonate. After they reproduce (by mitosis and cytokinesis), the daughter cells abandon the parent shell and make new shells of their own. The discarded shells of ancient foraminiferans make up extensive limestone deposits in various parts of the world, forming a layer hundreds to thousands of meters deep over millions of square kilometers of ocean bottom. Foraminiferan shells also make up much of the sand of some beaches. A single gram of such sand may contain as many as 50,000 foraminiferan shells and shell fragments.

The shells of individual foraminiferans are easily preserved as fossils in marine sediments. Each geological period has a distinctive assemblage of foraminiferan species. Because the shells of foraminiferan species have distinctive shapes (see Figure 26.11) and because they are so abundant, the remains of foraminiferans are especially valuable in classifying and dating sedimentary rocks. In addition, analyses of the chemical makeup of foraminiferan shells can be used to estimate the global temperatures prevalent at the time when the shells were formed.

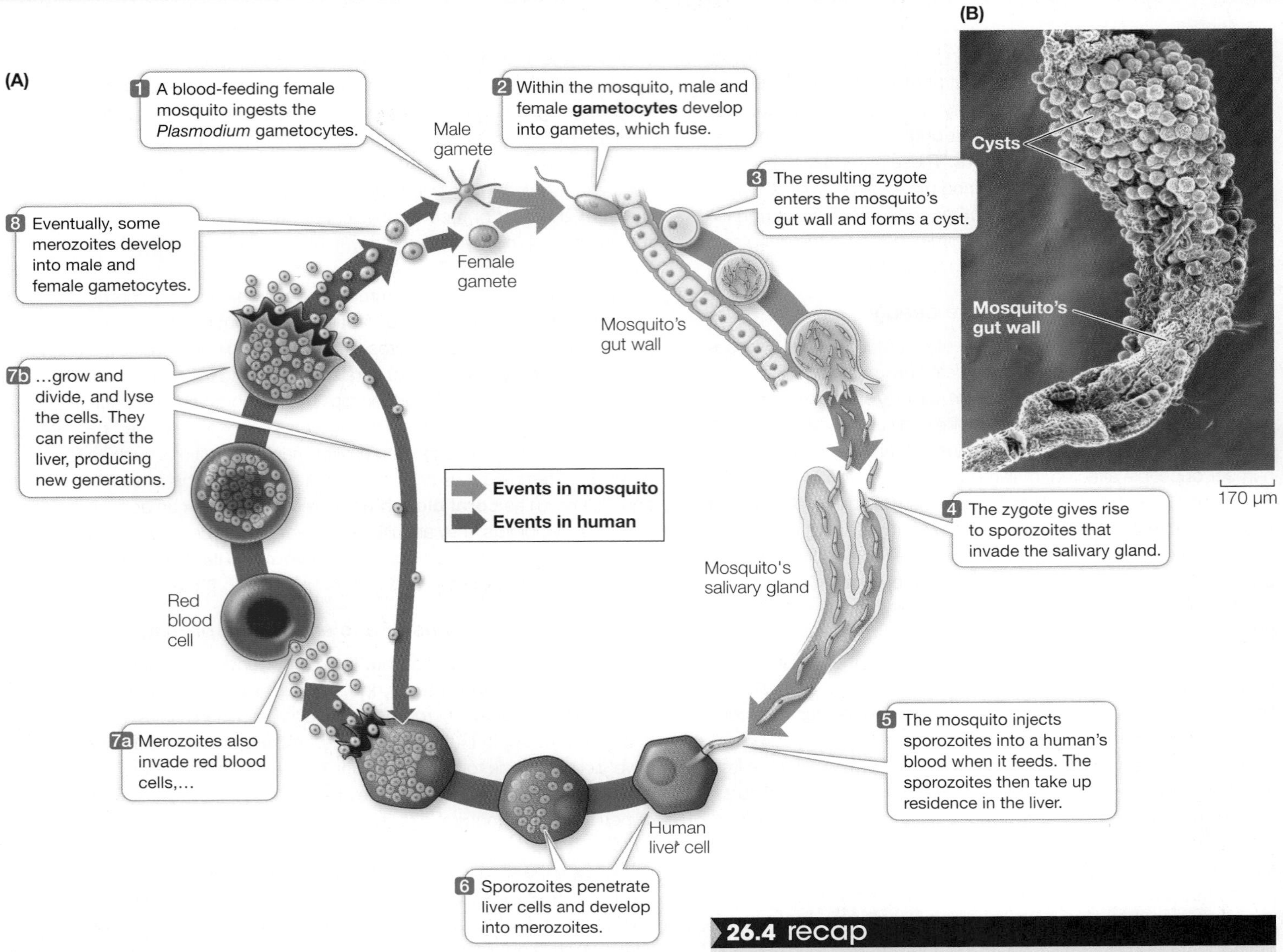

(A)

1 A blood-feeding female mosquito ingests the *Plasmodium* gametocytes.

2 Within the mosquito, male and female **gametocytes** develop into gametes, which fuse.

Male gamete

8 Eventually, some merozoites develop into male and female gametocytes.

3 The resulting zygote enters the mosquito's gut wall and forms a cyst.

Female gamete

Mosquito's gut wall

7b ...grow and divide, and lyse the cells. They can reinfect the liver, producing new generations.

Events in mosquito
Events in human

Red blood cell

4 The zygote gives rise to sporozoites that invade the salivary gland.

Mosquito's salivary gland

7a Merozoites also invade red blood cells,...

5 The mosquito injects sporozoites into a human's blood when it feeds. The sporozoites then take up residence in the liver.

Human liver cell

6 Sporozoites penetrate liver cells and develop into merozoites.

(B)

Cysts

Mosquito's gut wall

170 µm

Figure 26.20 Life Cycle of the Malarial Parasite   (A) Like many parasitic species, the apicomplexan *Plasmodium falciparum* has a complex life cycle, part of which is spent in mosquitoes of the genus *Anopheles* and part in humans. The sexual phase (gamete fusion) of this life cycle takes place in the insect, and the zygote is the only diploid stage. **(B)** Encysted *Plasmodium* zygotes (artificially colored blue) cover the stomach wall of a mosquito. Invasive sporozoites will hatch from the cysts and be transmitted to a human, in whom the parasite causes malaria.

 Animation 26.3 Life Cycle of the Malarial Parasite www.Life11e.com/a26.3

26.4 recap

Protists have many effects, both positive and negative, on their environment. Some species are important primary producers, many are endosymbionts, and some are pathogens. Protists are among the most important producers of fossil fuels, and they are important components of sedimentary rocks.

26.4 recap

learning outcomes

You should be able to:

- Describe the complex life cycle of a parasite that involves multiple hosts.
- Summarize the consequences of overproduction of some diatoms and dinoflagellates.
- Describe beneficial and harmful interactions among species that involve protists.
- Give examples of protists that are important to humans.

1. What is the role of female *Anopheles* mosquitoes in the transmission of malaria?

2. Explain the roles of dinoflagellates in the two very different phenomena of coral bleaching and red tides.

3. What are two ways in which diatoms are important to human society?

The next six chapters will explore the major evolutionary radiations of multicellular eukaryotes, along with the protist ancestors from which they arose. Chapters 27 and 28 will describe the origin and diversification of plants, Chapter 29 will present the fungi, and Chapters 30–32 will provide a brief overview of the animals.

Can Corals Reacquire Dinoflagellate Endosymbionts Lost to Bleaching?

experiment

Original Paper: Lewis, C. L. and M. A. Coffroth. 2004. The acquisition of exogenous algal symbionts by an octocoral after bleaching. *Science* 304: 1490–1492.

Some corals lose their chief nutritional source when their photosynthetic endosymbionts die, often as a result of changing environmental conditions. This experiment by Cynthia Lewis and Mary Alice Coffroth investigated the ability of corals to acquire new endosymbionts after bleaching.

HYPOTHESIS▶ Bleached corals can acquire new photosynthetic endosymbionts from their environment.

METHOD

1. Count numbers of *Symbiodinium*, a photosynthetic dinoflagellate, living symbiotically in samples of a coral (*Briareum* sp.).
2. Stimulate bleaching by maintaining all *Briareum* colonies in darkness for 12 weeks.
3. After 12 weeks of darkness, count numbers of *Symbiodinium* in the coral samples; then return all colonies to light.
4. In some of the bleached colonies (the experimental group), introduce *Symbiodinium* strain B211—dinoflagellates that contain a unique molecular marker. Do not expose the others (the control group) to strain B211. Maintain both groups in the light for 6 weeks.

RESULTS

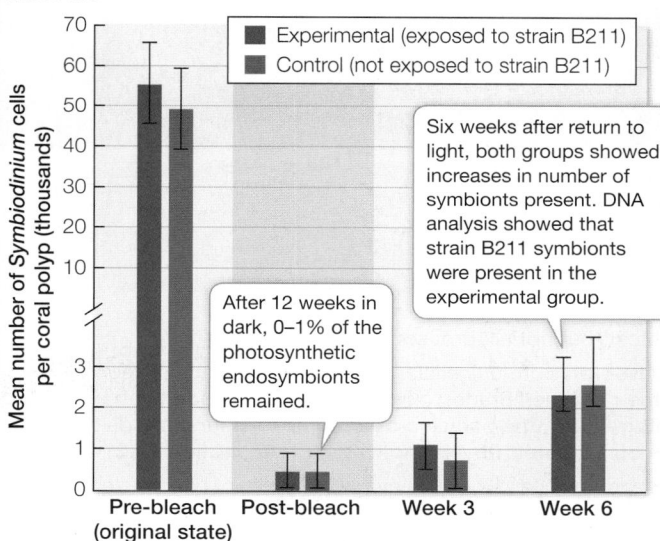

■ Experimental (exposed to strain B211)
■ Control (not exposed to strain B211)

Six weeks after return to light, both groups showed increases in number of symbionts present. DNA analysis showed that strain B211 symbionts were present in the experimental group.

After 12 weeks in dark, 0–1% of the photosynthetic endosymbionts remained.

Mean number of *Symbiodinium* cells per coral polyp (thousands)

Pre-bleach (original state)   Post-bleach   Week 3   Week 6

Pre-bleach
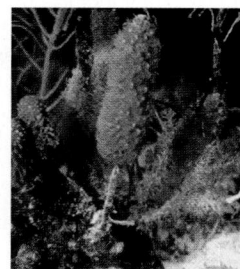

Post-bleach

CONCLUSION▶ Corals can acquire new endosymbionts from their environment following bleaching.

work with the data

The data in the table below come from DNA analyses of *Symbiodinium* strains found in the experimental and control colonies of corals (*Briareum*) before and after bleaching. *Symbiodinium* strain B211 (which was not present before bleaching) was introduced to the experimental colonies after bleaching. Use these data to answer the questions below.

	Symbiodinium strain present (% of colonies)			
	Non-B211	B211	None[a]	Colony died
Experimental colonies (strain B211 added)				
Pre-bleach	100	0	0	0
Post-bleach	58	0	42	0
Week 3	0	92	0	8
Week 6	8	58	8	25
Control colonies (no strain B211)				
Pre-bleach	100	0	0	0
Post-bleach	67	0	33	0
Week 3	67	0	33	0
Week 6	67	0	17	17

[a]Colonies remained alive but no *Symbiodinium* were detected.

QUESTIONS▶

1. Are new strains of *Symbiodinium* taken up only by coral colonies that have lost all their original endosymbionts?
2. Does the acquisition of a new *Symbiodinium* strain always result in survival of a recovering *Briareum* colony?
3. In week 3, only strain B211 was detected in the experimental colonies, but in week 6, non-B211 *Symbiodinium* were detected in 8 percent of the experimental colonies. Can you suggest an explanation for this observation?

investigatinglife

Red tides are harmful, but how are dinoflagellates beneficial to marine ecosystems?

The opening of this chapter discussed harmful red tides, or blooms of dinoflagellates. But not all dinoflagellate blooms produce problems for other species. Dinoflagellates are important components of many ecosystems, as we have seen throughout this chapter. Photosynthetic dinoflagellates also produce much of the atmospheric oxygen that most animals need to survive.

As we saw in Investigating Life: Can Corals Reacquire Dinoflagellate Endosymbionts Lost to Bleaching?, corals and many other species depend on symbiotic dinoflagellates for food. In addition, as photosynthetic organisms, free-living planktonic dinoflagellates are among the most important primary producers in aquatic food webs. They are a major component of the phytoplankton and provide an important food source for many species (see Key Concept 26.4).

Future directions

Some dinoflagellates produce a beautiful bioluminescence (Figure 26.21). Unlike the bioluminescent bacteria described at the start of Chapter 26, however, dinoflagellates cannot generate a steady bioluminescence, but produce bright flashes of light when disturbed, as people who swim in the ocean at night in certain regions often observe. What function do these flashes serve? Many light-emitting dinoflagellates are preyed on by other species, such as small crustaceans. When the dinoflagellates produce a bright flash, it functions like a "burglar alarm," and attracts secondary predators of the

Figure 26.21 **Light Up the Sea** Bioluminescent dinoflagellates flash as an outrigger disturbs the ocean surface off the island of Bali.

crustaceans. Experiments have shown that crustaceans reduce their feeding on dinoflagellates when they flash this signal. However, recent research shows another function of bioluminescence in toxic species of dinoflagellates. These toxic species produce a much lower intensity bioluminescence which may function as a warning of their toxicity to potential predators. So depending on the intensity and other details of the flash in different species, the bioluminescence can serve either to attract secondary predators or to warn off primary predators. Research on communication by bioluminescence continues as a rich field of investigation.

 Media Clip 26.8 Flashing Dinoflagellates
www.Life11e.com/mc26.8

Chapter Summary 26

▶ **26.1 Eukaryotes Acquired Features from Both Archaea and Bacteria**

- The term **protist** does not describe a formal taxonomic group. It is shorthand for "all eukaryotes that are not plants, animals, or fungi."
- Early events in the evolution of the eukaryotic cell included the loss of the firm cell wall and infolding of the cell membrane. Such infolding led to segregation of the genetic material in a membrane-enclosed nucleus. Review Focus: Key Figure 26.1
- Mitochondria evolved by endosymbiosis with a proteobacterium.
- Primary endosymbiosis of a eukaryote and a cyanobacterium gave rise to the first chloroplasts. **Secondary endosymbiosis** and **tertiary endosymbiosis** between chloroplast-containing eukaryotes and other eukaryotes gave rise to the distinctive chloroplasts of euglenids, dinoflagellates, and other groups. Review Figure 26.2, Animation 26.1

▶ **26.2 Major Lineages of Eukaryotes Diversified in the Precambrian**

- Most eukaryotes can be placed in one of eight major clades that diverged about 1.5 billion years ago: alveolates, stramenopiles, rhizarians, excavates, plants, amoebozoans, fungi, and animals. Review Figure 26.3

- Most, but not all, protists are unicellular.
- **Alveolates** are unicellular organisms with sacs (alveoli) beneath their cell membranes. Alveolate clades include the marine **dinoflagellates**, the parasitic **apicomplexans**, and the diverse, highly motile **ciliates**. Review Activity 26.1, Animation 26.2
- **Stramenopiles** typically have two flagella of unequal length, the longer one bearing rows of tubular hairs. Among the stramenopiles are the unicellular **diatoms**; the multicellular **brown algae**; and the nonphotosynthetic **oomycetes**, many of which are **saprobic**.
- **Rhizarians** are unicellular and aquatic. They include the **cercozoans**; the **foraminiferans**, which secrete shells of calcium carbonate; and the **radiolarians**, which have thin, stiff pseudopods and glassy endoskeletons.
- The **excavates** include parasitic as well as free-living species. The **diplomonads** and **parabasalids** lack typical mitochondria. **Heteroloboseans** have an amoeboid body form and a two-stage life cycle. **Euglenids** have anterior flagella; some are photosynthetic. The **kinetoplastids**, which include several human pathogens, have a single, large mitochondrion.
- The **amoebozoans** move by means of lobe-shaped pseudopods. A **lobosean** consists of a single amoeboid cell. **Plasmodial slime molds** are amoebozoans whose vegetative stage is a **coenocyte** that moves by cytoplasmic streaming. In **cellular slime molds**, the individual cells maintain their identity at all times but aggregate to form fruiting structures.

26.3 Protists Reproduce Sexually and Asexually

- Asexual reproduction gives rise to **clonal lineages** of organisms.
- **Conjugation** in *Paramecium* is a sexual process but not a reproductive one. **Review Figure 26.19**
- **Alternation of generations**, which includes a multicellular diploid stage and a multicellular haploid stage, is a feature of many multicellular protist life cycles (as well as those of some fungi and all land plants). The alternating generations may be **heteromorphic** or **isomorphic**.

26.4 Protists Are Critical Components of Many Ecosystems

- The diatoms are responsible for about one-fifth of the photosynthetic carbon fixation on Earth. They and other members of the phytoplankton are important **primary producers** in the marine environment. Ancient diatoms are a major source of today's petroleum and natural gas deposits.
- Some protists are pathogens of humans and other vertebrates. **Review Figure 26.20, Animation 26.3**
- Endosymbiotic relationships are common among microbial protists and often benefit both the endosymbionts and their protist or animal partners. **Review Investigating Life: Can Corals Reacquire Dinoflagellate Endosymbionts Lost to Bleaching?**

> Go to **LearningCurve** (in **LaunchPad**) for dynamic quizzing that helps you solidify your understanding of this chapter. **LearningCurve** adapts to your responses, giving you the practice you need to master each key concept.

Apply What You've Learned

Review

26.2 Ciliates are named for their numerous hair-like cilia.
26.2 Apicomplexans possess an apical complex.

Because closely related organisms are more likely to share characteristics than are more distantly related ones, biologists can use phylogenetic information to make predictions about organisms based on what is known about their relatives. This exercise examines such use of phylogenies in the alveolates. Below is a phylogenetic tree for some of the alveolates, along with a table listing some of the characteristics for selected species.

Species	Features
Theileria parva	Parasite of cattle, transmitted by ticks
Perkinsus marinus	Parasite of oysters; marine habitat
Crypthecodinium cohnii	Used to manufacture omega-3 fatty acids
Oxyrrhis marina	Omnivorous grazer; produces red tides
Pfiesteria shumwayae	Can be predatory on fish; can use chloroplasts of ingested green algae for photosynthesis

Questions

1. Based on this information and what you learned in the text, what life-history attribute was likely present in the common ancestor of the apicomplexans but not in other alveolates? Be as specific as possible.

2. Suppose a researcher examining the apicomplexans was considering which dinoflagellates to sample as an outgroup. Based on the phylogeny, is there a reason to favor choosing one particular genus of dinoflagellates over the others? If so, which genus? Explain your answer.

3. Suppose a researcher was studying the feeding mode of the genus Pfiesteria and wanted to determine if particular features evolved within this genus or were present in the ancestor of Pfiesteria. Based on the phylogeny, is there a reason to target sampling to a particular genus of dinoflagellate (outside *Pfiesteria*) over the others? If so, which genus? Explain your answer.

4. The red tides produced by *Oxyrrhis* and other dinoflagellates are not produced by apicomplexans. Similar red tides are produced by diatoms, which are distantly related to alveolates. Thus the production of red tides is convergent in these two groups. Describe another example from the text of convergent evolution in protists but outside alveolates.

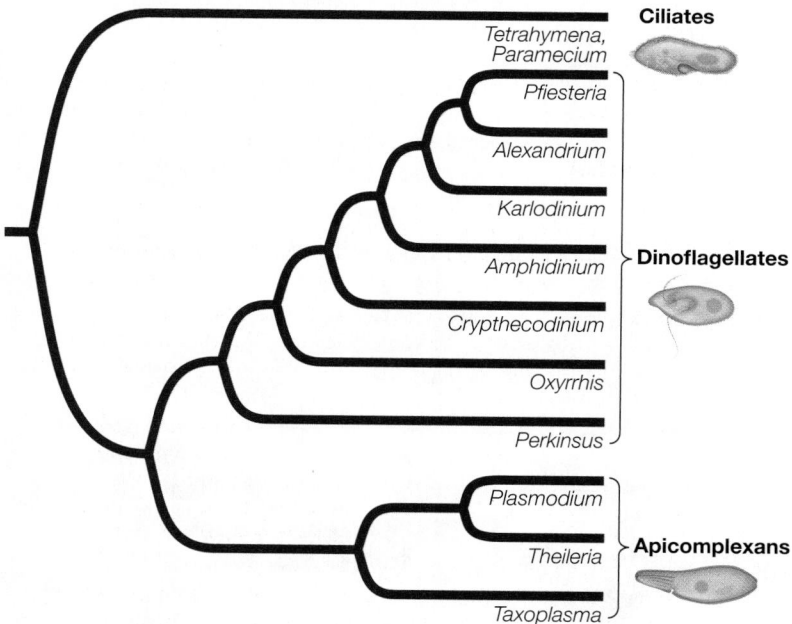

Ciliates — Tetrahymena, Paramecium

Dinoflagellates — Pfiesteria, Alexandrium, Karlodinium, Amphidinium, Crypthecodinium, Oxyrrhis, Perkinsus

Apicomplexans — Plasmodium, Theileria, Taxoplasma

Go to **LaunchPad** for the eBook, LearningCurve, animations, activities, flashcards, and additional resources and assignments.

27

Plants without Seeds: From Water to Land

Mosses, lycophytes, and ferns cover the trees and ground in this wet temperate rainforest.

> investigating**life**

A Toxic Spill of Ancient Fossil Algae

In the Gulf of Mexico, about 60 kilometers south of the Louisiana coast, the oil rig *Deepwater Horizon* was drilling an exploratory oil well in the seafloor beneath about 1,500 meters of water when, on April 20, 2010, an explosive blowout occurred and could not be contained. Over the next 3 months, almost 5 million barrels of petroleum flowed from the well into the Gulf, making this event the worst marine oil spill in history. The spill caused massive mortality among marine life, as well as considerable damage along the coast as the oil floated to the surface and washed ashore.

Why was oil to be found so deep beneath the Gulf, and what led geologists to expect to find oil there? Most people know that petroleum is a fossil fuel, meaning that it is derived from the ancient remains of once-living organisms. Fewer people know that most petroleum is derived largely from the remains of phytoplankton, including many species of green algae (as well as other microbial groups, as discussed in Chapter 26). These algae produce complex hydrocarbons through photo-

synthesis. They accumulate hydrocarbons both as an energy reserve and as a way to increase their buoyancy in water. When these algae die, they drop to the bottom of the ocean, and over many millions of years, their buried remains decompose into petroleum deposits.

Today there is great interest in using solar power to help meet human energy needs. But unicellular eukaryotes capitalized on this process first, and very long ago. Single-celled eukaryotes incorporated tiny solar energy converters into their cells about 1.5 billion years ago, when they formed partnerships with photosynthetic cyanobacteria. These endosymbionts—which over time would become the chloroplasts of modern plants—allowed many eukaryotes to use solar energy to drive the reactions that convert carbon dioxide into organic carbon compounds. Over many millions of years, the carbon compounds produced in the cells of marine algae accumulated in ocean sediments. Today humans are tapping that trapped solar energy in the form of petroleum and other fossil fuels.

Q&A Given that petroleum is derived naturally from green algae, can humans use green

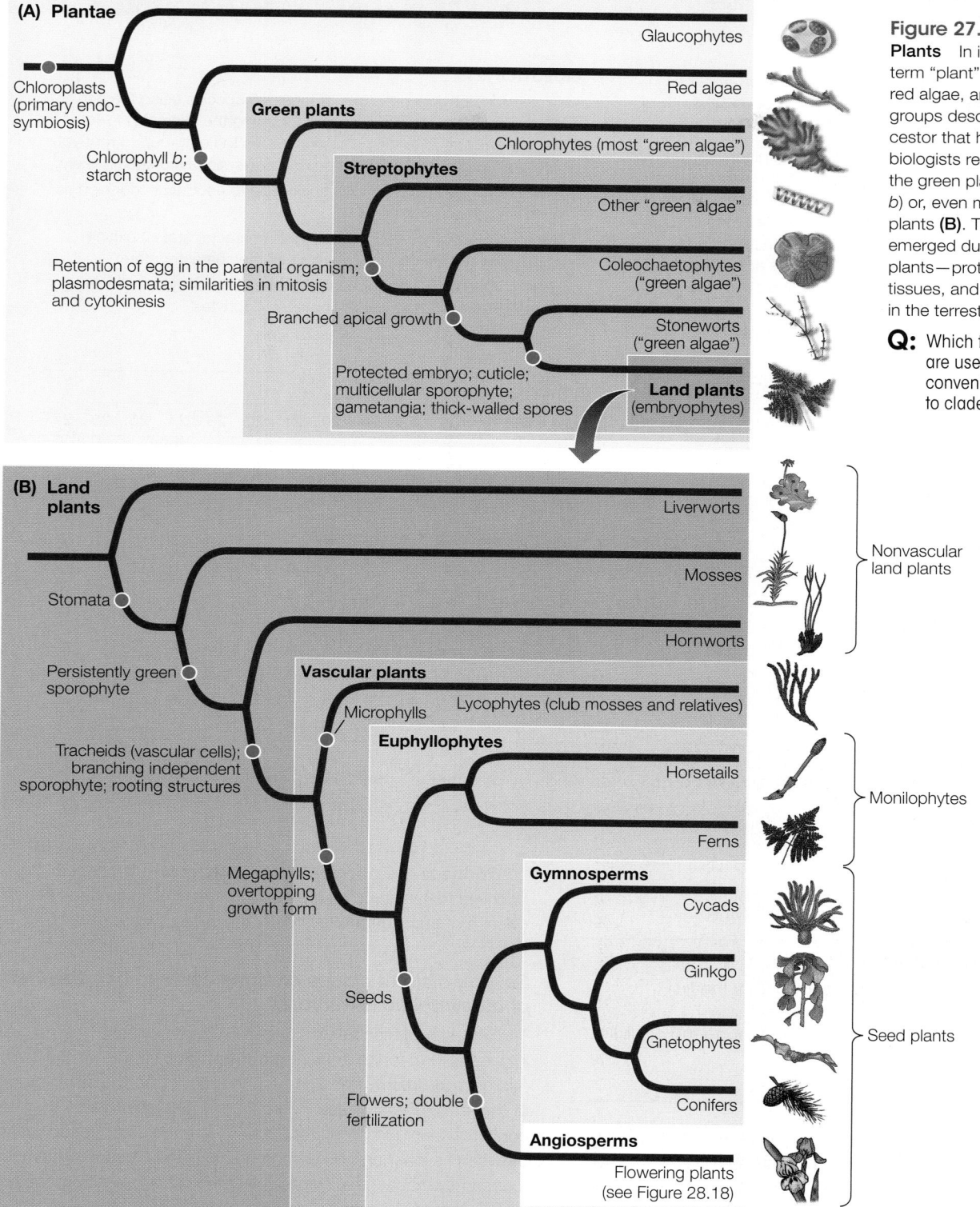

Figure 27.1 The Evolution of Plants In its broadest definition, the term "plant" includes the glaucophytes, red algae, and green plants—all the groups descended from a common ancestor that had chloroplasts **(A)**. Some biologists restrict the term "plant" to the green plants (those with chlorophyll *b*) or, even more narrowly, to the land plants **(B)**. Three key characteristics that emerged during the evolution of land plants—protected embryos, vascular tissues, and seeds—led to their success in the terrestrial environment.

Q: Which two names of plant groups are used in this figure only as convenience terms, and do not refer to clades?

key concept 27.1

Primary Endosymbiosis Produced the First Photosynthetic Eukaryotes

More than a billion years ago, when a cyanobacterium was first engulfed by an early eukaryote, the history of life was altered radically. The chloroplasts that resulted from primary endosymbiosis of this cyanobacterium (see Figure 26.2) were obviously important for the evolution of plants and other photosynthetic eukaryotes, but they were also critical to the evolution of all life on land. Until

work with the data

The Phylogeny of Land Plants

Original Paper: Qiu, Y.-L. et al. 2006. The deepest divergences in land plants inferred from phylogenetic evidence. *Proceedings of the National Academy of Sciences USA* 103: 15511–15516.

In addition to the morphological characters of land plants shown on the phylogeny in Figure 27.1, DNA sequences are widely used to study and reconstruct the evolutionary history of plants. These sequences are many tens of thousands of nucleotides long and have been collected from a large number of species. The full data set used by Yin-Long Qiu and his colleagues (available at treebase.org) includes DNA sequences from 67 genes. The table below provides sample sequences from a chloroplast gene that has been used to reconstruct the relationships of representative plant species; the table shows 27 nucleotide positions for 10 species.

QUESTIONS▶

1. Construct a phylogenetic tree of these 10 species using the parsimony method (see Key Concept 21.2 and the examples in Table 21.1 and Figure 21.5 for instructions). Use the outgroup to root your tree. Assume that all changes among nucleotides are equally likely.
2. How many changes (from one nucleotide to another) occur along each branch on your tree?
3. Which nucleotide positions (i.e., which character states) exhibit homoplasy (convergence or reversal of the character state)?
4. Which group on your tree represents the streptophytes? The land plants? The vascular plants? The euphyllophytes?

Species	Nucleotide position (character state)																										
	1	2	3	4	5	6	7	8	9	10	11	12	13	14	15	16	17	18	19	20	21	22	23	24	25	26	27
Outgroup (chlorophyte alga)	T	A	T	T	A	T	G	A	T	T	C	C	A	A	A	T	A	T	T	A	T	A	A	T	C	T	A
Stonewort	T	A	T	T	T	A	A	A	T	T	A	C	T	A	A	T	A	A	T	A	T	A	A	T	C	T	A
Liverwort	A	C	T	T	T	T	A	A	T	G	A	T	T	C	A	G	A	A	T	A	T	A	A	T	C	T	A
Moss	A	C	T	T	T	T	A	A	T	A	T	T	T	T	A	A	T	A	T	A	A	A	A	T	C	T	T
Hornwort	A	C	T	T	T	T	A	A	T	G	T	T	T	T	A	A	T	A	C	A	G	A	A	A	C	T	T
Lycophyte	A	C	T	C	C	C	G	G	T	G	T	T	C	T	G	A	T	A	C	A	A	G	A	C	C	C	T
Fern	C	C	T	C	C	G	A	G	C	G	T	T	C	T	T	A	G	A	T	A	A	G	G	A	C	C	T
Pine tree	A	C	C	C	C	G	C	G	C	G	T	T	C	T	G	A	T	G	C	G	A	G	G	A	T	C	T
Rice	A	C	C	C	C	G	C	G	C	G	T	T	C	T	G	A	T	G	C	G	A	G	G	A	T	A	T
Tobacco	A	C	C	A	C	G	C	G	C	G	T	T	C	T	G	A	T	G	C	G	A	G	G	A	T	A	T

A similar **work with the data** exercise may be assigned in **LaunchPad**.

---

photosynthetic plants were able to move onto land, there was very little on land to support multicellular animals or fungi, and almost all life was restricted to the oceans and fresh waters.

focus your learning

- The major clades of Plantae evolved from the first photosynthetic eukaryote.
- Key evolutionary innovations evolved among the aquatic Plantae.
- Land plants fall into ten major clades.

Primary endosymbiosis is a shared derived trait—a ***synapomorphy**—of the group known as **Plantae** (**Figure 27.1**). Although *Plantae* is Latin for "plants," in everyday language—and throughout this book—the unmodified common name "plants" is usually used to refer only to the land plants. However, the first several clades to branch off the tree of life after primary endosymbiosis are all aquatic. Most aquatic photosynthetic eukaryotes (other than those secondarily derived from land plants) are known by the common name **algae**. This name, however, is just a convenient way to refer to these groups, which are not all closely related. Many of the photosynthetic groups discussed in Chapter 26 (which acquired chloroplasts through secondary endosymbiosis) are also commonly called algae.

*connect the concepts   As noted in Key Concept 21.1, derived traits shared by a group of organisms that offer evidence of their common ancestry are called synapomorphies.

Several distinct clades of algae were among the first photosynthetic eukaryotes

The ancestor of Plantae was unicellular and may have been similar in general form to the modern **glaucophytes** (**Figure 27.2**). These microscopic freshwater algae are thought to be the sister group of the rest of Plantae (see Figure 27.1A). The chloroplast of glaucophytes is unique in containing a small amount of peptidoglycan between its inner and outer membranes—the same arrangement found in cyanobacteria. Peptidoglycan has been lost from the remaining photosynthetic eukaryotes.

In contrast to the glaucophytes, almost all **red algae** are multicellular (**Figure 27.3**). Their characteristic color is a result of the accessory photosynthetic pigment **phycoerythrin**, which is found in relatively large amounts in the chloroplasts of many red algae. In addition to phycoerythrin, red algal chloroplasts contain chlorophyll *a* as well as several other accessory pigments.

The red algae include species that grow in the shallowest tide pools as well as the photosynthesizers found deepest in the ocean

Glaucocystis   Chloroplasts

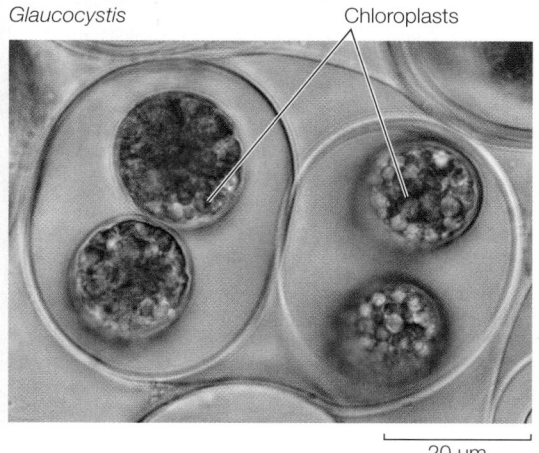

20 μm

Figure 27.2 Glaucophytes May Resemble Some of the Earliest Plantae The large chloroplasts of unicellular glaucophytes differ from chloroplasts of other Plantae in retaining a layer of peptidoglycan. This feature is thought to have been retained from the endosymbiotic cyanobacteria that gave rise to the chloroplasts of Plantae. The photograph shows a colony of two individuals, each with two chloroplasts.

(A) *Rhodymenia* sp.

1 cm

(B) *Calliarthron* sp.

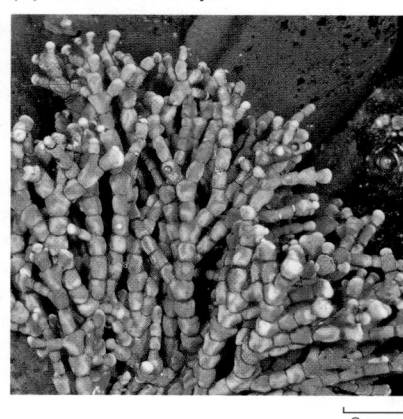

3 mm

Figure 27.3 Red Algae Contain a Red Accessory Photosynthetic Pigment **(A)** The rich red color of red algae is a result of the pigment phycoerythrin. **(B)** Coralline red alga is named for its coral-like appearance.

(as deep as 260 m if nutrient conditions are right and the water is clear enough to permit light to penetrate). A few red algae inhabit fresh water. Most grow attached to a substrate by a holdfast.

Despite their name, red algae don't always appear red in color. The ratio of two pigments—phycoerythrin (red) and chlorophyll *a* (green)—depends largely on the intensity of light that reaches the alga. In deep water, where light is dim, algae accumulate large amounts of phycoerythrin and appear red. But many species growing near the surface contain a higher concentration of chlorophyll *a* and thus appear bright green.

The remaining algal groups in Plantae are the various "green algae." Like land plants, the green algae contain both chlorophylls *a* and *b* and store their reserve of photosynthetic products as starch in chloroplasts. All the groups that share these features are commonly called **green plants** because both of their photosynthetic pigments are green.

The largest clade of "green algae" is the **chlorophytes**. There are more than 17,000 species of chlorophytes, most of which are aquatic (some are marine, though more are freshwater forms), although there are a few terrestrial forms that live in moist environments. Chlorophytes range in size from microscopic unicellular forms to multicellular forms many centimeters long and display an incredible variety of shapes and body forms. Surprisingly large and well-formed colonies of cells are found in some unicellular freshwater groups, such as the genus *Volvox* (**Figure 27.4A**). Certain cells in these colonies are specialized for reproduction. The cells in these colonies are not differentiated into specialized tissues and organs, as in land plants and animals, but they show vividly how the preliminary step of this great evolutionary innovation might have been taken.

Volvox is a colonial unicellular chlorophyte, but there are also many true multicellular species of chlorophytes. Some of these are

(A) *Volvox* sp.

Daughter colonies produced by reproductive cells   Somatic cells   Parent colony

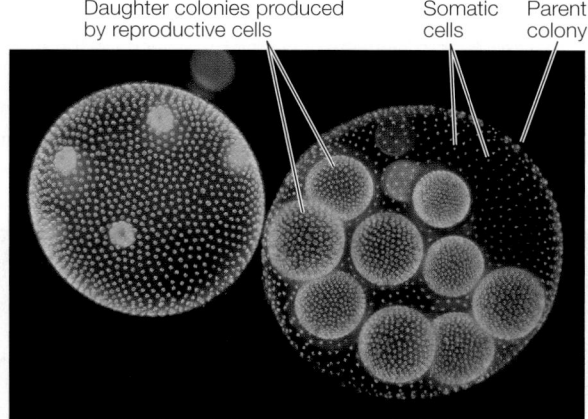

120 μm

(B) *Ulva latuca*

Figure 27.4 Chlorophytes Display a Wide Diversity of Forms **(A)** *Volvox* colonies are precisely spaced arrangements of individual cells. Specialized reproductive cells produce daughter colonies, which will eventually release new individuals. **(B)** Sea lettuce grows in marine waters and intertidal areas.

Can *Chlorella* Algae Be Grown in Municipal Sewage Wastewater for Biofuel Production?

experiment

Original Paper: Bhatnagar, A., M. Bhatnagar, S. Chinnasamy and K. C. Das. 2010. *Chlorella minutissima*—A promising fuel alga for cultivation in municipal wastewaters. *Applied Biochemistry and Biotechnology* 161: 523–536.

Growing algae for use in biofuel production requires an abundant source of freshwater that contains appropriate nutrients. Appropriate nutrients for growing algae are present in municipal wastewater, but other contaminants of wastewater (such as salts and detergents) could diminish algal growth. Ashish Bhatnagar and his colleagues at the University of Georgia conducted an experiment to see if *Chlorella minutissima* algae could be grown successfully in municipal wastewater.

HYPOTHESIS▶ *Chlorella minutissima* algae can be grown successfully in municipal wastewater, either with or without dilution from traditional growth medium.

METHOD

1. Grow replicate cultures of a standard inoculate of *Chlorella minutissima* algae in growth medium, as well as in 25%, 50%, 75%, and 100% municipal wastewater.
2. Control temperature and light conditions for all cultures.
3. Measure growth of the algae after 15 days by measuring the concentration of chlorophyll *a* present in each culture.

RESULTS

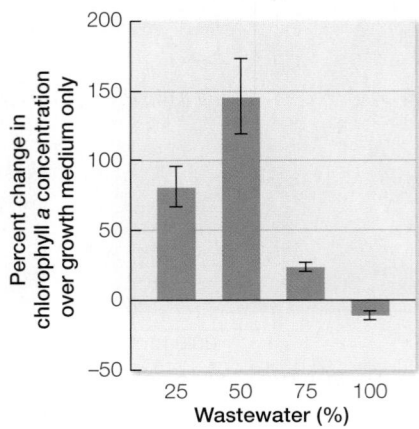

CONCLUSIONS▶ *Chlorella minutissima* algae can be grown effectively in municipal wastewater, and growth may even exceed production in standard growth medium.

Q: At what tested concentration of wastewater does the algae grow the best?

work with the data

After determining that municipal wastewater could be used to grow *Chlorella minutissima* for biofuel production, Ashish Bhatnagar and his colleagues explored the effects on *Chlorella* growth of various sources of carbon that were present in the wastewater. The table below shows three measures of algal growth in standard growth medium, as well as in the medium with added sources of carbon. Use these data to answer the following questions.

QUESTIONS▶

1. Calculate 95% confidence intervals for mean measures of chlorophyll *a* concentration in the various growth media (see Appendix B for

directions on calculating confidence intervals from standard errors). Which carbon source additives result in significantly higher or lower chlorophyll *a* concentrations (based on non-overlapping confidence intervals) compared with the growth medium only?

2. Plot the measures for chlorophyll *a* concentration, biomass, and cell count for each growth condition. Which two measures appear to be most strongly correlated? Can you think of reasons why the uncorrelated measure might show a different trend than the two correlated measures?

A similar **work with the data** exercise may be assigned in **LaunchPad**.

Measurement[a]	Growth medium only	Growth medium + propionate	Growth medium + acetate	Growth medium + citrate	Growth medium + sucrose	Growth medium + glucose
Chlorophyll *a* concentration (mg/L)	13.41 (1.24)	1.72 (0.03)	9.4 (0.82)	15.26 (0.59)	17.63 (0.86)	37.21 (3.76)
Biomass (mg/L)	899 (27.6)	120 (12.6)	630 (35.1)	1,007 (110.4)	1,164 (87.9)	2,456 (297.7)
Cell count (per 10^{-5} mL)	2.01 (0.17)	0.51 (0.02)	1.89 (0.08)	1.50 (0.07)	1.93 (0.17)	1.76 (0.18)

[a] Measures are means, with standard errors of mean shown in parentheses.

(A) *Coleochaete* sp.

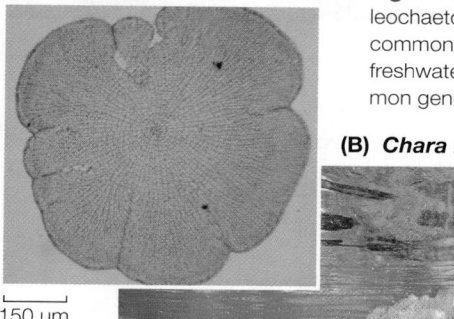

150 μm

Figure 27.5 The Closest Relatives of Land Plants **(A)** This species is a representative of the coleochaetophytes, the sister clade of stoneworts plus land plants. **(B)** The land plants probably evolved from a common ancestor shared with the stoneworts, an abundant group of multicellular green algae often found in freshwater pools and lakes (although a few species are found in marine environments). A species in the common genus *Chara* is shown here.

(B) *Chara rusbyana* (stonewort)

 Media Clip 27.1 **Reproductive Structures of Stoneworts**
www.Life11e.com/mc27.1

filamentous. Others, like species in the genus *Ulva* (**Figure 27.4B**), grow into thin, membranous sheets up to 30 centimeters in width.

Biologists are exploring mass production of several species of chlorophytes as a source of biofuels. One of the biggest limitations in using algae for biofuels relates to the cost of providing an appropriate growth medium for the algae. Growth media typically require fresh water and abundant sources of carbon, nitrogen, and phosphorus. These nutrients are already present in most municipal wasterwater supplies, and the removal of these nutrients by growing algae would help clean the water supply. **Investigating Life: Can *Chlorella* Algae Be Grown in Municipal Sewage Wastewater for Biofuel Production?** explores the poential of municipal wasterwater as a growth medium for algae used in biofuel production.

Two groups of green algae are the closest relatives of land plants

All green algae other than the chlorophytes form a group together with the land plants known as **streptophytes** (see Figure 27.1A). Several microscopic structural features, backed by clear evidence from molecular studies, indicate that the closest relatives of the land plants are two groups of aquatic green algae, the **coleochaetophytes** (**Figure 27.5A**) and the **stoneworts** (**Figure 27.5B**). Both of these multicellular algal groups retain their eggs in the parental organism, as land plants do. As in land plants, the cytoplasm of adjacent cells in these algal groups is connected through structures called plasmodesmata; they also share similarities in the details of mitosis and cytokinesis. Of these two groups, stoneworts are thought to be the sister group of land plants (see Figure 27.1A). The growth form of stoneworts is branching and apical (new growth occurs at the tips of branches), as in most land plants. Phylogenetic analysis of gene sequences has confirmed the close relationships of coleochaetophytes and stoneworts to the land plants.

There are ten major groups of land plants

One of the key synapomorphies of the **land plants** is development from an embryo that is protected by tissues of the parent plant. For

this reason, land plants are sometimes called **embryophytes** (*phyton*, "plant"). The green plants, the streptophytes, and the land plants have each been called "the plant kingdom" by different authorities; others take an even broader view and include red algae and glaucophytes as "plants." To avoid confusion in this chapter, we will use modifying terms (e.g., "land plants" or "green plants") to refer to the various clades of Plantae shown in Figure 27.1.

The land plants that exist today fall naturally into ten major clades (listed by their common names in the center column of **Table 27.1**). Members of seven of those clades possess well-developed vascular systems that transport materials throughout the plant body. We call these seven groups, collectively, the **vascular plants**, or **tracheophytes**, because they all possess fluid-conducting cells called **tracheids** (discussed in more detail in Chapter 34). The remaining three clades (liverworts, mosses, and hornworts) lack tracheids and are referred to collectively as **nonvascular land plants**. Note, however, that the three groups of nonvascular land plants do *not* form a clade (unlike the vascular plants, which *are* a clade).

▶ 27.1 recap

Primary endosymbiosis is a synapomorphy of the Plantae. The glaucophytes, the sister clade of the other Plantae, are unicellular algae that are similar to some of the earliest photosynthetic eukaryotes. The green plants contain chlorophyll *b* in addition to the chlorophyll *a* found in all Plantae.

learning outcomes

You should be able to:

- Plot the origin and distribution of key evolutionary innovations among aquatic Plantae.
- Describe the major clades of Plantae and their relationships to one another.

1. Explain the different possible uses of the term "plant."
2. Why doesn't the term "algae" designate a formal taxonomic group?
3. What are some of the key differences between glaucophytes, red algae, and the various clades of green algae?
4. What evidence supports the phylogenetic relationship between land plants and the various groups of aquatic green algae?

The green algal ancestors of the land plants lived at the margins of ponds or marshes, ringing them with a mat of dense green. It was from such a marginal habitat, which was sometimes wet and sometimes dry, that early plants made the transition onto land.

table 27.1 Classification of Land Plants

Group	Common name	Characteristics
NONVASCULAR LAND PLANTS		
Hepatophyta	Liverworts	No stomata; gametophyte flat or leafy
Bryophyta	Mosses	Filamentous stage; gametophyte leafy; sporophyte grows apically (at the tip)
Anthocerophyta	Hornworts	Embedded archegonia; sporophyte grows basally (i.e., from the ground)
VASCULAR PLANTS		
Lycopodiophyta	Lycophytes: Club mosses and allies	Microphylls in spirals; sporangia in leaf axils
Monilophyta	Horsetails, ferns	Simple leaves in whorls or frondlike compound leaves
Seed plants		
Gymnosperms		
Cycadophyta	Cycads	Compound leaves; swimming sperm; seeds on modified leaves
Ginkgophyta	Ginkgo	Deciduous; fan-shaped leaves; swimming sperm
Gnetophyta	Gnetophytes	Vessels in vascular tissue; opposite, simple leaves
Coniferophyta	Conifers	Seeds in cones; needle- or scalelike leaves
Angiosperms	Flowering plants	Endosperm; carpels; gametophytes much reduced; seeds contained within fruits

▶ key concept 27.2 Key Adaptations Permitted Plants to Colonize Land

How did the land plants arise? To address this question, we can compare land plants with their closest relatives among the green algae. The features that differ between the two groups include the adaptations that allowed the first land plants to survive in the terrestrial environment.

focus your learning

- Key innovations of Plantae facilitated their transition to land.
- Alternation of generations is a universal trait of the Plantae.

Adaptations to life on land distinguish land plants from green algae

Land plants first appeared in the terrestrial environment between 450 and 500 million years ago. How did they survive in an environment that differed so dramatically from the aquatic environment of their ancestors? While the water essential for life is everywhere in the aquatic environment, water is difficult to obtain and retain in the terrestrial environment.

No longer bathed in fluid, organisms on land faced potentially lethal desiccation (drying). Large terrestrial organisms had to develop ways to transport water to body parts distant from the source of the water. And whereas water provides aquatic organisms with support against gravity, a plant living on land must either have some other support system or sprawl unsupported on the ground. A land plant must also use different mechanisms for dispersing its gametes and progeny than its aquatic relatives, which can simply release them into the water.

Survival on land was facilitated by the evolution among plants of numerous adaptations, including:

- The *cuticle*, a membrane covered in waxes to retard water loss
- *Stomata*, small openings in leaves and stems that open and close to regulate gas exchange and water loss
- *Gametangia*, multicellular organs that enclose plant gametes and prevent them from drying out
- *Embryos*, young plants contained within a protective structure
- Certain *pigments* that afford protection against the mutagenic ultraviolet radiation that bathes the terrestrial environment
- Thick *spore walls* containing a polymer that protects the spores from desiccation and resists decay
- A *mutually beneficial association with fungi* (mycorrhizae) that promotes nutrient uptake from the soil

The cuticle may be the most important—and the earliest—of these features. Composed of several unique waxy lipids that coat the leaves and stems of land plants, the cuticle has several functions, the most obvious and important of which is to keep water from evaporating from the plant body.

As ancient plants colonized the land, they not only adapted to the terrestrial environment, they also modified it by contributing to the formation of soil. Acids secreted by plants helped break down rock, and the organic compounds produced by the breakdown of dead plants contributed nutrients to the soil. Such effects are repeated today wherever plants colonize and grow in new areas.

Life cycles of land plants feature alternation of generations

A universal feature of the life cycles of land plants is alternation of generations. Recall from Key Concept 26.3 the two hallmarks of alternation of generations:

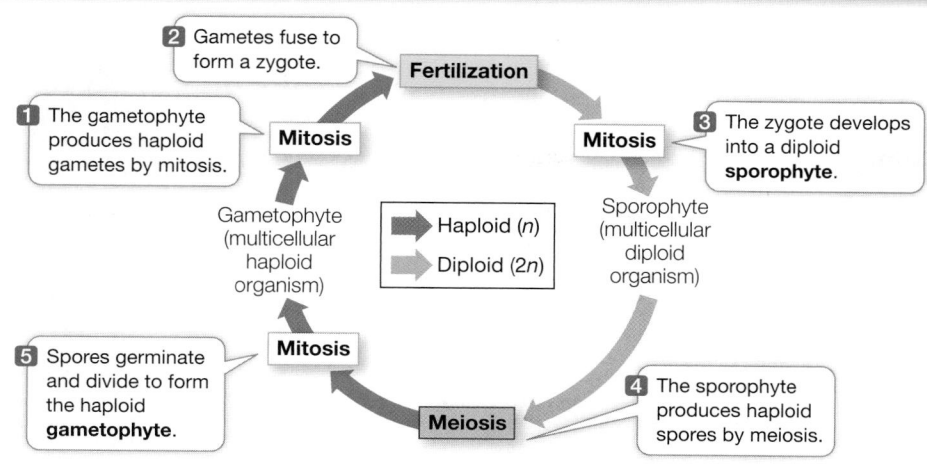

Figure 27.6 **Alternation of Generations in Land Plants**   A multicellular diploid sporophyte generation that produces spores by meiosis alternates with a multicellular haploid gametophyte generation that produces gametes by mitosis.

Q: What is a major difference in the haploid portion of the life cycle of plants and animals?

1. The life cycle includes both a multicellular diploid stage and a multicellular haploid stage.

2. Gametes are produced by mitosis, not by meiosis. Meiosis produces **spores** that develop into multicellular haploid organisms.

If we begin looking at the land plant life cycle at the single-cell stage—the diploid zygote—then the first phase of the cycle is the formation, by mitosis and cytokinesis, of a multicellular **embryo**, which eventually grows into a mature diploid plant. This multicellular diploid plant is called the **sporophyte** ("spore plant").

Cells contained within specialized reproductive organs of the sporophyte, called **sporangia** (singular *sporangium*), undergo meiosis to produce haploid, unicellular spores. By mitosis and cytokinesis, a spore develops into a haploid plant. This multicellular haploid plant, called the **gametophyte** ("gamete plant"), produces haploid gametes by mitosis. The fusion of two gametes (fertilization) forms a single diploid cell—the zygote—and the cycle is repeated (**Figure 27.6**).

The sporophyte generation extends from the zygote through the adult multicellular diploid plant and sporangium formation. In contrast, the gametophyte generation extends from the spore through the adult multicellular haploid plant to the gametes. The transitions between the generations are accomplished by fertilization and by meiosis. In all land plants, the sporophyte and the gametophyte differ genetically: the sporophyte has diploid cells, and the gametophyte has haploid cells.

There is a trend toward reduction of the gametophyte generation in plant evolution. In the nonvascular land plants, the gametophyte is larger, longer-lived, and more self-sufficient than the sporophyte. In those groups that appeared later in plant evolution, however, the sporophyte is the larger, more conspicuous, longer-lived, and more self-sufficient generation.

Nonvascular land plants live where water is readily available

The living species of nonvascular land plants are the liverworts, mosses, and hornworts. These three groups are thought to be similar in many ways to the earliest land plants. Most of these plants grow in dense mats, usually in moist habitats. Even the largest of these species are only about half a meter tall, and most are only a few centimeters tall or long. Why have they not evolved to be taller?

The probable answer is that they lack an efficient vascular system for transporting water and minerals from the soil to distant parts of the plant body.

The nonvascular land plants lack the true leaves, stems, and roots that characterize the vascular plants, although they have structures analogous to each. Their growth form allows water to move through the mats of plants by capillary action. They have leaf-like structures that readily catch and hold any water that splashes onto them. They are small enough that minerals can be distributed throughout their bodies by diffusion. As in most land plants, layers of maternal tissue protect their embryos from desiccation. Nonvascular land plants also have a cuticle, although it is often very thin (or even absent in some species) and thus is not highly effective in retarding water loss.

Most nonvascular land plants live on the soil or on vascular plants, but some grow on bare rock, on dead and fallen tree trunks, and even on buildings. Their ability to grow on such marginal surfaces results from a ***mutualistic association with fungi**. The earliest association of land plants with fungi dates back at least 460 million years. This mutualism probably facilitated the absorption of water and minerals, especially phosphorus, from the first soils.

---

*connect the concepts   Land plants of many groups have mutualistic associations with fungi, as described in Key Concept 29.2.

Nonvascular land plants are widely distributed over six continents and even exist (albeit very locally) on the coast of the seventh, Antarctica. Most are terrestrial. Although a few species live in fresh water, these aquatic species are descended from terrestrial ones. None live in the oceans.

The sporophytes of nonvascular land plants are dependent on the gametophytes

In the nonvascular land plants, the conspicuous green structure visible to the naked eye is the gametophyte. The gametophyte is photosynthetic and is therefore nutritionally independent; the sporophyte may or may not be photosynthetic, but it is always nutritionally dependent on the gametophyte and remains permanently attached to it.

Figure 27.7 illustrates the life cycle of a moss, which is typical of the life cycles of nonvascular land plants. A sporophyte produces

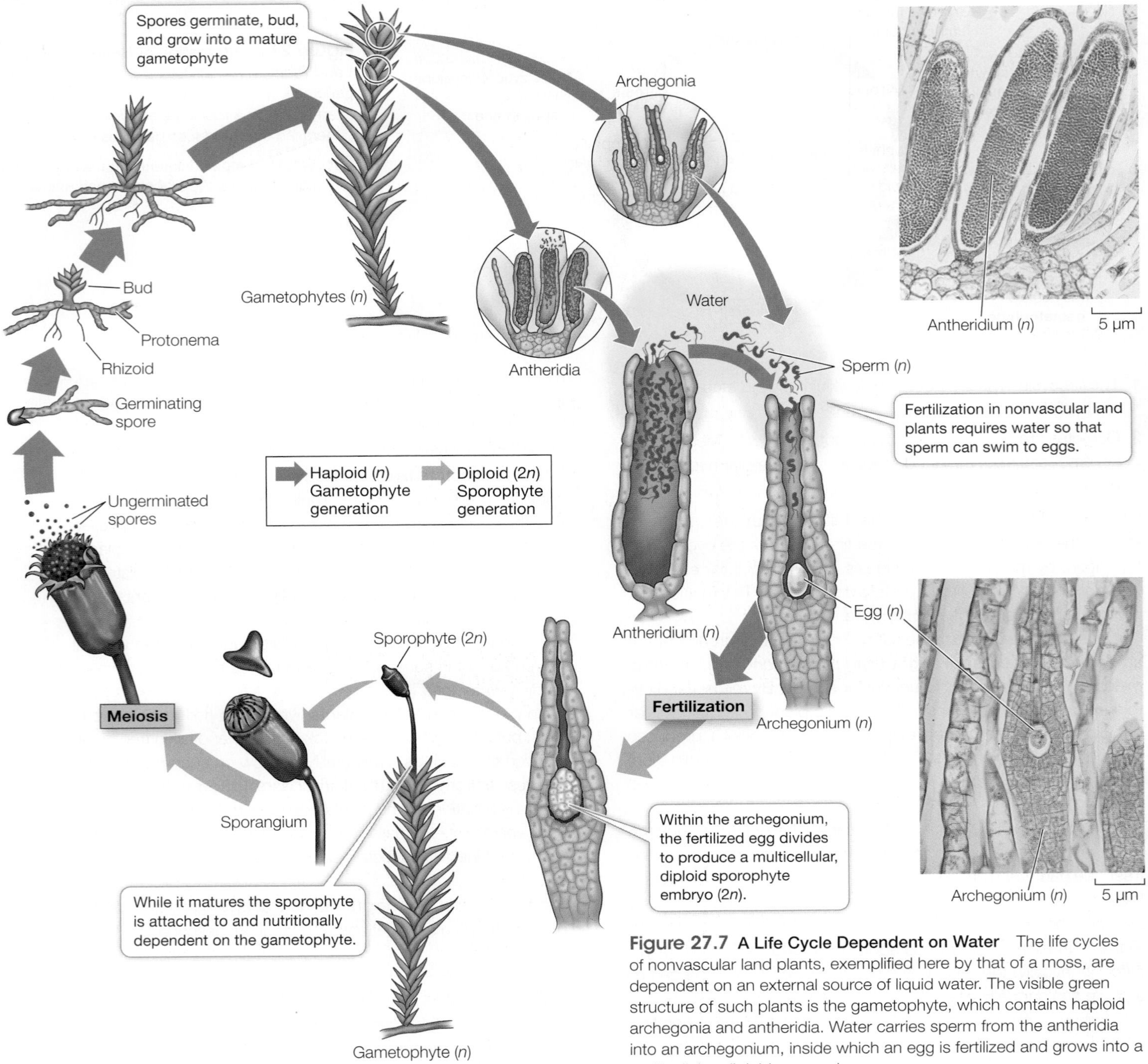

Spores germinate, bud, and grow into a mature gametophyte

Bud

Protonema

Rhizoid

Germinating spore

Ungerminated spores

Gametophytes (n)

Archegonia

Antheridia

Water

Antheridium (n)   5 μm

Sperm (n)

Fertilization in nonvascular land plants requires water so that sperm can swim to eggs.

Haploid (n) Gametophyte generation

Diploid (2n) Sporophyte generation

Antheridium (n)

Egg (n)

Meiosis

Sporophyte (2n)

Sporangium

While it matures the sporophyte is attached to and nutritionally dependent on the gametophyte.

Fertilization

Archegonium (n)

Within the archegonium, the fertilized egg divides to produce a multicellular, diploid sporophyte embryo (2n).

Archegonium (n)   5 μm

Gametophyte (n)

Figure 27.7 A Life Cycle Dependent on Water The life cycles of nonvascular land plants, exemplified here by that of a moss, are dependent on an external source of liquid water. The visible green structure of such plants is the gametophyte, which contains haploid archegonia and antheridia. Water carries sperm from the antheridia into an archegonium, inside which an egg is fertilized and grows into a multicellular, diploid sporangium.

Animation 27.1 Life Cycle of a Moss
www.Life11e.com/a27.1

Media Clip 27.2 Nonvascular Plant Reproduction
www.Life11e.com/mc27.2

unicellular haploid spores as products of meiosis within a sporangium. When a spore germinates, it gives rise to a multicellular haploid gametophyte whose cells contain chloroplasts and are thus photosynthetic. Eventually gametes form within specialized sex organs, called the **gametangia**. The **archegonium** is a multicellular, flask-shaped female sex organ with a long neck and a swollen base; each one produces a single egg. The **antheridium** is a male sex organ in which sperm, each bearing two flagella, are produced in large numbers. Archegonia and antheridia are produced on the same individual in many species, so each individual has both male and female reproductive structures. Adjacent individuals often fertilize

one another's gametes, however, which helps maintain genetic diversity in the population.

Once released from the antheridium, the sperm must swim or be splashed by raindrops to a nearby archegonium on the same or a neighboring plant—a constraint that reflects the aquatic origins of the nonvascular land plants' ancestors. The sperm are aided on their

These cups contain gemmae—small, lens-shaped outgrowths of the plant body, each capable of developing into a new plant.

The banana-like structures bear **archegonia**.

(A) *Lophocolea bidentata*   0.4 cm   (B) *Marchantia* sp.   0.3 cm   (C) *Marchantia* sp.   0.5 cm

Figure 27.8 Liverwort Diversity   **(A)** The gametophyte of a leafy liverwort. **(B)** The gametophytes of the thalloid liverwort *Marchantia* lie flat on the ground. **(C)** *Marchantia* gametophytes bearing archegonia.

journey by chemical attractants released by the egg or the archegonium. Before sperm can enter the archegonium, however, certain cells in the neck of the archegonium must break down, leaving a water-filled canal through which the sperm can swim to complete their journey. Notice that *all of these events require liquid water.*

Once sperm arrive at an egg, the nucleus of a sperm fuses with the egg nucleus to form a diploid zygote. Mitotic divisions of the zygote produce a multicellular, diploid sporophyte embryo. After the sporophyte grows out of the archegonium it produces a single sporangium, within which meiotic divisions produce spores and thus the next gametophyte generation.

Liverworts are the sister clade of the remaining land plants

There are about 9,000 species of **liverworts**. Most liverworts have leafy gametophytes (**Figure 27.8A**). Some have thalloid gametophytes: green, leaflike layers that lie flat on the ground (**Figure 27.8B and C**). The simplest liverwort gametophytes are flat plates of cells, a centimeter or so long, that produce antheridia or archegonia on their upper surfaces and rhizoids (rootlike filaments) on their lower surfaces.

Liverwort sporophytes are shorter than those of mosses and hornworts, rarely exceeding a few millimeters. The liverwort sporophyte has a stalk that raises the sporangium above the gametophyte. In most species, the stalk elongates by expansion of cells throughout its length. This elongation raises the sporangium above ground level, allowing the spores to be dispersed more widely. The sporangia of liverworts are simple: a globular sporangium wall surrounds a mass of spores. In some species of liverworts, spores are not released by the sporophyte until the surrounding sporangium wall rots. In other liverworts, however, the spores are thrown from the sporangium by structures that shorten and compress as they dry out. When the stress becomes sufficient, the compressed structure snaps back to its resting position, throwing spores in all directions.

Among the most familiar thalloid liverworts are species of the genus *Marchantia*. *Marchantia* is easily recognized by the characteristic

structures on which its male and female gametophytes bear their antheridia and archegonia (see Figure 27.8C). Like most liverworts, *Marchantia* also reproduces asexually by simple fragmentation of the gametophyte. In addition, *Marchantia* and some other liverworts and mosses reproduce asexually by means of gemmae (singular gemma), which are lens-shaped clumps of cells (see Figure 27.8B). In a few liverworts, the gemmae are held in structures called gemmae cups, which promote dispersal of the gemmae by raindrops.

▶ **Media Clip 27.3 Liverwort Life Cycle**
www.Life11e.com/mc27.3

Water and sugar transport mechanisms emerged in the mosses

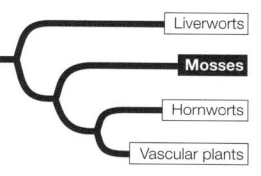

The most familiar of the nonvascular land plants are the **mosses**. These hardy little plants, of which there are about 15,000 species, are found in almost every terrestrial environment. They are often found on damp, cool ground, where they form thick mats (**Figure 27.9**). The mosses are the sister clade of the vascular plants plus the hornworts (see Figure 27.1B).

The mosses, along with the hornworts and vascular plants, share an advance over the liverworts in their adaptation to life on land: they have openings called **stomata**, which allow CO_2 to enter the plant body and allow water and O_2 to leave it. Stomata are a synapomorphy of mosses and all other land plants except liverworts.

In mosses, the gametophyte begins its development following spore germination as a branched, filamentous structure called a protonema (see Figure 27.7). Although the protonema looks a bit like a filamentous green alga, this structure is unique to the mosses. Some of the filaments contain chloroplasts and are photosynthetic; others, called rhizoids, are nonphotosynthetic and anchor the protonema to the substrate. After a period of linear growth, cells close to the tips of the photosynthetic filaments divide rapidly in three dimensions to form buds. The buds eventually develop a distinct tip, or apex, and produce the familiar leafy moss shoot with leaflike structures arranged spirally. These leafy shoots produce antheridia or archegonia (see Figure 27.7).

Some moss gametophytes are too large to transport enough water through their bodies solely by diffusion. Gametophytes and sporophytes of many mosses contain a type of cell called a hydroid, which dies and leaves a tiny channel through which water

(A)

Figure 27.9 Mosses Often Cover the Ground in Dense Mats **(A)** Dense layers of moss carpet a field of solidified volcanic lava in Iceland. **(B)** A close-up view of moss.

Gametophytes

Spore case

Sporophytes

(B) *Polytrichum commune*

Mosses of the genus *Sphagnum* (**Figure 27.10A**) often grow in cool, swampy places, where the plants begin to decompose in water after they die. Rapidly growing upper layers of moss compress the deeper-lying, decomposing layers. Partially decomposed plant matter is called **peat**. In some parts of the world, people derive the majority of their fuel from peat bogs (**Figure 27.10B**). *Sphagnum*-dominated peatlands cover a total area approximately half the size of the United States—more than 1 percent of Earth's surface. Millions of years ago, continued compression of peat composed primarily of other seedless plants gave rise to coal.

Hornworts have distinctive chloroplasts and stalkless sporophytes

The approximately 100 species of **hornworts** are so named because their sporophytes look like little horns (**Figure 27.11**). Hornworts appear at first glance to be liverworts with very simple gametophytes. Their gametophytes are flat plates of cells a few cells thick.

Liverworts
Mosses
Hornworts
Vascular plants

Hornworts have several characteristics that distinguish them from liverworts and mosses. First, the cells of hornworts each contain a single large, platelike chloroplast, whereas the cells of the other two groups contain numerous small, lens-shaped chloroplasts. Second, of the sporophytes in all three groups, those of the hornworts come closest to being capable of growth without a set limit. Liverwort and moss sporophytes have a stalk that stops growing as the sporangium matures, so elongation of the sporophyte is strictly limited. The hornwort sporophyte, however, has no stalk, and it is persistently green (a trait shared with vascular plants). A basal region of the sporangium remains capable of indefinite cell division, continuously producing new spore-bearing tissue above. The sporophytes of some hornworts growing in mild and continuously moist conditions can become as tall as 20 centimeters.

can travel. The hydroid is functionally similar to the tracheid, the characteristic water-conducting cell of the vascular plants, but it lacks lignin and the cell wall structure found in tracheids. The possession of hydroids and of a limited system for transport of sugar in some mosses shows that the term "nonvascular plant" is somewhat misleading when applied to these plants. Despite their simple system of internal transport, however, the mosses are not considered vascular plants because they lack tracheids or other components of xylem and phloem.

(A)

Sporophyte

Gametophyte

Sphagnum sp.

Figure 27.10 *Sphagnum* Moss **(A)** *Sphagnum* bogs are extremely dense growths of the moss shown here in a close-up view. **(B)** Workers mine a bog for peat, a fossil fuel formed from decomposing *Sphagnum* mosses.

(B)

The sporophytes of hornworts can reach 20 cm in height.

Gametophytes are flat plates a few cells thick.

Anthoceros sp.

Figure 27.11 Hornworts Get Their Name from Their Hornlike Sporophytes Unlike liverworts and mosses, hornworts have sporophytes that are persistently green. The hornworts share this trait with the vascular plants.

Eventually, however, the sporophyte's growth is limited by its simple transport system.

Hornworts have a symbiotic relationship with cyanobacteria that promotes their growth by providing them with access to nitrogen, which is often a limiting resource. Internal cavities of hornworts that are filled with mucilage can dry out, forming air-filled cavities through which cyanobacteria can enter. Cyanobacteria such as *Nostoc* convert atmospheric nitrogen gas into a form usable by their host plant; in return, the hornwort provides the symbiont with carbohydrates.

27.2 recap

The transition of plants to land required numerous adaptations, including the cuticle, stomata, gametangia, protected embryos, pigments, thick spore walls, and mutually beneficial associations with fungi. Nonvascular land plants rely on liquid water for reproduction.

learning outcomes

You should be able to:

- Describe one or more key innovations in land plants that were most important for their transition to a terrestrial environment, and justify your choice(s).
- Distinguish alternation of generations as seen in plants from the life cycles of animals.

1. Explain what is meant by "alternation of generations." Why do we not use this phrase for humans?
2. Describe key adaptations of plants to the terrestrial environment, and describe the distribution of those adaptations among the liverworts, mosses, and hornworts.

New features appeared in plants as they continued to adapt to the terrestrial environment. One of the most important of these was vascular tissues, the characteristic that defines the vascular plants.

key concept 27.3 Vascular Tissues Led to Rapid Diversification of Land Plants

The first plants possessing vascular tissues did not arise until tens of millions of years after the earliest nonvascular plants had colonized the land. But once vascular tissues arose, their ability to transport water and food throughout the plant body allowed the vascular plants to spread to new terrestrial environments and to diversify rapidly.

focus your learning

- Vascular tissues have several functions that allowed plants to colonize land.
- The precursors of vascular plants are extinct.
- Vascular plants evolved specialized structures and processes for transporting water, minerals, and the products of photosynthesis.
- Horsetails and ferns are more closely related than previously thought.
- Homospores produce a single type of spore, while heterospores produce two types: megaspores and microspores.

Vascular tissues transport water and dissolved materials

The key synapomorphy of the vascular plants is a well-developed vascular system containing two types of tissues that are specialized for the transport of materials from one part of the plant to another. One type of vascular tissue, the **xylem**, conducts water and minerals from the soil to aerial parts of the plant. Because some of its cell walls contain a stiffening substance called lignin, xylem also provides support against gravity in the terrestrial environment. The other type of vascular tissue, the **phloem**, conducts the products of photosynthesis from sites where they are produced or released to sites where they are used or stored. (We will discuss xylem and phloem in detail in Chapters 33 and 34.)

Although the vascular plants are an extraordinarily large and diverse group, one particular event was critical to their evolution. Sometime during the Paleozoic era, probably in the mid-Silurian (430 mya), a new cell type—the tracheid—evolved in sporophytes of the earliest vascular plants. The tracheid is the principal water-conducting element of the xylem in all vascular plants except the angiosperms (flowering plants) and gnetophytes—and tracheids persist even in these groups, along with a more specialized and efficient system derived from them.

The evolution of tracheids set the stage for the complete and permanent invasion of land by plants. First, these cells provided a pathway for the transport of water and mineral nutrients from a source of supply to regions of need in the plant body. Second, the cell walls of tracheids, stiffened by lignin, provided rigid structural support. This support is a crucial factor in a terrestrial environment because it allows plants to grow upward and thus compete for sunlight. A taller plant can intercept more direct sunlight (and thus conduct photosynthesis more readily) than a shorter plant, which may be shaded by the taller one. Increased height also improves the dispersal of spores.

Figure 27.12 Artist's Reconstruction of an Ancient Forest
Forests of the Carboniferous period were characterized by abundant vascular plants such as club mosses, ferns, and horsetails, some of which reached heights of 40 meters. Huge flying insects (see the opening of Chapter 24) thrived in these forests, which are the source of modern coal deposits.

The vascular plants featured another evolutionary novelty: a branching, independent sporophyte. A branching sporophyte body can produce more spores than an unbranched body, and it can develop in complex ways. The sporophyte of a vascular plant is nutritionally independent of the gametophyte at maturity. Among the vascular plants, the sporophyte is the large and obvious plant one normally notices in nature, in contrast to the relatively small, dependent sporophytes typical of most nonvascular land plants.

Vascular plants allowed herbivores to colonize the land

The initial absence of herbivores (plant-eating animals) on land helped make the first vascular plants successful. By the late Silurian period (about 425 mya), vascular plants were being preserved as fossils that we can study today. The proliferation of these plants made the terrestrial environment more hospitable to animals. Arthropods, vertebrates, and other animals moved onto land only after vascular plants became established there.

Trees of various kinds appeared in the Devonian period and dominated the landscape of the Carboniferous period (359–299 mya). Forests of lycophytes (club mosses) up to 40 meters tall, along with horsetails and tree ferns, flourished in the tropical swamps of what would become North America and Europe (**Figure 27.12**). Plant material from those forests sank into the swamps and was gradually covered by layers of sediment. Over millions of years, as

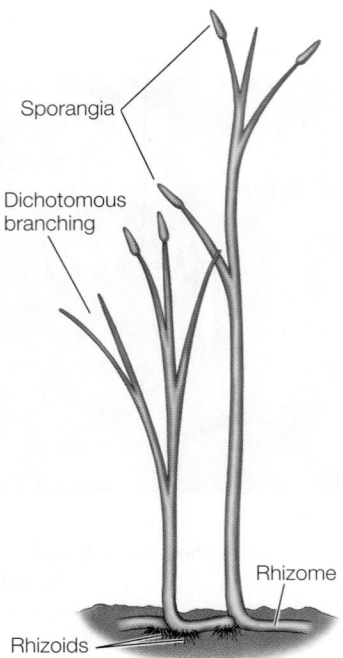

Figure 27.13 An Ancient Relative of the Vascular Plants
The extinct rhyniophyte *Aglaophyton major* lacked roots and leaves. It had a central column of xylem running through its stems but no true tracheids. A horizontal underground stem called a rhizome anchored the plant. The dichotomously branching aerial stems were less than 50 centimeters tall. Some stems were topped by sporangia.

the buried plant material was subjected to intense pressure and elevated temperatures, it was transformed into coal. Today that coal provides more than 40 percent of the electricity generated worldwide. The world's coal deposits, although huge, are not infinite, and humans are burning coal deposits at a far faster rate than they were produced.

In the subsequent Permian period, when the continents came together to form Pangaea, the continental interior became warmer and drier. The 200-million-year reign of the lycophyte–fern forests came to an end as they were replaced by forests of early gymnosperms.

The closest relatives of vascular plants lacked roots

The closest relatives of living vascular plants belonged to several extinct groups called **rhyniophytes** (**Figure 27.13**). The rhyniophytes were one of a very few types of land plants in the Silurian period. The landscape at that time probably consisted mostly of bare ground, with mats of nonvascular plants and stands of rhyniophytes in low-lying moist areas. Early versions of the structural features of the vascular plant groups appeared in the rhyniophytes of that time. These shared features strengthen the case for the origin of all vascular plants from a common nonvascular land plant ancestor.

Rhyniophytes did not have roots. They were apparently anchored in the soil by horizontal portions of stem called **rhizomes**, which bore water-absorbing unicellular filaments called **rhizoids**. These plants also bore aerial branches, and sporangia—homologous to the sporangia of mosses—were found at the tips of those branches. The branching pattern of rhyniophytes was **dichotomous**; that is,

Strobilus

Microphylls

(A) *Lycopodium digitatum*

Leaves
(in whorls)

Sporangia-bearing
structure

(B) *Equisetum pratense*

(C) *Marsilea mutica*

(D) *Dicksonia* sp.

Figure 27.14 Lycophytes and Monilophytes **(A)** Club mosses have microphylls arranged spirally on their stems. Strobili are visible at the tips of these stems. **(B)** Horsetails have a distinctive growth pattern in which the stem grows in segments above each whorl of leaves. These are fertile shoots with sporangia-bearing structures at the apex. **(C)** The floating leaves of a water fern. **(D)** Tree ferns dominate this forest in Oparara Basin Kahurangi National Park in New Zealand.

The sporangia of many club mosses are aggregated in conelike structures called **strobili** (singular *strobilus*; see Figure 27.14A), which are clusters of spore-bearing microphylls attached to the end of the stem. Other club mosses lack strobili and bear their sporangia on (or adjacent to) the upper surfaces of specialized microphylls.

the apex (tip) of the shoot divided to produce two equivalent new branches, with each pair of branches diverging at approximately the same angle from the original stem.

The lycophytes are sister to the other vascular plants

The club mosses and their relatives, the spike mosses and quillworts, are collectively called **lycophytes**. The lycophytes are the sister clade to the remaining vascular plants (see Figure 27.1B). There are relatively few (just over 1,200) surviving species of lycophytes.

The lycophytes have true roots that branch dichotomously. The arrangement of vascular tissue in their stems is simpler than in other vascular plants. They bear simple leaflike structures called **microphylls**, which are arranged spirally on the stem (**Figure 27.14A**). Growth in lycophytes comes entirely from apical cell division. Branching in the stems, which is also dichotomous, occurs by division of an apical cluster of dividing cells.

Horsetails and ferns constitute a clade

The horsetails and ferns were once thought to be only distantly related. From analysis of gene sequences we now know that they form a clade, the **monilophytes**. In the monilophytes—as in the seed plants, to which they are the sister clade (see Figure 27.1B)—the main stem is differentiated from side branches (including the leaves derived from these branches). This pattern contrasts with the dichotomous branching characteristic of the lycophytes and rhyniophytes, in which each split gives rise to two branches of similar size.

Today there are only about 15 species of **horsetails**, all in the genus *Equisetum*. The horsetails have reduced true leaves that form in distinct whorls (circles) around the stem (**Figure 27.14B**). Horsetails are sometimes called "scouring rushes" because rough silica deposits found in their cell walls make them useful for cleaning. They have true roots that branch irregularly. Horsetails have a large sporophyte and a small gametophyte, both independent of each other.

The first ferns appeared during the Devonian period; today this group comprises more than 12,000 species. Analyses of gene

Figure 27.15 Life Cycle of a Fern   The most conspicuous stage in the fern life cycle is the mature diploid sporophyte, shown at the bottom of this diagram. The inset shows sori on the underside of a fern leaf. Each sorus contains many spore-producing sporangia.

▶ **Activity 27.1 The Fern Life Cycle**
www.Life11e.com/ac27.1

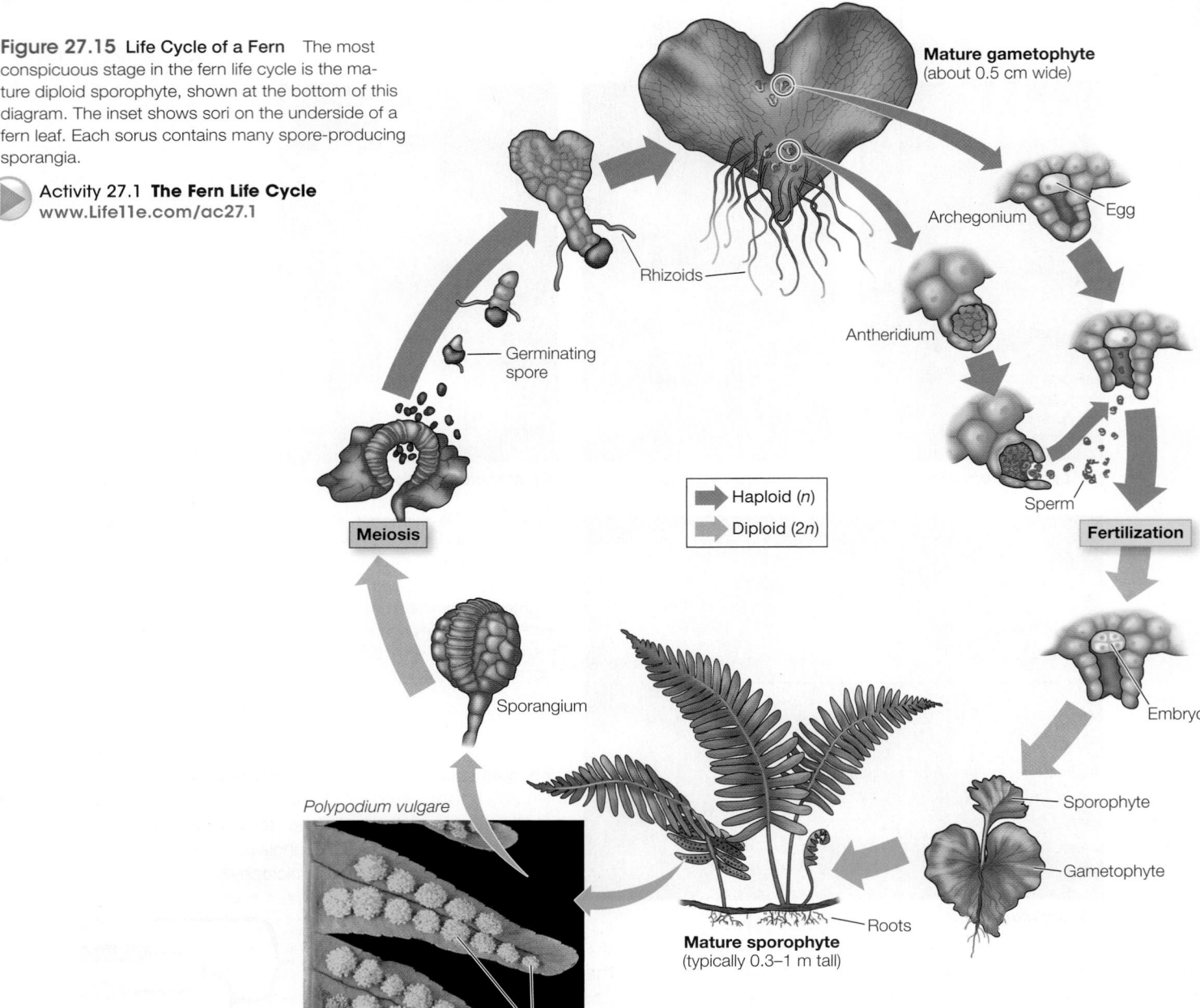

Mature gametophyte (about 0.5 cm wide)

Rhizoids

Germinating spore

Archegonium — Egg

Antheridium

Sperm

Fertilization

Meiosis

Haploid (*n*)
Diploid (*2n*)

Sporangium

Embryo

Polypodium vulgare

Sporophyte

Gametophyte

Roots

Mature sporophyte (typically 0.3–1 m tall)

Sori (clusters of sporangia)

sequences indicate that a few species traditionally allied with ferns may in fact be more closely related to horsetails than to ferns. Nonetheless, the majority of ferns form a monophyletic group.

Although most ferns are terrestrial, a few species live in shallow fresh water (**Figure 27.14C**). Terrestrial ferns are characterized by large leaves with branching vascular strands (**Figure 27.14D**). Some fern leaves become climbing organs and may grow to be as long as 30 meters.

In the alternating generations of a fern, the gametophyte is small, delicate, and short-lived, but the sporophyte can be very large and can sometimes survive for hundreds of years (**Figure 27.15**). Ferns require liquid water for the transport of the male gametes to the female gametes, so most ferns inhabit shaded, moist woodlands and swamps. The sporangia of ferns are typically borne on a stalk in clusters called **sori** (singular *sorus*). The sori are found on the

undersurfaces of the leaves, sometimes covering the entire undersurface and sometimes located at the edges.

The vascular plants branched out

Several features that were new to the vascular plants evolved in lycophytes and monilophytes. Roots probably had their evolutionary origins as branches, either of a rhizome or of the aboveground portion of a stem. These branches presumably penetrated the soil and branched further. The underground portions could anchor the plant firmly, and even in this primitive condition, they could absorb water and minerals.

The microphylls of lycophytes were probably the first leaflike structures to evolve among the vascular plants. Microphylls are usually small and only rarely have more than a single vascular strand, at least in existing species. Some biologists believe that microphylls

(A) Microphylls

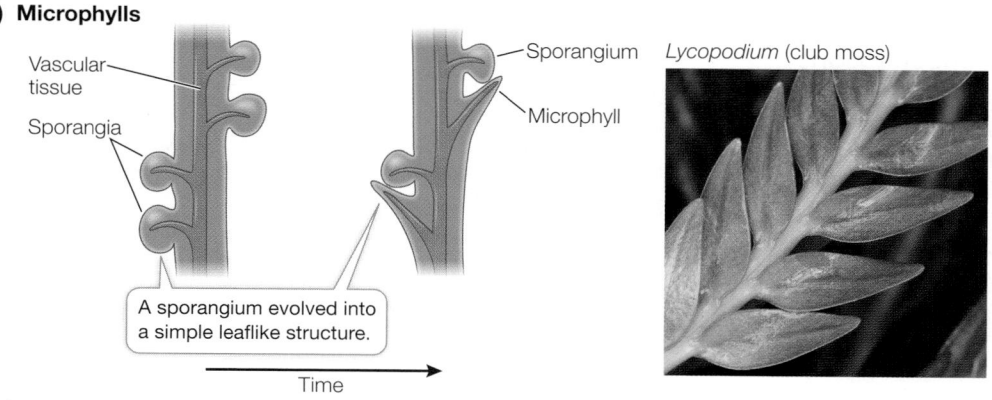

Lycopodium (club moss)

Figure 27.16 Evolution of Leaves
(A) Microphylls are thought to have evolved from sterile sporangia. **(B)** The megaphylls of monilophytes and seed plants may have arisen as photosynthetic tissue developed between branch pairs that were "left behind" as dominant branches overtopped them.

(B) Megaphylls

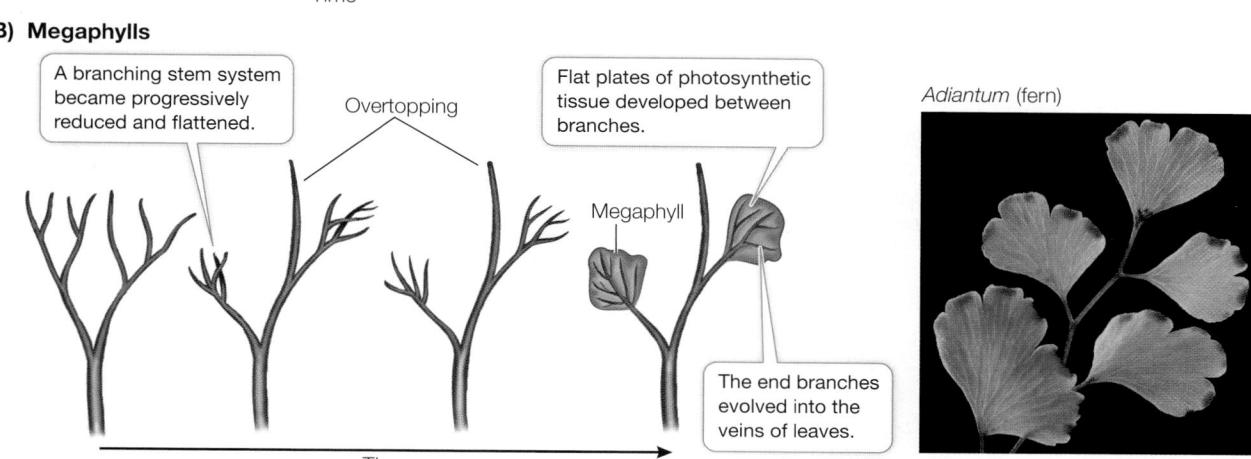

Adiantum (fern)

had their evolutionary origins as sterile sporangia (**Figure 27.16A**). A typical feature of this type of leaf is a vascular strand that departs from the vascular system of the stem in such a way that the structure of the stem's vascular system is scarcely disturbed. This pattern was evident even in the lycophyte trees of the Carboniferous period, many of which had microphylls many centimeters long.

The monilophytes and seed plants constitute a clade called the **euphyllophytes** (*eu*, "true," + *phyllon*, "leaf"). An important synapomorphy of the euphyllophytes is **overtopping**, a growth pattern in which one branch differentiates from and grows beyond the others (**Figure 27.16B**). Overtopping would have given these plants an advantage in the competition for light, enabling them to shade their dichotomously branching competitors. The overtopping growth of the euphyllophytes also allowed a new type of leaflike structure to evolve. This larger, more complex leaf is called a **megaphyll**. The megaphyll is thought to have arisen from the flattening of a portion of a branching stem system that exhibited overtopping growth. This change was followed by the development of photosynthetic tissue between the members of overtopped groups of branches, which had the advantage of increasing the photosynthetic surface area of those branches.

The first megaphylls, which were very small, appeared in the Devonian period. We might expect that evolution should have led swiftly to the appearance of more and larger megaphylls because of their greater photosynthetic capacity. However, it took some 50 million years, until the Carboniferous period, for large megaphylls to become common. Why should this have been so, especially given that other advances in plant structure were taking place during that time?

According to one theory, the high concentration of CO_2 in the atmosphere during the Devonian period reduced selection for the *****stomata** that allow a leaf to take up CO_2 for use in photosynthesis. With more CO_2 available, fewer stomata were needed. In the Devonian, larger leaves with a limited number of stomata would have absorbed heat from sunlight, but they would have been unable to lose heat fast enough by evaporation of water through their stomata. The resulting overheating would have been lethal. Recent research has supported this hypothesis, indicating that larger megaphylls evolved only as CO_2 concentrations dropped over millions of years (**Figure 27.17**).

*****connect the concepts**   As explained in Key Concept 35.3, stomata are pores in the epidermis of plants that open and close to regulate uptake of CO_2 into the plant and guard against excessive diffusion of water out of the plant. Various environmental factors trigger the opening and closing of the stomata, including light, the CO_2 level in photosynthetic cells of the plant, and water availability.

Heterospory appeared among the vascular plants

In the lineages of present-day, seedless vascular plants that are most similar to their ancestors, the gametophyte and the sporophyte are independent, and both are usually photosynthetic. The spores produced by the sporophyte are of a single type and develop into a single type of gametophyte that bears both female and male reproductive organs (see Figure 27.15). Such plants, which bear a single type of spore, are said to be **homosporous** (**Figure 27.18A**).

experiment

Figure 27.17 Atmospheric CO$_2$ Concentrations and the Evolution of Megaphylls

Original Paper: Osborne, C. P., D. J. Beerling, B. H. Lomax and W. G. Chaloner. 2004. Biophysical constraints on the origin of leaves inferred from the fossil record. *Proceedings of the National Academy of Sciences USA* 101: 10360–10362.

High concentrations of atmospheric CO$_2$ during the first part of the Devonian may have limited the evolution of leaf size. C. P. Osborne and colleagues compared the leaf sizes of fossil plants against estimates of CO$_2$ concentrations in the atmosphere at the time the plants were alive.

HYPOTHESIS▶ High atmospheric CO$_2$ concentrations during the early Devonian, and the resulting lack of selection for more stomata, kept leaf sizes small.

METHOD

1. Analyze 300 plant fossils from the Devonian and Carboniferous periods and measure the sizes of their leaves.
2. Compare the pattern of change in leaf size with that of the estimated change in atmospheric CO$_2$ concentrations over the same time span.

RESULTS

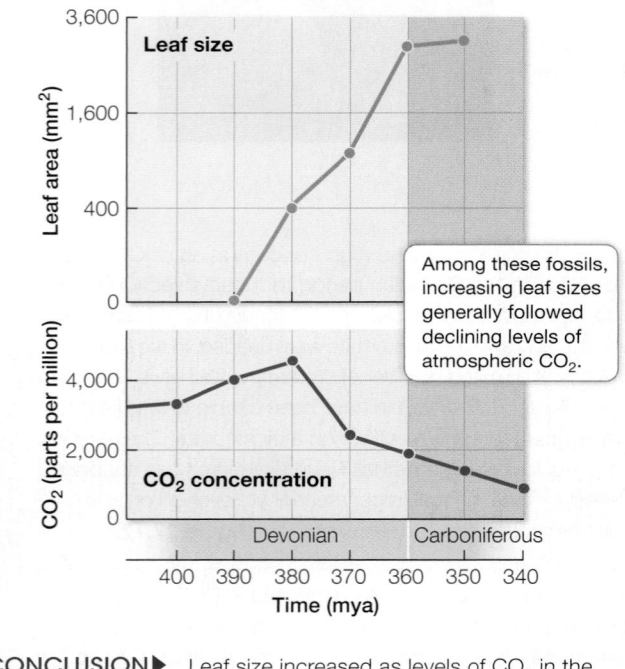

CONCLUSION▶ Leaf size increased as levels of CO$_2$ in the atmosphere decreased.

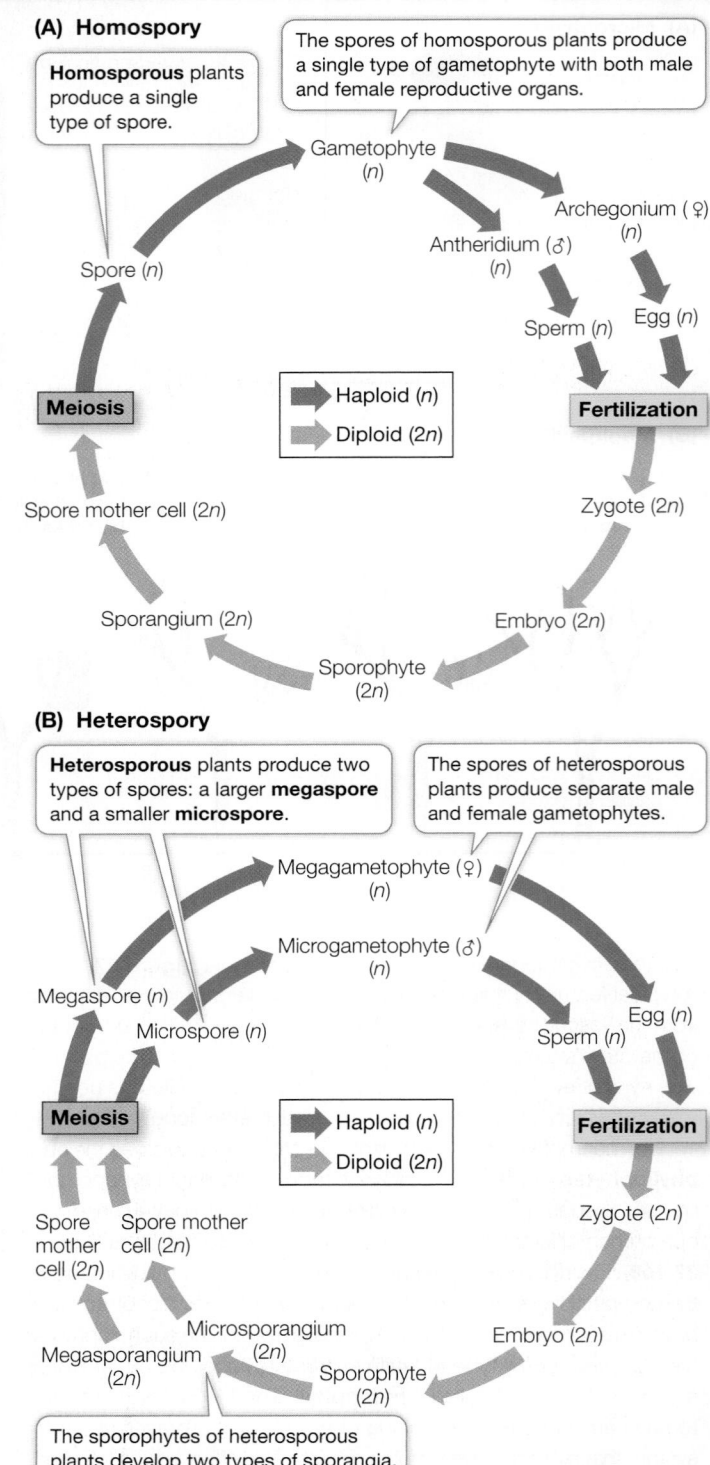

(A) Homospory

Homosporous plants produce a single type of spore.

The spores of homosporous plants produce a single type of gametophyte with both male and female reproductive organs.

(B) Heterospory

Heterosporous plants produce two types of spores: a larger **megaspore** and a smaller **microspore**.

The spores of heterosporous plants produce separate male and female gametophytes.

The sporophytes of heterosporous plants develop two types of sporangia.

Figure 27.18 Homospory and Heterospory **(A)** Homosporous plants bear a single type of spore. Each gametophyte has two types of sex organs, antheridia (male) and archegonia (female). **(B)** Heterosporous plants bear two types of spores that develop into distinctly male and female gametophytes.

Activity 27.2 **Homospory**
www.Life11e.com/ac27.2

Activity 27.3 **Heterospory**
www.Life11e.com/ac27.3

A system with two distinct types of spores evolved somewhat later. Plants of this type are said to be **heterosporous** (Figure 27.18B). In heterospory, one type of spore—the **megaspore**—develops into a specifically female gametophyte (a **megagametophyte**) that produces only eggs. The other type, the **microspore**, is smaller and develops into a male gametophyte (a **microgametophyte**) that produces only sperm. The sporophyte produces megaspores in small numbers in **megasporangia** and microspores in large numbers in **microsporangia**. Heterospory affects not only the spores and the gametophytes but also the sporophyte plant itself, which must develop two types of sporangia.

The earliest vascular plants were all homosporous, but heterospory evidently evolved several times independently among later groups of vascular plants. The fact that heterospory evolved repeatedly suggests that it affords selective advantages. Subsequent evolution in the land plants featured ever greater specialization of the heterosporous condition.

All of the vascular plant groups we have discussed thus far disperse by means of spores. The embryos of these seedless vascular plants develop directly into sporophytes, which either survive or die, depending on environmental conditions. The spores of some seedless plants may remain dormant and viable for long periods, but the embryos of seedless plants are relatively unprotected. Greater protection of the embryo evolved in the seed plants, which we will consider in the next chapter.

▶27.3 recap

Vascular plants are characterized by a vascular system specialized for the transport of materials from one part of the plant to another. A new type of cell, the tracheid, marked the origin of this group. Later evolutionary events included the appearance of roots, leaves, and heterospory.

learning outcomes

You should be able to:

- Explain the basic functions of vascular tissues.
- Describe important structures in vascular plants that probably first evolved in lycophytes.
- Compare possible selective advantages of heterospory over homospory.

1. How do the vascular tissues xylem and phloem serve the vascular plants?
2. Describe the evolution and distribution of different kinds of leaves and roots among the vascular plants.
3. Explain the concept of heterospory. How does heterospory provide selective advantages over homospory?

▶ investigating**life**

 Given that petroleum is derived naturally from green algae, can humans use green algae to produce oil commercially?

Scientists are developing new methods for growing green algae for the production of biofuels (fuels produced directly from living organisms, such as biodiesel). Some species of green algae can produce up to 60 percent of their dry weight in oil. So biofuels can certainly be produced from green algae, although the process is not yet commercially viable. Like conventional fossil fuels, biofuels release carbon dioxide into the atmosphere when burned. In the production of biofuels, however, algae remove carbon dioxide from the atmosphere, so the use of these fuels is more sustainable, and results in less accumulation of CO_2 in the atmosphere over time, than the use of fossil fuels.

Future directions

The primary commercial limitations to growing algae for biofuels include the need to establish efficient growing facilities, water needs, costs of fertilizers, costs and difficulties associated with harvest and refining, and labor expenses. As we explored in Investigating Life: Can *Chlorella* Algae Be Grown in Municipal Sewage Wastewater for Biofuel Production? considerable research is being conducted on ways to reduce these costs, such as growing algae for biofuels

in municipal wastewater. Many new methods for growing and harvesting algae are being developed (Figure 27.19). Once some of the technical difficulties have been overcome, it is possible that commercial production of biofuels from algae will provide a significant source of energy for humans.

Figure 27.19 Biodiesel from Algae In this algal cultivation system for biofuel production, algae are grown in tubes of clear plastic

Chapter Summary 27

▶ 27.1 Primary Endosymbiosis Produced the First Photosynthetic Eukaryotes

- Primary endosymbiosis gave rise to chloroplasts and the subsequent diversification of the **Plantae**. The descendants of the first photosynthetic eukaryote include **glaucophytes**, **red algae**, several groups of green algae, and **land plants**, all of which contain chlorophyll *a*. **Review Figure 27.1**

- **Green plants**, which include the green algae and the land plants, are characterized by the presence of chlorophyll *b* (in addition to chlorophyll *a*). **Review Figure 27.1**

- Land plants, also known as **embryophytes**, arose from an aquatic green algal ancestor related to today's **stoneworts**. Land plants develop from embryos that are protected by parental tissue. **Review Figure 27.1**

▶ 27.2 Key Adaptations Permitted Plants to Colonize Land

- The acquisition of a cuticle, **stomata**, **gametangia**, a protected **embryo**, protective pigments, thick spore walls with a protective polymer, and mutualistic associations with fungi were all adaptations of land plants to terrestrial life.

- All land plant life cycles feature alternation of generations, in which a multicellular diploid **sporophyte** alternates with a multicellular haploid **gametophyte**. **Review Figure 27.6**

- The **nonvascular land plants** comprise the **liverworts**, **mosses**, and **hornworts**. These groups lack specialized vascular tissues for the conduction of water or nutrients through the plant body.

- The life cycles of nonvascular land plants depend on liquid water. The sporophyte is usually smaller than the gametophyte and depends on it for water and nutrition. **Review Figure 27.7, Animation 27.1**

- Liverworts lack stomata, but they are present in mosses, hornworts, and vascular plants.

- Mosses have simple systems for the transport of water and sugars throughout the plant.

- Hornworts have a persistently green sporophyte, a characteristic shared with vascular plants. Hornworts also have a symbiotic relationship with cyanobacteria, which provide the hornworts with a usable form of nitrogen.

▶ 27.3 Vascular Tissues Led to Rapid Diversification of Land Plants

- The **vascular plants** have a vascular system consisting of **xylem** and **phloem** that conducts water, minerals, and products of photosynthesis through the plant body. The vascular system includes cells called **tracheids**.

- The **rhyniophytes**, the earliest known vascular plants, are known to us only in fossil form. They lacked true roots and leaves but possessed **rhizomes** and **rhizoids**. **Review Figure 27.13**

- The **lycophytes** (club mosses and relatives) have only small, simple leaflike structures (**microphylls**). **Monilophytes** (which include **horsetails** and ferns) have true leaves, and so together with seed plants are known as **euphyllophytes**.

- Unlike in nonvascular land plants, the diploid sporophyte is the more conspicuous life stage of lycophytes and monilophytes. **Review Figure 27.15, Activity 27.1**

- Microphylls probably evolved from sterile sporangia. **Megaphylls** (true leaves) may have resulted from the flattening and reduction of a portion of a stem system with **overtopping** growth. **Review Figure 27.16**

- The earliest-diverging groups of vascular plants are **homosporous**, but **heterospory**—the production of distinct **megaspores** and **microspores**—has evolved several times. Megaspores develop into female **megagametophytes**; microspores develop into male **microgametophytes**. **Review Figure 27.18, Activities 27.2, 27.3**

> Go to **LearningCurve** (in **LaunchPad**) for dynamic quizzing that helps you solidify your understanding of this chapter. **LearningCurve** adapts to your responses, giving you the practice you need to master each key concept.

Apply What You've Learned

Review

27.1 Land plants fall into ten major clades.

27.3 The precursors of vascular plants are extinct.

27.3 Horsetails and ferns are more closely related than previously thought.

Original Paper: Kenrick, P. and P. R. Crane. 1997. The origin and early evolution of plants on land. *Nature* 389: 33–39.

Together, phylogenetic analysis and the fossil record reinforce each other in deepening our understanding of the evolution of taxa. The figure below shows the evolution of early land plants, reflecting the fossil record and phylogenetic relationships. Use the figure to answer the questions that follow.

Questions

1. About how long ago and in what geological period did the last common ancestor of ferns and seed plants live?

2. Spike mosses are a group of lycophytes, some species of which are popular for cultivation. If you saw fossils of spike mosses in a geological stratum, would you expect to find plants closely related to the last common ancestor of ferns and seed plants in the same rock formation? Explain. If not, would you expect to find the last common ancestor of ferns and seed plants in a higher or lower stratum?

3. If you saw rhyniophyte fossils in a geological stratum, would you expect to find plants closely related to the last common ancestor of ferns and seed plants in the same stratum? Explain.

4. About how long ago and in what geological period did the last common ancestor of embryophytes live? What significant change in the life history of plants occurred in this lineage?

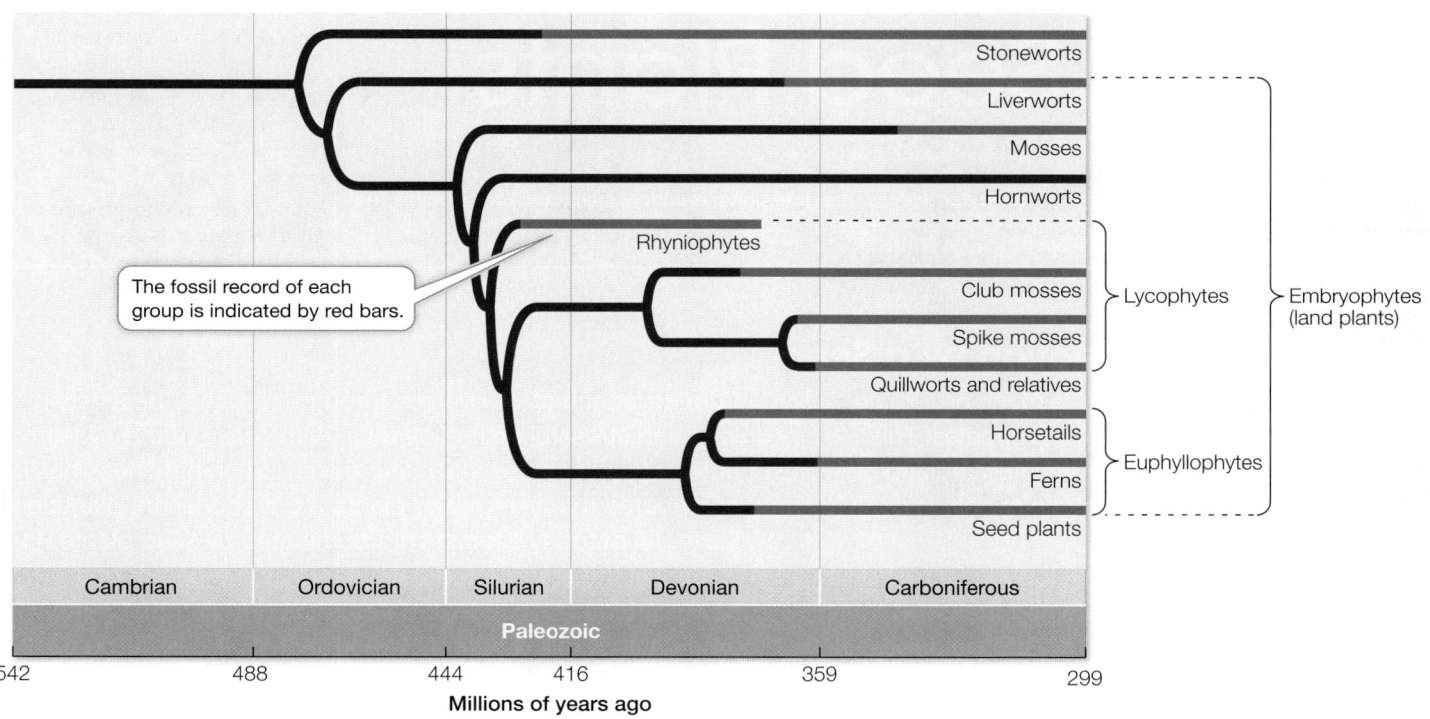

28

key concepts

28.1 Pollen, Seeds, and Wood Contributed to the Success of Seed Plants

28.2 Once Dominant Gymnosperms Still Thrive in Some Environments

28.3 Flowers and Fruits Led to Increased Diversification of Angiosperms

28.4 Plants Play Critical Roles in Terrestrial Ecosystems

The Evolution of Seed Plants

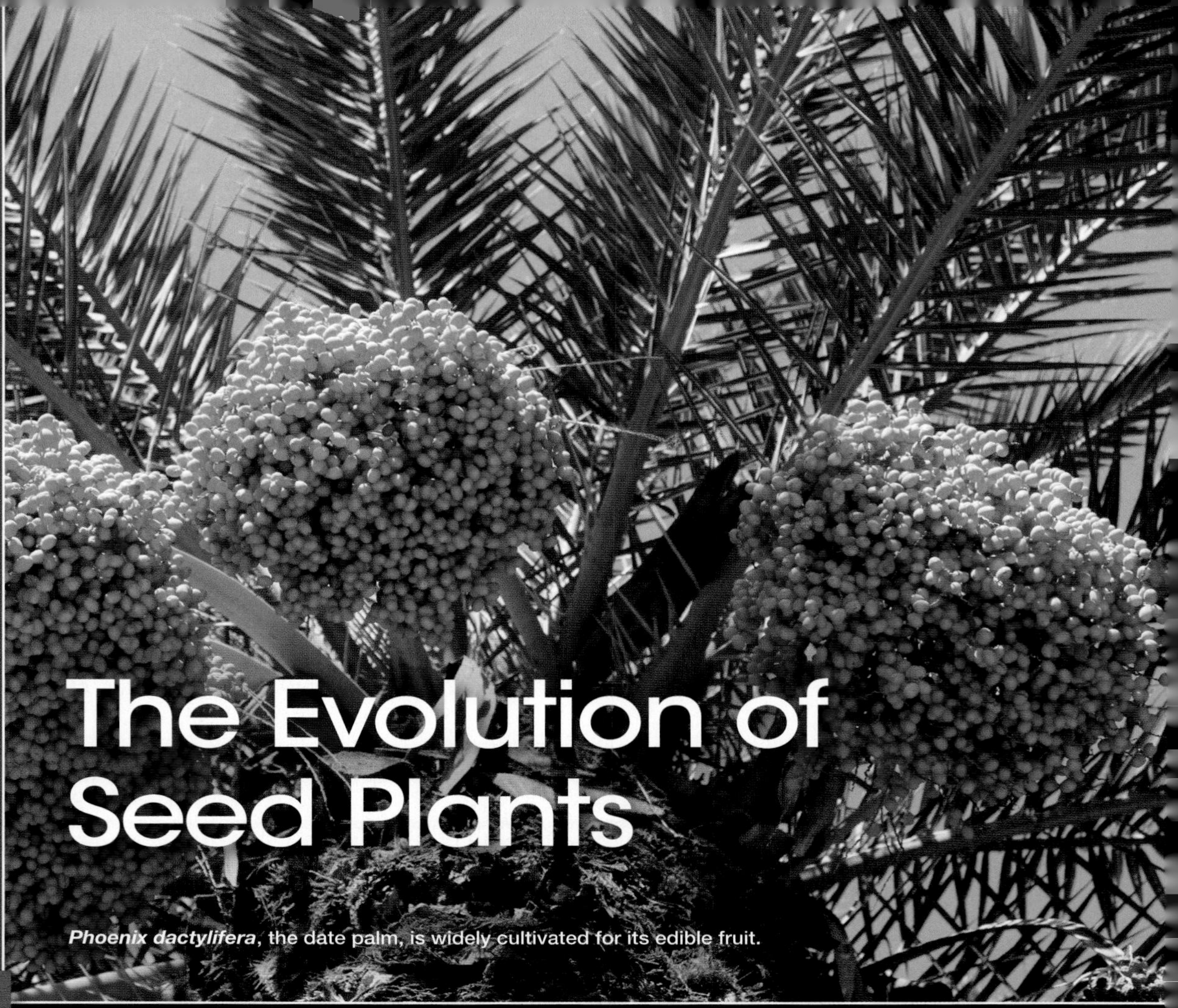

Phoenix dactylifera, the date palm, is widely cultivated for its edible fruit.

investigatinglife

Brought Back from Extinction by a Seed

The Judean date is a fruit that was once much prized, admired for its nutritional and medicinal properties. This fruit was the source of the “honey” in the biblical “land of milk and honey.” Today that ancient strain of date is gone. Or is it?

About 2,000 years ago, a seed developed in a fruit on a Judean date palm. The fruit that contained that seed found its way to a storeroom in the fortress Masada in Judea. In 73 C.E. a band of Jewish zealots involved in a revolt against Rome fled to this refuge with their families and provisions. The fortress was eventually sacked and left in ruins.

Twenty centuries later, archeologists working in Masada discovered the date seed that was long ago stored in the Masada fortress and confirmed its age. The previous record for seed survival and germination was 1,300 years, held by lotus seeds that recently germinated under the care of scientists in China. But botanist Elaine Solowey succeeded in making the 2,000-year-old date seed germinate! For 10 years the seedling has thrived, and is now mature and producing pollen.

Seeds are important structures for the evolutionary survival of plants. Each seed protects an enclosed plant embryo from environmental extremes through what may be a very long and stressful resting period—in the case of the Judean date, many centuries in a harsh desert. Such hardiness has contributed to making seed plants the predominant plants on Earth. All of today’s forests are dominated by seed plants.

So will the resurrected palm serve as the parent of a new population of Judean dates, thus bringing that genotype back from extinction? Unfortunately, it cannot do so alone, because date palms have separate sexes, and the single known individual is a male that can only produce pollen. However, the search is on for other ancient seeds of Judean dates from archeological sites, and it is hoped that one of those will grow into a female plant that can produce new fruit and seeds. Even if a female Judean date seed cannot be found, the male Judean date’s pollen can be used to introduce valuable genes back into other, still living, strains of dates.

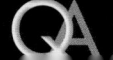

Q A How long can most seeds survive, and why is seed dormancy important?

28.1 Pollen, Seeds, and Wood Contributed to the Success of Seed Plants

By the late Devonian period, more than 360 million years ago, Earth was home to a great variety of land plants, many of which we discussed in Chapter 27. The land plants shared the hot, humid terrestrial environment with insects, spiders, centipedes, and early tetrapods. These plants and animals evolved together, each acting as agents of natural selection on the other.

focus your learning

- Seed plants are heterosporous.
- The seed is a well-protected resting stage.
- Secondary growth increases the diameter of stems and roots in many seed plants.

In the Devonian, a new innovation appeared when some plants developed extensively thickened woody stems. Among the first plants with this adaptation were seedless vascular plants called **progymnosperms**, all species of which are now extinct. The progymnosperms included many large trees.

Another innovation, the **seed**, arose in the seed plants. Seeds provide a secure and lasting structure that protects the dormant stage of the embryo. A plant embryo may safely wait within its seed (in some cases for many years, or even centuries) until conditions are right for germination.

The earliest fossil evidence of seed plants is found in late Devonian rocks. Like the progymnosperms, these now-extinct **seed ferns** were woody. They possessed fernlike foliage but had seeds attached to their leaves. By the end of the Permian, other groups of seed plants had become dominant (**Figure 28.1**). Today's living seed plants fall into two major groups, the **gymnosperms** (such as pines and cycads) and the hugely diverse group known as the **angiosperms** (flowering plants).

Features of the seed plant life cycle protect gametes and embryos

In Key Concept 27.2 we described a major trend in land plant evolution: the sporophyte became less dependent on the gametophyte, which became smaller in relation to the sporophyte. This trend continued with the seed plants, whose gametophyte generation is reduced even further than that of the ferns (**Figure 28.2**). The haploid seed plant gametophyte develops partly or entirely while attached to and nutritionally dependent on the diploid sporophyte.

Among the seed plants, only the earliest diverging groups of gymnosperms (including modern cycads and ginkgos) have swimming

Figure 28.1 The Fossil Record of Seed Plant Evolution Woody growth evolved in the seedless progymnosperms. The now-extinct seed ferns had woody growth, fernlike foliage, and seeds attached to their leaves. New lineages of seed plants arose during the Carboniferous, but the earliest known fossils of flowering plants are from near the Jurassic–Cretaceous boundary. The flowering plants have dominated most terrestrial environments through the Cenozoic era.

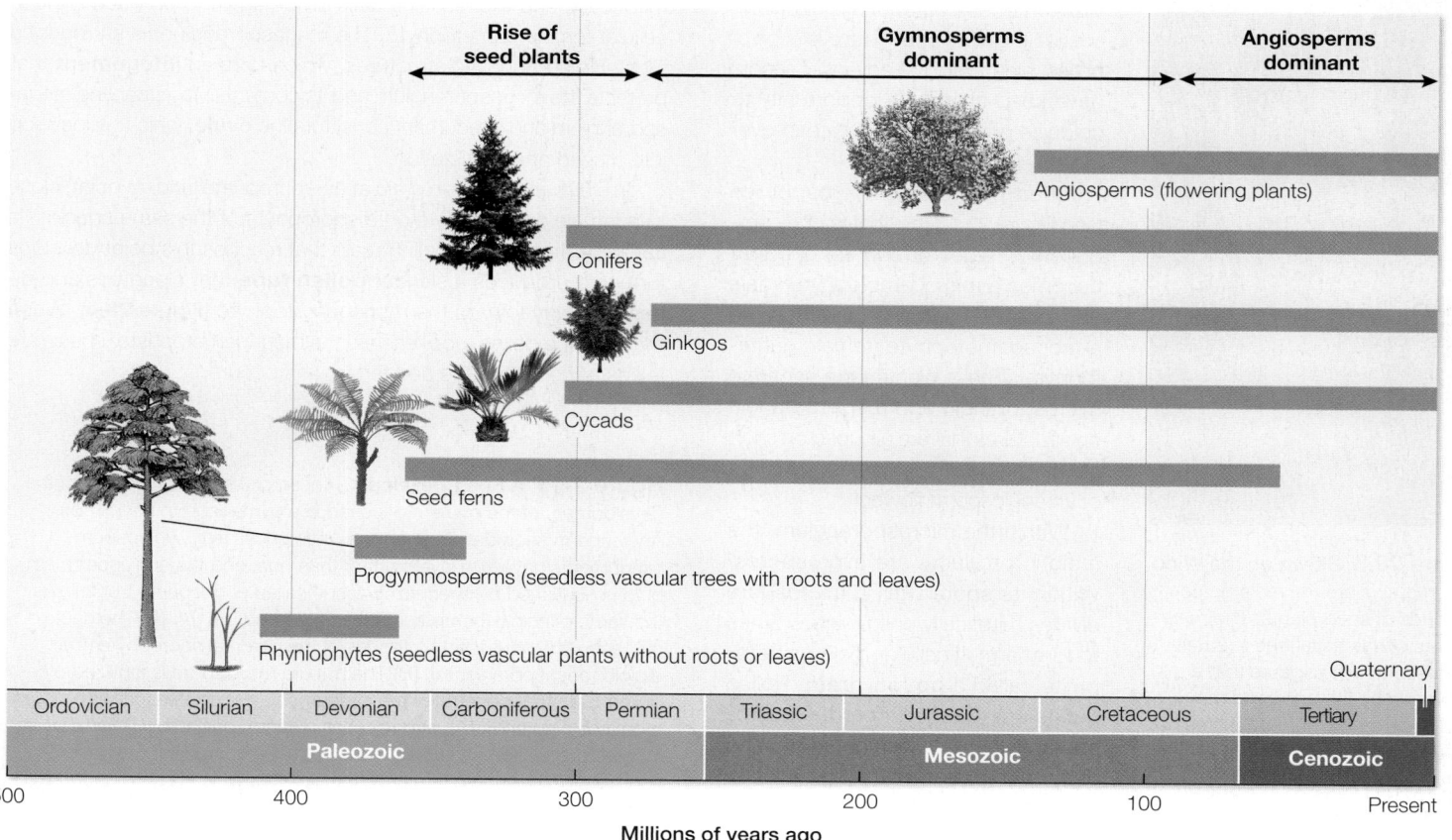

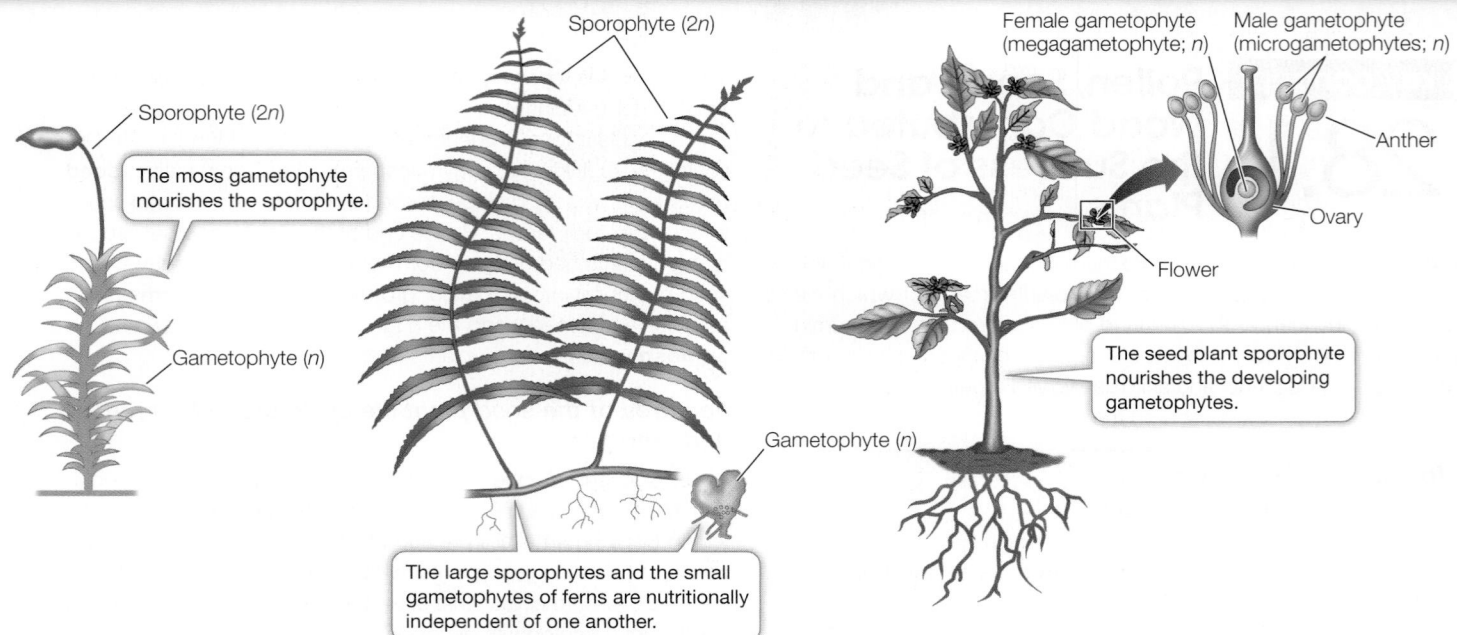

Figure 28.2 The Relationship between Sporophyte and Gametophyte
In the course of plant evolution, the gametophyte (brown) has been reduced and the sporophyte (blue) has become more prominent.

Q: When you see a growing moss, are you more likely to see the haploid or the diploid multicellular stage? Is your answer the same for a fern and for a flowering plant?

Corylus avellana

Figure 28.3 Blown on the Wind
Pollen grains are the male gametophytes of seed plants. The male flowers of this hazel tree release pollen-containing spores that are dispersed by the wind and may land near female gametophytes of other hazel plants.

Media Clip 28.1 Pollen Transfer by Wind
www.Life11e.com/mc28.1

sperm. Even in these groups, sperm is transferred via pollen grains, so fertilization does not require liquid water outside the plant body. The evolution of pollen, along with the advent of seeds, gave seed plants the opportunity to colonize drier areas and spread over the terrestrial environment.

Seed plants are heterosporous (see Figure 27.18B)—that is, they produce two types of spores, one that becomes a microgametophyte (male gametophyte) and one that becomes a megagametophyte (female gametophyte). These plants form separate microsporangia and megasporangia on structures that are grouped on short stems, such as the stamens and pistils of an angiosperm flower.

Within the microsporangium, the meiotic products are microspores. Within its spore wall, a microspore divides mitotically one or a few times to form a multicellular male gametophyte called a **pollen grain**. Pollen grains are released from the microsporangium to be distributed by wind or by an animal pollinator (**Figure 28.3**). As in seedless land plants, the spore wall that surrounds the pollen grain

contains sporopollenin, the most chemically resistant biological compound known, which protects the pollen grain against dehydration and chemical damage—another advantage in terms of survival in the terrestrial environment.

In contrast to the microspores, the megaspores of seed plants are not shed. Instead, they develop into female gametophytes within the megasporangia. These megagametophytes are dependent on the sporophyte for food and water.

In most seed plant species, only one of the meiotic products in a megasporangium survives. The surviving haploid nucleus divides mitotically, and the resulting cells divide again to produce a multicellular female gametophyte. The megasporangium is surrounded by sterile sporophytic structures, which form an **integument** that protects the megasporangium and its contents. Together, the megasporangium and integument constitute the **ovule**, which will develop into a seed after fertilization.

The arrival of a pollen grain at an appropriate landing point, close to a female gametophyte on a sporophyte of the same species, is called **pollination**. A pollen grain that reaches this point develops further. It produces a slender **pollen tube** that elongates and digests its way toward the megagametophyte (**Figure 28.4**). When the tip of the pollen tube reaches the megagametophyte, sperm are released from the tube and fertilization occurs.

Figure 28.5 A Seed Develops These cross sections diagram the development of the ovule into a seed in a gymnosperm (*Pinus* sp.). Angiosperm seed development has differences (e.g., angiosperm integuments have two layers rather than one, and the angiosperm embryo is nourished by specialized tissue called endosperm) but follows the same principle (compare Figures 28.8 and 28.16). **(A)** The haploid megaspore is nourished by tissues of the parental sporophyte (the diploid megasporangium). **(B)** The mature megaspore is fertilized by a pollen grain that penetrates the integument, germinates (grows a pollen tube; see Figure 28.4A), and releases a sperm nucleus. **(C)** Fertilization initiates production of a seed. A mature seed contains three generations: a diploid embryo (the new sporophyte), which is surrounded by haploid female gametophyte tissue that supplies nutrition, which is in turn surrounded by the seed coat (diploid parental sporophyte tissue).

(A) A growing pollen tube

Pollen grains

The **pollen tube** elongates on its way to the mega-gametophyte.

Figure 28.4 Pollination Is a Hallmark of the Seed Plants   In most seed plants, a pollen tube grows from the pollen grain to the female gametophyte, where sperm are released and fertilize the egg within the ovule. Once fertilization takes place, the ovule can develop into a seed (see Figure 28.5). **(A)** Scanning electron micrograph of a pollen tube growing in the flower of an angiosperm, the prairie gentian. **(B)** The process of pollination is diagrammed for a generalized angiosperm flower.

Activity 28.1 **Flower Morphology**
www.Life11e.com/ac28.1

(B) The process of pollination

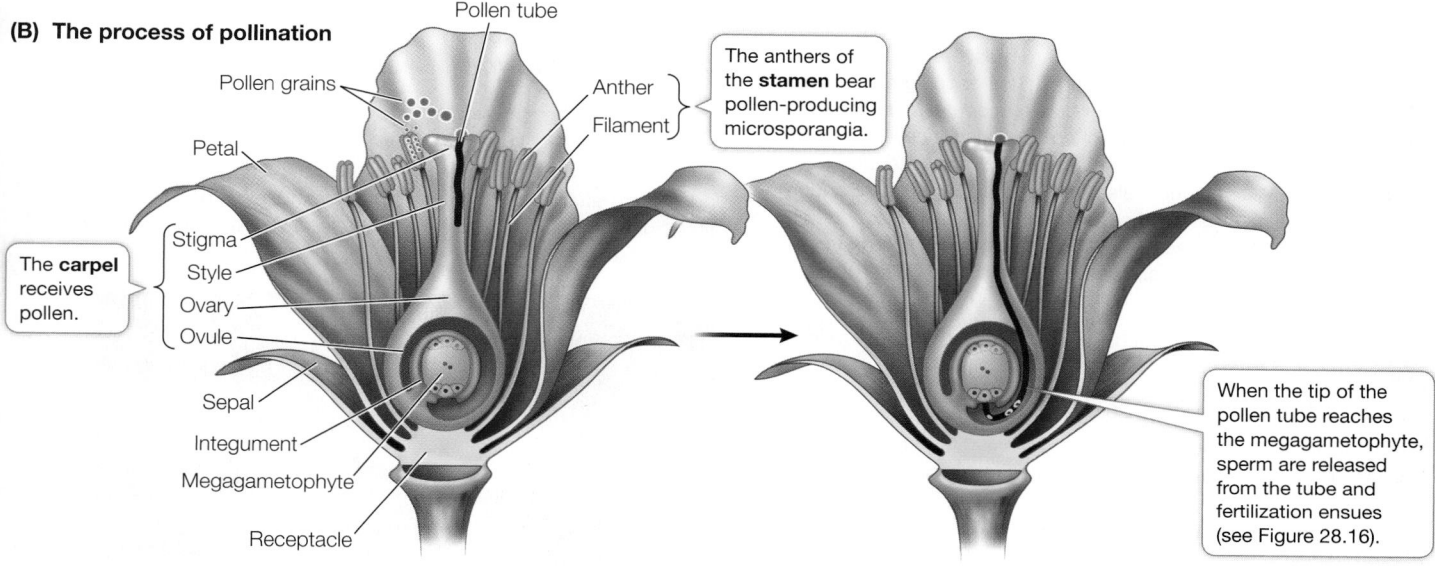

Pollen tube

Pollen grains

Petal

The **carpel** receives pollen. {
Stigma
Style
Ovary
Ovule

Sepal

Integument

Megagametophyte

Receptacle

Anther
Filament } The anthers of the **stamen** bear pollen-producing microsporangia.

When the tip of the pollen tube reaches the megagametophyte, sperm are released from the tube and fertilization ensues (see Figure 28.16).

The resulting diploid zygote divides repeatedly, forming an embryonic sporophyte. After a period of embryonic development, growth is temporarily suspended (the embryo enters a dormant stage). The end product at this stage is the multicellular seed.

The seed is a complex, well-protected package

A seed contains tissues from three generations (**Figure 28.5**). A seed coat develops from the integument—the tissues of the diploid sporophyte parent that surround the megasporangium. Within the megasporangium is haploid tissue from the female gametophyte,

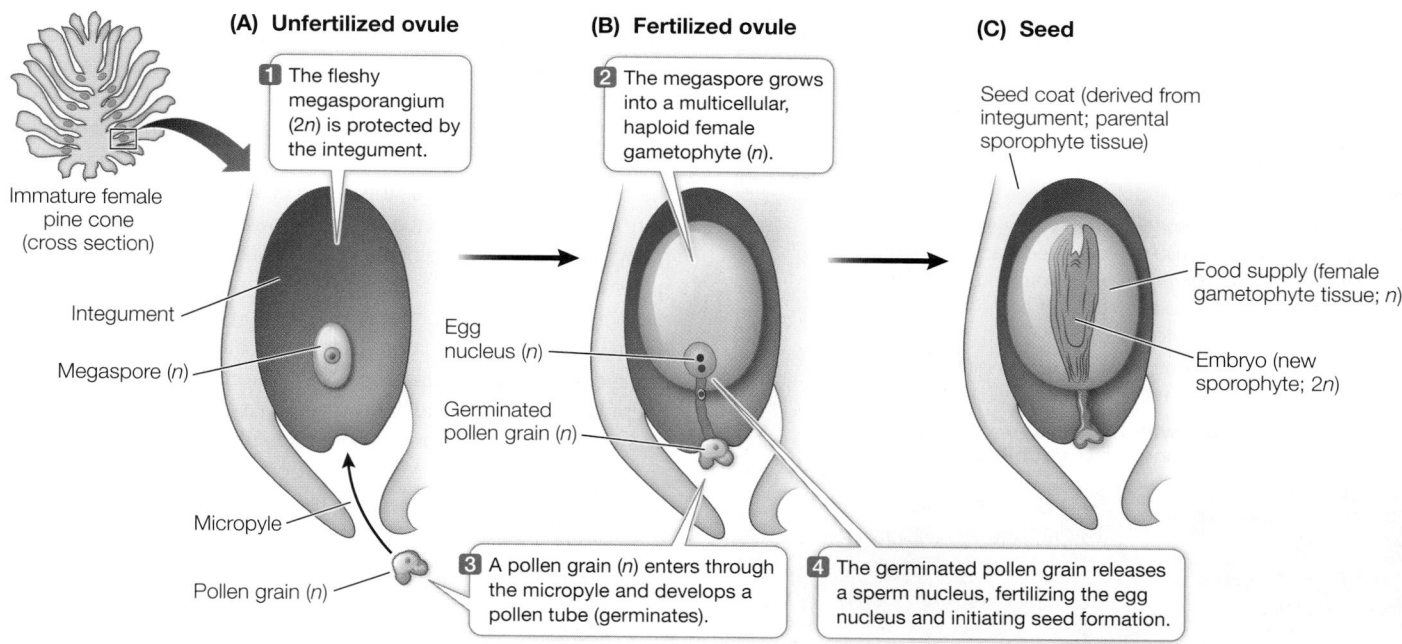

(A) Unfertilized ovule

Immature female pine cone (cross section)

Integument

Megaspore (n)

Micropyle

Pollen grain (n)

1 The fleshy megasporangium (2n) is protected by the integument.

(B) Fertilized ovule

Egg nucleus (n)

Germinated pollen grain (n)

2 The megaspore grows into a multicellular, haploid female gametophyte (n).

3 A pollen grain (n) enters through the micropyle and develops a pollen tube (germinates).

(C) Seed

Seed coat (derived from integument; parental sporophyte tissue)

Food supply (female gametophyte tissue; n)

Embryo (new sporophyte; 2n)

4 The germinated pollen grain releases a sperm nucleus, fertilizing the egg nucleus and initiating seed formation.

experiment

Original Papers: Beal, W. J. 1884. The vitality of seeds. *Proceedings of the Society for the Promotion of Agricultural Sciences* 5: 44–46.

Darlington, H. T. 1941. The sixty-year period for Dr. Beal's seed viability. *American Journal of Botany* 28: 271–273.

Kivilaan, A. and R. S. Bandurski. 1981. The one hundred-year period for Dr. Beal's seed viability experiment. *American Journal of Botany* 68: 1290–1292.

Telewski, F. W. and J. Zeevaart. 2002. The 120th year of the Beal seed viability study. *American Journal of Botany* 89: 1285–1288.

William Beal began an experiment in 1879 to measure the long-term viability of seeds of several common plants. This ongoing experiment has been continued by biologists for well over a century. For the first 40 years of the experiment, Beal checked seed viability every 5 years. H. T. Darlington took over the experiment in 1915 and extended the sampling period to 10-year intervals beginning in 1920. R. S. Bandurski took over the experiment when Darlington retired, and extended the sampling period to 20 years in 1980, a century after the experiment began. Results for three species of plants in years 50–100 of the experiment are shown here.

QUESTION▶   How long do seeds of common plants remain viable?

METHOD

1. Collect 1,000 seeds of each of 23 species of plants.
2. Divide the seeds into 20 lots of 50 seeds of each of the 23 species.
3. Mix each of the lots of seeds with sand, and place the mixtures in uncapped bottles.
4. Bury the bottles upside down (to prevent the entry of water) on a marked, sandy knoll.
5. At regular intervals, excavate a bottle and check the viability of its contents.

RESULTS

The table below shows the number of germinating seeds (of the original 50) from three of the species in years 50–100 of this ongoing experiment.

Species	Years after burial					
	50	60	70	80	90	100
Oenothera biennis (Evening primrose)	19	12	7	5	0	0
Rumex crispus (Curly dock)	26	2	7	1	0	0
Verbascum blattaria (Moth mullein)	31	34	37	35	10	21

CONCLUSION▶   Biologists have used this and similar experiments to estimate the maximum viability of seeds of many species of plants under near-natural environmental conditions. Explore how these data can be used to make predictions about seed viability in the work with the data exercise.

work with the data

Use the data presented in the preceding experiment to answer the following questions.

QUESTIONS▶

1. Calculate the percent of viable seeds for these three species in years 50–100 and graph seed survivorship as a function of time buried.
2. No seeds of the first two species were viable after 90 years of the experiment. Assume 100 percent seed viability at the start of the experiment (year 0), and predict from your graph the approximate year when you think the last of the *Verbascum blattaria* seeds will germinate.
3. What factors do you think might influence the differences among the species in long-term seed viability?

A similar **work with the data** exercise
may be assigned in **LaunchPad**.

which contains a supply of nutrients for the developing embryo. (This tissue is fairly extensive in most gymnosperm seeds. In angiosperm seeds it is greatly reduced, and nutrition for the embryo is supplied instead by a tissue called endosperm.) In the center of the seed is the third generation, the embryo of the new diploid sporophyte.

The seed is a well-protected resting stage. As we discussed in the opening of this chapter, the seeds of some species may remain dormant but stay viable (capable of growth and development) for many years, germinating only when conditions are favorable for the growth of the sporophyte. During the dormant stage, the seed coat protects the embryo from excessive drying and may also protect it against potential predators that would otherwise consume the embryo and its nutrient reserves. Many seed plants have structural adaptations that promote the dispersal of seeds by wind, water, or by animals. When the young sporophyte resumes growth, it draws on the food reserves in the seed. The possession of seeds is a major reason for the enormous evolutionary success of the seed plants, which are the dominant life forms of most modern terrestrial floras.

The germination of the Judean date described in the opening story is an extreme example of seed dormancy. How do we know how long most seeds remain viable? To find out, William J. Beal, a biologist at Michigan State University, decided to begin an experiment in 1879 that he could not hope to finish in his lifetime (**Investigating Life: William Beal's Seed Viability Study**). He prepared 20 lots of seeds for long-term storage. Each lot consisted of 50 seeds from each of 23 species of plants. He mixed each lot of seeds with sand and placed the mixture in an uncapped bottle, then buried all the bottles upside down (so they would stay dry) on a sandy knoll. The seeds experienced normal temperature fluctuations for Michigan. At regular intervals ever since, other biologists have excavated a bottle and checked the viability of the seeds it contained. The seeds of most species remained viable for decades, whereas others have remained viable for more than a century.

A change in stem anatomy enabled seed plants to grow to great heights

Fossils of the closest relatives of seed plants (progymnosperms) and the earliest seed plants (seed ferns) are found in late Devonian rocks (see Figure 28.1). These plants had thickened woody stems, developed through the proliferation of xylem. This type of growth, which increases the diameter of stems and roots in many modern seed plants, is called **secondary growth**. Its product is secondary xylem, or wood.

The younger portion of the wood produced by secondary growth is well adapted for water transport, but older wood becomes clogged with resins or other materials. Although no longer functional in transport, the older wood continues to provide support for the plant. This support allows woody plants to grow taller than other plants around them and thus capture more light for photosynthesis.

Not all seed plants are woody. In the course of seed plant evolution, many groups lost the woody growth habit; however, other advantageous attributes helped them become established in an astonishing variety of places.

28.1 recap

Today's living seed plants fall into two major groups, the gymnosperms and the angiosperms. Pollen grains, seeds, and wood are major evolutionary innovations of the seed plants. Protection of embryos is a hallmark of seed plants.

learning outcomes

You should be able to:

- Explain how the evolution of pollen enabled seed plants to thrive in terrestrial environments.
- Summarize the major advantages of having seeds.
- Explain how seed plants altered the terrestrial landscape.

1. Explain the importance of pollen in freeing seed plants from dependence on liquid water.
2. How do seeds and seed dormancy help plants survive in highly seasonal environments?
3. What part did the evolution of wood play in changing the terrestrial landscape?

The seed ferns have long been extinct, but the surviving seed plants have been remarkable successes. After the seed ferns, the gymnosperms were the next group of plants to dominate terrestrial environments.

28.2 Once Dominant Gymnosperms Still Thrive in Some Environments

The gymnosperms are seed plants that do not form flowers or fruits. Gymnosperms (which means "naked-seeded") are so named because their ovules and seeds, unlike those of angiosperms, are not protected by ovary or fruit tissue. Gymnosperms dominated the

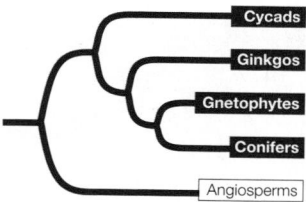

terrestrial world through the Mesozoic era but were replaced across large areas of Earth by flowering plants during the Cenozoic era (see Figure 28.1). Although there are now fewer than 1,200 living species of gymnosperms, these plants are still the primary forest trees in some areas.

focus your learning

- Some pines are fire-adapted.
- The reproductive structures of conifers are contained in male and female cones.

There are four major groups of living gymnosperms

The living gymnosperms can be divided into four major groups (see Figure 27.1):

1. **Cycads** are palmlike plants of the tropics and subtropics (**Figure 28.6A**). Of the present-day gymnosperms, the cycads are probably the earliest-diverging clade. There are about 300 species, some of which grow as tall as 20 meters. The tissues of many species are highly toxic to humans if ingested.

2. **Ginkgos**, common during the Mesozoic era, are represented today by a single species, *Ginkgo biloba*, the maidenhair tree (**Figure 28.6B**). There are both male and female maidenhair trees. The difference is determined by X and Y sex chromosomes, as in humans; few other plants have distinct sex chromosomes.

3. **Gnetophytes** number about 90 species in three very different genera, which share certain characteristics analogous to ones found in the angiosperms. One of the gnetophytes is *Welwitschia* (**Figure 28.6C**), a long-lived desert plant with straplike leaves that sprawl on the sand and can grow as long as 3 meters.

4. **Conifers** are by far the most abundant of the gymnosperms. There are about 700 species of these cone-bearing plants, including the pines and redwoods (**Figure 28.6D**).

With the exception of the gnetophytes, the living gymnosperm groups have only tracheids as water-conducting and support cells within the xylem. Most gymnosperms lack the vessel elements and fibers (cells specialized for water conduction and support, respectively) that are found in angiosperms. While the gymnosperm

(A) *Encephalartos* sp.

(B) *Ginkgo biloba*

(C) *Welwitschia mirabilis*

(D) *Pinus longaeva*

Figure 28.6 Diversity among the Gymnosperms **(A)** Many cycads have growth forms that resemble both ferns and palms, although cycads are not closely related to either group. **(B)** The characteristic broad leaves of the maidenhair tree. **(C)** The straplike leaves of *Welwitschia*, a gnetophyte, grow throughout the life of the plant, breaking and splitting as they grow. **(D)** Conifers dominate many types of landscapes in the Northern Hemisphere. Bristlecone pines such as these are the longest-lived individual trees known.

water-transport and support system may seem somewhat less efficient than that of the angiosperms, it serves some of the largest trees known. The coastal redwoods of California are the tallest gymnosperms, with some individuals growing to well over 100 meters tall.

During the Permian, as environments became warmer and dryer, the conifers and cycads flourished. Gymnosperm forests changed over time as the gymnosperm groups evolved. Gymnosperms dominated the Mesozoic era, during which the continents drifted apart and large dinosaurs lived. Gymnosperms were the principal trees in all forests until about 65 million years ago, and even today conifers are the dominant trees in many forests, especially at high latitudes and elevations. The oldest living single organism on Earth today is a gymnosperm in California—a bristlecone pine that germinated about 4,800 years ago, at about the time the ancient Egyptians were starting to develop writing (see Figure 28.6D).

Conifers have cones and lack swimming sperm

The great Douglas fir and cedar forests found in the northwestern United States and the massive boreal forests of pine, fir, and spruce of northern Eurasia and North America, as well as on the upper slopes of mountain ranges everywhere, rank among the great forests of the world. All these trees belong to one group of gymnosperms: the conifers, or cone-bearers.

Male and female **cones** contain the reproductive structures of conifers. The female (seed-bearing) cone is known as a **megastrobilus** (plural *megastrobili*). An example of a familiar

(A) Female cones (megastrobili) in *Pinus contorta*

Woody scale

Seed

Central axis

Cross section of a megastrobilus

Female cones, or megastrobili

(B) Male cones (microstrobili) in *Pinus contorta*

Herbaceous scale

Pollen-containing microspores

Central axis

Cross section of a microstrobilus

Male cones, or microstrobili

Figure 28.7 Female and Male Cones **(A)** The woody scales of female cones (megastrobili) are modified branches. **(B)** The herbaceous scales of male cones (microstrobili) are modified leaves.

microstrobilus to the female gametophyte inside a cone. A pollen tube provides the sperm with the means for the last stage of travel by elongating through maternal sporophytic tissue. When the pollen tube reaches the female gametophyte, it releases two sperm, one of which degenerates after the other unites with the egg. Union of sperm and egg results in a zygote. Mitotic divisions and further development of the zygote result in an embryo.

The megasporangium, in which the female gametophyte will form, is enclosed in a layer of sporophytic tissue—the integument—that will eventually develop into the seed coat that protects the embryo. The integument, the megasporangium inside it, and the tissue attaching it to the maternal sporophyte constitute the ovule. The pollen grain enters through a small opening in the integument at the tip of the ovule, the **micropyle**.

Most conifer ovules, which will develop into seeds after fertilization, are borne exposed on the upper surfaces of the scales of the megastrobilus. The only protection of the ovules comes from the scales, which are tightly pressed against one another within the cone. Some pines, such as the lodgepole pine, have tightly closed cones that are sealed with resin. Fire is needed to melt the resin and open the cones to release the seeds. These species are said to be fire-adapted, and fire is essential to their reproduction. A fire devastated lodgepole pine forests in Yellowstone National Park in 1988, but also released large numbers of seeds from cones. As a result, large numbers of lodgepole pine seedlings are now emerging in the burn area (**Figure 28.9**).

About half of all conifer species have soft, fleshy tissues that envelop their seeds. Some of these are fleshy, fruitlike cones, as in junipers. Others are fruitlike extensions of the seeds, called arils, as in yews. These tissues, although often mistaken for "berries," are not true fruits. As you will see in the next section, true fruits are the plant's ripened ovaries, which are absent in gymnosperms. Nonetheless, the fleshy tissues that surround the seeds of many conifers serve a similar purpose as that of the fruits of flowering plants, acting as an enticement for seed-dispersing animals. Animals eat these fleshy tissues and disperse the seeds in their feces, often depositing the seeds considerable distances away from the parent plant.

megastrobilus is the woody cone of pine trees. The seeds in a megastrobilus are protected by a tight cluster of woody scales, which are modifications of branches extending from a central axis (**Figure 28.7A**). The typically much smaller male (pollen-bearing) cone is known as a **microstrobilus**. The microstrobilus is typically herbaceous rather than woody, as its scales are composed of modified leaves, beneath which are the pollen-bearing microsporangia (**Figure 28.7B**).

The life cycle of a pine illustrates reproduction in gymnosperms (**Figure 28.8**). As in other seed plants, conifers have male gametophytes in the form of pollen grains, which frees the plants completely from their dependence on liquid water for fertilization. Wind assists conifer pollen grains in their first stage of travel from the

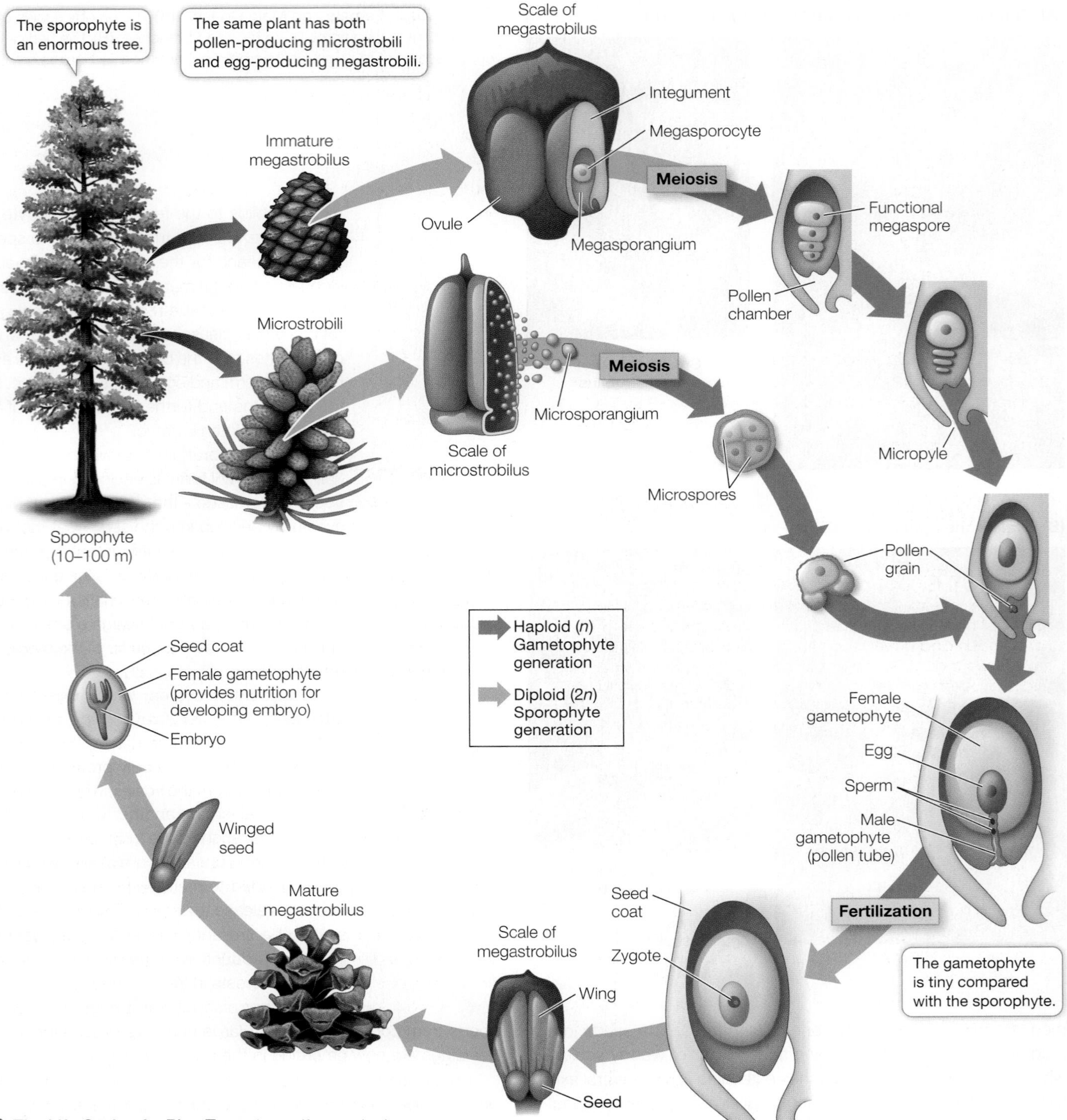

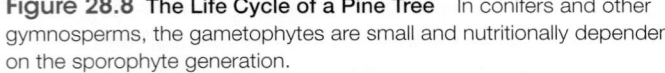

The sporophyte is an enormous tree.

The same plant has both pollen-producing microstrobili and egg-producing megastrobili.

Scale of megastrobilus

Immature megastrobilus

Integument

Megasporocyte

Meiosis

Functional megaspore

Ovule

Megasporangium

Pollen chamber

Microstrobili

Microsporangium

Meiosis

Micropyle

Scale of microstrobilus

Microspores

Sporophyte (10–100 m)

Pollen grain

Seed coat

Female gametophyte (provides nutrition for developing embryo)

Embryo

Haploid (*n*) Gametophyte generation

Diploid (2*n*) Sporophyte generation

Female gametophyte

Egg

Sperm

Male gametophyte (pollen tube)

Winged seed

Fertilization

Mature megastrobilus

Seed coat

Scale of megastrobilus

Zygote

The gametophyte is tiny compared with the sporophyte.

Wing

Seed

Figure 28.8 The Life Cycle of a Pine Tree   In conifers and other gymnosperms, the gametophytes are small and nutritionally dependent on the sporophyte generation.

 Animation 28.1 **Life Cycle of a Conifer**
www.Life11e.com/a28.1

 Activity 28.2 **Life Cycle of a Conifer**
www.Life11e.com/ac28.2

Figure 28.9 From Devastation, New Life A stand of lodgepole pines in Yellowstone National Park. The mature trees were destroyed by a forest fire in 1988. However, the fire released large numbers of seeds from cones, and now many young lodgepole pine trees are growing in the burn area.

28.2 recap

Living gymnosperms can be divided into four major groups: cycads, ginkgos, gnetophytes, and conifers. Although they have declined since the Mesozoic era, gymnosperms are still the dominant trees in some areas, especially at high elevations and latitudes. All gymnosperms are woody and have seeds that are not protected by ovaries, although some have fleshy, fruitlike cones or extensions of seeds that entice animals to disperse the seeds.

learning outcomes

You should be able to:

- Describe and contrast the roles of male and female gametophytes in reproduction.
- Summarize the functions of cones in conifers.
- Describe the role of fire in the life history of the lodgepole pine.

1. How is fire necessary for the survival of some gymnosperms?
2. Distinguish between the roles of the megagametophyte and the pollen grain in the life cycle of a conifer.
3. What is the function of the fleshy cones that surround the seeds of many gymnosperms?

Although gymnosperms still dominate the terrestrial landscape in some environments, angiosperms (the flowering plants) have become the dominant land plants across much of the Earth.

key concept 28.3 Flowers and Fruits Led to Increased Diversification of Angiosperms

The most obvious feature defining the angiosperms is the **flower**, which is their sexual structure. Production of **fruits** is also a synapomorphy (shared derived trait) of angiosperms. After fertilization, the ovary of a flower (together with the seeds it contains) develops into a fruit that protects the seeds and can promote seed dispersal. As you will see, both flowers and fruits gave angiosperms major reproductive advantages that led to their dominance of the terrestrial environment in the Cenozoic era (see Figure 28.1).

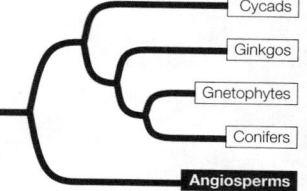

focus your learning

- In nearly all angiosperms, two male gametes, contained in a single microgametophyte, participate in fertilization.
- Fruits are often specialized to encourage dispersal by animals, wind, or water.
- All parts of a flower are modified leaves.
- Angiosperms and their animal pollinators have been coevolving for more than 150 million years.

Angiosperms have many shared derived traits

The name *angiosperm* ("enclosed seed") is derived from another distinctive trait of flowering plants that is related to the formation of fruits: the ovules and seeds are enclosed in a modified leaf called a **carpel**. Besides protecting the ovules and seeds, the carpel often interacts with incoming pollen to prevent self-pollination, thus favoring cross-pollination and increasing genetic diversity.

The female gametophyte of the angiosperms is even more reduced than that of the gymnosperms, usually consisting of only seven cells (see Figure 28.16). Thus the angiosperms represent the current extreme of the trend we have traced throughout the evolution of the vascular plants: the sporophyte generation becomes larger and more independent of the gametophyte, while the gametophyte generation becomes smaller and more dependent on the sporophyte.

The xylem of most angiosperms is distinguished by the presence of specialized water-transporting cells called **vessel elements**. These cells are larger in diameter than tracheids and connect with one another without obstruction, allowing easy water movement. A second distinctive cell type in angiosperm xylem is the **fiber**, which plays an important role in supporting the plant body. The phloem of angiosperms possesses its own unique cell type, called a companion cell. Like the gymnosperms, woody angiosperms exhibit secondary growth, increasing in diameter by producing secondary xylem and secondary phloem.

A more comprehensive list of angiosperm synapomorphies, then, includes the following (some of these traits will be discussed later in this chapter):

- Flowers
- Fruits
- Highly reduced female gametophytes
- Ovules and seeds enclosed in a carpel
- Germination of pollen on a stigma
- Double fertilization
- Endosperm (nutritive tissue for the embryo)
- Phloem with companion cells

(A) *Heracleum mantegazzianum*

Flowers

Umbel

Compound umbel

(B) *Zinnia elegans*

Ray flowers   Disc flowers

(C) *Sorghum halepense*

Anthers of spikelet   Stigma of spikelet

Figure 28.10 Inflorescences **(A)** The inflorescence of the giant hog-weed, a member of the carrot family, is a compound umbel. Each umbel bears flowers on stalks that arise from a common center. **(B)** Zinnias are members of the aster family; their inflorescence is a head. Within the head, each of the long, petal-like structures is a ray flower; the central portion of the head consists of dozens to hundreds of disc flowers. **(C)** Some grasses, such as Johnson grass, have inflorescences called spikes, which are composed of many smaller groups of flowers, or spikelets.

The majority of these traits bear directly on angiosperm reproduction, which is a large factor in the success of this dominant plant group.

The sexual structures of angiosperms are flowers

Flowers come in an astonishing variety of forms—just think of a few of the flowers you recognize. Flowers may be single, or they may be grouped together to form an **inflorescence**. Different families of flowering plants have characteristic types of inflorescences, such as the compound umbels of the carrot family (**Figure 28.10A**), the heads of the aster family (**Figure 28.10B**), and the spikes of many grasses (**Figure 28.10C**).

If you examine any familiar flower, you will notice that the outer parts look somewhat like leaves. In fact, all the parts of a flower *are* modified leaves. The diagram in Figure 28.4B represents a generalized flower (for which there is no exact counterpart in nature). The structures bearing microsporangia are called **stamens**. Each stamen is composed of a **filament** bearing an **anther** that contains the pollen-producing microsporangia. The structures bearing megasporangia are called carpels. A structure composed of one

carpel or two or more fused carpels is called a **pistil**. The swollen base of the pistil, is called the **ovary**, which contains one or more ovules. Each ovule contains a megasporangium surrounded by two protective integuments. The apical stalk of the pistil is the **style**, and the terminal surface that receives pollen grains is the **stigma**.

In addition, many flowers contain specialized sterile (non-spore-bearing) leaves. The inner ones are called **petals** (collectively, the corolla) and the outer ones **sepals** (collectively, the calyx). The corolla and calyx can be quite showy and often play roles in attracting animal pollinators to the flower. The calyx more commonly protects the immature flower in bud. From base to apex, these floral organs—sepals, petals, stamens, and carpels—are usually positioned in circular arrangements or whorls and attached to a central stalk.

The generalized flower in Figure 28.4B has both functional megasporangia and functional microsporangia; such flowers are referred to as **perfect** (or hermaphroditic). Many angiosperms produce two types of flowers, one with only megasporangia and the other with only microsporangia. Consequently, either the stamens or the carpels are nonfunctional or absent in a given flower, and the flower is referred to as **imperfect**.

Species such as corn or birch, in which both megasporangiate (female) and microsporangiate (male) flowers occur on the same plant, are said to be **monoecious** ("one-housed"—but, it must be added, one house with separate rooms). Complete separation of imperfect flowers occurs in some other angiosperm species, such as willows and date palms; in these species, an individual plant produces either flowers with stamens or flowers with carpels, but never both. Such species are said to be **dioecious** ("two-housed").

(A) *Nymphaea* sp.

(B) *Paphiopedilum* sp.

Figure 28.11 Flower Form and Evolution **(A)** A water lily shows the major features of early flowers: it is radially symmetrical, and the individual tepals, stamens, and carpels are separate, numerous, and attached at their bases. **(B)** Orchids such as this Venus slipper have a bilaterally symmetrical structure that evolved much later than radial flower symmetry.

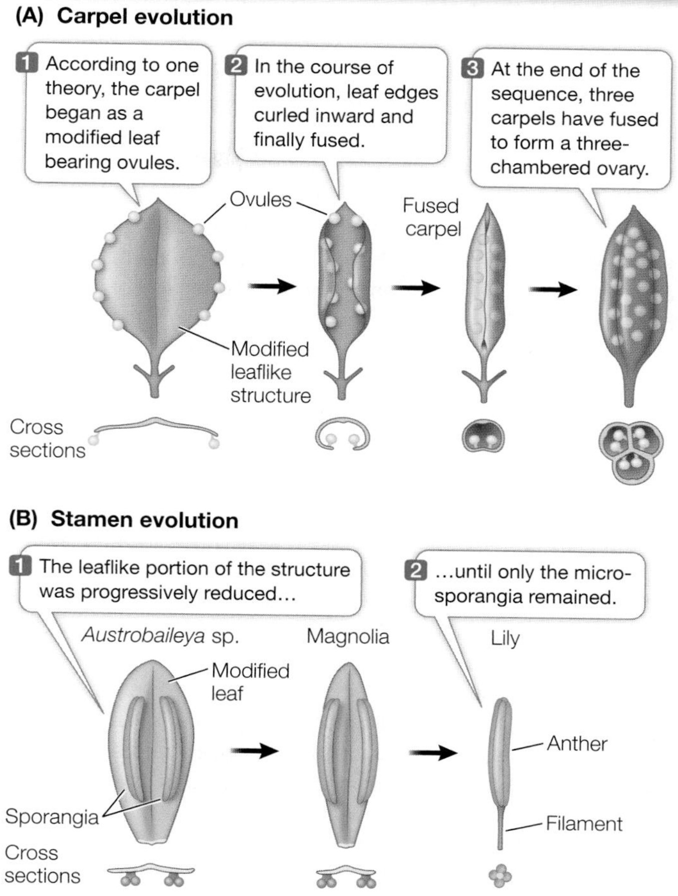

Figure 28.12 **Carpels and Stamens Evolved from Leaflike Structures** **(A)** Possible stages in the evolution of a carpel from a more leaflike structure. **(B)** The stamens of three modern plants show three possible stages in the evolution of that organ. (It is *not* implied that these species evolved from one another; their structures simply illustrate the possible stages.)

Flower structure has evolved over time

The flowers of the earliest-diverging clades of angiosperms have a large and variable number of tepals (undifferentiated sepals and petals), carpels, and stamens (**Figure 28.11A**). Evolutionary change within the angiosperms has included some striking modifications of this early condition: reductions in the number of each type of floral organ to a fixed number, differentiation of petals from sepals, and changes in symmetry from radial (as in a lily or magnolia) to bilateral (as in a sweet pea or orchid), often accompanied by an extensive fusion of parts (**Figure 28.11B**).

According to one hypothesis, the first carpels to evolve were leaves with marginal sporangia, folded but incompletely closed. Early in angiosperm evolution, carpels fused with one another, forming a single, multichambered ovary (**Figure 28.12A**). In some flowers, the other floral organs are attached at the top of the ovary rather than at the bottom as in Figure 28.4B. The stamens of the most ancient flowers may have been leaflike (**Figure 28.12B**), with little resemblance to the stamens of the generalized flower seen in Figure 28.4B.

Why do so many flowers have pistils with long styles and anthers with long filaments? Natural selection has favored length in both of these floral organs, probably because length increases the likelihood of successful pollination. Long filaments may bring the anthers into contact with insect bodies, or they may place the anthers in a better position to catch the wind. Similar arguments apply to long styles.

A perfect flower represents a compromise of sorts. On the one hand, by attracting a pollinating bird or insect, the plant is attending to both its female and male functions with a single flower type, whereas plants with imperfect flowers must create that attraction twice—once for each type of flower. On the other hand, the perfect flower can result in self-pollination, which is usually disadvantageous. Another potential problem is that the female and male functions might interfere with each other—for example, the stigma might be placed so as to make it difficult for pollinators to reach the anthers, thus reducing the export of pollen to other flowers.

Might there be a way around these problems? One solution is seen in the bush monkeyflower (*Mimulus aurantiacus*), which is pollinated by hummingbirds. Its flower has a stigma that initially serves as a screen, hiding the anthers (**Figure 28.13**). Once a hummingbird touches the stigma, however, one of the stigma's two lobes is retracted, so that subsequent hummingbird visitors pick up pollen from the previously screened anthers. Thus the first bird to visit the flower transfers pollen from another plant to the stigma. Later visitors

(A) Carpel evolution

1 According to one theory, the carpel began as a modified leaf bearing ovules.

2 In the course of evolution, leaf edges curled inward and finally fused.

3 At the end of the sequence, three carpels have fused to form a three-chambered ovary.

Ovules

Fused carpel

Modified leaflike structure

Cross sections

(B) Stamen evolution

1 The leaflike portion of the structure was progressively reduced…

2 …until only the microsporangia remained.

Austrobaileya sp.   Magnolia   Lily

Modified leaf

Sporangia

Anther

Filament

Cross sections

pick up pollen from the now-accessible anthers, fulfilling the flower's male function. **Figure 28.14** describes the experiment that revealed the function of this mechanism.

Angiosperms have coevolved with animals

Whereas most gymnosperms are pollinated by wind, most angiosperms are pollinated by animals. The many different **mutualistic* pollination relationships between plants and animals are vital to both parties.

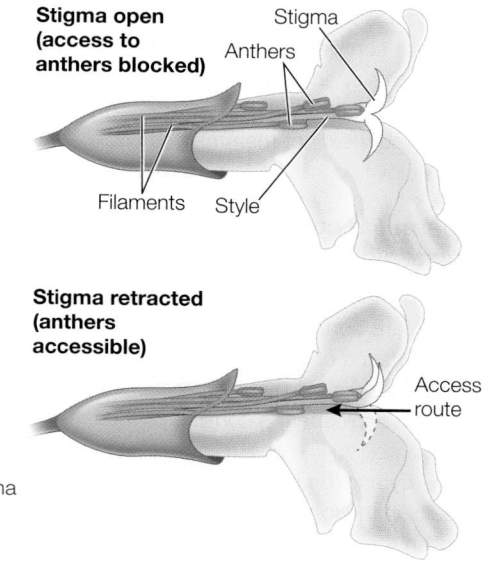

Stigma open (access to anthers blocked)

Stigma

Anthers

Filaments   Style

Stigma retracted (anthers accessible)

Access route

Figure 28.13 **An Unusual Way to Prevent Selfing** Both long stamens and long styles facilitate cross-pollination, but if these male and female structures are too close to each other, the likelihood of (disadvantageous) self-pollination increases. In *Mimulus aurantiacus*, the stigma is initially open, blocking access to the anthers. A hummingbird's touch as it deposits pollen on the stigma causes one lobe of the stigma to retract, creating a path to the anthers and allowing pollen dispersal by subsequent hummingbird visitors.

*connect the concepts   Mutualisms benefit both partners in a relationship, as described in Chapter 55.

Many flowers entice animals to visit them by providing food rewards. Pollen grains themselves sometimes serve as food for animals. In addition, some flowers produce a sugary fluid called nectar, and some of these flowers have specialized structures to store and distribute it. In the process of visiting flowers to obtain nectar or pollen, animals often carry pollen from one flower to another or from one plant to another. Thus, in their quest for food, the animals contribute to the genetic diversity of the plant population. Insects, especially bees, are among the most important pollinators. Other major pollinators include some species of birds and bats.

 Media Clip 28.2 Pollen Transfer by a Bat
www.Life11e.com/mc28.2

For more than 150 million years, angiosperms and their animal pollinators have coevolved in the terrestrial environment. The animals have affected the evolution of the plants, and the plants have affected the evolution of the animals. Flower structure has become incredibly diverse under these selection pressures. Some of the products of coevolution are highly specific. For example, the flowers of some yucca species are pollinated by only one species of yucca moth, and that moth may exclusively pollinate just one species of yucca. Such specific relationships provide plants with a reliable mechanism for transferring pollen only to members of their own species.

Most plant–pollinator interactions are much less specific. In most cases, many different animal species pollinate the same plant species, and the same animal species pollinates many different plant species. However, even these less specific interactions have developed some specialization. Bird-pollinated flowers are often red and odorless. Many insect-pollinated flowers have characteristic odors. Bees see ultraviolet light well, so many bee-pollinated flowers have nectar guides that are visible in this region of the light spectrum (**Figure 28.15**).

The fruits of some plants that are still around today originally evolved to attract large frugivorous animals—many of which went extinct in the Pleistocene. For example, the large round fruits of the osage orange tree attracted wooly mammoths, which ate the fruits and dispersed the seeds. After the extinction of the mammoths, osage orange trees survived in large part because early humans used the wood of this species for making bows for hunting, dispersing the fruits to new areas in the process. Some other species that depended on now-extinct mammals to disperse their seeds, however, may be slowly declining toward extinction.

The angiosperm life cycle produces diploid zygotes nourished by triploid endosperms

Like all seed plants, angiosperms are heterosporous. As you have seen, their ovules are contained within carpels rather than being exposed on the surfaces of scales, as in most gymnosperms. The male gametophytes, as in the gymnosperms, are pollen grains.

Pollination in the angiosperms consists of the arrival of a microgametophyte—a pollen grain—on a receptive surface in a flower (the stigma). As in the gymnosperms, pollination is the first in a series of events that results in the formation of a seed. The next event is the growth of a pollen tube extending to the megagametophyte. The

experiment

Figure 28.14   The Effect of Stigma Retraction in Monkeyflowers

Original Paper: Fetscher, A. E. 2001. Resolution of male-female conflict in an hermaphroditic flower. *Proceedings of the Royal Society B* 268: 525–528.

Elizabeth Fetscher's experiments showed that the unusual stigma retraction response to pollination in monkeyflowers (illustrated in Figure 28.13) enhances the dispersal of pollen to other monkeyflowers.

HYPOTHESIS▶   The stigma-retraction response in *M. aurantiacus* increases the likelihood that an individual flower's pollen will be exported to another flower once pollen from another flower has been deposited on its stigma.

METHOD

1. Set up three groups of monkeyflower arrays. Each array consists of one pollen-donor flower and multiple pollen-recipient flowers (with the anthers removed to prevent pollen donation).
2. In control arrays, the stigma of the pollen donor is allowed to function normally.
3. In one group of experimental arrays, the stigma of the pollen donor is permanently propped open (blocking access to the anthers).
4. In a second group of experimental arrays, the stigma of the pollen donor is artificially sealed closed (allowing access to the anthers).
5. Allow hummingbirds to visit the arrays, then count the pollen grains transferred from each donor flower to the recipient flowers in the same array.

RESULTS

Error bars indicate the standard errors of the mean.

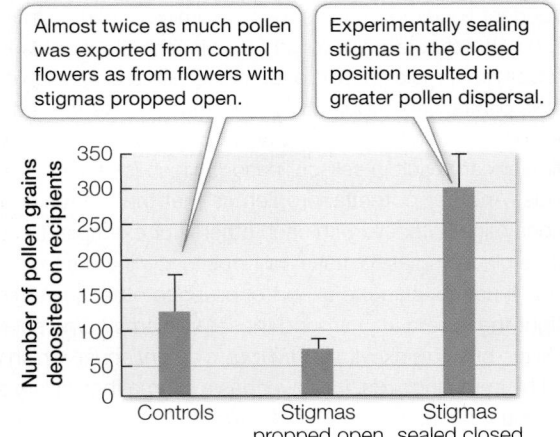

Almost twice as much pollen was exported from control flowers as from flowers with stigmas propped open.

Experimentally sealing stigmas in the closed position resulted in greater pollen dispersal.

CONCLUSION▶   The stigma-retraction response enhances the male function of the flower (dispersal of pollen) once the female function (receipt of pollen) has been performed.

third event is a fertilization process that, in detail, is unique to the angiosperms (**Figure 28.16**).

In nearly all angiosperms, *two* male gametes, contained in a single microgametophyte, participate in fertilization. The nucleus

Argentina anserina

Figure 28.15 See Like a Bee
To normal human vision, the petals of silverweed appear solid yellow. Ultraviolet photography reveals patterns that attract bees to the central region, where pollen and nectar are located.

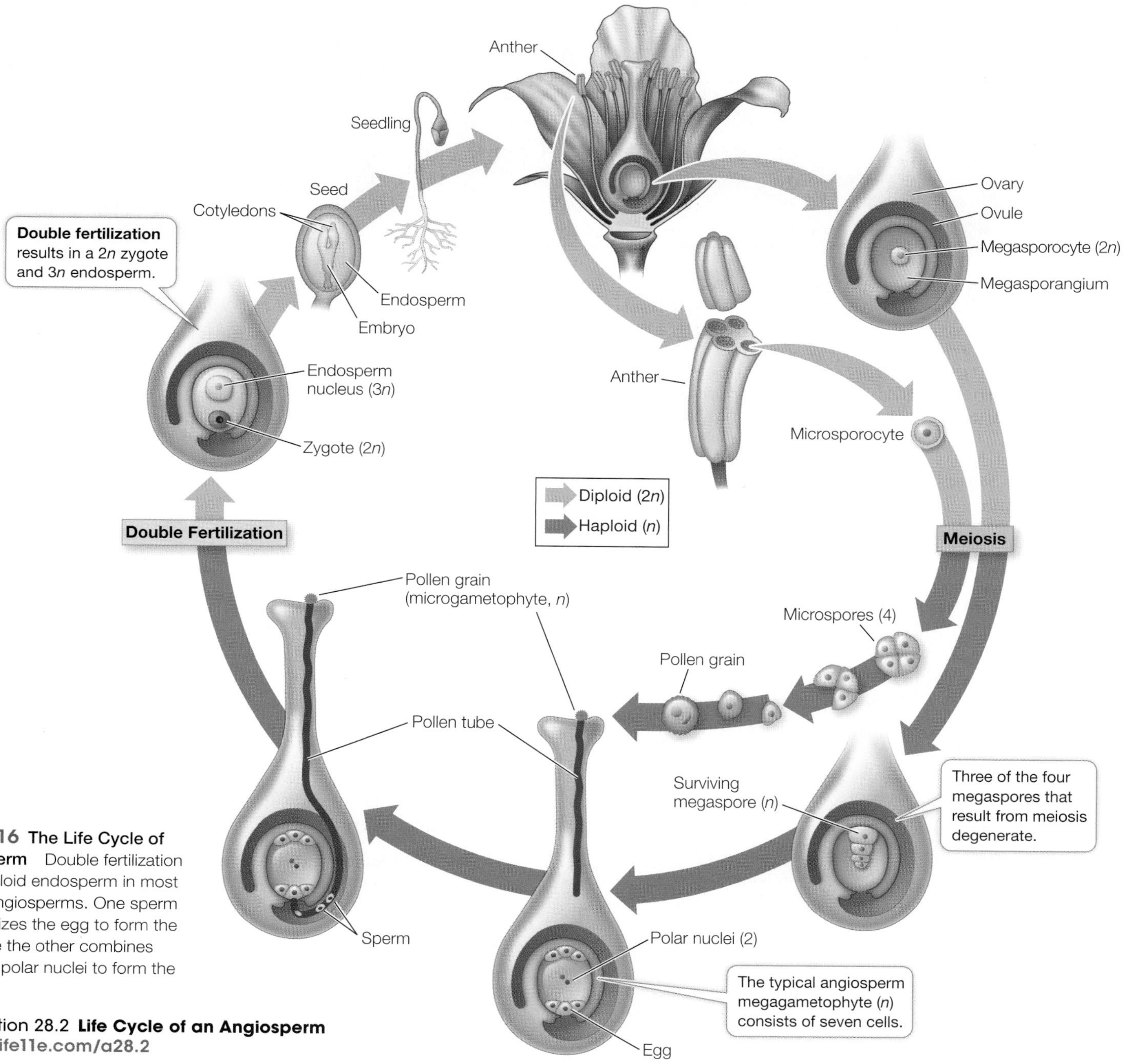

Flower of mature sporophyte

Anther

Seedling

Seed

Cotyledons

Double fertilization results in a 2*n* zygote and 3*n* endosperm.

Endosperm

Embryo

Ovary

Ovule

Megasporocyte (2*n*)

Megasporangium

Endosperm nucleus (3*n*)

Zygote (2*n*)

Anther

Microsporocyte

Double Fertilization

Diploid (2*n*)

Haploid (*n*)

Meiosis

Pollen grain (microgametophyte, *n*)

Microspores (4)

Pollen grain

Pollen tube

Surviving megaspore (*n*)

Three of the four megaspores that result from meiosis degenerate.

Figure 28.16 The Life Cycle of an Angiosperm Double fertilization results in triploid endosperm in most species of angiosperms. One sperm nucleus fertilizes the egg to form the zygote, while the other combines with the two polar nuclei to form the endosperm.

Sperm

Polar nuclei (2)

The typical angiosperm megagametophyte (*n*) consists of seven cells.

Egg

▶ Animation 28.2 **Life Cycle of an Angiosperm**
www.Life11e.com/a28.2

(A)

(B)

(C)

(D)

(E)

(F)

Figure 28.17 Fruits Come in Many Forms   **(A)** The single seeds inside the simple fruits of Bing cherries are dispersed by animals. **(B)** Each seed of the horse chestnut is covered by a hard, woody fruit that allows it to survive drought. Although such fruits are commonly called "nuts," this is a culinary rather than a biological term. **(C)** The highly re-duced simple fruits of dandelions are dispersed by wind. **(D)** A multiple fruit, the pineapple (*Ananas comosu*), has become one of the most economically significant fruit crops of the tropics. **(E)** An aggregate fruit (blackberry). **(F)** An accessory fruit (pear).

Q: The large incisors of rodents help these animals penetrate the shells of woody nuts, which they consume as food. But unlike many animals that eat fruits without digesting the seeds, rodents destroy the seeds when they eat them. So how do rodents aid in dispersing the seeds of nut-bearing plants?

of one sperm combines with that of the egg to produce a diploid zygote, the first cell of the sporophyte generation. In most angio-sperms, the other sperm nucleus combines with two other haploid nuclei of the female gametophyte to form a cell with a *triploid* (3*n*) nucleus. That cell, in turn, gives rise to triploid tissue, the **endo-sperm**, which nourishes the embryonic sporophyte during its early development. This process, in which two fertilization events take place, is known as **double fertilization**. In some angiosperms, additional haploid nuclei are incorporated to form even higher ploidy levels in the endosperm, or the second sperm fuses with only one haploid nucleus, resulting in diploid endosperm.

As Figure 28.16 shows, the zygote develops into an embryo, which consists of an embryonic axis (the "backbone" that will be-come a stem and a root) and one or two **cotyledons**, or "seed leaves." The cotyledons have different fates in different plants. In many, they serve as absorptive organs that take up and digest the endosperm. In others, they enlarge and become photosynthetic when the seed germinates. Often they play both roles.

The ovule develops into a seed containing the products of the double fertilization that characterizes angiosperms: the diploid zy-gote and a triploid endosperm (see Figure 28.16). The endosperm serves as storage tissue for starch or lipids, proteins, and other substances that will be needed by the developing embryo.

Fruits aid angiosperm seed dispersal

Fruits often aid in seed dispersal. Fruits may attach to or be eaten by an animal. The animal is then likely to move, after which the seeds may fall off or be defecated. Fruits are not necessarily fleshy. Fruits can also be hard and woody, or small and have modified structures that allow the seeds to be dispersed by wind or water (**Figure 28.17**).

A fruit may consist of only the mature ovary and its seeds, or it may include other parts of the flower or structures associated with it. A **simple fruit** is one that develops from a single carpel or several fused carpels, such as a plum or cherry. A raspberry is an example of an **aggregate fruit**—one that develops from several separate car-pels of a single flower. Pineapples and figs are examples of **multiple fruits**, formed from a cluster of flowers (an inflorescence). Fruits derived from parts in addition to the carpel and seeds are called **accessory fruits**—examples are apples, pears, and strawberries.

 Media Clip 28.3 **Flower and Fruit Formation**
www.Life11e.com/mc28.3

Recent analyses have revealed the phylogenetic relationships of angiosperms

Figure 28.18 shows the relationships among the major angio-sperm clades. Recent molecular and morphological analyses

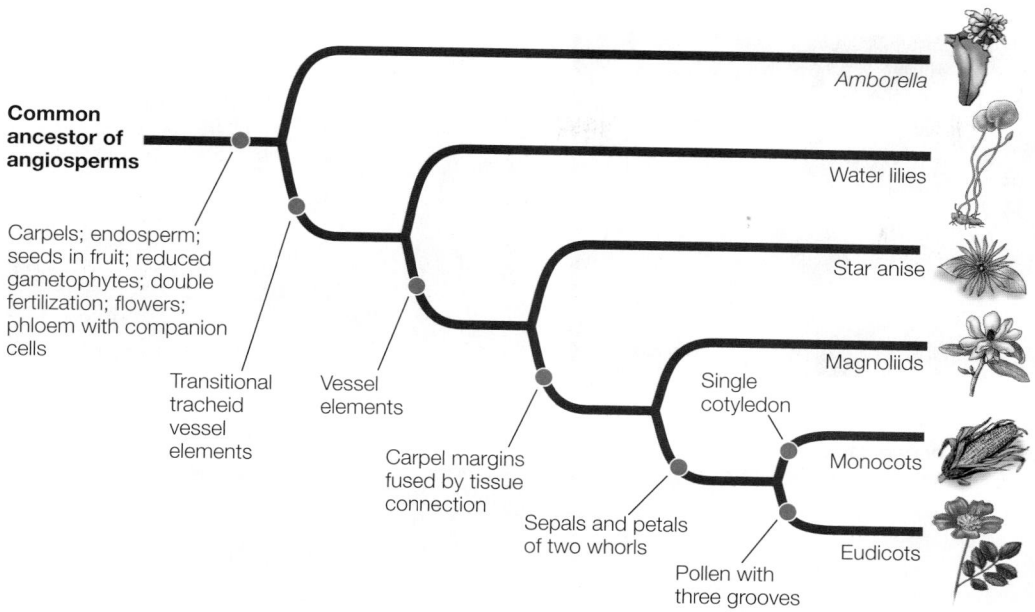

Figure 28.18 Evolutionary Relationships among the Angiosperms Recent analyses of many angiosperm genes have clarified the relationships among the major groups.

Common ancestor of angiosperms

Carpels; endosperm; seeds in fruit; reduced gametophytes; double fertilization; flowers; phloem with companion cells

Transitional tracheid vessel elements

Vessel elements

Carpel margins fused by tissue connection

Sepals and petals of two whorls

Pollen with three grooves

Single cotyledon

Amborella

Water lilies

Star anise

Magnoliids

Monocots

Eudicots

have supported the hypothesis that the sister group of remaining flowering plants is a single species of the genus *Amborella* (**Figure 28.19A**). This woody shrub with cream-colored flowers lives only on New Caledonia, an island in the South Pacific. Other early-branching angiosperm groups include the water lilies (**Figure 28.19B**), star anise and its relatives (**Figure 28.19C**), and the **magnoliids** (**Figure 28.19D, E**). The magnoliids include many familiar and useful plants, such as avocados, cinnamon, black pepper, and magnolias.

The two largest clades—the **monocots** and the **eudicots**—include the great majority of angiosperm species. The monocots are so called because they have a single embryonic cotyledon, whereas the eudicots have two.

Representatives of the two largest angiosperm clades are everywhere. The monocots (**Figure 28.20**) include grasses, cattails, lilies, orchids, and palms. The eudicots (**Figure 28.21**) include the vast majority of familiar seed plants, including most herbs (i.e., nonwoody

(A) *Amborella trichopoda*

(B) *Victoria amazonica*

Sterile stamens

Figure 28.19 Monocots and Eudicots Are Not the Only Surviving Angiosperms (A) *Amborella*, a shrub, is sister to the remaining extant angiosperms. Notice the sterile stamens on this female flower, which may serve to lure insects that are searching for pollen. (B) The water lily clade was the next to diverge after *Amborella*. (C) Star anise and its relatives belong to another early-diverging angiosperm clade. (D, E) The largest clade other than the monocots and eudicots is the magnoliid complex, which includes magnolias and the group known as "Dutchman's pipe."

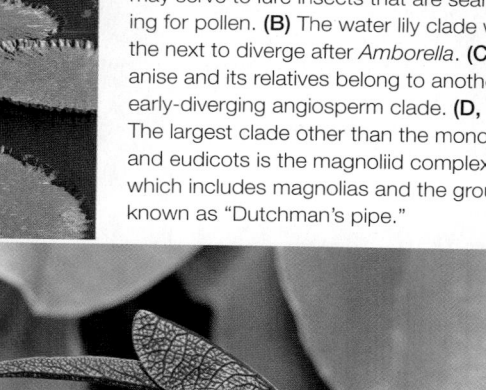

(C) *Illicium floridanum*

(D) *Magnolia* sp.

(E) *Aristolochia ringens*

Figure 28.20 Monocots
(A) Monocots include many popular garden flowers such as these tulips (pink and white) and daffodils (yellow). (B) Monocot grasses such as rice feed the world; wheat, sugarcane, and corn (maize) are also grasses. (C) Seagrasses such as this Neptune's grass form "meadows" in the shallow, sunlit waters of the world's oceans. (D) Palms are among the few monocot trees. Date palms like these are a major food source in some areas of the world.

(A) *Tulipa* sp. and *Narcissus* sp.

(B) *Oryza sativa*

(C) *Posidonia oceanica*

(D) *Phoenix dactylifera*

(A) *Malus* sp.

(B) *Banksia coccinea*

(C) *Escobaria vivipara*

(D) *Rafflesia arnoldii*

Figure 28.21 Eudicots (A) Eudicots include many trees, such as this crabapple tree. (B) Scarlet Banksia is a species of an Australian genus of eudicots that attracts a wide diversity of pollinators by producing large quantities of nectar. (C) Cacti comprise a large group of eudicots, with about 1,500 species in the Americas. Many, such as this spinystar cactus, bear large flowers for a brief period of each year. (D) *Rafflesia arnoldii*, found in the rainforests of Indonesia, bears the largest flower in the world. The flower lives as a parasite on tropical vines and has lost its leaf, stem, and even root structures. It smells like decaying meat, which attracts its fly pollinators.

plants), vines, trees, and shrubs. Among the eudicots are such diverse plants as oaks, willows, beans, snapdragons, roses, and sunflowers.

28.3 recap

Angiosperms are the dominant terrestrial plants of the Cenozoic era. The synapomorphies of angiosperms include flowers, fruit, carpels, double fertilization, and endosperm. Most angiosperms also possess distinctive cells in the xylem and phloem. The largest angiosperm clades are the monocots and the eudicots.

learning outcomes

You should be able to:

- Distinguish between the two different roles of sperm in angiosperm double fertilization.
- Distinguish between pollination and fertilization.
- Make inferences about the likely modes of seed dispersal based on the type of fruit the plant produces.
- Compare the functions of flowers, fruits, and seeds.

1. What are the respective roles of the two sperm in double fertilization in angiosperms?
2. Explain the difference between pollination and fertilization.
3. What are some of the differences between fruits that are dispersed by attracting animals to a food resource and those that are dispersed by attaching themselves to animals? Give examples of each.
4. What are the different functions of flowers, fruits, and seeds?

The remarkable diversity of the seed plants has been shaped by both biotic and abiotic components of the environments to which they have adapted. In turn, land plants—and seed plants in particular—shape their environments.

key concept 28.4 Plants Play Critical Roles in Terrestrial Ecosystems

Plants make profound contributions to ecosystem services—processes by which the environment maintains resources that benefit human society. Once life moved onto land, it was largely plants that shaped the terrestrial environment.

focus your learning

- Plants make profound contributions to ecosystem services.
- Many medicines are derived from seed plants.
- Most of the food eaten by humans comes from only 12 species of seed plants.

Plants produce oxygen and remove carbon dioxide from the atmosphere, and they play important roles in forming soils and renewing soil fertility. Plant roots help hold soil in place, providing protection against erosion by wind and water (**Figure 28.22**). Plants also moderate the local climate in various ways, such as by increasing humidity, providing shade, and blocking wind. All of these ecosystem services permit a great diversity of fungi and animals to exist on land.

Figure 28.22 Plants Prevent Erosion   When forest vegetation was cleared on these hillsides in Malaysia, landslides and extensive soil erosion quickly followed. Adjacent forested areas did not have landslides.

Seed plants have been sources of medicine since ancient times

Although we also use medicines derived from fungi, lichens, and actinobacteria, seed plants are the source of many of our medications. A few examples of medicines derived from plants are shown in **Table 28.1**. Even in synthetic pharmaceuticals, the chemical structures of the active ingredients are often based on the biochemistry of substances isolated from plants.

How are plant-based medicines discovered? Many were discovered over the millennia by people who lived alongside useful plants and discovered the properties of the plants through trial and error. Some plant-based medicines have come into widespread use through the work of ethnobotanists, who study how people use and view plants in their local environments. This work proceeds all over the globe today. An older example of this approach is the discovery of quinine as a treatment for malaria. In the sixteenth century, Spanish priests in Peru became aware that the native population used the bark of local *Cinchona* trees to treat fevers. The priests successfully used the bark to treat malaria. Word of the medicine spread to Europe, where it was put into use as early as 1631. The active ingredient of *Cinchona* bark—quinine—was identified in 1820, and quinine remained the standard malarial remedy well into the twentieth century.

table 28.1 Some Medicinal Plants and Their Products

Product	Plant source	Medical application
Atropine	Belladonna	Dilate pupils for eye examination
Bromelain	Pineapple stem	Control tissue inflammation
Digitalin	Foxglove	Strengthen heart muscle contraction
Ephedrine	*Ephedra*	Ease nasal congestion
Menthol	Japanese mint	Relieve coughing
Morphine	Opium poppy	Relieve pain
Quinine	*Cinchona* bark	Treat malaria
Taxol	Pacific yew	Treat ovarian and breast cancers
Tubocurarine	Curare plant	Muscle relaxant (used in surgery)
Vincristine	Periwinkle	Treat leukemia and lymphoma

Figure 28.23 Rice Feeds Much of the World's Human Population These rice fields, or "paddies," are in south China. Rice has been cultivated in this manner for thousands of years.

Since the mid-1900s, many plant-based medicines have been found by systematic testing of plants from all over the world. One example of a medicine discovered in this way is taxol, an important anti-cancer drug. Among the myriad plant samples that had been tested by 1962, extracts of the bark of the Pacific yew (*Taxus brevifolia*) showed anti-tumor activity in tests against rodent tumors. The active ingredient, taxol, was isolated in 1971 and tested against human cancers in 1977. After another 16 years, the U.S. Food and Drug Administration approved it for human use, and taxol is now widely used in treating breast and ovarian cancers as well as several other types of cancers.

Seed plants are our primary food source

Plants are primary producers. They trap energy and carbon by means of photosynthesis, making those resources available not only for their own needs, but also for the herbivores and omnivores that consume them, for the carnivores and omnivores that eat the herbivores, and for the prokaryotes and fungi that complete food webs. The earliest steps in human civilization involved cultivating angiosperms to provide a reliable food supply.

Today, twelve species of seed plants account for most of the food eaten by humans: rice, coconut, wheat, corn (maize), potato, sweet potato, cassava (also called tapioca or manioc), sugarcane, sugar beet, soybean, common bean (*Phaseolus vulgaris*), and banana.

Hundreds of other seed plants are cultivated for food, but none rank with these twelve in importance. Indeed, more than half of the world's human population derives the bulk of its food energy from the seeds of a single plant, *Oryza sativa*, better known as rice. Rice is particularly important in eastern Asia, where it has been cultivated for more than 8,000 years (**Figure 28.23**).

In addition to providing food directly for human consumption, plants also provide feed for agricultural production of animals. In particular, the grasses that grow in the world's rangelands are the primary forage for the grazing animals that humans use for meat and dairy products.

28.4 recap

Terrestrial ecosystems could not function without the foods and habitats provided by plants. Plants produce oxygen and remove carbon dioxide from the atmosphere, help form soils, and hold soils in place against erosion by wind and water. Plants also provide us with many important medicines and are the basis (directly or indirectly) of virtually all of our food.

learning outcomes

You should be able to:

- Describe how seed plants have been used for medicinal purposes.
- Summarize the ecological services angiosperms perform.
- Summarize the main plants humans use for food.

1. Name three plants that are used in medicine for humans, and give an example of how each is used.

2. A parcel of agricultural land that has been used to produce corn for decades no longer produces healthy crops. Tests indicate that the soil is depleted of nitrogen and organic matter. A farm extension agent recommends that the farmer grow soybeans and then rotate this crop with corn in future years. She also recommends that the farmer till the stubble from the crops after harvest into the soil. Why are these practices expected to help crop production?

3. What major group of plants provides (either directly or indirectly) the majority of calories consumed by humans? Explain your answer.

investigatinglife

 **How long can most seeds survive, and why is seed dormancy important?**

As you saw in Investigating Life: William Beale's Seed Viability Study and in the opening of this chapter, seeds of some plants can remain dormant but viable for decades, centuries, or in extreme cases, even millennia. Seed dormancy is especially important for the long-term survival of plant species that grow in unpredictable or highly fluctuating environments. If the appropriate conditions for seed growth and seedling maturation occur only rarely, then seeds need to be able to remain dormant in the soil until those favorable

conditions occur. Biologists refer to dormant seeds in a given area as the "seed bank." What conditions will cause the seeds to break their dormancy and sprout? Fire, heavy rain, appropriate temperatures, or disturbance by animals are some of the more common stimulants for breaking seed dormancy. Many seeds need to be consumed by an animal and pass through the acidic conditions of the animal's digestive system before the seeds will sprout. As anyone who has planted a new garden will know, soil disturbance will often produce the conditions needed for many dormant seeds to start growing. A new garden often requires persistent weeding to remove all the unwanted (by the gardener) plants that have been

Future directions

We've noted that humans depend on a relatively small number of plants for the majority of our food needs, although many more species of plants are grown and eaten for food. Traditionally, different locally adapted strains of crop plants have been grown by different farmers in different regions of the world. Agricultural companies now produce new varieties of crops that increase agricultural production. Unfortunately, a downside to these new strains is that farmers across the world have switched to growing relatively few varieties of important crop plants, which increases our global risk to disease and pests. If all farmers grow the same strain of wheat, for example, and that strain is susceptible to a particular pathogen or insect pest, what happens when the pathogen or pest spreads and all the susceptible wheat is lost?

To guard against devastating losses of important plants, many nations around the world have collaborated to develop a secure seed vault, which stores and protects the genetic diversity of crop plants (Figure 28.24). This seed vault is built into a sandstone mountain, surrounded by permafrost on the Arctic island of Spitsbergan, Norway. It is situated at a high enough elevation that it will remain above sea level even if all the polar ice caps were melted by global warming. Cooling systems maintain the vault at –18°C, and even if these systems fail, insulation and local cold weather would protect the seeds inside and maximize their viability. This seed vault complements several shorter-term seed depositories that are maintained in individual countries. Such seed depositories are used by biologists to develop new varieties of plant crops and to research the genetic variation of plant species from around the world.

Seed vaults
Airlock doors
Office and handling area
Sleeve to protect tunnel from erosion and climatic changes
Tunnel entrance
Bridge

Figure 28.24 The Global Seed Vault This vault is located over 100 meters inside a sandstone mountain on the Norwegian island of Spitsbergen, about 1,000 kilometers from the North Pole. The location was chosen for secure, long-term storage of seeds because the surrounding ground is perpetually frozen and the site is secure against earthquakes, floods, and other natural disasters.

Chapter Summary 28

▶ 28.1 Pollen, Seeds, and Wood Contributed to the Success of Seed Plants

- Fossils of woody **seed ferns** are the earliest evidence of seed plants. The surviving groups of seed plants are the **gymnosperms** and **angiosperms**. Review Figure 28.1
- All seed plants are heterosporous, and their gametophytes are much smaller than (and dependent on) their sporophytes. Review Figure 28.2
- Seed plants do not require liquid water for fertilization. **Pollen grains**, the microgametophytes of seed plants, are carried to a megagametophyte by wind or by animals.
- An **ovule** consists of the seed plant megagametophyte and the **integument** of sporophytic tissue that protects it.
- Following **pollination**, a **pollen tube** emerges from the pollen grain, elongates, and usually delivers gametes to the megagametophyte. Review Figure 28.4, Activity 28.1
- The ovule develops into a **seed** that contains an embryo (the new sporophyte generation). Seeds are well protected and are often capable of long periods of dormancy, germinating only when conditions are favorable. Review Figure 28.5

▶ 28.2 Once Dominant Gymnosperms Still Thrive in Some Environments

- Gymnosperms were the dominant terrestrial plants of the Mesozoic era. Although they declined in the Cenozoic era, they continue to thrive today in some areas—especially in high-elevation and high-latitude forests.
- The gymnosperms produce ovules and seeds that are not protected by ovary or fruit tissues.
- The major gymnosperm groups are the **cycads**, **ginkgos**, **gnetophytes**, and **conifers**. Review Figure 28.6
- The megaspores of conifers are produced in woody **cones** called **megastrobili**; the microspores are produced in herbaceous cones called **microstrobili**. Review Figures 28.7, 28.8, Activity 28.2, Animation 28.1

▶ 28.3 Flowers and Fruits Led to Increased Diversification of Angiosperms

- Angiosperms are the dominant plants of the Cenozoic era (the most recent 65 million years of Earth's history).
- **Flowers** and **fruits** are unique to the angiosperms, distinguishing them from the gymnosperms.

(continued)

Chapter Summary 28 (continued)

- The xylem of most angiosperms is more complex than that of the gymnosperms. It contains two specialized cell types: **vessel elements**, which function in water transport, and **fibers**, which play an important role in structural support.

- The ovules and seeds of angiosperms are enclosed in and protected by **carpels**.

- The floral organs, from the base to the apex of the flower, are the **sepals**, **petals**, **stamens**, and **pistil**. Stamens bear microsporangia in **anthers**. The pistil (consisting of one or more carpels) includes an **ovary** containing ovules. The **stigma** is the receptive surface of the pistil for pollen.

- A flower with both megasporangia and microsporangia is referred to as **perfect**; a flower with only one or the other is **imperfect**.

- A **monoecious** species has megasporangiate and microsporangiate flowers on the same plant. A **dioecious** species is one in which megasporangiate and microsporangiate flowers occur on different plants.

- The carpels and stamens of flowers probably evolved from leaflike structures. **Review Figure 28.12**

- Some plants with perfect flowers have adaptations to prevent self-fertilization. **Review Figure 28.13**

- Many angiosperms have coevolved with their animal pollinators.

- Angiosperms exhibit **double fertilization**, usually resulting in the production of a diploid zygote and triploid **endosperm**. **Review Figure 28.16, Animation 28.2**

- The oldest evolutionary split among the angiosperms is between the clade represented by the single species in the genus *Amborella* and all the remaining flowering plants. **Review Figure 28.18**

- The most species-rich angiosperm clades are the **monocots** and the **eudicots**.

▶ 28.4 Plants Play Critical Roles in Terrestrial Ecosystems

- Plants provide critical ecosystem services that affect soil, water, air quality, and climate.

- Plants provide humans with many important medicinal products.

- Plants are primary producers and as such are the foundation of terrestrial food webs.

> Go to **LearningCurve** (in **LaunchPad**) for dynamic quizzing that helps you solidify your understanding of this chapter. **LearningCurve** adapts to your responses, giving you the practice you need to master each key concept.

▶ Apply What You've Learned

Review

28.3 All parts of a flower are modified leaves.

28.3 Angiosperms and their animal pollinators have been coevolving for more than 150 million years.

Original Paper: Schemske, D. W. and H. D. Bradshaw, Jr. 1999. Pollinator preference and the evolution of floral traits in monkeyflowers (*Mimulus*). *Proceedings of the National Academy of Sciences USA* 96: 11910–11915.

Recent studies show that switches in pollinators and large changes in flower morphology can evolve quickly. In the photographs at right, the *Mimulus* species are excellent examples of such changes. Despite being sufficiently closely related that they hybridize readily in the lab, these species differ greatly in morphology and color. They also differ in their pollinators, with *M. cardinalis* pollinated exclusively by hummingbirds and *M. lewisii* pollinated primarily by bees and other insects. (Note that other species may visit the flowers, but not pollinate them).

M. lewisii

M. cardinalis

Questions

1. Given their pollinators, are the floral colors of the two species what you would expect? Explain your answer.

2. What differences are apparent between the flower corollas in the two species? Present a reasonable explanation for these differences, based on the pollinator differences.

3. Researchers found that variation at a single locus (called *yup*) was responsible for the color differences and that the *M. cardinalis* allele (*C*) is recessive, whereas the *M. lewisii* allele (*L*) is dominant. Thus *CC* genotypes are red, while both *LL* and *LC* genotypes are light pink. The researchers then examined the preferences of bees and hummingbirds for the different genotypes at the *yup* locus, keeping the rest of the genetic background of the plant the same. Their findings are shown at right. (Remember, these are visitation rates, not necessarily pollination rates.)

   Based on these results, how does variation at the *yup* locus affect visitation by bees and hummingbirds?

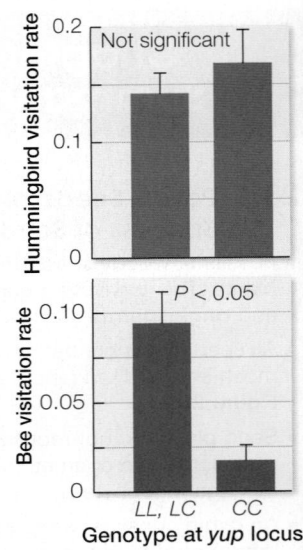

4. Do the results presented in Question 3 show a significant preference for red or pink flowers by birds and bees, respectively? Given your answer, why do you think most bird-pollinated flowers are red?

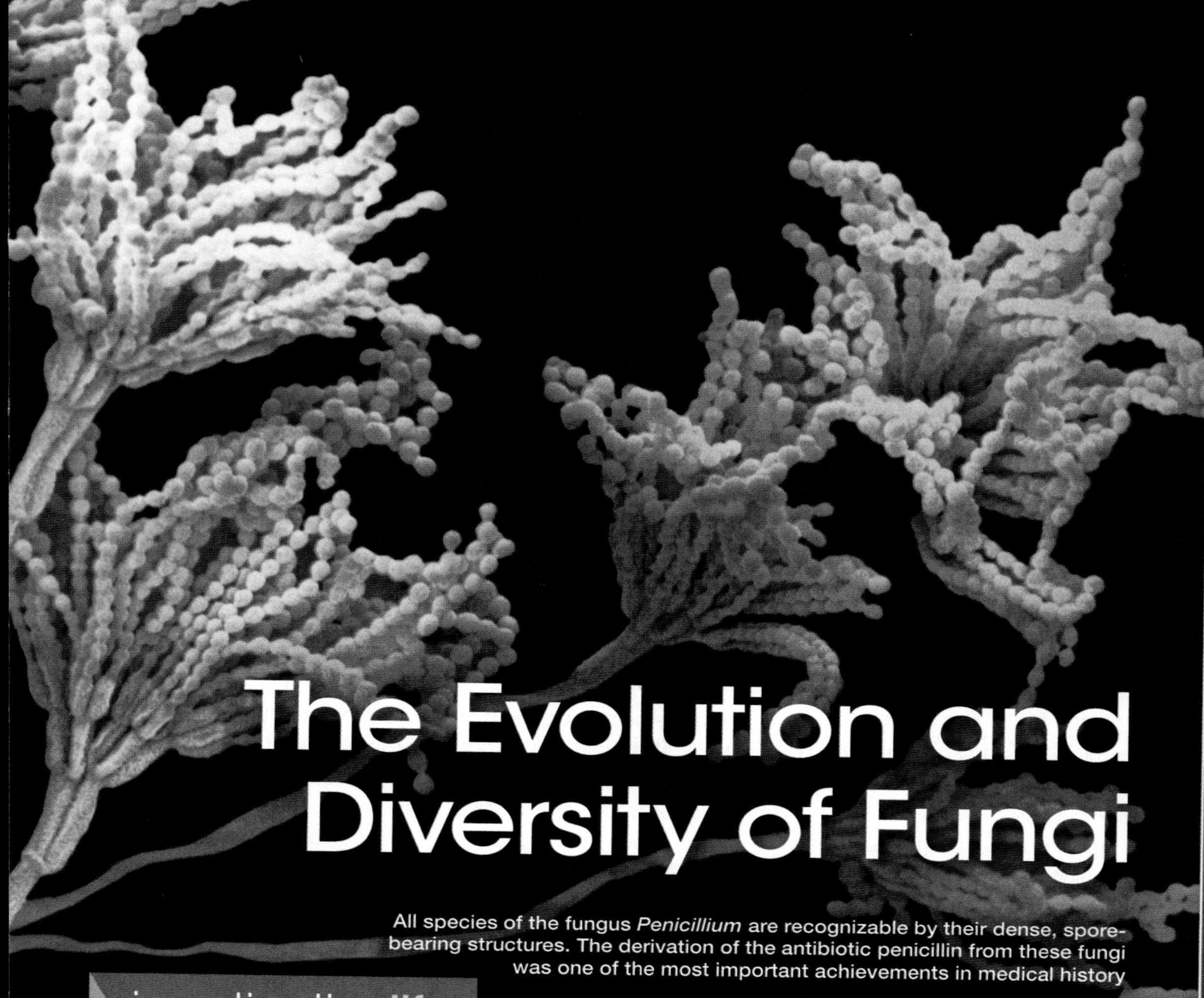

29

The Evolution and Diversity of Fungi

All species of the fungus *Penicillium* are recognizable by their dense, spore-bearing structures. The derivation of the antibiotic penicillin from these fungi was one of the most important achievements in medical history

key concepts

29.1 Fungi Digest Food Outside Their Bodies

29.2 Fungi Are Decomposers, Parasites, Predators, or Mutualists

29.3 Sex in Fungi Involves Multiple Mating Types

29.4 Fungi Have Many Practical Uses

investigatinglife

The Accidental Discovery of Antibiotics

Alexander Fleming was already a famous scientist in 1928, but his laboratory was often a mess. That year he was studying the properties of *Staphylococcus* bacteria, the agents of dangerous staph infections. In August he took a long vacation with his family. When he returned in early September, he found that some of his petri dishes of *Staphylococcus* had become infested with a fungus that killed many of the bacteria.

Many scientists would have sighed at the loss, thrown out the petri dishes, and started new cultures of bacteria. But when Fleming looked at the dishes, he saw something exciting. Around each colony of fungi was a ring within which all the bacteria were dead.

Fleming hypothesized that the bacteria-free rings around the fungal colonies were produced by a substance excreted from the fungi, which he initially called "mould juice." He identified the fungi as members of the genus *Penicillium* and eventually named the antibacterial substance they produced "penicillin." Fleming published his discovery in 1929, but initially it received very little attention.

Over the next decade, Fleming produced small quantities of penicillin for testing as an antibacterial agent. Some of the tests showed promise, but many were inconclusive, and eventually Fleming gave up on the research. But his tests had shown enough promise to attract the attention of several chemists, who worked out the practical problems of producing a stable form of the substance. Clinical trials of this stable form of penicillin were extremely successful, and by 1945 it was being produced and distributed as an antibiotic on a large scale. That same year, Fleming and two of the chemists, Howard Florey and Ernst Chain, received the Nobel Prize in Medicine for their work on penicillin.

The development of penicillin was one of the most important achievements in modern medicine. Until the introduction of modern antibiotics, the most widespread agents of human death included bacterial infections such as gangrene, tuberculosis, and syphilis. Penicillin proved to be highly effective in curing such infections, and its success led to the creation of the modern pharmaceutical industry. Soon many additional antibiotic compounds were isolated from other fungi or synthesized in the laboratory, leading to a "golden age" of human health.

QA How do fungi affect our daily lives?

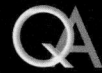

29.1 Fungi Digest Food Outside Their Bodies

Fungi are distinctive in the way they digest their food. They secrete digestive enzymes to break down large food molecules in the environment, then absorb the breakdown products through their cell membranes in a process known as **absorptive heterotrophy**. This mode of nutrition allows them to be successful in a wide variety of environments. Many fungi are **saprobes**, meaning they absorb nutrients from dead organic matter. Others are parasites or predators, feeding on or consuming other living organisms. Still others are mutualists, living in intimate associations with other organisms that benefit both partners.

focus your learning

- A mycelium, the body of a multicellular fungus, is made up of hyphae.
- Mycelia have enormous surface area-to-volume ratios.

Modern fungi evolved from a unicellular protist ancestor that had a flagellum, although most fungi are not motile and have now lost this structure. The probable common ancestor of the animals was also a flagellated protist much like the living choanoflagellates (see Figure 30.2). Current evidence, including the sequences of many genes, suggests that the fungi, choanoflagellates, and animals share a common ancestor not shared by other eukaryotes. These three lineages form a group known as the **opisthokonts** (**Figure 29.1**). A synapomorphy (shared derived trait) of the opisthokonts is a flagellum that, if present, is posterior, as in animal sperm. The flagella of all other eukaryotes attach at the front or sides of the cell, rather than at the rear.

Synapomorphies that distinguish the fungi as a group among the opisthokonts include absorptive heterotrophy and the presence of **chitin**, a nitrogen-containing structural polysaccharide, in their cell walls. The fungi represent one of the four independent evolutionary origins of large multicellular organisms (plants, brown algae, and animals are the other three).

Yeasts are unicellular, free-living fungi

Most fungi are multicellular, but single-celled species are found in most fungal groups. Unicellular, free-living fungi are referred to as **yeasts** (**Figure 29.2**). Some fungi have both a yeast life stage and

Saccharomyces cerevisiae

New cells have broken off from each of these yellow areas.

5 μm

Figure 29.2 Yeasts Unicellular, free-living fungi are known as yeasts. Many yeasts reproduce by budding—mitosis followed by asymmetrical cell division—as illustrated here.

a multicellular life stage. Thus the term "yeast" does not refer to a single taxonomic group, but rather to a lifestyle that has evolved multiple times. Yeasts live in liquid or moist environments and absorb nutrients directly across their cell surfaces.

The ease with which many yeasts can be cultured, combined with their rapid growth rates, has made them ideal *model organisms* for study in the laboratory. They present many of the same advantages to laboratory investigators as do many bacteria, but because they are eukaryotes, their genome structures and cells are much more like those of humans and other eukaryotes than are those of bacteria.

---

connect the concepts Why we select and study a variety of model organisms is discussed in Key Concept 17.3. Each model organism has distinctive advantages and disadvantages. The features of yeasts make them ideal for some investigations but impractical for others.

Multicellular fungi use hyphae to absorb nutrients

The body of a multicellular fungus is called a **mycelium** (plural *mycelia*). A mycelium is composed of a branching mass of individual tubular filaments called **hyphae** (singular *hypha*; **Figure 29.3A**), in which absorption of nutrients takes place primarily at each tip. The cell walls of the hyphae are greatly strengthened by microscopic fibrils of chitin. In some species of fungi, the hyphae are subdivided into cell-like compartments by incomplete cross-walls called **septa** (singular *septum*). These subdivided hyphae are referred to as **septate**. Septa do not completely close off compartments in the hyphae. Pores at the centers of the septa allow cytoplasm and organelles—sometimes even nuclei—to move in a controlled way between compartments (**Figure 29.3B**). In other species of fungi, the hyphae lack septa but may contain hundreds of nuclei. These multinucleate, undivided hyphae are referred to as **coenocytic**. The coenocytic condition results from repeated nuclear divisions without cytokinesis.

Certain modified hyphae, called **rhizoids**, anchor some fungi to their substrate (i.e., the dead organism or other matter on which they grow). These rhizoids are not homologous to the

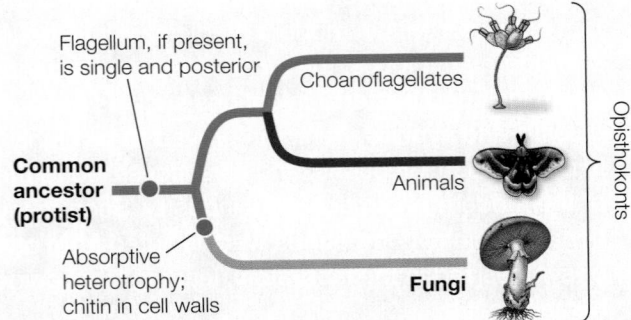

Flagellum, if present, is single and posterior

Choanoflagellates

Common ancestor (protist)

Animals

Absorptive heterotrophy; chitin in cell walls

Fungi

Opisthokonts

Figure 29.1 Fungi in Evolutionary Context Absorptive heterotrophy and the presence of chitin in their cell walls distinguish the fungi from other opisthokonts.

(A) Fungal hyphae in dry rot

Vessel in xylem   Fungal hyphae

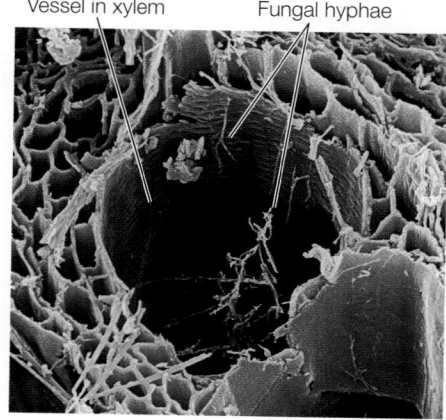

10 μm

(B) Anatomy of hyphae

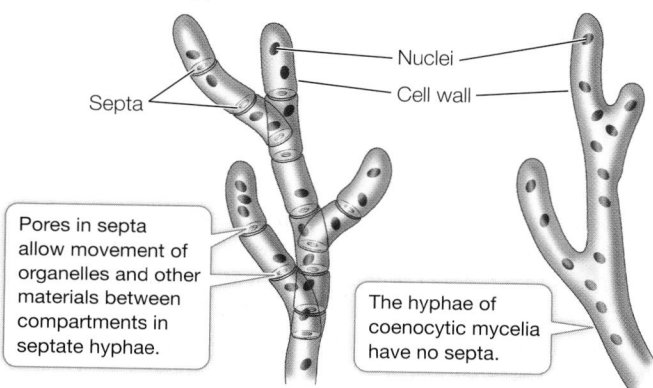

Nuclei

Cell wall

Septa

Pores in septa allow movement of organelles and other materials between compartments in septate hyphae.

The hyphae of coenocytic mycelia have no septa.

(C) The mycelium of a club fungus

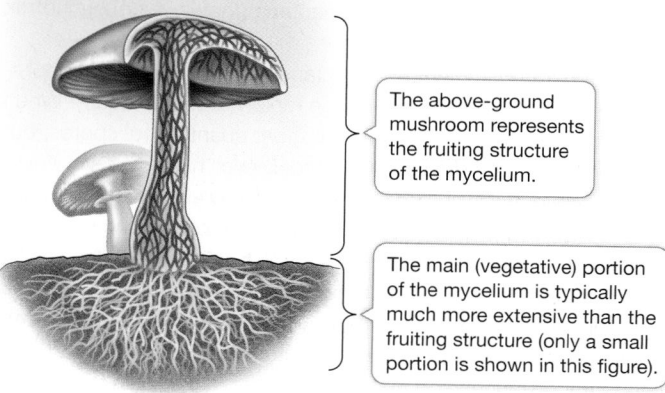

The above-ground mushroom represents the fruiting structure of the mycelium.

The main (vegetative) portion of the mycelium is typically much more extensive than the fruiting structure (only a small portion is shown in this figure).

Figure 29.3 Mycelia Are Made Up of Hyphae **(A)** The minute individual hyphae of fungal mycelia can penetrate small spaces. In this artificially colored micrograph, hyphae (yellow structures) of a dry-rot fungus are penetrating the xylem tissues of a log. **(B)** The hyphae of septate fungal species are divided into organelle-containing compartments by porous septa. The hyphae of coenocytic fungal species have no septa. **(C)** The fruiting structure of a club fungus is short-lived, but the filamentous, nutrient-absorbing mycelium can be long-lived and cover large areas.

mass is often far larger than the visible mushroom. The mycelium of one individual fungus discovered in Oregon covers almost 900 hectares underground and weighs considerably more than a blue whale (the largest animal). Aboveground, this individual is evident only as isolated clumps of mushrooms.

Fungi are in intimate contact with their environment

The filamentous hyphae of a fungus give it a unique relationship with its physical environment. The fungal mycelium has an enormous surface area-to-volume ratio compared with that of most large multicellular organisms. This large ratio allows the fungus to take in a great deal of water and mineral nutrients when soil moisture is high, which allows rapid growth in moist environments. The downside of the large surface area-to-volume ratio of the mycelium is its tendency to lose water rapidly in a dry environment. Thus fungi are most common in moist environments. You have probably observed the tendency of molds, toadstools, and other fungi to appear in damp places.

Another characteristic of some fungi is a tolerance for highly hypertonic environments (those with a solute concentration higher than their own; see Key Concept 6.3). Many fungi are more resilient than bacteria in hypertonic surroundings. Jelly, for example, does not support bacterial growth because its high sugar content makes it too hypertonic to those organisms, but it may eventually harbor mold colonies. Mold in the refrigerator illustrates yet another trait of many fungi: tolerance of temperature extremes. Many fungi grow in temperatures as low as –6°C, and some can tolerate temperatures higher than 50°C.

▶ 29.1 recap

Fungi, like animals, are opisthokonts. Fungi are distinguished from other opisthokonts by absorptive heterotrophy and by the presence of chitin in their cell walls. Unicellular fungi called yeasts absorb nutrients directly across their cell surfaces. The body form of multicellular fungi—a mycelium made up of rapidly growing hyphae—allows them to practice absorptive heterotrophy efficiently in a variety of moist environments.

learning outcomes
You should be able to:
- Explain how the structure of fungi facilitates their ability to decompose and consume other organisms.
- Make inferences about the presence and abundance of fungi given different environmental features.

1. How does fungal structure facilitate absorptive heterotrophy?
2. What are the advantages and disadvantages to multicellular fungi of the large surface area-to-volume ratio of the mycelium? How does the mycelium restrict the types of environments in which fungi are generally found?

rhizoids of plants, and they are not specialized to absorb nutrients and water.

Fungi can grow very rapidly when conditions are favorable. In some species, the total hyphal growth of a fungal mycelium (not the growth of an individual hypha) may exceed 1 kilometer a day! The hyphae may be widely dispersed to forage for nutrients over a large area, or they may clump together in a cottony mass to exploit a rich nutrient source. The familiar mushrooms you may notice growing in moist areas are spore-producing fruiting structures (**Figure 29.3C**). In the fungal species that produce these structures, the mycelial

Fungi are important components of healthy ecosystems. They interact with other organisms in many ways, some of which are harmful and some beneficial to those other organisms.

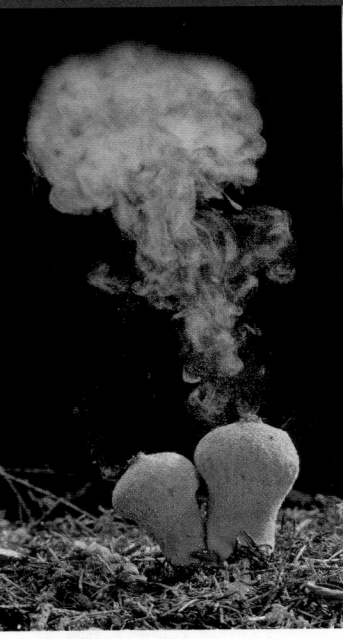

Figure 29.4 Spores Galore   Puffballs (a type of club fungus) disperse trillions of spores in great bursts. Few of the spores travel very far, however; some 99 percent of them fall within 100 meters of the parent puffball.

Lycoperdon perlatum

key concept

29.2 Fungi Are Decomposers, Parasites, Predators, or Mutualists

Without the fungi, our planet would be very different. Picture Earth with only a few stunted plants and watery environments choked with the remains of dead organisms. Fungi do much of Earth's garbage disposal. Fungi not only help clean up the landscape and form soil, but also play a key role in recycling mineral nutrients. Furthermore, the colonization of the terrestrial environment was made possible in large part by associations that fungi formed with land plants and other organisms.

focus your learning

- Were it not for saprobic fungi, Earth's carbon cycle would fail.
- A lichen is a complex mutualism between fungi and a photosynthetic microorganism.
- Mycorrhizae are mutualistic associations between fungi and the roots of vascular plants.

Saprobic fungi are critical to the planetary carbon cycle

Saprobic fungi, along with bacteria, are the major decomposers on Earth, contributing to the decay of nonliving organic matter and thus to recycling of the elements used by living things. In forests, for example, the mycelia of fungi secrete extracellular enzymes that degrade plant materials into simpler compounds that are than absorbed. Fungi are the principal decomposers of cellulose and lignin, the main components of plant cell walls (most bacteria cannot break down these materials). Other fungi produce enzymes that decompose keratin and thus break down animal structures such as hair and nails.

Were it not for the fungal decomposers, Earth's ***carbon cycle** would fail. Great quantities of carbon atoms would remain trapped forever on forest floors and elsewhere. Instead, those carbon atoms are returned to the atmosphere in the form of CO_2 by fungal respiration, where they are again available for photosynthesis by plants.

*connect the concepts   Earth's carbon cycle is discussed in detail in Key Concept 57.4. Fungi play a critical role in this cycle by decomposing other organisms and releasing carbon dioxide.

There was a time in Earth's history when populations of saprobic fungi declined dramatically. Vast tropical swamps existed during the Carboniferous period, as you saw in Chapter 24. When plants in these swamps died, they began to form peat. Peat formation led to acidification of the swamps. That acidity, in turn, drastically reduced the fungal population. The result? With the decomposers largely absent, large quantities of peat remained on the swamp floor and over time were converted into coal.

In contrast to their decline during the Carboniferous, fungi did very well at the end of the Permian, a quarter of a billion years ago,

when the aggregation of continents produced volcanic eruptions that triggered a global mass extinction. The fossil record shows that even as the maority of all multicellular species became extinct, fungi flourished—demonstrating both their hardiness and their role in recycling the elements in dead plants and animals.

Simple sugars and the breakdown products of complex polysaccharides are the favored source of carbon for saprobic fungi. Most fungi obtain nitrogen from proteins or the products of protein breakdown. Many fungi can use nitrate (NO_3^-) or ammonium (NH_4^+) ions as their sole source of nitrogen. No known fungus can get its nitrogen directly from inorganic nitrogen gas, however, as can some free-living and plant-associated bacteria (that is, fungi cannot fix nitrogen; see Key Concept 25.3).

What happens when a fungus faces a dwindling food supply? A common strategy is to reproduce rapidly and abundantly. When conditions are good, fungi produce great quantities of spores, but the rate of spore production is commonly even higher when nutrient supplies go down. The spores may then remain dormant until conditions improve, or they may be dispersed to areas where nutrient supplies are higher.

Not only are fungal spores abundant in number, but they are extremely tiny and easily spread by wind or water (**Figure 29.4**). These attributes virtually ensure that the spores will be scattered over great distances and that at least some of them will find conditions suitable for growth. The air we breathe contains as many as 10,000 fungal spores per cubic meter. No wonder we find fungi just about everywhere.

 Media Clip 29.1 Fungal Decomposers
www.Life11e.com/mc29.1

Some fungi engage in parasitic or predatory interactions

Whereas saprobic fungi obtain their energy, carbon, and nitrogen directly from dead organic matter, other species of fungi obtain their nutrition from parasitic—and even predatory—interactions.

(A) Parasitic mildew

Hyphae of fungal mycelium

This hypha is penetrating the leaf's interior through a stoma.

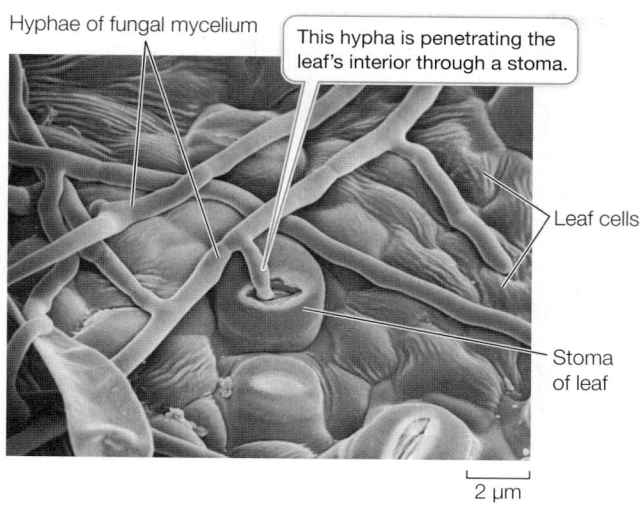

Leaf cells

Stoma of leaf

2 µm

Figure 29.5 Invading a Leaf   (A) Hyphae of the mildew *Phyllactinia guttata* growing on the surface of a hazel leaf. **(B)** Haustoria are fungal hyphae that push into the living cells of plants, from which they absorb nutrients.

(B) The parasitic action of haustoria

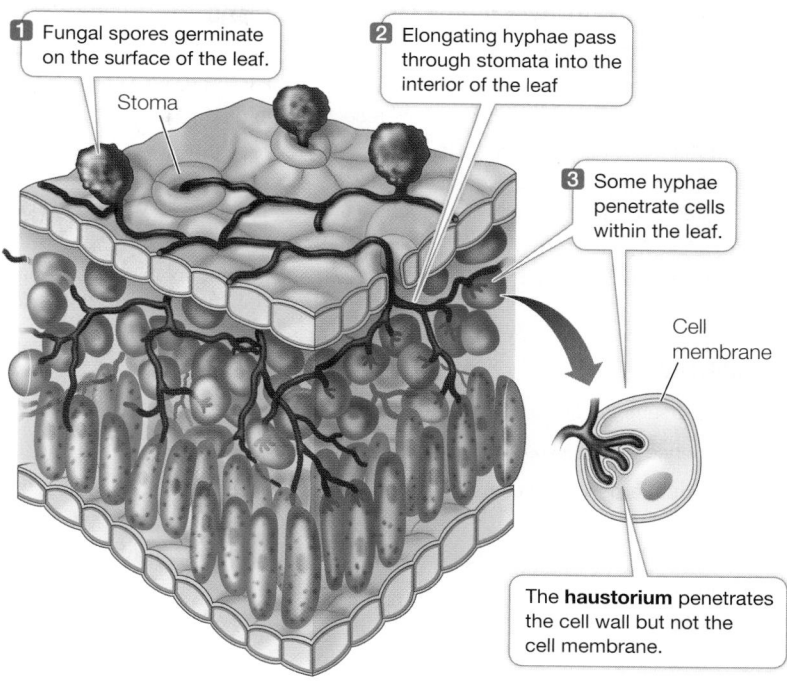

1 Fungal spores germinate on the surface of the leaf.

Stoma

2 Elongating hyphae pass through stomata into the interior of the leaf

3 Some hyphae penetrate cells within the leaf.

Cell membrane

The **haustorium** penetrates the cell wall but not the cell membrane.

PARASITIC FUNGI   Mycologists (biologists who study fungi) distinguish between two classes of parasitic fungi based on their degree of dependence on their host. **Facultative parasites** can grow on living organisms but can also grow independently (including on artificial media). **Obligate parasites** can grow only on a specific, living host. The fact that their growth depends on a living host shows that obligate parasites have specialized nutritional requirements.

Plants and insects are the most common hosts of parasitic fungi. The filamentous structure of fungal hyphae is especially well suited to a life of absorbing nutrients from living plants. The slender hyphae of a parasitic fungus can invade a plant through stomata, through wounds, or in some cases, by direct penetration of epidermal cell walls (**Figure 29.5A**). Once inside the plant, the hyphae branch out to expand the mycelium. Some hyphae produce **haustoria**, branching projections that push through cell walls into living plant cells, absorbing the nutrients within those cells. The haustoria do not break through the cell membranes inside the cell walls, but instead invaginate into the membranes, so that the cell membrane fits them like a glove (**Figure 29.5B**). Fruiting structures may form, either within the plant body or on its surface.

Some parasitic fungi live in a close physical (symbiotic) relationship with a plant host that is usually not lethal to the plant. Others are *pathogenic*, weakening or even killing the host from which they derive nutrition.

 Media Clip 29.2 Mind-Control Killer Fungi
www.Life11e.com/mc29.2

PATHOGENIC FUNGI   Although most human diseases are caused by bacteria or viruses, fungal pathogens are a major cause of death among people with compromised immune systems. Many people with AIDS die of fungal diseases, such as the pneumonia caused by *Pneumocystis jirovecii*. Even *Candida albicans* and certain other yeasts that are normally part of a healthy microbiome can cause severe diseases, such as esophagitis (which impairs swallowing), in individuals with AIDS and in individuals taking immunosuppressive drugs. Various fungi cause other, less threatening human diseases, such as ringworm and athlete's foot. The relatively close relationship between fungi and animals makes it difficult to find drugs that target unique features of fungal metabolism, without harming human patients. As a result, fungal diseases are a growing international health problem.

The worldwide decline of amphibian species has been linked to the spread of a chytrid fungus, *Batrachochytrium dendrobatidis* (or *Bd* for simplicity). In some areas of the world where this fungus has been present for millennia, amphibian populations appear to be tolerant of *Bd* and it does not cause widespread die-offs. But in other areas, such as western North America and Australia, introductions of *Bd* are implicated in the loss or decline of many frog and salamander populations, or the extinction of entire species. Some strains of *Bd* are native to southern Africa, where they do not appear to cause widespread amphibian declines. Biologists have hypothesized that the spread of *Bd* around the world may have been initiated in the 1930s with exports of the African clawed frog (*Xenopus laevis*), which was once widely used in human pregnancy tests. Recent studies by Erica Rosenblum and her colleagues have shown that Africa may indeed have been one source of the introductions, but that strains of *Bd* appear to have been introduced multiple times from several geographic areas (**Figure 29.6**).

Fungi are by far the most important plant pathogens, causing annual crop losses amounting to billions of dollars. Major fungal diseases of crop plants include black stem rust of wheat and other diseases of wheat, corn, and oats. The agent of black stem rust is *Puccinia graminis*, which has a complicated life cycle that involves two plant hosts (wheat and barberry). In an epidemic in 1935, *P. graminis* was responsible for the loss of about one-fourth of the wheat crop in Canada and the United States.

experiment

Figure 29.6 **What Is the Origin of Amphibian-Killing Chytrids in North America?**

Original Paper: Rosenblum, E. B. et al. 2013. Complex history of the amphibian-killing chytrid fungus revealed with genome resequencing data. *Proceedings of the National Academy of Sciences USA* 110: 9385–9390.

An investigation by Erica Rosenblum and colleagues of *Batrachochytrium dendrobatidis* isolates from around the world suggests that this pathogenic fungus was introduced to western North America several times, from different geographic localities.

HYPOTHESIS▶ Pathogenic forms of *Bd* that kill any amphibians were introduced into western North America from Africa, through importation of African clawed frogs beginning in the 1930s.

METHOD

Collect samples of *Bd* from throughout the world.

Sequence the genomes of the sampled chytrids.

Construct a phylogenetic tree by comparing the DNA sequences of the different samples of *Bd*.

Assess whether or not the resulting tree is consistent with the introduction of *Bd* from Africa to western North America.

RESULTS

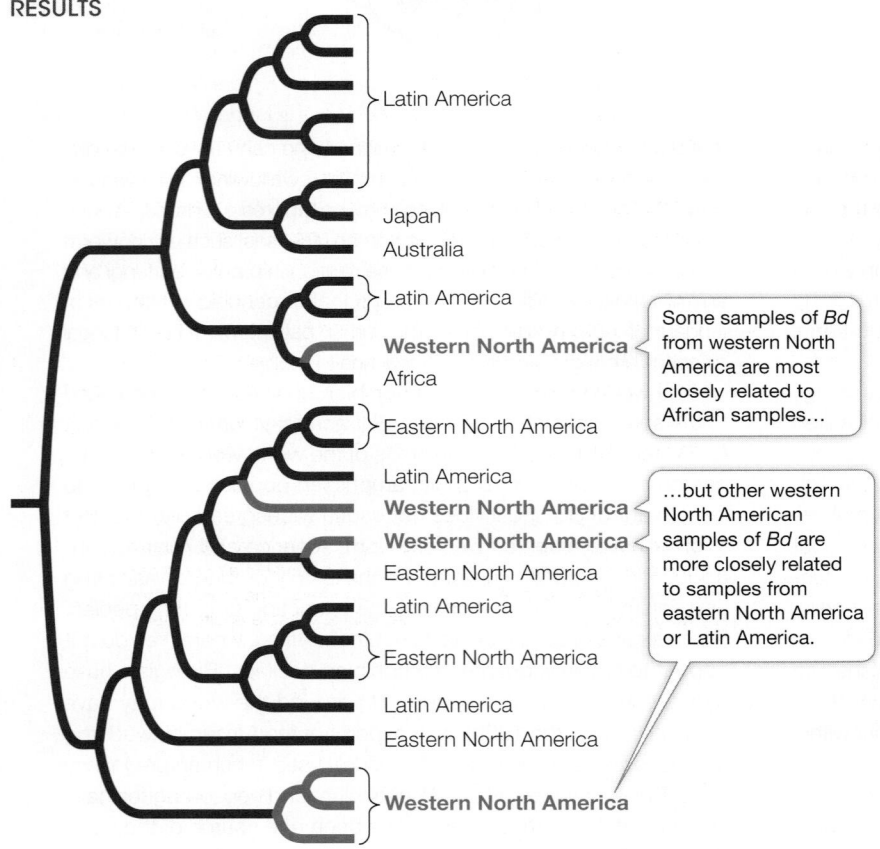

Latin America

Japan
Australia

Latin America

Western North America

Africa

> Some samples of *Bd* from western North America are most closely related to African samples…

Eastern North America

Latin America

Western North America

Western North America

Eastern North America

Latin America

> …but other western North American samples of *Bd* are more closely related to samples from eastern North America or Latin America.

Eastern North America

Latin America

Eastern North America

Western North America

CONCLUSION▶ Pathogenic strains of *Bd* appear to have been introduced to western North America from several different sources. At least one strain of *Bd* is consistent with an introduction from Africa, but other strains were more likely introduced from eastern North America or Latin America.

PREDATORY FUNGI Some fungi have adaptations that enable them to function as active predators, trapping nearby microscopic protists or animals. This feeding strategy is especially important in environments with lots of available carbon, but little nitrogen, such as wood. The most common predatory strategy seen in fungi is to secrete sticky substances from the hyphae so that passing organisms stick to them. The hyphae then quickly invade the trapped prey, growing and branching within it, spreading through its body, absorbing nutrients, and eventually killing it.

A more dramatic adaptation for predation is the constricting ring formed by some species of soil fungi (**Figure 29.7**). When nematodes (tiny roundworms) are present in the soil, these fungi form three-celled rings with a diameter that just fits a nematode. A nematode crawling through one of these rings stimulates the fungus, causing the cells of the ring to swell and trap the worm. Fungal hyphae quickly invade and digest the unlucky victim.

Mutualistic fungi engage in relationships that benefit both partners

Certain relationships between fungi and other organisms have nutritional consequences for both partners. Two relationships of this type are **symbiotic** (the partners live in close, permanent contact with each other) as well as **mutualistic** (the relationship benefits both partners).

LICHENS A **lichen** is not a single organism, but rather a meshwork of at least two radically different species: a fungus and a photosynthetic alga or cyano-bacterium. Recent studies suggest that an additional fungal species (a unicellular yeast) is also usually involved in this mutualism. Together the organisms that constitute a lichen can survive some of the harshest environments on Earth (although they are sensitive to poor air quality; see Key Concept 29.4). The biota of Antarctica, for example, features more than 100 times as many kinds of lichens as of plants. Relatively little experimental work has focused on lichens, perhaps because they grow so slowly—typically less than 1 centimeter in a year.

There are nearly 30,000 described "species" of lichens, each of which is assigned the name of its multicellular fungal component. These fungal components may constitute as many as 20 percent of all fungal species. Most of them are sac fungi (Ascomycota). Some of them are able to grow independently without a photosynthetic partner, but most have never been observed in nature other than in a lichen association. The photosynthetic component of a lichen is most

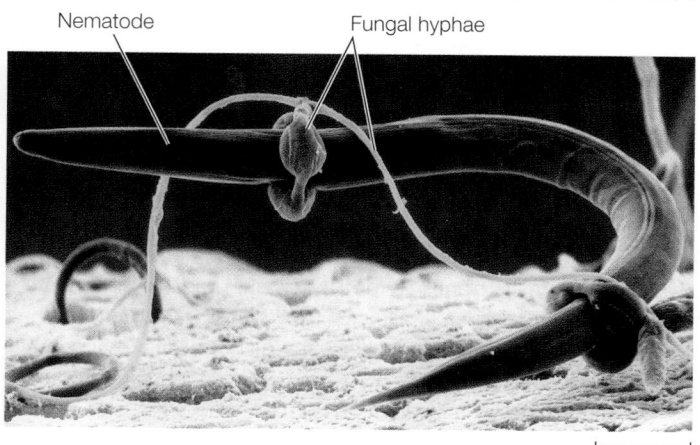

Figure 29.7 **Fungus as Predator** A nematode is trapped by hyphal rings of the soil-dwelling fungus *Arthrobotrys dactyloides*.

often a unicellular green alga, but it can be a cyanobacterium, or may even include both.

Lichens are found in all sorts of exposed habitats: on tree bark, on open soil, and on bare rock. Reindeer moss (not a moss at all, but the lichen *Cladonia subtenuis*) covers vast areas in Arctic, sub-Arctic, and boreal regions, where it is an important part of the diets of reindeer and other large mammals.

The body forms of lichens fall into three principal categories. **Crustose** (crustlike) lichens adhere tightly to their substrate (**Figure 29.8A**). **Foliose** (leafy) lichens are loosely attached and grow parallel to their substrate (**Figure 29.8B, right**). **Fruticose** lichens are highly

(A) Crustose lichen

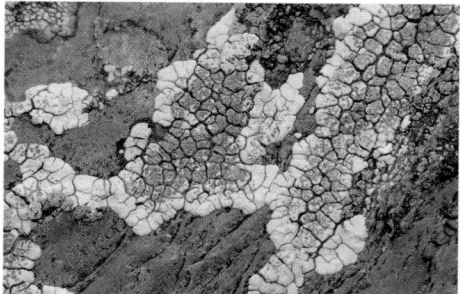

(B) Fruticose (left) and foliose (right) lichen

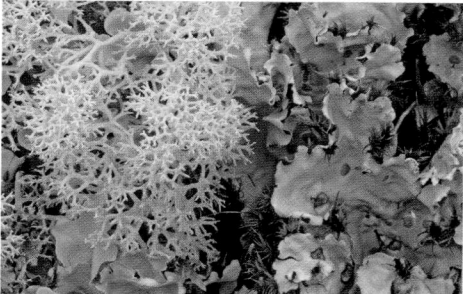

Figure 29.8 **Lichen Body Forms** The body forms of lichens fall into three principal categories. **(A)** Two crustose lichen species are growing together on this exposed rock surface. **(B)** The light green "shrubby" growth is a fruticose lichen. Foliose lichens have a leafy appearance.

branched and can grow upward like shrubs, or hang in long strands from tree branches or rocks (**Figure 29.8B, left**).

A cross section of a typical foliose lichen reveals a tight upper region of fungal hyphae, a layer of photosynthetic cyanobacteria or algae, a looser hyphal layer, and finally hyphal rhizoids that attach the entire structure to its substrate (**Figure 29.9**). The meshwork of fungal hyphae takes up mineral nutrients needed by the photosynthetic cells and also holds water tenaciously, providing a suitably moist environment. The fungus obtains fixed carbon from the photosynthetic products of the algal or cyanobacterial cells.

Within the lichen, fungal hyphae are tightly pressed against the photosynthetic cells and sometimes invade them without breaching the cell membrane (similar to the haustoria in parasitic fungi; see Figure 29.5). The photosynthetic cells not only survive these intrusions but continue to grow. Algal cells in a lichen "leak" photosynthetic products at a greater rate than do similar cells growing on their own, and photosynthetic cells taken from lichens grow more rapidly on their own than when associated with a fungus. On the basis of these observations, we could consider lichen fungi to be parasitic on their photosynthetic partners. In many places where lichens grow, however, the photosynthetic cells could not grow at all on their own.

Lichens can reproduce simply by fragmentation of the vegetative body (the **thallus**) or by means of specialized structures called **soredia** (singular *soredium*). Soredia consist of one or a few photosynthetic cells bound by fungal hyphae. They become detached from the lichen, are dispersed by air currents, and upon arriving at a favorable location, develop into a new lichen thallus (see Figure 29.9). Alternatively, the fungal partner may go through its sexual reproductive cycle, producing haploid spores. When these spores are discharged, however, they disperse alone, unaccompanied by the photosynthetic partner.

Lichens are often the first colonists on new areas of bare rock. They get most of the mineral nutrients they need from the air and rainwater, augmented by minerals absorbed from dust. A lichen begins to grow shortly after a rain, as it begins to dry. As it grows, the lichen acidifies its environment slightly, and this acidity contributes to

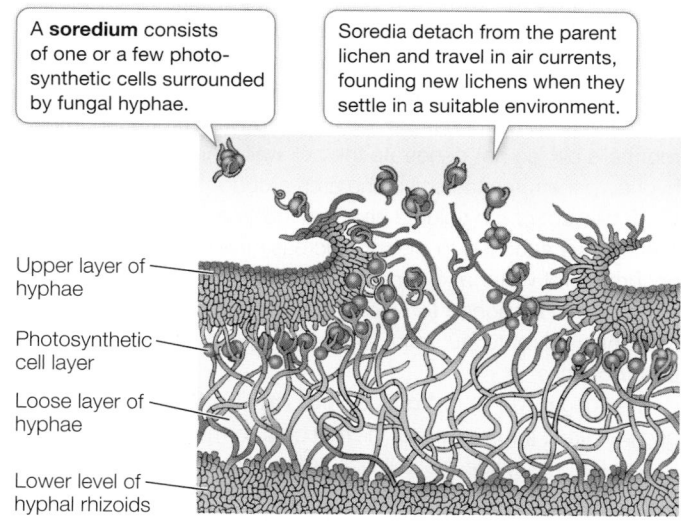

A **soredium** consists of one or a few photosynthetic cells surrounded by fungal hyphae.

Soredia detach from the parent lichen and travel in air currents, founding new lichens when they settle in a suitable environment.

Upper layer of hyphae

Photosynthetic cell layer

Loose layer of hyphae

Lower level of hyphal rhizoids

Figure 29.9 **Lichen Anatomy** Cross section showing the layers of a foliose lichen and the release of soredia.

(A) An ectomycorrhizal fungus

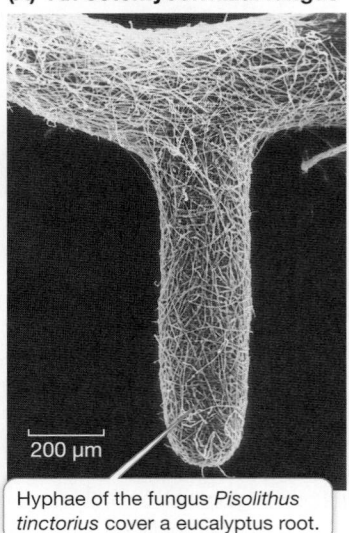

200 μm

Hyphae of the fungus *Pisolithus tinctorius* cover a eucalyptus root.

(B) An arbuscular mycorrhizal fungus

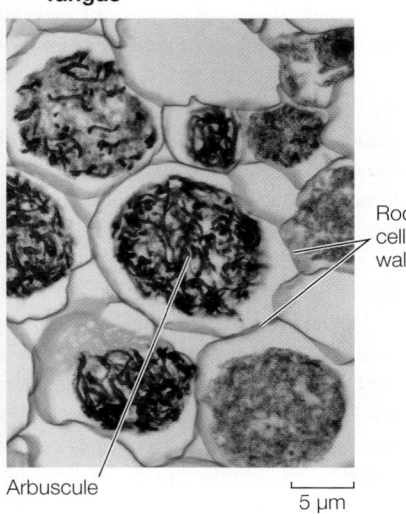

Root cell walls

Arbuscule

5 μm

Figure 29.10 Mycorrhizal Associations **(A)** Ectomycorrhizal fungi wrap themselves around a plant root, increasing the area available for absorption of water and minerals. **(B)** Hyphae of arbuscular mycorrhizal fungi infect the root internally and penetrate the root cell walls, branching within the cells and forming a treelike structure, the arbuscule. (The cell cytoplasm has been removed to better visualize the arbuscule.)

the slow breakdown of rocks, an early step in soil formation. With further drying, the lichen's photosynthesis ceases. The water content of the lichen may drop to less than 10 percent of its dry weight, at which point it becomes highly insensitive to extremes of temperature.

MYCORRHIZAE   Many vascular plants depend on a symbiotic association with fungi. This ancient association between plants and fungi was critical to the successful exploitation of the terrestrial environment by plants. Unassisted, the root hairs of many plants often do not take up enough water or minerals to sustain growth. However, the roots of such plants usually do become infected with fungi, forming an association called a **mycorrhiza** (plural *mycorrhizae*). There are two types of mycorrhizae, distinguished by whether or not the fungal hyphae penetrate the plant cell walls.

In **ectomycorrhizae**, the fungus wraps around the root tip, and its mass is often as great as that of the root itself (**Figure 29.10A**). The fungal hyphae penetrate the root and wrap around individual root cells but do not penetrate the cell walls. An extensive web of hyphae penetrates the soil in the area around the root, so that up to 25 percent of the volume near the root may be fungal hyphae. The hyphae attached to the root increase the surface area for the absorption of water and minerals, and the mass of hyphae in the soil acts like a sponge to hold water in the neighborhood of the root. Infected roots are short, swollen, and club-shaped, and they lack root hairs.

The fungal hyphae of *arbuscular mycorrhizae enter the root and penetrate the cell walls of the root cells, forming arbuscular (tree-like) structures inside the cell wall but outside the cell membrane. These structures, like the haustoria of parasitic fungi and the contact regions of fungal hyphae and photosynthetic cells in lichens, become the primary site of exchange between plant and fungus (**Figure 29.10B**). As in the ectomycorrhizae, the fungus forms a vast web of hyphae leading from the root surface into the surrounding soil.

*connect the concepts   Mycorrhizae can vastly increase the root surface area in contact with essential nutrients in the soil. The importance of arbuscular mycorrhizae in plant nutrition is described in detail in Key Concept 35.4.

The mycorrhizal association is important to both partners. The fungus obtains needed organic compounds, such as sugars and amino acids, from the plant. In return, the fungus, because of its very high surface area-to-volume ratio and its ability to penetrate the fine structure of the soil, greatly increases the plant's ability to absorb water and minerals (especially phosphorus). The fungus may also provide the plant with certain growth hormones and may protect it against attack by disease-causing microorganisms.

Plants that have active arbuscular mycorrhizae typically are a deeper green and may resist drought and temperature extremes better than plants of the same species that have little mycorrhizal development. Attempts to introduce some plant species to new areas have failed until some of soil from the native area (presumably containing the fungus necessary to establish mycorrhizae) was provided. Trees without ectomycorrhizae do not grow well in the absence of abundant nutrients and water, so the health of our forests depends on the presence of ectomycorrhizal fungi. Many agricultural crops require inoculation of seeds with appropriate mycorrhizal fungi prior to planting. Without these fungi, the plants are unlikely to grow well, or in some cases at all. Certain plants that live in nitrogen-poor habitats, such as cranberry bushes and orchids, invariably have mycorrhizae. Orchid seeds will not germinate in nature unless they are already infected by the fungus that will form their mycorrhizae. Plants that lack chlorophyll always have mycorrhizae, which they often share with the roots of green, photosynthetic plants. In effect, these plants without chlorophyll are feeding on nearby green plants, using the fungus as a bridge.

Endophytic fungi protect some plants from pathogens, herbivores, and stress

In a tropical rainforest, 10,000 or more fungal spores may land on a single leaf each day. Some are plant pathogens, some do not affect the plant at all, and some invade the plant in a beneficial way. Fungi that live within aboveground parts of plants without causing obvious deleterious symptoms are called **endophytic fungi**. Recent research has shown that endophytic fungi are abundant in plants in all terrestrial environments.

Among the grasses, individual plants with endophytic fungi are more resistant to pathogens and to insect and mammalian herbivores than are plants lacking endophytes. The fungi produce alkaloids (nitrogen-containing compounds) that are toxic to animals. The alkaloids do not harm the host plant. In fact, some plants produce alkaloids (such as nicotine) themselves. The fungal alkaloids also increase the ability of grasses to resist stress of various types, including drought (water shortage) and salty soils. Such resistance is beneficial for some crops, although it can produce toxicity for grazing animals.

The role, if any, of endophytic fungi in most broad-leaved plants is unclear. They may convey protection against pathogens, or they may simply occupy space within leaves without conferring any benefit, but also without doing harm. The benefit, in fact, might be all for the fungus.

29.2 recap

Fungi interact with other organisms in many ways, both harmful and beneficial. Saprobic fungi play critical roles in the recycling of elements required by living organisms. Lichens are mutualistic associations of fungi with algae or cyanobacteria. Mycorrhizae are associations of fungi and the roots of plants; they are essential for the survival of most plant species.

learning outcomes

You should be able to:

• Summarize why saprobic fungi are critical to Earth's carbon cycle.

• Summarize the nature and benefits of the lichen association.

• Summarize the nature and benefits of mycorrhizal associations.

1. What is the role of fungi in Earth's carbon cycle?

2. Describe the nature and benefits of the lichen association.

3. How does the mycorrhizal association benefit both the plant and the fungus?

Before molecular techniques clarified the phylogenetic relationships of fungi, one criterion used for assigning fungi to taxonomic groups was the nature of their life cycles—including the types of fruiting structures they produced. The next section will take a closer look at life cycles in the six major groups of fungi.

key concept 29.3 Sex in Fungi Involves Multiple Mating Types

Major fungal groups were originally defined by their structures and processes for sexual reproduction and also, to a lesser extent, by other morphological differences. Although fungal life cycles are even more diverse than was once realized, specific types of life cycles generally distinguish the six major groups of fungi: microsporidia, chytrids, zygospore fungi (Zygomycota), arbuscular mycorrhizal fungi (Glomeromycota), sac fungi (Ascomycota), and club fungi (Basidiomycota). **Figure 29.11** diagrams the evolutionary relationships of these groups as they are understood today.

focus your learning

• In most fungi, sexual reproduction involves genetically determined distinctions between two or more mating types.

• Microsporidia are parasitic fungi that lack true mitochondria.

• Chytrids have flagellated gametes and spores.

• Club fungi produce spectacular fruiting structures called basidiomata.

The chytrids and the zygospore fungi may not represent monophyletic groups, as they each consist of several distantly related lineages that retain some ancestral features. The clades that are thought to

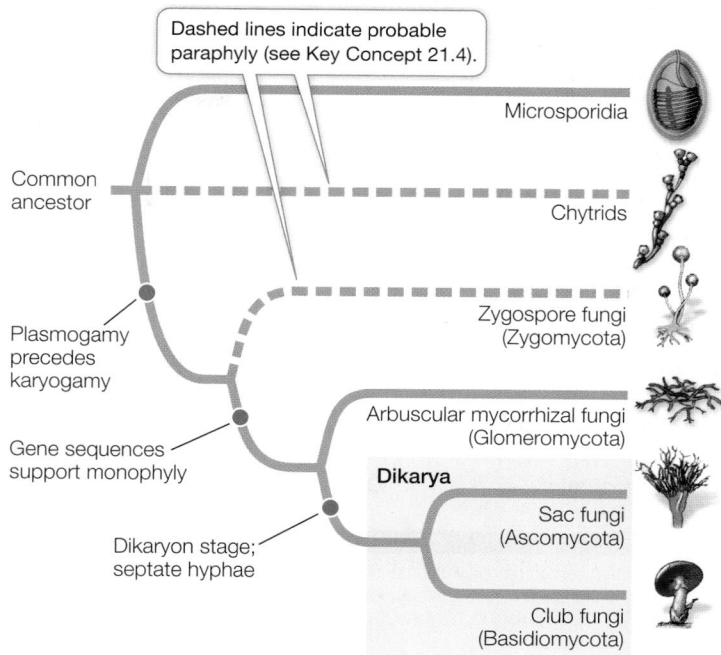

Figure 29.11 A Phylogeny of the Fungi Microsporidia are reduced, parasitic fungi whose relationships among the fungi are uncertain. They may be the sister group of most other fungi, or they may be more closely related to particular groups of chytrids or zygospore fungi. The dashed lines indicate that chytrids and zygospore fungi are thought to be paraphyletic; the relationships of the lineages within these two informal groups (see Table 29.1) are not yet well resolved. The sac fungi and club fungi together form the clade Dikarya.

 Activity 29.1 Fungal Phylogeny www.Life11e.com/ac29.1

be monophyletic within these two informal groupings are listed in **Table 29.1**. Recent evidence from DNA analyses has established the placement of the microsporidia among the fungi, the likely paraphyly of the chytrids and the zygospore fungi, the independence of arbuscular mycorrhizal fungi from the other fungal groups, and the monophyly of sac fungi and club fungi.

Fungi reproduce both sexually and asexually

Both asexual and sexual reproduction occur among the fungi (**Figure 29.12**). Asexual reproduction takes several forms:

• The production of (usually) haploid spores within sporangia

• The production of haploid spores (not enclosed in sporangia) at the tips of hyphae; in sac fungi, such spores are called **conidia** (Greek *konis*, "dust")

• Cell division by unicellular fungi—either a relatively equal division of one cell into two (*fission*) or an asymmetrical division in which a smaller daughter cell is produced (*budding*)

• Simple fragmentation of the mycelium

Sexual reproduction is rare (or even unknown) in some groups of fungi but common in others. Sexual reproduction may not occur in some species, or it may occur so rarely that biologists have never observed it. Species in which no sexual stage has been observed were once placed in a separate taxonomic group because

table 29.1 Classification of the Fungi

Group	Common name	Features
Microsporidia	Microsporidia	Intracellular parasites of animals; greatly reduced, among smallest eukaryotes known; polar tube used to infect hosts
Chytrids (paraphyletic)[a] Chytridiomycota Neocallimastigomycota Blastocladiomycota	Chytrids	Mostly aquatic and microscopic; zoospores and gametes have flagella
Zygomycota (paraphyletic)[a] Entomophthoromycotina Kickxellomycotina Mucoromycotina Zoopagomycotina	Zygospore fungi	Reproductive structure is a unicellular zygospore with many diploid nuclei; hyphae coenocytic; no fleshy fruiting body
Glomeromycota	Arbuscular mycorrhizal fungi	Form arbuscular mycorrhizae in plant roots; only asexual reproduction is known
Ascomycota	Sac fungi	Sexual reproductive saclike structure known as an ascus, which contains haploid ascospores; hyphae septate; dikaryon
Basidiomycota	Club fungi	Sexual reproductive structure is a basidium, a swollen cell at the tip of a specialized hypha that supports haploid basidiospores; hyphae septate; dikaryon

[a]The formally named groups within the chytrids and Zygomycota are each thought to be monophyletic, but their relationships to one another (and to microsporidia) are not yet well resolved.

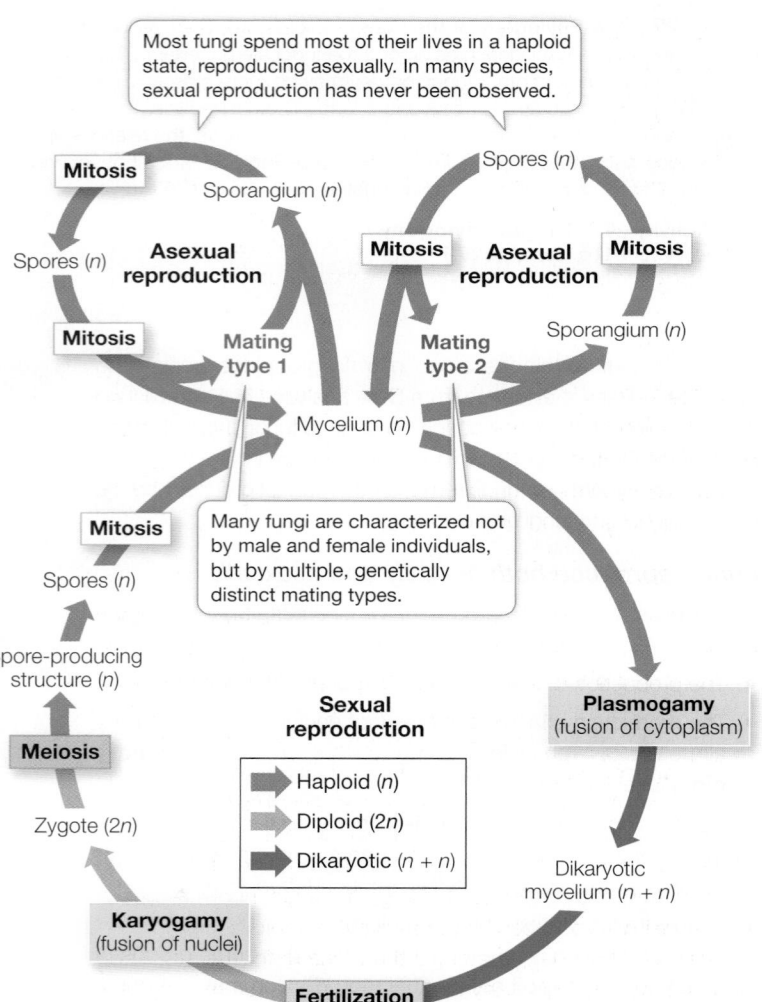

knowledge of the sexual life cycle was considered necessary for classifying fungi. Now, however, these species can be related to other species of fungi through analysis of their DNA sequences.

Sexual reproduction in most fungi features an interesting twist: There is no morphological distinction between female and male structures, or between female and male individuals, in most groups of fungi. Rather, there is a genetically determined distinction between two *or more* **mating types**. Individuals of the same mating type cannot mate with each other, but they can mate with individuals of another mating type within the same species, thus avoiding self-fertilization. Individuals of different mating types differ genetically but are often visually and physiologically indistinguishable.

Microsporidia are highly reduced, parasitic fungi

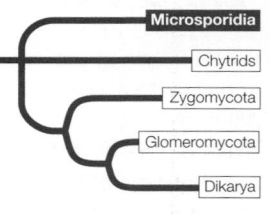

Microsporidia are unicellular parasitic fungi. They are among the smallest eukaryotes known, with infective spores that are only 1–40 micrometers (μm) in diameter. About 1,500 species have been described, but many more species are thought to exist. Their relationships among the eukaryotes have puzzled biologists for many decades

Microsporidia lack true mitochondria, although they have reduced structures, known as **mitosomes**, that are derived from mitochondria. Unlike mitochondria, however, mitosomes contain no DNA—the mitochondrial genome has been completely transferred to the nucleus. Because microsporidia lack mitochondria, biologists initially suspected that they represented an early lineage of eukaryotes that diverged before the endosymbiotic event from which mitochondria evolved. The presence of mitosomes, however, indicates that this hypothesis is incorrect. DNA sequence analysis, along with the fact that

Figure 29.12 A Generalized Fungal Life Cycle   Environmental conditions may determine which mode of reproduction—sexual or asexual—takes place at a given time.

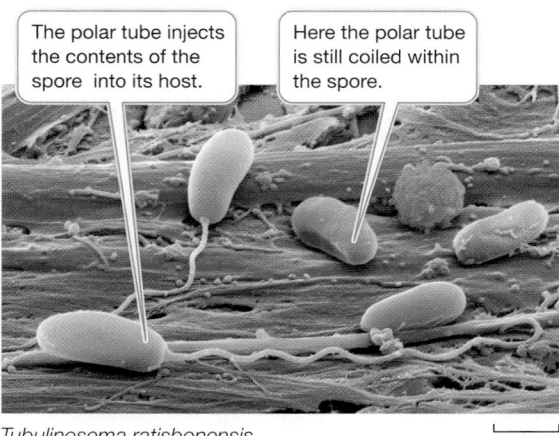

Figure 29.13 **Invasion of the Microsporidia** The spores of microsporidia grow polar tubes that transfer the contents of the spores into the host's cells. The species shown here infects many animals, including humans.

Tubulinosema ratisbonensis
20 µm

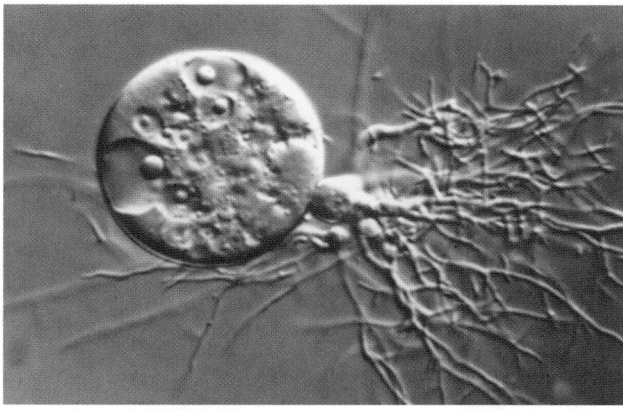

Chytriomyces hyalinus
25 µm

Figure 29.14 **A Chytrid** Branched rhizoids emerge from the sporangium of a mature chytrid.

their cell walls contain chitin, has confirmed that the microsporidia are in fact highly reduced, parasitic fungi, although their exact placement among the fungal lineages is still being investigated.

Microsporidia are obligate intracellular parasites of animals, especially of insects, crustaceans, and fish. Some species are known to infect mammals, including humans. Most infections by microsporidia cause chronic disease in the host, with effects that include weight loss, reduced fertility, and a shortened life span. The host cell is penetrated by a polar tube that grows from the microsporidian spore. The function of the polar tube is to inject the contents of the spore, the sporoplasm, into the host (**Figure 29.13**). The sporoplasm then replicates within the host cell and produces new infective spores. The life cycle of some species is complex and involves multiple hosts, whereas other species infect a single host. In some insects, parasitic microsporidia are transmitted vertically (i.e., from parent to offspring). Reproduction is thought to be strictly asexual in some microsporidians, but includes poorly understood asexual and sexual cycles in other species.

Most chytrids are aquatic

The **chytrids** (**Figure 29.14**) include several distinct lineages of mostly aquatic microorganisms once classified with the protists. However, morphological evidence (cell walls that consist primarily of chitin) and molecular evidence support their classification as early-diverging fungi. In this book we use the term "chytrid" to refer to all three of the formally named clades listed as chytrids in Table 29.1, but some mycologists use this term to refer to only one of those clades, the Chytridiomycota. There are fewer than 1,000 described species among the three groups of chytrids.

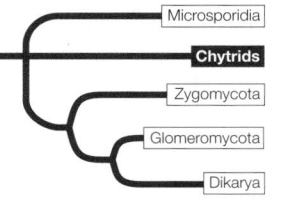

Chytrids reproduce both sexually and asexually. Like the animals, chytrids that reproduce sexually possess flagellated gametes. The retention of this trait reflects the aquatic environment in which fungi first evolved. Chytrids are the only fungi that include

species with flagella at any life cycle stage. Both the spores (called zoospores) and the gametes are flagellated (**Figure 29.15A**). There are two types of spores and gametes. Unlike in most fungi, one type of gamete is larger than the other, so the gametes can be distinguished as male (small gametes) and female (large gametes; see Figure 29.15A). Except for size, however, the two gametes are very similar.

The chytrids are diverse in form. Some are unicellular, others have rhizoids, and still others have coenocytic hyphae. They may be parasitic (on organisms such as algae, mosquito larvae, nematodes, and amphibians) or saprobic. Some have complex mutualistic relationships with foregut-fermenting animals such as cattle and deer. Many chytrids live in freshwater habitats or in moist soil, but some are marine.

Some fungal life cycles feature separate fusion of cytoplasms and nuclei

Most members of the remaining four groups of fungi are terrestrial. Although the terrestrial fungi grow in moist places, they do not have motile gametes, so liquid water is not required for fertilization. Instead, the cytoplasms of two individuals of different mating types fuse (a process called **plasmogamy**) before their nuclei fuse (a process called **karyogamy**; see Figure 29.12). Sexual species of terrestrial fungi include some zygospore fungi, sac fungi, and club fungi.

Zygospore fungi reproduce sexually when adjacent hyphae of two different mating types release chemical signals that cause them to grow toward each other. These hyphae produce gametangia, which are specialized cells for reproduction that are retained as part of the hyphae. In the gametangia, nuclei replicate without cell division, resulting in multiple haploid nuclei in both gametangia. The two gametangia then fuse to form a zygosporangium that contains many haploid nuclei of each mating type (**Figure 29.15B**). Haploid nuclei of different mating types then pair up to form multiple diploid nuclei within the zygosporangium. A

(A) Sexual life cycle of the Chytrids

Unlike other fungi, some chytrids produce flagellated male and female gametes from a multicellular haploid stage.

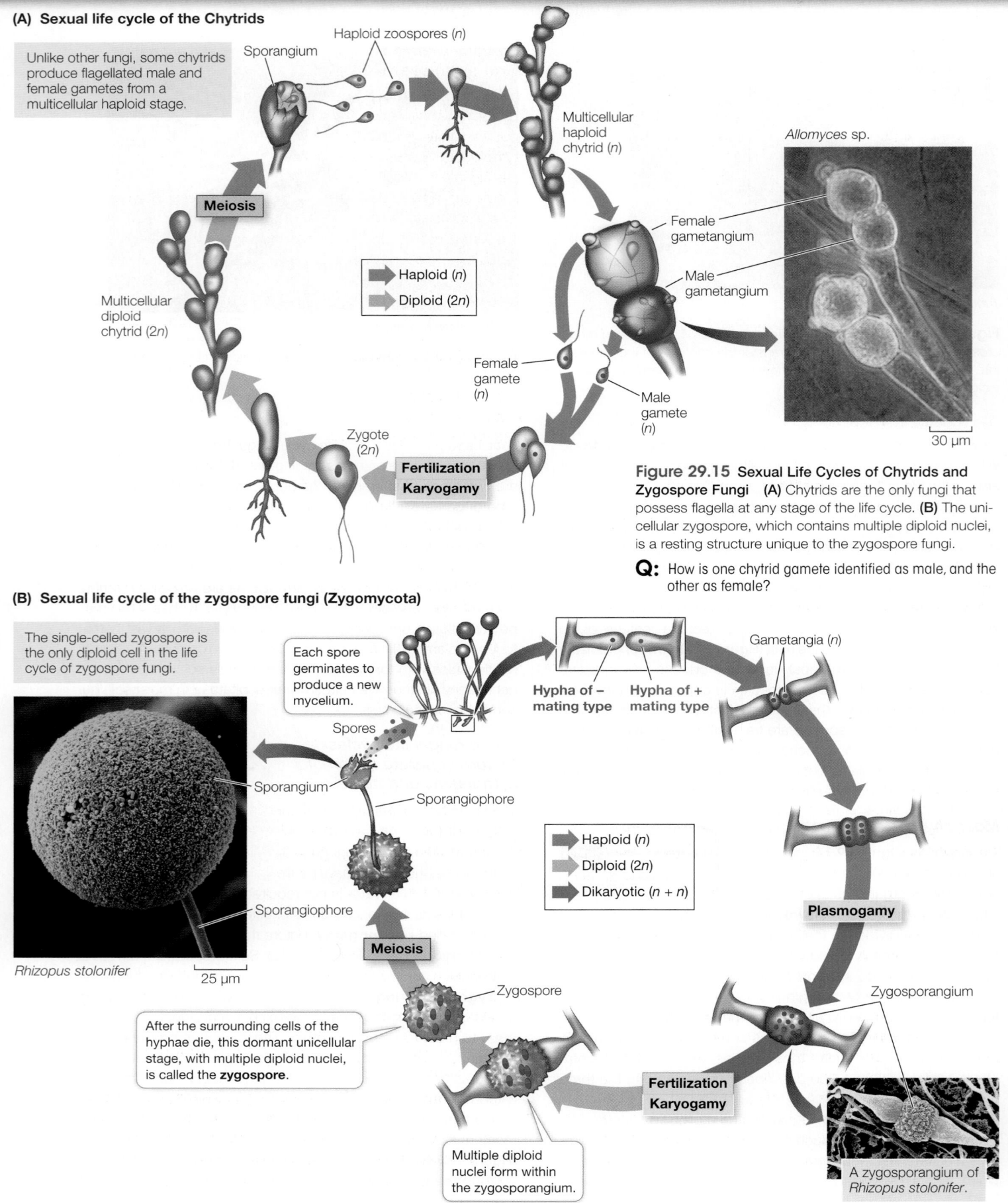

Sporangium

Haploid zoospores (n)

Multicellular haploid chytrid (n)

Allomyces sp.

Female gametangium

Male gametangium

Meiosis

Haploid (n)
Diploid (2n)

Multicellular diploid chytrid (2n)

Female gamete (n)

Male gamete (n)

Zygote (2n)

**Fertilization
Karyogamy**

30 μm

Figure 29.15 Sexual Life Cycles of Chytrids and Zygospore Fungi **(A)** Chytrids are the only fungi that possess flagella at any stage of the life cycle. **(B)** The unicellular zygospore, which contains multiple diploid nuclei, is a resting structure unique to the zygospore fungi.

Q: How is one chytrid gamete identified as male, and the other as female?

(B) Sexual life cycle of the zygospore fungi (Zygomycota)

The single-celled zygospore is the only diploid cell in the life cycle of zygospore fungi.

Each spore germinates to produce a new mycelium.

Hypha of – mating type
Hypha of + mating type

Gametangia (n)

Spores

Sporangium

Sporangiophore

Sporangiophore

Haploid (n)
Diploid (2n)
Dikaryotic (n + n)

Plasmogamy

Meiosis

Rhizopus stolonifer

25 μm

Zygospore

Zygosporangium

After the surrounding cells of the hyphae die, this dormant unicellular stage, with multiple diploid nuclei, is called the **zygospore**.

**Fertilization
Karyogamy**

Multiple diploid nuclei form within the zygosporangium.

A zygosporangium of *Rhizopus stolonifer*.

Figure 29.16 Zygospore Fungi Produce Sporangiophores
These transparent structures are sporangiophores produced by a zygospore fungus growing on decomposing animal dung. The sporangiophores grow toward the light and end in tiny sporangia, which the stalked sporangiophores can eject as far as 2 meters. Animals ingest sporangia and then disseminate the spores in their feces.

 Animation 29.1 **Life Cycle of a Zygospore Fungus**
www.Life11e.com/a29.1

thick, multilayered cell wall develops around the zygosporangium to form a well-protected resting stage that can remain dormant for months. In harsh environmental conditions, this resting stage may be the only cell that survives as the surrounding cells of the hyphae die. At this stage the single surviving cell is known as a **zygospore**, which is the basis of the name of the zygospore fungi. When environmental conditions improve, the nuclei within the zygospore undergo meiosis and one or more stalked **sporangiophores** sprout, each bearing a sporangium. Each sporangium contains the products of meiosis: haploid nuclei that are incorporated into spores. These spores disperse and germinate to form a new generation of haploid hyphae.

The zygospore fungi include four major lineages of terrestrial fungi that live on soil as saprobes, as parasites of insects and spiders, or as mutualists of other fungi and invertebrate animals. They produce no cells with flagella, and only one diploid cell—the zygospore—appears in the entire life cycle. Their hyphae are coenocytic. Most species do not form a fleshy fruiting structure. Instead, the hyphae spread in a radial pattern from the spore, with occasional stalked sporangiophores reaching up into the air (**Figure 29.16**).

More than 1,000 species of zygospore fungi have been described. One species you may have seen is *Rhizopus stolonifer*, the black bread mold. *Rhizopus* produces many stalked sporangiophores, each bearing a single sporangium containing hundreds of minute spores (see Figure 29.15B).

Arbuscular mycorrhizal fungi form symbioses with plants

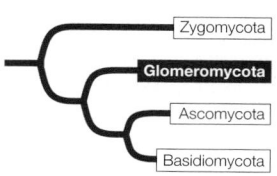

Arbuscular mycorrhizal fungi (Glomeromycota) are terrestrial fungi that associate with plant roots in a symbiotic, mutualistic relationship (see Figure 29.10B). Fewer than 200 species have been described, but 80 to 90 percent of all plants have associations with them. Molecular systematic studies have suggested that arbuscular mycorrhizal fungi are the sister group to the Dikarya (sac fungi and club fungi).

The hyphae of arbuscular mycorrhizal fungi are coenocytic. These fungi use glucose from their plant partners as their primary energy source, converting it into other, fungus-specific sugars that cannot return to the plant. Arbuscular mycorrhizal fungi are only known to reproduce asexually.

The dikaryotic condition is a synapomorphy of sac fungi and club fungi

In the two remaining groups of fungi—the sac fungi and the club fungi—some stages have a nuclear configuration other than the familiar haploid or diploid states (**Figure 29.17**). In these fungi, karyogamy (fusion of nuclei) occurs long after plasmogamy (fusion of cytoplasm), so that *two genetically different haploid nuclei coexist and divide within each cell of the mycelium*. This stage of the life cycle is called a **dikaryon** ("two nuclei"), and its ploidy is indicated as $n + n$. The dikaryon is a synapomorphy of the sac fungi and club fungi, which are placed together in a clade called **Dikarya**.

Eventually, specialized fruiting structures form, within which pairs of genetically dissimilar nuclei—one from each parent—fuse, giving rise to zygotes long after the original "mating." The diploid zygote nucleus then undergoes meiosis, producing four haploid nuclei. The mitotic descendants of those nuclei become spores, which germinate to give rise to the next haploid generation.

A life cycle with a dikaryotic stage has several unusual features. First, there are no gamete *cells*, only gamete *nuclei*. Second, the only true diploid structure is the zygote, although for a long period the genes of both parents are present in the dikaryon and can be expressed. In effect, the dikaryon is neither diploid (2n) nor haploid (n). Rather, it is dikaryotic (n + n). Therefore a harmful recessive mutation in one nucleus may be compensated for by a normal allele on the same chromosome in the other nucleus, and dikaryotic hyphae often have characteristics that are different from their n or 2n products. The dikaryotic condition is perhaps the most distinctive of the genetic peculiarities of the fungi.

The sexual reproductive structure of sac fungi is the ascus

The **sac fungi** (Ascomycota) are a large and diverse group of fungi found in marine, freshwater, and terrestrial habitats. There are approximately 64,000 known species, nearly half of which are the fungal partners in lichens. The hyphae of sac fungi are segmented by more or less regularly spaced septa. A pore in each septum permits

(A) Sexual life cycle of the sac fungi (Ascomycota)

In sac fungi, the products of meiosis are borne in a microscopic sac called an ascus. The fleshy fruiting bodies consist of both dikaryotic and haploid hyphae.

(B) Sexual life cycle of the club fungi (Basidiomycota)

In club fungi, the products of meiosis are borne on the surface of the gills on specialized hyphal tips called basidia. Fruiting bodies consist solely of dikaryotic hyphae, and the dikaryotic phase can last a long time.

Populations of club fungi may produce dozens to hundreds of different mating types.

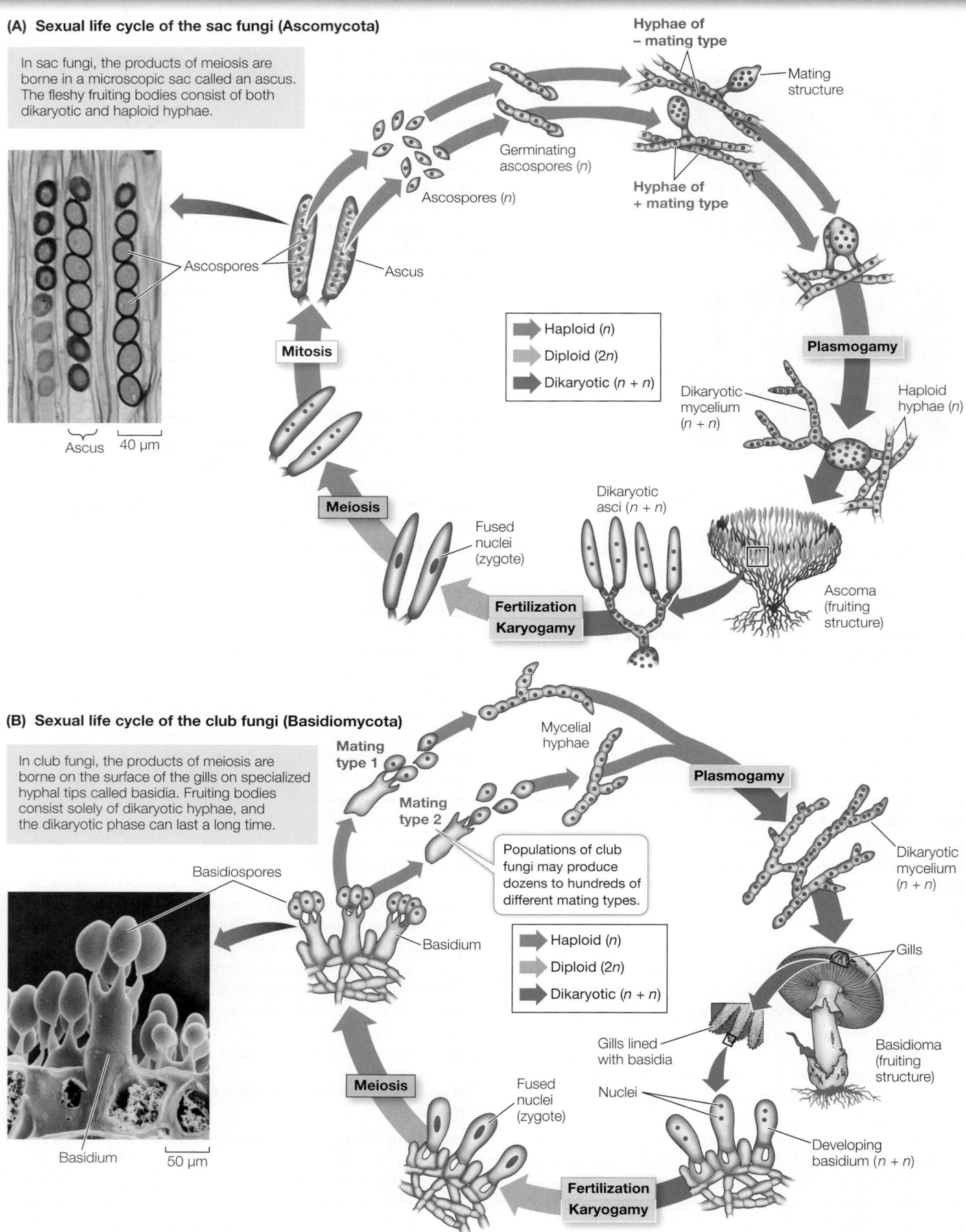

Figure 29.17 Sexual Life Cycles among the Dikarya   (A) In sac fungi, the products of meiosis are borne in a microscopic sac called an ascus. The fleshy fruiting structure, the ascoma, consists of both dikaryotic and haploid hyphae. **(B)** The basidium is the characteristic sexual reproductive structure of the club fungi. The fruiting structures, called basidiomata, consist solely of dikaryotic hyphae, and the dikaryotic phase can last a long time.

Q: How does the dikaryotic condition differ from diploidy?

Activity 29.2 Life Cycle of a Dikaryotic Fungus
www.Life11e.com/ac29.2

extensive movement of cytoplasm and organelles (including nuclei) from one segment to the next.

Sac fungi are distinguished by the production of sacs called **asci** (singular *ascus*), which at maturity contain sexually produced haploid **ascospores** (see Figure 29.17A). The ascus is the characteristic sexual reproductive structure of the sac fungi. In the past, the sac fungi were classified on the basis of whether or not the asci are contained within a specialized fruiting structure known as an ascoma (plural *ascomata*) and on differences in the morphology of that fruiting structure. DNA sequence analyses have resulted in a revision of these traditional groupings, however.

SAC FUNGUS YEASTS   Some species of sac fungi are unicellular yeasts. The 1,000 or so species in this group are among the most important domesticated fungi. Perhaps the best known is baker's, or brewer's, yeast (*Saccharomyces cerevisiae*; see Figure 29.2 and Key Concept 29.4), which metabolizes glucose obtained from its environment into ethanol and carbon dioxide by fermentation. Other sac fungus yeasts live on fruits such as figs and grapes and play an important role in the making of wine. Many others are associated with insects. In the guts of some insects, they provide enzymes that break down materials that are otherwise difficult for the insects to digest, especially cellulose.

Sac fungus yeasts reproduce asexually by budding. Sexual reproduction takes place when two adjacent haploid cells of dissimilar mating types fuse. In some species, the resulting zygote buds to form a diploid cell population. In others, the zygote nucleus undergoes meiosis immediately. When this happens, the entire cell becomes an ascus. Depending on whether the products of meiosis then undergo mitosis, a yeast ascus contains either eight or four ascospores, which germinate to become haploid cells. The sac fungus yeasts have lost the dikaryon stage.

FILAMENTOUS SAC FUNGI   Most sac fungi are filamentous species, such as the cup fungi (**Figure 29.18**), in which the ascomata are cup-shaped and can be as large as several centimeters across (although most are much smaller). The inner surfaces of the ascomata, which are covered with a mixture of specialized hyphae and asci, produce huge numbers of spores. The edible ascomata of some species, including morels and truffles, are regarded by humans as gourmet delicacies (and can sell at prices higher than gold). The underground ascomata of truffles have a strong odor that attracts mammals such as pigs, which then eat the fungi and disperse the spores. Humans sometimes take advantage of pigs' attraction to truffles and use pigs to locate truffles for human consumption.

The sexual reproductive cycle of filamentous sac fungi includes the formation of a dikaryon, although this stage is relatively brief compared with that in club fungi. Many filamentous sac fungi form multinucleate mating structures (see Figure 29.17A). Mating structures of two different mating types fuse and produce a dikaryotic mycelium, containing nuclei from both mating types. The dikaryotic mycelium often forms a cup-shaped ascoma, which bears the asci. Only after the formation of asci do the nuclei from the two mating types finally fuse. Both nuclear fusion and the subsequent meiosis that produces haploid ascospores take place within individual asci. The ascospores are ultimately released (sometimes shot off forcefully) by the ascus to begin the new haploid generation.

The sac fungi also include many of the filamentous fungi known as molds. **Molds** consist of filamentous hyphae that do not form large ascomata, although they can still produce asci and ascospores. Many molds are parasites of flowering plants. Chestnut blight and Dutch elm disease are both caused by molds. The chestnut blight fungus, which was introduced to the United States in the 1890s, had destroyed the American chestnut as a commercial species by 1940. Before the blight, this species accounted for more than half the trees in the eastern North American forests. Another familiar story is that of the American elm. Sometime before 1930, the Dutch elm disease fungus (first discovered in the Netherlands, but native to Asia) was introduced into the United States on infected elm logs from Europe. Spreading rapidly—especially via bark beetles—the fungus destroyed great numbers of American elm trees.

(A) *Cookeina tricholoma*

(B) *Morchella esculenta*

Figure 29.18 Sac Fungi   (A) These brilliant red cups are the ascomata of a cup fungus. **(B)** Morels, which have a sponge-like ascoma and a subtle flavor, are considered a culinary delicacy by humans.

Penicillium notatum

5 µm

Figure 29.19 Conidia Chains of conidia (yellow) are developing at the tips of specialized hyphae arising from a *Penicillium notatum* mold. This species is used to produce the antibiotic penicillin.

(A) *Amanita muscaria*

(B) *Laetiporus sulphureus*

Figure 29.20 Club Fungus Basidiomata **(A)** The fly agaric is a poisonous mushroom that contains hallucinogenic compounds. **(B)** A bracket fungus growing parasitically on a tree. Although this particular species is occasionally eaten when young, many similar-appearing bracket fungi are poisonous.

Other plant pathogens among the sac fungi include the powdery mildews that infect cereal crops, lilacs, and roses, among many other plants. Mildews can be a serious problem to farmers and gardeners, and a great deal of research has focused on ways to control these agricultural pests.

The filamentous sac fungi can also reproduce asexually by means of conidia that form at the tips of specialized hyphae (**Figure 29.19**). Small chains of conidia are produced by the millions and can survive for weeks in nature. The conidia are often what give molds their characteristic colors.

The basidium is the sexual reproductive structure of club fungi

Club fungi (Basidiomycota) produce some of the most spectacular fruiting structures found among the fungi. These fruiting structures, called **basidiomata** (singular *basidioma*), include mushrooms of all kinds, puffballs (see Figure 29.4), and the bracket fungi often encountered on trees and fallen logs in a damp forest. About 30,000 species of club fungi have been described. They include about 4,000 species of mushrooms, including both poisonous and edible species (**Figure 29.20A**). Bracket fungi (**Figure 29.20B**) play an important role in the carbon cycle by breaking down wood. They also do great damage to both cut lumber and timber stands. Some of the most economically damaging plant pathogens are club fungi, including the rust fungi and smut fungi that parasitize cereal grains. In contrast, other club fungi contribute to the survival of plants as fungal partners in ectomycorrhizae.

The hyphae of club fungi characteristically have septa with small, distinctive pores. As they grow, haploid hyphae of different mating types meet and fuse, forming dikaryotic hyphae, each cell of which contains two nuclei, one from each parent hypha. The dikaryotic mycelium grows and eventually, when triggered by rain or another environmental cue, produces a basidioma. The dikaryon stage may persist for years, or even centuries. This pattern contrasts with the life cycle of the sac fungi, in which the dikaryon is found only in the stages leading up to formation of the asci.

The **basidium** (plural *basidia*), a swollen cell at the tip of a specialized hypha, is the characteristic sexual reproductive structure of the club fungi (see Figure 29.17B). In mushroom-forming club fungi, the basidia typically form on specialized structures of the basidiomata known as gills. The basidium is the site of nuclear fusion and meiosis and thus plays the same role in the club fungi as the ascus does in the sac fungi and the zygosporangium does in the zygospore fungi.

After nuclei fuse in the basidium, the resulting diploid nucleus undergoes meiosis, and the four resulting haploid nuclei are incorporated into haploid **basidiospores**, which form on tiny stalks on the outside of the basidium. A single basidioma of the common bracket fungus *Ganoderma applanatum* can produce as many as 4.5 *trillion* basidiospores in one growing season. Basidiospores typically are forcibly discharged from their basidia and then germinate, and give rise to hyphae with haploid nuclei.

29.3 recap

Sexual reproduction is common in some groups of fungi but has never been observed in others. Many fungal species have two or more genetically distinct mating types. The sac fungi and club fungi share a dikaryotic condition, in which two genetically different haploid nuclei coexist in the same cell.

learning outcomes

You should be able to:

- Compare and contrast sexual reproduction in fungi with that of plants and animals.
- Summarize the parasitic life style of microsporidia.
- Infer the likely ancestral habitat of chytrids based on chytrid features.
- Compare and contrast the life cycles of zygospore fungi, sac fungi, and club fungi.

Fungi are of special interest to biologists because of the roles they play in interactions with other organisms, including humans. But they are also useful as tools for studying many kinds of biological problems and for finding solutions to those problems.

key concept 29.4
Fungi Have Many Practical Uses

We've briefly noted the important part that fungi play in the production of human foods and beverages. We have also described the diverse roles that fungi play in natural ecosystems, from decomposers to pathogens to mutualistic partners. These diverse ecological roles have led to the use of fungi in studies of environmental change and in remediation of environmental pollution. Many fungi are also important model organisms for laboratory investigations of basic biological process. Others, as you saw at the opening of this chapter, have given us treatments for human diseases.

focus your learning

- Baker's yeast, *Saccharomyces cerevisiae,* has many uses in food and drink.
- Some *Penicillium* and *Aspergillus* mold species are important in food production.
- Many fungi serve directly as a food source.
- Lichens are good biological indicators of air quality.
- Mycorrhizal fungi are important to reforestation efforts.

Fungi are important in producing food and drink

Grains from grasses provide most of the world's food supply for humans. But in most cases, we do not eat these grains directly as they are produced by the plants. Instead, we use them as a source of starch. To make the starch more pleasing and digestible for human consumption, we usually convert it to more complex and tasty forms of food and drink, often with the help of fungi.

Baker's (or brewer's) yeast (*Saccharomyces cerevisiae*) converts the starch from grain into ethanol. This process also forms carbon dioxide bubbles in bread dough, causing it to rise, which gives baked bread its light texture. The ethanol and carbon dioxide are baked away in bread making (which produces the pleasant aroma of baking bread). In contrast, the ethanol and carbon dioxide are retained when yeast is used to ferment grain into beer. The carbon dioxide gives beer its fizz, and the alcohol and yeast contribute to the taste and appeal of beer to those who enjoy it. Sugars, especially from fruit such as grapes, are also converted into alcohol and carbon dioxide by yeasts in the production of wine (although the carbon dioxide is not retained in most finished wine, as it is in beer). Many different strains of *S. cerevisiae* are used in wine production, which contributes to the distinctive nature of wine from different regions and wineries. Many other species of local, native yeasts are also used in producing distinctive local wines and beers. For example, fission yeast (*Schizosaccharomyces pombe*) was first isolated from African millet beer. Fission yeast takes its specific name (*pombe*) from the Swahili word for beer.

 Media Clip 29.3 **Time Lapse of Beer Fermentation**
www.Life11e.com/mc29.3

Brown molds of the genus *Aspergillus* are important in some human diets. *Aspergillus tamarii* acts on soybeans in the production of soy sauce, and *A. oryzae* is used in brewing the Japanese alcoholic beverage sake from rice. *Aspergillus niger* is the source of most commercial citric acid production. Citric acid gives food and soft drinks a tart taste and is also used as a food preservative. But some species of *Aspergillus* that grow on grains and on nuts such as peanuts and pecans produce extremely carcinogenic (cancer-inducing) compounds called aflatoxins. Aflatoxins can occur in high concentrations in foods such as peanut butter. In the United States and most other industrialized countries, moldy grain infected with *Aspergillus* is typically thrown out. In Africa, where food is scarcer, the grain is often eaten, moldy or not, and causes severe health problems, including high levels of certain cancers.

Penicillium is a genus of green molds, of which some species produce the antibiotic penicillin, as described in the beginning of this chapter. But several species of *Penicillium* are important for food production as well. For example, *P. camembertii* and *P. roqueforti* are the organisms responsible for the characteristic strong flavors of Camembert and Roquefort cheeses, respectively.

Many fungi serve directly as a human food source. Mushroom enthusiasts seek out the delicious fruiting structures of a wide variety of edible sac and club fungi. In the United States, relatively few species of mushrooms are grown commercially, and wild mushrooms are collected mostly for personal consumption. But in many parts of the world, a wide variety of wild mushrooms are collected for sale and consumption. Fungi used for food are not limited to fruiting bodies such as mushrooms, however. Various species of lichens are eaten in Arctic regions as well as in parts of North America and Asia. In southwestern China, for example, several species of lichens are used as a primary ingredient in cooking (**Figure 29.21**).

Fungi provide important weapons against diseases and pests

We started this chapter with the story of the discovery of penicillin. The discovery of antibiotics produced by fungi revolutionized medical treatment of bacterial diseases in humans and their domestic animals. Live strains of fungi are also used to combat various pest species of plants and animals.

In Africa, the parasitic plant witchweed (*Striga*) causes crop losses of about $7 billion every year. A group of Canadian biologists discovered that a strain of the mold *Fusarium oxysporum* could be

Figure 29.21 Some Lichens Are Edible In southwestern China, several species of lichens that grow on tree bark serve as a primary ingredient in the local cuisine.

applied to crops to control witchweed without harming the crop plants (**Figure 29.22**). Other strains of *Fusarium* that preferentially attack coca plants, the source of cocaine, have been proposed to combat illicit drug production. Still other fungi are used to attack various animal pests, such as termites and aphids, and even malaria-carrying mosquitoes.

Lichen diversity and abundance are indicators of air quality

Lichens can live in many harsh environments where few other species can survive, as you saw in Key Concept 29.2. In spite of their hardiness, however, lichens are highly sensitive to air pollution because they are unable to excrete any toxic substances they absorb. This sensitivity means that lichens are good biological indicators of air pollution levels. It also explains why they are not commonly found in heavily industrialized regions or in large cities.

Monitoring the diversity and abundance of lichens growing on trees is a practical and inexpensive system for gauging air quality

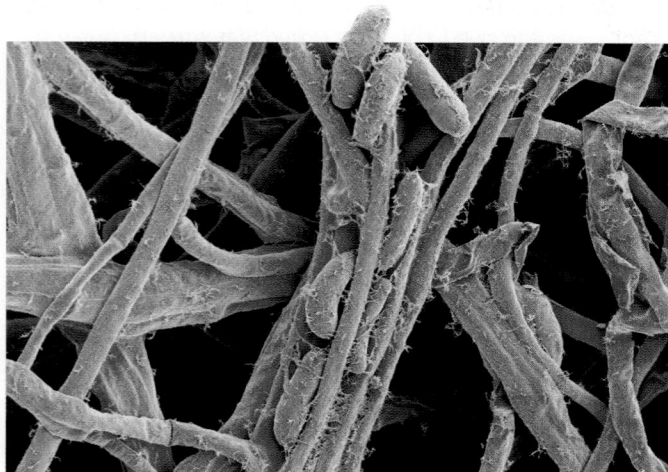

Figure 29.22 A Pathogenic Fungus Attacks a Parasitic Plant The fungus *Fusarium oxysporum* is a potent pathogen of witchweed (*Striga*), a parasitic plant that attacks crops. The fungal spores are shown in blue; fungal filaments are in tan. Both colors were added to enhance this scanning electron micrograph.

around cities. Maps of lichen diversity provide environmental biologists with a tool for tracking the distribution of air pollutants and their effects. Sensitive biological indicators of pollution, such as lichen growth, allow biologists to monitor air quality without the use of specialized equipment. Lichens are naturally distributed across the environment, and they can also provide a long-term measure of the effects of air pollution across many seasons and years.

Fungi record and help remediate environmental pollution

Each year, biologists deposit samples of many groups of organisms in the collections of natural history museums. These museum collections serve many purposes, one of which is to document changes in the biota of our planet over time.

Collections of fungi made over many decades or centuries provide a record of the environmental pollutants that were present when the fungi were growing. Biologists can analyze these historical samples to see how different sources of pollutants were affecting our environment before anyone thought to take direct measurements (**Investigating Life: Using Fungi to Study Environmental Contamination**). These long-term records are also useful for analyzing the effectiveness of cleanup efforts and regulatory programs for controlling environmental pollutants.

You have already seen that fungi are critical to the planetary carbon cycle because of their role in breaking down dead organic matter. Fungi are also used in remediation efforts to help clean up sites that have been polluted by oil spills or contaminated with toxic petroleum-derived hydrocarbons. Many herbicides, pesticides, and other synthetic hydrocarbons are broken down primarily through the action of fungi.

Reforestation may depend on mycorrhizal fungi

When a forest is cut down, it is not just the trees that are lost. A forest is an ecosystem that depends on the interaction of many species. As we have discussed, many plants depend on close relationships with mycorrhizal fungal partners. When trees are removed from a site, the populations of mycorrhizal fungi there decline rapidly. If we wish to restore the forest on the site, we cannot simply replant it with trees and other plants and expect them to survive. The mycorrhizal fungal community must be reestablished as well. For large forest restoration projects, a planned succession of plant growth and soil improvement is often necessary before forest trees can be replanted. As the community of soil fungi gradually recovers, trees that have been inoculated with appropriate mycorrhizal fungi in tree nurseries can be planted to reintroduce greater diversity to the soil fungal community.

Fungi are used as model organisms in laboratory studies

Much of what we know about many basic aspects of cell and molecular biology comes from the study of model organisms. Among the eukaryotes, some fungi have numerous advantages over model plant and animal systems for laboratory investigations.

Of particular importance as model organisms are several species of sac fungi: *Aspergillus nidulans* (a brown mold), *Neurospora crassa* (a red bread mold), *Saccharomyces cerevisiae* (baker's, or brewer's, yeast), and *Schizosaccharomyces pombe* (fission yeast).

experiment

Original Paper: Flegal, A. R., C. Gallon, S. Hibdon, Z. E. Kuspa and L. F. Laporte. 2010. Declining—but persistent—atmospheric contamination in central California from the resuspension of historic leaded gasoline emissions as recorded in the lace lichen (*Ramalina menziesii* Taylor) from 1892 to 2006. *Environmental Science and Technology* 44: 5613–5618.

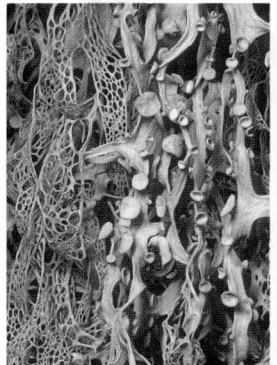

A. Russell Flegal and his colleagues analyzed museum samples of lace lichens (*Ramalina menziesii*) collected between 1892 and 2006 near San Francisco, California. Their goals were to determine the sources of lead contamination and to evaluate the effectiveness of efforts to reduce this contamination.

QUESTION▶　Is environmental lead contamination through time consistent with known sources of pollution, and have efforts to reduce lead pollution been effective?

METHOD

1. Obtain museum samples of lace lichen (*Ramalina menziesii*) collected near San Francisco.
2. Measure concentrations of lead (Pb) as well as the ratios of two lead isotopes, ^{206}Pb and ^{207}Pb.
3. Compare the ratio of the two isotopes in the lichen samples with that of the possible sources:
   a. Lead smelter, in operation from 1885 to 1971: ^{206}Pb/^{207}Pb ratio of 1.15–1.17
   b. Leaded gasoline, in use from 1930s to the early 1980s: ^{206}Pb/^{207}Pb ratio of 1.18–1.23
   c. Resuspension of historic lead contamination as atmospheric aerosols in recent decades: ^{206}Pb/^{207}Pb ratio of 1.16–1.19

RESULTS

Peak environmental lead contamination occurred in the 1960s to 1980s, with a rapid drop in the early 1980s after the closure of the smelter and the elimination of leaded gasoline. Average ^{206}Pb/^{207}Pb ratios of lead were below 1.17 before the 1930s, between 1.18 and 1.19 in the 1940s and 1950s, and >1.22 in the early 1980s. The ratios have declined since then.

CONCLUSIONS▶

1. The levels of environmental lead contamination, and ratios of lead isotopes, are consistent with the identified sources.
2. The closure of the lead smelter and elimination of leaded gasoline have significantly reduced environmental lead contamination.

work with the data

From the data below, plot lead concentration in the lichen samples against year of sample collection. Make a second plot, this one of ^{206}Pb/^{207}Pb ratio against year of sample collection.

Sample	Year collected	Lead concentration (μg of Pb/g lichen)	^{206}Pb/^{207}Pb ratio
1	1892	11.9	1.165
2	1894	4.0	1.155
3	1906	13.7	1.154
4	1907	22.9	1.157
5	1945	49.9	1.187
6	1957	34.2	1.185
7	1978	50.9	1.221
8	1982	10.0	1.215
9	1983	4.6	1.224
10	1987	1.0	1.198
11	1988	1.3	1.199
12	1995	1.9	1.202
13	2000	0.4	1.184
14	2006	1.8	1.184

QUESTIONS▶

1. Calculate trend lines for your plots by plotting the average for the each adjacent pair of observations, from 1894 to 2006. (In other words, plot the average for sample 1 and 2, then the average for sample 2 and 3, and so on, to produce a trend line for the data.)
2. Do your analyses support the conclusions of the study? How do you think you could improve the analysis?

A similar **work with the data** exercise may be assigned in **LaunchPad**.

▶ **29.4** recap

Fungi are important to humans in many ways. Some species are consumed directly as food, while others are important in food production. Fungi serve as important indicators of ecosystem health and are critical in reforestation and in pollution remediation efforts. Several species are important model organisms for studies of eukaryote cell and molecular biology. Fungi are widely used to combat diseases and pests.

learning outcomes

You should be able to:

- Describe the ways baker's yeast is used in food and drink.
- Give examples of uses of molds in food products.
- Summarize how fungi can be used for food.

(continued)

These species can be cultured in large numbers in small spaces, and they have short generation times, so that genetic investigations can be conducted in days rather than years. Furthermore, their genomes are relatively small and encode relatively few genes compared with those of most plants and animals, so it is easier to elucidate the functions of the fungal genes responsible for basic biological functions.

▶ 29.4 recap (continued)

- Assess the advantages of using lichens, over traditional measures, to monitor environmental conditions.
- Generate plans for land use that take advantage of the importance of mycorrhizal associations.

1. What are some of the ways in which fungi or fungal products contribute to the human food supply?

2. What are some advantages of using surveys of lichen diversity and museum collections of lichens to measure long-term changes in air quality, compared with direct measurements of atmospheric pollutants?

3. Can you develop a strategy for tree harvest that would ease the difficulty and expense of reforestation projects by retaining viable communities of mycorrhizal fungi?

4. Consider the data below for lichens at five survey sites. Predict the relative order of the sites with respect to their distance from the center of a large city. Other factors (besides distance to city center, such as prevailing wind direction) might affect your prediction. Can you think of two other major factors that might influence these results?

Site number	Number of lichen species	Tree branches covered in lichens (%)
1	5	38
2	1	2
3	3	15
4	8	75
5	13	100

▶ investigatinglife

 ## How do fungi affect our daily lives?

Beginning in the 1940s, antibiotics derived from fungi ushered in a "golden age" of freedom from bacterial infections. Today, however, many of these antibiotics are losing their effectiveness as pathogenic bacteria evolve resistance to them (Figure 29.23).

Most medical antibiotics are chemically modified forms of substances that are found naturally in fungi and other organisms. Fungi naturally produce antibiotic compounds to defend themselves against bacterial growth and to reduce competition from bacteria for nutritional resources. These naturally occurring compounds are usually chemically modified to increase their stability, improve their effectiveness, and facilitate synthetic production. From the late 1950s to the late 1990s, no new major classes of antibiotics were discovered. In recent years, however, three new classes of antibiotics have been synthesized based on information learned from naturally occurring, fungally derived antibiotics, leading to improved treatment of some formerly resistant strains of bacteria.

In addition to uses in human health, fungi are critical for production of human food and drink, as well as in agricultural applications. Fungi (especially yeasts) are important model organisms in biological research. Fungi are also sensitive indicators of environmental contamination, and long-term studies of fungi can inform us about changes in levels of environmental toxins over time.

Future directions

Biologists are developing applications to use fungi to combat nonbacterial diseases. One of the more unusual applications of fungi is in the war against malaria. Biologists have discovered that two species of fungi, *Beauveria bassiana* and *Metarhizium anisopliae*, can

kill malaria-causing mosquitoes when applied to mosquito netting. Mosquitoes have not yet shown evidence of developing resistance to these biological pathogens.

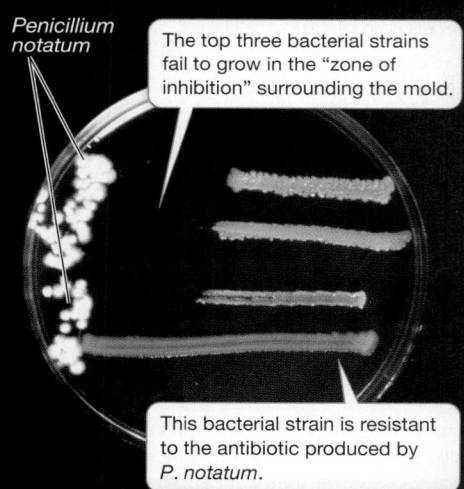

Penicillium notatum

The top three bacterial strains fail to grow in the "zone of inhibition" surrounding the mold.

This bacterial strain is resistant to the antibiotic produced by *P. notatum*.

Figure 29.23 Penicillin Resistance   In a petri dish similar to those in Alexander Fleming's lab, four strains of a pathogenic bacterium have been cultured along with *Penicillium* mold. One strain is resistant to the mold's antibiotic substance, as is evidenced by its growth up to the mold.

Chapter Summary 29

29.1 Fungi Digest Food Outside Their Bodies

- Fungi are distinguished from other **opisthokonts** by **absorptive heterotrophy** and by the presence of **chitin** in their cell walls. Review Figure 29.1
- Some fungi are **saprobes**, others are parasites or predators, and some are mutualists.
- **Yeasts** are unicellular, free-living fungi.
- The body of a multicellular fungus is a **mycelium**—a meshwork of filaments called **hyphae**. Hyphae may be **septate** (having **septa**) or **coenocytic** (multinucleate). Review Figure 29.3

29.2 Fungi Are Decomposers, Parasites, Predators, or Mutualists

- Saprobic fungi, which act as decomposers, make crucial contributions to the recycling of elements, especially carbon.
- Many fungi are parasites, harvesting nutrients from host cells by means of **haustoria**. Review Figure 29.5
- Certain fungi have relationships with other organisms that are **symbiotic** and **mutualistic**.
- Some fungi associate with unicellular green algae, cyanobacteria, or both to form **lichens**, which live on exposed surfaces of rocks, trees, and soil. Review Figure 29.8
- **Mycorrhizae** are mutualistic associations of fungi with plant roots. They improve a plant's ability to take up nutrients and water.
- **Endophytic fungi** live within plants and may provide their hosts with protection from herbivores and pathogens.

29.3 Sex in Fungi Involves Multiple Mating Types

- The microsporidia, chytrids, and zygospore fungi diversified early in fungal evolution. The arbuscular mycorrhizal fungi, sac fungi, and club fungi form a monophyletic group, and the latter two groups form the clade Dikarya. Review Figure 29.11, Table 29.1, Activity 29.1
- Many species of fungi reproduce both sexually and asexually. In many fungi, sexual reproduction occurs between individuals of different **mating types**. Review Figure 29.12
- The **microsporidia** are highly reduced unicellular fungi. They are obligate intracellular parasites of animals.
- The three distinct lineages of **chytrids** all include species with flagellated gametes. Review Figure 29.15A

- In the sexual reproduction of terrestrial fungi, **plasmogamy** (fusion of cytoplasm) precedes **karyogamy** (fusion of nuclei).
- **Zygospore fungi** have a resting stage known as a **zygospore**, which contains many diploid nuclei. Their spores are dispersed from simple stalked **sporangiophores**. Review Figure 29.15B, Animation 29.1
- **Arbuscular mycorrhizal fungi** form symbiotic associations with plant roots. They are only known to reproduce asexually. Their hyphae are coenocytic.
- In sac fungi and club fungi, a mycelium containing two genetically different haploid nuclei, called a **dikaryon**, is formed. The dikaryotic ($n + n$) condition is unique to the fungi. Review Figure 29.17, Activity 29.2
- **Sac fungi** have septate hyphae with large pores; their sexual reproductive structures are **asci**. Some sac fungi are unicellular yeasts. Many filamentous sac fungi produce fleshy fruiting structures called ascomata. The dikaryon stage in the sac fungus life cycle is relatively brief. Review Figure 29.17A
- **Club fungi** have septate hyphae with distinctive small pores. Their fruiting structures are called **basidiomata**, and their sexual reproductive structures are **basidia**. The dikaryon stage may last for years. Review Figure 29.17B

29.4 Fungi Have Many Practical Uses

- Some fungi are consumed as food by humans; other fungi are critical in baking, fermentation, and flavoring food.
- Fungi play important roles in cleaning up environmental pollutants such as synthetic petroleum-derived hydrocarbons.
- The diversity and abundance of lichen growth on trees is a sensitive indicator of air quality.
- Reforestation projects require restoration of the mycorrhizal fungal community.
- Several species of fungi are important model organisms.
- Fungi provide important weapons against diseases and pests.

> Go to **LearningCurve** (in **LaunchPad**) for dynamic quizzing that helps you solidify your understanding of this chapter. **LearningCurve** adapts to your responses, giving you the practice you need to master each key concept.

Apply What You've Learned

Review

29.2 Mycorrhizae are mutualistic associations between fungi and the roots of vascular plants.

Original Paper: Stinson, K. A. et al. 2006. Invasive plant suppresses the growth of native tree seedlings by disrupting belowground mutualisms. *PLoS Biology* 4: e140.

Garlic mustard (*Alliaria petiolata*) is a highly invasive plant of Eurasian origin that causes much damage to native plants where it invades in North America. The manner by which this plant outcompetes native plants is

unusual: it secretes chemicals that interfere with the mutualistic relationships the native plants have with arbuscular mycorrhizal fungi.

Some plants are more dependent than others on arbuscular mycorrhizal fungi. Researchers examined the effect of exposure to garlic mustard on several plant species and measured the effect that removal of arbuscular mycorrhizal fungi had on each plant. The figure on p. 634 shows the results of the study. Yellow points, herbaceous colonizers of disturbed forest edges; red points, herbaceous species that live in forest edges and in the gaps between patches of forest; blue points, woody colonizers of forest edges and gaps between patches of forest; green points, tree species of mature forests.

(continued)

Questions

1. What is the relationship between the effect of exposure to garlic mustard on a plant species and its dependence on arbuscular mycorrhizal fungi? Is this relationship consistent with garlic mustard reducing plant growth by interfering with the plant's fungal mutualists?

2. Which types of plants are most dependent on their mycorrhizal mutualists? Explain your answer. What other observations can be made about habitat and reliance on fungal mutualists?

3. Climate change is expected to lead to greater extremes in temperature and precipitation. What are the likely implications for plants with arbuscular mycorrhizal fungi as climate change becomes more pronounced? Would these aspects of climate change likely increase or decrease the negative effects of exposure to garlic mustard?

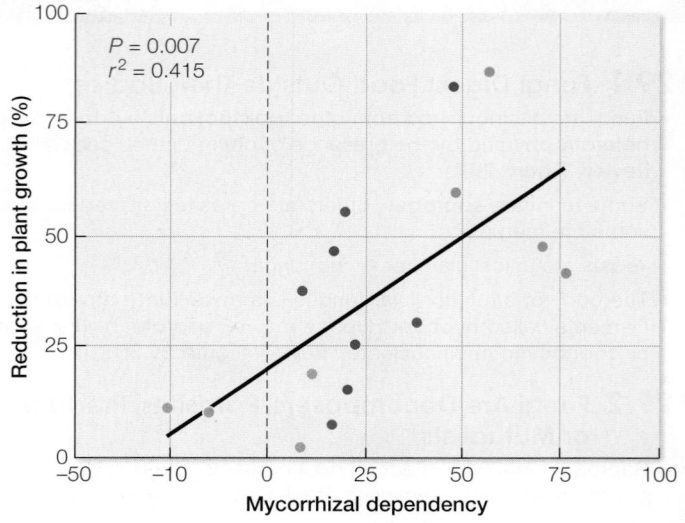

$P = 0.007$
$r^2 = 0.415$

Reduction in plant growth (%) — y-axis
Mycorrhizal dependency — x-axis

Go to **LaunchPad** for the eBook, LearningCurve, animations, activities, flashcards, and additional resources and assignments.

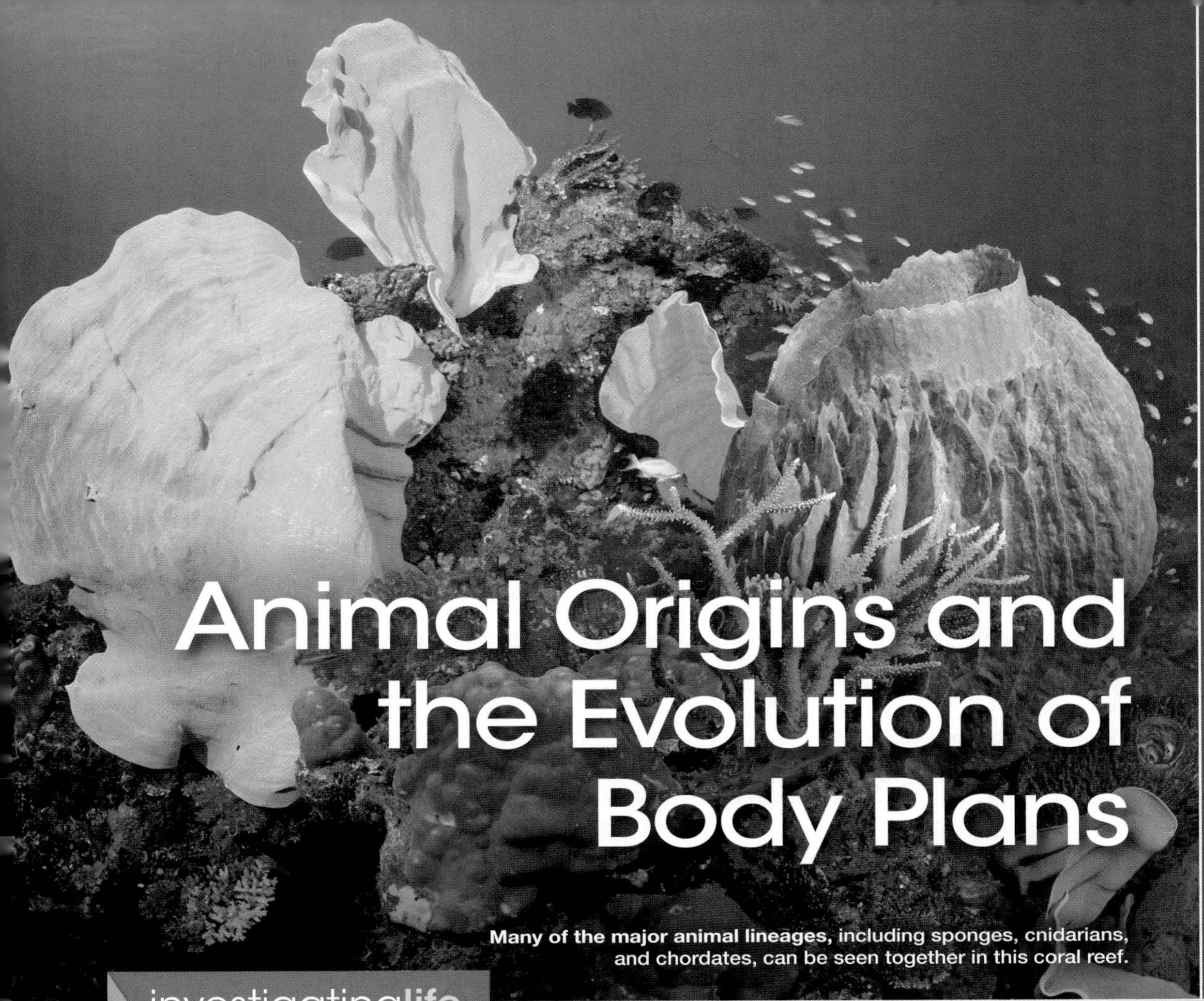

30

key concepts

30.1 Some Animal Characteristics Evolved More Than Once

30.2 Animals Diverged with Distinct Body Plans

30.3 Animals Use Diverse Forms of Movement to Feed

30.4 Animal Life Cycles Involve Trade-Offs

30.5 The Root of the Animal Tree Provides Clues to Early Animal Diversification

Animal Origins and the Evolution of Body Plans

Many of the major animal lineages, including sponges, cnidarians, and chordates, can be seen together in this coral reef.

investigatinglife

A Mysterious Animal Discovered in an Aquarium

In 1883 the zoologist Franz Schulze noticed something unusual in his Austrian laboratory: transparent, flattened organisms were crawling on the sides of his saltwater aquarium. These organisms, which Schulze had collected accidentally along with the sponges that were his primary interest, appeared to be animals, but they were unlike any animals previously described—especially since they continually changed shape as they moved.

Further examination revealed that the new organisms were indeed animals. Structurally, however, they were among the simplest animals that Schulze—or anyone else—had ever observed, being made up of only four types of cells. He named the new species *Trichoplax adhaerens*, which means "sticky hairy plate." For decades, however, most biologists dismissed Schulze's findings, insisting that the transparent organisms must be larval forms of other, well-known animals.

In the 1960s more detailed studies confirmed the distinctive nature of *Trichoplax*. Even then, this odd animal continued to be known almost exclusively from aquariums. Finally, in the early 2000s biologists located natural populations of *T. adhaerens* growing on hard surfaces in tropical and subtropical coastal regions. A few additional closely related species have been discovered (although most have not yet been formally named). Collectively, these species are known as placozoans (Greek, "flat animals").

The more biologists have studied *Trichoplax*, the odder this animal appears. It has the smallest genome of any animal studied to date. The mature stages lack body symmetry and have no mouth, gut, or nervous system. Is *Trichoplax* a relict representative of a group of animals that appeared early in animal evolution?

Which groups of organisms are at the root of the animal tree has been a subject of considerable investigation and debate. Several possible hypotheses of relationships have been explored with genomic analyses. The structural simplicity of *Trichoplax* is now considered by most biologists to be an evolutionary reversal from a more complex body form. Most genomic studies point to other groups of animals as forming the earliest split with the remaining species.

This chapter explores the earliest branches on the animal tree and shows how a few fundamental "body plans" have been modified to yield the remarkable variety of animal forms described in this and the following two chapters.

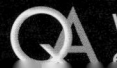

QA Which animal groups are involved in the earliest split in the animal tree?

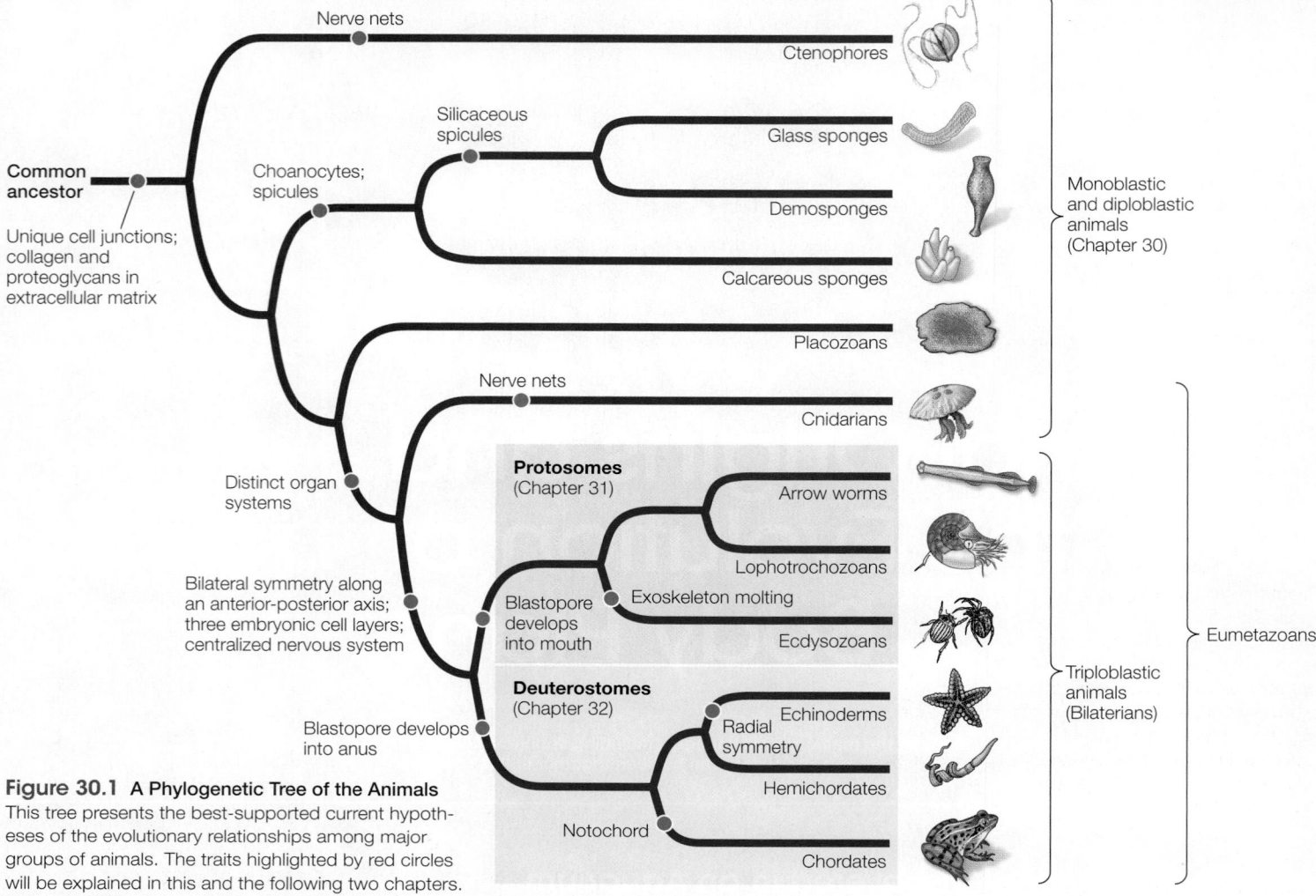

Figure 30.1 A Phylogenetic Tree of the Animals
This tree presents the best-supported current hypotheses of the evolutionary relationships among major groups of animals. The traits highlighted by red circles will be explained in this and the following two chapters.

Q: Based on this tree, which of the depicted traits evolved multiple times among animals, and in which lineages?

key concept

30.1 Some Animal Characteristics Evolved More Than Once

How do we recognize an organism as an animal? That may seem obvious for many familiar animals, but less so for groups such as sponges, which were once thought to be plants.

focus your learning

- Most animals are multicellular heterotrophs with internal digestion, muscular systems that allow movement, and nervous systems, although none of these features is completely diagnostic for all animals.
- A few microstructural features and genome sequences provide the best evidence of animal monophyly.
- Several patterns of embryonic cleavage exist among animals.
- Distinct layers of cells form during the early development of most animals.
- Bilaterians, named for their pattern of symmetry, include protostomes and deuterostomes.

Some of the general characteristics we associate with animals include:

- *Multicellularity.* In contrast to the bacteria, archaea, and most protists (see Chapters 25 and 26), all animals are multicellular. Animal life cycles feature complex patterns of development from a single-celled zygote into a multicellular adult.

- *Heterotrophic metabolism.* In contrast to most plants, all animals are heterotrophs. Animals are able to synthesize very few organic molecules from inorganic chemicals, so they must obtain the necessary organic molecules from their environment.

- *Internal digestion.* Although the fungi are also heterotrophs (see Chapter 29), animals and fungi digest their food differently. While fungi digest food outside their bodies, most animals use internal processes to break down materials from their environment into the organic molecules they need. Most animals ingest food into an internal **gut** that is continuous with the outside environment and in which digestion takes place.

- *Movement and nervous systems.* In contrast to the majority of plants and fungi, most animals can move their bodies. This movement is often coordinated through a well-developed nervous system, which also typically functions as a sensory system. Animals must move to find food or bring food to them.

(A) Colony of choanoflagellate protists

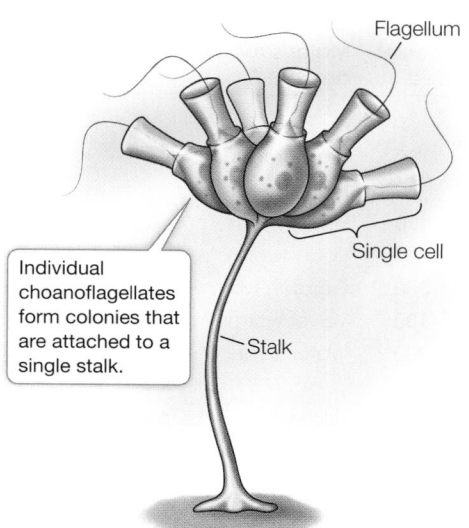

Flagellum

Single cell

Individual choanoflagellates form colonies that are attached to a single stalk.

Stalk

(B) Structure and function of sponge choanocytes

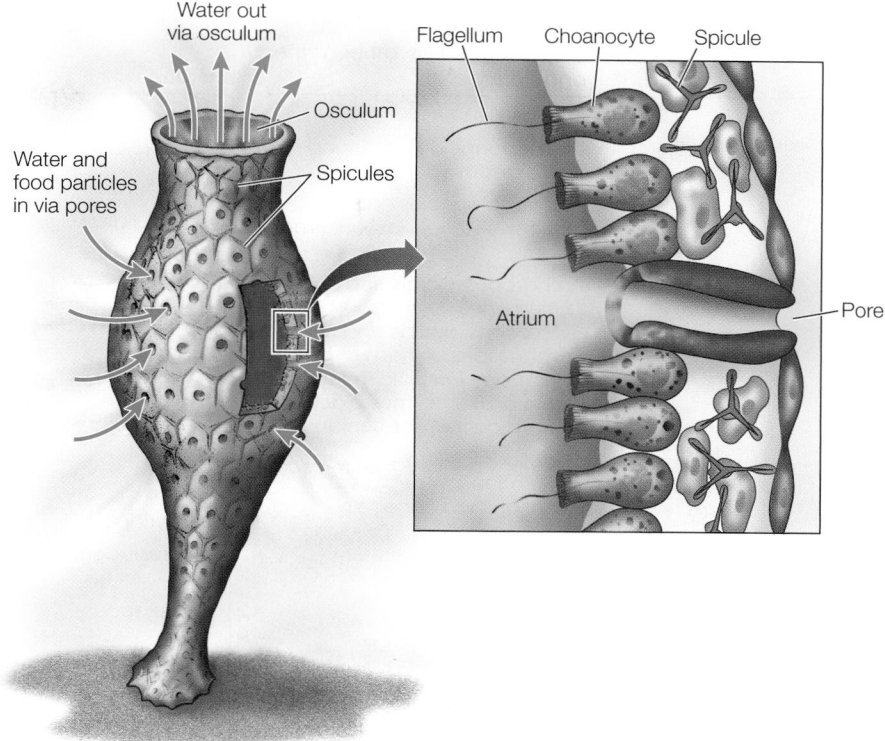

Water out via osculum

Water and food particles in via pores

Osculum

Spicules

Flagellum   Choanocyte   Spicule

Atrium

Pore

Figure 30.2 Choanocytes in Sponges Resemble Choanoflagellate Protists **(A)** The similarity of choanoflagellate protist colonies to sponge choanocytes supports an evolutionary link between this protist lineage and the animals. **(B)** A sponge moves food-containing water through its body by beating the flagella of its choanocytes. Water enters the sponge through small pores and passes into water canals or an open atrium, where the choanocytes capture food particles from the water. The spicules are supportive, skeletal structures.

Muscle tissue and nervous systems are unique to animals, and many animal body plans are specialized for movement and detection of prey.

Although these general features help us recognize animals, none is diagnostic for all animals. Some animals do not move, at least during certain life stages, and some plants and fungi do have limited movement. Some animals lack a nervous system, and the diffuse nerve nets (which lack a coordinating brain) of ctenophores and cnidarians are distinct and independently evolved from the **central nervous systems** of bilaterians. Some animals lack a gut, and many multicellular organisms are not animals. So on what basis do we group all animals together in a single clade?

Animal monophyly is supported by gene sequences and morphology

The most convincing evidence that all the organisms considered to be animals share a common ancestor comes from phylogenetic analyses of their gene sequences. Relatively few complete animal genomes are available, but more are being sequenced each year. Analyses of these genomes, as well as of many individual gene sequences, have shown that the animals are indeed monophyletic. The best-supported phylogenetic tree for the major animal groups is shown in **Figure 30.1**. **Table 30.1** summarizes the living members of those groups.

Although animals were considered to belong to a single clade long before gene sequencing became possible, surprisingly few morphological features are shared across all species of animals. Two morphological synapomorphies have been identified that distinguish the animals:

1. A common set of extracellular matrix molecules, including collagen and proteoglycans (see Figure 5.22)

2. Unique types of junctions between cells (tight junctions, desmosomes, and gap junctions; see Figure 6.7)

Although some animals in a few groups lack one or the other of these traits, it is believed that these traits were possessed by the ancestor of all animals and subsequently lost in those groups. Similarities among animals in the organization and function of Hox and other developmental genes (see Chapter 19) provide additional evidence of developmental mechanisms shared by a common animal ancestor.

The common ancestor of animals was likely a colonial flagellated protist similar to existing colonial choanoflagellates. Choanoflagellate colonies have clearly retained similarities to the multicellular sponges (**Figure 30.2**). Why did early animals begin to form multicellular colonies? One hypothesis is that multicellular colonies are more efficient than single cells are at capturing their prey. Experiments with living species of choanoflagellates show that they spontaneously form multicellular colonies in response to signaling compounds that are found on certain species of planktonic bacteria they eat (**Figure 30.3**).

One hypothesis of animal origins postulates a choanoflagellate-like lineage in which certain cells in the colony became specialized—some for movement, others for nutrition, others for reproduction, and so on. Once this functional specialization had begun, cells could have continued to differentiate. Coordination among groups of cells could have improved by means of specific regulatory and

table 30.1 Summary of Living Members of the Major Animal Groups

Group	Approximate number of living species described	Major subgroups, other names, and notes	Group	Approximate number of living species described	Major subgroups, other names, and notes
Ctenophores	250	Comb jellies	Ecdysozoans		
Sponges	8,500	Demosponges, glass sponges, calcareous sponges	Kinorhynchs	180	Mud dragons
			Loriciferans	30	Brush heads
Placozoans	2	Additional species have been discovered but not yet formally named	Priapulids	20	Penis worms
			Nematodes	25,000	Roundworms
Cnidarians	12,500	Anthozoans: Corals, sea anemones	Horsehair worms	350	Gordian worms
		Hydrozoans: Hydras and hydroids	Onychophorans	180	Velvet worms
		Scyphozoans: Jellyfish	Tardigrades	1,200	Water bears
		Myxozoans: Parasitic mucous animals; sometimes placed in group distinct from cnidarians	Arthropods		
			Chelicerates	114,000	Horseshoe crabs, pycnogonids, and arachnids (scorpions, harvestmen, spiders, mites, ticks)
Orthonectids	45	Microscopic wormlike parasites of marine invertebrates; relationships uncertain			
			Myriapods	12,000	Millipedes, centipedes
Rhombozoans	125	Tiny (0.5–7 mm) parasites of cephalopods; relationships uncertain	Crustaceans	67,000	Crabs, shrimps, lobsters and crayfish, barnacles, copepods
			Hexapods	1,020,000	Insects and their wingless relatives
PROTOSTOMES			**DEUTEROSTOMES**		
Arrow worms	180	Glass worms	Xenoturbellids	5	Secondarily simple marine worms; relationships uncertain
Lophotrochozoans					
Bryozoans	5,500	Moss animals	Acoels	400	Very small (mostly <2 mm) flattened marine worms; relationships uncertain
Entoprocts	170	Sessile aquatic animals, 0.1–7 mm long, superficially similar to bryozoans			
Flatworms	30,000	Free-living flatworms; flukes and tapeworms (all parasitic); monogeneans (ectoparasites of fish)	Echinoderms	7,500	Crinoids (sea lilies and feather stars), brittle stars, sea stars, sea daisies, sea urchins, sea cucumbers
Gastrotrichs	800	"Hairy backs"	Hemichordates	120	Acorn worms and pterobranchs
Rotifers and relatives	3,000	Rotifers, spiny-headed worms, and jaw worms	Tunicates	2,800	Sea squirts (ascidians), salps, and larvaceans
Ribbon worms	1,200	Proboscis worms	Lancelets	35	Cephalochordates
Phoronids	10	Sessile marine filter feeders	Vertebrates	65,000	Hagfish, lampreys, cartilaginous fish, ray-finned fish, coelacanths, lungfish, amphibians, reptiles (including birds), and mammals
Brachiopods	450	Lampshells			
Annelids	19,000	Polychaetes (generally marine; may not be monophyletic) Clitellates: earthworms, freshwater worms, leeches			
Mollusks	117,000	Monoplacophorans Chitons Bivalves: Clams, oysters, mussels Gastropods: Snails, slugs, limpets Cephalopods: Squid, octopuses, nautiloids			

signaling molecules that guided differentiation and migration of cells in developing embryos. Such coordinated groups of cells eventually could have evolved into the larger and more complex organisms that we call animals.

Nearly 80 percent of the 1.8 million named species of living organisms are animals, and millions of additional animal species await discovery (see Chapter 31 opening story). Evidence for the evolutionary relationships among animal groups can be found in fossils, in patterns of embryonic development, in the morphology and physiology of living animals, in the structure of animal proteins, and in gene sequences. Increasingly, studies of the phylogenetic relationships among major animal groups have come to depend on genomic sequence comparisons.

A few basic developmental patterns differentiate major animal groups

Differences in patterns of embryonic development have until recently provided many of the important clues to animal phylogeny. Analyses of gene sequences, however, are now showing that some developmental patterns are more evolutionarily variable than previously thought. Here we describe the basic developmental patterns that vary among the major animal clades.

The early cell divisions of an embryo are known as *cleavage. Several different patterns of cleavage exist among animals. Although these patterns can be useful for characterizing major animal groups, genomic analyses have shown that many changes have occurred in cleavage patterns throughout animal evolution.

---

*connect the concepts   The various patterns of embryonic cleavage and animal development are described in more detail in Key Concept 43.1.

Cleavage patterns are influenced by the configuration of the yolk, the acellular nutritive material that nourishes the growing embryo. The eggs of many animal groups contain a small amount of yolk that is evenly distributed throughout the egg cytoplasm. In some of these groups, the zygote and its descendant cells divide completely and evenly in a pattern known as **radial cleavage**. Radial cleavage is thought to be the ancestral condition for the bilaterian animals, as it is widely distributed among the major lineages. **Spiral cleavage**—a complicated permutation of radial cleavage—is found among many lophotrochozoans (a group that includes earthworms and clams). Lophotrochozoans with spiral cleavage are thus sometimes known as spiralians. The early branches of the ecdysozoans (molting animals, such as insects and nematodes) have radial cleavage, but most ecdysozoans have an idiosyncratic cleavage pattern that is neither radial nor spiral in organization (see Figure 43.3C). In reptiles, the presence of a large body of yolk within the fertilized egg creates an incomplete cleavage pattern in which the dividing cells form an embryo on top of the yolk mass.

Distinct layers of cells form during the early development of most animals. These cell layers differentiate into specific organs and organ systems as development continues. The embryos of **diploblastic** animals have two cell layers: an outer **ectoderm** and an inner **endoderm**. Embryos of **triploblastic** animals have, in addition to ectoderm and endoderm, a third distinct cell layer, **mesoderm**, which lies between the ectoderm and the endoderm (see Figure 43.6). The existence of three cell layers in embryos is a synapomorphy of the triploblastic animals (which form a clade), whereas the diploblastic animals (ctenophores, placozoans, and cnidarians, which are not a clade) exhibit the ancestral condition (see Figure 30.1). Some biologists also consider sponges to be diploblastic, but since they do not have clearly differentiated tissue

experiment

Figure 30.3   What Induces Choanoflagellates to Form Multicellular Colonies?

Original Paper: Alegado, R. A. et al. 2012. A bacterial sulfonolipid triggers multicellular development in the closest living relatives of animals. *eLife* 1: e00013.

The choanoflagellate *Salpingoeca rosetta* forms multicellular, rose-shaped colonies (rosettes) at low frequencies in natural conditions. The rosette colonies are more efficient than are single cells at capturing some prey species of planktonic bacteria. Rosanna Alegado and her collaborators investigated the chemical signals that induce the choanoflagellates to form multicellular colonies when environmental conditions favor the rosette formations.

HYPOTHESIS▶   Multicellular colonies of *S. rosetta* are induced by chemical compounds found on specific species of planktonic bacteria.

METHOD

1. Isolate individual species of bacteria from an environmental sample of *S. rosetta* and its associated bacterial community.

2. Use a combination of antibiotics to produce a bacteria-reduced culture line of *S. rosetta* that does not spontaneously form rosette colonies (the RCA line, for "rosette colonies absent").

3. Feed the RCA line with each of the isolated species of bacteria from Step 1. Record any rosette colony formation.

4. Isolate compounds that induce multicellular colony formation from rosette-producing bacteria identified in Step 3.

RESULTS

Alegado and colleagues tested various compounds from planktonic bacteria in the genus *Algoriphagus* and found that RIF-1, a sulfonolipid produced by the bacteria, is the signal that induces formation of multicellular colonies in *S. rosetta*.

In an environmental sample with many diverse bacteria, *S. rosetta* forms occasional multicellular colonies.

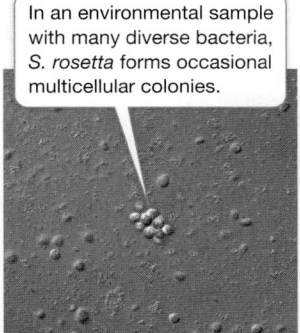

After treatment with antibiotics, no multicellular colonies were produced.

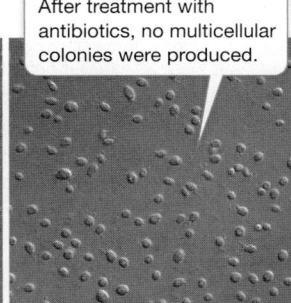

The addition of high concentrations of planktonic bacteria in the genus *Algoriphagus* induces the *S. rosetta* to form multicellular colonies.

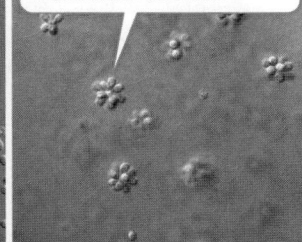

CONCLUSION▶   A chemical signal (RIF-1) from a particular genus of prey bacteria can induce the formation of multicellular colonies in some species of choanoflagellates.

---

types or embryonic cell layers, they are usually considered to be **monoblastic**.

During early development in many animals, in a process known as gastrulation, a hollow ball one cell thick indents to form a cup-shaped structure. The opening of the cavity formed by this indentation is called the blastopore (**Figure 30.4**). The process of gastrulation is detailed in Key Concept 43.3; the point to remember here is that the *overall pattern* of gastrulation immediately after formation of the blastopore divides the triploblastic animals into two major groups:

1. In the **protostomes** (Greek, "mouth first"), the mouth arises from the blastopore, and the anus forms later.

Blastopore

Figure 30.4 Gastrulation Illuminates Evolutionary Relationships The blastopore is clear in this scanning electron micrograph of a sea urchin gastrula. Because sea urchins (echinoderms) are deuterostomes, this blastopore will eventually become the anal end of the animal's gut.

2. In the **deuterostomes** ("mouth second"), the blastopore becomes the anus, and the mouth forms later.

Although the developmental patterns of animals are more varied than suggested by this simple dichotomy, sequencing data indicate that the protostomes and deuterostomes represent distinct animal clades. Together these two groups are known as the **bilaterians** (named for their usual bilateral symmetry), and they account for the vast majority of animal species.

30.1 recap

We generally recognize animals as multicellular heterotrophs with internal digestion, muscular systems that allow movement, and nervous systems. Animals are thought to be monophyletic because they share several derived traits, especially among their gene sequences. Major developmental differences also provide evidence of their evolutionary relationships.

learning outcomes

You should be able to:

- Explain the ways that animals differ from each of the other major groups.
- Contrast the major morphological and developmental features that distinguish major groups of animals.

1. What general features distinguish animals from other major groups of living organisms?
2. Differentiate among the members of each of the following sets of related terms:
   a. radial symmetry/bilateral symmetry
   b. protostome/deuterostome
   c. diploblastic/triploblastic

We will begin our exploration of animal diversity by discussing the general features of animal body plans. Later in this chapter we will describe several groups of animals that diverged before the origin of the bilaterians. We will devote Chapter 31 to the protostomes and Chapter 32 to the deuterostomes.

key concept
30.2 Animals Diverged with Distinct Body Plans

The general structure of an animal, the arrangement of its organ systems, and the integrated functioning of its parts are referred to as its **body plan**. As Chapter 19 described, the regulatory and signaling genes that govern the development of body symmetry, body cavities, segmentation, and appendages are widely shared among the different animal groups. Thus we might expect animals to share body plans.

focus your learning

- Bilateral symmetry is strongly associated with cephalization.
- The body plans of triploblastic animals can be divided into three types based on the presence and structure of an internal body cavity.
- Segmentation facilitates specialization of different body regions.
- Appendages that project from the body greatly enhance an animal's ability to move around.
- Nervous systems are important for sensing the environment, processing information, and coordinating movement.

Although animal body plans vary tremendously, they can be seen as variations on five key features:

1. The *symmetry* of the body
2. The structure of the *body cavity*
3. The *segmentation* of the body
4. *External appendages* that are used for sensing, chewing, locomotion, mating, and other functions
5. The development of the *nervous system*

Each of these features affects how an animal moves and interacts with its environment.

Most animals are symmetrical

The overall shape of an animal can be described by its **symmetry**. An animal is said to be symmetrical if it can be divided along at least one plane into similar halves. Animals that have no plane of symmetry are said to be asymmetrical. Placozoans and many sponges are asymmetrical, but most other animals have some kind of symmetry, which is governed by the expression of regulatory genes during development.

The simplest form of symmetry is **spherical symmetry**, in which body parts radiate out from a central point. An infinite number of planes passing through the central point can divide a spherically symmetrical organism into similar halves. Spherical symmetry is widespread among unicellular protists, but most animals possess other forms of symmetry.

In organisms with **radial symmetry**, body parts are arranged around one main axis at the body's center (**Figure 30.5A**). Ctenophores (comb jellies) are radially symmetrical, as are many cnidarians (such as sea anemones and jellyfish) and echinoderms. A

(A) Radial symmetry

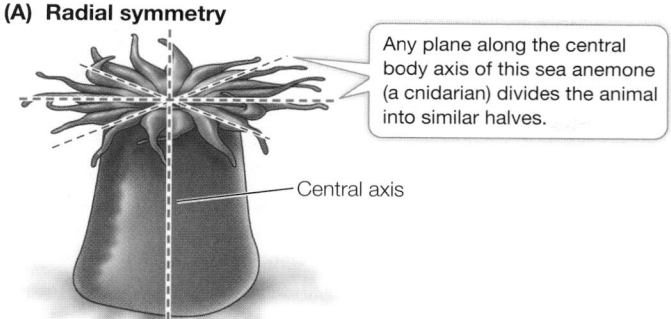

Any plane along the central body axis of this sea anemone (a cnidarian) divides the animal into similar halves.

Central axis

(B) Bilateral symmetry

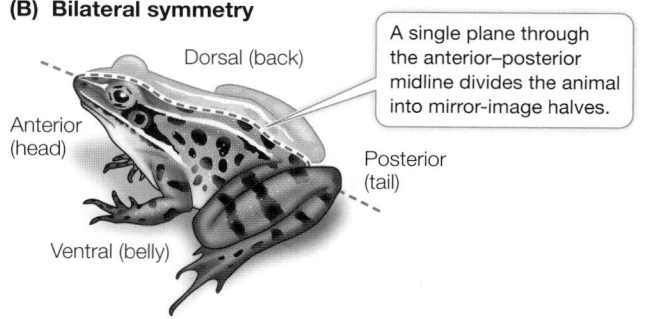

Dorsal (back)

A single plane through the anterior–posterior midline divides the animal into mirror-image halves.

Anterior (head)

Posterior (tail)

Ventral (belly)

Figure 30.5 Body Symmetry Most animals are either **(A)** radially or **(B)** bilaterally symmetrical.

perfectly radially symmetrical animal can be divided into similar halves by any plane that contains the main axis. However, most radially symmetrical animals—including the adults of echinoderms such as sea stars—are slightly modified, so that only some planes can divide them into identical halves. Some radially symmetrical animals are sessile (they remain fixed in one place) or drift with water currents. Others move about slowly but can move equally well in any direction.

Bilateral symmetry is characteristic of animals that have a distinct front end, which typically precedes the rest of the body as the animal moves. A bilaterally symmetrical animal can be divided into mirror-image (left and right) halves by a single plane that passes through the midline of its body. This plane runs from the front, or **anterior**, end of the body, to the rear, or **posterior**, end (**Figure 30.5B**). A plane at right angles to the midline divides the body into two dissimilar sides. The back of a bilaterally symmetrical animal is its **dorsal** surface; the underside is its **ventral** surface.

Bilateral symmetry is strongly correlated with **cephalization** (Greek *kephalos*, "head"), which is the concentration of sensory organs and nervous tissues at the anterior end of the animal. Cephalization has been favored by natural selection because the anterior end of a bilaterally symmetrical animal typically encounters new environments first.

The structure of the body cavity influences movement

The body plans of triploblastic animals can be divided into three types based on the presence and structure of an internal, fluid-filled **body cavity**.

1. **Acoelomate** animals such as flatworms lack an enclosed, fluid-filled body cavity. Instead, the space between the gut

(A) Acoelomate (flatworm)

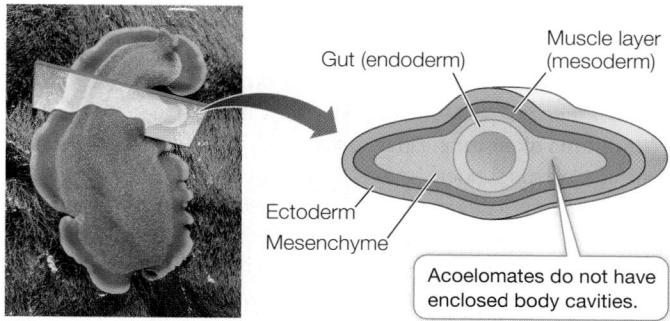

Gut (endoderm)

Muscle layer (mesoderm)

Ectoderm

Mesenchyme

Acoelomates do not have enclosed body cavities.

(B) Pseudocoelomate (roundworm)

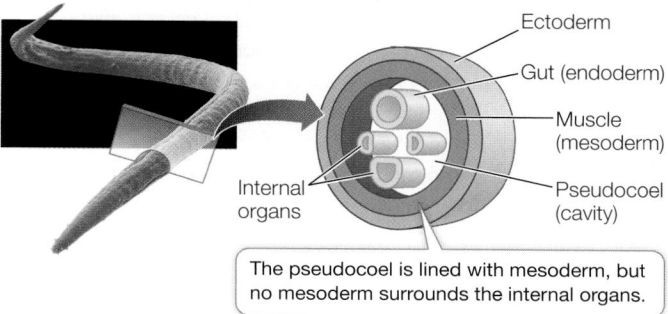

Ectoderm

Gut (endoderm)

Muscle (mesoderm)

Internal organs

Pseudocoel (cavity)

The pseudocoel is lined with mesoderm, but no mesoderm surrounds the internal organs.

(C) Coelomate (earthworm)

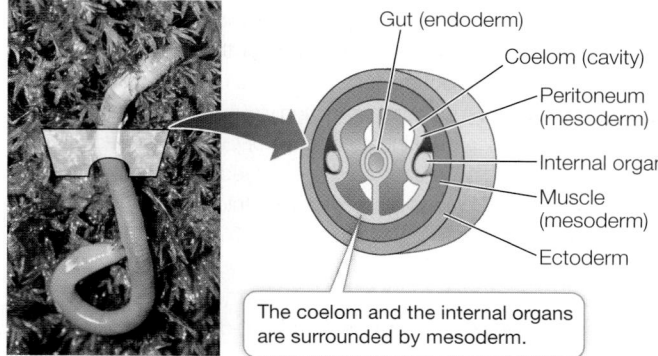

Gut (endoderm)

Coelom (cavity)

Peritoneum (mesoderm)

Internal organ

Muscle (mesoderm)

Ectoderm

The coelom and the internal organs are surrounded by mesoderm.

Figure 30.6 Animal Body Cavities **(A)** Acoelomates do not have enclosed body cavities. **(B)** Pseudocoelomates have a body cavity enclosed by mesoderm only on its outside. **(C)** Coelomates have a body cavity that is enclosed by mesoderm on both its inside and its outside.

 Activity 30.1 **Animal Body Cavities**
www.Life11e.com/ac30.1

(derived from endoderm) and the muscular body wall (derived from mesoderm) is filled with masses of cells called **mesenchyme** (**Figure 30.6A**). These animals typically move by beating cilia.

2. **Pseudocoelomate** animals have a body cavity called a pseudocoel, a fluid-filled space lying between the mesoderm and endoderm. Many of the internal organs are suspended in the pseudocoel, which is enclosed by muscles (mesoderm) only on its outside; there is no inner layer of mesoderm surrounding the internal organs (**Figure 30.6B**).

3. **Coelomate** animals have a body cavity, the coelom, that develops within the mesoderm. The coelom is lined with a thin

layer of tissue called the **peritoneum**, which also surrounds the internal organs. The coelom is thus completely enclosed by mesoderm (**Figure 30.6C**).

The structure of an animal's body cavity strongly influences the ways in which it can move. The body cavities of many animals function as **hydrostatic skeletons**. Fluids are relatively incompressible, so when the muscles surrounding a fluid-filled body cavity contract, fluids shift to another part of the cavity. If the body tissues around the cavity are flexible, fluids squeezed out of one region can cause some other region to expand. The moving fluids can thus move specific body parts. (You can see how a hydrostatic skeleton works by watching a snail emerge from its shell.) A coelomate animal has better control over the movement of the fluids in its body cavity than a pseudocoelomate animal does. An animal that has longitudinal muscles (running along the length of the body) as well as circular muscles (encircling the body cavity) has even greater control over its movement.

In terrestrial environments, the hydrostatic function of fluid-filled body cavities applies mostly to relatively small, soft-bodied organisms. Most larger animals (as well as many smaller ones) have hard skeletons that provide protection and facilitate movement. Muscles are attached to those firm structures, which may be inside the animal or on its outer surface (in the form of a shell or cuticle).

Segmentation improves control of movement

Segmentation—the division of the body into segments—is seen in many animal groups. Segmentation facilitates specialization of different body regions. It also allows an animal to alter the shape of its body in complex ways and to control its movements precisely. If an animal's body is segmented, muscles in each individual segment can change the shape of that segment independently of the others. In only a few segmented animals is the body cavity separated into discrete compartments, but even partly separated compartments allow better control of movement. As we will see in Chapters 31 and 32, segmentation occurs in several groups of protostomes and deuterostomes.

In many animals, such as annelids (earthworms and their relatives), similar body segments are repeated many times (**Figure**

30.7A). In other animals, including most arthropods, segments differ strikingly from one another (**Figure 30.7B**). As we'll describe in Chapter 31, the dramatic evolutionary radiation of the arthropods (including the insects, spiders, centipedes, and crustaceans) was based on modifications of a segmented body plan that features muscles attached to the inner surface of an external skeleton, including a variety of external appendages that move these animals. In some animals, distinct body segments are not apparent externally (as with the segmented vertebrae of vertebrates, including humans). Nonetheless, muscular segmentation is clearly visible in humans with well-defined, muscular bodies (**Figure 30.7C**).

Appendages have many uses

Getting around under their own power is important to many animals. It allows them to obtain food, avoid predators, and find mates. Even some species that are sessile as adults, such as sea anemones, have larval stages that use cilia to swim, thus increasing the animal's chances of finding a suitable habitat.

Appendages that project from the body greatly enhance an animal's ability to move around. Many echinoderms, including sea urchins and sea stars, have myriad tube feet that allow them to move slowly across the substrate (see Figure 32.3B). Animals whose appendages have become modified into specialized limbs are capable of better controlled, more rapid movement. The presence of jointed limbs has been a prominent factor in the evolutionary success of the arthropods and the vertebrates. In four independent instances—among the arthropod insects and among the vertebrate pterosaurs, birds, and bats—body plans emerged in which limbs were modified into wings, allowing these animals to use powered flight (see Figure 19.15).

Appendages also include many structures that are not used for locomotion. Many animals have antennae, which are specialized appendages used for sensing the environment. Other appendages (such as the claws and mouthparts of many arthropods) are adaptations for capturing prey or chewing food. In some species, appendages are used for reproductive purposes, such as sperm transfer or egg incubation.

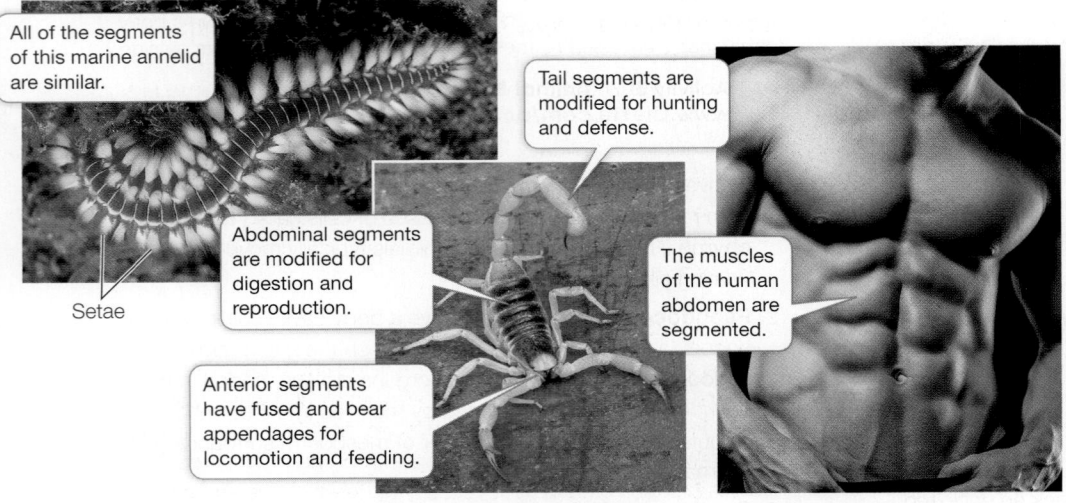

(A) *Hermodice carunculata*

All of the segments of this marine annelid are similar.

Setae

Abdominal segments are modified for digestion and reproduction.

Anterior segments have fused and bear appendages for locomotion and feeding.

Tail segments are modified for hunting and defense.

The muscles of the human abdomen are segmented.

(B) *Hadrurus arizonensis*

(C) *Homo sapiens*

Figure 30.7 Segmentation (**A**) All of the segments of this marine fireworm, an annelid, are similar. Its appendages are tipped with bristles (setae) that are used for locomotion and (in this species) for protection—the setae contain a noxious toxin. (**B**) Segmentation allows the evolution of differentiation among the segments. The segments of this scorpion, an arthropod, differ in their form, their function, and the appendages they bear. (**C**) Muscular segmentation is clearly visible in the abdomen of this body builder.

Nervous systems coordinate movement and allow sensory processing

The bilaterian animals have a well-coordinated central nervous system. More diffuse nervous systems, called **nerve nets**, are present in some other animals, such as ctenophores and cnidarians. Nervous systems appear to be completely absent in a few groups, such as sponges and placozoans.

The central nervous system of bilaterians coordinates the actions of muscles, which allows coordinated movement of appendages and body parts. This coordination of muscles permits highly effective and efficient movement on land, in water, or through the air. The central nervous system is also essential for the processing of sensory information gathered from a wide variety of sensory systems. Many animals have sensory systems for detecting light, for forming images of their environment (sight), for mechanical touch, for detecting movement, for detecting sounds (hearing), for detecting electric fields, and for chemical detection (e.g., taste and smell). These sensory systems allow animals to find food, and the ability of animals to move allows them to capture or collect food from their environment. These same abilities also allow most animals to move to avoid potential predators or to search for suitable mates. Most animals can also assess the suitability of different environments and move appropriately in response to that information.

 Media Clip 30.1 **Nervous Systems Lead to Efficient Predators** www.Life11e.com/mc30.1

Many of the modifications to the body plans of animals affect their ways of finding, capturing, and processing food. Evolutionary changes in symmetry, body cavities, appendages, segmentation, and sensory systems have played key roles in enabling animals to obtain food from their environment as well as helping them avoid becoming food for other animals.

key concept 30.3 Animals Use Diverse Forms of Movement to Feed

As we noted in Key Concept 30.1, animals are heterotrophs, or "other-feeders." Although some animals rely on photosynthetic endosymbionts to nourish them (see Investigating Life: Can Corals Reacquire Dinoflagellate Endosymbionts Lost to Bleaching? in Chapter 26), most animals must expend energy to obtain an outside source of nutrition, otherwise known as food.

focus your learning

• Plant eaters are called herbivores.
• Predators possess features that enable them to capture and subdue their prey.
• Parasites obtain nutrients by living on or within a host organism.

The need to locate food has favored the evolution of sensory structures that provide animals with detailed information about their environment as well as nervous systems that can receive, process, and coordinate that information. Furthermore, in order to acquire food, animals must either move through the environment to where food is located, or move the environment and the food it contains to them. Animals that move from one place to another are **motile**; animals that stay in one place are **sessile**.

The principal feeding strategies that animals use fall into five broad categories:

1. **Filter feeders** (or **suspension feeders**) strain small organisms from their environment.

2. **Herbivores** eat plants or parts of plants.

3. **Predators** capture and eat other animals.

4. **Parasites** live in or on other, generally much larger, organisms, from which they obtain energy and nutrients.

5. **Detritivores** feed on dead organic material.

Each of these strategies can be found in many different animal groups, and none of them is limited to a single group. Individuals of some species employ more than one of these feeding strategies, and some animals employ different feeding strategies at different stages of their life cycle. The constant and ongoing need to obtain food, the variety of nutrient sources available in a given environment, and the necessity of competing with other animals to obtain food means that a variety of feeding strategies can be found among all the major animal groups.

Filter feeders capture small prey

Air and water often contain small organisms and organic molecules that are potential food for animals. Moving air and water may carry those items to an animal that positions itself in a good location. Other animals can move through the environment, filtering out prey items as they move. In either case, filter feeders use some kind of straining device to filter the food from the environment.

Figure 30.8 Filter-Feeding Strategies **(A)** Sessile marine filter feeders such as this Christmas tree worm, a polychaete (annelid), allow the ocean currents to bring their food—plankton—to them. **(B)** The southern right whale is a motile filter feeder that strains small crustaceans such as krill from the oceans for its food.

(A) *Spirobranchus giganteus*

(B) *Eubalaena australis*

Many sessile aquatic animals rely on water currents to bring prey to them (**Figure 30.8A**). Some sessile filter feeders (such as sponges; see Figure 30.2) expend energy to move water past their food-capturing devices. Motile filter feeders move their bodies to the nutrient source. Flamingos, for example, use their serrated beaks to filter small organisms out of the muddy mixture they pick up as they wade through shallow water. Baleen whales—the largest animals that have ever lived—are filter feeders that strain tiny crustaceans from the water column as they swim (**Figure 30.8B**).

 Media Clip 30.2 Filter Feeders
www.Life11e.com/mc30.2

Herbivores eat plants

An individual plant has many different structures—leaves, wood, roots, sap, flowers, fruits, nectar, and seeds—that animals can consume. Not surprisingly, then, many different kinds of herbivores—animals that feed only on plants—may feed on a single kind of plant, consuming different parts of the plant or feeding on the same part in different ways. Whereas an individual animal that is captured by a predator is likely to die, herbivores often feed on plants without killing them.

Animals do not need to expend energy subduing and killing plants in order to feed on them. However, plant matter can be difficult to digest and can pose special challenges to terrestrial herbivores because the dominant land plants tend to have several different kinds of tissues, many of which are tough or fibrous. Herbivorous animals typically have long, *complex guts to accomplish the tasks involved in digesting plants. Animals also must expend energy to detoxify plants' defensive chemicals.

---

*connect the concepts   The structure and function of animal guts, including the differences in the guts of herbivores and carnivores, are presented and explained in Key Concept 50.2.

Predators and omnivores capture and subdue prey

Predators possess features that enable them to capture and subdue other animals (referred to as their **prey**). Many vertebrate predators have sensitive sensory organs that enable them to locate prey, as well as sharp teeth or claws that allow them to capture and subdue prey (**Figure 30.9**). Predators may stalk and pursue their prey or wait (often camouflaged) for their prey to come to them.

Omnivores ("all-devouring") are animals, such as raccoons and humans, that eat both plants and other animals. The diets of some omnivores differ at different life stages; many songbirds, for example, eat fruit or seeds as adults but feed insects to their young.

Parasites live in or on other organisms

Parasites obtain nutrients from another organism—a **host**—by living on or within the host. Some animal parasites consume parts of the host itself (such as ticks that suck body fluids); others highjack nutrients the host would otherwise consume (such as tapeworms that may live in the intestines of mammals). Most animal parasites are much smaller than their hosts, and many parasites can consume parts of their host without killing it. To set up residence within a host, a parasite must first overcome the host's defenses. Parasites often have complex life cycles that rely on multiple hosts, as you will see in Key Concept 30.4.

Parasites that live inside their hosts are called endoparasites, and these are often morphologically very simple. Endoparasites often function without a digestive system, absorbing their food directly from the host's gut or body tissues. Many flatworms are endoparasites of humans and other mammals, as we will describe in Chapter 31.

Parasites that live outside their hosts are called ectoparasites; they are generally more complex morphologically than endoparasites. Ectoparasites have digestive tracts and mouthparts that enable them to pierce the host's tissues or suck on the host's body fluids. Fleas and ticks are ectoparasitic arthropods that feed on many vertebrates, including humans.

Detritivores live on the remains of other organisms

Detritivores feed on the dead bodies or waste products of other organisms, organic matter known as **detritus**. Detritivores (sometimes called decomposers) perform an important ecosystem function by breaking down dead organic matter and returning the nutrients it contains to the environment in a form that can be used by other organisms. Detritivores are common in any soil with high organic content, as well as on the ocean floor. Well-known detritivores include earthworms and other annelids, millipedes, and many insects and crustaceans.

Charles Darwin became fascinated with earthworms and wrote a book called *The Formation of Vegetable Mould through the Action of Worms*. He was particularly impressed by the importance of earthworms in soil formation. Darwin conducted many interesting

(A)

(B)

Figure 30.9 Active and Sit-and-Wait Predators **(A)** The appendages (legs and wings) of the merlin along with its strong beak, are adaptations to the life of an active predatory hunter. **(B)** The crocodile typically sit motionless in the water, waiting in one spot for unsuspecting prey to walk within striking range.

experiments to establish how quickly earthworms break down organic matter and build up rich soils.

 Media Clip 30.3 Detritivores
www.Life11e.com/mc30.3

30.3 recap

Animals are heterotrophs that must expend energy to acquire food from their environment. Most animals either move through the environment to where food is located or move the environment and the food it contains to them.

learning outcomes

You should be able to:
- Summarize common adaptations of herbivores.
- Summarize features that predators use to capture and subdue their prey.
- Distinguish among predators, parasites, and filter feeders.

1. How can you distinguish among filter feeders, predators, and parasites—all of which may feed on other animals?
2. What adaptations are necessary for animals that eat plants? What adaptations are needed for a predatory lifestyle?

As an animal grows from a single-celled zygote into a larger, more complex adult, its body structure, its diet, and the environment in which it lives may all change. In the next section we will describe some animal life cycles and discuss why they are so varied.

key concept 30.4 Animal Life Cycles Involve Trade-Offs

The life cycle of an animal encompasses its embryonic development, birth, growth to maturity, reproduction, and death. During its life an individual animal ingests food, grows, interacts with other individuals of the same and other species, and reproduces.

focus your learning
- Direct development and metamorphosis each have advantages under different conditions.
- Life cycles often involve trade-offs.
- Animal colonies can provide some of the same specializations provided by multicellularity.

Many animal life cycles feature specialized life stages

In some groups of animals, newborns look much like miniature versions of the adults (a pattern called **direct development**). Newborns of most animal species, however, differ dramatically from adults. Many animal species have a life stage called a **larva** (plural *larvae*), which is an immature form that the animal takes early in its life before assuming an adult form. Some of the most striking life cycle changes are found among insects such as beetles, flies, moths, butterflies, and bees, which undergo radical change (called **complete metamorphosis**) between their larval and adult stages (**Figure 30.10**). In these animals, one stage may be specialized for feeding and the other for reproduction. Adults of many moth species, for example, do not eat. In some animal species, individuals eat during all life cycle stages, but what they eat changes with the stage. For example, butterfly larvae (known as caterpillars) eat leaves and flowers, whereas most adult butterflies eat only nectar. Having different life cycle stages that are specialized for different activities may increase the efficiency with which an animal performs those activities and reduce the competition for food between juveniles and adults.

Most animal life cycles have at least one dispersal stage

At some time during their lives, most animals move, or are moved, so few animals die exactly where they were born. Movement of organisms away from a parent organism or from an existing population is called **dispersal**.

Animals that are sessile as adults typically disperse as eggs or larvae. Most sessile marine animals discharge their eggs and sperm into the water, where fertilization takes place. A larva soon hatches and floats freely in the plankton, where it filters small food items from the water. Many animals that live on the seafloor, including polychaete worms and mollusks, have a radially symmetrical larval form known as a **trochophore** (**Figure 30.11A**). Other animals, such as crustaceans, have a bilaterally symmetrical larval form called a

(A) Larva   **(B) Pupa**   **(C) Adult**

Figure 30.10 A Life Cycle with Complete Metamorphosis **(A)** The larval stage (caterpillar) of the cecropia moth (*Hyalophora cecropia*) is specialized for feeding. **(B)** The pupa is the stage during which the transformation to the adult form occurs. **(C)** The adult moth is specialized for dispersal and reproduction and does not feed.

nauplius (**Figure 30.11B**). Both types of larvae feed for some time in the plankton and may travel long distances before settling on the ocean floor and transforming into adults.

Other animal species that are motile as adults disperse when they are mature. A caterpillar, for example, may spend its entire larval stage feeding on a single plant, but after it metamorphoses into a flying adult—a butterfly or moth—it may fly to and lay eggs on other plants located far from the one where it spent its caterpillar days. In some species, individuals disperse during several different life cycle stages.

Parasite life cycles facilitate dispersal and overcome host defenses

Animals that live as endoparasites are bathed in the nutritious tissues of their host or in the digested food that fills their host's digestive tract. Thus they may not need to exert much energy to obtain food, but to survive they must overcome the host's defenses. Furthermore, either they or their offspring must disperse to new hosts while their host is still living, because they die when their host dies.

The fertilized eggs of some parasites are voided with the host's feces and later ingested directly by other host individuals. Most parasite species, however, have complex life cycles involving one or more intermediate hosts and several larval stages (**Figure 30.12**). Some intermediate hosts transport individual parasites directly between other host species. Others house and support the parasite until another host ingests them. Complex life cycles may thus facilitate the transfer of individual parasites among hosts.

Some animals form colonies of genetically identical, physiologically integrated individuals

Most people tend to view the distinction between individuals and populations as clear-cut. In several groups of animals, however, asexual reproduction without fission leads to the formation of

colonies composed of many physiologically integrated individuals. At first appearance, these colonies may look much like a single integrated organism. The individuals in the colony are clonal copies of one another, so they are genetically homogeneous.

Coloniality has arisen several times among animal groups, with widely varying levels of integration and specialization among the individuals. In some species, colonies are composed of loosely connected but integrated individuals that all function alike (**Figure 30.13**). In other colonial species, the individuals may become specialized for different functions, just as different cell types in multicellular organisms do. The Portuguese man-of-war (a cnidarian; see Figure 30.20) is an example of such a colonial animal, as it is composed of many individuals of four different specialized body forms, all integrated and functioning together. The individuals in the colony are themselves multicellular, however, unlike the cells of a single multicellular organism.

(A) Trochophore

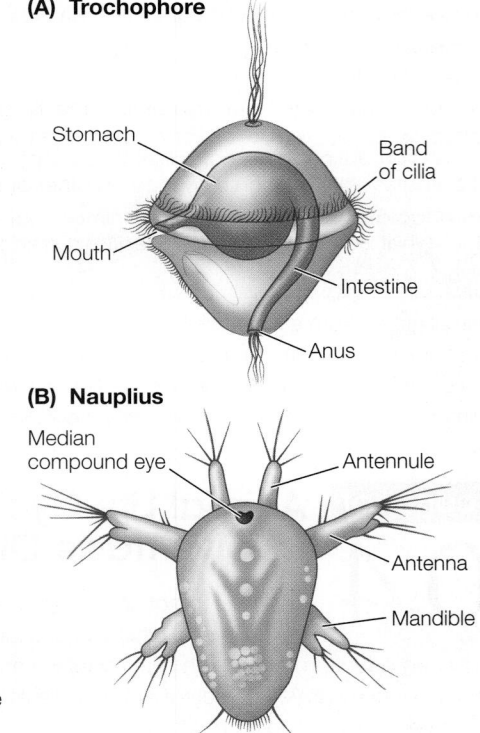

Stomach

Band of cilia

Mouth

Intestine

Anus

(B) Nauplius

Median compound eye

Antennule

Antenna

Mandible

Figure 30.11 Planktonic Larval Forms of Marine Animals **(A)** The trochophore ("wheel-bearer") is a distinctive larval form found in several marine animal clades with spiral cleavage, most notably the polychaete worms and the mollusks. **(B)** This nauplius larva will mature into a crustacean with a segmented body and jointed appendages.

Figure 30.12 Reaching a New Host by a Complex Route The broad fish tapeworm *Diphyllobothrium latum* must pass through the bodies of a copepod (a type of crustacean) and at least one fish before it can reinfect its primary host, a mammal. Such complex life cycles assist the parasite's colonization of new host individuals, but they also provide opportunities for humans to break the cycle with hygienic measures (such as thoroughly cooking food to kill the parasites).

No life cycle can maximize all benefits

A common saying, "a jack-of-all-trades is master of none," suggests why there are constraints on the evolution of life cycles. The characteristics an animal has in any one life cycle stage may improve its performance in one activity but reduce its performance in another—a situation known as a trade-off. An animal that is good at filtering small food items from the water, for example, is unlikely to be good at capturing large prey. Similarly, energy devoted to building protective structures such as shells cannot be used for growth.

Some major trade-offs can be seen in animal reproduction. Some animals produce large numbers of small eggs, each with a small energy store (**Figure 30.14A**). Other animals produce a small number of large eggs, each with a large energy store (**Figure 30.14B**). With a fixed amount of energy available for reproduction, a female animal can produce many small eggs or a

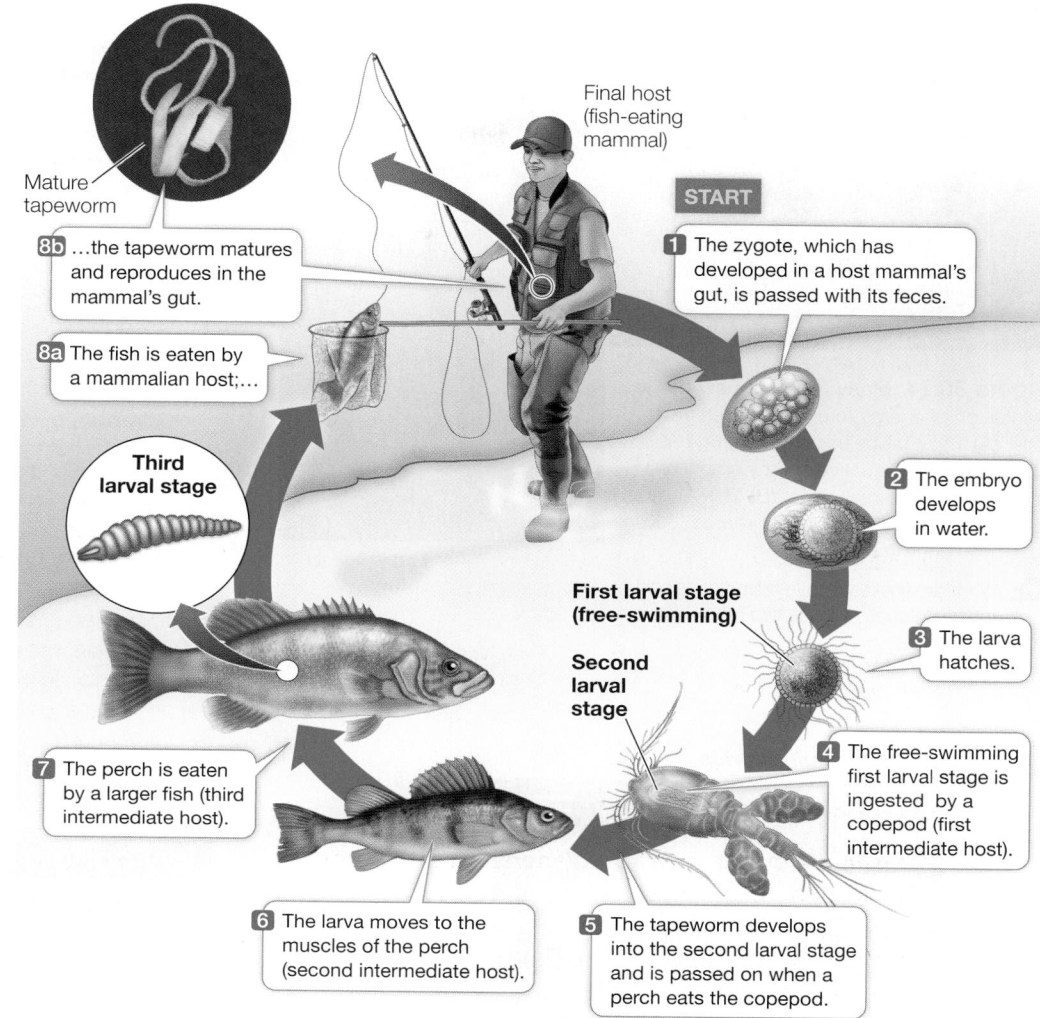

Final host (fish-eating mammal)

START

1 The zygote, which has developed in a host mammal's gut, is passed with its feces.

2 The embryo develops in water.

3 The larva hatches.

First larval stage (free-swimming)

Second larval stage

4 The free-swimming first larval stage is ingested by a copepod (first intermediate host).

5 The tapeworm develops into the second larval stage and is passed on when a perch eats the copepod.

6 The larva moves to the muscles of the perch (second intermediate host).

7 The perch is eaten by a larger fish (third intermediate host).

Third larval stage

8a The fish is eaten by a mammalian host;...

8b ...the tapeworm matures and reproduces in the mammal's gut.

Mature tapeworm

few large eggs, but she cannot produce many large eggs. Thus there is a trade-off between the number of offspring produced and the energy resources each offspring receives from its mother.

The larger the energy store in an egg, the longer an offspring can develop before it must either find its own food or be fed by its parents. Birds of all species lay relatively small numbers of relatively large eggs, but incubation periods vary. In some species, eggs hatch when the young are still helpless (**Figure 30.15A**). Such **altricial** young must be fed and cared for until they can feed themselves; parents can provide for only a small number of altricial offspring. In contrast, some bird species incubate their eggs longer, and the hatchlings are developed to the point that they are able to forage for themselves almost immediately (**Figure 30.15B**). The young of such species are called **precocial**.

Reteporella couchii

Figure 30.13 Colonial Animals This bryozoan colony consists of many asexually reproducing, genetically homogeneous, physiologically interacting individuals. The colony looks much like a single individual with many parts, but in fact is a group of individuals acting together.

(A) *Rana temporaria*

(B) *Aptenodytes patagonicus*

Figure 30.14 Many Small or Few Large
Allocation of energy to eggs requires trade-offs. **(A)** This European common frog has divided her reproductive energy among a large number of small eggs. **(B)** This king penguin has invested all of her reproductive energy in one large egg.

Q: What are the respective advantages and disadvantages of these two reproductive strategies?

(A) *Turdus migratorius*

(B) *Anser anser*

Figure 30.15 Helpless or Independent **(A)** The altricial young of the common robin are essentially helpless when they hatch. Their parents feed and care for them for several weeks. **(B)** Graylag goose hatchlings are precocial, ready to swim and feed independently almost immediately after hatching.

30.4 recap

Many animals have life cycle stages that differ from one another morphologically. In some animals, the larval form is a dispersal stage; in other animals, the adults are more likely to disperse than are larvae. In several groups of organisms, asexual reproduction without fission leads to coloniality.

learning outcomes

You should be able to:

- Using examples, show why trade-offs can constrain evolution.
- Compare and contrast multicellularity and coloniality.

1. Why can't a life cycle evolve that maximizes all aspects of feeding and reproduction simultaneously?
2. How does a colony of unicellular animals differ from a multicellular animal?

Variations in body symmetry, body cavity structure, life cycles, patterns of development, and survival strategies differentiate millions of animal species. In the remainder of this chapter and in Chapters 31 and 32, we will become acquainted with the major animal groups and learn how the general characteristics we have just described apply to each of them.

key concept
30.5 The Root of the Animal Tree Provides Clues to Early Animal Diversification

The bilaterians make up a large monophyletic group embracing all animals other than ctenophores, sponges, placozoans, cnidarians, and a few poorly known groups of parasitic animals (see Table 30.1). Some of the traits that support the monophyly of bilaterians are the presence of three distinct cell layers in embryos (triploblasty) and the presence of at least seven Hox genes (see Chapter 19). Although bilateral symmetry is often viewed as a synapomorphy of bilaterians (and the trait gives the group its name), some groups of cnidarians are also bilaterally symmetrical. Recent studies have shown that the genetic basis of bilateral symmetry in bilaterians and in those cnidarian groups that have this trait is similar, so this feature may have been present in the ancestor of both groups.

focus your learning

- Some major features common among animal groups appear to have evolved independently several times.
- In many respects, sponges are the simplest animals.
- Placozoan simplicity appears to be secondarily derived.
- Placozoans are difficult to observe in nature because of their small size and transparency.

(A) Generalized anatomy of a ctenophore

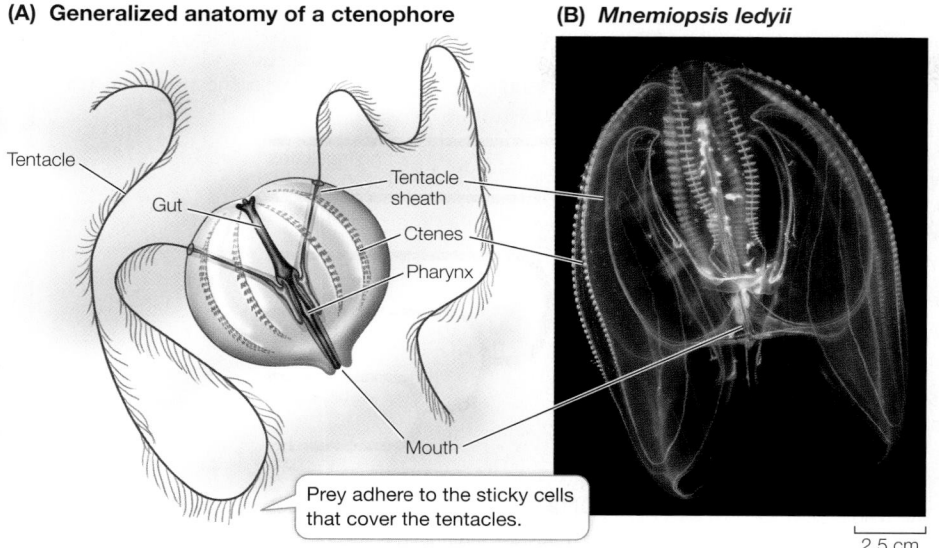

Tentacle

Gut

Tentacle
sheath

Ctenes

Pharynx

Mouth

Prey adhere to the sticky cells
that cover the tentacles.

(B) *Mnemiopsis ledyii*

2.5 cm

Figure 30.16 Comb Jellies Feed with Tentacles
(A) The body plan of a typical ctenophore. The long,
sticky tentacles sweep through the water, efficiently
harvesting small prey. **(B)** This comb jelly, photographed
in Sydney Harbor, Australia, has short tentacles. Comb
jellies get their name from the ctenes—rows or "combs"
of cilia used for movement.

Bilaterian animals can be divided into the two major clades mentioned earlier, the protostomes and the deuterostomes. These two groups, as well as the remaining animal groups we will discuss below, have been diversifying separately for at least 700 million years. We will describe the protostomes in Chapter 31 and the deuterostomes in Chapter 32.

The remainder of this chapter describes those animal groups that are not bilaterians. The ctenophores, sponges, and placozoans have weakly differentiated tissue layers or even undifferentiated layers, and two of these groups (sponges and placozoans) also lack a nervous system. The remaining animals (those that don't fall into any of the groups mentioned thus far) are called **eumetazoans**. Eumetazoans typically have some form of body symmetry, a gut, a nervous system, and tissues organized into distinct organs (although there have been secondary losses of some of these features in some eumetazoans).

 Activity 30.2 **Sponge and Diploblast Classificaton**
www.Life11e.com/ac30.2

Ctenophores are the sister group of all other animals

Ctenophores, also known as comb jellies, were until recently thought to be most closely related to the cnidarians (jellyfish, corals, and their relatives). But ctenophores lack most of the Hox genes and many other genes found in all other animals, and recent studies of their genomes provide strong evidence that ctenophores were the earliest lineage to split from the remaining animals (**Investigating Life: Reconstructing Animal Phylogeny from Protein-Coding Genes**). This position in the animal tree does not imply that ctenophores look anything like the ancestral animal, however. All living ctenophores are quite closely related to one another, and they have been evolving for as long as all other animal lineages—ever since animals last shared a common ancestor more

Ctenophores

Sponges

Placozoans

Cnidarians

Bilaterians
(protostomes and
deuterostomes)

than 700 million years ago. Thus many traits of modern ctenophores are highly derived and specialized.

Ctenophores have a radially symmetrical, diploblastic body plan. The two cell layers are separated by an inert, gelatinous extracellular matrix called **mesoglea**. Ctenophores, unlike sponges, have a **complete gut**: food enters through a mouth, and wastes are eliminated through two anal pores.

Ctenophores move by beating cilia rather than by muscular contractions. Most of the 250 known species have eight comblike rows of cilia-bearing plates, called **ctenes** (**Figure 30.16**). Recent studies of the many genes that encode nervous systems indicate that these genes were duplicated and specialized separately in ctenophores, cnidarians, and bilaterians. This finding suggests that nervous systems evolved independently in these three groups in response to similar selection pressures associated with multicellularity, movement, and prey capture.

The feeding tentacles of ctenophores are covered with cells that discharge adhesive material when they contact prey. After capturing its prey, a ctenophore retracts its tentacles to bring the food to its mouth. In some species, the entire surface of the body is coated with sticky mucus that captures prey. Most ctenophores eat small planktonic organisms, although some eat other ctenophores. They are common in open seas and can become abundant in coastal bays, where large populations of ctenophores may inhibit the growth of other organisms.

Ctenophore life cycles are uncomplicated. Gametes are released into the gut and discharged through the mouth or the anal pores. Fertilization takes place in open seawater. In nearly all species, the fertilized egg develops directly into a miniature ctenophore, which gradually grows into an adult.

 Media Clip 30.4 **Ctenophores**
www.Life11e.com/mc30.4

Sponges are loosely organized animals

In several respects, **sponges** are the simplest animals. For this reason, they were long thought to be the sister group of the remaining animals, until genome sequencing placed the ctenophores in that

Ctenophores

Sponges

Placozoans

Cnidarians

Bilaterians
(protostomes and
deuterostomes)

experiment

Original Paper: Dunn, C. W. et al. 2008. Broad phylogenomic sampling improves resolution of the Animal Tree of Life. *Nature* 452: 745–749.

Several breakthroughs in our understanding of animal phylogeny have occurred in recent years as the sequences of genes and proteins have been compared across species. Casey Dunn and his colleagues compared sequences from many different proteins across a wide variety of animal groups to reconstruct their phylogeny.

QUESTION▶ What is the earliest split in the animal tree? Are cteno-phores, sponges, or placozoans the sister group of other animals?

METHOD

1. Collect mRNA from major groups of animals and choanoflagellate outgroups. (See Figure 30.2 to see why choanoflagellates were used as an outgroup.)
2. Reverse transcribe the mRNA into cDNA.
3. Sequence the cDNA, translate the genes into the amino acid sequences of the proteins, and align sequences of homologous proteins.
4. Reconstruct the phylogeny of the animal groups from the aligned protein sequences.

RESULTS

The tree below represents just a sample of the major lineages examined by Dunn and colleagues.

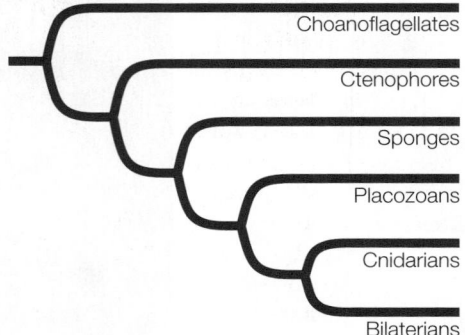

Choanoflagellates
Ctenophores
Sponges
Placozoans
Cnidarians
Bilaterians

CONCLUSION▶ The earliest split in the animal tree was between ctenophores and the remaining animals.

work with the data

The table below includes a small sample of the data collected and analyzed by Dunn and his colleagues. Use these representative data to reconstruct a tree of the represented species. In the original paper, Dunn and colleagues reported on 11,234 amino acid positions among 77 species of animals. Twenty-seven of these amino acid positions for 11 of those species are shown in the table. (See Table 3.2 for the one-letter amino acid abbreviations.)

QUESTIONS▶

1. Construct a phylogenetic tree of these 11 species using the parsimony method. Use the out-group (a choanoflagellate) to root your tree. Assume that all changes from one amino acid to another are equally likely. See Key Concept 21.2 for a review of the parsimony method of reconstructing phylogenetic trees.
2. How many character state changes (i.e., changes from one amino acid to another) occur along each branch on your tree?
3. Which characters support the ctenophores as the sister group of the remaining animals?
4. Which group on your tree represents the bilaterian animals? The protostomes? The deuterostomes?

Character state (amino acid at position)																										
Species	1	2	3	4	5	6	7	8	9	10	11	12	13	14	15	16	17	18	19	20	21	22	23	24	25	26	27
Clam	Y	S	T	G	L	H	E	N	Y	A	R	A	M	R	I	A	L	T	I	V	K	L	S	I	V	I	L
Earthworm	Y	A	T	G	L	H	E	N	Y	P	H	A	M	R	I	A	L	T	I	V	K	L	S	I	V	M	L
Tardigrade	Y	A	T	G	L	H	E	H	Y	K	R	A	M	R	V	A	T	S	I	V	R	L	N	L	V	L	L
Fruit Fly	F	A	T	G	L	H	E	N	Y	K	R	A	M	R	I	A	L	S	I	V	S	L	D	L	V	L	L
Sea Urchin	Y	A	T	G	L	L	E	N	Y	P	N	A	M	R	I	A	L	T	V	I	R	Q	N	L	T	V	K
Human	W	A	A	G	L	R	E	H	Y	P	K	A	I	R	I	S	V	T	V	I	R	Q	N	L	T	V	K
Chicken	W	A	A	G	L	R	E	H	Y	P	R	A	I	R	I	A	V	T	V	I	R	Q	N	L	T	V	K
Lancelet	Y	A	T	G	L	R	E	H	Y	P	K	A	M	R	I	A	V	T	V	I	R	L	N	L	T	V	K
Sponge	Y	G	L	S	L	R	P	N	F	P	K	S	M	S	V	A	L	T	V	I	R	Q	N	L	V	I	L
Ctenophore	Y	G	L	G	Q	D	P	N	F	P	K	S	M	S	V	A	L	T	V	I	R	Q	N	L	V	I	L
Choano-flagellate (outgroup)	Y	G	L	G	Q	D	P	N	F	P	K	S	F	S	V	A	L	T	V	I	R	Q	N	L	V	I	L

A similar **work with the data** exercise may be assigned in **LaunchPad**.

position. But some of the perceived complexity of ctenophores now appears to have evolved independently from other animal lineages, whereas sponges have retained more of the features of the ancestral animal. Although sponges have some specialized cells, they have no distinct embryonic cell layers and no true organs. Early naturalists classified sponges as plants because they were sessile and lacked body symmetry.

Sponges have hard skeletal elements called **spicules**, which may be small and simple or large and complex (see Figure 30.2B). Three major groups of sponges, which separated soon after the origin of sponges, are distinguished by their spicules. Members of two groups (glass sponges and demosponges) have skeletons composed of silicaceous spicules made of hydrated silicon dioxide (**Figure 30.17A and B**). These spicules are remarkable in having greater flexibility and toughness than synthetic glass rods of similar length. Members of the third group, the calcareous sponges, take their name from their calcium carbonate spicules (**Figure 30.17C**). There is some question about the monophyly of sponges, although genomic analyses that combine information from many genes support their monophyly.

The body plan of sponges of all three groups—even large ones, which may reach 1 meter or more in length—is an aggregation of cells built around a water canal system. Sponges bring water into their bodies by beating the flagella of their specialized feeding cells, called **choanocytes** (see Figure 30.2B). Water, along with any food particles it contains, enters the sponge by way of small pores and passes into the water canals or a central atrium, where the choanocytes capture food particles. (You may recall from Key Concept 30.1 that the choanocytes are similar in structure to protists known as choanoflagellates, which may indicate a retained ancestral similarity between the two groups.)

A skeleton of simple or branching spicules, often combined with a complex network of elastic fibers, supports the body of most sponges. Sponges also produce an extracellular matrix, composed of collagen, adhesive glycoproteins, and other molecules, that holds the cells together. Most species are filter feeders; a few species are predators that trap prey on hook-shaped spicules that protrude from the body surface.

Most of the 8,500 species of sponges are marine animals; only about 50 species live in fresh water. Sponges come in a wide variety of sizes and shapes that are adapted to different movement patterns of water. Sponges living in intertidal or shallow subtidal environments with strong wave action are firmly attached to the substrate. Most sponges that live in slowly flowing water are flattened and are oriented at right angles to the direction of current flow. They intercept water and the food items it contains as it flows past them.

Sponges reproduce both sexually and asexually. In most species, a single individual produces both eggs and sperm, but individuals do not self-fertilize. Water currents carry sperm from one individual to another. Sponges also reproduce asexually by budding and fragmentation.

(A) *Aplysina fistularis*

(B) *Euplectella aspergillum*

(C) *Sycon* sp.

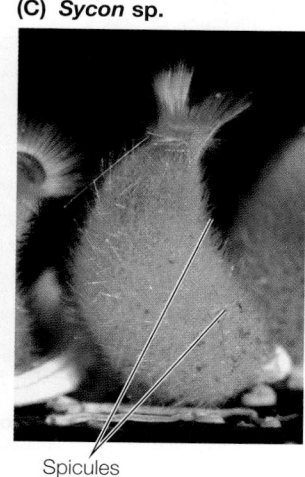

Spicules

Figure 30.17 Sponge Diversity   (A) The majority of sponge species are demosponges, such as these yellow tube sponges. The system of pores and water canals (see Figure 30.2B) that is typical of the sponge body plan is apparent in this photograph. **(B)** The supporting structures of both demosponges and glass sponges are silicaceous spicules, seen here in the skeleton of a glass sponge. **(C)** The spicules of calcareous sponges are made of calcium carbonate.

Placozoans are abundant but rarely observed

As discussed at the start of this chapter, **placozoans** are (like sponges) structurally very simple animals with only a few distinct cell types (**Figure 30.18A**). Individuals in the mature, asymmetrical life stage are usually observed adhering to surfaces (such as the glass of aquariums, where they were first discovered, or to rocks and other hard substrates in nature). Their structural simplicity—they have no mouth, gut, or nervous system—initially led to yet another hypothesis about the earliest split among the animals. Phylogenetic analyses and some aspects of the placozoans' structural simplicity, however, suggest that placozoan simplicity may be secondarily derived. They are generally considered to have a diploblastic body plan, with upper and lower surface layers that sandwich a layer of contractile fiber cells.

Recent studies have found that placozoans have a pelagic (open-ocean) life stage that is capable of swimming (**Figure 30.18B**), but the life history of placozoans is incompletely known. Most studies have focused on the larger adherent stages that are most easily observed in aquariums. The transparent nature and small size of placozoans make them very difficult to observe in nature. Nonetheless, it is known that placozoans can reproduce both asexually and sexually, although the details of their sexual reproduction are poorly understood. As we noted at the opening of this chapter, placozoans have been studied mainly in aquariums, where they appear after being inadvertently collected with other marine organisms, although we now know that pelagic-stage placozoans are abundant in warm seas around the world.

Ctenophores

Sponges

Placozoans

Cnidarians

Bilaterians (protostomes and deuterostomes)

(A)

(B)

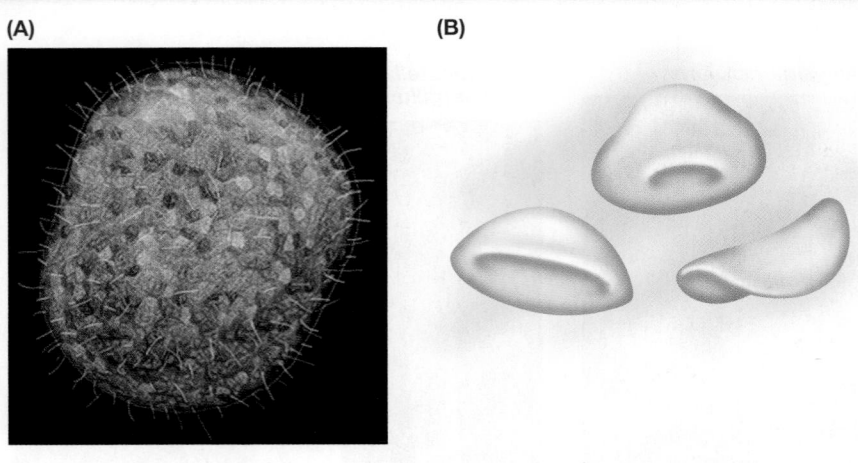

Figure 30.18 Placozoan Simplicity **(A)** As seen in this artist's rendition, adult placozoans are tiny (1–2 mm across), flattened, asymmetrical animals. **(B)** Recent studies have found a symmetrical, weakly swimming pelagic stage of placozoan to be abundant in many warm tropical and subtropical seas.

Cnidarians are specialized predators

The **cnidarians** (jellyfish, sea anemones, corals, and hydrozoans) make up the largest and most diverse group of non-bilaterian animals. The mouth of a cnidarian is connected to a blind sac called the **gastrovascular cavity** (a cnidarian thus does not have a complete gut). The gastrovascular cavity functions in digestion, circulation, and gas exchange, and it also acts as a hydrostatic skeleton. The single opening serves as both mouth and anus.

The life cycle of many cnidarians has two distinct stages, one sessile and the other motile (**Figure 30.19**), although one or the other of these stages is absent in some groups. In the sessile **polyp** stage, a cylindrical stalk attaches the animal to the substrate. The motile **medusa** (plural *medusae*) is a free-swimming stage shaped like a bell or an umbrella. It typically floats with its mouth and feeding tentacles facing downward.

Mature polyps produce medusae by asexual budding. Medusae then reproduce sexually, producing eggs or sperm by meiosis and releasing the gametes into the water. A fertilized egg develops into a

Figure 30.19 The Life Cycle of Most Cnidarians Has Two Stages The life cycle of a scyphozoan (jellyfish) exemplifies the typical cnidarian body forms: the sessile, asexual polyp and the motile, sexual medusa. Some species of cnidarians have life cycles that lack polyps or medusae.

▶ Animation 30.1 **Life Cycle of a Cnidarian**
www.Life11e.com/a30.1

free-swimming, ciliated larva called a **planula**, which eventually settles to the bottom and develops into a polyp.

Cnidarians are specialized predators adapted for capturing and subduing relatively large and complex prey. As we noted earlier in discussing ctenophores, recent genome studies suggest that the nerve nets found in cnidarians are largely independently derived from those found in ctenophores, as well as from the nervous systems found in bilaterians. The tentacles of cnidarians are covered with specialized cells that contain stinging organelles called **nematocysts**, which inject toxins into their prey (**Figure 30.20**). Some cnidarians, including many corals and anemones, gain additional nutrition from photosynthetic endosymbionts that live in their tissues.

Cnidarians have cells containing muscle fibers whose contractions enable the animals to move, as well as simple nerve nets that integrate the body's activities. Their bodies also contain specialized structural molecules (collagen, actin, and myosin). Yet cnidarians, like ctenophores, are largely made up of inert mesoglea. Most species have low metabolic rates and can survive in environments where they encounter prey only infrequently.

Of the roughly 12,500 living cnidarian species, all but a few live in the oceans (**Figure 30.21**). The smallest cnidarians can barely be seen without a microscope. One small group, known as myxozoans, consists of tiny parasites, usually with a two-host life cycle that includes a fish and an annelid worm or a bryozoan. The largest known jellyfish is 2.5 meters in diameter, and some colonial hydrozoans (which include the Portuguese man-of-war; see Figure 30.20) can reach lengths in excess of 30 meters. Here we describe the three

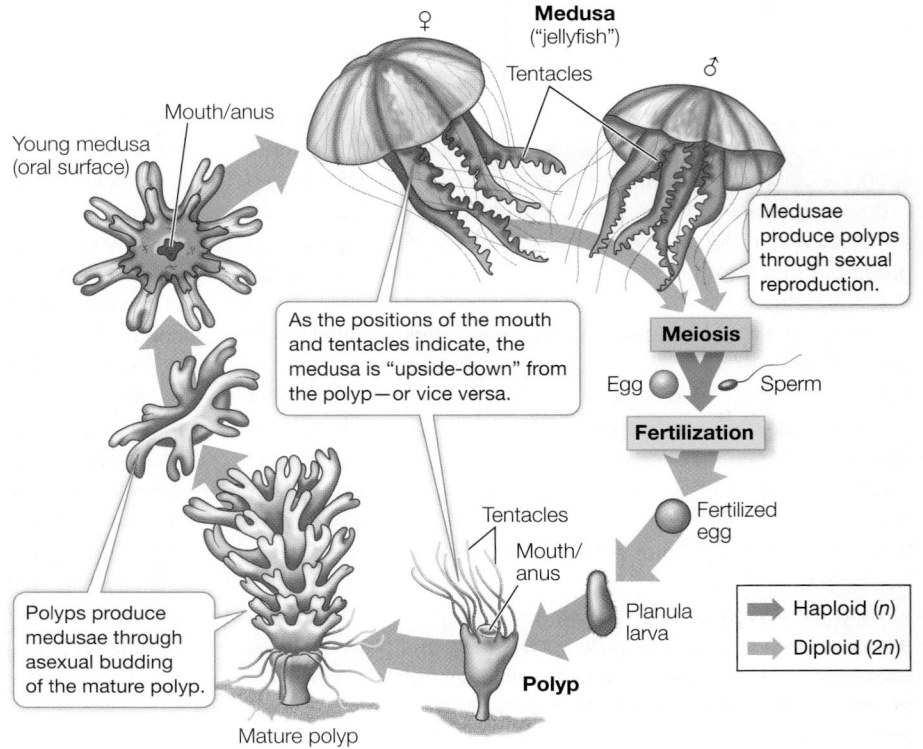

(A) Portuguese man-of-war
(Physalia physalis)

Media Clip 30.5 **Stunning Siphonophore: Colonial Hydrozoans**
www.Life11e.com/mc30.5

(B)

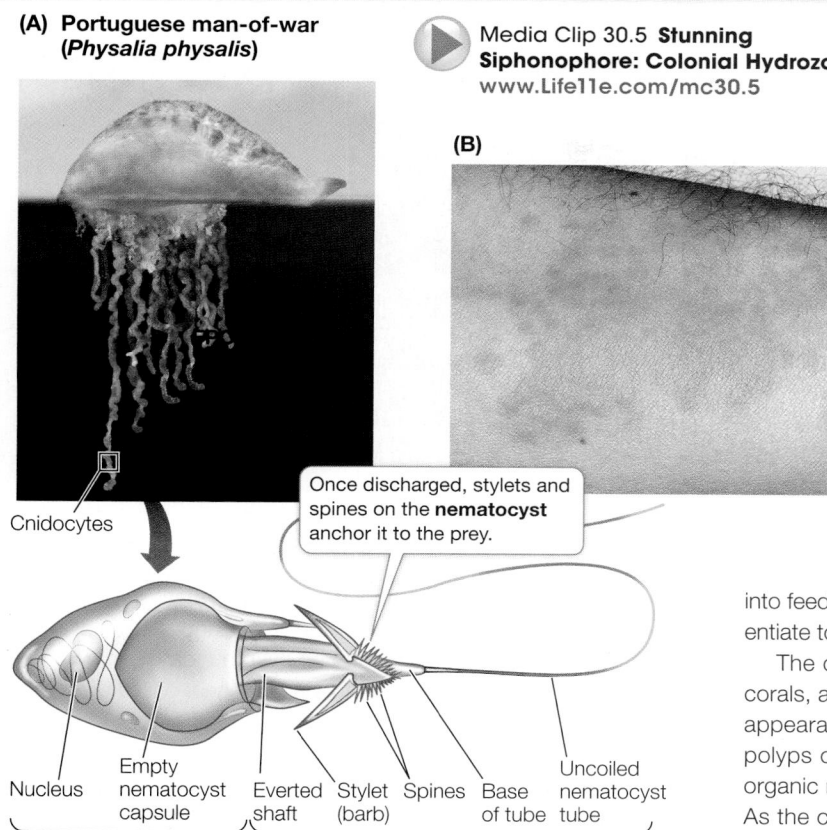

Cnidocytes

Once discharged, stylets and spines on the **nematocyst** anchor it to the prey.

Nucleus | Empty nematocyst capsule | Everted shaft | Stylet (barb) | Spines | Base of tube | Uncoiled nematocyst tube

Cnidocyte | **Nematocyst**

Figure 30.20 Nematocysts Are Potent Weapons
(A) The tentacles of the Portuguese man-of-war, a hydrozoan, are rife with specialized cells called cnidocytes. These cells contain stinging organelles called nematocysts, which inject toxins into their prey. The Portuguese man-of-war is a colonial organism, composed of many physiologically integrated individuals with specialized functions. **(B)** Nematocysts of the Portuguese man-of-war cause painful reactions when they come into contact with human skin.

waters. Sea pens (see Figure 30.21B), by contrast, are colonial. Each colony consists of two or more different kinds of polyps. The primary polyp has a lower portion anchored in the bottom sediment and a branched upper portion that projects above the substrate. Along the upper portion, the primary polyp produces smaller secondary polyps by budding. Some of these secondary polyps differentiate into feeding polyps; in some species, other secondary polyps differentiate to circulate water through the colony.

The common names of coral groups—brain corals, staghorn corals, and organ pipe corals, among others—often describe their appearance (**Figure 30.22A**). Corals are sessile and colonial. The polyps of most species form a skeleton by secreting a matrix of organic molecules on which they then deposit calcium carbonate. As the colony grows, old polyps die, but their calcium carbonate skeletons remain. Living corals form a layer on top of a growing bank of skeletal remains, eventually forming chains of islands and reefs. The Great Barrier Reef along the northeastern coast of Australia

clades of cnidarians that contain the most species: anthozoans, scyphozoans, and hydrozoans.

ANTHOZOANS   Members of the **anthozoan** clade include sea anemones, sea pens, and corals. Sea anemones (see Figure 30.21A), all of which are solitary, are widespread in both warm and cold ocean

(C) *Phyllorhiza punctata*

(D) *Obelia* sp.

(A) *Urticina crassicornis*

(B) *Virgularia gustaviana*

Figure 30.21 Diversity among Cnidarians
(A) Sea anemones are sessile, living attached to marine substrates. Water currents carry prey into the nematocyst-studded tentacles. **(B)** The sea pen is a colonial cnidarian that lives in soft bottom sediments and projects polyps above the substrate. **(C)** This jellyfish illustrates the complexity of a scyphozoan medusa. **(D)** Polyp colonies of this hydrozoan grow on rocks of intertidal pools.

Figure 30.22 Corals
(A) The descriptive common name of this Caribbean coral is "brain coral." **(B)** Many different coral species form the Great Barrier Reef off the coast of Australia.

(A) *Symphyllia* sp.

(B)

is a system of coral formations more than 2,000 kilometers long, which is about the distance from New York City to St. Louis (**Figure 30.22B**). A single coral reef in the Red Sea has been calculated to contain more material than all the buildings in the major cities of North America combined.

Corals flourish at shallow depths in clear, nutrient-poor tropical waters. They grow well in such environments because unicellular photosynthetic dinoflagellates live endosymbiotically within their cells. These dinoflagellates provide the corals with products of photosynthesis; the corals, in turn, provide the dinoflagellates with nutrients and a place to live. This endosymbiotic relationship explains why reef-forming corals are restricted to clear surface waters, where light levels are high enough to support photosynthesis.

Coral reefs throughout the world are threatened by rising CO_2 levels (which result in increased ocean temperatures and acidification of ocean waters through the formation of carbonic acid). Polluted runoff from development on adjacent shorelines is an additional threat to corals. Warmer temperatures lead to the loss of **coral endosymbionts* (known as coral bleaching), and acidification can cause coral skeletons to dissolve. An overabundance of nitrogen from fertilizer in runoff is advantageous to algae, which overgrow and eventually smother the corals.

---

*connect the concepts Investigating Life: Can Corals Reaquire Dinoflagellate Endosymbionts Lost to Bleaching? in Chapter 26 describes an experiment in which coral endosymbionts were lost and replaced by new endosymbionts in response to changing environmental conditions.

SCYPHOZOANS The several hundred species of **scyphozoans** are all marine. The mesoglea of their medusae is thick and firm, giving rise to their common name of jellyfish (or sea jellies). The medusa rather than the polyp dominates the life cycle of scyphozoans. An individual medusa is male or female, releasing eggs or sperm into the open sea. A fertilized egg develops into a small planula larva that quickly settles on a substrate and develops into a small polyp. This polyp feeds and grows and may produce additional polyps by budding. After a period of growth, the polyp begins to bud off small medusae, which feed, grow, and transform into adult medusae (see Figures 30.19 and 30.21C).

HYDROZOANS The polyp typically dominates the life cycle of **hydrozoans**, but some species have only medusae and others have only polyps. Most hydrozoans are colonial. A single planula larva eventually gives rise to a colony of many polyps, all interconnected and sharing a continuous gastrovascular cavity (**Figure 30.23**). Within

♀ **Medusa** ♂
Oral surface
Gonad

Medusae develop asexually within an enlarged polyp.

→ Haploid (*n*)
→ Diploid (*2n*)

Meiosis

Egg   Sperm

Fertilization

Eggs produced by medusae are fertilized in the open water by sperm produced by other medusae.

Fertilized egg

Planula larva

The polyps of the **hydrozoan** *Obelia* are interconnected and share a gastrovascular cavity.

Larvae settle on the substrate and grow into polyps.

Figure 30.23 Many Hydrozoans Are Colonial The polyps in a hydrozoan colony may differentiate to perform specialized tasks. In the species whose life cycle is diagrammed here, the medusa is the sexual reproductive stage, producing eggs and sperm in organs called gonads.

such a colony, some polyps have tentacles with many nematocysts; they capture prey for the colony. Other individuals lack tentacles and are unable to feed, but are specialized for the asexual production of medusae. Still others are fingerlike and defend the colony with their nematocysts.

Some small groups of parasitic animals may be the closest relatives of bilaterians

Two small groups of tiny marine parasites are listed in Table 30.1 but are not depicted in the phylogeny in Figure 30.1: the orthonectids and the rhombozoans. Recent genomic analyses suggest that these groups may be among the closest surviving relatives of the bilaterians, although their exact phylogenetic placement is uncertain. Both groups are highly reduced parasites that lack many of the structures that traditionally have been used to study animal relationships. As their genomes become more completely known, the relationships of these two groups to other animals should become clearer. Two other small groups (also listed in Table 30.1 but not shown in Figure 30.1) have been proposed as also falling just outside the bilaterians: the xenoturbellids and acoels. Some genomic analyses, however, suggest that these animals are actually highly specialized deuterostomes.

30.5 recap

Bilaterian animals are classified into two major clades, protostomes and deuterostomes. The non-bilaterian animals comprise the ctenophores, sponges, placozoans, cnidarians, and some small groups of parasitic animals.

learning outcomes

You should be able to:

- Apply knowledge of phylogenetic relationships among major groups of animals to describe the evolution and loss of major structural features.
- Apply information about the life histories of major animal groups to design methods for studying them.

1. What is the basis for the claim that animal nervous systems evolved independently in ctenophores, cnidarians, and bilaterians?

2. Why are sponges and placozoans considered to be animals even though they lack the complex body structures found in most other animal groups?

3. The discoveries that the pelagic stages of placozoans are abundant in warm seas and that the mature stages settle on smooth surfaces suggest how these organisms might be collected and surveyed. What sampling procedures might you use to discover whether placozoans occur at a particular location along a coast?

Activity 30.3 **Diversity Simulation**
www.Life11e.com/ac30.3

investigatinglife

Q&A   Which animal groups are involved in the earliest split in the animal tree?

Three major hypotheses have been proposed in recent decades about the earliest split in the animal tree. Given the structural simplicity of sponges and their similarity to some animal outgroups (especially choanoflagellates), for many years biologists considered sponges to form the sister group of all other animals. When placozoans were first described, their structural simplicity also led other biologists to suspect that they, rather than sponges, were the sister group of other animals. In contrast, the ctenophores share some superficial similarities with cnidarians and were traditionally considered to be more closely related to them, rather than splitting earlier at the base of the animal tree. However, as biologists collected complete genomes from all these lineages, it became clearer that the simplicity of sponges and placozoans derives from a mix of ancestral retained features and secondarily derived losses of complexity. For example, although placozoans, with only four cell types and no true organs, are structurally less complex than ctenophores, this structural

simplicity is now thought to represent an evolutionary reversal in the placozoan lineage. In contrast, some of the "advanced" features of ctenophores (such as their nervous systems) were gained through independent gene duplications and specializations compared with the nervous systems of cnidarians and bilaterians. As we explored in Investigating Life: Reconstructing Animal Phylogeny from Protein-Coding Genes, phylogenetic analyses of many genes now suggest that ctenophores are the sister group of all other animals.

Future directions

As biologists collect more genomes from new species of animals, it is becoming possible to reconstruct the various gene duplications and changes that have led to the major structural and functional differences among animal groups. Having several independent origins of nervous systems is a boon for understanding how nervous systems arise and evolve. The understanding and study of these genomes is just beginning, but already complete genomes are shedding considerable light on the origins of animal complexity.

Chapter Summary 30

▶ 30.1 Some Animal Characteristics Evolved More Than Once

- Animals share a set of derived traits not found in other groups of organisms. These traits include similarities in the sequences of many of their genes, the structure of their cell junctions, and the components of their extracellular matrix.

- Several features common to most living animals, including multi-cellularity, motility, and nervous systems, evolved independently in different lineages. In particular, the decentralized **nerve nets** of ctenophores and cnidarians evolved independently of the **central nervous systems** of bilaterians.

- Patterns of embryonic development provide clues to the evolutionary relationships among animals. **Diploblastic** animals, which include the ctenophores, placozoans, and cnidarians, develop two embryonic cell layers. The **monoblastic** sponges lack embryonic cell layer differentiation, whereas the **triploblastic** animals develop three cell layers. **Review Figure 30.1**

- Differences in their patterns of early development characterize two major triploblastic clades, the **protostomes** and the **deuterostomes**.

- **Sponges** are structurally simple animals that lack well-differentiated cell layers and true organs. They have skeletons made up of silicaceous or calcareous **spicules**. They create water currents and capture food with flagellated feeding cells called **choanocytes**. **Review Figure 30.2**

▶ 30.2 Animals Diverged with Distinct Body Plans

- Animal **body plans** can be described in terms of **symmetry**, **body cavity** structure, **segmentation**, types of appendages, and nervous system development.

- A few animals have no symmetry, but most animals have either **radial symmetry** or **bilateral symmetry**. **Review Figure 30.5**

- Most bilaterally symmetrical animals exhibit **cephalization**: the concentration of sensory organs and nervous tissues in an anterior head.

- On the basis of their body cavity structure, animals can be described as **acoelomates**, **pseudocoelomates**, or **coelomates**. **Review Figure 30.6, Activity 30.1**

- Segmentation, which takes many forms, improves control of movement, as do appendages. The development of a nervous system is important for the coordination of muscular movement and the processing of sensory information.

▶ 30.3 Animals Use Diverse Forms of Movement to Feed

- **Motile** animals can move to find food; **sessile** animals stay in one place but may expend energy to move the environment and the food it contains to them

- **Filter feeders** strain small organisms and organic molecules from their environment.

- **Herbivores** consume plants, usually without killing them.

- **Predators** have morphological features such as sharp teeth, beaks, and claws that enable them to capture and subdue animal **prey**.

- **Parasites** live in or on other organisms and obtain nutrition from those **host** individuals.

- **Detritivores** consume dead organic matter and return the nutrients it contains to the ecosystem.

▶ 30.4 Animal Life Cycles Involve Trade-Offs

- The stages of an animal's life cycle may be specialized for different activities. An immature stage whose morphology is dramatically different from that of the adult stage is called a **larva**.

- Most animal life cycles have at least one **dispersal** stage. Many sessile marine animals can be grouped by the presence of one of two distinct larval dispersal stages: **trochophore** or **nauplius**. **Review Figure 30.11**

- A characteristic of an animal or a life cycle stage may improve the animal's performance in one activity but reduce its performance in another, a situation known as a trade-off.

- Parasites have complex life cycles that may involve one or more hosts and several larval stages. **Review Figure 30.12**

- In some groups of animals, asexual reproduction without fission leads to the formation of colonies composed of many genetically homogeneous, physiologically integrated individuals.

▶ 30.5 The Root of the Animal Tree Provides Clues to Early Animal Diversification

- **Ctenophores** are radially symmetrical and have two cell layers separated by an inert extracellular matrix called **mesoglea**. Their weakly differentiated organ systems are independently derived from those in eumetazoans. **Review Figure 30.16**

- **Eumetazoans** have well-developed organ systems and include all animals except ctenophores, sponges, and placozoans. The triploblastic protostomes and deuterostomes belong to a large monophyletic group called **bilaterians**. **Review Figure 30.1, Investigating Life: Reconstructing Animal Phylogeny from Protein-Coding Genes, Activity 30.2**

- **Placozoans** are asymmetrical as adults. They have only a few cell types and lack true organs, although some features of their simplicity may be secondarily derived.

- The life cycle of most **cnidarians** has two distinct stages: a sessile **polyp** stage and a motile **medusa** stage that reproduces sexually. A fertilized egg develops into a free-swimming **planula** larva, which settles to the bottom and develops into a polyp. **Review Figures 30.19, 30.23, Animation 30.1**

See Activity 30.3

Go to **LearningCurve** (in **LaunchPad**) for dynamic quizzing that helps you solidify your understanding of this chapter. **LearningCurve** adapts to your responses, giving you the practice you need to master each key concept.

Apply What You've Learned

Review

30.4 Direct development and metamorphosis each have advantages under different conditions.

30.4 Life cycles often involve trade-offs.

Original Papers: Ma, X. and J. E. Purcell. 2005. Temperature, salinity, and prey effects on polyp versus medusa bud production by the invasive hydrozoan *Moerisia lyonsi*. *Marine Biology* 147: 225–234.

Schmich, J., Y. Kraus, D. De Vito, D. Graziussi, F. Boero and S. Piraino. 2007. Induction of reverse development in two marine Hydrozoans. *International Journal of Developmental Biology* 51: 45–56.

In many ways, *Turritopsis dohrnii* is similar to many other marine cnidarians: It's a small (5 mm in diameter) hydrozoan that drifts freely around warm ocean waters. It begins its life as a motile planula larva and grows into a sessile polyp when it finds a suitable substrate. Once it reaches adulthood, it develops into a motile, bell-shaped medusa capable of sexual reproduction. However, adults of *T. dohrnii* are capable of undergoing something strange: they can revert *back* to a younger version of themselves, rejuvenating and becoming "immortal jellies" (**Figure A**).

Figure A

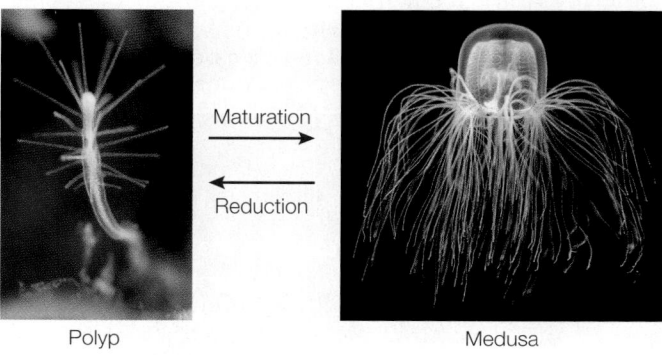

Polyp   Medusa

Individuals of *T. dohrnii* in the medusa stage undergo this reversal by reactivating genetic instructions from earlier in their life cycles to develop polyp cells. These polyp cells develop into sessile, asexual polyps, which then produce new motile, sexual medusae. Scientists curious about what triggers this unusual phenomenon hypothesized that some cnidarians may "throw their life cycle in reverse" if starving, injured, or facing stressors in their environment.

To test their hypothesis, scientists simulated unfavorable environmental conditions by incubating animals for 3 hours in different concentrations of cesium chloride (CsCl), which causes chemical disruptions in the function of various essential medusa cell types. Each treatment group included 30–50 specimens. All developmental assays, which categorize the life stages of specimens (medusa, reducing medusa, polyp), were performed multiple times to confirm the results. (A "reducing medusa" is a medusa that has started to revert back into a polyp.) Each treatment group showed a significant increase in the possibility of reverse development, compared with the control group (**Figure B**). Bars show 95% confidence intervals.

Figure B

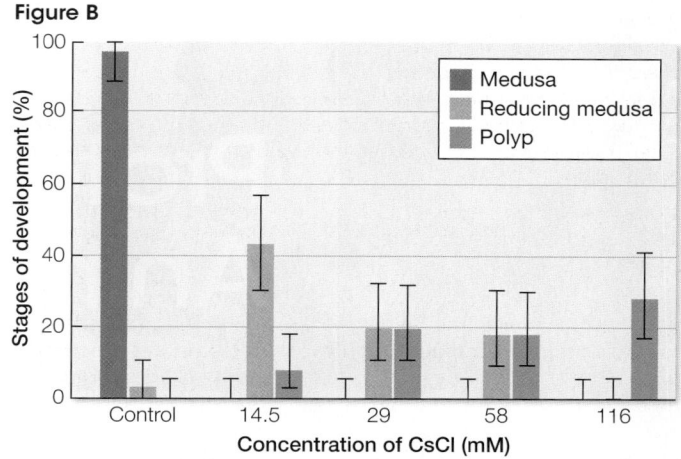

Questions

1. What conclusions can you draw from the data represented in the graph? What purpose do you think reverse development serves for some cnidarians?

2. Describe the reproductive and locomotive differences between medusa and polyp forms of cnidarians. Use this information to predict a possible trade-off that cnidarians face if they undergo reverse development.

3. In a different species of cnidarian (*Moerisia lyonsi*), individuals are also capable of maintaining polyp or medusa forms, depending on environmental conditions. One population of this species lives in an area of high food abundance, and the other lives in an area of low food abundance. Would you expect the two populations to look different? Explain your answer.

Go to **LaunchPad** for the eBook, LearningCurve, animations, activities, flashcards, and additional resources and assignments.

31

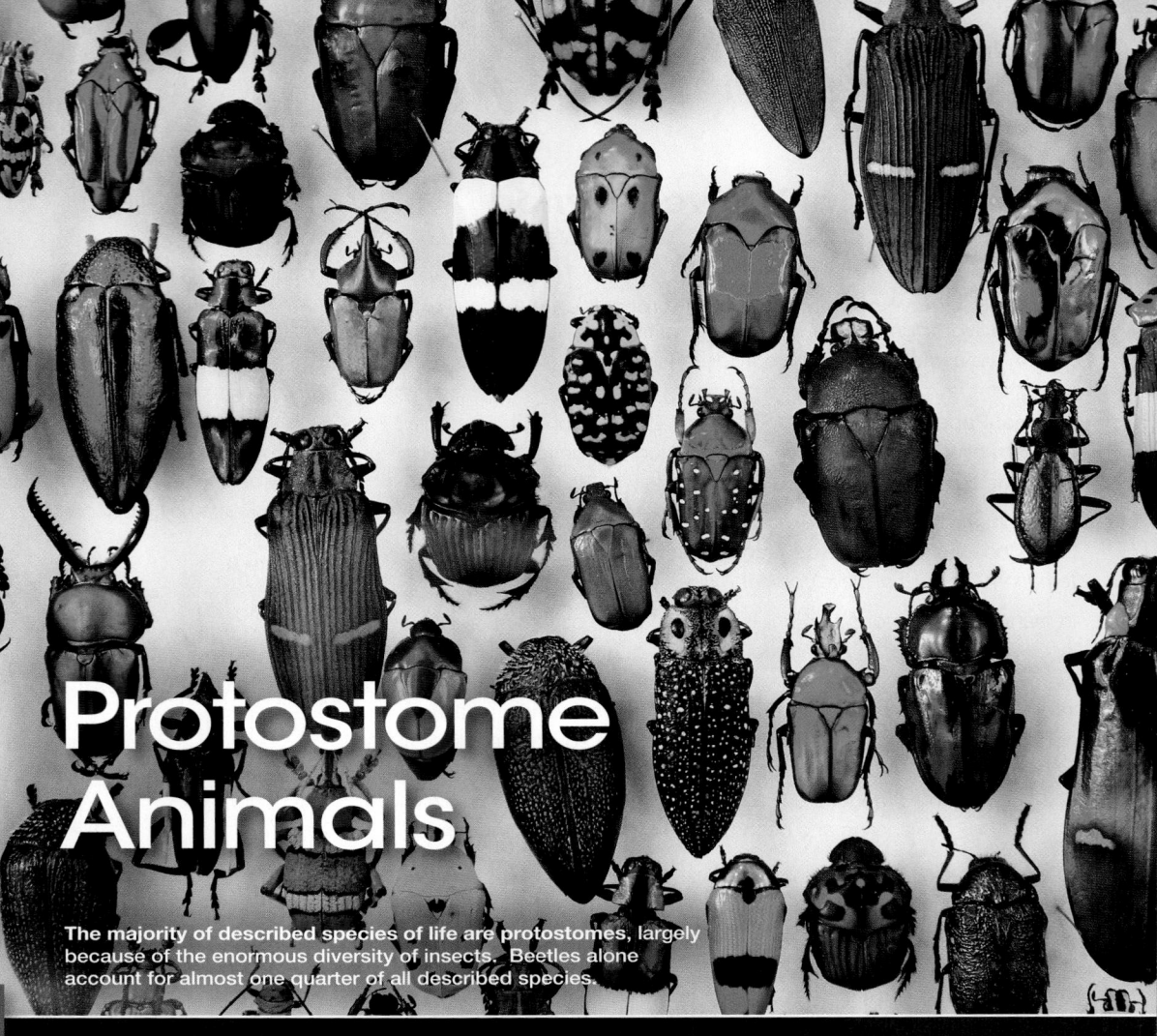

Protostome Animals

The majority of described species of life are protostomes, largely because of the enormous diversity of insects. Beetles alone account for almost one quarter of all described species.

investigatinglife

Exploring the Diversity of Life on Earth

Of the 1.8 million species of life that have been discovered and named by biologists, a large majority are protostomes. One group of protostomes, the insects, accounts for more than 1 million of these species, or more than half of all known species of living organisms. Although these numbers may seem incredibly large, they represent a relatively small fraction of the total protostome diversity that is thought to exist on Earth.

As recently as the 1980s, many biologists thought that about half of existing insect species had been described, but today they think that the number of described insect species may be a much smaller fraction of the total number of living species. Why did they change their minds?

A simple but important field study suggested that the number of existing insect species had been significantly underestimated. Knowing that the insects of tropical rainforests—the most species-rich habitat on Earth—were poorly known, entomologist Terry Erwin made a comprehensive sample of one group of insects, the beetles, in the canopies of a single species of tropical forest tree, *Luehea seemannii*, in Panama. Erwin fogged the canopies of large *L. seemannii*

trees with an insecticide and collected the insects that fell from the trees in collection nets. He collected about 1,200 species of beetles—many of them previously undescribed—from this one species of tree.

Erwin then used a set of assumptions to estimate the total number of insect species in tropical rainforests. His assumptions included estimates of the number of species of trees in these forests; the proportion of beetles that specialize on a specific species of host tree; the relative proportion of beetles to other insect groups; and the proportion of beetles that live in trees versus those that live in leaf litter on the ground. From this and similar studies, Erwin estimated that there may be 30 million or more species of insects on Earth. Although recent tests of Erwin's assumptions suggest that 30 million was an overestimate, it is clear that the vast majority of insect species remain to be discovered.

Erwin's pioneering study highlighted the fact that we live on a poorly known planet, most of whose species have yet to be named and described. Much of that undiscovered diversity occurs among several groups of protostomes.

Q&A Which groups of protostomes are thought to contain the most undiscovered species?

key concept 31.1

Protostomes Account for More Than Half of All Described Species

You may recall that the embryos of diploblastic animals (the cteno-phores, placozoans, and cnidarians, which we discussed in Chapter 30) have two cell layers: an outer ectoderm and an inner endoderm (see Key Concept 30.1). Sometime after the origin of the diploblastic animals, a third embryonic cell layer evolved: the mesoderm, which lies between the ectoderm and the endoderm. Mesoderm is found in the two major triploblastic animal clades, the protostomes and the deuterostomes. If we were to judge solely on the basis of numbers, both of species and of individuals, the protostomes would emerge as by far the more successful of the two groups.

focus your learning

- Ecdysozoans have an external covering—a cuticle—that they must shed as they grow.
- Arthropods have a rigid exoskeleton and have made use of a great variety of appendages.

As noted in Key Concept 30.1, the name "protostome" means "mouth first." In protostomes, the embryonic blastopore becomes the mouth as the animal develops. In contrast, in deuterostomes ("mouth second"), the blastopore becomes the anal opening of the gut. The protostomes are extremely varied, but they are all bilaterally symmetrical animals whose bodies exhibit two major derived traits:

1. An anterior brain that surrounds the entrance to the digestive tract

2. A ventral nervous system consisting of paired or fused longitudinal nerve cords

Other aspects of protostome body organization differ widely from group to group (**Table 31.1**). Before gene sequences were available for phylogenetic analysis, biologists considered the structure of the body cavity to be a critical feature in animal classification. But the results of genetic analyses have shown that body cavity forms have undergone considerable convergence in the course of protostome evolution. Although the common ancestor of the protostomes had a coelom, subsequent modifications of the coelom distinguish many protostome lineages. In some lineages (such as the flatworms and entoprocts), the coelom has been lost (that is, these groups reverted to an acoelomate state). Some lineages are characterized by a pseudocoel, a body cavity lined with mesoderm in which the internal organs are suspended (see Figure 30.6B). In two of the most prominent protostome clades, the coelom has been highly modified:

- Arthropods lost the ancestral condition of the coelom over the course of evolution. Their internal body cavity has become a **hemocoel**, or "blood chamber," in which fluid from an open circulatory system bathes the internal organs before returning to blood vessels.

- Most mollusks have an open circulatory system with some of the attributes of the hemocoel, but they retain vestiges of an enclosed coelom around their major organs.

table 31.1 Anatomical Characteristics of Some Major Protostome Groups

Group	Body cavity	Digestive tract	Circulatory system
Arrow worms	Coelom	Complete	None
Lophotrochozoans			
Bryozoans	Coelom	Complete	None
Entoprocts	None	Complete	None
Flatworms	None	Blind gut	None
Rotifers	Pseudocoel	Complete	None
Gastrotrichs	Pseudocoel	Complete	None
Ribbon worms	Coelom	Complete	Closed
Brachiopods	Coelom	Complete in most	Open
Phoronids	Coelom	Complete	Closed
Annelids	Coelom	Complete	Closed or open
Mollusks	Reduced coelom	Complete	Open except in cephalopods
Ecdysozoans			
Nematodes	Pseudocoel	Complete	None
Horsehair worms	Pseudocoel	Greatly reduced	None
Arthropods	Hemocoel	Complete	Open

The protostomes can be divided into two major clades—the lophotrochozoans and the ecdysozoans—largely on the basis of DNA sequence analysis (**Figure 31.1**).

 Activity 31.1 **Features of the Protostomes**
www.Life11e.com/ac31.1

Cilia-bearing lophophores and trochophores evolved among the lophotrochozoans

Lophotrochozoans derive their name from two different ciliated features: a feeding structure known as a lophophore and a free-living larval form known as a trochophore. Neither the lophophore nor the trochophore is universal to all lophotrochozoans, however.

Several distantly related groups of lophotrochozoans (including bryozoans, entoprocts, brachiopods, and phoronids) have a **lophophore**, a circular or U-shaped ring of ciliated, hollow tentacles around the mouth (**Figure 31.2**). This complex organ is used for both food collection and gas exchange. Biologists once grouped taxa that have lophophores together as "lophophorates," but it is now clear that they are not one another's closest relatives. The lophophore appears to have evolved independently at least twice, or else it is an ancestral feature of lophotrochozoans and has been lost in many groups. Nearly all animals with a lophophore are sessile as adults. They use the tentacles and cilia of the lophophore to capture small floating organisms from the water. Other sessile lophotrochozoans have less well developed tentacles that they use for the same purpose.

 Media Clip 31.1 **Feeding with a Lophophore**
www.Life11e.com/mc31.1

Figure 31.1 Phylogenetic Tree of Protostomes Two major lineages, the lophotrochozoans and the ecdysozoans, dominate the protostome tree. Some small groups are not included in this tree. The phylogenetic relationships shown here are supported mainly by genomic sequence data. Although genomic studies are contributing greatly to our knowledge of animal phylogeny, most species of protostomes have yet to be studied in detail.

▶ Activity 31.2 **Protostome Classification** www.Life11e.com/ac31.2

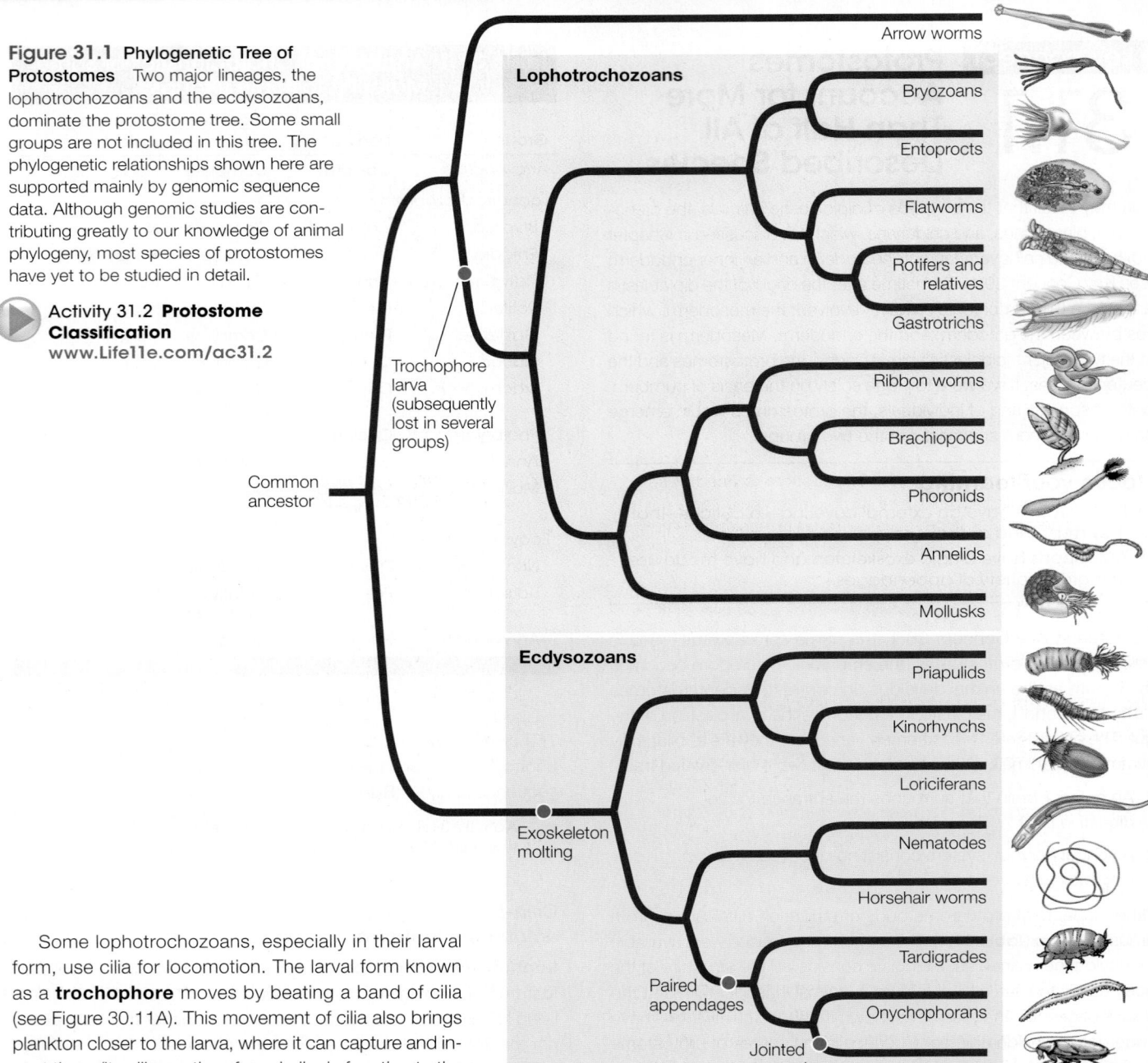

Some lophotrochozoans, especially in their larval form, use cilia for locomotion. The larval form known as a **trochophore** moves by beating a band of cilia (see Figure 30.11A). This movement of cilia also brings plankton closer to the larva, where it can capture and ingest them (its cilia are therefore similar in function to the cilia of the lophophore). Trochophore larvae are found among many of the major groups of lophotrochozoans, including the mollusks, annelids, ribbon worms, entoprocts, and bryozoans. This larval form was probably present in the common ancestor of lophotrochozoans but has been subsequently lost in several lineages.

Some lophotrochozoans (including flatworms, ribbon worms, annelids, and mollusks) exhibit spiral *cleavage in early development, in which layers of new cells spiral obliquely around the axis of the animal–vegetal pole of the early embryo. Some biologists group these taxa together as "spiralians," although phylogenetic analyses of gene sequences do not support monophyly of taxa with spiralian cleavage. Nonetheless, spiral cleavage may have been present in the lophotrochozoan ancestor and subsequently lost in several descendant lineages.

*connect the concepts As described in Key Concept 44.2, cleavage is the sequence of early cell divisions that transforms the diploid zygote into a mass of undifferentiated cells that will develop into the embryo. Different patterns of cleavage result in differential distributions of nutrients and cytoplasmic determinants within the egg cytoplasm.

Many lineages of lophotrochozoans have a wormlike body form, which means that they are bilaterally symmetrical, legless, soft-bodied, and at least several times longer than they are wide. A wormlike body form enables animals to burrow efficiently through marine sediment or soil. However, as you will see

Plumatella repens

> Bryozoans can oscillate, rotate, and retract their lophophore tentacles.

1 mm

Figure 31.2 Bryozoans Use the Lophophore to Feed The extended lophophore dominates the anatomy of the colonial bryozoans. This species inhabits fresh water, although most bryozoans are marine.

in Key Concept 31.2, the mollusks—the most familiar of the lophotrochozoans to many people—have a very different body organization.

Ecdysozoans must shed their cuticles

Ecdysozoans have an external covering, or **cuticle**, that is secreted by the underlying epidermis (the outermost cell layer). The cuticle provides these animals with both protection and support. Once formed, however, the cuticle cannot grow. How, then, can ecdysozoans increase in size? They do so by shedding, or **molting**, the cuticle and replacing it with a new, larger one.

 Media Clip 31.2 Molting a Cuticle
www.Life11e.com/mc31.2

A fossil Cambrian arthropod preserved in the process of molting shows that molting evolved more than 500 million years ago (**Figure 31.3A**). An increasingly rich array of molecular and genetic evidence, including a set of Hox genes shared by all ecdysozoans, suggests they have a single common ancestor. Thus molting of a cuticle is a trait that may have evolved only once during animal evolution.

Before an ecdysozoan molts, a new cuticle is already forming underneath the old one. Once the old cuticle is shed, the new one expands and hardens. Until it has hardened, though, the animal is vulnerable to its enemies, both because its outer surface is easy to penetrate and because an individual with a soft cuticle can move only slowly or not at all (**Figure 31.3B**).

The cuticles of some ecdysozoans, mainly arthropods, function as external skeletons, or **exoskeletons**. These exoskeletons are thickened by layers of protein and a strong, waterproof polysaccharide called **chitin**. An animal with a rigid, chitin-reinforced exoskeleton can neither move in a wormlike manner nor use cilia for locomotion. A hard exoskeleton also impedes the passage of

(A) Fossil evidence of molting

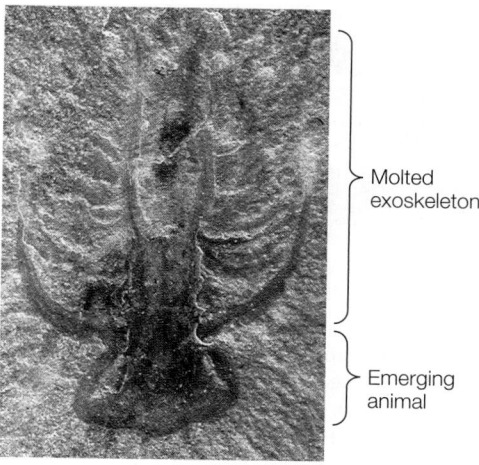

Molted exoskeleton

Emerging animal

(B) Whip scorpion (*Heterophrynus batesii*) undergoing molting

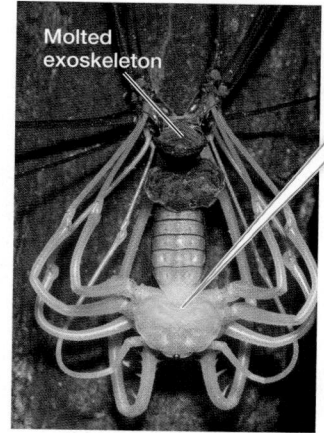

Molted exoskeleton

> The newly emerged whip scorpion's body is still soft and vulnerable.

Figure 31.3 Molting: Past and Present (A) This 500-million-year-old fossil from the Cambrian, an individual of a long-extinct arthropod species captured in the process of molting, shows that the molting process is an evolutionarily ancient trait. (B) This tailless whip scorpion has just emerged from its discarded exoskeleton. It will be highly vulnerable until its new cuticle has hardened.

oxygen and nutrients into the animal, presenting new challenges in other areas besides growth. Thus new mechanisms of locomotion and gas exchange evolved in those ecdysozoans with hard exoskeletons.

To move rapidly, an animal with a rigid exoskeleton must have body extensions that can be manipulated by muscles. Such appendages evolved in the late Precambrian, leading to the **arthropod** ("jointed foot") clade. Arthropod appendages exist in an amazing variety of forms. They serve many functions, including walking and swimming, gas exchange, food capture and manipulation, copulation, and sensory perception. Arthropods grasp food with their mouths and associated appendages and digest it internally. Their muscles are attached to the inside of the exoskeleton. Each segment has muscles that operate that segment and the appendages attached to it (**Figure 31.4**).

The arthropod exoskeleton has had a profound influence on the evolution of these animals. Encasement within a rigid body covering provides support for walking on dry land, and the waterproofing provided by chitin keeps the animal from dehydrating in dry air. Thus aquatic arthropods were, in short, excellent candidates to invade terrestrial environments. As you will see, they did so several times.

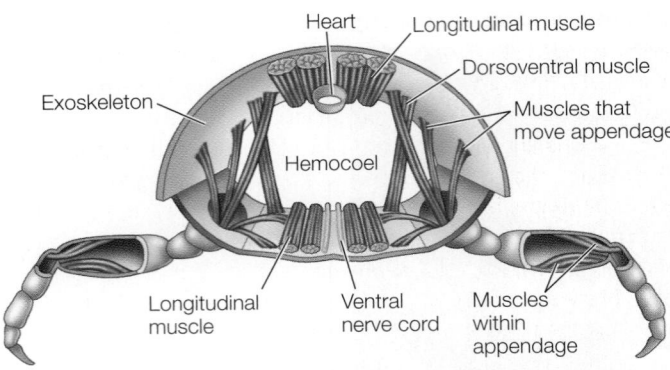

Figure 31.4 Arthropod Skeletons Are Rigid and Jointed This cross section through a thoracic segment of a generalized arthropod illustrates the arthropod body plan, which is characterized by a rigid exoskeleton with jointed appendages.

Arrow worms retain some ancestral developmental features

Nearly all triploblastic animal groups can be readily classified as either protostomes or deuterostomes, but the evolutionary relationships of one small group, the **arrow worms**, were debated for many years. The blastopore of arrow worms develops into an anus (as in deuterostomes), but they have ventral nerve cords and

spiral cleavage (as in many protostomes). Recent studies of gene sequences, however, clearly identify arrow worms as protostomes. There is still some question as to whether they are the closest relatives of the lophotrochozoans (as shown in Figure 31.1) or possibly the sister group of all other protostomes.

The arrow worm body is divided into three compartments: head, trunk, and tail (**Figure 31.5**). The body is transparent or translucent. Most arrow worms swim in the open sea. A few species live on the seafloor. Their abundance as fossils indicates that they were common more than 500 million years ago. The 180 or so known living species of arrow worms are small enough—ranging from 3 millimeters to 12 centimeters in length—that their gas exchange and waste excretion requirements are met by diffusion through the body surface. They lack a circulatory system; wastes and nutrients are moved around the body in the coelomic fluid, which is propelled by cilia that line the coelom.

Arrow worms are hermaphroditic; that is, each individual produces both male and female gametes. In most species, eggs are fertilized internally following elaborate courtship between two individuals, although some species can self-fertilize. Miniature adults hatch directly from the eggs; these animals have no distinct larval stage.

Arrow worms are stabilized in the water by means of one or two pairs of lateral fins and a tail fin. They are major predators of planktonic organisms in the open ocean, ranging in size from small protists to young fish as large as the arrow worms themselves. An arrow worm typically lies motionless in the water until water movement signals the approach of prey. The arrow worm then darts forward and uses the stiff spines adjacent to its mouth to grasp its prey.

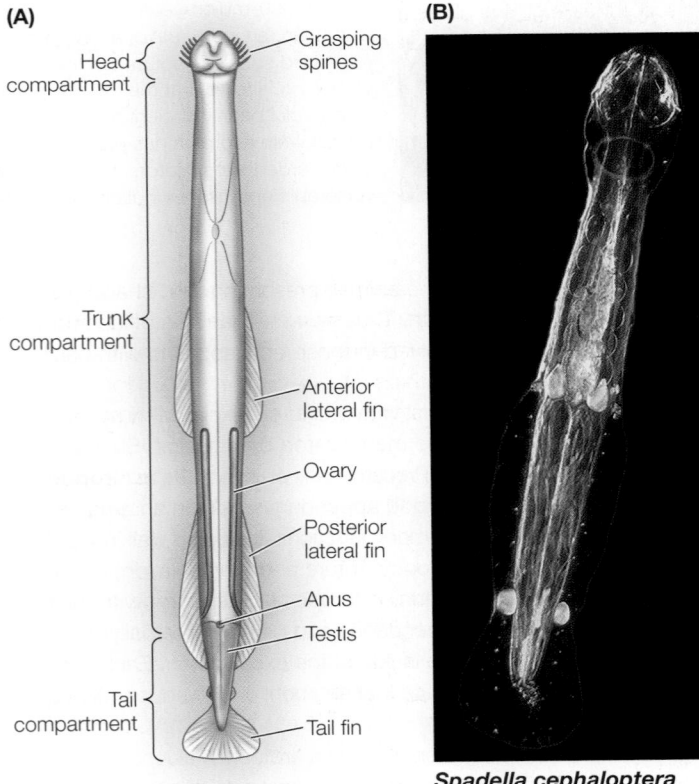

Spadella cephaloptera

Figure 31.5 An Arrow Worm Arrow worms have a three-part body organization. Their fins and grasping spines are adaptations for a predatory lifestyle. Individuals are hermaphroditic, producing both eggs in an ovary and sperm in a testis.

▶ 31.1 recap

The shared derived traits of protostomes include a blastopore that develops into a mouth (except in arrow worms), an anterior brain, and a ventral nervous system. Several lophotrochozoan groups are characterized by a filter-feeding ciliated structure known as a lophophore or by cilia-bearing larvae known as trochophores. Ecdysozoans, which have a body covering known as a cuticle, must molt periodically in order to grow.

learning outcomes

You should be able to:

- Synthesize the consequences of having a hard body covering versus a soft body for life as an animal.
- Describe major evolutionary innovations among protostomes that are related to life in aquatic versus terrestrial environments.

1. How does an animal's body covering influence the way it exchanges gases, feeds, and moves?
2. What features make arthropods well adapted for colonizing terrestrial environments?

In the next section we continue our survey of the protostomes with a more detailed look at the major groups of lophotrochozoans and the diverse body forms that are found among them.

key concept

31.2 Many Lophotrochozoans Have Ciliated Feeding Structures or Life Stages

Lophotrochozoans come in a variety of sizes and shapes, ranging from relatively simple animals with a blind gut (that is, a gut with only one opening) and no internal transport system to animals with a complete gut (having separate entrance and exit openings) and a complex internal transport system. They include some species-rich groups, such as flatworms, annelids, and mollusks. A number of these groups have wormlike bodies, but the lophotrochozoans encompass a wide variety of morphologies, including a few groups with external shells. Some lophotrochozoan groups have only recently been discovered by biologists.

focus your learning

- Flatworms have flat bodies because they lack specialized organs for transporting oxygen to their internal tissues.
- Most annelids lack a rigid external protective covering and instead have a thin, permeable body wall that serves as a surface for gas exchange.
- Modifications of the shell, mantle, and foot have allowed mollusks to diversify in a wide variety of environments.

Most bryozoans and entoprocts live in colonies

Most of the 5,500 known species of **bryozoans** ("moss animals") and 170 known species of **entoprocts** (meaning "anus inside") are colonial animals that live in a "house" made of material secreted by the external body wall. The colonial species are sessile, but the few solitary species can slowly move around in their environment. Almost all bryozoans and entoprocts are marine, although a few species occur in fresh or brackish water.

A bryozoan colony consists of many small (1–2 mm) individuals connected by strands of tissue along which nutrients can be moved. Bryozoan colonies can grow to contain more than 2 million individuals, all stemming from the asexual reproduction of the colony's founder. Rocks in coastal regions in many parts of the world are covered with luxuriant growths of bryozoans. Some bryozoans create miniature reefs in shallow waters. In some species, the individual colony members are differentially specialized for feeding, reproduction, defense, or support. Individual bryozoans in a colony are able to oscillate their lophophore to increase contact with prey. They can also retract it into their "house" (see Figure 31.2).

Bryozoans can reproduce sexually by releasing sperm into the water, which carries the sperm to other individuals. Eggs are fertilized internally; developing embryos are brooded before they exit as

larvae to seek suitable sites for attachment to the substrate. Entoprocts can also reproduce asexually. Some species of entoprocts release unfertilized eggs into the water for fertilization, whereas other species brood their developing young as bryozoans do.

Bryozoans and entoprocts differ in the placement of the anus. In bryozoans, the anus is located outside the ring of tentacles that make up the lophophore, whereas the anus of entoprocts is located in the center of this ring. The lophophores of the two groups also function differently: food particles are carried from the tips to the bases of the tentacles in bryozoans, but from the bases to the tips of the tentacles in entoprocts. Entoprocts lack a coelom, whereas bryozoans have a three-part coelom.

Flatworms, rotifers, and gastrotrichs are structurally diverse relatives

Flatworms, rotifers, gastrotrichs, and their close relatives are a structurally diverse group of organisms whose relationships to one another have been hypothesized only recently. Recent genomic studies show that this monophyletic lophotrochozoan group includes both acoelomate subgroups (e.g., the flatworms) and pseudocoelomate subgroups (e.g., the rotifers and gastrotrichs), and yet the closest relatives of this group—the bryozoans and entoprocts—are coelomate and acoelomate, respectively. Thus this group provides an example of the evolutionary convergence in body cavity form that we described in Key Concept 31.1.

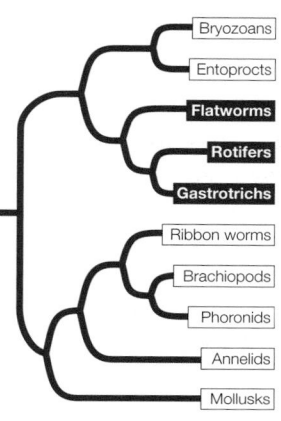

FLATWORMS   Flatworms lack specialized organs for transporting oxygen to their internal tissues. In the absence of a gas transport system, each cell must be near a body surface, a requirement met by the dorsoventrally flattened body form that gives these animals their common name. The digestive tract of a free-living flatworm consists of a mouth opening into a blind gut. The gut is often highly branched, forming intricate patterns that increase the surface area available for the absorption of nutrients. Some small free-living flatworms are cephalized, with a head bearing chemoreceptor organs, two simple eyes, and a tiny brain composed of anterior thickenings of the longitudinal nerve cords. Free-living flatworms glide over surfaces on a layer of mucus, powered by broad bands of cilia (**Figure 31.6A**).

Although many flatworms are free-living, most flatworm species are parasites. Of the parasitic species, most are endoparasites. There are also flatworms that feed externally on animal tissues (living or dead), and some graze on plants. A likely evolutionary transition was from feeding on dead organisms to feeding on the body surfaces of dying hosts to invading and consuming parts of healthy hosts.

Most of the 30,000 known species of living flatworms are tapeworms and flukes; members of these two groups are endoparasites, particularly of vertebrates (**Figure 31.6B**). Because they absorb digested food from the digestive tracts of their hosts, many endoparasitic flatworms lack digestive tracts of their own. Some

Figure 31.6 Flatworms Include Both Parasites and Free-Living Forms
(A) Some flatworm species, such as this Pacific marine flatworm, are free-living. **(B)** The fluke diagrammed here lives endoparasitically in the gut of sea urchins and is typical of endoparasitic flatworms. Because their hosts provide all the nutrition they need, these internal parasites do not require elaborate feeding or digestive organs and can devote most of their bodies to reproduction.

(A) *Eurylepta californica,* a free-living flatworm

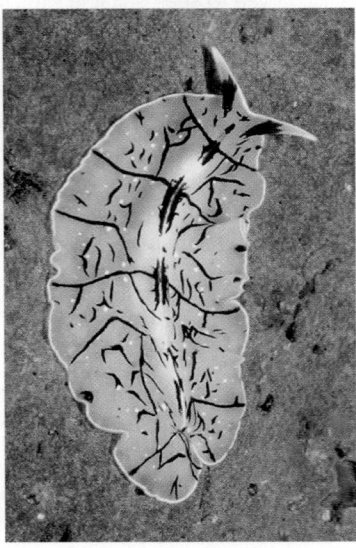

(B) Diagram of a typical parasitic flatworm

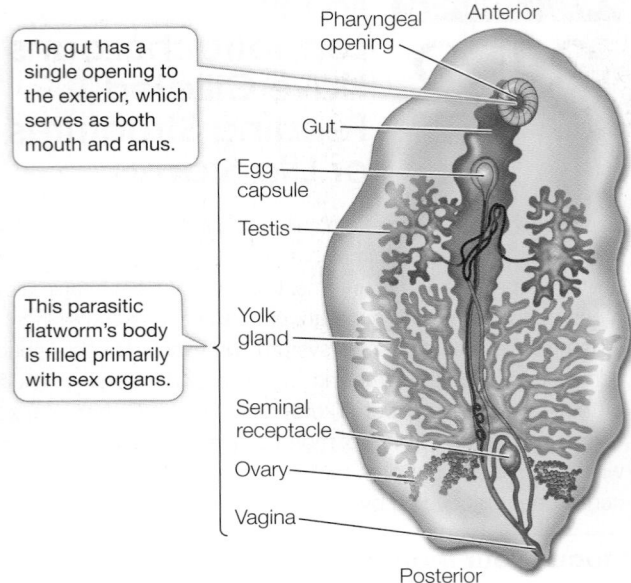

The gut has a single opening to the exterior, which serves as both mouth and anus.

This parasitic flatworm's body is filled primarily with sex organs.

Pharyngeal opening
Anterior
Gut
Egg capsule
Testis
Yolk gland
Seminal receptacle
Ovary
Vagina
Posterior

cause serious human diseases, such as schistosomiasis, which is common in parts of Asia, Africa, and South America. The species that causes this devastating disease has a complex life cycle involving both freshwater snails and mammals as hosts. Members of another flatworm group, the monogeneans, are ectoparasites of fishes and other aquatic vertebrates. The turbellarians include most of the free-living species.

ROTIFERS   Most species of **rotifers** are tiny (50–500 μm long)—smaller than some ciliate protists—but they have specialized internal organs (**Figure 31.7A and B**). A complete gut passes from an anterior mouth to a posterior anus; the body cavity is a pseudocoel that functions as a hydrostatic skeleton. Rotifers typically propel themselves through the water by means of rapidly beating cilia rather than by muscular contraction.

 Media Clip 31.3 Rotifer Feeding
www.Life11e.com/mc31.3

The most distinctive organ of rotifers is a conspicuous ciliated organ called the corona, which surmounts the head of many species. Coordinated beating of the cilia sweeps particles of organic matter from the water into the animal's mouth and down to a complicated structure called the mastax, in which food is ground into small pieces. By contracting muscles around the pseudocoel, a few rotifer species that prey on protists and small animals can protrude the mastax through the mouth and seize small objects with it.

Most of the known species of rotifers live in fresh water. Some species rest on the surfaces of mosses or lichens in a desiccated, inactive state until it rains. When rain falls, they absorb water and become mobile, feeding in the films of water that temporarily cover the plants. Most rotifers live no longer than a few weeks.

Both males and females are found in some species of rotifers, but only females are known among the bdelloid rotifers (the *b* in "bdelloid" is silent). Biologists have concluded that the bdelloid rotifers may have existed for tens of millions of years without regular sexual reproduction. Lack of genetic recombination generally leads to the buildup of deleterious mutations, so long-term *asexual

reproduction typically leads to extinction. Recent studies, however, have indicated that bdelloid rotifers may avoid this problem by picking up fragments of genes from their environment during the desiccation–rehydration cycle, which allows genetic recombination among individuals in the absence of direct sexual exchange.

---

*connect the concepts   The consequences of asexual reproduction, and the reasons that it commonly leads to a buildup of deleterious mutations, are discussed in Key Concept 20.5.

A few highly reduced lineages appear to have descended from the free-living rotifers. The spiny-headed worms are parasites with complex life cycles, often parasitizing several animal hosts (**Figure 31.7C**). The jaw worms are tiny marine organisms that glide between sand grains in shallow marine environments. Although spiny-headed worms and jaw worms are structurally quite distinct, molecular analyses have revealed that both groups are essentially highly modified rotifers.

GASTROTRICHS   The 800 known species of **gastrotrichs** (also called "hairy backs") are abundant, tiny (0.05–3 mm) animals that live in marine sediments, in fresh waters, and in the water films that surround grains of soil. Their transparent bodies have a flat ventral surface that is covered with cilia (**Figure 31.7D**). Most species are simultaneous hermaphrodites, with both male and female reproductive organs, although the male organs have been greatly reduced or lost in some species that reproduce asexually.

Ribbon worms have a long, protrusible feeding organ

Ribbon worms (nemerteans) have simple nervous and excretory systems similar to those of flatworms. Unlike flatworms, however, they have a closed circulatory system and a complete digestive tract with a mouth at one end and an anus at the other. Small ribbon worms move slowly by beating their cilia. Larger ones employ waves of muscle contraction to move over the surface of sediments or to burrow into them.

(A) Rotifer anatomy

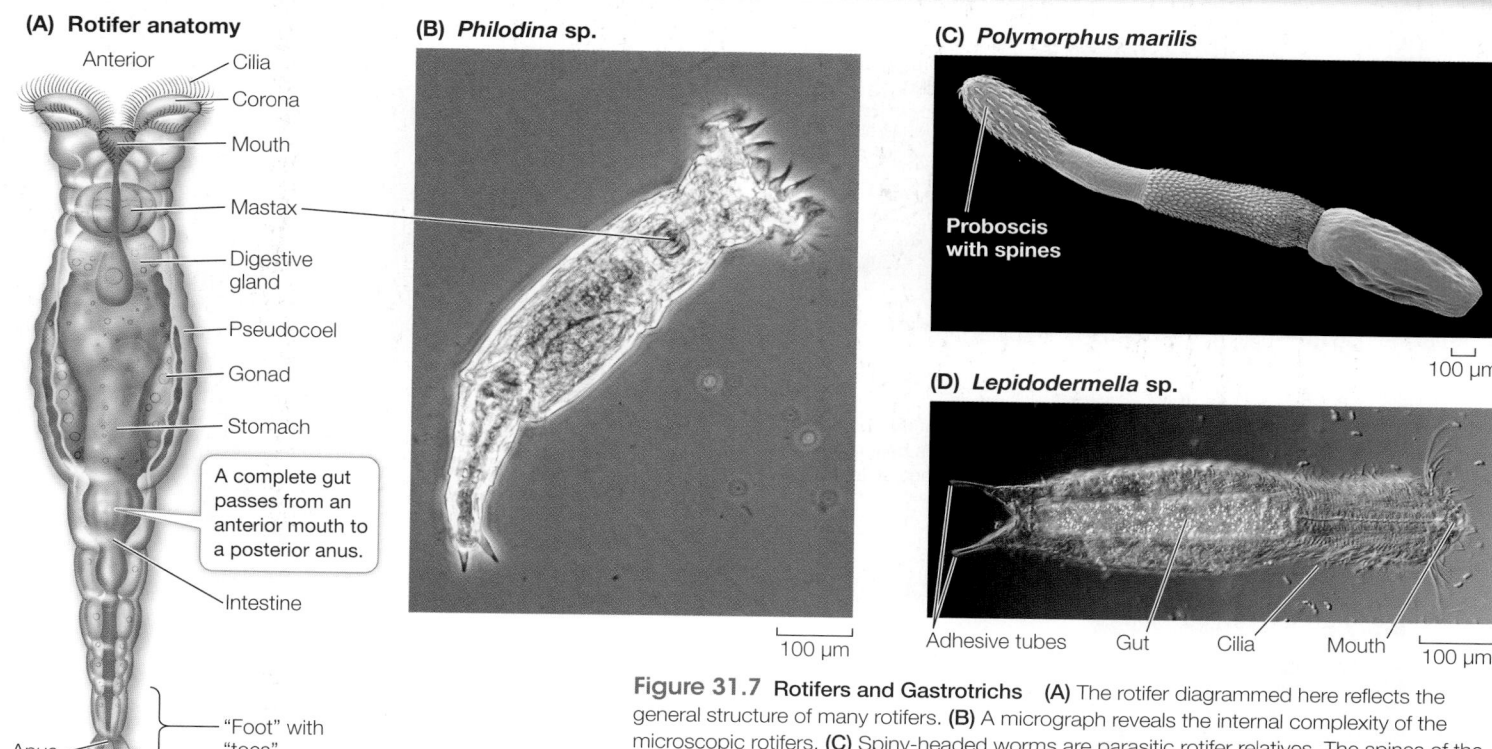

Anterior
Cilia
Corona
Mouth
Mastax
Digestive gland
Pseudocoel
Gonad
Stomach

A complete gut passes from an anterior mouth to a posterior anus.

Intestine

"Foot" with "toes"

Anus

Posterior

(B) Philodina sp.

100 µm

(C) Polymorphus marilis

Proboscis with spines

100 µm

(D) Lepidodermella sp.

Adhesive tubes   Gut   Cilia   Mouth

100 µm

Figure 31.7 Rotifers and Gastrotrichs **(A)** The rotifer diagrammed here reflects the general structure of many rotifers. **(B)** A micrograph reveals the internal complexity of the microscopic rotifers. **(C)** Spiny-headed worms are parasitic rotifer relatives. The spines of the proboscis anchor the animal to the organs of its host. **(D)** Gastrotrichs superficially resemble rotifers but have flattened ventral surfaces covered with cilia, as flatworms do.

Within the body of nearly all of the 1,200 known species of ribbon worms is a fluid-filled cavity called the rhynchocoel, within which lies a hollow, muscular **proboscis**. The proboscis, which is the worm's feeding organ, may extend much of the length of the body. Contraction of the muscles surrounding the rhynchocoel causes the proboscis to evert explosively through an anterior pore (**Figure 31.8A**). The proboscis may be armed with sharp stylets that pierce prey and discharge paralysis-causing toxins into the wound.

 Media Clip 31.4 Explosive Extrusion of Ribbon Worm Proboscis
www.Life11e.com/mc31.4

Most ribbon worm species are marine, although there are species that live in fresh water or on land. Most species are less than 20 centimeters long, but individuals of some species reach 20 meters or more. Some genera feature species that are conspicuous and brightly colored (**Figure 31.8B**). Recent molecular analyses suggest that ribbon worms may be most closely related to the brachiopods and phoronids.

Brachiopods and phoronids use lophophores to extract food from the water

Recall that the bryozoans and entoprocts use a lophophore to feed. Brachiopods and phoronids also feed using a lophophore, but this

[Cladogram listing: Bryozoans, Entoprocts, Flatworms, Rotifers, Gastrotrichs, **Ribbon worms**, Brachiopods, Phoronids, Annelids, Mollusks]

structure may have evolved separately in these groups. Although neither the brachiopods nor the phoronids are represented by many living species, the brachiopods (which have hard external shells and thus leave an excellent fossil record) are known to have been much more abundant in the past.

BRACHIOPODS Brachiopods (lampshells) are solitary marine animals. They have a rigid shell that is divided into two parts connected by a ligament (**Figure 31.9**). The two halves can be pulled shut to protect the soft body. Brachiopods superficially resemble bivalve mollusks, but shells have evolved independently in the two groups. The two halves of the brachiopod shell are dorsal and ventral, rather than lateral as in bivalves. The lophophore is located within the shell. The beating of cilia on the lophophore draws water into the slightly opened shell. Food is trapped in the lophophore and directed to a ridge, along which it is transferred to the mouth.

Most brachiopods are 4 to 6 centimeters long. They live attached to a solid substrate or embedded in soft sediments. Most species are attached by means of a short, flexible stalk that holds the animal above the substrate. Gases are exchanged across body surfaces, especially the tentacles of the lophophore. Most brachiopods release their gametes into the water, where they are fertilized. The larvae remain among the plankton for only a few days before they settle and develop into adults.

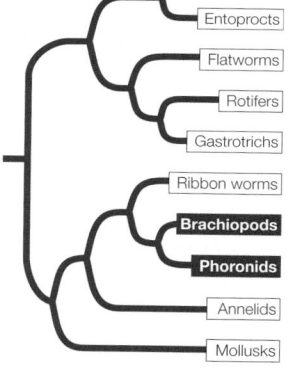

[Cladogram listing: Bryozoans, Entoprocts, Flatworms, Rotifers, Gastrotrichs, Ribbon worms, **Brachiopods**, **Phoronids**, Annelids, Mollusks]

(A)

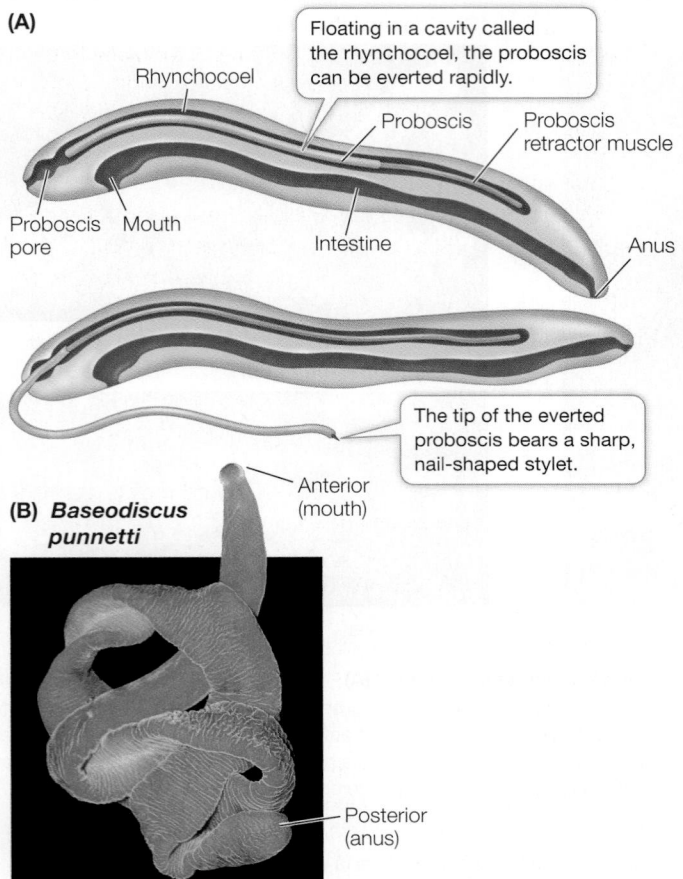

Rhynchocoel

Floating in a cavity called the rhynchocoel, the proboscis can be everted rapidly.

Proboscis

Proboscis retractor muscle

Proboscis pore

Mouth

Intestine

Anus

The tip of the everted proboscis bears a sharp, nail-shaped stylet.

Anterior (mouth)

(B) *Baseodiscus punnetti*

Posterior (anus)

Figure 31.8 Ribbon Worms **(A)** The proboscis is the ribbon worm's feeding organ. **(B)** This large marine nemertean is found in harbors and bays along the Pacific Coast of North America. Its proboscis is not everted in this photograph.

Brachiopods reached their peak abundance and diversity in Paleozoic and Mesozoic times. More than 26,000 fossil species have been described. Only about 450 species are known to survive, but they remain common in some marine environments.

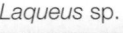

Laqueus sp.

Lophophore ring

Tentacles

Figure 31.9 A Brachiopod's Lophophore The lophophore of this North Pacific brachiopod can be seen between the valves of its shell.

(A) *Phoronis californica*

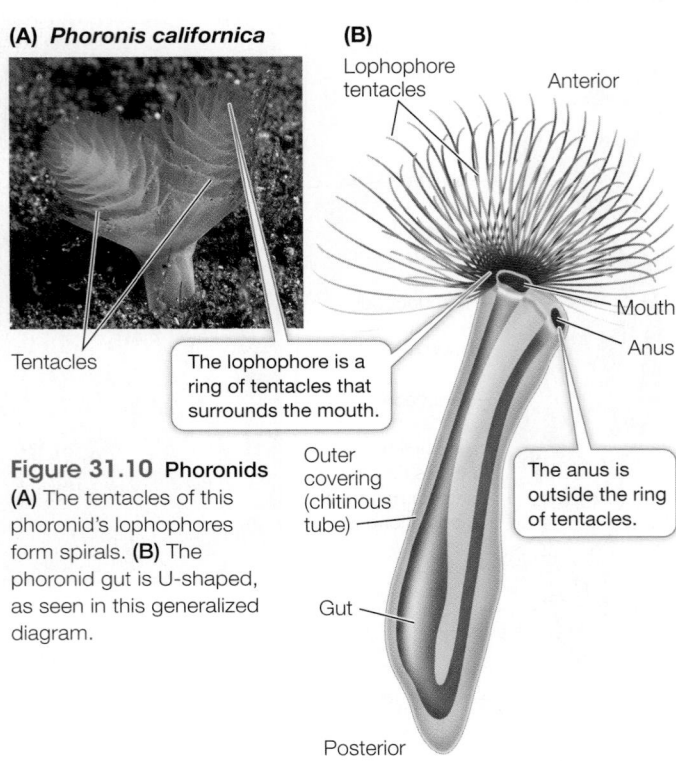

(B)

Lophophore tentacles

Anterior

Tentacles

The lophophore is a ring of tentacles that surrounds the mouth.

Mouth

Anus

Outer covering (chitinous tube)

The anus is outside the ring of tentacles.

Gut

Posterior

Figure 31.10 Phoronids **(A)** The tentacles of this phoronid's lophophores form spirals. **(B)** The phoronid gut is U-shaped, as seen in this generalized diagram.

PHORONIDS The ten known species of **phoronids** are small (5–25 cm long), sessile worms that live in muddy or sandy sediments or attached to rocky substrates. Phoronids are found in marine waters from the intertidal zone to about 400 meters deep. They secrete tubes made of chitin, within which they live, and have a U-shaped gut with the anus located outside the lophophore (**Figure 31.10**). Their cilia drive water into the top of the lophophore, and the water exits through the narrow spaces between the tentacles. Suspended food particles are caught and transported to the mouth by ciliary action. Some species release eggs into the water, where they are fertilized, but other species produce large eggs that are fertilized internally and retained in the parent's body, where they are brooded until they hatch.

Annelids have segmented bodies

The wormlike bodies of **annelids** are clearly segmented. As described in Key Concept 30.2, segmentation allows an animal to move different parts of its body independently, giving it much better control of its movement. The earliest segmented worms, preserved as fossils from the middle Cambrian, were burrowing marine annelids.

In most large annelids, the coelom in each segment is isolated from those in other segments (**Figure 31.11**). A separate nerve center called a ganglion (plural ganglia)

Bryozoans

Entoprocts

Flatworms

Rotifers

Gastrotrichs

Ribbon worms

Brachiopods

Phoronids

Annelids

Mollusks

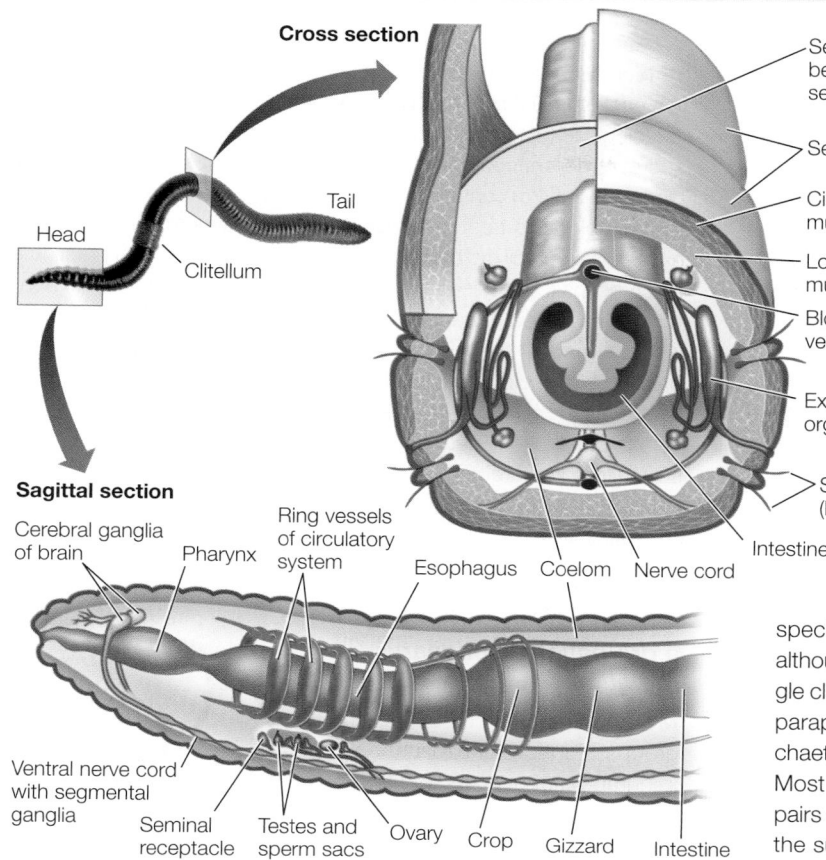

Cross section

- Septum between segments
- Segments
- Circular muscle
- Longitudinal muscle
- Blood vessel
- Excretory organ
- Setae (bristles)
- Intestine

Head — Clitellum — Tail

Sagittal section

- Cerebral ganglia of brain
- Pharynx
- Ring vessels of circulatory system
- Esophagus
- Coelom
- Nerve cord
- Ventral nerve cord with segmental ganglia
- Seminal receptacle
- Testes and sperm sacs
- Ovary
- Crop
- Gizzard
- Intestine

Figure 31.11 Annelids Have Many Body Segments The segmented structure of annelids such as this earthworm is apparent both externally and internally. Many organs are repeated serially.

controls each segment; nerve cords that connect the ganglia coordinate their functioning. Most annelids lack a rigid external protective covering; instead they have a thin, permeable body wall that serves as a general surface for gas exchange. Most annelids are thus restricted to moist environments because they lose body water rapidly in dry air. The approximately 19,000 described species live in marine, freshwater, and moist terrestrial environments.

POLYCHAETES   More than half of all annelid species are commonly known as **polychaetes** ("many hairs"), although this is a descriptive term rather than the name of a single clade. Recent molecular studies indicate that polychaetes are paraphyletic with respect to the remaining annelids. Most polychaetes are marine, and many live in burrows in soft sediments. Most of them have one or more pairs of eyes and one or more pairs of tentacles, with which they capture prey or filter food from the surrounding water, at the anterior end of the body (**Figure 31.12A**; see also Figure 30.7A). In some species, the body wall of most segments extends laterally as a series of thin outgrowths called parapodia. The parapodia function in gas exchange, and

(A) *Spirographis spallanzanii*

(B) *Riftia sp.*

(C) *Lumbricus terrestris*

(D) *Hirudo medicinalis*

Figure 31.12 Diversity among the Annelids **(A)** Fan worms, or "feather duster worms," are sessile marine polychaetes that grow in masses, filtering food from the water with their tentacles. This individual has been removed from its chitinous tube. **(B)** Pogonophorans live around hydrothermal vents deep in the ocean. Their tentacles can be seen protruding from their chitinous tubes. **(C)** Earthworms, like all oligochaetes, are hermaphroditic; when they copulate, each individual donates and receives sperm. **(D)** The medicinal leech has been a tool of physicians and healers for centuries. Even today, leeches have uses in clinical practice.

some species use them to move. Stiff bristles called setae protrude from each parapodium, forming temporary contact with the substrate and preventing the animal from slipping backward when its muscles contract.

Members of one polychaete clade, the **pogonophorans**, secrete tubes made of chitin and other substances, in which they live (**Figure 31.12B**). Pogonophorans have lost their digestive tract (they have no mouth or gut). So how do they obtain nutrition? Part of the answer is that pogonophorans can take up dissolved organic matter directly from the sediments in which they live or from the surrounding water. Much of their nutrition, however, is provided by endosymbiotic bacteria that the pogonophorans house in a specialized organ known as the trophosome. These bacteria oxidize hydrogen sulfide and other sulfur-containing compounds, fixing carbon from methane in the process. Uptake of the hydrogen sulfide, methane, and oxygen used by the bacteria is facilitated by hemoglobin in the pogonophorans' tentacles. It is this hemoglobin that gives the tentacles their red coloration.

Pogonophorans were not discovered until early in the twentieth century, when the first species were discovered on the seafloor at depths of up to a few hundred meters. In recent decades, deep-sea explorers have found them living many thousands of meters below the ocean surface. In these deep oceanic sediments, they may reach densities of many thousands per square meter. About 160 species have been described. The largest and most remarkable pogonophorans are 2 meters or more in length and live near deep-sea hydrothermal vents—volcanic openings in the seafloor through which hot, sulfide-rich water pours. The methane and hydrogen sulfide from these vents provide the raw materials for carbon fixation by the pogonophorans' endosymbiotic bacteria.

CLITELLATES The approximately 3,000 described species of **clitellates**, which form a well-supported clade within the annelids, are found in freshwater, marine, and terrestrial environments. The clitellates appear to be phylogenetically nested among various groups of polychaetes, although the exact relationships are not yet clear. There are two major groups of clitellates, the oligochaetes and the leeches.

Oligochaetes ("few hairs") have no parapodia, eyes, or anterior tentacles, and they have only four pairs of setae bundles per segment. Earthworms—the most familiar oligochaetes—burrow in and ingest soil, from which they extract food particles. All oligochaetes are hermaphroditic. Sperm are exchanged simultaneously between two copulating individuals (**Figure 31.12C**). Eggs and sperm are deposited in a cocoon outside the adult's body. Fertilization occurs within the cocoon after it is shed, and when development is complete, miniature worms emerge and immediately begin independent life.

Leeches, like oligochaetes, lack parapodia and tentacles. The coelom of leeches is not divided into compartments; the coelomic space is largely filled with undifferentiated tissue. Groups of segments at each end of the body are modified to form suckers, which serve as temporary anchors that aid the leech in its movement. With its posterior sucker attached to a substrate, the leech extends its body by contracting its circular muscles. The anterior sucker is then attached, the posterior one detached, and the leech shortens itself by contracting its longitudinal muscles. Leeches live in freshwater or terrestrial habitats.

Most leeches are ectoparasites that feed by making an incision in a host, from which blood flows. A leech can ingest so much blood in a single feeding that its body may enlarge severalfold. The leech secretes an anticoagulant into the wound that keeps the host's blood flowing. For centuries, medical practitioners employed leeches to draw blood to treat diseases they believed were caused by an excess of blood or by "bad blood." Although most leeching practices (such as inserting a leech in a person's throat to alleviate swollen tonsils) have been abandoned, *Hirudo medicinalis* (the medicinal leech; **Figure 31.12D**) is used today to reduce fluid pressure and prevent blood clotting in damaged tissues, to eliminate pools of coagulated blood, and to prevent scarring. The anticoagulants of certain other leech species also contain anesthetics and blood vessel dilators and are being studied for possible medical uses.

 Media Clip 31.5 **Leeches Feeding on Blood**
www.Life11e.com/mc31.5

Mollusks have undergone a dramatic evolutionary radiation

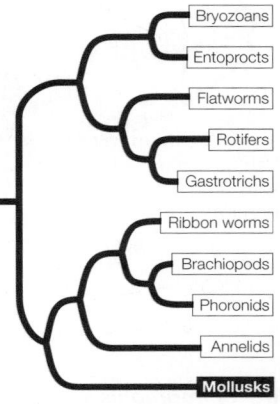

Mollusks are the most diverse group of lophotrochozoans, both in numbers of species and in the environments they occupy. Although the major groups of mollusks differ dramatically in morphology, they all share the same three major body components: a foot, a visceral mass, and a mantle (**Figure 31.13A**).

1. The molluscan foot is a large, muscular structure that originally was both an organ of locomotion and a support for the internal organs. In squids and octopuses, the foot has been modified to form arms and tentacles borne on a head with complex sensory organs. In other groups, such as clams, the foot is a burrowing organ. In some groups the foot is greatly reduced.

2. The heart and the digestive, excretory, and reproductive organs are concentrated in a centralized, internal **visceral mass**.

3. The **mantle** is a fold of tissue that covers the organs of the visceral mass. The mantle secretes the hard, calcareous shell that is typical of many mollusks.

In most mollusks, the mantle extends beyond the visceral mass to form a mantle cavity. Within this cavity lie gills that are used for gas exchange. When cilia on the gills beat, they create a current of water. The tissue of the gills, which is highly vascularized (contains many blood vessels), takes up oxygen from the water and releases carbon dioxide. Many mollusk species use their gills as filter-feeding devices, whereas others feed using a rasping structure known as a **radula** to scrape algae from rocks. In some mollusks, such as the marine cone snails, the radula has been modified into a drill or poison dart.

In all mollusks except cephalopods, the blood vessels do not form a closed system. Blood and other fluids empty into a large, fluid-filled hemocoel, through which fluids move and deliver oxygen

(A) Generalized molluscan body plan

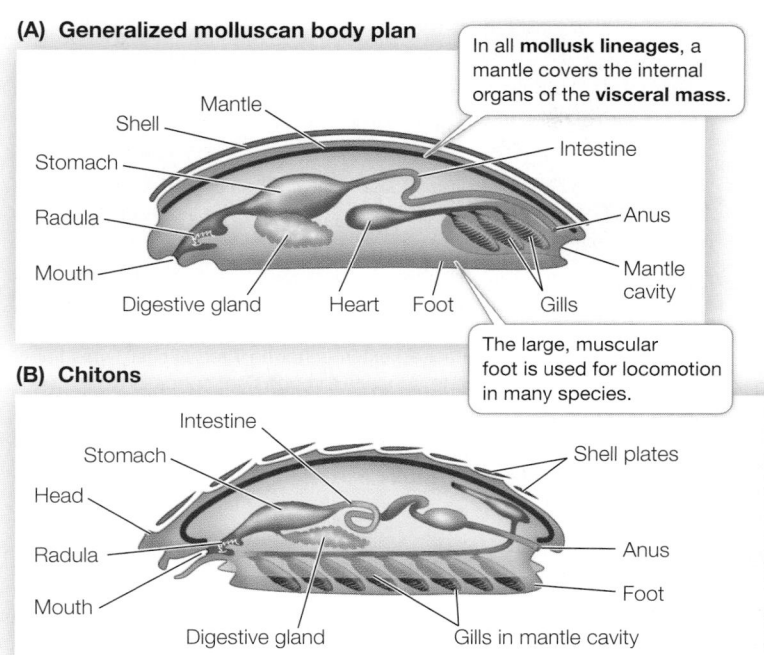

In all **mollusk lineages**, a mantle covers the internal organs of the **visceral mass**.

Mantle
Shell
Stomach
Radula
Mouth
Digestive gland
Heart
Foot
Intestine
Anus
Mantle cavity
Gills

The large, muscular foot is used for locomotion in many species.

Figure 31.13 Organization and Diversity of Molluscan Bodies **(A)** The major molluscan groups display different variations on a body plan that includes three major components: a foot, a visceral mass of internal organs, and a mantle. In many species, the mantle secretes a calcareous shell. **(B)** Chitons have eight overlapping calcareous plates surrounded by a girdle. **(C)** Most gastropods have a single dorsal shell, into which they can retreat for protection. **(D)** Bivalves get their name from their two hinged shells, which can be tightly closed. **(E)** Cephalopods are active predators; they use their arms and tentacles to capture prey. This squid has an internal shell but no external shell.

(B) Chitons

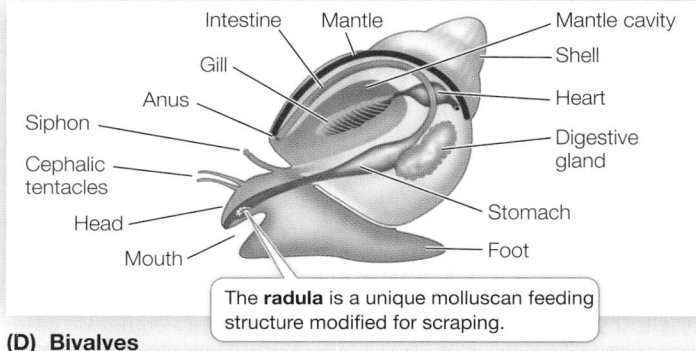

Intestine
Stomach
Head
Radula
Mouth
Digestive gland
Shell plates
Anus
Foot
Gills in mantle cavity

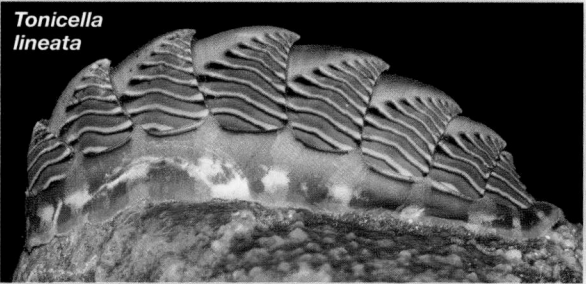

Tonicella lineata

(C) Gastropods

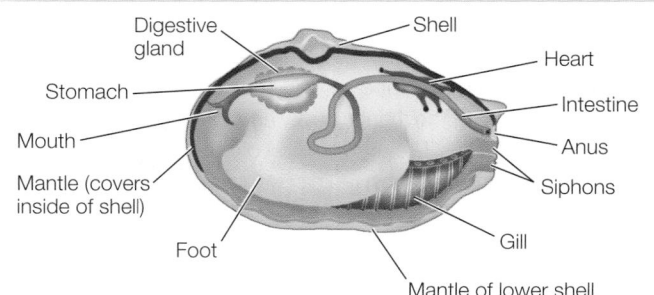

Intestine
Gill
Anus
Siphon
Cephalic tentacles
Head
Mouth
Mantle
Mantle cavity
Shell
Heart
Digestive gland
Stomach
Foot

The **radula** is a unique molluscan feeding structure modified for scraping.

Helix sp.

(D) Bivalves

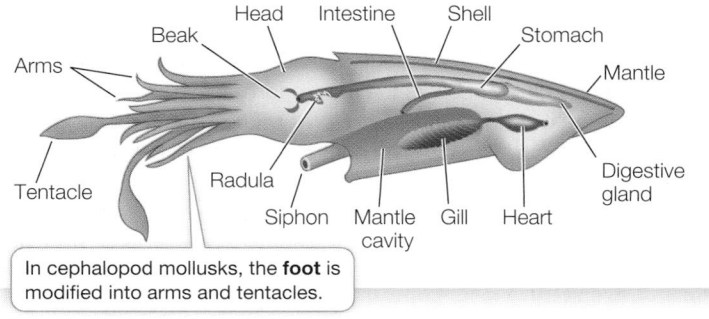

Digestive gland
Stomach
Mouth
Mantle (covers inside of shell)
Foot
Shell
Heart
Intestine
Anus
Siphons
Gill
Mantle of lower shell

Argopecten irradians

(E) Cephalopods

Head
Beak
Arms
Tentacle
Radula
Siphon
Mantle cavity
Intestine
Shell
Stomach
Mantle
Gill
Heart
Digestive gland

In cephalopod mollusks, the **foot** is modified into arms and tentacles.

Sepioteuthis lessoniana

(A) *Mexichromis macropus*

(B) *Deroceras* sp.

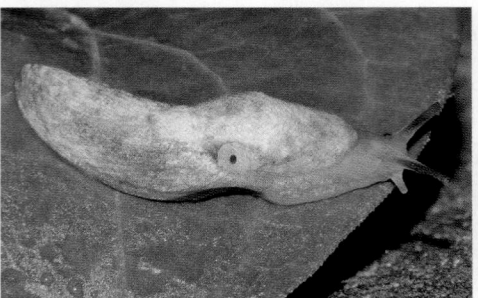

(C) *Octopus cyanea*

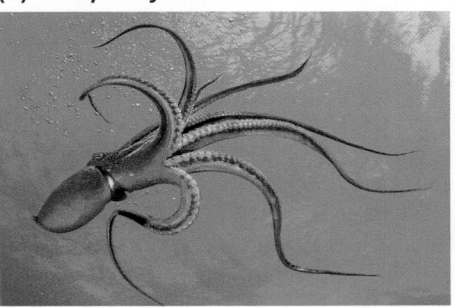

Figure 31.14 Mollusks in Some Groups Have Lost Their Shells **(A)** Nudibranchs ("naked gills"), also called sea slugs, are shell-less gastropods. This species is brightly colored, alerting potential predators of its toxicity. **(B)** Slugs are terrestrial, shell-less gastropods that feed on decomposing vegetation on the damp forest floor.

(C) Octopuses have neither an external nor an internal shell, which allows these cephalopods to squeeze through tight spaces.

 Media Clip 31.6 **Octopuses Can Pass through Small Openings**
www.Life11e.com/mc31.6

to the internal organs. Eventually the fluids reenter the blood vessels and are moved by a heart.

Monoplacophorans were the most abundant mollusks during the Cambrian period, 500 million years ago, but only a few species survive today. In monoplacophorans, in contrast to all other living mollusks, the gas exchange organs, muscles, and excretory pores are repeated over the length of the body.

The four major clades of living mollusks are the chitons, gastropods, bivalves, and cephalopods. Each of these groups is readily identifiable and distinct, even though they share variations on a common body plan.

CHITONS Eight overlapping calcareous plates, surrounded by a structure known as the girdle, protect the internal organs and muscular foot of **chitons** (**Figure 31.13B**). The chiton body is bilaterally symmetrical, and the internal organs, particularly the digestive and nervous systems, are relatively simple. Most chitons are marine omnivores that scrape algae, bryozoans, and other organisms from rocks with a sharp radula. An adult chiton spends most of its life clinging tightly to rock surfaces with its large, muscular, mucus-covered foot. It moves slowly by means of rippling waves of muscular contraction in the foot. Fertilization in most chitons takes place in the water, but in a few species fertilization is internal and embryos are brooded within the body. There are approximately 1,000 living species of chitons known.

GASTROPODS Gastropods (**Figure 31.13C**) are the most species-rich and widely distributed mollusks, with about 85,000 known living species. Snails, whelks, limpets, slugs, nudibranchs (sea slugs), and abalones are all gastropods. Most species move by gliding on the muscular foot, but in a few species—the sea butterflies and heteropods—the foot is a swimming organ with which the animal moves through open ocean waters.

Marine nudibranchs and terrestrial slugs are gastropods that have lost their protective shell over the course of evolution (**Figure 31.14**). Without a shell, these groups rely on other forms of protection from predation. The coloration of many nudibranchs is aposematic, meaning that it serves to warn potential predators of toxicity. Other nudibranch species and most terrestrial slugs exhibit camouflaged coloration.

Shelled gastropods have one-piece shells. The only mollusks that live in terrestrial environments—land snails and slugs—are gastropods. In these terrestrial species, the mantle tissue is modified into a highly vascularized lung.

BIVALVES Clams, oysters, scallops, and mussels are all familiar **bivalves**. The approximately 30,000 known species are found in both marine and freshwater environments. Bivalves have a hinged, two-part shell that extends over the sides of the body as well as the top (**Figure 31.13D**). Many clams use the foot to burrow into mud and sand. Bivalves feed by taking in water through an opening called an incurrent siphon and filtering food from the water with their large gills, which are also the main sites of gas exchange. Water and gametes exit through the excurrent siphon. Fertilization takes place in open water in most species.

CEPHALOPODS The **cephalopods**—squids, cuttlefish, octopuses, and nautiluses—first appeared near the beginning of the Cambrian period. By the Ordovician period a variety of types were present. Today there are about 800 described living species. In these mollusks the excurrent siphon is modified to allow the animal to control the volume of the mantle cavity and thereby bring in or expel water (**Figure 31.13E**). The modification of the mantle into a device for forcibly ejecting water from the cavity through the siphon enables these animals to move rapidly through the water by "jet propulsion." With their greatly enhanced mobility, cephalopods became the major predators in the open waters of the Devonian oceans. They remain important marine predators today.

As is typical of active, rapidly moving predators, cephalopods have a head with complex sensory organs—most notably eyes that are comparable to those of vertebrates in their ability to resolve images. The head is closely associated with a large, branched foot that bears the arms and/or tentacles and a siphon. Arms are distinguished by the presence of suckers along most of their length. Tentacles, in contrast, have suckers only near the tips or lack suckers altogether. Octopuses typically have eight arms and no tentacles, whereas squids and cuttlefishes have eight arms plus two tentacles. Cephalopods use their arms and tentacles to capture and subdue prey; octopuses also use their arms to move over the substrate. The large, muscular mantle provides a solid external supporting structure. The gills hang in the mantle cavity.

Many early cephalopods had an external chambered shell divided by partitions. The only surviving cephalopods with such shells are the nautiluses (genus *Nautilus*). The chambers inside nautilus shells are connected by a strand of tissue that runs through ducts in the partitions. Blood in this tissue carries water *from* the chambers and gases *into* the chambers, thus providing buoyancy. Most cephalopods retain an internal shell that functions for internal support and, in some species, is also chambered and buoyant. Octopuses have completely lost their shells, which allows them to compress their bodies through very small openings.

▶ 31.2 recap

Lophotrochozoans include animals with diverse body types. Wormlike forms include some flatworms, ribbon worms, phoronids, and annelids. There has been convergent evolution of lophophores (in bryozoans and entoprocts versus brachiopods and phoronids) and of external two-part shell coverings (in brachiopods versus bivalve mollusks).

learning outcomes

You should be able to:

- Associate major morphological features of protostomes with the environment and life history of the respective organisms.
- Provide examples demonstrating how modifications of existing structures can lead to increases in species diversity.

1. How do flatworms survive without an internal transport system?
2. Why are most annelids restricted to moist environments?
3. Mollusks have evolved a great diversity of body plans. Provide some examples that demonstrate how the basic body organization of mollusks has been modified to yield this diversity.

The second of the two major protostome clades, the ecdysozoans, contains the vast majority of Earth's animal species. What evolutionary innovations led to this massive diversity?

▶ key concept 31.3 Ecdysozoans Grow by Shedding Their Cuticles

Many ecdysozoans are wormlike in form, although others—the arthropods, onychophorans, and tardigrades—have limbs. In this section we will look at the two clades of wormlike ecdysozoans: the priapulids, kinorhynchs, and loriciferans in one clade and the nematodes and horsehair worms in the other. Key Concept 31.4 will be devoted to the most diverse ecdysozoans—the arthropods and their relatives—and the many forms their appendages take.

focus your learning

- An exterior covering known as the cuticle is regularly shed as an ecdysozoan grows.
- Nematodes are among the most abundant animals in many environments.
- Although many nematodes are free-living, others are parasitic or predatory.

As we noted in Key Concept 31.1, ecdysozoans can be identified by the presence of a body covering known as a cuticle, which provides these animals with both protection and support. The name of the group, Ecdysozoa, is derived from the Greek *ecdysis* ("to get out of") plus *zoa* ("animals"). Periodic shedding of the cuticle is necessary so that ecdysozoans can grow. In many ecdysozoans that have wormlike bodies, the cuticle is relatively thin and flexible; it offers the animal some protection but provides only modest body support. A thin cuticle allows the exchange of gases, minerals, and water across the body surface, but it restricts the animal to moist habitats. Many species of ecdysozoans with thin cuticles live in marine sediments from which they obtain food, either by ingesting sediments and extracting organic material from them or by capturing larger prey using a toothed **pharynx** (a muscular organ at the anterior end of the digestive tract). Some freshwater species absorb nutrients directly through their thin cuticles, as do parasitic species that live within their hosts (endoparasites). Many wormlike ecdysozoans are predators, eating protists and small animals.

Several marine ecdysozoan groups have relatively few species

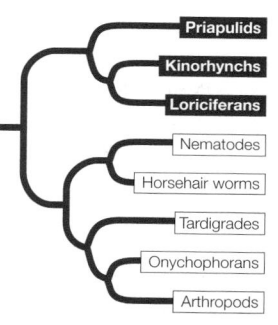

Members of several species-poor groups of wormlike marine ecdysozoans—the priapulids, kinorhynchs, and loriciferans—have relatively thin cuticles that are molted periodically as the animals grow to full size. Embryos of a fossil species related to these ecdysozoans have been discovered in sediments laid down in China about 500 million years ago. This remarkable discovery shows that the ancestors of these animals developed directly from an egg to the adult form, as most of their modern descendants do.

The 20 known species of **priapulids** are cylindrical, unsegmented, wormlike animals with a three-part body plan consisting of a proboscis, trunk, and caudal appendage ("tail"). It should be clear from their appearance why they were named after the Greek fertility god Priapus (**Figure 31.15A**). Priapulids range in length from 0.5 millimeters to 20 centimeters. They live in burrows in fine marine sediments and prey on soft-bodied invertebrates such as polychaetes, which they capture with a toothed, muscular pharynx that they evert through the mouth and then withdraw into the body together with the grasped prey. Fertilization is external, and most species have a larval form that also lives in the mud.

About 180 species of **kinorhynchs** have been described. They live in marine sands and muds and are virtually microscopic; no kinorhynchs are longer than 1 millimeter. Their bodies are

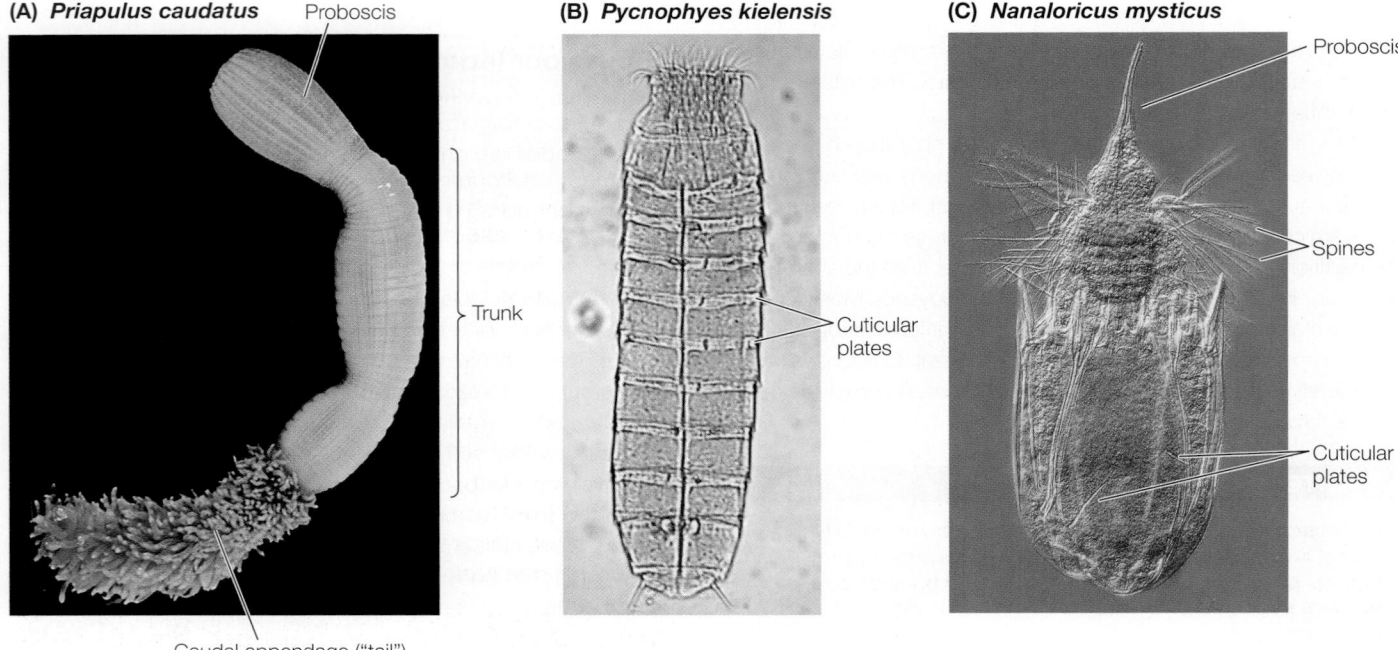

(A) *Priapulus caudatus* — Proboscis — Trunk — Caudal appendage ("tail")

(B) *Pycnophyes kielensis* — Cuticular plates

(C) *Nanaloricus mysticus* — Proboscis — Spines — Cuticular plates

Figure 31.15 Wormlike Marine Ecdysozoans Members of three ecdysozoan groups are marine bottom-dwellers. **(A)** Most priapulid species live in burrows on the ocean floor, extending the proboscis to feed. **(B)** Kinorhynchs are virtually microscopic. The cuticular plates that cover their bodies are molted periodically. **(C)** Six cuticular plates form a "corset" around the minute loriciferan body.

divided into 13 segments, each covered with a separate cuticular plate (**Figure 31.15B**). These plates are periodically molted during growth. Kinorhynchs feed by ingesting sediments through a

retractable proboscis (the group name means "movable snout"). They then digest the organic material found in the sediment, which may include living algae as well as dead matter. Kinorhynchs have no distinct larval stage; fertilized eggs develop directly into juveniles, which emerge from their egg cases with 11 of the 13 body segments already formed.

Loriciferans are also minute animals less than 1 millimeter long. They were not discovered until 1983. About 100 living species are known to exist, although only about 30 of these have been formally described to date. The body is divided into a head, neck, thorax, and abdomen and is covered by six plates, from which the loriciferans get their name (Latin *lorica*, "corset"). The plates around the base of the neck bear anterior-directed spines of unknown function (**Figure 31.15C**). Loriciferans live in coarse marine sediments. Little is known about what they eat, but some species apparently eat bacteria.

Nematodes and their relatives are abundant and diverse

Nematodes (roundworms) have a thick, multilayered cuticle that gives their unsegmented body its shape (**Figure 31.16**). As a nematode grows, it sheds its cuticle four times. Nematodes exchange oxygen and nutrients with their

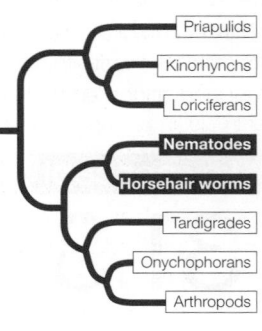

Priapulids
Kinorhynchs
Loriciferans
Nematodes
Horsehair worms
Tardigrades
Onychophorans
Arthropods

(A)

Brain
Pharynx
Ventral nerve
Excretory tube
Anus

> Nematodes shed their cuticle four times.

Cuticle
Dorsal nerve
Gut
Testis

> The large gut (maroon) and testis (blue) fill most of the body of a male *Trichinella spiralis*.

(B)

(C)

Figure 31.16 Nematodes **(A)** The body plan of *Trichinella spiralis*, which causes trichinosis, is typical of parasitic nematodes. **(B)** This polarized light micrograph shows a cyst of *T. spiralis* in the muscle tissue of a host. **(C)** This free-living nematode lives in freshwater environments.

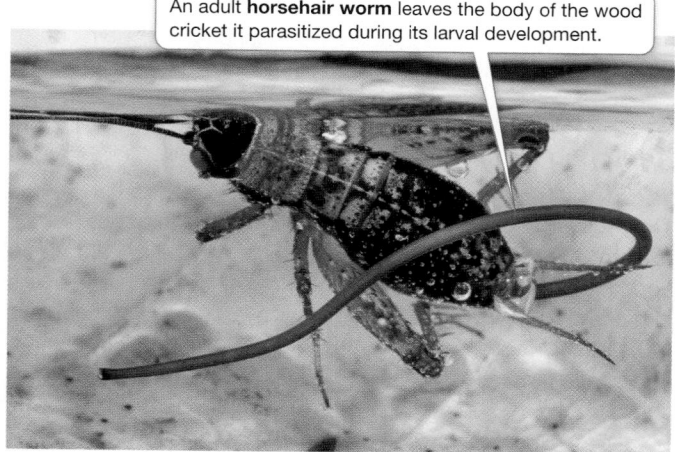

An adult **horsehair worm** leaves the body of the wood cricket it parasitized during its larval development.

Figure 31.17 Horsehair Worm Larvae Are Parasitic The larva of this horsehair worm (*Paragordius tricuspidatus*) can manipulate its host's behavior. The hatching worm causes the cricket to jump into water, where the worm emerges from the insect's body to continue its life cycle as a free-living adult. The cricket, having delivered its parasitic burden, drowns.

environment through both the cuticle and the gut wall, which is only one cell layer thick. Materials are moved through the gut by rhythmic contraction of a highly muscular pharynx. Nematodes move by contracting their longitudinal muscles.

Nematodes are probably the most abundant and universally distributed of all animal groups. Many nematodes are microscopic; the largest known nematode, which reaches a length of 9 meters, is a parasite in the placentas of sperm whales. About 25,000 species have been described, but the actual number of living species may be more than 1 million. Countless nematodes live as scavengers in the upper layers of the soil, on the bottoms of lakes and streams, and in marine sediments. The topsoil of rich farmland may contain from 3 to 9 billion nematodes per acre. A single rotting apple may contain as many as 90,000 individuals.

One soil-inhabiting nematode, *Caenorhabditis elegans*, serves as a model organism in the laboratories of geneticists and developmental biologists. It is ideal for such research because it is easy to cultivate, matures in 3 days, and has a fixed number of body cells. Its genome has been completely sequenced.

Many nematodes are predators, feeding on protists and small animals (including other roundworms). Most significant to humans, however, are the many species that parasitize plants and animals. The nematodes that parasitize humans (causing serious diseases such as trichinosis and elephantiasis), domestic animals, and economically important plants have been studied intensively in an effort to find ways of controlling them.

The structure of parasitic nematodes is similar to that of free-living species, but the life cycles of many parasitic species have special stages that facilitate the transfer of individuals among hosts. *Trichinella spiralis*, the species that causes the human disease trichinosis, has a relatively simple life cycle. A person may become infected by eating the flesh of an animal (usually a pig) that has *Trichinella* larvae encysted in its muscles (see Figure 31.16B). The larvae are activated in the person's digestive tract, emerge from their cysts, and attach to the intestinal wall, where they feed. Later they bore through the intestinal wall and are carried in the bloodstream to muscles, where they form new cysts. If present in great numbers, these cysts can cause severe pain or death.

About 350 species of the unsegmented **horsehair worms** have been described. As their name implies, these animals are extremely thin in diameter; horsehair worms range from a few millimeters up to 1 meter in length. Most adult worms live in fresh water, among leaf litter and algal mats near the edges of streams and ponds. A few species live in damp soil.

Horsehair worm larvae are endoparasites of freshwater crayfishes and of terrestrial and aquatic insects (**Figure 31.17**). An adult horsehair worm has no mouth, and its gut is greatly reduced and probably nonfunctional. Some species may feed only as larvae, absorbing nutrients from their hosts across the body wall. But other species continue to shed their cuticles and grow after they have left their hosts, suggesting that adult worms may also absorb nutrients from their environment.

31.3 recap

Priapulids, kinorhynchs, and loriciferans are relatively small, poorly known groups of wormlike marine ecdysozoans. Nematodes and horsehair worms have unsegmented wormlike bodies. Nematodes are among the most abundant and widely distributed animal groups.

learning outcomes

You should be able to:

- Compare cuticle thickness in the different ecdysozoans and explain the consequences of such variation.
- Explain why the nematode *Caenorhabditis elegans* is a widely used model system in laboratory research.
- Explain why nematode diversity is important to global ecosystems.
- Provide examples of nematodes that are human parasites.

1. How and why does cuticle thickness vary among the ecdysozoans?
2. Describe at least three ways in which nematodes have significant impacts on humans.

We will turn next to the animals that not only dominate the ecdysozoan clade but also constitute the most diverse group of animals on Earth.

key concept 31.4 Arthropods Are the Most Abundant and Diverse Group of Animals

Arthropods and their relatives are ecdysozoans with paired appendages. Arthropods are the most diverse group of animals in numbers of species (more than 1.2 million have been described, and many more remain to be discovered). Furthermore, the number of individual arthropods alive at any one time is estimated to be about 10^{18}, or 1 billion billion. Among the animals, only the nematodes are thought to exist in greater numbers.

focus your learning

- Several key features have contributed to the success of arthropods.
- Most pterygotes have two pairs of wings and go through complete or incomplete metamorphosis.
- Insects started to diversify about 450 million years ago, about the same time as land plants first appeared.
- Insects exchange gases via their tracheae.

Several key features have contributed to the success of the arthropods. As you have seen, their muscles are attached to the inside of their rigid exoskeletons. Their bodies are segmented, and each segment has muscles that operate that segment and the jointed appendages attached to it (see Figure 31.4). Jointed appendages permit complex movements, and different appendages are specialized for different functions. Encasement of the body within a rigid exoskeleton provides the animal with support for walking in the water or on dry land and provides some protection against predators. The waterproofing provided by chitin keeps the animal from dehydrating in dry air.

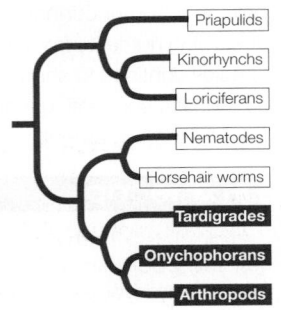

- Priapulids
- Kinorhynchs
- Loriciferans
- Nematodes
- Horsehair worms
- **Tardigrades**
- **Onychophorans**
- **Arthropods**

The four major arthropod groups living today are all species-rich: chelicerates (including the arachnids—spiders, scorpions, mites, and their relatives), myriapods (millipedes and centipedes), crustaceans (including shrimps, crabs, and barnacles), and hexapods (insects and their wingless relatives). The latter three groups are together known as mandibulates.

The jointed appendages of arthropods gave the clade its name, from the Greek words *arthron*, "joint," and *podos*, "foot" or "limb." Arthropods evolved from ancestors with simple, unjointed appendages. The exact forms of those ancestors are unknown, but some arthropod relatives with segmented bodies and unjointed appendages survive today. Before we describe the modern arthropods, we will discuss those arthropod relatives, as well as an early clade that went extinct but left an important fossil record.

Arthropod relatives have fleshy, unjointed appendages

The two living groups most closely related to the arthropods provide us with clues about the likely appearance of ancestral arthropod appendages. **Tardigrades** (water bears) have fleshy, unjointed legs and use their fluid-filled body cavities as hydrostatic skeletons (**Figure 31.18A**). Tardigrades are tiny (0.5–1.5 mm long) and lack both a circulatory system and gas exchange organs. The 1,200 known extant species live in marine sands and on temporary water films on plants. When these films dry out, the animals also lose water and shrink to small, barrel-shaped objects that can survive for at least a decade in a dormant state. Tardigrades have been found at densities as high as 2 million per square meter of moss.

Until fairly recently, biologists debated whether the **onychophorans** (velvet worms) were more closely related to annelids or to arthropods, but molecular evidence clearly links them to the latter. Indeed, with their soft, fleshy, unjointed, claw-bearing legs and elongate bodies, onychophorans may be similar in appearance to the ancestors of arthropods (**Figure 31.18B**). The 180 known species of onychophorans live in leaf litter in humid tropical environments. They have soft, segmented bodies that are covered by a thin, flexible cuticle that contains chitin. Like the tardigrades, they use their fluid-filled body cavities as hydrostatic skeletons. Fertilization is internal, and the large, yolky eggs are brooded within the body of the female.

Jointed appendages appeared in the trilobites

The **trilobites** flourished in Cambrian and Ordovician seas, but they disappeared in the great Permian extinction at the close of the Paleozoic era (251 million years ago). Because they had heavy exoskeletons that readily fossilized, they left behind an abundant record of their existence (**Figure 31.19**). About 10,000 species have been described.

The trilobites are the earliest known arthropods to have had jointed appendages. The body segmentation and appendages of trilobites followed a relatively simple, repetitive plan, but some of their appendages were modified for different functions. This specialization of appendages is a theme in the continuing evolution of the arthropods.

(A) *Paramacrobiotus tonolli*

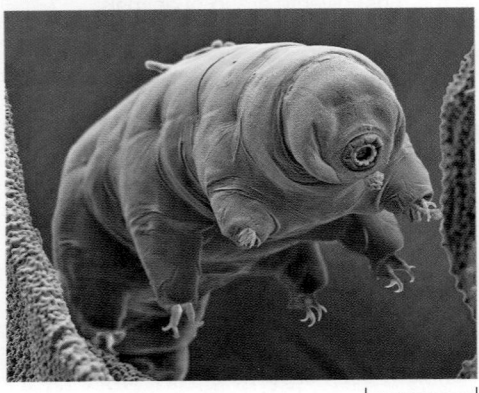

50 µm

(B) *Peripatus* sp.

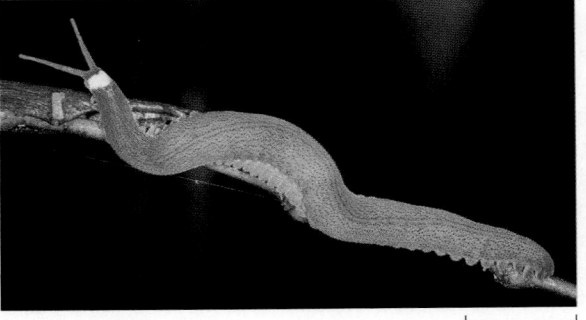

1 cm

Figure 31.18 Arthropod Relatives with Unjointed Appendages **(A)** Tardigrades (water bears) can be abundant on the wet surfaces of mosses and plants and in temporary pools of water. **(B)** Onychophorans (velvet worms) have unjointed legs and use the body cavity as a hydrostatic skeleton. They are sometimes referred to as "living fossils," meaning they are an ancient group that has changed very little over millennia.

Cheirurus ingricus

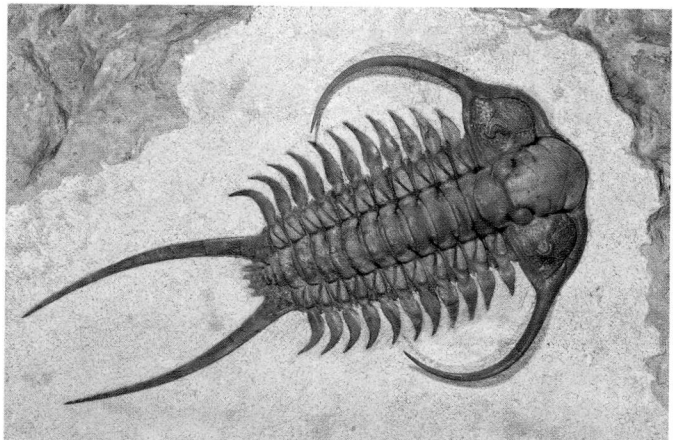

Figure 31.19 **A Trilobite Fossil**   The relatively simple, repetitive segments of the now-extinct trilobites are illustrated by a fossil trilobite from the shallow seas of the Ordovician period, some 450 million years ago.

(A) *Pseudopallene* **sp.**

(B) *Limulus polyphemus*

Figure 31.20 **Two Small Chelicerate Groups**   **(A)** Although they are not spiders, it is easy to see how sea spiders got their common name. **(B)** A spawning aggregation of horseshoe crabs. Horseshoe crabs, like the onychophorans (see Figure 31.18B), are an example of "living fossils."

Chelicerates have pointed, nonchewing mouthparts

In the **chelicerates**, the head bears two pairs of pointed appendages modified to form mouthparts, called chelicerae, that are used to grasp (rather than chew) prey. Chelicerates typically have a two-part body plan, with anterior segments fused to form a cephalothorax, and rear segments fused to form an abdomen. In some groups, such as mites and ticks, there is no clear distinction between these two body parts. Most chelicerates have four pairs of walking legs. The 114,000 described species are grouped into three major clades: pycnogonids, horseshoe crabs, and arachnids.

The pycnogonids, or sea spiders, make up a poorly known group of about 1,000 marine species (**Figure 31.20A**). Most are small, with leg spans less than 1 centimeter, but some deep-sea species have leg spans up to 60 centimeters. A few pycnogonids eat algae, but most are carnivorous, eating a variety of small invertebrates.

There are only four living species of horseshoe crabs, but many close relatives are known from fossils. Horseshoe crabs, which have changed very little morphologically over their long history, have a large horseshoe-shaped covering over most of the body. They are common in shallow waters along the eastern coast of North America and the southern and eastern coasts of Asia, where they scavenge and prey on bottom-dwelling animals. Periodically they

crawl into the intertidal zone in large numbers to mate and lay eggs (**Figure 31.20B**).

Arachnids are abundant in terrestrial environments. Most arachnids have a simple life cycle in which miniature adults hatch from internally fertilized eggs and begin independent lives almost immediately. Some arachnids retain their eggs during development and give birth to live young.

The most species-rich and abundant arachnids are the spiders, scorpions, harvestmen, mites, and ticks (**Figure 31.21**). More than 60,000 described species of mites and ticks live in soil, leaf litter,

(A) *Simaetha* **sp.**

(B) *Androctonus* **sp.**

(C) *Leiobunum rotundum*

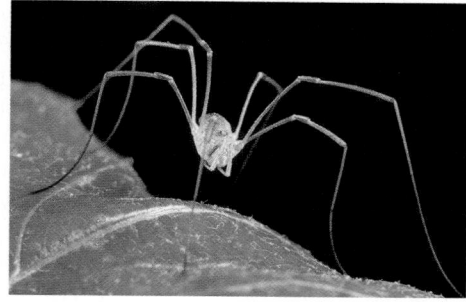

(D) *Trombidium* **sp.**

Figure 31.21 **Arachnid Diversity**   **(A)** Jumping spiders are active, visually oriented, diurnal predators of insects. **(B)** Scorpions are nocturnal predators of small animals. **(C)** Harvestmen, also called daddy longlegs, are scavengers. **(D)** Mites include many free-living species as well as blood-sucking ectoparasites.

(A) *Scolopendra hardwickei*

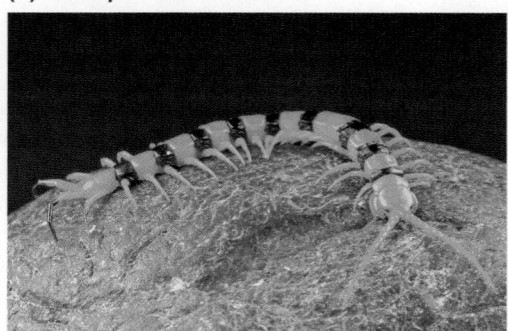

(B) *Motyxia monica*

Figure 31.22 Myriapods **(A)** Centipedes have modified appendages that function as poisonous fangs for capturing active prey. They have one pair of legs per segment. **(B)** Millipedes are scavengers and plant eaters; they have smaller jaws and legs than centipedes do, and they have two pairs of legs per segment.

mosses, and lichens, under bark, and as parasites of plants and animals. Mites are vectors for wheat and rye mosaic viruses; they cause mange in domestic animals and skin irritation in humans.

Spiders, of which about 50,000 species have been described, are important terrestrial predators with hollow chelicerae, which they use to inject venom into their prey. Some have excellent vision that enables them to chase and seize their prey. Others spin elaborate webs made of protein threads in which they snare prey. The threads are produced by modified abdominal appendages connected to internal glands that secrete the proteins, which solidify on contact with air. The webs of different groups of spiders are strikingly varied, and this variation enables the spiders to position their snares in many different environments for many different types of prey.

Mandibles and antennae characterize the remaining arthropod groups

The remaining three arthropod groups—the myriapods, crustaceans, and hexapods—have mouthparts that are mandibles, rather than chelicerae, so they are collectively called **mandibulates**. Mandibles can be used for chewing as well as for biting and holding food. Another distinctive characteristic of the mandibulates is the presence of sensory antennae on the head.

MYRIAPODS   The **myriapods** comprise the centipedes, millipedes, and their close relatives. Centipedes and millipedes have a well-formed head that bears the mandibles and antennae characteristic of mandibulates. Their distinguishing feature is a long, flexible, segmented trunk that bears many pairs of legs. Centipedes, which have one pair of legs per segment (**Figure 31.22A**), prey on insects and other small animals. In millipedes, two adjacent segments are fused so that each fused segment has two pairs of legs (**Figure 31.22B**). Millipedes scavenge and eat plants. More than 3,000 species of centipedes and 9,000 species of millipedes have been described; many more species probably remain unknown. Although most myriapods are less than a few centimeters long, some tropical species are ten times that size.

CRUSTACEANS   **Crustaceans** are the dominant marine arthropods today, and they are also common in fresh water and some terrestrial environments. The most familiar crustaceans are the shrimps, lobsters, crayfishes, and crabs (all decapods; **Figure 31.23A**) and the sow bugs (isopods; **Figure 31.23B**). Additional species-rich groups include the amphipods, ostracods, branchiopods (**Figure 31.23C**), and copepods (**Figure 31.23D**), all of which are found in freshwater and marine environments.

Barnacles are unusual crustaceans that are sessile as adults (**Figure 31.23E**). Adult barnacles look more like mollusks than like other crustaceans, but as the zoologist Louis Agassiz remarked more than a century ago, a barnacle is "nothing more than a little shrimp-like animal, standing on its head in a limestone house and kicking food into its mouth."

 Media Clip 31.7 Barnacles Feeding
www.Life11e.com/mc31.7

Most of the 67,000 described species of crustaceans have a body that is divided into three regions: head, thorax, and abdomen (**Figure 31.24**). The segments of the head are fused together, and the head bears five pairs of appendages. Each of the multiple thoracic and abdominal segments usually bears one pair of appendages. The appendages on different parts of the body are specialized for different functions, such as gas exchange, chewing, capturing food, sensing, walking, and swimming. In many species, a fold of the exoskeleton, the carapace, extends dorsally and laterally back from the head to cover and protect some of the other segments.

The fertilized eggs of most crustacean species are attached to the outside of the female's body, where they remain during their early development (see Figure 31.23D). At hatching, the young of some species are released as larvae; those of other species are released as juveniles that are similar in form to the adults. Still other species release eggs into the water or attach them to an object in the environment.

More than half of all described species are insects

During the Devonian period, more than 400 million years ago, some mandibulates colonized terrestrial environments. Of the several groups (including some crustacean isopods and decapods) that successfully colonized the land, none is more prominent today than the six-legged **hexapods**, which include the **insects** and their wingless relatives.

As you saw at the opening of this chapter, more than 1 million of the 1.8 million described living species on Earth are insects, but that is a small fraction of the number of species of insects that are thought to exist on Earth. One of Terry Erwin's experiments is shown in more detail in **Investigating Life: How Many Unknown Species?** This and similar experiments show that we have much to learn about the biodiversity of our own planet.

The wingless relatives of the insects—the springtails, two-pronged bristletails, and proturans—are probably the most similar of living forms to insect ancestors (**Figure 31.25**). These hexapods have a simple life cycle: they hatch from eggs as miniature adults.

(A) *Johngarthia lagostoma*

(B) *Oniscus asellus*

(C) *Triops longicaudatus*

(D) *Eudiaptomus gracilis*

(E) *Lepas* sp.

Figure 31.23 Crustacean Diversity   (A) This decapod crustacean, a land crab, is found only on four islands in the South Atlantic Ocean. **(B)** This pillbug, a terrestrial isopod, can roll into a tight ball when threatened. **(C)** This tadpole shrimp, a branchiopod, is common in seasonal pools of the southwestern United States. **(D)** This minute copepod is found in freshwater lakes throughout Europe. **(E)** Gooseneck barnacles attach to a substrate by their muscular stalks and feed by protruding and retracting their feeding appendages.

They differ from insects in having internal mouthparts. Springtails can be extremely abundant (up to 200,000 per m^2) in soil, leaf litter, and on vegetation and are the most abundant hexapods in the world in terms of number of individuals (as opposed to number of species).

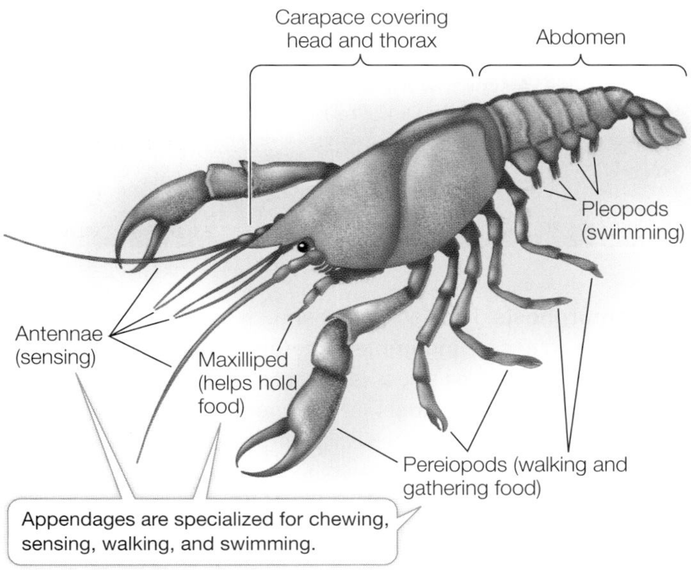

Like crustaceans, insects have a body with three regions: head, thorax, and abdomen. They have a single pair of antennae on the head and three pairs of legs attached to the thorax. In most groups of insects, the thorax also bears two pairs of wings. Unlike other arthropods, insects have no appendages growing from their abdominal segments (**Figure 31.26**). Insects are distinguished from springtails and other hexapods by their external mouthparts and by antennae that contain a motion-sensitive receptor called Johnston's organ. In addition, insects have a derived mechanism for gas exchange in air: a system of air sacs and tubular channels called tracheae (singular trachea) that extend

Carapace covering head and thorax

Abdomen

Pleopods (swimming)

Antennae (sensing)

Maxilliped (helps hold food)

Pereiopods (walking and gathering food)

Appendages are specialized for chewing, sensing, walking, and swimming.

Figure 31.24 Crustacean Body Plan   The bodies of crustaceans are divided into three regions: the head, thorax, and abdomen. Each body region bears specialized appendages. A shell-like carapace covers the head and thorax.

Q: In what ways is the crustacean body plan similar to and different from the insect body plan shown in Figure 31.26?

Tomocerus minor

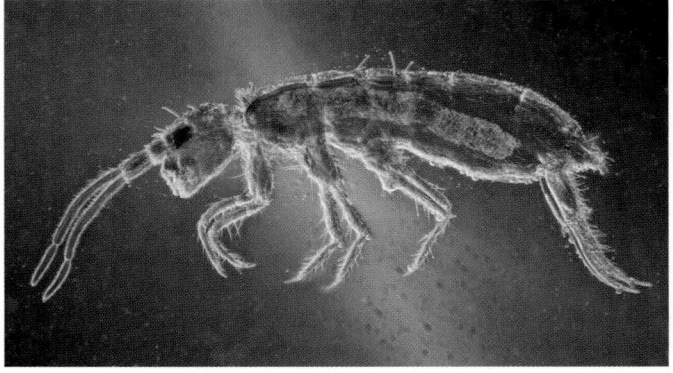

0.5 mm

Figure 31.25 Wingless Hexapods   The wingless hexapods, such as this springtail, have a simple life cycle. They hatch looking like miniature adults, then grow by successive molts of the cuticle.

experiment

Original Papers: Erwin, T. L. 1988. The tropical forest canopy: The heart of biotic diversity. In E. O. Wilson, ed., *Biodiversity*, pp. 123–129. National Academy Press, Washington, D.C.

Erwin, T. L. 1997. Biodiversity at its utmost: Tropical forest beetles. In M. L. Reaka-Kudla, D. E. Wilson and E. O. Wilson, eds., *Biodiversity II*, pp. 27–40. Joseph Henry Press, Washington, D.C.

Entomologist Terry Erwin decided to estimate the number of undescribed species of insects. He devised and executed the following experiment in the forests of Panama.

QUESTION▶   How many species of insects exist on Earth?

METHOD

1. Use insecticide fog to collect all beetles from *Luehea seemannii* trees.
2. Sort beetles into species based on morphological differences.
3. Calculate the number of host-specific species (those restricted to *L. seemannii*) and the number of generalists.
4. Estimate the number of tree species per hectare of Panamanian forest.
5. Calculate the percent of tree-dwelling versus ground-dwelling beetles.
6. Calculate the percent of beetles among all insect species.
7. Use the calculations above to extrapolate the total number of insect species on Earth.

RESULTS

Erwin estimated that there are more than 30 million insect species on Earth, only about 1 million of which have been described.

CONCLUSION▶   Insect diversity is far greater than indicated by the known number of species, and the number of undescribed insect species far exceeds the total number of described species on Earth.

work with the data

The data in the table were used by entomologist Terry Erwin to estimate the undescribed diversity of insects. Review the design of Erwin's experiment above and in the opening story of this chapter. Then use Erwin's data to answer the questions.

Approximate number of beetle species collected from *Luehea seemannii* trees	1,200
Estimated number of host-specific beetles in this sample	163
Number of tree species per hectare of Panamanian forest	70
Percent of beetle species living in tree canopy (as opposed to ground-dwelling species)	75%
Percent of beetles among all insect species	40%

QUESTIONS▶

1. From the data in the table, estimate the number of insect species in an average hectare of Panamanian forest. Assume that the data for beetles on *L. seemannii* are representative of the other tree species, and that all the species of beetles that are *not* host-specific were collected in the original sample. Remember to sum your estimates of the number of (a) host-specific beetle species in the forest canopy; (b) non-host-specific beetle species in the forest canopy; (c) beetle species on the forest floor; and (d) species of all insects other than beetles.

2. There are about 50,000 species of tropical forest trees. Assume that the data for beetles on *L. seemannii* are representative of other species of tropical trees and calculate the number of host-specific beetles found on these trees. Add an estimated 1 million species of non-host-specific beetles that are expected across different species of trees (including those in temperate regions). Estimate the number of ground-dwelling beetle species based on the percentage used in Question 1. Now use this information to estimate the number of insect species on Earth, based on the percentage of beetles among all insect species.

*A similar **work with the data** exercise may be assigned in **LaunchPad**.*

---

from external openings called spiracles inward to tissues throughout the body (see Figure 48.4).

Table 31.2 lists the major insect groups. Two groups—the jumping bristletails and silverfish—are wingless and have simple life cycles, like the springtails and other non-insect hexapods. The remaining groups are all pterygote insects. **Pterygotes** have two pairs of wings, except in some groups in which one or both pairs of wings have been secondarily lost. These secondarily wingless groups include the parasitic lice and fleas, some beetles, and the worker individuals of many ant species.

Hatchling pterygotes do not look like adults; they undergo substantial changes at each molt. The immature stages of insects between molts are called **instars**. A substantial change that occurs between one developmental stage and another is called

metamorphosis. If the changes between stages are gradual, an insect is said to have **incomplete metamorphosis**. If the change between at least some stages is dramatic, an insect is said to have **complete metamorphosis** (see Figure 30.10). In many insects with complete metamorphosis, different stages are specialized for different functions and use different food sources. In many species the larvae are specialized for feeding and growing, whereas the adults are specialized for reproduction and dispersal.

 Media Clip 31.8 Complete Metamorphosis www.Life11e.com/mc31.8

The adults of most flying insects have two pairs of stiff, membranous wings attached to the thorax. True flies, however, have one pair of wings and a pair of stabilizers called halteres (which

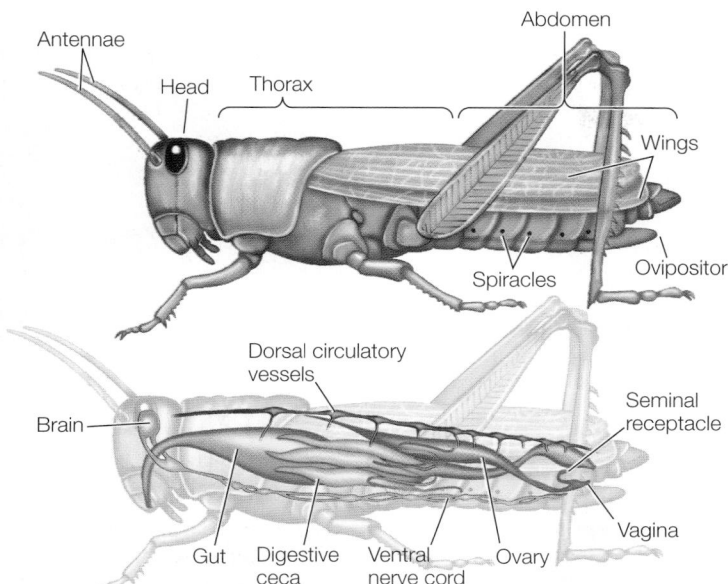

Figure 31.26 **Insect Body Plan**   Like those of crustaceans, the bodies of insects are divided into three regions: head, thorax, and abdomen. In insects, however, the thorax bears three pairs of legs and, in most groups, two pairs of wings. Unlike other arthropods, insects have no appendages growing from their abdominal segments.

Labels (top figure): Antennae, Head, Thorax, Abdomen, Wings, Ovipositor, Spiracles

Labels (bottom figure): Brain, Dorsal circulatory vessels, Seminal receptacle, Gut, Digestive ceca, Ventral nerve cord, Ovary, Vagina

table 31.2 The Major Insect Groups[a]

Group	Approximate number of described living species
Jumping bristletails (Archaeognatha)	550
Silverfish (Zygentoma)	575
PTERYGOTE (WINGED) INSECTS (PTERYGOTA)	
Mayflies (Ephemeroptera)	3,250
Dragonflies and damselflies (Odonata)	6,000
Neopterans (Neoptera)[b]	
Ice-crawlers (Grylloblattodea)	35
Gladiators (Mantophasmatodea)	15
Stoneflies (Plecoptera)	3,800
Webspinners (Embioptera)	500
Angel insects (Zoraptera)	40
Earwigs (Dermaptera)	2,000
Grasshoppers and crickets (Orthoptera)	24,000
Stick insects (Phasmida)	3,000
Cockroaches (Blattodea)	4,500
Termites (Isoptera)	3,000
Mantids (Mantodea)	2,500
Booklice and barklice (Psocoptera)	5,750
Thrips (Thysanoptera)	6,000
Lice (Phthiraptera)	5,000
True bugs, cicadas, aphids, leafhoppers (Hemiptera)	104,000
Holometabolous neopterans (Holometabola)[c]	
Ants, bees, wasps, sawflies (Hymenoptera)	150,000
Beetles (Coleoptera)	390,000
Strepsipterans (Strepsiptera)	650
Lacewings, ant lions, mantidflies (Neuroptera)	6,000
Dobsonflies, alderflies, fishflies (Megaloptera)	350
Snakeflies (Raphidoptera)	250
Scorpionflies (Mecoptera)	800
Fleas (Siphonaptera)	2,000
True flies (Diptera)	125,000
Caddisflies (Trichoptera)	12,000
Butterflies and moths (Lepidoptera)	180,000

[a]The hexapod relatives of insects include the springtails (Collembola; 3,000 spp.), two-pronged bristletails (Diplura; 600 spp.), and proturans (Protura; 10 spp.). All are wingless and have internal mouthparts.
[b]Neopteran insects can tuck their wings close to their bodies.
[c]Holometabolous insects are neopterans that undergo complete metamorphosis.

are derived from the second pair of wings). In winged beetles, one pair of wings—the forewings—forms heavy, hardened wing covers.

Two groups of pterygotes, the mayflies and dragonflies (**Figure 31.27A**), cannot fold their wings against their bodies. This is the ancestral condition for pterygote insects, and the mayflies and dragonflies are not closely related to one another. Members of both groups have predatory or herbivorous aquatic larvae that transform into flying adults after they crawl out of the water. Dragonflies (and their relatives the damselflies) are active predators as adults. In contrast, adult mayflies lack functional digestive tracts. Adult mayflies live only about a day, just long enough to mate and lay eggs.

All other pterygote insects—the **neopterans**—can tuck their wings out of the way upon landing and crawl into crevices and other tight places. Some neopteran groups undergo incomplete metamorphosis, so hatchlings of these insects are sufficiently similar in form to adults to be recognizable. Examples include the grasshoppers (**Figure 31.27B**), roaches, mantids, stick insects, termites, stoneflies, earwigs, thrips, true bugs (**Figure 31.27C**), aphids, cicadas, and leafhoppers. These groups acquire adult organ systems, such as wings and compound eyes, gradually through several juvenile instars.

More than 80 percent of all insects belong to a subgroup of the neopterans called the **holometabolous** insects (see Table 31.2), which undergo complete metamorphosis (**Figure 31.27D**). The many species of beetles account for almost half of this group (**Figure 31.27E**). Also included are lacewings and their relatives; caddisflies; butterflies and moths (**Figure 31.27F**); true flies (**Figure 31.27G**); and bees, wasps, and ants, some species of which display unique and highly specialized social behaviors (**Figure 31.27H**).

Molecular data suggest that insects began to diversify about 450 million years ago, about the time of the appearance of the first land plants. These early hexapods evolved in a terrestrial environment that lacked any other similar organisms, which in part accounts for their remarkable success. But the success of the insects is also due to their wings. Pterygote insects were the first animals in evolutionary history to achieve the ability to fly. Homologous genes control the development

(A) *Libellula luctuosa*   **(B)** *Poekilocerus pictus*   **(C)** *Dysdercus* sp.

(D) *Hydropsyche* sp.   **(E)** *Chryschus cobaltinus*   **(F)** *Papilio machaon*

(G) *Delia* sp.   **(H)** *Polistes gallicus*

Figure 31.27 Diverse Pterygotes   **(A)** Unlike most flying insects, a dragonfly cannot fold its wings over its back. **(B)** Orthopteran insects such as grasshoppers have incomplete metamorphosis: they undergo several molts, but the juvenile instars resemble small adults (incomplete metamorphosis). **(C)** Hemipterans such as this red cotton bug are known as "true" bugs. **(D–H)** Holometabolous insects undergo complete metamorphosis. **(D)** A larval caddisfly (bottom) emerges from its dark pupal case. **(E)** The beetles (Coleoptera) comprise the largest insect group; beetles such as this dogbane leaf beetle account for more than half of all holometabolous species. **(F)** Butterflies and moths are the lepidopterans, whose phases of complete metamorphosis are familiar to many (see Figure 30.10). **(G)** Blow-flies are among the "true" flies, the Dipterans. Adult blowflies feed on pollen or nectar, but lay their eggs on carrion, upon which the larvae feed. **(H)** These paper wasps are hymenopterans, a group in which most members display social behaviors.

of insect wings and crustacean appendages, suggesting that the insect wing evolved from a dorsal branch of a crustacean-like limb (**Figure 31.28**). The dorsal limb branch of crustaceans is used for gas exchange. Thus the insect wing probably evolved from a gill-like structure that had a gas exchange function.

Figure 31.28 The Origin of Insect Wings?   Insect wings may be derived from an ancestral appendage similar to that of modern crustaceans. **(A)** A diagram of the ancestral, multibranched arthropod limb. The uppermost dorsal branch may have been used for gas exchange. **(B, C)** The *pdm* gene, a Hox gene, is expressed throughout the dorsal limb segment of the thoracic limb of a crayfish **(B)** and in the wings of *Drosophila* **(C)**.

Q: If the hypothesized origin of insect wings is correct, then what was a possible function of the structure that would become a wing in flying insects?

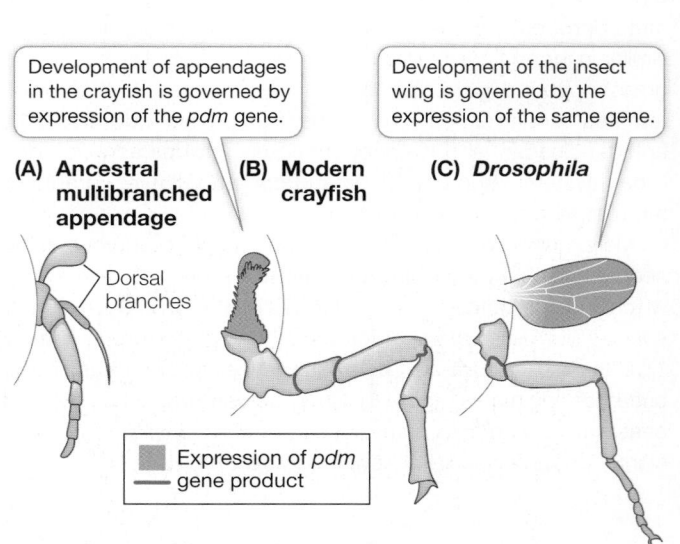

Development of appendages in the crayfish is governed by expression of the *pdm* gene.

Development of the insect wing is governed by the expression of the same gene.

(A) Ancestral multibranched appendage   **(B) Modern crayfish**   **(C)** *Drosophila*

Dorsal branches

Expression of *pdm* gene product

Flight opened up many new lifestyles and feeding opportunities that only the insects could exploit, such as pollination of (and coevolution with) flowering plants. Flight is almost certainly one of the reasons for the remarkable numbers of both insect species and individual insects, and for their unparalleled evolutionary success.

31.4 recap

All arthropods have segmented bodies. Muscles in each segment operate that segment and the appendages attached to it. Jointed, specialized appendages permit complex patterns of movement, including, in insects, the ability to fly. With flight, insects took advantage of new feeding and lifestyle opportunities, which contributed to the unparalleled evolutionary success of this group.

learning outcomes

You should be able to:

- Describe morphological innovations of arthropods, and their potential effects on arthropod diversity.
- Distinguish among different forms of metamorphosis in arthropods.
- Explain how the evolutionary innovations among arthropods are related to the distribution of major groups across the Earth's environments.

1. What features have contributed to making arthropods among the most abundant animals on Earth, both in number of species and in number of individuals?
2. Describe the difference between incomplete and complete metamorphosis.
3. There are more described and named species of insects than of all other species on Earth combined. However, only a very few insect species live in marine environments, and those species are restricted to the intertidal zone or the ocean surface. What factors may have contributed to the insects' lack of success in the oceans?

Key Features of Protostome Evolution

The protostomes encompass a staggering number of different body forms and lifestyles. The following aspects of protostome evolution have contributed to this enormous diversity:

- The evolution of *segmentation* permitted some groups of protostomes to move different parts of the body independently of one another. Species in some groups gradually evolved the ability to move rapidly over and through the substrate, through water, and through air.

- *Complex life cycles* with dramatic changes in form between one stage and another allow individuals of different stages to specialize on different resources.

- *Parasitism* has evolved repeatedly, and many protostome groups parasitize plants and animals.

- The evolution of *diverse feeding structures* allowed protostomes to specialize on many different food sources. Specialization on food sources undoubtedly contributed to reproductive isolation and further diversification.

- Predation was a major selection pressure favoring the development of *hard external body coverings* (exoskeletons and shells). Such coverings evolved independently in many lophotrochozoan and ecdysozoan groups. In addition to providing protection, these coverings became key elements in the development of new systems of locomotion.

- *Better locomotion* permitted prey to escape from predators, but also allowed predators to pursue their prey more effectively. Thus the evolution of animals has been, and continues to be, a complex "arms race" among predators and prey.

 Animation 31.1 **An Overview of the Protostomes**
www.Life11e.com/a31.1

Many major evolutionary trends among the protostomes are shared by the deuterostomes, which include the chordates, the group to which humans belong. We will turn to the deuterostomes in the next chapter.

 investigatinglife

Which groups of protostomes are thought to contain the most undiscovered species?

It is perhaps easier to list the groups of protostomes for which a nearly complete inventory of living species has been done than to list all the groups for which many new species remain to be described. Among the insects, the best-studied group, in terms of species, is the butterflies, which are widely collected and studied. There are still many species of other lepidopterans (such as moths), however, remaining to be discovered. As we explored in Investigating Life: How Many Unknown Species?, some other major insect groups, such as beetles, contain many undescribed species.

New species discovery and description rates remain high for almost all other major groups of protostomes. Second to the insects,

the nematodes. Although known nematode diversity is only about one-fortieth of known insect diversity (in terms of number of described species), the taxonomy of nematodes has been much more poorly studied than that of insects. Some biologists think there are likely to be species-specific parasitic nematodes specializing on most other species of multicellular organisms. If so, then there may be as many species of nematodes as there are of plants, fungi, and other animals combined.

Most of the other diverse groups of protostomes also contain many as yet undetected species, judging from the rate of new species descriptions. In particular, flatworms (especially the parasitic flukes and tapeworms), marine annelids, mollusks, crustaceans, myriapods, and chelicerates all contain large numbers of undescribed species.

(continued)

Future directions

Biologists have now placed almost all known species into a comprehensive tree of life (see Appendix A). This achievement allows the development of new technology for automated species discovery and identification. Once we can identify an organism, it is then possible to connect it to all the known information about that species.

It should soon be possible to construct a hand-held device that will isolate DNA from any sample, and then to sequence a series of target genes that can be used to place an unknown sample within the tree of life. This process results in either identification of a known species, or placement of an undescribed species with its closest relatives in the tree. This technology is expected to revolutionize the study of biodiversity and its applications.

Chapter Summary 31

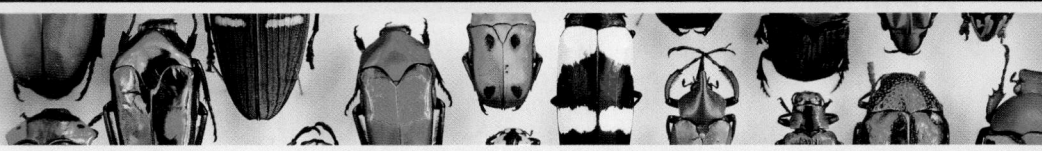

▶ 31.1 Protostomes Account for More Than Half of All Described Species

- Protostomes ("mouth first") are bilaterally symmetrical animals with an anterior brain that surrounds the entrance to the digestive tract and a ventral nervous system. The embryonic blastopore of protostomes develops into a mouth.

- There are two major clades of protostomes, the lophotrochozoans and the ecdysozoans. Review Figure 31.1, Table 31.1, Activities 31.1, 31.2

- **Lophotrochozoans** include a wide variety of body forms. Within this group, **lophophores** (complex organs for both food collection and gas exchange), free-living **trochophore** larvae, and spiral cleavage evolved. Some of these features were subsequently lost in some lineages (or evolved convergently).

- **Ecdysozoans** have a body covering known as the **cuticle**, which they must **molt** in order to grow. Some ecdysozoans, especially the **arthropods**, have a rigid cuticle reinforced with **chitin** that functions as an **exoskeleton**. Review Figure 31.4

- **Arrow worms** may be most closely related to lophotrochozoans, or they may be the sister group of all other protostomes. Review Figure 31.5

▶ 31.2 Many Lophotrochozoans Have Ciliated Feeding Structures or Life Stages

- Lophotrochozoans range from animals with a blind gut and no internal transport system to animals with complete digestive tracts and complex internal transport systems. Review Figure 31.6

- Most species of **bryozoans** and **entoprocts** live in colonies produced through asexual reproduction. Individuals of both groups feed using a lophophore.

- **Flatworms, rotifers, gastrotrichs,** and their close relatives form a structurally diverse clade of ciliated lophotrochozoans. Review Figure 31.7

- **Ribbon worms** feed using a long, protrusible proboscis. Review Figure 31.8

- The shelled **brachiopods** and wormlike **phoronids** use a lophophore to feed; this lophophore may have evolved independently of the lophophore in bryozoans and entoprocts. Review Figures 31.9, 31.10

- **Annelids** are a diverse group of segmented worms that live in moist terrestrial and aquatic environments. Review Figure 31.11

- **Mollusks** underwent a dramatic evolutionary radiation based on a body plan consisting of three major components: a **foot**, a **mantle**, and a **visceral mass**. The four major living molluscan clades—**chitons, bivalves, gastropods,** and **cephalopods**—demonstrate the diversity that evolved from this three-part body plan. Review Figure 31.13

▶ 31.3 Ecdysozoans Grow by Shedding Their Cuticles

- Several species-poor groups of wormlike marine ecdysozoans—**priapulids, kinorhynchs,** and **loriciferans**—have thin cuticles.

- **Nematodes** have a thick, multilayered cuticle. Nematodes are among the most abundant and universally distributed of all animal groups. Review Figure 31.16

- **Horsehair worms** are extremely thin; many are endoparasites as larvae.

▶ 31.4 Arthropods Are the Most Abundant and Diverse Group of Animals

- One major ecdysozoan clade, the arthropods, has evolved jointed, paired appendages that have a wide diversity of functions. Collectively, arthropods are the dominant animals on Earth in number of described species.

- Encasement within a rigid exoskeleton provides arthropods with support for walking as well as some protection from predators. The waterproofing provided by chitin keeps arthropods from dehydrating in dry air.

- Jointed appendages permit complex movement patterns. Each arthropod segment has muscles attached to the inside of the exoskeleton that operate that segment and attached appendages.

- The **onychophorans** and the **tardigrades** are arthropod relatives that have simple, unjointed appendages. **Trilobites,** the first arthropods known to have had jointed appendages, disappeared in the Permian mass extinction.

- **Chelicerates** have a two-part body and pointed mouthparts that grasp prey; most chelicerates have four pairs of walking legs.

- Mandibles and antennae are synapomorphies of the **mandibulates,** which include the myriapods, crustaceans, and hexapods.

- The bodies of **myriapods** have two regions: a head with mandibles and antennae, and a segmented trunk that bears many pairs of legs.

- **Crustaceans** have segmented bodies that are divided into three regions—head, thorax, and abdomen—with different, specialized appendages in each region. Review Figure 31.24

- **Hexapods**—insects and their relatives—are the dominant terrestrial arthropods. They have the same three body regions as crustaceans, but no appendages form in their abdominal segments. Review Figure 31.26, Table 31.2

- Wings and the ability to fly first evolved among the **pterygote** insects, allowing them to exploit new lifestyles. Review Figure 31.28

See Animation 31.1 for an overview of the protostomes.

Go to **LearningCurve** (in **LaunchPad**) for dynamic quizzing that helps you solidify your understanding of this chapter. **LearningCurve** adapts to your responses, giving you the practice you need to master each key concept.

Apply What You've Learned

Review

31.4 Insects exchange gases via their tracheae.

Original Papers: Braddy, S. J., M. Poschmann and O. E. Tetlie. 2008. Giant claw reveals the largest ever arthropod. *Biology Letters* 4: 106–109.

Kaiser A., C. J. Klok, J. J. Socha, W-K Lee, M. C. Quinlan and J. F. Harrison. 2007. Increase in tracheal investment with beetle size supports hypothesis of oxygen limitation on insect gigantism. *Proceedings of the National Academy of Sciences USA* 104: 13198–13203.

British scientists recently came upon something striking: a well-preserved fossil chelicera of a giant eurypterid sea scorpion dating back to the early Devonian period (**Figure A**). The claw belonged to the largest arthropod ever known on the planet, measuring nearly 2.5 meters in body length (longer than humans are tall). This new discovery was not the only representative of massive arthropods roaming Earth at that time; the Paleozoic fossil record is teeming with massive millipedes, monster cockroaches, and titanic trilobites. Why were these animals so large, and why don't we see supersized arthropods today?

To find out, we must look back to life in the Paleozoic era, approximately 542–251 million years ago. During this time, seed plants invaded land, land vertebrates became abundant, and winged insects took to the skies for the first time. According to geologic estimates, atmospheric oxygen in the late Paleozoic made up 35 percent of the air, as compared with 21 percent today. Chapter 24 described an experiment examining the relationship between atmospheric oxygen concentration and body size in fruit flies. Experimentally increasing O_2 concentrations resulted in fruit fly lineages that were significantly larger than control lineages, which lends support to the hypothesis that the increase in insect body size may have been due to higher oxygen concentrations. According to the fossil record, as O_2 concentrations declined at the end of the Permian period, so did the body size of arthropods.

Alexander Kaiser and his team wanted to know *why* oxygen might increase or limit growth. They knew that the larger an animal grows, the greater the distance required for oxygen to be transported to internal tissue. He thought that perhaps after a certain point the network of tracheae that supplies oxygen would start crowding other organs. To address this, the researchers compared mean values of tracheal density of the leg (TrD_{leg}) and the body (TrD_{tot}) against body length of living beetles from the literature, as shown in **Figure B**. They then extrapolated upper limits (dashed horizontal lines) of tracheal density in beetles based on the probable density at which tracheae would start to crowd other tissues.

Figure A

10 cm

Figure B

Upper limit of TrD_{leg}

TrD_{leg}

These are the 95% confidence limits for mean tracheal density.

Upper limit of TrD_{tot}

TrD_{tot}

These lines show the estimated relationship between tracheal density and body length.

Tracheal density, TrD (proportion of available volume)

Body length (cm)

Questions

1. Describe what is shown in the graph. Which is more limiting to the size of insects, the tracheal density of the leg or the tracheal density of the entire body? Explain your answer.

2. How do insects take in oxygen, and how does this compare with how humans and other vertebrates take in oxygen?

3. Based on the graph generated by Kaiser and his team, what size would you predict the largest living beetle could be today, given our current atmospheric oxygen levels? The largest living beetle is *Titanus giganteus*, which has a body length of approximately 17 cm. Does this fit your prediction?

Go to **LaunchPad** for the eBook, LearningCurve, animations, activities, flashcards, and additional resources and assignments.

32

key concepts

32.1 Deuterostomes Include Echinoderms, Hemichordates, and Chordates

32.2 Echinoderms and Hemichordates Are Restricted to Marine Environments

32.3 Chordates Have a Dorsal Nerve Cord and a Notochord

32.4 Life on Land Contributed to Vertebrate Diversification

32.5 Humans Evolved among the Primates

Deuterostome Animals

Grass snake (*Natrix natrix*) with eggs. Shelled eggs facilitated the colonization of drier terrestrial environments by deuterostomes.

investigatinglife

Which Came First, the Chicken or the Egg?

When people ponder the chicken and egg problem, it is supposedly a riddle without an answer. The first chicken must have hatched from an egg, but then didn't another chicken have to lay that egg? But if you understand evolutionary biology, then the answer to the question of whether the chicken or the egg came first is obvious, and it isn't even close. Shelled eggs originated hundreds of millions of years before the first chicken walked on Earth.

The lineage of birds that we now know as chickens is only a few million years old. If we followed that lineage back through time, the birds in it would look less like modern chickens with each preceding generation. But all birds lay shelled eggs (a phenomenon called oviparity, or "egg birth"), so we can logically infer that eggs existed long before the first chicken did. Birds are the only living group of dinosaurs, but fossil eggs for many groups of extinct dinosaurs have been discovered. The closest living relatives of birds are the crocodilians, and they also lay shelled eggs, as do turtles. Many (but not all) lizards and snakes also lay shelled eggs, and so do the earliest branching lineages of mammals, including the echidnas and duck-billed platypus. Other mammals modify the shelled

egg into structures that allow the embryo to develop within the mother, so that the young are born after "hatching" inside the mother (viviparity, or "live birth").

By comparing reproductive biology across species, biologists can trace the origin of the shelled egg to the origin of amniotes. Amniotes include all living mammals and reptiles (including birds). The group derives its name from the amnion, one of four membranes that surround the embryo within the shelled egg of amniotes.

Most amphibians also lay eggs, but their eggs have no shell. The soft, jellylike coating of amphibian eggs can easily dry, threatening the developing embryo. So amphibian eggs need to be deposited in water, or at least in moist places where desiccation is not a problem. This limitation kept the earliest terrestrial tetrapods from fully exploiting drier terrestrial environments. The shelled egg was thus an important evolutionary innovation that resulted in a great terrestrial explosion of the amniotes.

QA What are the advantages and disadvantages of egg laying

32.1 Deuterostomes Include Echinoderms, Hemichordates, and Chordates

It may surprise you to learn that both you and a sea urchin are deuterostomes. Adult sea stars, sea urchins, and sea cucumbers—the most familiar echinoderms—look so different from adult vertebrates (fishes, frogs, lizards, birds, and mammals) that it may be difficult to believe all these animals are closely related. The evidence that all deuterostomes share a common ancestor that is not shared with the protostomes includes early developmental patterns and phylogenetic analysis of gene sequences, factors that are not apparent in the forms of the adult animals.

focus your learning

- The best evidence that deuterostomes are a clade comes from phylogenetic analysis of gene sequences.
- There are three major deuterostome clades: echinoderms, hemichordates, and chordates.

Deuterostomes share early developmental patterns

Historically, the deuterostomes were distinguished by three early developmental patterns:

1. Radial cleavage

2. Development of the blastopore into the anus and formation of the mouth at the opposite end of the embryo from the blastopore (the pattern that gives the deuterostomes their name)

3. Development of a coelom from mesodermal pockets that bud off from the cavity of the gastrula rather than by splitting of the mesoderm, as occurs among protostomes

These distinctions, however, are not the strongest evidence for the monophyly of the deuterostomes. Radial cleavage is not exclusive to deuterostomes, and it is now thought to be the ancestral condition for all bilaterians. In fact, some of the groups now known to be protostomes were once thought to be deuterostomes because their developmental patterns are similar to those of echinoderms and chordates. The development of the blastopore into an anus does characterize the deuterostomes, but it may be the ancestral condition for bilaterians rather than a derived feature of deuterostomes. Today the strongest support for the shared evolutionary relationships of the deuterostomes comes from phylogenetic analyses of DNA sequences of many different genes.

Figure 32.1 Phylogeny of the Deuterostomes

The three principal groups of deuterostomes are the echinoderms, the hemichordates, and the chordates, which include the lancelets, tunicates, and vertebrates. The echinoderms and the vertebrates contain most of the described species.

 Activity 32.1 **Deuterostome Phylogeny**
www.Life11e.com/ac32.1

There are three major deuterostome clades

All deuterostomes are triploblastic, coelomate animals (see Figure 30.6C). Skeletal elements, where present, are internal rather than external. Some species have segmented bodies, but the segments are less obvious than those of annelids and arthropods. Although there are far fewer species of deuterostomes than of protostomes (see Table 30.1), we have a special interest in deuterostomes because we are members of that clade. The deuterostomes are also of interest because they include many large animals that strongly influence the characteristics of ecosystems. Many deuterostome species have been intensively studied in all fields of biology. Complex social behaviors are well developed among some deuterostomes (especially among the vertebrates).

The major groups of living deuterostomes comprise three distinct clades (**Figure 32.1**):

1. **Echinoderms**: sea stars (starfish), sea urchins, and their relatives

2. **Hemichordates**: acorn worms and pterobranchs

3. **Chordates**: tunicates, lancelets, and vertebrates

In addition, some recent genomic analyses suggest that two poorly known groups, the xenoturbellids and the acoels, may also be deuterostomes, although the exact position of these groups among animals is controversial.

 Animation 32.1 **An Overview of the Deuterostomes**
www.Life11e.com/a32.1

Fossils shed light on deuterostome ancestors

Scientists are learning much about the ancestors of modern deuterostomes from fossils discovered in 520-million-year-old rocks in China. Some of these early deuterostomes had skeletons similar to those of echinoderms, but unlike modern adult echinoderms, they had bilateral symmetry and a pharynx with slits through which water flowed. Another early deuterostome group, the yunnanozoans, was discovered in China's Yunnan Province. The well-preserved fossils reveal animals that had large mouths, six pairs of external gills, and

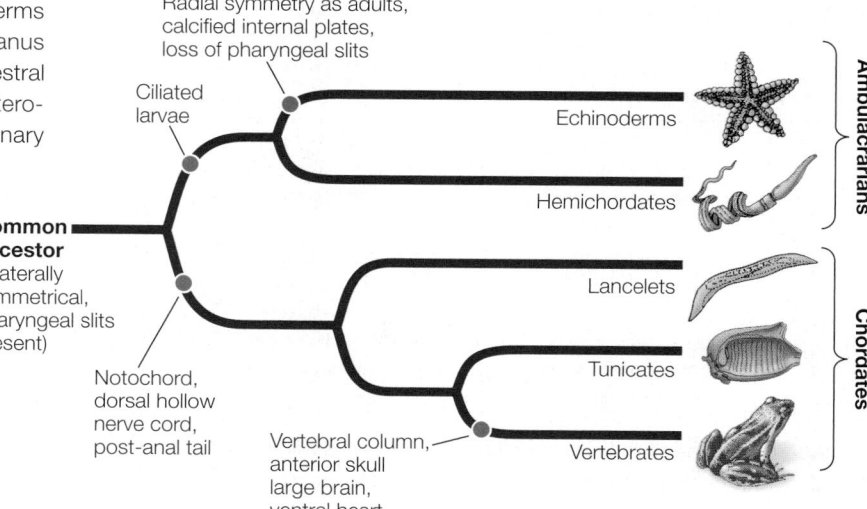

Radial symmetry as adults, calcified internal plates, loss of pharyngeal slits

Ciliated larvae

Common ancestor (bilaterally symmetrical, pharyngeal slits present)

Notochord, dorsal hollow nerve cord, post-anal tail

Vertebral column, anterior skull large brain, ventral heart

Echinoderms

Hemichordates

Lancelets

Tunicates

Vertebrates

Ambulacrarians

Chordates

Yunnanozoon lividum

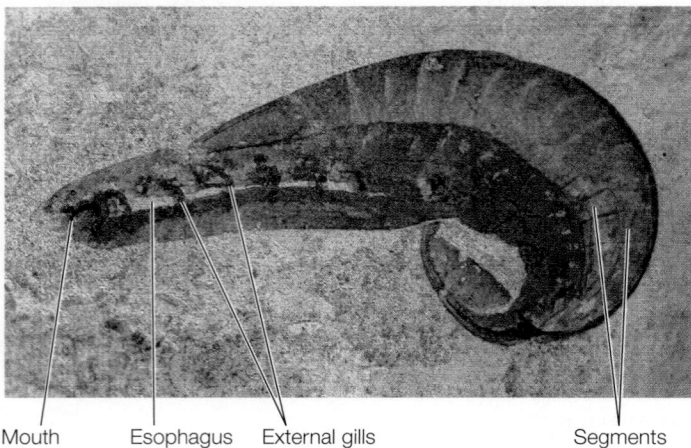

Mouth   Esophagus   External gills   Segments

Figure 32.2 Ancestral Deuterostomes Had External Gills The extinct yunnanozoans may be ancestral deuterostomes. This fossil, which dates from the Cambrian, shows the six pairs of external gills and segmented posterior body that characterized these animals.

a segmented posterior body section bearing a light cuticle (**Figure 32.2**). The features of these fossil animals, together with findings from phylogenetic analyses of living species, show that the earliest deuterostomes were bilaterally symmetrical, segmented animals with pharyngeal slits. The adult forms of the living echinoderms with their unique symmetry (in which the body parts are arranged along five radial axes) evolved much later. Other deuterostomes retained the ancestral bilateral symmetry.

32.1 recap

The three major clades of deuterostomes are the echinoderms, the hemichordates, and the chordates. The common ancestry of these groups is supported by early developmental similarities and by phylogenetic analyses of DNA sequences.

learning outcomes

You should be able to:

- Summarize the developmental and molecular characteristics that distinguish deuterostomes.
- Identify the major deuterostome clades and give examples of each.

1. What three developmental traits do all deuterostomes have in common? Why are some of these traits no longer considered to be strong evidence for the monophyly of deuterostomes?
2. What evidence provides strong support for the monophyly of deuterostomes?
3. What are the three major groups of deuterostomes? Place each of the following in the appropriate group: humans, acorn worms, rattlesnakes, tunas, and sea stars.

We will begin our survey of the deuterostomes with the echinoderms and hemichordates, the most distant of our relatives among the deuterostomes.

32.2 Echinoderms and Hemichordates Are Restricted to Marine Environments

About 13,000 species of echinoderms in 23 major groups have been described from their fossil remains. They are probably only a small fraction of the echinoderm species that have ever lived. Only 6 of the 23 major groups known from fossils are represented by species that survive today; many clades were lost during the periodic mass extinctions that have occurred throughout Earth's history. All of the 7,500 extant species of echinoderms and 120 species of hemichordates live in marine environments.

focus your learning

- Echinoderms are bilaterally symmetrical as larvae but pentaradially symmetrical as adults.
- Echinoderms use their tube feet in a variety of ways to capture prey.
- Hemichordates have a bilaterally symmetrical, wormlike body organized into three major parts: a proboscis, a collar, and a trunk.

The echinoderms and hemichordates (together known as **ambulacrarians**) have a bilaterally symmetrical, ciliated larva (**Figure 32.3A**). Adult hemichordates are also bilaterally symmetrical. Echinoderms, however, undergo a radical change in form as they develop into adults (**Figure 32.3B**), changing from a bilaterally symmetrical larva into an adult with **pentaradial symmetry** (symmetry in five or multiples of five). As is typical of animals with radial symmetry, echinoderms have no head, and most species move equally well in many directions. Rather than having an anterior–posterior (head–tail) and dorsal–ventral (back–belly) body organization, most echinoderms have an **oral** side containing the mouth and an opposite **aboral** side containing the anus.

Some recent genomic analyses suggest that two groups of small, highly reduced, soft-bodied marine organisms, the xenoturbellids and the acoels, may be the sister group of the ambulacrarians, although other studies suggest that they are the sister group of bilaterians. The five known species of **xenoturbellids** are wormlike organisms up to 20 centimeters long that feed on or parasitize marine mollusks. They have a very simple body plan, with almost no well-defined organ systems. The mostly tiny (<2 millimeter) **acoels** are also highly reduced, wormlike organisms that live as plankton, between grains of sediment, or on other organisms such as corals (**Figure 32.4**). They are among the simplest of bilaterian animals, with no gut, circulatory system, respiratory system, or excretory system. They feed through a mouth and form a vacuole around tiny food items. They are hermaphrodites, and their sperm form between the epidermis and digestive structure. There are about 400 known species of acoels.

Echinoderms have unique structural features

In addition to pentaradial symmetry, adult echinoderms have two unique structural

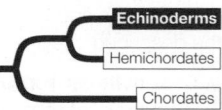

Echinoderms
Hemichordates
Chordates

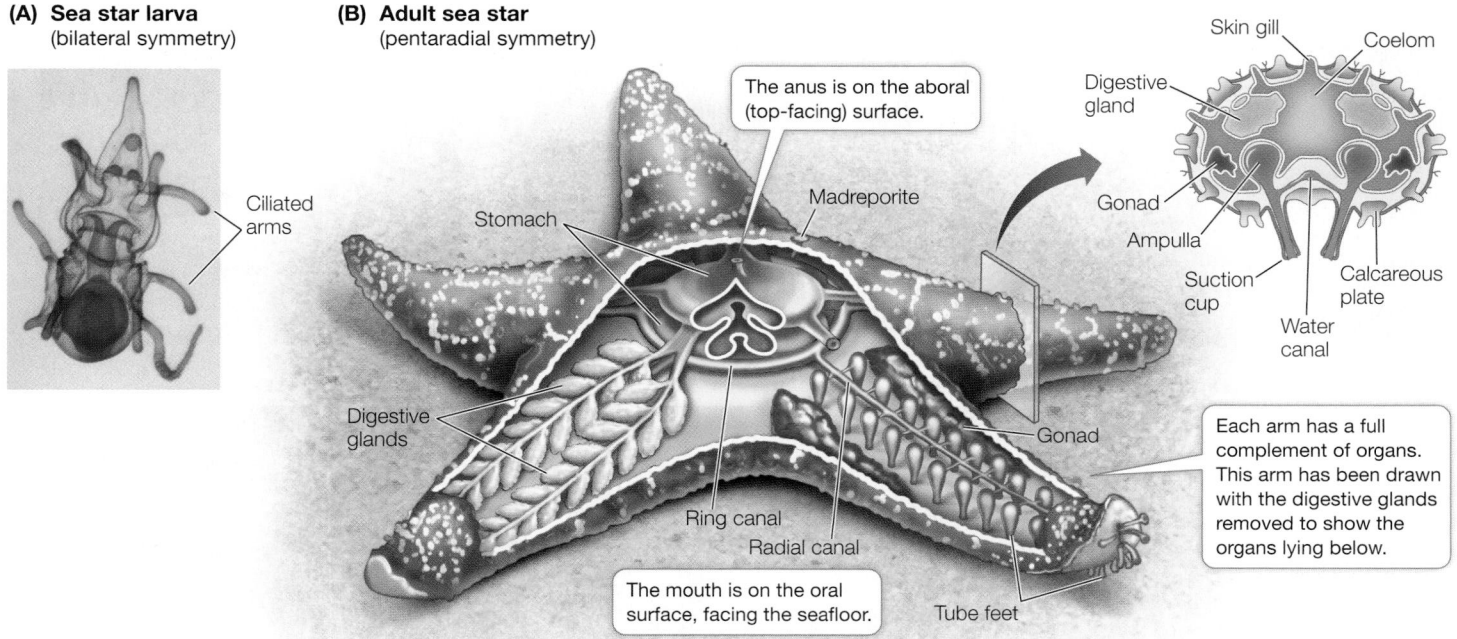

(A) Sea star larva
(bilateral symmetry)

Ciliated arms

(B) Adult sea star
(pentaradial symmetry)

The anus is on the aboral (top-facing) surface.

Madreporite

Stomach

Digestive glands

Ring canal

Radial canal

The mouth is on the oral surface, facing the seafloor.

Gonad

Tube feet

Each arm has a full complement of organs. This arm has been drawn with the digestive glands removed to show the organs lying below.

Skin gill

Coelom

Digestive gland

Gonad

Ampulla

Suction cup

Calcareous plate

Water canal

Figure 32.3 Echinoderms Are Bilaterally Symmetrical as Larvae but Radially Symmetrical as Adults **(A)** The ciliated larva of a sea star has bilateral symmetry. Hemichordates have a similar larval form. **(B)** An adult sea star displays the pentaradial symmetry of adult echinoderms. The canals and tube feet of the water vascular system, as well as the calcified internal skeleton, are shown in this diagram. The body's orientation is oral–aboral rather than anterior–posterior.

features. One is a system of calcified internal plates covered by thin layers of skin and some muscles. The calcified plates of most echinoderms are thick, and they fuse inside the entire body, forming an **internal skeleton**. The other unique feature of this group is a **water vascular system**, a network of water-filled canals leading to extensions called **tube feet** (see Figure 32.3B). This system functions in gas exchange, locomotion, and feeding. Seawater enters the system through a perforated structure called a madreporite. A calcified canal leads from the madreporite to the ring canal, which surrounds the esophagus (the tube leading from the mouth to the stomach). Radial canals branch off from the ring canal, extending through the arms (in species that have arms) and connecting with the tube feet. These structural innovations have been modified in many ways, resulting in a striking array of very different animals.

Members of one major extant echinoderm clade, the crinoids (sea lilies and feather stars), were more abundant and species-rich 300–500 million years ago than they are today. There are some 80 described living sea lily species, most of which are sessile organisms attached to the substrate by a stalk. Feather stars (**Figure 32.5A**) grasp the substrate with specialized flexible appendages that allow for limited movement. About 600 living species of feather stars have been described.

Although some crinoids are sessile, most surviving echinoderms are motile. The two main groups of motile echinoderms are the echinozoans (sea urchins and sea cucumbers) and the asterozoans (sea stars and brittle stars). Sea urchins are hemispherical in shape and lack arms (**Figure 32.5B**). They are covered with spines that are attached to the underlying skeleton with ball-and-socket joints. These joints enable the spines to be moved so they can converge toward a point that has been touched. The spines, which vary among species in size and shape, can be used for locomotion; a few produce toxic substances. They provide effective protection for the urchin, as many a scuba diver has found out the hard way. Sand dollars are flattened, disc-shaped relatives of sea urchins.

Sea cucumbers also lack arms, and their bodies are oriented in an atypical manner for an echinoderm (**Figure 32.5C**). The mouth is anterior and the anus is posterior (front and rear), in contrast to

Wamionoa sp.

Figure 32.4 Highly Reduced Acoels May Be Relatives of the Ambulacrarians Acoels (yellow) are seen here living on bubble coral (white). Acoels ("without coelom") feed by enveloping food particles in a vacuole within which nutrients are digested. These hermaphroditic animals can reproduce rapidly and may become problematic in saltwater aquariums.

(A) *Comanthina schlegeli*

(B) *Echinometra mathaei*

(C) *Bohadschia argus*

Expelled guts

Figure 32.5 Echinoderm Diversity
(A) The flexible arms of this variable bushy feather star (a crinoid) are clearly visible. **(B)** Sea urchins are important grazers on algae in the intertidal zones of the world's oceans. **(C)** This sea cucumber has expelled a portion of its viscera as a sticky defense against a potential predator. **(D)** Sea stars are important predators on bivalve mollusks such as mussels and clams. Suction tips on its tube feet allow a sea star to grasp both shells of a bivalve and pull them open. **(E)** The arms of a brittle star are composed of hard but jointed plates.

(D) *Marthasterias glacialis*

(E) *Ophiarachna incrassata*

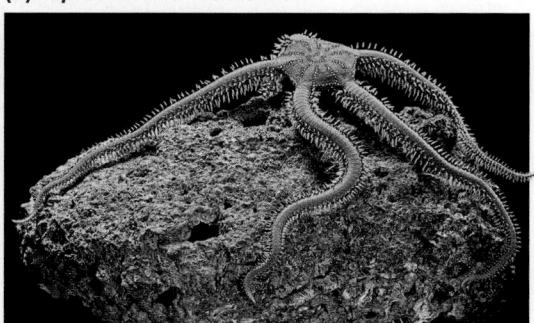

the oral–aboral (top and bottom) orientation of other echinoderms. Sea cucumbers can use most of their tube feet to move, but they use them primarily for attaching to the substrate.

Sea stars, popularly called starfish, are the most familiar echinoderms (**Figure 32.5D**). Their gonads and digestive organs are located in the arms, as seen in Figure 32.3B. Their tube feet serve as organs of locomotion, gas exchange, and attachment. Each tube foot of a sea star consists of an internal ampulla connected by a muscular tube to an external suction cup that can stick to the substrate. The tube foot is moved by expansion and contraction of the circular and longitudinal muscles of the tube. Brittle stars are similar in structure to sea stars, but their flexible arms are composed of jointed, hard plates (**Figure 32.5E**), their gonads and viscera are contained in the central disk, and their tube feet lack suction cups or ampullae.

Echinoderms use their tube feet in a great variety of ways to capture prey. Sea lilies, for example, feed by orienting their arms in passing water currents. Food particles then strike and stick to the tube feet, which are covered with mucus-secreting glands. The tube feet transfer these particles to grooves in the arms, where ciliary action carries the food to the mouth.

Most sea urchins capture phytoplankton with their tube feet or scrape algae from rocks with a complex rasping structure. Sea cucumbers capture food with their anterior tube feet, which are modified into large, feathery, sticky tentacles that can be protruded from the mouth. Periodically a sea cucumber withdraws the tentacles, wipes off the material that has adhered to them, and digests it.

Many sea stars use their tube feet to capture large prey such as polychaetes, gastropod and bivalve mollusks, small crustaceans such as crabs, and fishes. With hundreds of tube feet acting simultaneously, a sea star can grasp a bivalve in its arms, anchor the arms with its tube feet, and by steady contraction of the muscles in its arms, gradually exhaust the muscles the bivalve uses to keep its shell closed (see Figure 32.5D). To feed on the bivalve, the sea star can push its stomach out through its mouth and then through the narrow space between the two halves of the bivalve's shell. The sea star's stomach then secretes enzymes that digest the prey.

 Media Clip 32.1 Sea Star Hunting Bivalves
www.Life11e.com/mc32.1

Most of the 2,000 species of brittle stars ingest particles from the upper layers of sediments and assimilate the organic material from them, although some species filter suspended food particles from the water, and others capture small animals.

Hemichordates are wormlike marine deuterostomes

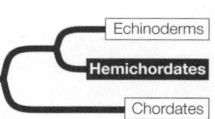

Hemichordates—acorn worms and pterobranchs—have a bilaterally symmetrical, wormlike body organized in three major parts: a proboscis, a collar (which bears the mouth), and a trunk (which contains the other body parts). The 90 known species of acorn worms range up to 2 meters long (**Figure 32.6A**). They live in burrows in muddy and sandy marine sediments. The digestive tract of an acorn worm consists of a mouth behind which are a muscular pharynx and an intestine. The pharynx opens to the outside through several pharyngeal slits through which water can exit. Highly vascularized tissue surrounding the pharyngeal slits serves as a gas exchange apparatus. Acorn worms respire by pumping water into the mouth and out through the pharyngeal slits. They capture prey with the large proboscis, which is coated with sticky mucus to which small organisms in the sediment stick. The mucus and its attached prey are conveyed

(A) *Saccoglossus kowalevskii*

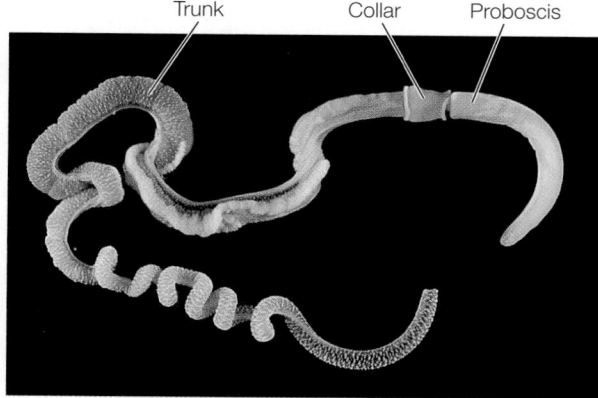

(B)

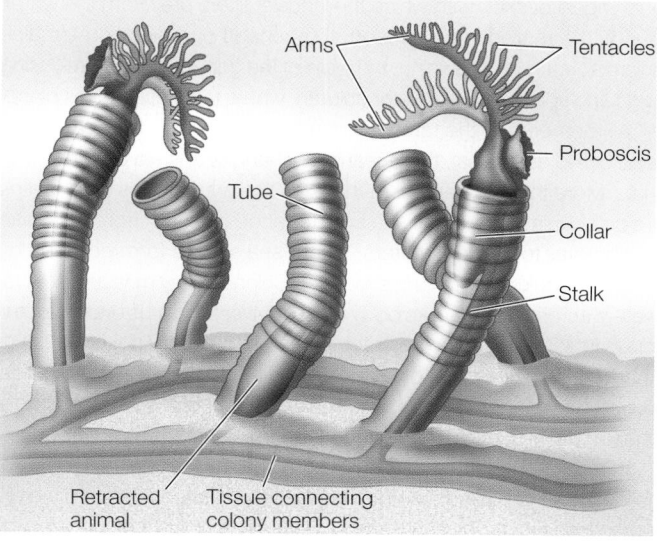

Figure 32.6 Hemichordates (A) The proboscis of an acorn worm is modified for burrowing. **(B)** Some pterobranch species form colonies.

by cilia to the mouth. In the esophagus, the food-laden mucus is compacted into a ropelike mass that is moved through the digestive tract by ciliary action.

The 30 living species of pterobranchs are sedentary marine animals up to 12 millimeters long that live in a tube secreted by the proboscis. Some species are solitary; others form colonies of individuals joined together (**Figure 32.6B**). Behind the proboscis is a collar with anywhere from one to nine pairs of arms. The arms bear long tentacles that capture prey and function in gas exchange.

32.2 recap

Adult echinoderms are characterized by pentaradial symmetry, an internal skeleton of calcified plates, and a unique water vascular system. Hemichordates have a bilaterally symmetrical body divided into three parts: proboscis, collar, and trunk. Both groups are restricted to marine environments.

32.2 recap

learning outcomes

You should be able to:

- Compare the body plans of larval and adult echinoderms.
- Summarize the ways that various echinoderms use their tube feet to feed.
- Contrast the ways that echinoderms and hemichordates feed.

1. How does the body form of echinoderm larvae differ from that of echinoderm adults?
2. Describe some of the ways that echinoderms use their tube feet to obtain food, and contrast this with the feeding mechanism of hemichordates.

Having described the deuterostome groups that are most distantly related to us, we will next turn our attention to the unique features that evolved in the chordates, a clade dominated by the vertebrates.

key concept

32.3 Chordates Have a Dorsal Nerve Cord and a Notochord

As you have seen, it is not obvious from examining adult animals that echinoderms and chordates share a common ancestor. The evolutionary relationships among some chordate groups are not immediately apparent either. The features that reveal all of these evolutionary relationships are seen primarily in the larvae—in other words, it is during the early developmental stages that these evolutionary relationships are evident.

focus your learning

- Chordates are characterized by a dorsal hollow nerve cord, a post-anal tail, and a dorsal supporting rod called a notochord.
- Vertebrates are distinguished by an additional dorsal supporting structure—the vertebral column—and other features.
- Hagfishes have weak circulatory systems and lack, or have reductions in, many otherwise shared vertebrate features.
- Lampreys and hagfishes may look superficially similar, but they differ greatly in their biology.
- Ray-finned fishes evolved swim bladders, which are organs of buoyancy.

There are three principal chordate clades: the **lancelets** (also called cephalochordates), the **tunicates** (also called urochordates), and the **vertebrates** (see Figure 32.1). Adult chordates vary greatly in form, but all chordates display the following derived structures at some stage in their development (**Figure 32.7**):

- A *dorsal hollow nerve cord*
- A *tail* that extends beyond the anus
- A dorsal supporting rod, the *notochord*

The **notochord** is the most distinctive derived chordate trait. It is composed of a core of large cells with turgid fluid-filled vacuoles,

(A)

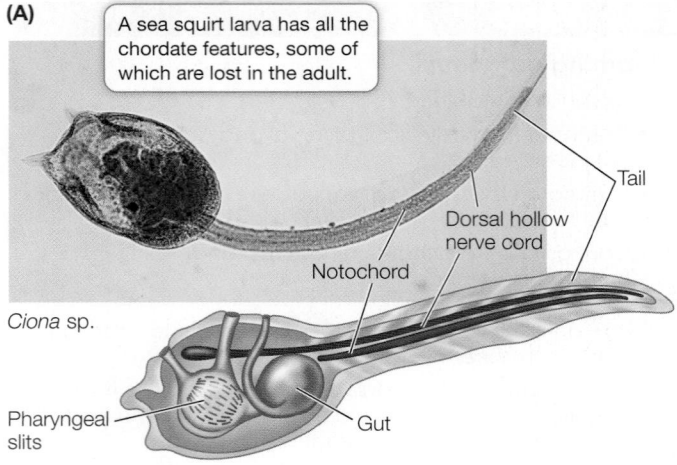

A sea squirt larva has all the chordate features, some of which are lost in the adult.

Ciona sp.

Tail

Dorsal hollow nerve cord

Notochord

Pharyngeal slits

Gut

(B)

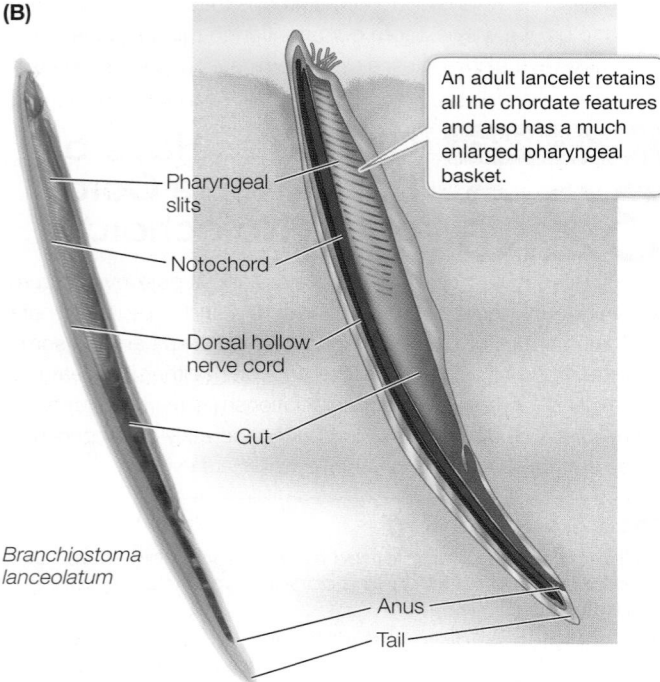

An adult lancelet retains all the chordate features and also has a much enlarged pharyngeal basket.

Pharyngeal slits

Notochord

Dorsal hollow nerve cord

Gut

Branchiostoma lanceolatum

Anus

Tail

Figure 32.7 Key Features May Be Most Apparent in Early Development **(A)** The tunicate larva (but not the adult) has all three key features of chordates: a dorsal hollow nerve cord, a post-anal tail, and a notochord. **(B)** All three chordate synapomorphies are retained in the adult lancelet.

which make it rigid but flexible. In the tunicates the notochord is lost during metamorphosis to the adult stage. In most vertebrate species it is replaced during development by skeletal structures that provide support for the body.

The pharyngeal slits found in the common ancestor of deuterostomes are present at some developmental stage in all chordates but are often lost or greatly modified in adults. In chordates, the pharyngeal slits are separated and supported by structural elements called pharyngeal arches. In tunicates and lancelets, the pharynx functions as a straining device to filter small food particles. In fishes and larval amphibians, some of the pharyngeal arches develop into gill arches, which support the respiratory gills and are often used as feeding structures as well. Developmentally, some pharyngeal arches also develop into elements of the vertebrate jaw, as well as parts

of the tongue, larynx, trachea, and middle ear of tetrapods (four-legged vertebrates). Some of the pharyngeal slits are modified in tetrapods to form the eustachian tube and middle ear chamber.

Adults of most lancelets and tunicates are sedentary

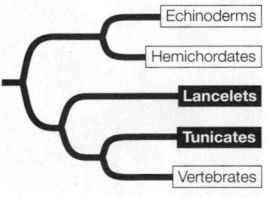

The 35 species of lancelets are small animals that rarely exceed 5 centimeters in length. The notochord, which provides body support, extends the entire length of the body throughout the animal's life (see Figure 32.7B). Lancelets are found in shallow marine and brackish waters worldwide. Most of the time they lie covered in sand with their head protruding above the sediment, but they can swim. The pharynx has been enlarged and modified to form a structure called a pharyngeal basket, with which the lancelet filters prey from the water. During the reproductive season, the gonads of males and females enlarge greatly. At spawning, the walls of the gonads rupture, releasing eggs and sperm into the water column, where fertilization takes place.

All members of the three major tunicate groups—the sea squirts (also called ascidians), thaliaceans, and larvaceans—are marine animals. More than 90 percent of the 2,800 known species of tunicates are sea squirts. Individual sea squirts range in length from less than 1 millimeter to 60 centimeters. Some sea squirts form colonies by asexual budding from a single founder. Colonies may measure several meters across. The baglike body of an adult sea squirt is enclosed in a tough tunic, which is the basis for the name "tunicate" (**Figure 32.8A**). The tunic is composed of proteins and a complex polysaccharide secreted by epidermal cells. The sea squirt pharynx is enlarged into a pharyngeal basket that filters prey from the water passing through it.

In addition to its pharyngeal slits, a sea squirt larva has a dorsal hollow nerve cord and a notochord that is restricted mostly to the tail region (see Figure 32.7A). Bands of muscle surround the notochord, which provide support for the body. After a short time swimming in the plankton, the larvae of most species settle on the seafloor and transform into sessile adults. The swimming, tadpolelike larvae suggest a close evolutionary relationship between tunicates and vertebrates (see Figure 21.6).

Thaliaceans (salps and their relatives) are tunicates that can live singly or in chainlike colonies up to several meters long (**Figure 32.8B**). They float in tropical and subtropical oceans at depths down to 1,500 meters. Larvaceans are solitary planktonic animals that retain the notochord and dorsal hollow nerve cord throughout their lives. Most larvaceans are less than 5 millimeters long, but some species that live near the bottom of deep ocean waters build delicate casings of mucus that may be more than a meter wide. They snare sinking organic particles (their primary food source) with elaborate filters built into their mucus "houses." When the old "house" gets clogged with excess debris, the animal builds a new one.

A dorsal supporting structure replaces the notochord in vertebrates

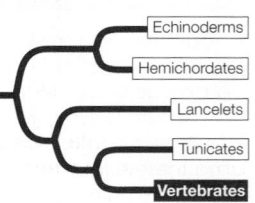

In one chordate group, the vertebrates, a new dorsal supporting structure evolved. This group takes its name from the jointed, dorsal **vertebral column**

(A) *Clavelina dellavallei*

(B) *Pegea sp.*

1 cm

1 cm

Figure 32.8 Adult Tunicates   (A) The transparent tunic and the pharyngeal basket are clearly visible in this sea squirt. **(B)** A chainlike colony of thaliaceans (salps) floats in tropical waters.

that typically replaces the notochord during early development as the primary supporting structure. The individual elements in the vertebral column are called vertebrae. Four other key features characterize the vertebrates as well (**Figure 32.9**):

- An anterior *skull* enclosing a large brain
- A rigid internal *skeleton* supported by the vertebral column
- Internal organs *suspended in a coelom*
- A well-developed *circulatory system*, driven by contractions of a ventral *heart*

These structural features can support large, active animals. The internal skeleton provides support for an extensive muscular system, which receives oxygen from the circulatory system and is controlled by the central nervous system. The evolution of these features allowed many vertebrates to become large, active predators, which in turn allowed the vertebrates to diversify widely (**Figure 32.10**).

All of the nonvertebrate deuterostomes live in marine environments. The lineage that led to the vertebrates is also thought to have evolved in the oceans, although probably in an estuarine environment (where fresh water meets salt water). The first vertebrates appeared in the Cambrian; since then they have radiated into marine, freshwater, terrestrial, and aerial environments worldwide. There are about 65,000 species of living vertebrates.

The phylogenetic relationships of jawless fishes are uncertain

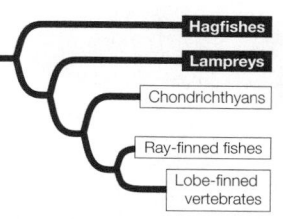

The **hagfishes** are thought by many to be the sister group to the remaining vertebrates (see Figure 32.10). Hagfishes (**Figure 32.11A**) have a weak circulatory system with three small accessory hearts (rather than a single, large heart); a partial cranium, or skull (containing a brain with no cerebrum or cerebellum, two main regions that characterize the brains of other vertebrates); and no jaws or stomach. They also lack separate, jointed vertebrae and have a skeleton composed of a firm but pliable material called cartilage. Thus some biologists do not consider hagfishes to be vertebrates and instead use the term "craniates" to refer collectively to the hagfishes and the vertebrates. Some analyses of gene sequences suggest, however, that hagfishes may be the sister group of the **lampreys** (**Figure 32.11B**); in this phylogenetic arrangement, the hagfishes and the lampreys are collectively called the cyclostomes ("circle mouths"). If in fact the hagfishes and lampreys do form a monophyletic group, then hagfishes must have secondarily lost many of the major vertebrate morphological features during their evolution.

The 80 known species of hagfishes are unusual marine animals that produce copious quantities of mucus as a defense.

The anterior skull of vertebrates contains the brain and many sensory organs.

Vertebral column

Spines (modified rays)

Dorsal fins

Rays (structures supporting fins)

Caudal fin

Heart

Liver

Gut

Gonad

Anal fin

Pelvic fin (paired appendage)

Cross section

Vertebrates have an internal skeleton, supported by a vertebral column that encloses the dorsal spinal cord…

…a well-developed circulatory system…

…and internal organs suspended in a large coelom.

Muscles

Spinal cord

Vertebra

Kidney

Gonad

Gut

Coelom

Peritoneum

Figure 32.9 The Vertebrate Body Plan   A ray-finned fish is used here to illustrate some of the structural elements common to all vertebrates. In addition to the paired pelvic fins, these fishes have paired pectoral fins on the sides of their bodies (not seen in this cutaway view).

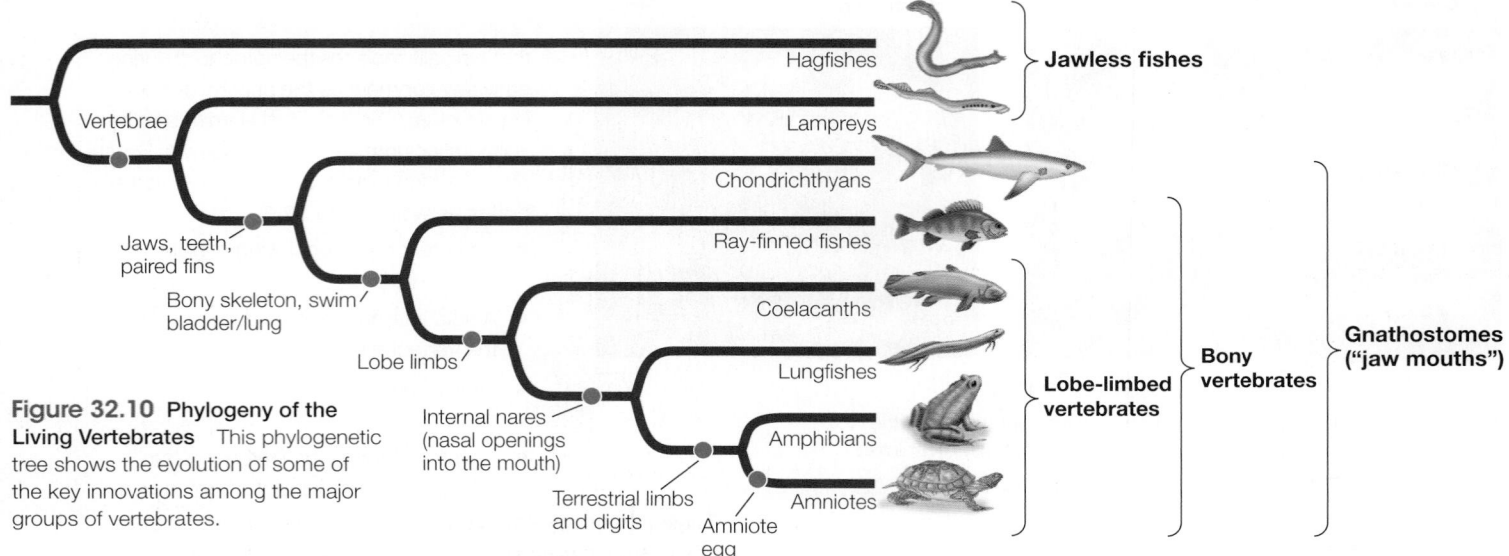

Figure 32.10 Phylogeny of the Living Vertebrates This phylogenetic tree shows the evolution of some of the key innovations among the major groups of vertebrates.

They are virtually blind and rely largely on the four pairs of sensory tentacles around the mouth to detect food. Although they have no jaws, hagfishes have a tonguelike structure equipped with toothlike rasps that they use to tear apart dead organisms and to capture their principal prey, polychaete worms. Hagfishes have direct development (no larvae), and individuals may actually change sex from year to year (from male to female and vice versa).

(A) Eptatretus stoutii

(B) Lampreta fluviatilis

Media Clip 32.2 **Hagfish Slime**
www.Life11e.com/mc32.2

Although the lampreys and hagfishes may look superficially similar (with elongate eel-like bodies and no paired fins), they differ greatly in their biology. Lampreys have a complete skull and distinct and separate (although rudimentary) vertebrae, all cartilaginous rather than bony. Lampreys undergo a complete metamorphosis from filter-feeding larvae, known as ammocoetes, which are morphologically similar to adult lancelets. The adults of many species of lampreys are parasitic, although several lineages of lampreys evolved to become nonfeeding as adults. These nonfeeding adults survive for only a few weeks after metamorphosis—just long enough to breed. In the species that are parasitic as adults, the round mouth is a rasping and sucking organ that is used to attach to prey and rasp at the flesh (see Figure 32.11B).

The nearly 50 species of lampreys either live permanently in fresh water or are anadromous—meaning they live in coastal salt water and move into fresh water to breed. Some species of lampreys are critically endangered because of recent habitat changes and losses.

Jaws and teeth improved feeding efficiency

Many kinds of jawless fishes were found in the seas, estuaries, and fresh waters of the Ordovician, Silurian, and Devonian periods, but hagfishes and lampreys are the only jawless fishes that survived beyond the Devonian. Late in the

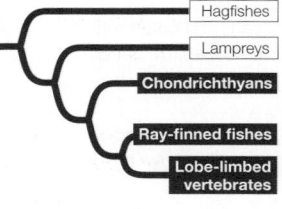

Figure 32.11 Modern Jawless Fishes (A) Hagfishes burrow in the ocean mud, from which they extract small prey. They also scavenge on dead or dying fishes. Hagfishes have degenerate eyes, which has led to their being called (inaccurately) "blind eels." (B) Many species of lampreys are ectoparasites that attach to the bodies of living fishes and use their large, jawless mouth to suck blood and flesh. This species lives in freshwater rivers, but some lampreys can live in marine environments.

(A)

Jawless fishes

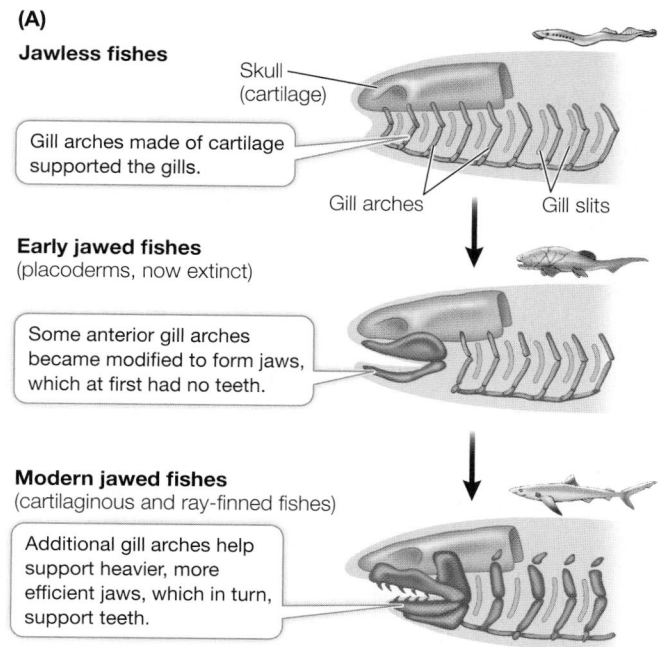

Skull (cartilage)

Gill arches made of cartilage supported the gills.

Gill arches   Gill slits

Early jawed fishes
(placoderms, now extinct)

Some anterior gill arches became modified to form jaws, which at first had no teeth.

Modern jawed fishes
(cartilaginous and ray-finned fishes)

Additional gill arches help support heavier, more efficient jaws, which in turn, support teeth.

(B)

Figure 32.12 Jaws and Teeth Increased Feeding Efficiency
(A) These diagrams illustrate one probable scenario for the evolution of jaws from the anterior gill arches of jawless fishes. **(B)** Jaws of the extinct giant shark *Carcharodon megalodon* display the teeth that indicate an extreme predatory lifestyle.

Ordovician, some fishes evolved jaws via modifications of the skeletal arches that supported the gills (**Figure 32.12A**). Those fishes and their descendants are referred to as **gnathostomes** (Greek, "jaw mouths"). Jaws greatly improved feeding efficiency, as an animal with jaws can grasp, subdue, and swallow large prey. Jawed fishes rapidly diversified during the Devonian period, eventually replacing the jawless fishes in dominance of the seas.

The earliest jaws were simple, but the evolution of teeth made feeding even more efficient (**Figure 32.12B**). In predators, teeth function crucially both in grasping and in breaking up prey. In both predators and herbivores, teeth enable an animal to chew both soft and hard body parts of their food organisms. Chewing also aids chemical digestion and improves an animal's ability to extract nutrients from its food. Vertebrates are remarkable in the diversity of their jaws and teeth, and in their physiological adaptations for *digestion of a wide variety of food items.

***connect the concepts**   The many physiological adaptations of vertebrates for digesting plants and animals are described in detail in Chapter 50.

Fins and swim bladders improved stability and control over locomotion

Most jawed fishes have a pair of pectoral fins just behind the gill slits and a pair of pelvic fins anterior to the anus (see Figures 32.9 and 32.13A). These paired fins stabilize the fish's position in water (and in some cases, help propel it). Median dorsal and anal fins also stabilize the fish, or may be used for propulsion in some species. In many fishes, the caudal (tail) fin helps propel the animal and enables it to turn rapidly.

Several groups of jawed fishes became abundant during the Devonian. Among them were the **chondrichthyans**—sharks, skates, and rays (about 1,000 living species) and chimaeras (40

living species). Like hagfishes and lampreys, these fishes have a skeleton composed entirely of cartilage. Their skin is flexible and leathery, sometimes bearing scales that give it the consistency of sandpaper. Sharks move forward by means of lateral (side-to-side) undulations of the body and caudal fin (**Figure 32.13A**). Skates and rays propel themselves by means of vertical undulating movements of their greatly enlarged pectoral fins (**Figure 32.13B**).

Most sharks are predators, but some feed by straining plankton from the water. Most skates and rays live on the ocean floor, where they feed on mollusks and other animals buried in the sediments. Nearly all chondrichthyans live in the oceans, but a few are estuarine or migrate into lakes and rivers. One group of stingrays is found in river systems of South America. The less familiar chimaeras (**Figure 32.13C**) live in deep-sea or cold waters.

One lineage of aquatic gnathostomes gave rise to the bony vertebrates, which soon split into two main lineages—the **ray-finned fishes** and the **lobe-limbed vertebrates**. Bony vertebrates have internal skeletons of calcified, rigid bone rather than flexible cartilage. In early bony vertebrates, gas-filled sacs supplemented the gas exchange function of the gills by giving the animals access to atmospheric oxygen. These features enabled these fishes to live where oxygen was periodically in short supply, as it often is in freshwater environments. In the ray-finned fishes, these lunglike sacs evolved into **swim bladders**, which are organs of buoyancy. By adjusting the amount of gas in its swim bladder, ray-finned fishes can control the depth at which it remains suspended in the water while expending very little energy to maintain its position.

The outer body surface of most species of ray-finned fishes is covered with thin, flat, lightweight scales that provide protection or enhance movement through the water. The gills open into a single chamber covered by a hard flap, called an operculum. Movement of the operculum increases the flow of water over the gills, where gas exchange takes place.

Ray-finned fishes began to diversify during the Mesozoic era and continued to radiate extensively throughout the Tertiary period. Today there are about 32,000 known living species, encompassing a

(A) *Charcharodon charcharis*

Dorsal fin

Caudal fin   Pelvic fin   Pectoral fin

(B) *Taeniura lymma*

Pectoral fins

(C) *Hydrolagus colliei*

Dorsal fin

Pectoral fin   Pelvic fin

Figure 32.13 Chondrichthyans **(A)** Most sharks are active marine predators, as epitomized by the great white shark seen here. **(B)** Skates and rays, represented here by a blue spotted fantail ray, feed on the ocean bottom. Their modified pectoral fins are used for propulsion; their other fins are greatly reduced. **(C)** A chimaera, or ratfish. Many of these deep-sea fishes possess modified dorsal fins that contain toxins.

remarkable variety of sizes, shapes, and lifestyles (**Figure 32.14**). The smallest are less than 1 centimeter long; the largest weigh as much as 900 kilograms. Ray-finned fishes exploit nearly all types of aquatic food sources. In the oceans they filter plankton from the water, rasp algae from rocks, eat corals and other soft-bodied colonial animals, dig animals from soft sediments, and prey on virtually all kinds of

(A) *Pygoplites diacanthus*

(B) *Phyllopteryx taeniolatus*

(C) *Ostracion meleagris*

(D) *Edriolychnus schmidti*

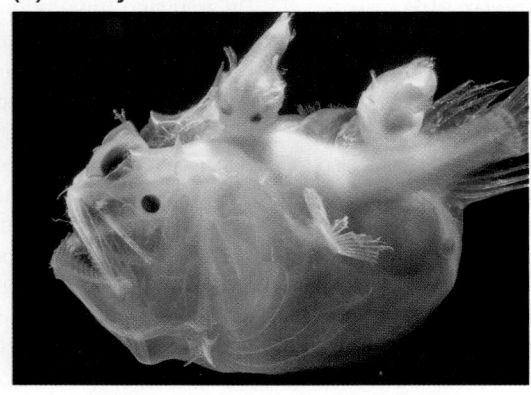

Figure 32.14 Diversity among the Ray-Finned Fishes **(A)** The brightly colored regal angelfish lives around Indo-Pacific coral reefs. Its deep, laterally compressed body and enlarged dorsal and anal fins increase its apparent size. **(B)** This slow-moving weedy seadragon relies on its highly modified body and fins to provide camouflage in swaying seaweed. **(C)** This spotted boxfish is a member of another large ray-fin clade, the serranids, which also includes the sea basses and groupers. **(D)** Deep-sea anglerfishes live below the level of light penetration. The large fish shown here is a female; the two smaller fishes are males, which fuse to the female for life.

other fishes. In fresh water they eat plankton, devour insects, eat fruits that fall into the water, and prey on other aquatic vertebrates and, occasionally, terrestrial vertebrates. Many ray-finned fishes are solitary, but in open water others form large aggregations called schools. Many species perform complicated behaviors to maintain schools, build nests, court mates, and care for their young.

Although ray-finned fishes can readily control their position in open water using their fins and swim bladder, their eggs tend to sink. Some species produce small eggs that are buoyant enough to complete their development in open water, but many marine fishes move to food-rich shallow waters to lay their eggs. That is why coastal waters and estuaries are so important in the life cycles of many marine fishes. Some ray-finned fishes, such as salmon, are anadromous, moving from the ocean to the fresh waters in which they breed.

▶ 32.3 recap

Chordates are characterized by a dorsal hollow nerve chord, a post-anal tail, and a dorsal supporting rod called a notochord at some point during the life cycle. Specialized structures for support (a vertebral column), locomotion (such as fins), and feeding (jaws and teeth) evolved among aquatic vertebrates.

learning outcomes

You should be able to:

- Recognize chordates and vertebrates, and distinguish them from other animals.
- Use examples to assess whether hagfishes should be considered vertebrates.
- Compare and contrast hagfishes and lampreys.
- Describe the modifications and losses of appendages in different lineages of vertebrates and how these modifications are related to function.

1. Describe the synapomorphies that characterize the chordates and the vertebrates, respectively.

2. How do the hagfishes differ from the lampreys in morphology? Why do some biologists contend that hagfishes are not vertebrates?

3. The body plan of most vertebrates is based on four appendages. What are the varied forms that these appendages take, and how are they used? In which lineages have two or more of these appendages been lost?

In the lobe-limbed vertebrates, the gas-filled sacs that gave rise to swim bladders in ray-finned fishes became specialized for another purpose: breathing air. That adaptation set the stage for the vertebrates to move onto the land.

▶ key concept
32.4 Life on Land Contributed to Vertebrate Diversification

The evolution of lunglike sacs in fishes set the stage for the vertebrate invasion of the land. Some early ray-finned fishes probably used those sacs to supplement their gills when oxygen levels in the water were low, as many groups of ray-finned fishes do today. But with their unjointed fins, those fishes could only flop around when out of water. Changes in the structure of the fins first allowed lobe-limbed vertebrates to support themselves better in shallow water, and later, to move better on land.

focus your learning

- Most modern amphibians are confined to moist environments because they rapidly lose water when their skin is exposed to air.
- Birds are a specialized group of theropods and are the only group of dinosaurs to survive the mass extinction event at the end of the Cretaceous.
- Feathers are modified scales, which probably evolved first for insulation and then for flight.
- Mammals increased dramatically in number, diversity, and size after non-avian dinosaurs disappeared during the mass extinction at the end of the Cretaceous.
- Mammals are distinguished by having sweat glands, mammary glands, hair, and a four-chambered heart.
- Movements of the continents affected the diversification of eutherians.
- Modifications of the amniote egg allowed the embryo to develop inside its mother's body.

Jointed limbs enhanced support and locomotion on land

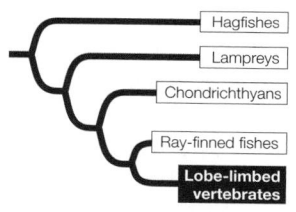

In the lobe-limbed vertebrates, the paired pelvic and pectoral fins developed into more muscular fins that were joined to the body by a single enlarged bone. The modern representatives of these lobe-limbed vertebrates include the coelacanths, lungfishes, and tetrapods.

The coelacanths flourished from the Devonian until about 65 million years ago, when they were thought to have become extinct. But in 1938 a commercial fisherman caught a living coelacanth off South Africa. Since that time, hundreds of individuals of this extraordinary fish, *Latimeria chalumnae*, have been collected (**Figure 32.15A**). A second species, *L. menadoensis*, was discovered in 1998 off the Indonesian island of Sulawesi. *Latimeria*, a predator of other fishes, reaches a length of about 1.8 meters and weighs up to 82 kilograms. Its skeleton is composed mostly of cartilage, not bone. The cartilaginous skeleton is a derived feature in this clade because it had bony ancestors.

 Media Clip 32.3 **Coelacanths in the Deep Seas**
www.Life11e.com/mc32.3

Lungfishes were important predators in shallow-water habitats in the Devonian, but most lineages died out. The six surviving species live in stagnant swamps and muddy waters in South America, Africa, and Australia (**Figure 32.15B**). Lungfishes have lungs derived from the lunglike sacs of their ancestors as well as gills. When ponds dry up, individuals of most species can burrow deep into the mud and survive for many months in an inactive state while breathing air.

It is believed that some early aquatic lobe-limbed vertebrates began to use terrestrial food sources, became more fully adapted to life on land, and eventually evolved to become ancestral **tetrapods**

(A) *Latimeria chalumnae*

(B) *Protopterus annectens*

Figure 32.15 The Closest Relatives of Tetrapods **(A)** The African coelacanth, discovered in deep waters of the Indian Ocean off the South African coast, represents one of two surviving species of a group that was once thought to be extinct. **(B)** All surviving lungfish species, such as this African lungfish, live in the Southern Hemisphere. **(C)** *Tiktaalik*, a fossil lobe-limbed vertebrate from the Devonian, is believed to represent a transitional species intermediate between the finned fishes and the limbed tetrapods.

(C) *Tiktaalik roseae*

Tiktaalik's pectoral fins show some of the skeletal structures of tetrapod limbs.

of which remained tied to moist environments, and the **amniotes**, many of which adapted to much drier conditions.

Amphibians usually require moist environments

Most modern amphibians are confined to moist environments because they lose water rapidly through the skin when exposed to dry air. In addition,

("four legs"). How was this transition from an animal that swam in water to one that walked on land accomplished? Early in 2006, scientists reported the discovery of a Devonian fossil lobe-limbed vertebrate, since then named *Tiktaalik*, which possessed intermediate appendages between the fins of fishes and the limbs of terrestrial tetrapods (**Figure 32.15C**). It appears that limbs capable of propping up a large fish and making the front-to-rear movements necessary for walking evolved while these animals still lived in water. These limbs appear to have functioned in holding the animals upright in shallow water, perhaps even allowing them to hold their head above the water's surface. These same structures were then co-opted for movement on land, at first probably for foraging on brief trips out of water.

Among the lobe-limbed vertebrates, limbs capable of movement on land evolved from the short, muscular fins of aquatic ancestors (**Figure 32.16**). The resulting four terrestrial limbs give the tetrapods their name. The basic skeletal elements of those limbs can be traced through major changes in limb form and function among the terrestrial vertebrates.

An early split in the tetrapod tree (see Figure 32.10) led to two main groups of terrestrial vertebrates: **amphibians**, most

Figure 32.16 Tetrapod Limbs Are Modified Fins The major skeletal elements of the tetrapod limb were already present in aquatic lobe-limbed fishes some 380 million years ago. The relative sizes and positions of these elements changed as lobe-limbed vertebrates moved to a terrestrial environment, where limbs were needed to support and move the animal's body on land.

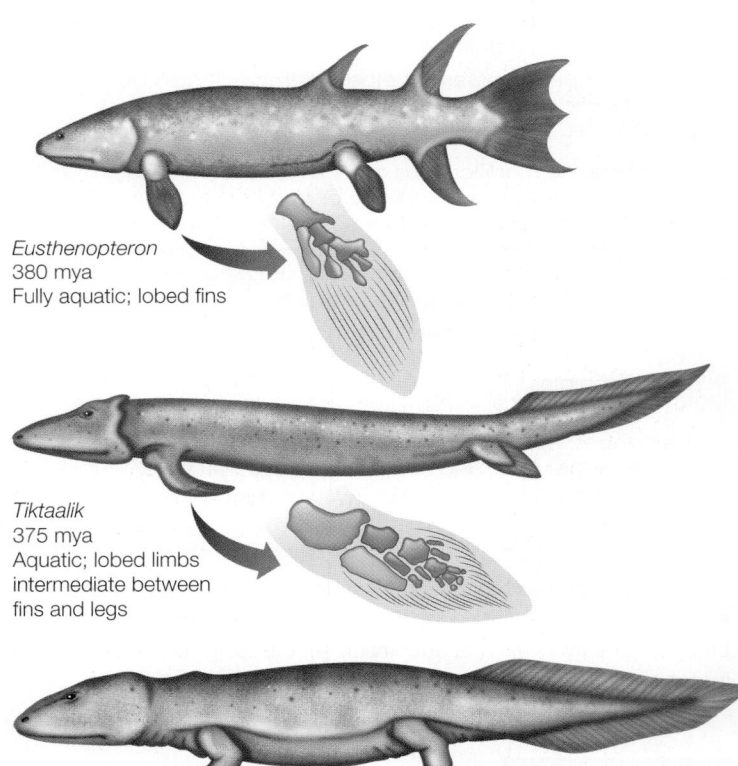

Eusthenopteron
380 mya
Fully aquatic; lobed fins

Tiktaalik
375 mya
Aquatic; lobed limbs intermediate between fins and legs

Acanthostega
365 mya
Semiterrestrial tetrapod

- Humerus
- Radius
- Ulna
- Distal elements of fin/wrist/hand

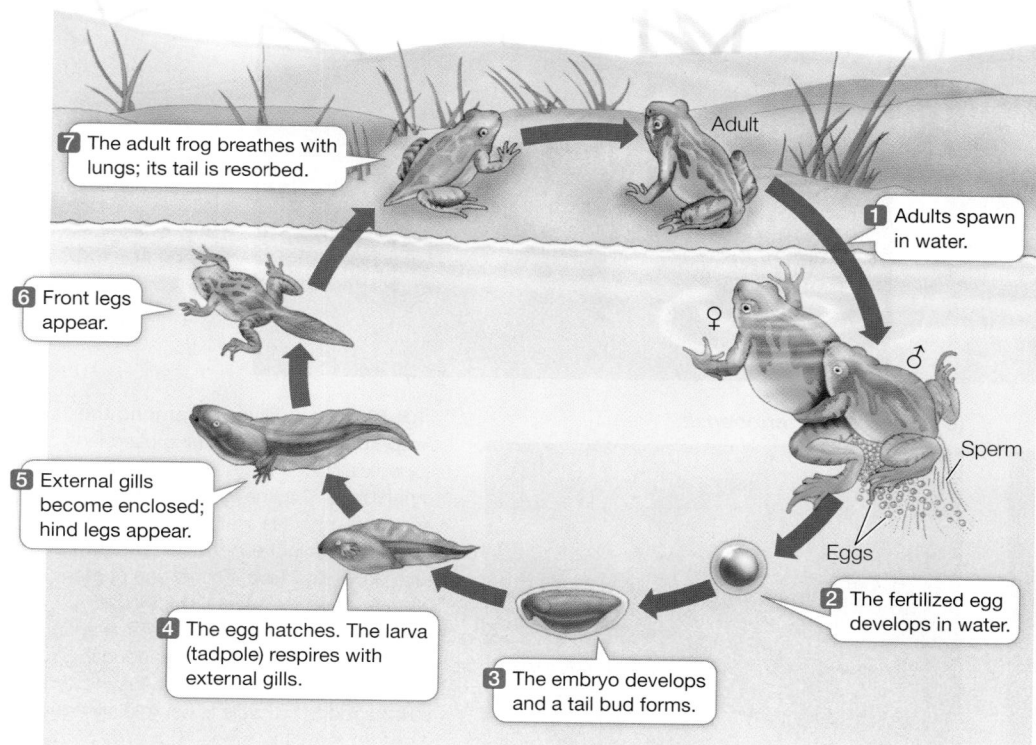

7 The adult frog breathes with lungs; its tail is resorbed.

6 Front legs appear.

5 External gills become enclosed; hind legs appear.

4 The egg hatches. The larva (tadpole) respires with external gills.

3 The embryo develops and a tail bud forms.

Adult

1 Adults spawn in water.

Sperm

Eggs

2 The fertilized egg develops in water.

Figure 32.17 In and Out of the Water
Many amphibian species have life cycles like the one diagrammed here, in which the early stages take place in water and the aquatic tadpole transforms into a terrestrial adult through metamorphosis. Some species of amphibians, however, have direct development (with no aquatic larval stage), and others are aquatic throughout life.

Q: What is a potential advantage of maintaining an aquatic tadpole stage, even though the adults are terrestrial?

▶ Animation 32.2 **Life Cycle of a Frog**
www.Life11e.com/a32.2

their eggs are enclosed within delicate gelatinous envelopes that cannot prevent water loss in dry conditions. In some amphibian species, adults live mostly on land but return to fresh water to lay and fertilize their eggs (**Figure 32.17**). The fertilized eggs give rise to larvae that live in water until they undergo metamorphosis to become terrestrial adults. However, many amphibians (especially those in tropical and subtropical areas) have evolved a wide variety of additional reproductive modes and types of parental care. Internal fertilization, for example, evolved several times among amphibian species. Many species develop directly into adultlike forms from fertilized eggs laid on land or carried by the parents. Other species of amphibians are entirely aquatic, never leaving the water at any stage of their lives, and many of these species retain a larval-like morphology.

The more than 7,000 known species of amphibians living today belong to three major groups: the wormlike, limbless, tropical, burrowing, or aquatic caecilians (**Figure 32.18A**), the tailless frogs and toads (collectively called anurans) (**Figure 32.18B**), and the tailed salamanders (**Figure 32.18C and D**).

Anurans are most diverse in wet tropical and warm temperate regions, although a few are found at very high latitudes. There are far more anurans than any other amphibians, with well over 6,000 described species and more being discovered every year. Some anurans have tough skins and other adaptations that enable them to live for long periods in deserts, whereas others live in moist terrestrial and arboreal environments. Some species are completely aquatic as adults. All anurans have a short vertebral column and a pelvic region that is modified for leaping, hopping, or propelling the body through water by kicking the hind legs.

The more than 600 described species of salamanders are most diverse in temperate regions of the Northern Hemisphere and in

cool, moist environments in the mountains of Central America, although a few species penetrate into tropical regions. Many salamanders live in rotting logs or moist soil. One major group has lost the lungs, and these species exchange gases entirely through the skin and mouth lining—body parts that all amphibians use, in addition to their lungs, for gas exchange. A completely aquatic lifestyle has evolved several times among the salamanders (see Figure 32.18D). These aquatic species have arisen through a developmental process known as **neoteny**, or the retention of juvenile traits (in this case, gills) by delayed somatic development. Most species of salamanders have internal fertilization, which is usually achieved through the transfer of a small, jellylike, sperm-embedded capsule called a spermatophore.

Many amphibians have complex social behaviors. Most male anurans utter loud, species-specific calls to attract females of their own species (and sometimes to defend breeding territories), and they compete for access to females that arrive at the breeding sites. Many amphibians lay large numbers of eggs, which they abandon once they are deposited and fertilized. Other amphibians lay only a few eggs, which are fertilized and then cared for. A few species of frogs, salamanders, and caecilians are viviparous, meaning that they give birth to well-developed young that have received nutrition from the female during gestation.

▶ Media Clip 32.4 **Answering a Mating Call**
www.Life11e.com/mc32.4

Amphibians are the focus of much attention today because populations of many species are declining rapidly, especially in mountainous regions of western North America, Central and South America, and northeastern Australia. Worldwide, about one-third of amphibian species are now threatened with extinction or have disappeared

(A) *Gymnopis multiplicata*

(B) *Bufo periglenes*

(C) *Gyrinophilus porphyriticus*

(D) *Eurycea waterlooensis*

Figure 32.18 Diversity among the Amphibians   **(A)** Burrowing caecilians superficially look more like worms than like amphibians They have internal fertilization and many species are viviparous. **(B)** Male golden toads in the cloud forest of Monteverde, Costa Rica. This is one of many amphibian species lost to extinction in the past few decades. **(C)** An adult spring salamander. This species has aquatic larvae but terrestrial adults. **(D)** This Austin blind salamander's life cycle is completely aquatic; it has no terrestrial stage. The eyes of this cave dweller are greatly reduced.

completely in the last few decades. Scientists are investigating several hypotheses to account for these *amphibian declines.

---

*connect the concepts   One of the major factors associated with amphibian declines is the global spread of a pathogenic chytrid fungus, as explored in Key Concept 29.2 and Figure 29.6.

Amniotes colonized dry environments

Several key innovations for conserving water contributed to the ability of the amniotes to exploit a wide range of terrestrial habitats. The **amniote egg** (which gives the group its name) is relatively impermeable to water and allows the embryo to develop in a contained aqueous environment (**Focus: Key Figure 32.19A**). Its leathery or brittle, calcium-impregnated shell retards evaporation of the fluids

▶ focus: key figure

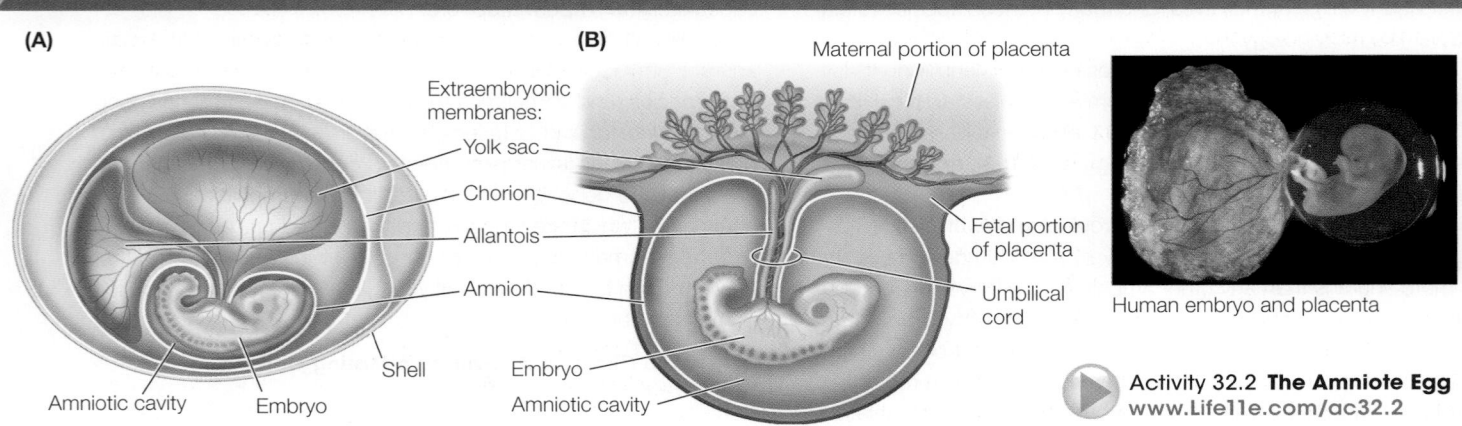

(A)

Extraembryonic membranes:
- Yolk sac
- Chorion
- Allantois
- Amnion

Shell
Amniotic cavity
Embryo

(B)

Maternal portion of placenta
Fetal portion of placenta
Umbilical cord
Embryo
Amniotic cavity

Human embryo and placenta

▶ Activity 32.2 **The Amniote Egg**
www.Life11e.com/ac32.2

Figure 32.19 The Amniote Egg   **(A)** The evolution of the amniote egg, with its water-retaining shell, four extraembryonic membranes, and embryo-nourishing yolk, was a major step in adaptation to the terrestrial environment. A chicken egg is shown here. **(B)** In most mammals, the developing embryo is retained inside the mother's body, with which it exchanges nutrients and wastes via the placenta.

Q: How were the various membranes in the amniote egg modified in mammals to support viviparity?

experiment

Original Papers: Pyron, R. A. and F. T. Burbrink. 2014. Early origin of viviparity and multiple reversions to oviparity in squamate reptiles. *Ecological Letters* 17: 13–21.

Wright, A. M., K. M. Lyons, M. C. Brandley and D. M. Hillis. 2015. Which came first: The lizard or the egg? Robustness in phylogenetic reconstruction of ancestral states. *Journal of Experimental Zoology (Molecular and Developmental Evolution)* 324B: 504–516.

The shelled amniote egg is a complex structure, and the transition from oviparity to viviparity requires extensive changes in reproductive morphology and physiology. For example, oviparous species have uterine shell glands that secrete calcium for eggshell construction. These shell glands are modified in viviparous species to provide calcium to the developing embryo. Viviparity is thought to have evolved from oviparity more than 100 times among lizards and snakes, but biologists wondered if the process can be reversed: does egg laying ever re-evolve in live-bearing lineages?

Oviparous: Pueblan milk snake

Viviparous: Blotched blue-tongue lizard

QUESTION▶ Can complex traits such as shelled eggs re-evolve once they have been lost?

METHOD

1. Reconstruct the phylogeny of lizards and snakes (squamates).
2. Collect data on the reproductive mode of more than 8,000 species of lizards and snakes.
3. Reconstruct the evolution of transitions between oviparity and viviparity on the phylogeny.

RESULTS

The number of reversals from viviparity to oviparity depends on the model of character change, but there is support for the re-evolution of egg laying from live birth in one group of lizards and in two groups of snakes.

CONCLUSION▶ Although transitions from oviparity to viviparity are much more common than the reverse, egg laying may have re-evolved in a few lineages of squamates.

investigating**life work with the data** follows on next page.

---

inside but permits passage of oxygen and carbon dioxide. The amniote egg also stores large quantities of food in the form of **yolk**, allowing the embryo to attain a relatively advanced state of development before it hatches. Within the shell are **extraembryonic membranes** that protect the embryo from desiccation and assist it with gas exchange and excretion of nitrogenous waste products of metabolism.

As we described in the opening of this chapter, modifications of the amniote egg in several different groups of amniotes allowed the embryo to develop inside (and exchange nutrients and wastes with) its mother's body (**Focus: Key Figure 32.19B**). For example, in most mammals the egg lost its shell entirely while the functions of the extraembryonic membranes were retained and expanded, giving rise to viviparity. Viviparity appears to have evolved many times among lizards and snakes. In **Investigating Life: Have Shelled Eggs Re-Evolved among Viviparous Reptile Lineages?**, we explore the possibility that shelled eggs could have re-evolved in some of these lineages.

Other innovations evolved in the organs of terrestrial adults. A tough, impermeable skin, covered with scales or modifications of scales such as hair and feathers, greatly reduced water loss. Adaptations of the kidneys (the vertebrate excretory organs), allowed amniotes to excrete concentrated urine, ridding the body of nitrogenous wastes without losing a large amount of water in the process (see Chapter 51).

During the Carboniferous, the amniotes split into two major groups: the **reptiles** and the lineage that eventually led to the **mammals (Figure 32.20).**

Reptiles adapted to life in many habitats

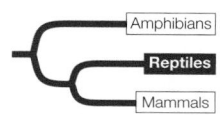

The lineage leading to modern reptiles began to diverge from other amniotes more than 300 million years ago. More than 19,000 species of reptiles exist today, more than half of which are birds. Birds are the only living representatives of the otherwise extinct dinosaurs, the dominant terrestrial vertebrates of the Mesozoic.

The **lepidosaurs** constitute the second most species-rich clade of living reptiles. This group is composed of the **squamates** (lizards, snakes, and amphisbaenians—the last a group of mostly legless, wormlike, burrowing reptiles with greatly reduced eyes) and the **tuataras**, which superficially resemble lizards but differ from them in possessing more rigid skulls and other anatomical features. Many species related to the tuataras lived during the Mesozoic era, but today only a single species survives (**Figure 32.21A**), and is restricted to a few islands off New Zealand.

The skin of a lepidosaur is covered with horny scales that greatly reduce loss of water from the body surface. These scales, however, make the skin unavailable as an organ of gas exchange. Gases are exchanged almost entirely via the lungs, which are proportionally much larger in surface area than those of amphibians. A lepidosaur forces air into and out of its lungs by bellows-like movements of its ribs. The three-chambered lepidosaur heart partially separates oxygenated blood from the lungs from deoxygenated blood returning from the body. With this type of heart, lepidosaurs can generate high blood pressure and can sustain a relatively high metabolism.

work with the data

Original Paper: Fenwick, A. M., H. W. Greene and C. L. Parkinson. 2012. The serpent and the egg: Unidirectional evolution of reproductive mode in vipers? *Journal of Zoological Systematics and Evolutionary Research* 50: 59–66.

The tree below shows a small portion of the phylogeny and reproductive data collected for squamates, from the New World pit vipers. Use this data sample to answer the questions.

QUESTIONS▶

1. Starting with a root state of oviparity, plot the minimum number of changes between oviparity and viviparity (in either direction) that is necessary to explain the distribution of reproductive states on the tree. What is the minimum number of changes required on this tree? Does that number include any reversals to oviparity?

2. Assume that the only changes that can occur are changes from oviparity to viviparity (do not allow any changes in the opposite direction). How many additional transitions in reproductive mode are necessary under this assumption compared with the result you obtained in Question 1?

A similar **work with the data** exercise may be assigned in **LaunchPad**.

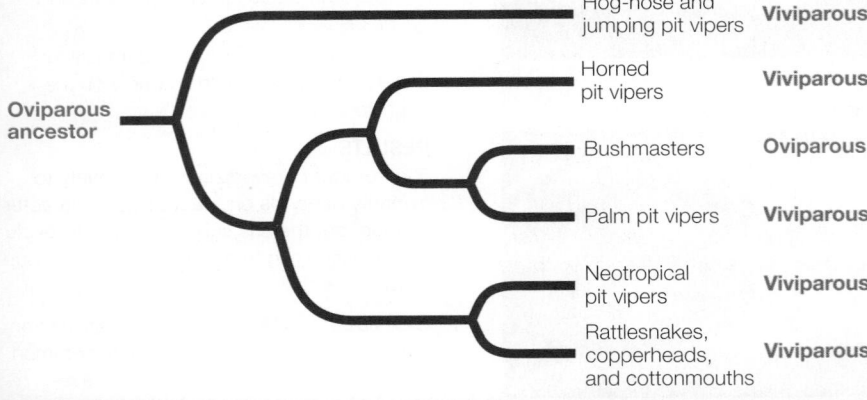

Oviparous ancestor	
Hog-nose and jumping pit vipers	**Viviparous**
Horned pit vipers	**Viviparous**
Bushmasters	**Oviparous**
Palm pit vipers	**Viviparous**
Neotropical pit vipers	**Viviparous**
Rattlesnakes, copperheads, and cottonmouths	**Viviparous**

Most lizards are insectivores, although some are herbivores and a few prey on other vertebrates. Most lizards walk on four limbs (**Figure 32.21B**), although limblessness has evolved repeatedly among the lizards, especially in burrowing and grassland species. The largest lizard is the predaceous Komodo dragon of the East Indies, which grows as long as 3 meters and can weigh more than 150 kilograms.

 Media Clip 32.5 **Komodo Dragons Bring Down Prey**
www.Life11e.com/mc32.5

The major group of limbless squamates is the snakes (**Figure 32.21C**). All snakes are carnivores, and many can swallow objects much larger than themselves. Several snake groups possess venom glands and the ability to inject venom rapidly into their prey.

The **turtles** comprise a reptilian group that has changed relatively little since the early Mesozoic. In these reptiles, dorsal and ventral bony plates form a shell into which the head and limbs can be withdrawn in many species (**Figure 32.21D**). The dorsal shell is

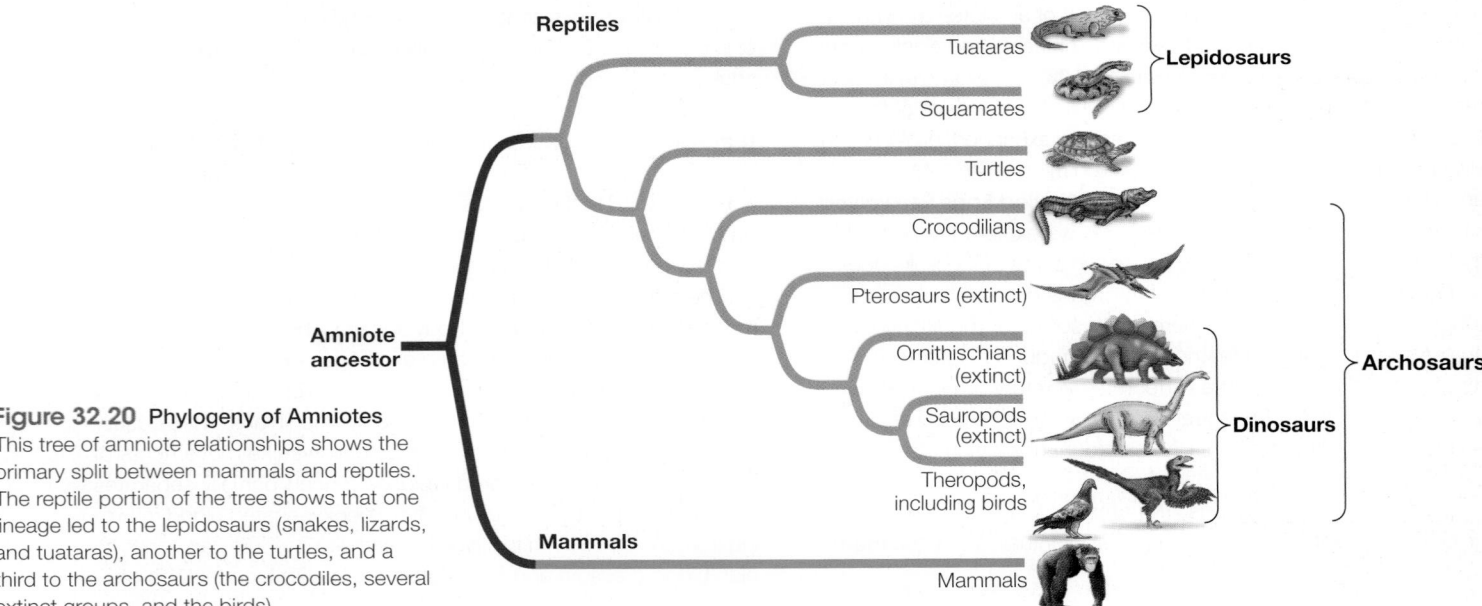

Figure 32.20 Phylogeny of Amniotes
This tree of amniote relationships shows the primary split between mammals and reptiles. The reptile portion of the tree shows that one lineage led to the lepidosaurs (snakes, lizards, and tuataras), another to the turtles, and a third to the archosaurs (the crocodiles, several extinct groups, and the birds).

(A) *Sphenodon punctatus*

(B) *Furcifer pardalis*

(C) *Orthriophis taeniura friesi*

(D) *Chelonoidis nigra abingdonii*

Figure 32.21 Reptilian Diversity **(A)** This tuatara represents the only surviving species in its lineage. **(B)** The panther chameleon is found in eastern and northern Madagascar. It uses its projectile tongue to capture insects. **(C)** The Taiwan beauty snake is a nonvenomous, semi-arboreal rodent eater. **(D)** Galápagos tortoises are the largest turtles and among the largest reptiles. They have been documented to live for more than 100 years in the wild.

a modification of the ribs. It is notable that pectoral girdles are inside the ribs of turtles, making them unlike any other vertebrates. Most turtles live in aquatic environments, but several groups, such as tortoises and box turtles, are terrestrial. Sea turtles spend their entire lives at sea except when they come ashore to lay eggs. Human exploitation of sea turtles and their eggs has resulted in worldwide declines of these species, all of which are now endangered. A few species of turtles are strict herbivores or carnivores, but most species are omnivores that eat a variety of aquatic and terrestrial plants and animals.

Crocodilians and birds share their ancestry with the dinosaurs

Another reptilian clade, the **archosaurs**, includes the crocodilians, pterosaurs, dinosaurs, and birds. Only the crocodilians and birds are represented by living species today. Modern **crocodilians**—crocodiles, caimans, gharials, and alligators—are confined to tropical and warm temperate environments (**Figure 32.22A**). All crocodilians are carnivorous; they eat vertebrates of all kinds, including large mammals. Crocodilians spend much of their time in water, but they lay their eggs in nests they build on land or on floating piles of vegetation. The eggs are warmed by heat generated by decaying organic matter that the female places in the nest. The female provides other forms of parental care as well: typically she guards the eggs until they hatch, and in some species she continues to guard and communicate with her offspring after they hatch.

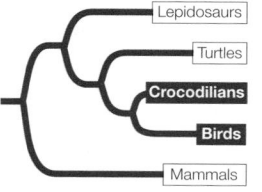

Lepidosaurs
Turtles
Crocodilians
Birds
Mammals

Dinosaurs rose to prominence about 215 million years ago and dominated terrestrial environments for about 150 million years. However, only one group of dinosaurs, the **birds**, survived the mass extinction at the end of the Cretaceous. During the Mesozoic, most terrestrial animals more than a meter long were dinosaurs. Many were agile and could run rapidly; they had special muscles that enabled the lungs to be filled and emptied while the limbs moved. We can infer the existence of such muscles in dinosaurs from the structure of the vertebral column in fossils. Some of the largest dinosaurs weighed as much as 70,000 kilograms.

Biologists have long accepted the phylogenetic position of birds among the reptiles, although birds clearly have many unique, derived morphological features. In addition to the strong morphological evidence for this placement, fossil and molecular data emerging over the last few decades have provided definitive supporting evidence. Birds are a specialized group of **theropods**, a clade of predatory dinosaurs that shared such traits as a bipedal stance, hollow bones, a furcula ("wishbone"), elongated metatarsals with three-fingered feet, elongated forelimbs with three fingers, and a pelvis that points backward. Modern birds are endothermic, meaning that they regulate their body temperatures by producing and retaining metabolic heat, rather than by absorbing heat from their external environment (see Key Concept 39.4). Although we cannot directly assess this physiological trait in extinct species, many fossil theropods share morphological traits that suggest they may have been endothermic as well.

The living bird species fall into two major groups that diverged about 80–90 million years ago from a flying ancestor. The few modern descendants of one lineage include a group of secondarily flightless and weakly flying birds, some of which are very large. This group, called the palaeognaths, includes the South and Central American tinamous and several large flightless birds of the southern continents—the rheas, emu, kiwis, cassowaries, and the world's largest birds, the ostriches (**Figure 32.22B**). The second lineage, the neognaths, has left a much larger number of descendants, most of which have retained the ability to fly.

(A) *Crocodylus porosus*

(B) *Casuarius casuarius*

Figure 32.22 Archosaurs The two surviving groups of archosaurs are very different. **(A)** The crocodilians live in tropical and warm temperate climates. This saltwater crocodile lives in saltwater and estuarine environments along Australia's coast. **(B)** Birds are the only other living archosaur group, represented here by the winged but flightless southern cassowary.

Feathers allowed birds to fly

Fossil theropods discovered in early Cretaceous deposits in Liaoning Province, in northeastern China, show that the scales of some small predatory dinosaurs were highly modified to form **feathers**. Initially these feathers were simply a body covering that probably provided insulation and enhanced coloration. But the feathers of

some later dinosaurs, such as *Microraptor gui*, were structurally similar to those of modern birds (**Figure 32.23A**).

Another theropod that was even more closely related to modern birds, *Archaeopteryx*, lived about 150 million years ago. *Archaeopteryx* had teeth (unlike modern birds), but it was covered with feathers that are virtually identical to those of birds (**Figure 32.23B**). It also had well-developed wings, a long tail, and a furcula to which some of the flight muscles were probably attached. *Archaeopteryx* had clawed fingers on its forelimbs, but it also had typical perching bird claws on its hindlimbs. It probably lived in trees and shrubs and used the fingers to assist it in clambering over branches. It probably glided or flew weakly. The descendants of *Archaeopteryx* and similar Mesozoic theropods were the modern birds, most of which are accomplished fliers.

 Media Clip 32.6 Falcons in Flight
www.Life11e.com/mc32.6

The evolution and specialization of feathers was a major force for diversification. Feathers are lightweight, but they are strong and structurally complex (**Figure 32.24**). The stiff central shaft of the

(A)

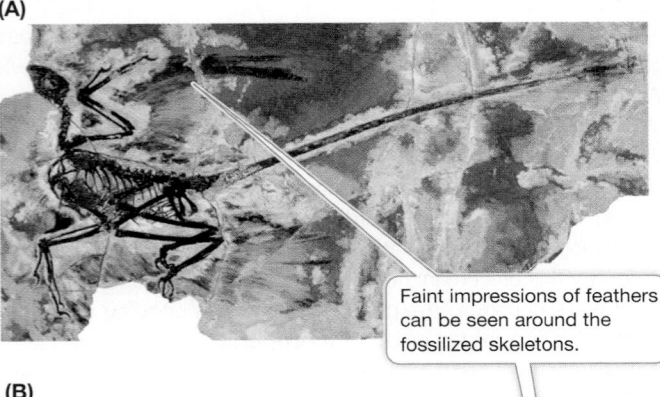

> Faint impressions of feathers can be seen around the fossilized skeletons.

(B)

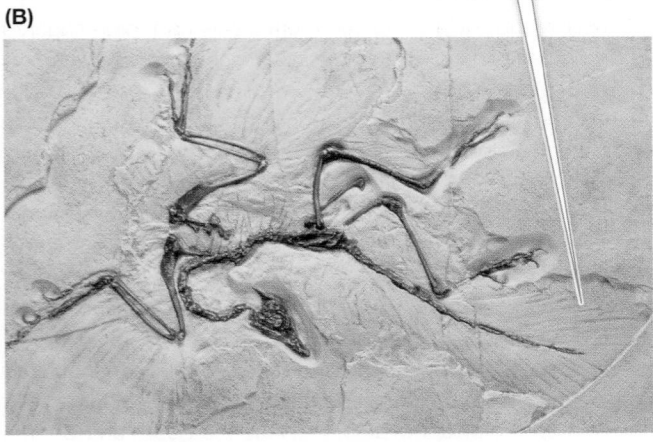

Figure 32.23 Mesozoic Bird Relatives Fossils support the evolution of birds from other theropods. **(A)** *Microraptor gui* was a feathered theropod from the early Cretaceous (about 140 mya). **(B)** *Archaeopteryx* was even more closely related to modern birds.

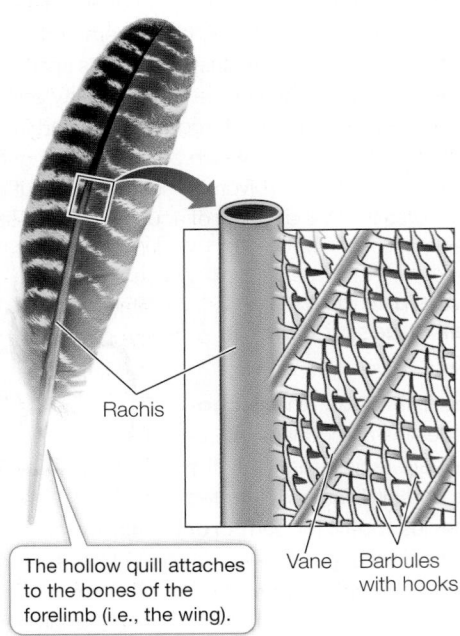

Rachis

> The hollow quill attaches to the bones of the forelimb (i.e., the wing).

Vane   Barbules with hooks

Figure 32.24 Feather Anatomy The flight feathers of birds are attached to the wing's skin by the hollow portion, or quill, of a stiff central shaft. The rachis is the solid portion of the shaft from which radiate fine branches (vanes) with interlocking hooks and barbules. Overall, this structure represents a major evolutionary innovation: a strong, lightweight surface that enables flight.

(A) *Sialia mexicana*

(B) *Trichoglossus haematodus*

(C) *Bubo virginianus*

(D) *Numenius americanus*

Figure 32.25 **Diversity among the Birds** **(A)** Perching, or passeriform, birds such as this western bluebird constitute the most species-rich of all bird groups. **(B)** Some 375 species of parrots, macaws, parakeets, and lorikeets such as the one shown here are another large bird group. **(C)** The great horned owl is a nighttime predator that can find prey using its sensitive auditory "sonar" system. **(D)** The long-billed curlew of North America is one of many species of wading birds.

flight feathers on a bird's wings arises from the skin of the forelimbs to create the flying surfaces. Other strong feathers sprout like a fan from the shortened tail and serve as stabilizers during flight. The feathers that cover the body, along with an underlying layer of down feathers, provide birds with insulation that helps them survive in virtually all of Earth's climates.

The bones of theropod dinosaurs, including birds, are hollow with internal struts that increase their strength. Hollow bones would have made early theropods lighter and more mobile; later they facilitated the evolution of flight. The sternum (breastbone) of flying birds forms a large, vertical keel to which the flight muscles are attached.

Flight is metabolically expensive. A flying bird consumes energy at a rate about 15–20 times faster than a running lizard of the same weight. Because birds have such high metabolic rates, they generate large amounts of heat. They control the rate of heat loss using their feathers, which may be held close to the body or elevated to alter the amount of insulation they provide. The lungs of birds allow air to flow through unidirectionally rather than pumping air in and out (see Key Concept 48.2). This flow-through structure of the lungs increases the efficiency of gas exchange and thereby supports a high metabolic rate.

There are about 10,000 species of living birds, which range in size from the 150-kilogram ostriches to a tiny hummingbird weighing only 2 grams (**Figure 32.25**). The teeth so prominent among other dinosaurs were secondarily lost in the ancestral birds, but birds nonetheless eat almost all types of animal and plant material. Insects and fruits are the most important dietary items for terrestrial species. Birds also eat seeds, nectar and pollen, leaves and buds, carrion, and other vertebrates. By eating the fruits and seeds of plants, birds serve as major agents of seed dispersal.

Mammals radiated after the extinction of non-avian dinosaurs

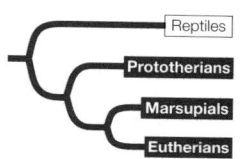

Small and medium-sized mammals coexisted with the large dinosaurs throughout most of the Mesozoic era, and most of the major groups of mammals that are alive today arose in the Cretaceous. After the non-avian dinosaurs disappeared during the mass extinction at the end of the Cretaceous, mammals increased dramatically in numbers, diversity, and size. Today mammals range in size from tiny shrews and bats weighing only about 2 grams to the blue whale, the largest animal on Earth, which measures up to 32 meters long and can weigh as much as 160,000 kilograms. Mammals have far fewer, but more highly differentiated, teeth than do fishes, amphibians, or reptiles. Differences among mammals in the number, type, and arrangement of teeth reflect their varied diets (see Figure 50.6).

Four key features distinguish the mammals:

1. *Sweat glands*, which secrete sweat that evaporates and thereby cools an animal

2. *Mammary glands*, which in females secrete a nutritive fluid (milk) on which newborn individuals feed

3. *Hair*, which provides a protective and insulating covering

4. A *four-chambered heart* that completely separates the oxygenated blood coming from the lungs from the deoxygenated blood returning from the body (this last characteristic is convergent with the archosaurs, including modern birds and crocodiles)

Mammalian eggs are fertilized within the female's body, and in nearly all mammalian groups the resulting embryos undergo a period of development inside the female's body in an organ called the uterus. In the uterus, the embryo is contained in an amniotic sac that is homologous to one of the four membranes found in the amniote egg (see Figure 32.19). The embryo is connected to the wall of the uterus by an organ called a placenta. The placenta allows for nutrient and gas exchange,

table 32.1 Major Groups of Living Mammals

Group	Number of described species	Examples
PROTOTHERIANS		
Monotremes (Monotremata)	5	Echidnas, duck-billed platypus
THERIANS		
Marsupials		
Diprotodonts (Diprotodontia)	146	Kangaroos, wallabies, possums, koala, wombats
New World opossums (Didelphimorphia)	93	Opossums
Carnivorous marsupials (Dasyuromorphia)	75	Quolls, dunnarts, numbat, Tasmanian devil
Omnivorous marsupials (Peramelemorphia)	24	Bandicoots and bilbies
Shrew opossums (Paucituberculata)	7	Andean rat opossums
Marsupial moles (Notoryctemorphia)	2	Southern and northern marsupial moles
Microbiothere (Microbiotherea)	1	Monito del monte
Eutherians		
Rodents (Rodentia)	2,337	Rats, mice, squirrels, woodchucks, ground squirrels, beaver, capybara
Bats (Chiroptera)	1,171	Fruit bats, echo-locating bats
Even-toed hoofed mammals and cetaceans (Cetartiodactyla)	469	Deer, sheep, goats, cattle, antelopes, giraffes, camels, swine, hippopotamus, whales, dolphins
Shrews, moles, and relatives (Soricomorpha)	428	Shrews, moles, solenodons
Primates (Primates)	396	Lemurs, monkeys, apes, humans
Carnivores (Carnivora)	284	Wolves, dogs, bears, cats, weasels, pinnipeds (seals, sea lions, walruses)
Rabbits and relatives (Lagomorpha)	92	Rabbits, hares, pikas
African insectivores (Afrosoricida)	50	Tenrecs, golden moles
Hedgehogs (Erinaceomorpha)	24	European hedgehog
Armadillos (Cingulata)	21	Giant armadillo, nine-banded armadillo
Tree shrews (Scandentia)	20	Pygmy tree shrew, pen-tailed tree shrew
Odd-toed hoofed mammals (Perissodactyla)	16	Horses, zebras, tapirs, rhinoceroses
Elephant shrews (Macroscelidea)	15	Elephant shrews, jumping shrews, sengis
Anteaters, sloths (Pilosa)	10	Anteaters, tamanduas, two- and three-toed sloths
Pangolins (Pholidota)	8	Asian and African pangolins
Hyraxes and relatives (Hyracoidea)	5	Hyraxes, dassies
Sirenians (Sirenia)	4	Manatees, dugongs
Elephants (Proboscidea)	3	African and Indian elephants
Colugos (Dermoptera)	2	Flying lemurs
Aardvark (Tubulidentata)	1	Aardvark

as well as waste elimination from the developing embryo, via the female's circulatory system. Most mammals develop a covering of hair (fur), which is luxuriant in some species but has been greatly reduced in others, including cetaceans (whales and dolphins) and humans. Thick layers of insulating fat (blubber) replace hair as a heat-retention mechanism in the cetaceans. Humans learned to use clothing for this purpose when they dispersed from warm tropical areas.

The approximately 5,700 species of living mammals are divided into two primary groups: the **prototherians** and the **therians** (**Table 32.1**). Members of the therian clade are further divided into the **marsupials** and the **eutherians**.

PROTOTHERIANS   Only five species of living prototherians are known, and they are found only in Australia and New Guinea. These mammals, the duck-billed platypus and four species of echidnas, differ from other mammals in laying shelled eggs and having sprawling legs (**Figure 32.26**). Prototherians supply milk for their young, but they have no nipples on their mammary glands; the milk simply oozes out and is lapped off the fur by the offspring.

MARSUPIALS   Females of most marsupial species have a ventral pouch in which they carry and feed their offspring (see Figure 32.27A). Gestation (pregnancy) in marsupials is brief; the young are born tiny but with well-developed forelimbs, with which they climb to the pouch. They attach to a nipple but cannot suck. The mother ejects milk into the tiny offspring until it grows large enough to suckle. Once her offspring have left the uterus, a female marsupial may become sexually receptive again. She can then carry fertilized eggs that are capable of initiating development to replace the offspring in her pouch should something happen to them.

(A) *Tachyglossus aculeatus*

(B) *Ornithorhynchus anatinus*

Figure 32.26 Prototherians   (A) The short-beaked echidna is one of four surviving species of echidnas. **(B)** The duck-billed platypus lives in freshwater streams in eastern Australia.

At one time marsupials were found on all continents, but the approximately 350 living species are now restricted to Australasia (**Figure 32.27A and B**) and the Americas (especially South America) (**Figure 32.27C**). Of the seven major groups of marsupials shown in Table 32.1, only the New World opossums, the shrew opossums, and the diminutive monito del monte are found in the Americas. Only one species, the Virginia opossum, is found in North America north of Mexico. Marsupials radiated to become herbivores, insectivores, and carnivores, but no marsupials live in the oceans. None can fly, although some arboreal (tree-dwelling) marsupials are gliders. The largest living marsupials are the kangaroos of Australia, which can weigh up to 90 kilograms. Much larger marsupials existed in Australia until humans exterminated them soon after reaching that continent about 40,000 years ago.

EUTHERIANS   The majority of mammals are eutherians ("true" therians). Eutherians are sometimes called placental mammals, but this name is inappropriate because marsupials also have placentas, although they are not as well developed as in eutherians. Eutherians are more developed at birth than are marsupials; no external pouch houses them after they are born.

The more than 5,300 species of living eutherians are divided into 20 major groups (see Table 32.1). The relationships of these groups to one another have been difficult to determine because most of the major groups diverged within a short time during an explosive adaptive radiation. Modern genomic analyses have elucidated these relationships, however (**Figure 32.28**). These studies have revealed that the major early splits in eutherian lineages are closely associated with the breakup of the continents during the Mesozoic (see Figure 24.14), after which the major groups of mammals radiated independently in Laurasia, Africa, and South America. The reconnection of South America and North America via the Panamanian land bridge about 3 million years ago resulted in a huge faunal exchange between those continents, which is particularly evident among the mammals. South American groups such as armadillos moved north into North America, and Laurasian groups such as carnivores and odd- and even-toed hoofed animals moved south into South America.

Eutherians are extremely varied in their form and ecology (**Figure 32.29**). The extinction of the non-avian dinosaurs at the end of the Cretaceous may have made it possible for them to diversify and radiate into a large range of ecological niches. Many eutherian species grew large, and some assumed the roles of dominant terrestrial predators previously occupied by the large dinosaurs. Among these predators, social hunting behavior evolved in several species, including members of the carnivore and primate clades.

The two most diverse groups of eutherians are the rodents and the bats, which together account for about two-thirds of the species. Rodents are traditionally defined by the unique morphology of their teeth,

(A) *Macropus giganteus*

(B) *Myrmecobius fasciatus*

(C) *Didelphis virginiana*

Figure 32.27 Diversity among the Marsupials
(A) Australia's eastern gray kangaroo is among the largest living marsupials. This female carries her young offspring in the characteristic marsupial pouch. **(B)** The diet of the banded anteater of western Australia consists almost entirely of termites. **(C)** The North American opossum is the only marsupial found north of Mexico.

Figure 32.28 Major Groups of Eutherians Diversified as the Continents Drifted Apart This phylogenetic tree shows the relationships among most of the major terrestrial eutherian groups, the location of the earliest fossils found for each group (also indicated in the color distinctions of the various branches), and the current distributions of the groups. The major splits in eutherian evolution correspond in large degree to the tectonic history of the major continents (see Figures 24.14 and 53.14).

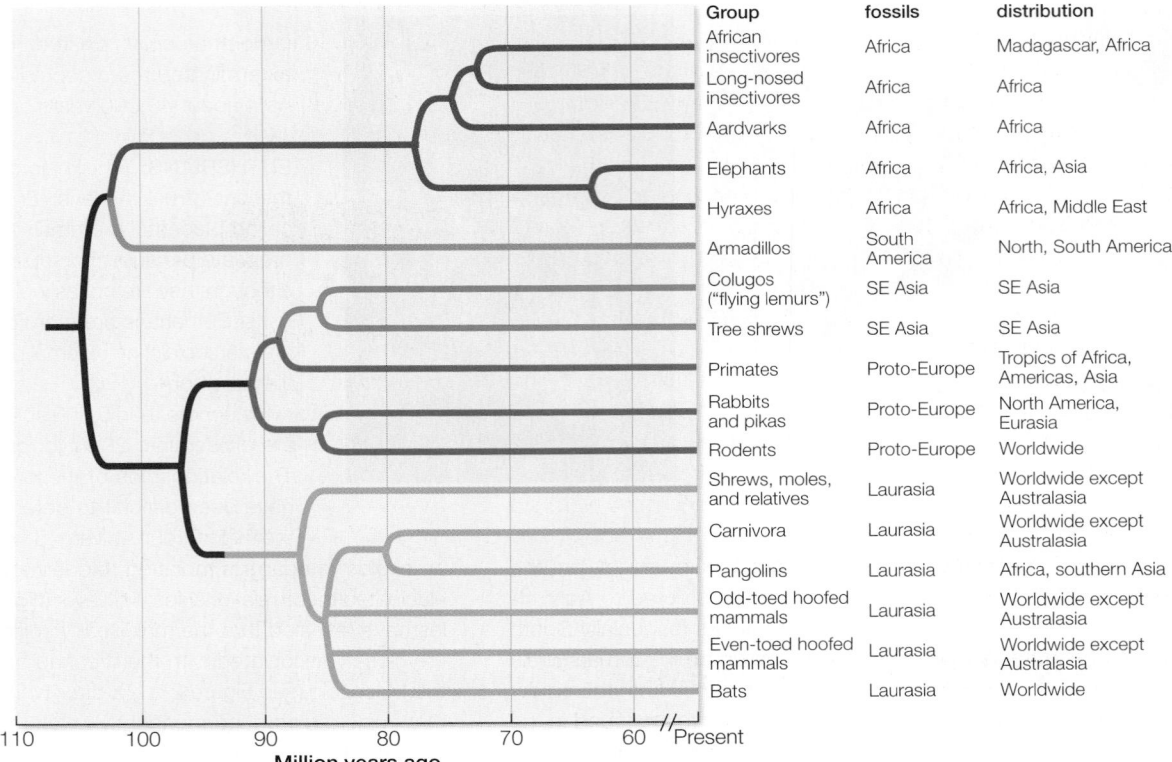

Group	Earliest fossils	Current native distribution
African insectivores	Africa	Madagascar, Africa
Long-nosed insectivores	Africa	Africa
Aardvarks	Africa	Africa
Elephants	Africa	Africa, Asia
Hyraxes	Africa	Africa, Middle East
Armadillos	South America	North, South America
Colugos ("flying lemurs")	SE Asia	SE Asia
Tree shrews	SE Asia	SE Asia
Primates	Proto-Europe	Tropics of Africa, Americas, Asia
Rabbits and pikas	Proto-Europe	North America, Eurasia
Rodents	Proto-Europe	Worldwide
Shrews, moles, and relatives	Laurasia	Worldwide except Australasia
Carnivora	Laurasia	Worldwide except Australasia
Pangolins	Laurasia	Africa, southern Asia
Odd-toed hoofed mammals	Laurasia	Worldwide except Australasia
Even-toed hoofed mammals	Laurasia	Worldwide except Australasia
Bats	Laurasia	Worldwide

110   100   90   80   70   60 //Present
Million years ago

(A) *Castor canadensis*

(B) *Plecotus townsendii*

(D) *Tursiops truncates*

(C)

which are adapted for gnawing through substances such as wood. The bats probably owe much of their success to the evolution of flight, which allows them to exploit a variety of food sources and colonize remote locations with relative ease.

 Media Clip 32.7 **Bats Feeding in Flight**
www.Life11e.com/mc32.7

Grazing and browsing by members of several eutherian groups helped transform the terrestrial landscape. Herds of grazing herbivores fed on open grasslands, whereas browsers fed on shrubs and trees. The effects of these herbivores on plant life favored the evolution of the spines, tough leaves, and

Figure 32.29 Diversity among the Eutherians **(A)** The North American beaver exhibits the gnawing teeth that characterize rodents. Almost half of all eutherians are rodents. **(B)** Flight evolved in the ancestor of bats. This Townsend's big-eared bat uses its enlarged ears to detect insect prey by echolocation. **(C)** Large hoofed mammals such as giraffes, zebras, and springboks are major herbivores of grasslands and savannahs. **(D)** Bottle nosed dolphins are cetaceans, a cetartiodactyl group that returned to the marine environment.

difficult-to-eat growth forms found in many plants. In turn, adaptations in the teeth and digestive systems of many herbivore lineages allowed these species to consume many plants despite such defenses—a striking example of coevolution. A large animal can survive on food of lower quality than a small animal can, and large size evolved in several groups of grazing and browsing mammals (see Figure 32.29C). The evolution of large herbivores, in turn, favored the evolution of large carnivores able to attack and overpower them.

Several lineages of terrestrial eutherians subsequently returned to the aquatic environments their ancestors had left behind (see Figure 32.29D). The completely aquatic cetaceans—whales and dolphins—evolved from even-toed hoofed ancestors (whales are closely related to the hippopotamuses). The seals, sea lions, and walruses also returned to the marine environment, and their limbs became modified into flippers. Weasel-like otters retain their limbs but have also returned to aquatic environments, colonizing both fresh and salt water. The manatees and dugongs colonized estuaries and shallow seas.

32.4 recap

The initial vertebrate colonization of dry land was facilitated by the evolution of lunglike sacs and jointed limbs. The amniotes also evolved impermeable body coverings, efficient kidneys, and the amniote egg, which resists desiccation.

learning outcomes

You should be able to:

- Explain how features of amphibians can make them especially vulnerable to environmental change.
- Summarize the evidence showing that birds are indeed reptiles.
- Describe how a mass extinction affected the diversification of mammals.
- Trace the evolution of endothermy, hair, and feathers across the phylogeny of amniotes.
- Describe how major changes in Earth's geology and climate have affected the evolution of tetrapods.

1. Amphibians have survived and prospered for millions of years, but today many species are disappearing and populations of others are in serious decline. What features of their life histories might make amphibians especially vulnerable to the kinds of environmental changes now happening on Earth?

2. How has the diversification of mammals been influenced by mass extinction events and continental drift?

3. In the not-too-distant past, the idea that birds were reptiles met with skepticism. Explain how fossils, morphology, and molecular evidence now support the position of birds among the reptiles.

4. Consider the phylogeny of the amniotes and the evolution of endothermy, hair, and feathers. Which of these traits likely evolved convergently in more than one lineage? What is a likely functional relationship among these traits? Why do many paleontologists think that several extinct theropods were endothermic?

The biology of one eutherian group—the primates—has been the subject of extensive research. The behavior, ecology, physiology, and molecular biology of the primates are of special interest to us because this lineage includes humans.

▶ key concept 32.5 Humans Evolved among the Primates

One lineage of small, arboreal, insectivorous eutherians underwent extensive evolutionary radiation to become the **primates** (Figure 32.30). Grasping limbs with opposable digits, an adaptation to arboreal life, are one of the major features that distinguish primates from other mammals.

focus your learning

- The two major groups of primates are the wet-nosed primates (lemurs, lorises, and galagos) and the dry-nosed primates (tarsiers, Old and New World monkeys, and apes).
- Possibly driven by an increasingly complex social life, larger brains evolved in one lineage of the genus *Homo*.

Two major lineages of primates split late in the Cretaceous

About 90 million years ago, late in the Cretaceous period, the primates split into two clades: the wet-nosed primates (strepsirrhines) and the dry-nosed primates (haplorhines). **Wet-nosed primates**—lemurs, lorises, and galagos—once lived on all continents, but today they are restricted to Africa, Madagascar, and tropical Asia. All mainland wet-nosed primate species are arboreal and nocturnal. On the island of Madagascar, however, the site of a remarkable radiation of lemurs, there are also diurnal and terrestrial species (Figure 32.31). Tarsiers were once considered a part of this lineage as well, although today we know that they are more closely related to **simians**—monkeys and apes—than to lemurs, lorises, and galagos.

The second primate lineage, the **dry-nosed primates**—tarsiers, New World monkeys, Old World monkeys, and apes—began to diversify shortly after the mass extinction event at the end of the Cretaceous, in Africa or Asia. New World monkeys diverged from Old World monkeys and apes slightly later, but early enough that they may have originated in Africa and reached South America when those two continents were still close to each other. New World monkeys now live only in South and Central America, and all of them are arboreal (Figure 32.32A). Many of them have a long, prehensile tail with which they can grasp branches. Many Old World monkeys are arboreal as well, but several species are terrestrial (Figure 32.32B). No Old World monkey has a prehensile tail.

About 35 million years ago, a lineage that led to the modern apes separated from the Old World monkeys. Between 22 and 5.5 million years ago, dozens of species of apes lived in Europe, Asia, and Africa. The Asian apes—gibbons and orangutans (Figure 32.33A and B)—descended from two of these ape lineages. Orangutans are the closest living sister group of the modern African apes: gorillas (Figure 32.33C), chimpanzees (Figure 32.33D), and humans.

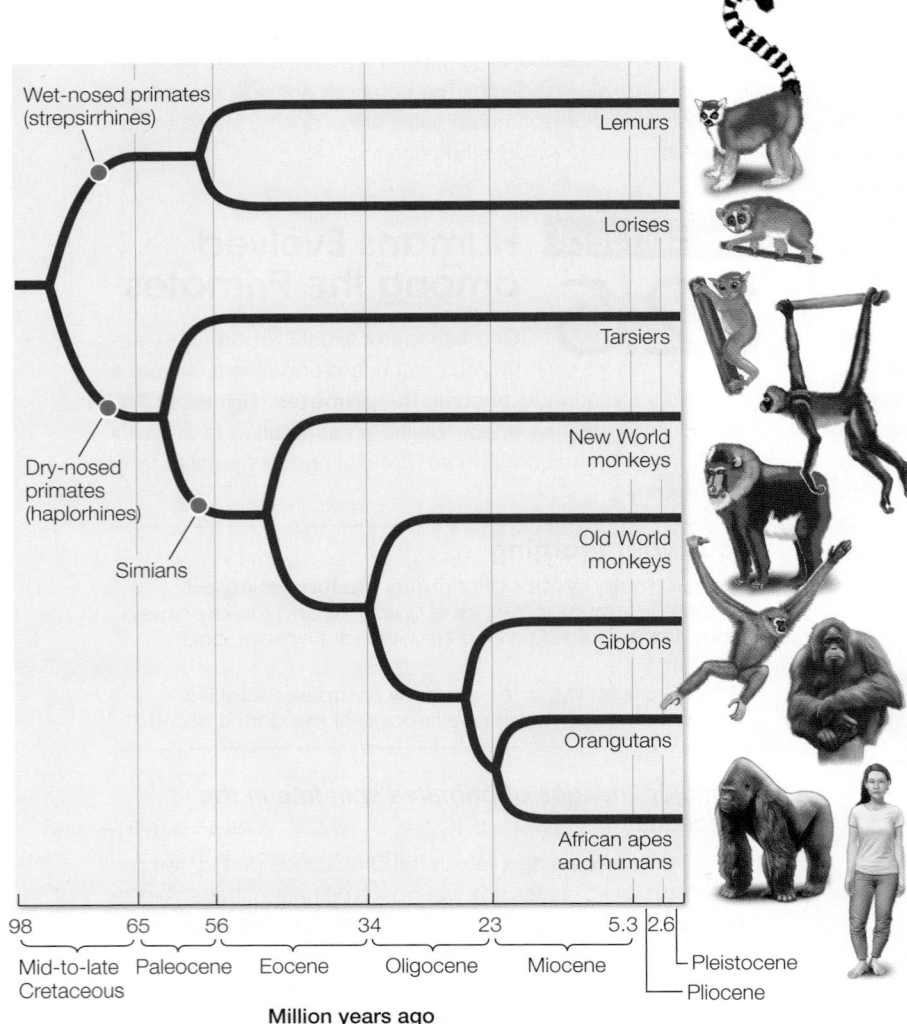

Wet-nosed primates (strepsirrhines)

Lemurs

Lorises

Dry-nosed primates (haplorhines)

Tarsiers

Simians

New World monkeys

Old World monkeys

Gibbons

Orangutans

African apes and humans

98	65	56		34	23		5.3	2.6

Mid-to-late Cretaceous | Paleocene | Eocene | Oligocene | Miocene | Pleistocene / Pliocene

Million years ago

Figure 32.30 Phylogeny of the Primates The phylogeny of primates is among the best studied of any major group of mammals. This tree is based on evidence from many genes, morphology, and fossils.

(A) *Ateles geoffroyi*

(B) *Mandrillus sphinx*

Varecia variegata variegata

Figure 32.31 A Wet-Nosed Primate The diademed sifaka is one of the many lemur species found in Madagascar, where it is part of a unique assemblage of endemic plants and animals. Sifakas live in groups of up to a dozen animals and defend their territories.

Bipedal locomotion evolved in human ancestors

About 6 million years ago in Africa, a lineage split occurred that would lead to the chimpanzees on the one hand and to the **hominin** clade, which includes modern humans and their extinct close relatives, on the other.

The earliest protohominins, known as ardipithecines, had distinct morphological adaptations for **bipedal locomotion** (walking on two legs). Bipedal locomotion frees the forelimbs to manipulate objects and to carry them while walking. It also elevates the eyes, enabling the animal to see over tall vegetation to spot predators and prey. Bipedal locomotion is also energetically more economical than quadrupedal locomotion (walking on four legs). All three advantages were probably important for the ardipithecines and their descendants, the australopithecines (**Figure 32.34**).

The first australopithecine skull was found in South Africa in 1924. Since then australopithecine fossils have been found at many sites in Africa. The most complete fossil skeleton yet found was discovered in Ethiopia in 1974. The skeleton, approximately 3.5 million years old, was that of a young female who has since become known to the world as "Lucy." Lucy was assigned to the species *Australopithecus afarensis*. Fossil remains of more than 100 *A. afarensis* individuals have since been discovered, and there have been recent discoveries of

Figure 32.32 Monkeys **(A)** The spider monkeys of Central America are typical of the New World monkeys, all of which are arboreal. Note the prehensile (gripping) tail. **(B)** Although many Old World monkeys are arboreal, none has a prehensile tail. Many Old World monkey species, like this mandrill, are thoroughly terrestrial.

(A) *Hylobates lar* **(B)** *Pongo pygmaeus* **(C)** *Gorilla gorilla*

(D) *Pan troglodytes*

Figure 32.33 Apes **(A)** The several genera of gibbons are all smaller in size than the other apes. Gibbons are found throughout Southeast Asia. **(B)** Orangutans are also native to Asia, living in the forests of Sumatra and Borneo. **(C)** Gorillas—the largest apes—are restricted to humid African forests. This male is a lowland gorilla. **(D)** Chimpanzees, our closest relatives, are found in forested regions of Africa.

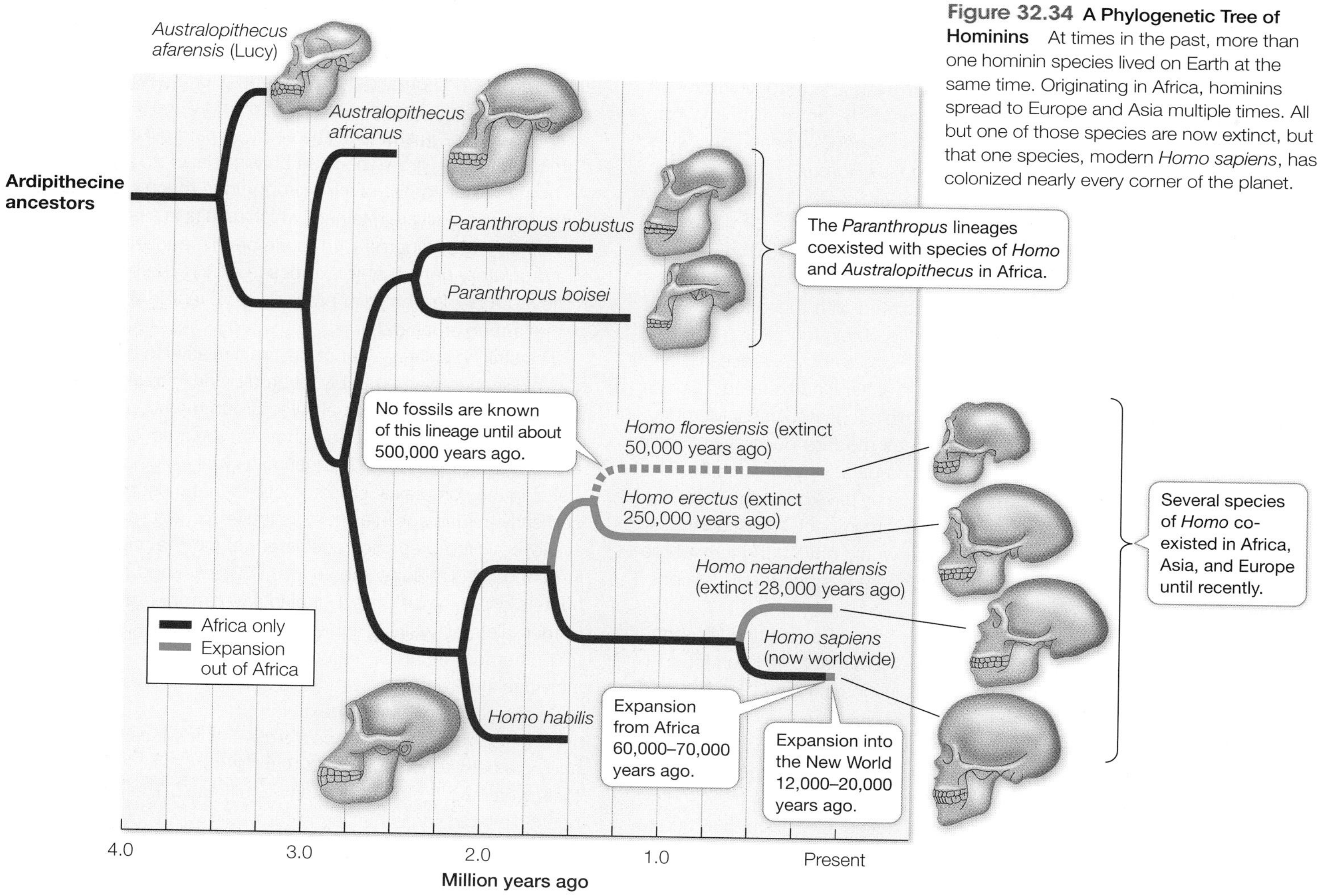

Figure 32.34 A Phylogenetic Tree of Hominins At times in the past, more than one hominin species lived on Earth at the same time. Originating in Africa, hominins spread to Europe and Asia multiple times. All but one of those species are now extinct, but that one species, modern *Homo sapiens*, has colonized nearly every corner of the planet.

Australopithecus afarensis (Lucy)

Australopithecus africanus

Ardipithecine ancestors

Paranthropus robustus

Paranthropus boisei

The *Paranthropus* lineages coexisted with species of *Homo* and *Australopithecus* in Africa.

No fossils are known of this lineage until about 500,000 years ago.

Homo floresiensis (extinct 50,000 years ago)

Homo erectus (extinct 250,000 years ago)

Homo neanderthalensis (extinct 28,000 years ago)

Homo sapiens (now worldwide)

Several species of *Homo* coexisted in Africa, Asia, and Europe until recently.

■ Africa only
■ Expansion out of Africa

Homo habilis

Expansion from Africa 60,000–70,000 years ago.

Expansion into the New World 12,000–20,000 years ago.

4.0   3.0   2.0   1.0   Present

Million years ago

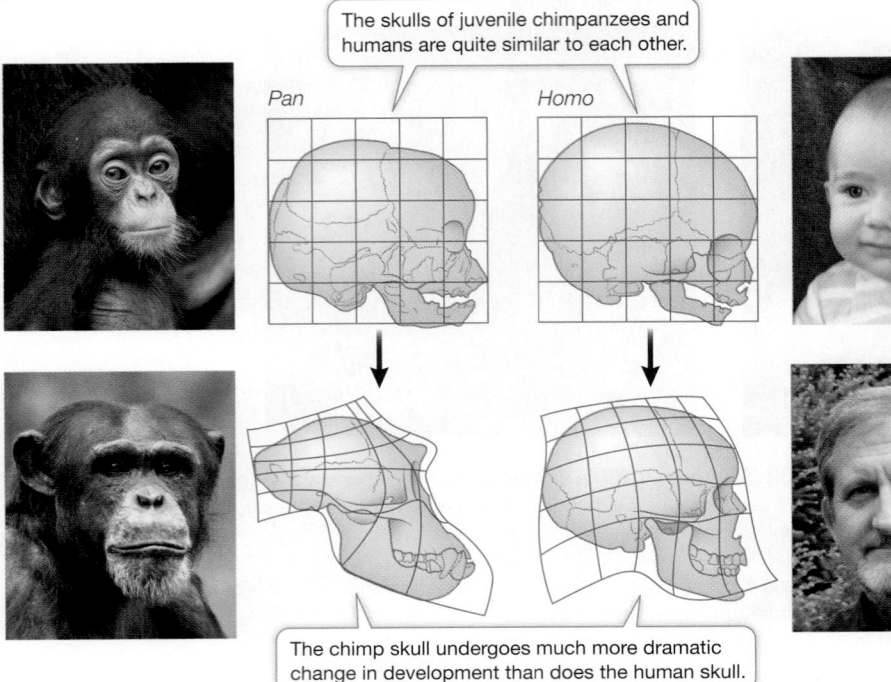

The skulls of juvenile chimpanzees and humans are quite similar to each other.

Pan   *Homo*

The chimp skull undergoes much more dramatic change in development than does the human skull.

Figure 32.35 Neoteny in the Evolution of Humans The skulls and heads of juvenile chimpanzees and humans start out relatively similar in shape. The grid over the skulls shows how the various bony elements change in their relative proportions over the course of maturation. The adult human skull retains a shape closer to its juvenile shape, resulting in a brain that is much larger relative to other parts of the skull (notably the jaw).

12,000 years ago, when a major volcanic eruption on Flores likely led to their extinction. Many anthropologists think that this small species, named *H. floresiensis*, was most closely related to *H. erectus*.

Human brains became larger as jaws became smaller

In another hominin lineage that diverged from *H. erectus* and *H. floresiensis*, the brain increased rapidly in size, and the jaw muscles, which were large and powerful in earlier hominins, dramatically decreased in size. These two changes were simultaneous, which suggests that they might have been developmentally linked. These changes are another example of evolution by neoteny (which, you may recall from our discussion of amphibians, is the retention of juvenile traits through delayed somatic development). Human and chimpanzee skulls are similar in shape at birth, but chimpanzee skulls undergo a dramatic change in shape as the animals mature (**Figure 32.35**). In particular, the jaw grows considerably in relation to the brain case. As human skulls grow, relative proportions much closer to those of the juvenile skull are retained, which results in a large brain case and small jaw compared with those of chimpanzees. A mutation in a regulatory gene that is expressed only in the head may have removed a barrier that had previously prevented this remodeling of the human cranium.

The striking enlargement of the brain relative to body size in the hominin lineage was probably favored by an increasingly complex social life. Any features that allowed group members to communicate more effectively with one another would have been valuable in cooperative hunting and gathering as well as for improving one's status in the complex social interactions that must have characterized early human societies, just as they do ours today.

Several *Homo* species coexisted during the mid-Pleistocene, from about 1.5 million to about 250,000 years ago. All were skilled hunters of large mammals, but plants were important components of their diets as well. During this period another distinctly human trait emerged: rituals and a concept of life after death. Deceased individuals were buried with tools and clothing, supplies for their presumed existence in the next world.

One species, *Homo neanderthalensis*, was widespread in Europe and Asia between about 500,000 and 28,000 years ago. Neanderthals were short, stocky, and powerfully built. Their massive skull housed a brain somewhat larger than our own. They manufactured

fossils of other australopithecine species that lived in Africa 4–5 million years ago.

Experts disagree over how many species are represented by australopithecine fossils, but it is clear that multiple species of hominins lived together over much of eastern Africa several million years ago. A lineage of larger species (weighing about 40 kilograms) is represented by *Paranthropus robustus* and *P. boisei*, both of which died out between 1 and 1.5 million years ago. A lineage of smaller australopithecines gave rise to the genus *Homo*.

Early members of the genus *Homo* lived contemporaneously with *Paranthropus* in Africa for about a million years. Some 2-million-year-old fossils of an extinct species called *H. habilis* were discovered in the Olduvai Gorge, Tanzania. Other fossils of *H. habilis* have been found in Kenya and Ethiopia. Associated with these fossils were tools that these early hominins used to obtain food.

Another extinct hominin species, *Homo erectus*, arose in Africa about 1.6 million years ago. Soon thereafter it had spread as far as eastern Asia, becoming the first hominin to leave Africa. Members of *H. erectus* were nearly as large as modern people, but their brains were smaller and they had comparatively thick skulls. The cranium, which had thick, bony walls, may have been an adaptation to protect the brain, ears, and eyes from impacts caused by a fall or a blow from a blunt object. What would have been the source of such blows? Fighting with other *H. erectus* individuals is a possible answer.

Homo erectus used fire for cooking and for hunting large animals, and made characteristic stone tools that have been found in many parts of Africa and Asia. Populations of *H. erectus* survived until at least 250,000 years ago, although more recent fossils may also be attributable to this species. In 2004 some 18,000-year-old fossil remains of a small *Homo* were found on the island of Flores in Indonesia. Since then, numerous additional fossils of this diminutive hominin have been found on Flores, dating from about 500,000 to

a variety of tools and hunted large mammals, which they probably ambushed and subdued in close combat. A closely related lineage, the Denisovans, is known from less complete fossils, primarily from from Asian caves.

Recently, biologists have been able to extract DNA from the bones of Neanderthals and Denisovans, and have compared the genomes of these species to those of modern humans. Early modern humans (*H. sapiens*) expanded out of Africa between 70,000 and 60,000 years ago. Then, about 35,000 years ago, *H. sapiens* moved into the range of *H. neanderthalensis* in Europe and western Asia, and into the range of the Denisovans in eastern Asia. Neanderthals abruptly disappeared about 28,000 years ago. Many anthropologists believe that Neanderthals and Denisovans were exterminated by those early modern humans. But comparisons of Neanderthal, Denisovan, and modern human genomes indicate that there was some limited interbreeding among these species while they occupied the same range. In humans with Eurasian ancestry, 1–4 percent of the genes in their genomes may be derived from Neanderthal ancestors. The genomes of some modern native people of Southeast Asia, Melanesia, and Australia contain even higher proportions of Denisovan genes. So all three of these lineages contributed in some degree to modern humans.

Early modern humans made and used a variety of sophisticated tools. They created the remarkable paintings of large mammals, many of them showing scenes of hunting, found in European caves. The animals they depicted were characteristic of the cold steppes and grasslands that occupied much of Europe during periods of glacial expansion. Early modern humans also spread across Asia, reaching North America perhaps as early as 20,000 years ago, although the date of their arrival in the Americas is still uncertain. Within a few thousand years, they had spread southward through North America to the southern tip of South America.

Humans developed complex language and culture

As our ancestors evolved larger brains, their behavioral capabilities increased, especially the capacity for language. Most animal communication consists of a limited number of signals, which refer mostly to immediate circumstances and are each used in a specific context. Human language is far richer in its symbolic character than other animal vocalizations. Our words can refer to past and future times and to distant places. We are capable of learning thousands of words, many of them referring to abstract concepts. We can rearrange words to form sentences with complex meanings.

The expanded mental abilities of humans enabled the development of a complex culture, in which knowledge and traditions are passed along from one generation to the next by teaching and observation. Cultures can change rapidly because genetic changes are not necessary for a cultural trait to spread through a population. Cultural norms, however, are not transferred automatically and must be deliberately taught to each generation.

Cultural transmission greatly facilitated the domestication of plants and animals and the resultant conversion of most human societies from ones in which food was obtained by hunting and gathering to ones in which pastoralism (herding large animals) and agriculture provided most of the food. The development of agriculture led to an increasingly sedentary life, the growth of cities, greatly expanded food supplies, rapid increases in the human population, and the appearance of occupational specializations, such as artisans, shamans, and teachers.

> **32.5 recap**

Grasping limbs with opposable digits distinguish primates from other mammals. Bipedal locomotion and large brains evolved in the primate ancestors that led to humans, which facilitated the development of complex language and culture.

learning outcomes

You should be able to:

- Summarize the diversity within the primates.
- Articulate the relationship between bigger brains, smaller jaws, and neoteny.

1. Describe the differences between Old World and New World monkeys.
2. Explain how neoteny resulted in the development of humans with relatively large brains and small jaws.

> ## investigating**life**

 ## What are the advantages and disadvantages of egg laying versus live birth?

Oviparity and viviparity involve some trade-offs of parental resources. In egg-laying species, all the nutrition for the developing embryo is supplied in the yolk. This means there is no continuing drain of parental resources as the embryo develops. One advantage of oviparity is that a female can often produce more offspring (since she devotes fewer resources to each one). Clutch size is often larger in oviparous species of lizards and snakes than in closely related viviparous species, although there are exceptions. Because each offspring in oviparous species receives a smaller parental investment compared with offspring of viviparous species, hatchlings tend to be smaller than the offspring of viviparous species.

Why is it sometimes advantageous for a species to commit more resources to fewer offspring? Developing embryos are retained in females in viviparous species, so females can control the temperature of development by behavioral thermoregulation. In high elevation and other cool environments, for example, females can bask in the sun during the day, and then retreat to cover at night. In addition, sometimes survival of offspring may depend on larger size and more parental investment. For example, in areas with short growing seasons and relatively cold temperatures, producing a

(continued)

few large offspring may yield significantly more descendants than producing a large number of small offspring that are less likely to survive. In contrast, in a favorable environment with abundant resources, producing many smaller offspring is likely to produce the most descendants. Indeed, there is a correlation between reproductive mode and elevational range in many groups of squamates, with oviparous species occurring at lower elevations and viviparous species at higher elevations.

Future directions

It is comparatively easy, from evolutionary and physiological points of view, for viviparity to evolve from oviparity in squamates. The membranes of the amniote egg can be modified to provide maternal nutrition, as well as gas and waste exchange (see Figure 32.19). Evolutionary reversal—from viviparity to oviparity—is more difficult, since complex modifications for shell deposition, independent growth, and hatching from the shell need to re-evolve. Nonetheless, a few candidate squamate lineages (including the huge venomous bushmasters of Central and South America) have been identified in which such reversals appear to have occurred. Biologists are focusing on these exceptional reversals to learn how lineages may redevelop structures and physiological processes that were lost in ancestors that lived millions of years earlier.

Chapter Summary 32

32.1 Deuterostomes Include Echinoderms, Hemichordates, and Chordates

- Deuterostomes vary greatly in adult form, but based on the distinctive patterns of early development they share and on phylogenetic analyses of their gene sequences, they are judged to be monophyletic.

- There are far fewer species of deuterostomes than of protostomes, but many deuterostomes are large and ecologically important.

- The deuterostomes comprise three major clades: the **echinoderms**, **hemichordates**, and **chordates**. **Review Figure 32.1, Activity 32.1, Animation 32.1**

32.2 Echinoderms and Hemichordates Are Restricted to Marine Environments

- Echinoderms and hemichordates, together called **ambulacrarians**, have bilaterally symmetrical, ciliated larvae. Adult echinoderms have **pentaradial symmetry** and an **oral–aboral** body orientation. **Review Figure 32.3**

- The **xenoturbellids** and **acoels** are reduced, soft-bodied wormlike marine animals with few distinct organ systems. Their relationships are uncertain, but recent analyses suggest that they may be the sister group of the ambulacrarians.

- Echinoderms have an **internal skeleton** of calcified plates and a unique **water vascular system** connected to extensions called **tube feet**. **Review Figure 32.3**

- Hemichordates are bilaterally symmetrical and have a three-part body divided into a proboscis, collar, and trunk. **Review Figure 32.6**

32.3 Chordates Have a Dorsal Nerve Cord and a Notochord

- Chordates fall into three principal clades: **lancelets**, **tunicates**, and **vertebrates**.

- At some stage in their development, all chordates have a dorsal hollow nerve cord, a post-anal tail, and a **notochord**. Lancelets have all three key chordate features as adults. Tunicates have these features as larvae but most species lose them as adults. **Review Figure 32.7**

- The vertebrate body is characterized by an internal skeleton, which is supported by a **vertebral column** that usually replaces the notochord. It is also characterized by internal organs suspended in a coelom, a ventral heart, and an anterior skull enclosing a large brain. **Review Figure 32.9**

- From estuarine ancestors, vertebrates diversified into many lineages of marine and freshwater fishes. One of these lineages, the lobe-limbed vertebrates, later radiated into terrestrial environments. **Review Figure 32.10**

- In the **gnathostomes**, jaws evolved from gill arches. Jaws enabled these vertebrates to grasp large prey and, together with teeth, allowed them to cut food into small pieces. **Review Figure 32.12**

- **Chondrichthyans** have skeletons of cartilage; almost all species are marine. The skeletons of **ray-finned fishes** are made of bone; these fishes have colonized all aquatic environments.

32.4 Life on Land Contributed to Vertebrate Diversification

- Lungs and jointed appendages enabled one lineage of **lobe-limbed vertebrates** to colonize the land. This lineage gave rise to the **tetrapods**. **Review Figure 32.16**

- The earliest split in the tetrapod tree is between the **amphibians** and the **amniotes** (**reptiles** and **mammals**).

- Most modern amphibians are confined to moist environments because their bodies and their eggs lose water rapidly. **Review Figure 32.17, Animation 32.2**

- An impermeable skin, efficient kidneys, and an **amniote egg** that could resist desiccation evolved in the amniotes. **Review Focus: Key Figure 32.19, Activity 32.2, Investigating Life: Have Shelled Eggs Re-Evolved among Viviparous Reptile Lineages?**

- The major living reptile groups are the **lepidosaurs** (tuataras, along with the **squamates**, which include lizards, snakes, and amphisbaenians), the **turtles**, and the **archosaurs** (crocodilians and **birds**). **Review Figure 32.20**

- Birds evolved from a group of active, predatory dinosaurs known as **theropods**. Feathers arose among the theropods, originally for insulation and to enhance coloration, but eventually developed into adaptations for flight in birds. **Review Figures 32.23, 32.24**

- Mammals are unique among animals in supplying their young with a nutritive fluid (milk) secreted by mammary glands. There are two primary mammalian clades: the **prototherians** (of which there are only five species) and the species-rich **therians**. The therian clade is further subdivided into the **marsupials** and the **eutherians**. Review Table 32.1

- Mammalian phylogeny is strongly associated with the breakup of the major continents during the Mesozoic. Major lineages of eutherians diversified in Laurasia, Africa, and South America. Review Figure 32.28

▶ **32.5 Humans Evolved among the Primates**

- Grasping limbs with opposable digits distinguish **primates** from other mammals. The **wet-nosed primate** clade includes the lemurs, lorises, and galagos; the **dry-nosed primate** clade includes tarsiers, monkeys, and apes. Review Figure 32.30

- The ancestors of **hominins** were terrestrial apes that developed efficient **bipedal locomotion**. Review Figure 32.34

- In the lineage leading to *Homo*, brains became larger as jaws became smaller; the two events appear to be developmentally linked and are an example of evolution via **neoteny**. Review Figure 32.35

See Activity 32.3 for a review of this chapter.

See Activity 32.4 for a review of the major groups of organisms.

Go to **LearningCurve** (in **LaunchPad**) for dynamic quizzing that helps you solidify your understanding of this chapter. **LearningCurve** adapts to your responses, giving you the practice you need to master each key concept.

▶ Apply What You've Learned

Review

32.4 Modifications of the amniote egg allowed the embryo to develop inside its mother's body.

Lambert, S. M. and J. J. Wiens. 2013. Evolution of viviparity: A phylogenetic test of the cold-climate hypothesis in phrynosomatid lizards. *Evolution* 67: 2614–2630.

Viviparity is estimated to have originated more than 100 times in squamate reptiles, making this an excellent group for the study of environmental conditions that favor live birth in ectothermic vertebrates. Fence lizards (*Sceloporus*) are a large, diverse clade that exhibits both oviparous and viviparous modes of reproduction. The "cold-climate" hypothesis suggests that viviparity is more likely to evolve in cooler environments. If this hypothesis is correct, then lizards in cooler environments (such as at higher elevations) should be more likely to evolve viviparity. Researchers at the University of Arizona tested this hypothesis using a combination of temperature records and phylogenetic comparative methods.

They constructed a phylogenetic tree that included (1) a time-calibrated phylogeny of *Sceloporus* species, (2) reproductive mode of *Sceloporus* species on the tips of the tree (solid = oviparous, open = viviparous), (3) node labels reflecting reproductive mode according to the likelihood of reconstructed ancestral states, and (4) color-coded lineages reflecting mean temperature at the warmest quarter of the year, since this is when lizards typically produce eggs or offspring.

Questions

1. What does the phylogenetic tree suggest about the evolution of viviparity as it relates to environmental conditions? According to the cold-climate hypothesis, why would viviparity provide an advantage in cold environments?

2. Describe at least one advantage and one disadvantage of both viviparous and oviparous modes of reproduction.

3. Viviparity evolved only once in extant lineages of mammals, more than 100 million years ago. Using the phylogeny provided on page 714, describe at least two logistical reasons why squamates offer a more suitable group to study the selective forces leading to viviparity.

4. The maternal manipulation hypothesis for the evolution of viviparity in squamate reptiles claims that gravid females can select optimal body temperatures for developing eggs through sun-seeking or sun-avoidance behaviors, and these temperatures are different from those available in external nests. Assume you are a research scientist with infinite resources. Design an experiment to test this hypothesis.

(continued)

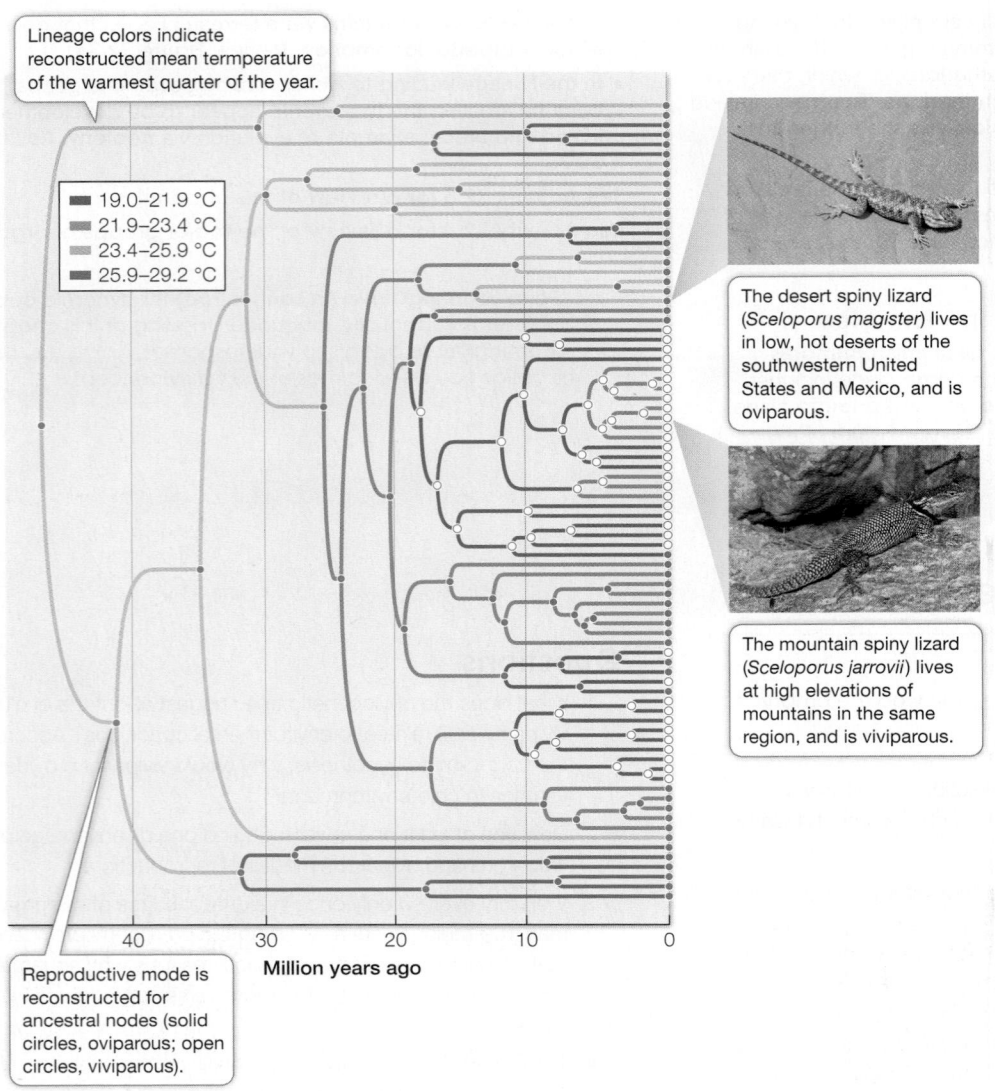

Lineage colors indicate reconstructed mean termperature of the warmest quarter of the year.

- 19.0–21.9 °C
- 21.9–23.4 °C
- 23.4–25.9 °C
- 25.9–29.2 °C

The desert spiny lizard (*Sceloporus magister*) lives in low, hot deserts of the southwestern United States and Mexico, and is oviparous.

The mountain spiny lizard (*Sceloporus jarrovii*) lives at high elevations of mountains in the same region, and is viviparous.

40   30   20   10   0

Million years ago

Reproductive mode is reconstructed for ancestral nodes (solid circles, oviparous; open circles, viviparous).

Go to **LaunchPad** for the eBook, LearningCurve, animations, activities, flashcards, and additional resources and assignments.

33

key concepts

33.1 The Plant Body Is Organized in a Distinctive Way

33.2 Plant Organs Are Made Up of Three Tissue Systems

33.3 Meristems Build a Continuously Growing Plant

33.4 Domestication Has Altered Plant Form

The Plant Body

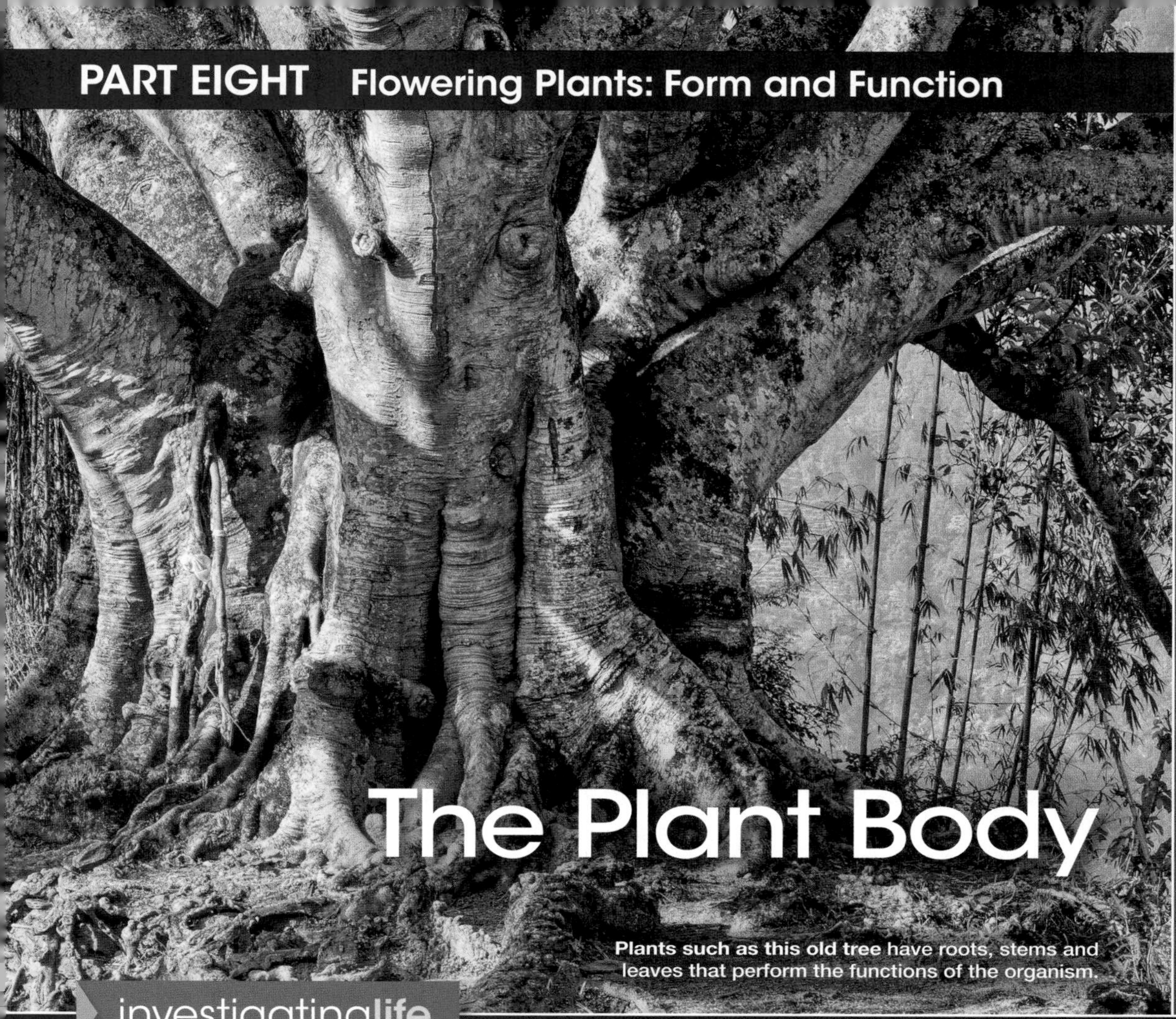

Plants such as this old tree have roots, stems and leaves that perform the functions of the organism.

investigatinglife

Bread of the Tropics

To North Americans and Europeans, it is tapioca. To Central and South Americans, it is yuca. To Africans, it is manioc or bananku. The roots of the cassava plant, *Manihot esculenta*, are important in the diets of more than 800 million people. Cassava is grown mostly by family farmers for their own consumption. The root of the plant is a store of starch, which can be hydrolyzed and used by the plant as the stems and leaves grow. Other crop plants such as rice and wheat apportion about 35 percent of the total carbon fixed in photosynthesis into their storage organs (the grains), whereas cassava apportions an astounding 80 percent into the root, as starch. For humans, this is a convenient and concentrated source of food energy. Indeed, cassava has been nicknamed the "bread of the tropics" because, just as wheat bread is the major starchy food in the Western world, cassava supplies the starch in tropical areas.

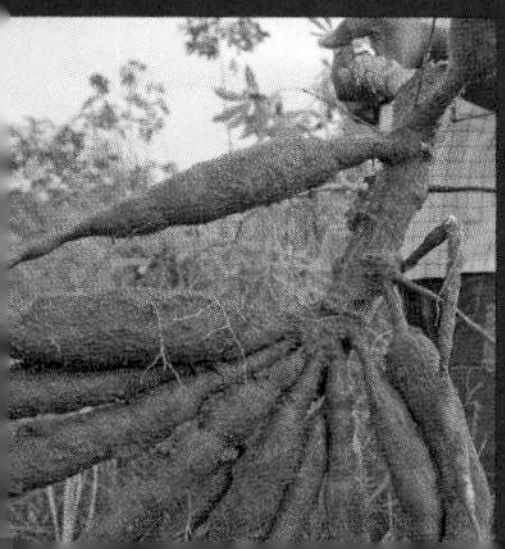

Most cassava plants are clones. The plants have wide adaptability and grow well in dry soils, in both hot and cool climates. Typically, the farmer breaks off some pieces of stem and plants them. Because most plant cells are totipotent (see Key Concept 19.1), some of the stem cells dedifferentiate and form roots, while others become growing shoots. A whole new plant develops, and the roots are ready to eat 6 months to 2 years later.

There are countless recipes for preparing cassava as food, but it is nutritionally incomplete. Although rich in carbohydrates, cassava is a relatively poor source of protein, a requirement for the human diet. Eating cassava presents other difficulties. Cassava roots and leaves contain cyanogenic compounds, which are converted to cyanide by enzymes released when the cells are broken open. Cyanide is highly toxic and potentially lethal because it blocks electron transport in the mitochondria. Therefore it is essential to soak, cook, or ferment cassava, to break down the cyanogenic compounds and eliminate the cyanide, before it can be eaten safely. In all probability, the plant uses cyanide production as a protection against predators that eat it.

We open with the topic of cassava because it offers a preview of a wide range of studies embraced by the discipline of plant physiology. Plant physiology is a broad subject, covering photosynthesis, transport, plant nutrition, regulation of growth and development, reproduction, and the interactions between plants and their environments.

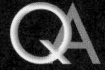

QA How might plant biologists improve the cassava plant for human use?

key concept 33.1 The Plant Body Is Organized in a Distinctive Way

Plants harvest energy from sunlight and collect water and mineral nutrients from the atmosphere and the soil. Because these resources are sometimes limited, plants must collect them from large areas, both above and below the ground. The plant is further challenged by its inability to move; a plant cannot, for example, relocate from a dry, shady location to one that is wet and sunny. The plant body plan responds to these challenges in two ways:

1. Stems, leaves, and roots enable a plant anchored to one spot to capture scarce resources effectively, both above and below the ground.

2. Plants can grow throughout their lifetimes, enabling them to respond to environmental cues. A plant can redirect its growth to exploit opportunities in its immediate environment; for example, it can extend its roots toward a water supply.

focus your learning

- Plant growth patterns reflect challenges imposed by scarce resources and an inability to move.
- Plant development is influenced by apical meristems, totipotency, vacuoles, and cell walls.
- Two basic patterns that develop early in plant embryogenesis are apical–basal polarity and radial symmetry.

Chapters 27 and 28 described how modern plants came from aquatic ancestors, giving rise to simple land plants and then to vascular plants. Despite their obvious differences in size and form, all vascular plants have essentially the same simple structural organization. This chapter describes the basic architecture of the largest group of vascular plants, the angiosperms (flowering plants), and shows how so much diversity can literally grow out of such a simple, basic form. There are more than 250,000 angiosperm species. A major distinguishing feature of angiosperms is flowers, which consist of modified leaves and stems and carry the organs for sexual reproduction. We will examine the structures and functions of flowers in detail in Chapter 37.

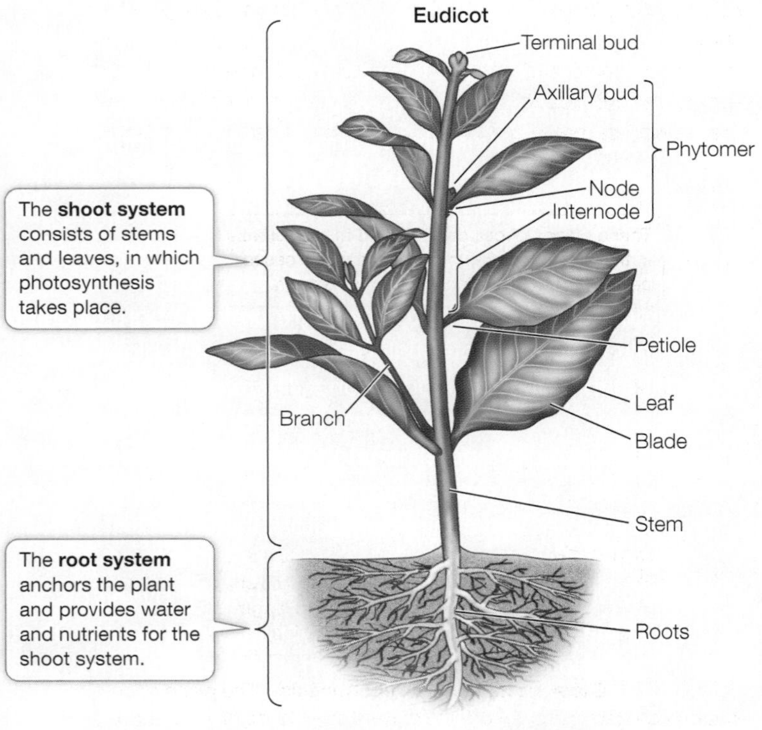

In this chapter we'll focus on the three kinds of **vegetative** (nonsexual) organs that angiosperms possess: roots, stems, and leaves. Each of these vegetative organs can be understood in terms of its structure. By structure we mean both the overall form of the organ (its morphology) and the arrangement of its cells and tissues (its anatomy). Plant organs are organized into two systems (**Figure 33.1**):

1. The **root system** anchors the plant in place, absorbs water and dissolved minerals, and stores the products of photosynthesis from the shoot system. The extreme branching of plant roots and their high surface area-to-volume ratios allow them to absorb water and mineral nutrients from the soil efficiently.

2. The **shoot system** of a plant consists of the stems, leaves, and flowers. Broadly speaking, the **leaves** are the chief organs of photosynthesis. The **stems** hold and display the leaves to the sun and provide connections for the transport of materials between roots and leaves.

Shoots and roots are composed of repeating modules called **phytomers**. Each phytomer in the shoot consists of a **node** carrying one or more leaves; an **internode**, which is the interval of stem between two nodes; and one or more **axillary buds**, each of which forms in the angle (axil) where a leaf meets the stem. A **bud** is an undeveloped shoot that can develop further to produce another leaf, a phytomer, a flower, or a flowering stem. The axillary buds (also called lateral buds) are distinguished from the bud at the end of a stem or branch, which is called a **terminal bud**. If it becomes active, an axillary bud can develop into a new branch, which is an extension of the shoot system. The arrangement of leaves along the stem (called phyllotaxy) is characteristic of plant species. Plant roots also have a modular construction. In the roots, each phytomer consists of a root segment between two branches.

Most angiosperms are either monocots or eudicots

As you saw in Key Concept 28.3, most angiosperms belong to one of two major clades. *Monocots* are generally narrow-leaved flowering plants such as grasses, lilies, and orchids. *Eudicots* are broad-leaved flowering plants such as soybeans, roses, sunflowers, and maples. These two clades, which account for 97 percent of flowering plant species, differ in several basic characteristics, as shown in **Figure 33.2**.

Plants develop differently than animals

Chapter 19 described four processes that govern development in all multicellular organisms: **determination** (the commitment of cells to their ultimate fates), **differentiation** (cell specialization), **morphogenesis** (the organization of cells into tissues and organs), and **growth** (increase in body size). In plants, these processes are

Eudicot

Terminal bud
Axillary bud
Phytomer
Node
Internode

The **shoot system** consists of stems and leaves, in which photosynthesis takes place.

Branch

Petiole
Leaf
Blade

Stem

The **root system** anchors the plant and provides water and nutrients for the shoot system.

Roots

Figure 33.1 Vegetative Organs and Systems　The basic plant body plan, with root and shoot systems, and the principal vegetative organs.

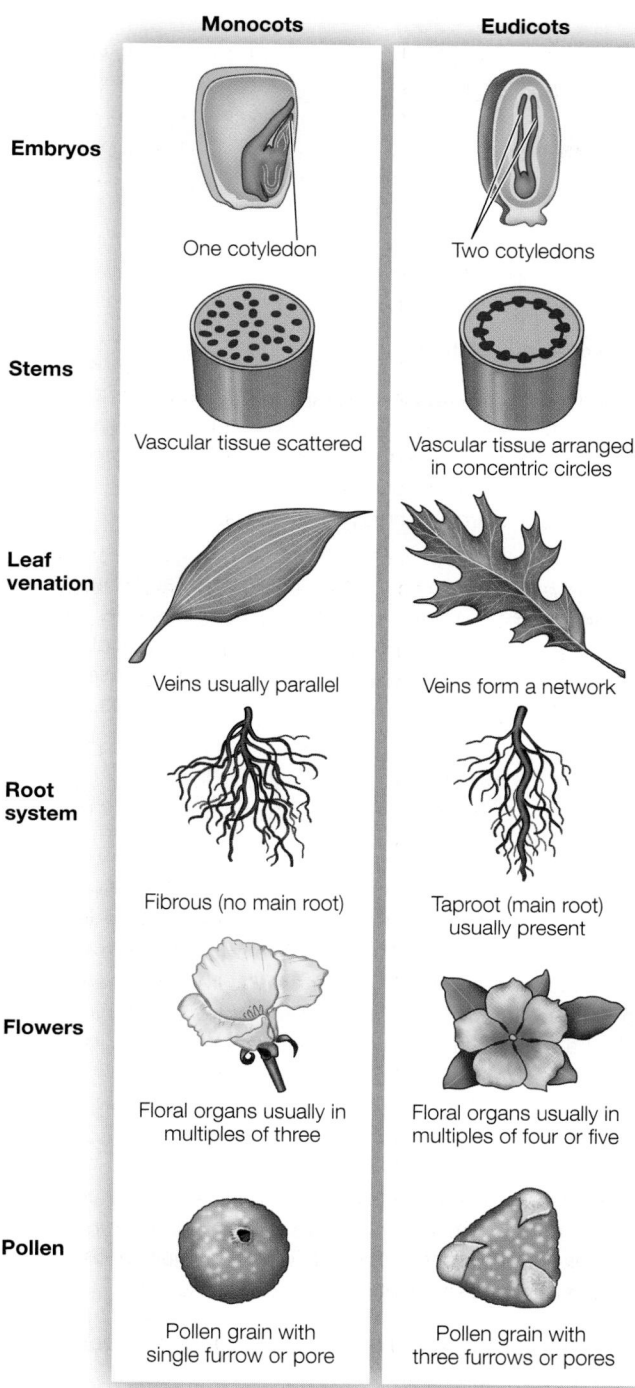

Monocots	Eudicots
Embryos One cotyledon	Two cotyledons
Stems Vascular tissue scattered	Vascular tissue arranged in concentric circles
Leaf venation Veins usually parallel	Veins form a network
Root system Fibrous (no main root)	Taproot (main root) usually present
Flowers Floral organs usually in multiples of three	Floral organs usually in multiples of four or five
Pollen Pollen grain with single furrow or pore	Pollen grain with three furrows or pores

Figure 33.2 Comparing the Two Major Angiosperm Clades There are several structural differences between monocots and eudicots.

influenced by four features: apical meristems, totipotency, vacuoles, and cell walls.

MERISTEMS   Animals have stem cells to replace tissues lost through damage or apoptosis. Hundreds of millions of new cells are produced every day in the human body just to maintain itself. Plants, too, experience losses, both from herbivory and programmed events (such as leaf drop in the fall). Plant tissues and organs are generated in plant **meristems**: regions of undifferentiated cells where cell division occurs. **Apical meristems** are found at the tips of shoots and roots and allow plants to continue growing throughout their lives.

TOTIPOTENCY   During normal animal development, only the early embryonic cells are *totipotent: they can differentiate into any type of cell in the body. In contrast, some differentiated plant cells can dedifferentiate and become totipotent. This means that a plant can readily repair damage wrought by the environment or herbivores.

---

*connect the concepts   Key Concept 19.1 describes distinctive aspects and functions of totipotency in plants.

VACUOLES   Mature plant cells usually contain a single **central vacuole**, which may account for up to 90 percent of a cell's volume (see Figure 5.13). The vacuole is a watery sac containing solutes, including enzymes, amino acids, and sugars produced by photosynthesis. Many of these solutes are pumped into the vacuole by transporter proteins located in the **tonoplast**, the vacuolar membrane. This active accumulation of solutes provides the osmotic force for water uptake into the vacuole, as we will explain in Key Concept 34.1. As the vacuole expands, it exerts turgor pressure on the cell wall. Turgor pressure keeps plants upright and is essential for plant growth (see Chapter 6).

CELL WALLS   Each plant cell is surrounded by a cell wall, which is interrupted by membrane-lined cytoplasmic channels called plasmodesmata (see Key Concept 5.4). This rigid extracellular matrix makes it impossible for cells to move from place to place, as they do in animal development. Instead, plant morphogenesis is controlled by the planes of cell division, which determine the direction in which a piece of tissue will grow (**Figure 33.3**). In addition, unequal cell division can occur when the cytoplasm contains differentiation signals that are localized in one part of a cell. Plant cell cytokinesis (the division of the cytoplasm) occurs along a cell plate laid down

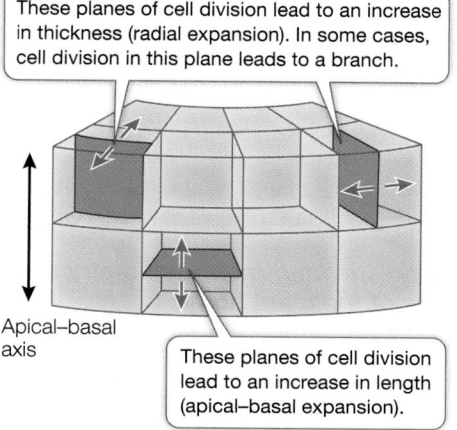

These planes of cell division lead to an increase in thickness (radial expansion). In some cases, cell division in this plane leads to a branch.

Apical–basal axis

These planes of cell division lead to an increase in length (apical–basal expansion).

Figure 33.3 Cytokinesis and Morphogenesis   The plane of cell division can determine the growth pattern of a plant's organs, as in this section of a shoot.

Q: How do you think the plane of cell division is controlled?

by membranous vesicles produced by the Golgi apparatus (see Figure 11.12B). Unlike animal cells, in which the location of cytokinesis depends on the location of the middle of the mitotic spindle, the location of the plant cell plate is determined earlier—as early as mitotic prophase.

One of the major ways that plants grow is by cell expansion. Some cells can increase in volume by 100,000 to 1,000,000 times! As a growing plant cell takes up water, it exerts turgor pressure on the cell wall, which resists cell expansion. For the cell to expand, the wall must expand too. Proteins appropriately called expansins reside in the cell wall and help loosen it by disrupting the noncovalent interactions between cellulose microfibrils and other polysaccharides in the cell wall. This is followed by the assembly of new polysaccharides and microfibrils, allowing the cell wall to grow:

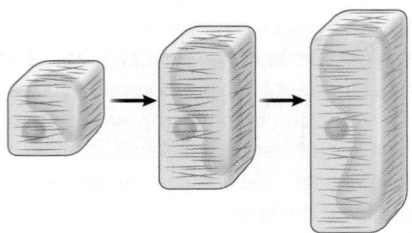

The wall of a growing plant cell is called the **primary cell wall**. When cell expansion stops, some types of plant cells deposit one or more additional cellulosic layers to form a thick, rigid **secondary cell wall** that is internal to the primary cell wall:

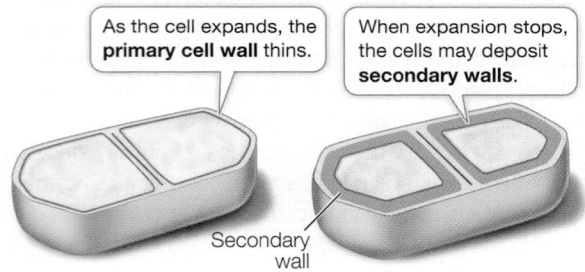

As the cell expands, the **primary cell wall** thins.

When expansion stops, the cells may deposit **secondary walls**.

Secondary wall

Secondary cell walls cannot expand. Instead they provide the mechanical support that allows some plants to produce large stems. The secondary wall contains layers of ordered cellulose microfibrils embedded in a remarkable substance called **lignin**. Lignin is a major component of wood; it is a complex, carbon-containing polymer. This matrix is strong, waterproof, and resistant to digestion by animals. After cellulose, lignin is the most abundant biological polymer on Earth, accounting for 20–35 percent of the dry weight of wood.

Apical–basal polarity and radial symmetry are characteristics of the plant body

Two basic patterns are established early in plant embryogenesis (embryo formation) (**Figure 33.4**):

1. *The apical–basal axis*: the arrangement of cells and tissues along the main axis from root to shoot

2. *The radial axis*: the concentric arrangement of the tissue systems (discussed in Key Concept 33.2)

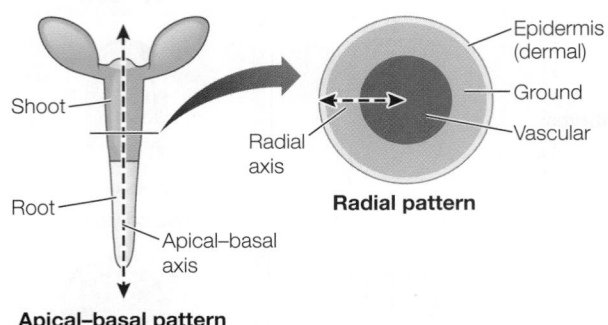

Figure 33.4 Two Patterns for Plant Morphogenesis The apical–basal pattern is the arrangement of cells and tissues along a main axis from root to shoot. The radial pattern determines the concentric arrangement of tissues as organs grow in thickness.

Both axes are best understood in developmental terms. We will focus here on embryogenesis in *Arabidopsis thaliana*. As you have seen in previous chapters, *Arabidopsis* is a model eudicot that has been studied extensively by plant physiologists and geneticists. You can follow the steps in plant embryogenesis in **Figure 33.5**.

The first step in the formation of an *Arabidopsis* embryo is a mitotic division of the zygote that gives rise to two daughter cells, an apical daughter cell and a basal daughter cell (see Figure 33.5, Step 1). An asymmetrical plane of cell division results in an uneven distribution of cytoplasm between these two cells, and this determines their different fates. Signals in the smaller, apical daughter cell induce it to produce the embryo proper, while the larger basal daughter cell produces the **suspensor**, a structure that connects the developing embryo to the storage material (see Figure 33.5, Step 2). This division not only establishes the apical–basal axis of the new plant but also determines its polarity (which end is the tip, or apex, and which is the base). A long, thin suspensor and a more spherical or globular embryo are distinguishable after just four mitotic divisions. The suspensor soon ceases to elongate.

In eudicots such as *Arabidopsis*, the initially globular embryo develops into the characteristic heart stage as the cotyledons start to grow (see Figure 33.5, Step 3). Further elongation of the cotyledons and of the main axis of the embryo gives rise to the torpedo stage, during which some of the internal tissues begin to differentiate (see Figure 33.5, Step 4). Between the cotyledons is the **shoot apical meristem**; at the other end of the axis is the **root apical meristem**. Each of these meristems contains undifferentiated cells that will continue to divide to give rise to the organs that will develop during the life of the plant.

As shown in Step 2 of Figure 33.5, the plant embryo is first a sphere and later a cylinder. The root and stem retain this cylindrical shape throughout the plant's life. You can see this most easily in the trunk (mature stem) of a tree. By the end of embryogenesis, the radial symmetry of the plant has been established (see Figure 33.4). The embryonic plant contains three tissue systems, arranged concentrically, which will give rise to the tissues of the adult plant body. We will discuss these tissue systems in the next section.

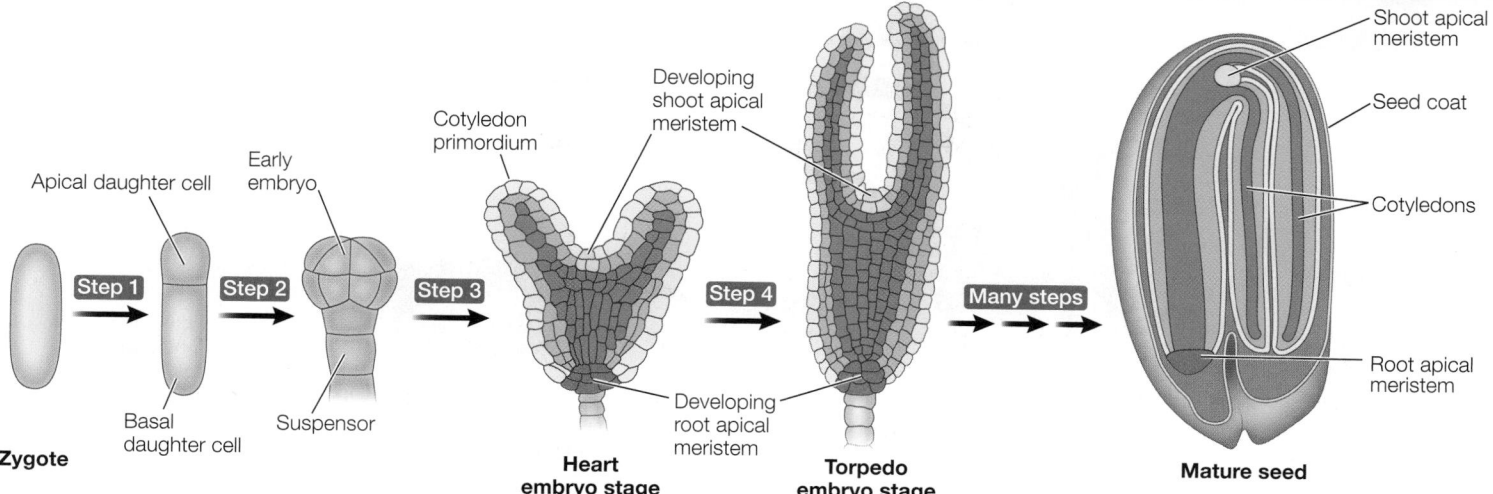

Figure 33.5 Plant Embryogenesis The basic body plant of the model eudicot *Arabidopsis thaliana* is established in several steps. By the heart stage, the three tissue systems are established: the dermal (tan), ground (light green), and vascular (blue) tissue systems. These tissue systems are shown in the mature seed, which develops after many more stages.

Q: How do you think the differences between the apical and basal daughter cells are established?

33.1 recap

The vegetative plant body consists of a root system and a shoot system. The plant body is modular, made up of repeated units called phytomers. Plants are characterized by apical meristems, totipotency, vacuoles, and cell walls. The tissues of the plant body develop along two axes: the apical–basal axis (which establishes a polar pattern from root to shoot) and the radial axis (which establishes a concentric pattern of vascular, ground, and dermal tissue systems). Most angiosperms are either monocots or eudicots, which differ in several basic ways.

learning outcomes

You should be able to:

- Identify ways in which plants have overcome the problems of scarce resources and an inability to move.
- Analyze major differences in plant and animal development.
- Examine the process by which a zygote develops into an embryo with an apical–basal axis.

1. How do plants explore their environment for resources even though they cannot move?
2. Describe the primary ways that plant development differs from animal development.
3. Explain how apical–basal and radial patterns develop.
4. A mutant strain of *Arabidopsis* lacks apical–basal polarity in the embryo. What would you expect the phenotype of the plant to be?

In the next two sections we will look more closely at the unique characteristics of the plant body, by following its development from a zygote into an adult.

key concept
33.2
Plant Organs Are Made Up of Three Tissue Systems

By the end of embryogenesis, both the apical–basal and radial axes have been established. Unlike complex animals that can have dozens of different tissues (for example, in humans there are three kinds of muscle tissue alone), plants have just three major tissue systems, each of which has specialized cells.

focus your learning

- Ground tissue forms most of the plant body and includes parenchyma, collenchyma, and sclerenchyma. The dermal and vascular systems have parenchyma and sclerenchyma.
- Vascular tissue consists of xylem and phloem, which are the plant's transport system.

As **Focus: Key Figure 33.6** illustrates, the three major plant tissue systems develop in a radial pattern in the embryo, establishing a concentric arrangement of dermal, ground, and vascular tissues. From these tissue systems will arise the roots, stems, and leaves of the adult vegetative plant body.

The dermal tissue system forms the outer covering of a plant

During plant development, the **dermal tissue system** develops in a plant's outermost layer of cells, called its **epidermis**. The stems and roots of woody plants develop a dermal tissue called **periderm**.

The epidermis grows to cover the expanding plant body. The cells of the epidermis are initially small and round and usually have a small central vacuole or none at all. Once cell division ceases in the epidermis of an organ, the epidermal cells expand. Some epidermal cells differentiate to form one of three specialized structures, about which you'll learn more later:

1. Stomatal guard cells, which form stomata (pores) for gas exchange in leaves

focus: key figure

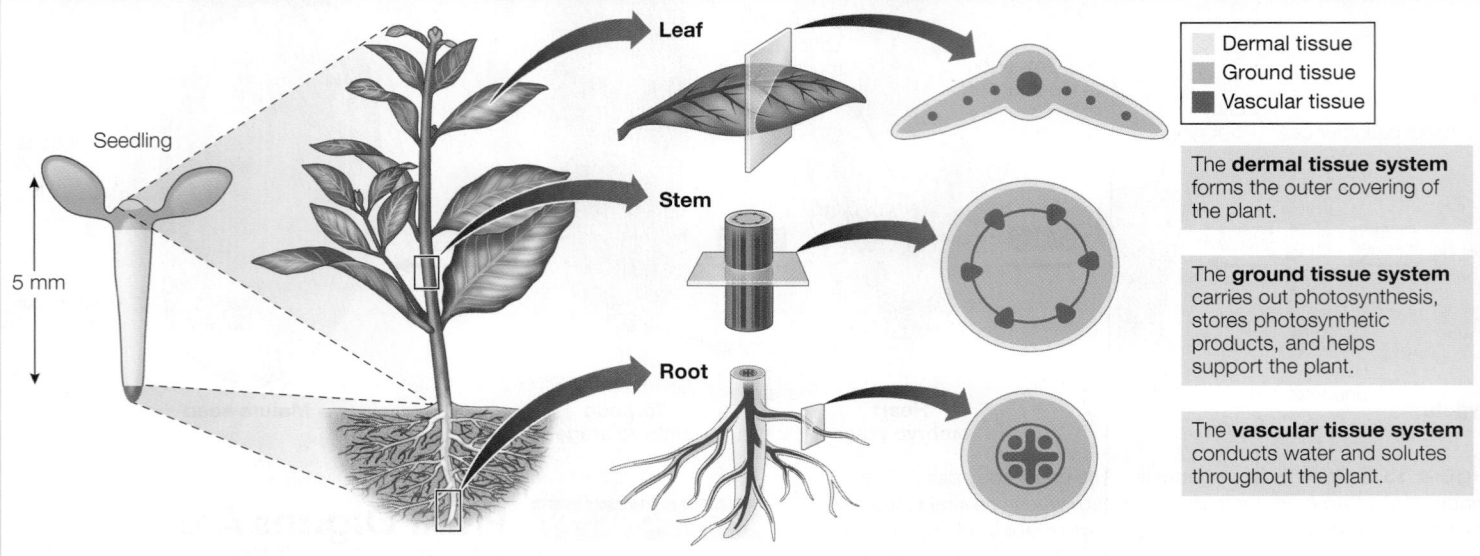

Figure 33.6 Three Tissue Systems Extend throughout the Plant Body
The arrangement shown here is typical of eudicots, but the three tissue systems are continuous in the bodies of all vascular plants.

Q: What is a major difference between cells in the ground tissue in the leaf compared to the root?

---

2. Trichomes, or leaf hairs, which provide protection against insects and damaging solar radiation, as well as secrete oils (such as those produced by mints)

3. Root hairs, which greatly increase root surface area, thus providing more surface for the uptake of water and mineral nutrients

The aboveground epidermal cells secrete a protective extracellular layer called a **cuticle**. The cuticle is made up of cutin (a polymer composed of long chains of fatty acids), a complex mixture of waxes, and cell wall polysaccharides. The cuticle limits water loss, reflects potentially damaging solar radiation, and serves as a barrier against pathogens.

The ground tissue system makes up most of the plant body

Virtually all the tissue lying between the dermal tissue and the vascular tissue in both shoots and roots is part of the **ground tissue system**. Therefore the ground tissues make up most of the plant body. Three main cell types are commonly seen in the ground tissue system: parenchyma, collenchyma, and sclerenchyma. Ground cell types are generally distinguished by the nature of their cell walls.

PARENCHYMA   The most common cell type in plants is the **parenchyma** cell (**Figure 33.7A**). Parenchyma cells often have large vacuoles and thin walls consisting only of a primary wall and the shared middle lamella. The **middle lamella** is a layer of pectin that cements adjacent plant cells together (see Figure 5.21). Parenchyma cells play important roles in photosynthesis (mainly in the leaves); proteins, starch, fats, or oils may be stored in parenchyma cells of the seeds and/or roots. Many retain the capacity to divide, as when a wound results in cell proliferation.

COLLENCHYMA   **Collenchyma** cells resemble parenchyma cells that have been modified to provide flexible support. They are generally elongated, and their primary walls are characteristically thick at the corners of the cells (**Figure 33.7B**). In these cells the primary wall thickens in part because of the deposition of pectins, but no secondary wall forms. Collenchyma cells provide support to leaf petioles, nonwoody stems, and growing organs. Tissue made of collenchyma cells is flexible, permitting stems and petioles to sway in the wind without snapping. The familiar "strings" in celery consist primarily of collenchyma cells.

SCLERENCHYMA   There are two types of sclerenchyma cells: elongated **fibers** and variously shaped **sclereids** (**Figure 33.7C**). Both cell types have thickened secondary walls that enable their major function: support. Many sclerenchyma cells undergo programmed cell death (apoptosis; see Key Concept 11.6) after developing the lignified secondary walls, and thus perform their supporting function when dead. Fibers are often organized into bundles and provide relatively rigid support to wood, bark, and other parts of the plant. Sclereids may pack together densely, as in a nut's shell or in some seed coats. Isolated clumps of sclereids, called stone cells, occur in pears and some other fruits and give them their characteristic gritty texture.

The vascular tissue system develops into the plant's transport system

The **vascular tissue system** is the plant's plumbing, or transport system—the distinguishing feature of vascular plants. Its two constituent tissues, the xylem and phloem, distribute materials throughout the plant. The **xylem** distributes water and mineral ions taken up by the roots to all the cells of the stems and leaves. **Phloem** performs a variety of functions, including transport, support, and storage. The phloem

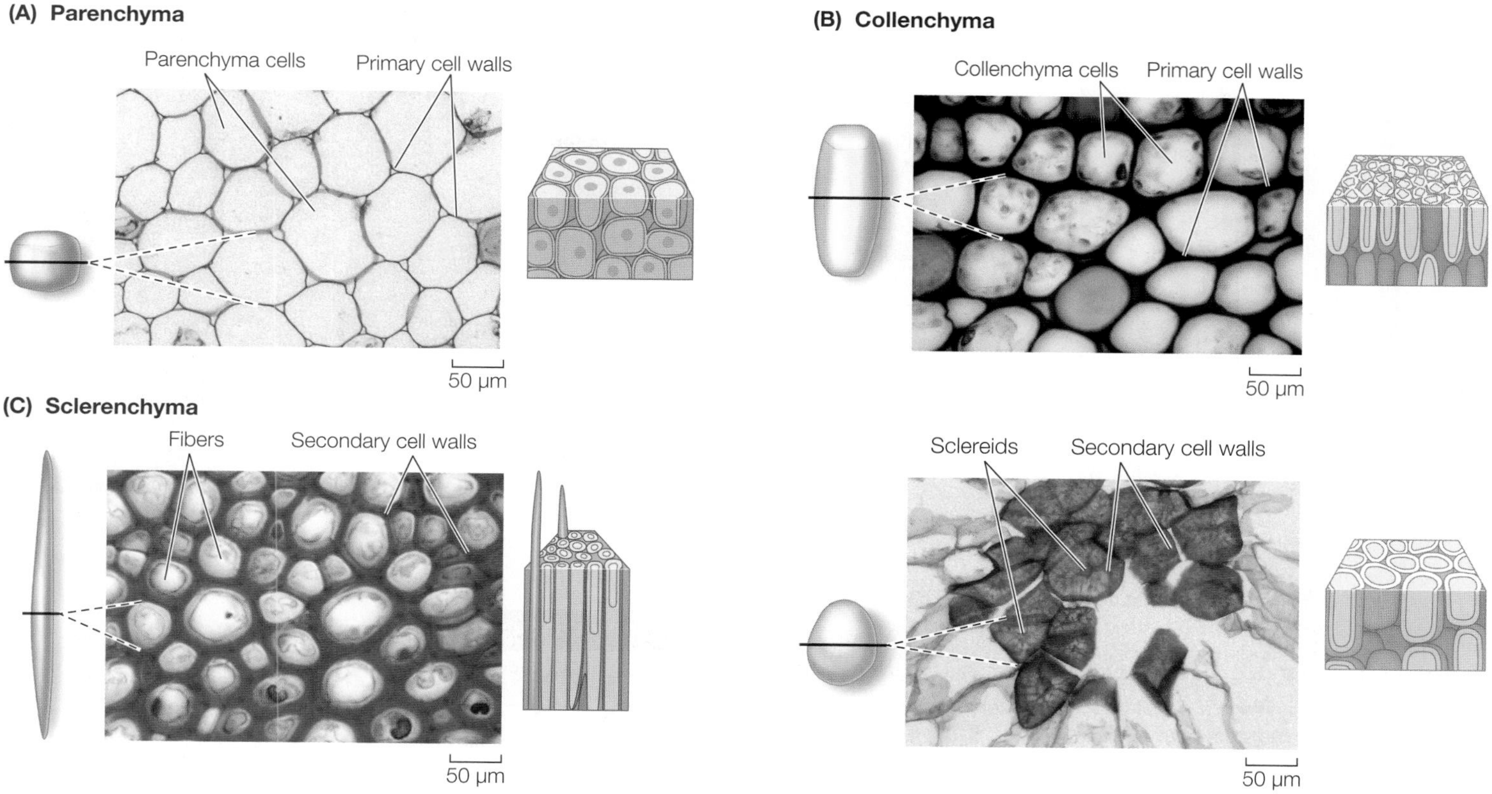

(A) Parenchyma

Parenchyma cells   Primary cell walls

50 µm

(B) Collenchyma

Collenchyma cells   Primary cell walls

50 µm

(C) Sclerenchyma

Fibers   Secondary cell walls

50 µm

Sclereids   Secondary cell walls

50 µm

Figure 33.7 Ground Tissue Cell Types **(A)** Parenchyma cells in the petiole of Coleus. Note the thin, uniform cell walls. **(B)** Collenchyma cells make up the outer cell layers of this spinach leaf vein. Their walls are thick at the corners of the cells and thin elsewhere. **(C)** The two types of sclerenchyma cells. Left: Fibers in a sunflower stem (*Helianthus*). The thick secondary walls are stained red. Right: Sclereids. The extremely thick secondary walls of sclereids are laid down in layers. They provide support and a hard texture to structures such as nuts and seeds.

plays a particularly critical role in supplying the energy-rich molecules and chemical building blocks needed by all living cells of the plant body. The phloem transports carbohydrates away from the sites of production, which are called **sources** (primarily leaves), and delivers them to sites of utilization or storage, called **sinks**. Sinks include growing tissues, storage organs, and developing flowers.

Let's take a closer look at the structures of the diverse cell types that make up these vascular tissues. In Chapter 34 you'll see how they transport water and materials throughout the plant body.

XYLEM TISSUE   Xylem tissue contains conducting cells called **tracheary elements**, which have secondary cell walls and undergo apoptosis before assuming their function of transporting water and dissolved minerals. There are two types of tracheary elements: tracheids and vessel elements. The spindle-shaped **tracheids** (**Figure 33.8A**) are evolutionarily more ancient than vessel elements and are the major cell type in the wood of gymnosperms (major groups of gymnosperms are covered in Key Concept 28.2). When tracheids die, their internal components disintegrate and pits remain between the cells. Pits are cavities in secondary walls, arranged such that pits of adjacent cells are situated across from each other. Water and minerals move rapidly from one cell to another because these substances need to cross only one thin extracellular barrier, the primary cell wall.

Flowering plants have evolved a water-conducting system made up of vessels, which are formed from individual cells called **vessel elements**. These cells are laid down end-to-end. Like tracheids, vessel elements have pits in their cell walls, but their pits are generally larger in diameter than those of tracheids. Before they undergo apoptosis, the end walls of vessel elements partially break down, forming a continuous hollow tube that functions as an open pipeline for water conduction (**Figure 33.8B**). In the course of angiosperm evolution, vessel elements have become shorter and wider, and their end walls have become less oblique and less obstructed. These adaptations have presumably increased the efficiency of water transport through the vessels. The xylem of many angiosperms includes both tracheids and vessels.

PHLOEM TISSUE   The transport cells of the phloem, unlike those of the mature xylem, are living cells. In flowering plants the characteristic cells of the phloem are called **sieve tube elements** (**Figure 33.8C**). Like vessel elements, these cells meet end-to-end. They form long sieve tubes, which transport carbohydrates and many other materials from their sources (usually leaves) to tissues that consume or store them (for example, roots).

Whereas the end walls of vessel elements break down to facilitate flow through the vessels, the end walls of sieve tube elements form tunnels through which carbohydrates and other needed substances flow. The end walls of sieve tube elements are called sieve plates. Although the sieve tube elements remain alive, some of their components, including the nucleus, ribosomes, and vacuole, break down during development. The sieve tube elements are, however, closely connected via plasmodesmata to **companion cells**—specialized parenchyma cells that retain

(A) Tracheids

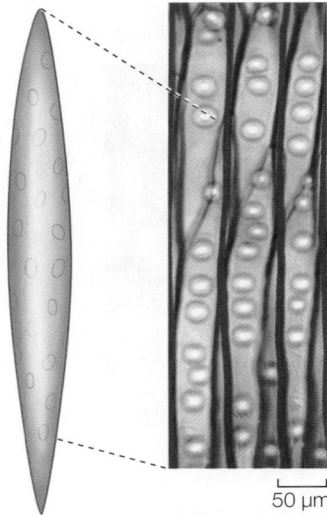

50 μm

(B) Vessel elements

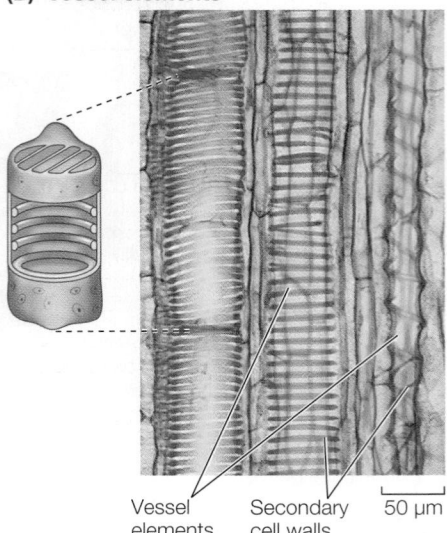

Vessel   Secondary   50 μm
elements   cell walls

(C) Sieve tube elements

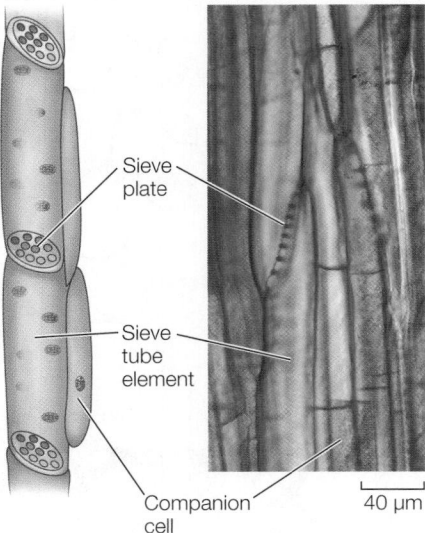

Sieve
plate

Sieve
tube
element

Companion   40 μm
cell

Figure 33.8 Vascular Tissue Cell Types   (A, B) Tracheary elements: **(A)** Tracheids in pinewood. The thick secondary walls are stained dark red. **(B)** Vessel elements in the stem of a squash. The secondary walls are stained red; note the different patterns of thickening, including rings and spirals. **(C)** Sieve tube elements and companion cells in the stem of a cucumber.

all their organelles and function as "life support systems" for the sieve tube elements.

▶ 33.2 recap

The three concentric tissue systems of the plant embryo—dermal, ground, and vascular—give rise to the tissues and organs of the adult plant. These tissue systems have unique combinations of specialized cells that carry out the various functions necessary for plant life.

learning outcomes

You should be able to:

- List and discuss the importance of the various functions of parenchyma.
- Define cell characteristics that make collenchyma useful as plant support structures.
- Compare collenchyma and sclerenchyma in terms of their ability to provide support to plants.
- Describe the water-conducting elements in plants, and compare these elements in gymnosperms and angiosperms.

1. Create a table comparing the three types of ground tissue cell types in terms of their structure and function.
2. What structural differences make tissues consisting of collenchyma cells more flexible than those consisting primarily of sclerenchyma cells?
3. Describe the differences between tracheids and vessel elements.

After the plant embryo has formed, it is encased in a seed coat and is ready to germinate. We will discuss aspects of seed germination in the chapters that follow. For now, let's consider the processes by which the embryo grows into a mature plant.

▶ key concept   Meristems Build a
33.3   Continuously Growing Plant

As noted earlier, plants and animals develop and function differently. While animals use their mobility to look for food, plants are sessile (rooted in one place) and instead grow toward scarce resources, both above (shoots toward the sun) and below the ground (roots toward water and minerals).

focus your learning

- Growth in plants can be either determinate or indeterminate, depending on the organ structure.
- Growth in terms of cell numbers occurs at meristems.
- Different apical meristems are where growth in cell numbers occurs and gives rise to leaves, stems, flowers, and roots.
- At the root meristem, zones of cell division, elongation, and maturation (differentiation) form the tissues of the root and root cap.
- The root consists of several tissue layers outside of the inner vascular tissues. These tissues have different arrangements in eudicot and monocot roots.
- Secondary growth in eudicots causes increase in diameter and forms wood and bark.

In most animals, growth is **determinate**: they stop growing when they reach the adult state. Some plant organs, such as leaves, flowers, and fruits, also show determinate growth. But shoots and roots keep growing; such open-ended growth is **indeterminate**.

How do plants increase in size?

Plant growth can occur in either of two ways:

1. **Primary growth** is characterized by cell division followed by cell enlargement. Primary growth results in the proliferation and lengthening of shoots and roots. All seed plants have a primary plant body, which consists of all the nonwoody parts of the

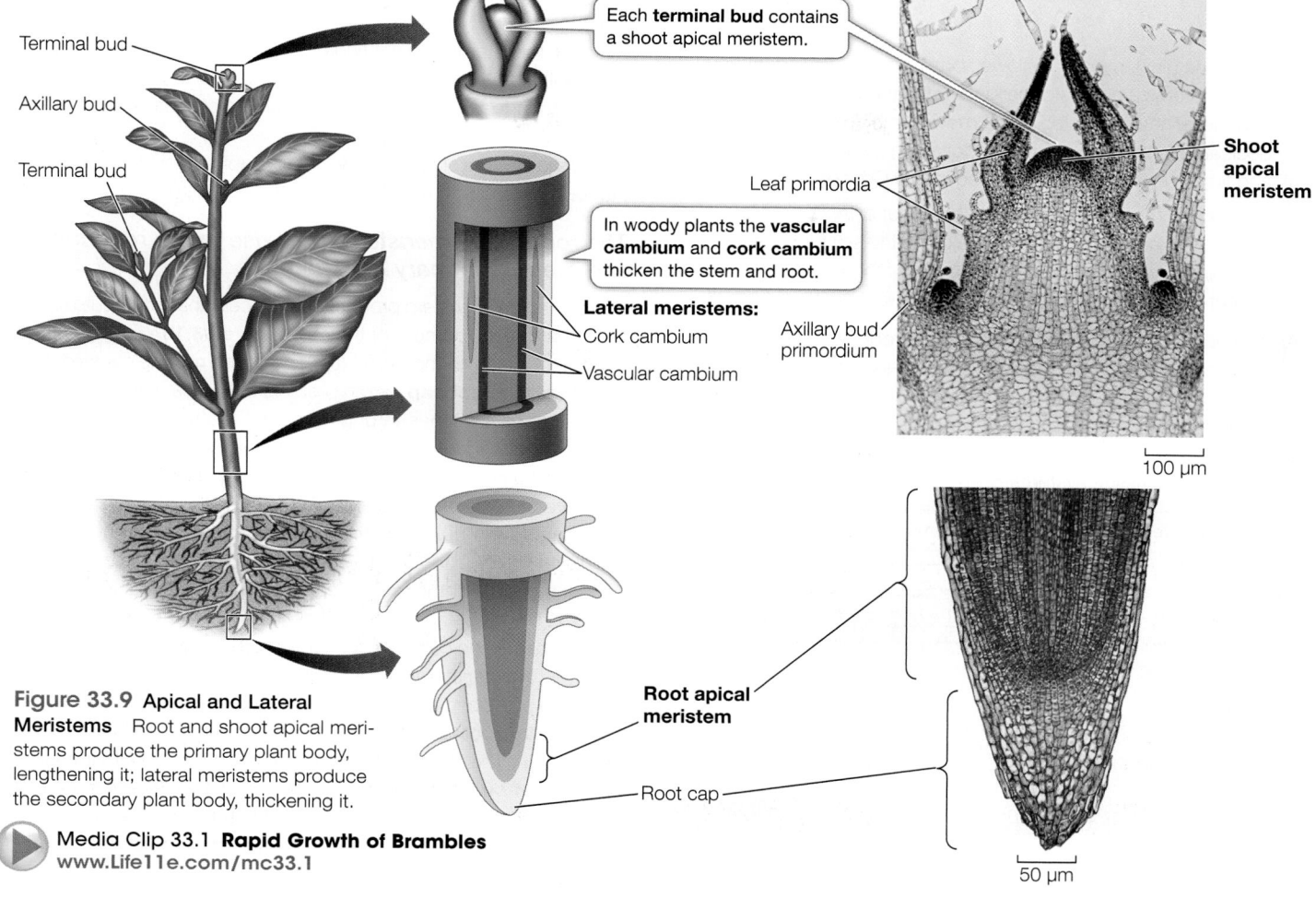

Terminal bud

Axillary bud

Terminal bud

Each **terminal bud** contains a shoot apical meristem.

Shoot apical meristem

Leaf primordia

In woody plants the **vascular cambium** and **cork cambium** thicken the stem and root.

Lateral meristems:
Cork cambium

Vascular cambium

Axillary bud primordium

100 µm

Root apical meristem

Root cap

50 µm

Figure 33.9 Apical and Lateral Meristems Root and shoot apical meristems produce the primary plant body, lengthening it; lateral meristems produce the secondary plant body, thickening it.

Media Clip 33.1 **Rapid Growth of Brambles**
www.Life11e.com/mc33.1

plant. Many herbaceous plants consist entirely of a primary plant body.

2. **Secondary growth** increases plant thickness. Woody plants, such as trees and shrubs, have a secondary plant body consisting of wood and bark. As the tissues of the secondary plant body are laid down, the stems and roots thicken.

Meristems generate the plant body

Meristems are localized regions of undifferentiated cells that are the sources of all new growth in the adult plant. Even before seed germination, the plant embryo has two meristems: a shoot apical meristem near the end of the embryonic shoot, and a root apical meristem at the end of the embryonic root (see Figure 33.5).

Meristematic cells are small and closely packed, with very small vacuoles and thin primary cell walls. The cells that perpetuate the meristems, called **initials**, are comparable to animal stem cells (discussed in Key Concept 19.1). When the initials divide, some of the daughter cells develop into new initials, and some differentiate into more specialized cells.

Several types of meristem contribute to the growth and development of the adult plant:

- Apical meristems in the root and shoot (**Figure 33.9**) orchestrate primary growth, ultimately giving rise to every cell in the primary plant body.

- When the initials of apical meristems divide, some of their daughter cells differentiate and become the **primary meristems**. Three kinds of primary meristem (see below) give rise to the three major tissue systems (dermal, ground, and vascular) that we described in Key Concept 33.2.

- **Lateral meristems** (also called secondary meristems) orchestrate secondary growth (see Figure 33.9). Two lateral meristems, vascular cambium and cork cambium, contribute to the secondary plant body.

Indeterminate primary growth originates in apical meristems

Because apical meristems can perpetuate themselves indefinitely, a stem or root can continue to lengthen and grow indefinitely; in other words, growth of the plant body as a whole is indeterminate. All plant organs arise ultimately from cell divisions in apical meristems, followed by cell expansion and differentiation. Several types of apical meristems play roles in organ formation:

- Shoot apical meristems supply the cells for new leaves and stems. In addition to the main stem of the plant, each branch has its own shoot apical meristem. Shoot apical meristems are also called **vegetative meristems**, because they give rise to vegetative tissues (leaves, stems, and roots).

- When the plant is ready to flower, one or more of its shoot apical meristems are transformed into **inflorescence meristems**, and these in turn develop **floral meristems**. (Chapter 37 is dedicated to the reproduction of flowering plants; inflorescence and floral meristems are discussed in Key Concept 37.2.)

- Root apical meristems supply the cells that extend roots, enabling the plant to penetrate and explore the soil for water and minerals. Each type of root (i.e., the taproot, a lateral root, or an adventitious root; see below) has its own root apical meristem.

Apical meristems in both the shoot and the root give rise to a set of primary meristems, which produce the tissues of the primary plant body. From the outside to the inside of the shoot or root, the primary meristems are the **protoderm**, the **ground meristem**, and the **procambium** (see Figure 33.10A). These meristems, in turn, give rise to the three tissue systems:

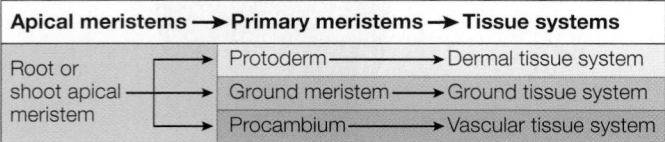

Apical meristems → Primary meristems → Tissue systems		
Root or shoot apical meristem	Protoderm → Dermal tissue system	
	Ground meristem → Ground tissue system	
	Procambium → Vascular tissue system	

(A) Root anatomy

Lateral root

Epidermis

Root hairs

Zone of cell maturation

Zone of cell elongation

Primary meristems:

Protoderm

Ground meristem

Procambium

Zone of cell division

New daughter cells are produced in the **root apical meristem**. Most daughter cells differentiate into the primary tissues of the root.

Some daughter cells become part of the **root cap**, which is constantly being eroded away.

Quiescent center

(B) Roots hairs (SEM)

1 mm

Figure 33.10 Tissues and Regions of the Root Tip (A) Extensive cell division creates the complex structure of the root. (B) Root hairs, seen with a scanning electron microscope (SEM).

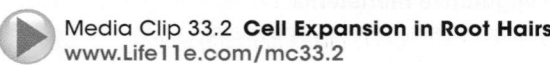
Media Clip 33.2 **Cell Expansion in Root Hairs**
www.Life11e.com/mc33.2

Because meristems can continue to produce new organs throughout the lifetime of the plant, the plant body is much more variable in form than the typical animal body, which produces each organ only once.

Let's look more closely at how the root apical meristem produces the root system.

The root apical meristem gives rise to the root cap and the root primary meristems

The root apical meristem produces all the cells that contribute to the growth and development of the root (**Figure 33.10A**). Some of the daughter cells from the apical (tip) end of the root apical meristem contribute to a **root cap**, which protects the delicate growing region of the root as it pushes through the soil. The root cap secretes a mucopolysaccharide (slime) that acts as a lubricant. Even so, the cells of the root cap are often damaged or scraped away and must therefore be replaced constantly. The root cap is also the structure that detects the pull of gravity and thus controls the downward growth of roots.

In the middle of the root apical meristem is a quiescent center, in which cell divisions are rare. The quiescent center cells may be in G1 phase of the cell cycle for 10 days and prevent the surrounding meristem cells from differentiating. The daughter cells produced above the quiescent center (that is, away from the root cap) become the three primary meristems.

The apical and primary meristems constitute the **zone of cell division**, the source of all the cells of the root's primary tissues. Just above this zone is the **zone of cell elongation**, where the newly formed cells are elongating and thus pushing the root farther into the soil. Above that zone is the **zone of maturation**, where the cells are differentiating, taking on specialized forms and functions. These three zones grade imperceptibly into one another; there is no abrupt border separating them.

The products of the root's primary meristems become root tissues

The products of the three primary meristems (the protoderm, ground meristem, and procambium) are the tissue systems of the mature root. The differing arrangement of the three tissue systems in the roots of eudicots and monocots is one of the ways in which the two clades are distinguished (**Figure 33.11**).

- The protoderm produces the epidermis, the outer layer of cells that is adapted for protection of the root and absorption of mineral ions and water. Many of the epidermal cells produce long, delicate **root hairs**, which vastly increase the surface area of the root (**Figure 33.10B**). Root hairs grow out among the soil particles, probing nooks and crannies and taking up water and minerals.

- The **cortex** is a region of ground tissue produced internal to the epidermis by the ground meristem. The cells of the multilayered cortex have various functions, including storage.

- The **endodermis** is the innermost layer of the cortex. Unlike other cortical cells, the endodermal cells contain a waterproof substance called suberin in their primary cell walls. The suberin forms a cylindrical ring around the inside of the endodermis,

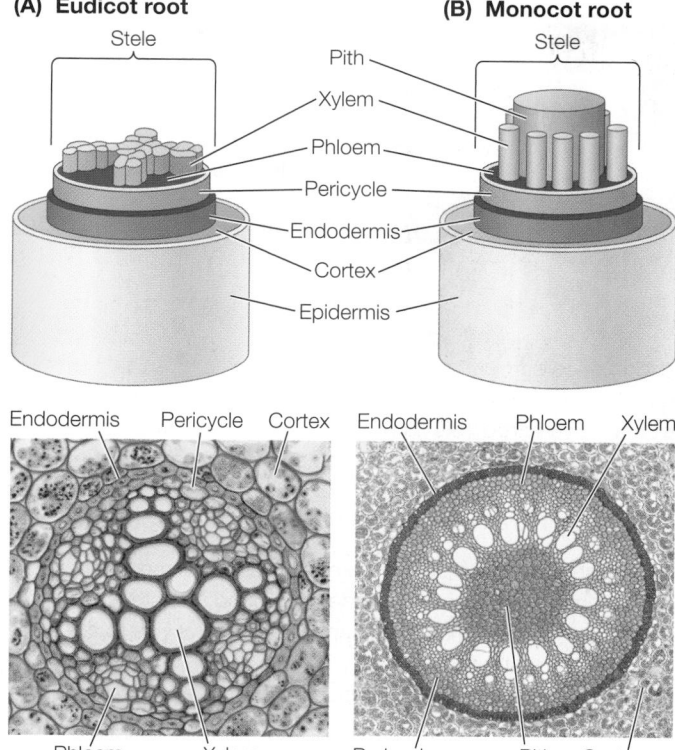

(A) Eudicot root

Stele

(B) Monocot root

Stele

Pith
Xylem
Phloem
Pericycle
Endodermis
Cortex
Epidermis

Endodermis   Pericycle   Cortex   Endodermis   Phloem   Xylem

Phloem   Xylem   Pericycle   Pith   Cortex

Figure 33.11 Products of the Root's Primary Meristems The protoderm gives rise to the outermost layer (epidermis). The ground meristem produces the cortex, the innermost layer of which is the endodermis. The primary vascular tissues of the root are found in the stele, which is the product of the procambium. The arrangement of tissues in the stele differs in the roots of **(A)** eudicots and **(B)** monocots. The photomicrographs show cross sections of the stele of a representative eudicot (the buttercup, *Ranunculus*) and a representative monocot (corn, *Zea mays*), showing the arrangement of the primary root tissues.

 Activity 33.1 **Eudicot Root**
www.Life11e.com/ac33.1

 Activity 33.2 **Monocot Root**
www.Life11e.com/ac33.2

which allows the endodermal cells to control the movement of water and dissolved mineral ions into and out of the vascular tissue system.

- Inner to the endodermis is the vascular cylinder, or **stele**, produced by the procambium. The stele consists of three tissues: pericycle, xylem, and phloem. The **pericycle** consists of one or more layers of relatively undifferentiated cells. It has three important functions: (1) It is the tissue within which lateral roots arise (**Figure 33.12**). (2) It can contribute to secondary growth by giving rise to lateral meristems that thicken the root. (3) Its cells contain membrane transport proteins that export nutrient ions into the cells of the xylem.

At the very center of the root of a eudicot lies the xylem. Seen in cross section, it typically has the shape of a star with a variable number of points (see Figure 33.11A). Between the points are bundles of phloem. In monocots, a region of parenchyma cells called the

pith typically lies in the center of the root, surrounded by xylem and phloem (see Figure 33.11B). Pith, which often stores carbohydrate reserves, is also found in the stems of both eudicots and monocots. Strips of pith of the stem of the papyrus plant *Cyperus papyrus* can be glued together and used to make a paperlike substance called by its species name. Papyrus dates from earlier than 2500 B.C.E. in Egypt, and was the writing surface of choice until about 300 C.E., when parchment from animal skins replaced it.

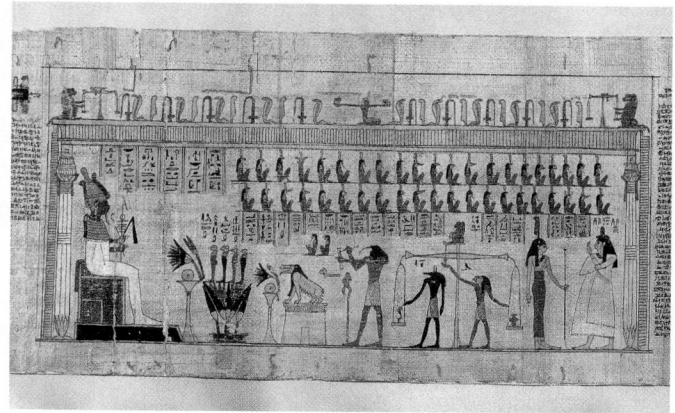

The root system anchors the plant and takes up water and dissolved minerals

Water and minerals enter most plants through the root system, which is located in the soil. Because light does not penetrate the soil, roots typically lack the capacity for photosynthesis. Although hidden from view, the root system is often larger than the visible shoot system. For example, the root system of a 4-month-old winter rye plant (*Secale cereale*) was found to be 130 times longer in total than the shoot system, with almost 13 million branches that had a cumulative length of more than 500 kilometers!

Angiosperm root systems develop from the embryonic root, called the **radicle**. From this common starting point, the root systems of monocots and eudicots develop differently. Following seed germination, the radicle of most eudicots develops as a primary root called the **taproot**, which extends downward by tip growth and outward by initiating **lateral roots**. The taproot and the lateral

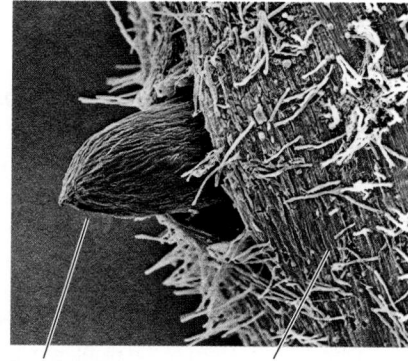

Developing lateral root   Epidermis

Figure 33.12 Lateral Root Anatomy SEM of the tip of a lateral root in a willow tree. Cells in the pericycle divide and the products differentiate, forming the tissues of a lateral root.

Q: How does totipotency relate to lateral root formation?

(A) Taproots

(B) Fibrous root system

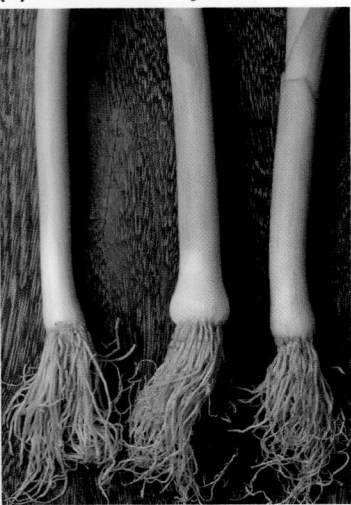

(C) Prop roots

Figure 33.13 Root Systems of Eudicots and Monocots **(A)** The taproot systems of eudicots, such as carrots and sugar beets, contrast with **(B)** the fibrous root system of a leek and **(C)** the adventitious prop roots of corn.

roots form a **taproot system**, which can take a variety of forms. For example, the taproot itself often functions as a nutrient storage organ, as in carrots (*Daucus carota*), sugar beets (*Beta vulgaris*), and sweet potato (*Ipomoea batatas*) (**Figure 33.13A**).

In contrast, the primary root of monocots (and some eudicots) is short-lived. Because they originate from the stem at ground level or just below, the roots of a typical monocot are called **adventitious** ("arriving from outside") **roots**, and they form a **fibrous root system** composed of numerous thin roots that are all roughly equal in length (**Figure 33.13B**). Many fibrous root systems have large surface areas for the absorption of water and minerals. A fibrous root system clings to soil very well. The fibrous root systems of grasses, for example, may protect steep hillsides where runoff from rain would otherwise cause erosion.

In some plants—corn, banyan, and pandanus trees, for example—adventitious roots grow down from above the ground and function as props to help support the shoot system (**Figure 33.13C**). These **prop roots** have evolved for different reasons in different species. For example, some monocots may develop prop roots because they are unable to support aboveground growth through the thickening of their stems. Pandanus trees often grow near coastal beaches, where their prop roots help provide support in very sandy soils. Banyans begin life as epiphytes (plants that grow on other plants) and then develop woody prop roots, which enable them to grow into huge trees.

The products of the stem's primary meristems become stem tissues

Now let's turn to the shoot, which comprises the stem and leaves. As seen in Figure 33.1, shoots and roots are composed of repeating modules called phytomers. A phytomer in the shoot system consists of a node with its attached leaf or leaves, the internode below the node, and axillary buds, each of which forms in the angle between a leaf and the stem. Shoots grow by adding new phytomers. Initially these form only on the primary stem of the plant, but if an axillary bud develops into a branch, it does so by producing new phytomers. The new phytomers originate from cells in the shoot apical meristems, which are present in the terminal buds of each branch and the main stem (see Figure 33.9).

The shoot apical meristem, like the root apical meristem, forms three primary meristems: protoderm, ground meristem, and procambium. These primary meristems, in turn, give rise to the three shoot tissue systems.

As the shoot grows and extends, bulges called **leaf primordia** develop on the sides of the shoot apical meristem at regular intervals (see Figure 33.9). These primordia are made up of primary meristematic tissues, which go on to develop into the mature tissues of the leaf. In addition, bud primordia form at the bases of the leaf primordia. These have the potential to become new apical meristems and initiate new shoots. The sites where leaf primordia form become nodes on the developing stem. The regions between the nodes (the internodes) lengthen initially via cell division in the primary meristematic tissues, and later by cell elongation in the mature stem tissues. The growing stem has no protective structure analogous to the root cap, but the leaf primordia can act as a protective covering for the shoot apical meristem.

The vascular systems of stems differ from those of roots. In a root, the vascular tissue lies deep in the interior, with the xylem at or near the center (see Figure 33.11). The vascular tissue of a young stem, however, is divided into discrete **vascular bundles** (**Figure 33.14**). Each vascular bundle contains both xylem and phloem. In eudicots the vascular bundles generally form a ring, whereas in monocots they are scattered throughout the stem.

In addition to the vascular tissues, the stem contains other important storage and supportive tissues. In eudicots the pith lies inside the ring of vascular bundles and also extends between them, forming regions called pith rays. To the outside lies the cortex, which may contain supportive collenchyma cells with thickened walls. The pith and cortex constitute the ground tissue system of the stem. The outermost cell layer of the young stem is the epidermis.

The stem supports leaves and flowers but can have other roles

The main role of stems is to elevate and support the photosynthetic organs (leaves) as well as the reproductive organs (flowers). But in some plants the stem does other things for the plant. For instance, the potato tuber that we eat comes from the ground but is not a root; it's an underground stem that acts as a storage organ for the plant.

Figure 33.14 Vascular Bundles in Stems **(A)** In herbaceous eudicot stems, the vascular bundles are arranged in a cylinder, with pith in the center and the cortex outside the cylinder. **(B)** A scattered arrangement of vascular bundles is typical of monocot stems.

 Activity 33.3 **Eudicot Stem**
www.Life11e.com/ac33.3

 Activity 33.4 **Monocot Stem**
www.Life11e.com/ac33.4

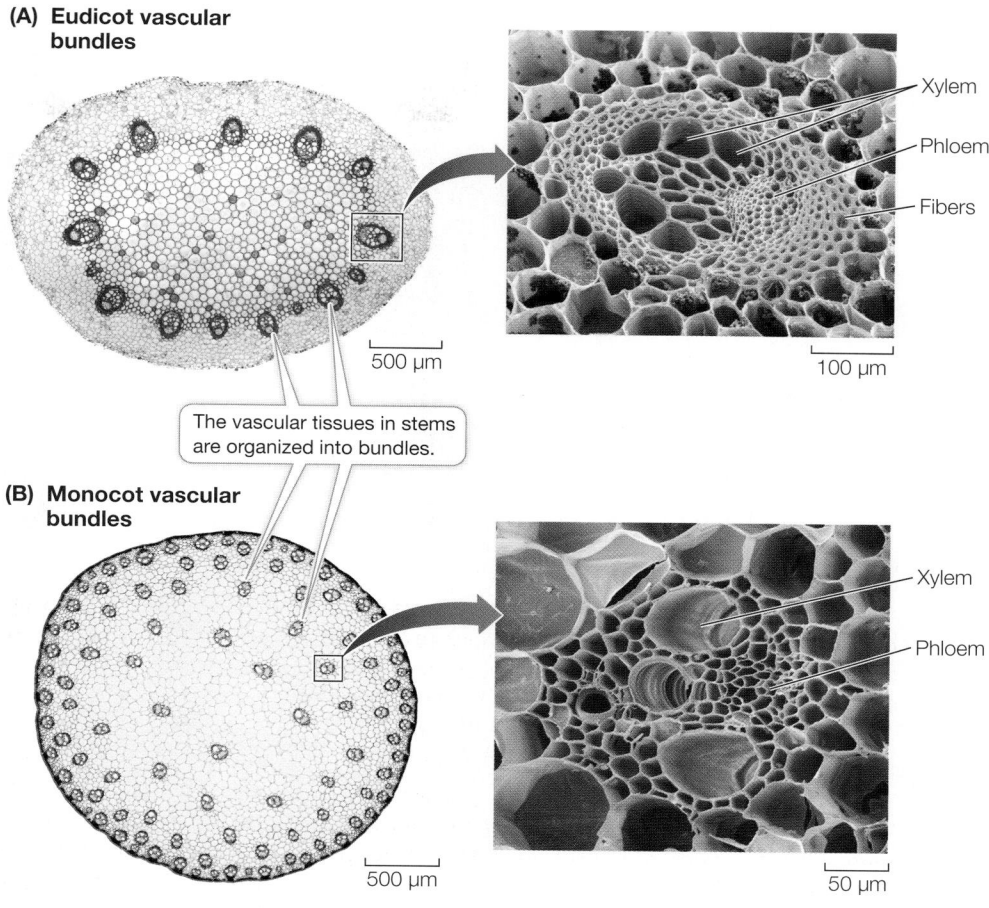

(A) Eudicot vascular bundles

500 μm

— Xylem
— Phloem
— Fibers

100 μm

The vascular tissues in stems are organized into bundles.

(B) Monocot vascular bundles

500 μm

— Xylem
— Phloem

50 μm

The "eyes" of a potato are depressions containing axillary buds—in other words, a sprouting potato is just a branching stem (**Figure 33.15A**). Many desert plants have enlarged, water-retaining stems (**Figure 33.15B**). The runners of strawberry plants are horizontal stems that develop adventitious roots some distance from the main stem (**Figure 33.15C**); if the links between the rooted portions are broken, independent plants can develop on each side of the break—a form of vegetative (asexual) reproduction (see Key Concept 37.3).

Leaves are determinate organs produced by shoot apical meristems

For most of its life, a plant produces leaves from apical meristems. Leaves originate from the edges of the apical meristem as initial cells that differentiate into leaf primordia. Leaves differ from stems in two important ways:

1. Unlike the growth of the stem, which is indeterminate, the growth of a leaf is determinate.

2. Whereas the tissues of the stem are arranged in a radial pattern, the *leaf, as a flat organ, has a distinct top side and bottom side.

(A) Branches

Tuber (modified stem)

(B) "Barrel" (enlarged stem)   Spines (modified leaves)

(C) Shoots   Runner (horizontal stem)

Figure 33.15 Modified Stems **(A)** A potato is a modified stem called a tuber; the sprouts that grow from its "eyes" are shoots, not roots. **(B)** The stem of this barrel cactus is enlarged to store water. Its highly modified leaves serve as thorny spines. Most of this plant's photosynthesis occurs in the stem. **(C)** The runners of strawberry are horizontal stems that produce roots and shoots at intervals. Rooted portions of the plant can live independently if the runner is cut.

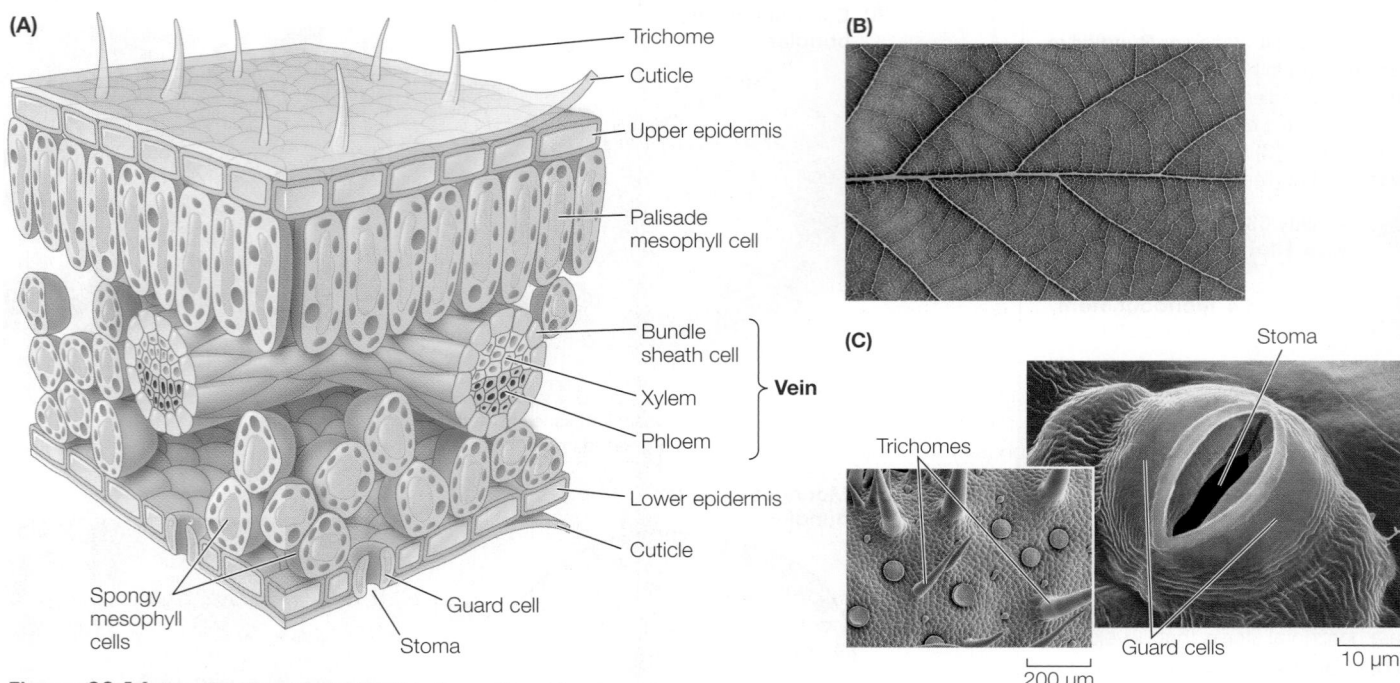

Figure 33.16 The Eudicot Leaf (A) This three-dimensional diagram shows a section of a eudicot leaf. (B) The network of fine veins in this maple leaf carries water to the mesophyll cells and carries photosynthetic products away from them. (C) Carbon dioxide enters the leaf through stomata like this one on the epidermis of a eudicot leaf. Trichomes are epidermal hairs that can break up airflow, reducing water loss from the leaf and repel water so that gas exchange can take place through the stomata.

 Activity 33.5 **Eudicot Leaf**
www.Life11e.com/ac33.5

---

*connect the concepts   The leaf is elegantly structured for the reactions of photosynthesis. For more detail on photosynthesis and the role of leaf tissues, see Key Concepts 10.1 and 10.4.

Leaf anatomy is beautifully adapted to carry out photosynthesis, and to support that process by exchanging the gases O_2 and CO_2 with the environment, while at the same time limiting evaporative water loss (**Figure 33.16**). Its extensive vascular system supplies the leaf with water and mineral nutrients and exports the products of photosynthesis to the rest of the plant. Figure 33.16A shows a typical eudicot leaf in three dimensions. The photosynthetic parenchyma tissue in leaves is the **mesophyll** (which means "middle of the leaf"). Most eudicot leaves have two zones of mesophyll: an upper layer of elongated cells called the palisade mesophyll, and a lower layer of irregularly shaped cells called the spongy mesophyll. Within the spongy mesophyll is a great deal of air space through which CO_2 can diffuse to photosynthesizing cells.

Vascular tissue branches extensively throughout the leaf, forming a network of veins (see Figure 33.16B). Veins extend to within a few cell diameters of all the cells of the leaf by the xylem, ensuring that the mesophyll cells are well supplied with water and minerals. The products of photosynthesis are loaded into the veins for export to

the rest of the plant. In addition to photosynthetic products, leaves make defensive molecules such as those you learned about in the opening investigation that produce cyanide in cassava. **Investigating Life: Understanding the Synthesis and Transport of Cyanogenic Glycosides** describes how it was determined that leaves produce these chemicals.

The epidermis covers the entire surface of the leaf. The epidermal cells secrete a waxy cuticle that is impermeable to water. Although this impermeability prevents excessive water loss, it also poses a problem: while the epidermis keeps water in the leaf, it also keeps out CO_2—the other raw material of photosynthesis.

The problem of balancing water retention and carbon dioxide availability is solved by guard cells, a system we'll examine in detail (Key Concept 34.3). Stomatal guard cells are modified epidermal cells that can change their shape, thereby opening or closing pores called **stomata** (singular *stoma*). The stomata serve as passageways between the environment and the leaf's interior (see Figure 33.16C). When the stomata are open, carbon dioxide can enter and oxygen can leave, but water can also be lost.

Many stems and roots undergo secondary growth

As you have seen, the roots and stems of some eudicots develop a secondary plant body, the tissues of which include wood and bark. These tissues are derived by secondary growth from the two lateral meristems:

- The **vascular cambium** is a cylindrical layer of tissue consisting predominantly of elongated cells that divide frequently. It supplies the cells of the secondary xylem and secondary phloem, which eventually become wood and bark, respectively.

- The **cork cambium** produces mainly waxy-walled protective cells. It supplies some of the cells that become bark.

experiment

Original Paper: Jørgensen, K. et al. 2005. Cassava plants with a depleted cyanogenic glycoside content in leaves and tubers. Distribution of cyanogenic glycosides, their site of synthesis and transport and blockage of biosynthesis by RNA interference technology. *Plant Physiology* 139: 363–374.

As described in the chapter opening, cassava is the most important root crop in the world. Its use as food is challenging, because the plant produces molecules that can be transformed into toxic cyanide when the tissues are disrupted. These cyanogenic glycosides (CGs) are important in defending the plant against animals that eat it, including humans. For the millions of people who eat cassava, careful processing is needed to avoid poisoning. Unfortunately, the processing methods lower the overall nutritional value of the food. Understanding how CGs are made and end up in the fleshy root is an important step in possibly using genetics to produce a crop that does not make CGs. At the Royal Veterinary and Agricultural University in Copenhagen, a team led by Birger Møller investigated the synthesis and transport of CGs in cassava plants.

HYPOTHESIS▶ Molecules that produce cyanide in the cassava are made in the leaves and transported to the underground root.

METHOD

Cassava plant

First petiole
2
1
Shoot apex
3
4
Third petiole
5
6
7

1 Leave stem intact and measure CG in plant shoot regions.

2 Remove phloem from stem below fifth leaf and then measure CG in plant parts.

RESULTS

Region	Relative CG concentration
Shoot apex	Moderate
First petiole	High
Third leaf	Low
Third petiole	Low

Region	Relative CG concentration
Fifth leaf	High
Stem above cut	Moderate
Stem below cut	Very Low

CONCLUSION▶ Cyanogenic glycosides are made in leaves and are transported through petioles to the stem, from which they are transported to the root.

work with the data

QUESTIONS▶

1. Cyanogenic glycosides in plant organs were measured in an intact stem of a cassava plant and in one in which the phloem had been cut below the fifth leaf. The results are shown in the graphs. What can you conclude about the source of CG?

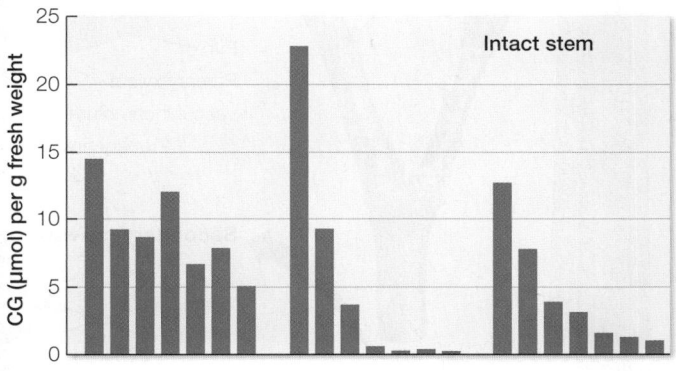

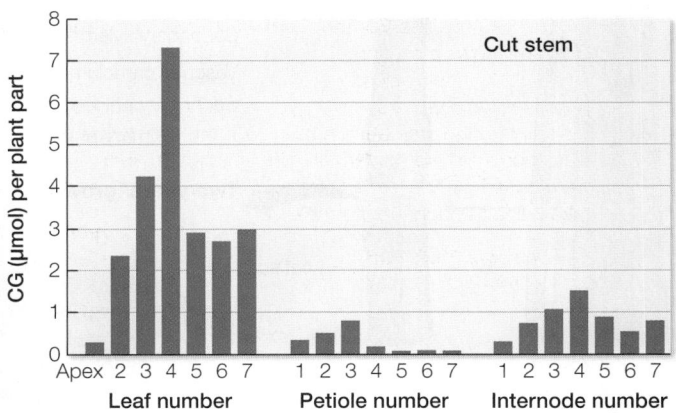

2. When a stem is girdled, it is cut carefully so that the epidermal, cortical, and outer vascular tissues (phloem) are removed. Using intact plants, the biologists girdled the stem of a cassava plant. After 2 days, they measured CG in various regions on the stem. The results for CG levels after girdling between the fifth and sixth nodes are shown in the table. What can you conclude about the source and transport of CG? Why was the level in the sixth leaf (below the girdle) so high?

Region	CG level (µmol)
Fifth leaf	59
Stem above cut	15
Stem below cut	0.2
Sixth leaf	55

A similar **work with the data** exercise may be assigned in **LaunchPad**.

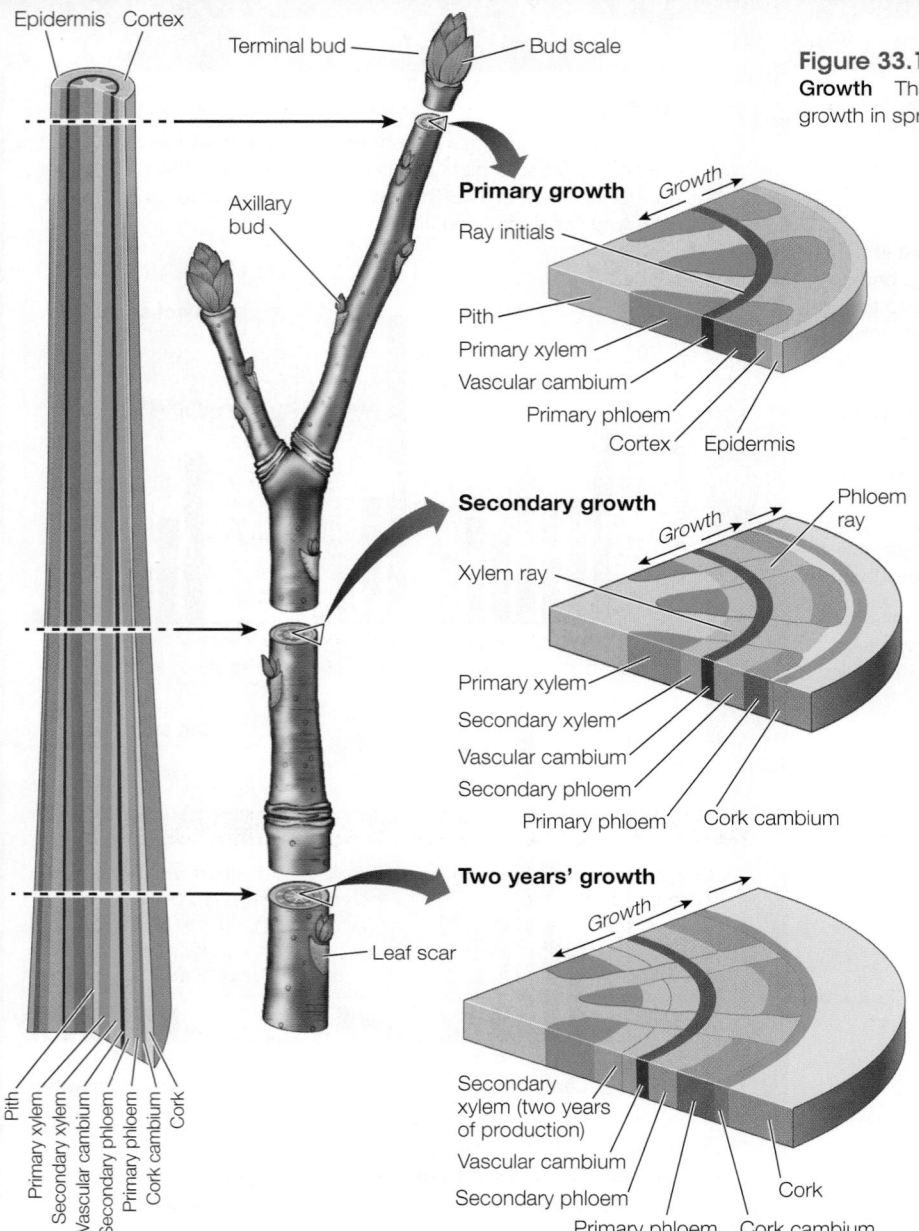

Epidermis Cortex

Terminal bud — — Bud scale

Axillary bud

Primary growth

Growth →

Ray initials

Pith

Primary xylem

Vascular cambium

Primary phloem

Cortex   Epidermis

Secondary growth

Growth →

Phloem ray

Xylem ray

Primary xylem

Secondary xylem

Vascular cambium

Secondary phloem

Primary phloem   Cork cambium

Two years' growth

Growth →

Leaf scar

Secondary xylem (two years of production)

Vascular cambium

Secondary phloem

Primary phloem   Cork cambium

Cork

Pith
Primary xylem
Secondary xylem
Vascular cambium
Secondary phloem
Primary phloem
Cork cambium
Cork

Figure 33.17 A Woody Twig Has Both Primary and Secondary Growth The apical meristems in this dormant twig will produce primary growth in spring. Lateral meristems are responsible for secondary growth.

the vascular bundles also divide, forming a continuous cylinder of vascular cambium running the length of the stem. This cylinder, in turn, gives rise to complete cylinders of secondary xylem (the **wood**) and secondary phloem, which contributes to the bark. It also produces vascular rays for lateral transport, a structure not found in primary xylem and phloem. Therefore the vascular cambium produces vessel elements, tracheids, parenchyma cells, and supportive fibers in the secondary xylem; and sieve tube elements, companion cells, fibers, and parenchyma cells in the secondary phloem.

▶ **Animation 33.1 Secondary Growth: The Vascular Cambium**
www.Life11e.com/a33.1

As secondary growth of stems or roots continues, the expanding vascular tissue stretches and breaks the epidermis and the outer layers of the cortex, which ultimately flake away. Before these dermal tissues are broken away, cells lying near the surface of the secondary phloem begin to divide, forming a cork cambium. This meristematic tissue produces layers of cork, a protective tissue composed of cells with thick walls waterproofed with suberin. The cork soon becomes the outermost tissue of the stem or root (see Figure 33.17). Without the activity of the cork cambium, the sloughing off of the outer primary tissues would expose the plant to potential damage, such as excessive water loss or invasion by microorganisms. Sometimes the cork cambium produces cells toward the inside as well as the outside; these cells constitute a tissue known as the phelloderm.

The cork cambium, cork, and phelloderm constitute the secondary dermal tissue called periderm. As the vascular cambium continues to produce secondary vascular tissue, these corky layers are lost, but the continuous formation of new cork cambia in the underlying secondary phloem gives rise to new corky layers. The periderm and the secondary phloem—that is, all the tissues external to the vascular cambium—constitute the **bark**.

When periderm forms on stems or roots, the underlying tissues still need to release carbon dioxide and take up oxygen for cellular respiration. Spongy regions in the periderm called **lenticels** can allow such gas exchange (**Figure 33.18**).

You may be curious about cork, which is used as stoppers in bottles of wine and in flasks in the chemistry lab. Cork is harvested from a tree, the cork oak (*Quercus suber*, see photo next page), which is grown for that purpose mainly in Portugal. The tree lives more than 100 years, and cork is first harvested when the tree is 25 years old. The outer bark is removed first, carefully exposing the cork layers, which are then removed without damaging the living tissues. The cork cambium then produces more cork, so a tree can be harvested every 10 years.

In cold parts of the world, each year, deciduous trees lose their leaves and have bare branches and twigs over the winter (**Figure 33.17**). The apical meristems of the twigs are enclosed in buds protected by bud scales. When the buds begin to grow in spring, the scales fall away, leaving scars that show where the bud was. These scars allow us to identify the parts of the twig from each year's growth. The dormant twig shown in Figure 33.17 is the product of both primary and secondary growth. The buds consist entirely of primary tissues.

The vascular cambium is initially a single layer of cells lying between the primary xylem and the primary phloem within the vascular bundles. The root or stem increases in diameter when the cells of the vascular cambium divide, producing secondary xylem cells toward the inside of the root or stem and producing secondary phloem cells toward the outside. In the stem, cells in the pith rays between

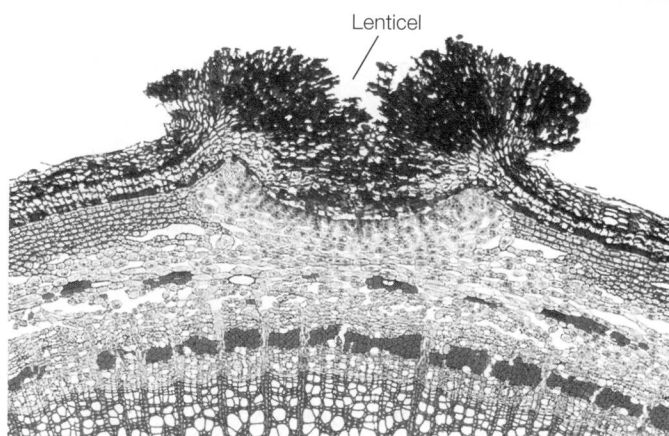

Figure 33.18 Lenticels Allow Gas Exchange through the Periderm The region of periderm that appears to be broken open is a lenticel in a year-old elderberry (*Sambucus*) twig; note the spongy tissue that constitutes the lenticel.

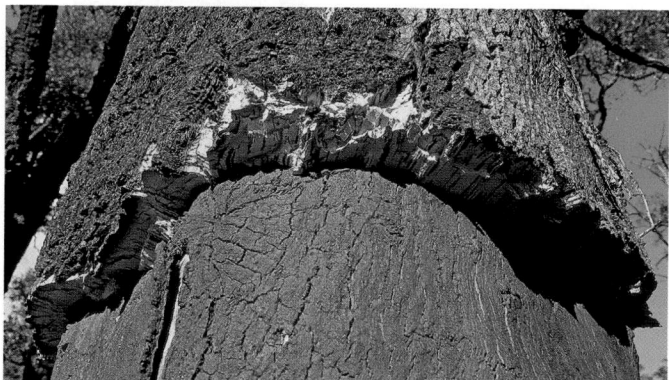

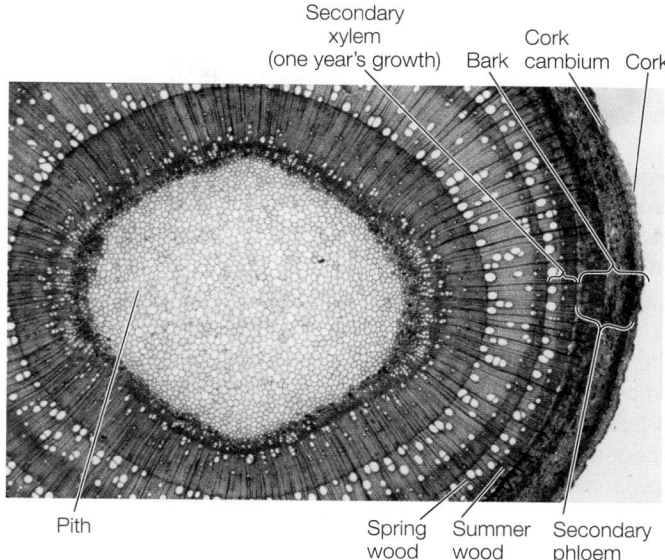

Figure 33.19 Annual Rings Rings of secondary xylem are the most noticeable feature of this cross section from a tree trunk.

Cross sections of most tree trunks (stems) in temperate-zone forests show annual rings of wood (**Figure 33.19**), which result from seasonal environmental conditions. In spring, when water is relatively plentiful, the tracheids or vessel elements produced by the vascular cambium tend to be large in diameter and relatively thin-walled. Such wood is well adapted for transporting water and minerals. As water becomes less available during the summer, narrower cells with thicker walls are produced, making this summer wood darker and perhaps more dense than the wood formed in spring. Thus each growing season is usually recorded in a tree trunk by a clearly visible annual ring. Trees in the moist tropics do not undergo seasonal growth, so they do not lay down such obvious regular rings. Variations in temperature or water supply can lead to the formation of more than one ring in a single year, but most commonly a single annual ring is formed. Next time you are in the woods and see a fallen tree, try to count the annual rings to gauge the age of a tree.

Only eudicots and other non-monocot angiosperms, along with many gymnosperms, have a vascular cambium and a cork cambium and thus undergo secondary growth. The few monocots that form thickened stems—palms, for example—do so without secondary growth. Palms have a very wide apical meristem that produces a wide stem, and dead leaf bases add to the diameter of the stem. All monocots grow in essentially this way, as do other angiosperms that lack secondary growth.

33.3 recap

Meristems are localized regions of cell division that are the sources of all new organs in the adult plant. Apical meristems are responsible for primary growth, which is associated with the lengthening and branching of shoots and roots. Lateral meristems increase plant thickness and form wood and bark in many eudicots.

learning outcomes

You should be able to:

- Describe characteristics of determinate and indeterminate growth in plants.
- Compare meristematic with non-meristematic cells, and explain the function of meristematic cells in plant growth.
- Analyze how apical meristems are able to produce different organs, including leaves, stems, flowers, and roots.
- Describe or illustrate a longitudinal section of the developing root, and explain the functions of special zones of cells from the tip upward.
- Explain how different types of meristem result in the various layers of the mature root.

1. How can an apical meristem be maintained for years while continuing to form leaves?
2. Define the primary meristems of the plant and how tissues and organs are formed from them.
3. What cells are derived from the root apical meristem, and what is the general process of root growth?
4. Compare primary and secondary growth in a tree trunk in terms of the meristems involved, the tissues formed, and the result.

The building of the plant body by meristems allows a plant to respond to its environment by redirecting its growth. Thus individual plants of the same species can vary greatly in form. What underlies this variation, and how have we humans used it to our advantage?

33.4 Domestication Has Altered Plant Form

You have seen in this chapter that a simple, modular plant body plan underlies the remarkable diversity of the flowering plants that cover our planet. Differences in plant form among species are not surprising, given the high levels of genetic diversity among plant species. However, members of the same species can also show remarkable diversity in form. From a genetic perspective, this suggests that minor differences in gene content or gene regulation can underlie dramatic differences in plant form.

---

focus your learning

- Genetic variation in the structure of plants is a valuable natural resource for crop plant evolution.

It is hard to believe that modern corn was domesticated from the wild grass teosinte, which still grows in the hills of Mexico (**Figure 33.20**). One of the most conspicuous differences is that teosinte, like other wild grasses, is highly branched, whereas domesticated corn has a single shoot. This morphological difference is due in large part to the activity of a single gene called *teosinte branched 1* (*tb1*). The protein product of *tb1* regulates the growth of axillary buds (see Figure 33.1). The allele of *tb1* in domesticated corn represses branching, whereas the allele in teosinte permits branching.

Even harder to believe is that a single species, * **Brassica oleracea** (**wild mustard**), is the ancestor of many familiar and morphologically diverse crops, including kale, broccoli, Brussels sprouts, and cabbage.

> ***connect the concepts** The diversity of plants derived from wild mustard illustrated in Figure 20.5 demonstrates how artificial selection—the purposeful selection of specific phenotypes by humans—can produce rapid evolutionary change.

Each of these familiar vegetable crops has the same basic body plan. Starting with morphologically diverse populations of the wild ancestor, humans selected and planted the seeds of variants with traits they found desirable. For example, Brussels sprout plants were selected for their enlarged axillary buds, cabbage plants were selected for their enlarged terminal buds and short internodes, and broccoli and cauliflower plants were selected for their large clusters of flower buds. Many generations of such artificial selection produced the crops that fill the produce section of the supermarket and the stands of farmers' markets.

Just as they were for ancient farmers, the genomes of plants are priceless resources today. The genetic variation in crop plants and their wild relatives can be used to improve our crop plants or adapt

Figure 33.20 Modern Corn Was Domesticated from the Wild Grass Teosinte Each teosinte plant has multiple branches. Beginning more than 8,000 years ago in Mexico, farmers favored plants with minimal branching. Reducing the number of branches results in fewer ears per plant but allows each ear to grow larger and produce more seeds.

Teosinte   Corn

them to changing conditions. The improvement of crop plants is a work in progress that is being carried out in plant breeding programs worldwide. These programs are now more important than ever. Increased human activity is dramatically changing our planet and leading to the extinction of more and more plant species. For this reason, various organizations around the world have developed seed banks, where seeds of diverse species, and variants within species, are stored.

33.4 recap

Crop domestication involves artificial selection of certain desirable traits found in wild plant populations. By understanding the basic body plan of plants, one can more easily understand the morphological relationship between a crop plant and its wild relatives.

---

learning outcome

You should be able to:

- Discuss reasons why wild plant genomes should be preserved.

---

1. Why are the seeds from wild relatives of crop plants valuable?

QA How might plant biologists improve the cassava plant for human use?

Many people depend on cassava roots for food. Unfortunately, the roots must undergo processing to remove sources of toxic cyanide before they are eaten. As you saw in Investigating Life: Understanding the Synthesis and Transport of Cyanogenic Glycosides, cyanide-forming molecules are formed in leaves and then transported to the roots through the vascular system. Molecular biologists have described the details of organ-specific synthesis of the machinery responsible for a major cyanogenic molecule. RNA interference (see Key Concept 18.4) has been used to block cassava leaves from making the toxic precursor, and as a result its concentration in the edible roots goes down by 99 percent compared with untreated plants. This is a promising way to reduce the requirements for processing of the crop.

Future directions

An international group of plant biologists and agricultural scientists has been set up to stimulate research on cassava. The BioCassava Plus Consortium has not only found ways to improve the nutritional quality of cassava, but has also sought ways to improve the plant's ability to grow in drought conditions. Cassava has been crossed with a treelike relative, *Manihot glaziovii*, and the resulting plants produce roots that are not only fleshy and edible but that grow deep into the soil, where they can tap into water supplies far below the surface in dry climates. The hope is that this new crop variant will improve conditions for people in drought-prone places such as sub-Saharan Africa, where tens of millions of people routinely experience famine.

Chapter Summary 33

33.1 The Plant Body Is Organized in a Distinctive Way

- The vegetative organs of flowering plants are roots, which form a **root system**, and stems, leaves, and flowers, which form a **shoot system**. Review Figure 33.1
- Plant development differs from animal development in that plants have apical meristems, cell walls, vacuoles, and totipotent cells.
- Plants have apical–basal and radial axes of symmetry. Review Figure 33.4

33.2 Plant Organs Are Made Up of Three Tissue Systems

- Three tissue systems, arranged concentrically, extend throughout the plant body: the vascular tissue, dermal tissue, and ground tissue systems. Review Focus: Key Figure 33.6
- The **dermal tissue system** protects the plant body surface. Dermal cells form the **epidermis** and, in woody plants, the **periderm**.
- The **ground tissue system** contains cells of three types. Some **parenchyma** cells carry out photosynthesis; others store starch. **Collenchyma** cells provide flexible support. Sclerenchyma cells include **fibers** and **sclereids**, which provide strength and mechanical support. Review Figure 33.7
- The **vascular tissue system** includes **xylem**, which conducts water and minerals absorbed by the roots, and **phloem**, which conducts the products of photosynthesis throughout the plant body.
- **Tracheary elements** include **tracheids** and **vessel elements**, which are the conducting cells of the xylem. **Sieve tube elements** are the conducting cells of the phloem. Review Figure 33.8

33.3 Meristems Build a Continuously Growing Plant

- All seed plants possess a primary plant body consisting of non-woody tissues. Woody plants also possess a secondary plant body consisting of wood and bark. **Apical meristems** generate the primary plant body, and **lateral meristems** generate the secondary plant body. Review Figure 33.9

- Apical meristems are responsible for **primary growth** (lengthening of roots and shoots). Apical meristems at the tips of shoots and roots give rise to three **primary meristems** (**protoderm**, **ground meristem**, and **procambium**), which in turn produce the three tissue systems of the primary plant body.
- The root apical meristem gives rise to the **root cap** and to three primary meristems. Root tips have overlapping **zones of cell division**, **cell elongation**, and **cell maturation**. Review Figure 33.10
- The vascular tissue of roots is contained within the **stele**. It is arranged differently in eudicot and monocot roots. Review Figures 33.11, 33.12, Activities 33.1, 33.2
- In nonwoody stems, the vascular tissue is divided into **vascular bundles**, each containing both xylem and phloem. Review Figure 33.14, Activities 33.3, 33.4
- Eudicot leaves have two zones of photosynthetic **mesophyll** that are supplied by veins with water and minerals. Veins also carry the products of photosynthesis to other parts of the plant body. A waxy **cuticle** limits water loss from the leaf. Guard cells control openings called **stomata** in the leaf that allow CO_2 to enter, but also allow some water to escape. Review Figure 33.16, Activity 33.5
- Two lateral meristems, the **vascular cambium** and **cork cambium**, are responsible for secondary growth. The vascular cambium produces secondary xylem (**wood**) and secondary phloem. The cork cambium produces a protective tissue called cork. Review Figure 33.17, Animation 33.1

33.4 Domestication Has Altered Plant Form

- The plant body plan is simple, yet it can be changed dramatically by minor differences in genes, as evidenced by the natural diversity of wild plants.
- Crop domestication involves artificial selection of certain desirable traits found in wild populations. Review Figure 33.20

Go to **LearningCurve** (in **LaunchPad**) for dynamic quizzing that helps you solidify your understanding of this chapter. **LearningCurve** adapts to your responses, giving you the practice you need to master each key concept.

▶ Apply What You've Learned

Review

33.3 Growth in plants can be either determinate or indeterminate, depending on the organ structure.

33.3 At the root meristem, zones of cell division, elongation, and maturation (differentiation) form the tissues of the root and root cap.

33.3 The root consists of several tissue layers outside of the inner vascular tissues. These tissues have different arrangements in eudicot and monocot roots.

33.3 Growth in terms of cell numbers occurs at meristems.

Original Paper: Doerner, P. 1998. Root development: Quiescent center not so mute after all. *Current Biology* 8: R42–R44.

Angiosperm root tissues are arranged in a cylinder with layers oriented from outside to inside and with a vascular cylinder composed of xylem and phloem in the center, as shown in **Figure A**.

These tissue layers originate in root apical meristems from a small, rapidly dividing group of cells called "initials," located just above the root cap. Like stem cells in animals, initials are generalized and can differentiate into all types of plant tissues. The small group of initials surrounds a smaller group (usually four to eight cells) called the quiescent center (QC), where cells divide very slowly, as shown in **Figure B**. For example, in corn (*Zea mays*), QC cells divide every 170 hours, while those in the zone of cell division divide every 10–16 hours.

To identify actively dividing cells, scientists incubate plant roots in radioactive thymidine, which enters and labels the DNA. Labeled cells are detected by a photographic technique. The table below shows simulated data for corn seeds incubated in radioactive thymidine for two hours and then removed to water containing non-radioactive thymidine. Numbers represent the number of labeled cells in various zones at different times. Results are shown in the table.

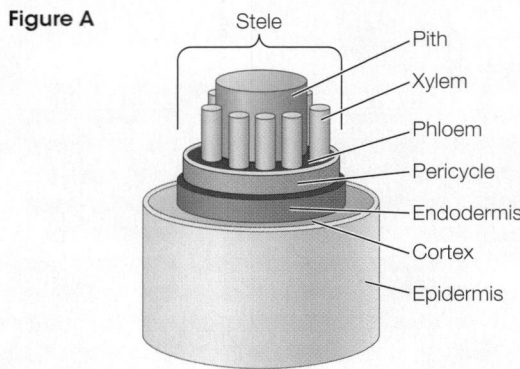

Figure A
Stele
Pith
Xylem
Phloem
Pericycle
Endodermis
Cortex
Epidermis

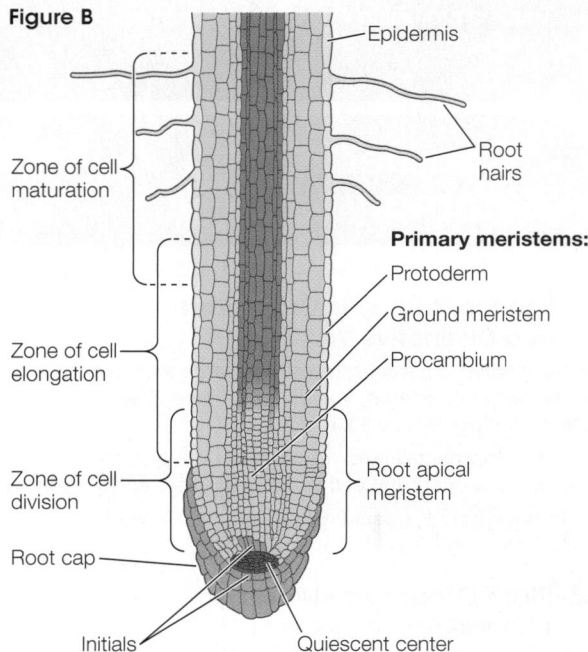

Figure B
Epidermis
Zone of cell maturation
Root hairs
Primary meristems:
Protoderm
Ground meristem
Procambium
Zone of cell elongation
Zone of cell division
Root apical meristem
Root cap
Initials
Quiescent center

Number of hours after transfer to nonradioactive thymidine	Radioactive cells/total cells			
	Quiescent center	Initials (zone of division)	Zone of elongation	Zone of maturation
0	1/4	75/95	0/30	0/19
12	1/4	85/90	12/32	0/23
36	1/4	55/110	15/31	11/21
72	1/4	3/98	3/29	28/30
144	1/4	0/108	0/30	29/31
192	1/4	0/100	0/27	30/31

Questions

1. Why does the number of labeled cells in the zones of division and elongation decrease over time?

2. Based on the distribution of labeled cells in the table, what can be inferred about the direction and pattern of root growth in this plant?

3. Plant cells, unlike animal cells, do not migrate during embryogenesis. Given this situation, how might the small group of initials just above the root cap result in separate layers of root tissue arranged in a cylinder?

4. Researchers have suggested that the QC serves as a "stem cell reservoir" that replenishes initials and ensures the continued existence of the root apical meristem. Does the pattern of labeling in the QC over time support this hypothesis? Explain your answer.

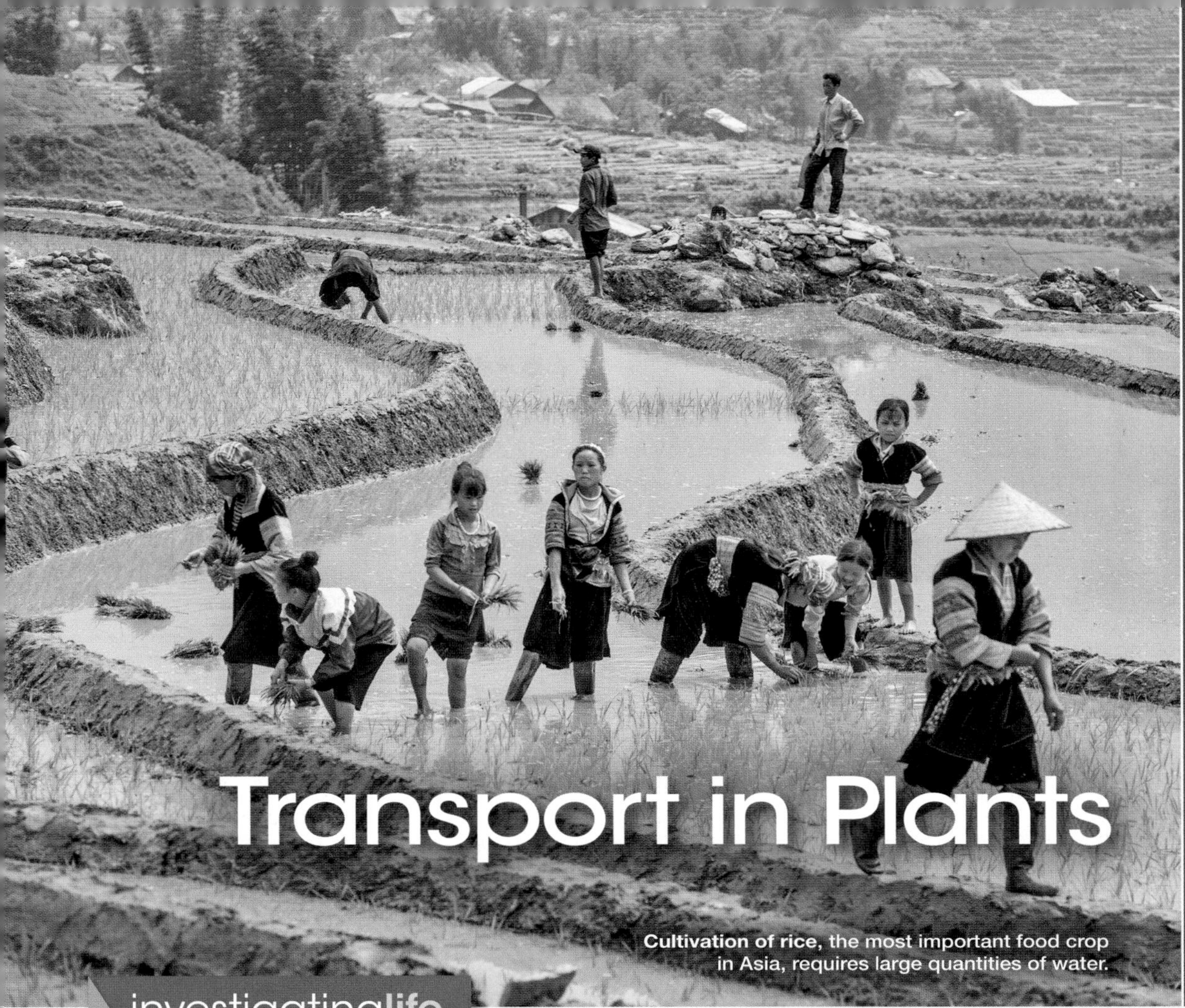

Cultivation of rice, the most important food crop in Asia, requires large quantities of water.

34 Transport in Plants

key concepts

34.1 Plants Acquire Water and Minerals from the Soil

34.2 Water and Minerals Are Transported in the Xylem

34.3 Stomata Control the Loss of Water and the Uptake of CO_2

34.4 Solutes Are Transported in the Phloem

investigatinglife

Thirsty Rice

Plants need a lot of water to grow. Crop plant cultivation consumes far more water than all other human activities combined. Meeting the demand is especially challenging in drought-prone regions, compelling us to understand how plants use water so that we can select or breed plants that use it more efficiently. Much of the mass that plants acquire as they grow is due to their net fixation of atmospheric CO_2 into carbohydrates through photosynthesis. The ratio of net photosynthetic carbon fixation into plant material to water uptake is known as a plant's water-use efficiency.

One of the least water-efficient of all crop plants is, unfortunately, one of our most important: rice. Rice plants use up to three times more water per unit of growth than crops such as wheat and corn. The precariousness of heavily water-dependent rice farming was dramatically demonstrated in eastern India between 1997 and 2003, when drought reduced rice production by more than 5 million tons—some farmers lost up to 50 percent of their crops.

A strain of rice requiring less water yet producing the same amount of grain would both make the world supply of rice less vulnerable to drought and help conserve water for other uses. A team of scientists began the quest for such a strain by studying the model organism *Arabidopsis thaliana*. They searched for mutations of *Arabidopsis* that had superior water-use efficiency. One genetic variant they studied has an extensive root system (indicating higher capacity for water uptake) and thick leaves with abundant photosynthetic tissue (indicating prolific photosynthesis). Molecular and physiological characterization linked the improved water usage of this *Arabidopsis* strain to a mutation in a single gene that codes for a transcription factor that the scientists called HARDY. This protein turns out to be a member of a family of transcription factors that plants use to regulate genes in response to biotic and abiotic stress. When biotechnologists isolated this gene (*HARDY*) and put it into rice plants, the transformed rice plants not only had higher water-use efficiency than wild-type rice plants but were more tolerant of dry soil as well.

Laboratories around the world are using *Arabidopsis* to isolate genes involved in water usage and other important physiological processes. The knowledge gained from these studies may lead to various improvements of crop plants.

What methods are used to reduce

key concept 34.1 Plants Acquire Water and Minerals from the Soil

Terrestrial plants must obtain both water and mineral nutrients from the soil, through their roots, and the water is transported through the plant in xylem tissue. Carbohydrates and other important materials mostly originate in leaves and circulate through the phloem (**Figure 34.1**). Water is required for photosynthesis in the leaves (see Key Concept 10.1), for transporting solutes between plant organs, for cooling the plant, and for developing the internal pressure that supports the plant body.

focus your learning

- Water potential is determined by solute potential and pressure potential.
- Membrane proteins form channels (aquaporins, ion channels) and act as pumps that help move materials across cell membranes.
- The two main pathways to get water from the soil into the xylem are the apoplast and the symplast.
- Movement of water and ions across a cell membrane can be impeded.
- The Casparian strip prevents water and ions in the apoplast from crossing the endodermis; to reach the xylem, they must enter the symplast.

The minerals that a plant needs are transported along with the water. Several steps in water and mineral transport will be considered in this chapter. In this section we will focus on the first part of the

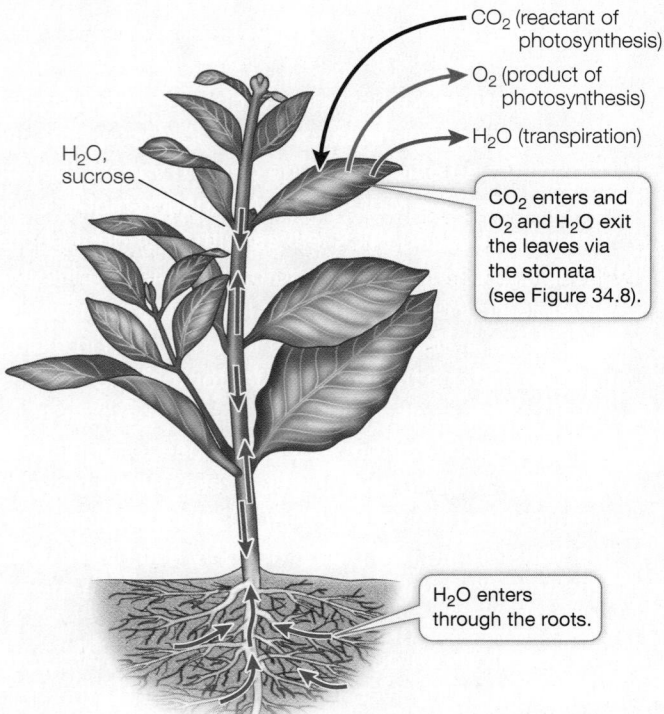

Figure 34.1 The Pathways of Water and Solutes in a Plant Water travels from the soil to the atmosphere, with only a small fraction used within the plant.

H_2O, sucrose

CO_2 (reactant of photosynthesis)

O_2 (product of photosynthesis)

H_2O (transpiration)

CO_2 enters and O_2 and H_2O exit the leaves via the stomata (see Figure 34.8).

H_2O enters through the roots.

Water potential differences govern the direction of water movement

Water moves into and out of plant cells. The direction of water movement is determined by **water potential** (Ψ, **psi**), defined as the tendency of water molecules to move in response to forces such as pressure or solute concentration gradients. In the case of plant roots in soil water, whenever water moves across a selectively permeable membrane by *osmosis, it moves toward the region of lower (more negative) water potential (**Figure 34.2A, left**). We measure water potential in megapascals (MPa), a unit of pressure. Atmospheric pressure, 1 atmosphere, is about 0.1 MPa, or 14.7 pounds per square inch; a typical pressure in an automobile tire is about 0.2 MPa.

---

*connect the concepts The key to water movement is osmosis. You can review the process of osmosis and the role of the cell membrane in Key Concept 6.3.

Water potential has two major components:

1. **Solute potential** (Ψ_s): As solutes are added to pure water, the concentration of free water is reduced, the tendency to take up water increases, and water potential decreases.

2. **Pressure potential** (Ψ_p): As plant cells take up water, they tend to swell. However, the presence of the cell wall provides resistance to swelling (see Figure 6.10). The result is an increase in pressure inside the cell (**turgor pressure**), which decreases the tendency of the cell to take up more water (increases the water potential) (**Figure 34.2A, right**).

A solution's water potential is the sum of its (usually negative) solute potential (Ψ_s) and its (usually positive) pressure potential (Ψ_p):

$$\Psi = \Psi_s + \Psi_p$$

By definition, the solute potential of pure water is zero (no solutes); because added solutes decrease water potential, solute potential is usually negative. Pressure potential is defined as zero when it equals atmospheric pressure. Pressure potential less than atmospheric pressure is negative; pressure potential greater than atmospheric pressure is positive.

In a plant cell immersed in pure water (**Figure 34.2B**), turgor pressure is comparable to the pressure potential exerted by the piston in Figure 34.2A. Water enters the cell by osmosis until the pressure potential exactly balances the solute potential and the water potential is zero. At this point the cell is **turgid**—that is, it has a significantly positive pressure potential. Plant cells are surrounded by water with dissolved solutes rather than pure water, and their water potential is dependent on the water potential in the soil. But because turgid cells have a positive pressure potential, there is no net movement of water into them. The physical structures of many plants are maintained by the positive pressure potential of the water in their cells. If the pressure potential drops (for example, if the plant does not have enough water), the plant wilts (**Figure 34.3**).

In living plant tissues, the movement of water from cell to cell follows a gradient of water potential. But over longer distances, in

(A) A theoretical illustration of water potential

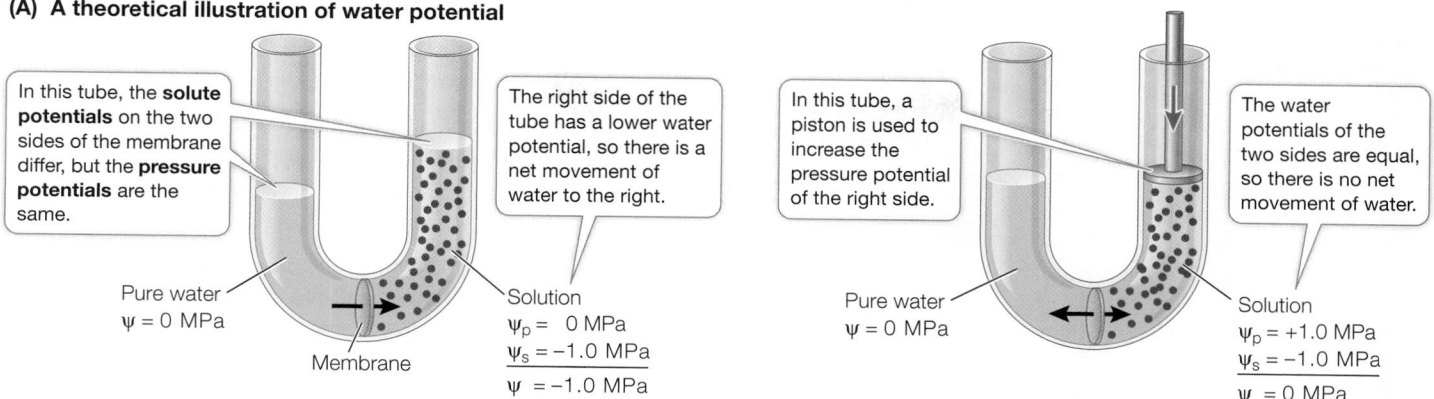

In this tube, the **solute potentials** on the two sides of the membrane differ, but the **pressure potentials** are the same.

The right side of the tube has a lower water potential, so there is a net movement of water to the right.

In this tube, a piston is used to increase the pressure potential of the right side.

The water potentials of the two sides are equal, so there is no net movement of water.

Pure water
$\psi = 0$ MPa

Membrane

Solution
$\psi_p = 0$ MPa
$\underline{\psi_s = -1.0}$ MPa
$\psi = -1.0$ MPa

Pure water
$\psi = 0$ MPa

Solution
$\psi_p = +1.0$ MPa
$\underline{\psi_s = -1.0}$ MPa
$\psi = 0$ MPa

(B) The effect of differences in water potential on a plant cell

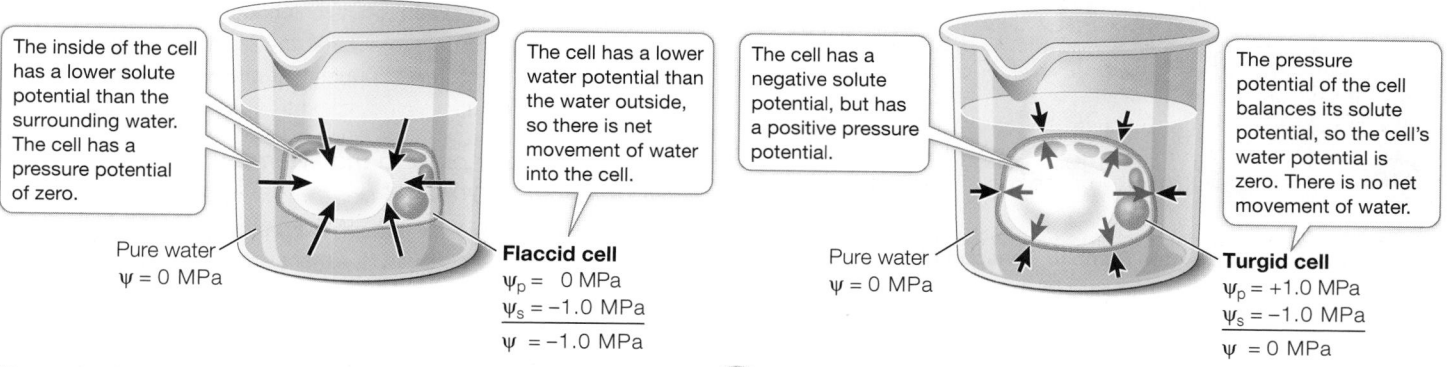

The inside of the cell has a lower solute potential than the surrounding water. The cell has a pressure potential of zero.

The cell has a lower water potential than the water outside, so there is net movement of water into the cell.

The cell has a negative solute potential, but has a positive pressure potential.

The pressure potential of the cell balances its solute potential, so the cell's water potential is zero. There is no net movement of water.

Pure water
$\psi = 0$ MPa

Flaccid cell
$\psi_p = 0$ MPa
$\underline{\psi_s = -1.0}$ MPa
$\psi = -1.0$ MPa

Pure water
$\psi = 0$ MPa

Turgid cell
$\psi_p = +1.0$ MPa
$\underline{\psi_s = -1.0}$ MPa
$\psi = 0$ MPa

Figure 34.2 Water Potential, Solute Potential, and Pressure Potential

Activity 34.1 Water Uptake in Plants
www.Life11e.com/ac34.1

unobstructed tubes such as xylem vessels, the flow of water and dissolved solutes is driven by a *gradient of pressure potential*, not a gradient of water potential. The movement of a solution from a region of higher pressure potential to a region of lower pressure potential is called **bulk flow**.

Water and ions move across the root cell's cell membrane

Now that you've seen how water can move into plant cells by osmosis, what about ions? And what is the role of the cell membrane? The movement of water and mineral ions across a root cell membrane can be impeded for two reasons:

1. The interior of the membrane is hydrophobic, whereas water and mineral ions are polar.

2. Some mineral ions must be moved against their concentration gradients.

However, as you saw in Chapter 6, membrane proteins assist with the movement of materials across membranes:

- *Aquaporins*. Aquaporins (see Key Concept 6.3) are located in both the cell membrane and the tonoplast (vacuolar membrane) of a plant cell. Aquaporins allow water to diffuse rapidly across these membranes. The number of aquaporins in a particular cell depends on that cell's need to obtain and retain water, and can vary with environmental conditions. The permeability of some aquaporins also can be regulated. Alterations in aquaporin abundance and permeability change the *rate* of osmosis across the membrane. Note that water movement

The cells of this plant have low **turgor pressure** and the plant is wilted.

The water potential of cells of this plant is zero because the negative solute potential is balanced by an equally positive pressure potential. The plant is upright because its cells are **turgid**.

Figure 34.3 A Wilted Plant A plant wilts when the pressure potential in its cells (the turgor pressure) is low.

Q: Sometimes a plant will wilt even if it is adequately watered because of an excess of fertilizer (dissolved ions in the soil). Why?

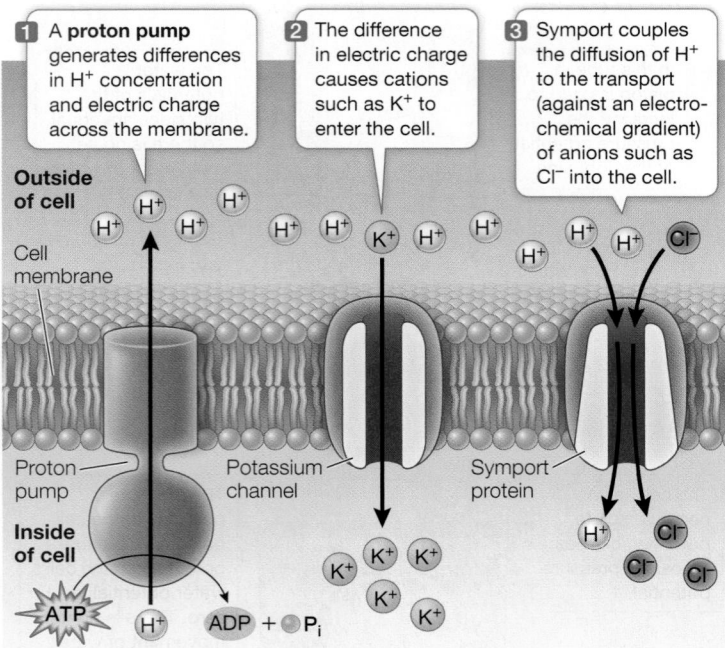

① A **proton pump** generates differences in H⁺ concentration and electric charge across the membrane.

② The difference in electric charge causes cations such as K⁺ to enter the cell.

③ Symport couples the diffusion of H⁺ to the transport (against an electrochemical gradient) of anions such as Cl⁻ into the cell.

Outside of cell

Cell membrane

Proton pump

Potassium channel

Symport protein

Inside of cell

ATP   H⁺   ADP + P$_i$

Figure 34.4 The Proton Pump in Transport of K⁺ and Cl⁻
The active transport of hydrogen ions (H⁺) out of the cell by the proton pump (1) drives the movement of both cations (2) and anions (3) into the cell.

through aquaporins is always passive: from a region of higher water potential to one of lower water potential.

- *Ion channels and pumps.* When the concentration of a charged ion in the soil is greater than that in the plant, transport proteins can move the ions into the plant by facilitated diffusion, which is a passive process (see Key Concept 6.3). The concentrations of most ions in the soil solution, however, are lower than those inside the plant. In these cases the plant must actively take up ions *against* their concentration gradients—a process that requires energy (see Key Concept 6.4).

Electric charge differences also play a role in the uptake of mineral ions. For example, a negatively charged ion that moves into a negatively charged compartment is moving against an *electrical gradient*, and this requires energy. Concentration and electrical gradients combine to form an *electrochemical gradient. Uptake against an electrochemical gradient involves active transport, which requires energy and specific transport proteins.

---

*connect the concepts   Electrochemical gradients are important in many biological systems. Learn more about electrochemical gradients, and their role in the animal nervous system, in Key Concept 44.2.

Unlike animals, plants do not have a sodium–potassium pump (described in Key Concept 6.4) to drive active transport. Rather, plants have a **proton pump**, which uses energy obtained from ATP to move protons out of the cell against a proton concentration

gradient (**Figure 34.4, Step 1**). Because protons (H⁺) are positively charged, their accumulation outside the cell has two results:

1. An electrical gradient is created, with the region outside the cell more positively charged than the inside.

2. A proton concentration gradient develops, with more protons outside the cell than inside.

Both the electrical gradient and the concentration gradient assist with the movement of other ions into the cell. Because the inside of the cell is more negative than the outside, cations (positively charged ions) such as potassium (K⁺) can move into the cell by facilitated diffusion through specific membrane channels (**Figure 34.4, Step 2**). In addition, the proton concentration gradient can be harnessed to drive secondary active transport, in which anions (negatively charged ions) such as chloride (Cl⁻) are moved into the cell. These ions can move against the electrochemical gradient because symport proteins couple their movement with that of H⁺ (**Figure 34.4, Step 3**).

Water and ions pass to the xylem by way of the apoplast and symplast

The journey from the soil through the roots to the xylem occurs primarily by one of two pathways, either separately or simultaneously: the fast lane (called the apoplast) and the slow(er) lane (called the symplast) (**Figure 34.5**):

1. The **apoplast** (Greek *apo*, "away from"; *plast*, "living material") consists of the cell walls, which lie outside the cell membranes, and the intercellular spaces (spaces between cells) that are common in many plant tissues. Typically, the apoplast occupies 5–20 percent of plant tissues by volume. The apoplast is a continuous meshwork through which water and dissolved substances can flow without ever having to cross a membrane. Movement of materials through the apoplast is thus unregulated and rapid.

2. The **symplast** (Greek *sym*, "together with") is the continuous cytoplasm of the living cells, which are connected by

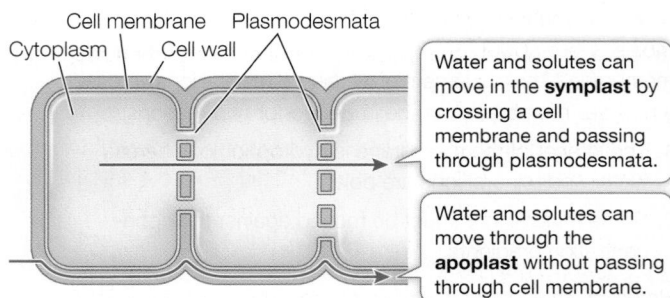

Cell membrane   Plasmodesmata
Cytoplasm   Cell wall

Water and solutes can move in the **symplast** by crossing a cell membrane and passing through plasmodesmata.

Water and solutes can move through the **apoplast** without passing through cell membrane.

Figure 34.5 Apoplast and Symplast   Plant cell walls and intercellular spaces constitute the apoplast. The symplast comprises the living cells connected by plasmodesmata. To enter the symplast, water and solutes must pass through a cell membrane. No such selective barrier limits movement through the apoplast.

Q: What type of animal cell junction is best compared with the Casparian strip? See Figure 6.7.

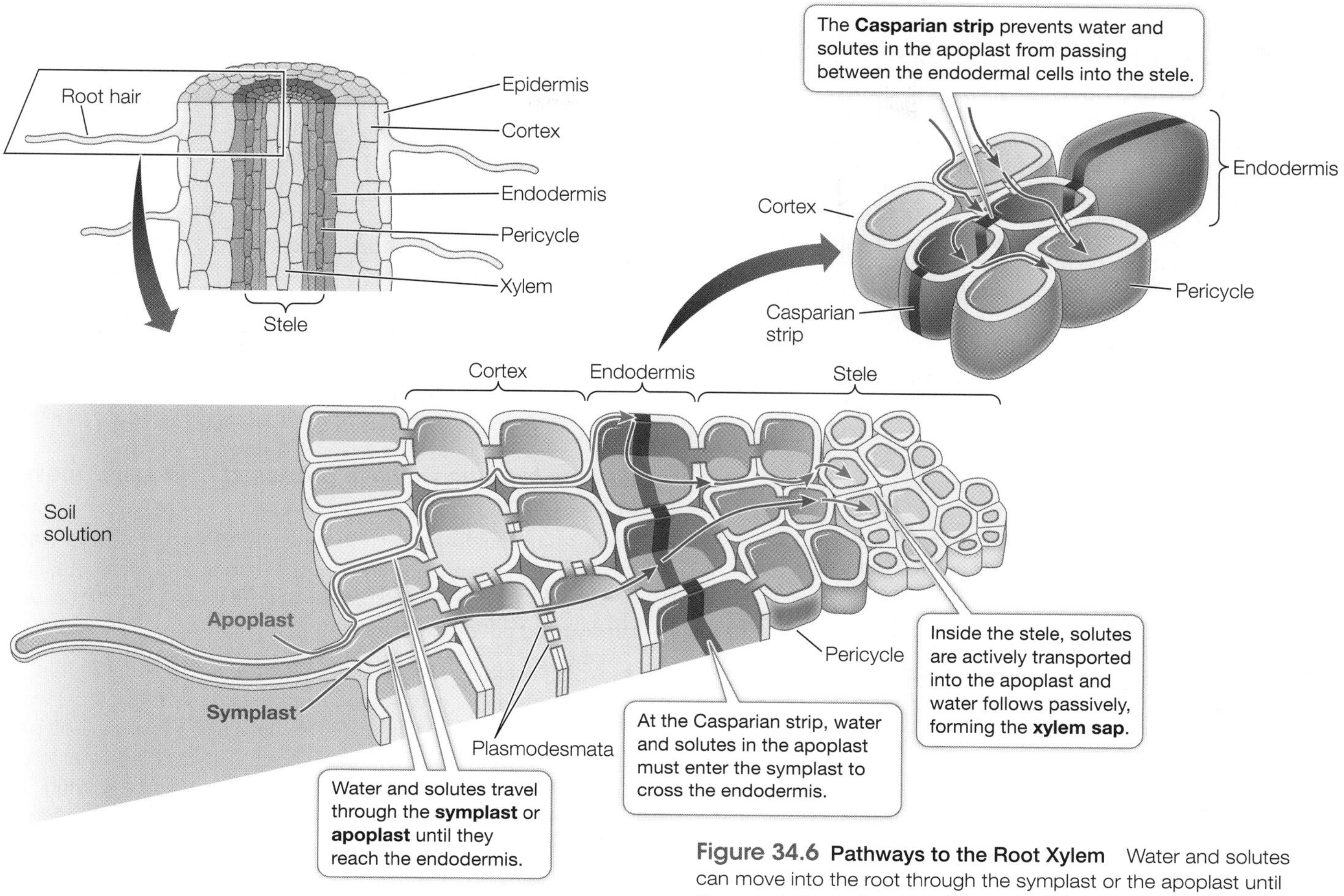

Root hair

Epidermis

Cortex

Endodermis

Pericycle

Xylem

Stele

The **Casparian strip** prevents water and solutes in the apoplast from passing between the endodermal cells into the stele.

Cortex

Endodermis

Pericycle

Casparian strip

Cortex   Endodermis   Stele

Soil solution

Apoplast

Symplast

Plasmodesmata

Pericycle

Water and solutes travel through the **symplast** or **apoplast** until they reach the endodermis.

At the Casparian strip, water and solutes in the apoplast must enter the symplast to cross the endodermis.

Inside the stele, solutes are actively transported into the apoplast and water follows passively, forming the **xylem sap**.

Figure 34.6 Pathways to the Root Xylem   Water and solutes can move into the root through the symplast or the apoplast until they reach the endodermis (shown in dark green); there the water and solutes must enter the symplast to bypass the Casparian strip (purple), a region of the endodermal cell wall that is impregnated with the water-repelling substance suberin. Inside the stele, the water and solutes enter the xylem (blue).

 Activity 34.2 **Apoplast and Symplast of the Root**
www.Life11e.com/ac34.2

plasmodesmata (see Figure 7.16B). Secretory cells such as those in flowers that make nectar have a lot of plasmodesmata. So do cells in the root tip, where the selectively permeable cell membranes of the root cells control access to the symplast. Movement of water and dissolved substances into the symplast is tightly regulated.

Water and minerals that pass from the soil solution through the apoplast can travel as far as the endodermis, the innermost layer of the root cortex (**Figure 34.6**; see Key Concept 33.3). The endodermis is distinguished from the rest of the ground tissue by the presence of the **Casparian strip**. This waxy, suberin-impregnated region of the endodermal cell wall forms a hydrophobic belt around each endodermal cell where it is in contact with other endodermal cells. The Casparian strip acts as a seal that prevents water and ions from moving through apoplastic spaces between the endodermal cells. Therefore all water and ions must enter the symplast in order to cross the endodermis into the stele, which contains the vascular tissues of the root. The materials pass from the endodermal cells to cells in the stele via plasmodesmata.

Once they have passed the endodermal barrier, water and minerals remain in the symplast until they reach parenchyma cells in the pericycle or xylem. These cells then actively export mineral ions into the apoplast of the stele. As the concentrations of mineral ions in the apoplast increase, its water potential becomes more negative. Consequently, water moves out of the cells and into the apoplast by osmosis. In other words, ions are transported actively, and water follows passively. The end result of transport is that water and minerals end up in the xylem, where they constitute the **xylem sap**. **Table 34.1** shows the composition of xylem sap.

table **34.1** Composition of Xylem and Phloem Fluids		
Substance	Xylem	Phloem
Ions, minerals (g/L)	0.2–4	1–5
Amino acids (g/L)	0.1–1	5–40
Sugars (g/L)	0	100–300
Solute potential, Ψ_s (MPa)	–0.02 to –0.2	–0.6 to –3

34.1 recap

Differences in water potential govern the osmotic flow of water from the soil into the plant stele; this is a passive process. Uptake of minerals from the soil that occurs along an electrochemical gradient is an active process requiring energy and membrane transport proteins. Water and minerals can move into the root vascular tissue through either the apoplast or the symplast, but must use the symplast pathway to reach the xylem.

learning outcomes

You should be able to:

• Describe or illustrate how solute potential and pressure potential affect the movement of water across the plant cell membrane.

• Describe how aquaporins, ion channels, and ion pumps affect the movement of water and solutes across the plant cell membrane.

• Discuss the roles of the apoplast and the symplast in moving water from the soil to the xylem.

1. What is the role of the cell wall in determining the direction of water movement and plant form?

2. A major component of root growth is osmotically driven cell expansion, which is achieved by the formation and swelling of the central vacuole. What role would you predict the expressions of aquaporins be in this cell expansion?

3. What are the differences between the apoplast and the symplast?

So far we've described the movement of water and minerals into plant roots and their entry into the root xylem. How does the xylem sap move once it is in the xylem?

key concept
34.2
Water and Minerals Are Transported in the Xylem

Once water has arrived in the xylem, it is all "uphill" from there. Before considering the ascent of water and minerals to the leaves, reacquaint yourself with the cells that make up the xylem: the tracheids and vessel elements (see Figure 33.8A and B). Recall that these xylem cells are dead and lack all cell contents. When fused end-to-end, they form long tubular "straws" of lignified cell walls called **xylem vessels**. These vessels provide both structural support and the rigidity needed to maintain a gradient of pressure.

focus your learning

• Xylem vessels must move large amounts of water upward for long distances.

• Experiments ruled out two possible mechanisms for movement of water upward through the xylem.

• The currently accepted mechanism for water transport through the xylem is the transpiration–cohesion–tension mechanism.

• There is evidence to support all parts of the transpiration–cohesion–tension mechanism.

Consider the magnitude of what the xylem accomplishes. A single maple tree 15 meters tall was estimated to have some 177,000 leaves, with a total leaf surface area of 675 square meters—about one and a half times the area of a basketball court. During a summer day, that tree loses 220 liters of *water *per hour* to the atmosphere by evaporation from the leaves. So to prevent wilting, the xylem needs to transport 220 liters of water up to 15 meters from the roots to the leaves every hour. (By comparison, a 50-gallon drum holds 189 liters.)

> ***connect the concepts**   As you learn about water movement in the xylem, keep in mind the properties of water described in Key Concept 2.4.

Various hypotheses have proposed how water moves in the xylem

Until the twentieth century, two mechanisms for moving water through the xylem were under consideration: upward pressure by living cells and capillary action. Both of these possibilities were largely ruled out by experiments:

• A simple experiment in 1893 ruled out the hypothesis that root cells might initiate a *pumping mechanism* to propel water upward. A tree was cut at its base and the sawed-off part was placed in a vat containing a solution of poison that killed living cells. The poison rose up the trunk, killing any living cells it encountered along the way. The experiment demonstrated that a living pump in the root is not necessary to push the xylem sap up a tree. Because the roots were absent, it was clear that they are not involved in movement through the xylem. Furthermore, when the poison sap reached the leaves, they died and all upward movement of the solution stopped, showing that living leaves are necessary for water to move in the xylem.

• Because of its surface tension (see Key Concept 2.4) and adhesive forces between water and its container, water will move up a narrow column by a mechanism called *capillary action*. You've seen this when you put a straw into a cup of water. Capillary action was ruled out as a primary mechanism for upward xylem sap transport when calculations showed that xylem vessels (at 100 micrometers [μm] in diameter) are too wide to get water to the top of a 15-meter tree in this fashion. In fact, the maximum height for a water column raised by capillary action alone in a 100-μm tube would be only 0.15 meters.

The transpiration–cohesion–tension mechanism accounts for xylem transport

The current model of xylem transport involves three processes (**Focus: Key Figure 34.7**):

1. *Transpiration* of water molecules from the leaves by evaporation

2. *Tension* (negative pressure) in the xylem sap resulting from transpiration from the leaves

3. *Cohesion* of water molecules in the xylem sap, from the leaves to the roots

The concentration of water vapor in the atmosphere is lower than that in the air spaces inside the leaf. Because of this difference, water vapor diffuses from the intercellular spaces of the leaf to the outside air, in a process called **transpiration**. Within the leaf, water evaporates from the moist walls of the mesophyll cells and enters the intercellular spaces. As water evaporates from the aqueous film coating each cell, the film shrinks back into tiny spaces in the cell walls, increasing the curvature of the water surface and thus increasing its surface tension. This increased *tension* (negative pressure potential) in the surface film draws more water into the walls from the cells, replacing that which was lost. The resulting tension in the mesophyll draws water from the xylem of the nearest vein into the apoplast surrounding the mesophyll cells. The removal of water from the veins, in turn, establishes tension on the entire column of water contained in the xylem. *Cohesion* between water molecules in the column prevents the column from breaking. So these three forces—transpiration, tension, and cohesion—operate together to draw water up the xylem, all the way from the roots to the leaves.

Each part of this theory is supported by evidence:

- The difference in water potential between the soil solution and the air is huge, on the order of -100 MPa. This difference should generate more than enough tension to pull a water column up a tree.

- There is a continuous column of water in the xylem, which is made possible by the cohesive forces holding water molecules together.

- Actual measurements of xylem pressures in cut stems show negative pressure potentials, indicating considerable tension in the xylem.

The transpiration–cohesion–tension mechanism accounts for the movement of water through the xylem. Dissolved mineral ions are carried along with the water to all of the plant's living tissues, where the ions are used for various cellular processes (see Chapter 35 for more on plant nutrition). In addition to promoting the transport of minerals, transpiration has an added benefit of cooling a plant's

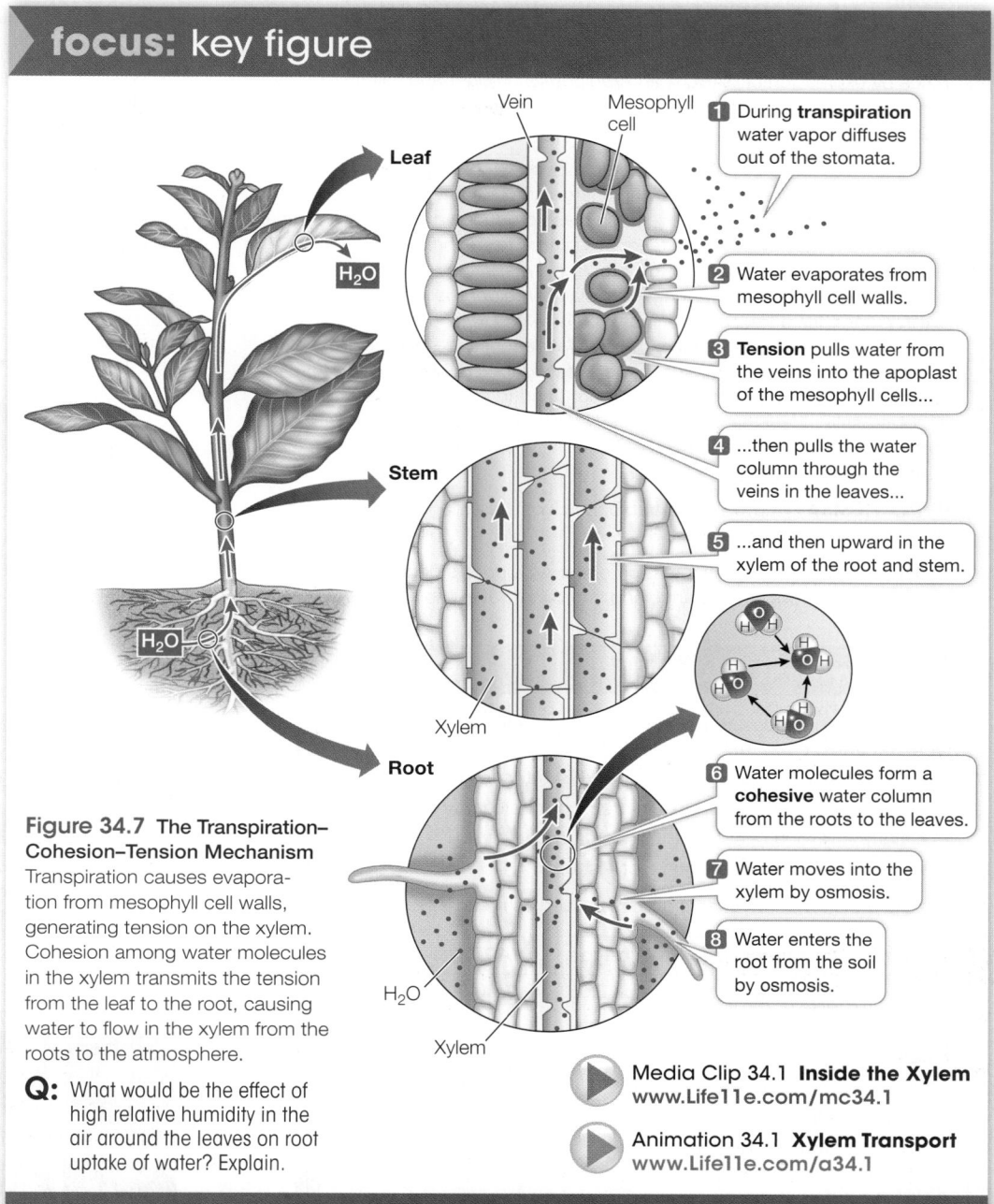

focus: key figure

1 During **transpiration** water vapor diffuses out of the stomata.

2 Water evaporates from mesophyll cell walls.

3 **Tension** pulls water from the veins into the apoplast of the mesophyll cells...

4 ...then pulls the water column through the veins in the leaves...

5 ...and then upward in the xylem of the root and stem.

6 Water molecules form a **cohesive** water column from the roots to the leaves.

7 Water moves into the xylem by osmosis.

8 Water enters the root from the soil by osmosis.

Figure 34.7 The Transpiration–Cohesion–Tension Mechanism Transpiration causes evaporation from mesophyll cell walls, generating tension on the xylem. Cohesion among water molecules in the xylem transmits the tension from the leaf to the root, causing water to flow in the xylem from the roots to the atmosphere.

Q: What would be the effect of high relative humidity in the air around the leaves on root uptake of water? Explain.

Media Clip 34.1 **Inside the Xylem**
www.Life11e.com/mc34.1

Animation 34.1 **Xylem Transport**
www.Life11e.com/a34.1

leaves. The evaporation of water from mesophyll cells consumes heat, thereby decreasing the leaf temperature. A farmer can hold a leaf between thumb and forefinger to estimate its temperature; if the leaf doesn't feel cool, that means transpiration is not occurring and it must be time to water.

As you saw in the opening investigation of this chapter, farmers strive to improve the water-use efficiency of their crops. The roles of the root system and transpiration are important. A gene from the model plant *Arabidopsis* has been shown to improve water-use efficiency by encoding a transcription factor that regulates genes involved in these processes (**Investigating Life: Improving Water-Use Efficiency in Rice**).

experiment

Original Paper: Karaba, A. et al. 2007. Improvement of water use efficiency in rice by expression of *HARDY*, an *Arabidopsis* drought and salt resistance gene. *Proceedings of the National Academy of Sciences USA* 104: 15270–15275.

The objective of improving rice production involves cultivating a large amount of plant material (biomass)—meaning a large amount of grain—while using less water. The relationship between biomass produced and water consumed is called water-use efficiency. A team at Virginia Polytechnic University led by Andrew Pereira approached this problem by isolating a gene that confers drought resistance from the model plant *Arabidopsis*. This gene, called *HARDY,* encodes a transcription factor in the AP/ERF family that activates transcription of genes involved in the plant's response to stresses. The team showed that *Arabidopsis* strains that expressed the HARDY transcription factor were much more resistant to drought and to an environment high in dissolved salts. They then inserted the *HARDY* gene into a high-expressing vector and made transgenic rice plants. Initial studies showed that the plants were drought-resistant.

HYPOTHESIS▶ A gene from *Arabidopsis* confers drought resistance and improves water-use efficiency.

CONCLUSION▶ High expression of the *Arabidopsis* gene *HARDY* in rice plants confers drought resistance and improved water-use efficiency.

work with the data

The investigators measured a number of parameters in both wild-type and transgenic *HARDY* rice:

- Water-use efficiency: g carbon fixed/kg water used
- Transpiration rate: water lost in g/cm^2 plant area/day
- Carbon fixation rate: g/cm^2 plant area/day
- Biomass accumulation: g in root + shoot

Results are shown in the table and expressed as mean ± SD.

Parameter	Wild-type	*HARDY*
Water-use efficiency	1.5 ± 0.06	3.0 ± 0.5
Transpiration rate	5.2 ± 0.2	4.0 ± 0.4
Carbon fixation rate	0.7 ± 0.06	1.3 ± 0.06
Biomass accumulation	7.5 ± 0.5	13.0 ± 0.6

QUESTIONS▶

1. Explain each result.
2. What statistical test would you use to evaluate the significance of any differences found?

A similar **work with the data** exercise may be assigned in **LaunchPad**.

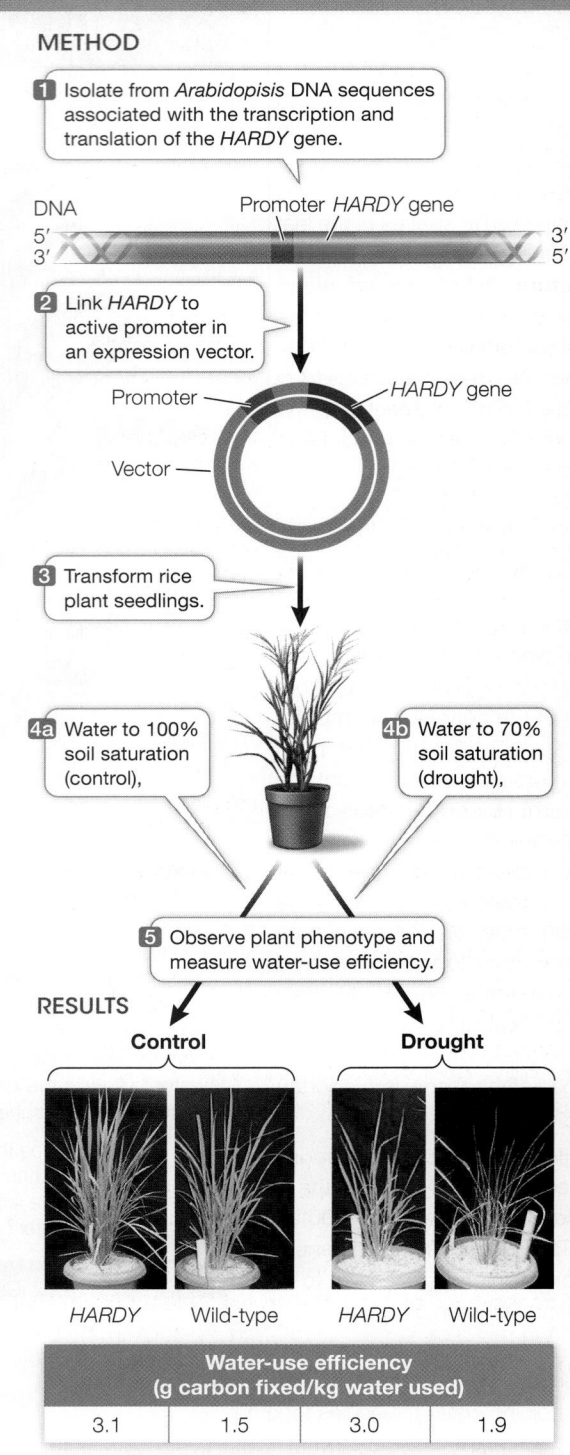

METHOD

1 Isolate from *Arabidopisis* DNA sequences associated with the transcription and translation of the *HARDY* gene.

DNA
Promoter *HARDY* gene
5′
3′
3′
5′

2 Link *HARDY* to active promoter in an expression vector.

Promoter
HARDY gene
Vector

3 Transform rice plant seedlings.

4a Water to 100% soil saturation (control),

4b Water to 70% soil saturation (drought),

5 Observe plant phenotype and measure water-use efficiency.

RESULTS

Control **Drought**

HARDY Wild-type *HARDY* Wild-type

Water-use efficiency (g carbon fixed/kg water used)			
3.1	1.5	3.0	1.9

34.2 recap

The transpiration–cohesion–tension mechanism explains the ascent of xylem sap. Transpiration draws water out of leaves, resulting in tension that pulls water from the xylem. Because of cohesion between water molecules, water is pulled passively through the xylem vessels in continuous columns, always toward a region with lower pressure potential.

learning outcomes

You should be able to:

- Solve problems relating to the transport of water by the xylem when given information on water potential.
- Describe how scientists ruled out past hypotheses for upward flow of water in plants.
- Discuss the interaction of the processes of transpiration, cohesion, and tension in the movement of water upward through the xylem.
- Provide evidence for the importance of each of the processes of transpiration, cohesion, and tension in the movement of water through the xylem.

1. What experiment ruled out the role of pressure from the roots in the upward flow of water in the xylem?

2. The table at right shows measurements of water potential (Ψ) in a 100-meter-tall tree and its surroundings:

Region	Ψ (MPa)
Soil water	–0.3
Xylem of root	–0.6
Xylem of trunk	–1.2
Inside of leaf	–2.0
Outside air	–58.5

Gravity exerts a force of –0.01 MPa per meter of height above ground.

a. Is the water potential in the leaf sufficiently low to draw water to the top of the tree?

b. Would transpiration continue if soil water potential decreased to –1.0 MPa?

c. What would you expect to happen to the xylem water potential if all of the stomata closed?

3. Explain why all three parts of the transpiration–cohesion–tension mechanism are necessary for water transfer through the xylem.

4. What types of evidence provide support for the transpiration–cohesion–tension mechanism of water transport through the xylem?

Although transpiration provides the driving force for the transport of water and minerals in the xylem, it also results in the loss of tremendous quantities of water from the plant. How plants control this loss will be the subject of the next section.

key concept 34.3 Stomata Control the Loss of Water and the Uptake of CO₂

The epidermis of leaves and stems secretes a waxy cuticle, which is impermeable to water and thus helps minimize the loss of water from transpiration. However, the cuticle is also impermeable to carbon dioxide. The cuticle poses a dilemma: how can the plant balance its need to retain water with its need to obtain CO_2 for photosynthesis?

focus your learning

- Stomata in leaves provide a balance between water loss and carbon dioxide diffusion into cells.
- Environmental factors affect the opening and closing of stomata; numbers of stomata can also change if the plant sheds leaves or grows new leaves.

Stomata control water loss and gas exchange

An elegant compromise has evolved in plants in the form of pores called **stomata** (singular *stoma*) in the epidermis of their leaves. A pair of specialized epidermal cells, called **guard cells**, controls the opening and closing of each stoma (**Figure 34.8A**). When the stomata are open, CO_2 can enter the leaf by diffusion—but water vapor diffuses out of the leaf at the same time. Closed stomata prevent water loss but also exclude CO_2 from the leaf.

Most plants open their stomata only when the light intensity is sufficient to maintain a moderate rate of photosynthesis. At night, when darkness precludes photosynthesis, their stomata are closed; no CO_2 is needed, and water is conserved. Even during the day, the stomata close if water is being lost at too rapid a rate.

Stomata are ancient structures; they have been found in plant fossils that are more than 400 million years old. For this reason they are thought to predate the evolution of leaves. Stomata are found in all vascular plants and in many nonvascular plants, including mosses (but not liverworts; see Chapter 27).

The stoma and guard cells seen in Figure 34.8A are typical of eudicots. Monocots typically have specialized epidermal cells associated with their guard cells. However, the principle of operation, which we will now describe in more detail, is the same for both monocot and eudicot stomata.

The guard cells control the size of the stomatal opening

The opening and closing of stomata are regulated by several environmental factors, including light, CO_2 levels, temperature, and water availability. Plants can also change the total number of stomata in response to longer-term changes in environmental conditions.

LIGHT AND CO_2 CONCENTRATION   Guard cells can respond to changes in light and CO_2 concentration in a matter of minutes by changing their solute potential. The absorption of light by a pigment in the guard cell's cell membrane activates a proton pump (see Figure 34.4), which actively transports H^+ out of the guard cells and into the apoplast of the surrounding epidermis. The resulting electrochemical gradient drives K^+ into the guard cells, where it accumulates (**Figure 34.8B**). Negatively charged chloride (Cl^-) ions and organic ions also move into and out of the guard cells along with the K^+ ions, maintaining electrical balance. The increased concentration of K^+ and other solutes inside the guard cells makes the solute potential of the guard cells more negative. Water then enters by osmosis, increasing the turgor pressure of the guard cells. The guard cells change their shape, becoming more turgid in response to the increase in pressure potential, so that a space—the

(A) An open stoma

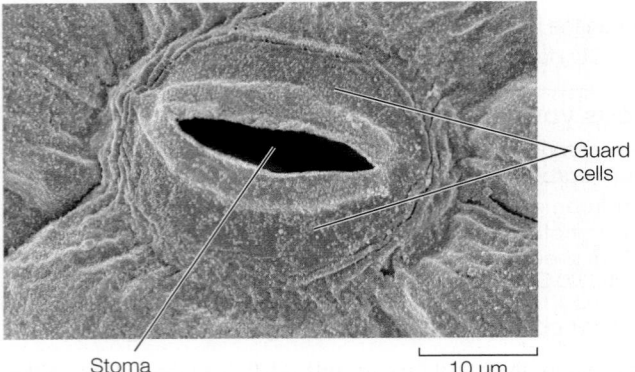

Guard cells

Stoma

10 µm

(B) Ion concentration effects on stomatal opening

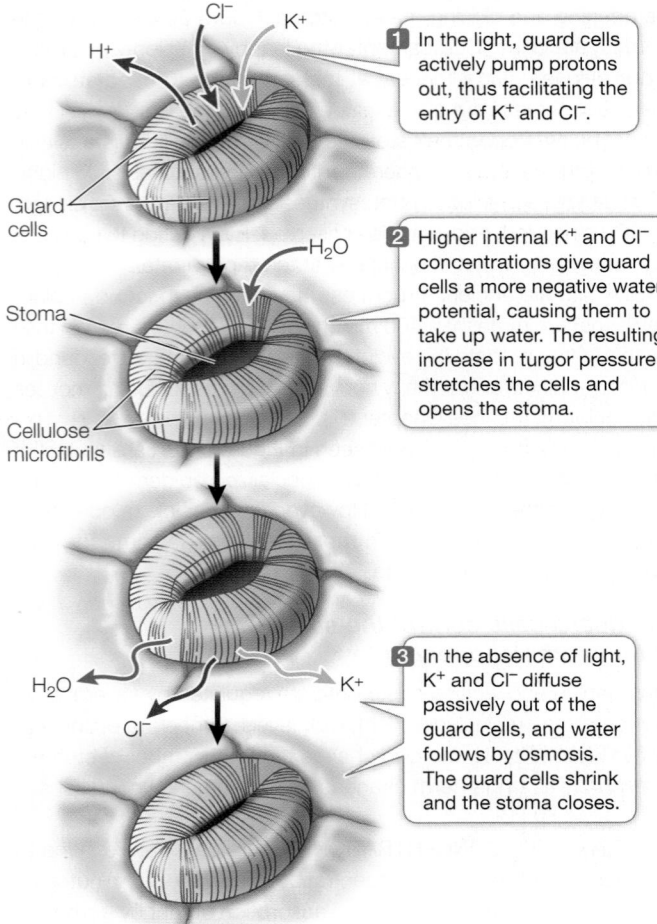

Cl⁻   K⁺

H⁺

1 In the light, guard cells actively pump protons out, thus facilitating the entry of K⁺ and Cl⁻.

Guard cells

H_2O

Stoma

2 Higher internal K⁺ and Cl⁻ concentrations give guard cells a more negative water potential, causing them to take up water. The resulting increase in turgor pressure stretches the cells and opens the stoma.

Cellulose microfibrils

H_2O

K⁺

Cl⁻

3 In the absence of light, K⁺ and Cl⁻ diffuse passively out of the guard cells, and water follows by osmosis. The guard cells shrink and the stoma closes.

Figure 34.8 Stomata **(A)** Scanning electron micrograph of an open stoma formed by two sausage-shaped guard cells. **(B)** Potassium ion concentrations affect the water potential of the guard cells, controlling the opening and closing of stomata. Negatively charged ions (e.g., Cl⁻) that accompany K⁺ maintain electrical balance and contribute to the changes in water potential that open and close the stomata.

stoma—appears between them. The stoma closes in the absence of light: the proton pump becomes less active, K⁺ ions diffuse passively out of the guard cells, water follows by osmosis, the pressure potential decreases, and the guard cells sag together and seal off

the stoma. Guard cell membranes are particularly rich in aquaporins, making guard cells well adapted for the rapid water movements involved in stomatal responses.

WATER Stomata also respond to water availability. Water stress is a common problem for plants, especially on a hot, windy day, when plants might close their stomata even when the sun is shining. The water potential of the leaf's mesophyll cells is the cue for this response. If the mesophyll becomes dehydrated, its cells release the hormone abscisic acid, which causes the stomata to close.

STOMATA NUMBER A plant can regulate water loss not just by the opening or closing of stomata, but by changing the number of stomata. The process of controlling the number of stomata takes place over days or weeks. Trees, for example, can reduce stomata numbers by shedding leaves, or by making new leaves that have few stomata.

> **34.3 recap**
>
> CO₂, which is needed for photosynthesis, enters leaves via tiny pores called stomata. Stomata also permit the loss of water by transpiration. Guard cells open or close stomata in response to a variety of environmental cues. Plants control their total numbers of stomata by shedding leaves or by altering the density of stomata on new leaves.
>
> **learning outcomes**
>
> You should be able to:
>
> - Analyze reasons why stomata do not always remain open, considering the processes of both photosynthesis and water conservation.
> - Analyze the importance of the proton pump and K⁺ ions in the actions of stomata, and list environmental changes that trigger the opening and closing of stomata.
>
> 1. What is the role of K⁺ ions in the functioning of guard cells?
> 2. Describe how water availability can affect stomatal function.

Stomata are normally open during daylight hours, allowing photosynthesis—the production of carbohydrates from CO₂ and water. In the next section we'll discuss how the products of photosynthesis are delivered to other parts of the plant, supporting plant growth.

key concept

34.4 Solutes Are Transported in the Phloem

Photosynthesis occurs primarily in the leaf (see Figure 10.1). The carbohydrate products of photosynthesis (mainly sucrose) diffuse to the nearest small vein (composed of xylem and phloem), where they are actively transported into sieve tube elements of the phloem. The movement of carbohydrates and other solutes through the phloem is called **translocation**.

focus your learning

- Phloem sap is transported, or translocated, from sources to sinks in plants.
- Translocation in plants is explained by the pressure flow model.
- Two steps in translocation, loading and unloading, involve active transport, which requires energy.

The products of photosynthesis are called **photosynthates**, and the content of the phloem is called the **phloem sap** (see Table 34.1). Translocation of phloem sap throughout the plant is by bulk flow, from sources to sinks:

- A **source** is an organ (such as a mature leaf or a storage root) that *produces*, by photosynthesis or by digestion of stored reserves, more sugars than it requires.
- A **sink** is an organ (such as a root, flower, developing fruit, or immature leaf) that *consumes* sugars for its own growth and storage needs of the plant.

Sources and sinks can change roles. For example, the storage roots of cassava (see Chapter 33) are sinks when they accumulate carbohydrates but are sources when their stored carbohydrates are mobilized to nourish other organs when the plant grows.

Sucrose and other solutes are carried in the phloem

Evidence that the phloem carries sugar and other solutes was first obtained in the 1600s when the Italian scientist Marcello Malpighi removed a ring of bark from the trunk of a tree—that is, he "girdled" the tree. The bark contained the phloem, while the xylem in the underlying wood remained intact. Over time, the bark in the region above the girdle swelled:

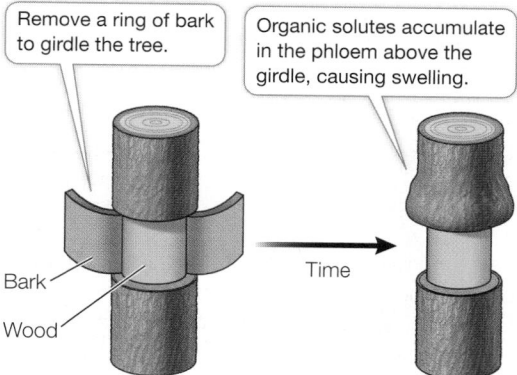

Remove a ring of bark to girdle the tree.

Organic solutes accumulate in the phloem above the girdle, causing swelling.

Bark

Wood

Time

Malpighi correctly concluded that a solution coming from the leaves above the girdle was trapped in the bark. Later the bark below the girdle died, presumably because it no longer received nutrients from the leaves. Eventually the roots, and then the entire tree, died—suggesting that sugar transport might occur in the phloem.

The cells that make up the phloem's conducting tubes are sieve tube elements (see Figure 33.8C). Like the vessel elements in the xylem, sieve tube elements meet end-to-end. However, unlike vessel elements, whose end walls are broken down as they mature, sieve tube elements retain their end walls. Communication between sieve tube elements is achieved by plasmodesmata in their end walls.

During sieve tube development, the diameter of these plasmodesmata increases 10- to 100-fold, resulting in pores that allow the flow of phloem sap between neighboring cells. Because the end walls of sieve tube elements look and function like sieves, they are called **sieve plates** (Figure 34.9).

What happens next is truly remarkable and makes sieve tube elements among the most unusual cell types in nature. As the holes in the sieve plates expand, most of the cell contents are lost, including the nucleus, Golgi apparatus, and most of the ribosomes and cytoskeleton. Despite this, sieve tube elements live for an entire growing season in deciduous trees, and for decades in some other plants. How can sieve tube elements live for so long with no nucleus? The answer is that each sieve tube element has one or more **companion cells** (see Figure 34.9). Companion cells are produced as daughter cells along with the sieve tube elements when parent cells divide. Numerous plasmodesmata link a companion cell with its neighboring sieve tube element. Companion cells retain all their organelles and provide all the components needed to maintain the sieve tube elements—they may be thought of as the "life support systems" of the sieve tube elements.

Plant biologists have used aphids to precisely analyze the contents of the phloem. Aphids are insects that feed on plants by drilling into sieve tube elements with a specialized organ, the stylet. The

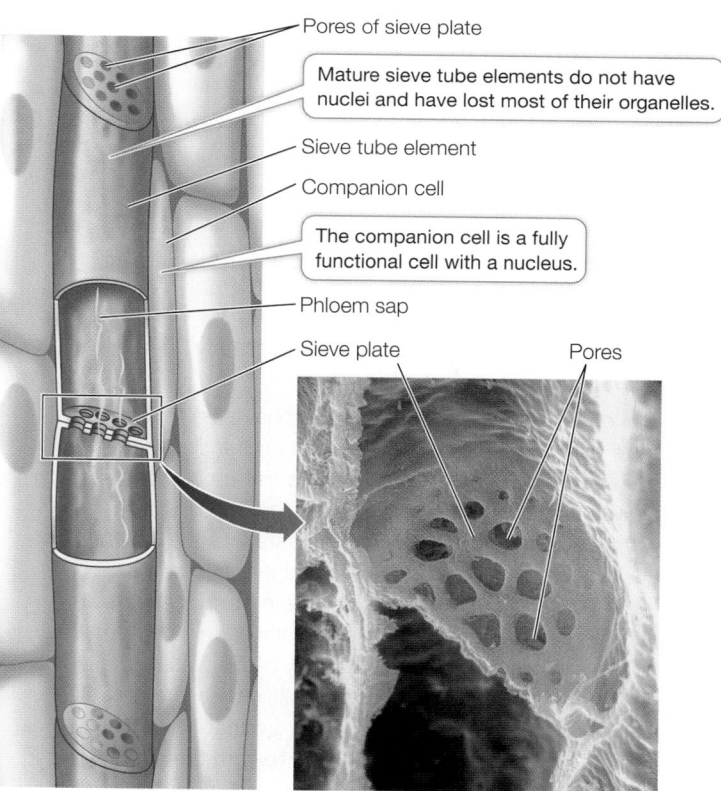

Pores of sieve plate

Mature sieve tube elements do not have nuclei and have lost most of their organelles.

Sieve tube element

Companion cell

The companion cell is a fully functional cell with a nucleus.

Phloem sap

Sieve plate

Pores

Pores

Dr. R. Kessel & Dr. G. Shih/Visuals Unlimited.

Figure 34.9 Sieve Tubes Individual sieve tube elements join together to form long tubes that transport carbohydrates and other nutrient molecules throughout the plant body in the phloem. Sieve plates form at the ends of each sieve tube element, and phloem sap passes through the pores in the sieve plate.

pressure potential in the sieve tube is higher than that outside the plant, so the phloem contents are forced through the stylet into the aphid's digestive tract. So great is the pressure that some of the liquid is forced through the insect's body and out its anus.

Sieve tube element

The aphid's stylet has successfully penetrated the sieve tube.

Longistigma caryae (aphid)   Sap droplet

Stylet

If an aphid is frozen in the act of feeding, its body can be chopped off the plant stem, leaving the stylet intact. Phloem sap continues to flow from the stylet for hours, and can be collected for analysis. These and other experiments led to several important observations:

● Sucrose makes up 90 percent of the phloem sap solutes. The phloem sap also contains hormones, proteins, small molecules such as amino acids, mineral nutrients, and viruses.

● The flow rate can be very high, as much as 150 centimeters per hour.

● Different sieve tube elements conduct their contents in different directions—for example, up or down the stem. Therefore the overall movement in the phloem is bidirectional.

● The movement of phloem sap requires living cells, in contrast to movement in the xylem.

The pressure flow model accounts for translocation in the phloem

The observations we just described have led to the **pressure flow model** as an explanation for translocation in the phloem (**Figure 34.10**). According to this model, sucrose is actively transported at a source into companion cells, from which it flows through plasmodesmata into the sieve tube elements. This gives those cells a higher sucrose concentration than the surrounding cells (a more negative solute potential), and water therefore enters the sieve tube elements from the xylem by osmosis. The entry of this water increases turgor pressure (causes a more positive pressure potential) at the source end of the sieve tube, so that the entire fluid content of the sieve tube is pushed toward the sink end of the tube—in other words, the sap moves in response to a pressure gradient. In the sink, the sucrose is unloaded both passively and by active transport, and water moves back into the xylem. In this way, the gradient of solute potential and pressure potential needed for the movement of phloem sap (translocation) is maintained.

Two steps in phloem translocation require metabolic energy:

1. **Loading**: Transport of sucrose and other solutes from sources into companion cells and then into the sieve tubes.
2. **Unloading**: Transport of solutes from the sieve tubes into sinks. The need for *****metabolic energy** is the reason why phloem transport, unlike xylem transport, requires living cells.

*****connect the concepts**   Interactions among biochemical pathways (such as photosynthesis, cellular respiration, and the Calvin cycle) that produce carbohydrates that fuel processes such as translocation are discussed in Key Concept 10.5.

Sucrose and other solutes can move from the mesophyll cells into the phloem by either an apolastic pathway or a symplastic pathway. In the apolastic pathway, sucrose and other solutes leave the mesophyll cells and enter the apoplast before they reach the sieve tube elements. Specific sugars and amino acids are then actively transported into cells of the phloem. Because the solutes cross at least one selectively permeable membrane in the apoplastic pathway, selective transport can be used to regulate which specific substances enter the phloem. In the symplastic pathway, the solutes remain within the symplast all the way from the mesophyll cells to the sieve tube elements. Because no membranes are crossed in the symplastic pathway, a mechanism that does not involve membrane transport is used to load sucrose into the phloem.

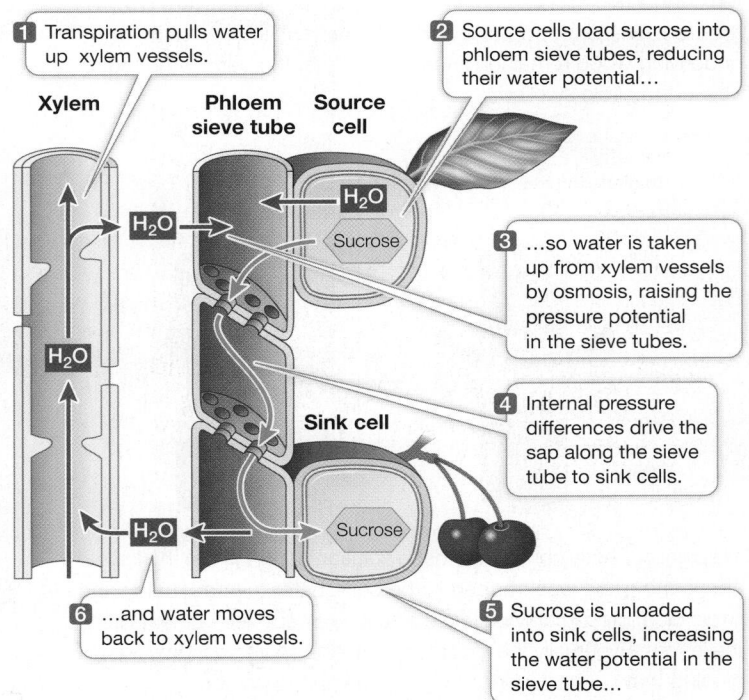

1 Transpiration pulls water up xylem vessels.

2 Source cells load sucrose into phloem sieve tubes, reducing their water potential...

Xylem   Phloem sieve tube   Source cell

H_2O

H_2O Sucrose

H_2O

3 ...so water is taken up from xylem vessels by osmosis, raising the pressure potential in the sieve tubes.

4 Internal pressure differences drive the sap along the sieve tube to sink cells.

Sink cell

Sucrose

6 ...and water moves back to xylem vessels.

5 Sucrose is unloaded into sink cells, increasing the water potential in the sieve tube...

Figure 34.10 The Pressure Flow Model   Water potential differences produce a pressure gradient and bulk flow of phloem sap from sources to sinks.

 Animation 34.2 The Pressure Flow Model
www.Life11e.com/a34.2

table 34.2 Mechanisms of Sap Flow in Plant Vascular Tissues

	Xylem	Phloem
Driving force for bulk flow	Transpiration from leaves	Active transport of sucrose at source and sink
Site of bulk flow	Nonliving vessel elements and tracheids	Living sieve tube elements
Pressure potential in sap	Negative (pull from top; tension)	Positive (push from source; pressure)

In sink regions, the solutes are actively transported *out* of the sieve tube elements and into the surrounding tissues. This unloading serves several purposes: it helps maintain the gradient of solute potential, and hence of pressure potential, in the sieve tubes; it helps build up high concentrations of carbohydrates in storage organs, such as developing fruits and seeds; and it moves nutrients to parts of the plant that are growing rapidly, such as shoots that form after the winter.

Table 34.2 contrasts the pressure flow model of translocation in the phloem with the transpiration–cohesion–tension model of xylem transport.

The control of phloem sap travel from source to sink has great importance for people. The parts of plants that we use for food, such as seeds (e.g., the rice in the opening story), fruits, and storage organs, are mostly sinks. Increasing the flow of sucrose into these organs as they develop can increase food production in crop plants.

34.4 recap

Carbohydrates produced by photosynthesis are translocated from source to sink through the phloem by a pressure flow mechanism.

learning outcomes

You should be able to:

- Define and identify sources and sinks in plants as they relate to the products of photosynthesis.
- Explain why energy and active transport are required to transport phloem sap and how the pressure flow model makes phloem sap flow possible.
- Define and compare the processes of loading and unloading in phloem sap transport.

1. Explain the difference between a source and a sink.
2. How does loading of sucrose at the source result in bulk flow toward the sink?

 **What methods are used to reduce water loss in agriculture?**

Ensuring the availability of water and its dissolved nutrients is a challenge to growing our crops. Historically, the approach of farmers has been to grow crops adapted to the local environment (e.g., heavy rainfall for rice) or adapt the environment for the plant (irrigation). With our changing climate resulting in droughts, a new approach—adapting the plant's genetic ability to the environment—has begun. The *HARDY* gene, described in Investigating Life: Improving Water-Use Efficiency in Rice, confers adaptation to drought in rice plants by upregulating the expression of clusters of genes involved in water-use efficiency, the ratio of plant biomass produced to water transpired. This is largely achieved through increased leaf thickness, with more mesophyll cells resulting in increased photosynthesis, as well as increased root growth, for greater uptake of water from the soil. The *HARDY* gene acts during development of leaves and roots, which you will recall happens throughout the life of the plant.

Future directions

Sorghum bicolor is an important grain crop worldwide, grown for human food, animal feed, and as a source for biofuels. It is especially valuable because it is both drought- and heat-tolerant, and so is important in hot, dry regions in Africa and South America. At the University of Illinois, scientists are trying to improve the water-use efficiency of sorghum by reducing transpiration in two ways: reduction of leaf stomata and shifting more photosynthesis to lower leaves, which are nearer to the ground and thus surrounded by humid air. Thousands of varieties of sorghum are stored in "seed banks." Samples of these seeds will be germinated and grown in the search for the two desired phenotypes. In addition, a deliberate molecular approach using transgenic plants will be attempted. It is hoped that the result will be a crop even better adapted to make the best use of limited water.

Chapter Summary 34

▶ **34.1 Plants Acquire Water and Minerals from the Soil**

- Water moves through biological membranes by osmosis, always moving toward regions with a more negative water potential. The **water potential** (Ψ) of a cell or solution is the sum of the **solute potential** (Ψ_s) and the **pressure potential** (Ψ_p). Review Figure 34.2, Activity 34.1

- **Turgid** plant cells have significant positive pressure potential because the rigid cell wall limits expansion of the cell. This positive pressure (**turgor pressure**) maintains the physical structure of many plant cells; if the pressure potential drops, the plant wilts.

- The movement of a solution due to a difference in pressure potential between two parts of a plant is called **bulk flow**.

- Aquaporins are channel proteins that facilitate movement of water molecules through biological membranes.

- Mineral uptake requires transport proteins. Some minerals enter the plant passively by facilitated diffusion; others enter by active transport. A **proton pump** provides energy for the active transport of many mineral ions across membranes in plants. Review Figure 34.4

- Water and minerals pass from the soil into the root by way of the **apoplast** and **symplast**, but must pass through the symplast to cross the endodermis and enter the xylem. The **Casparian strip** in the endodermis blocks movement of water and minerals through the apoplast. Review Figures 34.5, 34.6, Activity 34.2

▶ **34.2 Water and Minerals Are Transported in the Xylem**

- Experiments proved that neither a root pump nor capillary action can alone account for the ascent of xylem sap in trees.

- Water transport in the xylem results from the combined effects of **transpiration**, cohesion, and tension—the transpiration–cohesion–tension mechanism. Evaporation from the leaf produces tension in the mesophyll cells, which pulls a column of water—held together by cohesion—up through the xylem from the root. Review Focus: Key Figure 34.7, Animation 34.1

- Transport in the xylem is passive. It does not require the expenditure of energy by the plant.

- Water-use efficiency is the ratio of plant growth to water uptake.

▶ **34.3 Stomata Control the Loss of Water and the Uptake of CO₂**

- The waxy cuticle of plant epidermis is impermeable to both water and carbon dioxide. **Stomata** allow for carbon dioxide uptake (when open) while minimizing transpirational water loss (when closed).

- A pair of **guard cells** controls the size of the stomatal opening. A light-activated proton pump moves protons out of the guard cells to the walls of surrounding epidermal cells, setting up an electrochemical gradient that drives the transport of potassium ions into the guard cells. Water follows osmotically, swelling the guard cells and opening the stomata. Review Figure 34.8

- When threatened by dehydration, mesophyll cells release abscisic acid, which causes guard cells to close the stomata, even in the light.

▶ **34.4 Solutes Are Transported in the Phloem**

- Products of photosynthesis, as well as some minerals, are translocated through sieve tubes in the phloem by way of living sieve tube elements. Review Figure 34.9

- **Translocation** in the phloem can proceed in both directions in the stem. Translocation requires a supply of metabolic energy from living cells.

- Translocation in the phloem is explained by the **pressure flow model**: the difference in solute concentration between **sources** and **sinks** creates a difference in (positive) pressure potential along the sieve tubes, resulting in bulk flow. Review Figure 34.10, Table 34.2, Animation 34.2

> Go to **LearningCurve** (in **LaunchPad**) for dynamic quizzing that helps you solidify your understanding of this chapter. **LearningCurve** adapts to your responses, giving you the practice you need to master each key concept.

▶ Apply What You've Learned

Review

34.1 Water potential is determined by solute potential and pressure potential.

34.1 Movement of water and ions across a cell membrane can be impeded.

34.1 The two main pathways to get water from the soil into the xylem are the apoplast and the symplast.

Original Paper: Maurel, C., J. Reizer, J. I. Schroeder and M. J. Chrispeels. 1993. The vacuolar membrane protein γ-TIP creates water specific channels in *Xenopus* oocytes. *The EMBO Journal* 12: 2241–2247.

Most plant cells contain a large vacuole. In addition to the cell membrane, the vacuolar membrane (tonoplast) plays an important role in water relations between a plant cell and its environment. Changes in water potential in the environment can lead to changes in turgor pressure and therefore plant structure (see Key Concept 6.3 and Figure 6.10). The tonoplast contains a prominent protein called TIP (*tonoplast intrinsic protein*), which is related in sequence to the aquaporins of animal and bacterial cells.

Researchers investigated the possible aquaporin function of a tonoplast TIP from *Arabidopsis thaliana* by isolating the gene for the protein and expressing its mRNA in oocytes of the frog *Xenopus laevis*. These cells are large, and their size is easily measured in the light microscope. The oocyte cell membrane normally has low osmotic permeability. The researchers injected oocytes with the mRNA of the plant TIP, which contained the 5′ and 3′ untranslated sequences for protein synthesis in *Xenopus*. They injected other oocytes with an mRNA of a nontransport protein, and left yet other oocytes uninjected.

After injection, all of the oocytes were incubated in isotonic medium for 2 days to allow for protein synthesis and transport. Then the oocytes were placed in hypotonic medium to increase external water potential relative to the inside of the cells. The sizes of the oocytes were measured microscopically over 4 minutes and then compared. The results are shown in the graph.

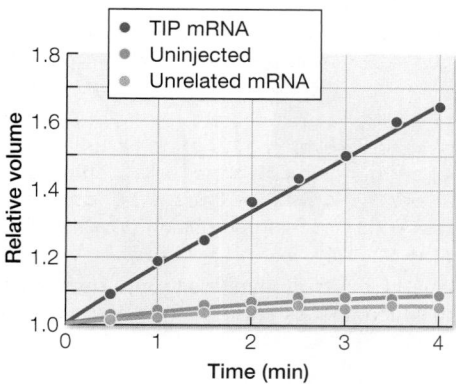

After 4 minutes in hypotonic medium, the oocytes containing TIP mRNA burst; those injected with unrelated mRNA and those that were uninjected did not burst.

The researchers then calculated the osmotic water permeability of the oocyte cell membranes during the experiment. Osmotic permeability coefficient was measured from the change in volume and surface area of the cells and water potentials inside and outside the cells. A higher osmotic coefficient indicates greater water movement across the membrane. The results are shown in the table below.

Condition	Osmotic water permeability (10^{-2} cm/sec)
TIP mRNA injected	1.25
Unrelated mRNA injected	0.15
Uninjected	0.15

Questions

1. From the graph, what can you conclude about the possible role of TIP? Why were the other two conditions important to your interpretation?

2. Why did the TIP-containing oocytes burst in the hypotonic environment? Would this have happened if the researchers had injected plant cells with TIP mRNA?

3. How do the data in the table relate to those in the figure?

4. Since the original discovery of the first plant aquaporin TIP, numerous related proteins have been found in plants. Some are in the cell membrane (plasma intrinsic proteins, or PIPs). If you examined a root in cross-section and stained it specifically for PIPs, where would you expect these proteins to be most prominent in the apoplastic and symplastic pathways?

5. In the growing root tip (see Figure 33.10), cells elongate considerably. This elongation is driven primarily by uptake of water by the cells and their prominent central vacuoles. How would you investigate the role of aquaporins in the uptake of water in these cells?

35

key concepts

35.1 Plants Require Nutrients

35.2 Plants Acquire Nutrients from the Soil

35.3 Soil Structure Affects Plant Nutrition

35.4 Soil Organisms Increase Nutrient Uptake by Plant Roots

35.5 Carnivorous and Parasitic Plants Obtain Nutrients in Unique Ways

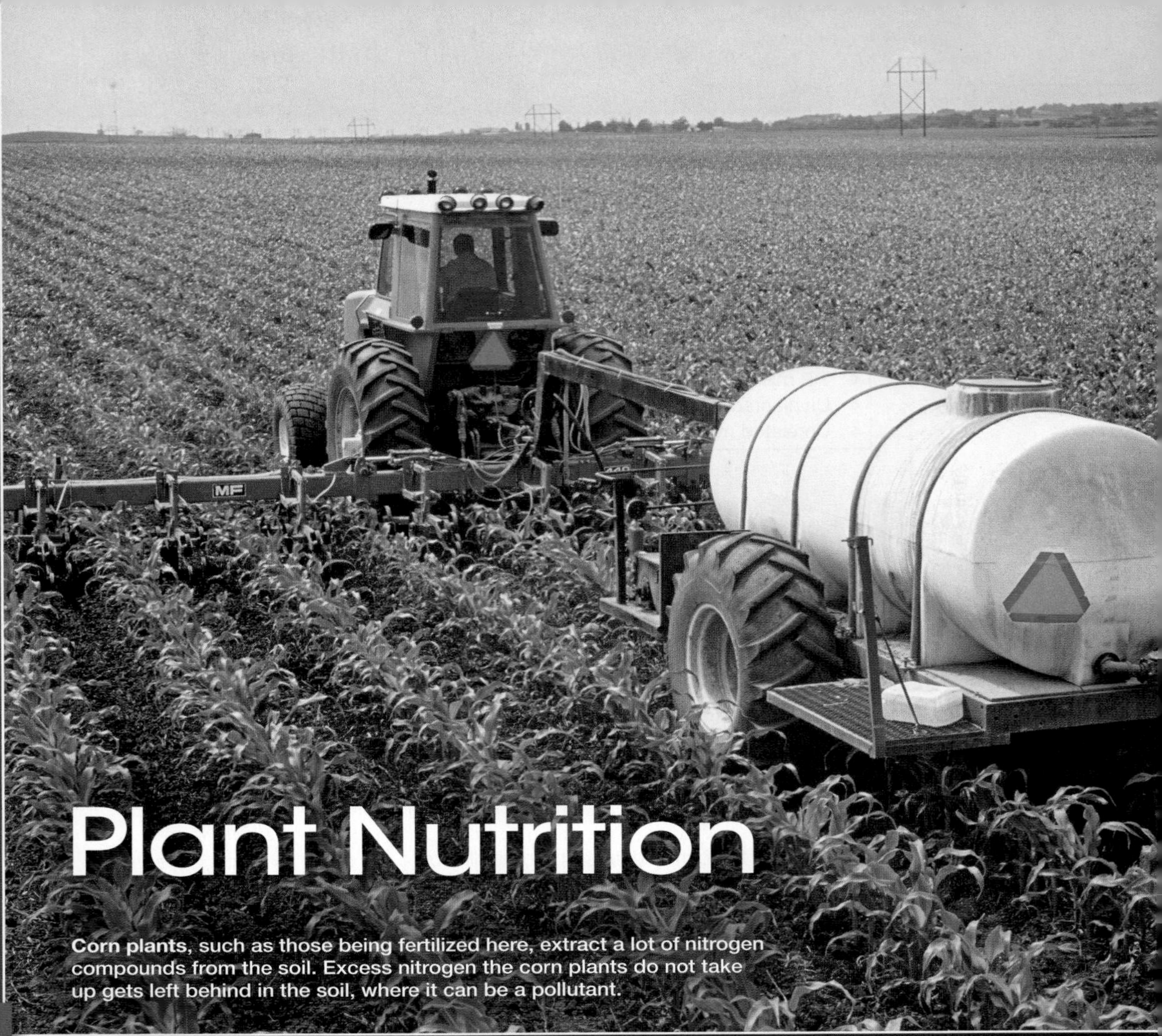

Plant Nutrition

Corn plants, such as those being fertilized here, extract a lot of nitrogen compounds from the soil. Excess nitrogen the corn plants do not take up gets left behind in the soil, where it can be a pollutant.

investigating**life**

Improving Plant Nutrition to Feed the World

Crops such as rice, wheat, and corn supply more than half of the human diet. Plants require good nutrition in order to grow and one of the nutrients often in short supply in soils is nitrogen. Organic farmers traditionally supply nitrogen by spreading animal waste (manure) over the field; as nature takes its course, the complex macromolecules in manure are broken down to ammonium (NH_4^+).

In 1917, Fritz Haber and Carl Bosch invented a way to make ammonia in the factory, revolutionizing crop plant nutrition. The Haber–Bosch process has been called one of the most important scientific feats of the twentieth century. Chemical fertilizer in the form of ammonium salts led to spectacular increases in food production. But nitrogen fertilizer is expensive in two ways. First, the Haber–Bosch process of manufacturing ammonium from hydrogen and nitrogen gases is very energy intensive and costly.

Second, nitrogen fertilizer is also environmentally expensive. When it rains excessively, nitrogen fertilizer can be lost from farm fields and end up in lakes, rivers, or groundwater. When nitrogen-laden rivers enter the sea, excessive growth of marine algae is likely to result. Eventually the algae die, and the organisms that decompose the algae use up so much oxygen in the water that there is not enough left to support marine life. Nitrogen fertilizer runoff has resulted in vast "dead zones" in waters near the mouths of major rivers, including the Mississippi River Delta in the Gulf of Mexico. An additional environmental cost is the conversion of some nitrogen fertilizer to nitrous oxide gas (N_2O), which contributes to global warming.

Scientists are working on several strategies by which nitrogen fertilizer might be used more efficiently. One strategy is to improve farming practices, by applying optimal rates of nitrogen to crops while reducing losses to the environment. The other strategy is to alter the genetics of crop plants to improve their uptake and assimilation of nitrogen. Many processes are involved in a plant's use of nitrogen from the soil, such as uptake into the roots, transport to other organs in the vascular system, and incorporation of nitrogen into organic molecules such as amino acids and nucleotides. Each process involves many genes, underscoring the complexity of inheritance.

How can nitrogen use efficiency

key concept
35.1
Plants Require Nutrients

Every living thing—and a plant is no exception—must obtain raw materials from its environment. These **nutrients** include the major ingredients of macromolecules: carbon, hydrogen, oxygen, and nitrogen. Plants are autotrophs, and obtain both carbon and oxygen from the atmosphere through the reactions of photosynthesis and cellular respiration (see Chapters 9 and 10). Hydrogen comes mainly from water, so it is plentiful when there is an adequate water supply. Nitrogen, as you will see later in this chapter, enters most plants from the soil. The activities of microorganisms are important in converting organic nitrogen and nitrogen gas into inorganic forms that are usable by plants.

focus your learning

- Plants receiving insufficient quantities of essential elements can be evaluated by observing symptomatic changes in physical features and/or growth.
- Experiments using hydroponic growth conditions have allowed scientists to determine the essential elements plants need for growth.

In addition to nitrogen, organisms require other **mineral nutrients**: inorganic elements that are used for various cellular processes. For example, proteins contain sulfur (S), nucleic acids contain phosphorus (P), chlorophyll contains magnesium (Mg), cytochromes contain iron (Fe), and cellular signaling can involve calcium (Ca). Most plants obtain these nutrients from the soil. Within the soil, minerals dissolve in water as ions, forming a solution—called the **soil solution**—that contacts the roots of plants.

Plants require nutrients in different amounts

A plant nutrient is called an **essential element** if the plant fails to complete its life cycle or grows abnormally when the element is absent or insufficient. Essential elements (**Table 35.1**) fall roughly into two categories—macronutrients and micronutrients—based on the amounts required by plants.

- A plant needs **macronutrients** in concentrations of at least 1 gram per kilogram of the plant's dry matter.
- A plant needs **micronutrients** in concentrations of less than 100 milligrams per kilogram of the plant's dry matter.

How do we know if a plant is getting enough of a particular nutrient?

Deficiency symptoms reveal inadequate nutrition

When a plant is deficient in an essential element, it displays characteristic **deficiency symptoms**. Table 35.1 and **Figure 35.1** describe some of the symptoms that help growers diagnose mineral nutrient deficiencies in plants. With proper diagnosis, the missing nutrient(s) can be provided in the form of a **fertilizer** (an added source of mineral nutrients).

Hydroponic experiments identified essential elements

The essential elements for plants were identified by growing plants **hydroponically**—that is, with their roots suspended in nutrient solutions instead of soil. Growing plants in this manner allows for greater control of nutrient availability than is possible in a complex medium such as soil.

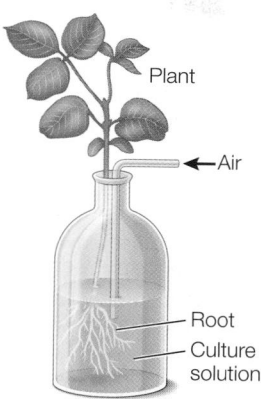

In the first experiments of this type, performed a century and a half ago, plants seemed to grow normally in solutions containing only calcium nitrate [$Ca(NO_3)_2$], magnesium sulfate ($MgSO_4$), and potassium phosphate (KH_2PO_4). A solution missing any of these compounds could not support normal growth. Tests with other compounds that included various combinations of these elements soon established the existence of six essential elements: calcium, nitrogen, magnesium, sulfur, potassium, and phosphorus. These are now known as the essential mineral macronutrients.

Iron was the first micronutrient to be clearly established as essential, in the 1840s. The most recent micronutrient to be listed as essential was nickel, in 1983. Identifying essential micronutrients proved to be more difficult than identifying macronutrients because of the small amounts involved. Sufficient amounts of micronutrients can be present in the environment used to grow plants or in the plants themselves. A seed may contain enough of a micronutrient to supply the embryo and the entire plant throughout its lifetime. There might even be enough left

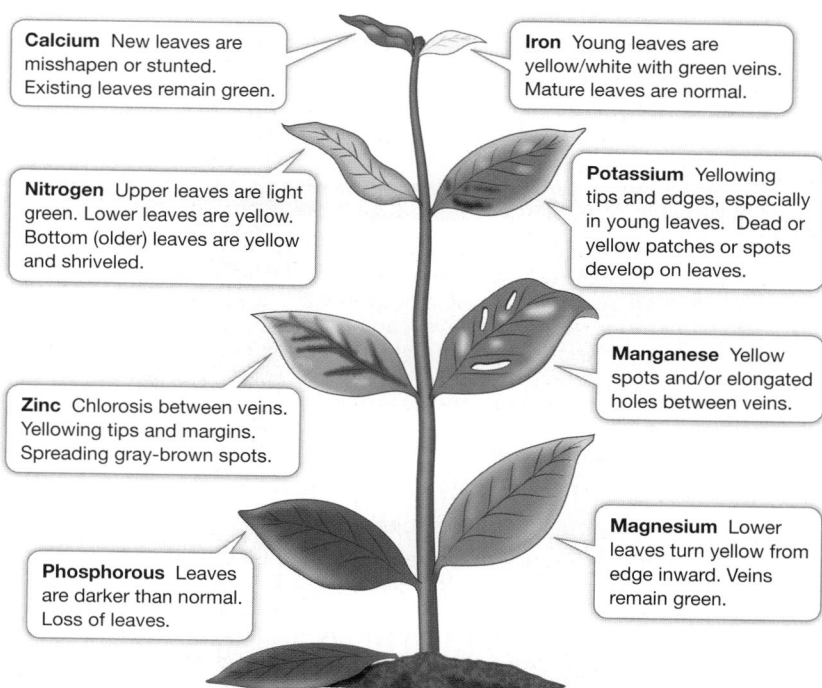

Figure 35.1 Mineral Nutrient Deficiency Symptoms Deficiencies of several nutrients are depicted in this diagram. Of course, it is very unlikely that a single plant would show all of these symptoms.

 Animation 35.1 **Nitrogen and Iron Deficiencies**
www.Life11e.com/a35.1

table 35.1 Mineral Elements Required by Plants

Element (abbreviation; absorbed form)	Typical amount in plant (g/kg dry wt)	Major functions	Deficiency symptoms
Macronutrients			
Nitrogen (N; NO_3^- and NH_4^+)	15	In proteins, nucleic acids	Oldest leaves turn yellow and die prematurely; plant is stunted
Phosphorus (P; $H_2PO_4^-$ and HPO_4^{2-})	2	In nucleic acids, ATP, phospholipids, and phosphoproteins	Plant is dark green with purple veins and is stunted
Potassium (K; K^+)	10	Enzyme activation; water balance; ion balance; stomatal opening	Older leaves have dead edges
Sulfur (S; SO_2^{4-})	1	In proteins and coenzymes	Young leaves are yellow to white with yellow veins
Calcium (Ca; Ca^{2+})	5	Affects the cytoskeleton, membranes, and many enzymes; second messenger	Growing points die back; young leaves are yellow and crinkly
Magnesium (Mg; Mg^{2+})	2	In chlorophyll; required by many enzymes; stabilizes ribosomes	Older leaves have yellow stripes between veins
Micronutrients			
Iron (Fe; Fe^{2+} and Fe^{3+})	0.1	In active site of many redox enzymes and electron carriers; chlorophyll synthesis	Young leaves are white or yellow
Chlorine (Cl; Cl^-)	0.1	Photosynthesis; ion balance	Leaf tips wilt; leaves turn yellow and die
Manganese (Mn; Mn^{2+})	0.05	Cofactor for many enzymes	Younger leaves are pale with green veins
Boron [B; $B(OH)_3$]	0.02	Required for proper cell wall formation and expansion	Poor growth of leaves and roots
Zinc (Zn; Zn^{2+})	0.02	Enzyme activation; auxin synthesis	Young leaves are abnormally small; older leaves have many dead spots
Copper (Cu; Cu^{2+})	0.006	Cofactor for some redox enzymes and electron carriers	New leaves are dark green, may have dead spots
Nickel (Ni; Ni^{2+})	0.001	Activation of the enzyme urease	Leaf tips die; deficiency is rare
Molybdenum (Mo; MoO_4^{2-})	0.0001	Cofactor of enzymes involved in nitrogen reduction	Leaves turn yellow between veins; older leaves die

over to pass on to third-generation plants. Because of such difficulties, nutrition experiments must be performed in tightly controlled laboratories with special air filters that exclude microscopic salt particles in the air, and must use only the purest available chemicals.

35.1 recap

Plants are autotrophs that obtain carbon and oxygen by photosynthesis, and mineral nutrients and water from the soil. Nutrients required by plants are classified as either macronutrients or micronutrients depending on the amount needed. Micronutrients are often needed in such minute amounts that only sophisticated chemical experiments can determine their essentiality.

learning outcomes

You should be able to:

- Identify nutrient deficiencies based on observations.
- Explain difficulties involved in testing nutrient requirements for plant growth.

1. Methods for determining whether a particular element is essential for plant growth were established a century and a half ago, yet the essentiality of some elements was discovered only recently. Why?

2. If the young leaves of a plant are yellow, what nutrient deficiencies may be suspected? What if the yellow leaves are the oldest ones?

You have seen that plants require a specific set of nutrients in large and small amounts for optimal growth and development. Let's look at how a plant finds and takes up these nutrients from its environment.

key concept

35.2 Plants Acquire Nutrients from the Soil

We can move from place to place to find the nutrients we need. But a plant cannot change its location (it is sessile), and so must obtain nutrients from its immediate environment. With the exception of carbon and oxygen, a plant's supply of nutrients is strictly local, and a plant may use up the water and mineral nutrients in its local environment as it grows. How does a plant cope with the problem of scarce nutrient supplies?

focus your learning

- Plants alter their direction of root growth, depending on the availability of dissolved nutrients in the soil.
- Plant cells use gene regulation to control the numbers of membrane transporters and enzyme regulation to control the rate of incorporation of nutrients into complex biomolecules.

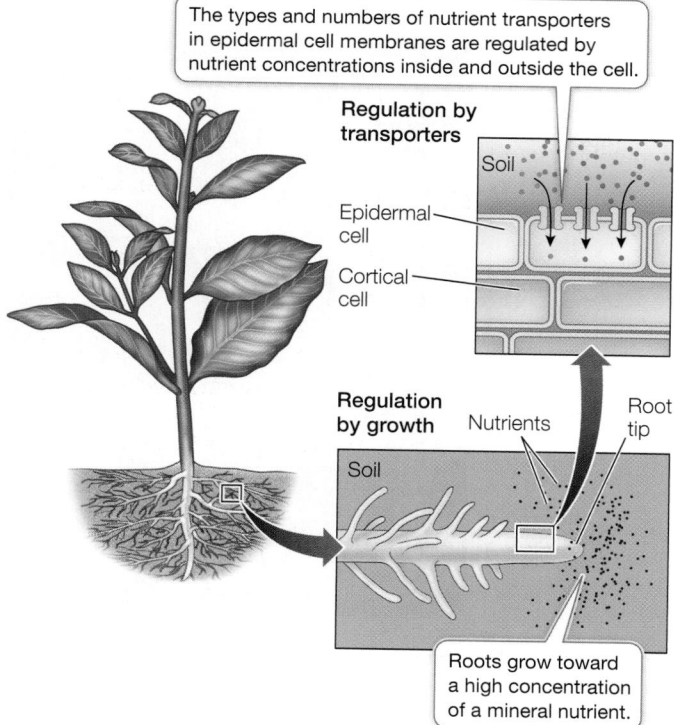

The types and numbers of nutrient transporters in epidermal cell membranes are regulated by nutrient concentrations inside and outside the cell.

Regulation by transporters

Soil

Epidermal cell

Cortical cell

Regulation by growth   Nutrients   Root tip

Soil

Roots grow toward a high concentration of a mineral nutrient.

Figure 35.2 Plants Regulate Their Nutrition   Plant roots branch and grow toward nutrients. Nutrients are taken up by transport proteins in the epidermal cell membrane. The number of transporters for a given nutrient can be regulated in response to nutrient availability.

Plants rely on growth to find nutrients

As you learned in Chapter 33, plants differ fundamentally from animals in that plants grow throughout their lifetimes. In fact, growth is a plant's version of movement. For example, roots obtain most of the mineral nutrients plants need. By growing through the soil, roots mine it for new sources of mineral nutrients and water. The growth of stems and leaves helps a plant secure light and carbon dioxide, which in turn allows the roots to continue their growth through the soil. Deficiencies in water or specific mineral nutrients can stimulate plants to grow more roots, to improve the plants' chances of finding the nutrients they need.

As it grows, a plant—or even a single root—must deal with a variable environment. Animal droppings create high local concentrations of nitrogen. A particle of calcium carbonate may make a tiny area of the soil alkaline, while dead organic matter may make a nearby area acidic. Such microenvironments encourage or discourage the proliferation of a root system and help direct its growth. A major effort is underway to identify the signals in the soil and signaling pathways in the root that result in growth toward a source of nutrients (**Figure 35.2**).

Nutrient uptake and assimilation are regulated

Nutrients must cross the cell membranes of cells in order to be used in the cell or be incorporated into larger molecules. Polar molecules, including mineral ions, cross the membrane via specialized transport systems. In Chapter 34 you saw how water and ions move into plant roots by way of the apoplast (through cell walls and intercellular spaces) or symplast (directly through cells). The Casparian strip

prevents water and ions from entering the xylem tissues of the roots (see Key Concept 34.1); therefore these nutrients must enter the symplast before they can be transported to other tissues. In most cases, ions are actively transported across the cell membrane of epidermal cells into the symplast because their concentrations in the soil solution are generally lower than their concentrations inside cells.

Plants have specialized transport systems for the uptake of specific ions (see Figure 35.2). For example, *Arabidopsis thaliana* has more than 50 genes that encode nitrate (NO_3^-) transporters, 6 genes encoding ammonium (NH_4^+) transporters, and at least 4 genes for phosphate (PO_4^{3-}) transporters. Nutrient uptake is highly regulated because the levels of ions inside cells must be maintained at constant levels. The genes for ion transporters are regulated at the transcriptional level by the amounts of each nutrient inside cells: low nutrient levels stimulate transcription, whereas high levels repress transcription. In addition, the transporter proteins themselves are regulated (for example, by phosphorylation) to control their ion transport activity.

The incorporation of nutrients into more complex molecules is also regulated according to the plant's needs. The enzymes involved in incorporating nitrate and ammonium into amino acids are regulated at the transcriptional and posttranscriptional levels to increase assimilation when available nitrogen is abundant. The uptake and assimilation of nitrogen are also stimulated by photosynthesis, and this ensures that the nitrogen status in the plant is coordinated with its carbon status.

▶ 35.2 recap

Both the uptake and assimilation of nutrients are regulated according to a plant's needs.

learning outcomes

You should be able to:

- Explain how growth is involved in a plant's search for nutrients.
- Describe how a plant is able to control the rate of movement of nutrients into a cell.

1. How do plants control the uptake and assimilation of mineral nutrients?
2. How does the ability to grow throughout their lifetimes allow plants to seek out nutrients?

Plants acquire many essential elements from the soil. As you will see in the next section, soils have complex structures that affect the availability of nutrients for plants.

▶ key concept 35.3   Soil Structure Affects Plant Nutrition

Most terrestrial plants grow in soil. Soils provide:

- Mechanical support
- Mineral nutrients and water from the soil solution
- O_2 for root respiration

Soils also harbor many bacteria and other organisms; some of these are beneficial to plant life, but others are harmful. Some soils contain toxic levels of metal ions such as cadmium, chromium, and lead.

focus your learning

- Soil can vary greatly in its composition of particles and nutrient availability (fertility). Most soils consist of three major horizons.
- Uptake of minerals that exist as cations in soil water is facilitated by a cation exchange process under slightly acidic pH conditions.
- The fertility of depleted and leached soils can be improved by adding organic and inorganic fertilizers.

Soils are modified by natural phenomena such as rain, temperature extremes, and the activities of plants and animals. They are also modified by the activities of people, particularly in agriculture. In this section you'll learn about the composition, structure, and formation of soils, as well as their role in plant nutrition.

Soil provides anchorage and nutrients for plants

Soils have living and nonliving components (**Figure 35.3**). The living components include plant roots as well as populations of bacteria, fungi, protists, and animals such as earthworms and insects. The nonliving portion of the soil includes rock fragments ranging in size from large stones to sand to silt, and finally to tiny particles of clay that are 2 micrometers (μm) or less in diameter. Soil also contains water and dissolved mineral nutrients, air spaces, and dead organic matter. The air spaces in soil contain O_2. Typical percentages of these components are:

Particles: 45%   Water: 25%   Air: 25%   Organisms: 5%

Although soils vary greatly, almost all of them have a soil profile consisting of several recognizable horizontal layers, called **horizons**, lying on top of one another. Soil scientists recognize three

Figure 35.4 A Soil Profile The A, B, and C horizons can sometimes be seen at construction sites such as this one in Massachusetts. The upper layer (the A horizon) is home to most of the living organisms in the soil.

major horizons—termed A, B, and C—in the profile of a typical soil (**Figure 35.4**).

- The **A horizon** is the **topsoil** that supports the plant's nutrient needs. It contains most of the soil's living and dead organic matter.
- The **B horizon** is the **subsoil**, which accumulates materials from the topsoil above it and from the parent rock below.
- The **C horizon** is the **parent rock**, also called bedrock, which is in the process of breaking down to form soil.

Soil fertility is a soil's ability to support plant growth. A topsoil's fertility is determined by several factors. Topsoils vary greatly in their proportions of sand, silt, and clay, and this influences their ability to support plant growth. For example, mineral nutrients tend to be **leached** from the upper soil horizons—dissolved in rain or irrigation water and carried to deeper horizons, where they are unavailable to plant roots. Dissolved minerals are readily leached from sandy soil because sand particles are relatively large and cannot hold water. Clay, by contrast, binds more water than sand does, and the charged surfaces of clay particles bind mineral ions that plant roots ultimately take up. But clay particles are tiny and pack tightly together, leaving little space for air. A **loam** is a soil that is an optimal mixture of sand, silt, and clay and thus has sufficient levels of air, water, and available nutrients for plants. Loams also usually contain organic matter. Most of the best topsoils for agriculture are loams.

In addition to mineral particles, soils contain dead organic matter, largely from plants. Soil organisms break down dead leaves and other plant organs on the ground into a substance called **humus**. This material is used as a food source by microbes that break down complex organic molecules and release simpler molecules into the soil solution. Humus also provides air spaces that increase O_2 availability to plant roots.

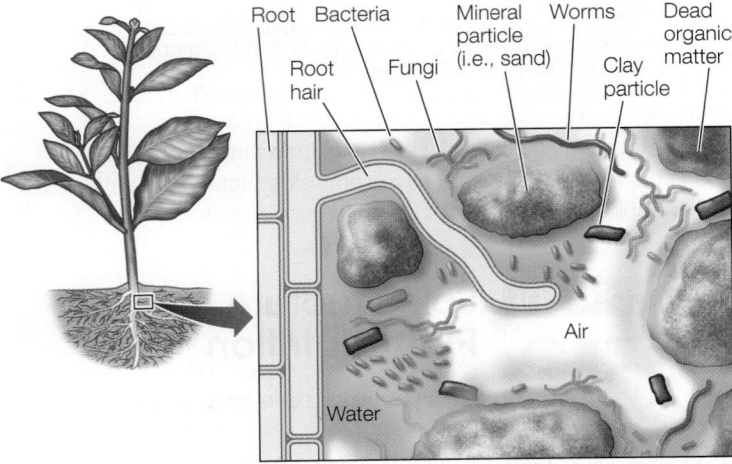

Figure 35.3 The Complexity of Soil Soils favorable for plant growth contain both clay and larger mineral particles, as well as water, air, and organic matter. Other organisms are also present.

Soils form through the weathering of rock

Rocks are broken down into soil particles—**weathered**—in two ways. First, there is *mechanical weathering*, which is the physical breakdown of materials by wetting, drying, and freezing. Second, there is *chemical weathering*, the alteration of the chemistry of the materials in the rocks. Several types of chemical weathering occur, all of which influence the availability of mineral nutrients:

- Oxidation by atmospheric oxygen
- Hydrolysis (reaction with water)
- Reaction with acids (particularly carbonic acid)

The parent rock and the weathering it undergoes determine the basic structure and chemical composition of a soil. However, a key soil characteristic for plants is the availability of nutrients, which must be dissolved in the soil solution for uptake by the plant. Chemical weathering often results in clay particles that are covered with negatively charged chemical groups, which bind positively charged mineral nutrients. How might roots obtain these mineral nutrients?

Soils are the source of plant nutrition

Humus and clay particles often carry negative charges. These particles form ionic attractions (see Key Concept 2.2) with the positively charged ions (cations) of many minerals that are important for plant nutrition, such as potassium (K^+), magnesium (Mg^{2+}), and calcium (Ca^{2+}). To become available to plants or other organisms, these cations must be detached from the clay particles. How can this breakage of ionic attraction happen?

Recall that the root surface is covered with root hair cells (see Figure 33.10). Transporters in the cell membranes of these cells actively pump protons (H^+) out of the cell. In addition, cellular respiration in the roots releases CO_2, which dissolves in the soil water and reacts with it to form carbonic acid. This acid ionizes to form bicarbonate and free protons:

$$CO_2 + H_2O \rightleftharpoons H_2CO_3 \rightleftharpoons H^+ + HCO_3^-$$

Proton-pumping by the root and ionization of carbonic acid both act to increase the proton concentration (lower the pH) in the soil surrounding the root. The protons bind more strongly to clay particles than do mineral cations; in essence, they trade places with the cations in a process called **cation exchange** (**Figure 35.5**). Cation exchange releases important cations into the soil solution, where they are available to be taken up by the roots. Soil fertility is determined in part by the soil's ability to provide nutrients in this manner.

Some soil particles, such as ones containing oxides of iron or aluminium, are positively charged under acid conditions and can exchange anions (Cl^-, for example) in a process similar to cation exchange. However, soil pH is rarely low enough for anion exchange to occur. As a result, important anions such as nitrate (NO_3^-) and sulfate (SO_4^{2-})—direct sources of nitrogen and sulfur, respectively—may leach rapidly from the A horizon.

As you've seen, soil fertility is affected by soil pH. The H^+ concentration affects the binding of cations and anions to soil particles, and

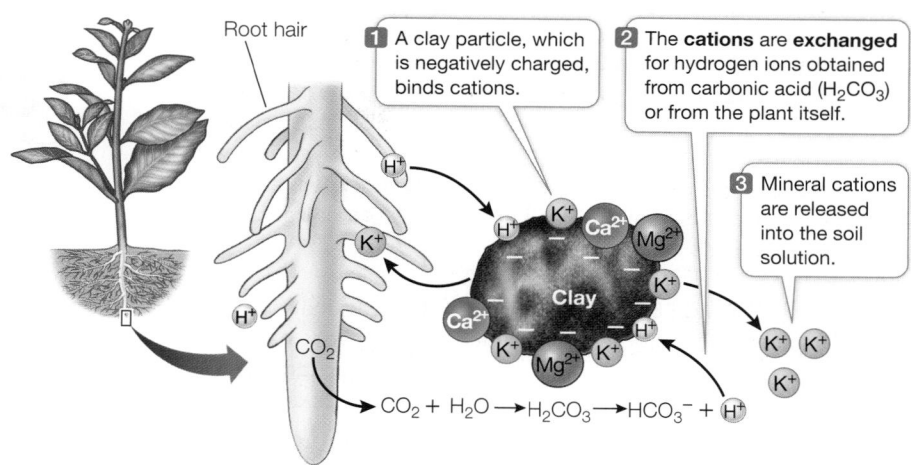

Figure 35.5 Cation Exchange Plants obtain mineral nutrients from the soil primarily in the form of positive ions; potassium (K^+) is the example shown here.

Q: Do negatively charged ions behave differently?

can affect the solubility of other nutrients, such as iron. In addition, soil pH affects the absorption of nutrients by plant roots. The pH level of a soil depends on its mineral and organic contents and can be altered by various factors, including rainfall, weathering, plant growth, and fertilizer applications. The optimal soil pH for most plants is in the range 6–7.5, but some plants, such as blueberries and cranberries, prefer pH levels of 4.5–5.

Fertilizers add nutrients to soil

Leaching and the harvesting of crops can deplete a soil of its nutrients, so that new crops grow poorly on that soil. Soil fertility can be restored or increased in various ways, including shifting agriculture to another location or applying organic or chemical fertilizers.

SHIFTING AGRICULTURE In the past, when the soil could no longer support a level of plant growth sufficient for agricultural purposes, people simply moved to another location. The nutrients in the soil of a field allowed to lie fallow will be replenished gradually through the addition of organic matter from the growth and death of plants naturally present, and by the weathering of the parent rock. Both processes take a long time, which is not a problem as long as a lot of land is available. Today, however, the food needs of a large human population are too great to allow land to be left vacant for a long time, and people are disinclined to move away from settled homesteads. As a consequence, chemical fertilizers are often used to improve soil fertility.

ORGANIC FERTILIZERS Microorganisms in the soil break down organic molecules into smaller, simpler molecules. These simpler molecules can dissolve in soil water and enter plant roots. For example, soil bacteria break down the proteins in dead leaves and produce ammonium ions (NH_4^+), which in turn are converted into nitrate (NO_3^-). Both ammonium and nitrate can be taken up and used by plants:

$$\text{Proteins in leaves} \xrightarrow{\text{Bacteria}} NH_4^+ \xrightarrow{\text{Bacteria}} NO_3^+$$

Farmers can increase the nutrient content of soil by adding organic materials such as compost (partially decomposed plant materials) or manure (waste from farm animals). Manure is a particularly good source of nitrogen. In either case, the addition of these **organic fertilizers** adds nutrients to the soil much more rapidly than weathering or the gradual addition of organic matter from natural vegetation. Organic fertilizers allow for a slow release of ions as the materials decompose.

INORGANIC FERTILIZERS   Organic fertilizers may act too slowly to restore fertility if a soil is to be used every year. As you read in the opening investigation of this chapter, **inorganic fertilizers** supply mineral nutrients in forms that can be taken up immediately by plants or that are rapidly converted to usable forms in the soil. Inorganic fertilizers are easily transported and handled, and allow farmers to control the amount of a particular nutrient that is supplied to each crop. Particular fertilizers are used in varying amounts, depending on the needs of the crop and the type of soil. For example, much higher amounts of nitrogen are applied to cornfields than to soybean fields. Inorganic fertilizers come in many forms. Common ones include ammonia (NH_3), urea (NH_2—CO—NH_2), and salts formed from positive and negative ions such as ammonium (NH_4^+), potassium (K^+), nitrate (NO_3^-), phosphate (PO_4^{3-}), and sulfate (SO_4^{2-}).

35.3 recap

Land plants live anchored in the soil and obtain water and mineral nutrients from it. Soils are complex in structure and vary in fertility. Farmers can add fertilizers to improve the nutrient contents of soils.

learning outcomes

You should be able to:

- Predict changes in a soil in response to environmental and land use changes.
- Explain how soil fertility is influenced by the process of cation exchange.
- Compare and contrast organic and inorganic fertilizers.

1. Soils are dynamic systems. What changes might result when land is subjected to heavy irrigation for agriculture after being relatively dry for many years? What changes in the soil might result when a deciduous forest is cut down and replaced by crops that are harvested each year?
2. How does the process of cation exchange enhance soil fertility?
3. What are the differences between organic and inorganic nitrogen fertilizers in terms of plant nutrition?

Thus far we have focused on the uptake of nutrients in the soil by plant roots. An understanding of how plants acquire nutrients from the soil would be incomplete, however, without taking into account the involvement of soil microbes, including fungi and bacteria. In the next section we will focus on the intimate interactions of plants with these organisms, which are essential to the success of most terrestrial plants.

key concept
35.4 Soil Organisms Increase Nutrient Uptake by Plant Roots

One gram of soil can contain 6,000–50,000 bacterial *species* and up to 200 *meters* of fungal hyphae (the long branching cells of fungi), although both are largely invisible to the naked eye. In Chapter 38 you will learn how plants prevent infection by harmful soil microbes. But plants actually encourage a few species of fungi and bacteria to infect their roots and even invade root cells. In this section we will describe the mutually beneficial relationships in which products are exchanged between the plants and these special soil microbes.

focus your learning

- Mutualistic relationships between soil organisms and plant roots are facilitated by chemical signal exchange.
- Although abundant in Earth's atmosphere, nitrogen is not a readily available nutrient for plants.
- Humans have used various methods, including crop rotation and use of natural and human-made fertilizers, to replenish soil nitrogen depleted by crop harvesting.

Plants send signals to soil organisms

In Chapter 29 we described mycorrhizae, the association of fungi with plant roots—an interaction that occurs in more than 90 percent of terrestrial plants. Our example in Chapter 29 was ectomycorrhizal fungi, which wrap themselves around a plant root (see Figure 29.10). In this chapter we examine arbuscular mycorrhizae, in which fungal hyphae penetrate root cells. We also describe here the close association between the roots of some plants and rhizobia, a group of nitrogen-fixing bacteria (bacteria that convert atmospheric N_2 into a more biologically useful form). You will see that ***mycorrhizal and rhizobial associations** are both initiated by signals sent by plant roots that attract the soil organisms, and that the development of these associations involves similar genes and cellular pathways.

> ***connect the concepts**   Mycorrhizal and rhizobial associations are examples of mutualism, one of the major types of interactions discussed in Key Concept 56.1.

FORMATION OF MYCORRHIZAL ASSOCIATIONS   You can follow the events in the formation of arbuscular mycorrhizae in **Figure 35.6A**. Plant roots produce molecules called **strigolactones** that stimulate rapid growth of fungal hyphae toward the root. In response, the fungi produce signals that stimulate expression of genes that promote symbioses with other organisms. The products of some of these genes give rise to the prepenetration apparatus (PPA), which guides the growth of the fungal hyphae into the root cortex. The sites of nutrient exchange between fungus and plant are the arbuscules, which form within root cortical cells. Despite the intimacy of this association, the plant and fungal cytoplasms never mix—they are separated by two membranes, the fungal cell membrane and the periarbuscular membrane (PAM), which is continuous with the plant cell membrane.

FORMATION OF NITROGEN-FIXING NODULES A group of plants called legumes (members of the plant family Fabaceae) can form symbioses with soil bacteria in several genera collectively known as rhizobia. The legume roots release flavonoids and other chemical signals that attract the rhizobia to the vicinity of the roots (**Figure 35.6B**). The flavonoids also trigger the transcription of bacterial *nod* genes, the products of which synthesize Nod (nodulation) factors. These and other factors, when secreted by the bacteria, cause cells in the root cortex to divide, leading to the formation of a primary nodule meristem. This meristem gives rise to the plant tissue that constitutes the **root nodule**. Bacteria enter the root via an infection thread, analogous to the PPA in mycorrhizal associations, and eventually reach cells inside the root nodule. There the bacteria are released into the cytoplasm of the nodule cells, enclosed in membrane vesicles similar to the PAM. Inside the vesicles, the bacteria differentiate into **bacteroids**—the form of the bacteria that can fix nitrogen.

Figure 35.6 Roots Send Signals for Colonization Plant roots send chemical signals to arbuscular mycorrhizal fungi **(A)** and nitrogen-fixing bacteria **(B)** to stimulate colonization. (PAM = periarbuscular membrane.)

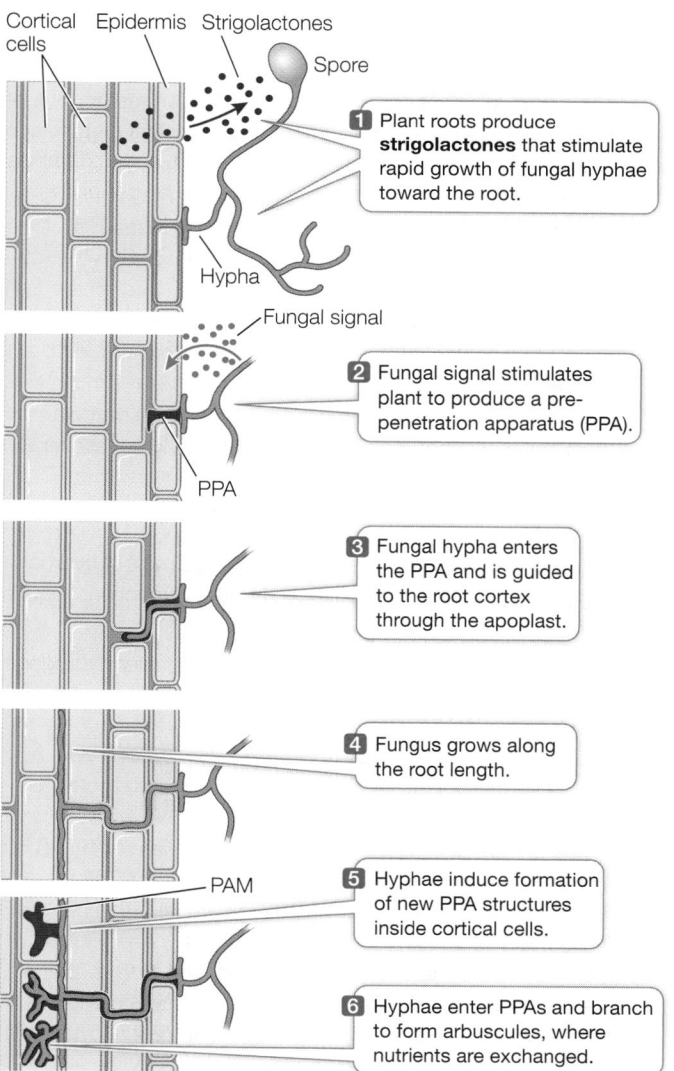

(A) Formation of arbuscular mycorrhizal fungi

1. Plant roots produce **strigolactones** that stimulate rapid growth of fungal hyphae toward the root.

2. Fungal signal stimulates plant to produce a pre-penetration apparatus (PPA).

3. Fungal hypha enters the PPA and is guided to the root cortex through the apoplast.

4. Fungus grows along the root length.

5. Hyphae induce formation of new PPA structures inside cortical cells.

6. Hyphae enter PPAs and branch to form arbuscules, where nutrients are exchanged.

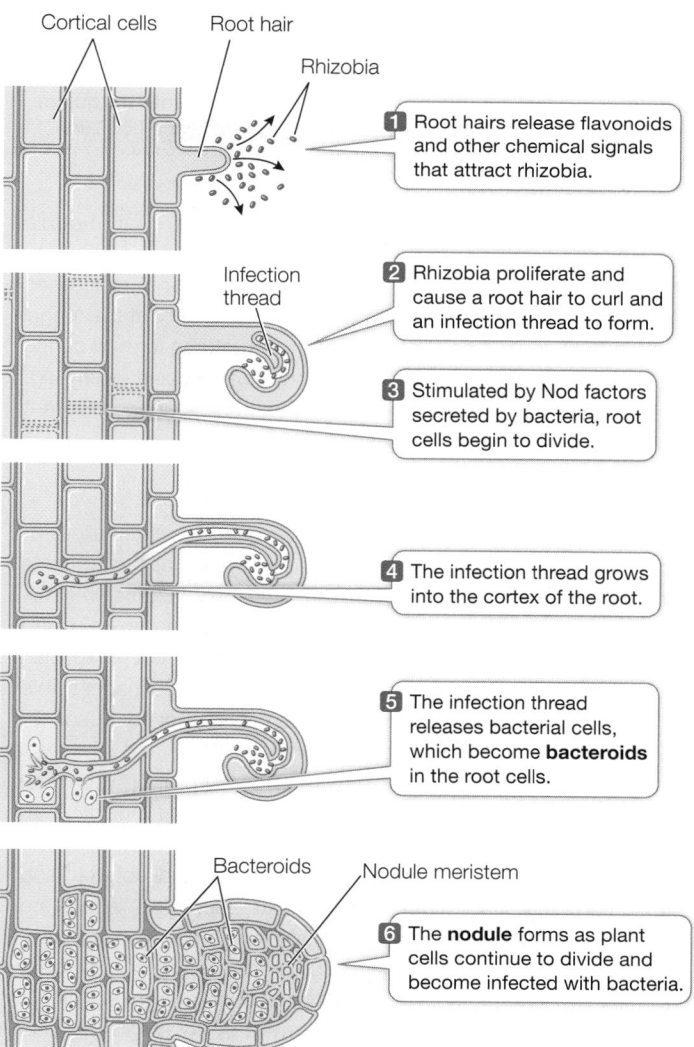

(B) Formation of a nitrogen-fixing root nodule

1. Root hairs release flavonoids and other chemical signals that attract rhizobia.

2. Rhizobia proliferate and cause a root hair to curl and an infection thread to form.

3. Stimulated by Nod factors secreted by bacteria, root cells begin to divide.

4. The infection thread grows into the cortex of the root.

5. The infection thread releases bacterial cells, which become **bacteroids** in the root cells.

6. The **nodule** forms as plant cells continue to divide and become infected with bacteria.

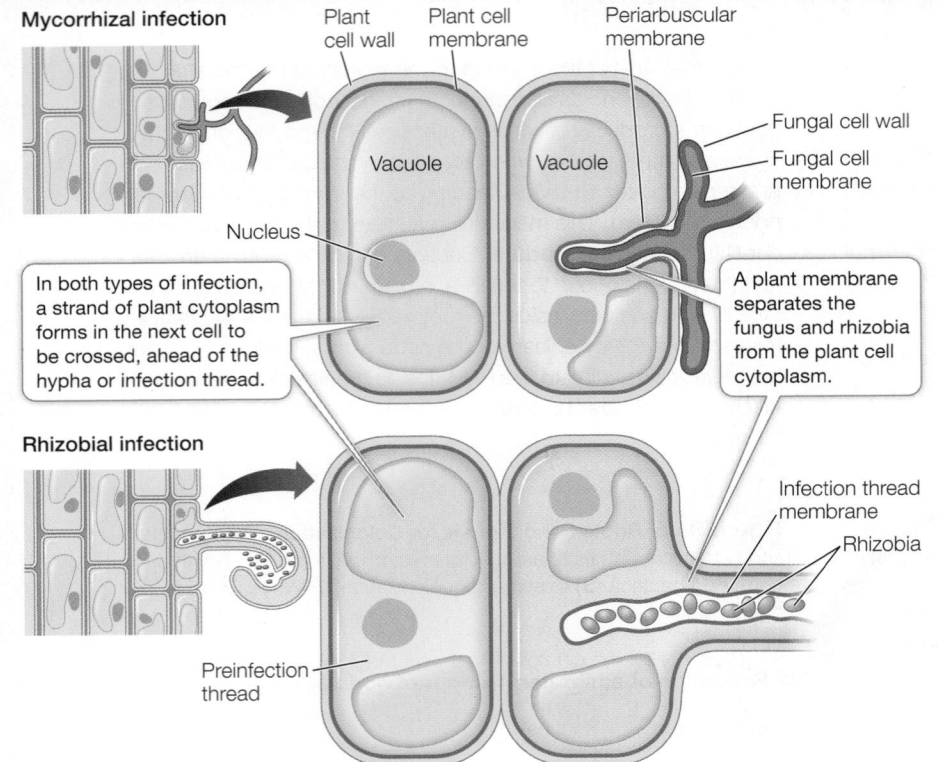

Mycorrhizal infection

Plant cell wall
Plant cell membrane
Periarbuscular membrane

Vacuole
Vacuole

Fungal cell wall
Fungal cell membrane

Nucleus

In both types of infection, a strand of plant cytoplasm forms in the next cell to be crossed, ahead of the hypha or infection thread.

A plant membrane separates the fungus and rhizobia from the plant cell cytoplasm.

Rhizobial infection

Infection thread membrane

Rhizobia

Preinfection thread

Figure 35.7 Intracellular Structures in Plant–Fungus and Plant–Rhizobium Symbioses Several steps in the development of mycorrhizae and nodules involve similar structures.

A COMMON MECHANISM Amazingly, it appears that nodule formation depends on some of the same genes and mechanisms that allow mycorrhizae to develop. For example, both processes involve folding in of the cell membrane to allow entry of the fungal hyphae or rhizobia. The similarities in the structures formed during the development of mycorrhizae and nodules are especially striking, considering that the symbioses involve members of two different kingdoms (fungi and bacteria) (**Figure 35.7**).

Mycorrhizae expand the root system

In many cases the roots of vascular plants cannot nutritionally support plant growth alone—they simply cannot reach all the nutrients available in the soil. Mycorrhizae expand the root surface area 10-fold to 1,000-fold, increasing the amount of soil that can be mined for nutrients. In addition, because fungal hyphae are much finer than root hairs, they can get into pores in the soil that are inaccessible to roots. In this way, mycorrhizae probe a vast quantity of soil for nutrients and deliver them into root cortical cells.

The primary nutrient that the plant obtains from a mycorrhizal interaction is phosphorus. In exchange, the fungus obtains an energy source: the products of photosynthesis. In fact, up to 20 percent of the photosynthate of terrestrial plants is directed to and consumed by arbuscular mycorrhizal fungi. Such associations are excellent examples of mutualism, an interaction between two species in which both species benefit. Mutualism is a type of symbiosis, in which two different species live in close contact for a significant portion of their life cycles. As is conveyed in **Investigating Life: Mycorrhizal Fungi Can Replace Fertilizer in Cassava Cultivation**,

the relationship between mycorrhizae and plant roots can be exploited by increasing the fungal population. Many species of fungus invade roots. Identification and genetic improvement of these species is now possible, which holds great promise in improving phosphorus nutrition in important crops.

Soil bacteria are essential in getting nitrogen from air to plant cells

The essential mineral nutrient most commonly in short supply, in both natural and agricultural situations, is nitrogen. This is surprising because elemental nitrogen (N_2) makes up almost four-fifths of Earth's atmosphere. However, plants cannot use N_2 directly as a nutrient. The triple bond linking the two nitrogen atoms is extremely stable, and a great deal of energy is required to break it; thus N_2 is a highly unreactive substance.

Some prokaryotes have an enzyme that enables them to convert N_2 into a more reactive and biologically useful form by a process called **nitrogen fixation**:

$$N_2 + 6 H^+ + 6 e^- \rightarrow 2 NH_3$$

This equation is simplified: actually, H_2 gas is a byproduct. **Nitrogen fixers**, including those present in root nodules, fix approximately 170 million metric tons of nitrogen per year. Humans use industrial methods to fix about 80 million metric tons per year. In addition, about 20 million metric tons per year are fixed in the atmosphere by nonbiological means such as lightning, volcanic eruptions, and forest fires. Rain brings these atmospherically formed products to the ground.

Two types of organisms can fix nitrogen:

1. Free-living organisms living in soil and water (e.g., *Azotobacter* bacteria and *Nostoc* cyanobacteria [sometimes called blue-green algae])
2. Symbiotic organisms living in other organisms (e.g., rhizobia in roots of legumes, and *Anabaena* cyanobacteria in aquatic ferns)

Nitrogenase catalyzes nitrogen fixation

Nitrogen fixation is the reduction (see Key Concept 9.1) of nitrogen gas. It proceeds by the stepwise addition of three pairs of hydrogen atoms to N_2 (**Figure 35.8**). In addition to N_2, these reactions require three things:

1. A strong reducing agent (e.g., ferredoxin) to transfer hydrogen atoms (protons and electrons) to N_2 and to the intermediate products of the reaction
2. A great deal of energy, which is supplied by ATP
3. The enzyme **nitrogenase**, which catalyzes the reaction

Nitrogenase is strongly inhibited by oxygen, and many nitrogen fixers are anaerobes that live in environments with little or no O_2. But rhizobia are *aerobic* and fix nitrogen in aerobic plant roots. How can nitrogenase function under these circumstances?

Plants typically house nitrogen-fixing bacteria in root nodules. Within a nodule, O_2 is maintained at a low level that is sufficient to support respiration, but not so high as to inactivate nitrogenase. This is possible because the cytoplasm of nodule cells contains

experiment

Original Paper: Ceballos, I., M. Ruiz, C. Fernandez, R. Pena, A. Rodriguez and I. Sanders. 2013. The *in vitro* mass-produced model mycorrhizal fungus, *Rhizophagus irregularis,* significantly increases yields of the globally important food security crop cassava. *PLoS ONE* 8(8): e70633.

Phosphorous is a plant macronutrient obtained from the soil. Unfortunately, in many tropical regions the phosphate level in the soil is inadequate to support the needs of crop production, forcing farmers to use inorganic phosphate fertilizer, which is a finite and expensive resource. The need for phosphate fertilizer poses an especially acute problem for farmers in poor, tropical regions of the world.

Mycorrhizal fungi very efficiently absorb phosphate ions from the soil. Plant roots with a symbiotic relationship with mycorrhizae obtain phosphorous, while the fungi obtain photosynthate (products of photosynthesis) from the plant. Ian Sanders and his team at the University of Lausanne in Switzerland, in collaboration with Alia Rodriguez and her team at the National University of Colombia in Bogota, investigated whether production of cassava could be increased by adding concentrated spores of mycorrhizae, thus lowering the need for phosphate fertilizer.

HYPOTHESIS▶ Adding arbuscular mycorrhizae to the soil can replace the need for phosphate fertilizer and increase food production in cassava.

METHOD

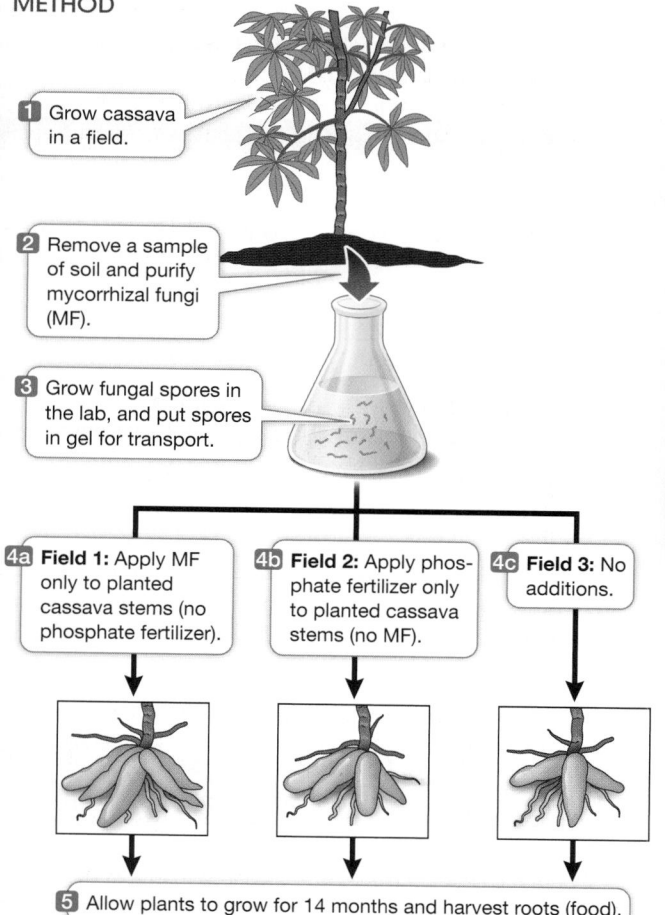

1 Grow cassava in a field.

2 Remove a sample of soil and purify mycorrhizal fungi (MF).

3 Grow fungal spores in the lab, and put spores in gel for transport.

4a **Field 1:** Apply MF only to planted cassava stems (no phosphate fertilizer).

4b **Field 2:** Apply phosphate fertilizer only to planted cassava stems (no MF).

4c **Field 3:** No additions.

5 Allow plants to grow for 14 months and harvest roots (food).

work with the data

QUESTIONS▶

1. Concentrated spores of the mycorrhizal fungus *Rhizophagus irregularis* were added to fields of cassava stems tied to stakes. Control fields did not receive the spores. The experiments took place at Yopal, Colombia, in a tropical region. At various times, developing roots were examined microscopically for the presence of fungal hyphae. The results are shown in **Table A**. Storage roots were harvested after 400 days.

Table A

Days after spores were added	Percent colonization
45	40
90	45
135	47
180	49
225	40
270	62
315	82

Plot the data as percent colonization versus time (days after spores added). When did the colonization of fungi increase? Why do you think there was a lag time?

2. In a large field experiment, plots of 300 plants each were treated either with or without fungal spores, and with or without phosphate fertilizer in an amount usually needed for optimal growth (100%) or half of that (50%). The results are shown in **Table B**, as fresh weight (g) of cassava roots, the crop (±SEM).

Table B

	No fertilizer	50% fertilizer	100% fertilizer
Fungal spores	38 (2)	40 (1)	43 (2)
No fungal spores	28 (2)	36 (3)	35 (2)

a. Did the addition of mycorrhizal spores affect the cassava production? What is the evidence for your answer?

b. What statistical test would you use to test for significance of the effects of fertilizer and added mycorrhizae?

A similar **work with the data** exercise may be assigned in **LaunchPad**.

RESULTS

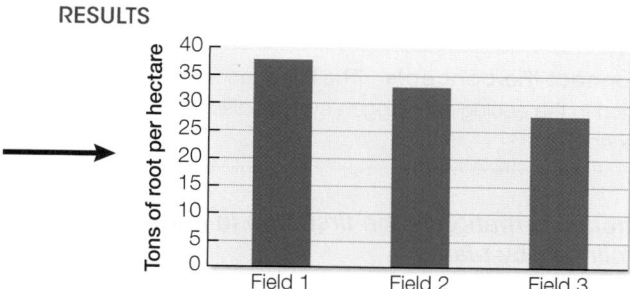

CONCLUSION▶ Adding purified mycorrhizal spores increases cassava production and reduces the need for phosphate fertilizer.

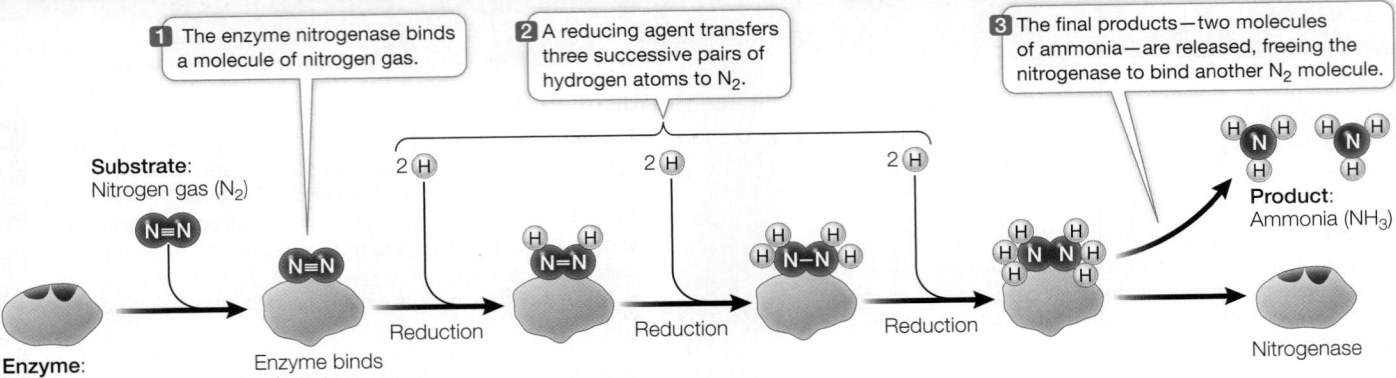

1 The enzyme nitrogenase binds a molecule of nitrogen gas.

2 A reducing agent transfers three successive pairs of hydrogen atoms to N_2.

3 The final products—two molecules of ammonia—are released, freeing the nitrogenase to bind another N_2 molecule.

Substrate: Nitrogen gas (N_2)

$N \equiv N$

Enzyme: Nitrogenase

Enzyme binds substrate

$N \equiv N$

Reduction

Reduction

Reduction

Product: Ammonia (NH_3)

Nitrogenase

Figure 35.8 Nitrogenase Fixes Nitrogen Throughout the chemical reactions of nitrogen fixation, the reactants are bound to the enzyme nitrogenase. A reducing agent transfers hydrogen atoms to nitrogen, and eventually the final product—ammonia—is released. This reaction requires a large input of energy: about 16 ATPs are consumed per reaction.

Q: Why does O_2 inhibit this reaction?

a plant-produced protein called **leghemoglobin**, which is an O_2 carrier. Leghemoglobin is a close relative of hemoglobin, the red, oxygen-carrying pigment of animals, and is thus an evolutionarily ancient molecule. Some plant nodules contain enough of it to be bright pink inside. Leghemoglobin, with its iron-containing heme groups, transports enough oxygen to the nitrogen-fixing bacteria to support their respiration, while keeping free oxygen concentrations low enough to protect nitrogenase.

Biological nitrogen fixation does not always meet agricultural needs

Crop rotation systems have been used for hundreds or thousands of years by many human civilizations. In these systems, each field is used to grow different crops in different years, with legumes (such as alfalfa, clover, peas, and beans) included in the rotation. The rotation may also include periods of grazing by farm animals. Because of their association with nitrogen-fixing bacteria, legumes can replace all or some of the nitrogen removed by grain crops such as wheat and corn. Even with these systems, however, bacterial *nitrogen fixation is not always sufficient to support the needs of agriculture. Some traditional farmers used to plant dead fish along with corn; the decaying fish released nitrogen that the developing corn could use. Today farmers use inorganic nitrogen fertilizers produced through industrial nitrogen fixation to meet the food needs of a rapidly expanding human population.

---

*connect the concepts The importance of nitrogen fixation in the ecological nitrogen cycle is discussed in Key Concept 57.4.

Ammonia formation is the first step in nitrogen assimilation by plants

Nitrogen fixation by free-living bacteria in the soil results in the formation of ammonia (NH_3), and most of it is rapidly ionized to form ammonium (NH_4^+). The latter can be taken up by plant roots and is then used as a source of amino groups ($-NH_2$) in the

formation of such molecules are amino acids and nucleotides. Soil bacteria called **nitrifiers** oxidize ammonia to nitrate ions (NO_3^-)—another form that plants can take up—by the process of **nitrification**. Soil pH affects which form of nitrogen is taken up by plants: nitrate ions are taken up preferentially under more basic conditions, and ammonium ions under more acidic ones. To use nitrate that has entered the roots, the plant must reduce it to ammonium in a process called **nitrate reduction**, which occurs in two enzyme-catalyzed steps. First, nitrate is converted to nitrite (NO_2^-) in the cytoplasm, and then nitrite is converted to ammonium in the plastids.

▶ 35.4 recap

Two mutualistic interactions with soil microbes are critical to the success of terrestrial plants. Fungi and plants form mycorrhizae, which greatly increase the soil volume that roots can scavenge for nutrients. Bacteria in soils and root nodules fix atmospheric nitrogen into forms that plants and ultimately animals can use.

---

learning outcomes

You should be able to:

- Describe the chemical signals exchanged between soil organisms and plant root cells.
- Justify the need for nitrogen-fixing organisms on Earth.
- Describe crop rotation and explain how it relates to plant nutrition.

---

1. What is exchanged between plants and fungi in mycorrhizae? Between leguminous plants and bacteria in nodules?

2. Earth's biosphere as we know it depends on the existence of a few species of nitrogen-fixing prokaryotes. What do you think might happen if one of these species were to become extinct? If all of them were to disappear?

3. A common cropping method is crop rotation, with corn or soybeans being rotated on a given plot of land. Explain this in terms of plant nutrition, especially available nitrogen in the soil.

Let's turn now to some special mechanisms for obtaining nutrients that have evolved in plant species with unusual lifestyles.

Most plants obtain their mineral nutrients from the soil solution. Carnivorous and parasitic plants are examples of plants that obtain nutrients from other sources.

focus your learning

- Some plants can obtain some of their nutrients by trapping and digesting insects.
- Chemical signaling between parasitic plants and their hosts is similar to chemical signaling observed in plant–bacteria and plant–fungus mutualistic relationships.

Carnivorous plants supplement their mineral nutrition

Some plants augment their nitrogen supply by capturing and digesting flies and other insects. There are about 500 of these **carnivorous plant** species, the best known of which are Venus flytraps (genus *Dionaea*) (**Figure 35.9A**), sundews (genus *Drosera*) (**Figure 35.9B**), and pitcher plants (genus *Sarracenia*).

Carnivorous plants are typically found in boggy habitats that are acidic and nutrient-deficient. To obtain extra nitrogen, these plants capture animals, digest their proteins, and absorb the amino acids. Pitcher plants have pitcher-shaped leaves that collect small amounts of rainwater. Insects and even small rodents are lured into the pitchers by bright colors or attractive scents and are prevented from leaving by stiff, downward-pointing hairs. The animals eventually die and are digested by a combination of plant enzymes and bacteria in the water. Sundews have leaves covered with hairs that secrete a sticky, sugary liquid. Insects become stuck to these hairs, and more hairs curve over to further entrap them. Enzymes secreted by the plant digest the insects. Venus flytraps have specialized leaves with two halves that fold together. When an insect touches special hairs called trigger hairs on the leaf, its two halves snap together. The spiny margins interlock and imprison the insect before it can escape. The leaf then secretes enzymes that digest its prey.

The closing of the Venus flytrap's leaf is one of the fastest movements in the plant world, requiring only 0.1 seconds. To find out how this happens, Dr. Lakshminarayanan Mahadevan and colleagues at Harvard University painted fluorescent dots on the surface of the flytrap's leaf surface and used high-speed cameras to record the trap snapping shut when its trigger hairs were touched. They then used computer analysis of the recorded dot movements to generate a mathematical model to help explain the movement. The researchers found that the first step is the osmosis-driven elongation of cells on the outer surface of the leaf. The expansion of only one side of the leaf causes it to snap from a convex into a concave shape, much like a contact lens flipping inside out.

Carnivorous plants photosynthesize and extract soil nutrients just like other plants, but eating insects helps them grow faster in their natural habitats. They use the additional nitrogen from the insects to make more proteins, chlorophyll, and other nitrogen-containing compounds.

Parasitic plants take advantage of other plants

Approximately 1 percent of flowering plant species derive some or all of their water, mineral nutrients, and sometimes even photosynthate from other plants. These **parasitic plants** have evolved absorptive organs called **haustoria**, which invade the host and tap into the vascular tissues in the root or stem.

Parasitic plants are divided into two broad classes based on their nutritional interactions with their hosts. **Hemiparasites** can still photosynthesize but derive water and mineral nutrients from the living bodies of other plants. Perhaps the most familiar hemiparasites are the several genera of mistletoes. Mistletoes are green and carry on some photosynthesis, but they parasitize other plants for water and mineral nutrients and may derive photosynthetic products from them as well. Dwarf mistletoe (*Arceuthobium americanum*) is a serious parasite in forests of the western United States, destroying more than 3 billion board feet of lumber per year.

Holoparasites are completely parasitic and do not perform photosynthesis. They are taxonomically and morphologically diverse. Some, such as members of the dodder family, are plantlike in appearance, with small leaf remnants and flowers (**Figure 35.10**). Some holoparasites do not have leaves or stems because they spend most of their life cycle underground and only break the surface to flower.

Several parasitic plant species lack many of the genes normally present in the chloroplast genome (which in turn is only a remnant of the genome of the original endosymbiont from which the chloroplast evolved; see Key Concepts 5.5 and 26.1). These genes, which are needed for photosynthesis, have been lost because there is no evolutionary pressure to retain them. Thus while the parasitic lifestyle can be viewed as a free ride, for some plants it is also a one-way ticket, with no possibility of return to self-sufficiency.

(A) *Dionaea muscipula*

(B) *Drosera scorpioides*

Figure 35.9 Carnivorous Plants   Some plants have adapted to nitrogen-poor environments by becoming carnivorous. **(A)** The Venus flytrap obtains nitrogen from the bodies of insects trapped inside the plant when its hinges snap shut. **(B)** Sundews trap insects on sticky hairs. Secreted enzymes will digest the carcass externally.

 Media Clip 35.1 **A Venus Flytrap "Snaps to It"**
www.Life11e.com/mc35.1

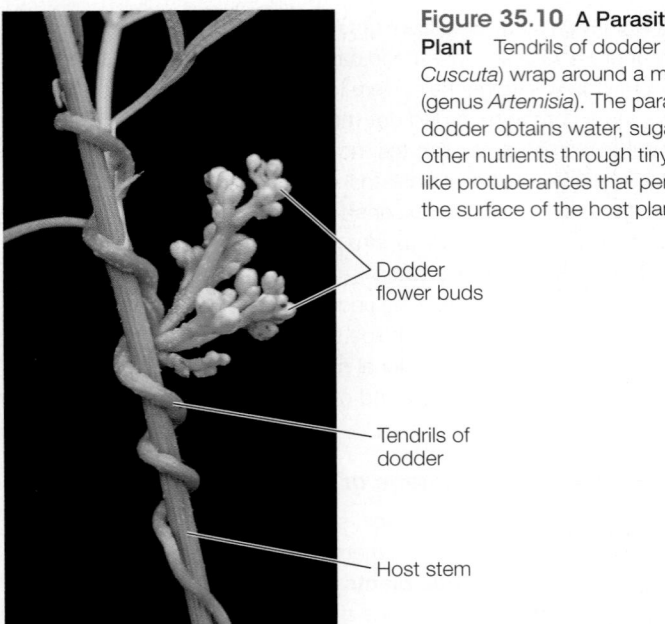

Figure 35.10 A Parasitic Plant Tendrils of dodder (genus *Cuscuta*) wrap around a mugwort (genus *Artemisia*). The parasitic dodder obtains water, sugars, and other nutrients through tiny, root-like protuberances that penetrate the surface of the host plant.

Dodder flower buds

Tendrils of dodder

Host stem

The plant–parasite relationship is similar to plant–fungus and plant–bacteria associations

Plant–bacteria and plant–fungus associations both involve recipro-cal signaling between the two species (see Figure 35.6). Parasitic plants also need to detect signals from nearby plants so they can grow toward them and obtain their nutrients, but obviously this is to the disadvantage of the potential host plant. In one interesting case, a parasitic plant has evolved the ability to recognize the chemical signals produced by plants to attract beneficial fungi.

The holoparasite *Striga* (witchweed) is a serious pest of cereal crops in Africa. Key Concept 35.4 described how arbuscular fungi are attracted to plant roots by compounds called strigolactones. One of these same molecules induces the seed germination of some parasitic plants, including *Striga*. Scientists strongly suspect that this is no coincidence. The mycorrhizal interaction is ancient (more than 400 million years old) and predates the evolution of parasitic plants. For this reason scientists hypothesize that a mechanism evolved in the ancestors of modern *Striga* to recognize a compound that was already produced by plants to attract soil microbes.

In *Striga* we thus find an example of "opportunistic evolution"—that is, the repurposing of preexisting processes rather than the develop-ment of new processes. This is not the first time you have encountered this phenomenon in this chapter. Recall that the formation of nodules by rhizobia uses some of the same mechanisms used by arbuscular fungi to establish residence inside plant cells (see Figure 35.6). This implies an evolutionary connection between the two symbioses.

35.5 recap

Carnivorous plants supplement their nutrition by extract-ing materials from animals. Rapid reflexes have evolved in some of these plants for trapping their prey. Parasitic plants, by contrast, get some or all of their sustenance from other plants. Holoparasites cannot function as auto-trophs, having lost chloroplast genes coding for photosyn-thetic machinery. At least one parasitic plant responds to the same signaling molecule that the host plant uses to attract beneficial fungi.

learning outcomes

You should be able to:

- Explain how carnivorous plants use nutrients from insects.
- Create a hypothesis to explain the evolution of a parasitic plant species.
- Analyze evidence that chemical signals involved in plant–bacteria and plant–fungi interactions represent an ancient process.

1. Describe the process by which carnivorous plants get nutrients from their prey. What is the primary nutrient they acquire?

2. Some mutant *Arabidopsis* plants that are very bushy (their shoots are more highly branched than wild-type plants) cannot make strigolactones because of a muta-tion in a gene necessary for strigolactone biosynthesis. If an investigator applies strigolactones to the plants, they grow normally. What does this experiment suggest about the role of strigolactones in plant growth? How does this add to the story of strigolactones as signals for arbuscules and parasitic plants?

3. Holoparasitic plants have lost many of the morphological and genetic traits necessary for an autotrophic lifestyle. From an evolutionary standpoint, how do you think this happened? (Hint: Think about selection pressures.)

investigatinglife

How can nitrogen use efficiency be improved?

We described in the opening of this chapter and in Investigating Life: Mycorrhizal Fungi Can Replace Fertilizer in Cassava Cultivation the possibility of improving plant nitrogen nutrition by manipulating the environment around the plant, in particular by adding fertilizers and mycorrhizae to the soil. Another approach is to improve the plant's ability to use this nutrient. Significant progress has been made in

using both conventional plant breeding (genetics) and biotechnol-ogy. The recent publication of the corn genome sequence will pro-vide useful information about genes involved in nitrogen use. For example, the seed company Pioneer Hi-Bred is developing a strain of corn that produces a more efficient version of the enzyme gluta-mine synthetase (GS). GS adds ammonia to glutamate to form glu-tamine, and therefore plays an important role in nitrogen assimilation in plants. The new corn strain uses up to 20 percent less nitrogen fertilizer to produce the same yields as other corn varieties.

Future directions

As we pointed out at the beginning of the chapter, the manufacture of nitrogen fertilizer by the Haber–Bosch process is expensive, both in terms of energy needed and economic costs. For too many farmers, nitrogen deficiency limits the growth of their crops. At Yale University, a team of chemists led by Patrick Holland is attempting to design a new compound that binds nitrogen from the atmosphere, mimicking the natural process of nitrogen fixation. In nature, nitrogen fixation is catalyzed by the enzyme nitrogenase. Nitrogenase has iron and sulfur atoms where catalysis takes place. Holland's group has made simple molecules in which N and S are bound, but not too tightly, so that atmospheric N_2 binds the complex. Eventually the group hopes to develop a synthetic catalyst to make ammonia on-site at farms, replacing industrial ammonia-generating processes that require high temperatures and pressures.

Chapter Summary 35

35.1 Plants Require Nutrients

- Plants are photosynthetic autotrophs that can produce all their organic molecules from carbon dioxide, water, and minerals, including a nitrogen source.
- **Mineral nutrients** are obtained from the **soil solution**.
- Plants require 14 **essential elements**, 6 of which are **macronutrients** and 8 of which are **micronutrients**. **Deficiency symptoms** suggest what essential element(s) a plant lacks. Review Figure 35.1, Table 35.1, Animation 35.1
- The essential elements were discovered by growing plants **hydroponically**, with their roots suspended in mineral solutions instead of soil.

35.2 Plants Acquire Nutrients from the Soil

- Root growth allows plants, which are sessile, to search for mineral resources.
- Plants can regulate the uptake of nutrients by increasing the number or activity of active transport proteins in root epidermal cells. Review Figure 35.2

35.3 Soil Structure Affects Plant Nutrition

- **Soils** contain water, air, and inorganic and organic substances. Soils have living (biotic) and nonliving (abiotic) components. Review Figure 35.3
- A soil typically consists of two or three horizontal zones called **horizons**. **Topsoil** forms the uppermost or **A horizon**. Topsoil tends to lose mineral nutrients through **leaching**. **Loams** are excellent agricultural topsoils, with a good balance of sand, silt, clay, and organic matter. Review Figure 35.4
- Soils form by mechanical and chemical **weathering** of rock. Chemical weathering imparts mineral nutrients to clay particles. Plant litter and other organic matter decompose to form **humus**. Plants obtain some mineral nutrients through **cation exchange** between the soil solution and the surface of clay particles. Review Figure 35.5
- Farmers use **fertilizers** to make up for deficiencies in soil mineral nutrient content.

35.4 Soil Organisms Increase Nutrient Uptake by Plant Roots

- In the earliest stages of mycorrhiza formation, the hyphae of arbuscular fungi grow toward **strigolactones**, compounds that are produced by the plant roots. Review Figure 35.6A

- Some **nitrogen fixers** live free in soil or water; others live symbiotically as **bacteroids** within plant roots. The formation of a **root nodule** requires interaction between the root system of a legume and rhizobia. Review Figure 35.6B
- Several steps in the formation of root nodules and arbuscules are similar and probably involve some of the same plant genes. Review Figure 35.7
- Mycorrhizae are symbiotic root–fungus associations that greatly increase a plant's absorption of water and minerals, especially phosphorus. They occur in more than 90 percent of terrestrial plant species. Review Investigating Life: Mycorrhizal Fungi Can Replace Fertilizer in Cassava Cultivation
- The arbuscules are the sites of nutrient exchange between the fungus and plant.
- In **nitrogen fixation**, nitrogen gas (N_2) is reduced to ammonia (NH_3) or ammonium ions (NH_4^+) in a reaction catalyzed by **nitrogenase**. Review Figure 35.8, Activity 35.1

35.5 Carnivorous and Parasitic Plants Obtain Nutrients in Unique Ways

- **Carnivorous plants** are autotrophs that supplement a low nitrogen supply by feeding on insects or other small animals.
- **Parasitic plants** draw on other plants to meet their needs, which may include minerals, water, or the products of photosynthesis.
- **Hemiparasites**, such as mistletoes, can still photosynthesize. **Holoparasites** cannot function as autotrophs because they have lost chloroplast genes that code for components of the photosynthetic apparatus (which they no longer need).
- A strigolactone—a compound in the same category of compounds plants use to attract mycorrhizal fungi—also induces the germination of some parasitic plants, including *Striga*. Scientists hypothesize that a mechanism evolved in the ancestors of modern *Striga* to recognize a compound that was already produced by plants to attract arbuscular fungi.

Go to Activity 35.1 for a concept review of this chapter.

Go to **LearningCurve** (in **LaunchPad**) for dynamic quizzing that helps you solidify your understanding of this chapter. **LearningCurve** adapts to your responses, giving you the practice you need to master each key concept.

Apply What You've Learned

Review

35.1 Plants receiving insufficient quantities of essential elements can be evaluated by observing symptomatic changes in physical features and/or growth.

35.1 Experiments using hydroponic growth conditions have allowed scientists to determine the essential elements plants need for growth.

Original Paper: Brown, P. H., R. M. Welch and E. E. Cary. 1987. Nickel: A micronutrient essential for higher plants. *Plant Physiology* 85: 801–803.

Both barley and wheat are cereal crops used for feeding humans and livestock, and the two have many similarities. Because of its characteristics when fermented, however, barley is favored over wheat for beer making. When you consider how much beer is brewed around the world, it's no wonder that researchers have spent considerable effort looking for ways to optimize barley growth.

Optimizing growth of any crop can be accomplished if the crop's nutritional requirements are known. In the case of barley, one research group investigated the element nickel as a possible essential micronutrient. Earlier studies where a single generation of plants was grown without Ni indicated that it was not a mineral requirement. They began by creating a stock growth solution containing extremely low amounts of nickel ions (less than 30 nanograms per liter [ng/L]). Then they divided this stock solution into three containers and added nickel sulfate ($NiSO_4$) to two of these. This step produced supplemented concentrations of 0, 0.6, and 1.0 micromolar (μM) nickel sulfate in the three final solutions. These were then used as experimental growth solutions to grow three groups of barley plants over three generations under different nickel concentrations.

The researchers harvested seeds from third-generation plants in each group. They measured the total mass and number of seeds produced by each plant. Results are shown in the table below.

Next, one portion of the seeds from each group was analyzed for nickel content. The other portion from each group was placed in nickel-free growth medium and tested for germination. The graph below shows seed germination success plotted against nickel content of the same seeds.

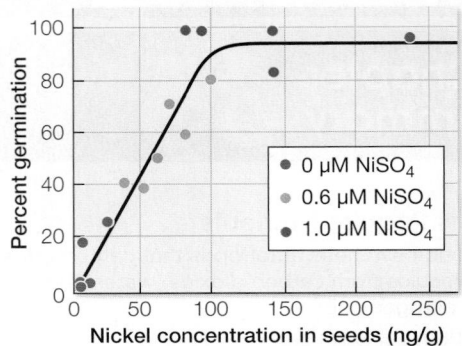

Questions

1. The researchers concluded that nickel is an essential micronutrient of barley. Do you agree with their conclusion? Explain your answer.

2. Analyze the data about the mass and number of seeds produced by the third-generation barley plants. What conclusion can you draw from these data? Is this consistent with your answer to Question 1 above? Explain.

3. Why didn't the researchers use soil to grow the barley plants? Will the experiments have to be repeated using soil as a growth medium, or can the results of this experiment be used to make predictions about plants grown in soil as well as in growth solutions?

Plant group	Supplemented nickel concentration in growth solution (μM)	Total seed mass per plant (grams dry weight)	Total number of seeds harvested per plant
1	0	7.3	175
2	0.6	7.5	179
3	1.0	8.4	195

Go to **LaunchPad** for the eBook, LearningCurve, animations, activities, flashcards, and additional resources and assignments.

36

Regulation of Plant Growth

The root, stem, and leaf develop from a plant embryo within a seed.

▶ investigating**life**

A Nobel Prize for a Plant Biologist

In their constant search for ways to help farmers produce more food for a growing population, biologists have developed cereal crops whose physiology allows them to produce more grain per plant (resulting in higher yields). The drawback of this approach is that the sheer weight of the load of seeds may cause the stem to bend over. The problem is made worse when fertilizer added to the soil stimulates plants to grow taller. Harvesting seeds on the ground is very difficult; think of how hard it would be to pick up seeds one by one, when some have already sprouted.

In 1945 the U.S. Army temporarily occupied Japan, which was defeated in World War II. During the war, Japan, an island nation with a limited amount of land suitable for farming, was blockaded and could not import food or other supplies. Food was rationed and many people were hungry, but there were no major famines in Japan during that period. How were the Japanese able to produce enough grain to feed their population? The answer to this question lay in the fields: the Japanese had bred genetic strains of rice and wheat with short, strong stems, growing in thickness rather than

height, that could bear high yields of grain without bending over. An agricultural advisor to the occupying American army sent samples of the grains to the United States.

A decade later, the American plant geneticist Norman Borlaug, who was working in Mexico at the time, began making genetic crosses between the Japanese wheat and other varieties that had genes conferring rapid growth, adaptability to varying climates, and resistance to fungal diseases. The results were "semi-dwarf" wheat varieties that gave record yields. The varieties were grown first in Mexico, and later in India and Pakistan during the 1960s. At about the same time and using a similar strategy, scientists in the Philippines developed semi-dwarf rice with equally spectacular results. People who had lived on the edge of starvation now produced enough food. Countries that had relied on food from other countries were now able to grow more than enough grain, and export the surplus. The development of these semi-dwarf grains began what was called the "Green Revolution." Borlaug was awarded the Nobel Peace Prize for his research on wheat, which is estimated to have saved a billion lives.

QA How is plant growth controlled, and what changes in growth patterns made the new strains of wheat and rice successful?

36.1   Plants Develop in Response to the Environment

As you learned in in Chapter 33, plants can't move to get food and so must seek out resources above and below the ground. A number of unique features enable plants to obtain the resources they need to grow and reproduce:

- *Meristems*. Plants have permanent collections of stem cells (undifferentiated, constantly dividing cells) that allow them to continue growing throughout their lifetimes.

- *Post-embryonic organ formation*. Unlike animals, plants can initiate development of new organs such as leaves and flowers throughout their lifetimes.

- *Differential growth*. Plants allocate resources to grow beneficial organs. For example, leaf growth allows plants to capture more sunlight or more roots permits greater water and nutrient uptake.

focus your learning

- Plants use several mechanisms to maintain dormancy; these mechanisms must be overcome before a seed can germinate.
- Seed germination begins with imbibition and ends with the emergence of the radicle, or embryonic root, in a process that requires plants to sense environmental changes.
- Plants respond to environmental cues during growth by the use of signal transduction pathways involving hormones and photoreceptors.
- Scientists use genetic screens to identify genes involved in the control of growth and development.

Plants must continuously monitor their ever-changing environments and redirect their growth appropriately. For example, the amount of available light changes from day to night and from season to season. In addition, other plants are often vying for the available light, and plants modulate their growth to compete with their neighbors for this precious resource. As you will see in this chapter, several mechanisms have evolved in plants that enable them to *sense changes in their environments and trigger appropriate growth responses.

> **connect the concepts**   This chapter is devoted to how plants regulate their growth in response to environmental signals. Chapter 38 is dedicated to how plants respond to environmental challenges such as pathogens, herbivory, and physical stresses.

Plant growth is regulated

The development of a plant—the series of progressive changes that take place throughout its life—is regulated in many ways. Key factors involved in regulating plant growth and development are:

- *Environmental cues*, such as day length, water availability, and various chemicals in the environment

- *Receptors* that allow a plant to sense environmental cues, such as photoreceptors that absorb light, and chemoreceptors that signal the presence of pathogens

- *Hormones*—chemical signals that mediate the effects of the environmental cues, including those sensed by receptors

- *Regulatory proteins and enzymes* that catalyze the biochemical reactions that control development

We will explore these *regulatory mechanisms in more detail later in this chapter. But first let's look at the initial steps of plant development—from seed to seedling—and the types of internal and external cues that guide them.

> **connect the concepts**   Much of growth regulation occurs via signal transduction pathways. Review these mechanisms, mostly from animal cells, in Chapter 7.

In early development, the seed germinates and forms a growing seedling

Chapter 37 will describe the events of plant reproduction and development that lead to the formation of seeds. Here we begin with the seed, the structure that contains the early embryo. Unlike most animal embryos, plant seeds may be held in "suspended animation," with the development of the embryo halted, for long periods. If development stops even when external conditions (such as water supply) are adequate for development, the seed is said to be **dormant**.

DORMANCY   Seed dormancy may last for weeks, months, or years. Plants use several mechanisms to maintain dormancy:

- *Exclusion of water or oxygen* from the embryo by an impermeable seed coat

- *Mechanical restraint* of the embryo by a tough seed coat

- *Chemical inhibition* of germination by growth regulators

- *Photodormancy*: some seeds need a period of light or dark before they can germinate

- *Thermodormancy*: some seeds need high or low temperatures to germinate

Dormancy can be broken by conditions that overcome these mechanisms. For example, the seed coat may be damaged by passage through an animal's digestive system, or heavy rains may wash away chemical inhibitors. There are some unusual methods to overcome dormancy. One example is the breaking of dormancy by components of smoke. *Emmenanthe penduliflora* is a common plant in dry chaparral of the southwestern United States, an area that is prone to wildfires.

Emmenanthe penduliflora seeds germinate after exposure to smoke.

These plants germinate rapidly after a fire. John Keeley of Occidental College in Los Angeles found that dormancy in seeds of this plant is broken not by heat but by smoke—in particular, by the nitrogen oxides found in smoke. Other molecules in smoke have been identified that regulate seed germination.

Plant biologists distinguish between seed *dormancy*, which prevents germination under conditions that are suitable for plant growth, and seed *quiescence*, which occurs when a seed fails to germinate because conditions are unfavorable for growth. Some seeds may remain quiescent, yet viable, for centuries; botanists have germinated a 1,300-year-old lotus seed recovered from a dry lake bed in China.

Seed dormancy and quiescence are common, so they must provide selective advantages for plants. Dormancy ensures that the seed will germinate at a time suitable for the plant to complete its life cycle. For example, some seeds require exposure to a long cold period (winter) before they germinate in the spring; this ensures that the plant has the entire growing season to mature and set new seeds. Dormancy and quiescence also help seeds survive droughts or long-distance dispersal, allowing plants to colonize new territory.

GERMINATION   Seeds begin to **germinate**, or sprout, when dormancy is broken and environmental conditions are satisfactory. The first step in germination is the uptake of water, called **imbibition** (from *imbibe*, "to drink in"). Before germination, a seed contains very little water: only 5–15 percent of its weight is water, compared with 80–95 percent in most other plant parts. Seeds also contain polar macromolecules, such as cellulose and starch, which attract and bind water molecules. Consequently a seed has a very negative water potential (see Chapter 34) and will take up water if the seed coat is permeable to water. The force exerted by imbibing seeds, which expand several-fold in volume, demonstrates the magnitude of their water potential; for example, imbibing cocklebur seeds can exert a pressure of up to 1,000 atmospheres.

As a seed takes up water, it undergoes metabolic changes: enzymes are activated upon hydration, RNA and then proteins are synthesized, the rate of cellular respiration increases, and other metabolic pathways are activated. In many seeds, cell division is not initiated during the early stages of germination. Instead, growth results solely from the expansion of small, preformed cells.

As germination proceeds, starch, proteins, and lipids that are stored in the seed are hydrolyzed to provide metabolic energy and chemical building

Figure 36.1 Patterns of Early Shoot Development **(A)** In grasses and some other monocots, growing shoots are protected by a coleoptile until they reach the soil surface. **(B)** In most eudicots, the growing point of the shoot is protected within the cotyledons.

blocks—carbohydrates, amino acids, and lipid monomers—for the growing embryo. These reserves are stored in the **cotyledons** (the first leaf or leaves of the embryo) or in the endosperm (the non-embryonic storage tissue of the seed). Germination is completed when the **radicle** (embryonic root) emerges from the seed coat. The plant is then called a **seedling**.

If the seed germinates underground, the new seedling must elongate rapidly (in the right direction!) and cope with a period of life in darkness or dim light. Photoreceptors that sense light and specialized cells that sense gravity direct this stage of development and prepare the seedling for growth in the light.

The pattern of early shoot development varies among the flowering plants. **Figure 36.1** shows the shoot development patterns of monocots and eudicots. In monocots, the growing shoot is protected by a sheath of cells called the **coleoptile** as it pushes its way through the soil. In eudicots, the shoot is protected by the cotyledons.

Several hormones and photoreceptors help regulate plant growth

The above description of the early stages of plant development illustrates some of the environmental cues that influence plant growth. A plant's responses to these cues involve signal transduction pathways. Various mechanisms are used by the plant to sense changes in the environment, and these mechanisms activate signal transduction pathways that result in the synthesis and activation of specific plant hormones. In turn, these hormones act as signals that trigger pathways resulting in changes in plant growth. In most cases these changes involve alterations in the expression of specific genes.

Hormones are chemical signals that act at very low concentrations at target sites often distant from where they are produced. Most plant hormones are very different from their counterparts in animals (**Table 36.1**). Each plant hormone plays multiple regulatory roles, and interactions among them can be complex. Several hormones regulate the growth and development of plants from seedling to adult

(A) Monocot shoot development (corn)

1 A **coleoptile** (a cylindrical sheath of cells) protects the early shoot as it grows to the soil surface.

First foliage leaf

Coleoptile

Coleoptile

2 After the shoot emerges from the soil, it continues to elongate, and the leaves emerge.

Primary root

(B) Eudicot shoot development (bean)

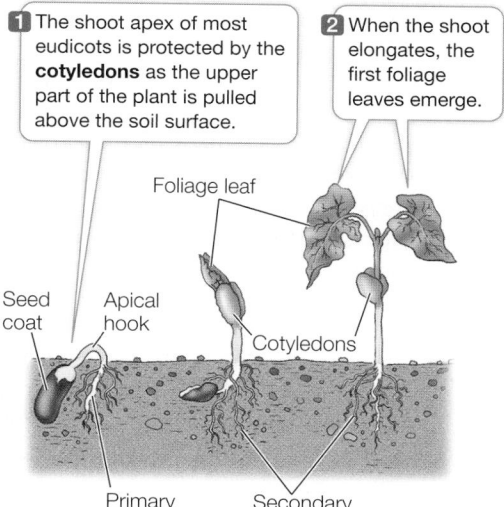

1 The shoot apex of most eudicots is protected by the **cotyledons** as the upper part of the plant is pulled above the soil surface.

2 When the shoot elongates, the first foliage leaves emerge.

Foliage leaf

Seed coat

Apical hook

Cotyledons

Primary root

Secondary roots

Activity 36.1 **Monocot Shoot Development**
www.Life11e.com/ac36.1

Activity 36.2 **Eudicot Shoot Development**
www.Life11e.com/ac36.2

(Table 36.2). Other hormones (for example, jasmonic acid and salicylic acid) are involved in the plant's defenses against herbivores and microorganisms, as we will discuss in Chapter 38.

Perhaps the most important environmental cue for a plant is light: the source of energy for photosynthesis. Plants have an abundance of **photoreceptors** that detect changes in the quality and direction of light as well as the timing of light availability (day length). Photoreceptors are often proteins associated with pigments. Light acts directly on photoreceptors, which in turn regulate developmental processes that need to be responsive to light, such as the many changes that occur as a seedling emerges from the soil.

Genetic screens have increased our understanding of plant signal transduction

In Chapter 19 we described how genetic studies can be used to identify the steps along a developmental pathway. Recall the reasoning behind these experiments:

- If a mutation in a specific gene disrupts a developmental process, then the product of that gene must be involved in that process.

In a similar way, genetic studies can be used to analyze pathways of receptor activation and signal transduction in plants:

- If proper signaling does not occur in a mutant plant, then the mutant gene must be involved in the signal transduction process.

Mapping the mutant gene and identifying its function is a starting point for understanding the signaling pathway. *Arabidopsis thaliana* has been a major model organism for plant biologists investigating signal transduction.

One technique for identifying the genes involved in a plant signal transduction pathway is illustrated in **Figure 36.2**. This technique, called a **genetic screen**, involves creating a large, random collection of mutant plants and identifying those individuals that are likely to have a defect in the pathway of interest. Plant genes can be randomly mutated in a variety of ways, including treatment with a chemical mutagen or the insertion of transposons (see Key Concept 17.2) randomly in the genome. After treatment, the plants are grown and then examined for a specific phenotype, usually a characteristic that is influenced by the pathway of interest. Once mutant plants have been selected, their genotypes are compared with those of wild-type plants. *Arabidopsis* mutants with altered developmental patterns have provided a wealth of new information about the mechanisms of hormone and receptor (particularly photoreceptor) action.

table 36.1 Comparison of Plant and Animal Hormones

Characteristic	Plant hormones	Animal hormones
Size, chemistry	Peptides, small organic molecules	Peptides, proteins, small molecules
Site of synthesis	Usually at many locations	Specialized glands or cells
Site of action	Local or distant	Usually distant, transported
Effects	Often diverse	Often specific
Regulation	By biochemical feedback	By central nervous system, ions, or feedback

research tools

Figure 36.2 A Genetic Screen   Genetics of the model plant *Arabidopsis thaliana* can be used to identify the steps of a signal transduction pathway. If a mutant strain does not respond to a hormone (in this case, ethylene), the corresponding wild-type gene must be essential for the pathway (in this case, ethylene response). This method has been instrumental to scientists in understanding plant growth regulation.

1 A phenotype of interest is chosen, in this case response to ethylene. Growth of wild-type seedlings in the dark is altered by ethylene (right).

No ethylene added   Ethylene added

2 Seeds are treated with a mutagen and then planted.

Many seeds suspended in a mutagen solution in a test tube

3 The treated plants are grown in the dark and then screened. One mutant shows insensitivity to the hormone (uninhibited growth).

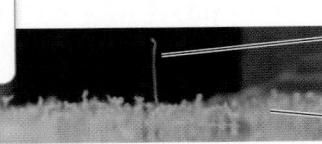

Ethylene-insensitive mutant

Ethylene-sensitive (wild-type) plants

4 The gene involved in ethylene signaling is mapped, cloned, and studied.

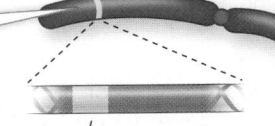

Gene required for ethylene response

Q: In undertaking a search for natural mutant strains for ethylene response, how would you modify the procedure?

table 36.2 Plant Growth Hormones

Hormone	Structure	Typical activities
Abscisic acid*		Maintains seed dormancy; closes stomata
Auxins (mainly indole-3-acetic acid)		Promote stem elongation, adventitious root initiation, and fruit growth; inhibit axillary bud outgrowth, leaf abscission, and root elongation
Brassinosteroids		Promote stem and pollen tube elongation; promote vascular tissue differentiation
Cytokinins		Inhibit leaf senescence; promote cell division and axillary bud outgrowth; affect root growth
Ethylene		Promotes fruit ripening, leaf abscission, and senescence; inhibits stem elongation and gravitropism
Gibberellins (e.g., gibberellic acid)		Promote seed germination, stem growth, and ovule and fruit development; break winter dormancy; mobilize nutrient reserves in grass seeds

*See Chapter 37.

36.1 recap

Early embryonic development in many plants occurs within seeds, which may remain dormant for a long period. Water uptake (imbibition) and germination occur when external cues in the environment are present. As in animals, development in plants is under the control of signals and is mediated by hormones. These signals activate pathways that may result in changes in gene expression. Genetic screens have been useful in describing signal transduction pathways in the model plant *Arabidopsis thaliana*.

learning outcomes

You should be able to:

- Evaluate the importance of seed dormancy, and contrast it (in importance and process) with seed quiescence.

- List the ways a seed can maintain dormancy, and describe with examples how each method might be overcome to achieve germination.

- Explore the physical and metabolic changes a seed must undergo during the early stages of germination.

- Evaluate the importance of photoreceptors in plant development and growth, and compare the action of photoreceptors with that of hormones.

- Draw a simple diagram to illustrate how a genetic screen can be used to identify the genes involved in the development of plant signal transduction pathways.

1. Under what conditions would it be advantageous for a species to have seed dormancy broken by fire?

2. What are the similarities and differences between hormone receptors and photoreceptors?

3. How would a genetic screen be used to identify a plant gene involved in seed dormancy?

You have now seen the early stages of plant development and growth, and how the environment influences these processes. We will now turn to the subject of plant hormones, which are central to the internal regulation of development. We will describe how hormones were discovered and what physiological effects they have on plants. We will emphasize how genetic screens and other methods have led to a deeper molecular understanding of the action of plant hormones.

key concept

36.2 Gibberellins and Auxin Have Diverse Effects but a Similar Mechanism of Action

The discovery of two key plant hormones exemplifies the experimental approaches that plant biologists have used to investigate the mechanisms of plant development.

focus your learning

- Gibberellins have multiple roles in plant growth and development.
- The first discovered function of auxin related to its role in phototropism.
- Polar transport of auxin distributes the hormone down the shoot; lateral transport distributes it from side to side.

Early in the twentieth century, two plant hormones were discovered: **gibberellins** (there are several active forms) and **auxin** (there are several forms, but one predominates). Initially, the discoveries came from observations of natural phenomena:

- *Gibberellins*: In rice plants, a disease caused by the fungus *Gibberella fujikori* resulted in plants that grew overly tall and spindly.

- *Auxin*: Biologists and indoor gardeners noted that seedlings would bend toward the light when placed near a light source.

Figure 36.3 Hormones Reverse a Mutant Phenotype
(A) The two mutant dwarf tomato plants in this photograph were the same size when the one on the right was treated with a gibberellin solution. **(B)** The short phenotype of this *Arabidopsis* mutant was reversed in the plant on the right by supplying auxin.

(A) Treatment with gibberellin

Twenty-two days after being sprayed with a dilute gibberellin solution, this plant reached the size of a nondwarf plant.

This untreated mutant plant did not make gibberellin and remained a dwarf.

(B) Treatment with auxin

Supplying auxin to this mutant plant made it grow.

This untreated mutant plant does not make auxin.

A chemical substance was then isolated that could cause each phenomenon:

- Gibberellic acid (see Table 36.2) made by the *G. fujikori* fungus caused rice plants to overgrow. Later it was found that plants make gibberellic acid as well, and that applying it to plants caused growth.

- Indole-3-acetic acid (see Table 36.2) applied asymmetrically to the growing tips of seedlings caused cell elongation on the side away from the light, which resulted in the shoot bending toward the light.

Finally, mutant plants that do not make each hormone exhibit a phenotype expected in the absence of the hormone, and adding the hormone reverses that phenotype (**Figure 36.3**):

- Tomato plants that do not make gibberellic acid are very short; supplying them with the hormone results in normal growth.

- *Arabidopsis thaliana* individuals that do not make auxin are also short; supplying them with that hormone reverses that phenotype.

Note that the phenotype involved—short stature, or dwarfism—is similar in both mutant plants even though different hormones are involved. This observation exemplifies a concept that is important to keep in mind when studying plant hormones: their *actions are not unique and specific*, as is the case with animal hormones (see Table 36.1).

The approaches outlined above—observation, hormone isolation, hormone treatment, and analysis of mutant plants—are just some of the methods used to identify plant hormones and understand their roles in plant development. Plant biologists have also studied hormones using chemical inhibitors and using plant transformation experiments that alter hormone levels or the plants' responses to hormones.

Gibberellins have many effects on plant growth and development

The functions of gibberellins can be inferred from the effects of experimentally decreasing gibberellins or blocking their action at various points in plant development. Such experiments reveal that gibberellins have multiple roles in regulating plant growth.

STEM ELONGATION   The effects of gibberellins on wild-type plants are not as dramatic as those seen on dwarf plants. However, gibberellins are indeed active in wild-type plants, because inhibitors of gibberellin synthesis cause a reduction in stem elongation. Such inhibitors can be put to practical uses. For example, plants such as chrysanthemums that are grown in greenhouses tend to get tall, but leggy plants do not appeal to consumers. Flower growers spray such plants with gibberellin synthesis inhibitors to control their height. Some wheat crops are similarly sprayed to keep them short, so they do not fall over when they produce grain; the result is chemically produced semi-dwarfs similar to the genetically produced varieties described in the opening story of this chapter. In some plants, such as cabbage, the normal growth habit is to be a squat, leafy head near the ground. When environmental signals are right, however, the plant "bolts," quickly producing a tall stem with flowers—a response that can be mediated by gibberellins.

FRUIT GROWTH   Gibberellins and other hormones regulate the growth of fruits. Grapevines that produce seedless grapes develop smaller fruits than varieties that produce seed-bearing grapes. Biologists wanting to explain this phenomenon removed seeds from immature seeded grapes and found that this prevented normal fruit growth, suggesting that the seeds are sources of a growth regulator. Biochemical studies showed that developing seeds produce gibberellins, which diffuse into the immature fruit tissue. Spraying young seedless grapes with a gibberellin solution causes them to grow as large as seeded ones, and this is now a standard commercial practice (**Figure 36.4**).

MOBILIZATION OF SEED RESERVES   Early in seed germination, hydrolytic enzymes are produced to break down stored reserves of starch, proteins, and lipids. Just after imbibition in germinating seeds of barley and other cereals, the embryo secretes gibberellins. The hormones diffuse through the endosperm to a surrounding tissue called the **aleurone layer**, which lies underneath the seed coat. The gibberellins trigger a cascade of events in the aleurone layer, causing it to synthesize and secrete enzymes that hydrolyze proteins and starch stored in the endosperm (**Figure 36.5**). These observations have practical importance: in the beer brewing industry, gibberellins are used to enhance the "malting" (germination) of

Figure 36.4 Gibberellins and Fruit Growth Spraying developing seedless grapes with gibberellins (right) increases their size compared with untreated fruit (left).

barley and the breakdown of its endosperm, producing sugar that is fermented to alcohol.

Another hormone, abscisic acid (ABA) (see Table 36.2) plays an antagonistic role with gibberellins in seed dormancy and germination. As we will describe in Key Concept 37.1, ABA levels are high in dormant seeds and fall off during germination. Thus ABA plays a role in maintaining seed dormancy, whereas gibberellins function to break dormancy and promote germination.

Auxin plays a role in differential plant growth

Auxin was first discovered for its role in the bending of plants toward light. Subsequent research has shown that auxin is involved in many other aspects of plant growth and development.

IDENTIFYING AUXIN AND ITS TRANSPORT Auxin (from the Latin "to increase") was discovered in the context of **phototropism**: a

response to light in which plant stems bend toward a light source. This was a familiar observation by biologists and home gardeners when the ever-curious Charles Darwin and his son Francis investigated this phenomenon and concluded that a signal was made in the shoot apex and diffused down the shoot in a polar (unidirectional) fashion, stimulating cell elongation.

 Animation 36.1 Tropisms
www.Life11e.com/a36.1

The Darwins worked with canarygrass (*Phalaris canariensis*) seedlings grown in the dark. While underground, a young grass seedling is protected by a coleoptile (see Figure 36.1). The coleoptiles of grasses are phototropic—they grow toward the light.

Light

To find the light-receptive region of the coleoptile, the Darwins "blindfolded" the coleoptiles of dark-grown canarygrass seedlings in various places and then illuminated them from one side (**Investigating Life: The Darwins' Phototropism Experiment**). The coleoptile grew toward the light whenever its tip was exposed. If the top millimeter or more of the coleoptile was covered, however, the coleoptile showed no phototropic response. The Darwins concluded that the

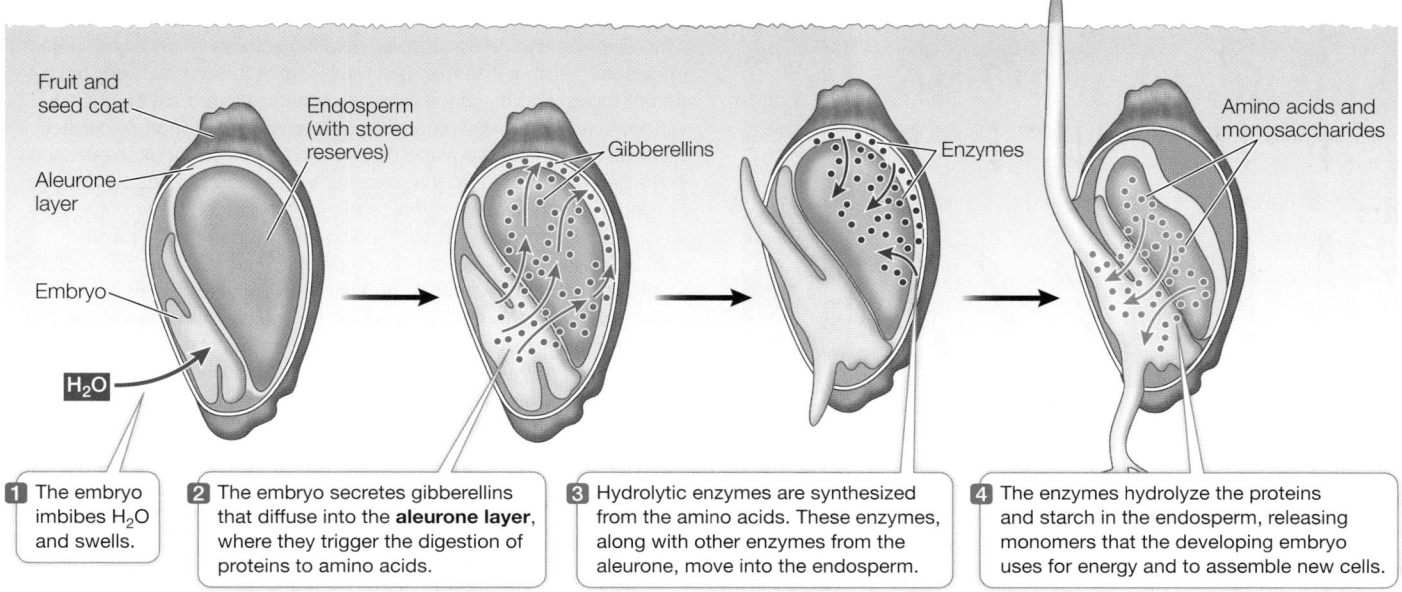

Fruit and seed coat

Endosperm (with stored reserves)

Aleurone layer

Embryo

H_2O

Gibberellins

Enzymes

Amino acids and monosaccharides

1 The embryo imbibes H_2O and swells.

2 The embryo secretes gibberellins that diffuse into the **aleurone layer**, where they trigger the digestion of proteins to amino acids.

3 Hydrolytic enzymes are synthesized from the amino acids. These enzymes, along with other enzymes from the aleurone, move into the endosperm.

4 The enzymes hydrolyze the proteins and starch in the endosperm, releasing monomers that the developing embryo uses for energy and to assemble new cells.

Figure 36.5 Embryos Mobilize Their Reserves During seed germination in cereal grasses, gibberellins trigger a cascade of events that result in the conversion of starch and protein reserves into monomers that can be used by the developing embryo.

Q: In brewing beer, stored starch in barley seeds is converted to maltose. How do you think this is done in the brewery?

 Activity 36.3 Events of Seed Germination
www.Life11e.com/ac36.3

experiment

Original Paper: Darwin, C. R. assisted by F. Darwin. 1880. *The Power of Movement in Plants*. Chapter IX: Sensitiveness of plants to light: Its transmitted effects. London, John Murray.

Charles Darwin and his son Francis wanted to know how plants bend toward the light. They grew canarygrass seedlings (coleoptiles) in the dark. To discover what part of the coleoptile responds to light, they covered up ("blindfolded") different regions of each coleoptile and then exposed the seedlings to light from one side. The Darwins discovered that the tip of the seedling senses the light and that growth occurs below the tip. Their observations led them to hypothesize the existence of a growth-promoting signal produced by the coleoptile tip.

HYPOTHESIS▶   Only part of the coleoptile senses the light that triggers phototropism.

METHOD

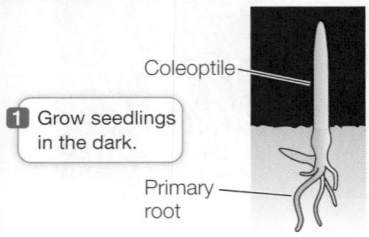

Coleoptile

1 Grow seedlings in the dark.

Primary root

2 "Blindfold" the seedlings in different places and expose to light on one side.

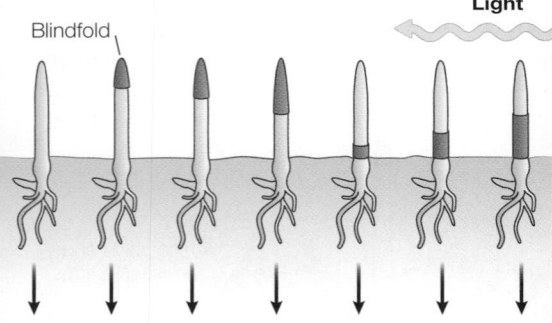

Blindfold

Light

RESULTS

3 Coleoptiles responded to light only when the tip was exposed.

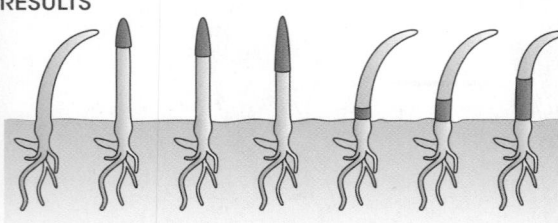

CONCLUSION▶   The coleoptile tip senses light and then transmits a growth signal to the tissue away from the light source.

work with the data

Fascinated by experiments showing that light induces plants to bend toward it and that this is due to increased growth on the "dark side" of the stem, Charles Darwin and his son Francis performed additional experiments on canarygrass (*Phalaris canariensis*) seedlings grown in the dark to learn more about how the seed germinates in the soil. Their conclusion that the light-sensing mechanism might reside in the coleoptile was validated when they repeated their experiments with the coleoptile tip cut off: bending did not occur. Because it was possible that the cut plants did not bend because they had been irreversibly damaged in some way, the Darwins repeated their initial experiment with undamaged seedlings whose coleoptile tips had been covered with a thin glass tube cap that was blacked out with India ink. In this case the plants did not bend. The Darwins' conclusion that the tip of the coleoptile senses the light and transmits a message to tissue below the tip to grow was initially controversial but ultimately led to the identification of auxin, the first plant growth hormone.

QUESTIONS▶

1. The figure shows a drawing by the Darwins of the bending of coleoptiles after 8 hours of light exposure. From which direction was the light shone?

Fig. 181.

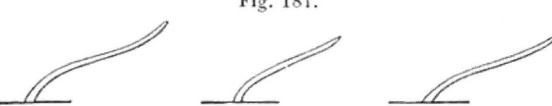

2. The Darwins reported:

   "Seven cotyledons [Note: The Darwins used the term "cotyledon" for what is now called a coleoptile.] had their tips cut off for lengths varying between 0.1 and 0.16 of an inch, and these, when left exposed all day to a lateral light, remained upright. In another set of 7 cotyledons, the tips were cut off for a length of only about 0.05 of an inch (1.27 mm) and these became bowed towards a lateral light, but not nearly so much as the many other seedlings in the same pots."

   What do these data indicate about the possible role of the tip and about the possibility that injury in cutting blocks the bending response?

3. The Darwins described their further experiments:

   "The summits of nine cotyledons, differing somewhat in height, were enclosed for rather less than half their lengths in uncoloured or transparent tubes; and these were then exposed before a south-west window on a bright day for 8 h. All of them became strongly curved towards the light, in the same degree as the many other free seedlings in the same pots; so that the glass-tubes certainly did not prevent the cotyledons from bending towards the light. Nineteen other cotyledons were, at the same time, similarly enclosed in tubes thickly painted with Indian ink. On five of them, the paint, to our surprise, contracted after exposure to the sunlight, and very narrow cracks were formed, through which a little light entered; and these five cases were rejected. Of the remaining 14 cotyledons, the lower halves of which had been fully exposed to the light for the whole time, 7 continued quite straight and upright; 1 was considerably bowed to the light, and 6 were slightly bowed, but with the exposed bases of most of them almost or quite straight."

   What do these data indicate about the role of the tip? Can you explain why there was slight bending in six of the coleoptiles that were covered with painted tubes?

A similar **work with the data** exercise may be assigned in **LaunchPad**.

tip contains the photoreceptor that responds to light. The actual bending toward the light, however, takes place in a growing region a few millimeters below the tip. Therefore, the Darwins reasoned, *some type of signal must travel from the tip of the coleoptile to the growing region.*

Others showed that placing the coleoptile tip (the source of the growth signal) on a decapitated coleoptile led to bending, even when a block of gelatin separated the tip and coleoptile.

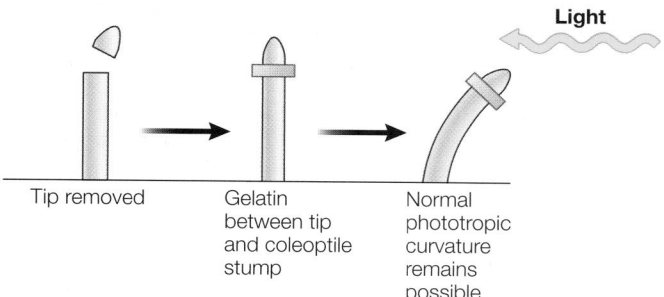

Light

| Tip removed | Gelatin between tip and coleoptile stump | Normal phototropic curvature remains possible |

In these experiments, the soluble growth signal moved from the coleoptile tip into the gelatin and then down into the decapitated coleoptile; later the growth signal was isolated from such gelatin blocks and identified as indole-3-acetic acid (see Table 36.2), the major form of auxin.

 Animation 36.2 **Went's Experiment**
www.Life11e.com/a36.2

MECHANISM OF AUXIN TRANSPORT   The movement of auxin down a coleoptile is an example of polar (apical-to-basal) transport. Polar transport of auxin (**Figure 36.6**) depends on four biochemical processes that we hope you recognize from earlier chapters:

1. *Diffusion across a cell membrane.* Polar molecules (in the chemical sense) diffuse across cell membranes less readily than nonpolar molecules (see Key Concept 6.3).

2. *Membrane protein asymmetry.* Active transport carriers (see Key Concept 6.4) for auxin are located only in the portion of the cell membrane at the basal (bottom) end of the cell.

3. *Proton pumping/chemiosmosis.* A proton pump (see Figure 34.4) moves H^+ from the cytoplasm to the cell wall, thereby increasing the intracellular pH and decreasing the pH in the cell wall. Proton pumping also sets up an electrochemical gradient (see Key Concept 9.3), which provides potential energy to drive the transport of auxin by the carriers mentioned above.

4. *Ionization of a weak acid.* The main form of auxin, indole-3-acetic acid, is a weak acid (see Key Concept 2.4):

$$A^- + H^+ \rightleftharpoons HA$$

When the pH is low, this reaction is driven to the right, and HA (non-ionized auxin) is the predominant form. When the pH is higher, there is more A^- (ionized auxin).

While polar auxin transport distributes the hormone along the *longitudinal* (up and down) axis of the plant, *lateral* (side-to-side) redistribution of auxin is responsible for directional plant growth. This was shown in early experiments that followed the Darwins' when

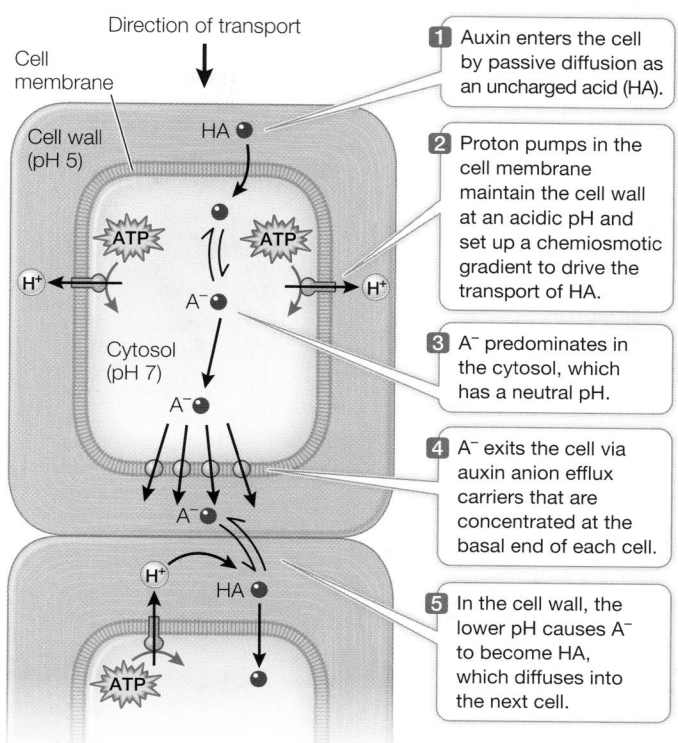

1 Auxin enters the cell by passive diffusion as an uncharged acid (HA).

2 Proton pumps in the cell membrane maintain the cell wall at an acidic pH and set up a chemiosmotic gradient to drive the transport of HA.

3 A^- predominates in the cytosol, which has a neutral pH.

4 A^- exits the cell via auxin anion efflux carriers that are concentrated at the basal end of each cell.

5 In the cell wall, the lower pH causes A^- to become HA, which diffuses into the next cell.

Figure 36.6 Polar Transport of Auxin   Proton pumps set up a chemiosmotic gradient directing ionized auxin (A^-) toward the basally placed active transport carriers for auxin, which leads to a net movement of auxin in a basal direction.

a coleoptile tip containing the growth hormone was placed asymmetrically on the decapitated coleoptile. The asymmetric distribution of growth hormone down the coleoptile resulted in excess growth on that side, and bending away from it, even in the absence of light:

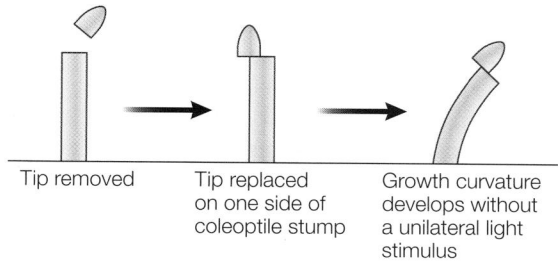

| Tip removed | Tip replaced on one side of coleoptile stump | Growth curvature develops without a unilateral light stimulus |

The redistribution of auxin to one side is carried out by auxin carrier proteins that move from the base of the cell to one side; because of this, auxin exits the cell only on that side of the cell, rather than at the base, and moves sideways within the tissue.

This lateral movement of auxin explains the bending of canarygrass seedlings toward light that the Darwins observed. When light strikes a canarygrass coleoptile on one side, auxin at the tip moves laterally toward the shaded side. The asymmetry thus established is maintained as polar transport moves auxin down the coleoptile, so that in the growing region below, the auxin concentration is highest on the shaded side. Cell elongation is thus stimulated on that side, causing the coleoptile to bend toward the light (**Figure 36.7A**).

(A) Phototropism

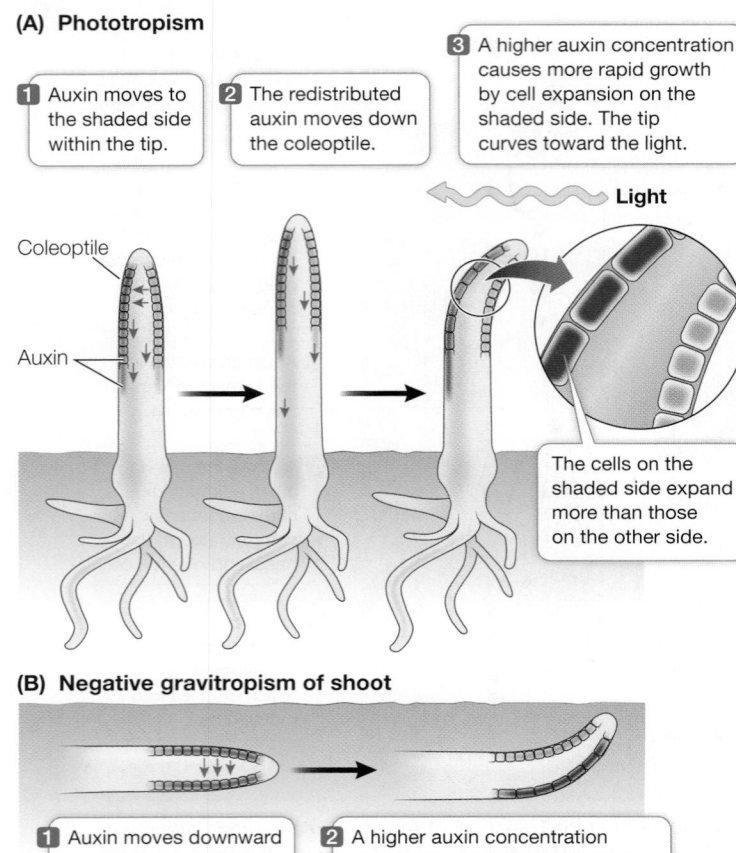

1 Auxin moves to the shaded side within the tip.

2 The redistributed auxin moves down the coleoptile.

3 A higher auxin concentration causes more rapid growth by cell expansion on the shaded side. The tip curves toward the light.

Light

Coleoptile

Auxin

The cells on the shaded side expand more than those on the other side.

(B) Negative gravitropism of shoot

1 Auxin moves downward in response to gravity.

2 A higher auxin concentration causes more rapid growth on the lower side. The tip curves upward.

Figure 36.7 Plants Respond to Light and Gravity Phototropism **(A)** and gravitropism **(B)** occur in shoot apices in response to a redistribution of auxin.

Light is not the only signal that can cause the redistribution of auxin. Auxin moves to the lower side of a shoot that has been tipped sideways, causing more rapid growth in the lower side and hence an upward bending of the shoot. Such growth in a direction determined by gravity is called **gravitropism** (**Figure 36.7B**). The upward gravitropic response of shoots is defined as negative gravitropism; that of roots, which bend downward, is positive gravitropism. Gravitropism in roots also involves differential growth caused by lateral movement of auxin, but the details of the mechanism differ between the root and the shoot.

Auxin affects plant growth in several ways

Like the gibberellins, auxin has many roles in plant development. Auxin affects the vegetative and reproductive growth of plants in several ways.

ROOT INITIATION   Cuttings from the shoots of some plants can produce roots and develop into entire new plants. For this to occur, parenchyma cells in the interior of the shoot must dedifferentiate and form an apical meristem of a new root. These changes are similar to those that take place in the pericycle of a root when a lateral root forms (see Figure 33.12). Shoot cuttings of many species can be made to develop roots by dipping the cut surfaces into an auxin solution. These observations suggest that in an intact plant,

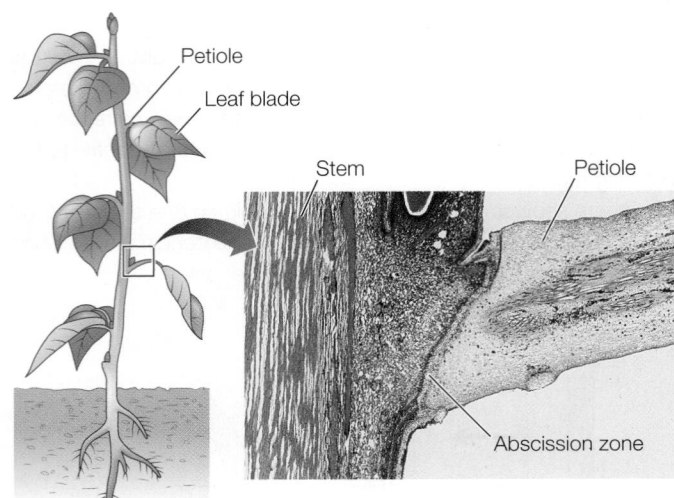

Petiole

Leaf blade

Stem   Petiole

Abscission zone

Figure 36.8 Changes Occur When a Leaf Is about to Fall The breakdown of cells in the abscission zone of the petiole causes the leaf to fall.

the plant's own auxin plays a role in the initiation of lateral roots. Commercial preparations that enhance the rooting of plant cuttings typically contain synthetic auxins.

LEAF ABSCISSION   Auxin inhibits the detachment of old leaves from stems. This detachment process, called **abscission**, is the cause of leaves falling from some trees in the autumn. Many leaves consist of a blade and a petiole that attaches the blade to the stem. Abscission results from the breakdown of a specific part of the petiole, the abscission zone (**Figure 36.8**). If the blade of a leaf is cut off, the petiole falls from the plant more rapidly than if the leaf had remained intact. If the cut surface is treated with an auxin solution, however, the petiole remains attached to the plant, often longer than an intact leaf would have. The timing of leaf abscission in nature appears to be determined in part by a decrease in the movement through the petiole of auxin produced in the blade.

APICAL DOMINANCE   Auxin helps maintain **apical dominance**, a phenomenon in which apical buds inhibit the growth of axillary buds (see Figure 33.1), resulting in the growth of a single main stem with minimal branching. Apical dominance can be demonstrated by an experiment with a young seedling. If the plant remains intact, the stem elongates and the axillary buds remain inactive. Removal of the apical bud—the major site of auxin production—results in growth of the axillary buds. If the cut surface of the stem is treated with auxin, however, the axillary buds do not grow. The apical buds of branches also exert apical dominance: the axillary buds on the branch are inactive unless the apex of the branch is removed. That is why gardeners prune shrubs to encourage branching.

FRUIT DEVELOPMENT   Fruit development normally depends on prior fertilization of the ovule (egg), but in many species treatment of an unfertilized ovary with auxin or gibberellins causes **parthenocarpy**—fruit formation without fertilization. Parthenocarpic fruits form spontaneously in some cultivated varieties of plants, including seedless grapes, bananas, and some cucumbers.

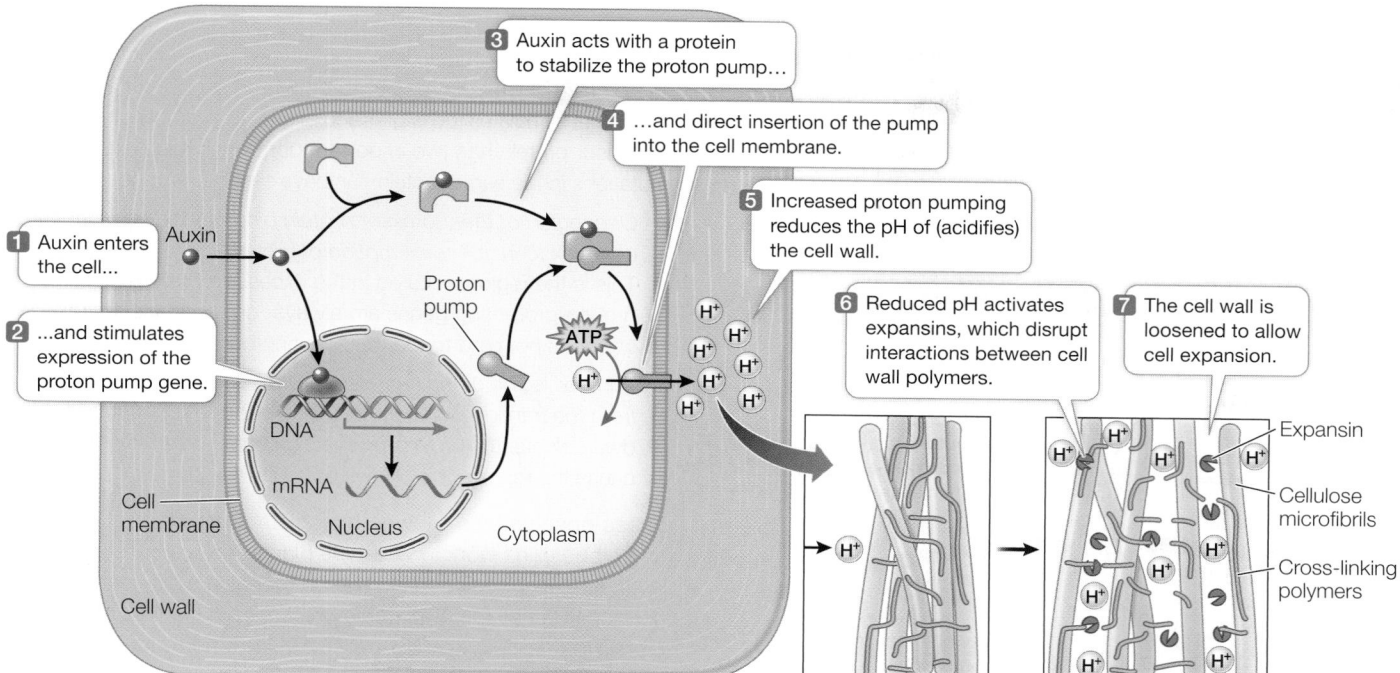

Figure 36.9 Auxin Increases Cell Expansion   The plant cell wall is an extensive network of cross-linked polymers. Auxin induces wall loosening by activating a proton pump that reduces pH in the cell wall.

Q: Where else in the plant cell are proton pumps important?

 Animation 36.3 Auxin Affects Cell Walls
www.Life11e.com/a36.3

CELL EXPANSION   Cell division followed by cell expansion is what causes plant growth. Because the plant cell wall normally prevents expansion of the cell contents inside the cell membrane (see Key Concept 33.1), the cell wall plays a key role in controlling the rate and direction of plant cell growth. Auxin acts on cell walls to regulate this process.

Cell expansion is driven mostly by the uptake of water, which enters the cytoplasm of the cell and accumulates in its central vacuole (see Key Concept 33.1). Growth of the vacuole accounts for most of the increase in volume of a growing cell, and the vacuole often makes up more than 90 percent of the volume of a mature cell. As the vacuole expands, it presses the cytoplasm against the cell wall, and the wall resists this force (the basis of turgor pressure). The cell wall is an extensively cross-linked network of polysaccharides and proteins, dominated by cellulose microfibrils. If the cell is to expand, some adjustments must be made in the wall structure to allow the wall to "give" under turgor pressure. Think of a balloon (the cell surrounded by a membrane) inside a box (the cell wall). How does the cell wall "box" loosen to allow expansion?

The **acid growth hypothesis** explains auxin-induced cell expansion (**Figure 36.9**). The hypothesis holds that protons (H⁺) are pumped from the cytoplasm into the cell wall, lowering the pH of the wall and activating enzymes called expansins that catalyze changes in the cell wall structure, thereby causing the polysaccharides to adhere to each other less strongly. Expansins loosen the cell wall, making it easier to stretch as the cell expands. Auxin has two roles in this process: to increase the synthesis of the proton pumps, and to guide their

insertion into the cell membrane. Auxin may also increase the activity of proton pump proteins already in the cell membrane by activating a protein kinase that phosphorylates the pump. Several lines of evidence support the acid growth hypothesis. For example, adding acid to the cell wall to lower the pH stimulates cell expansion even in the absence of auxin. Conversely, when a buffer is used to prevent the wall from becoming more acidic, auxin-induced cell expansion is blocked. The acid growth hypothesis applies to most species; in some plants, auxin stimulates secretion of new cell wall components quickly enough to account for even rapid changes in growth rate.

At the molecular level, auxin and gibberellins act similarly

The molecular mechanisms underlying both auxin and gibberellin action have been worked out with the help of genetic screens (see Figure 36.2). Biologists started by identifying mutant plants whose growth and development are insensitive to the hormones—that is, plants that are *not* affected by added hormone. Such mutant plants fall into two general categories:

1. *Excessively tall plants.* These plants resemble wild-type plants given an excess of hormone and grow no taller when given extra hormone. They grow tall even when treated with inhibitors of hormone synthesis. Their hormone response is always "on," even in the absence of the hormone. In such cases, it is presumed that the normal allele for the mutant gene codes for an inhibitor of the hormone signal transduction pathway. In wild-type plants, that pathway is "off," but in the mutant plants, the pathway is "on" and the plant grows tall.

2. *Dwarf plants.* These plants resemble dwarf plants that are deficient in hormone synthesis (see Figure 36.3), but they do not respond to added hormone. In these mutant plants the hormone response is always "off," regardless of the presence of the hormone.

focus: key figure

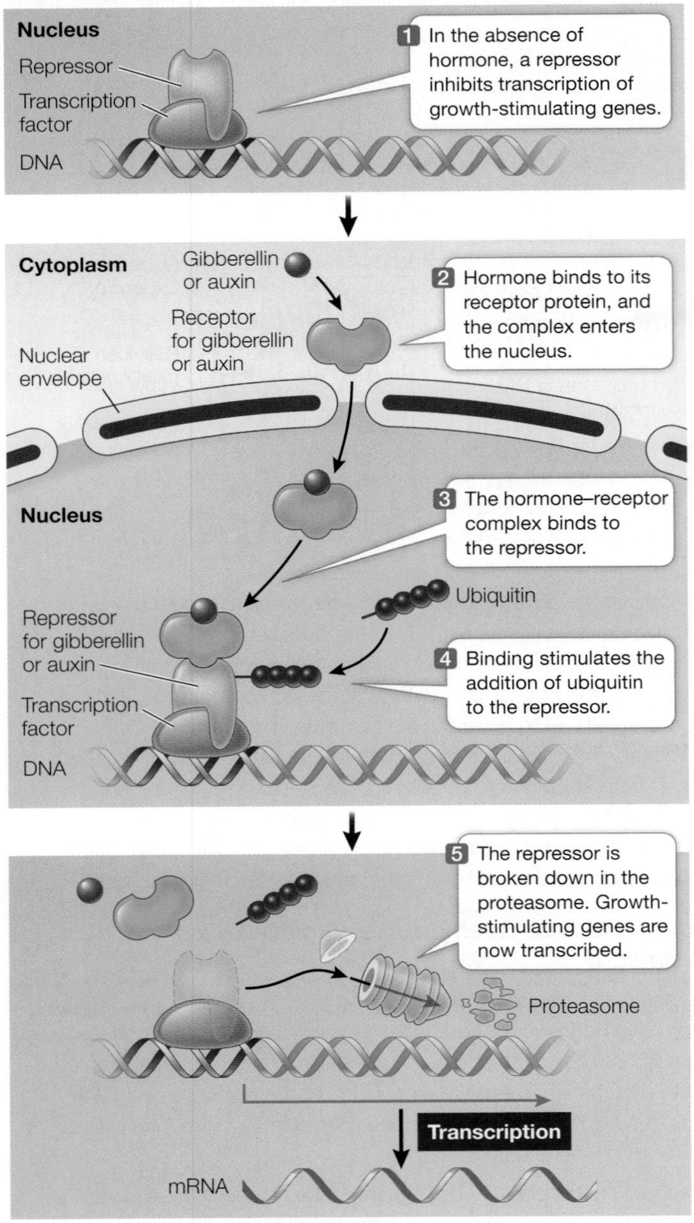

Nucleus

Repressor

Transcription factor

DNA

1 In the absence of hormone, a repressor inhibits transcription of growth-stimulating genes.

Cytoplasm   Gibberellin or auxin

Receptor for gibberellin or auxin

Nuclear envelope

2 Hormone binds to its receptor protein, and the complex enters the nucleus.

Nucleus

3 The hormone–receptor complex binds to the repressor.

Ubiquitin

Repressor for gibberellin or auxin

Transcription factor

DNA

4 Binding stimulates the addition of ubiquitin to the repressor.

5 The repressor is broken down in the proteasome. Growth-stimulating genes are now transcribed.

Proteasome

Transcription

mRNA

Figure 36.10 Gibberellins and Auxin Have Similar Signal Transduction Pathways   Although the specific proteins involved are different, both hormones act to stimulate gene transcription by inactivating a repressor protein.

Q: Are the molecules (receptor, repressor, transcription factor, ubiquitin, proteasome) the same or different in auxin versus gibberellin signaling?

 Media Clip 36.1 Gibberellin Binding to Its Receptor
www.Life11e.com/mc36.1

Remarkably, some mutations of both types turned out to affect *the same protein*, which turns out to be a repressor of a transcription factor that stimulates the expression of growth-promoting genes. The repressor protein has two important domains, which explains how mutations in the same protein can have seemingly opposite effects:

- *One region of the repressor protein binds to the transcription complex to inhibit transcription of growth-promoting genes.* This is the region mutated in the excessively tall plants: the growth-promoting genes are always "on" because the repressor does not bind to the transcription complex.

- *Another region of the repressor protein causes it to be removed from the transcription complex.* This is the region mutated in the dwarf plants: the growth-promoting genes are always "off" because the repressor is always bound to the complex.

These observations allowed biologists to figure out how auxin and gibberellins work in wild-type plants. The repressor proteins involved in responding to the two hormones are different, but both hormones *act by removing the repressor from the* * ***transcription complex*** (Focus: Key Figure 36.10). The hormones do this by binding to a receptor protein, which in turn binds to the repressor. Binding of the hormone–receptor complex stimulates polyubiquitination of the repressor, targeting it for breakdown in the proteasome (see Figure 16.20). The receptors contain or associate with a region called an F-box that facilitates protein–protein interactions necessary for polyubiquitination of a target protein. Whereas animal genomes have few F-box-containing proteins, plant genomes have hundreds, an indication that this type of gene regulation is common in plants.

---

*connect the concepts   Key Concept 16.2 explains how transcription factors bind to DNA and affect the rate of transcription in the regulation of gene expression.

▶36.2 recap

Gibberellins are plant hormones that affect stem growth, fruit size, seed germination, and many other aspects of plant development; the effects vary from species to species. Auxin regulates cell expansion and thus mediates phototropism and gravitropism; it also plays roles in apical dominance, leaf abscission, fruit development, and root initiation. The acid growth hypothesis explains auxin-induced cell wall loosening. The molecular mechanisms by which gibberellins and auxin convey their effects on plants are similar: both hormones stimulate gene transcription by inactivating a repressor protein.

learning outcomes

You should be able to:

- Give evidence for the role of auxin in plant growth and development.
- Give evidence for the role of gibberellins in various plant growth processes.
- Compare polar and lateral transport of auxin, as they relate to the bending of shoots toward the light.

36.2 recap

1. What would the result be, in terms of the synthesis of hydrolytic enzymes, if germinating barley seeds were treated with an inhibitor of mRNA translation? What would the result be if the seeds were treated with an inhibitor of proteasome function? What would be the result if the two inhibitors were applied to coleoptiles treated with auxin?

2. Explain why, even though auxin moves away from the lighted side of a coleoptile tip, the coleoptile bends toward the light.

3. Corn stunt spiroplasma (a bacterium) causes a significant reduction in the growth rate of infected corn plants. Diseased plants take on a dwarfed form. Since their appearance is reminiscent of a genetically dwarfed corn, you suspect that the bacterium may inhibit the synthesis of gibberellins by corn plants. Describe two experiments you might conduct to test this hypothesis, only one of which requires chemical measurement.

A corn plant infected with corn stunt spiroplasma

How can a single hormone, such as auxin or a gibberellin, have so many effects? As you have seen, a single signal transduction pathway may affect more than one gene. You will learn about other important plant hormones in the next section, and they too have multiple effects.

key concept 36.3   Other Plant Hormones Have Diverse Effects

The discoveries of gibberellins and auxin led to the isolation of additional classes of plant hormones (see Table 36.2). These hormones have multiple effects, and they interact with one another in many cases.

focus your learning

- Cytokinins are plant hormones that affect many plant growth functions related to cell division, growth, and senescence.
- The gaseous hormone ethylene has several functions related to senescence, including fruit ripening; it also affects auxin gradients during shoot development.
- Brassinosteroids, or plant steroid hormones, have a variety of functions that vary among plants.

Cytokinins are active from seed to senescence

Like bacteria and yeasts, plant cells such as parenchyma cells can be grown in a liquid or solidified growth medium containing sugars and salts. The cells will divide continuously until they run out of nutrients. In the early days of plant cell culturing, scientists experimented with many supplements to determine the optimal chemical environment for growth. The best supplement was coconut milk, the fluid that surrounds the developing embryo in coconut fruit. Investigators suspected that a molecule in the fluid must stimulate plant cell division.

A clue to the identity of the molecule came when Folke Skoog at the University of Wisconsin tested various pure substances that might substitute for coconut milk. DNA was among the substances tested, and it did not work; however, heating DNA at high pressure in an autoclave produced a mixture that strongly promoted plant cell division. A derivative of adenine called kinetin was identified as the active ingredient. Because it stimulated cell division (cytokinesis), it was called a **cytokinin**.

Kinetin does not exist in cells, but it gave scientists a hint as to what type of molecule might be the active ingredient in coconut milk. An adenine derivative called **zeatin** was extracted from corn endosperm, and since then more than 150 different cytokinins have been isolated, and most are derivatives of adenine.

Cytokinins (see Table 36.2) have several different effects, in many cases interacting with auxin:

- Adding an appropriate combination of auxin and cytokinins to a growth medium induces rapid proliferation of cultured plant cells.

- Cytokinins can cause certain light-requiring seeds to germinate even when kept in constant darkness.

- In plant cell cultures, a high cytokinin-to-auxin ratio promotes the formation of shoots; a low ratio promotes the formation of roots.

- Cytokinins usually inhibit the elongation of stems, but they cause lateral swelling of stems and roots (the fleshy roots of radishes are an extreme example).

- Cytokinins stimulate axillary buds to grow into branches; the auxin-to-cytokinin ratio controls the extent of branching (bushiness) of a plant. A lower ratio promotes more branching.

- Cytokinins delay the senescence of leaves. If leaf blades are detached from a plant and placed in water or a nutrient solution, they quickly turn yellow and show other signs of senescence. If instead they are placed in a solution containing a cytokinin, they remain green and senesce much more slowly. Roots contain abundant cytokinins, and cytokinin transport to the leaves delays senescence.

This leaf was not treated.

This leaf was treated with **cytokinin**, which delays senescence.

Ethylene is a gaseous hormone that hastens leaf senescence and fruit ripening

Whereas the cytokinins delay senescence, another plant hormone promotes it: the gas **ethylene** (see Table 36.2), which is sometimes called the senescence hormone. Ethylene can be produced by all parts of the plant, and like all plant hormones, it has several effects.

Back when streets were lit by gas rather than by electricity, leaves on trees near street lamps dropped earlier than those on trees farther from the lamps. We now know why: ethylene, a combustion product of the illuminating gas, caused the early abscission. Whereas auxin delays leaf abscission, ethylene strongly promotes it; thus the balance of auxin and ethylene controls abscission.

FRUIT RIPENING By promoting senescence, ethylene also speeds the ripening of fruit. As a fruit ripens, it loses chlorophyll and its cell walls break down; ethylene promotes both of these processes. Ethylene also causes an increase in its own production. Thus once ripening begins, more and more ethylene forms, and because it is a gas, it diffuses readily throughout the fruit and even to neighboring fruits on the same or other plants. The old saying "one rotten apple spoils the barrel" is true. That rotten apple is a rich source of ethylene, which speeds the ripening and subsequent rotting of the other fruit in a barrel or other confined space.

Farmers used to poke holes in developing figs to make them ripen faster. We now know that wounding causes an increase in ethylene production by the fruit, and that the raised ethylene level promotes ripening in many fruits, including apples, bananas, melons, apricots, and tomatoes. Today commercial shippers and storers of such fruit hasten ripening by adding ethylene to storage chambers. This use of ethylene is the single most important use of a natural plant hormone in agriculture and commerce.

These stored tomatoes were treated with ethylene to promote ripening.

These tomatoes were not treated with ethylene.

Ripening can also be delayed by the use of "scrubbers" and adsorbents that remove ethylene from the atmosphere in fruit storage chambers.

As flowers senesce, their petals may abscise, decreasing their value in the cut-flower industry. Growers and florists often immerse the cut stems of ethylene-sensitive flowers in dilute solutions of silver thiosulfate before sale. Silver salts inhibit ethylene action by interacting directly with the ethylene receptor—thus they delay senescence, keeping flowers "fresh" for longer. An alternative product, used to delay fruit ripening and preserve cut flowers, is 1-methylcyclopropene, a gas that also binds to ethylene receptors and blocks their function.

STEM GROWTH Although it is associated primarily with senescence, ethylene is active at other stages of plant development as well. Its effects on seedling development illustrate the interactions that occur among plant hormones. The hypocotyl of many eudicot seedlings forms an **apical hook** that protects the delicate shoot apex while the stem grows through the soil:

Apical hook

As in phototropic and gravitropic responses, the apical hook is maintained through an auxin gradient, which promotes the elongation of cells on the outer surface of the hook. Once the seedling breaks through the soil surface and is exposed to light, the auxin level on the inside of the hook increases and the hook unfolds, raising the shoot apex and the expanding leaves into the sun.

There is evidence that ethylene controls the formation of the auxin gradient during seedling development. Treatment of dark-grown seedlings of some species with ethylene results in what is called the "triple response": an exaggeration of the apical hook and a thickening and shortening of the hypocotyl and root (this response was exploited in a genetic screen for ethylene-response mutant plants; see Figure 36.2). It has been shown that ethylene affects both auxin synthesis and transport during apical hook development.

Brassinosteroids are plant steroid hormones

In animals, steroid hormones such as cortisol and estrogen are formed from cholesterol (see Figure 40.2B). Initially, biologists isolated a plant steroid hormone from the pollen of rape, a member of the Brassicaceae (mustard family). When applied to various plant tissues, this **brassinosteroid** (see Table 36.2) stimulated cell elongation, pollen tube elongation, and vascular tissue differentiation, but it inhibited root elongation. Since then, dozens of chemically related, growth-affecting brassinosteroids have been found in most plants.

Mutant plants that either do not make brassinosteroids or have defects in brassinosteroid reception and signal transduction are usually dwarf, infertile, and slow to develop. These effects can be reversed by adding small amounts of brassinosteroids, indicating that brassinosteroids are true hormones. These hormones have diverse effects, which vary among plants. Brassinosteroids can:

* promote xylem differentiation
* promote growth of pollen tubes during reproduction
* promote seed germination
* promote apical dominance and leaf senescence
* enhance cell elongation and cell division in shoots

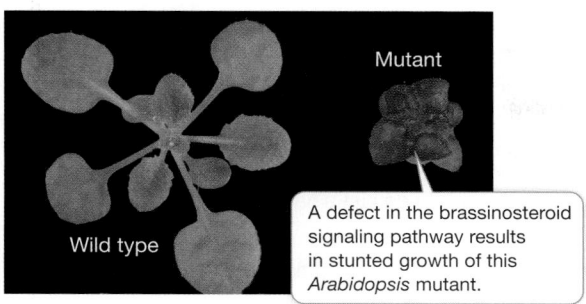

Mutant

A defect in the brassinosteroid signaling pathway results in stunted growth of this *Arabidopsis* mutant.

Wild type

The signaling pathway for these plant steroids differs sharply from those for steroid hormones in animals. In animals, steroids diffuse through the cell membrane and bind to receptors in the cytoplasm. In contrast, the receptor for brassinosteroids is an integral protein in the cell membrane.

36.3 recap

Cytokinins, ethylene, and brassinosteroids are hormones that work in concert with auxin and gibberellins to mediate plant development. They have various effects on different plants, ranging from stem growth to fruit ripening. They also interact in their effects with other hormones. Ethylene is remarkable in that it is a gas.

learning outcomes

You should be able to:

- Use information on the functions of cytokinins to predict how different aspects of plant growth will be affected by cytokinins and by various combinations of cytokinins and auxin.
- Describe the function of ethylene in fruit ripening, and explain its importance in the cut-flower and food-production industries.
- Identify some effects of brassinosteroids on plant growth.

1. A plant has an extremely bushy growth habit. What can you say about the likely auxin-to-cytokinin ratio in this plant?
2. Supermarkets sell plastic bags that are impregnated with activated charcoal, which binds gases. The bags are designed to keep fruit fresh. How do they work?
3. A plant shows rapid leaf senescence. Which hormone or hormones are most likely active in this plant? Which are most likely inactive?

A plant's response to light—the energy source for photosynthesis—is crucial to its survival. You saw how the Darwins' pioneering investigations of phototropism led to the discovery of auxin. Let's now look more closely at how plants sense and respond to light.

key concept
36.4 Photoreceptors Initiate Developmental Responses to Light

As we pointed out in Key Concept 36.1, plants respond to many different environmental cues, and light is possibly the most important of these cues. Much has been learned about the receptors that plants use to sense light, and we will focus on those receptors here.

focus your learning

- Plants have several blue-light receptor pigments.
- Phytochrome, a photoreceptor pigment, controls plant responses to red and far-red light, and exists in two forms, P_r and P_{fr}.

What are plant photoreceptors?

Plants respond to two aspects of light: (1) its *quality*—that is, the wavelengths of light that can be absorbed by molecules in the plant; and (2) its *quantity*—that is, the intensity and duration of light exposure.

In Chapter 10 you learned about the role of light in photosynthesis: how chlorophyll and other pigments absorb light at certain wavelengths (quality), and how light intensity affects photosynthetic rate (quantity). Here you'll learn how light affects plant development. Earlier in this chapter we described phototropism and how auxin mediates a plant stem's bending toward light. In addition to phototropism, light influences seed germination, shoot elongation, the initiation of flowering, and many other important aspects of plant development. Several photoreceptors take part in these processes. **Blue-light receptors** such as cryotochromes and phototropins mediate the effects of higher-intensity blue light, and phytochrome mediates the effects of red light.

Phototropins, cryptochromes, and zeaxanthin are blue-light receptors

Charles and Francis Darwin showed that the apical tip of a growing coleoptile receives light as a signal and then redistributes auxin to stimulate cell elongation below the tip on the shaded side. You may recall from Chapter 10 that an action spectrum involves exposing plants to different wavelengths of light to determine what wavelengths are most effective in driving a given process (e.g., photosynthesis). For photosynthesis, such studies showed that the most effective wavelengths are those absorbed by chlorophylls (see Figure 10.4). When an action spectrum was obtained for phototropism of coleoptiles, blue light (peak 436 nanometers [nm]) was found to be the most effective at inducing the coleoptile to curve (**Figure 36.11**). What is the blue-light-absorbing receptor/pigment? Biologists have used a genetic approach to answer this question, once again employing the model plant *Arabidopsis*.

Researchers recovered blue-light-insensitive mutant *Arabidopsis* plants from a genetic screen and identified the gene for a blue-light receptor protein located in the cell membrane called **phototropin**. Phototropin protein has a flavin mononucleotide associated with it that absorbs blue light, leading to a change in the shape of the protein. This change exposes an active site for a protein kinase, which in turn initiates a signal transduction cascade that ultimately results in the stimulation of cell elongation by auxin. Phototropin is also involved in chloroplast movements in relation to light, and participates with another type of blue-light receptor, the plastid pigment **zeaxanthin**, in the light-induced opening of stomata.

(A) The absorption spectrum for phototropin

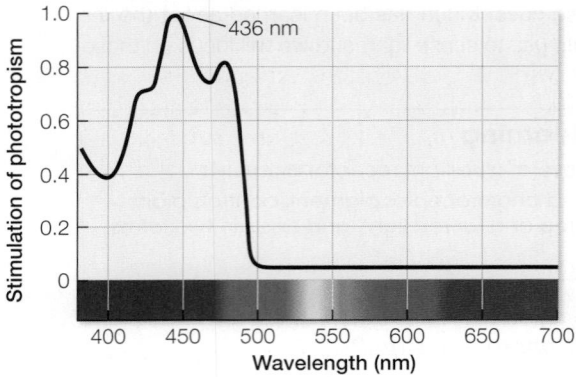

(B) Effect of blue light on coleoptiles

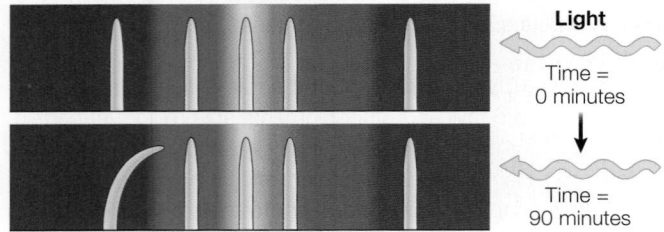

Light
Time =
0 minutes

Time =
90 minutes

Figure 36.11 Action Spectrum for Phototropism   **(A)** The action spectrum for bending of a coleoptile toward light is similar to the absorption spectrum for the receptor, phototropin. **(B)** After 90 minutes, only the coleoptiles exposed to blue light bend.

● Lettuce seeds spread on the soil will germinate only in response to light. Even just a flash of dim light will suffice.

● Adult cocklebur plants flower when they are exposed to long nights. If there is a brief light flash in the middle of the night, they do not flower.

Action spectra of the above processes show that they are induced by red light (650–680 nm). This indicates that plants must have a photoreceptor pigment that absorbs red light and initiates photomorphogenesis.

What is especially remarkable about these red light responses is that *they are reversible by far-red light* (710–740 nm). For example, if lettuce seeds are exposed to brief, alternating periods of red and far-red light in close succession, they respond only to the final exposure. If it is red, they germinate; if it is far-red, they remain dormant (**Figure 36.12**). This reversibility of the effects of red and far-red light regulates many other aspects of plant development, including flowering and seedling growth.

Another class of blue-light receptors is the **cryptochromes**, which absorb blue and ultraviolet light. These yellow pigments are located primarily in the plant cell nucleus and affect seedling development and flowering. The exact mechanism of cryptochrome action is not yet known. Strong blue light inhibits cell elongation through the action of cryptochromes, although the most rapid responses are mediated by phototropins.

Phytochromes mediate the effects of red and far-red light

Photomorphogenesis refers to a number of physiological and developmental events in plants that are controlled by light. For example:

● A bean seedling germinating below-ground has an elongated stem, a pale yellow, folded leaf, and a hook that protects the first leaves (see Figure 36.1 and p. 778)—it is **etiolated**. As the seedling reaches the surface of the soil, it undergoes several light-induced changes: the apical hook straightens, the rudimentary leaves unfold, and chlorophyll is made, so that photosynthesis can begin. Even very dim light will induce these changes.

experiment

Figure 36.12A   Sensitivity of Seeds to Red and Far-Red Light

Original Paper: Borthwick, H. A., S. B. Hendricks, M. W. Parker, E. H. Toole and V. K. Toole. 1952. A reversible photoreaction controlling seed germination. *Proceedings of the National Academy of Sciences USA* 38: 662–666.

Lettuce seeds will germinate if exposed to a brief period of light. An action spectrum indicated that red light was most effective in promoting germination, but far-red light would reverse the stimulation if presented right after the red light flash. Harry Borthwick and his colleagues asked what would be the effect of repeated alternating flashes of red and far-red light. In each case, the final exposure determined the germination response. This observation led to the conclusion that a single, photoreversible molecule was involved. That molecule turned out to be phytochrome.

HYPOTHESIS▶   The effects of red and far-red light on lettuce seed germination are mutually reversible.

METHOD

Expose lettuce seeds to alternate periods of red light R for 1 minute and far-red light FR for 4 minutes.

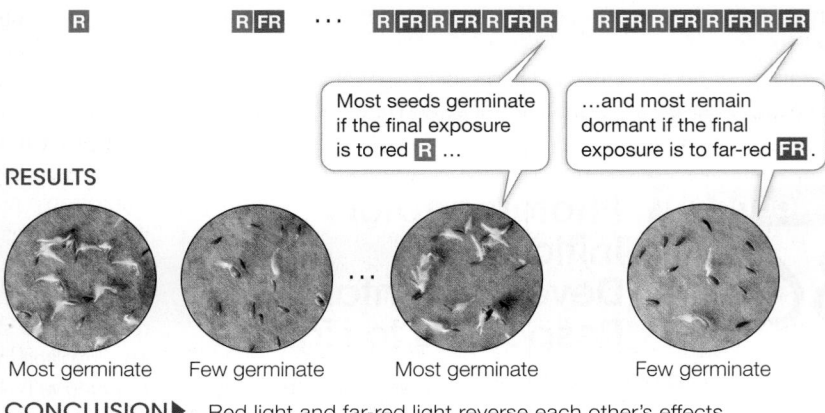

R

R FR   ···   R FR R FR R FR R   R FR R FR R FR R FR

Most seeds germinate if the final exposure is to red R …

…and most remain dormant if the final exposure is to far-red FR.

RESULTS

Most germinate   Few germinate   ···   Most germinate   Few germinate

CONCLUSION▶   Red light and far-red light reverse each other's effects.

The basis for the effects of red and far-red light resides in the photoreceptor pigment protein in the cytoplasm of plants called **phytochrome**. Phytochrome exists in two interconvertible "isoforms," or states. The molecule undergoes a conformational change upon absorbing light at particular wavelengths. The default or "ground" state, which absorbs principally red light, is called P_r. When P_r absorbs a photon of red light it is converted into P_{fr}. P_{fr} is the active form of phytochrome—the form that triggers important biological processes in various plants.

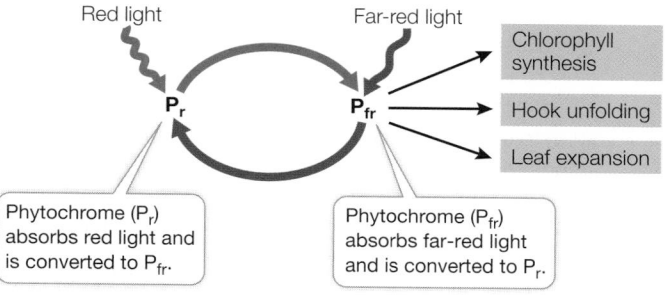

Phytochrome (P_r) absorbs red light and is converted to P_{fr}.

Phytochrome (P_{fr}) absorbs far-red light and is converted to P_r.

The part of phytochrome that absorbs red and far-red light is a covalently attached pigment called a chromophore (**Figure 36.13A**). The chromophore of P_r preferentially absorbs red light; when it does so, it changes conformation and the phytochrome is converted to the P_{fr} form. When the chromophore of P_{fr} absorbs far-red light, the phytochrome is converted back to the P_r form. If you know organic chemistry, this reaction is a familiar *cis-trans* isomerization.

The absorption spectra for P_r and P_{fr} correlate with their action spectra (**Figure 36.13B**). As you have seen, phytochrome affects seed germination, shoot development after etiolation, and flowering. In *Arabidopsis* there is a gene family that encodes five slightly different phytochromes, each functioning in different photomorphogenic responses.

For a plant in nature, the ratio of red to far-red light determines whether a phytochrome-mediated response will occur:

- In daylight the ratio is about 1.2:1; because there is more red than far-red light, the P_{fr} form predominates.

- In the shade of other plants, the ratio is as low as 0.13:1, and phytochrome is mostly in the P_r form. The low ratio of red to far-red light in the shade results from absorption of red light by

work with the data

Figure 36.12B Sensitivity of Seeds to Red and Far-Red Light

Original Paper: Borthwick, H. A. et al. 1952.

Harry Borthwick, Sterling Hendricks, and colleagues at the U.S. Department of Agriculture performed a series of landmark experiments that suggested the existence of a receptor for red light that determined seed germination. Borthwick was the son of a leading plant pathologist who followed in his father's footsteps and studied plant physiology. After it was discovered that plants responded to the length of day (photoperiodism; see Chapter 37), the importance of nonphotosynthetic responses to light as a developmental signal became a hot topic of research. For over a century, it was known that lettuce seeds required light to germinate. By placing lettuce seeds in an environment where other environmental variables such as temperature were controlled, Borthwick's group tested the light signaling on germination.

QUESTIONS▶

1. An action spectrum (see Figure 10.4) for seed germination was obtained by first soaking the seeds on wet filter paper in the dark for 16 hours for imbibition, and then exposing them to different wavelengths of light for 1 minute. Having been exposed to a possible germination signal, the seeds were returned to the dark and tested for germination after 2 days. The results are shown in **Table A** in terms of energy efficiency (how much light energy was required for 50% seed germination).

    a. Explain the energy efficiency measurement: what do high and low numbers mean?

    b. Plot the data as energy efficiency versus wavelength. What can you conclude about the most efficient wavelengths of light for germination?

2. There was evidence from other light signaling responses in plants that the light effects were reversible by exposure to different wavelengths of light. After the data in Table A were obtained, Borthwick's team decided to first expose groups of 200 seeds to light at 660 nm (red, R)

for 1 minute, and then some batches to light at 700 nm (far-red, FR) for 4 minutes. The seeds were returned to the dark and germination evaluated after 2 days. **Table B** shows the results.

a. What can you conclude about the nature of the light signal responder?

b. Why was there some germination in the "None" condition?

Table A

Wavelength (nm)	Energy required for 50% seed germination
560	35
570	25
580	15
590	10
600	8
620	6
640	4
660	3
680	4
690	45
700	80

Table B

Wavelengths used	Percent germination
None	8.5
R	98
FR	54
R then FR then R	100
R then FR then R then FR	43
R then FR then R then FR then R	99

A similar **work with the data** exercise may be assigned in **LaunchPad**.

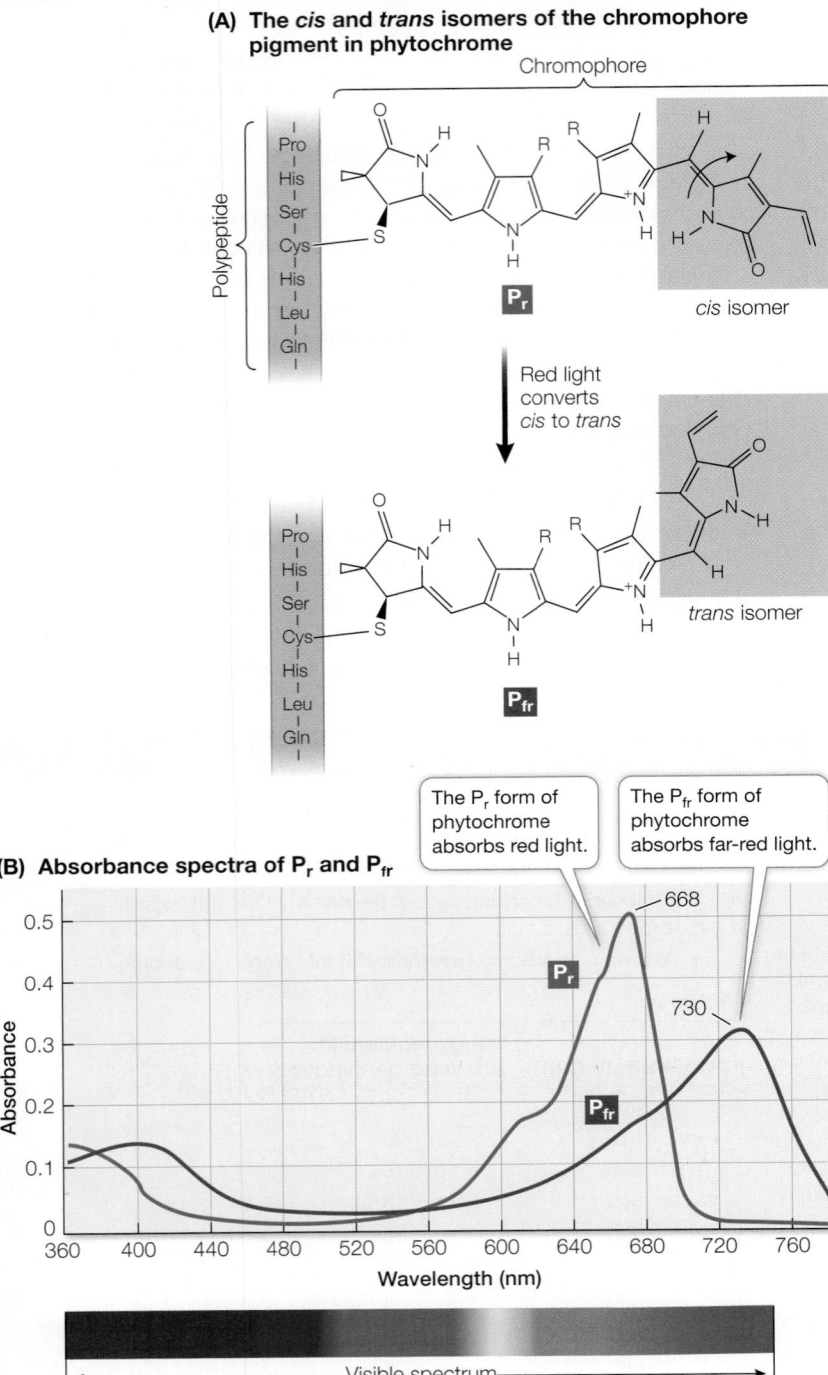

(A) The *cis* and *trans* isomers of the chromophore pigment in phytochrome

Red light converts *cis* to *trans*

(B) Absorbance spectra of P$_r$ and P$_{fr}$

The P$_r$ form of phytochrome absorbs red light.

The P$_{fr}$ form of phytochrome absorbs far-red light.

Figure 36.13 Phytochrome Exists in Two Forms   Absorption spectra of phytochrome reveal two interconvertible forms, corresponding to the *cis* and *trans* isomers of phytochrome's chromophore. The P$_r$ form absorbs red light; the P$_{fr}$ form absorbs far-red light.

chlorophyll in the leaves overhead, so less of the red light gets through to the plants below.

For some plants growing in the shade, the low ratio stimulates cell elongation and the plants grow upward toward the sun. The shade cast by other plants also prevents germination of seeds that require red light to germinate. The reflective properties of the soil can also affect the red to far-red ratio—and thus plant behavior. For example, cotton seedlings grow more slowly on soils (such as clay) that reflect more red than far-red light.

Phytochrome stimulates gene transcription

How does phytochrome, or more specifically P$_{fr}$, work? Phytochrome has two subunits (**Figure 36.14**), each composed of a protein chain and a chromophore. Gene transcription is stimulated when P$_r$ is converted to the P$_{fr}$ isoform. When P$_r$ absorbs red light, the chromophore changes shape, causing change in the conformation of the protein itself, from the P$_r$ form to the P$_{fr}$ form. Conversion to the P$_{fr}$ form exposes two important regions of the phytochrome protein, both of which affect transcriptional activity:

- Exposure of a *nuclear localization sequence* (see Figure 14.18) results in movement of P$_{fr}$ from the cytoplasm to the nucleus. Once in the nucleus, P$_{fr}$ binds to transcription factors and thereby stimulates expression of genes involved in photomorphogenesis.

- Exposure of a *protein kinase* domain causes P$_{fr}$ protein to phosphorylate itself and other proteins involved in red-light signal transduction, resulting in changes in the activity of transcription factors.

The effect of activating these transcription factors is quite large: in *Arabidopsis*, phytochrome affects an amazing 2,500 genes (10 percent of the entire genome!) by either increasing or decreasing their expression. Some of these genes are related to hormones. For example, when P$_{fr}$ is formed in seed germination, genes for gibberellin synthesis are activated and genes for gibberellin breakdown are repressed. As a result, gibberellins accumulate and seed reserves are mobilized.

Circadian rhythms are entrained by light reception

The timing and duration of biological activities in living organisms are governed in all eukaryotes and some prokaryotes by what is commonly called a "biological clock"—an oscillator within cells that alternates back and forth between two states at roughly 12-hour intervals. The major outward manifestations of this clock are known as **circadian rhythms** (Latin *circa*, "about," + *dies*, "day"). Think of your own life: in all probability you sleep at night and are awake during the day.

The circadian rhythms of animals are discussed in Key Concept 52.5. In plants, circadian rhythms influence, for example, the opening (during the day) and closing (at night) of stomata in *Arabidopsis*, and the raising toward the sun (during the day) and lowering (at night) of leaves in bean plants. From these two examples, it is obvious that circadian rhythms are ecologically useful adaptations, in that they relate the plant's physiology to its environment.

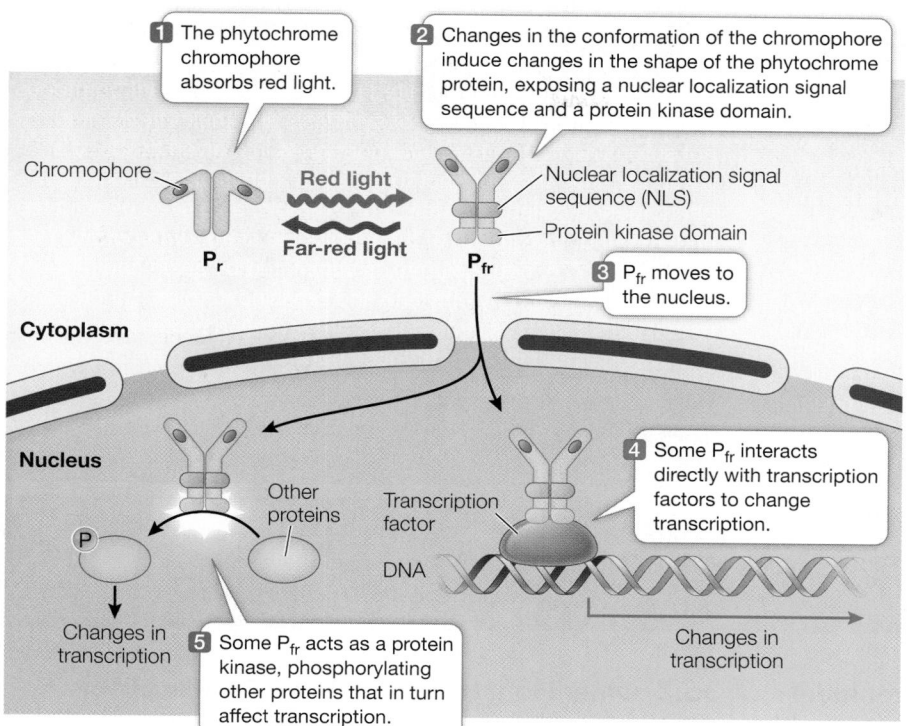

1 The phytochrome chromophore absorbs red light.

2 Changes in the conformation of the chromophore induce changes in the shape of the phytochrome protein, exposing a nuclear localization signal sequence and a protein kinase domain.

Chromophore

Red light

Far-red light

P_r

P_{fr}

Nuclear localization signal sequence (NLS)

Protein kinase domain

3 P_{fr} moves to the nucleus.

Cytoplasm

Nucleus

Other proteins

Transcription factor

DNA

4 Some P_{fr} interacts directly with transcription factors to change transcription.

P

Changes in transcription

5 Some P_{fr} acts as a protein kinase, phosphorylating other proteins that in turn affect transcription.

Changes in transcription

Figure 36.14 Phytochrome Stimulates Gene Transcription   Phytochrome is composed of two polypeptide chains, each with a chromophore. This pair of polypeptides undergoes a conformational change upon absorbing light. When phytochrome absorbs red light, it converts to the P_{fr} form, which activates transcription of phytochrome-responsive genes.

Two qualities characterize circadian rhythms, as well as other regular biological cycles: the **period** is the length of one cycle, and the **amplitude** is the magnitude of the change over the course of a cycle. The circadian rhythms of plants have several noteworthy characteristics:

- The period of a circadian rhythm is remarkably insensitive to temperature, although lowering the temperature may drastically reduce the amplitude.

- Circadian rhythms are highly persistent; they may continue for days, even in the absence of environmental cues, such as light–dark periods.

- Circadian rhythms can be entrained, within limits, by light–dark cycles that do not exactly correspond to 24 hours. That is, the period of a rhythm can be made to coincide (within limits) with that of the light–dark cycle to which the organism is exposed.

Consider what happens when a person abruptly moves across many time zones: what was the night becomes the day, and gradually the person's sleep–wakefulness circadian rhythm entrains to the new environmental cues. Similar entrainment occurs in plants adapting to day length as the seasons progress during the year. The action spectrum for plant entrainment indicates that phytochrome (and to a lesser extent, blue-light receptors) is very likely involved. At sundown phytochrome is mostly in the active P_{fr} form. But as the night progresses, P_{fr} gradually gets converted back to the inactive P_r form. By dawn phytochrome is mostly in the P_r state, but as daylight begins, it rapidly converts to P_{fr}. The switch to the P_{fr} state resets the plant's biological clock. However long the night, the clock is still reset at dawn every day. Thus while the total period measured by the clock is consistent, the clock adjusts to changes in day length over the course of the year.

36.4 recap

Light controls several physiological and developmental events in plants, a process called photomorphogenesis. Pigment photoreceptors such as phototropin, cryptochromes, and phytochrome mediate the effects of light on plant growth and development. Phytochrome exists in two interconvertible states; conversion from one state to the other is controlled by the ratio of red to far-red light. Circadian rhythms are influenced by light reception.

learning outcomes

You should be able to:

- Describe and evaluate evidence for the existence of blue-light receptors in plants.

- Explain the process and conditions under which conversion of the P_r and P_{fr} forms of phytochrome occur.

1. What is the evidence for blue-light receptors in plants?

2. Why does red light affect seed germination differently than far-red light does?

QA How is plant growth controlled, and what changes in growth patterns made the new strains of wheat and rice successful?

In normal wheat plants, gibberellins stimulate stem elongation. But in the semi-dwarf plants, a mutation affects the signal transduction mechanism for gibberellins so that the stem cells do not respond to it and growth is reduced. In rice, the mutation is in the gene for an enzyme in the biochemical pathway for the synthesis of gibberellins. Without the hormone, the stem does not elongate. The lives of countless people have been saved by intentional disruptions of hormone signaling. These high-yielding crops respond to added fertilizer with greatly increased growth and production of grain compared with conventional varieties of wheat and rice.

Future directions

Unlike animal hormones, which are made in one tissue and often have specific effects on a limited range of targets, plant hormones are widespread in their sites of synthesis and their effects. It is not surprising that with several hormones in a given tissue at a particular time, they interact to produce a physiological effect, a phenomenon called crosstalk. For example, as a seed germinates underground, the growth of the embryonic stem, or hypocotyl (see photo on p. 776, right column), is stimulated by auxin and gibberellins, and inhibited by ethylene and brassinosteroids. Studies on model plants such as *Arabidopsis* are revealing proteins common to the signaling pathways of these hormones that have a role in their interacting effects.

Chapter Summary 36

▶ 36.1 Plants Develop in Response to the Environment

- As sessile organisms, plants maximize their ability to grow by using meristems, forming new organs, and growing throughout life.
- The environment, receptors, hormones, and the plant's genome all regulate plant development.
- Seed **dormancy**, which has adaptive advantages, is maintained by a variety of mechanisms. In nature, dormancy is broken by, for example, abrasion, fire, leaching, and low temperatures. When dormancy ends and the seed **imbibes** water, it **germinates** and develops into a **seedling**. Review Figure 36.1, Activities 36.1, 36.2
- Plant hormones differ in structure and physiology from animal hormones. Review Table 36.1
- Plants have several hormones, each of which regulates multiple aspects of development. Interactions among these hormones are often complex. Review Table 36.2
- **Genetic screens** using the model organism *Arabidopsis thaliana* have contributed greatly to our understanding of signaling in plants. Review Figure 36.2

▶ 36.2 Gibberellins and Auxin Have Diverse Effects but a Similar Mechanism of Action

- Both **gibberellins** and **auxin** can induce growth in plants otherwise genetically destined to be dwarfs. Review Figure 36.3
- Gibberellins have many effects that vary among different plants, including cell elongation, fruit ripening, and mobilization of seed storage polymers. Review Figures 36.3–36.5, Activity 36.3
- Auxin was discovered in the context of stem and coleoptile growth, in particular **phototropism**. In the shoot, it is made in the growing tip and transported down to stimulate cell elongation. Review Investigating Life: The Darwins' Phototropism Experiment, Figures 36.6, 36.7, Animations 36.1, 36.2
- According to the **acid growth hypothesis**, auxin stimulates cell elongation through the release of protons into the cell wall (acidification of the cell wall). Review Figure 36.9, Animation 36.3
- Both auxin and gibberellins act through the breakdown of transcriptional repressors. Review Focus: Key Figure 36.10

▶ 36.3 Other Plant Hormones Have Diverse Effects

- **Cytokinins** are adenine derivatives that promote plant cell division, promote seed germination in some species, inhibit stem elongation, promote lateral swelling of stems and roots, stimulate the growth of axillary buds, promote the expansion of leaf tissue, and delay leaf senescence.
- A balance between auxin and **ethylene** controls leaf abscission. Ethylene promotes senescence and fruit ripening. It indirectly causes the formation of a protective **apical hook** in eudicot seedlings. In stems, it inhibits elongation, promotes lateral swelling, and causes a loss of gravitropic sensitivity.
- Dozens of different **brassinosteroids** affect cell elongation, pollen tube elongation, vascular tissue differentiation, and root elongation.

▶ 36.4 Photoreceptors Initiate Developmental Responses to Light

- **Phototropins** are blue-light photoreceptors for phototropism and chloroplast movements. **Zeaxanthin** acts in conjunction with the phototropins to mediate the light-induced opening of stomata. **Cryptochromes** are blue-light photoreceptors that control seedling development, stem elongation, and floral initiation.
- **Phytochrome** exists in the cytoplasm in two interconvertible forms, P_r and P_{fr}. The relative amounts of these two forms are a function of the ratio of red to far-red light. Phytochrome affects seedling growth, flowering, and etiolation. Review Figures 36.12, 36.13
- The phytochrome signal transduction pathway affects transcription in two different ways; the P_{fr} form interacts directly with some transcription factors, and influences transcription indirectly through interactions with protein kinases. Review Figure 36.14
- **Circadian rhythms** are activities that occur on a near-24-hour cycle. Light can entrain these activities through photoreceptors such as phytochrome.

Go to **LearningCurve** (in **LaunchPad**) for dynamic quizzing that helps you solidify your understanding of this chapter. **LearningCurve** adapts to your responses, giving you the practice you need to master each key concept.

Apply What You've Learned

Review

36.1 Plants respond to environmental cues during growth by the use of signal transduction pathways involving hormones and photoreceptors.

36.1 Scientists use genetic screens to identify genes involved in the control of growth and development.

Original Papers: Franklin, K. A., S. J. Davis, W. M. Stoddart, R. D. Vierstra and G. C. Whitelam. 2003. Mutant analyses define multiple roles for phytochrome C in *Arabidopsis* photomorphogenesis. *The Plant Cell* 15: 1981–1989.

Qin, M., R. Kuhn, S. Moran and P. H. Quail. 1997. Overexpressed phytochrome C has similar photosensory specificity to phytochrome B but a distinctive capacity to enhance primary leaf expansion. *The Plant Journal* 12: 1163–1172.

Sharrock, R. A. and T. Clack. 2002. Patterns of expression and normalized levels of the five *Arabidopsis* phytochromes. *Plant Physiology* 130: 442–456.

Phytochromes sense red (R) and far-red (FR) light and control many processes in plant growth and development, including chlorophyll synthesis, seed germination, and leaf expansion. The phytochrome photoreceptor has two parts: a chromophore that receives light and a protein involved in signal transduction for the response (see Figure 36.14). The model plant *Arabidopsis* has five distinct genes (*phyA*, *phyB*, *phyC*, *phyD*, and *phyE*) that control the formation of five types of phytochrome. These genes code for five slightly different proteins, the differential expression of which in different tissues at different times may underlie the varying responses of plants.

Elucidating the roles of the *phy* genes has involved both molecular and genetic approaches. In one molecular approach, a gene is coupled to a very active promoter and expressed at higher than normal levels (overexpressed) in developing *Arabidopsis* plants. For example, *phyA*, *phyB*, *phyC*, and *phyD* were separately overexpressed in *Arabidopsis*, and 18-day-old seedlings were examined with regard to leaf area. The table shows the results.

Gene expression	Leaf area (mm²)
Wild type	22
phyA overexpressed	21
phyB overexpressed	21
phyC overexpressed	31
phyD overexpressed	22

To complement the molecular approach, a genetic approach—a genetic screen for plants expressing altered leaf size phenotypes—was used to isolate mutants deficient in *phyC* (see Figure 36.2). All of the *phyC* mutants found using the genetic screen turned out to also be genetically deficient in *phyD*. This result prompted an investigation of plants deficient in different combinations of *phy* genes. These plants were analyzed with regard to leaf length (which increases due to elongation of the petiole that connects leaf to stem) and leaf area. The results are shown in **Figures A** and **B**.

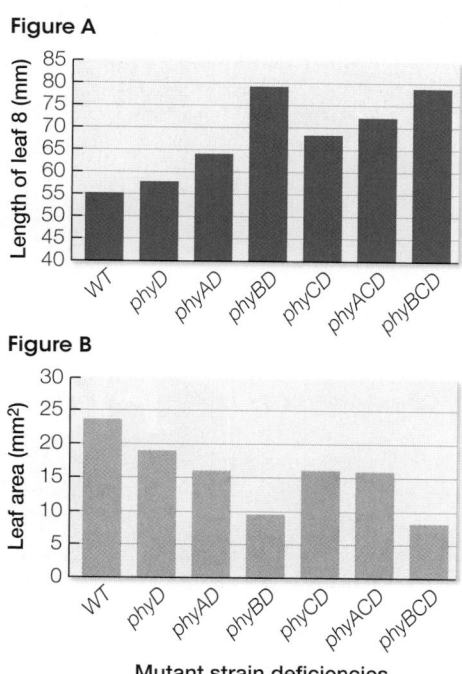

Figure A

Figure B

Mutant strain deficiencies

Questions

1. What do the data in the table indicate about the role of the four phytochrome genes in leaf development?

2. How would would a genetic screen for *phyC*-deficient mutants be done, and what phenotype would be examined in the screen?

3. Compare the results for *phyAD*, *phyCD*, and *phyACD* in Figure B. What do these results tell you about the function and necessity of the *phyA* and *phyC* gene in controlling *Arabidopsis* leaf area?

4. What do the data in Figures A and B suggest about the function of the gene *phyB* in leaf growth? How does its function relate to the function(s) of *phyA* and *phyC*?

37

key concepts

37.1 Most Angiosperms Reproduce Sexually

37.2 Hormones and Signaling Determine the Transition from the Vegetative to the Reproductive State

37.3 Angiosperms Can Reproduce Asexually

Reproduction in Flowering Plants

The poinsettia is a popular Christmas flower. But the red "flowers" are not flowers at all; instead, they are leaves (bracts).

investigatinglife

What Signals Flowering?

Dairy farmer Albert Ecke was fascinated by the red and green shrubs that grew all over southern California and were used by Mexican-Americans for red dye. The shrub, *Euphorbia pulcherrima*, got the name "poinsettia" from the man who first brought it to the United States: the first U.S. ambassador to Mexico, Joel Roberts Poinsett. In the early 1900s, Ecke started selling the plants at his farm in Hollywood. But two challenges stood in the way of his making this plant a commercial success.

First, although poinsettias generally bloomed in the fall and early winter in the mild climate of southern California, when Ecke grew them in fields in Hollywood, the formation of flowers was unreliable. Biologists later found that the flowering of poinsettias required at least a 14-hour night for several weeks. Any interruption in the long night—by passing cars or street lamps or lighting from nearby movie sets, for example—inhibited flowering. So Ecke's son Paul moved the growing operation south, to isolated fields far from Los Angeles. A second challenge was that the plants were tall and gangly. Although pretty, they were hard to transport and not attractive indoors. Paul Ecke found a variety of poinsettia that was much more compact. He propagated this variety and eventually developed it into the short potted plant that is so popular today. The Eckes decided that the time of flowering was just right for making the poinsettia a "Christmas flower," so they promoted this now portable plant as a holiday decoration, blanketing live television shows with free plants between Thanksgiving and Christmas. The campaign was successful: more than 100 million poinsettias are now sold in the United States during the Christmas season every year, making it the nation's best-selling potted plant.

Breeding more attractive flowering plants that are easier to grow is an ongoing part of floriculture, the industry involved with the production of floral crops. You may be surprised to learn that the brightly colored "flowers" of poinsettias are not flowers at all. The red parts of the plant that we most notice and appreciate are actually specialized leaves called bracts. The poinsettia has a single tiny yellow female flower, without petals, surrounded by male flowers.

 Q How did an understanding of angiosperm reproduction allow floriculturists to develop a commercially successful poinsettia?

key concept 37.1 Most Angiosperms Reproduce Sexually

Most angiosperms (flowering plants) have evolved to reproduce sexually because this strategy has the selective advantage of producing the genetic diversity that is the raw material for evolution. Sexual reproduction in angiosperms involves mitosis, meiosis, and the alternation of haploid and diploid generations (see Figure 11.14).

focus your learning

- The embryo sac is the female gametophyte and the pollen grain is the male gametophyte.
- A pollen grain germinates to form a pollen tube, which grows through the style to the embryo sac.
- Angiosperms prevent self-fertilization by physically separating male and female gametophytes and by genetic self-incompatibility.
- The embryo and endosperm develop within the seed, enclosed within a fruit derived from the ovary wall.

There are several important differences between sexual reproduction in angiosperms and in vertebrate animals:

- In most angiosperms, there are multicellular diploid (sporophyte) and haploid (gametophyte) life stages (alternation of generations); in animals, there is no multicellular haploid stage.
- In angiosperms, the cells that will form gametes are determined in the adult organism, usually in response to environmental conditions; in animals, the germ line cells are determined before birth.
- Meiosis in plants produces spores, after which mitosis produces gametes; in animals, meiosis usually produces gametes directly.

How does the flower function as a structure for sexual reproduction?

The plant life cycle typically involves the alternation of haploid and diploid generations (see Key Concept 11.4):

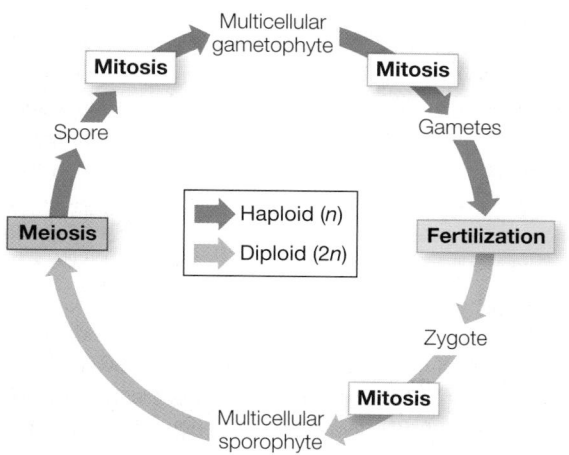

In angiosperms, the adult plant is a diploid sporophyte, and the male and/or female haploid gametophytes (structures that produce gametes) are contained in the flowers (see Key Concept 28.1 for a description of flower parts and floral evolution). A complete flower consists of four concentric groups of organs arising from modified leaves: the carpels, stamens, petals, and sepals.

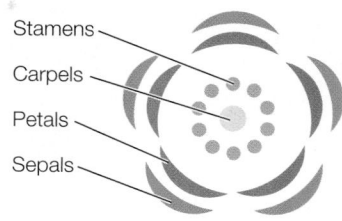

- The carpels are the female sex organs that contain the developing female gametophytes.
- The stamens are the male sex organs that contain the developing male gametophytes.

You learned in Chapter 19 how specific transcription factors control the differentiation of the meristem into the various organs of the flower. Most angiosperms have flowers with both stamens and carpels; such flowers are termed "perfect" (**Figure 37.1A**). "Imperfect" flowers, by contrast, are those with only male or only female sex organs. Male flowers have stamens but not carpels, and female flowers have carpels but not stamens. Some plants, such as corn, bear both male and female flowers on an individual plant; such species are called **monoecious** ("one house") (**Figure 37.1B**). In **dioecious** species, individual plants bear either male-only or female-only flowers; an example is holly (**Figure 37.1C**).

Flowering plants have microscopic gametophytes

You can follow the roles of male and female gametophytes in angiosperm reproduction in **Figure 37.2**. The gametophytes develop from haploid spores in the flower:

- Each female gametophyte (megagametophyte) is called an **embryo sac**, and it develops inside an ovule. One or more ovules are contained within the ovary, which is the lower part of the carpel.
- Male gametophytes (microgametophytes), which are called **pollen grains**, develop inside the anther, which is part of the stamen.

FEMALE GAMETOPHYTE Of the four haploid megaspores resulting from meiosis, three undergo apoptosis (programmed cell death). The remaining megaspore undergoes three mitotic divisions without cytokinesis, producing eight haploid nuclei, all initially contained within a single cell—three nuclei at one end, three at the other, and two in the middle. Subsequent cell wall formation leads to an elliptical, seven-celled megagametophyte with a total of eight nuclei:

- At one end of the megagametophyte are three small cells: the egg cell and two cells called synergids. The egg cell is the female gamete, and the synergids participate in fertilization by

(A) Perfect: lily (*Lilium* sp.)

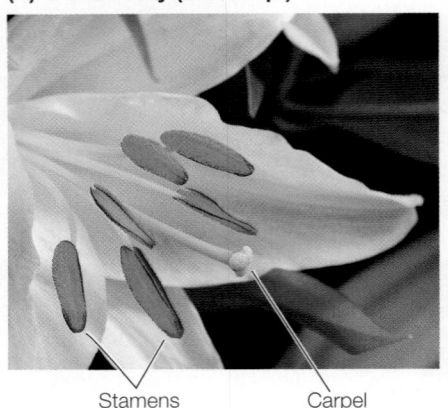

Stamens   Carpel

(B) Imperfect monoecious: corn

Male flower with stamens

Female flower with carpels

(C) Imperfect dioecious: American holly (*Ilex opaca*)

Female flower with carpels

Male flower with stamens

Figure 37.1 Perfect and Imperfect Flowers (A) A lily is an example of a perfect flower, meaning one that has both male and female sex organs. (B) Imperfect flowers are either male or female. Corn is a monoecious species: both types of imperfect flowers are borne on the same plant. (C) American holly is a dioecious species; some American holly plants bear male imperfect flowers whereas others bear female imperfect flowers.

attracting the pollen tube. The pollen tube enters one of the synergids before the sperm cells are released for fertilization.

- At the opposite end of the megagametophyte are three antipodal cells, which eventually degenerate.
- In the large central cell are two **polar nuclei**.

The megagametophyte, or embryo sac, is the entire seven-cell, eight-nucleus structure.

MALE GAMETOPHYTE The four haploid products of meiosis (the microspores) each develop a cell wall and undergo a single mitotic division, usually producing four two-celled pollen grains that are released into the environment. The two cells in a pollen grain have different roles:

- After pollination (see below) the generative cell divides by mitosis to form two sperm cells that participate in fertilization.
- The tube cell forms the elongating pollen tube that delivers the sperm to the embryo sac.

These events occur after the pollen grain is transferred to a stigma (part of the female reproductive organ)—a process called *****pollination**.

***connect the concepts** Flower structure evolved over time. Natural selection has favored floral features that have increased the likelihood of successful pollination. Key Concepts 28.3 and 56.4 describe how angiosperms coevolved with animal pollinators and how ecological communities are formed through interactions among plant and animal species.

A pollen tube delivers sperm cells to the embryo sac

When a functional pollen grain lands on the stigma of a compatible stigma, it germinates. A key event is water uptake by pollen from the stigma: pollen loses most of its water as it matures. Germination involves the development of a **pollen tube** (**Figure 37.3**). The pollen tube either traverses the tissue of the style (part of the

carpel; see Figure 37.2) or, if the style is hollow, grows on the inner surface of the style until it reaches an ovule. The growth rate of the pollen tube varies greatly among species, but it can be as fast as 1 centimeter an hour.

The growth of the pollen tube is guided in part by a chemical signal in the form of a small protein produced by the synergids within the ovule. If one synergid is destroyed, the ovule still attracts pollen tubes, but destruction of both synergids renders the ovule unable to attract pollen tubes, and fertilization does not occur.

Genetic screens (see Figure 36.2) of mutants in *Arabidopsis* have revealed several genes whose expression in synergids is needed for guiding the pollen tube. There is evidence that a small molecule made by the embryo sac—γ-aminobutyric acid, or GABA—is an attractant for pollen tube growth. This would be an amazing coincidence, since this molecule is also a signal in the vertebrate nervous system.

Many flowering plants control pollination or pollen tube growth to prevent inbreeding

You'll recall from discussions of Mendel's work (see Key Concept 12.1) that some plants can reproduce sexually by both cross-pollination and self-pollination. Self-pollination increases the chances of successful pollination but leads to homozygosity, which reduces genetic diversity. Because diversity is the raw material of evolution by natural selection, homozygosity can be selectively disadvantageous. Most plants have evolved mechanisms that prevent *****self-fertilization**. Angiosperms prevent self-fertilization in two main ways: (1) physical separation of male and female gametophytes and (2) genetic self-incompatibility.

***connect the concepts** As described in Key Concept 28.3, flowers known as perfect flowers have functioning megasporangia and microsporangia. Perfect flowers are thus capable of self-pollination, which is usually disadvantageous. Various mechanisms have evolved in perfect flowers to prevent "selfing"; one is described in Figure 28.13.

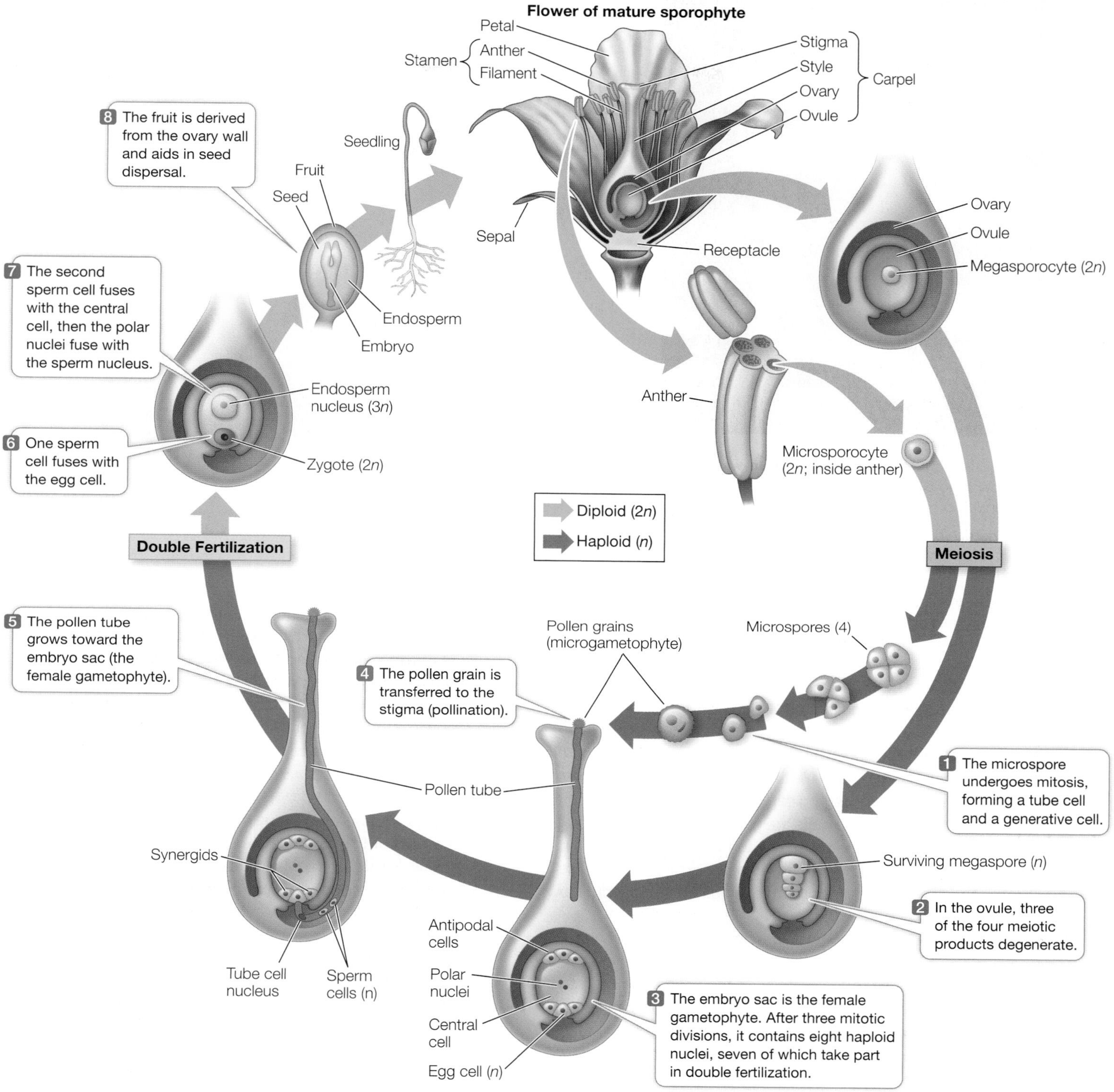

Flower of mature sporophyte

Petal
Stamen { Anther, Filament }
Stigma
Style
Ovary
Ovule
} Carpel

Sepal
Receptacle

Ovary
Ovule
Megasporocyte (2*n*)

8 The fruit is derived from the ovary wall and aids in seed dispersal.

Seedling
Fruit
Seed

Anther

Microsporocyte (2*n*; inside anther)

7 The second sperm cell fuses with the central cell, then the polar nuclei fuse with the sperm nucleus.

Endosperm
Embryo
Endosperm nucleus (3*n*)

Diploid (2*n*)
Haploid (*n*)

Meiosis

6 One sperm cell fuses with the egg cell.
Zygote (2*n*)

Double Fertilization

5 The pollen tube grows toward the embryo sac (the female gametophyte).

Pollen grains (microgametophyte)
Microspores (4)

4 The pollen grain is transferred to the stigma (pollination).

1 The microspore undergoes mitosis, forming a tube cell and a generative cell.

Pollen tube

Synergids

Surviving megaspore (*n*)

2 In the ovule, three of the four meiotic products degenerate.

Antipodal cells
Polar nuclei
Central cell
Egg cell (*n*)

Tube cell nucleus
Sperm cells (*n*)

3 The embryo sac is the female gametophyte. After three mitotic divisions, it contains eight haploid nuclei, seven of which take part in double fertilization.

Figure 37.2 Sexual Reproduction in Angiosperms The embryo sac is the female gametophyte; the pollen grain is the male gametophyte. The male and female cells meet and fuse within the embryo sac. Angiosperms have double fertilization, in which a zygote and an endosperm form from separate fusion events. The zygote forms by the fusion of one sperm cell with the egg cell. The endosperm forms after the other sperm cell fuses with the central cell, which contains two nuclei. The three nuclei fuse, forming a triploid cell.

 Activity 37.1 Sexual Reproduction in Angiosperms
www.Life11e.com/ac37.1

SEPARATION OF MALE AND FEMALE GAMETOPHYTES Self-fertilization is prevented in dioecious species, which bear only male or female flowers on a particular plant. Pollination in dioecious species is accomplished only when one plant pollinates another. In monoecious plants, which bear both male and female flowers on the same plant, the physical separation of the male and female flowers is often sufficient to prevent self-fertilization. Some monoecious species prevent self-fertilization by staggering the development of

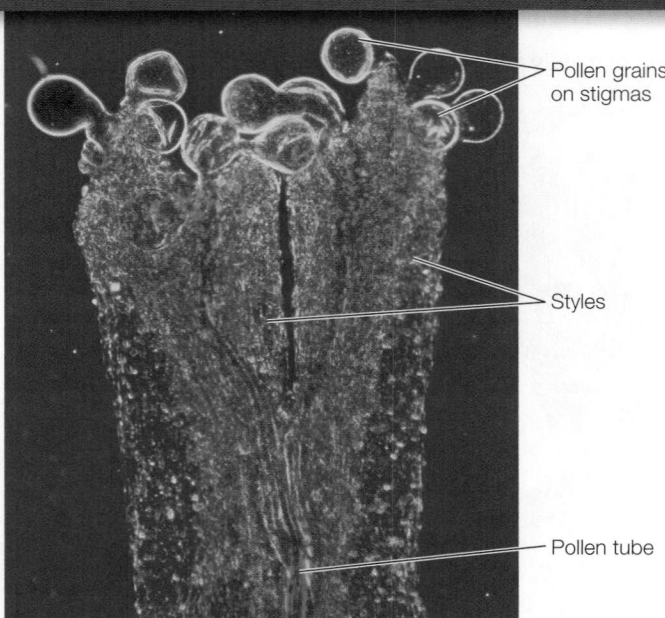

Figure 37.3 **Pollen Tubes Begin to Grow**   Staining pollen with a dye allows it to be seen through a microscope. These pollen grains have landed on the stigmas of a crocus.

Q: How do you think the pollen tube grows in length? What processes are involved in the pollen tube descending the solid tissue of the style?

 Media Clip 37.1 **Pollen Germination in Real Time**
www.Life11e.com/mc37.1

male and female flowers so they do not bloom at the same time, making these species functionally dioecious.

GENETIC SELF-INCOMPATIBILITY   A pollen grain that lands on the stigma of the same plant will fertilize the female gamete *only if the plant is self-compatible*, meaning capable of self-pollination. To prevent self-fertilization, many plants are genetically self-incompatible. **Self-incompatibility** depends on the ability of a plant to determine whether pollen is genetically similar or genetically different from itself. Rejection of "same-as-self" pollen prevents self-fertilization. How does it occur?

Self-incompatibility in many flowering plants is controlled by a cluster of tightly linked genes called the *S locus* (for *self-incompatibility*). The *S* locus encodes proteins in the pollen and style that interact during the recognition process. A self-incompatible species typically has many alleles of the *S* locus. The pollen phenotype may be determined by its own haploid genotype or by the diploid genotype of its parent plant. In either case, if the pollen expresses an allele that matches either of the alleles expressed in the recipient pistil, the pollen is rejected. Depending on the type of self-incompatibility system, the rejected pollen either fails to germinate or is prevented from growing through the style (**Figure 37.4**); either way, self-fertilization is prevented.

Angiosperms perform double fertilization

In most angiosperm species, the mature pollen grain consists of two cells: a large tube cell enclosing a much smaller generative cell. When a compatible pollen grain lands on the stigma of a plant of the same species, it germinates, and the pollen tube grows through the

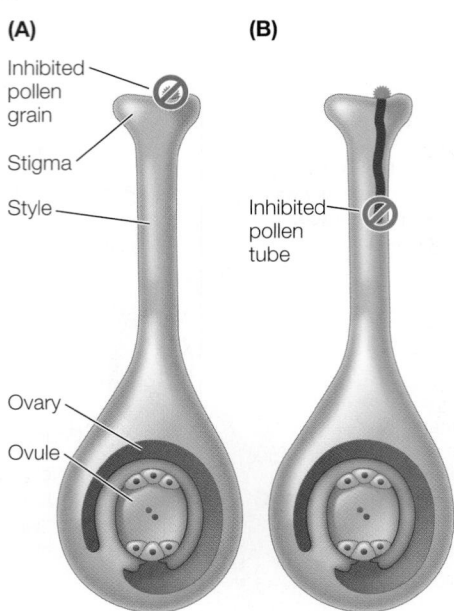

Figure 37.4 **Self-Incompatibility**   In a self-incompatible plant, pollen is rejected if it expresses an *S* allele that matches one of the *S* alleles of the stigma and style. Self pollen may **(A)** fail to germinate or **(B)** its pollen tube may die before reaching an ovule. In either case, the egg cannot be fertilized by a sperm from the same plant.

style tissue to the embryo sac. During this process, the generative cell undergoes one mitotic division and cytokinesis to produce two haploid **sperm cells** (**Figure 37.5, steps 1 and 2**).

Two fertilization events now occur. One of the two synergids degenerates when the pollen tube arrives and the two sperm cells are released into its remains (**Figure 37.5, step 3**). Each sperm cell then fuses with a different cell of the embryo sac (**Figure 37.5, steps 4 and 5**). One sperm cell fuses with the egg cell, and the two nuclei fuse, producing the diploid zygote. The other sperm cell fuses with the central cell, and its nucleus fuses with the two polar nuclei, forming a **triploid (3*n*) cell**. Immediately after fertilization the triploid nucleus undergoes rapid mitotic divisions to form a specialized nutritive tissue, the **endosperm**. In most species the endosperm nucleus initially divides without cytokinesis, forming a large, multinucleate cell, and cell walls form later between the nuclei. After the endosperm begins developing, the zygote undergoes mitotic division to form the new sporophyte embryo. The developing embryo uses the endosperm tissue as a source of nutrients, energy, and carbon-based anabolic building blocks. In some cases the endosperm persists until germination and is used as a source of nutrients by the developing seedling; this is prevalent in grains such as corn, wheat, and rice. The source of nutrients is important because the seedling often begins its development underground and cannot perform photosynthesis right away.

The remaining cells of the male and female gametophytes—the antipodal cells, the remaining synergid, and the pollen tube nucleus—degenerate as the embryo begins to develop.

Double fertilization, a hallmark of angiosperms, is so named because it involves two cell fusion events:

● One sperm cell fuses with the egg cell.

● The other sperm cell fuses with the central cell.

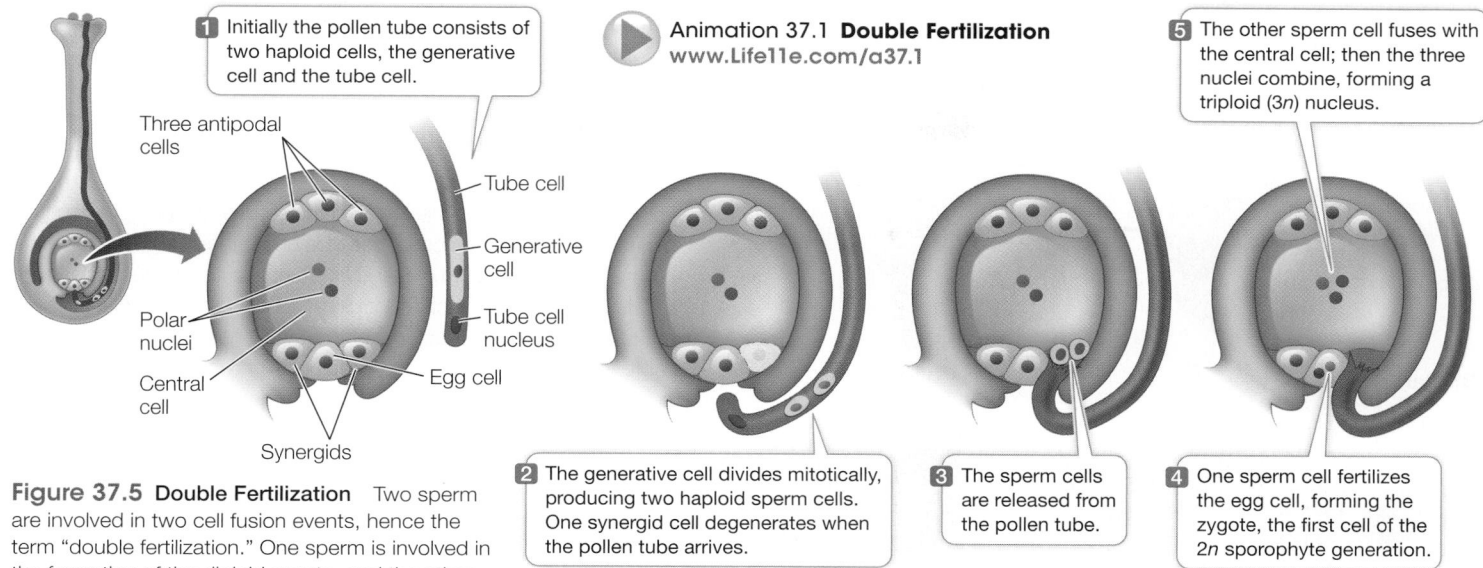

1 Initially the pollen tube consists of two haploid cells, the generative cell and the tube cell.

Animation 37.1 **Double Fertilization**
www.Life11e.com/a37.1

5 The other sperm cell fuses with the central cell; then the three nuclei combine, forming a triploid (3*n*) nucleus.

Three antipodal cells

Tube cell

Generative cell

Polar nuclei

Tube cell nucleus

Central cell

Egg cell

Synergids

2 The generative cell divides mitotically, producing two haploid sperm cells. One synergid cell degenerates when the pollen tube arrives.

3 The sperm cells are released from the pollen tube.

4 One sperm cell fertilizes the egg cell, forming the zygote, the first cell of the 2*n* sporophyte generation.

Figure 37.5 Double Fertilization   Two sperm are involved in two cell fusion events, hence the term "double fertilization." One sperm is involved in the formation of the diploid zygote, and the other results in the formation of the triploid endosperm cell that divides to form endosperm. Double fertilization is a characteristic feature of angiosperm reproduction.

The fusion of a sperm cell with the central cell to form the triploid endosperm nucleus is one of the defining characteristics of angiosperms.

Embryos develop within seeds contained in fruits

Fertilization initiates the highly coordinated growth and development of the embryo, endosperm, integuments, and carpel. The integuments—protective tissue layers surrounding the ovule—develop into the seed coat, and the ovary wall becomes the outer layers of the fruit that encloses the seed (see Figure 37.6A).

In Chapter 36 you learned about the events in plant embryonic development and its hormonal control. As seeds develop, they prepare for dispersal and dormancy by losing up to 95 percent of their water content. You can see this desiccation by comparing corn grains (e.g., popcorn) with ripe corn from the cob. A dry seed is still alive; it has protective proteins that keep its cells in a living state.

In angiosperms, the ovary—together with the seeds it contains—develops into a fruit after fertilization has occurred. Fruits have two main functions:

1. They aid in seed dispersal.

2. They protect the seed from damage by animals and infection by microbial pathogens.

A **fruit** may consist of only the mature ovary and seeds, or it may include other parts of the flower. Some species produce fleshy, edible fruits, such as peaches and tomatoes, whereas the fruits of other species are dry (**Figure 37.6**).

The diverse forms of fruits reflect the varied strategies plants use to disperse their progeny. Because plants cannot move, their progeny need some mechanism for separating themselves from their parents. Wide dispersal of progeny may not always be advantageous, however. If a plant has successfully grown and reproduced, its location is likely to be favorable for the next generation too. Some offspring do indeed stay near their parents. This is the case in many tree species, whose seeds simply fall to the ground. However, this strategy has several potential disadvantages. If the species is a perennial, the offspring that germinate near their parents will be competing with their parents for resources, which may be too limited to support a dense population. Furthermore, even though local conditions were good enough for the parent to produce at least some seeds, there is no guarantee that conditions will still be good the next year, or that they won't be better elsewhere. Thus in many cases seed dispersal is vital to a species' survival.

Many fruits help disperse seeds over substantial distances, increasing the probability that at least a few of the seeds will find suitable conditions for germination and growth to sexual maturity. Some wind-dispersed seeds have fruits with "wings," like those of the familiar maple. In other cases, the fruits and seeds are tiny, and they include feathery structures, such as in this thistle:

Other fruits, such as these burs, attach themselves to animals (or to your clothes and shoes):

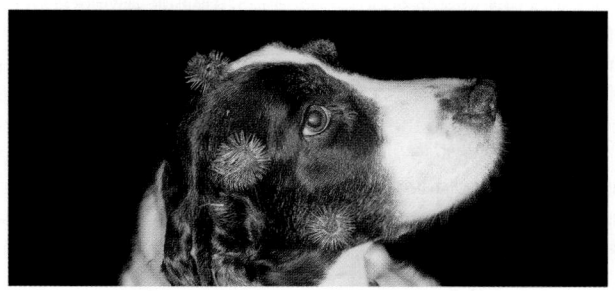

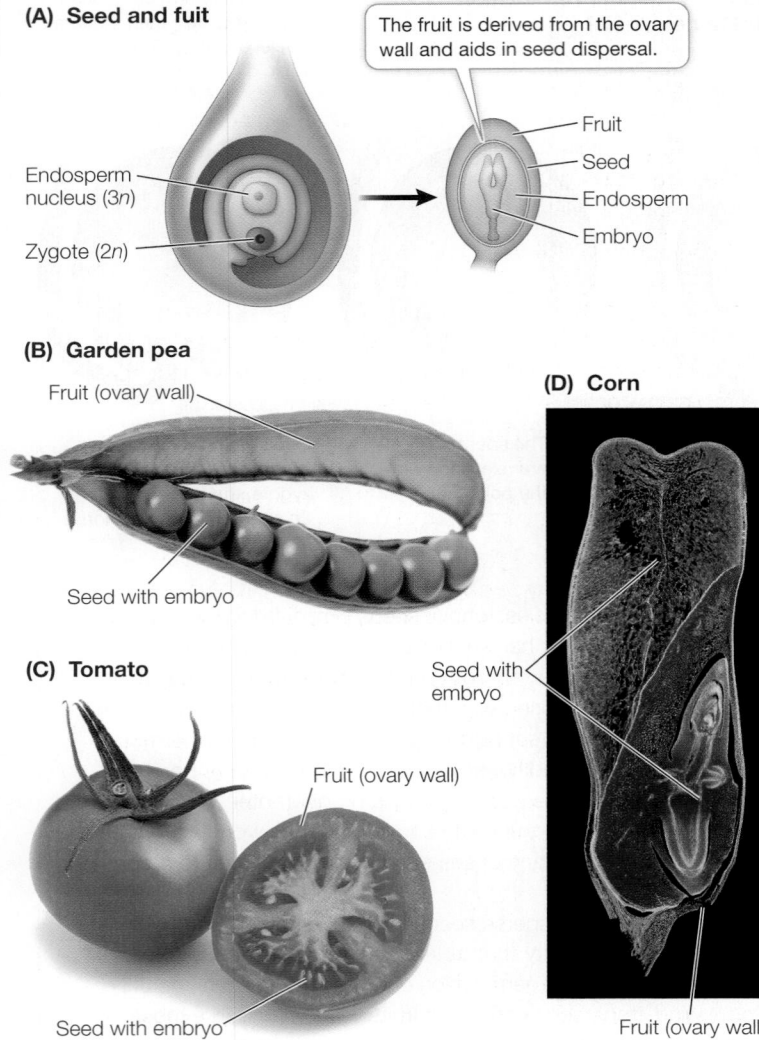

(A) Seed and fuit

The fruit is derived from the ovary wall and aids in seed dispersal.

Fruit
Seed
Endosperm
Embryo

Endosperm nucleus (3n)
Zygote (2n)

(B) Garden pea

Fruit (ovary wall)

Seed with embryo

(C) Tomato

Fruit (ovary wall)

Seed with embryo

(D) Corn

Seed with embryo

Fruit (ovary wall)

Figure 37.6 Angiosperm Fruits There are a variety of fruits, but all have a seed containing the embryo, surrounded by a fruit that comes from the wall of the ovary.

Water disperses some fruits; coconuts have been known to float thousands of miles between islands. Seeds swallowed whole by an animal along with fruits such as berries travel through the animal's digestive tract and are deposited some distance from the parent plant.

Biologists are beginning to understand the relationships between seed development (from ovules) and fruit development (from carpels). Some seedless fruits, such as varieties of watermelons and grapes, develop when fertilization occurs, but the embryo then aborts. In other cases, such as bananas, the fruit develops without fertilization. In most cases, however, fruit development does not occur in the absence of fertilization. Several years ago a farmer in Spain who grows sugar apples (*Annona squamosa*) noticed a seedless fruit and brought it to the attention of scientists. A single gene was subsequently identified whose mutated form results in this seedless phenotype. The wild-type version of this gene encodes a transcription factor important to the development of the outer tissues of the ovule. The gene is present in the genomes of all angiosperms

examined. In the future, therefore, it may be possible to produce other seedless fruits by engineering mutations in this gene.

Seed development is under hormonal control

Chapter 36 described the role of the gibberellin hormones in the mobilization of stored macromolecules in the seed endosperm during germination. The development of seeds is under the control of a different hormone, **abscisic acid** (**ABA**). (Unfortunately, its name is misleading, because it does not directly control leaf abscission.) Most plant tissues make ABA, and like other plant hormones, it has multiple effects (see Table 36.2). During early seed development the ABA level is usually low, and it rises as the seed matures. This increase stimulates the endosperm to synthesize seed storage proteins. It also stimulates the synthesis of proteins that prevent cell death as the seeds dry.

ABA also keeps the developing seed from germinating on the plant before it dries. Premature germination, termed **vivipary**, is undesirable in seed crops (such as wheat) because the grain is damaged if it starts to sprout. Viviparous seedlings are also unlikely to survive if they remain attached to the parent plant and are unable to establish themselves in the soil. Mutant strains of corn that are insensitive to ABA have viviparous seeds, indicating the importance of ABA in preventing precocious germination.

The general effect of ABA in preventing germination extends to seed dormancy. Seeds tend to stay dormant if their ABA levels are high and germinate when the levels go down. This usually occurs as dormancy is broken.

37.1 recap

Flowers contain the organs for sexual reproduction in angiosperms. Plants that use pollen for reproduction have several selective advantages. After fertilization, the flower develops into seed(s) and fruit. The selective advantages of seeds and fruits include long-term viability and multiple modes of dispersal.

learning outcomes

You should be able to:

- Compare the processes of male and female gamete formation.
- Describe the mechanisms that guide the growth of a pollen tube.
- Describe and compare two methods for preventing self-fertilization in angiosperms.
- Analyze the relationship between the diversity of fruits and their ability to disperse seeds.
- Relate fruit development to seed development.

1. How would plant reproduction be affected if the pollen tube failed to grow and develop properly?

2. How do the final products of meiosis in megagametophytes and microgametophytes differ, and how are these differences important to sexual reproduction?

3. Thompson Seedless grapes are produced from triploid plants. Explain why these grapes are seedless. Describe the role played by the flower in fruit formation when no seeds are being formed.

4. What is the selective advantage for the development of many diverse types of fruits?

We have now traced the sexual life cycle of angiosperms from the flower to the gametophytes, pollination, fertilization, and the dispersal of seeds. We discussed seed germination and seedling development in Chapter 36, and vegetative plant growth in Chapter 33. Next we look at the rest of the angiosperm life cycle—the transition from the vegetative to the flowering state—and how this transition is regulated.

key concept 37.2 Hormones and Signaling Determine the Transition from the Vegetative to the Reproductive State

Flowering is one of the major events in a plant's life. Flowering requires a reallocation of energy and materials away from making more non-reproductive plant parts (vegetative growth) to making flowers and gametes (reproductive growth). Once a plant is old enough, it can respond to internal or external signals to initiate reproduction. Flowering can happen right at maturity as part of a predetermined developmental program or in response to environmental cues such as light or temperature.

focus your learning

- Although in terms of flowering, many plants are classified as either short-day plants (SDPs) or long-day plants (LDPs), night length is actually the cue that controls flowering.
- Receptors for the photoperiodic signal for flowering are located in the leaf, and a signal travels to the apical meristem.
- The protein florigen converts a vegetative meristem into a reproductive meristem. Several genes are involved in the regulation and transport of florigen.
- Temperature or gibberellin can induce flowering in some plants.

Flowering occurs at specific places and specific times

Plants fall into three categories depending on when they mature and initiate flowering, and what happens after they flower:

1. **Annuals** complete their lives in one year. This class includes many crops important to the human diet, such as corn, wheat, rice, and soybean. When the environment is suitable, these plants grow rapidly, with little or no secondary growth. After flowering, they use most of their materials and energy to develop seeds and fruits, and the rest of the plant withers away.

2. **Biennials** take 2 years to complete their lives. They are much less common than annuals and include carrots, cabbage, and onions. Typically, biennials produce only vegetative growth during the first year and store carbohydrates in underground roots (carrot) and stems (celery). In the second year they use most of the stored carbohydrates to produce flowers and seeds rather than vegetative growth, and the plant dies after seeds form.

3. **Perennials** live 3 or more—sometimes many more—years. Maple trees can live up to 400 years. Perennials include many trees and shrubs, as well as wildflowers. Typically these plants flower every year but stay alive and keep growing for another season; the reproductive cycle repeats each year. However, some perennials (e.g., century plant) grow vegetatively for many years, flower once, and die.

No matter what type of life cycle they have, angiosperms all make the transition to flowering. The first visible sign of a transition to the flowering state may be a change in one or more apical meristems in the shoot system. As you saw in Chapter 33, meristems have a pool of undetermined cells. During vegetative growth, a shoot apical meristem continually produces leaves, axillary buds, and stem tissues (**Figure 37.7A**) in a kind of unrestricted growth called indeterminate growth (see Key Concept 33.3).

Flowers may appear singly or in an orderly cluster that constitutes an **inflorescence**. If a vegetative apical meristem becomes an **inflorescence meristem**, it stops making leaves and axillary buds and produces other structures: smaller leafy structures called bracts, as well as new meristems in the angles between the bracts and the stem (**Figure 37.7B**). These new meristems may also be inflorescence meristems, or they may be **floral meristems**, each of which gives rise to a flower.

Each floral meristem typically produces four consecutive whorls or spirals of organs—the sepals, petals, stamens, and carpels discussed earlier in the chapter—separated by very short internodes, keeping the flower compact (**Figure 37.7C**). In contrast to vegetative apical meristems and some inflorescence meristems, floral meristems are responsible for determinate growth—growth of limited duration.

A cascade of gene expression leads to flowering

The genes that determine the transition from shoot apical meristems to inflorescence meristems and from inflorescence meristems to floral meristems have been studied in model organisms such as *Arabidopsis*.

MERISTEM IDENTITY GENES Expression of two **meristem identity genes** initiates a cascade of further gene expression that leads to flower formation. These genes encode the transcription factors LEAFY and APETALA1, which together are necessary and sufficient for determining the transition to flowering. Evidence for the roles of these factors comes from both genetic and plant transformation experiments. For example, a mutant allele of the *APETALA1* gene leads to continued vegetative growth, even if conditions are suitable for flowering. However, if the wild-type *APETALA1* gene is coupled with a constitutive (always on) promoter and used to transform *Arabidopsis* plants, the plants will flower prematurely, regardless of environmental conditions. This is powerful evidence that APETALA1 plays a role in switching meristem cells from a vegetative to a reproductive fate (see Figure 37.7B).

FLORAL ORGAN IDENTITY GENES The products of the meristem identity genes trigger the expression of **floral organ identity genes**. They encode transcription factors that determine whether cells in the floral meristem will be sepals, petals, stamens, or carpels.

(A)

A vegetatively growing **apical meristem** continues to produce leaves and stem.

Vegetative apical meristem

Leaf

Figure 37.7 Flowering and the Apical Meristem A vegetative apical meristem **(A)** grows without producing flowers. Once the transition to the flowering state is made **(B)**, inflorescence meristems give rise to bracts and to floral meristems **(C)**, which become the flowers.

(B)

Inflorescence meristems give rise to floral meristems, bracts, and more inflorescence meristems.

Inflorescence meristem

Floral (or inflorescence) meristem

A **bract** is a modified, usually reduced, leaflike structure.

Meristem identity genes

(C)

A **floral meristem** gives rise to a flower.

Floral meristem

Floral organ identity genes

Petal

Stamen
Carpel
Sepal

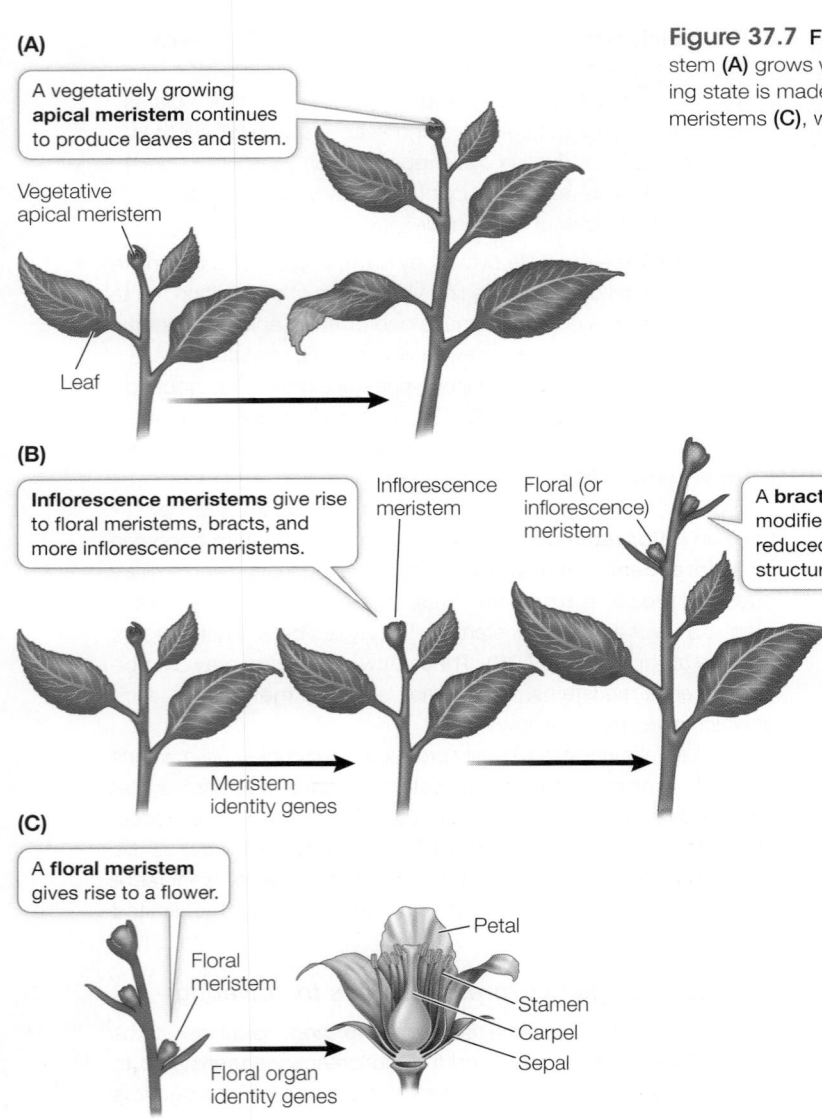

For example, *AGAMOUS* is a *class C gene that causes cells determined to be part of the flower to form stamens and carpels. Floral organ identity genes often become activated in response to external cues (day length, temperature) or internal cues (hormones). We begin with day length, or photoperiod.

---

*connect the concepts The roles of floral organ identity genes and their protein products are described in Key Concept 19.3. Class A genes are expressed in whorls 1 and 2 (which form sepals and petals, respectively). Class B genes are expressed in whorls 2 and 3 (which form petals and stamens). Class C genes are expressed in whorls 3 and 4 (which form stamens and carpels). See Figure 19.11.

Photoperiodic cues can initiate flowering

The study of how light affects the transition to flowering began with two observations in the early twentieth century:

- Normally, tobacco grows to about 1.5 meters tall before flowering in the summer, but a variety called Maryland Mammoth grows to an amazing 5 meters. Farmers in Virginia were frustrated because they could not easily get seeds from this luxuriant plant for successive crops. Instead of flowering, it continued to grow until the late fall frost killed it.

- Because of improvements in agricultural techniques, soybean yields became so great that it was hard for farmers to harvest all the plants at once. Hoping to stagger the harvests, farmers tried planting the seeds in groups several weeks apart, but all resulting plants nevertheless formed flowers and seeds at the same time.

The explanation for both of these observations was the same: the signal that set the plants' shoot apical meristems on the path to flowering was the length of daylight, or **photoperiod**. When soybeans experience days of a certain length, they flower, regardless of how "old" they are. Maryland Mammoth tobacco *can* flower, but it doesn't do so in Virginia because it dies when the weather there gets cold. Maryland Mammoth is now grown commercially in Florida. You saw another example of the role of photoperiod in the opening investigation of this chapter, dealing with poinsettias.

Scientists used greenhouse experiments to measure the day length required for different plant species to flower. Maryland Mammoth tobacco did not flower if exposed to more than 14 hours of light per day; flowering was only initiated once day length became shorter than 14 hours, as it does in December. Other plants (such as soybean and henbane) flowered only when the days were long (**Figure 37.8**). Control of an organism's responses by the length of day or night is called **photoperiodism**.

Plants vary in their responses to photoperiodic cues

Plants that flower in response to photoperiodic stimuli fall into two main classes:

1. **Short-day plants (SDPs)** flower only when the day is shorter than a critical maximum. They include coffee, morning glory, Maryland Mammoth tobacco, poinsettias, and chrysanthemums. Thus, for example, you see chrysanthemums in nurseries in the fall and poinsettias in winter.

2. **Long-day plants (LDPs)** flower only when the day is longer than a critical minimum. Spinach, lettuce, and clover are examples of LDPs. For example, spinach tends to flower in the summer, so for harvesting leaves, it is planted in early spring.

While there are variations on these two patterns, photoperiodic control of flowering serves an important role: it synchronizes the flowering of plants of the same species in a local population, and this promotes cross-pollination and successful reproduction.

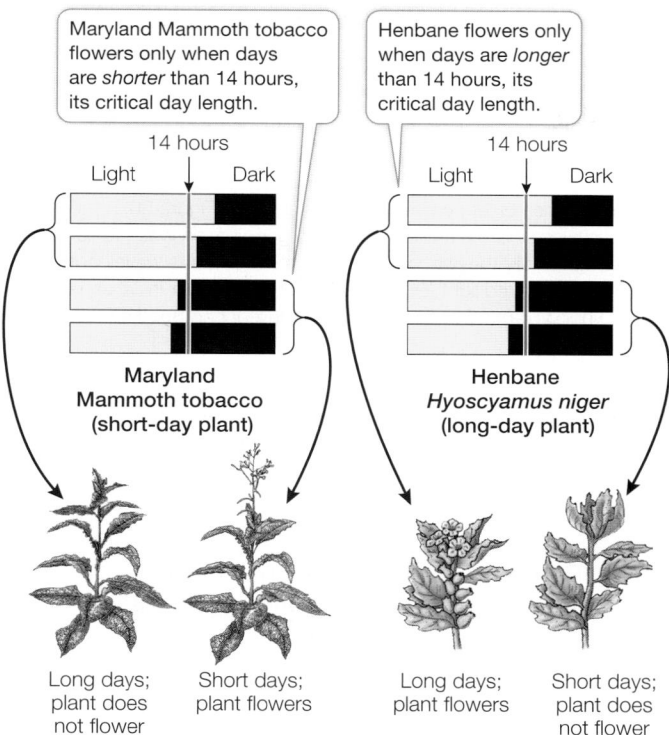

Figure 37.8 **Day Length and Flowering** Flowering of Maryland Mammoth tobacco is initiated when the days become shorter than a critical length. Maryland Mammoth tobacco is thus called a short-day plant. Henbane, a long-day plant, shows an inverse pattern of flowering.

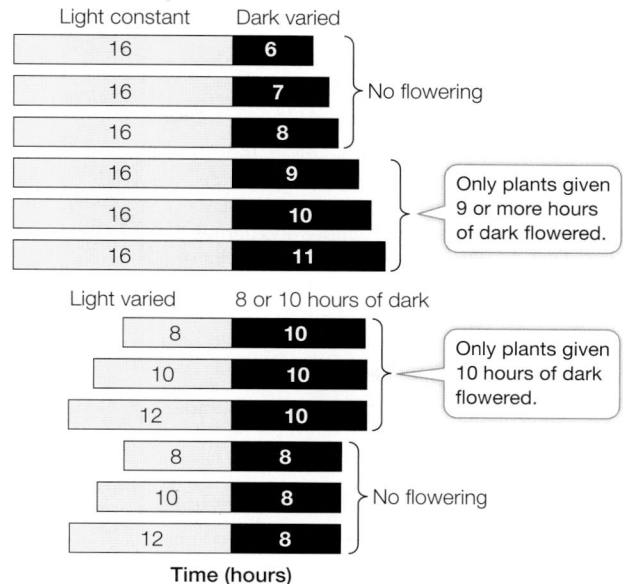

Figure 37.9 **Night Length and Flowering** Greenhouse experiments using cocklebur, a short-day plant, showed that night length, not day length, is the environmental cue that initiates flowering.

Night length is a key photoperiodic cue that determines flowering

The terms "short-day plant" and "long-day plant" imply that *day length* is the environmental cue that triggers flowering. Actually, the important cue is *night length*, as a series of greenhouse experiments confirmed (**Figure 37.9**). In a greenhouse, the overall length of a day or night can be varied, irrespective of the 24-hour natural cycle. For example, if cocklebur, an SDP, is exposed to several long periods of light (16 hours each), it will still flower as long as the dark period between them is 9 hours or longer. This 9-hour inductive dark period also induces flowering even if the light period varies from 8 hours to 12 hours.

Biologists noticed that when the inductive dark period was interrupted by a brief period of light, the flowering signal generated by the long night disappeared. It took several days of long nights for the plant to recover and initiate flowering. Interrupting the day with a dark period had no effect on flowering. A clue as to what occurred in the plant when the flash of light was given came when biologists determined the action spectrum for the wavelengths of light that were effective. As with lettuce seed germination (see Figure 36.12), red light was most effective at breaking the "night" stimulus, and its effect was reversible by far-red light (**Investigating Life: The Flowering Signal**). Later, the photoreceptor involved was identified as a phytochrome.

As described in Key Concept 36.4, phytochromes and a blue-light receptor function together to "entrain" a circadian rhythm in

plants. These photoreceptors cycle through active and inactive phases over repeated 24-hour periods. One gene whose expression follows a circadian rhythm is *CONSTANS* (*CO*), which encodes a transcriptional regulator that controls the expression of flowering genes.

Experiments with *Arabidopsis*, an LDP, have shown that photoperiodic flowering times are determined by interactions between photoreceptors and the CO protein. *CO* gene expression goes through regular 24-hour cycles with peak expression late in the day. On long days, this occurs in daylight, late in the afternoon, but on short days the peak expression occurs after dark. On long days, the active forms of phytochrome and a blue-light receptor activate pathways that stabilize the CO protein, which promotes flowering. This process does not occur on short days.

 Animation 37.2 **The Effect of Interrupted Days and Nights**
www.Life11e.com/a37.2

The flowering stimulus originates in a leaf

Early experiments indicated that reception of the photoperiodic stimulus occurs within the leaf. For example, in the LDP spinach, flowering occurred if the leaves were exposed to long-day periods of light, while the shoot apical meristem was masked to simulate short days. Flowering could *not* occur when the leaves were masked to simulate short days, while the bud was exposed to long-day periods of light.

These "masking" experiments were extended to SDP plants as well (**Figure 37.10**). Because the receptor of the stimulus (in the leaf) is physically separated from the tissue on which the stimulus acts (the bud meristem), the inference can be drawn that a systemic signal travels from the leaf through the plant's tissues to the bud

experiment

Original Paper: Hendricks, S. B. and H. W. Siegelman. 1967. Phytochrome and photoperiodism in plants. *Comparative Biochemistry* 27: 211–235.

Knowing that plants measure night duration, the question became whether the dark hours to which a plant is exposed must be continuous. Using SDPs and LDPs as test subjects, Karl Hamner and James Bonner demonstrated that this was the case by interrupting the night with short bursts of light. Sterling Hendricks and William Siegelman repeated the experiments using light of different wavelengths to gain information about the photoreceptor involved.

HYPOTHESIS▶   Red light participates in the photoperiodic timing mechanism.

METHOD

Grow plants under short-day conditions, but interrupt the night with light of different wavelengths.

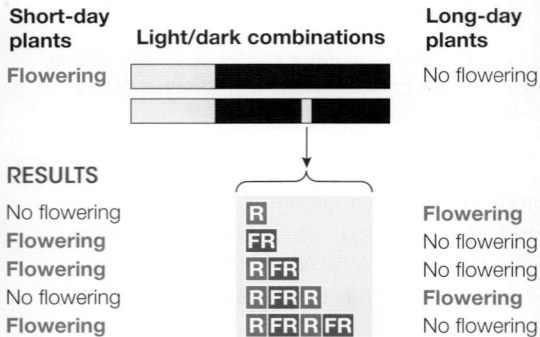

Short-day plants	Light/dark combinations	Long-day plants
Flowering		No flowering

RESULTS

No flowering	R	Flowering
Flowering	FR	No flowering
Flowering	R FR	No flowering
No flowering	R FR R	Flowering
Flowering	R FR R FR	No flowering

CONCLUSION▶   When plants are exposed to red (R) and far-red (FR) light in alternation, the final treatment determines the effect. Phytochrome is the photoreceptor.

 Activity 37.2 **Flowering and Day Length Simulation**
www.Life11e.com/ac37.2

work with the data

Original Paper: Borthwick, H. A., S. B. Hendricks and M. W. Parker. 1952. The reaction controlling floral initiation. *Proceedings of the National Academy of Sciences USA* 38: 929–934.

After it was established that the length of night, rather than day, was critical for flowering in some plants, the U.S. Department of Agriculture group that had worked on the phytochrome response of lettuce seed germination (see Figure 36.12) asked whether the signaling system that initiated lettuce seed germination was the same as that involved in flowering. Initially they showed that the most effective wavelength of light at inhibiting flowering was red light (660 nanometers [nm]). Their initial experiments were done on young plants, from which all but one leaf had been removed.

QUESTIONS▶

1. Plants were grown as seedlings in long-night conditions (12 h dark). After exposure to red light in the middle of the night, groups of four plants were exposed to light a second time and then allowed to resume growth in short-night conditions (8 h dark). After 12 days, the possible presence of a floral meristem was examined by dissection. **Table A** shows the results.

Table A

	Number of plants with a floral meristem after exposure to second light			
Wavelength (nm)	16 min	8 min	4 min	2 min
680	0	0	0	0
700	1	3	0	0
720	3	3	3	1

  a. Why were the plants grown in short-night conditions prior to evaluation of flowering?

  b. What can you conclude from the data in terms of the receptor and time needed?

2. Groups of four plants were exposed to red, far-red, or no light at the beginning of the dark period. That is, if there were a 7-hour dark period, the plants were exposed to light just after the day period ended (extending the day). The results are shown in **Table B**. Explain the data in terms of the phytochrome system.

Table B

	Number of plants with a floral meristem		
Length of dark period (h)	No light	660 nm	720 nm
7.0	—	—	4
7.5	—	—	4
8.0	0	0	4
8.5	4	0	4
9.0	4	4	—
9.5	4	4	—

A similar **work with the data** exercise may be assigned in **LaunchPad**.

meristem. Other evidence that a diffusible chemical travels from the leaf to the bud meristem signal includes the following:

- If a photoperiodically induced leaf is immediately removed from a plant after the inductive dark period, the plant does not flower. If, however, the induced leaf remains attached to the plant for several hours, the plant will flower. This result suggests that something is synthesized in the leaf in response to the inductive dark period, and then moves out of the leaf to induce flowering.

- If two cocklebur plants are grafted together, and if only one of the plants is exposed to inductive long nights, both plants flower.

- In several species, if an induced leaf from one species is grafted onto another, noninduced plant of a different species, the recipient plant flowers.

The transmissible signal was given a name, **florigen** ("flower inducing"), in 1937, but its chemical nature has been described only in the past decade.

experiment

Figure 37.10A The Flowering Signal Moves from Leaf to Bud

Original Paper: Hamner, K. C. and J. Bonner. 1938. Photoperiodism in relation to hormones as factors in floral initiation and development. *Botanical Gazette* 100: 388–431.

The receptors for photoperiod are in the leaf, but the transition to flowering occurs in the shoot apical meristem. To investigate whether there is a diffusible substance that travels from leaf to bud, Hamner and Bonner exposed only the leaf to the photoperiodic stimulus.

HYPOTHESIS▶ The leaves measure the photoperiod.

METHOD

Grow cocklebur plants (short-day) under long days and short nights. Mask a leaf on some plants and see if flowering occurs.

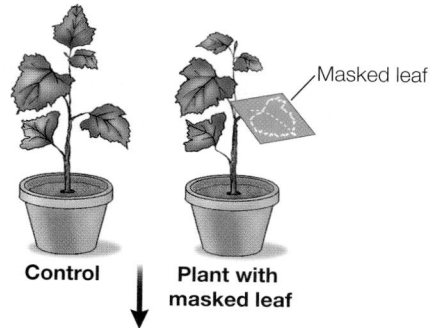

Control   Plant with masked leaf

Masked leaf

RESULTS

If even one leaf is masked for part of the day—thus shifting that leaf to short days and long nights—the plant will flower.

Burrs (fruit)

Masked leaf

CONCLUSION▶ The leaves measure the photoperiod. Therefore some signal must move from the induced leaf to the flowering parts of the plant.

work with the data

Figure 37.10B The Flowering Signal Moves from Leaf to Bud

Original Paper: Hamner, K. C. and J. Bonner. 1938.

In 1938 Karl Hamner at the University of Chicago was working on the role of plant nutrition in flowering. The plant he studied, cocklebur, is a short-day plant that requires 16 hours of darkness to flower. When the plants were kept in 6 hours darkness (16 hours light) in a greenhouse, they did not flower. One day Hamner came to the lab to find all the plants flowering. It turned out that there had been a power outage, and the plants had received a single inductive short day (long night). Realizing that this provided a simple system to study flowering, Hamner invited a major scientist in the field, James Bonner from Caltech, to join him for the summer. The two biologists carried out a series of experiments using the single inductive period that showed that flowering is induced by night length as opposed to day length (see Figure 37.9) and that the flowering signal is received by the leaf from which it travels to a bud, inducing flowering. A Russian plant physiologist, Mikhail Chailakhyan, named this signal florigen. More recently, the molecular nature of this signal was described.

QUESTIONS▶

1. Intact cocklebur plants (6) or plants with their leaves removed (6) were placed in the inductive (short-day) photoperiod. After 14 days, the researchers obtained the results in **Table A**. Based on these data, which part of the plant senses the photoperiod?

Table A

Treatment	Number of plants that flowered
No inductive period, intact plant	0
Inductive period, intact plant	6
Inductive period, leaves removed	0

2. Cocklebur plants were treated so that a single leaf was exposed to the inductive (short) photoperiod while the rest of the plant received the long photoperiod. The results obtained after 18 days are shown in **Table B**. What do these results indicate about the location of the receptor for flowering?

Table B

Treatment	Number of plants	Result
Untreated	6	Vegetative
Treated, one leaf	32	Flower

3. What do the data tell you about the signal generated by the plant in response to photoperiod and that induces flowering in the apical meristem?

A similar **work with the data** exercise may be assigned in **LaunchPad**.

Florigen is a small protein

The characterization of florigen was made possible by genetic and molecular studies of the model plant *Arabidopsis*. Three genes are involved in the signaling response for flowering (**Figure 37.11**).

1. **FT (FLOWERING LOCUS T)** *codes for florigen.* FT is a small (20 kilodalton [kDa]) protein that can travel through plasmodesmata. FT is synthesized in the phloem companion cells of the leaf and then diffuses into the adjacent sieve elements. It then is carried through the phloem to the apical meristem. If the *FT* gene is coupled to an active promoter and expressed at high levels in the shoot meristem, flowering is induced even in the absence of an appropriate photoperiodic stimulus.

2. **CO (CONSTANS)** *codes for the transcription factor that activates the synthesis of FT.* As described above, *CO* expression follows a circadian rhythm, and stabilization of the CO protein by photoreceptors allows it to function. Like *FT*, *CO* is

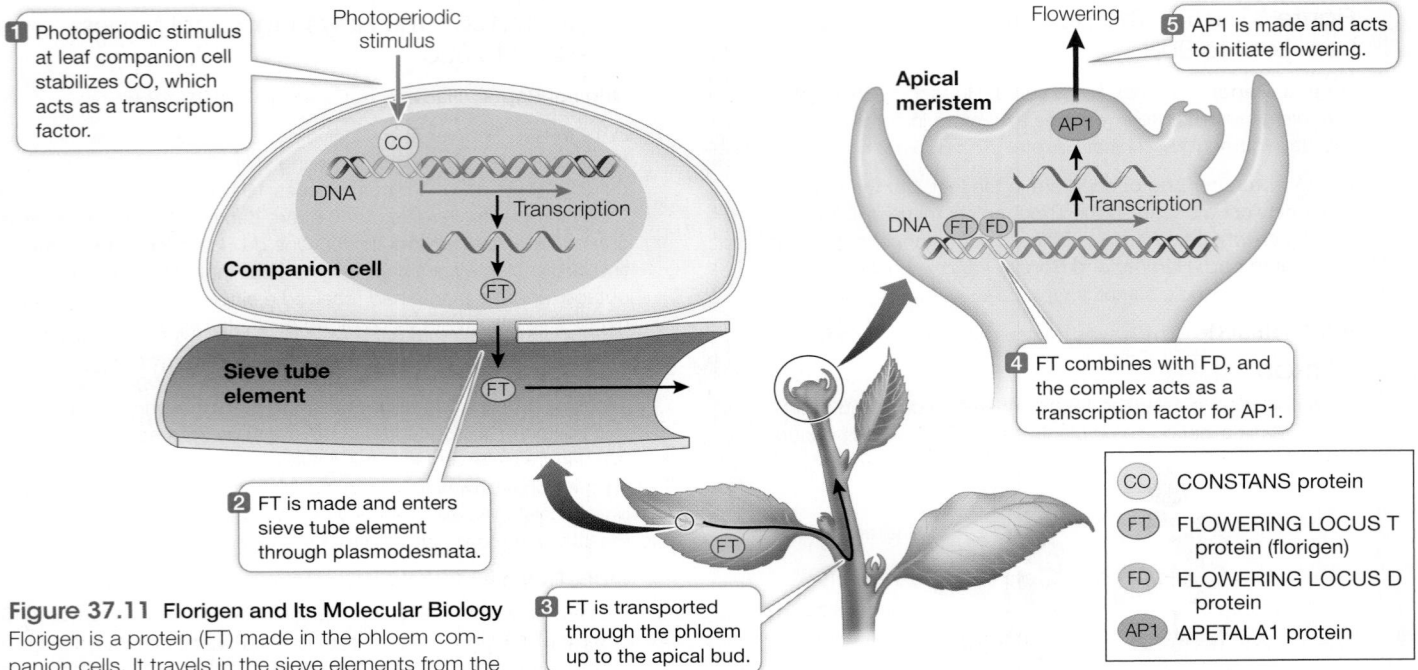

Figure 37.11 Florigen and Its Molecular Biology
Florigen is a protein (FT) made in the phloem companion cells. It travels in the sieve elements from the leaf to the bud meristem. There, florigen combines with another protein to stimulate transcription of genes that initiate flower formation.

expressed in leaf companion cells. If *CO* is experimentally overexpressed in the leaf, flowering is induced. However, if *CO* is overexpressed in the apical meristem, flowering is not induced, indicating that *CO* functions in the leaf.

3. **FD (FLOWERING LOCUS D)** *encodes a protein that binds to FT protein when it arrives in the apical meristem.* The FD protein is a transcription factor that, when bound to FT, activates promoters for meristem identity genes, such as *APETALA1* (see Figure 37.11). The expression of FD primes meristem cells to change from a vegetative fate to a reproductive fate once FT arrives.

Before FT was isolated, grafting experiments indicated that many different plant species could be induced to flower by the same chemical signal. Results of molecular experiments confirmed that the *FT* gene is involved in photoperiod signaling in many species:

- Transgenic plants (e.g., tobacco and tomato) that express the *Arabidopsis FT* gene at high levels flower regardless of day length.

- Transgenic *Arabidopsis* plants that express *FT* homologs from other plants (e.g., rice and tomato) flower regardless of day length.

How is this molecular system related to the photoperiodic stimulus? This has been described in the long-day plant *Arabidopsis*. Recall that the stimulus appears to be phytochrome, which exists in two forms, P_r and P_{fr}. At night, the P_{fr} form gradually is converted back to P_r. The P_r stimulates the breakdown of CO protein in proteasomes. So in the morning and during the day, CO protein levels go down. By the end of the day, P_r levels go down, allowing CO to accumulate.

So the key to flowering is a high level of the transcription factor CO, and this in turn is related to a low level of P_r. Now, think of a short night (long day): there is not a lot of dark time for all of the P_{fr} to be converted back to P_r, and a long day also causes more conversion of P_r to P_{fr}. The low level of P_r results in less breakdown of CO protein, and the high level of CO results in the transcription of genes for flowering, such as *FT*. To summarize:

Low P_r → more CO → more transcription of *FT* → flowering

We have considered the photoperiodic regulation of flowering, from photoreceptors in the leaf to florigen that travels from the induced leaf to the sites of flower formation. In some plants, however, flowering is induced by other stimuli. These additional stimuli can function with photoperiodism or independently of it.

Flowering can be induced by temperature or gibberellin

Whereas some plants use the environmental cue of day length to induce flowering, other plants use different mechanisms. These include temperature (another environmental cue) and gibberellin (an internal, hormonal cue).

TEMPERATURE In some plant species, notably certain cereal grains, the environmental signal for flowering is cold temperature, a phenomenon called **vernalization** (Latin *vernus*, "spring"). In both wheat and rye, we distinguish two categories of flowering behavior. Spring wheat, for example, is a typical annual plant: it is sown in the spring and flowers in the same year. Winter wheat is sown in the fall, grows to a seedling, overwinters (often covered by snow), and flowers the following summer. If winter wheat is not exposed to cold in its first year, it will not flower normally the next year.

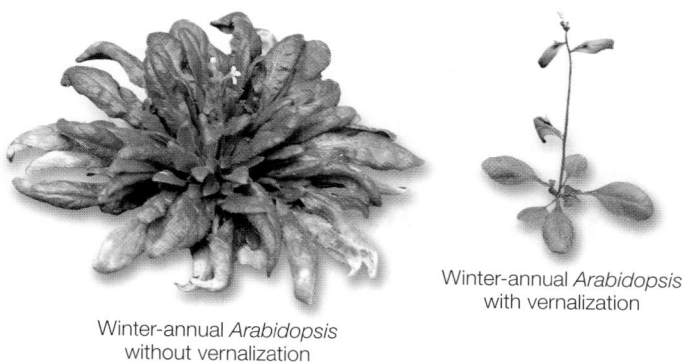

Winter-annual *Arabidopsis*
without vernalization

Winter-annual *Arabidopsis*
with vernalization

Figure 37.12 Vernalization A genetic strain of *Arabidopsis* (winter-annual *Arabidopsis*) requires vernalization for flowering. Without it, the plant is large and vegetative (left), but with the cold period it is smaller and flowers (right).

Studies of *Arabdipsosis* have shown how vernalization leads to flowering (**Figure 37.12**). In some strains of *Arabidopsis* a gene called *FLC* (*FLOWERING LOCUS C*) encodes a transcription factor that blocks the FT–FD florigen pathway (see Figure 37.11) by inhibiting expression of FT and FD. Cold temperature inhibits the synthesis of FLC protein, allowing the FT and FD proteins to be expressed, and flowering to proceed. Similar proteins control some steps in vernalization in cereals.

*Epigenetics plays an important role in the inhibition of *FLC* gene expression by cold temperature. Before vernalization, the chromatin at the promoter of the *FLC* gene is in a relaxed configuration, with histone protein acetylation lowering the ionic attraction of these proteins for DNA, which allows transcription (**Figure 37.13A**). During vernalization, a gene is expressed whose protein product is involved in the deacetylation of histones on the *FLC* gene. Deacetylation causes the chromatin to be more compact, which blocks *FLC* gene expression (**Figure 37.13B**).

 Media Clip 37.2 **How Plants Remember**
www.Life11e.com/mc37.2

*connect the concepts As described in Key Concept 16.4, epigenetic remodeling of chromatin involves loosening the nucleosome–DNA complex, opening up DNA for gene transcription. See Figure 16.15.

GIBBERELLIN *Arabidopsis* plants do not flower if they are genetically deficient in the hormone gibberellin, or if they are treated with an inhibitor of gibberellin synthesis. These observations implicate gibberellins in flowering. Direct application of gibberellins to *Arabidopsis* buds results in activation of the meristem identity gene *LEAFY*, which in turn promotes the transition to flowering.

Some plants do not require an environmental cue to flower

Several plant species and strains do not require a photoperiod or vernalization to flower, but instead flower on cue from an "internal clock." For example, flowering in some strains of tobacco will be initiated in the terminal bud when the stem has grown four phytomers in length (recall that stems are composed of repeating units called phytomers; see Figure 33.1). If such a bud with a single adjacent phytomer is removed and planted, the cutting will flower because the bud has already received the cue for flowering. But the rest of the shoot below the bud that has been removed will not flower because it is only three phytomers long. After it grows an additional phytomer, it will flower. These results suggest that there is something about the *position* of the bud (atop four phytomers of stem) that determines its transition to flowering.

The bud might "know" its position by the concentration of some substance that forms a positional gradient along the length of the plant. Such a gradient could be formed if the root makes a diffusible inhibitor of flowering whose concentration diminishes with plant height. When the plant reaches a certain height, the concentration of the inhibitor would become sufficiently low at the tip of the shoot to allow flowering. What this inhibitor might be is unclear, but there is evidence that it acts by decreasing the amount of FLC, allowing the FT–FD pathway to proceed (just as cold acts on FLC in vernalization). A positional gradient that acts on FLC would be consistent

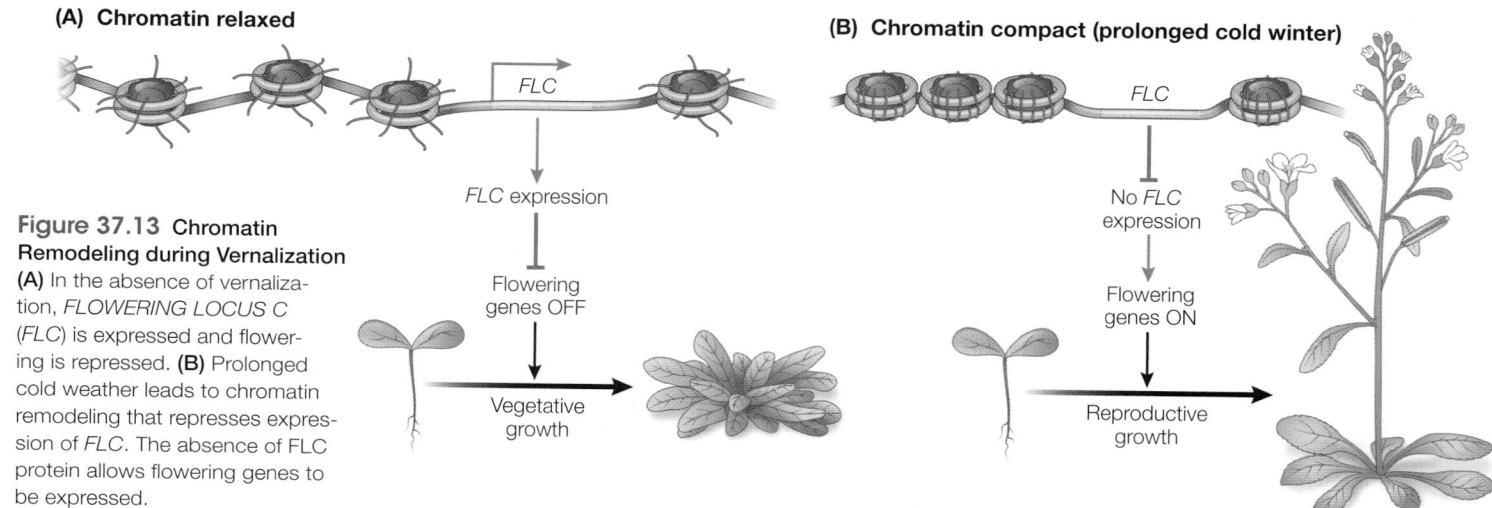

(A) Chromatin relaxed

FLC

FLC expression

Flowering genes OFF

Vegetative growth

Figure 37.13 Chromatin Remodeling during Vernalization **(A)** In the absence of vernalization, *FLOWERING LOCUS C* (*FLC*) is expressed and flowering is repressed. **(B)** Prolonged cold weather leads to chromatin remodeling that represses expression of *FLC*. The absence of FLC protein allows flowering genes to be expressed.

(B) Chromatin compact (prolonged cold winter)

FLC

No *FLC* expression

Flowering genes ON

Reproductive growth

with other mechanisms affecting flowering, which all converge on *LEAFY* and *APETALA1*:

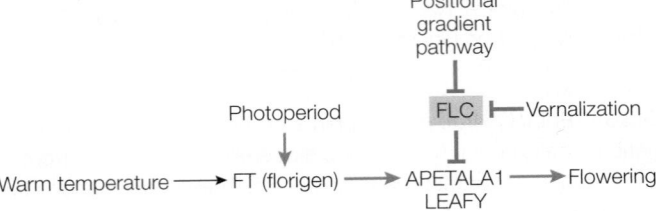

Positional gradient pathway

Photoperiod

Warm temperature → FT (florigen) → APETALA1 LEAFY → Flowering

FLC ⊣ Vernalization

37.2 recap

Flowering of some angiosperms is controlled by night length, a phenomenon called photoperiodism. Low temperatures can induce flowering in some species (vernalization). Some species flower when their stems have grown by a certain amount, independent of environmental cues. All pathways to flowering converge on the meristem identity genes. The photoperiodic flowering stimulus originates in the leaf, and a signaling pathway ends up in the expression of floral identity genes in the bud.

learning outcomes

You should be able to:

- Give evidence showing that night length, rather than day length, is the cue that triggers flowering.
- Describe evidence that a diffusible chemical travels from the leaf to the bud meristem to initiate flowering.
- Describe the three genes involved in florigen production and action, including where they are active, their functions, and their interactions.
- List factors other than photoperiodism and genetic triggering of florigen that can initiate flowering, and give evidence for each.

1. Describe the proteins and mutations that could be involved in the following observations:

   a. A mutant plant flowers without its normal inductive dark period. When a leaf from the mutant plant is grafted onto an unexposed wild-type plant, the recipient plant flowers.

   b. A mutant plant does not flower when exposed to the normal inductive dark period. When a leaf from a mutant plant that has been exposed to the inductive dark period is grafted onto an unexposed wild-type plant, the recipient plant flowers.

   c. A plant flowers only after exposure to cold.

   d. If a gene is experimentally overexpressed in the leaf, flowering is induced. Overexpression of the gene in the shoot apical meristem, however, does not induce flowering.

2. A horticulturist provides cut flowers for many occasions and often requires large numbers of flowers all blooming at the same time. Many of her flowers are normally summer-blooming, or long-day plants. How might she force these flowers to bloom during the winter?

3. Give at least two pieces of evidence supporting the presence of a transmissible substance that moves from leaves to bud meristems to initiate flowering.

You have seen how environmental factors interact with genes to control flowering in angiosperms. The function of flowers is sexual reproduction, which maintains beneficial genetic variation in a population. Many angiosperms, however, also benefit from being able to reproduce asexually.

▶ key concept 37.3   Angiosperms Can Reproduce Asexually

Although sexual reproduction takes up most of the space in this chapter, asexual reproduction accounts for many of the individual plants present on Earth. This fact suggests that in some circumstances asexual reproduction must be advantageous. In fact, many plants reproduce both asexually and sexually. An example is the iris plant, which has beautiful flowers for sexual reproduction but also a rhizome (an underground stem) from which new plants can arise asexually.

focus your learning

- Sexual and asexual reproduction provide separate and distinct advantages to plants.
- Asexual, or vegetative, reproduction in plants occurs by changes in vegetative (nonreproductive) organs.
- In apomixis, flowers produce clones; the technique has the potential to produce self-reproducing hybrids.

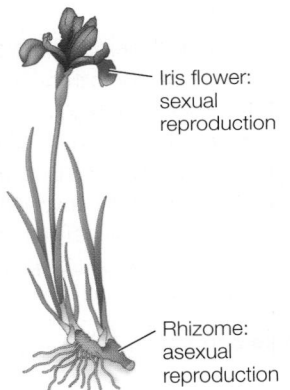

Iris flower: sexual reproduction

Rhizome: asexual reproduction

We have noted that genetic recombination is one of the advantages of sexual reproduction. Self-fertilization is a form of sexual reproduction, but offers fewer opportunities for genetic recombination than does cross-fertilization. A diploid, self-fertilizing plant that is heterozygous for a certain locus can produce both kinds of homozygotes for that locus plus the heterozygote among its progeny, but it cannot produce any progeny carrying alleles that it does not itself possess. Nevertheless many self-fertilizing plant species produce viable and vigorous offspring.

Asexual reproduction eliminates genetic recombination altogether. A plant that reproduces asexually produces progeny genetically identical to the parent. What, then, is the advantage of asexual reproduction? If a plant is well adapted to its environment, asexual reproduction allows it to pass on to all its progeny a superior combination of alleles, which might otherwise be separated by sexual recombination.

Plants reproduce asexually by several methods

Stems, leaves, and roots are considered vegetative organs and are distinguished from flowers, the reproductive parts of the plant. Asexual reproduction is often accomplished through the modification of a vegetative organ, which is why the term **vegetative reproduction** is sometimes used to describe asexual reproduction in plants. Often the stem is the organ that is modified for vegetative reproduction. Strawberries, for example, produce horizontal stems, called stolons or runners, which grow along the soil surface, form

(A) Rhizomes

These bamboo shoots all arise from the same underground stem.

(B) Bulbs

Each clove of garlic is a bulb that can give rise to a new plant.

(C) Plantlets

The plantlets forming on the margin of this *Kalanchoe* leaf will fall to the ground and become independent plants.

Figure 37.14 Vegetative Organs Modified for Reproduction
(A) The rhizomes of bamboo are underground stems that produce plants at intervals. **(B)** Bulbs are short stems with large leaves that store nutrients and can give rise to new plants. **(C)** In *Kalanchoe*, new plantlets can form on leaves.

roots at intervals, and establish potentially independent plants. Asexual reproduction by shoot tips is accomplished when the tips of upright branches sag to the ground and develop roots, as in blackberry and forsythia.

Some plants, such as potatoes, form enlarged fleshy tips of underground stems, called tubers, that can produce new plants (from the "eyes"). Rhizomes are horizontal underground stems that can give rise to new shoots. Bamboo is a striking example of a plant that reproduces vegetatively by means of rhizomes. A single bamboo plant can give rise to a stand—even a forest—of plants constituting a single, physically connected entity (**Figure 37.14A**).

Whereas stolons and rhizomes are horizontal stems, bulbs and corms are short, vertical, underground stems. Lilies and garlic form bulbs (**Figure 37.14B**), short stems with many fleshy, highly modified leaves that store nutrients. These storage leaves make up most of the bulb. They can give rise to new plants by dividing or by producing new bulbs from axillary buds. Crocuses, gladioli, and many other plants produce corms, underground stems that function very much as bulbs do. Corms are disclike and consist primarily of stem tissue; they lack the fleshy modified leaves that are characteristic of bulbs.

Stems are not the only vegetative organs modified for asexual reproduction. Leaves may also be the source of new plantlets, as in some succulent plants of the genus *Kalanchoe* (**Figure 37.14C**). Many kinds of angiosperms, ranging from grasses to trees such as aspens and poplars, form interconnected, genetically homogeneous populations by means of suckers—shoots produced by roots. What appears to be a whole stand of aspen trees, for example, may be a clone derived from a single tree by suckers. This is why the leaves of a whole stand of aspens typically turn yellow at the same time.

Plants that reproduce vegetatively often grow in physically unstable soils such as eroding hillsides. Plants with stolons or rhizomes, such as beach grasses, rushes, and sand verbena, are common pioneers on coastal sand dunes. Rapid vegetative reproduction enables these plants, once introduced, not only to multiply but also to survive burial by the shifting sand; in addition, the dunes are stabilized by the extensive network of rhizomes or stolons that develops. Vegetative reproduction is also common in some deserts, where the environment is often not suitable for seed germination and the establishment of seedlings.

Vegetative reproduction has a disadvantage

Vegetative reproduction is highly efficient in an environment that is stable over the long term. A change in the environment, however, can leave an asexually reproducing species at a disadvantage. A striking example is provided by the demise of the English elm, *Ulmus procera*, which was apparently introduced into England as a clone by the ancient Romans. This tree reproduces asexually by suckers and is incapable of sexual reproduction. In 1967 Dutch elm disease first struck the English elms. After two millennia of clonal growth, the population lacked genetic diversity, and no individuals carried genes that would protect them against the disease. Today the English elm is all but gone from England.

Vegetative reproduction is important in agriculture

One of the oldest methods of vegetative reproduction used in agriculture consists of simply making cuttings of stems, inserting them in soil, and waiting for them to form roots and become autonomous plants. The cuttings are usually encouraged to root by treatment with auxin, a plant hormone.

Woody plants can be propagated asexually by **grafting**: attaching a bud or a piece of stem from one plant to a root or root-bearing stem of another plant. The part of the resulting plant that comes from the root-bearing "host" is called the **stock**; the part grafted on is the **scion** (**Figure 37.15**). The vascular cambium of the scion associates with that of the stock, forming a continuous cambium that produces xylem and phloem. The cambium allows the transport of water and minerals to the scion and of photosynthate to the stock. Much of the fruit grown for market in the United States is produced on grafted trees, as are wine grapes.

Another method widely used for asexual plant propagation is **meristem culture**, in which pieces of shoot apical meristem are cultured on growth media to generate plantlets, which can then be planted in the field. This strategy is vital when uniformity is desired, as in forestry, or when virus-free plants are the goal, as with strawberries and potatoes.

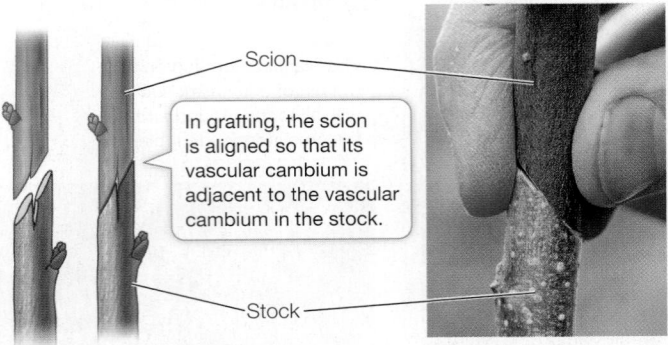

Figure 37.15 Grafting   Grafting—attaching a piece of a plant to the root or root-bearing stem of another plant—is a common horticultural technique. The "host" root or stem is the stock; the upper grafted piece is the scion. In the photo, a scion of one apple variety is being grafted onto a stock of another variety.

In apomixis, flowers are used in asexual reproduction

Some plants produce flowers but use them to reproduce asexually rather than sexually. Dandelions, blackberries, some citrus trees, and some other plants reproduce by the asexual production of seeds, which is called **apomixis**. In sexual reproduction, seeds form from the union of haploid gametes in the embryo sac. But in apomixis, one of two other things happens:

1. The megasporocyte in the ovule that is supposed to undergo meiosis fails to do so, resulting in a diploid egg cell, which then goes on to form an embryo and seed.

2. Diploid cells from the integument surrounding the embryo sac form a diploid embryo sac, and the sac goes on to form an embryo and seed.

In both cases, seed and fruit development proceed normally. But the genetic consequences are profound: *apomixis produces clones*.

Some agriculturally important plants are hybrids that are derived by crossing two genetically different varieties. Because these hybrids have two sets of chromosomes that are not homologous to one another, they cannot undergo meiosis and are sterile. This means they cannot form the seeds and fruits that are desired by people. In citrus, apomixis occurs naturally, which gets around this problem. Kentucky bluegrass, a mainstay of lawns, reproduces in this manner as well.

Many important crops, such as corn, are grown as hybrids because the progeny of a cross between two inbred, homozygous genetic strains are often superior to either of their parents, a phenomenon called hybrid vigor (see Key Concept 12.2). Unfortunately, once farmers have obtained a hybrid with desirable characteristics, they cannot use those plants for further crosses with themselves (selfing) to get more seeds for the next generation. You can imagine the genetic chaos when a hybrid, which is heterozygous at many of its loci (e.g., *AaBbCcDdEe*, etc.), is crossed with itself: there will be many new combinations of alleles (e.g., *AabbCCDdee*, etc.), resulting in highly variable progeny. The only way to reliably reproduce the hybrid is to maintain populations of the original parents to cross again each year. That is exactly what seed companies do. For farmers, it means buying new seeds every year—an expensive proposition.

If a hybrid carried a gene for apomixis, however, it could reproduce asexually, and its offspring would be genetically identical to itself. New hybrids could be developed that would be adapted to specific environments and could be propagated by the farmers on the spot. So an intensive search is on for genes for apomixis that could be introduced into desirable crops and allow them to be propagated indefinitely. Such a gene has been identified in corn, but the yield of the variety that contains it is low.

37.3 recap

Angiosperms may reproduce asexually by means of modified stems, roots, or leaves. Asexual reproduction is advantageous when a plant has a superior genotype well adapted to its environment, but it can decrease the genetic diversity of plant populations. Seeds and fruits, normally the result of sexual reproduction, can form asexually through apomixes.

learning outcomes

You should be able to:

- Compare and contrast sexual and asexual reproduction in plants, including both the end result and the advantages or disadvantages.
- List types of locations where vegetatively reproducing plants might occur, and give reasons for their occurrence in these locations.
- Describe how apomixis occurs and how this might be useful in agriculture.

1. What are the advantages and disadvantages of asexual reproduction in plants? What are the advantages and disadvantages of sexual reproduction?

2. In what types of situations might asexual reproduction in plants be adaptive?

3. The isolation of a mutation in the *Arabidopsis SWI1* gene that results in abnormal meiosis has offered insights into apomixis. How would you try to identify other genes or mutations that function along with the gene product of *SWI1*, with the goal of producing fully fertile apomictic plants?

You have seen how angiosperms reproduce sexually and asexually. A disadvantage of asexual reproduction is that its genetic inflexibility may leave a population unable to cope with new challenges. In the next chapter we will focus on the mechanisms that have evolved in plants to cope with biological and physical challenges in their environment.

investigatinglife

How did an understanding of angiosperm reproduction allow floriculturists to develop a commercially successful poinsettia?

Poinsettias are short-day plants, using the long nights of winter as a signal for flowering. In fact, the poinsettia's inductive dark period of 14 hours was among the first described in a flowering plant. To control photoperiod, the plants are grown in greenhouses, where the photoperiod is carefully regulated.

A second important factor in the poinsettia's commercial success was the plant's growth habit. In Mexico, the wild relatives of cultivated varieties grow up to 3 meters tall, with few branches (compare these plants with the 0.5-meter-tall plants in the photograph at the opening of this chapter). Paul Ecke found a variety that was much shorter and formed an attractive branching plant with axillary shoots. These compact poinsettias were initially propagated asexually by grafting to native plants.

Many new varieties of poinsettias have been generated by conventional sexual reproduction. You can look for the phenotypes of these varieties next holiday season. They include stiff stems, new colors (such as white), and longer lifetimes for the colorful bracts. How do you think these characteristics have been introduced into the cultivated varieties?

Future directions

Sequencing of the genomes of agriculturally important plants and genetic screening of mutant strains have led to an accumulation of knowledge of genes controlling the transition to flowering. Efforts are under way to use this understanding to breed crops that produce more seeds more often, by conventional means or biotechnology. This has enormous potential for human welfare. Consider legumes, which are in two general clades, most represented by peas and soybeans. The former is a long-day plant that responds to cold (vernalization), while the latter is more acclimated to warm climes and responds to short days to flower. Genes for the photoperiodic response and signal transduction for flowering have been identified for both species. For example, genetic changes in signaling could allow the plants to have more vegetative growth prior to flowering, which might lead to more floral meristems, more flowers, and more seeds.

Chapter Summary 37

37.1 Most Angiosperms Reproduce Sexually

- Sexual reproduction promotes genetic diversity in a population. The flower is an angiosperm's structure for sexual reproduction.

- Flowering plants have microscopic gametophytes. The megagametophyte is the **embryo sac**, which typically contains eight nuclei in a total of seven cells. The microgametophyte is the **pollen grain**, which usually contains two cells. Review Figure 37.2, Activity 37.1

- Following **pollination**, the pollen grain delivers **sperm cells** to the embryo sac by means of a **pollen tube**.

- Plants have both physical and genetic methods of preventing inbreeding. Physical separation of the gametophytes and genetic **self-incompatibility** prevent self-pollination. Review Figure 37.4

- Most angiosperms exhibit **double fertilization**: one sperm cell fertilizes the egg cell, forming a zygote, and the other sperm cell fertilizes the central cell, where its nucleus unites with the two **polar nuclei** to form a triploid **endosperm**. Review Figure 37.5, Animation 37.1

- Ovules develop into seeds, and the ovary wall and the enclosed seeds develop into a **fruit**. Review Figure 37.6

- The hormone **abscisic acid** (**ABA**) promotes seed development and dormancy.

37.2 Hormones and Signaling Determine the Transition from the Vegetative to the Reproductive State

- In **annuals** and **biennials**, flowering followed by seed formation usually leads to the death of the rest of the plant. **Perennials** live a long time and typically reproduce repeatedly.

- For a vegetatively growing plant to flower, an apical meristem in the shoot system must become an **inflorescence meristem**, which in turn must give rise to one or more **floral meristems**. These events are under the influence of **meristem identity genes** and **floral organ identity genes**. Review Figure 37.7

- Some plants flower in response to **photoperiod**. **Short-day plants** (**SDPs**) flower when the nights are longer than a critical night length specific to each species; **long-day plants** (**LDPs**) flower when the nights are shorter than a critical night length. Review Figure 37.9

- The mechanism of photoperiodic control involves phytochromes and a circadian rhythm. Review Investigating Life: The Flowering Signal, Animation 37.2, Activity 37.2

- A flowering signal, called **florigen**, is formed in a photoperiodically induced leaf and is translocated to the sites where flowers will form. Review Figures 37.10, 37.11

- In some angiosperm species, exposure to low temperatures—**vernalization**—is required for flowering; in others, internal signals (one of which is gibberellin in some plants) induce flowering. Review Figures 37.12, 37.13

37.3 Angiosperms Can Reproduce Asexually

- Asexual reproduction allows rapid multiplication of organisms that are well suited to their environment.

- **Vegetative reproduction** involves the modification of a vegetative organ—usually the stem—for reproduction. Review Figure 37.14

- Horticulturists often **graft** different plants together to take advantage of favorable properties of both **stock** and **scion**. Review Figure 37.15

- Some plant species produce seeds asexually by **apomixis**.

Go to **LearningCurve** (in **LaunchPad**) for dynamic quizzing that helps you solidify your understanding of this chapter. **LearningCurve** adapts to your responses, giving you the practice you need to master each key concept.

▶ Apply What You've Learned

Review

37.2 Receptors for the photoperiodic signal for flowering are located in the leaf, and a signal travels to the apical meristem.

37.2 Although in terms of flowering, plants are classified as either short-day plants (SDPs) or long-day plants (LDPs), night length is actually the cue that controls flowering.

Original Papers: Beveridge, C. A. and I. C. Murfet. 1996. The *gigas* mutant in pea is deficient in the floral stimulus. *Physiologia Plantarum* 96: 637–645.

Hecht, V. et al. 2011. The pea *GIGAS* gene is a *FLOWERING LOCUS T* homolog necessary for graft-transmissible specification of flowering but not for response to photoperiod. *The Plant Cell* 23: 147–161.

The garden pea (a long-day plant) is easy to graft, making it useful for studying the presence or absence of mobile signals. A mobile signal will diffuse from a wild-type root stock into a mutant graft that lacks the signal, restoring the signal in the graft. A nonmobile signal will not restore the signal. Grafting has been used to study genes involved in flowering in the pea.

The *gigas* (*gi*) allele in peas is a mutant version of an *FT* gene and encodes an FT-like protein. **Table A** compares the phenotypes of the wild type (WT) and the *gi* mutant at different long photoperiods. The results are given as the plant height (number of nodes) at first flowering. Presence of more nodes, or greater height, indicates delayed flowering (i.e., more vegetative growth before flowering occurs). "Veg." at 24 hours indicates that all growth was vegetative; flowering did not occur.

A grafting study was done to test whether the *gi* mutant was deficient in the ability to produce the floral stimulus molecule. Stems from 6-day-old plants were grafted, and plants were kept at a 24-hour photoperiod for 2–3 months. Grafts were either self (WT to WT, or *gi* to *gi*) or reciprocal (WT to *gi*, or *gi* to WT). Results are shown in **Table B**.

The graphs (at right) show the transcript levels (mRNA expression) of the WT allele (red circles) and the *gigas* allele (blue circles). Values are shown for both an open leaf and for the apex, or site of flowering. The tests were done on long-day plants.

Table A

Strain	Photoperiod (hours of light)		
	12	16	24
gi mutant	54	43	Veg.
WT	20	16	15

Table B

Graft partners (root/stem)	Node at first flowering
WT/WT	15
gi/*gi*	Veg.
gi/WT	17
WT/*gi*	16

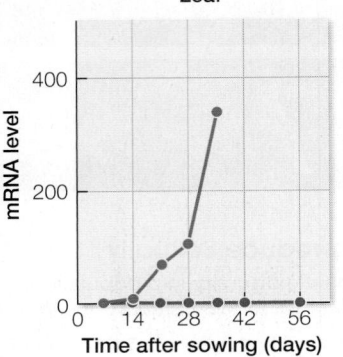
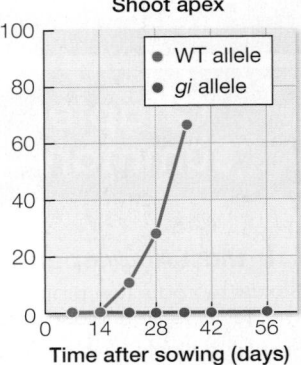

Questions

1. Based on the data in Table A, what is the effect of the *gi* mutant on flowering in the pea plant?

2. Based on the data in Table B, is it likely that the *gi* mutant is deficient in a mobile floral stimulus molecule? Provide evidence for your conclusion.

3. Develop a model describing the pathway of genetic control of flowering in long-day peas. Include the responses of both the normal (WT) *FT* gene and the mutant (*gigas*) *FT* gene.

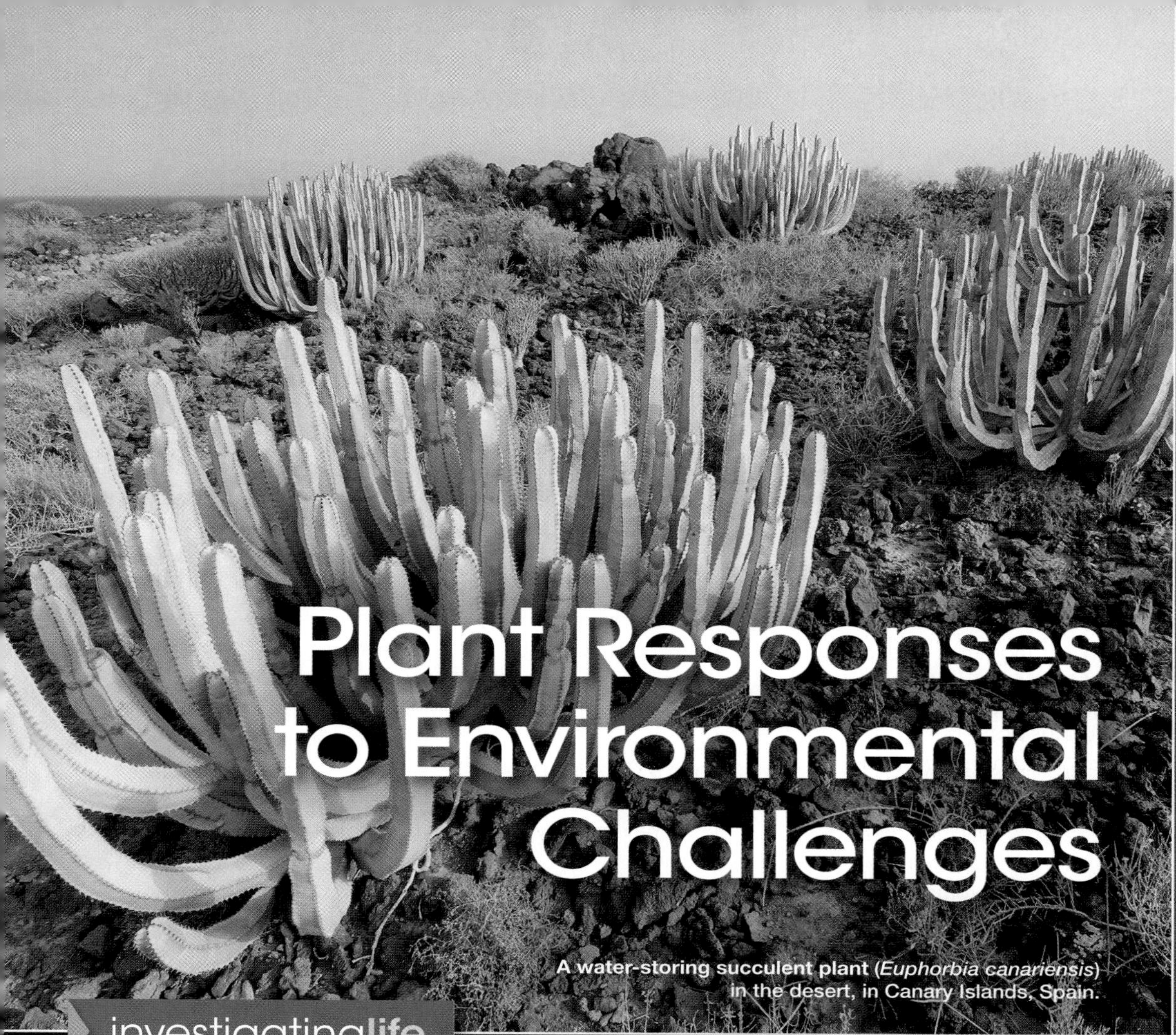

Plant Responses to Environmental Challenges

A water-storing succulent plant (*Euphorbia canariensis*) in the desert, in Canary Islands, Spain.

investigating life

Can Scientists Breed Disease-Resistant Wheat?

In 1998, William Wagoire, a plant geneticist in Uganda, was astounded to find red blotches (pustules) on the stems of wheat plants he was breeding. Wheat rust (*Puccinia graminis*), the fungus that causes the pustules, had supposedly been rendered almost extinct 25 years previously when a gene from rye plants was crossed into wheat, making it resistant to this fungal disease. This landmark achievement had protected the crop that provides one-third of the human diet from a mold that can devastate it.

Although two generations of farmers worldwide have never seen an epidemic of wheat rust, its history is well known. The ancient Romans invoked a deity, "Robigus," to ward the mold off their crops. Seventeenth-century colonists in Massachusetts almost starved because their wheat got infected with wheat rust. In 1917, an epidemic of wheat rust in the United States reduced the crop by one-third, leading to widespread panic.

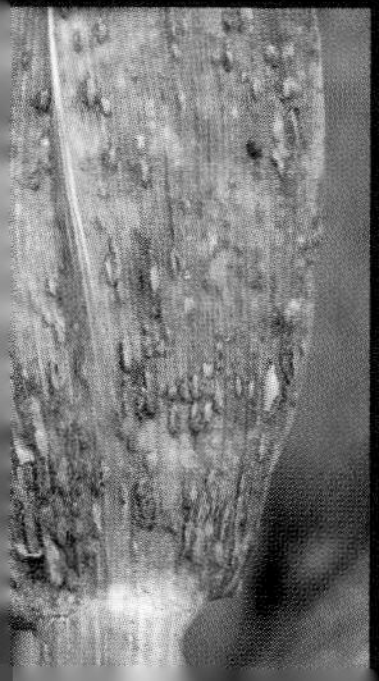

Wagoire used DNA markers to identify the strain of wheat rust he had found. Comparing it with known strains stored in a few laboratories, he found that his strain was new and unique.

Clearly, Ug99 (for Uganda 1999) had evolved a way to get around the resistance genes in modern wheat.

When wheat stem rust pustules burst, thousands of spores are released, any one of which can be carried by the wind to a susceptible plant. One hectare of an infected wheat field can release more than 10 billion spores. Carried by prevailing winds, the epidemic spread of Ug99 quickly crossed Africa. By 2001 it had infected wheat in Kenya; in 2003 it was in Ethiopia; in 2006 it crossed the Red Sea to Yemen; in 2009 it was in Iran. At the same time, winds carried the spores south to Zimbabwe and South Africa. Biologists fear that Ug99 could reach central Asia and Australia by prevailing winds; spores could reach North America by the "747 route," through international airplane travel. The Borlaug Global Rust Initiative has been set up to try to use knowledge of plant and fungal biology to stop the spread of this disease before it is too late. More than 1,000 scientists from hundreds of institutions are working together to study the pathogen and host, find genes for resistance, and breed wheat strains that have these genes.

QA **How can knowledge of plant and fungal biology be used to prevent the spread of**

key concept
38.1

Plants Respond to Pathogens with Constitutive and Induced Responses

Thousands of diseases can affect plants. Each is caused by a different strain of pathogen. Plant pathogens—which include bacteria, fungi, protists, nematodes, and viruses—are part of nature, and for that reason alone they merit our study in biology. For example, many diseases affect tomato plants, some of which may be familiar to you from growing tomatoes in your backyard or finding spoiled tomatoes in a neglected area of your kitchen (**Figure 38.1**).

---

focus your learning

- The plant immune system, which is triggered by elicitors, can produce either general or specific immunity.
- Several forms of systemic acquired resistance provide long-term immunity against various pathogens.
- Avirulence (*Avr*) gene products in pathogens interact with resistance (*R*) genes in plants to trigger immune responses.

Many plant diseases are named for the symptoms rather than the agent that causes them. For example, the condition called "blight," characterized by browning and death of plant tissues, can be caused by bacteria, fungi, or oomycetes. Just as medical schools have departments of pathology, many universities in agricultural regions have departments of plant pathology.

Successful infection by a pathogen can have significant effects on a plant, reducing photosynthesis and causing massive cell and tissue death. Like the responses of the human immune system (see Chapter 41), the responses by which plants fight off disease are varied and fascinating. Plants and pathogens have evolved together in a continuing "arms race": pathogens have evolved ways to attack plants, and plants have evolved ways to defend themselves against those attacks.

What determines the outcome of a battle between a plant and a pathogen? The key to success for the plant is to respond to information from the pathogen quickly and massively. Plants use both mechanical and chemical defenses in this effort. These defenses can either be:

- **Constitutive**, always present in the plant, or
- **Induced**, produced in reaction to damage or stress.

Physical barriers form constitutive defenses

A plant's first line of defense is its outer surfaces, which can prevent the entry of pathogens. As Chapter 33 described, the organs of a growing plant that are exposed to the outside environment are largely covered with cutin, suberin, and waxes. These substances not only prevent water loss by evaporation but can also prevent fungal spores and bacteria from entering the underlying tissues. Some fungi get around this defense by secreting enzymes that hydrolyze these substances, breaking them down to gain entry.

Much more important to the plant are the induced resistance mechanisms. Refer to these in **Focus: Key Figure 38.2** as you read the sections that follow.

Plants can seal off infected parts to limit damage

Whereas animals generally repair tissues that have been damaged by pathogens, plants do not. Instead, plants seal off and sacrifice damaged tissues so that the rest of the plant does not become infected. Plants have the option of discarding damaged tissues because most plants, unlike most animals, can replace damaged parts by growing new ones.

Before we look at the details of the defensive process, we consider a key response by plant cells to invasion by pathogens: the rapid

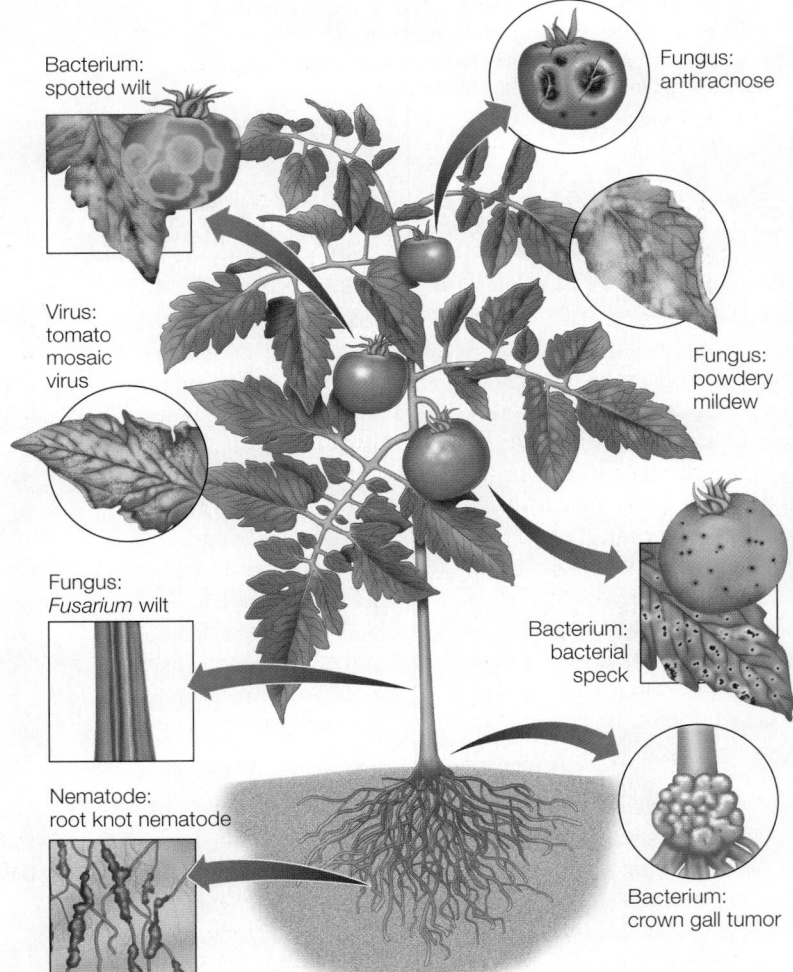

Bacterium: spotted wilt

Fungus: anthracnose

Virus: tomato mosaic virus

Fungus: powdery mildew

Fungus: *Fusarium* wilt

Bacterium: bacterial speck

Nematode: root knot nematode

Bacterium: crown gall tumor

Figure 38.1 Diseases of Tomato Plants   A wide variety of disease agents cause a variety of symptoms.

Q: As in animals, diseases in plants are caused by a wide variety of organisms. Can you name human diseases caused by the four types of agents described in this figure?

focus: key figure

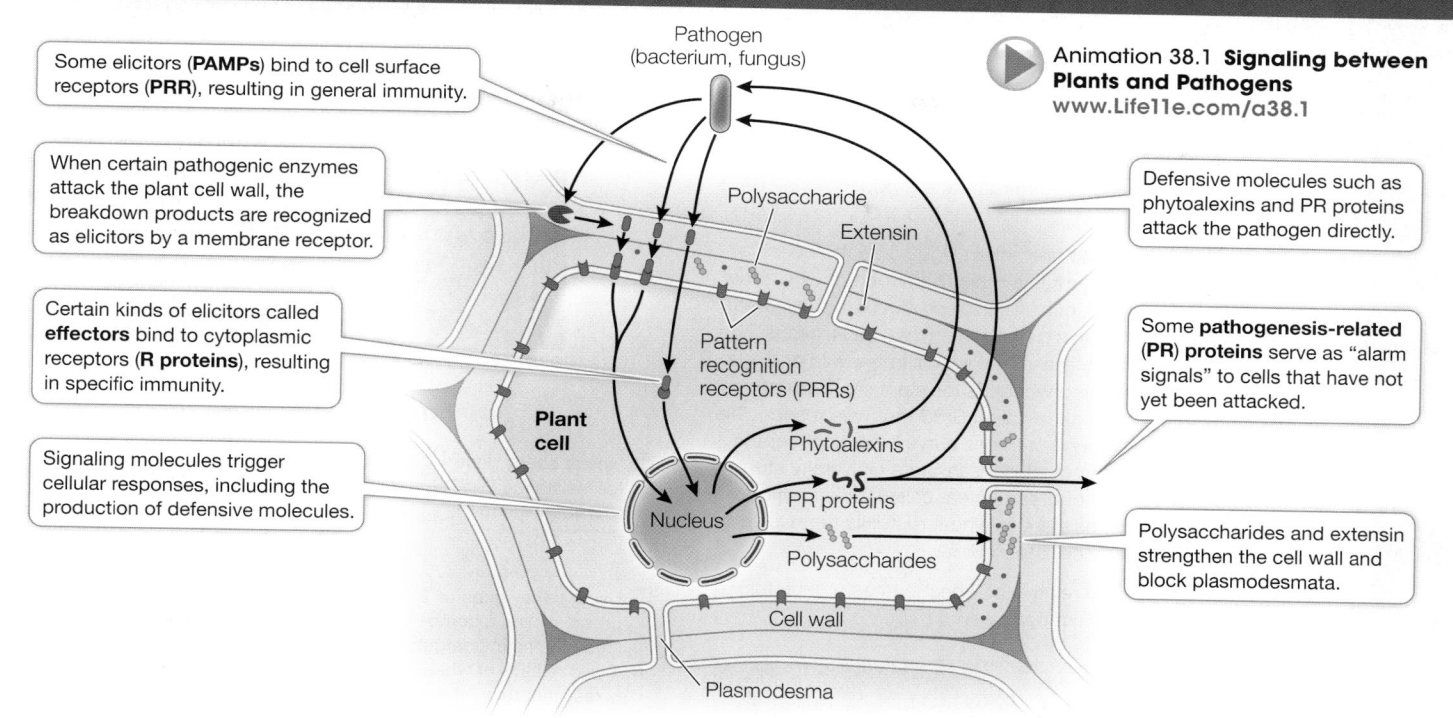

Some elicitors (**PAMPs**) bind to cell surface receptors (**PRR**), resulting in general immunity.

When certain pathogenic enzymes attack the plant cell wall, the breakdown products are recognized as elicitors by a membrane receptor.

Certain kinds of elicitors called **effectors** bind to cytoplasmic receptors (**R proteins**), resulting in specific immunity.

Signaling molecules trigger cellular responses, including the production of defensive molecules.

Pathogen (bacterium, fungus)

Polysaccharide

Extensin

Pattern recognition receptors (PRRs)

Plant cell

Phytoalexins

Nucleus

PR proteins

Polysaccharides

Cell wall

Plasmodesma

Animation 38.1 **Signaling between Plants and Pathogens**
www.Life11e.com/a38.1

Defensive molecules such as phytoalexins and PR proteins attack the pathogen directly.

Some **pathogenesis-related (PR) proteins** serve as "alarm signals" to cells that have not yet been attacked.

Polysaccharides and extensin strengthen the cell wall and block plasmodesmata.

Figure 38.2 Signaling between Plants and Pathogens   Molecular interactions between plants and pathogens are highly coevolved. The presence of a pathogen stimulates the plant to produce defensive molecules that work in many different ways.

Q: Would you expect these defense molecules to be expressed in the cells of roots, leaves, or both?

deposition of additional polymers on the inside of the cell wall. These macromolecules not only reinforce the mechanical barrier formed by the cell wall, but also block the plasmodesmata, limiting the ability of viral pathogens to move from cell to cell. The polysaccharides also serve as a base on which lignin, a polymer that provides strength and rigidity, may be laid down. Lignin enhances the mechanical barrier, and the toxicity of lignin precursor chemicals makes the cell inhospitable to some pathogens. These lignin building blocks are only one example of the toxic substances that plants use as chemical defenses.

Induced responses to pathogens are controlled by receptors. Plant pathogens cause the host plant to activate various chemical defense responses. A wide range of molecules called **elicitors** have been identified that trigger these defenses. These molecules vary in character, from peptides made by bacteria to cell wall fragments from fungi. Elicitors can also be derived from fragments of plant cell wall components broken down by pathogens.

The responses of plants to elicitors can be described in terms of the "plant immune system." Two forms of immunity are recognized (see Figure 38.2):

1. *General immunity* is triggered by general elicitors called pathogen associated molecular patterns (PAMPs). PAMPs are usually molecules that are produced by entire classes of pathogens, such as flagellin (found in bacterial flagella) or chitin (found in fungal cell walls). Thus, general immunity is an overall response rather than a response that is triggered by a specific pathogen

in a particular plant. PAMPs are recognized by transmembrane receptors called **pattern recognition receptors** (**PRRs**), which activate signaling pathways that lead to general immunity.

2. *Specific immunity* is triggered by specific elicitors called **effectors**. Effectors include a wide variety of specific pathogen-produced molecules that enter the plant cell. Once inside the cell, effectors bind to cytoplasmic receptors called R proteins that trigger the specific immunity response.

Understanding the pathways of pathogen infection is important for agricultural scientists, because pathogens can severely reduce crop production, as you saw in the opening investigation of this chapter. One approach to understanding is to describe the plant genes whose expression is changed after pathogen infection, as mutations in these genes may lead to resistance. Also, comparative genomics can be used: if a gene that confers resistance is found in one species, it may exist in close relatives, and may even be used to create a transgenic crop that is resistant to the pathogen. An example is described in **Investigating Life: A Gene for Resistance to Wheat Rust**.

General and specific immunity both involve multiple responses

Many of the signaling pathways associated with general and specific immunity are the same, although the latter is specific for particular pathogens in particular plants, and is much stronger than general

experiment

Original Papers: Sambasivam, P. et al. 2013. The gene *Sr33*, an ortholog of barley *Mla* genes, encodes resistance to wheat stem rust race Ug99. *Science* 341: 786–788.

Saintenac, A. et al. 2013. Identification of wheat gene *Sr35* that confers resistance to Ug99 stem rust race group. *Science* 341: 783–786.

Wheat rust, particularly the new strain Ug99 (see the opening investigation of this chapter), is a major pathogen of wheat that can cause significant reduction in grain production. Cultivated wheat is not resistant to this pathogen strain. But *Aegilops tauschii*, a species of goatgrass that grows in the wild and is a genetic relative of cultivated wheat, is resistant. An international research team isolated a gene called *Sr33* that was responsible for this resistance. The gene encodes a cytoplasmic receptor. Meanwhile, Eduard Akhunov at Kansas State University studied another primitive relative of bread wheat called *Triticum monococcum*, which also is resistant to wheat rust. His team isolated another gene for resistance, called *Sr35*.

HYPOTHESIS▶ A wild relative of wheat carries a gene for resistance to the wheat rust pathogen.

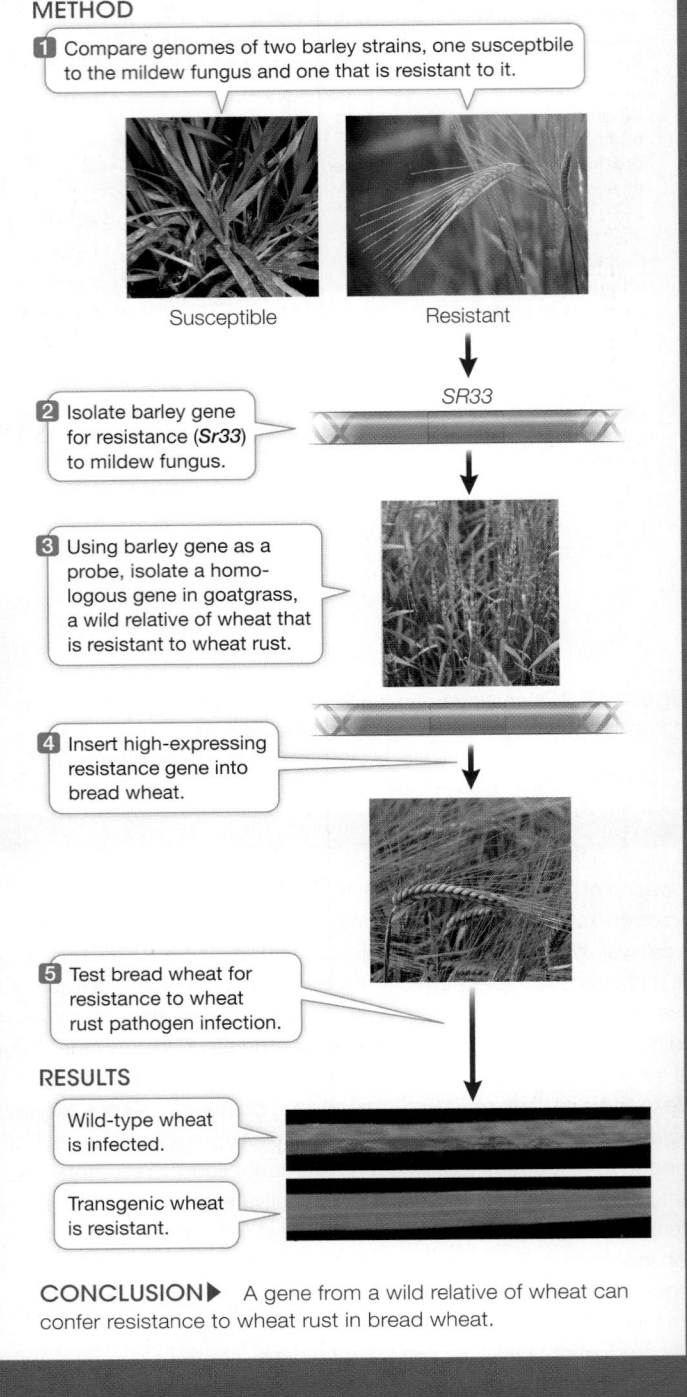

METHOD

1. Compare genomes of two barley strains, one susceptible to the mildew fungus and one that is resistant to it.

Susceptible Resistant

SR33

2. Isolate barley gene for resistance (*Sr33*) to mildew fungus.

3. Using barley gene as a probe, isolate a homologous gene in goatgrass, a wild relative of wheat that is resistant to wheat rust.

4. Insert high-expressing resistance gene into bread wheat.

5. Test bread wheat for resistance to wheat rust pathogen infection.

RESULTS

Wild-type wheat is infected.

Transgenic wheat is resistant.

CONCLUSION▶ A gene from a wild relative of wheat can confer resistance to wheat rust in bread wheat.

---

immunity. Both forms of immunity involve signaling pathways that are triggered by binding between the elicitors (either PAMPs or effectors) and their receptors. These pathways lead to various responses:

- *Formation of reactive oxygen and NO*: Receptor binding triggers the rapid production of nitric oxide (NO) and reactive oxygen species such as superoxide and hydrogen peroxide. These reactive molecules are toxic to some pathogens, and they are components of signal transduction pathways leading to local and systemic (plantwide) defenses.

- *Callose deposition*: The β-1,3-glucan polymer callose is deposited on the inside of the cell wall to strengthen the wall and seal off the cell.

- *Hormone signaling*: Some pathways result in the production of plant hormones, including salicylic acid and jasmonic acid. We will describe the roles of these hormones in immunity later in the chapter.

- *Changes in gene expression*: Signal transduction cascades lead to changes in gene expression. The upregulated genes include pathogenesis-related (PR) genes and genes encoding the production of antimicrobial substances called phytoalexins.

PHYTOALEXINS **Phytoalexins** are antibiotics produced by infected plants and are toxic to many fungi and bacteria. Most are small molecules, and each is made by only a few plant species. They are elicited by effectors of pathogenic bacteria and fungi and are made within hours of the onset of infection. Because their antimicrobial activity is nonspecific, phytoalexins can destroy many species of fungi and bacteria in addition to the one that originally triggered their production. Some phytoalexins can also kill the plant cells that produced them, thus sealing off the infection site (the hypersensitive response, discussed below).

A series of mutations in *Sr33* was generated by mutagenesis and a genetic screen was performed for susceptibility to wheat rust infection (**Figure A**). The white bars indicate the deleted DNA, and the map location of the *Sr33* gene is noted between the brackets.

Figure A

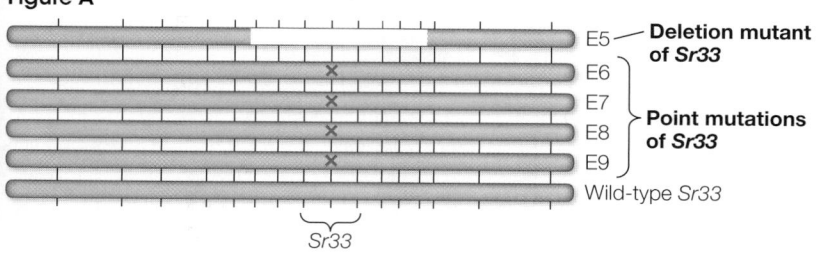

E5 — **Deletion mutant of *Sr33***

E6
E7
E8
E9 } **Point mutations of *Sr33***

Wild-type *Sr33*

Sr33

Figure B

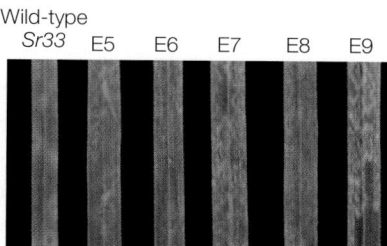

Wild-type *Sr33*   E5   E6   E7   E8   E9

QUESTIONS▶

1. The results for the wild-type *Sr33* and mutant E5–E9 *Sr33* are shown in **Figure B**. What can you conclude about the role of *Sr33* in rust resistance?

2. The bread wheat strain called Fielder was made transgenic for a high-expressing *Sr33* gene. It was then exposed to wheat rust. The results are shown in **Figure C**. What does this transgenic approach indicate about the role of *Sr33* in rust resistance? Compare your conclusion with that of Question 1.

3. *Triticum monococcum* with wild-type *Sr35* (**Figure D, left**) was compared with two strains of wheat with mutated *Sr35* genes (**Figure D, center and right**) with regard to infection by wheat rust. What can you conclude about the role of *Sr35* in resistance?

4. Cultivated Fielder bread wheat was made transgenic for a high-expressing *Sr35* gene. It was then exposed to the Ug99 strain of wheat rust. The results are shown in **Figure E**. What does this indicate about the *Sr35* gene?

Figure C

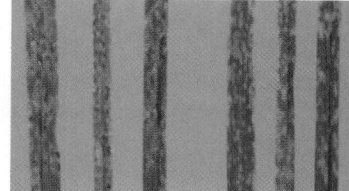

Fielder   Fielder + *Sr33*

Figure E

Sr35 transgenic   Fielder control

Figure D

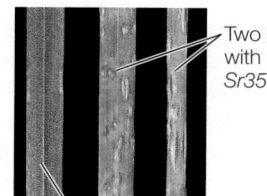

Two strains with mutated *Sr35* gene

Wild-type *Sr35*

A similar **work with the data** exercise may be assigned in **LaunchPad**.

An example of a phytoalexin is camalexin, made by the model organism *Arabidopsis thaliana*. Its production is induced by a conserved ***protein kinase cascade** that is triggered by receptor binding in either general or specific immunity. This protein kinase cascade results in the upregulation of genes encoding enzymes that convert the amino acid tryptophan to camalexin:

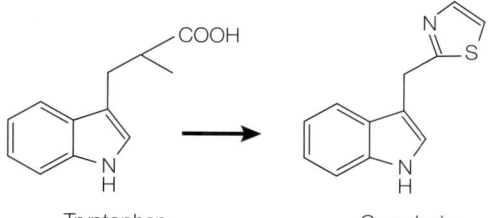

Tryptophan   Camalexin

Camalexin appears to work by disrupting the cell membranes of invading pathogens.

***connect the concepts**   Protein kinase cascades are common in signal transduction. See Key Concept 7.3.

PATHOGENESIS-RELATED PROTEINS   Plants produce several types of **pathogenesis-related (PR) proteins**. Some are enzymes that break down the cell walls of pathogens. Chitinase, for example, is a PR protein that breaks down chitin, which is found in many fungal cell walls. In some cases the breakdown products of the pathogen cell walls serve as elicitors that trigger further defensive responses.

Another class of PR proteins are the plant defensins, which are similar to defensins produced by animals (see Key Concept 41.2).

These small peptides bind to fungal membranes and are toxic to a wide range of fungal targets, but they are not toxic to plant or animal cells. Other PR proteins may serve as alarm signals to plant cells that have not yet been attacked. In general, PR proteins appear not to be rapid-response weapons; rather, they act more slowly, perhaps after other, faster responses have blunted the pathogen's attack.

Specific immunity is genetically determined

Pathogenic effectors are molecules made inside host plant cells by invaders. Often, these effectors block some aspect of *plant immunity. For example, many effectors are proteases that break down specific plant proteins involved in immune responses. So effectors enable pathogens to overcome general immunity and invade the plant, causing disease. The genes that encode effectors have evolved as part of the "arms race" between pathogens and their host plants.

---

*connect the concepts   The concepts of the plant immune response are similar to those of the animal immune response. See Chapter 41.

So the plants fight back. They have evolved intracellular receptors that recognize specific effectors. These receptors are called R (for *resistance*) proteins. When an R protein binds its ligand (a specific effector), it activates the signal transduction pathways of the specific immunity response. As we mentioned above, specific immunity is stronger than general immunity, and it enables the plant to prevent growth of the pathogen and remain healthy.

R proteins are encoded by **resistance (R) genes**. During the middle of the twentieth century, the plant pathologist Harold Henry Flor, at North Dakota State University, realized that there is a special relationship between R genes and certain pathogen genes named **Avirulence (Avr) genes**. Flor studied strains of the rust fungus (*Melampsora lini*) and the flax plant (*Linum usitatissimum*). He found that specific strains of rust fungus (with particular Avr genes) were able to cause disease only in some varieties of flax—those that *didn't* carry specific R genes. If a flax variety carried the right R gene, it was resistant to those strains of rust fungus. Thus, a particular Avr gene rendered the fungus *avirulent* on flax plants with the corresponding R gene. Flor named this the **gene-for-gene concept**. We now know that Avr genes encode effector molecules that bind to receptors encoded by the R genes, resulting in the specific immunity response (**Figure 38.3**). Hundreds of R genes and their corresponding Avr genes have been identified. A major goal of plant breeders for the past 50 years has been to breed new R genes into crops to make them more resistant to pathogens.

Specific immunity usually leads to the hypersensitive response

Many signaling pathways and plant responses are shared between general and specific immunity, although in the latter these responses are accelerated and amplified. Furthermore, specific immunity usually leads to a form of programmed cell death called the **hypersensitive response**. When this occurs, plant cells at and near the site of pathogen infection die, forming a necrotic lesion. This cell death

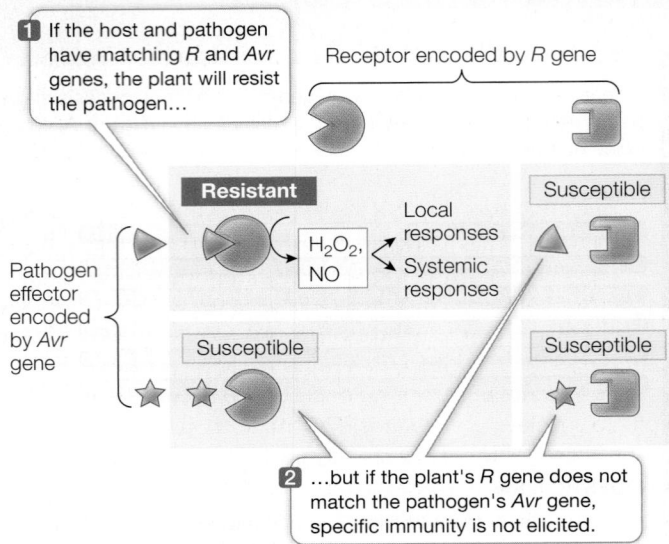

1 If the host and pathogen have matching *R* and *Avr* genes, the plant will resist the pathogen...

Receptor encoded by *R* gene

Pathogen effector encoded by *Avr* gene

Resistant

H_2O_2, NO

Local responses

Systemic responses

Susceptible

Susceptible

Susceptible

2 ...but if the plant's *R* gene does not match the pathogen's *Avr* gene, specific immunity is not elicited.

Figure 38.3 Gene-for-Gene Resistance If a gene in a pathogen that codes for an elicitor "matches" a gene in a plant that codes for a receptor, the receptor binds the elicitor, and a defensive response results.

Q: How do the elicitor and receptor recognize one another?

deprives the pathogen of nutrients and prevents the spread of the infection. The rest of the plant remains pathogen-free.

Necrotic lesion

Systemic acquired resistance is a form of long-term immunity

Thus far we have described the events that occur in plant cells at or near the site of invasion by a pathogen. In both general and specific immunity, the infected cells also send hormonal signals to the rest of the plant, stimulating a systemic response. **Systemic acquired resistance** is a general increase in the resistance of the entire plant to a wide range of pathogens. It is not limited to the pathogen that originally triggered it, or to the site of the original infection, and its effect may last as long as an entire growing season.

This defensive response is initiated by the plant hormone salicylic acid.

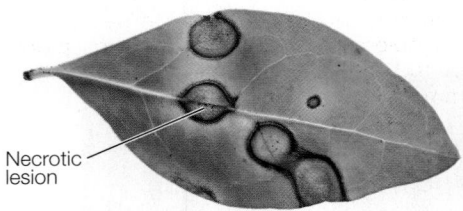

Salicylic acid

Salicylic acid production is triggered by receptor binding in both general and specific immunity; salicylic acid then functions as a signal that mediates various defense responses. For example, salicylic acid triggers the production of reactive oxygen species and the induction of genes that encode PR proteins. There are many classes of PR proteins, which function in various ways to protect plants against insect attack and against invasion by fungi, bacteria, viruses, and nematodes.

These responses occur at the site of infection and, to a lesser extent, throughout the plant. A derivative of salicylic acid can form a gas that can travel through the air, carrying the defense signal not only to other parts of the same plant but to other nearby plants. Plant-to-plant signaling is also done through mycorrhizae associated with roots. These beneficial fungi (see Key Concept 35.4) form a network that appears to transmit signals from an infected plant to its neighbors.

Another type of systemic acquired resistance is a more specific defense against viruses with RNA genomes. The plant uses its own enzymes to convert some of the single-strand RNA of the invading virus into double-strand RNA (dsRNA) and to chop that dsRNA into small interfering RNAs (siRNAs) (see Key Concept 16.5). Simultaneously, some of the viral RNA is transcribed, forming mRNAs that advance the infection. However, the siRNAs interact with another cellular component to degrade those mRNAs, blocking viral replication. The siRNAs spread quickly throughout the entire plant through plasmodesmata, providing systemic resistance.

38.1 recap

Plants have both constitutive and induced defense mechanisms to protect themselves against pathogens. General and specific immunity involve several common signaling pathways, but specific immunity is a stronger response that usually leads to hypersensitive cell death. Specific immunity involves reactions between genetically determined molecules of pathogen and host plant. Systemic acquired resistance provides a longer-lasting, more general immunity throughout the plant.

learning outcomes

You should be able to:

- Compare and contrast general immunity and specific immunity.
- Apply the concept of the genetic "arms race" between plants and their pathogens to the evolution of resistance genes and avirulence genes.
- Describe and discuss the influence of salicylic acid in systemic acquired resistance.

1. A tomato plant can be infected with the fungus *Cladosporium*. The host plant and pathogen can have various genes involved in the hypersensitive response. Complete the table below to show the effects of each fungal strain on tomato leaves:

Cladosporium genotype	Tomato genotype		
	R1R2	R3R4	R1R4
Avr1Avr2			
Avr2Avr3			
Avr1Avr4			

38.1 recap

2. Describe how a researcher could determine whether an induced response in a plant cell represents general immunity or specific immunity.

3. A researcher makes a small incision in a leaf of a healthy plant and applies salicylic acid to the cut. How might the researcher expect the plant to respond?

Not all biological threats to plants come from pathogens. Another threat comes from the many animals that eat plants.

key concept 38.2   Plants Have Mechanical and Chemical Defenses against Herbivores

Herbivores—animals that eat plants—depend on plants for energy and nutrients. Their foraging activities cause physical damage to plants, and they often spread disease among plants as well. While the majority of herbivores are insects (**Figure 38.4**), every major class of vertebrates includes at least a few herbivores. Plants cannot evade their consumers by running away, but they have many other ways of protecting themselves against herbivory.

focus your learning

- Plants can defend themselves against herbivores by mechanical methods.
- Secondary metabolites are plants' constitutive chemical defenses against herbivory.
- Herbivory induces plants to activate elicitors, which in turn activate signal transduction pathways.
- Signal transduction pathways involve membrane signaling, reactive oxygen species, and hormone signaling.
- The hormone jasmonic acid and its derivatives are key to the expression of many defenses against herbivory.

(A) *Locusta migratoria*

(B) *Manduca sexta*

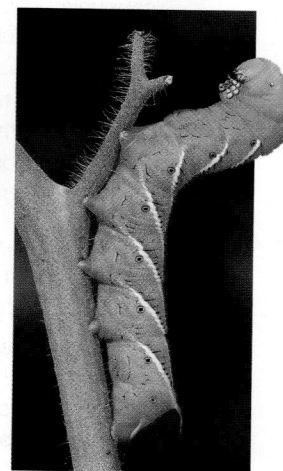

Figure 38.4 **Insect Herbivores**   The great majority of herbivores are insects. **(A)** Some herbivores, such as this locust, are generalists that will attack nearly any plant. **(B)** Others are specialists, like this tobacco hornworm, which feeds only on tobacco and related plants.

table 38.1 Secondary Metabolites Used in Defense

Class	Type	Role	Example
Nitrogen-containing	Alkaloids	Neurotoxin	Nicotine in tobacco
	Glycosides	Inhibit electron transport	Dhurrin in sorghum
	Nonprotein amino acids	Disrupt protein structure	Canavanine in jack bean
Ephedrine (an alkaloid)			
Nitrogen–sulfur-containing	Glucosinolates	Inhibit respiration	Methylglucosinolate in cabbage
Methylglucosinolide			
Phenolics	Coumarins	Block cell division	Umbelliferone in carrots
	Flavonoids	Phytoalexins	Capsidol in peppers
	Tannins	Inhibit enzymes	Gallotannin in oak trees
Umbelliferone			
Terpenes	Monoterpenes	Neurotoxins	Pyrethrin in chrysanthemums
	Diterpenes	Disrupt reproduction and muscle function	Gossypol in cotton
	Triterpenes	Inhibit transport	Digitalis in foxglove
	Sterols	Block animal hormones	Spinasterol in spinach
	Polyterpenes	Deter feeding	Latex in *Euphorbia*
Pyrethrin			

Plants have mechanical defenses against herbivores

Plants have both constitutive and induced mechanical defenses against herbivores. Constitutive anatomical barriers include trichomes (specialized hairs; see Figure 38.8) and thorns, spines that are specialized for defense. An example of an induced mechanical defense is the production of latex. Some plants, such as *Euphorbia* species, produce a thick, white aqueous suspension of cellular debris, oils, and resins called latex when they are injured by an *herbivore. Insects trapped by this sticky substance starve to death.

---

*connect the concepts In this chapter we discuss herbivory in the context of how a plant's physiology is affected by an herbivore. See also Key Concept 55.2, which discusses herbivory in the ecological context of species interactions.

Plants produce constitutive chemical defenses against herbivores

Plants attract, resist, and inhibit other organisms with a wide range of chemicals known as secondary metabolites. Primary metabolites are substances such as proteins, nucleic acids, carbohydrates, lipids, and their building blocks, which are produced and used by all living organisms, including plants. Primary metabolites are used in basic cellular processes such as photosynthesis, respiration, and nutrient uptake. **Secondary metabolites** are substances that are not used for basic cellular processes. Each is found in only certain organisms or groups of organisms.

The more than 10,000 known plant secondary metabolites range in molecular mass from about 70 to more than 380,000 daltons, but most have a low molecular mass (**Table 38.1**). The effects of defensive secondary metabolites on animals are diverse. Some act on the nervous systems of herbivorous insects, mollusks, or mammals. Others mimic the natural hormones of insects, causing some larvae to fail to develop into adults. Still others damage the digestive tracts of herbivores. Some secondary metabolites are toxic to fungal pathogens. Many secondary metabolites are not just used for defense; they have other roles in the plant. For instance, the phenolics called anthocyanins are toxic to insects and bacteria. But they also are pigmented and aid in attracting pollinators, accumulate in vacuoles and so contribute to osmotic pressure in cells, and even regulate dormancy in some plants.

Here is an example of how one of these metabolites works on pathogens. Canavanine is a secondary metabolite whose role is defensive and is based on its chemical structure. Canavanine is an amino acid that is not normally found in proteins, but is very similar to the amino acid arginine, which is found in almost all proteins.

NH₂ structure diagrams:

A seemingly slight chemical difference...

Arginine

...results in an inactive protein.

Canavanine

When an insect larva consumes canavanine-containing plant tissue, the canavanine is incorporated into the insect's proteins in some of the places where the mRNA codes for arginine, because the enzyme that charges the tRNA specific for arginine fails to discriminate accurately between the two amino acids (see Key Concept 14.5). The structure of canavanine, however, is different enough from that of arginine that some of the resulting proteins end up with a modified tertiary structure, and hence reduced biological activity. These defects in protein structure and function lead to developmental abnormalities that kill the insect.

Taking cues from nature, we have practical uses for many plant secondary metabolites or molecules made from them in the lab. Throughout human history, these have formed the basis of herbal medicine as plant extracts are used. As knowledge of chemistry accumulated, people purified the active ingredients of these plants. For instance, the foxglove plant gave us the cardiac drug digitalis.

Plants respond to herbivory with induced defenses

In Key Concept 38.1 we described the defenses that are induced in plants in response to pathogen attack. Plants also respond to wounding and herbivory with induced defenses involving signal transduction pathways. Less is known about the elicitors and receptors involved in these responses, but several classes of chemical elicitors have been identified. These elicitors are either derived from the herbivores themselves or are products of the digestion of plant tissues. For example, the enzymes that insects use to digest plant carbohydrates and lipids can elicit defensive responses. Some elicitors are produced when plant material passes through an herbivore's digestive tract; these elicitors are composed of a fatty acid derived from the plant and an amino acid derived from the insect gut. One such elicitor, produced by insects feeding on corn plants, has been named volicitin for its ability to induce production of volatile signals that can travel to other plant parts—and to neighboring corn plants—and stimulate their defense responses. In addition, many herbivorous insects lay their eggs on plants, and some components of the fluids secreted during egg laying have been identified as elicitors.

The signal transduction pathways that are activated by herbivory or wounding involve several key components, some of which are shared by the pathways that are induced by pathogens:

- *Membrane signaling*: The cell membrane is the part of the plant cell that is in contact with the environment. Within the first minute after an herbivore strikes, changes in the electric potential of the

cell membrane occur in the damaged area. As we will describe in our chapter on the animal nervous system (see Key Concept 44.2), such changes can be rapidly transmitted as a signal along the cell membrane. In the case of plants responding to herbivory, the continuity of the symplast (see Figure 34.5) ensures that the signal travels over much of the plant within 10 minutes.

- *Reactive oxygen species*: Both wounding and herbivory trigger the production of reactive oxygen species (such as superoxide and hydrogen peroxide), which act as signaling molecules in pathways that lead to changes in gene expression.

- *Hormone signaling*: Herbivory induces the production of several hormones that stimulate various plant responses. The most important of these is **jasmonic acid (jasmonate)**, which triggers systemic defenses against herbivores.

Jasmonic acid

 Media Clip 38.1 **Plants Can Hear Danger**
www.Life11e.com/mc38.1

Jasmonates trigger a range of responses to wounding and herbivory

When the plant senses an herbivore-produced elicitor, it makes jasmonate (the ionized form of jasmonic acid) and a variety of jasmonate derivatives. These molecules trigger many plant defenses both at the site of herbivore attack and throughout the plant (**Figure 38.5**). These defenses include the production of specific secondary metabolites and defensive proteins. Jasmonates induce changes in gene expression by binding a transcriptional inhibitor called a JAZ protein. After binding by jasmonate, the JAZ protein is targeted for degradation, and the previously inhibited genes can be expressed.

Protease inhibitors are an important group of defensive proteins that are synthesized in response to insect attack. Once inside an insect's gut, these inhibitors interfere with the digestion of proteins and thus stunt the insect's growth. Jasmonates can also "call for help" by triggering the formation of volatile compounds that attract insects that prey on the herbivores attacking the plant.

Why don't the defensive chemicals that are so toxic to herbivores and pathogens kill the plants that produce them? In some cases the defensive chemicals are directed at organs or systems that are not found in plants, such as the nervous, digestive, or endocrine systems of animals. In addition, plants that produce toxic defensive chemicals sequester them in vacuoles or store them as harmless precursors until used. Key Concept 56.2 describes these methods in more detail.

Plants don't always win the arms race

Milkweeds such as *Asclepias syriaca* store their defensive chemicals in latex in specialized tubes called **laticifers**, which run alongside the veins in the leaves. When damaged, a milkweed releases copious amounts of toxic latex from its laticifers. Field studies have shown that most insects that feed on neighboring plants of other species do not attack laticiferous plants, but there are exceptions. One population of beetles that feeds on *A. syriaca*

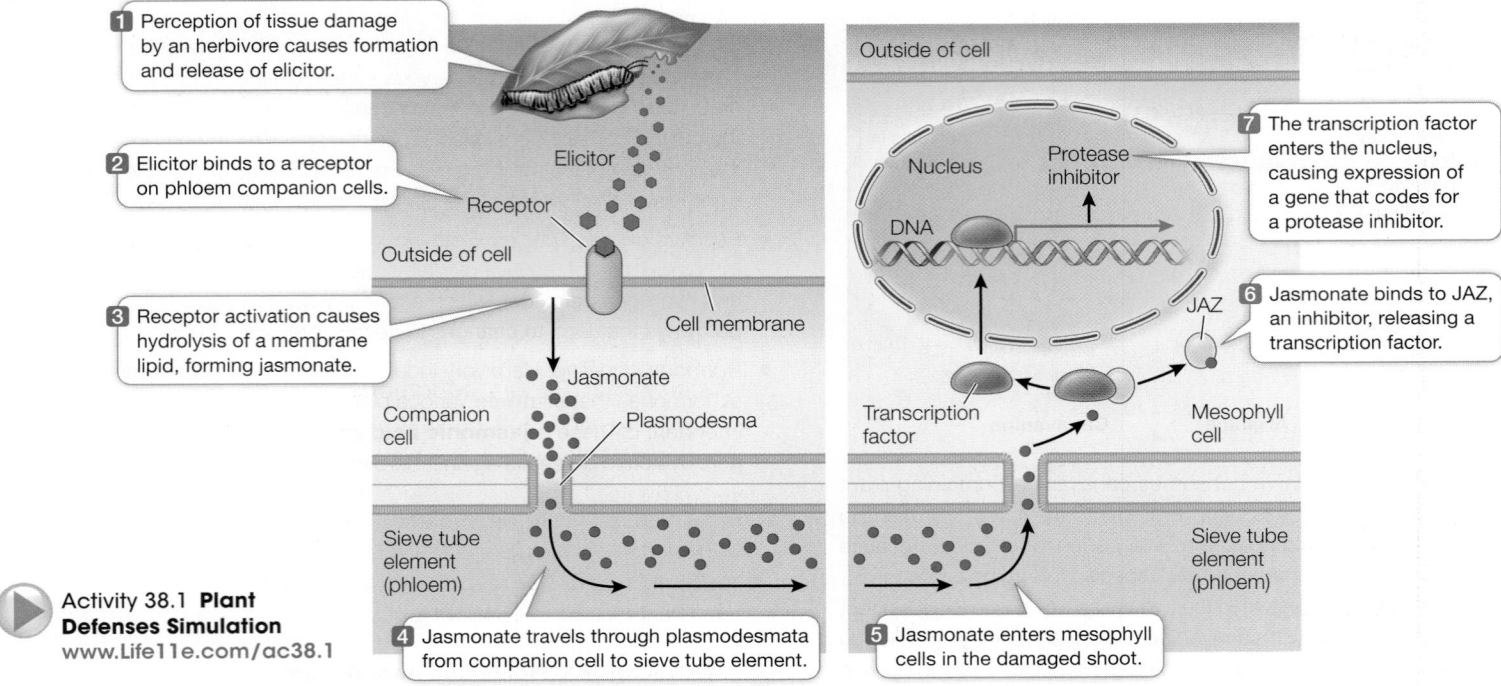

1 Perception of tissue damage by an herbivore causes formation and release of elicitor.

Elicitor

Receptor

Outside of cell

2 Elicitor binds to a receptor on phloem companion cells.

Cell membrane

3 Receptor activation causes hydrolysis of a membrane lipid, forming jasmonate.

Jasmonate

Companion cell

Plasmodesma

Sieve tube element (phloem)

Activity 38.1 **Plant Defenses Simulation**
www.Life11e.com/ac38.1

4 Jasmonate travels through plasmodesmata from companion cell to sieve tube element.

Outside of cell

Nucleus

Protease inhibitor

DNA

7 The transcription factor enters the nucleus, causing expression of a gene that codes for a protease inhibitor.

JAZ

6 Jasmonate binds to JAZ, an inhibitor, releasing a transcription factor.

Transcription factor

Mesophyll cell

Sieve tube element (phloem)

5 Jasmonate enters mesophyll cells in the damaged shoot.

Figure 38.5 A Signal Transduction Pathway for Induced Defenses The chain of events initiated by herbivory that leads to the production of a defensive chemical can consist of many steps. These steps may include the synthesis of one or two hormones, binding of receptors, gene activation, and finally, synthesis of defensive compounds.

exhibits a remarkable prefeeding behavior: these beetles cut a few veins in the leaves before settling down to dine. Cutting the veins causes massive latex leakage from the adjacent laticifers and interrupts the latex supply to a downstream portion of the leaf. The beetles then move to the relatively latex-free portion and eat their fill (**Figure 38.6A**).

(A)

(B)

Figure 38.6 Disarming a Plant's Defense and Applying It (A) This beetle is inactivating a milkweed's defense system by cutting its laticifer supply lines. **(B)** Collecting latex from a rubber tree in the Philippines.

When latex dries, it becomes rubbery. That's a good adjective, because we have exploited the sticky, moldy properties of dry latex from the tree *Hevea brasiliensis* to turn it into rubber for tires, shoes, adhesives, and condoms, to name a few products (**Figure 38.6B**). Because Japan occupied the regions of the world where rubber trees grow, during World War II U.S. and British chemists investigated ways to make synthetic rubber. They analyzed latex, determined its composition of hydrocarbons, and then made rubber in the lab. Today, half of the rubber used worldwide is synthetic.

38.2 recap

Many plants use secondary metabolites as constitutive defenses against herbivory. Other defenses are induced by herbivory through signal transduction pathways. The plant hormone jasmonate stimulates local and systemic responses to herbivores.

learning outcomes

You should be able to:

- Describe physical methods plants use to defend themselves against herbivores.
- Discuss the range of effects that secondary metabolites might have on organisms attacking plants.
- Compare constitutive and induced chemical defenses against herbivores.
- Give examples of types of elicitors known to trigger signal transduction pathways.
- Describe the key components of signal transduction pathways that are activated by herbivory or wounding.
- Describe and analyze the process by which jasmonate defends plants against herbivory.

1. Latex is harvested from certain trees to make rubber. After a region of the tree has been drained of latex, slugs are

table **38.2** Environmental Stresses on Plants

Condition	Effect on plants
Drought	Reduced water potential, dehydration
Flooding	Reduced O_2 and respiration
High temperature	Changes in membrane fluidity and in proteins
Low temperature	Changes in membrane fluidity, damage by ice crystals
Salinity	Reduced water potential, dehydration
Metal element toxicity	Disruption of metabolism

▶ **38.2 recap**

observed consuming the leaves in that region. What does this indicate about the role of latex in the tree?

2. Tobacco plants produce nicotine as a secondary metabolite. The plants also have induced defenses against herbivores. How would these defenses work together to protect the plant from herbivores?

3. Armyworm larvae feed on corn leaves. They deposit their frass, or waste, in crevices where corn leaves meet stalks. The frass releases chemicals that signal the plant that it is being attacked by armyworms, and the corn plant mounts a defense against armyworms.

   a. Does this represent a constitutive or an induced defense? Explain your answer.

   b. How would chemicals in armyworm frass affect the corn plant's response?

4. Explain how plants attacked by herbivores can use jasmonate to "call for help."

A plant's survival depends not only on successful defenses against pathogens and herbivores but also on coping with a sometimes hostile physical environment. Next we will consider how plants deal with environmental stresses.

▶ key concept

38.3 Plants Can Adapt to Environmental Stresses

Plants are threatened by many biologically harmful aspects of the physical environment, such as drought, waterlogged soils, and extreme temperatures (**Table 38.2**). Plants cope with environmental stresses through adaptation or acclimation.

- **Adaptation** is genetically encoded strategy for maximizing survival and reproduction in a particular environment. A plant may have structures or biochemical properties that aid in its survival in the face of environmental challenges. Adaptations are the result of evolution by natural selection.

- **Acclimation** is increased tolerance for environmental extremes because of prior exposure to them. An individual plant previously exposed to extreme cold, for example, may be more likely to survive the subsequent winter.

focus your learning

- Plants living in water-saturated environments show a variety of adaptations for obtaining oxygen.
- Most plant responses to drought are constitutive, but under extreme conditions plants rely on inducible responses.
- Only certain plants, called halophytes, are adapted to salty soils; they are able to take up and excrete excess salt.
- Phytoremediation takes advantage of hyperaccumulators (plants that take up heavy metals) to clean up toxic environments.

 Media Clip 38.2 Leaves for Every Environment
www.Life11e.com/mc38.2

Plants can adapt to dry conditions

Many plants, especially those living in deserts, must cope with extremely limited water supplies. A variety of anatomical and life-cycle adaptations allow plants to survive under these conditions. Many of these adaptations are ways to avoid, reduce, or cope with the inevitable water loss through transpiration that occurs during active photosynthesis. Other adaptations help plants tolerate the excessive light and heat that are often found in deserts.

DROUGHT AVOIDERS   Some desert plants have no special structural adaptations for water conservation. Instead, these desert annuals, called drought avoiders, simply evade periods of drought. Drought avoiders carry out their entire life cycle—from seed to seed—during a brief period in which rainfall has made the surrounding desert soil sufficiently moist for growth and reproduction (**Figure 38.7**). A different drought avoidance strategy is seen in some African and South American deciduous perennial plants, which shed their leaves in response to drought as a way to conserve water. These plants remain dormant until conditions are again favorable for growth, much as deciduous trees in temperate climates shed their leaves in fall and are dormant until the following spring.

LEAF STRUCTURES   Most desert plants are not drought avoiders, but rather grow in their dry environment year-round. Plants adapted

Figure 38.7 Desert Annuals Avoid Drought   The seeds of many desert annuals lie dormant for long periods, awaiting conditions appropriate for germination. When they do receive enough moisture to germinate, they grow and reproduce rapidly before the short wet season ends. During the long dry spells, only dormant seeds remain alive.

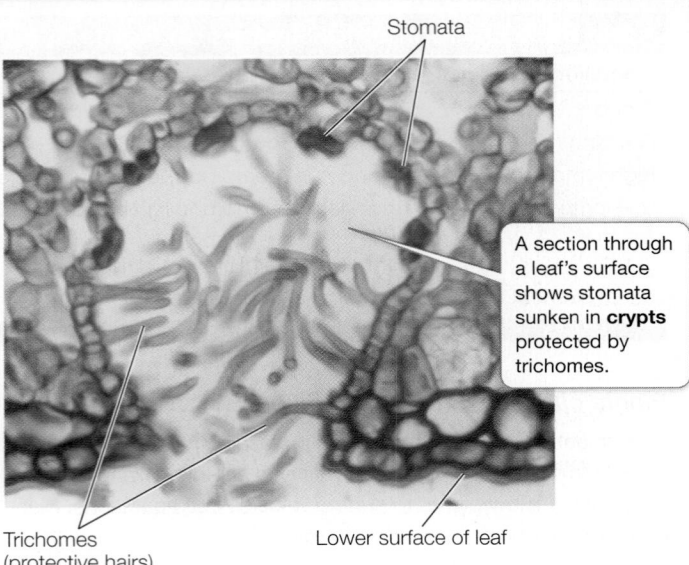

Stomata

A section through a leaf's surface shows stomata sunken in **crypts** protected by trichomes.

Trichomes (protective hairs)

Lower surface of leaf

Figure 38.8 Stomatal Crypts Stomata in the leaves of some xerophytes are located in sunken cavities called stomatal crypts. The trichomes (hairs) covering these crypts trap moist air. Note also the thick cuticle that prevents water loss.

to dry environments are called **xerophytes** (Greek *xeros*, "dry"). Three structural adaptations are found in the leaves of many xerophytes:

1. Specialized leaf anatomy that reduces water loss

2. A thick cuticle and a profusion of hairs over the leaf epidermis, which retard water loss

3. Trichomes that diffract and diffuse sunlight, thereby decreasing the intensity of light impinging on the leaves and the risk of damage to the photosynthetic apparatus by excess light

In some xerophytes the stomata are strategically located in sunken cavities below the leaf surface (known as **stomatal crypts**), where they are sheltered from the drying effects of air currents (**Figure 38.8**). Trichomes surrounding the stomata slow air currents further. Cacti and similar plants have spines rather than typical leaves, and photosynthesis is confined to the fleshy stems. The spines may help the plants cope with desert conditions by reducing leaf surface area and therefore, water loss. The spines may also deter herbivores.

WATER-STORING STRUCTURES **Succulence**—the possession of fleshy, water-storing leaves or stems—is another adaptation to dry environments (**Figure 38.9**). This adaptation allows plants to take up large amounts of water when it is available (such as after a brief thunderstorm) and then draw on the stored water during subsequent dry periods. Other adaptations of succulents include a reduced number of stomata and a variant form of photosynthesis (the CAM pathway; see Key Concept 10.4), both of which reduce water loss.

ROOT SYSTEMS THAT MAXIMIZE WATER UPTAKE Roots may also be adapted to dry environments. Cacti have shallow but extensive fibrous root systems that effectively intercept water at the soil surface following even light rains. Mesquite (*Prosopis*) (**Figure 38.10**) obtains water through taproots that grow to great depths, reaching water supplies far underground, as well as from condensation on its leaves.

Figure 38.9 Succulence The *Aloe* plant stores water in its fleshy leaves.

SOLUTE ACCUMULATION Xerophytes and other plants that must cope with inadequate water supplies may accumulate high concentrations of the amino acid proline or of secondary metabolites in their vacuoles. This solute accumulation lowers the water potential in the plant's cells below that of the soil, which allows the plant to take up water via osmosis. Plants living in saline environments share this and several other adaptations with xerophytes, as you will see shortly.

Some plants grow in saturated soils

For some plants, the environmental challenge is too much water, the opposite of that faced by xerophytes. Some plants live in environments so wet that the diffusion of oxygen to their roots is severely limited. These plants have shallow root systems that grow slowly; oxygen levels are likely to be highest near the surface of the soil, and slow growth decreases the roots' need for oxygen.

The root systems of some plants adapted to swampy environments, such as cypresses and some plants that grow in coastal mangrove habitats, have **pneumatophores**, which are extensions that grow out of the water and up into the air (**Figure 38.11A**).

Figure 38.10 Mining Water with Deep Taproots In Death Valley, California, the root of this mesquite tree must reach far beneath the dunes for its water supply.

(A)

Pneumatophores are root extensions that grow out of the water, under which the rest of the roots are submerged.

(B)

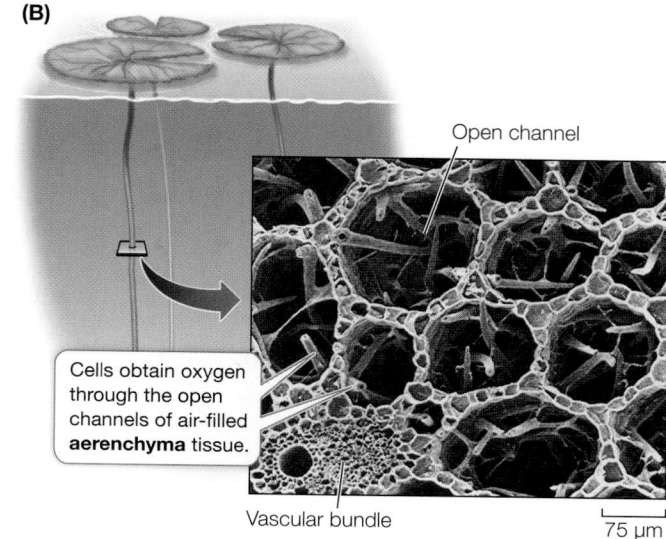

Cells obtain oxygen through the open channels of air-filled **aerenchyma** tissue.

Open channel

Vascular bundle

75 μm

Figure 38.11 Coming Up for Air **(A)** The roots of these mangroves obtain oxygen through pneumatophores. **(B)** This scanning electron micrograph of a cross section of a petiole of the yellow water lily shows the structure of the air-filled channels that make up aerenchyma tissue.

Pneumatophores contain lenticels (openings; see Figure 33.18) that allow oxygen to diffuse through them, aerating the submerged parts of the root system.

Many submerged or partly submerged aquatic plants have large air spaces in the leaf and stem parenchyma and in the petioles. Tissue containing such air spaces is called **aerenchyma** (**Figure 38.11B**). Aerenchyma stores oxygen produced by photosynthesis and permits its ready diffusion to parts of the plant where it is needed for cellular respiration. Aerenchyma also imparts buoyancy. Furthermore, because aerenchyma contains far fewer cells than most other plant tissues, metabolism in aerenchyma proceeds at a lower rate, so the need for oxygen is much reduced.

You may be familiar with rice paddies, where rice grows in a flooded environment. The strains of rice that thrive there—and provide food for millions—have hollow stems like straws, so that air can be supplied to submerged parts of the plant. Remarkably, these strains also have genes that respond to flooding due to heavy rains, by stimulating the plant to grow faster and keep the leaves above the water surface. These genes, appropriately called *SNORKEL-1* and *SNORKEL-2*, encode transcription factors that stimulate the expression of genes involved in cell elongation.

Plants can respond to drought stress

The adaptations of xerophytes for coping with dry environments are generally constitutive—they are always present—and under normal conditions they prevent the plants from experiencing drought stress. When conditions become so dry that even xerophytes are stressed, however, the plants turn to inducible responses. The same responses are found in many other plants, including those that are not adapted to grow in dry climates.

When the weather is abnormally dry, the water content of the soil is reduced and less water is available to plants. Water deficits in plant cells have two major biochemical effects:

1. A reduction in membrane integrity as the polar–nonpolar forces that orient the lipid bilayer are reduced

2. Changes in the three-dimensional structures of proteins

Plant growth is reduced when the structure of plant cells is compromised in these ways. Indeed, inadequate water supply is the single most important factor that limits production of our most important food crops.

When plants sense a water deficit in their ***roots**, a signaling pathway is set in motion that initiates several measures to conserve water and maintain cellular integrity. This pathway begins with the production of the hormone abscisic acid in the roots. This hormone travels from the roots to the shoot, where it results in stomatal closure and initiates gene transcription that leads to other physiological events that conserve water and cellular integrity (**Figure 38.12**).

---

*connect the concepts   Understanding the concept of water movement into plant roots by osmosis is essential to understanding how plants cope with water stresses. See Key Concept 34.1.

Many plant genes whose expression is altered by drought stress have been identified, largely through research using DNA microarrays, proteomics, and other molecular approaches (see Chapters 17 and 18). One group of proteins whose production is upregulated during drought stress is the *late embryogenesis abundant* (LEA; pronounced "lee-yuh") group of proteins. These hydrophobic proteins also accumulate in maturing seeds as they dry out (hence their name). LEA proteins bind to membrane proteins and other cellular proteins to stabilize them, preventing their aggregation during desiccation.

Plants can cope with temperature extremes

Temperatures that are too high or too low can stress plants and even kill them. Plant species differ in their sensitivity to heat and cold, but all plants have their limits. Any temperature extreme can damage cellular membranes.

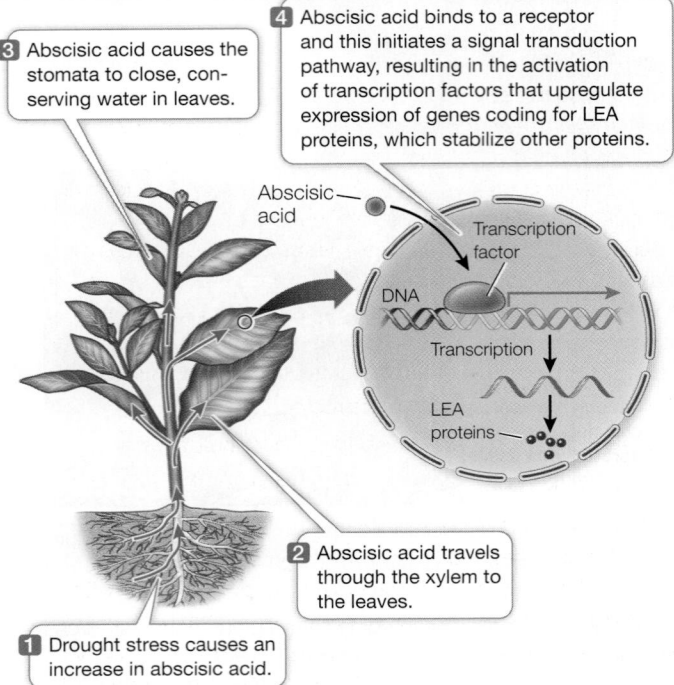

3 Abscisic acid causes the stomata to close, conserving water in leaves.

4 Abscisic acid binds to a receptor and this initiates a signal transduction pathway, resulting in the activation of transcription factors that upregulate expression of genes coding for LEA proteins, which stabilize other proteins.

Abscisic acid

Transcription factor

DNA

Transcription

LEA proteins

2 Abscisic acid travels through the xylem to the leaves.

1 Drought stress causes an increase in abscisic acid.

Figure 38.12 A Signaling Pathway in Response to Drought Stress Adaptation drought stress begins in the root with the production of the hormone abscisic acid.

Q: What genes do you think are stimulated by the LEA proteins?

- High temperatures destabilize membranes and denature many proteins, especially some of the enzymes of photosynthesis.
- Low temperatures cause membranes to lose their fluidity and alter their permeabilities to solutes.
- Freezing temperatures may cause ice crystals to form, damaging membranes.

Plants have both constitutive adaptations and inducible responses for coping with temperature extremes.

ANATOMICAL ADAPTATIONS Many plants living in hot environments have constitutive adaptations similar to those of xerophytes. These adaptations include hairs and spines that dissipate heat and leaf forms that intercept less direct sunlight.

HEAT SHOCK RESPONSE The plant inducible response to heat stress is similar to the response to drought stress in that new proteins are made, often under the direction of an abscisic acid–mediated signaling pathway. Within minutes of experimental exposure to raised temperatures (typically a 5°C–10°C increase), plants synthesize several kinds of **heat shock proteins**. Among these proteins are chaperonins, which help other proteins maintain their structures and avoid denaturation. Threshold temperatures for the production of heat shock proteins vary, but 38°C is sufficient to induce them in most plants.

COLD-HARDENING Low temperatures above the freezing point can cause chilling injury in many plants, including crops such as rice, corn, and cotton as well as tropical plants such as bananas. Many plant species can acclimate to cooler temperatures through a

process called **cold-hardening**, which requires repeated exposure to cool temperatures over many days. A key change during hardening is an increase in the proportion of unsaturated fatty acids in cell membranes, which allows them to retain their fluidity and function normally at cooler temperatures (see Figure 3.21). Plants have a greater ability to modify the degree of saturation of their membrane lipids than animals do. In addition, low temperatures induce the formation of proteins similar to heat shock proteins, which protect against chilling injury.

If ice crystals form within plant cells, they can kill the cells by puncturing organelles and cell membranes. Furthermore, the growth of ice crystals outside the cells can draw water from the cells and dehydrate them. Freeze-tolerant plants have a variety of adaptations to cope with these problems, including the production of antifreeze proteins that slow the growth of ice crystals.

Plants can adapt to salty soil

A number of toxic solutes are found in soils, but worldwide, no toxic substance restricts angiosperm growth more than ordinary salt (sodium chloride). Saline—salty—habitats support, at best, limited types of vegetation. Saline habitats are found in diverse locales, from hot, dry deserts to moist, cool coastal marshes. Along the seashore, saline environments are created by ocean spray. The ocean itself is a saline environment, as are estuaries, where fresh and salt water meet and mingle. Salinization of agricultural land is an increasing global problem (**Figure 38.13**). Even where crops are irrigated with fresh water, sodium ions from the water accumulate in the soil to ever-greater concentrations as the water evaporates.

Saline environments pose an osmotic challenge for plants. Because of its high salt concentration, a saline environment has an unusually negative soil water potential (see Figure 34.2). To obtain water from such an environment, a plant must have an even more negative soil water potential; otherwise water will diffuse out of its cells, and the plant will wilt and die. Plants in saline environments are also challenged by the potential toxicity of sodium, which inhibits enzymes and protein synthesis.

Figure 38.13 Salty Soil Accumulation of salt from irrigation water with inadequate drainage has caused this soil in central California to become unsuitable for most plant growth.

Figure 38.14 Excreting Salt This saltwater mangrove plant has special salt glands that excrete salt, which appears here as crystals on the leaves.

Halophytes—plants adapted to saline habitats—are found in a wide variety of flowering plant groups. Most halophytes share one adaptation: they take up sodium and, usually, chloride ions and transport those ions to their leaves. The accumulated ions are stored in the central vacuoles of leaf cells, away from more sensitive parts of the cells. Nonhalophytes accumulate relatively little sodium, even when placed in a saline environment; of the sodium that is absorbed by their roots, very little is transported to the shoot. The increased salt concentration in the tissues of halophytes lowers their water potential and allows them to take up water from their saline environment.

Some halophytes have other adaptations to life in saline environments. Some, for example, have **salt glands** in their leaves. These glands excrete salt, which collects on the leaf surface until it is removed by rain or wind (**Figure 38.14**). This adaptation, which reduces the danger of poisoning by accumulated salt, is found in some desert plants, such as *Frankenia palmeri*, and in some mangroves growing in seawater.

Salt glands can play multiple roles, as in the desert shrub *Atriplex halimus*. This shrub has glands that secrete salt into small bladders on the leaves. By lowering the water potential of the leaves, this salt not only helps them obtain water from the roots but also reduces their transpirational loss of water to the atmosphere.

The adaptations we have just discussed are specific to halophytes. Several other adaptations are shared by halophytes and xerophytes, including thick cuticles, succulence, and CAM photosynthesis.

Some plants can tolerate heavy metals

Salt is not the only toxic solute found in soils. High concentrations of some heavy metal ions, such as chromium, mercury, lead, and cadmium, are toxic to most plants; many of these ions are more toxic than sodium at equivalent concentrations.

Some places have a lot of heavy metals as a result of normal geological processes. In other places, acid rain leads to the release of toxic aluminum ions in the soil. Human activities, notably the mining of metallic ores, leave localized areas—known as tailings—with high concentrations of heavy metals and low concentrations of nutrients. Such sites are hostile to most plants, and seeds falling on them generally do not produce adult plants.

Some mine tailings are not completely barren. Investigations have shown that certain species of metal-tolerant plants accumulate concentrations of heavy metals that would kill most plants. More than 200 plant species have been identified as **hyperaccumulators** that store large quantities of metals such as arsenic (As), cadmium (Cd), nickel (Ni), aluminum (Al), and zinc (Zn).

Perhaps the best-studied hyperaccumulator is alpine pennycress (*Thlaspi caerulescens*). Before the advent of chemical analysis, miners would use the presence of this plant as an indicator of mineral-rich deposits. Studies of *Thlaspi* and other hyperaccumulators have revealed several adaptive mechanisms of hyperaccumulators:

- Increased ion transport into the roots
- Increased rates of translocation of ions to the leaves
- Accumulation of ions in vacuoles in the shoot
- Resistance to the ions' toxicity

Knowledge of these hyperaccumulation mechanisms and the genes underlying them has led to the emergence of **phytoremediation**, a form of bioremediation (see Key Concept 18.5) that uses plants to clean up environmental pollution. Some phytoremediation projects use natural hyperaccumulators, whereas others use genes from hyperaccumulators to create transgenic plants that grow more rapidly and are better adapted to a particular polluted environment. In either case, the plants are grown in the contaminated soil, where they act as natural "vacuum cleaners" by taking up the contaminants. The plants are then harvested and disposed of to remove the contaminants. Perhaps the most dramatic use of phytoremediation occurred after an accident at the nuclear power plant at Chernobyl, Ukraine (then part of the Soviet Union), in 1986, when sunflower plants were used to remove uranium from the nearby soil. These plants are now being used in Japan to soak up radioactive substances from the Fukushima Daiichi reactor accident in 2011. Sunflowers are now the international symbol of nuclear disarmament (**Figure 38.15**)!

Figure 38.15 Phytoremediation Plants that accumulate heavy metals can be used to clean up contaminated soils. Here, sunflowers are growing near a nuclear power plant in France.

38.3 recap

Plants that live in continually dry or water-saturated environments have structural adaptations to cope with those conditions. Mechanisms that protect plants from drought stress are initiated by a signaling pathway involving abscisic acid. Heat shock proteins help plants acclimate to high and low temperatures. Halophytes have several adaptations to saline habitats, most of which involve mechanisms that lower their water potential. Some plants can tolerate heavy-metal-rich soils that are toxic to most other plants.

learning outcomes

You should be able to:

- Analyze and compare different methods by which plants respond to saturated environments.
- Discuss factors that trigger inducible responses to drought in plants, and describe the possible responses.
- Describe, analyze, and evaluate methods by which plants are able to remove excess salt.
- Relate phytoremediation to hyperaccumulating plants, and explain how phytoremediation is applied to restore contaminated environments.

1. In the coming decades, climate change may have significant effects on the growth and productivity of plants, in particular the crops on which we depend for our food. What are the physiological effects, and possible genetic responses, of the following, in terms of plant breeding?
   a. In Pakistan, reduced rainfall causes a reduction in wheat yields.
   b. In the Mekong Delta of Vietnam, rising sea level inundates rice fields, causing a drastic reduction in yields.
2. Irrigation leads to increasing salinization of soil. An agricultural researcher wants to develop grain crops able to withstand higher salt concentrations. What type or types of genes might she add to the genomes of the crop plants, and how would these genes function to help the plant withstand salt?
3. One disadvantage of using hyperaccumulators to clean up toxic waste sites is that some hyperaccumulators grow slowly and produce little biomass.
   a. Why would this be a disadvantage?
   b. Suggest ways in which this disadvantage might be overcome.

investigatinglife

 How can knowledge of plant and fungal biology be used to prevent the spread of wheat rust?

Today, *Puccinia graminis* strain Ug99 continues its spread and threatens wheat crops worldwide. In the Green Revolution of the 1960s to 1980s, wheat plants were bred to have the gene complexes *Sr24* and *Sr31*, *R* genes that conferred resistance to existing strains of wheat rust fungus. The first strain of Ug99 found in Uganda had *Avr* genes that form elicitor proteins that do not bind to *Sr31* and therefore overcome part of the plant's resistance. Then, in 2006, a new genetic variant of Ug99 was found in Kenya with additional *Avr* genes that allow the pathogen to overcome the *Sr24* gene resistance as well. Because 90 percent of the wheat grown in the world has no resistance to this new strain of Ug99, an intensive search is under way for additional wheat resistance genes. In the regions where Ug99 infections are now present, occasional wheat plants grow that are resistant. Samples of both wheat plants and fungus are sent to laboratories, where, under high security and sterility, they are tested and examined for *Avr* and *R* genes. In addition, seeds from thousands of varieties of wheat, collected from all over the world, are being grown into seedlings and examined for *R* genes.

Future directions

The roles of the hormones jasmonic acid and salicylic acid in plant immunity are well described. It is now apparent that other hormones are involved as well. This is yet another example of crosstalk between signaling pathways. Auxins affect the architecture of the plant cell wall, leading to wall loosening. But this has a downside, making the plant's mechanical defenses less effective, and pathogens grow more readily in loose walls. Auxins can also interrupt other hormone signaling pathways, leading to reduced immunity. So there is a trade-off between growth and disease susceptibility. Perhaps most dramatically, certain bacteria have a complete biochemical pathway for the synthesis of auxins, so that when they enter a plant, auxins are secreted and overgrowth occurs. We are only beginning to understand the direct and indirect relations between hormones and plant immunity.

Chapter Summary 38

38.1 Plants Respond to Pathogens with Constitutive and Induced Responses

- Plants and pathogens have evolved together in a continuing "arms race": pathogens have evolved mechanisms for attacking plants, and plants have evolved mechanisms for defending themselves against those attacks.

- **Constitutive** defenses include plants' ability to strengthen their cell walls and block plasmodesmata when attacked, limiting the ability of viral pathogens to move from cell to cell.

- **Induced** defenses are triggered by a wide range of molecular **elicitors** and fall into two main categories: general immunity and specific immunity. Review Focus: Key Figure 38.2, Animation 38.1

- There is an ongoing search for genes involved in plant defense. Review Investigating Life: A Gene for Resistance to Wheat Rust

- The **gene-for-gene concept** depends on a match between a plant's **resistance** (*R*) genes and a pathogen's **Avirulence** (*Avr*) **genes**. Review Figure 38.3

- In the **hypersensitive response** to infection by bacteria or fungi, cells produce two kinds of defensive molecules: **phytoalexins** and **pathogenesis-related (PR) proteins**. Some cells around the infected area die, sealing off the pathogens and the damage they have caused.

- The hypersensitive response is often followed by **systemic acquired resistance**, in which salicylic acid activates further synthesis of defensive compounds.

- Plants use RNA interference to develop specific immunity to invading RNA viruses.

38.2 Plants Have Mechanical and Chemical Defenses against Herbivores

- Some plants produce **secondary metabolites** as defenses against herbivores. Review Table 38.1

- Hormones, including **jasmonates**, participate in signal transduction pathways leading to the production of defensive compounds. Review Figure 38.5, Activity 38.1

- Plants protect themselves against their own toxic defensive chemicals by sequestering them in vacuoles or storing them as harmless precursors until used.

38.3 Plants Can Adapt to Environmental Stresses

- Plants cope with environmental stresses by **adaptation** (genetically encoded resistance) or **acclimation** (increased tolerance). Review Table 38.2

- **Xerophytes** are plants that are adapted to dry environments.

- Some xerophytic adaptations are structural, including thickened cuticles, specialized trichomes, **stomatal crypts**, **succulence**, and long taproots.

- Some plants accumulate solutes, making their water potential lower so they can tolerate drought.

- Adaptations to water-saturated habitats include **pneumatophores**, extensions of roots that allow oxygen uptake from the air, and **aerenchyma**, tissue in which oxygen can be stored and used to supply the rest of the plant.

- A signaling pathway involving abscisic acid initiates a plant's response to drought stress. Review Figure 38.12

- Membranes and proteins can be damaged by extremely high or low temperatures. Plants respond to extreme temperatures by producing **heat shock proteins**.

- Some plants undergo **cold-hardening**, an acclimation process that includes changes in membrane lipids and production of heat shock proteins.

- Some plants resist freezing by producing antifreeze proteins.

- Most **halophytes** accumulate salt. Some have **salt glands** that excrete salt to the leaf surface.

- Some plants living in soils that are rich in heavy metals are **hyperaccumulators** that take up large amounts of those metals into their tissues.

- **Phytoremediation** is the use of hyperaccumulating plants or their genes to clean up environmental pollution.

See Activity 38.2 for a concept review of this chapter.

> Go to **LearningCurve** (in **LaunchPad**) for dynamic quizzing that helps you solidify your understanding of this chapter. **LearningCurve** adapts to your responses, giving you the practice you need to master each key concept.

▷ Apply What You've Learned

Review

38.2 Secondary metabolites are plants' constitutive chemical defenses against herbivory.

38.2 Herbivory induces plants to activate elicitors, which in turn activate signal transduction pathways.

38.2 The hormone jasmonic acid and its derivatives are key to the expression of many defenses against herbivory.

Original Paper: Halitschke, R., U. Schittko, G. Pohnert, W. Boland, and I. T. Baldwin. 2001. Molecular interactions between the specialist herbivore *Manduca sexta* and its natural host *Nicotiana attenuata*. *Plant Physiology* 125: 711–717.

Herbivores feed on the tobacco plant *Nicotiana attenuata*, which defends itself by producing secondary metabolites, most notably nicotine. Researchers did a series of experiments comparing two different types of responses in *Nicotiana*: response to a wound (mechanical damage), such as that done by animals, and response to herbivory by the caterpillar of the tobacco hornworm, *Manduca sexta*. They questioned whether the plant showed differences in its chemical defenses to these two types of predation.

The researchers exposed leaves on tobacco plants to physical wounds that might occur when a large animal chews the leaves. They also exposed the leaves to wounds along with hornworm secretions called FACs. This simulated the second condition, attack by a hornworm.

In one experiment, researchers measured nicotine production under both conditions and compared the results to those from untreated controls, as shown in **Figure A**.

In a second experiment, the researchers repeated the above two experimental conditions, but after a short period measured the concentration of the signaling molecule jasmonic acid (see Figure 38.5). The results are shown in **Figure B**.

The hornworm is able to eat tobacco without ill effects. One way it might be able to do this is by suppressing nicotine production in the plants. The hormone ethylene (see Figure 36.2) suppresses nicotine production. Thus, the researchers measured ethylene in the two conditions of predation, as shown in **Figure C**.

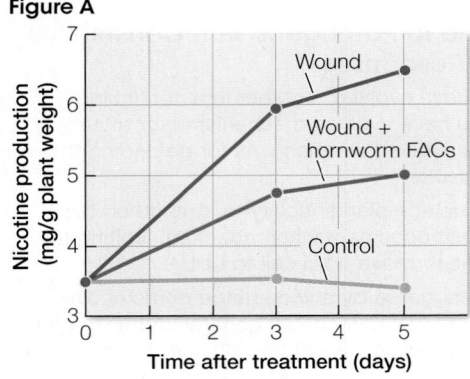
Figure A

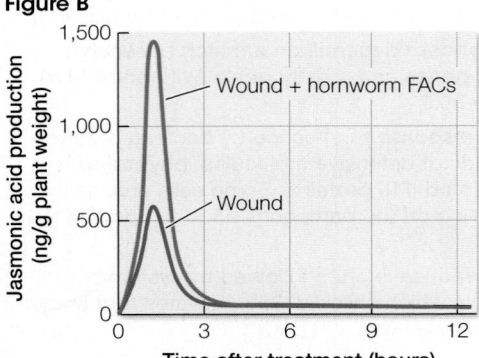
Figure B

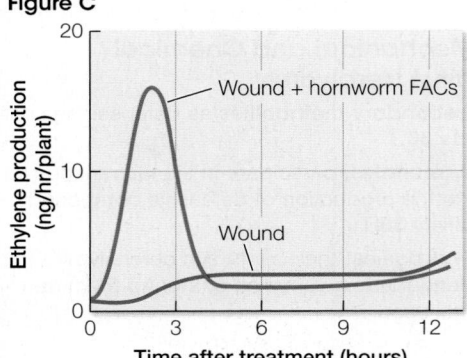
Figure C

Questions

1. Compare nicotine production in the two types of herbivory. What might account for the difference? Is nicotine production a constitutive or an induced response? Explain your answer.

2. What is the role of jasmonic acid in the plant's response to herbivory? How does it differ from nicotine in its effects?

3. In the graph illustrating jasmonic acid production, compare the plant's response to the wound and to the wound + hornworm FACs. What does the difference in these responses suggest about the importance of the FACs produced by the caterpillar?

4. Compare the production of jasmonic acid and ethylene under the two experimental conditions. What might explain the difference in these two responses?

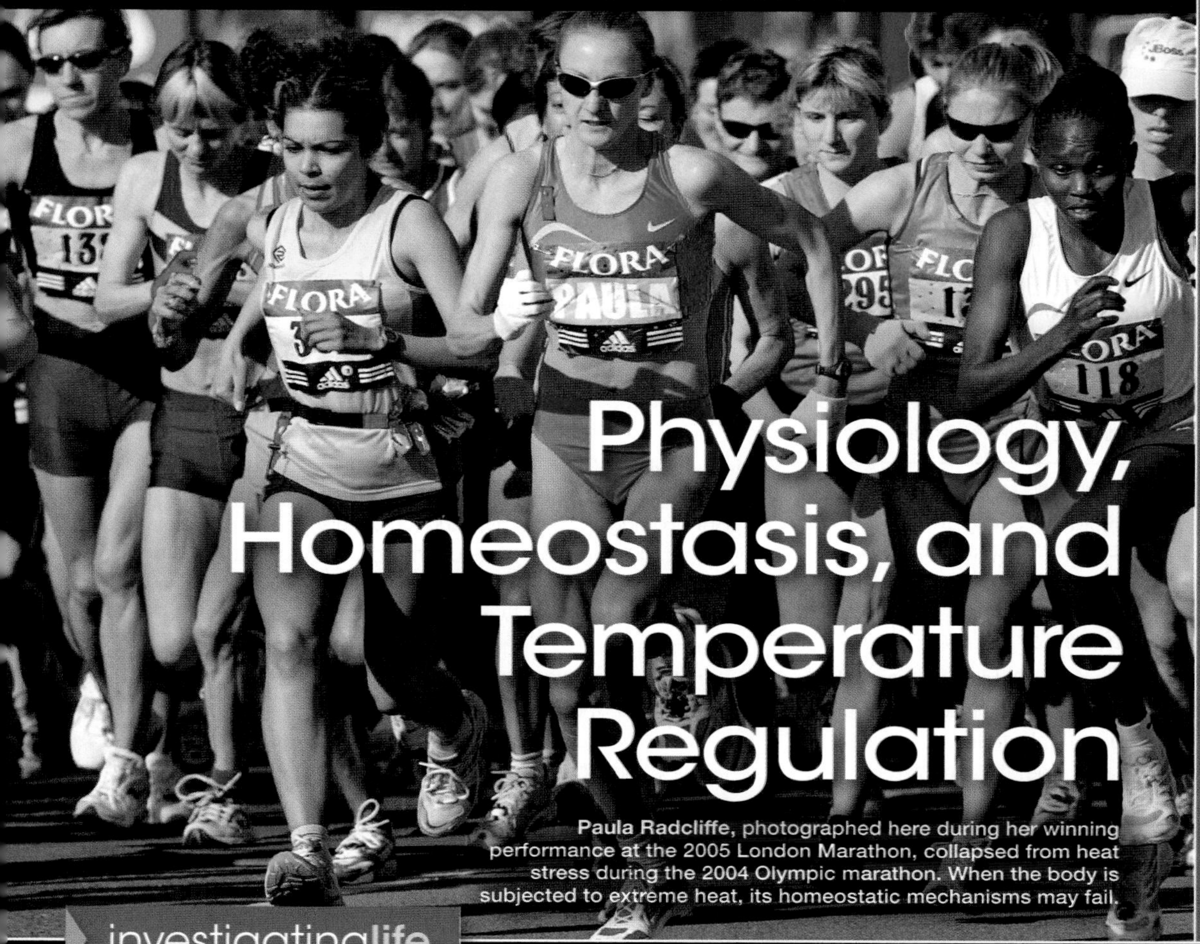

39 Physiology, Homeostasis, and Temperature Regulation

Paula Radcliffe, photographed here during her winning performance at the 2005 London Marathon, collapsed from heat stress during the 2004 Olympic marathon. When the body is subjected to extreme heat, its homeostatic mechanisms may fail.

key concepts

39.1 Animals Are Composed of Organs Built from Four Types of Tissues

39.2 Physiological Systems Maintain Homeostasis of the Internal Environment

39.3 Biological Processes Are Temperature-Sensitive

39.4 Body Temperature Depends on the Balance between Heat In and Heat Out of the Body

39.5 Body Temperature Is Regulated through Adaptations for Heat Production and Heat Loss

investigatinglife

Heat Limits Physical Performance

The 2008 New York City Marathon took place on a cold, clear, windy day in November. For the third time, the first-place woman in this 41-kilometer race was world record holder Paula Radcliffe. Radcliffe had also been expected to win the women's marathon in the 2004 Olympics. But that race took place on an extremely hot (high of 34°C), humid day in Athens. Overcome by heat stress, Radcliffe collapsed 6 kilometers from the finish line.

Elite runners generally have their best times when temperatures are below 10°C. The 2012 Boston Marathon coincided with an unseasonable April heat wave, with temperatures exceeding 27°C. During the race, 120 runners were rushed to hospitals with severe heat stress.

When a person's internal body temperature rises above 40°C, major organs begin to fail, a condition known as heat stroke. Every year some athletes suffer heat stroke, which leads to death in a high percentage of cases. Soldiers in desert environments are at extreme risk of heat stroke, as are workers in many occupations, including firefighting, agriculture, and construction.

Heat stroke is a particular danger for those who are active in the heat because working muscles generate heat. The blood carries that heat out of the muscles and distributes it throughout the body, raising the temperature of the body's internal tissues. Some of the heated blood flows to the skin, where heat can be lost to the environment, but humans are subject to the problems faced by all mammals in losing excess heat. First, their normal internal temperatures are not far from the environmental temperatures that cause heat stress, so they don't have much of a safety zone. Second, most mammalian skin surfaces are covered with an insulating layer of fur—great for conserving body heat in cold environments, but an impediment to heat loss in warm ones. Even human skin is an insulator.

A major evolutionary adaptation in mammals for heat loss is efficient heat-loss portals of non-furred areas such as the nose, tongue, and footpads. In these areas, specialized blood vessels can open up and act like radiators to disperse heat (conversely, these portals can close down to conserve heat). Humans are not furred, but our evolutionary ancestors were, and we retain these mammalian blood vessel adaptations in our hands, feet, and face (which is why we blush).

QA How can we increase heat loss from the body to protect against heat stress?

39.1 Animals Are Composed of Organs Built from Four Types of Tissues

Animals are amazing in all they do to survive, thrive, and propagate. They live in environments that could not support single cells, they extract energy from the environment, they build complex molecules and structures, they alter their environments, and they reproduce. All of these feats and more depend on complex body components that have evolved since the first appearance of multicellular organisms on this planet. What does it take to be an animal?

focus your learning

- Multicellularity enabled the evolution of organisms that are much larger than single-celled organisms can be.
- Organs are composed of four tissue types.

How would you build a multicellular animal?

This might seem like a silly question; even bioengineers don't build animals. But as you begin to study animal form and function, or animal physiology, it is instructive to think of all of the problems animals must solve to survive. For example: they must:

- extract energy and nutrients from the environment.
- build all of the internal structures they need.
- eliminate toxins and metabolic waste products.
- sense the environment and respond to it in various ways, including movement.
- maintain constant conditions in their internal environments.
- reproduce.

Single- and multicellular organisms have evolved to conquer these challenges. In fact, single-celled organisms are remarkably abundant and thrive in widely diverse environments. Multicellular organisms have evolved complex body systems to conquer the same challenges. Given the success of single-celled organisms, why have multicellular animals evolved? How is it advantageous to be multicellular? Multicellularity confers two advantages—large size and the opportunity for cells to specialize.

There are advantages and challenges to getting bigger

Advantages to being bigger include being able to prey on other organisms and also being able to resist or work against forces in the environment, such as wave forces in reefs. There is a problem, however, that limits how big single-celled organisms can get. All exchanges with the environment, such as uptake of nutrients and elimination of wastes, occur across the cell membrane, so the extent of these exchanges is limited by the surface area of the cell membrane. *Cell surface area increases with the square of the linear dimension of the cell, but the needs of the cell increase as a function of the volume of the cell, which is a function of the cube of the linear dimension. Thus there is a size limit beyond which single-celled organisms cannot satisfy their needs.

*connect the concepts   Figure 5.2 illustrates how cell volume grows faster than cell surface area as cells get bigger.

Cell specialization is essential for multicellular animals to grow bigger. If a multicellular animal consists of only a few layers of cells, those cells can carry out their own exchanges directly with the environment. A good example is a sponge (**Figure 39.1**). Bigger organisms have cells that are not in contact with the external environment. The needs of those cells must be supplied by an *internal environment*—extracellular fluid that bathes all the cells of the animal's body. Different cells and groups of cells must be specialized to contribute to the maintenance of that internal environment. Thus each cell contributes to the internal environment, and each cell is supported by that internal environment.

Cell specializations have tremendous adaptive value for even the simplest multicellular animals. Some cells can be specialized for movement, some for intake of food, and some for reproduction. But evolving a specialized function usually comes at the expense of other cellular functions, which may be lost or at least compromised. The specialized cells of an organism thus engage in a kind of collaborative enterprise by providing services for all of the cells of the organism in return for different services provided by other cells. Thus the cells specialized for movement do not have to also capture and process food, and the reproductive cells do not have to move the whole animal to find a mate. For example, male sex cells are motile cells that are good at seeking out nonmotile female sex cells, but the male sex cells have lost the ability to do many other cell functions such as acquire and process food.

There are four tissue types

Specialized cells rarely function alone, but rather as groups of similar cells called **tissues**. There are many different kinds of cells, but they are all classified as one of four tissue types: epithelial, muscle, connective, and nervous (**Focus: Key Figure 39.2**):

1. **Epithelial tissues** are sheets of cells that create barriers between different compartments and frequently have secretory functions.

2. **Muscle tissues** contract to generate forces and movement.

Figure 39.1 No Cell Too Far The sponge *Leucosolenia* is a tube about 2 cm. long with an outer and an inner layer of cells and a jelly-like matrix. There are pores between the cells of the tube. The inner cells have flagella that create a current of water in through the pores and out through the opening at the top. The individual cells of this animal can obtain nutrients and oxygen directly from the seawater.

focus: key figure

Creating Compartments: Epithelial Tissues

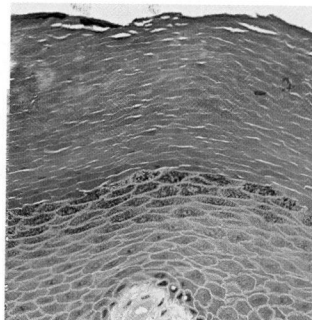

Stratified squamous epithelium: The outer layers of skin (epidermis) separate internal from external environment and protect the body.

40 μm

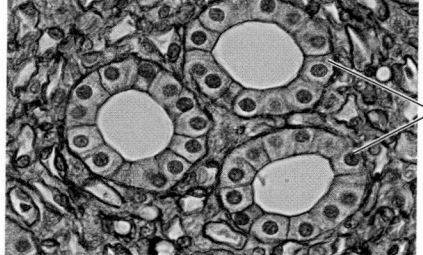

Cubodial epithelial cells

Cubodial epithelium: Makes up tubules and ducts and can have secretory and absorptive functions.

25 μm

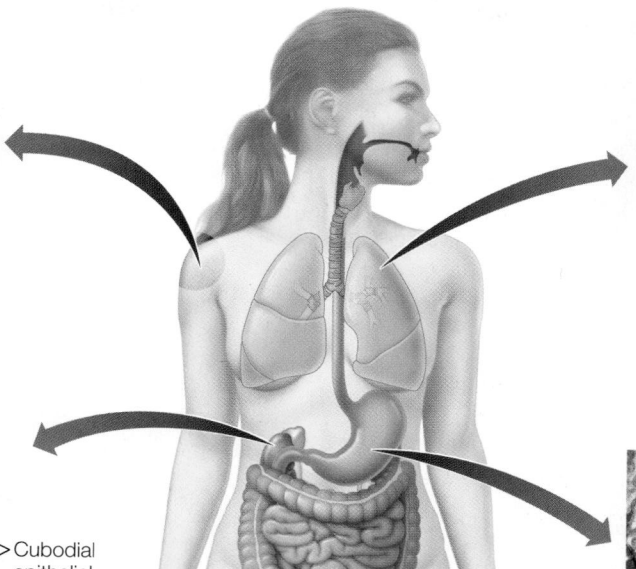

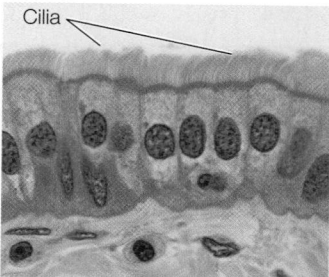

Cilia

Columnar epithelium: Lines many internal organs such as the airways of the lungs and the small intestine. Can have secretory and absorptive functions. Enables movements of substances between body compartments.

20 μm

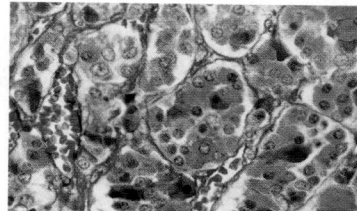

Secretory cells: The stomach lining includes epithelial cells that secrete digestive juices and acid. Other secretory epithelial cells are found in the salivary, sweat, and mammary glands, and in the pancreas and anterior pituitary.

Generating Forces: Muscle Tissues

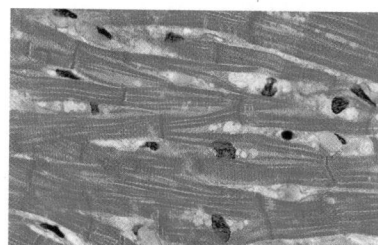

Cardiac muscle: Responsible for contractions of the heart.

15 μm

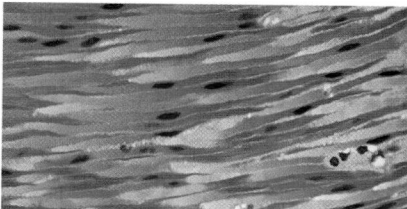

Smooth muscle: Provides motility to internal organs such as the digestive tract. Controls diameter of blood vessels.

30 μm

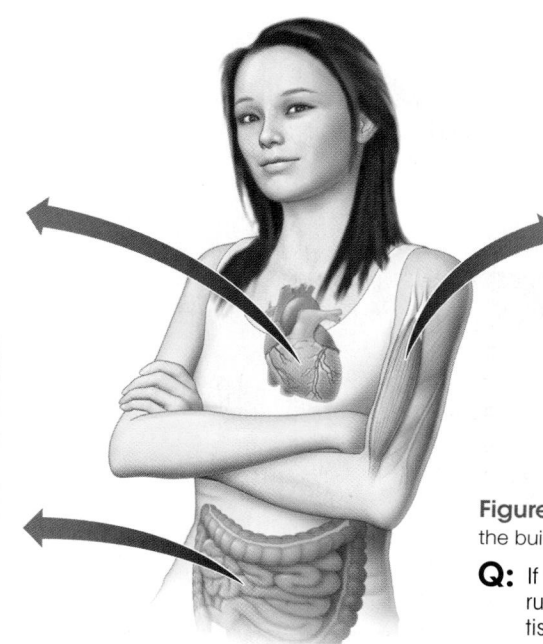

Skeletal muscle: Responsible for voluntary movements of the body.

15 μm

Figure 39.2 How to Build an Animal   Tissues are the building blocks.

Q: If we consider the functions of skeletal muscle in running, what are roles played by the three other tissue types in this process?

(continued)

focus: key figure (continued)

Providing Support: Connective Tissues

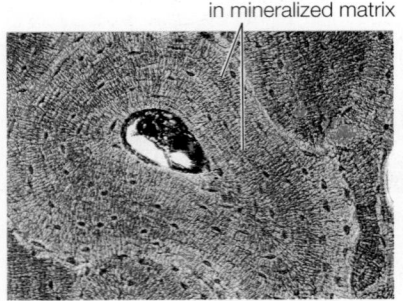

Layers of bone cells in mineralized matrix

Bones: Provide support structures for the body that make it possible for muscles to generate movements.

250 µm

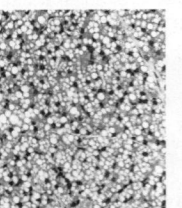

Brown fat 80 µm

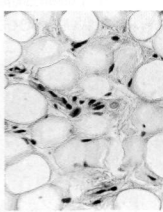

White fat 80 µm

Adipose tissue: White fat cushions and supports organs, provides thermal insulation, and stores energy. Brown fat produces heat.

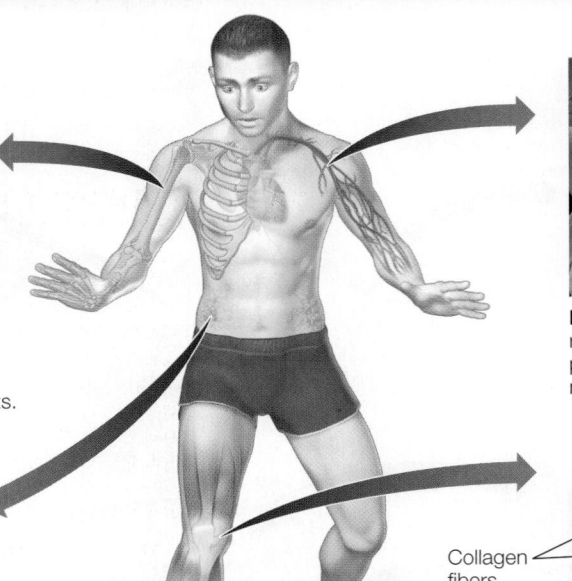

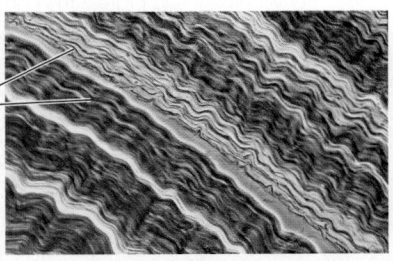

White blood cell   Red blood cells

Blood cells: Red blood cells carry respiratory gases. White blood cells protect against foreign substances and microorganisms invading the body.

15 µm

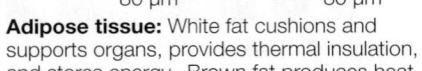

Collagen fibers

Ligaments and tendons: Connect bones to bones and muscles to bones.

Processing Information: Nervous Tissues

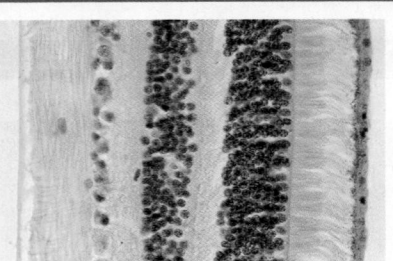

Sensors: These cells of the retina encode information about the external environment.

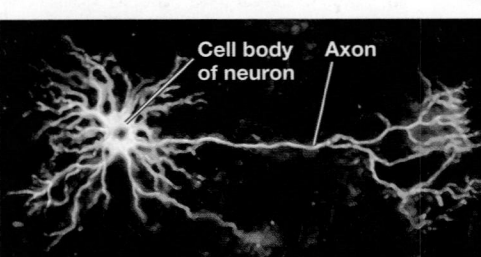

Cell body of neuron   Axon

Neurons: Communicate information from sensors to the central nervous system, stores and integrates information, communicates commands to muscles and glands.

20 µm

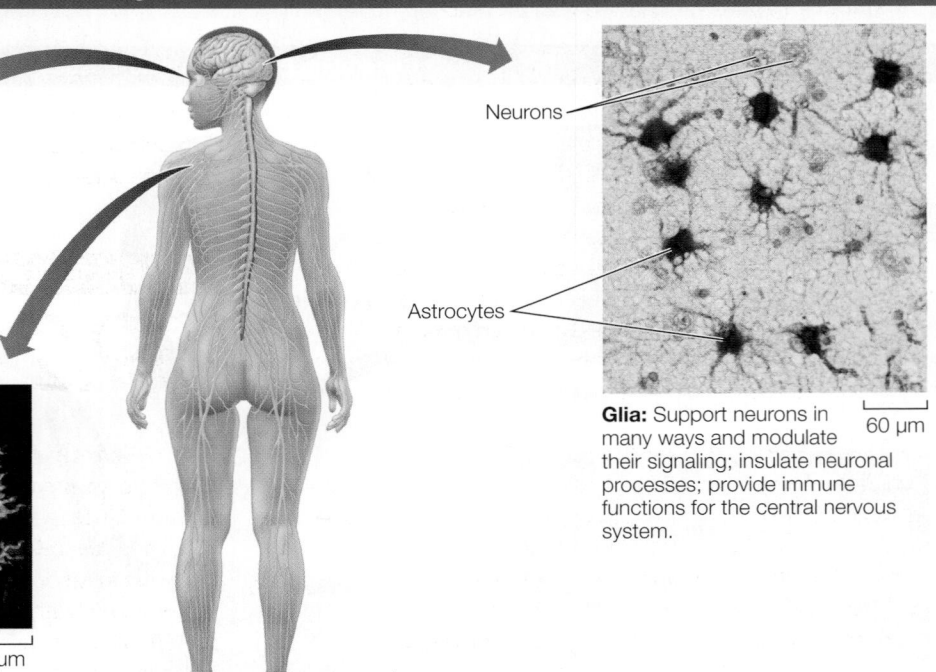

Neurons

Astrocytes

Glia: Support neurons in many ways and modulate their signaling; insulate neuronal processes; provide immune functions for the central nervous system.

60 µm

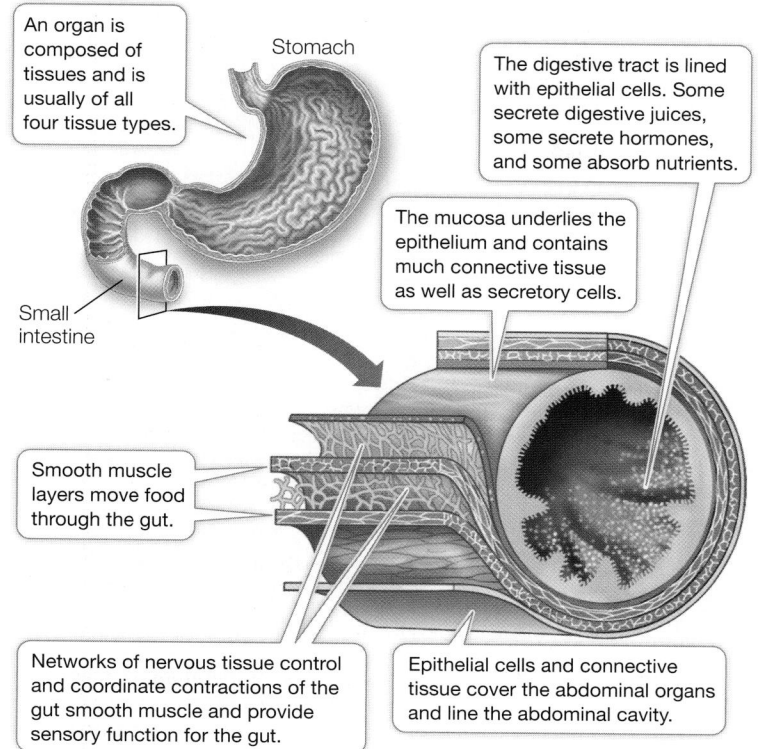

An organ is composed of tissues and is usually of all four tissue types.

Stomach

The digestive tract is lined with epithelial cells. Some secrete digestive juices, some secrete hormones, and some absorb nutrients.

The mucosa underlies the epithelium and contains much connective tissue as well as secretory cells.

Small intestine

Smooth muscle layers move food through the gut.

Networks of nervous tissue control and coordinate contractions of the gut smooth muscle and provide sensory function for the gut.

Epithelial cells and connective tissue cover the abdominal organs and line the abdominal cavity.

Figure 39.3 Tissues Form Organs The organs of the human digestive system, such as the stomach and small intestine, are made up of all four tissue types.

 Activity 39.1 Tissues and Cell Types
www.Life11e.com/ac39.1

3. **Connective tissues** provide structure and support.

4. **Nervous tissues** convey and process information.

These are very general groupings that reflect developmental origins, but within each group there are various specializations that accommodate different functions. For example, connective tissue includes bones, cartilage, fat, and blood. Epithelial tissue includes the outer layers of skin (epidermis) and many types of secretory cells that produce and release substances internally, such as hormones, and externally, such as sweat and digestive juices. You might think of digestive juices as being secreted internally, but the lumen of the digestive tract is continuous with the external environment. You will learn a lot about the diversity of functions accomplished through specializations of these four tissue types in the chapters that follow.

Organs are built from multiple tissues

Organs are internal structures that carry out specific functions. An organ is composed of an outer covering, or epithelium, and one or more other kinds of tissues. Indeed, most organs include all four tissue types. The gut (digestive tract) is a good example (**Figure 39.3**). Its inner surface is lined with a sheet of columnar epithelial cells. Different types of epithelial cells in this lining secrete hormones or digestive juices or absorb nutrients from the gut. Beneath the epithelial lining is a layer of connective tissue called the mucosa. Within this connective tissue are blood vessels, neurons, and glands (clusters of secretory epithelial cells). Concentric layers of smooth muscle tissue enable the gut to contract to mix food with digestive juices. A network of neurons between the muscle layers controls these movements.

An individual organ is usually part of an **organ system**, a group of organs that work together to carry out certain functions. The stomach, small intestine, liver, and pancreas, for example, are parts of the digestive system. The structure of an animal involves an organizational hierarchy. Cells form tissues, tissues form organs, organs work together as organ systems, and all work together as a multicellular organism.

39.1 recap

There are limits to how large cells can be because metabolic needs are related to cell volume, and abilities to satisfy those needs are determined by cell surface area. Cell volume grows faster than cell surface area. Multicellularity and cell specialization enabled organisms to grow larger. Groups of similar cells form tissues, and multicellular animals are composed of four tissue types: epithelial, connective, nervous, and muscle. Individual organs include multiple tissue types and are part of organ systems.

learning outcomes

You should be able to:

- Demonstrate an understanding of the factors that limit cell size.
- Describe the advantages of multicellularity.
- Explain structure–function relationships within an organ.

1. What limits the size of single cells?
2. What are the advantages of being a multicellular organism rather than a single-celled organism?
3. Provide an example of how multiple tissue types are involved in the construction of an organ. What does each tissue type contribute to the function of that organ?

Subsequent chapters will describe each of the organ systems in much greater detail. The remainder of this chapter focuses on the mechanisms of control and regulation of the internal environment, using temperature as the example.

key concept 39.2 Physiological Systems Maintain Homeostasis of the Internal Environment

The evolution of multicellularity made it possible to enclose and therefore create an internal fluid environment that bathed all cells of the organism. Since that internal environment provided for all of the needs of the individual cells, individual cell lines could become specialized as to what they contributed to the internal environment. Just as with people, specialization in one function usually goes hand in hand with losses in other functions. But those functional losses in each cell line were compensated for by the constancy of the internal environment. The evolution of organs composed of multiple tissues and the evolution of informational systems to control organ activities improved the abilities of animals to maintain a constant internal environment. In essence, the evolution of physiological systems to maintain various aspects of the internal environment made it possible for multicellular animals to become larger, and more complex, and allowed them to occupy many different environments.

focus your learning

- The needs of cells in the multicellular animal are served through exchanges with the internal environment, which consists of the extracellular fluid.
- Homeostasis of the internal environment is maintained through control and regulation of activities of organs and organ systems.

The internal environment is the extracellular fluid that serves all the needs of the cells of the body

Individual cells get their nutrients from the surrounding extracellular fluid (ECF) and dump their waste products into it. A human is about 60 percent water. Two-thirds of that water is intracellular fluid and one-third is the ECF. About 20 percent of the ECF (about 3 liters) is the **blood plasma** that circulates in our blood vessels. The remaining 80 percent (about 11 liters) is the **interstitial fluid** that bathes every cell of the body (**Figure 39.4**). Water and small molecules freely exchange between the interstitial fluid and the blood plasma. Thus nutrients move into the interstitial fluid from the blood plasma, and wastes move into the blood plasma from the interstitial fluid. As long as the composition and physical conditions of the interstitial fluid are held within certain limits, cells are protected from the changes and harsh conditions of the external environment. A stable internal environment makes it possible for an animal to occupy habitats that would kill its cells if they were directly exposed to the external conditions. How is the internal environment kept constant?

The constant composition of the internal environment is always being challenged by the external environment and by the metabolic activities of the cells of the body. Organisms must maintain their internal environment in a state of **homeostasis**—a narrow range of stable and optimal physical and biochemical conditions.

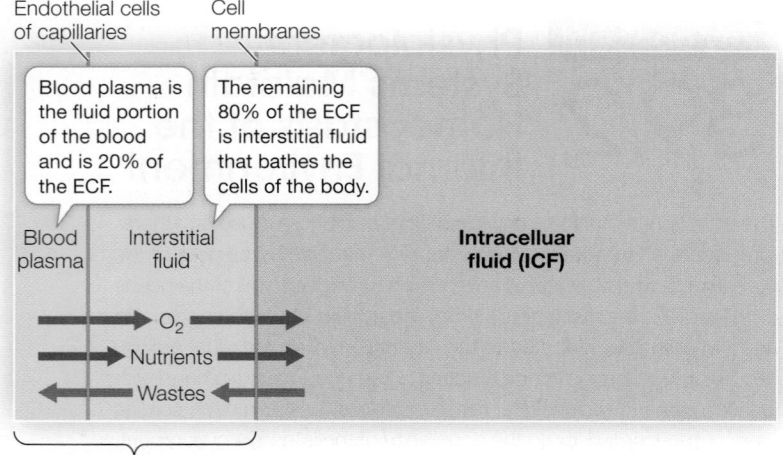

Endothelial cells of capillaries   Cell membranes

Blood plasma is the fluid portion of the blood and is 20% of the ECF.

The remaining 80% of the ECF is interstitial fluid that bathes the cells of the body.

Blood plasma   Interstitial fluid   Intracelluar fluid (ICF)

O₂
Nutrients
Wastes

Extracellular fluid (ECF)

Figure 39.4 The Internal Environment   The extracellular fluid (ECF) is the "internal environment," and it is about 1.3 of total body water. The ECF is about 20% blood plasma and 80% interstitial fluid. The composition and physiological state of the ECF must remain stable within narrow limits, and maintaining that stability is the job of the body's organ systems.

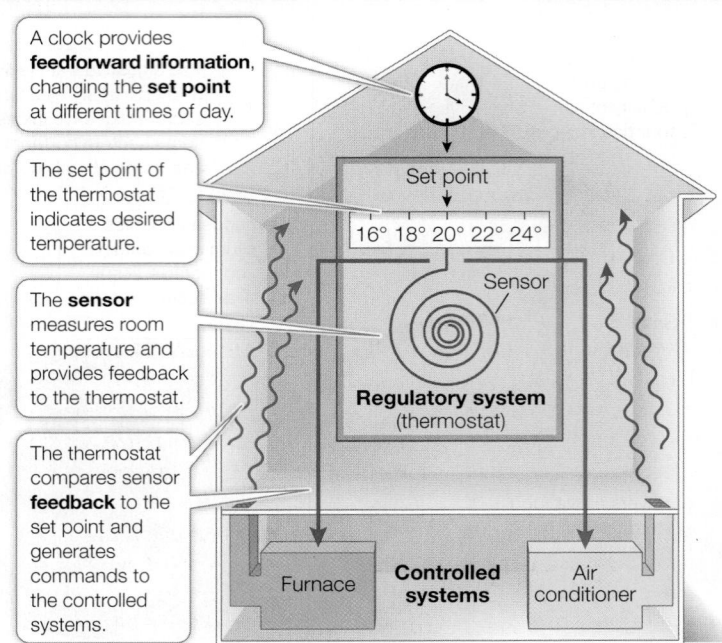

A clock provides **feedforward information**, changing the **set point** at different times of day.

The set point of the thermostat indicates desired temperature.

The **sensor** measures room temperature and provides feedback to the thermostat.

The thermostat compares sensor **feedback** to the set point and generates commands to the controlled systems.

Set point

16° 18° 20° 22° 24°

Sensor

Regulatory system (thermostat)

Furnace   **Controlled systems**   Air conditioner

Figure 39.5 A Thermostat Regulates Temperature   A thermostat regulates the temperature of a room by turning the furnace or air conditioner on or off in response to the difference between feedback information (room temperature) and set points that are programmed into the thermostat.

If a physiological system fails to function properly, homeostasis is compromised, and cells are damaged and can die. To avoid the loss of homeostasis, organ systems are controlled and regulated in response to changes in both the external and internal environments. The maintenance of homeostasis is a central theme of physiology.

Physiological systems are regulated to maintain homeostasis

The activities of physiological systems are *controlled*—speeded up or slowed down—by actions of the nervous and endocrine systems. But to *regulate* these systems and maintain homeostasis, information is required. As an analogy, think of the thermostat that controls the furnace and air conditioner to regulate the temperature of a house (**Figure 39.5**). The desired temperature is a **set point**, or reference point on the thermostat. The thermostat acts as a **comparator** by sensing the current temperature in the house and comparing that value to the set point. Thus the sensing of the air temperature is **feedback** information. Any difference between the set point and feedback information results in an **error signal**. The error signal is converted into commands to the furnace or air conditioner, turning them on or off.

Regulatory systems such as a building thermostat or one of the many physiological regulatory systems described in this book obtain, process, and integrate information. Regulatory systems use that information to issue commands to **effectors** such as muscles or glands that effect changes in the internal environment. Effectors are also called **controlled systems** because their activities are controlled by the neural or hormonal signals from their respective regulatory systems. Important components of any regulatory system are the sensors such as light-, temperature-, and pressure-sensitive cells that provide feedback information to be compared with internal set points.

Negative feedback is information used to counteract the influence that created an error signal. Whatever force is pushing the system away from its set point must be "negated." In our thermostat analogy, an air temperature below the set point causes the furnace to be turned on, which then reverses the direction of change in the air temperature.

Although not as common as negative feedback, **positive feedback** also exists in physiological systems. Rather than returning a system to a set point, positive feedback amplifies a response (i.e., it increases the deviation from the set point). An example is sexual behavior, in which a little stimulation causes more behavior, which causes more stimulation, and so on. Positive feedback responses tend to reach a limit and terminate rapidly. The birth process is a good example. Contractions of the uterus push the baby into the birth canal, and stretching of the birth canal stimulates more and stronger contractions until the baby is delivered, at which time contractions cease.

Feedforward information is another feature of regulatory systems. Its function is to change the set point in anticipation of a change in conditions. The timer on a thermostat provides feedforward information by changing the system's temperature set point, usually lowering it in the evening and raising it in the morning. Hearing the words "on your mark" before a race is feedforward information that raises your heart rate in anticipation of running. Feedforward information anticipates a change in the internal environment before that change occurs.

 Activity 39.2 **System Simulation**
www.Life11e.com/ac39.2

39.2 recap

The internal environment provides for the needs of all the cells that make up a complex multicellular animal. Organs and organ systems control the composition of the internal environment so as to maintain homeostasis. The activities of organs and organ systems are regulated. Regulation requires information about optimal conditions (set points) and existing conditions (feedback) to create commands to effector cells and organs.

learning outcomes

You should be able to:

- Explain why homeostasis of the internal environment in a multicellular animal is critical to the animal's survival.
- Use knowledge about fluid compartments in the human body to perform analyses.
- Differentiate between negative feedback, positive feedback, and feedforward control mechanisms.

1. Why is maintenance of the internal environment considered to be the primary focus of physiology?
2. Explain the differences between negative and positive feedback and feedforward control mechanisms.
3. Based on your body mass, estimate how much water is in your body. What is your intracellular fluid volume, your extracellular fluid volume, and your blood plasma volume?

Principles of control and regulation help organize our thinking about physiological systems. Once we understand how a system works, we can then ask how it is regulated. Part Nine of this book describes various physiological systems, how they function, and how they are regulated. We begin by extending our example of a house thermoregulatory system to that of the mammalian body.

▶ key concept 39.3 Biological Processes Are Temperature-Sensitive

Temperatures of environments where organisms live vary enormously, from the boiling hot springs of Yellowstone National Park to the interior of Antarctica, where the temperature can fall below −80°C. But in general, cells can function over only a narrow range of temperatures. The evolution of thermoregulatory adaptations enables maintenance of suitable thermal conditions in the internal environment.

focus your learning

- Cells can survive only within a small range of the temperatures that occur on Earth.
- Q_{10} values are measures of temperature sensitivity of biochemical reactions or physiological processes.
- Temperature acclimatization allows animals to shift their metabolic rates as they experience seasonal changes in external temperatures.
- Metabolic acclimatization enables animals to adapt to seasonal temperature changes.

Temperature affects living systems

If cells cool below the freezing point of their body fluids, ice crystals form and damage cell structures. Some animals have adaptations, such as antifreeze molecules in their blood that help them resist freezing; others can survive freezing (**Figure 39.6**). Generally, however, cells must remain above freezing to stay alive. The upper temperature limit for survival in most cells is about 45°C because of **protein denaturation**. Some specialized algae, however, can grow in hot springs at 70°C, and some archaea live at near 100°C.

> ***connect the concepts** Protein denaturation by heat is due to effects on secondary and tertiary structure of the protein as described in Key Concept 3.2.

Although the survival limits for most cells fall in the range of 0°C–40°C, most species have much narrower thermal limits for optimal function. To stay within those limits in spite of environmental conditions, animals have **thermoregulatory adaptations** that enable them to tolerate extreme conditions or to control their body temperature in spite of environmental conditions. These adaptations may determine the thermal tolerances of species and thereby determine their geographic ranges. When environments change rapidly, as is happening globally at present, animals may find themselves in situations that exceed their thermal tolerances. A good example is the reef-building corals discussed in the opening of Chapter 1.

Q_{10} is a measure of temperature sensitivity

Even between 0°C and 40°C, changes in body temperature create problems for animals. Most ***biochemical reactions** and the physiological processes they are involved in are

Figure 39.6 Frozen Frogs   Wood frogs (*Rana sylvatica*) live in the northeastern United States, across most of Canada, and up into northern Alaska. They freeze solid in the winter and survive. During the fall and early winter, urea and glucose accumulate in the frogs' cells and act as intracellular cryoprotectants (like antifreeze in your car radiator). Also, much intracellular water is redistributed to the extracellular fluid, which can freeze without damaging cells.

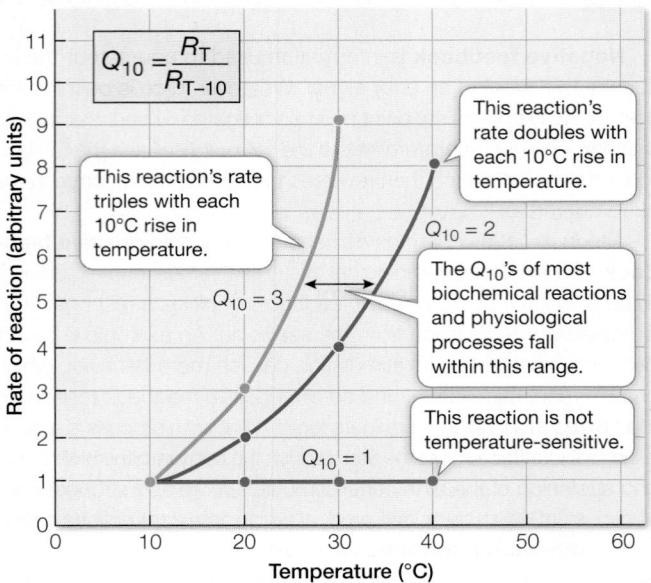

Figure 39.7 Q_{10} and Reaction Rate   The larger the Q_{10} of a reaction or process, the faster its rate rises in response to an increase in temperature.

temperature-sensitive, going faster at higher temperatures. The temperature sensitivity of a reaction or process can be described in terms of **Q_{10}**, a factor calculated by dividing the rate of a process or reaction at a certain temperature, R_T, by the rate of that same process or reaction at a temperature 10°C lower, R_{T-10}:

$$Q_{10} = \frac{R_T}{R_{T-10}}$$

---

*connect the concepts   As discussed in Key Concept 8.5, increases in temperature increase reaction rates because reactant molecules have more kinetic energy. Enzymes that control rates of biochemical reactions in metabolic pathways are influenced by temperature as well.

Q_{10} can be measured for a simple biochemical reaction or for a complex physiological process, such as rate of oxygen consumption. If a reaction or process is not temperature-sensitive, it has a Q_{10} of 1. Most biological Q_{10} values are between 2 and 3. A Q_{10} of 2 means that the reaction rate doubles as temperature increases by 10°C, and a Q_{10} of 3 indicates a tripling of the rate over a 10°C temperature range (**Figure 39.7**).

Notice that the Q_{10} values (except for $Q_{10} = 1$) plotted in Figure 39.7 produce curves rather than straight lines. This is because temperature increases in additive intervals (10, 20, 30, etc.) but reaction rates increase in a multiplicative fashion (2, 4, 8, 16, 32, etc.). Such curvilinear plots are common in biological data.

Changes in body temperature can disrupt an animal's physiology because not all of the biochemical reactions that constitute the metabolism of an animal have the same Q_{10}. These biochemical reactions are linked together in complex networks: the products of one reaction are the reactants for other reactions. Because different reactions have different Q_{10}'s, changes in tissue temperature will shift the rates of some reactions more than others, disrupting the overall network. Therefore, to maintain homeostasis, organisms compensate for or prevent changes in body temperature.

Animals acclimatize to seasonal temperatures

The body temperatures of some animals (especially aquatic animals) are coupled to environmental temperature. The body temperature of a fish in a pond, for example, is the same as the water temperature, which might range from 4°C in winter to 24°C in summer. If we bring that fish into the laboratory in the summer and measure the rates of any of its physiological or biochemical processes such as oxygen consumption, we will demonstrate a Q_{10} relationship. On the basis of the Q_{10} relationship, we would predict that the fish's oxygen consumption—a measure of its metabolic rate—would be lower in the winter pond because the water temperature is lower. However, if we bring that fish back into the laboratory in the winter and measure its metabolic rate at winter pond temperature, we find that rate to be higher than we predicted. The fish's biochemistry and physiology will have **acclimatized** to the seasonal change in water temperature so that it can remain active at winter temperatures. What could be the mechanism of this acclimatization? Many organisms express multiple forms of enzymes (**isozymes**) that have different temperature optima. In winter, organisms express the forms that show maximum substrate affinity at low temperatures, and in summer they express the forms that have maximum substrate affinity at high temperature. Another mechanism of temperature acclimatization is a change in the composition of cell membranes designed to maintain optimum fluidity despite changes in temperature (see Key Concept 6.1). Acclimatization in animals essentially means that their metabolic functions are less sensitive to long-term changes in their environment than to short-term changes.

Small changes in temperature can have large physiological effects

The story opening this chapter described a world-class marathon runner failing to finish an important race because of heat stress. As

experiment

Original Paper: Grahn, D. A., V. H. Cao, C. M. Nguyen, M. T. Liu and H. C. Heller. 2012. Work volume and strength training responses to resistive exercise improve with periodic heat extraction from the palm. *Journal of Strength and Conditioning Research* 26(9): 2558–2569.

At the beginning of the chapter you learned that mammals dissipate excess heat through their non-hairy skin areas. Those skin areas are small, but they have special high-volume blood vessels that transfer heat from the body core to the skin. Humans have these same blood vessel adaptations. In this experiment, heat was extracted from the palm of a hand, and muscle work capacity was measured.

HYPOTHESIS▶ Extracting excess heat from the body will increase the capacity of muscles to do work.

METHOD

1. Recruit seventeen healthy recreational athletes between the ages of 19 and 23 years old.

2. Instruct subjects in proper bench press technique, and determine each subject's 1 repetition maximum (the maximum weight they could lift once).

3. Divide subjects into two groups (controls with no cooling and treatment with palmar cooling). Have each group complete 5 trials of 6 sets of bench presses, until muscle failure, with the weight set to 50% of the subject's 1 repetition maximum. Between sets allow subjects to rest for 3 minutes, with the experimental group receiving palmar cooling and the control group receiving no cooling. Separate trials by at least 3 days. Record the total numbers of bench presses per trial.

4. Repeat the experiment with control and treatment groups switched.

RESULTS

Only results for first and last trials of each experiment are shown in the table.

CONCLUSION▶ By working through the work with the data exercise, you will be able to conclude whether the hypothesis is supported or not supported by this experiment.

investigatinglife work with the data follows on next page.

	Number of bench presses			
	Experiment 1		Experiment 2	
	First trial	Last trial	First trial	Last trial
Subject	Control group (no cooling)		Treatment group (palmar cooling)	
1	34	46	45	50
2	50	62	58	74
3	66	69	69	73
4	53	68	80	90
5	58	73	79	107
6	61	63	67	77
7	37	38	43	76
8	50	53	56	72
9	76	72	76	98
	Treatment group		Control group	
10	66	92	89	96
11	87	97	89	94
12	70	91	80	87
13	65	86	71	99
14	57	90	84	106
15	61	87	88	107
16	45	69	75	89
17	49	73	74	75

many of us have experienced, our ability to do hard work is reduced when we get overheated. Because high tissue temperatures can damage cells, shutting off a muscle's ability to do work may serve a protective function.

Pyruvate kinase is an enzyme critical for production of ATP in muscles; it catalyzes the last step in glycolysis, producing pyruvate that can enter the mitochondria and be metabolized to produce ATP (see Figure 9.12). Without ATP, muscles fatigue. Muscle pyruvate kinase inactivates around 40°C, which shuts off muscle function and thereby prevents thermal damage to the muscle. The thermal inactivation of muscle pyruvate kinase is probably a protective adaptation to prevent thermal damage to the muscle. You may thus wonder, if we could facilitate the extraction of heat from muscle, could we increase its capacity to do work? We consider this question in **Investigating Life: Can the Work Capacity of Muscle Be Increased by Extracting Heat from the Palms of the Hands?**

▶ 39.3 recap

Cells can survive only within a narrow range of temperatures, but even changes within that range can be disruptive because different physiological processes have different temperature sensitivities.

learning outcomes

You should be able to:

- Explain why animals' body temperatures must be maintained within narrow ranges.
- Plot a given Q_{10} value for a biochemical process.
- Explain how isozymes may be involved in seasonal temperature acclimatization.

1. Explain how fluctuations in body temperature can disrupt complex physiological processes.

2. Plot a $Q_{10} = 2.5$ curve for a physiological process.

3. What characteristic of isozymes might explain their role in metabolic seasonal temperature acclimatization?

Now that you know how animals are affected by temperature, we next take a look at the adaptations that allow animals to survive wide ranges of thermal environments and wide ranges of metabolic activities.

work with the data

This experiment involves a simple outcome measure: number of bench presses in 6 sets done to muscle failure. The experimental manipulation is heat extraction from a palm between sets. The control condition is an equivalent period of rest between sets without palmar cooling. The main question is whether there is a difference in the work volume when cooled versus when not cooled. However, there are some complicating factors: (1) the subjects varied considerably in their initial work capacity, and (2) since the treatments were sequential there could be an order effect. To minimize the order effect, the treatments were randomized as to palmar cooling or no cooling. Your analysis must first standardize the data, then test for an order effect, and finally test for a treatment effect.

How can you standardize the data? Because the subjects began the experiment at different ability levels, the absolute increase in repetitions cannot be used as a comparative outcome measure. An increase of 5 repetitions is more impressive for a subject who began at 25 than for one who began at 50—a 20 percent increase in contrast to a 10 percent increase. For each individual and each treatment it is necessary to convert the raw data to percent increase.

How can you test for a treatment effect? Each subject was cooled in one treatment and not cooled in the other treatment. To see if there is a difference in the percent increase due to palmar cooling, you can average the percent increases for all subjects in their cooling treatments and for all subjects in their control treatments. Using a paired t-test, you can assess whether palmar cooling had a significant effect on the outcome.

QUESTIONS▶

1. Because this experiment involved repetitions of a physical activity, you would expect there to be a conditioning effect whether the subjects were cooled or not, and you might expect the improvements to be greater in the second treatment than in the first. To test for an order effect, compare the percent increases for all subjects in treatment 1 to the percent increases for all subjects in treatment 2. For statistical analysis use a paired t-test (see Appendix B). What do you conclude about a possible order effect in your collection of data?

2. Each subject was cooled in one treatment and not cooled in the other treatment. To see if there is a difference in the percent increase due to palmar cooling, average the percent increases for all subjects in their cooling treatments and for all subjects in their control treatments. Using a paired t-test, you can assess whether palmar cooling had a significant effect on the outcome. What do you conclude about a treatment effect in your experimental results?

3. Is the hypothesis supported or disproven? Explain.

A similar **work with the data** exercise
may be assigned in **LaunchPad**.

---

key concept 39.4 Body Temperature Depends on the Balance between Heat In and Heat Out of the Body

Body temperature is a measure of the heat content of the body. Animals produce heat internally through their metabolism, and they continuously exchange heat with their environment. The balance between heat gain and heat loss determines the animal's body temperature. Thermoregulatory adaptations enable animals to modulate the various paths of heat exchange between themselves and their environment.

focus your learning

- Endotherms can increase metabolic heat production to balance increased heat loss to the environment, whereas ectotherms rely largely on behavior to control their heat exchange with the environment.
- The four avenues of heat exchange between an animal and its environment are radiation, convection, conduction, and evaporation.
- A countercurrent heat exchange system in some highly active fishes conserves heat generated by muscle activity.

Animals can be classified by their thermoregulatory characteristics

One way of classifying animals is according to the constancy of their body temperature. Animals that maintain a constant body temperature are called **homeotherms**, and animals that experience a fluctuating body temperature are called **poikilotherms**. Another classification is based on the source of heat that predominantly determines body temperature. **Endotherms** such as birds and mammals have the ability to vary their metabolic heat production to compensate for the loss of heat to the environment. **Ectotherms** such as invertebrates, fishes, amphibians, and non-avian reptiles are largely dependent on environmental sources of heat. We use the terminology of ectotherm and endotherm in this book, but no classification scheme is perfect. Many ectotherms can generate considerable internal heat to raise their body temperature, and endotherms that hibernate can periodically appear to give up the ability to generate internal heat. To accommodate these departures from a strict dichotomy, we add the term **heterotherms**—organisms that act like ectotherms some of the time and like endotherms at other times.

Endotherms produce substantial amounts of metabolic heat

*Transfers of energy** in biological systems are inefficient. In both ectotherms and endotherms, in every transfer of energy—from food

(A)

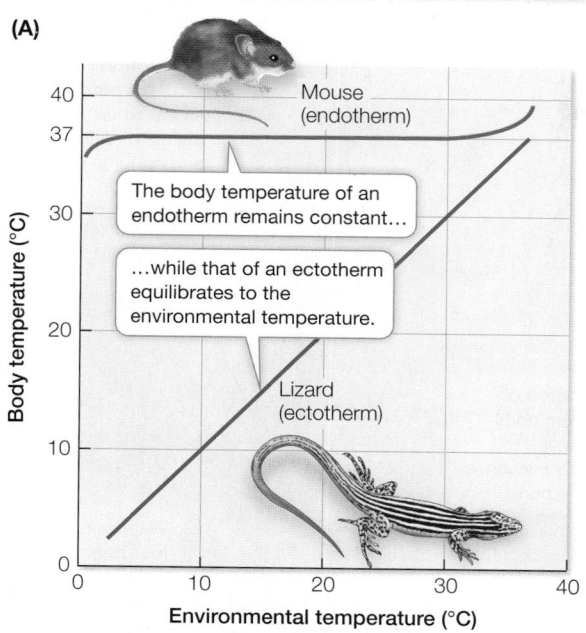

(B)

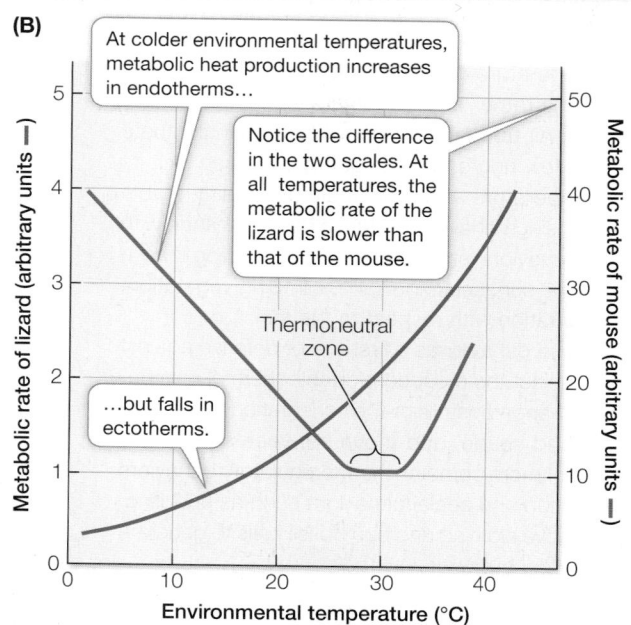

Figure 39.8 Ectotherms and Endotherms React Differently to Environmental Temperatures **(A)** At the same environmental temperature, an ectotherm and an endotherm of approximately the same body size (here, a lizard and a mouse) have different body temperatures. **(B)** The metabolic rates of the lizard and mouse react in opposite ways to cooler temperatures.

Q: In the plot of body temperature versus environmental temperature, why does the mouse curve fall on the far left side and rise on the far right side? In the plot of metabolic rate versus environmental temperature, why does the mouse's metabolic rate rise at higher temperatures?

molecules to ATP, from ATP to biological work—some of the energy is lost as heat. Working muscles produce heat, as do all of the metabolic activities of all tissues.

---

*connect the concepts As discussed in Key Concept 8.1, transfers of energy are governed by the first and second laws of thermodynamics, which explain why heat is always a by-product of metabolism.

Why do endotherms produce more heat than ectotherms? The surprising answer is that the cells of endotherms are less efficient at using energy than are the cells of ectotherms. The cells of endotherms are more "leaky" to ions than are the cells of ectotherms. Therefore Na^+ ions are constantly diffusing into the cells, and K^+ ions are constantly diffusing out. Even an endotherm at rest must spend considerable amounts of energy to transport Na^+ out of the cells and transport K^+ back in. Because of their constant need to actively transport ions, endotherms expend more energy than do ectotherms just to maintain the ion concentration gradients across their cell membranes. This situation is analogous to a leaky boat: the faster water comes in (i.e., the faster ions diffuse down their concentration gradients), the more metabolic energy has to be expended to bail the water out (i.e., pump ions back up their concentration gradients). Since endotherms expend more energy than ectotherms do to maintain ion concentration gradients, they produce more internal heat.

We can speculate that a mutation resulting in seemingly faulty or leaky ion channels may underlie the evolution of endothermy. Such a mutation in a small ectotherm would have increased its energy expenditure and therefore its heat production. Increased heat production would have enabled the animal to be active earlier in the

morning or for a longer time after sunset. Being active in twilight, and eventually at night, would open up a new world of ecological opportunities—a world in which there was less competition from similar-sized ectotherms and less danger of predation.

Two major differences between endotherms and ectotherms are (1) their resting metabolic rates—the sum total of all energy expenditures in their bodies when at rest—and (2) their responses to changes in environmental temperature.

Ectotherms and endotherms respond differently to changes in environmental temperature

How do similar-sized ectotherms and endotherms respond to changes in environmental temperature? Let's compare a lizard and a mouse. We put each animal in a closed chamber and measure its body temperature and its resting metabolic rate as we change the temperature of the chamber from 40°C to 0°C. The body temperature of the lizard equilibrates with that of the chamber, whereas the body temperature of the mouse remains stable over most of this range (**Figure 39.8A**). The metabolic rate of the lizard (already much lower than that of the mouse when both are at warm temperatures) decreases as the temperature drops (**Figure 39.8B**). In contrast, the mouse's metabolic rate increases as the chamber temperature falls below 25°C. This increase in metabolism produces enough heat to prevent the mouse's body temperature from falling. In other words, the mouse can regulate its body temperature by increasing its metabolic rate and thereby maintain its physiological functions at an optimal level; the lizard cannot.

This experiment might lead us to conclude that the ectotherm cannot regulate its body temperature, but observations of the lizard in nature do not support this conclusion. In nature, the lizard can regulate its body temperature quite well, although it does so by behavioral mechanisms rather than by altering its internal metabolic heat production. Air temperature in the desert can fluctuate by 40°C or more in a few hours; the lizard, however, can limit its variation in body temperature and even maintain a fairly stable body temperature by using behavior to alter its heat exchange with the environment. Its behavioral strategies include spending time in a burrow, basking in the sun, seeking shade, climbing vegetation, and changing its orientation with respect to the sun.

Endotherms also use behavior as a first line of defense against thermal stress. They select the most comfortable thermal environment possible. They change posture or their orientation to the sun, move between sun and shade, and move between still air and moving air. Examples of more complex thermoregulatory behaviors include nest construction and social behaviors such as huddling. Humans put on or remove clothing and burn fossil fuels to generate the energy to heat or cool buildings.

Energy budgets reflect adaptations for regulating body temperature

Both ectotherms and endotherms influence their body temperatures by altering four avenues of heat exchange between their bodies and the environment (**Figure 39.9**):

1. **Radiation** Heat moves from warmer objects to cooler ones via the exchange of infrared radiation (what you feel when you stand in front of a fire).

2. **Convection** Heat exchanges with a surrounding medium such as air or water that flows over a surface (the wind-chill factor).

3. **Conduction** Heat flows directly between two objects at different temperatures when they come into contact (e.g., an icepack on a sprained ankle).

4. **Evaporation** Heat is transferred away from a surface when water evaporates on that surface (the effect of sweating).

The total balance of heat production and heat exchange can be expressed as an **energy budget**, based on the simple fact that if the body temperature of an animal is to remain constant, the heat entering the animal must equal the heat leaving it. The heat coming in is usually from metabolism and radiation (R_{abs}, for radiation absorbed). Heat leaves the body via the four mechanisms listed above—radiation emitted (R_{out}), convection, conduction, and evaporation. The energy budget takes the mathematical form

$$\underbrace{\text{heat}_{in}}_{\text{metabolism} + R_{abs}} = \underbrace{\text{heat}_{out}}_{R_{out} + \text{convection} + \text{conduction} + \text{evaporation}}$$

The energy budget is a useful concept because any adaptation that influences the ability of an animal to control its temperature must affect one or more components of the energy budget. The energy budget gives us the ability to quantify and compare the thermal adaptations of animals. One interesting observation is that all of the components on the right side of the energy budget equation—that is, the heat-loss side—depend on the surface temperature of the animal. In fact, if the environmental temperature is above skin surface temperature, convection and conduction are avenues of

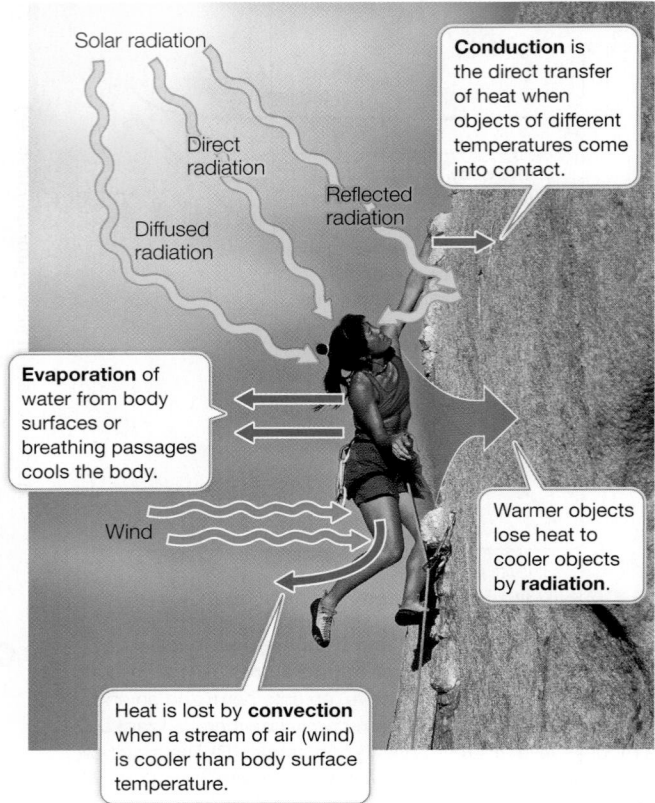

Conduction is the direct transfer of heat when objects of different temperatures come into contact.

Evaporation of water from body surfaces or breathing passages cools the body.

Warmer objects lose heat to cooler objects by radiation.

Heat is lost by convection when a stream of air (wind) is cooler than body surface temperature.

Solar radiation
Direct radiation
Diffused radiation
Reflected radiation
Wind

Figure 39.9 Animals Exchange Heat with the Environment An animal's body temperature is determined by the balance between internal heat production and four avenues of heat exchange with the environment: radiation, convection, conduction, and evaporation.

heat gain rather than loss. One way surface temperature can be controlled is by altering the flow of blood to the skin.

Both ectotherms and endotherms control blood flow to the skin

Heat is mostly moved around the internal environment by blood flow. Heat produced in muscles during exercise is transported out of the muscle and to the heart in the blood. That heat is then distributed around the body by the blood, raising body temperature. Blood flow to the skin enables internal heat to be lost to the environment through radiation, convection, conduction, and evaporation, thus bringing the body temperature back toward normal. When body temperature is too low or the environment is too cold, the blood vessels supplying the skin constrict, reducing heat loss to the environment.

The ability to control blood flow to the skin can be an important adaptation for an ectotherm such as the marine iguana (a reptile) of the Galápagos archipelago (**Figure 39.10**). The Galápagos are volcanic islands that lie on the equator but are bathed by cold ocean currents. Iguanas bask on hot black lava rocks on the shore, and periodically enter the cold ocean water to feed on seaweed. While the iguanas are feeding, they lose heat to the cold ocean water. This cooling lowers their metabolism, making them slower, more vulnerable to predators, and incapable of efficient digestion. The iguanas therefore alternate between feeding in the cold seawater

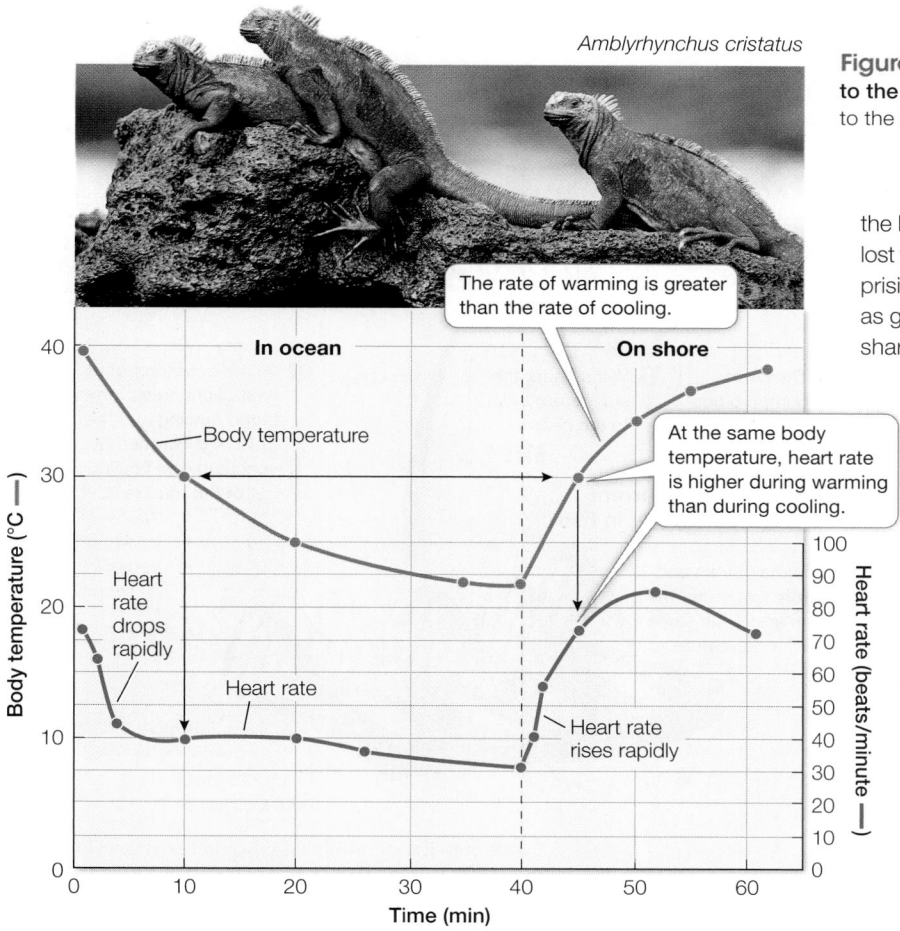

Amblyrhynchus cristatus

Figure 39.10 Some Ectotherms Regulate Blood Flow to the Skin   Galápagos marine iguanas control blood flow to the skin to alter their heating and cooling rates.

the blood picks up from metabolically active muscles is lost to the water flowing across its gills. It is therefore surprising that some large, rapidly swimming fishes, such as giant bluefin tuna (see Figure 1.15A) and great white sharks, can maintain swimming muscle temperatures 10°C–15°C higher than the surrounding water. The heat comes from their powerful swimming muscles, and the ability of these "hot" fishes to conserve that heat is based on the remarkable arrangement of their blood vessels.

In the typical ("cold") fish circulatory system, oxygenated blood from the gills collects in a large blood vessel, the aorta, that runs through the center of the fish, distributing blood to all organs and muscles (**Figure 39.11A**). "Hot" fishes have a small aorta, and most of their oxygenated blood is transported in large side vessels just under the skin (**Figure 39.11B**). The cold blood leaving the gills is therefore kept close to the surface of the "hot" fishes as it flows posteriorly to the swimming muscles. The cold blood flows into the muscle mass through small vessels branching off the large side vessels. These many small vessels run parallel to the vessels carrying warm blood from the swimming muscles back toward the heart. Because the vessels carrying the cold blood into the muscles are in close contact with the vessels carrying warm blood out of the muscles, heat flows from the warm to the cold blood by conduction. Therefore the heat produced by the active muscles is retained in the muscle mass.

Because heat is exchanged between blood vessels carrying blood in opposite directions, this adaptation is called **countercurrent heat exchange** (**Figure 39.11C**). By keeping heat within the muscles, these fishes can maintain an internal body temperature considerably higher than the water temperature. Each 10°C rise in muscle temperature increases the fish's power output almost threefold, giving it a faster and more powerful swimming capability.

and basking on the hot rocks. It is advantageous for iguanas to retain body heat as long as possible while swimming and to warm up as fast as possible when basking. They accomplish these results by changing their heart rate and thus the rate of blood flow to their skin and around their bodies.

What about furred mammals? Fur acts as insulation to keep body heat in, making it possible for mammals to function in cold environments. When they are active, however, mammals must get rid of excess heat, and it does little good to transport that heat to the skin under the fur. Thus, as mentioned at the opening of this chapter, mammals have specialized blood vessels for transporting heat to their hairless skin surfaces. Heat loss from these areas is tightly controlled by the opening and closing of these blood vessels. When you are cold, the blood flow to your hands and feet decreases and they feel cold, but when you exercise, the blood vessels serving these skin areas dilate increasing the blood flow and therefore the heat dissipation.

 Media Clip 39.1 **Thermoregulation in Animals**
www.Life11e.com/mc39.1

Some fishes conserve metabolic heat

The muscles of active fishes produce substantial amounts of metabolic heat, but most fishes cannot retain that heat. Blood pumped from the fish heart goes directly to the gills, where it comes very close to the surrounding water to exchange oxygen and carbon dioxide. Any heat

Some ectotherms regulate metabolic heat production

By definition, ectotherms are largely dependent on environmental sources of heat for thermoregulation, but some ectotherms raise their body temperatures by producing metabolic heat. For example, the powerful flight muscles of many insects must reach 35°C–40°C before the insects can fly, and they must maintain these high temperatures during flight. Such insects warm up to fly by contracting their flight muscles isometrically in a manner similar to shivering in mammals. The heat-producing ability of insect muscle can be quite remarkable. An impressive case is a species of scarab beetle that lives mostly underground in mountains north of Los Angeles, California. These beetles come aboveground to mate, with the males flying in search

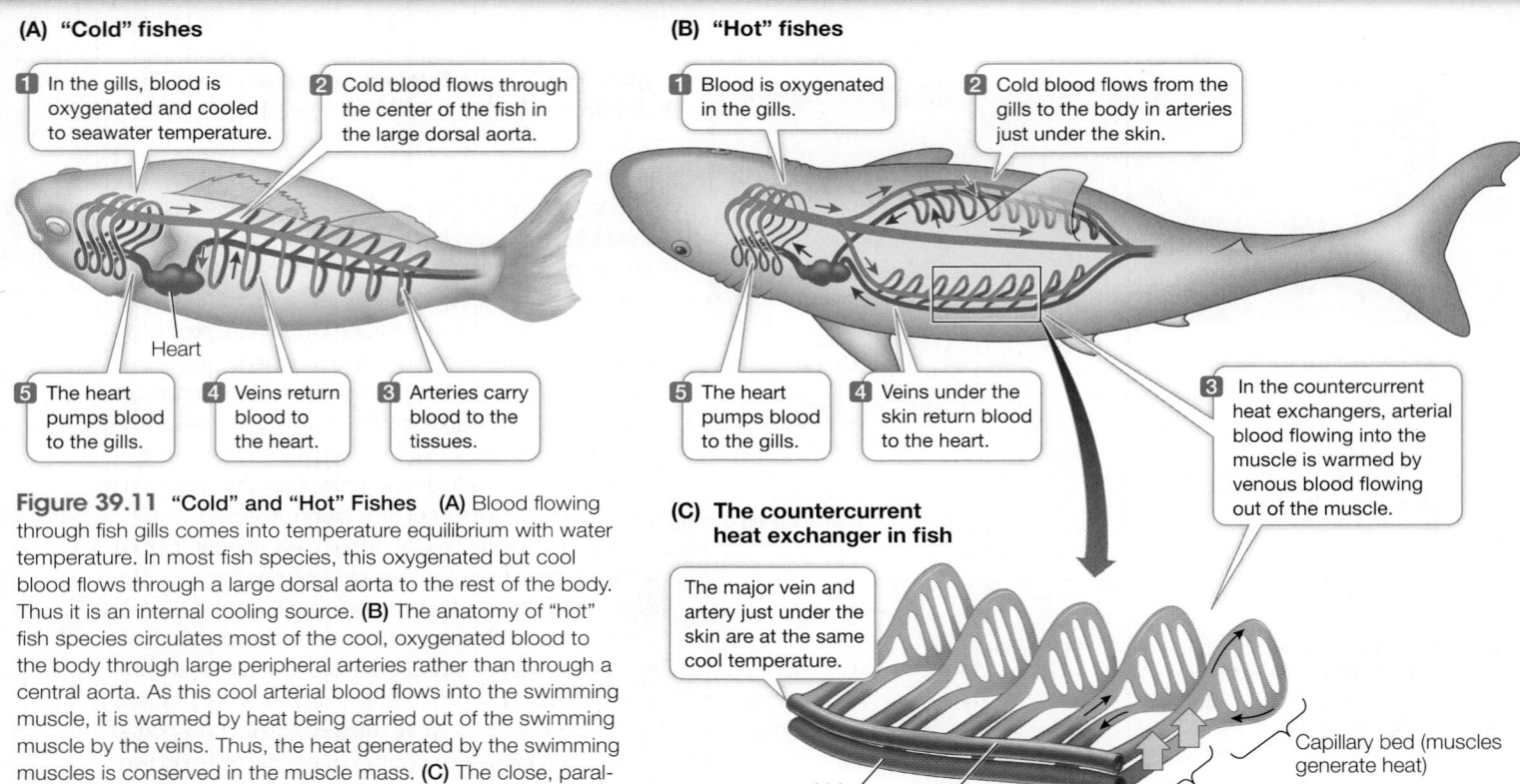

(A) "Cold" fishes

1 In the gills, blood is oxygenated and cooled to seawater temperature.

2 Cold blood flows through the center of the fish in the large dorsal aorta.

Heart

5 The heart pumps blood to the gills.

4 Veins return blood to the heart.

3 Arteries carry blood to the tissues.

(B) "Hot" fishes

1 Blood is oxygenated in the gills.

2 Cold blood flows from the gills to the body in arteries just under the skin.

5 The heart pumps blood to the gills.

4 Veins under the skin return blood to the heart.

3 In the countercurrent heat exchangers, arterial blood flowing into the muscle is warmed by venous blood flowing out of the muscle.

(C) The countercurrent heat exchanger in fish

The major vein and artery just under the skin are at the same cool temperature.

Vein   Artery

Capillary bed (muscles generate heat)

Countercurrent heat exchanger

Figure 39.11 "Cold" and "Hot" Fishes **(A)** Blood flowing through fish gills comes into temperature equilibrium with water temperature. In most fish species, this oxygenated but cool blood flows through a large dorsal aorta to the rest of the body. Thus it is an internal cooling source. **(B)** The anatomy of "hot" fish species circulates most of the cool, oxygenated blood to the body through large peripheral arteries rather than through a central aorta. As this cool arterial blood flows into the swimming muscle, it is warmed by heat being carried out of the swimming muscle by the veins. Thus, the heat generated by the swimming muscles is conserved in the muscle mass. **(C)** The close, parallel arrangement of the arteries carrying blood into the muscles and the veins carrying blood out of the muscle makes efficient countercurrent heat exchange possible.

of females. They undertake this mating ritual only at night, in winter, during snowstorms.

Honey bees regulate temperature as a group. They live in large colonies consisting mostly of female worker bees that maintain the hive and rear the larval offspring of the single queen bee. During winter, worker bees cluster around the brood (eggs and larvae). They adjust their individual metabolic heat production and density of clustering so that the brood temperature remains remarkably constant, at about 34°C, even as the outside air temperature drops below freezing.

39.4 recap

Endotherms use metabolic heat production to maintain constant high body temperatures. Body temperatures of ectotherms are more dependent on environmental heat sources. Both ectotherms and endotherms use behavior to control their body temperature. Heat exchange between an animal and its environment occurs via radiation, convection, conduction, and evaporation, all of which depend on body surface temperature. Some ectotherms use metabolic heat production episodically to raise body temperature.

learning outcomes

You should be able to:

- Explain responses to environmental temperature by an ectotherm, an endotherm, and a heterotherm.
- Describe the paths of heat exchange between an animal and its environment.
- Explain the principle of countercurrent heat exchange.

39.4 recap

1. Why is the body temperature of an ectotherm that is exposed to variations in environmental temperature more constant in nature than in the laboratory?

2. For each path of heat exchange, what determines whether heat is gained or lost by the animal?

3. Compared to most fishes, explain the circulatory system adaptations that enable "hot" fishes to raise the temperature of their swimming muscles.

Endotherms respond to changes in environmental temperature by changing their rates of metabolic heat production. They also have other adaptations for controlling their rates of heat exchange with their environments. How do they regulate these various avenues of heat exchange to achieve a constant internal body temperature?

key concept

39.5 Body Temperature Is Regulated through Adaptations for Heat Production and Heat Loss

As explained in Key Concept 39.2, there is a difference between control and regulation. Think of driving your car. You control the speed of the car with the accelerator and the brake, but you regulate the speed of your car by comparing the speed limit (set point) with

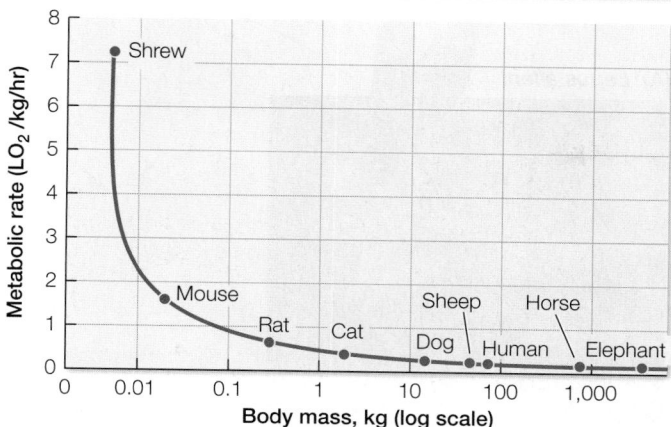

Figure 39.12 The Mouse-to-Elephant Curve   On a weight-specific basis, the basal metabolic rate of small endotherms is much greater than that of larger endotherms. This classic illustration, originally published in the 1930s, plots O_2 consumption per kilogram of body mass (a measure of metabolic rate) against a logarithmic plot of body mass.

Q: Only 9 species are shown on this curve, but hundreds of species from the tropics to the artic fit on this curve as well. What does that fact allow you to conclude about whether or not basal metabolic rate is an adaptation to climate?

your speedometer (feedback) to generate command signals to your foot that is operating the accelerator and brake. For understanding how body temperature is regulated, you must understand the adaptations for heat production and heat loss, how they are controlled, and what information is used to modulate the control signals to achieve a regulated body temperature.

focus your learning

- The basal metabolic rate (BMR) of an endotherm is the lowest metabolic rate necessary for biochemical and physiological processes of a resting animal.
- Endotherms produce and conserve metabolic heat to offset heat loss in cold environments.
- Mammalian body temperature is controlled by a regulatory center in the hypothalamus that uses hypothalamic temperature as the feedback information.

Basal heat production rates of endotherms correlate with body size

Physiologists can determine an animal's metabolic rate and therefore its rate of heat production by measuring its consumption of O_2 or production of CO_2. Within a narrow range of environmental temperatures, called the **thermoneutral zone** (see Figure 39.9B), the metabolic rates of endotherms (birds and mammals) are at low levels and independent of environmental temperature. The metabolic rate of a resting endotherm at a temperature within its thermoneutral zone is its **basal metabolic rate** (**BMR**). BMR is usually measured in animals that are quiet but awake and not using energy for digestion, reproduction, or growth. Thus BMR is the rate at which a resting animal is consuming just enough energy to carry out its minimal body functions.

As you might expect, the BMR of an elephant is greater than that of a mouse. After all, the elephant is more than 100,000 times larger than the mouse. However, the BMR of the elephant is only about

7,000 times greater than that of the mouse. That means that a gram of mouse tissue uses energy at a rate 15 times greater than a gram of elephant tissue (**Figure 39.12**). Across all of the endotherms, BMR per gram of tissue increases as animals get smaller.

Why should this disproportionate difference exist? There are several possible reasons. As animals get bigger, they have a smaller surface area-to-volume ratio (see Figure 5.2). Since heat production is related to the volume (i.e., mass) of the animal, but its capacity to dissipate heat is related to its surface area, it has been reasoned that larger animals evolved lower metabolic rates to avoid overheating. However, this explanation alone is insufficient because the relationship between body mass and basal metabolic rate holds for even very small organisms and for ectotherms, in which overheating is not a problem. Other hypotheses have also been proposed. For example, a larger animal has a greater proportion of support tissues (e.g., skin and bone), which are not as metabolically active as other tissue types. The real explanation is probably a mixture of different factors, but the relationship holds over a very broad range of species.

For an endotherm, a metabolic rate versus environmental temperature curve represents the integrated response of all the animal's thermoregulatory adaptations (**Figure 39.13**). The thermoneutral zone is bounded by a **lower critical temperature** and an **upper critical temperature**. When the environmental temperature

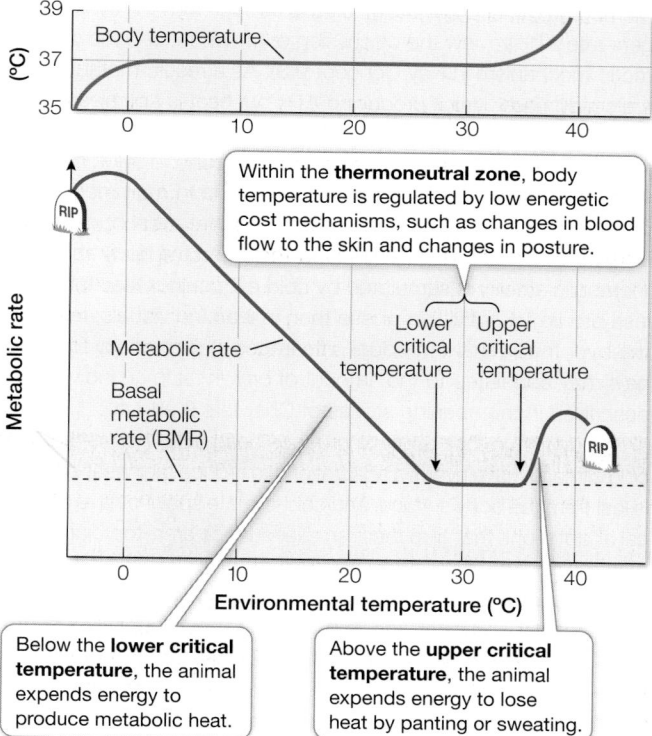

Figure 39.13 Environmental Temperature and Mammalian Metabolic Rates   A plot of an endotherm's metabolic rate versus environmental temperature represents the integrated response of all of its thermoregulatory mechanisms. Outside the thermoneutral zone, maintaining a constant body temperature requires expending energy. Outside extreme limits (0°C and 40°C in this instance), the animal cannot maintain its body temperature and dies.

is within the thermoneutral zone, an endotherm's thermoregulatory responses do not require much energy and could be considered passive; such responses include changing posture, fluffing fur or feathers, and altering blood flow to the skin. Outside its thermoneutral zone, however, an endotherm's thermoregulatory responses are active and require metabolic energy.

Endotherms respond to cold by producing heat and adapt to cold by reducing heat loss

When environmental temperatures fall below the lower critical temperature, endotherms increase metabolic heat production to compensate for heat loss. Mammals can accomplish this by shivering and/or nonshivering heat production. Birds use only shivering heat production. Shivering uses the contractions of skeletal muscles to convert ATP to ADP, with the energy from this process released as heat. Shivering muscles pull against each other so that little movement other than a tremor results. "Shivering heat production" is perhaps too narrow a term, however; increased muscle tone and increased body movements also contribute to increased heat production in cold environments.

Most nonshivering heat production occurs in specialized adipose tissue called **brown fat**. This tissue looks brown because of its abundant mitochondria and rich blood supply. In brown fat cells, a protein called **thermogenin** uncouples proton movement from ATP production, allowing protons to leak across the inner mitochondrial membrane rather than having to pass through the ATP synthase and generate ATP (review the discussion of brown fat and the chemiosmotic mechanism in Key Concept 9.3). As a result, metabolic fuels are consumed without producing ATP, but heat is still released.

Pads of brown fat are found in newborns of many mammalian species. Brown fat is also commonly found in adult mammals that are small and acclimatized to cold, and in mammals that hibernate. Recently it has been discovered that adult humans have small amounts of brown fat distributed around the body and that its metabolic activity is stimulated by cold exposure. One study found less brown fat activity in obese than in lean individuals, leading to the hypothesis that individual differences in propensity for weight gain may be related to the amount of brown fat in an individual, as described in the opening stories of Chapters 9 and 40.

In addition to their ability to produce heat, endotherms that live in cold climates have evolved adaptations to reduce their heat loss. Heat is lost from the body surface, and cold-climate species have anatomical adaptations that give them smaller surface area-to-volume ratios than their warm-climate relatives (**Figure 39.14**). These adaptations include rounder body shapes and shorter appendages.

The most common and important means of decreasing heat loss is to increase thermal insulation. Animals adapted to cold climates have much thicker layers of fur, feathers, and/or fat than do their warm-climate relatives. Fur and feathers are good insulators because they trap a layer of still, warm air close to the skin surface. If that air is displaced by water, insulation is drastically reduced. In many species, oil secretions spread through fur or feathers by grooming are critical for resisting wetting and maintaining a high level of insulation. In terms of the energy budget equation presented in Key Concept 39.4, remember that heat exchange between the animal and its environment is a function of the animal's surface

(A) *Lepus alleni*

(B) *Lepus arcticus*

Figure 39.14 Adaptations to Cold and Hot Climates **(A)** The antelope jackrabbit is found in the Sonoran Desert of Arizona. Its large ears serve as heat exchangers, passing heat from the animal's blood to the surrounding air. **(B)** The thick fur of the Arctic hare provides insulation, and its rounded body shape lowers its surface area-to-volume ratio. The ears and extremities are smaller than those of its warm-climate relatives, so less heat is lost to the environment.

temperature. For a body area covered by fur or feathers, surface temperature is that of the outer surface of the fur or feathers.

The ability to decrease blood flow to the noninsulated skin is an important thermoregulatory adaptation for cold-climate endotherms. Constriction of blood vessels in the skin, and especially in the appendages, greatly improves an animal's ability to conserve heat. Countercurrent heat exchange like that you saw in "hot" fishes is also an important adaptation in the appendages of endotherms. Warm blood flowing out to the paw of a wolf, the hoof of a caribou, or the foot of a bird parallels the flow of the cooler blood returning to the body core. Heat is transferred from the outgoing to the returning blood, thus retaining heat in the animal's core.

 Activity 39.3 Thermoregulation in an Endotherm www.Life11e.com/ac39.3

Evaporation of water can dissipate heat, but at a cost

At the upper end of the thermoneutral zone, an endotherm dissipates heat through adaptations that don't require much energy: increasing blood flow to the skin, seeking shade and cool breezes, and decreasing activity. But when the environmental temperature exceeds the upper critical temperature, these adaptations are not sufficient and overheating becomes a problem. For exercising animals (including athletes), overheating can occur even at low environmental temperatures. Large mammals, especially those in hot habitats such as elephants, rhinoceroses, and water buffaloes, have little or no insulating fur and seek out water to wallow in when the air temperature is high. Having water in contact with the skin greatly increases heat loss because the heat-absorbing capacity of water is much greater than that of air.

Evaporation from external or internal body surfaces through sweating or panting can also cool an endotherm. A gram of water absorbs about 580 calories of heat when it evaporates. If this evaporation occurs on the skin, most of the heat comes from the skin. However, sweat or saliva that falls off the body provides no cooling. Thus when the need for heat loss is greatest, water from the internal environment can be squandered with no cooling benefit. Water is heavy, so animals do not carry an excess supply of it, and many hot environments are also arid. In habitats that are both hot and dry, sweating and panting are cooling adaptations of last resort.

Sweating and panting are *active* processes that require expending metabolic energy. That is why the metabolic rate increases when the upper critical temperature is exceeded (see Figure 39.16). A sweating or panting animal is generating heat in the process of dissipating heat, which can be a losing battle.

The mammalian thermostat uses feedback information

The thermoregulatory mechanisms and adaptations of endotherms are controlled by neural regulatory systems that integrate information from environmental and internal sources and then issue commands to the effectors that alter the heat content of the body. These regulatory systems are similar in principle in birds and mammals but differ in many details. Here we focus on the nervous system thermostats of mammals.

The major thermoregulatory integrative center of mammals is at the base of the brain in a structure called the **hypothalamus (Figure 39.15)**. The hypothalamus is a key player in many regulatory systems of vertebrates. Experiments demonstrating its role have shown that slight cooling of the hypothalamus stimulates constriction of skin blood vessels and that stronger cooling increases metabolic heat production. As a result, cooling of the hypothalamus in an unchanging, thermoneutral environment will cause body temperature to rise. Conversely, hypothalamic heating causes the overall body temperature to fall **(Figure 39.16)**.

In mammals, the temperature of the hypothalamus itself is the major feedback signal. The hypothalamus generates set points for various thermoregulatory responses. When the temperature of the hypothalamus exceeds or drops below those set points, thermoregulatory responses are controlled to reverse the direction of temperature change (see Figure 39.15). The system integrates other sources of information in addition to hypothalamic temperature. For example, temperature sensors in the skin register environmental temperature. A change in skin temperature is feedforward information that shifts hypothalamic set points; the set point for metabolic heat production is higher when the skin is cold and lower when the skin is warm.

Hypothalamic set points are higher during wakefulness than during sleep, and they are higher during the active part of the daily cycle than the inactive part, even if the animal is awake at both times. Even when an endotherm is kept under constant environmental conditions, its body temperature displays a daily cycle of changes

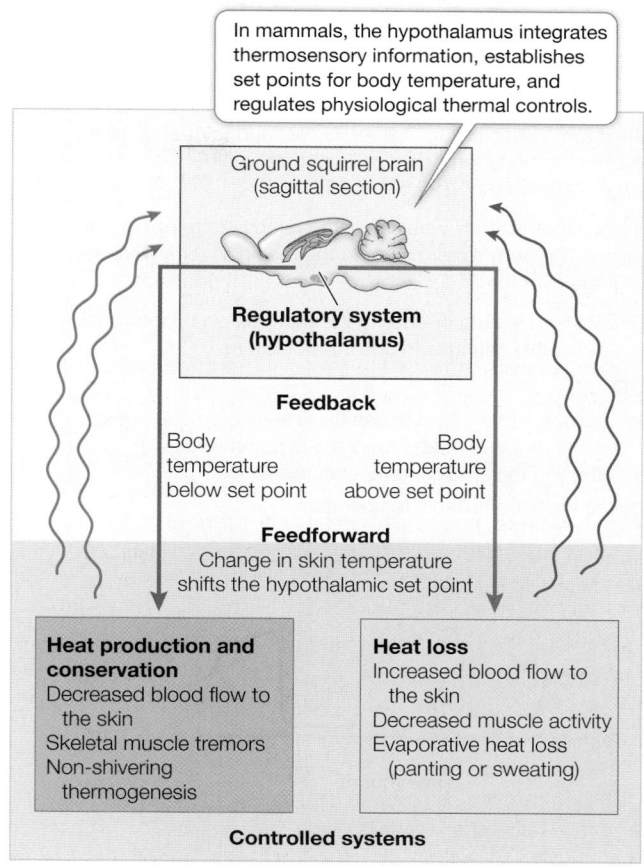

In mammals, the hypothalamus integrates thermosensory information, establishes set points for body temperature, and regulates physiological thermal controls.

Figure 39.15 The Mammalian Thermostat Like the home thermostat in Figure 39.5, the mammalian hypothalamus is the regulatory system that controls the body's heating and cooling mechanisms.

 Animation 39.1 **The Hypothalamus**
www.Life11e.com/a39.1

in set point. This kind of cycle, a ***circadian rhythm**, is controlled by an internal biological clock.

***connect the concepts** Key Concept 52.5 explains how a region of the hypothalamus called the suprachiasmatic nucleus generates endogenous daily cycles called circadian rhythms that influence many physiological processes, including temperature regulation.

Some animals conserve energy by turning down the thermostat

Hypothermia is a below-normal body temperature. It can result from starvation (lack of metabolic fuel), exposure to extreme cold, serious illness, or anesthesia. In each of these cases, the drop in body temperature is *unregulated*. However, many birds and mammals undergo *regulated* hypothermia to survive periods of cold and food scarcity.

Hummingbirds, for example, are very small endotherms with a high metabolic rate, and going even a single day without food could exhaust their metabolic reserves. Hummingbirds and other small endotherms extend the period over which they can survive without food by dropping their body temperature by 10°C–20°C during

experiment

Figure 39.16 The Hypothalamus Regulates Body Temperature

Original Paper: Heller, H. C., Colliver, G. W., and P. Anand. 1974. CNS regulation of body temperature in euthermic hibernators. *American Journal of Physiology*. 227:576–582.

A mammal's hypothalamus was subjected directly to temperature manipulation. The body's responses to the manipulations were as expected if the hypothalamus is the mammalian "thermostat."

HYPOTHESIS▶ Heating or cooling the mammalian hypothalamus results in predictable changes in body temperature.

METHOD

1. Implant a probe into the hypothalamus of a living ground squirrel's brain. Use the probe to heat or cool the hypothalamus directly (i.e., without affecting the ambient temperature).
2. Manipulate the hypothalamic temperature T_H.

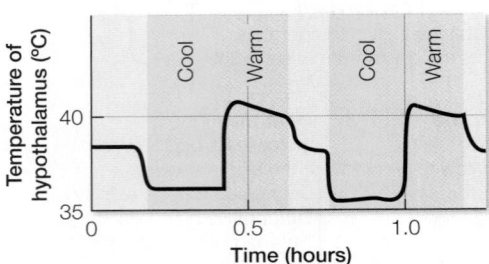

3. Measure the animal's metabolic rate and body temperature throughout the period of hypothalamic manipulation.

RESULTS

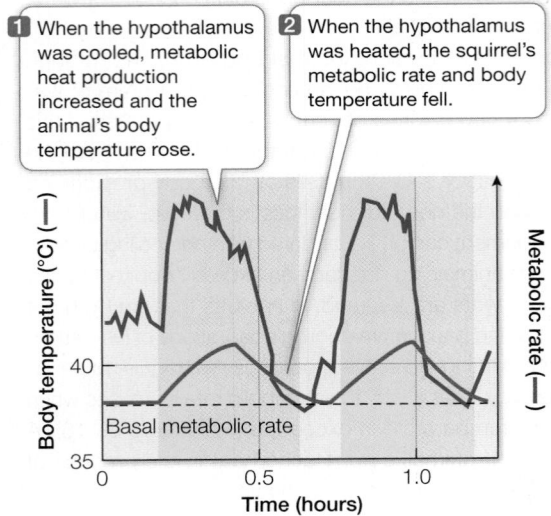

1 When the hypothalamus was cooled, metabolic heat production increased and the animal's body temperature rose.

2 When the hypothalamus was heated, the squirrel's metabolic rate and body temperature fell.

CONCLUSION▶ The ground squirrel's hypothalamus acts as a thermostat. When cooled, it activates metabolic heat production; when warmed, it suppresses metabolic heat production and favors heat loss.

---

the portion of day when they are normally inactive, thus lowering their metabolic rate and conserving energy. Daily bouts of regulated hypothermia are called **daily torpor**.

Regulated hypothermia that lasts for days or even weeks, during which the body temperature falls close to environmental temperature, is called **hibernation** (Figure 39.17). Many species of mammals, including bats, bears, marmots, and ground squirrels, hibernate, but only one species of bird (the common poorwill) has been shown to

hibernate. The metabolic rate needed to sustain a hibernating animal may be only one-fiftieth its basal metabolic rate, and many hibernating animals maintain body temperatures close to the freezing point. Arousal from hibernation occurs when the hypothalamic set point returns to the normal level. The ability of animals to enter daily torpor or deep hibernation to reduce their thermoregulatory set point so dramatically probably evolved as an extension of the set point decrease that accompanies sleep in all mammals and birds.

39.5 recap

Within the thermoneutral zone, an endotherm controls its body temperature by passive means. Above or below the thermoneutral zone, an endotherm must expend considerable metabolic energy to control body temperature. Thermoregulatory responses in mammals are regulated by the hypothalamus.

learning outcomes

You should be able to:

- Describe the thermoneutral zone and the upper and lower critical temperatures and how they relate to the basal metabolic rate of an endotherm.
- Interpret a plot of metabolic rate versus environmental temperature for an endotherm.
- Explain differences in physical features of endotherms of the same or similar species that live in climates of different temperature extremes.
- Describe how the mammalian brain thermostat uses feedback and feedforward information to regulate body temperature.
- Explain the variable features of the mammalian thermoregulatory system.

1. What determines the upper and lower boundaries of the thermoneutral zone?

2. In the metabolic rate versus ambient temperature curve for mammals shown in Figure 39.13, the segment of the curve below the lower critical temperature is described by the equation $MR = K(T_b - T_a)$. In this equation, what does K stand for? Why does this equation predict that the curve it describes will meet the x axis at a temperature that equals body temperature?

3. Explain why arctic species of mammals tend to be bigger and stockier than closely related species in warmer environments.

4. What is the role of feedforward information in the mammalian thermoregulatory system?

5. Explain why the mammalian thermoregulatory system has been described as a proportional thermostat with an adjustable set point.

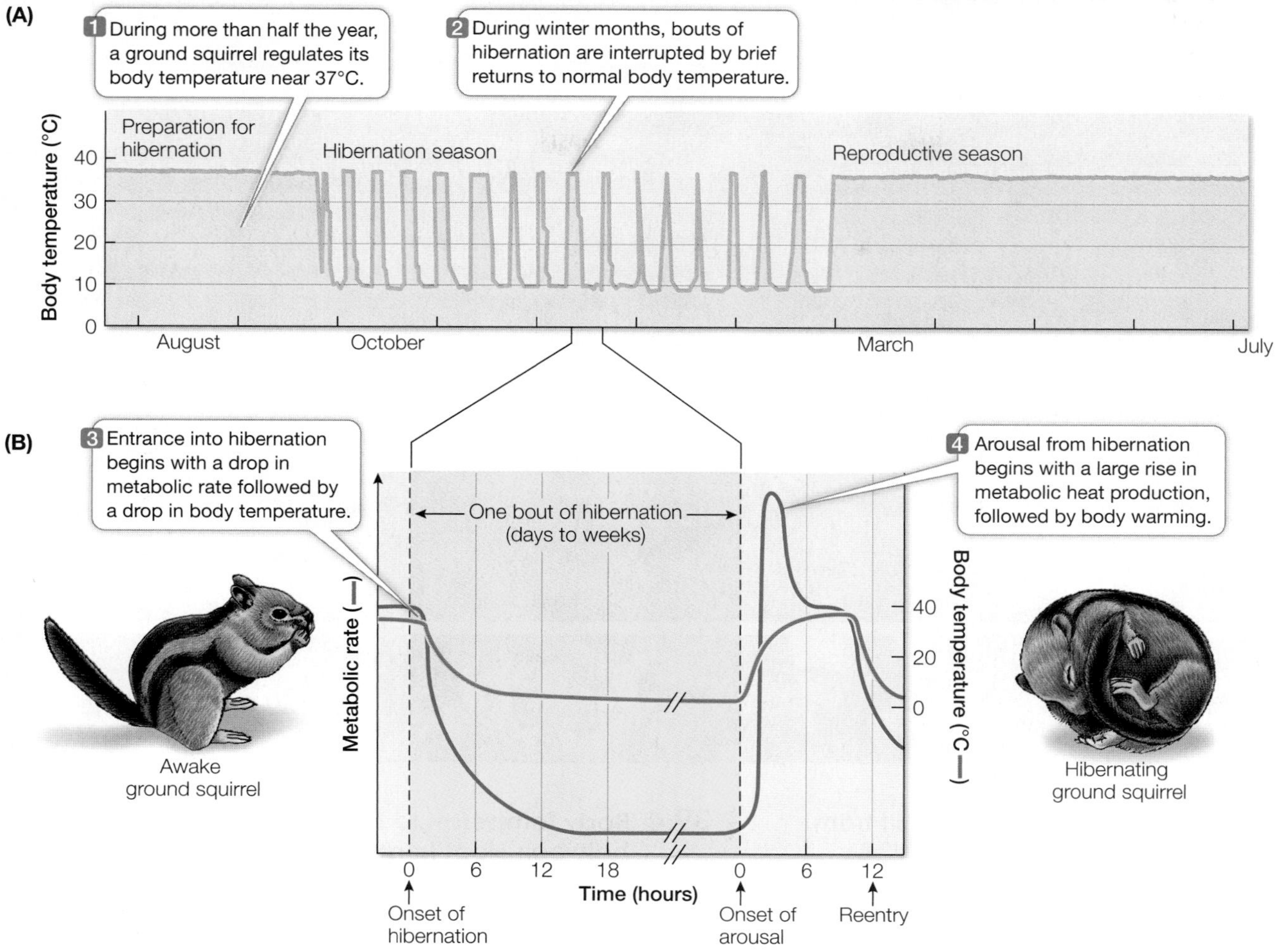

(A)

1 During more than half the year, a ground squirrel regulates its body temperature near 37°C.

2 During winter months, bouts of hibernation are interrupted by brief returns to normal body temperature.

Preparation for hibernation

Hibernation season

Reproductive season

Body temperature (°C): 40, 30, 20, 10, 0

August   October   March   July

(B)

3 Entrance into hibernation begins with a drop in metabolic rate followed by a drop in body temperature.

One bout of hibernation (days to weeks)

4 Arousal from hibernation begins with a large rise in metabolic heat production, followed by body warming.

Metabolic rate (—)

Body temperature (°C) (—): 40, 20, 0

Awake ground squirrel

Hibernating ground squirrel

Time (hours): 0 6 12 18 0 6 12

Onset of hibernation   Onset of arousal   Reentry

Figure 39.17 Hibernation Patterns in a Ground Squirrel **(A)** During most of the year, the ground squirrel regulates its body temperature around 37°C. During winter months, however, these animals hibernate in underground burrows, living off stored fuel (in the form of either fat or cached food). **(B)** The metabolic demands for these stored fuels are decreased by bouts of torpor, during which body temperature drops close to that of the environment for long periods of time.

investigatinglife

How can we increase heat loss from the body to protect against heat stress?

In the opening story to this chapter you saw that body temperature is a critical factor in limiting physical performance. In Key Concept 39.4 you learned that body surface temperature is a critical factor in each path of heat loss. The relevant body surface temperature for furred mammals is mostly the surface of the fur. For us it is mostly the surface of our clothing. We can take off our clothes, but furred mammals cannot take off their fur. Mammals have evolved special blood vessel adaptations in their non-hairy skin that can accommodate large volumes of blood and therefore support high levels of heat loss. The experiment in Investigating Life: Can the Work Capacity of Muscle Be Increased by Extracting Heat from the Palms of the Hands? used a technology to extract large amounts of heat from the palm of one hand and showed a resulting improvement in ability to do muscular work

How does this heat-loss adaptation, and the technology designed to amplify it, work?

Blood flows from the heart through large arteries, then through small arteries, and then through the tiny capillaries that nourish the tissues and carry away waste products of metabolism, and then into veins that carry the blood back to the heart. The circulatory adaptations of the "heat portals" in non-hairy mammalian skin are an exception to this pattern. In the non-hairy skin, gated shunts deliver arterial blood directly to veins, bypassing the capillaries. These veins in non-hairy skin form networks that can accommodate a large volume of blood when the shunts are open. Based on this knowledge, biologists at Stanford University developed the "rapid-cooling" technology used in the experiment in Investigating Life: Can the Work Capacity of Muscle Be Increased by Extracting Heat from the Palms of the Hands? The palm of the hand (non-hairy skin) is placed in contact with a cooled surface and a mild vacuum is used to pull more blood into the large, heat-exchanging blood vessels. When this device is used, body temperature rises more slowly during exercise and cools more rapidly during rest after exercise.

As expected, this rapid cooling technology greatly increased endurance for exercise in the heat, but an unexpected discovery was enhanced athletic performance. Because muscle fatigue is partly due to increased muscle temperature, enhanced cooling reduces fatigue and increases exercise capacity, which in turn can lead to conditioning gains. In one study, first-year college students in a conditioning program doubled their rate of physical conditioning when they used the cooling technology. After 6 weeks of training, some men and women in the study achieved more than 900 push-ups or hundreds of pull-ups in a 45-minute workout session.

Future directions

In this chapter you learned that virtually all biological processes are affected by temperature. Therefore, if we have a new technology that can rapidly and efficiently alter the heat content of the

body, we should find many applications for it. What could some of those applications be besides improving athletic performance? Consider a few medical applications. Active thermoregulatory responses such as vasoconstriction, shivering, and sweating depend on control by the nervous system, but all of these mechanisms are inactivated by anesthesia and patients can become seriously hypothermic. Their body temperatures could be stabilized by heat exchange through their natural heat portals. Some medical conditions render patients terribly temperature-sensitive. An example is multiple sclerosis. Individuals with MS can suffer serious increases in their symptoms if their body temperature rises even a little bit. They may lose vision, balance, the ability to walk, or the ability to think clearly. Could they be protected from the effects of heat by cooling their natural heat portals? Evidence suggests that some cancer treatments could be augmented by the induction of hyperthermia and that the consequences of stroke or heart attack could be lessened by timely induction of hypothermia.

Chapter Summary 39

39.1 Animals Are Composed of Organs Built from Four Types of Tissues

- Advantages of multicellularity include the ability to grow bigger and the opportunity for cells specialize.
- The cells of the body are organized into assemblages called **tissues**.
- Although there are many cell types, there are only four tissue types: **epithelial**, **muscle**, **connective**, and **nervous** tissues. Review Focus: Key Figure 39.2
- **Organs** are made up of tissues, and most organs contain all four tissue types. Organs are grouped into **organ systems**. Review Figure 39.3, Activity 39.1

39.2 Physiological Systems Maintain Homeostasis of the Internal Environment

- Multicellular animals provide for the needs of all their cells by maintaining a stable internal environment. That environment consists of two extracellular fluid compartments: the **interstitial fluid** and the **blood plasma**. Review Figure 39.4
- Regulation of physiological systems is mostly through **negative feedback**. **Feedforward information** functions to change **set points**. Review Figure 39.5, Activity 39.2

39.3 Biological Processes Are Temperature-Sensitive

- Life is possible only within a narrow range of environmental temperatures. Q_{10} is a measure of the sensitivity of a life process to temperature. A Q_{10} of 2 means that the reaction rate of that process doubles as temperature increases by 10°C. Review Figure 39.7
- Animals can **acclimatize** to seasonal changes in temperature through biochemical and physiological adaptations.
- Even small changes in body temperature can have major physiological effects. Review Investigating Life: Can the Work Capacity of Muscle Be Increased by Extracting Heat from the Palms of the Hands?

39.4 Body Temperature Depends on the Balance between Heat In and Heat Out of the Body

- The body temperature of **ectotherms** is determined primarily by external sources of heat. **Endotherms** can regulate their body temperature by varying their rates of metabolic heat production. Review Figure 39.8
- Behavior is an important thermoregulatory adaptation.
- The four avenues of heat exchange with the environment are **radiation**, **convection**, **conduction**, and **evaporation**. The balance between heat production and heat exchange can be expressed as an **energy budget**. Review Figure 39.9
- Control of blood flow to the skin is an important means of temperature regulation. Review Figure 39.10
- Circulatory system adaptations such as **countercurrent heat exchange** can conserve metabolic heat. Review Figure 39.11

39.5 Body Temperature Is Regulated through Adaptations for Heat Production and Heat Loss

- Within the **thermoneutral zone**, resting endotherms have a **basal metabolic rate (BMR)** that correlates with body size. Review Figures 39.12, 39.13, Activity 39.3
- In mammals, control of body temperature relies on commands from a regulatory center in the **hypothalamus**. This thermostat uses its own temperature as negative feedback information and skin temperature as feedforward information. Review Figure 39.15, Animation 39.1
- **Daily torpor** and **hibernation** are regulated lowerings of body temperature. Review Figure 39.17

Go to **LearningCurve** (in **LaunchPad**) for dynamic quizzing that helps you solidify your understanding of this chapter. **LearningCurve** adapts to your responses, giving you the practice you need to master each key concept.

Apply What You've Learned

Review

39.4 The four avenues of heat exchange between an animal and its environment are radiation, convection, conduction, and evaporation.

39.5 Endotherms produce and conserve metabolic heat to offset heat loss in cold environments.

39.5 The basal metabolic rate (BMR) of an endotherm is the lowest metabolic rate necessary for biochemical and physiological processes of a resting animal.

Original Paper: Karpovich, S. A., Ø. Tøien, C. L. Buck and B. M. Barnes. 2009. Energetics of arousal episodes in hibernating arctic ground squirrels. *Journal of Comparative Physiology* 179: 691–700.

During winter, hibernating small mammals alternate between 1- to 3-week torpor bouts at low body temperature and 1- to 2-day returns to normal mammalian body temperature. These intermittent arousals from torpor include three phases: warming, maintenance of normal body temperature, and recooling. This pattern is repeated in hibernating small mammals regardless of the environmental temperature. Does the environmental temperature make a difference in metabolic rate as the animals transition through these three phases?

Researchers set up hibernation chambers for arctic ground squirrels (*Spermophilus kennicottii*). In nature, their winter burrows can reach −10°C. The laboratory chambers were held at either +2°C or −12°C. All squirrels entered repeated bouts of torpor. During periods of arousal, body temperatures (T_b) and metabolic rates of individual squirrels were measured. Recordings of two individual squirrels are shown in the figure at right. They are representative of the squirrels assigned to the two groups.

The table reports the mean values during the rewarming phase for metabolic rate and body temperature plus or minus (±) the standard error for the two groups of squirrels at the different ambient temperatures. An asterisk (*) indicates that the mean values were significantly different from each other.

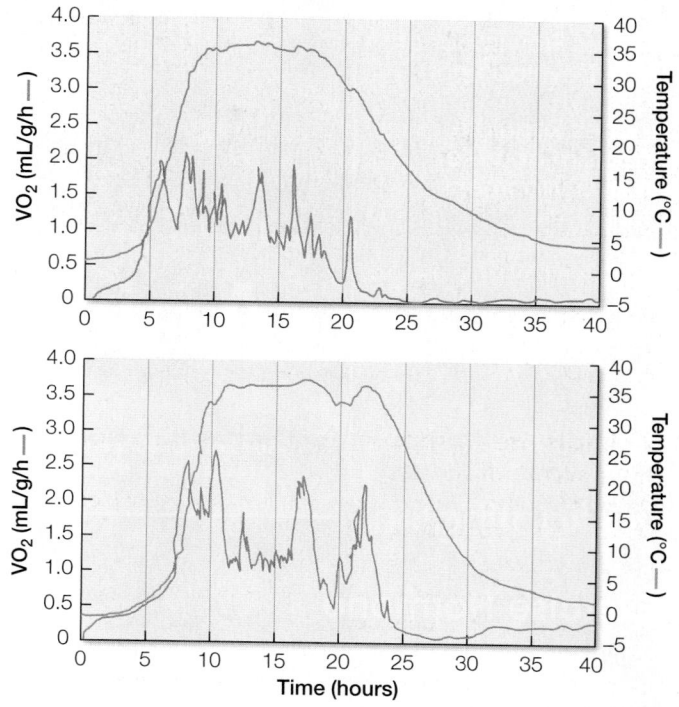

Environmental temperature (°C)	Starting body temperature (°C)	Time to reach T_b = 30°C (h)	Peak metabolic rate (mL O_2/g/h)	Total O_2 consumption from initiation until T_b = 30°C (mL O_2/g)	Time to peak metabolic rate (h)
+2	2.37 ± 0.38	5.65 ± 0.51	2.65 ± 0.22	5.42 ± 0.28	4.25 ± 0.30
−12	−1.44 ± 0.40*	8.20 ± 0.89*	3.40 ± 0.18*	7.71 ± 0.34*	7.08 ± 0.65*

Questions

1. Did environmental temperature significantly affect the time to reach a T_b of 30°C? Did the initial body temperature affect the rate of rewarming? Explain your answer.

2. Using the explanation in the chapter for an energy budget, explain why it took significantly longer for one group of squirrels to reach normal body temperatures and why their total O_2 consumption was significantly different from that of the squirrels in the other group.

3. During the warming period, the subjects engaged in limited gross or large body movement. Describe the mechanisms and processes these squirrels used to raise their body temperatures.

40

key concepts

40.1 Hormones Circulate Around the Body and Affect Target Cells

40.2 The Endocrine System and Nervous System Work Together

40.3 Hormones Play Important Roles in Development

40.4 Hormones Regulate Metabolism and the Internal Environment

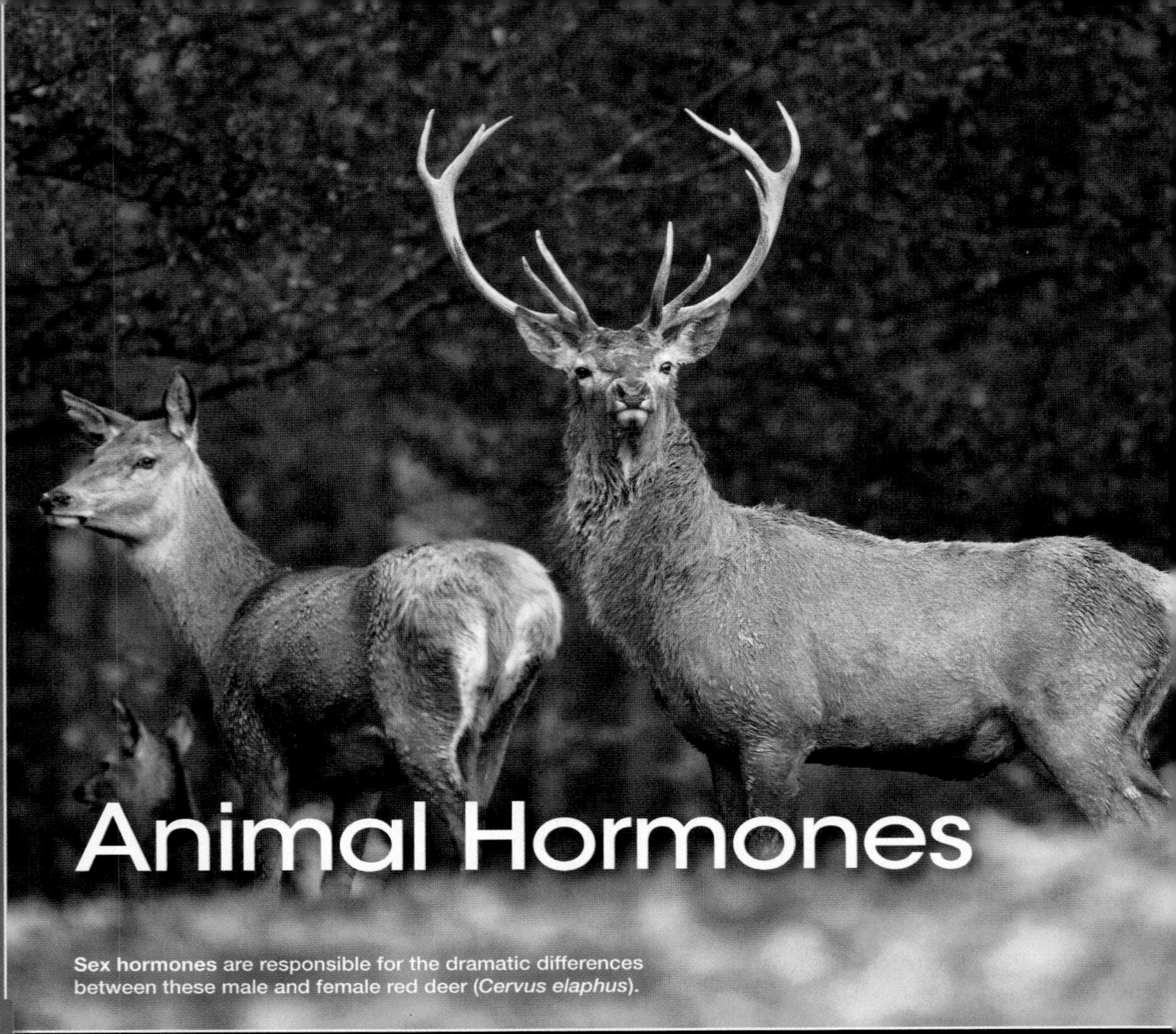

Animal Hormones

Sex hormones are responsible for the dramatic differences between these male and female red deer (*Cervus elaphus*).

investigatinglife

The Exercise Hormone

Some active people eat a lot and never put on weight. Do they stay thin simply because exercising burns excess calories that would be stored as fat? There are two kinds of fat. "White" fat stores lipids, and "brown" fat metabolizes lipids to produce heat without producing ATP. Brown fat is present in cold-acclimated rodents, hibernators, and newborn humans, but it was not thought to be present in adult humans. Recently, imaging techniques revealed brown fat activity in cold-exposed adults. What is interesting is that the amount of brown fat is inversely proportional to total body mass—lean people have more brown fat and obese people have less. Perhaps the excess calories burned by brown fat contribute to low body mass.

A recently discovered signaling molecule may explain the difference in individual propensities to put on weight. Because this molecule, which was named irisin, circulates in the blood, it was designated a hormone. Irisin was discovered in a strain of mice bred for increased exercise endurance capacity, and it was shown that the irisin molecule is produced in and released from active muscles.

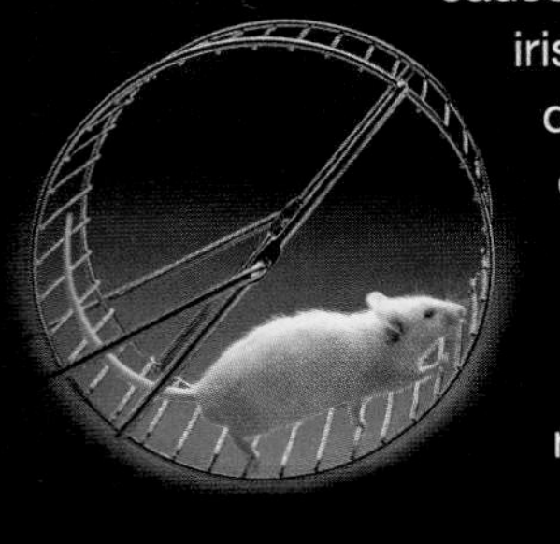

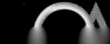

Training causes numerous structural and metabolic changes in muscle. Training also improves many other aspects of health. How are these changes mediated? Analysis of fat tissue from the super athletic mice showed a remarkable finding: their white fat had properties of brown fat. Moreover, this "browning" of white fat was triggered by irisin. That is, active muscles in mice signal white fat to change its properties to become metabolically active, burn more calories, and produce more heat. So in addition to improving the condition of muscles, exercise also causes muscles to "talk" to white fat, telling it to "shape up."

Lots of questions followed the discovery of irisin in mice. Is it produced in humans during exercise? Does it have effects on tissues other than fat? Might other benefits of exercise be attributed to it? More research is under way.

This chapter covers the remarkable roles that hormones play in regulating physiological systems. A hormone is a chemical message that circulates in the blood and activates distant target cells. The power of hormones to initiate physiological change is reflected in the name irisin. Iris was the messenger of the Greek gods, traveling the world with the speed of the wind.

In what ways could irisin mediate the

key concept 40.1 Hormones Circulate Around the Body and Affect Target Cells

In multicellular animals, physiological regulatory systems require information and cell-to-cell communication. Most intercellular communication is by means of chemical signals that bind to receptors, as described in Chapter 7. Some types of chemical signals discussed in other chapters include growth factors, morphogens, and cytokines. In future chapters you will learn about chemical signals used by the nervous system called neurotransmitters. Each of these examples involves a chemical message secreted by cells and received by other cells (target cells) that have appropriate receptors. These four classes of chemical signals are named according to the types of actions they have: growth factors influence cell division, morphogens influence developmental processes, cytokines control immune system cells, and neurotransmitters enable nerve cells to communicate with each other and with target cells such as muscles. In this chapter we cover another broad class of chemical signals called **endocrines**, which are secreted by epithelial cells directly into the extracellular fluid (ECF; see Key Concept 39.2). From the ECF, endocrines can diffuse locally and also into the blood, where they circulate throughout the body.

focus your learning

- Hormones affect target cells that have appropriate receptors.
- Hormones stimulate signal transduction pathways in target cells either by binding to specific receptors on cell surfaces or by diffusing through the cell membrane to bind to internal receptors.
- Different target cells can respond to the same hormone in different ways depending on the signal transduction pathway stimulated.
- Intercellular chemical signaling mechanisms evolved early in the evolutionary history of multicellular animals.

There are multiple chemical communication systems in the body

Some analogies might help differentiate how chemical signaling in the immune, nervous, and endocrine informational systems works. The immune system (the topic of Chapter 42) operates like an army of private security guards. Immune system cells make their rounds of the body, and if they detect a security breach, they sound their alarms—cytokines—that activate the body's defenses. The nervous system (see Chapters 44–46) operates like a landline telephone system, with a central integration and command center that sends signals along specific wires to specific receivers. The endocrine system is more like a radio, television, or cell-phone system, broadcasting signals that can be picked up by anyone who has an appropriate receiver that is turned on and tuned in. Some endocrine signals are weak like those of a walkie-talkie and are picked up only locally, but others are strong like those of a major broadcast station and reach distant parts of the body. In all

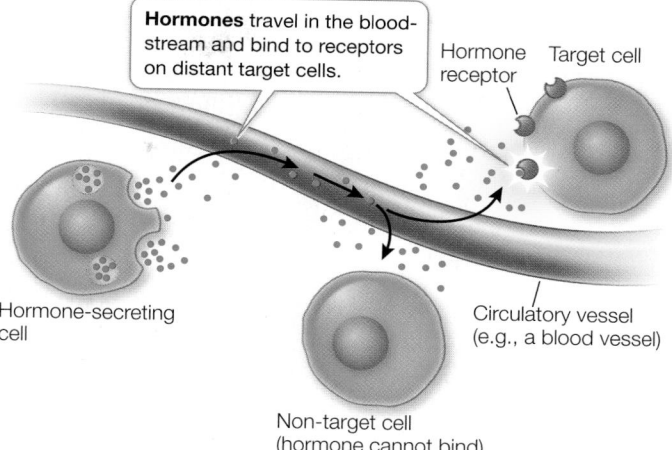

(A) Signals that act at a distance

Hormones travel in the blood-stream and bind to receptors on distant target cells.

Hormone receptor   Target cell

Hormone-secreting cell

Non-target cell (hormone cannot bind)

Circulatory vessel (e.g., a blood vessel)

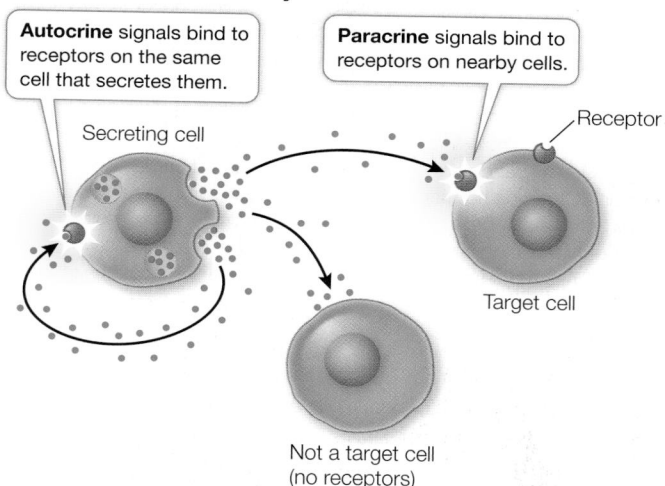

(B) Signals that act locally

Autocrine signals bind to receptors on the same cell that secretes them.

Paracrine signals bind to receptors on nearby cells.

Secreting cell

Receptor

Target cell

Not a target cell (no receptors)

Figure 40.1 Chemical Signaling Systems **(A)** Hormones are distributed throughout the body in the bloodstream. **(B)** Paracrines and autocrines diffuse locally in the extracellular fluid. Paracrines influence nearby cells, and autocrines influence the same cells that release them.

cases, the signals are received only by cells that have appropriate receptors, and the responses are determined by the receiving cell's internal signal transduction machinery.

Endocrine signaling can act locally or at a distance

Endocrine signals that enter the blood are called **hormones** (from a Greek word meaning to excite, arouse, or stimulate). Hormones activate target cells wherever they occur in the body (**Figure 40.1A**). You probably are familiar with several hormones, such as testosterone, estrogen, adrenaline (also called epinephrine), and insulin. Those molecules are well known, as are their sites of release and their actions.

Some endocrine signals are released in such tiny quantities, or are so rapidly inactivated by enzymes or taken up so efficiently by local cells, that they never diffuse into the blood in sufficient amounts to act on distant cells (**Figure 40.1B**). Because these

(A) Protein hormone

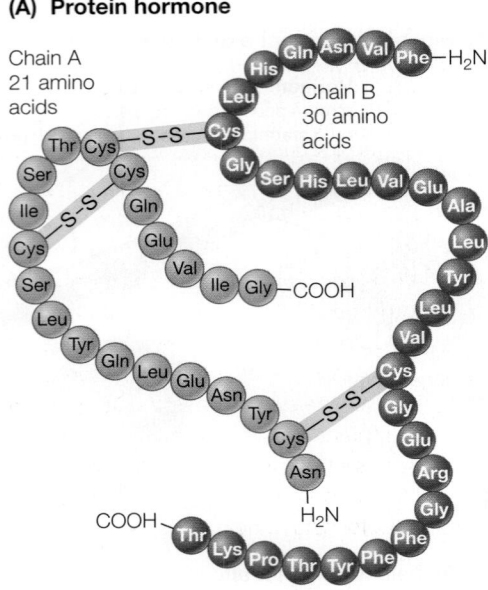

Chain A
21 amino
acids

Chain B
30 amino
acids

Human insulin

(B) Steroid hormones

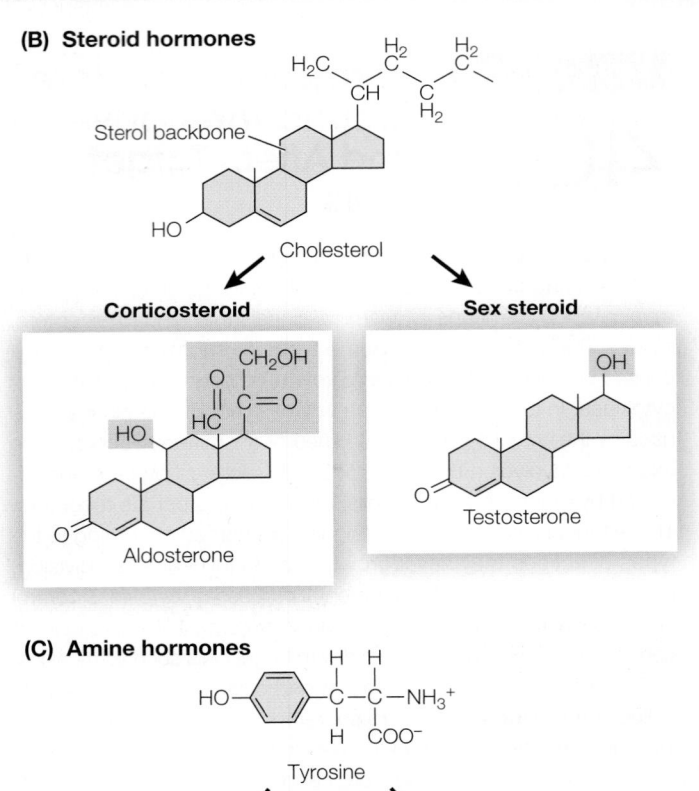

Sterol backbone

Cholesterol

Corticosteroid

Sex steroid

Aldosterone

Testosterone

(C) Amine hormones

Tyrosine

Epinephrine

Thyroxine

Figure 40.2 Three Classes of Hormones **(A)** The largest hormone molecules are peptides and proteins. An example of a protein hormone is the pancreatic hormone insulin. **(B)** Steroid hormones are modified from cholesterol molecules. They include the corticosteroids produced by the adrenal gland and the sex steroids produced primarily by the gonads. **(C)** Amine hormones are tiny molecules synthesized from a single amino acid. Both epinephrine and thyroxine are synthesized from tyrosine units, but thyroxine is lipid-soluble and epinephrine is water-soluble. Their modes of release and transport and the locations of their receptors differ accordingly.

signals affect only target cells near their release site, they are called **paracrines** (*para*, "near"). An example of a paracrine signal is histamine, one of the mediators of inflammation. The most local action an endocrine signal can have is when it binds to receptors on or in the same cell that secreted it. An **autocrine** signal is so named because it influences the cell that secreted it. Hormones and paracrines can have autocrine functions as a means of providing negative feedback to control their own rates of secretion.

Some endocrine cells exist as single cells within a tissue. Hormones of the digestive tract, for example, are secreted by isolated endocrine cells in the walls of the stomach and small intestine. Many hormones are secreted by aggregations of endocrine cells in secretory organs called **endocrine glands**. A single endocrine gland may secrete multiple hormones. Be sure you don't confuse endocrine glands with **exocrine glands**. Endocrine glands secrete hormones within the body, whereas exocrine glands secrete substances through ducts to the outside of the body (sweat glands and salivary glands are examples).

Two other categories of chemical signals are neuroendocrines or **neurohormones** that are secreted by nerve cells and enter the circulation, and the **pheromones** that are released from the body of an animal to influence the behavior and physiology of other animals in its vicinity. We will discuss pheromones in Chapter 52.

Hormones are divided into three chemical groups

There is enormous diversity in the chemical structure of hormones, but by and large they can be classified into three groups:

1. **Peptide** and **protein hormones** make up the majority of hormones. A peptide consists of 2 or more amino acids, and a protein is defined as having 50 or more amino acids. Insulin is therefore a small protein hormone (**Figure 40.2A**). Peptide and protein hormones are water-soluble and thus easily transported in the blood. They can be packaged in vesicles within the cells that make them, and then released by exocytosis.

2. **Steroid hormones** (such as estrogen and testosterone) are synthesized from cholesterol (**Figure 40.2B**), are lipid-soluble, and pass easily through cell membranes. Steroid hormones diffuse out of the cells that make them and are usually bound to carrier molecules in the blood.

3. **Amine hormones** are mostly synthesized from the amino acid tyrosine (thyroxine is one example; **Figure 40.2C**). Some amine hormones are water-soluble and others are lipid-soluble; their modes of release differ accordingly.

Hormone action is mediated by receptors on or within their target cells

Water-soluble hormones cannot pass readily through cell membranes; instead they bind to receptors on the surfaces of target cells.

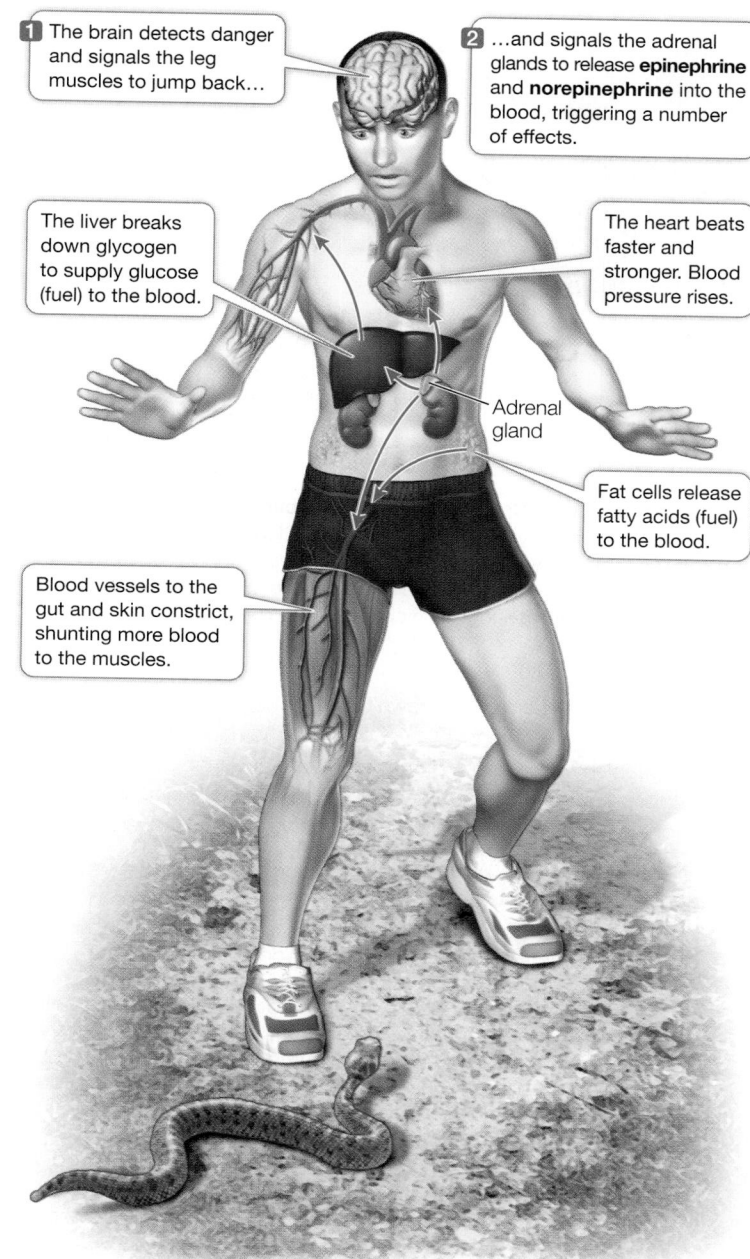

1 The brain detects danger and signals the leg muscles to jump back...

2 ...and signals the adrenal glands to release **epinephrine** and **norepinephrine** into the blood, triggering a number of effects.

The liver breaks down glycogen to supply glucose (fuel) to the blood.

The heart beats faster and stronger. Blood pressure rises.

Adrenal gland

Fat cells release fatty acids (fuel) to the blood.

Blood vessels to the gut and skin constrict, shunting more blood to the muscles.

Figure 40.3 The Fight-or-Flight Response   The brain of a person suddenly faced with a threatening situation sends a signal to the adrenal glands, which almost instantaneously release the hormones epinephrine and norepinephrine. Epinephrine and norepinephrine circulate around the body and induce the various components of the fight-or-flight response in different tissues.

These receptors are large transmembrane glycoprotein complexes with three domains:

1. A binding domain that projects outside the cell membrane

2. A transmembrane domain that anchors the receptor in the membrane

3. A cytoplasmic domain that extends into the cytoplasm of the cell

When a hormone binds to the binding domain, the cytoplasmic domain initiates the target cell's response, usually through second messengers. Second messengers activate a cascade of intracellular events, eventually activating protein kinases or protein phosphatases (see Figures 7.6 and 7.7). In most cases these protein kinases and phosphatases activate or inactivate enzymes in the cytoplasm that lead to the cell's response, but the signaling cascade initiated by the receptor can also generate signals that enter the nucleus and alter gene expression (see Figure 7.10).

Lipid-soluble hormones can diffuse through cell membranes, and therefore their receptors are usually inside cells, in either the cytoplasm or the nucleus. There are also some membrane-bound receptors for lipid-soluble hormones, but they are the exception rather than the rule. In most cases, the complex formed by the lipid-soluble hormone and its receptor acts by altering gene expression in the cell's nucleus (see Figure 7.8).

Hormone action depends on the nature of the target cell and its receptors

Wherever a hormone encounters a cell with an appropriate receptor, it can bind to that receptor and trigger a response. The nature of the response depends on the signal transduction pathways of the responding cell. Therefore the same hormone can cause different responses in different types of cells.

Consider the amine hormone **epinephrine**, and the closely related amine hormone as well as neurotransmitter **norepinephrine**. Both of these molecules are involved in the well-known **fight-or-flight response** (**Figure 40.3**). Suppose you are walking in the forest and almost step on a rattlesnake. You jump back, your heart starts to thump, and protective reactions are set in motion. The jump and the heart thumping are driven by your rapidly responding sympathetic nervous system (see Figure 46.8) which uses norepinephrine to signal target cells. Simultaneously with these muscular responses, your sympathetic nervous system stimulates endocrine cells in the adrenal glands just above your kidneys to secrete both epinephrine and norepinephrine. These hormones rapidly diffuse into your blood and circulate around your body, activating the many components of the fight-or-flight response.

Epinephrine and norepinephrine bind to receptors in your heart, causing a faster and stronger heartbeat. Your heart is now pumping more blood. Epinephrine also binds to receptors in certain blood vessels. By causing constriction of blood vessels supplying your skin, kidneys, and digestive tract (digesting lunch can wait!), the hormone diverts more blood to the muscles needed for your escape from danger.

Epinephrine binds to cells in the liver, stimulating them to break down glycogen and release glucose into the blood as a quick energy supply (see Figure 7.15). In fatty tissue, epinephrine stimulates the breakdown of fats to yield fatty acids—another source of energy. These are just some of the actions triggered by one hormone. In each case the cellular response depends on the cell's receptors, its associated intracellular signaling cascade, and the mechanisms activated by that signaling cascade. All of the diverse cellular re-

Fishes
Prolactin is required for osmoregulation in freshwater species. In saltwater species that return to fresh water to spawn (e.g., salmon), prolactin production in adults may play a role in generating the drive to return to natal streams.

Amphibians
In some species, prolactin creates a "water drive" that returns adults to breeding locations. Stimulates oviduct development and production of egg jelly in females. In some species, controls development of sexual characteristics.

Mammals
In females, prolactin stimulates growth of the mammary glands and milk production. In humans, it is responsible for the sensation of sexual gratification as well as the male refractory period following sexual intercourse.

Birds
In some species, prolactin stimulates nesting activity, incubation behavior, and parental care in both sexes. In pigeons and doves it stimulates production of crop milk to nurture the young nestlings.

Figure 40.4 **Prolactin's Structure Is Conserved, but Its Functions Have Evolved**
The hormone prolactin is found in all vertebrate groups and has a long evolutionary history. Its probable function in early vertebrates was in regulating the body's salt and water balance (osmoregulation). It maintains this function in some species, and has evolved in others to control a number of physiological processes, most of which are associated with reproduction.

The list of hormones known to exist is long and growing longer (as shown by the recent discovery of irisin). To make the subject manageable, we will focus primarily on the endocrine system of mammals. **Figure 40.5** presents the human endocrine glands and most of the hormones that are discussed in this and subsequent chapters.

sponses activated by epinephrine and norepinephrine increase your chances of surviving a dangerous situation.

Hormone structure is conserved through evolution, but functions change

Intercellular chemical signaling was critical for the evolution of multicellularity, so hormones have a long evolutionary history. The least complex of the multicellular animals—the sponges—do not have nervous systems, but they do have intercellular chemical communication. Studying the evolution of hormonal signaling reveals an interesting generalization: the signal molecules themselves are highly conserved. We find the same chemical compounds over broad groups of organisms, although their functions may differ. As organisms have evolved to occupy different environments and have different lifestyles, the same hormone–receptor systems have diversified to serve different functions. A good example of this evolutionary diversification is the hormone prolactin, described in **Figure 40.4**. The name prolactin reflects the role this hormone plays in mammals to stimulate the growth of mammary glands and the production of milk. Mammary glands are unique to mammals, but we find prolactin in all vertebrate groups—and in all cases it is involved in some way in reproduction.

40.1 recap

Endocrine signals initiate responses in target cells that vary depending on the signal transduction pathways of the responding cell. Autocrines and paracrines act locally. Hormones are chemical signals released by endocrine cells into the extracellular fluid, where they diffuse into the blood and travel to distant target cells. The receptors for water-soluble hormones are on the surfaces of target cells; receptors for most lipid-soluble hormones are inside the target cells. The chemical structures of signal molecules have been highly conserved over evolutionary time, but the chemical compounds serve different functions in different organisms.

learning outcomes

You should be able to:
- Describe the structural feature of hormones that determines whether they act on surface or cytoplasmic receptors.
- Explain how different target cells in the body can respond differently to a single hormone.
- Discuss scientific thinking about the evolutionary history of hormones in multicellular animals.

1. Why are the effects of epinephrine more rapid than the effects of estrogen?
2. Why is the presence of a particular hormone in two species not an indication of the closeness of their evolutionary relationship?
3. How can a single hormone have two different effects in the same species?

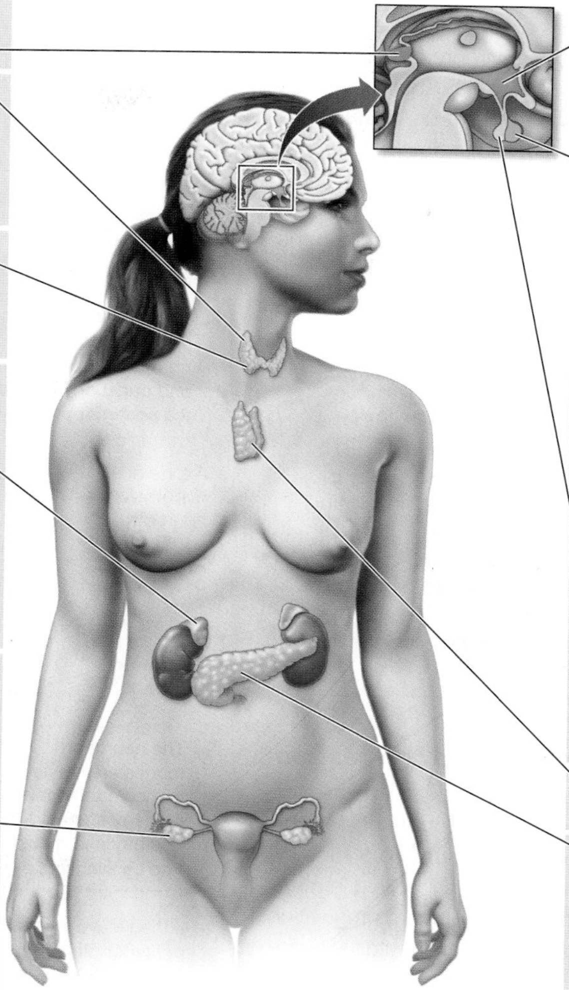

Pineal gland
Melatonin: helps entrain daily rhythms

Thyroid gland (see Figures 40.11 and 40.13)
Thyroxine (T₃ and T₄): increases cell metabolism; essential for growth and neural development
Calcitonin: stimulates incorporation of calcium into bone

Parathyroid glands (on posterior surface of thyroid; see Figure 40.13)
Parathyroid hormone (PTH): stimulates release of calcium from bone and absorption of calcium by gut and kidney

Adrenal gland (see Figure 40.16)
Cortex
Cortisol: mediates metabolic responses to stress
Aldosterone: involved in salt and water balance
Sex steroids: in small amounts

Medulla
Epinephrine (adrenaline) and *norepinephrine* (noradrenaline): stimulate immediate fight-or-flight reactions

Gonads (see Chapter 42)
Testes (male)
Testosterone: development and maintenance of male sexual characteristics

Ovaries (female)
Estrogens: development and maintenance of female sexual characteristics
Progesterone: supports pregnancy

Other organs include cells that produce and secrete hormones

Organ	Hormone
Adipose tissue	Leptin
Heart	Atrial natriuretic peptide
Kidney	Erythropoietin
Stomach	Gastrin, ghrelin
Intestine	Secretin, cholecystokinin
Liver	Somatomedins, insulin-like growth factors
Skeletal muscle	Irisin

Hypothalamus (see Figure 40.6)
Release and release-inhibiting neuro-hormones control the anterior pituitary; *ADH* and *oxytocin* are transported to and released from the posterior pituitary

Anterior pituitary (see Figure 40.7)
Thyrotropin (TSH): activates the thyroid gland
Follicle-stimulating hormone (FSH): in females, stimulates maturation of ovarian follicles; in males, stimulates spermatogenesis
Luteinizing hormone (LH): in females, triggers ovulation and ovarian production of estrogens and progesterone; in males, stimulates production of testosterone
Adrenocorticotropin (ACTH): stimulates adrenal cortex to secrete cortisol
Growth hormone (GH): stimulates protein synthesis and growth
Prolactin: stimulates milk production
Melanocyte-stimulating hormone (MSH): stimulates production of the pigment melanin
Endorphins and enkephalins: pain control

Posterior pituitary (see Figure 40.6)
Receives and releases two hypothalamic hormones:
Oxytocin: stimulates contraction of uterus, flow of milk, interindividual bonding
Antidiuretic hormone (ADH; also known as vasopressin): promotes water conservation by kidneys

Thymus (diminishes in adults)
Thymosin: activates immune system T cells

Pancreas (islets of Langerhans)
Insulin: stimulates cells to take up and use glucose
Glucagon: stimulates liver to release glucose
Somatostatin: slows release of insulin and glucagon and digestive tract functions

Figure 40.5 The Endocrine System of Humans Cells that produce and secrete hormones may be organized into discrete endocrine glands, or they may be embedded in the tissues of other organs, such as the digestive tract or kidneys. The hypothalamus is part of the brain, but it includes cells that secrete neurohormones into the extracellular fluid.

 Activity 40.1 The Human Endocrine Glands
www.Life11e.com/ac40.1

The advantage of hormone signals is that they can originate from a localized source but reach cells in all areas of the body and coordinate their activities. The disadvantage of hormones is that they are relatively slow in delivering and in terminating their messages. Thus they are not good for communicating rapidly changing information. In contrast, the nervous system sends messages to specific target cells, and its messages are rapid and can be turned off quickly. In addition, the nervous system has access to a great diversity of both external and internal information that indicates the state of the body and threats to homeostasis. The different characteristics of each of these informational systems make it advantageous for them to work together.

key concept
40.2 The Endocrine System and Nervous System Work Together

The nervous system and the endocrine system control and regulate virtually all physiological and behavioral mechanisms that contribute to homeostasis, survival, and reproductive success. Challenges to an organism usually trigger both physiological and behavioral responses, as we saw with respect to thermoregulation in Chapter 39 and also with the fight-or-flight reactions shown in Figure 40.3. Therefore the endocrine and nervous systems must work together in an integrated fashion, and indeed they do. The nervous system receives an enormous amount of information from the body and from the environment. That information is critical for maintenance of homeostasis, survival, and reproductive success. The nervous system controls many tissue and organ functions as well as behavior. It is therefore not surprising that the nervous system regulates many endocrine functions. Conversely, hormones circulate to the nervous system, providing feedback information used in regulation. Hormones reaching the brain also stimulate behaviors that contribute to the health and well being of an organism.

focus your learning

- Oxytocin and vasopressin are produced in the hypothalamus and secreted from nerve terminals in the posterior pituitary.
- The anterior pituitary produces four tropic hormones plus growth hormone, prolactin, and melanocyte stimulating hormone.
- The release of hormones from the anterior pituitary is controlled by hypothalamic hormones transported to the pituitary in portal vessels.
- Regulation involves short-loop and long-loop negative feedback.

The pituitary is an interface between the nervous and endocrine systems

The **pituitary gland** sits in a depression at the bottom of the skull, just over the back of the roof of the mouth (**Figure 40.6**). It is attached by a stalk to the hypothalamus, which is involved in many physiological regulatory systems such as that of thermoregulation (see Key Concept 39.5). Through its close connection with the hypothalamus, the pituitary serves as the interface between the nervous system and the endocrine system and is involved in the hormonal control of many physiological processes.

The pituitary has two parts with different developmental origins. The **anterior pituitary** originates as an outpocketing of the roof of the embryonic mouth cavity. The **posterior pituitary** originates as an outpocketing of the floor of the developing brain. Thus the anterior pituitary originates from gut epithelial tissue and the posterior pituitary from neural tissue. Both parts interact with the nervous system but in different ways. The anterior pituitary is controlled by hypothalamic neurohormones that reach the anterior pituitary via the blood. The posterior pituitary contains long extensions of hypothalamic neurons that release their neurohormones in the posterior pituitary.

 Animation 40.1 **The Hypothalamic–Pituitary–Endocrine Axis**
www.Life11e.com/a40.1

THE POSTERIOR PITUITARY The long hypothalamic neuron extensions into the posterior pituitary are called **axons**. The terminals of these axons in the posterior pituitary contain vesicles of neurohormones—either antidiuretic hormone or oxytocin. When these hypothalamic neurons are activated, they fire electrical signals (***action potentials**) that travel down the axon and cause the release of the neurohormones (see Figure 40.6).

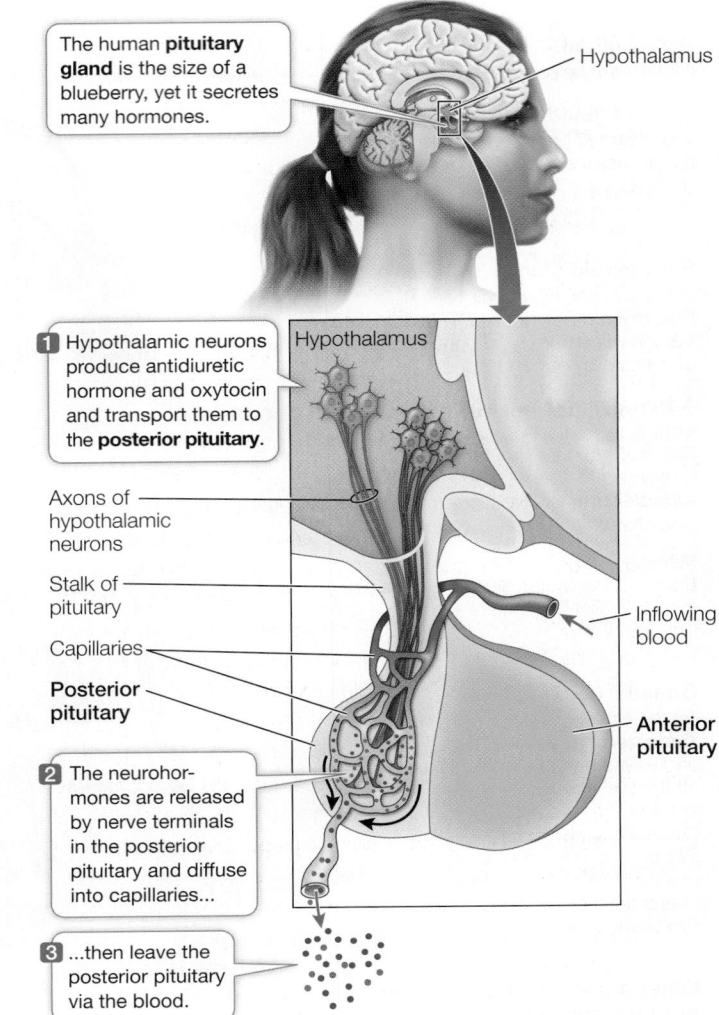

The human **pituitary gland** is the size of a blueberry, yet it secretes many hormones.

Hypothalamus

1 Hypothalamic neurons produce antidiuretic hormone and oxytocin and transport them to the **posterior pituitary**.

Hypothalamus

Axons of hypothalamic neurons

Stalk of pituitary

Capillaries

Posterior pituitary

2 The neurohormones are released by nerve terminals in the posterior pituitary and diffuse into capillaries...

3 ...then leave the posterior pituitary via the blood.

Inflowing blood

Anterior pituitary

Figure 40.6 The Posterior Pituitary Releases Neurohormones
Neurons in the hypothalamus produce two peptide neurohormones—antidiuretic hormone and oxytocin—which are stored in and released by nerve endings in the posterior pituitary.

*connect the concepts As described in Key Concept 45.2, action potentials are sudden and transient electric signals generated by voltage-gated ion channels in the cell membrane of the axon. These signals travel rapidly along axons, and their arrival at the axon terminals initiates the release of neurotransmitters.

The main action of **antidiuretic hormone (ADH)** in mammals and birds is to increase the amount of water conserved by the kidneys. When ADH secretion is high, the kidneys produce only a small volume of highly concentrated urine. When ADH secretion is low, the kidneys produce a large volume of dilute urine. The posterior pituitary increases its release of ADH when blood pressure falls or the blood becomes too salty. ADH is also known as **vasopressin** because at high concentrations it causes the constriction of peripheral blood vessels as a means of elevating blood pressure.

The hormone **oxytocin** is released from the posterior pituitary when a woman is about to give birth. Oxytocin stimulates contractions of the uterine muscles, resulting in the delivery of the baby. Oxytocin also brings about the flow of milk from the mother's breasts. The baby's suckling stimulates neurons in the mother's brain that cause the secretion of oxytocin. Even the crying of her baby can cause a nursing mother to secrete oxytocin, resulting in the release of breast milk—a good example of how the nervous system integrates information that regulates hormonally mediated processes.

THE ANTERIOR PITUITARY The anterior pituitary produces and releases into the circulation four peptide and protein hormones that act as **tropic hormones**, meaning they control the activities of other endocrine glands. These four tropic hormones are thyrotropin (thyroid-stimulating hormone), luteinizing hormone (LH), follicle-stimulating hormone (FSH), and adrenocorticotropin (ACTH). Each is produced by a different type of pituitary cell. We will say more about these tropic hormones when we describe their target glands—the thyroid, testes, ovaries, and adrenal cortex—later in this chapter and in Chapter 42. Other peptide and protein hormones produced by the anterior pituitary are prolactin (see Figure 40.4), growth hormone, enkephalins and endorphins, and melanocyte stimulating hormone (MSH).

Growth hormone (**GH**) acts on a wide variety of tissues to promote growth and development. One of its important effects is to stimulate cells to take up amino acids. Growth hormone also promotes growth by stimulating the liver to produce chemical signals that stimulate the growth of bone and cartilage. Overproduction of GH in children causes gigantism; affected individuals may grow to nearly 8 feet tall. Underproduction of GH results in pituitary dwarfism, in which individuals fail to reach normal adult height.

Endorphins and **enkephalins** are the body's natural painkillers. In the brain, these molecules act as neurotransmitters in pathways that control pain. Their production in the anterior pituitary is normally quite small and probably has little significant effect.

Melanocyte-stimulating hormone (**MSH**) stimulates the production of the pigment melanin in skin and hair. MSH is produced by a thin layer of cells between the anterior and posterior pituitary that is sometimes called the intermediate lobe of the pituitary.

Hypothalamic neurohormones control the anterior pituitary

Hypothalamic neurons do not extend into the anterior pituitary as they do into the posterior pituitary. So how do the neurohormones produced by hypothalamic neurons reach the anterior pituitary cells they control? The pituitary is connected to the brain by the pituitary stalk (**Focus: Key Figure 40.7**). Within that stalk are hypothalamic–pituitary **portal blood vessels**. The hypothalamic cells secrete neurohormones at the bottom of the hypothalamus where the stalk begins. The neurohormones diffuse into the portal blood vessels that begin there, and are carried in the blood to where the portal vessels end in the anterior pituitary. Hypothalamic neurohormones are released in such tiny amounts that they were only hypothesized to exist before the technology developed to isolate and characterize their structure.

In the 1960s two large teams of scientists led by Roger Guillemin and Andrew Schally initiated the search for these hypothalamic secretions. Massive numbers of hypothalami from pigs and sheep were collected from slaughterhouses and shipped to laboratories. One extraction effort began with the hypothalami from 270,000 sheep and yielded only 1 mg of purified **thyrotropin-releasing hormone** (**TRH**). TRH was the first hypothalamic release-stimulating hormone to be isolated and characterized. It turned out to be a simple tripeptide consisting of glutamine, histidine, and proline. It causes certain anterior pituitary cells to release the tropic hormone thyrotropin, which in turn stimulates the activity of the thyroid gland.

Soon after discovering TRH, Guillemin's and Schally's teams identified **gonadotropin-releasing hormone** (**GnRH**), which stimulates certain anterior pituitary cells to release the tropic hormones that control the activity of the gonads (the ovaries and the testes). For these discoveries, Guillemin and Schally shared the 1977 Nobel Prize in Medicine with Rosalyn Yalow, who invented the technique of radioimmunoassay, which made possible the measurement of miniscule amounts of specific molecules such as releasing hormones.

Many other hypothalamic neurohormones, including both releasing and release-inhibiting hormones, are now known. The major hypothalamic neurohormones that control anterior pituitary function are listed in Figure 40.7.

Negative feedback loops regulate hormone secretion

In addition to being controlled by hypothalamic releasing and release-inhibiting hormones, the endocrine cells of the anterior pituitary are also under direct and indirect negative feedback control by the hormones of the target glands they stimulate (see Figure 40.7). For example, cortisol, produced by the adrenal gland in response to ACTH secreted by the anterior pituitary, reaches the pituitary in the circulating blood and inhibits further release of ACTH. Cortisol also acts as a negative feedback signal to the hypothalamus, inhibiting the release of corticotropin-releasing hormone (CRH). Because cortisol is the final hormone released in this chain of interacting structures called the hypothalamic–pituitary–adrenal axis, its feedback actions are called long-loop negative feedback. In this system, the tropic hormone ACTH also exerts negative feedback control on the hypothalamic cells that produce the corresponding releasing hormone (CRH). This action is called short-loop negative feedback because of the proximity of the pituitary and the hypothalamus.

Hormones influence the nervous system

Hormones have more influences on the nervous system than simply being negative feedback signals controlling their own secretion. For example, the posterior pituitary hormone oxytocin, which plays a role in the birth process and in stimulating the flow of milk from the breasts, also has a strong behavioral effect in promoting bonding (see the story that opens Chapter 7). If oxytocin release is experimentally blocked, mammalian mothers, from rats to sheep, will reject their newborn offspring, but if a virgin rat is given a dose of oxytocin, she will adopt strange pups as if they were her own. Oxytocin promotes pair bonding in a variety of animals. In humans, oxytocin secretion rises with intimate sexual contact, and it has been nicknamed the "cuddle hormone." Experiments using games involving financial exchanges have shown that puffs of oxytocin applied to the nostrils raise the level of trust between players. Thus circulating hormones can have

focus: key figure

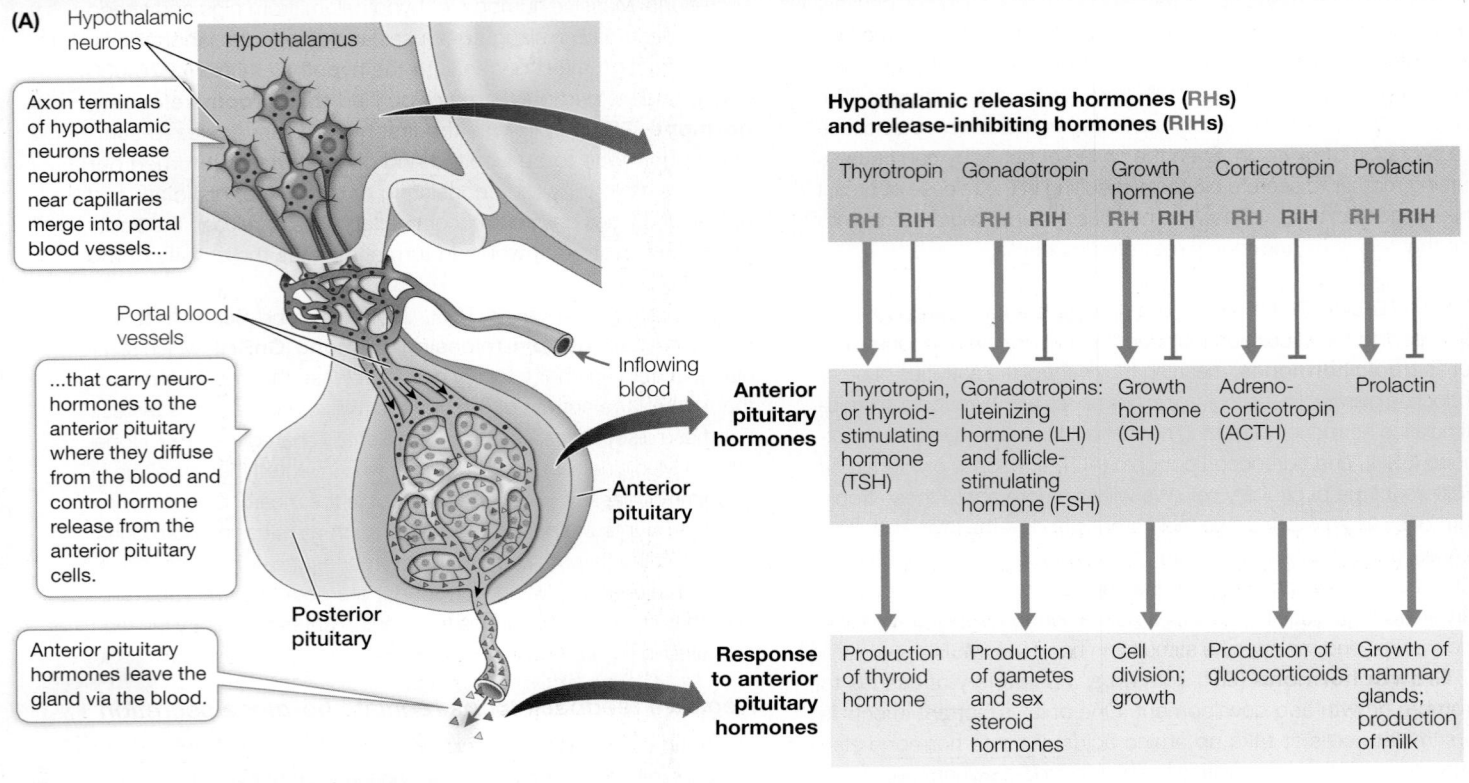

(A)

Hypothalamic neurons

Hypothalamus

Axon terminals of hypothalamic neurons release neurohormones near capillaries merge into portal blood vessels...

Portal blood vessels

...that carry neuro-hormones to the anterior pituitary where they diffuse from the blood and control hormone release from the anterior pituitary cells.

Inflowing blood

Anterior pituitary

Posterior pituitary

Anterior pituitary hormones leave the gland via the blood.

Hypothalamic releasing hormones (RHs) and release-inhibiting hormones (RIHs)

	Thyrotropin		Gonadotropin		Growth hormone		Corticotropin		Prolactin	
	RH	RIH	RH	RIH	RH	RIH	RH	RIH	RH	RIH
Anterior pituitary hormones	Thyrotropin, or thyroid-stimulating hormone (TSH)		Gonadotropins: luteinizing hormone (LH) and follicle-stimulating hormone (FSH)		Growth hormone (GH)		Adreno-corticotropin (ACTH)		Prolactin	
Response to anterior pituitary hormones	Production of thyroid hormone		Production of gametes and sex steroid hormones		Cell division; growth		Production of glucocorticoids		Growth of mammary glands; production of milk	

(B)

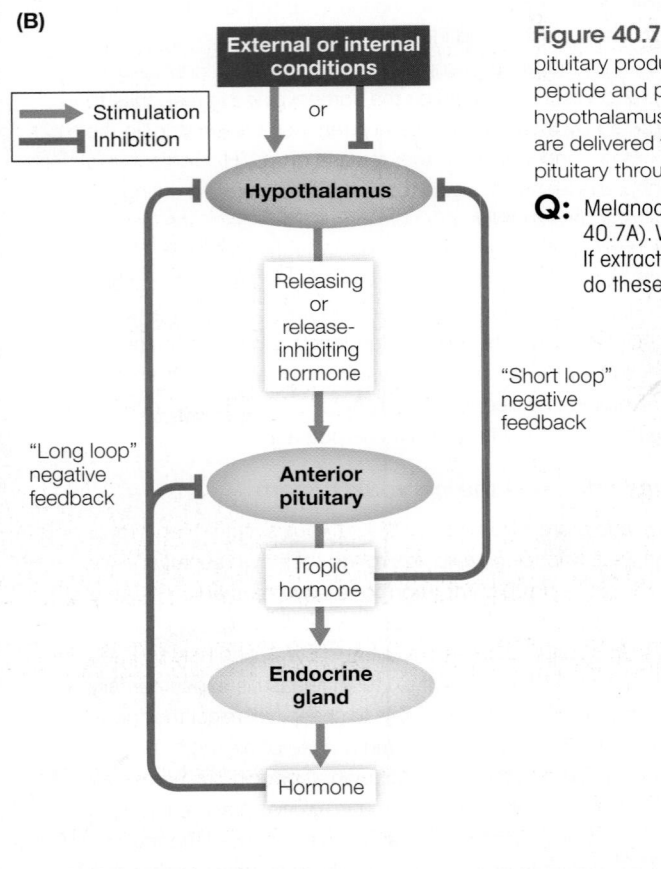

→ Stimulation
⊣ Inhibition

External or internal conditions

or

Hypothalamus

Releasing or release-inhibiting hormone

"Long loop" negative feedback

"Short loop" negative feedback

Anterior pituitary

Tropic hormone

Endocrine gland

Hormone

Figure 40.7 The Anterior Pituitary Is Controlled by the Hypothalamus Cells of the anterior pituitary produce tropic hormones that control other endocrine glands, as well as several other peptide and protein hormones. These cells are controlled by neurohormones produced in the hypothalamus and are called releasing or release-inhibiting hormones. These neurohormones are delivered through portal blood vessels that run between the hypothalamus and the anterior pituitary through the pituitary stalk.

Q: Melanocyte stimulating hormone (MSH) is an anterior pituitary hormone (not shown in Figure 40.7A). When anterior pituitary cells are grown in culture they release MSH into the medium. If extracts of the hypothalamus are added to the medium, the release of MSH decreases. What do these results tell you about the hypothalamic control over MSH producing cells?

a strong influence on behavior. Another example is the effects of sex hormones, which we will discuss in Key Concept 53.3.

Recent investigations involving the hormone irisin that was presented at the beginning of this chapter reveal that this same molecule may mediate effects of exercise on the brain. Many experiments have shown that exercise improves cognitive functions. Beneficial effects of exercise have been shown especially for a part of the brain called the hippocampus that is involved in learning and memory. These effects include the production of new brain cells, increased blood flow, and changes in the structures supporting communication between brain cells. It had been established that one molecular factor involved in the beneficial effects of exercise on the hippocampus is BDNF, which stands for brain-derived neurotrophic factor. How does exercise mediate the release of BDNF in the brain? **Investigating Life: How Could Irisin Mediate the Benefits of Exercise on Cognition?** explores this interesting question.

experiment

Original Paper: Wrann, C. D. et al. 2013. Exercise induces hippocampal BDNF through a PGC-1α/FNDC5 pathway. *Cell Metabolism* 18: 649–659.

Irisin secreted by exercising muscle is cleaved from a protein called FNDC5, and FNDC5 expression is controlled by the transcription factor PGC-1α. Irisin then circulates in the blood, but because of the blood–brain barrier (see Key Concept 44.1), it is unlikely to reach brain structures such as the hippocampus where it could enhance cognition. However, could irisin also be expressed in the brain? If so, exercise should result in similar changes in expression of FNDC5 and its transcription factor PGC-1α in the hippocampus as it does in skeletal muscle. First, we need to show in our experimental system and with our assays that exercise induces expression of PGC-1α and FNDC5 in exercising muscle. Then we can use the same experimental system to investigate expression in the hippocampus.

HYPOTHESIS▶ Endurance exercise results in increased expression of PGC-1α and FNDC5 in skeletal muscle.

METHOD

1. Have one group of mice exercise on running wheels for 30 days and a control group that does not. *N* = 10 for each group.
2. At the end of 30 days, measure PGC-1α and FNDC5 mRNA levels in leg muscle of both groups of mice.
3. Plot the data for each variable (FNDC5 and PGC-1α) for the exercised and control mice, and use an un-paired *t*-test (see Appendix B) to evaluate the significance of the results.

RESULTS Error bars indicate ± 1 standard deviation.

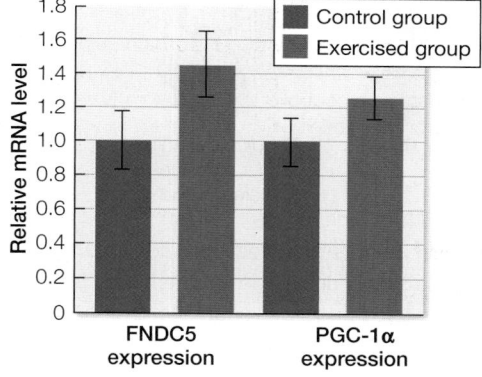

CONCLUSION▶ Relative levels of FNDC5 mRNA were significantly higher in the exercised than in the non-exercised mice (*P* < 0.001). Relative levels of PGC-1α mRNA also were significantly higher in the exercised than in the non-exercised mice (*P* < 0.01).

work with the data

Similar measurements of expression of FNDC5 and PGC-1α mRNAs were made from the hippocampus of the exercised and the non-exercised mice. Again, *N* = 10 for each group. The results are shown in the table.

FNDC5		PGC-1α	
Control	Exercised	Control	Exercised
1.02	1.33	1.22	1.23
1.10	1.24	0.84	1.09
0.92	1.68	1.06	1.35
0.93	1.30	1.26	1.18
1.21	1.34	1.04	1.40
0.93	1.33	1.12	1.18
0.96	1.50	1.04	1.46
1.01	1.29	1.09	1.19
1.00	1.44	1.02	1.54
0.97	1.41	0.54	1.16

QUESTIONS▶

1. Plot the data as was done for the leg skeletal muscle results. Then you can apply the same analysis methods to these data that were applied to the leg muscle data in the preceding experiment. Did exercise have a significant effect on FNDC5 and PGC-1α expression in the hippocampus?

2. The neurotrophic factor BDNF is a secreted protein that plays important roles in neural changes associated with learning and memory. How would you determine if the exercise protocol induced a significant increase in BDNF expression in the hippocampus, assuming you have an assay for BDNF?

3. How could you determine if a noted increase in the expression of FNDC5 in the hippocampus is sufficient for an increased expression of BDNF in the hippocampus? Assume you can culture hippocampal neurons so that the expression of FNDC5 is stimulated or suppressed.

A similar **work with the data** exercise may be assigned in **LaunchPad**.

40.2 recap

The pituitary is the interface between the nervous system and the endocrine system. The posterior pituitary releases two neurohormones, antidiuretic hormone (ADH) and oxytocin. The anterior pituitary, under the control of other neurohormones from the hypothalamus, releases four tropic hormones that control other endocrine glands. In addition, the anterior pituitary produces and releases five other hormones that act directly on non-endocrine tissues. The hypothalamus and pituitary are under negative feedback control by the hormones of their target glands.

learning outcomes

You should be able to:

- Describe the relationships between the hypothalamus and the pituitary gland.
- Explain developmental origins of the anterior pituitary and the posterior pituitary.
- Explain how a negative feedback loop controls hormone secretion.

(continued)

40.2 recap (continued)

1. Provide a developmental explanation of why posterior pituitary hormones are released from neurons and anterior pituitary hormones are released from epithelial cells.
2. How do hypothalamic neurons control the release of hormones from the anterior pituitary?
3. Explain what is meant by negative feedback in the hypothalamic control of endocrine function.

The hypothalamic and pituitary hormones and some of the endocrine glands they control are easy to conceptualize because they share common features in how they are organized and function. But the entire list of hormones is highly diverse, making generalizations difficult. We cannot consider all of the hormones in this chapter. Instead, we focus on hormones that have some common functions—one of which is the coordination of developmental processes.

> **key concept**
> # 40.3 Hormones Play Important Roles in Development

The development of an animal from a fertilized egg requires the coordination of processes throughout the body. Imagine the problems that would arise if growth of the body were not symmetrical, or if the growth of organ systems did not match the demands placed on them by the growing body. Because hormones circulate in the blood to all parts of the body, they are ideal signals for integrating and coordinating the many diverse processes involved in development.

focus your learning

- Classic experiments revealed that hormones control insect metamorphosis.
- Vertebrate sexual development is controlled by steroid hormones.
- Human puberty results from a change in sensitivity of the hypothalamus to negative feedback.

Insect development consists of either complete or incomplete metamorphosis

The changes that occur in a developing insect can be dramatic and involve major changes in body form. Insects such as bees and butterflies undergo a developmental process called **complete metamorphosis**. Starting out as a caterpillar (a larva), a butterfly feeds voraciously and goes through a series of growth stages called **instars**, which are separated by molts. The last instar builds a cocoon and enters into a quiescent, transformative state called a pupa. The tissues of the pupa are extensively remodeled to form the adult body, which is entirely different from that of the larva. The larval butterfly crawls and eats vegetation; the adult butterfly flies and eats flower nectar. Many other species of insects undergo **incomplete metamorphosis**, which involves less extreme changes of body form. These insects also go through a series of instars, getting bigger with each molt, but the last juvenile instar

simply molts into the adult form with all of the essential adult body parts. Molting involves shedding the skin, or in many cases the rigid exoskeletons, and therefore requires coordination of cellular processes in the entire epidermis of the animal. Both complete and incomplete *metamorphosis require integrated processes in all parts of the insect body.

> *connect the concepts   Insect metamorphosis is described in more detail at the end of Key Concept 32.4.

Experiments on insect development revealed hormonal signaling systems

Pioneering experiments of hormonal action and control in insects were done on the bloodsucking bug *Rhodnius prolixus*, which undergoes incomplete metamorphosis. Newly hatched *Rhodnius* lack certain adult features and molt five times before developing into a mature adult. A blood meal triggers each episode of molting and growth. *Rhodnius* is an amazingly hardy experimental animal—it survives for quite a long time after its head is cut off. If decapitated within an hour after a blood meal, *Rhodnius* can survive for up to a year, but it never molts. If decapitated a week after its blood meal, however, it does molt into an adult.

  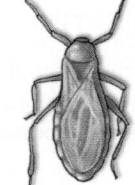

Juvenile *Rhodius*   *Rhodius* decapitated 1 hour after meal (remains a juvenile)   *Rhodius* decapitated 1 week after meal (molts into adult)

This observation suggested that molting was triggered by some substance slowly diffusing from the head. To test this hypothesis, two decapitated bugs—one soon after a blood meal and one a week after a blood meal—were connected together with a glass tube that allowed body fluids to cross between the two bugs. Both molted into adults proving that a diffusible substance from the bug's head triggered molting.

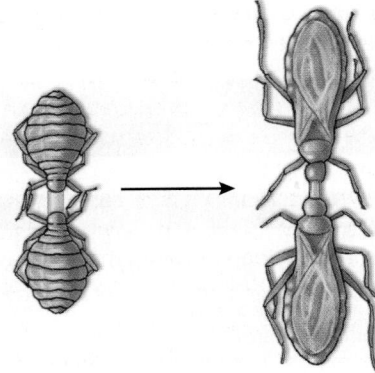

Since those pioneering experiments on *Rhodnius*, the chemical nature of the diffusible substances (hormones) controlling molting in arthropods have been identified and their functions described.

Three hormones regulate molting and maturation in arthropods

An arthropod's nervous system receives various types of information about the environment (e.g., day length, temperature, social cues, and nutrition) that help determine the optimal timing for the stages of growth and development. When conditions are right, the brain signals the **prothoracic gland** to produce the hormones that orchestrate physiological processes involved in development and molting.

PTTH AND ECDYSONE   Two hormones, **prothoracicotropic hormone (PTTH)** and **ecdysone**, work in sequence to regulate molting in arthropods. Cells in the brain produce PTTH, which is why it has also been called "brain hormone." PTTH is transported to and stored in paired structures called the **corpora cardiaca** attached to the brain (**Figure 40.8**). After appropriate stimulation (which for *Rhodnius* is a blood meal), PTTH is released and diffuses through the extracellular fluid to an endocrine gland, the prothoracic gland. PTTH stimulates the prothoracic gland to secrete the hormone ecdysone. Ecdysone diffuses to target tissues and stimulates molting.

Ecdysone is a steroid hormone (see Figure 40.2B) and thus is related to the vertebrate hormones estrogen and testosterone (which also play roles in controlling growth and development). Ecdysone is lipid-soluble and readily passes through the cell membrane of its target cells (mostly cells of the epidermis). In the target cells, ecdysone binds to a receptor that induces expression of genes encoding enzymes involved in digesting the old cuticle and secreting a new one.

JUVENILE HORMONE   What determines whether an instar molts into another juvenile form or matures into an adult? **Juvenile hormone** is a molecule that is released continuously from the **corpora allata** (structures that are attached to the *corpora cardiaca*, which release PTTH). As long as juvenile hormone is present, *Rhodnius* molts into another juvenile instar. Normally *Rhodnius* stops producing juvenile hormone during the fifth instar and then molts into an adult.

The role of juvenile hormone is more complex in insects that undergo complete metamorphosis. An example of complete metamorphosis is provided by the silkworm moth *Hyalophora cecropia* (see Figure 40.8). As long as juvenile hormone is present in high concentrations, larvae molt into larger larvae. When the level of juvenile hormone falls, larvae spin cocoons and molt into pupae. Because no juvenile hormone is produced in pupae, they molt into adults. Many modern pesticides use juvenile hormone analogs to prevent larvae developing into adults.

These cases of insect development show how just a few hormones can coordinate major developmental effects that involve very different cell types and tissues in the body. The hormonal signals reach all parts of the body simultaneously, and depending on the receptors on different cell types and the internal mechanisms driven by those receptors, the developmental processes are orchestrated.

Sex steroids control sexual development

Hormones also control many developmental processes in vertebrates. One example already mentioned in Key Concept 40.2 is the influence that the anterior pituitary growth hormone has on growth.

Figure 40.8 Hormonal Control of Metamorphosis   Three hormones control molting and metamorphosis in the silkworm moth *Hyalophora cecropia*.

 Animation 40.2 Complete Metamorphosis www.Life11e.com/a40.2

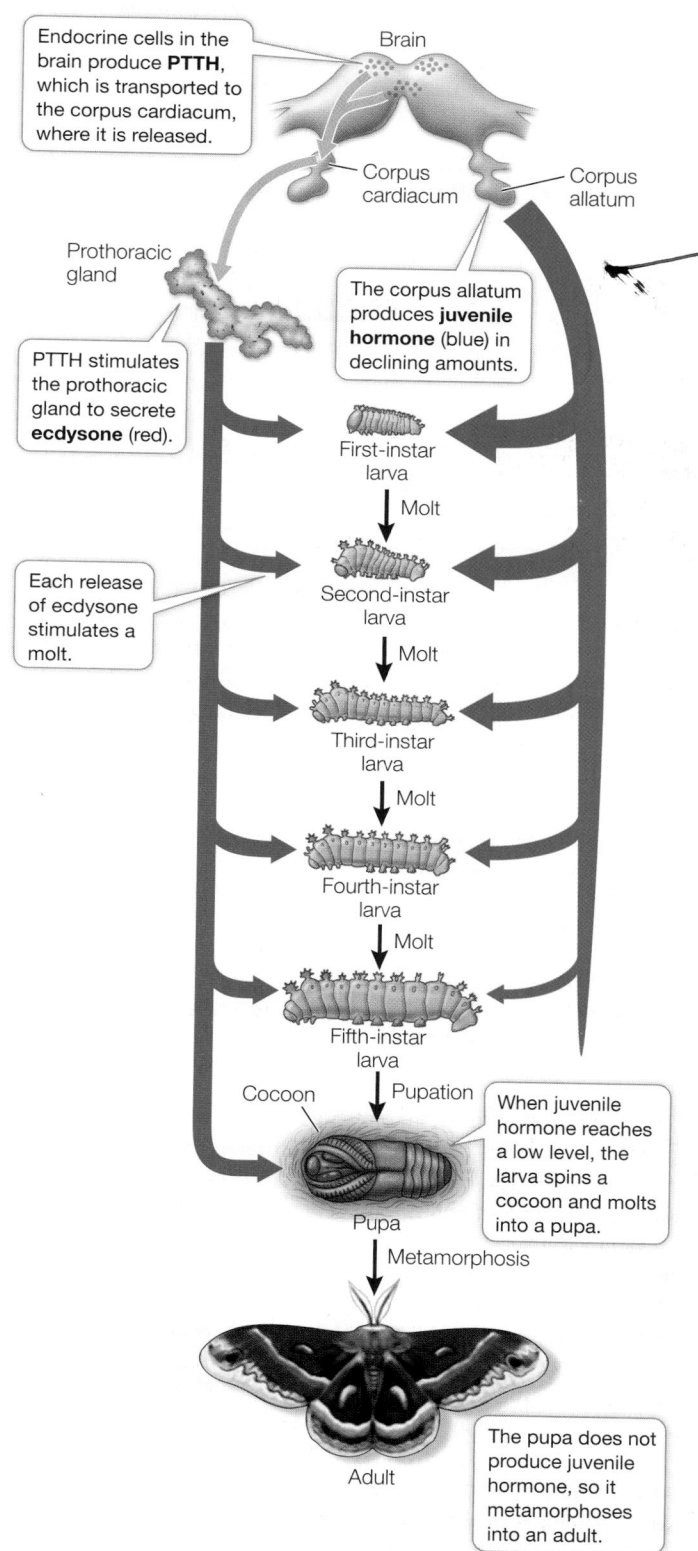

Endocrine cells in the brain produce **PTTH**, which is transported to the corpus cardiacum, where it is released.

Brain

Corpus cardiacum

Corpus allatum

Prothoracic gland

The corpus allatum produces **juvenile hormone** (blue) in declining amounts.

PTTH stimulates the prothoracic gland to secrete **ecdysone** (red).

First-instar larva

Molt

Each release of ecdysone stimulates a molt.

Second-instar larva

Molt

Third-instar larva

Molt

Fourth-instar larva

Molt

Fifth-instar larva

Cocoon   Pupation

When juvenile hormone reaches a low level, the larva spins a cocoon and molts into a pupa.

Pupa

Metamorphosis

The pupa does not produce juvenile hormone, so it metamorphoses into an adult.

Adult

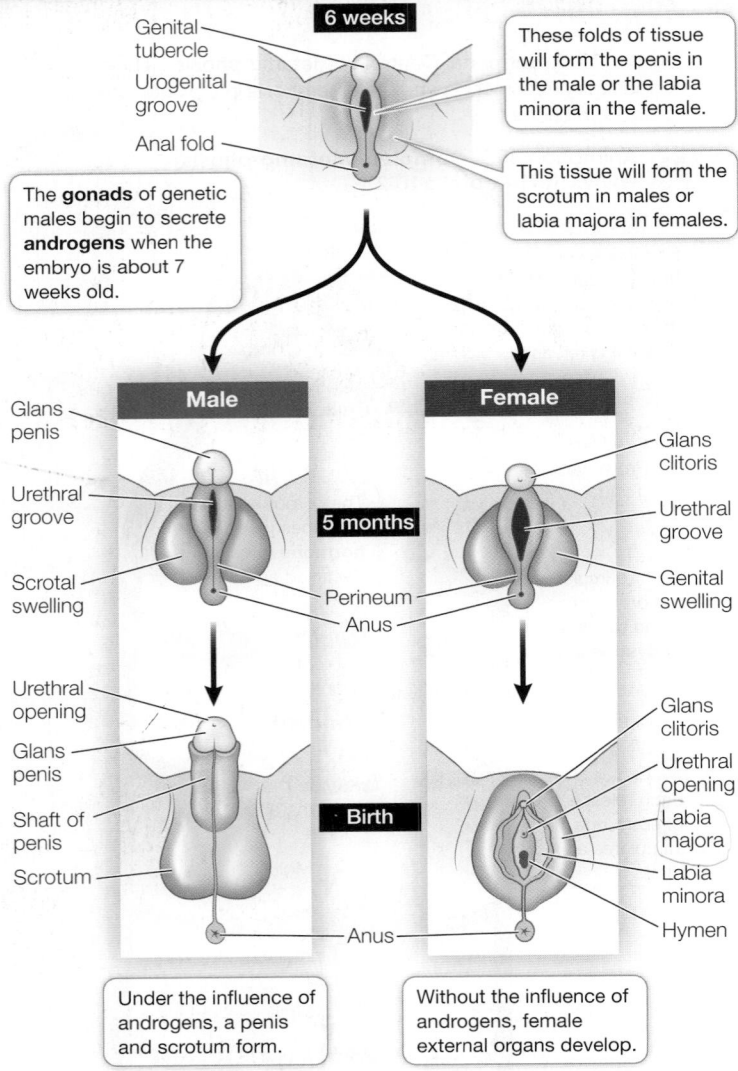

6 weeks

Genital tubercle

Urogenital groove

Anal fold

These folds of tissue will form the penis in the male or the labia minora in the female.

This tissue will form the scrotum in males or labia majora in females.

The **gonads** of genetic males begin to secrete **androgens** when the embryo is about 7 weeks old.

Male

Glans penis

Urethral groove

Scrotal swelling

5 months

Perineum

Anus

Female

Glans clitoris

Urethral groove

Genital swelling

Urethral opening

Glans penis

Shaft of penis

Scrotum

Birth

Anus

Glans clitoris

Urethral opening

Labia majora

Labia minora

Hymen

Under the influence of androgens, a penis and scrotum form.

Without the influence of androgens, female external organs develop.

Figure 40.9 Sex Steroids Direct the Development of Human Sex Organs The external sex organs of early human embryos are undifferentiated. Androgens promote the development of male sex organs. In the absence of androgens, female sex organs form.

Q: What do you think the effect of a mutation that results in nonfunctional androgen receptors would be?

Other examples are the hormones involved in sex development. The **gonads**—the testes of the male and the ovaries of the female—produce hormones as well as sperm and ova. The male steroid hormones are collectively called **androgens**, and the dominant hormone is testosterone. The female steroids are **estrogens** and **progesterone**. The dominant estrogen is estradiol, which is synthesized from testosterone. Males and females both synthesize testosterone, but females have an enzyme (aromatase) that converts testosterone to estradiol.

 Media Clip 40.1 The Testosterone Factor
www.Life11e.com/mc40.1

PHENOTYPIC SEX DETERMINATION The sex steroids determine whether a mammalian embryo develops into a phenotypic female or

male. In humans, the gonads of an early embryo are undifferentiated. Beginning in about the seventh week of development, the expression of genes on the Y chromosome of an XY individual normally causes the undifferentiated gonads to produce androgens. In response to androgens, the reproductive system develops the male phenotype. If no Y chromosome is present (i.e., the individual is genotype XX), androgens are not produced at this time and female structures develop (**Figure 40.09**). After birth, the sex steroids control the maturation of the reproductive organs and the development and maintenance of secondary sexual characteristics, such as breasts and facial hair.

Phenotypic sex determination provides a good example of the importance of receptors in hormonal control mechanisms. Individuals are occasionally born who because of a mutation do not have functional androgen receptors—a condition called androgen insensitivity syndrome. These individuals have an XY genotype and internal testes, but they develop the external body form and genitalia of a female.

PUBERTY In juvenile humans, the sex steroids are produced at low levels. At puberty (around age 12 or 13), sex steroid production increases rapidly. Why does this sudden increase occur? In both juvenile and adult humans, the activities of the gonads are controlled by the anterior pituitary tropic hormones **luteinizing hormone** (**LH**) and **follicle-stimulating hormone** (**FSH**), which together are the **gonadotropins**. The production of gonadotropins by the anterior pituitary is under the control of gonadotropin-releasing hormone (GnRH)

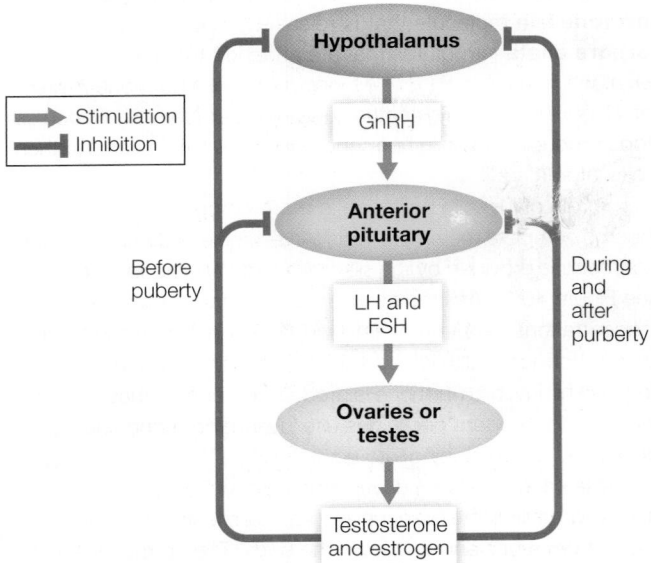

Stimulation

Inhibition

Hypothalamus

GnRH

Anterior pituitary

Before puberty

LH and FSH

During and after purberty

Ovaries or testes

Testosterone and estrogen

Figure 40.10 The Trigger for Puberty In a prepubertal individual, the gonads produce sex steroids in small quantities, and because the hypothalamus is highly sensitive to the negative feedback effects of those sex steroids, GnRH production is suppressed. A sudden decrease in the sensitivity of the hypothalamus to this feedback causes increased production of GnRH and the onset of puberty.

produced by the hypothalamus (**Figure 40.10**). The production of GnRH is controlled by negative feedback from circulating sex steroids. Before puberty, the hypothalamus is very sensitive to that negative feedback signal, even though the circulating levels of sex steroids are low, and only low levels of GnRH are released. Puberty is initiated by a reduction in the sensitivity of hypothalamic GnRH-producing cells to negative feedback. As a result, GnRH production increases, stimulating increased production of gonadotropins and hence increased production of sex steroids.

In females, increasing levels of LH and FSH at puberty stimulate the ovaries to increase their production of the female sex hormones. The increased circulating levels of these hormones stimulate the devel-opment of the traits of a sexually mature woman: enlarged breasts, vagina, and uterus; broadened hips; increased subcutaneous fat; pubic hair; onset of menstrual and ovarian cycles (see Figure 42.11), and behavioral changes. In males, an increasing level of LH stimulates groups of cells in the testes to increase their synthesis of testosterone, which in turn initiates the physiological, anatomical, and behavioral changes associated with adolescence. The voice deepens, hair begins to grow on the face and body, and the testes and penis grow larger. Testosterone also stimulates bone and skeletal muscle growth. FSH in males stimulates production of sperm.

The roles that sex steroids play in adult sexual behavior and reproduction will be described in Chapter 42.

40.3 recap

Hormones play major roles in development. Because they circulate throughout the body, they can simultaneously affect many cells and tissues, but in different ways depending on the receptors and intracellular pathways in those cells. Thus hormones can coordinate the complex sets of responses involved in development. Arthropod development involves complete or incomplete metamorphosis controlled by ecdysone, juvenile hormone, and prothoracicotropic hormone (PTTH). Human gonadal and phenotypic sex development is controlled by steroid hormones.

learning outcomes

You should be able to:

- Analyze observations made during experiments of arthropod development.
- Explain how an individual can have a male genotype but female body characteristics and genitalia.
- Explain the role of negative feedback in the timing of puberty in humans.

1. Why did decapitation of *Rhodnius* prevent molting when done 1 hour after feeding but not when done 1 week after feeding?

2. How can a person with an XY genotype develop with the phenotype of a female?

3. What is the role of negative feedback in the timing of puberty?

A common feature of the roles of hormones in development is that they coordinate the execution of events programmed into the genome. These functions have evolved to be rather stable in the face of changes in the internal and external environments. In contrast, hormones involved in maintaining homeostasis of the internal environment have evolved to be very sensitive to changes in the internal or external environment that could compromise homeostasis. We will look at some of these hormonal regulatory mechanisms in the next section.

key concept 40.4 Hormones Regulate Metabolism and the Internal Environment

Hormones participate in regulation in all physiological systems. In this section we will examine a few major examples of hormonal action that are critical in regulating the internal environment. For simplicity we will focus on humans, but these systems are similar in all mammals.

focus your learning

- The thyroid gland secretes the hormone thyroxine, which has multiple effects on metabolism, and calcitonin, which reduces blood calcium levels.
- Calcitriol increases blood calcium.
- Clusters of endocrine cells called islets of Langerhans in the pancreas secrete the hormones insulin, glucagon, and somatostatin.
- The adrenal gland consists of the adrenal medulla, which secretes epinephrine and norepinephrine, and the adrenal cortex, which secretes steroid hormones.

Thyroxine stimulates many metabolic processes

Thyroxine is produced and secreted by the **thyroid gland**. This gland wraps around the front of the windpipe (trachea) and expands into a lobe on either side (see Figures 40.5 and 40.12). There are two cell types in the thyroid gland, each of which produces a specific hormone. **Thyroxine** is produced by epithelial cells that make up round, colloid-containing structures called follicles (**Figure 40.11A**). **Calcitonin** is produced by cells in the spaces between the follicles and is involved in blood calcium regulation (which we will describe shortly).

Thyroxine, a crucial signal in the regulation of cellular energy metabolism, begins as the glycoprotein thyroglobulin, which is synthesized by the follicle cells and packaged in secretory vesicles. The follicle cells actively take up iodide from the blood and move it into the lumen of the follicle (**Figure 40.11B**). Each thyroglobulin molecule contains about 100 tyrosine units. When the secretory vesicles release thyroglobulin into the lumen of the follicle, they also release an enzyme that catalyzes the iodination of the tyrosine units in the thyroglobulin. When the thyroid gland is stimulated to release thyroxine, the follicle cells take up thyroglobulin from the follicle by endocytosis. These bits of thyroglobulin are then cleaved to form smaller molecules consisting of only two tyrosine units, and these molecules leave the follicle cells and enter the blood. If these molecules are iodinated at

(A) Cross section of a thyroid gland

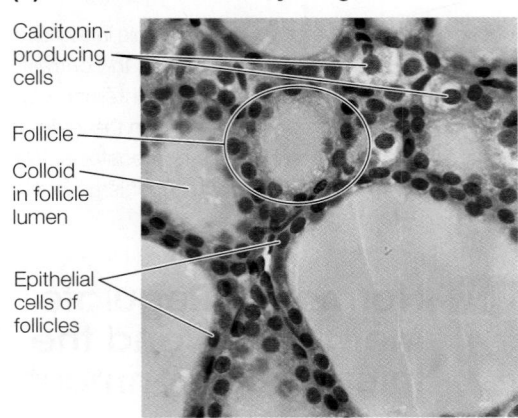

Calcitonin-producing cells

Follicle

Colloid in follicle lumen

Epithelial cells of follicles

Figure 40.11 The Thyroid Gland Consists of Many Follicles
(A) Cross section through a thyroid gland, showing numerous follicles bounded by epithelial cells. Calcitonin-secreting cells are located in the spaces between the follicles.
(B) The epithelial cells of the follicle synthesize thyroglobulin and secrete it into the lumen of the follicle, where it is iodinated and stored until it is processed by the epithelial cells to generate T_3 and T_4.

(B) Production of T_3 and T_4 hormones by the thyroid gland

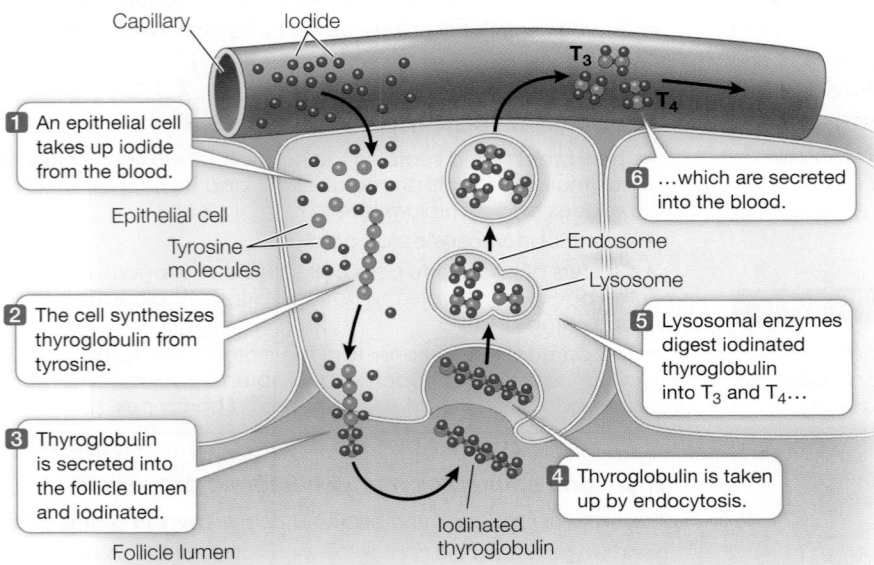

Capillary   Iodide

T_3
T_4

1 An epithelial cell takes up iodide from the blood.

6 ...which are secreted into the blood.

Epithelial cell

Tyrosine molecules

Endosome

Lysosome

2 The cell synthesizes thyroglobulin from tyrosine.

5 Lysosomal enzymes digest iodinated thyroglobulin into T_3 and T_4...

3 Thyroglobulin is secreted into the follicle lumen and iodinated.

4 Thyroglobulin is taken up by endocytosis.

Iodinated thyroglobulin

Follicle lumen

the maximum of four sites on the tyrosine units, the hormone is tetraiodothyronine, or T_4:

and if they are iodinated at only three sites, they are triiodothyronine, or T_3:

The thyroid usually releases about ten times as much T_4 as T_3. However, T_3 is a more active hormone than T_4, so when you read about the effects of thyroxine, keep in mind that the actions discussed are primarily those of T_3. The difference in the activities of T_3 and T_4 makes it possible to control the effects of thyroxine in different tissues. Within target cells, T_4 can be converted to T_3 by an enzyme called deiodinase. A different deiodinase can convert T_4 into an inactive hormone called reverse T_3. Deiodinase can also inactivate T_3 by converting it into T_2 or T_1. Thus each target cell can set a unique sensitivity to thyroid hormones using these enzymes to control the conversion of T_4 to T_3 or to reverse T_3.

TSH AND TRH REGULATE THYROXINE PRODUCTION The tropic hormone **thyroid-stimulating hormone** (**TSH**, also known as **thyrotropin**), produced by the anterior pituitary, activates the thyroxine-producing follicle cells in the thyroid. Thyrotropin-releasing hormone (TRH), produced in the hypothalamus and transported to the anterior pituitary through the portal blood vessels, activates the TSH-producing pituitary cells. The hypothalamus uses environmental information, such as temperature or day length, to determine whether to increase or decrease its secretion of TRH. This sequence of steps is regulated by a negative feedback loop (see Figure 40.7B). Circulating thyroxine inhibits the response of pituitary cells to TRH, so less TSH is released when thyroxine levels are high, and more TSH is released when thyroxine levels are low. Circulating thyroxine also exerts negative feedback on the production and release of TRH by the hypothalamus.

Because thyroxine is lipid-soluble, it enters cells readily and binds to receptors in the nucleus. When combined with thyroxine, these receptors (which are found in most cells of the body) stimulate the transcription of numerous genes whose products are transport proteins, structural proteins, and enzymes involved in metabolic pathways; thus thyroxine elevates the metabolic rates of most cells and tissues. Exposure to cold for several days leads to an increased release of thyroxine, an increased conversion of T_4 to T_3, and therefore an increased basal metabolic rate (see Key Concept 39.5).

During development and growth, thyroxine promotes amino acid uptake and protein synthesis. Insufficient thyroxine in a human fetus or growing child greatly retards physical and mental development, resulting in a condition known as cretinism.

GOITER A **goiter** is an enlarged thyroid gland (**Figure 40.12**) that can be associated with either hyperthyroidism (excess production of thyroxine) or hypothyroidism (thyroxine deficiency). The negative feedback loop whereby thyroxine controls TSH release helps explain how two seemingly opposite conditions can result in the same symptom.

1. The most common cause of *hyperthyroid* goiter is Graves' disease, an autoimmune disease involving an antibody to the TSH receptor. This antibody binds to and activates the TSH receptors

Figure 40.12 A Hypothyroid Goiter In this condition, dietary iodide deficiency leads to a lack of functional thyroxine, resulting in the oversynthesis of thyroglobulin and subsequent enlarged follicles.

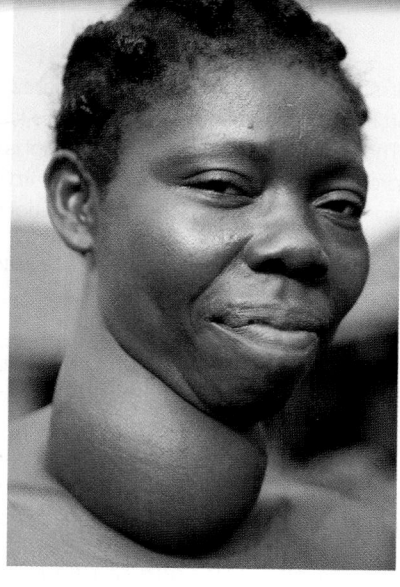

on the follicle cells, causing uncontrolled production and release of thyroxine. Blood levels of TSH are low due to negative feedback from high levels of thyroxine, but the thyroid remains maximally stimulated and grows bigger. People with hyperthyroidism have high metabolic rates, usually feel hot, and may develop a buildup of fat behind the eyeballs that causes their eyes to bulge.

2. *Hypothyroid* goiter results when there is not enough circulating thyroxine to turn off TSH production. The most common cause is a deficiency of dietary iodide, without which the follicle cells cannot make thyroxine. Without sufficient thyroxine, TSH levels remain high and the thyroid continues to produce large amounts of thyroglobulin. Because sufficient iodine is not available, however, the thyroglobulin is poorly iodinated. When it is broken down by the follicle cells, it produces little functional T_3 or T_4. TSH levels remain high and stimulate more and more synthesis of thyroglobulin, and the thyroid gets bigger. The symptoms of hypothyroidism are low metabolism, intolerance of cold, and general physical and mental sluggishness.

Goiter affects about 5 percent of the world's population. The addition of iodide to table salt has greatly reduced the incidence of hypothyroid goiter in industrialized nations, but the condition is still common in other parts of the world and is a leading cause of intellectual impairment.

Three hormones regulate blood calcium concentrations

The regulation of calcium concentration in the blood is crucial, and shifts in blood calcium concentration above or below a narrow range can cause serious problems. When blood calcium falls below this range, the nervous system becomes overly excited, resulting in muscle spasms and even seizures. When blood calcium rises above this range, the nervous system becomes depressed and muscles— including the heart—weaken. Regulation of blood calcium is difficult because only about 0.1 percent of the calcium

in the body is located in the extracellular fluid. About 1 percent is in cells, and almost 99 percent is in the bones. Therefore the body must maintain a tiny pool of calcium in the blood at a precise concentration, and that tiny pool can be influenced greatly by relatively small shifts in the much larger pools of calcium in the cells and bones.

The body has multiple mechanisms for changing blood calcium levels, including:

- Deposition or absorption of bone
- Excretion or retention of calcium by the kidneys
- Absorption of calcium from the digestive tract

These mechanisms are controlled by three hormones: calcitonin, parathyroid hormone, and calcitriol (synthesized from vitamin D).

 Animation 40.3 Hormonal Regulation of Calcium www.Life11e.com/a40.3

CALCITONIN REDUCES BLOOD CALCIUM **Calcitonin** is released by the thyroid and lowers the concentration of calcium in the blood, mainly by regulating bone turnover (**Figure 40.13**). Bone is continuously remodeled through a dynamic process that involves both resorption of old bone and synthesis of new bone, as we will discuss in Key

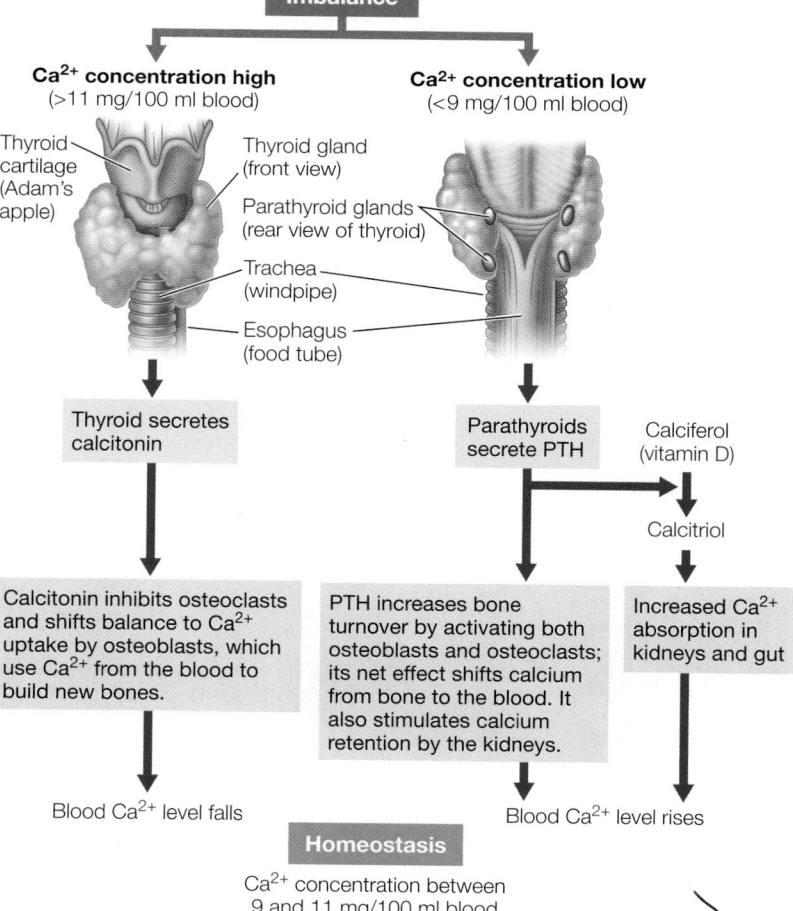

Figure 40.13 Hormonal Regulation of Calcium Calcitonin, parathyroid hormone (PTH), and calcitriol (the active form of calciferol, or vitamin D) regulate Ca^{2+} levels in the blood.

Concept 47.3. Cells called osteoclasts break down bone and release calcium into the blood, and cells called osteoblasts take up calcium from the blood and deposit it in new bone. Calcitonin decreases the activity of osteoclasts and thereby favors removal of calcium from the blood and its deposition in bone by osteoblasts. The turnover of bone in adult humans is not very high, so calcitonin does not play a major role in calcium homeostasis in adults. It is probably more important in young individuals whose bones are actively growing.

PARATHYROID HORMONE INCREASES BLOOD CALCIUM   The **parathyroid glands** are four tiny structures embedded in the posterior surface of the thyroid gland (see Figure 40.13). Their single hormone product, **parathyroid hormone** (**PTH**, also called parathormone), is the most important hormone in the regulation of blood calcium levels. Circulating calcium activates receptors in the cell membrane of the parathyroid cells. When these receptors are active, they inhibit the synthesis and release of PTH. A fall in blood calcium removes this inhibition and triggers the synthesis and release of PTH. PTH stimulates bone turnover by actions on both osteoclasts and osteoblasts. The end result of these actions of PTH is a net increase of calcium in the blood. PTH also maintains blood concentration of calcium by stimulating the kidneys to reabsorb it rather than excrete it in the urine.

CALCITRIOL INCREASES BLOOD CALCIUM   It had long been known that fragile bones were common among people living at high latitudes, where winter days are short and the winter diet often lacks fish, dairy products, and fresh vegetables. Since the condition could be reversed by taking cod-liver oil, it was assumed that a vitamin deficiency was involved. That vitamin was named vitamin D, but when its chemical identity was established, this molecule turned out not to be a vitamin at all.

A vitamin is a substance that the body requires in small quantities but cannot synthesize for itself and must therefore obtain from food (or from supplements such as vitamin pills). However, vitamin D—now more accurately named calciferol—is synthesized naturally from cholesterol when skin cells receive ultraviolet light. Calciferol is not an active hormone, but through actions of the liver and kidneys

it is converted into the active form called **calcitriol**, which circulates in the blood and acts on distant cells (and therefore *is* a hormone). The conversion of calciferol to calcitriol is activated by PTH. Calcitriol promotes the absorption of calcium from food in the gut. Thus the combined actions of PTH and calcitriol raise blood calcium levels.

PTH lowers blood phosphate levels

Bones are made of phosphate as well as calcium. When PTH stimulates the release of calcium from bone, it also releases phosphate. Normal blood concentrations of calcium and phosphate are just below the levels at which they precipitate out of solution as calcium phosphate salts. Even a small rise in concentrations may cause precipitation that leads to maladies such as kidney stones and calcium deposits in the arteries (hardening of the arteries). To reduce this risk, PTH acts on the kidneys to increase the elimination of phosphate via the urine.

Insulin and glucagon regulate blood glucose concentrations

Before the 1920s, the disease diabetes mellitus was fatal. Characterized by weakness, lethargy, and a dramatic loss of body mass, this condition was known to be connected somehow with the pancreas—a large gland located just below the stomach (see Figure 40.5)—and with abnormal glucose metabolism. The exact links, however, were not clear.

Today we know there are two forms of diabetes mellitus. The form that was mostly seen prior to the 1920s usually occurred in young people and was called juvenile diabetes. We now call it type I diabetes, and it is caused by a lack of the protein hormone **insulin**. Glucose enters cells by diffusion, but cell membranes are not very permeable to glucose. Glucose transporter proteins in cell membranes facilitate the movement of glucose into cells, and the glucose transporters called GLUT4 that are most common in muscle and adipose tissue are controlled by insulin. When insulin binds to its receptor on the cell membrane, it causes these glucose transporters to move from cytoplasmic vesicles to the cell membrane, thus making the cell more permeable to glucose (**Figure 40.14**). When insulin is not present, these transporters are returned to the cytoplasmic pool through endocytosis.

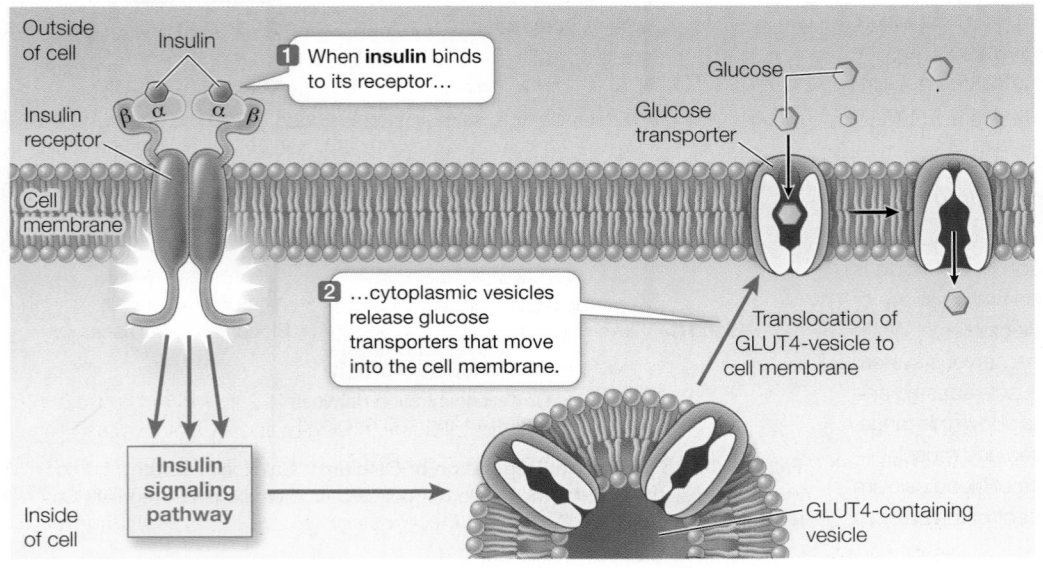

Figure 40.14 Glucose Transporters Are Controlled by Insulin   Most tissues of the body depend on insulin to promote the uptake of glucose into cells. Glucose enters cells by diffusion, but that diffusion is facilitated by passive glucose transporters. When insulin binds to receptors on the cell surface, it stimulates the transfer of one type of glucose transporter (GLUT4) from intracellular sequestration to the cell membranes. When insulin levels fall, the glucose transporters are again sequestered intracellularly.

experiment

Figure 40.15A   Effects of Exercise on Glucose Metabolism

Original Paper: Zhang, Y. et al. 2014. Irisin stimulates browning of white adipocytes through mitogen-activated protein kinase p38 MAP kinase and ERK MAP kinase signaling. *Diabetes* 63: 514–525.

Exercise has been shown to have benefits for individuals with type II diabetes and for obese individuals. Imaging studies have revealed an inverse correlation between brown fat abundance and body mass—lean people have more brown fat and obese people have less. Given that irisin has been shown to have a "browning" effect on white fat, making it more metabolically active, might irisin mediate the beneficial effects of exercise on glucose metabolism and body mass?

HYPOTHESIS▶   Irisin can mediate exercise-induced improvements in glucose metabolism.

METHOD

1. For 2 weeks, treat mice with daily injections of irisin (experimentals) to simulate effects of exercise. Treat control mice with injections of saline.

2. After 2 weeks, sacrifice mice and collect fat tissues to analyze for expression of genes for uncoupling protein (UCP1, an indication of irisin's "browning" effect) and betatrophin (βt, a peptide that stimulates proliferation of insulin-secreting cells in the pancreas).

3. Determine the average expression for the two genes under control and experimental conditions. Determine the standard deviations. Do *t*-tests to determine if the treatment (irisin injections) had a significant effect on the expression of genes for UCP1 and for βt.

RESULTS

The table shows gene expression (in arbitrary units) in white adipose tissue (simulated data based on published means and SDs). The bar graph shows the mean expression levels of the two genes ± 1 SD.

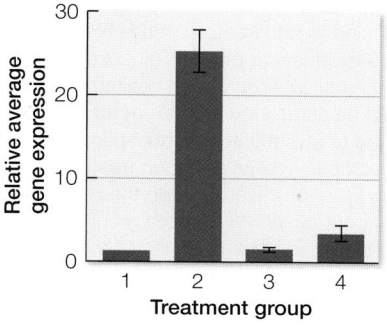

CONCLUSIONS▶

1. The treatment with irisin resulted in a significant increase ($P < 0.001$) in expression of UCP1 in the white fat, demonstrating the known effect of irisin.

2. The treatment with irisin also resulted in a significant increase in the expression of βt ($P < 0.001$), which could stimulate insulin production in the pancreas.

Gene	Treatment group	Relative gene expression of individual								Mean	SD
		1	2	3	4	5	6	7	8		
UCP1	Control (1)	1.42	1.42	1.40	1.42	1.38	1.41	1.40	1.41	1.41	0.01
UCP1	Experimental (2)	23.46	21.68	28.2	26.36	25.77	22.99	28.72	26.13	25.41	2.50
βt	Control (3)	1.36	1.30	1.33	1.40	2.01	1.43	1.46	1.55	1.48	0.23
βt	Experimental (4)	3.62	3.56	3.40	3.03	3.56	4.65	3.41	1.83	3.38	0.78

Figure 40.15B **work with the data** follows on next page.

The other form of diabetes, which is seen in epidemic proportions in the United States today, is called adult onset or type II diabetes. Type II diabetes is associated with obesity, high carbohydrate consumption, and lack of exercise. It is caused either by low production of insulin or by insensitivity to insulin. More than 90 percent of diabetes cases seen in the United States today are type II diabetes, and the numbers are rising alarmingly. Statistics from the Centers for Disease Control and Prevention show that the number of new cases of type II diabetes reported in the United States in 1991 was less than 600,000 and had been fairly stable since 1980. Between 1991 and 2012, the annual number of new cases tripled. Diabetes is currently the seventh leading cause of death in the United States, and it is a common contributor to blindness, kidney failure, stroke, heart disease, and amputations resulting from ulcerations. This disease is largely preventable and even reversible with lifestyle changes.

In both type I and type II diabetes, glucose entry into cells is impaired, resulting in so much glucose accumulating in the blood that it starts to spill over into the urine. A high concentration of glucose in the blood increases urine output by two mechanisms. First, it causes water to move from cells into the blood by osmosis, and this increase in blood volume results in increased urine production. Second, the increased glucose in the tubules of the kidneys pulls

more water into the urine by osmosis. Diabetic individuals thus lack metabolic fuel, and they can also become dehydrated. Because glucose uptake by muscle and adipose tissue is impaired in the absence of insulin, muscle cells must depend on fat and protein for fuel and adipose tissue cannot replenish its stores of triglycerides. If the condition is not treated, the body wastes away.

For centuries, the prospects for people suffering with type I diabetes were bleak. A change came almost overnight in 1921, when the physician Frederick Banting and a medical student, Charles Best, at the University of Toronto, discovered they could reduce the symptoms of diabetes by injecting an extract prepared from pancreatic tissue. The active component of this extract was found to be insulin, a small protein consisting of just 51 amino acids. In the United States today, insulin replacement therapy using manufactured insulin allows more than 1.5 million people with type I diabetes to lead almost normal lives.

Much evidence now shows that type II diabetes can be treated with a combination of dietary changes and weight loss, and **Figure 40.15** looks at the intriguing question of whether the hormone irisin (introduced in the opening of this chapter) can mediate the beneficial effects of exercise on glucose metabolism in type II diabetes patients.

work with the data

Figure 40.15B Effects of Exercise on Glucose Metabolism

Original Paper: Zhang et al. 2014.

Figure 40.15A showed that elevated irisin levels could stimulate pancreatic function in ways that could benefit individuals with type II diabetes. Betatrophin had been shown to stimulate proliferation of insulin-producing cells in the pancreas, increase release of insulin, and improve the ability of individuals to process glucose from dietary sources. The question then arose whether these health benefits associated with betatrophin could be stimulated by irisin treatment. Additional experiments were done to investigate this question. Researchers kept mice on a high-fat diet for 10 weeks and then treated half of the mice with daily injections of irisin for 2 weeks while the other half received daily saline injections (control). $N = 8$ for both groups. The table shows body mass before and after the two-week injection treatment, fasting insulin levels, and blood glucose levels. All values are means ± 1 standard deviation. Use an unpaired t-test to compare the body mass and fasting insulin of the two groups.

Group	Body mass (g) before	Body mass (g) after	Fasting insulin (ng/mL)	Blood glucose (mg/dL)
Control	40.1 ± 4.4	40.9 ± 4.8	4.3 ± 1.0	150 ± 5
Irisin treated	43.7 ± 2.7	40.2 ± 2.8	2.1 ± 0.5	130 ± 7

QUESTIONS▶

1. In the first experiment, why was it important to measure a known marker of brown fat function, namely UCP1, in the same experiment in which the expression of a new candidate gene, namely βt was being measured as a response to irisin treatment?

2. In the second experiment, why were the mice first placed on a high-fat diet?

3. Type II diabetes is frequently associated with obesity and is characterized by insulin resistance. As a result, insulin levels can be high while glucose metabolism is still impaired. What evidence was produced supporting the hypothesis that raised irisin levels could benefit obese individuals or individuals with type II diabetes?

A similar **work with the data** exercise may be assigned in **LaunchPad**.

ISLETS OF LANGERHANS Insulin is produced in clusters of endocrine cells in the pancreas. These clusters are called **islets of Langerhans** after the German medical student who discovered them. They contain three types of cells, each of which produces a specific hormone:

1. Beta (β) cells produce and secrete insulin.
2. Alpha (α) cells produce and secrete **glucagon**, a hormone that has effects mostly opposite from those of insulin.
3. Delta (δ) cells produce the hormone **somatostatin**.

The rest of the pancreas is made up of exocrine tissue, which produces enzymes and other secretions that travel through ducts to the gut, where they participate in digestion.

After a meal, the concentration of glucose in the blood rises, stimulating the β cells of the islets to release insulin. Insulin causes target cells throughout the body to use circulating glucose as fuel and convert it into storage products such as glycogen and fat. When the gut is empty of food, blood glucose concentration falls and the islets stop releasing insulin. As a result, most cells shift to using glycogen and fat rather than glucose as fuel. If blood glucose concentration falls substantially below normal, the islet α cells release glucagon, which stimulates the liver to break down stored glycogen and release glucose into the blood. We will discuss these actions in greater detail in Key Concept 50.4.

SOMATOSTATIN Somatostatin is released from the δ cells of the pancreas in response to rapid increases of glucose and amino acids in the blood. This hormone has paracrine functions within the islets, where it inhibits the release of both insulin and glucagon. Outside the pancreas it acts as a hormone, slowing the digestive activities of the gut and extending the period during which nutrients are absorbed. Somatostatin is also produced in very small amounts by cells in the hypothalamus. Hypothalamic somatostatin, also called **growth**

hormone inhibiting hormone (**GHIH**), is transported in the portal blood vessels to the anterior pituitary, where it inhibits the release of growth hormone and thyrotropin.

The adrenal gland is two glands in one

An **adrenal gland** sits above each kidney, just below the middle of your back. Functionally and anatomically, each adrenal gland consists of a gland within a gland (**Figure 40.16**). The core, or **adrenal**

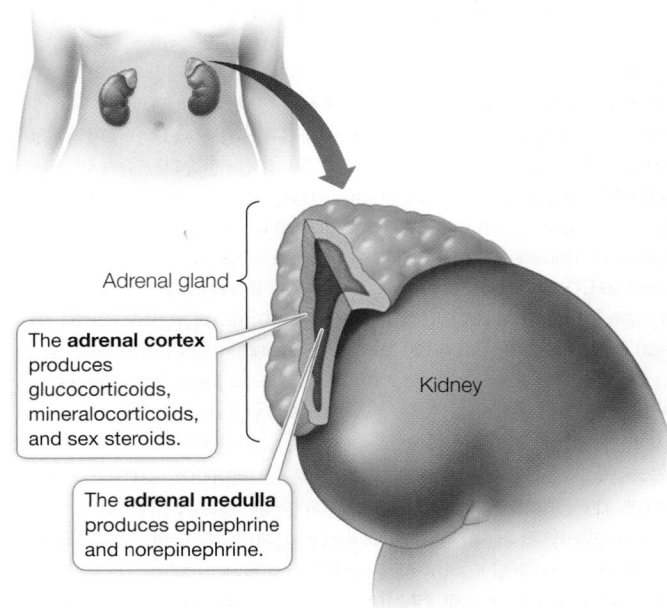

Adrenal gland

The **adrenal cortex** produces glucocorticoids, mineralocorticoids, and sex steroids.

The **adrenal medulla** produces epinephrine and norepinephrine.

Kidney

Figure 40.16 The Adrenal Is Really Two Glands An adrenal gland, consisting of an outer cortex and an inner medulla, sits above each kidney. The medulla and the cortex produce different hormones.

medulla, produces epinephrine and, to a lesser degree, norepinephrine. The medulla develops from nervous tissue and is under the control of the nervous system. Surrounding the medulla is the **adrenal cortex**, which produces steroid hormones. The cortex is under hormonal control, largely by a tropic hormone produced by cells in the anterior pituitary.

THE ADRENAL CORTEX The cells of the adrenal cortex use cholesterol to produce three classes of steroid hormones (see Figure 40.2B), collectively called **corticosteroids**:

1. *Mineralocorticoids* influence the salt and water balance of the extracellular fluid.

2. *Glucocorticoids* influence blood glucose concentrations as well as other aspects of fat, protein, and carbohydrate metabolism.

3. *Sex steroids* play roles in sexual development, sexual behavior, and anabolism (tissue building).

In adult humans, the adrenal cortex secretes only negligible amounts of sex steroids. The major producers of sex steroids are the gonads, as we saw in Key Concept 40.3.

Aldosterone, the primary mineralocorticoid, stimulates the kidneys to conserve sodium and excrete potassium, as we will discuss in Chapter 51. If the adrenal glands are removed from an animal, sodium must be added to its diet, or its sodium will be depleted and it will die.

The main glucocorticoid in humans is **cortisol**, which is critical for mediating the body's metabolic responses to stress. Within minutes of a stressful stimulus (one provoking fear or anger, for example), blood cortisol levels begin to rise. This response is much slower than the neurally mediated epinephrine and norepinephrine response to stress, but it lasts longer. Whereas epinephrine and norepinephrine cause sudden increases in blood pressure, cortisol causes a sustained increase in blood pressure. Cortisol stimulates tissues that are not critical for escaping the danger or threat to switch from using blood glucose for energy to using fats and proteins. This action preserves circulating glucose to fuel the actions of the nervous system and skeletal muscles necessary to meet the threat. Cortisol also raises blood pressure, to increase delivery of oxygen to the muscles to prepare them for flight or fight, and inhibits the immune system. Dealing with the immediate stressor is more important than feeling sick, having allergic reactions, or healing wounds. This explains why cortisol and drugs that mimic its action are useful for reducing inflammation and allergic responses. Cortisol decreases activity of the gut and reproductive systems—you can digest lunch and reproduce later if you escape the threat. You can see that actions of cortisol are adaptive in responding to acute stress, but they can cause problems if sustained over time, as sometimes occurs in modern society due to social, economic, and job-related stressors.

Cortisol release is controlled by **adrenocorticotropic hormone** (**ACTH**) from the anterior pituitary. ACTH release is controlled in turn by **corticotropin-releasing hormone** (**CRH**) from the hypothalamus (see Figure 40.7) . The action of ACTH on the adrenal cortex is to stimulate the synthesis of cortisol. Like other steroid hormones, cortisol is not stored in vesicles and therefore is available for immediate release. As cortisol or other steroid hormones diffuse into the blood, they combine with carrier proteins, and their release from

these proteins can have a long time course, thus stretching out their actions. Also, many of their actions stimulate gene expression in target cells, which also takes time but has a long-lasting effect.

Turning off the stress responses activated by cortisol is as important as turning them on. A study of stress in rats showed that old rats could turn on these stress responses as effectively as young rats, but they had lost the ability to turn them off as rapidly. As a result, they suffered from the well-known consequences of stress seen in humans: digestive system problems, cardiovascular problems, strokes, impaired immune system function, and increased susceptibility to cancers and other diseases. Acute stress responses are turned off by negative feedback from cortisol on both the ACTH-secreting cells of the anterior pituitary and the CRH-secreting cells of the hypothalamus. With chronic or prolonged stress, these control mechanisms become insufficient and cortisol must exert negative feedback through another brain region, the hippocampus. Prolonged exposure to cortisol, however, causes trauma to and loss of hippocampal cells, resulting in the decreased ability to turn off the stress response.

Loss of adrenal cortical function results in Addison's disease, which is characterized by fatigue, muscle weakness, digestive problems, low blood sodium (hyponatremia), low blood pressure, and salt hunger. Without replacement hormone treatment, death is likely. An interesting symptom of Addison's disease is darkening of many skin areas that normally are not exposed to the sun—for instance, the lining of the cheeks. You can understand the cause of this symptom by thinking about the source of ACTH and the control of its secretion. In the hypothalamic–pituitary–endocrine gland axis (see Figure 40.7), an end hormone (in this case cortisol) frequently serves as a negative feedback signal to the pituitary and also the hypothalamus. If the hormones of the adrenal cortex are absent, the negative feedback is also gone, and the system will call for more ACTH. The production of ACTH by pituitary cells involves the expression of a polypeptide called proopiomelanocortin which is then cleaved to produce multiple peptide hormones. These include ACTH, MSH, endorphins, and enkephalins. The excess production of MSH is the cause of the skin darkening in Addison's disease.

THE ADRENAL MEDULLA Stress causes the adrenal medulla to release epinephrine and norepinephrine that arouse the body to action. As we saw earlier in this chapter, epinephrine and norepinephrine increase heart rate and blood pressure and divert blood flow to active muscles and away from the gut and skin. Only about one-fifth as much norepinephrine is secreted from the adrenal medulla as epinephrine, but it has similar functions. Norepinephrine is also used by the nervous system as a neurotransmitter, and it is involved in controlling many physiological processes.

Epinephrine and norepinephrine are both water-soluble, and both bind to the same set of receptors on the surfaces of target cells. These **adrenergic receptors** are of two general types, α-adrenergic and β-adrenergic. The α-adrenergic receptors respond more strongly to norepinephrine than to epinephrine, whereas β-adrenergic receptors respond about equally to both epinephrine and norepinephrine. Because of this difference in receptor affinities, it is possible for drugs to blunt the flight-or-fight responses without disrupting physiological regulatory processes. Such drugs are called "beta blockers" because, by inhibiting β-adrenergic receptors, they can reduce the fight-or-flight response to epinephrine without disrupting the physiological regulatory functions of norepinephrine mediated through the α-adrenergic receptors. Beta blockers are commonly prescribed to reduce symptoms of anxiety such as dry mouth and elevated heart rate (palpitations).

Many chemicals may act as hormones

We have discussed the major mammalian endocrine glands and their hormones in this chapter, but many more hormones exist. As we discuss the organ systems of the body in the chapters that follow, we will frequently describe hormones that their tissues produce as well as hormones that control their functions.

The pineal gland has a daily cycle of melatonin release

Humans like most vertebrate species have a **pineal gland** located between the cerebral hemispheres. The pineal has a daily rhythm of production and secretion of the hormone **melatonin**. The functions of the pineal and melatonin have to do with coordinating daily rhythms of physiological functions and, in many species, seasonal changes such as reproduction timing, migration, and preparations for environmental stress such as cold weather.

▶ 40.4 recap

The major endocrine glands of mammals include the hypothalamus, pituitary gland, thyroid gland, parathyroid glands, pancreas, adrenal glands, gonads, and pineal gland. Each of these glands secretes and responds to hormones that play crucial roles in controlling physiology and development. Thyroxine is a lipid-soluble hormone that influences the expression of genes involved in regulating cell metabolism. Calcitonin and parathyroid hormone (PTH) regulate blood calcium levels, mainly by regulating the remodeling of bone. Blood glucose levels are controlled by insulin when blood glucose levels rise and largely by the lack of insulin when blood glucose levels fall. Glucagon is released in the event of serious hypoglycemia, and it stimulates breakdown of glycogen to raise blood glucose levels. The adrenal cortex produces steroid hormones, the most important of which is cortisol, which is involved in stress responses. The adrenal medulla produces epinephrine and norepinephrine.

learning outcomes

You should be able to:

- Explain the effects of disruptive changes in normal thyroxine levels on humans.
- Describe calcitriol and its effects on calcium in the blood.
- Describe how insulin and glucagon regulate blood sugar levels.
- Apply concepts about hormonal control of stress to explain why prolonged stress can have negative effects on the body.

1. Describe how both hypothyroidism and hyperthyroidism can result in goiter.
2. How does vitamin D contribute to raising blood calcium levels, and why is it not a vitamin?
3. How does insulin control the rate of glucose uptake by cells?
4. Explain how chronic stress can lead to stress-related diseases.

▶ investigating life

 **In what ways could irisin mediate the health benefits of exercise?**

The initial discovery of irisin and its functions revealed that it induces brown fat–like changes in white fat, including expression of the gene that codes for uncoupling protein (UCP1), which enables brown fat cells to metabolize lipids without producing ATP. As a result, metabolism in the brown fat cells is not inhibited by the buildup of ATP or the lack of ADP, and the brown fat can simply metabolize excess energy and generate heat.

Exercise does more than promote weight loss and toned muscles. It also improves cognition and glucose metabolism. Improving glucose metabolism can help reverse the changes associated with type II diabetes—elevated blood glucose, elevated blood insulin, loss of sensitivity of the pancreas to glucose, and loss of sensitivity of body tissues to insulin. Investigating Life: How Could Irisin Mediate the Benefits of Exercise on Cognition? and Figure 40.15 present experiments showing that irisin is a mediator of benefits of exercise on cognition and glucose metabolism. It seems likely that irisin is a major integrating factor in a large suite of healthful benefits of exercise

Future directions

The experiments explored in Investigating Life: How Could Irisin Mediate the Benefits of Exercise on Cognition? and Figure 40.15 suggest the interesting possibility that irisin could be developed as a therapeutic that might be useful in treating the national epidemic of obesity and type II diabetes. It might also prove to be beneficial in cognitive declines due to conditions such as stroke and Alzheimer's disease. Before such treatments can be developed, however, research is needed on such questions as: Are there potential negative side effects of abnormal elevations of irisin? Is there the potential for abuse of irisin, similar to what has been seen with other growth-promoting agents such as growth hormone and testosterone? It would seem that the safest way to benefit from the actions of irisin would be to use the natural approach of regular exercise to promote its normal production and release.

Chapter Summary 40

▶ 40.1 Hormones Circulate Around the Body and Affect Target Cells

- Endocrine cells secrete chemical signals that induce responses in other cells that have receptors for those molecules. In some cases endocrine cells are aggregated into **endocrine glands**.

- **Hormones** are endocrine signals that are secreted from a cell, circulate in the blood, and bind to target cells distant from the secreting cell. Review Figure 40.1

- Hormones fall into three general categories: **protein** and **peptide hormones**, **steroid hormones**, and **amine hormones**. Peptide and protein hormones and some amine hormones are water-soluble; steroid hormones and some amine hormones are lipid-soluble. Review Figure 40.2

- Receptors for water-soluble hormones are located on the cell surface. Receptors for most lipid-soluble hormones are inside the cell.

- The same hormone can cause different responses in different target cells. Review Figure 40.3

- The chemical structures of hormones are highly conserved. Through evolution, however, hormones acquire different functions in different animal groups. Review Figure 40.4

- In humans, the major endocrine glands are distributed around the body. Review Figure 40.5, Activity 40.1

▶ 40.2 The Endocrine System and Nervous System Work Together

- The **pituitary gland** is the interface between the nervous and endocrine systems. The **anterior pituitary** develops from embryonic mouth tissue; the **posterior pituitary** develops from the developing brain. Review Figure 40.6, Focus: Key Figure 40.7

- The posterior pituitary secretes two **neurohormones: antidiuretic hormone (ADH)** and **oxytocin**. The anterior pituitary secretes **tropic hormones** (thyrotropin, adrenocorticotropin, luteinizing hormone, and follicle-stimulating hormone) as well as **growth hormone**, prolactin, **endorphins, enkephalins,** and **melanocyte stimulating hormone.**

- The anterior pituitary is controlled by neurohormones produced by cells in the hypothalamus and transported through **portal blood vessels** to the anterior pituitary. See Animation 40.1

- Hormone release is controlled in part by negative feedback loops.

- The hormone irisin may mediate effects of exercise on the brain. Review Investigating Life: How Could Irisin Mediate the Benefits of Exercise on Cognition?

▶ 40.3 Hormones Play Important Roles in Development

- Pioneering experiments in illustrating hormonal action showed that two hormones, PTTH and ecdysone, control molting in arthropods. A third hormone, **juvenile hormone**, prevents maturation. Review Figure 40.8, Animation 40.2

- Sex hormones (**androgens** in males, **estrogens** and **progesterone** in females) control sexual development, secondary sexual characteristics, and reproductive functions. Review Figures 40.9, 40.10

▶ 40.4 Hormones Regulate Metabolism and the Internal Environment

- The **thyroid gland** is controlled by **thyroid-stimulating hormone** (**TSH**, also called **thyrotropin**) and secretes **thyroxine**, which controls cell metabolism. Review Figure 40.11

- The level of calcium in the blood is regulated by three hormones. **Calcitonin** from the thyroid lowers blood calcium by promoting bone deposition. **Parathyroid hormone** (**PTH**) raises blood calcium by promoting bone turnover and decreasing calcium excretion. **Calcitriol** promotes calcium absorption from the digestive tract. Review Figure 40.13, Animation 40.3

- The pancreas secretes three hormones. **Insulin** stimulates glucose uptake by cells and lowers blood glucose. **Glucagon** raises blood glucose, and **somatostatin** slows the rate of nutrient processing. Review Figure 40.14

- The **adrenal gland** has two portions, one within the other. The inner portion, the **adrenal medulla**, releases epinephrine and norepinephrine in response to stress. The outer portion, the **adrenal cortex**, produces three classes of **corticosteroids**: glucocorticoids, mineralocorticoids, and small amounts of sex steroids. Review Figure 40.16

- **Aldosterone** is a mineralocorticoid that stimulates the kidneys to conserve sodium and excrete potassium. **Cortisol** is a glucocorticoid that is released in response to stressful stimuli but acts more slowly than the hormones of the adrenal medulla.

- The **pineal gland** releases **melatonin**, a hormone involved in controlling biological rhythms.

See Activity 40.2 for a concept review of this chapter.

Go to **LearningCurve** (in **LaunchPad**) for dynamic quizzing that helps you solidify your understanding of this chapter. **LearningCurve** adapts to your responses, giving you the practice you need to master each key concept.

Apply What You've Learned

Review

40.1 Hormones affect target cells that have appropriate receptors.

40.1 Different target cells can respond to the same hormone in different ways depending on the signal transduction pathway stimulated.

Original Paper: Corbett, M. C. and A. B. Richards. 1994. Intraocular adrenaline maintains mydriasis during cataract surgery. *British Journal of Ophthalmology* 78: 95–98.

The opening in the iris forms the pupil of the eye. The iris is composed of two arrangements of smooth muscle fibers. Constrictor muscles surrounding the pupil are arranged in concentric circles, like tree rings. Peripheral to the constrictor fibers are the dilator fibers, which are arranged like the spokes on a bicycle wheel. When the constrictor muscles contract, the pupil becomes very small, and when the dilators contract, the pupil becomes large.

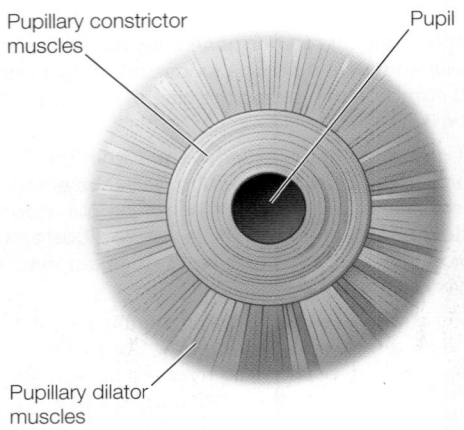

Pupillary constrictor muscles

Pupil

Pupillary dilator muscles

Aging and exposure to UV light can cause the lens of the eye to become cloudy and nontransparent—a condition known as a cataract. To restore sight, a surgeon must remove the cataract through the pupil. To facilitate this surgery it is important to dilate the pupil as much as possible and to keep it dilated as long as possible. Pupillary dilation is facilitated by the sympathetic nervous system and the adrenal medulla releasing epinephrine (adrenaline) and norepinephrine.

Could epinephrine be used to maintain pupil dilation during cataract surgeries? To find out, researchers studied the effects of adding epinephrine to the irrigation fluid injected into the eye during cataract-removal surgery. Twenty-seven patients received irrigation fluid containing epinephrine at a concentration of 1:1,000,000. Forty-three patients received irrigation fluid without epinephrine. The table below shows the percentage of patients with pupil diameter less than 5 mm at various stages during cataract surgery.

Stage of surgery	Without epinephrine (%)	With epinephrine (%)
Start	0	0
After lens removal	16	0
After soft tissue removal	21	7
10 min after surgery	9	4
20 min after surgery	21	4
30 min after surgery	21	9

Questions

1. In this experiment, which part of the iris is being affected by the epinephrine injection? Explain your answer based on the experimental results of the two surgeries.

2. Both parts of the iris are composed of smooth muscle tissue. Propose a hypothesis to explain why the epinephrine produced the observed experimental effect.

3. Epinephrine is a hormone released during the normal fight-or-flight response in mammals. Name some of the other tissue and organ changes that occur during this response, and discuss why they respond so differently to epinephrine than the smooth muscle in the iris does.

4. The drug atropine, a known acetylcholine receptor inhibitor, is also used to dilate the pupil. It affects the pupillary constrictor muscle. What can you conclude about how atropine causes pupillary dilation? What are the implications of this for how the autonomic nervous system controls pupil diameter?

Go to **LaunchPad** for the eBook, LearningCurve, animations, activities, flashcards, and additional resources and assignments.

41

Immunology: Animal Defense Systems

Soldiers being vaccinated against smallpox in the 19th century. Using vaccines continues to be important in protecting an army as it is exposed to pathogens in the environment.

▶ investigating**life**

Vaccines and Immunity

Five tiny vials in deep freezers, three in Bethesda, Maryland, one in Atlanta, Georgia, and one in Siberia, are all that is left of the smallpox virus. Smallpox killed more than 300 million people in the twentieth century, but the disease made its last appearance in 1978, thanks to vaccination. A vaccine is usually an inactive form of a pathogen or toxin that provokes the immune system to produce antibodies: specific proteins directed against the target. The immune system destroys whatever is bound to the antibodies.

The eradication of smallpox was a spectacular international accomplishment. So why hasn't vaccination for other potentially lethal diseases such as the flu (influenza) been widely accepted by the public? More than one-third of Americans refuse flu shots, and a significant number of parents refuse vaccination for their children.

Unfortunately, those who refuse vaccination may harm people other than themselves. A vaccination program can control or eradicate a disease only if a high percentage (typically above 80 percent) of people are vaccinated, thus disrupting the chain of infection from person to person. A very high level of vaccination results in "herd immunity," a state in which even those who cannot be vaccinated or who have weak immune

systems are protected from infection. This protection is lost if the vaccination rate falls below the level needed for herd immunity. Those who are old or sick, and infants whose immune systems have not yet fully developed, are most at risk.

The best way to develop herd immunity in a population is compulsory vaccination. In many countries, vaccination is a prerequisite for school enrollment and military enlistment and is required during epidemics. For example, during the periodic smallpox epidemics during the twentieth century in the United States, doctors accompanied by police would go into neighborhoods where the disease raged, vaccinating all those who were uninfected and removing infected people to quarantine. You can imagine the reaction of parents whose children were taken away, often to die. Henning Jacobson was arrested when he refused vaccination during a smallpox epidemic. He took his case to the U.S. Supreme Court, which ruled in 1905 that while personal freedom is important, each state is entitled to protect its citizens. This provided a legal framework for compulsory vaccination that continues to this day. But opposition and court challenges continue.

Q:A What are the mechanisms and implications of long-lasting immunity?

key concept

41.1 Animals Use Innate and Adaptive Mechanisms for Defense

Animals have several ways of defending themselves against **pathogens**—harmful organisms and viruses that can cause disease. These defense systems are based on the distinction between *self*—the animal's own molecules—and *nonself*, or foreign, molecules.

focus your learning

- The two general types of defense mechanisms are innate defenses and adaptive defenses.
- Many animal groups have Toll-like receptors (TLRs) that participate in innate defense responses.
- All white blood cells originate from multipotent stem cells in the bone marrow.
- In mammals, the major immune system proteins include antibodies, MHC proteins, T cell receptors, and cytokines.

The defensive response involves three phases:

1. *Recognition phase.* The organism must be able to discriminate between self and nonself.

2. *Activation phase.* The recognition event leads to a mobilization of cells and molecules to fight the invader.

3. *Effector phase.* The mobilized cells and molecules destroy the invader.

There are two general types of defense mechanisms:

1. **Innate defenses**, *or nonspecific defenses, are inherited mechanisms that provide the first line of defense against pathogens*. Innate defenses typically act very rapidly and include barriers such as the skin, molecules that are toxic to invaders, and phagocytic cells that ingest invaders. (Recall from Key Concept 6.5 that phagocytosis is a form of endocytosis, in which a cell engulfs a large particle or another cell.) The innate immune system recognizes broad classes of organisms and molecules, such as viruses, bacteria, fungi, protists, worms, and various toxins.

2. **Adaptive defenses** *are aimed at specific pathogens and are activated by the innate immune system*. For example, the adaptive immune system can make an antibody protein that will recognize, bind to, and aid in the destruction of a specific pathogen, if that specific pathogen ever enters the body. Adaptive defenses are typically slower to develop than the innate defenses and longer-lasting.

Immunity occurs when an organism has sufficient defenses to successfully avoid the effects of biological invasion by a pathogen.

Innate defenses evolved before adaptive defenses

All animals have innate defenses against their enemies. For example, the arthropod *Tachypleus tridentatus*—the Japanese horseshoe crab—first appeared in the fossil record about 400 million years ago. It relies only on innate defenses. These defenses include barriers, defensive cells, and defensive molecules.

- Barriers include physical, chemical, and biological mechanisms for resisting infections. The horseshoe crab has the hard exoskeleton that is characteristic of arthropods. This shell acts to protect the crab from invasion by pathogens.

Barrier: The first line of innate defense

- Cells involved in innate defenses include phagocytes that bind to microbial pathogens, ingest them by endocytosis, and destroy them by hydrolysis. Amebocytes in the blood of the horseshoe crab fulfill this defensive role.

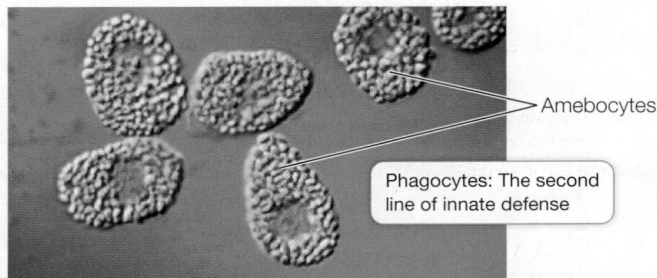

Amebocytes

Phagocytes: The second line of innate defense

- Molecules that are toxic to invading pathogens are important in innate defense. The horseshoe crab has a wide array of such molecules that are released from cells in its blood. These molecules include peptides that disrupt the bacterial cell membrane, rendering it permeable; and peptides that bind to bacterial surfaces and cross-link them.

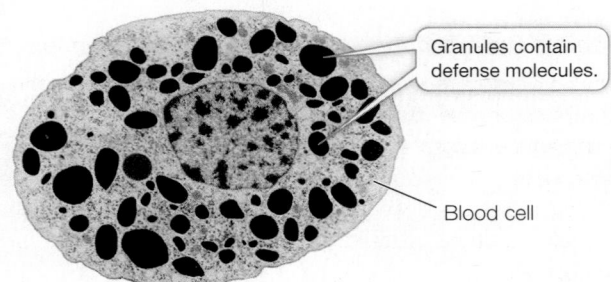

Granules contain defense molecules.

Blood cell

Studies of innate immunity, along with genome sequencing, have revealed that the recognition and activation phases of innate immunity evolved very early in animals. For example, animals as diverse as humans and fruit flies share a class of receptors, called **Toll-like receptors** (**TLRs**), that participate in innate defense responses. These receptors recognize nonself molecules called *pathogen-associated molecular patterns (PAMPs). In vertebrates, each Toll-like receptor recognizes and binds to a specific molecule that is found in a broad class of pathogens, such as a component of the

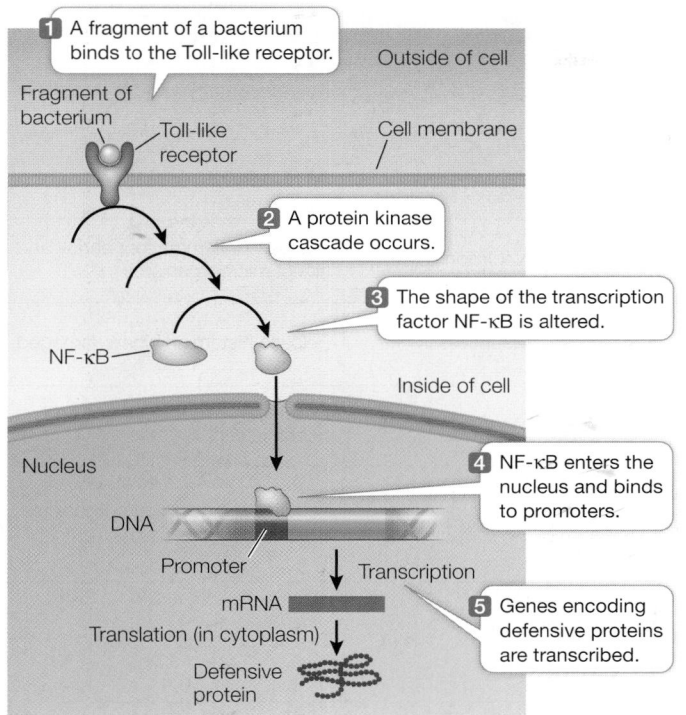

1 A fragment of a bacterium binds to the Toll-like receptor.

Outside of cell

Fragment of bacterium

Toll-like receptor

Cell membrane

2 A protein kinase cascade occurs.

3 The shape of the transcription factor NF-κB is altered.

NF-κB

Inside of cell

Nucleus

4 NF-κB enters the nucleus and binds to promoters.

DNA

Promoter

Transcription

mRNA

5 Genes encoding defensive proteins are transcribed.

Translation (in cytoplasm)

Defensive protein

Figure 41.1 Cell Signaling and Defense   Binding of a pathogenic molecule or fragment to the Toll-like receptor initiates a signal transduction pathway that results in the transcription of genes whose products are involved in adaptive and innate defenses.

bacterial cell wall. Binding triggers a signal transduction pathway that ends with the expression of genes for anti-pathogen molecules (**Figure 41.1**). This pathway exists in some form in many animal groups, including humans.

---

* **connect the concepts**   The recognition of PAMPs is widespread in immunity and is therefore an evolutionarily ancient mechanism. Plants have an immune response to invading pathogens that also involves PAMPs. See Key Concept 38.1.

Mammals have both innate and adaptive defenses

Mammals have both kinds of defense mechanisms, and these mechanisms are the focus of this chapter. In mammals, the innate and adaptive mechanisms operate together as a coordinated defense system. **Table 41.1** gives an overview of these defenses during the course of an infection. Innate immunity is the body's first line of defense because the adaptive defenses often require days or even weeks to become effective.

Blood and lymph tissues play important roles in defense

The components of the mammalian defense system are dispersed throughout the body and interact with almost all of its other tissues and organs. The lymphoid tissues, which include the thymus, bone marrow, spleen, and lymph nodes, are essential parts of the defense system (**Figure 41.2**). The blood and lymph are complex systems with nondefensive functions that will be discussed in Chapter 49. They each have central roles in defense as well.

table 41.1 Innate and Adaptive Immune Responses to an Infection

Response time	System	Mechanisms
Early (0–4 hr)	Innate, nonspecific (first line)	Barrier (skin and lining of organs) Dryness, low pH Mucus Lysozyme, defensins
Middle (> 4–96 hr)	Innate, nonspecific (second line)	Inflammation Phagocytosis Natural killer cells Complement system Interferons
Late (> 96 hr)	Adaptive, specific	Humoral immunity (antibodies from B cells) Cellular immunity (T cells)

The blood and lymph both consist of liquids in which cells are suspended:

- **Blood plasma** is a yellowish solution containing ions, small molecule solutes, and soluble proteins. Suspended in the plasma are red blood cells, white blood cells, and platelets

Activity 41.1 **The Human Defense System** www.Life11e.com/ac41.1

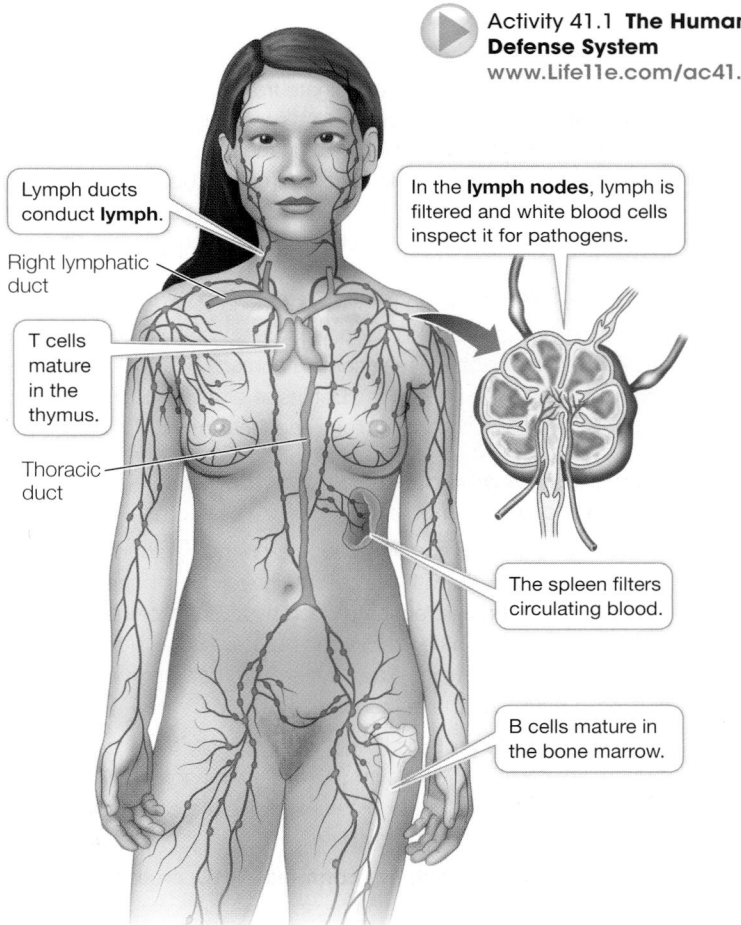

Lymph ducts conduct **lymph**.

Right lymphatic duct

T cells mature in the thymus.

Thoracic duct

In the **lymph nodes**, lymph is filtered and white blood cells inspect it for pathogens.

The spleen filters circulating blood.

B cells mature in the bone marrow.

Figure 41.2 The Human Lymphatic System   A network of ducts and vessels collects lymph from body tissues and carries it toward the heart, where it mixes with blood to be pumped back to the tissues. Other lymphoid tissues, including the thymus, spleen, and bone marrow, are also essential to the body's defense system.

(cell fragments essential to blood clotting). Whereas red blood cells are normally confined to the closed circulatory system (the heart, arteries, capillaries, and veins), white blood cells and platelets are also found in the lymph.

- **Lymph** is a fluid that is derived from the blood (but lacking red blood cells) and other tissues and accumulates in intercellular spaces throughout the body. From these spaces, the lymph moves slowly into the vessels of the lymphatic system. Tiny lymph capillaries conduct this fluid to larger ducts that eventually join together, forming two large vessels, the thoracic duct and right lymphatic duct, which join a major vein (the left subclavian vein) near the heart. By this system of vessels, the lymph is eventually returned to the blood and the circulatory system.

At many sites along the lymph vessels are small, roundish structures called **lymph nodes**, which contain a type of white blood cell called a **lymphocyte**. As lymph passes through a lymph node, the lymphocytes encounter foreign cells and molecules that have entered the body, and if they are recognized as nonself, an immune response is initiated.

White blood cells play many defensive roles

One milliliter of human blood typically contains about *5 billion* red blood cells and *7 million* of the larger **white blood cells** (also called leukocytes). All of these cells originate from multipotent stem cells (constantly dividing undifferentiated cells that can form several different cell types; see Key Concept 19.1) in the bone marrow. There are two major families of white blood cells: lymphocytes and phagocytes (**Figure 41.3**). Lymphocytes include the B cells and T cells; they are smaller than other white blood cells and are not phagocytic. **Phagocytes** include most of the other cells shown in Figure 41.3, and as their name suggests, they are phagocytic. Each kind of white blood cell has specialized functions. Some phagocytes are also referred to collectively as granulocytes because they contain numerous granules (vesicles containing defensive molecules). Defensive proteins and signals play fundamental roles in the interactions and functioning of these cells.

Immune system proteins bind pathogens or signal other cells

The cells that defend mammalian bodies work together, interacting with one another and with the cells of invading pathogens. These cell–cell interactions are accomplished by a variety of key proteins, including receptors, other cell surface proteins, and signaling molecules. Four of the major players are listed here, and will be discussed in more detail later in the chapter.

1. **Antibodies** are proteins that bind specifically to certain substances identified by the immune system as nonself. The specific molecules that bind to antibodies are called **antigens**. Antibodies are produced by B cells.

2. **Major histocompatibility complex** (**MHC**) proteins are used to display antigens on the surfaces of self cells, so that the antigens can be detected by the T cells of the immune system. MHC proteins also function as important self-identifying labels.

3. **T cell receptors** are integral membrane proteins on the surfaces of T cells. They recognize and bind to antigens presented by the MHC proteins on the surfaces of other cells.

Type of cell	Function
Basophils (I, A)	Release histamine
Eosinophils (A)	Kill antibody-coated parasites
Neutrophils (I)	Stimulate inflammation; engulf and digest microorganisms
Mast cells (I)	Release histamine when damaged
Monocytes (I, A)	Develop into macrophages and dendritic cells
Macrophages (I, A)	Engulf and digest microorganisms; activate T cells
Dendritic cells (A)	Present antigens to T cells
Natural killer cells (I)	Attack and lyse virus-infected or cancerous body cells
B lymphocytes (A)	Differentiate to form antibody-producing cells and memory cells
T lymphocytes (A)	Kill virus-infected cells or cancer cells; regulate activities of other white blood cells

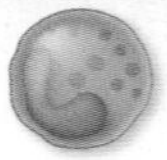

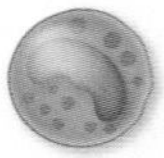

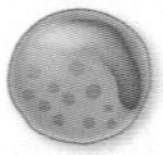

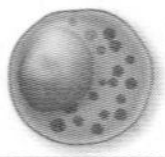

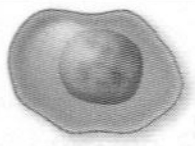

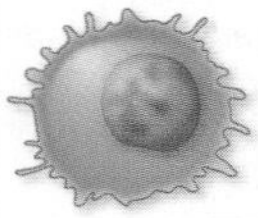

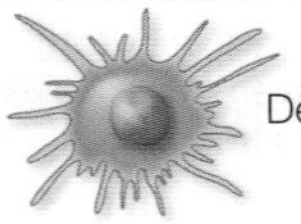

Figure 41.3 White Blood Cells　White blood cells have key roles in both innate (I) and adaptive (A) immunity. The lymphocytes are the B cells and T cells; the other cell types are phagocytes.

 Activity 41.2 Cells of the Immune System
www.Life11e.com/ac41.2

4. **Cytokines** are soluble signaling proteins released by many cell types. They bind to cell surface receptors and alter the behavior of their target cells.

▶ 41.1 recap

All animals have innate defenses against pathogens, and vertebrates have innate and adaptive defenses. Both kinds of mechanisms are based on the ability to differentiate self from nonself. Innate defenses target a broad range of molecules and organisms, whereas adaptive defenses target specific pathogens. Both innate and adaptive defenses involve specialized white blood cells.

learning outcomes

You should be able to:

- Compare various types of defense mechanisms.
- Describe the role of the Toll-like receptor (TLR) pathway in innate immunity.

- Infer what types of immune-related molecules and genes a particular animal would have.

1. Describe three characteristics of innate immunity and adaptive immunity that distinguish them from each other.
2. Some people have a rare genetic disease that results in defective signaling in the TLR pathway. What would be the phenotype of such individuals?
3. If you compared the genomes of an insect and a human with regard to the presence of genes involved in immunity, what might you find?

The outcome of an infectious disease—the life or death of the host—often depends on the success of both rapid, innate responses and long-lasting, adaptive responses to invading pathogens. We will turn now to the innate defenses that protect vertebrates from disease.

key concept 41.2 Innate Defenses Are Nonspecific

Innate defenses are general protection mechanisms that attempt to stop pathogens from invading the body or to quickly eliminate those that do manage to invade. They are genetically programmed (innate) and "ready to go," in contrast to adaptive responses, which take time to develop after a pathogen or toxin has been recognized as nonself.

focus your learning

- Innate immunity is applied to any potential harmful invader.
- Physical barriers, such as skin, are the first line of innate defense.
- Inflammation involves recruitment of cells and defensive molecules to an area damaged by a pathogen or other injury.

Innate immunity is sometimes described as nonspecific immunity because it is applied indiscriminately to any potentially harmful invader. There are two lines of innate defense (**Figure 41.4**). The first line of innate defense is encountered by a potential pathogen as soon as it lands on the surface of an animal. Consider a pathogenic bacterium that lands on human skin. The challenges faced by the bacterium just to reach its target are formidable:

- The *physical barrier of the skin*: Bacteria rarely penetrate intact skin, which has about 30 cell layers; by the same token, broken skin increases the risk of infection.
- The *saltiness and dryness of skin*: This environment may not be hospitable to the growth of the bacterium.
- The *presence of normal flora*: Bacteria and fungi that normally live and sometimes reproduce in great numbers on our body surfaces without causing disease will compete with pathogens for space and nutrients.

If a pathogen lands inside the nose or another internal organ, it faces other innate defenses:

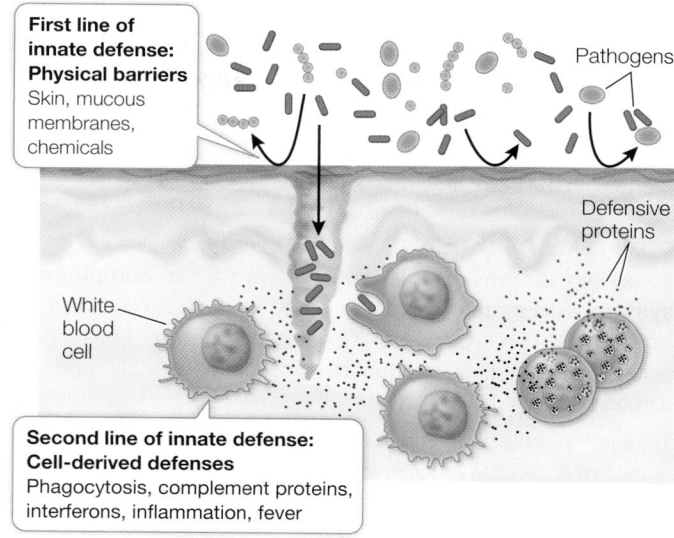

First line of innate defense: Physical barriers Skin, mucous membranes, chemicals

Pathogens

Defensive proteins

White blood cell

Second line of innate defense: Cell-derived defenses Phagocytosis, complement proteins, interferons, inflammation, fever

Figure 41.4 Innate Immunity   Physical barriers, cells, and proteins (complement and interferons) provide nonspecific defenses against invading pathogens.

 Media Clip 41.1 The Chase Is On: Phagocyte versus Bacteria www.Life11e.com/mc41.1

- **Mucus** is a slippery secretion produced by mucous membranes found at the inner surfaces of the nose (as well as the digestive, respiratory, and urogenital systems). Mucus traps microorganisms so they can be removed by the beating of cilia (see Figure 5.17), which continuously move the mucus and its trapped debris away.

- **Lysozyme** is an enzyme made by cells of the mucous membranes that attacks the cell walls of many bacteria, causing them to lyse (burst open).

- **Defensins**, also made by mucous membranes, are peptides of 18–45 amino acids that contain hydrophobic domains. They are toxic to a wide range of pathogens, including bacteria, microbial eukaryotes, and enveloped viruses. Defensins insert themselves into the cell membranes of these organisms and make the membranes permeable, thus killing the invaders. Defensins are also produced in phagocytes, where they kill pathogens ingested by phagocytosis. Plants also produce *defensins in response to pathogen exposure.

---

*connect the concepts   Key Concept 38.1 describes constitutive and cellular responses to pathogens in plants. Among these are the pathogenesis-related (PR) proteins, which are similar to defensins produced by animals. These small peptides bind to fungal membranes and are toxic to a wide range of fungal targets, but they are not toxic to plant or animal cells. Other PR proteins may serve as alarm signals to plant cells that have not yet been attacked.

Harsh conditions in an animal's internal environment can also kill pathogens. For example, gastric juice in the stomach is a deadly environment for many bacteria because of the hydrochloric acid and proteases that are secreted into it.

Specialized proteins and cells participate in innate immunity

Several proteins are produced by the body either before an infection occurs or in response to invasion by pathogens. Two important groups are the complement and interferon proteins.

COMPLEMENT PROTEINS   Vertebrate blood contains more than 20 different proteins that make up the antimicrobial **complement system**. This system can be activated by various mechanisms, including both innate and adaptive defense responses. The proteins act in a characteristic sequence, or cascade, with each protein activating the next:

1. First, the proteins attach to specific components on the surface of a microbe or to an antibody that has already bound to the microbe's surface. In either case, binding helps phagocytes recognize and destroy the microbe.

2. Then, complement proteins activate the inflammatory response (see Figure 41.5) and attract phagocytes to the site of infection.

3. Finally, complement proteins lyse invading cells (such as bacteria).

INTERFERONS   When a cell is infected by a pathogen, it produces small amounts of signaling proteins called **interferons** that increase the resistance of neighboring cells to infection. Interferons are a class of cytokines and have been found in many vertebrates. Various molecules, including double-strand (viral) RNA, induce the production of interferons. Thus interferons are particularly important as a defense against viruses. Interferons bind to receptors on the cell membranes of uninfected cells, stimulating a signaling pathway that inhibits viral reproduction if the cells are subsequently infected. In addition, interferons stimulate the cells to hydrolyze bacterial or viral proteins to peptides, an initial step in adaptive immunity (see Key Concept 41.3).

PHAGOCYTES   Some phagocytes travel freely in the circulatory and lymphatic systems; others can move out of blood vessels and adhere to certain tissues. Pathogenic cells, viruses, or fragments of these invaders are recognized by phagocytes, which then ingest them by phagocytosis.

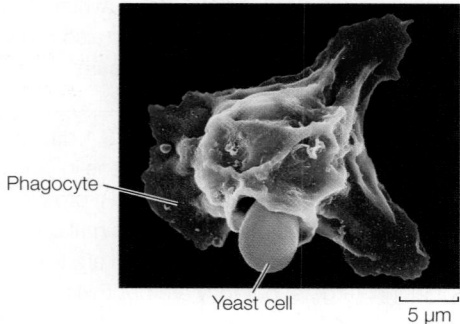

Phagocyte

Yeast cell   5 μm

Defensins, nitric oxide, and reactive oxygen intermediates inside these phagocytes then kill the pathogens.

NATURAL KILLER CELLS   One class of lymphocytes, known as **natural killer cells**, can distinguish virus-infected cells and

some tumor cells from their normal counterparts and initiate apoptosis (programmed cell death) of these target cells. In addition to this innate defense action, natural killer cells interact with the adaptive defense mechanisms by lysing antibody-labeled target cells.

DENDRITIC CELLS   These phagocytes occur in tissues exposed to the environment, such as skin and the linings of the digestive and respiratory tracts, as well as in blood. They act as messengers between the innate and adaptive immune systems. They can endocytose microbes, viruses, and even remnants of the virus-infected host cells. Once inside a dendritic cell, these particles are digested to fragments, and the dendritic cell "presents" an antigenic fragment on its surface, along with class II MHC proteins (see Key Concept 41.5). In addition, the dendritic cell secretes signals that activate cells of the adaptive immune system.

Inflammation is a coordinated innate response to infection or injury

When mammalian tissue is damaged because of infection or injury, the body responds with **inflammation**. This response can happen almost anywhere in the body, internally as well as on the surface. Inflammation is an important phenomenon: it isolates the damaged area to stop the spread of the damage; it recruits cells and molecules to the damaged location to kill the invader; and it promotes healing. The first responders to tissue damage are **mast cells**, which adhere to the skin and the linings of organs and release numerous chemical signals, including:

● **Tumor necrosis factor**, a cytokine protein that kills target cells and activates immune cells.

● **Prostaglandins**, fatty acid derivatives involved in various responses, including the widening of blood vessels. Prostaglandins interact with nerve endings and are partly responsible for the pain caused by inflammation.

● **Histamine**, an amino acid derivative that leads to the important process of dilation of blood vessels, as well as itchy, watery eyes and rashes seen with some types of allergic reactions.

The redness and heat of inflammation result from the dilation and leakiness of blood vessels in the infected or injured area (**Figure 41.5**). Phagocytes enter the inflamed area, where they engulf the invaders and dead tissue cells. Phagocytes are responsible for most of the healing associated with inflammation. They produce several cytokines, which (among other functions) can signal the brain to produce a fever. This rise in body temperature accelerates lymphocyte production and phagocytosis, thereby speeding the immune response. In some cases, pathogens are temperature-sensitive and their growth is inhibited. The pain of inflammation results from increased pressure due to swelling, the action of leaked enzymes on nerve endings, and the action of prostaglandins, which increase the sensitivity of the nerve endings to pain.

Following inflammation, pus may accumulate. Pus is a mixture of leaked fluid and dead cells: bacteria, neutrophils (the most abundant white blood cells—see Figure 41.3), and damaged body cells. Pus is a normal result of inflammation and is gradually consumed and further digested by macrophages.

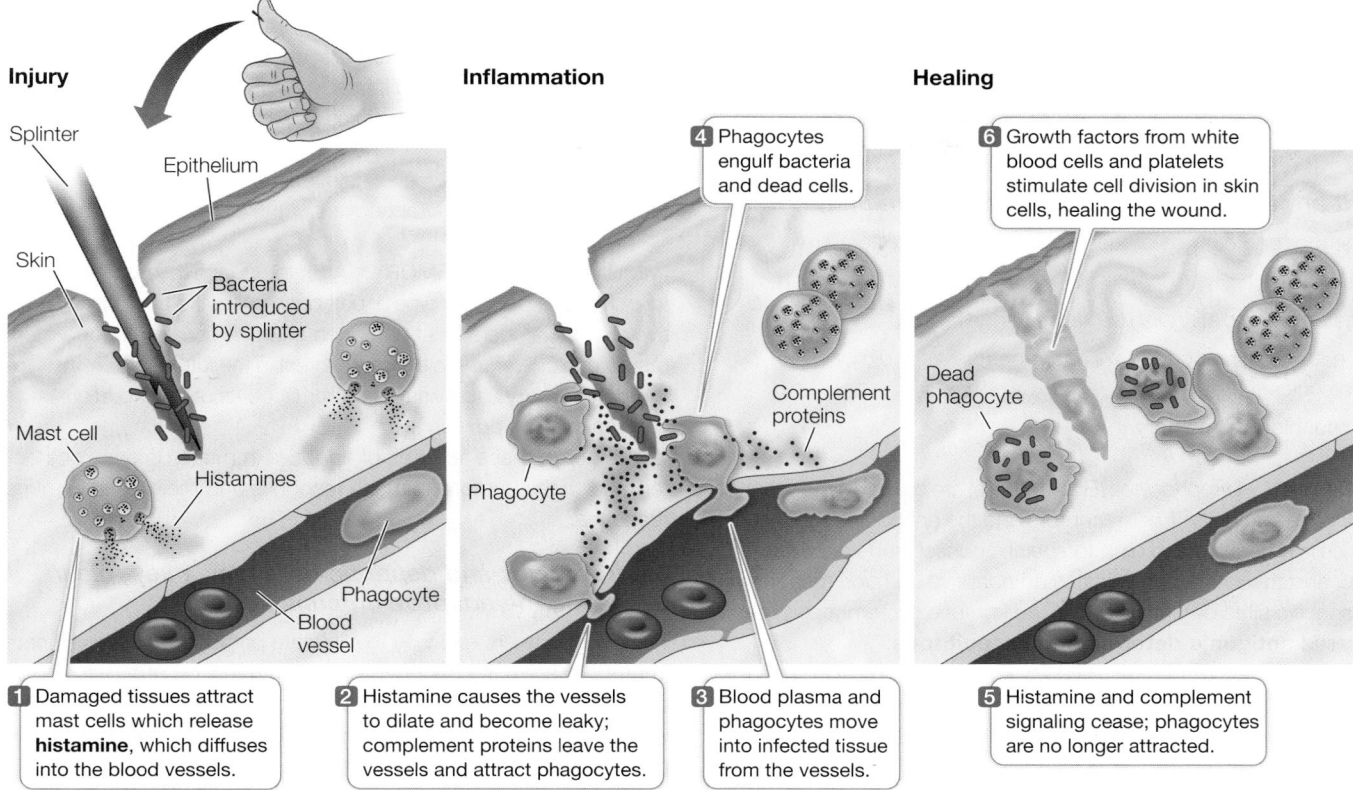

Injury

Splinter

Epithelium

Skin

Bacteria introduced by splinter

Mast cell

Histamines

Phagocyte

Blood vessel

1 Damaged tissues attract mast cells which release **histamine**, which diffuses into the blood vessels.

Inflammation

4 Phagocytes engulf bacteria and dead cells.

Complement proteins

Phagocyte

2 Histamine causes the vessels to dilate and become leaky; complement proteins leave the vessels and attract phagocytes.

3 Blood plasma and phagocytes move into infected tissue from the vessels.

Healing

6 Growth factors from white blood cells and platelets stimulate cell division in skin cells, healing the wound.

Dead phagocyte

5 Histamine and complement signaling cease; phagocytes are no longer attracted.

Figure 41.5 Interactions of Cells and Chemical Signals Result in Inflammation   Histamine and other signals are released from mast cells to initiate the inflammatory response. The chemical signals associated with inflammation attract phagocytes, which digest the pathogens and damaged cells.

▶ Activity 41.3 **Inflammatory Response**
www.Life11e.com/ac41.3

Q: Antihistamines are a class of drugs widely used to treat people with inflammation. How do you think they work?

41.2 recap

Innate immunity is the first line of defense against pathogens. Innate immunity includes physical barriers such as the skin, and cellular responses involving the recognition of self and nonself molecules. Recognition of nonself molecules by white blood cells leads to coordinated responses such as the production of defensive proteins and inflammation.

learning outcomes

You should be able to:

• Summarize the roles of physical barriers, chemical defenses, and cellular defenses in innate immunity.

• Explain how the innate defense system defends against a pathogen.

• List the steps in the inflammatory response.

1. A pathogenic bacterium lands on your skin. Outline the innate defenses that come into play.

2. Your immune system does not mount an innate defense response to a speck of dust that lands on your skin, but it may mount an innate response to a bacterium. Why?

3. When a splinter of wood permeates the skin, the skin swells up. Outline the events by which this occurs.

Often the innate immune system, with its nonspecific defenses, is adequate to prevent or fight off a pathogenic infection. But in many cases this system works together with adaptive immunity, which detects and responds to specific pathogens. We will now turn to the development and functioning of adaptive immunity.

▶ key concept 41.3 Adaptive Defenses Are Specific

More than a century ago, experiments showed that guinea pigs that were infected with bacteria that cause diphtheria produced a chemical in their blood that can protect other guinea pigs from infection. This observation led to the concept of adaptive immunity and to the understanding that factors promoting immunity are found in blood.

focus your learning

• The humoral and cellular immune responses work simultaneously and cooperatively.

• Memory cells, the key to immunological memory, retain the capacity to divide to produce effector and more memory cells.

• Vaccinations initiate a primary immune response, generating memory cells without causing illness.

In this section we outline the main features of the adaptive immune system. We will consider the two major types of adaptive responses: the humoral immune response, which produces antibodies; and the cellular immune response, which destroys infected cells.

What are the key features of adaptive immunity?

Four important features of the adaptive immune system are:

1. Specificity
2. The ability to distinguish self from nonself
3. The ability to respond to an enormous diversity of nonself molecules
4. Immunological memory

SPECIFICITY   Lymphocytes (B and T cells) are crucial components of adaptive immunity. T cell receptors and the antibodies produced by B cells recognize and bind to specific nonself substances (antigens), and this interaction initiates an adaptive immune response. The specific sites on antigens that the immune system recognizes are called **antigenic determinants**, or **epitopes**:

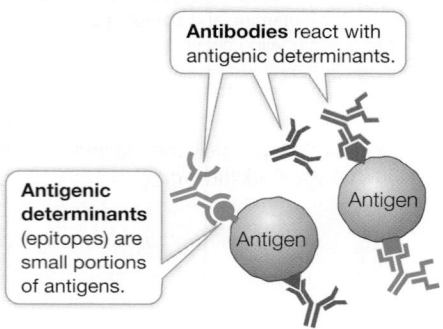

Antibodies react with antigenic determinants.

Antigenic determinants (epitopes) are small portions of antigens.

Antigen

Antigen

An antigenic determinant is a specific portion of a large molecule, such as a certain sequence of amino acids that may be present in a protein. Antigens are usually proteins or polysaccharides, and there can be multiple antigens on a single invading bacterium. A single antigenic molecule can have multiple, different antigenic determinants. The host animal responds to the presence of an antigen with highly specific defenses involving T cell receptors and antibodies. These receptors and soluble proteins bind to the antigenic determinants. Each T cell and each antibody is specific for a single antigenic determinant. For the remainder of the chapter, we will refer to antigenic determinants simply as "antigens."

DISTINGUISHING SELF FROM NONSELF   You have seen how the innate immune system distinguishes between self and nonself molecules. The adaptive immune system has another set of mechanisms for distinguishing self from nonself. The human body contains tens of thousands of different molecules, each with a specific three-dimensional structure capable of generating immune responses. Thus every cell in the body has a tremendous number of antigens. A crucial requirement of an individual's adaptive immune system is that it recognize the body's own antigens and not attack them.

DIVERSITY   Challenges to the immune system are numerous. Pathogens take many forms: viruses, bacteria, protists, fungi, and

multicellular parasites. Furthermore, each pathogenic species usually exists as many subtly different genetic strains, and each strain possesses multiple surface features. Estimates vary, but a reasonable guess is that humans can respond specifically to 10 million different antigens. Upon recognizing an antigen, the adaptive immune system responds by activating lymphocytes of the appropriate specificity.

IMMUNOLOGICAL MEMORY   After the adaptive immune system responds to a particular type of pathogen once, the adaptive immune system "remembers" that pathogen and can usually respond more rapidly and powerfully to the same threat in the future. This **immunological memory** usually saves us from repeats of childhood diseases such as chicken pox.

All four of these features of adaptive immune defense characterize both the humoral immune response and the cellular immune response.

Macrophages and dendritic cells play a key role in activating the adaptive immune system

As you have seen, one way to eliminate pathogens with the innate immune system is phagocytosis. After ingestion of a pathogenic organism or infected host cell, phagocytic cells display fragments of the pathogen on their cell surfaces. These fragments function as antigens, and **antigen presentation** is one way that components of the innate immune system communicate with the adaptive immune system. Macrophages and dendritic cells play a key role in activating the adaptive immune system. After engulfing pathogens or infected host cells, these cells migrate to lymph nodes, where they present antigen to immature (previously unexposed) T cells. In addition, the antigen-presenting cells secrete cytokines and other signals that stimulate the activation and differentiation of the T cells.

Two types of adaptive immune responses interact

The adaptive immune system mounts two types of responses against invaders: the humoral immune response and the cellular immune response. These two responses work simultaneously and cooperatively, sharing many mechanisms. We will use the example of a viral infection in an overview of these two types of responses (**Focus: Key Figure 41.6**). These responses also occur when bacteria or other pathogens infect and grow inside host cells.

B cells that make antibodies are the workhorses of the **humoral immune response**, and **cytotoxic T (T$_C$) cells** are the workhorses of the **cellular immune response**. There are three phases in both types of adaptive immune response:

1. *Recognition phase*: In both cellular and humoral immunity, recognition occurs when an antigen is inserted into the cell membrane of an antigen-presenting cell, with the unique antigen structure protruding from the cell membrane. In addition to dendritic cells and macrophages, developing B cells can also perform phagocytosis and function as antigen-presenting cells. A virus-infected host cell can be engulfed by an antigen-presenting cell, or the free viral particles may be engulfed. The antigen on the surface of the antigen-presenting cell is recognized by a **T-helper (T$_H$) cell** bearing a T cell receptor protein that is specific for the antigen. In both humoral and cellular immunity, binding initiates the activation phase.

focus: key figure

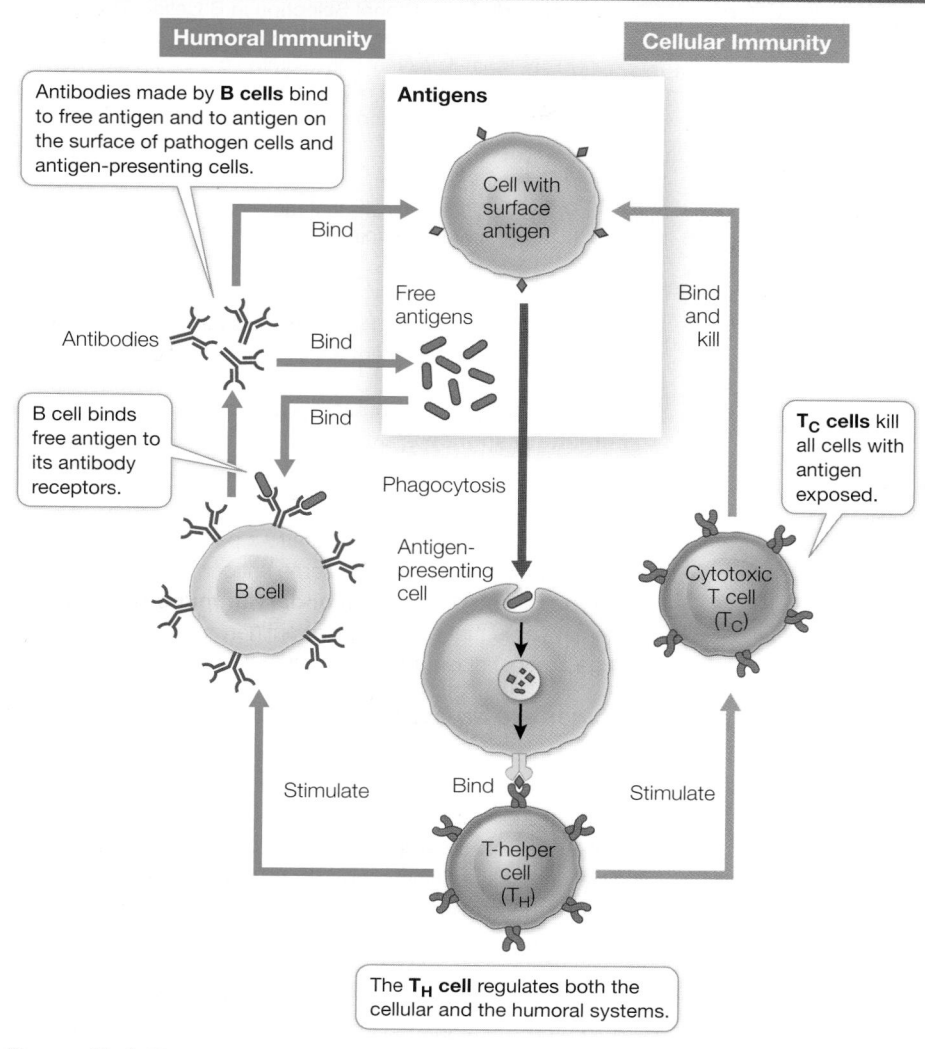

Humoral Immunity

Cellular Immunity

Antibodies made by **B cells** bind to free antigen and to antigen on the surface of pathogen cells and antigen-presenting cells.

Antigens

Cell with surface antigen

Bind

Bind and kill

Antibodies

Bind

Free antigens

T$_C$ cells kill all cells with antigen exposed.

B cell binds free antigen to its antibody receptors.

Bind

Phagocytosis

B cell

Antigen-presenting cell

Cytotoxic T cell (T$_C$)

Stimulate

Bind

Stimulate

T-helper cell (T$_H$)

The **T$_H$ cell** regulates both the cellular and the humoral systems.

Figure 41.6 The Adaptive Immune System Humoral immunity involves the production of antibodies by B cells. Cellular immunity involves the activation of cytotoxic T cells that bind to cells expressing the antigen. For further details, see Figure 41.13.

Q: In AIDS, a virus infects T$_H$ cells and kills them. What would be the effects on adaptive immunity?

2. *Activation phase*: When the T$_H$ cell recognizes an antigen on an antigen-presenting cell, it propagates and releases cytokines that stimulate B cells and T$_C$ cells bearing receptors to the same antigen to divide. The results are a clone of B cells that function in humoral immunity and a clone of T$_C$ cells that function in cellular immunity.

3. *Effector phase*: In the humoral immune response, cells of the B clone produce antibodies that bind to viral particles and/or virus-infected cells. The bound antibodies attract phagocytes and complement proteins that engulf and destroy the virus and the virus-infected cells. In cellular immunity, cells of the T$_C$ clone bind to virus-infected cells and destroy them.

 **Animation 41.1 Pregnancy Test**
www.Life11e.com/a41.1

Adaptive immunity develops as a result of clonal selection

Before the humoral and cellular immune responses can take place, the body needs to generate a vast diversity of lymphocytes that have the ability to bind different antigens. How does this tremendous diversity arise? As we will discuss in Key Concept 41.4, this diversity is generated primarily by DNA changes—chromosomal rearrangements and other mutations—that occur just after the B and T cells are formed in the bone marrow. Millions of different B cells develop, each of which can produce only one kind of antibody. Similarly, there are millions of different T cells, each with one specific kind of T cell receptor. Thus the adaptive immune system is "predeveloped"—*all of the machinery available to respond to an immense diversity of antigens is already there, even before the antigens are ever encountered*.

As we have described, when a pathogen enters the vertebrate body it stimulates the innate immune system. In addition to triggering its own defensive responses, the innate immune system triggers adaptive defensive responses via specific antigens that are presented on the surfaces of antigen-presenting cells, particularly dendritic cells. This triggers the proliferation of lymphocytes (B and T cells) that are specific for those particular antigens. How does this proliferation occur? The answer lies in the process of **clonal selection**: *antigen binding "selects" a particular B or T cell for proliferation*. When an antigen fits the surface receptor on a B or T cell and binds to it, that cell is activated. It divides to form a clone of cells (a genetically identical group derived from a single cell), all of which recognize and react to the same antigen. This process is illustrated for B cells in **Figure 41.7**. Binding and activation select a particular lymphocyte, while proliferation generates the clone, hence the term "clonal selection."

Clonal deletion helps the immune system distinguish self from nonself

Normally the body is tolerant of its own molecules. One way that the immune system does this is through the process of **clonal deletion**. This occurs during the early differentiation of T and B cells, when these cells encounter self antigens. Any immature B or T cell that shows the potential to mount an immune response against self antigens undergoes apoptosis within a short time.

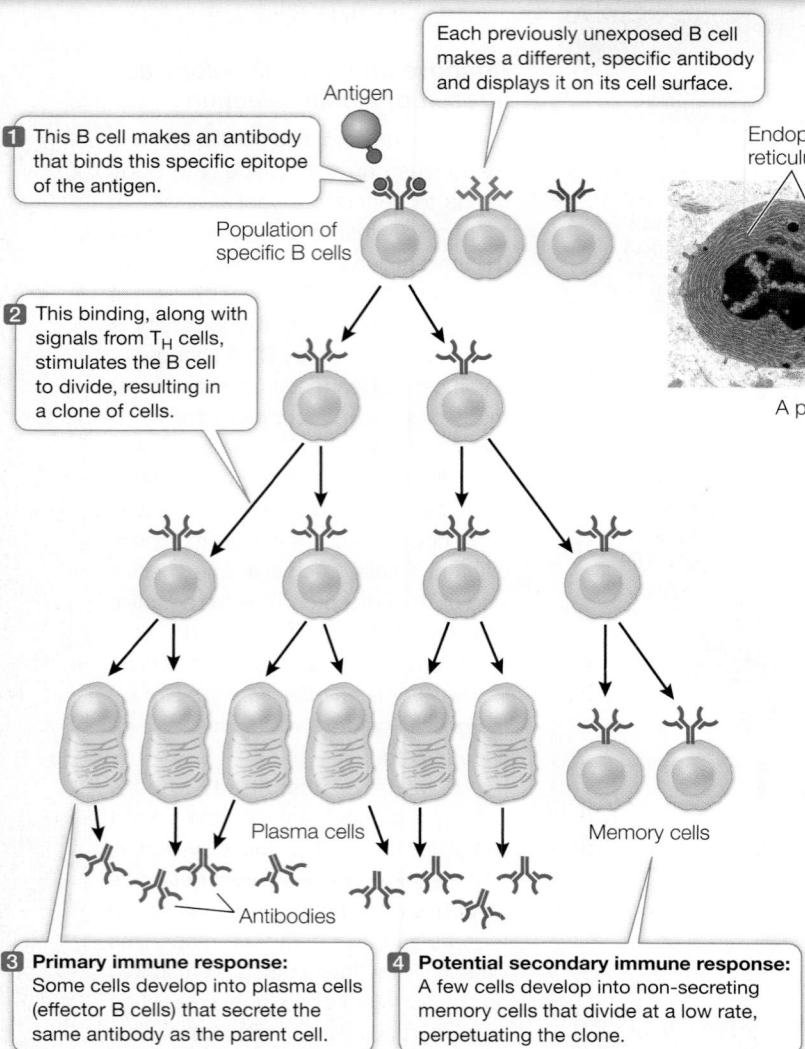

Each previously unexposed B cell makes a different, specific antibody and displays it on its cell surface.

Antigen

1 This B cell makes an antibody that binds this specific epitope of the antigen.

Population of specific B cells

2 This binding, along with signals from T_H cells, stimulates the B cell to divide, resulting in a clone of cells.

Endoplasmic reticulum

A plasma cell

Plasma cells

Memory cells

Antibodies

3 **Primary immune response:** Some cells develop into plasma cells (effector B cells) that secrete the same antibody as the parent cell.

4 **Potential secondary immune response:** A few cells develop into non-secreting memory cells that divide at a low rate, perpetuating the clone.

Figure 41.7 Clonal Selection in B Cells   The binding of an antigen to a specific receptor on the surface of a B cell stimulates that cell to divide, producing a clone of genetically identical cells to fight that invader. Plasma cells have extensive endoplasmic reticulum for synthesizing antibodies.

Q: You may never have been exposed to Ebola virus. Do you have cells that make antibodies against this virus? Explain.

Effector cells and memory cells can respond to an antigen in two different ways:

1. When the body first encounters a particular antigen, a **primary immune response** is activated, in which the previously unexposed lymphocytes that recognize that antigen proliferate to produce clones of effector and memory cells.

2. After a primary immune response to a particular antigen, subsequent encounters with the same antigen will trigger a much more rapid and powerful **secondary immune response**. The memory cells that bind with that antigen proliferate, launching a huge army of plasma cells and effector T cells.

Vaccines are an application of immunological memory

Thanks to immunological memory, exposure to many diseases (including childhood diseases such as chicken pox) provides a natural immunity to those diseases. Furthermore, it is possible to provide artificial immunity against many life-threatening diseases by **vaccination**: the introduction of antigen into the body in a form that does not cause disease.

Vaccination initiates a primary immune response, generating memory cells without making the person ill. Later, if a pathogen carrying the same antigen attacks, specific memory cells already exist. They recognize the antigen and quickly overwhelm the invaders with a massive production of lymphocytes and antibodies.

Immunological memory results in a secondary immune response

The first time a vertebrate animal is exposed to a particular antigen there is a time lag (usually several days) before the B cell–produced antibody molecules and T cells specific to that antigen slowly increase. But for years afterward—sometimes for life—the immune system "remembers" that particular antigen, allowing the body to mount a faster response the next time it encounters the antigen. How does this happen?

The answer lies in the fact that activated lymphocytes divide and differentiate to produce *two types* of daughter cells: effector cells and memory cells.

1. **Effector cells** carry out the attack on the antigen. Effector B cells, called **plasma cells**, secrete antibodies. Effector T cells release cytokines and other molecules that initiate reactions that destroy nonself or altered cells. Effector cells live only a few days.

2. **Memory cells** (see Figure 41.7) are long-lived cells that retain the ability to start dividing on short notice to produce more effector and more memory cells. Memory B and T cells may survive in the body for decades, rarely dividing.

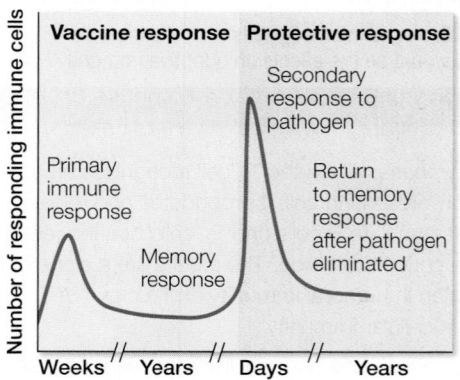

Because of mutation, an antigenic determinant on a pathogen can change over time. If the change is significant in terms of three-dimensional structure, a vaccine directed against an antigen from a pathogen may become ineffective. You are familiar with this with regard to influenza. More often than not, the annual flu season features a genetic strain of the flu virus that is not sufficiently

experiment

Original Paper: Miller, M. et al. 2013. Neutralizing antibodies against previously encountered influenza virus strains increase over time: A longitudinal analysis. *Science Translational Medicine* 5: 1–9 (August 14 issue).

The Framingham Heart Study was begun in 1948 with the objective of finding out the causes of heart disease and stroke. About 5,000 people ages 30–62 and living in the town Framingham, Massachusetts, have been subjected every 2 years to extensive medical exams in an attempt to relate lifestyle factors to heart disease. Subsequent generations have been added to the study ever since. Not only has this study led to major findings about the causes of heart disease, but it has provided a wealth of data on other medical conditions as well. Peter Palese of the Icahn School of Medicine at Mount Sinai Hospital in New York City measured antibodies in blood to flu virus strains from 40 people in the Framingham study, born between 1917 and 1952. They had been exposed to major flu outbreaks in 1957, 1968, and 1977, as well as to the typical flu that comes up every year. Blood samples were examined between 1987 and 2008 to see if these people still had antibodies against the three major flu outbreak strains, and whether these antibody amounts increased over time as people fought off the new, less virulent strains that arise every year.

HYPOTHESIS▶ Exposure to new strains of flu stimulates antibody production against strains to which an individual was exposed previously.

CONCLUSION▶ People exposed to flu virus develop long-lasting memory cells that make antibodies that have broad specificity to other flu strains.

investigatinglife work with the data follows on next page.

METHOD

1 Obtain blood samples from people born between 1917 and 1952 who had been exposed to Asian flu (1957), Hong Kong flu (1968), and Russian flu (1977).

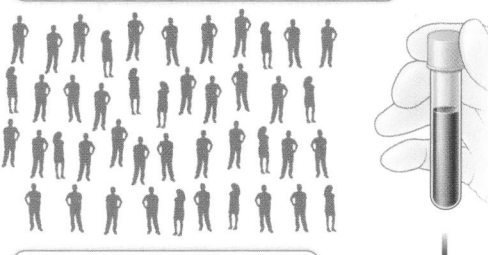

2 Measure antibodies against the three strains in blood samples taken between 1987 and 2008.

RESULTS

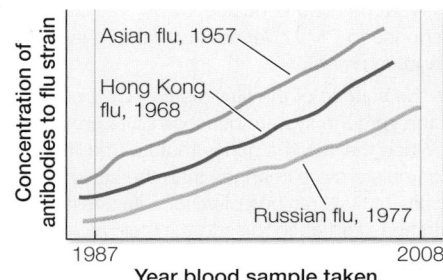

Asian flu, 1957
Hong Kong flu, 1968
Russian flu, 1977

Concentration of antibodies to flu strain

Year blood sample taken
1987 — 2008

---

related to the previous season's vaccine; so a new vaccine, with new B cell clones, must be made. A remarkable recent finding is that memory cells can last a long time and help in a response to new, related antigens. People who were exposed to a particular strain of flu retain antibody-making memory cells against that strain for decades and develop a rapid immune response to a new, genetically different strain that appears years later. In the case of flu, the memory cells are making antibodies that bind to not just the first strain of flu they responded to, but to later ones as well. Characterizing these antibodies provides a clue to developing a vaccine that is broadly effective and obviates the need to develop new ones (**Investigating Life: What Are the Mechanisms and Implications of Long-Lasting Immunity?**).

As you saw in the opening story, vaccination has completely or almost completely wiped out some deadly diseases in industrialized countries.

▶ 41.3 recap

The adaptive immune system reacts against nonself molecules called antigen[s]. The system generates amazing diversity in both antibodies (produced by B ce[lls]) and in T cell receptors. In the primary immune response, B cells and T cells tha[t] recognize a particular antigen proliferate by clonal selection. Immunological memory prepares the body for a much stronger secondary immune response[.]

learning outcomes

You should be able to:

- Distinguish between the recognition, activation, and effector phases in adaptive immunity.
- Make and explain inferences regarding rates of infection in the context of immunological memory.
- Summarize how vaccines provide long-lasting immunity to pathogens.

---

1. A fellow passenger sneezes on the train and the common cold virus enters y[our] nose. Outline how your body responds, noting the kinds and stages of the ad[ap]tive immune response.

2. In 2009 the H1N1 strain of influenza spread around the world. People who ha[d] been alive at the time of the 1918 epidemic had low rates of infection. Explai[n] in terms of immunological memory.

3. How do vaccines make use of immunological memory?

work with the data

QUESTIONS ▶

1. Blood samples were taken from 1987 to 2008, and the people's reactivity against various flu strains was tested. The results for antibodies against the Russian flu strain of 1977 in numerous individuals are shown in **Figure A**. Note that these people were exposed to many flu strains after 1977. The data shown are typical for the way specific immune reactions are measured. The titer is presented as a reciprocal. For example, a titer of 2 means that undiluted serum sample gave an antibody reaction (titer 1), and so did a dilution of 1 part serum and 1 part water (dilution of 1/2, reciprocal = 2). If the sample was diluted to a titer of 4 (1 part serum, 3 parts water), there was no reaction because the concentration of antibody was too low. The higher the titer number, the more the blood serum could be diluted to still react with the antigen; that is, a higher titer number means that there were more antibodies in the original serum sample. What can you conclude about the titer of anti-Russian flu strain over the years 1987–2008?

2. The flu virus mutates rapidly. In contrast, another virus to which people are exposed, cytomegalovirus (CMV), does not mutate rapidly. Antibody titer against CMV were measured, and the data are shown in **Figure B**. What can you conclude from these data? Compare the antibody response to CMV over time with that against the Russian flu virus strain and comment.

3. A protein on the surface of the flu virus called hemagluttinin (HA) is exposed to the immune system because of its prominent location (**Figure C**). When the flu virus genes mutate, the "head" of the protein tends to accumulate more changes than the stalk. Thus the amino acid sequences of stalk regions of various flu strains are similar. The investigators measured antibodies to the stalk region that were either broadly reactive to various flu strains, or not broadly reactive. The results are shown in the table. What can you conclude about the broadly reactive antibodies? How does this provide a guide for the development of a broadly based flu vaccine?

Figure A

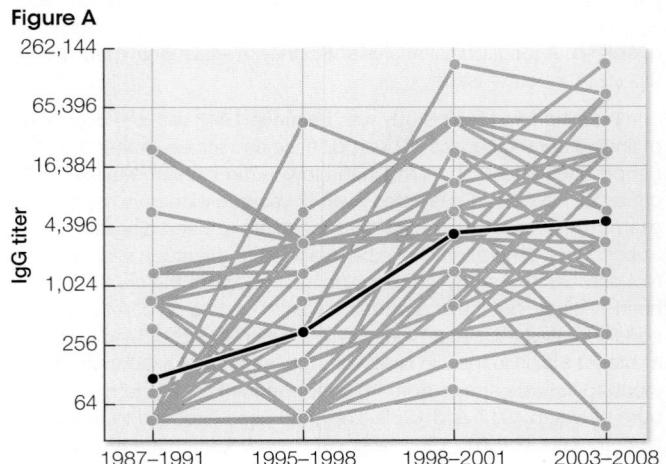

Figure B

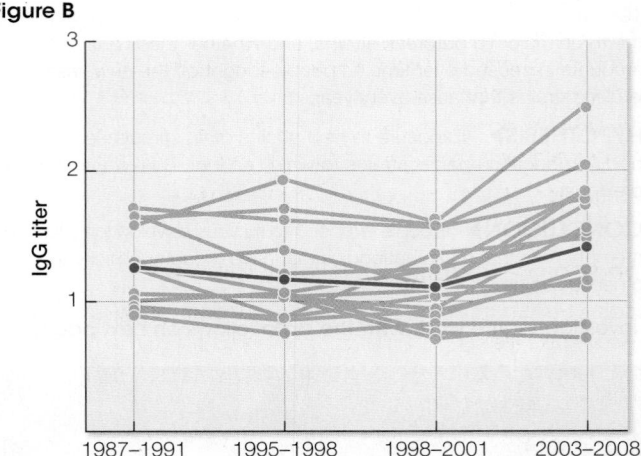

Figure C

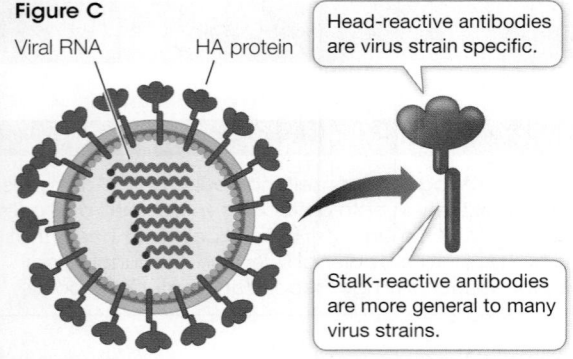

Viral RNA   HA protein

Head-reactive antibodies are virus strain specific.

Stalk-reactive antibodies are more general to many virus strains.

Type of antibodies to stalk region	Titer
Broadly reactive	4,080
Not broadly reactive	1,024

A similar **work with the data** exercise may be assigned in **LaunchPad**.

The general features of the adaptive system are that it is pre-made to respond specifically to virtually any antigen, and it involves an initial primary response to the antigen and a more powerful secondary response when the same antigen is encountered in the future. As we noted, the adaptive system has a humoral response with B cells making antibodies and a cellular response with T cells expressing specific receptors. We now turn to these two types of adaptive response, beginning with the humoral system.

▶key concept The Humoral Adaptive Response Involves Antibodies

41.4

Every day in the human body, billions of B cells survive the test of clonal deletion and are released from the bone marrow into the circulatory system. B cells are the basis for the humoral immune response.

focus your learning

- Antibody proteins are immunoglobulins with four polypeptide chains—two identical light chains and two identical heavy chains, each with constant and variable regions.
- IgG is the most common of the five classes of immunoglobulins.
- DNA rearrangements and other mutations generate the diversity of immunoglobulins.

A B cell begins as an unexposed B cell with a receptor protein on its cell surface that is specific for a particular antigen. The cell is activated by antigen binding to this receptor, and after stimulation by a T_H cell, it gives rise to a clone of plasma cells that make antibodies as well as to a smaller number of memory cells (see Figure 41.7). The stimulation occurs after the B cell presents antigen to a T_H cell with a receptor that can recognize the antigen. The T_H cell then secretes cytokines that stimulate the B cell to divide.

Activated plasma cells (effector B cells) can synthesize and secrete large amounts of antibody proteins—up to 2,000 molecules per second! *All the plasma cells arising from a given B cell produce antibodies that are specific for the antigen that originally bound to the parent B cell.* Thus antibody specificity is maintained as B cells proliferate.

 Animation 41.2 Humoral Immune Response
www.Life11e.com/a41.2

Antibody protein structure reflects function

Antibodies belong to a class of proteins called **immunoglobulins**. There are several types of immunoglobulins, but all contain a tetramer consisting of four polypeptide chains: two "heavy chains" and two "light chains" (**Figure 41.8**). In each immunoglobulin molecule, the two light chains are identical, and the two heavy chains are identical. Disulfide bonds hold the chains together, characteristically in a Y-like shape.

Each of the four polypeptide chains has a constant region and a variable region (designated C and V in Figure 41.8A):

- The amino acid sequences of the **constant regions** are similar among the immunoglobulins. They determine the destination and function—the class—of each immunoglobulin.

- The amino acid sequences of the **variable regions** are different for each specific immunoglobulin. Their three-dimensional antigen-binding sites are determined by their secondary structures and are responsible for antibody specificity.

Antigen–antibody binding is noncovalent. The attractive forces involved in holding antigen and antibody together include hydrogen bonding, ionic attractions, hydrophobic interactions, and van der Waals forces (see Key Concept 2.2). These attractive forces are generated from particular amino acids at the binding site on both molecules. So for an antibody protein, the sequence of amino acids

Figure 41.8 The Structure of an Immunoglobulin   Four polypeptide chains (two light, two heavy) make up an immunoglobulin molecule. Both diagrammatic **(A)** and space-filling **(B)** representations of immunoglobulin are shown here.

 Activity 41.4 Immunoglobulin Structure
www.Life11e.com/ac41.4

at the binding site is crucial to its specificity. Generating antibody diversity is thus a molecular genetics challenge, as we will soon discuss.

The two antigen-binding sites on each immunoglobulin molecule are identical, making the antibody bivalent (*bi*, "two," + *valent*, "binding"), meaning the molecule can bind two antigen molecules at once. Bivalency, in addition to the presence of multiple epitopes on the surfaces of many antigens (including large proteins, viruses, and bacteria, as you saw in the case of the influenza virus explored in Investigating Life: What Are the Mechanisms and Implications of Long-Lasting Immunity?), permits antibodies to form large complexes with the antigens. These complexes are easy targets for ingestion and breakdown by phagocytes.

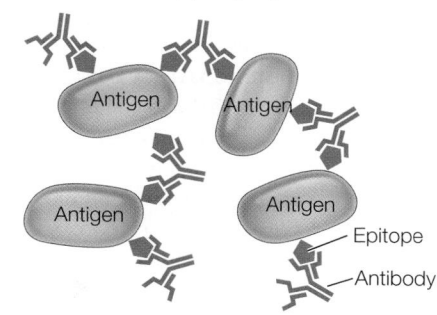

(A) General structure of an immunoglobulin

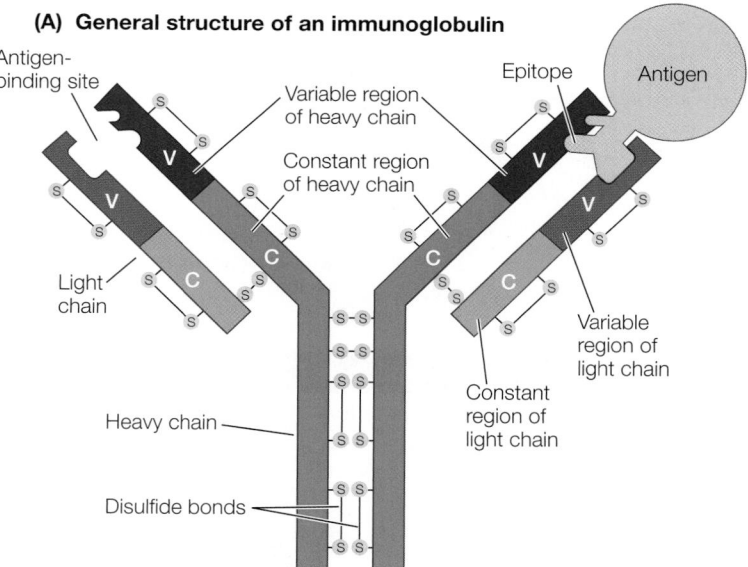

(B) Space-filling model of an immunoglobulin

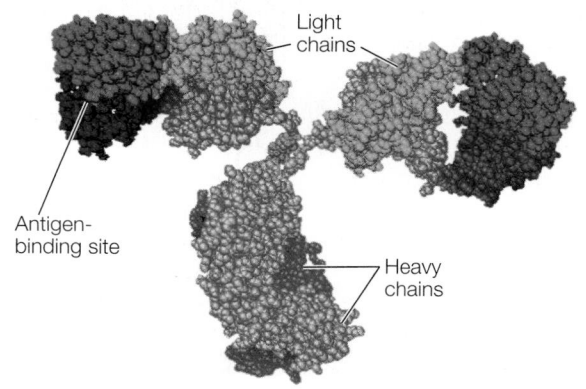

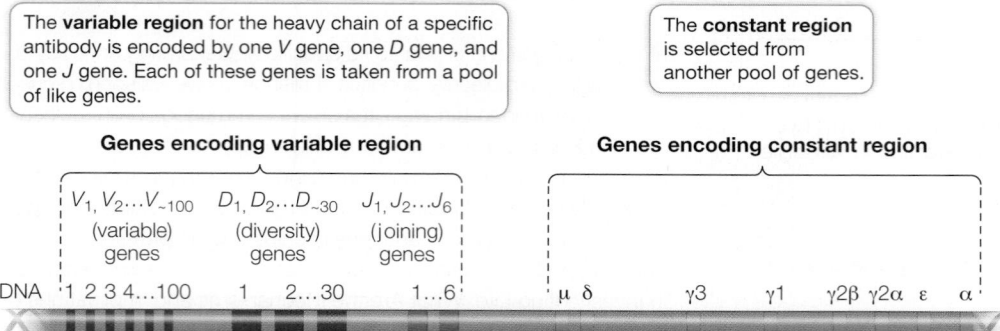

The **variable region** for the heavy chain of a specific antibody is encoded by one *V* gene, one *D* gene, and one *J* gene. Each of these genes is taken from a pool of like genes.

The **constant region** is selected from another pool of genes.

Genes encoding variable region

$V_1, V_2...V_{\sim100}$ $D_1, D_2...D_{\sim30}$ $J_1, J_2...J_6$
(variable)   (diversity)   (joining)
genes   genes   genes

Genes encoding constant region

DNA 1 2 3 4...100   1   2...30   1...6 μ δ $\gamma 3$ $\gamma 1$ $\gamma 2\beta$ $\gamma 2\alpha$ ε α

Figure 41.9 Supergene   Mouse immunoglobulin heavy chains have four domains, each of which is coded for by one of several possible genes selected from a cluster of similar genes. The immunoglobulin protein has one domain from each cluster.

There are five classes of immunoglobulins

While the variable regions are responsible for the specificity of an immunoglobulin, the constant regions of the heavy chain determine the class of the immunoglobulin. The five immunoglobulin classes are described in **Table 41.2**. The most abundant class is IgG; these soluble antibody proteins make up about 80 percent of the total immunoglobulin content of the bloodstream. They are made in greatest quantity during a secondary immune response. IgG molecules defend the body in several ways. For example, after some IgG molecules bind to antigens, they become attached by their heavy chains to macrophages. This attachment permits the macrophages to destroy the antigens by phagocytosis.

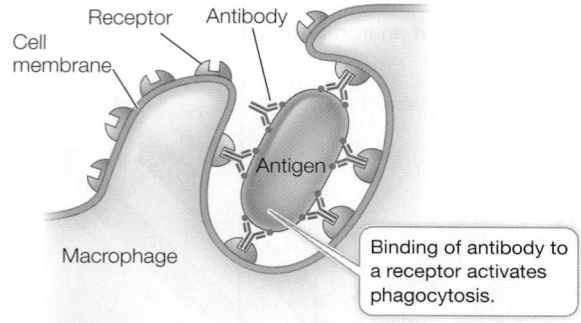

Receptor   Antibody
Cell membrane
Antigen
Macrophage

Binding of antibody to a receptor activates phagocytosis.

Immunoglobulin diversity results from DNA rearrangements and other mutations

Each mature B cell makes antibodies targeted to only one single antigen. And there are millions of possible antigens to which a human can be exposed. How can the genome encode enough different antibodies to protect the body against all the possible pathogens? Although there are millions of possible amino acid sequences in immunoglobulins, there are not millions of different immunoglobulin genes. It turns out that instead of a single gene encoding each complete immunoglobulin, the genome of the differentiating B cell has multiple different coding regions for each domain of the protein, and diversity is generated by putting together different combinations of these regions. Shuffling of this genetic deck generates the enormous immunological diversity that characterizes each individual mammal.

Each gene encoding an immunoglobulin chain is in reality a "supergene" assembled by means of genetic recombination from several clusters of smaller genes scattered along part of a chromosome (**Figure 41.9**). Every cell in the body has hundreds of immunoglobulin genes located in separate clusters that are potentially capable of participating in the synthesis of both the variable and constant regions of immunoglobulin chains. In most body cells

table **41.2** Antibody Classes				
Class	General structure		Location	Function
IgG	Monomer		Free in blood plasma; about 80 percent of circulating antibodies	Most abundant antibody in primary and secondary immune responses; crosses placenta and provides passive immunization to fetus
IgM	Pentamer		Surface of B cell; free in blood plasma	Antigen receptor on B cell membrane; first class of antibodies released by B cells during primary response
IgD	Monomer		Surface of B cell	Cell surface receptor of mature B cell; important in B cell activation
IgA	Dimer		Saliva, tears, milk, and other body secretions	Protects mucosal surfaces; prevents attachment of pathogens to epithelial cells
IgE	Monomer		Secreted by plasma cells in skin and tissues lining gastrointestinal and respiratory tracts	Binds to mast cells and basophils to sensitize them to subsequent binding of antigen, which triggers release of histamine that contributes to inflammation and some allergic responses

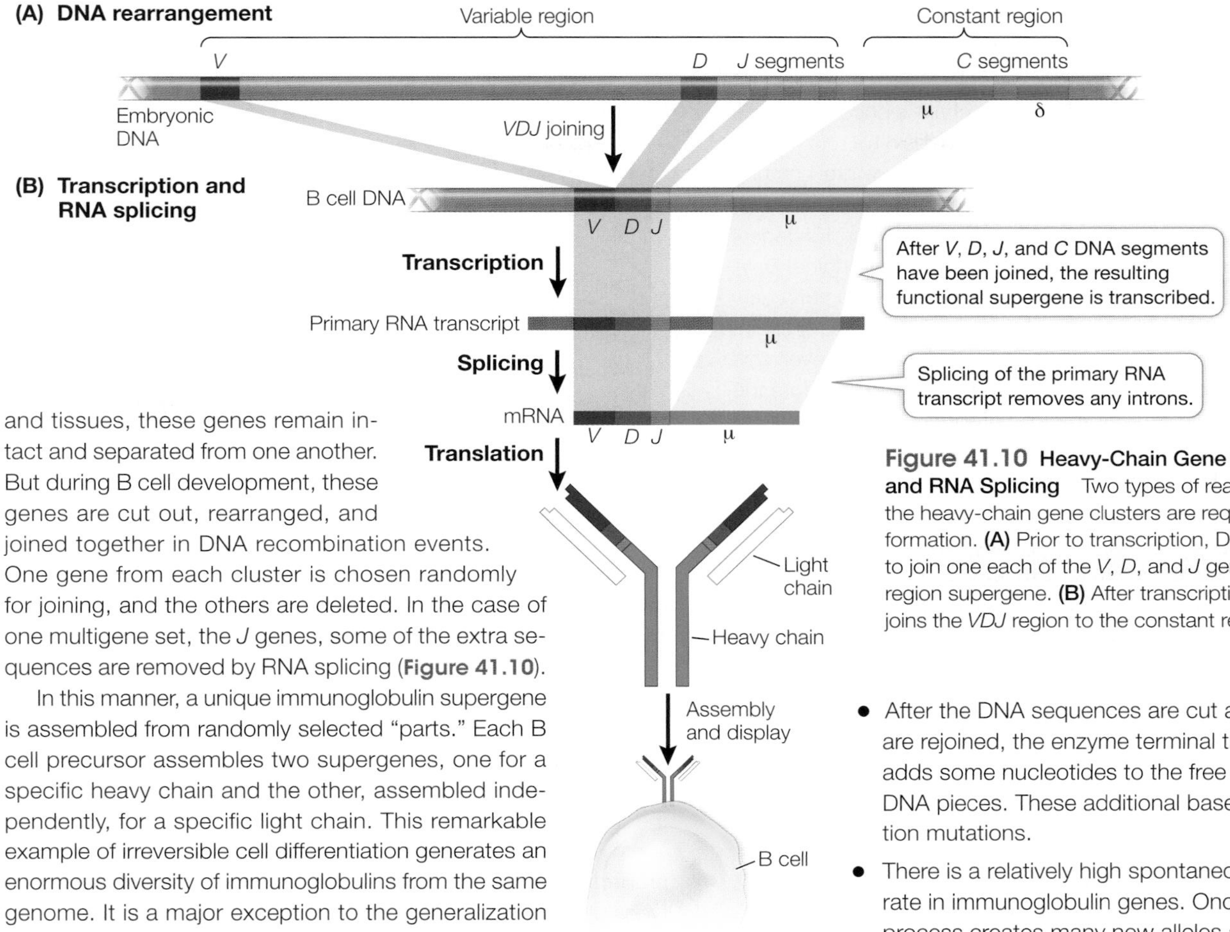

(A) DNA rearrangement

Variable region

Constant region

V   D   J segments   C segments

Embryonic DNA

μ   δ

VDJ joining

(B) Transcription and RNA splicing

B cell DNA

V   D   J   μ

Transcription

Primary RNA transcript   μ

After V, D, J, and C DNA segments have been joined, the resulting functional supergene is transcribed.

Splicing

mRNA   V   D J   μ

Splicing of the primary RNA transcript removes any introns.

Translation

Light chain

Heavy chain

Assembly and display

B cell

and tissues, these genes remain intact and separated from one another. But during B cell development, these genes are cut out, rearranged, and joined together in DNA recombination events. One gene from each cluster is chosen randomly for joining, and the others are deleted. In the case of one multigene set, the J genes, some of the extra sequences are removed by RNA splicing (**Figure 41.10**).

In this manner, a unique immunoglobulin supergene is assembled from randomly selected "parts." Each B cell precursor assembles two supergenes, one for a specific heavy chain and the other, assembled independently, for a specific light chain. This remarkable example of irreversible cell differentiation generates an enormous diversity of immunoglobulins from the same genome. It is a major exception to the generalization that all somatic cells derived from the fertilized egg have identical DNA.

Figure 41.9 illustrates the gene families that encode the constant and variable regions of the heavy chain in mice. Multiple genes encode each of the three parts of the variable region: 100 V, 30 D, and 6 J genes. Each B cell randomly selects one gene from each of these clusters to make the final coding sequence (VDJ) of the heavy-chain variable region. So the number of different heavy chains that can be made through this random recombination process is quite large:

$$100\ V \times 30\ D \times 6\ J = 18{,}000 \text{ possible combinations}$$

Now consider that the light chains are similarly constructed, with a similar amount of diversity made possible by random recombination. If we assume that the degree of potential light-chain diversity is the same as that for heavy-chain diversity, the number of possible combinations of light- and heavy-chain variable regions is:

18,000 different light chains × 18,000 different heavy chains = 324 million possibilities!

Other mechanisms generate even more diversity:

- When the DNA sequences that encode the V, D, and J regions are rearranged so that they are next to one another, the recombination event is not precise, and errors occur at the junctions. This imprecise recombination can create ***frame-shift mutations**, generating new codons at the junctions, with resulting amino acid changes.

Figure 41.10 Heavy-Chain Gene Recombination and RNA Splicing   Two types of rearrangement in the heavy-chain gene clusters are required for antibody formation. **(A)** Prior to transcription, DNA is rearranged to join one each of the V, D, and J genes into a variable region supergene. **(B)** After transcription, RNA splicing joins the VDJ region to the constant region.

- After the DNA sequences are cut and before they are rejoined, the enzyme terminal transferase often adds some nucleotides to the free ends of the DNA pieces. These additional bases create insertion mutations.

- There is a relatively high spontaneous mutation rate in immunoglobulin genes. Once again, this process creates many new alleles and adds to antibody diversity.

---

*connect the concepts   When they occur in the coding regions of proteins, point mutations can cause silent, missense, nonsense, or frame-shift mutations (see Figure 15.2). Key Concept 17.2 discusses how some sequences of DNA can move within a genome.

When we include these possibilities with the millions of combinations that can be made by random DNA rearrangements, it is not surprising that the immune system can mount a response to almost any natural or artificial substance.

Once the DNA rearrangements are completed, each supergene is transcribed and then translated to produce an immunoglobulin light chain or heavy chain. These chains combine to form an active immunoglobulin protein.

▶ Animation 41.3 **A B Cell Builds an Antibody**
www.Life11e.com/a41.3

The constant region is involved in immunoglobulin class switching

Table 41.2 describes the different classes of immunoglobulins and their functions. Generally, a B cell makes only one class at a time. But **class switching** can occur, in which a B cell changes the immunoglobulin class it synthesizes. For example, a B cell making IgM can switch to making IgG.

Early in its life, a B cell produces IgM molecules, which are the receptors responsible for its recognition of a specific antigen. At this time, the constant region of the heavy chain is encoded by the first constant region gene, the μ gene (see Figures 41.9 and 41.10). If the B cell later becomes a plasma cell during a humoral immune response, another deletion occurs in the cell's DNA, positioning the variable region genes (consisting of the same *V*, *D*, and *J* genes) next to a constant region gene farther away on the original DNA molecule. Such a DNA deletion results in the production of a new immunoglobulin with a different constant region of the heavy chain, and therefore a different function (see Table 41.2). However, this immunoglobulin has the same variable regions—and therefore the same antigen specificity—as the IgM produced by the parent B cell. The new immunoglobulin protein falls into one of the other four classes (IgA, IgD, IgE, or IgG), depending on which of the constant region genes is placed adjacent to the variable region genes.

41.4 recap

The humoral immune response is based on the synthesis by B cells of specific immunoglobulins. The specificity of an immunoglobulin derives from the amino acid sequence of its variable regions. B cells can make millions of immunoglobulins with different specificities by rearranging the genes that encode the variable regions of the heavy and light chains.

learning outcomes

You should be able to:

- Compare the antigen-binding site of an antibody with other molecules such as enzymes that bind other molecules.
- Describe the roles of molecules such as immunoglobulins in humoral immunity and of T cell receptors in cellular immunity.
- Compare the diversity of specific antibodies in an individual with the diversity of other biological molecules.

1. Describe the part of an antibody molecule that interacts with an antigen. How is it similar to the active site of an enzyme? How does it differ?

2. Compare immunoglobulins and T cell receptors with respect to structure and function.

3. Discuss the diversity of antibody specificities in an individual in relation to the diversity of enzymes. Does every cell in an animal contain genetic information for all of that organism's enzymes? For all of the immunoglobulins? Explain your answer.

By making antibodies, B cells are the major players in the humoral immune response. We will now turn to the cellular immune response, and the T cells responsible for it.

key concept

41.5 The Cellular Adaptive Response Involves T Cells and Receptors

Two types of effector T cells (T-helper cells and cytotoxic T cells) are involved in the cellular immune response. They work along with

proteins of the major histocompatibility complex (the MHC proteins), which present antigens on the surfaces of cells and contribute to the immune system's tolerance for the body's own cells.

focus your learning

- In the effector phase of the humoral response, T-helper (T_H) cells activate previously unexposed B cells with the same specificity to produce antibodies.
- Regulatory T cells (Tregs), which recognize self antigens, help ensure that the immune system does not attack self cells and molecules indiscriminately.
- Two classes of effector T cells (cytotoxic T cells and T-helper cells) are involved in the cellular adaptive response.
- Diversity in MHC molecules is very high in humans.

Like B cells, T cells possess specific membrane receptors. The T cell receptor is not an immunoglobulin, but a glycoprotein with a molecular weight of about half that of an IgG. It is made up of two polypeptide chains, each encoded by a separate gene (**Figure 41.11**). The two chains have distinct regions with constant and variable amino acid sequences. As in the immunoglobulins, the variable regions provide the site for specific binding to antigens. But there is one major difference: whereas an antibody can bind to an antigen whether it is present on the surface of a cell or not, a T cell receptor binds only to an antigen displayed by an MHC protein on the surface of an antigen-presenting or target cell.

When a T cell is activated by contact with a specific antigen, it proliferates and forms a clone. Its descendants form clones of two types of effector T cells:

1. Cytotoxic T cells, or T_C cells, recognize virus-infected or mutated cells and kill them by inducing lysis.

2. T-helper cells (T_H cells, also called helper T cells) assist both the cellular and the humoral immune responses.

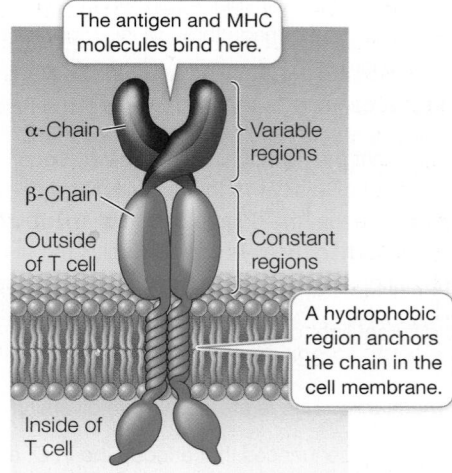

The antigen and MHC molecules bind here.

α-Chain

β-Chain

Outside of T cell

Variable regions

Constant regions

A hydrophobic region anchors the chain in the cell membrane.

Inside of T cell

Figure 41.11 A T Cell Receptor The receptors on T lymphocytes are smaller than those on B lymphocytes, but their two polypeptides contain both variable and constant regions. As with the B cell receptors, the constant regions fix the receptor in the cell membrane, while the variable regions establish the specificity for binding to antigen.

Animation 41.4 **Cellular Immune Response**
www.Life11e.com/a41.4

MHC proteins present antigen to T cells, causing recognition

Both T_H and T_C cells express T cell receptors that bind to antigen on the cell surface. But the response of each cell type to binding is quite different. T_H binding results in activation of the adaptive immune response, whereas T_C binding results in death of the cell carrying the antigen. MHC proteins form complexes with antigens on cell surfaces and assist with recognition by the T cells, so that the appropriate type of T cell binds.

The MHC proteins are cell membrane glycoproteins. Two types of MHC proteins function to present antigens to the two different types of T lymphocytes:

1. **Class I MHC proteins** are present on the surface of every nucleated cell in the mammalian body. They present antigens to T_C cells. These antigens can be fragments of virus proteins in virus-infected cells or abnormal proteins made by cancer cells as a result of somatic mutations. A T_C cell with the appropriate T cell receptor binds to the MHC-antigen complex. To ensure binding, the T_C cell also has a cell surface receptor protein called CD8 that recognizes and binds to MHC I.

2. **Class II MHC proteins** are on the surfaces of macrophages, B cells, and dendritic cells. They present antigens to T_H cells. The three cell types ingest antigens and break them down; one of the fragments then binds to MHC II for presentation (**Figure 41.12**). A T_H cell with the appropriate T cell receptor binds to the MHC-antigen complex. To ensure binding, the T_H cell also has a cell surface receptor protein called CD4 that recognizes and binds to MHC II.

In humans there are three genetic loci for class I MHC proteins and three for class II MHC proteins. Each of these six loci has as many as 100 different alleles. With so many possible allele combinations, it is not surprising that different people are very likely to have different MHC genotypes. This is why it can be difficult to find a good "match" for organ donations. MHC proteins are "self" markers. To accomplish its role in antigen presentation, an MHC protein has an antigen-binding site that can hold a peptide of about 10–20 amino acids.

Information on MHC proteins, the cellular origins of antigens, and T lymphocytes is summarized in **Table 41.3**.

T-helper cells and MHC II proteins contribute to the humoral immune response

When a T_H cell survives the selection processes and binds to an antigen-presenting cell, it releases cytokines that activate the T_H cell to proliferate, producing a clone of T_H cells with the same

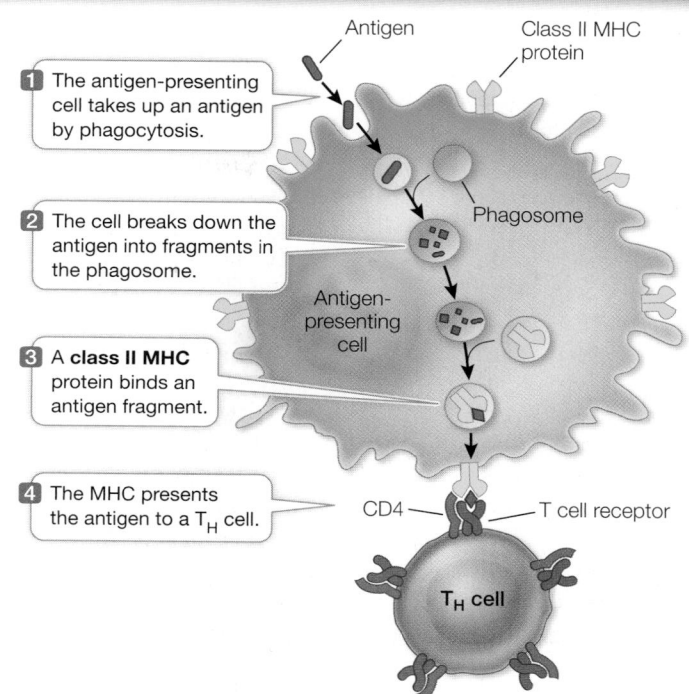

1 The antigen-presenting cell takes up an antigen by phagocytosis.

2 The cell breaks down the antigen into fragments in the phagosome.

3 A **class II MHC** protein binds an antigen fragment.

4 The MHC presents the antigen to a T_H cell.

Antigen

Class II MHC protein

Phagosome

Antigen-presenting cell

CD4

T cell receptor

T_H cell

Figure 41.12 **Macrophages Are Antigen-Presenting Cells**
A fragment of an antigen is displayed by MHC II on the surface of a macrophage. T cell receptors on a specific T_H cell can then bind to and interact further with the antigen–MHC II complex.

specificity. The steps to this point constitute the activation phase of the humoral immune response, and they occur in the lymphoid tissues. Next comes the effector phase, in which the T_H cells activate previously unexposed B cells with the same specificity to produce antibodies.

B cells are also antigen-presenting cells. B cells take up antigens bound to their surface immunoglobulin receptors by endocytosis, break them down, and display antigenic fragments on class II MHC proteins. When a T_H cell binds to the displayed antigen–MHC II complex, it releases cytokines that cause the B cell to produce a clone of plasma cells and memory cells (**Figure 41.13A**). Finally, the plasma cells secrete antibodies, completing the effector phase of the humoral immune response.

Cytotoxic T cells and MHC I proteins contribute to the cellular immune response

Class I MHC proteins play a role in the cellular immune response that is similar to the role played by class II MHC proteins in the humoral immune response. In a virus-infected or mutated cell, foreign or abnormal proteins or peptide fragments combine with class I MHC molecules. The resulting complex is displayed on the cell surface and

table **41.3** The Interaction between T Cells and Antigen-Presenting Cells				
Presenting cell type	Antigen presented	MHC class	T cell type	T cell surface protein
Any cell	Intracellular protein fragment	Class I	Cytotoxic T cell (T_C)	CD8
Macrophages, dendritic cells, and B cells	Fragments from extracellular proteins	Class II	Helper T cell (T_H)	CD4

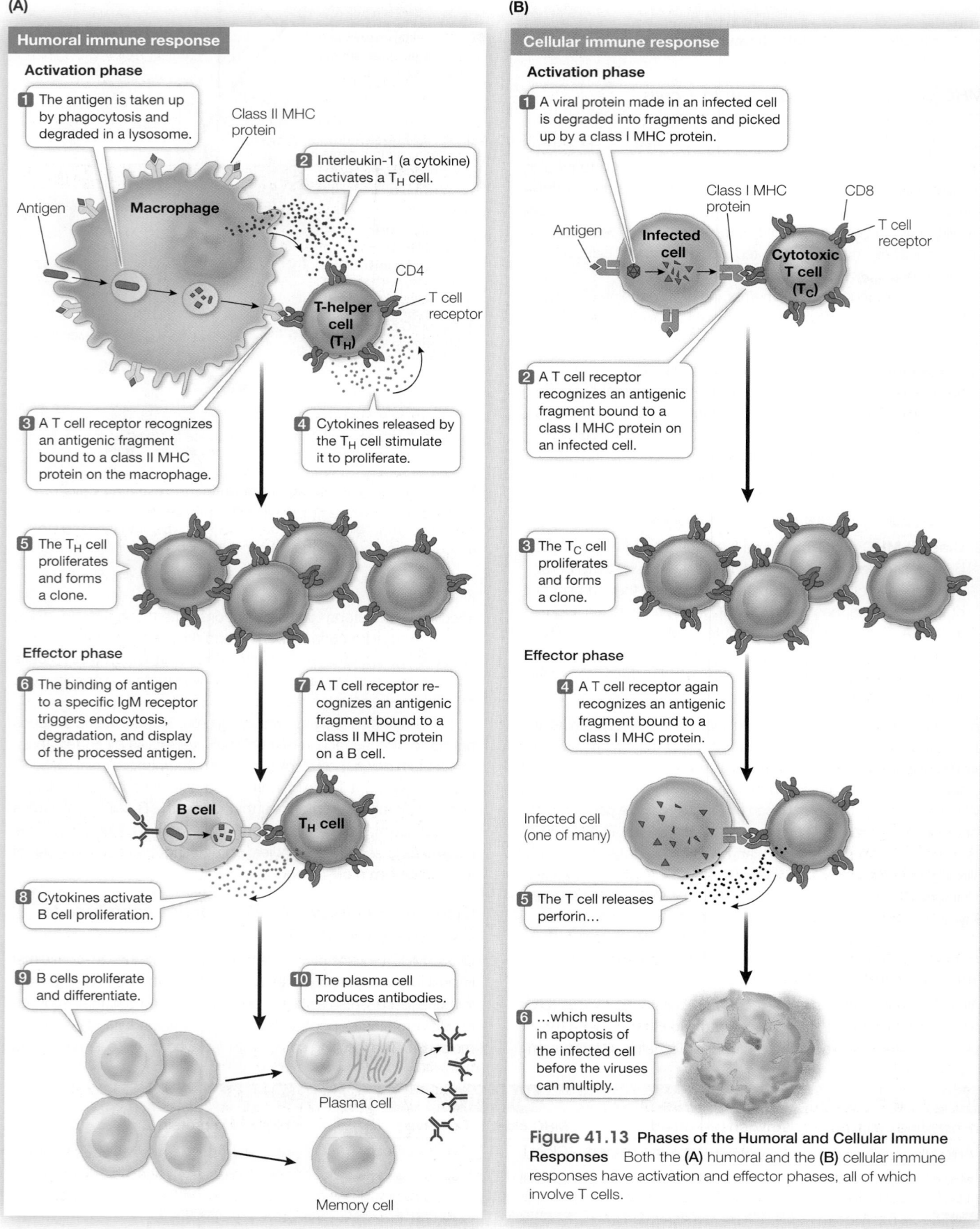

(A)

Humoral immune response

Activation phase

1 The antigen is taken up by phagocytosis and degraded in a lysosome.

Class II MHC protein

2 Interleukin-1 (a cytokine) activates a T_H cell.

Antigen

Macrophage

CD4

T cell receptor

T-helper cell (T_H)

3 A T cell receptor recognizes an antigenic fragment bound to a class II MHC protein on the macrophage.

4 Cytokines released by the T_H cell stimulate it to proliferate.

5 The T_H cell proliferates and forms a clone.

Effector phase

6 The binding of antigen to a specific IgM receptor triggers endocytosis, degradation, and display of the processed antigen.

7 A T cell receptor recognizes an antigenic fragment bound to a class II MHC protein on a B cell.

B cell

T_H cell

8 Cytokines activate B cell proliferation.

9 B cells proliferate and differentiate.

10 The plasma cell produces antibodies.

Plasma cell

Memory cell

(B)

Cellular immune response

Activation phase

1 A viral protein made in an infected cell is degraded into fragments and picked up by a class I MHC protein.

Class I MHC protein

CD8

Antigen

Infected cell

Cytotoxic T cell (T_C)

T cell receptor

2 A T cell receptor recognizes an antigenic fragment bound to a class I MHC protein on an infected cell.

3 The T_C cell proliferates and forms a clone.

Effector phase

4 A T cell receptor again recognizes an antigenic fragment bound to a class I MHC protein.

Infected cell (one of many)

5 The T cell releases perforin...

6 ...which results in apoptosis of the infected cell before the viruses can multiply.

Figure 41.13 Phases of the Humoral and Cellular Immune Responses Both the **(A)** humoral and the **(B)** cellular immune responses have activation and effector phases, all of which involve T cells.

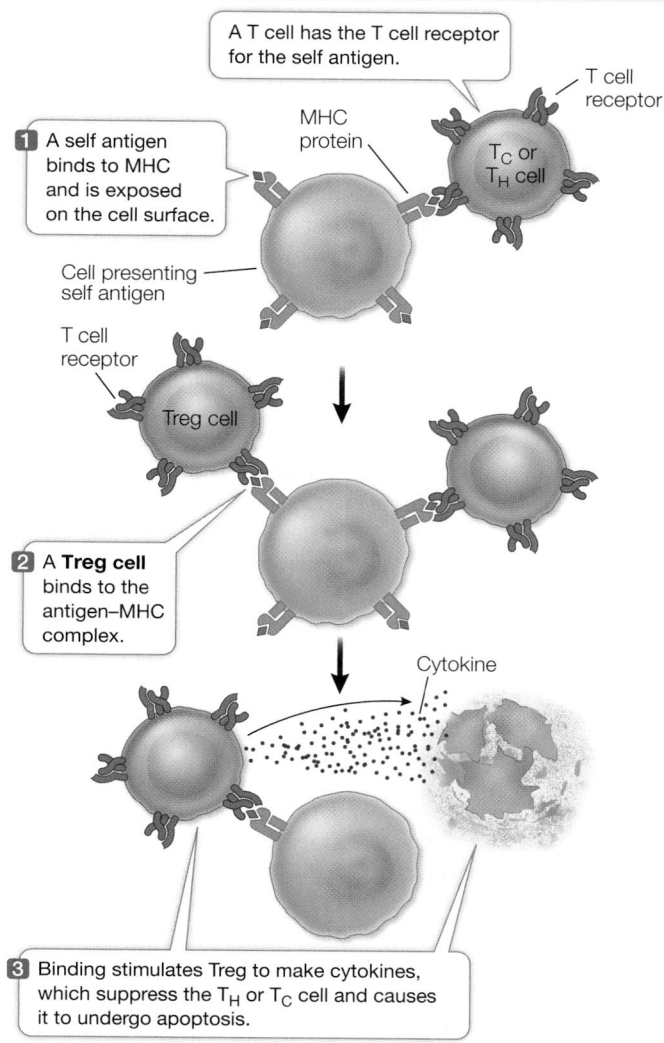

A T cell has the T cell receptor for the self antigen.

T cell receptor

MHC protein

T$_C$ or T$_H$ cell

1 A self antigen binds to MHC and is exposed on the cell surface.

Cell presenting self antigen

T cell receptor

Treg cell

2 A **Treg cell** binds to the antigen–MHC complex.

Cytokine

3 Binding stimulates Treg to make cytokines, which suppress the T$_H$ or T$_C$ cell and causes it to undergo apoptosis.

Figure 41.14 Tregs and Tolerance   A special class of T cells called regulatory T cells (Tregs) inhibits the activation of the immune system in response to self antigens.

presented to T$_C$ cells. When a T$_C$ cell recognizes and binds to this antigen–MHC I complex, it is activated to proliferate (**Figure 41.13B**).

In the effector phase of the cellular immune response, T$_C$ cells recognize and bind to cells bearing the same antigen–MHC I complex. These bound T$_C$ cells produce a substance called perforin, which ultimately results in apoptosis of the bound target cell. In addition, the T$_C$ cells can bind to a specific receptor (called Fas) on the target cell that initiates apoptosis in that cell. These two mechanisms work in concert to eliminate the antigen-containing host cell. Because T$_C$ cells recognize MHC proteins complexed with nonself antigens, they help rid the body of its own virus-infected or cancer cells.

Regulatory T cells suppress the humoral and cellular immune responses

A third class of T cells called **regulatory T cells** (**Tregs**) ensures that the immune system does not attack self cells and molecules indiscriminately. Like T$_H$ and T$_C$ cells, Tregs mature in the thymus gland, carry T cell receptors, and become activated if they bind to

antigen–MHC complexes. But Tregs are different in one important way: the antigens that Tregs recognize are *self antigens*. The activation of Tregs causes them to secrete cytokines, which blocks the activation of T cells that are bound to the same antigen-presenting cell (**Figure 41.14**). Thus Tregs constitute another mechanism for distinguishing self from nonself.

41.5 recap

The cellular immune response acts against virus-infected or mutated body cells. Specific receptors on T cells bind to antigen–MHC complexes displayed on cell surfaces. During development, T cells are selected that recognize MHC proteins. T-helper cells bind antigens on MHC II proteins and contribute to the humoral immune response. Cytotoxic T cells bind antigens on MHC I proteins and contribute to the cellular immune response. Tregs suppress immune responses to self antigens.

learning outcomes

You should be able to:

- Summarize the role of MHC proteins in the humoral immune response.
- Provide reasonable speculation regarding how drugs could generate immunosupression by targeting T-helper cells, and predict the side effects of these drugs.
- Describe how Tregs suppress the cellular and humoral immune systems.
- Predict consequences of inappropriate levels of activity of Tregs.

1. MHC proteins almost always differ among unrelated people, so organ transplants usually provoke a cellular immune response by the host, causing rejection. People receiving transplants are given a drug called cyclosporine which inhibits T cell development. How do you think cyclosporine works?

2. When a virus infects a cell, the immune system mounts an adaptive response to this infection. What are the roles of the T cell receptor and MHC protein in this response?

3. Cancer cells often express proteins that the body has never seen, so they are targeted by the cellular adaptive immune system. In some cancers, however, Tregs are increased in numbers and activity. Why would this indicate a poor prognosis?

Given the numerous and complex cellular interactions that activate the immune system and generate antibody diversity, you may have perceived many points at which the immune system could fail. We will now turn to several situations in which one or more components of this complex system malfunction.

key concept

41.6 Malfunctions in Immunity Can Be Harmful

Sometimes the immune system overreacts, as in an allergic response. Sometimes it attacks self antigens, as in autoimmune disease. And sometimes it fails to act, or acts weakly, as in immune deficiency disease.

Initial response: sensitization

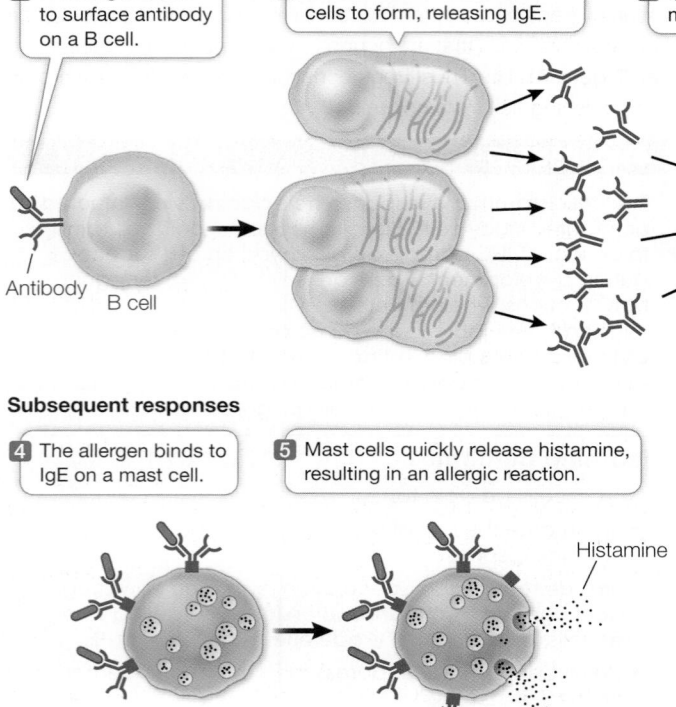

1 An allergen binds to surface antibody on a B cell.

2 This causes a clone of plasma cells to form, releasing IgE.

3 IgE binds to receptors on mast cells or basophils.

Receptor

Antibody   B cell

Mast cell

Subsequent responses

4 The allergen binds to IgE on a mast cell.

5 Mast cells quickly release histamine, resulting in an allergic reaction.

Histamine

Figure 41.15 An Allergic Reaction   An allergen is an antigen that stimulates B cells to make large amounts of IgE antibodies, which bind to mast cells and basophils. When the body encounters the allergen again, these cells produce large amounts of histamine, which has harmful physiological effects.

focus your learning

- Allergic reactions, which can be immediate or delayed, result from an over-stimulation of the immune response.
- Allergies can be treated by desensitization.
- Autoimmunity occurs when T cells bind to antigen–MHC complexes that carry self antigens.
- The study of HIV has led to the development of treatments for HIV.

The immune system mounts an excessive response in allergic reactions

An **allergic reaction** arises when the human immune system overreacts to (is hypersensitive to) a dose of antigen. Although the antigen itself may present no danger to the host, the inappropriate immune response may produce inflammation and other symptoms, which can cause serious illness or even death. Allergic reactions are the most familiar examples of this phenomenon. Allergic reactions may involve immediate hypersensitivity or delayed hypersensitivity.

IMMEDIATE HYPERSENSITIVITY   **Immediate hypersensitivity** arises when an allergic individual is exposed to an antigen (in this case referred to as an allergen) from the environment, such as a

food, pollen, or the venom of an insect. In response to the allergen, the individual makes large amounts of IgE. When this happens, mast cells in tissues and basophils in the blood bind the constant end of the IgE. If that individual is exposed to the same allergen again, binding of the allergen to the IgE causes the mast cells and basophils to rapidly release a large amount of histamine (**Figure 41.15**). This results in symptoms such as dilation of blood vessels, inflammation, and difficulty breathing. If not treated with antihistamines, a severe allergic reaction can lead to death. It is not known why some people produce excessive amounts of IgE in response to allergens. There is some evidence for genetic factors predisposing people to allergic responses.

Allergy to pollen can be treated using a process called desensitization. The process involves injecting small amounts of the allergen (typically just an extract of the offending plant tissue) into the skin—enough to stimulate IgG production but not enough to stimulate IgE production. The next time the person is exposed to the allergen, IgG binds to it, tying it up before IgE can bind it and exert its harmful effects.

Desensitization does not work well for food allergens because the IgE response to those substances is so strong that even a small amount of antigen provokes it. The best approach for those with food allergies is to avoid foods containing the allergens. This can be difficult, but food labels listing all the ingredients are helpful. Molecular biologists are beginning to identify the antigens that act as allergens, with the hope of developing vaccines or genetically modified foods that lack the allergenic epitopes.

DELAYED HYPERSENSITIVITY   **Delayed hypersensitivity** is an allergic reaction that does not begin until hours after exposure to an antigen. In this case the antigen is taken up by antigen-presenting cells, and a T cell response is initiated. A T_H cell produces a clone of cells that secrete various cytokines, which cause such reactions as inflammation and rash. These events take time (hence the term "delayed"). An example is the rash that develops after exposure to poison ivy.

Autoimmune diseases are caused by reactions against self antigens

Errors in the selection of T cells in the thymus can result in T cells that bind to antigen–MHC complexes that carry self antigens. Although the precise origin of **autoimmunity** is not known, there are several hypotheses:

- *Failure of negative selection*. A clone of lymphocytes making antibodies against self antigens that should have been destroyed by clonal deletion is not destroyed.

- *Molecular mimicry*. T cells that recognize a nonself antigen, such as a virus, also recognize something on a self antigen that has a similar structure.

Autoimmunity does not always result in disease, but several autoimmune diseases are common:

- People with *systemic lupus erythematosis* (SLE) have antibodies to many cellular components, including DNA and nuclear proteins released from dying cells. These antinuclear antibodies can cause serious damage when they bind to normal tissue antigens and form large circulating antigen–antibody complexes, which become stuck in tissues and provoke inflammation.

- People with *rheumatoid arthritis* have difficulty in shutting down a T cell response to self antigens. These patients may have low activity of CTLA4, an inhibitory protein that blocks T cells from reacting to self antigens. This results in inflammation of the joints and other tissue damage.

- *Hashimoto's thyroiditis* is the most common autoimmune disease in women over 50. Immune cells attack thyroid tissue, resulting in fatigue, depression, weight gain, and other symptoms.

- *Insulin-dependent diabetes mellitus*, or type I diabetes, occurs most often in children. It is caused by an immune reaction against several proteins in the cells of the pancreas that manufacture the protein hormone insulin. This reaction kills the insulin-producing cells, so people with type I diabetes must take insulin daily in order to survive.

AIDS is an immune deficiency disorder

There are several inherited and acquired immune deficiency disorders. In some individuals, T or B cells never form; in others, B cells lose the ability to give rise to plasma cells. In either case, the affected individual is unable to mount an adaptive immune response and thus lacks a major line of defense against pathogens. The T_H cell is perhaps the most central component of the immune system because of its essential roles in both the humoral and cellular immune responses (see Figure 41.6). This cell is the target of **human immunodeficiency virus** (**HIV**), the retrovirus that results in **acquired immune deficiency syndrome** (**AIDS**).

HIV can be transmitted from person to person in body fluids containing the virus (such as blood, semen, or vaginal fluid). The recipient tissue is either blood (by transfusion) or a mucous membrane lining an organ (the mucus contains a high concentration of lymphocytes). HIV initially infects macrophages, T_H cells, and antigen-presenting dendritic cells in the blood and tissues. At first there is an immune response to the viral infection, and T_H cells are activated. But because HIV infects the T_H cells, they are killed both by HIV itself and by T_C cells that lyse infected T_H cells. Consequently T_H cell numbers decline after the first month or so of infection. Meanwhile, the extensive production of HIV by infected cells activates the humoral immune system. Antibodies bind to HIV, and the complexes are removed by phagocytes. The HIV level in blood goes down. There is still a low level of infection, however, because of the depletion of T_H cells (**Figure 41.16**). This process reaches a low, steady-state level called the "set point." This point varies among individuals and

is a strong predictor of the rate of progression of the disease. For most people it takes 8–10 years without treatment for the more severe manifestations of AIDS to develop. In some it can take as little as a year; in others, 20 years.

During this dormant period, people carrying HIV generally feel fine, and their T_H cell levels are adequate for them to mount immune responses. Eventually, however, the virus destroys the T_H cells, and their numbers fall to the point where the infected person is susceptible to infections that the T_H cells would normally eliminate. These infections result in conditions such as Kaposi's sarcoma, a skin tumor caused by a herpes virus; pneumonia caused by the fungus *Pneumocystis jirovecii*; and lymphoma tumors caused by the Epstein–Barr virus. These conditions result from opportunistic infections because the pathogens take advantage of the crippled immune system of the host. They lead to death within a year or two.

HIV has been intensively studied. This has resulted in the development of drugs targeted to HIV proteins, such as the reverse transcriptase that makes cDNA from the viral RNA, and the viral protease that cuts the large precursor viral protein into its final active proteins. Treatment with combinations of such drugs has had spectacular success. Getting AIDS before the 1990s was a death sentence, with few sufferers surviving beyond a year or two. Today the survival of a treated, infected person is decades longer. Unfortunately, like many medical treatments, HIV drugs are not available to all who need them—particularly in poor regions of the world where AIDS is prevalent. Since AIDS is caused by a viral infection, a vaccine would seem an appropriate preventive agent. Unfortunately, despite a huge international effort that is ongoing, a vaccine against HIV has not been developed, probably because the virus "hides out" inside cells and is prone to a high mutation rate. There are about 1.2 million deaths per year worldwide from AIDS.

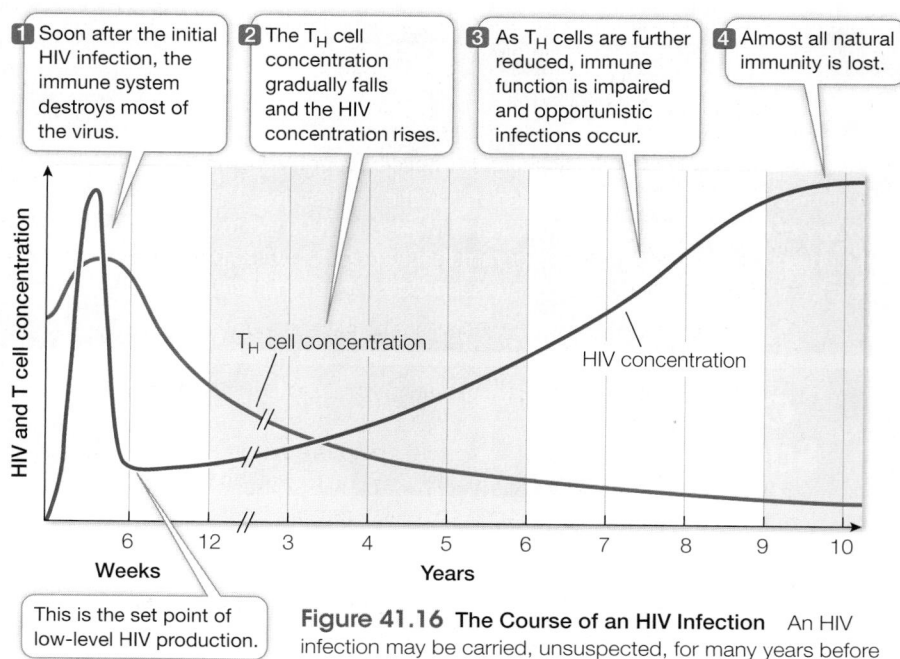

1 Soon after the initial HIV infection, the immune system destroys most of the virus.

2 The T_H cell concentration gradually falls and the HIV concentration rises.

3 As T_H cells are further reduced, immune function is impaired and opportunistic infections occur.

4 Almost all natural immunity is lost.

T_H cell concentration

HIV concentration

HIV and T cell concentration

Weeks   6   12

Years   3   4   5   6   7   8   9   10

This is the set point of low-level HIV production.

Figure 41.16 The Course of an HIV Infection   An HIV infection may be carried, unsuspected, for many years before the onset of symptoms.

41.6 recap

Failures of the immune system include allergic reactions (caused by hypersensitivity to antigens), autoimmune diseases (caused by reactions against self antigens), and immune deficiency disorders.

learning outcomes

You should be able to:

- Summarize the responses made during immediate hypersensitivity.
- Describe how desensitization can be used to minimize allergic reactions.
- Summarize the hypotheses that explain how autoimmunity can occur.
- Outline experiments to test the effectiveness of an HIV vaccine.

1. Some people are allergic to peanuts. Outline the events that characterize the immune reactions involved.
2. There has been recent progress in treating peanut allergy by desensitization. How does this work?
3. A new cancer treatment involves promoting autoimmunity by designing a drug to inhibit CTLA4, a protein on T cells that normally blocks them from mounting an autoimmune response. How does this treatment work? What might be its side effects?
4. Development of an effective HIV vaccine requires that a person being vaccinated develop both cellular and humoral immunity against HIV. What experiments would you perform to test whether both types of immunity developed in people given a potential new vaccine?

investigatinglife

Q&A What are the mechanisms and implications of long-lasting immunity?

You have seen in this chapter that lasting immunity against a pathogen can be achieved in two ways, both of which involve exposure to an antigen associated with the pathogen and the resulting production of memory cells. The first way is natural: In Investigating Life: What Are the Mechanisms and Implications of Long Lasting Immunity?, we showed that people exposed to virulent strains of flu viruses retain their immunity, and even increase it when exposed to less harmful flu strains. The second way is artificial: As we described in the opening story, vaccines have been spectacularly successful at inducing lasting immunity and even eradicating diseases when enough people are vaccinated. Why, then, do people refuse vaccination? There appear to be several reasons. One is complacency—the threat of disease may seem remote. For example, measles, which used to kill thousands of children every year in the United States and still does in poor countries, is no longer a highly visible threat to public health. Second, some people believe that vaccines, although exhaustively tested and proven safe, are actually unsafe and cause disease. The internet is full of such assertions. Third, false alarms have led people to dismiss vaccination advisories. The discovery of the H1N1 flu virus (swine flu) in Mexico in 2009 led to a high alert and a mass vaccination program that turned out not to be necessary. In addition, some people are suspicious of governmental programs in general. In Pakistan, a polio vaccination program was used as a ruse by officials to gain access to the home of the international terrorist Osama bin Laden. Vaccination is a scientific success in terms of bolstering immunity and eradicating disease. But like any technology, its acceptance is a political issue.

Future directions

Following activation of the immune system, many effector cells are mobilized. Most eventually die, but some survive as memory cells. Researchers are investigating how this happens. At the University of Wisconsin, Marulasiddappa Suresh and his colleagues have shown that a single transcription factor called FoxO1 plays an important role in the effector-to-memory-cell transition. This molecule is not needed for the production of effector cells, but if it is not there, memory cells do not form. The genes regulated by FoxO1 are being characterized and may hold the key to the memory cell. Meanwhile, at Yale University, Susan Kaech and her colleagues have shown that an intercellular signaling molecule, interleukin-7 (IL-7), binds to effector cells and results in changes that may allow memory cell survival. Specifically, the target cells produce a membrane channel that allows rapid uptake of glycerol, the backbone of fats. Production of fats in the memory cell may provide the energy needed for long-term survival. Understanding how memory cells form and survive may provide valuable information for the production of better vaccines.

Chapter Summary 41

▶ 41.1 Animals Use Innate and Adaptive Mechanisms for Defense

- Animal defenses against **pathogens** are based on the body's ability to distinguish between self and nonself.

- **Innate** (nonspecific) **defenses** are inherited mechanisms that protect the body from many kinds of pathogens. They typically act rapidly.
- **Adaptive** (specific) **defenses** respond to specific pathogens. They develop more slowly than innate defenses but are longer-lasting.

- Innate defenses evolved before adaptive defenses, and in animals there is a common signaling pathway that sets innate defenses in motion. **Review Figure 41.1**

- Many defenses are implemented by cells and proteins carried in the blood plasma and **lymph. Review Figure 41.2, Activity 41.1**

- **White blood cells** fall into two broad groups. **Phagocytes** engulf pathogens by phagocytosis. **Lymphocytes**, which include B cells and T cells, participate in adaptive responses. **Review Figure 41.3, Activity 41.2**

41.2 Innate Defenses Are Nonspecific

- An animal's innate defenses include physical barriers such as the skin, and competing resident microorganisms known as normal flora. **Review Figure 41.4**

- The **complement system** consists of more than 20 different antimicrobial proteins that act to alter membrane permeability and kill targeted cells.

- Circulating defensive cells, such as phagocytes and **natural killer cells**, eliminate invaders.

- **Inflammation** involves activation of several types of cells and proteins that act against invading pathogens. **Mast cells** release **histamine**, which causes blood vessels to dilate and become "leaky." **Review Figure 41.5, Activity 41.3**

41.3 Adaptive Defenses Are Specific

- The adaptive immune response recognizes specific **antigens**, responds to an enormous diversity of **antigenic determinants** (**epitopes**), distinguishes self from nonself, and remembers the antigens it has encountered. **Review Animation 41.1**

- Each antibody and each T cell is specific for a single antigenic determinant. **T cell receptors** and antibodies bind to antigens and initiate an adaptive immune response.

- The **humoral immune response** is directed against pathogens in the blood, lymph, and tissue fluids. The **cellular immune response** is directed against an antigen established within a host cell. Both responses are mediated by antigenic fragments being presented on a cell surface. **Review Focus: Key Figure 41.6**

- **Clonal selection** accounts for the specificity and diversity of the immune response and for **immunological memory**. **Review Figure 41.7**

- An activated B or T lymphocyte produces **effector cells** that attack the antigen, and **memory cells** that are long-lived and rarely divide. Effector B cells are called **plasma cells** and secrete specific **antibodies**.

- **Vaccination** is inoculation with modified pathogens or antigens that provoke an immune response but are not pathogenic. **Review Investigating Life: What Are the Mechanisms and Implications of Long-Lasting Immunity?**

41.4 The Humoral Adaptive Response Involves Antibodies

See Animation 41.2

- B cells are the basis of the humoral immune response. Unexposed B cells are activated by binding of antigen and by stimulation by T_H cells with the same specificity, and then form plasma cells. These cells synthesize and secrete specific antibodies.

- An antibody is an **immunoglobulin**, a tetramer of four polypeptides: two identical light chains and two identical heavy chains, each consisting of a **constant region** and a **variable region**. **Review Figure 41.8, Activity 41.4**

- The variable regions determine the specificity of an immunoglobulin, and the constant regions of the heavy chain determine its class. There are five classes of immunoglobulins with different body locations and functions. **Review Table 41.2**

- B cell genomes undergo random recombination of genes coding for regions of the immunoglobulin polypeptide chains so that each cell can produce a specific antibody protein. The immunoglobulin chains derive from "supergenes" that are constructed from different combinations of V, D, J, and C genes. This DNA rearrangement and rejoining yields millions of different immunoglobulin chains. **Review Figures 41.9, 41.10, Animation 41.3**

- Once a B cell becomes a plasma cell, it may undergo **class switching**, in which a deletion of one or more constant region genes results in the production of an immunoglobulin with a different constant region and a different function.

41.5 The Cellular Adaptive Response Involves T Cells and Receptors

See Animation 41.4

- T cells are the effectors of the cellular immune response. T cell receptors are somewhat similar in structure to the immunoglobulins, having variable and constant regions. **Review Figure 41.11**

- The genes of the major histocompatibility complex (MHC) encode membrane proteins that bind antigenic fragments and present them to T cells. **Review Figures 41.12, 41.13**

- There are three types of T cells. **Cytotoxic T cells** (T_C cells) recognize and kill virus-infected or mutated cells. **T-helper cells** (T_H cells) assist both the cellular and humoral immune responses. **Regulatory T cells** (**Tregs**) inhibit the other T cells from mounting an immune response to self antigens. **Review Figure 41.14**

41.6 Malfunctions in Immunity Can Be Harmful

- An **allergic reaction** is an inappropriate immune response caused by **immediate hypersensitivity** or **delayed hypersensitivity** to certain antigens. **Review Figure 41.15**

- Autoimmune diseases result when the immune system produces B and T cells that attack self antigens.

- Immune deficiency disorders result from failure of some part of the immune system. **Acquired immune deficiency syndrome** (**AIDS**) is a disorder that arises from depletion of the T_H cells as a result of infection with **human immunodeficiency virus** (**HIV**). **Review Figure 41.16**

Apply What You've Learned

Review

41.1 All white blood cells originate from multipotent stem cells in the bone marrow.

41.5 Two classes of effector T cells (cytotoxic T cells and T-helper cells) are involved in the cellular adaptive response.

41.5 Diversity in MHC molecules is very high in humans.

Original Papers: Candotti, F. et al. 2012. Gene therapy for adenosine deaminase-deficient severe combined immune deficiency: Clinical comparison of retroviral vectors and treatment plans. *Blood* 120: 3635–3646.

Gaspar, H. B., A. Alessandro, P. Fulvio, C. Fabio, M. S. Hershfield and L. D. Notarangelo. 2009. How I treat ADA deficiency. *Blood* 114: 3524–3532.

Hershfield, M. S. et al. 1987. Treatment of adenosine deaminase deficiency with polyethylene glycol-modified adenosine deaminase. *New England Journal of Medicine* 316: 589–596.

Severe combined immunodeficiency (SCID) is a very rare inherited condition in which individuals show high susceptibility to viral, fungal, and bacterial infection. Left untreated, it is fatal. SCID can be caused by mutations in any of several genes, one of which codes for the enzyme adenosine deaminase (ADA). ADA is produced in all cells, but the highest levels are found in lymphocytes.

Inside lymphocytes, when DNA is metabolized, toxic deoxyadenosine is produced. When ADA is present, deoxyadenosine is converted to a nontoxic molecule. When ADA is not present, deoxyadenosine accumulates and causes cell death, especially of T cells.

There are currently three treatments for ADA-SCID. In enzyme replacement therapy (ERT), patients are injected with ADA, usually weekly. This restores normal levels of ADA (6–12 μmol/h/mL) and deoxyadenosine (less than 10 percent of adenine nucleotides) in blood and immune function. The table below shows the results of ERT in one patient.

Weeks after initial treatment	Dosage (mL/kg)	Plasma ADA levels (μmol/h/mL)	Percent deoxyadenosine compared with all adenine nucleotides
0	0.1	0	44
1	0.1	2	28
2	0.2	3	19
3	0.2	7	11
4	0.2	8	5
5	0.2	5	4
6	0.2	4	3

The second treatment method is hematopoietic stem cell transplantation (HSCT), in which stem cells from a donor give rise to lymphocytes that carry the normal ADA gene. These are transplanted into the patient. The greatest obstacle in HSCT is finding a donor whose cell surface markers (in this case, human leukocyte antigens, a type of MHC protein) are similar to those of the patient. To understand the significance of the matching process, 87 patients underwent HSCT with donors having various levels of "matching," and were studied for long-term survival (**Figure A**).

Figure A

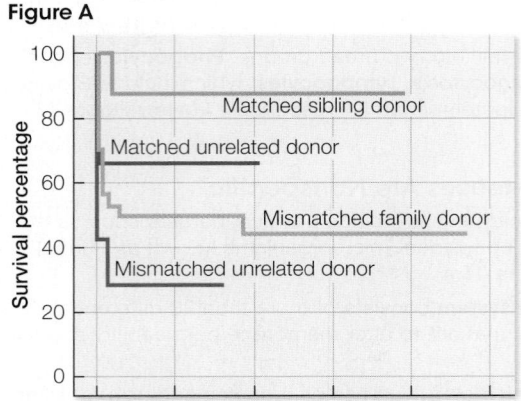

The third treatment for ADA-SCID, gene therapy, can be used when a matching stem cell donor is not available. Blood stem cells are removed from an ADA patient and then exposed in the lab to a viral vector that inserts the correct gene sequence for ADA into the cells. These modified cells are then returned to the patient. **Figure B** shows the results for one patient. The normal range of ADA activity is at least 60 nmol/10^8 cells/min.

Figure B

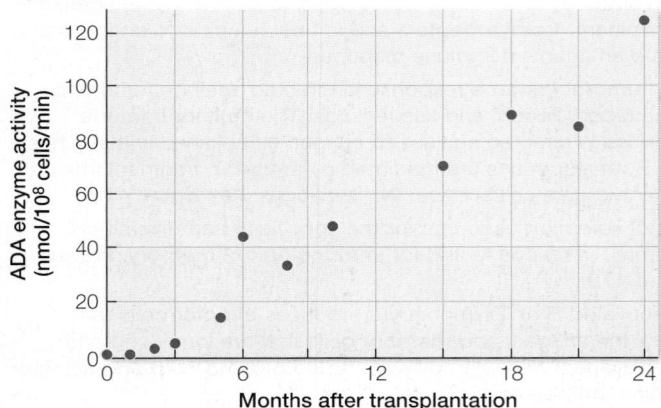

Questions

1. According to the table, how many weeks of ERT therapy were needed to restore blood ADA and deoxyadenosine levels to normal? Why were weekly injections necessary? Explain your answer.

2. Summarize the survival of the patients in Figure A from each type of matched donor. Which donor category would most closely approximate what would happen with an untreated patient? Explain your answer.

3. Most patients who are successfully treated with gene therapy also receive ERT. Explain how this would be advantageous.

This Blue-footed booby (*Sula nebouxii*) pair is in a courtship dance.

42

Animal Reproduction

investigatinglife

No Time to Waste

Female tammar wallabies become sexually receptive and mate soon after they give birth. The resulting embryo can be born any time from a month to a year later. Humans have a set 9-month pregnancy, so this highly variable duration of pregnancy seems unusual and leads to questions of "why" and "how." Knowing more about the development of the young wallaby, called a joey, helps. The joey is very immature when born. It crawls into its mother's pouch (the marsupium) where it suckles and grows for up to 10 months when it is weaned.

Answering the "why" question is not difficult. Reproductive success is the currency of evolution. Reproductive success depends on how many offspring a female can produce and on the chances that those offspring survive, reproduce, and pass on their genes to the following generation. Mating soon after giving birth (as the wallaby does) can maximize the number of offspring a female produces, but it does not necessarily maximize the survival and success of those offspring. Survival of offspring in many spe-cies depends on the ability of the mother to provide maternal care. Mammalian mothers produce milk to nourish their young, and that has an enormous energy cost. Therefore the females of many mammalian species do not become fertile and sexually receptive until they have weaned their young. But, this is not true for the tammar wallaby.

The large energy investment that a tammar wallaby mother makes to successfully rear a single offspring means that giving birth while still having one joey in the pouch lowers the chances of having either of those young survive. The adaptation of this species is developmental arrest, or diapause. The fertilized egg begins to develop into an embryo, but at an early stage of development its cell divisions cease, its metabolism decreases, and it remains in the uterus without implanting in the uterine lining as long as the previous offspring is suckling. If that offspring dies or is weaned, the quiescent embryo in the uterus is activated, implants in the uterine lining, and continues its development. Having an embryo ready to go when conditions are right is a way of maximizing reproductive success, but the "how" question remains.

Q A What signal controls the breaking of diapause in the tammar wallaby?

42.1 Asexual Reproduction Is Efficient but Limits Genetic Variability

When we think of reproduction, we think of sex, but many species can reproduce asexually—the offspring are exact genetic copies of the parents. Most of the animal species that reproduce asexually are invertebrates living in rather constant environments. However, examples of asexual reproduction do occur in vertebrates. Reproduction by female sharks living in aquaria with no males have been reported. The natural habitat of these fish is the open ocean where potential mates might not be readily available. Asexual reproduction is efficient because no time or energy is wasted on mating and every member of the population can convert resources into offspring.

focus your learning

- Asexual reproduction does not generate genetic diversity.
- Budding and regeneration are forms of asexual reproduction.
- Parthenogenesis is the development of offspring from unfertilized eggs.

You can demonstrate to yourself the efficiency of asexual reproduction with a thought exercise. Start with two females, one reproducing sexually and one asexually. Assume that each can have two offspring, but in the case of the sexually reproducing female, one will be a male. Also, assume that in each subsequent generation all of the females (mothers and daughters) will have two offspring each. In the first generation, the asexually reproducing female will have two daughters who can then each have two female offspring. Thus in the next generation there will be three females having two offspring each. The sexually reproducing female will give rise to only one female, so in the next generation there will be two females reproducing having two offspring each—only one of which is female. Run this thought exercise through four or five cycles and compare the numbers of offspring produced.

Asexual reproduction limits genetic diversity

A disadvantage of asexual reproduction is that the offspring are genetically identical to one another and to their parents. Thus asexual reproduction does not generate genetic diversity. Genetic diversity is the raw material that enables natural selection to shape adaptations in response to environmental change. When environmental changes occur, lack of genetic diversity can be disadvantageous to a population. Asexual reproduction is, however, a good way to preserve a successful genotype in a stable environment. Aphids are an example of a species that has it both ways (**Figure 42.1**). In the spring, the aphids that hatch on rose bushes from overwintering eggs are all females. Each of those aphids reproduces asexually to produce many more females—all genetically identical. This is a time when there is plenty of food and the environment is stable. When the conditions change due to drought or season, the females still reproduce asexually, but some of the offspring develop into males. The ensuing sexual reproduction introduces genetic variation into the eggs that will remain dormant until favorable conditions return.

Three common modes of asexual reproduction are budding, regeneration, and parthenogenesis.

Budding and regeneration produce new individuals by mitosis

Many simple multicellular animals produce offspring by **budding**. New individuals form as outgrowths or buds from the bodies of older animals. A bud grows by mitotic cell division, and the cells differentiate before the bud breaks away from the parent (**Figure 42.2A**). The bud is genetically identical to the parent, and it may grow as large as the parent before it becomes independent.

Regeneration is usually thought of as the replacement of damaged tissues or lost limbs, but in some cases pieces of an organism can regenerate complete individuals. Echinoderms, for example, have remarkable abilities to regenerate. If sea stars are cut into pieces, each piece that includes an arm and a portion of the central disc can grow into a new animal (**Figure 42.2B**). In the early 1900s oyster fishermen in Narragansett Bay tried to eliminate the sea stars that were preying on their oysters. Whenever they encountered sea stars, they cut them up with their knives and threw them back into the water. As a

Second instar

Third instar

First instar

Summer Cycle
(many generations)

Fourth instar

Live birth

Hatchling

Winter Cycle
(one generation)

Fall sexual reproductive (female)

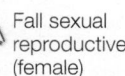

Egg

Female   Male

Figure 42.1 Aphids Are Sexual and Asexual   When environmental conditions are favorable, such as in a rose garden in the summer, aphids make the most of resources by rapidly reproducing asexually. They are actually making identical copies of themselves. But when environmental conditions are unsure, such as in the fall, aphids reproduce sexually and thereby increase their genetic diversity in the face of changing, unpredictable conditions.

(A) *Hydra littoralis*

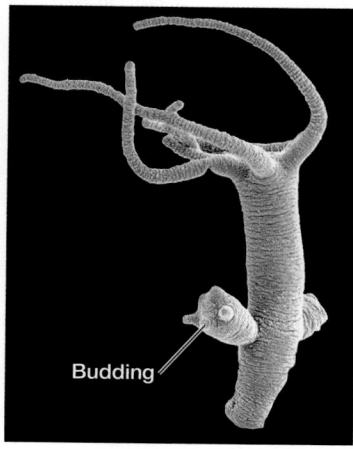

Budding

(B) *Linkia multifora*

Regeneration

Figure 42.2 Two Forms of Asexual Reproduction
(A) Budding: a new individual forms as an outgrowth from an adult hydra. **(B)** Regeneration: a single severed arm and a piece of the central disc of a mature sea star can regenerate into an entire animal.

result, the sea star population increased explosively. A variant of regeneration called fission involves the animal splitting in half with each half growing into a whole individual. An example is sea anemones.

Regeneration can occur when an animal is broken by an outside force such as wave action in the intertidal zone. In some cases, breakage occurs in the absence of external forces. Some species of segmented marine worms develop segments with rudimentary heads bearing sensory organs. The segments then break apart and each one forms a new worm.

Parthenogenesis is the development of unfertilized eggs

Not all eggs must be fertilized to develop. A common mode of asexual reproduction in arthropods (including aphids) is the development of offspring from unfertilized eggs. This phenomenon, called **parthenogenesis**, also occurs in some species of fish, amphibians, and reptiles. Most species that reproduce parthenogenetically also engage in sexual reproduction or at least sexual behavior at other times. In some species, parthenogenesis is part of the mechanism that determines sex. In honeybees (as well as in most ants and wasps), males develop from unfertilized eggs and are haploid, whereas females develop from fertilized eggs and are diploid.

Parthenogenetic reproduction in some species requires sexual behavior even though sperm are not involved and eggs are not fertilized. David Crews and his students at the University of Texas extensively investigated one such case, that of parthenogenetic reproduction in a species of whiptail lizard. There are no males of this species. Females can act as males, engaging in all aspects of courtship display and mating, although no sperm are produced or transferred (**Figure 42.3**). Whether a specific female acts as a female or as a male depends on cyclical hormonal states. When estrogen levels are high, she acts as a female. When her progesterone levels peak, she acts as a male. The stimulation resulting from the sexual activity triggers the release of eggs from the ovaries of the acting female.

(A) Sexual behavior in parthenogenetic whiptail lizards

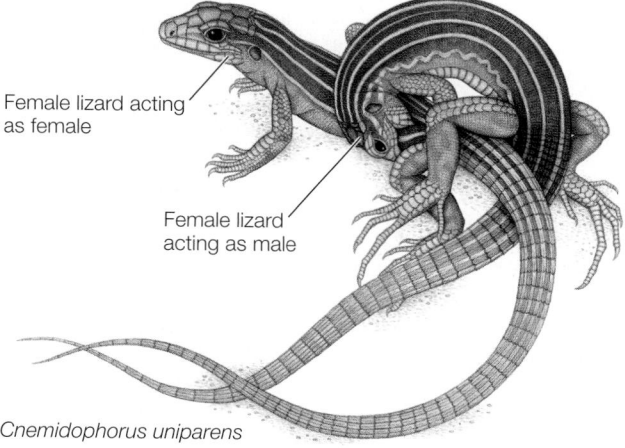

Female lizard acting as female

Female lizard acting as male

Cnemidophorus uniparens

(B) Sex role in whiptail lizards is under hormonal control

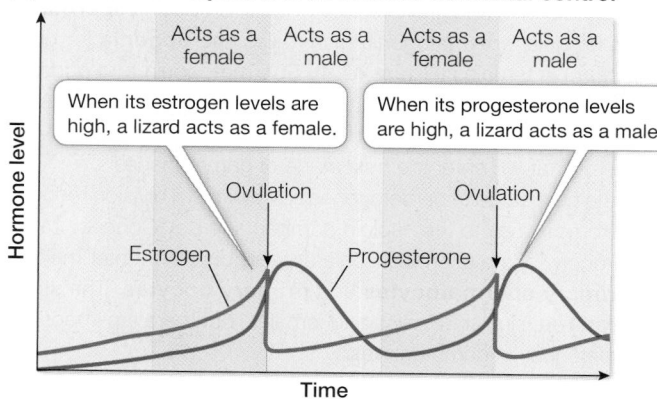

Acts as a female   Acts as a male   Acts as a female   Acts as a male

When its estrogen levels are high, a lizard acts as a female.

When its progesterone levels are high, a lizard acts as a male.

Hormone level

Ovulation

Ovulation

Estrogen

Progesterone

Time

Figure 42.3 Asexual Reproduction May Require Sexual Behavior   **(A)** Parthenogenetic whiptail lizards are all females, but they take turns acting the male role in reproductive behavior. The stimulation from sexual behavior is necessary for ovulation to occur. **(B)** The cycling of ovarian hormones determines the role an individual whiptail plays.

Q: What selective advantages could accrue to a bisexual species becoming unisexual? What disadvantages?

42.1 recap

Many species of animals, especially invertebrates, can reproduce asexually through budding, regeneration, or parthenogenesis. Asexual reproduction is an efficient use of resources but does not produce genetic diversity.

learning outcomes

You should be able to:

• Describe circumstances that make asexual reproduction disadvantageous.

• Describe the characteristics of the different types of asexual reproduction.

(continued)

42.1 recap (continued)

- Explain conditions that would favor parthenogenesis over other forms of asexual reproduction.

1. What is similar about budding and regeneration but *not* parthenogenesis?

2. Is global climate change likely to have a more negative impact on a species that reproduces asexually or one that reproduces sexually?

3. The Komodo dragon is a very large, sexually breeding species of lizard that lives on the Indonesian island of Komodo. A young Komodo dragon named Flora was raised alone for many years in the London Zoo. One year after reaching sexual maturity, she laid a clutch of eggs that produced several baby dragons, all males. How did Flora reproduce? What advantage might this ability have for an island-dwelling species? Why is it significant that the offspring were all male?

Asexual reproduction is an efficient way to use resources. Since most animals reproduce sexually, however, the genetic diversity produced by sexual reproduction must confer a tremendous advantage.

key concept

42.2 Sexual Reproduction Involves the Union of Haploid Egg and Sperm

Given the inefficiency of sexual reproduction, its prevalence is somewhat surprising. In most sexually reproducing species, the sex ratio is 1:1 and therefore half the population (males) cannot produce offspring. And mating behaviors involve costs and risks. Costs include time and energy spent finding, attracting, and competing for a mate, as well as the "opportunity costs" of detracting from other activities such as feeding and caring for existing offspring. Risks include increased exposure to predation and the potential for physical damage. Despite these disadvantages, most eukaryotic organisms reproduce sexually. Thus it would seem that the production of genetic diversity is an evolutionary advantage that can overwhelm the cost of sex (see Key Concepts 20.2 and 20.4).

focus your learning

- In spermatogenesis, four haploid sperm are produced from each primary spermatocyte.
- Oogenesis involves asymmetrical division of cytoplasm.
- Specific recognition molecules mediate interactions between sperm and eggs in sea urchins.
- Organisms in aquatic environments generally have external fertilization and engage in spawning to bring eggs and sperm together.
- Mechanisms of internal fertilization are highly diverse.
- Hermaphrodites can function both as male and as female.

Sexual reproduction requires the joining of two haploid sex cells to form a diploid individual. These haploid cells, or **gametes**, are produced through gametogenesis, a process that involves meiotic cell divisions. Two events in meiosis contribute to genetic diversity: crossing over between homologous chromosomes and the independent assortment of chromosomes (see Key Concepts 11.5 and 12.1). Mating behavior itself also contributes to genetic diversity. The genetic variation among the gametes of a single individual and the genetic variation between any two parents produce an enormous potential for genetic variation between any two offspring of a sexually reproducing pair of individuals.

Sexual reproduction in animals consists of three fundamental steps:

1. **Gametogenesis**: making gametes
2. **Spawning** or **mating**: bringing gametes together
3. **Fertilization**: fusing gametes

The process of gametogenesis is similar across sexually reproducing animal species. Processes of fertilization are also quite similar in widely different species. Therefore, while our discussion of gametogenesis will focus generally on mammals, and our discussion of fertilization will feature sea urchins, the facts would not be dramatically different were we to consider many other animal groups. Adaptations for spawning and mating, in contrast, show incredible anatomical, physiological, and behavioral diversity across species.

Gametogenesis produces eggs and sperm

Gametogenesis occurs in the gonads: male **testes** (singular *testis*) and female **ovaries** (singular *ovary*). The tiny male gametes, **sperm**, move by beating their flagella. The larger female gametes, **eggs** or **ova** (singular *ovum*), are nonmotile.

Gametes are produced from **germ cells**, which have their origin in early cell divisions of the embryo and remain distinct from all the other cells of the body (the somatic cells). Germ cells are sequestered in the body of the embryo until its gonads begin to form. The germ cells then migrate to the developing gonads, where they take up residence and proliferate by mitosis, producing **spermatogonia** (singular *spermatogonium*) in males and **oogonia** (singular *oogonium*) in females (**Figure 42.4**). Spermatogonia and oogonia are diploid, multiply by mitosis, and are stem cells. They are self-regenerating, and they also produce progeny—spermatocytes and oocytes—that will enter the next stage of gametogenesis.

In the next stage of gametogenesis, meiotic cell division reduces the chromosomes to the haploid number (see Key Concept 11.5). The progeny of the spermatogonia and oogonia that enter meiosis are **primary spermatocytes** and **primary oocytes**. The steps of meiosis are similar in males and females, but there are important sex differences in gametogenesis.

SPERMATOGENESIS   The initial proliferation of male germ cells into spermatogonia proceeds by mitosis in the embryo. But throughout the male life span, spermatogonia continue to divide by mitosis, with one daughter cell retaining the self-regenerating capacity of a spermatogonium while the other daughter cell becomes a primary spermatocyte. As illustrated in Figure 42.4A, primary spermatocytes then undergo the first meiotic division—the reduction division—to form two haploid **secondary spermatocytes**.

(A) Spermatogenesis

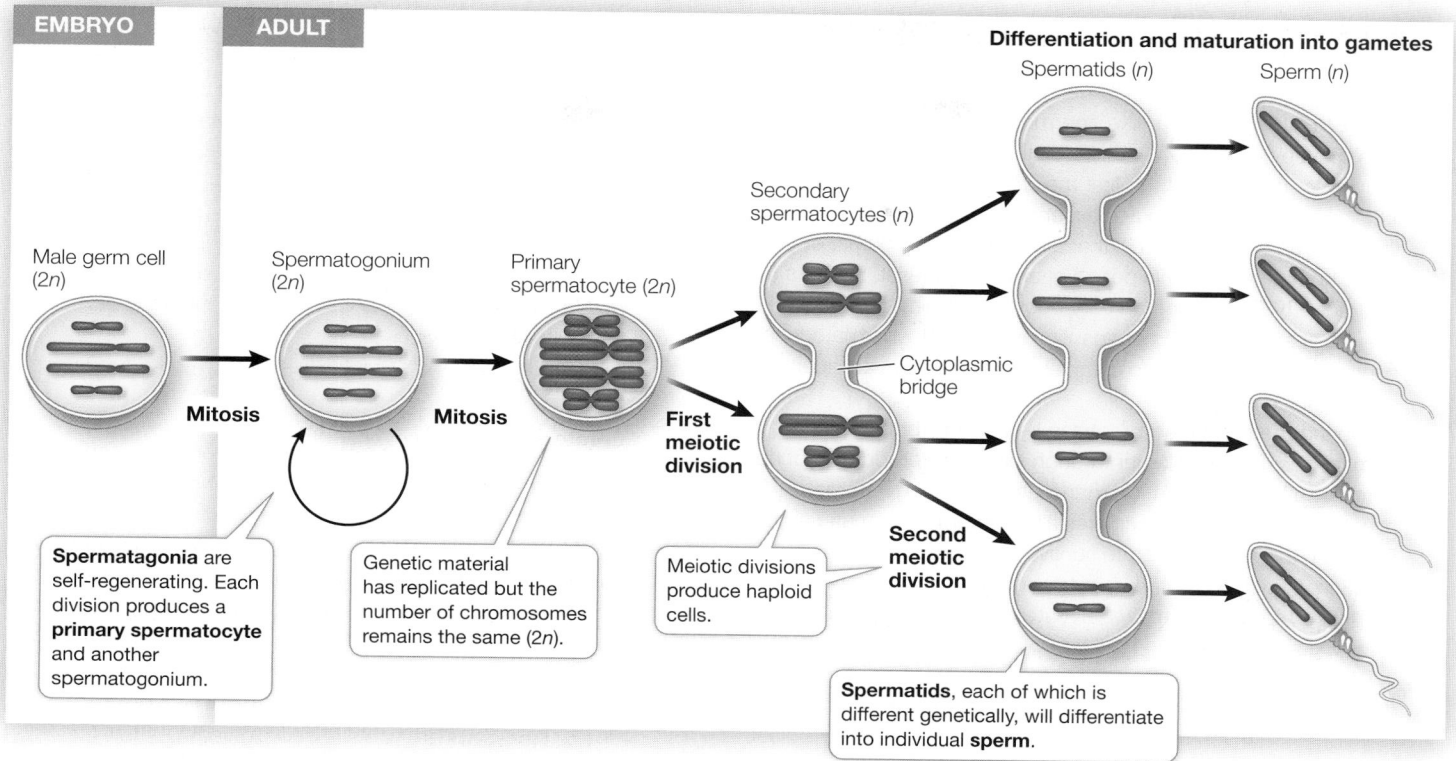

(B) Oogenesis

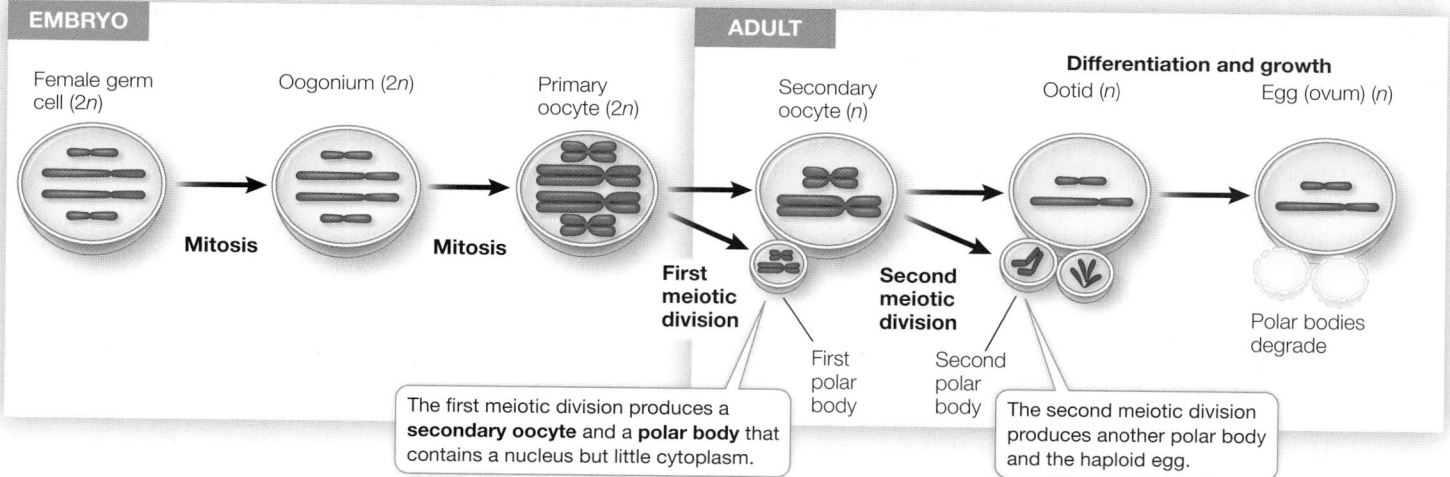

Figure 42.4 Gametogenesis   Male and female germ cells proliferate by mitosis and produce diploid spermatogonia and oogonia that mature into primary spermatocytes and oocytes before entering meiosis. **(A)** Spermatogonia continue to divide by mitosis in adults, producing a steady supply of spermatocytes that divide meiotically to produce haploid spermatids, which differentiate into sperm. In many species, the progeny of spermatocytes remain in contact through cytoplasmic bridges until the sperm mature. **(B)** In mammals, oogonia cease division in the embryo, and primary oocytes remain arrested in prophase I of meiosis until they are ovulated and fertilized. Each oocyte will produce one haploid ootid, which matures into an egg.

The second meiotic division produces four haploid **spermatids** for each primary spermatocyte that enters meiosis. In mammals, the progeny of primary spermatocytes remain connected by cytoplasmic bridges after each division.

One reason that mammalian spermatocytes remain in cytoplasmic contact throughout their development is the asymmetry of sex chromosomes in males. Half the secondary spermatocytes receive an X chromosome, the other half a Y chromosome. The Y chromosome contains fewer genes than the X chromosome, and some of the products of genes found only on the X chromosome are essential for spermatocyte development. By remaining in cytoplasmic contact, all four spermatocytes can share the gene products of the X chromosomes, although only half of them have an X chromosome.

A spermatid bears little resemblance to a mature sperm. Through further differentiation the spermatid becomes compact, streamlined, and develops a flagellum to become motile. We will look at the production of human sperm in Key Concept 42.3.

OOGENESIS   Oogonia, like spermatogonia, proliferate through mitosis (see Figure 42.4B). The resulting primary oocytes immediately enter prophase of the first meiotic division. In many species, including humans, the oocyte experiences developmental arrest at this point and may remain in that state for days, months, or years. In the human female, this period of arrest is at least 10 years (i.e., until puberty), and some primary oocytes remain in prophase I for over 50 years (i.e., until menopause). In contrast, spermatogenesis continues, uninterrupted, to completion once the primary spermatocyte has differentiated.

During the prolonged prophase I, or shortly before it ends, the primary oocyte grows larger through increased production of ribosomes, RNA, cytoplasmic organelles, and energy stores. At this point the primary oocyte acquires all the energy, raw materials, and RNA that the egg will need to survive its first cell divisions after fertilization. In fact, the nutrients in the egg must maintain the embryo until it is either nourished by the maternal circulatory system or can feed on its own.

When a primary oocyte resumes meiosis, its nucleus completes the first meiotic division near the surface of the cell. The daughter cells of this division receive grossly unequal shares of cytoplasm. This asymmetry represents another major difference from spermatogenesis, in which cytoplasm is apportioned equally. The daughter cell that receives almost all the cytoplasm becomes the **secondary oocyte**, and the one that receives almost none forms the **first polar body** (see Figure 42.4B).

The second meiotic division—that of the large, secondary oocyte—is accompanied by asymmetrical division of the cytoplasm. One daughter cell forms the large, haploid **ootid** that eventually differentiates into a mature egg, and the other forms the **second polar body**. In most species the polar bodies rapidly degenerate, so the end result of oogenesis is only one mature egg for each primary oocyte that entered meiosis. However, that egg is a large, well-provisioned cell. In a few species the polar bodies have evolved specialized functions. In some species that reproduce parthenogenetically, a polar body can fuse with the egg to restore the diploid number of chromosomes. In some insect species the polar bodies form organs with specialized functions. An example is scale insects in which cells derived from the polar bodies form a container for endosymbiotic bacteria that the insects require for life.

A second period of arrested development occurs after the first meiotic division forms the secondary oocyte. The egg may be expelled from the ovary in this condition. In many species, including humans, the second meiotic division is not completed until the egg is fertilized by a sperm.

Fertilization is the union of sperm and egg

The union of the haploid sperm and the haploid egg (fertilization) creates a single diploid cell, a **zygote**, which will develop into an embryo. Fertilization does more than just restore the full genetic complement of the animal. The processes associated with fertilization help the egg and sperm get together, prevent the union of the sperm and egg of different species, and guarantee that only one sperm will enter and activate the egg. Fertilization involves a complex series of events:

1. The sperm and the egg chemically recognize each other.
2. The sperm is activated, enabling it to gain access to the cell membrane of the egg.
3. The cell membrane of the egg fuses with the cell membrane of a single sperm.
4. The egg blocks entry of additional sperm.
5. The egg is metabolically activated and stimulated to start development.
6. The egg and sperm nuclei fuse to create the diploid nucleus of the zygote.

SPECIFICITY IN SPERM–EGG INTERACTIONS   Specific recognition molecules mediate interactions between sperm and eggs. These molecules ensure that the activities of sperm are directed toward eggs and not other cells, and they help prevent eggs from being fertilized by sperm from the wrong species. The latter function is particularly important in aquatic species that release eggs and sperm into the surrounding water where the eggs can readily be exposed to sperm of other species. The sea urchin is a good example of such a species, and sea urchin fertilization has been well studied.

Sea urchin eggs release chemical attractants that increase the motility of sperm and cause them to swim toward the egg. These chemical attractants are species-specific. For example, eggs of one species of sea urchin release a specific peptide consisting of 14 amino acids. This peptide binds to receptors present on sperm of the same species. The sperm respond by increasing their mitochondrial respiration and motility. Before exposure to the peptide, the sperm swim in tight little circles, but after binding to the peptide, they swim energetically up the concentration gradient of the peptide until they reach the egg that is releasing it.

When sperm reach an egg, they must get through two protective layers before they can fuse with the egg cell membrane. The eggs of sea urchins are covered with a jelly coat that surrounds a proteinaceous **vitelline envelope** (Figure 42.5A). The success of a sperm's assault on these protective layers depends on a membrane-enclosed structure at the front of the sperm head called an **acrosome**.

The acrosome contains enzymes and other proteins. When a sperm makes contact with an egg of its own species, substances in the jelly coat trigger an acrosomal reaction, which begins with the breakdown of the cell membrane covering the sperm head and the underlying acrosomal membrane (Figure 42.5B). The acrosomal enzymes are released and digest a hole through the jelly coat.

As a result of the polymerization of actin triggered by the acrosomal reaction, an acrosomal process extends out of the head of the sperm. The acrosomal process is coated with species-specific recognition molecules called **bindin**, and there are bindin receptors on the vitelline envelope of the egg. The interaction of bindin with their receptors enables the sperm to contact the egg cell membrane. That contact results in fusion of the sperm and egg cell membranes and the formation of a fertilization cone that engulfs the sperm head, bringing it into the egg cytoplasm. The sperm mitochondria, which largely constitute the midpiece of the sperm, are also drawn into the egg cytoplasm, but they degrade and disappear; this means that

(A) A sea urchin egg

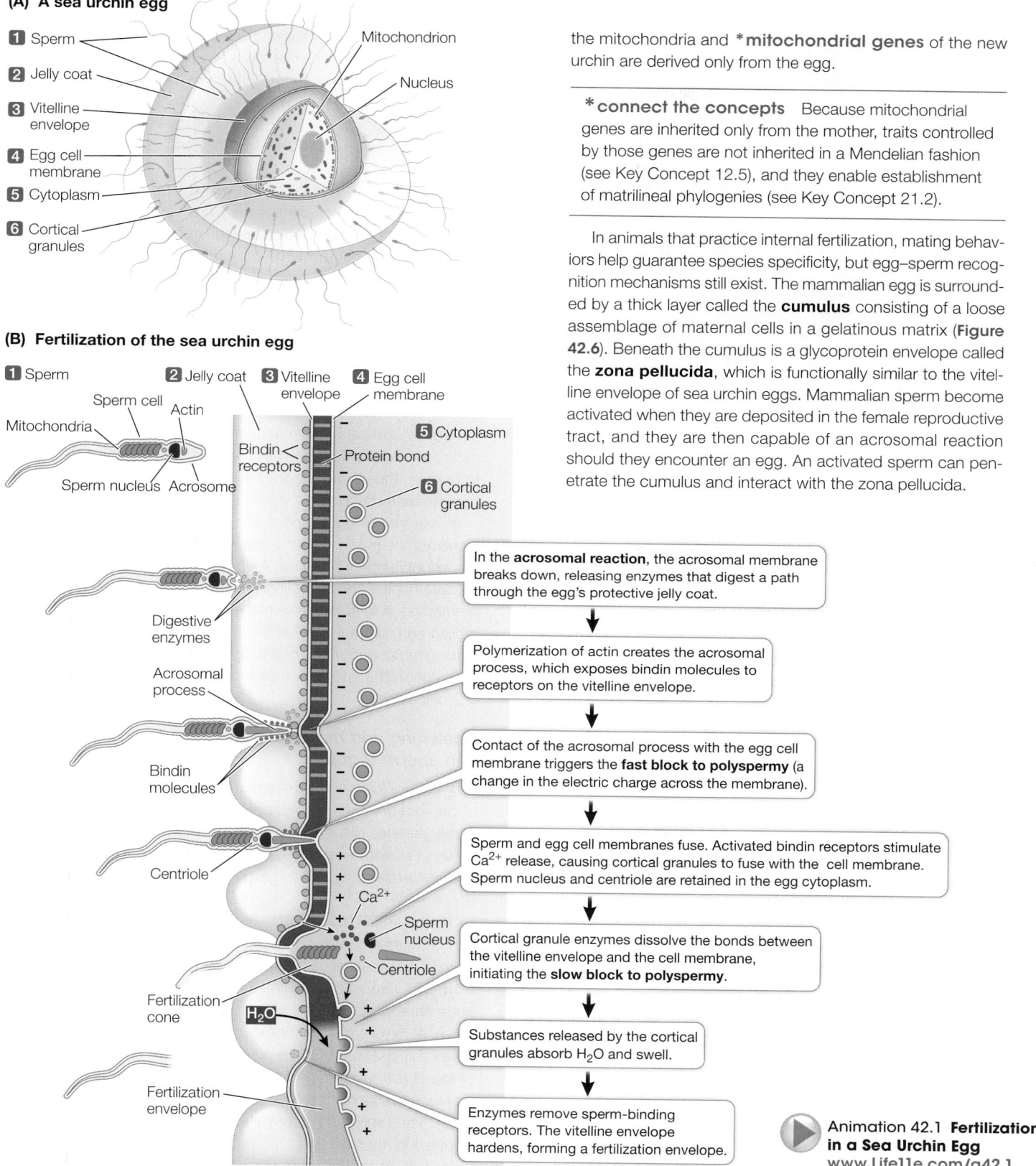

1 Sperm

2 Jelly coat

3 Vitelline envelope

4 Egg cell membrane

5 Cytoplasm

6 Cortical granules

Mitochondrion

Nucleus

(B) Fertilization of the sea urchin egg

1 Sperm   2 Jelly coat 3 Vitelline envelope   4 Egg cell membrane

Sperm cell   Actin

Mitochondria

Sperm nucleus   Acrosome

Bindin receptors

Protein bond

5 Cytoplasm

6 Cortical granules

Digestive enzymes

Acrosomal process

Bindin molecules

Centriole

Ca²⁺

Sperm nucleus

Centriole

Fertilization cone

H₂O

Fertilization envelope

In the **acrosomal reaction**, the acrosomal membrane breaks down, releasing enzymes that digest a path through the egg's protective jelly coat.

Polymerization of actin creates the acrosomal process, which exposes bindin molecules to receptors on the vitelline envelope.

Contact of the acrosomal process with the egg cell membrane triggers the **fast block to polyspermy** (a change in the electric charge across the membrane).

Sperm and egg cell membranes fuse. Activated bindin receptors stimulate Ca²⁺ release, causing cortical granules to fuse with the cell membrane. Sperm nucleus and centriole are retained in the egg cytoplasm.

Cortical granule enzymes dissolve the bonds between the vitelline envelope and the cell membrane, initiating the **slow block to polyspermy**.

Substances released by the cortical granules absorb H₂O and swell.

Enzymes remove sperm-binding receptors. The vitelline envelope hardens, forming a fertilization envelope.

▶ Animation 42.1 **Fertilization in a Sea Urchin Egg**
www.Life11e.com/a42.1

the mitochondria and ***mitochondrial genes** of the new urchin are derived only from the egg.

> ***connect the concepts** Because mitochondrial genes are inherited only from the mother, traits controlled by those genes are not inherited in a Mendelian fashion (see Key Concept 12.5), and they enable establishment of matrilineal phylogenies (see Key Concept 21.2).

In animals that practice internal fertilization, mating behaviors help guarantee species specificity, but egg–sperm recognition mechanisms still exist. The mammalian egg is surrounded by a thick layer called the **cumulus** consisting of a loose assemblage of maternal cells in a gelatinous matrix (**Figure 42.6**). Beneath the cumulus is a glycoprotein envelope called the **zona pellucida**, which is functionally similar to the vitelline envelope of sea urchin eggs. Mammalian sperm become activated when they are deposited in the female reproductive tract, and they are then capable of an acrosomal reaction should they encounter an egg. An activated sperm can penetrate the cumulus and interact with the zona pellucida.

Figure 42.5 Fertilization of the Sea Urchin Egg **(A)** Sea urchin eggs are protected by a jelly layer and a proteinaceous vitelline envelope. Sperm must penetrate both to reach the egg cell membrane. Many sperm attach to the vitelline envelope, but only the first one to reach the egg cell membrane achieves fertilization. The numbers match structures with the events shown in (B). **(B)** The acrosomal reaction allows a sea urchin sperm to recognize an egg of the same species and pass through its protective layers. Enzymes from the egg's cortical granules trigger the slow block to polyspermy.

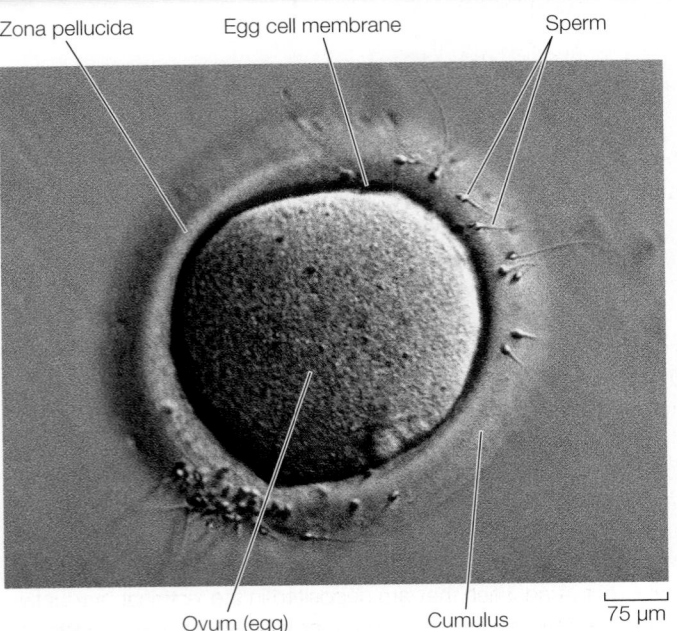

Zona pellucida   Egg cell membrane   Sperm

Ovum (egg)   Cumulus   75 μm

Figure 42.6 Barriers to Sperm   This human egg, like other mammalian eggs, is surrounded by the cumulus and zona pellucida. Sperm must penetrate both to fertilize the egg. Only one sperm will penetrate the zona pellucida and fuse with the egg cell membrane.

Q: What would be the consequences of two sperm fertilizing the egg?

Unlike the jelly coat of sea urchin eggs, the cumulus of mammalian eggs does not trigger the acrosomal reaction. When sperm make contact with the zona pellucida, a species-specific glycoprotein binds to recognition molecules on the head of the sperm. This binding triggers the acrosomal reaction, releasing acrosomal enzymes that digest a path through the zona pellucida. When the sperm head reaches the egg cell membrane, other proteins facilitate its adhesion to and fusion with the egg cell membrane.

BLOCKS TO POLYSPERMY   The fusion of the sperm and egg cell membranes and the entry of the sperm into the egg initiate a programmed sequence of events. The first responses to sperm entry are **blocks to polyspermy**: mechanisms that prevent more than one sperm from entering the egg. Survival of the embryo is unlikely if more than one sperm enters the egg. The sperm contributes a haploid set of chromosomes, so an extra set would disrupt the mitotic division of the egg and subsequent cells. In addition, in most mammals the sperm contributes a centriole to the fertilized egg, and the *centriole forms the centrosome which is critical for organizing the mitotic spindle. Having two centrosomes would disrupt mitosis of the fertilized egg.

---

*connect the concepts   The roles of the centriole and the centrosome in mitosis are discussed in Key Concept 11.3.

Blocks to polyspermy have been studied extensively in sea urchin eggs, which can be fertilized in a dish of seawater. Within seconds after the sperm membrane contacts the egg membrane, an influx of sodium ions changes the electric charge difference across the egg cell membrane. This fast block to polyspermy prevents the fusion

of any other sperm with the egg cell membrane, but it is transient. The change in membrane electric charge lasts only about a minute, but that is enough time to allow a slower block to sperm entry to develop.

The slow block to polyspermy involves converting the vitelline envelope to a physical barrier that sperm cannot penetrate. Before fertilization, the vitelline envelope is bonded to the egg cell membrane. Just under the cell membrane are vesicles called cortical granules (see Figure 42.5) which contain enzymes and other proteins.

The sea urchin egg, like all animal cells, sequesters calcium in its endoplasmic reticulum. Sperm entry into the egg stimulates the release of calcium ions from the endoplasmic reticulum and into the egg cytosol. This increase in cytosolic calcium causes the egg's cortical granules to fuse with the cell membrane and release their contents. Cortical granule enzymes break the bonds between the vitelline envelope and the cell membrane, and other proteins released from the cortical granules attract water into the space between them. As a result, the vitelline envelope rises to form a fertilization envelope. Cortical granule enzymes also degrade sperm-binding molecules on the surface of the fertilization envelope and cause it to harden, thus preventing additional sperm from contacting the egg cell membrane.

In mammalian eggs, sperm entry does not cause a rapid change in membrane potential, but it does trigger a release of calcium from the endoplasmic reticulum. As in the sea urchin egg, increased calcium causes the cortical granules to fuse egg with the egg cell membrane. A fertilization envelope does not form around the mammalian egg, but the cortical granule enzymes destroy the sperm-binding molecules in the zona pellucida. The rise in cytosolic calcium also signals the egg to complete meiosis. The stage is set for the first cell division.

Spawning and mating behaviors get eggs and sperm together

As you have just seen, sexual reproduction requires the production of haploid gametes (gametogenesis) and the joining together of those gametes to form a diploid zygote (fertilization). Spawning and mating behaviors get eggs and sperm close enough together that fertilization can occur. Fertilization can occur externally or internally.

EXTERNAL FERTILIZATION   In an aquatic environment, animals can bring their gametes together by simply releasing them into the water. This practice, called spawning, results in **external fertilization**. Many aquatic animals are not very mobile, but they produce huge numbers of gametes that can travel far from the point of release. A female oyster, for example, will release millions of eggs when she spawns, and the number of sperm produced by a male oyster is astronomical.

Numbers alone, however, do not guarantee that gametes will meet. The reproductive activities of the males and females of a population must be synchronized, since released gametes have a limited life span. Seasonal breeders may use day length, changes in temperature, or changes in weather to time the production and release of their gametes. Mutual stimulation is also important. Release of gametes into the water by one individual can stimulate others to spawn.

Behavior can play an important role in bringing gametes together even when fertilization is external. Many species travel great distances to congregate with potential mates and release their gametes at the same time in a suitable environment. Many salmon are an extreme example. They hatch and develop in freshwater streams and then migrate to the ocean, where they remain for years. When they are mature and ready to spawn, they travel hundreds of miles back to and then up the stream to the region where they hatched. Males and females expend great amounts of energy to swim up the streams to the spawning grounds, where they pair up, prepare a depression in the streambed gravel, and together release their sperm and eggs. As the gametes drift down into the gravel, fertilization occurs.

INTERNAL FERTILIZATION   Terrestrial animals cannot simply release their gametes into the environment. Sperm can move only through liquid, and delicate gametes released into air would dry out and die. Most terrestrial animals avoid these problems by **internal fertilization**, the release of sperm into the female reproductive tract. Some aquatic animals also practice internal fertilization, but it is ubiquitous in terrestrial animals.

Animals have evolved an astonishing diversity of behavioral and anatomical adaptations for internal fertilization. The reproductive organs together are the animal's **genitalia**. Gametogenesis occurs in the gonads, which are the **primary sex organs**. All additional components of an animal's genitalia are called **accessory sex organs**. The internal accessory sex organs include a variety of glands, tubules, ducts, and other structures. External accessory sex organs are referred to as the external genitalia. In males of many species, the external genitalia includes a **penis** that enables the male to deposit sperm in the female's **vagina**, the entry to her reproductive tract.

Copulation is the physical joining of male and female accessory sex organs. Most male insects copulate and transfer sperm to the female's vagina through a penis. The external genetalia of insects often have species-specific shapes that match in a lock-and-key fashion to ensure a tight, secure fit between the mating pair during the prolonged period of sperm transfer. In some insect species in which females mate with more than one male, the male external genitalia includes elaborate structures that can scoop sperm deposited by other males out of the female's reproductive tract, replacing it with his own.

Transfer of sperm in internal fertilization can also be indirect. Males of many invertebrate species (e.g., mites and scorpions) and a few vertebrates (e.g., salamanders) deposit spermatophores—packets of sperm protected from desiccation—in the environment. When a female mite encounters a spermatophore from a potential mate, she straddles it and opens a pair of plates in her abdomen so that the tip of the spermatophore enters her reproductive tract and allows the sperm to enter.

Male squid and spiders play a more active role in spermatophore transfer. The male spider secretes a drop containing sperm onto a bit of web, then uses a special structure on his foreleg to pick up the sperm-containing web and insert it through the female's genital opening. Some male squid use one specialized tentacle to pick up a spermatophore and insert it into the female's genital opening.

Some individuals can function as both male and female

In most species, gametes are produced by individuals that are either male or female. Species that have separate male and female members are referred to as **dioecious** (Greek, "two houses"). In some species, however, a single individual may produce both sperm and eggs. Such species are **monoecious** ("one house"), or **hermaphroditic**.

Almost all invertebrate groups contain some hermaphroditic species. An earthworm is an example of a simultaneous hermaphrodite, meaning an individual is both male and female at the same time. When two earthworms mate, they exchange sperm, and as a result the eggs of each are fertilized (see Figure 32.12C). Some vertebrates are sequential hermaphrodites, meaning an individual may function as a male or a female at different times in its life. An example is the anemone fish, or clown fish, a species that lives in small groups within large sea anemones (**Figure 42.7**). All anemone fish are born male. The largest fish in a group becomes a functional female. If that fish is removed from the group, the next-largest male becomes a female. The second-largest fish in the group is the only male in breeding condition. The reverse situation also exists in some species where a large dominant male controls a harem of females. If the male is removed, the largest female becomes the male in the group.

What is the evolutionary advantage of hermaphroditism? Some simultaneous hermaphrodites, such as parasitic tapeworms, have a low probability of meeting a potential mate—an individual tapeworm may be the only one in its host. Tapeworms can fertilize their own eggs, but most simultaneous hermaphrodites must mate with another individual. Because every simultaneous hermaphrodite is both male and female, however, the probability of encountering a possible mate doubles. In some sequential hermaphrodites, siblings are all male or all female at the same time, thus reducing the incidence of inbreeding.

Amphiprion sp.

Figure 42.7 When Size Determines Sex   Anemonefish (also known as clownfish) live in groups of about a dozen centered on a single sea anemone. All anemone fish are born male; the largest fish in the group becomes a functional female. Thus a fish may function as a male and as a female at different times in its life.

(A) *Bungarus fasciatus*

(B) *Merops apiaster*

Figure 42.8 The Shelled Egg   The terrestrial environment is desiccating. Therefore, the shelled egg was a major evolutionary step that allowed reptiles to reproduce in the terrestrial environment. **(A)** A female banded krait has deposited her eggs in the sand. **(B)** The shelled egg requires that sperm meet egg before the shell forms. Therefore reptiles must practice internal fertilization, as these European bee-eaters are doing.

The evolution of vertebrate reproductive systems parallels the move to land

The earliest vertebrates evolved in aquatic environments. The closest living relatives of those earliest vertebrates are modern-day fishes. They remain exclusively aquatic, and most practice external fertilization. The most primitive fishes, the lampreys and hagfish, simply release their gametes into the environment. In most fishes, however, mating behaviors bring females and males into close proximity at the time of gamete release. In sharks and rays, fins have evolved into claspers that hold the male and female together and enable sperm to be transferred directly into the female reproductive tract.

Amphibians were the first vertebrates to live in terrestrial environments. They dealt with the challenge of a dry environment by returning to water to reproduce, as most amphibians still do today.

Reptiles were the first vertebrate group to solve the problem of reproduction in the terrestrial environment (**Figure 42.8**). Their solution was the amniote egg (see Figure 32.19A). A hard or leathery shell protects the embryo and impedes water loss while allowing the diffusion of oxygen into the egg and carbon dioxide out of the egg. The eggshell creates an obvious problem for fertilization. Sperm cannot penetrate the shell, so they must reach the egg before the shell forms. Hence internal fertilization and the evolution of accessory sex organs were necessary for the evolution of the amniote egg.

Male snakes and lizards have paired hemipenes (hemipenis, sing.), which can be engorged with blood and thereby extruded from the male's body. Only one hemipenis is inserted into the female's reproductive tract. It is usually rough or spiny at the end to achieve a secure hold while sperm are transferred down a groove on its surface. Retractor muscles pull the hemipenis back into the male's body when mating is completed. Some evolutionarily ancient bird species have erectile penises that channel sperm along a groove into the female's reproductive tract. Birds with more recent evolutionary origins, however, do not have erectile penises; instead, the male and female simply bring their genital openings (cloacae, sing. cloaca) close together to transfer sperm. Usually this involves the male standing on the female's back (see Figure 42.8B).

All mammals practice internal fertilization. Then, all except the prototherian mammals retain the developing embryo for some time in the female reproductive tract. Prototherian mammals (the monotremes; see Figure 32.26) lay eggs. The other mammals (the therians) vary enormously as to the developmental stage of their offspring at the time of birth. As you saw at the beginning of the chapter, the tammar wallaby is extremely immature at birth. By comparison, a horse foal can stand up soon after birth and is able to run within hours.

Animals with internal fertilization are distinguished by where the embryo develops

Two patterns of care and nurture of the embryo have evolved in animals: oviparity (egg laying) and viviparity (live bearing).

Oviparous animals lay eggs in the environment, and their embryos develop outside the mother's body. Oviparous terrestrial animals such as insects and reptiles—including birds— protect their eggs from desiccation with waterproof membranes or shells. Oviparity is possible because eggs are stocked with abundant nutrients to supply the needs of the embryo. Some oviparous animals engage in various forms of parental behavior to protect their eggs, but until the eggs hatch, the embryos depend entirely on the nutrients stored in the egg.

Viviparous animals retain the embryo within the mother's body during its early developmental stages. The embryo is nourished and its wastes removed through exchanges with the mother's tissues. Viviparity occurs in many vertebrate groups, but with the exception of the prototherians (the monotremes), viviparity is universal in mammals. Therian mammals have a specialized portion of the female reproductive tract, the **uterus** or womb that contains the embryo until it is born. The early embryo arriving in the uterus from the Fallopian tube implants into the lining of the uterus. Implantation involves interactions between the tissues of the embryo and the uterus to form a **placenta**. Intimate contact in the placenta between the blood of the embryo and the blood of the mother facilitates the exchange of nutrients and wastes.

In most non-mammalian viviparous animals, such as garter snakes and the well-known aquarium fish the guppy, fertilized eggs are retained in the mother's body until they hatch. These embryos still receive nutrition from stores in the egg, so this reproductive adaptation is called **ovoviviparity**.

42.2 recap

Sexual reproduction involves gametogenesis, mating, and fertilization. Fertilization can be external or internal and involves mechanisms for ensuring that only one sperm from the right species enters the egg. Oviparous female animals lay eggs in the environment, and their embryos develop outside the mother's body. Viviparous female animals retain the developing embryo in the uterus until it is born.

learning outcomes

You should be able to:

- Explain the connection between sex chromosome asymmetry and spermatocytes remaining in cytoplasmic contact.
- Compare oogenesis and spermatogenesis.
- Summarize the interactions between sperm and eggs in sea urchins.
- Summarize ways aquatic animals enhance the likelihood that sperm and eggs of the same species will meet.
- Identify and describe conditions that favor the evolution of hermaphroditism.

1. Crossing over is the reciprocal exchange of segments in homologous chromatids. Why does it occur more frequently in human oocytes than in human spermatocytes?

2. If a mutation in a mammal eliminated the maintenance of cytoplasmic bridges between spermatids, would you expect all offspring sired by that mammal to be male or female? Explain your answer.

3. In animals that spawn, what mechanisms ensure that an egg is fertilized by sperm of the same species, and by only one sperm?

4. What are three situations in which hermaphroditism would be advantageous?

Now that we have covered some of the general aspects of gametogenesis and fertilization and have briefly discussed the great diversity of mating systems, we will next consider the human reproductive systems in detail.

key concept 42.3 Male Sex Organs Produce and May Deliver Sperm

This section describes the structures and functions of the male reproductive system in mammals, using humans as the prime example. You will also learn about the hormonal regulation of the male reproductive system. The discussion covers:

- *The primary sex organs*, testes, that produce sperm and serve endocrine functions.
- *The accessory sex organs*, including the ducts through which the sperm pass, the glands that produce and deliver secretions into those ducts, and the external genitalia.
- *Secondary sexual characteristics* that result in male appearance but are not directly involved in reproduction.

focus your learning

- Semen contains sperm and a mixture of molecules that support the sperm and facilitate fertilization.
- Sexual stimulation triggers responses in the nervous system and penile arterioles that result in erection.
- Spermatogenesis and the maintenance of male secondary sexual characteristics depend on testosterone.

Semen is the product of the male reproductive system

Semen contains sperm and a complex mixture of fluids and molecules (seminal fluid) that support the sperm and facilitate fertilization. Sperm make up less than 5 percent of the volume of the semen. Sperm are produced in the testes, the paired male gonads, and the seminal fluid is produced by accessory glands.

Figure 42.9 illustrates the human male reproductive organs. The testes of most mammals are located outside the body cavity in a pouch of skin called the **scrotum**. Why should the testes be located outside the body cavity? The optimal temperature for spermatogenesis in most mammals is slightly lower than the normal body temperature. The scrotum keeps the testes at this optimal temperature. Muscles in the scrotum contract in a cold environment, bringing the testes closer to the warmth of the body; in a hot environment the muscles relax, cooling the testes by suspending them farther from the body.

Spermatogenesis takes place within the **seminiferous tubules** tightly coiled in each testis (**Figure 42.10A**). Between the seminiferous tubules are clusters of **Leydig cells** that produce testosterone (**Figure 42.10B**). Spermatogonia reside in the outer regions of the seminiferous tubules, just under the basement membrane. Moving inward toward the lumen of the tubule are germ cells in successive stages of spermatogenesis (**Figure 42.10C**). The germ cells are intimately associated with **Sertoli cells** that provide nutrients for the developing sperm and secrete several things including hormones and growth factors that are necessary for spermatogenesis (which is why Sertoli cells are sometimes called "nurse cells"). Tight junctions (see Key Concept 6.2) between Sertoli cells create a blood–sperm barrier that prevents many harmful substances that can be circulating in the blood from getting to the developing sperm.

When the second meiotic division is complete, each primary spermatocyte has produced four spermatids (see Figure 42.4A). The spermatids develop into spermatozoa as they migrate toward the lumen of the seminiferous tubule. The nucleus becomes compact, and the surrounding cytoplasm is lost. A flagellum—the sperm tail—develops. The mitochondria that provide the energy for sperm motility become condensed into a midpiece between the head and tail. An acrosome forms over the nucleus in the head of the sperm. Immature sperm are shed into the lumen of the seminiferous tubule.

From the seminiferous tubules, sperm move into the **epididymis** (see Figure 42.9A), where they mature, become motile, and are stored. The epididymis connects to the **urethra** via the **vas deferens** (plural *vasa deferentia*) and the **ejaculatory duct**. The urethra originates in the bladder, runs through the penis, and opens to the outside of the body at the tip of the penis. It serves as the common final duct for urine and semen.

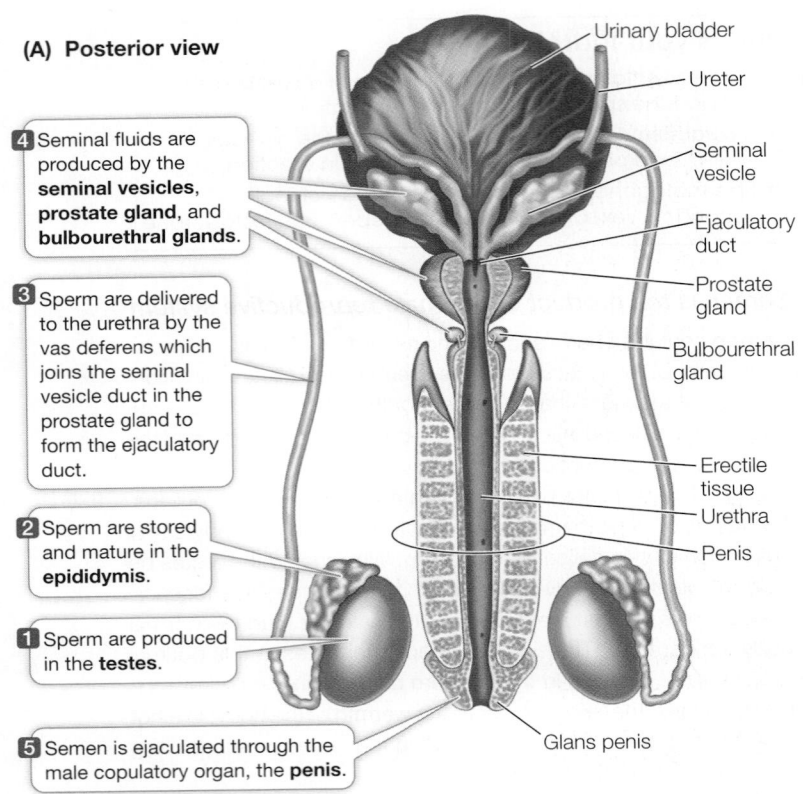

(A) Posterior view

4 Seminal fluids are produced by the **seminal vesicles**, **prostate gland**, and **bulbourethral glands**.

3 Sperm are delivered to the urethra by the vas deferens which joins the seminal vesicle duct in the prostate gland to form the ejaculatory duct.

2 Sperm are stored and mature in the **epididymis**.

1 Sperm are produced in the **testes**.

5 Semen is ejaculated through the male copulatory organ, the **penis**.

Urinary bladder
Ureter
Seminal vesicle
Ejaculatory duct
Prostate gland
Bulbourethral gland
Erectile tissue
Urethra
Penis
Glans penis

(B) Side view

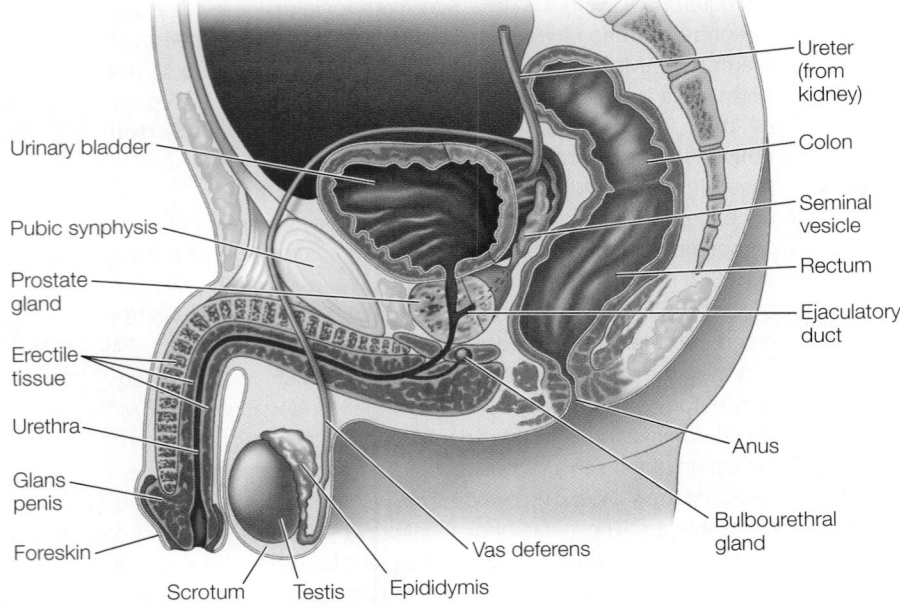

Ureter (from kidney)
Colon
Seminal vesicle
Rectum
Ejaculatory duct
Anus
Bulbourethral gland

Urinary bladder
Pubic synphysis
Prostate gland
Erectile tissue
Urethra
Glans penis
Foreskin
Scrotum   Testis   Epididymis
Vas deferens

Figure 42.9 Reproductive Tract of the Human Male   The male reproductive organs are shown **(A)** from the rear and **(B)** from the side.

Q: What problem might an enlarged prostate gland produce?

 Activity 42.1 The Human Male Reproductive Tract
www.Life11e.com/ac42.1

The components of the semen other than sperm come from three accessory glands. About 60 percent of the volume of semen is secreted by the paired **seminal vesicles** that empty into the vas deferens just before it joins the urethra. Seminal fluid is thick because it contains mucus and fibrinogen, a protein also found in the blood, where it can polymerize to form blood clots. Seminal fluid also contains the monosaccharide fructose, an energy source for the sperm.

The **prostate gland** contributes about 30 percent of the volume of the semen. Prostate fluid is alkaline, so it neutralizes the acidity in the male and female reproductive tracts and makes these environments more hospitable to sperm. The prostate also secretes a clotting enzyme that causes fibrinogen from the seminal vesicles to convert the semen into a clotted, gelatinous mass, facilitating the semen's propulsion into the upper regions of the female reproductive tract. Another prostate enzyme, fibrinolysin, then dissolves the clotted semen and liberates the sperm. Prostaglandins—hormones produced by the prostate—stimulate contractions of the female reproductive tract.

The **bulbourethral glands** produce a small volume of an alkaline, mucoid secretion that helps neutralize acidity in the urethra and lubricate it to facilitate the passage of semen during the climax of sexual intercourse.

The penis and the scrotum are the male external genitalia

The shaft of the penis is covered with normal skin, but the highly sensitive tip, the **glans penis**, is covered with thinner, more sensitive skin that is especially responsive to sexual stimulation. A fold of skin called the foreskin covers the glans of the human penis. The procedure known as circumcision removes a portion of the foreskin.

Sexual stimulation triggers responses in the nervous system that result in penile **erection**. Nerve endings release a neurotransmitter that causes the endothelial cells lining the penile blood vessels to release a gaseous neurotransmitter, nitric oxide (NO). NO diffuses into the smooth muscle cells that control the diameter of the penile arteries and stimulates them to produce the second messenger cGMP (see Figure 7.13). Increased cGMP in these muscle cells causes them to relax; the arteries dilate and carry more blood into the penis. Increased blood flow swells the shafts of spongy erectile tissue located along the length of the penis (see Figure 42.9). The enlargement of these blood-filled cavities compresses the vessels that carry blood out of the penis, and the erectile tissue becomes engorged with blood. The penis becomes hard and erect, facilitating its insertion into the vagina. Many species of mammals (though not humans) have a bone in the penis, but these species still depend on erectile tissue for copulation.

At the climax of copulation, 2 to 6 milliliters of semen are propelled through the vasa deferentia and the urethra in two steps, emission and ejaculation. During emission, rhythmic contractions of smooth muscles in the vasa deferentia and accessory glands move the semen into the urethra at the base of the penis. Ejaculation is caused by contractions of other muscles at the base of the penis surrounding the urethra. These contractions force the coagulum of semen through the urethra and out of the

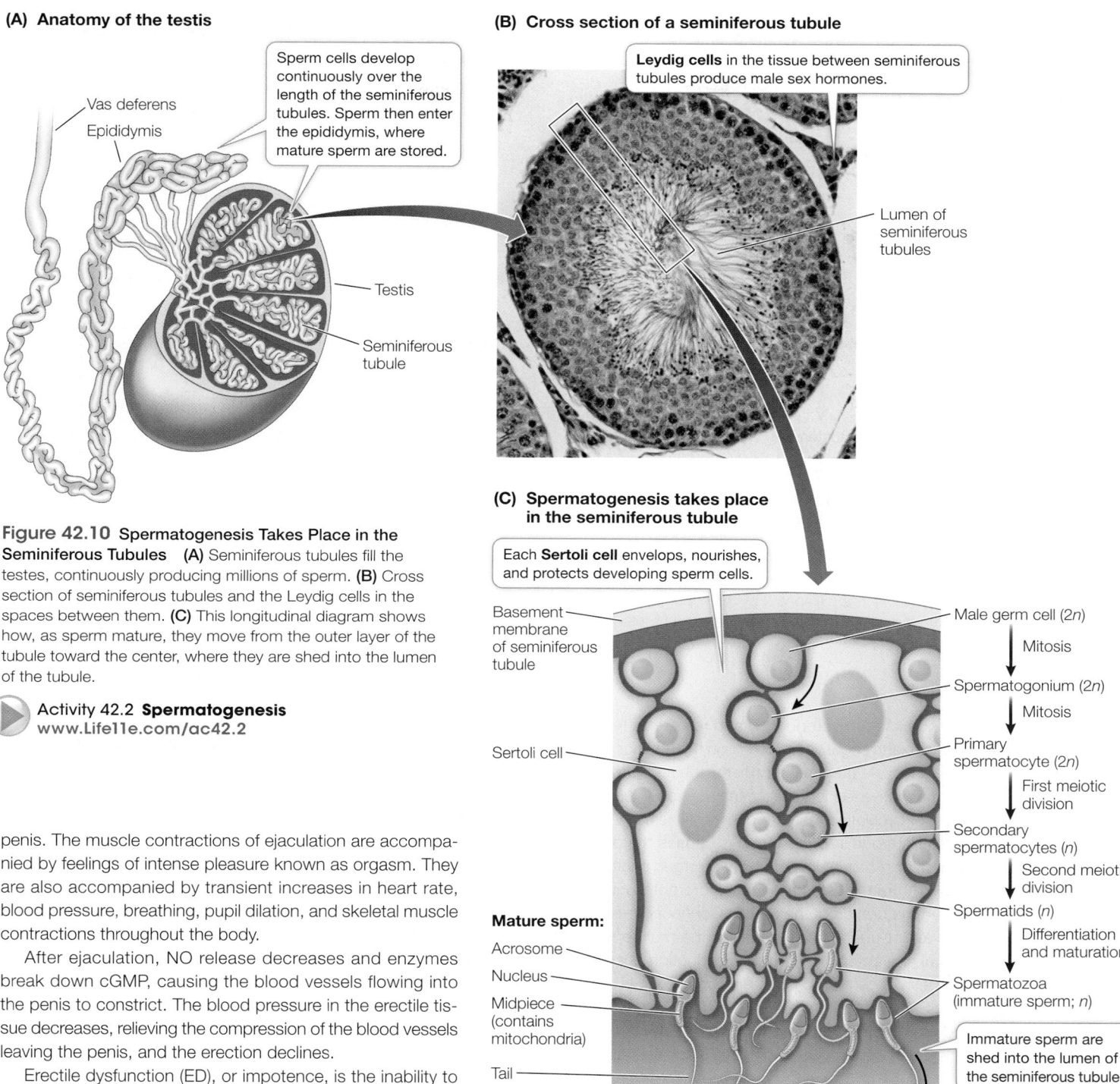

(A) Anatomy of the testis

Vas deferens

Epididymis

Sperm cells develop continuously over the length of the seminiferous tubules. Sperm then enter the epididymis, where mature sperm are stored.

Testis

Seminiferous tubule

Figure 42.10 Spermatogenesis Takes Place in the Seminiferous Tubules **(A)** Seminiferous tubules fill the testes, continuously producing millions of sperm. **(B)** Cross section of seminiferous tubules and the Leydig cells in the spaces between them. **(C)** This longitudinal diagram shows how, as sperm mature, they move from the outer layer of the tubule toward the center, where they are shed into the lumen of the tubule.

▶ Activity 42.2 **Spermatogenesis**
www.Life11e.com/ac42.2

(B) Cross section of a seminiferous tubule

Leydig cells in the tissue between seminiferous tubules produce male sex hormones.

Lumen of seminiferous tubules

(C) Spermatogenesis takes place in the seminiferous tubule

Each **Sertoli cell** envelops, nourishes, and protects developing sperm cells.

Basement membrane of seminiferous tubule

Sertoli cell

Mature sperm:

Acrosome

Nucleus

Midpiece (contains mitochondria)

Tail

Lumen

Male germ cell (2n)

⬇ Mitosis

Spermatogonium (2n)

⬇ Mitosis

Primary spermatocyte (2n)

⬇ First meiotic division

Secondary spermatocytes (n)

⬇ Second meiotic division

Spermatids (n)

⬇ Differentiation and maturation

Spermatozoa (immature sperm; n)

Immature sperm are shed into the lumen of the seminiferous tubule.

penis. The muscle contractions of ejaculation are accompanied by feelings of intense pleasure known as orgasm. They are also accompanied by transient increases in heart rate, blood pressure, breathing, pupil dilation, and skeletal muscle contractions throughout the body.

After ejaculation, NO release decreases and enzymes break down cGMP, causing the blood vessels flowing into the penis to constrict. The blood pressure in the erectile tissue decreases, relieving the compression of the blood vessels leaving the penis, and the erection declines.

Erectile dysfunction (ED), or impotence, is the inability to achieve or sustain an erection. ED may have different causes, including cardiovascular disease. Drugs used to treat ED act by inhibiting the enzyme that breaks down cGMP, thus enhancing the effect of NO released in the penis, which improves the ability to achieve and maintain an erection.

Male sexual function is controlled by hormones

Spermatogenesis and maintenance of male secondary sexual characteristics such as facial hair and a deep voice depend on testosterone produced by the Leydig cells of the testes. As described in

Key Concept 40.3, increased production of testosterone at puberty results from an increased release of gonadotropin-releasing hormone (GnRH) by the hypothalamus. GnRH stimulates anterior pituitary cells to secrete luteinizing hormone (LH) and follicle-stimulating hormone (FSH) (**Figure 42.11**). Higher levels of LH stimulate the

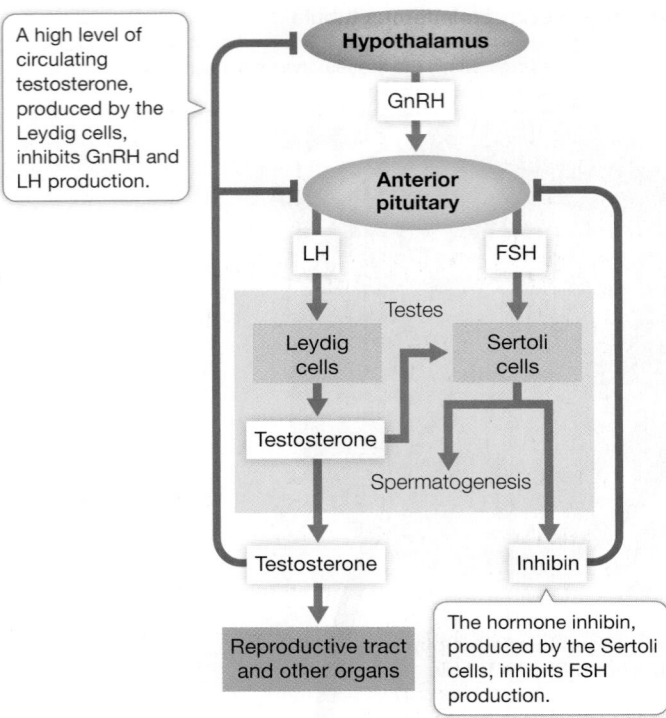

A high level of circulating testosterone, produced by the Leydig cells, inhibits GnRH and LH production.

The hormone inhibin, produced by the Sertoli cells, inhibits FSH production.

Figure 42.11 Male Reproductive Hormones   The male reproductive system is under hormonal control by the hypothalamus and the anterior pituitary. Red lines indicate inhibition; green lines indicate stimulation and secretion.

Leydig cells to increase their production and release of testosterone. Testosterone exerts negative feedback on the anterior pituitary and the hypothalamus. At the time of puberty, the sensitivity of the hypothalamus to negative feedback from testosterone declines; as a result, the level of circulating testosterone increases.

Increased testosterone in pubertal boys causes the development of pubic and facial hair, a deeper voice, enlarged genitals, and an increased growth rate. Testosterone also promotes increased muscle mass and maturation of the testes. Continued production of testosterone after puberty is essential for the maintenance of secondary sexual characteristics and the production of sperm.

Spermatogenesis is controlled by the influence of FSH and testosterone on the Sertoli cells in the seminiferous tubules. The Sertoli cells also produce a hormone called inhibin that exerts negative feedback on the anterior pituitary cells producing and secreting FSH.

42.3 recap

Sperm are produced in the testes and are suspended in the seminal fluids produced by accessory glands. The resulting semen is ejaculated into the female reproductive tract during copulation. Prior to copulation, sexual stimulation causes the penis to become erect by increasing the blood flow into the erectile tissues. The male reproductive system is under hormonal control by the hypothalamus and anterior pituitary, with negative feedback coming from hormones produced by the testes.

42.3 recap

learning outcomes

You should be able to:

- Summarize where the different components of semen are produced, and describe their functions.
- Describe the anatomical and functional relationships between Leydig cells and Sertoli cells.
- Describe the roles of NO and cGMP in the male sexual response.
- Summarize the hormonal changes that occur during puberty in males and how they lead to the development of male secondary sexual characteristics.

1. What makes ejaculated semen thick and gelatinous, and how do sperm escape the gel to move up the female reproductive tract?
2. What are the roles of Sertoli cells in spermatogenesis?
3. cGMP is broken down by an enzyme called a phosphodiesterase. How does an inhibitor of that enzyme treat erectile dysfunction?
4. What is the significance of the fact that the testes of a prepubertal male are responsive to FSH and LH?

Whereas the male reproductive system produces sperm continuously throughout reproductive life, the female system produces eggs that can be fertilized by sperm periodically. We now turn to the cyclical processes by which the female reproductive system produces and releases eggs and prepares the female reproductive tract to receive a developing embryo should fertilization occur.

key concept 42.4 Female Sex Organs Produce Eggs and Nurture Embryos

Figure 42.12 illustrates the human female reproductive system. Eggs are produced in and released from the ovaries located on each side of the lower abdominal cavity. When an egg leaves the ovary (**ovulation**), it enters the abdominal cavity, but it does not go far. The ovaries are close to the openings of the **oviducts** (also known as the Fallopian tubes). The openings are surrounded by undulating, fringed tissues called fimbria that sweep the egg into the oviduct. Cilia lining the oviduct propel the egg toward the **uterus**, a muscular, thick-walled cavity shaped in humans like an upside-down pear. The uterus is where the embryo will develop if the egg is fertilized. At its bottom, the uterus narrows into a region called the **cervix**, which opens into the vagina.

focus your learning

- The menstrual (uterine) cycle refers to the buildup and breakdown of the endometrial lining of the uterus.
- If pregnancy occurs, human chorionic gonadotropin (hCG) stimulates the corpus luteum to continue producing estrogen and progesterone.
- Mechanical and hormonal stimuli are involved in labor.

- The ovarian and uterine cycles are coordinated and timed by the same hormones that initiate sexual maturation.

Externally, the vagina is enclosed by two sets of skin folds. The inner, more delicate folds are the labia minora; the outer, thicker folds are the labia majora. At the anterior tip of the labia minora is the clitoris, a small bulb of highly sensitive erectile tissue that has the same developmental origins as the male glans penis. Clitoral tissues, homologous to the erectile tissues of the penis, extend deep under the labia minora and the walls of the vagina. The labia minora and the clitoral tissues become engorged with blood in response to sexual stimulation. Just as stimulation of the glans penis during sexual intercourse culminates in orgasm in the male, stimulation of the clitoral tissues results in orgasm in the female.

The external opening of an infant's vagina is usually, but not always, partly covered by a thin membrane, the hymen. Eventually the hymen can be torn by vigorous physical activity or by first sexual intercourse; it can sometimes make first intercourse difficult or painful for the woman.

Sperm deposited in the vagina swim and are propelled by contractions of the female reproductive tract through the cervical opening, across the uterus, and most of the way up an oviduct, where fertilization takes place. Before meeting a sperm, the egg is still a secondary oocyte. Fertilization stimulates the egg to complete its second meiotic division. Following that division, the haploid nuclei of the sperm and the egg can fuse to produce a diploid zygote nucleus.

The *zygote undergoes its first few cell divisions to become a **blastocyst** while still in the oviduct. The blastocyst moves down the oviduct to the uterus, where it attaches itself to the epithelial lining of the uterus—the **endometrium**. Once attached, the blastocyst implants in the endometrium and interacts with it to form the placenta, as you will see in Chapter 42. The mother's body nurtures the embryo through the placenta, which also produces hormones that sustain pregnancy.

---

*connect the concepts   The processes by which the zygote forms, develops into a blastocyst, and becomes implanted in the uterus are illustrated in Key Concept 44.2.

The maturation of eggs and ovulation are cyclical events in the ovaries. Changes in the uterus are also critical processes. These cycles in humans average about 28 days, but they can vary considerably in an individual and among different individuals. In anticipation of receiving a blastocyst, the endometrium thickens and develops lots of blood vessels. The endometrium regresses if a blastocyst does not arrive within a certain window of time. Thus female reproductive functions consist of two linked cycles: an **ovarian cycle** that produces eggs and hormones; and a **uterine**, or **menstrual**, **cycle** that prepares the endometrium for the arrival of a blastocyst (**Focus: Key Figure 42.13**). The two cycles must be synchronized so that a blastocyst arrives in the uterus at the optimal time to implant in the endometrium and continue its development.

(A) Front view

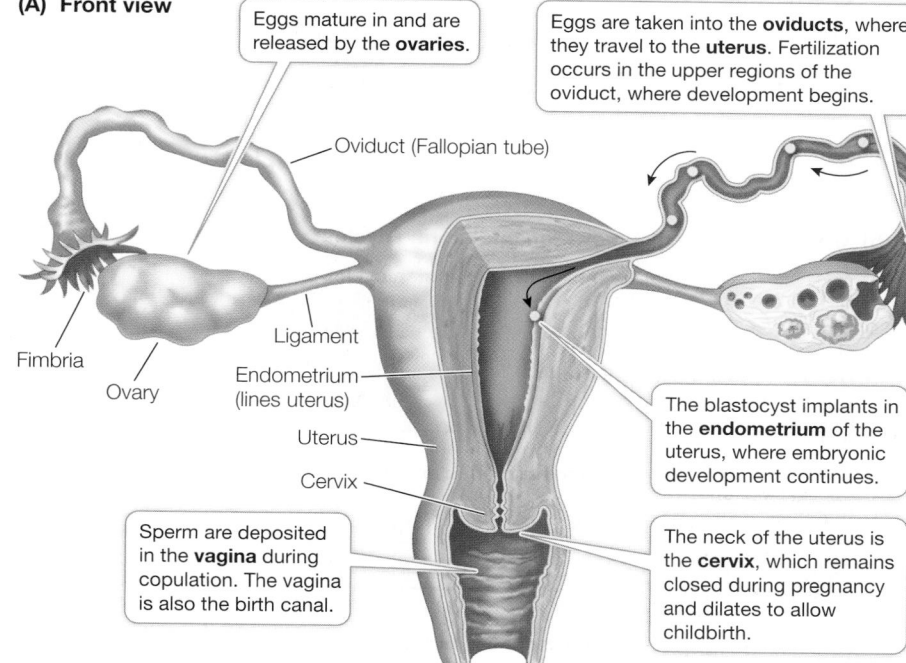

Eggs mature in and are released by the **ovaries**.

Eggs are taken into the **oviducts**, where they travel to the **uterus**. Fertilization occurs in the upper regions of the oviduct, where development begins.

Oviduct (Fallopian tube)

Fimbria

Ligament

Ovary

Endometrium (lines uterus)

Uterus

Cervix

Sperm are deposited in the **vagina** during copulation. The vagina is also the birth canal.

The blastocyst implants in the **endometrium** of the uterus, where embryonic development continues.

The neck of the uterus is the **cervix**, which remains closed during pregnancy and dilates to allow childbirth.

(B) Side view

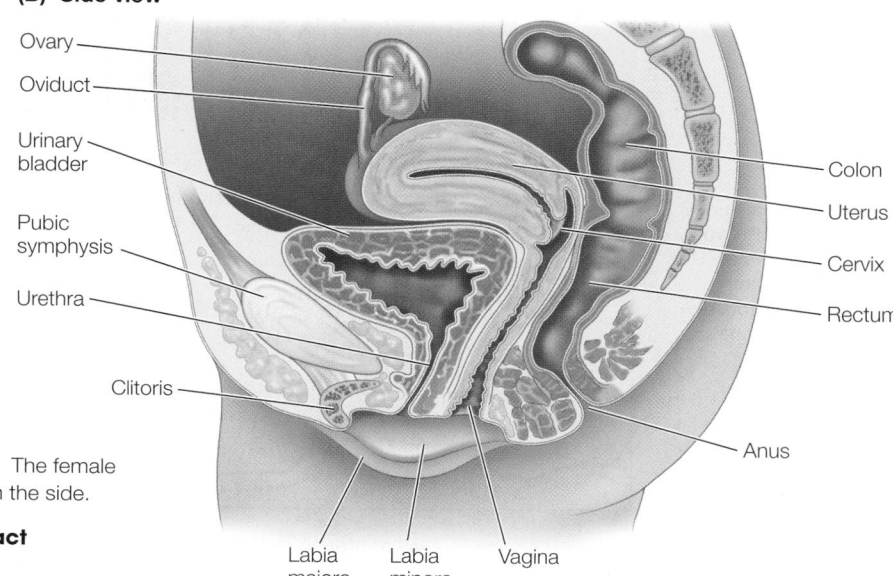

Ovary

Oviduct

Urinary bladder

Pubic symphysis

Urethra

Clitoris

Colon

Uterus

Cervix

Rectum

Anus

Labia majora

Labia minora

Vagina

Figure 42.12 Reproductive Tract of the Human Female   The female reproductive organs are shown **(A)** from the front and **(B)** from the side.

 Activity 42.3 **The Human Female Reproductive Tract**
www.Life11e.com/ac42.3

focus: key figure

(A) Changes in the ovary during the ovarian cycle

1 About once a month between puberty and menopause, 6–12 primary oocytes begin to mature. A primary oocyte and its surrounding cells constitute a **follicle**.

2 The surrounding follicular cells nourish the developing oocyte and release estrogen.

3 After 1 week, usually only one primary oocyte continues to develop. A meiotic division just before ovulation creates the secondary oocyte (*n*).

Primary oocyte (2*n*)

Ligament (holds ovary in place in the abdomen)

Ovary

Ruptured follicle

6 If pregnancy does not occur, the corpus luteum degenerates.

4 At ovulation, the follicle ruptures, releasing an oocyte.

5 The remaining follicle cells form the **corpus luteum**, which produces progesterone and estrogen.

Mature mammalian follicle

Primary oocyte   Granulosa cells   Thecal cells

(B) Coordinated events in ovarian and uterine cycles

FSH stimulates the development of follicles; the LH surge causes ovulation and then the development of the corpus luteum.

Events in ovary (ovarian cycle)

Oocyte maturation   Developing follicle   Ovulation (day 14)   Corpus luteum   Developing oocyte

Estrogen and progesterone stimulate the development of the endometrium in preparation for pregnancy.

Ovarian hormones and the uterine cycle

Estrogen

Progesterone

Endometrium of uterus   Bleeding and sloughing (menstruation)   Highly proliferated and vascularized endometrium

Thickness of endometrium

FSH and LH are under control of GnRH from the hypothalamus and the ovarian hormones estrogen and progesterone.

Gonadotropins (from anterior pituitary)

Estrogen inhibits LH and FSH release   Estrogen stimulates LH and FSH release   Estrogen inhibits LH and FSH release

Luteinizing hormone (LH)

LH surge triggers ovulation.

Follicle-stimulating hormone (FSH)

0   7   14   21   28
Day of uterine cycle

Figure 42.13 The Ovarian and Uterine Cycles **(A)** The ovarian cycle progresses from the development of a follicle to ovulation and then to growth and finally to degeneration of the corpus luteum. **(B)** During a woman's ovarian and uterine cycles, coordinated changes occur in the development of the ovarian follicle, the release of female sex steroids, and the endometrial lining of the uterus. The release of hypothalamic GnRH stimulates the release of pituitary LH and FSH which in turn control the development of the follicle and eventually ovulation. The cycles begin with the onset of menstruation; ovulation is at midcycle (yellow bar). **(C)** The ovarian and uterine cycles are under a complex series of positive and negative feedback controls involving several hormones.

Q: Hormonal contraceptive pills contain estrogen and progesterone, which portion of the ovarian cycle do they simulate and how do they prevent ovulation?

Animation 42.2 The Ovarian and Uterine Cycles
www.Life11e.com/a42.2

(C) Hormones control the ovarian and uterine cycles

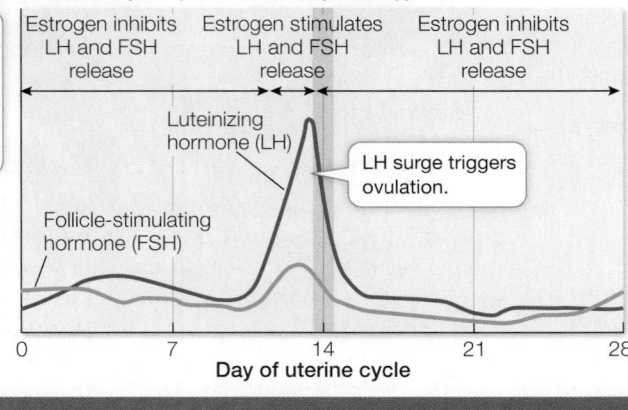

Hypothalamus

GnRH

Anterior pituitary

LH/FSH

Ovary

Estrogen and progesterone

Uterus

Positive feedback occurs during days 12 through 14.

Negative feedback occurs throughout most of the cycle.

Ovarian cycles produce mature eggs

A newborn baby girl has about a million primary oocytes in each ovary. By the time she reaches puberty, she has only about 200,000; the rest have degenerated. During a woman's fertile years, her ovaries go through about 450 ovarian cycles. During each cycle, 10 to 20 primary oocytes begin to mature, but usually only one matures completely and is ovulated; the others degenerate (see Figure 42.13A). At around the age of 50, a woman reaches **menopause**—the end of fertility—and may have few oocytes left in each ovary. For more than 50 years, the gradual age-related decline in oocytes was taken as evidence that oogenesis in humans is limited to prenatal life. However, in recent years oogonial stem cells (OSCs) have been isolated from adult ovaries of mice and of humans. These OSCs have been shown to be capable of division and production of oocytes. Thus menopause may not be the absolute end of fertility.

A primary oocyte surrounded by a layer of ovarian cells is the functional unit of the ovary—the **follicle** (see the photo in Figure 42.13A). The follicle cells surrounding the oocyte supply it with nutrients and growth factors. The follicle cells are also the main site of production of the female hormones estrogen and progesterone. Usually only one follicle matures completely and reaches the stage of ovulation, releasing its egg at midcycle. Events that lead up to and follow ovulation are illustrated in Figure 42.13B. Follicle maturation occurs in the first half of the cycle, therefore called the follicular phase. The midcycle point at which the follicle ruptures to release the oocyte is marked by a vertical yellow bar. The notable hormonal event leading up to ovulation is the gradual and then escalating increase in estrogen, which peaks just prior to midcycle (see Figure 42.13B).

The second half of the ovarian cycle is called the luteal phase because the follicle cells left in the ovary after the egg is ovulated develop into an endocrine gland—the **corpus luteum** (yellow body)—secreting estrogen and progesterone. If the egg is not fertilized and does not embed into the endometrium, the corpus luteum becomes inactive and degenerates in 12 to 14 days.

 Media Clip 42.1 **Human Ovulation**
www.Life11e.com/mc42.1

The uterine cycle prepares an environment for a fertilized egg

The uterine cycle parallels the ovarian cycle and consists of a buildup and then a breakdown of the endometrium (see Figure 42.13B). The breakdown of the endometrium, or **menstruation** (Latin *menses*, "months"), marks the beginning of each uterine and ovarian cycle. About 5 days after menstruation begins, the endometrium starts to thicken in preparation for receiving a blastocyst. The uterus attains its maximal state of preparedness about 5 days after ovulation and remains in that state for another 9 days. If a blastocyst has not arrived by that time, the endometrium breaks down and the sloughed-off tissue, including blood, flows from the body through the vagina as a menstrual discharge.

The uterine cycles of most mammals other than humans do not include menstruation; instead, the uterine lining typically is reabsorbed. In these species, the most obvious correlate of the ovarian cycle is a state of sexual receptivity called **estrus** ("heat") around the time of ovulation. You may be aware of the bloody discharge that occurs in dogs at the time of estrus. This discharge is not the same as menstruation—in fact it is exactly the opposite. Bleeding in dogs occurs during the *proliferation* of the uterine lining, which occurs just before ovulation. When the female mammal comes into estrus, she actively solicits male attention and may be aggressive to other females. Humans are unusual among mammals in that females are potentially sexually receptive throughout their ovarian cycles and at all seasons of the year.

Hormones control and coordinate the ovarian and uterine cycles

The ovarian and uterine cycles are coordinated and timed by the same hormones that initiate sexual maturation. Gonadotropins (FSH and LH) secreted by the anterior pituitary are the central elements of this control (see Figure 42.13B). Before puberty (i.e., before about 11 years of age), the secretion of FSH and LH is low and the ovaries are inactive. At puberty the hypothalamus increases its release of GnRH, stimulating the anterior pituitary to secrete FSH and LH. In response to FSH and LH, ovarian tissue grows and produces estrogen. The rise in estrogen causes the maturation of the accessory sex organs and the development of female secondary sexual characteristics. Between puberty and menopause, interactions of GnRH, gonadotropins, and sex steroids control the ovarian and uterine cycles.

A few days before menstruation begins, the anterior pituitary begins to increase its secretion of FSH and LH. In response, several follicles begin to mature in the ovaries, and these follicles steadily increase their production of estrogen. After about a week, all but one of the follicles wither away.

For the first 12 days of a 28-day ovarian cycle, estrogen exerts positive rather than negative feedback control on the pituitary. As a result, a surge of LH and a lesser surge of FSH occur (see Figure 42.13B). The LH surge triggers the mature follicle to rupture and release its egg, and it stimulates the cells of the ruptured follicle to develop into a corpus luteum.

Estrogen and progesterone secreted by the corpus luteum following ovulation are crucial to growth and maintenance of the endometrium. These sex steroids also exert negative feedback control on the pituitary, inhibiting gonadotropin release and thus preventing new follicles from maturing.

If the egg is not fertilized, the corpus luteum degenerates on about day 26 of the cycle. Without production of progesterone by the corpus luteum, the endometrium sloughs off and menstruation occurs. The decrease in circulating steroids also releases the hypothalamus and pituitary from negative feedback control, so GnRH, FSH, and LH all begin to increase. The increase in these hormones induces the next round of follicle development, and the ovarian cycle begins again.

FSH receptors determine which follicle ovulates

Early in the ovarian cycle several follicles begin to develop, but only one reaches full maturity. What determines which one will survive? There are two types of follicle cells, arranged in two layers. The inner layer immediately surrounding the ovum is made up of **granulosa cells**, and the outer layer consists of **thecal cells**. Thecal cells are stimulated by LH to produce testosterone (similar to Leydig cells in the male). Testosterone diffuses into the granulosa cells, where the enzyme aromatase converts the testosterone to estrogen. Estrogen, along with FSH, stimulates the growth and maturation of the granulosa cells (similar to the Sertoli cells in the male). The estrogen plays two important roles in the selection of the follicle that will ovulate: (1) estrogen stimulates the granulosa cells to express more FSH and LH receptors, and (2) estrogen entering the circulation feeds back on the pituitary to decrease the production of FSH. The granulosa cells also produce inhibin (similar to Sertoli cells in the male), and inhibin also decreases the production of FSH. As FSH and LH levels fall, the follicle with the most FSH and LH receptors survives and matures while the others regress.

In pregnancy, hormones from the extraembryonic membranes take over

If the egg is fertilized and a blastocyst arrives in the uterus and implants in the endometrium, a new hormone comes into play. A layer of cells covering the blastocyst begins to secrete **human chorionic gonadotropin**, or **hCG** (Figure 42.14). This gonadotropin, a molecule similar to LH, stimulates the corpus luteum to continue to produce estrogen and progesterone to support the growth and maintenance of the endometrium and thereby prevent menstruation. Because it is present only in the blood of pregnant women, the presence of hCG is the basis for pregnancy testing. Pregnancy tests use an antibody to detect hCG in urine; they take only minutes and can be done at home.

Blastocyst and endometrial tissues form the placenta, which nourishes the embryo. The placenta also replaces the corpus luteum as the major producer of estrogen and progesterone. Continued high levels of estrogen and progesterone prevent the pituitary from secreting gonadotropins; thus the ovarian cycle ceases for the duration of pregnancy. This mechanism underlies the action of birth control pills, which contain synthetic hormones resembling estrogen and progesterone that exert negative feedback control on the hypothalamus and pituitary.

Breast feeding delays the return of the ovarian cycle

Following birth, estrogen and progesterone levels drop because of the loss of the placenta, and therefore negative feedback on the

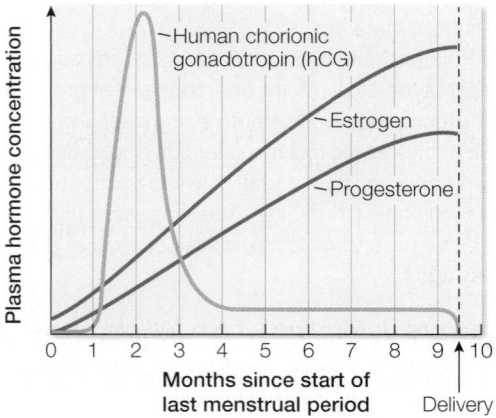

Figure 42.14 Pregnancy and Childbirth When a fertilized ovum implants in the uterus, cells surrounding it produce human chorionic gonadotropin, which acts like LH and keeps the corpus luteum functioning as an endocrine gland. The ovarian and uterine cycles are put on hold for the duration of pregnancy.

hypothalamus and pituitary imposed by these hormones during pregnancy is gone. Nevertheless, women who breast feed their infants may not have their ovarian and menstrual cycles return for some time—a phenomenon called lactational delay. In Figure 40.4 you learned that the pituitary hormone prolactin stimulates the growth of the mammary glands and the production of milk. Prolactin release is stimulated by the suckling of the infant. Even the sight or sound of the infant can stimulate prolactin release in the mother. Investigations have now shown that in addition to stimulating the mammary glands, prolactin also inhibits GnRH release and LH release, possibly explaining lactational delay. Going back to the case of the tammar wallaby in the opening to this chapter, could prolactin release stimulated by the suckling of the joey in the pouch be responsible for the delayed implantation seen in this species? **Investigating Life: The Control of Diapause in the Tammar Wallaby** examines this question.

Childbirth is triggered by hormonal and mechanical stimuli

Throughout pregnancy, the muscles of the uterine wall periodically undergo slow, weak, rhythmic contractions called Braxton Hicks contractions. These contractions become stronger during the third trimester of pregnancy and are sometimes called false labor. The contractions of true labor mark the beginning of childbirth. Both hormonal and mechanical stimuli contribute to the onset of labor (**Figure 42.15A**).

Progesterone inhibits and estrogen stimulates contractions of uterine muscle. Toward the end of the third trimester, the estrogen–progesterone ratio shifts in favor of estrogen. The onset of labor is marked by increased secretion of the hormone oxytocin by the posterior pituitaries of both mother and fetus. Oxytocin is a powerful stimulant of uterine muscle contraction. Manufactured oxytocin is used to induce labor when that is necessary.

Mechanical stimuli come from the stretching of the uterus by the fully grown fetus and the pressure of the fetal head on the cervix. These mechanical stimuli increase the release of oxytocin by the mother's posterior pituitary, which in turn increases the activity of uterine muscle, causing even more pressure on the cervix. This positive feedback loop converts the weak, slow, rhythmic Braxton Hicks contractions into stronger labor contractions.

In the early stage of labor, hormonal changes and pressure created by the contractions cause the cervix to dilate (expand) until it is large enough to allow the baby to pass through. Gradually the contractions become more frequent and more intense. This stage of labor lasts an average of 12 to 15 hours in a first pregnancy, but is usually 8 hours or less in subsequent ones.

The second stage of labor begins when the cervix is fully dilated to a diameter of about 10 cm (**Figure 42.15B**). The baby's head can now move into the vagina. Passage of the fetus through the vagina is assisted by the mother's bearing down ("pushing") with her abdominal and other muscles. Once the head and shoulders of the baby clear the cervix, the rest of its body eases out rapidly, but it is still connected to the placenta by the umbilical cord. Once the baby clears the birth canal, it starts breathing and is independent of its mother's circulation. The umbilical cord may then be clamped and cut. The segment still attached to the baby dries up and sloughs

experiment

Original Paper: Hinds, L. A. and C. H. Tyndale-Biscoe. 2013. Daily prolactin pulse inhibits the corpus luteum during lactational quiescence in the marsupial, *Macropus eugneii*. *Reproduction, Fertility, and Development* 25: 456–461.

In Chapter 40 (see Figure 40.4) we discussed how the hormone prolactin (PRL) has evolved to serve many physiological processes in different vertebrate groups, but these various processes all have some connection to reproduction. Since prolactin is involved in controlling lactation in mammals, could prolactin be the signal in the tammar wallaby that maintains embryonic diapause as long as suckling continues?

HYPOTHESIS▶ Pulses of prolactin in lactating female tammar wallabies prevent new embryos from developing as long as previous offspring are suckling.

METHOD

The hormone progesterone (P) prepares the uterine lining for implantation of an embryo and maintains the uterine lining during pregnancy. Thus measurements of blood levels of P reveal when diapause is broken in tammar wallabies. A stimulus known to break diapause in these animals is the removal of the prior young from the pouch (RYP). The RYP method was used to study the effects of prolactin on maintenance of diapause.

1. Remove young from the pouches of three groups of five wallabies.
2. Following RYP administer the following treatments:

   Group A: saline injection for 8 days following RYP (controls)

   Group B: PRL injection every 24 hr for 8 days following RYP

   Group C: PRL injection every 48 hr for 8 days following RYP
3. Take daily blood samples to measure P levels.

RESULTS

Data are approximate means of the actual measurements, progesterone concentrations (P) expressed in picograms per milliliter (pg/mL).

Days following RYP	P level (picograms/mL)		
	Group A	Group B	Group C
0	100	125	100
1	150	175	113
2	150	125	125
3	180	150	135
4	300	150	175
5	450	125	215
6	175	125	240
7	200	150	325
8	175	16	210
9	175	175	240
10	225	250	335
11	200	450	260
12	250	350	200
13	225	225	225
14	300	250	230
15	350	200	210
16	360	225	225

CONCLUSION▶ When tammar wallabies were treated with daily saline injections (controls), they broke diapause 4 to 5 days following RYP. However, when they were treated with daily PRL injections, they broke diapause 11 or 12 days following RYP. Hence daily PRL pulses can prolong diapause and delay implantation of a new blastocyst.

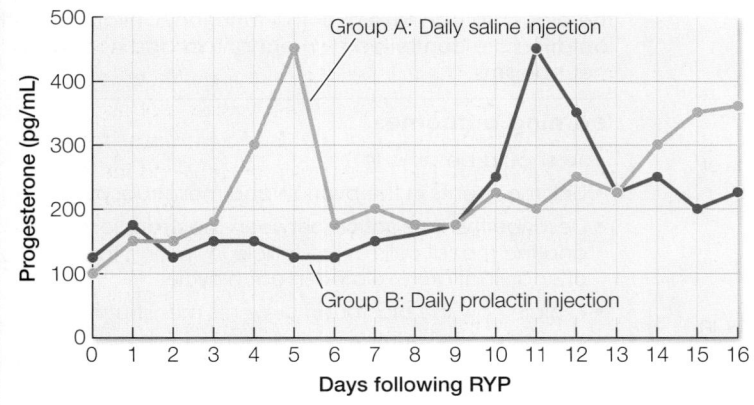

work with the data

Following the experiment showing that daily pulses of PRL can maintain diapause, the investigators asked whether pulses of PRL every 2 days could maintain diapause. The data are shown in the table as Group C. Plotting the data will show if there is a difference in the response to PRL when pulses were given every other day following RYP.

QUESTION▶

1. Plot the data as was done for Groups A and B. Was the 48-hr PRL pulse effective in preventing the breaking of diapause?

A similar **work with the data** exercise may be assigned in **LaunchPad**.

off in a few days, leaving behind its distinctive signature, the belly button—more properly called the umbilicus. The third stage of labor, the detachment and expulsion of the placenta and fetal membranes, takes from a few minutes to an hour, and may be accompanied by uterine contractions. A baby's suckling at the breast immediately following birth stimulates additional secretion of oxytocin, which augments uterine contractions, reduces the size of the uterus, and helps stop bleeding.

(A) Signals stimulating labor and birth

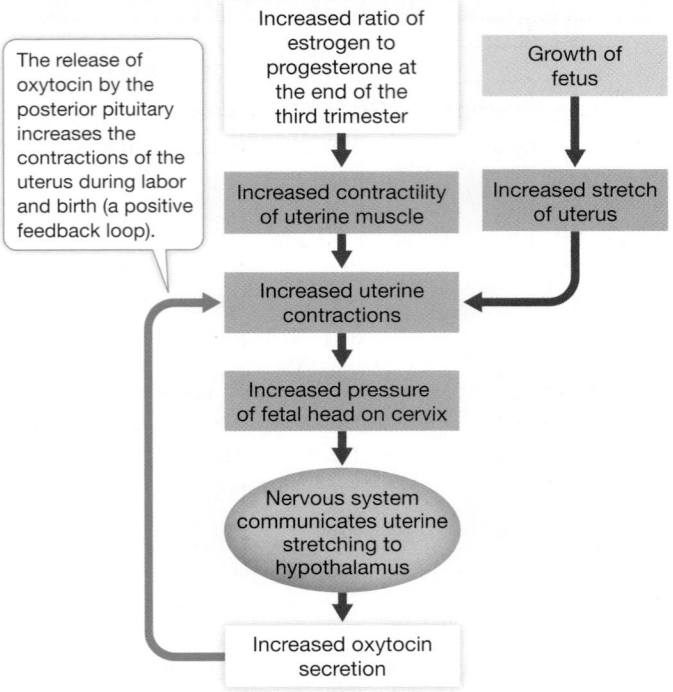

The release of oxytocin by the posterior pituitary increases the contractions of the uterus during labor and birth (a positive feedback loop).

Increased ratio of estrogen to progesterone at the end of the third trimester

Growth of fetus

Increased contractility of uterine muscle

Increased stretch of uterus

Increased uterine contractions

Increased pressure of fetal head on cervix

Nervous system communicates uterine stretching to hypothalamus

Increased oxytocin secretion

(B) Delivery

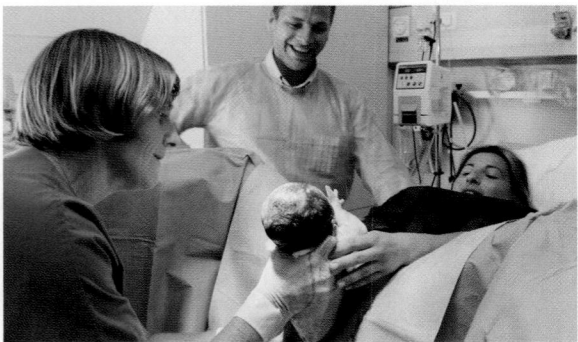

Figure 42.15 Labor and Delivery **(A)** Both mechanical and hormonal signals are involved in stimulating the uterine contractions of labor and delivery. **(B)** A new person comes into the world headfirst.

Birth as well as reproduction is timed

Reproduction in many species is seasonal. This seasonality in many species of vertebrates involves mechanisms of photoperiodism. Briefly, changes in day length enable animals to anticipate the best time of year for the rearing of young and therefore time their reproduction and birth accordingly. How do animals measure day length? As we will discuss in more detail in Chapter 52, animals have endogenous daily rhythms called circadian (*circa*, "day," + *dia*, "length") rhythms. As day length changes, sunrise and sunset occur at different phases of the circadian rhythm. The release of melatonin from the pineal gland is controlled by the circadian rhythm but is inhibited by light. Therefore as days get longer, the nightly pulse of melatonin gets shorter, and conversely as days get shorter, the melatonin pulse gets longer. Day length determines when many species come into reproductive condition and mate.

That is not true for the tammar wallaby that was introduced at the beginning of this chapter. The wallaby mates at any time of year soon after it gives birth. However, diapause in the wallaby is sensitive to night length—that is, to the duration of the melatonin pulse. A non-lactating wallaby will break diapause when night length reaches 12 hr. Similarly, when non-lactating wallabies are held at a night length of 9 hr but given melatonin injections 2.5 hr before sunset, they break diapause. These results support the hypothesis that short daily melatonin pulses prolong diapause and that long daily melatonin pulses are a signal to break diapause. The significance for the nocturnal wallaby is that birth tends to be at the time of year when nights are longest.

Humans are not seasonal breeders, but there is a timing component of childbirth. Most women go into labor at night. However, it is quite common for labor to cease when women arrive in the brightly lit hospital. Bright light interrupts the nightly melatonin pulse in humans. The work with the data exercise (Circadian Timing, Hormone Release, and Labor) explores whether melatonin is involved in the timing of human births.

42.4 recap

The reproductive system of women produces oocytes (eggs) and hormones, and these functions are controlled by hypothalamic and anterior pituitary hormones. The follicle is the functional unit of the ovary. Usually one follicle matures and ovulates each month as controlled by a cyclical feedback interplay between ovarian hormones and the hypothalamic–pituitary axis. These hormonal cycles also coordinate the uterine, or menstrual, cycle so that the endometrium is prepared to receive a blastocyst should fertilization occur. A blastocyst implanted in the uterus produces hCG, which suspends the ovarian and uterine cycles by sustaining the ovarian production of progesterone and estrogen until the placenta can take over this function. Childbirth and lactation are controlled by mechanical and hormonal mechanisms.

learning outcomes

You should be able to:

- Link the events in the ovarian and menstrual cycles.
- Describe the interaction between the granulosa cells and the thecal cells of the follicle in limiting the number of eggs that are ovulated in each cycle.
- Explain how the blastocyst prevents menstruation.
- Explain how mechanical and hormonal stimuli interact during labor.

1. What events in the ovary are responsible for the onset of menses, and what prevents menses if pregnancy occurs?
2. What are the roles of thecal cells and granulosa cells in the ovarian cycle, and how do they limit the number of eggs that are ovulated?
3. How does positive feedback between hormonal and mechanical processes play a role in childbirth?

Understanding the physiology of human reproduction has led to numerous methods and technologies for controlling it, either to prevent unwanted pregnancies or to overcome infertility.

key concept
42.5
Fertility Can Be Controlled

Sexual issues and sexual behavior are dominant aspects of our society, and reproductive technologies have had huge impacts on our sexual and reproductive lives.

focus your learning

- Birth control techniques are designed to prevent fertilization and/or implantation.
- Assisted reproductive technologies enable couples who are infertile to have children.

Humans use a variety of methods to control fertility

According to a recent study, almost half of the more than 6 million pregnancies that occur in the United States each year are unintended. For women of college age, a single act of unprotected intercourse in the 2 days prior to ovulation carries a chance of conception as high as 50 percent.

The only failure-proof methods of preventing pregnancy are either complete abstinence from sexual activity or surgical removal of the gonads. Those options are not acceptable to most people, and they turn to other methods to prevent pregnancy. Many of these methods prevent fertilization or implantation (conception) and are therefore referred to as **contraception**. **Table 42.1** lists some of the most commonly used contraceptive methods and their relative failure rates; note that these methods vary enormously in their effectiveness. Most methods are used by the woman, although some are used by the man.

Once a fertilized egg is successfully implanted in the uterus, any termination of the pregnancy is called an **abortion**. A spontaneous abortion is the medical term for what is commonly called a miscarriage. Spontaneous abortions frequently occur early in pregnancy and are usually the result of either a chromosomal abnormality in the fetus or a breakdown in the process of implantation. Many spontaneous abortions occur before the woman even realizes she is pregnant.

Abortions that result from medical intervention may be performed either for therapeutic purposes or for fertility control. A therapeutic abortion may be necessary to protect the health of the mother, or it may be performed because prenatal testing reveals that the fetus has a severe defect. In a medical abortion, the cervix is dilated and some of the endometrium, along with the implanted fetus, is removed. When performed in the first trimester of a pregnancy, a medical abortion carries less risk of death to the mother than a full-term pregnancy. The risk rises after the first 12 weeks of pregnancy but remains less than that of a full-term pregnancy through the second trimester.

Reproductive technologies help solve problems of infertility

About 15 percent of couples in the United States are infertile—that is, they can't have children. The reasons for infertility are many, and are about equally distributed between men and women. Several technologies have been developed to overcome barriers to both conceiving and bearing a child. The simplest of these is **artificial insemination**, in which the physician places sperm in the woman's reproductive tract. This technique is useful if the male's sperm count is low, if his sperm lack motility, or if problems in the female's reproductive tract prevent the normal movement of sperm up the oviducts. Artificial insemination is also widely used in the production of domesticated animals such as cattle.

Assisted reproductive technologies, or **ARTs**, involve procedures that remove unfertilized eggs from the ovary, combine them with sperm outside the body, and then place fertilized eggs or egg–sperm mixtures in the appropriate location in the woman's reproductive tract for development to take place. The first successful ART was in vitro fertilization (IVF). In IVF the woman is treated with hormones that stimulate many follicles in her ovaries to mature. Eggs are collected from these follicles, and sperm are collected from the intended father. Eggs and sperm are combined in a culture medium outside the body, where fertilization takes place. The resulting embryos can be injected into the mother's uterus in the blastocyst stage or kept frozen for implantation later. The first "test tube baby" resulting from IVF was born in England in 1978. Since then, millions of babies have resulted from this ART.

A major cause of IVF failure is failure of sperm to gain access to the egg cell membrane (see Figure 42.5). To solve this problem, methods have been developed to inject a sperm cell directly into the cytoplasm of an egg. In intracytoplasmic sperm injection (ICSI), an egg is held in place by suction applied through a polished glass pipette. A slender, sharp pipette is then used to penetrate the egg and inject a sperm (**Figure 42.16**). This ART was used successfully for the first time in 1992 by researchers in Belgium; now thousands of these procedures are performed in U.S. clinics each year, with a success rate of about 25 percent.

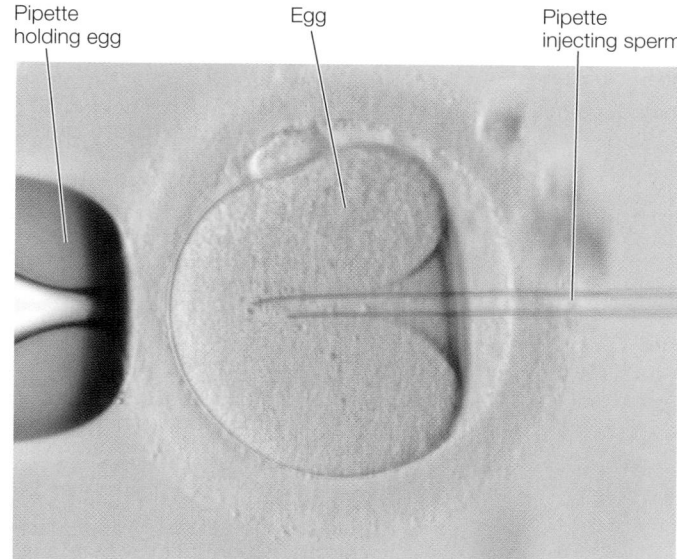

Pipette holding egg   Egg   Pipette injecting sperm

Figure 42.16 Intracytoplasmic Sperm Injection In this procedure, a sperm is injected directly into a mature egg cell. The fertilized egg is then placed in the female reproductive tract, where it can implant and develop into a fetus.

table 42.1 Methods of Contraception

Method	Mode of action	Failure rate[a]	Comments
Unprotected	No form of birth control.	85	High risk of pregnancy, especially for women 15–30.
Nontechnological methods			
Rhythm method	The couple abstains from intercourse between days 10 and 20 of the ovarian cycle (peak fertility).	15–35	High failure rate due to miscalculation and/or variation of individual cycles.
Coitus interruptus	The man withdraws his penis prior to ejaculation with the intention of not depositing sperm into the vagina.	20–40	Requires self-control, especially by the man. Very high failure rate.
Barrier methods[b]			
Condom (male or female)	A sheath of impermeable material (often latex) is fitted over the erect penis or into the vagina. Semen is trapped in the condom, so no sperm are deposited in the vagina.	3–20	If fitted correctly, an intact condom can prevent pregnancy. Failure can be due to material failure or improper use. Condoms are the only contraceptive method that protects against sexually transmitted diseases (STDs), including HIV (AIDS).
Spermicidal jellies	Applied deep inside the vagina, these chemical compounds kill or immobilize sperm.	25	Used alone, spermicidal compounds have a high failure rate.
Diaphragms, cervical caps, cervical sponges containing spermacide	Inserted by the woman prior to intercourse, these devices work by blocking the cervix so that sperm cannot pass into the uterus.	15–30	Approximately the same failure rate as condom use by men, but do not protect against STDs. Can be used in conjunction with spermicidal jelly for extra protection.
Hormone-based contraceptives			
Oral hormones ("the pill")	A daily pill for women containing a combination of synthetic estrogens and progesterone. These hormones mimic pregnancy to the extent that the ovarian cycle and ovulation are suspended. The uterine cycle can continue by including a week of non-hormone administration every 21–28 days.	0–3	Requires medical consultation and prescription. Taken correctly, oral contraceptives are extremely effective. In the United States, more than 12 million women use them each year; they are sometimes prescribed to treat menstrual disorders.
Non-orally administered hormones	Making use of same hormonal actions as the pill, these methods include long-acting injections, patches that release hormones transdermally (through the skin), and a hormone-containing vaginal ring. Implantable progestin releasing have a failure rate of <1 and can last up to 3 years.	<5	Same as oral hormones. A slightly lower failure rate because the woman does not have to remember to take a daily pill.
Progestin-only pill (Plan B[b])	An oral contraceptive meant to be taken within 72 hours after unprotected sex. A high dose of progestin in two pills prevents ovulation in the same manner birth control pills do.	5–40	Not an "abortion pill," this drug will not terminate an existing pregnancy. Currently available to women over 17 without a prescription. Failure rate varies widely depending on when taken.
Implantation blockers			
Intrauterine device (IUD)	A medical professional inserts a small plastic or metal device into the uterus. The resulting inflammation reaction (see Chapter 40) releases prostaglandins, which prevent implantation of the fertilized egg.	<1	A highly effective contraceptive, it is the most widely used birth control device in China (and hence the world). With medical monitoring, can remain in place for several years.
Contragestational pill	Mifepristone or RU-486 is a drug that blocks the action of progesterone that is necessary to maintain the endometrium during implantation and pregnancy.	0.5–6	Prevents implantation when taken up to several days after unprotected intercourse. Can terminate a pregnancy up to the time of the first missed menstrual period. In the United States, available from specialized providers.
Sterilization			
Vasectomy	The vasa deferentia (see Figure 42.9) are cut and tied off so that sperm can no longer pass into the urethra. Sperm continue to be produced but are reabsorbed by the man's body. Male hormone levels and sexual responses are not affected.	0–0.15	A simple surgical procedure performed under local anesthetic in a doctor's office. Although theoretically it can be reversed, vasectomy should be considered permanent.
Tubal ligation	The oviducts (see Figure 42.12A) are tied off so that eggs cannot reach the uterus and sperm cannot reach the egg. As with vasectomy, hormone levels and sexual responses are not affected.	0–0.05	This surgical procedure is somewhat more complex than vasectomy. It is often performed in conjunction with childbirth when a woman has decided that her family is complete.

[a]Failure rate refers to the number of pregnancies per 100 women per year.
[b]All of these barrier methods are routinely available without medical prescription.

IVF, coupled with techniques of genetic analysis, can eliminate the risk that adults who are carriers of genetic diseases will produce affected children. It is now possible to take a cell from a human embryo at the 4- or 8-cell stage (see Figure 43.3) without damaging its developmental potential. The sampled cell can be subjected to molecular analysis to determine whether it carries the harmful gene. This procedure, called preimplantation genetic diagnosis (PGD), makes it possible to determine whether an embryo produced by IVF carries the genetic defect of concern.

42.5 recap

Controlling fertility is an important aspect of modern human life. Decreasing the probability of pregnancy is achieved through methods that prevent sperm and egg from meeting and from preventing implantation. Pregnancies can be facilitated through technologies that extract eggs from the ovary and fertilize them in vitro before reintroducing them into the female reproductive tract.

learning outcomes

You should be able to:

• Describe the relative advantages and disadvantages of different contraceptive methods.

42.5 recap

• Compare contraception and contragestational techniques.
• Explain how procedures that use genetic analysis and in vitro fertilization (IVF) can help carriers of genetic disease avoid producing afflicted children.

1. Which method of contraception is the only one to offer protection against sexually transmitted diseases (STDs)? Explain your answer.
2. Both the contraceptive pill ("the pill") and the contragestational pill (RU-486) involve progesterone signaling. How do their mechanisms of action differ?
3. How can IVF be used to prevent transmission of a genetic disease?

The fertilized egg of a sexually reproducing organism is a single cell containing all the genetic information needed to create a new organism. Chapter 19 introduced some of the molecular aspects of the process of development in multicellular animals. Chapter 43 will describe the physiological and anatomical events of animal development.

investigating**life**

What signal controls the breaking of diapause in the tammar wallaby?

The experiment presented in Investigating Life: The Control of Diapause in the Tammar Wallaby exploited the fact that removal of the young joey from the pouch of its mother (RYP) stimulated the blastocyst in her uterus to break diapause, implant, and continue development. What diapause-breaking signal might be involved in conjunction with RYP? Researchers hypothesized that RYP resulted in cessation of prolactin release stimulated by suckling of the young, and that the absence of those prolactin pulses initiated the breaking of diapause. The test of that hypothesis was to apply RYP to groups of mother wallabies and test whether applications of prolactin could prevent the breaking of diapause. The results showed that injections of prolactin every 24 hr for 8 days following RYP delayed the progesterone rise that indicates the breaking of diapause. Prolactin injections at 48-hr or longer intervals were less effective or ineffective in maintaining diapause. The conclusion was thus reached that prolactin pulses stimulated by suckling are a

signal for the maintenance of diapause, and that the removal of that signal results in the breaking of diapause. Another signal, melatonin, could be involved in the breaking of diapause. The prolactin experiments were all done at the time of year when night length was appropriate for the breaking of diapause.

Future directions

Reproductive success depends on many factors, including ecological, physiological, and socio-behavioral factors. It is therefore not surprising that evolutionary adaptations have exploited many kinds of information to control the timing of reproduction in various species. While these controlling factors are diverse, in vertebrates they all work through the hypothalamic, pituitary, and gonadal regulatory mechanisms. Therefore whatever we can learn about the physiology of reproduction in one vertebrate species may also apply to other species, or at least inform us as to what questions we should ask of other species, including humans. The answers to these questions can help us preserve endangered species, improve domesticated species, solve reproductive problems in humans, and even help control human population growth.

Chapter Summary 42

42.1 Asexual Reproduction Is Efficient but Limits Genetic Variability

- Asexual reproduction produces offspring that are genetically identical to their parent and to one another; it produces no genetic diversity.
- Means of asexual reproduction include **budding**, **regeneration**, and **parthenogenesis**. Review Figures 42.1–42.3

42.2 Sexual Reproduction Involves Union of Haploid Egg and Sperm

- Sexual reproduction involves three basic steps: **gametogenesis**, **spawning** or **mating**, and **fertilization**.
- Gametogenesis and fertilization are similar in all animals, but spawning and mating include a great variety of anatomical, physiological, and behavioral adaptations.
- Gametogenesis occurs in **testes** and **ovaries**. In spermatogenesis (the production of **sperm**) and oogenesis (the production of eggs), the **germ cells** proliferate mitotically, undergo meiosis, and mature into gametes.
- Each **primary spermatocyte** can produce four haploid sperm through the two divisions of meiosis. Review Figure 42.4A
- **Primary oocytes** immediately enter prophase of the first meiotic division, and in many species, including humans, their development is arrested at this point. Each **oogonium** produces only one egg. Review Figure 42.4B
- Fertilization involves sperm activation, species-specific binding of sperm to egg, the acrosomal reaction, digestion of a path through the protective coverings of the egg, and fusion of sperm and egg cell membranes. Fusion of these two membranes triggers **blocks to polyspermy**, which prevent additional sperm from entering the egg and, in mammals, signal the egg to complete meiosis and begin development. Review Figure 42.5, Animation 42.1
- **External fertilization** is common in aquatic species. **Internal fertilization** is necessary in terrestrial species and usually involves **copulation**.
- **Hermaphroditic**, or **monoecious**, species have both male and female reproductive systems in the same individual, either sequentially or simultaneously. **Dioecious** species have separate male and female individuals.
- Animals can be classified as **oviparous** or **viviparous**, depending on whether the early stages of development occur outside or inside the mother's body.

42.3 Male Sex Organs Produce and May Deliver Sperm

- Males produce **semen** consisting of sperm suspended in seminal fluid (which nourishes the sperm and facilitates fertilization).
- Sperm are generated in the **seminiferous tubules** of the testes, mature in the **epididymis**, and are delivered to the **urethra** through the **vasa deferentia**. Other components of semen are produced in the **seminal vesicles**, **prostate gland**, and **bulbourethral glands**. Review Figure 42.9, Activities 42.1, 42.2

- All components of the semen join in the urethra at the base of the **penis** and are ejaculated through the erect penis by muscle contractions at the climax of copulation.
- Spermatogenesis depends on testosterone secreted by the **Leydig cells** of the testes, which are under the control of hormones produced in the anterior pituitary and the hypothalamus. The production of these hormones is controlled by negative feedback from testosterone and from inhibin, a hormone produced by the **Sertoli cells** of the testes. Review Figures 42.10, 42.11

42.4 Female Sex Organs Produce Eggs and Nurture Embryos

- Primary oocytes in the ovary are surrounded by granulosa and thecal cells that make up a follicle. Each month a small number of follicles respond to pituitary FSH by growing, producing estrogen, and promoting development of the oocyte.
- Usually in each cycle only one follicle fully matures and releases its oocyte, which is picked up by the opening of the oviducts. Sperm deposited in the **vagina** during copulation move up through the **cervix** and **uterus** into the oviducts. Fertilization occurs in the upper regions of the oviducts. Review Figure 42.12, Activity 42.3
- The maturation and release of eggs constitute an **ovarian cycle**. The **uterine**, or **menstrual**, **cycle** prepares the uterus for receipt of a blastocyst. If no blastocyst is implanted, the lining of the uterus sloughs off in the process of **menstruation**. Review Focus: Key Figure 42.13, Animation 42.2
- Both the ovarian and the uterine cycles are under the control of hypothalamic and pituitary hormones, which in turn are under the feedback control of estrogen and progesterone. Review Figure 42.14
- Childbirth is initiated by hormonal and mechanical stimuli that increase the contractility of uterine muscle. Review Figures 42.14, 42.15A

42.5 Fertility Can Be Controlled

- Methods of **contraception** include abstention from copulation and the use of technologies that decrease the probability of fertilization. Review Table 42.1
- **Assisted reproductive technologies** (**ARTs**) have been developed to increase fertility.

Go to **LearningCurve** (in **LaunchPad**) for dynamic quizzing that helps you solidify your understanding of this chapter. **LearningCurve** adapts to your responses, giving you the practice you need to master each key concept.

Apply What You've Learned

Review

42.4 The ovarian and uterine cycles are coordinated and timed by the same hormones that initiate sexual maturation.

Original Paper: Johnston, S. D., P. O'Callaghan, K. Nilsson, G. Tzipori and J. D. Curlewis. 2004. Semen-induced luteal phase and identification of a LH surge in the koala (*Phascolarctos cinereus*). *Reproduction* 128: 629–634.

Many female mammals, including humans, are spontaneous ovulators, meaning they ovulate in a cyclical pattern; the release of eggs requires no external stimulus. Other female mammals, including camels, rabbits, and koalas, are reflex ovulators; in these females, sexual activity induces ovulation. The females come into estrus but do not ovulate and progress to the luteal phase unless copulation occurs. What is the ovulation "trigger" in these mammals? Is it a pheromone, the physical stimulus of coitus, or a chemical signal in the seminal fluid? Researchers studied female koalas to find the answer.

The goal of Experiment 1 was to determine whether the presence of a male or the physical act of mating stimulates ovulation. Female koalas were separated into two groups of four. On day 2 of estrus, Group A females were paired with males, and mating occurred. Group B females were housed next to males but without physical contact. At intervals following initiation of exposure to males, the researchers took blood samples from the females and measured luteinizing hormone (LH) as the indicator of ovulation. The results of the blood samples indicated that all four females in Group A showed 25 nanogram per milliliter (ng/mL) peaks of LH between 24 and 32 hours postcoitus, and all gave birth 35 days postcoitus. During the same time interval, no females in Group B showed any change in LH hormone levels, and none gave birth.

In Experiment 2, female koalas were separated into four groups of nine. On day 2 of estrus, Group A females underwent an artificial insemination procedure but received 1 mL of 0.9% sterile saline. Group B females were artificially inseminated with 1 mL of koala semen. Group C females had their urogenital sinus artificially stimulated with a glass rod designed to mimic the action of a male koala's penis during mating. Group D females had the same treatment as females in Group C, followed by artificial insemination with 1 mL of koala semen. At 14 and 28 days after the treatments, the researchers took blood samples from the females and measured levels of progesterone as an indicator of the postovulatory luteal phase. The mean values of blood progesterone levels were calculated for each group and are shown in the figure below.

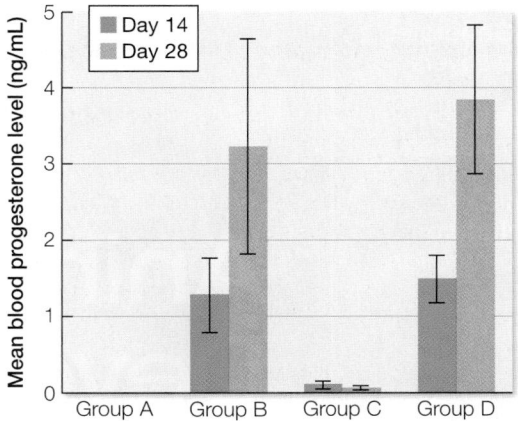

Questions

1. Describe the series of events in a female koala's body that would produce the results in Experiment 1. What potential triggers of ovulation could be eliminated through this experiment?

2. Based on the results of Experiment 2, what appears to be the necessary and sufficient stimulus for reflex ovulation in the koala? What potential triggers of ovulation could be eliminated through this experiment?

3. Propose a possible evolutionary advantage of reflex versus spontaneous ovulation.

Go to **LaunchPad** for the eBook, LearningCurve, animations, activities, flashcards, and additional resources and assignments.

43

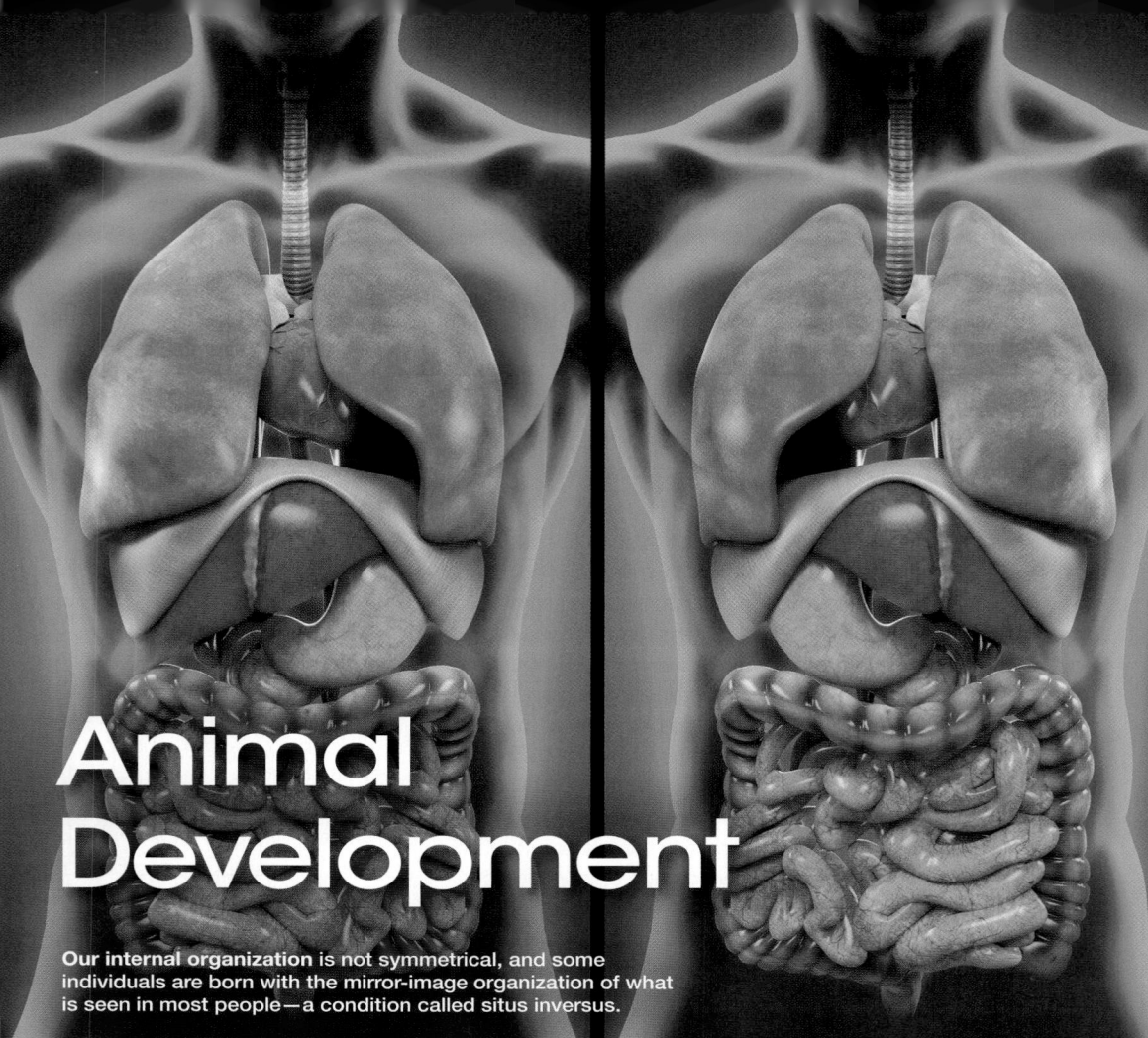

Animal Development

Our internal organization is not symmetrical, and some individuals are born with the mirror-image organization of what is seen in most people—a condition called situs inversus.

investigating life

Go With the Flow

Place your hand over your heart. Point to your appendix, then your pancreas. Surely your hand went to the left side of your chest, then to the right side of your lower abdomen, and lastly to the left side of your upper abdomen. Vertebrates are bilaterally symmetrical, but our symmetry is not absolute. Some internal organs are oriented to the left and some to the right. In about 1 out of every 7,000 people, these orientations are reversed, a condition known as situs inversus ("location inverted"). The difference arises from events very early in the development of the embryo. Most people with situs inversus lead normal lives and are unaware of their condition until they have diagnostic imaging or surgery.

You will learn in this chapter that a mammalian embryo goes from a single layer of cells to the next stage, with two layers of cells, by forming a midline slit, and that cells migrate through that slit to form a second cell layer. Below where the inward movement of cells starts is an area called the node. Each cell at the node has a primary cilium. The primary cilia of some of these cells are nonmotile and probably serve sensory functions. The primary cilia of other nodal cells

are motile, and they beat in a clockwise motion that moves extracellular fluid across the surface of the node. That fluid movement is from right to left. Why?

Imagine that the movement of a cilium describes a clock face oriented so that 12 o'clock is closest to the embryo's future head. The cilia protrude from the cell surfaces at an angle pointing posteriorly. Thus when the cilia rotate through the 6 o'clock position, they are closer to the cell surface and experience greater resistance than at the 12 o'clock position. As a result, the circular beating of cilia at this critical spot—the node—creates left–right flow and therefore left–right positional information.

Why do we hypothesize that the information generated by beating nodal cilia is responsible for left–right asymmetrical patterns? In mutant strains of mice that have no cilia or only nonmotile cilia in the node, about half of the mice have reversed organ symmetry. Similarly, about half of the people with Kartagener's syndrome, a spectrum of rare genetic disorders in humans characterized by nonmotile cilia, have situs inversus.

key concept 43.1 Fertilization Activates Development

Fertilization is the joining of sperm and egg to produce a diploid zygote. You might think of fertilization as the event that begins development. But keep two things in mind: (1) in asexually reproducing animals, development proceeds without fertilization, and (2) in animals where fertilization does occur, it is preceded by critical events in the maturing egg that influence subsequent development. Thus in studying fertilization we are asking how it activates multicellular development in sexually reproducing animals.

focus your learning

- The egg and sperm contribute different materials to the zygote.
- Sperm entry converts the symmetry of the frog egg from radial to bilateral.
- The movement of cytoplasm, proteins, and organelles following fertilization changes the distribution of critical signaling factors.

Fertilization does more than restore a full diploid complement of maternal and paternal genes. The fusion of sperm and egg cell membranes:

- triggers ion fluxes across the egg membrane.
- creates blocks to the entry of additional sperm into the egg.
- changes the pH of the egg cytoplasm.
- increases egg metabolism and stimulates protein synthesis.
- initiates the rapid series of cell divisions that produce a multicellular embryo.

We described the mechanisms of fertilization in Key Concept 42.2. Here we will take a closer look at the cellular and molecular interactions of sperm and egg that initiate the first steps of development.

The sperm and the egg make different contributions to the zygote

In most species, eggs are much larger than sperm. Egg cytoplasm is well stocked with organelles, nutrients, and a variety of molecules, including cytoplasmic determinants such as transcription factors and mRNAs (see Key Concept 19.2). Cytoplasmic determinants in the egg play important roles in setting up the signaling cascades that orchestrate the major events of development: determination, differentiation, morphogenesis, and growth. Nearly everything the embryo needs during its early stages of development comes from the mother, including its mitochondria (and therefore all of its mitochondrial DNA).

In addition to its haploid nucleus, the sperm makes one other crucial contribution to the zygote in most species—the centriole. The centriole contributes to the zygote's centrosome, which organizes the mitotic spindles for subsequent cell divisions (see Figure 11.9). Centrioles are also the origin of the microtubules of the primary cilia, which are important in cell signaling, as you will see when we return to the opening question about how embryos know left from right.

Fertilization sets the stage for determination

Unique attributes of amphibian eggs make them ideal models for illustrating how rearrangements of egg cytoplasm set the stage for determination. Remember from Key Concept 19.1 that determination sets the developmental fates of cells, and it precedes differentiation—the expression of different cell properties. The molecules in the cytoplasm of the amphibian egg are not homogeneously distributed. The entry of the sperm into the egg stimulates rearrangements of the egg cytoplasm that introduce additional organization to the egg. Sperm entry establishes the polarity of the zygote, and informational molecules in the egg cytoplasm are organized with respect to that polarity. Therefore, when cell divisions begin, these informational molecules—which guide subsequent development—are not divided equally among daughter cells.

Rearrangement of egg cytoplasm following fertilization is easily observed in some frog species because of pigments in the cytoplasm. The nutrients in an unfertilized frog egg are dense yolk granules that are concentrated by gravity in the lower half of the egg, called the **vegetal hemisphere**. The haploid nucleus of the egg is located at the opposite end, in the **animal hemisphere**. The outermost (cortical) cytoplasm of the animal hemisphere is more pigmented than the underlying cytoplasm. The vegetal hemisphere is not pigmented. Because of these differences, it is easy to observe how the cytoplasm rearranges when the egg is fertilized.

The unfertilized frog egg is radially symmetrical. You can turn it on its vegetal–animal pole axis, and all sides are the same. Sperm-binding sites are localized on the surface of the animal hemisphere. When a sperm binds to and enters the egg, the egg's radial symmetry is converted to bilateral symmetry and a dorsal–ventral axis is created. Cortical cytoplasm rotates toward the site of sperm entry (**Figure 43.1**). This rotation brings the animal and vegetal regions of cytoplasm into contact with each other, producing a band of lightly pigmented cytoplasm on the side opposite the site of sperm entry. This band, called the **gray crescent**, marks the location of important developmental events in some species of amphibians.

The centriole that was the sperm's contribution to the egg initiates the cytoplasmic reorganization that coincides with the appearance

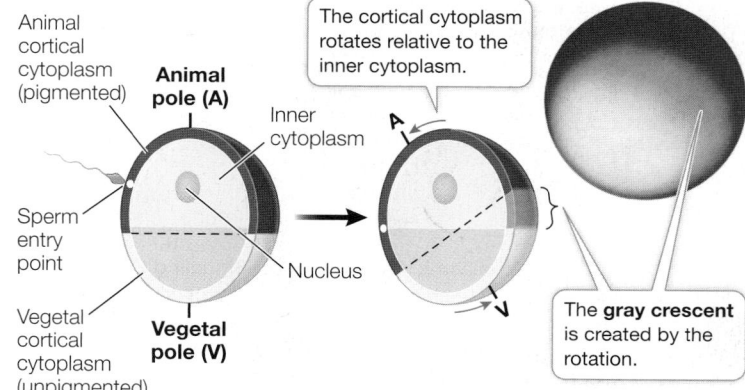

Figure 43.1 The Gray Crescent In amphibian eggs, cortical rotation and rearrangement of the cytoplasm after fertilization create the gray crescent opposite the point of sperm entry. These events are important for specifying the body axes and other important events in later development.

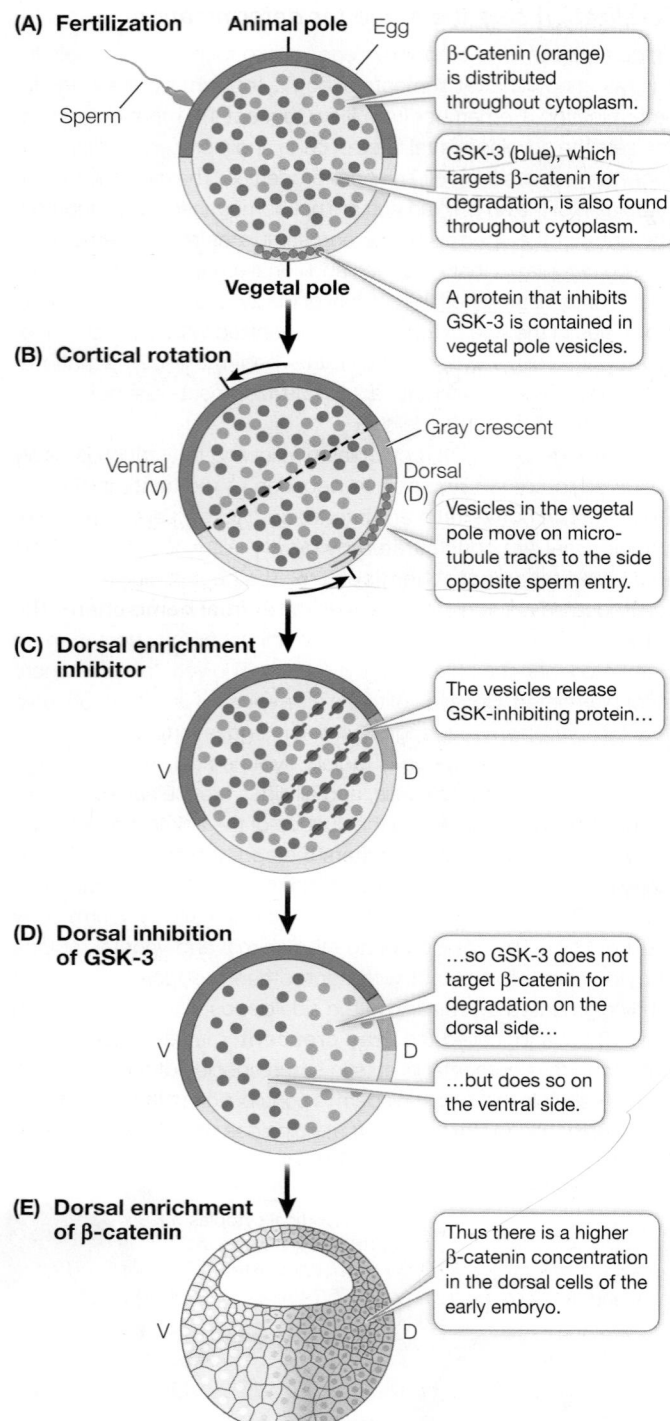

(A) Fertilization

Animal pole

Sperm

Egg

β-Catenin (orange) is distributed throughout cytoplasm.

GSK-3 (blue), which targets β-catenin for degradation, is also found throughout cytoplasm.

Vegetal pole

A protein that inhibits GSK-3 is contained in vegetal pole vesicles.

(B) Cortical rotation

Gray crescent

Ventral (V)

Dorsal (D)

Vesicles in the vegetal pole move on microtubule tracks to the side opposite sperm entry.

(C) Dorsal enrichment inhibitor

The vesicles release GSK-inhibiting protein...

V   D

(D) Dorsal inhibition of GSK-3

...so GSK-3 does not target β-catenin for degradation on the dorsal side...

...but does so on the ventral side.

V   D

(E) Dorsal enrichment of β-catenin

Thus there is a higher β-catenin concentration in the dorsal cells of the early embryo.

V   D

Figure 43.2 Cytoplasmic Factors Set Up Signaling Cascades
Cytoplasmic movement changes the distributions of critical developmental signals. In the frog fertilized egg, the interaction of the protein kinase, the GSK-3 inhibitor, and the protein β-catenin results in differential concentrations of β-catenin in blastomeres following cleavage thereby specifying the dorsal–ventral axis of the embryo.

of the gray crescent. The centriole organizes the microtubules in the vegetal hemisphere cytoplasm into a parallel array that guides the movement of the cortical cytoplasm as well as proteins and organelles.

The movement of cytoplasm, proteins, and organelles changes the distribution of critical signals. A key transcription factor in early development is β-catenin produced from maternal mRNA (mRNA produced and stored in the egg while it was maturing in the ovary). β-Catenin is found throughout the egg cytoplasm (**Figure 43.2**). Also present throughout the egg cytoplasm is a protein kinase, glycogen synthase kinase-3 (GSK-3), that phosphorylates β-catenin and thereby targets it for degradation. An inhibitor of GSK-3 is localized in the vegetal cortex of the egg. After sperm entry, this inhibitor moves along microtubules to the gray crescent, where it prevents the degradation of β-catenin. As a result, the concentration of β-catenin is higher on the dorsal than on the ventral side of the developing embryo, setting up regional differences within the egg cytoplasm that have developmental consequences.

> **43.1 recap**
>
> The egg is stocked with nutrients and informational molecules that power and direct the early stages of development. Fertilization activates the egg and stimulates rearrangement of the cytoplasm, setting up the body axes and organizing positional information that will control determination and differentiation.
>
> **learning outcomes**
>
> You should be able to:
> - Compare the contributions of the sperm and the egg to the zygote.
> - Explain how sperm entry converts frog egg symmetry from radial to bilateral.
> - Explain how the distribution of β-catenin in the egg cytoplasm is altered following fertilization and why this alteration is important.
>
> 1. Apart from its nucleus, what is the major contribution of the sperm to subsequent cell signaling in the embryo?
> 2. How is the gray crescent generated in the fertilized frog egg?
> 3. How does β-catenin become more concentrated on the dorsal side of the amphibian embryo?

The uneven distribution of informational molecules in the cytoplasm of the fertilized egg is essential to later events in development. In the next section you will see how these informational molecules end up in different cells of the developing embryo.

> **key concept**
>
> **43.2**
>
> ## Mitosis Divides Up the Early Embryo
>
> β-Catenin plays a major role in the cell–cell signaling cascade that begins the process of cell determination and the formation of the embryo. But cell–cell signaling requires more than one cell. The single-celled zygote must become a multicellular embryo.

(A) Complete cleavage (frog)

Animal pole

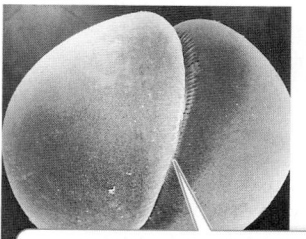

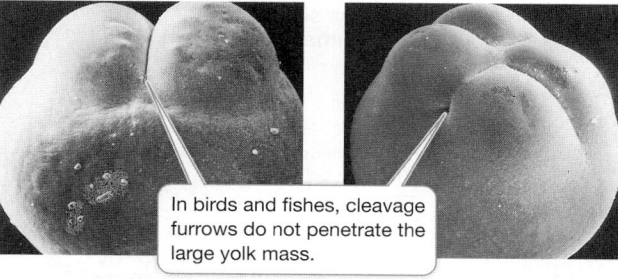
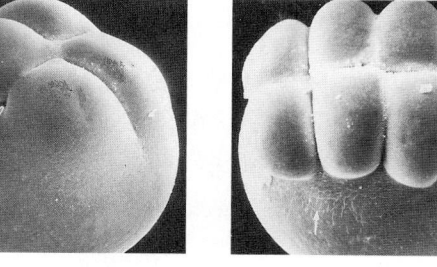
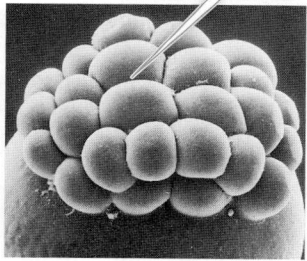

The first **cleavage** passes through the poles and site of sperm entry and bisects the gray crescent.

The second cleavage is at right angles to the first.

The third cleavage is at right angles (horizontal) to the first two and separating the animal and vegetal hemispheres. Vegetal hemisphere cells are larger.

The embryo forms as a **blastodisc** that sits on top of the yolk mass.

(B) Incomplete cleavage (zebrafish)

In birds and fishes, cleavage furrows do not penetrate the large yolk mass.

(C) Superficial cleavage (fruit fly)

1 Mitosis (nuclear division) occurs without cell division.

3 The nuclei migrate to the edge of the cell membrane.

Nucleus

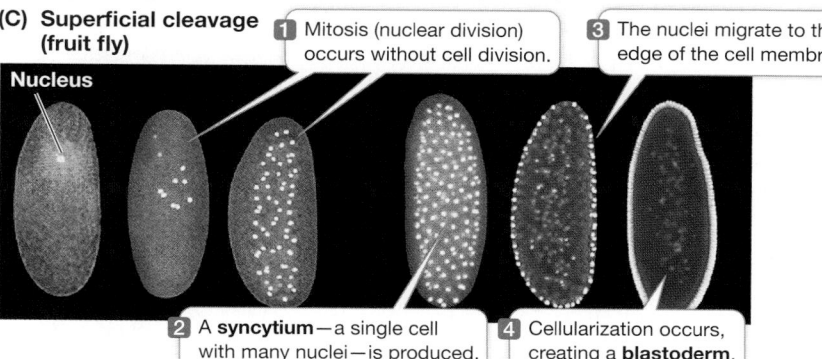

2 A **syncytium**—a single cell with many nuclei—is produced.

4 Cellularization occurs, creating a **blastoderm**.

Figure 43.3 Some Patterns of Cleavage Differences in patterns of early embryonic development reflect differences in the way the egg cytoplasm is organized. **(A)** The frog is a model organism representing complete cleavage in these scanning electron micrographs (SEMs). **(B)** SEMs of zebrafish embryos illustrate incomplete cleavage, in which the large yolk mass limits the planes of cleavage. **(C)** Nuclear staining reveals the syncytial nuclei characteristic of the early embryo of a fruit fly. These nuclei migrate to the periphery. Cleavage furrows then move inward to separate the nuclei into individual cells, forming the blastoderm.

focus your learning

- The three main types of cleavage are complete, incomplete, and superficial.
- Mammalian cleavage is unusual because of the requirements associated with the connection between the embryonic and maternal circulatory systems.
- Determination can arise from informational molecules present in the egg or by information received by their environment and neighboring cells.
- Twins can form in several ways.
- Reproductive germ cells are determined very early in cleavage.

Cleavage produces a multicellular embryo

Cleavage is the cell divisions that transform the diploid zygote into a mass of undifferentiated cells that develop as the embryo. Zygote cytoplasm is not homogeneous, and these first cell divisions result in the differential distribution of nutrients and cytoplasmic determinants in the early embryo.

In most animals, cleavage proceeds with rapid DNA replication and mitosis but with no cell growth and little gene expression. The embryo becomes a solid ball of smaller and smaller cells. Eventually this ball forms a central fluid-filled cavity called a **blastocoel**, and the embryo is called a **blastula**. Its individual cells are called **blastomeres**. The pattern of cleavage in different species influences the form of their blastulas.

- **Complete cleavage** occurs in eggs that have little yolk such as the eggs of mammals. Early cleavage furrows divide the egg completely. The amphibian egg undergoes complete cleavage, but vegetal pole yolk causes unequal cytoplasm division and animal hemisphere blastomeres are smaller than those in the vegetal hemisphere (**Figure 43.3A**).

- **Incomplete cleavage** occurs in eggs with a lot of yolk and the cleavage furrows do not penetrate it completely. **Discoidal cleavage** is a type of incomplete cleavage common in fishes as well as in birds and other reptiles, in which the embryo forms as a disc of cells, or **blastodisc**, that sits on top of the dense yolk mass (**Figure 43.3B**).

- **Superficial cleavage** is a variation of incomplete cleavage that occurs in insects such as the fruit fly (*Drosophila*). Early in development, cycles of mitosis occur without cell division, producing a **syncytium**—a single cell with many nuclei

(A) Mammalian rotational cleavage

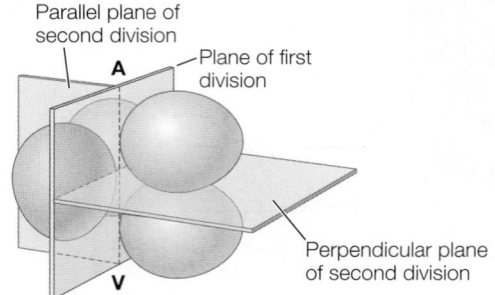

Parallel plane of second division

A

Plane of first division

Perpendicular plane of second division

V

Figure 43.4 Becoming a Blastocyst **(A)** Mammals have rotational, complete cleavage, in which the plane of the first cleavage is parallel to the animal–vegetal (A–V) axis, but the second cell division involves two planes (beige) at right angles to each other. **(B)** Scanning electron micrographs (color added) of early cleavage (leading to the formation of the blastocyst) in a human embryo. The cells' outer surfaces are covered with cilia (bright yellow). The small spheres, or "blebs," of cytoplasmic material, prominent at the 8-cell stage, disintegrate as cleavage progresses. **(C)** Seen in cross section under a light microscope, a mammalian blastocyst consists of an inner cell mass adjacent to a fluid-filled blastocoel and surrounded by trophoblast cells.

(B) Early cleavage in human embryo

8-Cell stage   Bleb   16-Cell stage   Compacted cells

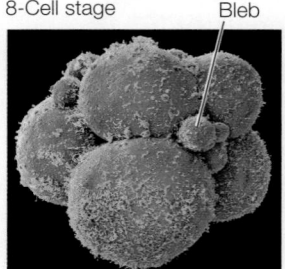
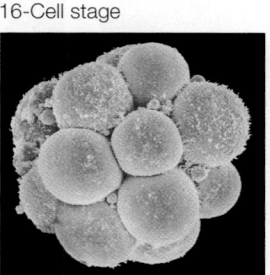
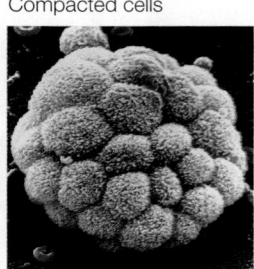

(C) Cross section of a mammalian blastocyst

Trophoblast (outer cells)

The inner cell mass will form the embryo.

Blastocoel

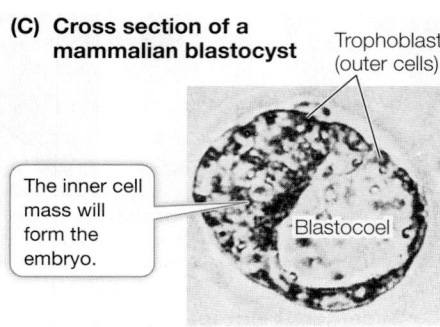

(**Figure 43.3C**). The nuclei eventually migrate to the periphery of the egg, after which the cell membrane of the egg grows inward, creating a **blastoderm** by partitioning the nuclei into individual cells surrounding a core of yolk.

The positions of the mitotic spindles during cleavage are not random but are defined by *cytoplasmic determinants produced from the maternal genome and stored in the egg. The orientation of the mitotic spindles can determine the planes of cleavage and the arrangement of the blastomeres.

---

*connect the concepts   As described in Key Concept 19.2, cytoplasmic determinants are regulatory substances that are distributed unequally in the egg cytoplasm and that influence polarity and cell fate during embryogenesis.

In complete cleavage, if the mitotic spindles of successive cell divisions form parallel or perpendicular to the animal–vegetal axis of the zygote, a pattern of **radial cleavage** occurs. The first two cell divisions are parallel to the animal–vegetal axis, and the third is perpendicular to it (see Figure 43.3A). **Spiral cleavage** results when the mitotic spindles are at oblique angles to the animal–vegetal axis. In spiral cleavage, each new cell layer is shifted to the left or right, depending on the orientation of the mitotic spindles. Most mollusks have spiral cleavage, reflected in some species by a coiling shell pattern (as seen in snails).

Cleavage in mammals is unique

Several features of early cell divisions in placental mammals (eutherians) are very different from those seen in other animal groups. First, this process in mammals is very slow. Cell divisions are 12 to 24 hours apart, compared with tens of minutes to a few hours in non-mammalian species. Also, the cell divisions of mammalian blastomeres are not in synchrony with each other. Because the blastomeres do not undergo mitosis at the same time, the number of cells in the embryo does not increase in the regular progression (2, 4, 8, 16, 32, etc.) typical of other species. The slowness of mammalian cleavage means that genes expressed during cleavage can play roles in cleavage. In animals such as sea urchins and frogs where cleavage progresses rapidly, very little if any gene transcription occurs in the blastomeres. Instead, cleavage is directed by molecules that were present in the egg before fertilization.

The pattern of mammalian cleavage is unique and is called **rotational cleavage**. The first cell division is parallel to the animal–vegetal axis as in radial cleavage, but in the second cell division, the two blastomeres divide at right angles to one other. One blastomere divides parallel to the animal–vegetal axis, while the other divides perpendicular to this axis (**Figure 43.4A**). As in other animals that have complete cleavage, the early cell divisions in a mammalian zygote produce a loosely associated ball of cells. After the 8-cell stage, however, the behavior of the mammalian blastomeres changes. They change shape to maximize their surface contact with one another, form tight junctions (see Figure 6.7), and become a compact mass of cells (**Figure 43.4B**).

Soon after the transition to the 32-cell stage, the cells of the mammalian embryo separate into two groups. The **inner cell mass** will develop as the embryo, while the surrounding outer cells become an encompassing sac called the **trophoblast**. Trophoblast cells secrete fluid, creating a cavity—the blastocoel—with the inner cell mass at one end. At this stage the mammalian embryo is called a **blastocyst**, distinguishing it from the blastulas of other animal groups (**Figure 43.4C**). The pluripotent cells of the inner cell mass are known as **embryonic stem cells** and are the subject of much research because of their therapeutic potential (see Key Concept 19.1).

Why is mammalian cleavage so different? A key factor is that mammalian eggs contain little or no yolk and must derive all nutrients from the mother. To support the developing embryo, a connection develops between the circulatory systems of the embryo and the mother. As you will see later in this chapter, the structures that provide this connection are the placenta and the umbilical cord. Thus the blastocyst of placental mammals must produce both the embryo (from the inner cell mass) and its support structures (from the trophoblast).

Fertilization in mammals occurs in the upper reaches of the oviduct, and cleavage occurs as the zygote travels down the oviduct to the uterus (**Figure 43.5**). When the blastocyst arrives in the uterus, the trophoblast adheres to the lining of the uterus (the **endometrium**), beginning the process of **implantation**. In humans, implantation begins about 6 days after fertilization and is aided by adhesion molecules and enzymes secreted by the trophoblast.

As the blastocyst moves down the oviduct to the uterus, it must not embed itself in the oviduct (Fallopian tube) wall, or the result will be an ectopic, or tubal, pregnancy—a very dangerous condition. Early implantation is prevented by the zona pellucida that surrounded the egg (see Figure 42.6) and remains around the cleaving ball of cells. At about the time the blastocyst reaches the uterus, it hatches from the zona pellucida so implantation can occur.

The fates of blastomeres depend on the cytoplasm they receive during cleavage

Cleavage results in a repackaging of the egg cytoplasm into a large number of small cells surrounding the fluid-filled blastocoel. Except in mammals, there is little gene expression during early cleavage divisions. Cells in different regions of the blastula do, however, possess different complements of the nutrients and cytoplasmic determinants that were present in the egg. For example, Figure 43.2 illustrated the processes by which β-catenin becomes localized in the region of the amphibian zygote that will become the dorsal side of the embryo.

The blastocoel prevents cells from different regions of the blastula from coming into contact and interacting, but that will soon change. During the next stage of development, the cells of the blastula move around and come into new associations with one another, communicate instructions to one another, and begin to differentiate. In many animals, these movements of the blastomeres are so regular and well orchestrated that it is possible to label specific blastomeres with a dye, thus producing **fate maps** that identify the tissues and organs formed from each blastomere's progeny (**Figure 43.6**).

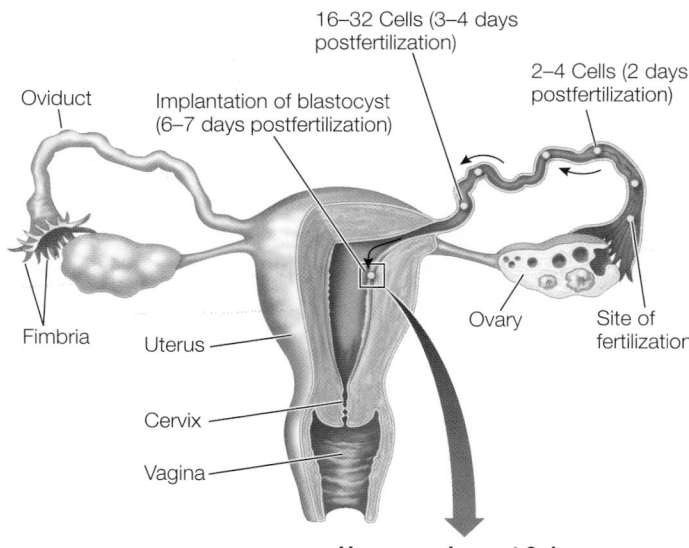

16–32 Cells (3–4 days postfertilization)

2–4 Cells (2 days postfertilization)

Oviduct

Implantation of blastocyst (6–7 days postfertilization)

Fimbria

Uterus

Cervix

Vagina

Ovary

Site of fertilization

Human embryo at 9 days

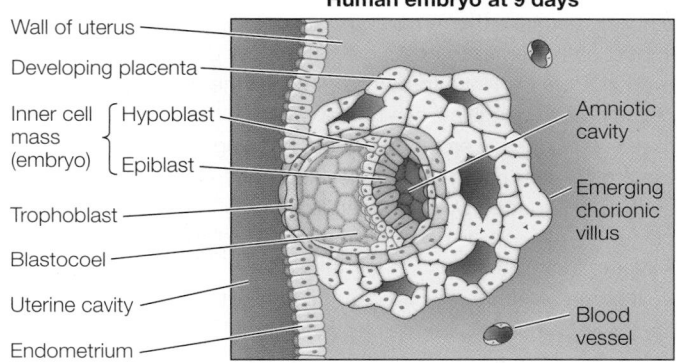

Wall of uterus

Developing placenta

Inner cell mass (embryo) { Hypoblast / Epiblast

Trophoblast

Blastocoel

Uterine cavity

Endometrium

Amniotic cavity

Emerging chorionic villus

Blood vessel

Figure 43.5 A Human Blastocyst at Implantation   Adhesion molecules and proteolytic enzymes secreted by trophoblast cells allow the blastocyst to burrow into the endometrium. Once the blastocyst is implanted in the wall of the uterus, the trophoblast cells send out numerous projections—the chorionic villi—which increase the embryo's area of contact with the mother's bloodstream. The inner cell mass divides into two embryonic tissues, the hypoblast and the epiblast. The epiblast splits to form the amniotic cavity.

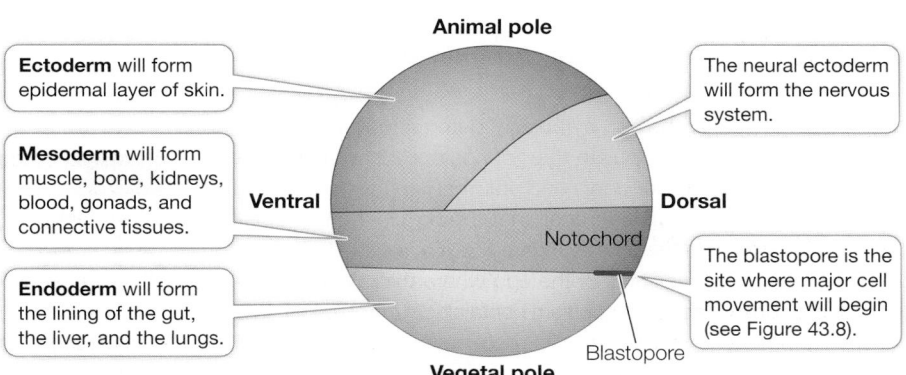

Animal pole

Ectoderm will form epidermal layer of skin.

The neural ectoderm will form the nervous system.

Mesoderm will form muscle, bone, kidneys, blood, gonads, and connective tissues.

Ventral

Dorsal

Notochord

Endoderm will form the lining of the gut, the liver, and the lungs.

The blastopore is the site where major cell movement will begin (see Figure 43.8).

Blastopore

Vegetal pole

Figure 43.6 Fate Map of a Typical Amphibian Blastula

Blastomeres become **determined**—committed to specific fates—at different times in different species. In some species, such as roundworms, the fates of blastomeres are restricted as early as the two-cell stage. This early determination depends on informational molecules that were originally present in the egg cytoplasm and is called **autonomous specification**. If one of these blastomeres is experimentally removed, a particular portion of the embryo will not form. The resulting developmental pattern has been called **mosaic development** because each blastomere seems to contribute a specific set of "tiles" to the final "mosaic" that is the adult animal. Mosaic development is the common form of development in invertebrates.

The fates of blastomeres in most vertebrate embryos are determined by information they receive from their environment and their neighboring cells. This mechanism of determination is called **regulated specification**, and it gives rise to **regulative development**. Unlike in mosaic development, the loss of some cells during cleavage in regulative development does not affect the developing embryo, because the remaining cells compensate for the loss. A subset of the remaining cells may add a cell division, or they may even change their fate to compensate for the cells lost. Because development is regulative in humans, a single blastomere can be removed from early embryos without harming the remaining blastomeres or disrupting normal development. Cells removed from embryos produced by in vitro fertilization can be used for preimplantation genetic diagnosis to ensure that healthy embryos are selected for implantation in the mother.

If some blastomeres can change their fate to compensate for the loss of other cells during cleavage and blastula formation, can those cells form an entire embryo? Yes, in species with regulative development, they can. If during cleavage or early blastula formation, the blastomeres are physically separated into two groups, both groups can produce complete embryos. Since the two embryos come from the same zygote, they will be monozygotic twins—genetically identical.

Non-identical twins occur when two separate eggs are fertilized by two separate sperm. Thus, although identical twins are always the same sex, non-identical twins have a 50 percent chance of being the same sex (that is, the same as two non-twin siblings). If genetic or environmental factors cause the blastula or blastocyst to split partially, the result is twins who are conjoined at some point on their bodies and usually share some of their organs and limbs. Conjoined twins occur in about 1 out of 50,000 human pregnancies.

Reproductive germ cells are determined early in cleavage

Molecules present in the egg cytoplasm determine which lineage of cells will eventually populate the gonads and become the reproductive stem cells—oogonia and spermatogonia. In fruit flies, at the ninth nuclear division (recall that the egg is a multinucleate syncytium at this stage; see Figure 43.3C), a group of nuclei migrate to the posterior pole of the egg where they become surrounded by **pole plasm**—cytoplasm containing a complex mixture of fibrils, mitochondria, and specific proteins and mRNAs. As the cellularization of the blastoderm proceeds, the nuclei within the pole plasm give rise to the lineage of cells that will eventually migrate to the gonads (when they form) and produce germ cells (eggs and sperm).

As in fruit flies, the germ cell lineage in frogs starts with a special type of cytoplasm—the germ cell plasm—localized to one part of the egg. As a result of cleavage, the germ cell plasm becomes enclosed within some of the cells in the vegetal hemisphere; descendants of these cells will eventually migrate to the gonads once those structures form. The components of germ cell plasm have not been fully characterized, but one hypothesis is that they include inhibitors of transcription and translation that prevent these cells from differentiating into anything other than germ cells.

▶ 43.2 recap

Cleavage divides up the cytoplasm of the zygote such that different blastomeres contain different combinations of informational molecules. The amount of nutrients stored in the egg influences the pattern of cell cleavage that produces the blastula. Blastulation in mammals is different compared to other vertebrates. The cells that will give rise to germ cell lineages are set aside very early in development.

learning outcomes

You should be able to:

- Compare the three main types of cleavage.
- Explain why cleavage in mammals is different from other vertebrates.
- Compare and contrast regulated and autonomous specification.
- Compare and contrast identical and non-identical twins.
- Compare the timing of determination in reproductive germ cells with that in other types of cells in the embryo.

1. What characteristic of the eggs of different species relates to whether cleavage is complete or incomplete?
2. Why can genes expressed during cleavage play roles in development in mammals but not in sea urchins and frogs?
3. Why can identical twins arise from a single egg in mammals but not in most invertebrates?
4. Why is there a longer time between determination and differentiation in germ cells versus skin cells?

Of the next stage of development—gastrulation—the developmental biologist Lewis Wolpert once said, "It is not birth, marriage, or death, but gastrulation which is the most important time in your life." During gastrulation, cell movements create new cell-to-cell contacts, which in turn set up signaling cascades that initiate the differentiation of cells and tissues and set the stage for the emergence of the body plan.

▶ key concept 43.3 Gastrulation Generates Multiple Tissue Layers

The blastula is typically a fluid-filled ball of cells. How does this simple ball of cells become an embryo made up of multiple tissue layers with head and tail ends and

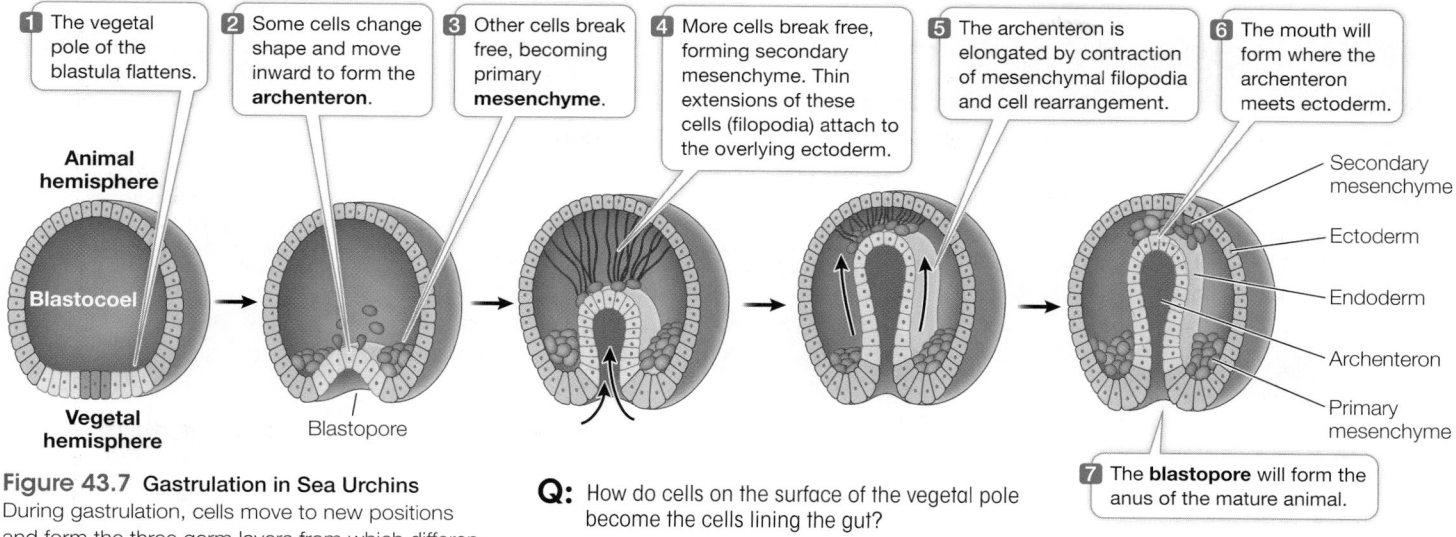

1 The vegetal pole of the blastula flattens.

2 Some cells change shape and move inward to form the **archenteron**.

3 Other cells break free, becoming primary **mesenchyme**.

4 More cells break free, forming secondary mesenchyme. Thin extensions of these cells (filopodia) attach to the overlying ectoderm.

5 The archenteron is elongated by contraction of mesenchymal filopodia and cell rearrangement.

6 The mouth will form where the archenteron meets ectoderm.

7 The **blastopore** will form the anus of the mature animal.

Animal hemisphere

Blastocoel

Vegetal hemisphere

Blastopore

Secondary mesenchyme

Ectoderm

Endoderm

Archenteron

Primary mesenchyme

Figure 43.7 Gastrulation in Sea Urchins
During gastrulation, cells move to new positions and form the three germ layers from which differentiated tissues develop.

Q: How do cells on the surface of the vegetal pole become the cells lining the gut?

dorsal and ventral sides? **Gastrulation** is the process whereby the blastula is transformed by massive movements of cells into an embryo with multiple tissue layers and distinct body axes. The resulting spatial relationships between tissues make possible the inductive interactions between cells that trigger differentiation and organ formation.

focus your learning

- Cell movements during gastrulation result in three germ layers and new tissue interactions.
- Gastrulation in frogs is more complex than in sea urchins, owing to frog blastulas having more yolk and more cell layers.
- The dorsal lip of the blastopore is the primary organizer in the amphibian embryo.
- Showing that a protein is an inductive signal requires that it be both necessary and sufficient for the proposed effect.
- In amphibians, organizer cells moving anteriorly from the dorsal lip of the blastopore initiate the formation of different organs along the anterior–posterior axis.
- Inductive tissue interactions can suppress as well as activate.
- Because bird and reptile eggs have a large yolk mass, the embryos have a flattened blastodisc, and display a pattern of gastrulation very different from that of amphibians.

In the triploblastic animals (see Key Concept 30.1), three **germ layers** (also called cell layers or tissue layers, and not to be confused with germ cells) form during gastrulation:

1. The **endoderm** is the innermost germ layer, created as some blastomeres move to the inside of the embryo. The endoderm gives rise to the epithelial lining of the digestive tract, respiratory tract, pancreas, thyroid, and liver.

2. The **ectoderm** is the outer germ layer, formed from those cells remaining on the outside of the embryo. The ectoderm contributes to the nervous system, including the eyes and ears; and

to the epidermal layer of the skin and structures derived from skin, such as hair, feathers, nails or claws, sweat glands, oil glands, and even teeth and other tissues of the mouth.

3. The **mesoderm** is the middle layer and is made up of cells that migrate between the endoderm and the ectoderm. The mesoderm contributes tissues to many organs, including the heart, blood vessels, muscles, and bones.

The three germ layers are illustrated for a very early embryo in the fate map shown in Figure 43.6.

Some of the most interesting and important challenges in animal development have dealt with two related questions: what directs the cell movements of gastrulation, and what is responsible for the resulting patterns of cell differentiation and organ formation? Scientists have made significant progress in answering both these questions at the molecular level. In the following discussion we will begin with sea urchin gastrulation because it is the simplest to conceptualize in spatial terms. We will then describe the more complex pattern of gastrulation in frogs, and then the still more complex patterns in reptiles (including birds), and mammals.

 Animation 43.1 **Gastrulation**
www.Life11e.com/a43.1

Invagination at the vegetal pole initiates gastrulation in the sea urchin

The sea urchin blastula is a hollow ball of cells only one cell layer thick. At the end of blastulation the rate of mitosis slows, and the beginning of gastrulation is marked by a flattening of the vegetal hemisphere (**Figure 43.7**). Some cells at the vegetal pole break away from neighboring cells and migrate into the cavity to become **mesenchyme**—cells of the middle germ layer, the mesoderm. Mesenchymal cells are not organized in tightly packed sheets or tubes like epithelial cells are; they act as independent units, migrating into and among the other tissue layers. These early mesenchymal cells will contribute to the skeletal structures of the sea urchin.

focus: key figure

Figure 43.8 Gastrulation in the Frog Embryo
Yellow, blue, green, and red in this diagram are matched to those colors in Figure 43.6, the frog fate map.

Q: Epiboly results in the formation of which tissues in the adult frog?

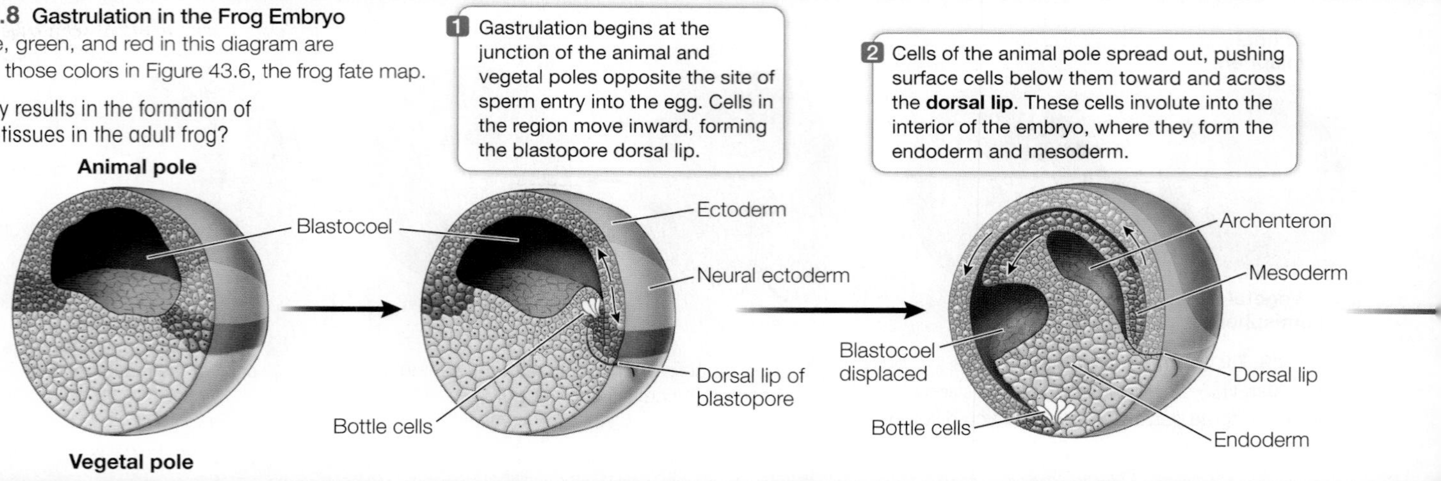

1 Gastrulation begins at the junction of the animal and vegetal poles opposite the site of sperm entry into the egg. Cells in the region move inward, forming the blastopore dorsal lip.

2 Cells of the animal pole spread out, pushing surface cells below them toward and across the **dorsal lip**. These cells involute into the interior of the embryo, where they form the endoderm and mesoderm.

Animal pole — Blastocoel — Ectoderm — Neural ectoderm — Dorsal lip of blastopore — Bottle cells — Vegetal pole — Blastocoel displaced — Bottle cells — Archenteron — Mesoderm — Dorsal lip — Endoderm

The flattening at the vegetal pole results from changes in the shape of individual blastomeres. These cells, which are originally rather cuboidal, become wedge-shaped, with smaller outer edges and larger inner edges. As a result, the vegetal pole bulges inward, or invaginates, as if someone were poking a finger into a hollow ball (see Figure 43.7). The invaginating cells become endoderm and form the primitive gut, called the **archenteron**. The archenteron begins as a short stubby invagination, but its cells extend, flatten, interdigitate, and migrate over one another to form a long thin tube. This process is called **convergent extension**. As the archenteron gets closer to the opposite side of the blastocoel, more mescenchymal cells migrate out of it into the blastocoel. These cells remain attached to the tip of the archenteron and send out extensions called filopodia that adhere to the overlying ectoderm. The filopodia contract and pull the archenteron toward the overlying ectoderm. The mouth of the animal will form where the archenteron makes contact with this overlying ectoderm. The opening created by the invagination of the vegetal pole is called the **blastopore**, and it will become the anus of the animal.

We should recall at this point that where the mouth and anus form during gastrulation distinguishes the two major groups of animals that have bilateral symmetry, the protostomes and the deuterostomes (see Key Concept 30.1). If the blastopore becomes the mouth, the animals are classified as protostomes ("mouth first"), and if the blastopore becomes the anus, the animals are classified as deuterostomes ("mouth second").

What mechanisms control the various cell movements of sea urchin gastrulation? The immediate answer is that specific properties of particular blastomeres change. For example, some vegetal cells change shape and bulge into the blastocoel, and these cells become mesenchyme. Once they lose contact with their neighboring cells on the surface of the blastula, they send out filopodia that then move along an extracellular matrix of proteins laid down by the cells lining the blastocoel.

A deeper understanding of gastrulation requires that we discover the molecular mechanisms whereby different blastomeres develop different properties. Cleavage systematically divides the cytoplasm of the egg. The sea urchin blastula at the 64-cell stage is radially symmetrical, but it has polarity, as described in Key Concept 19.2. It consists of tiers of cells. As in the frog blastula, the top is the animal pole and the bottom the vegetal pole.

If different tiers of blastula cells are separated experimentally, they show different developmental potentials; only cells from the vegetal pole are capable of initiating the development of a complete larva. It has been proposed that these differences are due to uneven distribution of various transcriptional regulatory proteins in the egg cytoplasm. As cleavage progresses, these proteins end up in different groups of cells. Therefore specific sets of genes are activated in different cells, determining their different developmental capacities.

Next we will turn to gastrulation in the frog and to the key signaling molecules involved.

Frog gastrulation begins at the gray crescent

Amphibian blastulas have considerable yolk and are more than one cell layer thick; gastrulation is therefore more complex in amphibians than in sea urchins. Variation is considerable across different species of amphibians, so this brief account combines results from studies done on different species to produce a generalized picture of amphibian development.

Amphibian gastrulation begins when certain cells in the gray crescent region (see Figure 43.1) change their shapes and cell-adhesion properties. These cells bulge inward toward the blastocoel while they remain attached to the outer surface of the blastula by slender necks; because of their shape, they are called bottle cells. Bottle cells mark the spot where the **dorsal lip** of the blastopore will form (**Focus: Key Figure 43.8**).

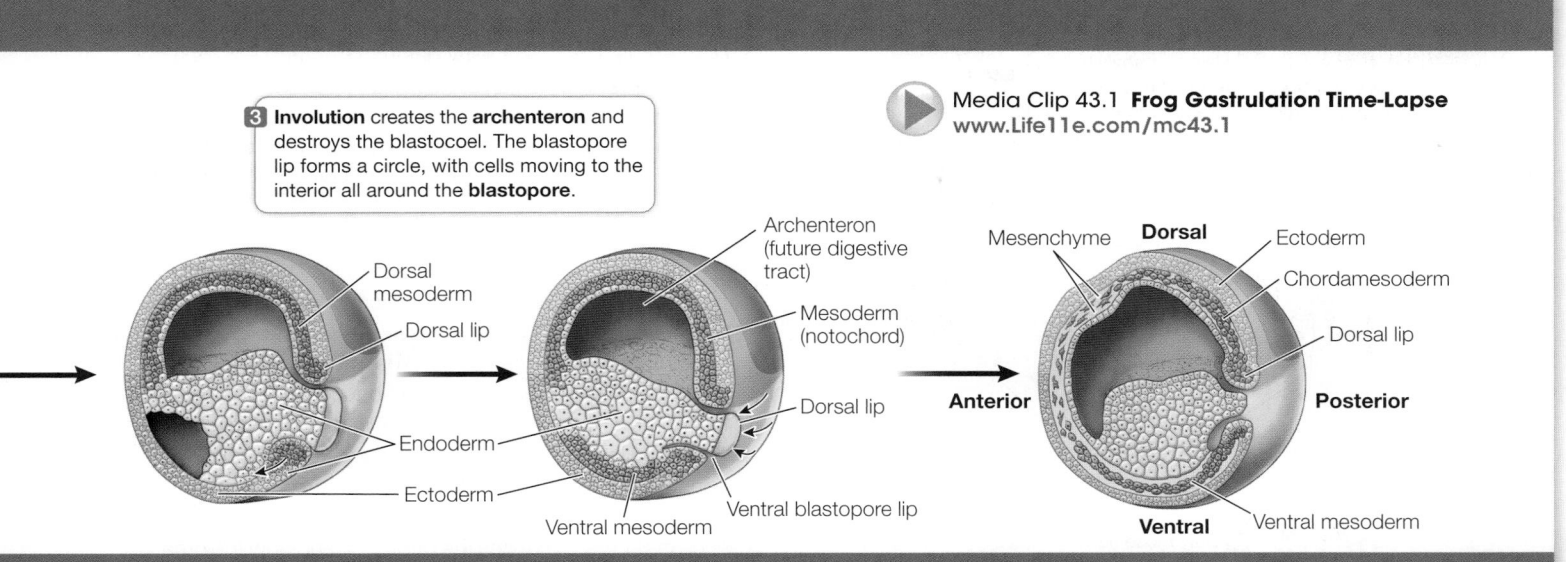

③ **Involution** creates the **archenteron** and destroys the blastocoel. The blastopore lip forms a circle, with cells moving to the interior all around the **blastopore**.

Media Clip 43.1 **Frog Gastrulation Time-Lapse**
www.Life11e.com/mc43.1

As the bottle cells move inward, the dorsal lip is created, and a sheet of cells moves over it into the blastocoel. This process is called **involution**. One group of involuting cells is the prospective endoderm; these cells form the primitive gut, or archenteron. Another group will move between the endoderm and the outermost cells to form mesoderm. As with the extension of the archenteron in the sea urchin, these cell rearrangements involve convergent extension. The cells elongate in the direction of growth, but they also intercalate (move in between each other). As gastrulation proceeds, cells from the animal hemisphere flatten and move toward the site of involution in a process called **epiboly**. The blastopore lip widens and spreads ventrally, and eventually forms a complete circle surrounding a "plug" of yolk-rich cells. As cells continue to move inward through the blastopore, the archenteron grows, gradually displacing the blastocoel.

As gastrulation comes to an end, the amphibian embryo consists of three germ layers: ectoderm on the outside, endoderm on the inside, and mesoderm in between. The embryo also has a dorsal–ventral and anterior–posterior organization. Most important, the fates of specific regions of the endoderm, mesoderm, and ectoderm have been determined. The beautiful experiments revealing how determination takes place in the amphibian embryo are an old but exciting story.

The dorsal lip of the blastopore organizes the formation of the amphibian embryo

In the early 1900s the German biologist Hans Spemann was studying the development of salamander eggs. He was interested in finding out whether the nuclei of blastomeres remain capable of directing the development of complete embryos or whether these nuclei lose some developmental potential. With great patience and dexterity, he formed loops from single hairs taken from a baby (in fact, his daughter) and tied them around fertilized eggs along the

plane of the first cell division, effectively dividing the eggs in half, with the nucleus restricted to one side. The side containing the nucleus went through cell divisions and developed into a salamander; the other half simply degenerated. Up until the 16-cell stage, if one nucleus escaped to the other side of the constriction, twin salamanders could develop. Thus each of the nuclei of the blastula (at least up to the 16-cell stage) was capable of directing and supporting development of the whole organism (regulative development).

As often happens in science, Spemann's bisection experiments revealed a new phenomenon. Sometimes the half of the blastula receiving an escaped nucleus did not develop. When his loops bisected the gray crescent, both halves of the zygote developed into a complete embryo. When he tied the loops so the gray crescent was on only one side of the constriction, however, only that half of the zygote developed into a complete embryo (**Figure 43.9**). When the half lacking gray crescent material received a nucleus, it underwent cell division but formed only a clump of undifferentiated cells that Spemann called a "belly piece." Spemann hypothesized that cytoplasmic factors unequally distributed in the fertilized egg were necessary for gastrulation and the development of a normal salamander.

To further test the hypothesis that cells receiving different complements of cytoplasmic factors had different developmental fates, Spemann transplanted pieces of early gastrulas to various locations on other gastrulas. Guided by fate maps (see Figure 43.6), he was able to take a piece of ectoderm he knew would develop into the epidermis of the skin and transplant it to a region that normally becomes part of the nervous system, and vice versa.

When he performed these transplants in early gastrulas—when the blastopore was just beginning to form—the transplanted pieces always developed into tissues that were appropriate for the location where they were placed. Transplanted cells destined to become epidermis in their original location developed into nervous system tissue, and transplanted cells destined to become nervous system

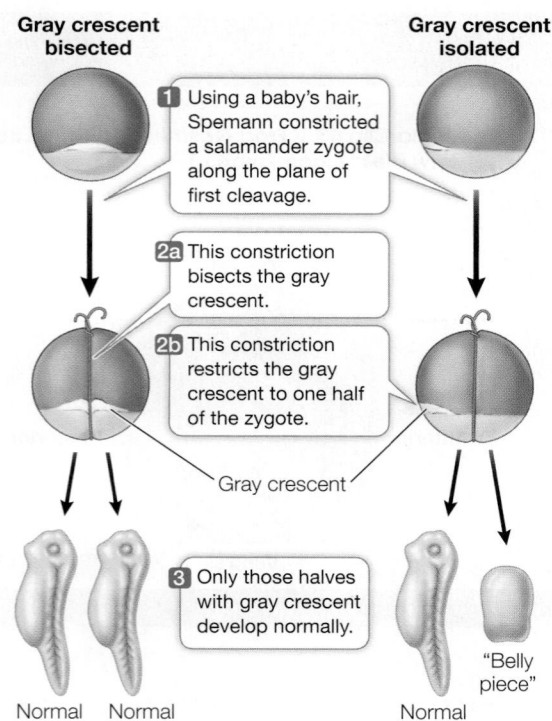

Gray crescent
bisected

Gray crescent
isolated

1 Using a baby's hair, Spemann constricted a salamander zygote along the plane of first cleavage.

2a This constriction bisects the gray crescent.

2b This constriction restricts the gray crescent to one half of the zygote.

Gray crescent

3 Only those halves with gray crescent develop normally.

Normal   Normal

Normal

"Belly piece"

Figure 43.9 Gastrulation and the Gray Crescent Spemann's research revealed that gastrulation and subsequent normal development in salamanders depend on cytoplasmic determinants localized in the region of the blastula that corresponds to the gray crescent region in the fertilized egg.

tissue in their original location developed into host epidermis. Thus Spemann learned that the fates of the transplanted cells had not been determined before the transplantation.

In late gastrulas, however, the same experiment yielded opposite results. Transplanted cells destined to become epidermis in their original location produced patches of skin cells in the host nervous system, and the transplanted cells from regions that would develop into nervous system tissue produced nervous tissue in the skin of the recipient. At some point during gastrulation, the fates of the embryonic cells had become determined.

Spemann's next experiment, done with his student Hilde Mangold, produced momentous results: they transplanted the dorsal lip of the blastopore (**Figure 43.10**). When this small piece of tissue was transplanted into the presumptive belly area of another gastrula, it stimulated a second site of gastrulation—and a second complete embryo formed belly-to-belly with the original embryo. Because the dorsal lip of the blastopore was apparently capable of inducing the host tissue to form an entire embryo, Spemann and Mangold called the dorsal lip tissue the **primary embryonic organizer**, or simply the **organizer**. For almost 100 years, what has come to be called the Spermann organizer has been an active area of research.

Transcription factors and growth factors underlie the organizer's actions

With the advent of modern molecular methods, the primary embryonic organizer has been studied intensively to discover the molecular mechanisms involved in its action. The distribution of the transcription factor β-catenin in the late blastula corresponds to the location of the organizer in the early gastrula, so β-catenin is a candidate for the initiator of organizer activity. To prove that a protein is an inductive signal, it has to be shown that it is both *necessary* and *sufficient* for the proposed effect. In other words, the effect should not occur if the candidate protein is not present (necessity), and the candidate protein should be capable of inducing the effect where it would otherwise not occur (sufficiency).

The criteria of necessity and sufficiency have been satisfied for β-catenin. If β-catenin mRNA transcripts are depleted by injections of antisense RNA into the egg (see Key Concept 18.4), gastrulation does not occur. If β-catenin is experimentally overexpressed in another region of the blastula, it can induce a second axis of embryo formation, as the transplanted dorsal lip did in the Spemann–Mangold experiments. Thus β-catenin appears to be both necessary and sufficient for the formation of the primary embryonic organizer—but it is only one component of a complex signaling process. How the presence of β-catenin creates the organizer, and how the organizer then induces the beginnings of the body plan, involves a complex series of interactions between transcription factors and growth factors that control gene expression.

Properties of organizer cells change as they migrate from the dorsal lip

Organizer cells begin the process of formation of the dorsal lip of the blastopore. Specifically, these cells are at the center of the dorsal lip and involute, moving forward on the midline (i.e., the middle of the anterior–posterior axis). The first organizer cells to enter the embryo migrate anteriorly to become the head endoderm and head mesoderm. Here they induce neighboring cells to participate in making structures of the head. Organizer cells that involute later in gastrulation induce structures of the trunk, and the last of the organizer cells to move inward from the dorsal lip induce structures of the tail. How do the organizer cells change to enable them to induce head, trunk, or tail structures?

Inductive tissue interactions can suppress as well as activate. Goosecoid is a transcription factor expressed by early organizer cells. Goosecoid activates genes encoding certain signaling factors. As the early organizer cells move forward in the blastocoel, they come into contact with new populations of cells that produce several different growth factors. For head structures to form, certain of these growth factors have to be suppressed. The most anterior organizer cells, under the influence of Goosecoid, produce and release antagonists to those growth factors.

The induction of trunk structures requires suppression of a different set of growth factors. In organizer cells that involute later than the head organizers, Goosecoid is no longer the dominant transcription factor, and these cells express different growth factor antagonists. The induction of tail structures requires still different activities of the organizer cells that involute last. Thus the organizer cells express appropriate sets of growth factor antagonists at the right times to achieve different patterns of differentiation on the anterior–posterior axis.

The initiation of nervous system development also involves a suppressive tissue interaction. For a long time it was thought that the involuting organizer cells actively induced the overlying ectoderm

experiment

Figure 43.10 The Dorsal Lip Induces Embryonic Organization

Original Paper: Spemann, H. and H. Mangold. 1924. *Roux' Arch. Entw. Mech.* 100: 599–638. Viktor Hamburger's translation appeared in *Foundations of Experimental Embryology,* 1964, (B. H. Willier and J. M. Oppenheimer, eds.), pp. 146–184.

In a classic experiment, Hans Spemann and Hilde Mangold transplanted the dorsal blastopore lip mesoderm of an early gastrula stage salamander embryo. The results showed that the cells of this embryonic region, which they dubbed "the organizer," could direct the formation of an entire embryo.

HYPOTHESIS▶ The early dorsal blastopore lip organizes cell differentiation in amphibian embryos.

METHOD

1. Excise a patch of mesoderm tissue from above the dorsal blastopore lip of an early gastrula stage salamander embryo (the donor).

2. Transplant the donor tissue onto a recipient embryo at the same stage. The donor tissue is transplanted onto a region of ectoderm that should become epidermis (skin).

CONCLUSION▶ The cells of the dorsal blastopore lip can induce other cells to change their developmental fates.

 Animation 43.2 The Primary Embryonic Organizer
www.Life11e.com/a43.2

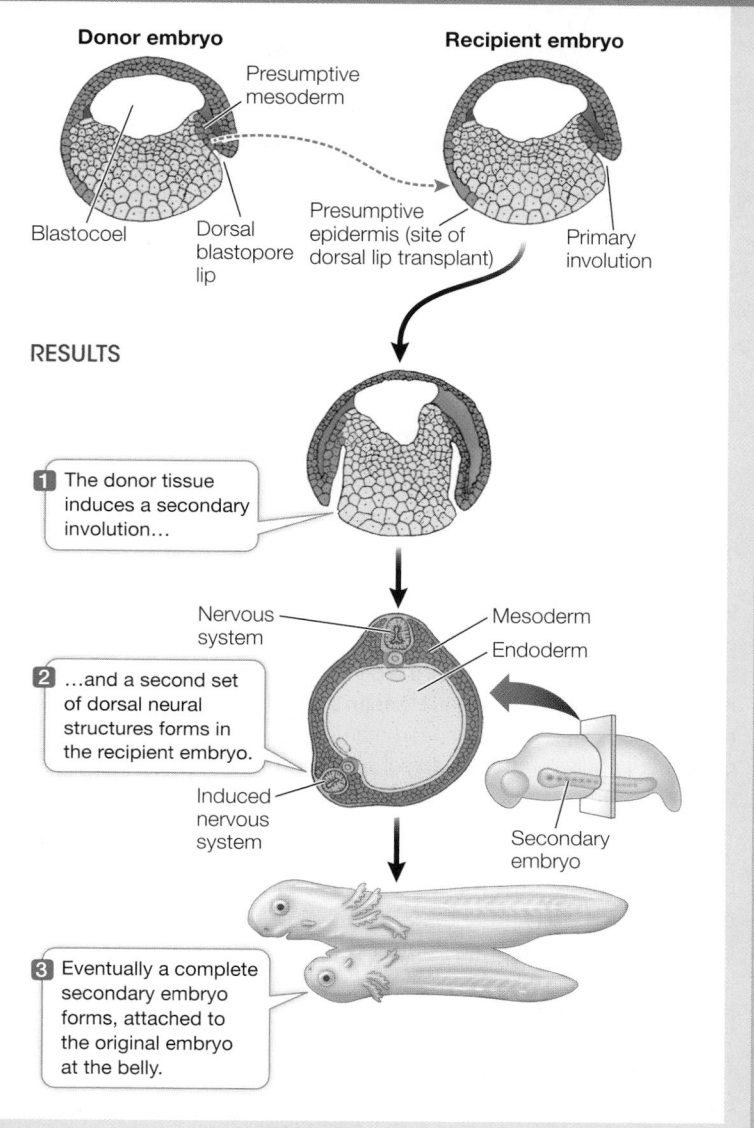

Donor embryo — Presumptive mesoderm — Blastocoel — Dorsal blastopore lip

Recipient embryo — Presumptive epidermis (site of dorsal lip transplant) — Primary involution

RESULTS

1 The donor tissue induces a secondary involution…

2 …and a second set of dorsal neural structures forms in the recipient embryo.

Nervous system — Mesoderm — Endoderm — Induced nervous system — Secondary embryo

3 Eventually a complete secondary embryo forms, attached to the original embryo at the belly.

---

to form nervous tissue rather than becoming epidermis. We now know, however, that epidermis is not the default state of the dorsal ectoderm. Rather, the underlying mesoderm secretes factors called BMP proteins that induce the ectoderm to become epidermis. The role of the involuting organizer cells is to block that induction, allowing the overlying ectodermal cells to follow what is really their default pathway—differentiation into nervous tissue.

The amount of yolk influences gastrulation

The eggs of reptiles, including birds, contain a mass of yolk, and the blastulas of these groups develop as a disc of cells on top of the yolk (see Figure 43.3B). We will use the chicken egg to show how gastrulation proceeds in a flat disc of cells rather than in a ball of cells.

Cleavage in the chick results in a flat, circular layer of cells called a blastodisc (**Figure 43.11**). Between the blastodisc and the yolk

mass is a fluid-filled space. Some cells from the blastodisc break free and move into this space. These cells come together to form a continuous layer called the **hypoblast**, which will later contribute to extraembryonic membranes that will support and nourish the developing embryo. The overlying cells make up the **epiblast**, from which the embryo will form. Thus the avian blastula is a flattened structure consisting of an upper epiblast and a lower hypoblast, which are joined at the margins of the blastodisc. The blastocoel is the fluid-filled space between the epiblast and hypoblast.

Gastrulation begins with a thickening in the posterior region of the epiblast, caused by the movement of cells toward the midline and then forward along the midline (see Figure 43.12). The result is a midline ridge called the **primitive streak**. A depression called the primitive groove forms along the length of the primitive streak. The primitive groove functions as the blastopore, and cells migrate through

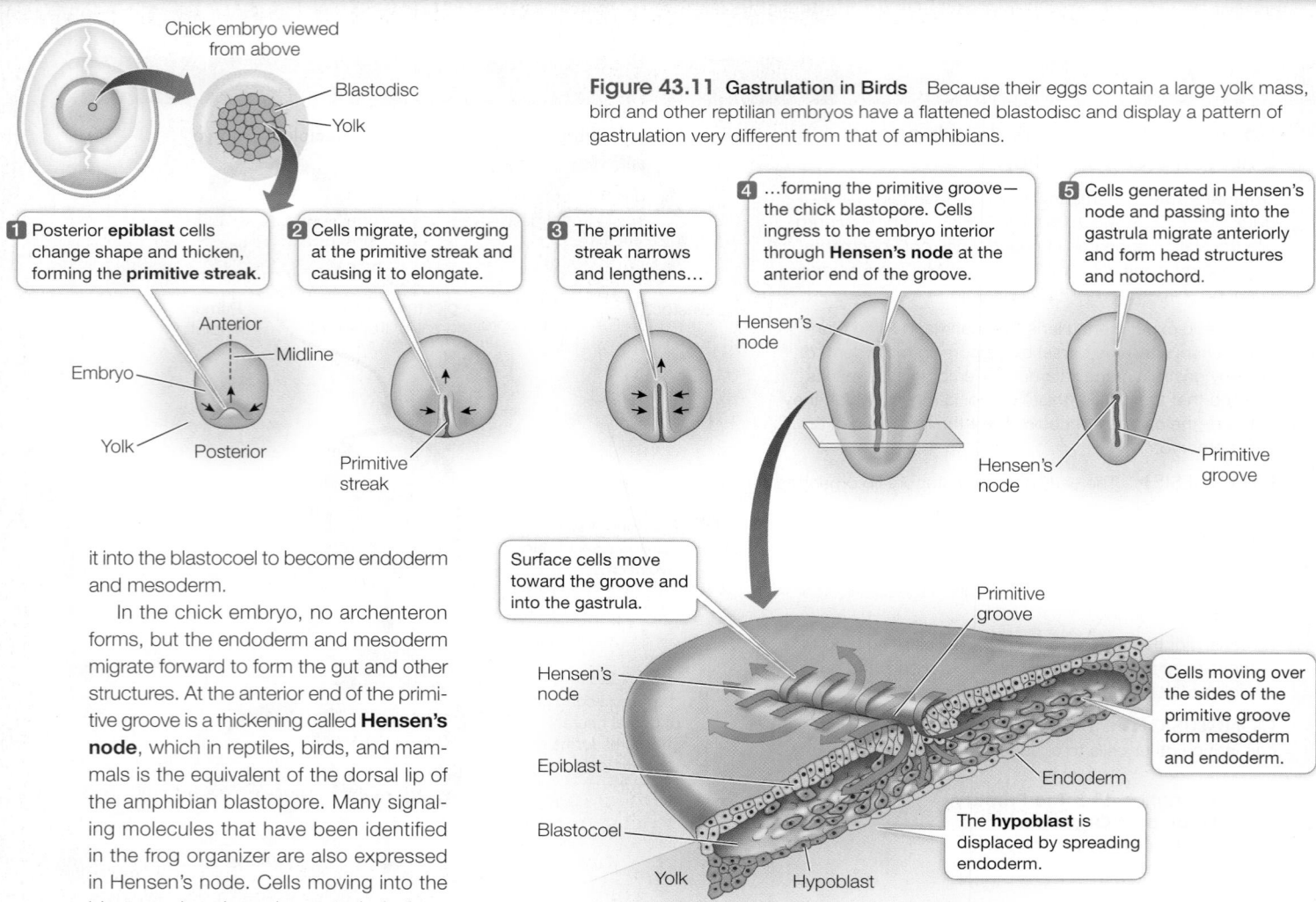

Chick embryo viewed from above

Blastodisc
Yolk

Figure 43.11 Gastrulation in Birds Because their eggs contain a large yolk mass, bird and other reptilian embryos have a flattened blastodisc and display a pattern of gastrulation very different from that of amphibians.

1 Posterior **epiblast** cells change shape and thicken, forming the **primitive streak**.

2 Cells migrate, converging at the primitive streak and causing it to elongate.

3 The primitive streak narrows and lengthens…

4 …forming the primitive groove—the chick blastopore. Cells ingress to the embryo interior through **Hensen's node** at the anterior end of the groove.

5 Cells generated in Hensen's node and passing into the gastrula migrate anteriorly and form head structures and notochord.

Anterior
Midline
Embryo
Yolk
Posterior
Primitive streak

Hensen's node

Hensen's node

Hensen's node

Primitive groove

Surface cells move toward the groove and into the gastrula.

Primitive groove

Hensen's node

Cells moving over the sides of the primitive groove form mesoderm and endoderm.

Epiblast

Endoderm

Blastocoel

The **hypoblast** is displaced by spreading endoderm.

Yolk

Hypoblast

Cross section through chick embryo

it into the blastocoel to become endoderm and mesoderm.

In the chick embryo, no archenteron forms, but the endoderm and mesoderm migrate forward to form the gut and other structures. At the anterior end of the primitive groove is a thickening called **Hensen's node**, which in reptiles, birds, and mammals is the equivalent of the dorsal lip of the amphibian blastopore. Many signaling molecules that have been identified in the frog organizer are also expressed in Hensen's node. Cells moving into the blastocoel and moving anteriorly from Hensen's node become the notochord and organize the chick embryo in a manner similar to that of the frog embryo.

Gastrulation in mammals is similar to avian gastrulation

Mammalian embryos (with the exception of monotremes) derive their nourishment from the maternal circulation, and therefore mammalian eggs do not have large amounts of yolk constraining their cleavage and early development. Nevertheless, mammals evolved from reptilian ancestors, so it is not surprising that mammals, birds, and reptiles share certain patterns of early development. Earlier we described the development of the mammalian inner cell mass (the equivalent of the avian blastodisc) and the outer trophoblast.

As in avian development, in placental mammals the inner cell mass splits into an upper layer called the epiblast and a lower layer called the hypoblast. The embryo forms from the epiblast, while the hypoblast contributes to the extraembryonic membranes that will encase the developing embryo and help form the placenta (see Figure 43.5). The epiblast also contributes to the extraembryonic membranes; specifically, it splits off an upper layer of cells that will form the amnion. The amnion will grow to surround the developing embryo as a membranous sac filled with amniotic fluid. Gastrulation occurs in the mammalian epiblast just as it does in the avian epiblast. A primitive groove forms, and epiblast cells migrate through the groove to become layers

of endoderm and mesoderm. At the top of the groove is the **node** which is homologous with Hensen's node in birds.

How is bilateral symmetry broken?

Gastrulation results in the anterior–posterior and dorsal–ventral patterning of the embryo, but all of the mechanisms discussed produce bilateral symmetry. At the opening of this chapter we introduced an early mechanism that breaks that symmetry to produce the well-known left–right asymmetries of the body. The heart is on the left; the major lobe of the liver is on the right; the pancreas and spleen are on the left; the appendix is on the right; the major blood vessel leaving the heart, the aorta, bends to the left; the right lung has three lobes and the left lung has two; and so on. At the very beginning of gastrulation, the location where the involution of cells begins in a mammal is marked by a node of cells. The ventral surface of the node has ciliated cells. Some of these cilia are motile and others are not. The motile cilia create a leftward flow of extracellular fluid over the node. The experiments in **Investigating Life: Does the Direction of Nodal Flow Influence the Development of Left–Right Asymmetry in the Mouse Embryo?** investigate whether that directional flow is both necessary and sufficient to initiate signaling processes that break bilateral symmetry of the embryo.

Does the Direction of Nodal Flow Influence the Development of Left–Right Asymmetry in the Mouse Embryo?

experiment

Original Paper: Nonaka, S., H. Shiratori and H. Hamata. 2002. Determination of left–right patterning of the mouse embryo by artificial nodal flow. *Nature* 418: 96–99.

HYPOTHESIS▶ The leftward flow of extracellular fluid created by the beating of nodal primary cilia is the stimulus for breaking bilateral symmetry in organ development the mouse.

METHOD

1. Mount wild-type mouse embryos at different stages of development (presomite and 1, 2, or 3 somites; see Figure 43.13) in a culture medium–filled chamber with their nodal ends pointed upward.
2. Artificially pump culture medium through the chamber either from the left (normal) or from the right (reversed) and at slow or fast flow rates.
3. After 4 days assess the embryos as to the direction of looping (normal or reversed) of their developing heart tubes.

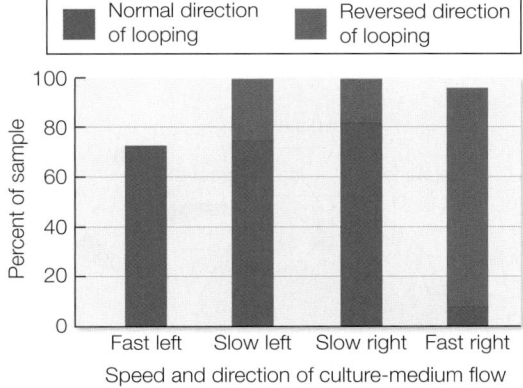

RESULTS

Since there were different numbers of embryos in different conditions, and some embryos did not survive, the data are displayed as percent of the total sample under each condition that showed normal or reversed direction of looping of their heart tubes. For example, in the slow left column there were 16 total embryos: 12 had normal direction of looping and 4 had reversed looping. The graph displays how heart-looping patterns were affected in the presomite stage by the direction of culture-medium flow.

Genotype, stage	Speed and direction of culture-medium flow[a]			
	Fast left	Slow left	Slow right	Fast right
Wild type, presomite	9N, 0R, 10T	12N, 4R, 16T	13N, 3R, 16T	2N, 21R, 24T

[a]N = normal heart looping; R = reversed looping; T = total embryos in sample.

CONCLUSIONS▶ When extracellular fluid moved across the node from right to left, the direction of heart-tube looping was normal. If this signal was weak (slow left flow), or if a weak reversed signal was applied (slow right flow), there was a small effect on normal direction of heart-tube looping, but when the reversed signal was strong, the heart-tube looping was in the reversed direction. These results support the conclusion that nodal flow is an important signal in breaking bilateral symmetry in the mouse embryo.

work with the data

The experiment illustrated in the graph was done on presomite embryos—before any somites had formed. Developmental signals usually have a time window of effectiveness given that developmental events are sequential and different events must be coordinated. Is there a time window within which nodal flow determines left–right symmetry in the embryo? To investigate the timing of sensitivity to nodal flow, another experiment was conducted using only fast right flow on early embryos at different stages of development.

Genotype, stage	Results
Wild type, presomite	2N, 21R, 24T
Wild type, 1-somite	9N, 0R, 12T
Wild type, 2-somite	22N, 0R, 22T
Wild type, 3-somite	14N, 0R, 14T

QUESTIONS▶

1. Plot the data as the percent of responses on the y axis and developmental stage on the x axis.
2. Why did this experiment not include left flow as a stimulus?
3. Do these data provide evidence for a sensitive period for nodal flow determining left–right asymmetry?
4. What do you conclude from the results on the 1-, 2-, and 3-somite wild-type embryos?

A similar **work with the data** exercise may be assigned in **LaunchPad**.

43.3 recap

The cell movements of gastrulation convert the blastula into an embryo with three tissue layers. New contacts between cells set up inductive signaling interactions that determine cell fates. Dorsal lip tissue is the source of organizer cells that induce development of preliminary head, trunk, and tail structures.

learning outcomes

You should be able to:

- Compare and contrast gastrulation in frogs and sea urchins.
- Explain the experiment that revealed the role of the dorsal lip of the amphibian blastopore in embryonic development.
- Describe the cell movements that occur during gastrulation in sea urchins and amphibians.
- Summarize how organizer cells achieve a pattern by appropriately expressing growth factor antagonists.
- Explain the meaning of necessity and sufficiency in investigating potential inductive signals.
- Compare gastrulation in reptiles and amphibians.

(continued)

1. The sea urchin blastula develops into a radially symmetrical larva, but the frog blastula develops into a bilateral embryo. How does this difference relate to processes of gastrulation in these two types of animals?

2. What is meant by satisfying the criteria of necessity and sufficiency, and how did the Spemann–Mangold experiments use these criteria to demonstrate the role of the dorsal lip of the blastopore as the primary embryonic organizer?

3. How do the cells of the primary organizer induce different structures on the anterior–posterior axis as they migrate from the dorsal lip to the head region of the embryo?

4. How does gastrulation in reptiles compare with that in amphibians?

We have described how the fertilized egg develops into an embryo with three germ layers and how cellular signals trigger different patterns of differentiation. In the next section we will describe how organs and organ systems develop.

▶ key concept 43.4 Organs Develop from the Three Germ Layers

Gastrulation produces an embryo with three germ layers that are positioned to influence one another through inductive tissue interactions. During the next phase of development, called **organogenesis**, organs and organ systems develop simultaneously and in coordination with each other. In the chordates (see Key Concept 32.1), an early process of organogenesis is **neurulation**, the initiation of the nervous system. We will examine neurulation in the amphibian embryo, but it occurs in a similar fashion in reptiles (including birds), and in mammals.

focus your learning

- The chordamesoderm forms the notochord and initiates neurulation.
- Segmented blocks of cells, known as somites, produce cells that develop into the vertebrae, ribs, trunk muscles, and limbs.
- Hox genes control differentiation on the anterior–posterior axis of vertebrates.
- Genetic and environmental factors can lead to neural tube defects.

The organizer sets the stage for organogenesis

One group of cells that passes over the dorsal lip of the blastopore moves anteriorly and becomes the endodermal lining of the digestive tract. Another group of cells that passes over the dorsal lip on the midline becomes **chordamesoderm**, so named because it forms a rod of mesoderm—the **notochord**—that extends down the center of the embryo. These cells also have important organizer functions (see Figure 43.8). The notochord gives structural support to the developing embryo and in vertebrates is replaced by the vertebral column. The organizing capacity of the chordamesoderm enables the overlying ectoderm to become neural ectoderm.

Neurulation involves the formation of an internal neural tube from an external sheet of cells. The first signs of neurulation are flattening and thickening of the ectoderm overlying the notochord; this thickened area forms the neural plate (**Figure 43.12A**). The edges of the neural plate that run in an anterior–posterior direction continue to thicken to form ridges or folds. Between these neural folds, a groove forms and deepens as the folds roll over it to converge on the midline. The folds fuse, forming a cylinder, the **neural tube**, and a continuous overlying layer of epidermal ectoderm (**Figure 43.12 B–D**).

Cells from the most lateral portions of the neural plate do not become part of the neural tube, but disassociate from it and come to lie between the neural tube and the overlying epidermis. These **neural crest cells** migrate outward to lead the development of the connections between the central nervous system (brain and spinal cord) and the rest of the body. Neural crest cells give rise to many

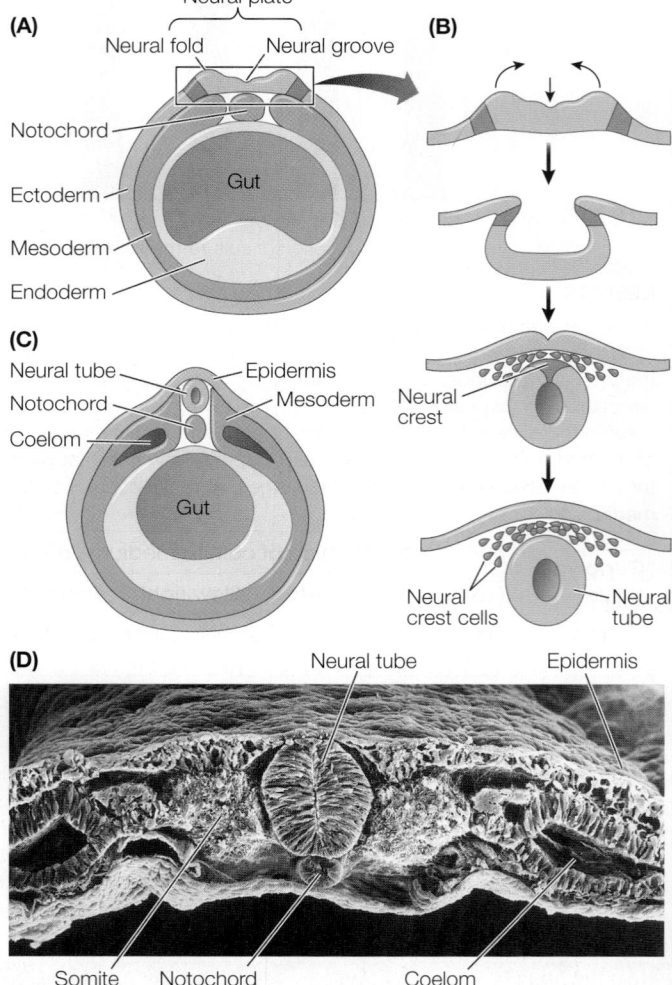

Figure 43.12 Neurulation in a Vertebrate **(A)** At the start of neurulation, the ectoderm of the neural plate (green) is flat. **(B)** The neural plate invaginates and folds, forming a tube. **(C, D)** The completely formed neural tube seen (C) in diagrammatic form and (D) in a scanning electron micrograph of a chick embryo.

(A)

2-Day chick embryo

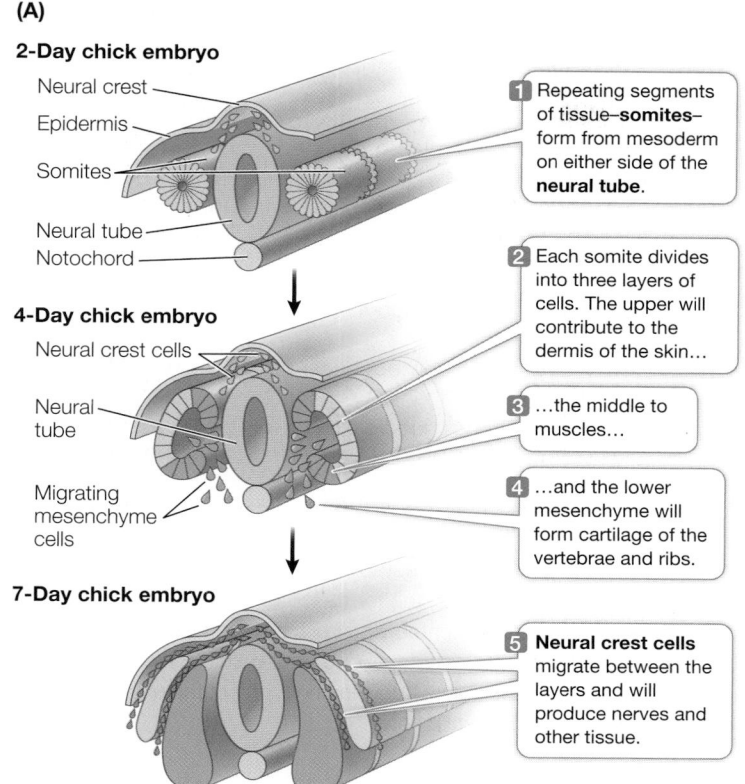

Neural crest

Epidermis

Somites

Neural tube

Notochord

1 Repeating segments of tissue–**somites**–form from mesoderm on either side of the **neural tube**.

4-Day chick embryo

Neural crest cells

Neural tube

Migrating mesenchyme cells

2 Each somite divides into three layers of cells. The upper will contribute to the dermis of the skin...

3 ...the middle to muscles...

4 ...and the lower mesenchyme will form cartilage of the vertebrae and ribs.

7-Day chick embryo

5 **Neural crest cells** migrate between the layers and will produce nerves and other tissue.

(B)

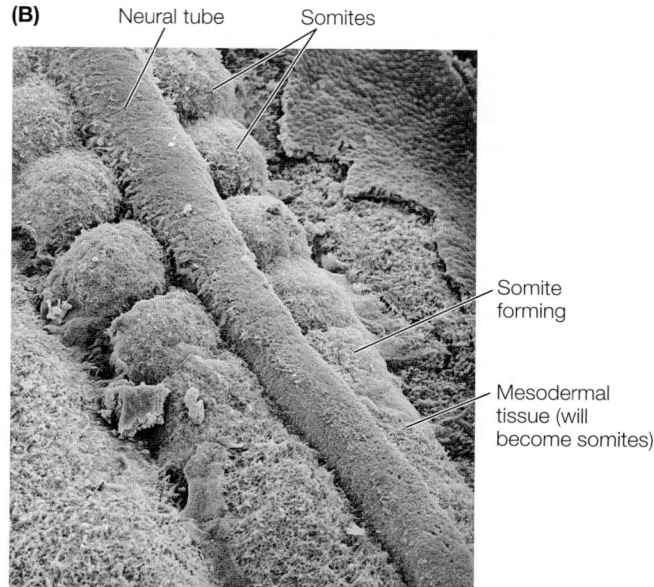

Neural tube   Somites

Somite forming

Mesodermal tissue (will become somites)

Figure 43.13 Developing Body Segmentation (A) Repeating blocks of tissue called somites form on either side of the neural tube. Muscle, cartilage, bone, and the inner layer of the skin form from the somites. (B) In this scanning electron micrograph of somite formation in a chick embryo, the overlying ectoderm has been removed and the neural tube and somites are seen from above.

diverse structures including jaws, skull, face, pigment cells, glands, smooth muscle, and many others.

The neural tube develops bulges at the anterior end, which become the major divisions of the brain; the rest of the tube becomes the spinal cord. In humans, failure of the neural folds to fuse in this posterior region results in spina bifida, a birth defect in which the spinal cord is exposed because the vertebrae do not fuse. If the folds fail to fuse at the anterior end, an infant can develop without a forebrain (a condition called anencephaly). Although several genetic factors can cause these defects, other factors are environmental, including maternal diet. The incidence of neural tube defects in the United States in the early 1900s was as high as 1 in 300 live births; today it is less than 1 in 1,000. A major factor in this improvement has been the inclusion of folic acid (a B vitamin, also known as folate) in the mother's diet. It is essential for pregnant women to ingest sufficient folic acid.

Body segmentation is an early feature of vertebrate development

The vertebrate body plan, like that of arthropods, consists of repeating segments that are modified during development. These segments are most evident as the repeating patterns of vertebrae, ribs, nerves, and muscles along the anterior–posterior axis.

As the neural tube forms, mesodermal tissues gather along the sides of the notochord to form separate, segmented blocks of cells called **somites** (Figure 43.13). Somites produce cells that will become the vertebrae, ribs, muscles of the trunk, and limbs.

Nerves that connect the brain and spinal cord with tissues and organs throughout the body are also arranged segmentally. The somites help guide the organization of these peripheral nerves, but the nerves are not of mesodermal origin. As you saw above, when the neural tube fuses, the neural crest cells break loose and migrate inward between the epidermis and the somites and through the somites. These neural crest cells have diverse fates, including the development of peripheral nerves.

As development progresses, the different segments of the body change. Regions of the spinal cord differ, regions of the vertebral column differ in that some vertebrae grow ribs of various sizes and others do not, forelegs arise in the anterior part of the embryo, and hind legs arise in the posterior region.

Hox genes control differentiation along the anterior–posterior axis

How is mesoderm in the anterior part of a mouse embryo programmed to produce forelegs rather than hind legs? In Key Concept 19.4, you saw how homeotic genes control body segmentation in *Drosophila*. You also learned that all homeotic genes contain a DNA sequence called the **homeobox**. In vertebrates, the homeotic genes that control differentiation along the anterior–posterior body axis are called **Hox genes** (see Key Concept 19.3).

In mammals, four Hox gene complexes reside on different chromosomes in clusters of about ten genes each. Remarkably, the temporal and spatial expression of these genes generally follows the same pattern as their linear order on their chromosome. That is, the Hox genes closest to the 3' end of each gene complex are expressed first and in the anterior of the embryo. The Hox genes at the 5' end of the gene complex are expressed later and in a more posterior part of the embryo. As a result, different segments of the

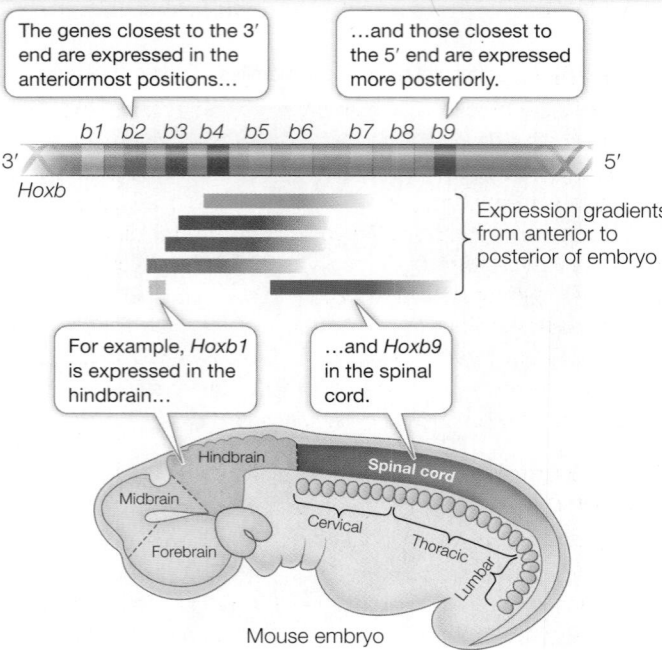

The genes closest to the 3' end are expressed in the anteriormost positions...

...and those closest to the 5' end are expressed more posteriorly.

b1 b2 b3 b4 b5 b6   b7 b8 b9

3'   5'
Hoxb

Expression gradients from anterior to posterior of embryo

For example, *Hoxb1* is expressed in the hindbrain...

...and *Hoxb9* in the spinal cord.

Hindbrain   Spinal cord
Midbrain
            Cervical   Thoracic
Forebrain   Lumbar

Mouse embryo

Figure 43.14 Hox Genes Control Body Segmentation Hox genes are expressed along the anterior–posterior axis of the embryo in the same order as their arrangement between the 3' and 5' ends of the gene complex. As a result of gene duplication during evolution, vertebrates have four copies of the Hox gene complex shown.

embryo receive different combinations of Hox gene products, which serve as transcription factors (**Figure 43.14**; see also Figure 19.15).

Whereas Hox genes give cells information about their position on the anterior–posterior body axis, other genes provide information about their dorsal–ventral position. Tissues in each segment of the body differentiate according to their dorsal–ventral location. The notochord provides many of these signals.

After body segmentation develops, the formation of organs and organ systems progresses rapidly. The development of an organ involves extensive inductive interactions. An example is the ability of the chordamesoderm to induce overlying ectoderm to become neural ectoderm, and form the neural tube (see Figure 43.12). These inductive interactions are a current focus of study for developmental biologists.

▶ 43.4 recap

Gastrulation sets up tissue interactions that initiate organogenesis. Neurulation is initiated by organizer mesoderm that forms the notochord. Neurulation results in the formation of the central nervous system, and associated neural crest cells form the peripheral nerves. Suites of Hox genes are expressed sequentially in vertebrates, but in overlapping patterns providing a variety of signaling molecule patterns that govern differentiation on the anterior–posterior axis.

learning outcomes

You should be able to:

• Describe how nervous system development begins with neurulation.

▶ 43.4 recap

• Describe the formation of somites and their role in development.
• Explain how genetic and environmental factors can influence neural tube formation.
• Explain how a limited number of Hox genes can organize complex anterior–posterior patterning of the vertebrate embryo.

1. How does the ectoderm overlying the notochord respond to chordamesoderm signals to form a neural tube as well as neural crest cells?
2. What might be a consequence of a failure in somite formation in the embryo?
3. How do Hox genes generate a diversity of signals sufficient to control differentiation of tissues and organs on the anterior–posterior axis?

You have seen how the basic structure of the developing embryo arises, through the establishment of the anterior–posterior and dorsal–ventral axes, the formation of the neural tube, and the emergence of a segmented body plan. In the next section we will examine the developmental events that result in the formation of structures that support the developing embryo: the extraembryonic membranes and the placenta.

▶ key concept
43.5 Extraembryonic Membranes Nurture Avian and Mammalian Embryos

There is more to a developing reptile or mammal than the embryo itself. The embryos of these vertebrates are surrounded by several **extraembryonic membranes** that originate from the embryo but are not part of it. Extraembryonic membranes function in nutrition, gas exchange, and waste removal. In mammals they interact with tissues of the mother to form the placenta. The evolutionary relationships between the extraembryonic membranes of birds and mammals were discussed in Key Concept 32.4.

focus your learning

• The four extraembryonic membranes in birds are the yolk sac, the allantoic membrane, the amnion, and the chorion.
• The placenta forms from the embryonic chorion and tissues of maternal uterine wall.
• During the first trimester, the embryo is most sensitive to environmental effects.

Birds develop four extraembryonic membranes

The chicken provides a good example of how extraembryonic membranes form from the germ layers created during gastrulation. In the chick, four membranes form—the yolk sac, the allantoic membrane, the amnion, and the chorion. The **yolk sac** is the first

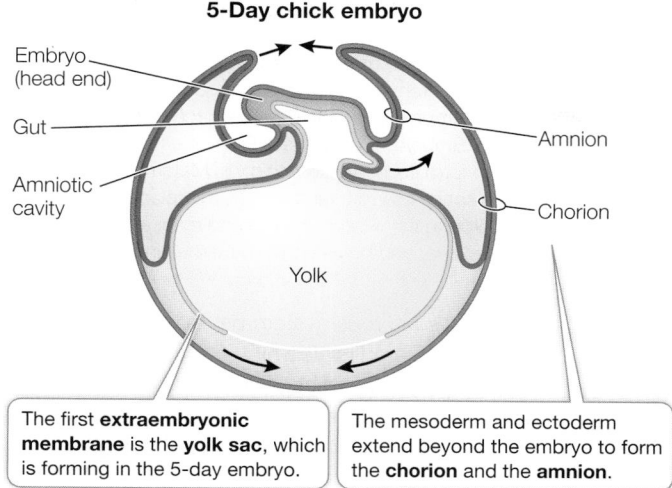

5-Day chick embryo

Embryo (head end)

Gut

Amniotic cavity

Amnion

Chorion

Yolk

The first **extraembryonic membrane** is the **yolk sac**, which is forming in the 5-day embryo.

The mesoderm and ectoderm extend beyond the embryo to form the **chorion** and the **amnion**.

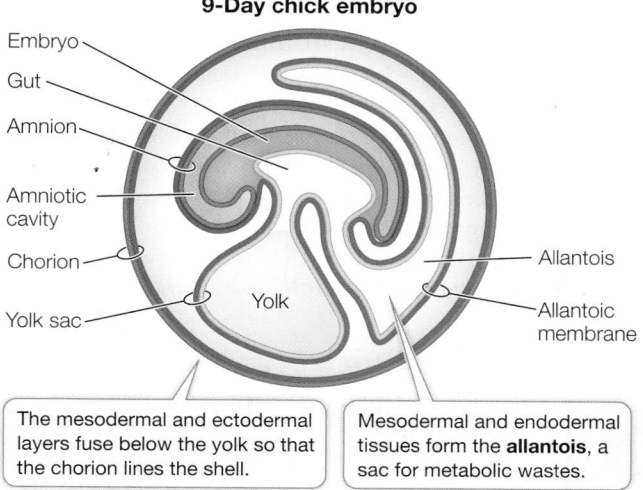

9-Day chick embryo

Embryo

Gut

Amnion

Amniotic cavity

Chorion

Yolk sac

Yolk

Allantois

Allantoic membrane

The mesodermal and ectodermal layers fuse below the yolk so that the chorion lines the shell.

Mesodermal and endodermal tissues form the **allantois**, a sac for metabolic wastes.

Figure 43.15 The Extraembryonic Membranes In birds (and other reptiles) and mammals, the embryo constructs four extraembryonic membranes. In birds, the yolk sac encloses the yolk, and the amnion and chorion enclose the embryo. Fluids secreted by the amnion fill the amniotic cavity, providing an aqueous environment for the embryo. The chorion, along with the allantoic membrane, mediates gas exchange between the embryo and its environment. The allantois stores the embryo's waste products. (See also Figure 32.19.)

  Activity 43.1 Extraembryonic Membranes
www.Life11e.com/ac43.1

to form, and it does so by extension of the hypoblast layer along with some adjacent mesoderm. The yolk sac grows to enclose the entire body of yolk in the egg (**Figure 43.15**). It constricts at the top to create a tube that is continuous with the gut of the embryo. However, yolk does not pass through this tube. Yolk is digested by the cells of the yolk sac, and the nutrients are transported to the embryo through blood vessels that form from mesoderm and line the outer surface of the yolk sac. The **allantoic membrane** is also an outgrowth of the extraembryonic endoderm plus adjacent mesoderm. It forms the **allantois**, a sac for storage of metabolic wastes.

Ectoderm and mesoderm combine and extend beyond the limits of the embryo to form the other extraembryonic membranes. Two layers of cells extend all along the inside of the eggshell, both over the embryo and below the yolk sac. Where they meet, they fuse, forming two membranes, the inner **amnion** and the outer **chorion**. The amnion surrounds the embryo, forming the amniotic cavity. The amnion secretes fluid into the cavity, providing a protective environment for the embryo. The outer membrane, the chorion, forms a continuous membrane just under the eggshell (see Figure 43.16) that limits water loss from the egg. The exchange of O_2 and CO_2 between the embryo and the environment requires diffusion of those gases across the chorion and the allantois.

The mammalian placenta forms from extraembryonic membranes

In placental mammals, the first extraembryonic membrane to form is the trophoblast (see Figures 43.4C and 43.5). When the blastocyst reaches the uterus and hatches from its encapsulating zona pellucida, trophoblast cells interact directly with the endometrium. Adhesion molecules expressed on the surfaces of these cells attach them to the uterine wall. By secreting proteolytic enzymes, the trophoblast burrows into the endometrium, beginning the process of implantation. Eventually the entire trophoblast is within the wall of the uterus. The trophoblast cells send out numerous projections, or villi, to increase the surface area of contact with maternal blood.

Meanwhile, the hypoblast cells proliferate to form what in the bird would be the yolk sac. But there is virtually no yolk in eggs of placental mammals; instead the yolk sac contributes mesodermal tissues that interact with trophoblast tissues to form the chorion. The chorion, along with tissues of the uterine wall, produces the **placenta**, the organ that exchanges nutrients, respiratory gases, and metabolic wastes between the mother and the embryo (**Figure 43.16**).

At the same time the yolk sac is forming from the hypoblast, the epiblast produces the amnion, which grows to enclose the entire embryo in a fluid-filled amniotic cavity. The rupturing of the amnion and chorion and the loss of **amniotic fluid** (the "water breaks") herald onset of labor in humans.

An allantois also develops in mammals, but its importance depends on how well nitrogenous wastes can be removed by the blood vessels of the placenta. The human allantoic sac is small because the human placenta deals effectively with the fetal nitrogenous wastes. In contrast, the pig placenta is less effective at clearing nitrogenous wastes from the fetus, so the pig's allantoic sac is large. In humans and other placental mammals, allantoic tissues contribute to the formation of the umbilical cord, by which the embryo is attached to the chorionic placenta (see Figure 32.19B). It is through the blood vessels of the umbilical cord that nutrients and oxygen from the mother reach the developing fetus, and wastes, including carbon dioxide and urea, are removed.

Human gestation is divided into trimesters

In humans, **gestation**, or pregnancy, lasts about 266 days, or 9 months. Gestation is shorter in smaller mammals (21 days in mice) and longer in larger mammals (600 days in elephants). The events of human gestation (**Figure 43.17**) can be divided into three **trimesters**

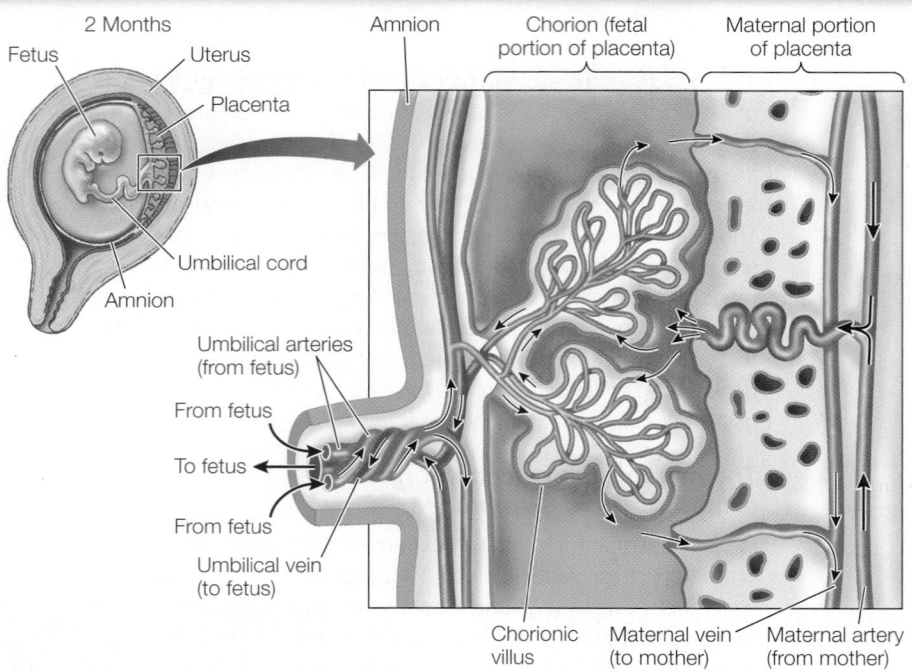

2 Months
Fetus
Uterus
Placenta
Umbilical cord
Amnion
Amnion
Chorion (fetal portion of placenta)
Maternal portion of placenta
Umbilical arteries (from fetus)
From fetus
To fetus
From fetus
Umbilical vein (to fetus)
Chorionic villus
Maternal vein (to mother)
Maternal artery (from mother)

Figure 43.16 The Mammalian Placenta In humans and most other mammals, nutrients and wastes are exchanged between maternal and fetal blood in the placenta, which forms from the chorion and tissues of the uterine wall. The embryo is attached to the placenta by the umbilical cord. Embryonic blood vessels invade the placental tissue to form fingerlike chorionic villi. Maternal blood flows into the spaces surrounding the villi, and placental blood flows through the villi so nutrients and respiratory gases can be exchanged between the maternal and fetal blood.

the fetus's nervous system develops rapidly. Fetal movements are first felt by the mother early in the second trimester and become progressively stronger and more coordinated.

The fetus grows rapidly during the third trimester. As this final stage approaches its end (see Figure 43.17C), the internal organs mature. The digestive system begins to function, the liver stores glycogen, the kidneys produce urine, and the brain undergoes cycles of sleep and waking. A human infant is born when the last of its critical organs—the lungs—mature.

Although the first-trimester embryo is the most susceptible to adverse effects of drugs, chemicals, and diseases, the potential for serious effects from exposure to environmental factors exists throughout pregnancy. Protein malnutrition and exposure to alcohol and cigarette smoke are examples of factors that can result in low birth weight, mental retardation, and other developmental complications.

of roughly 3 months each. The first trimester (see Figure 43.17A) is a time of rapid cell division and tissue differentiation. Because signal transduction cascades that determine sequences of developmental processes are in their early stages, the first trimester is the period when the embryo is most sensitive to damage from radiation, drugs, chemicals, and pathogens that can cause birth defects. An embryo can be damaged before the mother even realizes she is pregnant. A classic case is that of thalidomide, a drug widely prescribed in Europe in the late 1950s to treat nausea. Women who took this drug in the fourth and fifth weeks of pregnancy, when the embryo's limbs are beginning to form, gave birth to children with missing or severely malformed arms and legs.

During the second trimester the fetus grows rapidly; its limbs elongate, and its fingers, toes, and facial features become well formed (see Figure 43.17B). Eyebrows and fingernails grow, and

Figure 43.17 Stages of Human Development (A) At 4 weeks, most of the embryo's organ systems have been formed and the heart is beating. The body structures of this 8-week-old embryo are forming rapidly. (B) At 4 months, the fetus has fully formed limbs with fingers and toes and moves freely within the amniotic cavity. (C) This fetus is well along in its ninth month. Soon its lungs will be mature enough to trigger the onset of contractions and birth.

(A) First trimester

4 Weeks

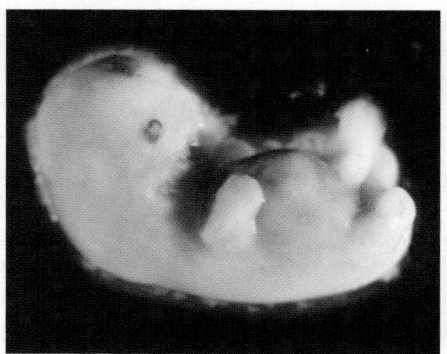

Actual length ~0.4 cm (4 mm)

8 Weeks

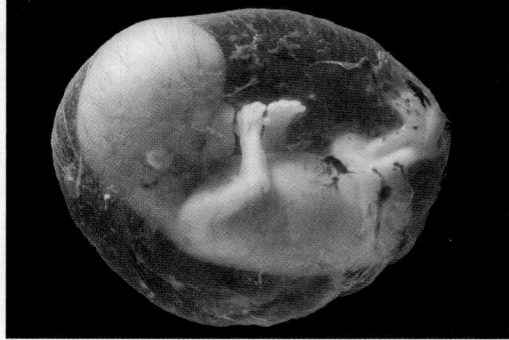

Actual length ~3 cm

(B) Second trimester

4 Months

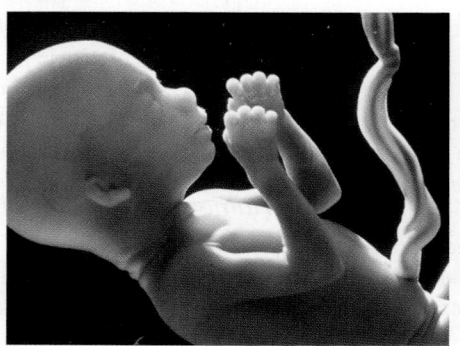

Actual length ~10 cm

(C) Third trimester

9 Months

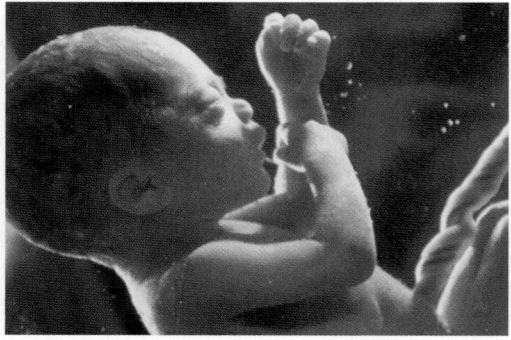

Actual length ~40 cm

43.5 recap

The extraembryonic membranes of reptiles and mammals sustain the growing embryo. In reptiles these membranes surround the embryo within the shelled egg. In mammals the extraembryonic membranes form the placenta, an organ that exchanges nutrients, respiratory gases, and metabolic wastes between the mother and the embryo. Human gestation lasts 9 months. At the end of the first trimester, the fetus is very small but most of its organs have begun to form. In the second trimester, limbs elongate and the fetus moves. By the end of the third trimester, most organs have begun to function.

learning outcomes

You should be able to:

- Describe the formation of the yolk sac, amnion, chorion, and allantois, using the chick as an example.
- Summarize how the placenta forms.
- Explain why a first-trimester embryo is particularly sensitive to environmental risks for birth defects.

1. Which germ layers participate in the development of the chick's yolk sac, allantois, chorion, and amnion?
2. What is the origin of the placenta in mammalian development?
3. Why is a first-trimester embryo particularly sensitive to environmental risks?

investigatinglife

How does the embryo know left from right?

At the beginning of gastrulation in the mammal when midline cells of the epiblast begin to move inward to create the primitive groove, a region called the node forms at the anterior end of the groove. Motile cilia at the node create a flow of extracellular fluid across the node. Non-motile nodal cilia are believed to sense the direction of fluid flow across the node. There are two hypotheses about how these cilia function, one chemical and the other mechanical. Both sensory mechanisms are used by cilia elsewhere in the body. In the nose, cilia of the olfactory cells sense chemicals. In the ear, cilia are bent by sound waves, which stimulates ion channels to open. The leftward flow of fluid across the node imposes mechanical forces on the nonmotile cilia that differ on the two sides of the node. Other research, however, supports the possibility of chemical signaling. Research on the mouse node shows that the beating of the cilia causes proteins of a certain size range to form a concentration gradient across the node. More important, it was discovered that nodal cells secrete small vesicles that are swept to the left side of the

node by the flow and burst when they contact the nonmotile cilia of surrounding cells. The contents of these vesicles could be the chemical signals that initiate left–right asymmetry in gene expression and development.

Future directions

A major player in the determination of left–right asymmetry is the primary cilium of cells in the node that forms during gastrulation. However, just about every cell in the body has a primary cilium. Up until fairly recently the primary cilia were thought to be vestigial structures or simply organelles to sequester the centriole and thereby inhibit cell division. We now know that the primary cilia of cells function widely in mechanisms of cell signaling and in transduction of physical and chemical stimuli. Just about every organ in the body uses primary cilia for essential functions. Primary cilia have been implicated in many medical conditions such as cancer, kidney disease, obesity, blindness, and developmental disorders. In the near future we will be seeing lots of new discoveries about the functions of primary cilia.

Chapter Summary 43

43.1 Fertilization Activates Development

- The sperm and the egg contribute differentially to the zygote. The sperm contributes a haploid nucleus and, in most species, a centriole. The egg contributes a haploid nucleus, nutrients, ribosomes, mitochondria, mRNAs, and proteins.
- In amphibians, the cytoplasmic contents of the unfertilized egg are not distributed homogeneously, and they are rearranged after fertilization to set up the major axes of the future embryo. The nutrient molecules are generally found in the **vegetal hemisphere**,

whereas the nucleus is found in the **animal hemisphere**. Review Figures 43.1, 43.2

43.2 Mitosis Divides Up the Early Embryo

- The protein β-catenin activates a signaling cascade that induces the primary embryonic organizer and sets up the anterior–posterior body axis. Review Figures 43.2
- **Cleavage** is a period of rapid cell division. Except in mammals, little if any gene expression occurs during cleavage. Cleavage can be complete or incomplete, and the pattern of cell divisions

(continued)

depends on the orientation of the mitotic spindles. The result of cleavage is a ball or mass of cells called a **blastula**. Review Figure 43.3

- Early cell divisions in mammals are unique in being slow and allowing for gene expression early in the process. These cell divisions produce a **blastocyst** composed of an **inner cell mass** that becomes the embryo and an outer cell mass that develops as the **trophoblast**. At the time of **implantation**, the trophoblast secretes molecules that help the blastocyst implant in the uterine wall. Review Figures 43.4, 43.5

- A **fate map** can be created by labeling specific **blastomeres** and observing what tissues and organs are formed by their progeny later in development. Review Figure 43.6

- Some species undergo **mosaic development**, in which the fate of each cell is determined during early divisions. Other species, including vertebrates, undergo **regulative development**, in which remaining cells can compensate for cells lost in early cleavages.

43.3 Gastrulation Generates Multiple Tissue Layers

- **Gastrulation** involves massive cell movements that produce three **germ layers** and place cells from various regions of the blastula into new associations with one another. Review Figure 43.7, Animation 43.1

- The initial step of sea urchin and amphibian gastrulation is inward movement of certain blastomeres. The site of inward movement becomes the **blastopore**. Cells that move into the blastula become the **endoderm** and **mesoderm**; cells remaining on the outside become the **ectoderm**. Cytoplasmic factors in the vegetal pole cells are essential to initiate development. Review Figure 43.7, Focus: Key Figure 43.8

- The **dorsal lip** of the amphibian blastopore is a critical site for cell determination. It has been called the **primary embryonic organizer**, or **organizer**, because it induces determination in cells that pass over it during gastrulation. Review Figures 43.8–43.10, Animation 43.2

- Gastrulation in reptiles, including birds, differs from that in sea urchins and frogs because the large amount of yolk causes the blastula to form a flattened disc of cells. Review Figure 43.11

- Although their eggs have no yolk, placental mammals have a pattern of gastrulation similar to that of reptiles and birds.

- The direction of extracellular fluid flow over the ventral surface of the node breaks bilateral symmetry in the mammalian gastrula. Review Investigating Life: Does the Direction of Nodal Flow Influence the Development of Left–Right Asymmetry in the Mouse Embryo?

43.4 Organs Develop from the Three Germ Layers

- Gastrulation is followed by **organogenesis**, the process where tissues interact to form organs and organ systems.

- In the formation of the vertebrate nervous system, one group of cells that migrates over the blastopore lip is determined to become the **notochord**. The notochord organizes the overlying ectoderm to thicken, form parallel ridges, and fold in on itself to form a **neural tube** below the epidermal ectoderm. The nervous system develops from this neural tube. Review Figure 43.12

- The notochord and **neural crest cells** participate in the segmental organization of mesoderm into structures called **somites** along the body axis. Rudimentary organs and organ systems form during these stages. Review Figure 43.13

- In vertebrates, **Hox genes** determine the pattern of anterior–posterior differentiation along the body axis. Other genes, such as *sonic hedgehog*, contribute to dorsal–ventral differentiation. Review Figure 43.14

43.5 Extraembryonic Membranes Nurture Avian and Mammalian Embryos

- The embryos of reptiles and mammals are protected and nurtured by four **extraembryonic membranes**. In reptiles the **yolk sac** surrounds the yolk and provides nutrients to the embryo, the **chorion** lines the eggshell and participates in gas exchange, the **amnion** surrounds the embryo and encloses it in an aqueous environment, and the **allantois** stores metabolic wastes. Review Figure 43.15, Activity 43.1

- In mammals the chorion and the trophoblast cells interact with the maternal uterus to form a **placenta**, which provides the embryo with nutrients and gas exchange. The amnion encloses the embryo in an aqueous environment. Review Figure 43.16

- Human pregnancy, or **gestation**, lasts 9 months. The embryo forms in the first **trimester**; during this time, it is most vulnerable to environmental factors that can lead to birth defects. During the second and third trimesters the fetus grows, external features form, and the organ systems mature. Review Figure 43.17

Go to **LearningCurve** (in **LaunchPad**) for dynamic quizzing that helps you solidify your understanding of this chapter. **LearningCurve** adapts to your responses, giving you the practice you need to master each key concept.

Apply What You've Learned

Review

43.3 Cell movements during gastrulation result in three germ layers and new tissue interactions.

43.3 Inductive tissue interactions can suppress as well as activate.

43.4 Segmented blocks of cells, known as somites, produce cells that develop into the vertebrae, ribs, trunk muscles, and limbs.

43.4 Genetic and environmental factors can lead to neural tube defects.

43.5 During the first trimester, the embryo is most sensitive to environmental effects.

Original Paper: Finnell, R. H., J. D. Waes, J. D. Eudy and T. H. Rosenquist. 2002. Molecular basis of environmentally induced birth defects. *Annual Review of Pharmacology and Toxicology* 42: 181–208.

Both genetic and environmental factors—alone or acting together—can affect the incidence of birth defects in humans. Chemicals in the environment sometimes mimic or interfere with signals involved in inductive tissue interactions, with detrimental results for fetal development. The table below shows the effects of some of these agents.

Questions

1. An infant has various bone, cardiovascular, and muscular system defects. Is this consistent with a vitamin A deficiency? Explain your answer.

2. Suggest a reasonable hypothesis for why vitamin A has different effects on ectoderm-derived tissues than on mesoderm-derived tissues.

3. Suppose there is a genetic variant that causes faster progression through G1 phase of the cell cycle. Would you expect individuals with this variant to be more susceptible or less susceptible to valproic acid during early development?

4. A woman takes valproic acid before she realizes she is pregnant. Would supplements of folic acid be beneficial for this woman's pregnancy? Explain your answer. Why would effects of valproic acid occur so early in the pregnancy?

5. Folate receptor alpha is a protein that is involved in bringing the biologically active form of folic acid into cells. Explain how variants of this protein could affect the frequency of neural tube defects during pregnancy.

Chemical/agent	Effects on fetus	Comments
Folic acid	Deficiency increases the risk of neural tube defects.	Genetic variants at folate receptor alpha affect the risk of neural tube defects.
Vitamin A	Excess increases mesenchyme defects. Deficiency increases ectoderm defects.	Vitamin A is not synthetized by humans; it must be acquired from the diet.
Valproic acid	Exposure increases the risk of neural tube and other neural defects, craniofacial abnormalities, and skeletal defects.	Valproic acid is used to treat bipolar disorder and epilepsy. It appears to act by impeding the cell cycle during G1 phase.

Go to **LaunchPad** for the eBook, LearningCurve, animations, activities, flashcards, and additional resources and assignments.

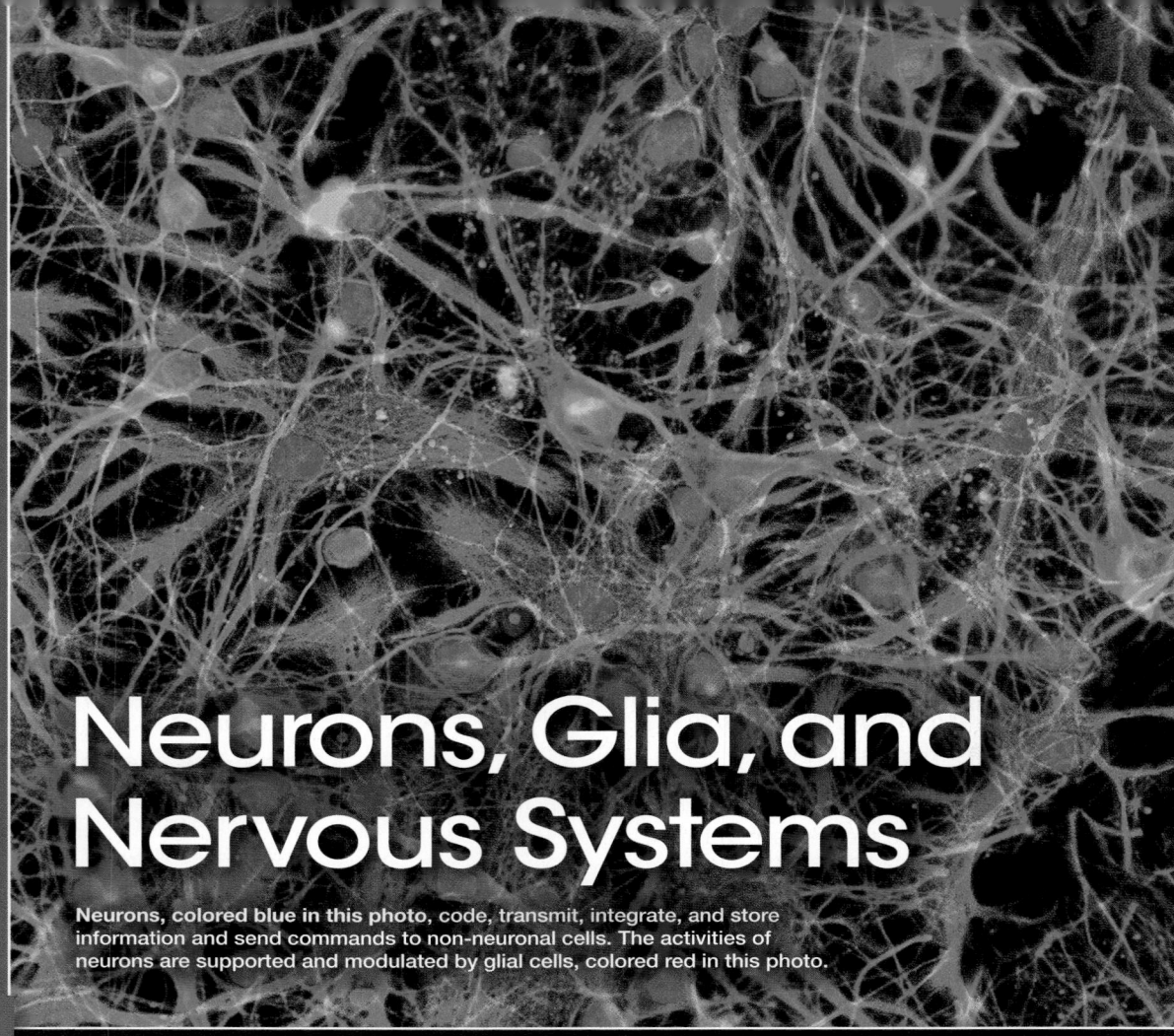

44

Neurons, Glia, and Nervous Systems

Neurons, colored blue in this photo, code, transmit, integrate, and store information and send commands to non-neuronal cells. The activities of neurons are supported and modulated by glial cells, colored red in this photo.

▶ investigatinglife

Balancing the Brain

Your brain enables you to learn the material in this chapter—to read the words, understand the illustrations, and store information to allow you to answer questions. You must study to master this chapter, but if you do you will learn how the brain receives and processes information.

Imagine what it would be like if you could not learn any of this material—no matter how much you poured over the book. This is the situation faced by individuals with a condition known as Down syndrome, which affects 1 out of every 700 children born in the United States.

Individuals with Down syndrome are born with three copies of most of chromosome 21, one of the smallest of the human chromosomes (containing about 225 confirmed genes). Having an extra copy of these genes causes numerous developmental and functional problems, including learning disability. Can we understand the cause of this disability and perhaps find a way to remedy it?

One productive way of investigating the causes and possible treatments for a human disease or deficit is to develop an animal model. Through genetic engineering, researchers created a "Down syndrome mouse" that has most of the same genes triplicated as those in humans with Down syndrome. These mice cannot perform in mouse memory tests as normal mice do. The mouse Down syndrome model revealed that excess inhibition in the brain causes the learning disability. What is meant by "excess inhibition in the brain"? We tend to think mostly about the excitatory actions of the brain, likening the brain to a puppet master who pulls the strings that activate the muscles and organs of the body. In fact the brain is more like an orchestra conductor, making some sections louder, some softer, speeding up, slowing down. The brain must constantly maintain a delicate balance of excitation and inhibition, acting on some signals and ignoring others. In the brains of Down syndrome model mice there is consistently too much inhibition. When inhibition is experimentally reduced with certain drugs, the mice are able to learn. Research like this is a first step toward bringing potential therapies for humans into clinical trials in the hope of improving the cognitive functions of individuals with Down syndrome.

QA Can learning be restored in mouse models of Down syndrome?

key concept
44.1 Neurons and Glia Are Unique Cells of Nervous Systems

Nervous systems are informational systems. They encode, process, and store a wide variety of information from the external and internal environments, and they use that information to control and regulate the physiology and behavior of the organism. The nervous systems of all animals can carry out these functions because of the properties of a unique class of cells: nerve cells, or **neurons**. Vertebrate nervous systems also include another unique class of cells called glial cells or **glia**.

focus your learning

- The structure of a neuron defines its functions in receiving, integrating, and communicating information.
- Specific types of glial cells have unique roles within the nervous system to support, insulate, and nurture neurons and to affect neuronal signaling.

There are many types of neurons, which vary in size, shape, and function, but they all generate and conduct electric signals. Glia do not generate and conduct electric signals, and there are two major types—macroglia and microglia. **Macroglia** support and modulate the functions of neurons. **Microglia** are phagocytic cells derived from blood-forming stem cells. Microglia are motile, are activated by foreign matter or damage to neural tissue, and are the first and major immune defense mechanism in the nervous system.

Nervous systems in all but the simplest invertebrates are divided into central and peripheral divisions. The central division includes dense assemblages of neurons that process and store information and generate commands. In lower invertebrates these assemblages of neurons are called ganglia. With increased complexity, some large ganglia of invertebrates are designated as brains. In vertebrates the **central nervous system** (**CNS**) includes the brain and spinal cord. The **peripheral nervous system** (**PNS**) provides communication between the central division and all of the rest of the body.

Vertebrate neurons and macroglia originate in the embryonic neural tube

Neurons and macroglia have a common origin in the neural tube of the early vertebrate embryo (see Figure 43.14). The neural tube contains rapidly dividing neuronal stem cells. Initially the divisions of these neural stem cells are symmetrical—all of the progeny are stem cells. Eventually, however, the cell divisions become asymmetrical, with one daughter cell remaining a stem cell and the other daughter cell becoming either a neural progenitor cell or a macroglial progenitor cell—a neuroblast or a gliablast. These progenitor cells produce the neurons and glia that together form the various structures of the nervous system.

The structure of neurons reflects their functions

Neurons vary enormously in structure and appearance, but they have a basic structure that includes four regions (**Figure 44.1**).

1. A **cell body** contains the nucleus and most of the cell's organelles.

2. Shrublike projections called **dendrites** (Greek *dendron*, "tree") may extend from the cell body. Dendrites receive information from other neurons or sensory cells and bring it to the cell body.

3. In most neurons, one projection—the **axon**—is much longer than the others. Axons carry information from the cell body to target cells.

4. At the target cell, the axon divides into a spray of fine nerve endings. The tips of these tiny nerve endings have swellings called **axon terminals**.

Neurons have a wide variety of forms that reflect different functionalities (**Figure 44.2**). Neurons with few dendrites receive information from specific and limited sources, whereas neurons with large arrays of dendrites can collect and integrate information from a wide range of sources. Some neurons communicate over very short distances, but others communicate over extremely long distances and therefore have very long axons. For example, axons from neurons in your spinal cord control the muscles in your toes.

Regardless of form and function, all neurons process and communicate information through changes in the electric potential across their membranes. As in all cells, the electric charge on the inside of the cell membrane of a neuron is slightly negative in comparison with the outside. The electric potential across the cell membranes of neurons can change in response to specific stimuli, and these electric potential changes can travel along the cell membrane. In most neurons, small changes in membrane electric potential generate large, rapidly reversed changes in membrane potential that are called **action potentials** (**APs**). Axons conduct APs over long distances.

In summary, a neuron receives specific stimuli (e.g., light, sound, pressure, chemicals) that cause changes in the electric potentials

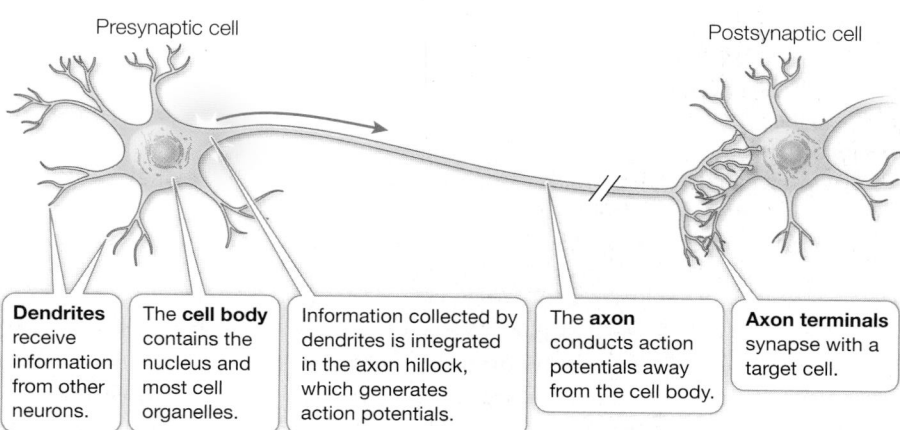

Figure 44.1 A Generalized Neuron The diagram shows the features typical of most neurons. The forms of these features, including the length of the single axon and the density and branching patterns of the dendrites, vary greatly across the many different types of neurons.

Presynaptic cell   Postsynaptic cell

| **Dendrites** receive information from other neurons. | The **cell body** contains the nucleus and most cell organelles. | Information collected by dendrites is integrated in the axon hillock, which generates action potentials. | The **axon** conducts action potentials away from the cell body. | **Axon terminals** synapse with a target cell. |

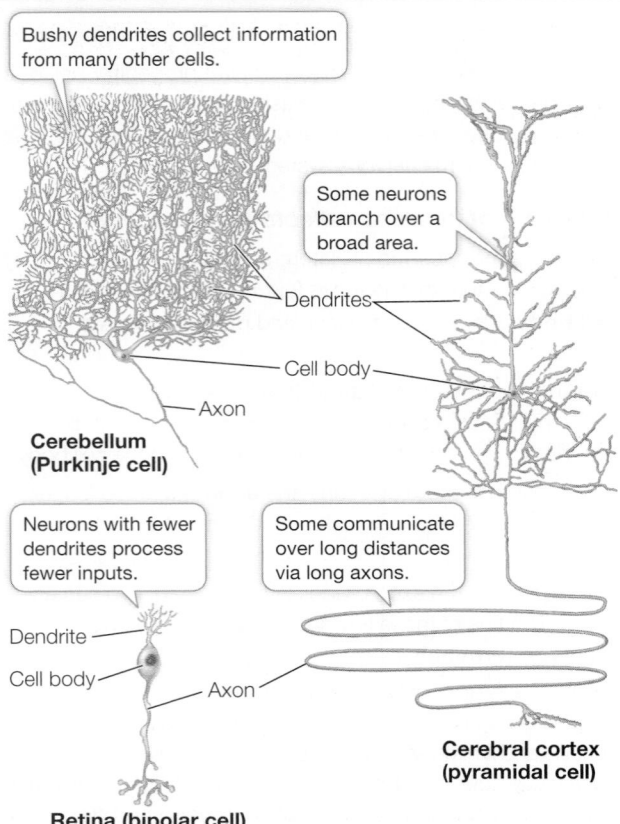

Bushy dendrites collect information from many other cells.

Some neurons branch over a broad area.

Dendrites

Cell body

Axon

Cerebellum (Purkinje cell)

Neurons with fewer dendrites process fewer inputs.

Some communicate over long distances via long axons.

Dendrite

Cell body

Axon

Cerebral cortex (pyramidal cell)

Retina (bipolar cell)

Figure 44.2 Neurons Have Many and Varied Forms The morphological differences in neurons from different parts of the mammalian nervous system are related to their specific functional adaptations. The small sample here shows two neuronal types from the human brain (a Purkinje cell and a pyramidal cell) and a sensory neuron from the retina of the human eye.

Q: Which of these neurons is likely to receive a greater number of inputs?

across the cell membranes of its dendrites. These electric signals from dendrites spread to and are averaged by the cell body. Changes in the electric potential of the membrane of the cell body at the base of its axon can cause an AP that is conducted along the axon to its terminals. APs can travel at speeds up to 100 meters per second (360 kilometers per hour), making it possible for an individual to sense, process, and act on information very quickly.

What happens when the AP reaches the axon terminals? Axon terminals come extremely close to the membrane of the target cell, which can be another neuron, a muscle cell, or a secretory cell. Structures called **synapses** form at these locations. The synapse transfers the information conveyed by the AP from the presynaptic cell to the postsynaptic cell (see Figure 44.1).

Synapses can be either electrical or chemical. Electrical synapses allow the AP to pass directly between two neurons. In vertebrates most synapses are chemical. At chemical synapses a space about 25 nanometers wide (about 1/2000th the width of a human hair) separates the presynaptic and postsynaptic membranes. An AP arriving at an axon terminal causes it to release chemical messenger molecules called **neurotransmitters**. Neurotransmitters diffuse across the synaptic space and bind to receptors on the cell membrane of the postsynaptic (target) cell. The binding of the

(A) Schwann cells

Myelin-producing Schwann cells

Site and direction of myelin growth

Nodes of Ranvier

Nucleus of Schwann cell

(B) Cross section of a myelinated axon Mitochondria

Axon

Multiple layers of **myelin** insulate the axon.

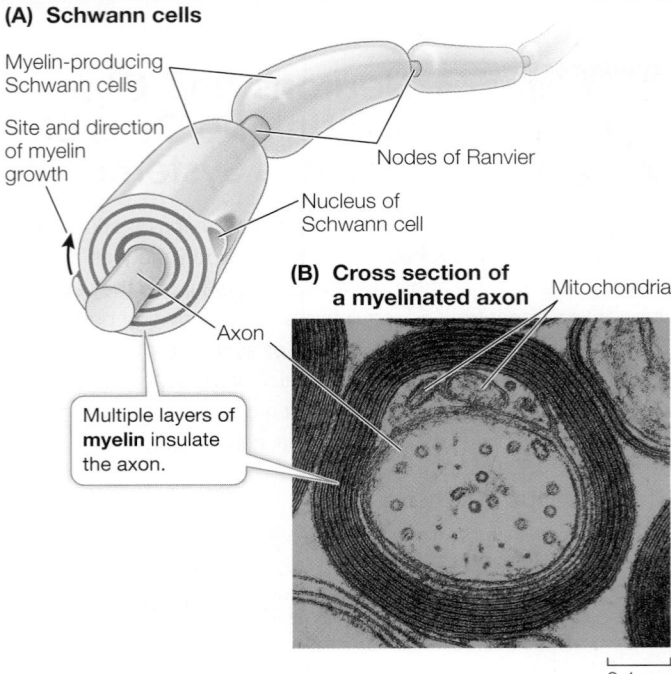

0.1 µm

Figure 44.3 Glial Cells Insulate Axons **(A)** Schwann cells produce layers of myelin, a type of cell membrane that provides electrical insulation to the axon. At the intervals between Schwann cells—the nodes of Ranvier—the axon is exposed. Action potentials travel along the axon by "jumping" from node to node, as described in Key Concept 44.2. **(B)** A myelinated axon, seen in cross section through an electron microscope.

neurotransmitter to its receptor alters the activity of the postsynaptic neuron. Some neurotransmitter–receptor combinations inhibit activity of the postsynaptic neuron, and other neurotransmitter–receptor combinations excite it. Neurons integrate information by summing excitatory and inhibitory inputs.

Glia are the "silent partners" of neurons

The human brain has at least as many glial cells as neurons. The ratio differs in different brain areas, and in some areas is considerably greater than 1 to 1. A neurobiologist once said that "flashy neurons get all of the attention, but glial cells do most of the brain's work and are the cause of many of its diseases." It has been easier to study the functions of neurons because their APs can be observed. The mostly silent glia are more difficult to study, and therefore we know much less about them. Our knowledge of glia will grow enormously in the years to come.

Like neurons, glia come in several forms and have diverse functions. As mentioned above, microglia are macrophages (see Figure 41.3) and provide immune defense responses for the nervous system. Macroglia include Schwann cells, oligodendrocytes, and astrocytes. In the brain and spinal cord, **oligodendrocytes** wrap around the axons of neurons, covering them with concentric layers of cell membrane. You can think of this wrapping as being insulation, like the insulation on electrical wires. It prevents electric current from leaking out of the axon. Outside the brain and spinal cord, **Schwann cells** provide this same function for the peripheral nerves that communicate between the brain and spinal cord and all parts of the body (**Figure 44.3**). **Myelin** is the wrapping produced by oligodendrocytes and Schwann cells, and it gives many parts of the nervous system a glistening white appearance. Not all axons are myelinated, but those that are conduct APs more rapidly than can axons that are not myelinated, for reasons described in Key Concept 44.2.

Diseases that affect myelin can be devastating because they impair conduction of APs. The most common of these demyelinating diseases is multiple sclerosis (MS)—literally "multiple scars"—which occurs in about 1 in 700 people in the United States. The cause of MS is not known, but it involves inflammatory autoimmune damage to the myelin in the brain and spinal cord. The symptoms of the disease depend on where in the nervous system the myelin is damaged. Motor impairment is common. An example of a demyelinating disease that attacks myelin outside the brain and spinal cord is Guillain–Barre syndrome, which is usually the result of a severe infection. Environmental factors such as pesticide exposure can also damage myelin. There are no known cures for demyelinating diseases.

The third type of macroglia, **astrocytes** (so named because they look like stars), contribute to the **blood–brain barrier** that protects the brain from toxic chemicals in the blood. Blood vessels throughout the body are very permeable to many chemicals, including toxic ones, that would reach the brain if this barrier did not exist. Astrocytes help form the blood–brain barrier by surrounding the smallest, most permeable blood vessels in the brain. The barrier is not perfect, however. Because it consists of cell membranes, it is permeable to fat-soluble substances such as anesthetics and alcohol (which explains why these substances have such rapid and marked effects on the nervous system).

In addition to their role in the blood–brain barrier, astrocytes have several known functions at the synapse:

- They can take up neurotransmitter that has been released into the synapse and thereby control communication between the pre- and postsynaptic cells.

- They can supply neurons with nutrients. Neurons have no energy reserves, but astrocytes store glycogen that they can break down to supply the neurons with fuel.

- They have signaling properties. Even though most astrocytes do not generate APs , they do release neurotransmitters that can alter the activities of neurons.

- They aid in the repair and regeneration of neurons.

- They make contact with both blood vessels and neurons and can therefore signal changes in the composition of the blood.

Astrocytes play crucial yet poorly understood roles in modulating synapse activity. The projections of a single astrocyte may make contact with more than 100,000 synapses. The contact of the astrocyte with the neuronal components of the synapse is so intimate that it has inspired the concept of the **tripartite synapse**—the idea that a synapse includes not only the pre- and postsynaptic neurons but also connections from astrocytes.

Although astrocytes do not generate APs , they do communicate with each other. They are connected to each other through electrical synapses, and the electric signals traveling across these synapses cause changes in the Ca^{2+} content of the postsynaptic astrocyte. When Ca^{2+}-sensitive dyes are applied to neural tissue, Ca^{2+} waves can be seen traveling through extensive networks of astrocytes. The functions of these Ca^{2+} waves are not understood.

44.1 recap

Nervous systems have two unique types of cells: neurons and glia. There are many types of neurons, but they can all generate and conduct electric signals, and most generate APs. A neuron has four regions: cell body, dendrites, axon, and axon terminals. Neurons communicate with target cells at synapses, which can be chemical or electrical. Although glia do not generate APs, they have a variety of functions, including contributing to the blood–brain barrier, insulating axons, and modulating synaptic activity.

learning outcomes

You should be able to:

- Describe the parts of a neuron and explain their functions.
- Explain how the functions of axons and dendrites make it possible to communicate information across the synapse.
- Describe the different functions of glial cells.

1. Draw two neurons, including a synapse between them, and label all of the parts. At a chemical synapse, how does a signal from one neuron cross the synaptic space to another neuron?

2. Explain how a neuron integrates information received from several other neurons.

3. Explain how astrocytes modulate synaptic activity.

The one feature common to neurons is that they process information in the form of changes in the electric potential across their cell membranes. In the next section we will focus on how electric signals are generated and transmitted by neurons.

key concept 44.2 Neurons Generate and Transmit Electric Signals

Various ion pumps create ion concentration gradients across animal cell membranes. For example, Na^+ concentrations are higher outside cells and K^+ concentrations are higher inside. The positive charges of these ions are balanced by negatively charged ions both inside and outside the cell. But, across the cell membrane there is an electric charge difference, with the inside of the cell being negative relative to the outside. This electrical charge is due to the cell membrane being differentially permeable to certain ions. For neurons at rest, their membranes are mostly permeable to K^+, and that permeability is primarily responsible for the electrical charge across their membranes.

focus your learning

- The membrane at rest is close to the equilibrium potential for K^+ due to the selective permeability to K^+ through "leak" channels open at rest.
- When gated ion channels are activated by electrical, chemical, or mechanical stimuli, permeability of the membrane to the respective ions changes, resulting in a change in membrane potential.
- Potential changes, through both inhibitory and excitatory inputs, are summed and result in the membrane potential deviating from the resting potential.
- Rapid activation and inactivation of voltage-gated Na^+ channels are responsible for action potentials.

(continued)

- Action potentials are all-or-none, self-regenerating events that are conducted rapidly and unidirectionally along axons. Myelination of axons results in saltatory conduction, which is faster than conduction in nonmyelinated axons.

Because the concentration of K⁺ is higher inside the cell than in the extracellular fluid, K⁺ diffuses out of the cell down its concentration gradient. But when K⁺ leaks out of the cell, it leaves behind an unbalanced negative electric charge that tends to pull K⁺ back into the cell. An equilibrium is reached when the tendency for K⁺ to diffuse out is countered by the electric charge pulling K⁺ back in. The resulting charge difference across the membrane is called the **membrane potential**, with the inside of the cell negative relative to the outside. Membrane potentials exist in all cells.

Membrane potentials exist in all cells. In neurons the steady-state membrane potential is called the **resting potential**. Stimuli that cause changes in the permeability of the cell membrane cause local changes in that membrane potential. These local changes can generate the large electrical signals called action potentials (APs) that are conducted along the axons and convey information about the stimulus that caused the initial small change in membrane potential. The AP is a sudden, large, transient change in membrane potential generated by sudden changes in permeability to Na⁺ ions. Before describing the mechanisms of these permeability changes and the resulting APs in detail, a review of some simple concepts of electricity may be useful.

Simple electrical concepts underlie neuronal function

Voltage (electric potential difference) is a force that causes electrically charged particles to move between two points. Voltage is to the flow of electrically charged particles as pressure is to the flow of

water. If the negative and positive poles of a battery are connected by a wire, an electric current will flow through the wire because there is a voltage difference between the two poles. This flow of electric current can be used to do work, just as a current of water can be used to do work.

In wires, electric current is carried by electrons, but in solutions and across cell membranes, electric current is carried by ions. The major ions that carry electric charges across the cell membranes of neurons are sodium (Na⁺), potassium (K⁺), calcium (Ca²⁺), and chloride (Cl⁻). Recall that ions with opposite charges attract one another and those with like charges repel one another. How do these basic principles of bioelectricity establish the resting potential of the neuronal cell membrane? And how is the flow of ions through membrane channels turned on and off to generate APs?

Activities of neurons are recorded as changes in membrane potential

We record electrical events in a cell using electrodes. **Figure 44.4** shows how this technique is applied across the cell membrane of a neuron to measure the resting potential, which is usually between –60 and –70 millivolts (mV). The minus sign indicates that the inside of the cell is electrically negative compared with the outside.

As explained above, the resting potential is largely due to a balance between the tendency of K⁺ ions to diffuse down their concentration gradient and the electrical potential that holds them back. This resting potential provides a means for a neuron to respond to stimuli. If some determinant of this balance of forces is altered, the membrane potential will change. The concentration gradient of the different ions inside and outside of the neuron may not change, but if the permeability of the membrane to ions changes, the membrane

research tools

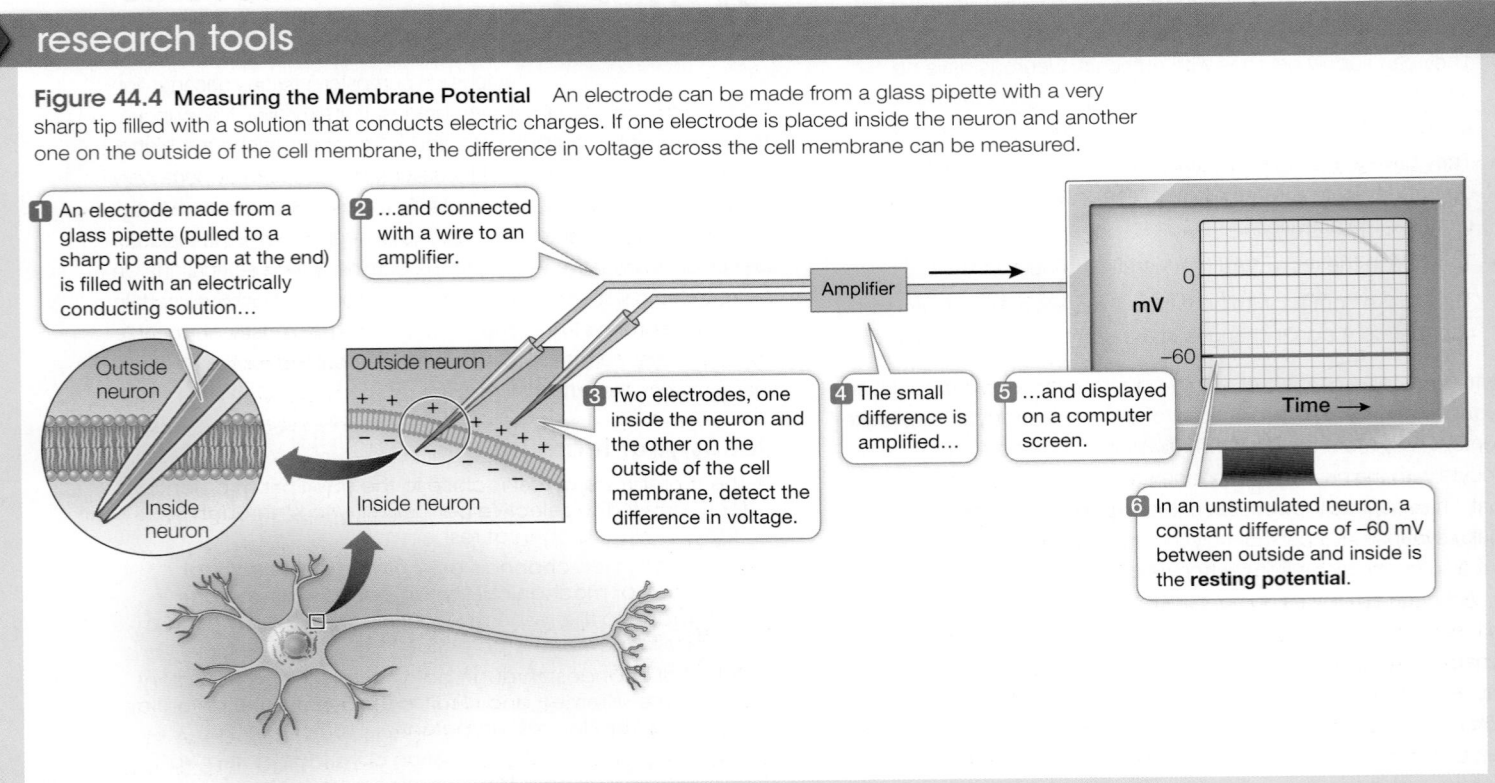

Figure 44.4 Measuring the Membrane Potential An electrode can be made from a glass pipette with a very sharp tip filled with a solution that conducts electric charges. If one electrode is placed inside the neuron and another one on the outside of the cell membrane, the difference in voltage across the cell membrane can be measured.

1 An electrode made from a glass pipette (pulled to a sharp tip and open at the end) is filled with an electrically conducting solution...

2 ...and connected with a wire to an amplifier.

Outside neuron
Inside neuron

Outside neuron
Inside neuron

3 Two electrodes, one inside the neuron and the other on the outside of the cell membrane, detect the difference in voltage.

4 The small difference is amplified...

5 ...and displayed on a computer screen.

Amplifier

mV
0
–60
Time →

6 In an unstimulated neuron, a constant difference of –60 mV between outside and inside is the **resting potential**.

potential will change. Therefore, any chemical or physical stimulus that changes the permeability of the cell membrane to ions will cause a change in the cell's membrane potential. For example, what would happen to the cell membrane potential if there was a sudden increase in the permeability of the membrane to Na⁺ ions. Na⁺ ions are more abundant outside the cell than inside. Also, the inside of the resting cell is negatively charged which attracts the positively charged N⁺ ions. Thus, a sudden large increase in permeability of the membrane to Na⁺ would cause a change of the membrane potential in the positive direction. If the change in Na⁺ permeability were transient, the membrane potential would rapidly return to its resting, negative value. Such a rapid change in membrane potential is an AP.

 Animation 44.1 The Resting Membrane Potential
www.Life11e.com/a44.1

Ion transporters and channels generate membrane potentials

The cell membranes of neurons, like those of all other cells, are lipid bilayers that are impermeable to ions but contain many protein molecules that serve as *ion transporters and ion channels. Ion transporters and channels are responsible for the distribution of charges across the membrane that determine membrane potential.

---

*connect the concepts   The passive and active mechanisms that transport materials across cell membranes are discussed in Key Concepts 6.3 and 6.4. The passive mechanisms are membrane proteins that form channels (see Figure 6.10), and the active mechanisms (which require ATP) are membrane proteins engaged in active transport (see Figure 6.14).

Ion transporters require energy to move ions against their concentration or electrical gradients and are therefore called ion pumps. A major ion transporter in the cell membranes of neurons (and all other cells) is the **sodium–potassium pump**, so called because it actively expels Na⁺ ions from inside the cell, exchanging them for K⁺ ions from outside the cell (see Figure 6.14). The Na⁺–K⁺ pump is also known as sodium–potassium ATPase, a term emphasizing that it is an enzyme complex requiring ATP to do its work. The Na⁺–K⁺ pump keeps the concentration of K⁺ inside the cell greater than the K⁺ concentration of the extracellular fluid, and the concentration of Na⁺ inside the cell less than that of the extracellular fluid. The concentration differences established by this active transporter mean that K⁺ would diffuse out of the cell and Na⁺ would diffuse in if the ions could cross the lipid bilayer. How do these concentration gradients relate to the electrical gradients we discussed above?

Ion channels permit the diffusion of ions across membranes. These channels are water-filled pores formed by proteins that span the lipid bilayer and are generally selective, allowing some types of ions to pass through more easily than others (see Figure 6.11). Thus there are potassium channels, sodium channels, chloride channels, and calcium channels, and there are different kinds of channels for each ion. Ions can diffuse through these channels in either direction. The direction and magnitude of the net movement of ions through a channel depend on the concentration gradient of that ion type across the cell membrane, as well as on the voltage difference

across that membrane. These two motive forces acting on an ion are termed its **electrochemical gradient**. Although the electrochemical gradient drives the movement of ions through channels, that movement is modified by gates that open and close the channels.

Potassium channels are the most common open, or leak, channels in the cell membranes of resting (nonstimulated) neurons. As a consequence, resting neurons are more permeable to K⁺ than to any other ion. Thus, open potassium channels are largely responsible for the resting membrane potential. Because the potassium channels make the cell membrane permeable to K⁺, and because the Na⁺–K⁺ pump keeps the concentration of K⁺ inside the cell much higher than that outside the cell, K⁺ tends to diffuse down its electrochemical gradient, out of the cell, through the channels. But, if K⁺ ions left the cell they would leave behind unbalanced negative charges, generating an electric potential across the membrane that tends to pull K⁺ back into the cell.

The membrane potential at which the net diffusion of K⁺ out of the cell ceases (that is, the point at which K⁺ diffusion out due to the concentration gradient is balanced by its inward movement due to the negative electric potential) is the **potassium equilibrium potential**, or E_K. The value of E_K can be calculated from the concentrations of K⁺ on the two sides of the membrane using the **Nernst equation** (**Figure 44.5A**). This equation, developed in the late 1800s, shows that the existence of ion channels in neuronal membranes was hypothesized long before their specific structures and properties were discovered.

In the late 1940s, A. L. Hodgkin and A. F. Huxley at the University of Cambridge set out to study the electrical properties of axonal membranes. With the techniques available at that time, the necessary measurements could be made only if you had a very large axon to work with. Such an axon exists in nature—the huge neuron that controls the escape response of squid. Hodgkin and Huxley used electrodes to measure the voltage across the cell membrane of this large axon, as seen in Figure 44.5, and to pass electric current into it to change its membrane potential. They also changed the concentrations of Na⁺ and K⁺ both inside and outside the squid axon and measured the resulting changes in membrane potential. On the basis of their many careful experiments, Hodgkin and Huxley developed virtually all of our basic concepts about the electrical properties of neurons, and they shared a Nobel Prize in 1963.

The resting potential of a neuron is less negative than the E_K calculated from the Nernst equation. This means that the resting potential is not due solely to leak K⁺ channels. The neuronal membrane is slightly permeable to other ions, especially Na⁺ and Cl⁻, and movements of these ions influence the resting potential. A different equation takes into account (1) all of the ions that can cross the membrane and (2) the relative permeability of the membrane to those ions. This equation, called the **Goldman equation**, predicts the membrane potential more accurately than does the Nernst equation (**Figure 44.5B**).

Ion channels and their properties can be studied directly

Because Hodgkin and Huxley were working long before there were laboratory techniques enabling the investigation of ion channels, they could only hypothesize ion channel properties. These hypotheses

research tools

Figure 44.5A Using the Nernst Equation The Nernst equation calculates membrane potential when only one type of ion can cross a membrane that separates solutions with different concentrations of that ion.

1. Measure concentrations of ions inside and outside a neuron.

   To measure the concentration of ions in a neuron, the neuron (and its axon) must be big. Squid have giant neurons that control their escape response (see Figure 44.13C). It is possible to sample the cytoplasm of these axons, which are about 1 mm in diameter.

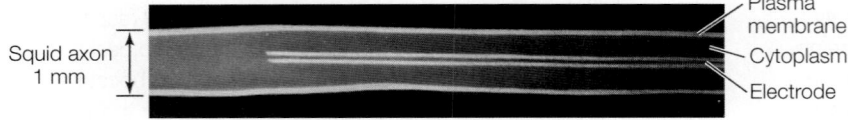

2. Use the Nernst equation to calculate what the membrane potential would be if it were permeable to each of the ions that are differently concentrated on the two sides of the membrane: Na^+, K^+, Ca^{2+}, and Cl^-.

   The Nernst equation predicts the membrane potential resulting from membrane permeability to a single type of ion that differs in concentration on the two sides of the membrane. The equation is written

   $$E_{ion} = 2.3 \frac{RT}{zF} \log \frac{[ion]_o}{[ion]_i}$$

   where E is the equilibrium (resting) membrane potential (the voltage across the membrane in mV), R is the universal gas constant, T is the absolute temperature, z is the charge on the ion, and F is the Faraday constant. The subscripts o and i indicate the ion concentrations outside and inside the cell, respectively.

   At this point you could just "plug and play," but do you understand this equation?

   A concentration difference of ions across a membrane creates a chemical force that pushes the ions across the membrane by diffusion; however, the resulting unbalanced electric charges will pull the ions back the other way. At equilibrium, the two forces are in balance and there is no net movement of ions across the membrane.

   The chemical energy pushing the ions will equal $2.3\ RT \log [ion]o/[ion]i$

   The electric energy pulling the ions will equal zEF. So, at equilibrium:

   $$zEF = 2.3\ RT \log \frac{[ion]_o}{[ion]_i}$$

   Rearranging the equation to solve for E, we get the Nernst equation:

   $$E_{ion} = 2.3 \frac{RT}{zF} \log \frac{[ion]_o}{[ion]_i}$$

   We can simplify the equation by picking a temperature—let's use "room temperature," or 20°C—and solving for $2.3\ RT/F$. At 20°C, $2.3\ RT/F$ equals 58. Thus:

   $$E_{ion} = 58/z \log \frac{[ion]_o}{[ion]_i}$$

   Measuring ion concentrations in squid giant axon cytoplasm and in seawater, then solving the Nernst equation for each ion, we find:

Ion	Ion concentration (mM)		Predicted membrane potential (mV)
	in squid axon	in seawater	
K^+	400	20	−75
Na^+	50	460	+56
Ca^{2+}	0.5	10	+38
Cl^-	50	560	−60

3. Since the measured membrane potential is −66 mV, it is clear that the resting potential of the axon is due to permeability of the membrane to more than just one type of ion.

could not be tested until the late 1970s, when B. Sakmann and E. Neher developed a technique called **patch clamping**, for which they shared the Nobel Prize in 1991. Patch clamping, described in **Figure 44.6**, is widely used by neurobiologists, enabling them to record in real time the tiny electric currents caused by the openings and closings of single ion channels.

Gated ion channels alter membrane potential

The ion channels called leak channels are always open, but other ion channels in the cell membranes of neurons behave as if they contain "gates"; they are open under some conditions and closed under other conditions. **Voltage-gated channels** open or close in response to a change in the voltage across the cell membrane. **Chemically gated channels** open or close depending on the presence or absence of a specific molecule that binds to the channel protein, or to a separate receptor that in turn alters the channel protein. **Mechanically gated channels** open or close in response to mechanical force applied to the cell membrane. Gated channels play important roles in neuronal function.

Openings and closings of gated channels alter the membrane potential. Above, you considered what would happen if the permeability of the membrane to Na^+ suddenly increased. Such a change in local membrane permeability is generated by the opening of Na^+ channels. Na^+ ions would move down their electrochemical gradient until the membrane approaches the equilibrium potential for Na^+ (E_{Na}). Therefore, the inside of the cell becomes less negative. When the inside of a neuron becomes less negative (or more positive) in comparison to its resting condition, its cell membrane is **depolarized (Figure 44.7)**. The opposite change in the membrane potential occurs if gated K^+ channels open. Opening of these channels increases the permeability of the cell membrane to K^+, and the membrane potential becomes even more negative than when only the leak K^+ channels were open, and it is said to be **hyperpolarized**.

The openings and closings of ion channels that result in changes in the voltage across the cell membrane are the basic mechanisms by which neurons respond to stimuli, be they electrical, chemical, or mechanical. How do such local changes in membrane potential get communicated to other parts of the cell?

A local change in membrane potential causes a flow of ions that spreads the change in membrane potential to adjacent regions of the membrane. For example, when Na^+ channels in an axon open making that region of the axon positive, a rapid

work with the data

Figure 44.5B Equilibrium Membrane Potential: The Goldman Equation

Original Papers: Goldman, D. E. 1943. Potential, impedence and rectification in membranes. *Journal of General Physiology* 27: 37–60.

Hodgkin, A. L. and B. Katz. 1949. The effect of sodium ions on the electrical activity of the giant axon of the squid. *Journal of Physiology* 108: 37–77.

Figure 44.5A presented the Nernst equation, by which the membrane potential for a single ion can be determined. But you also saw (at the end of Figure 44.5A) that the equilibrium membrane potential is the product of more than one ion. The Goldman equation (frequently called the Goldman, Hodgkin, Katz equation) calculates the equilibrium membrane potential by taking into account all of the ions that can diffuse across a given membrane and the relative permeabilities of the membrane to those ions. The ions involved in mammalian neurons here are K^+, Na^+, and Cl^-, and the Goldman equation is

$$V_m = \frac{RT}{F}\ln\left(\frac{p_K\left[K^+\right]_o + p_{Na}\left[Na^+\right]_o + p_{Cl}\left[Cl^-\right]_i}{p_K\left[K^+\right]_i + p_{Na}\left[Na^+\right]_i + p_{Cl}\left[Cl^-\right]_o}\right)$$

Relative permeabilities (*p*) are expressed as ratios. The membrane's permeability to potassium ions is the highest, so $p_K = 1.0$. Then $p_{Na} = 0.05$ and $p_{Cl} = 0.44$. Bracketed elements refer to the inside and outside ion concentrations, as in the Nernst equation.

QUESTION ▶

1. The table gives the intra- and extracellular ion concentrations for a mammalian neuron. Use these values and the Goldman equation to calculate the membrane potential. Refer to Figure 44.5 for a comparison with calculations based on the Nernst equation. (Hint: In redrafting the equation, you can substitute "2.3*RT/F* log" for "*RT/F* ln.")

	Ion concentration (mM)	
	Intracellular	Extracellular
K^+	140	5
Na^+	10	145
Cl^-	20	110

A similar **work with the data** exercise may be assigned in **LaunchPad**.

research tools

Figure 44.6 Patch Clamping   The patch clamp electrode is a glass micropipette with a polished tip and filled with an electrically conductive solution that has the same composition as extracellular fluids. When this pipette/electrode is positioned against the membrane of a cell and slight suction is applied, a seal forms. If a single ion channel (or a few ion channels) are within the patch of membrane bounded by the seal, the openings and closings of individual channels can be recorded by the electrode. If the pipette is retracted, it can tear the patched membrane away from the cell, and the activities of the ion channels in the patch can continue to be recorded.

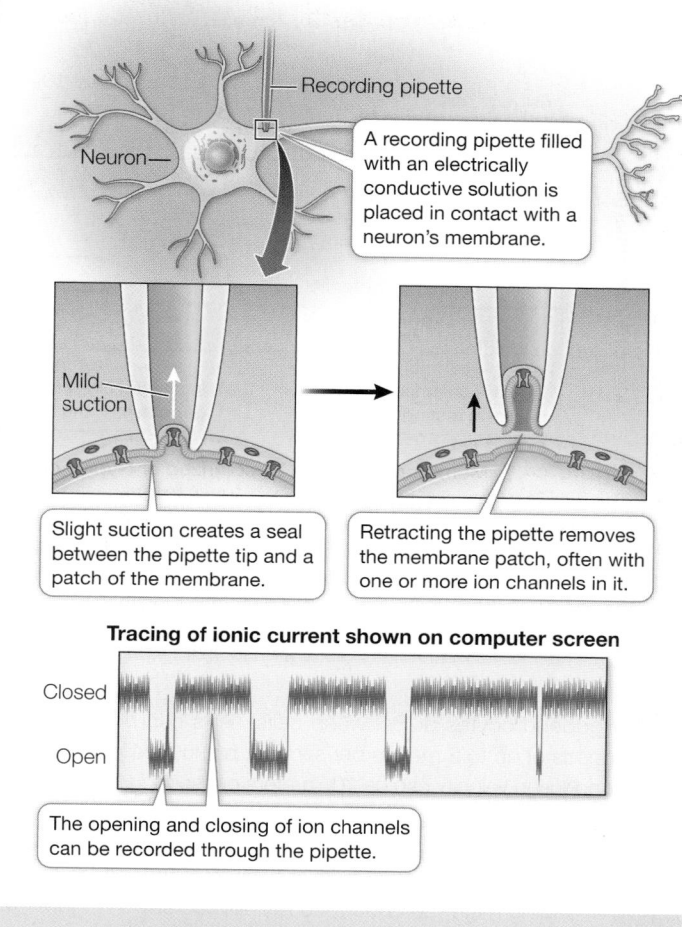

Recording pipette

A recording pipette filled with an electrically conductive solution is placed in contact with a neuron's membrane.

Neuron

Mild suction

Slight suction creates a seal between the pipette tip and a patch of the membrane.

Retracting the pipette removes the membrane patch, often with one or more ion channels in it.

Tracing of ionic current shown on computer screen

Closed

Open

The opening and closing of ion channels can be recorded through the pipette.

---

flow of ionic electric current in the axoplasm causes a decay in that electrical potential change, and it does not spread very far.

Graded changes in membrane potential can integrate information

The flow of ionic electric current along cell membranes can only extend over a short distance, but it causes a graded change in the local membrane potential. A **graded membrane potential** is a change from the resting potential that is proportional to the magnitude of a stimulus. That stimulus can be chemical or mechanical. Graded potentials are a means of integrating stimuli because the membrane can respond with proportional amounts of depolarization

or hyperpolarization to each stimulus, and those changes in membrane potential are summed. In the next chapter you will learn how graded potentials play important roles in sensory systems. However, the spread of graded potentials can only be local and cannot be transmitted down long axons. Therefore axons code information as discrete APs that travel along their membranes. Graded potentials, however, play an important role in the generation of APs.

Sudden changes in Na^+ and K^+ channels generate action potentials

APs are sudden, transient, large changes in membrane potential. In unmyelinated axons (those not wrapped in myelin by

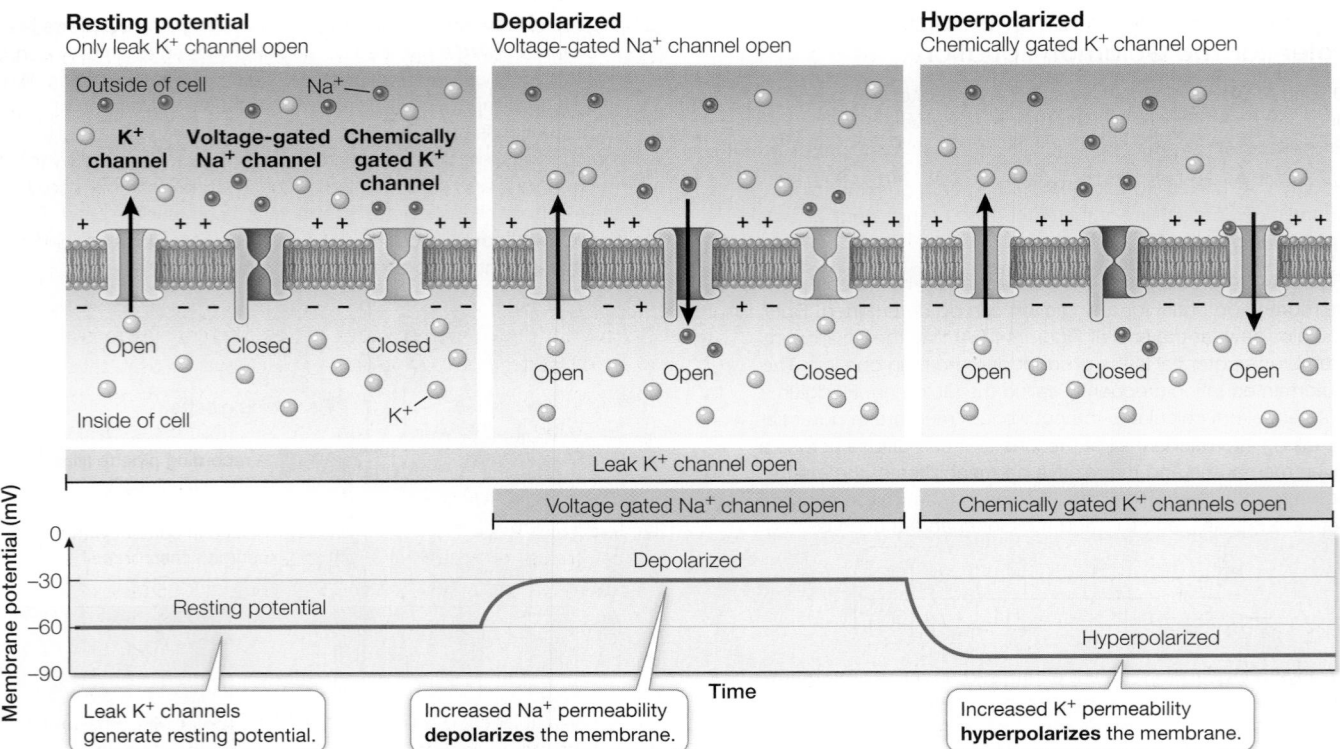

Figure 44.7 Membranes Can Be Depolarized or Hyperpolarized
The resting potential is produced mostly by leak K⁺ channels with a small contribution from leak Na⁺ channels (not shown here). A shift from the resting potential to a less negative membrane potential caused by increased Na⁺ permeability is called depolarization. Hyperpolarization occurs when the membrane potential becomes more negative as when the permeability of the membrane to K⁺ increases.

oligodendrocytes or Schwann cells), APs can be conducted at speeds of up to 2 meters per second, but in myelinated axons the conduction velocity can be 100 meters per second. Think of running the 100-meter dash—the world record is slightly under 10 seconds.

If we place the tips of a pair of electrodes on either side of the cell membrane of a resting axon and measure the voltage difference, the reading might be about –60 mV, as you saw in Figure 44.4. If these electrodes are in place when an AP travels down the axon, they register a rapid change in membrane potential, from –60 mV to about +50 mV. The membrane potential then rapidly returns to its resting level of –60 mV (**Figure 44.8**).

The AP is generated by the openings and closings of voltage-gated Na⁺ and K⁺ channels in the cell membrane of the axon. At the resting potential, most of these channels are closed except for the leak K⁺ channels (balloon 1 in Figure 44.8). A slight depolarization of the membrane can cause some voltage-gated Na⁺ channels to open. Specifically, if a neuron is stimulated sufficiently to cause the membrane of its cell body to depolarize slightly, that graded potential spreads by local current flow to the **axon hillock**, the region of the cell body at the base of the axon (see Figure 44.1). Voltage-gated Na⁺ channels are concentrated in the axon hillock. A slight depolarization of the cell membrane in this area causes some of these voltage-gated channels to open briefly—for less than a millisecond (balloon 2 in Figure 44.8). When these channels open,

the increased Na⁺ permeability depolarizes the membrane even more, causing more Na⁺ channels to open—a positive feedback effect. When the membrane is depolarized about 5–10 mV above the resting potential, a **threshold** is reached; a large number of sodium channels open (balloon 3 in Figure 44.8), and the membrane potential becomes positive—an AP. The rising phase of the AP halts abruptly in 1–2 milliseconds, and the membrane potential rapidly becomes negative once again.

What causes the axon to return to resting potential? There are two contributing factors: the voltage-gated Na⁺ channels close, and voltage-gated K⁺ channels open (balloon 4 in Figure 44.8). Voltage-gated K⁺ channels open in response to the same depolarization that triggered the Na⁺ channels, but they do so more slowly and they stay open longer, allowing K⁺ to carry excess positive charges out of the axon. As a result, the membrane potential returns to a negative value and usually becomes even more negative than the resting potential until the voltage-gated K⁺ channels close (balloon 5 in Figure 44.8).

Another feature of the voltage-gated Na⁺ channels is that once they open and close, they have a **refractory period** of 1–2 milliseconds during which they cannot open again. This property can be explained by the channels having two gates, an **activation gate** and an **inactivation gate** (see Figure 44.8). Under resting conditions, the activation gate is closed and the inactivation gate is open. Depolarization of the membrane to the threshold level causes both gates to change state, but the activation gate responds faster. As a result, the channel is open for a brief time between the opening of the activation gate and the closing of the inactivation gate. Inactivation gates remain closed for 1–2 milliseconds before they spontaneously open again, thus explaining the refractory period. By the time the inactivation gate reopens, the activation gate is closed, and the membrane is poised to generate another AP. Another contribution to the refractory period is the duration of the opening of the voltage-gated K⁺ channels, as you saw above. The dip in the membrane potential following an AP is called the **after-hyperpolarization** or **undershoot**.

The difference in the concentration of Na⁺ across the cell membrane and the negative resting potential constitute the "battery" that drives APs. How rapidly does the battery run down? It might seem that a substantial number of ions would have to cross the membrane for the membrane potential to change from –60 mV to

+50 mV and back to –60 mV again. In fact, only a vanishingly small number of Na⁺ ions move through the channels during the passage of an active potential. Thus, the effect of a single action potential on the concentration gradients of Na⁺ and K⁺ is very small, and the sodium–potassium pump can keep the "battery" charged, even when the neuron is generating many APs every second.

Action potentials are conducted along axons without loss of signal

An AP is conducted over long distances with no loss of signal. If we place two pairs of electrodes at two different locations along an axon, we can record an AP at those two locations as it travels along the axon (**Figure 44.9A**). The magnitude of the AP does not change between the two recording sites. This constancy is possible because an AP is an *all-or-none, self-regenerating event*.

- An AP is *all-or-none* because of the interaction between the voltage-gated Na⁺ channels and the membrane potential. If the membrane is depolarized slightly, some voltage-gated Na⁺ channels open. Some sodium ions cross the

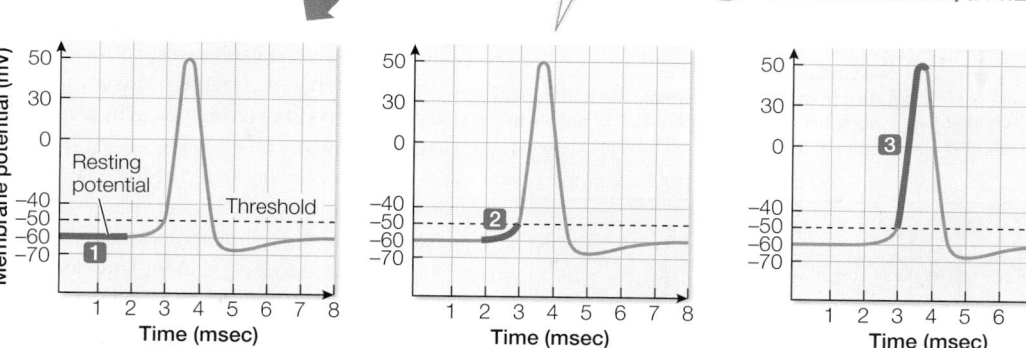

Tracing of membrane potential change through time shows the action potential.

Figure 44.8 The Course of an Action Potential Action potentials result from rapid changes in voltage-gated Na⁺ and K⁺ channels.

Animation 44.2 **The Action Potential**
www.Life11e.com/a44.2

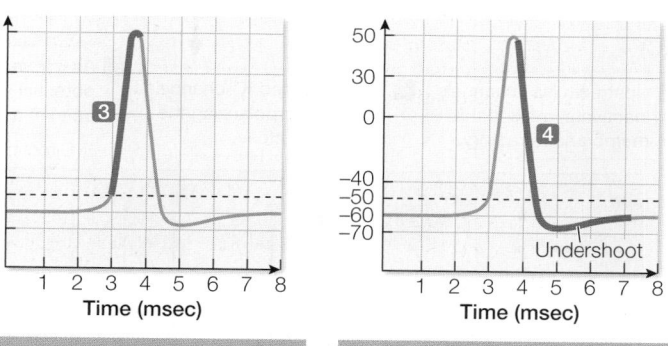

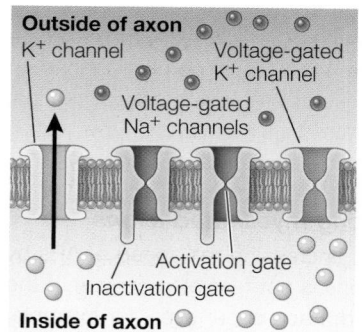

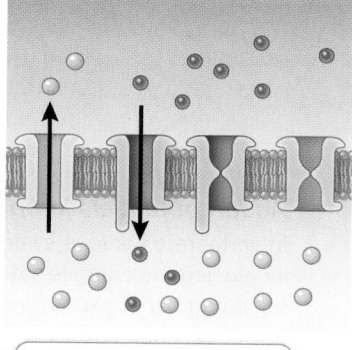

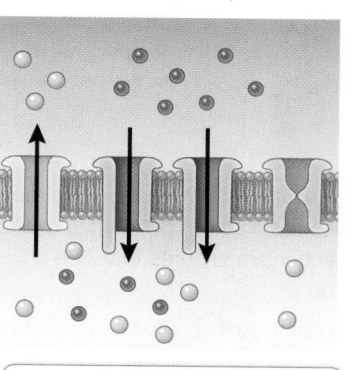

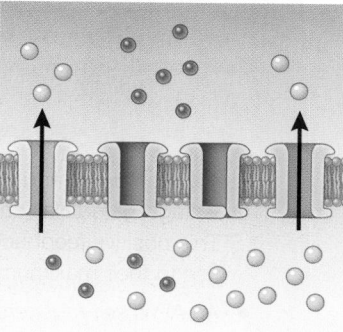

1 Leaky K⁺ channels create the resting potential. Gated channels are closed.

2 Some voltage-gated Na⁺ channels open, depolarizing the cell to threshold.

3 Additional voltage-gated Na⁺ channel activation gates open, causing a rapid spike of depolarization—an action potential.

4 Na⁺ channel inactivation gates close; gated K⁺ channels open, repolarizing and even hyperpolarizing the cell.

5 All gated channels close. The cell returns to its resting potential. Na⁺ inactivation gates reopen.

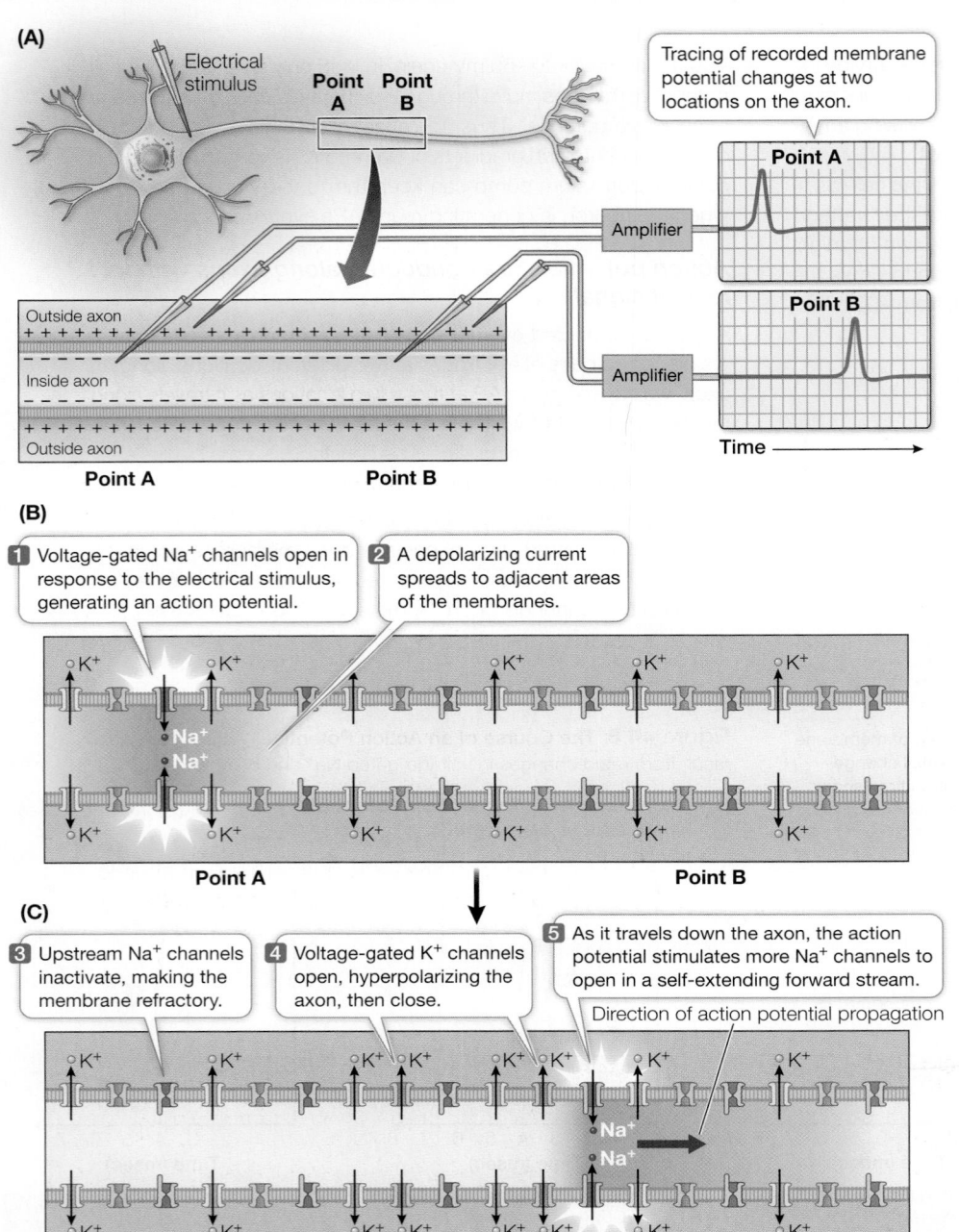

(A)

Electrical stimulus

Point A   Point B

Tracing of recorded membrane potential changes at two locations on the axon.

Point A

Amplifier

Point B

Amplifier

Time

Outside axon
+ + + + + +
Inside axon
- - - - - -
Outside axon
+ + + + + + + +

Point A   Point B

(B)

1 Voltage-gated Na⁺ channels open in response to the electrical stimulus, generating an action potential.

2 A depolarizing current spreads to adjacent areas of the membranes.

Point A   Point B

(C)

3 Upstream Na⁺ channels inactivate, making the membrane refractory.

4 Voltage-gated K⁺ channels open, hyperpolarizing the axon, then close.

5 As it travels down the axon, the action potential stimulates more Na⁺ channels to open in a self-extending forward stream.

Direction of action potential propagation

Point A   Point B

Figure 44.9 Action Potentials Travel along Axons (A) There is no loss of signal as an action potential travels along an axon. (B) When an action potential is stimulated in one region of membrane, ionic current flows to and depolarizes adjacent areas of membrane. (C) The advancing wave of depolarization causes more Na⁺ channels to open, and the action potential is generated anew in the next section of membrane. Meanwhile, in the region where the action potential has just fired, the Na⁺ channels are inactivated and the voltage-gated K⁺ channels are still open, rendering this section of the axon incapable of generating an action potential. Hence the action potential cannot "back up," but moves continuously forward, regenerating itself as it goes.

We can use an electrode to stimulate an axon, causing it to depolarize and to fire an AP that is then conducted along the axon. **Figure 44.9B** shows the changes in the ion channels in the membrane that are responsible for conducting the AP along the axon without a reduction in amplitude. Normally an AP is propagated in only one direction—away from the cell body. It cannot reverse itself because the voltage-gated Na⁺ channels in the region of the membrane it came from are in their refractory period (**Figure 44.9C**).

APs are not conducted at the same speed in all axons. They travel faster in large-diameter axons than in small-diameter axons because the resistance to ionic current flow decreases as an axon's diameter gets bigger. They travel faster in myelinated than in nonmyelinated axons because they can move down the axon in short "jumps" as described in the next section (**Figure 44.10**). Invertebrates depend on increased axon diameter for fast conduction, but vertebrates mostly depend on myelination of axons to increase conduction velocity.

cell membrane and depolarize it even more, opening more voltage-gated Na⁺ channels, and so on, generating an AP. This positive feedback mechanism ensures that APs always rise to their maximum value.

- An AP is *self-regenerating* because it spreads by local current flow to adjacent regions of the cell membrane. The resulting depolarization brings those neighboring areas of membrane to threshold. Therefore, when an AP occurs at one location on an axon, it stimulates the adjacent region of axon to generate an AP, and so on down the length of the axon. Although we say that the AP travels down the axon, what is really happening is that new APs are being created sequentially.

Action potentials jump along myelinated axons

In vertebrate nervous systems, increasing the speed of APs by increasing the diameter of axons is not feasible because of the huge number of axons involved. Each of our eyes, for example, has about a million axons connecting it to the brain. These axons conduct APs at about the same speed as does the squid giant axon—about 20 meters per second—yet the diameter of each is 200 times smaller than the squid axon's diameter. Imagine having optic tracts 200 times bigger. A different way of increasing conduction velocity of axons has evolved in vertebrates, and that adaptation is myelination.

Figure 44.10 Saltatory Action Potentials Action potentials "jump" from node to node in myelinated axons, allowing faster transmission of information.

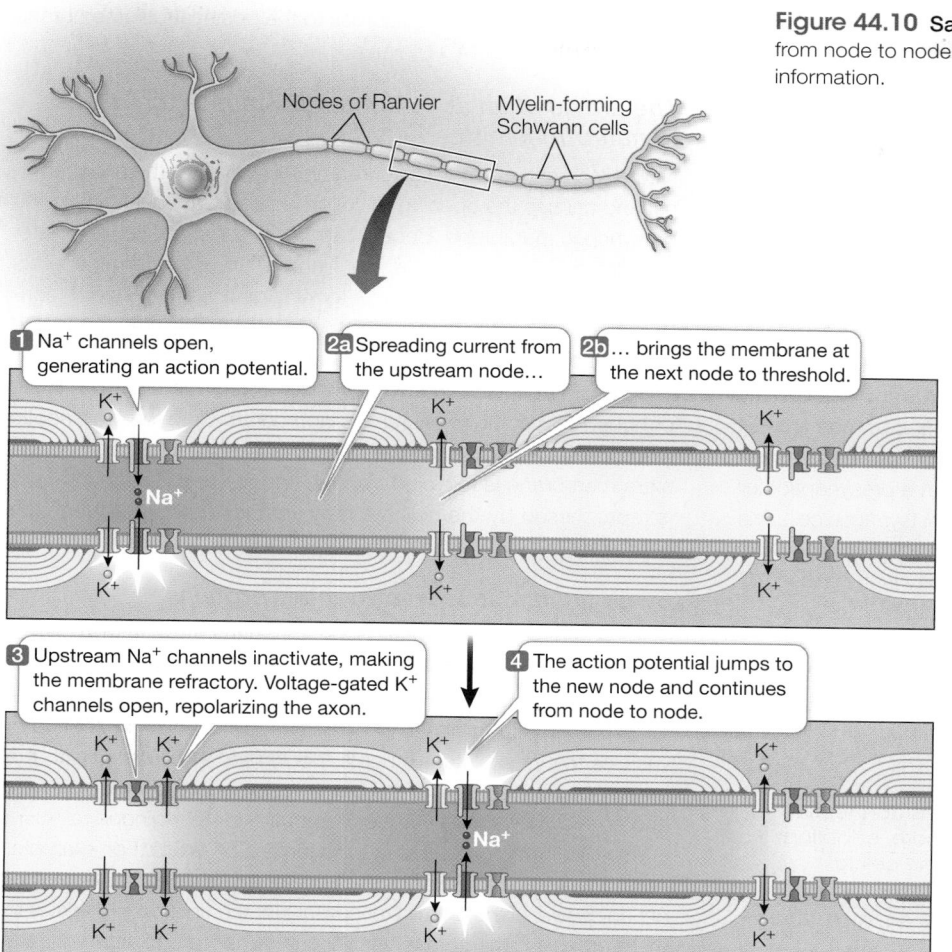

1 Na⁺ channels open, generating an action potential.

2a Spreading current from the upstream node…

2b … brings the membrane at the next node to threshold.

3 Upstream Na⁺ channels inactivate, making the membrane refractory. Voltage-gated K⁺ channels open, repolarizing the axon.

4 The action potential jumps to the new node and continues from node to node.

Nodes of Ranvier

Myelin-forming Schwann cells

When glia wrap around axons, they cover the axons with concentric layers of myelin (see Figure 44.3). However, they leave regularly spaced gaps called **nodes of Ranvier**, where the axon is not covered (see Figure 44.10). Underneath the myelin sheaths, there are no Na⁺ or K⁺ channels, therefore, APs cannot propagate under the myelin sheath. However, an AP firing at a node of Ranvier creates a local electrical field inside the axon that spreads almost instantaneously to the next node of Ranvier. The resulting depolarization of that node triggers another AP, and so on down the axon. Thus the APs appear to jump from node to node, and their conduction down the axon is very fast.

The speed of conduction is increased in these myelin-wrapped axons because electric current flows much faster through the cytoplasm than ion channels can open and close. This form of rapid impulse propagation is called **saltatory conduction** (Latin *saltare*, "to jump").

44.2 recap

Neurons have membrane potentials due to ion concentration differences across their membranes and the different permeabilities of the membrane to those ions created by ion channels. Leak channels generate the resting potential, and opening and closing of gated channels can create graded changes in membrane potentials depending on how many of which channel types open or close. Voltage-gated channels have threshold potentials for opening. If some voltage-gated Na⁺ channels open, they create a local depolarization that brings other Na⁺ channels to threshold—a positive feedback mechanism that generates APs. Sudden closure of Na⁺ channels and the opening of K⁺ channels terminate the AP. APs are rapid, all-or-none changes in membrane potential that are conducted along axons from the cell body to the axon terminals. In myelinated axons, APs jump between nodes of Ranvier and conduction is more rapid than in nonmyelinated axons.

learning outcomes

You should be able to:

- Demonstrate an understanding of how a membrane potential can be calculated based on intra- and extracellular ion concentration differences and relative permeabilities of the membrane to those ions.

- Describe how graded membrane potentials enable a neuron to integrate various inputs.

- Explain what changes in ion channels in the axonal membrane are responsible for different components of the action potential.

- Explain how action potential conduction velocity is related to myelination and axon diameter, and why conduction of action potentials is unidirectional.

1. Why does the Goldman equation produce a more accurate calculation of membrane resting potential than the Nernst equation?

2. How do graded membrane potentials enable the activity in the axon of a neuron to integrate the various dendritic inputs to that neuron?

3. Why is an action potential self-regenerating?

4. A clinical nerve conduction velocity test electrically stimulates a nerve at one location and records the muscle response at a more distal location. Why is this test used when a physician suspects a demyelinating disease?

Having described how APs are generated and transmitted along axons, we will next address the question of what happens when an AP reaches the axon terminal. How is its signal communicated to the next cell—which could be another neuron, a muscle cell, or a secretory cell?

key concept 44.3 Neurons Communicate with Other Cells

Neurons communicate with each other and with other cells at synapses. In **electrical synapses**, which are common among invertebrates, the AP spreads directly from presynaptic to postsynaptic cell. The most common type of synapse in the vertebrate nervous system is the **chemical synapse**, in which neurotransmitters released from a presynaptic cell induce changes in a membrane potential of a postsynaptic cell. This section begins with a discussion of the synapses between neurons and muscle cells. We then consider the diversity in synapses between neurons and how they integrate information.

focus your learning

- Action potentials arriving at a chemical synapse such as the neuromuscular junction trigger an influx of Ca^{2+} that activates the molecular machinery responsible for vesicular release of neurotransmitter.
- Gap junctions are comprised of connexins connecting electrically-coupled neurons through a pore, enabling ions and small molecules to cross from one cell to another.
- The action of a neurotransmitter depends on the nature of the postsynaptic receptor.
- Neurotransmitter action can be terminated by diffusion, reuptake, or degradation.
- Agonists and antagonists are drugs that target specific receptors.

The neuromuscular junction is a model chemical synapse

Neuromuscular junctions are synapses between neurons and skeletal muscle cells. They are excellent models for how chemical synaptic transmission works. Neurons that control muscles are called motor neurons. Like other neurons, a motor neuron has only one axon, but that axon can branch into numerous axon terminals that form many synapses with muscle cells. At each axon terminal an enlarged knob or buttonlike structure contains membrane-bound vesicles filled with neurotransmitter molecules. The neurotransmitter used by all vertebrate neuromuscular synapses is **acetylcholine (ACh)**. ACh is released by exocytosis when the membrane of a vesicle containing this neurotransmitter fuses with the presynaptic membrane of the axon terminal.

The postsynaptic membrane of the neuromuscular junction is a modified part of the muscle cell membrane called a **motor end plate**. It appears as a depression in the muscle cell membrane, and the terminals of the motor neuron sit in the depression. The space between the presynaptic membrane and the postsynaptic membrane is the **synaptic cleft**, which in chemical synapses is about 20–40 nanometers wide. ACh released into the cleft by the presynaptic cell diffuses across to the postsynaptic membrane (**Focus: Key Figure 44.11**).

The arrival of an action potential causes the release of neurotransmitter

Neurotransmitter is released when an AP arrives at the axon terminal and causes the opening of voltage-gated Ca^{2+} channels in the presynaptic membrane. Because the Ca^{2+} concentration is greater outside the cell than inside, Ca^{2+} enters the axon terminal. This increase in Ca^{2+} inside the axon terminal causes the vesicles containing neurotransmitter to fuse with the presynaptic membrane and empty their contents into the synaptic cleft.

In neuromuscular synapses, vesicle fusion and emptying is all-or-none. The vesicle membrane is incorporated into the presynaptic membrane, which actually gets larger as a result—at least until the extra membrane is recycled through endocytosis. The membrane is reprocessed by the cell into new vesicles that are refilled with neurotransmitter.

Synaptic functions involve many proteins

Our description of the release of neurotransmitter from the presynaptic membrane may seem simple, but it involves hundreds of proteins that are responsible for various aspects of the process: vesicle formation, transport of neurotransmitter into vesicles, anchoring of vesicles to cytoskeletal elements, docking of the vesicles with the presynaptic membrane, fusion of the vesicular and cell membranes, and endocytosis of the vesicle membrane for recycling.

Some of these proteins are the targets of toxins. For example, botulinum and tetanus toxins from bacteria of the genus *Clostridium* act on several of the proteins necessary for the docking of vesicles to the presynaptic membrane, resulting in diseases that are frequently fatal. Botulinum toxin impairs muscle contraction, whereas tetanus toxin causes uncontrolled muscle contraction. Poisons can become medicines. Botulinum toxin (in the form of Botox) is used therapeutically to subdue muscle spasms and cosmetically to reduce wrinkles.

The postsynaptic membrane responds to neurotransmitter

When acetylcholine is released at a synapse, some of it diffuses across the synaptic cleft and binds to ACh receptors on the postsynaptic membrane. The postsynaptic membrane of the motor end plate is highly folded. ACh receptors are on the crests of the folds, and voltage-gated cation channels are at the bottoms of the folds and in the surrounding muscle cell membrane (see Figure 44.11). The ACh receptors are gated channels that allow both Na^+ and K^+ to flow through, but since the electrochemical gradients favor a net influx of Na^+, the response of the motor end plate to ACh is to depolarize. That graded potential reflects the number of receptors activated. If sufficient numbers of receptors are activated, the membrane depolarization spreads to the depths of the folds of the motor end plate membrane and to the surrounding muscle cell membrane, which contain voltage-gated Na^+ channels.

The spreading depolarization of the motor end plate activates the voltage-gated Na^+ channels and causes the firing of an AP. This AP is then conducted throughout the muscle cell's system of membranes, causing the cell to contract. We will discuss the contraction of muscle cells in greater detail in Key Concept 47.10.

focus: key figure

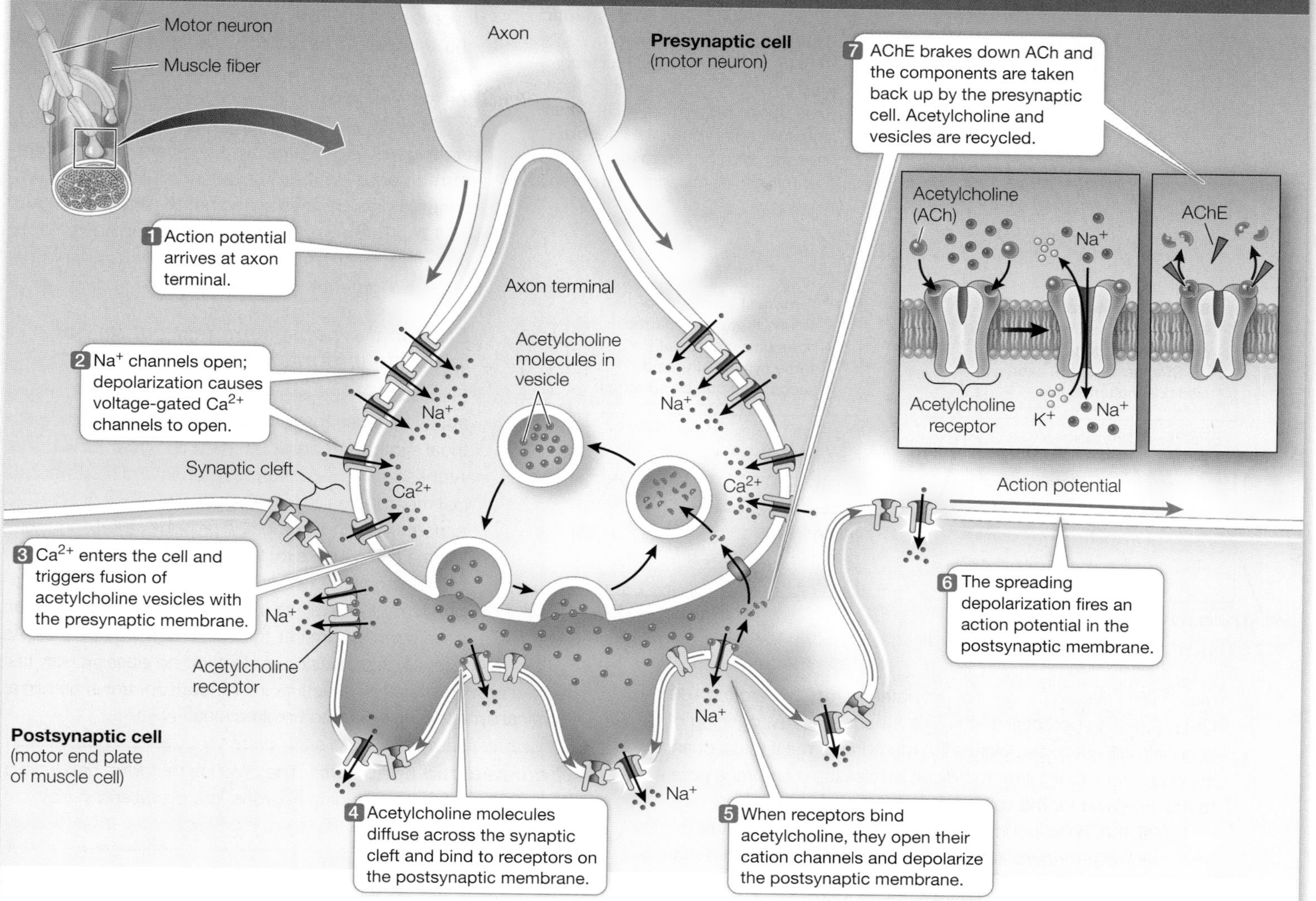

Motor neuron
Muscle fiber

Axon

Presynaptic cell
(motor neuron)

7 AChE brakes down ACh and the components are taken back up by the presynaptic cell. Acetylcholine and vesicles are recycled.

Acetylcholine
(ACh)

Na^+

AChE

1 Action potential arrives at axon terminal.

Axon terminal

Acetylcholine molecules in vesicle

Acetylcholine
receptor

K^+ Na^+

2 Na^+ channels open; depolarization causes voltage-gated Ca^{2+} channels to open.

Na^+

Na^+

Action potential

Synaptic cleft

Ca^{2+}

Ca^{2+}

3 Ca^{2+} enters the cell and triggers fusion of acetylcholine vesicles with the presynaptic membrane.

Na^+

6 The spreading depolarization fires an action potential in the postsynaptic membrane.

Acetylcholine
receptor

Na^+

Postsynaptic cell
(motor end plate of muscle cell)

4 Acetylcholine molecules diffuse across the synaptic cleft and bind to receptors on the postsynaptic membrane.

Na^+

5 When receptors bind acetylcholine, they open their cation channels and depolarize the postsynaptic membrane.

Figure 44.11 Chemical Synaptic Transmission Begins with the Arrival of an Action Potential The neuromuscular junction is a typical chemical synapse. Events shown here are similar for other neurotransmitters at other synapses. Both chemically gated and voltage gated ACh receptors are found in the motor end plate. These are non-selective cation channels, but are more permeable to Na^+ than to K^+ (inset). When one of the chemically gated receptors binds ACh, its channel pore opens and the net increase in Na^+ permeability of the postsynaptic membrane depolarizes it. The depolarization spreads beyond the motor endplate activating voltage gated Na^+ channels in the muscle cell membrane, depolarizing it and initiating an action potential. The enzyme acetylcholinesterase (AChE) breaks down ACh in the synapse, closing the chemically gated channels. The breakdown products (acetate and choline) are taken up by the presynaptic membrane and resynthesized into more ACh.

Q: If this synapse were between neurons, and the neurotransmitter were inhibitory rather than excitatory, what chemically gated channels would you expect to be on the postsynaptic membrane?

Media Clip 44.1 **Put Some ACh Into It!**
www.Life11e.com/mc44.1

Animation 44.3 **Synaptic Transmission**
www.Life11e.com/a44.3

How much neurotransmitter is enough? Neither a single ACh molecule nor the contents of an entire vesicle (about 10,000 ACh molecules) will bring the membrane of a muscle cell to threshold. However, a single AP in an axon terminal releases the contents of about 100 vesicles—more than enough to fire an AP in the muscle cell and cause it to contract.

 Activity 44.1 Neurons and Synapses Simulation
www.Life11e.com/ac44.1

Synapses can be excitatory or inhibitory

The synapses between vertebrate motor neurons and muscle cells are always excitatory. The motor end plates always respond to ACh with a graded potential that is less negative than the resting potential (depolarization). However, synapses between neurons are frequently inhibitory; such a synapse causes hyperpolarization of the postsynaptic membrane or simply makes it less likely that membrane potential will reach threshold for the voltage gated Na^+ channels. For example, there are more Cl^- ions outside than inside the cell. However, in a

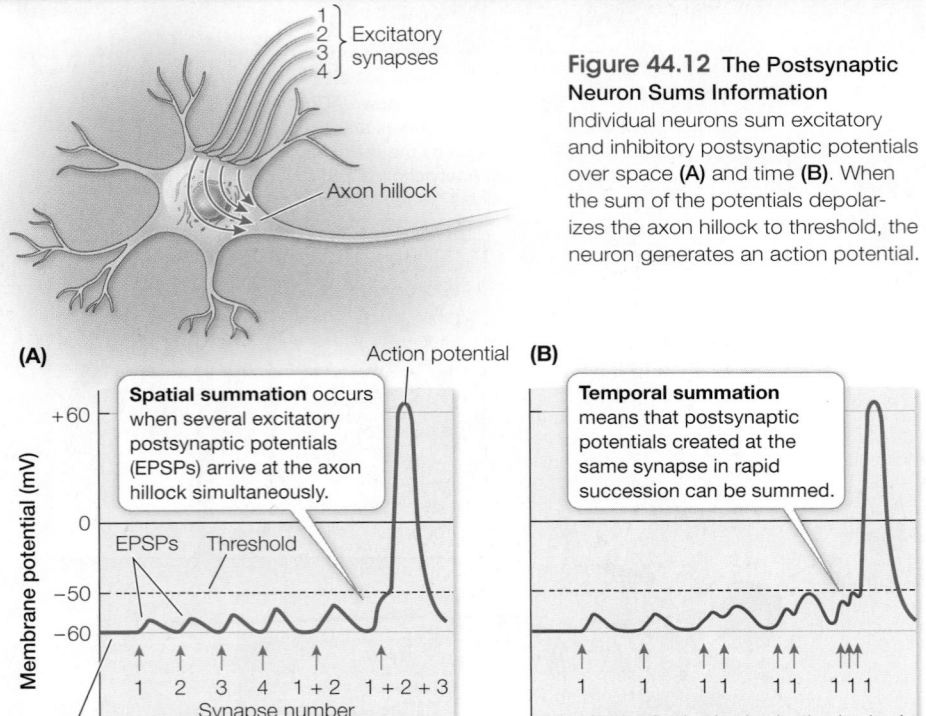

Figure 44.12 The Postsynaptic Neuron Sums Information
Individual neurons sum excitatory and inhibitory postsynaptic potentials over space (**A**) and time (**B**). When the sum of the potentials depolarizes the axon hillock to threshold, the neuron generates an action potential.

decrease in strength as they spread from the site of the synapse, a synapse at the tip of a dendrite has less influence than a synapse on the cell body, near the axon hillock.

Excitatory and inhibitory postsynaptic potentials are summed over space and over time. **Spatial summation** adds up the simultaneous influences of synapses at different sites on the postsynaptic cell (**Figure 44.12A**). **Temporal summation** adds up postsynaptic potentials generated at the same site in a rapid sequence (**Figure 44.12B**).

Electrical synapses are fast but do not integrate information well

Electrical synapses are different from chemical synapses because they couple neurons electrically. Electrical synapses contain numerous *****gap junctions**. At these synapses, the presynaptic and postsynaptic cell membranes are separated by a space of only 2–3 nanometers, and membrane proteins called connexins link the two neurons by forming pores that connect the cytoplasm of the two cells (see Figure 7.16A). Electrical current as well as ions and small molecules can pass directly from cell to cell through these pores. As a result, transmission at electrical synapses is very fast and can proceed in either direction. In comparison, transmission at chemical synapses is slow and unidirectional.

***connect the concepts**   The proteins that form gap junctions between cells, including neurons, are discussed in Key Concept 6.2.

The properties of electrical synapses make them especially well adapted for processes that have to be fast–for example, escape from danger. They are also effective means of achieving synchronous activity of populations of cells. Electrical synapses are less common in the nervous systems of vertebrates than are chemical synapses. The major reason is that they offer fewer means for integration of information. Electrical synapses are almost exclusively excitatory whereas chemical synapses are both excitatory and inhibitory. Electrical synapses are not well suited for temporal summation of inputs. Also, because of their relative simplicity of structure and function, electrical synapses offer fewer possible mechanisms for plasticity which is the foundation of learning and memory. One advantage they do have is the ability to synchronize large numbers of cells.

mammalian neuron the equilibrium potential for Cl^- is around –60mV. Given a resting potential that is also around –60 mV, opening of Cl^- channels will not hyperpolarize the membrane, but if these channels are open, it will make it more difficult to raise the membrane potential to the threshold for the voltage gated Na^+ channels.

Recall that most neurons have many dendrites. Axon terminals from many other neurons can form synapses with those dendrites and with the cell body. The axon terminals of different presynaptic neurons can store and release different neurotransmitters, and the cell membrane of the dendrites and cell body of a postsynaptic neuron can have receptors for a variety of neurotransmitters. The mix of synaptic activity impinging on a neuron will cause it to have a graded membrane potential that can be either more positive or more negative than its resting potential.

The postsynaptic neuron sums excitatory and inhibitory input

What determines when an individual neuron will fire an AP? As you just learned, the sum of excitatory and inhibitory postsynaptic potentials creates a graded membrane potential in the postsynaptic cell body. This summation ability is the major mechanism by which the nervous system integrates information. A neuron may receive 1,000 or more synaptic inputs, but it has only one output: APs in a single axon. At any one time, the information from all of the active inputs is translated into the rate at which that neuron generates APs in its axon.

For most neurons, summation takes place in the axon hillock at the base of the axon. The cell membrane of the axon hillock is not insulated by glia and has many voltage-gated Na^+ channels. Excitatory and inhibitory postsynaptic potentials from synapses anywhere on the dendrites or the cell body spread to the axon hillock by local current flow. If the resulting graded potential depolarizes the axon hillock to threshold, it fires an AP. Because postsynaptic potentials

The action of a neurotransmitter depends on the receptor to which it binds

More than 100 neurotransmitters are now recognized, and more will surely be discovered. Acetylcholine (ACh) is an important neurotransmitter because it is the substance the nervous system uses in signaling muscles to contract. ACh also plays roles in certain synapses between neurons in the brain, but it accounts for only a small percent of the total neurotransmitter content of the brain.

The workhorse neurotransmitters of the brain are three simple amino acids: glutamate, which is usually excitatory, and glycine and γ-aminobutyric acid (GABA), which are usually inhibitory. The

experiment

Original Paper: Colas, D., B. Chuluun, D. Warrier, M. Blank, D. Z. Wetmore, P. Buckmaster, C. C. Garner and H. C. Heller. 2013. Short-term treatment with the GABA antagonist pentylenetetrazole produces a sustained procognitive benefit in a mouse model of Down's syndrome. *British Journal of Pharmacology* 169: 963–973.

If the learning disability in the mouse model of Down syndrome (DS) is due to overinhibition, then reducing the activity of the major inhibitory neurotransmitter, GABA, might improve learning in these mice.

HYPOTHESIS▶ The inability of DS model mice to learn and remember experiences as well as wild-type (WT) mice is due to overactive GABA signaling in their brains. If GABA signaling is reduced, the ability of mice to perform learning tasks should improve.

METHOD

1. Keep DS and WT mice of the same age on a 12-hour:12-hour light–dark cycle. At the same time each morning, for 2 weeks, inject the mice in the experimental group with the drug pentylenetetrazole (PTZ), which blocks the action of GABA. Use three different doses. Inject the control mice with saline. The doses are measured as mg PTZ/kg body weight of mouse.

2. Following completion of the two-week drug-dosing protocol, evaluate the mice using the Novel Object Recognition (NOR) test. For training, place each mouse in a 4-square-foot arena that contains two objects such as chessmen or small bottles. Allow the mouse to explore the objects for 10 min and then return it to its home cage. An acceptable baseline is when the mouse spends the same amount of time exploring each object. Test again 1 day later, placing the mouse back in the arena but with one of the objects changed (the novel object). If the mouse spends more time exploring the novel object, that is evidence that it remembered the original object from the day before.

3. Quantify the mouse's behavior as a NOR preference (time spent exploring object A/time spent exploring objects A and B × 100%). For the testing trial, A is the novel object. A NOR preference of 50% means the mouse spent equal amounts of time with each object, and therefore did not remember the objects it explored previously. A NOR preference significantly greater than 50% means the mouse spent more time exploring the novel object and is evidence that the mouse remembered the objects it saw before.

RESULTS

Data are NOR preference pecentages ± 1 SEM (standard error of the mean). An asterisk (*) indicates $P < 0.01$ for difference between training (blue) and testing (orange) results as determined by a *t*-test: **, $P < 0.001$.

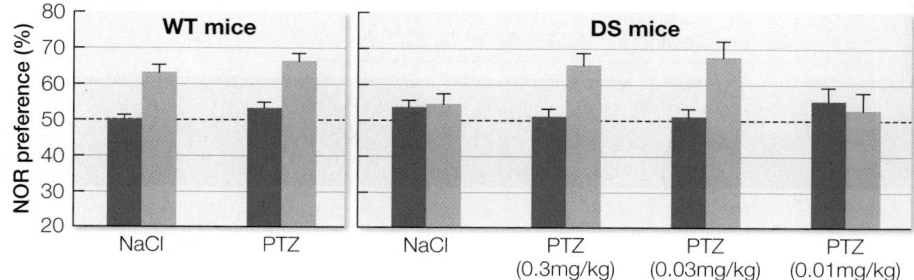

CONCLUSION▶

The saline treatments (controls) show that the WT mice can recognize the novel object 24 hr after training and that the DS mice cannot. PTZ treatment had no effect on the WT mice. PTZ treatment at 0.3 and 0.03 mg/kg improved the ability of the DS mice to remember the familiar objects and therefore recognize the novel object, whereas treatment at 0.01 mg/kg did not.

investigatinglife work with the data follows on next page.

---

integration of information at the cellular level is a balance between excitation and inhibition, so it is understandable that excessive inhibition such as you saw in the Down syndrome mice described at the start of this chapter could impair an animal's ability to process and integrate new information. Indeed, when GABA's inhibitory action in these mice was reduced by a drug that blocks the GABA receptor, the learning ability of the mice improved, as described in **Investigating Life: Can Learning Be Restored in Mouse Models of Down Syndrome?**

Another important group of neurotransmitters in the brain is the monoamines, which are derivatives of amino acids. They include dopamine and norepinephrine (derivatives of tyrosine) and serotonin (a derivative of tryptophan). Peptides also function as neurotransmitters; for example, endorphins and enkephalins are the body's opiates and modulate the sensation of pain. Another peptide, substance P, transmits pain sensations. Even a gas, nitric oxide, is used by neurons as an intercellular messenger (see Figure 7.13).

Neurotransmission is complex in part because each neurotransmitter has multiple receptor types. ACh, for example, has two receptor types: nicotinic receptors and muscarinic receptors. All of the ACh receptors in skeletal muscle are nicotinic. Both types of ACh receptors are found in the brain and spinal cord, where nicotinic

receptors tend to be excitatory and muscarinic receptors tend to be inhibitory. ACh actions can differ outside the brain and spinal cord as well. ACh acting through nicotinic receptors causes the smooth muscle of the gut to increase its motility, but ACh acting through muscarinic receptors causes cardiac muscle to hyperpolarize and therefore to slow down. There are many more examples of neurotransmitters that have different effects in different tissues, but the important thing to remember is that the action of a neurotransmitter depends on the receptor to which it binds. In addition, turning neurotransmitter action off is as important as turning it on.

 Activity 44.2 **Neurotransmitters**
www.Life11e.com/ac44.2

To turn off responses, synapses must be cleared of neurotransmitter

If released neurotransmitter molecules simply remained in the synaptic cleft, the postsynaptic membrane would become saturated and receptors would be constantly activated. The postsynaptic cell would remain hyperpolarized or depolarized and would be unresponsive additional APs arriving at the presynaptic cell. The more rapidly neurons can respond to input, the more information they can process in a

work with the data

All of the PTZ injections in the experiment described above were given to the mice early in the light phase of their day. This is the time when these nocturnal animals are normally sleeping. The researchers therefore asked whether the same results could be obtained if the treatments were delivered during the dark phase of the day, when the mice are usually active and exploring. They obtained the following data:

Group	Treatment	
	Saline	PTZ 0.3 mg/kg
WT Mice Training	55.0 ± 3.1	55.4 ± 2.4
DS Mice Training	51.9 ± 2.4	50.7 ± 2.0
WT Mice Testing	65.8 ± 5.3*	67.4 ± 4.9*
DS Mice Testing	53.7 ± 4.7	51.3 ± 6.9

QUESTIONS▶

1. How does the learning ability of the WT and DS mice compare during the dark phase of the day, and does this result differ from when the training and testing are done during the light phase?

2. Describe any difference between the effect of the PTZ treatment on the performance of the WT and DS mice on the NOR task during the light and dark phases of the day.

A similar **work with the data** *exercise may be assigned in* **LaunchPad**.

---

given amount of time. Thus neurotransmitter must be cleared from the synaptic cleft shortly after it is released by the axon terminal.

Neurotransmitter action can be terminated in several ways. First, enzymes can destroy the neurotransmitter. Acetylcholine, for example, is rapidly destroyed by the enzyme acetylcholinesterase (AChE), which is present in the synaptic cleft in close association with ACh receptors on the postsynaptic membrane (see Figure 44.12). When AChE is inhibited, ACh lingers in the synaptic cleft, causing spastic (contracted) muscle paralysis and usually resulting in death. Some of the most deadly nerve gases developed for chemical warfare work by inhibiting AChE. Some agricultural insecticides, such as malathion, also inhibit AChE and can poison farm workers if used without safety precautions.

Neurotransmitter can also simply diffuse away from the cleft or be taken back up by the presynaptic membrane to be recycled. Glial cells also take up neurotransmitters. An antidepressant drug commonly prescribed under the brand name Prozac slows the reuptake of the neurotransmitter serotonin, thus enhancing serotonin's activity at the synapse.

The diversity of receptors makes drug specificity possible

Many drugs used to treat the nervous system act by modulating specific synaptic interactions. Drugs that mimic or potentiate the effect of a neurotransmitter are called **agonists**; those that block the actions of neurotransmitters are called **antagonists**. For example, morphine is an agonist at the endorphin receptor and therefore blocks pain. Propranolol, a widely used beta blocker, is an antagonist of receptors in the fight-or-flight response and therefore decreases panic attacks and anxiety. A major emphasis in neurobiology is to identify neurotransmitter receptor subtypes and design drugs that selectively bind to them to have highly specific effects on nervous system activity.

44.3 recap

Chemical synapses involve the release of neurotransmitter molecules stored in vesicles in the presynaptic terminal. APs reaching that terminal cause the fusion of vesicles with the presynaptic membrane, releasing neurotransmitter that can then bind to receptors on the postsynaptic membrane and influence its membrane potential. At excitatory synapses, the neurotransmitter causes depolarization of the postsynaptic membrane, and at inhibitory synapses the neurotransmitter hyperpolarizes or stabilizes the postsynaptic membrane potential. There is a great diversity of neurotransmitters and their receptors. At electrical synapses the AP moves directly between pre- and postsynaptic cells.

learning outcomes

You should be able to:

- Describe the events at a neuromuscular synapse, beginning with the action potential in the presynaptic neuron and ending with the response of the postsynaptic cell.
- Compare the characteristics of electrical versus chemical synapses.
- Discuss how the properties of transmission differ for electrical and chemical synapses, and why chemical synapses present a greater variety of postsynaptic responses than do electrical synapses.
- Discuss how the properties of chemical synapses present a variety of postsynaptic responses.
- Describe the reasons why synaptic transmission is typically transitory.

1. What are the two types of cation channels in the motor end plate, and what are their functions?

2. How can the same neurotransmitter be excitatory to some postsynaptic neurons and inhibitory to others?

3. Sarin is a colorless, odorless liquid that is highly lethal. It irreversibly inhibits the enzyme acetylcholinesterase. Explain why victims of sarin poisoning generally die of asphyxiation.

4. Explain why parts of nervous systems that integrate complex information use chemical rather than electrical synapses.

Now that you understand how neurons communicate with each other, we will explore how they are organized into circuits and systems that process sensory information and generate responses to that information.

key concept 44.4 Neurons and Glia Form Information-Processing Circuits

Nervous systems can process information because their neurons are organized into **neural networks**. These networks include three functional categories of neurons that are responsible for input, output, and integration:

1. **Afferent neurons** carry sensory information into the nervous system. That information comes from specialized sensory cells that transduce (convert) various kinds of sensory stimuli (e.g., light, heat, pressure) into APs.

2. **Efferent neurons** carry commands to physiological and behavioral effectors such as muscles and glands.

3. **Interneurons** integrate and store information and communicate between afferent and efferent neurons.

focus your learning

- From invertebrates to vertebrates, nervous systems have evolved with increasing complexity from simple nerve nets to clusters of neurons (ganglia) to the development of anterior brain structures.
- The knee-jerk reflex involves a simple circuit with input from sensory neurons stimulating interneurons and motor neurons in the spinal cord that trigger the quadriceps to contract.

Nervous systems range in complexity

Simple animals such as cnidarians (sea anemones, for example) process information with a limited number of simple neural networks that do little more than provide direct lines of communication from sensory cells to effectors; there is little or no integration or processing of signals (**Figure 44.13A**). The cnidarian's **nerve net** is most developed around the tentacles and the oral opening, where it facilitates detection of food or danger and causes tentacles to extend or retract.

Animals that are more complex and move about in search of food and mates must process and integrate larger amounts of information. Even earthworms fit this description, and their increased need for information processing is met by higher numbers of neurons organized into clusters of neuronal cell bodies called **ganglia** (singular *ganglion*). Ganglia serving different functions may be distributed around the body, as in earthworms (**Figure 44.13B**). In animals that are bilaterally symmetrical, ganglia frequently come in pairs, one on each side of the body (**Figure 44.13C**). Also, as animals increase in complexity, some ganglia may become enlarged or fused together at the anterior end, forming a **brain**. Small nervous systems of invertebrates can be remarkably complex. Consider the nervous systems of spiders that have programmed within them the thousands of precise movements necessary to construct an intricate web without prior experience or opportunities to learn the specific web architecture of their species.

In vertebrates, most cells of the nervous system are found in the central nervous system (CNS). The CNS includes the brain and the spinal cord, which are the sites of most information processing, storage, and retrieval (**Figure 44.13D**). Information is transmitted from sensory cells to the CNS and from the CNS to cells that respond (effectors) via neurons that extend or reside outside the brain and the spinal cord. These sensory and effector neurons and their supporting cells are the peripheral nervous system.

(A) Sea anemone

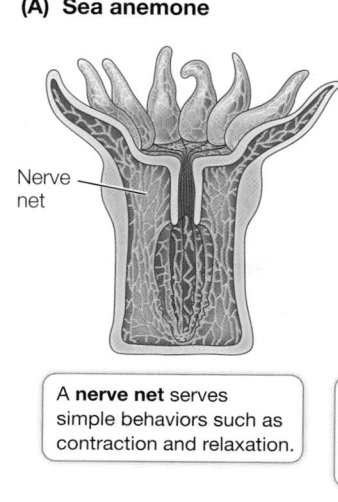

Nerve net

A **nerve net** serves simple behaviors such as contraction and relaxation.

(B) Earthworm

Segmental nerve

Ganglion in ventral nerve cord

Anterior ganglia

In the earthworm, **ganglia** in each segment coordinate movement, and anterior ganglia control more complex behavior.

(C) Squid

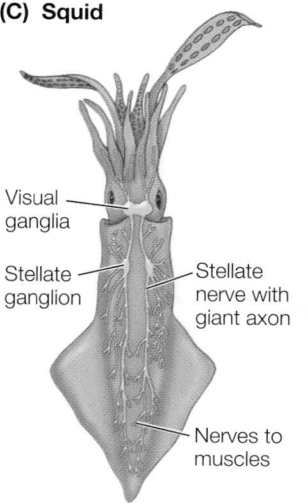

Visual ganglia

Stellate ganglion

Stellate nerve with giant axon

Nerves to muscles

In squid, more complex behaviors are served by collections of neurons in specialized ganglia that process and integrate information.

(D) Human

The human brain and spinal cord are the central nervous system…

…which communicates to the cells and organs of the body via the peripheral nervous system.

Figure 44.13 Nervous Systems Vary in Size and Complexity In animal species with increasingly complex sensory and behavioral abilities, information processing is increasingly centralized in ganglia (collections of neurons) or in a brain.

Reflexes are controlled by simple circuits involving sensory neurons, interneurons, and effectors

A reflex is a behavioral or physiological response that does not require conscious information processing—it is autonomic, meaning an involuntary or unconscious reaction. Examples are salivating at the sight or smell of food, and the jerk of your lower leg when a physician taps your knee with a small rubber hammer. The knee-jerk reflex involves a very simple neural circuit involving just a few neurons that connect with each other in the spinal cord. Thus it is called a **spinal reflex**. A cross section of the human spinal cord reveals a central area of gray matter in the shape of a butterfly, surrounded by areas of white matter (see Figure 44.14). In the nervous system, **gray matter** is rich in neuronal cell bodies, and **white matter** contains myelinated axons. The gray matter of the spinal cord contains the cell bodies of the spinal neurons; the white matter contains myelinated axons that conduct information up and down the spinal cord.

Spinal nerves extend from the spinal cord at regular intervals on each side. Each spinal nerve has two roots, one connecting with the dorsal horn of the gray matter, the other with the ventral horn. The afferent (sensory) axons in a spinal nerve enter the spinal cord through the dorsal root, and the efferent (motor) axons leave through the ventral root. The neural circuit of the knee-jerk reflex is diagrammed in **Figure 44.14**. The tap of the physician's rubber hammer stretches the tendon going over the knee. That tendon attaches the muscle of the upper leg to bone in the lower leg. Stretching the tendon stretches muscle fibers in the upper leg, and stretch receptors in that muscle transduce the physical stimulus into APs. The APs are then conducted by a sensory neuron into the dorsal horn of the spinal cord. That sensory neuron goes all the way to the ventral horn and synapses onto a motor neuron, which fires APs. The axon of that motor neuron travels out through the ventral horn of the spinal cord and extends all the way to the same muscle that initially was stretched, causing that muscle to contract.

What is the function of the spinal knee-jerk reflex? This simple circuit senses increased load on a muscle and adjusts its level of contraction to match the load. Most spinal reflexes are more complex. For example, limb movement is controlled by antagonistic sets of muscles that work against each other. When one member of an antagonistic set of muscles contracts, it bends (flexes) the limb; it is therefore called a flexor. The antagonist muscle, the extensor, straightens (extends) the limb. For a limb to move, one muscle of the pair must relax while the other contracts. Thus sensory input that activates the motor neuron of one muscle also inhibits its antagonist. This coordination is achieved by an interneuron between the sensory neuron and the motor neuron of the antagonist muscle (see Figure 44.15). Thus the reciprocal inhibition of antagonistic muscles involves at least two synapses.

The withdrawal reflex is an example of a polysynaptic spinal reflex that involves many interneurons. When you step on a tack, you immediately pull back your foot, but you don't fall over because many other muscles change your posture. The tack stimulates pain receptors in the foot, and the sensory neurons transmit APs into the dorsal horn of the spinal cord on the same side of the body. In the dorsal horn, these neurons synapse with a variety of interneurons. Some send the pain information through their axons to the brain, resulting in the conscious sensation of pain. But even before the brain is aware of the pain, other interneurons coordinate the withdrawal of the foot with actions of other muscles to shift your weight onto the other leg. Thus a rather complex suite of movements is coordinated by a network of interneurons in the spinal cord. By extension, you can appreciate how much more complex the neural networks are that enable you to execute complex movements in time with music and coordinated with another person—in other words, to dance.

The vertebrate brain is the seat of behavioral complexity

Vertebrates differ greatly in their behavioral complexity and in their physiological specializations, and their neural networks reflect this diversity. **Figure 44.15** shows the brains of four vertebrate species of similar body mass drawn to the same scale.

The human nervous system contains an estimated 10^{11} neurons. A given neuron in the brain can have 1,000 or more synapses. Thus the human brain can contain 10^{14} synapses (10^{11} neurons × 10^3 synapses per neuron). Then there are the glia. A single astrocyte might participate

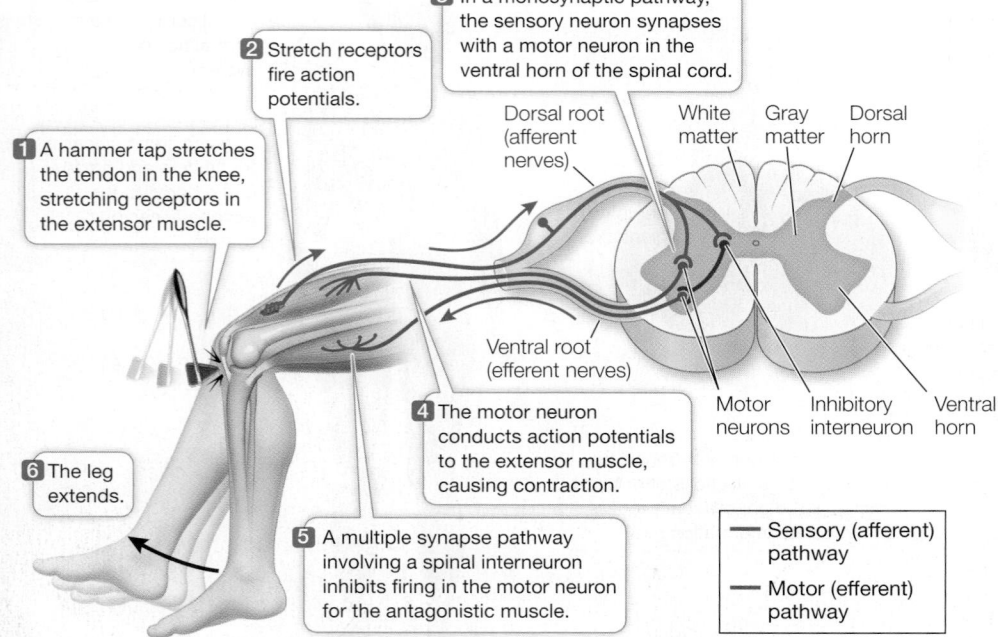

2 Stretch receptors fire action potentials.

3 In a monosynaptic pathway, the sensory neuron synapses with a motor neuron in the ventral horn of the spinal cord.

1 A hammer tap stretches the tendon in the knee, stretching receptors in the extensor muscle.

Dorsal root (afferent nerves)   White matter   Gray matter   Dorsal horn

6 The leg extends.

Ventral root (efferent nerves)

Motor neurons   Inhibitory interneuron   Ventral horn

4 The motor neuron conducts action potentials to the extensor muscle, causing contraction.

5 A multiple synapse pathway involving a spinal interneuron inhibits firing in the motor neuron for the antagonistic muscle.

— Sensory (afferent) pathway
— Motor (efferent) pathway

Figure 44.14 A Neural Network in the Spinal Cord Generates the Knee-Jerk Reflex Sensory (afferent) information enters the spinal cord through the dorsal horns, and motor (efferent) output leaves it via the ventral horns. Information travels to the brain in white matter tracts. Interneurons make connections within the spinal cord that result in a complex, coordinated behavior pattern.

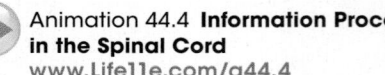

Animation 44.4 **Information Processing in the Spinal Cord**
www.Life11e.com/a44.4

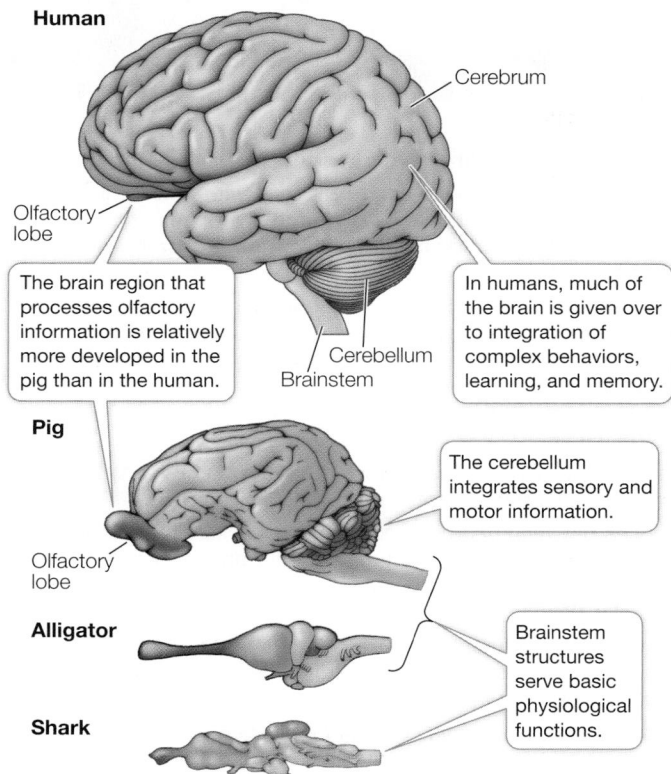

Human

Cerebrum

Olfactory lobe

The brain region that processes olfactory information is relatively more developed in the pig than in the human.

Cerebellum
Brainstem

In humans, much of the brain is given over to integration of complex behaviors, learning, and memory.

Pig

The cerebellum integrates sensory and motor information.

Olfactory lobe

Alligator

Brainstem structures serve basic physiological functions.

Shark

Figure 44.15 Brains Vary in Size and Complexity   The brains of four vertebrate species—all of which may have a similar body mass—show immense differences. Note that the brainstem, which is involved in physiological regulation and stereotypic behavior, differs less among these species than does the cerebrum, which is responsible for complex behavior and learning.

Q: What sensory capability do you think is well developed in the pig?

in 100,000 synapses while at the same time monitoring signals in the extracellular fluid and the blood. In addition, synapses are not constant but can be highly plastic. They can increase or decrease in number and size, and they can become more or less sensitive.

This astronomical number of neurons and synapses is divided into thousands of distinct but interacting networks that function in parallel. The possible number of informational networks in the brain is almost infinite, and therein lies the incredible ability of the human brain to process information, to learn, to do complex tasks, to remember, and to have emotions.

44.4 recap

Nervous systems are composed of neural networks that include afferent neurons, interneurons, and efferent neurons. Nervous systems range in complexity from simple nerve nets to the human brain. The spinal knee-jerk reflex is an example of a simple neural network, but networks that control more complex behavior are much more complicated.

learning outcomes

You should be able to:

- Distinguish the central from the peripheral nervous system and recognize how these systems range in complexity among different species.
- Analyze a simple circuit and map the communication between afferent and efferent neurons.

1. What are ganglia, and why are they concentrated in the anterior region of many invertebrates?
2. Draw or describe the flow of information from stimulus to response in the knee-jerk reflex circuit.
3. In the knee-jerk reflex circuit, how can the same stimulus cause contraction in one muscle and relaxation in another?
4. When considering fish, then reptiles, then mammals, which part of the vertebrate brain increases most in size?

investigatinglife

Can learning be restored in mouse models of Down syndrome?

Some neurotransmitter–receptor combinations excite activity in the postsynaptic neuron, depolarizing it and making it likely to fire an AP. Other neurotransmitter–receptor combinations inhibit the responses of the postsynaptic cell. Neurons integrate information by summing excitatory and inhibitory synaptic inputs. Over-inhibition in the nervous system can arise when too much inhibitory neurotransmitter (such as GABA) is released, or if postsynaptic cells have too many receptors for inhibitory neurotransmitters. Drugs that decrease the synthesis and release of the inhibitory neurotransmitter can reduce the inhibition, as can drugs that block the receptors for such neurotransmitters. In the case of the Down syndrome model mice, drugs such as pentylenetetrazole, which blocked the GABA receptors (which function also as Cl^- channels), reduced the level of inhibition in the mouse nervous system and increased the ability of the mice to learn and form memories.

Future directions

Going from experiments on the effects of drugs on mice to a therapy for humans is a difficult path. To protect patients, the Federal Food and Drug Administration (FDA) has a rigorous procedure for approving drugs for human use. After a candidate drug is shown to be effective and safe in animal models, four stages of approval must be passed before a drug is allowed on the market. In Phase 0, the first in-human trials are conducted on a small group of volunteers to study how the body absorbs, distributes, metabolizes, and excretes the drug, and whether it has any obvious side effects. In Phase 1, trials are conducted in a small group of volunteers to evaluate safety and dosage, and again to look for side effects. Phase 2 involves a larger group of subjects who have the disease that the drug is expected to treat, to see if it is effective and safe. The comparison is with a placebo. In Phase 3, trials are conducted with a large group of patients to assess the drug's effectiveness and safety, and to compare it with other commonly used treatments. Each of these phases must be reviewed by

Chapter Summary 44

44.1 Neurons and Glia Are Unique Cells of Nervous Systems

- The cells of the nervous systems include many types of **neurons** and **glia**.

- All neurons can generate and conduct ionic electrical signals, and most can generate **action potentials (APs)**. Glia support and modulate the activities of neurons but do not generate APs .

- A neuron generally receives information via its **dendrites**, of which there can be many, and transmits information via its single **axon**, which ends in **axon terminals**. Review Figures 44.1, 44.2

- Where neurons and their target cells meet, information is transmitted across specialized junctions called **synapses**.

- Glia include **Schwann cells** and **oligodendrocytes**, both of which generate **myelin** sheaths on axons. Review Figure 44.3

- Glia also include **astrocytes**, which support neurons metabolically, modulate synaptic signaling, and contribute to the **blood-brain barrier**.

44.2 Neurons Generate and Transmit Electric Signals

- Neurons have an electric charge difference across their cell membranes. This **membrane potential** is generated by ion gradients (due to ion transporters) and ion channels. When a neuron is not stimulated, its membrane potential is referred to as the **resting potential**. Review Figure 44.4, Animation 44.1

- The **sodium-potassium pump** concentrates K^+ on the inside of a neuron and Na^+ on the outside. Leak K^+ channels in the cell membrane allow K^+ to diffuse out leaving behind unbalanced negative charges, creating the resting potential. Review Figure 44.5

- **Patch clamping** allows the study of single ion channels. Review Figure 44.6

- The resting potential is perturbed when ion channels open or close, changing the permeability of the cell membrane to charged ions. Through this mechanism, the cell membrane can become **depolarized** or **hyperpolarized** and therefore have a **graded membrane potential** response to input. Review Figure 44.7

- An AP is a rapid transient reversal in charge across a portion of the cell membrane resulting from the sequential opening and closing of voltage-gated Na^+ and K^+ channels. These changes in voltage-gated channels occur when the cell membrane depolarizes to a **threshold** level. Review Figure 44.8, Animation 44.2

- APs are all-or-none, self-regenerating events. They are conducted down axons because local current flow depolarizes adjacent regions of membrane and brings them to threshold. Review Figure 44.9

- In myelinated axons, APs appear to jump between **nodes of Ranvier**, areas of axonal cell membrane that are not covered by myelin. Review Figure 44.10

44.3 Neurons Communicate with Other Cells

- Neurons communicate with each other and with other cell types by transmitting information over **electrical synapses** or by the transmission of molecular signals called **neurotransmitters** over **chemical synapses**.

- The **neuromuscular junction** is a well-studied chemical synapse between a motor neuron and a skeletal muscle cell. Its neurotransmitter is **acetylcholine (ACh)**, which causes depolarization of the postsynaptic membrane when it binds to its receptor at the **motor end plate**. Review Focus: Key Figure 44.11, Animation 44.3

- When an AP reaches an axon terminal, it causes the release of neurotransmitters, which diffuse across the **synaptic cleft** and bind to receptors on the postsynaptic membrane. See Activity 44.1

- Synapses between neurons can be either excitatory or inhibitory. A postsynaptic neuron integrates information by summing excitatory and inhibitory postsynaptic potentials both spatially and temporally. Review Figure 44.12

- There are many different neurotransmitters and even more types of receptors. The action of a neurotransmitter depends on the type of receptor to which it binds. See Activity 44.2

- Neurotransmitter molecules cannot be allowed to accumulate in a synapse but must be cleared to turn off responses in the postsynaptic cell. This may be done by enzymatic degradation, simple diffusion, or reuptake of the neurotransmitter.

44.4 Neurons and Glia Form Information-Processing Circuits

- In vertebrates, the brain and spinal cord form the **central nervous system (CNS)**, which communicates with the rest of the body via the **peripheral nervous system (PNS)**. The CNS increases in complexity from invertebrates to vertebrates and from fish to mammals. Review Figures 44.14, 44.15

- **Neural networks** include **afferent neurons** and **efferent neurons**, generally connected through **interneurons**.

- A **spinal reflex** is an example of a simple neural network that integrates information and controls a response. Review Figure 44.14, Animation 44.4

Go to **LearningCurve** (in **LaunchPad**) for dynamic quizzing that helps you solidify your understanding of this chapter. **LearningCurve** adapts to your responses, giving you the practice you need to master each key concept.

Apply What You've Learned

Review

44.2 When gated ion channels are activated by electrical, chemical, or mechanical stimuli, permeability of the membrane to the respective ions changes, resulting in a change in membrane potential.

44.2 Rapid activation and inactivation of voltage-gated Na⁺ channels are responsible for action potentials.

44.3 Agonists and antagonists are drugs that target specific receptors.

Original Paper: Adamantidis, A. R., F. Zhang, A. M. Aravanis, K. Deisseroth and L. de Lecea. 2007. Neural substrates of awakening probed with optogenetic control of hypocretin neurons. *Nature* 450: 420–424.

A common question in neurobiology is whether or not neurons in a particular brain region control a specific physiological or behavioral response in the animal. Recording neural activity with electrodes may show a correlation between activity of the neurons and the response, but that does not prove that those neurons *control* the response. The best evidence for whether specific neurons control a specific response is obtained by stimulating the activity of the neurons and observing—and thus proving—that the stimulation results in the response.

Dr. Karl Deisseroth's lab at Stanford University developed a means of activating specific neurons with light. They called their method "optogenetics." Algae have light-sensitive ion channels in their membranes. The scientists isolated the gene for one of these ion channels, rhodopsin (ChR2), and developed methods to modify the gene so that when it was introduced by viral infection into a specific brain area, it would be expressed only in the neurons of interest. Using this technique, the researchers introduced ChR2 into a specific group of cells in mouse brains that produce the neurotransmitter hypocretin. They also implanted optical fibers into the brains of these mice so they could shine light on these neurons and activate them. They hypothesized that these hypocretin-releasing neurons coordinated wakefulness.

Using electrodes, the scientists recorded the action potentials of the hypocretin neurons when they were stimulated with light. **Figure A** shows the response of a neuron to 1 second of light, and **Figure B** shows the response to 15-millisecond pulses of light. Each figure superimposes two trials.

Figure A

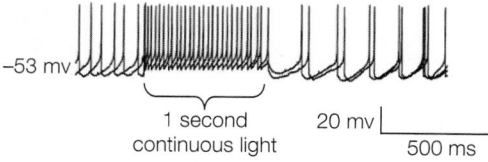

−53 mv

1 second continuous light   20 mv   500 ms

Figure B

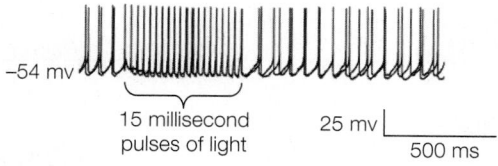

−54 mv

15 millisecond pulses of light   25 mv   500 ms

The researchers recorded sleep and wakefulness in the mice, and they stimulated the hypocretin neurons when the mice were asleep. The control mice received the same viral infection treatment, but the virus did not include the ChR2 gene. The time between the stimulation and the animal awakening (sleep-to-wake latency) was determined for different stimulation frequencies. They repeated the same experiment in mice that were expressing the ChR2 gene but were injected with saline or high and low does of a drug that blocked the hypocretin receptor (HctR). **Figure C** shows the results of 15-ms light pulses over 10 seconds (1-30 Hz) and continuous light illumination for 10 seconds (ON). Asterisks indicate significant differences from controls.

Figure C

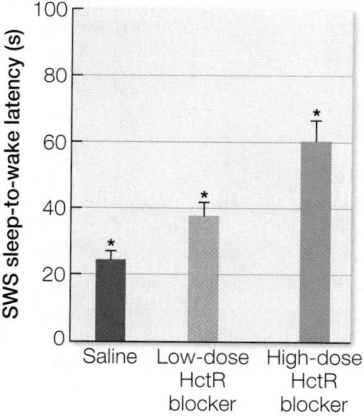

Questions

1. Describe the effect of 1 second of light on the membrane potential and the firing rate of the hypocretin neurons. What do these results suggest about the nature of the ChR2 protein and its response to light?

2. What is the evidence that light stimulation has a direct effect on the activity of the hypocretin neurons that are expressing ChR2?

3. Describe the effect of light stimulation on the behavior of the mice. Did the frequency of stimulation matter?

4. What is the evidence that the effects of the stimulation were due to release of hypocretin by the stimulated neurons?

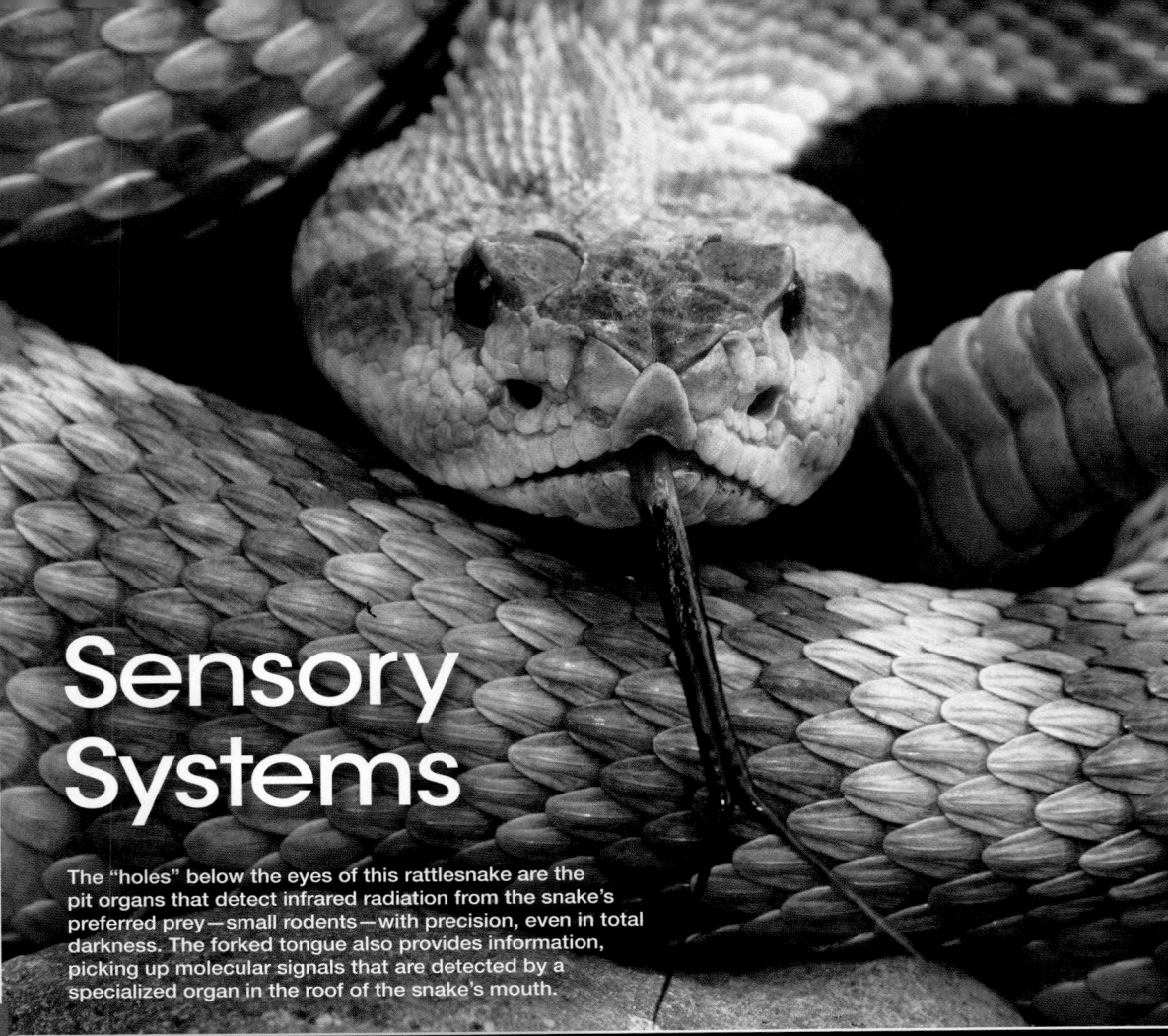

Sensory Systems

The "holes" below the eyes of this rattlesnake are the pit organs that detect infrared radiation from the snake's preferred prey—small rodents—with precision, even in total darkness. The forked tongue also provides information, picking up molecular signals that are detected by a specialized organ in the roof of the snake's mouth.

▶ **investigatinglife**

Seeing in the Dark

A rattlesnake can see to strike a running rodent in complete darkness. How can this be, when "seeing" means using the eyes to detect light waves, and "complete darkness" means no light? It is possible because these definitions are based on human capabilities. What we call "light" is actually only a small portion (red, orange, yellow, green, blue, indigo, and violet) of the spectrum of electromagnetic radiation. Other animals see wavelengths humans cannot. Some insects, for example, perceive patterns on flowers that reflect ultraviolet wavelengths invisible to humans (see Figure 28.15). Similarly, rattlesnakes "see" infrared wavelengths that we cannot (although at high enough levels of intensity, humans feel infrared wavelengths as heat).

It is not the snake's eyes that perceive infrared light. Pit vipers, rattlesnakes, and their relatives have pit organs located between the nostril and the eye on each side of the skull that contain high densities of infrared-sensitive neurons. The two pits are positioned in such a way that sensory receptor cells in the pits receive directional information. Information from the pit organs goes to the same region of the brain as information from the eyes, so rattlesnakes actually do "see" the

world in a range of electromagnetic radiation that is different from the human visual spectrum.

Our definition of silence is as human-oriented as our definition of darkness. "Sound" is actually pressure waves in the environment, and many animals are sensitive to pressure waves with frequencies, or pitches, we cannot hear. Elephants communicate using sound waves that are below human hearing range; such long waves travel great distances, an advantage to large animals that roam over extensive areas. Bats emit incredibly loud, brief sound pulses that are above our range of hearing. A flying bat hears echoes of these pulses bouncing off objects in the environment and can therefore avoid obstacles and catch prey in total darkness.

"Reality" is what our eyes see, our ears hear, our noses smell, and what we touch and taste. Humans sense only a limited range of the information available. Animals with different ranges of sensitivity process different sources of information and perceive the world quite differently than we do. A challenge for neurobiologists is to understand how neurons are adapted to detect different types of information in the environment.

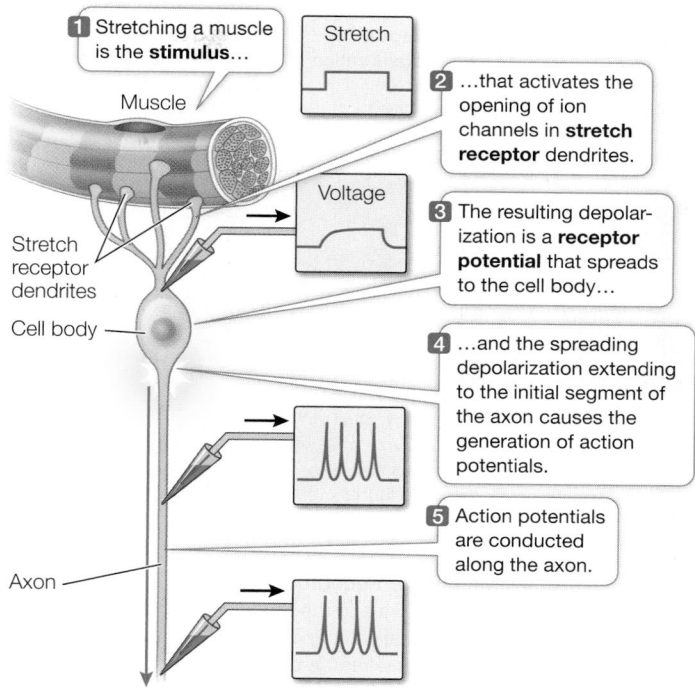

key concept 45.1 Sensory Receptor Cells Convert Stimuli into Action Potentials

Sensory receptor cells, usually called **sensors** or **receptors**, transduce (convert) physical and chemical stimuli such as light and sound waves, pressure (e.g., touch), and odorant and taste molecules into neural signals. These signals are then transmitted to the central nervous system (CNS) for processing and interpretation. The first step in this process of **sensory transduction** is a change in the membrane potential of the receptor cell in response to a specific type of stimulus.

focus your learning

- Specific stimuli cause changes in membrane potentials of sensory cells.
- Sensory receptors act directly or indirectly on ion channels in the cell membrane.
- Responses to sensory information depend on the neural circuits they activate.
- Receptors can adapt to constant stimulation.

Sensory transduction involves changes in membrane potentials

Sensory transduction typically begins with a **receptor protein** that opens or closes ion channels in response to a specific stimulus such as heat, light, chemicals, mechanical force, or electric fields. The resulting change in membrane permeability to ions alters the receptor cell's membrane potential. A change in the membrane potential of a receptor cell in response to a stimulus is a **receptor potential**. Receptor potentials are graded membrane potentials that spread over only short distances. To signal over long distances in the nervous system, receptor potentials must generate action potentials, which they can do in two ways:

1. The receptor potential may trigger action potentials in the receptor cell itself.
2. The receptor potential may cause the receptor cell to release neurotransmitters that induce a postsynaptic neuron to generate action potentials.

A good model of how a receptor cell generates action potentials is the **stretch receptor** of a crayfish (**Figure 45.1**). Stretching the muscle to which the stretch receptor is attached causes receptor potentials. These receptor potentials spread to the base of the cell's axon, where they generate action potentials that travel down the axon to the CNS. The rate at which action potentials are fired depends on the magnitude of the receptor potential; that magnitude, in turn, depends on how much the muscle is stretched.

In a receptor cell that does not fire action potentials (such as the photoreceptors in the vertebrate eye), the spreading receptor potential reaches a presynaptic patch of cell membrane and induces the release of a neurotransmitter. The intensity of the stimulus influences how much neurotransmitter is released. That neurotransmitter binds to receptor proteins on an associated sensory neuron, altering its membrane potential and causing it to

Figure 45.1 Stimulating a Sensory Cell Produces a Receptor Potential Signal transduction in the stretch receptor of a crayfish can be investigated by measuring the membrane potential at different places on the stretch receptor neuron while stretching the muscle innervated by that sensory neuron.

increase or decrease its rate of firing action potentials. In a few cases, this second cell also does not generate action potentials, but simply changes the rate at which it releases neurotransmitter onto another neuron. Eventually, however, the stimulation of a sensory cell is always coded as a change in firing of action potentials in a sensory circuit.

Sensory receptor proteins act on ion channels

Sensory receptor proteins respond to stimuli by directly or indirectly opening or closing ion channels in the sensory cell (**Figure 45.2**), leading either to an action potential or to the release of neurotransmitter. Those sensory receptor proteins that act directly on membrane potential are either ion channels themselves or directly affect the opening of an associated ion channel. Examples are receptors that respond to physical force (mechanoreceptors) and those that respond to temperature (thermoreceptors). Electroreceptors do not have receptor proteins, but their cell membranes are sensitive to voltage changes and respond to depolarization by opening voltage-gated Ca^{2+} channels, and the Ca^{2+} triggers neurotransmitter release. Sensory receptor proteins that influence ion channels indirectly usually do so by activating G proteins and second messengers, as described in Key Concepts 7.2 and 7.3. Examples are most chemoreceptors and photoreceptors.

A recently discovered superfamily of receptor proteins is the focus of considerable research in an effort to identify the molecular mechanisms by which infrared signals are detected. Transient receptor potential (TRP) channel proteins generate a large initial response

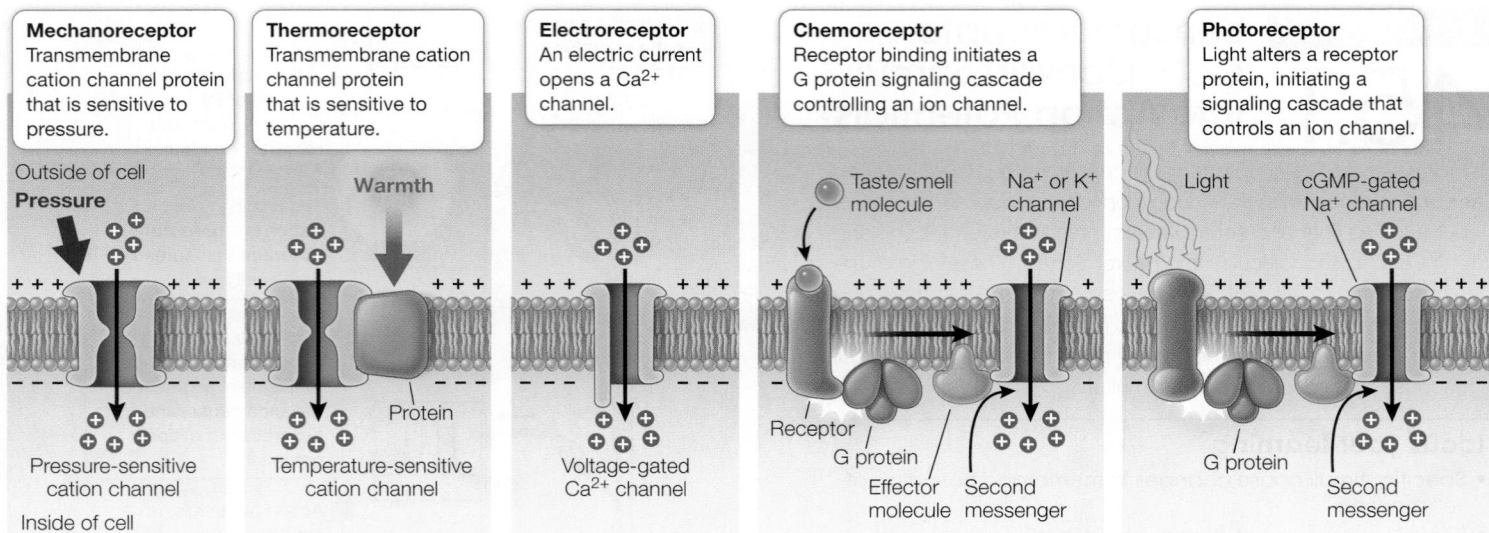

Mechanoreceptor
Transmembrane cation channel protein that is sensitive to pressure.

Outside of cell
Pressure

Pressure-sensitive cation channel

Inside of cell

Thermoreceptor
Transmembrane cation channel protein that is sensitive to temperature.

Warmth

Protein

Temperature-sensitive cation channel

Electroreceptor
An electric current opens a Ca^{2+} channel.

Voltage-gated Ca^{2+} channel

Chemoreceptor
Receptor binding initiates a G protein signaling cascade controlling an ion channel.

Taste/smell molecule

Na^+ or K^+ channel

Receptor
G protein
Effector molecule
Second messenger

Photoreceptor
Light alters a receptor protein, initiating a signaling cascade that controls an ion channel.

Light

cGMP-gated Na^+ channel

G protein

Second messenger

Figure 45.2 Sensory Cell Membrane Receptor Proteins Respond to Stimuli Mechanoreceptors and thermoreceptors have transmembrane proteins that are also cation channels. Changes in their respective stimuli, pressure or heat, alter the permeability of the channels. Electroreceptors are ion channels responsive to electric currents. Chemoreceptors and photoreceptors influence ion channels through G proteins and second messengers.

to a stimulus and then gradually fade even as the stimulus remains constant. Seven classes of TRP channels have been identified, along with about 28 different molecular structures. Each type of TRP channel is specialized to respond to a different stimulus, such as stretch, pressure, temperature, light, and certain chemicals. Some respond to both physical and chemical stimuli, such as one that responds to heat, capsaicin (an active ingredient in chili peppers), and wasabi. Could a heat-sensitive TRP channel be responsible for the infrared-detection function of the pit organs of rattlesnakes? You will pursue this question in **Investigating Life: How Do Pit Vipers "See" in the Dark?**

Sensation depends on which neurons receive action potentials from sensory cells

All sensory systems process information in the form of action potentials. But the sensations we perceive—heat, pressure, light, smell, sound—differ because the messages from different kinds of sensory cells arrive at different places in the CNS. Action potentials arriving in the visual cortex of the brain are interpreted as light, whereas those that arrive in the olfactory bulb are perceived as smells.

A small patch of skin on your arm contains some sensory receptor cells that increase their firing rates when the skin is warmed and others that increase their activity when the skin is cooled. Other sensory cells in the same patch of skin respond to touch, irritants such as insect bites, and painful stimuli. These receptor cells transmit their messages through axons that enter the CNS at the spinal cord. The synapses made by those axons in the spinal cord and the subsequent pathways of transmission determine whether the stimulation of the skin on your arm is perceived as warmth, cold, touch, itch, or pain; even though the action potentials

carried by all of these sensory axons are the same, the connectivity of each axon is specific for a given sensory modality. The *intensity* of a given sensation is coded by the frequency of the action potentials.

Some sensory receptors transmit information about the body's internal conditions of which we are not consciously aware. The brain continuously receives information about body temperature, the concentrations of carbon dioxide and oxygen in the blood, arterial pressure, muscle tension, and the position of the limbs. All of this information is important for homeostasis but does not necessarily result in conscious sensation.

Some sensory receptor cells are assembled with other types of cells into **sensory organs**, such as eyes, ears, and noses, that enhance the ability of sensory cells to collect, filter, and amplify stimuli. **Sensory systems** include the sensory cells, their associated organs, and the neural networks that process the information.

Many receptors adapt to repeated stimulation

Some sensory receptor cells give gradually diminishing responses to maintained or repeated stimulation. This phenomenon, known as **sensory adaptation**, enables an animal to ignore background or unchanging conditions while remaining sensitive to changes and new information. (Note that this use of the term "adaptation" is different from its application in an evolutionary context.) When you get dressed, you feel each item of clothing touch your skin, but the sensation of clothes touching your skin is not constantly on your mind throughout the day. You are immediately aware of new sensations, however, such as your shoe coming untied or someone touching your back.

Some sensory receptor cells adapt very little or very slowly; examples are some types of pain receptors and the mechanoreceptors that control balance. You do not want to ignore pain, which usually is signaling that something is wrong in your body, and to maintain equilibrium you must continuously know (albeit unconsciously) the tensions and forces on all of your joints and muscles.

experiment

Original Paper: Gracheva, E. O. et al. 2010. Molecular basis of infrared detection by snakes. *Nature* 464: 1006–1012.

Neurons of the pit organ have their cell bodies in a neural ganglion in the head—the trigeminal ganglion (TG). Sensory neurons conducting information from the body of the snake have their cell bodies in neural ganglia associated with the spinal cord—the dorsal root ganglia (DRGs). Any receptor protein that is involved in sensing heat by the pit organ or by temperature sensors in the skin should be expressed in the TG or the DRGs, respectively.

HYPOTHESIS▶ Different transient receptor potential (TRP) channel proteins responsive to heat are expressed in the TG and the DRGs, and they have different properties.

METHOD

Many neurons in the TG of rattlesnakes innervate the pit organs, and therefore the RNA transcripts for any protein that is abundant in the pit organ nerve endings should be highly expressed in the TG. Similarly, RNA transcripts for sensory receptor proteins in the skin of the body should be highly expressed in the DRGs. Whole genome RNA sequencing (RNA-seq) was used to create the RNA profiles of the TG and the DRGs of the rattlesnake. Of particular interest were specific TRP RNA reads (specific RNA sequences) that were differentially expressed in the TG and the DRGs. If there are more reads for a certain TRP channel in the TG versus the DRGs, that TRP channel might play a role in the pit organ's ability to sense infrared radiation. Because the range of read values was so large, the data are plotted as logarithmic values (log–log plot).

RESULTS

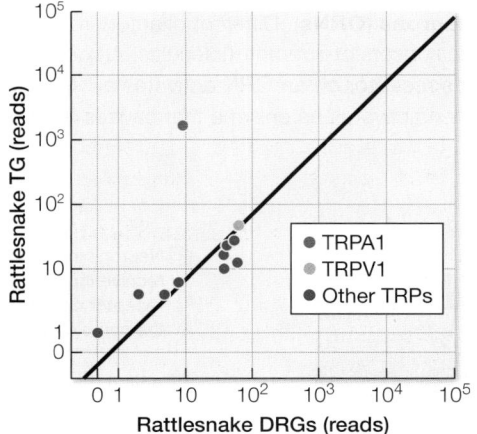

CONCLUSIONS▶

1. TRPV1 and "Other TRPs" are expressed equally in the TG and the DRGs, so they probably are not responsible for the unique infrared-sensing properties of the pit organ.
2. TRPA1 is expressed more in the TG than in the DRGs. This channel could therefore serve a unique function in the pit organ.

work with the data

The data from the experiment comparing TRP channel expression in the TG and the DRGs support the conclusion that the TRPA1 channel could be the infrared (IR) sensor in the pit organ. Another piece of evidence would be finding that the *TRPA1* gene is not highly expressed in the TG of snakes without pit organs, or that it is highly expressed in snakes not closely related to rattlesnakes but that have pit organs. Accordingly, the investigators performed RNA-seq analysis of the TG and the DRGs of rat snakes (no pits) and pythons (pits).

RESULTS

(Simulated data, arbitrary units)

	Rat snake		Python	
Channel	TG	DRGs	TG	DRGs
TRPA1	105	110	900,000	50
TRPV1	85	90	15	9

QUESTIONS▶

1. Compare the relative expression of TRPV1 and TRPA1 in the TG and the DRGs of the rat snake (no pits) and the python (pits). What is the major difference?
2. Explain how these results do or do not support the conclusion that the TRPA1 channel is the IR sensor in the pit organ.
3. What functional data on the TRPA1 channel would further support the conclusion that TRPA1 is the IR sensor in the pit organ?

A similar **work with the data** exercise may be assigned in **LaunchPad**.

Sensory receptor cells have receptor proteins that respond to specific stimuli from the external or internal environment by directly or indirectly opening or closing ion channels, causing the generation of action potentials or neurotransmitter release in sensory neurons. How the resulting information is processed and interpreted depends on the neural circuitry that is activated.

learning outcomes

You should be able to:

- Describe the roles of receptor potentials and action potentials in sensory systems.
- Explain how the same membrane event, an action potential, can result in sensations of heat, light, sound, touch, or taste.
- Describe the various ways that environmental stimuli can generate receptor potentials.
- Explain the functional significance of sensory adaptation.

---

1. In the crayfish stretch receptor, what creates the receptor potential? Where and how does it stimulate an action potential?

(continued)

45.1 recap (continued)

2. How are we able to perceive action potentials—all of which are essentially the same—as different sensations?

3. Peppers have a distinctive flavor, but they also can be hot if they contain the molecule capsaicin. Explain the difference in how the flavor versus the hotness of a pepper is sensed.

4. Why is adaptation or the lack thereof important in different sensory modalities?

focus your learning

- Many different chemoreceptors can be activated in different combinations and to different degrees, enabling discrimination of an enormous number of odorant and pheromone molecules.
- Five classes of taste chemoreceptors have different mechanisms of membrane depolarization.

Now that you have a general view of how sensory systems code and process information, you will next learn how sensory systems gather and filter stimuli, transduce specific stimuli into action potentials, and transmit action potentials to the CNS, which perceives them as different types of information.

key concept 45.2 Chemoreceptors Respond to Specific Molecules

A colony of corals responds to a small amount of meat extract in seawater by extending its coral polyps and their tentacles in search of food; a solution of a single type of amino acid can stimulate this response. Conversely, a small amount of seawater in which corals were crushed will stimulate a defensive retraction of the coral polyps. Humans also react strongly to certain chemical stimuli. When you smell freshly baked bread you salivate and feel hungry, and when you smell rotting meat you feel nauseated.

All animals receive information about chemical stimuli through **chemoreceptors**, which are receptor proteins that bind to specific molecules—their **ligands**—and are responsible for smell and taste. Chemoreceptors are also responsible for monitoring some aspects of the internal environment, such as the level of carbon dioxide in the blood.

Olfaction is the sense of smell

The sense of smell, **olfaction**, depends on chemoreceptors. In vertebrates the olfactory receptors are neurons embedded in a layer of epithelial tissue in the uppermost region of the nasal cavity (**Figure 45.3**). The dendrites of these neurons project as olfactory cilia on the surface of the nasal epithelium, and their axons extend through holes in the overlying bone into the **olfactory bulb** (the olfactory integration area of the brain). A protective layer of mucus covers the nasal cavity epithelium. Molecules from the environment must diffuse through this mucus to reach the receptor proteins on the olfactory cilia. When you have a cold, the amount of mucus in your nose increases, and the epithelium swells. With this in mind, you can easily understand why respiratory infections can cause you to lose your sense of smell.

An **odorant** is a molecule in the environment that binds to and activates an olfactory receptor protein on the cilia of **olfactory receptor neurons** (**ORNs**). Different olfactory receptor proteins bind specific subsets of odorant molecules. An odorant molecule binding to its receptor on an ORN activates a *G protein. The G protein then activates an enzyme that causes an increase of a

Figure 45.3 Olfactory Receptors Communicate Information Directly to the Brain   The receptor cells of the human olfactory system are embedded in epithelial tissues lining the nasal cavity and send their axons to the olfactory bulb of the brain.

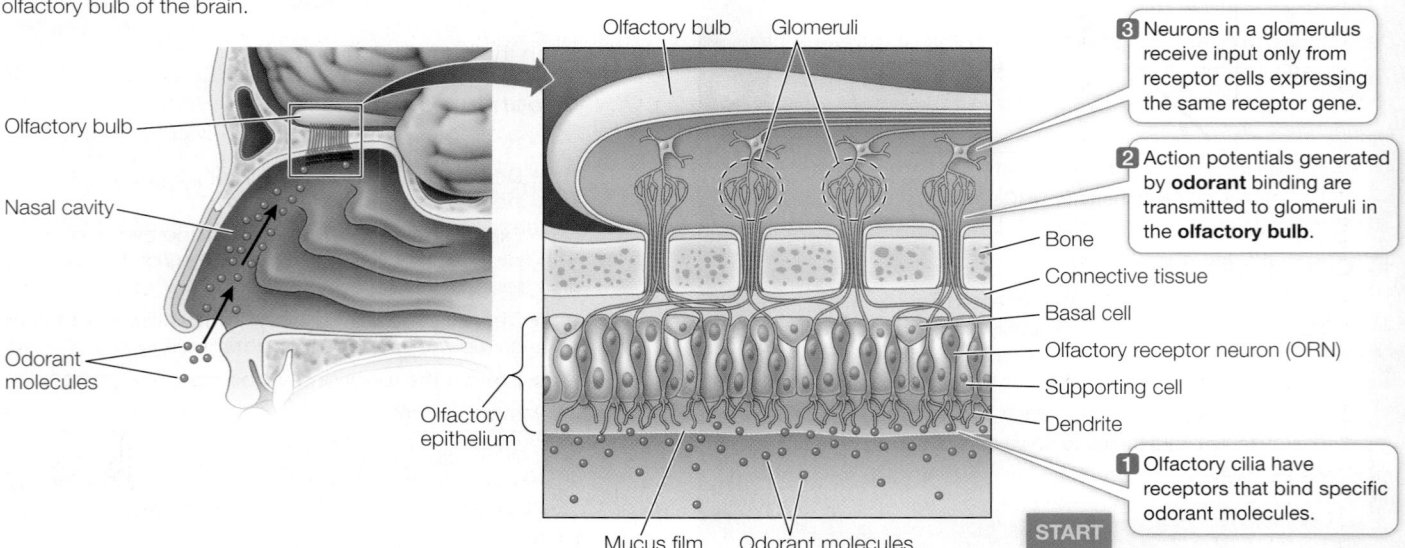

3 Neurons in a glomerulus receive input only from receptor cells expressing the same receptor gene.

2 Action potentials generated by **odorant** binding are transmitted to glomeruli in the **olfactory bulb**.

1 Olfactory cilia have receptors that bind specific odorant molecules.

START

(A) Release of pheromone by female

The female moth releases a **pheromone** from a gland at the tip of her abdomen. The pheromone can travel thousands of meters downwind.

(B) Detection of pheromone by male

A male moth detects this pheromone in the air passing over his antennae, which are covered with chemo-sensitive hairs.

Figure 45.4 Pheromones Can Communicate over Great Distances Mating in silkworm moths of the genus *Bombyx* is coordinated by a pheromone called bombykol.

Q: Molecules have different diffusion coefficients. *Bombykol* has a high diffusion coefficient. How do you think its diffusion coefficient would compare with that of a territory-marking pheromone, and why might they be different?

second messenger (cAMP in vertebrates) in the cytoplasm. The second messenger binds to and opens cation channels in the ORN's cell membrane, causing an influx of Na^+ into the ORN—which then depolarizes to threshold and fires action potentials. An interesting feature of ORNs is that they are continuously re-generated. Because they are embedded in nasal epithelium that, like other epithelial linings, is regularly shed, ORNs have to be constantly replaced.

---

*connect the concepts Key Concept 7.2 describes how G protein-coupled receptors transmit signals from a variety of stimuli outside a cell to its interior. Ligand binding of the receptor activates the G protein on the cytoplasmic surface of the membrane, which—if bound by GTP—will activate an effector.

The olfactory world has an enormous number of odors and a correspondingly large number of olfactory receptor proteins. In the 1990s Linda Buck and Richard Axel discovered a family of about 1,000 genes in mice (about 3 percent of the mouse genome) that code for olfactory receptor proteins. Each receptor protein that is expressed is found in a limited number of ORNs in the olfactory epithelium, and each ORN expresses just one receptor type. The investigators matched specific gene products with the odorants they detect. For their discoveries of the molecular nature of the olfactory system, Buck and Axel received a Nobel Prize in 2004.

Olfactory sensitivity enables discrimination of many more odorants than there are olfactory receptors. An odorant molecule can be quite complex, and different regions of that molecule may bind to different receptor proteins. The next stage of processing olfactory information is in the olfactory bulb, where axons from ORNs expressing the same receptor protein cluster together on olfactory bulb neurons, forming structures called glomeruli (see Figure 45.3). A complex odorant molecule can activate a unique combination of glomeruli in the olfactory bulb, so an olfactory system with hundreds of different receptor proteins can discriminate an astronomically large number of smells. The more odorant molecules that bind to ORNs, the greater the frequency of action potentials and thus the greater the intensity of the perceived smell.

Humans have a sensitive olfactory system, but in comparison with most mammals we depend far more on vision than on olfaction. The nasal epithelium of a typical dog is 15–20 times larger than a human's and has about *300 million* odorant receptors, compared with about 6 million in the average human. For some scents, the threshold sensitivity of the dog is 100,000 times lower; a dog's nose reveals a huge amount of information not available to people.

Some chemoreceptors detect pheromones

A specialized type of chemical signal used for communication among conspecifics (individuals of the same species) is called a **pheromone**. Individual animals secrete pheromone molecules into the environment, triggering behavioral responses in other individuals of the species. Pheromones may communicate alarm signals, mark food trails, or define territories, among many other uses. Their function in mating and mate attraction especially illustrates the remarkable sensitivity of chemosensory systems, and the silkworm moth *Bombyx mori* is an extensively studied example.

To attract a mate, the female silkworm moth releases a pheromone called bombykol from a gland at the tip of her abdomen (**Figure 45.4A**). The male silkworm moth has about 10,000 receptors for this molecule on each of his feathery antennae (**Figure 45.4B**). A single molecule of bombykol may be sufficient to generate action potentials in the antennal nerve, which transmits the signal to the male's CNS. The extreme sensitivity of the male bombykol receptors ensures that the sexual message sent by a female moth is likely to reach any male within a huge downwind area. The rate of firing in the male's sensory nerves is proportional to the concentration of bombykol in the air, so by flying upwind he can follow an airborne concentration gradient to "home in" on the signaling female.

The vomeronasal organ contains chemoreceptors

The **vomeronasal organ** (**VNO**) is a small, paired tubular structure embedded in the nasal epithelium of amphibians, reptiles, and many mammals (although not humans). In mammals the VNO is located on the septum dividing the two nostrils (see Figure 52.3) and has a pore that opens into the nasal cavity. The VNO pulsates and draws a sample of nasal fluid over the chemoreceptors embedded in its walls. The information from these chemoreceptors goes to an accessory olfactory bulb in the brain, and from there to brain regions involved in sexual and other instinctive behaviors. The VNO of a rodent detects pheromones that signal the sex, strain, identity, and sexual receptivity of another individual.

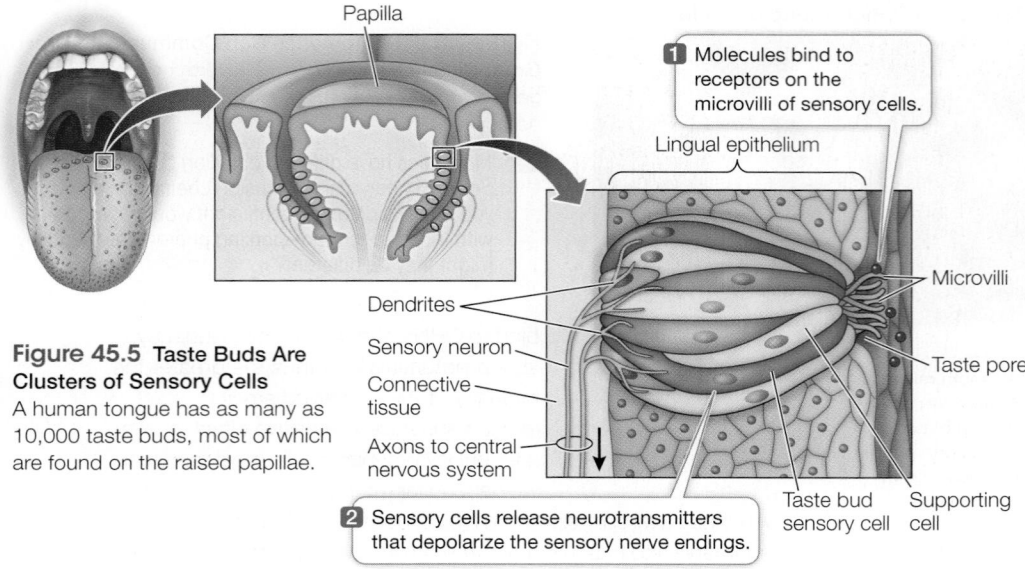

Papilla

1 Molecules bind to receptors on the microvilli of sensory cells.

Lingual epithelium

Microvilli

Dendrites

Sensory neuron

Connective tissue

Taste pore

Axons to central nervous system

2 Sensory cells release neurotransmitters that depolarize the sensory nerve endings.

Taste bud sensory cell   Supporting cell

Figure 45.5 Taste Buds Are Clusters of Sensory Cells
A human tongue has as many as 10,000 taste buds, most of which are found on the raised papillae.

In snakes the VNO opens into the roof of the mouth cavity. Each time the snake's forked tongue darts in and out, the forks fit into the VNO openings and present the chemoreceptors located there with a sample of molecules from the surrounding air (see the chapter opening photo). Thus the snake uses its tongue to smell its environment, not to taste it. Why doesn't the snake simply use the flow of air to and from its lungs, as we do, to smell the environment? In snakes, air flows to and from the lungs slowly (and can even stop entirely for long periods of time), but the tongue can dart in and out rapidly. It is a quick source of olfactory information used in social communication and hunting prey.

Gustation is the sense of taste

In humans and other vertebrates, the sense of taste, **gustation**, depends on clusters of chemoreceptors called **taste buds**. The taste buds of terrestrial vertebrates are confined to the mouth cavity, but some fish have taste buds in the skin that enhance their ability to sense their environment. Some fish living in murky water are very sensitive to small amounts of amino acids in the water and can find food without the use of vision. The duck-billed platypus, a prototherian mammal (see Figure 32.26B), has similar talents as a result of taste buds on the sensitive skin of its bill.

The human tongue has 5,000–10,000 taste buds embedded in the epithelium. Most of them are found on the sides of the papillae (**Figure 45.5**). (Look at your tongue in a mirror—the papillae make it look fuzzy.) The outer surface of a taste bud has a pore that exposes the tips of the sensory receptor cells. Microvilli (tiny hairlike projections) increase the surface area of these cells where their tips converge at the pore. These chemoreceptor cells generate action potentials and release neurotransmitter at their bases, where they form synapses with sensory neurons that convey the signals to the CNS.

The tongue does a lot of hard work, so its epithelium, along with cells of its taste buds, are shed and replaced at a rapid rate. Individual taste bud cells last about 10 days before they are replaced, but the sensory neurons associated with them live on, constantly forming new synapses as new taste buds form.

Gustation begins with the membrane receptor proteins of the microvilli of the taste bud sensory cells (see Figure 45.5). Humans

perceive five taste classes: sweet, salty, sour, bitter, and umami. However, taste buds can distinguish among a variety of sweet-tasting molecules and a variety of bitter-tasting molecules. Umami is a savory, meaty taste that is stimulated by amino acids, including monosodium glutamate (MSG), a commonly used flavor enhancer. Sweet, bitter, and umami tastes are the functions of a small family of G protein-coupled TRP channels. Other TRP channels related to the ones you saw associated with heat sensation in Investigating Life: How Do Pit Vipers "See" in the Dark? are responsible for spicy/hot tastes. Still other TRP channels that act as cold sensors are activated by minty, cool-tasting foods. Saltiness receptor proteins allow Na^+ to diffuse into taste bud sensory cells through open Na^+ channels, depolarizing the sensory cell. Sourness receptors are activated by a direct effect of H^+ ions on Na^+ channels.

Various selective pressures led to the evolution of diversity in gustatory receptors. Receptors for bitter tastes may have evolved as protective mechanisms enabling animals to detect toxic plant compounds such as quinine, caffeine, and nicotine. Because many such toxic compounds evolved in plants as protections against herbivorous predators, having a variety of receptors in those herbivores is a product of the continuous "arms race" between predators and prey (see Key Concept 55.1). Also, a large number of molecules in food could indicate nutritional value, so a variety of receptors is of value. The diversity of sweet receptors helps explain why it has been possible to invent many chemically distinct artificial sweeteners.

Regardless of the mechanism of taste transduction by the receptors of the taste buds, all these cells release neurotransmitter onto sensory neurons. These neurons then generate action potentials that are conducted to the CNS, where they are interpreted as specific taste sensations. The full complexity of the chemosensitivity that enables us to enjoy the subtle flavors of food comes from the combined activation of gustatory and olfactory receptors, which is why you may lose your sense of taste when you have a cold.

▶ 45.2 recap

All animals receive information about chemical stimuli through chemoreceptors, which have diverse structures and bind to a tremendous variety of ligands. Chemoreceptors are the basis of the sensations of olfaction and gustation and the reception of pheromones. They also monitor some aspects of an animal's internal environment.

learning outcomes

You should be able to:

- Describe how a large but limited number of olfactory receptor proteins can discriminate a much larger number of odorant molecules, as well as their concentrations.

45.2 recap

- Explain how the five classes of taste receptors generate action potentials in their sensory neurons.

1. Different odorants have different cell surface receptors, but these receptors activate the same intracellular G protein signaling pathway. How does the olfactory system discriminate different odors?

2. How does the olfactory system discriminate between the whiff of a skunk, a repulsive odor of skunk, and the smell of a skunk cabbage?

3. The concentration thresholds for sensing bitter tastes are lower than for salt or sweet. Offer an evolutionary explanation for why this is so.

4. With only five classes of taste receptors, how can we discriminate so many different tastes?

You have now seen how chemoreceptors give rise to the sensations of smell and taste, and how some animals use chemoreception to communicate with others of their species. Next we describe the sensory cells that respond to mechanical forces, including the vibrations you perceive as sound.

key concept 45.3 Mechanoreceptors Respond to Physical Forces

The cell membranes of **mechanoreceptors** have cation channels that open when the membrane is disturbed by physical forces. The intensity of the physical force is registered as a graded receptor potential, which in turn leads to either the release of neurotransmitter or the generation of action potentials. The amount of neurotransmitter release or the rate of action potentials tells the CNS the strength of the stimulus to the mechanoreceptor. A diversity of mechanoreceptor cells has evolved in different sensory systems. The functions of these cells range from interpreting skin sensations to sensing blood pressure to hearing and maintaining balance.

focus your learning

- A variety of pressure-sensitive receptors enable multiple modes of mechanosensation.
- Stretch receptors in muscle spindles signal for contraction, and stretch receptors in tendons signal for relaxation.
- As the frequency of the sound stimulus decreases, the stimulated region of the basilar membrane moves from basal to apical.
- The ear converts sound waves in air to pressure waves in fluid that bend hair cells to create auditory sensation.

Many different receptor cells respond to touch and pressure

Human skin (and that of other mammals) is packed with diverse mechanoreceptors that generate varied sensations (**Figure 45.6**). The most important tactile receptors, found in both hairy and non-hairy skin, are **Merkel's discs**, which adapt rather slowly and provide continuous information about anything touching the skin. **Meissner's corpuscles**, found primarily in non-hairy skin, are very sensitive but adapt rapidly; they provide information about *changes* in things touching the skin. The rapid adaptation of Meissner's corpuscles is why you roll a small object between your fingers (rather than holding it still) to discern its shape and texture: as you roll it, the object continually stimulates Meissner's corpuscles.

Deeper in the skin, **Ruffini endings** adapt slowly and are good at providing information about vibrating stimuli of low frequencies, while **Pacinian corpuscles**, which adapt rapidly, provide information about vibrating stimuli of higher frequencies. Even deeper in the skin, the dendrites of sensory neurons wrap around hair follicles. When the surface hairs are displaced, those neurons are stimulated.

The density of tactile mechanoreceptor cells varies across the body's surface. By touching the skin with two toothpicks simultaneously (when a person's eyes are closed!), you can determine how far apart two stimuli have to be before the person can tell whether the sensations are produced by one toothpick or by two. On back skin, stimuli have to be relatively far apart before they are perceived as two discrete stimuli. But when this same "two-point spatial discrimination test" is applied to the lips or fingertips, a person can identify two stimuli as separate even when they are quite close together, because the receptor density is much greater in these regions.

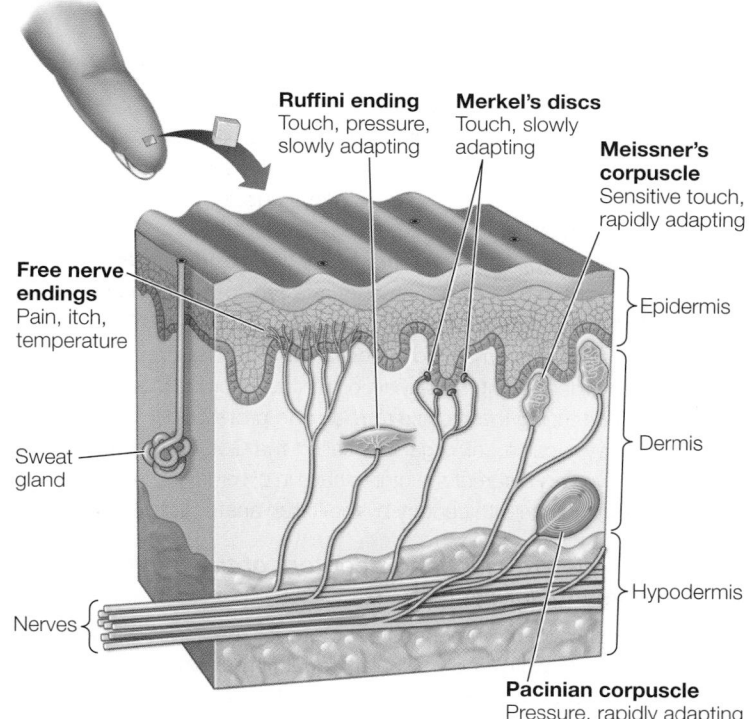

Ruffini ending
Touch, pressure, slowly adapting

Merkel's discs
Touch, slowly adapting

Meissner's corpuscle
Sensitive touch, rapidly adapting

Free nerve endings
Pain, itch, temperature

Epidermis

Sweat gland

Dermis

Nerves

Hypodermis

Pacinian corpuscle
Pressure, rapidly adapting

Figure 45.6 The Skin Feels Many Sensations Even a very small patch of skin contains a variety of sensory cells, making the skin a multimodal receptor that can sense temperature, pressure, texture, pain, touch, and itch.

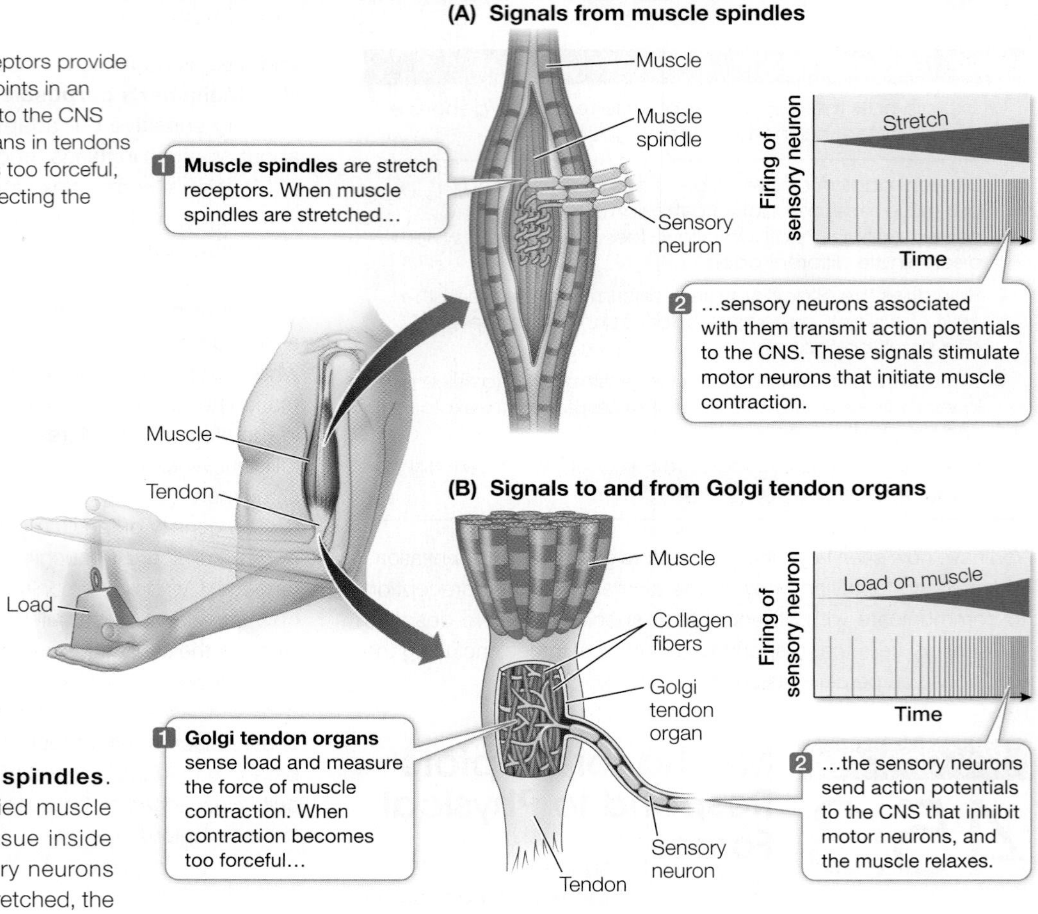

Figure 45.7 Stretch Receptors Stretch receptors provide information about the stresses on muscles and joints in an animal's limbs. **(A)** Signals from muscle spindles to the CNS initiate muscle contraction. **(B)** Golgi tendon organs in tendons and ligaments inhibit a contraction that becomes too forceful, triggering a reduction in muscle tension and protecting the muscle from tearing.

(A) Signals from muscle spindles

Muscle

Muscle spindle

Sensory neuron

1 **Muscle spindles** are stretch receptors. When muscle spindles are stretched…

Stretch

Firing of sensory neuron

Time

2 …sensory neurons associated with them transmit action potentials to the CNS. These signals stimulate motor neurons that initiate muscle contraction.

Muscle

Tendon

Load

(B) Signals to and from Golgi tendon organs

Muscle

Collagen fibers

Golgi tendon organ

Sensory neuron

Tendon

1 **Golgi tendon organs** sense load and measure the force of muscle contraction. When contraction becomes too forceful…

Load on muscle

Firing of sensory neuron

Time

2 …the sensory neurons send action potentials to the CNS that inhibit motor neurons, and the muscle relaxes.

Mechanoreceptors are also found in muscles, tendons, and ligaments

An animal receives information from mechanoreceptors about the position of its limbs and the stresses on its muscles and joints. These mechanoreceptors supply information continuously to the CNS, and this information is essential for postural control and the coordination of movements.

The mechanoreceptors in skeletal muscle are the **muscle spindles**. These stretch receptors are modified muscle cells embedded in connective tissue inside muscles and innervated by sensory neurons (**Figure 45.7A**). When a muscle is stretched, the muscle spindles are stretched as well, signaling the spindle neurons to transmit action potentials to the CNS. Figure 45.1 showed how crayfish stretch receptors transduce physical force into action potentials; the actions of muscle spindles are similar. The CNS uses the information from muscle spindles to adjust the strength of the muscle contraction so that it matches the load put on the muscle; thus a person can hold a mug steady while it is being filled.

Another type of mechanoreceptor, the **Golgi tendon organ**, is found in tendons and ligaments and provides information about the force generated by a contracting muscle (**Figure 45.7B**). When a contraction becomes too forceful, action potentials from the Golgi tendon organ inhibit the spinal cord motor neurons innervating that muscle, causing the muscle to relax and protecting it from tearing. (You may recall a cell organelle called the Golgi apparatus. What these two very different structures have in common is their discovery in the late nineteenth century by the Italian anatomist Camillo Golgi.)

Hair cells are mechanoreceptors of the auditory and vestibular systems

Hair cells are the mechanoreceptors for the vertebrate auditory (sound-perceiving) and vestibular (equilibrium-maintaining) systems. Both of these systems are housed in the complex structures of the vertebrate ear. **Stereocilia**—fingerlike extensions of the cell membrane stiffened by cross-linked actin filaments—project from the surface of each hair cell like a set of organ pipes (**Figure 45.8**). Stereocilia bend in response to waves of pressure; bending of the stereocilia in one direction depolarizes the hair cell, and bending in the other direction hyperpolarizes it.

Measurements with microelectrodes have shown that the bending of stereocilia creates local electric currents near their tips, indicating that ion channels near the tips must be opening or closing. Electron microscope images reveal minute filaments that connect the tip of each stereocilium to its taller neighbor. It is hypothesized that these filaments are fine molecular attachments to the ion channels, and that they act like springs that open the channels. If the taller neighboring stereocilium is bent away, the spring tightens and the ion channel is opened. If the taller neighbor bends toward its shorter neighbor, the spring is relaxed and the channel closes (see Figure 45.8).

Auditory systems use hair cells to sense sound waves

The stimuli that animals perceive as sounds are pressure waves. Auditory systems use mechanoreceptors to convert pressure waves into receptor potentials. Auditory systems include special structures that gather sound waves, direct them to the sensory organ, and amplify their effect on the mechanoreceptors. A good example of an auditory system is the human ear (**Focus: Key Figure 45.9**), which can be divided into three major areas: the outer, middle, and inner ear (see Figure 45.9A).

OUTER EAR   The outer ear consists of the pinnae (singular, pinna) and the auditory canal. The pinnae collect sound waves and direct

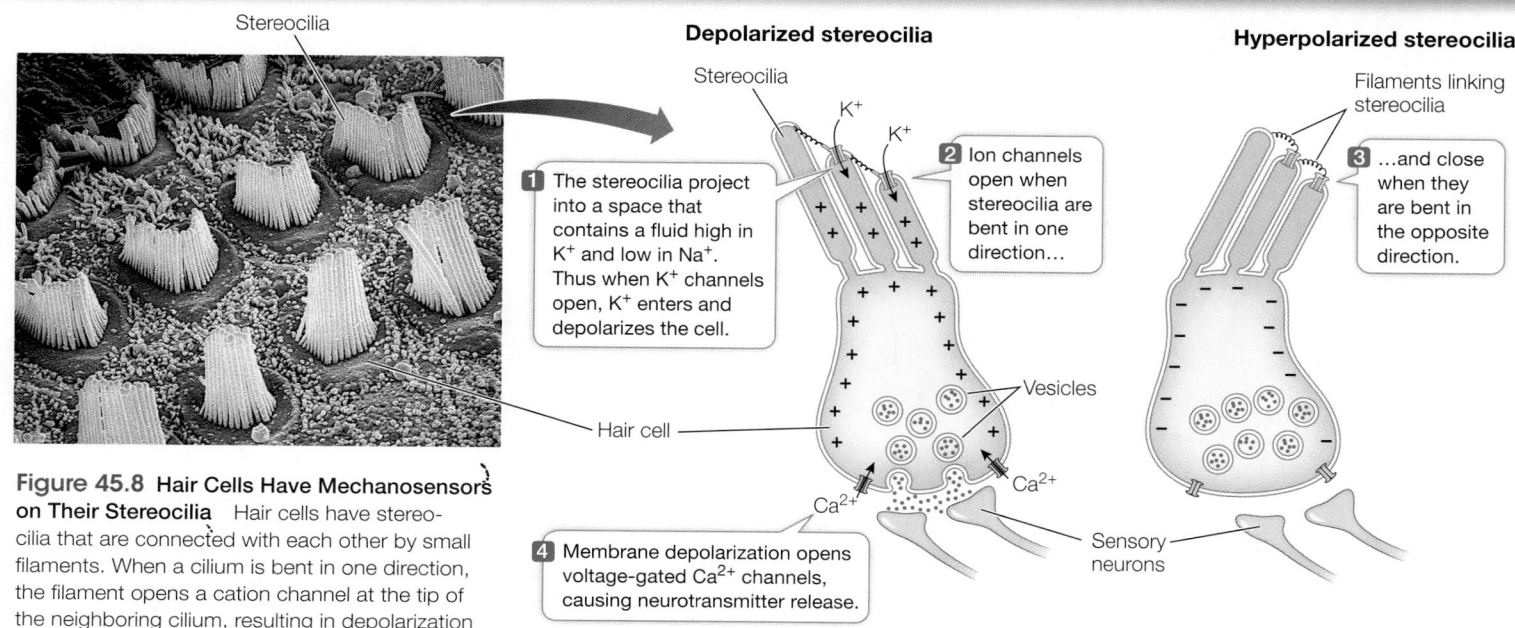

Figure 45.8 **Hair Cells Have Mechanosensors on Their Stereocilia** Hair cells have stereocilia that are connected with each other by small filaments. When a cilium is bent in one direction, the filament opens a cation channel at the tip of the neighboring cilium, resulting in depolarization of membrane potential. When depolarized, the hair cell releases neurotransmitter onto a sensory neuron. Movement of the hair cell in the opposite direction closes the cation channel and the cell hyperpolarizes, causing a reduction in its release of neurotransmitter.

them into the auditory canals; watch a dog change the orientation of its ears to focus on a particular sound to get the idea of the role pinnae play in hearing. Covering the end of the auditory canal is the **tympanic membrane** (commonly called the eardrum) which vibrates in response to pressure waves traveling down the canal, thus converting the pressure waves to physical forces in the middle ear.

MIDDLE EAR On the other side of the tympanic membrane is the middle ear, an air-filled cavity connected to the throat at the back of the mouth through the **eustachian tube** (also called the pharyngotympanic tube). Because the eustachian tube is also filled with air, pressure can equilibrate between the middle ear and the environment. When you have a cold, the tube can be blocked by mucus or tissue swelling and you may have difficulty equilibrating the pressure in the middle ear with the outside air pressure (as you have to do when changing altitude in an airplane).

The middle ear contains the **ossicles**, three delicate bones individually named the malleus ("hammer"), incus ("anvil"), and stapes ("stirrup") (see Figure 45.9B). The ossicles transmit the vibrations of the tympanic membrane to another flexible membrane, the **oval window**. The ossicles act as a lever (like a hammer pulling out a nail), translating a large movement of the tympanic membrane into a smaller movement of the oval window, but a movement of greater force. Because the oval window is much smaller than the tympanic membrane, the pressure the stapes transmits to the oval window is more than 20 times greater than the pressure exerted by a sound wave on the tympanic membrane.

Behind the oval window lies the fluid-filled inner ear. Movements of the oval window impart pressure changes to that enclosed fluid. These pressure waves are transduced into action potentials.

INNER EAR The inner ear is a bony structure consisting of two sets of canals. One is the organ of balance, the **vestibular system**, and the other is the organ of hearing, the **cochlea**. The cochlea ("snail" or "spiral shell") is a long, tapered, coiled structure. A cross section of the cochlea reveals that it is composed of three parallel canals separated by two membranes, the vestibular membrane and the basilar membrane (see Figure 45.9C). Sitting on the basilar membrane is the **organ of Corti**, which transduces pressure waves into action potentials. The organ of Corti contains hair cells with stereocilia (see Figure 45.8). The tips of the hair cells are embedded in a gelatinous overhanging shelf called the tectorial membrane (see Figure 45.9D).

Stereocilia are not motile, but because their tips are attached to the more rigid tectorial membrane, stereocilia bend when the basilar membrane flexes. As shown in Figure 45.8, when the stereocilia are bent in one direction, the hair cell is depolarized by an inflow of K^+ ions. This effect of opening K^+ channels happens because the fluid in the cochlear canal has a higher K^+ concentration than does the intracellular fluid of the hair cell. The response of the hair cell is a graded membrane potential. The hair cells do not fire action potentials, but the changes in their membrane potential alter the rate at which the hair cells release neurotransmitter onto the sensory neurons whose axons make up the auditory nerve and transmit action potentials to the brain.

The vestibular and tympanic canals are separate until they reach the distal end of the cochlea (the end farthest from the oval window), where they join; thus they form one continuous canal that turns back on itself (see Figure 45.10). Just as the oval window is a flexible membrane at the beginning of the vestibular canal, the **round window** is a flexible membrane at the end of the tympanic canal. When the oval window vibrates, the waves of fluid pressure create traveling waves of flexion in the basilar membrane.

Flexion of the basilar membrane is perceived as sound

What causes the basilar membrane to flex, and how does this mechanism distinguish sounds of different frequencies? Air is highly

focus: key figure

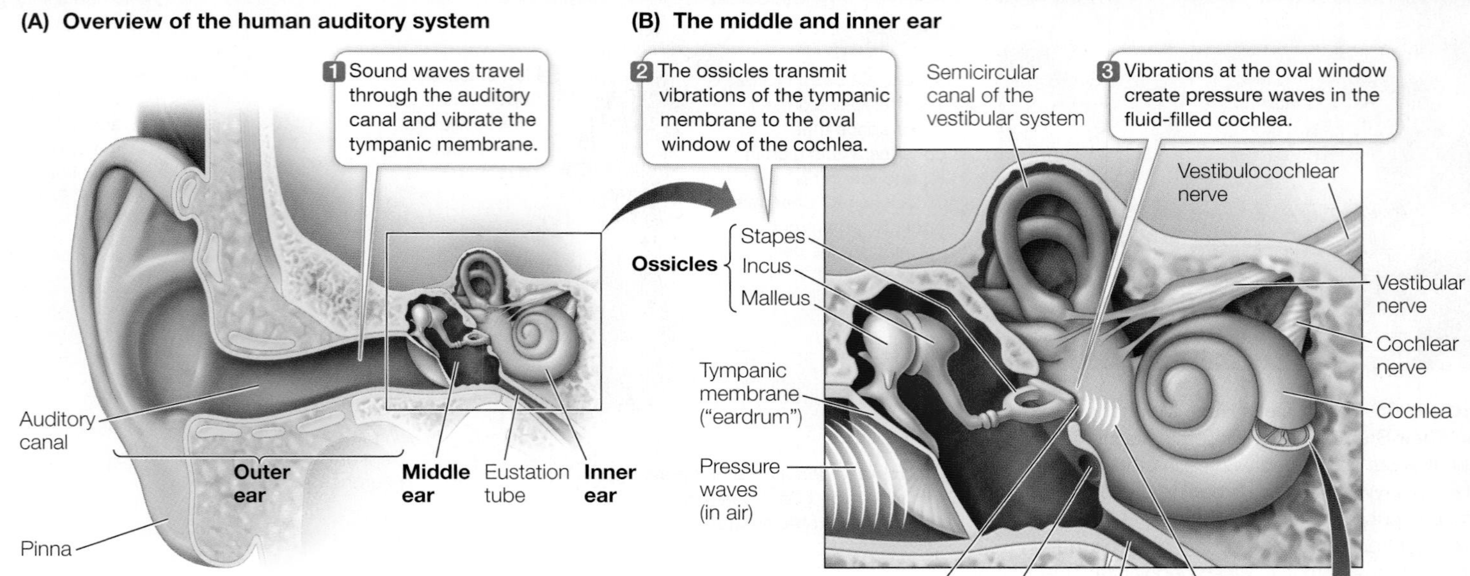

(A) Overview of the human auditory system

1 Sound waves travel through the auditory canal and vibrate the tympanic membrane.

Auditory canal

Outer ear   **Middle ear**   Eustation tube   **Inner ear**

Pinna

Figure 45.9 Structures of the Human Ear   **(A)** The pinnae direct sound waves down the auditory canal to impinge on the tympanic membrane. The tympanic membrane mechanically transmits these pressure waves into movements of the ossicles in the middle ear. **(B)** The ossicles transmit their movement into pressure waves in the fluid of the cochlea at the oval window. **(C)** The cochlea is divided into fluid-filled canals; pressure waves from the ossicles cause the membranes between the canals to flex. **(D)** Flexing of the basilar membrane bends stereocilia on hair cells in the organ of Corti.

Q: How can the ears detect directionality of a sound source?

 Activity 45.1 Structures of the Human Ear
www.Life11e.com/ac45.1

(B) The middle and inner ear

2 The ossicles transmit vibrations of the tympanic membrane to the oval window of the cochlea.

Semicircular canal of the vestibular system

3 Vibrations at the oval window create pressure waves in the fluid-filled cochlea.

Vestibulocochlear nerve

Ossicles — Stapes / Incus / Malleus

Tympanic membrane ("eardrum")

Pressure waves (in air)

Oval window (under stapes)   Round window   Eustachian tube   Pressure waves (in air)

Vestibular nerve
Cochlear nerve
Cochlea

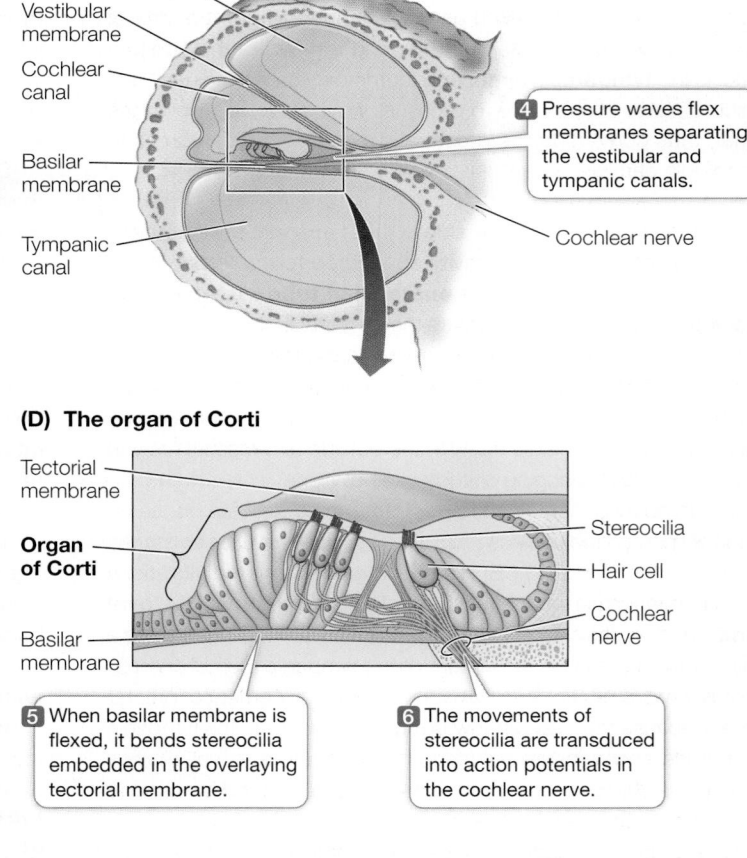

(C) Cross section of the cochlea

Vestibular canal
Vestibular membrane
Cochlear canal
Basilar membrane
Tympanic canal

4 Pressure waves flex membranes separating the vestibular and tympanic canals.

Cochlear nerve

(D) The organ of Corti

Tectorial membrane
Organ of Corti
Basilar membrane

Stereocilia
Hair cell
Cochlear nerve

5 When basilar membrane is flexed, it bends stereocilia embedded in the overlaying tectorial membrane.

6 The movements of stereocilia are transduced into action potentials in the cochlear nerve.

compressible but fluids are not; therefore a pressure wave can travel through air without displacing much air, whereas a pressure wave in fluid displaces that fluid. When the stapes pushes on the oval window, the fluid in the vestibular canal is displaced. If the movement of the oval window occurs slowly, the cochlear fluid pressure wave travels down the vestibular canal, around the bend, and back through the tympanic canal (**Figure 45.10**). At the end of the tympanic canal, the displacement pressure is dissipated by the outward bulging of the round window.

The basilar membrane is not uniform—it is thicker and stiffer at its base and wider and thinner at its apical end. Pressure waves in the cochlear fluid have different frequencies and set up different patterns of traveling waves. High-frequency waves cause maximal flexion at the basal end of the basilar membrane, whereas low-frequency pressure waves result in maximal flexion at the apical end (see Figure 45.10). Thus different pitches of sound flex the basilar membrane at different locations and activate different sets of hair cells. Action potentials stimulated by the mechanoreceptors at different positions along the organ of Corti travel along the cochlear nerve and are transmitted to different regions of the brain's auditory cortex by the vestibulocochlear nerve.

Figure 45.10 Sensing Pressure Waves in the Inner Ear
Pressure waves of different frequencies flex the basilar membrane at different locations. Information about sound frequency is specified by which hair cells are activated. For simplicity, this representation illustrates the cochlea as uncoiled.

Various types of damage can result in hearing loss

There are two general types of acquired hearing loss, or deafness. Conduction deafness is caused by the loss of function of the tympanic membrane and/or the ossicles of the middle ear. Repeated infections of the middle ear can cause scarring of the tympanic membrane and stiffening of the connections between the ossicles. The consequence is less efficient conduction of sound waves from the tympanic membrane to the oval window. With increasing age, the ossicles inevitably stiffen, resulting in a gradual loss of the ability to hear high-frequency sounds.

Nerve deafness is caused by damage to the inner ear or the auditory pathways. A common cause of nerve deafness is damage to the hair cells of the delicate organ of Corti by exposure to loud sounds such as jet engines, pneumatic drills, or highly amplified music. Consistent exposure to sounds above 85 decibels can damage hearing; this damage is cumulative and irreversible. Even using earphones can put you at risk for hearing loss because they generate high-pressure sound waves close to the tympanic membrane. Personal stereo earphones can reach 120 decibels, and people commonly use them at 100 decibels (equivalent to being at a rock concert).

The vestibular system uses hair cells to detect forces of gravity and momentum

Hair cells in the vestibular system of the inner ear detect the position and movement of the head—information essential for maintaining equilibrium (balance). Information from the vestibular system is also crucial for the control of eye movements. When you look at something, you can move your head while staying focused on the object because of your vestibulo-ocular reflex.

In the mammalian inner ear, the vestibular system consists of three bony **semicircular canals** and a bony chamber called the **vestibule**. Within each canal is a membranous semicircular duct, and within the vestibule are the membranous saccule and the utricle. The ducts and the saccule and utricle are filled with fluid. In the semicircular ducts, the fluid shifts when the head changes position (**Figure 45.11A**). Since the three semicircular canals have different orientations, the fluid in their ducts responds differentially to the direction of movement. Projecting into the base of each duct is a **cupula**, a gelatinous swelling enclosing a cluster of hair cell stereocilia. When the shifting fluid pushes on the cupula, it bends the stereocilia and causes a graded potential in the hair cell membranes.

The stereocilia in the saccule and utricle are bent in a different way. These stereocilia are embedded in gelatinous membranes that contain **otoliths** ("ear stones") which are crystals of calcium carbonate. When the head changes position or when it accelerates or decelerates, gravitational forces are exerted on the otoliths and the stereocilia bend (**Figure 45.11B**).

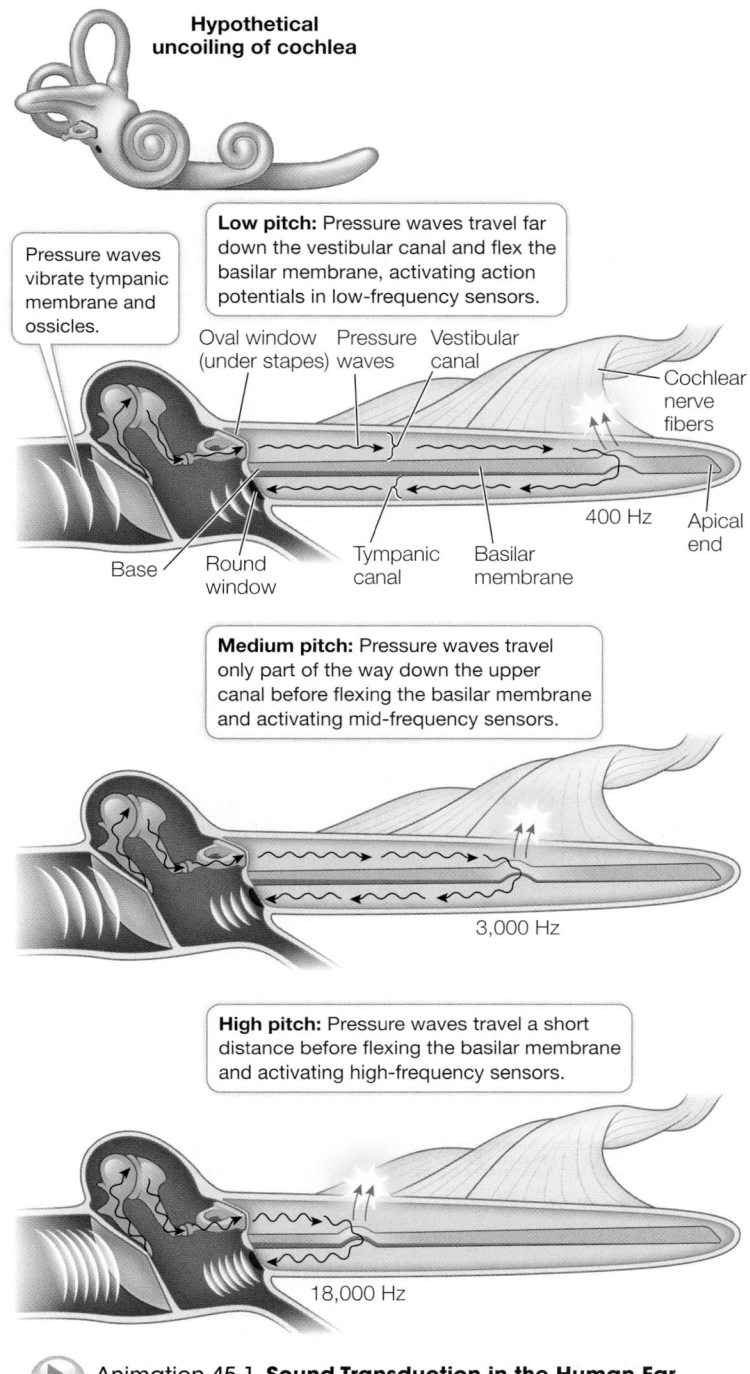

Hypothetical uncoiling of cochlea

Pressure waves vibrate tympanic membrane and ossicles.

Low pitch: Pressure waves travel far down the vestibular canal and flex the basilar membrane, activating action potentials in low-frequency sensors.

Oval window (under stapes) Pressure waves Vestibular canal

Cochlear nerve fibers

Base Round window Tympanic canal Basilar membrane 400 Hz Apical end

Medium pitch: Pressure waves travel only part of the way down the upper canal before flexing the basilar membrane and activating mid-frequency sensors.

3,000 Hz

High pitch: Pressure waves travel a short distance before flexing the basilar membrane and activating high-frequency sensors.

18,000 Hz

 Animation 45.1 **Sound Transduction in the Human Ear**
www.Life11e.com/a45.1.

As in the cochlea, the hair cells of the vestibular system do not fire action potentials, but they release neurotransmitter at synapses with sensory neurons, which in turn fire action potentials.

Motion sickness is a common malady experienced as seasickness or car sickness. This condition occurs when the vestibulular system and the visual system produce conflicting information. The vestibular system may be indicating movement when the visual system is not, or vice versa.

 Activity 45.2 **Mechanoreceptors Simulation**
www.Life11e.com/ac45.2

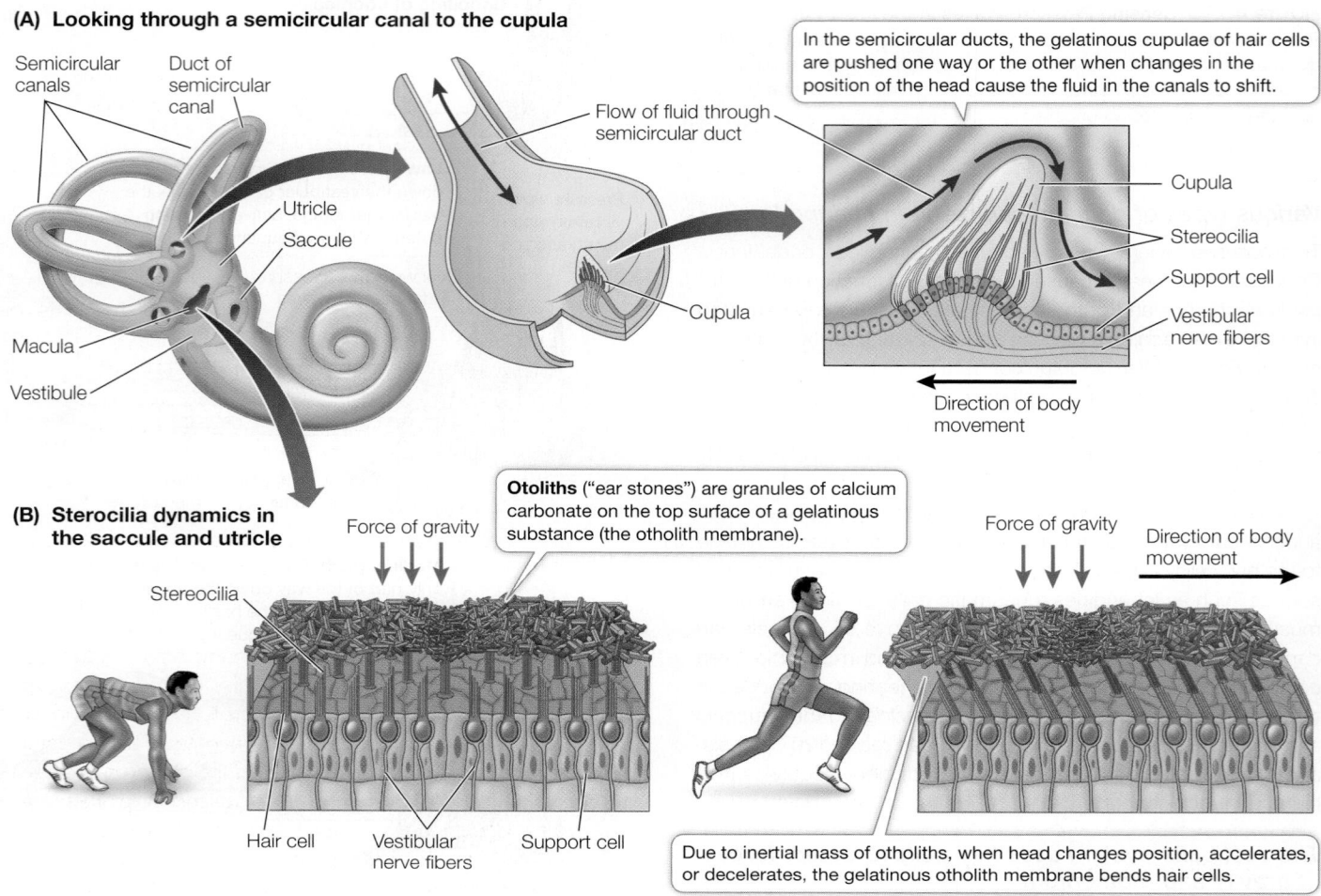

(A) Looking through a semicircular canal to the cupula

Semicircular canals

Duct of semicircular canal

Utricle

Saccule

Macula

Vestibule

Flow of fluid through semicircular duct

Cupula

In the semicircular ducts, the gelatinous cupulae of hair cells are pushed one way or the other when changes in the position of the head cause the fluid in the canals to shift.

Cupula

Stereocilia

Support cell

Vestibular nerve fibers

Direction of body movement

(B) Sterocilia dynamics in the saccule and utricle

Force of gravity

Stereocilia

Otoliths ("ear stones") are granules of calcium carbonate on the top surface of a gelatinous substance (the otolith membrane).

Force of gravity

Direction of body movement

Hair cell

Vestibular nerve fibers

Support cell

Due to inertial mass of otoliths, when head changes position, accelerates, or decelerates, the gelatinous otolith membrane bends hair cells.

Figure 45.11 Organs of Equilibrium The vestibular system consists of bony chambers and fluid-filled canals. **(A)** Each semicircular duct has a cupula containing stereocilia. When fluid moves against the cupula, the stereocilia bend. **(B)** In the saccule and utricle, stereocilia are bent by gravitational forces on the otoliths.

45.3 recap

Sensations that derive from mechanoreceptors include touch, tickle, pressure, joint position, muscle load, hearing, and equilibrium. The membranes of mechanoreceptors have ion channels that are opened by distortion of the membrane, resulting in graded receptor potentials that cause the cell to release neurotransmitter or to fire action potentials. Hair cells involved in hearing and equilibrium have stereocilia that are bent by pressure waves, resulting in receptor potentials.

learning outcomes

You should be able to:

- Describe the various touch sensors and their unique properties.
- Describe how tension in a muscle is sensed and adjusted for load.
- Explain how sound waves of different frequencies are heard as sounds of different pitches.

1. Why are there different types of mechanoreceptor cells that have different rates of adaptation?

2. Explain the role of specific mechanoreceptors when you are (a) holding a glass that is being filled and (b) lifting heavy furniture.

3. How do different frequencies of pressure waves in the air result in action potentials being fired in different acoustic neurons?

We next turn to vision, perhaps the most elaborate of the senses.

key concept 45.4 Photoreceptors Respond to Light

Sensitivity to light—**photosensitivity**—confers on the simplest animals the ability to orient to the sun and sky, and it gives more complex animals rapid and extremely detailed information about objects in their environment. Photosensitivity is ubiquitous in the animal kingdom, and the molecular basis for that sensitivity is a family of visual

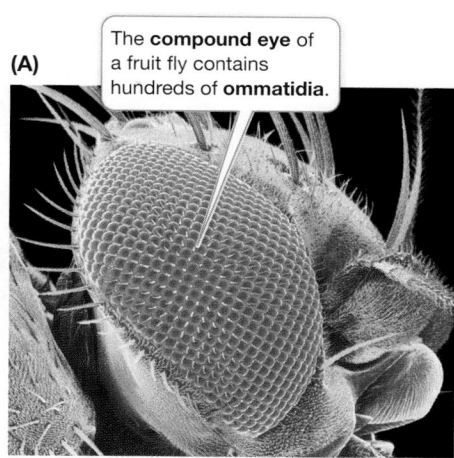

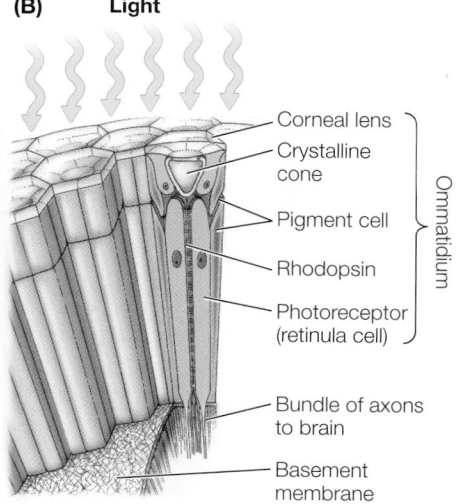

Figure 45.12 Ommatidia: The Functional Units of Insect Eyes **(A)** The micrograph shows the compound eye of a fruit fly. **(B)** The rhodopsin-containing retinula cells are the photoreceptors in ommatidia.

The **compound eye** of a fruit fly contains hundreds of **ommatidia**.

(A)

(B)   Light

Corneal lens
Crystalline cone
Pigment cell
Rhodopsin
Photoreceptor (retinula cell)
Ommatidium

Bundle of axons to brain
Basement membrane

pigments that have been evolutionarily conserved.

focus your learning

- Rhodopsin transduces light energy into changes in membrane potential.
- Vertebrate rods are sensitive to light with high sensitivity; cones are sensitive to color.
- Receptive fields of photoreceptors send their output to ganglion cells along with lateral processing by horizontal and amacrine cells, enabling motion and contrast detection.

In this section you will learn about the structures of eyes, the organs that gather light energy and focus it onto **photoreceptor cells**. Then you will learn how visual pigment molecules of photoreceptors respond when stimulated by light energy and the molecular mechanism that transforms that light energy into action potentials. You will also learn about color vision and how photoreceptor information is processed in the neural layers of the vertebrate eye.

 Animation 45.2 **Photosensitivity**
www.Life11e.com/a45.2

Invertebrates have a variety of visual systems

Photoreceptors and visual pigments are incorporated into a variety of visual systems, from simple to complex. Flatworms obtain directional information about light from photoreceptor cells that are organized into **eye cups**. The eye cups are paired, bilateral structures, each partly shielded from light by a layer of pigmented cells lining the cup.

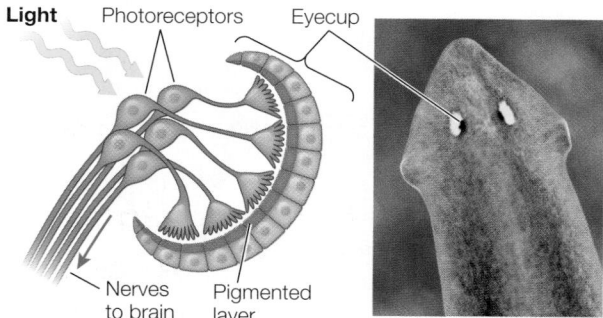

Light   Photoreceptors   Eyecup

Nerves to brain   Pigmented layer

The photoreceptors on the two sides of the animal are unequally stimulated unless the animal is facing directly toward or away from a light source. The flatworm generally uses directional information from the eye cups to move away from light.

Arthropods have **compound eyes** that provide them with information about patterns in the environment. These eyes are called compound because each eye consists of many optical units called **ommatidia** (singular *ommatidium*), each with its own narrow-angle lens (**Figure 45.12**). In contrast, a vertebrate eye consists of just one optical unit with a wide-angle lens. The number of ommatidia in a compound eye varies from only a few in some ants to 800 in fruit flies and to 30,000 in some dragonflies.

Each ommatidium has a lens that directs light onto photoreceptor cells. Flies, for example, have eight elongated photoreceptors in each ommatidium. The inner borders of the photoreceptors are covered with microvilli that contain a photosensitive pigment that traps light. Axons from the photoreceptors send the light information to the nervous system. Since each ommatidium of a compound eye is directed at a slightly different part of the visual world, only a low-resolution (pixillated) image can be communicated from the compound eye to the CNS.

Image-forming eyes evolved independently in vertebrates and cephalopods

Both vertebrates and cephalopod mollusks (such as squid and octopus) have eyes with exceptional abilities to form detailed images of the visual world. Like cameras, both of these eye types focus inverted images on an internal surface that is sensitive to light. Considering that they evolved completely independently of each other, their degree of similarity is remarkable (**Figure 45.13**).

The vertebrate eye (see Figure 45.13A) is a spherical, fluid-filled structure bounded by a tough connective tissue layer called the **sclera**. The sclera at the front of the eye forms the transparent **cornea**, through which light passes to enter the eye. Just inside the cornea is the pigmented **iris**, which gives the eye its color. The iris controls the amount of light that reaches the photoreceptor cells at the back of the eye, just as the diaphragm of a camera controls the exposure. The central opening of the iris is the **pupil**. The iris is under neural control. In bright light, the iris constricts and the pupil is very small. As light levels fall, the iris opens and the pupil enlarges.

Behind the iris is the crystalline protein **lens** that makes fine adjustments in the focus of images falling on the photosensitive layer—the **retina**—at the back of the eye. The cornea and the gel-like mass (vitreous humor) within the eye bend light rays passing through them

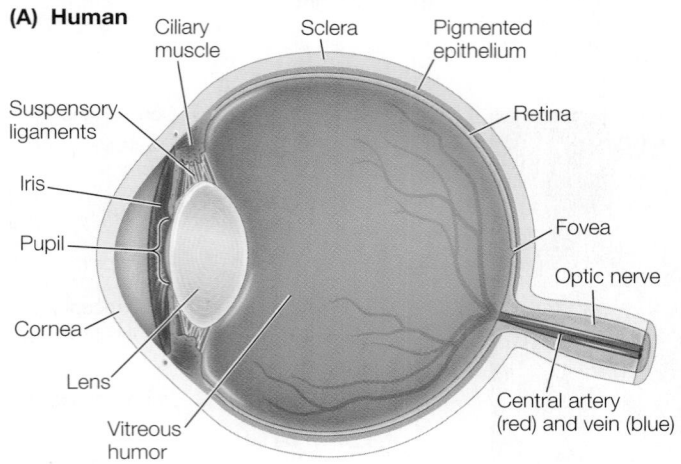

(A) Human

- Ciliary muscle
- Sclera
- Pigmented epithelium
- Suspensory ligaments
- Retina
- Iris
- Pupil
- Fovea
- Optic nerve
- Cornea
- Lens
- Vitreous humor
- Central artery (red) and vein (blue)

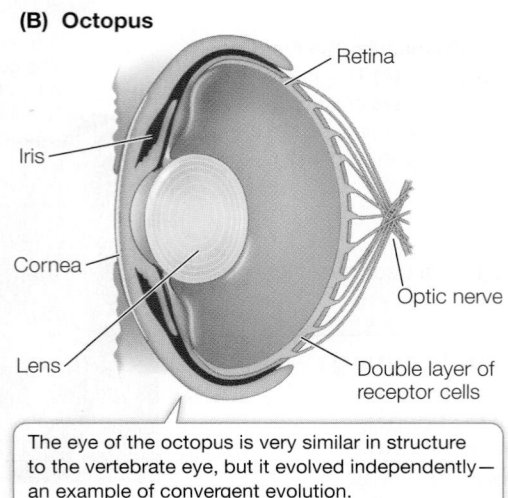

(B) Octopus

- Retina
- Iris
- Cornea
- Lens
- Optic nerve
- Double layer of receptor cells

The eye of the octopus is very similar in structure to the vertebrate eye, but it evolved independently—an example of convergent evolution.

Figure 45.13 Convergent Evolution of Eyes The lenses of vertebrate (A) and cephalopod (B) eyes focus images on layers of photoreceptor cells.

Q: Humans commonly experience "floaters" in their eyes—tissue fragments that float around in the field of vision. Where in the structure of the eye would you expect floaters to exist?

so that they are focused on the retina. The lens makes fine adjustments to the focus and allows the eye to accommodate—that is, to focus on objects at various locations in the near visual field. To focus a camera on objects close at hand, you adjust the distance between the lens and the internal surface sensitive to light. Fish, amphibians, and non-avian reptiles accommodate in a similar manner, moving the lenses of their eyes closer to or farther from their retinas. Mammals and birds use a different method; they alter the shape of the lens.

The mammalian lens is contained in a connective tissue sheath that tends to keep it in a spherical shape, but the sheath is attached to suspensory ligaments that pull the lens into a flatter shape. Circular ciliary muscles counteract the pull of the suspensory ligaments, permitting the lens to round up. When the ciliary muscles are at rest, the flatter lens has the correct optical properties to focus distant images on the retina. Contracting the ciliary muscles rounds up the lens, changing its light-bending properties to bring close images into focus (**Figure 45.14**).

Lenses become less elastic with age, so we lose the ability to focus on objects close at hand without the help of corrective lenses. Most people over the age of 45 need the assistance of reading glasses or bifocal lenses.

▶ Activity 45.3 **Structure of the Human Eye**
www.Life11e.com/ac45.3

The vertebrate retina receives and processes visual information

During embryonic development, neural tissue grows out from the brain to form the retina. In addition to a layer of photoreceptor cells, the retina includes other cells arranged

Figure 45.14 Staying in Focus Mammals and birds focus their eyes by changing the shape of the lens depending on the eye's distance from the object of focus.

in layers that process visual information from the photoreceptors (see Figure 45.20). Light must pass through all the layers of retinal cells before being captured by photosensors. In humans and other day-active animals, the light that is not captured is absorbed by a black-pigmented epithelial tissue layer behind the retina. In contrast, nocturnal animals such as deer and raccoons have an iridescent reflective layer behind their retinas, which maximizes the capture of photons by reflecting them back onto the photoreceptors. This is why a deer in the headlights or your pet dog or cat in a flash photo appears to have bright white eyes. In contrast, humans do not have a white reflective layer in the retina, so flash photographs often have a "red-eye effect," caused by the flash being reflected by the abundant blood vessels in the retina.

Opsins are the universal photoreceptor molecule in animals

Photosensitivity in animals depends on the ability of **opsin** molecules to absorb photons of light and activate a G protein. Opsins are proteins that are not themselves photosensitive, but they contain covalently bound functional groups called 11-*cis*-retinal (**Figure 45.15**). Retinal absorbs photons of light, and the color

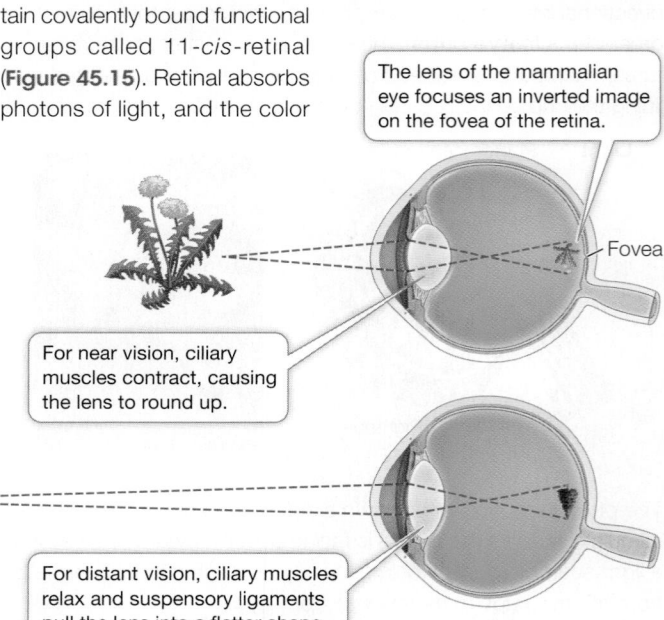

The lens of the mammalian eye focuses an inverted image on the fovea of the retina.

Fovea

For near vision, ciliary muscles contract, causing the lens to round up.

For distant vision, ciliary muscles relax and suspensory ligaments pull the lens into a flatter shape.

of light it is receives depends on the structure of its opsin. There are several different opsins, and the most common one that gives humans sensitivity to low levels of light, but not color, is called **rhodopsin**.

When 11-*cis*-retinal absorbs a photon of light energy, it changes into a different isomer of retinal, called all-*trans*-retinal. This change puts a strain on the bonds between retinal and opsin, changing the conformation of opsin and signaling the detection of light. In vertebrate eyes, the retinal and the opsin eventually separate from each other (a process called bleaching), which causes the opsin to lose its photosensitivity. A series of enzymatic reactions returns the all-*trans*-retinal to the 11-*cis* isomer, which then recombines with opsin so that it once again becomes photosensitive.

How does the conformational change of opsin transduce light into a cellular response? After retinal is converted from the 11-*cis* to the all-*trans* form, its interactions with opsin pass through several unstable intermediate stages. One of these stages triggers a cascade of reactions involving a G protein signaling mechanism that results in the alteration of membrane potential that is the photoreceptor cell's response to light.

Rod and cone cells are the photoreceptors of the vertebrate retina

The photoreceptors of the vertebrate retina are two types of modified neurons called **rod cells** and **cone cells** based on their shapes (**Figure 45.16**). Rod cells are highly light-sensitive and perceive shades of gray in dim light. Cone cells function at high light levels and are responsible for high-acuity color vision. The human retina has about 5 million cones and 100 million rods, but the density of each varies across the retina.

ROD CELLS   Rod cells do not produce action potentials, but they release neurotransmitter from their bases, where they form synapses with the next neurons in the visual pathway. Each rod cell has an outer segment, an inner segment, and a synaptic terminal (see Figure 45.16). The highly specialized outer segment contains stacks of discs made up of cell membranes densely packed with rhodopsin. These discs capture photons of light. The inner segment contains the cell nucleus, mitochondria, and other organelles. The synaptic terminal is where the rod cell communicates with other neurons.

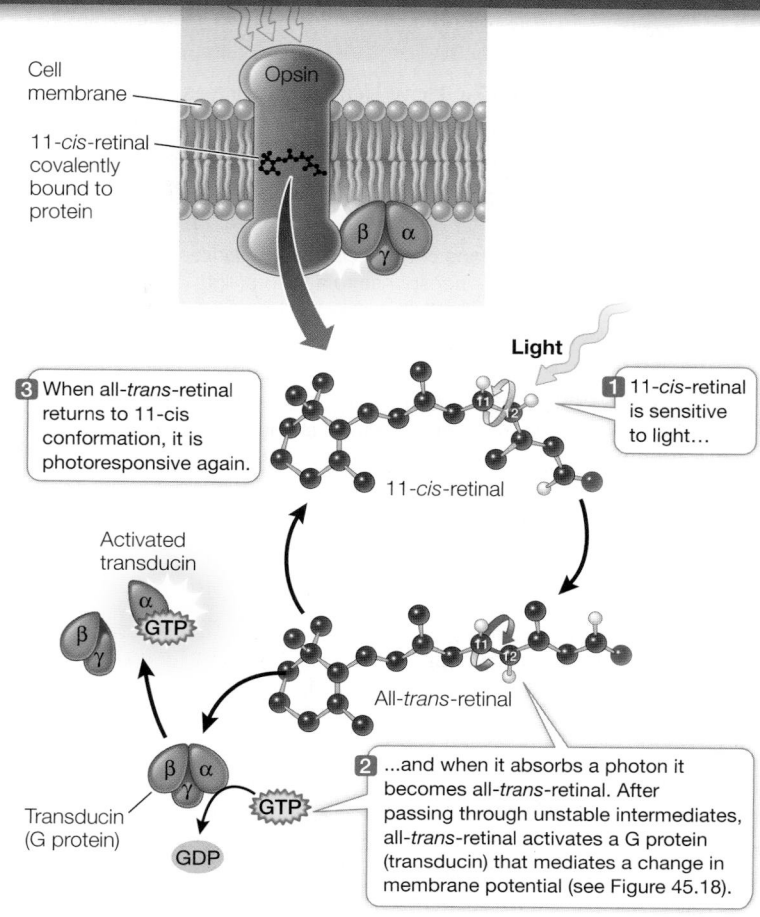

3 When all-*trans*-retinal returns to 11-cis conformation, it is photoresponsive again.

Light

1 11-*cis*-retinal is sensitive to light...

Cell membrane

Opsin

11-*cis*-retinal covalently bound to protein

11-*cis*-retinal

Activated transducin

All-*trans*-retinal

2 ...and when it absorbs a photon it becomes all-*trans*-retinal. After passing through unstable intermediates, all-*trans*-retinal activates a G protein (transducin) that mediates a change in membrane potential (see Figure 45.18).

Transducin (G protein)

Figure 45.15 Light Changes the Conformation of Rhodopsin The light-absorbing molecule 11-*cis*-retinal bonds with the protein opsin to form the vertebrate visual pigment rhodopsin.

To see how a rod cell responds to light, we can penetrate a single rod cell with an electrode and record its membrane potential in the dark and in the light, as shown in **Figure 45.17**. From what you have learned about other types of sensory receptors, you might expect that stimulation of the rod cell by light would make its membrane potential less negative, but the opposite is true—it becomes more negative, that is, it becomes hyperpolarized.

Rod cell outer segments have cGMP gated sodium channels. In the dark, abundant cGMP in the cell keeps these channels open, creating the dark current. When light is flashed on the dark-adapted rod cell, these Na^+ channels close in proportion to the intensity of the light, and the cell's membrane potential becomes more negative—it hyperpolarizes (see Figure 45.17). The rate of neurotransmitter release from the base of the cell changes as membrane potential changes. As the rod cell hyperpolarizes, its release of neurotransmitter decreases.

How does the absorption of light by rhodopsin hyperpolarize the rod cell? As illustrated

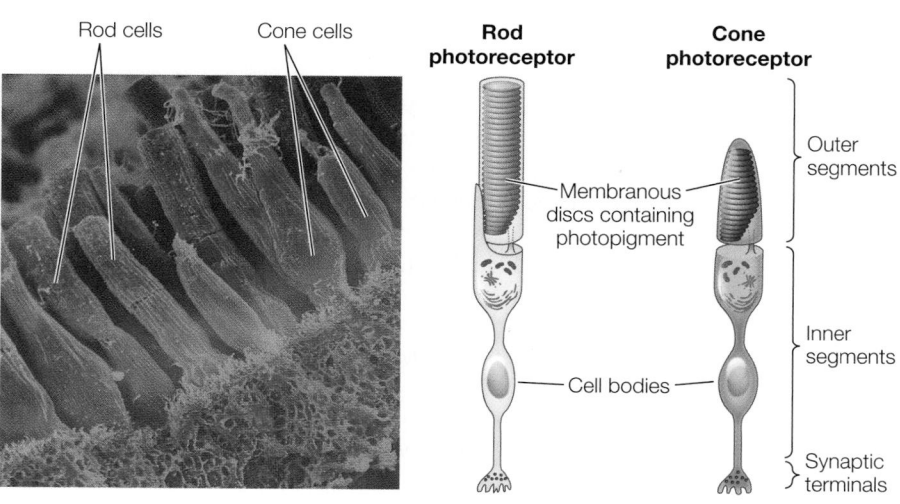

Rod cells   Cone cells

Rod photoreceptor   **Cone photoreceptor**

Membranous discs containing photopigment

Outer segments

Inner segments

Cell bodies

Synaptic terminals

Figure 45.16 Rods and Cones   This scanning electron micrograph of photoreceptors in the retina of a mud puppy (an amphibian) shows cylindrical rods and tapered cones.

experiment

Figure 45.17   How Light Stimulates Rod Cells

Original Paper: Baylor, D. A. and B. J. Nunn. 1986. Electrical properties of the light-sensitive conductance of rods of the salamander *Ambystoma tigrinum*. *Journal of Physiology* 371: 115–145.

The cell membrane of a rod cell hyperpolarizes—becomes more negative—in response to a flash of light. Rod cells do not fire action potentials, but in response to the absorption of light energy, the neuron undergoes a change in membrane potential.

HYPOTHESIS▶   When a rod cell absorbs photons (light energy), its membrane potential changes in proportion to the strength of the light stimulus.

METHOD

1. Record membrane potentials from the inner segment of a rod cell.
2. Stimulate the rod cells with light flashes of varying intensity and record the results.

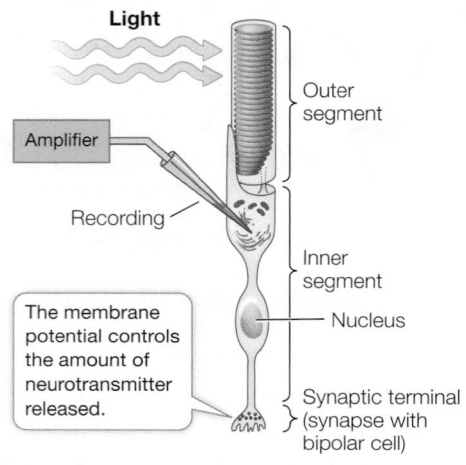

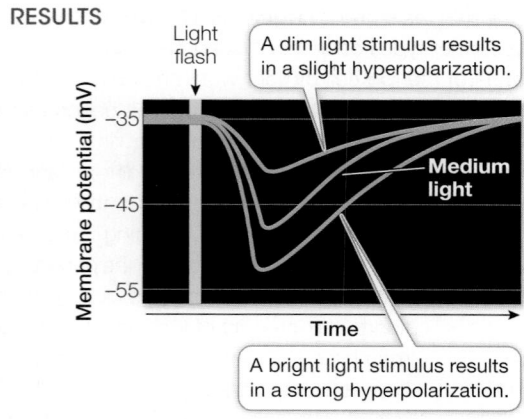

RESULTS

A dim light stimulus results in a slight hyperpolarization.

Medium light

A bright light stimulus results in a strong hyperpolarization.

CONCLUSION▶   The membrane potential of rod cells is depolarized in the dark and hyperpolarizes (becomes more negative) in response to light.

in Figure 45.15, when rhodopsin is excited by light it undergoes conformational changes that initiate a signaling cascade. The dark-adapted rod cell has open Na^+ channels—the depolarizing dark current

(**Figure 45.18**). Light photons excite rhodopsin, which activates a G protein called transducin. Activated transducin activates a phosphodiesterase (PDE). Activated PDE converts cyclic GMP (cGMP) to GMP, which causes the cGMP gated Na^+ channels to close, changing the balance between the inward flow of Na^+ and the outward flow of K^+ and causing the cell to hyperpolarize.

This cascade may seem like a roundabout way of doing business, but its advantage is its enormous amplification ability. Each molecule of light-excited rhodopsin can activate several hundred transducin molecules, thus activating a large number of PDE molecules. The catalytic capacity of PDE is great—one PDE molecule can hydrolyze several hundred molecules of cGMP per second. The bottom line is that a single photon of light can result in the closure of a huge number of Na^+ channels.

CONE CELLS   Cone cells are responsible for the high-acuity color vision of day-active vertebrates such as humans. It is therefore logical that the highest density of cone cells is in the area of the retina that receives light from the center of the visual field, a region called the **fovea** (see Figure 45.13A). The human fovea has about 160,000 cones per square millimeter. But humans are not the champions of high-acuity vision; a hawk's fovea has almost twice that number of cones, making the hawk's vision much sharper than ours. Birds also have *two* foveae in each eye; one receives light from straight ahead, the other from a more lateral field of vision. The forward-looking foveae make binocular vision possible, while the lateral-looking foveae provide high-acuity vision. Birds use both sets of foveae by frequently turning their heads slightly; they cannot move their eyes in the sockets as humans can.

Cones have low sensitivity to light and contribute little to night vision. Night vision depends mostly on rod cells, which is why vision in dim light is mostly in shades of gray and acuity is low. You may have trouble seeing a small object at night when you are looking straight at it—that is, when its image is falling on your fovea. If you look a little to the side, so that the image falls on a rod-rich area of your retina, you see the object better. Astronomers looking for faint objects in the sky learned this trick a long time ago. The retinas of nocturnal animals, such as flying squirrels, contain a high percentage of rods. By contrast, some animals that are active only during the day (such as chipmunks) have mostly cones in their retinas.

The human retina has three kinds of cone cells, each containing slightly different opsin molecules that differ in the wavelengths of light they absorb best. Although the same 11-*cis*-retinal group is the light-absorber in all three kinds of cones, its molecular interactions with opsin determine the spectral sensitivity of the cone cell as a whole (**Figure 45.19**). Those that absorb maximally around 419 nanometers (nm) are called S-cones for short wavelength (blue range). Those that absorb maximally around 531 nm are called M-cones for medium wavelength (green range), and those that absorb maximally around 559 nm are called L-cones for long wavelength (yellow-red range). Because different wavelengths of light are differentially absorbed by the different cone cell visual pigments, the brain interprets the relative inputs from the different classes as a full range of color. Color blindness in humans results from the absence or dysfunction of one or more of the three classes of cone cells. Some mammals have only one or two classes of cone cells, whereas birds have four.

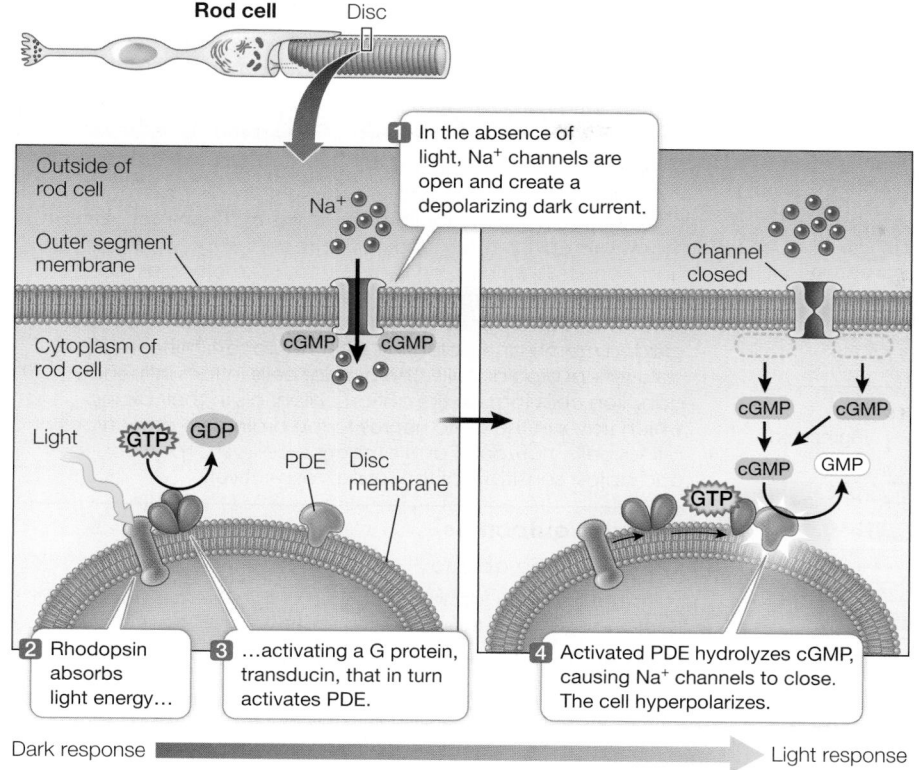

Figure 45.18 **Light Absorption Closes Sodium Channels**   The absorption of light by rhodopsin initiates a signaling cascade that hyperpolarizes the rod cell.

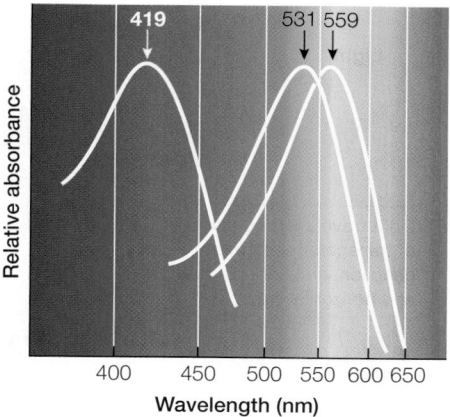

Figure 45.19 **Absorption Spectra of Cone Cells**   The three kinds of cone cells contain slightly different visual pigments that absorb different wavelengths of light.

Information flows through layers of neurons in the retina

The human retina is organized into layers of neurons that receive visual information and process it before sending it to the brain (**Figure 45.20**). Closest to the lens (and thus to light input) is a layer of **ganglion cells**; a central layer contains three neuronal types, **bipolar cells**, **horizontal cells**, and **amacrine cells**; and at the "rear" of the retina lie the photoreceptors (rods and cones). The layers of cells between the photoreceptors and the ganglion cells process information about the visual field.

GANGLION AND BIPOLAR CELLS   From our discussion of rod cells, you know that the photoreceptor cells at the back of the retina hyperpolarize in response to light and do not generate action potentials. The ganglion cells at the front of the retina do, however, fire action potentials. The axons of ganglion cells form the **optic nerves** that travel to the brain.

The ganglion cells are connected to the photoreceptors by bipolar cells. Changes in the membrane potential of rods and cones in response to light alter the rates at which the rods and cones release neurotransmitter at their synapses with the bipolar cells. In response to this neurotransmitter, the membrane potentials of the bipolar cells change, altering the rate at which they release neurotransmitter onto ganglion cells. The neurotransmitter release from the bipolar cells determines the rate at which ganglion cells fire action potentials. Thus the direct flow of information in the retina is from photoreceptor to bipolar cell to ganglion cell. The ganglion cells send the information to the brain via the optic nerves.

Each human eye contains about 1.2 million ganglion cells but more than 100 million rods and cones. Therefore there must be convergence of information as it passes from the photoreceptors to the ganglion cells. A given bipolar cell can receive input from multiple rods or multiple cones, but not from both. The relationship between photoreceptors, bipolar cells, and ganglion cells depends on their location on the retina. In the fovea, a ganglion cell may receive input from as few as five photoreceptors, but in the periphery of the retina, a ganglion cell may receive input from thousands of photoreceptors. Visual acuity is a reflection of these quantitative relationships.

The patch of photoreceptors that communicates with a ganglion cell forms a circular receptive field. When light falls on a receptive field, its ganglion cell can be either excited or inhibited. As mentioned above, each ganglion cell sends an axon to the brain in the optic nerves. Thus the information coming from the retina to the brain is about the pattern of patches of light and dark falling on the retina. We'll return to a discussion of how the brain constructs the visual field from the input from receptive fields in Chapter 46.

HORIZONTAL AND AMACRINE CELLS   The other two cell layers, the horizontal cells and the amacrine cells, consist of interneurons that communicate laterally across the retina. Horizontal cells form synapses with neighboring photoreceptors and bipolar cells. Thus light falling on one photoreceptor can influence the sensitivity of its neighbors to light. This lateral flow of information enables the retina to sharpen the perception of contrast between light and dark patterns. Amacrine cells form local interconnections between bipolar cells and ganglion cells. Some amacrine cells are highly sensitive to changing illumination or to motion. Others assist in adjusting the sensitivity of the eyes according to the overall level of light falling on the retina. When background light levels change, amacrine cell connections to the ganglion cells adjust the range of intensities to which they are sensitive. Thus even with large changes in background illumination such as going from outdoors into a

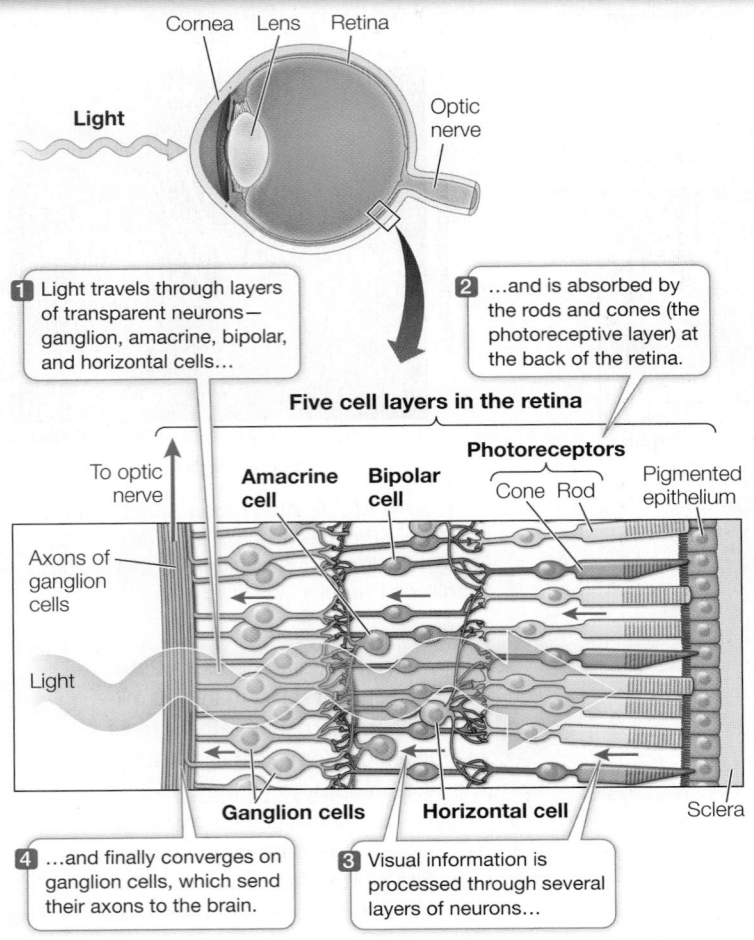

Light — Cornea   Lens   Retina   Optic nerve

1 Light travels through layers of transparent neurons—ganglion, amacrine, bipolar, and horizontal cells...

2 ...and is absorbed by the rods and cones (the photoreceptive layer) at the back of the retina.

Five cell layers in the retina

To optic nerve   **Amacrine cell**   **Bipolar cell**   **Photoreceptors** Cone Rod   Pigmented epithelium

Axons of ganglion cells

Light

Ganglion cells   **Horizontal cell**   Sclera

4 ...and finally converges on ganglion cells, which send their axons to the brain.

3 Visual information is processed through several layers of neurons...

Figure 45.20 The Human Retina   Five layers of neurons receive and process visual information. The rods and cones are photoreceptors. The other four layers are the ganglion cells, the bipolar cells, the horizontal cells, and the amacrine cells, all of which are involved in transmitting light signals received by the photoreceptors to the brain.

 Media Clip 45.1 Into the Eye
www.Life11e.com/mc45.1

 Activity 45.4 Structure of the Human Retina
www.Life11e.com/ac45.4

of magnitude, modulation by amacrine cells can rapidly shift that functional range to accommodate a wide range of light intensities.

45.4 recap

Vertebrates have image-forming eyes in which a lens focuses incoming light onto the photosensitive retina. Opsin molecules are responsible for light sensitivity in all animals, and in mammals opsins are contained in receptor cells called rods and cones. Slight differences in the opsin molecules result in differences in spectral sensitivity, which is the basis for color vision. Rods and cones do not fire action potentials, but their graded membrane potentials alter neurotransmitter release onto retinal bipolar cells. The bipolar cells in turn influence ganglion cells that do fire action potentials in their axons, which travel in the optic nerves to the brain. Two other types of retinal cells, horizontal and amacrine cells, sharpen contrast and adjust sensitivity of the retina, respectively.

learning outcomes

You should be able to:

- Trace the steps for how a photon of light is transduced into a change in membrane potential.
- Compare the properties and locations of rod and cone cells in the retina.
- Describe the structural implications for movement detection and visual acuity.
- Explain ganglion cell receptive fields in terms of the lateral processing afforded by the horizontal and amacrine cells.

1. Explain why a compound eye is better at detecting movement than at forming images.
2. How do photons of light change the receptor potential of rod cells?
3. Explain how rods and the various cones differ in their spectral sensitivity.
4. If you are looking straight ahead and someone holds up a colored card on your right or left side, why can't you tell what the color is?
5. Give two reasons why your vision is impaired when you come from bright outside light into a dimly lit room.

building, the eyes are still sensitive to small, rapid changes in the pattern of light falling on the retina. The light reflected from this page can vary by nine orders of magnitude (1,000,000,000-fold) between a sunny outdoor environment and a moonless night. Even though our eyes function best over a light-intensity range of only two orders

Knowing the path of information from sensory receptor cells to the CNS still does not tell us how that information is processed by the brain. What does the eye tell the brain, for example, in response to a pattern of light falling on the retina? In Chapter 46 you will learn how the mammalian brain reassembles sensory information into our perception of the world.

 investigating life

How do pit vipers "see" in the dark?

The pit organs of rattlesnakes are the infrared detectors that signal the presence of warm-blooded prey. In this chapter we considered how the pit organs transduce infrared radiation into neural information. Infrared is a portion of the electromagnetic radiation spectrum, as is visible light. So could the receptor cells in the pit organs be

modified photoreceptors? Or have they evolved from the types of sensory neurons in the skin that sense temperature? The temperature-sensitive neuron theory seems more likely since there is no rhodopsin in the pit organs. But how did such neurons acquire an acute sensitivity to minute changes in temperature of the pit organ caused by distant infrared emitters? The experiments covered in Investigating Life: How Do Pit Vipers "See" in the Dark? identified a particular TRP channel that is highly expressed in the trigeminal ganglia but not in the dorsal root ganglia of rattlesnakes. These experiments also revealed that the TRPA1 channel is not highly expressed in the TG or DRGs of non-pit snakes, but is highly expressed in a distantly related pit snake, the python. In the online exercise to Investigating Life: How Do Pit Vipers "See" in the Dark? it was determined that expression of the *TRPA1* gene in frog oocytes causes them to have a steep temperature sensitivity in the range of 36°C to 42°C. Thus we can conclude that the TRPA1 channel has evolved along with the pit organ to enable pit vipers such as the rattlesnake to detect warm-blooded prey.

Future directions

The identification of specific genes that convey specific sensory capabilities—for example, identification of the different opsin genes for color vision and the *TRPA1* gene for infrared detection—raise the possibility that genetic engineering efforts could be directed toward expanding the sensory capabilities of animals, and even humans. What would it be like to be able to see in the infrared and the ultraviolet ranges or to hear infrasound and ultrasound? Would the human brain be able to process and integrate that huge amount of additional information? How would our conscious perception of reality be changed?

Chapter Summary 45

45.1 Sensory Receptor Cells Convert Stimuli into Action Potentials

- Sensory receptor cells, also known as **sensors** or **receptors**, transduce information about an animal's external and internal environment into action potentials that the brain perceives as different forms of sensory information.

- **Receptor potentials** can spread to regions of the cell membrane that generate action potentials. Some sensors do not fire action potentials but release neurotransmitter onto sensory neurons that do fire action potentials. **Review Figure 45.1**

- Sensors have **receptor proteins** that cause ion channels to open or close, affecting the receptor cell's membrane potential. Some receptors open ion channels physically through forces such as pressure or stretch. Other receptors act through signal transduction pathways to generate receptor potentials. A large family of genes called transient receptor potential (TRP) genes is responsible for many modalities of sensory transduction. **Review Figure 45.2, Investigating Life: How Do Pit Vipers "See" in the Dark?**

- The interpretation of action potentials as particular sensations depends on which neurons in the central nervous system receive them.

- **Adaptation** enables the nervous system to ignore irrelevant or continuous stimuli while remaining responsive to relevant or new stimuli.

45.2 Chemoreceptors Respond to Specific Molecules

- **Chemoreceptors** are responsible for **olfaction**, **gustation**, and the sensing of **pheromones**.

- Mammalian **olfactory receptor neurons (ORNS)** project directly to the **olfactory bulb** of the brain. ORNs for the same **odorant** project to the same area of the olfactory bulb.

- Each ORN expresses one receptor protein that can bind a specific type of molecule or ion. Binding causes a second messenger to open ion channels, which creates an action potential. **Review Figure 45.3**

- Pheromones are chemicals that communicate information among individuals of the same species.

- In vertebrates, **taste buds** in the mouth cavity are responsible for gustation. The five basic tastes are sweet, salty, sour, bitter, and umami. **Review Figure 45.5**

45.3 Mechanoreceptors Respond to Physical Forces

- The skin contains a variety of **mechanoreceptors** that respond to touch and pressure. The density of mechanoreceptors in any skin area determines the sensitivity of that area. **Review Figure 45.6**

- Stretch receptors in **muscle spindles** and in **Golgi tendon organs** inform the CNS of the positions of and loads on parts of the body. **Review Figure 45.7**

- **Hair cells** are mechanoreceptors of the auditory and vestibular systems. Physical bending of the hair cells' **stereocilia** alters their receptor proteins and therefore their membrane potentials. **Review Figure 45.8**

- In mammalian auditory systems, ear pinnae collect and direct sound waves to the **tympanic membrane**, which vibrates in response to sound waves. The movements of the tympanic membrane are amplified through a chain of **ossicles** that conduct the vibrations to the **oval window**. Movements of the oval window create pressure waves in the fluid-filled **cochlea**. **Review Focus: Key Figure 45.9, Activity 45.1**

- The basilar membrane running down the center of the cochlea is distorted by pressure waves at specific locations that depend on the frequency of the wave. These distortions cause hair cells in the **organ of Corti** to bend and to release neurotransmitter, generating action potentials in the cochlear nerve that are transmitted to the auditory cortex of the brain. **Review Figure 45.10, Animation 45.1**

- Hair cells are also the mechanoreceptors of the organs of equilibrium in the mammalian vestibular system, which include the semicircular canals and the saccule and utricle. **Review Figure 45.11, Activity 45.2**

45.4 Photoreceptors Respond to Light

- Visual systems range from the simple **eye cups** of flatworms, which sense the direction of a light source, to the **compound eyes** of arthropods, which detect shapes and patterns, to the image-forming eyes of vertebrates and cephalopods. **Review Figures 45.12, 45.13**

- Vertebrate and cephalopod eyes focus detailed images of the visual field onto dense arrays of photoreceptors that transduce the visual image into neural signals. **Review Figures 45.13, 45.14, Activity 45.3**

- **Photosensitivity** in animals depends on the absorption of photons of light by the visual pigment **opsin**, which contains a light absorbing group called 11-*cis*-retinal. Absorption of light is the first step in a cascade of intracellular events leading to a change in the membrane potential of the **photoreceptor cell**. Review Figures 45.15, 45.18, Animation 45.2

- Vertebrates have two types of photoreceptors, **rod cells** and **cone cells**. Rod cells are more sensitive to light and are responsible for dim light vision. Cone cells are less sensitive to light but are responsible for high-acuity and color vision.

- Photoreceptors do not fire action potentials. Light hyperpolarizes rod cells, and their release of neurotransmitter decreases. Review Figure 45.17

- **Rhodopsin** is the visual pigment of rod cells. The visual pigments of cone cells have three different opsin components, which gives them different spectral sensitivities. Review Figure 45.19

- The vertebrate **retina** consists of layers of neurons lining the back of the eye. The light-absorbing photoreceptor cells are at the rear of the retina. The axons of the **ganglion cells** are bundled together in the **optic nerves**. Between the photoreceptors and the ganglion cells are neurons that process information from the photoreceptors. Review Figure 45.20, Activity 45.4

> Go to **LearningCurve** (in **LaunchPad**) for dynamic quizzing that helps you solidify your understanding of this chapter. **LearningCurve** adapts to your responses, giving you the practice you need to master each key concept.

▷ Apply What You've Learned

Review

45.3 The ear converts sound waves in air to pressure waves in fluid that bend hair cells to create auditory sensation.

Original Paper: Corcoran, A. J. and W. E. Conner. 2012. Sonar jamming in the field: Effectiveness and behavior of a unique prey defense. *Journal of Experimental Biology* 215: 4278–4287.

Bats prey on nocturnal moths using echolocation. A bat pursuing prey emits ultrasonic signals that reflect off flying moths and echo back to the bat, enabling it to home in on the prey. Moths can "hear" ultrasonic bat signals through paired auditory organs (tympana) that respond to both low- and high-intensity sound waves. The moths have evolved evasive behaviors to evade capture.

Some moths use rapid changes in flight behavior, such as turning away or rapid dives, to evade capture. Others use unique tymbal organs, located in the thorax, to produce ultrasonic clicks. Researchers hypothesized that these clicks serve as echolocation-jamming mechanisms. To test this hypothesis they used tiger moths (*Bertholdia trigona*). One group of moths (*N* = 41) had their tymbal organs punctured, making them incapable of producing sounds (silenced), and another group (*N* = 38) were left with intact tymbal organs (clicking). Both groups were released into bat feeding areas. The fates of the released moths were documented with video and audio recordings. Results of the experiment are shown in the figures.

In **Figure A**, numbers over the bars show the number of attacks when silenced or clicking moths exhibited a particular evasive behavior. Statistically, there was no difference between behaviors or between silenced moths and clicking moths.

In **Figure B**, numbers over the bars represent the number of captures that were made when silenced or clicking moths exhibited a particular evasive behavior. In all three categories the difference between silenced moths and clicking moths was statistically significant.

Questions

1. According to the data in Figure A, were moths with disrupted tymbal organs still capable of responding to the bats with evasive behaviors? Explain your answer. Why is this important for the validity of this experiment?

2. According to the data in Figure B, what effect did disruption of the tymbal organs have on the number of captures? How do these results support the hypothesis that the tymbal organ in *B. trigona* provides an echolocation-jamming function?

3. The sound pulses produced by the bats (*Myotis* sp.) for echolocation in this study have a frequency of around 50 Khz (50,000 cycles per second). The echo from the moth has the same frequency. The upper limit of human hearing is 20,000 Hz. How do you think the basilar membrane in the bat's cochlea differs in structure from the human basilar membrane?

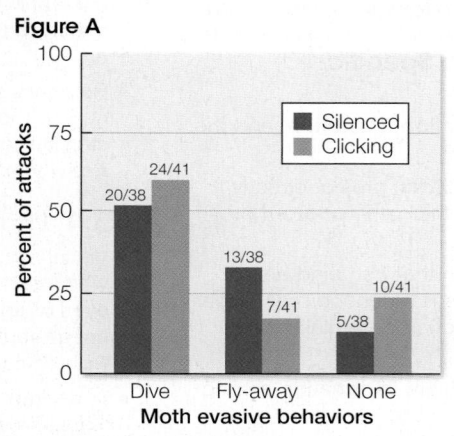

Figure A

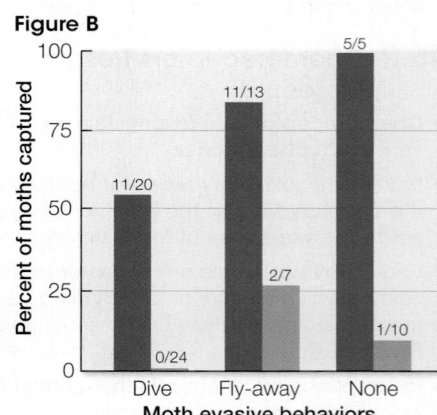

Figure B

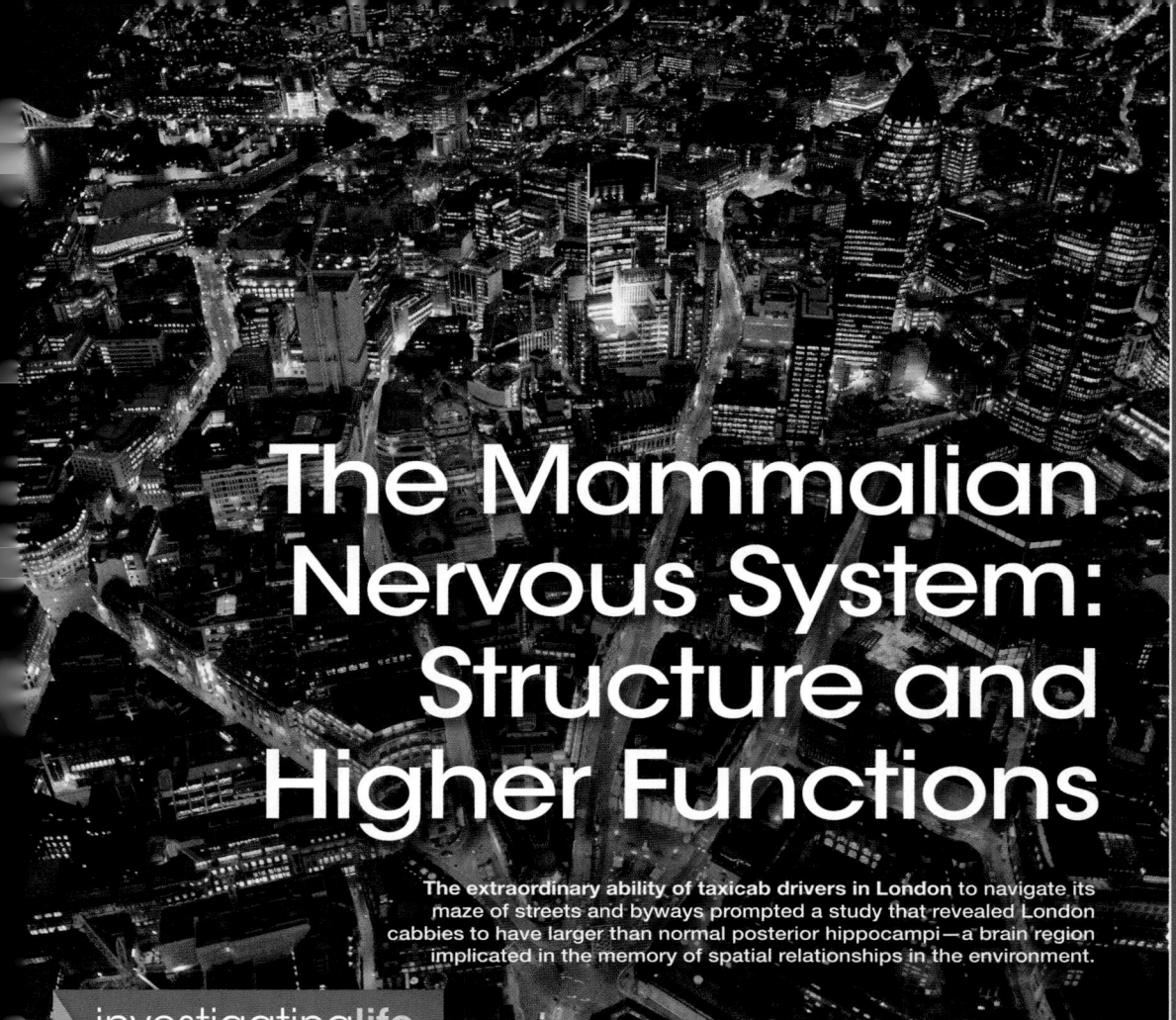

The Mammalian Nervous System: Structure and Higher Functions

The extraordinary ability of taxicab drivers in London to navigate its maze of streets and byways prompted a study that revealed London cabbies to have larger than normal posterior hippocampi—a brain region implicated in the memory of spatial relationships in the environment.

▶ investigating**life**

The Brains of Taxi Drivers

Compare Google maps of London and New York City at the same scale. In which city do you think it is easier to drive a taxi? Eleanor Maguire at University College London was so impressed with the navigational abilities of London taxi drivers that she investigated whether there was anything "special" about their brains. Using magnetic resonance imaging, Maguire and her colleagues examined the brains of taxi drivers with varying numbers of years of experience and compared them with each other and with the brains of control subjects who were not taxi drivers. The studies revealed significant differences in the sizes of a brain region called the hippocampus.

The hippocampus is involved in learning and memory. The posterior hippocampus in particular is implicated in the memory of spatial relationships among objects in the environment. Maguire found that the posterior hippocampi of taxi drivers were larger than those of control subjects and that, among the cab drivers themselves, there was a positive correlation between the size of the posterior hippocampus and years of driving experience.

In Matt Wilson's lab at the Massachusetts Institute of Technology, researchers record the activity of hippocampal neurons of rats running in a maze. Specific neurons called "place cells" fire action potentials only when the rat is at a particular location in the maze. Thus when the rat runs the maze from beginning to end, a sequence of cell firing is recorded. Sometimes when the rat is resting, the same sequence is recorded but is speeded up about 20 times, or the reverse sequence is recorded. It seems as if the recordings are reflecting what the resting rat is thinking—where it has been, where it is going to go. In a maze with a choice point (go right or go left), the sequence recorded when the rat is resting at the start position frequently predicts which direction it will turn.

The hippocampus is important for acquiring memories and consolidating them into long-term memory. We also know that sleep is important for learning. Could something be happening in the hippocampus during sleep to promote memory?

Can recordings of place cells reveal

key concept
46.1
Functions Are Localized in the Nervous System

The organization of the mammalian nervous system can be described anatomically and functionally. In anatomical terms, all vertebrate nervous systems consist of three parts: a brain, a spinal cord, and a set of peripheral nerves that reach all parts of the body. As discussed in Key Concept 44.4, the brain and spinal cord are the **central nervous system**, or **CNS**, and the neurons that connect the CNS to all the tissues and sensors of the body are the **peripheral nervous system**, or **PNS**. An additional division of the nervous system exists in the gut; we will discuss this **enteric nervous system** in Chapter 50.

focus your learning

- Conscious sensory and unconscious physiological information flows to the central nervous system via afferent neurons, resulting in efferent outputs to voluntary muscles (i.e., movement) and in involuntary commands over autonomic responses.
- From anterior to posterior, the embryonic neural tube gives rise to the cerebrum, thalamus/hypothalamus, midbrain structures, and cerebellum/pons/medulla, respectively.
- Specific areas within the five lobes of the cerebral cortex have sensory, motor, and associative functions.
- Humans are called the "big-brain primates," but factors other than size contribute to brain function in humans.

Recall from Key Concept 45.1 that a neuron is an electrically excitable cell that communicates via an axon. When used in the context of a nervous system, the term **nerve** refers to a bundle of axons in the PNS that carry information about many things simultaneously. Some axons in a nerve may be carrying information to the CNS while other axons in the same nerve are carrying information from the CNS to the body's organs. A discussion of the functional organization of the nervous system refers to these paths of information flow. In this chapter we will divide the anatomy of the mammalian brain, spinal cord, and PNS into smaller, discrete functional units.

Functional organization is based on flow and type of information

Figure 46.1 illustrates the major avenues of information flow through the human nervous system. The white boxes explain the four divisions of the PNS; two of these bring information from the periphery to the CNS, and two transmit information from the CNS to the periphery.

1. The **afferent** portion of the PNS (Figure 46.1, left) carries information from sensory receptor cells to the CNS. We are *conscious* of much of this information (e.g., light, sound, skin temperature, limb position), but we are usually *unconscious* of the information involved in physiological regulation (e.g., blood pressure, deep body temperature, blood oxygen levels).

2. The **efferent** portion of the PNS (Figure 46.1, right) carries information from the CNS to the muscles and glands of the body. Efferent pathways are divided into a *voluntary* division that executes our conscious movements; and an *involuntary*, or *autonomic*, division that controls physiological functions.

In addition to the neural information it receives from the PNS, the CNS receives chemical information from hormones circulating in the blood. In turn, neurohormones released by neurons enter the circulation and affect neurons and other cells distant from the site of release (see Key Concept 40.1).

The anatomical organization of the CNS emerges during development

Early in the development of a vertebrate embryo, a tube of neural tissue forms (see Key Concept 43.4). At its anterior end, this neural tube forms three swellings that become the **hindbrain**, **midbrain**, and **forebrain**. The rest of the neural tube becomes the spinal cord. Peripheral nerves sprout from the midbrain and hindbrain (the cranial nerves) and from the spinal cord (the spinal nerves). From these early stages we see the linear axis of information flow in the nervous system. Although the developing brain will fold and become a complex structure, the information flow in the adult nervous system will follow paths that emerge from the simple, linear neural tube.

Each of these three regions of the embryonic brain develops into several structures in the adult brain (**Figure 46.2**). From the embryonic midbrain come structures that integrate information from the different senses and coordinate motor responses. From the hindbrain come the **medulla**, the **pons**, and the **cerebellum**. The medulla is continuous with the spinal cord, the pons is anterior to the medulla, and the cerebellum is a dorsal outgrowth of the pons. The medulla and pons contain distinct groups of neurons involved in controlling physiological functions such as breathing, circulation, and basic motor patterns such as swallowing and vomiting. All

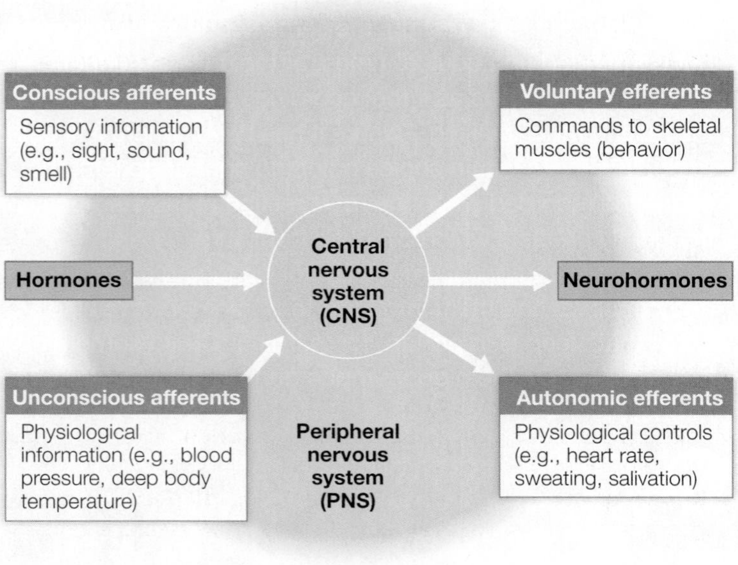

Conscious afferents
Sensory information (e.g., sight, sound, smell)

Hormones

Unconscious afferents
Physiological information (e.g., blood pressure, deep body temperature)

Central nervous system (CNS)

Peripheral nervous system (PNS)

Voluntary efferents
Commands to skeletal muscles (behavior)

Neurohormones

Autonomic efferents
Physiological controls (e.g., heart rate, sweating, salivation)

Figure 46.1 Organization of the Nervous System   The peripheral nervous system carries information to (afferent) and from (efferent) the central nervous system (center circle in the diagram). The CNS also receives hormonal inputs and produces hormonal outputs (see Key Concept 40.2).

Q: How do hormonal inputs reach the CNS and hormonal outputs leave the CNS?

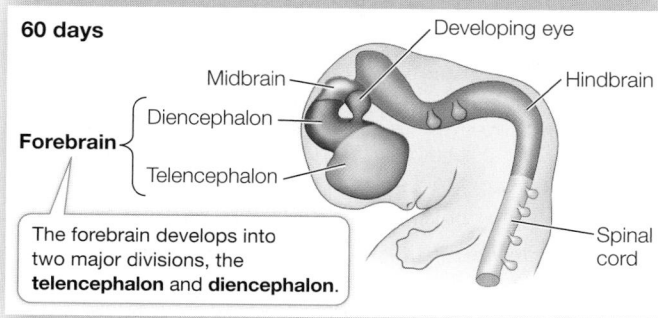

30 days

The brain develops from the anterior embryonic neural tube.

Neural tube
Midbrain
Forebrain
Optic vesicle
Cranial and spinal ganglia

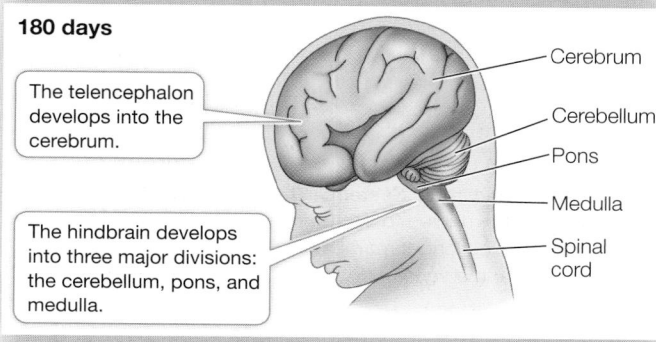

60 days

Forebrain { Diencephalon, Telencephalon }

The forebrain develops into two major divisions, the **telencephalon** and **diencephalon**.

Developing eye
Midbrain
Hindbrain
Spinal cord

180 days

The telencephalon develops into the cerebrum.

The hindbrain develops into three major divisions: the cerebellum, pons, and medulla.

Cerebrum
Cerebellum
Pons
Medulla
Spinal cord

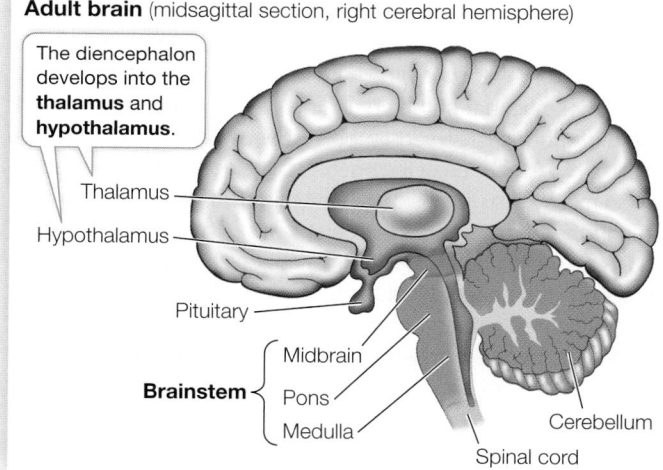

Adult brain (midsagittal section, right cerebral hemisphere)

The diencephalon develops into the **thalamus** and **hypothalamus**.

Thalamus
Hypothalamus
Pituitary

Brainstem { Midbrain, Pons, Medulla }

Cerebellum
Spinal cord

Figure 46.2 Development of the Central Nervous System In vertebrate embryos, the anterior end of the hollow neural tube differentiates into forebrain, midbrain, and hindbrain. Each of these regions develops into several structures of the adult brain. The remainder of the neural tube becomes the spinal cord.

information traveling between the spinal cord and higher brain areas must pass through the pons, the medulla, and the midbrain, which are collectively known as the **brainstem**.

One function of the cerebellum is coordinating muscle activity and maintaining balance. It is like the director of a movie; the cerebellum receives a "script" of the commands going to the muscles from higher brain areas, and it receives information about the actual performance coming up the spinal cord from the "actors"—the joints and muscles. The cerebellum compares the "script" with the performance and refines motor commands accordingly. Damage to the cerebellum results in loss of fine motor control and coordination.

The embryonic forebrain develops a central region called the **diencephalon** and a surrounding structure called the **telencephalon**. The diencephalon is the core of the forebrain and consists of an upper structure, the **thalamus**, and a lower structure, the **hypothalamus**. The thalamus is the final relay station for sensory information going to the telencephalon. The hypothalamus receives a lot of physiological information of which we are not conscious, and it uses that information to regulate many physiological functions and biological drives. A major function of the hypothalamus is to control the pituitary gland (see Key Concept 40.2).

The embryonic telencephalon gives rise to the **cerebrum**, consisting of the left and right **cerebral hemispheres**. The outer layer of the cerebrum is the **cerebral cortex**, a thin layer rich in cell bodies. If we compare vertebrate groups from fish through amphibians, reptiles, and mammals, the cerebrum increases in size, complexity, and importance—an evolutionary trend called telencephalization (see Figure 44.15). In humans, the cerebrum is by far the largest part of the brain and plays major roles in sensory perception, learning, memory, and conscious behavior.

The spinal cord transmits and processes information

The spinal cord conveys information to and from the brain. However, the spinal cord is more than an information pipe. As you saw in Key Concept 44.4, the spinal cord carries out integrative functions as well. The knee-jerk reflex (see Figure 44.14) is an example of a circuit between the PNS and the spinal cord that controls a simple behavioral function. That simple circuit can be built on to control more complex behaviors such as the withdrawal reflex, which involves readjusting tension in many muscles on both sides of the body to coordinate movement and maintain balance.

The brainstem carries out many autonomic functions

Swallowing, salivating, breathing, eye movements, blood pressure regulation, and gut activity are only a few of the many autonomic functions that are localized in the medulla, the pons, and the midbrain. To carry out these functions, the brainstem has its own components of the PNS, the 12 paired **cranial nerves**. You encountered the olfactory nerve, the optic nerve, and the auditory nerve (cranial nerves I, II, and VIII) in Chapter 45. Another one, cranial nerve X, is called the vagus ("wandering") nerve because it travels into the body cavity and communicates with many of the organs, including the heart and the gut. You will encounter the vagus nerve in subsequent chapters.

Anatomically distinct groups of neurons in the CNS that work together as a unit are called a **nucleus** (not to be confused with the nucleus of a single cell). Axons from the neurons of many brainstem nuclei extend to various regions of the brain and modulate brain activity; for example, brainstem nuclei are involved in keeping the higher brain areas awake or allowing them to sleep. All of the sensory information coming up the neural axis from the spinal cord passes through the brainstem on its way to the forebrain, and many of these ascending neuronal tracts give off branches to the awake-promoting nuclei in the brainstem. Because the neuronal circuitry in this part of the brain is so complicated and because activity in these ascending sensory pathways can promote wakefulness, the core of the brainstem has been termed the **reticular-activating system** ("reticular" means netlike). Damage to the brain or spinal cord below the reticular-activating system can result in paralysis but leave sleep–wake cycle behavior normal. Damage above the level of the reticular-activating system can interrupt ascending wake-promoting activity resulting in a coma.

The core of the forebrain controls physiological drives, instincts, and emotions

As mentioned above, the diencephalon consists of the thalamus and the hypothalamus. The thalamus communicates sensory information to the cerebral cortex; the hypothalamus receives information about physiological conditions in the body and regulates many homeostatic functions. Key Concept 39.5 describes how the hypothalamus is involved in regulating body temperature, and Key Concept 40.3 discusses the intimate association between the hypothalamus and the pituitary gland in the control of many homeostatic functions.

The forebrain of all vertebrates includes a phylogenetically old, complex set of structures called the **limbic system** (**Figure 46.3**). The limbic system is responsible for some basic instincts and drives such as hunger and thirst, sexual behavior, and emotions. Within the limbic system there are areas that, when stimulated with small electric currents, can cause intense sensations of pleasure, pain, or rage.

Pleasure and pain centers in the limbic system are believed to play roles in learning and in physiological drives. One component of the limbic system—the **amygdala**—is involved in fear and fear memory. If a certain portion of the amygdala is damaged or chemically blocked, an animal that becomes tame and cannot learn to be afraid of a situation that would normally induce a strong fear reaction. The amygdala is involved in post-traumatic stress disorder (PTSD). Another part of the limbic system, the **hippocampus**, is involved in the transfer of certain types of short-term memory to long-term memory, as discussed in the chapter opener and in **Investigating Life: Place Cells Reveal Processes of Memory Consolidation during Sleep**.

The cerebrum is responsible for complex behavior and consciousness

The cerebrum is the dominant structure in the mammalian brain. In humans it is so large that it covers all other parts of the brain except the cerebellum (**Figure 46.4A**). Layers of cells at the surface of the cerebrum form the cerebral cortex, which is about 4 millimeters thick and is folded into ridges (**gyri**; singular *gyrus*) and valleys (**sulci**; singular *sulcus*). If the cortex were flattened out, it would be about 1 square meter, but the foldings, or **convolutions**, enable that extensive surface area of cortex to fit within the skull.

As we explore the infoldings of the cerebral cortex and other parts of the brain, we will occasionally mention an individual whose brain was damaged by an accident or other unfortunate event. Until recently the study of such individuals has been the main source of

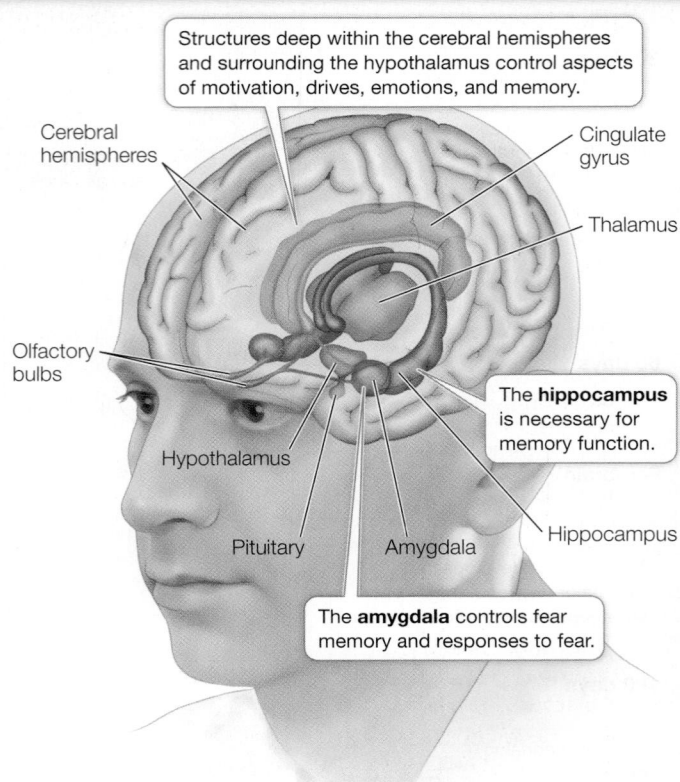

Figure 46.3 The Limbic System   The evolutionarily primitive parts of the forebrain are referred to as the limbic system. The hippocampus is involved in forming long-term memory. The amygdala triggers fear emotions and fear memories.

functional information about the human brain, but new imaging technologies such as positron emission tomography (PET) and magnetic resonance imaging (MRI) are providing a wealth of new information and opportunities to study the human brain.

A curious feature of the human nervous system is that the left side of the body is served (in both sensory and motor aspects) mostly by the right side of the brain, and the right side of the body is served mostly by the left side of the brain. Thus sensory input from the right hand goes to the left cerebral hemisphere, and sensory input from the left hand goes to the right cerebral hemisphere. The exception is the head, where the left side is controlled by the left cerebral hemisphere and the right side by the right cerebral hemisphere. The two hemispheres are not symmetrical with respect to all functions. Language abilities, for example, reside predominantly in the left hemisphere in most people.

Different regions of the cerebral cortex have specific functions (**Figure 46.4B**). Some of those functions are easily defined, such as receiving and processing sensory information or generating motor commands, but in most humans most of the cortex is involved in higher-order information processing that is less easy to define. These latter areas are given the general name of **association cortex**, so named because they integrate, or *associate*, information from different sensory modalities and from memory.

To understand the cerebral cortex, it helps to have an anatomical road map. Viewed from the left side, the left cerebral hemisphere looks like a boxing glove for the right hand with the fingers pointing

experiment

Original Paper: Lee, A. K. and M. A. Wilson. 2002. Memory of sequential experience in the hippocampus during slow wave sleep. *Neuron* 36: 1183–1194.

Multiple hippocampal cells were recorded while a rat ran a maze. Individual cells fired only when the rat was at particular locations in the maze; thus the rat's progress through the maze can be represented by a specific sequence of firing of these "place cells." Since memory is consolidated during sleep, are the sequences of place cells expressed during the maze-running experience replayed and processed into long-term memory during sleep?

HYPOTHESIS▶ Memory of the maze-running experience is represented by a sequence of firing of hippocampal place cells. This sequence is repeated during sleep as part of the process of memory consolidation.

METHOD

1. Record from multiple hippocampal neurons while a rat is running through a maze.

2. Identify sequences of cell firing that reoccur whenever the rat runs the maze.

3. Continue to record the neuronal activity during sleep and determine if similar sequences are occurring with a probability greater than expected from random activity in that group of cells.

CONCLUSIONS▶

1. Sequences of hippocampal-cell firing that correspond to locations in the maze were highly repeatable and probably represent the coding of that place information in the hippocampus.

2. During sleep, bursts of cell firing sometimes occurred that correspond wholly or partially to the sequences recorded when the rat was running in the maze.

investigating**life work with the data** follows on next page.

RESULTS

1 After 30 runs through the maze, a group of nine place cells are identified that fire in the same order correlating with position in the maze.

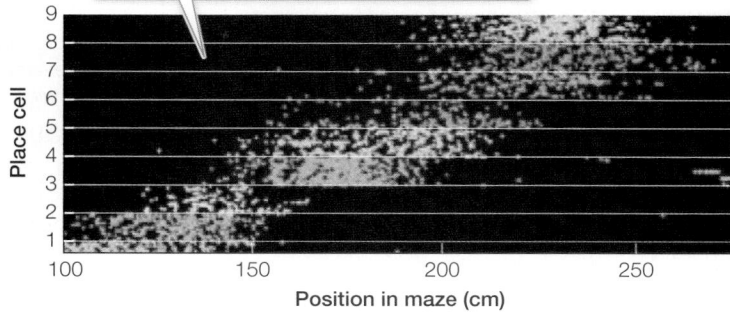

2 The peak firing rate of each neuron can be calculated as a function of position in the maze. When the neurons are arranged in the temporal order of their peak firing rates, you can think of each neuron as being a "letter" in a "word" that codes spatial information about the maze.

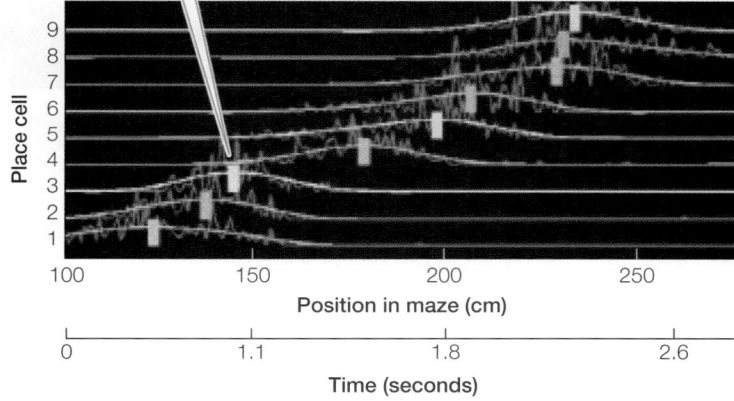

3 Subsequent recordings while the animal is awake, at rest, or asleep are analyzed for the occurrence of the same or similar "words"—complete or partial sequences of place-cell activity in the same order as the base sequence. Thus a similar word to a nine-letter (nine-neuronal) sequence might be a sequence of seven cells firing in the same order as in the original nine-letter sequence.

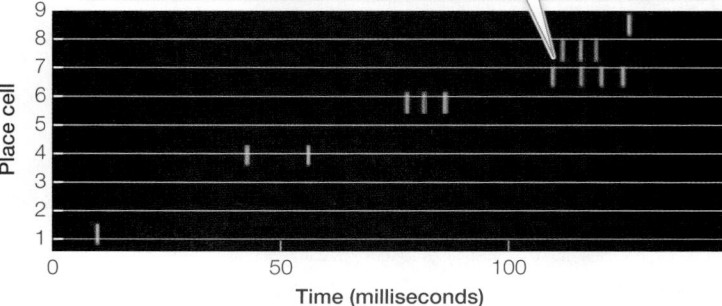

work with the data

To conclude that a particular sequence of cell firing observed during sleep is a significant event—meaning that it could be a replay of the waking experience—we have to compare it with the probability that it would occur by random chance. And we need to consider whether an observed sequence that includes some but not all of the components of the baseline "word" is nevertheless a significant event.

What is the minimum number of elements that would have to match the baseline sequence for it to be considered a significant event with $P < 0.01$? To find out, we need to calculate the probabilities of the recorded neurons firing randomly and creating the sequence matches. First, let's consider the probability of a recurrence by chance of the exact nine-"letter" baseline sequence being recorded in a single burst of that group of nine cells. The total number of possible combinations of those nine letters can be calculated as the factorial 9!:

$$9! = 9 \times 8 \times 7 \times 6 \times 5 \times 4 \times 3 \times 2 \times 1 = 362,880$$

In other words, the probability of repeating the entire nine-letter sequence would be 1 out of 362,880. That repeat would have an extremely low probability of occurring and would therefore be a highly significant event. But what if the recorded sequence has only six of the nine cells firing in the proper order? How do you determine the

probability of that? First, calculate the probability of a combination of six cells out of the nine. So for the first cell you have nine choices, for the second you have eight choices, and so forth. Then you have to calculate the probability of that combination of six cells being in the proper sequence, and that will be 1 out of the total number of possible arrangements of those six cells, or 6!. Then what do you do for probabilities of successive events? You multiply the probabilities. For example, if you flip a coin, the probability of heads is 0.5. If you flip it twice, the probability of getting heads both times is 0.5 × 0.5, or 0.25.

QUESTIONS▶

1. What is the probability of the sequence 1, 4, 6, 7, 8, 9 out of a random firing of nine cells?
2. What is the probability of a sequence of four letters occurring out of a random firing of nine cells?
3. Are both of the recordings from Questions 1 and 2 low-probability events?
4. What do you note about the time frame of the sequence during sleep versus during wakefulness?

A similar **work with the data** exercise
may be assigned in **LaunchPad**.

forward, the thumb pointing out, and the wrist at the rear. The "thumb" area is the **temporal lobe**, the fingers the **frontal lobe**, the back of the hand the **parietal lobe**, and the wrist the **occipital lobe** (see Figure 46.4A). What you can't see from the external view is the **insular cortex**, which is folded in between the thumb an finger areas of the boxing glove (between the frontal/parietal an the temporal lobes). The right cerebral hemisphere shows a mirr image of this arrangement. We will look at each lobe separately.

(A) The lobes of the cerebrum

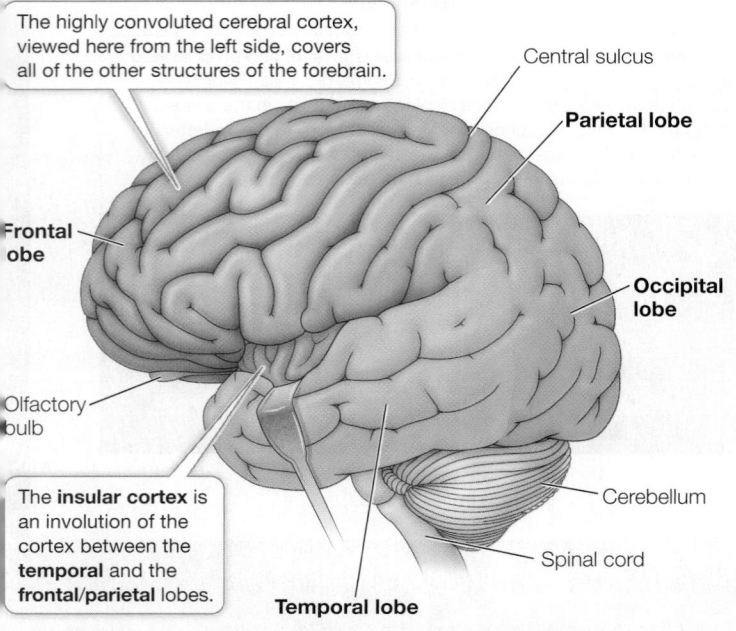

The highly convoluted cerebral cortex, viewed here from the left side, covers all of the other structures of the forebrain.

Central sulcus
Parietal lobe
Occipital lobe
Frontal lobe
Olfactory bulb
The **insular cortex** is an involution of the cortex between the **temporal** and the **frontal/parietal** lobes.
Cerebellum
Spinal cord
Temporal lobe

(B) Functions of selected regions of the cerebral cortex

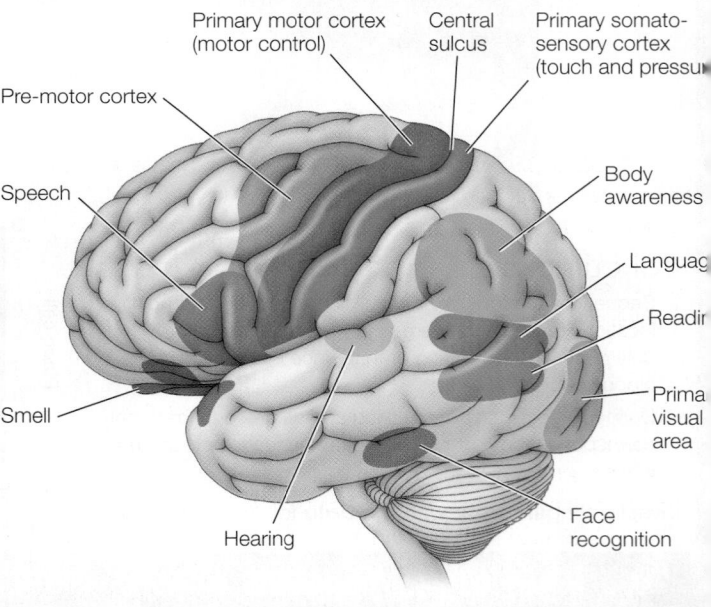

Primary motor cortex (motor control)
Central sulcus
Primary somato-sensory cortex (touch and pressu
Pre-motor cortex
Body awareness
Speech
Languag
Readir
Smell
Prima visual area
Hearing
Face recognition

Figure 46.4 The Human Cerebrum (A) Each cerebral hemisphere is divided into frontal, temporal, parietal, occipital, and insular lobes. (B) Different functions are localized in particular areas of the four cerebral lobes.

▶ Activity 46.1 **The Human Cerebru**
www.Life11e.com/ac46.1

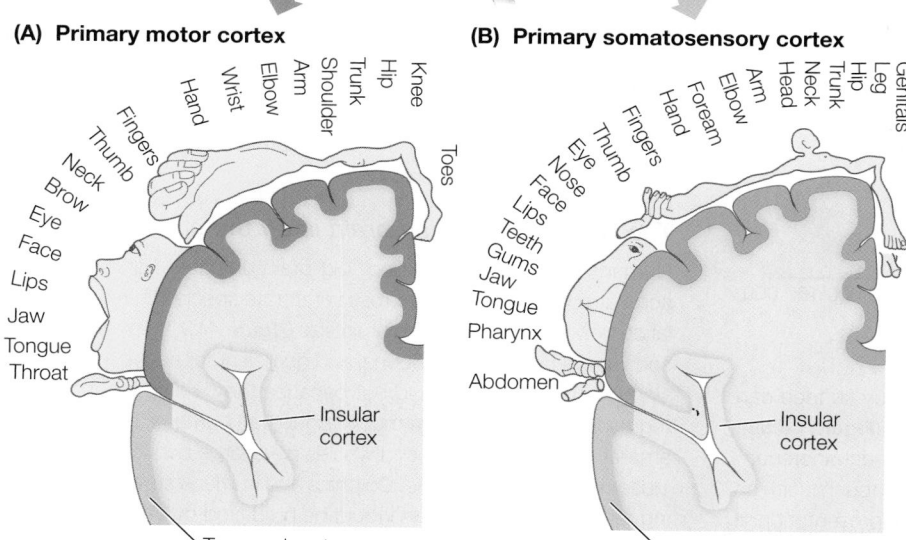

The left side of the cerebral cortex communicates with the right side of the body, and vice versa.

Motor   Somatosensory

Figure 46.5 The Body Is Represented in Primary Motor and Primary Somatosensory Cortexes Neurons in the primary motor cortex **(A)** control muscles in specific parts of the body, while neurons in the primary somatosensory cortex **(B)** receive information from specific parts of the body. The locations of these neurons within each cortex correspond to "maps" on which regions of the body are represented in proportion to the amount of innervation they receive.

(A) Primary motor cortex

Hand, Wrist, Elbow, Arm, Shoulder, Trunk, Hip, Knee, Toes, Fingers, Thumb, Neck, Brow, Eye, Face, Lips, Jaw, Tongue, Throat

Insular cortex

Temporal cortex

(B) Primary somatosensory cortex

Foream, Arm, Elbow, Hand, Fingers, Head, Neck, Trunk, Hip, Leg, Genitals, Thumb, Eye, Nose, Face, Lips, Teeth, Gums, Jaw, Tongue, Pharynx, Abdomen

Insular cortex

Temporal cortex

THE TEMPORAL LOBE The upper region of the temporal lobe receives and processes auditory information. The association areas of this lobe are involved in recognizing, identifying, and naming objects. Damage to the temporal lobe results in disorders called agnosias, in which the individual is aware of an object but cannot identify it.

Damage to a specific region of the temporal lobe results in the inability to recognize faces. Even old acquaintances cannot be identified by facial features, although they may be identified by other attributes such as voice, body features, and posture. Damage to other association areas of the temporal lobe can cause deficits in understanding spoken language, although speaking, reading, and writing abilities may be intact.

THE FRONTAL LOBE The frontal and parietal lobes are separated by a deep valley called the central sulcus. A strip of the frontal lobe cortex just in front of the central sulcus is called the **primary motor cortex** (see Figure 46.4B). The neurons in this region control muscles in specific parts of the body; the parts of the body map onto the primary motor cortex, with the head represented in the lower outside region and the legs and feet in the top region near the midline. The graphic drawing of body parts overlying the diagram of the cortex is called a homunculus, and its distortion represents disproportionate innervation. Parts of the body with fine motor control, such as the face and hands, have disproportionate representation **(Figure 46.5A)**. Electrical stimulation of neurons in different regions of the primary motor cortex causes specific muscles of the body to twitch. Just anterior to the primary motor cortex is the **pre-motor cortex**, which is involved in planning more complex and coordinated movements.

The association functions of the frontal lobe are diverse and best described as having to do with feeling and planning.

They are said to have **executive function** and they contribute significantly to personality. People with frontal lobe damage have drastic alterations of personality and difficulty planning future events. A dramatic case of frontal lobe damage is that of Phineas Gage, who in 1848 was an industrious and responsible young railroad construction foreman. Then a blasting accident shot a meter-long, 3-centimeter-wide iron tamping rod through his brain. The rod entered Gage's head below his left eye, passed through his frontal lobe, and exited the top of his head **(Figure 46.6)**.

Remarkably, Gage survived, but he had a different personality. In the years following his recovery, he was quarrelsome, impatient, obstinate, and used profane language, which he did not do before. He lost his railroad job and earned money by telling his story and exhibiting his scars (and the tamping iron). Several years later, however, he took a job as a stage coach driver in Chile—a challenging job indicating considerable behavioral recovery. He died of a seizure in 1860, at the age of 38. If you are in Boston, you can pay him a visit—his skull, death mask, and the tamping iron are on display in the Warren Anatomical Museum of Harvard Medical School.

THE PARIETAL LOBE The strip of parietal lobe cortex just behind the central sulcus is the **primary somatosensory cortex** (see Figure 46.4B). This area receives touch and pressure information relayed from the body through the thalamus.

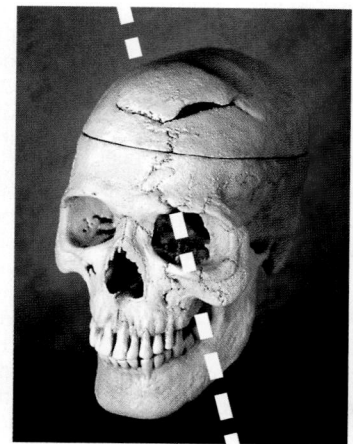

Figure 46.6 A Mind-Altering Experience Phineas Gage miraculously survived a nineteenth-century railroad construction accident that blew an iron rod through his brain. His personality, however, was permanently altered from that of a responsible foreman to a sometimes quarrelsome drifter.

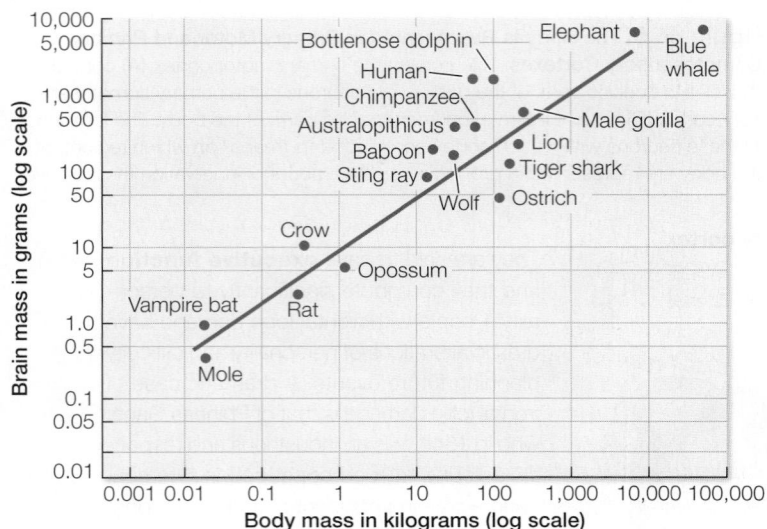

Figure 46.7 Evolution of the Human Brain Brain size scales to body size across a wide range of vertebrates. The higher primates have larger brains than predicted by the correlation, and humans stand outside this relationship with much bigger brains. The increase in brain size in humans is mostly due to an increase in the cerebral cortex. The human brain is also highly convoluted, and more of it is devoted to associative functions.

and marine mammals, receives information about social interactions. Thus in mammals the insular cortex appears to integrate physiological information from all over the body to create a sensation of how the body "feels," and in the higher mammals it may extend that function into a sense of self.

The size of the human brain is off the curve

Humans are sometimes called "big-brain primates," and that is an accurate characterization. Across vertebrate species there is a correlation between body size and brain size (**Figure 46.7**). Higher primates such as chimpanzees, baboons, and gorillas all fall above this regression line, but humans stand out because they are so far above the regression line. Gorillas are much larger than humans, but they have smaller brains. Elephants and whales have large brains, but they fall closer to the regression line. Dolphins and humans stand out as having much larger brains than would be predicted by their body sizes.

The correlation of brain size to body size does not tell the whole story of human brain evolution, however. In Figure 44.15, which compares the brains of four vertebrates, we see that the forebrain is larger than other brain regions, and in mammals this is seen as an elaboration of the cerebral cortex. If we look just at mammals, another feature is the degree of convolution of the cortex. Since the cortex is a layered, two-dimensional array of neurons, the area of cortex is increased by convolutions, which are greatest in humans. And finally, the percent of the cortex that is association cortex (i.e., devoted to the integration of information) is by far the greatest in humans. It is these evolutionary changes, primarily in the cortex, that provide the resources for the intellectual capacity of humans—a topic to which we will return at the end of the chapter.

As with the primary motor cortex, the entire body surface can be mapped onto the primary somatosensory cortex (**Figure 46.5B**). Areas of the body that have a high density of tactile mechanoreceptors and are capable of making fine discriminations in touch (such as the lips and fingers) have disproportionately large representation. If a very small area of the primary somatosensory cortex is stimulated electrically, the subject reports feeling specific sensations, such as touch, in a localized part of the body.

A major association function of the parietal lobe is attending to complex stimuli. Damage to the right parietal lobe causes a condition called contralateral neglect syndrome, in which the individual tends to ignore stimuli from the left side of the body or the left visual field. Such individuals have difficulty performing complex tasks, such as dressing the left side of the body; an afflicted man may not be able to shave the left side of his face. When asked to copy simple drawings, a person who exhibits this syndrome can do well with the right side of the drawing but not the left.

The parietal cortex is not symmetrical with respect to its role in attention. Damage to the left parietal cortex does not cause the same degree of neglect of the right side of the body. You will see similar asymmetries in cortical function later in the chapter when we discuss language.

THE OCCIPITAL LOBE The occipital lobe receives and processes visual information. The association areas of the occipital cortex are essential for making sense of the visual world and translating visual experience into language. Some deficits resulting from damage to these areas are specific. In one case, a woman with limited damage was unable to see motion. Her vision was intact, but she could see a waterfall only as a still image, and an approaching car only as a series of a stationary object at different distances.

THE INSULAR CORTEX The insular cortex, buried deep in the forebrain, receives a great variety of afferent information. The posterior regions receive somatosensory information (e.g., touch, pain, temperature); the middle region combines that with autonomic regulatory information and drives (e.g., hunger, thirst, sex); and the anterior region, which is well developed in higher primates, elephants,

46.1 recap

The central nervous system communicates with the rest of the body through the peripheral nervous system. We are conscious of some sensory information coming into the CNS, but we are not conscious of other afferent information used in physiological regulation. The brain develops from the anterior region of the embryonic neural tube. The forebrain gives rise to the cerebrum, which is divided into five lobes. Specific functions such as somatosensation, vision, hearing, motor control, smell, and speech are localized in specific cortical areas. At the center of the cerebrum is the limbic system which contributes to the processing of emotions, inborn responses, and some types of memory. Underneath the cerebrum is the hypothalamus, responsible for many autonomic functions, and the thalamus, which directs ascending information to the cortex. The midbrain and hindbrain constitute the brainstem, within which are located many nuclei that support autonomic functions, including sleep and wakefulness. Evolution of the human brain has resulted in a greatly increased cerebral cortex devoted to integration of information.

46.1 recap

learning outcomes

You should be able to:

- Describe and classify the types of afferent information to and efferent information from the nervous system.
- Describe the developmental relationships between the major brain regions, starting with the forebrain, midbrain, and hindbrain and ending with structures such as the amygdala, hypothalamus, primary motor cortex, reticular-activating system, and other components of the adult brain.
- Sketch the five lobes of the cerebrum and identify the primary sensory and motor areas.
- Characterize the human brain in comparison to other mammals.

1. Given your understanding of afferent and efferent systems operating in the nervous system, give examples of brain–body feedback loops that influence each of the following:

   a. Voluntary behavior through a conscious sensory stimulus

   b. Involuntary responses through a conscious sensory stimulus

   c. Involuntary responses through an unconscious stimulus

2. In the developmental and lethal disorder anencephaly, the anterior-most region of the developing neural tube fails to close. Explain what impact this has on development of specific brain structures.

3. What is significant about the location of the reading and language areas of the cerebral cortex as shown in Figure 46.4B?

4. Key Concept 45.3 described the two-point spatial discrimination test. How do the typical results of that test relate to Figure 46.5B?

5. The brain size of humans is significantly above the regression line for brain size as a function of body size in most other mammals. Why is this observation an understatement of the difference in cerebral cortical capacity in humans as compared with other mammals?

Having briefly described the structure and function of different regions of the nervous system, we will now explore some examples of how information is processed by the neural circuitry in some specific brain regions.

key concept

46.2   Nervous System Functions Rely on Neural Circuits

Specific functions are localized in specific parts of the nervous system and depend on the neural circuits, or networks, in those structures. A major focus of modern neuroscience is to understand how the various functions of the nervous system, ranging from simple reflexes to complex learning and memory, are accomplished by the interactions of neurons in circuits. Two extensively studied examples of how neural networks process information are the autonomic nervous system (an output pathway) and the visual system (an input pathway).

focus your learning

- To control involuntary autonomic responses, the sympathetic and parasympathetic divisions of the autonomic nervous system have distinct circuitry and typically induce opposite functional responses from target organs and tissues.
- Visual information is transduced in the retina via center-surround receptive fields, resulting in patterns of action potential firing in ganglion cells that are transmitted to the visual cortex via the thalamus.
- Binocular neurons in the visual cortex receive input from both eyes and thereby enable depth perception in humans.

Pathways of the autonomic nervous system control involuntary physiological functions

The **autonomic nervous system**, or **ANS**, includes CNS and PNS components and controls many involuntary functions, such as heart rate, blood flow, sweating, and digestive activities. Its control of diverse organs and tissues is crucial to homeostasis. The ANS has two divisions, **sympathetic** and **parasympathetic**, that work in opposition to each other in their effects on most organs (**Figure 46.8**). The sympathetic and parasympathetic divisions are easily distinguished by their anatomy, neurotransmitters, and actions.

The best-known functions of the ANS are those of the sympathetic division that produce the fight-or-flight response: increasing heart rate, blood pressure, and cardiac output and which prepares the body for emergencies (see Figure 40.3). In contrast, the parasympathetic division slows the heart and lowers blood pressure; its actions have been characterized as "rest and digest." It is tempting to think of the sympathetic division as speeding things up and the parasympathetic division as slowing things down, but it is not that simple; for example, the sympathetic division slows down the digestive system whereas the parasympathetic division accelerates it.

Whether sympathetic or parasympathetic, every autonomic efferent pathway begins with a cholinergic neuron (that is, a neuron that uses acetylcholine as its neurotransmitter) that has its cell body in the brainstem or spinal cord. These cells are called preganglionic neurons because the second neuron in the pathway with which they synapse resides in a collection of neurons outside the CNS called a **ganglion** (plural *ganglia*). The second neuron is called a postganglionic neuron because its axon extends out from the ganglion. The axon of the postganglionic neuron synapses with cells in the target organs (see Figure 46.8).

The postganglionic neurons of the sympathetic division mostly use norepinephrine (also known as noradrenaline) as their neurotransmitter. In contrast, the postganglionic neurons of the parasympathetic division are mostly cholinergic. In organs that receive both sympathetic and parasympathetic input, the target cells respond in an opposite manner to norepinephrine and to acetylcholine. This happens, for example, in a region of the heart called the pacemaker, which initiates the heartbeat. Stimulating the sympathetic nerve to the heart or dripping norepinephrine onto pacemaker cells increases their firing rate and causes the heart to beat faster. In contrast, stimulating the parasympathetic

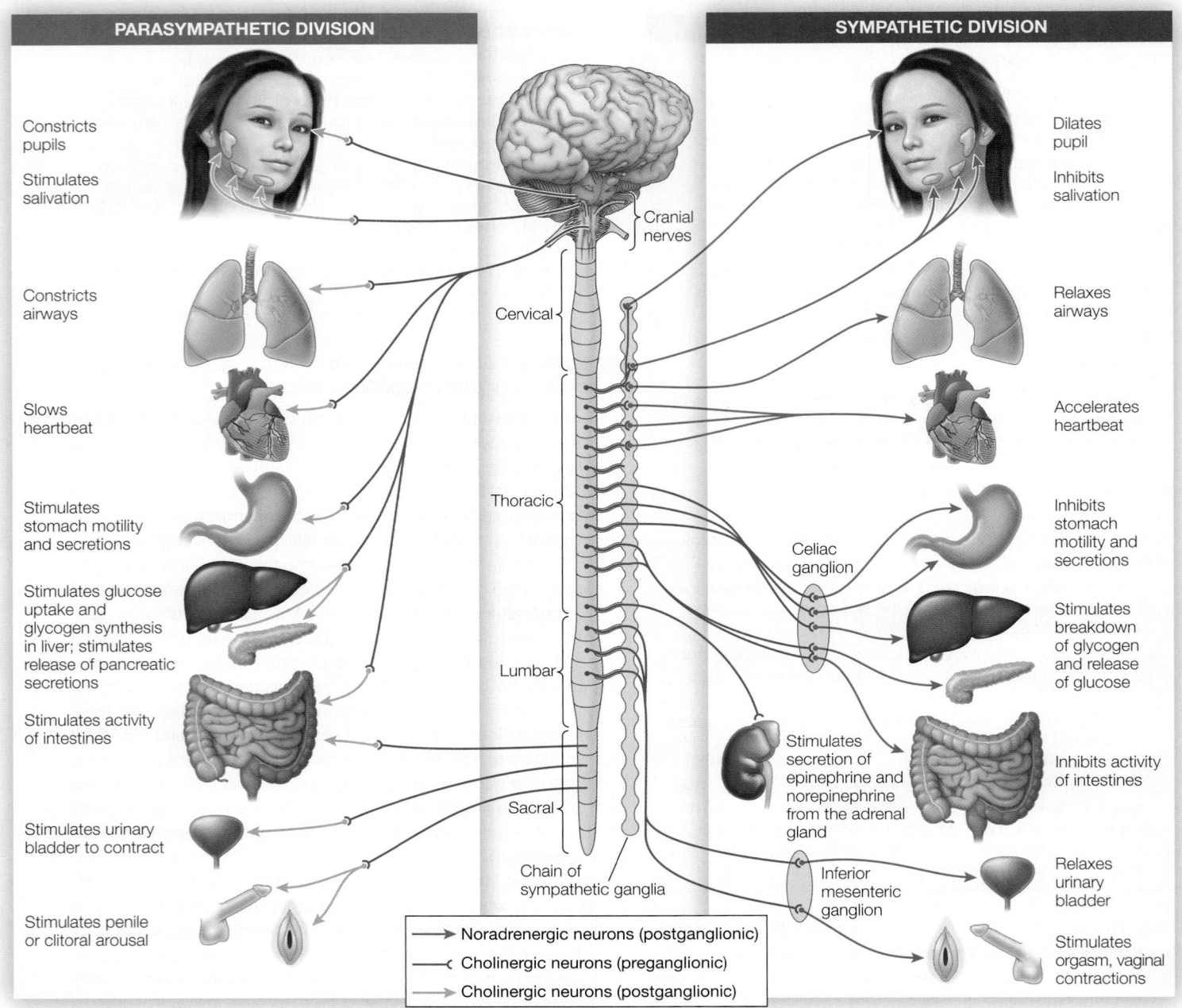

PARASYMPATHETIC DIVISION

Constricts pupils

Stimulates salivation

Constricts airways

Slows heartbeat

Stimulates stomach motility and secretions

Stimulates glucose uptake and glycogen synthesis in liver; stimulates release of pancreatic secretions

Stimulates activity of intestines

Stimulates urinary bladder to contract

Stimulates penile or clitoral arousal

SYMPATHETIC DIVISION

Dilates pupil

Inhibits salivation

Relaxes airways

Accelerates heartbeat

Inhibits stomach motility and secretions

Stimulates breakdown of glycogen and release of glucose

Inhibits activity of intestines

Relaxes urinary bladder

Stimulates orgasm, vaginal contractions

Cranial nerves

Cervical

Thoracic

Lumbar

Sacral

Chain of sympathetic ganglia

Celiac ganglion

Stimulates secretion of epinephrine and norepinephrine from the adrenal gland

Inferior mesenteric ganglion

→ Noradrenergic neurons (postganglionic)
⤙ Cholinergic neurons (preganglionic)
→ Cholinergic neurons (postganglionic)

Figure 46.8 The Autonomic Nervous System   The autonomic nervous system is divided into the sympathetic and parasympathetic divisions. The two divisions work in opposition to each other in their effects on most organs; one results in an increase and the other a decrease in activity.

Q: Why does a person get an extremely dry mouth when speaking in public?

nerve to the heart or dripping acetylcholine onto pacemaker cells decreases their firing rate and causes the heart to beat more slowly (see Figure 49.7).

The sympathetic and parasympathetic divisions of the ANS can also be distinguished by anatomy. The preganglionic neurons of the parasympathetic division come from the cranial nerves of the brainstem and the sacral (lowest) region of the spinal cord; those of the sympathetic division come from the thoracic and lumbar regions of the spinal cord (see Figure 46.8). Most of the ganglia of the sympathetic division are lined up and interconnected in two chains, one on either side of the spinal cord. The parasympathetic ganglia are close to the target organs and are not interconnected.

The ANS is an important link between the CNS and many physiological functions. Its control of diverse organs and tissues is crucial to homeostasis. Despite the complexity of the ANS, work by neurobiologists and physiologists over many decades has made it possible to understand its functions in terms of neuronal properties and circuits.

The visual system is an example of information integration by the cerebral cortex

The visual system is one of the most-studied input pathways to the CNS. Key Concept 45.4 described how light falling on the retina produces signals that are transmitted through the cellular circuits of the retina, resulting in action potentials in the optic nerve. But how does the brain use this information to represent the visual world?

RETINAL RECEPTIVE FIELDS   Key Concept 45.4 described how a retinal ganglion cell collects information from several photoreceptors—an example of "convergence of information." Each ganglion cell is communicating to the brain something more than simply the presence or absence or intensity of light falling on a portion of the retina.

The question of what information the retinal ganglion cell extracts from the photoreceptors was addressed in classic experiments by Stephen Kuffler, then at Johns Hopkins University. He used electrodes to record the activity of single ganglion cells of cat eyes while stimulating their retinas with spots of light (**Figure 46.9**). Kuffler's experiments revealed that each ganglion cell has a well-defined **receptive field** composed of a group of photoreceptor cells that receive light from a small area of the entire visual field. Stimulating these photoreceptors with light activates the ganglion cell, which sends action potentials to the thalamus and on to the visual cortex (the area of the occipital lobe where visual information is processed; see Figure 46.4). Information from many photoreceptors is therefore communicated to the brain as a single message. Individual photoreceptors may contribute to the receptive fields of multiple ganglion cells, so that receptive fields overlap.

The receptive fields of most ganglion cells are circular, but whether a spot of light falling on a receptive field excites or inhibits its ganglion cell depends both on the nature of the receptive field and on where the spot of light falls on it. Receptive fields have a center and a concentric surround, and can be either "on-center" or "off-center." Light falling on the center of an on-center receptive field excites the ganglion cell, and light falling on the center of an off-center receptive field inhibits the ganglion cell. Light falling on the surround has the opposite effect: the surround for an on-center receptive field inhibits the ganglion cell, and the surround for an off-center field is excitatory. Thus the activity of the ganglion cell reflects how much of the light stimulus is on the center and how much is on the surround of its receptive field (see Figure 46.9).

Center effects are always stronger than surround effects. Thus a small dot of light directly on the center of a receptive field has the maximum effect, and a larger light stimulus illuminating the center and parts of the surround has a smaller effect. A uniform patch of light falling equally on the center and surround has very little effect on the firing rate of the ganglion cell for that receptive field. The center and surround almost exactly cancel one another, even though the area of the surround is much larger.

 Activity 46.2 **Visual Receptive Fields Simulation**
www.Life11e.com/ac46.2

As is seen in Figure 46.9, photoreceptors synapse onto bipolar cells and bipolar cells onto ganglion cells. This pattern of connectivity describes the relationship between the photoreceptors in the center of a receptive field. The photoreceptors in the surround area modify communication between the center photoreceptors and their bipolar cells through the lateral connections of horizontal cells and amacrine cells. Thus the receptive field of a ganglion cell results from a pattern of synapses between photoreceptors, horizontal cells, amacrine cells, and bipolar cells. A general lesson to learn from this seemingly confusing chain of events is that inhibition can be as important as excitation in neural circuits.

In summary, the neural circuitry of the retina results in the generation of signals in the axons of the optic nerve that communicate simple information about the contrasting patterns of light and dark falling on different parts of the retina. But once the action potentials in the optic nerve reach their destinations, how does the brain integrate them to construct visual images of the outside world?

 Animation 46.1 **Information Processing in the Retina**
www.Life11e.com/a46.1

RECEPTIVE FIELDS OF CELLS IN THE VISUAL CORTEX   The axons of the optic nerves terminate in a region of the thalamus that is a relay station receiving information from both the right and left eyes. From the thalamus, the information encoded in the activity of axons in the optic nerves is relayed to the visual cortex in the occipital lobes at the back of the brain. In the 1960s David Hubel and Torsten Wiesel of Harvard University studied the activity of neurons in the visual cortex by shining spots and bars of light on retinas while recording the activities of single cells in the cortex. They found that many neurons in the visual cortex, like retinal ganglion cells, have receptive fields with regions that are mutually antagonistic.

Hubel and Weisel discovered that neurons in the visual cortex respond selectively to bars of light of different orientations falling on the retina, and in some cases to movement of those bars of light in different directions. The concept that emerges from these experiments is that the brain assembles a mental image of the visual world by analyzing edges in patterns of light falling on the retina. Each retina sends a million axons to the brain, but there are *hundreds of millions* of neurons in the visual cortex. The action potentials from one retinal ganglion cell are received via the thalamus by hundreds of cortical neurons, each responsive to a different combination of orientation, position, color, and movement of contrasting lines in the patterns of light and dark falling on the retina.

Three-dimensional vision results from cortical cells receiving input from both eyes

How do we perceive objects in three dimensions? The short answer is that our two eyes see overlapping, yet slightly different, visual fields—that is, humans have **binocular vision**. A person who is blind in one eye has difficulty discriminating distances. Animals whose eyes are on the sides of the head rather than facing front have minimal overlap in their fields of vision and, as a result, poor depth vision except for a narrow field straight ahead; however, they can see predators creeping up from all sides.

The story of how the brain integrates information from two eyes begins with the paths of the optic nerves. The two optic nerves run

experiment

Figure 46.9 What Does the Eye Tell the Brain?

Original Paper: Kuffler, S. W. 1953. Discharge patterns and functional organization of mammalian retina. *Journal of Neurophysiology* 16: 37–68.

Stephen Kuffler's experiments recorded the activity of single ganglion cells in the eyes of cats. These groundbreaking experiments revealed the existence of a circular receptive field for each of the retina's ganglion cells. Signals from photoreceptor cells in a receptive field are either excitatory or inhibitory to the ganglion cell, which sends action potentials via the optic nerve to the brain.

HYPOTHESIS▶ Retinal ganglion cells are excited or inhibited by light and dark stimuli falling on local areas of the retina.

METHOD

1. Place electrodes next to an individual retinal ganglion cell.
2. Stimulate the retina with different combinations of light and dark stimuli and record the responses of the ganglion cell.
3. Continue recording and move the stimuli around the retina to find the area of sensitivity—the receptive field—for a specific ganglion cell.

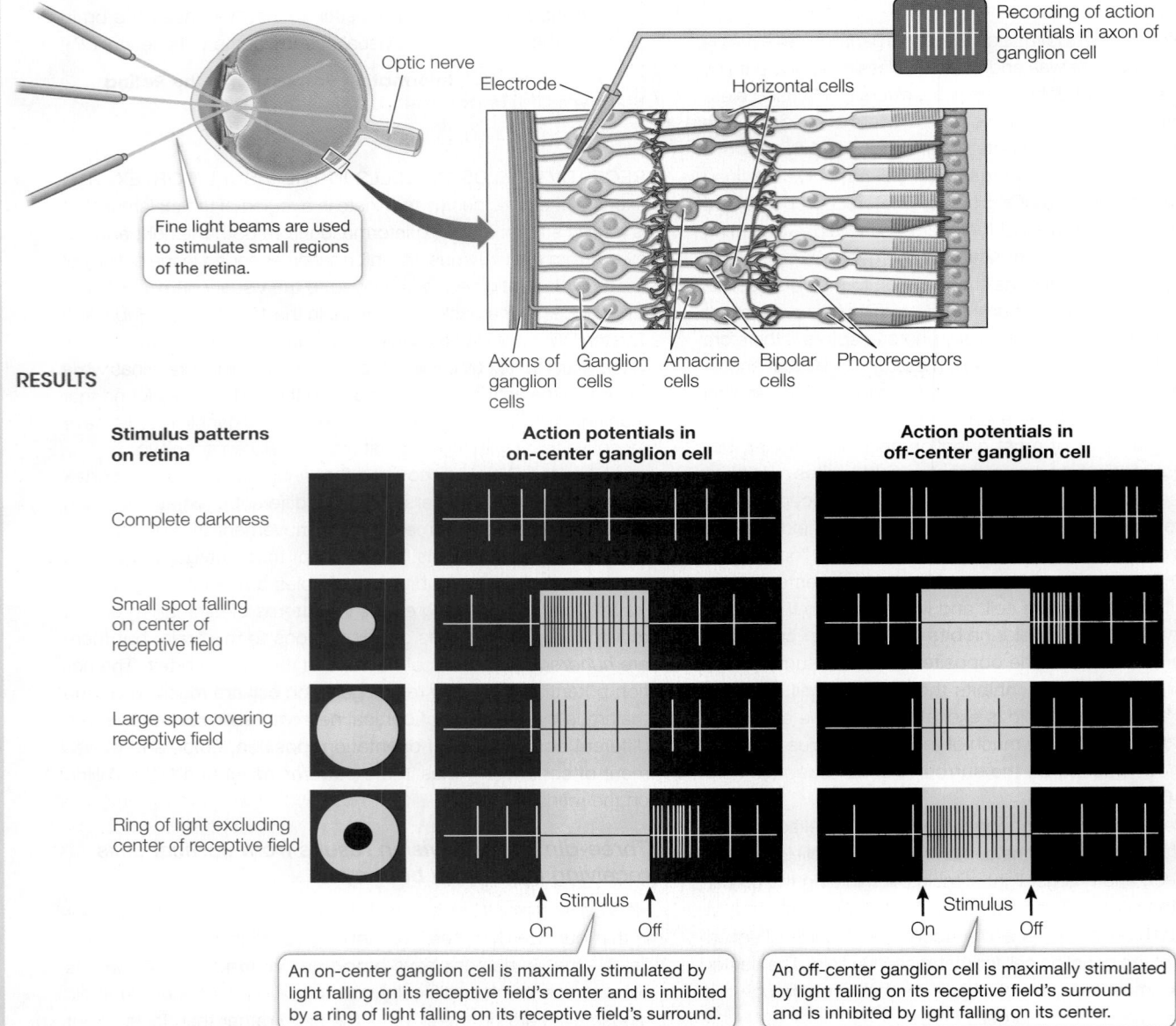

CONCLUSION▶ Ganglion cells use a center-surround dichotomy to encode patterns of contrast between light and dark.

Q: Why is there a blind spot in each eye?

> **focus:** key figure

(A) Visual field

(B) Binocular cells

(C) Binocular disparity

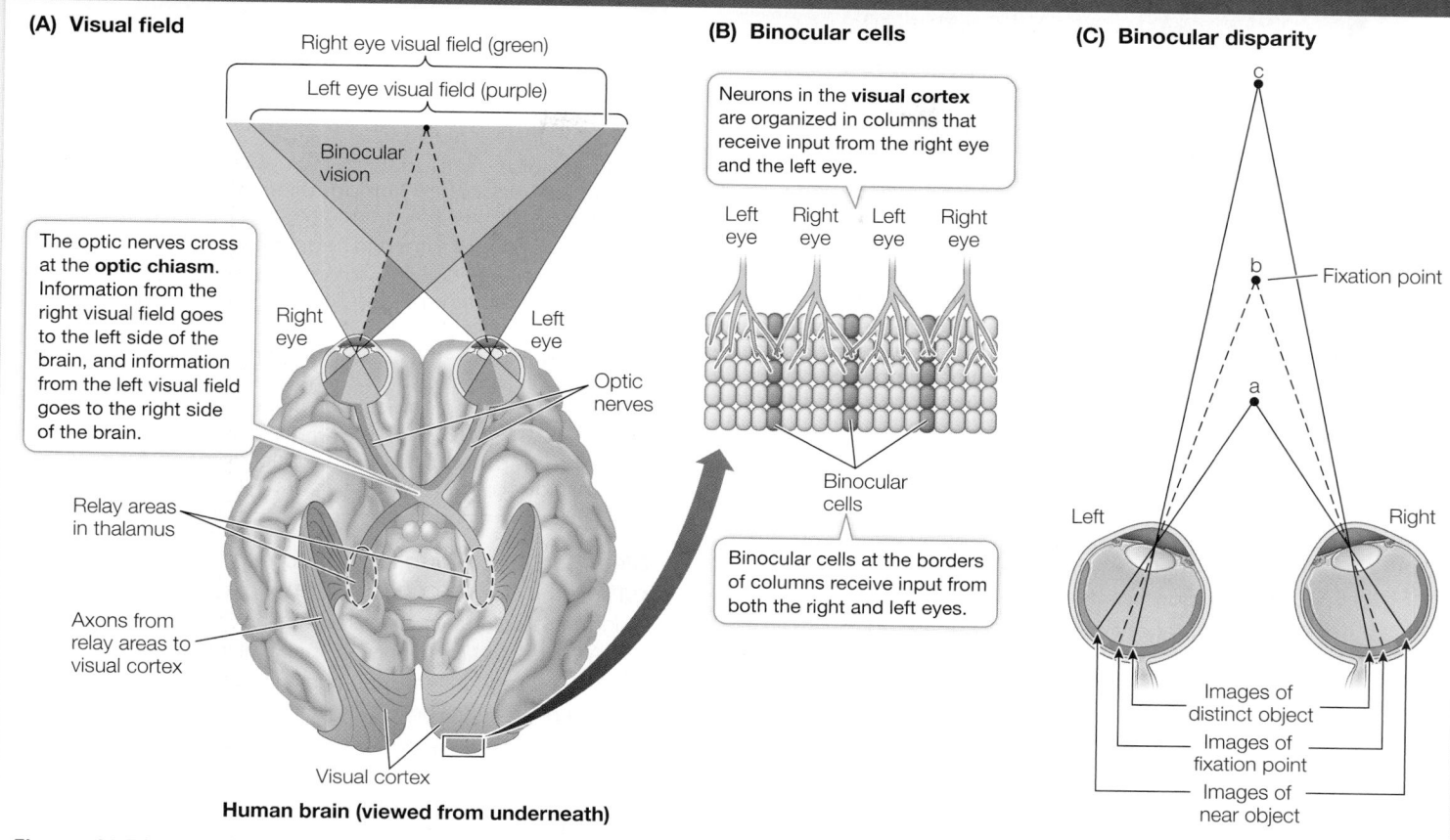

Human brain (viewed from underneath)

Figure 46.10 How Do We See in 3-D? **(A)** Our eyes have largely overlapping visual fields, and both eyes transmit information to both sides of the brain. However, because the axons from the inner sides of the retinas cross at the optic chiasm, information from the visual space to the left of straight ahead goes to the right side of the brain, and information from the visual space to the right of straight ahead goes to the left side of the brain. **(B)** Between these left and right eye columns of neurons are neurons that receive input from both eyes. These are the binocular cells.

(C) Light from objects that are different distances away will fall at different locations on the retinas. These disparities between where the image of the fixation point and the images of other objects fall on the retina means that different binocular cells will receive input from different retinal axons depending on how far away the object is. These disparity sensitive cells enable depth perception and 3-D vision.

Q: What would be the consequences of an anterior–posterior cut of the optic chiasm on midline?

along the underside of the brain, join just under the hypothalamus, and then separate again (**Focus: Key Figure 46.10**). The place where they join is called the **optic chiasm**. Axons from the half of each retina closest to the nose cross in the optic chiasm and go to the opposite side of the brain. The axons from the outer half of each retina do not cross over at the optic chiasm; axons from the outer left retina go to the left side of the brain, and vice versa for axons from the outer right retina.

The functional consequence of the optic chiasm is that all of the visual information from the left side of your field of vision when you are looking straight ahead goes to the right side of your brain, and all of the visual information from the right side of your field of vision goes to the left side of your brain. However, there is a very large area of overlap between the visual fields of the two eyes. These relationships are shown in purple and green in Figure 46.10.

Because of the overlap between the visual fields of the two eyes, there can be neurons in the visual cortex that receive input from both the left and the right eye. These are called binocular cells, and

they can detect disparity between where the image of an object falls on the two retinas. What is disparity? Hold your finger out in front of you and look at it closing one eye and then the other. Your finger appears to jump back and forth because its image falls on a different position on each retina. Repeat the exercise with a distant object. It doesn't jump back and forth as much because there is less disparity in the positions of the image on the two retinas. Certain binocular cells respond optimally to a stimulus falling on both retinas with a particular disparity. Which set of binocular cells is stimulated depends on how far away the stimulus is.

When we look at something, we can detect its shape, color, depth, and movement. Where does all this information come together? Is there a single cell that fires only when a red sports car drives by? The answer to that is probably no. Visual experience comes from simultaneous activity in many different cells. In addition, most visual experiences are enhanced by information from the other senses and from memory, which helps explain why about 75 percent of the cerebral cortex is association cortex.

46.2 recap

Information in the nervous system is processed by cellular interactions in neural networks. The opposing actions of the sympathetic and parasympathetic divisions of the ANS can be understood in terms of neural pathways consisting of just two neurons. Vision begins with retinal photoreceptors organized into circular retinal receptive fields that influence the activity of retinal ganglion cells. Information from ganglion cells is transmitted in the optic nerves to the thalamus and relayed from there to the visual cortex. The left visual field from both eyes is represented in the right visual cortex, and vice versa for the right visual field. Receptive fields of cortical cells represent input from multiple ganglion cells, and binocular, disparity-sensitive cortical cells enable depth perception.

learning outcomes

You should be able to:

- Predict the outcome on effector tissues when specific peripheral nerves of the sympathetic or parasympathetic divisions of the autonomic nervous system are damaged or stimulated.
- Explain how receptive fields are generated in the retina through the interactions between photoreceptors, bipolar cells, and ganglion cells, and through the modulation of their signaling by horizontal and amacrine cells.
- Understand how neural connections between the eyes and the visual cortex produce three-dimensional vision.

1. Explain how a knife wound to left side of the neck can result in pupillary constriction in the left eye.
2. Why can a small spot of light on the retina cause a larger response in a retinal ganglion cell than a large spot of light falling on the same area of retina?
3. What changes in vision would occur if the optic chiasm were cut right on the midline?

By studying the neural circuitry of the visual system and the ANS, you have gained some understanding of how information reaches the CNS and how the CNS controls various functions of the body. But what about the higher functions of the mammalian CNS—the complex functions between input and output, such as language, learning, memory, and dreams?

key concept 46.3 Higher Brain Functions Involve Integration of Multiple Systems

The higher brain functions discussed in the remaining pages of this chapter are undeniably complex. Nevertheless, neuroscientists, using a wide range of techniques, are making considerable progress in understanding these functions in terms of their cellular, molecular, and network properties. The following discussion will address several aspects of brain and behavior that present challenges to neuroscientists: sleep and dreaming, learning and memory, language use, and consciousness.

focus your learning

- During wakefulness and REM sleep (dreaming), certain brainstem nuclei provide excitatory input that activates the thalamus and cortex, while in non-REM sleep these nuclei are less active.
- Sleep states are divided into slow-wave sleep (non-REM) and REM sleep, and humans alternate between these states four or five times each night.
- Sleep deprivation impairs cognitive functions.
- Language functions are lateralized (typically within the left cerebral hemisphere). Language ability involves parts of the occipital, parietal, temporal, and frontal lobes.
- The cellular correlate for learning and memory is the modification of synaptic strength. The hippocampus is important for acquiring and consolidating declarative memories.
- In humans and higher mammals, the insular lobe integrates physiological information to generate a conscious awareness of the self.

Sleep and dreaming are reflected in electrical patterns in the cerebral cortex

A dominant feature of behavior is the daily cycle of sleep and waking. All birds and mammals, probably all other vertebrates, and also many invertebrates, sleep. We humans spend one-third of our lives sleeping, yet we do not know why or how. We do know, however, that we need to sleep. Loss of sleep impairs alertness and performance. Many people in our society—certainly most college students—are chronically sleep-deprived. Accidents and serious mistakes that endanger lives can be attributed to impaired alertness caused by lack of sleep. Insomnia (difficulty in falling or staying asleep) is one of the most common medical complaints.

THE ELECTROENCEPHALOGRAM   A common tool of sleep researchers is the **electroencephalogram**, or **EEG**. Rather than recording the activity of single neurons, the EEG characterizes activity in huge numbers of neurons. EEG electrodes are much larger than the fine electrodes used to detect single cell activity. Placed at different locations on the head and scalp (**Figure 46.11A**), EEG electrodes record fluctuations in the electric potential differences between electrodes over time. These differences reflect the electrical activity of the neurons in the brain regions under the electrodes, primarily regions of the cerebral cortex. Usually the electrical activity of one or more skeletal muscles is also recorded; this record is called an electromyogram (EMG). Movements of the eyes are recorded as an electrooculogram (EOG).

EEG, EMG, and EOG patterns reveal the transition from being awake to being asleep. They also reveal that there are different states of sleep. In mammals other than humans, two major sleep states are easily distinguished: **slow-wave sleep** and **rapid eye movement (REM) sleep**. Slow-wave sleep gets its name from the high-amplitude, slow-frequency waves in the EEG. REM sleep gets its name from jerky movements of the eyeballs that occur during this state. In humans, sleep states are characterized as non-REM sleep and REM sleep. Human non-REM sleep is divided into three stages.

When you fall asleep, you first enter stage 1 non-REM sleep, which then progresses through stage 2 to stage 3 (**Figure 46.11B**). Stage 3 is deep, restorative, slow-wave sleep. This first full cycle of

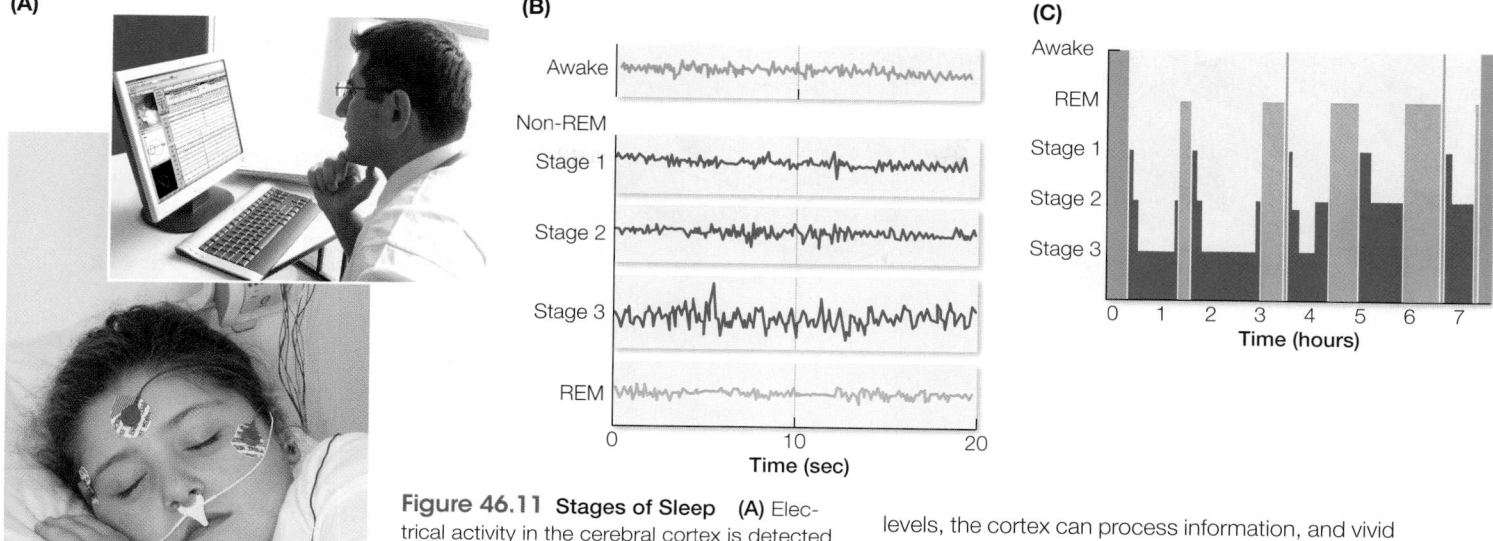

Figure 46.11 Stages of Sleep   (A) Electrical activity in the cerebral cortex is detected by electrodes that are placed on the head and record changes in voltage between the electrodes through time. (B) The resulting record is an electroencephalogram, or EEG. (C) Humans cycle through different stages of sleep throughout the night.

non-REM sleep is followed by an episode of REM sleep. Throughout the night you experience four or five cycles of non-REM and REM sleep (**Figure 46.11C**). About 80 percent of your sleep is non-REM sleep. The most vivid dreams and nightmares occur during the 20 percent of sleep that is REM sleep.

CELLULAR CHANGES DURING SLEEP   When we are awake, several nuclei (groups of neurons) in the brainstem reticular-activating system are continuously active. Axons from neurons in these nuclei extend to the thalamus and throughout the cerebral cortex, where they release depolarizing neurotransmitters (acetylcholine, norepinephrine, and serotonin). These broadly distributed neurotransmitters keep the resting potential of the neurons of the thalamus and cortex close to threshold and sensitive to synaptic inputs, thereby maintaining the responsiveness of the brain that characterizes being awake.

With the onset of sleep, activity in these brainstem nuclei decreases, and their axon terminals release less neurotransmitter. With the withdrawal of the depolarizing neurotransmitters, the resting potentials of the cells of the thalamus and cortex become more negative (hyperpolarized), and the cells are less sensitive to excitatory synaptic input. Their processing of information is inhibited, and consciousness is lost. You have probably experienced this as "nodding off." As you enter non-REM sleep, cells of the cortex begin to fire action potentials in bursts. The synchronization of these bursts over broad areas of cerebral cortex results in the EEG slow-wave pattern that characterizes deep non-REM sleep.

At the transition from non-REM to REM sleep, dramatic changes occur. Some of the brainstem nuclei that were inactive during non-REM sleep become active again, causing a general depolarization of cortical neurons. Thus in REM sleep the synchronized bursts of firing cease, and the EEG resembles that of the awake brain. Because the resting potentials of the neurons return to near threshold

levels, the cortex can process information, and vivid dreams occur.

So why don't we act out our dreams? During REM sleep the brain inhibits both afferent (sensory) and efferent (motor) pathways; we are paralyzed during REM sleep. Limb twitches and the jerky eye movements are motor signals breaking through the inhibition. The bizarre nature of dreams may be due to the lack of sensory feedback to the cortex from the body and the outside world. In other words, a functioning cortex is out of touch with reality. The function of muscle paralysis during REM sleep may be to prevent the acting out of dreams.

Knowing the cellular mechanisms of sleep has not yet led to an understanding of its function. Many questions remain. Why do we have two sleep states with very different neurophysiological characteristics? Why does non-REM sleep always occur first? Why do the two states cycle during the rest phase? We know sleep is essential for life, but we don't know why. One set of hypotheses is that sleep is necessary for the maintenance and repair of neural connections and for the neural changes involved in learning and memory—and possibly forgetting. These hypotheses are supported by many experiments showing that performance of a learned task or recall of declarative information on the day following training is impaired if sleep is prevented, and is best following a good night's sleep.

Language abilities are localized in the left cerebral hemisphere

No aspect of brain function is as integrally related to human consciousness and intellect as language. Therefore brain mechanisms that underlie the acquisition and use of language are extremely interesting to neuroscientists. A curious observation about language ability is that it resides in one cerebral hemisphere—which in 97 percent of people is the left hemisphere. This phenomenon is referred to as the **lateralization** of language functions.

Fascinating research on this subject was conducted by Roger Sperry and his colleagues at the California Institute of Technology. The two cerebral hemispheres are connected by a tract of white matter called the corpus callosum. In one severe form of epilepsy, bursts of action potentials causing seizures travel between hemispheres via the corpus callosum. Cutting the tract eliminates the problem, and patients function well following surgery. However,

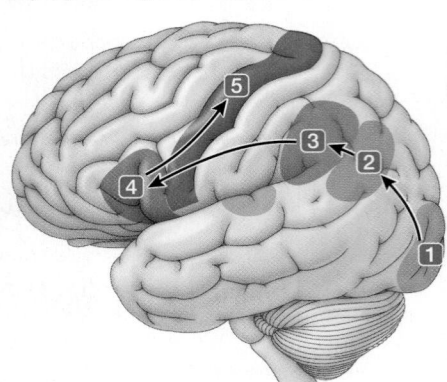

(A) Repeating a heard word

Broca's area

Primary motor cortex (motor control)

Wernicke's area

Angular gyrus

Primary visual area

Speech

Hearing

(B) Speaking a written word

Figure 46.12 Language Areas of the Cortex Different regions of the left cerebral cortex participate in the processes of (A) repeating a word that is heard and (B) speaking a written word.

Activity 46.3 **Language Areas of the Cortex** www.Life11e.com/ac46.3

these "split-brain" subjects display interesting deficits in language ability.

After the surgery, if an object is shown in the right visual field and the left eye is closed (see Figure 46.10), the patient can describe it verbally and in writing. If the object is shown in the left visual field and the right eye is closed, the patient cannot describe it either verbally or in writing, but can use his or her left hand to point to a picture of the object. Without the connecting tissue between the two hemispheres, knowledge or experience of the right hemisphere can no longer be expressed in language.

Individuals who have suffered damage to the left hemisphere frequently suffer from some form of **aphasia**, a deficit in the ability to use or understand words. Studies of such individuals have identified several language areas in the left hemisphere.

- **Broca's area**, located in the frontal lobe just in front of the primary motor cortex, is essential for speech. Damage to Broca's area results in halting, slow, poorly articulated speech or even

complete loss of speech, but the patient can still read and understand language.

- **Wernicke's area**, located in the temporal lobe close to its border with the occipital lobe, is more involved with sensory than with motor aspects of language. Damage to Wernicke's area can cause a person to lose the ability to speak sensibly while retaining the abilities to form the sounds of normal speech and to imitate its cadence. Such a patient cannot understand spoken or written language.

- The **angular gyrus**, located near Wernicke's area, is involved in integrating spoken and written language.

Normal language ability depends on the flow of information among various areas of the left cerebral cortex. Input from spoken language travels from the auditory cortex to Wernicke's area (**Figure 46.12A**). Input from written language travels from the visual cortex to the angular gyrus to Wernicke's area (**Figure 46.12B**). Commands to speak are formulated in Wernicke's area and travel to Broca's area and from there to the primary motor cortex. Damage to any one of those areas or the pathways between them can result in aphasia. Using modern methods of brain imaging such as functional magnetic resonance imaging (fMRI) or positron emission tomography (PET), it is possible to see the metabolic activity in different brain areas when the brain is using language (**Figure 46.13**).

Some learning and memory can be localized to specific brain areas

Learning is the modification of behavior by experience. Memory is the ability of the nervous system to retain what is learned and experienced. Even very simple animals can learn and remember, but these two abilities are most highly developed in humans. Consider the amount of information associated with

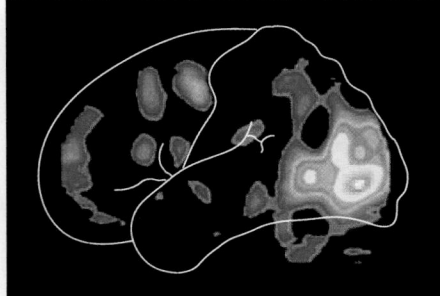

Listening to words

Passively viewing words

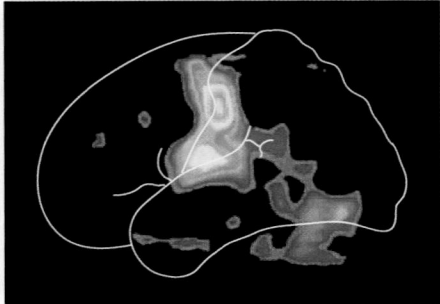

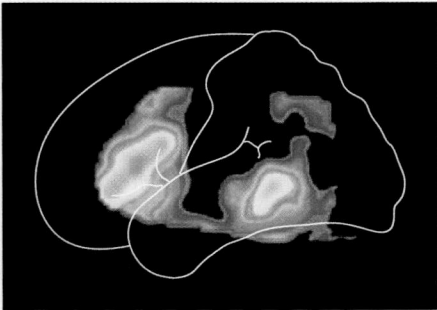

Speaking words

Generating word associations

Figure 46.13 Imaging Techniques Reveal Active Parts of the Brain Positron emission tomography (PET) scanning reveals the brain regions activated by different aspects of language use. Radioactively labeled glucose is given to the subject. Brain areas take up radioactivity in proportion to their metabolic use of glucose. The PET scan visualizes levels of radioactivity in specific brain regions when a particular activity is performed. The red and white areas are the most active.

learning a language, and then the much greater amount of information that language enables our brains to store and to process. The capacity of memory and the rate at which memories can be retrieved are remarkable features of the human nervous system.

Learning that leads to long-term memory and modification of behavior must involve long-lasting synaptic changes. A phenomenon that may explain how long-term synaptic changes might arise is **long-term potentiation**, or **LTP**. LTP results from high-frequency electrical stimulation of certain identifiable circuits that makes these circuits more sensitive to subsequent stimulation. In contrast, continuous, repetitive, low-level stimulation of these same circuits reduces their responsiveness, a phenomenon that has been called **long-term depression** (**LTD**). LTP and LTD may be fundamental cellular or molecular mechanisms involved in learning and memory.

Above the level of the synapse, memory involves interactions between several brain areas, as we discussed in the opener to this chapter and in Investigating Life: Places Cells Reveal Processes of Memory Consolidation during Sleep. Some of the first insights into memory processes came from surgical treatment of patients with severe seizures. That treatment involves destruction of the excessively active brain area triggering the seizures. To find the right area, the surgery is done under local anesthesia, with the patient remaining conscious. As different regions of the brain are electrically stimulated with electrodes, the patient reports the resulting sensations. Stimulation of some regions of the association cortex elicits recall of vivid memories. Such observations provided the first evidence that specific areas in the brain are associated with specific memories and that memory can be attributed to networks of neurons. Destroying a small area of the brain does not completely erase a memory, however, so it is postulated that memory is a function distributed over many brain regions and can be stimulated via many different routes.

You experience several forms of memory everyday. You have **immediate memory** for events that are happening now. Immediate memory is almost perfectly photographic but lasts only seconds. **Short-term memory** contains less information but lasts longer—on the order of 10–15 minutes. When you are introduced to a group of several new people, you probably will have forgotten their names in an hour or so if you have not written them down, used them in a conversation, or made a conscious effort to repeat them. Repetition, use, or reinforcement by something that gets your attention (a title such as "President," for example) facilitates the transfer of short-term memory to **long-term memory**, which can last for days, months, years, or a lifetime. There is also intermediate-term memory, which lasts about a day.

Knowledge about neural mechanisms for the transfer of short-term memory to long-term memory—memory consolidation—has come from observations of persons who have lost parts of the limbic system, notably the hippocampus. A famous case is that of the man identified as H.M., whose hippocampus on both sides of the brain was removed to control his severe epilepsy. After the surgery, H.M. was unable to transfer information to long-term memory. If someone was introduced to him, had a conversation with him, and then left the room for several minutes, when that person returned, H.M did not recognize him—it was as if the conversation had never taken place. Up until his death 55 years later, H.M. remembered events that happened before his surgery but could not remember postsurgery events for more than 10–15 minutes.

The studies of H.M., other such patients, and experiments on animals and even London taxi drivers show that the hippocampus plays an essential role in the acquisition of **declarative memories**—those that involve people, places, things, and events. The study featured in the opener to this chapter revealed the mapping of spatial information by neural networks in the rat, and the replay of that information when the rat is resting before or after running a maze. Subsequent experiments revealed the replay of the same spatial information during sleep, possibly revealing a process that is consolidating the memory and transferring it into long-term storage (see Investigating Life: Place Cells Reveal Processes of Memory Consolidation during Sleep).

 Media Clip 46.1 **The Man with No Short-Term Memory**
www.Life11e.com/mc46.1

We still cannot answer the question "What is consciousness?"

This chapter has only scratched the surface of the organization and functions of the human brain. Even with all of our knowledge of the human brain, and with all of the sophisticated new research tools, we still cannot answer the question "What is consciousness?"

The word "consciousness" is used in everyday language to refer to being awake in contrast to being asleep or in a coma. Here we are referring to being mentally aware of yourself, your environment, and events going on around you in such a way that you can plan for future events and make decisions based on experience, evidence, value systems, and predicted consequences. Speculations about consciousness have been the realm of philosophers, but we are getting closer to a neurobiological understanding.

One requirement for conscious experience is a perception of self that can be integrated with information from the physical and social environment and information from past experience. The basis for a perception of self derives from the huge amount of somatosensory and visceral information that comes from all parts of the body. This information is used for motor control and for homeostatic regulation. It enables animals to find food, seek mates, seek warmth, avoid cold, avoid danger, and so on. Different types of this afferent information go to appropriate control and regulatory systems in the brainstem and forebrain. But these diverse sources of information also go to the insular cortex. The insular lobe appears to integrate physiological information from all over the body to create a sensation of how the body "feels." Thus when an animal's actions restore homeostasis, it "feels" better, and this is motivation to do the right thing for well-being.

In humans and the great apes, the insular cortex is greatly expanded and its circuitry communicates with parts of the brain involved in planning and decision making. In imaging studies, the insular lobe is activated by a great diversity of situations that involve strong feelings such as pleasure, disgust, humor, pain, lust, craving, humiliation, guilt, or empathy. Damage to the insular lobe causes apathy, loss of the ability to enjoy music, loss of sexual responsiveness, and even loss of the ability to distinguish good food from spoiled food. Humans and the few other species (great apes, elephants, and some marine mammals) that have an expanded insular lobe are the only species that can recognize themselves in a mirror. Could it be that this very discrete part of our brains and its circuitry are the neurobiological basis for self-awareness?

46.3 recap

Complex functions of the nervous system can be understood in terms of the properties of neurons and neural networks. Language functions are lateralized to the left cerebral hemisphere. So, since information about objects seen in the left visual field is processed in the right visual cortex, cannot be described in words, but they can be drawn. Imaging shows the integration of different brain areas in the hearing, reading, and speaking of words. REM and non-REM sleep and awake states involve very different activities in many areas of the brain. Sleep states have a regular organization of expression during the daily rest phase. At the cellular level, learning and memory involve changes in synaptic strength, and memory processes involve interactions of neural networks from different brain areas. Declarative memories are acquired through the hippocampus. Consciousness may involve creation of a perception of the self in the insular cortex.

learning outcomes

You should be able to:

- Recognize the characteristic EEGs associated with wakefulness, non-REM sleep, and REM sleep, and infer which brain areas are active during these distinct phases of the sleep–wake cycle.
- Describe the flow of information when speaking a written or a heard word, and specify which areas of the brain are involved in speech processing.
- Connect the modification of synaptic strength in place cells to the development of a spatial map in the hippocampus.
- Connect evolutionary changes in brain structures to the development of self-awareness.

---

1. What is the significance of paralysis of skeletal muscles during REM sleep?
2. If a person with a severed corpus callosum is shown an object in his left visual field, how will he be able to communicate to an observer what that object is?
3. What does the case of H.M. tell us about the brain mechanisms underlying the acquisition of declarative memories?
4. What is unique about the insular lobes of higher mammals that may explain the evolution of consciousness?

investigatinglife

Can recordings of place cells reveal processes of memory consolidation during sleep?

Unfortunately we cannot give a rat an exam to ask what it remembers. However, we can look in other brain areas for electrophysiological patterns of activity that correlate with the hippocampal place-cell patterns. Wilson and colleagues described the fast replay of the hippocampal place-cell patterns as a unique EEG signal called "ripples." They observed that these ripples were tightly coupled to slower EEG brainwaves that spread to cortical areas of the brain. When they recorded in some of those other areas—the frontal cortex and the visual cortex—they found similar patterns of firing (i.e., ripples) in those areas that were synchronized to the hippocampal ripples. They hypothesized that the ripples represent memory transcripts that are transferred to and stored in areas of the cortex.

Future directions

The neurophysiological studies of the acquisition and consolidation of declarative memories bring us closer to understanding the physical nature of a memory. The synaptic strengthening known as LTP (long-term potentiation) lasts a long time in contrast to action potentials, but not nearly as long as a memory that can virtually last a lifetime. We can think of a memory as a piece of information that is stored in the brain. We have used the word "information" a lot in this chapter, but we really do not know what "information" is in the brain. It is not simply action potentials; they are brief, transient events. It is not patterns of synapses; they are constantly being altered. It must involve patterns of connectivity in and between neural networks in the brain; but what maintains those networks over years? We have much to learn about how our brains work.

Chapter Summary 46

46.1 Functions Are Localized in the Nervous System

- The brain and spinal cord make up the **central nervous system (CNS)**; the cranial and spinal nerves make up the **peripheral nervous system (PNS)**.

- The nervous system can be modeled conceptually in terms of the direction of information flow and whether we are conscious of the information. The **afferent** component of the PNS carries information from sensory cell receptors to the CNS, and the **efferent** component of the PNS carries information from the CNS to the target tissues and organs that are being controlled. **Review Figure 46.1**

- The vertebrate nervous system develops from a hollow dorsal neural tube. The brain forms from three swellings at the anterior end of the neural tube, which become the **hindbrain**, the **midbrain**, and the **forebrain**. The developing forebrain consists of the **telencephalon** and the **diencephalon**. The telencephalon develops into the **cerebrum (cerebral hemispheres)**. The underlying diencephalon develops into the **thalamus** and **hypothalamus**. The midbrain and hindbrain develop into the **brainstem** and the **cerebellum**. **Review Figure 46.2**

- The spinal cord communicates information between the brain and the rest of the body.

- The **reticular-activating system** is a complex network that directs incoming information to appropriate brainstem **nuclei** that control autonomic functions, and transmits information to the forebrain that results in conscious sensation. The reticular-activating system controls the level of arousal of the nervous system, including sleep and wakefulness.

- The **limbic system** is a complex set of phylogenetically old forebrain structures involved in emotions, mood, physiological drives (such as hunger and thirst), instincts, and memory. **Review Figure 46.3**

- The cerebral hemispheres are the dominant structures of the human brain. Their surfaces are layers of neurons called the **cerebral cortex**. The cerebral hemispheres can be divided into the **temporal**, **frontal**, **parietal**, **occipital**, and **insular lobes**. Many motor functions are localized in parts of the frontal lobe. Information from many sensory receptors projects to a region of the parietal lobe. Visual information projects to the occipital lobe, and auditory information projects to a region of the temporal lobe. The insular cortex integrates much information about the physiological state of the body, but it is also activated by psychological and emotional conditions and may be responsible for the sense of self. **Review Figures 46.4, 46.5, Activity 46.1**

46.2 Nervous System Functions Rely on Neural Circuits

- The **autonomic nervous system (ANS)** controls the physiological function of organs and organ systems. Its **sympathetic** and **parasympathetic** divisions are characterized by their anatomy, neurotransmitters, and effects on target tissues. **Review Figure 46.8**

- The neural network of vision involves patterns of light falling on **receptive fields** in the retina. Receptive fields have a center and a surround, which have opposing effects on ganglion cell firing. **Review Figure 46.10, Activity 46.2, Animation 46.1**

- Information from retinal ganglion cells is communicated via the optic nerve to the thalamus and then to the visual cortex. The visual cortex seems to assemble an image of the visual world by analyzing edges of patterns of light.

- **Binocular vision** is possible because information from both eyes is communicated to binocular cells in the visual cortex. These cells interpret distance by measuring the disparity between where the same stimulus falls on the two retinas. **Review Focus: Key Figure 46.10**

46.3 Higher Brain Functions Involve Integration of Multiple Systems

- Humans have a daily cycle of sleep and waking. Sleep can be divided into **rapid eye movement (REM) sleep** and non-REM sleep. Deep non-REM sleep is known as **slow-wave sleep** because of its characteristic EEG patterns. **Review Figure 46.11**

- Language abilities are localized mostly in the left cerebral hemisphere, a phenomenon known as **lateralization**. Different areas of the left hemisphere—including **Broca's area**, **Wernicke's area**, and the **angular gyrus**—are responsible for different aspects of language. **Review Figures 46.12, 46.13, Activity 46.3**

- Some learning and memory processes can be localized to specific brain areas. Long-lasting changes in synaptic properties referred to as **long-term potentiation (LTP)** and **long-term depression (LDP)** may be involved in learning and memory.

- Complex memories can be elicited by stimulating small regions of association cortex. Damage to the hippocampus can destroy the ability to form long-term **declarative memory**.

- A sense of the physiological state of the body may be created in the insular cortex from visceral afferent information. Evolution of this integrative function in humans and other higher primates could be the basis for conscious experience.

See Activity 46.4 for a concept review of this chapter.

Go to **LearningCurve** (in **LaunchPad**) for dynamic quizzing that helps you solidify your understanding of this chapter. **LearningCurve** adapts to your responses, giving you the practice you need to master each key concept.

Apply What You've Learned

Review

46.1 Specific areas within the five lobes of the cerebral cortex have sensory, motor, and associative functions.

46.3 Sleep deprivation impairs cognitive functions.

46.3 Language functions are lateralized (typically within the left cerebral hemisphere). Language ability involves parts of the occipital, parietal, temporal, and frontal lobes.

Original Paper: Chee, M. W. L., L. Y. M. Chuah, V. Venkatraman, W. Y. Chan, P. Philip and D. F. Dinges. 2006. Functional imaging of working memory following normal sleep and after 24 and 35 h of sleep deprivation: Correlations of fronto-parietal activation with performance. *NeuroImage* 31: 419–428.

Insufficient sleep has negative consequences. There are the physiological ones: lack of energy, difficulty staying awake, and even high blood pressure, diabetes, and stroke. And what about cognitive abilities—functions such as memory and alertness? Looking for answers, researchers ran tests on three groups of subjects. In the first group, subjects were rested and wakeful. This "RW" group was compared with subjects who were sleep-deprived for 24 hours ("SD24") and subjects who were sleep-deprived for 35 hours ("SD35"). The researchers looked at both alertness and working memory (the ability to hold new information in memory for a short time to enable cognitive processing). They also performed brain scans, using magnetic resonance imaging (MRI), to see what specific areas of the brain were affected by sleep deprivation.

In one test (referred to as the LTR test), subjects looked at a picture of four capital letters for 0.5 seconds. Three seconds later they were shown a lowercase letter for 1.5 seconds and were to press a button if the lowercase letter matched one of the capital letters they had just seen.

The second test (referred to as the PLUS test) required more cognitive processing than the LTR test. In the PLUS test, subjects looked at a picture of two capital letters for 0.5 seconds. Three seconds later they were shown a lowercase letter for 1.5 seconds and were to press a button if this was the next letter, in alphabetical sequence, after either of the two capital letters.

The effect of sleep deprivation on working memory was measured by the accuracy of responses of sleep-deprived (SD) subjects compared with rested and wakeful (RW) subjects. The alertness of the SD and RW subjects was measured by their reaction times in milliseconds. The results are shown in the table. Asterisks (*) indicate significant differences between RW and SD subjects.

Using MRI, the researchers measured activation of different brain regions during the testing. Two areas with the most significant differences between RW and SD subjects are shown in the graph. Plus signs (+) indicate significant differences between RW and SD24 subjects; hashtags (#) indicate significant differences between RW and SD35 subjects.

Accuracy (%)	RW	SD24	SD35	Reaction time (ms)	RW	SD24	SD35
LTR	95.2	86.5*	85.0*	LTR	745	792*	769
PLUS	95.0	86.6*	84.3*	PLUS	698	747*	746*

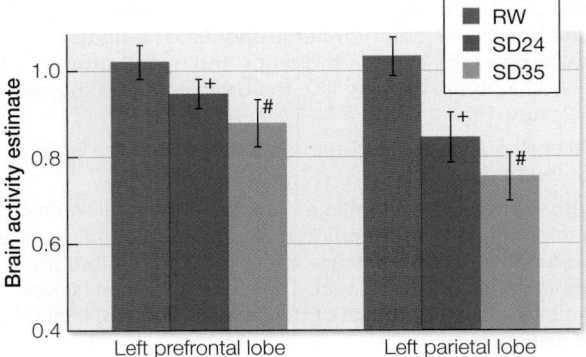

Questions

1. Based on the data in the table, justify the conclusion that sleep deprivation impairs working memory.

2. One interpretation of the reaction-time data in the table is that they reflect decreased alertness after sleep deprivation. What information would lead to this conclusion? What is another way to explain the results?

3. Based on the information in the text, explain why the areas of the cerebral cortex shown in the graph were more affected by sleep deprivation than most other areas of the cerebral cortex examined in this study.

4. After looking at the MRI data, the researchers concluded that sleep deprivation has a stronger effect on the parietal lobe than on the frontal lobe. How do the data support that conclusion? What statistical comparison would you use to test this conclusion?

Go to **LaunchPad** for the eBook, LearningCurve, animations, activities, flashcards, and additional resources and assignments.

Musculoskeletal Systems

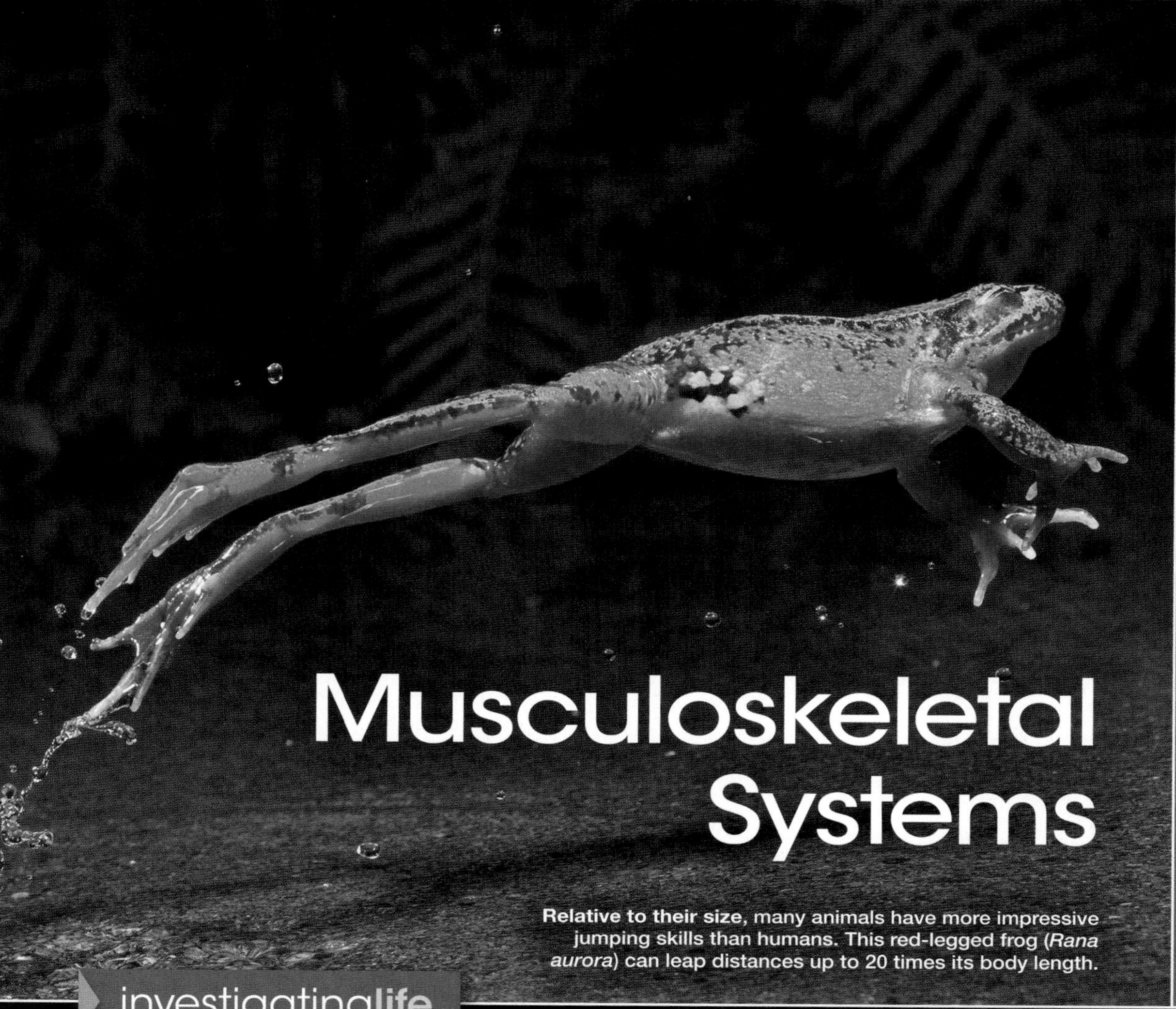

Relative to their size, many animals have more impressive jumping skills than humans. This red-legged frog (*Rana aurora*) can leap distances up to 20 times its body length.

▶ investigating**life**

Champion Jumpers

The women's long jump Olympic record is 7.4 meters, set in 1988 by Jackie Joyner-Kersee. Another world-record long jump was set two years earlier by Rosie the Ribeter, who jumped 6.5 meters. Rosie was a frog competing in the Calaveras County Jumping Frog Contest. Rosie's jump is actually more impressive—it was 20 times her body length, whereas Jackie's was 5 times her body length. And Rosie jumped from a sitting, crouched position, while Jackie's jump was propelled by a sprint.

Both jumps were powered by skeletal muscle. The cellular mechanisms of muscle contraction are the same in the frog and the human, so why is the frog's jump more impressive? The answer involves the concept of *leverage*, which depends on the muscles and skeletal elements working together. Muscles pull on bones that are connected at joints to make levers. With a lever you can use the same force to move a large mass a small distance or a small mass a large distance. The ratio of leg length to body mass is greater for the frog than for the human. Thus the frog's long legs can move its small body mass a long distance.

The power for the frog's jump comes from a single contraction of its leg muscles (there is no sprint), and those muscles contract only a very short distance compared to the length of the jump. One species of frog, *Rana pipiens*, has a body length of only 11 centimeters. It has much longer legs, but muscles that power its jump are only about 3.5 centimeters long and shorten by just 0.75 centimeters (21%) when the frog jumps. Once again leverage comes into play. The thigh muscles are attached to the lower leg bones close to the knee joint, thus they can move the opposite ends of those bones over a longer distance and more quickly than if they were attached farther away from the joint. There is a trade-off, however, in terms of force and speed. To use a shovel, you have one hand near the end of the shaft and the other hand farther down the shaft. With hands close together, you can move the shovel blade fast, but not pick up a heavy load of dirt. With hands farther apart, you can't move the blade as fast, but you can pick up a heavier load.

 **What adaptations optimize the jumping muscles of the frog?**

47.1 Interactions of Actin and Myosin Cause Muscles to Contract

Most behavior and many physiological actions of animals, such as beating of the heart and moving of food through the digestive tract, depend on muscle contraction. Wherever tissues contract, muscle cells are responsible. As introduced in Key Concept 39.1 and shown in Figure 39.2, there are three types of vertebrate muscle:

1. **Skeletal muscle** is responsible for all voluntary movements, such as running or playing a piano. It is also involved in many involuntary actions, including breathing, shivering, and maintaining posture.
2. **Cardiac muscle** is responsible for the beating of the heart.
3. **Smooth muscle** creates movement in many hollow internal organs, such as the digestive system, bladder, and blood vessels, and is under the control of the autonomic (involuntary) nervous system.

focus your learning

- When skeletal muscle contracts, the sarcomeres shorten and the band pattern changes.
- Muscle contraction is due to repeated cross bridge forming, breaking, and reforming between the actin and myosin filaments causing them to slide past each other.
- Action potentials arriving at a neuromuscular junction trigger action potentials in the muscle cell membrane, resulting in release of Ca^{2+} from the sarcoplasmic reticulum.
- Cardiac muscle has similarities and differences with skeletal muscle that are important for their respective functions.

Skeletal and smooth muscle occur in all animal phyla, but cardiac muscle is unique to vertebrates. All three muscle types share the same underlying molecular mechanisms for generating force. We will use vertebrate skeletal muscle as our primary example. Later in this section we will discuss the differences in cardiac and smooth muscle that adapt them to their particular functions, and at the end of Key Concept 47.2 we will describe some interesting specializations of insect muscle.

Sliding filaments of actin and myosin cause skeletal muscle to contract

Skeletal muscle is also called striated muscle—striated because of its striped appearance (**Figure 47.1**; also see Figure 39.2) and skeletal because its attachments to skeletal elements make most movements of the body possible. Skeletal muscle cells, called **muscle fibers**, are large and long and have many nuclei.

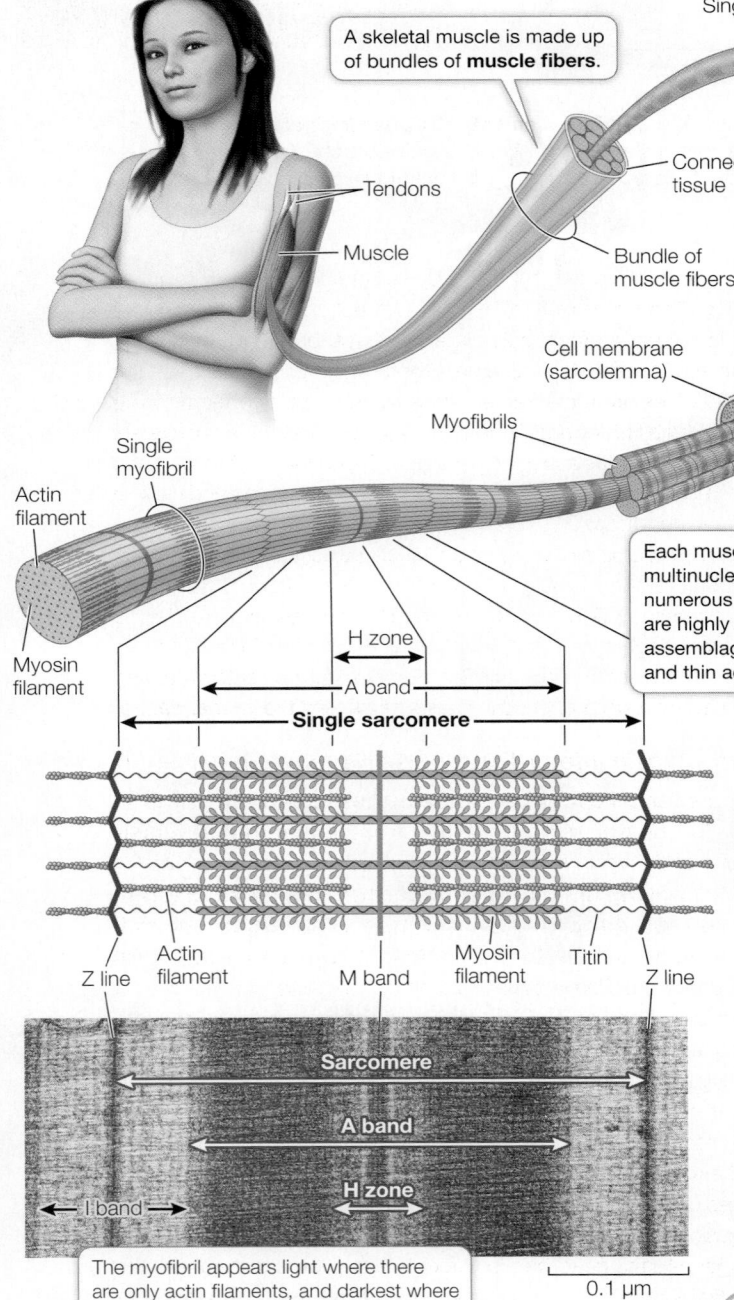

A skeletal muscle is made up of bundles of **muscle fibers**.

Single muscle fiber (cell)

Tendons

Muscle

Connective tissue

Bundle of muscle fibers

Nucleus

Cell membrane (sarcolemma)

Myofibrils

Single myofibril

Actin filament

Myosin filament

Each muscle fiber is a multinucleate cell containing numerous **myofibrils**, which are highly ordered assemblages of thick myosin and thin actin filaments.

H zone

A band

Single sarcomere

Sarcomeres are the units of contraction.

Z line

Actin filament

M band

Myosin filament

Titin

Z line

Sarcomere

A band

H zone

I band

The myofibril appears light where there are only actin filaments, and darkest where there are both actin and myosin filaments.

0.1 µm

Figure 47.1 The Structure of Skeletal Muscle   A skeletal muscle is made up of bundles of muscle fibers. Each muscle fiber is a multinucleate cell containing numerous myofibrils, which are highly ordered assemblages of thick myosin and thin actin filaments. The arrangement of the actin and myosin filaments gives skeletal muscle fibers their characteristic striated appearance.

 Activity 47.1 The Structure of a Sarcomere
www.Life11e.com/ac47.1

These multinucleate cells form in development through the fusion of many individual embryonic muscle cells called myoblasts. A specific muscle such as your biceps (which bends your arm) is composed of hundreds or thousands of muscle fibers bundled together by connective tissue.

Muscle contraction is due to the interaction between the contractile proteins **actin** and **myosin**. Within muscle cells, *actin and myosin molecules are organized into filaments. Actin filaments are also called thin filaments, and myosin filaments are called thick filaments. The two kinds of filaments lie parallel to each other. When muscle contraction is triggered, the actin and myosin filaments slide past each other in a telescoping fashion.

---

*connect the concepts   Key Concept 5.3 discussed how microfilaments made of polymerized actin are elements of the cytoskeleton that are involved in maintaining cell shape and cell movements in all eukaryotic cells.

What is the relationship between a skeletal muscle fiber and the actin and myosin filaments responsible for its contraction? Each muscle fiber (cell) is packed with **myofibrils**—bundles of thin actin and thick myosin filaments arranged in orderly fashion. In most regions of the myofibril, each thick myosin filament is surrounded by six thin actin filaments, and each thin actin filament sits within a triangle of three thick myosin filaments.

A longitudinal view of a myofibril reveals why skeletal muscle appears striated. The myofibril consists of repeating units called **sarcomeres**. Each sarcomere is made of overlapping filaments of actin and myosin, which create a distinct banding pattern (see Figure 47.1). Before the molecular nature of the muscle banding pattern was known, the bands were given names that are still used today. Each sarcomere is bounded by Z lines, which anchor the thin actin filaments. Centered in the sarcomere is the A band, which contains all the myosin filaments. The H zone and the I band, which appear light, are regions where actin and myosin filaments do not overlap in the relaxed muscle. The dark stripe within the H zone is called the M band; it contains proteins that hold the myosin filaments in their regular arrangement.

The bundles of myosin filaments are held in a centered position within the sarcomere by a protein called **titin** (see Figure 47.2). Titin is the largest protein in the body; it runs the full length of the sarcomere from Z line to Z line. Each titin molecule runs right through a myosin bundle. Between the ends of the myosin bundles and the Z lines, titin molecules are very stretchable, like bungee cords. In a relaxed skeletal muscle, resistance to stretch is mostly due to the elasticity of the titin molecules.

As the muscle contracts, the sarcomeres shorten and the band pattern changes. The H zone and the I band become much narrower, and the Z lines move toward the A band as if the actin filaments were sliding into the H zone, the region occupied by the myosin filaments (**Figure 47.2**). In the mid-1950s this observation

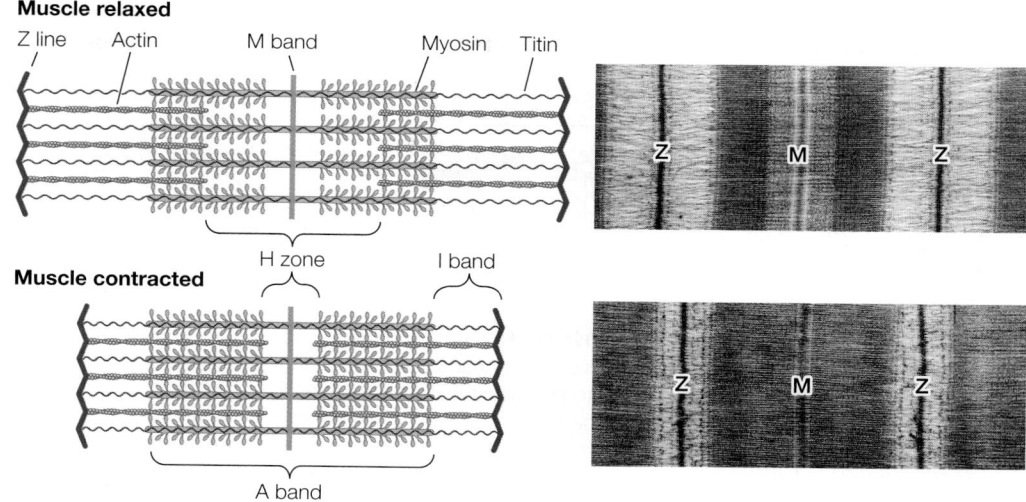

Figure 47.2 **Sliding Filaments**   The banding pattern of the sarcomere changes as it shortens, as seen in electron micrographs. These observations led to the sliding filament model of muscle contraction.

independently led two teams of British biologists to propose the **sliding filament model** of muscle contraction.

It is not uncommon in science for critical breakthroughs to be made simultaneously in different laboratories, but in this case the coincidences are remarkable. The leaders of the two teams were named Hugh Huxley and Andrew Huxley—but they were not related. Andrew Huxley was a faculty member at Cambridge University in the early 1950s, and Hugh Huxley was a graduate student working in a different laboratory. Both were interested in the structure of muscle and the mechanism of contraction. Hugh Huxley moved to MIT as a postdoctoral fellow where he worked with Jean Hanson using electron microscopy. Andrew Huxley, joined by Rolf Niedergerke, used a new form of light microscopy that he and Huxley developed. The two groups proposed the sliding filament model at the same time, and both papers were published in the same issue of the journal *Nature*.

Actin–myosin interactions cause filaments to slide

To understand the mechanism causing the actin and myosin filaments to slide past each other, we must first examine the structures of actin and myosin (**Figure 47.3**). A myosin molecule consists of two long polypeptide chains coiled together, each ending in a large globular head. A myosin filament is made up of many myosin molecules arranged in parallel, with their heads projecting sideways at each end of the filament like a bunch of golf clubs.

An actin filament consists of actin monomers polymerized into long chains that look like two strands of beads twisted together. Twisting around the actin chains is another protein, **tropomyosin**, and attached to tropomyosin at intervals are molecules of **troponin**. We'll discuss these two proteins in more detail later in this section.

The myosin heads can bind specific sites on actin, to form crossbridges between the myosin and the actin filaments. Moreover, when a myosin head binds to an actin filament, the head's conformation changes. As the head bends, it exerts a tiny force that causes the actin filament to move 5–10 nanometers relative to the myosin filament. When the myosin heads are bound to actin, they can bind and

▶ Animation 47.1 **Molecular Mechanisms of Muscle Contraction**
www.Life11e.com/a47.1

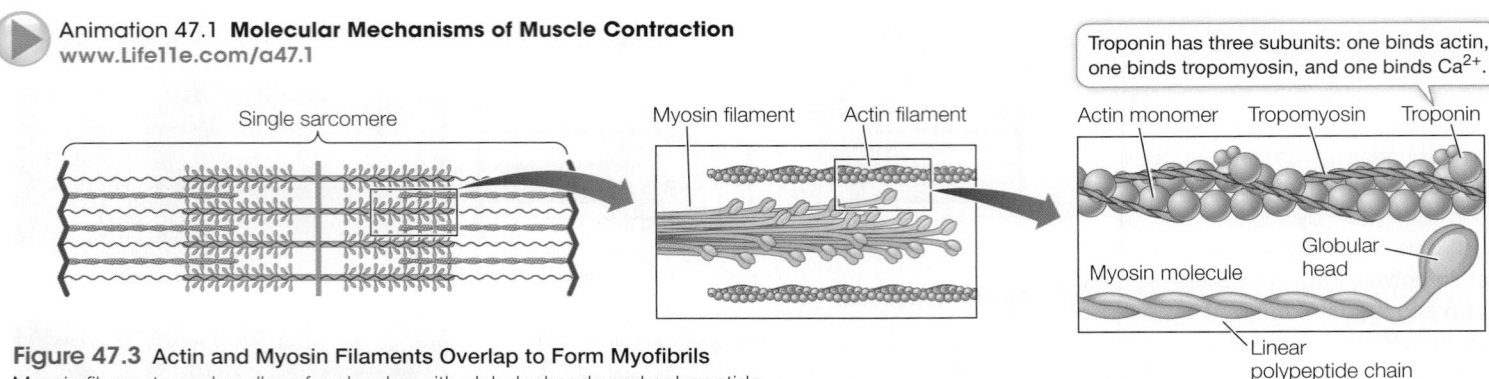

> Troponin has three subunits: one binds actin, one binds tropomyosin, and one binds Ca²⁺.

Figure 47.3 Actin and Myosin Filaments Overlap to Form Myofibrils
Myosin filaments are bundles of molecules with globular heads and polypeptide tails; the protein titin holds these filaments centered within the sarcomeres. Actin filaments consist of two chains of actin monomers twisted together. They are wrapped by chains of the polypeptide tropomyosin and are studded at intervals with another protein, troponin.

hydrolyze ATP. The energy released when this happens changes the conformation of the myosin head, causing it to release the actin and return to its extended position, from which it can bind to actin again.

Together these details explain the cycle of events that cause the actin and myosin filaments to slide past each other and shorten the sarcomere. They also explain rigor mortis—the stiffening of muscles soon after death. ATP binding causes myosin to release from actin, so when ATP production stops with death, myosin cannot release and the muscles stay contracted. Eventually, however, the proteins lose their integrity and the muscles soften. The timing of these events helps a medical examiner estimate the time of death.

We have been discussing the cycle of contraction in terms of a single myosin head. Remember that each myosin filament has many myosin heads at both ends and is surrounded by six actin filaments; thus the contraction of the sarcomere involves a great many cycles of interaction between actin and myosin molecules. That is why when a single myosin head breaks its contact with actin, the actin filaments do not slip backward.

Actin–myosin interactions are controlled by calcium ions

Like neurons, muscle cells are excitable—that is, their cell membranes generate and conduct action potentials. Action potentials are initiated in skeletal muscle fibers by motor neurons arriving at **neuromuscular junctions** (see Key Concept 44.3). The axon terminals of motor neurons are generally highly branched and form synapses with hundreds of muscle fibers (**Figure 47.4**). A motor neuron and all of the fibers with which it synapses constitute a **motor unit**. The fibers contract simultaneously when the unit's motor neuron fires. One muscle can consist of many motor units. Thus there are two ways to increase a muscle's strength of contraction—increase the firing rate of an individual motor neuron, or recruit more motor neurons.

When an action potential arrives at a neuromuscular junction, the neurotransmitter acetylcholine is released from the motor neuron terminals, diffuses across the synaptic cleft, binds to receptors in the postsynaptic membrane, and causes ion channels in the motor end plate to open (see Figure 44.11). Most of the ions that flow through these channels are Na⁺, and therefore the motor end plate is depolarized. The depolarization spreads to the surrounding cell

membrane of the muscle fiber, which contains voltage-gated sodium channels. When threshold is reached, the cell membrane fires an action potential that is conducted rapidly to all points on the surface of the muscle fiber.

An action potential in a muscle fiber also travels deep within the cell. The cell membrane is continuous with a system of **T tubules** that descend into the muscle fiber cytoplasm (also called the **sarcoplasm**). The "T" stands for "transverse." The contents of the T tubules are ouside the cell. The T tubules are like the indentations created in a water-filled balloon by poking your finger into it. The action potential that spreads over the cell membrane also spreads through this system of T tubules (**Figure 47.5**).

The T tubules come very close to the endoplasmic reticulum (ER) of the muscle cell. In muscle cells the ER is called the **sarcoplasmic reticulum**, and it is a closed compartment surrounding every myofibril. Calcium pumps in the sarcoplasmic reticulum take up Ca²⁺ from the sarcoplasm. Therefore when the muscle fiber is at rest, there is a high concentration of Ca²⁺ in the sarcoplasmic reticulum and a very low concentration of Ca²⁺ in the sarcoplasm.

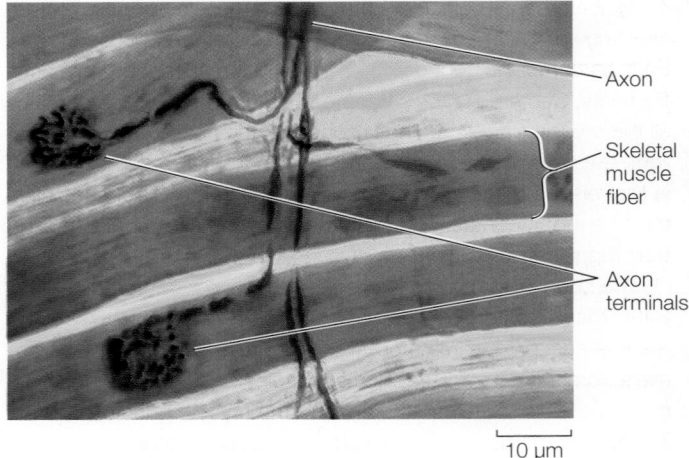

Figure 47.4 The Neuromuscular Junction Axons branching from a single motor neuron end in terminals that innervate multiple skeletal muscle fibers.

Q: Is there a relationship between the number of fibers that a motor neuron innervates and fine motor control, or between the number of fibers innervated and the exertion of force?

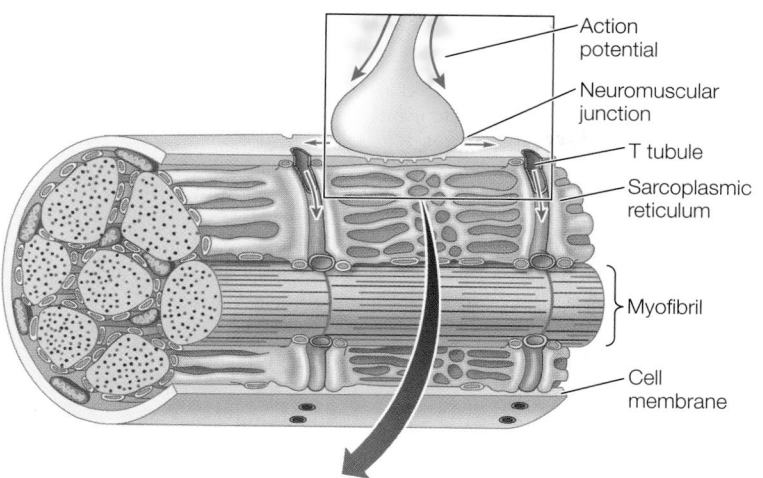

Action potential

Neuromuscular junction

T tubule

Sarcoplasmic reticulum

Myofibril

Cell membrane

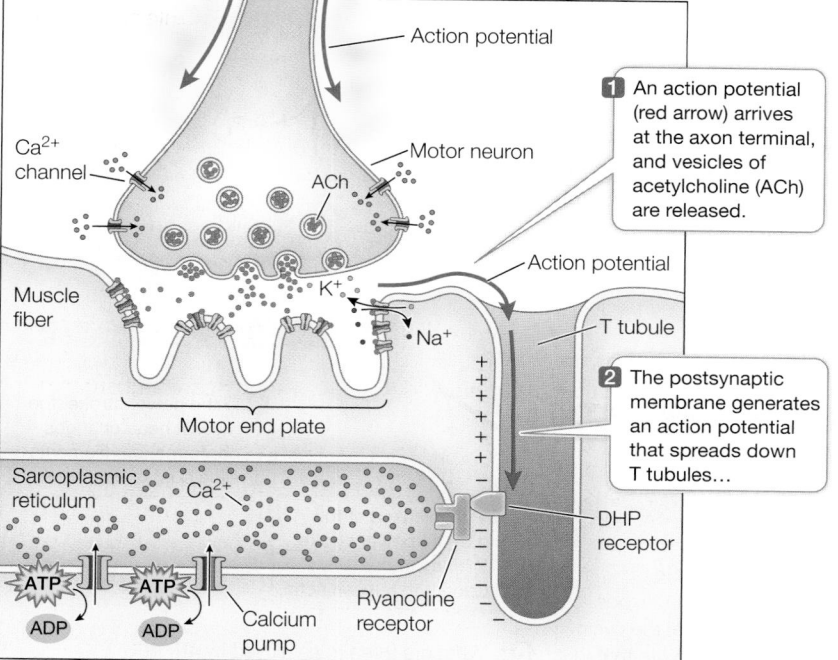

Action potential

Ca^{2+} channel

Motor neuron

ACh

Muscle fiber

K$^+$

Na$^+$

Motor end plate

Sarcoplasmic reticulum

Ca^{2+}

ATP

ADP

ATP

ADP

Calcium pump

Action potential

1 An action potential (red arrow) arrives at the axon terminal, and vesicles of acetylcholine (ACh) are released.

T tubule

2 The postsynaptic membrane generates an action potential that spreads down T tubules...

DHP receptor

Ryanodine receptor

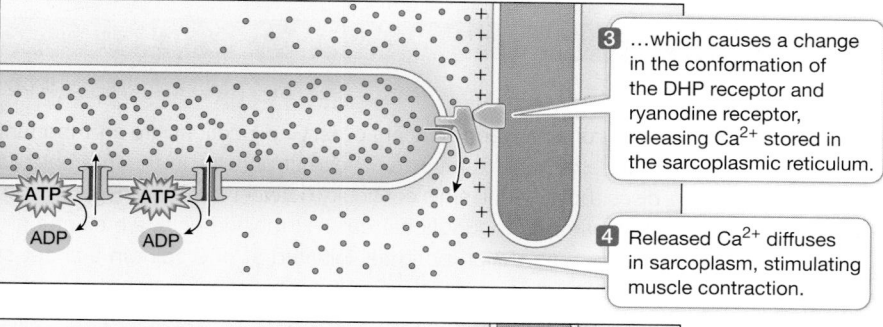

ATP

ADP

ATP

ADP

3 ...which causes a change in the conformation of the DHP receptor and ryanodine receptor, releasing Ca^{2+} stored in the sarcoplasmic reticulum.

4 Released Ca^{2+} diffuses in sarcoplasm, stimulating muscle contraction.

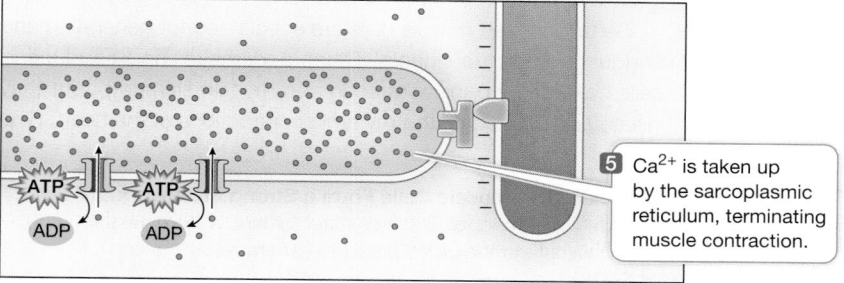

ATP

ADP

ATP

ADP

5 Ca^{2+} is taken up by the sarcoplasmic reticulum, terminating muscle contraction.

Spanning the space between the membranes of the T tubules and the membranes of the sarcoplasmic reticulum are two proteins. One protein, the dihydropyridine (DHP) receptor, is located in the T tubule membrane; it is voltage-sensitive and changes its conformation in response to an action potential spreading through the T tubule system. The other protein, the ryanodine receptor, is located in the sarcoplasmic reticulum membrane and is a Ca^{2+} channel. These two proteins are physically connected. When the DHP receptor changes its conformation due to an action potential, it causes Ca^{2+} to flow through the ryanodine receptor from the sarcoplasmic reticulum to the sarcoplasm. Ca^{2+} ions diffusing through the sarcoplasm surrounding the actin and myosin filaments trigger the interaction of actin and myosin and the sliding of the filaments. How do the Ca^{2+} ions do this?

An actin filament, as you have seen, is a helical arrangement of actin monomers. Twisted around the actin filament are two strands of the protein tropomyosin (**Focus: Key Figure 47.6**; see also Figure 47.3). At regular intervals, the filament also includes a globular protein, troponin. The troponin molecule has three subunits: one binds actin, one binds tropomyosin, and one binds Ca^{2+}.

When the muscle is at rest, the tropomyosin strands are positioned so that they block the myosin-binding sites on the actin filaments. When Ca^{2+} is released into the sarcoplasm, it binds to troponin, changing its conformation. Because the troponin is bound to the tropomyosin, this conformational change twists the tropomyosin enough to expose the actin–myosin binding sites, and the cycle of making and breaking actin–myosin bonds is initiated. The filaments pull past each other, and the muscle fiber contracts. When the calcium pumps in the sarcoplasmic reticulum membranes remove the Ca^{2+} ions from the sarcoplasm, the conformation of the tropomyosin reverts and again blocks the binding of myosin heads to actin, and the muscle fiber returns to its resting condition. Figure 47.6 summarizes this cycle.

Cardiac muscle is similar to and different from skeletal muscle

Like skeletal muscle, cardiac muscle appears striated because of the regular arrangement of actin and myosin filaments into sarcomeres (**Figure 47.7**). The difference between cardiac and skeletal muscle is that cardiac muscle cells are smaller and have only one nucleus each (uninucleate). Cardiac muscle cells branch, and the branches of adjoining cells interdigitate, forming a meshwork resistant to tearing. As a result, the heart walls can withstand high pressures while pumping blood, without the danger of developing leaks. Adding to the strength of cardiac muscle are **intercalated discs** that provide strong mechanical

Figure 47.5 T Tubules Spread Action Potentials into the Fiber   An action potential at the neuromuscular junction spreads throughout the muscle fiber via a network of T tubules, triggering the release of Ca^{2+} from the sarcoplasmic reticulum.

 Activity 47.2 The Neuromuscular Junction
www.Life11e.com/ac47.2

focus: key figure

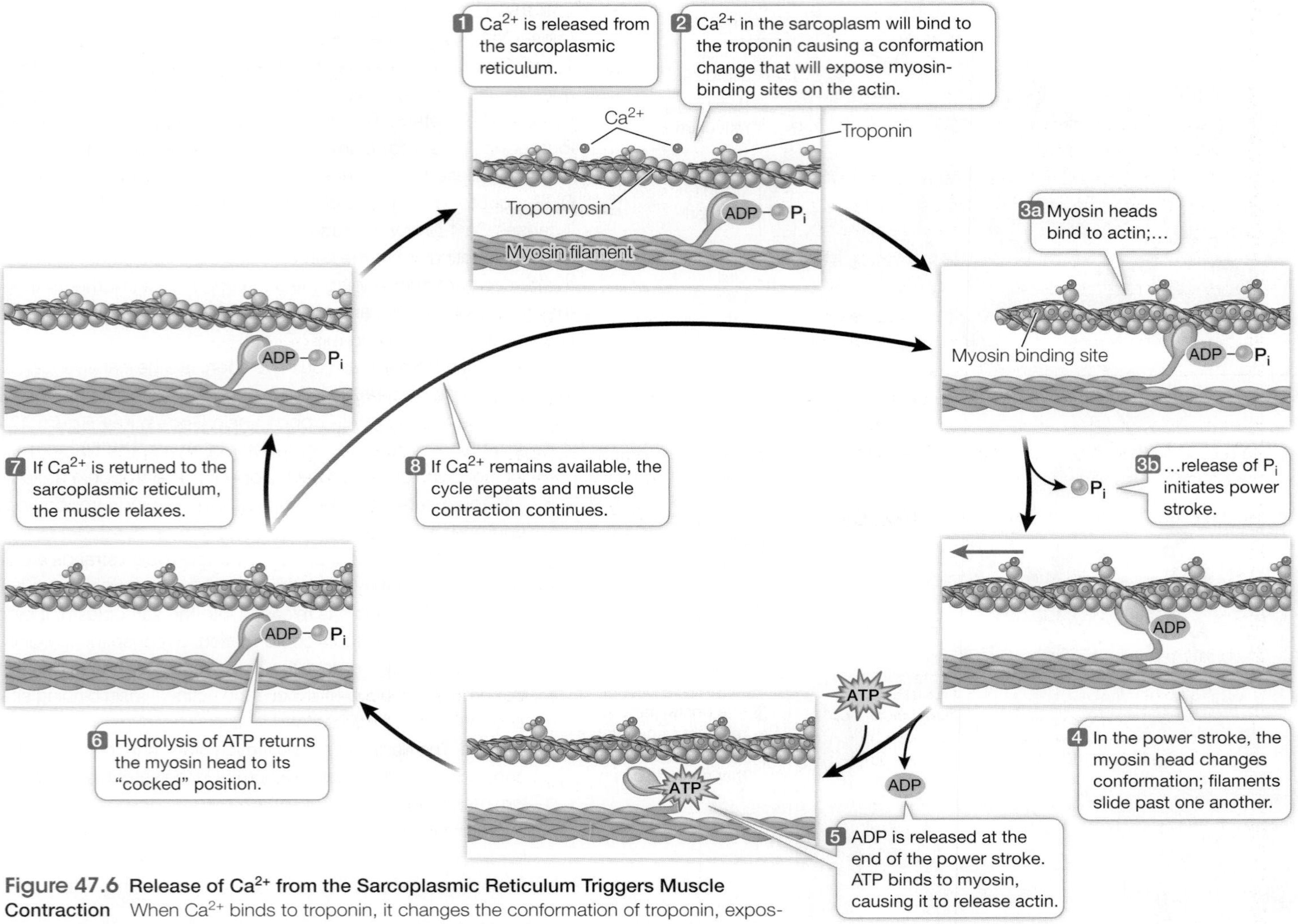

1 Ca²⁺ is released from the sarcoplasmic reticulum.

2 Ca²⁺ in the sarcoplasm will bind to the troponin causing a conformation change that will expose myosin-binding sites on the actin.

Ca²⁺

Troponin

Tropomyosin

ADP — Pᵢ

Myosin filament

3a Myosin heads bind to actin;…

Myosin binding site

ADP — Pᵢ

3b …release of Pᵢ initiates power stroke.

Pᵢ

7 If Ca²⁺ is returned to the sarcoplasmic reticulum, the muscle relaxes.

8 If Ca²⁺ remains available, the cycle repeats and muscle contraction continues.

ADP — Pᵢ

ADP

4 In the power stroke, the myosin head changes conformation; filaments slide past one another.

6 Hydrolysis of ATP returns the myosin head to its "cocked" position.

ADP — Pᵢ

ATP

ATP

ADP

5 ADP is released at the end of the power stroke. ATP binds to myosin, causing it to release actin.

Figure 47.6 Release of Ca²⁺ from the Sarcoplasmic Reticulum Triggers Muscle Contraction When Ca²⁺ binds to troponin, it changes the conformation of troponin, exposing myosin-binding sites on the actin. As long as binding sites and ATP are available, the cycle of actin and myosin interactions continues and the filaments slide past each other. Note that in the relaxed condition, the myosin heads are energized.

Q: What are three roles played by ATP in skeletal muscle contraction?

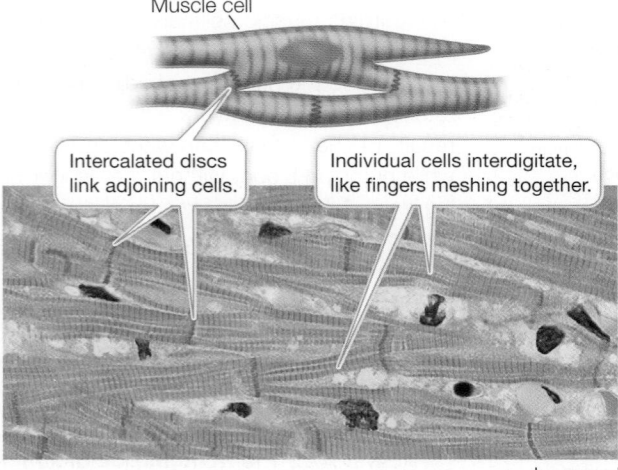

Muscle cell

Intercalated discs link adjoining cells.

Individual cells interdigitate, like fingers meshing together.

15 μm

adhesions between adjacent cells. Gap junctions are an important feature of cardiac muscle. These structures in the intercalated discs allow cytoplasmic continuity between cells (see Figure 7.16). Because of gap junctions, cardiac muscle cells are electrically coupled. An action potential initiated at one point in a sheet of cardiac muscle spreads rapidly, causing a large number of cardiac muscle cells to contract simultaneously.

Certain cardiac muscle cells are specialized for generating and conducting electric signals. These pacemaker and conducting cells have a low density of actin and myosin filaments, but they initiate and coordinate the rhythmic contractions of the heart. (The

Figure 47.7 Cardiac Muscle Cells Form a Strong Meshwork Cardiac muscle cells branch and interdigitate, forming a tear-resistant mesh that can withstand the blood pressure generated by the contraction of the heart muscle.

molecular basis for this pacemaking function will be covered in Key Concept 49.3.) Pacemaker cells make the vertebrate heartbeat autorhythmic, meaning it is generated by the heart muscle itself. A heart removed from a vertebrate can continue to beat with no input from the nervous system. Input from the autonomic nervous system modifies the *rate* of the pacemaker cells, but it is not essential for their continued rhythmic function.

The mechanism of excitation–contraction coupling in cardiac muscle cells is different from that in skeletal muscle cells. The T tubules are larger, and the DHP receptor proteins in the T tubules are Ca^{2+} channels. These T tubule proteins are not physically connected with the ryanodine receptors in the sarcoplasmic reticulum. Instead, when an action potential spreads down the T tubules, it causes the voltage-gated DHP channels to open, allowing extracellular Ca^{2+} to flow into the sarcoplasm. That increased sarcoplasmic Ca^{2+} binds to the ryanodine receptors and opens the calcium channels in the sarcoplasmic reticulum. The resulting huge rise in sarcoplasmic Ca^{2+} concentration stimulates fiber contraction. This mechanism is called Ca^{2+}-induced Ca^{2+} release.

 Media Clip 47.1 Be Still My Beating Stem Cell Heart
www.Life11e.com/mc47.1

Smooth muscle causes slow contractions of many internal organs

Smooth muscle provides the contractile force for most of our internal organs, which are under the control of the autonomic nervous system. Smooth muscle moves food through the digestive tract, controls the flow of blood through blood vessels, and empties the urinary bladder. Smooth muscle also generates "goose bumps" on the skin and causes the hair on the back of a dog or cat to stand erect when the animal is being aggressive or defensive. Structurally, smooth muscle cells are the simplest muscle cells. They are smaller than skeletal muscle cells, usually long and spindle-shaped, and each has a single nucleus. They are "smooth" because the actin and myosin filaments are not as regularly arranged as they are in skeletal and cardiac muscle, and so do not produce the striated appearance.

Some smooth muscle tissue, such as that from the wall of the digestive tract, is arranged in sheets, and individual cells in a sheet are in electrical contact with one another through gap junctions, as they are in cardiac muscle. As a result, an action potential generated in one smooth muscle cell spreads to all the cells in that sheet of tissue, and the sheet contracts in a coordinated fashion.

The cell membranes of smooth muscle cells are sensitive to stretch, with important consequences. For example, if the wall of the digestive tract is stretched in one location (as by a mouthful of

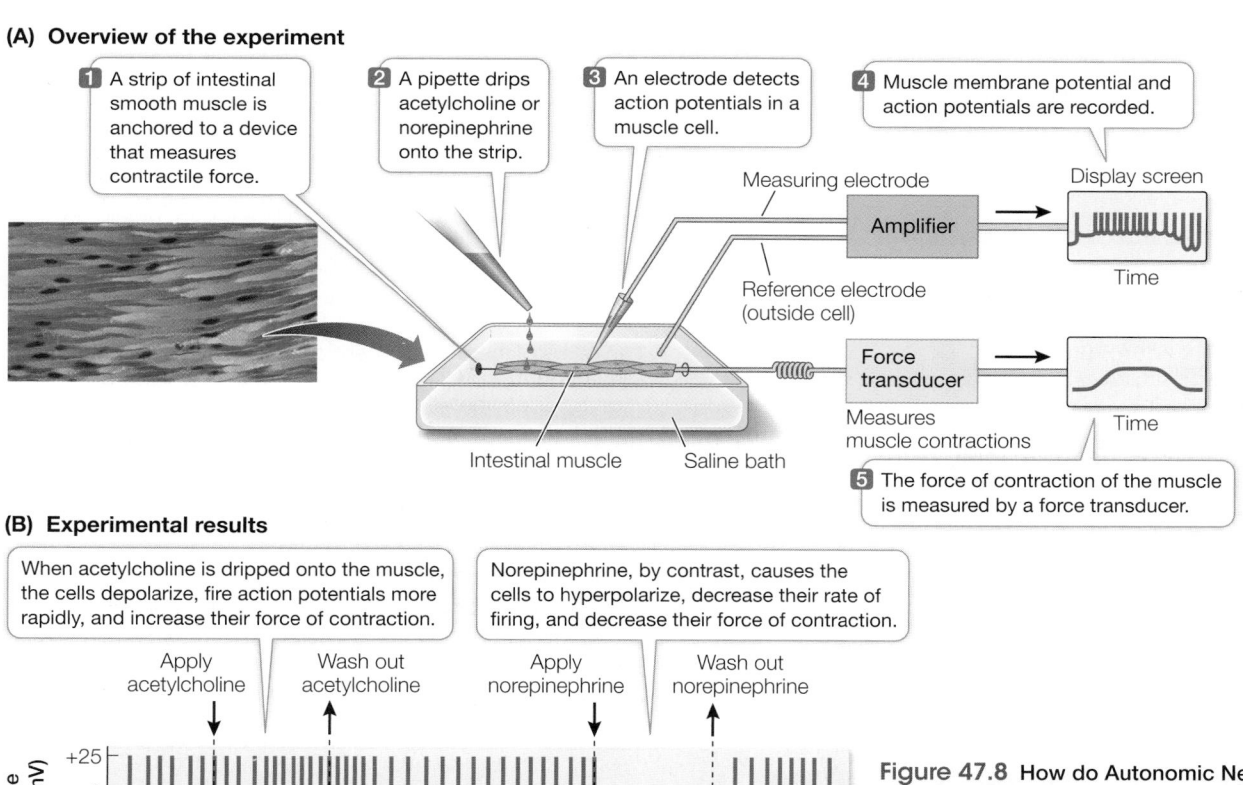

(A) Overview of the experiment

1 A strip of intestinal smooth muscle is anchored to a device that measures contractile force.

2 A pipette drips acetylcholine or norepinephrine onto the strip.

3 An electrode detects action potentials in a muscle cell.

4 Muscle membrane potential and action potentials are recorded.

Measuring electrode
Reference electrode (outside cell)
Amplifier → Display screen
Time
Intestinal muscle   Saline bath
Force transducer
Measures muscle contractions
Time

5 The force of contraction of the muscle is measured by a force transducer.

(B) Experimental results

When acetylcholine is dripped onto the muscle, the cells depolarize, fire action potentials more rapidly, and increase their force of contraction.

Norepinephrine, by contrast, causes the cells to hyperpolarize, decrease their rate of firing, and decrease their force of contraction.

Apply acetylcholine   Wash out acetylcholine   Apply norepinephrine   Wash out norepinephrine

Membrane potential (mV)
+25
0
−25
−50

Force
Muscle contracts
Muscle relaxes

Time

Figure 47.8 How do Autonomic Nervous System Neurotransmitters Control the Firing Rate of Smooth Muscle Cells? **(A)** When a strip of intestinal smooth muscle is mounted in an incubation bath so that its generation of action potentials and its force of contraction can be recorded, it is possible to test the effects of sympathetic and parasympathetic neurotransmitters on that tissue. **(B)** Acetylcholine depolarizes (makes less negative) the membranes and increases action potentials; norepinephrine hyperpolarizes (makes more negative) the membranes and decreases action potentials.

food passing down the esophagus to the stomach), the membranes of the stretched cells depolarize, reach threshold, and fire action potentials, which cause the cells to contract. Thus smooth muscle contracts after being stretched, and the harder it is stretched, the stronger it contracts. This behavior of smooth muscle is important for moving food through the digestive system.

The walls of blood vessels—except capillaries—are smooth muscle. This is especially true on the arterial side where the blood is under higher pressure. Changes in vascular smooth muscle tone are responsible for controlling the distribution of blood in the body.

The neural influences on smooth muscle come from the two divisions of the autonomic nervous system (see Figure 46.8). The neurotransmitters of the sympathetic and parasympathetic postganglionic cells alter the membrane potential of smooth muscle cells. For example, in the digestive tract, acetylcholine causes smooth muscle cells to depolarize, making them more likely to fire action potentials and contract. Antagonistically, norepinephrine causes these muscle cells to hyperpolarize and thus be less likely to fire action potentials and contract (**Figure 47.8B**). In contrast, norepinephrine acting through G protein-coupled receptors causes the smooth muscle in arteries serving the gut to contract. Remember that the action of the neurotransmitters depends on the receptors in the target tissues. Sympathetic activity is high in a fight-or-flight situation; in an emergency you don't need to digest your lunch, but it could help to shunt blood to the tissues critical for survival.

Smooth muscle cell contraction is not controlled by the troponin–tropomyosin mechanism, but calcium still plays a critical role. A Ca^{2+} influx into the sarcoplasm of a smooth muscle cell can be stimulated by action potentials, hormones, or stretching. The Ca^{2+} that enters the sarcoplasm combines with a protein called calmodulin. The Ca^{2+}–calmodulin complex activates an enzyme called myosin kinase that phosphorylates myosin heads. When the myosin heads in smooth muscle are phosphorylated, they undergo cycles of binding and releasing actin, causing muscle contraction. As Ca^{2+} is removed from the sarcoplasm, it dissociates from calmodulin, and the activity of myosin kinase falls. An additional enzyme, myosin phosphatase, dephosphorylates the myosin to help reduce actin–myosin interactions (**Figure 47.9**).

 Animation 47.2 Smooth Muscle Action
www.Life11e.com/a47.2

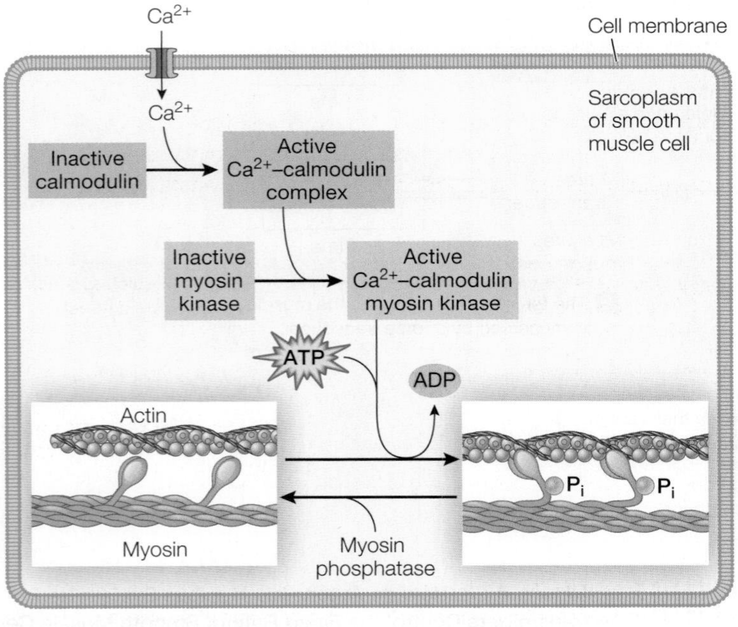

47.1 recap

The three types of muscle are skeletal, cardiac, and smooth. The contractile ability of muscle derives from interactions between actin and myosin filaments. Contraction depends on control by Ca^{2+} in the sarcoplasm. Tropomyosin and troponin are controlling elements in skeletal and cardiac muscle. Calmodulin is the controlling element in smooth muscle.

learning outcomes

You should be able to:

- Predict changes in sarcomere structure and appearance as muscle contraction occurs.
- Describe the molecular events underlying the sliding filament model of muscle contraction.
- Explain how ATP is involved in muscle contraction.
- Beginning with an action potential in a motor neuron, describe the sequence of events at the molecular level that leads to skeletal muscle contraction.
- Justify the need for gap junctions in cardiac and smooth muscle but not in skeletal muscle.

1. In terms of the arrangement of actin and myosin filaments in the sarcomere, why do the H zone and I band get smaller when the skeletal muscle contracts, and why does the A band stay the same?
2. What roles do the three subunits of troponin play in the control of skeletal muscle contraction?
3. Why does rigor mortis—the stiffening of muscles after death—occur first in muscles with the least glycogen content?
4. The insecticide Malathion destroys the enzyme acetylcholinesterase. Accidental exposure can cause extreme rigid muscle paralysis. Why?
5. Why are gap junctions important to the function of cardiac and smooth muscle, but not skeletal muscle?

Figure 47.9 The Role of Ca^{2+} in Smooth Muscle Contraction
When a smooth muscle cell is depolarized by a neurotransmitter, Ca^{2+} enters the sarcoplasm and binds to calmodulin, which in turn activates an enzyme that phosphorylates the myosin heads, causing them to bind to actin. As long as the myosin remains phosphorylated, actin and myosin go through cycles of binding and release. Thus in smooth muscle the Ca^{2+}-mediated change is on myosin, whereas in skeletal and cardiac muscle it is on the actin–tropomyosin filament.

Now that you understand how muscles generate force, let's look at what determines the characteristics of a muscle, its performance, and how individual muscles can change their characteristics with regular use and conditioning.

key concept 47.2 Many Factors Affect Muscle Performance

The functions that different muscles perform place different demands on them. Some muscles, such as postural muscles, must sustain a load continuously over long periods of time. Muscles used in walking or running must vary their strength of contraction over a wide range of positions. In contrast, the jumping muscles of the frog must exert maximum force instantaneously. The muscles that control your fingers generally do not have to sustain long, strong contractions, but they must be able to undergo quick yet graded contractions for fine manual dexterity. What is "quick" for humans, however, doesn't begin to compare with insect flight muscles that can contract as fast as 1,000 times per second. How are different muscles adapted for specific functions and demands?

focus your learning

- A twitch is a unit of contraction in one muscle fiber initiated by a single action potential.
- Many twitches can be summed in response to many action potentials to produce a large and/or sustained force.
- Slow- and fast-twitch muscle fiber types differ in the rate at which they develop and maintain tension.
- Three systems supply ATP to muscles for contraction, and together they determine the work capacity and endurance of a muscle.

The strength of a muscle contraction depends on how many fibers are contracting and at what rate

In skeletal muscle, the arrival of an action potential at a neuromuscular junction causes an action potential in a muscle fiber. The spread of that action potential through the muscle fiber's T tubule system causes a minimum unit of contraction, called a **twitch**. A twitch can be measured in terms of the tension, or force, it generates (**Figure 47.10A**). A single action potential stimulates a single twitch, but the ultimate force generated by an action potential varies enormously depending on how many muscle fibers it reaches. The level of tension an entire muscle generates depends on two factors: the number of motor units activated, and the frequency at which the motor units fire.

In muscles responsible for fine movements, such as those of the fingers, a motor neuron may innervate only one or a few muscle fibers, but in a muscle that produces large forces, such as the biceps, a motor neuron innervates a large number of muscle fibers.

At the level of a muscle fiber, a single action potential stimulates a single twitch. If action potentials reaching the muscle fiber are adequately separated in time, each twitch is a discrete, all-or-none phenomenon. If action potentials are fired more rapidly, however, new twitches are triggered before the myofibrils have a chance to return to their resting condition. As a result, the twitches sum, and the tension generated by the fiber increases and becomes more sustained. Thus an individual muscle fiber can show a graded response to increased levels of stimulation by its motor neuron.

Twitches sum at high levels of stimulation because the calcium pumps in the sarcoplasmic reticulum (see Figure 47.5) are unable to clear the Ca^{2+} ions from the sarcoplasm between action potentials. Eventually a stimulation frequency can be reached that results in the continuous presence of Ca^{2+} in the sarcoplasm at high enough levels to cause continuous activation of the contractile machinery—a condition known as **tetanus (Figure 47.10B)**. (Do not confuse this condition with the disease tetanus, which is caused by a bacterial toxin and is characterized by spastic contractions of skeletal muscles.)

How long a muscle fiber can maintain a tetanic contraction depends on its supply of ATP. Eventually the contracting fiber will become fatigued and be unable to sustain the contraction. It may seem paradoxical that the *lack* of ATP causes fatigue, since the action of ATP is to break actin–myosin bonds. But remember that the energy released from the hydrolysis of ATP "re-cocks" the myosin heads, allowing them to cycle through another power stroke. When a muscle is contracting against a load, the cycle of making and breaking actin–myosin bonds must continue in order to prevent the load from stretching the muscle. The situation is like rowing a boat upstream. You cannot maintain your position relative to the stream bank by just holding the oars out against the current; you have to keep rowing. Likewise, actin–myosin bonds have to keep cycling to maintain tension in the muscle.

Many muscles of the body maintain a low level of tension even when the body is at rest. For example, the muscles of the neck, trunk, and limbs that maintain our posture against the pull of gravity are always working, even when we are standing or sitting still. Muscle tone comes from the activity of a small but changing number of motor units in a muscle; at any one time, some of the muscle's fibers are contracting and others are relaxed. The nervous system is constantly readjusting muscle tone.

Muscle fiber types determine endurance and strength

Not all skeletal muscle fibers are alike, and a single muscle often contains more than one type of fiber. The two major types of skeletal muscle fibers express different genes for their myosin molecules, and these myosin variants have different rates of ATPase activity. Those

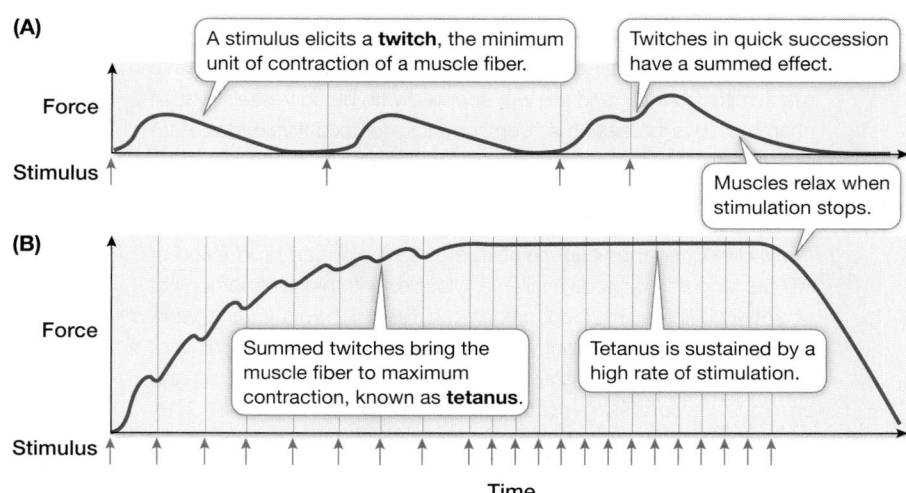

(A)

A stimulus elicits a **twitch**, the minimum unit of contraction of a muscle fiber.

Twitches in quick succession have a summed effect.

Muscles relax when stimulation stops.

Force

Stimulus

(B)

Summed twitches bring the muscle fiber to maximum contraction, known as **tetanus**.

Tetanus is sustained by a high rate of stimulation.

Force

Stimulus

Time

Figure 47.10 Twitches and Tetanus **(A)** Action potentials from a motor neuron cause a muscle fiber to twitch. Twitches in quick succession can be summed. **(B)** Summation of many twitches can bring the muscle fiber to the maximum level of contraction, known as tetanus.

(A) Force sustained by fast- and slow-twitch fibers

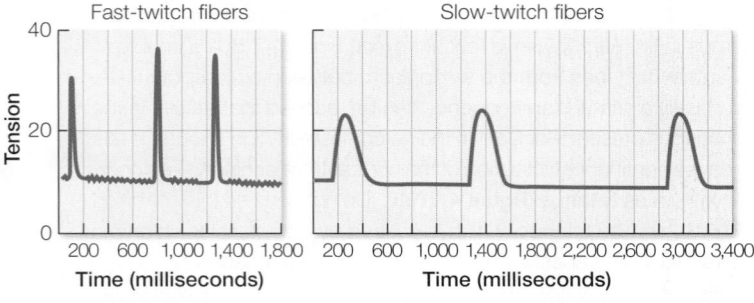

(B) Fiber type and performace

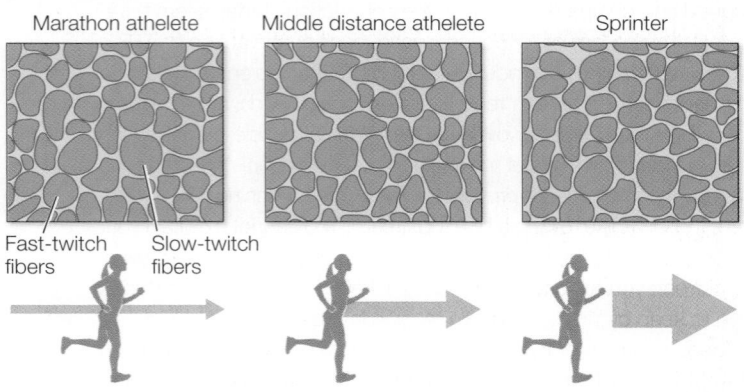

Figure 47.11 Fast- and Slow-Twitch Muscle Fibers **(A)** Fast-twitch fibers generate a large force quickly but fatigue rapidly. Slow-twitch fibers generate smaller forces, and do so more slowly, but can sustain them for a longer time. **(B)** The skeletal muscles in these three panels show slow-twitch fibers as red and fast-twitch fibers as blue. Arrows represent the duration (length) and speed (thickness) of running.

Q: What would you expect the muscle fiber distribution to be in a champion weight lifter?

with high ATPase activity can recycle their actin–myosin cross-bridges rapidly and are therefore called **fast-twitch fibers**; those with lower ATPase activity are called **slow-twitch fibers**, and they develop tension more slowly but can maintain it longer (**Figure 47.11A**).

Slow-twitch fibers are also called oxidative or red muscle because they contain especially large amounts of **myoglobin** (an oxygen-binding protein similar to the hemoglobin in red blood cells), have many mitochondria, and are well supplied with blood vessels. These characteristics increase the fibers' capacity for oxidative metabolism and result in their red appearance. The maximum tension a slow-twitch fiber produces is low and develops slowly but is highly resistant to fatigue. Slow-twitch fibers have substantial reserves of fuel, mostly lipids, so they can maintain steady, prolonged production of ATP as long as oxygen is available. Muscles with high proportions of slow-twitch fibers are good for long-term aerobic work (that is, work that requires oxygen). Long-distance runners, swimmers, cyclists, and other athletes whose activities require endurance have leg and arm muscles consisting mostly of slow-twitch fibers (**Figure 47.11B**).

Fast-twitch fibers, also called glycolytic or white muscle, have few mitochondria, little or no myoglobin, and fewer blood vessels; thus they look pale. Their major energy source is glycolysis, and therefore their major stored fuel is glycogen. Fast-twitch glycolytic fibers can develop maximum tension more rapidly than slow-twitch

fibers can, and that maximum tension is greater. However, fast-twitch fibers fatigue rapidly (see Figure 47.11A). The myosin of these fibers puts the energy of ATP to work very rapidly, but the fibers cannot replenish ATP quickly enough to sustain contraction for a long time. Fast-twitch fibers are especially good for short-term work that requires maximum strength, such as sprinting (see Figure 47.11B).

The types of fibers that make up a muscle influence the performance properties of that muscle, and different muscles have different fiber compositions depending on their function. Postural muscles that maintain continuous contractions are mostly composed of slow-twitch fibers. An example is the soleus muscle that runs up the back of the leg from the heel. Its contraction extends the foot and is therefore used in walking, but its continuous contraction is required for standing. If the soleus muscle fatigued, we would fall forward. We can walk or stand for a long period of time because the soleus muscle is resistant to fatigue. In contrast, a muscle that is used mostly for short-term work, such as the biceps, has a higher percentage of fast-twitch fibers than does the soleus. We can use our biceps to pick up a heavy weight, but we cannot hold that weight in a given position for a long period of time.

Can you change the fiber composition of your muscles to optimize your performance in a particular activity? To a limited extent, you can alter the properties of your muscle fibers through training. There are fast-twitch fibers that are somewhat oxidative and therefore intermediate in their properties between slow-twitch and fast-twitch fibers. These intermediate fibers can become more oxidative with endurance training and more glycolytic with strength training. However, the most important determinant of your muscle fiber types is your genetic heritage. There is some truth to the statement that champions are born, not made. A person born with a high proportion of fast-twitch fibers in her legs is unlikely to become a champion marathon runner, and a person born with a high proportion of slow-twitch fibers in her legs is unlikely to become a champion sprinter.

A muscle has an optimal length for generating maximum tension

Two parts of a pull-up exercise are especially difficult. When your arms are fully extended, it is hard to start the pull-up; and when your chin has almost reached the bar, the last little bit is difficult. The structure of the sarcomere is the reason.

When a muscle changes length the sarcomeres also change length, and the spatial relationships between the actin and myosin filaments changes (**Figure 47.12**). If there is no actin–myosin overlap, cross-bridges cannot form and the sarcomere can't generate force. This is almost the situation when you are hanging from the pull-up bar. At the other extreme, when the muscle is fully contracted, the actin filaments are overlapping and the myosin filaments are bumping up against the Z lines, so additional force cannot be generated. With these considerations, you can ask how well a specific muscle—such as the frog jumping muscle—will function given a range of sarcomere lengths. In **Investigating Life: What Is the Optimal Resting Position for the Jumping Muscle of the Frog?**, we examine experiments done to identify the optimal sarcomere length for a forceful frog jump. As you review these experiments, think about how they apply to deciding on the best starting position for a sprint.

experiment

Original Paper: Lutz, G. J. and L. C. Rome. 1994. Built for jumping: The design of the frog muscular system. *Science* 263: 370–372.

The force propelling a frog's jump should be greatest when there is optimal overlap between the actin and myosin filaments and not when the muscle is stretched. The following experiment attempted to determine the resting sarcomere length that allows a frog to achieve a maximal jump.

QUESTION▶ If you want to get the maximum force out of a slingshot, you stretch it as far as you can. Would the frog's jump be greater if the jumping muscle were stretched when the frog was crouched before the jump?

METHOD

1. Through anatomical studies measure sarcomere length at different resting positions observed in frogs before they jump and also measure the changes in muscle length and sarcomere length during jumps.
2. Excise frog jumping muscle from a frog and mount it in a tissue culture chamber so that the muscle can be held at different lengths and electrically stimulated to contract. Attach one end of the muscle to a force transducer that measures the force generated by the contraction. With this experimental set-up you can determine the relationship between sarcomere length and maximum force the muscle can generate.
3. Compare the sarcomere length measurements with the forces measured in muscles held at different lengths.
4. Measure velocity of contraction of the muscle under different load conditions.

RESULTS

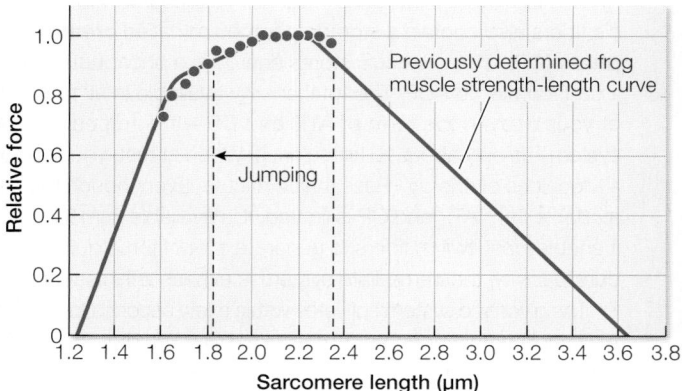

1. With methods not described here, it was determined that the jumping muscle shortened from 33.6 mm to 26.1 mm during a jump, and the resting sarcomere lengths varied from 2.34 μm to 1.82 μm.
2. The relative forces generated by isometric contraction at different sarcomere lengths corresponded well with a previously determined strength-length curve (data points shown in the graph).

CONCLUSION▶ During the jump, the sarcomeres of the frog shortened over the optimal range for generation of force.

work with the data

The resting position of the sarcomeres and the extent of their shortening during a jump maximize the force that the frog's muscle can generate to power the jump. But in physical terms, power is a force applied over a specific time. Therefore the velocity of contraction (measured as muscle length/second [ML/s]) will influence the maximum power the muscle can generate, and the velocity of contraction depends on the load. If the load is too great, the contraction is isometric (velocity = 0). If the load is negligible, the velocity of contraction can be maximal. **Figure A** shows the relationship between force and velocity of contraction in a frog jumping muscle under different loads. The mean jump velocity is indicated. **Figure B** shows the change in muscle length (mm) during a jump (milliseconds).

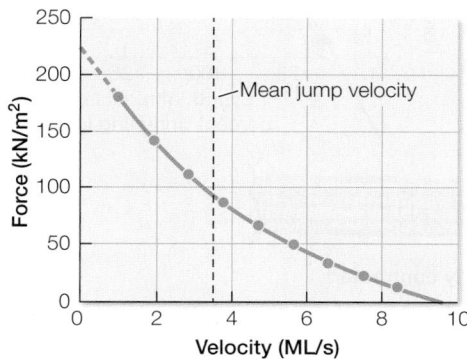

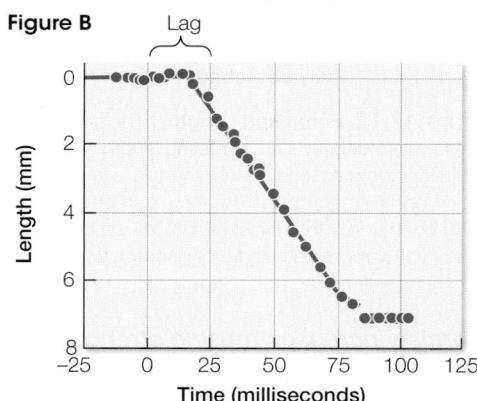

QUESTIONS▶

1. Power is work/time and work is force × displacement. Displacement/time = velocity. Therefore power = force × velocity. Using the data in Figure A, plot a curve that shows power as a function of contraction velocity for the frog muscle.
2. What is the velocity of contraction of the frog jumping muscle in muscle lengths/sec? Use the data in Figure B.
3. Is the speed and force of the frog muscle optimal for generating the most powerful jump?

A similar **work with the data** exercise may be assigned in **LaunchPad**.

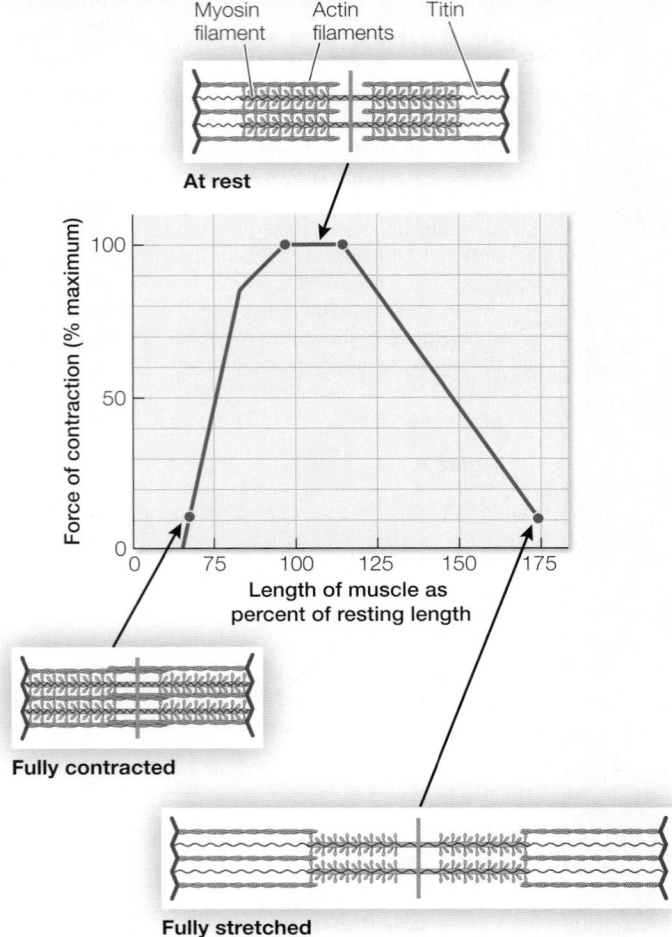

At rest

Fully contracted

Fully stretched

Figure 47.12 Force and Length   The amount of force a sarcomere can generate depends on its resting length. When a muscle is stretched and its sarcomeres lengthened, there is less overlap between the actin and myosin filaments, and less force can be produced. When a muscle is fully contracted, the myosin filaments are bumping up against the Z lines of the sarcomeres, so further contraction is not possible.

Exercise increases muscle strength and endurance

Different types of exercise produce different physical conditioning responses. In general, anaerobic activities such as weight lifting, increase strength, and aerobic activities such as jogging, increase endurance. Strength is the maximum force a muscle can exert, and endurance is work capacity or how long a given workload can be sustained. What are the physiological bases for these differences?

Strength is a function of the volume of the muscles: the more actin and myosin filaments in a muscle fiber, and the more muscle fibers in a muscle, the more tension it can produce. When athletes undertake strength training, they use weights or exercises such as pull-ups to repeatedly contract specific muscles under heavy loads. Repetitions are usually done until the muscle is completely fatigued. Such stress on a muscle does minor tissue damage—hence the soreness the day or two after a hard workout—but it also induces the formation of new actin and myosin filaments in existing muscle fibers. The muscle fibers, and hence the muscles, get bigger and stronger. In extreme cases, and after serious muscle damage, new muscle fibers can also be produced from stem cells called satellite

cells in the muscle. In general, however, the major effect of strength training is to produce bigger, rather than more, muscle fibers.

Aerobic exercise has a completely different effect on muscles: it enhances their oxidative capacity. This effect comes from increases in the number of mitochondria, in enzymes involved in energy use, and in the density of capillaries that deliver oxygen to the muscle. Myoglobin also increases in skeletal muscle cells. Myoglobin is similar to hemoglobin, but it has a higher affinity for oxygen. Therefore myoglobin accepts oxygen from the blood, facilitates the diffusion of oxygen throughout the muscle, and provides a store of oxygen for use when oxygen delivery by the blood is insufficient. By increasing the capacity of muscle to use oxygen to produce ATP, aerobic training increases the length of time that a given workload can be sustained.

Muscle ATP supply limits performance

Muscles have three systems for supplying the ATP they need for contraction:

1. The *immediate system* uses preformed ATP and creatine phosphate.

2. The *glycolytic system* metabolizes carbohydrates to lactate and pyruvate.

3. The *oxidative system* metabolizes carbohydrates or fats all the way to H_2O and CO_2.

The capacity of these three systems and the rates at which they can produce ATP determine both work capacity and endurance of a muscle (**Figure 47.13**).

ATP is present in muscles in very small amounts. However, muscle fibers also contain a storage compound called **creatine phosphate** (**CP**). This molecule stores energy in a phosphate bond that it can transfer to ADP. The total energy available in all the muscles of your body in the form of ATP and CP—the immediate energy system—is only about 10 kilocalories. When at rest, you metabolize a kilocalorie of energy in less than a minute. Even though the energy available from ATP and CP is limited, it is available immediately, and it enables fast-twitch fibers to generate a lot of force quickly. During burst activity, the immediate system is exhausted in seconds.

The glycolytic system activates within a few seconds to replace the ATP depleted at the onset of muscle activity. The glycolytic enzymes are located in the cytoplasm of the muscle fiber, and therefore the ATP they generate is rapidly available to the myosin filaments. However, as noted in Chapter 9, glycolysis alone is an inefficient way to produce ATP, and it leads to the accumulation of lactic acid, which slows the process. Thus the glycolytic system and the immediate system together provide the energy needed by active muscles for less than a minute (see Figure 47.13). Fortunately, oxidative metabolism becomes fully active in about a minute, producing relatively huge amounts of ATP because it can completely metabolize carbohydrates and fats to CO_2 and water. However, it requires many reactions, and it takes place in the mitochondria, so O_2 and substrate must get into the mitochondria, and the formed ATP must get from the mitochondria to the myosin filaments in the muscle. These processes are not instantaneous, so the rate at which oxidative metabolism can make ATP available to do work is slower than the rate at which the other two systems can supply ATP.

(A) Three systems that supply ATP

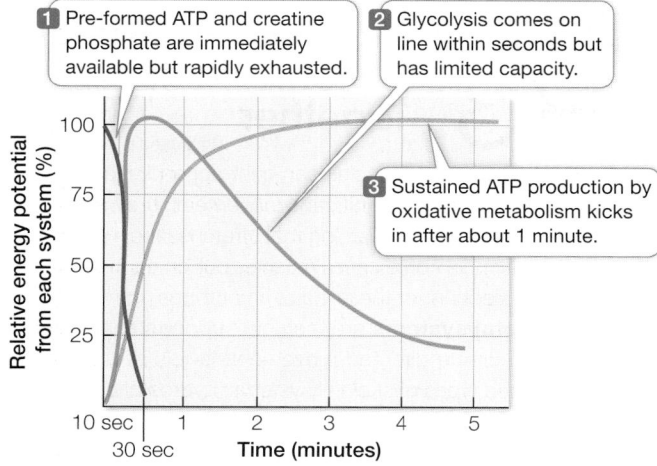

1 Pre-formed ATP and creatine phosphate are immediately available but rapidly exhausted.

2 Glycolysis comes on line within seconds but has limited capacity.

3 Sustained ATP production by oxidative metabolism kicks in after about 1 minute.

(B) Timing of ATP replenishment systems

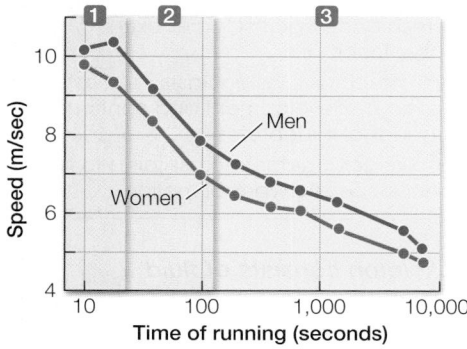

Figure 47.13 Supplying Fuel for High Performance **(A)** Muscles have three systems for obtaining the ATP they need for contraction during exertion such as running. **(B)** Looking at a plot of world-record times for running events of different durations, you can see that the performance of the athletes corresponds to the time courses of the three energy systems.

Q: Figure 47.13B shows how maximum speed decreased as the duration of the running events got longer. Explain why there are three different slopes for this curve.

The ***fuel resources** available to the muscles influences how long someone can sustain a high level of aerobic exercise. From the circulating blood, muscle receives glucose and free fatty acids that it metabolizes to generate ATP. At high levels of aerobic exercise, however, most of the fuel used by muscles to produce ATP comes from the reserve of glycogen stored in the muscle itself. Depletion of muscle glycogen contributes to fatigue.

***connect the concepts** The work performance curves for different activities reflect the complexity of the biochemical processes by which different fuel sources are metabolized to product ATP. Glycolysis supports anaerobic activities and it only entails ten enzymatic steps, so it can come on fast (see Key Concept 9.2). For aerobic activity the citric acid cycle and the respiratory chain reactions add more steps and limit the rate of ATP production.

The rate at which muscle glycogen is replenished depends on diet. Glycogen is quickly replenished with a high-carbohydrate diet,

work with the data

Does Heat Cause Muscle Fatigue?

Original Paper: Grahn, D. A., V. H. Cao, C. M. Nguyen, M. T. Lieu and H. C. Heller. 2012. Work volume and strength training responses to resistive exercise improve with periodic heat extraction from the palm. *Journal of Strength and Conditioning Research* 26: 2558–2569.

Physical conditioning requires repeated intense physical activity, and the capacity of such workouts is limited by muscle fatigue. Because metabolic heat production raises the temperature of muscles during workouts, it is possible that the rise in temperature contributes to muscle fatigue and limits the capacity of workouts. To test this idea, investigators used rapid-cooling technology to extract heat from subjects during 3-minute rests between ten sets of pull-ups twice a week. In each of the ten sets of pull-ups, the subject worked until muscle failure—the inability to complete an additional pull-up. The control condition was 3-minute rests without cooling. Each subject was his own control; the subjects were randomly assigned to begin with 6 weeks of training with cooling or 6 weeks of training without cooling, followed by a reversal of the treatments. The results are shown in the figure.

QUESTIONS▶

1. What do these data indicate about the possible role of muscle temperature in muscle fatigue?

2. What do these data indicate about the relationship between workout capacity and physical conditioning effects?

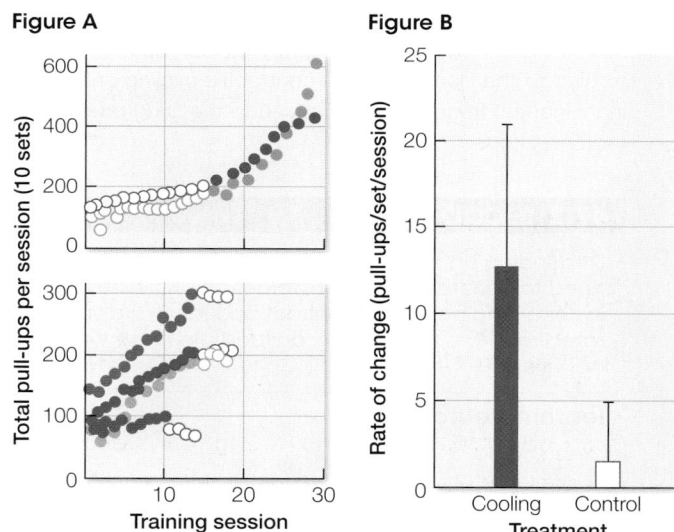

The Effect of Cooling on Workout Capacity **(A)** The results for individual subjects. Open symbols indicate control treatment, closed symbols indicate cooling treatment; different colors indicate different subjects. **(B)** The slopes of the lines in Figure A show the rates of increase in workout capacity over the course of the experiment. Mean ± standard deviation of rate of change in pull-ups per set during the two treatment phases ($N = 6$, $P < 0.001$ paired *t*-tests).

A similar **work with the data** exercise may be assigned in **LaunchPad**.

but slowly replenished with a high-fat diet. A method used to maximize muscle glycogen is called carbo-loading. About 1 week before an athletic event, athletes typically exercise for 3–5 days at a level that depletes muscle glycogen. Then, 2–3 days before the event, they taper down their level of training and eat a diet rich in complex carbohydrates. The result can be glycogen supercompensation, in which the restoration of muscle glycogen stores "overshoots" and reaches above-normal levels.

Insect muscle has the greatest rate of cycling

Insect flight muscle can produce a wingbeat frequency of up to 1,000 cycles per second. Since neuronal action potentials last 1–3 milliseconds, that number of cycles per second would exceed the capacity of motor neurons, let alone the mechanism of cycling of striated muscle contraction/relaxation. The extremely fast wingbeat of a hummingbird may be only about 50 cycles per second. How do insects do it?

Vertebrate skeletal muscle and much of invertebrate skeletal muscle is called "synchronous" because the cycling of the contractile mechanism is linked to the firing of the motor neurons. Insect flight muscle does not contract in tandem with the firing of motor neurons, however, and is thus called "asynchronous" muscle. The firing of action potentials in the insect flight motor neurons is not particularly fast, but it does cause depolarization of the muscle cell membrane, the spreading of an action potential throughout the membrane, and the release of Ca^{2+} from the sarcoplasmic reticulum. However, once the asynchronous muscle fiber is stimulated, its cycle of contraction/relaxation proceeds at its own characteristic frequency as long as Ca^{2+} is available to bind to the troponin. Thus contractile cycling and the resulting wingbeat frequency are not tied to the firing rate of the flight motor neurons.

47.2 recap

Depending on the function a muscle serves, it may need to generate maximum force rapidly, sustain activity for a long period, or contract and relax at a very rapid rate. Properties of muscles can facilitate these types of activities.

learning outcomes

You should be able to:

- Explain the factors that determine how much force or tension is generated by a muscle.
- Apply knowledge about muscle fiber types to explain how they contribute to performance abilities in aerobic versus anaerobic work.

1. In terms of the relationships between motor neurons and skeletal muscles, how can the nervous system control the amount of force generated by a muscle?
2. Why do postural muscles have a higher percentage of slow-twitch fibers than do muscles used for sprinting?
3. Apply knowledge about the three systems used by muscles as sources of ATP to explain observations about human physical performance.

Regardless of how much force a muscle can generate, how long it can sustain a workload, or how fast it can contract and relax, a muscle needs something to pull on; otherwise it would just be a lump of pulsating, quivering tissue. Let's look now at how muscle and skeletal elements work together to produce movement.

---

key concept 47.3   Muscles and Skeletal Systems Work Together

Muscles can contract and exert force, or they can relax. To create significant movement, they must have something to pull on and something that stretches the muscle back to a longer position. In some cases muscles pull on each other, as in the trunk of an elephant or the arms of an octopus. In most cases, however, **skeletal systems** are the rigid supports against which muscles pull to create directed movement. In this section we will examine the three types of skeletal systems: hydrostatic skeletons, exoskeletons, and endoskeletons.

focus your learning

- Three types of skeletons have evolved in animals, and each provides resistance against forces exerted by muscles to enable locomotion.
- Three cell types, osteoblasts, osteoclasts, and osteocytes are responsible for the development and continuous remodeling of bones of vertebrates.
- Bones and muscles act together across joints to form levers that produce specific types of motion.

A hydrostatic skeleton consists of fluid in a muscular cavity

Cnidarians, annelids, and other soft-bodied invertebrates have **hydrostatic skeletons** consisting of a volume of fluid enclosed in a body cavity surrounded by muscle (see Key Concept 30.2). When muscles oriented in one direction contract, the fluid-filled body cavity bulges out in a perpendicular direction.

An earthworm uses its hydrostatic skeleton to crawl (**Figure 47.14**). The earthworm's body cavity is divided into many separate

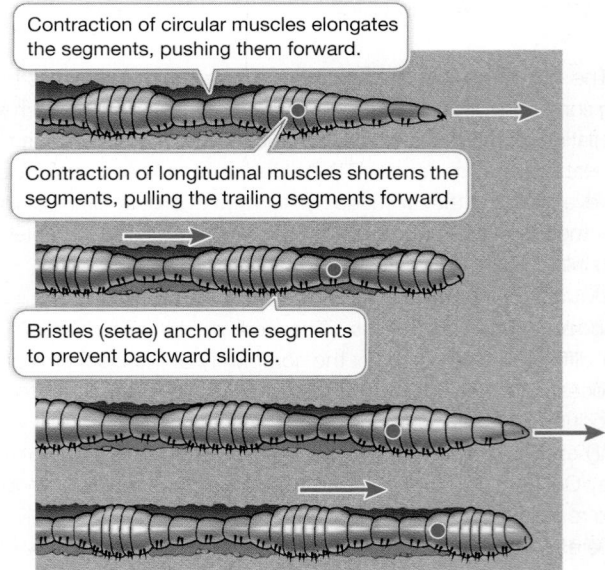

Contraction of circular muscles elongates the segments, pushing them forward.

Contraction of longitudinal muscles shortens the segments, pulling the trailing segments forward.

Bristles (setae) anchor the segments to prevent backward sliding.

Figure 47.14 A Hydrostatic Skeleton Alternating waves of muscle contraction move the earthworm through the soil. The red dot enables you to follow the changes in one segment as the worm moves forward.

segments filled with extracellular fluid. The body wall surrounding each segment has two muscle layers: a circular layer and a longitudinal layer. If the circular muscles in a segment contract, that segment constricts and elongates. If the longitudinal muscles in a segment contract, it shortens and bulges outward. Alternating contractions of the earthworm's circular and longitudinal muscles create waves of narrowing and widening, lengthening and shortening, that travel down the body. Bulging, shortened segments serve as anchors as long, narrow segments project forward and longitudinal contractions pull other segments forward. Bristles help the widest parts of the body to hold firm against the substrate, so the body moves forward.

Exoskeletons are rigid outer structures

An **exoskeleton** is a hardened, rigid outer surface to which muscles can be attached. Contractions of the muscles cause jointed segments of the exoskeleton to move relative to each other. The simplest example of an exoskeleton is the shell of a mollusk. Some marine mollusks, such as clams, have shells composed of protein strengthened by crystals of calcium carbonate (a rock-hard material). These shells can be massive, affording significant protection against predators. Some terrestrial mollusks (and snails) generally lack the hard mineral component and are much lighter.

The most complex exoskeletons are found among the arthropods (see Key Concept 31.1). A type of exoskeleton called a **cuticle** covers the outer surfaces of the arthropod body including its appendages. It is made up of **chitin** (a nitrogen-containing polysaccharide) secreted by a layer of cells just below the exoskeleton. Chitin stiffens and hardens the cuticle everywhere except at the joints, where flexibility must be retained. Muscles attached to the inner surfaces of the arthropod cuticle move its parts around the joints (see Figure 31.4). Growth of animals with exoskeletons is accommodated by molting (see Figure 31.3). Soft-shelled crabs, a gourmet delicacy, are crabs caught while they are molting.

Vertebrate endoskeletons consist of cartilage and bone

The **endoskeleton** of vertebrates is an internal scaffolding. Muscles are attached to it and pull against it. Endoskeletons are composed of rodlike, platelike, and tubelike bones connected to one another at a variety of joints that allow a wide range of movements. An advantage of endoskeletons over the exoskeletons of arthropods is that bones in the body can grow without the animal shedding its skeleton.

The adult human skeleton consists of 206 bones, some of which are shown in **Figure 47.15**. It can be divided into an axial skeleton, which includes the skull, vertebral column, sternum, and ribs; and an appendicular skeleton, which includes the pectoral girdle, pelvic girdle, and bones of the arms, legs, hands, and feet.

The vertebrate endoskeleton consists of two kinds of connective tissue, cartilage and bone, which are produced by two kinds of connective tissue cells that produce extensive extracellular matrices. **Cartilage** cells produce a matrix that is a tough, rubbery mixture of polysaccharides and proteins—mainly fibrous collagen. Collagen fibers run in all directions like reinforcing cords through the gel-like matrix and give it the well-known strength and resiliency of gristle in meat. Cartilage is found in parts of the endoskeleton where both

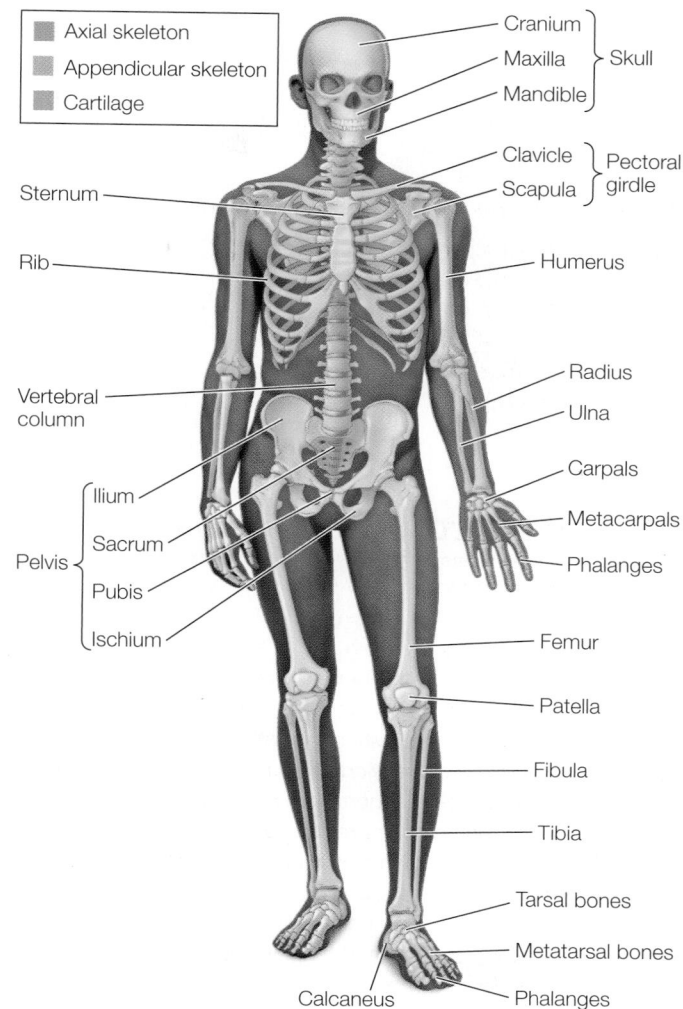

Figure 47.15 The Human Endoskeleton   Cartilage and bone make up the internal skeleton of a human being.

stiffness and resiliency are required, such as on the surfaces of joints where bones move against one another. Cartilage is also the supportive tissue in stiff but flexible structures such as the larynx (voice box), nose, and ear pinnae. Sharks and rays are called cartilaginous fishes because their skeletons are composed entirely of cartilage. In most other vertebrates, cartilage is the principal component of the embryonic skeleton, but during development most of it is gradually replaced by bone.

Bone also contains collagen fibers, but it gets its rigidity and hardness from an extracellular matrix of insoluble calcium phosphate crystals. Bone serves as a reservoir of calcium for the rest of the body and is in dynamic equilibrium with soluble calcium in the extracellular fluids of the body. This equilibrium is under the control of calcitonin and parathyroid hormone (see Figure 40.13). If too much calcium is taken from the skeleton, the bones are significantly weakened.

The living cells of bone—osteoblasts, osteocytes, and osteoclasts—are responsible for the constant dynamic remodeling of bone (**Figure 47.16**). **Osteoblasts** lay down new matrix material on bone surfaces. These cells gradually become surrounded by matrix and eventually become enclosed within the bone, at which point they cease laying down matrix but continue to exist within

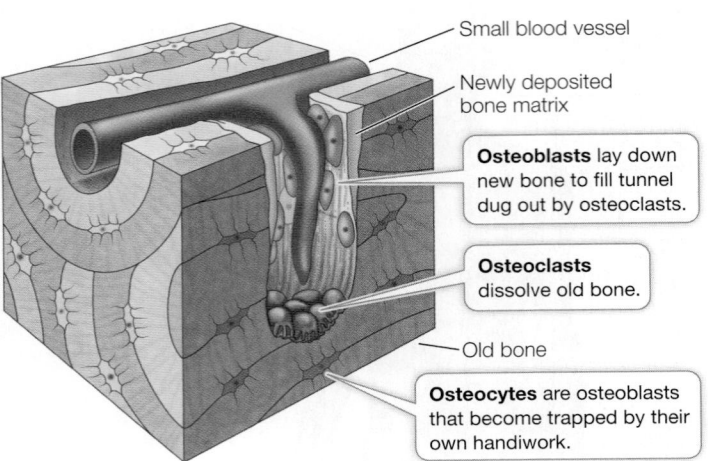

Figure 47.16 Bone Is Living Tissue Bones are constantly being remodeled by osteoblasts, which lay down bone, and osteoclasts, which break down bone.

small lacunae (cavities) in the bone. In this state they are called **osteocytes**. Despite the vast amounts of matrix between them, osteocytes remain in contact with one another through long cellular extensions that run through tiny channels in the bone. Communication between osteocytes is important in controlling the activities of the cells that are laying down or removing bone.

The cells that break down bone and release calcium into the extracellular fluid are called **osteoclasts**. They are derived from the same cell lineage that produces white blood cells. Osteoclasts erode bone, forming cavities and tunnels. Osteoblasts follow osteoclasts, depositing new bone. Thus the interplay of osteoblasts and osteoclasts constantly replaces and remodels the bones, allowing a bone to recover from damage and adjust to the forces placed on it.

A dramatic finding from the early days of manned space missions was that the bones of astronauts who spent long periods in zero gravity decalcified. Conversely, in athletes, certain bones thicken during training. Both thickening and thinning of bones are experienced by anyone who has had a leg in a cast for a long time: the bones of the uninjured leg carry the person's weight and thicken while the bones of the inactive leg in the cast thin. We now know that there are multiple mechanisms in bone that transduce physical stress into activation of various cell signaling mechanisms that can control the bone remodeling responses to those stressors.

Because of the positive effects of physical stress on bone deposition, weight-bearing exercise is effective in preventing and treating osteoporosis, which is the loss of bone density (and hence strength). More than 25 million people in the United States suffer from this debilitating condition. Although osteoporosis is most commonly a problem for postmenopausal women, it can occur in younger people as a result of malnutrition. For example, the condition known as female athlete triad includes eating disorders, cessation of menstrual cycling, and osteoporosis. These are interactive conditions in which the eating disorder and excessive training lead to malnutrition that can result in endocrine disruption

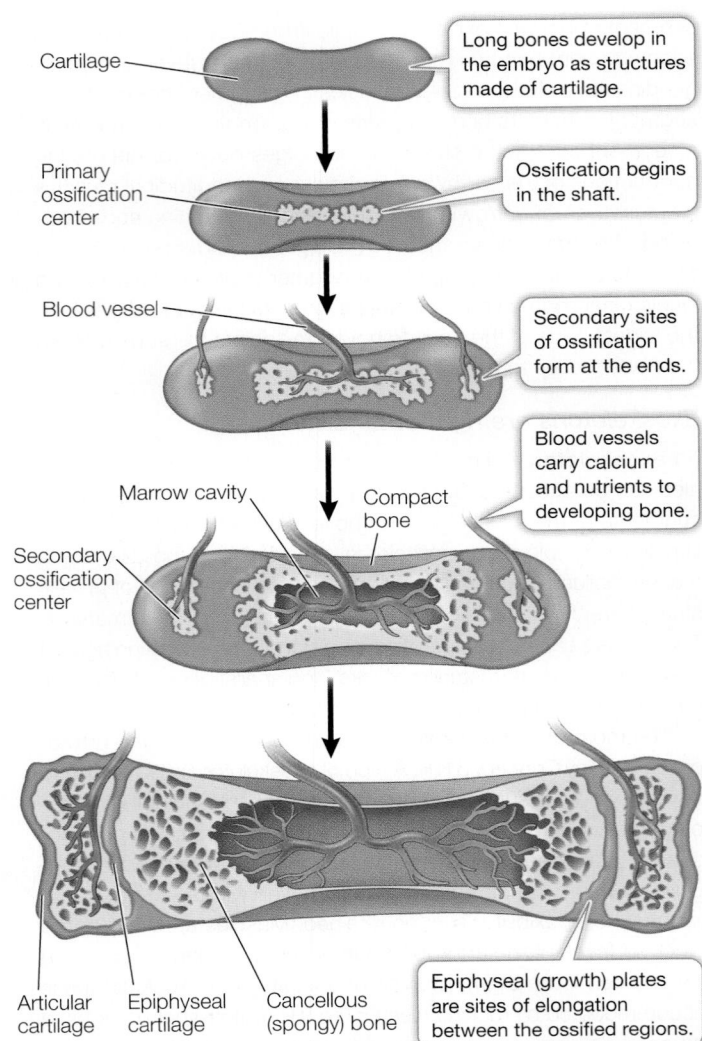

Figure 47.17 The Growth of Long Bones In the long bones of human limbs, ossification occurs first at the centers and later at each end.

and osteoporosis. Excessive training and malnutrition can lead to bone loss in males as well.

Bones develop from connective tissues

Bones are divided into two types on the basis of how they develop. **Membranous bone** forms on a scaffold of connective tissue membrane. **Cartilage bone** forms first as a cartilaginous structure resembling the future mature bone, then gradually hardens, or ossifies, to become bone. The outer bones of the skull are membranous bones; the bones of the limbs are cartilage bones.

Cartilage bones can grow throughout the ossification process. The long bones of the legs and arms, for example, ossify first at the centers and later at each end (**Figure 47.17**). The cartilage between these ossification centers forms **epiphyseal plates**. Growth can continue until these areas of ossification join, eliminating the epiphyseal plates. The membranous bones forming the skull cap grow until their edges meet. The large soft spot on the top of a baby's head, called a fontanel, is a point where the skull bones have not

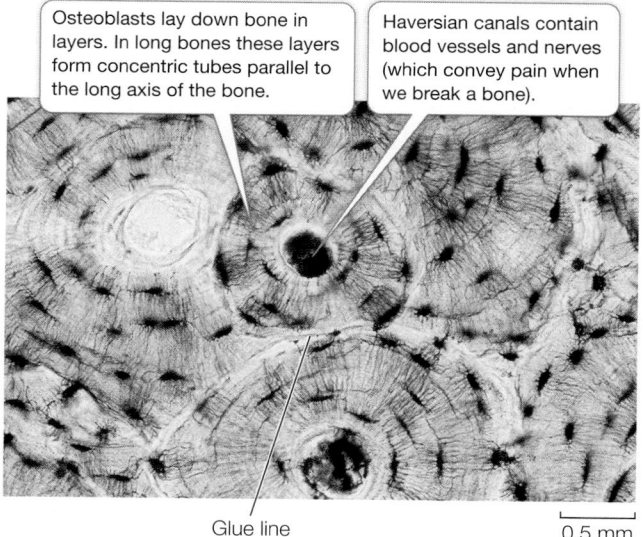

Osteoblasts lay down bone in layers. In long bones these layers form concentric tubes parallel to the long axis of the bone.

Haversian canals contain blood vessels and nerves (which convey pain when we break a bone).

Glue line   0.5 mm

Figure 47.18 Most Compact Bone Is Composed of Haversian Systems A micrograph of a section of a long bone shows Haversian systems with their central canals. Glue lines separate Haversian systems. Osteocytes reside in small spaces and are in communication by means of thin cell processes running through small channels.

yet joined. There are multiple fontanels in the neonate where the skull bones eventually join.

The structure of bone may be **compact** (solid and hard) or **cancellous** (having numerous internal cavities that make it appear spongy, although it is rigid). The architecture of a specific bone depends on its position and function, but most bones have both compact and cancellous regions. The shafts of the long bones of the limbs, for example, are cylinders of compact bone surrounding central cavities that contain fatty bone marrow called yellow bone marrow. The rigid, tubelike shaft of compact bone can withstand compression and bending forces. Architects and nature alike use hollow tubes as lightweight structural elements.

The ends of the long bones are cancellous (see Figure 47.17). Cancellous bone is lightweight because of its numerous cavities. These cavities contain the red bone marrow responsible for red blood cell formation as well as other cellular elements of the blood. Cancellous bone is also strong because its internal meshwork constitutes a support system. It can withstand considerable forces of compression.

Most compact bone in mammals is called Haversian bone because it is composed of structural units called **Haversian systems** (**Figure 47.18**). Each Haversian system is a set of thin, concentric bony cylinders that surround a narrow canal, the Haversian canal, that contains blood vessels and nerves. Osteocytes are arranged in small spaces between the concentric circles of bone surrounding the canals. Adjacent Haversian systems are separated by boundaries called glue lines. Haversian bone is resistant to fracturing because cracks tend to stop at glue lines.

Bones that have a common joint can work as a lever

Muscles and bones work together around **joints**, where two or more bones come together. Different kinds of joints allow motion in different directions (**Figure 47.19**), but muscles can exert force

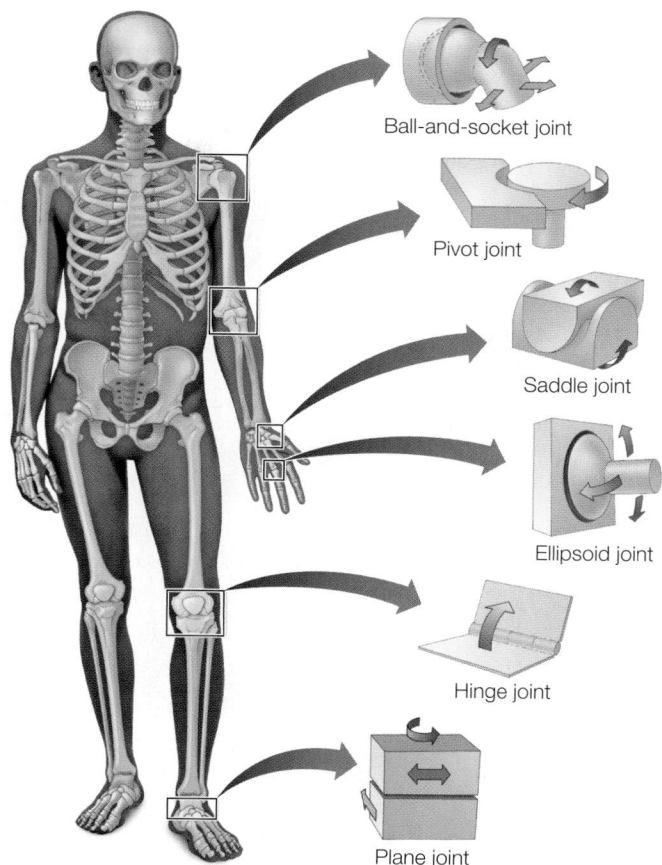

Ball-and-socket joint

Pivot joint

Saddle joint

Ellipsoid joint

Hinge joint

Plane joint

Figure 47.19 Types of Joint Motions The designs of joints are similar to mechanical counterparts and enable a variety of movements.

 **Activity 47.3 Joints**
www.Life11e.com/ac47.3

in only one direction. Therefore muscles create movement around joints by working in antagonistic pairs: when one muscle contracts, the other relaxes. When both contract, the joint becomes rigid (which is important for maintaining posture, for example).

With respect to a particular joint, such as the knee, we refer to the muscle that bends, or flexes, the joint as the **flexor**, and the muscle that straightens, or extends, the joint as the **extensor**. The bones that meet at the joint are held together by **ligaments**, which are flexible bands of connective tissue. Other straps of connective tissue, called **tendons**, attach the muscles to the bones. In **Figure 47.20**, which shows the knee joint, you can see a tendon that connects a flexor muscle in the upper leg (biceps femoris) to a bone in the lower leg. You can also see a tendon connecting the extensor muscle (quadriceps) to the kneecap (patella) and the patellar ligament that connects the patella to the lower leg bone. It is the patellar ligament that is tapped when testing the knee-jerk reflex (see Figure 44.14).

Bones constitute a system of levers that are moved around joints by the muscles. Levers have three parts: load arm, force arm, and fulcrum. The joint is the fulcrum, and the length of the force arm relative to the load arm depends on where the muscle attaches to the bone. Three classes of levers are defined based on the location

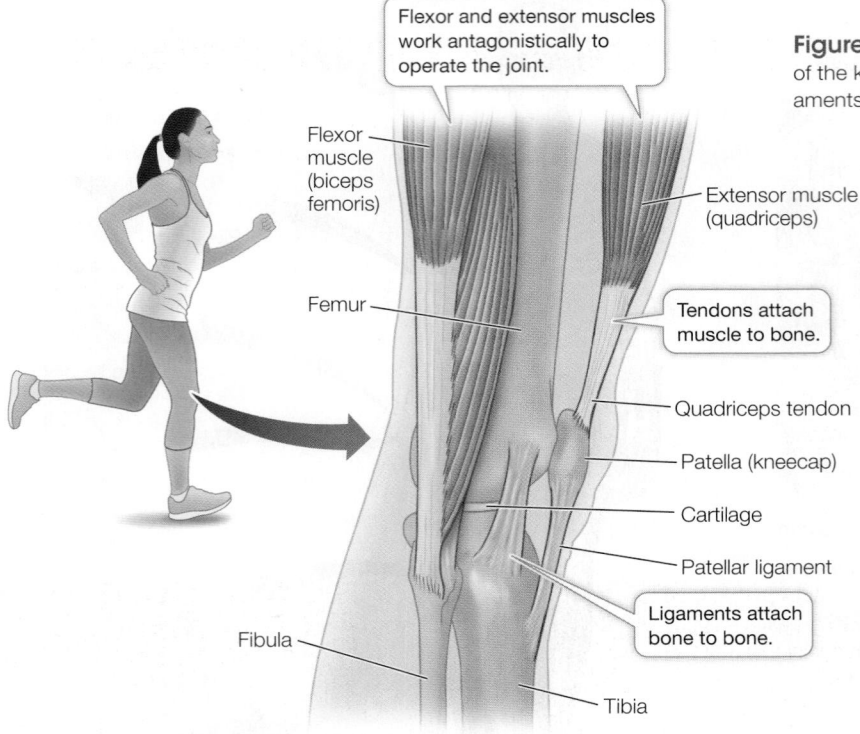

Flexor and extensor muscles work antagonistically to operate the joint.

Flexor muscle (biceps femoris)

Extensor muscle (quadriceps)

Femur

Tendons attach muscle to bone.

Quadriceps tendon

Patella (kneecap)

Cartilage

Patellar ligament

Ligaments attach bone to bone.

Fibula

Tibia

Figure 47.20 Joints, Ligaments, and Tendons A side view of the knee shows the interactions of muscle, bone, cartilage, ligaments, and tendons at this crucial and vulnerable human joint.

of the fulcrum (**Figure 47.21**). In a class 1 lever, the fulcrum is between the load and force arms, as in a see-saw. An example is the joint between the skull and the spinal cord. A class 2 lever is like a wheelbarrow, with the load in the middle. An example is the ankle joint and the muscle in the back of the leg that enables you to walk and stand on tiptoe. In a class 3 lever, the force arm is between the fulcrum and the load. An example is your elbow or your knee. The length ratio of the load and force arms determines whether a particular lever can exert a lot of force over a short distance or is better at translating force into large or fast movements. Think of the powerful jaws of carnivores that can easily crack bones. In contrast, the force arm of the lower legs of Rosie the Ribeter and Jackie Joyner-Kersee, featured in the opener to this chapter, is short relative to the load arm, so they can jump fast and high.

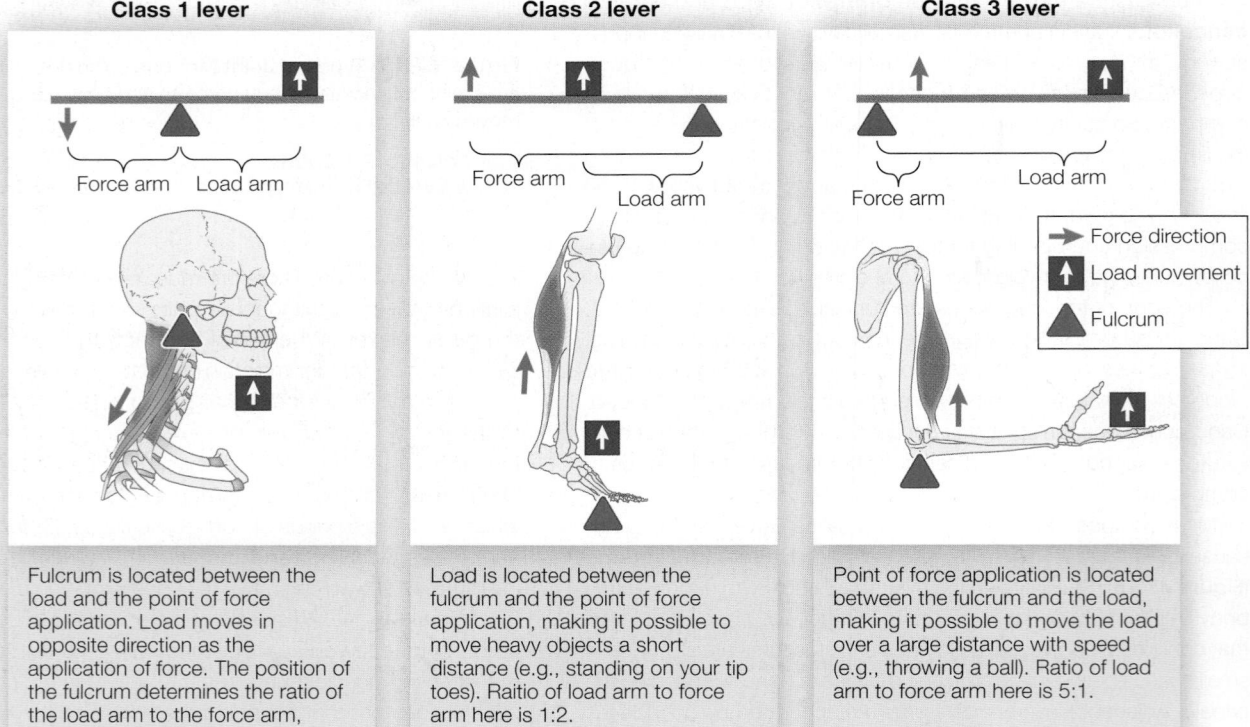

Class 1 lever

Force arm | Load arm

Fulcrum is located between the load and the point of force application. Load moves in opposite direction as the application of force. The position of the fulcrum determines the ratio of the load arm to the force arm, which in this case is 1:1.

Class 2 lever

Force arm

Load arm

Load is located between the fulcrum and the point of force application, making it possible to move heavy objects a short distance (e.g., standing on your tip toes). Raitio of load arm to force arm here is 1:2.

Class 3 lever

Force arm | Load arm

→ Force direction
↑ Load movement
▲ Fulcrum

Point of force application is located between the fulcrum and the load, making it possible to move the load over a large distance with speed (e.g., throwing a ball). Ratio of load arm to force arm here is 5:1.

Figure 47.21 Bones and Joints Work as Levers The three classes of levers are defined by the relationship between the fulcrum and the load and force arms. For each class of lever, the load and force arm ratio determines whether force or speed of movement is maximized.

Q: What class of lever is the knee joint?

▶ 47.3 recap

Muscles can only contract and relax; to achieve organized movement, they must pull against rigid structures—other muscles, hydrostatic skeletons, exoskeletons, or endoskeletons.

learning outcomes

You should be able to:

- Describe features of hydrostatic skeletons, exoskeletons, and endoskeletons.
- Explain the functional relationships among osteoblasts, osteoclasts, and osteocytes.
- Explain the structural features of a lever of an endoskeletal system that give the lever its functional characteristics.

1. Why is molting a universal characteristic of arthropods but not of vertebrates?

2. What changes in the activities of bone cells might explain the loss of bone mass in astronauts who spend long periods at the International Space Station, and what mechanism in the bone might be responsible?

3. What is the significance of the fact that the ratio of the force arm to the load arm is much greater in the jaw joint than in the elbow joint?

▶ investigating life

 ### What adaptations optimize the jumping muscles of the frog?

Rosie the Ribeter was an exceptional jumper judged by the length of time her record remains unbeaten, but frogs in general are remarkable jumpers. This ability is not due to the molecular structure of their muscles, as the arrangement of sarcomeres is very similar across a wide cross section of animals, and in all cases, the sarcomeres and therefore the muscles only contract by about 25–30 percent of their length. So how can a muscle only a few centimeters long produce a jump of many meters? One answer, which is considered in the experiments described in Investigating Life: What Is the Optimal Resting Position for the Jumping Muscle of the Frog?, involves maximizing force and power. First, you saw that the resting posture of the frog maintains the sarcomere length where it can generate the maximum force when it contracts. Then by using the measured force/velocity curve, you saw that the velocity of the jump was at the level that maximized the power of the jump. This rather fast velocity of contraction is possible because the jumping muscle contains mostly fast-twitch fibers. As mentioned in the opening story, and further explained in Key Concept 47.3, the legs operate as class 3 levers to move the body mass. Because the legs of the frog are long, the load arms of those levers are long in comparison with the force arms—that is, the length ratio of the load and force

arms is high. A frog's long legs and big feet have another advantage as well. Muscles must have something to exert force against to maximize power, and once the frog leaves the ground, its muscles can no longer generate power. Long legs and feet ensure that as the jumping muscles contract, the feet remain in contact with the ground for a longer time, thus maximizing the power of the jump.

Future directions

Understanding how muscles and skeletal structures work together to generate movements of different types is essential in the growing field of robotics. Robots that repeat the same movement over and over are quite common in manufacturing, but more and more robotic systems are being created to do more variable and complex tasks that mimic and frequently augment the abilities of humans and other animals. For example, robots have been developed to travel over rough terrain, even while carrying heavy loads. Robots have been developed to do household chores such as washing dishes and folding laundry. Robotic systems are improving the functionality of prosthetic limbs for amputees. And melding of robotic systems with normal human functionality is leading to technologies called "exoskeletons" that greatly enhance the physical abilities of humans. The starting point in all of these engineering efforts is a thorough characterization of how the biological musculoskeletal system works.

Chapter Summary 47

47.1 Interactions of Actin and Myosin Cause Muscles to Contract

- **Skeletal muscle** consists of bundles of **muscle fibers**. Each skeletal muscle fiber is a large cell containing multiple nuclei.

- Skeletal muscles contain numerous **myofibrils**, which are bundles of **actin** and **myosin** filaments. The regular, overlapping arrangement of the actin and myosin filaments into **sarcomeres** gives skeletal muscle its striated appearance. Review Figure 47.1, Activity 47.1

- Observations of the changes in the banding patterns of sarcomeres led to the development of **sliding filament model** of muscle contraction. Review Figure 47.2

- The molecular mechanism of muscle contraction involves the binding of the globular heads of myosin molecules to actin. Review Figure 47.3, Animation 47.1

- A single motor neuron and all the fibers it activates constitute a **motor unit**. Each nerve ending of the motor neuron forms a synapse with the muscle cell membrane. Action potentials spread across the muscle cell membrane and through the **T tubules**, causing Ca^{2+} to be released from the **sarcoplasmic reticulum**. Review Figure 47.5, Activity 47.2

- Ca^{2+} binds to **troponin** and changes its conformation, pulling the **tropomyosin** strands away from the myosin-binding sites on the actin filament. The muscle fiber continues to contract until the Ca^{2+} is returned to the sarcoplasmic reticulum. Review Focus: Key Figure 47.6

- **Cardiac muscle** cells are striated, uninucleate, branching, and electrically connected by gap junctions, so that action potentials spread rapidly throughout sheets of cardiac muscle and cause coordinated contractions.

- **Smooth muscle** provides contractile force for internal organs. Smooth muscle cells respond to stretch and to neurotransmitters from the autonomic nervous system. Review Figures 47.8, 47.9, Animation 47.2

47.2 Many Factors Affect Muscle Performance

- In skeletal muscle, a single action potential causes a minimum unit of contraction called a **twitch**. Twitches occurring in rapid succession can be summed to achieve sustained contraction, which is known as **tetanus**. Review Figure 47.10

- **Slow-twitch fibers** facilitate extended, aerobic work; **fast-twitch fibers** generate maximum forces for short periods of time. The ratio of slow-twitch to fast-twitch fibers in the muscles of an individual is largely genetically determined. Review Figure 47.11

- The force that a muscle fiber can produce depends on its initial state of extension or contraction. Review Figure 47.12, Investigating Life: What Is the Optimal Resting Position for the Jumping Muscle of the Frog?

- Anaerobic exercise stimulates the enlargement of muscle fibers through production of new microfilaments. Aerobic exercise stimulates greater oxidative capacity of muscle fibers.

- Muscle performance depends on a supply of ATP. Review Figure 47.13

47.3 Muscles and Skeletal Systems Work Together

- **Skeletal systems** provide supports against which muscles can pull.

- **Hydrostatic skeletons** are fluid-filled body cavities that can be squeezed by muscles. Review Figure 47.14

- **Exoskeletons** are hardened outer surfaces to which internal muscles are attached.

- **Endoskeletons** are internal systems of rigid rodlike, platelike, and tubelike supports, consisting of **bone** and **cartilage** to which muscles are attached. Review Figure 47.15

- Bone is continually remodeled by **osteoblasts**, which lay down new bone, and **osteoclasts**, which erode bone. Review Figure 47.16

- Bones develop from connective tissue membranes (**membranous bone**) or from cartilage (**cartilage bone**) through ossification. Review Figure 47.17

- Bone can be **compact** (solid and hard) or **cancellous** (containing numerous internal spaces). Most of the compact bone of mammals is composed of **Haversian systems**. Review Figure 47.18

- **Joints** enable muscles to power movements in different directions. Muscles and bones work together around joints as systems of levers. Review Figures 47.19, 47.21, Activity 47.3

- **Tendons** connect muscles to bones; **ligaments** connect bones to one another. Review Figure 47.20

> Go to **LearningCurve** (in **LaunchPad**) for dynamic quizzing that helps you solidify your understanding of this chapter. **LearningCurve** adapts to your responses, giving you the practice you need to master each key concept.

Apply What You've Learned

Review

47.2 Three systems supply ATP to muscles for contraction, and together they determine the work capacity and endurance of a muscle.

Original Paper: Coyle, E. F., A. R. Coggan, M. K. Hemmert and J. L. Ivy. 1986. Muscle glycogen utilization during prolonged strenuous exercise when fed carbohydrates. *Journal of Applied Physiology* 61: 165–172.

An athlete trained for endurance can sustain repeated muscle contractions far longer than someone without such training. But even trained athletes have a limit on how long they can exercise before their muscles become fatigued. One factor necessary to sustain muscle activity is the type and amount of metabolic fuel used. In this study, researchers investigated whether administering carbohydrates to athletes as they exercised would affect their time to fatigue.

Seven male, endurance-trained cyclists performed two strenuous cycling trials until they fatigued. At 20-minute intervals throughout one trial, subjects ingested a placebo (flavored water). On another day, the athletes performed the same trial, this time ingesting a glucose solution at 20-minute intervals. The researchers measured time to fatigue, muscle glycogen content in the vastus lateralis (thigh muscle), blood glucose (**Figure A**), carbohydrate oxidation rate (**Figure B**), and blood free fatty acid (FFA) concentration (**Figure C**). Results are shown in the table and graphs. Glycogen levels were measured in millimoles (mM) of glucosyl units per kilogram of muscle tissue. Values are means ± SE, and means are compared with a *t*-test. In the figures, significant differences are indicated with an asterisk.

Treatment	Muscle glycogen at 0 hours	Muscle glycogen at 2 hours	Muscle glycogen at 3 hours	Muscle glycogen at 4 hours	Time to fatigue (hrs)
Placebo	163 ± 13	60 ± 9	37 ± 5	No data	3.02 ± 0.19
Glucose solution	173 ± 11	71 ± 11	44 ± 12	39 ± 8	4.02 ± 0.33
	$P = 0.27$	$P = 0.22$	$P = 0.30$	$P = 0.01$	

Questions

1. In these two trial groups, what were the effects of strenuous aerobic exercise on blood glucose, muscle glycogen, and carbohydrate utilization?

2. Did glucose feedings have an effect on time to fatigue? Did this effect involve maintenance of blood glucose and a slowing of muscle glycogen depletion?

3. Consider one more piece of data: comparing the muscle glycogen levels for the glucose-fed group between hour 3 and point of fatigue, the *P*-value is 0.63. What do you conclude about the effect of glucose feeding on carbohydrate metabolism in this

Figure A

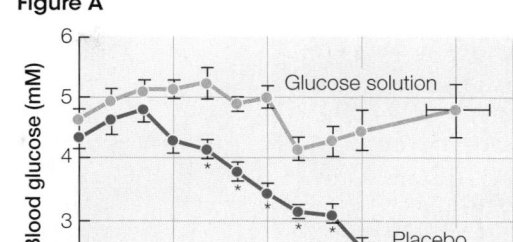

Figure B

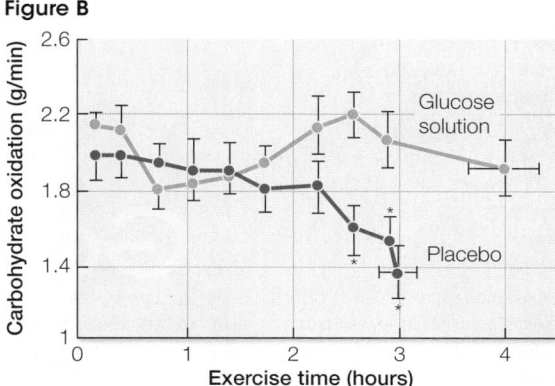

Figure C

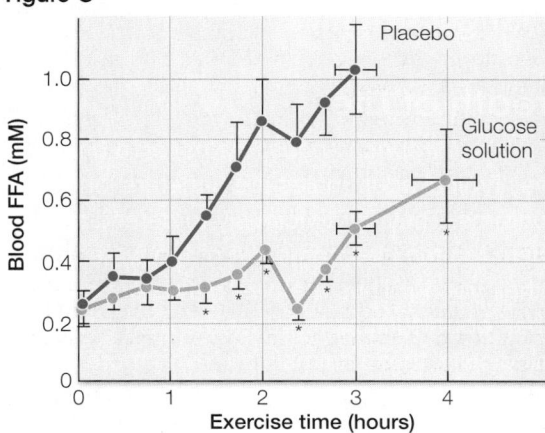

experiment? How might the placebo group have compensated for declining carbohydrate availability?

4. Using the data on blood FFA levels in the two groups, develop an explanation for the overall effects of the glucose feeding on the energy metabolism of the subjects.

5. Suppose you have a friend training to run a marathon. Based on this study, suggest something your friend might do to improve his or her performance and ability to complete the long-distance run.

Go to **LaunchPad** for the eBook, LearningCurve, animations, activities, flashcards, and additional resources and assignments.

48

Gas Exchange

Northern Elephant Seals (*Miirounga angustirostris*) are champion breath-hold divers. Diving repeatedly, they may only spend 20% of their time at the surface.

▶ investigating**life**

The Breath of Life

Elephant seals (*Mirounga angustirostris*) on a beach are impressive. Typical adult males are 5 meters long and weigh 2,500 kilograms. They fight viciously to defend areas of beach where the females come out of the water to give birth and rear their young. Thus the beach masters are really defending their opportunities to father the next generation.

Physiologically, these animals are even more impressive in the water. Female elephant seals, half the size of the males and therefore easier and safer to manage for experiments, have been fitted with instruments to record the depth and duration of their dives after they leave the beach where they gave birth. They spend up to 7 months at sea, and during that entire time they dive continually—day and night—to feed on eels, skates, rays, squid, and other prey. Most dives are 300–600 meters deep and last 20 minutes, but dives lasting over an hour and exceeding 2,000 meters have been recorded. The seals spend only 4–5 minutes at the surface between dives. Thus for the 80 percent of their lives that they spend at sea, they are breathing only about 20 percent of the time—12 minutes per hour!

The best human divers are perhaps the pearl divers of Japan and Korea. They typically dive repeatedly during their workday to about 20 meters, and their dives last about a minute. Record breath-hold dives for humans resting face-down in a pool are about 10 minutes, but that is a one-time event that cannot be repeated for many hours. The elephant seals are not resting—they are descending and ascending great distances and pursuing prey.

How can these behemoths spend so much time not breathing? How can they survive such crushing depths? How can they not get the bends, which is a danger for human divers? How much oxygen do these animals need to carry out their normal behavior? How much oxygen can they store in their bodies during the short times they spend at the surface? And what adaptations do they have to make those oxygen stores last as long as they do? To answer these questions, elephant seals and other diving animals have been studied in unique laboratory facilities and in the field.

 **What adaptations make seals champion breath-hold divers?**

key concept

48.1 Respiratory Gas Exchange Is Governed by Physical Factors

The **respiratory gases** that animals must exchange are oxygen (O_2) and carbon dioxide (CO_2). Cells need to obtain O_2 from the environment to produce an adequate supply of ATP by cellular respiration (see Chapter 9). CO_2 is an end product of cellular respiration, and it must be removed from the body to prevent toxic effects. Gas exchange systems of animals consist of (1) specialized body surface areas where these gases can move between the body and the external environment, (2) mechanisms that ventilate the environmental side of those surfaces with air or water, and (3) mechanisms that circulate extracellular fluids on the internal sides.

focus your learning

- Respiratory gases diffuse from areas of higher partial pressure to areas of lower partial pressure.
- Fick's law of diffusion describes the factors that determine the rate of diffusion of a gas between two locations.
- Rising temperature decreases oxygen solubility in water and increases the metabolism of aquatic ectotherms.
- With increasing altitude, oxygen partial pressure decreases, which decreases the rate of oxygen diffusion into the respiratory systems of animals.

Diffusion is the only means by which respiratory gases are exchanged between an animal's internal body fluids and the outside medium (air or water). There are no active transport mechanisms to move respiratory gases across biological membranes. Because diffusion is a physical process, knowing what physical factors influence rates of diffusion helps us understand the diverse adaptations of gas exchange systems. (You may want to review the coverage of diffusion in Key Concept 6.3.)

Diffusion of gases is driven by partial pressure differences

Diffusion is due to the random motion of molecules, and it results in a net movement of molecules from areas of higher concentration to areas of lower concentration. In other words, diffusion evens out the distribution of molecules. Diffusion of respiratory gases occurs either in water or in air. Concentrations of gases in water or air vary with pressure because gases are compressible, a relationship described in physics by Boyle's Law. For example, there are twice as many gas molecules in a liter of gas at 2 atmospheres of pressure as there are in a liter of gas at 1 atmosphere of pressure. And if that gas is in contact with a liquid such as water, twice as many gas molecules will enter into solution when the gas pressure is 2 atmospheres than when it is 1 atmosphere.

The concentrations of different gases in a mixture are described as the **partial pressures** of those gases. To calculate the partial pressure of a gas such as oxygen in a mixture of gases such as air, we have to know the total pressure. The total pressure of air inhaled by air-breathing animals is the atmospheric pressure. At sea level, atmospheric pressure is about 760 millimeters of mercury (mm Hg),

depending on the weather. Because dry air is 20.9 percent O_2, the partial pressure of oxygen (P_{O_2}) at sea level is 20.9 percent of 760 mm Hg, or about 159 mm Hg. If two gas mixtures are separated by a membrane permeable to O_2, O_2 will diffuse from the mixture where its partial pressure is higher to the mixture where its partial pressure is lower.

To calculate the concentration of O_2 in the environment of a water breather, we have to know two things: the partial pressure of O_2 in the air in contact with the water, and the solubility of O_2 in water. The amount of a gas that dissolves in a liquid depends both on its partial pressure in the gas phase in contact with the liquid *and* on its solubility in that liquid. The *diffusion* of a gas between the gas phase and the liquid phase is a function of its partial pressures in those two phases; the gas diffuses from the phase with the higher partial pressure to the phase with the lower partial pressure until equilibrium is reached—the point at which the partial pressures in the two phases are equal. However, the *amount* of the gas that can be contained in the liquid depends on the solubility of that gas in that liquid. Furthermore, the solubility of a gas in a particular liquid can vary widely depending on conditions. What follows is a practical illustration of these facts.

Solubility of a gas in a liquid, such as oxygen in water, is a function of temperature—solubility is higher at low temperatures. Think of opening a bottle of warm soda in comparison to a bottle of cold soda. There is more gas out of solution—more fizz—in the warm soda. So if we have similar containers of water in equilibrium with air, but at different temperatures, the concentrations of oxygen in these water containers will be different (less O_2 in the warmer one), but the partial pressures of oxygen will be the same. Thus for water-breathing animals, the warmer the water is, the less O_2 there is per liter of water. The important point is that for a gas in solution, its concentration is not the same as its partial pressure, but partial pressures are what drive diffusion. Thus in our continuing discussions of respiratory gas exchange, we will always use partial pressure rather than concentration when referring to the diffusion of respiratory gases.

Fick's law applies to all systems of gas exchange

Whether in air or water, the diffusion rates of respiratory gases depend on their partial pressure gradients and on other factors that can be described quantitatively with a simple equation called **Fick's law of diffusion**. All environmental variables that limit respiratory gas exchange and all adaptations that maximize respiratory gas exchange are reflected in one or more components of this equation. Fick's law is written as

$$Q = DA\frac{P_1 - P_2}{L}$$

where

- Q is the rate at which a gas such as O_2 diffuses between two locations.
- D is the diffusion coefficient, which is a characteristic of the diffusing substance, the medium, and the temperature. For example, perfume has a higher D than motor oil vapor, and all substances diffuse faster at higher temperatures and faster in air than in water. Temperature is not expressed explicitly in

(A) *Maiazoon orsaki* **(B)** *Callyspongia plicifera* **(C)** *Triturus* sp.   Gills

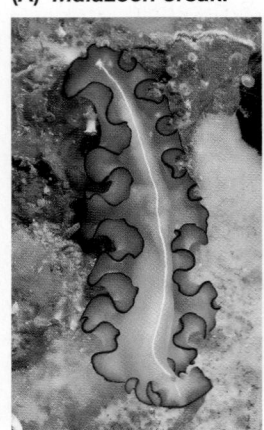

Channels

Figure 48.1 Keeping in Touch with the Medium **(A)** No cell in the leaflike body of this marine flatworm is more than a millimeter away from seawater. **(B)** Sponges have body walls perforated by many channels, which allow water to flow between the outside world and a central cavity. No cell in the sponge is more than a millimeter away from seawater. **(C)** A feathery fringe of gills on this aquatic salamander provides a large surface area for gas exchange. Blood circulating through the gills comes into close contact with the respiratory medium.

Fick's law because the diffusion coefficient is usually determined at room temperature (about 20°C).

- A is the area across which the gas is diffusing.
- P_1 and P_2 are the partial pressures of the gas at the two locations.
- L is the path length, or distance, between the two locations.
- $(P_1 - P_2)/L$ is a partial pressure gradient.

The strict dependence of animals on diffusion for gas exchange with their environments has selected for various adaptations that maximize Q, many of which we will describe in this chapter. Animals can maximize D for respiratory gases by using air rather than water as their gas exchange medium whenever possible. All other adaptations for maximizing respiratory gas exchange must influence the surface area (A) for gas exchange or the partial pressure gradient across that surface area.

Air is a better respiratory medium than water

The slow diffusion of O_2 molecules in water affects both air- and water-breathing animals. Eukaryotic cells carry out cellular respiration in their mitochondria, which are located in the cytoplasm—an aqueous medium. Cells are bathed in extracellular fluid—also an aqueous medium. In addition, all respiratory surfaces must be protected from drying out by a thin film of fluid through which O_2 must diffuse. Even in air-breathing animals, the slow rate of O_2 diffusion in water limits the efficiency of O_2 distribution from gas exchange surfaces to the sites of cellular respiration.

Diffusion of O_2 in water is so slow that even animal cells with low rates of metabolism cannot function more than a few millimeters away from a good source of environmental O_2. Therefore there are severe size and shape limits on the many species of invertebrates that lack internal systems for distributing O_2. Most of these species are very small, but some, such as marine flatworms, have grown larger by evolving a flat, thin body with a large external surface area (**Figure 48.1A**). Another strategy is seen in sponges, which have bodies made of cells that surround water channels through which the external medium flows (**Figure 48.1B**) (see also Figure 30.2B). A critical factor enabling larger, more complex animal bodies has been the evolution of specialized respiratory systems with large surface areas such as gills that are highly permeable to respiratory gases (**Figure 48.1C**).

O_2 can be obtained more easily from air than from water for several reasons:

- *The O_2 content of air is much higher than the O_2 content of an equal volume of water.* The maximum O_2 content of a bubbling stream in equilibrium with air is less than 10 milliliters (mL) of O_2 per liter of water. The O_2 content of the air over the stream is about 200 mL of O_2 per liter of air.
- *O_2 diffuses about 8,000 times more rapidly in air than in water.* That is why the O_2 content of a stagnant pond can be zero only a few millimeters below the surface.
- *An animal has to work (expend energy) to ventilate its gas exchange surfaces with water or air.* More energy is required to move water than air because water is 800 times denser than air and about 50 times more viscous.

You can appreciate how important these facts were for the evolutionary transition of life to the terrestrial environment, because they meant that there were fewer constraints on the evolution of higher metabolic rates.

High temperatures create respiratory problems for aquatic animals

Animals that use water for their respiratory exchange medium are in a double bind when environmental temperatures rise. Most water-breathing animals are ectotherms—their body temperatures are closely tied to the temperature of the water around them. As the water temperature rises, an ectotherm's body temperature and metabolic rate rise (see Figure 39.8). Thus water breathers need more O_2 as the water gets warmer (**Figure 48.2**). But as mentioned above, warm water holds less dissolved gas than cold water does. In addition, since a water-breathing animal performs work to move water across its gas exchange surfaces, it must expend more energy to breathe as water temperature rises. Therefore, as water temperature goes up, a water-breathing animal must extract more and more O_2 from an environment that is increasingly O_2 deficient, and a lower percentage of that O_2 is available to support activities other than breathing.

O_2 availability decreases with altitude

Just as a rise in water temperature reduces the supply of O_2 available to water-breathing animals, an increase in altitude reduces the O_2 supply for air breathers. At all altitudes, O_2 makes up 20.9 percent of the dry air; however, as you go up in altitude, the total

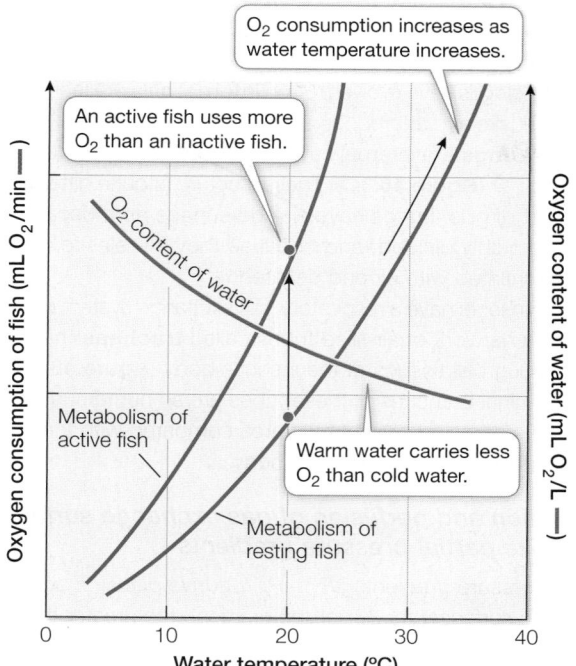

O₂ consumption increases as water temperature increases.

An active fish uses more O₂ than an inactive fish.

O₂ content of water

Metabolism of active fish

Warm water carries less O₂ than cold water.

Metabolism of resting fish

Oxygen consumption of fish (mL O₂/min —)

Oxygen content of water (mL O₂/L —)

Water temperature (°C)

Figure 48.2 The Double Bind of Water Breathers   Fishes need more O_2 when the water is warmer, but warm water carries less O_2 than cold water.

amount of gas per unit of volume decreases, as reflected in the atmospheric pressure. For example, at 5,800 meters, atmospheric pressure is only half what it is at sea level, so the P_{O_2} at that altitude is only about 80 mm Hg. At the summit of Mount Everest (8,850 m), P_{O_2} is only about 50 mm Hg—roughly one-third what it is at sea level.

Because the movement of O_2 across respiratory gas exchange surfaces and into the body depends on diffusion, its rate of movement depends on the P_{O_2} difference between the air and the body fluids. Therefore the drastically reduced P_{O_2} in the air at high altitudes constrains O_2 uptake. Because of this, mountain climbers attempting peaks such as Mount Everest or K2 usually breathe O_2 from pressurized bottles.

CO₂ is lost by diffusion

Respiratory gas exchange is a two-way process: CO_2 diffuses out of the body as O_2 diffuses in. The direction and rate of diffusion of the respiratory gases across the exchange surfaces depend on the partial pressure gradients of the gases. The partial pressure gradients of O_2 and CO_2 across these exchange surfaces are quite different. The amount of CO_2 in the atmosphere is extremely low (0.03%), so for air-breathing animals there is always a large concentration gradient for diffusion of CO_2 from the body to the environment. Whereas the partial pressure gradient for O_2 decreases with increasing altitude, the gradient driving CO_2 out of the body hardly changes. The partial pressure of carbon dioxide (P_{CO_2}) in the atmosphere is close to zero both at sea level and atop Mount Everest.

In general, getting rid of CO_2 is not a problem for water-breathing animals because CO_2 is much more soluble in water than is O_2. Even in stagnant water, where the P_{CO_2} is higher than in moving water, the lack of O_2 becomes a problem for an animal long before CO_2 exchange difficulties arise.

48.1 recap

Respiratory gases are exchanged only by diffusion. All factors influencing respiratory gas exchanges are components of Fick's law of diffusion. Air is a better respiratory medium than water because a given volume of air has more O_2 than the same volume of water. O_2 diffuses faster in air than in water, and less work is required to move air over respiratory exchange surfaces. CO_2 exchange with the external environment occurs readily because of a large concentration gradient from the body to the environment and because CO_2 is highly soluble in water.

learning outcomes

You should be able to:

- Analyze changes in partial pressure of a gas at different altitudes and at different depths under water.
- Apply Fick's law to explain animal adaptations that influence respiratory gas exchange.
- Describe three factors that require a fish to increase its gill ventilation in response to an increase in its environmental temperature.
- Explain how altitude influences respiratory gas exchange in terms of partial pressures of gases.

1. What is the partial pressure of oxygen in inhaled air at sea level; at 2,000 meters above sea level, where atmospheric pressure is 80 percent that of sea level; and while breathing air at the same pressure as the environment at 10 meters (= an additional 1 atmosphere of pressure) underwater?

2. Flatworms (see Figure 48.1A) are invertebrates that do not have a respiratory system. What two components of Fick's law enable them to satisfy their need for oxygen?

3. The O_2 content of water decreases with an increase in water temperature, but the gill ventilation rate of a fish increases much more than is necessary to match that decrease in O_2 availability. Why?

4. In terms of Fick's law, what is the difference between the ability to exchange O_2 and CO_2 at sea level and at high altitude?

Now that we have discussed the physical factors that influence diffusion rates of respiratory gases between animals and their environments, let's look at some of the adaptations that have evolved for maximizing respiratory gas exchange.

> **key concept**
> **48.2**
> ## Enhancing Diffusion Maximizes Respiratory Gas Exchange

Adaptations to maximize respiratory gas exchange are reflected in the values of Fick's law of diffusion $Q = DA(P_1 - P_2)/L$. Such adaptations have to do one or more of the following:

- Increase the surface area for gas exchange (A)
- Maximize the partial pressure difference driving diffusion ($P_1 - P_2$)
- Minimize the diffusion path length (L)
- Minimize the diffusion that takes place in an aqueous medium (maximize D)

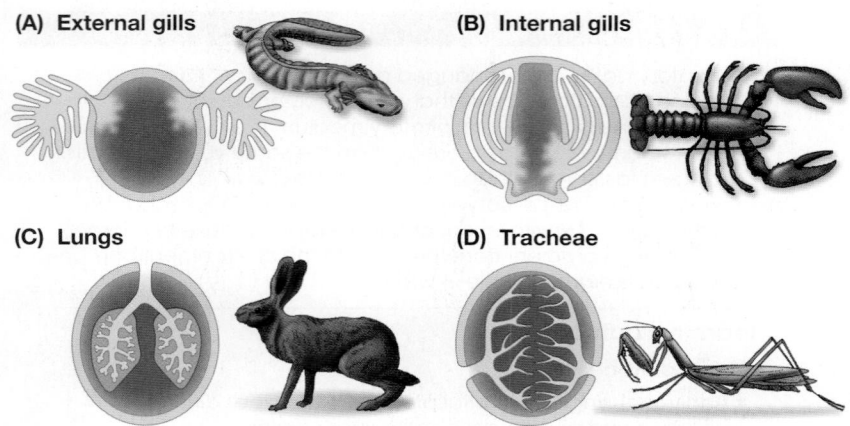

(A) External gills

(B) Internal gills

(C) Lungs

(D) Tracheae

Figure 48.3 Gas Exchange Systems Large surface areas (tan in these diagrams) for the diffusion of respiratory gases are common features of animals. External **(A)** and internal **(B)** gills are adaptations for gas exchange with water. Lungs **(C)** and tracheae **(D)** are organs for gas exchange with air.

With these factors in mind, we will look at four different respiratory systems: those of insects, fishes, birds, and mammals.

focus your learning

- In insects, efficient gas exchange is facilitated by an internal network of air passages.
- In fishes, countercurrent ventilation and perfusion facilitate efficient gas exchange.
- In birds, anterior and posterior air sacs act as bellows to maintain a continuous unidirectional flow of air through the respiratory system.
- Tidal volume, inspiratory reserve volume, and expiratory reserve volume can be measured directly, and residual volume can be determined using the helium dilution method.

Respiratory organs have large surface areas

A variety of anatomical adaptations maximize the specialized body surface areas (*A*) for respiratory gas exchange. Water-breathing animals generally have gills, and air-breathing animals have tracheae or lungs. Gills are highly branched and folded extensions of the body surface that provide a large surface area for gas exchange (**Figure 48.3A**; see also Figure 48.1C). Because gills consist of thin, delicate tissues, they minimize the path length (*L*) traversed by diffusing molecules of O_2 and CO_2. **External gills** are found in larval amphibians and in the larvae of many insects. External gills are vulnerable to damage and are tempting

morsels for predators, so in many animals protective body cavities for gills have evolved. Such **internal gills** are found in most mollusks and arthropods and in all fishes (**Figure 48.3B**).

Lungs are internal cavities for respiratory gas exchange with air (**Figure 48.3C**). Their structure is quite different from that of gills. Lungs have a large surface area because they are highly divided, and because they are elastic, they can be inflated with air and deflated.

Insects have a respiratory gas exchange system consisting of a network of air-filled tubes called **tracheae** that branch through all tissues of the insect's body (**Figure 48.3D**). The terminal branches of these tubes are so numerous that they have an enormous surface area compared with the external surface area of the insect's body.

Ventilation and perfusion of gas exchange surfaces maximize partial pressure gradients

Partial pressure gradients [$(P_1 - P_2)/L$] drive diffusion across gas exchange surfaces; the larger the gradient, the greater the rate of gas exchange. These gradients can be maximized in several ways:

- *Minimizing path length*: Very thin tissues in gills and lungs reduce the diffusion path length (*L*).
- *Ventilating*: Actively moving the external medium over the gas exchange surfaces (i.e., breathing) regularly exposes those surfaces to fresh respiratory medium containing maximum O_2 and minimum CO_2 concentrations. This maximizes the partial pressure gradients.
- *Perfusing*: Actively moving the internal medium (e.g., blood) over the internal side of the exchange surfaces transports CO_2 to those surfaces and O_2 away from them, thus maximizing the partial pressure gradients driving diffusion.

Insects have airways throughout their bodies

The tracheal system that enables insects to exchange respiratory gases extends to all tissues in the insect body. Thus respiratory

Figure 48.4 The Tracheal Gas Exchange System of Insects **(A)** In insects, respiratory gases diffuse through a system of air sacs and tubes (tracheae) that open to the external environment through holes called spiracles. **(B)** The spiracles of a hawkmoth larva run down its sides. **(C)** A scanning electron micrograph shows an insect trachea dividing into smaller tracheoles and still finer air capillaries.

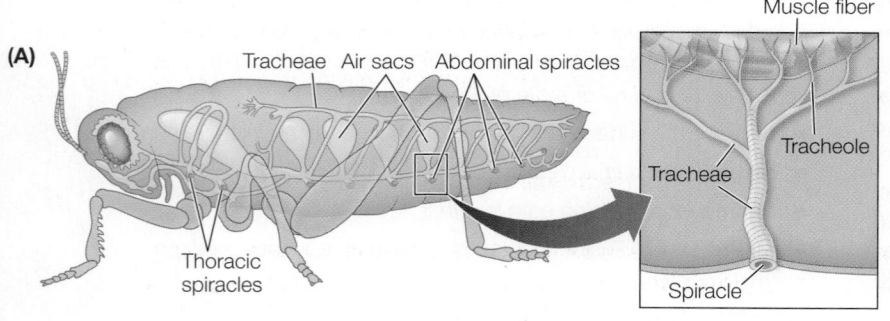

(A)

Tracheae   Air sacs   Abdominal spiracles

Muscle fiber

Tracheae

Tracheole

Thoracic spiracles

Spiracle

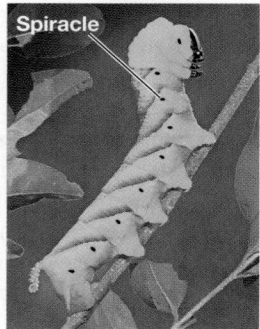

(B) *Acherontia atropos*

Spiracle

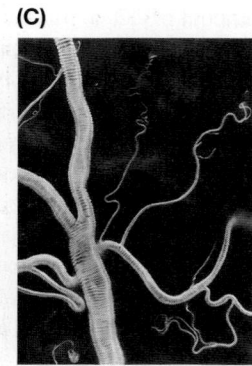

(C)

gases diffuse through air most of the way to and from every cell. The insect respiratory system communicates with the outside environment through gated openings called spiracles in the sides of the abdomen and thorax (**Figure 48.4A and B**). The spiracles open to allow gas exchange and then close to decrease water loss. They open into tubes called tracheae that branch into even finer tubes, or tracheoles, that end in tiny air capillaries—the actual gas exchange surfaces (**Figure 48.4C**).

Fish gills use countercurrent flow to maximize gas exchange

The internal gills of fish are supported by gill arches that lie between the mouth cavity and the protective opercular flaps on the sides of the fish just behind the eyes (**Figure 48.5A**). Water flows unidirectionally into the fish's mouth, over the gills, and out from under the opercular flaps. Thus there is a constant one-way flow of oxygenated water over the gills, maximizing the P_{O_2} on the external gill surfaces. On the internal side of the gill membranes, the circulation of blood minimizes the P_{O_2} by sweeping O_2 away as rapidly as it diffuses across.

Gills have an enormous surface area for gas exchange because they are so highly subdivided. Each gill consists of hundreds of ribbonlike gill filaments (**Figure 48.5B**). The upper and lower flat surfaces of each gill filament are covered with rows of evenly spaced folds, or lamellae (**Figure 48.5C**). The lamellae are the actual gas exchange surfaces. Because the lamellae are exceedingly thin, the path length (L) for diffusion of gases between blood and water is minimized. The surfaces of the lamellae consist of highly flattened epithelial cells, so the water and the fish's red blood cells are separated by little more than 1–2 micrometers.

The flow of blood perfusing the inner surfaces of the lamellae, like the flow of water over the gills, is unidirectional. **Afferent** blood vessels bring deoxygenated blood to the gills, while **efferent** blood vessels take oxygenated blood away from the gills (see Figure 48.5C). Blood flows through the lamellae in the direction opposite to the flow of water over the lamellae. This **countercurrent flow** optimizes the P_{O_2} gradient between water and blood, making gas exchange more efficient than it would be in a system using concurrent (parallel) flow (**Figure 48.6**).

Some fishes, including anchovies, tuna, and certain sharks, ventilate their gills by swimming almost constantly with their mouths open. Most fishes, however, ventilate their gills by means of a two-pump mechanism. The closing and contracting of the mouth cavity pushes water over the gills, and the expansion of the opercular cavity prior to opening of the opercular flaps pulls water over the gills.

These adaptations for maximizing the surface area (A) for diffusion, minimizing the path length (L) for diffusion, and maximizing the P_{O_2} gradient allow fishes to extract an adequate supply of O_2 from meager environmental sources.

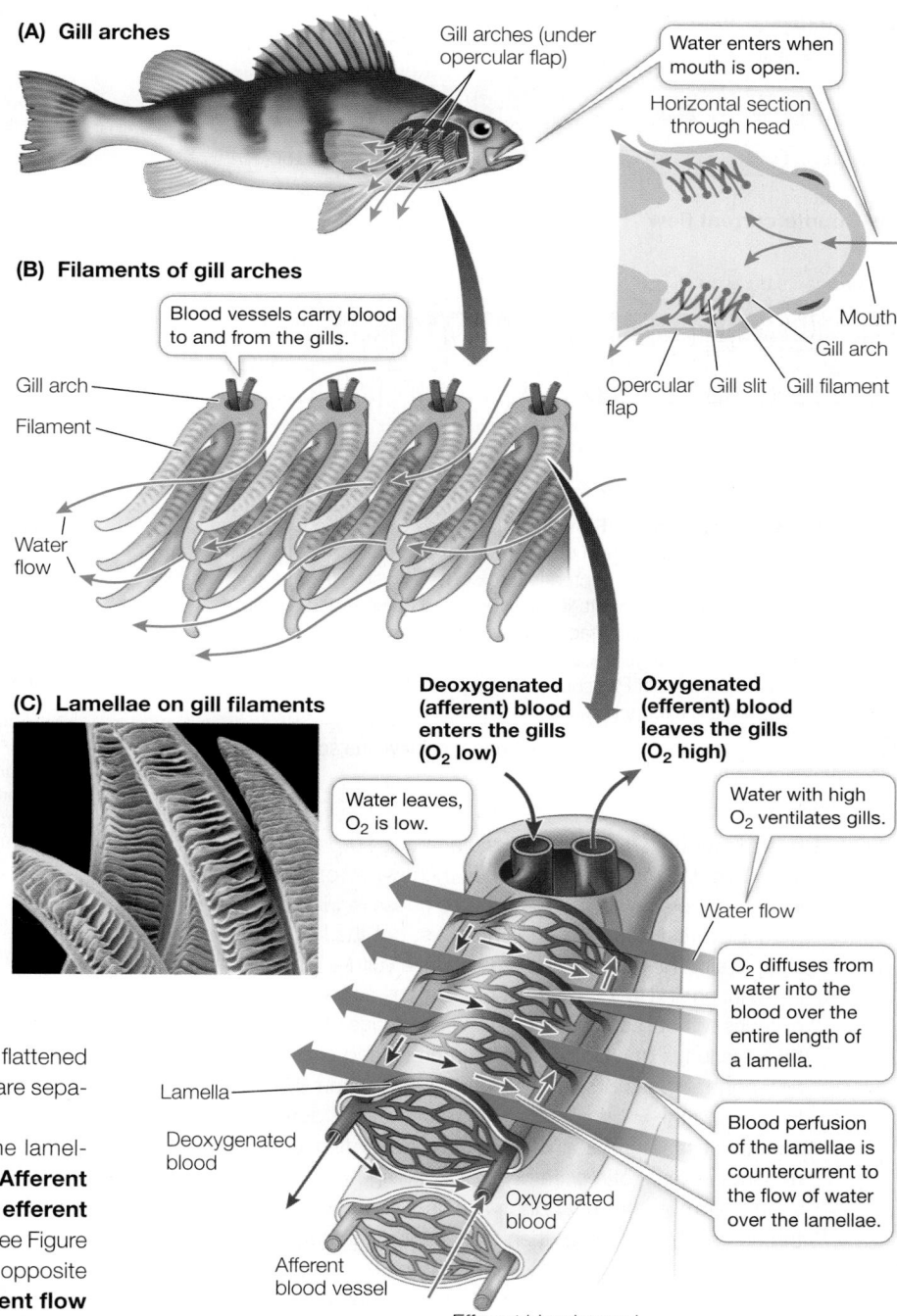

(A) Gill arches

Gill arches (under opercular flap)

Water enters when mouth is open.

Horizontal section through head

Mouth

Opercular flap Gill slit Gill filament

Gill arch

(B) Filaments of gill arches

Blood vessels carry blood to and from the gills.

Gill arch

Filament

Water flow

Deoxygenated (afferent) blood enters the gills (O_2 low)

Oxygenated (efferent) blood leaves the gills (O_2 high)

(C) Lamellae on gill filaments

Water leaves, O_2 is low.

Water with high O_2 ventilates gills.

Water flow

O_2 diffuses from water into the blood over the entire length of a lamella.

Blood perfusion of the lamellae is countercurrent to the flow of water over the lamellae.

Lamella

Deoxygenated blood

Oxygenated blood

Afferent blood vessel

Efferent blood vessel

Figure 48.5 Fish Gills **(A)** Water flows unidirectionally over the gills of a fish, in through the mouth and out through the opercular opening. **(B)** Gill filaments have a large surface area and thin tissues. **(C)** Gill filaments are lined with lamellae, which are the gas exchange surfaces. Blood flows through the lamellae countercurrent (left to right in this drawing; small blue and red arrows) to the direction of the flow of water across the gills (right to left in this drawing; large blue arrows).

Birds use unidirectional ventilation to maximize gas exchange

Birds are remarkable for their ability to sustain high levels of activity for a long time—for example, on long-distance flights—even at high altitudes where mammals cannot even survive. The first team

(A) Concurrent flow

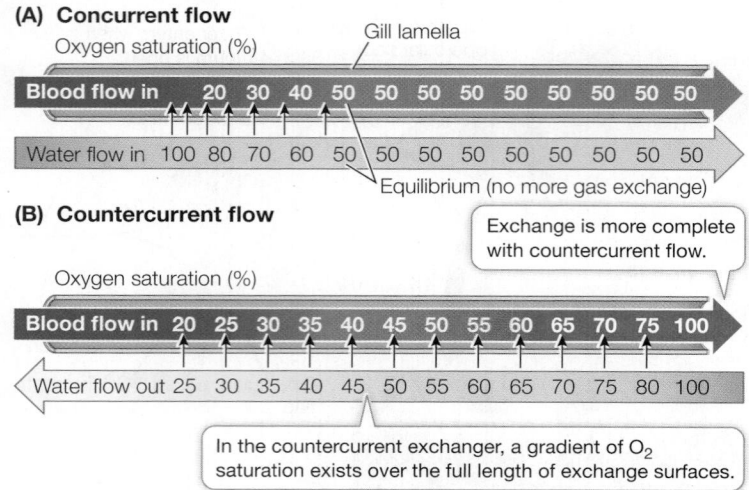

(B) Countercurrent flow

Exchange is more complete with countercurrent flow.

In the countercurrent exchanger, a gradient of O_2 saturation exists over the full length of exchange surfaces.

Figure 48.6 Countercurrent Flow Enables More Complete Exchange In these models of concurrent and countercurrent gas exchange, the numbers represent the O_2 saturation percentages of blood and water. **(A)** In a concurrent exchanger, the saturation percentages of blood and water reach equilibrium halfway across the exchange surface. **(B)** A countercurrent exchanger allows more complete gas exchange because the water always has a higher P_{O_2} than the blood; thus a gradient of P_{O_2} supports diffusion of O_2 into the blood across the entire length of the exchanger.

Q: What could make the concurrent exchanger achieve the same maximum O_2 saturation as the countercurrent exchanger?

to climb Mount Everest (8,850 meters) was surprised to see birds flying over the mountain when they themselves could barely move without supplemental O_2. Bar-headed geese regularly migrate over Mount Everest and surrounding peaks, but the highest recorded flight of a bird is from a Ruppell's griffon, a vulture, that was sucked into a jet engine at 11,278 meters. Humans cannot survive at such altitudes without supplemental O_2. Yet the lungs of a bird are smaller than the lungs of a similar-sized mammal, and bird lungs expand and contract less during a breathing cycle than do mammalian lungs. And the volume of bird lungs decreases during inhalation and increases during exhalation. How do birds accomplish such remarkable feats of respiratory gas exchange?

The structure of bird lungs allows air to flow unidirectionally through the lungs, rather than bidirectionally through all the same airways, as it does in tidally ventilating mammals (discussed below). Because mammalian lungs are never completely emptied of air during exhalation, there is always some lung volume that is not ventilated with fresh air. The air remaining in lungs and airways after exhalation is called **dead space**. Bird lungs, by contrast, have very little dead space, and the fresh incoming air is not mixed with stale air. In this way, a high P_{O_2} gradient is maintained.

Air sacs are an important and unique feature of the avian respiratory system, and they occupy much of the body cavity of the bird (**Figure 48.7A**). They can be divided into a group of anterior air sacs and a group of posterior air sacs. The air sacs are interconnected with each other, with the lungs, and with air spaces in some of the bones. The air sacs are not gas exchange surfaces; rather, they act as bellows to maintain a unidirectional flow of air through the lungs.

As in other air-breathing vertebrates, air enters and leaves a bird's gas exchange system through the **trachea** (commonly known

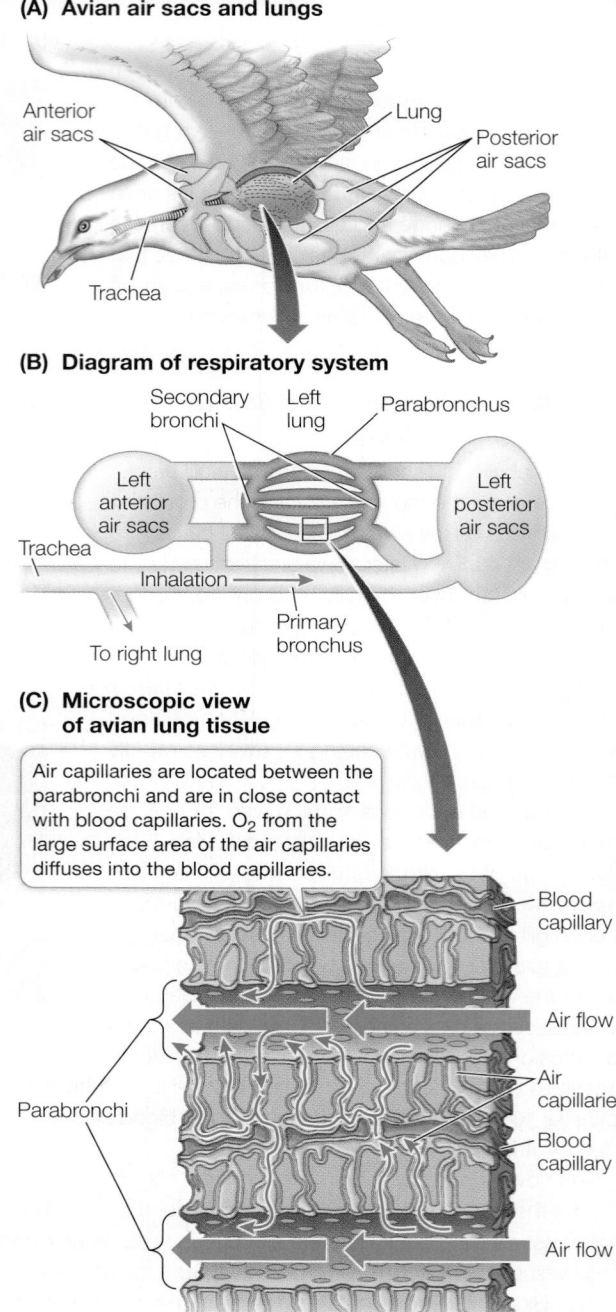

(A) Avian air sacs and lungs

(B) Diagram of respiratory system

(C) Microscopic view of avian lung tissue

Air capillaries are located between the parabronchi and are in close contact with blood capillaries. O_2 from the large surface area of the air capillaries diffuses into the blood capillaries.

Figure 48.7 The Respiratory System of a Bird **(A)** Air sacs in the body cavity and air spaces in the bones are unique to birds. **(B)** Air comes in through the trachea, which divides into a left and a right primary bronchus. The primary bronchi go to the posterior air sacs. Secondary bronchi connect the posterior air sacs to the lungs. Air flows unidirectionally through the lungs in numerous parabronchi and flows in bronchi from the lungs to the anterior air sacs, then out through the trachea. **(C)** Air capillaries running between the parabronchi are the site of gas exchange.

as the windpipe, and not to be confused with the air-conducting tracheae of insects) (**Figure 48.7B**). The trachea divides into two smaller airways, the **primary bronchi** (singular *bronchus*). The primary bronchi extend all the way to the posterior air sacs and also

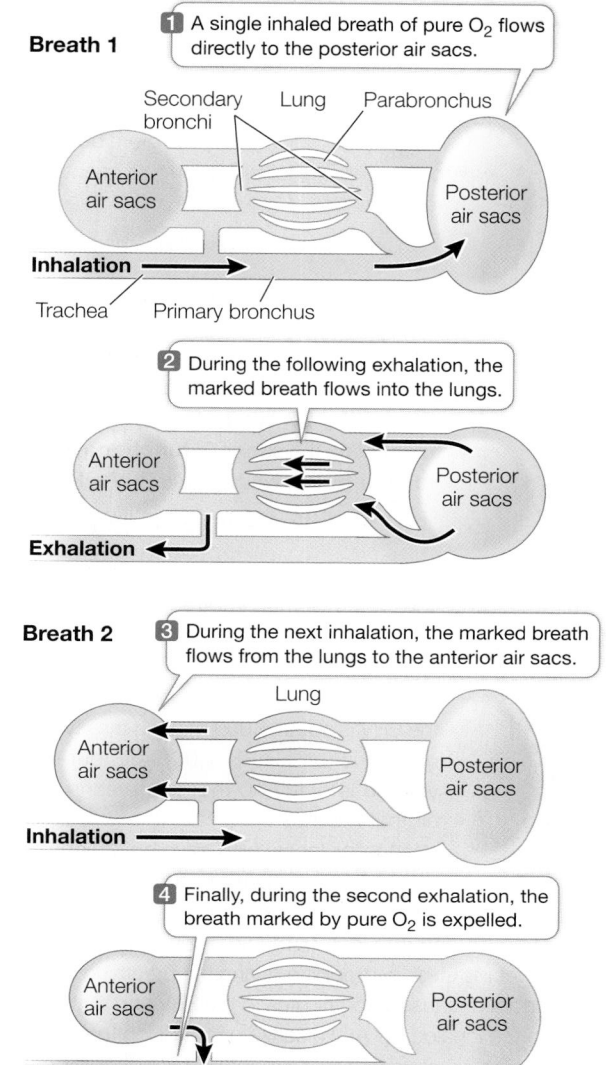

Breath 1

1 A single inhaled breath of pure O_2 flows directly to the posterior air sacs.

Secondary bronchi   Lung   Parabronchus

Anterior air sacs

Posterior air sacs

Inhalation

Trachea   Primary bronchus

2 During the following exhalation, the marked breath flows into the lungs.

Anterior air sacs

Posterior air sacs

Exhalation

Breath 2

3 During the next inhalation, the marked breath flows from the lungs to the anterior air sacs.

Lung

Anterior air sacs

Posterior air sacs

Inhalation

4 Finally, during the second exhalation, the breath marked by pure O_2 is expelled.

Anterior air sacs

Posterior air sacs

Exhalation

Figure 48.8 An Experiment to Describe Airflow in Bird Respiratory System   Oxygen sensors placed in different locations in the bird respiratory system trace the flow of a single breath of pure oxygen (blue) through the system. A single breath remains in the system for two breathing cycles.

Animation 48.1 Airflow in Birds
www.Life11e.com/a48.1

branch into **secondary bronchi**. The posterior air sacs also have connections to the secondary bronchi. Secondary bronchi divide into tubelike **parabronchi** that run parallel to one another through the lungs.

Branching off the parabronchi are numerous tiny air capillaries (**Figure 48.7C**). Air flows through the parabronchi and diffuses into the air capillaries, which are the gas exchange surfaces. The air capillaries are so numerous that they provide an enormous surface area for gas exchange. The blood capillaries are in close contact with the air capillaries, and oxygen from the air diffuses into the blood. The parabronchi coalesce into larger bronchi that take the

air out of the lungs and into anterior air sacs and then back to the trachea. Thus the anatomy of a bird's airways allows air to flow unidirectionally through the lungs: trachea, bronchi, posterior air sacs, parabronchi, bronchi, anterior air sacs, trachea.

The puzzle of how birds breathe was solved by placing small O_2 sensors at different locations in birds' air sacs and airways. The birds were then allowed to inhale pure O_2 for just a single breath, which made it possible to track that particular breath by recording when each of the O_2 sensors registered a peak value. The experiment showed that a single breath remains in a bird's respiratory system for two cycles of inhalation and exhalation. During inhalation, the air sacs expand and air flows through the trachea into the posterior air sacs. At the same time, the lungs are contracting as air moves from the lungs into the anterior air sacs. During exhalation, the air sacs contract. The posterior air sacs push air into the lungs, which are expanding, and the anterior air sacs push air into the trachea and out of the respiratory system. This bellowslike action of the air sacs maintains a continuous, unidirectional flow of fresh air through the lungs (**Figure 48.8**).

The advantages of the bird respiratory system are similar to those of fish gills. The air sacs keep fresh air flowing unidirectionally over the gas exchange surfaces without interruption. Thus those gas exchange surfaces are constantly exposed to a P_{O_2} that is very close to that of the ambient air. Even at high altitude where the P_{O_2} of the ambient air may be only slightly above that of the blood, O_2 can still diffuse from air to blood.

Tidal ventilation produces dead space that limits gas exchange efficiency

Lungs evolved in early lungfishes as outpocketings of the digestive tract. Although lung structure has evolved considerably, lungs remain dead-end sacs in all air-breathing vertebrates except birds (and likely their extinct reptilian ancestors). Ventilation of dead-end sacs with common passageways for inflow and outflow cannot be constant and unidirectional but must be **tidal**: air flows in and exhaled gases flow out by the same route. Since the lungs and airways can never be completely emptied of air, they always contain dead space. We can easily measure the volumes of air exchanged during breathing, but we have to use an indirect method to measure the dead space contained in the lungs and airways. Measures of dead space are important in assessing lung health and disease.

A flowmeter measures the volume of air breathed in and out (**Figure 48.9**). Using a human as an example, the amount of air that moves either in or out per breath when at rest is called the **tidal volume (TV)** (about 500 mL for an average human adult). When we breathe in as much as possible, the additional volume is the **inspiratory reserve volume (IRV)**. Conversely, if we forcefully exhale as much air as possible, the additional amount of air expelled is the **expiratory reserve volume (ERV)**. The maximum capacity for air exchange in one breath, or the **vital capacity (VC)**, is the sum of TV + IRV + ERV. The vital capacity of an athlete is generally greater than that of a nonathlete, and vital capacity decreases with age because of stiffening of the lung tissue.

Vital capacity is not the entire lung capacity because of the dead space, also called the **residual volume (RV)**. We can't measure RV directly with the flowmeter, but we can measure it indirectly using

research tools

Figure 48.9A Measuring Lung Ventilation   A flowmeter is a device that measures the volume of air a person breathes through a mouthpiece. The combined tidal volume, inspiratory reserve volume, and expiratory reserve volume are the lungs' vital capacity. The vital capacity plus the residual volume are the total lung capacity.

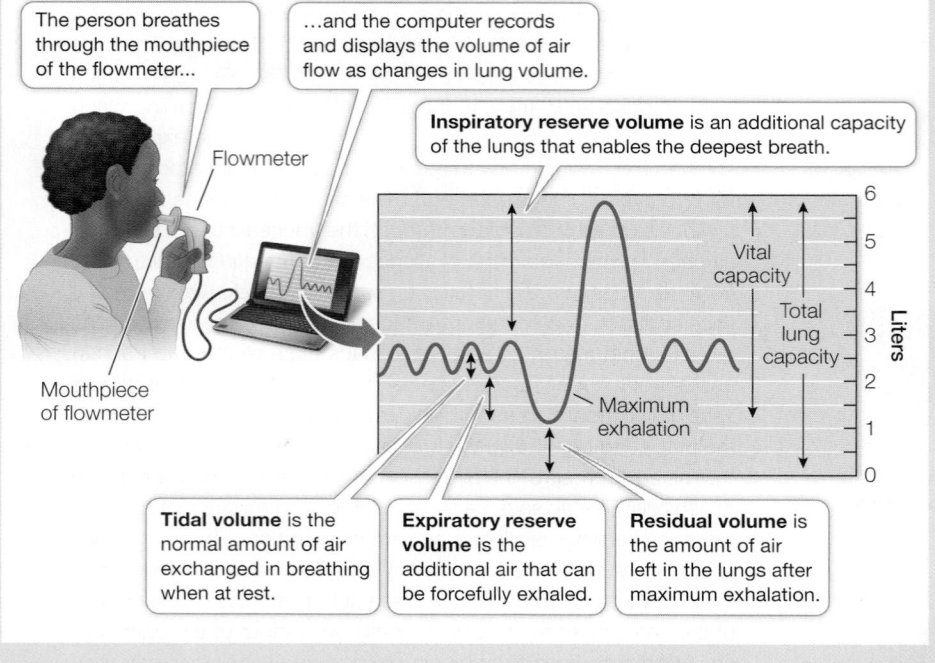

The person breathes through the mouthpiece of the flowmeter...

...and the computer records and displays the volume of air flow as changes in lung volume.

Flowmeter

Mouthpiece of flowmeter

Inspiratory reserve volume is an additional capacity of the lungs that enables the deepest breath.

Vital capacity

Total lung capacity

Maximum exhalation

Liters

Tidal volume is the normal amount of air exchanged in breathing when at rest.

Expiratory reserve volume is the additional air that can be forcefully exhaled.

Residual volume is the amount of air left in the lungs after maximum exhalation.

the helium dilution method. The flowmeter is attached to a closed reservoir of air of a known volume (V_{res}) so that a person can breathe in and out from that reservoir. If this reservoir has a known amount of helium (He) in it, and a subject breathes the air in that reservoir through several breathing cycles, the helium will become distributed in the reservoir and in the subject's respiratory system. Helium is

not absorbed into the body from the lungs, so the total amount of helium at the beginning of a test (He_i) will equal the total amount at the end of the test (He_f)—it will just be distributed over a larger volume and therefore be less concentrated. The following formula quantifies that relationship:

Initial amount of He = final amount of He

We start the subject breathing from the reservoir at an end-tidal breath, and we end the breathing from the reservoir at an end-tidal breath. Since the V_{res} does not change, the decrease in helium concentration [He] enables us to calculate the subject's **functional residual volume** (**FRV**), which is the ERV + RV.

$$V_{res} \times [He]_i = (V_{res} + FRV) \times [He]_f$$

Since we *can* measure the ERV with the flowmeter, we can subtract the ERV from the FRV to obtain the RV.

Why is the RV important? Referring to Figure 48.9, you will see that for a normal person the ERV is about 1,000 mL and the RV is 1,000 mL. Thus the FRV is 2,000 mL, but the tidal volume is only 500 mL; this means that the air that reaches the alveoli (the actual gas exchange surfaces) with each breath consists of only 500 mL of fresh air diluted by 2,000 mL of stale air. The maximum P_{O_2} in this mixed air is much below the P_{O_2} of the outside air, and because of the tidal ventilation pattern, the P_{O_2} in the alveoli is steadily dropping during the breathing cycle. The RV is important because it contributes to the FRV and to the dilution of the O_2 in the inhaled air. Any disease or condition that increases the RV (such as emphysema or pulmonary fibrosis) compromises a patient's respiratory ability. Similarly, considering the mixing of fresh air with the FRV, you can understand why reductions in tidal volume can be a problem—and therefore why patients recovering from surgery are encouraged to breathe deeply, even if it hurts.

Small residual volume prevents the bends in seals

In the opening of this chapter we discussed the remarkable abilities of elephant seals to dive to great depths for long times and then spend only minutes breathing at the surface. We asked how elephant seals avoid getting the bends, a very painful condition that develops in scuba divers who surface too rapidly after spending time at depths. The pressure of the water on the body increases by 1 atmosphere for every 10 meters of depth. Body tissues are not compressible, but air in the respiratory system is, so scuba divers breathe air through a regulator that equalizes the air pressure in their lungs to the pressure of the water on the body. As the total air pressure goes up, the partial pressures of the component gases go up, increasing their rate of diffusion into the tissues. Eighty percent of the air is nitrogen, so the amount of nitrogen dissolved in tissues increases. If the diver comes up too rapidly, the dissolved nitrogen comes out of solution in the tissues, creating small bubbles that block blood vessels, damage tissues, cause pain, and can even

work with the data

Figure 48.9B   Measuring Lung Ventilation

A patient complains of always being short of breath. Using a flowmeter, the physician determines that the patient's tidal volume is 0.3 L and his expiratory reserve volume is 1.5 L The physician decides to use the Fick dilution method to measure the patient's residual volume. She attaches the flowmeter to a 30-liter air reservoir containing 5% helium. Using a two-way valve, she can have the patient breathe room air or the air in the reservoir. At the end of a tidal exhalation, the physician turns the valve and lets the patient take 10 breaths from the reservoir. She then switches the patient to room air at the end of a tidal exhalation. She determines that the volume of the resevoir is still 30 L, but the helium concentration in the reservoir is now 4.4%.

QUESTIONS ▶

1. What is the patient's residual volume?
2. Why do you think the patient complains of always being short of breath?

A similar **work with the data** exercise may be assigned in **LaunchPad**.

result in death. To avoid the bends, divers use decompression tables that tell them how to ascend depending on how deep and how long their dives have been. For example, if a diver has spent 1 hour at a depth of 60 meters, he would have to spend 3 hours gradually ascending to shallower and shallower depths to allow the nitrogen dissolved in his tissues to diffuse out through the respiratory system rather than form bubbles in his tissues. Why does a seal who makes repeated dives not face the same consequences? The seal exhales before diving and the residual air in its respiratory system gets compressed into its airways, which are not respiratory exchange surfaces. So, when seals dive they do not have compressible air in their lungs that can be forced into tissues at depth and then come out of solution upon resurfacing.

48.2 recap

The major adaptations that increase efficiency of respiratory gas exchange are large surface areas for exchange and maximized partial pressure gradients across those surfaces. Fish and avian respiratory systems have continuous and unidirectional ventilation of their respiratory exchange surfaces. Mammals have tidal ventilation, resulting in residual volumes of O_2 that dilute the incoming fresh air.

learning outcomes

You should be able to:

- Explain the concept and adaptive significance of countercurrent exchange, using the fish gills as the example.
- Describe how the avian respiratory system maximizes the partial pressure gradient for O_2 uptake.
- Determine the total lung capacity (including residual volume) of a human.

1. Given constant rates of ventilation and perfusion, how can the directionality of blood and water flow on opposite sides of fish gill membranes influence the maximum O_2 exchange that can occur?
2. How do birds maintain a constant and unidirectional flow of air through their lungs?
3. Find the functional residual volume of a person who yields the following data while breathing from a flowmeter that has a volume of 30 liters (L) and an initial gas mixture of 20 percent O_2, 75 percent nitrogen (N), and 5 percent helium (He). Beginning with an inhalation, the person inhales and exhales 20 times. The final concentration of He in the flowmeter is 4.6 percent.

Despite their limitations, mammalian lungs serve the respiratory needs of mammals well. Offsetting the inefficiencies of tidal breathing, mammalian lungs have an enormous surface area and a very short path length for diffusion. Next we will look at the human respiratory system as an example.

key concept 48.3   Humans Have Tidal Respiration

The respiratory exchange surfaces in mammalian lungs are the very thin-walled alveoli that cluster around a huge number of finely branching terminal airways. To understand how these respiratory surfaces are ventilated and perfused, we will first examine the functional anatomy of the whole respiratory system (**Focus: Key Figure 48.10**).

focus your learning

- Mucus traps particulates in the inhaled air and surfactants decrease the work of ventilation.
- Inhalation and exhalation of air result from changes in the volume and the pressure in the thoracic cavity.

Lungs are ventilated through a branching system of airways

Air enters the lungs through the oral cavity or through the nasal passages, which join together in the **pharynx** (see Figure 48.10A). Below the pharynx, the esophagus conducts food to the stomach, and the trachea conducts air to the lungs. At the beginning of this airway is the **larynx**, or voice box, which houses the vocal cords. The larynx is the lump that you can see or feel on the front of your neck. The trachea is about 2 centimeters in diameter. C-shaped bands of cartilage prevent the thin walls of the trachea from collapsing as air pressure changes during the breathing cycle. If you run your fingers down the front of your neck just below your larynx, you can feel a few of these bands of cartilage.

The trachea branches into two **bronchi**, one leading to each lung. The bronchi branch repeatedly to generate a treelike structure of progressively smaller airways extending to all regions of the lungs. After four branchings, the cartilage supports disappear, marking the transition to **bronchioles**. After about 16 branchings, the bronchioles are less than a millimeter in diameter, and tiny, thin-walled air sacs called **alveoli** begin to appear. Alveoli are the sites of gas exchange. After the first alveoli there are about six more branchings of the airways that end in clusters of alveoli (see Figure 48.10B). Because the airways conduct air only to and from the alveoli and do not themselves participate in gas exchange, their volume is dead space.

Human lungs have about 300 million alveoli. Although each alveolus is very small, their combined surface area for diffusion of respiratory gases is about 70 square meters—about one-fourth the size of a basketball court. The walls of each alveolus are made of very thin cells. Wrapped around the alveoli are networks of capillaries that also have very thin walls. Thus where capillary meets alveolus, the diffusion path length between air and blood is less than 2 micrometers (see Figure 48.10C).

Diseases of the bronchioles and alveoli are the third leading cause of death in the United States as of 2014. Among these diseases, the most lethal is emphysema, a condition in which inflammation damages and eventually destroys the walls of the alveoli. As a result, the lungs have fewer but larger alveoli, the lungs lose elasticity, and the RV increases. The principal cause of emphysema is smoking.

Respiratory tract secretions aid ventilation

Mammalian lungs produce two secretions that do not directly influence their gas exchange but do affect the process of ventilation: mucus and surfactant.

Many cells lining the airways produce sticky mucus that captures bits of dirt and microorganisms that are inhaled. Other cells lining the airways have cilia that beat continually and sweep the mucus,

focus: key figure

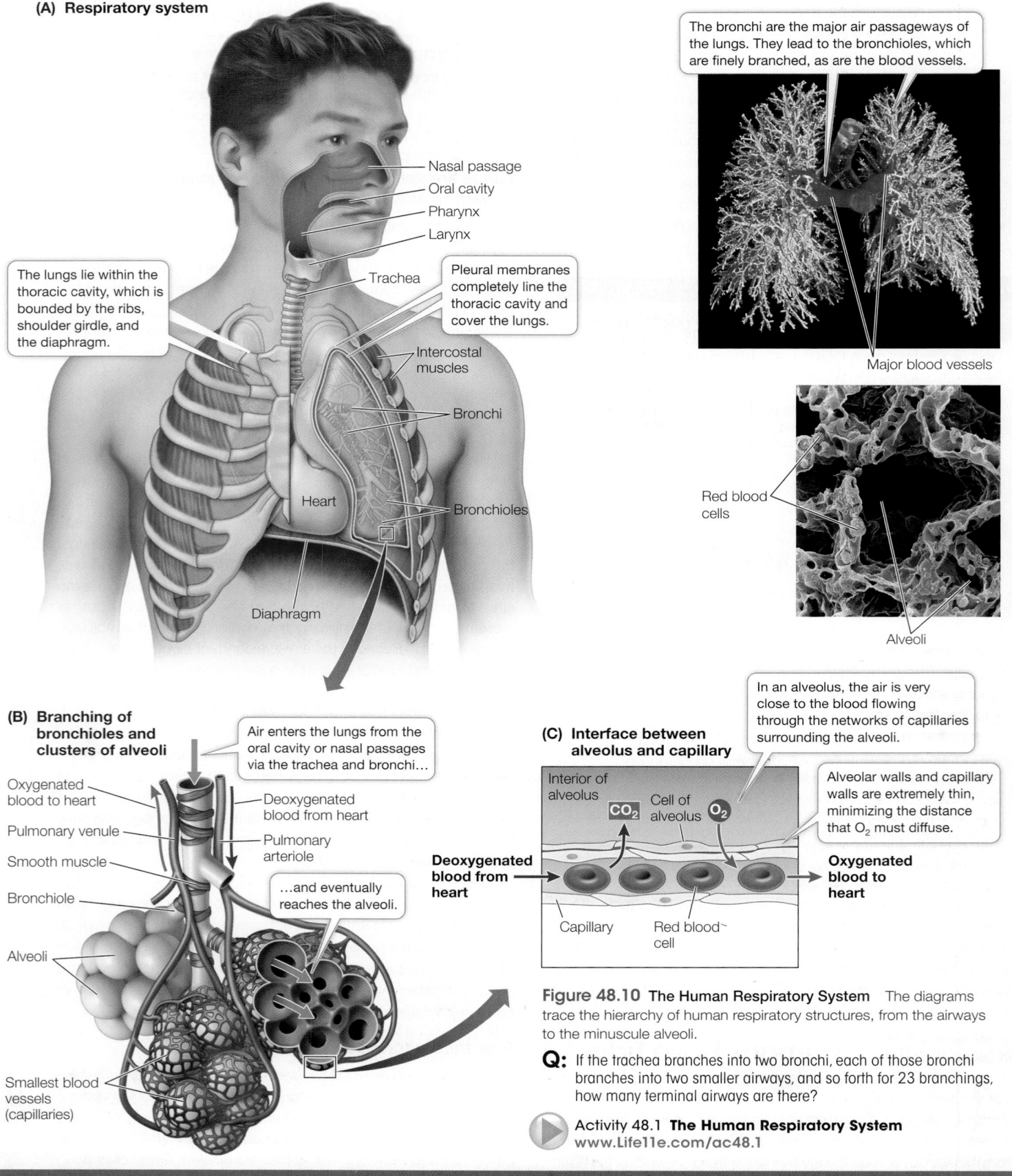

(A) Respiratory system

Nasal passage
Oral cavity
Pharynx
Larynx

The lungs lie within the thoracic cavity, which is bounded by the ribs, shoulder girdle, and the diaphragm.

Trachea

Pleural membranes completely line the thoracic cavity and cover the lungs.

Intercostal muscles

Bronchi

Heart

Bronchioles

Diaphragm

The bronchi are the major air passageways of the lungs. They lead to the bronchioles, which are finely branched, as are the blood vessels.

Major blood vessels

Red blood cells

Alveoli

(B) Branching of bronchioles and clusters of alveoli

Air enters the lungs from the oral cavity or nasal passages via the trachea and bronchi…

Oxygenated blood to heart

Deoxygenated blood from heart

Pulmonary venule

Pulmonary arteriole

Smooth muscle

Bronchiole

…and eventually reaches the alveoli.

Alveoli

Smallest blood vessels (capillaries)

(C) Interface between alveolus and capillary

In an alveolus, the air is very close to the blood flowing through the networks of capillaries surrounding the alveoli.

Interior of alveolus

Cell of alveolus

CO₂   O₂

Alveolar walls and capillary walls are extremely thin, minimizing the distance that O₂ must diffuse.

Deoxygenated blood from heart

Oxygenated blood to heart

Capillary

Red blood cell

Figure 48.10 The Human Respiratory System   The diagrams trace the hierarchy of human respiratory structures, from the airways to the minuscule alveoli.

Q: If the trachea branches into two bronchi, each of those bronchi branches into two smaller airways, and so forth for 23 branchings, how many terminal airways are there?

Activity 48.1 The Human Respiratory System
www.Life11e.com/ac48.1

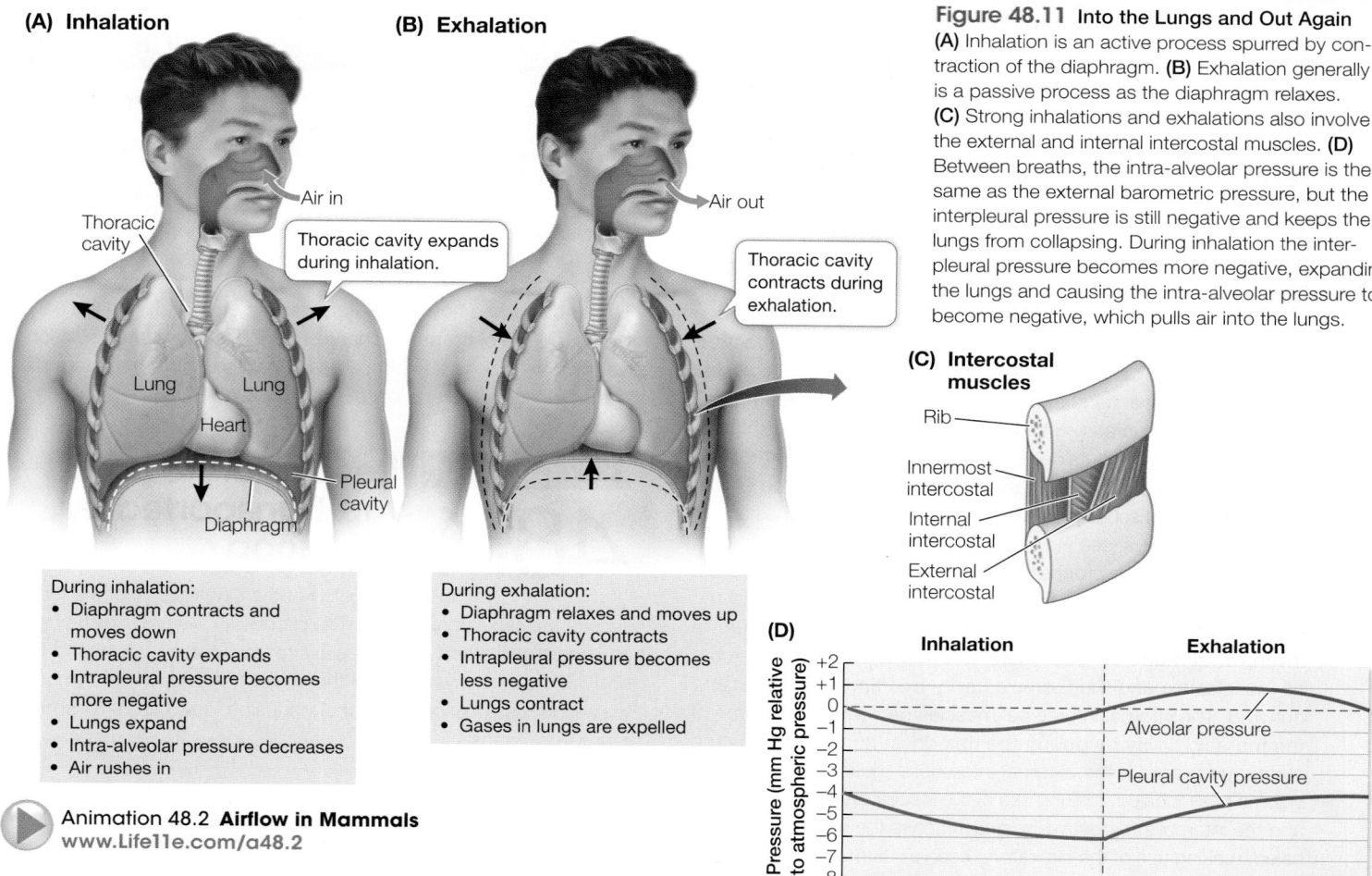

(A) Inhalation

Thoracic cavity

Air in

Thoracic cavity expands during inhalation.

Lung   Lung

Heart

Pleural cavity

Diaphragm

During inhalation:
- Diaphragm contracts and moves down
- Thoracic cavity expands
- Intrapleural pressure becomes more negative
- Lungs expand
- Intra-alveolar pressure decreases
- Air rushes in

▶ Animation 48.2 **Airflow in Mammals**
www.Life11e.com/a48.2

(B) Exhalation

Air out

Thoracic cavity contracts during exhalation.

During exhalation:
- Diaphragm relaxes and moves up
- Thoracic cavity contracts
- Intrapleural pressure becomes less negative
- Lungs contract
- Gases in lungs are expelled

Figure 48.11 Into the Lungs and Out Again
(A) Inhalation is an active process spurred by contraction of the diaphragm. **(B)** Exhalation generally is a passive process as the diaphragm relaxes. **(C)** Strong inhalations and exhalations also involve the external and internal intercostal muscles. **(D)** Between breaths, the intra-alveolar pressure is the same as the external barometric pressure, but the interpleural pressure is still negative and keeps the lungs from collapsing. During inhalation the interpleural pressure becomes more negative, expanding the lungs and causing the intra-alveolar pressure to become negative, which pulls air into the lungs.

(C) Intercostal muscles

Rib

Innermost intercostal

Internal intercostal

External intercostal

(D)

Inhalation — Exhalation

Pressure (mm Hg relative to atmospheric pressure)

Alveolar pressure

Pleural cavity pressure

with its trapped debris, up toward the pharynx, where it is swallowed or spit out. This phenomenon, called the mucus escalator, can be adversely affected by inhaled pollutants. Smoking one cigarette can immobilize the cilia of the airways for hours. A smoker's cough results from the need to clear the obstructing mucus from the airways when the mucus escalator is out of order.

A **surfactant** is a substance that reduces the surface tension of a liquid. **Surface tension** gives the surface of a liquid the properties of an elastic membrane, and it is why certain insects, such as water-striders, can walk on water. As discussed in Key Concept 2.4, surface tension is the result of chemical forces of attraction between water molecules. The attractive forces working on the water molecules at the surface pull from below and from the sides but not from above. This imbalance of forces creates surface tension. The thin film of fluid covering the air-facing surfaces of the alveoli has surface tension that contributes to the lungs' elasticity. To inflate the lungs, enough force has to be generated to overcome both the elasticity of the lung tissue and the surface tension in the alveoli.

Lung surfactant is a fatty, detergent-like substance that is critical for reducing the work necessary to inflate the lungs. Certain cells in the alveoli release surfactant molecules. If a baby is born more than a month prematurely, these cells may not have developed the ability to produce surfactant. A baby with this condition, known as respiratory distress syndrome, will have great difficulty breathing and may die from exhaustion and lack of O_2. Common treatments for premature babies have been to put them on respirators to assist their breathing

and to give them hormones to speed lung development. A newer approach is to apply surfactant to the lungs via an aerosol.

Lungs are ventilated by pressure changes in the thoracic cavity

Human lungs are suspended in the **thoracic cavity**, a closed compartment bounded on the bottom by a sheet of muscle called the **diaphragm** (see Figure 48.10A). Each lung is covered by a continuous sheet of tissue called the **pleural membrane** that also lines the thoracic cavity adjacent to the lung. A thin film of fluid between the pleural membranes of the lung and the thoracic cavity lubricates the inner surfaces of the pleural membranes so they can slip and slide against each other. As with surface tension, there are forces of attraction between the molecules of fluid covering the pleural membranes. As a result, it is difficult to pull the pleural membranes apart. Think of two wet panes of glass or two wet microscope slides; you can slide them past each other, but it is difficult to separate them. While the inner surfaces of the pleural membranes are "stuck" to each other by surface tension, they can move relative to each other during breathing movements.

Inhalation and exhalation involve changes in the volume of the thoracic cavity (**Figure 48.11**). Because the pleural membranes covering the cavity wall and the lung surface are stuck to each other by surface tension, any attempt to increase the volume of the thoracic cavity increases the tension between the pleural membranes. Even between breaths, there is tension between the pleural membranes because the rib cage is pulling outward and the elasticity of the lung tissue is pulling inward. This slight negative pressure keeps the alveoli partly inflated even at the end of an exhalation. If the thoracic cavity is punctured—by a knife wound, for example—air can leak into the space between the pleural membranes and cause the lung to deflate. If the wound is not sealed, breathing movements pull air in between the pleural membranes rather than into the lung (a "sucking chest wound"), and there is no ventilation of the alveoli in that lung—a condition called "collapsed lung."

At rest, inhalation is initiated by contraction of the muscular diaphragm (see Figure 48.11A). As the domed diaphragm contracts, it pulls down, expanding the thoracic cavity and pulling on the pleural membranes. Since the pleural membranes cannot separate, they pull on the lungs increasing their volume, and air rushes in through the trachea from the outside. Exhalation begins when contraction of the diaphragm ceases. As the diaphragm relaxes, the elastic recoil of the lung tissues pulls the diaphragm up and pushes air out through the airways (see Figure 48.11B). When a person is at rest, inhalation is an active process and exhalation is a passive process.

The diaphragm is not the only muscle that can change the volume of the thoracic cavity. Between the ribs are two sets of **intercostal muscles** (see Figure 48.11C). The external intercostal muscles expand the thoracic cavity by lifting the ribs up and outward. The internal intercostal muscles decrease the volume of the thoracic cavity by pulling the ribs down and inward. During strenuous exercise, the external intercostal muscles increase the volume of air inhaled, making use of the inspiratory reserve volume, and the internal intercostal muscles increase the amount of air exhaled, making use of the expiratory reserve volume. The abdominal muscles can also aid in breathing. When they contract, they cause the abdominal contents to push up on the diaphragm and thereby contribute to the expiratory reserve volume.

Remember that ventilation and perfusion work together to maximize the partial pressure gradients across the alveolar membranes (see Figure 48.11D). Ventilation delivers O_2 to the environmental side of the exchange surface, where it diffuses across and is swept away by the perfusing blood, which carries it to the tissues. The reverse is true for CO_2. Perfusion delivers CO_2 to the exchange surface, where it diffuses across and is swept away by ventilation.

▶ 48.3 recap

The mammalian respiratory system consists of a highly branching system of airways that lead to alveoli—the gas exchange surfaces. Respiratory muscles ventilate the alveoli by creating pressure differences between the lungs and the outside air. CO_2 and O_2 are exchanged across thin capillary and alveoli walls by diffusion.

learning outcomes

You should be able to:

- Explain the roles of surface tension and surfactant in breathing.
- Explain how changes in volume and pressure in the thoracic cavity are involved in ventilation.
- Describe the changes (in arbitrary units) in tension between the pleural membranes and the pressure in the alveoli, and air flow during a breathing cycle.

▶ 48.3 recap

1. Why do some premature newborns have difficulty breathing?
2. Why does a puncture wound of the chest cause a collapsed lung?
3. In the breathing cycle, what is the alveolar pressure when the tension between the pleural membranes is maximal? Explain your answer.

Having discussed how respiratory gases get to and from the environmental side of the gas exchange membranes through ventilation, we will now look at how these gases get to and from the internal side of those membranes through perfusion.

▶ key concept
48.4 Respiratory Gases Are Transported by the Blood

Perfusion of the lungs is one of the functions of the circulatory system. The circulatory system uses a pump (the heart) and a network of vessels to transport blood around the body. Circulatory systems are the subject of Chapter 49, so here we will discuss only one aspect of perfusion: how blood transports respiratory gases.

focus your learning

- The protein hemoglobin in red blood cells binds, transports, and releases O_2 to the tissues of the body.
- Factors influencing the affinity of hemoglobin for oxygen include its chemical composition, the P_{O_2} in the blood plasma, the blood pH, and the presence of 2,3-bisphosphoglyceric acid.
- Most of the carbon dioxide produced by body cells is transported from tissues to alveoli in the form of bicarbonate ions dissolved in blood plasma.

The liquid part of blood, the plasma, carries some O_2 in solution, but its ability to transport this nonpolar molecule is limited. The blood plasma of a human can contain in solution about 3 mL of O_2 per liter of plasma, which is inadequate to support even basal metabolism. However, the blood of vertebrates and many invertebrates, contains molecules that bind and release O_2 and thus augment its transport capacity. These molecules pick up O_2 where P_{O_2} is high and release it where P_{O_2} is lower. There are many O_2 transport molecules in the animal kingdom, but in vertebrates this role is played by hemoglobin, a protein contained in red blood cells (RBCs). The percent of the blood volume consisting of RBCs is called the **hematocrit**. The hematocrit of human males averages 45 percent and that of females 40 percent. Hemoglobin increases the capacity of blood to carry 60 times more oxygen than it could carry in solution, making high rates of metabolism possible.

Red blood cells are produced in the bone marrow

In humans, about 25 percent of the cells in the body are RBCs. They are produced in the bone marrow at a rate of about 2 million per second, and they circulate around the body for 100–120 days before they are destroyed. Each circuit around the body takes about a minute when at rest. RBCs are among the smallest cells of the body. They have to pass through the tiniest blood vessels, the capillaries. To be so small yet carry a lot of hemoglobin (at least 25% of their volume), RBCs lose their nuclei and most cell organelles prior to being released into the circulation. Because they have no mitochondria, they depend on *glycolysis for energy.

---

*connect the concepts   In Key Concept 9.1, glycolysis is described as a sequence of reactions that by the sixth step has converted one molecule of glucose to two molecules of 1,3-bisphosphoglycerate and by the tenth step to two molecules of pyruvate. However, some of the 1,3-bisphosphoglycerate can be enzymatically converted to 2,3-bisphosphoglycerate which plays an important role in regulating the oxygen-binding properties of hemoglobin, as you will learn below.

Hemoglobin combines reversibly with O_2

RBCs contain enormous numbers of hemoglobin molecules. **Hemoglobin** is a protein consisting of four polypeptide subunits, each of which surrounds a heme group—an iron-containing ring structure that can reversibly bind a molecule of O_2. Thus each hemoglobin molecule can bind and release up to four O_2 molecules, enabling the blood to carry a large amount of O_2 to the body's tissues.

Hemoglobin's ability to pick up or release O_2 depends on the P_{O_2} in its environment. When the P_{O_2} of the blood plasma is high, as it usually is in the lung capillaries, each hemoglobin molecule can carry its maximum load of four O_2 molecules. As the blood circulates through the rest of the body, it releases some of the O_2 it carries when it encounters lower P_{O_2} values in tissues.

The relationship between P_{O_2} and the amount of O_2 bound to hemoglobin is not linear but S-shaped (sigmoidal). The hemoglobin–oxygen binding/dissociation curve reflects interactions between the four subunits of the hemoglobin molecule. At low P_{O_2} values, only one subunit will bind an O_2 molecule (**Figure 48.12A**). When it does so, the shape of that subunit changes, altering the quaternary structure of the entire hemoglobin molecule. That structural change makes it easier for the other subunits to bind an O_2 molecule; that is, their O_2 affinity is increased. Therefore a smaller increase in P_{O_2} is necessary to get the hemoglobin molecule to bind a second O_2 molecule (that is, to become 50% saturated) than was necessary to get it to bind one O_2 molecule (to become 25% saturated). This change in affinity is reflected in the increased steepness of the O_2-binding curve. The influence of O_2 binding/dissociation by one subunit on the O_2 affinity of the other subunits is called **positive cooperativity**.

Once the third O_2 molecule is bound, the relationship seems to change, as a larger increase in P_{O_2} is required for the hemoglobin to reach 100 percent saturation. This upper bend of the sigmoid curve is due to a probability phenomenon. The closer we get to having all subunits occupied, the less likely it is that any particular O_2 molecule will find a place to bind. Therefore it takes a relatively greater P_{O_2} to achieve 100 percent saturation.

The O_2-binding/dissociation properties of hemoglobin help get O_2 to the tissues that need it most (**Figure 48.12B**). In the lungs, where the P_{O_2} is about 100 mm Hg, hemoglobin is 100 percent

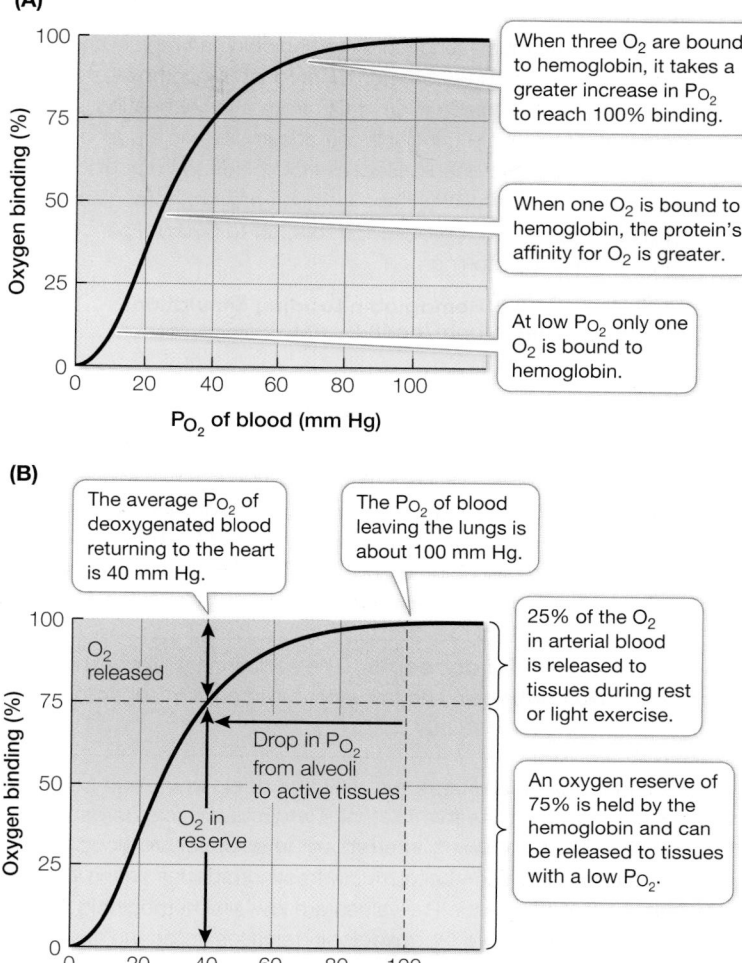

Figure 48.12 Binding of O_2 to Hemoglobin Depends on P_{O_2} **(A)** The sigmoidal shape of the hemoglobin–oxygen binding/dissociation curve reflects positive cooperativity among hemoglobin's subunits. **(B)** Hemoglobin in blood leaving the lungs is 100 percent saturated (four O_2 molecules are bound to each hemoglobin molecule). Most hemoglobin molecules drop only one of their four O_2 molecules as they circulate through the body and are still 75 percent saturated when the blood returns to the lungs. The steep portion of this O_2-binding/dissociation curve comes into play when tissue P_{O_2} falls below 40 mm Hg, which is a normal mixed venous blood P_{O_2}.

saturated. The P_{O_2} in blood returning to the heart from the body (at rest) is usually about 40 mm Hg. You can see that at this P_{O_2} the hemoglobin is still about 75 percent saturated. This means that as the blood circulates around the body, it releases only about one in four of the O_2 molecules it carries. This system seems inefficient, but because the hemoglobin keeps 75 percent of its O_2 in reserve, it can meet the demands of highly active tissues.

When a tissue becomes starved of O_2 and its local P_{O_2} falls below 40 mm Hg, the hemoglobin flowing through that tissue is on the steep portion of its binding/dissociation curve. That means relatively small decreases in P_{O_2} below 40 mm Hg will result in the release of lots of O_2 to the tissue. Thus hemoglobin is very effective in making O_2 available to tissues precisely when and where it is needed most.

The O_2 transport function of hemoglobin can rapidly and tragically be disrupted by a common by-product of incomplete combustion: carbon monoxide (CO). If CO from a faulty heating system, engine exhaust, or burning charcoal accumulates in a closed space, the results can be deadly. Because CO binds to hemoglobin with a 240-fold higher affinity than O_2, it can prevent hemoglobin from transporting O_2. In the United States, up to 500 people die each year from CO poisoning.

 Activity 48.2 Hemoglobin Loading Simulation
www.Life11e.com/ac48.2

Myoglobin holds an O_2 reserve

Muscle cells have their own O_2-binding molecule, **myoglobin**. Myoglobin consists of just one polypeptide chain associated with an iron-containing ring structure that can bind one O_2 molecule. *Myoglobin has a higher affinity for O_2 than hemoglobin does, so it picks up and holds O_2 at P_{O_2} values at which hemoglobin is releasing its bound O_2 (**Figure 48.13**).

---

*connect the concepts   The evolutionary relationships between myoglobin and the various hemoglobin subunits is shown in Figure 23.10.

Myoglobin facilitates the diffusion of O_2 in muscle cells and provides an O_2 reserve for times when metabolic demands are high and blood flow is interrupted. Interruption of blood flow in muscles is common because contracting muscles squeeze blood vessels. When tissue P_{O_2} values are low and hemoglobin can no longer supply more O_2, myoglobin releases its bound O_2. Diving

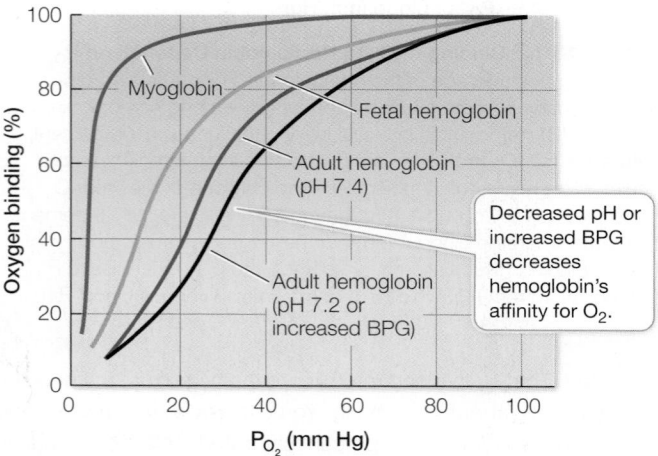

Figure 48.13 Oxygen-Binding Adaptations   Myoglobin and the different hemoglobins have different O_2-binding properties adapted for different circumstances. Fetal hemoglobin, for example, has a higher affinity for O_2 than does adult hemoglobin, facilitating O_2 transfer in the placenta. When high metabolism lowers the pH of the blood, or low O_2 increases bisphosphoglyceric acid (BPG), hemoglobin releases more of its O_2.

Q: How would the breathing of some carbon monoxide affect the hemoglobin–oxygen binding curve?

 **Activity 48.3 Oxygen-Binding Curves**
www.Life11e.com/ac48.3

mammals such as seals have high concentrations of myoglobin in their muscles, which is one reason they can stay under water for so long (**Investigating Life: Seals Are Champion Breath-Hold Divers**). Even in non-diving animals, muscles called on for extended periods of work frequently have more myoglobin than do muscles that are used for short, intermittent periods, as noted in Key Concept 47.2.

Hemoglobin's affinity for O_2 is variable

Various factors influence the O_2-binding/dissociation properties of hemoglobin, thereby influencing O_2 delivery to tissues. Three of these factors are the chemical composition of the hemoglobin, the blood pH, and the presence of 2,3-bisphosphoglycerate (BPG) in RBCs.

HEMOGLOBIN COMPOSITION   There is more than one type of hemoglobin because the chemical composition of the polypeptide chains that form the hemoglobin molecule varies. The normal hemoglobin of adult humans has two each of two kinds of polypeptide chains—two α-globin chains and two β-globin chains. This normal adult hemoglobin has the O_2-binding characteristics shown in Figure 48.13.

Before birth, the human fetus has a different form of hemoglobin, consisting of two α-globin and two γ-globin chains. The functional difference between fetal and adult hemoglobin is that fetal hemoglobin has a higher affinity for O_2. Therefore the fetal hemoglobin–oxygen binding/dissociation curve is shifted to the left compared with the adult curve (see Figure 48.13). You can see from these curves that if both types of hemoglobin are at the same P_{O_2} (as they are in the placenta), fetal hemoglobin will pick up O_2 that the adult hemoglobin releases. This difference in O_2 affinities enables the efficient transfer of O_2 from the mother's blood to the fetus's blood.

HEMOGLOBIN AND pH   The O_2-binding properties of hemoglobin are also influenced by physiological conditions. The influence of pH on the function of hemoglobin is known as the **Bohr effect**. As blood passes through metabolically active tissue such as exercising muscle, it picks up acidic metabolites. As a result, blood pH falls. The excess H^+ ions bind preferentially to deoxygenated hemoglobin and decrease its affinity for O_2, and the O_2-binding/dissociation curve of hemoglobin shifts to the right (see Figure 48.13). This shift means the hemoglobin will release more O_2 in tissues where pH is low—another way that O_2 is supplied where and when it is most needed.

2,3-BISPHOSPHOGLYCERATE   1,3-BPG is an intermediate in glycolysis (see p. 177), and therefore an important source of energy for cells. However, 1,3-BPG can be converted to 2,3-BPG by an enzyme in the RBC. 2,3-BPG is an important regulator of hemoglobin function. Like excess H^+, 2,3-BPG reversibly combines with deoxygenated hemoglobin and lowers its affinity for O_2. The result is that at any P_{O_2}, hemoglobin releases more of its bound O_2 than it otherwise would. In other words, 2,3-BPG shifts the O_2-binding/dissociation curve of mammalian hemoglobin to the right.

When humans go to high altitudes, or when they cease being sedentary and begin to exercise, the level of 2,3-BPG in their RBCs goes up, making it easier for hemoglobin to deliver more O_2 to tissues. During pregnancy, a woman has about a 30 percent increase

experiment

Original Paper: Castellini, M. A., G. L. Kooyman and P. J. Ponganis. 1992. Metabolic rates of freely diving Weddell seals: Correlations with oxygen stores, swim velocity, and diving duration. *Journal of Experimental Biology* 165: 181–194.

Weddell seals live near the edges of the ice in Antarctica where they feed by diving in the surrounding ocean. The ice-covered ocean presents a unique opportunity to study the diving and breathing behaviors of these seals if they have to return to the same breathing hole in the ice.

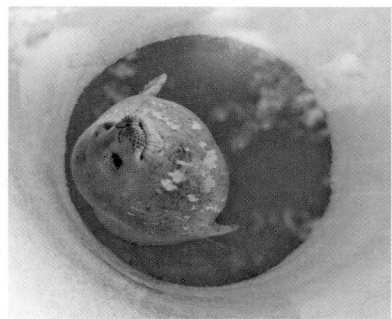

QUESTION▶ How do metabolic requirements of a freely diving seal correlate with its oxygen stores?

METHOD

1. Transport Weddell seals several kilometers away from the ice edge, but still on the extensive Antarctic ice sheet.
2. Fit the seals with distance and depth gauges to calculate the metabolic costs of individual dives.
3. Make a hole in the ice and release the seals so they can return to the sea through the hole.
4. Allow the seals to dive and feed naturally.
5. When seals return to the hole to breathe, measure O_2 consumption and CO_2 release and determine each seal's metabolic rate (MR).

RESULTS

Most dives were only a few minutes long, but many were over 30 minutes and the longest was 82 minutes. MR was also measured when the seals rested and slept. Seals frequently do not breathe during sleep. The longest sleep episode without breathing was 13.5 minutes.

Condition	Mean MR (mL O_2/min kg) (± SD)
Rest	4.1 (0.3)
Sleep	3.5 (0.6)
All dives	4.5 (0.9)
Short dives	5.0 (0.9)
Long dives	3.4 (0.6)

CONCLUSIONS▶

1. There were no significant differences in MR during rest, sleep, and the mean for all dives taken together. The relatively consistent measurements of MR indicate a very low metabolic cost of diving and activities during the dives. Going from resting to diving resulted in only a 10% increase in MR. (In humans, by contrast, going from resting to standing involves a 16% increase in MR.)
2. MRs during long dives were lower than at rest, but MRs during short dives (<15 minutes) were significantly higher than during long dives, indicating the capacity of the seals to conserve O_2 supplies to extend dive duration.

work with the data

How do the seal's oxygen stores compare with its oxygen demands? To calculate the oxygen reserves of a Weddell seal, we have to assess the amount of oxygen in its blood and in the myoglobin in its muscles at the beginning of the dive. The amount of oxygen in its lungs is irrelevant since seals exhale before diving, so there is little oxygen available in its alveoli during the dive.

The problem requires calculating the amount of O_2 in the arterial and venous blood and myoglobin at the beginning of the dive that is available to support the seal's MR.

Assume the following:

- Average body mass: 355 kg
- Blood volume: 14.8% of mass (7% in humans)
- Hematocrit: 58% (human averages 42%)
- Hemoglobin content of red blood cells: 23.7%
- O_2 holding capacity of hemoglobin: 1.34 mL/g
- Blood: 34% is arterial, 95% saturated at beginning of dive, can desaturate to 20%

Available O_2 in arterial blood is calculated as follows:

Total arterial blood = 355 kg × 0.148 × 0.34 = 17.9 kg

Total arterial hemoglobin = 17.9 kg × 0.58 × 0.24 = 2.49 kg

Maximum O_2 content = 3.34 L

Arterial O_2 available = 3.34 × (95% − 20%) = 2.5 L

QUESTIONS▶

1. Calculate the available O_2 in the venous blood. Assume venous blood is 66% of total blood volume, and it is 90% saturated at the beginning of the dive and can desaturate to 0%.
2. Calculate the available O_2 in myoglobin. Assume 33% of body mass is muscle, myoglobin content of seal muscle is 45 g/kg (20–25 g/kg for humans), and O_2 holding capacity of myoglobin is 1.34 mL O_2/g.
3. What are the total O_2 reserves at the beginning of the dive?
4. Given the MR values found in the experiment, what do you predict would be the maximum duration of no breathing during sleep?
5. Using the table of MR values for all dives in the results portion of the experiment, predict the maximum duration of a Weddell seal dive.

A similar **work with the data** exercise may be assigned in **LaunchPad**.

---

in 2,3-BPG, which makes more O_2 available in the placenta to be picked up by the fetal hemoglobin. In addition, fetal hemoglobin has a left-shifted O_2-binding/dissociation curve because its γ-globin chains have a lower affinity for 2,3-BPG than do the β-globin chains of adult hemoglobin.

Most CO_2 is transported as bicarbonate ions in the blood

Delivering O_2 to tissues is only half the respiratory function of blood. Blood also carries CO_2, a metabolic waste product, away from tissues (**Figure 48.14**). CO_2 is highly soluble and readily diffuses

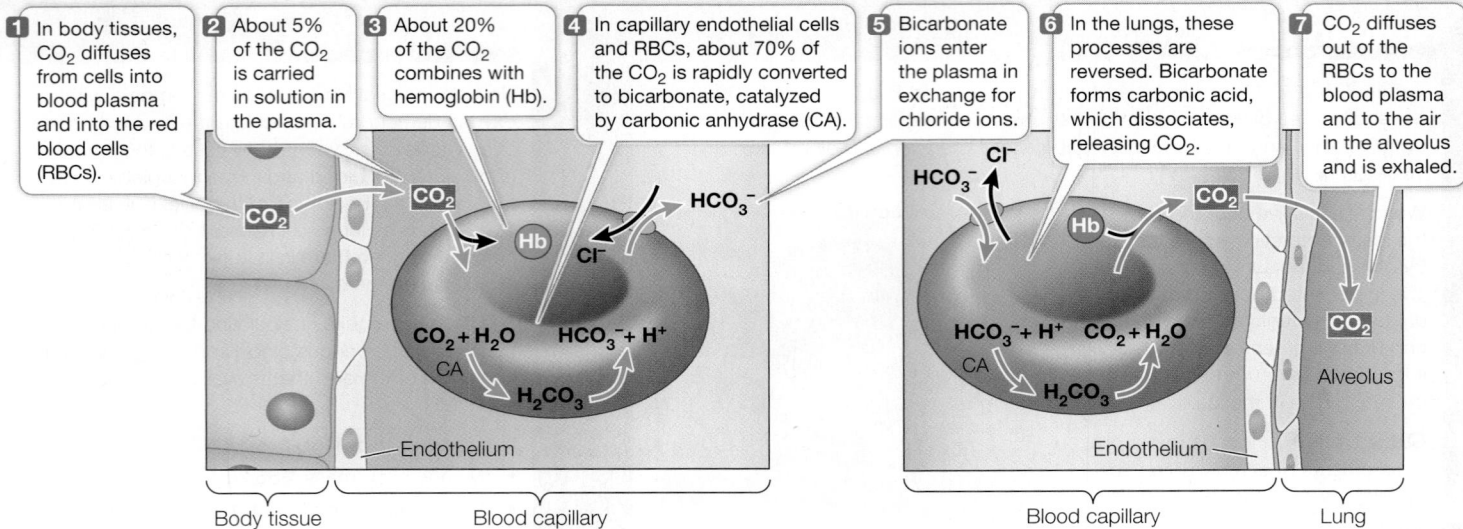

① In body tissues, CO_2 diffuses from cells into blood plasma and into the red blood cells (RBCs).

② About 5% of the CO_2 is carried in solution in the plasma.

③ About 20% of the CO_2 combines with hemoglobin (Hb).

④ In capillary endothelial cells and RBCs, about 70% of the CO_2 is rapidly converted to bicarbonate, catalyzed by carbonic anhydrase (CA).

⑤ Bicarbonate ions enter the plasma in exchange for chloride ions.

⑥ In the lungs, these processes are reversed. Bicarbonate forms carbonic acid, which dissociates, releasing CO_2.

⑦ CO_2 diffuses out of the RBCs to the blood plasma and to the air in the alveolus and is exhaled.

Figure 48.14 Carbon Dioxide Is Transported as Bicarbonate Ions The left panel represents the circulation through the tissues of the body where carbonic anhydrase (CA) in capillary endothelial cells and in RBCs facilitates conversion of CO_2 produced in the tissues to bicarbonate ions in the blood. The right panel shows the process reversing in the lungs as CO_2 diffuses from blood to alveolar air and is exhaled.

through cell membranes, moving from its site of production in the tissues into the blood, where P_{CO_2} is lower. However, only about 5 percent of the CO_2 carried by the blood is dissolved CO_2. About 20 percent of the CO_2 carried to the lungs in the blood is in chemical combination with hemoglobin. And most CO_2 produced by the tissues is transported to the lungs in the form of bicarbonate ions (HCO_3^-). CO_2 diffusing from cells is converted to HCO_3^-, transported to the lungs, and then converted back to CO_2 to be exhaled.

When CO_2 dissolves in water, some of it slowly reacts with the water molecules to form carbonic acid (H_2CO_3), some of which then dissociates into a proton (H^+) and a bicarbonate ion (HCO_3^-). This reversible reaction is expressed as follows:

$$CO_2 + H_2O \rightleftharpoons H_2CO_3 \rightleftharpoons H^+ + HCO_3^-$$

In the extracellular fluid, the reaction between CO_2 and H_2O proceeds slowly. But it is a different story in the endothelial cells of the capillaries and in the RBCs, where the enzyme **carbonic anhydrase** speeds up the conversion of CO_2 to H_2CO_3. The newly formed H_2CO_3 dissociates, and the resulting bicarbonate ions enter the plasma in exchange for Cl^- (see Figure 48.14). By converting CO_2 to H_2CO_3, carbonic anhydrase reduces the P_{CO_2} in these cells and in the plasma, facilitating the diffusion of CO_2 from tissue cells to endothelial cells, plasma, and RBCs.

In the lungs, the reactions involving CO_2 and bicarbonate ions are reversed. Remember that an enzyme such as carbonic anhydrase only speeds up a reversible reaction; it does not determine its direction. The direction is determined by concentrations of reactants and products. Ventilation keeps the P_{CO_2} in the alveoli low, so CO_2 diffuses from the blood plasma into the alveoli, lowering the P_{CO_2} in the blood, which favors the conversion of HCO_3^- into CO_2. Also, the oxygenation of hemoglobin in the alveoli decreases its affinity for CO_2, a phenomenon known as the Haldane effect.

48.4 recap

O_2 is transported from the lungs to the body's tissues in reversible combination with hemoglobin. Each hemoglobin molecule can reversibly combine with four O_2 molecules; the percent saturation of the binding sites is a function of the P_{O_2} in the hemoglobin's environment. The hemoglobin–oxygen binding/dissociation curve is shifted by the concentration of H^+ and the level of 2,3-BPG in the blood. CO_2 is transported by the blood mostly in the form of HCO_3^-.

learning outcomes

You should be able to:

• Explain the adaptive advantage of hemoglobin remaining 75 percent saturated with O_2 in mixed venous blood.

• Explain how 2,3-BPG and pH alter the O_2 binding properties of hemoglobin, and discuss the functional significance of these changes.

• Explain how the same enzyme, carbonic anhydrase, catalyzes both the loading of CO_2 from cells to the blood and the off-loading of CO_2 from the blood to the alveoli.

1. The P_{O_2} of atmospheric air at sea level is about 159 mm Hg. What is the functional significance of the fact that human hemoglobin fully saturates with O_2 at a P_{O_2} of 100 mm Hg and returns to the heart in venous blood with a P_{O_2} of about 40 mm Hg, reflecting 75 percent saturation?

2. When red blood cells are stored in the blood bank, they will eventually use their 1,3-BPG for energy. What will be the consequence for their ability to function after transfusion into a patient?

3. Why do HCO_3^- ions leave the red blood cells in systemic venous blood but enter red blood cells in the alveolar circulation?

We must breathe every minute of our lives, but most of us usually don't worry about it, or even think about it very often. In the next section we will examine how the regular breathing cycle is generated and controlled by the central nervous system.

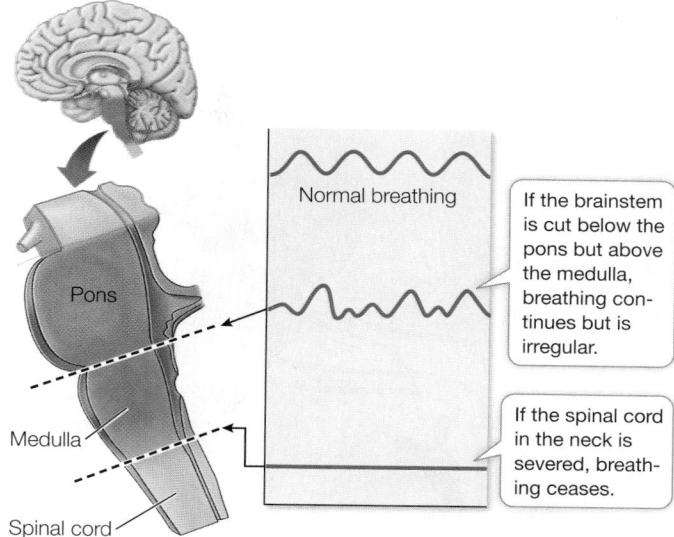

key concept

48.5

Breathing Is Homeostatically Regulated

Breathing is an involuntary function of the central nervous system (CNS). The breathing pattern is adjusted around other activities (such as speech and eating), and breathing rates change to match the metabolic demands of the body. How is this accomplished?

focus your learning

- The breathing pattern of inhalation followed by exhalation is controlled by a neural circuit in the brainstem.
- Breathing is regulated by feedback mechanisms involving CO_2 sensitive cells in the brain and O_2 sensitive cells in the great arteries.
- Brain chemosensitive cells sense CO_2 levels through changes in hydrogen ion concentration, whereas nodes in the aorta and the carotid arteries are sensitive to blood O_2 levels.

The basic breathing rhythm is an involuntary function that is modified by feedback information that conveys the body's need for an adequate O_2 supply and CO_2 elimination.

Breathing is controlled in the brainstem

Breathing is driven by a rhythm-generating neural circuit in the brainstem, but where? Breathing ceases if the spinal cord is severed in the neck region, showing that breathing is generated in the brain. If the brainstem is cut just above the pons, breathing is normal. If the brainstem is cut between the pons and the medulla (the segment of the brainstem just above the spinal cord), the breathing pattern becomes irregular (**Figure 48.15**). Electrophysiological recordings reveal a group of respiratory motor neurons in the dorsal medulla that increase their firing rates just before an inhalation begins. The axons of these neurons leave the CNS in the neck region to form the phrenic nerve, which innervates the diaphragm. As more and more of these neurons fire—and fire faster and faster—the diaphragm contracts. All of a sudden the neurons stop firing, the diaphragm relaxes, and exhalation begins. Exhalation is usually a passive process that depends on the elastic recoil of the lung tissues. Another group of respiratory motor neurons is found in the ventral medulla that becomes active when breathing demand is high. These motor neurons communicate through thoracic spinal nerves to the intercostal muscles. By expanding and contracting the rib cage, the intercostal muscles increase both the inhalation and the exhalation volumes.

Neurons in the lower region of the pons help regularize the basic respiratory rhythm. Still higher brain areas modify breathing to accommodate speech, ingestion of food, coughing, and emotional states.

Regulating breathing requires feedback

When breathing or metabolism changes, the P_{O_2} and P_{CO_2} in the blood change as well. It is reasonable to surmise that the blood levels of one or both of these gases provide feedback information to the breathing rhythm generator in the medulla. Humans and other mammals are remarkably insensitive to falling levels of O_2 in arterial

Figure 48.15 Breathing Is Controlled in the Brainstem   Basic breathing rhythm is generated in the medulla and is modified by neurons in or above the pons.

▶ Media Clip 48.1 **Sea Bed Hunting on One Breath**
www.Life11e.com/mc48.1

blood but are extremely sensitive to increases in CO_2. That is, arterial P_{O_2} can deviate considerably from normal without causing much of an increase in ventilation rate, but even a small rise in arterial P_{CO_2} causes a large increase in ventilation (**Figure 48.16**). This relationship

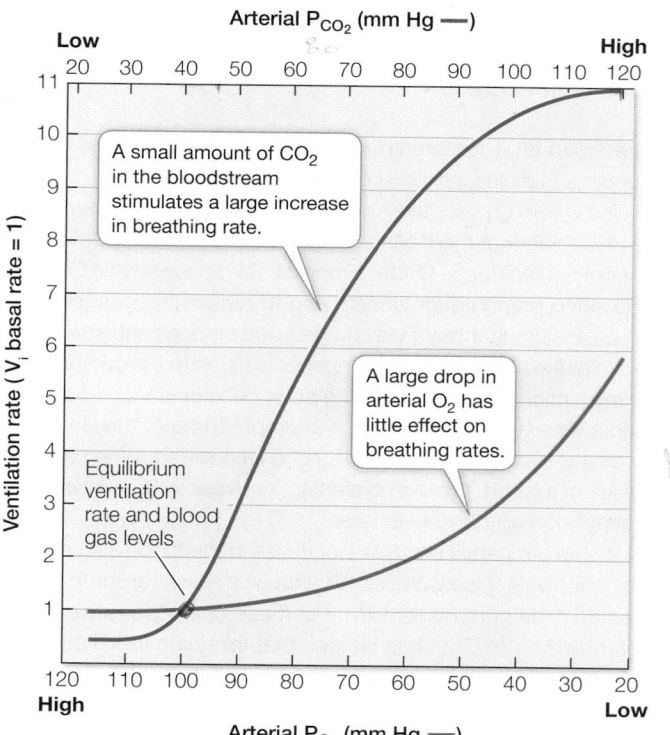

Figure 48.16 Carbon Dioxide Affects Breathing Rate
The breathing mechanism is more sensitive to increased levels of CO_2 in arterial blood than to decreased amounts of O_2.

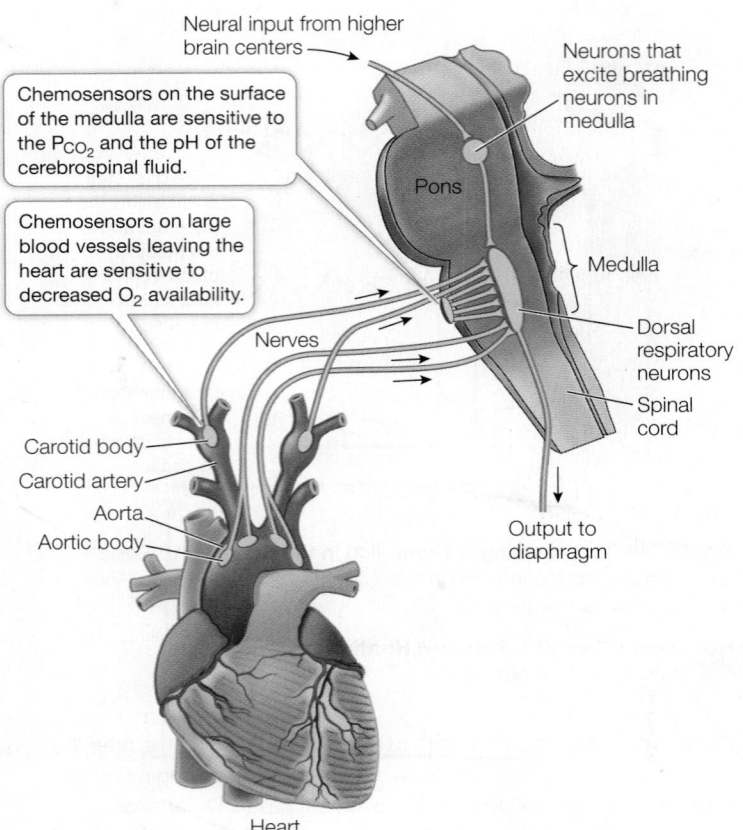

Neural input from higher brain centers

Chemosensors on the surface of the medulla are sensitive to the P_{CO_2} and the pH of the cerebrospinal fluid.

Chemosensors on large blood vessels leaving the heart are sensitive to decreased O_2 availability.

Neurons that excite breathing neurons in medulla

Pons

Medulla

Nerves

Dorsal respiratory neurons

Spinal cord

Carotid body
Carotid artery
Aorta
Aortic body

Output to diaphragm

Heart

Figure 48.17 Feedback Information Controls Breathing
The body uses feedback information from chemosensors in the great arteries leaving the heart and on the ventral surface of the medullary brainstem to match breathing rate to metabolic demand.

is reversed for water-breathing animals, in which O_2 is the primary feedback stimulus for gill ventilation.

The lack of O_2 sensitivity in humans was tragically demonstrated by an accident at the Kennedy Space Center in 1981 prior to the first flight of the Space Shuttle Program. The space shuttle *Columbia* was being prepared for launch, and to reduce the danger of fire, the engine compartment was purged with nitrogen and sealed. Unaware of the nitrogen, workers opened a hatch to the compartment for inspection. The first worker to enter became unconscious, and others followed him in a rescue attempt. Three of these workers died, and two others who were stricken recovered. When exposed to lack of oxygen, there is no alerting, emergency response—there is simply loss of consciousness.

Where are partial pressures of gases in the blood sensed? The major site of P_{CO_2} sensitivity is an area on the ventral surface of the medulla. The primary sensitivity of these chemosensitive cells is not, however, to CO_2. It is H^+ ions that stimulate these cells. The H^+ ion concentration in the environment of these cells is a direct reflection of the P_{CO_2} of the blood. When the P_{CO_2} of the blood is higher than that of the extracellular fluid in this area, CO_2 diffuses

out of the blood. This CO_2 interacts with H_2O in the plasma to form carbonic acid (H_2CO_3), which dissociates into H^+ ions and HCO_3^- ions (see Figure 48.14, which describes this same reaction in the RBCs). It is the H^+ ions produced by the dissociation of carbonic acid that stimulate the chemosensitive cells that increase respiratory gas exchange. Thus even though we measure blood P_{CO_2} as the stimulus for breathing, the major stimulus is an increase in H^+ ion concentration (lowered pH).

We do have some physiological sensitivity to blood P_{O_2} in nodes of neural tissue on the large blood vessels leaving the heart: the aorta and the carotid arteries (**Figure 48.17**). These **carotid bodies** and **aortic bodies** are chemosensors. If the blood supply to these structures decreases, or if the blood P_{O_2} falls dramatically, the chemosensors are activated and send nerve impulses to the breathing control center. Although we are not very sensitive to changes in blood P_{O_2}, the carotid and aortic bodies can stimulate increases in breathing during exposure to high altitudes or when blood volume or blood pressure is very low. Also, there is a synergism between CO_2 and O_2 sensing. When blood P_{CO_2} increases, there is an increased sensitivity to low O_2, and vice versa.

48.5 recap

The rhythmic contractions of the respiratory muscles that drive breathing are generated by neurons in the brainstem. Breathing is regulated predominantly in response to the P_{CO_2} of the blood. Respiratory chemosensors are cells on the ventral surface of the medulla that are sensitive to the blood P_{CO_2}, and the carotid and aortic bodies on the large arteries leaving the heart that are sensitive to the blood P_{O_2} or to a decrease in the blood they are receiving.

learning outcomes
You should be able to:
- Explain the roles of the dorsal and ventral groups of respiratory neurons in the medulla.
- Apply knowledge about the physiological processes underlying breathing to explain the body's response to intense exercise and to decreases in P_{O_2}.
- Describe where chemosensitive cells are located and how they participate in negative feedback control of breathing.

1. In terms of respiratory motor neurons and respiratory muscle groups, what changes occur in transitioning from breathing at rest to breathing that uses more of the lung's vital capacity?
2. When you go rapidly to high altitude, you may experience a breathing pattern in which periods of no breathing alternate with periods of fast breathing. Explain that response in terms of diffusion gradients for O_2 and CO_2 and the feedback signals regulating breathing.
3. Cells in the carotid and aortic bodies have high metabolic rates, and they are sensitive either to a decrease in blood flow or to a decrease in the P_{O_2} of the blood. How are these two facts related?

investigating**life**

What adaptations make seals champion breath-hold divers?

Seals have remarkable abilities to dive deep and long and to spend only minutes at the surface to breathe between dives. In the chapter opener about elephant seals, we posed several questions that apply to all seals: Why aren't seals crushed by the water pressure at extreme depths? Why don't they get the bends? What are their O_2 demands? And what are their O_2 reserves? The answers to the first two questions are related. Only air-filled structures of the body are compressible, and the seals exhale at the beginning of a dive. Their residual volume is restricted to their airways, where gas exchanges do not occur. Thus the high pressure of the small amount of gas in their respiratory system does not drive dissolved nitrogen into their tissues. Investigating Life: Seals Are Champion Breath-Hold Divers presented metabolic rate (MR) measurements for Weddell seals resting, sleeping, and diving. No significant differences in MR for various activities were observed, leading to the conclusion that the metabolic costs of diving are quite modest. In fact, MRs during dives were only about 10 percent higher than at rest—not much different than for humans going from sitting to standing.

How is seal activity supported by such a low metabolic expenditure? And given those low MRs, how do seals maintain O_2 reserves sufficient to support them during dives? The calculations made in the work with the data exercise in Investigating Life: Seals Are Champion Breath-Hold Divers showed that the O_2 reserve of a sleeping seal would permit it to not breathe for a maximum of 12.8 minutes, which corresponds well with the maximum episode of sleep observed—13.5 minutes. Using the MR value for all dives, calculations show that the seals' O_2 reserves would only support dives that were under 10 minutes in duration. This means that additional adaptations enable the seal to remain active underwater for long periods at an average MR similar to what they experience during sleep.

Future directions

Recording devices attached to seals diving in nature reveal adaptations that minimize their O_2 demands during a dive. Because they exhale at the beginning of a dive, they are negatively buoyant, meaning they can drift down to depths like a falling leaf. These recordings also answered another question. Since the seals at sea dive continually day and night, when do they sleep? They spend much of their passive descent time sleeping, thus lowering their MR even more. Yet more metabolic savings are needed to enable long dives. One clue comes from measures of the seals' heart rates during dives. Their heart rates are very much reduced, a phenomenon called diving bradycardia. We will discuss diving bradycardia in the next chapter and see how it relates to seals' ability to drastically lower their MR while diving.

Chapter Summary 48

48.1 Respiratory Gas Exchange Is Governed by Physical Factors

- Most cells require a constant supply of O_2 and continuous removal of CO_2. These **respiratory gases** are exchanged between an animal's body fluids and its environment by diffusion.
- **Fick's law of diffusion** shows how various physical factors influence the diffusion rate of gases. Adaptations to maximize respiratory gas exchange influence one or more variables of Fick's law.
- In water-breathing animals, gas exchange is limited by the low diffusion rate and low amount of O_2 in water. If water temperature rises, water-breathing animals face a double bind: while O_2 in water decreases, the amount of work required to move water over the gas exchange surfaces increases. **Review Figure 48.2**
- In air, O_2 **partial pressure** (P_{O_2}) decreases with increasing altitude.

48.2 Enhancing Diffusion Maximizes Respiratory Gas Exchange

- Adaptations to maximize gas exchange include increasing the surface area for gas exchange and maximizing partial pressure gradients across those exchange surfaces. Partial pressure gradients are established both by ventilating the outer surface with the respiratory medium, and perfusing the inner surface with blood.
- **External** and **internal gills** are adaptations for gas exchange with water. **Lungs** and **tracheae** are adaptations for gas exchange with air. **Review Figures 48.3, 48.4**
- The gills of fish have large gas exchange surface areas that are ventilated continuously and unidirectionally with water. The **countercurrent flow** of blood helps increase the efficiency of gas exchange. **Review Figures 48.5, 48.6**

- The gas exchange system of birds includes **air sacs** that communicate with the lungs but are not used for gas exchange. Air flows unidirectionally through bird lungs; gases are exchanged in air capillaries that run between **parabronchi**. **Review Figure 48.7**
- Each breath of air remains in a bird's respiratory system for two breathing cycles. The air sacs work as bellows to supply the air capillaries with a continuous unidirectional flow of fresh air. **Review Figure 48.8, Animation 48.1**
- In all air-breathing vertebrates except birds, breathing is **tidal**. Although the volume of air exchanged with each breath can vary considerably in tidal breathing, the inhaled air is always mixed with stale air. **Review Figure 48.9**

48.3 Humans Have Tidal Respiration

- In mammalian lungs, the gas exchange surface area provided by the millions of **alveoli** is enormous, and the diffusion path length between the air and perfusing blood is short. **Surface tension** in the alveoli would make inflation of the lungs difficult if the alveoli did not produce **surfactant**. **Review Focus: Key Figure 48.10, Activity 48.1**
- Inhalation occurs when contractions of the **diaphragm** increase volume and reduce pressure in the **thoracic cavity**, thereby pulling on the **pleural membranes**. Relaxation of the diaphragm increases pressure in the thoracic cavity and results in exhalation. **Review Figure 48.11, Animation 48.2**
- During periods of heavy metabolic demands such as strenuous exercise, the **intercostal muscles**, located between the ribs, increase the volume of air inhaled and exhaled.

(continued)

Chapter Summary 48 (continued)

▶ 48.4 Respiratory Gases Are Transported by the Blood

- O_2 is reversibly bound to **hemoglobin** in red blood cells. Each hemoglobin molecule can carry a maximum of four O_2 molecules. Because of **positive cooperativity**, hemoglobin's affinity for O_2 depends on the P_{O_2} to which the hemoglobin is exposed. Therefore hemoglobin picks up O_2 as it flows through respiratory exchange structures and gives up O_2 in metabolically active tissues. **Review Figure 48.12, Activity 48.2**

- **Myoglobin** serves as an O_2 reserve in muscle.

- There is more than one type of hemoglobin. Fetal hemoglobin has a higher affinity for O_2 than does adult hemoglobin, allowing fetal blood to pick up O_2 from the maternal blood in the placenta. **Review Figure 48.13, Activity 48.3**

- CO_2 is transported in the blood principally as bicarbonate ions (HCO_3^-). **Review Figure 48.14**

▶ 48.5 Breathing Is Homeostatically Regulated

- The basic breathing rhythm is an involuntary function generated by neurons in the medulla and modulated by higher brain centers. The most important feedback stimulus for breathing is the level of CO_2 in the blood. **Review Figures 48.15, 48.16**

- The breathing rhythm is sensitive to feedback from chemosensors on the ventral surface of the medulla and in the **carotid** and **aortic bodies** on the large vessels leaving the heart. **Review Figure 48.17**

See **Activity 48.4** for a concept review of this chapter.

Go to **LearningCurve** (in **LaunchPad**) for dynamic quizzing that helps you solidify your understanding of this chapter. **LearningCurve** adapts to your responses, giving you the practice you need to master each key concept.

▶ Apply What You've Learned

Review

48.1 Respiratory gases diffuse from areas of higher partial pressure to areas of lower partial pressure.

48.1 Rising temperature decreases oxygen solubility in water and increases the metabolism of aquatic ectotherms.

48.2 In insects, efficient gas exchange is facilitated by an internal network of air passages.

Original Paper: Hetz, S. K. and T. J. Bradley. 2005. Insects breathe discontinuously to avoid oxygen toxicity. *Nature* 433: 516–519.

Insects' air-delivery system consists of tracheae that branch throughout the body, facilitating the exchange of O_2 and CO_2 in and out of tissues. The finest of these tubes are less than a micrometer in diameter. Tracheal openings on the body's surface are guarded by valve-like spiracles that can be opened, closed, or "fluttered."

To study tracheal function, microscopic plastic tubes were surgically inserted into the tracheae of the pupal stages of *Attacus atlas* moths. The microtubes were attached to sensors that recorded the rate of CO_2 released from the tracheae and the intratracheal concentrations of O_2. Pupae were placed in chambers with normal atmospheric levels of gases, and the spiracle behavior, rate of CO_2 release, and O_2 concentrations in the tracheae were recorded (**Figure A**). Kilopascals (kPa) was the measure of gas pressure.

To test how the spiracles performed in differing atmospheric conditions and what effects this might have on the rate of CO_2 release and O_2 concentrations, researchers exposed pupae to differing levels of atmospheric O_2 (labeled red line in graphs). They then measured levels of CO_2 release and O_2 concentrations during the fluttering stage (**Figure B**).

Questions

1. If 1 kPa = 7.5 mm Hg, calculate the change in P_{O_2} in units of mm Hg in a typical open–close–fluttering cycle. What is the approximate mm Hg P_{O_2} in the atmosphere and inside the tracheal tubes in each of the three experiments shown in Figure B?

Figure A

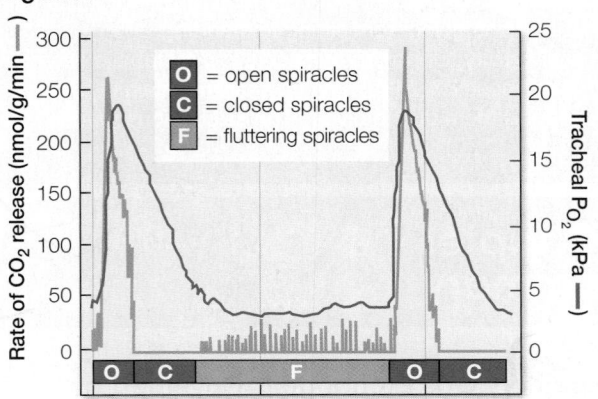

Figure B

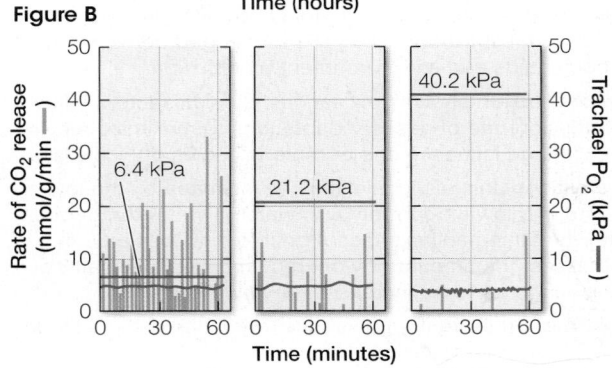

2. Based on your understanding of the purpose of insect tracheae, explain why the P_{O_2} rises to such a high level inside the tracheae when the spiracle is open, and then decreases over time. What effect does the fluttering of spiracles have on P_{O_2}?

3. One function of the fluttering of spiracles might be to prevent the P_{O_2} from reaching a toxic level. Based on the data in Figure B, is this a viable hypothesis? Explain your answer.

Go to **LaunchPad** for the eBook, LearningCurve, animations, activities, flashcards, and additional resources and assignments.

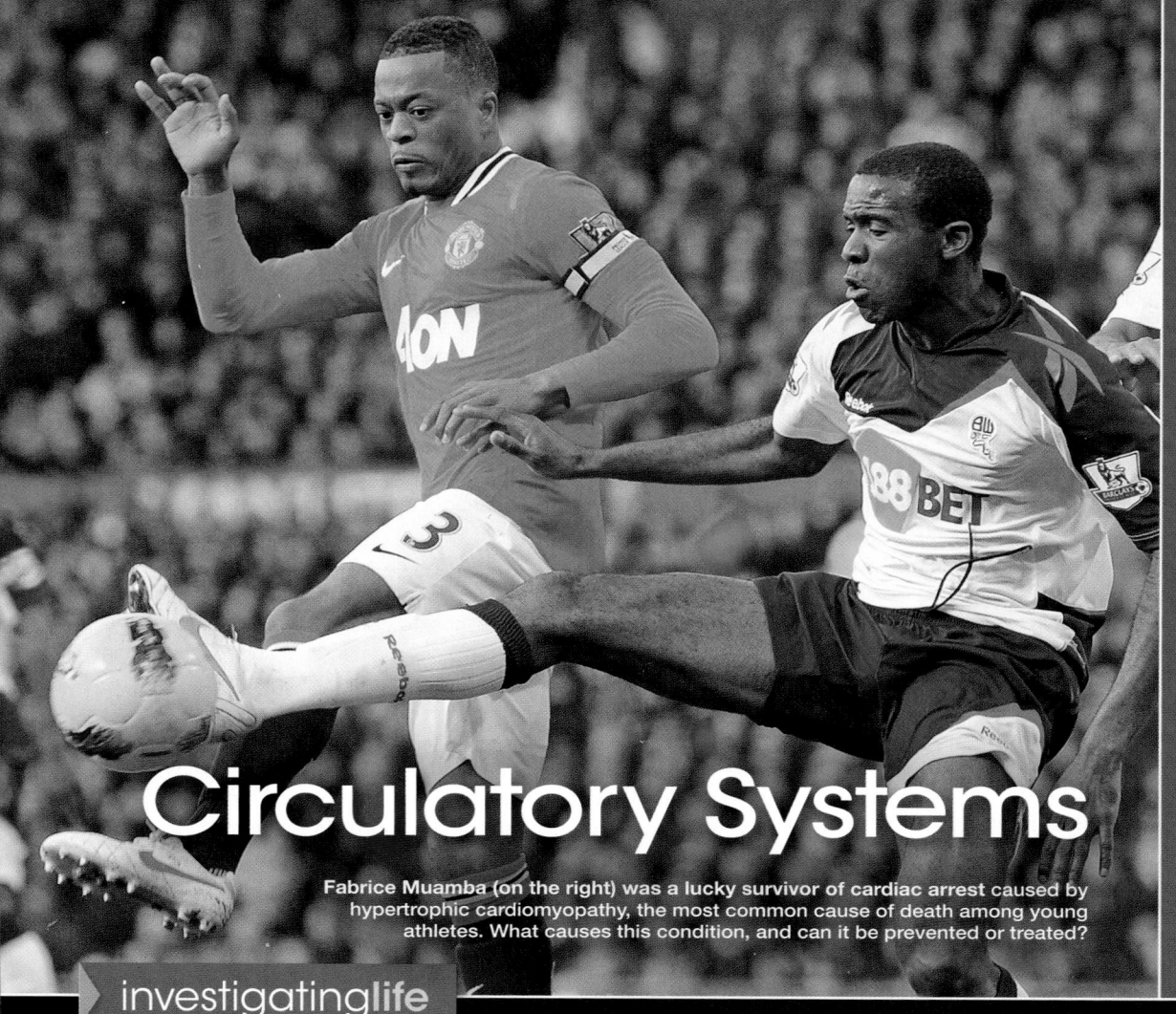

49

Circulatory Systems

Fabrice Muamba (on the right) was a lucky survivor of cardiac arrest caused by hypertrophic cardiomyopathy, the most common cause of death among young athletes. What causes this condition, and can it be prevented or treated?

investigatinglife

Athletes with Big Hearts

Fabrice Muamba was a much-loved and admired professional soccer player. On March 17, 2012, during a championship quarterfinal match, 23-year-old Fabrice suddenly collapsed on the field. He was immediately attended to by team physicians and a cardiologist who happened to be a spectator at the match. Fabrice received numerous defibrillator shocks, cardiopulmonary resuscitation, and was rushed to a hospital, but still his heart stopped for 78 minutes. Remarkably he survived, but there is a long list of young, fit athletes who have died from cardiac arrest during physical exertion. Sudden cardiac arrest is the most common cause of death for athletes. In most cases the cause is not a heart attack, but a condition known as hypertrophic cardiomyopathy (HCM).

Your heart is a muscular pump. At rest it beats an average of 60–70 times per minute. With each beat, it circulates about 70 milliliters of blood through your body. Without adding the increases due to work or exercise, that is 300 liters per hour, 7,200 liters per day, 2.6 million liters per year—no time-outs.

Heart failure is the leading cause of death in the United States, accounting for some one-fourth (about 600,000) of the deaths each year. Heart failure most commonly results from blockage of the blood vessels that supply the heart muscle with blood, and its risk increases with age. The most common cause of heart failure in athletes is not blocked vessels but gene mutations that affect the contractile proteins of the heart. About 1,000 such mutations have been identified, and they exist in about 1 in 500 people. Most people with one of these mutations live normally, so why are they so dangerous in athletes? The mutations decrease the efficiency of the contractions of the heart muscle, so the heart has to work much harder. Especially in athletes, the heart gets bigger to compensate for its inefficiency. The walls of the heart can get so thick that they impede the blood from leaving the heart, and the thick walls can interfere with the signals that coordinate the contractions of the heart muscle. The result is cardiac arrest. Assessment of HCM risk is improving, and many afflicted athletes are choosing to retire rather than taking a chance of dying from cardiac arrest.

QA Are there potential therapies that could treat HCM?

key concept

49.1 Circulatory Systems Serve Many Functions

A **circulatory system** consists of a muscular pump (the heart), a fluid (blood), and a series of conduits (blood vessels) through which the fluid can be pumped throughout the body. Heart, blood, and vessels are also known collectively as a **cardiovascular system** (Greek *kardia*, "heart," + Latin *vasculum*, "vessel"). The function of a circulatory system is to transport things around the body. Preceding chapters discussed how circulatory systems transport heat, hormones, respiratory gases, blood cells, platelets, and cells and molecules of the immune system. Succeeding chapters will add nutrients and waste products to that list. In this section we will describe the general types of circulatory systems found in animals.

focus your learning

- A circulatory system functions to transport materials around the body.
- Small or thin aquatic organisms may not have circulatory systems, and the needs of their cells are served by direct exchange with the environment.
- Circulatory systems use a muscular heart to distribute extracellular fluid throughout the body to transport nutrients, respiratory gases, wastes, and other materials.
- Closed circulatory systems allow more rapid flow and differential control of blood flow to different tissues.

Some animals do not have a circulatory system

Single-celled organisms serve all of their needs through direct exchanges with the environment. Such organisms are found mostly in aquatic or very moist terrestrial environments. Similarly, many multicellular aquatic organisms are small or thin enough that all of their cells are close to the external environment. Such species may not have a circulatory system because nutrients, respiratory gases, and wastes can diffuse directly between the cells of their bodies and the environment.

The cells of some larger aquatic multicellular animals without a circulatory system are served by highly branched central cavities called **gastrovascular systems** that bring the external environment into the animal. All the cells of a sponge are in contact with, or very close to, the water that surrounds the animal and circulates through its central cavity (see Figure 48.1B). Very small animals without a circulatory system can maintain high levels of metabolic activity, but larger animals without a circulatory system such as sponges, jellyfishes, and flatworms tend to be inactive, slow, or even sedentary. Large, active animals require circulatory systems.

Circulatory systems can be open or closed

The cells of large, mobile animals are supported by extracellular fluid. All requirements of cells—oxygen, fuel, nutrients, essential molecules—come from that fluid, and the waste products of cell metabolism go into it. Circulatory systems have muscular chambers, or **hearts**, that move the extracellular fluid through the body. In open circulatory systems, extracellular fluid is the same as the fluid in the circulatory system and is called **hemolymph**. This fluid leaves the vessels of the circulatory system, percolates between cells and through tissues, and then flows back into the heart or vessels of the circulatory system to be pumped out again. In contrast, closed circulatory systems completely contain the circulating fluid (blood) in a continuous system of vessels. Blood cells and large molecules stay within the system, but water and low-molecular-weight solutes leak out of the smallest vessels, the capillaries, which are highly permeable.

In animals with a closed circulatory system, **extracellular fluid** refers to both the fluid in the circulatory system and the fluid outside it. The fluid in the circulatory system is the blood plasma; the *extracellular fluid outside the circulatory system is the **interstitial fluid**. A 70-kilogram person has a total extracellular fluid volume of about 14 liters. Less than a quarter of it—about 3 liters—is the blood plasma.

---

*connect the concepts   Key Concept 39.2 and Figure 39.4 describe the different extracellular fluid compartments that constitute the internal environment that supports all of the metabolic needs of the cells of the body.

Open circulatory systems move extracellular fluid

Open circulatory systems are found in arthropods, mollusks, and some other invertebrate groups. In these systems a heart moves the hemolymph through vessels leading to different regions of the body. The fluid leaves the vessels to filter through the tissues before returning to the heart. In the arthropod shown in **Figure 49.1A**, the fluid returns directly to the heart through openings called ostia. Ostia have valves that allow hemolymph to enter the relaxed heart but prevent it from flowing in the reverse direction when the heart contracts. In a mollusk such as a clam, open vessels collect hemolymph from different regions of the body and return it to the heart (**Figure 49.1B**).

Lest you think that open circulatory systems are inefficient and can support only sluggish lifestyles such as those of mollusks, remember that crabs scuttling along the beach, yellow jackets buzzing around your picnic, and scorpions dashing across the desert all have open circulatory systems.

Closed circulatory systems circulate blood through a system of blood vessels

In **closed circulatory systems**, blood vessels separate blood from interstitial fluid. Blood is pumped through this vascular system by one or more hearts, and cellular and macromolecular components of the blood never leave the vessels. Closed circulatory systems characterize vertebrates and some invertebrate groups, among them annelids.

A simple example of a closed circulatory system is that of the earthworm (**Figure 49.1C**). One large ventral blood vessel carries blood from the worm's anterior end to its posterior end. Smaller vessels branch off and transport the blood to even smaller vessels serving the tissues in each body segment. In the smallest vessels, respiratory gases, nutrients, and metabolic wastes diffuse between the blood and interstitial fluid. The blood then flows from these vessels into larger vessels that lead into one large dorsal vessel that carries the blood from the posterior to the anterior end of the body. Five pairs of muscular vessels connect the large dorsal and ventral vessels in the anterior end, thus completing the circuit.

(A) Arthropod circulation

A muscular, tubular heart pumps hemolymph to different body regions. The hemolymph percolates through tissues and re-enters the heart through openings called ostia.

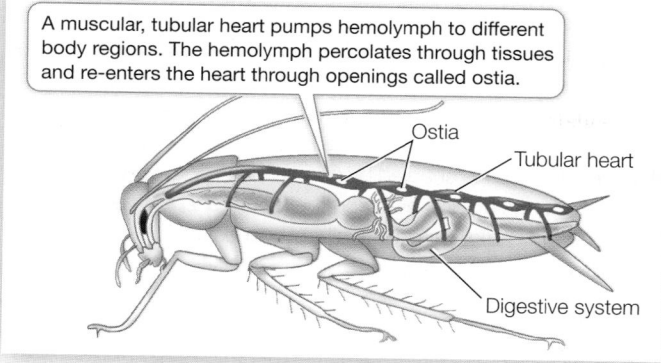

Ostia

Tubular heart

Digestive system

(B) Mollusk circulation

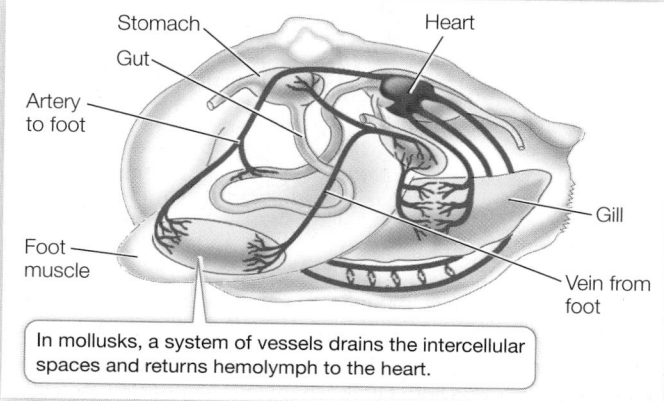

Stomach

Gut

Heart

Artery to foot

Foot muscle

Gill

Vein from foot

In mollusks, a system of vessels drains the intercellular spaces and returns hemolymph to the heart.

(C) Annelid worm circulation

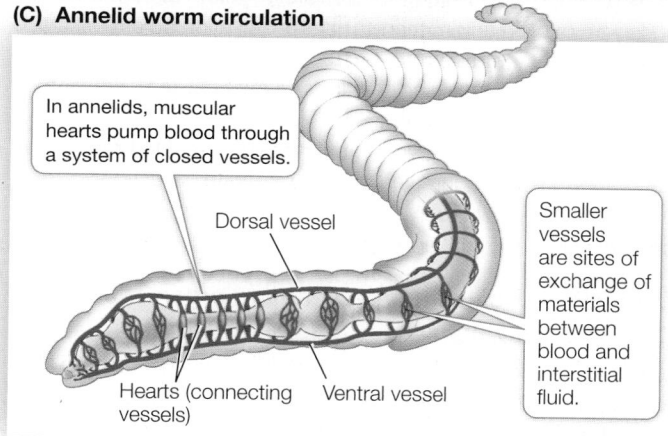

In annelids, muscular hearts pump blood through a system of closed vessels.

Dorsal vessel

Smaller vessels are sites of exchange of materials between blood and interstitial fluid.

Hearts (connecting vessels)

Ventral vessel

Figure 49.1 Circulatory Systems Arthropods, illustrated here by an insect (A), and mollusks such as clams (B) have an open circulatory system. Hemolymph is pumped by a tubular heart and directed to different regions of the body through vessels that open into intercellular spaces. (C) Annelids such as earthworms have a closed circulatory system, in which the cellular and macromolecular elements of the blood are confined in a system of vessels, and the blood is pumped through those vessels by one or more muscular hearts. Earthworms exchange respiratory gases across their skin. The circulation distributes those gases between the skin and the deeper tissues.

The dorsal vessel and the five connecting vessels serve as hearts for the earthworm; their contractions keep the blood circulating. The direction of circulation is determined by one-way valves in the dorsal vessel.

Closed circulatory systems have several advantages compared with open systems:

- Fluid can flow more rapidly through vessels than through intercellular spaces and can therefore transport things to and from tissues more rapidly.
- By changing the diameter (and hence the resistance) of specific vessels, closed systems can control the flow of blood to selective tissues and organs to match their needs.
- Specialized cells and large molecules that aid in transporting hormones, nutrients, and oxygen can be kept in the vessels but can drop their cargo in the tissues where it is needed.

With all of these "advantages" of closed circulatory systems, you might wonder how some species with open circulatory systems can sustain such high levels of activity. In the case of insects, the answer is clear: insects do not depend on their circulatory systems for respiratory gas exchange (see Figure 48.4).

49.1 recap

Circulatory systems consist of a pump and an open or closed set of vessels through which a fluid transports oxygen, nutrients, wastes, and a variety of other substances. Flow to and from tissues in closed systems of vessels can be more rapid and can be directed to specific tissues according to their needs.

learning outcomes

You should be able to:

- Describe the diverse functions of circulatory systems.
- Explain why some animals can function without a circulatory system.
- Describe the differences among hemolymph, blood plasma, and interstitial fluid.
- Identify features that make a closed circulatory system more efficient than an open circulatory system.

1. Describe three ways that circulatory systems support exercising muscles.
2. A sponge is neither small nor thin. How can it survive without a circulatory system?
3. Explain the importance of extracellular fluid in large mobile animals.
4. How does a closed circulatory system facilitate the fight-or-flight response?

Our overview of the open and closed systems found among invertebrates introduced some basic concepts about circulatory systems. Next we will turn to describing the closed circulatory systems of vertebrates.

key concept
49.2 Vertebrate Circulatory Systems Evolved from Single to Double Circuits

Vertebrates have a closed circulatory system and a heart with two or more chambers. When a heart chamber contracts, it squeezes the blood, putting it under pressure. Blood then flows out of the heart and into vessels, where pressure is lower. Resistance to flow in the vessels dissipates the pressure imparted to the blood by the heart. One-way valves prevent backflow of blood as the heart cycles between contraction and relaxation. Comparisons of the circulatory systems of different vertebrates reveals that as circulatory systems become more complex, the blood that flows to the gas exchange organs (gills or lungs; see Figure 48.3) becomes increasingly separated from the blood that flows to the rest of the body.

focus your learning

- The circulatory system in fishes is organized as a single circuit in which blood is pumped by a four-chambered heart through the gills and then to the other organs of the body.
- The lungfish is a transitional form in the evolution of a single blood-flow circuit into separate pulmonary and systemic circuits in other vertebrates.
- Ectothermic reptiles have the ability to control blood flow to the pulmonary circuit, enabling them to conserve energy when they are not breathing.
- Separation of pulmonary and systemic circulations maximizes oxygen transport to body cells and gas exchange in respiratory tissues, and allows the two circuits to operate at different pressures.

In fishes, the phylogenetically oldest vertebrates, blood is pumped from the heart to the gills and then to the tissues of the body and back to the heart—a single circuit. In birds and mammals, blood is pumped from the heart to the lungs and back to the heart in a **pulmonary circuit**, and then from the heart to the rest of the body and back to the heart in a **systemic circuit**. In amphibians and ectothermic reptiles we see various adaptations that incompletely separate the blood flow into pulmonary and systemic circuits.

Both pulmonary and systemic circuits begin with vessels called **arteries** that carry blood away from the heart. Arteries branch into smaller **arterioles** that feed blood into capillary beds. **Capillaries** are tiny, thin-walled vessels where materials are exchanged between blood and the interstitial fluid. Small vessels called **venules** drain capillary beds. The venules join to form the larger **veins** that ultimately deliver blood back to the heart.

We can trace the evolutionary history of vertebrate circulatory systems by comparing the circulatory systems of fishes, lungfishes, amphibians, reptiles (including crocodilians), and mammals.

Circulation in fishes is a single circuit

The fish heart has four chambers that are connected in series. Blood returning from all parts of the body collects in a **sinus venosus** that feeds into the muscular **atrium**. The atrium pumps blood into the more muscular chamber, the **ventricle**. Contraction of the ventricle pushes blood into the last chamber, the **bulbus arteriosus**, a highly elastic chamber. The pressure imparted to the blood by the ventricle stretches the bulbus arteriosus, and its elastic recoil dampens the blood pressure oscillations generated by the beating of the heart. The arterial blood leaving the bulbus arteriosus under pressure flows through the gills, where respiratory gases are exchanged. Blood

leaving the gills collects in a large dorsal artery, the **aorta**, which distributes blood to smaller arteries and arterioles leading to all the organs and tissues of the body. In the tissues, blood flows through beds of tiny capillaries, collects in venules and veins, and eventually returns to the sinus venosus of the heart. The unidirectional flow of blood in this circuit is enabled by one-way valves between the sinus venosus and the atrium, between the atrium and the ventricle, and between the ventricle and the bulbus arteriosus.

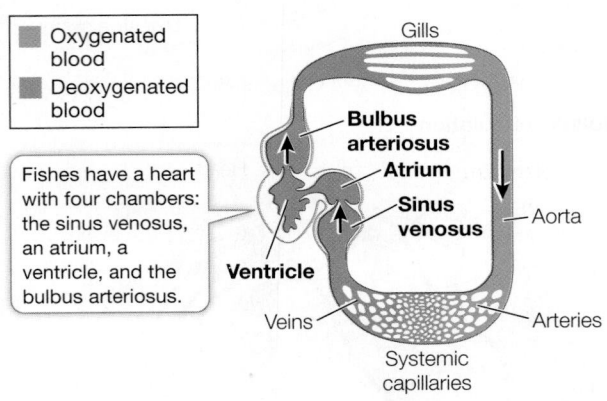

Most of the pressure imparted to the blood by the contraction of the ventricle is dissipated as a result of resistance to flow in the many narrow spaces in the gill lamellae (see Figure 48.5). Therefore blood leaving the gills and entering the aorta is under low pressure, limiting the capacity of the fish circulatory system to supply the tissues with oxygen and nutrients. Yet this limitation on arterial blood pressure does not seem to limit swimming performance. Some species, such as tuna and marlin, can swim at remarkably high rates of speed for long distances.

The evolutionary transition from breathing water to breathing air had important consequences for the vertebrate circulatory system. An example of how the system changed to serve a primitive lung can be seen in the African lungfishes.

Lungfishes evolved a gas-breathing organ

Lungfishes are exposed periodically to water with low oxygen content or to situations in which their aquatic environment dries up. The adaptation that deals with these conditions is an outpocketing of the gut that serves as a lung. The lung contains many thin-walled blood vessels, so blood flowing through those vessels can pick up oxygen from air gulped into the lung.

How does the lungfish circulatory system take advantage of this new organ? In fishes, the gills are arranged on supportive gill arches (see Figure 48.5A and B). Blood flows into the gill arch in an afferent arteriole and leaves in an efferent arteriole. In lungfishes, the blood vessels in the posterior pair of gill arteries have been modified into a low-resistance conduit for blood to the lung, and a separate pulmonary vessel carries oxygenated blood from the lung back to the heart. In addition, blood from the heart flows directly into the dorsal aorta through two anterior gill arches that have lost their gill filaments. A few of the gill arches retain gill filaments, so the African lungfishes can exchange respiratory gases with either air or water.

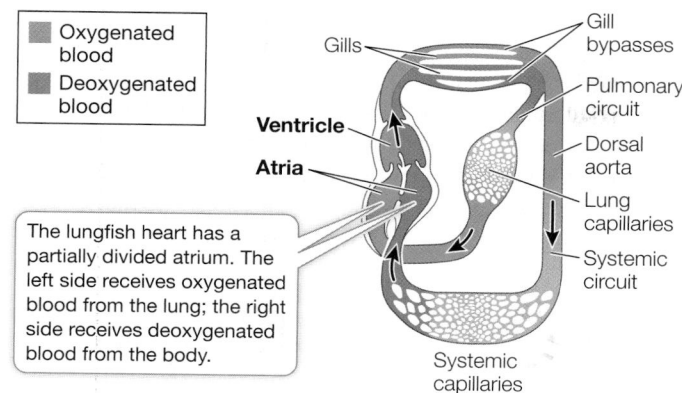

Oxygenated blood

Deoxygenated blood

Gills

Ventricle

Atria

Gill bypasses

Pulmonary circuit

Dorsal aorta

Lung capillaries

Systemic circuit

Systemic capillaries

The lungfish heart has a partially divided atrium. The left side receives oxygenated blood from the lung; right side receives deoxygenated blood from the body.

The lungfish heart partially separates its flow of blood into pulmonary and systemic circuits; it has a partially divided atrium. The left side receives oxygenated blood from the lung, and the right side receives deoxygenated blood from the body via the sinus venosus. These two bloodstreams stay mostly separate as they flow through the ventricle and the bulbus arteriosus. As a result, oxygenated blood goes mostly to the anterior gill arteries leading to the dorsal aorta, and deoxygenated blood goes mostly to the other gill arches that have functional gill filaments. The blood leaving those posterior gill arches flows to the dorsal aorta as well as to the lung. Thus the dorsal aorta can receive oxygenated blood either from the lung (via the anterior gill arches) or from the gill filaments of the posterior gill arches.

We can conclude that the lungfish lung evolved as a means of supplementing oxygen uptake from the gills. When the water is oxygenated, the lungfish can obtain oxygen through its gills; but in oxygen-depleted water, it can depend on getting oxygen from its lung. Associated modifications of the lungfish vascular system set the stage for the evolution of separate pulmonary and systemic circulations in higher vertebrates.

Amphibians have partial separation of systemic and pulmonary circulation

In adult amphibians, a single ventricle pumps blood to the lungs and the rest of the body, but two atria receive blood returning to the heart. The left atrium receives oxygenated blood from the lungs, and the right atrium receives deoxygenated blood from the body.

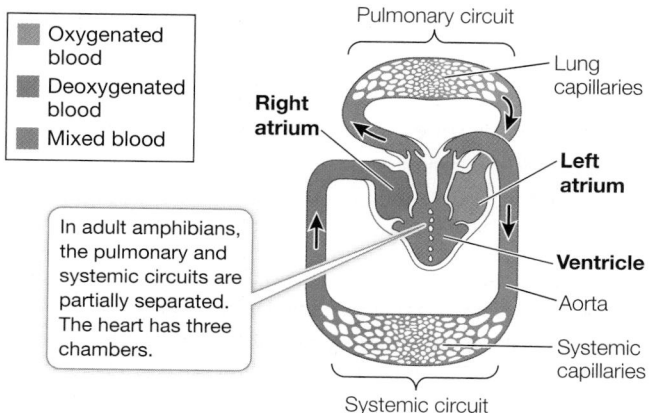

Oxygenated blood

Deoxygenated blood

Mixed blood

Right atrium

Left atrium

Ventricle

Aorta

Systemic capillaries

Pulmonary circuit

Lung capillaries

Systemic circuit

In adult amphibians, the pulmonary and systemic circuits are partially separated. The heart has three chambers.

Because both atria deliver blood to the same ventricle, the oxygenated and deoxygenated blood could mix, in which case blood going to the tissues would not carry a full load of oxygen. Mixing is limited, however, because anatomical features of the ventricle direct the flow of deoxygenated blood from the right atrium primarily to the pulmonary circuit and the flow of oxygenated blood from the left atrium primarily to the aorta. Partial separation of pulmonary and systemic circulation has the advantage of allowing blood destined for the tissues to sidestep the large pressure drop that occurs in the gas exchange organ. Blood leaving the amphibian heart for the tissues moves directly to the aorta, and hence to the body, at a higher pressure than if it had first flowed through the lungs as it does in the fishes.

Amphibians have another adaptation for oxygenating their blood: they can pick up a considerable amount of oxygen in blood flowing through small blood vessels in their skin.

Reptiles have exquisite control of pulmonary and systemic circulation

As described in Chapter 32, the reptiles include turtles, snakes, lizards, crocodilians, and birds (see Figures 32.21 and 32.22). Crocodilians and birds have circulatory systems with two completely separated ventricles, creating a four-chambered heart. All other reptiles have ventricles that are not completely separated into left and right chambers.

Consider the behavior, ecology, and physiology of ectothermic reptiles (i.e., excluding birds). Many are active, powerful, fast animals, but their activity comes in bursts that are interspersed with long periods of inactivity. At these times the animals' metabolic rates are much lower than the resting metabolic rates of the endothermic birds and mammals. So enormous is the range of metabolic demand in ectothermic reptiles that they do not need to breathe continuously. Some species are accomplished divers and spend long periods under water, where they cannot breathe air.

When these animals are not breathing, it would be a waste of energy for them to pump blood through their lungs. Thus they pump blood to the lungs and the rest of the body when they are breathing, but when they are not breathing, they can bypass the pulmonary circuit and pump most of the blood to the body. How do they do this?

In ectothermic reptiles with a three-chambered heart—that is, the turtles, snakes, and lizards—the ventricle is partially divided into left and right halves by a septum. Oxygenated blood from the lungs enters the left side of the ventricle through the left atrium. Deoxygenated blood from the body enters the right side of the ventricle through the right atrium. These species have two aortas, left and right. The left aorta is positioned so that it receives oxygenated blood from the left side of the ventricle. The right aorta, however, is positioned so that it can receive blood from either the right or left side of the ventricle.

When the animal is breathing air, the resistance in the pulmonary circuit is lower than the resistance in the systemic circuit, so blood from the right side of the ventricle tends to flow into the pulmonary artery rather than the right aorta. When the animal is not breathing, pulmonary vessels constrict, resistance in the pulmonary circuit goes up, and blood from the right side of the ventricle tends to flow into the right aorta. As a result, blood from both sides of the ventricle flows through the two aortas to the systemic circuit.

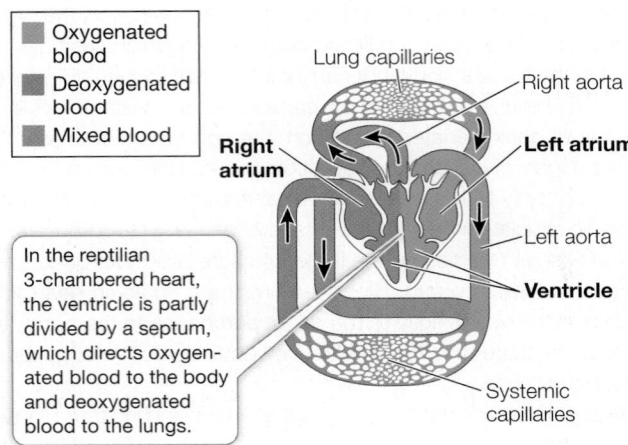

In the reptilian 3-chambered heart, the ventricle is partly divided by a septum, which directs oxygenated blood to the body and deoxygenated blood to the lungs.

- Oxygenated and deoxygenated blood cannot mix; therefore the systemic circuit always receives blood with the highest oxygen content.
- Respiratory gas exchange is maximized because the blood with the lowest oxygen content and highest CO_2 content is sent to the lungs.
- Separate systemic and pulmonary circuits can operate at different pressures.

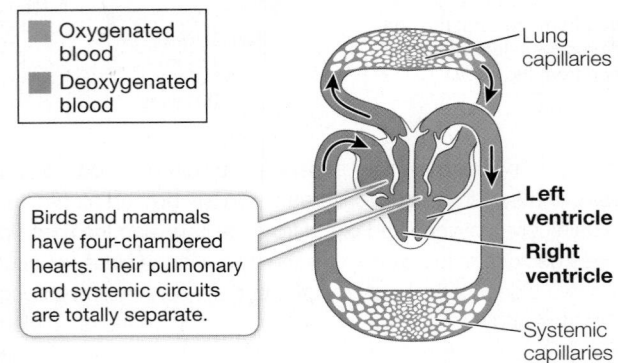

Birds and mammals have four-chambered hearts. Their pulmonary and systemic circuits are totally separate.

Crocodilians, like birds, have two completely separated ventricles. Unlike birds, they have two aortas, one originating in each ventricle. But there is a connection between the two aortas just as they leave the heart, and this connection enables them to alter the proportions of blood going to their pulmonary and systemic circuits. When a crocodile or alligator is breathing and resistance in the pulmonary circuit is low, backpressure from the stronger left ventricle closes the valve between the right ventricle and the right aorta, forcing all of the blood from the right ventricle to flow into the pulmonary circuit. When the animal stops breathing, pulmonary vessels constrict, resistance in the pulmonary circuit rises, and pressure in the right ventricle overcomes the backpressure in the right aorta from the left ventricle. As a result, the blood from the right ventricle flows into the right aorta. This ability of all ectothermic reptiles to direct blood to their pulmonary or systemic circuits is highly adaptive for their lifestyle of intermittent breathing.

The tissues of birds and mammals have high nutrient demands and thus a very high density of blood vessels, requiring the heart to generate a high blood pressure to perfuse all the vessels of the systemic circuit. The pulmonary circuit of these animals receives a blood flow equal to that of the systemic circuit, but the lungs have far fewer blood vessels. Thus the pulmonary circuit of birds and mammals can function at lower pressures, and the four-chambered heart makes that possible.

 Activity 49.1 Vertebrate Circulatory Systems
www.Life11e.com/ac49.1

49.2 recap

The closed circulatory system of vertebrates has evolved from a single circuit system in fishes to separate pulmonary and systemic circuits in birds and mammals. A key step was the evolution of a primitive lung in a lineage of air-breathing fishes. Further adaptations increasingly separated the flows of oxygenated and deoxygenated blood in separate circuits, as seen in amphibians and reptiles. Ectothermic reptiles retained the ability to direct blood to either the pulmonary or systemic circuit, but mammals and birds have complete division of those two circuits, allowing the circuits to function at different pressures.

learning outcomes

You should be able to:

- Understand the single circuit of the circulatory system of fishes, which are phylogenetically the earliest vertebrates.
- Describe how a lungfish can distribute oxygen to its tissues in both water and air environments.
- Explain why the circulatory system of ectothermic reptiles is unique in its control of pulmonary and systemic circulation.

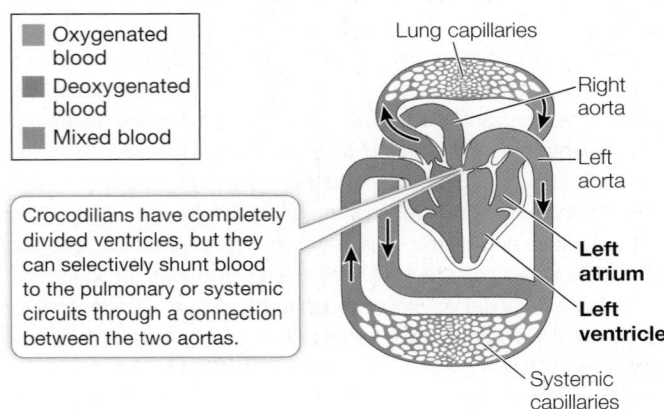

Crocodilians have completely divided ventricles, but they can selectively shunt blood to the pulmonary or systemic circuits through a connection between the two aortas.

Birds and mammals have fully separated pulmonary and systemic circuits

The four-chambered hearts of birds and mammals have completely separate pulmonary and systemic circuits. Separate circuits have several advantages for these active animals with continuously high metabolic rates:

- Apply knowledge about four-chambered hearts and fully separate pulmonary and systemic circulatory systems to explain the consequences of heart defects

1. Trace the flow of blood in circulatory system of a fish.
2. Lungfish brooding in stagnant water can increase the oxygen content in the water immediately surrounding their eggs. How is this possible?
3. Explain how and why ectothermic reptiles shunt blood from the pulmonary circuit when not breathing.
4. If the ductus arteriosus in a human fetus does not close at birth, what are the consequences?

We will turn now to the structure and function of the mammalian heart, focusing in particular on human cardiac function.

key concept 49.3 Heart Function Depends on Properties of Cardiac Muscle

In this section we describe in detail the structure and function of the mammalian heart, using the human heart as the example. We first trace the paths of blood flow through the heart and the body, and then we examine the unique electrical properties of cardiac muscle that result in the rhythmic contractions of the heart.

focus your learning

- Blood flow in mammalian hearts goes from right atrium to right ventricle to lungs and from left atrium to left ventricle to the systemic circulation. Valves prevent backflow.
- In each cardiac cycle, both atria contract first, followed by both ventricles contracting, which is then followed by relaxation.
- Action potentials in cardiac muscle cells depend on the properties of the ion channels in their membranes.

The mammalian heart, and therefore the human heart, has four chambers—a right and a left atrium and a right and a left ventricle (**Figure 49.2**). The right atrium receives blood from the systemic circuit, and the left atrium receives blood from the pulmonary circuit. The right ventricle pumps blood through the pulmonary circuit, and the left ventricle pumps blood through the systemic circuit.

One-way valves between the atria and ventricles, the **atrioventricular (AV) valves**, prevent backflow of blood into the atria when the ventricles contract. The right AV valve is called the **tricuspid valve** because it has three leaves. The left AV valve is called the **bicuspid valve** because it has two leaves. The bicuspid valve is also called the **mitral valve** because it has the shape of a religious headdress called a miter. There are also one-way valves between the ventricles and the arteries leaving the heart. The **pulmonary valve** goes to the lungs, and the **aortic valve** goes to the aorta. These two valves are also called semi-lunar valves because their separate leaves are shaped like half moons.

Blood flows from right heart to lungs to left heart to body

The **superior** (upper) **vena cava** and the **inferior** (lower) **vena cava** (see Figure 49.2) are the large veins that return deoxygenated blood to the right atrium from the upper and lower body, respectively. The veins of the heart itself also drain into the right atrium. From the right atrium, the blood flows through the tricuspid valve into the right ventricle. The filling of the ventricle is mostly passive flow while the heart is relaxed between beats. At the end of this phase of passive ventricular filling, the atrium contracts and adds a little more blood to the ventricular volume. The right ventricle then contracts, causing the tricuspid valve to close and pumping the blood into the **pulmonary artery** leading to the lungs.

Pulmonary veins return the oxygenated blood from the lungs to the left atrium, from where the blood enters the left ventricle through the bicuspid valve. As on the right side of the heart, most left ventricular filling is passive, and is only topped off when the atria contract.

The walls of the ventricles are powerful muscles that contract around the blood with a wringing motion starting from the bottom. When pressure in the left ventricle is high enough to push open the aortic valve, blood rushes into the aorta to begin its circulation throughout the body. In Figure 49.2 observe that the walls of the left ventricle are thicker than those of the right ventricle. The left ventricle has to propel blood through many more kilometers of blood vessels than does the right ventricle, and must therefore push against more resistance, even though both ventricles pump the same volume of blood.

It is the left ventricle that creates the condition of hypertrophic cardiomyopathy (HCM) that you learned about at the beginning of this chapter. Mutations that reduce the efficiency of contraction of the cardiac muscle result in the work load of the left ventricle being greater than normal, so the ventricular muscle compensates by getting bigger (hypertrophy). When the heart is repeatedly stressed by high levels of activity, as is common in athletes, the hypertrophy can become extreme (**Figure 49.3**). The consequences can be blockage of blood flowing into the aorta and disruption of the signals that coordinate the contractions of the ventricular muscles.

Both sides of the heart contract at the same time. Contraction of the two atria, followed by contraction of the two ventricles and then relaxation, is the **cardiac cycle**. The cardiac cycle is divided into two phases: **systole** (pronounced sís-toll-ee), when the ventricles contract, and **diastole** (die-ás-toll-ee), when the ventricles relax (**Focus: Key Figure 49.4**). At the very end of diastole (step 1 in Figure 49.4), just before the ventricles contract, the atria contract and top off the volume of blood in the ventricles.

The sounds of the cardiac cycle—the "lub-dup" heard through a stethoscope—are created by the heart valves slamming shut. The closing and opening of these valves are simple mechanical events resulting from pressure differences on the two sides of the valves. As the ventricles begin to contract (step 2 in Figure 49.4), the pressure in them rises above the pressure in the atria, so the AV valves close ("lub"). When the ventricles begin to relax (step 4 in Figure 49.4), the high pressure in the aorta and pulmonary artery closes the aortic and pulmonary valves ("dup").

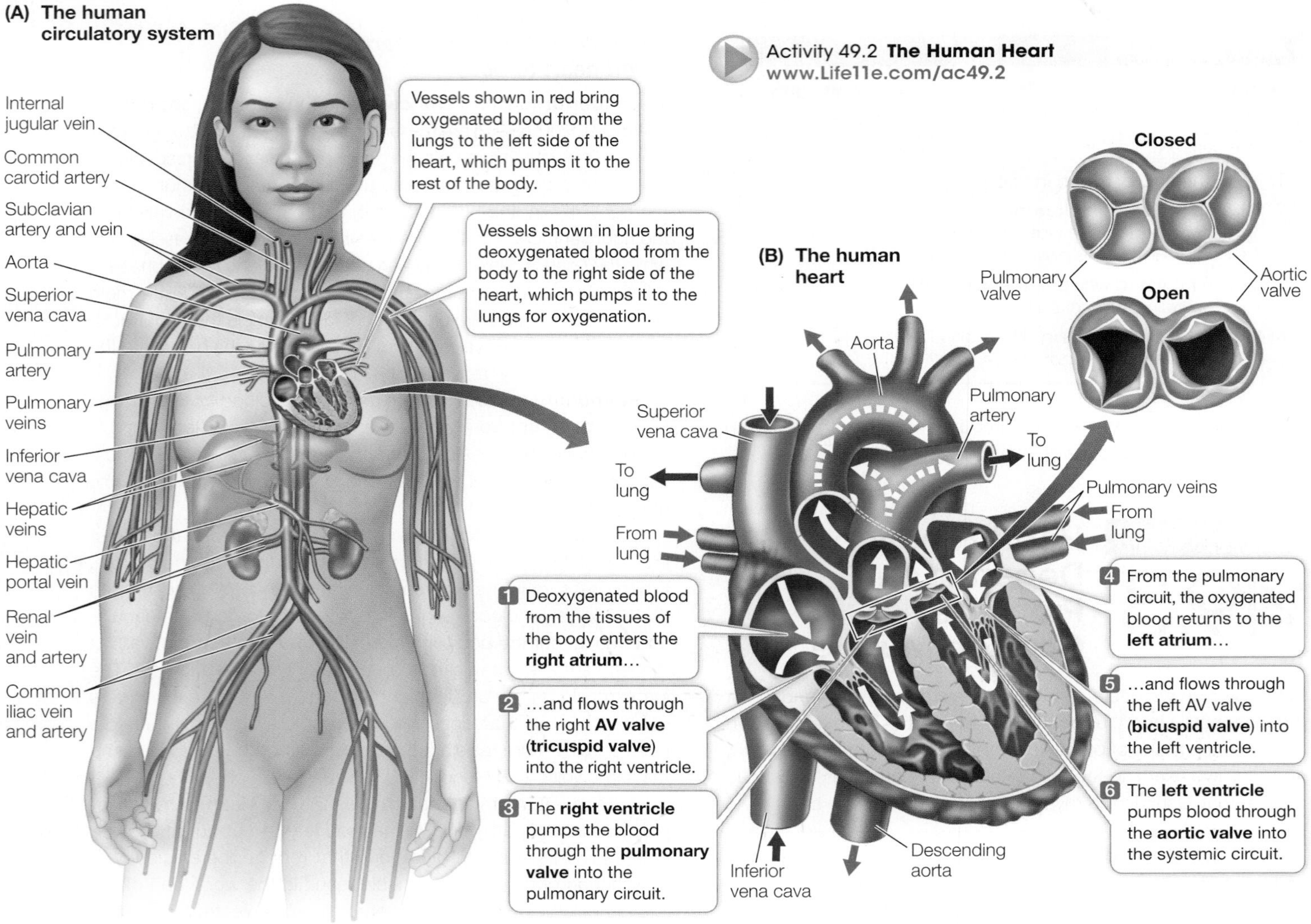

(A) The human circulatory system

Internal jugular vein

Common carotid artery

Subclavian artery and vein

Aorta

Superior vena cava

Pulmonary artery

Pulmonary veins

Inferior vena cava

Hepatic veins

Hepatic portal vein

Renal vein and artery

Common iliac vein and artery

Vessels shown in red bring oxygenated blood from the lungs to the left side of the heart, which pumps it to the rest of the body.

Vessels shown in blue bring deoxygenated blood from the body to the right side of the heart, which pumps it to the lungs for oxygenation.

Activity 49.2 **The Human Heart**
www.Life11e.com/ac49.2

(B) The human heart

Closed

Pulmonary valve   Aortic valve

Open

Aorta

Superior vena cava

Pulmonary artery

To lung

To lung

From lung

From lung

Pulmonary veins

From lung

1 Deoxygenated blood from the tissues of the body enters the **right atrium**…

2 …and flows through the right **AV valve** (**tricuspid valve**) into the right ventricle.

3 The **right ventricle** pumps the blood through the **pulmonary valve** into the pulmonary circuit.

4 From the pulmonary circuit, the oxygenated blood returns to the **left atrium**…

5 …and flows through the left AV valve (**bicuspid valve**) into the left ventricle.

6 The **left ventricle** pumps blood through the **aortic valve** into the systemic circuit.

Inferior vena cava

Descending aorta

Figure 49.2 The Human Heart and Circulation **(A)** The major blood vessels of the body either carry blood away from the heart (arteries) or return blood to the heart (veins). **(B)** In the human heart, blood flows from right side of the heart to the lungs to the left side of the heart and then out to the body. The atrioventricular valves prevent blood from flowing back into the atria when the ventricles contract. The pulmonary and aortic valves prevent blood from flowing back into the ventricles from the arteries when the ventricles relax.

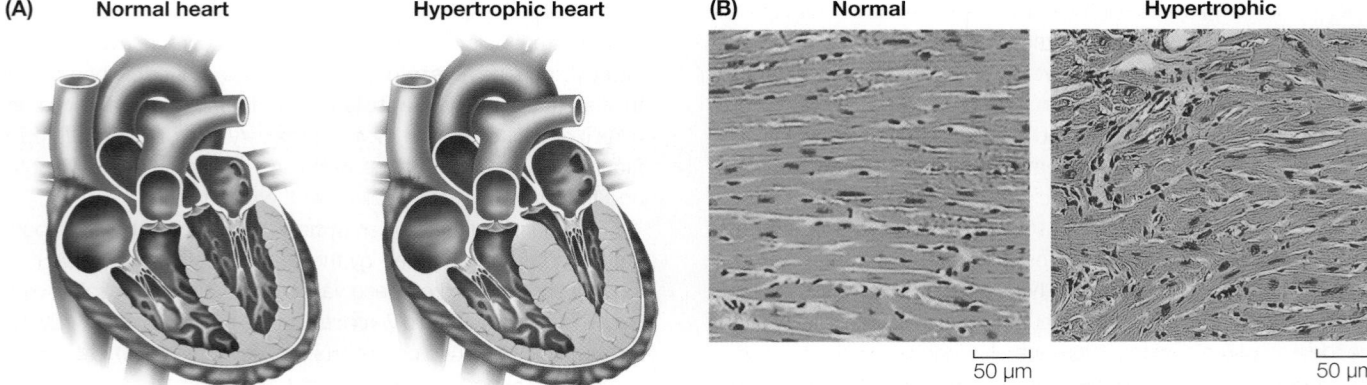

(A)   Normal heart   Hypertrophic heart

(B)   Normal   Hypertrophic

50 μm   50 μm

Figure 49.3 Hypertrophic Cardiomyopathy **(A)** Comparison of a normal and a hypertrophic heart. **(B)** Micrographs of the ventricular heart muscle fibers of a normal and a hypertrophic heart. The organization of the muscle fibers is irregular causing inefficient contractions.

focus: key figure

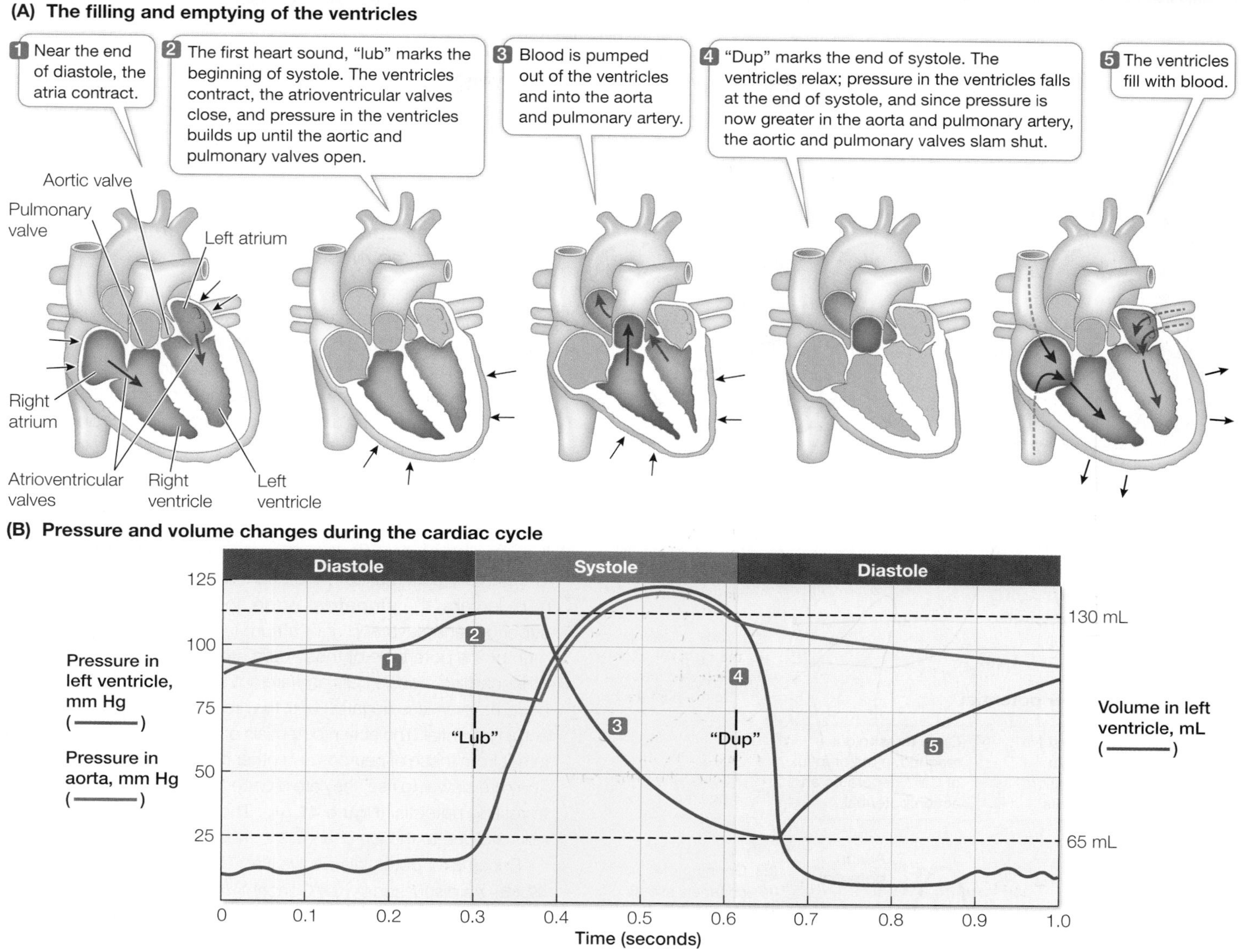

(A) The filling and emptying of the ventricles

1 Near the end of diastole, the atria contract.

2 The first heart sound, "lub" marks the beginning of systole. The ventricles contract, the atrioventricular valves close, and pressure in the ventricles builds up until the aortic and pulmonary valves open.

3 Blood is pumped out of the ventricles and into the aorta and pulmonary artery.

4 "Dup" marks the end of systole. The ventricles relax; pressure in the ventricles falls at the end of systole, and since pressure is now greater in the aorta and pulmonary artery, the aortic and pulmonary valves slam shut.

5 The ventricles fill with blood.

Aortic valve
Pulmonary valve
Left atrium
Right atrium
Atrioventricular valves
Right ventricle
Left ventricle

(B) Pressure and volume changes during the cardiac cycle

Diastole | Systole | Diastole

Pressure in left ventricle, mm Hg (———)

Pressure in aorta, mm Hg (———)

Volume in left ventricle, mL (———)

130 mL

65 mL

"Lub" "Dup"

Time (seconds)

Figure 49.4 The Cardiac Cycle (A) The rhythmic contraction (systole) and relaxation (diastole) of the ventricles is called the cardiac cycle. **(B)** During diastole, the ventricles fill and pressure is low. During systole, the ventricular pressure rises and the ventricles empty. The representation here shows pressure and volume changes for the left ventricle only. The dashed lines represent the maximum and minimum volumes of blood in the left ventricle.

Q: If these curves were for the right ventricle instead of the left ventricle, how would they differ and how would they be the same?

 Animation 49.1 The Cardiac Cycle
www.Life11e.com/a49.1

Defective valves that do not close completely produce turbulent blood flow and the sounds known as heart murmurs. For example, if an AV valve does not close completely, blood will flow back into the atrium with a "whoosh" at the beginning of systole.

Blood pressure changes associated with the cardiac cycle can be measured in the large artery in your arm by using an inflatable pressure cuff and a pressure gauge, together called a sphygmomanometer, and a stethoscope (**Figure 49.5**). This method measures the minimum pressure necessary to compress an artery so blood does not flow through it at all (the systolic pressure) and the maximum pressure that allows continuous flow through the artery (the diastolic pressure). A conventional blood pressure reading is expressed as the systolic pressure placed over the diastolic pressure. Healthy values for a young adult might be 120 millimeters of mercury (mm Hg) during systole and 70 mm Hg during diastole, or 120/70. Modern blood pressure measuring devices such as you might encounter in a drugstore are automated so that the stethoscope is unnecessary.

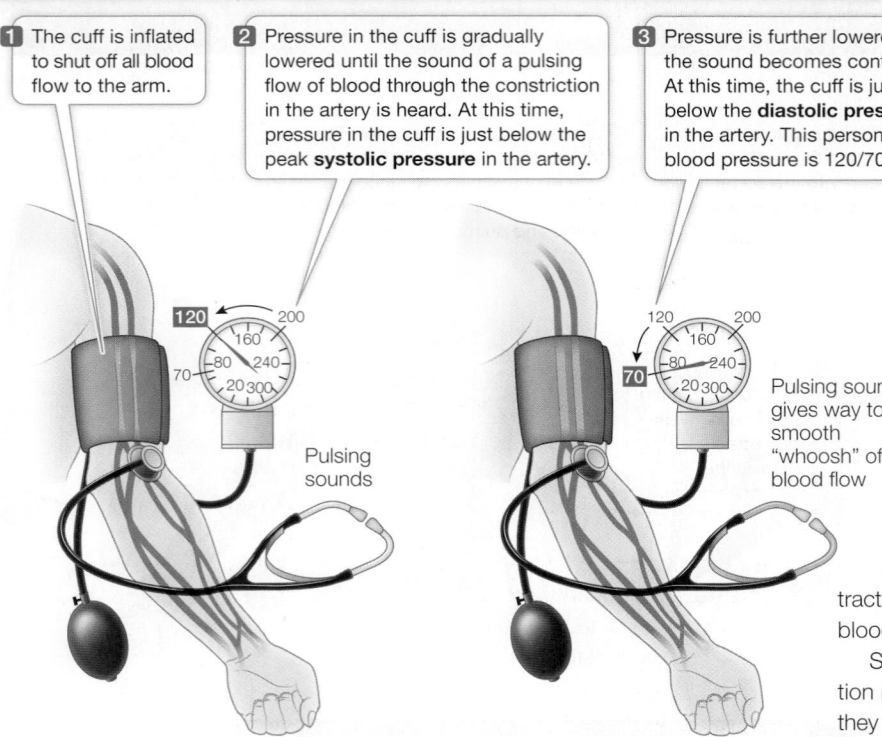

1 The cuff is inflated to shut off all blood flow to the arm.

2 Pressure in the cuff is gradually lowered until the sound of a pulsing flow of blood through the constriction in the artery is heard. At this time, pressure in the cuff is just below the peak **systolic pressure** in the artery.

3 Pressure is further lowered until the sound becomes continuous. At this time, the cuff is just below the **diastolic pressure** in the artery. This person's blood pressure is 120/70.

Pulsing sounds

Pulsing sound gives way to smooth "whoosh" of blood flow

Figure 49.5 Measuring Blood Pressure Blood pressure in the major artery of the arm can be measured with a device called a sphygmomanometer, which combines an inflatable cuff and a pressure gauge. A stethoscope or an automatic sensor is used to detect sounds created by the blood vessels in response to changes in pressure during the cardiac cycle.

The heartbeat originates in the cardiac muscle

Cardiac muscle has unique adaptations that enable it to function as a pump. Cardiac muscle cells are in electrical contact with one another through gap junctions that enable action potentials to spread rapidly from cell to cell. Because a spreading action potential stimulates contraction, large groups of cardiac muscle cells contract in unison. This coordinated contraction is essential for pumping blood effectively.

Some cardiac muscle cells are **pacemaker cells** that initiate action potentials without stimulation from the nervous system. When they fire action potentials, they stimulate neighboring cells to contract. The primary pacemaker of the heart is a group of modified cardiac muscle cells, the **sinoatrial node**, located at the junction of the superior vena cava and right atrium (see Figure 49.8). The resting membrane potentials of these cells are less negative than those of other cardiac muscle cells and are not stable; instead they gradually become even less negative until they reach threshold for initiating an action potential. The action potentials of pacemaker cells are very different from those of neurons and other muscle cells (see Figure 44.8). They are slower to rise; they are broader; and they are slower to return to resting potential (**Figure 49.6A**). These properties of pacemaker cells are due to the ion channels in their membranes.

Pacemaker potentials involve Na+, Ca2+, and K+ channels (**Figure 49.6B**). As discussed in Key Concept 44.2, when Na+ or Ca2+ channels open, positive charges flow into the cell and the membrane potential becomes less negative. When K+ channels open, positive charges flow out of the cell and the membrane potential becomes more negative. Because the Na+ channels of pacemaker cells are open more of the time than are those of other cardiac muscle cells, the pacemaker resting potential is less negative. The action potential of pacemaker cells is due to voltage-gated Ca2+ channels rather than voltage-gated Na+ channels as in neurons, skeletal muscle, and other cardiac muscle cells. These Ca2+ channels open and close more slowly than voltage-gated Na+ channels, explaining the shape of pacemaker action potentials.

The unstable resting potential of pacemaker cells is due to the behavior of cation channels. As in neurons and skeletal muscle

(A) The pacemaker potential

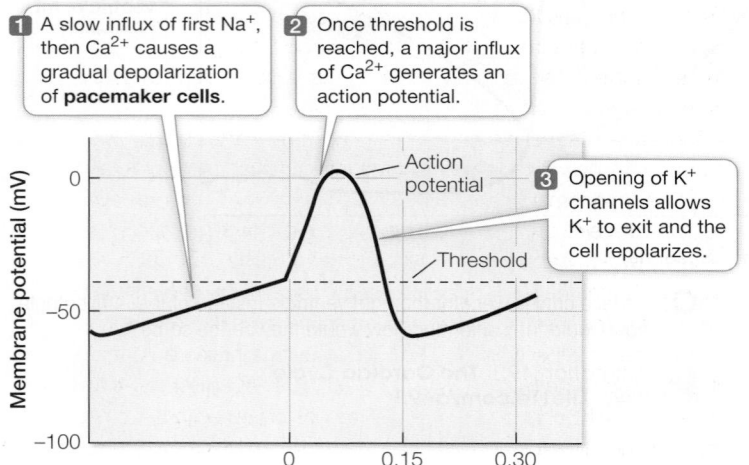

1 A slow influx of first Na+, then Ca2+ causes a gradual depolarization of **pacemaker cells**.

2 Once threshold is reached, a major influx of Ca2+ generates an action potential.

3 Opening of K+ channels allows K+ to exit and the cell repolarizes.

Action potential

Threshold

(B) Ion channel activity of pacemaker cells

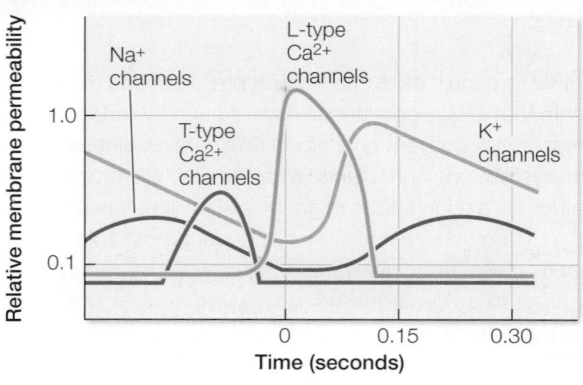

Na+ channels

T-type Ca2+ channels

L-type Ca2+ channels

K+ channels

Time (seconds)

Figure 49.6 The Pacemaker Potential (A) The resting potential of the pacemaker cells of the sinoatrial node gradually drifts upward between action potentials. The pacemaker action potential is slow to rise and is broad. These characteristics are due to properties of Na+, K+, and two types of Ca2+ channels. (B) At rest, sinoatrial pacemaker cells are more permeable to Na+ than are neurons or other muscle cells. During rest, the K+ channels gradually close, but voltage-gated T-type Ca2+ channels open. When a threshold is reached, the L-type Ca2+ channels open, generating the action potential. Opening of the K+ channels helps the membrane potential return to its most negative value.

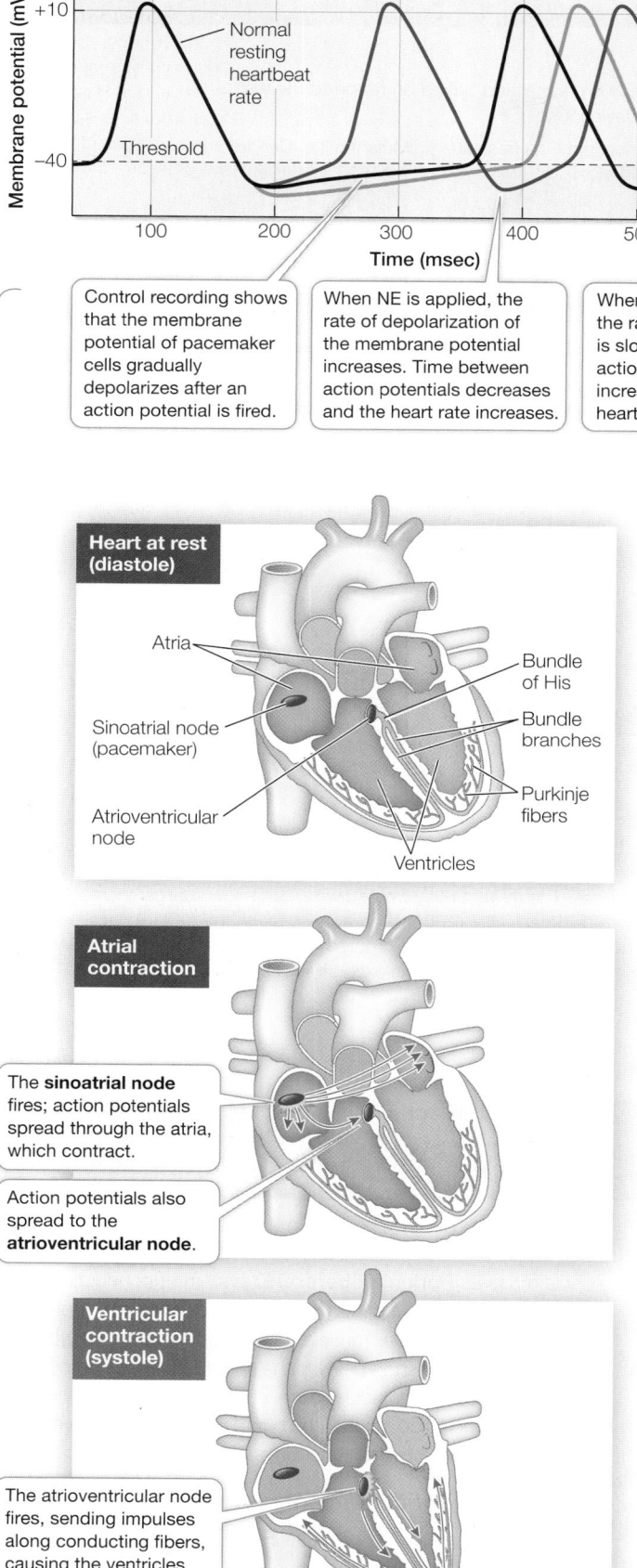

Figure 49.7 The Autonomic Nervous System Controls Heart Rate   Neurotransmitters from the two divisions of the autonomic nervous system either speed up (norepinephrine, NE) or slow down (acetylcholine, ACh) the rate at which the pacemaker membrane potential drifts upward, thereby controlling the rate at which pacemaker cells fire action potentials.

Control recording shows that the membrane potential of pacemaker cells gradually depolarizes after an action potential is fired.

When NE is applied, the rate of depolarization of the membrane potential increases. Time between action potentials decreases and the heart rate increases.

When ACh is applied, the rate of depolarization is slower. Time between action potentials increases and the heart rate slows down.

Heart at rest (diastole)

Atria

Sinoatrial node (pacemaker)

Atrioventricular node

Bundle of His

Bundle branches

Purkinje fibers

Ventricles

Atrial contraction

The **sinoatrial node** fires; action potentials spread through the atria, which contract.

Action potentials also spread to the **atrioventricular node**.

Ventricular contraction (systole)

The atrioventricular node fires, sending impulses along conducting fibers, causing the ventricles to contract.

cells, the rise of the action potential is followed by opening of voltage-gated K^+ channels. The increased permeability of the cells to K^+ brings the membrane potential back to a negative level. But, that restoration of a negative membrane potential causes a unique class of voltage-gated Na^+ channels to open. At the same time, the voltage-gated K^+ channels that restored the negative membrane potential are gradually closing. The opening Na^+ channels and closing K^+ channels cause the membrane potential to drift upwards (become less negative) (see Figure 49.6B).

The gradual rise in membrane potential closes the channels that allow Na^+ to move into the cell, but as the membrane becomes less negative, some Ca^{2+} channels open, causing the membrane potential to continue its gradual rise. These Ca^{2+} channels are called T-types because they open transiently. Eventually the rising membrane potential reaches threshold for the major voltage-gated Ca^{2+} channels to open. These Ca^{2+} channels are called L-types because their opening is long-lasting. It is the opening of these L-type Ca^{2+} channels that generates the pacemaker action potential. The intricate interaction of these ion channels through their effects on membrane potential causes the rhythmic generation of action potentials that characterizes pacemaker cells.

The autonomic nervous system controls the heartbeat (speeds it up or slows it down) by influencing the rate at which the membrane potentials of pacemaker cells drift upward (**Figure 49.7**). Norepinephrine (NE) released onto pacemaker cells by sympathetic nerves increases the permeability of the Na^+ channels and the Ca^{2+} channels. The result is that the resting potential of the pacemaker cells drifts up more rapidly, the interval between action potentials is decreased, and the heart beats faster. Conversely, the parasympathetic neurotransmitter acetylcholine (ACh) has opposite effects. ACh increases the permeability of K^+ channels so that the membrane potential becomes even more negative following an action potential and rises more slowly. ACh also decreases the permeability of the Ca^{2+} channels so that the rate of rise of the membrane potential slows, the interval between pacemaker action potentials lengthens, and the heart slows down.

A conduction system coordinates the contraction of heart muscle

A normal heartbeat begins with an action potential in the sinoatrial node (**Figure 49.8**). This action potential spreads rapidly throughout

Figure 49.8 The Heartbeat   Pacemaker cells in the sinoatrial node initiate the heartbeat by firing action potentials that spread through the electrically coupled atrial muscle. The atrial action potential spreads to the atrioventricular node which, with a delay, conducts it through the bundle of His and Purkinje fibers to the cells of the ventricles.

experiment

Original Paper: Jiang, J., H. Wakimoto, J. B. Seidman and C. E. Seidman. 2013. Allele-specific silencing of mutant *Myh6* allele in mice suppresses hypertrophic cardiomyopathy. *Science* 342: 111–114.

Hypertrophic cardiomyopathy (HCM) is caused by mutations in myosin genes. A similar mutation has been engineered into mice, and they develop HCM. In addition, it was shown that treatment of these mice with cyclosporine A (CsA) exacerbates their development of HCM. Since the HCM model mice are heterozygous for the mutated gene (as are most afflicted humans), and since there are multiple myosin genes, the question arose as to whether or not the selective silencing of the mutant gene would allow the normal alleles to correct the disorganization of the cardiac muscle due to the mutant allele.

HYPOTHESES▶

1. Mice genetically engineered to carry a mutation in the myosin gene most commonly responsible for HCM in humans will show the HCM phenotype, and it will be exacerbated by treatment with CsA.
2. Silencing expression of the mutant gene will prevent the development of the HCM phenotype.

METHOD

1. Using genetic engineering techniques, generate a mouse model that carries a mutation in one of its myosin genes. The heterozygotes are viable, but the homozygotes are not.
2. Using the wild type mice as a comparison, measure left ventricular wall thickness in mutant mice with and without treatment with CsA to exacerbate the development of left ventricular wall hyperthrophy.
3. Develop an interference RNA (RNAi) that will silence expression of the mutant gene in vitro.
4. Transfect mice heterozygous for the mutation with the RNAi construct and at a later time assess them for hypertrophy of their left ventricular walls. Some mice are also treated with CsA, which exacerbates the development of hypertrophy in the mutant mice.
5. Calculate the means and SDs for each cohort of mice. Then do non-paired *t*-tests to see if there are significant differences.

RESULTS

Data in the graph are simulated based on reported means and standard deviations.

Mutant mice, not CsA treated

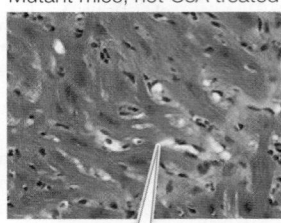

Mutant mice, CsA treated

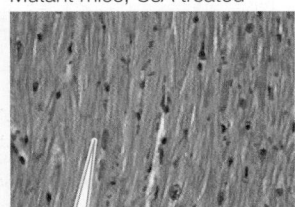

In untreated mice, cardiac muscle cells are disorganized.

Treated mice have normal cardiac muscle cells.

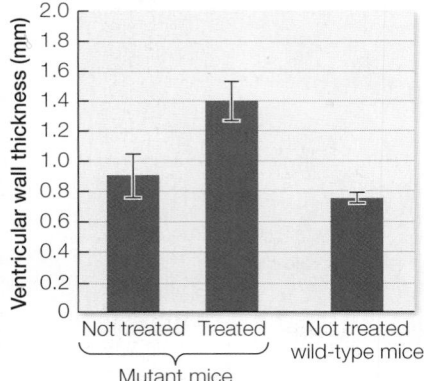

CONCLUSION▶

The mice carrying the mutation show a significant ($P < 0.003$) increase in left ventricular wall thickness. The mutant mice treated with CsA show a larger increase in left ventricular wall thickness ($P < 0.001$) in comparison with the mutant mice not CsA treated. Thus the mutation has the expected effect of promoting the development of HCM symptoms, and the effect is exacerbated by treatment with CsA.

---

the electrically coupled cells of the atria, causing them to contract in unison. Because there are no gap junctions between the cells of the atria and those of the ventricles, the action potential does not spread directly to the ventricles. Therefore the ventricles do not contract in unison with the atria.

How does the action potential move from the atria to the ventricles? Situated at the junction of the atria and the ventricles is a nodule of modified cardiac muscle cells—the **atrioventricular node**—which is stimulated by the depolarization of the atria. With a slight delay, it generates action potentials that are conducted to the ventricles via the **bundle of His**, which are modified cardiac muscle fibers that do not contract but do conduct action potentials. These fibers course down the septum between the left and right ventricles and divide into right and left bundle branches. The branches of conducting fibers run to the tips of the ventricles and then spread throughout the ventricular muscle mass as **Purkinje fibers**. This system of conducting fibers ensures that the cardiac action potential spreads rapidly and evenly throughout the ventricular muscle mass, starting at the very bottom of the ventricles. The short delay in the spread of the action potential imposed by the atrioventricular node ensures that the atria contract before the ventricles do, so that the blood passes progressively from the atria to the ventricles to the arteries.

HCM can disrupt conduction

Returning to the case of HCM introduced at the beginning of this chapter, consider in Figure 49.8 how HCM can impair the function of the left ventricle. First, in many cases the septum between the ventricles is greatly modified by the hypertrophy, and this can disrupt

work with the data

The second hypothesis posed in the experiment above is that treatment of the mutant mice with interference RNA (RNAi) targeted to the mutant myosin gene will reduce the development of the HCM phenotype. The data are listed in the table. All values are for left ventricular wall thickness (mm) of individual mice and are simulated based on the reported means and SDs.

CsA treated	Not CsA treated	RNAi and CsA treated	RNAi but not CsA treated
1.60	0.96	0.87	0.63
1.34	0.98	0.82	0.68
1.50	1.01	0.87	0.85
1.39	0.89	0.91	0.65
1.50	0.86	0.74	0.68
1.44	0.91	0.83	
1.42		0.92	

QUESTIONS▶

1. Comparing mutant mice that did not receive CsA, did the treatment with RNAi significantly reduce their left ventricular wall thickness?
2. Comparing mutant mice that did receive CsA, did the treatment with RNAi significantly reduce their left ventricular wall thickness?

A similar **work with the data** exercise may be assigned in LaunchPad.

the conducting fibers that run down through the septum. Second, the thickening of the septum near the aorta can block the entrance to the aorta and therefore impede blood flow and the emptying of the ventricle. How can these effects be prevented? In **Investigating Life: Silencing Mutant Myosin Genes** we describe a recent study on mice that have been genetically engineered to carry mutations of a gene for one of the myosin chains of cardiac muscle that is associated with a large percentage of human HCM cases. These mice develop the pathologies that are seen in humans who inherit the genes causing HCM. The investigators found that by using *RNAi to silence the mutated genes, they could prevent the development of the HCM pathology.

*connect the concepts The use of RNAi to block gene expression at the mRNA level is discussed in Key Concept 18.4.

Electrical properties of ventricular muscles sustain heart contraction

Electrical properties of ventricular muscle fibers allow them to contract for about 300 milliseconds—much longer than skeletal muscle fibers. As in neuronal and skeletal muscle action potentials, the rising phase of the ventricular muscle cell action potentials is due to the opening of voltage-gated Na+ channels. Unlike neurons and skeletal muscle

fibers, however, ventricular muscle cells remain depolarized for a long time. This extended plateau of the action potential is due to sustained opening of voltage-gated Ca2+ channels (**Figure 49.9**). Like other muscle, cardiac muscle is stimulated to contract when Ca2+ is available to bind with troponin (see Figure 47.6). As long as Ca2+ remains in the sarcoplasm, the ventricular muscle cells continue to contract.

To terminate systole and allow the ventricles to fill again, Ca2+ must be rapidly cleared from the sarcoplasm of the ventricular cells. Ca2+ pumps in the sarcoplasmic reticulum membrane and the cell membrane actively transport Ca2+ ions out of the sarcoplasm and into the sarcoplasmic reticulum or into the intersitial fluid. Thus Ca2+

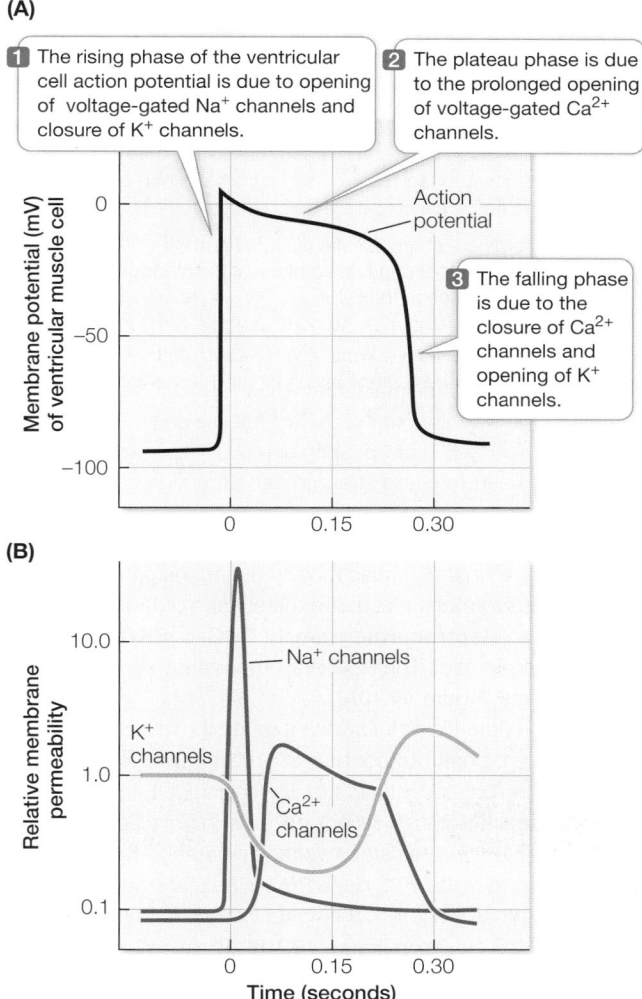

Figure 49.9 The Action Potential of Ventricular Muscle Fibers (A) The three phases of the action potential of ventricular muscle fibers are due to the opening and closing of voltage-gated channels. (B) At the initiation of the action potential, voltage-gated Na+ channels open rapidly but briefly. At the same time, but more slowly, K+ channels are closing and Ca2+ channels are opening. The open Ca2+ channels sustain the depolarization. Repolarization occurs when the Ca2+ channels close, and the slow opening of the K+ channels also contributes to the repolarization.

Q: Why is the broadening of the ventricular muscle action potential important?

(A) ECG set-up

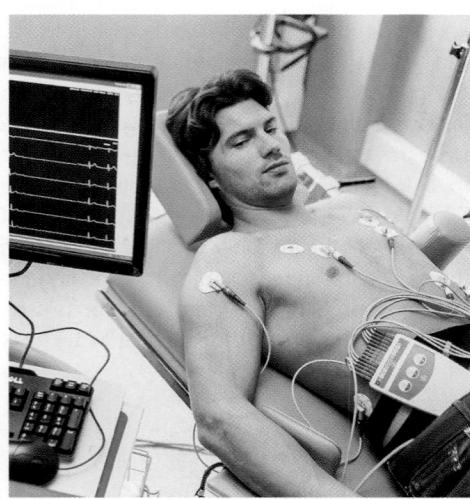

(B) Typical ECG recording with corresponding arterial pressures and heart sounds

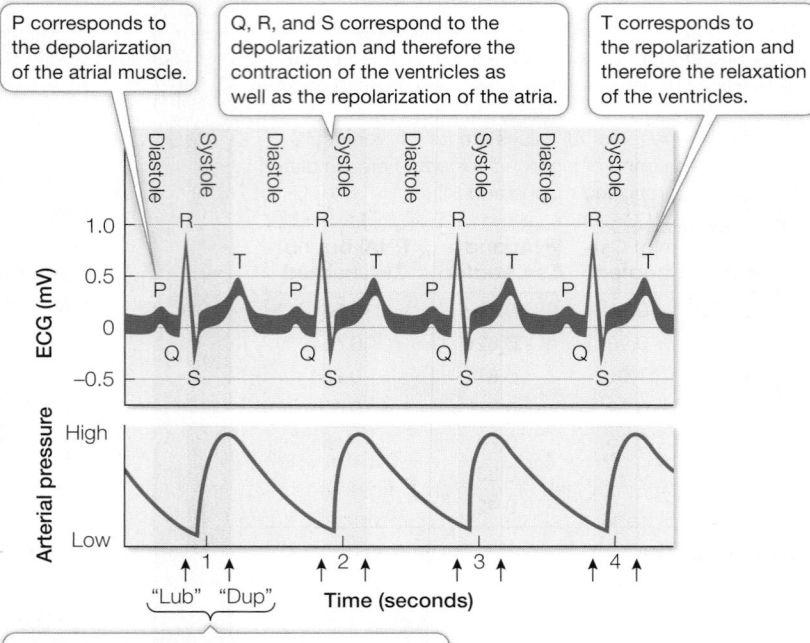

P corresponds to the depolarization of the atrial muscle.

Q, R, and S correspond to the depolarization and therefore the contraction of the ventricles as well as the repolarization of the atria.

T corresponds to the repolarization and therefore the relaxation of the ventricles.

The sounds heard through a stethoscope occur at the beginning and end of ventricular systole.

Figure 49.10 The Electrocardiogram (A) An electrocardiogram (abbreviated as ECG or EKG) is used to monitor heart function. Electrodes attached to the person record an ECG that is amplified and displayed on a monitor. **(B)** Variations from the normal pattern shown here can be used to diagnose heart problems.

Q: Abnormal ECG patterns are characteristic of HCM hearts. If there were an occasional or a long-lasting block of the AV node, what would you expect to see in the ECG?

in the sarcoplasm is maintained at a low level until the next action potential triggers another round of Ca^{2+} release and muscle contraction. The rate of cycling of Ca^{2+} into and out of the sarcoplasm puts limits on the heart rate and strength of contraction of the ventricle.

The ECG records the electrical activity of the heart

Electrical events in the cardiac muscle during the cardiac cycle are recorded by electrodes placed on the surface of the body. Such a recording is an **electrocardiogram**, or **ECG**, or **EKG** reflecting the Greek spelling (*kardia*). The ECG is an important tool for diagnosing heart problems (**Figure 49.10A**).

The action potentials that sweep through the muscles of the atria and ventricles before they contract are massive, localized electrical events causing electric currents to flow throughout the body. Electrodes placed at different locations on the skin detect those currents at different times and register a voltage difference between them. The appearance of the ECG depends on the placement of the electrodes. Electrodes placed on the right wrist and left ankle produced the normal ECG shown in **Figure 49.10B**. The wave patterns of the ECG are designated P, Q, R, S, and T, each letter representing a particular event in the cardiac muscle, as shown in the figure.

> **49.3 recap**
>
> The mammalian heart has two atria and two ventricles. The right atrium and ventricle serve the pulmonary circuit, and the left atrium and ventricle serve the systemic circuit. Valves that are open and shut by pressure differences prevent the backflow of blood. Modified cardiac muscle tissue in the right atrium functions to spontaneously generate pacemaker action potentials. Other modified

cardiac muscle tissue between the atria and ventricles and throughout the ventricles conducts those signals and coordinates the heart contraction. Broad action potentials in ventricular muscle reflect Ca^{2+} cycling in the ventricular muscle cells and make sustained contractions possible.

learning outcomes

You should be able to:

- Trace the path of blood through both sides of the heart, naming the major blood vessels and heart valves.
- Describe how heart valves function.
- Describe the cardiac cycle in terms of left ventricular volume and pressure, aortic pressure, and heart sounds.
- Explain how calcium ion movement into and out of the sarcoplasm of cardiac muscle cells influences the rate and strength of contraction.

1. Valves of the heart can become stiff and calcified, limiting the flow of blood through them in a condition called stenosis. What would be the consequences of aortic valve stenosis and of pulmonary valve stenosis?

2. Digitalis is a drug that causes an increase in Ca^{2+} concentration in the cardiac muscle cells. Why is it used to treat patients with a weakened heart?

3. Heart valve stenosis impedes blood flow, whereas heart valve prolapse allows blood to flow backward (regurgitation). Both of these conditions can be detected as heart murmurs when a stethoscope is used to listen to heart sounds. When in the cardiac cycle could a heart murmur indicate aortic valve stenosis, and when could it indicate aortic valve prolapse? Explain your answer.

Next we will consider the composition of the blood and the characteristics of the vessels through which blood circulates around the body, illustrating once again how structure serves function. We will also consider the role of the lymphatic vessels that return interstitial fluid to the blood.

key concept

49.4 Circulatory System Functions Depend on Blood and Blood Vessels

Blood is a connective tissue, and like all connective tissues, it consists of cells suspended in an extracellular matrix. There are a variety of blood cell types with different functions, and the extracellular matrix has a complex composition that is highly regulated. The unusual feature of blood is that the extracellular matrix is a liquid, so blood is a fluid tissue that circulates in a system of closed vessels through the bodies of vertebrates and invertebrates that have closed circulatory systems. The properties of the different classes of vessels reflect their functions.

focus your learning

- Red blood cells are generated in bone marrow and, until they are destroyed in the spleen, they transport respiratory gases.
- Platelets and blood proteins are responsible for blood clotting.
- Smooth muscle cells in arteries and arterioles control blood flow to specific tissues and also control blood pressure.

- Extracellular fluid exchange between the blood and the interstitial space occurs in capillary beds and depends on hydrostatic pressure pushing water out of the capillaries and osmotic pressure pulling water back into the capillaries.
- One-way valves and muscle contraction help return venous blood to the heart.

The cells of the blood can be separated from the fluid matrix, called blood plasma, by centrifugation (**Figure 49.11**). If a sample of blood is spun in a centrifuge, all the cells move to the bottom of the tube, leaving the clear, straw-colored blood plasma on top. The packed-cell volume, or hematocrit, is the percentage of the blood volume made up by red blood cells (RBCs). Normal hematocrit is about 42 percent for women and 46 percent for men, but these values can vary considerably. They are usually higher, for example, in people who live and work at high elevations, because their exposure to low oxygen stimulates the production of more RBCs. Between the blood plasma and the packed RBCs there is a very thin layer—less than 1 percent of the blood volume—called the buffy coat that contains most of the white blood cells and platelets. **Platelets** are pinched-off fragments of a class of bone marrow cells. Here we consider the RBCs and the platelets. White blood cells, or leukocytes, are cells of the immune system, discussed in Chapter 41.

Red blood cells transport respiratory gases

As is obvious from the hematocrit, most blood cells are RBCs—also called **erythrocytes**. Mature RBCs are biconcave, flexible discs

Figure 49.11 The Composition of Blood   Blood consists of a complex aqueous solution (the blood plasma) and numerous cell types and cell fragments. The hematocrit (arrow) is a measure of the volume of RBCs as a percentage of total blood volume. At the surface of the packed cells is the thin buffy coat that contains leukocytes and platelets.

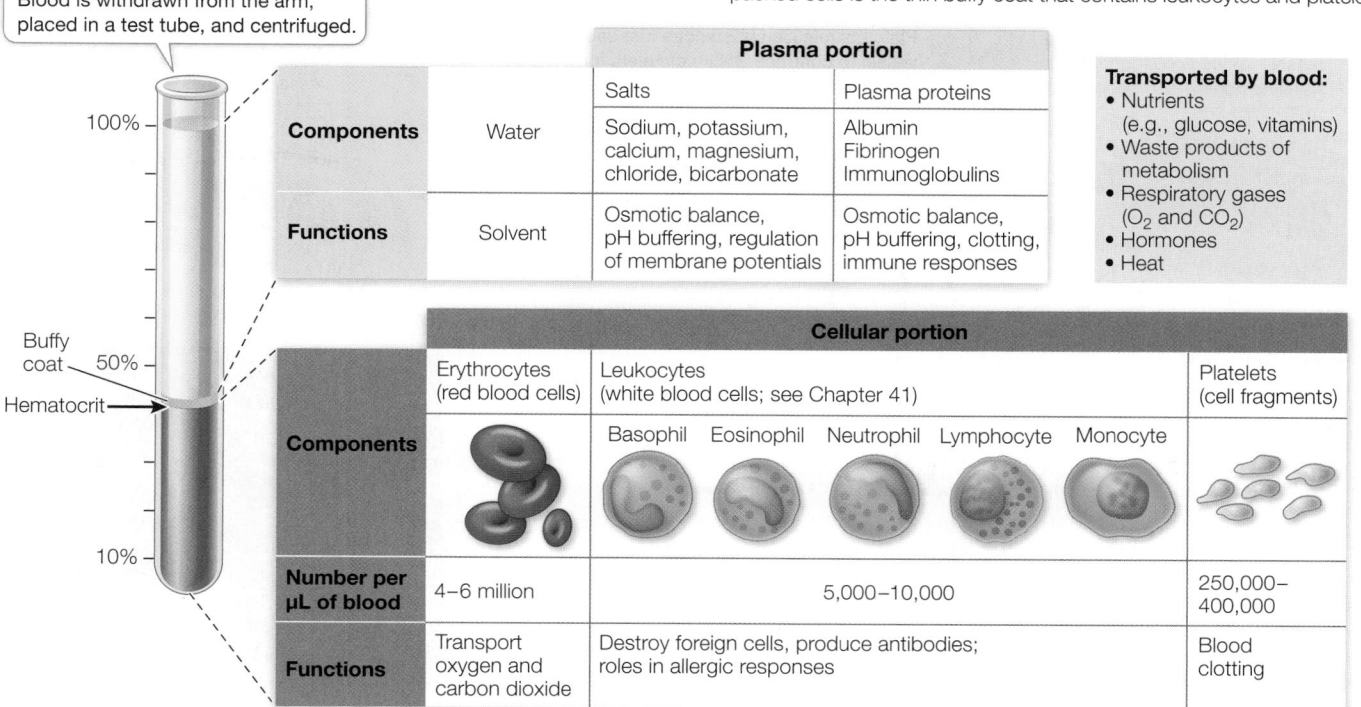

Blood is withdrawn from the arm, placed in a test tube, and centrifuged.

Plasma portion			
Components	Water	Salts	Plasma proteins
		Sodium, potassium, calcium, magnesium, chloride, bicarbonate	Albumin Fibrinogen Immunoglobulins
Functions	Solvent	Osmotic balance, pH buffering, regulation of membrane potentials	Osmotic balance, pH buffering, clotting, immune responses

Transported by blood:
- Nutrients (e.g., glucose, vitamins)
- Waste products of metabolism
- Respiratory gases (O_2 and CO_2)
- Hormones
- Heat

Cellular portion			
	Erythrocytes (red blood cells)	Leukocytes (white blood cells; see Chapter 41)	Platelets (cell fragments)
Components		Basophil Eosinophil Neutrophil Lymphocyte Monocyte	
Number per µL of blood	4–6 million	5,000–10,000	250,000–400,000
Functions	Transport oxygen and carbon dioxide	Destroy foreign cells, produce antibodies; roles in allergic responses	Blood clotting

100%

Buffy coat   50%

Hematocrit

10%

that lack nuclei but are packed with hemoglobin. Their function is to transport respiratory gases (see Key Concept 48.4). The shape of RBCs gives them a large surface area for gas exchange, and their flexibility enables them to squeeze through narrow capillaries. Men have 4.5–6.0 million red blood cells per microliter of blood, and women have 3.5–5.0 million.

RBCs, as well as all the other cellular components of blood, are generated by stem cells in the bone marrow, particularly in the ribs, breastbone, pelvis, vertebrae, and the long bones of the limbs. RBC production is controlled by a hormone, **erythropoietin** (**EPO**), released by cells in the kidneys in response to insufficient oxygen—**hypoxia**. Many tissues respond to hypoxia by expressing a transcription factor called hypoxia-inducible factor 1 (HIF-1). When the kidneys become hypoxic and express HIF-1, one of the actions of the transcription factor is to activate the gene encoding erythropoietin. Increased circulating erythropoietin extends the lives of mature RBCs and stimulates production of new RBCs in the bone marrow.

Under normal conditions, your bone marrow produces about 2 million RBCs every second. Developing RBCs divide many times while still in the bone marrow, and during this time they produce hemoglobin. When hemoglobin makes up about 25 percent of the volume of the immature RBC, the nucleus is extruded from the cell and other cell organelles are also lost. This process is almost complete when the newly mature RBC squeezes between the endothelial cells of blood vessels in the bone marrow and enters the circulation. Loss of nuclei from the RBCs occurs in most mammalian species, but the RBCs of a few mammals and of all other vertebrates are nucleated.

Each RBC circulates for about 120 days. As it gets older, its membrane becomes less flexible and more fragile, so older RBCs are more likely to rupture as they bend to fit through narrow capillaries.

RBCs are particularly squeezed in the **spleen**, an organ that sits near the stomach in the upper left side of the abdominal cavity. The spleen has many sinuses (cavities) that serve as reservoirs for RBCs. To get into the sinuses, however, the RBCs must squeeze between spleen cells. When old RBCs are ruptured by this squeezing, their remnants are taken up and degraded by macrophages (a class of white blood cells that ingest debris and foreign materials; see Key Concept 41.3).

Platelets are essential for blood clotting

Besides producing erythrocytes and leukocytes, the bone marrow stem cells described in Key Concept 41.1 also produce cells called **megakaryocytes**. Megakaryocytes are large cells that remain in the bone marrow and release platelets into the circulation. A platelet is a tiny fragment of a cell without cell organelles, but it is packed with enzymes and chemicals necessary for its function: sealing leaks in blood vessels and initiating **blood clotting** (Figure 49.12).

Damage to a blood vessel exposes collagen fibers. An encounter with collagen fibers activates platelets. The platelet swells, becomes irregularly shaped and sticky, and releases chemicals that activate other platelets and initiate the clotting of blood. The sticky platelets also form a plug at the damaged site.

Blood clotting requires many steps and many clotting factors, most of which are circulating in the blood in an inactive form. The absence of any one of these proteins can impair clotting, resulting in excessive bleeding. Because the liver produces most of the clotting factors, liver diseases such as hepatitis and cirrhosis can result in excessive bleeding. People with hemophilia experience uncontrolled bleeding because of a genetic inability to produce a clotting factor.

Blood clotting factors participate in a cascade of chemical activations of other substances circulating in the blood. The cascade

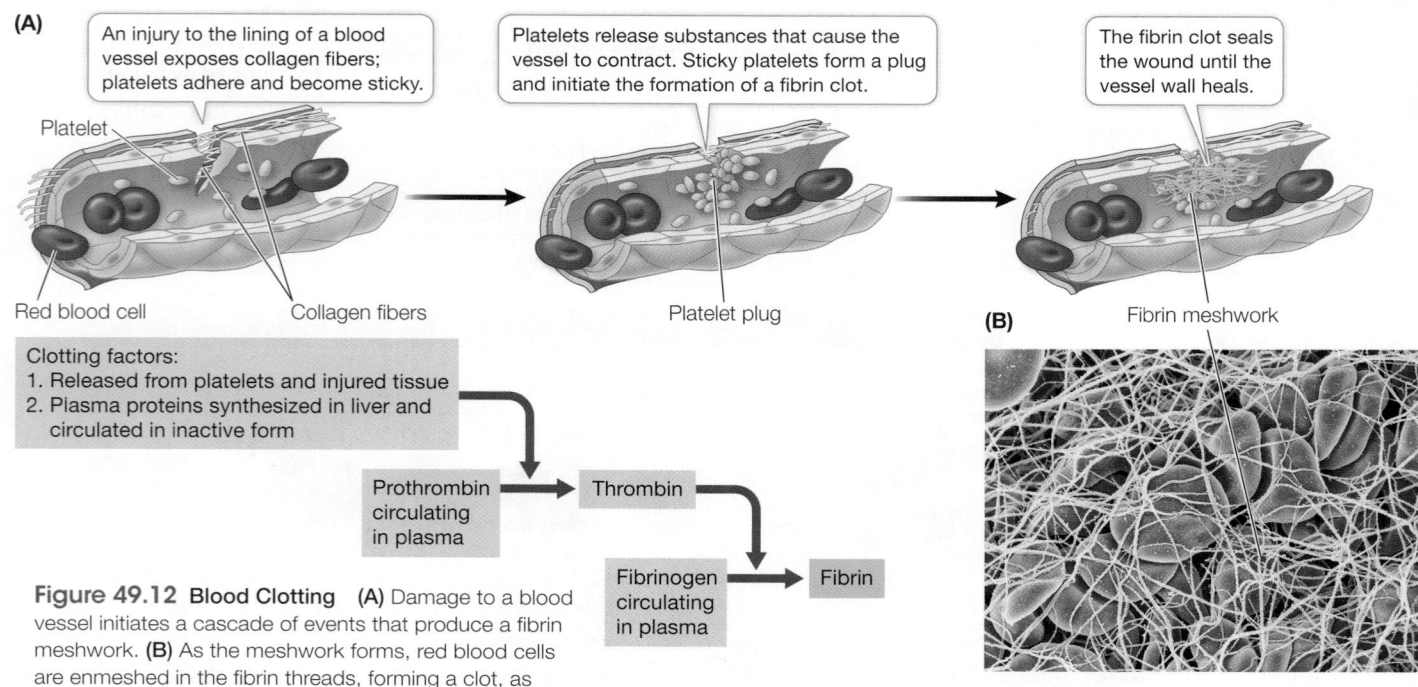

Figure 49.12 Blood Clotting (A) Damage to a blood vessel initiates a cascade of events that produce a fibrin meshwork. (B) As the meshwork forms, red blood cells are enmeshed in the fibrin threads, forming a clot, as shown in this color-enhanced electron micrograph.

(A) An injury to the lining of a blood vessel exposes collagen fibers; platelets adhere and become sticky.

Platelets release substances that cause the vessel to contract. Sticky platelets form a plug and initiate the formation of a fibrin clot.

The fibrin clot seals the wound until the vessel wall heals.

Platelet

Red blood cell   Collagen fibers

Platelet plug

(B)   Fibrin meshwork

Clotting factors:
1. Released from platelets and injured tissue
2. Plasma proteins synthesized in liver and circulated in inactive form

Prothrombin circulating in plasma → Thrombin

Fibrinogen circulating in plasma → Fibrin

begins with damage to a blood vessel or other tissue that exposes the blood to proteins such as collagen that are normally separated from the blood by endothelial cells lining the blood vessels. This exposure activates platelets and begins the clotting factor cascade. The end result of this cascade is to convert an inactive circulating enzyme, **prothrombin**, to its active form, **thrombin**. Thrombin cleaves molecules of **fibrinogen**, a plasma protein, forming insoluble threads of **fibrin**. The fibrin threads form the meshwork that binds platelets, seals the vessel, and provides a scaffold for the formation of scar tissue (see Figure 49.12).

Arteries withstand high pressure, arterioles control blood flow

The walls of the large arteries have many extracellular collagen and elastin fibers that enable them to withstand the high blood pressures generated by the heart (**Figure 49.13A**). These elastic tissues have another important function: as you saw with the bulbus arteriosus in fish, they are stretched during systole, and thereby store some of the energy imparted to the blood by the heart. Elastic recoil during diastole returns this energy to the blood by squeezing it and pushing it forward. As a result, even though pressure in the arteries pulsates

with the beating of the heart, the flow of blood is smoother than it would be through a system of rigid pipes.

Contraction of smooth muscle cells in the walls of the arteries constricts the vessels, and relaxation of the smooth muscle dilates them. When the diameter of a vessel changes, its resistance to blood flow changes, and the amount of blood flowing through it changes as a result. Neural and hormonal mechanisms act on smooth muscle cells in the walls of the arteries and arterioles, controlling the flow of blood through these vessels. The arterioles are referred to as resistance vessels because their resistance can vary to control the blood flow to specific tissues.

Materials are exchanged in capillary beds by filtration, osmosis, and diffusion

Beds of capillaries lie between arterioles and venules (**Figure 49.13B**). Most cells are not more than a few cell diameters away from a capillary. The needs of cells are served by the exchange of materials between blood and interstitial fluid across the capillary walls. Capillary walls are thin and permeable to water and many solutes. Also, blood flows slowly through capillaries, allowing time for exchange.

It may seem strange that blood flows through the large arteries rapidly at high pressures, but when it reaches the small capillaries the pressure and rate of flow decrease (**Figure 49.13C**). When you constrict the opening of a garden hose, the pressure in the hose and the velocity of the water spraying out increase. But keep in mind that the arteries branch into many arterioles that serve huge numbers of capillaries. Even though each capillary has a diameter so small that red blood cells pass through in single file, there are so many capillaries that their total cross-sectional area is much greater than that of any other class of vessel. As a result, all the capillaries together have a much greater capacity for blood than do the arterioles.

Capillary walls consist of a single layer of endothelial cells (**Figure 49.14**). Capillaries are permeable to water, some ions, and some small molecules

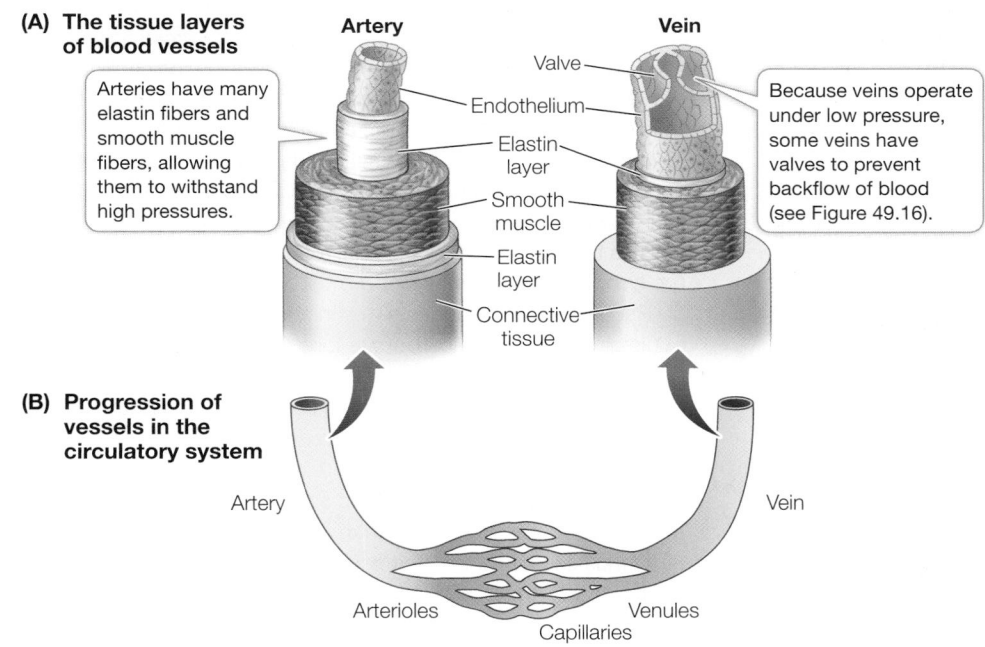

(A) The tissue layers of blood vessels

Arteries have many elastin fibers and smooth muscle fibers, allowing them to withstand high pressures.

Artery

Vein

Valve

Endothelium

Elastin layer

Smooth muscle

Elastin layer

Connective tissue

Because veins operate under low pressure, some veins have valves to prevent backflow of blood (see Figure 49.16).

(B) Progression of vessels in the circulatory system

Artery

Vein

Arterioles

Venules

Capillaries

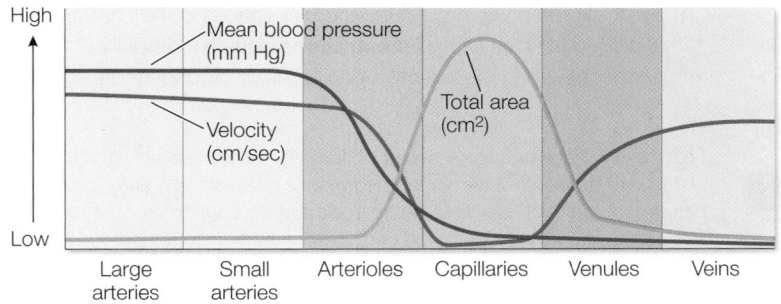

(C) Characteristics of classes of vessels

High

Low

Mean blood pressure (mm Hg)

Total area (cm²)

Velocity (cm/sec)

Large arteries

Small arteries

Arterioles

Capillaries

Venules

Veins

Figure 49.13 Anatomy of Blood Vessels **(A)** The different anatomical characteristics of arteries and veins match their functions. **(B)** Blood from the arterial system feeds into capillary beds where exchanges with the interstitial fluid occur. The venous system returns the blood to the heart. **(C)** The cross-sectional area encompassed by each vessel type is graphed along with the pressure and velocity of the blood within it.

 Activity 49.3 **Structure of Blood Vessels**
www.Life11e.com/ac49.3

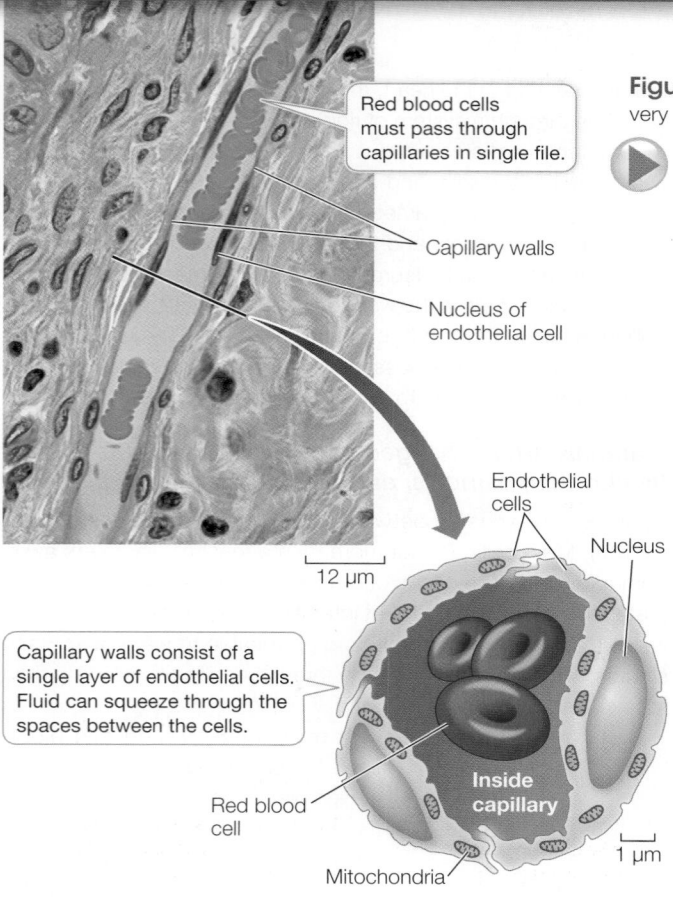

Red blood cells must pass through capillaries in single file.

Capillary walls

Nucleus of endothelial cell

12 µm

Capillary walls consist of a single layer of endothelial cells. Fluid can squeeze through the spaces between the cells.

Endothelial cells

Nucleus

Red blood cell

Inside capillary

Mitochondria

1 µm

Figure 49.14 A Narrow Lane   Capillaries have a very small diameter, and blood flows through them slowly.

▶ **Media Clip 49.1 Capillary Flow: A Tight Squeeze**
www.Life11e.com/mc49.1

but not to large molecules such as proteins, and certainly not to red blood cells. At the arterial (high pressure) end, blood pressure squeezes water and small solutes out through spaces between the cells of the capillary walls into the surrounding intercellular space.

Why don't water and small-molecular-weight solutes collect in the intercellular spaces? How is the blood volume maintained if fluid is continuously leaking out of the capillaries? An answer to such questions was put forth more than 100 years ago by the physiologist E. H. Starling. Starling suggested that water movement across capillary walls is a result of two opposing forces, which are now known as **Starling's forces**:

1. Blood hydrostatic pressure squeezes water and small solutes out of the capillaries.

2. Blood osmotic pressure pulls water back into the capillaries.

Blood hydrostatic pressure is high at the arterial end of a capillary bed and steadily drops as blood moves toward the venous end (**Figure 49.15**). Blood osmotic pressure is due to the large protein molecules that cannot leave the capillaries, and it is relatively constant along the capillaries. As long as the blood hydrostatic pressure is above the blood osmotic pressure, fluid leaves the capillaries. At the venule end of most capillaries, blood hydrostatic pressure falls below blood osmotic pressure, so fluid returns to the capillaries. The actual numbers for a normal capillary bed in a resting person suggest that there would be a *slight* net loss of fluid to the intercellular spaces. This loss, about 4 liters per day, percolates between cells as the interstitial fluid before it returns to the venous blood via the **lymphatic system**, which we will discuss later in this chapter.

Recent research suggests that in addition to large protein molecules in the blood, bicarbonate ions (HCO_3^-) may also contribute significantly to pulling water back into the capillaries. The CO_2 produced by cellular metabolism diffuses into the endothelial cells lining the capillaries, where it is converted into HCO_3^- and released into the plasma. When an individual is at rest, the increasing HCO_3^- concentration can cause the blood osmotic pressure at the venous end to be 30 mm Hg higher than at the arterial end, and during strenuous exercise this difference can be much higher. Thus it appears that CO_2 and HCO_3^- are major factors that pull water back into the capillaries.

All capillaries are permeable to O_2, CO_2, and small ions. Lipid-soluble substances readily pass through the capillary walls. Water and small solutes pass through intercellular spaces in the capillary wall, and through holes called **fenestrations** in the endothelial cells. More permeable and less selective capillaries are found in the digestive

(A)

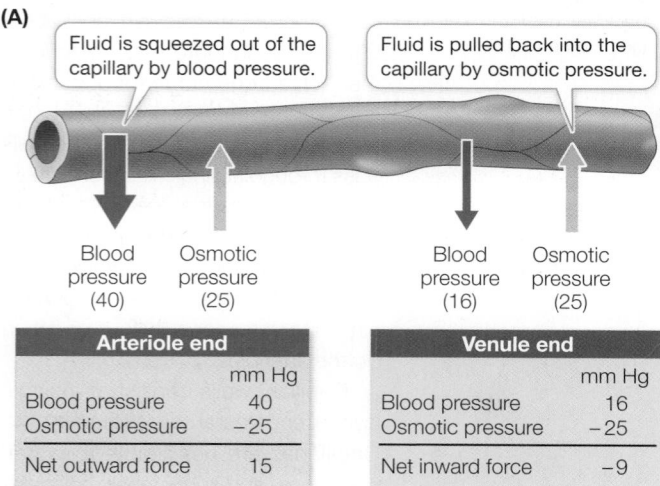

Fluid is squeezed out of the capillary by blood pressure.

Fluid is pulled back into the capillary by osmotic pressure.

Blood pressure (40)   Osmotic pressure (25)

Blood pressure (16)   Osmotic pressure (25)

Arteriole end	
	mm Hg
Blood pressure	40
Osmotic pressure	−25
Net outward force	15

Venule end	
	mm Hg
Blood pressure	16
Osmotic pressure	−25
Net inward force	−9

(B)

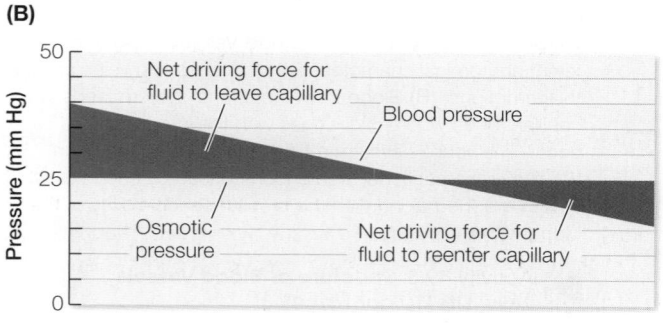

Net driving force for fluid to leave capillary

Blood pressure

Osmotic pressure

Net driving force for fluid to reenter capillary

Pressure (mm Hg) — 50, 25, 0

Figure 49.15 Starling's Forces   Starling's model explains fluid balance in capillary beds. **(A)** When blood hydrostatic pressure is greater than blood osmotic pressure, fluid leaves the capillaries; when blood hydrostatic pressure falls below blood osmotic pressure, fluid returns to the capillaries. **(B)** The balance of these two forces changes over the capillary bed as blood hydrostatic pressure falls.

tract, where nutrients are absorbed, and in the kidneys, where wastes are filtered. Capillary endothelial cells also contain transporters that can facilitate the passage of specific molecules, such as glucose and lactate. Overall, permeability varies widely in different capillary beds—an important consideration in the design and delivery of drugs.

The capillaries of the brain are a special case, being rather impermeable and wrapped by glia. Not much can pass through them other than lipid-soluble substances (including alcohol and anesthetics). This high selectivity of brain capillaries is known as the **blood–brain barrier**. Even in the brain, however, there are specific regions where the capillaries are more permeable, enabling the brain to detect non-lipid-soluble hormones.

Blood flows back to the heart through veins

The pressure of the blood flowing from capillaries to venules is low and insufficient to propel blood back to the heart. The walls of veins are more expandable than the walls of arteries, and blood tends to accumulate in veins. As much as 60 percent of your total blood volume may be in your veins when you are resting. Because of their high capacity to stretch and store blood, veins are called capacitance vessels.

Blood flow through veins that are above the level of the heart is assisted by gravity. Below the level of the heart, however, venous return is against gravity. The most important force propelling blood from these regions is the squeezing of the veins by the contractions of surrounding skeletal muscles. As muscles contract, the veins are compressed and blood is squeezed through them. Blood flow may be temporarily obstructed during a prolonged muscle contraction, but when muscles relax, blood is free to move again. One-way valves in the veins of the extremities prevent backflow of blood. Thus whenever a vein is squeezed, blood is propelled forward toward the heart (**Figure 49.16**).

In a resting person, gravity causes blood accumulation in the veins of the lower body and exerts backpressure on the capillary beds. This backpressure shifts the balance between blood hydrostatic pressure and osmotic pressure, causing increased loss of fluid to the intercellular spaces. That is why your feet swell during a long airline flight.

Because of the one-way valves in the veins of the legs, the contractions of leg muscles act as auxiliary vascular pumps when an animal walks or runs and facilitate the return of blood to the heart from the lower body. As a greater volume of blood is returned to the heart, the heart contracts more forcefully and its pumping action is enhanced. The heartbeat gets stronger because of a property of cardiac muscle cells described by the **Frank–Starling law**: if the cardiac muscle cells are stretched, as they are when the volume of returning blood increases, they contract more forcefully.

The actions of breathing also help return venous blood to the heart. The muscles involved in inhalation create negative pressure that pulls air into the lungs (see Figure 48.11), and this negative pressure also pulls blood toward the chest, increasing venous return to the right atrium. In addition, some of the largest veins closest to the heart contain smooth muscle that contracts at the onset of exercise. Contraction of veins can rapidly increase venous return and stimulate the heart in accord with the Frank–Starling law, increasing cardiac output.

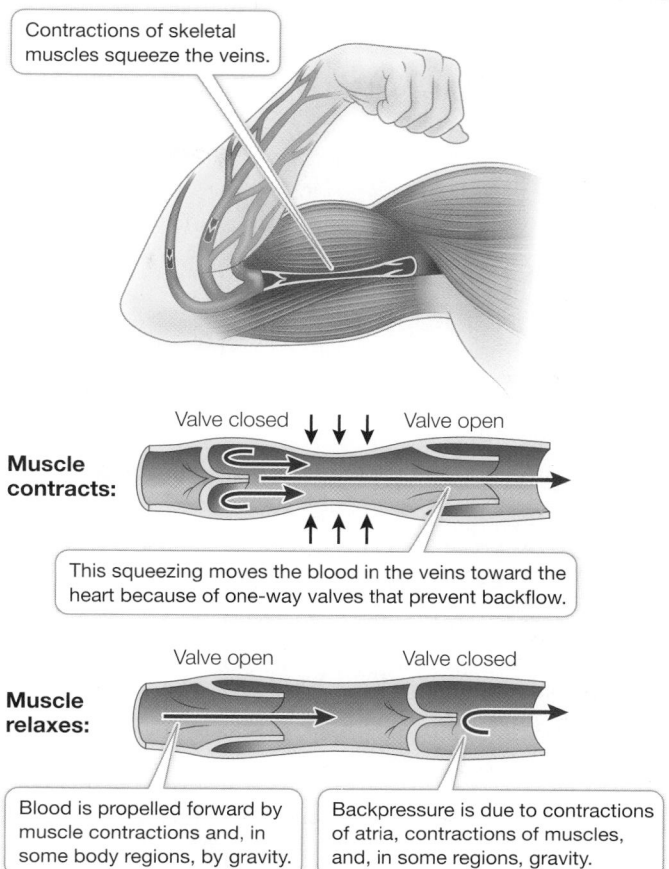

Contractions of skeletal muscles squeeze the veins.

Valve closed ↓ ↓ ↓   Valve open

Muscle contracts:

This squeezing moves the blood in the veins toward the heart because of one-way valves that prevent backflow.

Valve open   Valve closed

Muscle relaxes:

Blood is propelled forward by muscle contractions and, in some body regions, by gravity.

Backpressure is due to contractions of atria, contractions of muscles, and, in some regions, gravity.

Figure 49.16 One-Way Flow   Veins of the extremities have valves that prevent blood from flowing backward, and contractions of skeletal muscle help move blood toward the heart.

Lymphatic vessels return interstitial fluid to the blood

The interstitial fluid contains water and other small molecules, but no RBCs, and less protein than found in plasma. A separate system of vessels—the *lymphatic system*—returns interstitial fluid to the blood. Each capillary bed contains at least one blind-ended lymph capillary.

> ∗ **connect the concepts**   The lymphatic system is introduced in conjunction with the immune system. See Key Concept 41.1.

Interstitial fluid that enters the lymphatic vessels is called **lymph**. Fine lymphatic capillaries merge into progressively larger vessels and ultimately into the right and the left **lymphatic ducts**—that empty into large veins at the base of the neck (see Figure 41.2). The left lymphatic duct, also known as the **thoracic duct**, carries most of the lymph from the lower part of the body and is larger than the right lymphatic duct. Lymphatic vessels, like veins, have one-way valves that keep the lymph flowing toward the lymphatic ducts. Therefore lymph, like blood, is propelled toward the heart by skeletal muscle contractions and breathing movements. **Lymph nodes** along the major lymphatic vessels are a major site of lymphocyte production and of the phagocytic action that removes microorganisms and other foreign materials from the circulation.

(A) Healthy

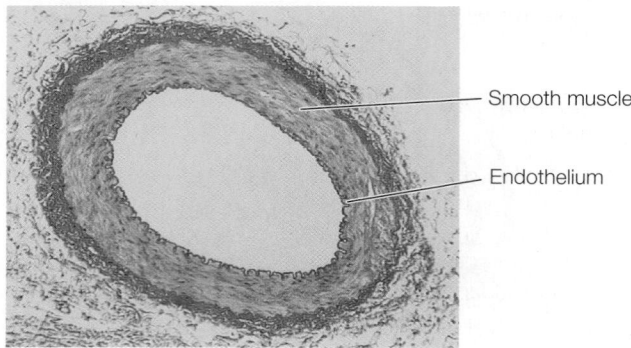

Smooth muscle

Endothelium

(B) Atherosclerotic

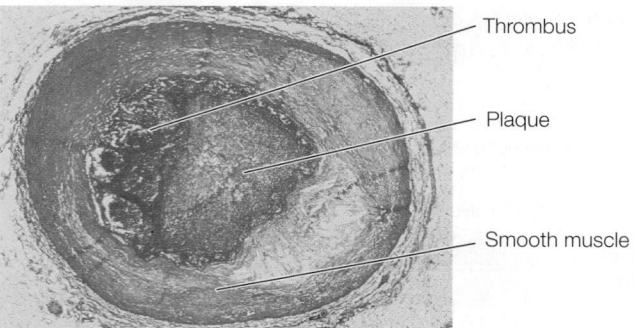

Thrombus

Plaque

Smooth muscle

Figure 49.17 Atherosclerotic Plaque   (A) A healthy, clear artery. **(B)** An atherosclerotic artery, clogged with plaque and a thrombus.

Vascular disease is a killer

As mentioned at the beginning of this chapter, cardiovascular disease is responsible for about one-fourth of all deaths each year in the United States, and the same is true in Europe. The immediate cause of most of these deaths is not a defect in heart muscle as in cases of HCM, but is heart attack or stroke—both of which are usually the result of a disease called **arteriosclerosis** ("hardening of the arteries") that begins many years before symptoms are detected.

Healthy arteries have a smooth internal lining of endothelial cells (**Figure 49.17A**) that can be damaged by chronic high blood pressure, smoking, a high-fat diet, or microorganisms. Deposits called **plaque** begin to form at sites of endothelial damage, initiating an early stage of arteriosclerosis called **atherosclerosis**. First, the damaged endothelial cells attract certain white blood cells to the site. These cells are then joined by smooth muscle cells migrating from the deeper layers of the arterial wall. Lipids, especially cholesterol, are deposited in these cells, so that the developing plaque becomes fatty. Fibrous connective tissue made by the invading smooth muscle cells in the plaque, along with deposits of calcium, gradually make the artery wall less elastic—hence "hardening of the arteries." Plaque deposits narrow the artery and causes turbulence in the blood flow. Blood platelets stick to the plaque (see Figure 49.12) and initiate formation of an intravascular blood clot, a **thrombus** that can block the artery (**Figure 49.17B**).

The blood supply to the heart muscle flows through the **coronary arteries**, which are highly susceptible to atherosclerosis. As these arteries narrow, blood flow to the heart muscle decreases, causing the symptoms of chest pain and shortness of breath during mild exertion. A person with atherosclerosis is at high risk of forming a thrombus in a

coronary artery. This condition, called **coronary thrombosis**, can totally block the vessel, causing a **myocardial infarction** (heart attack).

A piece of a thrombus that breaks loose, called an **embolus**, is likely to travel to and become lodged in a vessel of smaller diameter, blocking its flow (an **embolism**). Arteries already narrowed by plaque formation are likely places for an embolism. An embolism in an artery in the brain causes the cells fed by that artery to die. This event is a **stroke**. The specific damage resulting from a stroke, such as memory loss, speech impairment, or paralysis, depends on the location of the blocked artery in the brain.

Important risk factors for developing atherosclerosis are your genetic predisposition and your age. Environmental risk factors also play a large role, however. These include high-fat and high-cholesterol diets, smoking, and a sedentary lifestyle. Certain untreated medical conditions such as hypertension (high blood pressure), obesity, and diabetes are also risk factors for atherosclerosis. For those who have a genetic predisposition to atherosclerosis, it is even more important to minimize environmental risk factors. Changes in diet and behavior and treatment of predisposing medical conditions can prevent and reverse early atherosclerosis and help fend off this silent killer.

49.4 recap

Blood is a fluid tissue with cellular components that play roles in transport of respiratory gases, immune system function, and blood clotting. The properties of the arteries, arterioles, capillaries, venules, and veins reflect their functions. Exchanges between the blood and interstitial fluids occur in the smallest of those vessels, the capillaries. These exchanges are governed by Starling's forces and the movement of HCO_3^- into the blood. Lymphatic vessels return interstitial fluid to the blood.

learning outcomes

You should be able to:

- Describe the features, life cycle, and function of a red blood cell.
- Describe the blood clotting process.
- Explain the role of arterioles in the control of blood flow.
- Explain the forces that cause water to leave and reenter the capillaries.
- Describe the factors that affect the return of blood in veins to the heart.

1. Why do athletes train at high altitudes?
2. Why is the blood clotting process called a cascade?
3. Why are arterioles called resistance vessels and veins called capacitance vessels, and what do these terms imply about function?
4. In advanced starvation, the body breaks down its blood plasma proteins for energy. At this stage of disease, fluid accumulates in the abdomen and the extremities. What is the connection between these two consequences of starvation?

Every tissue in the body requires an adequate flow of oxygen-saturated blood. Blood flow depends on the maintenance of an appropriate blood pressure, and the distribution of blood flow throughout the body depends on control of the resistance in the blood vessels supplying different tissues.

key concept
49.5
The Circulation Is Controlled by Hormonal and Neural Signals

When we investigate how a physiological process is regulated, we start by identifying the critical components of that process, how they can be controlled, and the information used to govern that control. Because blood flow depends on pressure, we can identify the mean arterial pressure as a critical variable of the circulatory system.

focus your learning

- Local blood flow is controlled to meet local metabolic need by autoregulatory responses of arteriole smooth muscles.
- Mean arterial pressure is regulated through neural and hormonal control of smooth muscles in arterioles.
- Baroreceptors in the walls of large arteries continuously convey blood pressure information to the cardiovascular regulatory center in the brain, which controls the neural and hormonal mechanisms that influence mean arterial pressure.

Pressure in the aorta oscillates between systole and diastole, and the duration of diastole is usually twice as long as that of systole (see Figure 49.4). Therefore a simple way of estimating the mean arterial pressure (MAP) is to take the diastolic pressure and add to it one-third of the difference between systolic and diastolic pressure. However, to understand the physiological mechanisms that control MAP, we have to consider the factors that determine it. MAP is determined by the cardiac output (CO) and the resistance to flow in the blood vessels, or total peripheral resistance (TPR):

$$MAP = CO \times TPR$$

Since CO is equal to the heart rate (HR) times how much blood the heart pumps with each beat (stroke volume [SV]), the critical relationships can be expressed as:

$$MAP = HR \times SV \times TPR$$

HR, SV, and TPR are controlled by neural and hormonal mechanisms at both the local and systemic levels.

At the local level, each tissue controls its own blood flow through **autoregulatory mechanisms** that alter the diameter and therefore the resistance of the arterioles serving it. The collective autoregulatory actions in the arterioles in all tissues of the body influence TPR and therefore MAP. If many arterioles suddenly dilate, TPR goes down and MAP falls. If many arterioles constrict, TPR goes up and MAP goes up. Therefore changes in MAP provide information about changing needs of the body. In addition, as blood flows through capillary beds, its composition changes—its CO_2 content goes up and its O_2 content goes down. Thus blood composition also provides information the body uses to regulate the circulatory system.

The nervous and endocrine systems respond to changes in MAP and blood composition by changing breathing rate, heart rate, stroke volume, and peripheral resistance to match the metabolic needs of the body.

Activity 49.4 **Blood Pressure and Heart Rate Simulation**
www.Life11e.com/ac49.4

Autoregulation matches local blood flow to local need

The amount of blood that flows through a capillary bed is controlled by the smooth muscle of the arteries and arterioles feeding that bed. **Figure 49.18** illustrates the flow of blood in a typical capillary bed. Blood flows into the bed from an arteriole. Smooth muscle "cuffs," or **precapillary sphincters**, on the arteriole can shut off the supply of blood to the capillary bed. When the precapillary sphincters are relaxed and the arteriole is open, blood flow through the capillaries increases.

Autoregulation depends on the sensitivity of the smooth muscle to its local chemical environment. Low O_2 concentrations and high CO_2 concentrations cause the smooth muscle to relax, thus increasing the supply of blood, which brings in more O_2 and carries away CO_2—a response known as hyperemia, which means "excess blood." Increases in other by-products of metabolism, such as lactic acid, hydrogen ions, potassium, and adenosine (all of which increase in exercising muscle), also promote hyperemia. Hence activities that increase the metabolism of a tissue also induce hyperemia in that tissue.

Arterial pressure is regulated by hormonal and neural mechanisms

Control and regulation of the circulatory system begins with the local autoregulatory mechanisms that alter the resistance of arteries and arterioles feeding capillary beds. The demands of the capillary beds influence MAP and blood composition. Both of these provide

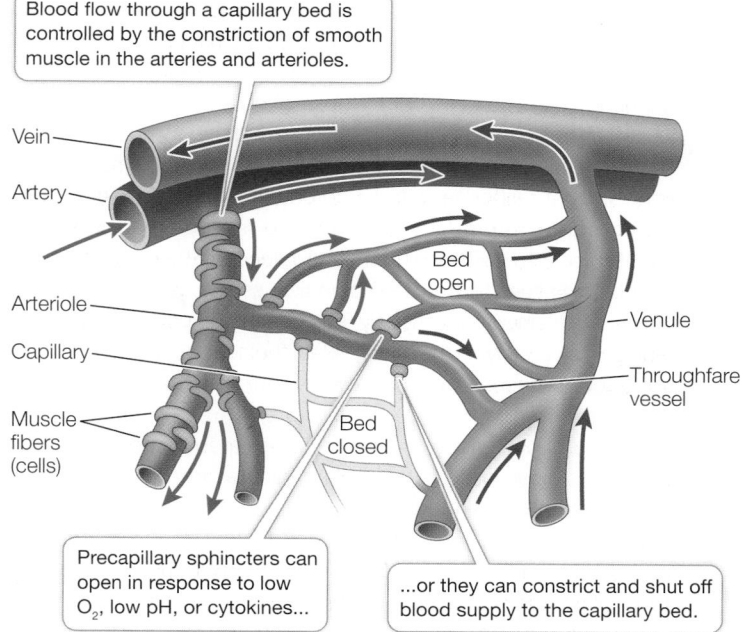

Blood flow through a capillary bed is controlled by the constriction of smooth muscle in the arteries and arterioles.

Vein

Artery

Arteriole

Capillary

Muscle fibers (cells)

Bed open

Venule

Throughfare vessel

Bed closed

Precapillary sphincters can open in response to low O_2, low pH, or cytokines...

...or they can constrict and shut off blood supply to the capillary bed.

Figure 49.18 Local Control of Blood Flow Low O_2 concentrations or high levels of metabolic by-products cause the smooth muscle of the arteries and arterioles to relax, thus increasing the supply of blood to the capillary bed.

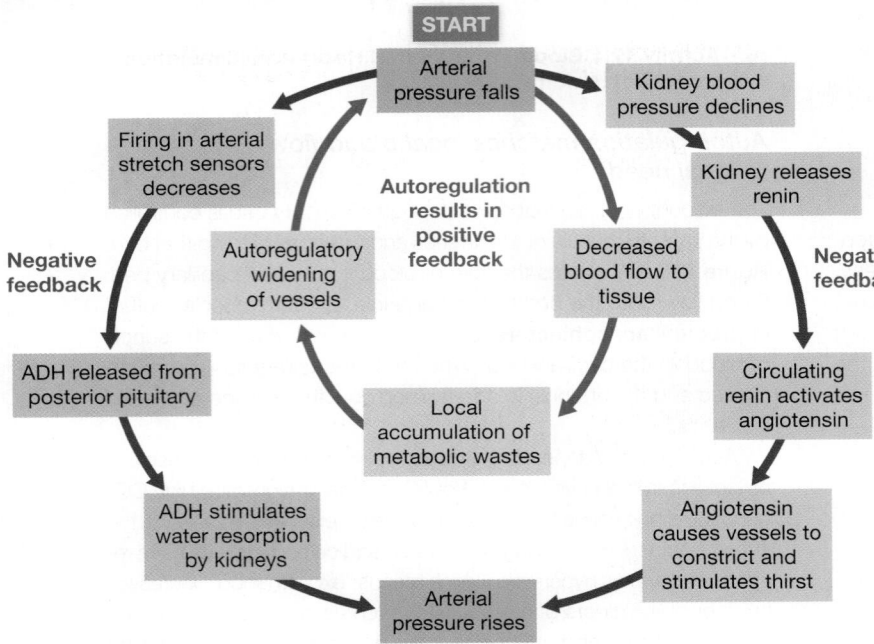

Figure 49.19 Control of Blood Pressure through Local and Systemic Mechanisms A drop in arterial pressure reduces blood flow to tissues, resulting in local accumulation of metabolic wastes. This change in the extracellular environment stimulates autoregulatory opening of the arterioles. A fall in central blood pressure is prevented by negative feedback mechanisms (including the release of antidiuretic hormone, ADH) that constrict arterioles in less essential tissues and stimulate maintenance of blood volume and blood pressure.

information for the control of endocrine and neural responses that act to return blood pressure and composition to normal. Thus circulatory functions are matched to the regional and overall needs of the body.

Arteries and arterioles are innervated by the autonomic nervous system, particularly the sympathetic division. The sympathetic postganglionic neurotransmitter norepinephrine binds to receptors in smooth muscle in blood vessels in the gut and other tissues not essential for "fight or flight" and causes these vessels to constrict, resulting in reduced blood flow through them and an elevation in MAP. As we discussed earlier in this chapter, increased sympathetic activity increases heart rate, and by increasing the strength of the cardiac muscle contraction, it also increases stroke volume.

Hormones also play a role in regulating arterial pressure. Epinephrine has actions similar to those of norepinephrine and is released from the adrenal medulla during massive sympathetic activation stimulated by a fall in arterial pressure or by activation of the fight-or-flight response to a dangerous threat. Another hormone, **angiotensin**, is produced when blood pressure in the kidneys falls (**Figure 49.19**). These hormones influence arterioles located in peripheral tissues (extremities) or in tissues whose functions need not be maintained continuously (such as the digestive system). By reducing blood flow in those arterioles, the hormones increase central blood pressure and blood flow to essential organs such as the heart, brain, and kidneys.

The autonomic nervous system activity that controls heart rate and constriction of blood vessels originates in a cardiovascular control center in the medulla. Many inputs converge on this central integrative network and influence the commands it issues via parasympathetic and sympathetic nerves (**Figure 49.20**). Of special importance is incoming information about changes in blood pressure from **baroreceptors** (stretch receptors) in the walls of the large arteries leading to the brain—the aorta and the carotid arteries.

Increased activity in baroreceptors of the large arteries signals rising blood pressure and inhibits sympathetic nervous system

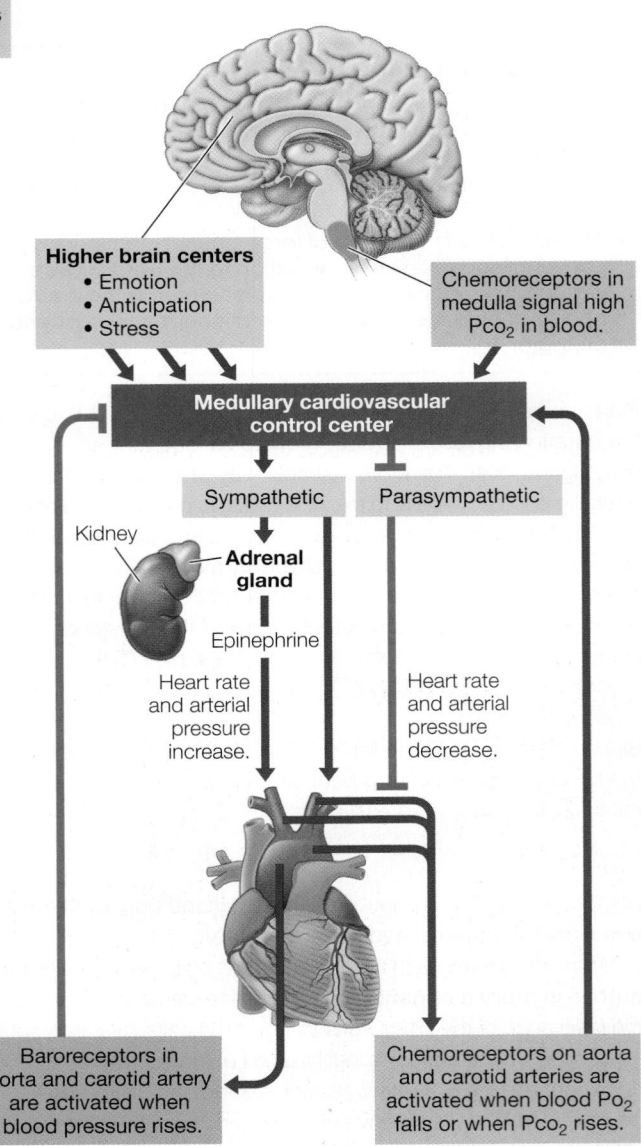

Figure 49.20 Regulating Cardiac Output The autonomic nervous system controls heart rate in response to information about blood pressure and blood composition originating in baroreceptors and chemoreceptors shown at the bottom of the figure. Information from these sensors goes to the cardiovascular control center in the medulla, where it is integrated with other information. The medullary center generates responses in the sympathetic and parasympathetic nervous systems that control cardiac output.

signaling to arteries and arterioles while increasing parasympathetic signaling to the heart's pacemaker. As a result, the heart slows and arterioles in peripheral tissues dilate, reducing blood pressure. If pressure in the large arteries falls, the activity of the baroreceptors decreases, stimulating sympathetic output to the arteries and arterioles while reducing parasympathetic output to the heart's pacemaker. As a result, the heart beats faster and the arterioles in peripheral tissues constrict, increasing blood pressure.

Another hormone that helps stabilize blood pressure is **antidiuretic hormone** (**ADH**, also called **vasopressin**), which is secreted by the posterior pituitary in response to a fall in the activity of the baroreceptors, signaling a fall in arterial pressure. ADH causes the kidneys to reabsorb more water and thereby maintain blood volume and increase blood pressure (see Figure 51.14). Increased activity of the baroreceptors inhibits the release of ADH, and as a result the kidneys excrete more water, reducing blood volume and contributing to a fall in arterial pressure (see Figure 49.19).

Other information that causes the cardiovascular control center to increase heart rate and blood pressure comes from *chemoreceptors in the medulla, aorta, and the carotid arteries.

---

*connect the concept   Key Concept 48.5 discusses the mechanisms for sensing CO_2 and O_2 concentrations in the blood. The medullary chemoreceptors are activated by increases in CO_2 levels and decreases in pH, and the carotid and aortic bodies are activated by falling O_2 levels. Chemoreceptors send signals to the cardiovascular regulatory center as well as to the respiratory regulatory center.

49.5 recap

The delivery of blood to tissues is controlled locally by autoregulatory mechanisms that dilate or constrict arterioles. These local actions are translated into alterations in central blood pressure and composition that are detected by neural and hormonal mechanisms, which then mediate corrective cardiovascular adjustments.

learning outcomes

You should be able to:

- Describe mechanisms underlying autoregulation of local blood flow.
- Describe the role of the autonomic nervous system in regulating MAP.
- Draw a diagram showing how an increase or a decrease in the firing rate of baroreceptors causes a change in MAP.
- Apply knowledge about local and systemic controls of cardiovascular functions to describe interactions between the two.

1. Why is autoregulation in Figure 49.19 described as a positive feedback loop?
2. Distinguish between the roles of the two divisions of the autonomic nervous system (ANS) in regulating MAP?
3. Baroreceptor neurons have a range of firing rates depending on their degree of stretch, and at rest they are firing at the midpoint of that range. Why is this adaptive?
4. Explain how local autoregulatory controls of blood flow can stimulate hormonal adjustments of MAP.

---

investigatinglife

 Are there potential therapies that could treat HCM?

Any organ of your body can shut down for a while and you can recover—with one exception, your heart. Cardiac arrest lasting only minutes can result in death. But why should the heart of a healthy, physically fit, top-performing athlete stop? Just as physical conditioning results in making skeletal muscles bigger and stronger, it does the same for the heart. The heart responds to increased demand by getting bigger and stronger. However, if an individual has a large number of mutations in genes responsible for proteins involved in contractions of cardiac muscle, that muscle will be less efficient. To compensate for that inefficiency, the heart gets even bigger than a normal heart would when repeatedly challenged by excessive exercise. Especially in the walls of the left ventricle, hypertrophy can have two serious consequences: (1) The opening of the aorta is at the top of the heart next to the septum that divides the two ventricles. Enlargement of that septum can block the entrance to the aorta, so the cardiac output falls. (2) The bundle of His and the left and right bundle branches that convey the pacemaker potential to the ventricular muscle course down through the septum.

Hypertrophy of the septum can interfere with conduction of these critical signals. If advanced conditions of HCM are discovered, there are surgical procedures that remove some of the heart muscle.

Future directions

The mutations responsible for the development of HCM are not rare, but most people who carry them live their entire lives without even knowing they are at risk for HCM. Why? Most people do not place the excessive demands on their hearts that athletes do. So what can be done to reduce the risk of death by HCM? Preventative measures such as genetic screening could allow carriers of the mutations to make decisions about whether or not to pursue extreme physical activity. Physical examinations including ECGs, sonograms, and even magnetic resonance imaging (MRI) would enable diagnosis and recommendations to decrease levels of physical exertion. The work presented in Investigating Life: Silencing Mutant Myosin Genes suggests a potential future treatment through gene therapy. If safe reagents could be developed to silence the mutant genes, and those reagents were applied before severe HCM developed, they could have a protective benefit.

Chapter Summary 49

49.1 Circulatory Systems Serve Many Functions

- The metabolic needs of the cells of many small animals are met by direct exchange of materials with the external medium. The metabolic needs of the cells of larger animals are met by a **circulatory system** that transports nutrients, respiratory gases, and metabolic wastes throughout the body.

- In **open circulatory systems**, extracellular fluid called **hemolymph** leaves vessels and percolates through tissues before returning to the heart. In **closed circulatory systems**, blood—a portion of the extracellular fluid—is contained in a system of vessels. Closed circulatory systems have the ability to selectively direct blood, and therefore hormones and nutrients, to specific tissues. **Review Figure 49.1**

49.2 Vertebrate Circulatory Systems Evolved from Single to Double Circuits

- The circulatory system of vertebrates consists of a heart and a closed system of vessels containing blood that is separate from the interstitial fluid. **Arteries** and **arterioles** carry blood from the heart; **capillaries** are the site of exchange between blood and interstitial fluid; **venules** and **veins** carry blood back to the heart.

- The vertebrate circulatory system evolved from a single circuit in fishes to partially or completely separate pulmonary and systemic circuits in amphibians, reptiles, and mammals.

- In the single-circuit system of fishes, blood flow is unidirectional and is propelled by one-way valves between the **sinus venosus** and the **atrium**, between the atrium and the **ventricle**, and between the ventricle and the **bulbus arteriosus**.

- In birds and mammals, blood circulates through two completely separate circuits. The **pulmonary circuit** transports blood between the heart and lungs, and the **systemic circuit** transports oxygen-rich blood between the heart and tissues. **See Activity 49.1**

49.3 Heart Function Depends on Properties of Cardiac Muscle

- The mammalian heart has four chambers. Valves in the heart prevent the backflow of blood. **Review Figure 49.2, Activity 49.2**

- The **cardiac cycle** has two phases: **systole** refers to the contraction phase, and **diastole** refers to the relaxation phase. The sequential heart sounds ("lub-dup") are made by the closing of the heart valves. **Review Focus: Key Figure 49.4, Animation 49.1**

- Blood pressure can be measured using a sphygmomanometer and a stethoscope. **Review Figure 49.5**

- **Pacemaker cells** of the sinoatrial node set the heart rate as a result of the properties of their ion channels. The autonomic nervous system controls heart rate: sympathetic activity increases heart rate, and parasympathetic activity decreases it by altering the rate of depolarization of the pacemaker cell resting membrane potentials following the termination of systole. **Review Figures 49.6, 49.7**

- The **sinoatrial node** controls the cardiac cycle by initiating a wave of depolarization in the atria, which is conducted to the ventricles through a system consisting of the **atrioventricular node**, **bundle of His**, and Purkinje fibers. **Review Figure 49.8**

- Sustained contraction of ventricular muscle cells is due to long-duration action potentials that are generated by voltage-gated Na^+ and Ca^{2+} channels. **Review Figure 49.9**

- An **electrocardiogram** (**ECG** or **EKG**) records electrical events caused by the depolarizations and repolarizations of the cardiac muscles. **Review Figure 49.10**

49.4 Circulatory System Functions Depend on Blood and Blood Vessels

- Blood consists of a **plasma** portion (water, salts, and proteins) and a cellular portion (**erythrocytes** or red blood cells, platelets, and white blood cells). All of the cellular components are produced from stem cells in the bone marrow. **Review Figure 49.11**

- Erythrocytes transport oxygen. Their production in the bone marrow is stimulated by **erythropoietin**, which is produced in response to **hypoxia** (low oxygen levels) in the tissues.

- **Platelets**, along with circulating proteins, are involved in **blood clotting**, which results in a meshwork of **fibrin** threads that help seal damaged vessels. **Review Figure 49.12**

- Abundant smooth muscle cells allow vessels to change their diameter, altering their resistance and thus blood flow. Arteries have elastic fibers that enable them to withstand high pressures. **Review Figure 49.13, Activity 49.3**

- Capillary beds are the site of exchange of materials between blood and tissue fluid.

- **Starling's forces** suggest that blood volume is maintained in the capillary beds by an exchange of fluids driven by both blood hydrostatic pressure and osmotic pressure. **Review Figure 49.15**

- Bicarbonate ions in the blood plasma contribute to the osmotic forces that draw water back into capillaries.

- The ability of a specific molecule to cross a capillary wall depends on the architecture of the capillary and the chemical characteristics of the molecule.

- Veins have a high capacity for storing blood. Aided by gravity, by contractions of skeletal muscle, and by the actions of breathing, they return blood to the heart. **Review Figure 49.16**

- The **Frank–Starling law** describes forces that increase cardiac output, such as stretch of the cardiac muscles cells caused by increased venous return.

- The **lymphatic system** returns the interstitial fluid to the blood.

49.5 The Circulation Is Controlled by Hormonal and Neural Signals

- Blood flow through capillary beds is controlled by local **autoregulatory mechanisms**, hormones, and the autonomic nervous system. **Review Figure 49.18, Activity 49.4**

- Blood pressure is controlled in part by the hormones **ADH** and **angiotensin**, which stimulate contraction of blood vessels. **Review Figure 49.19**

- Heart rate is controlled by the autonomic nervous system, which responds to information about blood pressure and blood composition that is integrated by regulatory centers in the medulla. **Review Figure 49.20**

Go to **LearningCurve** (in **LaunchPad**) for dynamic quizzing that helps you solidify your understanding of this chapter. **LearningCurve** adapts to your responses, giving you the practice you need to master each key concept.

Apply What You've Learned

Review

49.2 Ectothermic reptiles have the ability to control blood flow to the pulmonary circuit, enabling them to conserve energy when they are not breathing.

Original Paper: Eme, J., J. Gwalthney, T. Owerkowicz, J. M. Blank and J. W. Hicks. 2010. Turning crocodilian hearts into bird hearts: Growth rates are similar for alligators with and without right-to-left cardiac shunt. *Journal of Experimental Biology* 213: 2673–2680.

Crocodilian hearts are four-chambered, just like those of birds and mammals. But crocodiles have two aortas—one from each ventricle—and the flow into each is controlled by a valve. The two aortas are connected by a shunt after they leave the heart. Thus, the right ventricle of the crocodile heart can pump blood to the lungs or to the rest of the body. If the crocodile is breathing, the resistance in the pulmonary circuit is low, and blood flows through the pulmonary artery because high backpressure from the left ventricle exerted through the shunt keeps the right aortic valve shut. When the animal is not breathing, the pulmonary resistance increases, and blood from the right ventricle flows into the right aorta. Researchers tried to demonstrate an adaptive advantage to the aortic shunt in crocodiles.

Working with young American alligators (5–7 months of age), the researchers surgically closed the right aorta near the right ventricle so that blood from the right ventricle could not flow through it to the body but had to flow through the lungs, as in an avian or mammalian circulatory system. The researchers performed sham operations on a similar group of animals, as controls. Sham operations were carried out exactly as the experimental operations, except that the aortas were not actually closed. After the crocodiles recovered from surgery, the researchers divided them into two experimental groups and two control groups that would be either sedentary (no exercise) or that would run on a treadmill to exhaustion three times a week. Over the next 21 months, the researchers measured each animal's body mass and body length every 2 weeks. Animals were provided enough food to satisfy their hunger at each feeding.

Data for each group were subjected to least squares analysis. Bars represent mean values determined for each group, and error bars are the standard errors of the means. The probabilities that the means in each comparison are due to chance alone are shown above the graphs.

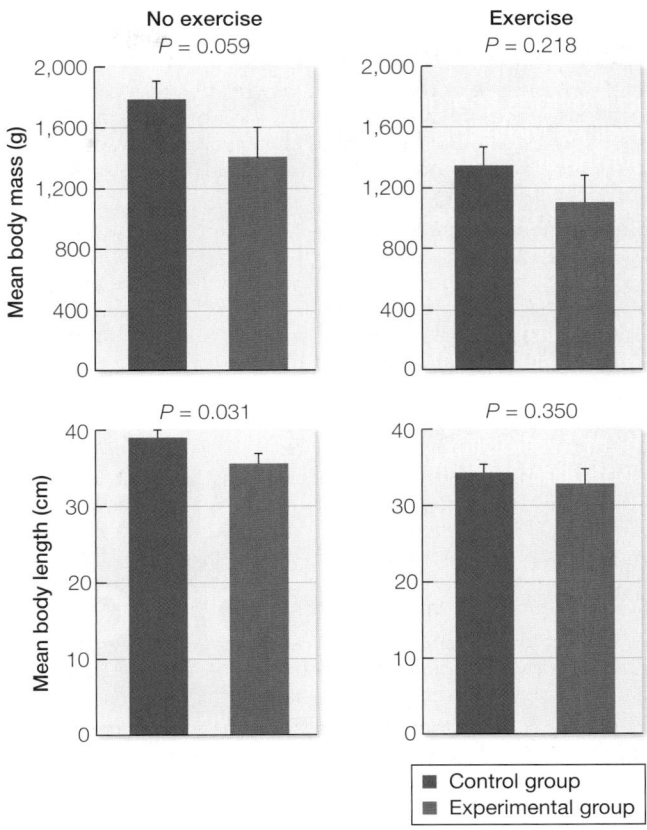

Questions

1. Create a diagram like that on p. 1048 showing the path of blood through a crocodilian heart. Mark the diagram to show the surgery done by the research group. On the diagram, show how the surgery changed the blood flow through the crocodilian heart. Write an explanation to accompany your figure. Then explain why the surgery made the crocodilian heart similar to a bird heart.

2. For an alligator, what advantage is gained by being able to shunt blood away from the lungs?

3. What do the data regarding body mass and body length suggest about the effects of blocking blood flowing from the right ventricle to the body? Does this provide evidence in support of the adaptive advantage of having a pulmonary bypass shunt? Explain your answer.

4. The animals in this study were maintained in captivity and supplied with food. How might these conditions limit the potential of the experiment to show an adaptive advantage of the aortic shunt?

Go to **LaunchPad** for the eBook, LearningCurve, animations, activities, flashcards, and additional resources and assignments.

50

Nutrition, Digestion, and Absorption

In **Pima populations**, historic occurrences of food deprivation may have imposed selection for genes that improve the efficiency of managing the energy obtained from food. With modern diets and lifestyles, these "thrifty genes" can contribute to obesity and diabetes.

▶ investigating**life**

Thrifty Phenotypes

For thousands of years the Pima of southwestern North America were hunters and gatherers, supplementing their diet with subsistence agriculture. Their environment was arid, so they developed sophisticated irrigation systems; even so, they frequently encountered drought and subsequent starvation. Today most individuals of the ethnic Pima population in North America are clinically obese. As a population they are one of the heaviest in the world.

Obesity contributes to health problems such as diabetes, high blood pressure, and heart disease. Diabetes incidence in the Pima is seven times the national average. Two-thirds of Pima adults over the age of 40 are diabetic, and diabetes is occurring in younger individuals than previously seen. What has caused such a radical health change in the population? Two interacting factors are involved: genetics and lifestyle.

Geneticists hypothesize that recurring episodes of starvation produce strong selective pressure for "thrifty genes"—particular alleles of the genes involved in digestion, absorption, and energy storage that result in greater-than-average efficiency in converting food into fat. Thrifty genes would carry a strong selective advantage when food is scarce. The Pima display a "thrifty" phenotype. They have low resting metabolic rates and convert food into fat readily. Insulin facilitates conversion of

dietary sugar into fat. For many Pima, consuming a standard amount of glucose causes their insulin levels to rise three times higher than it does in people of European ancestry.

Modern lifestyle also plays a role in the Pima obesity epidemic. Instead of eating their traditional diet, the Pima now eat a high-calorie, high-fat Western diet, and they engage in less physical activity than their ancestors did. A population of Pima living in the Sierra Madre of northern Mexico are genetically the same as the Arizona population. However, they eat traditional foods and live a traditional lifestyle that involves much physical activity. Obesity and diabetes are not prevalent among the Mexico Pima.

A high-calorie diet and sedentary lifestyle affect not just the Pima but contribute to the overall increase in obesity throughout the U.S. population. Recently an additional factor that might contribute to obesity and diabetes has been discovered—the type of food that is eaten. The composition of the diet influences the populations of gut microbes, and those microbes influence outcomes of digestion. Apparently calories are not all equal; their impact depends on their origin.

Q **A** If the nutritional values were the same, could different foods have different impacts on health?

key concept
50.1 Food Provides Energy As Well As Materials for Biosynthesis

Animals are **heterotrophs**—they derive their nutrition from eating other organisms. In contrast, **autotrophs** (most plants, some bacteria, some archaea, and some protists) can use solar energy or inorganic chemical energy to synthesize all of their components. Of course, some ***plants** are also heterotrophs.

***connect the concepts**   Carnivorous and parasitic plants are discussed in Key Concept 35.5.

---

focus your learning

- Fats, carbohydrates, and proteins in food provide energy.
- Energy is stored in animal bodies as glycogen and fat.
- Some small organic molecules required for biosynthesis, including essential amino acids and fatty acids, must come from food.
- Vitamins are organic compounds that the animal cannot synthesize but are required for healthy cell function.
- Animals require a variety of mineral nutrients.

Directly and indirectly, heterotrophs take advantage of—indeed, depend on—the organic synthesis carried out by autotrophs and have evolved an enormous diversity of adaptations to exploit this resource (**Figure 50.1**). In this section we describe how animals use food, be it plants or other animals, to obtain energy and building blocks of complex molecules. We also consider the need for special mineral nutrients and organic molecules and the diseases that result when they are lacking in the diet.

Energy needs and expenditures can be measured

Energy—the capacity to do work—comes in different forms, including electric, heat, chemical, and nuclear energy. As discussed in Chapter 8, a calorie (note the small *c*) is a unit of heat energy; specifically, it is the amount of heat necessary to raise the temperature of 1 gram of water 1°C. Because this is such a tiny amount of energy, physiologists commonly use the **kilocalorie** (**kcal**) as a unit of measure (1 kcal = 1,000 calories). Nutritionists also use the kilocalorie as a standard unit of energy, but they traditionally refer to it as the **Calorie** (**Cal**), which is capitalized to distinguish it from the single calorie.

Just about any food container you pick up in the United States carries the label "Nutrition Facts," which includes the item "Calories." How do Calories (or kcal) relate to the discussion in Chapter 9 about how energy in the chemical bonds of food molecules is transferred to the high-energy phosphate bonds of adenosine triphosphate (ATP) and is then used to do cellular work? And why do we use heat energy as a measure of nutrition?

The reason is found in the ***laws of thermodynamics**, which tell us that energy cannot be created or destroyed, but can be converted from one form to another.

(A)

(B)

Figure 50.1 Heterotrophs Get Energy from Autotrophs
(A) Herbivores get their energy directly from autotrophs. The large herbivores of the African grasslands must consume huge amounts of plant matter to fulfill their nutritional needs. (B) A carnivore's energy is indirectly obtained from autotrophs, since the energy stored in a prey animal was originally obtained from autotrophs.

***connect the concepts**   The laws of thermodynamics and their application to biological energetics are described in Key Concept 8.1.

However, every energy conversion is inefficient; a large portion of the original energy always ends up as heat. Whether we are using the energy in glucose to make ATP or are using that ATP to power muscle contraction or ion transport, most of the available chemical energy is lost as heat. The bottom line is that if an animal is not growing, not doing any external work, and not changing its body temperature, the heat it loses to the environment is a measure of its total energy expenditure, or metabolism.

An animal's energy needs must be met by the ingestion, digestion, and assimilation of food. The basal energy expenditure (basal metabolic rate, BMR) of a human is 1,300–1,500 Cal/day for an adult female and 1,600–1,800 Cal/day for an adult male. Any physical activity adds to the BMR. For a person doing sedentary work, about 30 percent of the total Calories expended are used for skeletal muscle activity; for a person doing heavy physical labor, more than 95 percent of caloric expenditure is for skeletal muscle activity.

The components of food that provide energy are fats, carbohydrates, and proteins. Fats yield 9.5 Cal/gram, carbohydrates 4.2

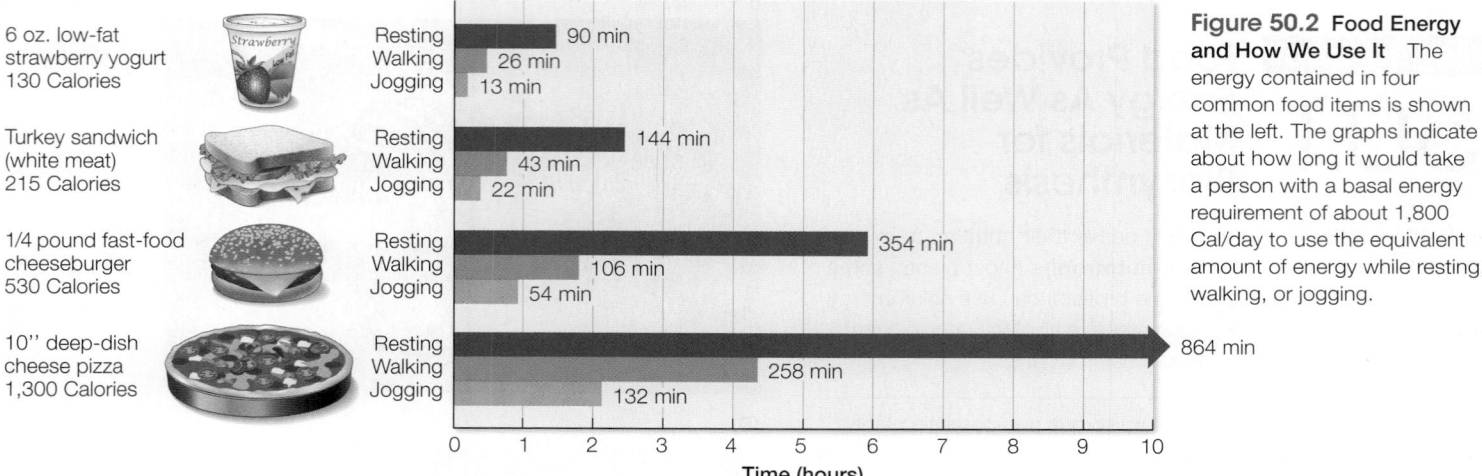

6 oz. low-fat strawberry yogurt 130 Calories
Resting 90 min
Walking 26 min
Jogging 13 min

Turkey sandwich (white meat) 215 Calories
Resting 144 min
Walking 43 min
Jogging 22 min

1/4 pound fast-food cheeseburger 530 Calories
Resting 354 min
Walking 106 min
Jogging 54 min

10" deep-dish cheese pizza 1,300 Calories
Resting 864 min
Walking 258 min
Jogging 132 min

Time (hours)

Figure 50.2 Food Energy and How We Use It The energy contained in four common food items is shown at the left. The graphs indicate about how long it would take a person with a basal energy requirement of about 1,800 Cal/day to use the equivalent amount of energy while resting, walking, or jogging.

Cal/gram, and proteins about 4.1 Cal/gram. **Figure 50.2** shows some equivalencies of food, energy, and energy consumption.

Even though the units calorie, kilocalorie, and Calorie remain in popular use, most scientists now use the International System of Units (ISU). In this system the basic unit of energy is the joule: 1 joule = 0.239 calories, and the measure of energy use is 1 joule/second = 1 watt. You are familiar with light bulb ratings, so think about that when you convert kcal/day into watts. The 1,700 Cal/day energy expenditure of the average man converts to 82 watts (note that a watt includes the time dimension, so it is a rate of energy use).

Thus it is possible to quantify the caloric value of any food an animal eats. It is also possible to quantify the caloric expenditure of any activity or behavior an animal performs. By comparing calories consumed with calories expended, we can construct **energy budgets** that allow ecologists and evolutionary biologists to apply a *cost–benefit analysis to feeding behavior.

---

*connect the concepts   Any behavior involves energetic costs, opportunity costs, and risk costs and can be measured in terms of its effect on reproductive success. Key Concept 52.4 explores costs and benefits of various behaviors, illustrating how measures of metabolism permit quantitative analysis of evolutionary adaptations.

Sources of energy are stored in the body

Although cells of the body use energy continuously, most animals do not eat continuously and so must store fuel molecules that can be released as needed between meals. Carbohydrates are stored in the liver and in muscle cells as glycogen, but the total glycogen stored represents only about a day's basal energy requirement. Fat is the most important form of stored energy in the bodies of animals. Not only does fat have more energy per gram than glycogen, but it can be stored with little associated water, making it more compact and less expensive to carry. Migrating birds store energy as fat to fuel their long flights;

if they had to store the same amount of energy as glycogen, they would be too heavy to fly. Proteins are not used as energy storage compounds, but the body's proteins can be metabolized as an energy source of last resort.

If an animal takes in too little food to meet its energy requirements, it starts to metabolize some of the molecules of its own body. This "self-consumption" begins with the energy storage compounds glycogen and fat. Once fat reserves are seriously depleted, the body increases its metabolism of proteins for energy (**Figure 50.3A**). The first proteins to be sacrificed are those of the blood plasma. The loss of plasma proteins decreases the osmotic concentration of the plasma, resulting in increased loss of fluid from the blood to the interstitial spaces (edema). Accumulation of fluid in the extremities and abdomen

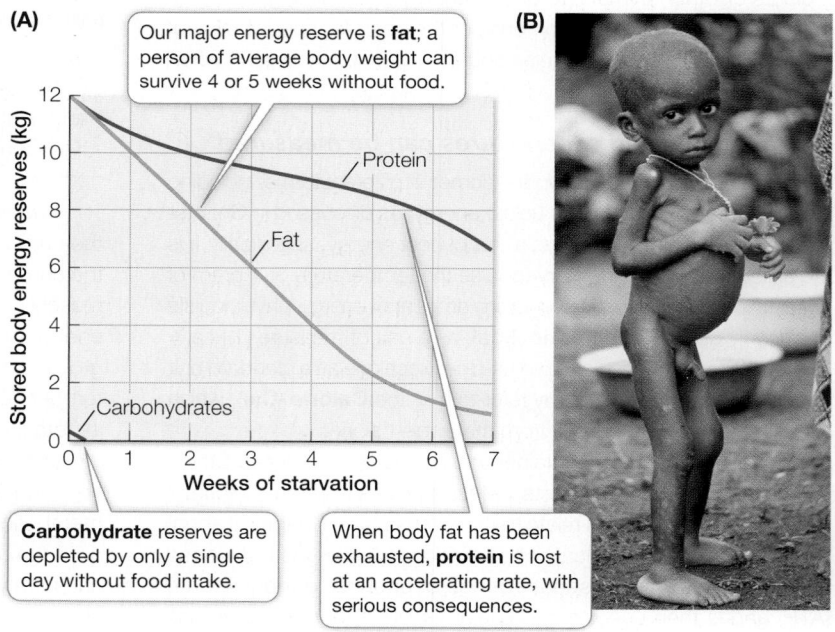

(A)

Our major energy reserve is **fat**; a person of average body weight can survive 4 or 5 weeks without food.

Protein

Fat

Carbohydrates

Stored body energy reserves (kg)

Weeks of starvation

Carbohydrate reserves are depleted by only a single day without food intake.

When body fat has been exhausted, **protein** is lost at an accelerating rate, with serious consequences.

(B)

Figure 50.3 The Course of Starvation (A) When a person is subjected to undernutrition, the body's energy reserves are depleted. (B) The swollen abdomen, hands, and feet of this boy are due to edema. Along with his spindly limbs, these are symptoms of kwashiorkor, a disease resulting from the body breaking down blood proteins and muscle tissue to fuel metabolism.

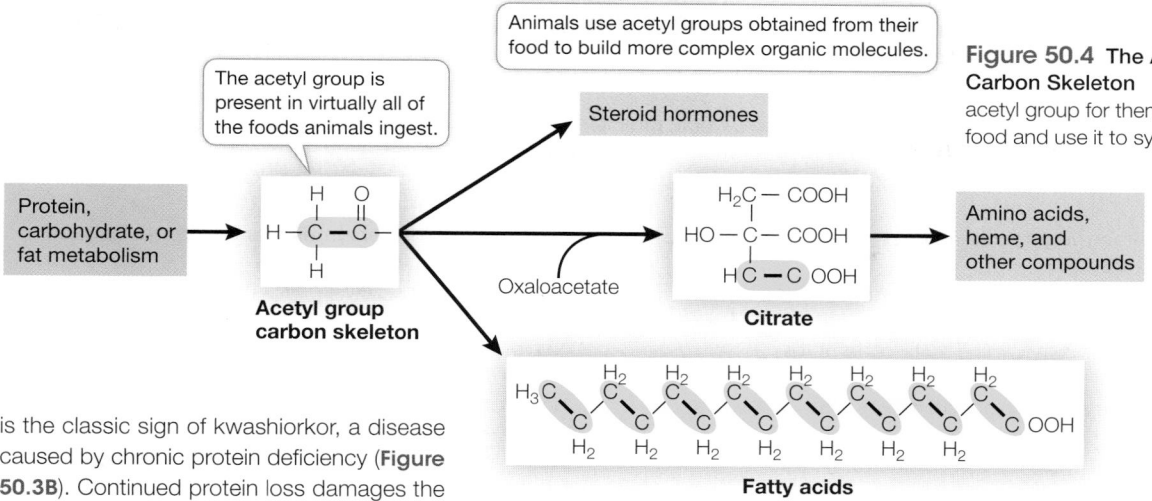

Figure 50.4 **The Acetyl Group Is an Acquired Carbon Skeleton** Animals cannot synthesize the acetyl group for themselves, but they ingest it in their food and use it to synthesize a wide variety of molecules.

is the classic sign of kwashiorkor, a disease caused by chronic protein deficiency (**Figure 50.3B**). Continued protein loss damages the body's organs, eventually leading to death.

When an animal consistently takes in more food than it needs to meet its energy requirements, the excess nutrients are stored as increased body mass. First glycogen reserves build up; then additional dietary carbohydrates, fats, and proteins are converted to body fat. In some species, such as hibernators, seasonal overnutrition is an important adaptation for surviving periods when food is not available. In humans, however, overnutrition can be a serious health hazard, increasing the risk of high blood pressure, heart attack, diabetes, and other disorders, as seen in the Pima that were discussed in the chapter opener.

Food provides carbon skeletons for biosynthesis

Every animal must take in certain organic molecules that it cannot synthesize for itself but needs to form the building blocks of its own complex organic molecules. The acetyl group (CH_3CO-) is one such required building block, supplying the **carbon skeleton** of larger organic molecules (**Figure 50.4**). Animals cannot synthesize acetyl groups from carbon, oxygen, and hydrogen molecules but must obtain them from food. Acetyl groups can be derived from the metabolism of almost any food, but they originate in plants.

Acetyl groups are never in short supply for an adequately nourished animal. However, some groups supplying carbon skeletons can be deficient in an animal's diet even if caloric intake is adequate. One such group includes certain amino acids, the building blocks of proteins. Animals can synthesize some of their own amino acids using carbon skeletons from acetyl or other groups and transferring to them amine groups ($-NH_2$) derived from other amino acids. However, most animals cannot synthesize all the amino acids they need and thus must obtain certain **essential amino acids** from food. If an animal does not take in enough of even one of its essential amino acids, its protein synthesis is impaired and its capacity to maintain enzymatic and transport functions is challenged.

Essential amino acids vary by species. Most researchers agree that adult humans must obtain eight essential amino acids from their food: isoleucine, leucine, lysine, methionine, phenylalanine, threonine, tryptophan, and valine. All eight are available in milk, eggs, meat, and soybean products, but most plant foods do not contain adequate quantities of all eight, so a strict vegetarian diet carries a risk of protein malnutrition. A **complementary diet** of plant foods, however, supplies all eight essential amino acids (**Figure 50.5**). In

general, grains (such as rice, wheat, and corn) are complemented by legumes (such as beans and peas). Long before the chemical basis for complementarity was understood, societies with little access to meat developed complementary diets. Many Central and South American peoples traditionally eat beans with corn, and the native peoples of North America complemented their beans with squash.

Human infants are thought to require four additional amino acids in their diets: histidine, tyrosine, cysteine, and arginine. Also, some amino acids are required by individuals with certain metabolic disorders who cannot synthesize them adequately. For example, individuals with the genetic disease phenylketonuria lack the enzyme for converting phenylalanine to tyrosine (see Key Concept 15.2) and must obtain tyrosine from their diets. They must keep their dietary intake of phenylalanine low to prevent its accumulation to toxic levels.

Why are dietary proteins completely digested to their constituent amino acids before being used by the body? Wouldn't it be more energy-efficient to reuse some dietary proteins directly? There are several reasons why ingested proteins are not used "as is":

● Macromolecules such as proteins are not readily absorbed by the cells of the gut, but their constituent monomers (such as amino acids) are readily absorbed.

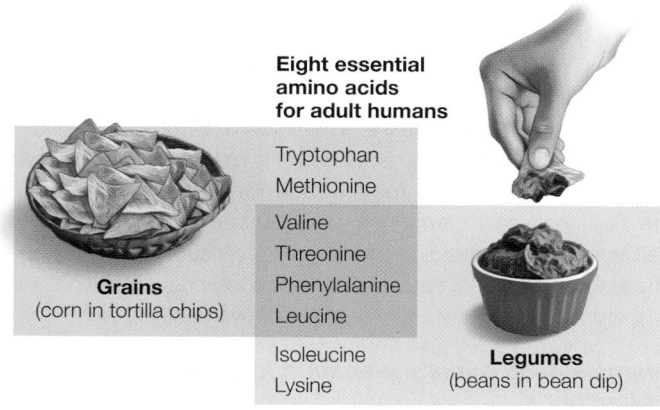

Eight essential amino acids for adult humans

Tryptophan
Methionine
Valine
Threonine
Phenylalanine
Leucine
Isoleucine
Lysine

Grains
(corn in tortilla chips)

Legumes
(beans in bean dip)

Figure 50.5 **A Strategy for Vegetarians** By combining cereal grains with legumes, an adult vegetarian can obtain all eight essential amino acids.

table 50.1 Mineral Elements Required by Animals

Element	Source in human diet	Major functions
MACRONUTRIENTS		
Calcium (Ca)	Dairy foods, eggs, green leafy vegetables, whole grains, legumes, nuts, meat	Found in bones and teeth; blood clotting; nerve and muscle action; enzyme activation
Chlorine (Cl)	Table salt (NaCl), meat, eggs, vegetables, dairy foods	Water balance; digestion (as HCl); principal negative ion in extracellular fluid
Magnesium (Mg)	Green vegetables, meat, whole grains, nuts, milk, legumes	Required by many enzymes; found in bones and teeth
Phosphorus (P)	Dairy, eggs, meat, whole grains, legumes, nuts	Component of nucleic acids, ATP, and phospholipids; bone formation; buffers; metabolism of sugars
Potassium (K)	Meat, whole grains, fruits, vegetables	Nerve and muscle action; protein synthesis; principal positive ion in cells
Sodium (Na)	Table salt, dairy foods, meat, eggs	Nerve and muscle action; water balance; principal positive ion in extracellular fluid
Sulfur (S)	Meat, eggs, dairy foods, nuts, legumes	Found in proteins and coenzymes; detoxification of harmful substances
MICRONUTRIENTS		
Chromium (Cr)	Meat, dairy, whole grains, legumes, yeast	Glucose metabolism
Cobalt (Co)	Meat, tap water	Found in vitamin B_{12}; formation of red blood cells
Copper (Cu)	Liver, meat, fish, shellfish, legumes, whole grains, nuts	Found in active site of many redox enzymes and electron carriers; production of hemoglobin; bone formation
Fluorine (F)	Most municipal water supplies	Found in teeth; helps prevent tooth decay
Iodine (I)	Fish, shellfish, iodized salt	Found in thyroid hormones
Iron (Fe)	Liver, meat, green vegetables, eggs, whole grains, legumes, nuts	Found in active sites of many redox enzymes and electron carriers, hemoglobin, and myoglobin
Manganese (Mn)	Organ meats, whole grains, legumes, nuts, tea, coffee	Activates many enzymes
Molybdenum (Mo)	Organ meats, dairy, whole grains, green vegetables, legumes	Found in some enzymes
Selenium (Se)	Meat, seafood, whole grains, eggs, milk, garlic	Fat metabolism
Zinc (Zn)	Liver, fish, shellfish, and many other foods	Found in some enzymes and some transcription factors; insulin physiology

- Protein structure and function are highly species-specific. A protein that functions optimally in one species might not function well in another.
- Foreign proteins entering the body directly from the gut would be recognized as invaders and would be attacked by the immune system.

Humans can synthesize almost all the lipids required by the body using acetyl groups obtained from food (see Figure 50.4), but we must have a dietary source of two **essential fatty acids**—linoleic acid and alpha-linolenic acid—that we cannot synthesize. Linoleic acid is needed by mammals to synthesize other unsaturated fatty acids, such as arachidonic acid, which is a component of several signaling molecules, including prostaglandins. Essential fatty acids are also necessary components of membrane phospholipids. A deficiency of linoleic acid can lead to problems such as infertility and impaired lactation, but because it is commonly present in vegetable oils, a deficiency is unlikely in an adequately nourished individual.

Animals need mineral elements for a variety of functions

Principal mineral elements that animals require are listed in **Table 50.1**. Elements required in large amounts are called **macronutrients**; those required in only tiny amounts (generally less than 100

mg/day) are called **micronutrients**. Some micronutrients are required in such minute amounts that deficiencies are never observed, but they are nevertheless essential elements.

Calcium is an example of a macronutrient. It is the fifth most abundant element in the body; a 70-kg person contains about 1.2 kg of calcium. Calcium phosphate is the principal structural material in bones and teeth. Muscle contraction, neural function, and many other intracellular functions in animals require *****calcium ions** (Ca^{2+}). The turnover of calcium in the extracellular fluid is high, as bones are constantly being remodeled and calcium is constantly entering and leaving cells. Calcium is lost from the body in urine, sweat, and feces, so it must be replaced regularly. Adult humans require 800–1,000 mg of calcium per day in their diet.

***connect the concepts**   Many physiological processes are affected by Ca^{2+}; therefore it is essential that the blood levels stay between 9 and 11 mg/100 mL. This is only 0.1 percent of the Ca^{2+} in the body. About 1 percent is intracellular and 99 percent is in the bones, thus it is a challenge to maintain tiny, but critical, blood levels in the face of such huge reservoirs. Hormonal regulation of blood calcium is discussed in Key Concept 40.4, and the cellular mechanisms are discussed in Key Concept 47.1.

---

table **50.2** Vitamins in the Human Diet

Vitamin	Source	Function	Deficiency symptoms
WATER-SOLUBLE			
B_1 (thiamin)	Liver, legumes, whole grains	Coenzyme in cellular respiration	Beriberi, loss of appetite, fatigue
B_2 (riboflavin)	Dairy, meat, eggs, green leafy vegetables	Coenzyme in FAD	Lesions in corners of mouth, eye irritation, skin disorders
B_3 (niacin)	Meat, fowl, liver, yeast	Coenzyme in NAD and NADP	Pellagra, skin disorders, diarrhea, mental disorders
B_5 (pantothenic acid)	Liver, eggs, yeast	Found in acetyl CoA	Adrenal problems, reproductive problems
B_6 (pyridoxine)	Liver, whole grains, dairy foods	Coenzyme in amino acid metabolism	Anemia, slow growth, skin problems, convulsions
B_7 (biotin)	Liver, yeast, bacteria in gut	Found in coenzymes	Skin problems, loss of hair
B_{12} (cobalamin)	Liver, meat, dairy foods, eggs	Formation of nucleic acids, proteins, red blood cells	Pernicious anemia
Folic acid	Vegetables, eggs, liver, whole grains	Coenzyme in formation of heme and nucleotides	Anemia
C (ascorbic acid)	Citrus fruits, tomatoes, potatoes	Formation of connective tissues; antioxidant	Scurvy, slow healing, poor bone growth
FAT-SOLUBLE			
A (retinol)	Fruits, vegetables, liver, dairy	Found in visual pigments	Night blindness
D (calciferol)	Fortified milk, fish oils, sunshine	Absorption of calcium and phosphate	Rickets
E (tocopherol)	Meat, dairy foods, whole grains	Muscle maintenance, antioxidant	Anemia
K (menadione)	Intestinal bacteria, liver	Blood clotting	Blood clotting problems

Iron is an example of a micronutrient. It is found throughout the body because it is the oxygen-binding atom in hemoglobin and myoglobin and is a component of enzymes in the respiratory chain. Nevertheless, the total amount of iron in a 70-kg person is only about 4 g, and since iron is recycled efficiently in the body and is not lost in the urine, we require only about 15 mg per day in our food. Despite the small amount required, insufficient iron is the most common mineral nutrient deficiency in the world today. Iron deficiency leads to anemia, a condition that renders individuals weak and tired all the time.

 Activity 50.1 Mineral Elements Required by Animals
www.Life11e.com/ac50.1

Animals must obtain vitamins from food

Like essential amino acids and fatty acids, **vitamins** are carbon compounds that an animal requires for growth and metabolism but cannot synthesize for itself. Most vitamins function as coenzymes or parts of coenzymes (see Key Concept 8.4).

Each species has its own vitamin requirements. Primates, for example, require vitamin C (ascorbic acid). While most mammals can make their own ascorbic acid, primates (including humans) cannot, so for primates, ascorbic acid is a vitamin. If we do not get vitamin C in our food, we develop scurvy, a disease characterized by bleeding gums, loss of teeth, subcutaneous hemorrhages, and slow wound healing. Scurvy was a frequently fatal problem for sailors on long voyages until late in the eighteenth century, when a Scottish physician, James Lind, discovered that the disease could be prevented if the sailors ate fresh greens and citrus fruit. The British Admiralty made limes standard provisions for its ships (and British sailors have been called "limeys" ever since). When the active ingredient in limes was isolated, it was named ascorbic ("without scurvy") acid.

Humans require 13 vitamins; these are divided into two groups, water-soluble and fat-soluble (**Table 50.2**). When water-soluble vitamins are ingested in excess of bodily needs, they are simply eliminated in the urine. (This is the fate of much of the large doses of vitamin C that people take.) Fat-soluble vitamins, however, can accumulate in body fat and may build up to toxic levels in the liver if taken in excess.

The fat-soluble vitamin D (calciferol), which is essential for absorbing and metabolizing calcium, is a special case because the body can synthesize it. (As noted in Key Concept 40.4, vitamin D is by definition a hormone.) Certain lipids present in the human body can be converted into vitamin D by the action of ultraviolet light on the skin. Thus vitamin D must be obtained in the diet by individuals with inadequate exposure to the sun.

The need for vitamin D may have been an important factor in the evolution of skin color. For humans living in equatorial and low latitudes, dark skin pigmentation is adaptive, as it is a protection against the damaging effects of ultraviolet radiation. These peoples generally expose extensive areas of skin to the sun on a regular basis, so their skin synthesizes adequate amounts of vitamin D. Most races that adapted to life in the higher latitudes lost this dark skin pigmentation, probably because lighter skin facilitates vitamin D production in the relatively small areas of skin exposed to sunlight during the short days of winter. The dark-skinned Inuit peoples of the Arctic are an exception to the correlation between latitude and skin pigmentation, but the Inuit obtain ample vitamin D from the large amounts of animal fat (especially whale blubber) and fish oils in their diet.

Activity 50.2 **Vitamins in the Human Diet**
www.Life11e.com/ac50.2

Nutrient deficiencies result in diseases

The lack of an essential nutrient in the diet produces a state of **malnutrition**, and chronic malnutrition leads to a characteristic **deficiency disease** (see Table 50.2). We discussed kwashiorkor (protein deficiency) and scurvy (vitamin C deficiency). Another deficiency disease, beriberi, was directly involved in the discovery of vitamins.

Beriberi, which means "extreme weakness," became prevalent in Asia in the nineteenth century when it became standard practice to mill rice to a white polish and discard the hulls present in brown rice. A critical observation was that chickens and pigeons developed beriberi-like symptoms when they were fed only polished rice. In 1912 Casimir Funk, a Polish scientist working in England, cured pigeons of beriberi by feeding them discarded rice hulls.

At the time of Funk's discovery, all diseases were thought either to be caused by microorganisms or to be inherited. Funk suggested that beriberi and some other diseases are dietary in origin and result from deficiencies in specific substances. Funk coined the term "vitamines" from "vital amines" because he mistakenly thought that all these substances vital for life were compounds with amine groups. In 1926 thiamin (vitamin B_1)—the substance lost in the rice milling process—was the first vitamin to be isolated in pure form.

Deficiency diseases can also result from an inability to absorb or process an essential nutrient even if it is present in the diet. Vitamin B_{12} (cobalamin), for example, is present in all foods of animal origin. Since plants neither use nor produce vitamin B_{12}, a strictly vegetarian diet (not supplemented with dairy products or vitamin pills) can lead to a B_{12} deficiency disease called pernicious anemia, characterized by a failure of red blood cells to mature. The most common cause of pernicious anemia, however, is not a lack of vitamin B_{12} in the diet but an inability to absorb it. Normally cells in the stomach lining secrete a peptide called intrinsic factor that binds to vitamin B_{12} and makes it absorbable by the small intestine. Conditions that damage the stomach lining, such as alcoholism or gastritis, can thus lead to pernicious anemia.

Inadequate mineral nutrition can also lead to deficiency diseases. Examples are hypothyroidism and goiter resulting from iodine deficiency (see Key Concept 40.4), and anemia resulting from iron deficiency. Iodine deficiency is almost unheard of in the developed world because we add iodide to salt. However, it is still a major health problem in large segments of the human population.

50.1 recap

As heterotrophs, animals must obtain the energy and molecular building blocks for biosynthesis from their food. Energy can come from the metabolism of carbohydrates, fats, and proteins. Molecular building blocks include carbon skeletons, vitamins, and minerals. Lack of essential nutrients results in deficiency diseases.

learning outcomes

You should be able to:

- Describe energy storage adaptations in animal species.
- Discuss challenges related to obtaining all essential amino acids from the diet.

50.1 recap

- Discuss examples of mineral macro- and micronutrients and their functions.
- Compare water-soluble and fat-soluble vitamins and their effects on an animal's body.

1. Gray whales migrate each year from their summer range off Alaska to their winter range off Mexico. During their migrations and their stay in Mexican waters, they eat very little, if anything. What is their energy source over this long time, and what are the advantages of this energy source?

2. Inuit peoples do not eat polar bear liver because it contains so much vitamin A that consuming it produces a serious disease called hypervitaminosis A. Why isn't there a disease called hypervitaminosis C?

3. Anemia is lack of the micronutrient iron. The incidence of anemia is greater in premenopausal women than in men. Why?

4. A comatose patient is fed a nutrient mix through a food tube into the stomach, but a person with a nonfunctional gastrointestinal system will receive a nutrient mix via blood vessels. Mixes introduced into the stomach include proteins, but mixes delivered into blood vessels do not. Explain why this is, and what must take the place of proteins in the mix delivered into blood vessels.

We have surveyed the essential elements of nutrition in animals. Next we will look at various methods and adaptations by which animals obtain the food they need, and the mechanisms they use to extract nutrients from their food.

key concept 50.2 Diverse Adaptations Support Ingestion and Digestion of Food

Heterotrophic organisms can be classified by how they acquire their nutrition. **Saprobes** (also called saprotrophs) are organisms that absorb nutrients from dead organic matter. Most saprobes are either protists or fungi. **Detritivores**, or **decomposers**, such as earthworms and crabs, actively feed on dead organic material. Animals that feed on living organisms are **predators**: **herbivores** prey on plants, **carnivores** prey on animals, and **omnivores** prey on both. **Filter feeders**, such as clams and blue whales, prey on small organisms by filtering them from the aquatic environment. **Fluid feeders** include mosquitoes, aphids, leeches, and hummingbirds. The anatomical adaptations that enable a species to exploit a particular source of nutrition are usually obvious, but physiological and biochemical adaptations are also important.

focus your learning

- Heterotrophs display a wide diversity of adaptations for acquiring and processing food.
- Digestion occurs in a body cavity where secreted enzymes break down large molecules into small molecules that cells can absorb.
- Heterotrophs rely on symbiotic bacteria residing in their digestive system to carry out essential tasks in digestion.

The food of herbivores is often low in energy and hard to digest

Most vegetation is coarse, difficult to break down, and has low energy content. Therefore herbivores spend a great deal of time feeding and processing their food. Many have striking adaptations for feeding, such as the trunk (a flexible, gripping nose) of the elephant or the huge bill of the fruit-eating toucan, which can be half as long as its body. Many types of grinding, rasping, cutting, and shredding mouthparts have evolved in invertebrates for ingesting plant material, and the teeth of herbivorous vertebrates have been shaped by selection to tear, crush, and grind coarse plant matter.

The elephant provides an example of the importance of teeth. Elephants have six chewing teeth per quadrant, which appear one at a time. The first is present at birth, and subsequent ones appear at the back of the jaw and move forward as the elephant ages. The first falls out at 2–3 years of age, the second at 4–6 years of age, and so forth until the last, and biggest, one (20 cm long and 4 kg) is worn out sometime past 60 years of age. With the loss of that last tooth, the elephant starves and dies.

The digestive processes of herbivores can be quite specialized. An example is the koala, which almost exclusively eats leaves of eucalyptus trees. These leaves are very fibrous, low in usable energy and protein, and high in toxic chemicals. The koala has strong jaws for grinding the leaves, a very long gut for fermenting them, enzymes in its liver for detoxifying chemicals in the leaves, and a low metabolic rate (i.e., it expends little energy) to compensate for low energy intake.

Carnivores must find, capture, and kill prey

The predatory behaviors of many carnivores are legendary—the hunting skills of hawks, wolves, and tigers, for example. Carnivores have evolved stealth, speed, power, large jaws, sharp teeth, and strong gripping appendages. They also have evolved remarkable means of detecting prey. Bats use echolocation, pit vipers sense infrared radiation from the warm bodies of their prey, and certain fishes detect electric fields created in the water by their prey. There are many fascinating examples of adaptations for capturing prey, such as the immobilizing venom of many snakes, the long sticky tongues of chameleons, and the webs of spiders.

Some predators digest their prey externally. For example, a spider injects insect prey with digestive enzymes and then sucks out the liquefied contents, leaving behind the empty exoskeletons frequently seen in old spider webs. The majority of animals, however, digest their food internally. For many, the process of digestion begins with the physical breaking down of the food items by the teeth.

Vertebrate species have distinctive teeth

Teeth are adapted for the acquisition and initial processing of specific types of foods. Because they are among the hardest structures of the body, an animal's teeth remain in the environment long after it dies. Paleontologists use teeth to identify animals that lived in the distant past and to deduce their feeding behavior.

In general, teeth of all vertebrates have the same, three-layered structure (**Figure 50.6A**). An extremely hard material called **enamel**, composed principally of calcium phosphate, covers the crown of the tooth. Both the crown and root contain a layer of bony material called **dentine**, inside of which is a **pulp cavity** containing blood vessels, nerves, and the cells that produce the dentine.

There is a great deal of homology in the dentition of mammals, but the shapes and organization of mammalian teeth are adaptations to different diets (**Figure 50.6B**). In general, incisors are used

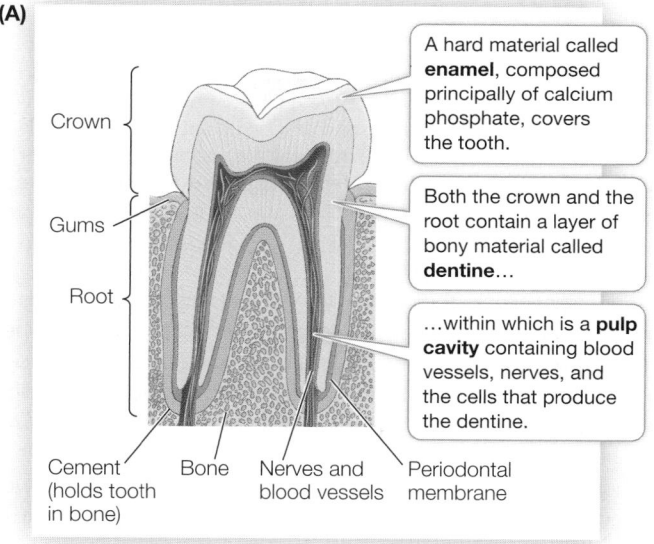

(A)

Crown

Gums

Root

A hard material called **enamel**, composed principally of calcium phosphate, covers the tooth.

Both the crown and the root contain a layer of bony material called **dentine**...

...within which is a **pulp cavity** containing blood vessels, nerves, and the cells that produce the dentine.

Cement (holds tooth in bone)   Bone   Nerves and blood vessels   Periodontal membrane

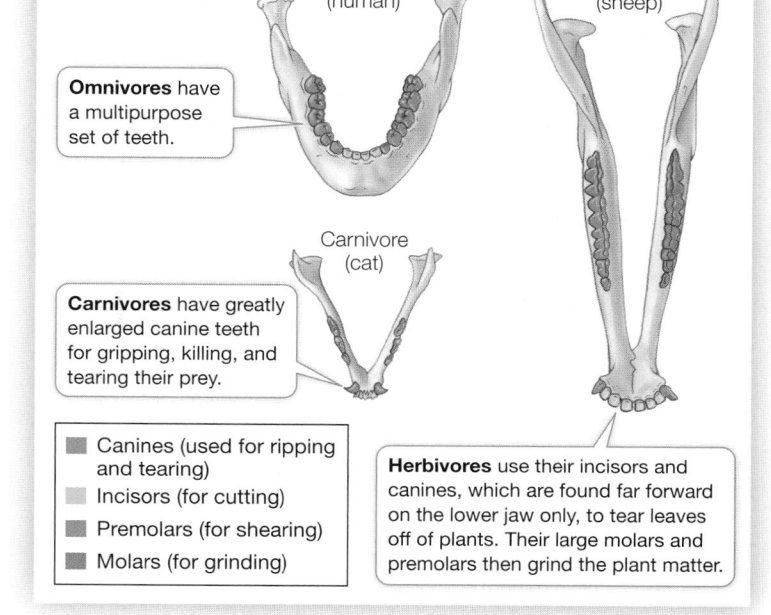

(B)

Omnivore (human)

Herbivore (sheep)

Omnivores have a multipurpose set of teeth.

Carnivore (cat)

Carnivores have greatly enlarged canine teeth for gripping, killing, and tearing their prey.

■ Canines (used for ripping and tearing)
□ Incisors (for cutting)
■ Premolars (for shearing)
■ Molars (for grinding)

Herbivores use their incisors and canines, which are found far forward on the lower jaw only, to tear leaves off of plants. Their large molars and premolars then grind the plant matter.

Figure 50.6 Mammalian Teeth **(A)** A mammalian tooth has three layers: enamel, dentine, and a pulp cavity. **(B)** The teeth of different mammalian species are specialized for different diets. This illustration depicts the teeth of the lower jaw, viewed from above.

▶ Activity 50.3 **Mammalian Teeth**
www.Life11e.com/ac50.3

for cutting, chopping, and gnawing; canines are used for stabbing, gripping, and ripping; and molars and premolars (the cheek teeth) are used for shearing, crushing, and grinding. The highly varied diet of humans is reflected in our multipurpose set of teeth, as is common among omnivores.

Digestion usually begins in a body cavity

Animals take food into a body cavity that is continuous with the outside environment. They secrete digestive enzymes into that cavity, and the enzymes break down the food into nutrient molecules that can be absorbed by the cells lining the cavity.

The simplest digestive system is found in the simplest animals such as sponges. Water flows from the environment through the body of the sponge in water channels, and individual cells capture food particles from the water (see Figure 30.2B). A slightly more specialized digestive system is a **gastrovascular cavity**, which connects to the outside world through a single opening. Cnidarians, such as jellyfishes, capture prey using stinging nematocysts and use their tentacles to cram the prey into their gastrovascular cavity. Enzymes in the gastrovascular cavity partially digest the prey. Cells lining the cavity take in small food particles by endocytosis. The vesicles created by endocytosis then fuse with lysosomes containing digestive enzymes, and intracellular digestion completes the breakdown of the food. Nutrients are released to the cytoplasm as the vesicles break down.

Tubular guts have an opening at each end

Animals classified as bilaterians (see Key Concept 30.1) have a tubular digestive system called a **gastrointestinal system** or more simply **gut**. The gut begins with a mouth that takes in food. The food is moved through the length of the gut as it is sequentially disassembled or digested, and the smallest units of the digested food are absorbed across the gut wall. Solid digestive wastes are eliminated through an **anus**. Different regions in the tubular gut are specialized for particular functions (**Figure 50.7**). These functions must be coordinated so they occur in the proper sequence, and they must be regulated so they occur slowly enough to allow complete digestion but quickly enough to supply the energy needs of the animal.

In the mouth cavity, food may be physically fragmented by teeth (in many vertebrates), by a **radula** (in snails), or by **mandibles** (in many arthropods). In most birds, food is ground by small stones in an early, muscular portion of the gut called the **gizzard**. Some animals, such as snakes, simply ingest whole prey with little or no fragmentation. Teeth in these species are angled backward to keep the prey moving in only one direction while being slowly swallowed. **Stomachs** and **crops** are storage chambers that enable animals to ingest relatively large amounts of food when it is available, and then digest it gradually. In these storage chambers, food may be further fragmented and mixed, and in most vertebrates it is an important site of digestion. Food delivered into the next section of the gut, the **intestine**, is in small particles, well mixed, and usually partially digested.

Most digestion occurs in the intestine, and nutrients, water, and ions are absorbed across its walls. Glands secrete digestive enzymes into the intestine, and other enzymes are produced and secreted by cells lining the intestine. The final segment of the intestine recovers water and ions and stores undigested wastes, or **feces**, so they can be released to the environment at an appropriate time or place. A muscular **rectum** near the anus assists in expelling feces.

Earthworm

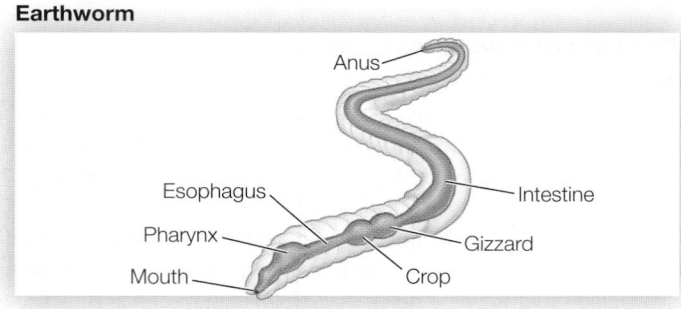

Cockroach

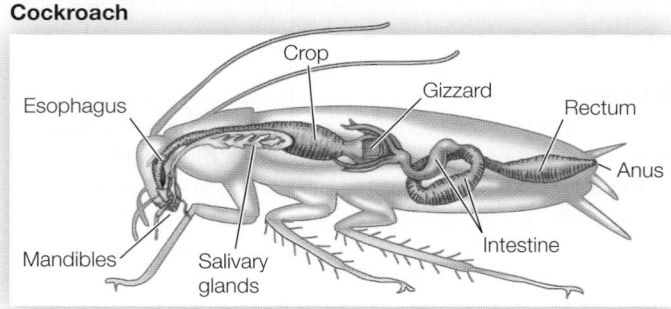

Rabbit

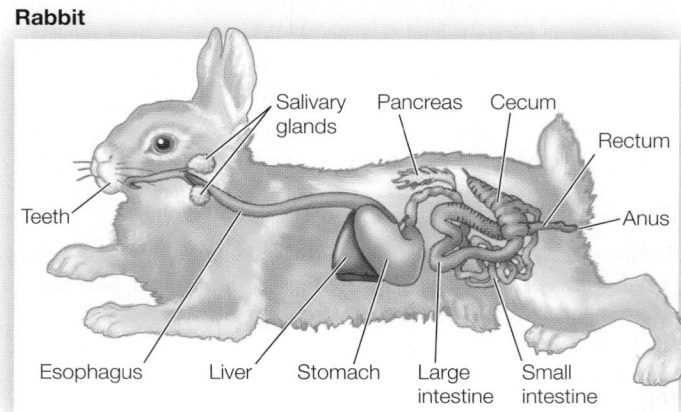

Figure 50.7 Compartments for Digestion and Absorption   Most invertebrates and all vertebrates have a tubular gut that begins with a mouth where food is taken in, and ends in an anus where waste is eliminated. Between these two structures are specialized regions for digestion and nutrient absorption; the structures in these regions are adapted to different diets and vary from species to species.

In many animals, the parts of the gut that absorb nutrients have greater surface areas than would be expected of a simple tube. The simplest way to increase the surface area of a tube without changing its diameter is to produce an infolding. Such an infolding of the gut, called a typhlosole, is seen in earthworms (**Figure 50.8A**). Sharks, which have large stomachs but short intestines, have a unique adaptation called a spiral valve (**Figure 50.8B**). The lower region of the intestine is enlarged and has an internal structure that forces the food to pass through it in a spiral fashion (like going down a spiral staircase). The walls of the spiral present a large surface area for absorption of nutrients. This spiral passage, however, will not accommodate large chunks of food or undigestible matter, so food remains in the stomach for a long time to be broken down, and some larger, undigestible items do not leave the stomach except by regurgitation.

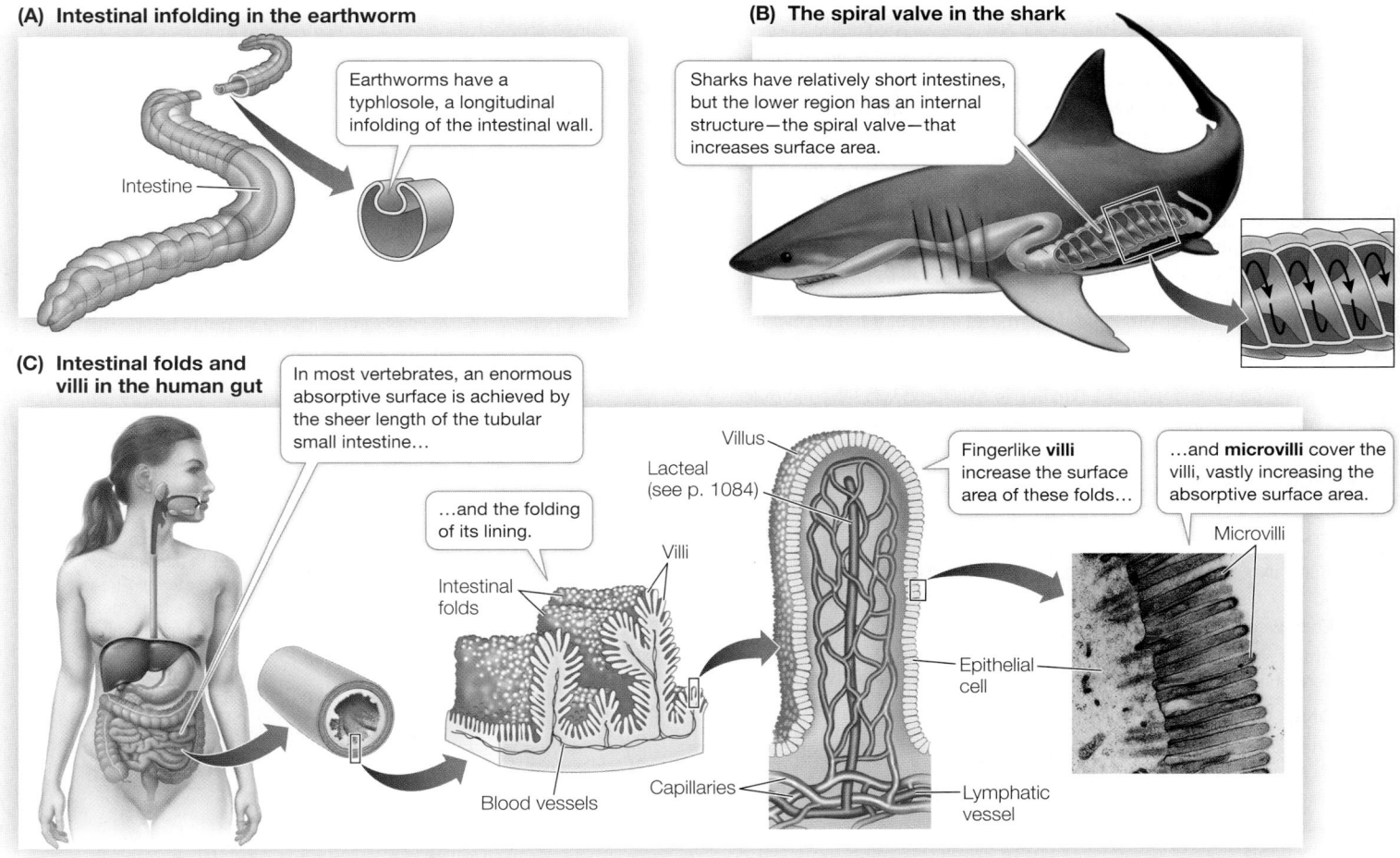

(A) Intestinal infolding in the earthworm

Earthworms have a typhlosole, a longitudinal infolding of the intestinal wall.

Intestine

(B) The spiral valve in the shark

Sharks have relatively short intestines, but the lower region has an internal structure—the spiral valve—that increases surface area.

(C) Intestinal folds and villi in the human gut

In most vertebrates, an enormous absorptive surface is achieved by the sheer length of the tubular small intestine…

…and the folding of its lining.

Lacteal (see p. 1084)

Villus

Fingerlike **villi** increase the surface area of these folds…

…and **microvilli** cover the villi, vastly increasing the absorptive surface area.

Microvilli

Intestinal folds

Villi

Epithelial cell

Blood vessels

Capillaries

Lymphatic vessel

Figure 50.8 Intestinal Surface Area and Nutrient Absorption
Maximizing the surface area of the gut increases an animal's ability to absorb nutrients.

In humans, as in most other vertebrates, the wall of the intestine is highly folded, with the individual folds bearing legions of tiny fingerlike projections called **villi** (**Figure 50.8C**). The cells that line the surfaces of the villi, in turn, have microscopic projections called **microvilli**. The microvilli give the intestine an enormous internal surface area for absorbing nutrients.

Digestive enzymes break down complex food molecules

Protein, carbohydrate, and fat macromolecules are broken down into their simplest monomeric units by hydrolytic enzymes produced at different locations in the digestive tract. Many are secreted into the lumen of the gut, and others remain associated with the membranes of the microvilli. All of these enzymes cleave the chemical bonds of macromolecules through hydrolysis, a reaction that adds a water molecule (see Figure 3.4B). Digestive enzymes are classified according to the substances they hydrolyze: **proteases** break the bonds between adjacent amino acids in proteins; **carbohydrases** hydrolyze carbohydrates; **peptidases** break down peptides; **lipases**, fats; and **nucleases**, nucleic acids. We will discuss the digestive enzymes of humans in detail in Key Concept 50.3.

The gut microbiome contributes to digestion

The gut is colonized by a huge population of microorganisms, mostly bacteria. Collectively the foreign species associated with an organism are called its **microbiota**, and their cumulative genomes are called the organism's **microbiome**. The microbiomes of the skin are different from the microbiomes of the mouth, gut, respiratory tract, and urogenital tracts. The microbiota of the gut obtain their nutrition from the food passing through the gut while contributing to the host's digestive processes. For example, members of the leech genus *Hirudo* produce no enzymes that can digest the proteins in the blood they suck from vertebrates; instead they depend on bacteria to perform this service. The resulting amino acids are subsequently used by both the leech and the bacteria. The microbiota of the human gut also provide important services in digestion, prevent the establishment of harmful microorganisms, and even produce some vitamins (vitamin K and biotin). It is estimated that the human body consists of 10^{13} cells, but our guts contain probably ten times that number of unicellular organisms representing at least 500 different species (see Figure 25.19). Recent studies have revealed that gut microbiomes are not constant, but are influenced by diet and other factors, including medications we ingest. The effects of different microbiomes may go far beyond simply facilitating digestion (**Investigating Life: How Does the Gut Microbiome Contribute to Obesity and Metabolic Disease?**).

investigatinglife

How Does the Gut Microbiome Contribute to Obesity and Metabolic Disease?

experiment

Original Paper: Caesar, R., V. Tremaroli, P. Kovatcheva-Datchary, P. D. Cani and F. Bäckhed. 2015. Crosstalk between gut microbiota and dietary lipids aggravates WAT inflammation through TLR signaling. *Cell* 22: 658–668.

Fats differ in fatty acid composition, and fatty acids differ in the number of single or double bonds between their carbon atoms (see Key Concept 3.4). Saturated fatty acids have no double bonds, and unsaturated fatty acids do. Diets rich in saturated fatty acids are implicated in several health concerns, but what is it about saturated fatty acids that is unhealthy? Recent research examines the effect of diet on the gut microbiome, and in turn, the effect of the microbiome on health. Does the amount of saturated fat in the diet influence the composition of the microbiome, and does the microbiome influence the health consequences of a diet rich in saturated fats? We examine the first question in the experiment below, and the second question in an optional online companion exercise.

HYPOTHESIS▶ The type of fat in the diet can influence the composition of the gut microbiome.

METHOD

1. Raise two groups of 15 mice on isocaloric diets rich in saturated fats (lard) or unsaturated fats (fish oil).
2. Each week, record the mice's body weight.
3. At the end of 11 weeks, compare the mice's (1) microbiomes, by sequencing the ribosomal RNA genes, and (2) body weight gain.

CONCLUSIONS▶

1. The fat composition of the diet influenced the composition of the microbiome.
2. Mice on the diet containing saturated fats gained more weight and gained weight faster than mice on the isocaloric diet containing unsaturated fats.

RESULTS

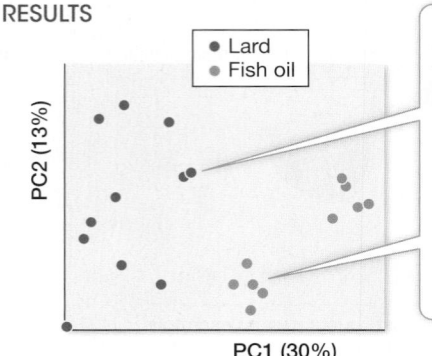

Using a statistical test called principle components (PC) analysis, the microbiomes of the two groups of mice were compared by sequencing. The ribosomal RNA genes of the two microbiomes were shown to be significantly different.

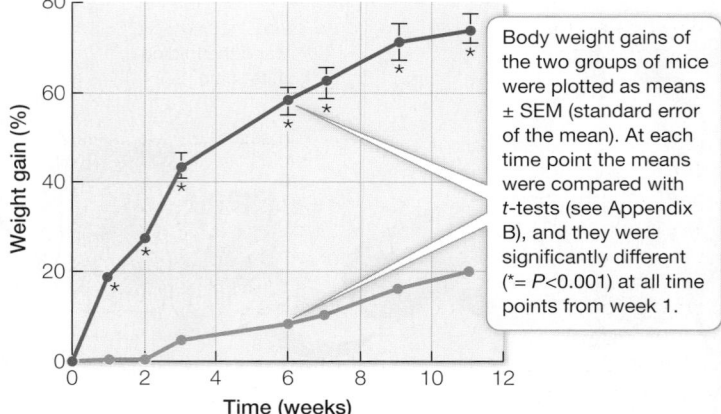

Body weight gains of the two groups of mice were plotted as means ± SEM (standard error of the mean). At each time point the means were compared with *t*-tests (see Appendix B), and they were significantly different (*= *P*<0.001) at all time points from week 1.

work with the data

The researchers hypothesized that mice consuming more saturated fats will have a higher probability of becoming obese and developing symptoms of diabetes. To test for dietary effects on the development of metabolic disease, the researchers collected additional data from the two cohorts of mice: fasting blood glucose levels and fasting blood insulin levels. They also did an insulin tolerance test (effect of a dose of insulin on blood glucose level). Data are reported as means (± SEM).

QUESTIONS▶

1. Using a *t*-test, determine whether the difference in the fasting blood glucose levels for the lard-fed and fish-oil-fed mice is significant.
2. Using a *t*-test, determine whether the difference in the fasting blood insulin levels for the lard-fed and fish-oil-fed mice is significant.
3. Plot the glucose over time data. Did the two groups of mice differ in their response to the same dose of insulin? At which time points?
4. What conclusions can you make about the effect of the type of dietary fat on carbohydrate metabolism?

A similar **work with the data** exercise may be assigned in **LaunchPad**.

Fasting blood glucose (mM)		Fasting blood insulin (ng/mL)	
Lard (N = 14)	Fish oil (N = 14)	Lard (N = 10)	Fish oil (N = 14)
11 (0.22)	10.32 (0.23)	3.96 (0.36)	1.09 (0.17)

Time after insulin injection (min)	Blood glucose (% of the time 0 level)	
	Lard	Fish oil
0	100	100
30	59.8 (3.05)	49.8 (4.09)
60	54.5 (3.51)	37.2 (4.07)
90	63.1 (3.91)	39.1 (4.09)
120	75.6 (3.06)	56.9 (3.31)

50.2 recap

Heterotrophs have diverse adaptations for acquiring food. Once captured or ingested, food is digested extracellularly by secreted enzymes to release nutrients that are absorbed into the animal's body, usually via a tubular gut. The gut microbiome plays a role in digestion.

learning outcomes

You should be able to:

- Compare the feeding and digestive adaptations of different species.
- Relate structural features of animal guts to their functions and adaptive significance in digestion.
- Describe digestive enzymes that are present in animal guts.
- Explain the importance of the microbiome in the digestion of food.

1. Why do herbivores typically spend more time feeding than do carnivores?
2. What do the actions of all digestive enzymes have in common?
3. Why is it common for individuals to have digestive problems after antibiotic therapy?
4. When people with celiac disease eat food containing gluten, their immune system damages their intestinal villi. If undiagnosed, such people are likely to experience bloating, abdominal pain, diarrhea, fatty stool, loss of body weight, and iron deficiency. Why do these symptoms occur?

Once ingested by an animal, food may be fragmented and moved into the gut for digestion by hydrolytic enzymes. The processes of digestion release nutrients that are absorbed into the animal's body. Next we will focus on how those processes occur in vertebrates.

key concept 50.3 The Vertebrate Gastrointestinal System Is a Disassembly Line

Digestion in vertebrates occurs in the gastrointestinal system, which includes a tubular gut running from mouth to anus and several accessory structures that produce secretions that play important roles in digestion (**Figure 50.9**). In this section we consider three important processes of this system: the movement of food through it, the sequential steps of digestion, and the absorption of nutrients. We use as our primary example a typical vertebrate, the human.

focus your learning

- The structure of the vertebrate gut supports its digestive function.
- A variety of chemical and physical processes take place in different parts of the digestive system to efficiently break down food into forms that the body can use.
- Herbivores do not produce the cellulases needed to digest cellulose in their plant diet, and thus require microbiota to carry out this task.

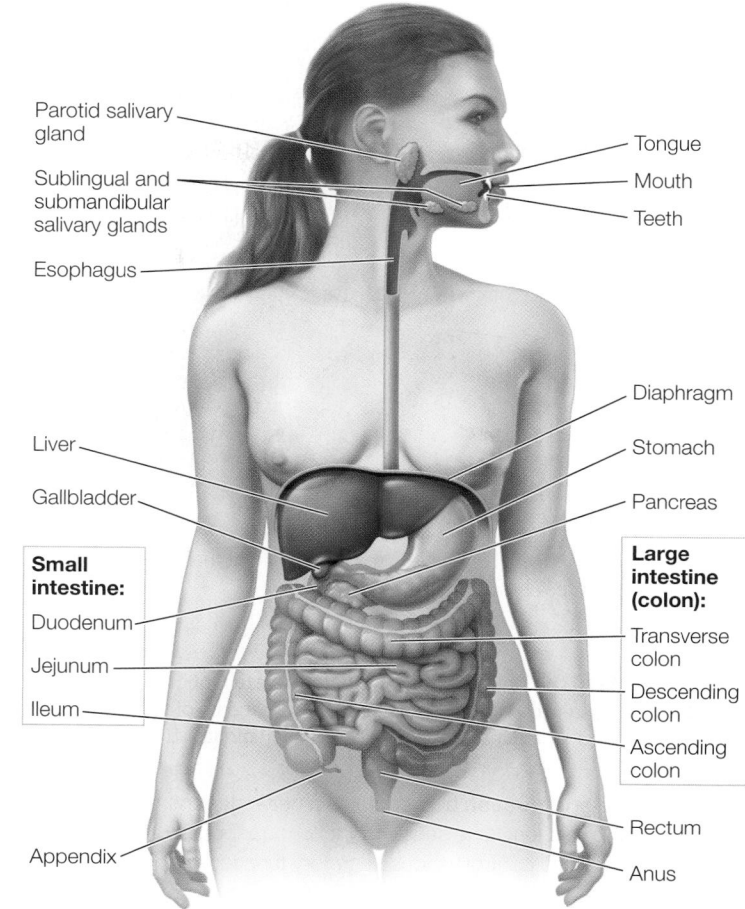

Figure 50.9 The Human Digestive System Different compartments in the long tubular gut specialize in digesting food, absorbing nutrients, and storing and expelling wastes. Accessory organs contribute secretions containing enzymes and other molecules.

 Activity 50.4 **The Human Digestive System**
www.Life11e.com/ac50.4

The vertebrate gut consists of concentric tissue layers

Tissues of the vertebrate gut are arranged in concentric layers that have a similar organization throughout its length (**Figure 50.10**). The **mucosa** lines the internal cavity, or **lumen**, of the gut. The mucosa consists of delicate epithelial cells with underlying connective tissue. Some cells of this mucosal epithelium secrete mucus to lubricate and protect the walls of the gut; some secrete digestive enzymes; and some secrete hormones. Mucosal epithelial cells in the stomach secrete hydrochloric acid, and as we noted in Key Concept 50.1, some secrete intrinsic factor to aid the absorption of vitamin B_{12}. In some regions of the gut, nutrients are absorbed by mucosal epithelial cells. The apical cell membranes of these absorptive cells have microvilli that increase the surface area over which absorption can take place (see Figure 50.8C).

At the base of the mucosa there is a thin layer of smooth muscle cells with random orientations. The function of this cell layer is to move the mucosal epithelium to maximize its contact with the gut contents. Just under the mucosa is the submucosal tissue layer where blood and lymph vessels pick up nutrients to transport to the rest of the

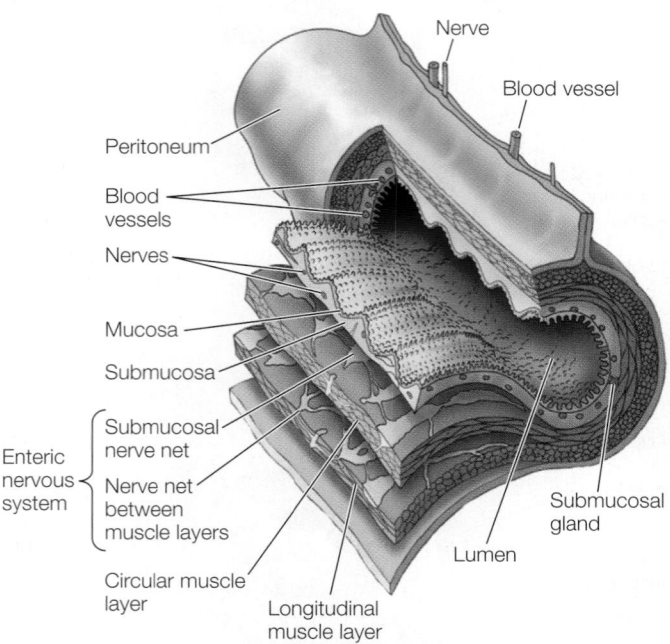

Figure 50.10 Tissue Layers of the Vertebrate Gut The organization of tissue layers is the same in all compartments of the gut, but specialized adaptations of specific tissues characterize different regions.

body. The **submucosa** also contains a network of nerves; the neurons in this network have sensory functions (responsible for stomach aches) and also control various secretory functions of the gut.

External to the submucosa are two layers of smooth muscle responsible for large movements of the gut. Innermost is the circular muscle layer, with its cells oriented around the gut. Outermost is the longitudinal muscle layer, with its cells oriented along the length of the gut (see Figure 50.10). The circular muscles constrict the gut, and the longitudinal muscles shorten it. Between the two layers of smooth muscle is another nerve network that controls and coordinates the movements of the gut. The coordinated activity of the two smooth muscle layers mixes contents of the gut and moves it continuously toward the anus.

 Media Clip 50.1 **Following Food from Mouth to Gut** www.Life11e.com/mc50.1

Nerve nets in the submucosa and between the smooth muscle layers are called the **enteric nervous system**, and they are unusual. The cell bodies of most peripheral nerves are in the central nervous system (CNS)—in the brain or spinal cord. The only exceptions are the postganglionic cells of the autonomic nervous system and the cells of the enteric nervous system. In addition, most neurons of the peripheral nervous system either receive synapses from neurons in the CNS or contribute synapses to neurons in the CNS. In contrast, most of the neurons in the enteric nervous system form synapses only with other neurons in their network. Thus they are responsible for communication within the gut. The CNS can influence activity in the enteric nervous system and receive information from it, but the gut truly has "a mind of its own."

A tissue membrane called the **peritoneum** surrounds the gut and most of the organs of the abdominal cavity as well as lining the wall of the cavity. The peritoneum includes connective and epithelial tissues that secrete a lubricating fluid that enables the organs to easily slide against each other in the body cavity.

Gut motility moves food through the gut and aids digestion

In humans and most other mammals, food is chewed in the mouth and mixed with saliva. Periodically the tongue pushes a bolus (mass) of the chewed food toward the throat. By making contact with the soft palate at the back of the mouth cavity, the food bolus initiates swallowing, which is a complex series of reflexes. Swallowing propels the food through the pharynx (where the mouth cavity and nasal passages join) and into the **esophagus** (food tube). To prevent food from entering the trachea (windpipe), the larynx (voice box) closes, and a flap of tissue called the **epiglottis** covers the entrance to the larynx (**Figure 50.11A**).

Once a bolus of food enters the esophagus, it is moved toward the stomach both by gravity and by waves of muscle contraction called **peristalsis** (**Figure 50.11B**). The muscle of the upper region of the esophagus is striated (i.e., skeletal muscle) and is controlled

Figure 50.11 Swallowing and Peristalsis **(A)** Food pushed to the back of the mouth triggers the swallowing reflex. **(B)** Once a food bolus enters the esophagus, peristalsis propels it from mouth to anus by coordinated actions of the circular and longitudinal muscle layers of the gut.

(A) Swallowing

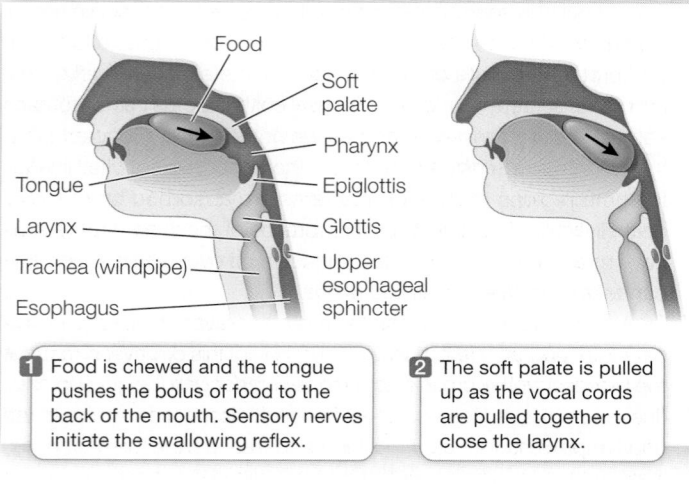

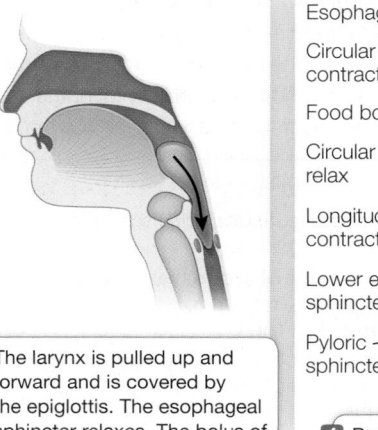

1 Food is chewed and the tongue pushes the bolus of food to the back of the mouth. Sensory nerves initiate the swallowing reflex.

2 The soft palate is pulled up as the vocal cords are pulled together to close the larynx.

3 The larynx is pulled up and forward and is covered by the epiglottis. The esophageal sphincter relaxes. The bolus of food enters the esophagus.

(B) Peristalsis

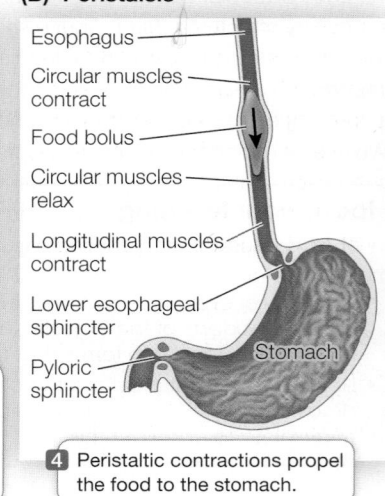

4 Peristaltic contractions propel the food to the stomach.

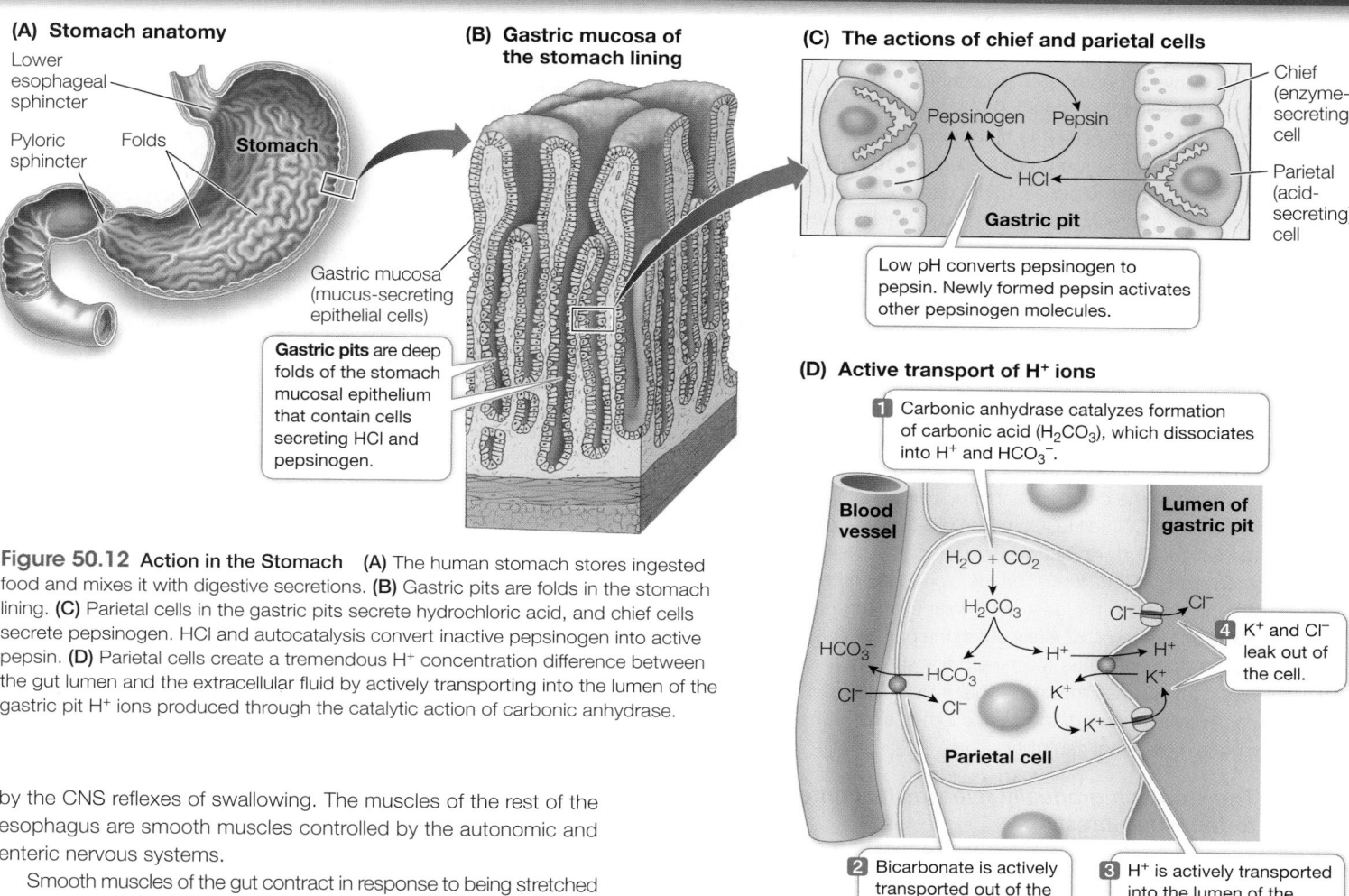

(A) Stomach anatomy

Lower esophageal sphincter

Pyloric sphincter   Folds   **Stomach**

Gastric mucosa (mucus-secreting epithelial cells)

Gastric pits are deep folds of the stomach mucosal epithelium that contain cells secreting HCl and pepsinogen.

(B) Gastric mucosa of the stomach lining

(C) The actions of chief and parietal cells

Pepsinogen   Pepsin

HCl

Gastric pit

Chief (enzyme-secreting) cell

Parietal (acid-secreting) cell

Low pH converts pepsinogen to pepsin. Newly formed pepsin activates other pepsinogen molecules.

(D) Active transport of H⁺ ions

1 Carbonic anhydrase catalyzes formation of carbonic acid (H_2CO_3), which dissociates into H^+ and HCO_3^-.

Blood vessel

Lumen of gastric pit

$H_2O + CO_2$

H_2CO_3

Cl^- Cl^-

HCO_3^- H^+ H^+

Cl^- HCO_3^- K^+ K^+

Cl^- K^+

4 K^+ and Cl^- leak out of the cell.

Parietal cell

2 Bicarbonate is actively transported out of the blood side of the cell in exchange for Cl^-.

3 H^+ is actively transported into the lumen of the gastric pit in exchange for K^+.

Figure 50.12 Action in the Stomach   (A) The human stomach stores ingested food and mixes it with digestive secretions. **(B)** Gastric pits are folds in the stomach lining. **(C)** Parietal cells in the gastric pits secrete hydrochloric acid, and chief cells secrete pepsinogen. HCl and autocatalysis convert inactive pepsinogen into active pepsin. **(D)** Parietal cells create a tremendous H^+ concentration difference between the gut lumen and the extracellular fluid by actively transporting into the lumen of the gastric pit H^+ ions produced through the catalytic action of carbonic anhydrase.

by the CNS reflexes of swallowing. The muscles of the rest of the esophagus are smooth muscles controlled by the autonomic and enteric nervous systems.

Smooth muscles of the gut contract in response to being stretched (see Figure 47.8). When a bolus of food reaches the smooth muscle region of the esophagus and stretches it, the muscle responds by contracting, thus pushing food toward the stomach. Why doesn't the contraction of the esophageal smooth muscle push food back toward the mouth? The nerve net between the two smooth muscle layers coordinates the muscles so that when a region of the gut smooth muscle contracts, the circular smooth muscle just beyond it relaxes. This action, controlled by the enteric nervous system, is called an anticipatory wave of relaxation. The contractions of smooth muscles behind the relaxing area push food into it—in the direction of the anus. The resulting stretch of that region stimulates its circular smooth muscle to contract as the anticipatory wave of relaxation causes the smooth muscle of the next region to relax. The peristaltic waves of contraction and relaxation of the gut smooth muscle move food along the gut from the mouth to the anus.

At the junction of the esophagus and stomach is an esophageal **sphincter**, a ring of circular smooth muscle that is normally constricted. Waves of peristalsis cause the sphincter to relax enough to let food pass from the esophagus into the stomach. Sphincter muscles are found throughout the digestive tract: the pyloric sphincter governs the passage of stomach contents into the small intestine; the ileocaecal sphincter controls the flow of food between the small and large intestines; and the anal sphincter relaxes to allow defecation.

Some movements of the stomach are not as coordinated as the peristaltic movements of the esophagus and serve to move the contents of the stomach around and mix them with the digestive juices. However, peristaltic waves of contraction reorganize in the

lower region of the stomach and push the semidigested material up against the pyloric sphincter which controls the opening to the small intestine. In the small intestine, peristalitic activity continues to move the gut contents toward the large intestine.

Chemical digestion begins in the mouth and the stomach

Salivary glands secrete the enzyme amylase into the mouth where it is mixed with the food being chewed. Amylase hydrolyzes bonds between glucose monomers that make up carbohydrate molecules. The action of amylase is what makes a chewed piece of bread or cracker taste slightly sweet if you hold it in your mouth long enough.

The main role of the stomach is to store food and deliver it gradually to the small intestine so that digestion can occur more slowly than ingestion. The stomach produces secretions that facilitate digestion and also kill potential pathogens that are ingestsed with food. **Gastric pits** in the stomach walls (**Figure 50.12A and B**) have three types of secretory cells: (1) mucus secreting cells, (2) **parietal cells** that secrete hydrocholoric acid (HCl), and (3) **chief cells** that secrete the inactive proteolytic enzyme pepsinogen (**Figure 50.12C**). The low pH of the stomach juices has multiple functions. It kills most microorganisms that are ingested with food; it denatures proteins, making them more accessible to the hydrolytic actions of pepsin;

and it converts pepsinogen to pepsin by cleaving away a sequence of amino acids that masks the active site of the enzyme. Newly activated pepsin activates other pepsinogen molecules, creating a positive feedback process called **autocatalysis**. Acid and the proteolytic enzyme pepsin could damage the stomach walls, but the mucus-secreting cells provide a protective coating for the walls of the gastric pits and stomach.

The production of hydrochloric acid (HCl) by the parietal cells of the gastric pits is quite impressive. It amounts to about 2 liters per day—enough to bring the pH of the stomach contents below 1, which is the same as battery acid and ten times more acidic than pure lemon juice. This means that across their cell membranes, parietal cells create a H^+ ion concentration difference of 3 million-fold. Such a feat of transport is not seen anywhere else in the body. How do the gastric pits do it? Enzymes and transporters are involved.

The enzyme carbonic anhydrase in parietal cells catalyzes the hydration of CO_2 to H_2CO_3, which dissociates into H^+ and bicarbonate ion (HCO_3^-). An antiporter transport protein (see Figure 6.13) exchanges HCO_3^- for Cl^- on the blood side of the gastric pits, and an antiporter on the gastric pit side exchanges H^+ for K^+ (**Figure 50.12D**). However, K^+ leaks out of the parietal cell down its concentration gradient. Thus the inward transport of K^+ acts like an endless conveyer belt moving H^+ out into the stomach lumen. Cl^- also passively leaks out of the gastric lumen side of the parietal cells to maintain electrical neutrality.

The stomach gradually releases its contents to the small intestine

The acidic, fluid mixture of gastric juice and partly digested food in the stomach is called **chyme**. A few substances can be absorbed across the stomach wall, including alcohol (hence its rapid effects), aspirin, and caffeine, but even these substances are absorbed in rather small quantities from the stomach; most absorption takes place in the small intestine. At the bottom of the stomach, peristaltic contractions push the chyme up against the pyloric sphincter. Associated waves of anticipatory relaxation cause the pyloric sphincter to relax briefly so that little squirts of chyme enter the small intestine. The human stomach empties gradually over a period of approximately 4 hours. This slow introduction of food into the small intestine enables it to work on a little material at a time.

Most chemical digestion occurs in the small intestine

In the **small intestine**, digestion of carbohydrates and proteins continues, and digestion of fats and absorption of nutrients begin. The small intestine takes its name from its diameter; it is in fact a very large organ, about 6 meters long in an adult human. Given its length and the folds, villi, and microvilli of its epithelial cells, its inner surface area is roughly the size of a tennis court. Across this surface the small intestine absorbs the nutrient molecules derived from food.

The small intestine of humans has three sections. The initial section (about 25 cm long) is called the **duodenum** and is the site of most digestion; the **jejunum** and the **ileum** (together about 600 cm) carry out 90 percent of the absorption of nutrients (see Figure 50.9).

Digestion in the small intestine involves many specialized enzymes, as well as several other secretions. Two accessory organs that are not part of the digestive tract—the liver and the

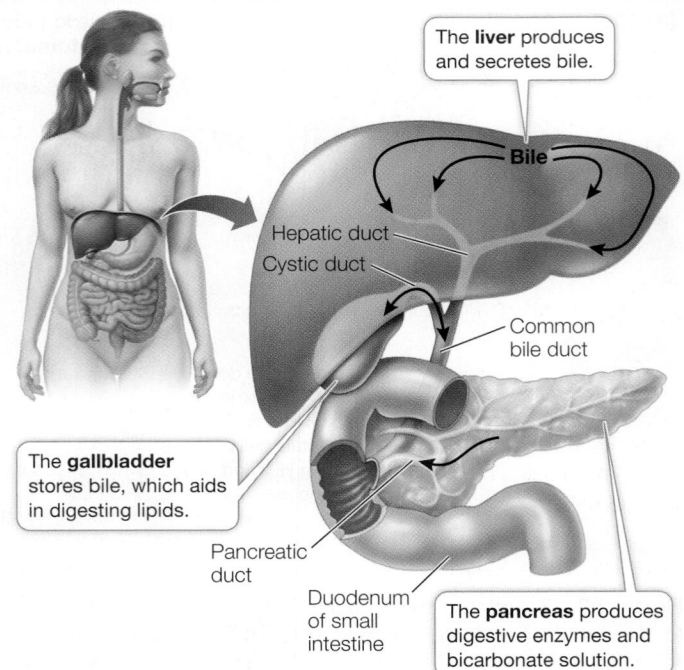

Figure 50.13 Ducts of the Gallbladder and Pancreas Bile produced in the liver flows from the liver via the hepatic duct. Branching off this duct is the gallbladder, which stores bile. Below the gallbladder, the hepatic duct is called the common bile duct and is joined by the pancreatic duct before entering the duodenum.

pancreas—produce most of these secretions and deliver them to the lumen of the intestine through ducts.

LIVER   The **liver** synthesizes **bile** from cholesterol. Bile also includes other substances, such as phospholipids and bilirubin (the breakdown product of hemoglobin). Bile flows from the liver through the hepatic duct. A side branch off the hepatic duct called the cystic duct goes to the **gallbladder**, where bile is stored. Below this junction, the hepatic duct is called the common bile duct. Before it reaches the duodenum, the common bile duct is joined by the pancreatic duct (**Figure 50.13**).

Fat entering the duodenum stimulates cells of the duodenal epithelium to release the hormone **cholecystokinin** (**CCK**), which stimulates the walls of the gallbladder to contract rhythmically. As a result, bile is squeezed out of the gallbladder and through the cystic duct to the common bile duct. Peristalsis moves the bile down the bile duct, and a small sphincter at the junction of the common bile duct with the duodenum relaxes in response to waves of peristalsis and allows squirts of bile to enter the duodenal lumen.

To understand the role of bile in fat digestion, think of an oil-and-vinegar salad dressing. Oil is hydrophobic and tends to aggregate in large globules. For that reason, many salad dressings include an emulsifier—something that prevents oil droplets from aggregating. Mayonnaise, for example, is oil and vinegar with egg yolk added as an emulsifier. Bile emulsifies fats in the chyme. One end of each bile molecule is lipophilic (soluble in fat), and the other end is hydrophilic (soluble in water). The lipophilic ends of bile molecules merge with the fat droplets, leaving their hydrophilic ends sticking out. As a result, bile prevents small fat droplets from congealing together into

Figure 50.14 Digesting Fats **(A)** Dietary fats are broken up by bile into small micelles that present a large surface area to lipases. **(B)** The products of fat digestion are absorbed by intestinal mucosal cells, where they are resynthesized into triglycerides and exported to lymphatic vessels.

Q: A small percentage of people who have had their gallbladders removed must thereafter eat a low-fat diet, or they will experience diarrhea. Why?

 Animation 50.1 The Digestion and Absorption of Fats
www.Life11e.com/a50.1

large fat droplets, and thereby the surface area of the fats exposed to the fat-digesting enzymes (lipases) is greatly enlarged. The very small fat particles that result from emulsification by bile are called **micelles (Figure 50.14A)**.

PANCREAS The **pancreas** is a large gland that lies just behind and below the stomach (see Figures 50.9 and 50.13). It is both an endocrine gland (secreting hormones into the circulation; see Key Concept 40.1) and an exocrine gland (secreting digestive juices through the pancreatic duct to the gut lumen). The exocrine tissues of the pancreas produce a host of digestive enzymes, including lipases, amylases, proteases, and nucleases (**Table 50.3**). As in the stomach, the protease enzymes are released in inactive forms called **zymogens**; if proteases were not in this inactive state, they would digest the pancreas and its ducts before ever reaching the duodenum. Once in the duodenum, the zymogen trypsinogen is activated by the enzyme enterokinase (secreted by cells lining the duodenum) to produce the active protease **trypsin**. Trypsin cleaves other zymogens, releasing other proteases as well as more active trypsin.

The mixture of zymogens produced by the pancreas can be dangerous if the pancreatic duct is blocked or if the pancreas is injured by infection or physical trauma such as a blow to the abdomen. A few activated trypsin molecules can initiate a chain reaction of enzyme activity that digests the tissues of the pancreas (a condition called pancreatitis), destroying both its endocrine and exocrine functions.

The ducts of the pancreas produce a secretion rich in bicarbonate ions (HCO_3^-). Bicarbonate ions are alkaline (basic) and neutralize the acidic pH of the chyme that enters the duodenum from the stomach. Intestinal enzymes function best at a neutral or slightly alkaline pH. The production of bicarbonate and its secretion into the pancreatic ducts is similar to the production of H^+ ions and their secretion into the gastric pits—but with opposite spatial orientations. In both types of secretory cells, carbonic anhydrase catalyzes the hydration of CO_2 to produce H_2CO_3 that dissociates into H^+ and HCO_3^- ions. Whereas in the gastric pit the H^+ is secreted into the lumen of the pit, in the pancreatic duct it is the HCO_3^- that is secreted into the lumen, and the H^+ is transported into the circulation.

Nutrients are absorbed in the small intestine

The final step in digesting proteins and carbohydrates and absorbing their components occurs among the intestinal microvilli. Mucosal epithelial cells produce peptidases that cleave small peptides into absorbable amino acids. These epithelial cells also produce the

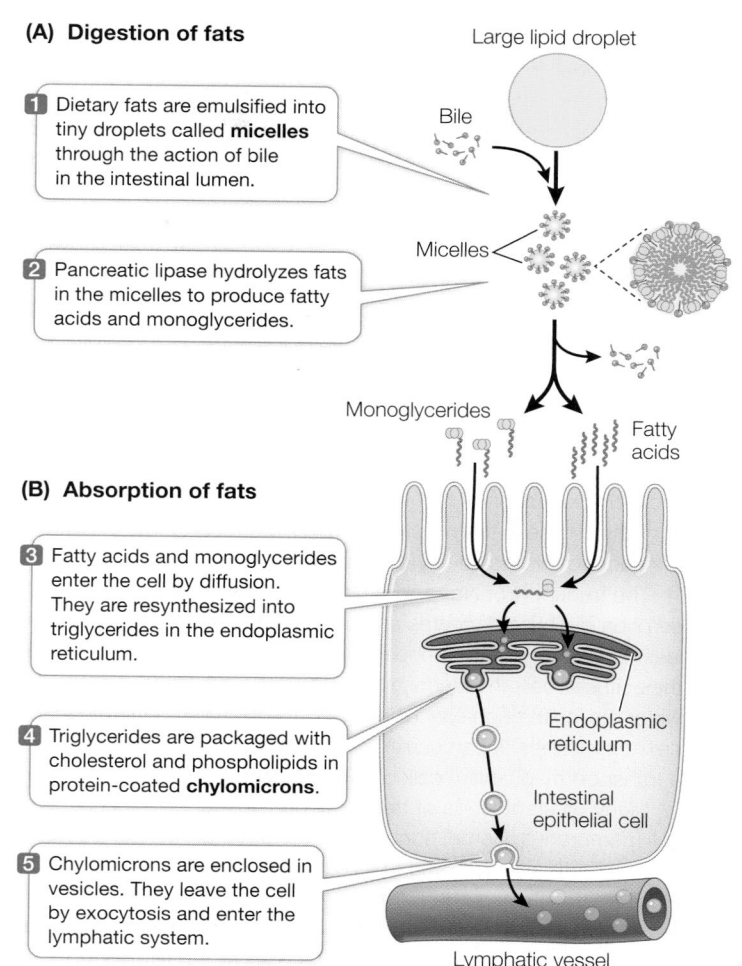

(A) Digestion of fats

1 Dietary fats are emulsified into tiny droplets called **micelles** through the action of bile in the intestinal lumen.

2 Pancreatic lipase hydrolyzes fats in the micelles to produce fatty acids and monoglycerides.

Large lipid droplet

Bile

Micelles

Monoglycerides

Fatty acids

(B) Absorption of fats

3 Fatty acids and monoglycerides enter the cell by diffusion. They are resynthesized into triglycerides in the endoplasmic reticulum.

4 Triglycerides are packaged with cholesterol and phospholipids in protein-coated **chylomicrons**.

5 Chylomicrons are enclosed in vesicles. They leave the cell by exocytosis and enter the lymphatic system.

Endoplasmic reticulum

Intestinal epithelial cell

Lymphatic vessel

table 50.3 Major Digestive Enzymes of Humans

Source/enzyme	Action
SALIVARY GLANDS	
Salivary amylase	Starch → Maltose
STOMACH	
Pepsin	Proteins → Peptides; autocatalysis
PANCREAS	
Pancreatic amylase	Starch → Maltose
Lipase	Fats → Fatty acids and glycerol
Nuclease	Nucleic acids → Nucleotides
Trypsin	Proteins → Peptides; zymogen activation
Chymotrypsin	Proteins → Peptides
Carboxypeptidase	Peptides → Shorter peptides and amino acids
SMALL INTESTINE	
Aminopeptidase	Peptides → Shorter peptides and amino acids
Dipeptidase	Dipeptides → Amino acids
Enterokinase	Trypsinogen → Trypsin
Nuclease	Nucleic acids → Nucleotides
Maltase	Maltose → Glucose
Lactase	Lactose → Galactose and glucose
Sucrase	Sucrose → Fructose and glucose

enzymes maltase, lactase, and sucrase that cleave the common disaccharides into absorbable monosaccharides—glucose, galactose, and fructose. There is also some lipase activity for fat digestion.

Many humans stop producing the enzyme lactase in childhood and thereafter have difficulty digesting lactose (the sugar in milk). Lactose is a disaccharide and cannot be absorbed without being cleaved into its constituents, glucose and galactose. Unabsorbed lactose is metabolized by bacteria in the large intestine, causing gas, diarrhea, and abdominal cramps.

The mechanisms by which cells of the intestinal epithelium absorb nutrients and inorganic ions are diverse and include diffusion, facilitated diffusion, active transport, and secondary active transport. Many inorganic ions such as sodium, calcium, and iron are actively transported by these cells. For example, active Na^+ transporters exist on the basal and lateral sides of the epithelial cells. They maintain a low concentration of Na^+ in those cells so that Na^+ can diffuse in from the chyme in the intestinal lumen. About 30 grams of Na^+ are transported this way every day, and Cl^- follows.

The transport of Na^+ and other ions is important for water absorption because it creates an osmotic concentration gradient. At least 7–8 liters of water per day move through the spaces between the epithelial cells in response to this osmotic gradient. Because the water moves through spaces between the cells and not through the cells themselves, it can carry with it nutrients that are in solution—a transport mechanism called **solvent drag** or **bulk transport**.

Many different kinds of transport proteins function in epithelial cell membranes. Some, such as the transport protein for fructose, only facilitate diffusion, and that requires a concentration gradient. Once fructose enters the cell it is converted to glucose, so the concentration of fructose in the cell is always low and the concentration gradient is maintained. Transport proteins known as symporters (see Figure 6.13) exploit the concentration gradient of Na^+ between the inside and outside of the cell that is maintained by the Na^+–K^+ pump common to all cells (see Figure 6.14). Symporters combine the transport of Na^+ and another molecule, such as glucose, galactose, or an amino acid. As Na^+ moves down its concentration gradient into the cell, the "hitchhiking" molecules are carried along with it.

The absorption of the products of fat digestion is relatively simple. Triglycerides are hydrolyzed to diglycerides, monoglycerides, and fatty acids, all of which are lipid-soluble and thus able to pass through the cell membranes of the microvilli. In the intestinal epithelial cells, these molecules are resynthesized into triglycerides, combined with cholesterol and phospholipids, and coated with protein to form water-soluble **chylomicrons** (**Figure 50.14B**). Rather than enter the blood directly, chylomicrons pass into blind-ended lymph vessels called **lacteals** that are inside each villus (see Figure 50.8C). They then flow through the lymphatic system, entering the bloodstream through the thoracic ducts at the base of the neck. After a meal rich in fats, chylomicrons can be so abundant in the blood that they give the blood plasma a milky appearance. Chylomicrons deliver their triglyceride and cholesterol cargo as they circulate through tissues.

The bile molecules that emulsify fats are not absorbed along with the monoglycerides, diglycerides, and the fatty acids, but are shuttled back and forth between the gut contents and the microvilli. In the ileum, bile is actively reabsorbed and returned to the liver via the bloodstream.

Absorbed nutrients go to the liver

Blood leaving the digestive tract flows to the liver in the **hepatic portal vein**. This large vein delivers the blood to small spaces called sinusoids between groups of liver cells. These cells absorb the nutrients coming from the digestive tract and either store them or convert them to molecules the body needs. Glucose, galactose, and fructose are used to synthesize glycogen. Amino acids are used to build proteins. Lipids from the chylomicrons are either stored as triglycerides or used to make lipoproteins, which are released by the liver and carry the triglycerides and cholesterol to other tissues (as we will discuss in Key Concept 50.4).

Water and ions are absorbed in the large intestine

The motility of the small intestine gradually pushes its contents through another sphincter—the ileocaecal sphincter—and into the large intestine, or **colon**. Most of the available nutrients have been removed from the chyme that enters the colon, but it contains a lot of water and inorganic ions. A major type of motility in the colon is called segmentation movements. These movements are mostly due to contractions of circular muscles that repeatedly divide the colon into separate segments. Functionally, segmentation movements cause the chyme to slosh around in the colon and promote absorption of ions and water by increasing contact between the chyme and the epithelium of the colon. As ions and water are absorbed, the chyme becomes a semisolid mass that forms feces. Absorption of too much water from the colon can cause hard feces and constipation. The opposite condition, diarrhea, results if too little water is absorbed. Excessive diarrhea caused by diseases such as cholera can produce such rapid loss of water and electrolytes that death can occur in hours. Simple rehydration therapy with electrolytes can save lives.

Fecal matter is stored in the descending colon and in the rectum until defecation occurs. This is usually once a day and is preceded by strong peristaltic activity. The distension of the walls of the rectum by fecal matter initiates a parasympathetic reflex, causing the rectal muscle to contract and the internal anal sphincter to relax. In addition, there is an external anal sphincter that is under conscious control so that defecation is not entirely an involuntary act.

Herbivores rely on their microbiota to digest cellulose

Cellulose is the primary component of plant cell walls, and is therefore the principal component of the food of herbivores. Most herbivores, however, cannot produce cellulases, the enzymes that break down cellulose. (Exceptions include earthworms, shipworms, and the silverfish that eat books and stored papers.) From termites to cattle, herbivores rely on the microbiota of their digestive tracts to digest cellulose.

Ruminants (cud chewers) such as cattle ferment their plant-based food in special stomachs before digestion. Their stomachs are large, four-chambered organs (**Figure 50.15**). The first two chambers, the **rumen** and the **reticulum**, are packed with endosymbiotic microorganisms that break down cellulose by fermentation. The ruminant periodically regurgitates the contents of the rumen (the cud) into the mouth for more chewing. When swallowed again, the vegetal fibers present more surface area to the microorganisms. The

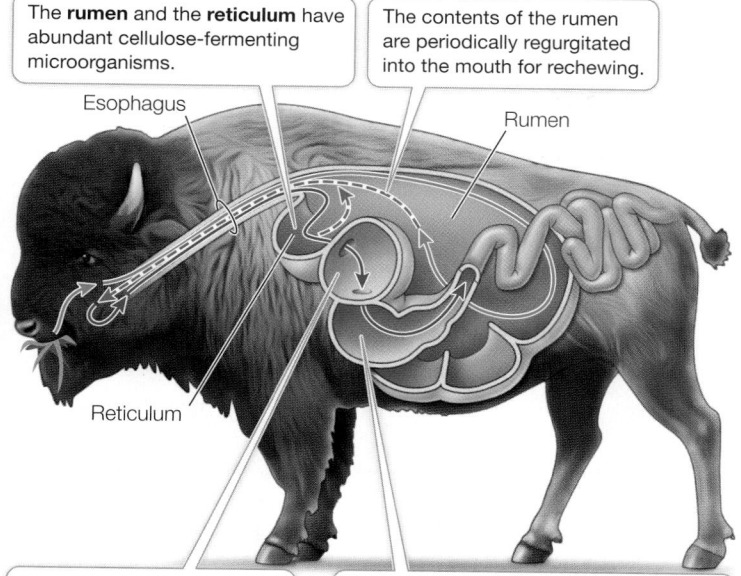

The **rumen** and the **reticulum** have abundant cellulose-fermenting microorganisms.

The contents of the rumen are periodically regurgitated into the mouth for rechewing.

Esophagus

Rumen

Reticulum

The mixture of fermented food and microorganisms passes through the **omasum**, where it is concentrated by water absorption.

The **abomasum** is the "true" stomach, secreting HCl and proteases. The microorganisms are killed by the HCl, digested by the proteases, and passed on to the small intestine for further digestion.

Figure 50.15 A Ruminant's Stomach   Bison, like their relatives domestic cattle, have a specialized stomach with four compartments that enables them to obtain energy from coarse plant material through bacterial fermentation of the otherwise indigestible plant material. The bacteria themselves become an important source of nutrition.

Q: Functionally, why do the rumen and reticulum come before the true stomach, the abomasum?

microorganisms metabolize cellulose and other nutrients to simple fatty acids, which are nutrients for their host.

Enormous numbers of microorganisms leave the rumen along with the partially digested food. This mass is concentrated by water absorption in the **omasum** before it enters the true stomach, the **abomasum**, where the microorganisms are killed by secreted hydrochloric acid, partially digested by proteases, and passed on to the small intestine for further digestion and absorption. A cow derives more than 100 grams of protein per day from digestion of its microbiota. The rate of multiplication of microorganisms in the rumen offsets their loss, so a well-balanced, mutually beneficial relationship is maintained.

Some mammalian herbivores have a microbial fermentation chamber called a **cecum** extending from the large intestine. An example is the rabbit (see Figure 50.7). Since the cecum empties into the large intestine, absorption of nutrients produced by the microorganisms in the cecum is inefficient because of the limited surface area of the large intestine. Such species frequently produce two kinds of feces—ones that are pure waste and ones that contain cecal material. In a behavior known as **coprophagy**, these species reingest the cecal feces directly from the anus so they can digest and absorb the nutrients that would otherwise be lost. In humans the cecum is small and ends in the vestigial **appendix** which serves no digestive function.

> **50.3 recap**

The vertebrate gastrointestinal system is a tubular gut that is adapted to ingest food, fragment it, digest it, and absorb nutrients. Peristalsis moves food through the gut. Segmentation movements mix the gut contents. Enzymatic digestion followed by absorption of nutrients occur mostly in the small intestine; water and ions are absorbed in the large intestine.

learning outcomes

You should be able to:

- Explain the mechanisms that move food through the vertebrate digestive system.
- Describe the role of acid and how it is produced in the stomach.
- Discuss autocatalysis and its role in digestion.
- Explain how bile assists in digestion.
- Explain observations about the presence of microorganisms found in the digestive tracts of herbivores.
- Compare the roles of microbiota in the digestive systems of different vertebrates.

1. Explain why in the weightlessness of space, food still travels from the mouth to the stomach.
2. When some biochemists ground up a piece of stomach wall and stimulated it with the hormone that causes the stomach to secrete HCl, no change in the pH of the solution occurred. Why not?.
3. How does bile assist in the digestion of fats, and why aren't equivalent molecules necessary to facilitate transport of lipids in the lymph and blood?
4. Explain why the gut microbiota is a significant source of nutrition for ruminants but not for humans.
5. Explain why in the ruminant digestive systyem HC1 is secreted in the abomasum but not in the rumen, the reticulum, or the omasum.
6. Why would the autocatalytic property of pepsin foster a positive feedback mechanism of protein hydrolysis in the stomach?

The steps included in the ingestion and digestion of food—from fragmentation in the mouth to the digestive processes in the stomach and intestines—make the nutrients in food available for absorption and ultimately for metabolism. We will look next at how the processes of digestion are controlled and how nutrients are handled by the body once food has been digested.

> **key concept**

50.4 Nutrient Availability Is Controlled and Regulated

As we indicated earlier, the vertebrate gut is a disassembly line, and as with a standard assembly line, the control and coordination of the sequential processes of digestion are critical. Both neuronal and hormonal controls govern these processes. Once the products of digestion are absorbed, their availability to the cells of the body must also be controlled and regulated.

focus your learning

- Hormones control actions of the digestive tract and its accessory organs.
- The liver interconverts and stores fuel molecules.
- The liver controls fat metabolism through its production of lipoproteins, which move fats between storage and production sites.
- The pancreas secretes hormones that help control glucose levels in the blood.
- Body mass and food intake behavior are influenced by stimulatory and inhibitory hormones that activate neurons in the arcuate nucleus region of the hypothalamus in the brain.

Your salivation at the sight or smell of food is an autonomic reflex, as is swallowing. Other autonomic reflexes include the release of digestive juices and increased stomach motility stimulated by good food smells, as well as gagging at the smell of rotting meat. These reflexes are called the "cephalic phase" of control of digestive processes because they involve the CNS. Most of the control of digestive processes, however, involves neural and hormonal mechanisms that are intrinsic to the gut.

Neuronal messages travel from one region of the digestive tract to another in the enteric nervous system. One function of the gut's nervous system is coordinating the movement of food through the gut. Of course, this intrinsic nervous system communicates information to the CNS and receives input from the CNS, but its most important role is to coordinate actions throughout the digestive tract. In spite of this marvelous intrinsic nervous system, however, much of the control and regulation of the digestive system and nutrient management involves hormonal mechanisms.

Hormones control many digestive functions

Several hormones control the activities of the digestive tract and its accessory organs (**Figure 50.16**). The first hormone ever discovered came from the duodenum; it was called **secretin** because it causes the pancreas to secrete digestive juices. We now know that secretin is only one of several hormones that control pancreatic secretion; specifically, secretin stimulates the cells of the pancreatic ducts to secrete a solution rich in bicarbonate ions.

The stimulus that causes the duodenum to release secretin is low pH caused by the arrival of acidic chyme from the stomach. Similarly, the presence of fats and proteins in the chyme stimulates the release of cholecystokinin (CCK), the hormone that stimulates the gallbladder to release bile. CCK also stimulates the pancreas to release digestive enzymes. Both CCK and secretin also have negative feedback effects on the stomach. They slow the movements of the stomach, thus slowing the delivery of chyme into the small intestine and allowing more complete digestion in the duodenum.

The presence of food in the stomach stimulates cells in the lower region of the stomach to secrete a hormone called **gastrin**. Gastrin returns to the stomach in the blood and stimulates the secretion of digestive juices and also increases the motility of the stomach. Gastrin release begins to be inhibited when the pH of the stomach contents falls below 3—another example of negative feedback.

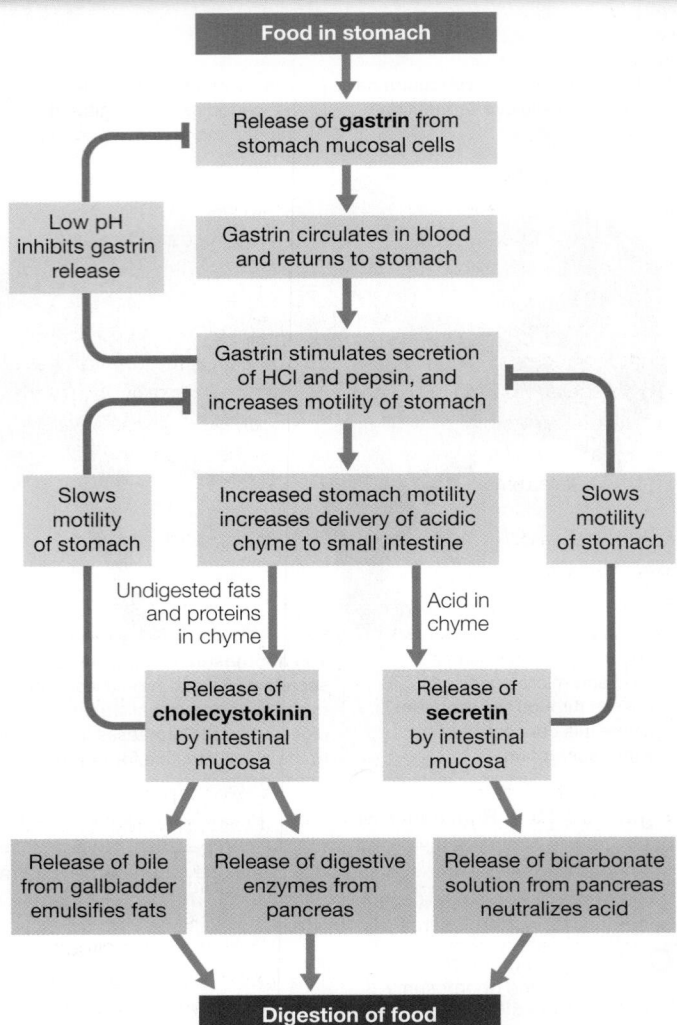

Figure 50.16 Hormones Control Digestion The hormones gastrin, cholecystokinin (CCK), and secretin are involved in feedback loops that control the sequential processing of food in the digestive tract. Red lines indicate inhibitory actions; green lines indicate stimulatory actions.

Q: Why do secretin and CCK stimulate the pancreas but inhibit the stomach?

The liver stores and releases the molecules that fuel metabolism

Most animals do not eat continuously, so they can be either in an **absorptive state** (food in the gut) or in a **postabsorptive state** (no food in the gut). Nutrient requirements for energy metabolism and biosynthesis are continuous, however. Thus nutrient traffic must be controlled so that reserves accumulate in the liver, muscle, and adipose (fat) tissue while the animal is in the absorptive state and are then used efficiently during the postabsorptive state. When fuel molecules are abundant in the blood, the liver stores them in the form of glycogen and fats. The liver also synthesizes blood plasma proteins from circulating amino acids. When levels of fuel molecules in the blood decline, the liver taps its reserves and delivers nutrients into the blood.

The liver has an enormous capacity to interconvert fuel molecules. Liver cells convert monosaccharides into either glycogen or fats during the absorptive state and then can reverse the process, returning glucose to the blood during the postabsorptive state. The liver can also convert certain amino acids and some other molecules, such as pyruvate and lactate, into glucose—the process of gluconeogenesis (see Figure 9.14). Gluconeogenesis provides an indirect pathway for exercising muscle to contribute to blood glucose levels. At high levels of aerobic activity, muscle cells break down their stores of glycogen to provide glucose for metabolic fuel. The glucose released from muscle glycogen, however, has to be used in the cells that produce it. Glucose cannot leave muscle cells as it can leave liver cells. However, when the activity of the muscle becomes anaerobic, pyruvate and lactate build up, leave muscle cells, and enter the circulation. Circulating pyruvate and lactate are taken up by the liver and converted to glucose that can then move out of the liver cells, circulate in the blood, and be taken up by muscle cells.

The liver produces lipoproteins: the good, the bad, and the ugly

The liver is the major controller of fat metabolism through its production of lipoproteins. A **lipoprotein** is a particle made up of a core of hydrophobic fat and cholesterol with a covering of hydrophilic protein that allows it to be suspended in water. Lipoproteins are the most abundant fuel reserve in the body, and they move fats from sites of absorption or synthesis to sites of storage, and from sites of storage to sites of use. You saw in Key Concept 50.3 how bile in the intestine solves the problem of processing hydrophobic fats in an aqueous medium. Transport of fats in the circulatory system presents the same problem, and lipoproteins provide the solution.

The chylomicrons produced by the mucosal cells of the intestine are the largest lipoprotein particles in the blood. As the circulation carries chylomicrons through muscle, adipose tissue, and liver, lipoprotein lipases attached to the endothelial cells of capillaries break down the chylomicrons' triglycerides so that the component fatty acids, monoglycerides, and glycerol can be taken up by the cells. The chylomicron remnants have lipid cores with a lot of cholesterol, and they are taken up by liver cells.

Lipoproteins different from chylomicrons are synthesized in the liver. These lipoproteins can be classified according to their density. Fat has a low density (it floats on water) and protein has a high density, so the greater the fat-to-protein ratio in the lipoprotein, the lower its density.

- **High-density lipoproteins (HDLs)** transport lipids and cholesterol from tissues to the liver where the cholesterol can be used to synthesize bile. HDL consists of about 50 percent protein, 35 percent lipids, and 15 percent cholesterol. These are the "good" lipoproteins, and their levels are higher in people who exercise and are fit.

- **Low-density lipoproteins (LDLs)** transport cholesterol around the body for use in biosynthesis and for storage. LDL consists of about 25 percent protein, 25 percent lipids, and 50 percent cholesterol. These are the "bad" lipoproteins associated with a high risk for cardiovascular disease.

- **Very low-density lipoproteins (VLDLs)** contain mostly triglycerides, which are taken up by fat cells in adipose tissues around the body. VLDL consists of about 2 percent protein, 94 percent lipids, and 3 percent cholesterol. These are the "ugly" lipoproteins, as they are associated with obesity and diabetes as well as a high risk for cardiovascular disease.

focus: key figure

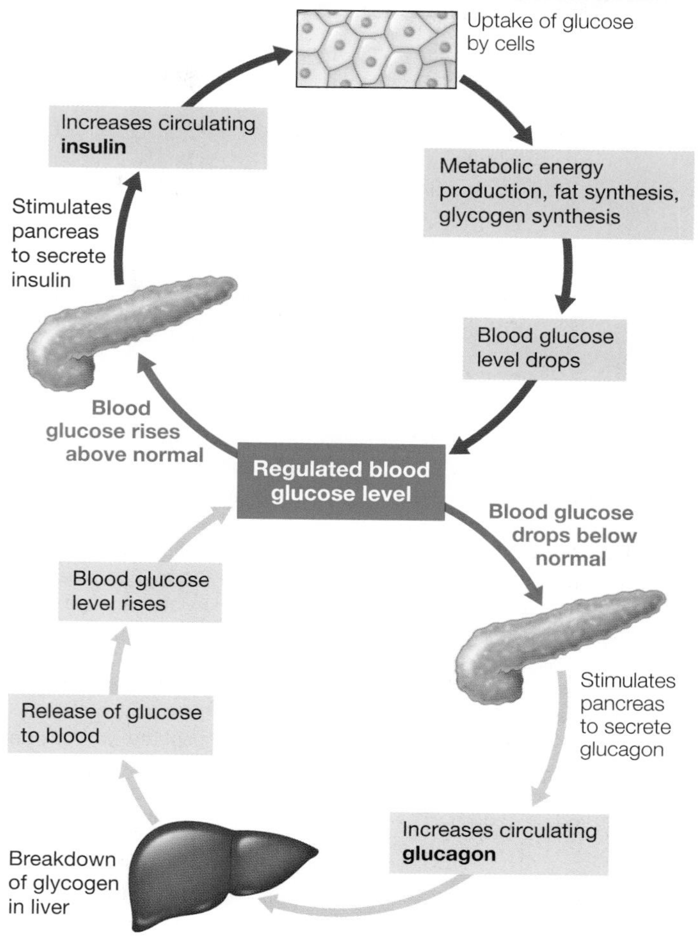

Figure 50.17 Regulating Glucose Levels in the Blood   The hormone insulin (blue) acts to promote glucose utilization and storage and thereby lower blood glucose. The hormone glucagon (brown) acts on the liver to breakdown glycogen and release glucose into the blood.

Q: The nervous system depends on a constant supply of glucose, yet none of the actions of insulin or glucagon in this figure refer to supplying glucose to the nervous system. Why?

▶ Animation 50.2 **Insulin and Glucose Regulation**
www.Life11e.com/a50.2

Insulin and glucagon control fuel metabolism

During the absorptive state, blood glucose levels rise as carbohydrates are digested and absorbed (**Focus: Key Figure 50.17**). During this time, the pancreas releases the hormone **insulin** which plays a major role in directing glucose to where it will be used or stored. In Key Concept 50.3 we mentioned that the pancreas has both an exocrine function (secretions involved in digestion) and endocrine functions. The endocrine functions are located in clusters of cells in the pancreas called the islets of Langerhans. One class of islet cells, the beta cells, produces and releases insulin. Another class, the alpha cells, produces and releases the hormone **glucagon** (discussed below). The actions of insulin vary in different tissues, but they are all aimed at promoting the use of glucose for metabolic fuel and directing the excess glucose into storage as either glycogen or fat.

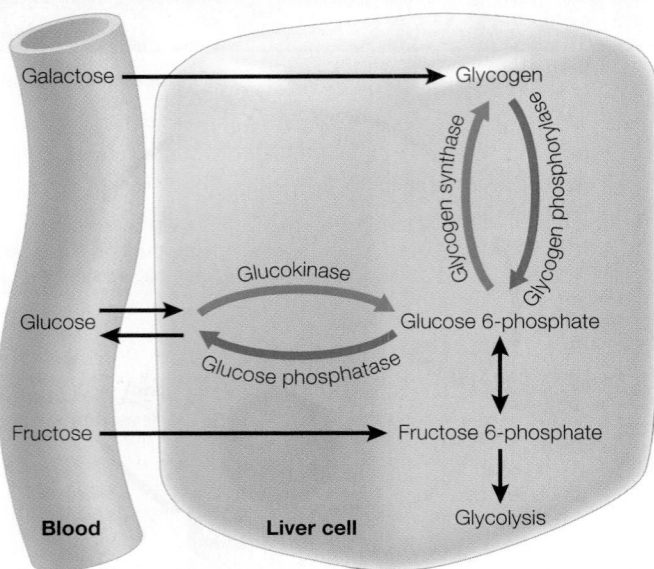

Figure 50.18 Insulin Controls Glucose Traffic in the Liver   Glucose freely enters and leaves liver cells by diffusion. But, when insulin is present (green arrows), glucose is phosphorylated and cannot leave the cell. The glucose 6-phosphate either enters glycolysis or is synthesized into glycogen. Insulin activates the enzymes of glycogen synthesis. Without insulin (red arrows) glucokinase is inhibited and glycogen phosphorylase and glucose phosphatase are activated. Thus glycogen is broken down to glucose phosphate, which is dephosphorylated so that it can leave the cell and enter the blood. Other monosaccharides—galactose and fructose—can also diffuse into the cell and be converted to glucose through these pathways.

Glucose enters cells by diffusion. This diffusion is facilitated by transporters, but they are not active transporters—they require a concentration gradient, which is why it is important to regulate blood glucose levels so there is always an adequate glucose concentration gradient across the cell membranes. There are several kinds of glucose transporters, and those in skeletal muscle and adipose tissues are normally sequestered in cytoplasmic vesicles until insulin binds its receptors on the cell surface and triggers the insertion of transporters into the cell membrane.

Insulin plays many roles in controlling how cells use the glucose they take up from the circulation. In adipose cells, insulin inhibits lipase and promotes fat synthesis from glucose. In the liver, insulin activates an enzyme, glucokinase, that phosphorylates glucose as it enters the liver cell so it cannot diffuse back out again, thereby enhancing the overall diffusion of glucose into the cells. At the same time, insulin inhibits the enzyme glucose phosphatase that enables glucose to leave the cell. Insulin also activates glycogen synthase, an enzyme in liver cells that catalyzes the incorporation of glucose into glycogen, and activates enzymes that increase the flow of glucose into glycolysis (**Figure 50.18**).

To maintain blood glucose levels during the postabsorptive state, liver cells break down their stored glycogen, releasing glucose into the blood. The multiple processes that make it possible for the liver to release glucose depend on a fall in blood insulin levels (see Figure 50.18). Falling insulin inhibits the enzyme responsible for glycogen synthesis and activates the enzyme that breaks down glycogen. Also, with less insulin, the enzyme that phosphorylates glucose is inhibited and activity of glucose phosphatase is increased. The result is a

breakdown of glycogen and the return of glucose to the blood. Another consequence of the fall in insulin levels is the increased activity of lipases in the liver and adipose tissue, releasing fatty acids to the blood. Most cells preferentially use fatty acids as their metabolic fuel during the postabsorptive state. *Overall, the most important control of fuel metabolism in the postabsorptive state is the lack of insulin.*

One tissue that does not switch fuel sources when an animal is postabsorptive is the nervous system. The cells of the nervous system require a constant supply of glucose and can use other fuels only to a very limited extent. Most neurons do not require insulin to absorb glucose from the blood, but they do need an adequate glucose concentration gradient to drive the facilitated diffusion of glucose across their cell membranes. Therefore it is critical that blood glucose levels are maintained when an animal is postabsorptive. The overall dependence of neural tissues on glucose, and their requirement for constant blood glucose levels, are the reasons it is so important for other cells of the body to shift to fat metabolism during the postabsorptive state.

The metabolism of fuel molecules during the postabsorptive state is mostly controlled by the lack of insulin, but if blood glucose falls below a certain level, glucagon is released. Glucagon's effect is opposite that of insulin: it stimulates liver cells to break down glycogen and to carry out gluconeogenesis. Thus, under the influence of glucagon, the liver produces glucose and releases it into the blood. Note that under conditions that stimulate glucagon release, the effects of low insulin are already in play—low glycogen synthase activity, low glucokinase activity, and high glucose phosphatase activity.

The brain plays a major role in regulating food intake

Multiple brain areas and signals are involved in the regulation of food intake. Long ago it was discovered that damaging a region in the center of the rat hypothalamus resulted in the rats increasing their food intake and becoming obese. Damage to the lateral hypothalamus, however, led to decreased food intake and the rats became thin. In both cases the rats eventually reached a new equilibrium body weight; thus it appeared that a capacity for regulation remained, but the set point was altered.

We now know that another region of the hypothalamus, the **arcuate nucleus**, plays an important role in integrating a variety of feedback signals that influence food intake and body mass. Cells within the arcuate nucleus send axons to the ventromedial and dorsal hypothalamus, as well as to other brain areas that influence food intake and metabolism. One group of arcuate neurons projects to brain areas that inhibit food intake, while the other projects to brain areas that stimulate food intake. But what stimulates or inhibits the activity of the arcuate neurons?

Several factors have been identified that reflect the body's energy balance. Three of these are the proteins insulin, leptin, and ghrelin. Insulin, as detailed earlier, is released when blood glucose levels are high, and it induces a decline in food intake. **Leptin** (Greek *leptos*, "thin") is released by fat cells in proportion to how much lipid they contain. Evidence that leptin is a satiety signal is described in **Figure 50.19**. **Ghrelin** is released by the stomach when it is empty; its levels rise before meals and fall after meals. In the arcuate nucleus, insulin and leptin activate the neurons that inhibit feeding and inhibit

experiment

Figure 50.19 A Single-Gene Mutation Leads to Obesity in Mice

Original Papers: Coleman, D. L. 1973. Effects of parabiosis of obese with diabetes and normal mice. *Diabetologia* 9: 294–297.

Coleman, D. L. and K. P. Hummel. 1969. Effects of parabiosis of normal with genetically diabetic mice. *American Journal of Physiology* 217: 1298–1304.

In mice the *Ob* gene codes for the protein leptin, a satiety factor that signals the brain when enough food has been consumed. The recessive *ob* allele is a loss-of-function allele, so *ob/ob* mice do not produce leptin; they do not experience satiety and become obese. The *Db* gene encodes the leptin receptor, so mice homozygous for the recessive loss-of-function allele *db*, even if they produce leptin, cannot use it and so become obese.

HYPOTHESIS▶ Mice that cannot produce the satiety signal protein leptin will not become obese if they are able to obtain leptin from an outside source.

METHOD

1. Create two strains of genetically obese laboratory mice, one that lacks functional leptin (genotype *ob/ob*) and one that lacks the receptor for leptin (genotype *db/db*).
2. Create parabiotic pairs by surgically joining the circulatory systems of a non-obese (wild-type) mouse with a partner from one of the obese strains.
3. Allow mice to feed at will.

Parabiotic pair

Wild-type mouse   Genetically obese mouse
(*Ob/–* and *Db/–*)   (either *ob/ob* or *db/db*)

RESULTS

Parabiotic *ob/ob* mice obtain leptin from the wild-type (WT) partner and lose fat. Parabiotic *db/db* mice remain obese because they lack the leptin receptor and thus the leptin they obtain from their partner has no effect.

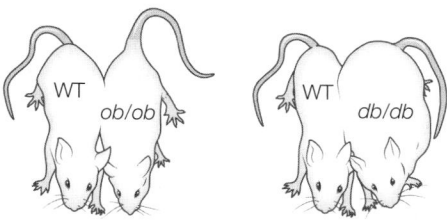

CONCLUSION▶ The protein leptin is a satiety signal that acts to prevent overeating and resultant obesity.

 Activity 50.5 **Parabiotic Mice Simulation**
www.Life11e.com/ac50.5

---

the neurons that stimulate feeding. Ghrelin has the opposite effect on these two groups of neurons.

An integrative signal that could be playing a central role in regulation of feeding is the enzyme AMP-activated protein kinase (AMPK). When most cells are nutrient deprived, they produce AMPK, which stimulates the oxidation of substrates to replenish ATP. Fasting increases AMPK levels in the hypothalamus, and feeding reduces them. Insulin and leptin decrease AMPK activity in the hypothalamus, and ghrelin increases it. Thus AMPK could be a common pathway for various signals controlling food intake.

▶50.4 recap

The major factors controlling gut function are autonomic reflexes, the enteric nervous system, and the hormones gastrin, secretin, and cholecystokinin. Insulin is the major hormonal controller of fuel metabolism through its effects on liver, muscle, and adipose tissues. The hypothalamus controls food intake by generating sensations of hunger and satiety influenced by feedback from blood glucose and hormones, including insulin, leptin, and ghrelin.

learning outcomes

You should be able to:

- Explain how hormones control the rate of delivery of substrate to the small intestine and the appropriate secretions of digestive juices.
- Describe metabolic interconversions that take place in liver and muscle to meet the body's demands for glucose.
- Differentiate between the classes of lipoproteins.
- Compare the effects of insulin on activities in skeletal muscle cells, adipose cells, and liver cells.
- Cite evidence that supports the role of leptin as a satiety signal and ghrelin as a hunger signal.
- Interpret results from experiments designed to elucidate the mechanisms underlying regulation of body mass and food intake behaviors.

---

1. For various reasons patients may be fed through a tube going into their stomachs or tubes going all the way to their jejunum, the middle region of their small intestines. How would the feeding formulas used in the jejunal tube have to differ from the one applied through the gastric tube?
2. How do the three classes of lipoproteins differ in structure and function?
3. How are the effects of high and low insulin levels the same and different in liver and muscle?
4. In the experiments in which the lateral or the ventromedial hypothalamus was lesioned, what result enabled the conclusion that the regulation of body mass was altered rather than just the ability to produce satiety or hunger by the nutritional state of the body?
5. When exercising muscles do not have an adequate supply of oxygen, they continue to produce small amounts of ATP through glycolysis, but they also produce pyruvate and lactate. Explain how this pyruvate and lactate can help support continuing glycolysis.
6. What evidence supports the hypothesis that leptin influences satiety?

If the nutritional values were the same, could different foods have different impacts on health?

The Pima of Arizona were among the first communities to be examined with a genome-wide survey. Many candidate genes were shown to be associated with obesity, but weakly so. The conclusion was that there are genetic contributions to the propensity to develop obesity, but except for a few rare cases, obesity is a multigenic trait that is highly influenced by environment. One aspect of the environment is physical activity, but another one is diet. The caloric content of the diet is obviously an important factor, but what about the composition of the diet? The experiments described in Investigating Life: How Does the Gut Microbiome Contribute to Obesity and Metabolic Disease? show that the composition of the diet can influence the composition of the gut microbiome, and the gut microbiome can influence the nutritional and other health effects of the diet. Thus, apart from differences in caloric intake, the consequences of the

Pima switching from a traditional diet to a high-fat Western diet could have resulted in changes in their microbiomes that have contributed to obesity and diabetes.

Future directions

It would be of interest to compare the microbiomes of the Arizona and the Mexican Pima peoples, and to compare those with the microbiomes of other populations on similar and different diets. If a particular microbiome profile is associated with the obesity–diabetes phenotype, it might be possible to change that microbiome through dietary change to promote better health. Just as we now do genomic screening to identify health risks, we might consider microbiome screening. Whereas we cannot change our genetic makeup, we could change our microbiome either through diet, or even therapeutically by uses of antibiotics and microbiome transplants. A new branch of medicine could develop around the testing, diagnosis, and therapy of the gut microbiome.

Chapter Summary 50

50.1 Food Provides Energy As Well As Materials for Biosynthesis

- Animals are **heterotrophs** that derive their energy and molecular building blocks, directly or indirectly, from **autotrophs**.

- Carbohydrates, fats, and proteins in food supply animals with energy. A measure of the energy content of food is the **kilocalorie (kcal)**. Excess caloric intake is stored as glycogen and fat. **Review Figure 50.2**

- For many animals, food provides essential **carbon skeletons** that they cannot synthesize themselves. **Review Figure 50.4**

- Most researchers consider 8 amino acids to be essential for adult humans; some believe that infants require as many as 12 **essential amino acids** in their diet. **Macronutrients** are mineral elements needed in large quantities; **micronutrients** are needed in small amounts. **Review Figure 50.5, Table 50.1, Activity 50.1**

- **Vitamins** are organic molecules that must be obtained in food. **Review Table 50.2, Activity 50.2**

- **Malnutrition** results when any essential nutrient is lacking from the diet. Chronic malnutrition causes **deficiency disease**.

50.2 Diverse Adaptations Support Ingestion and Digestion of Food

- Animals can be characterized by how they acquire nutrients: **saprobes** and **detritivores**, or **decomposers**, depend on dead organic matter, **filter feeders** strain the aquatic environment for small food items, **herbivores** eat plants, and **carnivores** eat other animals. Behavioral and anatomical adaptations reflect these feeding strategies. **See Activity 50.3**

- Digestion involves the breakdown of complex food molecules into monomers that can be absorbed and used by cells. In most animals, digestion takes place in a tubular gut. **Review Figure 50.7**

- Absorptive areas of the vertebrate gut are characterized by a large surface area produced by extensive folding and numerous **villi** and **microvilli**. **Review Figure 50.8**

- Hydrolytic enzymes break down proteins, carbohydrates, and fats into their monomeric units.

- The microbiome of the gut participates in digestive processes, makes certain nutrients available for absorption, and is sensitive to composition of diet, ingestion of drugs, and other factors. **Review Investigating Life: How Does the Gut Microbiome Contribute to Obesity and Metabolic Disease?**

50.3 The Vertebrate Gastrointestinal System Is a Disassembly Line

- The vertebrate gut can be divided into several compartments with different functions. **Review Figure 50.9, Activity 50.4**

- The cells and tissues of the vertebrate gut are organized in the same way throughout its length. The innermost tissue layer, the **mucosa**, is the secretory and absorptive surface. The **submucosa** contains blood and lymph vessels and a nerve network that is sensory and also controls gut secretions. External to the submucosa are two smooth muscle layers. Between the two muscle layers is another nerve network that controls the movements of the gut. **Review Figure 50.10**

- Swallowing is a reflex that pushes a bolus of food into the **esophagus**. **Peristalsis** and segmentation movements of the gut move the bolus down the esophagus and through the entire length of the gut. **Sphincters** block the gut at certain locations, but they relax as a wave of peristalsis approaches. **Review Figure 50.11**

- Digestion begins in the mouth, where amylase is secreted with the saliva. Digestion of protein begins in the **stomach**, where parietal cells secrete HCl and chief cells secrete pepsinogen, which becomes pepsin when activated by low pH and **autocatalysis**. The mucosa also secretes mucus, which protects the tissues of the gut. **Review Figure 50.12**

- In the **duodenum**, pancreatic enzymes carry out most of the digestion of food. **Bile** from the liver and **gallbladder** emulsifies fats into **micelles**. Bicarbonate ions from the **pancreas** neutralize the pH of the **chyme** entering from the stomach, thereby producing an environment conducive to the actions of pancreatic enzymes such as **trypsin**. **Review Figure 50.13, Table 50.3**

- Final enzymatic cleavage of polypeptides and disaccharides occurs among the microvilli of the intestinal mucosa. Amino acids, monosaccharides, and inorganic ions are absorbed by the microvilli. Specific transport proteins are sometimes involved. Symporters often power the absorption of nutrients.

- Fats broken down by **lipases** are absorbed mostly as monoglycerides and fatty acids and are resynthesized into triglycerides within the gut epithelium. The triglycerides are combined with cholesterol and phospholipids and coated with protein to form **chylomicrons**, which pass out of the mucosal cells and into lymphatic vessels in the submucosa. Review Figure 50.14, Animation 50.1

- Water and ions are absorbed in the large intestine as waste matter and consolidated into **feces**, which are periodically eliminated.

- Microorganisms in some compartments of the gut digest materials that their host cannot. Review Figure 50.15

▶ 50.4 Nutrient Availability Is Controlled and Regulated

- Autonomic reflexes coordinate activity of the digestive tract, which has an intrinsic nervous system that can act independently of the CNS.

- The actions of the stomach and small intestine are largely controlled by the hormones **gastrin**, **secretin**, and **cholecystokinin (CCK)**. Review Figure 50.16

- The liver plays a central role in directing the storage and release of fuel molecules. In the **absorptive state**, the liver takes up and stores fats and carbohydrates, converting monosaccharides to glycogen or fats. The liver also takes up amino acids and uses them to produce blood plasma proteins, and can engage in gluconeogenesis.

- Fat and cholesterol are shipped out of the liver as **low-density lipoproteins (LDLs)**. **High-density lipoproteins (HDLs)** act as acceptors of cholesterol and bring fat and cholesterol back to the liver.

- **Insulin** largely controls fuel metabolism during the absorptive state and promotes glucose uptake as well as glycogen and fat synthesis. In the **postabsorptive state**, lack of insulin blocks the uptake and use of glucose by most cells of the body except neurons. If blood glucose levels fall, **glucagon** secretion increases, stimulating the liver to break down glycogen and release glucose to the blood. Review Focus: Key Figure 50.17, Figure 50.18, Animation 50.2

- Food intake is governed by sensations of hunger and satiety, which are determined by brain mechanisms responding to feedback signals such as insulin, **leptin**, and **ghrelin**. Review Figure 50.19, Activity 50.5

Go to **LearningCurve** (in **LaunchPad**) for dynamic quizzing that helps you solidify your understanding of this chapter. **LearningCurve** adapts to your responses, giving you the practice you need to master each key concept.

▶ Apply What You've Learned

Review

50.1 Fats, carbohydrates, and proteins in food provide energy.

50.2 Heterotrophs display a wide diversity of adaptations for acquiring and processing food.

50.2 Heterotrophs rely on symbiotic bacteria residing in their digestive systems to carry out essential tasks in digestion.

Original Paper: Sommer, F. et al. 2016. The gut microbiota modulates energy metabolism in the hibernating brown bear *Ursus arctos. Cell Reports* 14: 1655–1661.

Winter in the Northern Hemisphere can be a harsh period of low temperatures, deep snow cover, and limited food. The brown bear (*Ursus arctos*) survives the winter by hibernating. Bears do not eat during hibernation; but prior to the onset of winter, they overeat—and can double their body mass. The extra calories stored in their bodies as fat provide the necessary energy to sustain them through the winter months.

In recent years, information has been accumulating about the importance of the gut microbiota in the digestion of food and in many other aspects of nutrition. Researchers asked whether the gut microbiome of the bear was different when the bear was fattening and when it was fasting. They studied blood and fecal samples from 16 free-ranging brown bears during hibernation (February or March) and following hibernation (June). Fecal samples were analyzed for composition of the microbe communities, and the blood was tested for triglycerides and other nutrients. The results shown in **Figure A** show that there are differences in the composition of the microbiome and in blood levels of triglycerides, lactate, and succinate between summer and winter. Data on the

Figure A

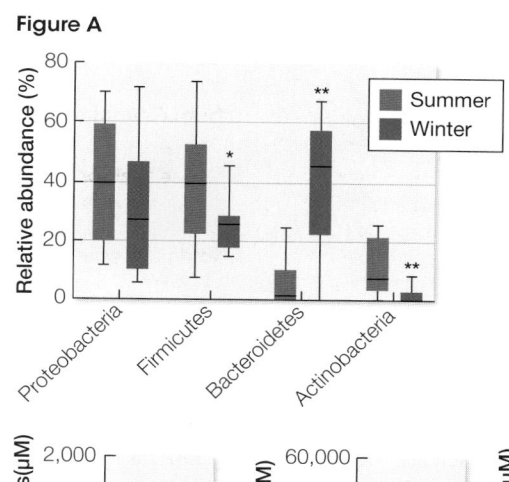

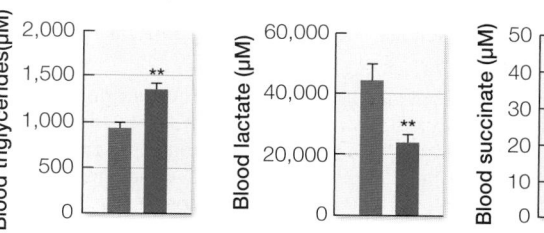

composition of the microbiome are presented as the means, the twenty-fifth to seventy-fifth percentiles of the data (colored bars), and the total range of the data (vertical lines). The graphs on blood levels of nutrients are presented as the means ± 1 SEM (* = $P < 0.05$; ** = $P < 0.01$).

(continued)

The observed seasonal differences in the microbiota could result simply from the presence or absence of food in the gut. To test whether summer and winter microbiota could have different influences on nutritional status, the researchers examined the effects of these microbiota when introduced into germ-free mice. Germ-free mice have no gut microbiota and do not gain mass when placed on a high-fat diet, as normal mice do. Germ-free mice were given fecal transplants from summer or winter bears. Two weeks after the transplants, the mice were analyzed for changes in body mass, fat content, and blood triglyceride levels. The results are shown in **Figure B**. Data are presented as means $\pm$ SEM (* = $P < 0.05$).

Figure B

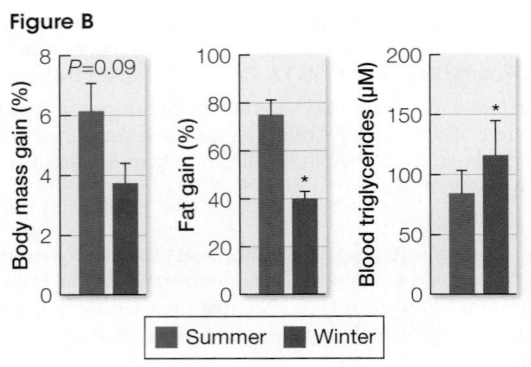

Questions

1. How do the data on seasonal changes in blood triglycerides, lactate, and succinate indicate changes in the bears' metabolism between summer and winter? (Keep in mind that lactate is a product of glycolysis, and succinate is a citric acid cycle intermediate.)

2. Do the data indicate significant seasonal changes in the gut microbiome of the bears? Propose hypotheses that would help explain any changes.

3. What do the data indicate about how the energy balance of the germ-free mice is influenced when they are infected with microbes isolated from hibernating bears? Is there a difference between the outcomes from summer versus winter microbiota? Construct a hypothesis that explains your answer.

4. Taken together, what do all of these data suggest about digestive adaptations of brown bears that aid their survival in their native habitat?

Go to **LaunchPad** for the eBook, LearningCurve, animations, activities, flashcards, and additional resources and assignments.

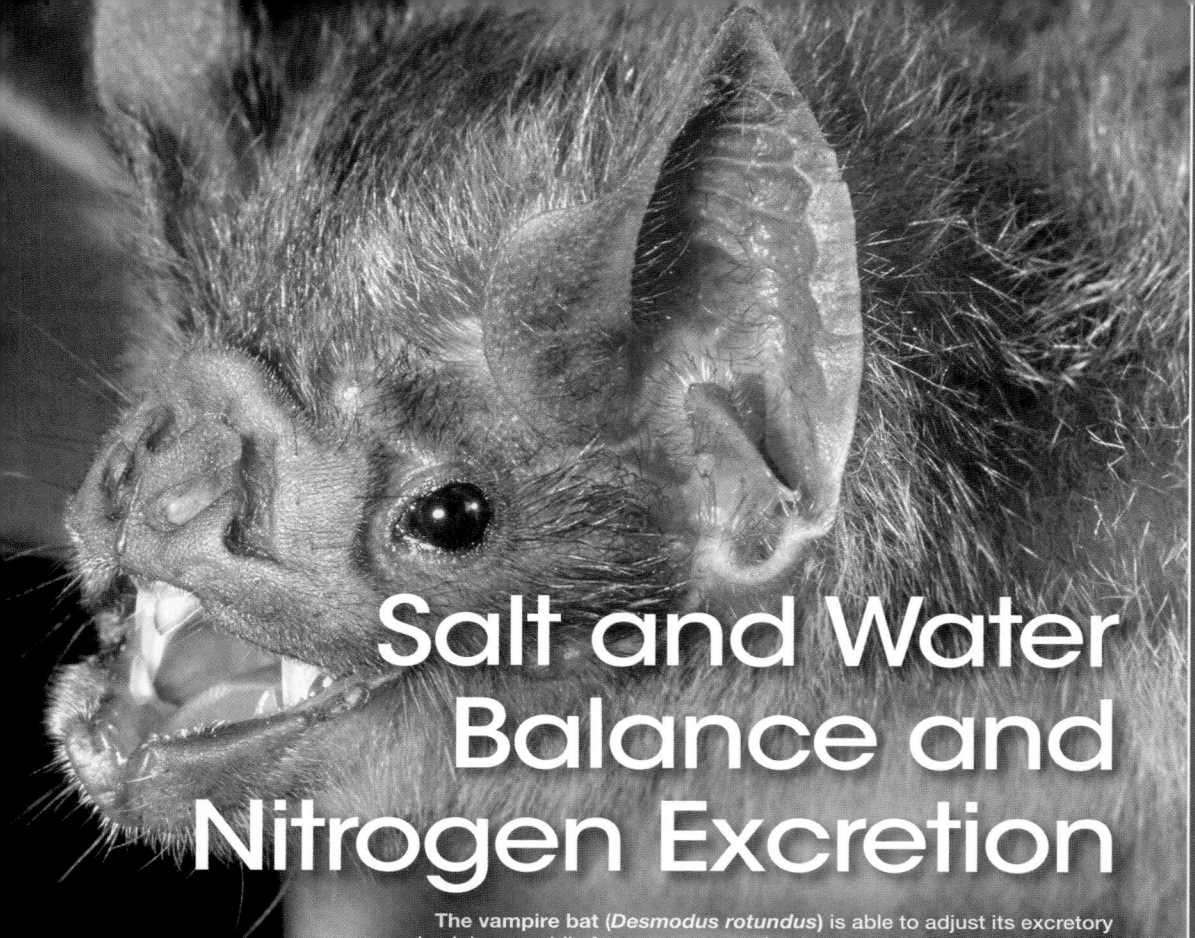

Salt and Water Balance and Nitrogen Excretion

The vampire bat (*Desmodus rotundus*) is able to adjust its excretory physiology rapidly from water-excreting to water-conserving, depending on whether it is ingesting or digesting its blood meal.

51

▶ investigating**life**

How Vampire Bats Use Blood as Fast Food

Blood, sweat, and tears have the composition of extracellular fluid and taste salty. The composition and volume of the extracellular fluid are regulated by the excretory system. The challenges to an animal's excretory system depend on its environment and lifestyle. Some desert animals rarely encounter free water, so they must conserve water by excreting their excess ions and nitrogenous wastes with as little water loss as possible. Animals living in fresh water have the opposite challenge: water continuously enters their bodies by osmosis and with their food, so they must excrete water and conserve ions. Vertebrate animals living in the ocean face a challenge similar to that of the desert dwellers; they lose body water by osmosis and therefore need to conserve water and excrete ions.

The physiological mechanisms animals have to maintain salt and water balance are similar, but they are used in different ways to solve the problems unique to each species. Consider vampire bats, which feed on the blood of animals such as goats and cattle. The bats use their sharp incisor teeth to make a small incision (usually on the ankle or leg of

a sleeping victim) and then lap up the blood. A vampire bat feeds only once a day, and has to do so quickly, before the victim awakens and shakes it off.

Blood contains nutritious protein but consists mostly of water. To meet its daily energy needs, a vampire bat must take in a huge volume of blood relative to its body size, and it has to be able to fly back to its roost. To maximize protein intake and still be able to fly, the bat gets rid of water fast. The warm fluid flowing down the victim's leg is not blood!

Back in the roost, the bat's challenge is reversed. To digest a high-protein meal, the bat must excrete a lot of nitrogenous waste with minimal loss of water, as it may not get more water until its next meal. So within a few hours, the bat's excretory system has to switch from rapidly excreting water to conserving water, in spite of the need to excrete nitrogenous wastes in the urine. The vampire bat's excretory system rapidly switches from producing lots of dilute urine to producing small amounts of very concentrated urine.

Q How does the excretory system of vampire

key concept
51.1
Excretory Systems Regulate Osmotic and Ionic Concentrations

Homeostasis of the extracellular fluid (the blood plasma and interstitial fluid; see Key Concept 39.2) is critical for several reasons:

- The solute concentration of the extracellular fluid determines the water balance of the cells of the body.
- The specific ionic composition of the extracellular fluid influences many cell functions.
- The health of cells requires the elimination of nitrogenous wastes.

The problems that have to be solved to maintain homeostasis of the extracellular fluid depend on the environment in which a species lives (salt water, fresh water, or terrestrial) and its lifestyle, as you saw in the case of the vampire bat in the opening story. Animals depend on **excretory systems** to maintain the volume, concentration, and composition of their extracellular fluids, and to excrete wastes.

focus your learning

- Animal excretory systems use filtration, secretion, and reabsorption to regulate osmolarity and ionic composition of the extracellular fluid.
- Aquatic animals either conform to the osmolarity of the environment or expend energy to maintain a constant internal osmolarity.

Osmosis causes water to enter or to leave cells

The volume of a cell depends on whether it takes up water from or loses water to the extracellular fluid. The movement of water across cell membranes via osmosis depends on differences in solute concentration on the two sides of the membrane and on the permeability of the membrane (see Key Concept 6.3). If the solute concentration of the extracellular fluid is less than that of the cytoplasm, water moves into the cells, causing them to swell and possibly burst (see Figure 6.10). If the solute concentration of the extracellular fluid is greater than that of the cytoplasm, the cells lose water and shrink. Thus the solute concentration of the extracellular fluid affects both the volume and the solute concentration of the cells.

Animal physiologists use the term **osmolarity** in discussing osmosis. The osmolarity of a solution is the number of *moles of osmotically active solutes per liter of solvent. Thus a 1-molar solution of glucose is also a 1-osmolar (1 osmole per liter) solution, but a 1-molar solution of sodium chloride (NaCl) is a 2-osmolar solution, because each NaCl molecule dissociates into two osmotically active ions.

---

*connect the concepts   As described in Key Concept 2.4, a mole is the amount of a substance in grams that is numerically equal to its molecular weight.

Excretory systems control extracellular fluid osmolarity and composition

Excretory systems control the osmolarity and composition of the extracellular fluids by excreting solutes that are present in excess (such as NaCl when we eat lots of salty food) and conserving solutes that are valuable or in short supply (such as glucose and amino acids). Excretory systems also eliminate the toxic waste products of protein metabolism. The output of the excretory system is **urine**.

Three basic processes are common to a wide variety of animal excretory systems: filtration, secretion, and reabsorption. Filtered extracellular fluid contains no cells or large molecules, such as proteins. In animals with a closed circulatory system, the blood plasma is filtered from capillaries into associated tubules. The walls of the capillaries and of the tubules are the filter, and the filtration is driven by blood pressure. As the filtrate flows through the tubules, its composition and concentration are modified through processes of secretion and reabsorption to form the urine that leaves the body.

In all of the discussions that follow about the movement of water across membranes, *it is important to remember that there are no mechanisms for the active transport of water*. The movement of water is due either to a pressure difference (filtration) or to a difference in solute concentration (osmosis). Water molecules move down a pressure gradient or up a solute concentration gradient.

Aquatic invertebrates are either ionic conformers or regulators

Most invertebrates that live in seawater conform to the osmotic concentration of their environment over a fairly wide range of salinities and are therefore called **osmoconformers** (**Figure 51.1**). The

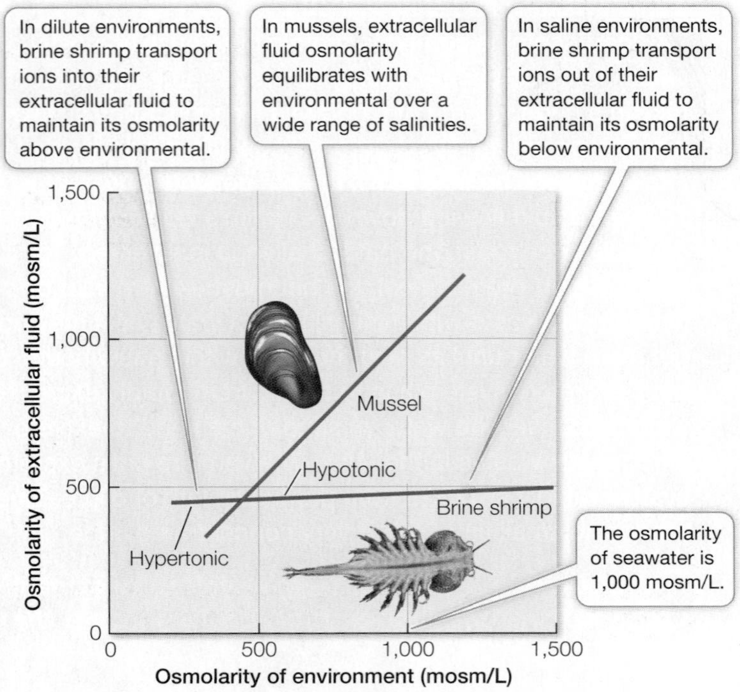

In dilute environments, brine shrimp transport ions into their extracellular fluid to maintain its osmolarity above environmental.

In mussels, extracellular fluid osmolarity equilibrates with environmental over a wide range of salinities.

In saline environments, brine shrimp transport ions out of their extracellular fluid to maintain its osmolarity below environmental.

The osmolarity of seawater is 1,000 mosm/L.

Figure 51.1 Some Marine Invertebrates Osmoregulate   Some aquatic animals, such as mussels, experience an extreme range of salt concentrations in their environment and are osmoconformers over much of that range. Other aquatic animals, such as brine shrimp, are osmoregulators in that they maintain a relatively constant osmolarity of their extracellular fluids as environmental salinity varies.

osmolarity of seawater in the open ocean is about 1,000 milliosmoles/liter (mosm/L), but it can vary quite a bit in estuaries where it is diluted by an influx of fresh water or in evaporating tide pools as the salt gets concentrated. Osmoconformity can result in considerable energetic savings, as it costs metabolic energy to move ions across membranes to achieve osmotic regulation.

Osmoconformity is not an option for freshwater animals. No cell can do without ions or nutrients, so freshwater animals always have to expend energy to conserve salts and excrete water.

Some marine invertebrates maintain a rather constant osmolarity of their extracellular fluids as the osmolarity of the environment changes, and these animals are therefore called **osmoregulators** (see Figure 51.1). Osmoregulators in the marine environment mostly maintain their osmotic concentration considerably below that of the environment and are therefore engaged in **hypotonic regulation**. Occasionally, however, seawater is diluted by an influx of fresh water, as in estuaries, and animals must maintain their osmotic concentrations above that of the environment—a process called **hypertonic regulation**.

The brine shrimp *Artemia* illustrated in Figure 51.1 is an osmoregulator with an enormous range of tolerances. *Artemia* are found in huge numbers in extremely salty environments, such as Utah's Great Salt Lake and in coastal evaporation ponds where salt is concentrated for commercial purposes (see Figure 25.17) and can reach an osmolarity of 2,500 mosm/L. No animal could survive with internal osmolarities that high; such a solute concentration would cause proteins to denature. *Artemia* are able to exploit these environments because of their ability to regulate hypotonically by actively transporting NaCl from their extracellular fluid out across their gill membranes to the environment. *Artemia* cannot survive in fresh water, but they can live in dilute seawater by reversing the direction of transport of NaCl across their gill membranes to maintain the osmolarity of their extracellular fluids above that of the environment, thus becoming hypertonic regulators.

Invertebrates that are osmoconformers still have to be ionic regulators with respect to certain ions. Although the concentrations of Na^+ and Cl^- in their extracellular fluids may be the same as in seawater, many other ions are regulated at different levels. For example, the concentration of K^+ is usually higher, and the concentrations of Mg^{2+} and of SO_4^{2+} are usually lower, in the extracellular fluids than in the seawater. Active transporters maintain the regulated ions in the extracellular fluid at optimal concentrations.

Vertebrates are osmoregulators and ionic regulators

All aquatic vertebrates, with two exceptions, regulate the osmolarity of their extracellular fluids at around 300 mosm/L. In doing so, they are selective in which ions they conserve and which ions they excrete; thus they are ionic regulators and osmoregulators. One exception is the hagfish, a primitive jawless fish and a very ancient vertebrate group (see Figure 32.11). Hagfishes are osmoconformers as well as ionic conformers for most of the ions found in seawater. The other exception is members of the class Chondrichthyes (cartilaginous fishes, the sharks and rays). The cartilaginous fishes retain in their extracellular fluid two organic solutes, urea and trimethylamine oxide (TMAO), that are products of protein metabolism. As a result, their extracellular fluid is slightly hypertonic to the seawater and they gain water by osmosis.

Terrestrial vertebrates obtain their salts mostly from food, and they regulate the ionic composition of their extracellular fluids by conserving some ions and excreting others. For example, herbivores have to conserve Na^+ because most plants have low concentrations of Na^+. In contrast, birds that feed on marine animals must excrete the excess sodium they ingest with their food. Such birds, which include penguins and gulls, excrete excess salt through nasal salt glands. These glands use secondary active transport of Cl^- ions (with Na^+ and some water following passively) into a series of canals and ducts to produce a concentrated solution of NaCl that empties into the nasal cavity (**Figure 51.2**). These birds can be seen frequently sneezing or

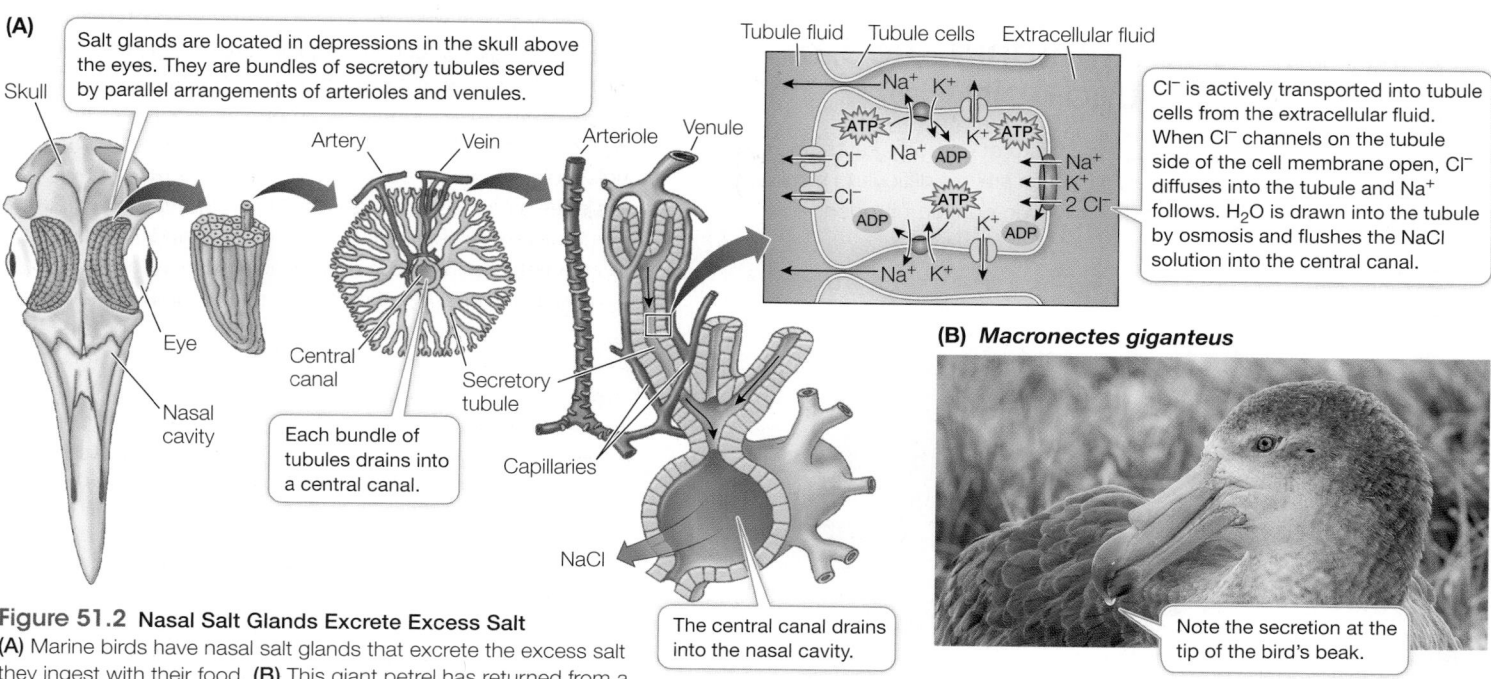

(A)

Salt glands are located in depressions in the skull above the eyes. They are bundles of secretory tubules served by parallel arrangements of arterioles and venules.

Skull

Artery   Vein   Arteriole   Venule

Eye

Central canal

Secretory tubule

Each bundle of tubules drains into a central canal.

Nasal cavity

Capillaries

NaCl

Tubule fluid   Tubule cells   Extracellular fluid

Na^+ K^+
ATP
Cl^- Na^+ ADP K^+ ATP
Cl^- ATP Na^+ K^+ 2 Cl^-
ADP K^+ ADP
Na^+ K^+

Cl^- is actively transported into tubule cells from the extracellular fluid. When Cl^- channels on the tubule side of the cell membrane open, Cl^- diffuses into the tubule and Na^+ follows. H_2O is drawn into the tubule by osmosis and flushes the NaCl solution into the central canal.

(B) *Macronectes giganteus*

Note the secretion at the tip of the bird's beak.

The central canal drains into the nasal cavity.

Figure 51.2 Nasal Salt Glands Excrete Excess Salt
(A) Marine birds have nasal salt glands that excrete the excess salt they ingest with their food. **(B)** This giant petrel has returned from a feeding trip at sea and is secreting salt through its nasal salt gland.

shaking their heads to get rid of the salty droplets excreted from their nasal salt glands.

51.1 recap

Excretory systems control water and salt balance and the excretion of nitrogenous waste products through three mechanisms: filtration of body fluids to form urine, active secretion of substances into the urine, and active reabsorption of substances from the urine.

learning outcomes

You should be able to:

- Describe the roles of filtration, secretion, and reabsorption in the excretory systems of animals with closed circulatory systems.
- Given data about the extracellular composition and environment of a particular species, determine its osmoregulatory characteristics.

1. What is the significance of a closed circulatory system for osmoregulation in a freshwater vertebrate?
2. Why is it important for an animal like *Artemia* that lives in the intertidal zone to be able to osmoregulate, and how does *Artemia* do that at the extremes of salinities in its environment?
3. How does water get into the nasal salt gland ducts of marine birds?

In addition to maintaining salt and water balance, animals must eliminate the waste products of metabolism from their extracellular fluids. The major problem is nitrogen. When nitrogen-containing molecules are broken down by metabolism, the end product can be toxic.

key concept 51.2 Animals Excrete Nitrogen as Ammonia, Urea, or Uric Acid

The end products of the metabolism of carbohydrates and fats are water and carbon dioxide, which are not difficult to eliminate. Proteins and nucleic acids, however, contain nitrogen, so their metabolism produces nitrogenous wastes in addition to water and carbon dioxide.

focus your learning

- Nitrogenous waste products have different advantages in different environments.
- Excreting uric acid conserves water.

Ammonia is toxic

The most common nitrogenous waste is **ammonia** (NH_3). Because it is highly toxic, ammonia is either excreted continuously to prevent its accumulation or is detoxified by conversion into **urea** or **uric acid** (**Figure 51.3**). Ammonia is highly soluble in water and diffuses rapidly, so its continuous excretion is relatively simple for many aquatic animals that continuously lose ammonia from their blood to the environment by diffusion across their gill membranes. Animals that excrete ammonia, such as aquatic invertebrates and bony fishes, are called **ammonotelic**.

Ammonia can also be lost by diffusion across the alveolar membranes of air-breathing animals, but with tidal ventilation, the buildup of ammonia in the extracellular fluids would be greater and therefore toxic. Therefore these animals detoxify ammonia by converting it into urea or uric acid.

Urea is highly soluble in water

Mammals, amphibians, cartilaginous fishes (sharks and rays), and hagfishes excrete urea as their principal nitrogenous waste product and are therefore called **ureotelic**. Urea is quite soluble in water, but its excretion can result in a large loss of water that many animals can ill afford. Mammals have somewhat solved this problem through the evolution of excretory systems that can produce urine with a high concentration of urea, thereby decreasing the water loss associated with excretion of urea. Also, recall from Key Concept 51.1 that sharks and rays solve the water loss problem of living in sea water by retaining high concentrations of urea and TMAO in their extracellular fluid, so that it is hyperosmotic to seawater.

Uric acid is not very soluble in water

A single molecule of uric acid eliminates twice the amount of nitrogen as does a molecule of urea (see Figure 51.3), but uric acid is not very soluble in water. Therefore it tends to precipitate out of solution and form a colloidal suspension in the urine and is excreted as a semisolid (for example, the whitish material in bird droppings). Because of the precipitation

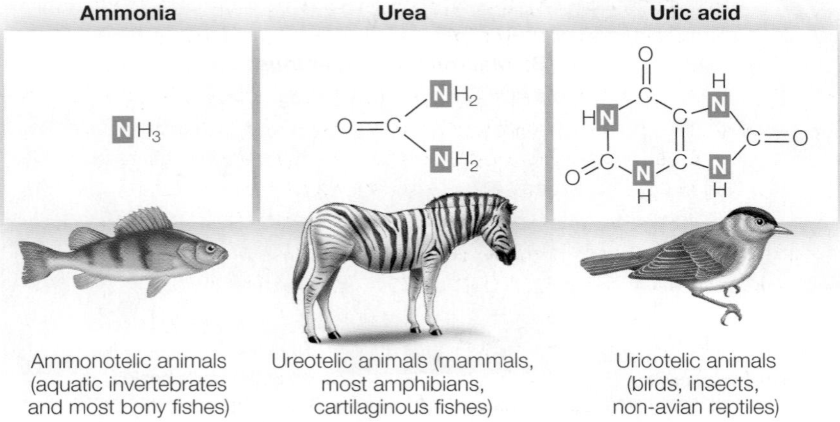

Figure 51.3 Nitrogenous Waste Products of Metabolism The metabolism of proteins and nucleic acids produces nitrogenous wastes. Many aquatic animals, including most fishes, excrete nitrogenous wastes as ammonia, which is highly diffusible and soluble in an aqueous environment. Most terrestrial animals and some aquatic animals excrete either urea or uric acid. Urea is more soluble in water and is the major nitrogenous excretory product for mammals, amphibians, and some fishes. Uric acid is not very soluble in water and is the major nitrogenous excretory product for reptiles (including birds), insects, and some amphibians.

Ammonia — NH_3 — Ammonotelic animals (aquatic invertebrates and most bony fishes)

Urea — Ureotelic animals (mammals, most amphibians, cartilaginous fishes)

Uric acid — Uricotelic animals (birds, insects, non-avian reptiles)

of the uric acid out of solution in the excretory system, water can be reabsorbed from the urine and thereby reduce the water loss necessary to excrete the nitrogenious waste. Animals that excrete uric acid are **uricotelic**, and they include insects, reptiles (including birds), and some amphibians. Many uricotelic species live in very arid environments.

Most species produce more than one nitrogenous waste

Humans are ureotelic, but we also excrete uric acid. The uric acid in human urine comes largely from the metabolism of nucleic acids. If uric acid levels in the extracellular fluid rise too high, uric acid can precipitate in joints, causing the age-old malady gout. Because solubility goes down with temperature, uric acid usually precipitates first in the extremities, especially the big toe. Pain in the big toe is a telltale symptom of gout. The risk for developing gout goes up with excessive consumption of protein, beer, and spirits (less so with wine) and in general with obesity and insulin resistance. Therefore in the past it was called "the disease of rich men." Gout has been around for a very long time; the first documented case was in Egypt around 2500 B.C.E.

Humans can also excrete ammonia, which is an important mechanism for regulating the pH of the extracellular fluids. As you will see later in this chapter, excreted ammonia buffers the urine and enables the excretion of excess hydrogen ions.

Species that live in different habitats at different developmental stages may use more than one mechanism of nitrogen excretion. The tadpoles of frogs and toads, for example, excrete ammonia across their gill membranes, but adult frogs and toads generally excrete urea. Some adult amphibians that live in arid habitats excrete uric acid.

51.2 recap

Ammonia is a common metabolic waste product of nitrogen-containing molecules. Most aquatic animals excrete ammonia by diffusion into the water. Terrestrial animals and some aquatic animals detoxify ammonia by converting it to urea or uric acid.

learning outcomes

You should be able to:

- Compare the properties of ammonia, urea, and uric acid as nitrogenous waste products.
- Describe the relative advantages and disadvantages of excreting nitrogenous waste as urea or as uric acid.

1. When amino acids are broken down, a first step is removal of the amine group, forming ammonia. What happens to that ammonia in a fish and in a mammal? Explain why this happens.

2. Why do sharks and rays produce urea rather than uric acid to raise the osmotic concentration of their extracellular fluids?

3. For terrestrial animals, what is the advantage of excreting uric acid as the nitrogenous waste product?

Animals exhibit a variety of adaptations for dealing with the challenges of salt and water balance in different environments. All of these adaptations, however, are based on two basic mechanisms—namely, filtration and tubular processing of the filtrate to conserve some solutes and excrete others.

key concept 51.3 Invertebrate Excretory Systems Use Filtration, Secretion, and Reabsorption

Freshwater and terrestrial invertebrates have a wide variety of adaptations for maintaining salt and water balance and excreting nitrogen. Here we explore three examples of invertebrate excretory systems: protonephridia, metanephridia, and Malpighian tubules. Each of these systems produces an extract of interstitial fluid lacking large molecules. They then change the solute composition (ions and small molecules) of that fluid to form an excretory product.

focus your learning

- Protonephridia of flatworms process filtered extracellular fluid by secretion and reabsorption of solutes.
- Metanephridia of annelids filter blood from their closed circulatory systems and process the resulting filtrate by secretion and reabsorption in a system of tubules.
- Malpighian tubules use active transport of solutes, including uric acid, to pull extracellular fluid into the tubules osmotically; precipitation of the uric acid enables reabsorption of water.

Protonephridia of flatworms excrete water and conserve salts

Many free-living flatworms, such as *Planaria*, live in fresh water. These animals excrete water through an elaborate network of tubules running throughout their bodies. The tubules end in flame cells, so called because each cell has a tuft of cilia projecting into the tubule (**Figure 51.4**). The beating of the cilia gives the appearance of a flickering flame. A flame cell and a tubule together form a **protonephridium** (plural *protonephridia*; Greek *proto*, "before," + *nephros*, "kidney").

Extracellular fluid enters the tubules by filtration. The beating of the cilia causes a slight negative pressure in the tubule, and movements of the animal create positive pressure in the extracellular fluid. This pressure difference causes extracellular fluid to be filtered through tiny spaces between tubule cells. The filtrate flows toward the animal's excretory pore, and along the way the cells of the tubules modify the composition of the fluid by reabsorption and secretion of specific ions and molecules. Because more ions are reabsorbed than are secreted, the urine that leaves the flatworm's body is less concentrated than the extracellular fluid. Thus the protonephridium conserves ions and excretes water and wastes.

Metanephridia of annelids process coelomic fluid

Filtration of body fluids and modification of urine by tubules are highly developed processes in annelids such as the earthworm. Annelids

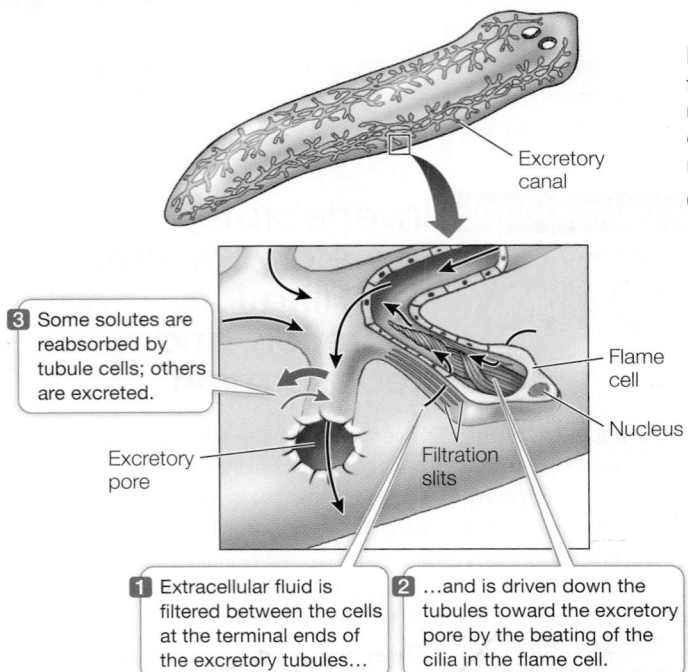

Figure 51.4 Protonephridia in Flatworms   The protonephridia of the freshwater flatworm *Planaria* consist of tubules ending in flame cells. In the region of the flame cells, body fluid is filtered between the tubule cells. The composition of the filtrate is modified as it flows down the tubule. Black arrows indicate direction of flow of tubular fluid.

Q: Do you expect the fluid coming out of the excretory pore to be hypertonic or hypotonic to the flatworm's interstitial fluid?

3 Some solutes are reabsorbed by tubule cells; others are excreted.

Excretory canal

Flame cell

Nucleus

Excretory pore

Filtration slits

1 Extracellular fluid is filtered between the cells at the terminal ends of the excretory tubules...

2 ...and is driven down the tubules toward the excretory pore by the beating of the cilia in the flame cell.

are segmented, and in each segment they have a fluid-filled body cavity called a coelom (see Figure 31.11). Annelids have a closed circulatory system through which blood is pumped under pressure. The pressure causes the blood to be filtered across the thin, permeable capillary walls into the coelom. Some waste products, such as ammonia, diffuse directly from the tissues into the coelom. Where does this coelomic fluid go?

Each segment of the earthworm contains a pair of **metanephridia** (singular *metanephridium*; Greek *meta*, "akin to"). Each metanephridium begins as a ciliated, funnel-like opening called a nephrostome. The nephrostome resides in one segment and continues as a tubule in the next segment. The tubule ends in a pore, called a nephridiopore, that opens to the outside of the animal (**Figure 51.5**). Coelomic fluid is swept into the metanephridia through the ciliated nephrostomes. As the fluid passes through the tubules, their cells actively reabsorb certain molecules from it and actively secrete other molecules into it. What leaves the animal through the nephridiopores is a dilute urine containing nitrogenous wastes and other solutes.

Malpighian tubules of insects use active transport to excrete wastes

Insects can excrete nitrogenous wastes with very little loss of water and can therefore live in the driest habitats on Earth. The insect excretory system consists of **Malpighian tubules**. An individual insect has from 2 to more than 100 of these blind-ended tubules that open into the gut between the midgut and hindgut (**Figure 51.6**).

Insects have an open circulatory system and therefore cannot use a pressure difference to filter extracellular fluids into the Malpighian tubules. Instead, the cells of the tubules actively transport uric acid, potassium ions, and sodium ions from the extracellular fluid into the tubules. The high concentration of solutes in the tubules causes water to follow osmotically, which flushes the tubule contents toward the gut.

The epithelial cells of the hindgut and rectum actively transport sodium and potassium ions from the gut contents back into the extracellular fluid. This local transport of salts creates an osmotic gradient that pulls water out of the rectal contents. As its concentration increases, the uric acid forms a colloidal suspension, freeing even more water to be reabsorbed. Remaining in the rectum is the uric acid mixed with other wastes; this semisolid matter is what the insect excretes. If you ever park your car under a tree being visited by bees, you will find

Capillaries   Bladder   Coelomic cavity

Metanephridium

Collecting tubules

Nephridiopore

Urine

Nephrostome

3 ...producing a dilute urine that is excreted through the nephridiopore.

2 The tubule cells of the metanephridium alter the composition of the fluid as it flows through the tubule...

1 Coelomic fluid is swept into the **metanephridium** by cilia surrounding the nephrostome.

Figure 51.5 Metanephridia in Earthworms   The metanephridia of annelids are arranged segmentally. The cross section at the left end shows a pair of metanephridia. Three longitudinal sections (right) show only one metanephridium of the two in each segment. Coelomic fluid enters the nephrostome and flows through tubules leading to the nephridiopore. A close association of the tubules and blood capillaries facilitates the active exchange of substances between the blood and the tubular fluid.

 Activity 51.1 Annelid Metanephridia
www.Life11e.com/ac51.1

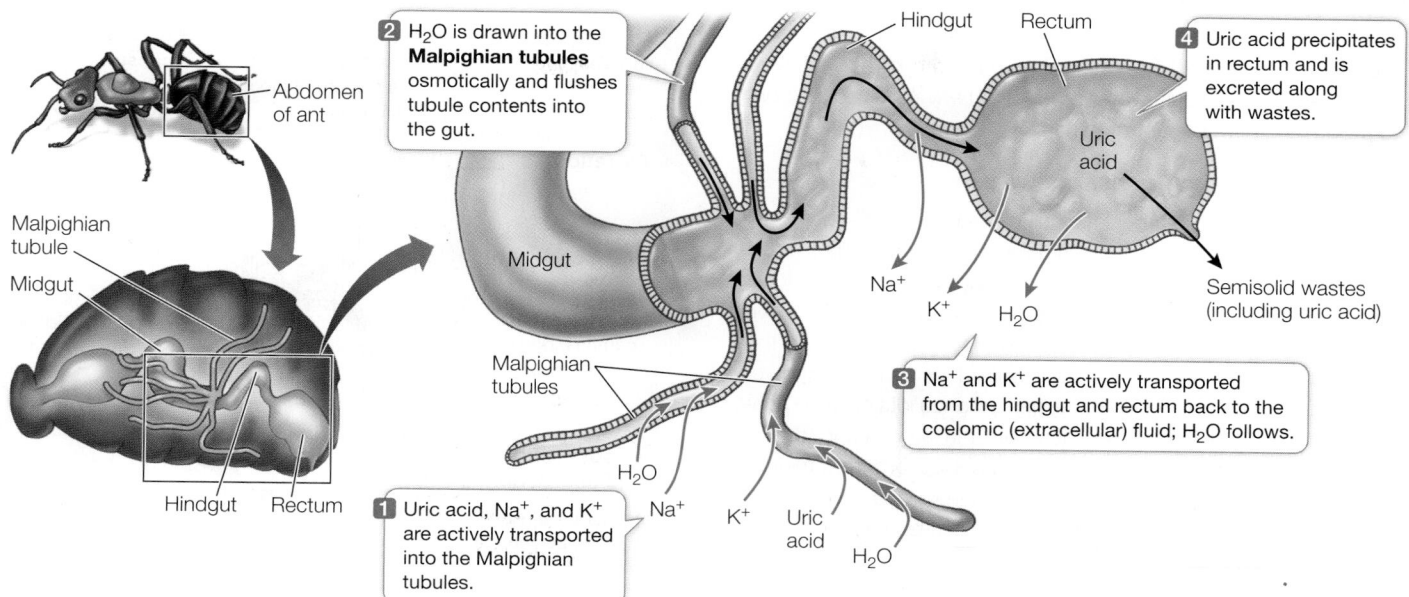

Figure 51.6 Malpighian Tubules in Insects   The blind, thin-walled Malpighian tubules are attached to the junction of the insect's midgut and hindgut and project into the spaces containing extracellular fluid. This system makes it possible to excrete wastes with very little loss of water.

Q: Why are Na⁺ and K⁺ ions transported into the Malpighian tubule but then transported out of the hindgut and rectum?

the little excretory droplets from the bees all over the car. The Malpighian tubule system is a highly effective mechanism for excreting nitrogenous wastes and some salts without giving up much water.

51.3 recap

Protonephridia and metanephridia work by creating a filtrate of the body fluids that is modified by the secretion and reabsorption of specific substances before being excreted. Insect Malpighian tubules actively secrete uric acid and other solutes into closed tubules.

learning outcomes

You should be able to:

- Describe the structure and function of a protonephridium.
- Diagram the process of urine formation by a metanephridium.
- Explain how Malpighian tubules enable insects to live in very arid habitats.

1. How do protonephridia enable flatworms to live in fresh water?
2. What are the relationships among blood plasma, coelomic fluid, and urine in an annelid worm?
3. How does the insect excretory system conserve water?

Having described how several invertebrate groups handle nitrogen excretion, we will next consider the nephron—the basic unit of the vertebrate excretory system—and how it evolved to be able to respond to a variety of salt- and water-balance challenges and maintain a relatively constant internal environment.

> **key concept**

51.4 The Nephron Is the Basic Functional Unit of Vertebrate Excretory Systems

The main excretory organ of vertebrates is the **kidney**, and the functional unit of the kidney is the **nephron**, which has a blood vessel component and a tubule component. The vascular component begins with a knot of capillaries that are highly permeable and filter the blood into the tubule component. Processes of secretion and reabsorption in the tubules produce urine, which differs in composition from the extracellular fluids of the body.

focus your learning

- Reptiles can be more independent of water sources than amphibians can.
- The nephron produces urine through filtration of blood plasma in the glomerulus and tubular processing of the filtrate by reabsorption and secretion of solutes.
- Glomerular filtration involves anatomical features of glomerular capillaries and cells of Bowman's capsule.
- The liquid that enters Bowman's capsule is called the renal filtrate and is further processed into urine as it moves through the renal tubules of the nephron.

The evolution of vertebrates is thought to have begun with a marine ancestor that moved into a freshwater habitat. The excretory system of this vertebrate ancestor would have evolved to excrete large quantities of water while conserving salts. But if the early vertebrate excretory system evolved in response to the need to excrete water, how did subsequent vertebrate lineages adapt to environments where water must be conserved and salts excreted? The answer to this question differs among vertebrate groups. Even among marine fishes, the excretory adaptations of the bony fishes differ from those of the cartilaginous fishes. Reptiles, including birds, and mammals have excretory systems that conserve water. The reptiles achieve this mainly by being uricotelic and producing a semisolid excretory product that contains little water. Mammals, in contrast, are ureotelic; they excrete a liquid waste product but have evolved the ability to produce highly concentrated urine.

Marine fishes must conserve water

Marine bony fishes osmoregulate their extracellular fluids to maintain them at one-third to one-half the osmolarity of seawater. Thus marine bony fishes must conserve water, which they are constantly losing through osmosis, and must excrete excess solutes. Marine bony fishes cannot produce urine that is more concentrated than their extracellular fluids, so they minimize water loss by producing very little urine. In contrast, freshwater fishes produce lots of dilute urine.

How do marine bony fishes deal with the large salt loads they ingest with food? Marine bony fishes do not absorb from their gut some of the ions they take in, especially divalent ions such as Mg^{2+} and SO_4^{2-}. NaCl, the major salt ingested, is actively excreted across the gill membranes. As mentioned earlier, bony fishes can lose their nitrogenous waste, ammonia, by diffusion across their gill membranes.

Sharks and rays are osmoconformers but not ionic conformers. As mentioned earlier, they raise the osmolarity of their body fluids by retaining urea and TMAO making their extracellular fluids hyperosmotic to seawater. These species have adapted to a concentration of urea in the body fluids that would be toxic to other vertebrates. Sharks and rays still have the problem of excreting the large amount of salts they take in with their food. They solve this problem by having a gland in the rectum that actively secretes NaCl by a mechanism similar to that of the nasal salt glands of seabirds.

Terrestrial amphibians and reptiles must avoid desiccation

Most amphibians live in or near fresh water, and they usually stay in humid habitats when they do venture from the water. Like freshwater fishes, most amphibians produce large amounts of dilute urine and conserve salts. Some amphibians, however, have adapted to habitats that require water conservation.

Amphibians living in dry terrestrial environments have skin with a reduced permeability to water. Some secrete a waxy substance over the skin for waterproofing. Several species of frogs that live in arid regions of Australia burrow into the ground and remain there during long dry periods. They enter **estivation**, a state of very low metabolic activity and therefore low water turnover. When it rains, the frogs come out of estivation, feed, and reproduce. Their most interesting adaptation is an enormous urinary bladder. Before entering estivation, they fill the bladder with dilute urine, which can amount to one-third of their body weight. This dilute urine serves as a water reservoir that is gradually reabsorbed into the blood during the long period of estivation. Aboriginal peoples have learned to locate these buried frogs and use them as an emergency source of water.

Reptiles occupy habitats ranging from aquatic to extremely hot and dry. In fact, snakes, lizards, and birds are among the most prominent members of many desert faunas. Three major adaptations have freed reptiles from the close association with water that is necessary for most amphibians (see Key Concept 33.4):

1. Reptiles are amniotes that do not need fresh water to reproduce because they employ internal fertilization and lay eggs with shells that retard evaporative water loss.

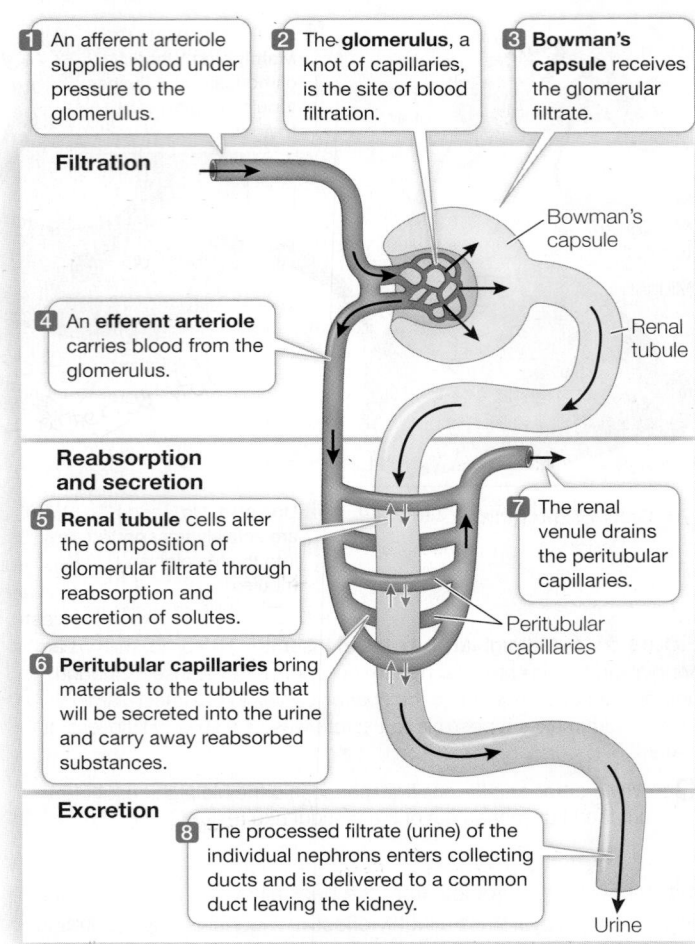

1 An afferent arteriole supplies blood under pressure to the glomerulus.

2 The **glomerulus**, a knot of capillaries, is the site of blood filtration.

3 **Bowman's capsule** receives the glomerular filtrate.

Filtration

Bowman's capsule

Renal tubule

4 An **efferent arteriole** carries blood from the glomerulus.

Reabsorption and secretion

5 **Renal tubule** cells alter the composition of glomerular filtrate through reabsorption and secretion of solutes.

7 The renal venule drains the peritubular capillaries.

Peritubular capillaries

6 **Peritubular capillaries** bring materials to the tubules that will be secreted into the urine and carry away reabsorbed substances.

Excretion

8 The processed filtrate (urine) of the individual nephrons enters collecting ducts and is delivered to a common duct leaving the kidney.

Urine

Figure 51.7 The Vertebrate Nephron   The vertebrate nephron consists of a renal tubule closely associated with two capillary beds, the glomerulus and the peritubular capillaries.

 **Activity 51.2 The Vertebrate Nephron**
www.Life11e.com/ac51.2

2. Reptiles have a dry epidermis (skin) that retards evaporative water loss.

3. Reptiles excrete nitrogenous wastes as uric acid semisolids, losing little water in the process.

Mammals can produce highly concentrated urine

Mammals occupy diverse habitats, many of which present special excretory system challenges. The most challenging environments are those in which water is severely limited. Mammals have a variety of adaptations to conserve water, but chief among them is the ability to produce urine that is more concentrated than their extracellular fluids. They are able to concentrate their urine because of adaptations of their kidneys that we will explore in detail in Key Concept 51.5. To understand how these adaptations work, you must first know the structure and function of the vertebrate nephron.

The nephron is the functional unit of the vertebrate kidney

Urine formation in vertebrate nephrons involves three main processes (**Figure 51.7**):

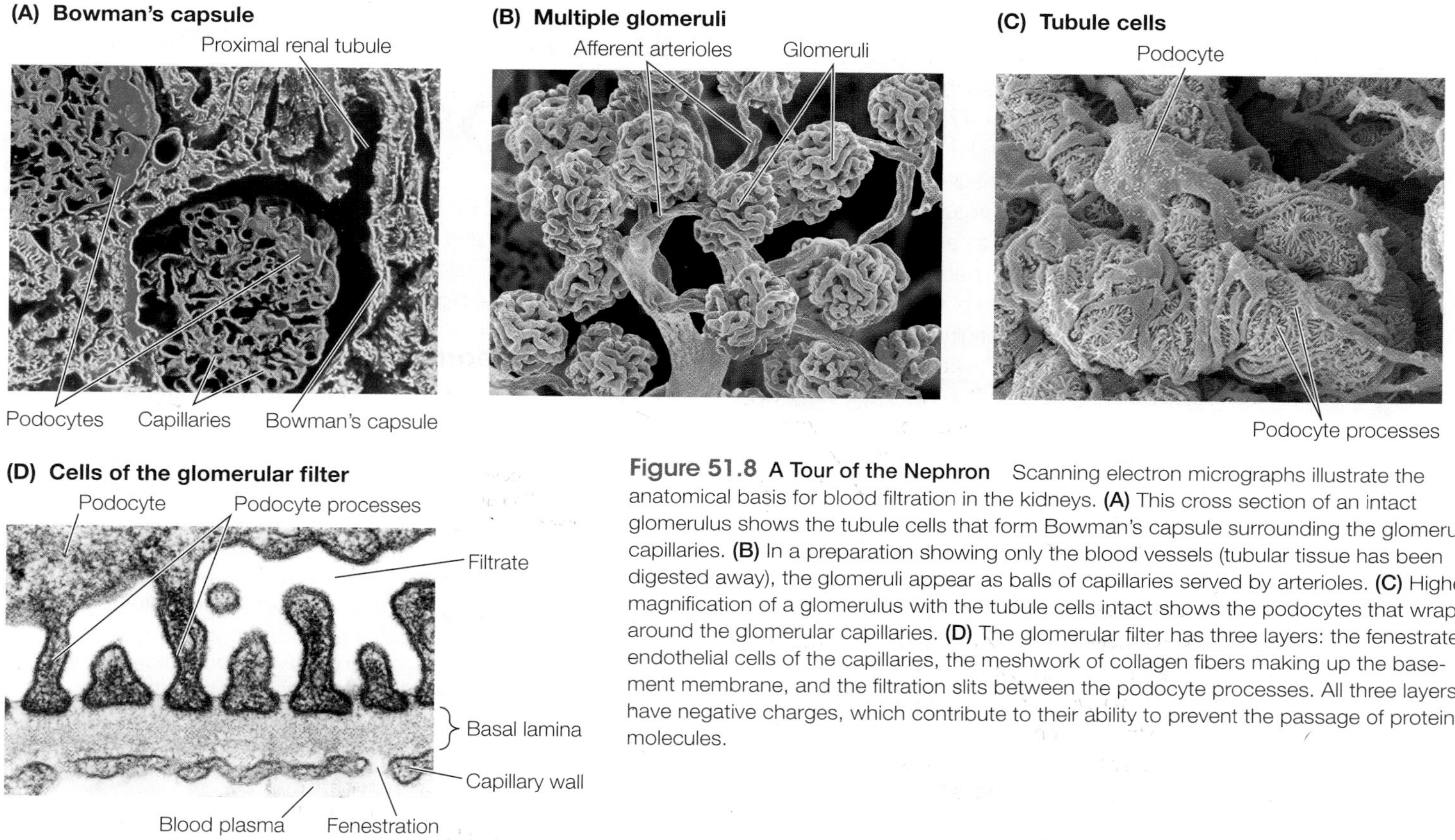

(A) Bowman's capsule

Proximal renal tubule

Podocytes   Capillaries   Bowman's capsule

(B) Multiple glomeruli

Afferent arterioles   Glomeruli

(C) Tubule cells

Podocyte

Podocyte processes

(D) Cells of the glomerular filter

Podocyte   Podocyte processes

Filtrate

Basal lamina

Capillary wall

Blood plasma   Fenestration

Figure 51.8 A Tour of the Nephron   Scanning electron micrographs illustrate the anatomical basis for blood filtration in the kidneys. **(A)** This cross section of an intact glomerulus shows the tubule cells that form Bowman's capsule surrounding the glomerular capillaries. **(B)** In a preparation showing only the blood vessels (tubular tissue has been digested away), the glomeruli appear as balls of capillaries served by arterioles. **(C)** Higher magnification of a glomerulus with the tubule cells intact shows the podocytes that wrap around the glomerular capillaries. **(D)** The glomerular filter has three layers: the fenestrated endothelial cells of the capillaries, the meshwork of collagen fibers making up the basement membrane, and the filtration slits between the podocyte processes. All three layers have negative charges, which contribute to their ability to prevent the passage of protein molecules.

1. *Filtration*. Each nephron has a dense bed of capillaries called a **glomerulus** (plural *glomeruli*). The glomerulus is highly permeable to water, ions, and small molecules but impermeable to large molecules. Blood pressure drives the movement of water and small-molecular-weight solutes out of the glomerular capillaries.

2. *Tubular reabsorption*. The filtrate from the glomerulus flows into the **renal tubule**. Cells in the renal tubule modify the filtrate by reabsorbing specific ions, nutrients, and water, returning these to the blood, and leaving behind and concentrating excess ions and waste products such as urea.

3. *Tubular secretion*. The filtrate in the renal tubule is further modified by tubule cells transporting substances into the tubule. These are substances that the body needs to excrete.

Blood enters the glomerular capillaries via an **afferent arteriole** and leaves the glomerulus in an **efferent arteriole**. This short vessel is called an arteriole because it feeds another capillary bed, called the **peritubular capillaries**. These capillaries intimately surround the renal tubules and deliver substances to the renal tubules that the cells of the tubules secrete into the urine. The peritubular capillaries also carry away substances that the tubule cells reabsorb from the urine.

Blood is filtered into Bowman's capsule

The renal tubule begins with **Bowman's capsule** (see Figure 51.7), which encloses the glomerulus (**Figure 51.8A and B**). The glomerulus appears to be pushed into Bowman's capsule much like a fist

pushed into an inflated balloon. The cells of the capsule that are in direct contact with the glomerular capillaries are called **podocytes** (**Figure 51.8C**). These highly specialized cells have numerous extensions, each with hundreds of fine processes. The podocytes wrap around the capillaries so that their processes interdigitate and intimately cover the capillaries.

The glomerulus filters the blood, producing a fluid (the renal filtrate) that lacks cells and large molecules. The walls of the capillaries, the basal lamina of the capillary endothelium, and the podocytes of Bowman's capsule all participate in filtration. Fenestrations in the walls of the capillaries (see Key Concept 49.4) allow water and many solute molecules, but not red blood cells, to pass through. The meshwork of the basal lamina and the spaces between the processes of the podocytes are even finer and prevent large molecules from leaving the capillaries (**Figure 51.8D**). The arterial pressure of the blood entering the permeable capillaries causes the filtration of water and small molecules out of the glomerular capillaries and into the Bowman's capsule. The glomerular filtration rate is high because blood pressure in the glomerular capillaries is high, and because the capillaries of the glomerulus, along with their covering of podocytes, are more permeable to water than are other capillary beds in the body.

You will recall that marine bony fishes produce very little urine. The explanation for this is that their kidneys have evolved to have few if any glomeruli, whereas the kidneys of freshwater fishes have many glomeruli.

The renal tubules convert glomerular filtrate to urine

The composition of the filtrate that enters the renal tubule is similar to that of the blood plasma, except the filtrate lacks high-molecular-weight solutes such as proteins. Reabsorption and secretion cause the composition of this fluid to change as it passes down the renal tubule. Cells of the tubule actively reabsorb certain molecules from the tubule fluid (which are returned to the blood flowing through the peritubular capillaries). For example, glucose and amino acids are reabsorbed. Most NaCl is reabsorbed. Other substances in the blood of the peritubular capillaries are actively secreted into the tubule fluid. An example is para-aminohippuric acid (PAH), which is produced in the liver from benzoic acid, a common food preservative. Because of the actions of the renal tubules, the excreted urine is very different from the original glomerular filtrate.

51.4 recap

The kidney is the major excretory organ of vertebrates. Its functional unit is the nephron, which includes a glomerulus that filters blood and a renal tubule that secretes and reabsorbs solutes, modifying the filtrate to produce urine. As seen in freshwater fishes, the nephron evolved as a mechanism for excreting excess water while conserving valuable solutes. Vertebrates have various adaptations for conserving water. Marine bony fishes produce very little urine but lose nitrogenous waste as ammonia across their gills and also excrete salts across their gills. Marine cartilaginous fishes produce urea and retain it, so their extracellular fluids are hyperosmotic to the environment and therefore take up water by osmosis. Many terrestrial vertebrates conserve water by excreting nitrogenous waste as uric acid, by minimizing evaporative water loss, and by active secretion of salts. Mammals have the ability to conserve water by producing urine that is more concentrated than their extracellular fluid.

learning outcomes

You should be able to:

- Relate structural and behavioral features of amphibians and reptiles to their abilities to minimize water loss.
- Describe the organization of blood vessels in the nephron.
- Explain the mechanism of filtration of blood plasma from the glomerular capillaries into Bowman's capsule.
- Explain why the renal filtrate changes in composition as it passes through the nephron.

1. What is the connection between the need to conserve water and the evolution of internal fertilization?
2. What are the roles of the afferent and efferent arterioles in the nephron?
3. What are the anatomical components of the filtration mechanism in the glomerulus?
4. How does the composition of the urine become different from the composition of the blood?

The adaptations that enable the mammalian kidney to produce urine more concentrated than extracellular fluids were important steps in vertebrate evolution, and they were largely achieved through changes in the structure and regional functions of the renal tubules. These changes converted a kidney that evolved to excrete water into an organ that conserves water.

key concept 51.5 The Mammalian Kidney Can Produce Concentrated Urine

Mammals have high body temperatures and high metabolic rates, and therefore have the potential for a high rate of water loss and a high level of nitrogenous waste production. Having an excretory system that minimizes water loss made it possible for these highly active species to occupy arid habitats.

focus your learning

- Nephrons are organized in the kidney so that the glomeruli and major portions of the renal tubules (called convoluted tubules) are in the outer region, or cortex, and straight sections of tubules (called loops of Henle) and collecting ducts are in the inner region, or medulla.
- Most of the filtrate is reabsorbed isosmotically in the proximal convoluted tubule.
- Loops of Henle are countercurrent multipliers that create a concentration gradient in the extracellular fluid.
- As the tubular fluid passes down the collecting duct, water can be reabsorbed into the extracellular fluid to produce a more concentrated urine in the tubule.
- The mammalian kidney functions in conjunction with the lungs to regulate the pH of the extracellular fluid.

 Media Clip 51.1 **Inside the Bladder**
www.Life11e.com/mc51.1

Kidneys produce urine and the bladder stores it

Mammalian excretory systems are similar, so we will use that of humans as our example. Humans have two kidneys at the back of the upper region of the abdominal cavity (**Figure 51.9A**). Each kidney filters blood, processes the filtrate into urine, and releases that urine into a duct called the **ureter**. The ureter of each kidney leads to the **urinary bladder**, where the urine is stored until it is excreted through the **urethra**, a short tube that opens to the outside of the body.

Two sphincter muscles surrounding the base of the urethra control urination. One of these sphincters is a smooth muscle and is controlled by the autonomic nervous system. As the bladder fills, stretch receptors in the walls of the bladder trigger a spinal reflex that relaxes this sphincter. This reflex is the only control of urination in infants, hence their frequent "accidents." The other sphincter is a skeletal muscle and is controlled by the voluntary nervous system. When the bladder is *very* full, only deliberate conscious effort prevents urination. Toilet training of children teaches them to control this sphincter.

Nephrons have a regular arrangement in the kidney

The kidney is shaped like a kidney bean. When sliced along its long axis on the midline, its key anatomical features are revealed (**Figure 51.9B**). The ureter and the **renal artery** and **renal vein** enter the kidney on its concave (punched-in) side. Inside the kidney, the ureter is continuous with structures that envelop kidney tissues called **renal pyramids**. The renal pyramids make up the internal core, or

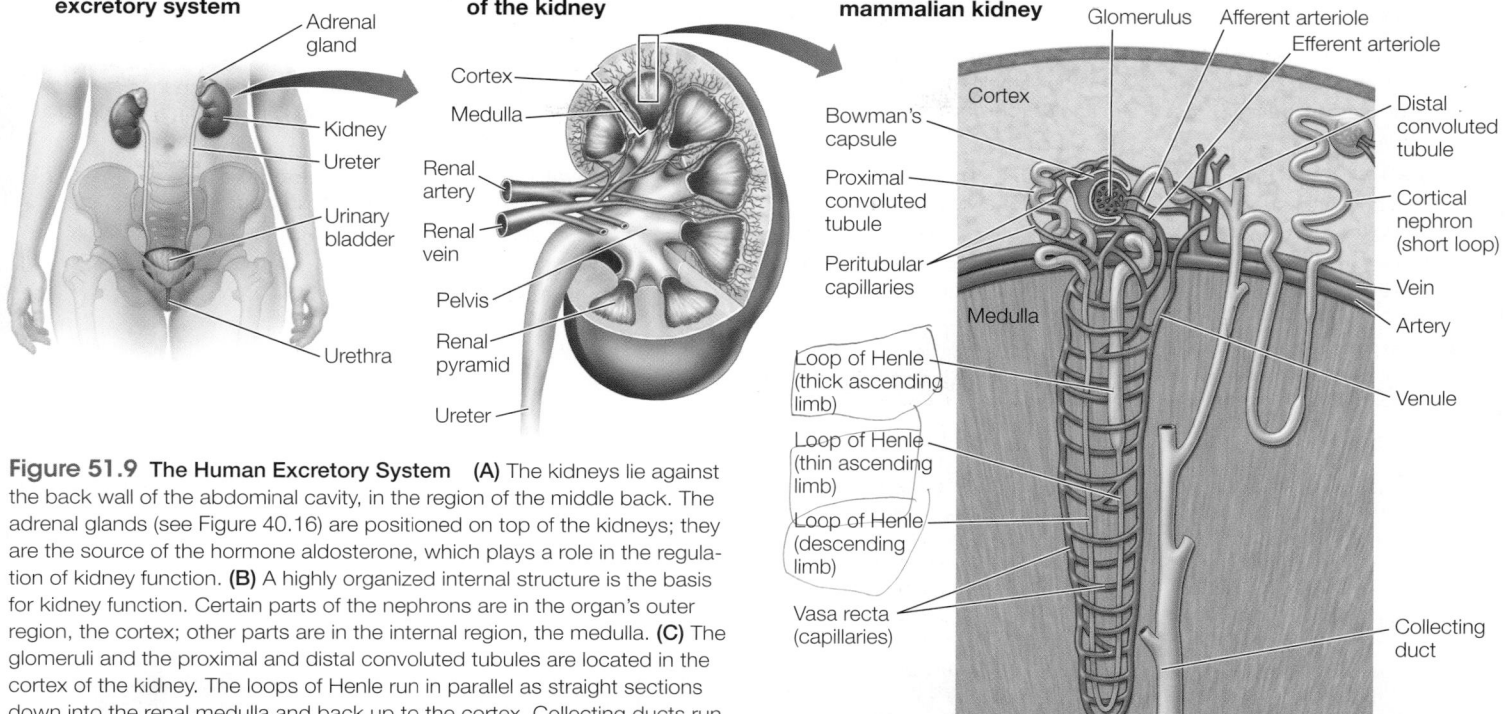

(A) Organs of the excretory system
- Adrenal gland
- Kidney
- Ureter
- Urinary bladder
- Urethra

(B) Cross section of the kidney
- Cortex
- Medulla
- Renal artery
- Renal vein
- Pelvis
- Renal pyramid
- Ureter

(C) Nephrons in a mammalian kidney
- Glomerulus
- Afferent arteriole
- Efferent arteriole
- Cortex
- Bowman's capsule
- Proximal convoluted tubule
- Peritubular capillaries
- Medulla
- Distal convoluted tubule
- Cortical nephron (short loop)
- Vein
- Artery
- Loop of Henle (thick ascending limb)
- Venule
- Loop of Henle (thin ascending limb)
- Loop of Henle (descending limb)
- Vasa recta (capillaries)
- Collecting duct

Figure 51.9 The Human Excretory System **(A)** The kidneys lie against the back wall of the abdominal cavity, in the region of the middle back. The adrenal glands (see Figure 40.16) are positioned on top of the kidneys; they are the source of the hormone aldosterone, which plays a role in the regulation of kidney function. **(B)** A highly organized internal structure is the basis for kidney function. Certain parts of the nephrons are in the organ's outer region, the cortex; other parts are in the internal region, the medulla. **(C)** The glomeruli and the proximal and distal convoluted tubules are located in the cortex of the kidney. The loops of Henle run in parallel as straight sections down into the renal medulla and back up to the cortex. Collecting ducts run from the cortex to the inner surface of the medulla, where they open into the ureter. The vasa recta are peritubular capillaries that parallel the loops of Henle and are permeable to salts and water.

Q: How do you think the osmolarity of the blood in the vasa recta compares with the osmolarity of the interstitial fluid at different levels of the medulla?

 Activity 51.3 The Human Excretory System
www.Life11e.com/ac51.3

medulla, of the kidney. The medulla is covered by an outer layer, or **cortex**, that has a granular appearance. Between the cortex and the medulla, the renal artery divides into the many arterioles that serve the nephrons. In this same region, the renal vein collects blood from the many venules that drain the peritubular capillaries.

The organization of nephrons within the kidney is very regular. All of the glomeruli with their Bowman's capsules are located in the cortex. The initial segments of the renal tubules that come from the Bowman's capsules are called the **proximal convoluted tubules**—"proximal" because they are closest to the glomerulus, and "convoluted" because they are twisted (**Figure 51.9C**). All of the proximal convoluted tubules are located in the cortex.

At the point at which the proximal convoluted tubule enters the medulla, it becomes thin, straight, and descends directly down toward the tip of a renal pyramid. In the medulla the tubule makes a hairpin turn and ascends back to the cortex, forming what is called a **loop of Henle**. Some nephrons have longer loops of Henle than others. Some 20–30 percent of human nephrons that have glomeruli deep in the cortex (i.e., near the border with the medulla) have long loops of Henle that go deep into the medulla. Nephrons that have glomeruli farther up in the cortex generally have short loops of Henle that descend only a short distance into the medulla. As you will see, the long loops are the critical

adaptation of the mammalian nephron that enables the kidney to concentrate the urine.

The ascending limb of the loop of Henle becomes the **distal convoluted tubule** when it reaches the cortex—"distal" because it is farther from the glomerulus. The distal convoluted tubules of many nephrons join a common **collecting duct** in the cortex. The collecting ducts descend back down through the renal pyramid, parallel to and past the tips of the loops of Henle, and empty into a funnel-shaped structure called the pelvis. Divisions of the pelvis that surround each renal pyramid join together to leave the kidney as the ureter (see Figure 51.9B).

The organization of the blood vessels of the kidney closely parallels the organization of the nephrons (see Figure 51.9C). Smaller arteries branch from the renal artery and radiate into the cortex, forming the afferent arterioles that carry blood to each glomerulus. Each glomerulus is drained by an efferent arteriole that gives rise to the peritubular capillaries, most of which surround the proximal and distal convoluted tubules. The intimate associations of the glomerular and peritubular capillaries with the renal tubules enable exchanges between the blood and the specialized regions of the tubules.

Some of the peritubular capillaries run into the medulla in parallel with the loops of Henle and the collecting ducts, forming a vascular network called the **vasa recta**. All of the peritubular capillaries from a nephron join back together into a venule that joins with venules from other nephrons and eventually leads to the renal vein. As you will see, the concentrating ability of the mammalian kidney depends on water reabsorption in the renal medulla, and the vasa recta are the avenue by which that water gets out of the renal medulla and back into the circulation.

experiment

Original Paper: Busch, C. 1988. Consumption of blood, renal function and utilization of free water by the vampire bat, *Desmodus rotundus*. *Comparative Biochemistry and Physiology Part A: Physiology* 90: 141–146.

You learned at the beginning of the chapter that vampire bats feed once a day. After sunset they fly out, search for a victim, and then return to their roost. This nightly excursion generally takes less than 2 hours. Therefore in a very short time these small animals must take in a large enough volume of their liquid blood diet to supply their energy and water needs for a 24-hour period. How do they do it?

HYPOTHESIS▶   Vampire bats maximize their food intake by rapidly excreting water, but they then conserve water by producing a highly concentrated urine while digesting their protein meal.

METHOD

1. Offer a colony of captive vampire bats a beef blood meal once a day for 2 hours, to mimic the natural once-a-day feeding behavior of the bats.
2. Do not allow bats access to free water.
3. Measure the daily amount of blood consumed and the time course of subsequent urine volume and concentration.

RESULTS

The table shows the average body mass and average amount of blood consumed by nine vampire bats over a period of 5–8 days.

Average body mass (g)	Average amount of blood consumed (g)
25.0	19.7
26.3	11.3
26.4	13.6
28.6	16.8
29.6	15.7
30.3	17.8
32.4	24.7
36.0	24.4
36.2	18.6

CONCLUSION▶   In a short daily feeding period, vampire bats consume on average 60% of their body mass.

work with the data

To consume 60% of body mass and still be able to fly, the bats must rapidly excrete a large percentage of the water content of their food. The time course of urine flow and its concentration was measured during and after daily feeding for three vampire bats for 1 week and averaged. Create a double plot of the flow and concentration data as a function of time. (—, no data reported.)

QUESTIONS▶

1. If you averaged the urine flow rates over the first hour following the beginning of the meal, and you assumed that the total blood intake was 60% of body mass and body mass was 30 g, what percentage of the ingested volume would be excreted in that first hour?

2. Considering that the normal osmolarity of interstitial fluid in a mammal is 300 mosm, what is the maximum concentrating factor in these vampire bats?

3. Three data points are missing in the urine osmolarity curve. Would it be reasonable to plot these data by connecting the 90-min sample and the 330-min sample with a straight line? Why or why not?

A similar **work with the data** *exercise may be assigned in* **LaunchPad**.

Time after beginning to feed (min)	Average urine flow rate (g/g body mass/hr)	Average urine osmolarity (osm)
10	0.13	0.5
20	0.24	0.25
45	0.16	0.40
90	0.05	0.70
150	0.01	—
210	0.01	—
270	0.005	—
330	0.003	3.3
390	0.02	3.4

Most of the glomerular filtrate is reabsorbed by the proximal convoluted tubule

Most of the water and solutes filtered by the glomerulus are reabsorbed and do not appear in the urine. We can reach this conclusion by comparing the rate of filtration by the glomeruli with the rate of urine production. The kidneys receive about 1 liter of blood per minute, or about 1,500 liters of blood per day. How much of this huge volume is filtered out of the glomeruli? The answer is about 12 percent. This is still a large volume—180 liters per day! We normally urinate less than 2 liters per day, so about 99 percent of the fluid volume that is filtered out of the glomerulus is returned to the blood. Where and how is this enormous amount of fluid reabsorbed?

The proximal convoluted tubule (PCT) is responsible for most of the reabsorption of water and solutes from the glomerular filtrate. The cells of this section of the renal tubule have many microvilli that increase their apical (facing into the tubule) surface area for reabsorption, and they have many mitochondria—an indication that they are metabolically active. PCT cells actively transport Na^+ (with Cl^- following) and other solutes, such as glucose and amino acids, out of the tubule fluid.

Almost all glucose and amino acid molecules that are filtered from the blood are actively reabsorbed by PCT cells and transported into the extracellular fluid. The active transport of solutes from the proximal tubule into the interstitial fluid causes water to follow osmotically. The water and solutes moved into the interstitial fluid are taken up by the peritubular capillaries and returned to the venous blood. These processes accomplish the reabsorption of more than 75 percent of the fluid that initially enters the nephron.

Despite the bulk reabsorption of water and solutes by the PCT, the overall osmolarity of the fluid flowing through the PCT does not change. Thus the process that is occurring in the PCT is called isosmotic reabsorption. The fluid that enters the loop of Henle has the same osmolarity as the blood plasma, although its composition is different. How then does the kidney produce urine that is more concentrated than the blood plasma?

The loop of Henle creates a concentration gradient in the renal medulla

Humans can produce urine that is four times more concentrated than their blood plasma. The vampire bat you encountered at the beginning of this chapter can produce urine that is 15 times more concentrated than its blood plasma (**Investigating Life: How Can Vampire Bats Use Blood as Fast Food?**). The concentrating ability of the mammalian kidney arises from a **countercurrent multiplier** mechanism made possible by the anatomical arrangement of the loops of Henle. The term "countercurrent" refers to the opposing directions of fluid flow in the descending and ascending limbs. The term "multiplier" refers to the ability of this system to create a solute concentration gradient in the renal medulla.

The loops of Henle do not themselves produce concentrated urine; rather, they increase the osmolarity of the extracellular fluid in the medulla in a graduated way. In humans, for example, the extracellular fluid at the top of the medulla bordering the cortex will be about 300 mosm/L (the concentration of blood plasma). But at the bottom of the medulla, where the loops of Henle make their hairpin turns, the extracellular fluid can be 1,200 mosm/L (**Focus: Key Figure 51.10**). How do the loops produce this effect?

The cells that make up the different segments of the loop of Henle differ anatomically and functionally. Cells of the descending limb and the initial cells of the ascending limb are thin, with no microvilli and few mitochondria. They are not specialized for transport. Partway up the ascending limb, the cells become specialized for active transport. These cells are thick and have many mitochondria. Accordingly, the segments of the loop of Henle are named the thin descending limb, the thin ascending limb, and the thick ascending limb (see Figure 51.9C).

The countercurrent multiplier mechanism is best understood by first considering events occurring in the thick ascending limb (see Figure 51.10, note 1). The cells of the thick ascending limb reabsorb Na^+ and Cl^- from the tubule fluid and move it into the interstitial fluid. (In the following discussion, we will distinguish between the two components of extracellular fluid—the blood plasma and the interstitial fluid.) The thick ascending limb is not permeable to water, so the reabsorption of Na^+ and Cl^- from the tubular fluid raises the concentration of those solutes in the surrounding interstitial fluid and decreases the concentration of the tubular fluid entering the distal convoluted tubule.

The thin descending limb, in contrast, is highly permeable to water but not very permeable to Na^+ and Cl^-. Since the local interstitial fluid has been made more concentrated by the Na^+ and Cl^- reabsorbed from the neighboring thick ascending limb, water is withdrawn osmotically from the fluid in the descending limb. Therefore the fluid in the descending limb becomes more concentrated as it flows toward the hairpin turn at the bottom of the renal medulla (see Figure 51.10, note 2).

The thin ascending limb, like the thick ascending limb, is not permeable to water. It is, however, permeable to Na^+ and Cl^-. As the concentrated tubule fluid flows up the thin ascending limb, it is more concentrated than the surrounding interstitial fluid, so Na^+ and Cl^- diffuse out. When the tubule fluid reaches the thick ascending limb, active transport continues to move Na^+ and Cl^- from the tubule fluid to the interstitial fluid.

Because of this countercurrent multiplier mechanism, the tubule fluid reaching the distal convoluted tubule is less concentrated than the blood plasma (see Figure 51.10, note 3), and the solutes that have been left behind in the renal medulla have created a concentration gradient in the interstitial fluid of the medulla (indicated by the background color gradient in Figure 51.10).

You may wonder why the blood flow through the medulla does not wash out the concentration gradient established by the loops of Henle. The parallel arrangement of the descending and ascending peritubular capillaries in the medulla —the vasa recta—helps preserve the concentration gradient in the medulla. These capillaries are permeable to both salts and water. Therefore as blood flows down the descending limb of the vasa recta into the increasingly concentrated interstitial fluid of the medulla, it loses water and gains solutes. As blood flows up from the bottom of the medulla in the ascending limb of the vasa recta, the opposite happens (water is gained and solutes are lost) because now the blood is more concentrated than the surrounding interstitial fluid (see Figure 51.10, notes 4–6). The dynamics of this countercurrent exchange of salts and water between the blood in the vasa recta and the interstitial fluids result in little net change in the composition of the interstitial fluid in the medulla.

Water permeability of kidney tubules depends on water channels

We have noted that some tubule regions, such as the PCT, are highly permeable to water whereas others, such as the thick ascending limb of the loop of Henle, are impermeable to water. What causes these differences in water permeability in different regions of the

focus: key figure

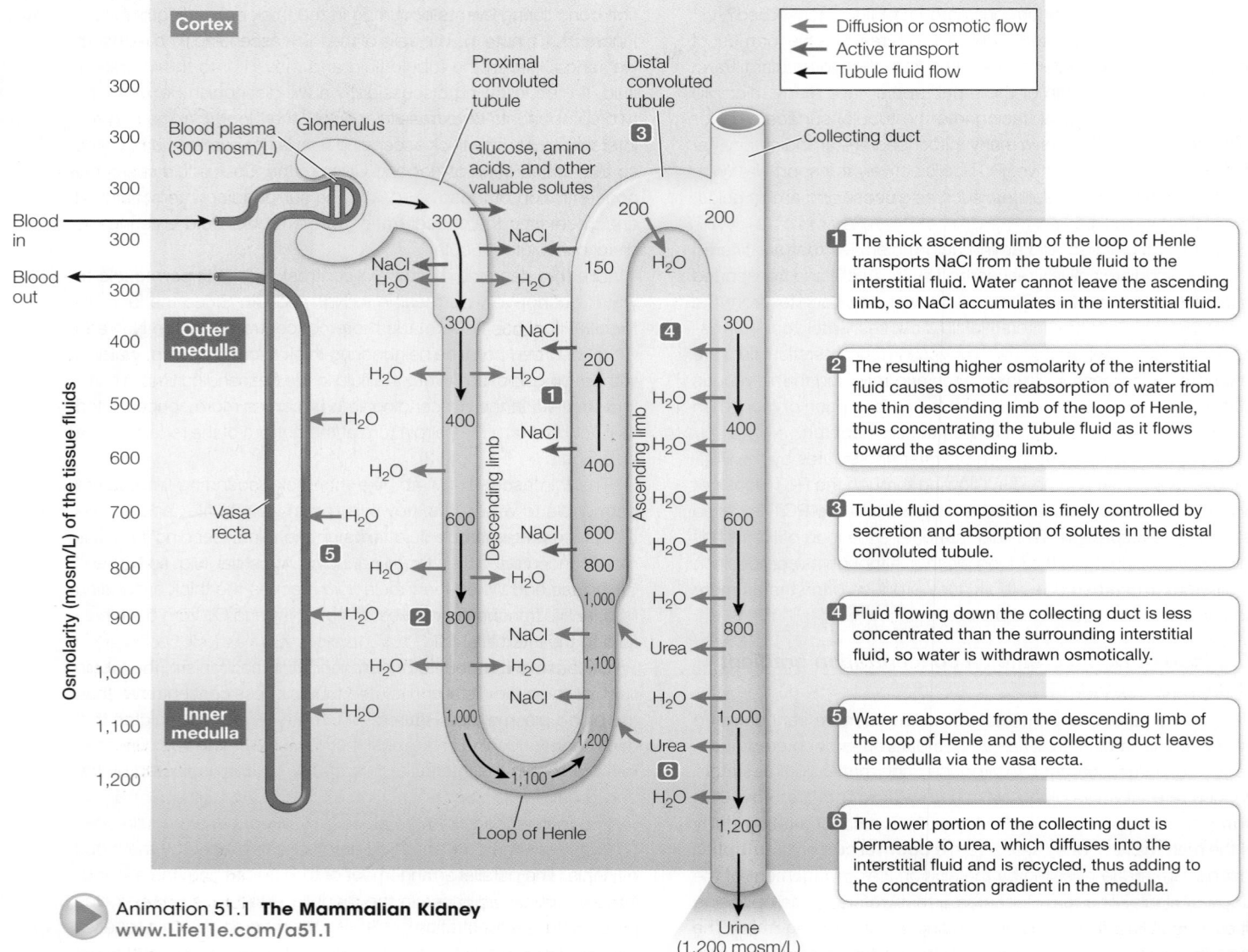

1 The thick ascending limb of the loop of Henle transports NaCl from the tubule fluid to the interstitial fluid. Water cannot leave the ascending limb, so NaCl accumulates in the interstitial fluid.

2 The resulting higher osmolarity of the interstitial fluid causes osmotic reabsorption of water from the thin descending limb of the loop of Henle, thus concentrating the tubule fluid as it flows toward the ascending limb.

3 Tubule fluid composition is finely controlled by secretion and absorption of solutes in the distal convoluted tubule.

4 Fluid flowing down the collecting duct is less concentrated than the surrounding interstitial fluid, so water is withdrawn osmotically.

5 Water reabsorbed from the descending limb of the loop of Henle and the collecting duct leaves the medulla via the vasa recta.

6 The lower portion of the collecting duct is permeable to urea, which diffuses into the interstitial fluid and is recycled, thus adding to the concentration gradient in the medulla.

▶ Animation 51.1 **The Mammalian Kidney**
www.Life11e.com/a51.1

Figure 51.10 Concentrating the Urine A countercurrent multiplier mechanism enables the mammalian kidney to produce urine that is far more concentrated than the blood plasma. The composition—but not the concentration—of the filtrate is changed by the proximal convoluted tubule, which reabsorbs valuable molecules (including NaCl). Bulk reabsorption of water follows osmotically. The urine concentration process begins in the thick ascending limb of the loop of Henle, which reabsorbs NaCl but is impermeable to H_2O. This reabsorbed NaCl creates an osmotic differential that pulls water out of the thin descending limb; NaCl also can diffuse into the thin descending limb, making the fluid in that limb more concentrated. Through this continuous process, a concentration gradient in the interstitial fluid of the renal medulla is created as indicated by the gradient of color in the medulla. As urine in the collecting duct passes through this concentration gradient, it can lose water osmotically, becomes more concentrated as indicated by the shading of color, and can become almost as concentrated as the interstitial fluid at the bottom of the renal medulla.

Q: What would be the effect of a drug called a "loop diuretic" that blocks the Na^+ transport in the thick ascending limb of the loop of Henle?

nephron? ***Aquaporins** are a class of membrane proteins that form water channels. Regions of the nephron that are highly permeable to water have greater numbers of aquaporins. Thus aquaporins are abundant in kidney PCT cells and in descending limbs of the loops of Henle, but not in the ascending limbs of the loop of Henle.

*connect the concepts As explained in Key Concept 6.3, water does not cross cell membranes easily, but aquaporins are protein channels that make cells permeable to water.

As an interesting evolutionary note, aquaporins are also important in maintaining water balance in amphibians. Many amphibians can gain water from a moist substrate because they have aquaporins in the epithelial cells of their belly skin. Thus water can cross their skin into the interstitial fluid by osmosis.

The distal convoluted tubule fine-tunes the composition of the urine

The first portion of the distal convoluted tubule is similar to the thick ascending limb of the loop of Henle. Na^+ and Cl^- are transported out of the tubule fluid, and water cannot follow. As a result, the tubule fluid becomes even more dilute. The later sections of the distal convoluted tubule, however, can be permeable to water, and therefore water can be osmotically drawn from the hypotonic tubule fluid into the interstitial fluid. As the tubule fluid flows through the distal tubule to the collecting duct, it equilibrates with the osmolarity of the blood plasma.

An important function of the distal tubule is the fine-tuning of the ionic composition of the urine. Even though bulk reabsorption of substances such as calcium, phosphate, bicarbonate, and potassium occurs in the proximal convoluted tubule, changes in the concentrations of these substances occur in the distal convoluted tubule. In the case of potassium, for example, if a person is potassium depleted, this ion is reabsorbed in the distal convoluted tubule, but if a person has an abundance of potassium, this ion is secreted in the distal convoluted tubule. As you will see in Key Concept 51.6, this exchange of K^+ is controlled by the hormone aldosterone. Another example is reabsorption of Ca^{2+} in the distal convoluted tubule, which is controlled by the actions of vitamin D. The fine-tuning of urine composition continues in the collecting duct. As you can imagine, the list of ion transporters in the distal convoluted tubule is large.

Urine is concentrated in the collecting duct

The tubule fluid entering the collecting duct is at about the same solute *concentration* as the blood plasma, but its solute *composition* is considerably different from that of the plasma. The major solute in the tubule fluid is now urea, since salts were reabsorbed earlier in the nephron. As the tubule fluid flows down the collecting duct, it loses water osmotically to the interstitial fluid, and that water returns to the circulatory system via the vasa recta (see Figure 51.10, note 4).

The concentration gradient established in the renal medulla by the countercurrent multiplier actions of the loops of Henle creates the osmotic potential that withdraws water from the collecting ducts. The collecting ducts begin in the renal cortex and run through the renal medulla before emptying into the ureter at the tips of the renal pyramids. During this journey, the solute concentration of the surrounding interstitial fluid increases, and more and more water can be absorbed from the urine in the collecting duct. By the time the urine reaches the ureter, it can become greatly concentrated, with urea as the major solute.

As water is withdrawn from the collecting duct, some urea also leaks out into the medullary interstitial fluid, adding to its osmotic potential. This urea diffuses back into the loop of Henle and is returned to the collecting duct. The recycling of urea in the renal medulla contributes significantly to the concentration gradient and

therefore the ability of the kidney to concentrate the urine in the collecting duct. The ability of a mammal to concentrate its urine is determined by the maximum concentration gradient it can establish in its renal medulla.

The kidneys help regulate acid–base balance

Besides regulating salt and water balance and excreting nitrogenous wastes, the kidneys have another important role: they regulate the hydrogen ion concentration (the pH) of the extracellular fluids. pH is a critical variable because it influences the structure and function of proteins.

One way to minimize pH changes in a chemical solution is to add a buffer—a substance that can either absorb or release hydrogen ions (see Key Concept 2.4). The major buffer in the blood is bicarbonate ions (HCO_3^-; see Figure 48.14) that are formed from the dissociation of carbonic acid, which in turn is formed by the hydration of CO_2 according to the following equilibrium reaction:

$$CO_2 + H_2O \rightleftharpoons H_2CO_3 \rightleftharpoons H^+ + HCO_3^-$$

From this equation, you can see that if excess hydrogen ions are added to this reaction mixture, the reaction will move to the left and absorb the excess H^+. If hydrogen ions are removed from the reaction mixture, however, the reaction will move to the right and supply more H^+.

The HCO_3^- buffer system is important for controlling the pH of the blood, and therefore of the interstitial fluids as well, because the reaction can be pushed to the right and pulled to the left physiologically. The lungs control the levels of CO_2 in the blood, thus altering the acid portion of the reaction. CO_2 is considered the acid portion of the reaction because if you add additional CO_2, the reaction shifts to the right, producing more H^+ ions. The kidneys control the base portion of the reaction by removing H^+ from the blood and returning HCO_3^- to the blood. How does this occur?

HCO_3^- is filtered in the glomerulus and is therefore present in the tubule fluid. As illustrated in **Figure 51.11A**, tubule cells transport H^+ into the tubule fluid in exchange for Na^+. In the tubule, the excreted H^+ combines with the filtered HCO_3^- to produce H_2CO_3 that then disassociates into H_2O and CO_2. The CO_2 diffuses into the tubule cells, where in the presence of the enzyme carbonic anhydrase it produces HCO_3^- that is transported out of the basal end of the cell into the interstitial fluid and thence to the blood. Thus for each H^+ secreted into the tubule fluid, a HCO_3^- ion is released into the blood.

Another mechanism for H^+ secretion and HCO_3^- reabsorption involves ammonium ions (NH_4^+). The metabolism of glutamine in tubule cells produces NH_4^+ and HCO_3^- (**Figure 51.11B**). The HCO_3^- is reabsorbed into the interstitial fluid. The NH_4^+ is transported into the tubule fluid and combines with Cl^-, which is excreted in the urine. This process results in the addition of a new HCO_3^- ion to the blood. The NH_4^+ is transported into the tubules by means of an NH_4^+ transporter.

Kidney failure is treated with dialysis

Loss of kidney function (renal failure) results in the retention of salts and water (hence high blood pressure), retention of urea (uremic poisoning), and a decreasing pH (acidosis). A person who suffers complete renal failure will die within 2 weeks if not treated. A drastic but

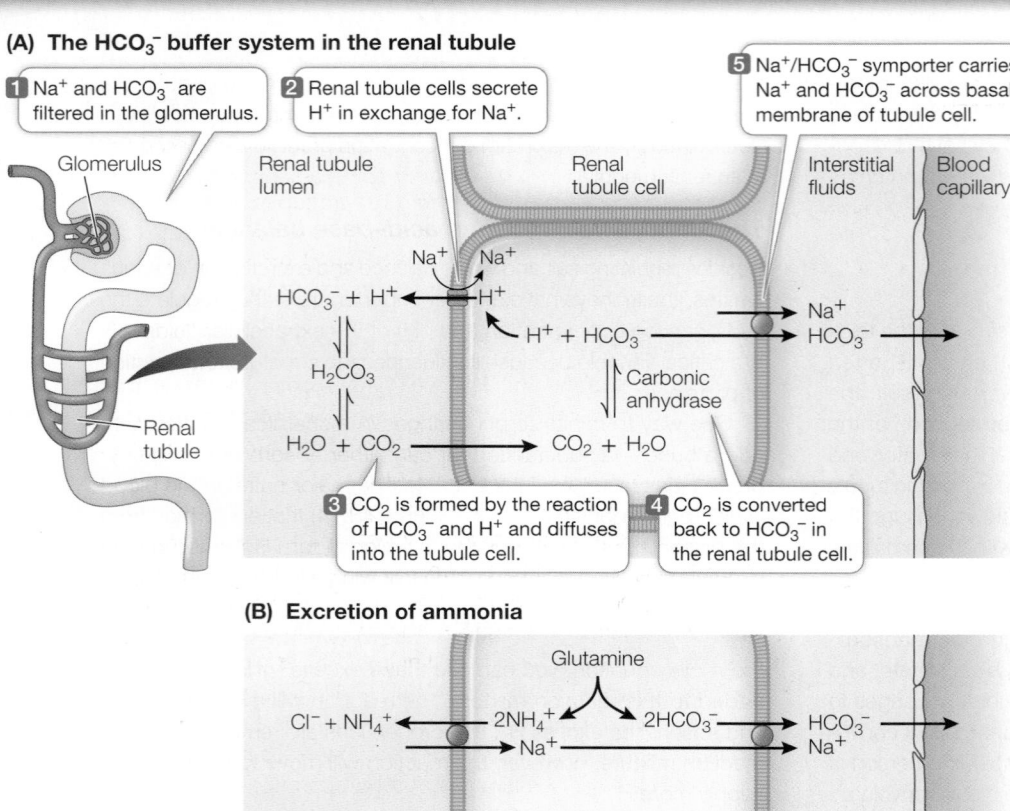

(A) The HCO$_3^-$ buffer system in the renal tubule

1 Na$^+$ and HCO$_3^-$ are filtered in the glomerulus.

2 Renal tubule cells secrete H$^+$ in exchange for Na$^+$.

5 Na$^+$/HCO$_3^-$ symporter carries Na$^+$ and HCO$_3^-$ across basal membrane of tubule cell.

Glomerulus

Renal tubule lumen

Renal tubule cell

Interstitial fluids

Blood capillary

HCO$_3^-$ + H$^+$

Na$^+$ Na$^+$

H$^+$

H$_2$CO$_3$

H$^+$ + HCO$_3^-$

Na$^+$
HCO$_3^-$

Carbonic anhydrase

Renal tubule

H$_2$O + CO$_2$

CO$_2$ + H$_2$O

3 CO$_2$ is formed by the reaction of HCO$_3^-$ and H$^+$ and diffuses into the tubule cell.

4 CO$_2$ is converted back to HCO$_3^-$ in the renal tubule cell.

(B) Excretion of ammonia

Glutamine

Cl$^-$ + NH$_4^+$ 2NH$_4^+$ 2HCO$_3^-$ HCO$_3^-$

Na$^+$ Na$^+$

Figure 51.11 The Kidney Excretes Acids and Conserves Bases **(A)** Bicarbonate ions are filtered out of the blood at the glomerulus, and renal tubule cells secrete hydrogen ions into the tubule fluid. In the renal tubule, the filtered bicarbonate buffers the secreted hydrogen ions and keeps the urine from becoming too acidic. The CO$_2$ formed by the reaction of bicarbonate and hydrogen ions is converted back to bicarbonate by the renal tubule cells and transported back into the interstitial fluid. **(B)** Excretion of ammonium ions (NH$_4^+$) by renal tubule cells is also important for acid–base balance.

highly successful treatment is kidney transplant, but it is usually necessary to sustain a patient for considerable time while waiting for a kidney to become available. Therefore artificial kidneys, or renal dialysis machines, are essential modes of treatment.

In a dialysis machine, the patient's blood flows through many small channels made of semipermeable membranes (**Figure 51.12**). A dialysis solution flows on the other side of these membranes, through which small molecules can diffuse. Molecules and ions diffuse from an area of higher concentration to an area of lower concentration, so the composition of the dialysis fluid is crucial. The concentrations of the molecules or ions that need to be conserved must be at the same concentration in the dialysis fluid as they are in the blood. The concentrations of molecules and ions that need to be removed from the blood are zero in the dialysis fluid. The total osmotic potential of the dialysis fluid must equal that of the blood plasma.

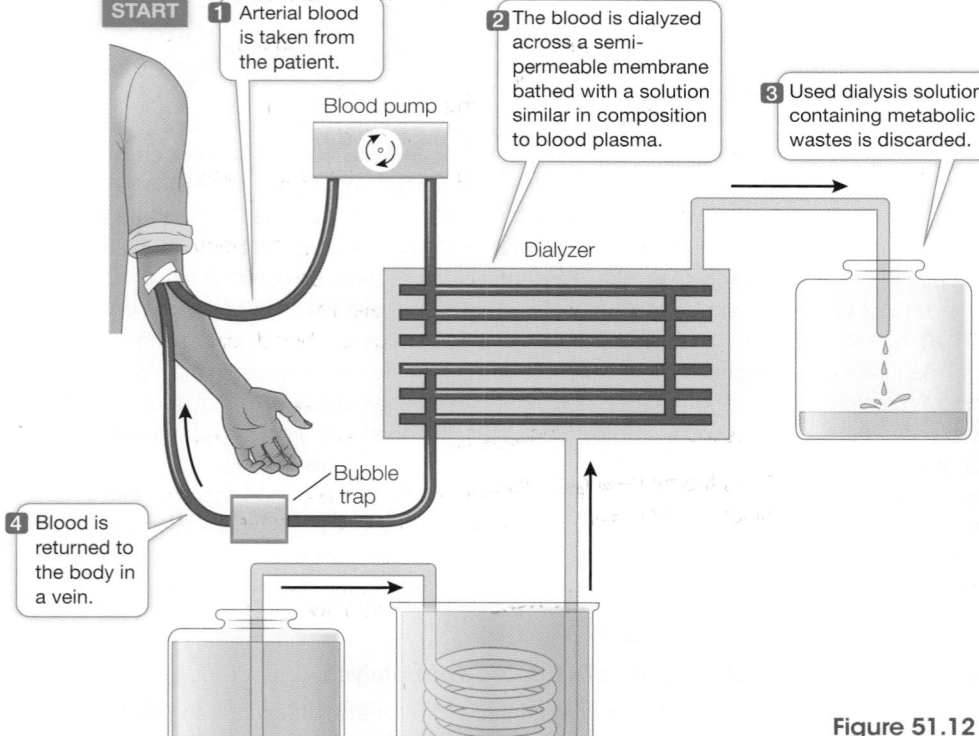

START

1 Arterial blood is taken from the patient.

2 The blood is dialyzed across a semipermeable membrane bathed with a solution similar in composition to blood plasma.

3 Used dialysis solution containing metabolic wastes is discarded.

Blood pump

Dialyzer

4 Blood is returned to the body in a vein.

Bubble trap

Fresh dialysis solution   Constant-temperature bath

Figure 51.12 Renal Dialysis Patients with kidney failure can have their blood cleansed of wastes by renal dialysis machines. Blood flows through channels of semipermeable membranes that allow diffusion of waste molecules from the blood to a dialysis fluid.

About 500 mL of the patient's blood is in the dialysis machine at any one time, and the unit processes several hundred milliliters of blood per minute. A patient with no kidney function must be on the dialysis machine for 4–6 hours three times a week.

51.5 recap

The anatomical organization of nephrons makes it possible for the mammalian kidney to produce a urine more concentrated than the blood, thereby conserving water to maintain extracellular fluid volume. Bulk reabsorption of salts, other valuable solutes, and water takes place in the proximal convoluted tubule. The loops of Henle act as a countercurrent multiplier, creating a concentration gradient of the interstitial fluids in the renal medulla. Collecting ducts run through the renal medulla and lose water osmotically to the surrounding interstitial fluids, concentrating the urine. The renal tubules contribute to acid–base balance by excreting H^+ ions.

learning outcomes

You should be able to:

- Describe the internal organization of the mammalian kidney.
- Describe the mechanism whereby most of the glomerular filtrate is reabsorbed in the proximal convoluted tubule.
- Describe how the loop of Henle maintains a concentration gradient in the extracellular fluid of the medulla of the kidney.
- Describe how the mammalian excretory system is involved in regulating pH of the extracellular fluid.

1. If you could measure the concentration of mitochondria in mammalian kidney tissues, where would you find the highest concentration, the lowest concentration, and an intermediate concentration? Explain your answer.
2. How do the loops of Henle act as a countercurrent multiplier?
3. If you compared a water rat and a desert rat of the same size, how would you expect their kidneys to differ in anatomical structure? Explain your answer.
4. How does the active secretion of H^+ ions by the renal tubule cells result in an increase in the bicarbonate concentration in the blood?

The kidneys contribute to homeostasis in several ways, including regulating extracellular fluid volume, maintaining the osmotic concentration and ionic composition of the extracellular fluid, and regulating pH. As you will see next, the kidneys also play a major role in regulating blood pressure.

key concept 51.6 Kidney Function Is Regulated

Several regulatory mechanisms act on the kidneys to maintain blood pressure, blood osmolarity, and blood composition. We will discuss these mechanisms separately, but keep in mind that they are always working together.

focus your learning

- Circulating renin initiates activation of angiotensin, which acts in several ways to increase blood pressure.
- The kidney autoregulates its vascular resistance to maintain constancy of glomerular filtration rate.
- Antidiuretic hormone (ADH) controls urine concentration in response to blood pressure and blood osmolarity.

Glomerular filtration rate is regulated

If the kidneys stop filtering blood, they cannot accomplish any of their functions. The maintenance of a constant **glomerular filtration rate** (**GFR**) depends on an adequate blood supply to the kidneys at an adequate blood pressure. Renal arteries usually deliver blood to the kidneys at high pressure because they are early branches off the aorta. In addition, autoregulatory mechanisms ensure adequate blood supply and blood pressure for kidney function regardless of what is happening elsewhere in the body. The kidney's autoregulatory adjustments compensate for decreases in cardiac output or decreases in blood pressure so that the GFR remains constant.

One autoregulatory mechanism is the dilation (expansion) of the afferent renal arterioles when blood pressure falls. This dilation decreases the resistance in the arterioles and helps maintain blood pressure in the glomerulus. If arteriole dilation does not keep the GFR from falling, the kidney releases an enzyme, **renin**, into the blood. Renin converts a circulating protein, angiotensinogen, into angiotensin I, which is then acted on by angiotensin-converting enzyme (ACE) in the lungs and other tissues to form the active hormone angiotensin II, or simply **angiotensin** (**Figure 51.13**). Angiotensin has several effects that help restore the GFR to normal:

- It constricts the efferent renal arterioles, raising the resistance for blood leaving the glomerulus. Like putting a finger over the end of a garden hose, this restriction of drainage elevates blood pressure in the glomerular capillaries.
- It constricts peripheral blood vessels all over the body, an action that elevates blood pressure.
- It stimulates the adrenal cortex to release the hormone **aldosterone**. Aldosterone stimulates sodium reabsorption by the kidney, making its reabsorption of water more effective. Enhanced water reabsorption helps maintain blood volume and therefore blood pressure.
- It acts on the brain to stimulate thirst. Increased water intake in response to thirst increases blood volume and blood pressure.

Thus the renin–angiotensin–aldosterone system, or RAAS, coordinates many responses to maintain blood pressure and kidney function.

Regulation of GFR uses feedback information from the distal tubule

A remarkable anatomical feature of the nephron is that where its renal tubule returns to the cortex and becomes the distal convoluted tubule, it makes contact with the afferent and efferent arterioles of its glomerulus. At this location the cells of the tubule are modified to form a structure called the macula densa (see Figure 51.13), and

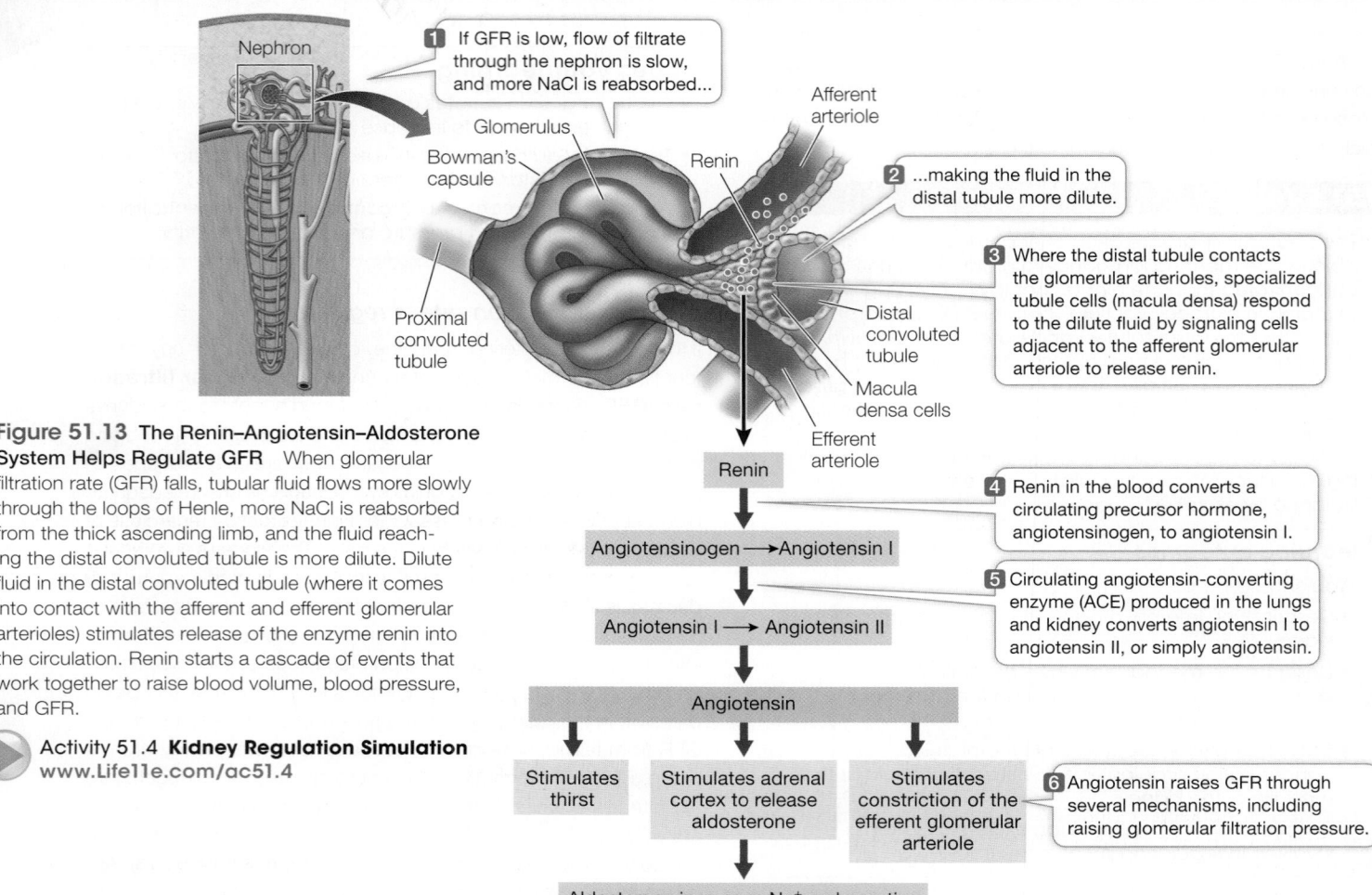

Figure 51.13 The Renin–Angiotensin–Aldosterone System Helps Regulate GFR When glomerular filtration rate (GFR) falls, tubular fluid flows more slowly through the loops of Henle, more NaCl is reabsorbed from the thick ascending limb, and the fluid reaching the distal convoluted tubule is more dilute. Dilute fluid in the distal convoluted tubule (where it comes into contact with the afferent and efferent glomerular arterioles) stimulates release of the enzyme renin into the circulation. Renin starts a cascade of events that work together to raise blood volume, blood pressure, and GFR.

▶ Activity 51.4 **Kidney Regulation Simulation**
www.Life11e.com/ac51.4

the arteriole cells are also modified and are called juxtaglomerular cells. The macula densa cells sense the concentration of NaCl in the fluid entering the distal convoluted tubule. If GFR is high, flow through the loop of Henle is high and the cells of the thick ascending limb cannot reabsorb all of the NaCl passing through it. If GFR is low, flow through the loop of Henle is slower and the cells of the thick ascending limb can reabsorb more of the NaCl. If the NaCl level in the distal convoluted tubule drops too low, the macula densa cells signal the juxtaglomerular cells to release renin and trigger the RAAS response. Thus the concentration of NaCl in the fluid passing over the macula densa is a function of GFR and is information that enables fine control of the RAAS system.

Blood osmolarity and blood pressure are regulated by ADH

Cells in the hypothalamus can stimulate the release of a hormone called **antidiuretic hormone** (**ADH**, also called vasopressin) from the posterior pituitary. ADH stimulates cells of the collecting duct to insert aquaporins (water channels) into their cell membranes. The aquaporins increase the permeability of these membranes to water, and therefore more water is reabsorbed from the collecting duct fluid into the interstitial fluid of the renal medulla. The higher the circulating levels of ADH, the greater the number of aquaporins. Various factors can stimulate or inhibit the release of ADH. Of key importance to

kidney function are (1) osmoreceptors that monitor blood osmolarity and (2) stretch receptors that monitor blood pressure (**Figure 51.14**).

Osmoreceptor neurons in the hypothalamus are activated by a rise in blood osmolarity, and they increase the release of ADH. ADH helps regulate blood osmolarity by controlling water reabsorption. The osmoreceptors also stimulate thirst. The resulting water retention and water intake dilute the blood as they expand blood volume.

Stretch receptors in the walls of the aorta and the carotid arteries (see Figure 49.19) that detect an increase in blood pressure will *inhibit* the release of ADH. With less circulating ADH, less water is reabsorbed, which increases urine volume and decreases blood volume, hence lowering blood pressure.

If blood pressure falls, as when you lose blood volume through hemorrhage or excessive evaporative water loss, activity of the stretch receptors in the aorta and carotid arteries decreases. Input via cranial nerves to the hypothalamus from these receptors inhibits the release of ADH, so when the firing rates of these stretch receptors fall, ADH release increases. More ADH results in more efficient water reabsorption and therefore a protection of blood volume and blood pressure.

Alcohol inhibits ADH release, explaining why excessive beer drinking leads to excessive urination and dehydration, which contributes to the symptoms of a hangover.

As already mentioned, the presence of aquaporins in cell membranes determines their water permeability. Aquaporins play an

Blood osmolarity

Blood pressure

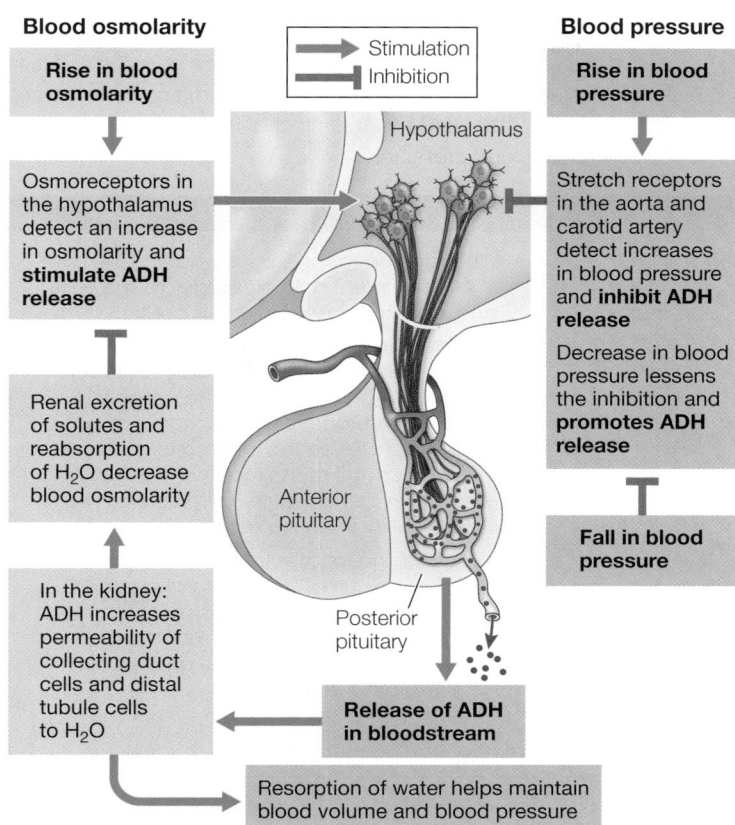

Stimulation
Inhibition

Rise in blood osmolarity

Osmoreceptors in the hypothalamus detect an increase in osmolarity and **stimulate ADH release**

Renal excretion of solutes and reabsorption of H_2O decrease blood osmolarity

In the kidney: ADH increases permeability of collecting duct cells and distal tubule cells to H_2O

Hypothalamus

Anterior pituitary

Posterior pituitary

Release of ADH in bloodstream

Resorption of water helps maintain blood volume and blood pressure

Rise in blood pressure

Stretch receptors in the aorta and carotid artery detect increases in blood pressure and **inhibit ADH release**

Decrease in blood pressure lessens the inhibition and **promotes ADH release**

Fall in blood pressure

Figure 51.14 Antidiuretic Hormone Increases Blood Pressure and Promotes Water Reabsorption ADH is produced by neurons in the hypothalamus and released from nerve endings in the posterior pituitary. The release of ADH is stimulated by hypothalamic osmoreceptors and inhibited by stretch receptors in the great arteries. Red lines indicate inhibitory actions; green lines show stimulatory actions.

important and unique role in the collecting duct. Several members of the aquaporin family of water channels are found in the cell membranes of the collecting duct cells. At least two aquaporins (AQP-3 and AQP-4) are localized in the basolateral membranes (facing the blood vessels). A different aquaporin, AQP-2, is found in the apical cell membranes (facing into the tubule). The presence of AQP-2 in these membranes is controlled by ADH (**Figure 51.15**).

When ADH levels are low, such as when a person is well hydrated, most of the AQP-2 protein is sequestered in the membranes of intracellular vesicles, and the collecting duct permeability is low. A rise in ADH levels stimulates the insertion of these vesicles along with their AQP-2 channels into the apical cell membranes. As a result, the membranes become more permeable to water. Water that enters the collecting duct cells passes into the interstitial fluid through the aquaporins in the basolateral membranes. ADH also stimulates the synthesis of new AQP-2 proteins. Thus circulating ADH controls the number of AQP-2 water channels in the cell membranes of the collecting duct cells, and therefore the permeability of the collecting duct to water.

The heart produces a hormone that helps lower blood pressure

You may not think of the heart as an endocrine organ, but it is. When blood volume is high, blood pressure is high, putting strain

Figure 51.15 ADH Induces Insertion of Aquaporins into Cell Membranes

Original Paper: Nielsen, S., C. L. Chou, D. Marples, E. I. Christensen, B. K. Kishore and M. A. Knepper. 1995. Vasopressin increases water permeability of kidney collecting duct by inducing translocation of aquaporin-CD water channels to plasma membrane. *Proceedings of the National Academy of Sciences USA* 92: 1013–1017.

Aquaporin proteins make some regions of renal tubules permeable to water. One aquaporin, AQP-2, is responsible for the permeability of the collecting duct cells. M. A. Knepper and colleagues did an experiment to find out how antidiuretic hormone acts on these proteins to control the level of permeability in renal cells.

HYPOTHESIS▶ Antidiuretic hormone (ADH) controls the location of aquaporin proteins.

METHOD

1. Isolate collecting ducts from rat kidney.
2. Use immunochemical staining to localize the AQP-2 aquaporins in collecting duct cells both with and without the presence of ADH. Also localize the aquaporins after ADH is applied and then washed away.

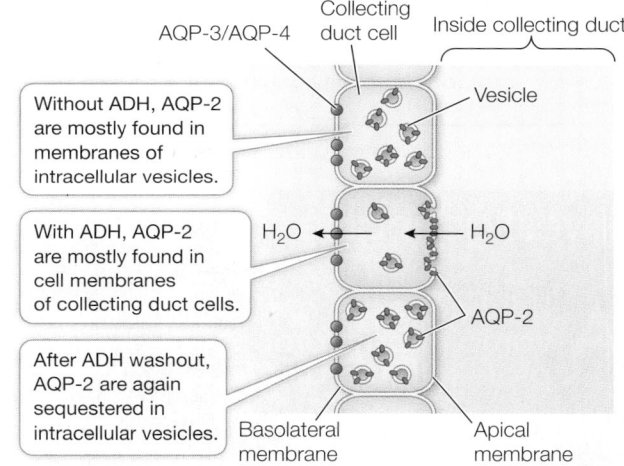

Without ADH, AQP-2 are mostly found in membranes of intracellular vesicles.

With ADH, AQP-2 are mostly found in cell membranes of collecting duct cells.

After ADH washout, AQP-2 are again sequestered in intracellular vesicles.

AQP-3/AQP-4
Collecting duct cell
Inside collecting duct
Vesicle
H_2O
H_2O
AQP-2
Basolateral membrane
Apical membrane

3. Measure the water permeability of the collecting duct cells under the same three conditions.

RESULTS

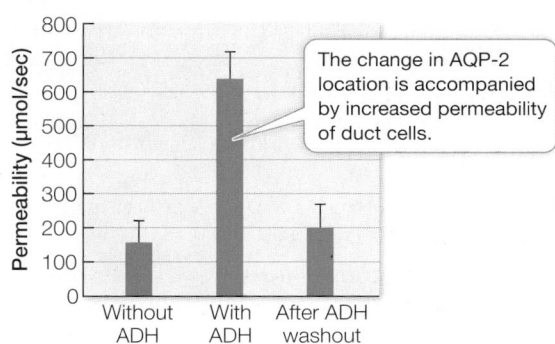

The change in AQP-2 location is accompanied by increased permeability of duct cells.

CONCLUSION▶ In the absence of ADH, AQP-2 is sequestered intracellularly. When ADH is present, AQP-2 channels are inserted into cell membranes, making the cells more permeable to water.

on the heart. Under these conditions, the increased venous return stretches the atria of the heart. When the atrial muscle fibers are over-stretched, they release a peptide hormone called **atrial natriuretic peptide (ANP)**. This peptide hormone enters the circulation, and in the kidney it decreases the reabsorption of sodium. If less sodium is reabsorbed, less water is reabsorbed, and more passes into the urine. Thus ANP has the effect of lowering blood volume and therefore blood pressure.

51.6 recap

Glomerular filtration is essential for kidney function and is sustained by autoregulatory mechanisms. These autoregulatory mechanisms include changes in dilation and constriction of the glomerular afferent arterioles and hormonal control over constriction of the efferent glomerular arteriole mediated by the renin–angiotensin–aldosterone system (RAAS). Angiotensin controls the release of aldosterone, which controls sodium reabsorption. Sensors that monitor blood pressure and blood osmolarity stimulate or inhibit the release of ADH, which controls the water permeability of the collecting duct.

learning outcomes

You should be able to:

- Describe how renin, angiotensin, and aldosterone affect glomerular filtration rate and blood pressure.
- Given an event that lowers blood pressure, predict the response of the renin–angiotensin–aldosterone system.
- Describe ADH's role in regulating kidney function in response to changes in blood osmolarity or blood pressure.

1. What are the mechanisms whereby the afferent and efferent arterioles regulate the GFR, and how are those responses controlled?

2. Aldosterone stimulates activity of the Na^+/K^+ exchangers in the distal convoluted tubule and the initial segments of the collecting ducts. Aldosterone release is triggered by angiotensin. Given this information, what is the significance of the fact that the $Na+/K+$ exchanger in the tubule cells secretes two K^+ ions for every three Na^+ ions it reabsorbs?

3. If you drank a lot of salty chicken broth that had an osmotic concentration of 300 mosm/L, what two mechanisms would play roles in the response of your kidneys to excrete the excess fluid volume and excess Na^+? Explain your answer.

investigatinglife

How does the excretory system of vampire bats process their high-protein liquid diet?

As we described in the opening story, vampire bats conserve water by producing urine that can be as much as 15 times more concentrated than their blood. For comparison, the maximum urine concentration of humans is about four times that of the blood.

Recall from Key Concept 51.5 that the mammalian body's ability to produce concentrated urine depends on the concentration gradient set up in the renal medulla by the loops of Henle. An important adaptation for increasing the concentration gradient is to increase the lengths of the loops of Henle relative to overall kidney size. Some tiny desert gerbils, for example, have such extremely long loops of Henle that the renal pyramid (each of the rodent kidneys has only one, in contrast to humans) extends far out of the concave surface of the kidney and into the ureter (Figure 51.16). The large concentration gradient that results draws most of the water out of the urine as it passes down the collecting duct. Desert gerbils are so effective in conserving water that they can survive on the water released by the metabolism of their food.

The vampire bat kidney is also particularly good at concentrating urea. Within minutes after feeding, the urea concentration in its urine can increase by more than 500-fold.

(A) Desert gerbil

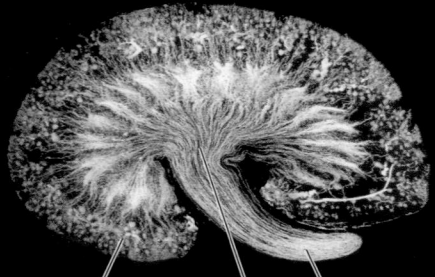

(B) Laboratory rat

Figure 51.16 The Ability to Concentrate The ability of the mammalian kidney to concentrate urine depends on the lengths of its loops of Henle relative to the overall size of the kidney. **(A)** The kidney of a desert gerbil has a single renal pyramid with loops of Henle so long that the pyramid extends far into the ureter (ureter not shown). **(B)** An ordinary laboratory rat has much shorter loops of Henle.

Q: Do you think the vampire bat has loops of Henle resembling those of the laboratory rat or the desert gerbil?

Renal cortex   Renal medulla   Loops of Henle (many)   Loops of Henle   Renal medulla

Future directions

Understanding the diverse mechanisms whereby physiological systems are controlled and regulated can contribute to the discovery and development of new pharmaceuticals. The renal system is an example. Many diseases involve problems with salt and water balance. A common problem is water retention. A large number of drugs have been developed that increase the excretion of water in urine, such as occurs naturally in vampire bats. These drugs are called diuretics, and since they act through different mechanisms,

they have different effects on Na^+, K^+, and Ca^{2+} levels in the blood, which can cause serious side effects. Excessive loss of K^+, for example, can lead to heart arrhythmia, muscle weakness, and paralysis. Excessive retention of K^+ can also lead to heart arrhythmia, along with muscle cramps and paralysis. Loss of too much Na^+ can result in central nervous system deficits. Too much Ca^{2+} has many negative side effects, ranging from depression and fatigue to vomiting and gout. Thus it is important to understand how drugs work on physiological systems, to avoid negative side effects.

Chapter Summary 51

▶ 51.1 Excretory Systems Regulate Osmotic and Ionic Concentrations

- **Excretory systems** maintain the **osmolarity** and volume of the extracellular fluids and eliminate the waste products of nitrogen metabolism through the processes of filtration, reabsorption, and secretion. **Urine** is the output of excretory systems.

- There is no active transport of water, so water must be moved across membranes by a difference in either osmolarity or pressure.

- Water enters and leaves cells by osmosis. To achieve cellular water balance, animals must maintain the osmolarity of their extracellular fluids within an acceptable range.

- Marine animals can be **osmoconformers** or **osmoregulators**. Freshwater animals must be osmoregulators and must continually excrete water and conserve salts. Terrestrial animals are osmoregulators, but the nature of their regulation depends on environment and lifestyle. Review Figure 51.1

- Apart from regulating osmolarity of cells and extracellular fluids, animals must also regulate their ionic composition by conserving some ions and secreting others. Salt glands are adaptations for secretion of NaCl. Review Figure 51.2

▶ 51.2 Animals Excrete Nitrogen as Ammonia, Urea, or Uric Acid

- Aquatic animals that breathe water can eliminate nitrogenous wastes such as **ammonia** by diffusion across their gill membranes. Terrestrial animals and some aquatic animals must detoxify ammonia by converting it to **urea** or **uric acid** before excretion. Review Figure 51.3

- Depending on the form in which they excrete their nitrogenous wastes, animals are classified as **ammonotelic**, **ureotelic**, or **uricotelic**.

▶ 51.3 Invertebrate Excretory Systems Use Filtration, Secretion, and Reabsorption

- The **protonephridia** of flatworms consist of flame cells and excretory tubules. Extracellular fluid is filtered into the tubules, which process the filtrate to produce dilute urine. Review Figure 51.4

- In annelid worms, blood pressure causes filtration of the blood across capillary walls. The filtrate enters the coelomic cavity, where it is taken up by **metanephridia**, which alter the composition of the filtrate by active transport mechanisms. Review Figure 51.5, Activity 51.1

- The **Malpighian tubules** of insects receive ions and nitrogenous wastes by active transport across the tubule cells. Water follows by osmosis. Ions and water are reabsorbed from the rectum, so the insect excretes semisolid wastes. Review Figure 51.6

▶ 51.4 The Nephron Is the Basic Functional Unit of Vertebrate Excretory Systems

- Marine bony fishes produce little urine. Cartilaginous fishes retain urea and TMAO, so the osmolarity of their body fluids remains close to that of seawater.

- Reptiles, including birds, have skin with low water permeability and excrete nitrogenous wastes as uric acid in a semisolid form.

- Mammals produce urine that is more concentrated than their extracellular fluids.

- The **nephron**, the functional unit of the vertebrate **kidney**, consists of a **glomerulus**, in which blood is filtered, a **renal tubule**, which use processes of active secretion and reabsorption to convert the glomerular filtrate into urine to be excreted, and a system of **peritubular capillaries**, which surround the tubule and support its functions of secretion and reabsorption. Review Figures 51.7, Activity 51.2

▶ 51.5 The Mammalian Kidney Can Produce Concentrated Urine

- The concentrating ability of the mammalian kidney is a function of its anatomy, which enables countercurrent exchange.

- The glomeruli and the **proximal** and **distal convoluted tubules** are located in the **cortex** of the kidney. Certain molecules are actively reabsorbed from the glomerular filtrate by the tubule cells, and other molecules are actively secreted. Straight sections of renal tubules called **loops of Henle** and **collecting ducts** are arranged in parallel in the **medulla** of the kidney. Review Figure 51.9, Activity 51.3

- Salts, water, and valuable molecules such as glucose and amino acids are reabsorbed in the proximal convoluted tubule without the renal filtrate becoming more concentrated, although its composition changes.

- The loops of Henle create a concentration gradient in the interstitial fluid of the renal medulla by a **countercurrent multiplier** mechanism. Urine flowing down the collecting ducts to the **ureter** is concentrated by the osmotic reabsorption of water caused by the concentration gradient in the surrounding interstitial fluid. Review Focus: Key Figure 51.10, Animation 51.1

- Hydrogen ions secreted by the renal tubules are buffered in the urine by bicarbonate and other chemical buffering systems. Review Figure 51.11

(continued)

Chapter Summary 51 (continued)

▶ 51.6 Kidney Function Is Regulated

- Kidney function in mammals is controlled by autoregulatory mechanisms that maintain a constant high **glomerular filtration rate (GFR)** even if blood pressure varies.

- The RAAS is an important autoregulatory mechanism. **Renin** is released by the kidney when blood pressure falls. Renin activates **angiotensin**, which causes the constriction of efferent glomerular arterioles and peripheral blood vessels, causes the release of **aldosterone** (which enhances water reabsorption), and stimulates thirst. Review Figure 51.13, Activity 51.4

- Changes in blood pressure and osmolarity influence the release of **antidiuretic hormone (ADH)**, which controls the permeability of the collecting duct to water and therefore the amount of water

that is reabsorbed from the urine. ADH stimulates the expression of and controls the intracellular location of aquaporins, which serve as water channels in the membranes of collecting duct cells. Review Figures 51.14, 51.15

- When the volume of blood returning to the heart increases and stretches the atrial walls, **atrial natriuretic peptide (ANP)** is released, which causes increased excretion of salt and water.

See Activity 51.5 for a review of the major human organ systems.

> Go to **LearningCurve** (in **LaunchPad**) for dynamic quizzing that helps you solidify your understanding of this chapter. **LearningCurve** adapts to your responses, giving you the practice you need to master each key concept.

▶ Apply What You've Learned

Review

51.5 As the tubular fluid passes down the collecting duct, water can be reabsorbed into the extracellular fluid to produce a more concentrated urine in the tubule.

Original Paper: Schmidt-Nielsen, B. and R. O'Dell. 1961. Structure and concentrating mechanism in the mammalian kidney. *American Journal of Physiology* 200: 1119–1124.

The ability of a mammalian kidney to produce concentrated urine is directly related to the countercurrent multiplier mechanism found in the loops of Henle. Because of the way this mechanism works, it could be hypothesized that the longer the loops of Henle, the more concentrated the urine can become. Since bigger kidneys should have the longest loops, a large mammal, such as an elephant, should be able to produce much more concentrated urine than a small mammal, like a shrew. However, some of the mammals that live in the driest conditions on Earth are very small. How can these small mammals conserve water by producing highly concentrated urine with such tiny kidneys?

To account for body size, scientists refined their hypothesis by saying that it wasn't just the total length of the loops of Henle, but the "relative" length compared with the rest of the kidney. The loops of Henle are found only in the medulla of the kidney, so the loops can be only as long as the medulla is thick. To normalize the data across different-sized mammals, a new measurement called the relative medullary thickness (RMT) was conceived. The RMT is equal to the thickness of the medulla divided by width × length × thickness of the whole kidney.

On further study, it was also noticed that in many species not all of the loops of Henle extend the full length of the medulla. Some are shorter than others. This presented a second hypothesis, namely that the ability to concentrate urine could be related to the number of long loops of Henle compared with short loops of Henle. In other words, the higher the percent of long loops to short, the higher the possible concentration of urine.

To gather evidence in support of one hypothesis over the other, scientists performed many measurements on various-sized mammals from different environments. The scientists deprived the mammals of water to find the limit of each mammal's ability to concentrate urine. A test called the freezing point depression (FPD) was used to measure urine concentration. Fresh water freezes at 0°C, but the more solutes that are added to the water, the lower the temperature must be to get the solution to freeze. So the FPD of urine samples reveals their concentrations. The results of all of these tests are seen in the table.

Animal	Percent long loops of Henle	RMT	FPD (°C)
Human	14	3	2.6
Pig	3	1.6	2
Dog	100	4.3	4.85
Cat	100	4.8	5.8
Beaver	0	1.3	0.96
Lab rat	28	5.8	4.85
Sand rat	100	10.7	9.2
Kangaroo rat	27	8.5	10.4
Jerboa	33	9.3	12

Questions

1. Is the RMT a good predictor of an animal's ability to concentrate urine? To answer this, plot the RMT versus FPD.

2. Are animals with a higher percent of long loops of Henle more capable of producing concentrated urine? To answer this, plot the percent long loops of Henle versus FPD.

3. Is the percent of long loops of Henle or RMT the better predictor of an animal's ability to concentrate urine? Explain your answer.

> Go to **LaunchPad** for the eBook, LearningCurve, animations, activities, flashcards, and additional resources and assignments.

52

Animal Behavior

The male songbird claims and defends his territory through song and attracts a mate through song, and the female selects her mate based on the quality and vigor of his song. Birdsong involves genetic determinants, learning, physiological conditions, and social interactions.

▶ investigating life

An Instinct to Learn

A springtime joy is hearing birds sing. Why do they do it? Male songbirds use species-specific song to compete with other males for territory. They also sing to attract mates. Females may not sing, but they recognize the songs of their own species. Males of most species, such as the white-crowned sparrow (*Zonotrichia leucophrys*), must learn their song, but *what* they can learn is influenced by their genes, and when they can learn it is influenced by their physiology. A hatchling in the nest hears his father and other white-crowned sparrows singing. He also hears the songs of many other bird species. He does not sing until he approaches sexual maturity almost a year later, and when he does, he sings his father's type of song and not that of other species. His first attempts at song are poor, but by trial and error he gradually matches his song to the one stored in his memory, even though other species are singing in his environment. His singing talent depends on two critical conditions: he must hear his species-specific song in the first 2 months of life, and he must be able to hear himself as he develops his song. These are both learning experiences, but the bird's genes play an important role too.

Male white-crowned sparrows raised alone in the laboratory and exposed to recordings of unrelated species will not learn any of those song templates even if they are played over and over. But if in that cacophony of songs, a bird's species-specific song is played just a few times, he will sing that song the following spring. Thus male white-crowned sparrows must learn their song, but a genetic predisposition ensures that they learn the song of their own species.

An important reason why males learn only their species-specific song is that the females also listen to their father's songs when they are nestlings. When they mature, they choose mates that sing like their fathers did. But how does one female chose her specific mate? There is variation in male songs, and some components of songs seem sexier than others. Also, the way the songs are performed may differ among individuals and in different social contexts. As in humans, variability surely plays a role in mate choice.

Q&A How does social context influence the learning and expression of birdsong?

key concept
52.1
Ethology Led to Modern Behavioral Biology

Humans have studied animal behavior since prehistoric times. Understanding the habits of potential prey, as well as those of their predators, was of great value to hunters. Appreciation of behavioral traits led to the domestication of animal species. Accounts of animal behaviors such as seasonal appearances and disappearances, mating displays, aggression, prey capture, parental care, and communication are found throughout recorded history. Yet the scientific study of animal behavior did not truly get under way until the early 1900s.

focus your learning

- Ethology is the study of natural behaviors in natural species.
- Behaviors have proximate and ultimate causes.

Conditioned reflexes are a simple behavioral mechanism

In the late 1800s the Russian physiologist Ivan Pavlov used dogs to study the neural control of digestive juice secretion. Of course his dogs salivated when they smelled food, but Pavlov observed that they also salivated whenever the technician who routinely fed the dogs entered the room—even when no food was present. Following up this observation, Pavlov substituted a sound stimulus for the technician; a metronome ticked while the dogs were fed. After several trials, the dogs salivated when they heard the metronome, even if no food was offered.

Salivation in response to the sight, smell, or taste of food is a natural reflex response to a stimulus, but salivation in response to a sound was a learned response. The pairing of a sound with the experience of receiving food conditioned the dogs' nervous systems to generate a response, which Pavlov dubbed the **conditioned reflex (Figure 52.1)**. The natural stimulus—food in this case—became known as the unconditioned stimulus, and the unnatural stimulus was termed the conditioned stimulus. Pavlov received a Nobel prize in 1904 for his work showing that a simple behavior controlled by the nervous system could be modified through experience. This work stimulated much new research because Pavlov had developed an experimental model of learning.

Going beyond the conditioning of autonomic reflexes, the psychologist B. F. Skinner showed that many behaviors could become a conditioned response to a stimulus if a reward was associated with the action and the stimulus. A rat, for example, could be conditioned to press a lever in response to a stimulus if it got a reward when it pressed the lever. Because the animal was conditioned to perform an *operation* on its environment, this experimental protocol was called **operant conditioning** and became another model of learning.

The experimental approaches to behavior initiated by Pavlov and Skinner stimulated a lot of research on animal behavior, but the focus of that research was mostly controlled laboratory experiments on learning and memory in only a few species—mostly the white rat. This field of animal behavior became known as **behaviorism.**

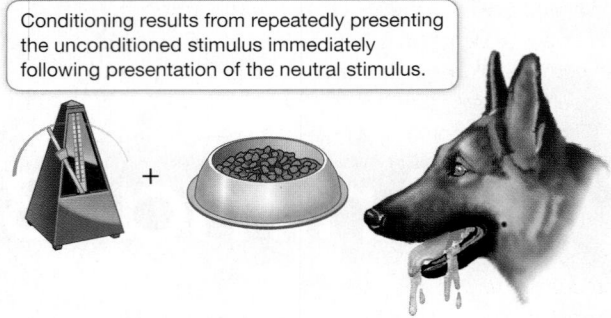

(A) Before conditioning

Food is an unconditioned stimulus that produces an unconditioned response (salivation).

Sound is a neutral stimulus that produces no response.

(B) Conditioning

Conditioning results from repeatedly presenting the unconditioned stimulus immediately following presentation of the neutral stimulus.

+

(C) After conditioning

The neutral stimulus has become a conditioned stimulus that by itself produces the **conditioned reflex** (salivation).

Figure 52.1 The Conditioned Reflex **(A)** Prior to conditioning, there is no response to the neutral stimulus. Ivan Pavlov discovered that when a normal response is paired with an artificial or neutral stimulus **(B)**, an animal learns to produce the response even when only the artificial stimulus is presented **(C)**.

Ethologists focused on the behavior of animals in their natural environment

An alternative approach to the study of animal behavior arose at the same time as behaviorism and focused on the various behaviors of animals in their natural environment. This field of study became known as **ethology** (Greek *ethos*, "character," + *logos*, "study"). Ethologists were interested in a wide variety of species, their evolutionary relationships, and the ways in which their behaviors were adapted to their environments. The leaders of the ethology movement were Karl von Frisch, who discovered the dance language of honey bees; Konrad Lorenz, who discovered that the strong bond between parent and offspring develops during a "critical period" following birth; and Niko Tinbergen, who studied inborn patterns of behavior commonly known as instincts. These three scientists shared the Nobel Prize in 1973 for "their discoveries concerning organization and elicitation of individual and social behavior patterns." Their work laid the foundation for modern research on animal behavior.

Ethologists were mainly interested in species-specific or instinctive behaviors that therefore had to have genetic components. Behaviors were thought to be genetically determined if they:

- are performed without learning.
- are stereotypic (that is, they are performed the same way each time).
- cannot be modified by learning.

The ethologists called such behaviors **fixed action patterns**.

Ethologists performed **deprivation experiments** to demonstrate the genetic determination of a behavior. They raised animals in an environment devoid of opportunities to learn their species-specific behavior, and then looked for the expression of that behavior. The web spinning of a spider is an example of a natural deprivation experiment. The parents of a young spider die before it hatches, and in a seasonal environment it has no model webs to copy when it spins its first web, which requires thousands of stereotyped sequential movements. Yet a young spider creates a perfect, species-specific web the first and every time it spins a web. Thus the information for the web spinning behavior has to be genetically programmed into its nervous system.

Fixed action patterns are usually responses to specific stimuli. The ethologists carefully characterized such stimuli, which they called **releasers**. In general, releasers are simple subsets of the information available in the environment. For example, Tinbergen studied the begging behavior of gull chicks. Adult gulls have a red dot on their lower bill. When a parent returns to the nest to feed its chicks, the chicks peck on the red dot, which stimulates the parent to regurgitate food (**Figure 52.2A**). Experimenters investigated what stimulated the chicks to peck their parents' bills. Models of gull heads of different shapes and colors were tested (**Figure 52.2B**), as were models of a beak without a head. The results showed that the red dot was necessary for the release of chick pecking behavior. In fact, a pencil with a red eraser elicited a more robust pecking response than an accurate model of a gull head without a red dot.

Ethologists probed the causes of behavior

The ethologists demonstrated the genetic basis for fixed action patterns by interbreeding closely related species. Konrad Lorenz studied the courtship behaviors of different species of dabbling ducks. Some of these species, such as mallards, teals, pintails, and gadwalls, are closely related and can interbreed, but they rarely do so in nature. Each male duck performs a courtship display consisting of a precise series of movements that is typical of his species. A female is not likely to accept him unless the entire display is successfully and correctly completed.

Lorenz crossbred these duck species and found that the hybrid offspring expressed some elements of each parent's courtship display, but in novel combinations. Furthermore, Lorenz observed that hybrids sometimes exhibited display elements that were not in the repertoire of either parent species but were seen in other dabbling duck species. Lorenz's interbreeding studies demonstrated that the stereotypic motor patterns of the courtship displays are inherited. The observation that females were not interested in males performing hybrid displays was evidence that sexual selection had shaped these genetically determined behaviors to be *reproductive isolating mechanisms*.

(A) *Larus marinus*

(B)

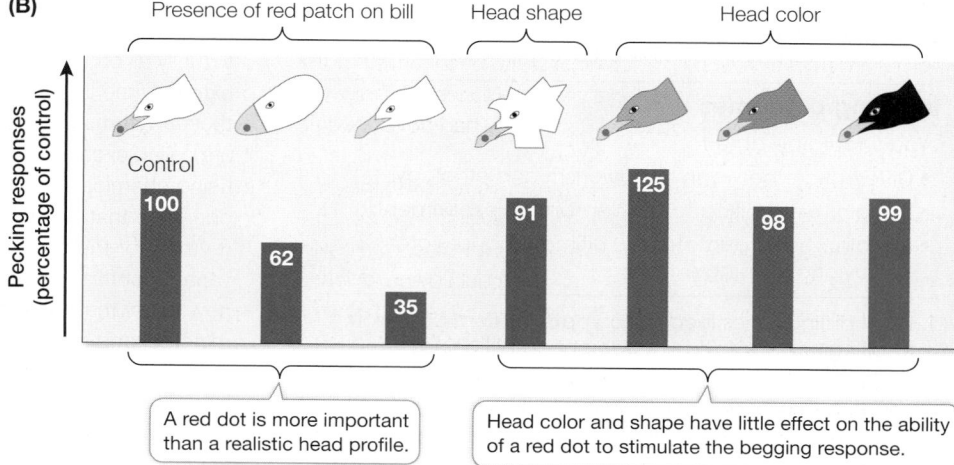

Figure 52.2 Releasing a Fixed Action Pattern   (A) In many gull species, chicks instinctively peck at the red dot on the parent's lower bill, a behavior that induces the parent to regurgitate food into the chick's mouth. **(B)** Tinbergen's work showed that the red dot on the parent's lower bill is the critical component that releases the pecking response.

Q: Why do you think sign stimuli are rather simple components of all of the sensory information available?

*connect the concepts The evolution of populations into different species occurs when hybrids of the two populations have lower fitness. In those cases, selection favors mechanisms that prevent hybridization, and behavioral reproductive isolating mechanisms are the most common, as described in Key Concept 23.4.

The ethologists laid the foundation for the application of modern biological methods to the study of animal behavior. Tinbergen outlined the challenges for investigators as four questions:

1. *Causation*: What is the mechanism underlying the behavior, and how can the relationship between mechanisms and behavior be modified by learning?

2. *Development*: What experiences are necessary for a behavior to be displayed, and how does the behavior change with age?

3. *Function*: How does the behavior affect the animal's chances for survival and reproduction?

4. *Evolution*: How does the behavior compare with similar behaviors in related species, and how might it have evolved?

The first two questions refer to the **proximate causes** of behavior: the immediate genetic, physiological, neurological, and developmental mechanisms that determine how an individual is behaving at a particular time. The third and fourth questions refer to the **ultimate causes** of behavior: the evolutionary processes that produced the animal's capacity and tendency to behave in particular ways. In the sections that follow, we will describe many experiments on animal behavior. For each one, ask yourself which of Tinbergen's four questions it addresses and whether it focuses on proximate or ultimate causes of behavior.

52.1 recap

Early scientific studies of animal behavior took two approaches. Behaviorists focused on the study of conditioned behavior in a few species of laboratory animals and asked questions about learning. Ethologists studied genetically determined behavior in many species in their natural environments and asked evolutionary questions.

learning outcomes

You should be able to:

- Differentiate between behaviorism and ethology.
- Explain the significance of deprivation experiments.
- Explain what proximate and ultimate causes of behavior are, using examples.

1. What distinguishes fixed action patterns from behaviors resulting from Pavlovian or operant conditioning?

2. What variables might account for a genetically determined behavior not being expressed in a deprivation experiment?

3. What are proximate and ultimate causes of mating-display behavior?

The work of the ethologists left no doubt that behavior can be genetically determined, but how? Genes code for proteins, whereas behaviors are highly complex traits involving sensory input and intricate patterns of control over responses to that input. Is it reasonable to think that a single gene can have a specific effect on a behavior?

key concept 52.2 Behavior Can Be Genetically Determined

The work of the ethologists showed the strong genetic basis for most animal behaviors. Of course, *most* animals are insects and other invertebrates, and even among the vertebrates *most* are not mammals, let alone primates. So even though the controversy over the relative influences of nature versus nurture in human behavior has raged for many years, the dominant role played by genes in shaping behavior is amply documented by research on other species. Genetically determined complex behaviors such as the web spinning of spiders or the mating behaviors of dabbling ducks mentioned above must surely depend on many genes. Yet with the rapid development of genomic analytical technologies, comparisons of genomes of animals with different behavioral phenotypes are producing many examples of single gene alterations that produce behavioral differences.

focus your learning

- Even single genes can influence complex behaviors.
- Breeding and gene knockout experiments reveal genetic determinants of behavior.

Single gene mutations can alter behavioral phenotypes

Are you a morning person or a night person? Some people are early risers and function best in the morning, and others have their best work hours late at night. It is possible that this big difference in the behavior of individuals could be due to differences in a single gene. This story begins with fruit flies, *Drosophila*. Fruit flies, like virtually all organisms, have daily rhythms that continue even if all external time cues are taken away. But without external time cues, these rhythms do not span exactly 24 hours, which is why they are called circadian (*circa*, "about," + *dies*, " a day"). Investigators using chemicals to randomly mutate genes in *Drosophila* found one gene that controlled the duration, or period, of the circadian rhythms of the flies. This gene was named *per* for period, and variants of this gene caused short or long circadian periods. We now know that there are many more genes involved in circadian rhythms, and that these genes are highly conserved in organisms ranging from *Drosophila* to humans. One variant of the *per* gene in humans results in familial advanced phase sleep syndrome. The affected individuals have short circadian rhythms and tend to go to sleep and wake up earlier than others.

That single genes can influence even complex behaviors should not be surprising. Many behaviors depend on a particular signaling molecule or a particular receptor molecule, and a change in the gene for that signal or receptor could therefore dramatically change the behavior. For example, mutation in a gene for an olfactory receptor

(A) The mouse VNO

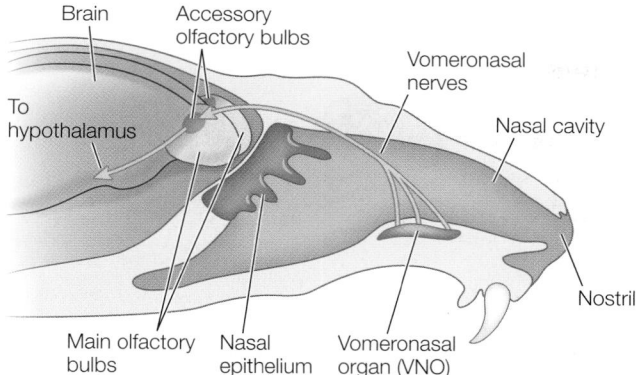

(B)

Figure 52.3 **The Mouse Vomeronasal Organ Identifies Sex**
(A) The mouse VNO is located adjacent to the nasal passages. It contains pheromone receptors whose input travels to a specific region of the olfactory bulb (the accessory olfactory bulb). Projections from the accessory olfactory bulb go only to areas of the brain involved in sexual behavior, whereas projection from the main olfactory bulb go to many different areas of the brain (not shown). **(B)** In male mice, information from the VNO is crucial in identifying gender and thus a potential sexual partner.

could result in the ability to sense a different odorant. Even changes in the expression pattern of a gene can alter a complex behavior. As we discussed at the beginning of Chapter 7, prairie voles are highly monogamous, and both sexes give much parental care to their young. Montane voles do not form pair bonds, are promiscuous, and provide a shorter period of maternal care than prairie voles. The difference between these two species lies in the expression of the receptors for the neuropeptides vasopressin (males) and oxytocin (females): there are far fewer receptors for these neuropeptides in the brains of montane voles. We'll return to the social systems of voles in Key Concept 52.6.

Knockout experiments reveal the roles of specific genes

As you may recall from Chapter 18, a knockout experiment is a genetic technique in which one of an organism's genes is made inoperable ("knocked out") to see what effect its elimination has on the organism's phenotype. Knocking out genes involved in sensory pathways can have pronounced effects on behavior. One example is a gene for a specific olfactory receptor in mice.

As was discussed in Key Concept 45.2, mice have two olfactory organs: the nasal olfactory epithelium common to all mammals, and a small organ in the nasal passages called the vomeronasal organ, or VNO (**Figure 52.3**). Catherine Dulac at Harvard University discovered that pheromone receptors were expressed in that organ. (Pheromones are signaling molecules that are released into the environment and are used for communication between individuals of the same species.) Dulac hypothesized that when sex pheromones produced by female mice bind to the receptors in the male's VNO, they stimulate mating behavior. To test this hypothesis, Dulac created a genetically engineered male mouse in which a gene for VNO receptor signaling was knocked out. Contrary to

the hypothesis, the knockout males in fact did pursue and mate with females placed in their cages. However, they also pursued and tried to mate with *males* placed in their cages. Normally a male mouse reacts aggressively to a strange male, but the knockout male could not discriminate between males and females placed in his cage. Thus properly functioning VNO receptors appear to be essential not for sexual attraction, but for sex identification. You can imagine how selection working on this one gene could modify the intensity of male–male aggression and lead to changes in social behavior.

Gene cascades can control complex behavioral phenotypes

Male courtship behavior in the fruit fly (*Drosophila*) is stereotypic, species-specific, and requires no learning—a classic fixed action pattern. When a male encounters a potential mate, he follows her, taps her body with his foreleg, extends and vibrates one wing, and licks her genitals (**Figure 52.4A**). The development of this complex male behavior is under the control of a single gene, called *fruitless* (*fru*), and the development of male anatomy is under the control of another gene, called *doublesex* (*dsx*). In both males and females, these two genes are part of gene expression cascades that differ in the splicing of the mRNA in the two sexes. Those splicing differences result in different *dsx* and *fru* gene products in males and females (**Figure 52.4B**). The female version of the Dsx protein controls the development of female anatomy, and the expression of *fru* in the male nervous system results in the organization of the neural circuitry controlling male sexual behavior.

There are two take-home lessons from this example. First, genes that control aspects of behavior, like other genes, are generally embedded in gene cascades that offer opportunities for simple genetic changes to alter the phenotype of even complex behaviors. Second, certain genes, such as *dsx* and *fru*, influence a range of other genes that contribute to complex behaviors. Modifications in any one of those genes or its expression can alter behavior. Thus even though no behavior is coded for by a single gene, alterations in single genes can influence behavior in ways that affect an animal's fitness.

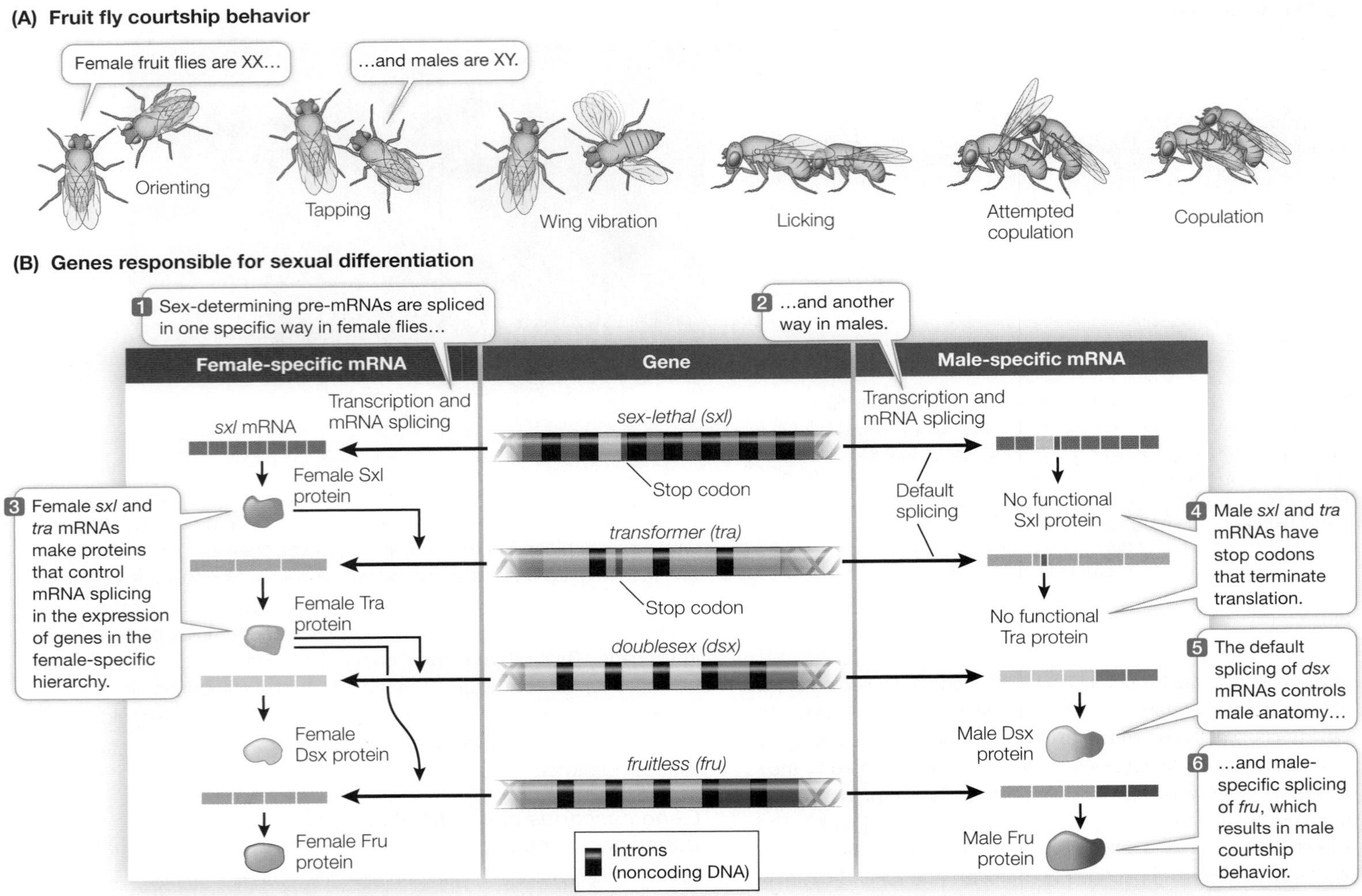

(A) Fruit fly courtship behavior

Female fruit flies are XX...

...and males are XY.

Orienting

Tapping

Wing vibration

Licking

Attempted copulation

Copulation

(B) Genes responsible for sexual differentiation

1 Sex-determining pre-mRNAs are spliced in one specific way in female flies...

2 ...and another way in males.

Female-specific mRNA	Gene	Male-specific mRNA

Transcription and mRNA splicing

sxl mRNA

Transcription and mRNA splicing

sex-lethal (sxl)

Stop codon

Female Sxl protein

3 Female *sxl* and *tra* mRNAs make proteins that control mRNA splicing in the expression of genes in the female-specific hierarchy.

Default splicing

No functional Sxl protein

4 Male *sxl* and *tra* mRNAs have stop codons that terminate translation.

transformer (tra)

Stop codon

Female Tra protein

No functional Tra protein

5 The default splicing of *dsx* mRNAs controls male anatomy...

doublesex (dsx)

Female Dsx protein

Male Dsx protein

6 ...and male-specific splicing of *fru*, which results in male courtship behavior.

fruitless (fru)

Introns (noncoding DNA)

Female Fru protein

Male Fru protein

Figure 52.4 The *fruitless* Gene **(A)** Male fruit flies display stereotypic, species-specific courtship behavior. **(B)** Sexual differentiation in *Drosophila* is controlled by a cascade of genes, including the *fru* gene, whose expression results in male sexual behavior.

52.2 recap

There are strong genetic determinants of behavior in most species. Gene manipulations such as mutations and knockouts show that a single gene can play a large role in the development and expression of a behavior. Although most behaviors are controlled by multiple genes, those genes frequently function in signaling cascades, so a single gene modification can influence even a complex behavior controlled by many genes.

learning outcomes

You should be able to:

- Give examples of a single gene influencing a behavior.
- Describe different kinds of evidence supporting the claim that single genes can influence behavior.
- Explain the evolutionary benefit of a gene expression cascade that influences behavior.

52.2 recap

1. A method has recently been developed to increase the expression of oxytocin and vasopressin genes in montane voles. What difference might that treatment make in the behavior of these animals? If the expected result is not seen, what could be the reason?

2. If you interbred two individuals from separate populations that bred true for a trait that differed between the two, and all of the offspring resembled one of the parents, what would you conclude about the genetic basis for that behavior?

3. What could be an evolutionary reason for a species-specific mating behavior being controlled by a gene expression cascade, such as in *Drosophila*?

How can the genetic cascades that underlie complex behaviors be programmed to respond selectively to specific sets of stimuli? How can their expression be limited to appropriate times in an animal's life? The answers to these questions can be found by studying how behaviors develop over the life span.

52.3 Behavior Can Be Studied Developmentally

The emergence of behavior as an animal develops and matures depends on the development of the nervous system as well as on the growth and maturation of other body systems. But even with anatomical and physiological competence, specific behaviors may not be expressed. Behaviors that are adaptive at one stage in an animal's life may not be adaptive at other stages. Behaviors typical of juvenile animals, such as begging for food, may disappear and new behavior patterns of a mature individual, such as courtship displays, appear.

focus your learning

- Hormones influence the development of certain behaviors and their expression at a later time.
- Complex behaviors can involve both genetic determinants and learning.

Hormones can determine behavioral potential and timing

Hormones can determine the development of a behavioral potential at an early age and the expression of that behavior at a later age. An excellent example of this is sexual behavior in rats (**Figure 52.5**). Normally, adult male and female rats exhibit different patterns of sexual behavior: females adopt a sexually receptive posture, called lordosis, in the presence of males, and males copulate with receptive females. Neither sex, however, expresses these behaviors until the animals have reached adulthood. Experiments in which newborn and adult rats were neutered (to remove the influence of sex steroids naturally produced by their gonads) and artificially treated with hormones led to the following conclusions:

- Development of male sexual behavior requires the brain of the newborn rat be exposed to testosterone. Female sexual behavior development does not require exposure to estrogen.

- Testosterone masculinizes the nervous systems of both genetic males and genetic females.

- Exposure to sex steroids in adulthood is necessary for the expression of sexual behavior, but testosterone produces male sexual behavior only in adult rats whose brains were masculinized when they were newborns, and estrogen produces female sexual behavior only in adult rats whose brains were not masculinized when they were newborns.

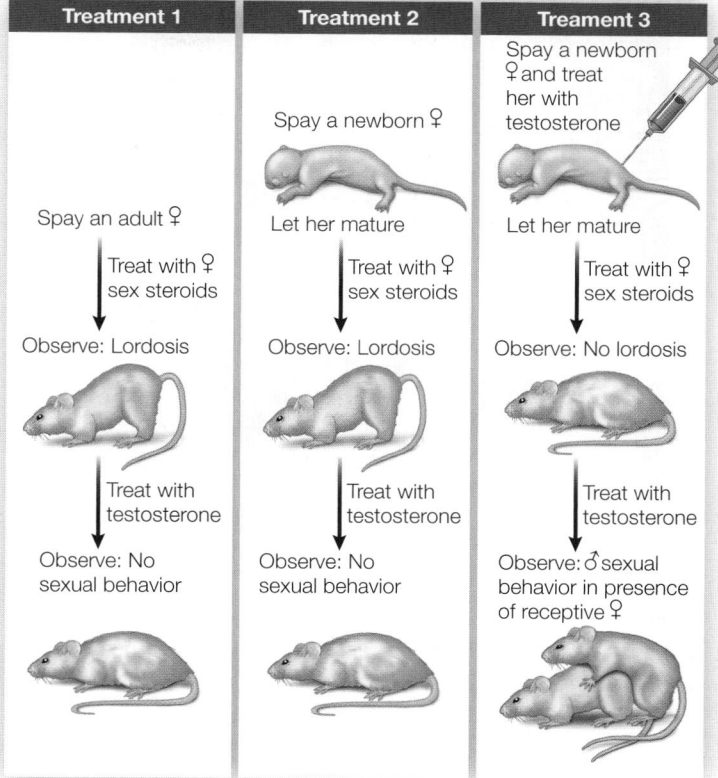

(A) Sex hormone manipulations of female rats

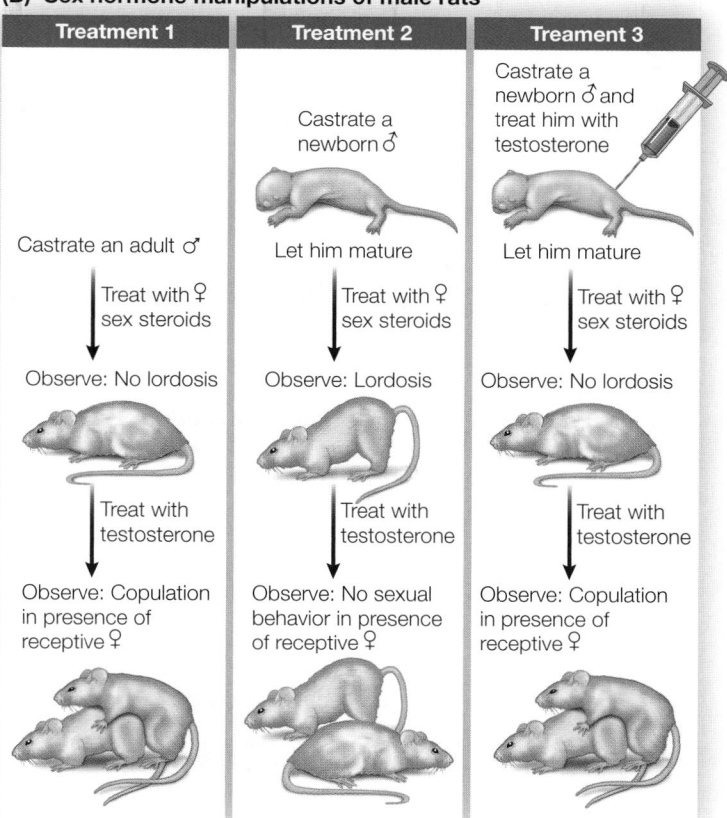

(B) Sex hormone manipulations of male rats

Figure 52.5 Hormonal Control of Sexual Behavior Experimental hormone treatments of rats demonstrated that the sex steroids present during early development determine which sexual behavior patterns develop, whereas the sex steroids present in adulthood control the expression of those behavior patterns. The elimination of sex steroids during early postnatal development reveals that the default pattern of behavioral development is female.

(A) *Anser anser*

(B) *Aptenodytes forsteri*

Figure 52.6 Imprinting Helps Parents and Offspring Recognize Each Other   **(A)** Greylag geese that imprinted on Konrad Lorenz as hatchlings followed him everywhere he went. **(B)** Imprinting allows a male emperor penguin to find his own chick among many others.

Thus the sex steroids that are present at birth determine which pattern of behavior develops, and the sex steroids that are present in adulthood determine when that pattern is expressed.

Some behaviors can be acquired only at certain times

Responsiveness to simple releasers is sufficient for certain behaviors such as begging behavior in gull chicks, but more complex information that cannot be genetically programmed is required for other behaviors. An example is parent–offspring recognition. When animals live in close proximity to other individuals, as in a herd or a nesting colony, it is important for parent and offspring to learn each other's identity soon after birth so they will be able to find each other in a crowded situation. In many such cases, a parent–offspring bond is formed by **imprinting**. What characterizes imprinting is that an animal learns a specific set of stimuli during a limited time called a **critical period**, also known as a **sensitive period**.

Konrad Lorenz demonstrated that young greylag geese (*Anser anser*) imprint on their parents between 12 and 16 hours after hatching. By positioning himself to be present during this critical period, Lorenz succeeded in imprinting goslings on himself. The imprinted goslings followed him around as if he were their parent (**Figure 52.6A**). In a subsequent experiment his assistants wore boots with different patterns on them. The goslings imprinted on the boots, and even in a situation that mixed different groups of goslings, they always sorted themselves out by following their "parental" boots.

Imprinting requires only a brief exposure, but its effects are strong and long-lasting. Emperor penguins (*Aptenodytes forsteri*) reproduce during the coldest, darkest time of year in Antarctica. The parents walk up to 150 kilometers inland to form a dense colony, where the female lays her egg. She then walks back to the ocean to feed while her mate incubates the egg. By the time she returns, the chick has hatched. She then takes over its care and feeding, and the father walks back to the ocean to feed. Generally he is away so long that the mother must leave to find food to avoid starvation. Thus after being away for weeks, the father must find his chick in a crowded, milling colony of chicks, all calling for their parents (**Figure 52.6B**). Yet he can unerringly locate his own offspring by recognizing its call, which he imprinted on before he left to feed.

The critical or sensitive period for imprinting may be determined by a brief hormonal state. For example, if a mother goat does not nuzzle and lick her newborn within 10 minutes after its birth, she will not recognize it as her own offspring later. For goats, the sensitive period is associated with peaking levels of the hormone oxytocin in the mother's circulatory system at the time she gives birth and at the same time she is sensing the olfactory cues emanating from her newborn kid. A female goat rendered incapable of smelling before giving birth is unable to differentiate between her own kid and other kids after giving birth.

Birdsong learning involves genetics, imprinting, development, and social interactions

The study of white-crowned sparrow singing at the beginning of this chapter is a classic case of how genetics, physiology, and experience come together in the development and expression of behavior. Genetics endows the birds with a neural template that must be matched by experience during early development. That type of learning—imprinting—is limited to a sensitive or critical period by processes of developmental physiology. Months later, the onset of sexual maturation motivates the birds to sing, initiating a second phase of learning that requires the birds to use auditory experience to develop the motor patterns required for the full expression of species-specific song (**Figure 52.7**). At the end of this learning phase, the song becomes permanent, or crystalized. Thus the overall process of transmission of a complex species-specific behavior necessary for reproductive success is well adapted to preserving constancy of the behavior. But variation is an important component of evolution.

Variation in birdsong is evident in nature. There are subtle differences between the songs of any two males, including those in neighboring territories. Regionally, white-crowned sparrows have dialects, and a bird migrating into an area that has a different dialect can adopt that song pattern. There are even observations of white-crowned sparrows singing the songs of closely related species. In an interesting laboratory experiment, white-crowned sparrows were raised in isolation but exposed to the sight and song of a male of a closely related species while recordings of white-crowned sparrow song played on speakers. When those birds matured, they sang the song of the closely related species. The ability to incorporate variation into an individual bird's song may relate to social interactions such as fitting into a regional norm, displaying individual qualities to a potential mate, and maintaining a bond between mates.

To investigate questions about the adaptive significance of variations in the male singing behavior, it is necessary to take female choice into consideration. Sarah Woolley and Allison Doupe at the

experiment

Original Paper: Woolley, S. C. and A. J. Doupe. 2008. Social context-induced song variation affects female behavior and gene expression. *PLoS Biology* 6(3): e62.

When male zebra finches (*Taeniopygia guttata*) sing alone, they improvise (undirected song), but when they sing in the presence of a female, they sing a stereotyped (directed) song. Woolley and Doupe asked whether the quality of the song influenced the behavior of the female.

HYPOTHESIS▶ Female zebra finches have a preference for directed over undirected song, and can identify the song of their mates.

METHOD

1. Record males singing alone and singing in the presence of their mates.

2. Place a female in the center chamber of a three-chamber apparatus; the female should be able to move between the chambers. Play recordings of her mate's directed song in one end-chamber and of her mate's undirected song in the other end-chamber.

3. Record the amount of time the female spends in the different chambers.

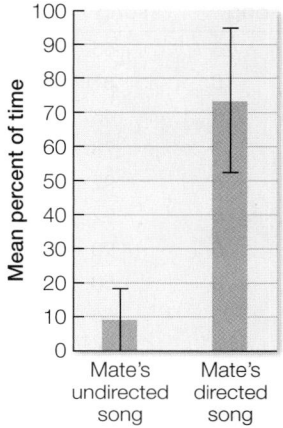

Recording of mate's undirected song

Recording of mate's directed song

The female zebra finch can choose either compartment.

RESULTS

Females spent a higher percentage of time in the chamber in which their mate's directed song was played. The data show results for 17 females with means and standard deviations. The *P*-value from a paired *t*-test is <0.001.

CONCLUSION▶ Females can discriminate between their mate's directed and undirected songs and are more attracted to the directed song.

investigatinglife work with the data
follows on next page.

[Bar chart: Mean percent of time (y-axis, 0 to 100). Mate's undirected song ≈ 10; Mate's directed song ≈ 73]

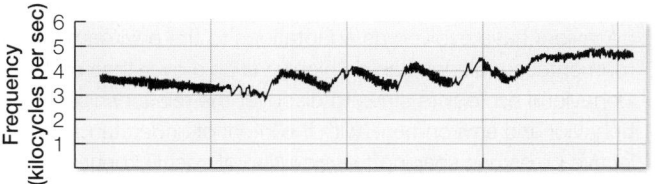

A) Control or wild bird

[Sonogram: Frequency (kilocycles per sec), 1–6]

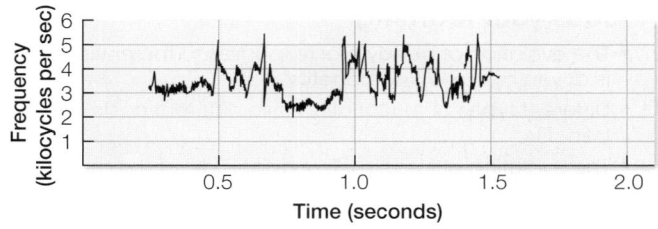

B) Isolated hand-reared bird

[Sonogram: Frequency (kilocycles per sec), 1–6]

C) Deafened bird

[Sonogram: Frequency (kilocycles per sec), 1–6]

Time (seconds): 0.5, 1.0, 1.5, 2.0

University of California, San Francisco did that with zebra finches (*Taeniopygia guttata*), a species that has become a valuable model system for studying the neurobiology of birdsong. When male zebra finches sing in isolation, their song is variable ("undirected"), almost as if they are improvising and trying out new motifs. In the presence of females, however, they sing "directed" song (directed to a female) that is less variable. Woolley and Doupe conducted experiments to see which songs female zebra finches preferred (**Investigating Life: Practice Makes Perfect**). The females preferred directed song, and if mated, they preferred the directed song of their mate. Thus the ability to incorporate variations into the species-specific song may be important to establish individuality and influence mate choice, but the nature of the singing depends on the social context.

Figure 52.7 Sensitive Periods for Song Learning **(A)** Sonogram showing the species-specific song of an adult male white-crowned sparrow (*Zonotrichia leucophrys*). **(B)** Song of an adult male raised in isolation (never having heard the song as a nestling). **(C)** Song of an adult male that heard the song as a nestling but was deafened prior to ever singing himself. Experiments showed that the bird must first acquire a song memory by hearing the song as a nestling, and must then be able to hear himself as he attempts to match his singing to that song memory.

work with the data

Since the experiment showed that females could distinguish between the directed and undirected songs of their mates, the next question was whether they could distinguish between the directed song of their mates and the directed song of other males. A second experiment used the same protocol as did the first one, but the two songs that were played to each female were the directed song of her mate and the directed song of an unfamiliar male. The results are shown in the table.

Percent time spent close to:	
Unfamiliar directed song	Mate's directed song
25	55
15	81
8	43
7	78
6	92
6	92
5	90
5	50
5	86
4	85
4	84
3	83
3	81
3	80
2	73
1	72
1	97
0	98

QUESTIONS▶

1. If you hypothesized that the female's preference for her mate's directed song was a factor in pair bonding rather than just a mating stimulus, what data might you collect to test this hypothesis?

2. Plot the data as bar graphs. Show the means and standard deviations, and do a paired *t*-test to determine the significance level. Can females distinguish their mate's directed song from the directed song of an unfamiliar male?

3. What do these results suggest about the mating system of the zebra finch?

4. What hypotheses can you suggest about the function of undirected song?

A similar **work with the data** exercise may be assigned in **LaunchPad**.

The timing and expression of birdsong are under hormonal control

Both male and female songbirds hear their species-specific song as nestlings, but only the males of many species sing as adults, and most do so only in spring. Hormones underlie both the difference in song expression between male and female songbirds and the timing of song expression. When investigators injected adult female songbirds with testosterone in spring, those females sang their species-specific song just as males did. Females as well as males form a memory of their species-specific song when they are nestlings, and they have the physical capacity to sing, but under normal circumstances they lack the necessary hormonal stimulation.

How does testosterone cause a songbird to sing? Each spring an increase in circulating testosterone levels causes certain parts of the male's brain necessary for learning and developing song to grow larger. Individual neurons in those regions of the brain increase in size and grow longer extensions, and the number of neurons in those regions increases. Thus hormones can control behavior by changing brain structure as well as brain function, both developmentally and in response to environmental cues.

▶52.3 recap

Genetics can determine what behaviors can be learned, and hormones can determine the sensitive period when they can be learned as well as the motivational states that trigger their expression. Imprinting ensures the ability of individuals to recognize other individuals or species at a later time. Some capacity for variation of genetically determined and imprinted behaviors enables behavior to be shaped by social context.

learning outcomes

You should be able to:

- Explain how hormones can affect the timing and expression of behavior.
- Describe the evidence supporting the claim that genetics are important in bird singing behavior.
- Explain the evolutionary benefit of variability in the singing behavior of a bird species.

1. Why is female considered to be the default pathway for the development of sexual behavior in rats?

2. Explain why a songbird deafened in adulthood can sing its song, but a songbird deafened before reaching sexual maturity cannot.

3. Under what conditions might variability in a species-specific behavior be adaptive or not be adaptive?

Complex behaviors are the product of interactions of genetic, physiological, and environmental factors. Many genes are involved in shaping behavior, and therefore there are multiple opportunities for selection to favor behavioral modifications. Questions about how changes in behavior adapt animals to environmental conditions are the province of an evolution-based field called **behavioral ecology**.

key concept 52.4 Selective Pressures Shape Behavior

Most of the behaviors we have discussed so far involve interactions between individuals. The individual also interacts with its environment, and many behaviors have evolved as adaptations to the environment. Environmental conditions can be highly variable over time and space. Behavioral ecologists strive to discover the relationships between behavior and environment, with the intent of understanding the selective pressures shaping the evolution of specific behaviors. Here we offer just a few examples.

focus your learning

- The evolution of behavioral responses to the environment is driven by costs and benefits.
- Different types of territoriality have different costs and benefits.
- The evolution of foraging behavior is driven by costs and benefits.

Animals are faced with many choices

Animal behavior is largely a sequence of choices: where and when to move, where to build a nest, what to eat, when to fight and when to flee, with whom to associate, with whom to mate. Making wrong choices reduces fitness. Behavioral ecologists seek to discover what information animals use to make behavioral choices and how that information relates to the environmental conditions that influence fitness.

Where an animal lives is referred to as its **habitat**. In most cases the habitat provides not only a protected nest site, but also food and access to mates. The environmental cues animals use to select their habitat may be quite simple. For example, seabirds select cliffs or offshore rocks for nesting, and both of those sites offer protection from predators. Animals with very specialized food requirements select habitats where those foods are abundant. The general hypothesis that guides behavioral ecologists is that the cues animals use to select habitats are reliable predictors of conditions suitable for future survival and reproduction.

For many species, the presence of **conspecifics**—other members of the same species—can be a valuable cue. Observing conspecifics can provide animals with information about the quality of a habitat. After all, you can't argue with success. During the breeding season, European collared flycatchers (*Ficedula albicollis*) are nosy neighbors, regularly visiting the nests of conspecifics. Researchers hypothesized that this behavior allows the flycatchers to assess the quality of the habitat by seeing how well their neighbors are faring. To test this hypothesis, they created some areas with supersized broods—normally an indication of abundant food—by taking young birds from some nests and adding them to nests in another area. The next year, flycatchers preferentially settled in the areas where broods had been artificially enlarged.

Behaviors have costs and benefits

A cost–benefit approach can be used to investigate the relationships between behavior, environment, and fitness. A **cost–benefit approach** assumes that an animal has only a limited amount of time and energy, and therefore cannot afford to engage in behaviors that cost more to perform than they return in benefits. A cost–benefit approach provides a framework that can be used to construct hypotheses and design experiments to investigate why behavior patterns evolve as they do.

The benefits of a behavior are measured in terms of the enhancement in fitness an animal accrues by performing the behavior. The cost of a behavior typically has three components:

1. **Energetic cost** is the energy the animal expends performing the behavior.
2. **Risk cost** is the increased chance of being injured or killed.
3. **Opportunity cost** is the benefit the animal forgoes by not being able to perform other behaviors at the same time.

Territorial behavior carries significant costs

Territorial behavior is a good subject for cost–benefit analysis. Territoriality is aggressive behavior that actively denies other animals access to a habitat or resource. Optimal habitats and resources may be in short supply, so conspecifics have to compete for them. Many animals—usually males—defend all-purpose territories that provide a nest site, food, and access to mates. The territory holder stakes out his boundaries by engaging in aggressive interactions with neighbors, and must then patrol those boundaries constantly and respond to trespassers. These aggressive interactions usually consist of highly stereotypic, species-specific displays such as birdsong. Through territorial behavior, the male obtains the resources he needs for reproductive success, but he also pays a price.

Territorial displays require considerable expenditure of energy, they make a male more vulnerable to predation, and they detract from the time he has for feeding or engaging in parental behavior. Michael Moore and Catherine Marler at Arizona State University performed an experiment to estimate the costs incurred by male Yarrow's spiny lizards (*Sceloporus jarrovii*) when defending a territory. These lizards defend territories that include the home ranges of several females. Their territorial behavior is normally most intense during September and October when circulating testosterone levels of the males are high and the females are most receptive to mating. The researchers varied the intensity of the lizards' territorial behavior by implanting testosterone capsules in some males in summer, when they are not normally highly territorial (**Figure 52.8**).

Testosterone-treated males spent more time patrolling their territories, performed more displays, and expended about one-third more energy than control males (energetic cost). They had less time to feed (opportunity cost), captured fewer insects, stored less energy, and had a higher death rate (risk costs). In summer, when females are not normally receptive, these high costs of vigorous territorial defense outweigh the reproductive benefits of territoriality. Thus natural selection has favored seasonal variation in the level of the hormone controlling territorial behavior in this species.

 Animation 52.1 **The Costs of Defending a Territory**
www.Life11e.com/a52.1

The cost–benefit approach explains the diversity of territorial behaviors seen in different species. Even if a resource is essential to an animal, if it cannot be defended economically, the animal will not engage in territorial behavior. Food is essential for all animals, but if the food is widely distributed in space or fluctuating in availability, there is no benefit to balance the high costs of trying to defend it. For example, the open ocean where seabirds feed cannot be defended. But safe nest sites on islands or rocky cliffs are in short supply, and they can be defended. The territories of seabirds may be no larger than the distance the birds can reach while sitting on their nests (**Figure 52.9A**).

In some cases the resource that is defended is the female herself. Elephant seals spend most of their lives at sea, but females come to land at traditional beach sites to give birth to their pups. Male elephant seals arrive at these sites ahead of time and stake out territories through vigorous fighting (**Figure 52.9B**). When the females arrive on the beaches, they have to enter the territories of the males. As long as the male territory holder can fend off challengers, he will be able to mate with all the females using his piece of the beach.

One unusual form of male territorial behavior arises in situations in which neither food, nest sites, nor females are defended. A **lek** is an area where males gather for the purpose of engaging in intense

experiment

Figure 52.8   The Costs of Defending a Territory

Original Paper: Marler, C. A. and M. C. Moore. 1988. Evolutionary costs of aggression revealed by testosterone manipulations in free-living male lizards. *Behavioral Ecology and Sociobiology* 23: 21–26.

By using testosterone implants to increase territorial behavior, Michael Moore and Catherine Marler measured the costs to male Yarrow's spiny lizards (*Sceloporus jarrovii*) of defending a territory during the summer, when they do not normally do so.

HYPOTHESIS▶   Yarrow's spiny lizards do not defend a territory during summer because the energetic costs of territorial behavior in that season outweigh the benefits.

METHOD

1. During the summer, when female lizards are not sexually receptive, insert testosterone capsules under the skin of some males; other males received empty capsules to serve as controls.

2. Observe the patterns of territorial behavior and the survival rate of the two groups of males.

RESULTS

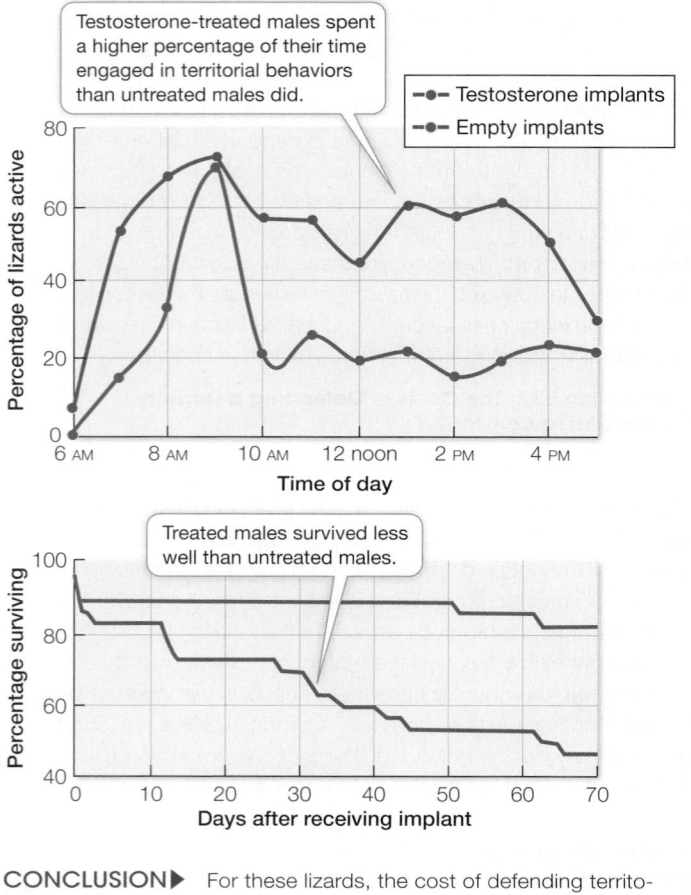

Testosterone-treated males spent a higher percentage of their time engaged in territorial behaviors than untreated males did.

Legend:
–•– Testosterone implants
–•– Empty implants

Y-axis: Percentage of lizards active (0, 20, 40, 60, 80)
X-axis: Time of day (6 AM, 8 AM, 10 AM, 12 noon, 2 PM, 4 PM)

Treated males survived less well than untreated males.

Y-axis: Percentage surviving (40, 60, 80, 100)
X-axis: Days after receiving implant (0, 10, 20, 30, 40, 50, 60, 70)

CONCLUSION▶   For these lizards, the cost of defending territories during summer significantly reduces their survival rate.

displays of their territorial prowess aimed at impressing females and winning the opportunity to mate. Even though space is not limited, each male defends a small piece of real estate on which he performs a display (**Figure 52.9C**). Those territories closest to the

(A) *Thalassarche melanophris*

(B) *Mirounga angustirostris*

(C) *Centrocercus urophasianus*

Figure 52.9 Animals Defend Territories of Different Sizes
(A) The nesting territories of many seabirds consist of only as much space as the birds can defend without leaving the nest. **(B)** Male elephant seals fight vigorously to defend areas of beach where females haul out of the water to give birth to their pups. **(C)** Male greater sage-grouse gather at a lek in Colorado to perform displays aimed at impressing females and winning the opportunity to mate.

center of the lek are the prime sites, and males compete intensely for those locations. The females visit the lek, observe the males, and generally mate with the males holding the prime sites. The benefit of this system to the female is that she is inseminated by a successful competitor, and therefore her offspring will carry the genes that contributed to his success. This is another example of sexual selection (see Key Concept 20.2). The costs of lekking to males are high, as they engage in continuous, intense territorial behavior that precludes eating, drinking, and sleeping until they are displaced. The benefit is the chance to maximize their fitness by mating with many females.

Foraging behavior has costs and benefits

Searching for food (foraging) involves many decisions: how much time to spend in each location, what is edible, and which potential foods should be eaten and which should be ignored. A cost–benefit approach to feeding behavior has resulted in an **optimal foraging theory**, which aims at identifying the fitness value of feeding choices. The primary benefit of foraging is nutrition: energy, minerals, and vitamins

(A) *Ara chloropterus*

(B)

Figure 52.10 Herbivores Seek Out Unusual Sources of Minerals
(A) Red-and-green macaws of the Amazon jungle obtain essential
minerals by eating dried clay. **(B)** Heliconian butterflies obtain needed
salts by drinking secretions from the skin and nostrils of a caiman.

(see Key Concept 50.1). The costs of foraging are similar to those of
other behaviors: energy expended, time lost from other activities that
could enhance fitness, and the risk of increased exposure to predators.

 Animation 52.2 **Foraging Behavior**
www.Life11e.com/a52.2

Animals frequently make choices among food items that may
differ not only in terms of energy content, but also abundance or
ease of acquisition and processing. Optimal foraging theory predicts
that in such situations, animals will make choices that will maximize
the rate at which they obtain energy. The more rapidly a foraging
animal satisfies its energetic requirements, the lower the opportunity
costs and risk costs of foraging.

The energy maximization hypothesis considers food items in
terms of the energy they provide, but animals have nutrient requirements in addition to energy that can play a role in shaping their foraging behavior. Essential minerals, for example, are in short supply
in some animals' diets, and those animals may incur large energetic
costs and risks to obtain them (**Figure 52.10**). Some foods may also
be sought for their medicinal value. Chimpanzees, for example,
have been observed eating the pith of the plant *Vernonia amygda-
lina*. The pith contains small quantities of a secondary metabolite
(vernonioside B1) that is toxic to chimps at high concentrations, but
at low concentrations can kill their intestinal parasites. Chimps that
consume this plant material have fewer parasites.

52.4 recap

Many behaviors are shaped by environmental conditions
that influence the decisions animals must make. Decisions can result in benefits that increase reproductive
fitness, but they also involve costs, including energetic
costs, opportunity costs, and risk costs. Cost–benefit analysis can be applied to territorial and foraging behaviors.

learning outcomes

You should be able to:

- Describe how a cost–benefit approach is applied to
  investigating animal behavior.
- Describe different types of territorial behaviors and the
  cost and benefit factors that shape their evolution.
- Explain why aggressive territorial behavior is associated
  with some essential resources, but not others.
- Explain how foraging involves trade-offs between nutrients acquired and the costs of energy required, opportunity loss, and risk.

1. Male redwing blackbirds prefer to build nests in marshes,
   but they feed in upland meadows. What costs and benefits most likely shaped these behaviors?

2. Cichlid fish have dominant males that have colorful markings and defend territories. There are also nondominant
   males that are plain and do not defend territories but
   occasionally dash in when a female is spawning to try to
   fertilize some of her eggs. What are the costs and benefits
   of these two male phenotypes?

3. Elephant seals defend a piece of beach as a mating territory, whereas prairie-chickens defend small pieces of grassland as a mating territory. Male elephant seals are much
   larger than the females and viciously fight with one another
   over their territories. Male prairie-chickens are not much
   larger than females; they have elaborate displays, and they
   don't fight viciously with neighboring males. What selective
   pressures have shaped the behaviors of these two different
   kinds of animals that defend mating territories?

4. Consider a species that has a choice of two potential
   sources of food. One is calorie-rich, but is dispersed in a
   sparse environment. The other is calorie-poor but occurs
   in dense patches in a protected environment. Which
   types of selective pressures would likely shape the food
   preference of this species?

Behavioral ecologists are interested in understanding how the natural environment influences the fitness value of behavioral choices—
the ultimate causes of those behaviors, in Tinbergen's terms. Other
behavioral biologists focus on the physiological mechanisms that
underlie the proximate causes of behavior.

>key concept
52.5
Behavior Can Be Studied Mechanistically

Control of behavior involves the nervous
and endocrine systems. Execution of behavior involves the musculoskeletal system as well as other effector mechanisms, such as those
that produce secretions, color changes, electrical impulses, sound,
and even light. We have already considered many of the physiological
systems that are involved in these processes, including hormones,

reproductive systems, nervous systems, sensory systems, and feeding mechanisms. The field of behavioral physiology, which encompasses aspects of all of these systems, is enormous, so here we will dig deeper into just three different phenomena studied by behavioral physiologists: the timing of behavior, navigation, and communication.

focus your learning

- Animals have genetically determined daily and annual rhythms that enable them to anticipate future conditions.
- Animals can navigate using landmarks or time and direction information, and some use true bicoordinate navigation.
- Animals communicate using light, sound, mechanical, and chemical signals.

Biological rhythms coordinate behavior with environmental cycles

Earth turns on its axis once every 24 hours, generating daily cycles of light and dark, temperature, humidity, and tides. In addition, Earth is tilted on its axis, so the light–dark cycle changes as Earth revolves around the sun. These daily and seasonal cycles profoundly influence the physiology and behavior of animals. Animals tend to be active either during the day (diurnal) or at night (nocturnal) and have sensory capabilities appropriate to this distinction. Therefore it is adaptive to organize behavior on a cycle that corresponds with the environmental cycle of light and dark. Similarly, a behavior that is adaptive at one time of year (such as midsummer) may not be adaptive at another time (midwinter). Thus it is important for animals to organize their behavior with respect to time of the day or year and to be able to anticipate those times.

CIRCADIAN RHYTHMS   Experimental animals kept under constant conditions with no daily time cues such as light and dark, temperature cycles, or episodic availability of food and water still have daily cycles of activities such as locomotor activity, sleeping, eating, drinking, learning, and just about anything else that can be measured. The persistence of these daily cycles in the absence of environmental time cues suggests that animals have an internal clock. As we mentioned above, because these daily cycles are not exactly 24 hours long, they are known as **circadian rhythms**.

Biological rhythm can be viewed as a series of cycles, and the length of one of those cycles is the period of the rhythm. Any point in the cycle is a phase of that cycle. When two rhythms completely match, they are in phase, and if a rhythm is shifted (as in the resetting of a clock), it is phase-advanced or phase-delayed. Because the period of a circadian rhythm is not exactly 24 hours, it must be phase-advanced or phase-delayed each day to remain in phase with the daily cycle of the environment. In other words, the rhythm has to be **entrained** to the environmental cycle of light and dark.

The circadian rhythm of an animal held under constant conditions will run according to its natural period—it will be **free-running**. If the period is less than 24 hours, the animal will begin its activity a little earlier each day (**Focus: Key Figure 52.11**). The period of the free-running circadian rhythm is under genetic control. Different species may have different average periods, and within a species, mutations can lead to different period lengths.

Under natural conditions, environmental time cues, such as the onset of light or dark, entrain the free-running rhythm to the light–dark cycle of the environment. In the laboratory it is possible to entrain the circadian rhythms of free-running animals with short pulses of light or dark administered every 24 hours (see the bottom panel of Figure 52.11).

In mammals, the master circadian "clock" consists of two clusters of neurons just above the optic chiasm (the area of the brain where the optic nerves come together). These structures are called the **suprachiasmatic nuclei** (**SCN**). If they are destroyed, the animal becomes arrhythmic (loses its circadian rhythm) and is just as likely to eat, drink, sleep, or wake at any time of day.

A notable study by Martin Ralph and his colleagues, then at the University of Virginia, demonstrated that the SCN is the source of circadian rhythms. When the SCNs of adult hamsters with typical 24-hour rhythms were destroyed, the animals became arrhythmic. After several weeks of this arrhythmic behavior, the researchers transplanted SCN tissue from hamster fetuses bred for an atypical (mutant) short-day rhythmicity into the original hamsters' brains. The experiment produced two remarkable results (**Figure 52.12**). First, circadian rhythms were restored by the transplanted SCN tissue, demonstrating that the SCN is sufficient to generate circadian rhythms—a unique case of a behavior being restored by a neural transplant. Second, the restored circadian rhythms had the period length of the *donor* strain, demonstrating that the specific phenotype of the behavior was a property of the donor neural tissue, and thus wholly generated by the SCN.

The molecular mechanism of the circadian clock involves negative feedback loops. Although there are several genes involved, including the *per* gene discussed in Key Concept 52.2, we can generalize about the mechanism by saying that when certain "clock genes" are expressed in SCN cells, the mRNA enters the cytoplasm, where it is translated. The resulting proteins combine, and the dimer returns to the nucleus as a transcription factor that shuts off the expression of the clock genes until the transcription factor is degraded. The period of this cycle is about a day. These findings show that it is possible to understand circadian rhythms of behavior at all levels, from the molecular rhythm generators to the environmental stimuli that entrain them to the daily cycle of light and dark.

CIRCANNUAL RHYTHMS   Seasonal changes in the environment present challenges to many species. Most animals reproduce most successfully if they time their reproductive behavior to coincide with the most favorable time of year for the survival of their offspring. Many species require considerable advance preparation for reproduction. Migratory animals must arrive on their breeding grounds at the right time, and animals that have specialized structures used in mating displays, such as the antlers of deer, moose, and caribou, must grow these structures before the breeding season arrives.

For many species, a change in day length—the *photoperiod*—is a reliable indicator of seasonal changes to come. For others, however, change in day length is not a reliable seasonal cue. Hibernators, for example, spend long months in dark burrows underground but must be physiologically prepared to breed almost as soon as they emerge in the spring. A bird overwintering near the equator cannot use changes in photoperiod as a cue to time its migration to its temperate-zone breeding grounds. When held under constant laboratory conditions, such animals show endogenous

focus: key figure

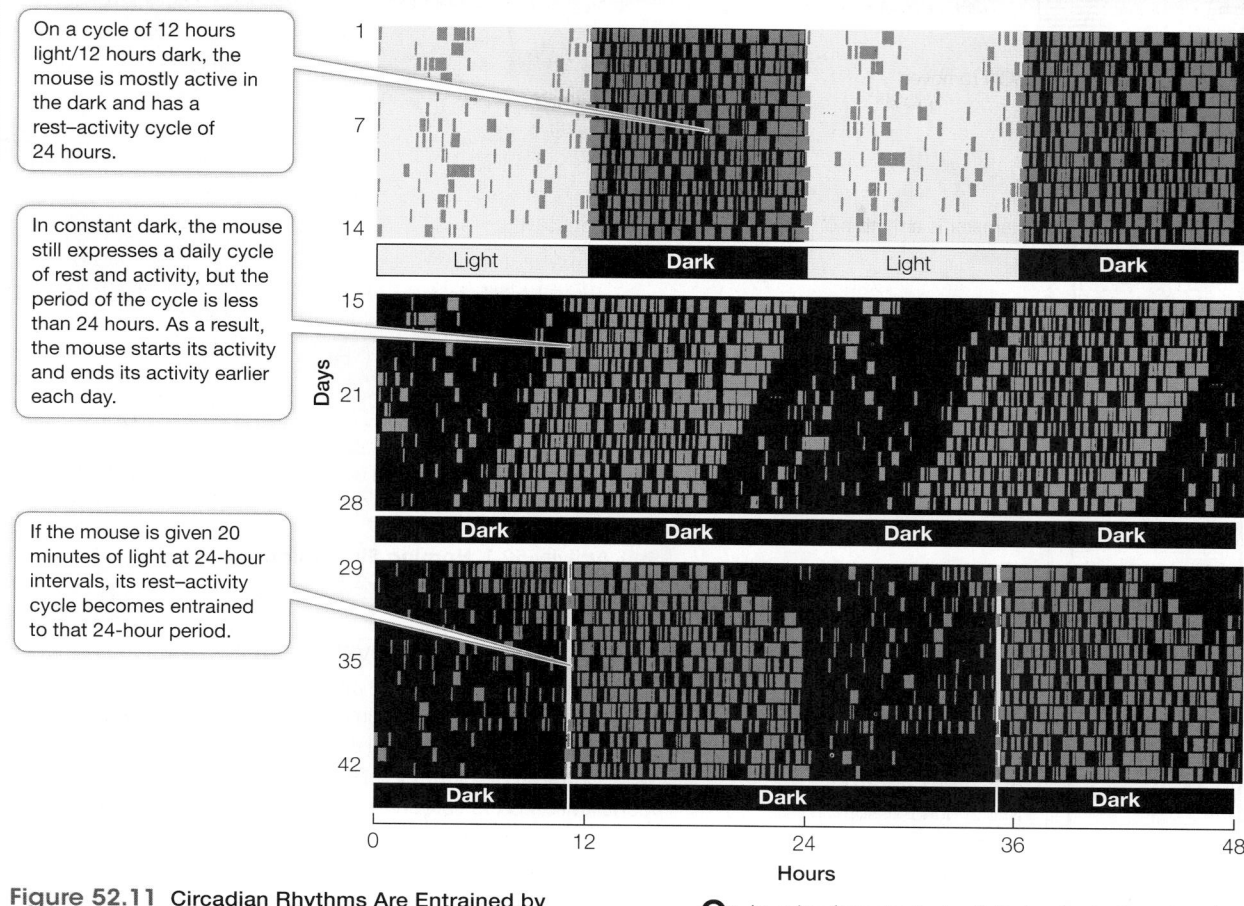

On a cycle of 12 hours light/12 hours dark, the mouse is mostly active in the dark and has a rest–activity cycle of 24 hours.

In constant dark, the mouse still expresses a daily cycle of rest and activity, but the period of the cycle is less than 24 hours. As a result, the mouse starts its activity and ends its activity earlier each day.

If the mouse is given 20 minutes of light at 24-hour intervals, its rest–activity cycle becomes entrained to that 24-hour period.

Figure 52.11 Circadian Rhythms Are Entrained by Environmental Cues The activity–rest cycle of a laboratory mouse (a nocturnal animal) responds to the light–dark cycle under which it is kept. The gray bars indicate times when the mouse is running on an activity wheel. Two days of activity are represented on each horizontal line. Thus, day 1 is 0–24 and day 2 is 24–48. Then on the next line day 2 is replotted as 0–24 and day 3 is 24–48, and so on. This double plotting is merely to make the pattern easier to see.

Q: In going from the last activity tracing in the second panel to the light-pulse experiment in the third panel, do this transition and its effects on the circadian clock seem more similar to what you experience if you are recovering from jet lag after traveling from San Francisco to New York, or in the opposite direction?

 Animation 52.3 **Circadian Rhythms**
www.Life11e.com/a52.3

circannual rhythms that keep track of the time of year. Unlike for circadian rhythms, the neural basis for circannual rhythms is unknown.

---

*connect the concepts Photoperiodism is also a common phenomenon in plants that have to anticipate seasonal changes, and many do so by being sensitive to the length of the night. See Key Concept 38.2.

Animals must find their way around their environment

To locate suitable habitats, find food and mates, and avoid predators and bad weather, an animal needs to be able to find its way around its environment. Within its local habitat, an animal can orient to landmarks. But what if its destination is a considerable distance away?

PILOTING: ORIENTATION BY LANDMARKS Most animals find their way by knowing and remembering the structure of their

environment, a form of navigation called **piloting**. Gray whales, for example, migrate seasonally between the Bering Sea and the coastal lagoons of Mexico (**Figure 52.13**). They find their way in part by following the western coast of North America. The rules are simple: keep the land on your right when traveling north and on your left when traveling south. Coastlines, mountain chains, rivers, water currents, and wind patterns can all serve as piloting cues for animals. But many remarkable cases of long-distance orientation and movement cannot be explained by piloting.

HOMING: RETURN TO A SPECIFIC LOCATION The ability to return to a nest site, burrow, or other specific location is called **homing**. Homing can be accomplished by piloting in a known environment, but some animals that travel long distances through unfamiliar territory perform much more sophisticated homing. The ability of pigeons to return to their home loft even after being transported to remote sites is well known. How do they find their way home?

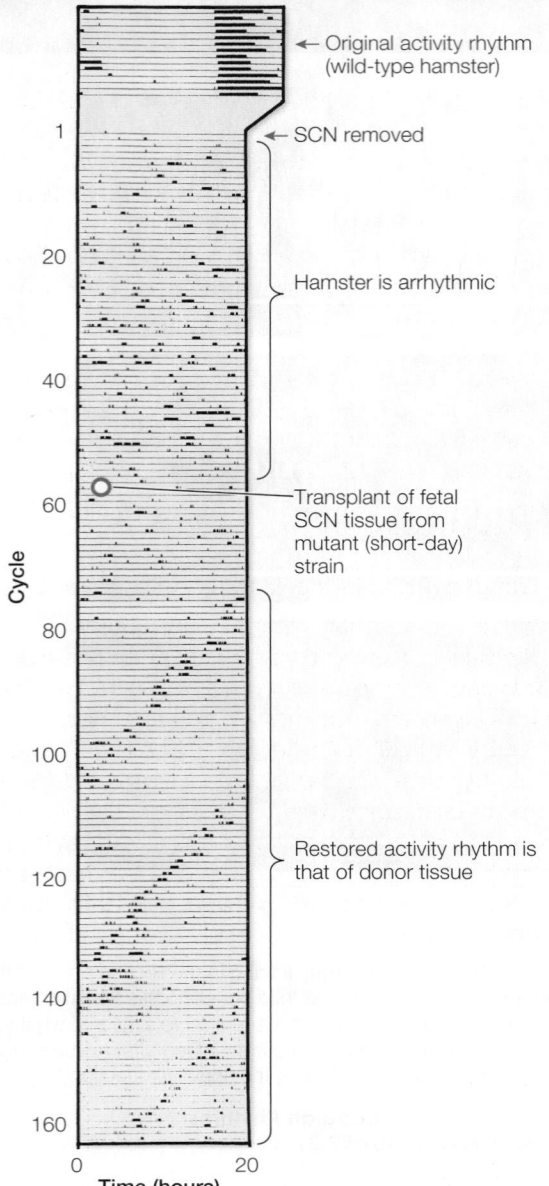

Figure 52.12 **The Brain Clock Can Be Transplanted**   In this experiment, the activity rhythm of a wild-type (i.e., genetically typical) hamster was measured; this animal had a circadian period of 24.5 hours (top of bar). After its SCN was removed (lesioned), the hamster became arrhythmic. SCN tissue from a fetal "short-day" hamster (a mutant strain with a 19-hour circadian period) was then transplanted into the region where the lesion was made in the wild-type hamster. The transplanted tissues restored circadian rhythm in the lesioned hamster, but the restored rhythm had the period of the donor animal.

Experiments have shown that pigeons use the sun as a compass, but they can still find their way home when the sun is not visible. Other experiments have shown that pigeons equipped with frosted contact lenses can find their way home, suggesting that visual cues are not essential. Most amazing has been the demonstration that pigeons can detect Earth's magnetic field and orient to it much as a human orients with a compass. Taken together, the studies of homing by pigeons suggest that they can use multiple, redundant sources of directional information and can switch among those sources depending on the circumstances.

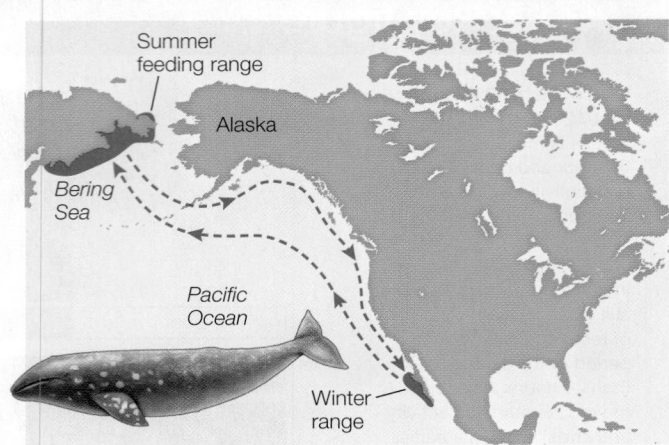

Figure 52.13 **Piloting**   Gray whales (*Eschrichtius robustus*) migrate south in winter from the Bering Sea to the coast of Baja California by piloting, in part by following the western coast of North America.

▶ Activity 52.1 **Homing Simulation**
www.Life11e.com/ac52.1

MIGRATION: NAVIGATION OVER GREAT DISTANCES   For as long as humans have inhabited high latitudes, they have been aware that entire populations of animals, especially birds, disappear and reappear seasonally. Not until the early nineteenth century, however, were patterns of migration traced by marking individual birds with identification bands around their legs. Only when individuals could be unmistakably identified was it possible to show that the same birds and their offspring returned to the same breeding grounds year after year, and that these same birds could be found during the nonbreeding season at locations hundreds or even thousands of kilometers from their breeding grounds.

Many homing and migrating species take direct routes to their destinations through environments they have never experienced because they use mechanisms of navigation other than piloting. Humans use two major forms of navigation:

1. *Distance–direction navigation* requires knowing in what direction and how far away the destination is. With a compass to determine direction and a means of measuring distance, humans navigate.

2. *Bicoordinate navigation*, also known as true navigation, requires knowing the latitude and longitude (the map coordinates) of both the current position and the destination, as well as a compass to determine direction.

Many animals seem to have a "compass sense" that allows them to use environmental cues to determine direction, and some seem to have a "map sense" that allows them to determine their position.

The behavior of some animals suggests that they are capable of bicoordinate navigation. Gray-headed albatrosses (*Thalassarche chrysostoma*), for example, breed on oceanic islands in the Southern Hemisphere. When a young albatross leaves its parents' nest, it flies widely over the southern oceans for 8–9 years. After those wandering years, it reaches reproductive maturity and flies back to the island where it was raised, where it mates and builds a nest (**Figure 52.14**). How can the bird find a tiny island in an enormous ocean after years of wandering? We don't know the answer, but

(A)

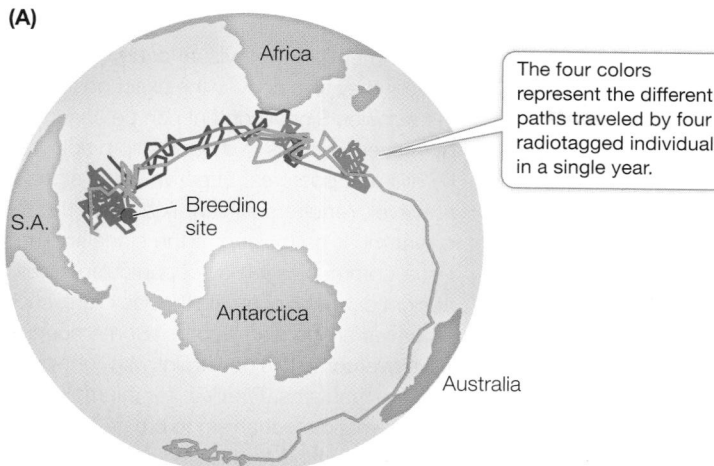

The four colors represent the different paths traveled by four radiotagged individuals in a single year.

(B) *Thalassarche chrysostoma*

Figure 52.14 Coming Home **(A)** Gray-headed albatrosses are born on islands in the subantarctic oceans. Young birds roam widely over the southern oceans for 8–9 years. **(B)** Once they reach maturity, the birds return to the island where they were hatched to mate and raise their own young.

the albatross provides strong circumstantial evidence that an animal can be capable of bicoordinate navigation, as were sailors before the days of global positioning satellites.

The ability to locate a position by calculating the angles between celestial objects such as the sun and stars and the horizon at specific times of day is called celestial navigation. During the day, the sun can serve as a compass if you know what time it is, and animals can use their circadian clocks for that information. This capacity has been demonstrated by "clock-shifting" experiments such as the one shown in **Figure 52.15**. Similar experiments have shown that many animal species can orient by means of a time-compensated solar compass.

 Animation 52.4 Time-Compensated Solar Compass
www.Life11e.com/a52.4

Many animals are normally nocturnal; in addition, many diurnal bird species migrate at night and thus cannot use the sun to determine direction. The stars offer two sources of information about direction: moving constellations and a fixed point. The positions of constellations (like that of the sun) change because Earth is rotating. With a

star map and a clock, direction can be determined using any constellation. But one point that does not change position during the night is the point directly over the axis on which Earth turns. In the Northern Hemisphere, the star Polaris—the "North Star"—lies in that region of the sky and reliably indicates north.

Stephen Emlen at Cornell University showed that birds can learn to use the stars for orientation. As the time of year approaches when young birds would normally migrate to their winter range, young captive birds become more active and orient their activity in the direction they would fly. How do they know that direction? If these birds are raised in a planetarium with a natural star pattern, but one that does not rotate, the birds do not learn to orient, and their premigratory activity is random. However, if the planetarium sky rotates, and even if it rotates around a different point than the North Star, the birds orient their premigratory activity as if the fixed point in the sky were north.

Animals use multiple modalities to communicate

As individual animals interact, they exchange information; therefore animal behaviors can evolve into systems of information exchange, or **communication**. The behaviors of individuals may become elaborated into communication signals, but only if the transmission of information benefits both the sender and the receiver. To understand why these conditions must be met, consider male courtship displays, which can be quite bizarre, time- and energy-consuming, and involve risk. Nevertheless, natural selection will favor them if they increase the males' probability of mating and passing on their genes. The displays can produce that result if they convey information to the females (the receivers) about the qualities of the males as potential fathers.

Animals communicate using a variety of sensory modalities that vary in the nature of the signal produced, the specificity of the information conveyed, the speed and persistence of the signal, and its suitability in different environments. Behavioral physiologists interested in communication must take into consideration the sensory and motor characteristics of their study animals, the physics of the communication modalities they use, and the environment in which the communication takes place.

CHEMICAL SIGNALS Because of the diversity of their molecular structures, **pheromones** can communicate very specific, information-rich messages (see Key Concept 45.2). Pheromones are effective day and night, and they can cover a broad range of transmission distances. Pheromones used in different types of communication vary in their volatility (ease of vaporization) and diffusibility; these chemical properties are functions of the nature and size of the pheromone molecule. Pheromones that act as alarm signals, for example, are highly volatile and diffusible, so their message spreads rapidly but disappears rapidly. Territory-marking and trail-marking pheromones have low volatility and diffusibility and stay effective for a long time, so they can convey directional information. Sex pheromones, such as that of the gypsy moth (see Figure 45.4), are intermediate in these properties, so they can spread a long distance but do not disappear rapidly.

Pheromones are an effective way to exchange species-specific information, and because the recipient must have the proper receptor molecule to detect the pheromone, it is not a signal that is easily intercepted by predators. Pheromonal signals cannot be changed rapidly, but they can convey static, complex information. Mammals that mark

experiment

Figure 52.15 A Time-Compensated Solar Compass

Original Paper: Kramer, G. 1952. Experiments on bird orientation. *Ibis* 94: 265–285.

Experiments show that pigeons use the sun to establish directions for navigation and finding food. "Clock-shifting" experiments demonstrate that the birds' circadian clocks factor into their ability to judge direction correctly based on the sun's position.

HYPOTHESIS ▶ Pigeons determine compass direction from the position of the sun with respect to their internal circadian clocks.

METHOD

1. Place a pigeon in a circular cage from which it can see the sun and sky, but not the horizon or any other visual cue.
2. Surround the cage with multiple food bins but place food only in the southernmost bin, thus training the bird to look for food in the south. (Rotating the cage but always placing the food in the southernmost bin confirms that the bird is navigating to find south.)
3. Place the trained pigeon in a room with a controlled light cycle for 2 weeks. Turn the lights on at midnight and off at noon to phase-advance its circadian rhythm by 6 hours (i.e., 6 A.M. feels like noon to the bird).
4. Return the pigeon to the circular cage under natural light and observe its food-seeking behavior.

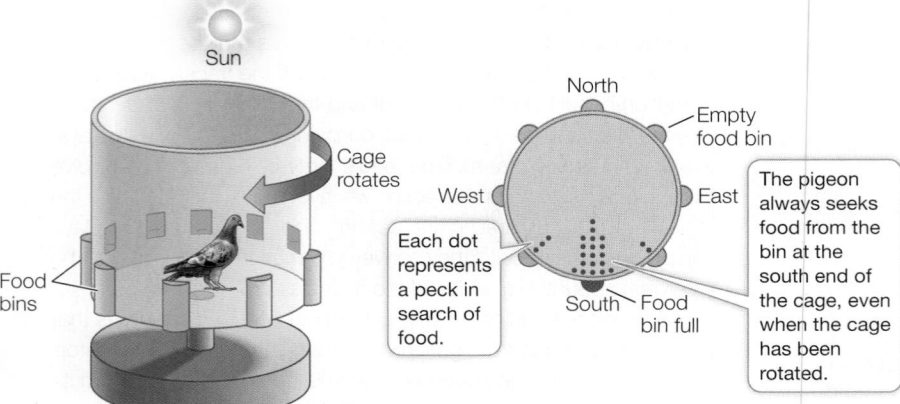

RESULTS

A 6-hour shift in the circadian clock results in a 90-degree error in the pigeon's orientation.

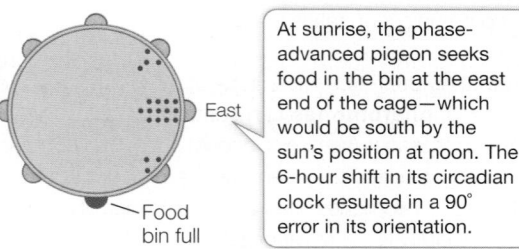

At sunrise, the phase-advanced pigeon seeks food in the bin at the east end of the cage—which would be south by the sun's position at noon. The 6-hour shift in its circadian clock resulted in a 90° error in its orientation.

CONCLUSION ▶ Pigeons have the ability to determine direction using time of day and the position of the sun as a compass.

their territories with pheromones reveal a great deal of information about themselves: species, individual identity, reproductive status, size (indicated by the height of the marking), and how recently the animal has been in the area (indicated by the strength of the scent).

VISUAL SIGNALS Visual signals offer the advantage of rapid delivery of information over considerable distances (depending on the environment and the visual acuity of the receiver); they also convey the exact position of the signaler. Signal content can be enhanced by movements (as in a courtship display) or by different postures. Effective visual signals, however, require sufficient light, and the receiver must be looking at the signaler. Thus visual communication is not particularly useful at night or in environments that lack light, such as caves and ocean depths. Some species have overcome this constraint with light-emitting mechanisms. Fireflies, for example, use an enzymatic mechanism to create flashes of light. By emitting flashes in species-specific patterns, fireflies advertise for mates at night.

Another drawback of visual signals is that they can be intercepted by other species. There are predatory firefly species, for example, that mimic the flash pattern of females of other species. A male that approaches the mimicking "female" becomes a meal rather than a mate. Thus deception can be part of animal communication systems, just as it is part of human communication.

ACOUSTIC SIGNALS Sound cannot convey complex information as rapidly as visual signals can. But acoustic signals, unlike visual signals, can be used at night and in dark environments. They are not hindered by objects that would interfere with visual signals, so they can be transmitted in complex environments such as forests. They are often better than visual signals at getting the attention of a receiver because the receiver does not have to be looking at the signaler for the message to be received. Sounds are also useful for communicating over long distances. Even though the intensity of a sound decreases with distance from the source, loud sounds can transmit information over much longer distances than visual signals can. The complex songs of humpback whales, when produced at ocean depths of about 1,000 meters, can be heard hundreds of kilometers away, allowing these whales to locate one another across vast expanses of ocean. Humpback whale populations that feed during the summer in the north Pacific Ocean migrate thousands of kilometers to winter breeding areas in either the eastern Pacific Ocean, the mid-Pacific Ocean (Hawaii), or the western Pacific Ocean (see Figure 54.4). Long-distance communication among individuals spread over such enormous expanses of open ocean is important if they are to congregate in rather limited breeding areas at the same time of year.

The information content of acoustic signals can be increased by varying their frequency, as you can see in the sonograms of the species-specific song of white-crowned sparrows shown in Figure 52.7, and as you practice in your own speech. However, acoustic signals place the signaler at risk for detection by predators. This danger can be minimized by adjustments of frequency and signal structure that decrease the directional information the receiver can extract from the signal. Alarm calls tend to be pure tones (a single frequency) without much temporal structure (starts and stops). It is very difficult to localize such calls. By contrast, territorial calls tend to cover a broad frequency range and have temporal structure. These calls are easy to localize. The frequencies and structures of acoustic signals are also adapted to specific habitats. Different vegetation types, for example, have different sound-absorbing properties: pure tones at lower frequencies carry better in forests, and more complex calls at higher frequencies carry well in open habitats.

MECHANOSENSORY SIGNALS   Animals in close contact with one another can communicate by touch. A classic case of mechano-sensory communication is the dance of honey bees (*Apis* spp.), first described by Karl von Frisch. Honey bees have a spectacular ability to navigate and can accurately communicate the location of food sources as far away as 10 kilometers. When a forager bee finds food, she returns to the hive and communicates her discovery to her hivemates by performing a vigorous **waggle dance** in the dark hive on the vertical surface of the honeycomb. Other bees follow the dancer and receive her message.

The waggle dance conveys information about both the distance and the direction of the food source. The dancing bee repeatedly traces out a figure-eight pattern as she runs on the honeycomb. She alternates half-circles to the left and right with vigorous wagging of her abdomen in the short, straight run between turns (**Figure 52.16**). Bees use the sun as their compass, and the angle of the straight run indicates the direction of the food source relative to the position of the sun projected down to the horizon. Even under cloudy conditions, the forager can provide directions to the food source because she can see polarized light. The bee's circadian clock allows the dancer to adjust her dance to take into account the sun's movement during her return flight. The clocks of the recruits enable them to adjust their flight direction to accommodate the sun's movement.

The distance to the food source is communicated by the duration of the waggle portion of the dance. The farther away the food source is, the longer the duration of each waggle run. When food is close to the hive, the waggle portion of the dance becomes so short that it appeared to von Frisch that it was a different dance, which he called a round dance. Thus the honey bee has a dance language that communicates both the distance and the direction to a food source.

When challenged to prove that the bees were not simply using an odor trail to find the indicated food source, von Frisch responded with a very common sense observation. Bees returning from a new food source fly around barriers such as buildings, but the recruits going out to the food source fly over the barriers in a "beeline," which would be impossible if they were following an odor trail. Careful observation is still one of the best tools for studying behavior.

COMMUNICATION IN MULTIPLE SENSORY MODALITIES   Avoiding ambiguity is a high priority in any signaling system. Signal specificity is enhanced if multiple sensory modalities are used. Courtship behavior in fruit flies, for example, involves visual, tactile, chemical,

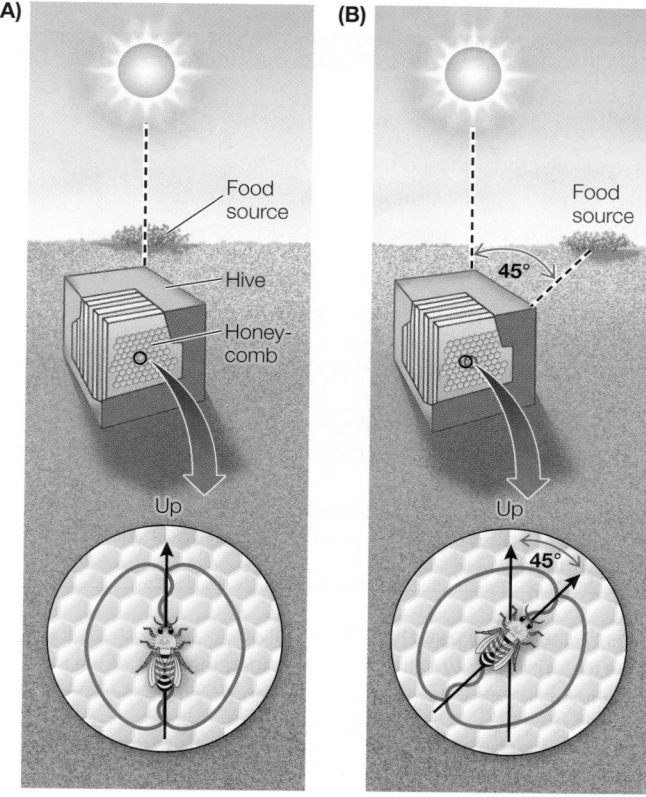

Figure 52.16   The Honey Bee Waggle Dance   (A) A honey bee (*Apis mellifera*) runs straight up on the vertical surface of the honeycomb in the dark hive while wagging her abdomen to tell her hivemates that there is a food source in the direction of the sun. (B) When her waggle runs are at an angle from the vertical, the other bees know that the same angle separates the direction of the food source from the direction of the sun.

 Activity 52.2 Honey Bee Dance Communication
www.Life11e.com/ac52.2

and acoustic signals (see Figure 52.4A). The male fruit fly orients toward the female's line of vision (visual signal) and taps her body with his foreleg (tactile signal). Upon detecting pheromones in her cuticle (chemical signal), the male begins to vibrate one wing, producing a species-specific courtship song (acoustic signal). The male then extends his mouthparts to taste the female's genitalia (chemical and tactile signals); if she is receptive, he initiates copulation. If at any point sensory feedback indicates to either the male or the female that their pairing is inappropriate, the courtship abruptly ends.

52.5 recap

Biological rhythms allow an animal to anticipate changes in its environment. In mammals, a circadian clock located in the suprachiasmatic nuclei generates a rhythm that is entrainable by environmental information. The navigational abilities of animals range from simple piloting by landmarks to distance–direction and bicoordinate navigation. Behaviors may evolve into communication signals if the transmission of information benefits both the sender and the receiver.

(continued)

52.5 recap (continued)

learning outcomes

You should be able to:

- Describe characteristics of circadian rhythms.
- Explain the concept of a time-compensated solar compass.
- Describe the different forms of animal communication and their advantages and disadvantages.
- Evaluate the effectiveness of various forms of animal communication in specific circumstances.

1. What is a free-running circadian rhythm, and how is it entrained by light to the 24-hour day?
2. Why is it important for honey bees to have a time-compensated solar compass, and how does that mechanism influence the waggle dance?
3. Which forms of communication are most efficient for inter-individual communication and for broader population-level communication? Explain your answer.

When behaviors involve interactions between multiple individuals, we have to consider how natural selection operates on all of the interacting parties, whether they are sending signals or receiving them. This concern becomes particularly relevant in the case of social behaviors.

key concept 52.6 Social Interactions Shape the Evolution of Behavior

Our discussion of behavior has focused mostly on behavior in individual animals. We've covered, for example, fixed action patterns, deprivation experiments, genetic determinants, development, and physiological mechanisms. Yet most behavior takes place in a social context. Social behavior and its evolution became a field of study in its own right in 1975, with the publication of E. O. Wilson's landmark book *Sociobiology*.

focus your learning

- Reproductive success of genetically related individuals contributes to an animal's inclusive fitness.
- Eusocial systems include nonreproductive members.
- The evolution of social behavior can be analyzed using a cost-benefit approach.

Mating, parenting, group behaviors such as migration, and group living such as in animal colonies all involve interactions among individuals, and selection operates on all participants in these interactions.

Mating systems evolve to maximize fitness

We begin consideration of this enormous field of social behavior with mating behavior involving a single male and a single female. Because natural selection is always operating to maximize the reproductive success of both the male and the female of the species, an amazing array of mating systems has evolved.

MONOGAMY AND PROMISCUITY At the start of Chapter 7 you learned about the mating behavior of two species of voles. Prairie voles (*Microtus ochrogaster*) are monogamous, forming strong pair

bonds that can last for life, and both parents participate in rearing the young. In contrast, montane voles (*M. montanus*) are promiscuous: males mate with many females, and the young are raised by the female alone. Behavioral physiologists have explained the proximate mechanisms behind these stark behavioral differences in terms of the release of neurohormones and the distribution of the receptors for those hormones in the brains of the two species. The ultimate question—and the one asked here—is why two such different mating systems evolved in two species that are so closely related.

We begin with the premise that there is an asymmetry in the contributions of male and female animals to their offspring at the time of fertilization. Females produce a limited number of eggs, and each egg is generously stocked with resources. Males produce an almost infinite number of sperm, which contain next to no resources. So the energetic and opportunity costs of reproduction are greater for the female than for the male. In mammals this asymmetry increases throughout gestation as the female bears most of the costs. By the time of birth or hatching, the female's investment in the young is much greater than the male's investment, and the main way for the female to maximize her fitness is to make sure her young are healthy and survive to pass on her genes.

The male has different options for maximizing his fitness. He can simply move on after inseminating the female and seek additional mates as a means of maximizing his reproductive success—as in the case of the montane vole. Or he can stay with the female he inseminated, protect her, and help care for their young—as in the case of the prairie vole. Which strategy maximizes male fitness in these species depends on a number of factors that are influenced by the environment, such as the likelihood that a female and her offspring will survive without a male's help, and a male's likelihood of finding another fertile female. Thus sociobiologists seek to quantify these factors in nature as a means of explaining observed differences in mating systems.

POLYGYNY Mating systems in which a male has more than one mate generally involve a male that manages to sequester a group of females from other males. The **polygynous** male increases his fitness by increasing the number of females in his group. As you saw in Key Concept 52.4, male elephant seals accomplish this by protecting an area of beach where females give birth. Male baboons do so by herding females. Male red-winged blackbirds may acquire more than one mate by defending high-quality nesting territories where females prefer to build their nests. Since sex ratios in all these species are close to 50:50, a large differential in male fitness is established, with some males having high reproductive success while many males have none. Thus selection favors males that are successful in competing with other males to obtain and protect access to many females. In general, bigger, stronger males are the winners, and sexual dimorphism in body size evolves. The elephant seal is an extreme example: males may weigh more than three times as much as females. When species with polygynous mating systems are compared, there is a strong correlation between the number of females a male controls and the degree of sexual dimorphism.

Why do females participate in these polygynous mating systems? Why doesn't a female seek out a nice, kind, noncompetitive male? In some cases, such as the female elephant seal, she has no choice.

If a female red-winged blackbird wants to nest in an optimal territory, she will have to share the attentions of the territory owner with other females. Even if the female has a choice of mates, she is likely to maximize her fitness by mating with a male that is strong and dominant enough to control a number of females. Why? If her mate is a dominant male, her male offspring are likely to have their father's traits, become dominant males, and give her more grandchildren. Nonrandom mate choice by females based on male characteristics produces *sexual selection. An extreme result of sexual selection is the lek mating system (see Figure 52.9C), in which the *only* thing a male offers a female is the display of his dominance over other males.

---

*connect the concepts   Charles Darwin recognized the phenomenon of sexual selection leading to the evolution of traits that could compromise the health and survival of the individual. See Key Concept 21.2.

POLYANDRY   Mating systems in which one female mates with multiple males are relatively rare, but **polyandry** is seen in some birds and a few mammal species in which paternal care for the young can have a large effect on fitness. An example of a polyandrous species is the golden lion tamarin (*Leontopithecus rosalia*), a primate native to Brazil's tropical rainforests (**Figure 52.17**). Tamarins are small—adults weigh less than 1 kilogram and face high predation pressure. Females usually give birth to twins, and newborns constitute a higher percentage of maternal weight than is typical of other primates. The young also grow more rapidly than other primates, so nursing costs are high. For all these reasons, young tamarins cared for by their mother alone are unlikely to survive.

What can a male tamarin do to help guarantee his reproductive success? Watching out for predators is one obvious contribution; gathering food for the female and her young is another. Like other primate parents, tamarins carry their young most of the time, but most other primates have single offspring. When tamarin mothers are carrying twins, they spend 92 percent of the time resting, compared with 58 percent of the time when they are not carrying young. Resting is not compatible with foraging and filling the mother's high energy requirements. When a male is present, however, he carries the young about one-third of the time, so the mother has much more time for foraging and feeding.

If one male tamarin is helpful in protecting and raising young, then two should be even more helpful. Some females can attract a second mate by being sexually receptive to him. Neither male can be sure that any eventual offspring are his, so it is in the best interest of both to help in their rearing. Of the tamarin families observed in field studies, only 22 percent had one male and one female, whereas 61 percent had multiple males and one female.

Fitness can include more than your own offspring

As humans, we readily understand the concept of extended family—brothers, sisters, aunts, uncles, nieces, nephews. Extended families are a common form of social organization in other species as well, and members of these families may cooperate in territory defense, predator avoidance, foraging, and rearing of young. If behavior is favored when it increases the fitness of the individual performing it, then how can we explain the evolution of social behaviors that do not lead to the performer having more offspring and that may even appear to be **altruistic**—benefiting another individual at a cost to the performer?

An individual's fitness is increased by having offspring because those offspring carry the parent's genes into the next generation. Fitness gained by producing offspring is referred to as **direct fitness**. However, an individual's genes are carried into the next generation by more than his or her own offspring. In diploid organisms, two offspring of the same parents share, on average, 50 percent of the same alleles, and an individual is likely to share 25 percent of its alleles with its siblings' offspring (nieces or nephews). Therefore, by helping parents and other relatives raise their offspring, an individual increases the transmission of those shared alleles to the next generation. **Inclusive fitness** is the individual's direct fitness plus its **indirect fitness**: the reproductive success of the individual's relatives, to the extent that those relatives share the individual's alleles.

The maximization of inclusive fitness is the mechanism driving **kin selection**, selection for behaviors that increase the reproductive success of relatives even when they come at a cost to the performer. One example is "helping at the nest" behavior, which was studied extensively in Florida scrub-jays (*Aphelocoma coerulescens*). Scrub-jay pairs mate for life and establish large territories, which they defend aggressively. The mating pair may be assisted in rearing their young by three to five helpers (**Figure 52.18**). The helpers guard against predators, feed the young, clean the nest, and fly with fledglings. Why are these birds helping others rather than rearing their own young? A long-term study established a number of important facts:

- The helpers are prior offspring of the mating pair and are usually 1–3 years old.
- Young birds that attempt to breed have almost zero reproductive success.
- Mating pairs with helpers have approximately three times the reproductive success of those without helpers.

Leontopithecus rosalia

Figure 52.17 Polyandry in a Small Primate   The endangered golden lion tamarins of Brazil are small primates whose unique life history has given rise to polyandry in some groups, with males playing a major role in rearing the young.

Aphelocoma coerulescens

Figure 52.18 Helpers at the Nest Young Florida scrub-jays often forego reproduction in their first few years of adulthood to help their parents raise their siblings. These young birds help their parents feed the nestlings, defend the territory, and protect the nest from predators.

These results support the conclusion that helper scrub-jays are maximizing their inclusive fitness by helping their parents raise siblings until they are mature enough to have a reasonable probability of successfully raising their own offspring.

The concept of kin selection was formalized by W. D. Hamilton in what has become known as **Hamilton's rule**. He argued that, for an apparent altruistic behavior to be adaptive, the fitness benefit of that act to the recipient times the degree of relatedness between the performer and the recipient has to be greater than the cost to the performer. This relationship was clearly stated years before by the eminent geneticist J. B. S. Haldane, who said during an argument about altruism that he would not be willing to risk his life to save his brother, but for two brothers or eight cousins, he would consider it.

Eusociality is the extreme result of kin selection

Hamilton's rule can be applied to explain **eusociality**: social groups that include nonreproductive members (that is, members that as individuals do not reproduce). The most obvious examples of eusociality occur among the Hymenoptera, an insect group that includes wasps, bees, and ants. In a honey bee colony, for example, the thousands of individuals in the colony are sterile females.

 Media Clip 52.1 **Social Shrimps**
www.Life11e.com/mc52.1

The key to understanding the evolution of eusociality in hymenopterans is their sex determination mechanism, **haplodiploidy**, in which diploid individuals are female and haploid individuals are male. The queen carries a lifetime supply of sperm obtained during her single mating flight, and she controls whether her eggs are fertilized or not. An unfertilized egg develops into a haploid male; a fertilized egg develops into a diploid female. The queen's daughters share all of their father's genes and, on average, half of their mother's genes. As a result, the sterile female workers in the hive—all sisters—share, on average, 75 percent of their alleles (**Figure 52.19**). The basic unit of reproductive success in eusocial insects is the creation of new colonies around a new queen. If a new queen developed from a worker's offspring, the worker would share fewer genes with the new queen than if the new queen developed from a sister.

Haplodiploidy is not essential for the evolution of eusociality. Nearly all eusocial animals construct elaborate nests or burrow systems within which their offspring are reared. Such a structure represents an enormous investment of resources. Naked mole-rats are eusocial mammals that live in elaborate underground tunnel systems (**Figure 52.20**). A colony includes 70–80 individuals but only 1 reproductive female and a few reproductive males. The other colony members are sterile workers that dig and maintain the tunnels, guard against intruders, harvest food (tubers), and use their feces to feed the queen and her offspring. Individuals attempting to found new colonies have a high risk of failing or being captured by predators. When chances of individual reproductive success are

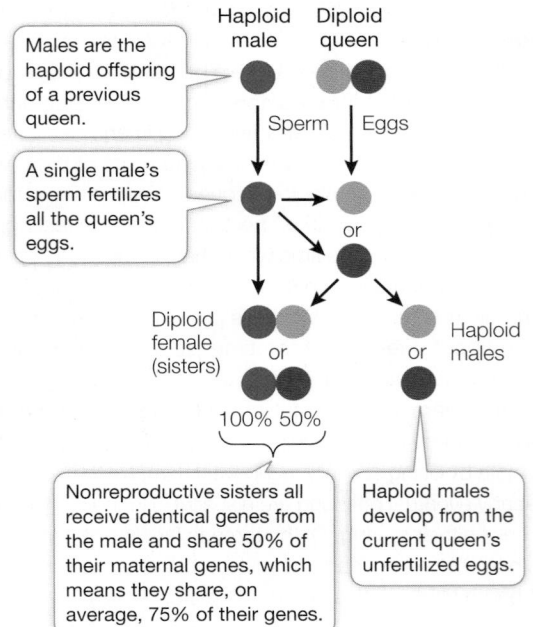

Figure 52.19 Favoring Sisters over Daughters Female honey bees are diploid and males are haploid. Thus if a female worker bee were to reproduce, she would share approximately 50 percent of her genes with her daughters. However, she shares an average of 75 percent of her genes with her sisters. In terms of inclusive fitness, then, a sister is more valuable than a daughter for this species.

Heterocephalus glaber

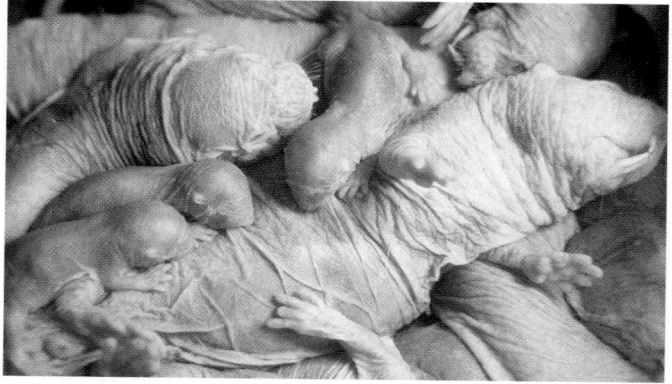

Figure 52.20 A Eusocial Mammal   Naked mole-rats live in a large colony with one reproductive female and a few reproductive males. Their home is an elaborate tunnel system excavated by the colony over time.

practically zero, an individual can best maximize its inclusive fitness by staying with and helping maintain the colony.

Group living has benefits and costs

Apart from their direct influences on reproductive success, social systems can contribute to survival in many ways, but they can also involve costs. Thus the cost–benefit approach of behavioral ecology is relevant to understanding the evolution of social behavior.

An obvious example of a benefit of group living is improved foraging efficiency. By hunting in packs, African wild dogs (see Figure 58.17) employ cooperative strategies that enable them to bring down larger prey than could a single dog. The larger the pack, the greater the hunting success rate. Once the prey is killed, the presence of conspecifics also reduces the risk that the wild dogs will lose their prey to larger scavengers, such as hyenas.

Living in a group can also reduce the risk of individuals becoming prey themselves. Many small birds forage in flocks. To test the hypothesis that flocking provides protection against predators, R. E. Kenward released a trained goshawk (*Accipiter gentilis*) near wild common wood-pigeons (*Columba palumbus*) in England. The hawk was most successful when it attacked solitary pigeons. Its success in capturing a pigeon in a flock decreased as the number of pigeons in the flock increased (**Figure 52.21A**). The larger the flock, the sooner some individual in the flock spotted the hawk and flew away. This escape behavior stimulated other individuals in the flock to take flight as well.

Alarm calling is another means of reducing predation risk, but the caller incurs a risk cost by calling attention to itself. Belding's ground squirrels (*Urocitellus beldingi*) live in large colonies in open meadows. When one squirrel announces the presence of a predator with loud, sharp barks, all the nearby squirrels dive into their burrows (**Figure 52.21B**). Paul Sherman of Cornell University showed that callers double their risk of being preyed on—so why do they do it? Research by Sherman and by others has shown that this altruistic behavior is a product of kin selection. In this polygynous species, males establish large territories in the spring that include the territories of several females, whom they inseminate. The females then drive off the males. Female offspring settle near their mothers, so neighboring females in a colony tend to be sisters, and they defend each other's young. Sherman showed that males are less likely to give alarm calls than females, and that females are more likely to give alarm calls when related individuals are nearby.

Social behavior has many costs as well as benefits. Foraging in a group may reduce the amount of food available to each individual, and the foraging individuals may interfere with one another's foraging activities. Individuals living in groups may face more competition for mates, as well as for food, than solitary individuals would. A large group may actually attract the attention of predators. And living at high population densities can increase the risk of disease transmission. The study of disease transmission in wild animal populations is a relatively new field, but such studies have made it apparent

(A)

Goshawk

Wood pigeon

The more pigeons in the flock, the sooner the hawk is spotted...

...and the lower the hawk's attack success.

Hawk's distance (meters) when spotted by pigeons (●)

Hawk's attack success (%) (●)

80

60

40

20

80

60

40

20

1   2–10   11–50   >50
Number of pigeons in flock

(B) *Urocitellus beldingi*

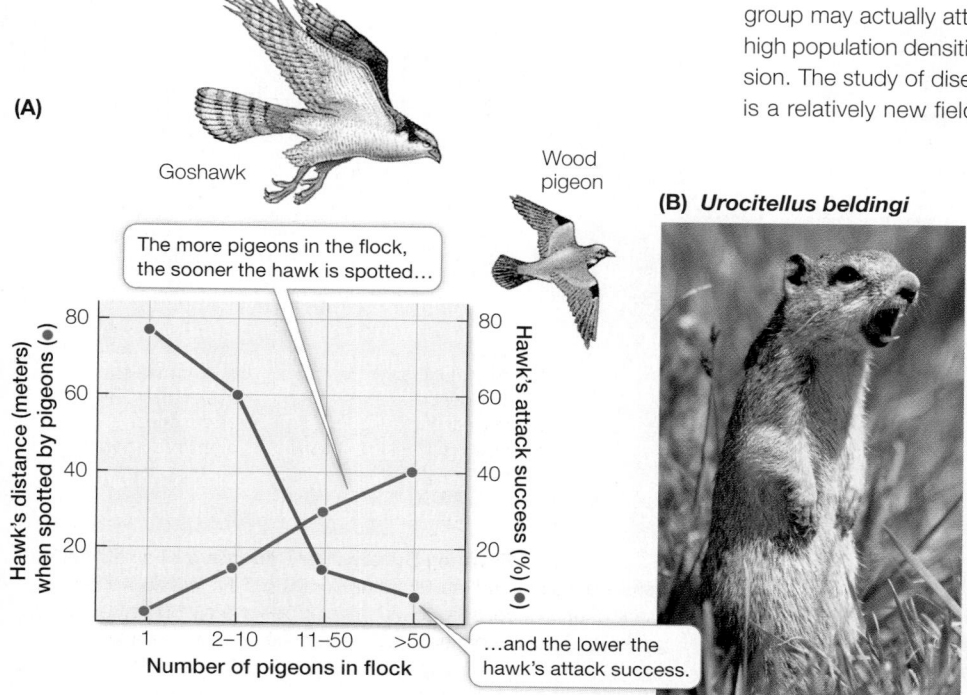

Figure 52.21 Group Living Provides Protection from Predators   Animals that live in groups can spread the cost of looking out for predators. **(A)** The larger the number of common wood-pigeons in a flock, the greater the chances that one of the pigeons will spot a predatory goshawk before it attacks, and the lower the chances that the hawk will capture one of the pigeons. **(B)** A male Belding's ground squirrel gives an alarm call upon spotting a predator. Although this behavior increases his individual risk of becoming prey, he increases the survival chances of many of his close relatives.

that species living in social groups are more prone to outbreaks of disease than are solitary species.

Can the concepts of sociobiology be applied to humans?

With the publication of E. O. Wilson's *Sociobiology*, applications of the concepts of evolutionary genetics to human behavior were controversial. The intensity of the debate stemmed from false impressions that sociobiological approaches resembled previously discredited pseudoscientific movements such as social Darwinism, biological determinism, and eugenics, all of which have been used as rationales for racism and discrimination. However, sociobiology is an objective science, questions of "what is" should not be conflated with questions of "what ought to be." For example, we can demonstrate the genetic basis of sexual dimorphism in body size and muscle mass in humans, and we can compare this dimorphism with that in other mammals. However, any attempt to use such data as a political or legal defense for the cultural practice of polygamy would not be science and should not be confused with science.

The fact that our biochemistry, our cell biology, physiology, and anatomy are shaped by our genes is beyond argument. However, it is also clear that these genetically shaped characteristics are also influenced by factors such as environment, nutrition, social interactions, and culture. Why should it be different for behavior? Studies of identical twins reared apart have produced evidence for inheritance of uncanny similarities in behavioral propensities. Studies of isolated human cultures around the world have also revealed remarkable similarities in social organization. None of these studies, however, would challenge the dominant role of learning and culture in the shaping of human behavior.

52.6 recap

Social behavior can be understood by asking how it contributes to the fitness of all the individuals involved. Asymmetry between the sexes in parental investment is a key factor in the evolution of mating systems. According to the theory of kin selection, an individual can increase its fitness by helping related individuals with whom it shares alleles. In extreme cases, kin selection has given rise to eusociality.

learning outcomes

You should be able to:
- Apply concepts of fitness to explain the evolution of social behaviors related to reproduction.
- Explain how haplodiploidy relates to eusociality.
- Identify costs and benefits of social behaviors.

1. Explain the selective pressures that favor an individual being a helper at the nest rather than engaging in reproduction.
2. What is the relationship between haplodiploid sex determination and the evolution of eusociality in ants, wasps, and bees?
3. Many species of small birds respond to the presence of a much larger predator by joining together to mob the predator and drive it off. What do you see as the risks and benefits of this behavior, and what hypotheses could you propose about the individuals taking part in an incidence of mobbing behavior?

investigating life

 How does social context influence the learning and expression of birdsong?

Birdsong is the classic example of how genetic determinants, development, physiology, and learning can interact to generate and control even complex behaviors. As you saw with the white-crowned sparrow, the nestlings must experience their species-specific song if they are going to sing it (males) and respond to it (females) as adults. But there is a genetically determined template of what they can and will learn. With physiological maturation, they begin to sing poorly and must hear themselves and practice to be able to sing their species-specific songs. Social context also plays a role in learning small variations of dialects in song that may play a role in population as well as individual identity. Social context also influences what and when singing occurs. Singing intensifies in the presence of competitors and also potential or current mates. The experiments by Woolley and Doupe on zebra finches showed that even though there were only small differences between songs when males sang alone or in the presence of a female, these differences were clearly recognized by females and influenced their behaviors. Studies of birdsong emphasize the importance of considering the

evolution of a behavior from the angles of all of the different individuals involved, and at the various levels of biological organization that contribute to the behavior.

Future directions

Language is considered a unique characteristic of humans. Many animals use simple acoustical signals to communicate, but the fine motor control necessary to produce precise, fine-tuned, rapid complex vocalizations is found only in humans and songbirds. Whereas birdsong is recognized to be elaborate but highly stereotyped territorial and mating signals, recent studies like those of Woolley and Doupe are showing that small variations in song can convey more detailed information about the identity of the individual, its intentions, and the social context in which it is behaving. Can birdsong be used to convey more complex information? Can birdsong be a form of simple language? If so, we have much more to learn about the cognitive abilities of birds. Language is used to construct our views of ourselves and of our position in our physical and social environments, and to be able to project those views into the past and into the future—the essence of consciousness. Could birds be better models for human cognitive processes than the mammals

Chapter Summary 52

▶ 52.1 Ethology Led to Modern Behavioral Biology

- Ivan Pavlov's discovery of **conditioned reflexes** and B. F. Skinner's research on **operant conditioning** as a model for learning led to an approach called behaviorism that mainly carried out laboratory experiments on rats and a few other animal models. Review Figure 52.1

- **Ethology** focuses on both the **proximate causes** of behavior (the immediate cause of the behavior, and how the behavior develops) and the **ultimate causes** (how the behavior affects the animal's evolutionary fitness).

- A major focus of the ethologists was **fixed action patterns** and their **releasers**. They performed **deprivation experiments** as well as crossbreeding experiments to demonstrate that certain behaviors are genetically determined. Review Figure 52.2

▶ 52.2 Behavior Can Be Genetically Determined

- Genetic manipulations such as gene knockout experiments can reveal the roles of specific genes underlying a behavioral phenotype. Review Figure 52.3

- Most behaviors are complex traits involving many genes that function in cascades, offering many points for a change in a single gene to influence behavior. Review Figure 52.4

▶ 52.3 Behavior Can Be Studied Developmentally

- Hormones can determine the pattern of behavior that develops and the timing of its expression. Review Figure 52.5

- **Imprinting** is a process by which an animal learns a specific set of stimuli during a limited **critical** or **sensitive period**. That critical period may be determined by hormones.

- The development and expression of song in white-crowned sparrows involve a genetic predisposition to learn the species-specific song, a critical period for imprinting of a song memory, and hormonally controlled timing of song expression. Social interactions may also play a role. Review Figure 52.7

- The ability to incorporate variations into species-specific behaviors may be important in establishing individuality and influencing mate choice. Review Investigating Life: Practice Makes Perfect

▶ 52.4 Selective Pressures Shape Behavior

- An animal's behavior involves a series of choices that influence its fitness. To make these choices, animals use environmental cues that are reliable predictors of the potential effects of their choice on their fitness.

- The **cost–benefit approach** can be used to investigate the fitness value of specific behaviors. The cost of a behavior typically has three components: **energetic cost**, **risk cost**, and **opportunity cost**. Review Figure 52.8, Animation 52.1

- According to **optimal foraging theory**, animals should practice feeding behaviors that maximize their energetic gain at the least cost. Review Animation 52.2

▶ 52.5 Behavior Can Be Studied Mechanistically

- **Circadian rhythms** control the daily cycle of behavior. Without environmental time cues, circadian rhythms free-run with a period that is genetically programmed. Circadian rhythms are normally **entrained** by the environmental light–dark cycle. Review Focus: Key Figure 52.11, Animation 52.3

- Circannual rhythms help organisms time their reproductive behavior to coincide with favorable times of year.

- Forms of navigation used by animals to find their way in the environment include **piloting** (orienting to landmarks), homing, distance–direction navigation, and bicoordinate navigation. Navigation mechanisms include celestial navigation and a time-compensated solar compass. Review Figures 52.13–52.15, Animation 52.4, Activity 52.1

- The behaviors of individuals may become **communication** signals if the transmission of information benefits both the sender and the receiver. Review Figure 52.16, Activity 52.2

- Chemical communication signals (pheromones) can be highly specific and have different time courses. Visual signals can convey complex messages rapidly, but only if the recipient can see the sender. Acoustic signals can travel over long distances, do not require a focused recipient, and can be modified to reveal or conceal directional information. Tactile signals can convey complex messages when animals are in close proximity.

▶ 52.6 Social Interactions Shape the Evolution of Behavior

- Differences in parental investment can lead to different mating strategies for males and females.

- **Polygynous** mating systems, in which one male controls and mates with many females, can result in great variation in male reproductive success.

- **Polyandry**—a female mating with multiple males—can evolve in circumstances in which males make substantial contributions to the survival of offspring.

- The fitness an individual gains by producing offspring (**direct fitness**) plus the fitness it gains by increasing the reproductive success of relatives with whom it shares alleles (**indirect fitness**) is called **inclusive fitness**. **Kin selection** may favor **altruistic** behavior toward relatives, despite its cost to the performer, if it increases the performer's inclusive fitness.

- As a result of **haplodiploidy**, the sex determination mechanism of hymenopteran insects, nonreproductive female workers (sisters) share more alleles with one another than reproductive females share with their own offspring. Review Figure 52.19

- Haplodiploidy has probably facilitated the evolution of **eusocial** behavior in the hymenopterans through kin selection. Eusociality has also arisen in diploid species in which chances of individual reproductive success are extremely low.

- Group living confers benefits such as greater foraging efficiency and protection from predators, but it also has costs, such as increased competition for food and ease of transmission of diseases.

See Activity 52.3 for a concept review of this chapter.

Go to **LearningCurve** (in **LaunchPad**) for dynamic quizzing that helps you solidify your understanding of this chapter. **LearningCurve** adapts to your responses, giving you the practice you need to master each key concept.

▶ Apply What You've Learned

Review

52.1 Behaviors have proximate and ultimate causes.

52.3 Hormones influence the development of certain behaviors and their expression at a later time.

52.4 The evolution of behavioral responses to the environment is driven by costs and benefits.

Original Paper: Engel, K. C. et al. 2016. A hormone-related female anti-aphrodisiac signals temporary infertility and causes sexual abstinence to synchronize parental care. *Nature Communications* 7: 1–10.

Burying beetles are called nature's undertakers because they bury the remains of small dead animals. These animal bodies serve as food for the beetles' young. A mating pair of beetles typically meets over a fresh animal carcass, copulates, and buries the carcass. The female lays fertilized eggs in the soil above the carcass. Some 50–60 hours later, the eggs hatch and both parents spend the next 3 days eating from the carcass and regurgitating the partially digested material to feed their young. After 3 days, the larvae feed themselves.

Researchers used these beetles to study trade-offs in time and energy investments between brood care and production of new broods. Both activities are demanding, so how do the beetles control their apportionment of energy? The researchers provided female beetles with mating partners and carcasses to stimulate egg laying. There were three treatment groups: (1) no larvae group: just as the eggs hatched, the new hatchlings were removed and the parents were presented with a new carcass; (2) new larvae group: just after their eggs hatched, the parents were left with 10 larvae and given a new carcass; and (3) old larvae group: 4 days after their eggs hatched and they had cared for their hatchlings, the parents were presented with a new carcass. The researchers measured the number of females in each group that produced eggs. The results are shown in **Figure A**.

The researchers measured juvenile hormone (JH) levels present at various times in the breeding cycles of females with and without larvae. Juvenile hormone is known to suppress the fertility of the females. The results are shown in **Figure B**.

The researchers also measured the effects of a possible pheromone, methyl geranate, on the beetles' sexual behavior. Methyl geranate is a volatile compound that shares the same biosynthetic pathway with juvenile hormone. The researchers recorded whether or not males copulated with females and the amount of methyl geranate emitted by those females. **Figure C** summarizes these results. The data points represent meetings resulting in copulation or not.

Questions

1. What do the data in Figure A indicate about energy trade-offs made by female burying beetles? How does this relate to the concept of evolution shaping behavior to maximize fitness? Explain your answer.

Figure A

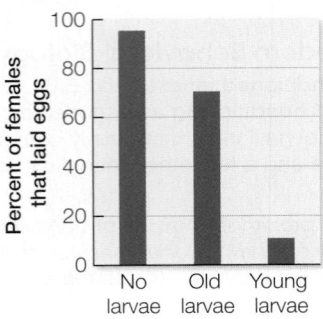

Figure B

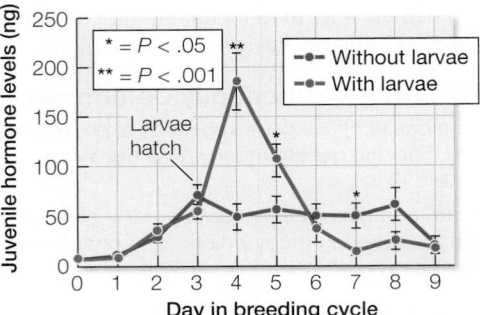

Figure C

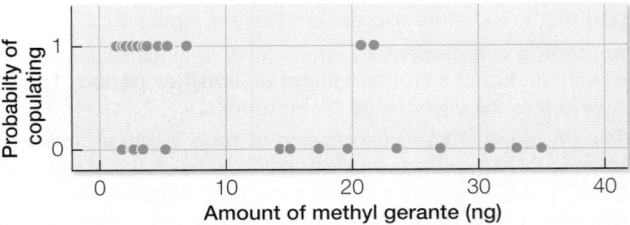

2. Recall that Tinbergen posed two questions about the proximal causes of behavior and two questions as to the ultimate causes of behavior. For the investigation in Question 1 above, decide which question or questions apply. Do these question(s) deal with proximate or ultimate causes of behavior in burying beetles? Explain your answer.

3. What do the data in the Figures B and C indicate about how hormones and pheromones influence burying beetle behavior? What is the biological significance of these results?

4. Refer again to the four questions posed by Tinbergen. For the investigations producing the data in Figures B and C, which question or questions apply? Do these question(s) deal with proximate or ultimate causes of behavior in burying beetles? Explain your answer.

5. Give examples of a cost and a benefit of the parental care behavior of burying beetles.

Go to **LaunchPad** for the eBook, LearningCurve, animations, activities, flashcards, and additional resources and assignments.

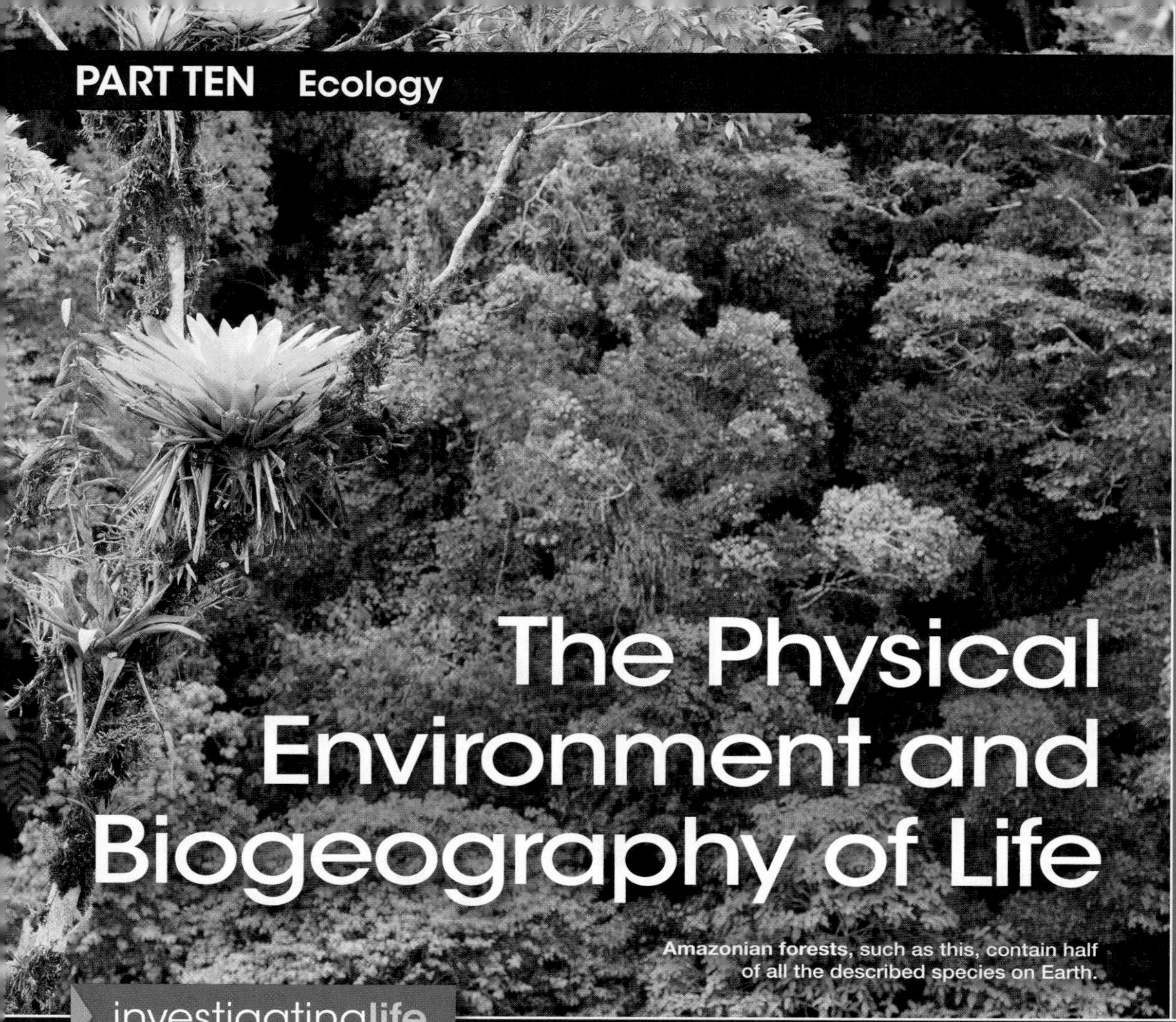

53

The Physical Environment and Biogeography of Life

Amazonian forests, such as this, contain half of all the described species on Earth.

investigatinglife

The Largest Experiment on Earth

Most people know that there are many species on Earth, and that new species are being discovered every day, but not everyone knows that roughly half of all species worldwide reside in one place: the tropical rainforests of the Amazon Basin. The statistics are staggering: an estimated 390 billion trees representing 16,000 species grow in Amazonia, one in five bird species globally resides there, and one-fifth of all fresh water falls on its slopes and valleys. The Amazon Basin is by far the largest watershed on Earth, with thousands of tributaries leading to the Amazon River and then out to sea. To date, 2,200 freshwater fish species have been described, which is more than all the fish species in the Atlantic Ocean.

It is reasonable to assume, then, that when the Amazon Basin is threatened by human activities, so, too, is global biodiversity. By far, the main destructive force is deforestation, which began about 50 years ago. As roads pushed their way into Amazonia's forests, more rainforest was logged and converted to agriculture and settlements. It is estimated that nearly 20 percent of the Amazon Basin has been clear-cut. Because soil fertility is generally poor in tropical rainforests, the cleared land is often useful only for a decade or less,

resulting in abandonment and new cycles of deforestation and fragmentation.

How does habitat fragmentation affect the species living in the Amazon Basin? Fragmentation makes habitats smaller, and it isolates populations. Deforestation forces species to sustain themselves in a smaller area or move to more suitable areas. In the late 1970s, some ecologists asked a deceptively simple question about Amazon deforestation: What is the minimum area needed to maintain species diversity within a rainforest fragment? They wanted to know whether the species remaining in the forest fragments could maintain themselves there, whether the sizes of the forest fragments mattered to this maintenance, and whether species would venture from one fragment to another. These ecologists conducted a large and long-running experiment in the Amazon forests near Manaus, Brazil. As you will see in Key Concept 53.5, the size of the fragments and the ability of species to move across the deforested landscape are key to maintaining the extraordinary diversity left in the Amazon Basin.

 How do geographic area and isolation

53.1 Ecology Is the Study of the Interrelationships among Organisms and the Environment

Whether you notice ants collecting crumbs on a sidewalk or watch a video of polar bears hunting on sea ice, you are studying ecology. From simple experiences such as these, most people know that organisms on Earth are interconnected with one another and with their environment. **Ecology** is the exploration of these interconnections and is most simply defined as the study of the interrelationships among organisms and the physical environment. Ecologists study these interactions at different scales and with different methods, but they all agree that ecology is a scientific endeavor, first and foremost.

focus your learning

- Ecology is the study of the interrelationships among organisms and the physical environment.
- Ecology uses observations, experiments, and models to test theories at multiple spatial and temporal scales.

"Ecology" is sometimes equated with "environmentalism," but the two terms are different. Ecology is a science that generates knowledge about interactions in the natural world. **Environmentalism** is the use of ecological knowledge, along with economics, ethics, and many other considerations, to inform both personal decisions and public policy related to stewardship of the natural world. So while an ecologist might study the number of bluefin tuna in the world's oceans, the environmentalist would focus on the social, economic, and political capital needed to enforce a sustainable harvest or, alternatively, to ban bluefin tuna fishing altogether.

Ecology is a relatively new branch of the biological sciences; in fact, it did not have a formal name until 150 years ago when Ernst Haeckel, a German biologist, constructed the new word "ecology," from the Greek root *oikos*, or "household."

Modern ecology has a more "use-inspired" focus

Although it was recognized in Darwin's day that humans could affect the natural world, and that human populations were undergoing unprecedented growth, for the most part early ecologists believed they were studying pristine systems largely untouched by human activities. Today most ecologists have a much greater awareness of ways in which humans shape the natural world, be it in the form of global climate change, the introduction of novel species to new locations, or the logging of Earth's tropical forests. Because of the dominant role that humans play in almost every ecological system on Earth, a "use-inspired" motivation often shapes ecologists' research and teaching. They understand that scientific knowledge of ecology greatly improves our ability to grow food for ourselves reliably and sustainably, to manage pests and diseases safely and effectively, and to deal with natural disasters such as floods and fires. The greater our understanding of ecological connections, the more likely it is that we can accomplish these things without causing a cascade of unanticipated consequences for ourselves and the other life on Earth.

Ecology is studied at many levels of organization

Ecology is studied at many levels of organization, from molecules to the entire globe. Some ecologists work at the level of genes to understand how individuals might adapt to a changing environment, while other researchers seek to understand how temperature might modify nutrient cycling at the ecosystem level. The work of ecologists focuses on many scales or levels of organization: individuals, populations, communities, ecosystems, landscapes, or the global biosphere (**Figure 53.1**).

As prior chapters have conveyed, a **population** is a group of individuals of the same species within a given area that have the

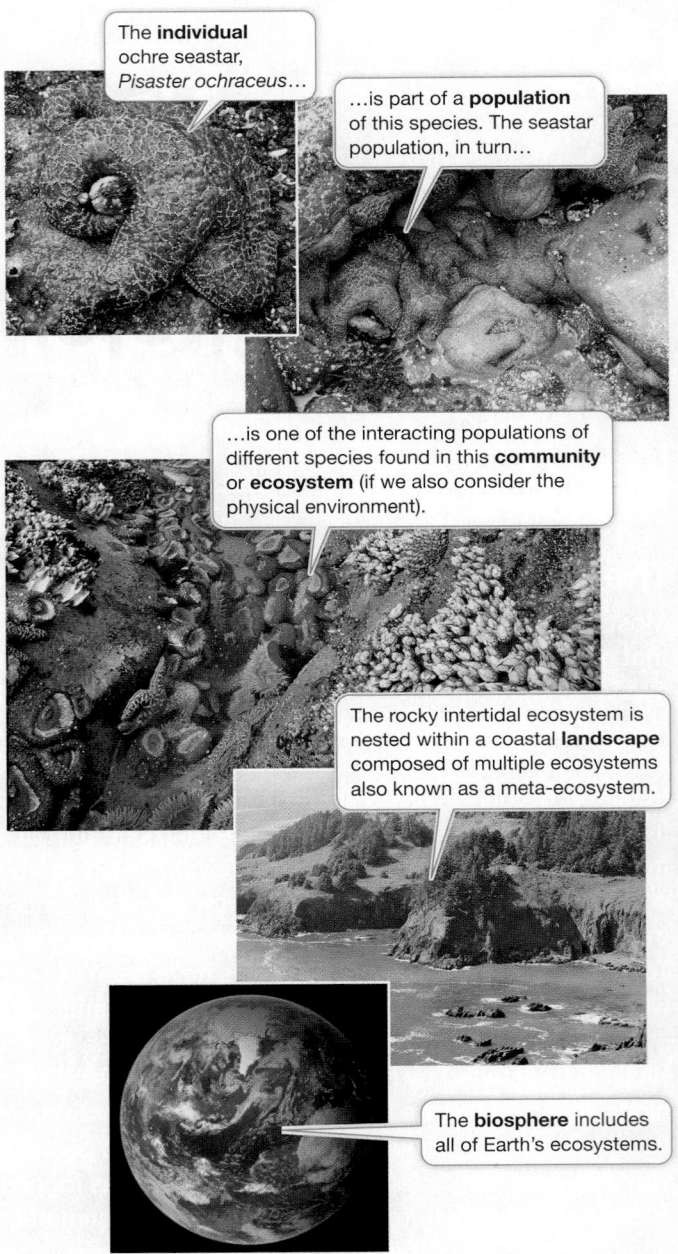

The **individual** ochre seastar, *Pisaster ochraceus*…

…is part of a **population** of this species. The seastar population, in turn…

…is one of the interacting populations of different species found in this **community** or **ecosystem** (if we also consider the physical environment).

The rocky intertidal ecosystem is nested within a coastal **landscape** composed of multiple ecosystems also known as a meta-ecosystem.

The **biosphere** includes all of Earth's ecosystems.

Figure 53.1 Scale Is Important to Ecological Systems   These photos from the Oregon coast of the United States illustrate how ecological systems can be studied at a number of scales, from the individual to the biosphere.

potential to interbreed and interact with one another. For example, the individual seastars in Figure 53.1 make up a population at this particular rocky shore location. The seastar population in turn interacts with other species, including its prey, sessile mussels, and its competitors, whelks and other seastar species. These species form a **community**, which is an assemblage of interacting species living together at the same place and time.

Ecological studies often encompass both the living, or **biotic**, and the nonliving, or **abiotic**, components of the environment. The biotic components of an organism's environment are other organisms, so ecology includes the study of interactions within and among species. The abiotic components of an organism's environment are the myriad physical and chemical characteristics of the system. For example, in the rocky shore community, tides are a key abiotic factor because they limit where seastars can feed on mussels and compete with whelks.

The term **ecosystem** is often used when describing a community of organisms in which their physical environment is explicitly taken into account. Ecosystem ecology particularly focuses on the movement of energy and nutrients through a group of organisms. The movement of energy and nutrients, as well as organisms, can affect higher levels of organization, including **landscapes**, geographic areas that include multiple ecosystems (sometimes referred to as meta-ecosystems). Ultimately, meta-ecosystems are linked to one another at larger geographic scales to form the **biosphere**, consisting of all living organisms on Earth plus their respective environments.

Answering ecological questions requires observations, experiments, and models

Just as in other fields in biology, ecologists use a mixture of observations, experiments, and models to test ecological theories (see Key Concept 1.2). Ecologists are unique, though, in using a wide range of venues and scales in their experiments, including short laboratory experiments, long-term field manipulations, and even decades-long ecosystem-wide experiments, such as the Biological Dynamics of Forest Fragments Project described in Key Concept 53.5. However, at large spatial scales or over long periods of time, well-replicated and controlled experiments are usually not possible. For example, when studying the ecological effects of climate change on a particular species, ecologists might have to combine observations of the temperature range of that species with short-term experiments on the effects of warming on the species' distribution and then use models to project how distributions could change with continued warming. Often this type of research requires multiple techniques and collaborative efforts to be successful.

> **53.1 recap**

Ecology is the scientific study of the interrelationships among organisms and the environment. Ecology is studied at multiple levels of organization, from individuals to the biosphere. Ecologists use a mixture of observations, experiments, and models to test ecological theories.

learning outcomes

You should be able to:

- Distinguish between ecology and environmentalism, and include examples of each.

> **53.1 recap**

- Describe the methods of study used by ecologists, and identify how and why ecological studies might differ from studies in other scientific fields.

1. How does ecology differ from environmentalism? Give a reason why ecology is a useful scientific enterprise.

2. Using observations, experiments, and models at different levels of organization, describe how you might test the hypothesis that the widespread death of the ochre sea star (*Pisaster ochraceus*) (shown in Figure 53.1) is the result of a disease.

We will begin our study of ecology in the spirit of the discipline in Haeckel's time, focusing on variations in the distribution and abundance of organisms worldwide and the factors that determine these patterns. Our first step will be to examine attributes of the physical environment, which is the ultimate determinant for where organisms can live. We'll begin with climate.

> key concept
># 53.2   Global Climate Is a Fundamental Component of the Physical Environment

The terms "weather" and "climate" both refer to atmospheric conditions—temperature, humidity, precipitation, and wind direction and velocity—but they consider different time scales. **Weather** is the short-term state of atmospheric conditions at a particular place and time, whereas **climate** refers to the average atmospheric conditions, and the extent of their variation, at a particular place over a longer time (years to millennia).

focus your learning

- "Weather" and "climate" refer to atmospheric conditions at different time scales.
- Earth's spherical shape creates latitudinal variation in solar radiation absorption and thus differences in temperature and precipitation; the proportion of land to ocean is also important.
- Earth's rotation around its axis deflects air and water and creates prevailing winds and ocean currents.
- The tilt of Earth's axis and its orbit create seasons.

Climate is what you expect given past conditions; weather is what you get on any one day. For example, a warming climate will cause glaciers to melt over decades or centuries (see Figure 1.16), whereas a single large snowstorm is unlikely to thwart that retreat. In addition, the responses of organisms to weather are usually short-term—seeking shelter from a sudden rainstorm, for example, or shivering to keep warm when the temperature drops. Climate, by contrast, acts as a sort of filter for organisms, ultimately determining their distribution and abundance. If organisms cannot tolerate the climate of a particular place, they will not be found there. What are the processes that create global climate patterns?

Solar radiation drives global climate patterns

The energy that drives global climate patterns ultimately originates from the sun. Of the solar radiation that hits Earth, about 30

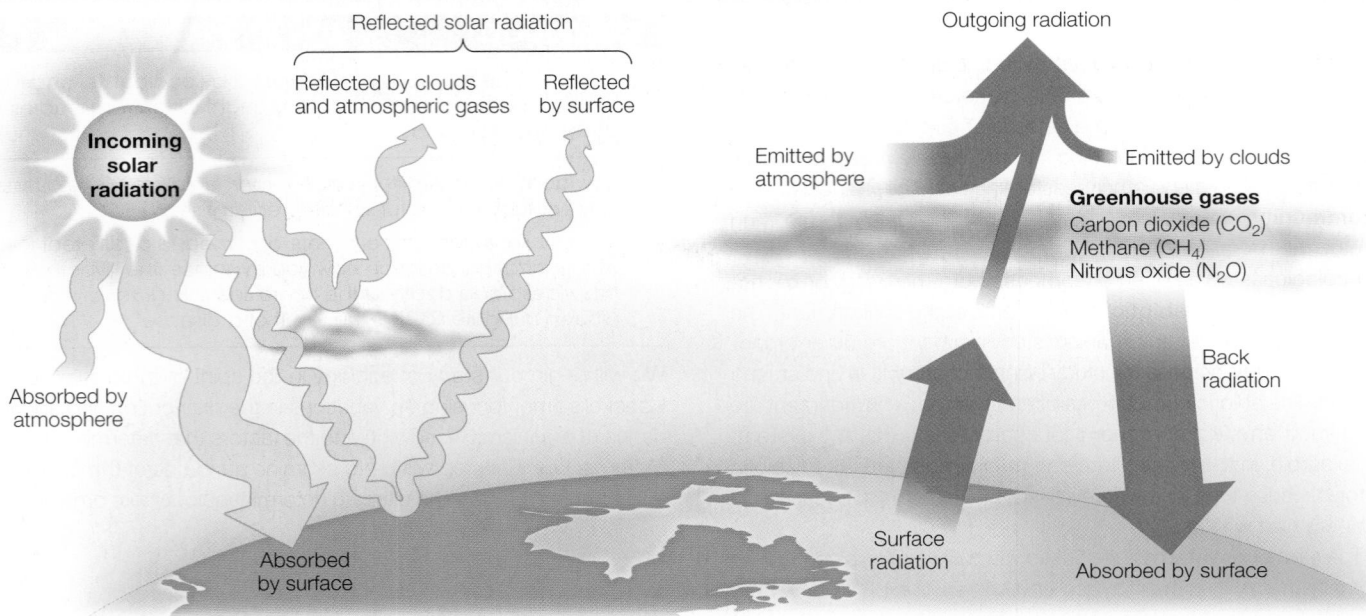

Figure 53.2 Radiant Energy Warms the Planet Solar energy input (yellow arrows) is absorbed by Earth's atmosphere and surface. Much of this energy is radiated from Earth's surface in the form of heat (orange arrows). Of that absorbed at the surface, most is prevented from escaping back into space by greenhouse gases in the atmosphere. The widths of the arrows are roughly proportional to the sizes of the energy fluxes.

Q: What effect do greenhouse gases have on Earth's energy balance?

 Animation 53.1 Radiant Energy Balance
www.Life11e.com/a53.1

percent is reflected back to space by clouds, atmospheric gases, and Earth's surface (**Figure 53.2**). The rest is absorbed either by the atmosphere (20%) or the surface of Earth (50%). The **atmosphere**, a thin layer of gases surrounding Earth composed of nitrogen gas (N_2, 78%), oxygen (O_2, 21%), argon (0.9%), carbon dioxide (CO_2, 0.037%), and traces of hydrogen, helium, ozone, methane (CH_4), and many other gases, moderates Earth's surface temperatures by trapping heat energy. If Earth had no atmosphere, its average surface temperature would be about −18°C, rather than its actual +17°C. Carbon dioxide, methane, nitrous oxide (N_2O), water vapor, and certain other gases in the atmosphere are known as **greenhouse gases** because they allow sunlight into Earth's atmosphere but trap heat radiating back out toward space (see Figure 53.2).

The amount of radiation received on Earth varies with latitude and with the seasons. These differences are driven by some well-known characteristics of Earth: it is a sphere that orbits the sun, it spins on a tilted axis, and it has both land and water. As you will see, it is the combination of these factors that determines the climate we experience on Earth.

Earth is a sphere, creating latitudinal variation in global temperature and precipitation

Because Earth is a sphere, the intensity of solar radiation hitting its surface varies with latitude. This latitudinal variation in solar energy depends primarily on the angle of the sun's rays striking Earth. At high latitudes (i.e., areas toward the North and South poles), incoming solar energy is distributed over a larger area (and thus is less intense) than at the equator, where sunlight strikes the surface perpendicularly (**Figure 53.3**). Moreover, when the sun's radiation comes in at an angle, it must pass through more of Earth's atmosphere, resulting in more of its energy being absorbed or reflected before reaching Earth's surface. Differences in solar energy input explain why the average global temperature varies with latitude, with warmer temperatures at the equator and colder temperatures at the poles (**Figure 53.4A**). The average air temperature over the course of a year decreases about 0.76°C for every degree of latitude (about 110 km) at sea level.

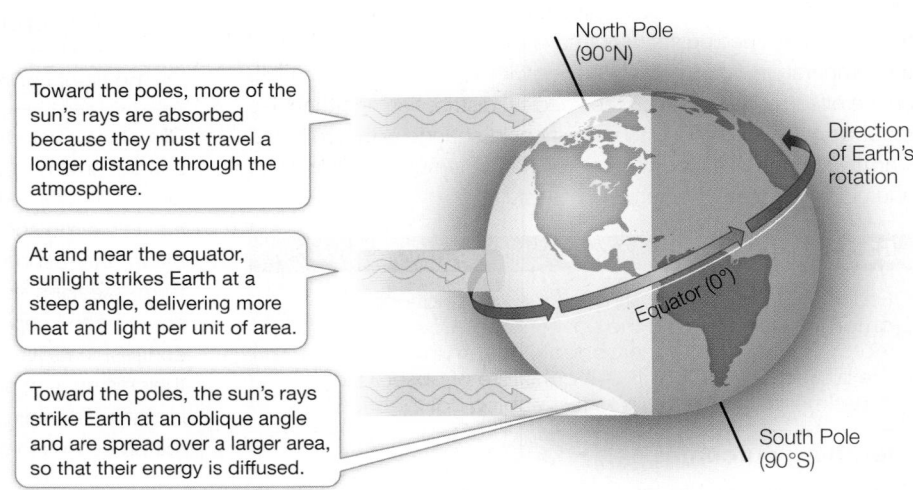

Toward the poles, more of the sun's rays are absorbed because they must travel a longer distance through the atmosphere.

At and near the equator, sunlight strikes Earth at a steep angle, delivering more heat and light per unit of area.

Toward the poles, the sun's rays strike Earth at an oblique angle and are spread over a larger area, so that their energy is diffused.

Figure 53.3 Solar Energy Input Varies with Latitude The angle of incoming sunlight affects the intensity of solar energy that reaches a given unit of Earth's surface.

(A) Average annual air temperature

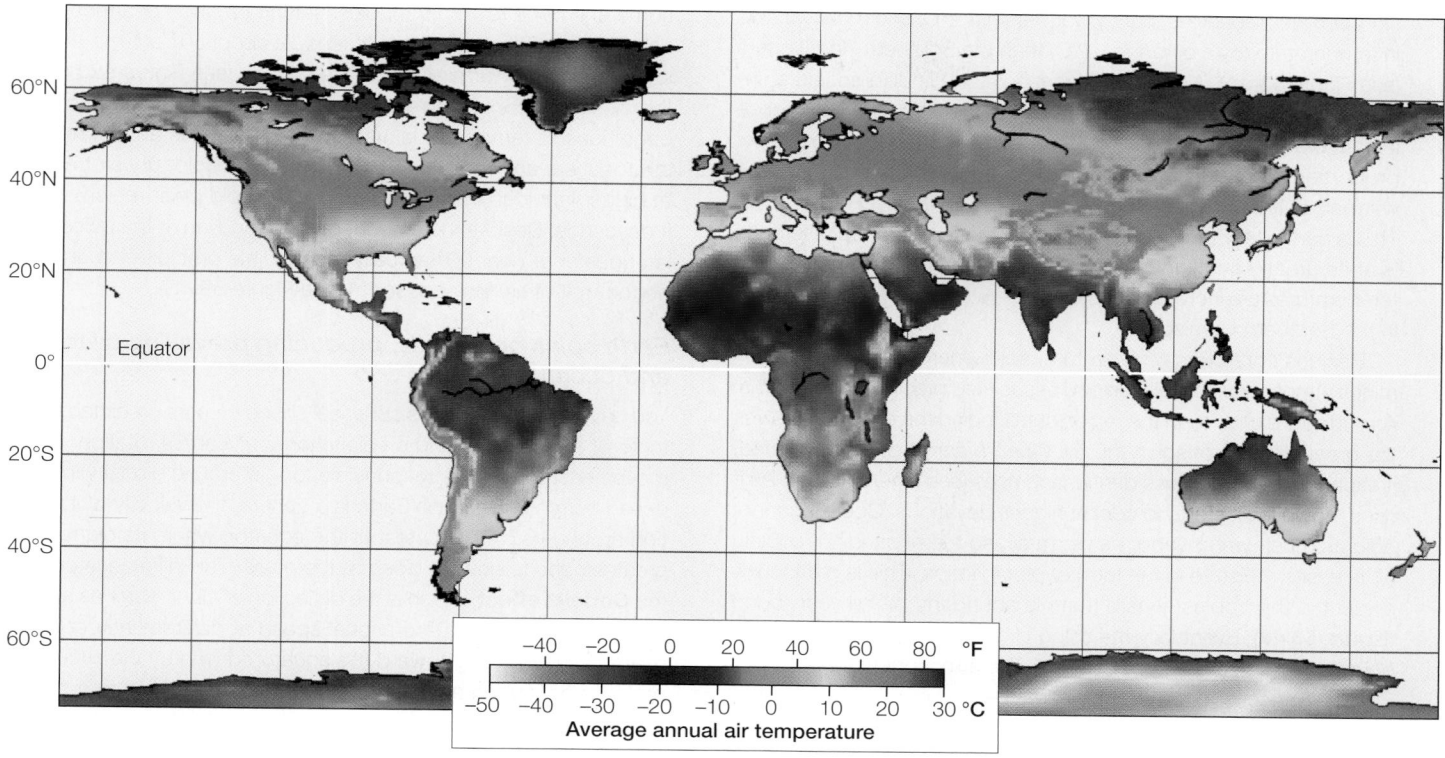

(B) Average annual precipitation

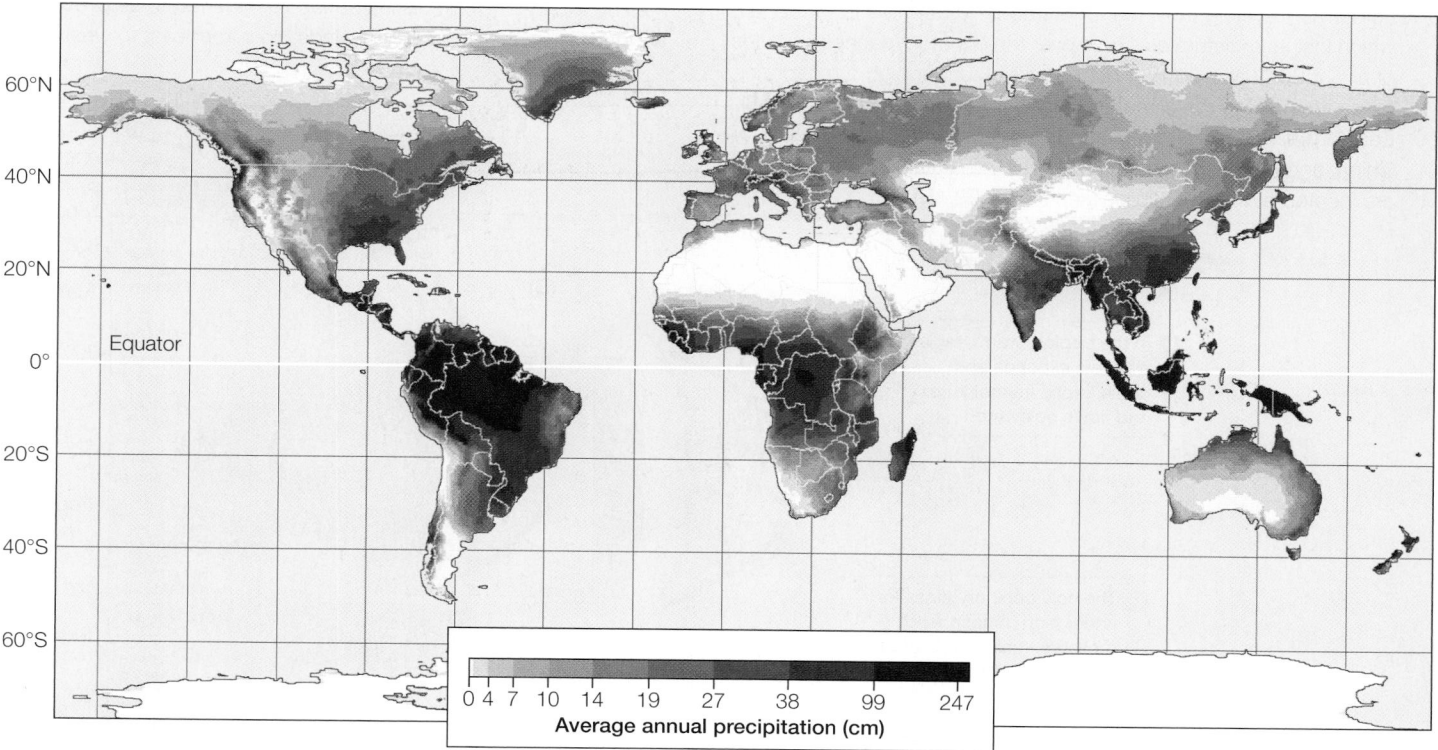

Figure 53.4 Global Temperature and Precipitation Vary with
Latitude   Average annual air temperature **(A)** and precipitation
(B) vary with Earth's latitude but can be modified by topography
(i.e., Earth's shape and surface features) (see Figure 53.9).

In addition to there being a latitudinal variation in temperature, the Northern Hemisphere is slightly warmer (by ~2°C) and more variable in temperature (by a of range ~7°C) than the Southern Hemisphere (see Figure 53.4A). Although the reasons for this difference are un-certain, one hypothesis is that because land heats up the air above it more than does the ocean (which absorbs more heat), the greater proportion of land to ocean in the Northern Hemisphere creates warmer air temperatures than those in the Southern Hemisphere. This is particularly true in summer, when the temperature difference is at its greatest. In light of global warming, it is worth noting that temperatures are increasing faster in the Northern Hemisphere than in the Southern Hemisphere.

Besides controlling temperature, solar energy input also deter-mines atmospheric circulation and associated precipitation patterns. As you saw earlier, air in the region surrounding the equator receives the greatest input of solar energy. When a parcel of air is warmed, it expands, becomes less dense, and rises (see Figure 53.5A). As it rises, however, it starts to cool at higher elevations. Cool air cannot hold as much water vapor as warm air, so the expanding, cooling air releases moisture in the form of precipitation. This is particularly evident in the tropics, where there is abundant rainfall year-round (**Figure 53.4B**). Eventually the rising air at the equator reaches the boundary between the troposphere, the atmospheric layer above Earth's surface, and the stratosphere, the next layer up. The strato-sphere is warmer than the troposphere at this boundary, causing the tropical air to stop rising and eventually move either north or south as newly warmed air rises to replace it (**Figure 53.5A**). Eventually this air starts to cool as it exchanges heat with the atmosphere and meets colder air traveling from higher latitudes. When this cool, dry air, which lost its moisture as it rose over the equator, reaches latitudes of about 30°N and 30°S, it begins to sink under high atmospheric pressure, making the climate at these latitudes dry (**Figure 53.5B**). Earth's great deserts—including the Sahara of Africa, the Gobi of China, and the deserts of Australia and the American Southwest—are located at these latitudes (see Figure 53.4B).

While some of the descending air flows back toward the equator, some of it flows toward the poles, setting up another atmospheric cell of air circulation at higher latitudes. At about 60° latitude, air rises again due to its warming at Earth's surface (see Figure 53.5B). As this air rises, it cools and releases precipitation, creating wet climate conditions at temperate latitudes. Eventually this air reaches the stratosphere and moves north or south. The cold, dry air traveling at higher latitudes eventually reaches the cold polar region, where it descends. Despite the amount of snow and ice at the poles, high latitudes at or near 90° actually receive little precipitation and can be thought of as "polar deserts" (see Figure 53.4B).

Earth spins on an axis, producing prevailing winds and ocean currents

You have seen that Earth's spherical shape creates predictable pat-terns of temperature and precipitation, but Earth's rotation, which moves east to west, is responsible for generating global wind and ocean currents. Because Earth is a sphere, the velocity of its rota-tion around its axis is fastest at the equator, where its diameter is greatest, and is slowest close to the poles. This difference creates the **Coriolis effect**, which is the deflection of air or water as a result of differences in Earth's rotational speed at different latitudes. For example, air traveling toward the equator (driven by the circulation patterns described above and in Figure 53.5) moves at a slower speed than that of the planet beneath it and is thus deflected to the west. Conversely, air traveling toward either pole is moving faster than that of the surface beneath it and is deflected to the east. This interaction of Earth's rotation and north–south air mass movement sets up a pattern of circulating surface air referred to as **prevailing winds** (**Figure 53.6**). Prevailing winds blow from east to west in the

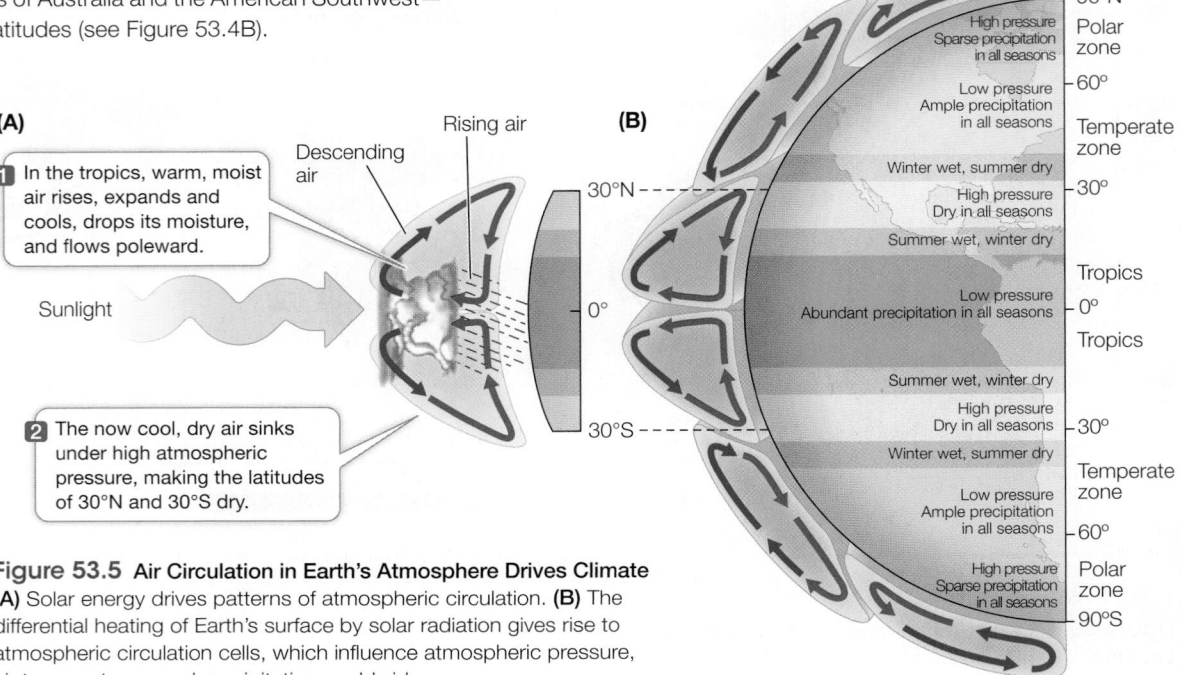

Figure 53.5 Air Circulation in Earth's Atmosphere Drives Climate
(A) Solar energy drives patterns of atmospheric circulation. **(B)** The differential heating of Earth's surface by solar radiation gives rise to atmospheric circulation cells, which influence atmospheric pressure, air temperatures, and precipitation worldwide.

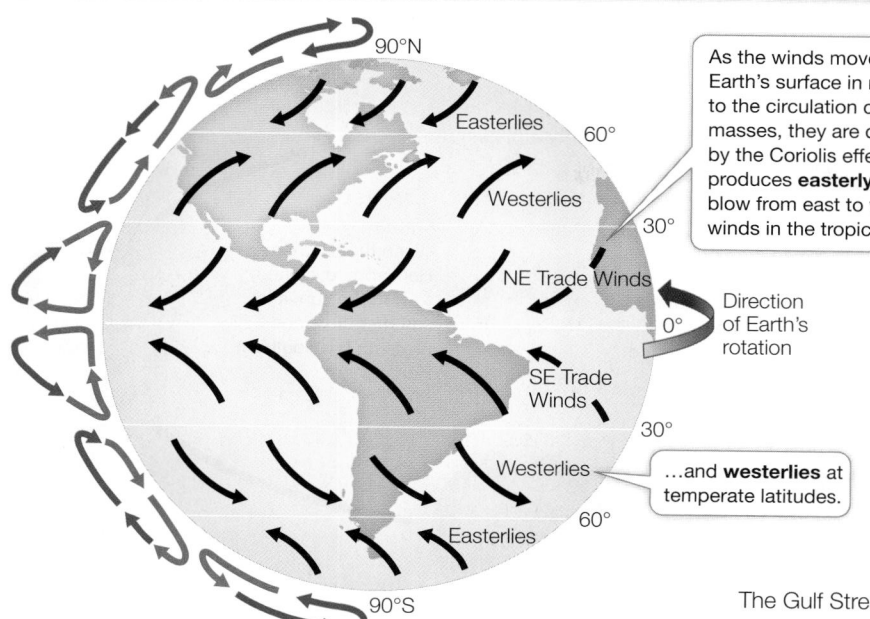

As the winds move across Earth's surface in response to the circulation of air masses, they are deflected by the Coriolis effect, which produces **easterly** (winds blow from east to west) trade winds in the tropics...

...and **westerlies** at temperate latitudes.

Figure 53.6 Prevailing Winds Created by Earth's Rotation and Atmospheric Circulation   The deflection of air as a result of differences in Earth's rotational speed at different latitudes (also known as the Coriolis effect) coupled with atmospheric circulation of air masses combine to create a global pattern of prevailing surface winds across the planet.

tropics (the **trade winds**); from west to east in mid-latitudes (the **westerlies**); and from east to west again above 60°N or 60°S latitude (the **easterlies**).

How, then, are the circulation patterns of ocean currents generated? Ocean currents are driven by prevailing winds, which move water by means of frictional drag. The trade winds, for example,

cause currents to converge at the equator and move westward until they encounter a continent (**Figure 53.7**). At that point, the strong Equatorial Countercurrent brings some of the water back eastward. The remaining water divides, some moving northward and some southward along continental shores. These patterns of water movement set up rotating circulation patterns called **gyres** (Greek *gyros*, "spiral"), which rotate clockwise in the Northern Hemisphere and counterclockwise in the Southern Hemisphere.

Because ocean currents transport heat, they have a tremendous effect on Earth's climates. The poleward movement of warm water from the tropics transfers large amounts of heat to high latitudes. The Gulf Stream, for example, carries warm water from the tropical Atlantic Ocean (including the Gulf of Mexico) north across the Atlantic to northern Europe, making Europe's climate considerably milder than that of corresponding latitudes in North America. Similarly, currents flowing toward the equator from high latitudes bring cool, wet winters to some western coastal regions that are otherwise warm and dry.

The tilt of Earth's axis and its orbit result in seasons

The tilt of Earth's axis, coupled with Earth's orbit around the sun, is responsible for seasonal changes in climate. Earth's axis is tilted at an angle of 23.5 degrees, changing the amount of sunlight a

Figure 53.7 Oceanic Circulation Created by Winds and Continents   The surface currents of the ocean are driven by the prevailing winds (see Figure 53.6), Earth's rotation, and the placement of the continents. Red arrows indicate warm surface water, and blue arrows indicate cold surface water.

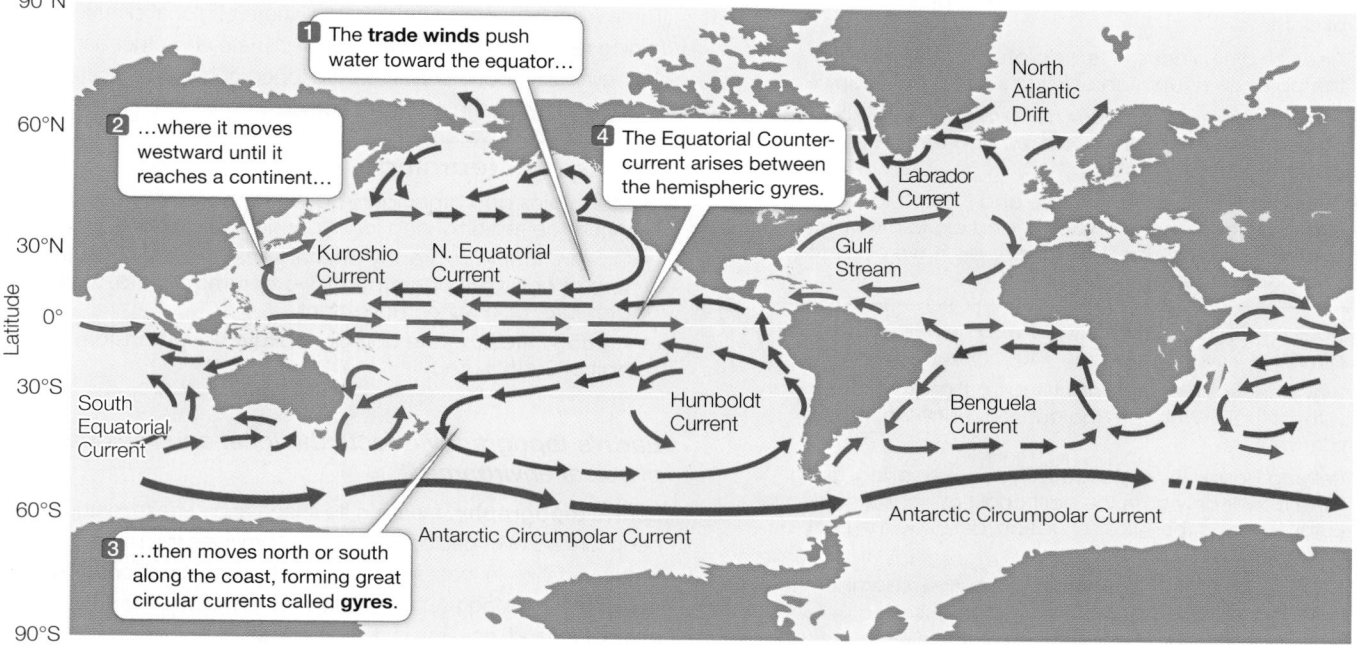

1 The **trade winds** push water toward the equator...

2 ...where it moves westward until it reaches a continent...

4 The Equatorial Countercurrent arises between the hemispheric gyres.

3 ...then moves north or south along the coast, forming great circular currents called **gyres**.

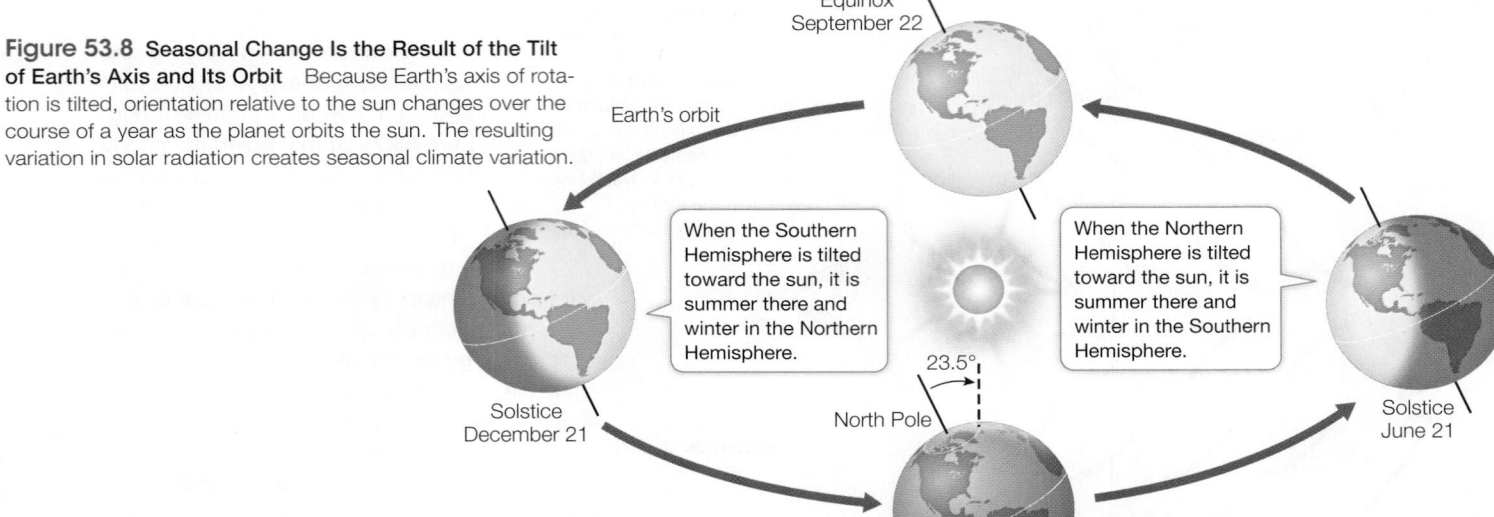

Figure 53.8 Seasonal Change Is the Result of the Tilt of Earth's Axis and Its Orbit Because Earth's axis of rotation is tilted, orientation relative to the sun changes over the course of a year as the planet orbits the sun. The resulting variation in solar radiation creates seasonal climate variation.

Equinox September 22

Earth's orbit

When the Southern Hemisphere is tilted toward the sun, it is summer there and winter in the Northern Hemisphere.

When the Northern Hemisphere is tilted toward the sun, it is summer there and winter in the Southern Hemisphere.

Solstice December 21

23.5°

North Pole

Solstice June 21

Equinox March 20

particular region receives over the course of a year as Earth orbits the sun (**Figure 53.8**). This tilt causes seasonal variation in temperature and day length. Higher latitudes experience greater seasonal variation than lower latitudes do. Around the equator, day length and seasonal temperatures change only slightly over the course of the year, although there are seasonal shifts in precipitation patterns.

53.2 recap

Climate and weather both refer to atmospheric conditions, but climate represents conditions over years whereas weather includes conditions over a span of days. Earth is a sphere, has an atmosphere, and rotates around its axis, which creates latitudinal variation in solar radiation, temperature and precipitation, and prevailing winds and ocean currents. Earth's tilt and orbit around the sun result in the seasons.

learning outcomes

You should be able to:

- Compare weather and climate, and give examples of both.
- Analyze differences in temperature at the equator and the poles as a function of Earth's spherical shape.
- Relate Earth's ocean currents to prevailing wind patterns, and explain how ocean currents affect climate.
- Connect the tilt of Earth's axis and its orbit to the existence of seasons on Earth, and explain why the Northern and Southern Hemispheres have opposite seasonal patterns.

1. Does a month of extremely hot weather tell us that Earth's climate is warming? Explain your answer.

2. Why does solar energy input into Earth's atmosphere vary with latitude? What are the consequences for global climate?

3. Refer to Figures 53.6 and 53.7. If you were sailing, using wind power only, from the east coast of North America to Europe across the Atlantic Ocean, describe the best route given winds and currents.

4. Explain why Earth's Northern and Southern Hemispheres have opposite seasons.

We have considered large, global patterns in climate, but we know that the physical environment can vary within certain regions or even locally. Just travel across a large mountain range and you can experience lush forests on one side and bone-dry deserts on the other. We next turn to what controls smaller, more regional and local variation in the physical environment, both on land and in water.

key concept 53.3 Topography, Vegetation, and Humans Modify the Physical Environment

Three main factors can modify the pattern of global climate, determining regional and local variation in climate and other aspects of the physical environment: (1) Earth's topography, (2) vegetation, and more recently, (3) the effects of humans.

focus your learning

- Mountains and other land-based topographic features influence regional and local physical conditions.
- Ocean-floor topography affects water depth, which in turn affects light penetration, water temperature, water pressure, and water movement.
- Human infrastructure and agriculture have transformed half of Earth's area.

Earth's topography affects the local and regional physical environment

Earth's **topography**, meaning its shape and surface features, has been formed over geological time and provides the basis for much of the variation in local and regional physical conditions. Here we consider how topography modifies the physical environment with respect to land, oceans, rivers, and lakes.

(A) Rain shadow

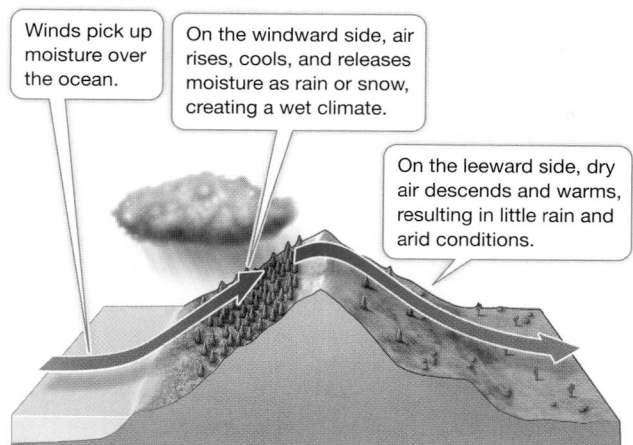

Winds pick up moisture over the ocean.

On the windward side, air rises, cools, and releases moisture as rain or snow, creating a wet climate.

On the leeward side, dry air descends and warms, resulting in little rain and arid conditions.

(B) Temperature inversion

Animation 53.2 **Rain Shadow**
www.Life11e.com/a53.2

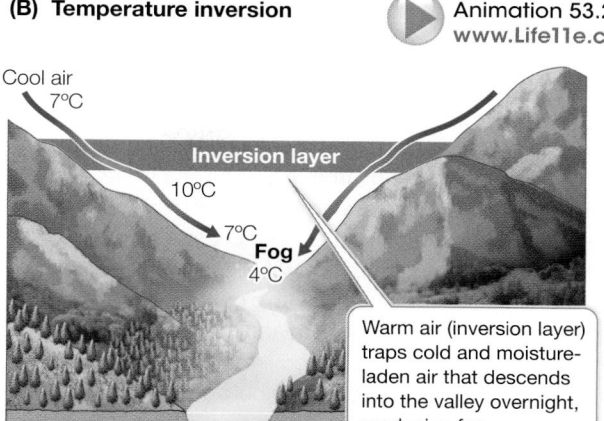

Cool air
7°C

Inversion layer

10°C

7°C

Fog
4°C

Warm air (inversion layer) traps cold and moisture-laden air that descends into the valley overnight, producing fog.

(C) Aquatic zones

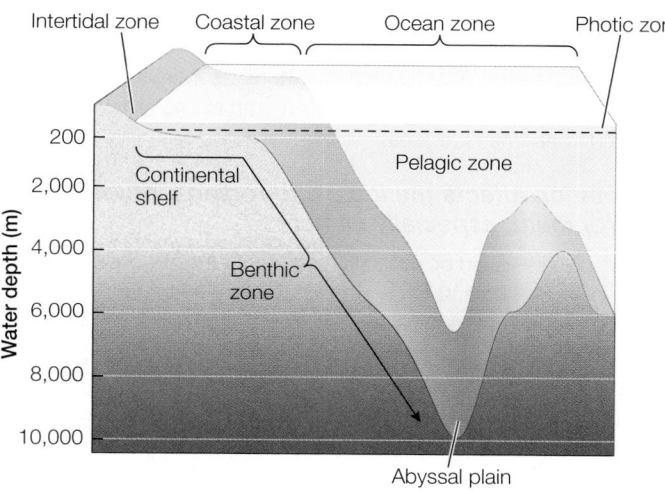

Intertidal zone   Coastal zone   Ocean zone   Photic zone

Water depth (m)

200

2,000

4,000

6,000

8,000

10,000

Continental shelf

Pelagic zone

Benthic zone

Abyssal plain

(D) Ocean upwelling

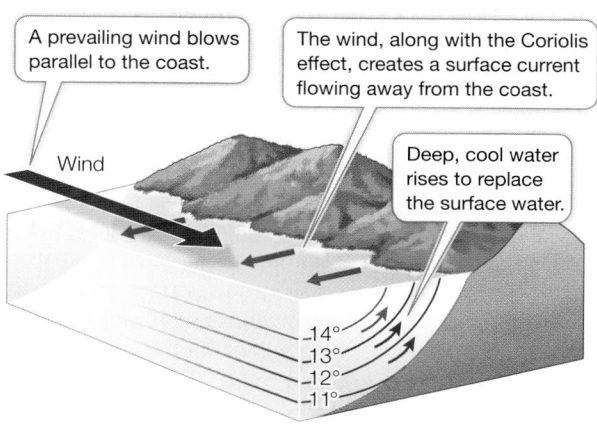

A prevailing wind blows parallel to the coast.

The wind, along with the Coriolis effect, creates a surface current flowing away from the coast.

Wind

Deep, cool water rises to replace the surface water.

14°
13°
12°
11°

Figure 53.9 **Effects of Earth's Topography on the Physical Environment**   Earth's topography can affect both land and oceans. Mountains and valleys create **(A)** rain shadows and **(B)** temperature inversions. Ocean topography affects water depth, which **(C)** creates aquatic zones that vary in light, temperature, and water pressure, and **(D)** along with wind, produces ocean upwelling.

Q: What kind of weather conditions would you expect to produce temperature inversions? How about ocean upwelling?

LAND   The topography of land plays a large role in the regional and local physical environment, including climate. Mountains create elevational gradients in temperature, precipitation and sunlight. For example, mountains show progressively colder temperatures and greater precipitation at higher elevations, creating different environmental conditions over relatively short distances. When a mountain range occurs adjacent to an ocean, a **rain shadow** is often created, where one side of the mountain has a wet climate and the other side has a dry climate (**Figure 53.9A**). Rain shadows occur when winds deliver moist air evaporated from the ocean to the windward side of the mountain, where the air rises, cools, and releases rain or snow. On the leeward side of the mountain (that is, opposite from the direction of the winds), the now-dry air descends, warms, and produces arid conditions. The High Desert, on the leeward side of the Cascade Range in the Pacific Northwest of the United States, is one such area.

Other topographic features, such as canyons and valleys, also modify local climate in significant ways. Valleys can experience **temperature inversions**, extremes in temperature that produce morning fog and intense afternoon heat. Temperature inversions occur when the valley concentrates heat from solar radiation during the day. As that heat rises, it forms an inversion layer that traps cold, dense, and moisture-laden air that descends into the valley overnight and produces fog (**Figure 53.9B**).

OCEANS   Beneath the oceans, which make up 70 percent of Earth's surface, dramatic differences in topography influence the marine environment. If you could drain the oceans, you would notice major differences in topography, ranging from the Hawaiian island Mauna Loa, the tallest "mountain" on Earth at 9,170 meters high from base to peak, to the Marianas Trench, the deepest "valley" at 10,994 meters deep. Variations in water depth affect light penetration, water temperature, water pressure, and water movement (i.e., waves and tides). These physical discontinuities create distinct zones that are identified principally by their physical conditions and the characteristic biota they contain, similar to biomes on land (see Key Concept 53.4). The depth of an ocean basin varies from the shoreline to the relatively shallow continental shelf and to the deepest part of the ocean, sometimes known as the abyssal plain (**Figure 53.9C**). Water depth affects how much light is available to sustain the photosynthetic organisms that form the base of the marine food chain. In both marine and freshwater environments, roughly 90 percent of all aquatic life is found in the **photic zone**, the layer of water reached by enough sunlight to support photosynthesis.

Other zones that vary in environmental conditions include the **coastal zone**, which extends from the shoreline to the edge of the continental shelf, and the **ocean zone**, which extends beyond the coastal zone (see Figure 53.9C). The coastal zone is diverse and highly productive, supporting high densities of pelagic, or floating, plankton and fish. On the benthic or bottom of the ocean, the coastal zone has a variety of living and nonliving habitats such as rocky reefs, soft sediment flats, coral reefs, and kelp beds. The portion of the benthic coastal zone lying between the high and low tide levels is the **intertidal zone**, where tidal movements create conditions of highly variable light and temperature, alternately exposing organisms to air and water. Intertidal zones can occur on the open coast as sandy or rocky beaches, or in more protected bays as estuaries. **Estuaries** form where rivers meet the sea, creating variability in salinity, sediment, and light conditions.

In some coastal zones, where the continental shelf is steep and prevailing winds are parallel to the coastline, upwelling can affect ocean and climate conditions. **Upwelling** is a process in which offshore winds in combination with the Coriolis effect push warmer surface waters away from the shore, allowing deeper, colder, and nutrient-rich bottom water to rise to the surface (**Figure 53.9D**). Upwelling affects local and regional climate on the coast by creating cooler and moister conditions. In addition, upwelled waters support high rates of primary production by phytoplankton, which in turn support dense consumer populations. Most of the world's great fisheries are concentrated in upwelling zones.

As we move away from the coastline, most of the ocean's physical environment is made up of vast amounts of water that is, with the exception of the photic zone (which can extend to 200 meters in some regions), mostly dark, cold, and under tremendous pressure from the weight of the water above. Little photosynthesis can take place, and many of the organisms inhabiting these regions subsist on decaying organic matter that sinks down from the photic zone. Deep-ocean trenches and rift valleys support hydrothermal vent ecosystems sustained by chemoautotrophic prokaryotes that can metabolize the nutrients without the aid of sunlight (see Key Concept 25.3).

RIVERS AND LAKES   In contrast to the vast oceans, freshwater lakes and rivers make up less than 3 percent of Earth's surface, but they are home to about 10 percent of all aquatic species. Earth's topography influences freshwater environments just as it does in the oceans—through the depth of the water—but it also influences the degree and direction of movement of fresh water. A river is water that flows downhill from its point of origin (the source) over the land surface as a result of gravity until it reaches either the ocean or a depression in Earth's surface, where it forms a lake or pond. The headwaters (those close to the source) tend to be cool, fast-flowing, and well oxygenated. As a river flows downstream, it widens, it slows down, it warms, and sediment accumulates, reducing light penetration and oxygen levels.

The Amazon River, mentioned in the chapter opening, offers a dramatic example of river dynamics. The Amazon is one of the longest rivers on Earth, at approximately 7,000 km. Eventually this source water reaches the Atlantic Ocean, where discharge is the largest of any river on Earth—the Amazon's mouth is 240 km wide and is referred to as "the River Sea."

Vegetation affects the local and regional physical environment, especially climate

Earth's climate would be very different without the influence of organisms, including humans. Organisms can "engineer" the environment in multiple ways that affect the surrounding climate and physical conditions. The most obvious influence comes from vegetation, especially forests, which can have large effects on temperature and precipitation at local and even regional scales. For example, measurements of energy exchange in tropical forests that have been converted to pastureland show that forests make the climate cooler and moister (**Figure 53.10**). While forests absorb heat by reflecting less sunlight

Figure 53.10 Deforestation Can Affect Climate   Evapotranspiration by trees reduces air temperatures and increases precipitation. When land is deforested and the trees are gone, the cooling effects of the albedo (reflection of solar radiation) and of convective heat loss are not enough to outweigh the warming effects of reduced evapotranspiration. The widths of the arrows are roughly proportional to the sizes of the fluxes.

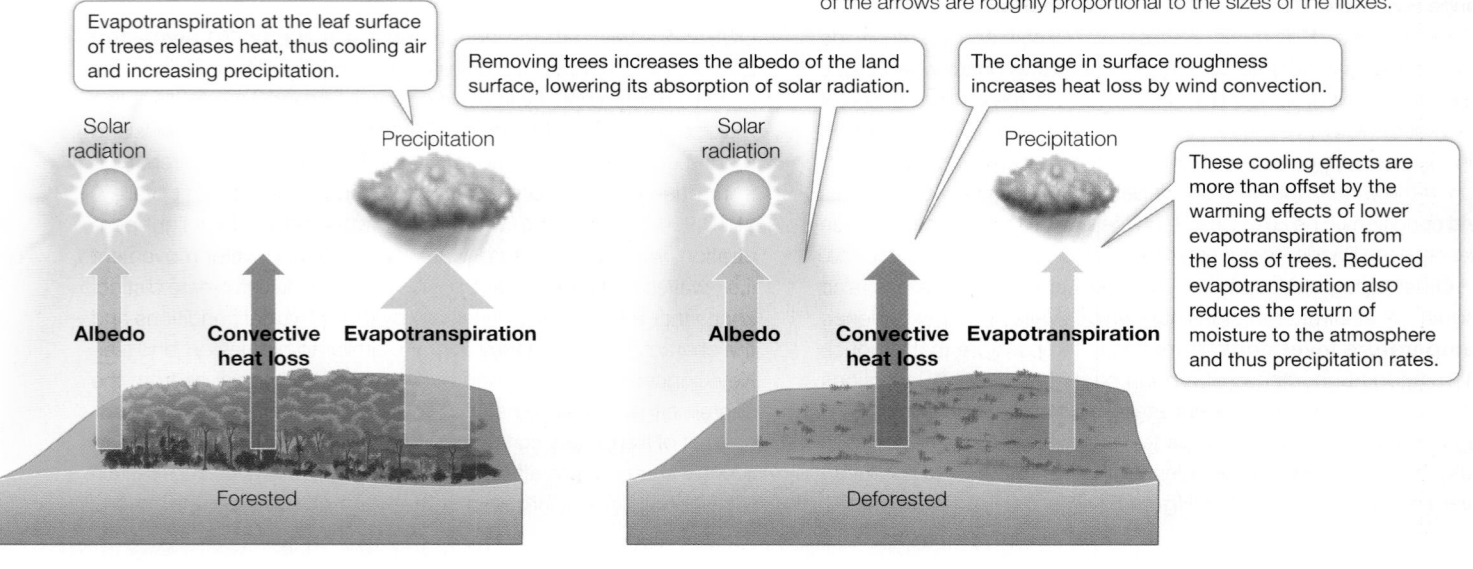

Figure 53.11 Cities Can Create Heat Islands and Country Breezes As this hypothetical urban heat island profile shows, cities and suburbs typically have higher temperatures than rural, park, or agricultural areas. Country breezes form when warm air from cities and suburbs is replaced by cool air from rural areas.

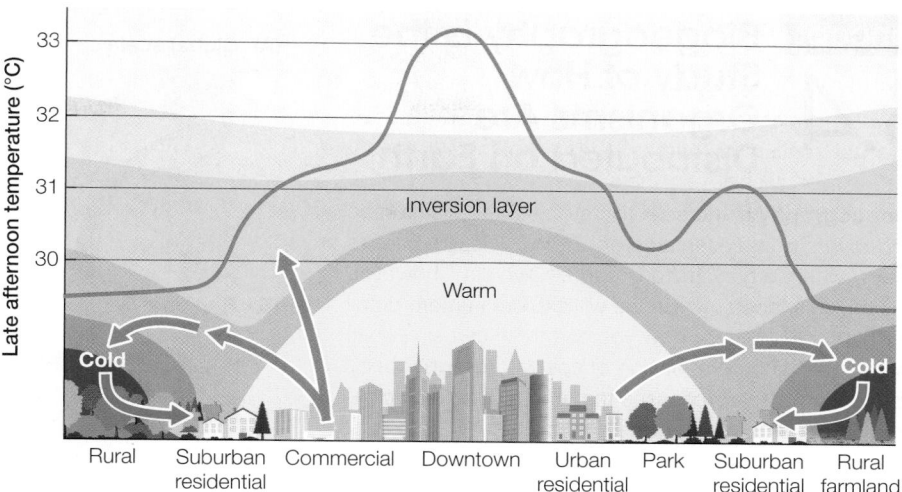

(reflection of solar radiation is known as the **albedo effect**) and decrease convective heat loss by acting as a wind screen, this is more than balanced by the cooling effect produced by **evapotranspiration**: the evaporative transfer of heat and water from the surfaces of plants into the atmosphere, which reduces air temperature and increases moisture. When forests are cut and replaced by vegetation such as grasses, the albedo effect and convective heat loss intensify while evapotranspiration decreases, causing overall air temperatures to rise and precipitation to decrease, creating more arid conditions. The climate changes associated with deforestation have been seen in the Amazon Basin, where massive forest clearing makes way for cattle ranching.

Oceans have their own "forests"—coastal areas dominated by seaweeds such as giant kelps or seagrasses that create calmer waters, promote sedimentation, and provide habitat for fish and invertebrates. In warmer coastal waters, corals generate complex reef structures that support ecosystems rivaling rainforests in diversity. In Chapter 56 you will learn more about organisms that engineer their environment, but for now let's turn to the final topic in this section: the effects of humans on climate and the physical environment.

Humans have transformed their physical environment, including urban climate

It is estimated that humans have transformed more than 50 percent of Earth's land surface, replacing it with infrastructure and agriculture needed to support our burgeoning population. You have already seen how the conversion of forestland to pastureland can change regional climate, and in Chapter 57 you will learn more about how humans have changed global climate through the burning of fossil fuels. Structures such as dams, bridges, and nuclear power plants can dramatically transform rivers and lakes. Urban areas, in particular, modify not only the immediate physical environment but also the surrounding climate. A survey of 60 U.S. cities in 2014 showed that urban summer temperatures were on average 1.5°C hotter than rural summer temperatures, with the top 10 cities having an average urban–rural temperature differential of 5°C. The cause of this so-called urban heat island effect is simple: concrete, asphalt, and even the dark roofs of buildings absorb heat from solar radiation and radiate that heat

in the evening hours (**Figure 53.11**). Cities also produce more heat through the burning of fossil fuels from cars, factories, and buildings. Parks can serve to cool the surrounding air through the evapotranspiration of trees and other vegetation. In addition, the temperature differential between cities and the surrounding countryside can create "country breezes," air movement generated as rising warm air from the city is replaced by cold surface air from surrounding rural areas.

53.3 recap

Earth's topography is responsible for rain shadows and temperature inversions, producing variation in temperature and precipitation. In the oceans, topographical variability produces differences in water depth, which affects light penetration, water temperature, water pressure, and water movement. Vegetation, especially forests, affects climate through the process of evapotranspiration. Humans have transformed urban climate; cities are on average hotter than rural areas.

learning outcomes

You should be able to:

- Explain and compare the causes and effects of rain shadows and temperature inversions.
- Explain why ocean-floor topography is important in determining the marine environment.
- Describe and contrast ways in which agriculture and human infrastructure (such as urbanization) may transform local and regional climates.

1. What are some of the major ways that Earth's topography affects land and ocean physical conditions?

2. Why are cities warmer than rural areas, and what are country breezes? How and why could the temperature within a city be changed with the establishment of a large forested park?

With an understanding of how Earth's physical environment, including climate, is determined at a variety of spatial scales, we turn next to biogeography: how the physical environment and other geographic and historical processes determine the global distribution of organisms on Earth.

<key concept>

53.4 Biogeography Is the Study of How Organisms Are Distributed on Earth

Biogeography is the scientific study of the distribution and diversity of organisms on Earth. "Distribution" is simply where organisms are found. "Diversity," however, has multiple facets, which we will explore in more detail in Chapter 56, but for now you can think of it simply as the number and composition of species or higher taxa. For centuries, the processes that control these biogeographic patterns have puzzled naturalists. Why do species vary in their distributions around Earth, and what controls this variability?

focus your learning

- Biomes cross continents, occurring where climatic conditions are similar. Earth's major biomes include tropical rainforests, deserts, temperate grasslands, temperate deciduous forests, temperate evergreen forests, boreal forests, and tundra.
- Alfred Russel Wallace observed distributional differences in biota on the Malay Archipelago that led to his theory of biogeography.
- Hypotheses to explain the causes of latitudinal variations in diversity include species diversification rate, species diversification time, and productivity.

Patterns of biogeography are interconnected across a hierarchy of spatial and temporal scales

The patterns of biogeography play out over global, regional, landscape, and local spatial scales (**Focus: Key Figure 53.12**). Spatial scales are interconnected in a hierarchical way, with patterns at one scale setting the conditions for patterns at other scales.

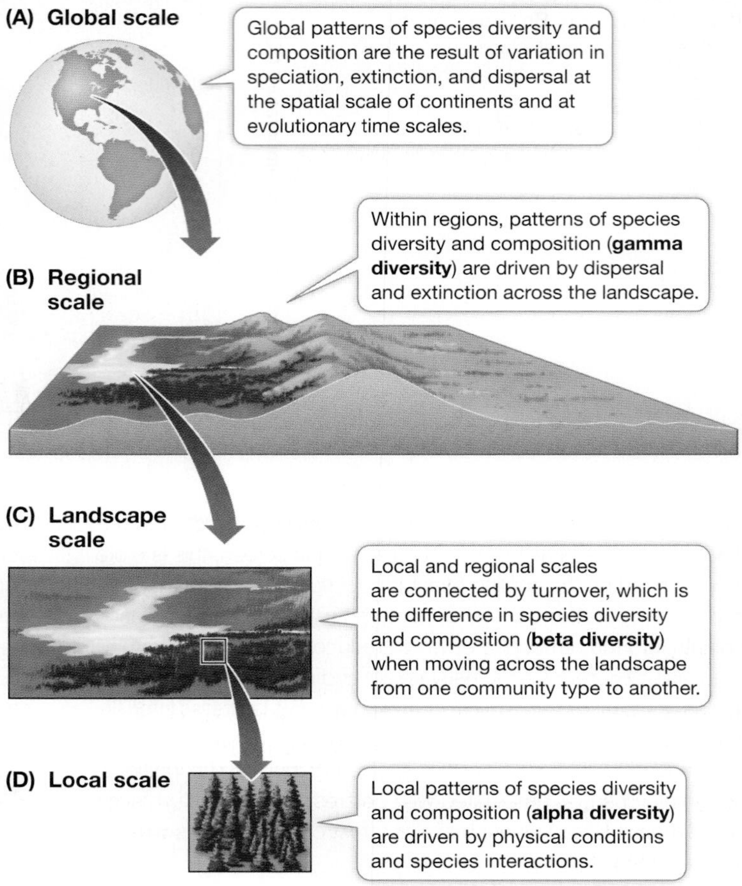

At the global scale, species are isolated from one another by continents or oceans, and for long periods of time. Thus the rates of three processes—speciation, extinction, and dispersal—help determine differences in biogeography across the globe (see Figure 53.12A). At the regional scale, large-scale climate conditions are relatively uniform and species are limited by their ability to disperse, or move between locations (see Figure 53.12B). All the species that are limited to a region are part of the **regional species pool** (also called **gamma diversity**). At the landscape scale, there are multiple community types—for example, lakes, rivers, forests, and prairies—arranged in a spatially heterogeneous way across the region (see Figure 53.12C). The change in the number of species from one community to another across the landscape is known as **beta diversity**. Finally, the local scale (also called **alpha diversity**) typically encompasses a single community—a group of species assembled together as the result of the physical environment, biotic interactions with other species, and the regional species pool (see Figure 53.12D).

focus: key figure

(A) **Global scale**

Global patterns of species diversity and composition are the result of variation in speciation, extinction, and dispersal at the spatial scale of continents and at evolutionary time scales.

(B) **Regional scale**

Within regions, patterns of species diversity and composition (**gamma diversity**) are driven by dispersal and extinction across the landscape.

(C) **Landscape scale**

Local and regional scales are connected by turnover, which is the difference in species diversity and composition (**beta diversity**) when moving across the landscape from one community type to another.

(D) **Local scale**

Local patterns of species diversity and composition (**alpha diversity**) are driven by physical conditions and species interactions.

Figure 53.12 Biogeography Depends on Interconnected Spatial and Temporal Scales The arrows represent the relationship between, and processes important to, species diversity and composition at **(A)** global, **(B)** regional, **(C)** landscape, and **(D)** local scales.

Q: At what scales is the regional species pool important to species diversity and composition?

In the remainder of this chapter we will consider the distributional and diversity patterns of organisms at global and regional biogeographic scales. Chapter 56 will consider the causes and consequences of diversity at community and local scales.

Terrestrial biomes reflect global patterns of temperature and precipitation

Biomes are groupings of ecologically similar organisms shaped by the environment in which they are found. The classification of biomes is most often and easily applied to terrestrial systems. Ecologists classify terrestrial biomes principally by the growth forms of their dominant plants, which reflect the evolution of those plants under annual patterns of temperature and precipitation.

The same classic types of biomes may be widely separated, occurring on different continents, depending in large part on the presence of suitable climatic conditions (**Figure 53.13**). For example, the desert biome occurs in such distant locations as Arizona in the

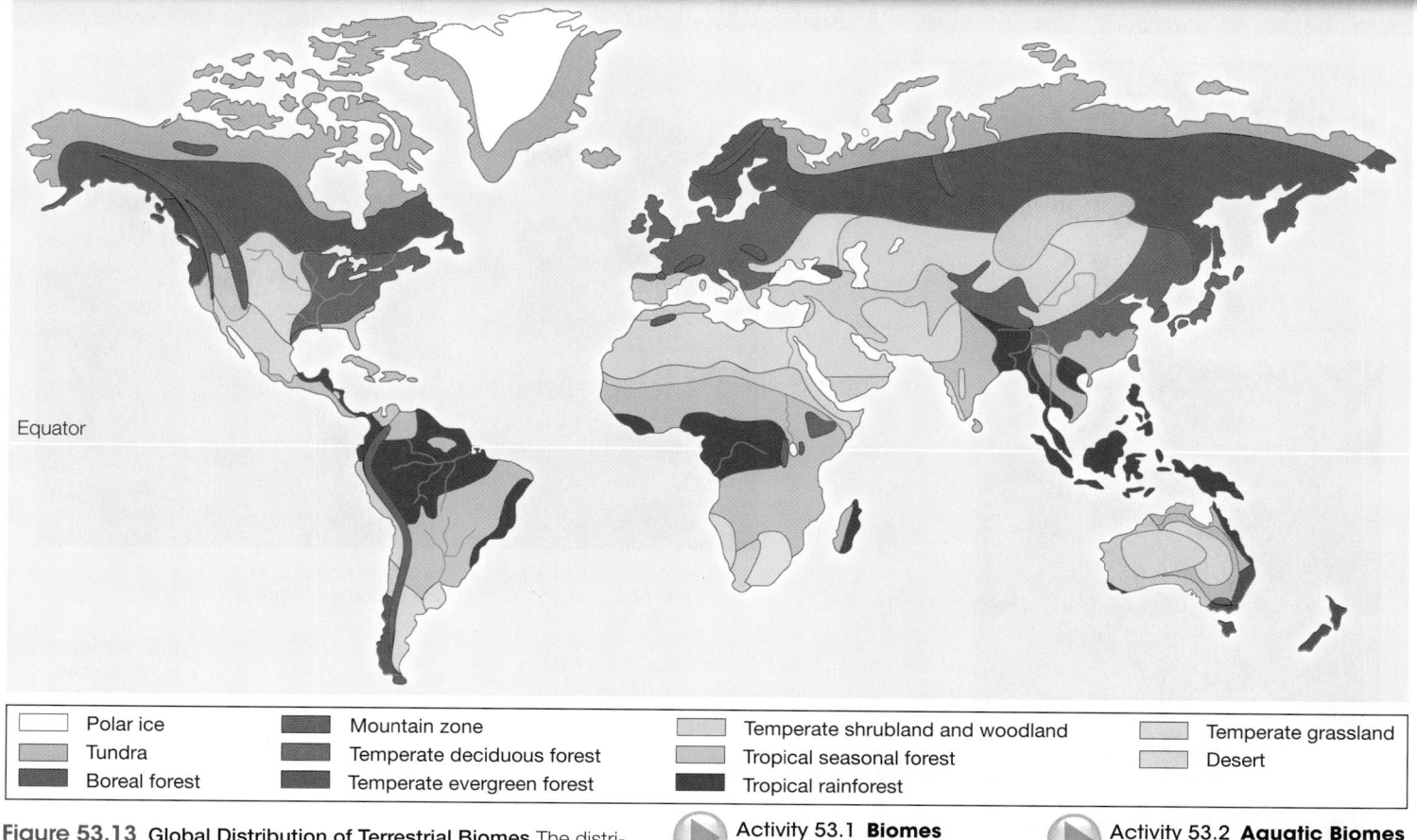

☐ Polar ice	■ Mountain zone	☐ Temperate shrubland and woodland	☐ Temperate grassland
☐ Tundra	■ Temperate deciduous forest	☐ Tropical seasonal forest	☐ Desert
■ Boreal forest	■ Temperate evergreen forest	■ Tropical rainforest	

Figure 53.13 Global Distribution of Terrestrial Biomes The distribution of terrestrial biomes is determined primarily by annual patterns of temperature and precipitation.

▶ **Activity 53.1 Biomes**
www.Life11e.com/ac53.1

▶ **Activity 53.2 Aquatic Biomes**
www.Life11e.com/ac53.2

Q: Does the Northern or Southern Hemisphere have a greater number of biomes? What do you think might determine this difference?

North American Southwest and the Namib Desert in Africa; both locations are extremely dry and dominated by succulent plants such as cacti and by drought-tolerant shrubs and grasses. Why use plants as a way of classifying biomes? Because plants are immobile, they must adapt to environmental conditions in order to be successful over long periods of time. Thus *the growth forms of plants strongly reflect their environment in ways that can be compared around the world. Furthermore, by providing three-dimensional structure, by modifying physical conditions near the ground, and by providing food, dominant plants strongly influence the organisms living there.

---

*connect the concepts   Plants growing in a particular type of environment (e.g., the desert) may display similar growth forms (e.g., succulence). See Key Concept 38.3.

In the following pages we briefly describe seven terrestrial biomes (a subset of those shown in Figure 53.13). For each, a plot of the seasonal patterns of temperature and precipitation—known as a **Walter climate diagram**—is included. Devised by the German biogeographers Heinrich Walter and Helmut Lieth, this graphic technique plots temperature and precipitation data in

a simple way that visualizes a "growing season"—those months when average temperatures are above freezing (0°C) and when average precipitation is sufficient for plant growth. Walter climate diagrams are predicated on the "rule of thumb" that plant growth requires temperatures above 0°C and at least 2 mm of precipitation for each 1°C rise in temperature above 0°C. They have two y-axis scales, one for temperature and one for precipitation; these axes align 0 mm of precipitation with 0°C of temperature. The x axis shows 12 months, with the summer solstice placed in the center of the axis.

TROPICAL RAINFOREST   The tropical rainforest biome is found in equatorial regions where rainfall and temperatures are high year round. With no season unsuitable for growth, it is the most productive and species-rich of all biomes. In some areas, there are up to 500 species of trees per square kilometer. Although historically these forests covered 12 percent of Earth's surface, and are now reduced through deforestation, they are home to more than half of all known species. These forests are home to many epiphytes—plants that grow on other plants, deriving their nutrients and moisture from air and water rather than soil.

Tropical rainforests provide humans with a range of products, including fruits, nuts, medicines, fuel, pulp, and furniture wood. Rainforests, however, are currently being cut down or converted to agriculture at a rate of almost 1 percent per year. In some cases, rainforests are recovering but the soils are often nutrient-poor, hindering regrowth.

TROPICAL RAINFOREST

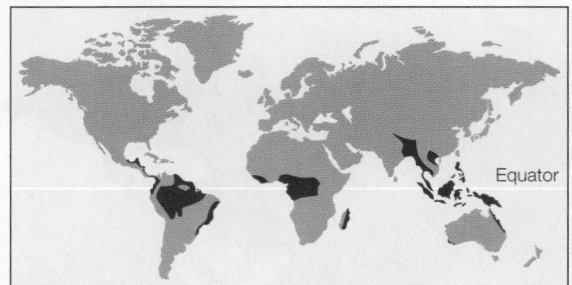

There are over 200 bird species in this tropical rainforest at Lambir Hills National Park, on the island of Borneo, Malaysia.

Bromeliads commonly grow on trees in tropical rainforests, such as this Ceibo tree in Ecuador, but can also be found in desert biomes.

Yanganbi, D.R.C.
0°
Average annual temperature 24.6°C
Total annual precipitation 1,828 mm

Note that the scale changes above 100 mm of precipitation.

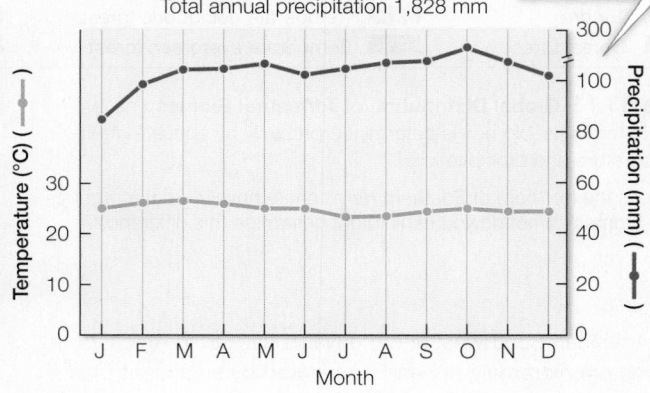

Mountain gorilla (*Gorilla gorilla beringei*) are critically endangered species. This family lives in the tropical rainforest of Volcanoes National Park, Rwanda.

A **work with the data** exercise that accompanies this figure may be assigned in **LaunchPad**.

DESERT The desert biome is concentrated in two belts, centered around 30°N and 30°S latitude (where warm, dry air sinks under high atmospheric pressure; see Figures 53.4 and 53.5). The driest of these regions, where rains rarely fall, are far from the oceans, as in the center of Australia and the middle of the Sahara in Africa.

Desert plants have several structural and physiological adaptations that help them conserve water, as described in Key Concept 38.3. Small desert animals are inactive during the hottest part of the day, remaining in underground burrows. Desert mammals have physiological adaptations for conserving water, including a reduced number of sweat glands and kidneys that produce highly concentrated urine. Many desert animals require no water beyond what they can extract from the carbohydrates in their food.

Humans have used deserts for livestock grazing and agriculture for centuries. Deserts can be irrigated from deep wells or distant mountains, but such efforts typically fail as a result of salinization, the buildup of salts from the evaporation of irrigation water.

TEMPERATE GRASSLAND Temperate grasslands are found in many parts of the world, all of which are relatively dry for much of the year. Most grasslands, such as the pampas of Argentina, the veldt of South Africa, and the Great Plains of North America, have hot summers and relatively cold winters. In some grasslands, most of the precipitation falls in winter (as in California grasslands); in others, the majority falls in summer (as in the Great Plains and the Russian steppe).

Grassland vegetation is structurally simple but rich in species of perennial grasses and forbs (herbaceous plants). Grassland plants support herds of large grazing mammals and are adapted to grazing and to fire. They store much of their energy underground and resprout quickly after being burned or grazed. There are comparatively

DESERT

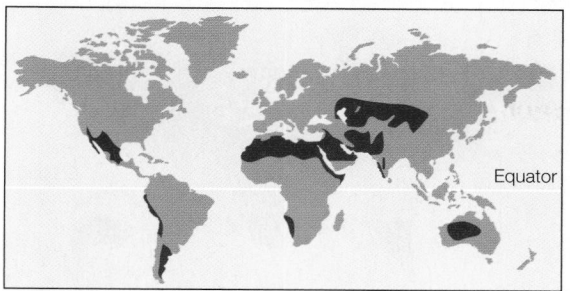

A desert biome in Namibia, Africa, is home to the gemsbuck, a large antelope in the *Oryx* genus.

Sonoran desert in bloom with saguaro (*Carnegiea gigantean*) and cholla (*Cylindropuntia* sp.) cacti and poppies at Organ Pipe Cactus National Monument, Arizona, USA.

When the precipitation curve falls below the temperature curve (yellow area), water availability limits plant growth.

Ouargla, Algeria
31°N
Average annual temperature 22.3°C
Total annual precipitation 39 mm

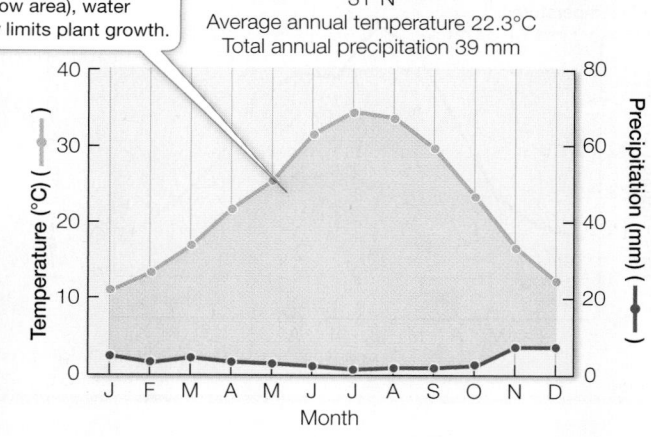

A desert tortoise (*Gopherus agassizii*) in its native habitat, the Mohave Desert, California, USA.

few trees in temperate grasslands because trees cannot survive the periodic fires or the dry conditions.

The topsoil of grasslands is usually rich and deep, and thus exceptionally well suited to growing crops such as corn and wheat. As a consequence, most of the world's temperate grasslands have been turned over to agriculture and no longer exist in their natural state.

TEMPERATE DECIDUOUS FOREST   The temperate deciduous forest biome is found in eastern North America, eastern Asia, and Europe. Temperatures in these regions fluctuate dramatically between summer and winter, although precipitation is fairly evenly distributed throughout the year. The deciduous trees that dominate these forests lose their leaves during the cold winters and produce new leaves during the warm, moist summers.

The temperate deciduous forests have many more species than boreal forest ecosystems. Those with the highest species richness occur in the southern Appalachian Mountains of the United States and eastern China and Japan—areas that were not covered by glaciers during the Pleistocene. Many plant genera are shared among the three geographically separate regions where this biome is found.

Although many animals are permanent residents of deciduous forests, some (including many birds) migrate to find food resources and escape the winter cold. Others that remain through the winter hibernate (see Key Concept 39.5), often in underground burrows. Many insects pass the winter in a state of diapause (suspended development), the onset of which is triggered by the decreasing hours of daylight—a reliable predictor of winter.

TEMPERATE GRASSLAND

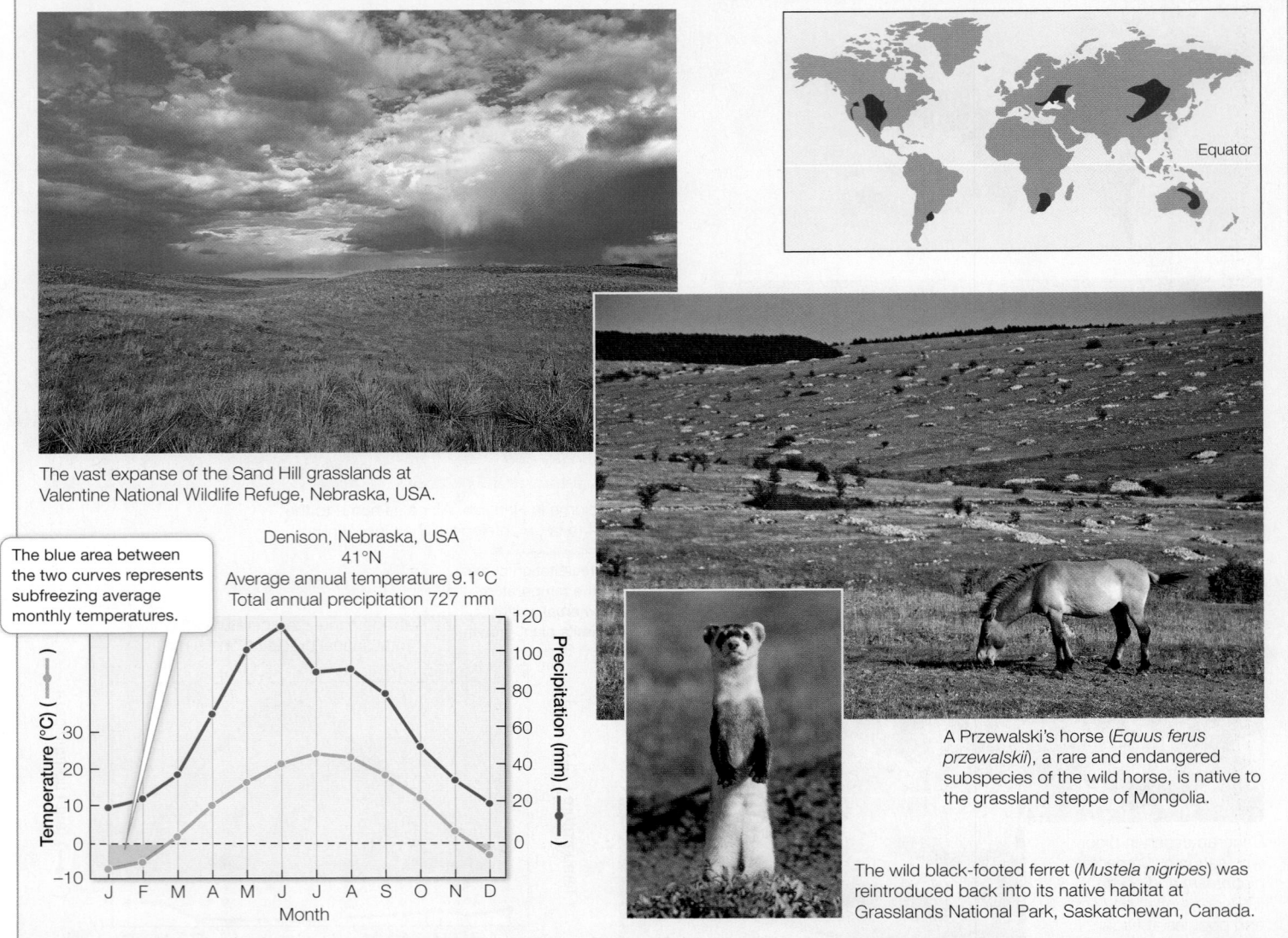

The vast expanse of the Sand Hill grasslands at Valentine National Wildlife Refuge, Nebraska, USA.

The blue area between the two curves represents subfreezing average monthly temperatures.

Denison, Nebraska, USA
41°N
Average annual temperature 9.1°C
Total annual precipitation 727 mm

A Przewalski's horse (*Equus ferus przewalskii*), a rare and endangered subspecies of the wild horse, is native to the grassland steppe of Mongolia.

The wild black-footed ferret (*Mustela nigripes*) was reintroduced back into its native habitat at Grasslands National Park, Saskatchewan, Canada.

BOREAL FOREST AND TEMPERATE EVERGREEN FOREST   The boreal forest biome (also known as taiga) occurs just above 50°N but below Arctic tundra, and at elevations below alpine tundra on temperate-zone mountains. Winters in the boreal forest are long and very cold; summers are short, although often relatively warm. The boreal forests of the Northern Hemisphere are dominated by coniferous trees such as spruces and firs. The short summers favor evergreen leaves, which are ready to photosynthesize as soon as temperatures warm. In winter, downward-drooping limbs allow the trees to shed snow easily.

The dominant mammals of the boreal forest, such as moose and hares, eat leaves, but the seeds in conifer cones support a variety of rodents, birds, and insects. Many small mammals hibernate in winter, but voles, lemmings, and mice remain active under the snowpack, serving as food for predators such as foxes and owls.

The temperate evergreen forest biome occurs along the coasts of continents in both hemispheres at middle to high latitudes, where winters are mild and wet and summers are cool and dry. In the Northern Hemisphere, the dominant trees in temperate evergreen

forests are conifers, some of which are the world's most massive tree species (including the giant sequoia and coast redwood). In the Southern Hemisphere, the dominant trees are southern beeches (*Nothofagus*), some of which are evergreen.

TUNDRA   The tundra biome is found at high latitudes (above 65°) characterized by low temperatures and a short growing season. This biome is underlain by permafrost—soil permeated with permanently frozen water. The top few centimeters of the soil thaw during the short summers, when the sun may be above the horizon 24 hours a day. Thus even though there is little precipitation near the poles, the soil in lowland Arctic tundra is wet because water cannot drain through the permafrost. As a result, trees cannot grow in the tundra biome because their roots cannot penetrate the permafrost. Instead, the tundra biome is characterized by sedges, forbs, grasses, and low-growing shrubs such as heaths, willows, and birches. Lichens and mosses are also important vegetation.

Tundra plants have several structural and physiological adaptations that help them conserve heat, as described in Key Concept

TEMPERATE DECIDUOUS FOREST

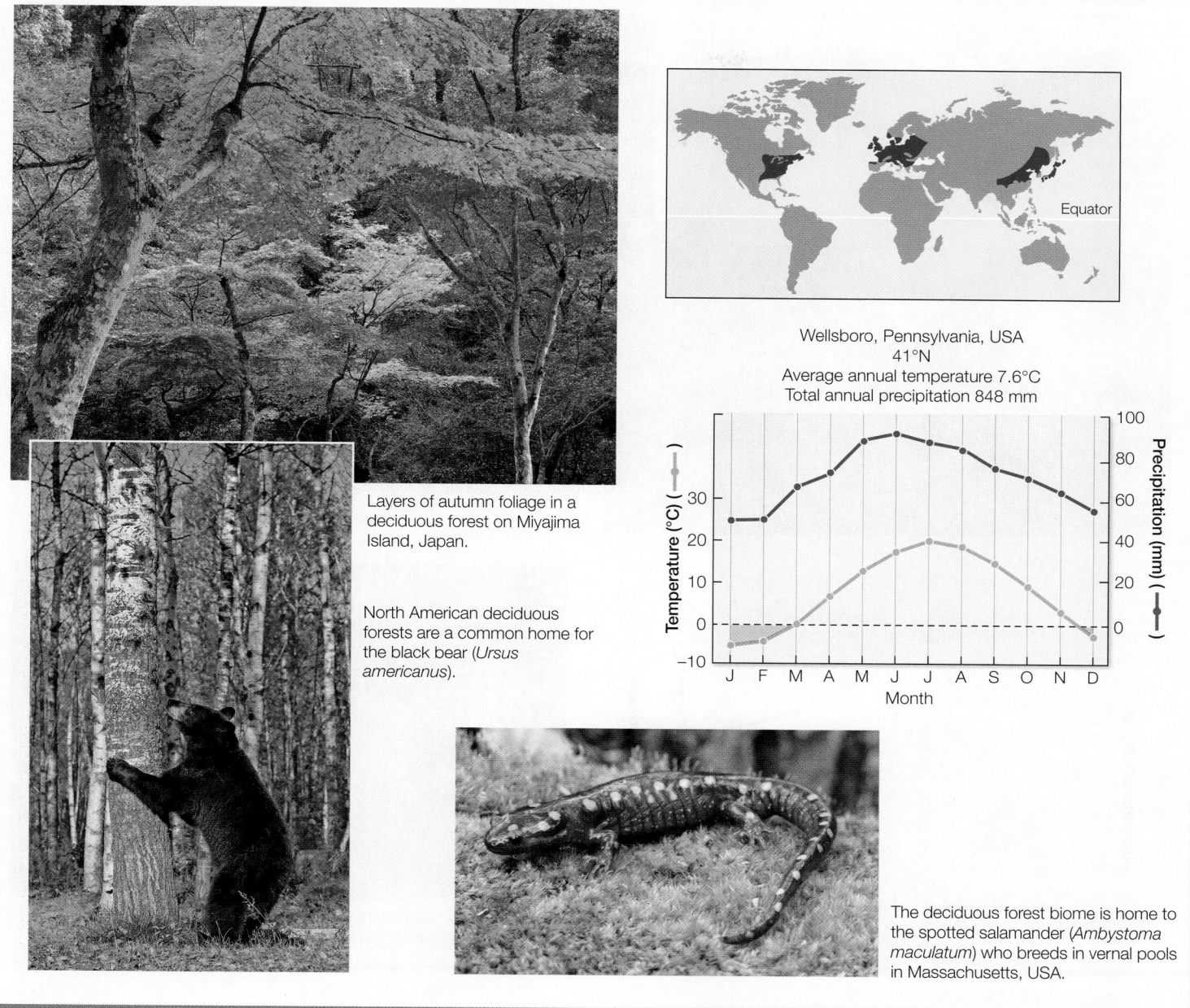

Layers of autumn foliage in a deciduous forest on Miyajima Island, Japan.

North American deciduous forests are a common home for the black bear (*Ursus americanus*).

Wellsboro, Pennsylvania, USA
41°N
Average annual temperature 7.6°C
Total annual precipitation 848 mm

The deciduous forest biome is home to the spotted salamander (*Ambystoma maculatum*) who breeds in vernal pools in Massachusetts, USA.

38.3. Most animals are either summer migrants or are dormant for much of the year. Resident birds and mammals, such as the willow ptarmigan (*Lagopus lagopus*) and Arctic fox (*Vulpes lagopus*), have thick fur or feathers that may change color with the seasons, from brown in summer to white in winter.

Biogeographic regions reflect evolutionary isolation

Climate interacts with local abiotic features to influence where and how organisms live, but these are not the only factors determining where organisms are found. Evolutionary history—where and when species originated and diverged—is key to determining their biogeography.

Until European naturalists traveled the globe in the nineteenth century, they had no way of knowing how organisms were distributed in other parts of the world. Alfred Russel Wallace, who along with Charles Darwin advanced the idea that natural selection could account for the evolution of life on Earth (see Key Concept 20.1), was one of those global travelers. Wallace spent seven years in the Malay Archipelago, where he noticed some remarkable patterns in the distributions of species. For example, he described the dramatically different species that inhabited the adjacent islands of Bali and Lombok. He pointed out that the differences could not be explained by the physical environment, because Bali and Lombok are only 24 kilometers apart.

Wallace saw that, based on the distributions of plant and animal species, he could draw a line that divided the Malay Archipelago into two distinct halves. He correctly deduced that the dramatic differences in flora and fauna were related to the depth of the channel separating Bali and Lombok. This channel is so deep that it would have remained full of water, and thus would have been a barrier

BOREAL FOREST and TEMPERATE EVERGREEN FOREST

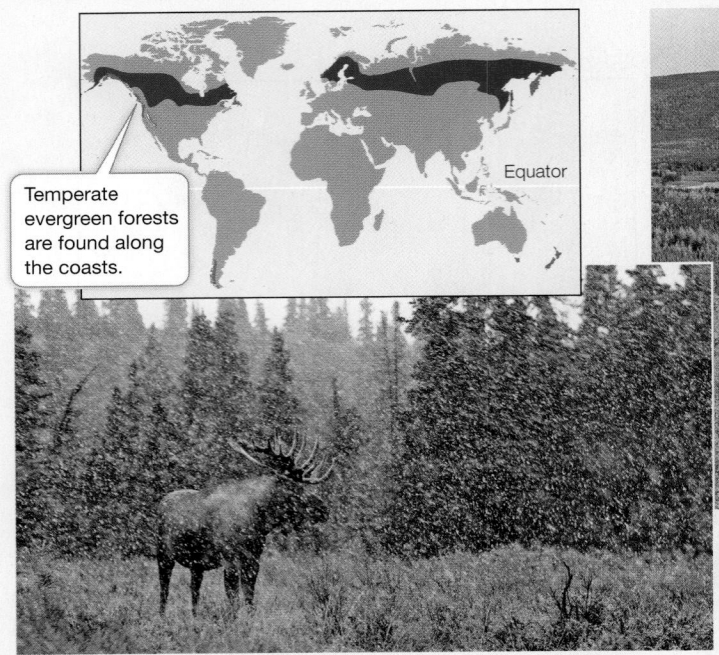

Temperate evergreen forests are found along the coasts.

Equator

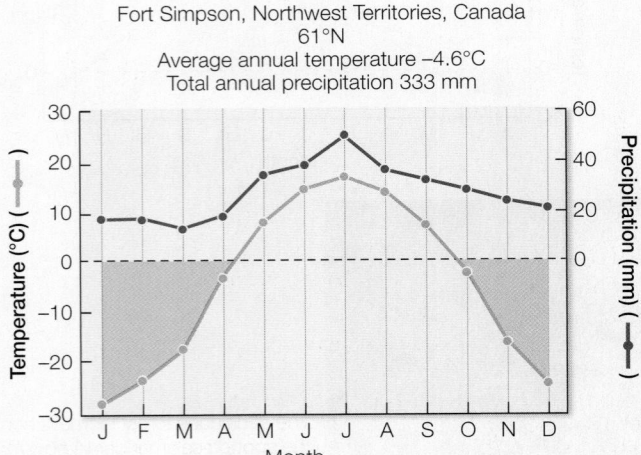

The boreal forests of Alaska are home to moose (*Alces americanus*), the largest North American terrestrial herbivore.

A boreal forest scene in Gros Morne National Park, Newfoundland, Canada.

Fort Simpson, Northwest Territories, Canada
61°N
Average annual temperature −4.6°C
Total annual precipitation 333 mm

A great gray owl (*Strix nebulosa*) is camouflaged against a boreal forest tree in Finland.

to the movement of terrestrial animals, even during the glaciations of the Pleistocene epoch, when sea level dropped more than 100 meters and Bali and the islands to the west were connected to the Asian mainland.

With these insights, Wallace established the conceptual foundations of biogeography. In *The Geographical Distribution of Animals*, published in 1876, he detailed the factors known at the time that influence the distributions of animals, including past glaciation, land bridges, deep ocean channels, and mountain ranges. He earned some measure of scientific immortality in that the Malay discontinuity that first piqued his curiosity is known to this day as "Wallace's line" (see Figure 53.14 on page 1160).

The biotas of different parts of the world differ enough to allow us to divide Earth into many continental-scale areas called **biogeographic regions**, each containing characteristic assemblages of species. The boundaries of the biogeographic regions in Figure 53.14 were originally proposed by Wallace, and represent assemblages of species that change dramatically, often over short distances. A major process controlling the formation of these biogeographic regions is ***continental drift**. For example, we now know that over the course of the Triassic and Jurassic periods, the supercontinent Pangaea separated into two great land masses, Laurasia and Gondwana (see Figure 24.14), which subsequently separated into the continents we know today. After the land masses broke up, the descendants of the organisms widely disturbed across Pangaea evolved independently, forming new species and new species assemblages. The legacy of these continental movements can be found in several existing taxonomic groups and in the fossil record. For example, the modern southern beeches—trees of the genus *Nothofagus*—are found in both the Neotropical and the Australasian biogeographic regions. Evidence of fossilized *Nothofagus* pollen from 55 to 34 million years ago has also been found in Australia, New Zealand, western Antarctica, and South America, suggesting that beeches originated in

TUNDRA

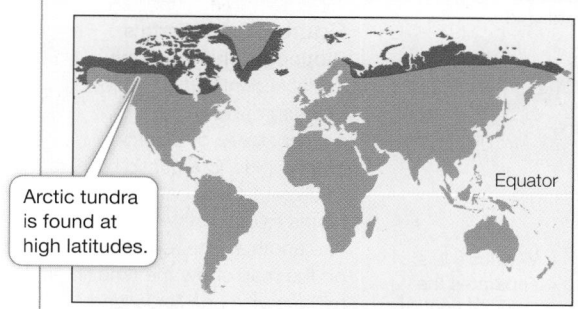

Arctic tundra is found at high latitudes.

Equator

A brilliant autumn tundra scene at Thorofare Pass in Denali National Park and Preserve, Alaska, USA.

The muskox (*Ovibos moschatus*) was introduced to the tundra biome of Dovrefjell National Park, Norway, in 1932.

The arctic fox (*Vulpes lagopus*) has a white coat that serves as camouflage in the winter.

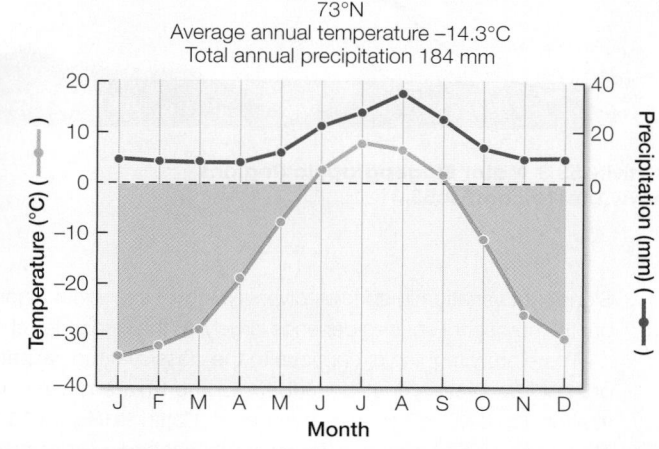

Olenek, Russia
73°N
Average annual temperature −14.3°C
Total annual precipitation 184 mm

The evolutionary separation of species can be attributed to two basic processes, vicariance and dispersal. **Vicariance** occurs when a physical barrier prevents dispersal and divides a species into two or more discontinuous populations. Dispersal occurs when the members of a species cross an existing barrier and establish a new population elsewhere.

Gondwana during the Cretaceous period and were geographically separated by the breakup of that land mass 100 million years ago (see Figure 53.14).

*connect the concept Earth's land masses have undergone continental drift over the millennia as a result of plate tectonics, shaping the major patterns of biogeography today and in the past. See Key Concepts 24.2 and 24.3.

Given that the processes of vicariance and dispersal both influence distribution patterns, how can biogeographers determine the role of each process when reconstructing the evolutionary history of a particular species? As you saw in Chapter 21, taxonomists have developed powerful molecular methods of reconstructing the phylogenetic relationships among organisms that can be used to understand how organisms came to occupy their present-day distributions. Phylogenetic trees can be used to discover whether the distribution of an ancestral species was influenced by a vicariant event, such as continental drift or a change in sea level, or is simply the result of a dispersal event.

Media Clip 53.1 **Rafting to Madagascar**
www.Life11e.com/mc53.1

Diversity varies with latitude and longitude

About 200 years ago, the German explorer and naturalist Alexander von Humboldt spent 5 years traveling around Latin America. He remarked in the account of his voyages that "the nearer we approach the tropics, the greater the increase in the variety of structure, grace of form, and mixture of colors, as also in perpetual youth and vigour of organic life." Humboldt and other nineteenth-century scientific explorers such as Wallace and Darwin became keenly aware of this pattern as they collected thousands of species in the tropics and compared them with their more meager collections from Europe. As more observations and collections have been made over the last 200 years, the latitudinal gradient in species diversity has been established for a wide variety of taxa, including some groups of birds, mammals, flowering plants, and insects such as swallowtail butterflies (**Figure 53.15A**).

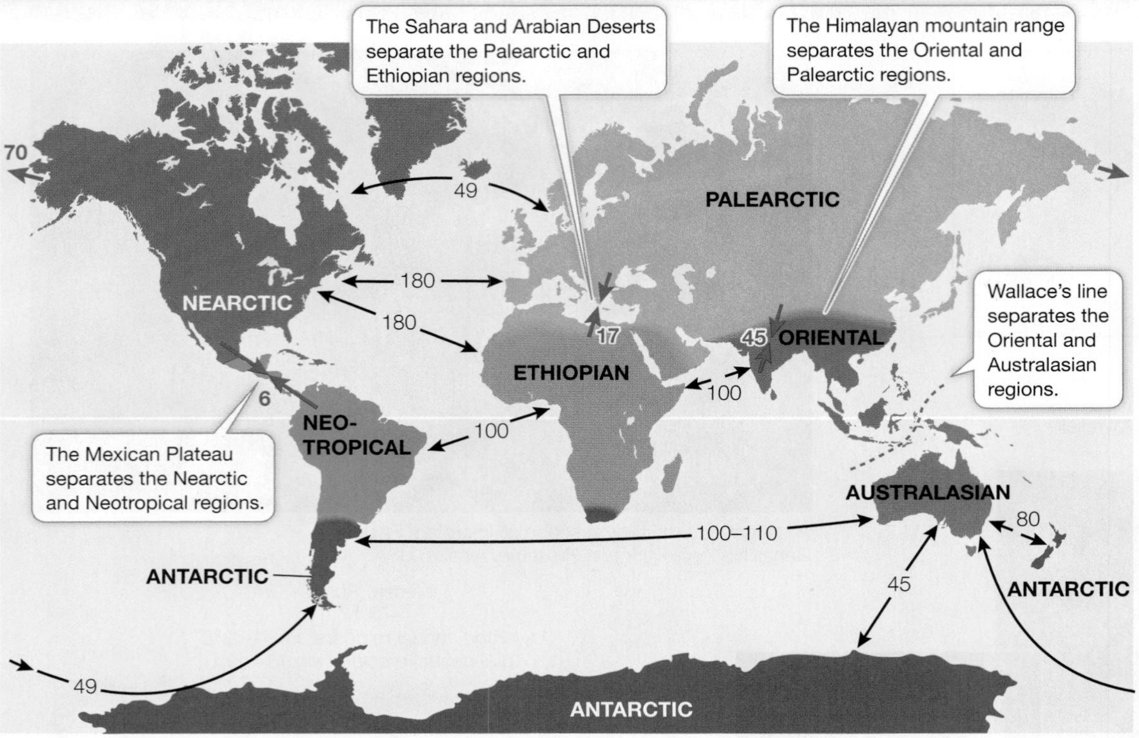

The Sahara and Arabian Deserts separate the Palearctic and Ethiopian regions.

The Himalayan mountain range separates the Oriental and Palearctic regions.

PALEARCTIC

NEARCTIC

180

180

Wallace's line separates the Oriental and Australasian regions.

45 ORIENTAL

17

ETHIOPIAN

100

The Mexican Plateau separates the Nearctic and Neotropical regions.

6

NEO-
TROPICAL

100

AUSTRALASIAN

80

100–110

45

ANTARCTIC

ANTARCTIC

49

ANTARCTIC

Figure 53.14 Earth's Biogeographic Regions Wallace identified six major biogeographic regions that are separated by climatic, topographic, or aquatic barriers to dispersal that cause their biotas to differ strikingly from one another. The red arrows on the map show the time (in millions of years) since land masses came together. Black arrows show the time since land masses separated. These areas roughly correspond to Earth's major tectonic plates.

Q: As land masses separated, would you expect speciation to increase? Explain.

 Activity 53.3 **Major Biogeographic Regions**
www.Life11e.com/ac53.3

Significant variation in species diversity is also seen with longitude, but the number of species depends largely on the longitudes chosen.

There are intriguing exceptions to the classic latitudinal patterns of species diversity just described. Some taxa show a positive relationship between species number and latitude (**Figure 53.15B**). That is, among some taxa—for example, seabirds—species number increases at higher latitudes. This pattern of seabird diversity correlates with marine *****net primary productivity**, which is substantially higher in temperate and polar oceans than in the tropics.

*****connect the concepts**   Net primary production (NPP) is the amount of biomass incorporated into the tissues of primary producers and is the base of all food chains. Oceanic NPP is highest at mid–latitudes. See Key Concept 57.2.

We consider productivity differences along with other explanations of latitudinal gradients in diversity in the next section.

Geographic variation in diversity is explained by multiple factors

Although most ecologists agree that latitudinal gradients in diversity exist, there is much less consensus as to why they exist. Dozens of hypotheses have been proposed to explain these patterns, but they are hard to evaluate—in part because there are multiple and confounding differences in such factors as geographic area, climate, and productivity with latitude. More important, because these factors operate at large spatial scales and over evolutionary time, it is impossible to conduct experiments to isolate the factors and separate correlation from causation.

Here we present three well-known hypotheses that have been advanced to account for latitudinal gradients in diversity. Corroborative evidence can be found for each of these hypotheses, varying with taxon, locality, and scale, which suggests that none of these hypotheses is mutually exclusive. Multiple factors, working at multiple spatial scales, are likely responsible for common biogeographic patterns.

SPECIES DIVERSIFICATION RATE   This hypothesis proposes that in the tropics the rate of speciation is higher and the rate of extinctions is lower, resulting in an overall higher species diversity than in temperate or polar regions. Two factors could account for a higher speciation rate in the tropics: a larger geographic area (this is the widest part of the planet, so it encompasses the greatest amount of area) and a warm and stable climate. The rationale is that a larger geographic area with a consistent climate should promote speciation because species will have larger geographic ranges, giving them a greater chance of reproductive isolation (see Key Concept 22.2). In addition, these large and stable areas should reduce extinctions. The net result should be an increase in the number of species in the tropics.

SPECIES DIVERSIFICATION TIME   This hypothesis proposes that the amount of time over which speciation has taken place is greatest in the tropics—in other words, that the tropics have a longer evolutionary history than temperate or polar regions. First championed by Wallace in 1878, this hypothesis proposes that tropical regions, because they are more climatically stable over time, have more time to diversify than temperate or polar regions, where severe climate conditions (such as ice ages) could have decreased speciation and increased extinction. Thus even if the rates of speciation and extinction are the same worldwide, the tropics should have accumulated more species over time simply because of the lack of dramatic changes in climate.

(A) Global swallowtail butterfly diversity

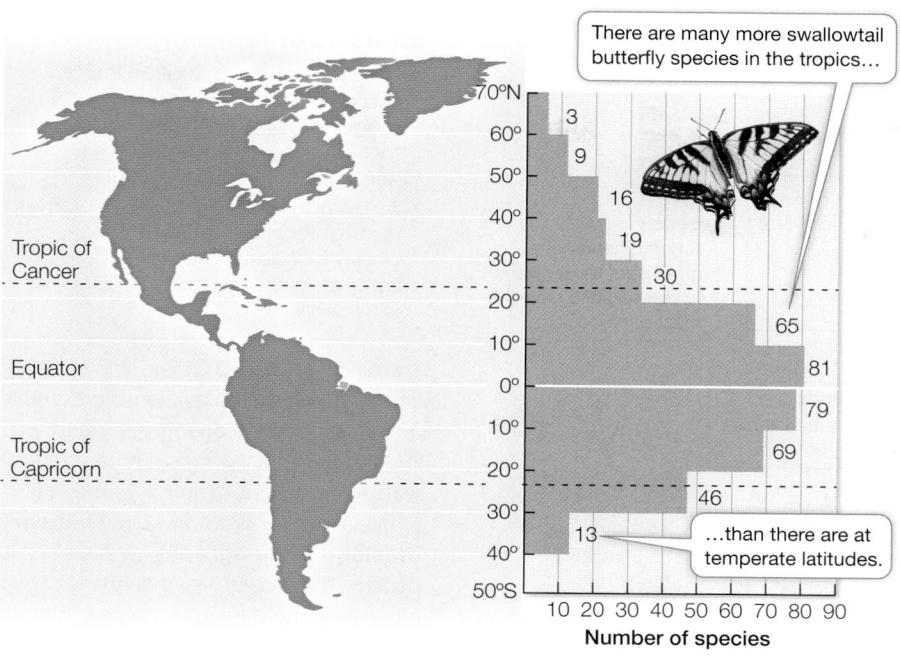

There are many more swallowtail butterfly species in the tropics...

...than there are at temperate latitudes.

Number of species

(B) Global seabird diversity

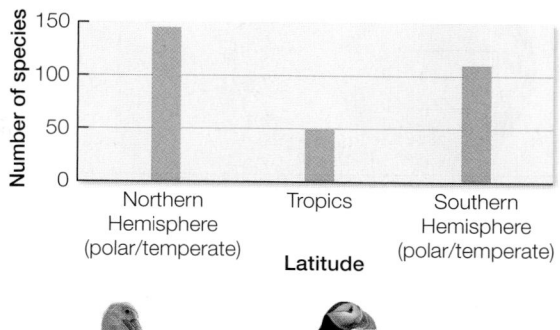

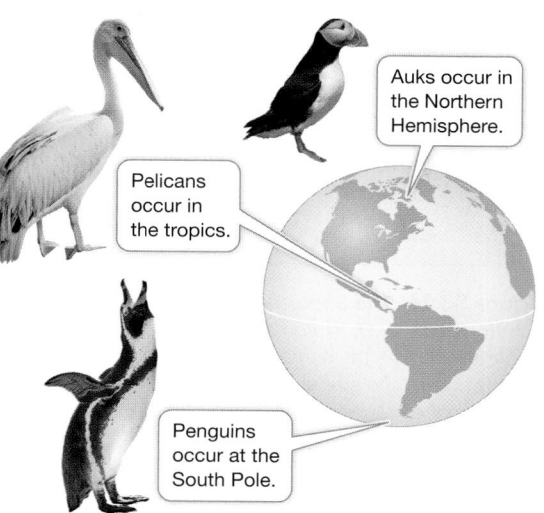

Auks occur in the Northern Hemisphere.

Pelicans occur in the tropics.

Penguins occur at the South Pole.

Figure 53.15 Latitudinal Gradients in Diversity   (A) Among swallowtail butterflies (Papilionidae), species richness decreases with latitude both north and south of the equator. **(B)** Seabirds, by contrast, show a positive relationship between species diversity and latitude: the number of seabird species increases at higher latitudes.

PRODUCTIVITY This hypothesis proposes that species diversification is promoted by higher productivity, which allows species more resources and thus decreased risk of species extinction due to competition. The productivity hypothesis might explain the positive relationship between seabird diversity and increasing latitude, given that ocean productivity is generally higher at temperate and polar latitudes. As you will see in Chapter 56, the role of productivity can more easily be explored at smaller spatial scales, where experiments can give us a better idea of its role in promoting species diversity.

53.4 recap

Biogeographic patterns are interconnected across a hierarchy of scales ranging from global to local. Three strong biogeographic patterns can be seen: (1) biomes or groupings of ecologically similar organisms are shaped by the environment, (2) diversity varies from continent to continent, forming biogeographic regions, and (3) diversity varies with latitude.

learning outcomes

You should be able to:

- Describe the principle characteristics of each of the seven major biomes featured in this section, including climate, latitude and/or continent(s), and representative plants and animals.
- Use the observations and insights of Alfred Russel Wallace to explain how biogeography might influence the distribution of plants and animals.

53.4 recap

- Compare and contrast the hypotheses posed for latitudinal patterns of species diversity, including species diversification rate, species diversification time, and productivity, giving examples of situations that could be used to explain each one.

1. Describe the information used to characterize biomes. What are the differences and similarities between the rainforest biome and the temperate evergreeen forest biome?
2. Refer to Figure 53.14. Why did Wallace's line help explain what controls the formation of biogeographic regions?
3. Over Earth's history, ice sheets extended into temperate latitudes. Does this fact better support the species diversification rate or the species diversification time hypothesis proposed to explain latitudinal variation in diversity? Explain.

As you have seen, an important aspect of biogeography is the relationship between diversity and geographic area. We have considered this relationship at global scales. We turn now to smaller regional scales, to understand in more detail the importance of geographic area to patterns of biogeography.

key concept 53.5 Geographic Area and Humans Affect Regional Species Diversity

Biogeographers have repeatedly documented what is termed the **species–area relationship**, in which species diversity increases with

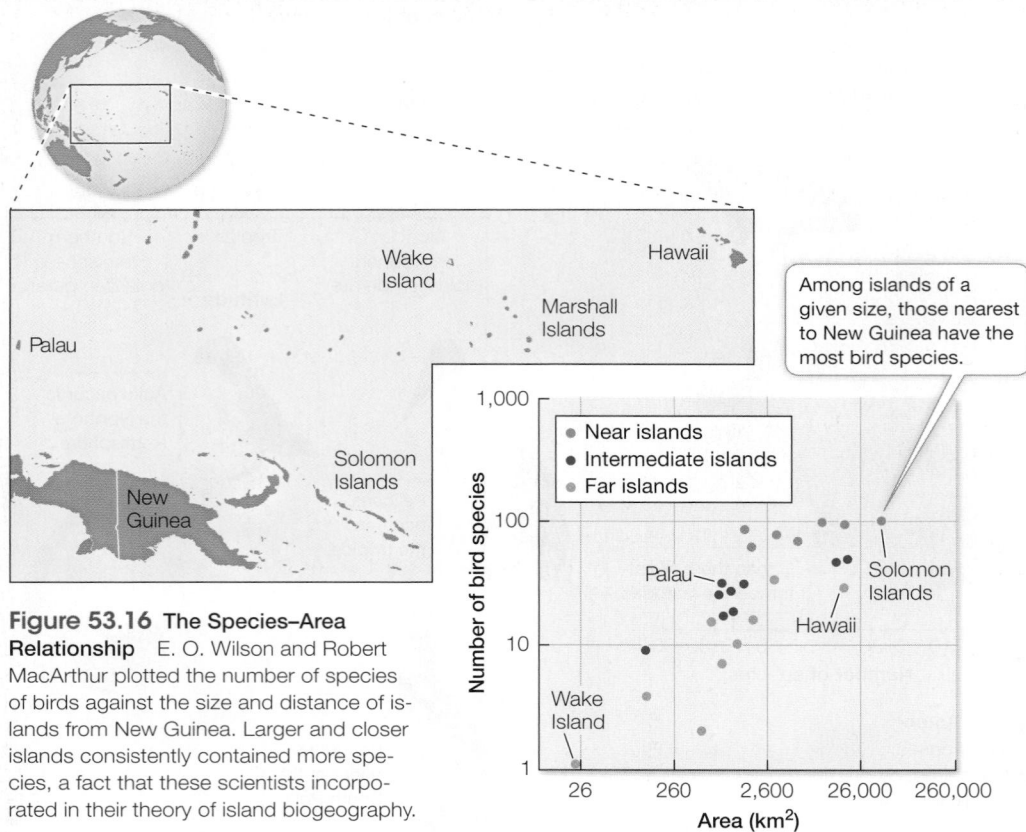

Figure 53.16 The Species–Area Relationship E. O. Wilson and Robert MacArthur plotted the number of species of birds against the size and distance of islands from New Guinea. Larger and closer islands consistently contained more species, a fact that these scientists incorporated in their theory of island biogeography.

focus your learning

- The species–area relationship is explained by the theory of island biogeography, which considers the balance between species immigration and extinction.
- As humans fragment Earth's biomes through agriculture, deforestation, and urbanization, remaining habitats become more isolated, or islandlike.

The biologist Edward O. Wilson was struck by the species–area relationship from his research on ant biogeography and was looking for explanations. He teamed up with Robert MacArthur, a gifted mathematical ecologist, to develop the **theory of island biogeography** to explain this pattern. They based their theory on just two processes: the immigration of new species to an island and the extinction of species already present on that island (**Figure 53.17**). The premise of island biogeography is that the number of species on an island represents a balance between the rate at which species immigrate to and colonize the island and the rate at which resident species become locally extinct.

increasing area. The relationship has most often been measured for islands or islandlike habitats—any isolated area surrounded by a "sea" of dissimilar habitat. In addition to showing a pattern of increasing species diversity with increasing area, islands also show a pattern of decreasing species number with distance from the source of those species, or the species pool (**Figure 53.16**). How do island area and distance act together to produce species–area relationships?

The rate of immigration is determined in part by the number of species in the regional species pool. In the case of oceanic islands, the species pool comprises all the species on the nearest mainland location. Not all species that reach the island will persist there, however. The more species there are on an island, the greater the likelihood that some will go extinct, either due to chance or the inability to coexist with other species on the island. The number of species that will be sustained on any particular island is determined by the balance of the immigration rate and the extinction rate for that island. Even though species will come and go over time (i.e., immigrate and become extinct), that turnover will always result in some equilibrium number of species, assuming a constant species pool and natural extinction events.

The actual number of species on an island will ultimately depend on two factors:

1. *The size (area) of the island*. The smaller the island, the fewer resources it provides, the greater the potential for competition, and the higher the extinction rate will be (see Figure 53.17). Larger islands provide greater resources and can sustain larger populations (which tend to have lower extinction rates than small populations).

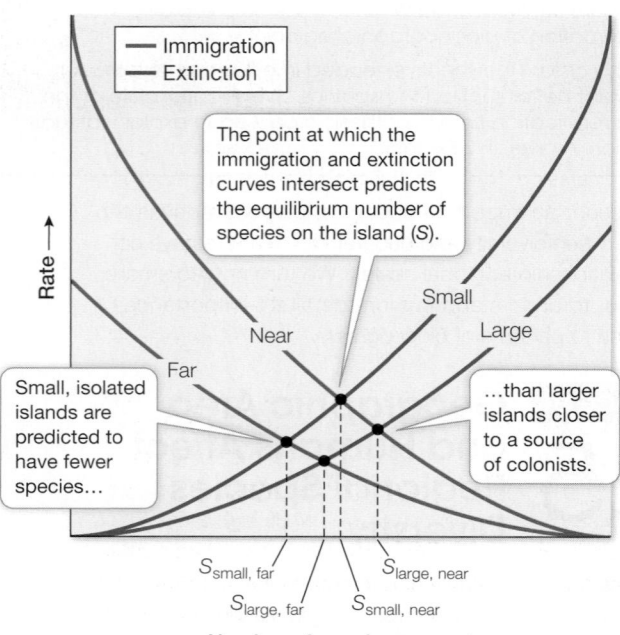

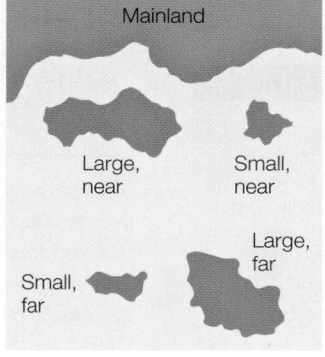

Figure 53.17 The Equilibrium Theory of Island Biogeography MacArthur and Wilson's theory emphasized the balance between species immigration rates and species extinction rates for islands of different sizes and at different distances from a source of colonizing species.

 Activity 53.4 Biogeography Simulation www.Life11e.com/ac53.4

experiment

Original Paper: Ferraz, G. et al. 2003. Rates of species loss from Amazonian forest fragments. *Proceedings of the National Academy of Sciences USA* 100: 14069–14073.

Thomas Lovejoy and his colleagues asked what was the minimum area needed to maintain species diversity in the rainforest fragments created by logging near Manaus, Brazil. They conducted an experiment, starting in 1979, that took advantage of an existing Brazilian law that required landowners who cut rainforest to leave half of it untouched. In 2003, Ferraz and colleagues reported on one particular aspect of the experiment: the number of forest understory birds living in different-sized forest fragments surrounded by deforested land.

HYPOTHESIS▶ The number of understory bird species will decline faster after the isolation of the fragment, and in smaller rainforest fragments compared with larger rainforest fragments.

METHOD

1. Starting in 1979, fragments of three sizes (1, 10, and 100 hectares) were established 80 kilometers north of Manaus. The fragments were separated from continuous forest by at least 100 meters of cleared land.

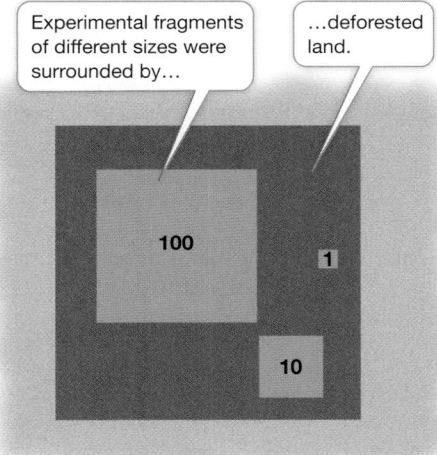

Experimental fragments of different sizes were surrounded by…

…deforested land.

100

1

10

investigatinglife work with the data follows on next page.

2. Birds were captured in mist nests placed in the fragments prior to deforestation and then for 12 years after fragment establishment. The mist nets were placed in roughly the same locations throughout the study.

3. Each newly captured bird was recorded for species, given an individually numbered band, and released back into the rainforest from which it was caught.

4. The data gathered were used in species-loss curves for each fragment, and a scaling factor (i.e., a number that scales, or multiplies, by some quantity) was estimated to determine the time it takes to lose half of the bird species (t_{50}) in fragments of different sizes.

RESULTS

The average initial number of species was 83 for the 1-hectare fragment, 92 for the 10-hectare fragment, and 113 for the 100-hectare fragment. Despite finding a total of 164 understory bird species in the fragments over the course of 13 years, there was a significant drop in the number of species over time. The t_{50} was shortest in the 1-hectare fragments and longest in the 100-hectare fragments. Fragments of 100 hectares were predicted to lose half of their species in 12 years.

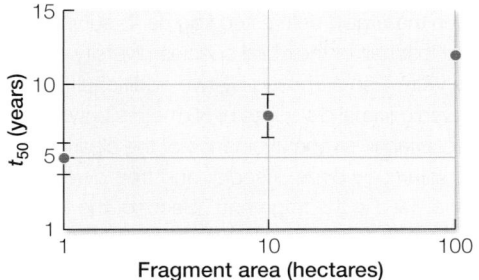

Fragment area (hectares)

CONCLUSION▶ The number of understory bird species declined after the fragmentation of tropical rainforest habitat by deforestation. Smaller rainforest fragments lost species faster than larger forest fragments. Even the largest fragments were predicted to lose half of their bird species in 12 years.

 Animation 53.4 Edge Effects
www.Life11e.com/a53.4

---

2. *Distance of the island from the species pool.* The farther the island is from the source of immigrants, the lower the immigration rate—the rate at which new species arrive—will be (see Figure 53.17).

Between 1966 and 1969, Wilson and his student Daniel Simberloff conducted an experiment to test the theory of island biogeography, using arthropods living on mangrove islands in the Florida Keys. You can view this experiment and the results online.

 Animation 53.3 Island Biogeography in the Florida Keys
www.Life11e.com/a53.3

Humans exert a powerful influence on biogeographic patterns

The theory of island biogeography has provided surprising insight into the role human's play in regional biogeographic patterns. As you

saw in Key Concepts 53.3 and 53.4, Earth's biomes are becoming smaller and more fragmented by local- and regional-scale forces such as agriculture, deforestation, and urbanization. They are being transformed into islandlike habitats—isolated patches of suitable habitat such as parks or forest fragments, surrounded by extensive areas of unsuitable habitat similar to the cattle pastures of Amazonia described at the opening of this chapter.

The fragmentation of the Amazon rainforest led Thomas Lovejoy and his colleagues to initiate, in 1979, one of the largest and longest-running ecological experiments ever conducted to investigate the effects of tropical forest fragmentation on species diversity (**Investigating Life: The Largest Experiment on Earth**). The Biological Dynamics of Forest Fragments Project (BDFFP) took advantage of an existing Brazilian law that required landowners who cut rainforest to leave half of it untouched. Lovejoy was able to survey the species diversity in different-sized fragments (either 1, 10, or 100 hectares [ha] in size

work with the data

The data in the graph shown in the experiment can be used to calculate a scaling factor for the time it takes to lose half of the bird species within a fragment. The scaling factor shows that to increase the t_{50} (i.e., the time it takes fragments to lose half their species) by 10-fold, fragment area would need to increase 1,000-fold. Here you will use this scaling factor and other data provided in the graph to determine the area needed for the conservation of understory birds in the Amazon forests outside Manaus, Brazil.

QUESTIONS▶

1. Graph the initial numbers of bird species by the size of the fragment area. Do these fragments follow the species–area relationship?
2. Suppose you are consulting for the Brazilian government on the conservation of rainforests near Manaus. Assuming that the t_{50} for 1 hectare is 5 years, what is the minimum size that a rainforest fragment in the Manaus area needs to be to ensure that half of the bird species remain 50 years after the deforestation event that isolated them? How about a century after deforestation?
3. Suppose it takes 100 years for rainforests to fully recover after a deforestation event in the Manaus region. Given that the average size of a rainforest fragment in this area is no bigger than 1,000 hectares, will the number of species decline by half before the recovery of rainforest occurs around the largest fragments?

A similar **work with the data** exercise
may be assigned in **LaunchPad**.

---

surrounded by deforested land). Among a number of results, his group found that, after more than 30 years of isolation, even the largest deforested fragments surveyed (100 ha) had lost half of their bird species diversity. The researchers were also surprised by two additional findings. First, even minimal distances of 80 meters between fragments resulted in strong avoidance of the clearings, and thus isolation, by birds, insects, and tree-dwelling mammals that lived in the fragments. Second, the fragmentation exposed species within a fragment to a variety of potential hazards, including extreme heat, fires, hunting, predators, diseases, and invasive species. These so-called **edge effects** not only reduce the immigration rate from one fragment to another, thus contributing to a smaller species pool, but they effectively reduce the size of the fragment, thus increasing the extinction rate, by making the edge of the fragment much less hospitable habitat.

53.5 recap

Species–area relationships, in which species diversity increases with increasing area, have been shown for islands and islandlike habitats. Smaller islands have higher extinction rates and lower immigration rates compared with larger islands. Earth's biomes are becoming smaller and more fragmented by human activities, leading to lower diversity in the remaining habitats.

learning outcomes

You should be able to:

- Explain the species–area relationship and use a conceptual model to describe it.
- Describe specific ways in which habitat fragmentation negatively affects species diversity, using the Amazon rainforest as an example.

1. Using Figure 53.17, explain how and why the small and far islands have lower species numbers than the large and near islands.
2. Why is habitat fragmentation by human activities threatening species diversity? Besides ceasing the activity altogether, what can be done to reduce this threat?

---

investigating**life**

How do geographic area and isolation affect the biogeography of life on Earth?

At the opening of the chapter, we asked how geographic area and isolation might affect the biogeography of life on Earth. You have seen that this question is highly relevant at different spatial scales—from whole continents to small islands—and also over different time scales—from millions of years to months. Because the spatial and temporal scales at which biogeographic patterns manifest themselves are so wide and interconnected, it can be hard to test theories of biogeography with experiments that are large and long enough to be meaningful. The Biological Dynamics of Forest Fragments Project (BDFFP), an experiment conducted in the most species-rich region of the world, and at arguably the largest scale ever attempted, offers support for theories proposed nearly 200 years ago by explorers such as Humboldt, Wallace, and Darwin

who hypothesized that the distribution of species on Earth is highly influenced by the geographic area and isolation those species have experienced. The question remains, in the case of the Amazon Basin, whether species will ultimately adapt to the changing biogeographic landscape or become a casualty of an extinction event that would be unrivaled in human history.

Future directions

One of the major findings of the BDFFP is the importance of connectivity in maintaining species diversity. Management and permitting of forest clearing are focusing on creative ways to connect forest fragments by taking advantage of existing landscape features to increase connectivity of disturbed habitat. For example, in Amazonia strong laws exist to prohibit clearing forest along rivers and steep slopes, providing strips of land that can serve to connect fragments around newly cleared land.

Chapter Summary 53

▶ 53.1 Ecology Is the Study of the Interrelationships among Organisms and the Environment

- **Ecology** differs from **environmentalism**, which involves the use of ecological knowledge to inform our stewardship of natural resources.
- Ecologists have a use-inspired focus to their research, motivated by a desire to improve Earth's ecological problems.
- Ecology is studied at multiple levels of organization, ranging from individuals to the biosphere. Review Figure 53.1

▶ 53.2 Global Climate Is a Fundamental Component of the Physical Environment

- **Climate** and **weather** both refer to atmospheric conditions, but climate represents conditions over years to millennia whereas weather represents conditions over days to weeks.
- Solar radiation drives global climate. About 30 percent of the solar radiation that hits Earth is reflected back to space. The rest is absorbed either by the **atmosphere** (20%) or by Earth's surface (50%). **Greenhouse gases** in the atmosphere allow sunlight into Earth's atmosphere but trap heat radiating back out toward space. Review Figure 53.2, Animation 53.1
- Latitudinal differences in solar energy input drive patterns of atmospheric circulation. Review Figures 53.3, 53.5
- The interaction of Earth's rotation and north–south air mass movements generate **prevailing winds**, which in turn drive ocean currents. Review Figures 53.6, 53.7
- The seasons are a consequence of Earth's tilt and orbit around the sun. Review Figure 53.8

▶ 53.3 Topography, Vegetation, and Humans Modify the Physical Environment

- On land, Earth's topography creates **rain shadows** and **temperature inversions**, producing variation in temperature and precipitation Review Figure 53.9A and B, Animation 53.2
- The topography of the ocean floor produces variation in water depth, which affects light penetration and water temperature, pressure, and movement, creating aquatic zones and **ocean upwelling**. Review Figure 53.9C and D

- Vegetation, especially forests, affects climate through the process of **evapotranspiration**. Review Figure 53.10

▶ 53.4 Biogeography Is the Study of How Organisms Are Distributed on Earth

- Patterns of biogeography are interconnected across a hierarchy of scales, ranging from global to regional to landscape to local. Review Focus: Key Figure 53.12
- **Biomes** are groups of ecologically similar dominant plants that are shaped by patterns of temperature and precipitation. Review Figure 53.13, Activities 53.1, 53.2
- Earth's diversity varies from continent to continent, forming **biogeographic regions** and reflecting evolutionary isolation as a consequence of Earth's geological history. Review Figure 53.14, Activity 53.3
- Earth's diversity varies with latitude. Hypotheses explaining this variation rely on multiple factors, including geographic area, time, climate, and productivity. Review Figure 53.15

▶ 53.5 Geographic Area and Humans Affect Regional Species Diversity

- **Species–area relationships**, in which species diversity increases with area and distance, have been shown for islands and area fragments. Review Figure 53.16
- The theory of **island biogeography** states that the number of species on an island is a balance between immigration and extinction rates. Smaller islands have lower species diversity because they have higher extinction rates and lower immigration rates compared with larger islands. Review Figure 53.17, Animations 53.3, 53.4, Activity 53.4

> Go to **LearningCurve** (in **LaunchPad**) for dynamic quizzing that helps you solidify your understanding of this chapter. **LearningCurve** adapts to your responses, giving you the practice you need to master each key concept.

▶ Apply What You've Learned

Review

53.4 Hypotheses to explain the causes of latitudinal variations in diversity include species diversification rate, species diversification time, and productivity.

53.5 The species–area relationship is explained by the theory of island biogeography, which considers the balance between species immigration and extinction.

Original Paper: Holbrook, S. J. et al. 2015. Reef fishes in biodiversity hotspots are at greatest risk from loss of coral species. *PLoS ONE* 10: e0124054.

As you learned in Chapter 1, climate change and human activities pose substantial threats to tropical coral reefs. These species-rich ecosystems make up less than 1 percent of the coastal ocean but contain more than 25 percent of all marine species.

A group of researchers studied the species diversity of reef fishes, which depend on coral reefs for food and habitat. They found a positive relationship between fish species diversity and coral species diversity. Coral reefs support a variety of fish species, including both generalists and specialists, but as coral diversity increases, the number of fish specialists—those that depend on one or a very few coral species—also increases. The researchers hypothesized that if reefs were degraded,

(continued)

locations with the highest fish diversity would suffer the greatest proportional declines, because of the loss of certain fish species that specialize on certain coral habitats. They reasoned that if coral species disappeared, their associated fish species would also disappear.

The researchers set up identical experiments in three geographic locations across the Indo-Pacific, as represented in **Figure A**. Kimbe Bay, in Papua New Guinea (PNG), has the greatest number of fish species (approximately 1,600 species). Moorea Island, in French Polynesia, has the lowest number of fish species. Lizard Island, in the Great Barrier Reef, has an intermediate number of fish species, 10–15 percent lower than PNG.

At each of the three reef sites across the Indo-Pacific, the researchers experimentally manipulated three levels of coral diversity: high diversity (six coral species), medium diversity (three coral species), and low diversity (one coral species) within separate plots. They used the same coral species in all the plots and counted fish species four times during a 12-month period.

Results for the final count are shown in **Figure B**. Researchers calculated an "index of sensitivity" to evaluate the relationship between the number of coral species in the experimental plots and the number of fish species that inhabited the plots. The zero line indicates no difference in fish species diversity; positive values indicate proportionally greater decreases in fish species present with declining coral species in the plots. That is, fish species from the Kimbe Bay site were more sensitive to changes in the number of coral species than were fish species from lower diversity areas, and they were more likely to suffer local extinctions.

Figure A

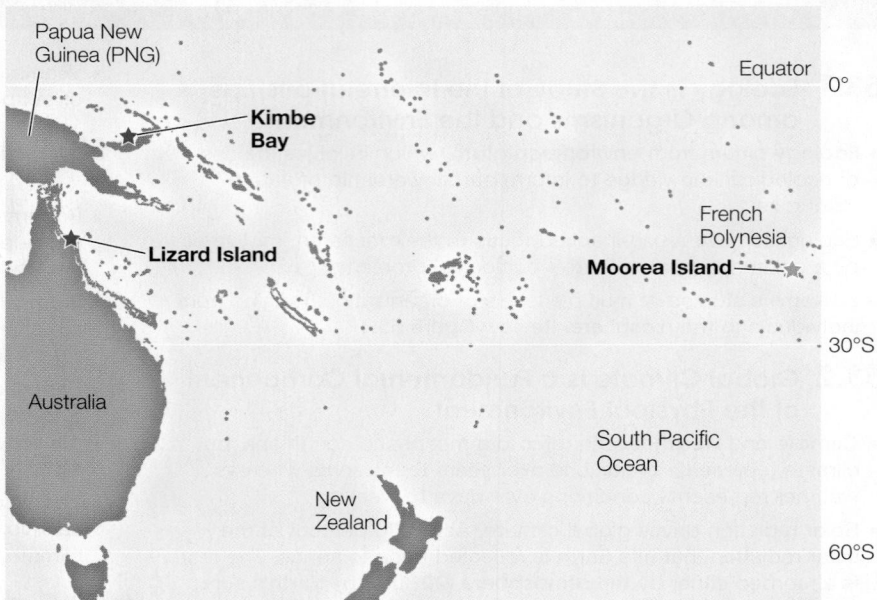

Figure B

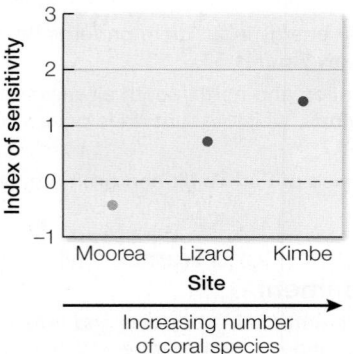

Questions

1. What do the results of this experiment suggest about the relationship between the number of fish species and the species diversity of corals on the reefs they inhabit? Explain how this insight might be used to preserve or maintain coral reef ecosystems currently under threat.

2. Based on the map, at which sites do you think the species diversity of corals and fish would be most affected by the process of dispersal, or immigration, of species to the site(s), and which would be least affected?

3. What major factors could have resulted in the development of the high coral reef species diversity at the Kimbe Bay site? Why do the other sites have lower species diversity?

4. As climate change progresses, ocean waters are becoming both warmer and more acidic, causing the decline in corals worldwide. What effect would these changes in coral reef species diversity likely have on fish species diversity in the three areas? Explain possible reasons for these changes and what variations might occur in each area.

Go to **LaunchPad** for the eBook, LearningCurve, animations, activities, flashcards, and additional resources and assignments.

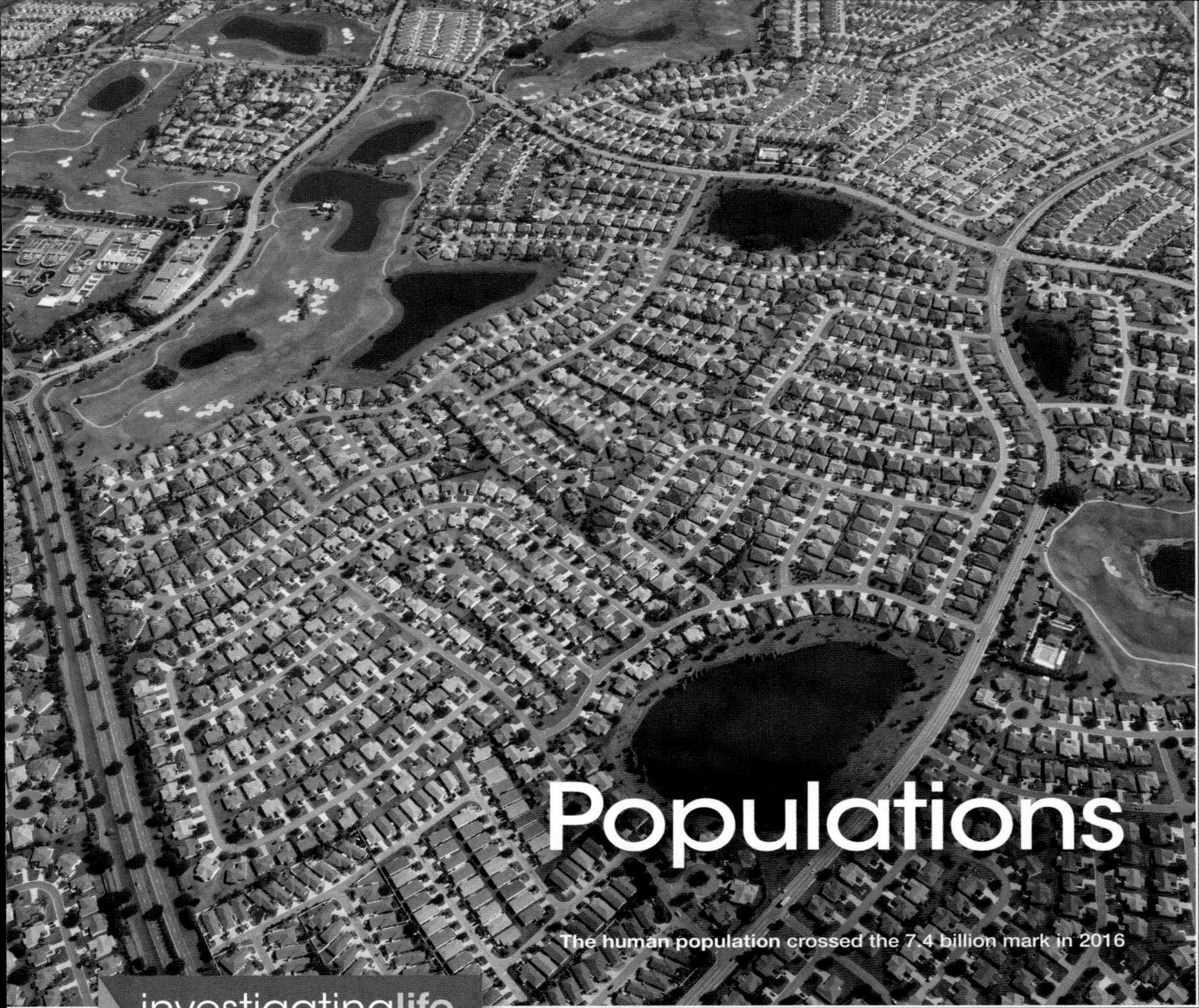

54

Populations

The human population crossed the 7.4 billion mark in 2016

> ## investigatinglife

Earth's Human Carrying Capacity

In 1798 Thomas Robert Malthus, in his *Essay on the Principle of Population*, pointed out that the human population was growing exponentially but its food supply was not, and argued that at some point, famine and death would be the ultimate fate of the human race. Malthus could not have anticipated the technological innovations over the next 200 years that would greatly enhance our food production and health care. Today, however, the size of the human population is once again a serious concern as we confront the effects of our contributions to climate change, pollution, habitat destruction, and species extinctions.

For thousands of years, Earth's capacity to support human populations was low because of the relative inefficiency with which we could obtain food and water. The development of social systems and communication, the domestication of plants and animals, ever-increasing crop and livestock yields due to ongoing technological advances, and our increasing proficiency at managing diseases all contributed to unprecedented growth of the human population. It took more than 200,000 years for the human population to reach 1 billion people, which happened in the early nineteenth century. Today, a mere 200 years later, the planet is home to roughly 7.4 billion human beings.

Many believe that human populations will (if they have not already) reach some maximum size (known as the population's carrying capacity) above which continued population growth is unsustainable. Food production is critical to human population growth and currently uses large amounts of land, water, fertilizers, and energy. Housing and quality health care require advances in infrastructure and drug production, and these too are highly energy-intensive. In addition, if Earth's ecosystems continue to be degraded, human populations may experience a general breakdown of Earth's life support systems.

Luckily, population growth rate has slowed from its high after World War II, but with a base of 7.4 billion, even minimal population growth rate means millions more individuals each year. As you will see later in the chapter, population biologists have used census data and standard population growth models to predict human population sizes into the future.

Q **A** What is the human population growth projected to be in the next century, and how will that affect population size?

Media Clip 54.1 **World Population Growth**
www.Life11e.com/mc54.1

key concept

54.1 Populations Show Dynamic Variation in Size over Space and Time

Species, whether they are humans, yeast, or giant redwood trees, are divided into groups of individuals, or populations. A **population** is a group of individuals of the same species within a given area that have the potential to interbreed and interact with one another. As you might expect, populations vary in their **population size**—the numbers of individuals within a population. Well before ecology became a distinct biological discipline, humans were interested in **population dynamics**, or the patterns and processes of population change over space and time. Growing crops, raising livestock, and controlling pests all involve paying attention to population dynamics. To this day, game wardens, park managers, and conservation biologists aim to maintain stable populations of fish, wildlife, and threatened or endangered species in the face of change. Here we consider population dynamics in more detail.

---

focus your learning

- Populations are dynamic; they vary in size, range, and dispersion patterns, and they vary over time.
- Three main factors affect population dynamics: the physical environment, biological interactions, and dispersal.
- Ecologists estimate population size and extent using a variety of techniques.

Part of the dynamism that populations show is a consequence of their distribution across the landscape (**Figure 54.1A**). For example, distribution maps of the herbaceous perennial *Clematis fremontii* show that it is patchily distributed across Missouri, Kansas and Nebraska. Here populations are restricted to dry, rocky meadows formed on particular limestone outcrops within the region. Populations, such as those of *Clematis*, rarely occur in isolation from one another but are usually connected through dispersal. **Dispersal** is simply the movement of individuals into (**immigration**) or out of (**emigration**) an existing population. A group of geographically isolated populations linked together by dispersal is known as a **metapopulation**. For example, a cluster of meadows might be considered a metapopulation if *Clematis* seeds from one meadow had the potential to disperse to another meadow. At larger spatial scales, the entire **geographic range** or distribution of a species might consist of one or multiple metapopulations, depending on the extent of the area occupied by a species.

Species differ greatly in their geographic ranges—they can be as wide ranging as the acorn barnacle *Semibalanus balanoides*, which extends along both North American coasts and the European Atlantic coast. However, even though dispersal links populations, most species have fairly restrictive geographic ranges such as we see for *Clematis*. We call such species **endemic** because they occur in one particular location and nowhere else on Earth. Thus *Clematis* can be said to be endemic to the central United States because its geographic range is restricted to that region.

Within a population, the spatial arrangement of individuals may vary in what are known as **dispersion patterns**. Ecologists recognize three basic dispersion patterns: (1) **regular** (or **uniform**) **dispersion**, characterized by evenly spaced individuals, (2) **random dispersion**, in which individuals are randomly spaced, and (3) **clumped dispersion**, in which individuals tend to be clustered in groups (**Figure 54.1B**). As we will consider later in this section, multiple processes such as

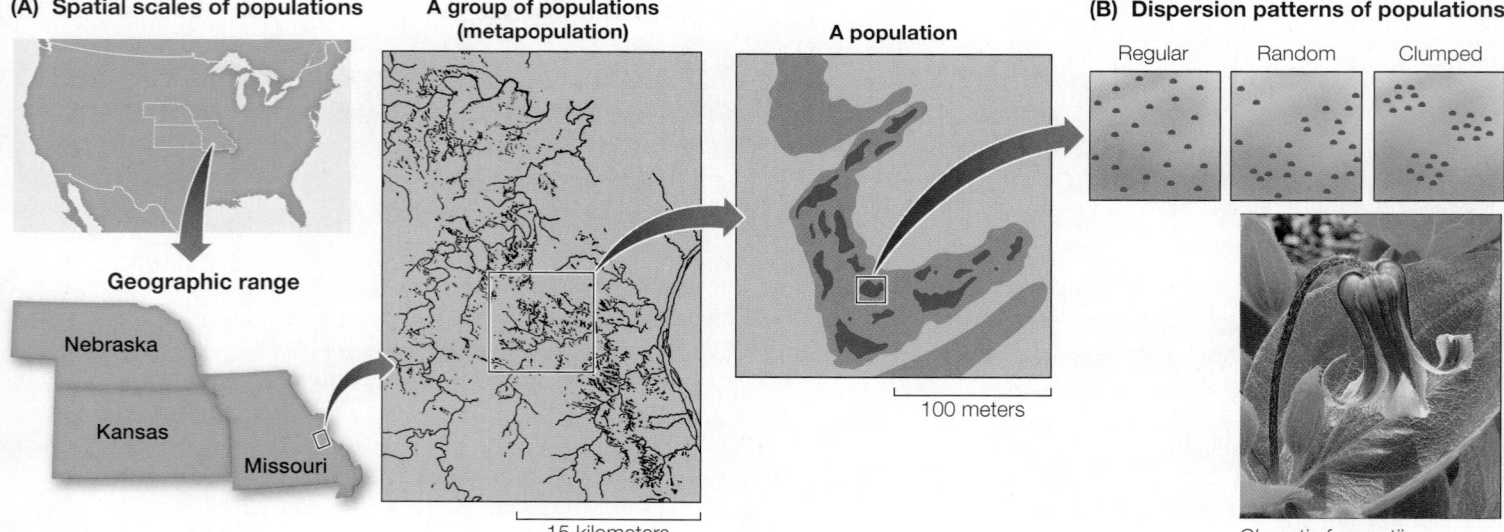

Figure 54.1 Many Populations Have Patchy Distributions
The distribution and abundance of the herbaceous perennial *Clematis fremontii* are patchy over different spatial scales. **(A)** Populations occur within limestone meadows. A group of populations makes up a metapopulation, and multiple metapopulations make up the geographic range (in this case, Missouri, Kansas, and Nebraska). **(B)** Individuals within a population show one of three different dispersion patterns.

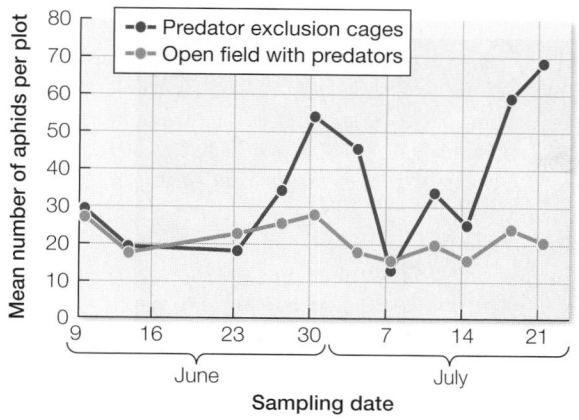

Figure 54.2 Population Fluctuations of Aphids and Their Predators A population of soybean aphids (*Aphis glycines*) fluctuates more when its insect predator, the insidious flower bug (*Orius insidiosus*), is excluded than when it is present. Without the predator, soybean aphids cost soybean growers hundreds of millions of dollars in damage per year.

resource availability, species interactions, and dispersal can shape the dispersion patterns we see in nature.

Populations can vary over time as well. For example, some populations may show "boom and bust" cycles, such as we see for bacteria, yeast, and phytoplankton, in which the population size can vary over just a few minutes to days. Species that show wide fluctuations in population size tend to have reproductive strategies that allow them to respond rapidly to changing resources. They may also have predators that are able to take advantage of these outbursts in growth. For example, aphids engage in parthenogenetic reproduction—a type of asexual reproduction resulting in genetically identical individuals (see Key Concept 42.1)—allowing them to become very abundant very quickly. If overcrowding occurs and food resources become limiting, populations can crash, leaving behind a few individuals that may ultimately serve to increase the population size through rapid clonal reproduction (**Figure 54.2**). Predatory insects such as the insidious flower bug can dampen these large swings in aphid population size, resulting in decreased resource limitation. Thus extrinsic factors such as predation can provide checks on populations, thereby limiting outbreaks, or on the other end, local extinctions.

Population dynamics are controlled by the physical environment, biological interactions, and dispersal

Even though we know that populations fluctuate in space and time, determining the factors responsible for this variation can be challenging. Ecologists recognize three main factors that affect population dynamics and that act simultaneously to set limits for populations and species: the physical environment, biological interactions, and dispersal.

As you saw in Chapter 53, the distribution and abundance of species are highly dependent on the physical environment; factors such as climate, topography, and even human infrastructure set physiological limits for species. Thus it makes sense that the physical features of the environment strongly limit a population's geographic extent and size through time. You saw this for *Clematis*, whose populations are restricted to dry, rocky limestone soils, resulting in a patchy distribution pattern.

Likewise, many types of biological interactions affect population dynamics. Interbreeding among individuals within populations affects their size and extent—in fact, many populations can be defined by the genetic similarity of the individuals within a given area (see Key Concept 20.4). For example, populations of wild endangered

salmon, which look nearly identical to hatchery-raised salmon, can be assessed for population size using genetic techniques that allow differentiation among populations. In addition, because individuals within populations share similar resources, **intraspecific competition** (competition for shared resources by individuals of the same species) can be critical in controlling population size, as you will see in Key Concept 54.2. ***Interspecific interactions** (interactions among individuals of different species) are another important factor in controlling population dynamics.

---

*connect the concepts Interspecific interactions such as competition, predation, and positive interactions (i.e., mutualisms and commensalisms) often set the upper limits on the number of individuals within populations. See Chapter 55.

Finally, dispersal can be a critical factor controlling population dynamics. Dispersal comes in many forms. It can be active when movement is controlled by the individual, such as you might see with elephants; passive when movement is controlled by the physical environment, as in corals and dandelions; or facilitated by active agents, such as birds or bats dispersing seed-laden fruits (**Figure 54.3**). Dispersal is a mechanism that can reduce competition for resources, either with a parent or another individual within the population. In addition, dispersal allows individuals to escape a harsh physical environment. Thus dispersal serves as a mechanism to reduce the probability of local extinctions of populations by spreading the risk among multiple populations that are geographically separated (see Key Concept 54.4).

A specific type of dispersal is migration. **Migration** typically occurs in response to seasonal variation in resources, involves round-trip movement, and includes the whole population. For example, North Pacific populations of the humpback whale (*Megaptera novaeangliae*) migrate more than 3,000 miles between their winter breeding grounds in the south (Mexico, Hawaii, and Japan) and their summer feeding grounds in the north (Northeast Pacific coast and Gulf of Alaska) (**Figure 54.4**). Scientists determine whale migration patterns and population numbers by taking photographs and collecting genetic material to follow and compare individuals in the feeding and breeding grounds. A 2006 survey of North Pacific humpback whales determined that close to 20,000 individuals in five separate populations make the migration every year. Remarkably, there were only 1,400 North Pacific humpback whales in 1966, when commercial whaling of these populations was banned.

As you can see, whale biologists use various techniques to estimate the population sizes of humpback whales. Let's consider in

Figure 54.3 Different Dispersal Mechanisms of Organisms **(A)** Elephants actively disperse, **(B, C)** corals and dandelions disperse passively, and **(D)** seeds within fruits are typically dispersed by an active agent, in this case, a toco toucan.

more detail the variety of approaches ecologists use to measure population size.

Ecologists use a variety of approaches to estimate population size and extent

Accurately measuring the size and extent of a population is harder than you might think. One challenge is determining the boundary of the population. If the area is self-contained and small enough, you might be able to simply count all the individuals of a particular species to get the population size. Biologists performed this type of count, called a **full census**, on the African elephant population of Samburu and Buffalo Springs National Reserves in Kenya. By monitoring the elephants for 21 months, the biologists learned to recognize each of the 760 individuals in the population, primarily by their unique and distinctive ear markings. But populations are often too large and too mobile for a full census; often the extent of the range of a population is not perfectly known. In these cases, ecologists instead determine

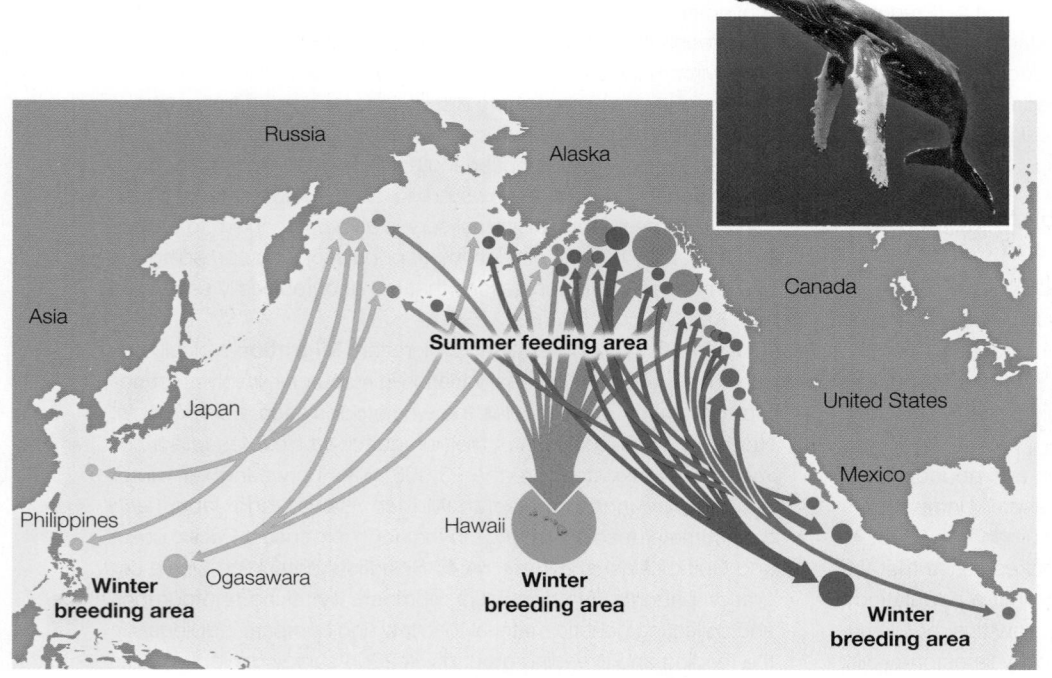

Figure 54.4 Migration of North Pacific Humpback Whales   North Pacific populations of the humpback whale (*Megaptera novaeangliae*) migrate between their winter breeding grounds off Mexico, Hawaii, and Japan and their summer feeding grounds in the Gulf of Alaska and Northeast Pacific coast. A 2006 survey of North Pacific humpback whales determined that close to 20,000 individuals migrate in five separate populations (represented by different colored arrows).

the **population density**, or the number of individuals within a given area (or volume, for organisms living in water), and then extrapolate from these samples to estimate the total population size.

As you might imagine, estimating population densities is easiest for sessile organisms. Investigators need only count the individuals in a sample of representative locations and extrapolate the counts to the entire geographic range of the population. Individuals may be counted within a measured area called a **quadrat** or along a **transect**—a line drawn across an area within the range of the population (often designated by a tape measure marked at regular intervals). By making repeated counts with either of these methods, investigators can use these surveys to make reasonably good estimates of the size of a population.

Counting mobile organisms is more difficult because individuals move into and out of sampling areas, as you saw in the humpback whale example. In such cases, investigators may use the **mark–recapture method** (**Figure 54.5**). They begin by capturing, marking, and then releasing a number of individuals. Later, after the marked individuals have had time to mix with unmarked individuals in the population (but before enough time has elapsed for births, deaths, and individual movement to affect the

research tools

Figure 54.5A The Mark–Recapture Method The method described here is used to estimate animal population sizes for highly mobile species (such as *Ixodes scapularis*, the black-legged tick). Once a sampling area has been determined, investigators capture, mark, and then release the animals back into the population. The proportion of marked individuals recaptured in a second sample is assumed to be the same as the proportion of the total individuals in that sample to the population size within the area.

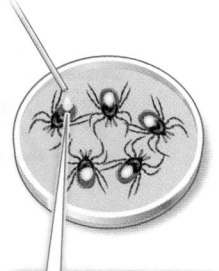 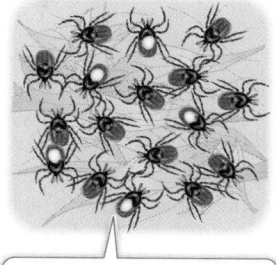 

1 Capture a random sample of individuals from the population of interest. Mark each captured individual.

2 Release the marked individuals and allow an appropriate amount of time for them to merge completely with unmarked individuals in the population.

3 Capture a second random sample of individuals. Determine both the total number of individuals captured and the number of marked individuals in this sample.

4 Estimate the total population size *N* using the equation

$$N = \frac{n_1 \times n_2}{M}$$

where
n_1 = the total number of individuals in the first sample (captured, marked, and released)
n_2 = the total number of individuals in the second sample
M = the number of marked individuals recaptured in second sample

Animation 54.1
The Mark–Recapture Method
www.Life11e.com/a54.1

Figure 54.5B **work with the data** follows below.

work with the data

Figure 54.5B Monitoring Tick Populations

Original Paper: Falco, R. C. and O. Fish. 1988. Prevalence of *Ixodes dammini* near the homes of Lyme disease patients in Westchester County, New York. *American Journal of Epidemiology* 127: 826–830.

Lyme disease is a chronic and debilitating condition caused by spirochete bacteria of the genus *Borrelia*, which infect humans by way of the bite of an intermediate host, the black-legged tick (*Ixodes scapularis*), also known as the deer tick. (In 1993 *I. dammini* and *I. scapularis* were found to be one species.) The incidence of Lyme disease has increased dramatically in the past 20 years, particularly in the northeastern United States. In order to assess the risk of exposure to this disease in Westchester County, New York, investigators measured the abundance of deer ticks in suburban lawns near wooded areas using the mark–recapture method described in Figure 54.5A. (Ticks are typically collected by dragging a white cloth along the ground; the ticks latch onto the cloth in much the same way they would to a passing leg.) By drag-sampling one representative lawn, the researchers collected the data shown in the table.

	Original capture event	Second capture event (3 weeks later)
Adult ticks captured	180	33
No. of marked ticks	180[a]	8

[a]All ticks captured in the first event were marked with acrylic paint and released.

QUESTIONS▶

1. Refer to Figure 54.5A. Using the equation and other information described in that figure, estimate the total number of adult ticks in the sampled lawn from the data table above.

2. The lawn was approximately 700 m² in size. What is the approximate density of ticks per square meter?

3. What do you think might be the implications of this study for residents of this neighborhood?

A similar **work with the data** exercise may be assigned in **LaunchPad**.

population size significantly), another sample of individuals is captured. This sample is then used to obtain an estimate of the total size of the population in the sampling area. The extrapolation is achieved by applying the equation described in Figure 54.5A, which assumes that the *proportion* of marked individuals in the second sample (i.e., individuals that were captured and marked in the first sample) is about the same as the proportion of individuals in the sampling area that were captured in the first sample.

More recently, ecologists have used DNA analysis to determine how many unique individuals might be contained within a population. In some cases, instead of capturing individuals and collecting DNA, ecologists collect the next best thing—tissues, fur, or feces left behind by organisms in their natural habitat to be analyzed for DNA signatures. In a novel twist on sample collection, some ecologists have even trained dogs to locate feces of some hard to find species such as orca whales. With enough samples, the extent and size of populations can then be determined for species that are hard to survey in the wild.

54.1 recap

Species are divided into groups of individuals, or populations, that can vary spatially across their geographic range. All populations change in size and spatial extent due to the physical environment, biological interactions, and dispersal. Ecologists use a variety of approaches to estimate population sizes, including full censuses, surveys using quadrats or transects, mark–recapture methods, and DNA analyses.

learning outcomes

You should be able to:

- Use examples to illustrate how species are divided into populations, metapopulations, and geographic ranges.
- Describe and compare the three main factors that control population dynamics over time.
- List and describe major approaches that ecologists use to estimate population size and extent.
- Given a description of a population, choose and justify the most appropriate method for estimating its population size and/or extent.

1. Refer to Figure 54.4. Draw a diagram showing which populations of humpback whales interact on their summer feeding grounds. Do you think all five populations make up one metapopulation during the summer? Explain.

2. Refer to Figure 54.2. What type of species interaction appears to have caused the aphid population to fluctuate in size? What type of species interaction keeps the population size from fluctuating?

3. What are some of the ways in which population size and geographic extent of organisms can be measured?

4. A researcher wants to determine how many leopard frog tadpoles are present in a pond in the spring. Later, after the tadpoles have metamorphosed, she wants to compare this number with the size of the frog population. What method should the researcher use to determine the population sizes? Explain your answer.

Quantifying population sizes in space and time provides valuable information on the dynamics of a species, but those observations alone cannot explain how, when, and why populations change in size. In order to understand and predict population sizes through time, ecologists characterize population growth, or the change in population size over time, using quantitative methods.

key concept 54.2 Population Growth Describes the Change in Population Size over Time

It is clear from the previous section that populations are constantly changing in size. **Population growth** describes the change in population size over time. For example, you read that North Pacific humpback whales experienced dramatic declines in population size from commercial whaling, but when hunting was banned, the populations rebounded from 1,400 whales in 1966 to nearly 20,000 in 2006. Measuring how a population changes over time requires following its **demography**: births, deaths, immigration, and emigration.

focus your learning

- Population size and growth rate are measured by the numbers of births, deaths, and migrants into and out of the population over time.
- Population growth can be limited by density-dependent and density-independent factors.
- Life tables help predict how age-related survival and reproduction affect population growth.

Births increase and deaths decrease population size over time

Over any given interval of time, the size of a population increases by the number of individuals added to the population by births and by immigration (the movement of individuals into the population from elsewhere) and decreases by the number of individuals lost from the population by deaths and by emigration (individuals leaving the population to go elsewhere). This relationship is expressed mathematically as

$$N_t = N_0 + (B - D) + (I - E) \tag{54.1}$$

where N_t = the population size at time t, N_0 = the population size at time 0, B = the number of individuals born between time 0 and time t, D = the number that died between time 0 and time t, I = the number that immigrated between time 0 and time t, and E = the number that emigrated between time 0 and time t.

As you saw earlier, population size rarely remains the same but instead changes over time. We can use Equation 54.1 to estimate how population size changes over the interval of time 0 to t and get a **population growth rate**, or the rate of change in population size over time. This can be expressed mathematically as

$$\Delta N = (B - D) + (I - E), \text{ or the change in } N (\Delta N)$$
$$\text{for the time interval 0 to } t$$

To understand population growth rate from births and deaths alone, ecologists often assume a "closed system" that does not include immigration and emigration. This simplifies the equation to

$$\Delta N = (B - D) \qquad (54.2)$$

The change in population size can then be calculated if one knows the number of births (B) and deaths (D) that have occurred over a given period of time. The birth and death numbers will naturally depend on the size of the population and the time interval considered. As a result we need to convert B and D into rates that will allow us to estimate population growth for any change in population size over time. Therefore we can further express B and D as

$$B = bN_0$$

where B is the product of the **per capita birth rate** (**b**) (i.e., the number of births per individual per unit of time) and N_0 (population size at time 0). Likewise, the death rate can be calculated as

$$D = dN_0$$

where D is the product of the **per capita death rate** (**d**) (i.e., the number of deaths per individual per unit of time) and N_0 (population size at time 0).

The term "per capita" literally translates to "per head" and is meant to indicate births and deaths per individual in the population. The change in the total number of births and deaths can then be calculated by multiplying the per capita birth and death rates by the population size at N_0. Here's what that looks like mathematically:

$$\Delta N = bN_0 - dN_0 \text{ or } \Delta N = (b - d)N_0$$

The difference between the per capita birth and death rate ($b - d$) is the **per capita growth rate** and is symbolized by **r**. Substituting r for ($b - d$), we get

$$\Delta N = rN_0 \qquad (54.3)$$

This simple model reflects why populations change in size. A population will increase in size if the per capita birth rate exceeds its per capita death rate; that is, if $b > d$, then $r > 0$. Likewise, the population will decline in size if $b < d$, or $r < 0$. If $b = d$, then $r = 0$, and the population will not change in size.

Ecologists use differential calculus to express the change in population size (ΔN) over very short, instantaneous periods of time (dN/dt). In this way, they are assuming that population growth is continuous, meaning the time interval is infinitely small and the growth curve will be smooth. This is expressed by

$$dN/dt = rN_0 \qquad (54.4)$$

Once we have a value for r, and we know population size, we can use it to characterize patterns of growth in populations through time. One pattern of growth that occurs when populations are not limited by resources is exponential growth, which we discuss next.

All populations have the potential for exponential growth

You know from Equation 54.4 that a population will grow, over some time period, as long as the per capita growth rate, r, is greater than zero (that is, when $b > d$). During this time period, the population will add a number of individuals that is r times its initial size. As the number of individuals in a population (N_0) increases, the number

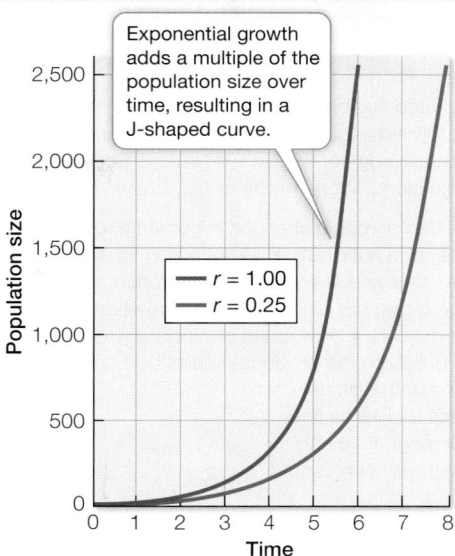

Figure 54.6 Exponential Population Growth Rate   Exponential population growth occurs when the size of a population changes by a constant proportion over time. This graph compares the exponential growth of two populations that vary in their values of r.

Q: Which curve, blue or red, has the faster population growth rate?

 Activity 54.1 Exponential Population Growth Simulation www.Life11e.com/ac54.1

of new individuals added per unit of time accelerates—that is, rN_0 increases—even though the per capita growth rate (r) remains constant (**Figure 54.6**). This multiplicative pattern is known as **exponential growth**. It is also clear from Figure 54.6 that the value of r can make a big difference in the rate of growth of the population—a population with $r = 1.00$ will have a much faster rate of growth than a population with $r = 0.25$, even though both populations will display an exponential growth pattern.

Equation 54.4 gives us the exponential growth of a population, but it does not allow us to predict the size of a population at a later time. Using calculus, we can integrate Equation 54.4 and get

$$N_t = N_0 e^{rt} \qquad (54.5)$$

where N_t = the population size at time t, N_0 = the population size at time 0, e = a constant, the base of the natural log (e = 2.718), r = per capita growth rate, and t = the time interval between time 0 and time t. Knowing the starting population size and the per capita growth rate, we can use Equation 54.5 to forecast the population size at some later time, assuming exponential growth. This is similar to the equation banks use to calculate compound interest on a loan.

One application of projecting population sizes into the future comes in the form of predicting human population growth. The human population increased relatively slowly until 1825, when it was 1 billion, and now stands at 7.4 billion (**Figure 54.7**). United Nations data show that world population growth rates have fallen from a high of 2.30 percent (note that the percent population growth rate is $r \times 100$) in the early 1960s to 1.18 percent today (**Investigating Life: Will the Global Human Population Growth Rate Decline?**). Projections indicate that human population growth rates will continue to decline to 0.50 percent in 2050 and to 0.10 percent in 2100, leading to a "best estimate" population size of 9.1 billion by 2050 and 11 billion by 2100. These projections in human

Original Paper: United Nations, Department of Economic and Social Affairs, Population Division. 2015. World Population Prospects: The 2015 Revision, Methodology of the United Nations Population Estimates and Projections, Working Paper No. ESA/P/WP.242.

Understanding the demographic changes in human populations over the coming years is critical in anticipating and planning for such challenges as food production, energy and water consumption, emerging diseases, and climate change facing our planet. The 2015 revision of World Population Prospects is the U.N.'s 24th round of official population estimates and projections. It builds on the previous reports by incorporating results from the most up-to-date national population censuses as well as findings from recent demographic and health surveys that have been carried out around the world.

HYPOTHESIS▶ World population growth rates have declined since 1950 and will continue to decline in the next century.

METHOD

1. Census and demographic survey data on survivorship, fertility, and international migration were gathered for 201 countries over the period 1950 to 2015.

2. The demographic data were used to establish parameters for life tables (similar to Table 54.1) in order to determine human population growth rates from 1950 to 2015.

3. To project the change in population growth rate values up to 2100, demographic data were used to make assumptions on future trends

in human survivorship, fertility, and international migration. Because the future is uncertain, several different projections were produced to understand their possibility of variability and arrive at a "best estimate" of the change in population growth.

RESULTS

Population growth rates fluctuated after 1950 but dropped from a high of 2.30% in the early 1960s to 1.18% in 2015. Projections indicate that human population growth rates will continue to decline to 0.50% in 2050 and to 0.10% in 2100. Note that the percentage population growth rate is equivalent to $r \times 100$.

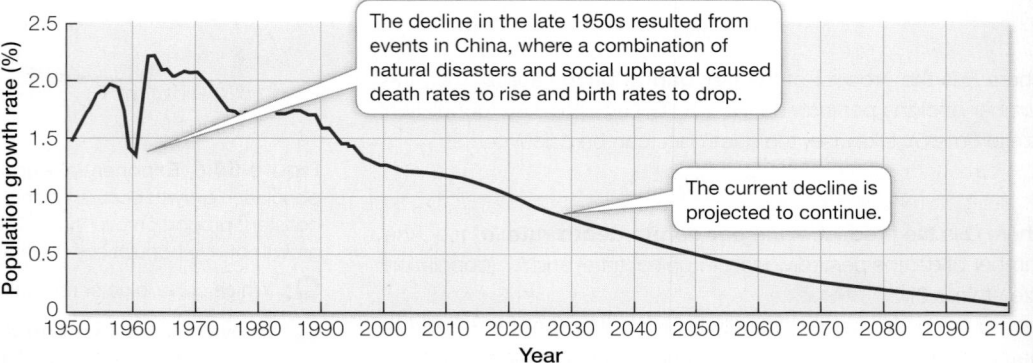

The decline in the late 1950s resulted from events in China, where a combination of natural disasters and social upheaval caused death rates to rise and birth rates to drop.

The current decline is projected to continue.

CONCLUSION▶ Population growth rates have dropped substantially in the last 65 years and are projected to continue this trend into 2100.

Here you will use the population growth rate from above to project the human population size on Earth in the future.

QUESTIONS▶

1. Using Equation 54.5, the population size in 2015 (7.4 billion), and the population growth rate for 2015 (1.18%, above), estimate what the world's population size would be in 2100.

2. Projections indicate that if the human population growth rate declines to 0.50% in 2050, the population size will be 9.1 billion by 2050 and 10 billion by 2080. Using the projected population size and growth rate in 2080, project the human population size in 2100.

3. Suppose the carrying capacity of Earth is 12 billion people. Using Equation 54.7, project the human population size in 2100, given the 2015 data and 2080 data.

4. Given your calculations in Question 3, will Earth's carrying capacity for humans be reached under either scenario by 2100? Which parameter (population growth rate or carrying capacity) is projected to most reduce population size by 2100?

A similar **work with the data** exercise may be assigned in **LaunchPad**.

population size beg the question: will world resources be able to sustain such large populations into the future? Is 11 billion above the maximum number of humans that can be supported sustainably on Earth? We will consider these questions at the end of the chapter.

Some populations may grow at rates close to their maximum, depending on the circumstances. During the 39 years following the ban on commercial whaling, North Pacific humpback whale populations grew exponentially at a rate of 5–7 percent per year. Since the whales were released from hunting, they have had ample habitat, abundant food, and no predators, so there appears to be little limiting their population growth at this point. However, changes in ocean conditions and thus food resources, entanglement in fishing gear, ship strikes, whale-watching harassment, and harvesting for scientific purposes all pose potential threats to humpback whale population recovery. In addition, there is some limit to the population sizes whales can attain, given intraspecific competition for food resources and breeding areas.

We will turn next to what happens to population growth when limiting resources affect the number of births and deaths within a population.

Logistic growth occurs as a population approaches its carrying capacity

What would happen if all the offspring produced by a population survived and reproduced? The prospects are alarming. In 1911 L. O. Howard, then chief entomologist of the U.S. Department of Agriculture, estimated that if all the offspring of the common house fly (*Musca domestica*) were to survive, a pair of flies beginning to reproduce on April 15 would produce a population of 5,598,720,000,000 adults by September 10 of the same year. Given such amazing reproductive capacities, it is clear that forces exist to limit the growth of fly populations (and populations of every other organism).

No real population can maintain exponential growth for long periods of time. As a population increases in density, the resources it requires—such as food, nest sites, and shelter—become depleted, causing birth rates to drop and death rates to rise. In other words, birth rates and death rates are dependent on the density of the population. Populations may initially grow rapidly when the population is small, but as populations increase in size, density-dependent birth and death rates cause r to decline toward zero, a pattern of population growth called **logistic growth**. Eventually when $r = 0$, the population stops changing in size because births equal deaths. Thus instead of continuing to increase exponentially, population growth slows down and the population eventually levels off at some size known as the **carrying capacity** or **K** (Focus: Key Figure 54.8). *K* can be thought of as the number of individuals that any particular environment can sustain indefinitely. If the number of individuals in the population exceeds *K*, then *r* will be negative until population size reaches *K* once again. Likewise, if the number of individuals falls below *K*, then *r* will be positive until population size reaches *K*.

To model logistic growth, the exponential growth equation is modified by adding a term, $(K - N_0)/K$, which represents the fraction of the carrying capacity (*K*) that is available for population growth. As long as the population size is less than the carrying capacity (i.e., $N_0 < K$), only a fraction of the available resources are being used. As the population size approaches the carrying capacity, however, the fraction of resources available for any new individual becomes smaller. The logistic growth equation is simply the exponential growth rate rN_0 multiplied by $(K - N_0)/K$ to get

$$\frac{dN}{dt} = rN_0 \left(\frac{K - N_0}{K} \right) \quad (54.6)$$

Population growth should stop when $N_0 = K$ because at that point, $K - N_0 = 0$, so $(K - N_0)/K = 0$, and thus $dN/dt = 0$ and the population will remain at a constant size.

Just as you saw with Equation 54.5, we can integrate Equation 54.6 to allow us to predict the population size at some later time, assuming logistic growth. When we do this we get

$$N_t = \frac{K}{\left[1 + \left((K - N_0)/N_0 \right) \right] e^{-rt}} \quad (54.7)$$

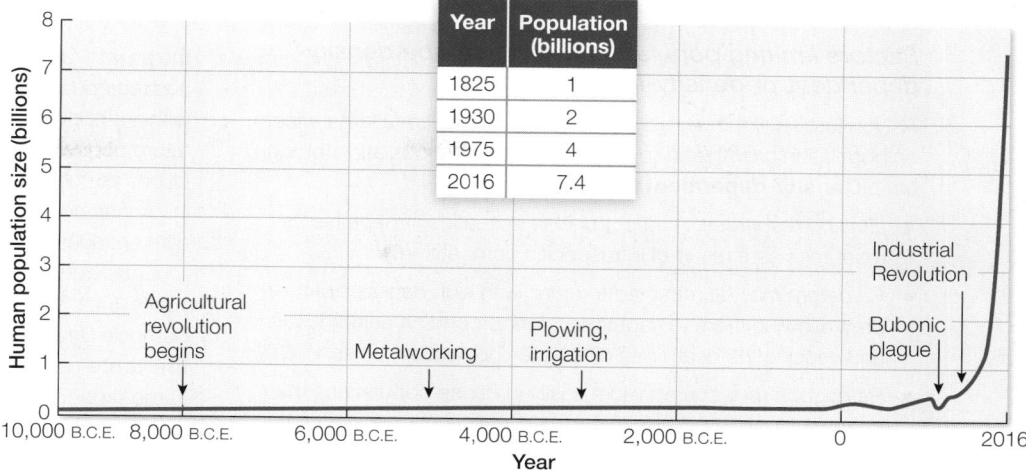

Year	Population (billions)
1825	1
1930	2
1975	4
2016	7.4

Figure 54.7 Explosive Growth of the Human Population   Human population growth was relatively slow until 1825, but advances in technology since then have allowed it to grow at an exponential rate.

where N_t = the population size at time t, N_0 = the population size at time 0, e = a constant, the base of the natural log (e = 2.718), r = per capita growth rate, K = carrying capacity, and t = the time interval between time 0 and time t. In this case, knowing the starting population size, the per capita growth rate, and the carrying capacity of the population, we can use Equation 54.7 to forecast population size at some later time, assuming logistic growth.

 Activity 54.2 Population Growth
www.Life11e.com/ac54.2

> ## focus: key figure

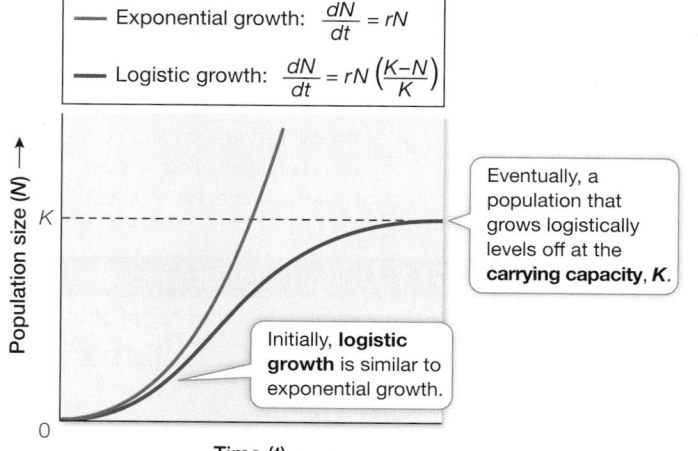

Figure 54.8 Population Growth Models: Exponential versus Logistic Growth   Exponential population growth (red J-shaped curve) is much faster than logistic population growth (blue S-shaped curve), which is limited by the density or carrying capacity of the population.

Q: How does exponential growth differ from logistic growth?

 Activity 54.3 Logistic Population Growth Simulation
www.Life11e.com/ac54.3

Factors limiting population growth can be density-dependent or density-independent

As you just saw, the density of a population can have large effects on population growth and ultimately the size of a population through time. **Density-dependent** factors can include:

- *Limiting resources*. As a population increases, it depletes its resources as a result of intraspecific competition.

- *Predators* may be attracted to areas with high densities of their prey, allowing them to capture a larger proportion of individuals and causing the death rate of the prey population to rise.

- *Pathogens* may spread more easily in dense populations than in populations with fewer individuals, resulting in a rise in the death rate.

It is important to note that density does not always cause a decline in population growth. Some populations grow better, albeit up to some limit, when population densities are higher rather than lower; this is known as the **Allee effect** (named after the ecologist W. C. Allee, who first described it). Sometimes individuals survive better in a group than on their own, as you saw with the predator avoidance strategies used by wood-pigeons and ground squirrels living within groups (see Figure 52.21).

Moreover, not all factors that change population size act in a density-dependent manner. A period of extreme cold, or an exceptionally strong hurricane, may kill a large number of individuals in a population regardless of the population's density; such an event is **density-independent** and may be a significant factor in depressing growth and keeping populations from reaching their carrying capacity.

Life tables keep track of demographic events

Up to this point we have assumed that individuals within a population do not vary in their birth and death rates. This is a big assumption given that we know real populations are made up of individuals of different ages, sizes, and sexes, which vary in their capacity to reproduce or survive. For example, a newborn whale or human cannot reproduce immediately, but must mature to reproductive age. Likewise, the likelihood of death will vary depending on the age of the individual. Ecologists use a form of accounting, known as a **life table**, to keep track of how demography will affect the growth rate of the population. Life tables summarize how survival and reproductive rates vary with the age, size, or sex of the individuals within a population. These summaries can then be used to predict future population trends and develop strategies for managing populations of commercial or ecological value. They are also used by life insurance companies to determine how much to charge people of different ages for insurance policies.

Let's consider an example of a life table from the literature. **Table 54.1** shows a cohort life table using data from the acorn barnacle *Balanus glandula* on the shorelines of Scotland. A **cohort life table** uses a cohort—a group of individuals born within the same time frame—from the larger population to estimate growth rate. The two columns on the left show the number of individuals surviving and the number of offspring produced at different ages (x) through time. As the number of individuals within the cohort die, N_x decreases from 1 million barnacles to only 2 barnacles after 8 years. The proportion of individuals that survive, known as **survivorship** (l_x), can be calculated by simply dividing N_x by N_0, or the number of barnacles originally born into the cohort (represented by age 0). In addition, we can calculate the **fecundity** (m_x), or the mean number of barnacle offspring produced per surviving adult barnacle per age class, by dividing the total number of offspring ($N_{x\ offspring}$) by the number of individuals (N_x) that produce those offspring. Multiplying survivorship (l_x) by fecundity (m_x) gives us the number of offspring produced for individuals within a particular age class within the population. The sum of these values for all the age classes gives us the **net reproductive rate**, R_0, which is simply the mean number of offspring produced per individual in the cohort, adjusted for survival:

$$R_0 = \text{sum } (l_x m_x) \qquad (54.8)$$

If R_0 is greater than 1.0, there is a net increase in offspring produced each generation, and assuming the birth and death rates do not change over time, the population should increase exponentially. If R_0 is less than 1.0, and individuals are not replaced as they die, the population declines eventually to extinction. If R_0 is 1.0, then the births and deaths balance out and the population will not change in size.

table 54.1 Cohort Life Table for the Barnacle *Balanus glandula*[a]

Age (x)	Number of individuals (N_x)	Number of offspring ($N_{x\ offspring}$)	Survivorship (l_x)	Fecundity (m_x)	$l_x m_x$	$x\ l_x m_x$
0	1,000,000	0	1	0	0	0
1	62	285,200	0.000062	4,600	0.285	0.285
2	34	295,800	0.000034	8,700	0.296	0.592
3	20	232,000	0.000020	11,600	0.232	0.696
4	15	190,500	0.000015	12,700	0.191	0.764
5	11	139,700	0.000011	12,700	0.140	0.700
6	6	76,200	0.000006	12,700	0.076	0.456
7	2	25,400	0.000002	12,700	0.025	0.175
8	2	25,400	0.000002	12,700	0.025	0.200

1 To estimate the per capita growth rate, r, sum all l_x by m_x values to get the **net reproductive rate, R_0.**
$R_0 = \text{sum } (l_x m_x) = 1.27$

2 Then, sum all l_x by m_x by x values and divide by R_0 to get the generation time, G.
$G = \text{sum } (x\ l_x m_x)/R_0 = 3.05$

3 Finally, divide the natural log of R_0 by G to get the intrinsic rate of increase, r.
$r = (\ln R_0)/G = 0.08$

[a]The life history of barnacles involves releasing larvae into the water column, where they feed and undergo a series of larval stages. Eventually they settle back on the rock and metamorphose into juvenile barnacles, which grow into reproductive adults.

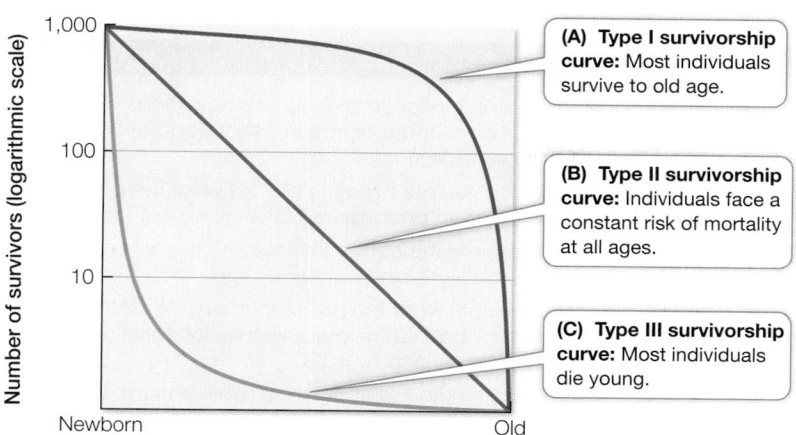

Figure 54.9 **Survivorship Curves**   Ecologists recognize three general types of survivorship curves. Notice that the number of survivors has been plotted on a logarithmic scale. Three species provide real-world examples of the three types of life histories.

Q: What type of survivorship curve do humans have?

 Activity 54.4 **Age Structure and Survivorship Simulation**
www.Life11e.com/ac54.4

We can use R_0 to estimate the per capita growth rate, r, of a cohort by scaling R_0 to account for the generation time of the cohort. The **generation time**, **G**, is the average age of the parents of all the offspring produced within the cohort (see Table 54.1 for equation). To estimate r, we simply divide the natural log (ln) of R_0 by G and get

$$r = (\ln R_0)/G \qquad (54.9)$$

Cohort life tables follow individuals from birth to death as a function of calendar year or life stage (e.g., eggs, larvae, pupae, and adults in insects). This is relatively easy to do if the organisms are easily followed—for example, if they are sessile and short-lived as we saw in the barnacle example. For organisms that are harder to track, a **static life table** can be constructed by sampling a population at a single slice of time. Here the numbers of individuals and their reproduction for different age classes are recorded at one time and used in a similar way to those data in Table 54.1 to calculate the net reproductive rate and per capita growth rate of the population.

The construction of life tables has allowed ecologists to observe common life history patterns, reflecting common solutions to ecological challenges, across a tremendous diversity of organisms. For example, the number of individuals surviving through each life stage (survivorship, l_x) can be taken from a life table and plotted graphically to construct a **survivorship curve**. Typically, a survivorship curve is constructed for a hypothetical cohort, usually of 1,000 individuals, by plotting the numbers of individuals expected to survive to reach each age category on a logarithmic scale.

Survivorship curves tend to take one of three general shapes:

1. Species with **type I survivorship curves** experience high overall survivorship through adulthood but steep declines late in life (the curve is concave) (**Figure 54.9A**). Species with this type of survivorship curve (humans, elephants, whales, and many other large mammals) typically have low reproduction rates but provide parental care to their offspring, which reduces the risk of death in early stages of development.

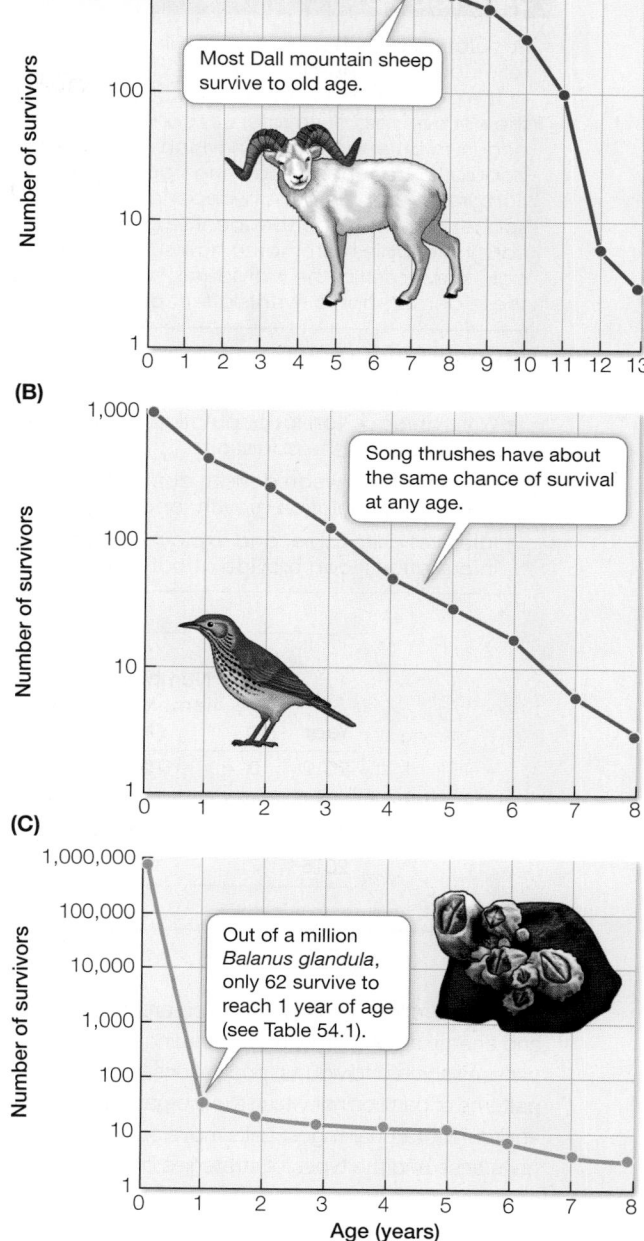

2. Species with **type II survivorship curves** face a constant risk of mortality at all ages (the curve is linear) (**Figure 54.9B**). Many birds, fish, and plants display this pattern.

3. Species with **type III survivorship curves** experience low survivorship early in life and higher survivorship once they reach maturity (the curve is convex) (**Figure 54.9C**). Species with this type of survivorship curve (most insects; marine invertebrates, including the barnacle in Table 54.1; and annual plants) tend to produce many offspring but provide little or no parental care.

Species survivorship curves help us classify how mortality plays out in populations over time, but many species have patterns of survivorship somewhere in between these extremes. For example, some organisms show a "stair-stepped" curve, with bouts of mortality during vulnerable periods of their life, followed by periods of relatively low mortality.

54.2 recap

Population growth is the change in population size over time, as reflected by the number of births, deaths, and migrants in and out of the population. Exponential (multiplicative) growth occurs when the per capita growth rate, r, is constant over time. Logistic growth occurs in an environment in which resources are limited and a population reaches its carrying capacity, K. Factors limiting population growth can be density-dependent (e.g., competition for food or space) or density-independent (e.g., physical environmental conditions). Life tables summarize how survival and fecundity vary with the age, size, or sex of the individuals in a population, and predict how these demographic events affect population growth.

learning outcomes

You should be able to:

- Write an equation for population size at time t, and explain each of the terms of the equation.
- Distinguish between density-dependent and density-independent factors in population growth, and give examples of each.
- Interpret a life table, and explain and calculate the types of information it can provide about a population.

1. Write an equation that describes how births, deaths, immigration, and emigration affect population size. Define each variable.

2. Suppose you are asked to predict what the population size of an endangered bird will be in 2035.

   a. Complete the cohort life table below, and calculate the R_0, G, and r of this cohort.

   b. Calculate what the population size, N_t, of this cohort will be in 2035, assuming exponential population growth and $N_0 = 120$.

   c. To add more realism to these calculations, suppose that K (carrying capacity) = 300 for this cohort. What will the population size be, assuming logistic growth, in 2035?

3. Plot the number of birds surviving with age for the cohort in Question 2. What type of survivorship curve does this cohort have?

4. Describe some density-dependent and density-independent factors that could control the population growth of the bird cohort in Question 2.

Year	Age (x)	Number of female birds (N_x)	Number of female offspring ($N_{x\ \text{offspring}}$)	Survivorship (l_x)	Fecundity (m_x)	$l_x m_x$	$x\, l_x m_x$
2012	0	100	0				
2013	1	50	75				
2014	2	40	80				
2015	3	30	60				

Life tables reflect the ways in which organisms partition their time and energy among growth, maintenance, and reproduction. Comparisons across individuals, populations, and species reveal different patterns of partitioning, which allow organisms to cope with different environmental challenges. Let's move on to discuss how life history traits arise and the types of strategies organisms employ.

key concept 54.3 Life History Is the Lifetime Pattern of Growth, Reproduction, and Survival

We know as humans that individuals vary widely in when they mature, when and how many children they have, and when they die. This lifetime pattern of growth, reproduction, and survival is known as an individual's **life history**. In addition to the variation we see in life history patterns at the individual level, all species have a common **life history strategy** that helps define the average timing and nature of important life history events (**Figure 54.10**). Life history strategies are shaped by the way individuals within and among species allocate resources to growth, reproduction, and survival based on genetic and environmental factors. For example, in the

case of humans today, the average age of puberty is 11 years, the average woman has one child at a time (except in the rare cases of multiple births), and the average life span is 70 years. These life history statistics are a product of the environmental and genetic constraints placed on humans in the twenty-first century. But they also belie the fact that human life history traits have changed dramatically from historical times and vary widely among populations around the world. Thus, as you will see, life history patterns not only vary within and among species but also through time.

focus your learning

- Life history strategies arise within and among species as a result of environmental and genetic constraints, resulting in trade-offs.
- Life history strategies vary across a continuum from maximum population growth (r-strategists) to minimum population growth (K-strategists).

Life history strategies can vary at species and population levels

Life history strategies can vary among species, even for species that are taxonomically closely related, as a consequence of environmental differences. For example, on Mount Kenya two species of unusual plants in the genus *Lobelia* live in different types of habitats

(**Figure 54.11**). One species, *Lobelia telekii*, lives on the dry rocky slopes of the mountain, where it reproduces only once (a strategy called **semelparity**), produces vast numbers of small seeds, and has a short life span. In the moist valley bottoms, another species, *Lobelia keniensis*, reproduces more than once (a strategy called **iteroparity**), produces fewer but larger seeds per reproductive event, and lives longer. Semelparity is typical of organisms whose probability of dying remains high even when they reach adulthood; it is seen in some fish, many insects, and all annual plants. In contrast, iteroparity is typical of organisms with long life spans. For example, because environmental conditions within the nests of social insects such as honey bees and ants are remarkably stable, iteroparity is the rule among these species; some queens live 10 years or longer and reproduce over their entire adult lives.

Life history strategies can also vary within species at the population level. For example, some populations of guppies (*Poecilia reticulata*) in Trinidad live in streams where they are attacked and eaten by larger fish (**Figure 54.12**). But some streams have waterfalls that predatory fish are unable to negotiate. Guppies that live in the predator-free areas upstream from those waterfalls have lower death rates compared with guppies below the falls. To see whether the risk of being eaten by a predator influenced the life history strategies of

these guppies, David Reznick and his colleagues collected guppies from high-predation and low-predation sites and raised them in the laboratory. Some guppies from each group were provided with plentiful food, and others with limited food, to simulate the variation the fish would typically encounter in their home streams. In the laboratory, where no predators were present, guppies from high-predation sites matured earlier, reproduced more frequently, and produced more offspring in each brood than did guppies from low-predation

Figure 54.10 A Life History Strategy
The timing and nature of life history events shape the overall life cycle of an organism. The life history strategy of this frog species reflects how environmental and genetic factors affect strategies to optimize growth, reproduction, and survival.

How large to grow?
When to begin reproducing?
How many offspring, and of what size?
How fast to grow?
How often to breed?
Whether to care for offspring?
How fast to grow and develop?
How long to live?
When to metamorphose?

(A) *Lobelia telekii*

(B) *Lobelia keniensis*

Figure 54.11 Related Species with Vastly Different Life History Strategies   On Mount Kenya, Africa, **(A)** *Lobelia telekii* lives on the dry rocky slopes of the mountain, where it reproduces once (is semelparous), produces many seeds, and has a short life span. In the moist valley bottoms, **(B)** *Lobelia keniensis* reproduces multiple times (is iteroparous), produces fewer but larger seeds, and lives longer.

Figure 54.12 Predation Affects the Life History of Guppies   In Trinidad, populations of guppies (*Poecilia reticulata*) from high-predation stream sites below the falls matured earlier, reproduced more frequently, and produced more offspring per brood compared with guppies from low-predation stream sites above the falls, no matter how many resources the guppies were given.

Q: Are the life history strategies of the two guppy populations, either from high-predation streams and low-predation streams, likely genetically determined? Why?

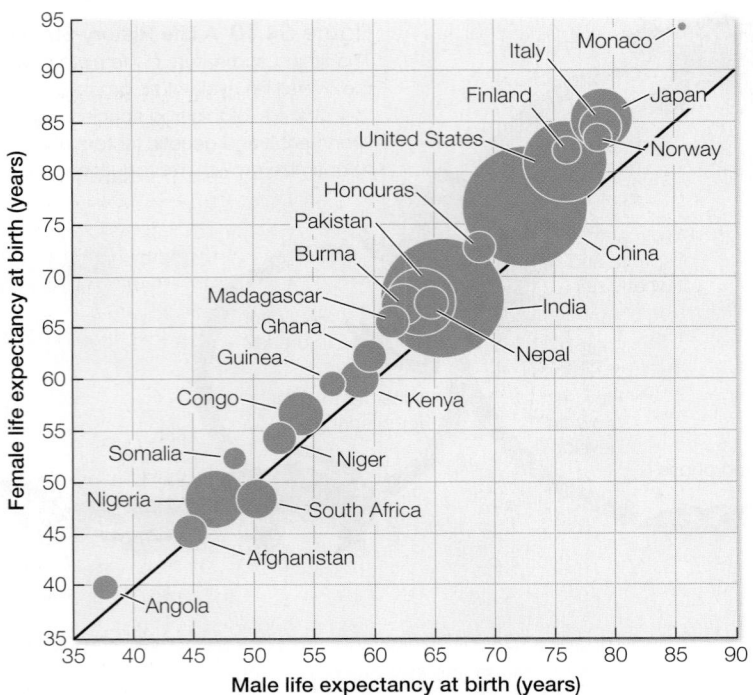

Figure 54.13 Human Life Expectancy around the World
Comparison of male and female life expectancy at birth for a variety of countries in 2016. The solid black line corresponds to equal female and male life expectancy. The sizes of the bubbles are linearly proportional to the population of the country.

sites, no matter how much food the guppies received. The investigators concluded that predation had genetically selected for early and frequent reproduction in the guppies living below the falls.

Perhaps the best-studied example of how life history traits can vary not only among populations, but also among individuals, is the case of human life span. Human life span varies widely around the world and between the sexes (**Figure 54.13**). Today, humans live on average to 87 years in Japan, Hong Kong, and many European countries, and to 35–50 years in some African countries

(e.g., Sierra Leone, Democratic Republic of the Congo, and Somalia) where HIV infection, famine, and war are common. In the United States, average life span is 80 years, with women outliving men by 5 years. Research has shown that the factors contributing to lower life expectancy rates in men include higher rates of homicide, accidental death, and parasitic and infectious diseases.

For the most part, the large variability in life span we see in humans can be attributed to environmental factors. If you lived during the Bronze Age, your life expectancy was 28 years, and it was not much better even in the early nineteenth century (about 35 years). Reduction in infant mortality has accounted for most of the increase in life expectancy, but since the end of the 1950s the mortality rate of those older than 80 years has decreased by 1.5 percent each year. Much of the increase in life expectancy that we enjoy today is the result of modern advances in human health and nutrition, which begs the question: Is there a limit to human life expectancy? One study, which plotted human life-span data over time, shows a remarkable linear relationship with no leveling off yet, that might indicate an upper limit (**Figure 54.14**). In fact, because life span has increased by 2.5 years every decade for the last 150 years, one possible scenario is that 60 years from now the average person's life span will reach 100 years! While some limit on human lifespan is mostly certain, these predictions suggest that human life expectancy is a highly flexible life history trait indeed.

Life history strategies arise from constraints on growth, reproduction, and survival

As it should be clear by now, the fitness of a population or species is highly dependent on its life history strategy in relationship to its surrounding environment. Thus it makes sense that genes that maximize growth, reproduction, and survival would be favored by natural selection, resulting in an optimal life history strategy for that species or population. However, there are ***constraints on optimality**, resulting in **trade-offs** in life history traits that put limits on the possible life history outcomes for species.

---

*connect the concepts Constraints on optimality include genetic variation for evolution to act on, as well as the mediating effects of the physical and biological environment. Review Key Concept 21.5.

Trade-offs include allocation of resources to such traits as growth versus reproduction or number of offspring versus their size. Energy devoted to one life history trait can reduce the amount available for others. For example, the growth of Douglas fir trees (*Pseudotsuga*

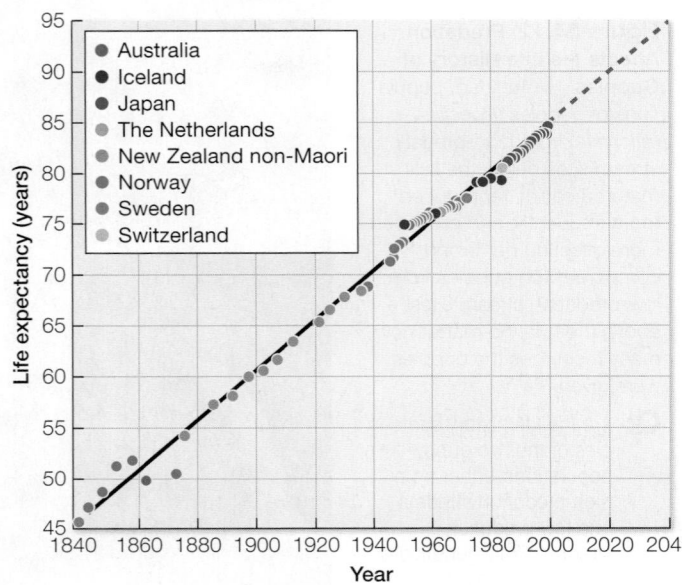

Figure 54.14 Human Life Expectancy—Past, Present, and Future Recorded human female life expectancy from 1840 to the present. The linear trend in life expectancy is depicted by the bold line, and the extrapolated trend by the dashed gray line. Based on the trend, in 2040 the average woman might have a life span of 95 years.

Q: Compare the differences in female life expectancy between 1840–1940 and 1940–2040. Is the rate of change in life expectancy predicted to increase?

Level	*r*-strategists	*K*-strategists
Habitat	Tolerate unpredictable conditions by using a wide variety of resources	Thrive in predictable environments but compete for high-quality resources
Reproductive strategy	Typically reproduce once (semelparity), but greater allocation to reproduction than growth results in large numbers of offspring	Reproduce more than once (iteroparity), but greater allocation to growth than reproduction results in fewer offspring
Survivorship	Short life span, density-independent mortality, typical of a type III survivorship curve (see Figure 54.9C)	Long life span, density-dependent mortality, typical of a type I or II survivorship curve (see Figure 54.9A, B)
Population growth	Period of exponential growth (*r*) followed by periodic or seasonal population declines	Slowly rising population growth that stabilizes at or near carrying capacity (*K*)
Examples		

Figure 54.15 Two Life History Strategy Extremes Species whose life histories are geared to achieve the maximum possible rate of population growth are referred to as *r*-strategists; those whose population dynamics are bounded by carrying capacity are *K*-strategists. The life histories of most species fall along a continuum between these two strategies.

menziesii) measured using annual growth ring size is negatively correlated with cone production. The more cones the tree produces, the narrower the width of the rings, suggesting that the allocation to reproduction takes energy from growth. Another trade-off that can have dramatic consequences for population growth is that of survival and reproduction. An example of this occurs in the giant Pacific octopus (*Enteroctopus dofleini*), which lives only 3–5 years and dies soon after significant parental investment in its thousands of offspring.

As you have seen, much of the variation in life history traits appears to result from an evolutionary response to environmental variability. Generally, unpredictable environments are associated with greater allocation to reproducing quickly, shorter life spans, and correspondingly high per capita growth rates. Conversely, predictable environments, where organisms must compete for high-quality resources, are associated with lower fecundity, longer life spans, and correspondingly lower per capita growth rates. Species whose life history strategies allow for high population growth rates are called **r-strategists**, and species whose life history strategies allow them to persist at or near the carrying capacity (*K*) of their environment are called **K-strategists** (**Figure 54.15**). Keep in mind, however, that these categories are extremes; most species fall along a continuum between these two strategies.

For *r*-strategists, life is uncertain. Individuals tend to reproduce only once and to produce large numbers of offspring, most of which are expected to die early in life (that is, they have a type III survivorship curve; see Figure 54.9C). They can generally use a wide variety of resources and tolerate a wide range of conditions. *K*-strategists are adapted to predictable conditions, are long-lived, and reproduce several times; their smaller numbers of offspring have a high probability of surviving to adulthood (that is, they have a type I or II survivorship curve; see Figure 54.9A, B). *K*-strategists tend to be good at enduring competition for resources, but this sometimes leads to trade-offs in reproduction.

That life history trade-offs can evolve is suggested by genetic correlations among suites of life history traits. Such genetic correlations imply either simultaneous selection on two or more life history traits

or *linkages among the genes that code for those traits. Across *Drosophila melanogaster* strains, for example, a high per capita rate of growth is correlated with the ability to reproduce under starvation conditions and with the ability to develop on a variety of media in the laboratory—both of which are consistent with the *r* strategy of tolerating a wide range of resources and conditions.

---

*connect the concepts   Linked genes are inherited together. Genes can be linked if they occur on the same chromosome. Review Key Concept 12.4.

54.3 recap

An organism's life history is characterized by its lifetime pattern of growth, reproduction, and survival. All species have a characteristic life history strategy, which arises from genetic and environmental constraints on life history traits. Two extremes of life history strategies include species with high population growth rates (*r*-strategists) and species that persist at or near their carrying capacity (*K*-strategists). Most species fall along a continuum between these two strategies.

learning outcomes

You should be able to:

- Distinguish between the relative contributions of environment and genetics to life history traits in specific situations.
- Describe how and why life history trade-offs might arise in a species or population.
- Distinguish between *r*-strategists and *K*-strategists, and identify specific situations in which evolution might favor one strategy over the other.

1. Refer to Figure 54.14.

   a. Does the change in life expectancy for humans over the past 175 years suggest a genetic or an environmental cause?

   b. Given your answer, are the large differences in human life span for Japan versus Angola, shown in Figure 54.13, likely the result of genetic or environmental factors? Explain.

54.3 recap

2. Why are species unable to maximize growth, reproduction, and survival? In other words, why are there life history trade-offs?

3. Refer to Figures 54.11 and 54.12. What life history strategies characterize *Lobelia telekii* versus *L. keniensis*? How about guppies in high-predation versus low-predation stream sites?

For millennia, humans have tried to increase populations of desirable or useful species and reduce populations of species they consider undesirable. Such efforts are most successful if they are based on knowledge of how those populations grow and what determines their densities.

key concept

54.4 Population Biology Can Be Used in Conserving and Managing Populations

If we want to manage other species—that is, to increase or decrease their populations—we need to understand their life histories and population dynamics. We turn now to how populations can be managed to our benefit.

focus your learning

- Appropriate management and conservation of wild or commercial populations requires knowledge of their life history strategies and population dynamics.
- Successful control of undesirable species, including biological control, relies on an understanding of population dynamics.
- Thorough understanding of metapopulation dynamics can help prevent species extinction.

Management plans must take life history strategies into account

Knowing the life history strategy of a species can be helpful in managing populations of commercial value. The black rockfish (*Sebastes melanops*), an important game fish that lives off the Pacific coast of North America, provides one such example. Rockfish have an indeterminate growth pattern—they continue to grow throughout their lives. As in

Figure 54.16 Overharvesting Can Reduce Fish Populations Harvests of cod and haddock on Georges Bank (map inset) have led to dramatic population declines of both species. Fishing regulations on haddock have allowed its stock to rebound enough to once again support a commercial fishery.

many other animals, the number of eggs a female rockfish produces is proportional to her size, so larger females produce more eggs than smaller females. In addition, older, larger females are better able to provision the eggs they produce with oil droplets, which provide energy to the newly hatched larvae, giving them a head start in life. Larvae from eggs with larger oil droplets, produced by larger females, grow faster and survive better than do larvae from eggs with smaller oil droplets. These life history traits have important implications for the management of rockfish populations.

Because fishermen prefer to catch big fish, intensive fishing off the Oregon coast from 1996 to 1999 reduced the average age of female rockfish from 9.5 to 6.5 years. Thus the females reproducing in 1999 were, on average, smaller than the females reproducing in 1996. This change decreased the average number of eggs produced by females and reduced the average growth rate of larvae by about 50 percent. This reduction in reproductive ability was linked to a decrease in the ability of the rockfish population to recover from intensive fishing. Maintaining productive populations of rockfish may require regulating the number or size of fish caught, the fishing season, and/or setting aside no-fishing zones where some females can be protected from fishing and allowed to grow to large sizes.

Management plans must be guided by the principles of population dynamics

If we look at a logistic growth curve (see Figure 54.8), we can see that the number of births tends to be highest when a population is well below its carrying capacity. Therefore if we wish to maximize the number of individuals that can be harvested from a population, we should manage the population so that it is far enough below the carrying capacity to have a high birth rate. Hunting and fishing regulations are established with this objective in mind.

Fish can, however, be overharvested, as illustrated by the story of the black rockfish. Many fish populations have been greatly reduced because so many individuals were harvested that the few surviving reproductive adults could not maintain the population. For example, Georges Bank, off the northeast coast of North America—a

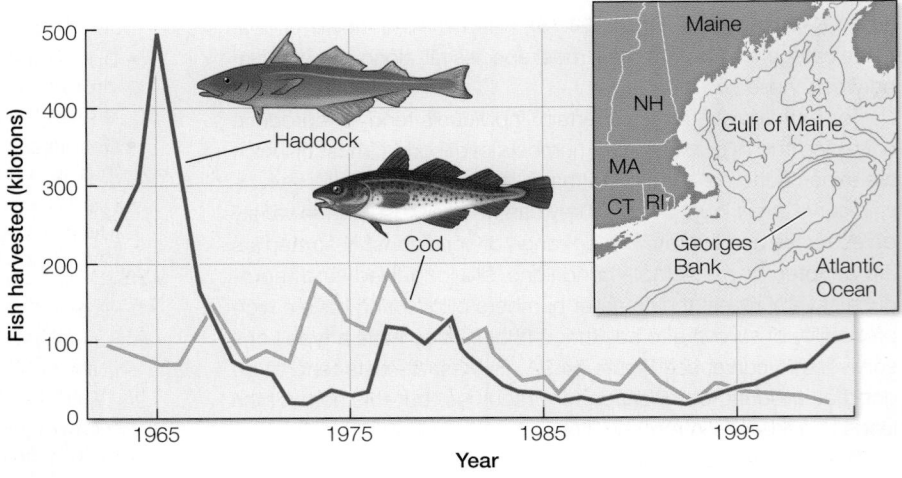

source of cod, haddock, and other prime food fish—was exploited so heavily during the twentieth century that many fish stocks have been reduced to levels insufficient to support a commercial fishery (**Figure 54.16**). With fishing restrictions the haddock population has rebounded, but the cod population has not.

Many rapidly reproducing species can recover if *****overharvesting** is stopped, but recovery is more difficult for slowly reproducing species. Just as they did with the humpback whale, twentieth-century whalers hunted the blue whale (*Balaenoptera musculus*), Earth's largest animal, nearly to extinction. These whales reproduce very slowly: they live up to 10 years before becoming reproductively mature, produce only one offspring at a time, and have long intervals between births. As opposed to humpback whales, however, blue whale populations have recovered more slowly.

***connect the concepts**   A study of fisheries worldwide shows a dramatic increase in harvest rate, leading to collapses of fisheries. But the effects of fish harvesting vary widely across the globe, in part because of current fisheries' restrictions to recover once-overfished species. See Key Concept 58.2 and Figure 58.8.

Whether we want to manage the sizes of populations of desirable species for sustainable harvesting or of undesirable species for control purposes, the same principles apply. **Biological control** is the use of natural enemies (predators, parasites, or pathogens) to reduce the population density of an economically damaging species. In many cases the target species is a pest only because it has been introduced to a new area. Natural enemies used for biological control are often obtained from the native region of the pest species. Biological control became popular in the nineteenth century after an outbreak of cottony-cushion scale, an Australian insect that attacks citrus, appeared in citrus groves in California. A predaceous ladybeetle and a parasitic fly were then introduced from Australia. Within a year of their release, these insects brought the scales under control.

Sometimes, however, introduced natural enemies not only fail to have any effect on the pest they were imported to control but also, freed of their own enemies, become pests themselves. This fact underlies the horror story of the cane toad (*Rhinella marina*) in Australia. This Central American toad was introduced to control cane beetles attacking Australian sugarcane fields. But Australian cane beetles stay high on the upper stalks of the plants; the toads could not reach that high, and thus had no effect on the beetle population. Unfortunately, they had massive effects on other species.

All stages of the *R. marina* life cycle are poisonous, and Australian reptiles (including snakes and lizards) and mammals that eat them usually die. With no enemies to limit their population growth, cane toads grow fast and outcompete native amphibian species for resources. The toads have spread from northern Australia down the east coast, where they threaten native frog species by preying on them as well as by competing with them. The Australian government is forced to spend millions of dollars in attempting to reduce their numbers.

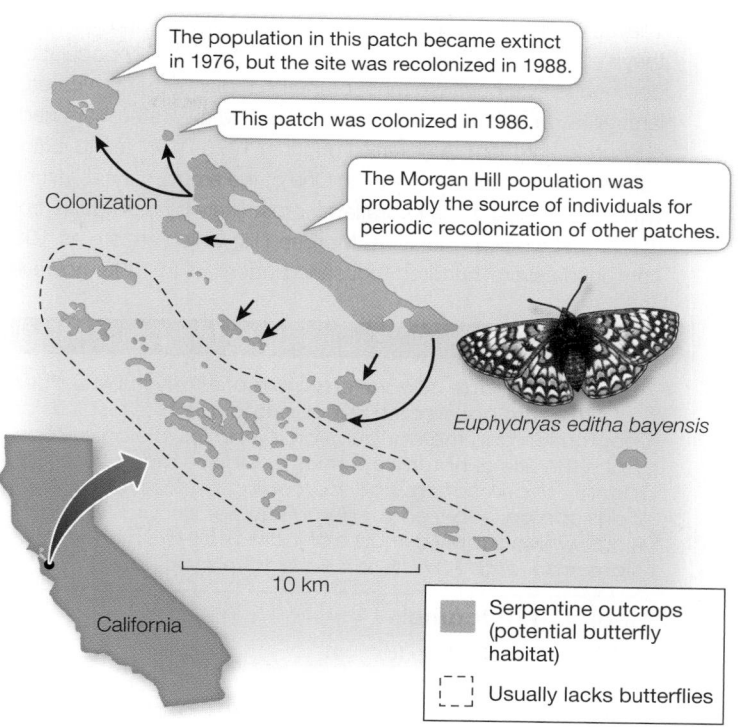

Euphydryas editha bayensis

Figure 54.17 A Metapopulation of Checkerspots   The San Francisco Bay Area metapopulation of Edith's checkerspot butterfly is divided into several populations confined to patches of suitable habitat on serpentine rock outcrops that contain the food plants of the species. Arrows indicate colonization events.

 Activity 54.5 **Metapopulation Simulation**
www.Life11e.com/ac54.5

Knowledge of metapopulation dynamics helps us conserve species

Ecologists have used information on metapopulation dynamics to conserve endangered species. We know from mathematical models that the risk of extinction for a species can depend on the number and size of its populations within a metapopulation, and on the rates of dispersal among them. In some metapopulations there are **source populations**, which serve as a net source of individuals for **sink populations**, which receive more immigrants than they produce. Thus, although some individual populations may undergo severe decline as a consequence of natural or human-caused factors, source populations within a metapopulation can serve to rescue them.

A dramatic example of such metapopulation dynamics is provided by Edith's checkerspot butterfly (*Euphydryas editha bayensis*). This endangered subspecies feeds on only two species of annual plants (California plantain and purple owl's clover) that are endemic to outcrops of serpentine rock on hills south of San Francisco. In 1960 Paul Ehrlich and his colleagues at Stanford University began studying a population of this butterfly in the nearby Jasper Ridge Biological Preserve and discovered that it was one of several populations within a large, very fragmented metapopulation (**Figure 54.17**).

They followed all the populations over several years and found that they varied in size. During drought years, the butterfly populations declined in size because most of the host plants that the caterpillars relied on died early in the spring, before the caterpillars had a chance to eat them. A severe drought in 1975–1977 led to extinctions of some of the populations. One of the empty habitat patches was repopulated a few years later, most likely by individuals from the largest single population, Morgan Hill, which as late as 1989 contained several hundred thousand butterflies. In 1998, however, the Morgan Hill population, which had historically been the largest in the metapopulation, went extinct. Ehrlich and his colleagues examined 70 years of climate data for the region and concluded that increasing climate variation accounted for the extinction. They also concluded that without a large, stable source population to provide emigrants for recolonization, as the Morgan Hill population did during the 1970s drought, it is unlikely that any of the other populations, which are sink populations, will persist without human intervention.

▶ 54.4 recap

Efforts to manage populations are more likely to be successful if they are based on an understanding of life histories and population and metapopulation dynamics. Biological control is the use of natural enemies to reduce the population density of an economically damaging species. Metapopulations can conserve endangered species if source populations, which supply a source of individuals to sink populations, can rescue those populations from extinction.

learning outcomes

You should be able to:

* Specify possible density-dependent approaches to controlling a population of undesirable organisms, and discuss the different results each approach would have.

* Analyze the importance of source and sink populations in a metapopulation in terms of their function in conserving a species from extinction.

---

1. Describe an effective strategy for recovering the population size of a desirable species that has a *K*-strategist life history.
2. Refer to Figure 54.17.
    a. Describe factors responsible for the metapopulation size of the Edith's checkerspot butterfly in the San Francisco Bay area.
    b. Given the 1998 extinction of the Morgan Hill population, what is the likely fate of the metapopulation? Explain your answer.

▶ investigatinglife

 **What is the human population growth projected to be in the next century, and how will that affect population size?**

You have seen that this question is highly dependent on the factors that ultimately control the number of births and deaths, or population growth, of humans over time. If the population growth rate projections are correct, there will be roughly 11 billion humans in 2100. Is 11 billion above the carrying capacity of humans? Estimates of the carrying capacity published in 1994 varied 10-fold: from less than 3 billion to 44 billion, with the median estimated at 12 billion. As opposed to the carrying capacities of other species, human carrying capacity depends on natural constraints important to all species, but also on individual and collective choices concerning the distribution of material goods, the use of technology, and the right to political, economic, and cultural freedoms. The number of people that can live on Earth will depend in part on how many will own cars, eat meat, or have air conditioning. For example, estimates suggest that if the carrying capacity values for the population in the United States were applied worldwide, Earth's carrying capacity would be limited to 1.2 billion. Compare this with the estimated values for India, where the worldwide carrying capacity could be extrapolated to 14 billion. Given that the current population size on Earth is 7.4 billion, it is clear that the vast majority of humans do not, and will not, experience the quality of life inherent to many in the United States.

Future directions

Some have argued that the number of people that can ultimately live on Earth will depend on the ability of humans to continue to expand their carrying capacity through technological innovations and greater efficiencies in food production and energy generation. These technological innovations will require that we increase food production and natural resource extraction without a heavy reliance on fossil fuels, which have a limited supply and adverse effects on the climate. Ultimately, to believe that no ceiling to human population size is inevitable defies what we know about natural populations.

Chapter Summary 54

▶ **54.1 Populations Show Dynamic Variation in Size over Space and Time**

- Species are divided into groups of individuals, or **populations**, that can vary spatially across their **geographic range**. Review Figure 54.1A

- A group of geographically isolated populations linked together by **dispersal** is known as a **metapopulation**. A species might consist of one or multiple metapopulations, depending on its geographic range. Review Figures 54.1, 54.4

- All populations fluctuate in size over time due to the physical environment, biological interactions, and dispersal. **Review Figure 54.2**

- Ecologists use a variety of approaches to estimate population sizes, including **full censuses**, surveys using **quadrats** or **transects**, **mark–recapture methods**, and DNA analysis. Review Figure 54.5, Animation 54.1

▶ **54.2 Population Growth Describes the Change in Population Size over Time**

- **Population growth** is the change in population size over time, as reflected by the number of births, deaths, and migrants in and out of the population. **Review Equation 54.1**

- **Exponential growth** occurs when the rate of change in population size is constant over time. **Review Figure 54.6, Activity 54.1**

- **Logistic growth** occurs when resources are limited, slowing growth to a population's maximum size (carrying capacity). Review Focus: Key Figure 54.8, Activities 54.2, 54.3

- Factors limiting population growth can be **density-dependent** when **intraspecific competition** between individuals affects population growth. **Density-independent** factors are those that act independently of the population density, such as physical conditions.

- **Life tables** summarize how survival and fecundity vary with the age, size, or sex of individuals within a population, and incorporate how these demographic events affect population growth. **Review Table 54.1**

- Species have different types of **survivorship curves** that reflect mortality patterns over time. **Review Figure 54.9, Activity 54.4**

▶ **54.3 Life History Is the Lifetime Pattern of Growth, Reproduction, and Survival**

- **Life history strategies** are characterized by lifetime patterns of growth, reproduction, and survival, and can vary at the species and population levels. **Review Figure 54.10**

- Life history **trade-offs** arise from genetic and environmental constraints on optimal growth, reproduction, and survival.

- Extremes in life history strategies of species allow for high growth rates (**r-strategists**) or for persistence at or near the carrying capacity (**K-strategists**), but most species fall along a continuum between these two strategies. **Review Figure 54.15**

▶ **54.4 Population Biology Can Be Used in Conserving and Managing Populations**

- Efforts to manage populations are more likely to be successful if they are based on an understanding of life histories and population and metapopulation dynamics.

- **Biological control** is the use of natural enemies to reduce the population density of an economically damaging species.

- Metapopulations can conserve endangered species if **source populations**, which serve as a source of individuals to **sink populations**, can rescue those populations from extinction. **Review Figure 54.17, Activity 54.5**

> Go to **LearningCurve** (in **LaunchPad**) for dynamic quizzing that helps you solidify your understanding of this chapter. **LearningCurve** adapts to your responses, giving you the practice you need to master each key concept.

▶ Apply What You've Learned

Review

54.2 Life tables help predict how age-related survival and reproduction affect population growth.

54.4 Appropriate management and conservation of wild or commercial populations requires knowledge of their life history strategies and population dynamics.

Original Paper: Simpfendorfer, C. A. 2005. Demographic models: Life tables, matrix models and rebound potential. In J. A. Musick, and R. Bonfil, eds. *Management techniques for elasmobranch fishes*. FAO Fisheries Technical Paper 474: 143–153.

Many temperate shark species grow slowly, have a long life span, mature late, and produce few young. By contrast, the tropical Australian sharpnose shark is fast-growing, short-lived, and matures early. Because of its small size, it is not routinely fished, but it is frequently caught as bycatch in gillnet fisheries for other species.

Management decisions about endangered shark populations require knowledge of how their population growth rates respond to mortality events such as predation, disease, and fishing (including bycatch mortality). Life tables can be used to predict how age-related survival and reproduction affect population growth rates of shark populations.

The table is a static life table for the Australian sharpnose shark. It was constructed from shark catch data during a single fishing season

(continued)

(only female individuals were used in the analysis). The table is composed of survivorship (l_x: the number of females that have survived within each age class) and fecundity (m_x: the number of females born per adult female shark within each age class). Note that survivorship and fecundity at age 0 are for individuals just born into the population.

Age (x)	Survivorship of females (l_x)	Fecundity of female offspring (m_x)	$l_x m_x$	$x\,l_x m_x$
0	1	0		
1	0.32	1.98		
2	0.19	2.59		
3	0.11	2.81		
4	0.06	2.89		
5	0.03	2.92		
6	0.02	2.93		
7	0.01	2.93		
8	0.005	2.93		
9	0.003	2.93		
10	0.002	0		

Questions

1. Complete the fourth column of the table by calculating $l_x m_x$ for each age class (x) of the population. Then, calculate the net reproductive rate (R_0) for the population, and explain what this value represents.

2. Given the value of R_0 for the population, can you determine whether the shark population is increasing, decreasing, or not changing in size? Explain your answer.

3. Complete the final column of the life table by calculating $x\,l_x m_x$ for each age class (x) of the population. Then, calculate the generation time, G, for the population.

4. Using previous calculations of R_0 and G, calculate an estimate of r, the per capita growth rate of the population.

5. Consider the life history differences between the Australian sharpnose shark versus the larger temperate sharks. Which shark populations do you think will have a higher r value? Given fishing pressures and life history characteristics, which shark populations do you think have a greater chance of being overfished? Explain your answer.

Go to **LaunchPad** for the eBook, LearningCurve, animations, activities, flashcards, and additional resources and assignments.

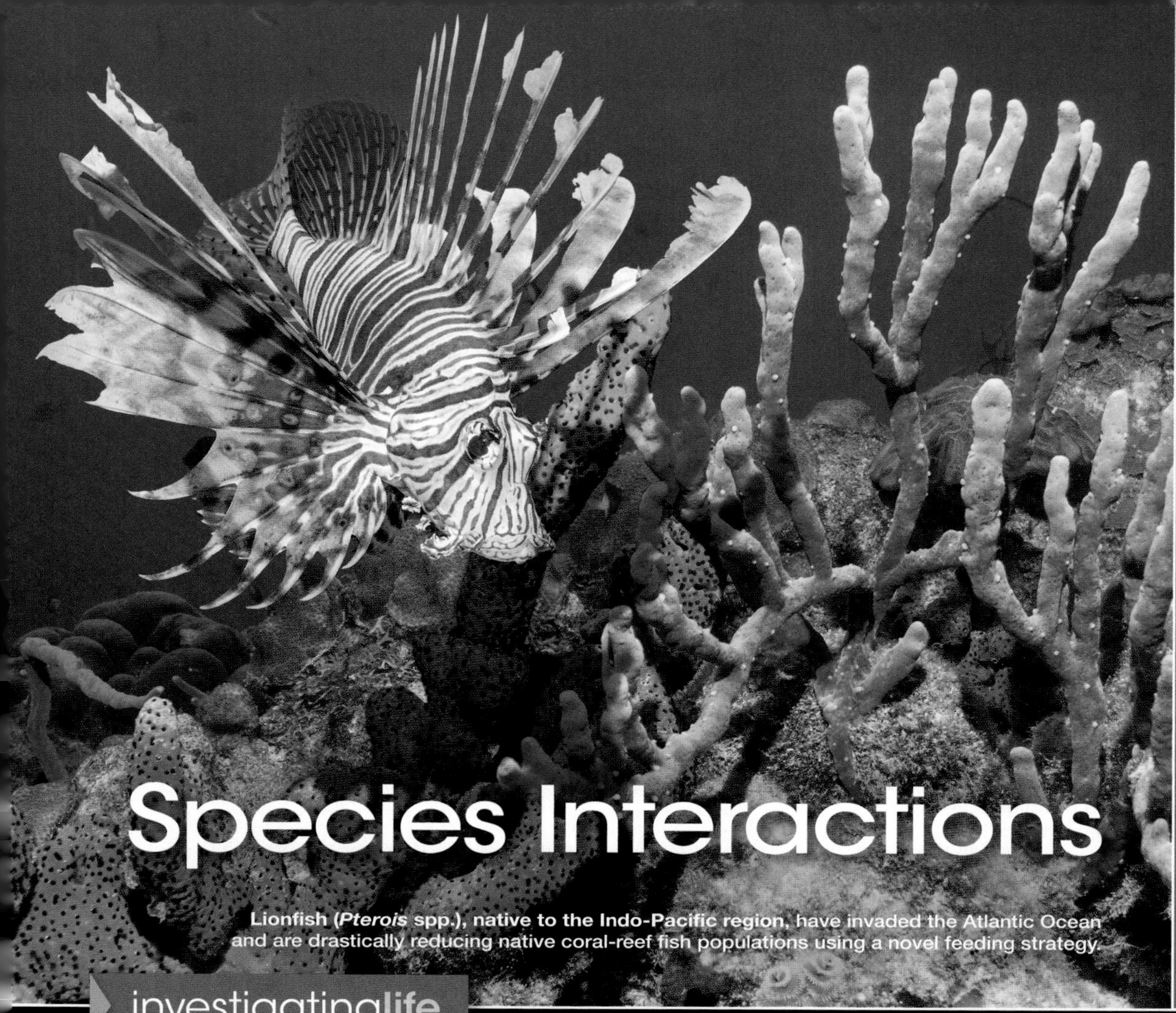

Species Interactions

Lionfish (*Pterois* spp.), native to the Indo-Pacific region, have invaded the Atlantic Ocean and are drastically reducing native coral-reef fish populations using a novel feeding strategy.

> ## investigatinglife

The Lionfish King

The coral reefs surrounding the Bahamas have endured the effects of human activity for centuries, but for the last two decades they have faced a threat unlike any other—the invasion of two species of beautiful but highly destructive lionfishes (*Pterois volitans* and *P. miles*). Lionfishes, first spotted off the coast of Florida in the mid-1980s, are native to the Indo-Pacific region and are popular aquarium fishes. How they came to their new habitats is uncertain, but DNA evidence suggests that they were released into the Atlantic Ocean by unsuspecting aquarium owners. Over the last two decades, lionfish populations have thrived, successfully invading the Western Atlantic, the Caribbean and Gulf of Mexico, and now as far south as Brazil. The invasion has progressed to the point that the density of lionfish in the Atlantic is greater than in their native Pacific range; in some locations, 450 lionfish per hectare (2.5 acres) have been recorded.

The invasion of marine habitats by fish predators is extremely rare. What, then, allows lionfishes to be so successful? Part of their success has to do with their ability to tolerate a variety of tropical and subtropical temperatures and salinities—from the warm waters of shallow reefs, mangroves, and sea-grass beds to the colder waters of deep

reefs and offshore currents. But the real story lies in their interactions with other species—they are dominant predators and competitors in their introduced range. A survey of lionfish populations at nine sites off the coast of the Bahamas showed that their abundance increased 2.5-fold from 2008 to 2010, despite the presence of 16 species of similar native competitor fish (which declined by 44% during that period). Gut analyses of lionfish showed that they are voracious generalist predators; the increase in lionfish abundance coincided with a 65 percent decline in small-bodied coral-reef fish representing 42 different species. Other studies have shown similar results, reinforcing the notion that lionfishes may be among the most damaging marine invaders known.

Explaining the success and devastating impact of these invaders requires an understanding of the ecology and evolution of species interactions, the subject of this chapter. We will return to the example of the lionfishes' extraordinary success in capturing their prey at the end of this chapter.

 Why are small coral-reef fish species so vulnerable to lionfish predation, and what can be done about it?

key concept

55.1 Species Interactions Vary in Direction and Strength across a Continuum

One of life's certainties is that every species on Earth will encounter and interact with other species. These interactions have consequences that affect not only individuals but populations, and ultimately the distribution and abundance of species. Over the long term, they can lead to evolutionary change in one or more of the interacting species. For ease of description, we will focus on two-way relationships, such as one species eating another or two species competing for the same limiting resources. In reality, species interact with multiple species as part of an integrated ecological community. We will explore these community dynamics in Chapter 56.

focus your learning

- Species interactions are classified by the effect (positive, negative, or neutral) they have on one another.
- Broad categories of species interactions are predation, competition, positive interactions, and amensalism.
- Species interactions are often dynamic and asymmetrical; that is, species have unequal effects on one another.
- Some species interactions can affect the survival or reproduction of individuals in interacting populations, resulting in genetic and evolutionary changes.

Ecologists characterize species interactions based on the effects species have on each other—positive (+), negative (–), or neutral (0)—and whether that effect is **trophic** (feeding) or not (**Focus: Key Figure 55.1**). In addition, species interactions may include a **symbiosis**, a condition in which one of the species lives in or on the other species. As you will see, symbioses can occur whether the interaction is positive or negative. Four broad categories into which species interactions fall are predation, competition, positive interactions, and amensalism.

1. **Predation** is a trophic interaction in which an individual of one species (a **predator**) kills and/or consumes individuals (or parts of individuals) of another species (its **prey**). Predation includes **carnivory**, in which the predator and prey are both animals; **herbivory**, in which the predator is an animal and the prey is a plant or algae; and **parasitism**, in which the predator (a **parasite**) lives symbiotically on or in the prey (its **host**) and consumes only certain tissues without necessarily killing the host. Some parasites are pathogens that cause disease in their hosts.

2. **Competition** is a nontrophic interaction in which two or more species overlap in the use of at least some of the same required limiting resources, negatively affecting their growth, reproduction, and/or survival. Competition can occur under a variety of circumstances: between

focus: key figure

(A)

Type of interaction	Effect on species 1	Effect on species 2
Predation (predator–prey)	+	–
Carnivory (animal–animal)	+	–
Herbivory (animal–plant)	+	–
Parasitism (parasite–host)	+	–
Competition	–	–
Positive interactions (facilitations)	+	0 or +
Mutualism	+	+
Commensalism	+	0
Amensalism	–	0

(B)

Parasitism, Predation, Mutualism
The buffalo's hide is infested with parasitic ticks. Oxpecker birds eat the ticks, to the mutual benefit of the birds and the buffalo.

Herbivory
The African buffalo feeds on the grasses of the savanna.

Amensalism, Commensalism
The large mammal unwittingly destroys insects and their nests. The white cattle egrets feed on insects disturbed by the buffalo's passage.

Predation
Carnivores such as wolves hunt and kill herbivorous mammals.

Competition
The grizzly bear is attempting to take over the wolves' kill.

Figure 55.1 Types of Species Interactions Interactions among species can be grouped into categories based on whether their influence on each of the interacting species is positive (+), negative (–), or neutral (0). Species interactions are an everyday occurrence within communities.

Q: Are any of the species interactions featured in the photos trophic or symbiotic? If so, which ones?

 Activity 55.1 Ecological Interactions
www.Life11e.com/ac55.1

predators that depend on the same prey species (such as you saw in the lionfish example), between herbivores that feed on the same host plant, or between pathogenic microbes attacking the same host. The limiting resource need not be food; species may compete for water, space, nesting sites, or even (in the case of plants) sunlight.

3. **Positive interactions** (also known as **facilitations**) are trophic or nontrophic interactions in which at least one species benefits from the interaction and none are harmed. Positive interactions include **mutualisms**, in which both species benefit from the interaction, sometimes in a highly dependent and symbiotic manner. **Commensalism** is a type of positive interaction in which one participant benefits but the other is unaffected; it includes a wide range of interactions that typically encompass the provisioning of food and habitat that improve conditions for the commensal organism.

4. **Amensalisms** occur when one participant is harmed but the other is not affected. An example is a herd of elephants moving through a forest and crushing plants; the elephants are not affected by this event, but the plants clearly are. Another example includes small understory plants that grow under trees. The understory plants are deprived of light but likely have no effect on the trees.

Species interactions are not always clear-cut

Although ecologists find it useful to classify interactions among species into a few basic categories, the boundaries between groupings are not always clear. In reality, the interaction categories described in this section are part of a continuum in how strongly each species affects the other (**Figure 55.2**). The continuum is a consequence of the variable strength in species interactions and their asymmetrical nature. For example, competition between the native Eurasian red squirrel (*Sciurus vulgaris*) and the non-native eastern gray squirrel (*S. carolinensis*) in the United Kingdom is highly asymmetrical, with the gray squirrel having a much stronger effect on the red squirrel than vice versa. This asymmetry is the result of the gray squirrel's ability to better obtain limiting food resources, particularly oak acorns, causing

Effects of species 1 on species 2

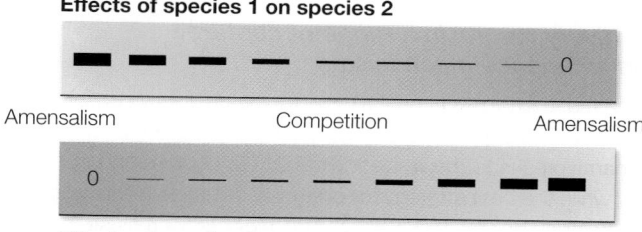

Amensalism   Competition   Amensalism

Effects of species 2 on species 1

Figure 55.2 A Continuum of Competitive Effects Species interactions, such as competition, vary in their strength (represented by the thickness of the bars) and symmetry depending on the effects of one species on the other species. Amensalism occurs when one species suffers a negative effect from the interaction but the other species is not affected.

Q: Suppose the Eurasian red squirrel described in the text is species 1 and the eastern gray squirrel is species 2. Indicate which pairs of interactions between species 1 and 2 would best represent their relationship.

Amphiprion percula

Figure 55.3 Interactions between Species Are Not Always Clear-Cut Ecologists long believed that the relationship between sea anemones and anemonefish was a commensalism: that the fish, by living among the anemone's stinging tentacles, gained protection from its predators. But it could also be considered a mutualism, if the fish's feces provide the anemone with beneficial nutrients.

the red squirrel's population sizes to decline. The red squirrel is also declining, because of a fatal disease spread by the gray squirrel.

Species interactions can be difficult to classify because of the different ways in which interactions affect particular species. For example, sea anemones in the Pacific Ocean sting and eat small fish, but a select few fish species (mostly in the genus *Amphiprion*) live within the tentacles of sea anemones and are unaffected by their stings. Safe from its predators, an anemonefish moves freely among the stinging tentacles to scavenge the leavings of the anemone (**Figure 55.3**).

Anemonefish must acclimate to the anemone's venom, and the anemone, in turn, must acclimate to the fish. The acclimation process appears to involve a change in the mucus coat of the fish; wiping off the mucus of an acclimated fish results in immediate stinging, whereas anemones do not sting fish with intact mucus. The benefits of this relationship to the anemonefish are clear: it escapes its own predators by hiding behind the anemone's stinging tentacles, and it has no need to forage widely for food. But does the anemone benefit from the association? By defecating while in residence, the anemonefish may provide nitrogen-rich nutrients to the anemone. However, the fish may occasionally steal the anemone's prey, which has a negative effect on the anemone's fitness. Thus the net outcome of the interaction depends on the ecological circumstances under which it occurs, including the influence of other species and the existing physical conditions.

Some interactions result in evolutionary change in the species involved

All types of interactions have the potential to influence the population sizes of interacting species. By contributing to the differential survival or reproduction of individuals with different traits, they can also alter genotype frequencies within the interacting populations over time. Thus these interactions have both ecological consequences, as when they affect the distribution and abundance of a species, and evolutionary consequences, as when they lead to evolutionary change. Darwin observed that evolutionary change occurs not only in response to physical conditions, as described in Chapter 53, but also in response to interactions among species. In his introduction to *On the Origin of Species*, Darwin pointed out that woodpeckers have feet, tails, beaks, and tongues "admirably adapted to catch insects under the bark of trees" as a result of their long-standing interactions with their insect prey.

While abiotic factors also act as agents of selection, they differ in a fundamental way from biotic agents of selection in that they do not themselves undergo change as a result of the interaction. Snow and ice do not become more deadly as a result of encountering cold-resistant organisms, but predators can, over evolutionary time, become swifter, more powerful, or more efficient at capturing their prey. In response, prey species may become swifter, tougher, less conspicuous, or more poisonous, all of which decrease their likelihood of being consumed. This back and forth evolutionary response between interacting species is known as **coevolution**.

The types of interactions most likely to lead to coevolution are those that occur predictably and with high frequency over time and that have a strong effect on the fitness of species. Thus species involved in amensal and commensal interactions are less likely to experience coevolutionary change as a result of the interaction than are species involved in predatory, competitive, and mutualistic interactions.

55.1 recap

Species interactions may be classified into four basic categories: predation (a trophic interaction in which a predator kills and/or consumes its prey, or part of its prey), competition (a nontrophic interaction in which species use some of the same limiting resources and have negative effects on one another), positive interactions (a trophic or nontrophic interaction in which at least one species benefits from the interaction and none are harmed) or amensalism (a nontrophic interaction in which one species is harmed but the other is not affected). Species interactions are not always clear-cut but instead fall along a continuum in which species vary in how strongly they affect one other. Some species interactions result in evolutionary change.

learning outcomes

You should be able to:

- Use examples to distinguish among predation, competition, positive interactions (mutualism and commensalism), and amensalism.
- Give examples of species interactions in which the responses of two species are asymmetrical, and evaluate possible reasons for the asymmetry.
- Compare biotic and abiotic factors as agents of natural selection and evolution.

1. Classify each of the following species interactions as carnivory, herbivory, parasitism, competition, amensalism, mutualism, or commensalism.

   a. Leafcutter ants harvest and transport leaf fragments to their nests, where the vegetation will support fungus, which the ants consume.

   b. Elephant seals lie on rocks, crushing and detaching algae underneath them.

   c. Head lice inhabit the human scalp and feed on blood.

   d. Wolves and cougars both prey on the same deer population, which is limiting to both predators.

2. Name some conditions that can modify or change the outcome of a two species interaction.

3. Do you think the interactions between lionfishes and their fish prey in the Atlantic Ocean have resulted in evolutionary change for the prey? Explain.

Trophic interactions are critical to life on Earth. As a result, the variety of ways that heterotrophic organisms acquire food plays a critical role in the success of every species on Earth. In the next section we will consider how predation influences the ecology and evolution of both predator and prey species.

▶ key concept 55.2   Predation Is a Trophic Interaction in which Predators Benefit and Prey Are Harmed

Predator–prey interactions are probably the most familiar, and perhaps the most dramatic, type of species interaction. Even though carnivory, herbivory, and parasitism all involve the simple interaction of one species eating another species, they often defy categorization. For example, a carnivore might sometimes function like an herbivore by eating berries and roots, or an insect might be considered an herbivore or a parasite if it spends its entire life feeding on a single tree. In addition, predators can differ considerably in the ways they affect their prey's population dynamics if the prey species is not killed in the process. Despite these complications, we will explore predation by considering carnivory, herbivory, and parasitism separately. We will end by drawing on some generalities regarding population and community consequences of predation.

focus your learning

- Prey have evolved many strategies to avoid being captured, including escape, avoiding detection, chemical defenses, warning signals, and mimicry.
- Plants and herbivores have developed reciprocal adaptations to counter their effects on one another.
- Parasites are abundant and have highly varied effects on their hosts.
- Coupled predator–prey cycles greatly influence the populations of both species and the communities in which they reside.

Carnivory results in a range of capture and avoidance mechanisms

Given that carnivores almost always kill the prey they consume, the interests of both predator and prey are at stake: the best strategy for carnivores is to maximize their ability to capture and consume prey while the best strategy for prey is to minimize the likelihood of being eaten. We first consider the strategies carnivores use to obtain food and then examine how prey avoid being eaten.

What strategies do carnivores use to obtain food? The success of carnivores depends on balancing the cost of pursuing, capturing, and handling prey against the energetic benefit of consuming them, as discussed in Key Concept 52.4. At one extreme, carnivores use strength and swiftness to actively search for and capture high-quality prey. The "active pursuit" strategy is used by predators of all sizes: both Orca whales pursuing gray whales and foxes pursuing birds are fast, powerful predators and are equipped with strong jaws (**Figure 55.4A**). At the other extreme, an ability to look inconspicuous allows the stealthy predator to ambush prey unlucky enough to pass its way (**Figure 55.4B**). Various other predation strategies have evolved in predators that are smaller than their prey. For example, the jaws of some snakes open wide enough to allow them to swallow prey larger than their head. The tiny short-tailed shrew, among the smallest mammalian predators, produces venomous saliva that paralyzes earthworms and snails but also prey much larger than itself, including mice and small birds.

Prey species have in turn developed a variety of defenses against predators. Some escape from predators simply by flying or running away. Others have morphological defenses; tough skin, shells, spines, or hair can foil even a determined predator (**Figure 55.5A**). Various other defenses that prey have evolved to avoid capture are described below.

AVOIDING DETECTION   Prey species can often escape predators by hiding. One form of hiding is camouflage, also known as **crypsis**. Crypsis allows species to resemble objects that their predators consider inedible. The katydid in **Figure 55.5B**, for example, looks very much like a dead leaf, even down to the likeness of a spot of fungal decay.

Because the vision of many types of predators is adapted to spot moving prey, many prey species simply stop moving if they are being pursued. "Playing possum," a term that is sometimes applied to this strategy, refers to the ability of opposums to simulate death.

CHEMICAL DEFENSES   Many animals use chemical defenses to escape or repel their predators. Chemical defenses are generally used by animal prey that are small, weak, sessile, or otherwise unprotected. Many insects produce sprays, oozes, or froths when attacked. Bombardier beetles, for example, eject a hot (100°C) noxious chemical spray from the tip of their abdomen that can be fatal to attacking insects.

 Media Clip 55.1 **Bombardier Beetle Sprays Its Enemies**
www.Life11e.com/mc55.1

 Activity 55.2 **Coevolution Simulation**
www.Life11e.com/ac55.2

 Activity 55.3 **Predator-Prey System Simulation**
www.Life11e.com/ac55.3

(A) *Vulpes vulpes fulva*

Figure 55.4 **Active Searcher or Stealthy Ambusher** Predators can be **(A)** active searchers such as the swift and strong fox or **(B)** stealthy ambushers such as the seemingly inconspicuous wide mouth frog.

(B) *Ceratophrys* sp.

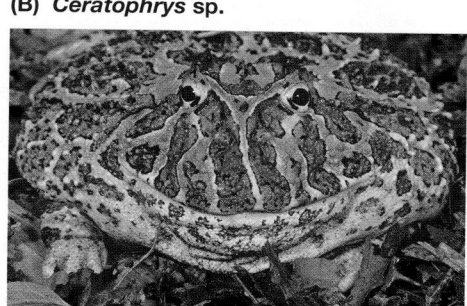

Predators may evolve adaptions to overcome their prey's chemical defenses, as you saw in the case of the rough-skinned newt and the garter snakes that have become insensitive to the newt's protective toxin (see Figure 21.20). Some predators ingest and sequester their prey's defensive chemicals and use them as defenses against their own predators. For example, some sea slug species incorporate toxic chemicals from their sponge prey, whereas others, which feed

(A) *Hystrix cristata*

(B) *Typophyllum* sp.

Figure 55.5 Adaptations to Avoid Being Eaten   **(A)** Porcupines have a coat of sharp spines, or quills, that protect them from predators. **(B)** Birds searching for insect prey are likely to bypass a katydid that looks like a partially eaten leaf. **(C)** Nudibranchs (sea slugs) are mollusks without protective shells that may possess stinging nematocysts acquired from their hydrozoan prey. **(D)** Poison dart frogs of Central and South America sequester highly toxic chemicals in their brightly colored skin.

(C) *Chromodoris* sp.

(D) *Dendrobates reticulatus*

(A) Batesian mimics

This harmless blenny...

...closely resembles a venomous related species.

Petroscirtes breviceps   *Meiacanthus grammistes*

Figure 55.6 Truth in Labeling? **(A)** Batesian mimics are vulnerable species that gain protection by mimicking the warning signals of dangerous species. The appearance of the harmless blenny species *Petroscirtes breviceps* closely resembles that of the fanged striped blenny, which possesses a pair of grooved fangs with associated venom glands. **(B)** The shared warning coloration of Müllerian mimics is an honest advertisement of their toxicity. As caterpillars, all of the longwing butterflies (genus *Heliconius*) of South America feed on toxic passionflower plants and incorporate the toxins into their adult bodies. The *Heliconius* species living together in a particular region have similar warning coloration.

(B) Müllerian mimics

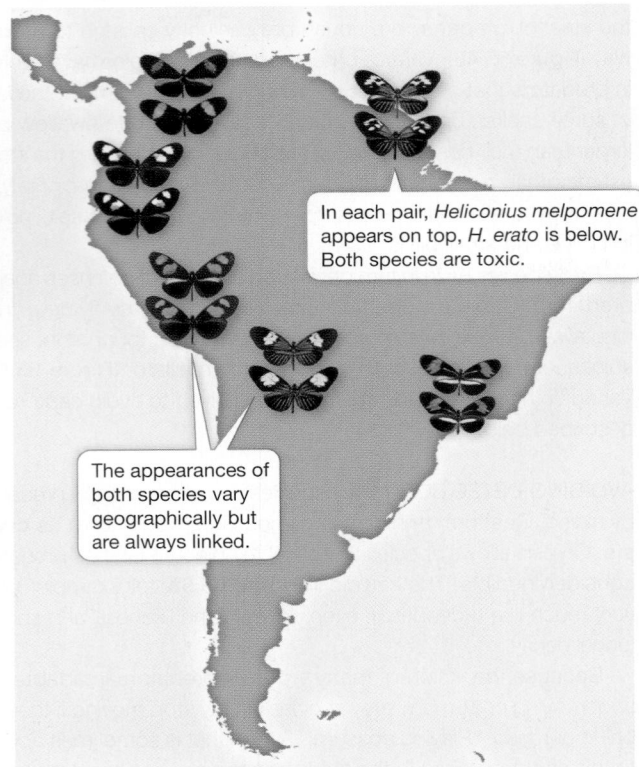

In each pair, *Heliconius melpomene* appears on top, *H. erato* is below. Both species are toxic.

The appearances of both species vary geographically but are always linked.

on hydrozoans, incorporate the stinging cells into their own bodies (**Figure 55.5C**). Likewise, poison dart frogs from Central and South America sequester toxins from their prey, which include ants, mites, and other small invertebrates (**Figure 55.5D**).

WARNING SIGNALS   Some prey species that defend themselves with toxic chemicals advertise that fact with a **warning signal**. Warning signals may be visual (many toxic species are brightly colored) or acoustical (the rattlesnake's warning rattle, for example), depending on what sensory cues their predators use to find prey.

Many toxic prey, such as the nudibranchs and frogs in Figures 55.5C and D, sport bright colors or striking patterns to protect themselves against visually orienting predators. Such warning coloration increases the probability that a predator will learn to recognize and avoid the toxic species. Some vertebrate predators that rely on visual cues can learn quickly to associate certain color patterns with an unpleasant feeding experience.

MIMICRY SYSTEMS   There are two general types of mimicry systems that have evolved repeatedly in nature: Batesian mimicry and Müllerian mimicry. In **Batesian mimicry**, a benign, edible species (the mimic) closely resembles a dangerous, toxic species (the model). The mimic benefits from the avoidance behavior of predators of the model (**Figure 55.6A**). In **Müllerian mimicry**, two or more species converge on a common warning signal; all benefit from providing a stronger recognition signal to predators. Many of the Neotropical longwing butterflies (*Heliconius*), which as caterpillars feed on toxic passionflower plants and incorporate the plant toxins into their bodies, are Müllerian mimics. *Heliconius* species living together in a particular geographic region are likely to have similar coloration and share a common warning pattern (**Figure 55.6B**). Genome sequencing of *Heliconius* Müllerian mimics has identified one gene, *optix*, that codes for a transcription factor that can, by changing gene expression patterns, create the same color patterns in *Heliconius* species that are not very closely related.

BEHAVORIAL MECHANISMS   Prey often use behavioral mechanisms to avoid predation. For example, as you saw in Key Concept 52.6 and Figure 52.21, group behaviors such as flocking and alarm vocalizations can provide protection from predators. In the lionfish example introduced in the chapter opening story, the evolutionary change in behavior that predator and prey may experience has not yet had time to develop. In **Investigating Life: The Lionfish King** we describe research showing that the small reef-fish prey of lionfishes are likely experiencing a novel predatory behavior not previously encountered, and thus may be defenseless in the face of lionfish predation. Lionfishes slowly approach their prey and direct jets of water toward them. Confused or distracted by the jets, the prey are often taken by headfirst capture. The highly efficient predatory behavior of lionfishes doubtless contributes to their success as invasive species in tropical Western Atlantic and Caribbean coral reefs.

Herbivory is a widespread but specialized interaction

Herbivory is a widespread interaction, with the vast majority of herbivores being insects. Of those herbivorous insects, more than 90 percent are specialists that feed on just one or a few, often taxonomically related, plant species. Generalist herbivores, in contrast, feed on as many as hundreds of unrelated plant species. Vertebrate herbivores are usually generalists; a cow grazing in a pasture, for example, can consume many different plant species in a single afternoon. There are exceptions to this pattern, however. Australian koalas famously feed exclusively on the foliage of eucalyptus trees, and the diet of giant pandas is made up almost entirely of bamboo.

Herbivores, particularly insects, generally consume only parts of their food plants and usually do not kill them. In most natural

experiment

Original Paper: Albins, M. A. and P. J. Lyons. 2012. Invasive red lionfish *Pterois volitans* blow directed jets of water at prey fish. *Marine Ecology Progress Series* 448: 1–5.

Mark Albins and his colleagues asked what feeding behaviors the invasive lionfish *Pterois volitans* might use to be a successful predator on Atlantic and Caribbean coral-reef fish. They conducted field observations of lionfish feeding on small reef fish and noticed that lionfish typically face their prey, flaring their pectoral fins and moving very slowly within striking distance of the prey. During the approach, lionfish produced a strong, pulsed jet of water from their mouths and directed it toward the prey. To investigate this novel feeding behavior, Albins and his colleagues conducted feeding trials on lionfish and native goby prey.

HYPOTHESIS▶ The novel feeding method of lionfish—in which jets of water are directed at their prey—disorients prey, allowing for easier headfirst capture.

METHOD

1. Using hand nets, capture lionfish and two species of native goby prey (*Coryphopterus glaucofraenum* and *Gnatholepis thompsoni*) from coral reefs and sea-grass beds in the Bahamas.
2. Acclimate lionfish and gobies to a flow-through aquarium and film their feeding behavior using a digital video camera.
3. Present lionfish with gobies in two ways: (1) release lionfish into an open aquarium setting containing the gobies, and (2) release lionfish into an aquarium that holds the gobies in transparent containers so the lionfish can see but not consume them.
4. Use a glass pipette to release food-grade dye in front of the lionfish to visualize the movement of water around its mouth.
5. Examine digital videos of all feeding trials to measure characteristics of water jets and response of the prey.

RESULTS

In all of the open aquarium trials, lionfish produced directed water jets at goby prey. Seven of the eight trials resulted in headfirst strikes. In the container trials, the lionfish produced an average of 18 water-jet pulses before striking at the prey. The maximum distance from the mouth of the lionfish that a visible jet-front reached was 10 cm.

Goby in container

Dye

Glass pipette

The lionfish produces jets of water (indicated by the spread of the dye) to disorient its goby fish prey.

CONCLUSION▶ What would you conclude about the feeding behavior of lionfishes and the responses of prey from the Atlantic Ocean versus the Pacific Ocean? Explore this question in the following work with the data exercise.

work with the data

In addition to the aquarium experiments, the researchers conducted field observations of the feeding behavior of lionfishes in different locations—their native Pacific Ocean and their non-native Atlantic Ocean. They found that, in the Atlantic Ocean, most successful captures by lionfishes did not require them to blow jets of water at their prey (only 18 percent of the captures involved this behavior). However, in the Pacific Ocean, they found that 56 percent of successful captures by lionfishes involved them blowing jets of water at their prey. The researchers also reported that an extensive literature search showed that blowing jets of water at prey appears to be unique to lionfishes.

QUESTIONS▶

1. Refer to Method step 3 in the experiment. Why do you suppose the researchers conducted lionfish aquarium feeding trials with free-swimming fish prey? Alternatively, what did the containers containing the fish prey afford the researchers?
2. Given what you know about the feeding behavior of lionfishes, would you expect them to produce more or fewer water jets to capture fish prey in the Atlantic Ocean compared with the Pacific Ocean? Would the distance the water jet had to travel be shorter or longer for the Atlantic Ocean prey compared with the Pacific Ocean prey?
3. Based on all the lionfish feeding behavior information presented here, what is the most plausible hypothesis to explain why lionfishes are such effective predators on coral-reef fishes in the Atlantic Ocean compared with the Pacific Ocean?

 Media Clip 55.2 **Lionfish "Strike" Again!**
www.Life11e.com/mc55.2

A similar **work with the data** exercise may be assigned in **LaunchPad**.

ecosystems, insects rarely remove more than a small percentage of the plant biomass. For that reason, some ecologists have questioned the ability of herbivores to exert strong selection pressure on plant traits. Mortality is not, however, the only form of selection that leads to evolutionary change; herbivores can reduce plant fitness if the plants they attack produce fewer offspring.

PLANT DEFENSES AGAINST HERBIVORES Plant chemistry is one of the principal defense mechanisms against herbivores. Plants in the mustard family (Brassicaceae) offer just one example of the strategy of producing secondary metabolites as a mechanism to reduce herbivory. The amazing variety of secondary metabolites produced by plants to defend themselves against herbivores is considered in Key Concept 38.2.

Some plants and algae protect themselves by being physically difficult to eat. For example, thorns and spines are highly effective deterrents to browsing vertebrate herbivores. Coralline algae contain calcium carbonate within their tissues, which deters many marine herbivores. An exception is some sea urchin species that have powerful articulated jaws and teeth and long coiling guts that can process less nutritious chalky tissue.

RECIPROCAL ADAPTATIONS IN HERBIVORES AND PLANTS   A spectacular variety of adaptations to plant defenses has evolved in herbivores. Many herbivores circumvent plant defenses by behavioral means. For example, the secondary metabolites produced by a plant called St. John's wort (*Hypericum perforatum*) require exposure to sunlight for optimal toxicity, so some insects that feed on this plant roll its leaves into a light-impervious cylinder and feed in comfort in the dark. Large herbivores such as deer and horses graze on a wide variety of plant species, minimizing their exposure to any particular defensive chemical. Long-lived and with relatively good memories, they can learn to avoid plants with an unpleasant taste.

Unlike large mammalian herbivores, caterpillars and many other insect herbivores may spend their entire lives feeding on a single individual plant. Such exclusive diets are associated with highly specialized detoxification systems. The diamondback moth caterpillar eats plants in the cabbage family, which are rich in toxic mustard oil glycosides. In its gut is an enzyme that breaks down the glycosides into harmless by-products.

Some herbivores take resistance a step further by storing, or sequestering, plant toxins in specialized organs or tissues that are insensitive to those toxins. This strategy also makes the sequestered chemicals available for defense against the herbivores' own enemies. The caterpillar of the monarch butterfly, for example, is insensitive to the neurotoxic glycosides in its milkweed host plants, but most of its enemies, including insect-eating birds, cannot tolerate these compounds.

Most parasites specialize on hosts, and most hosts house many species of parasites

Parasitism is an interaction in which a predator (a parasite) lives on or in the prey (its host) and consumes only certain tissues. Many parasite–host relationships are symbiotic. Effects of parasite–host interactions can vary dramatically in magnitude, from having weak effects to causing the death of the host. While parasites typically feed on only one or a few host species, hosts typically house many species of parasites (**Figure 55.7**). Parasites can also have their own parasites. This specialization helps explain why there are so many parasite species; it is estimated that roughly 50 percent of the species on Earth are parasites.

Parasites can be classified as **microparasites**, which include viruses, bacteria, and protists, or **macroparasites**, which are larger and include such groups as worms and insects.

MICROPARASITES   Microparasites are many orders of magnitude smaller than their hosts and generally live and reproduce inside their hosts. Multiple generations may reside within a single individual,

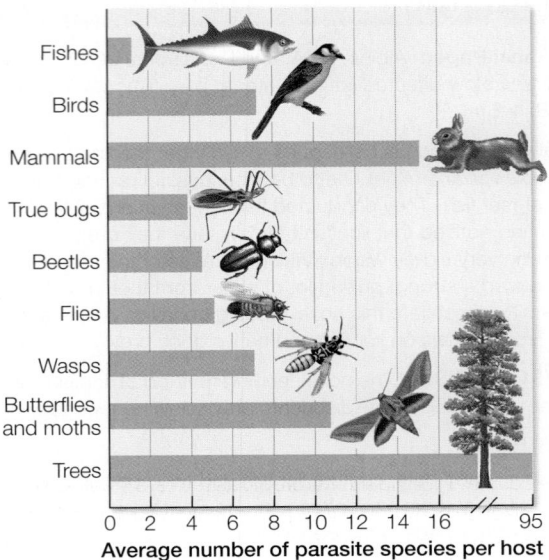

Figure 55.7 Why Are There So Many Parasites?   It is estimated that roughly 50 percent of the species on Earth are parasites. Most parasites specialize on host species, most host species have multiple parasite species, and most parasites have their own parasites. This specialization may help explain why there are so many species of parasites.

and a host may harbor thousands or millions of them. Many microparasites, in the process of acquiring nutrients at the expense of their host, cause symptoms of disease and thus are considered pathogens. Host mechanisms that can limit pathogen proliferation include the immune system in animals (see Chapter 41) and biochemical defenses in plants (see Key Concept 38.3).

Just as with other predator–prey interactions, hosts and pathogens can place strong selective pressure on one another. If a pathogen strain is to persist in a host population, the pathogens must continually infect new host individuals. A less deadly strain that kills a smaller proportion of host individuals may be able to infect a larger number of new hosts. Thus pathogen and host may reach a state of coexistence as increased host resistance (ability to withstand the effects of a pathogen) and decreased pathogen virulence (ability to cause disease) evolve. Yet new virulent strains may also arise, reminding us that evolution is constantly occurring.

An example of the complex interactions between host and pathogen comes from Australia, where the myxoma virus was introduced to control the exploding population of European rabbits (*Oryctolagus cuniculus*), which were devastating the rangeland vegetation used by cattle and sheep. The virus, which is transmitted by mosquitos, causes infections and blindness in rabbits, eventually leading to starvation and death. After the introduction of the virus in 1950, 99.8 percent of infected rabbits died. Millions of rabbits died over the following decades, but the rabbits slowly evolved resistance to the virus. The virus is still used to control rabbit populations, but this requires introducing new, more lethal strains of the virus to ultimately be effective.

MACROPARASITES   While microparasites generally live and reproduce inside their hosts, macroparasites are not always intimately

(A) *Cyamus* sp.

(B) *Macaca fuscata*

Figure 55.8 Ectoparasites Hang On **(A)** Ectoparasites such as these whale lice on the body of a gray whale are flattened and equipped with strong claws for gripping the body of their host. Each individual whale has its own population of lice. **(B)** Grooming behaviors among primates are believed to have evolved in response to ectoparasites. Japanese macaques form social groups in which this behavior plays a significant role.

associated with their hosts. Macroparasites rarely cause the same kinds of disease symptoms that pathogenic microparasites cause, but they may nevertheless affect host survival and reproduction and can thereby act as agents of selection on their hosts. **Ectoparasites** are macroparasites that live outside the bodies of their hosts. **Endoparasites**, such as the tapeworms described in Key Concept 30.4, are macroparasites that spend at least part of their life cycle inside the bodies of their hosts.

Some ectoparasites, such as leeches and mosquitoes, are only casually associated with their hosts, interacting with them just long enough to eat their fill and then moving on. Ectoparasites that spend their entire lives on their hosts have several attributes that keep them attached to their hosts. Whale lice, which are generally found in skin lesions, nostrils, and eyes of marine mammals, have claws on the tips of their legs that pierce the skin of the whale or dolphin host (**Figure 55.8A**). Most whale lice are associated with a single species of whale and spend their entire lives on one individual whale. The lice feed on algae or flaking skin on the whale's body, causing minor skin damage but no significant health risks.

Most hosts actively work to rid themselves of their ectoparasites. Grooming behavior—an important component of the social interactions of many primates—may have evolved in response to ectoparasites. The Japanese macaque (*Macaca fuscata*), for example, is prone to infestation by two species of lice, which tend to lay their multitudinous eggs on the outer surfaces of their host's back, arms, and legs. To keep louse populations in check, macaques form and maintain social bonds that ensure the consistent presence of grooming partners (**Figure 55.8B**). Some biologists believe that hairlessness in humans is an evolutionary response to ectoparasites.

Predator populations can cycle with their prey populations

The interaction between predators and prey, including parasites and hosts, not only influences the adaptations that species develop but also their population dynamics. Some predator–prey populations are linked and oscillate in response to one another. What controls such patterns of cycling?

Consider the classic example of predator–prey cycling in the Canada lynx (*Lynx canadensis*) and the snowshoe hare (*Lepus americanus*), which makes up a large portion of the lynx diet. Long-term records from Hudson's Bay Company, a Canadian retailer with a long history in the fur-trading business, show that the numbers

of lynx and hares trapped over time reflect a regular cycle in the population sizes of each species, with peak abundances occurring every 10 years but falling to low abundances in the intervening years (**Figure 55.9**). In addition, the peak in the lynx population lags behind that of the hare population by 2–3 years. Models of predator–prey interactions suggest that this cycling is the result of a tight coupling between prey and predator: as prey increase in abundance, predators do as well, causing prey numbers to decline as a result of increased mortality. When prey decline, so do predators, causing both predator and prey populations to decrease. At some point, the prey population starts to increase again due to the lack of predation, which then allows the predator population, which lags behind, to increase as well. However, in the case of lynx and hare populations, it is likely more complicated. A long-term study by Charles Krebs and colleagues showed that if lynx are absent or food is added for the hares, predator and prey populations continue to cycle, suggesting that the underlying cause of the cycling may be controlled by environmental factors such as large-scale variability in climate.

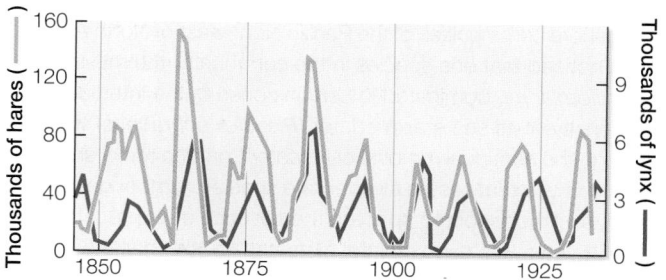

Figure 55.9 Lynx and Hare Population Cycles Historical trapping data from the Hudson's Bay Company show that numbers of both hares and lynx fluctuate in a 10-year cycle that is offset by 2–3 years.

Q: Does the peak number of hares typically occur before or after the peak number of lynx? Why is this so?

(A) Sea stars of the Pacific Northwest

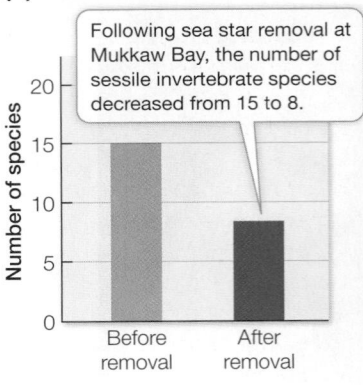

> Following sea star removal at Mukkaw Bay, the number of sessile invertebrate species decreased from 15 to 8.

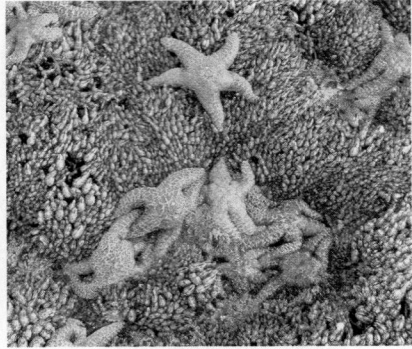

(B) Sea stars of New Zealand

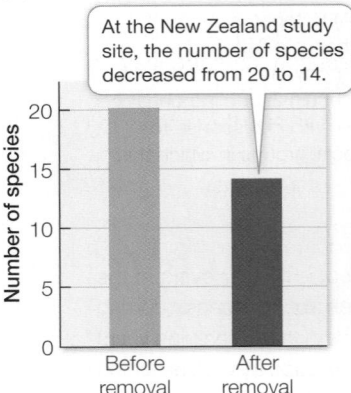

> At the New Zealand study site, the number of species decreased from 20 to 14.

Figure 55.10 Sea Star Predation on Mussels Has Community-Wide Effects Along the coasts of the Pacific Northwest of North America and New Zealand, sea stars preferentially feed on mussels, creating bare space for other sessile invertebrates such as acorn and gooseneck barnacles. Experiments by Robert Paine demonstrated that when sea stars were removed from the rocky shore, mussels outcompeted other sessile invertebrates for space, resulting in a decline in the number of invertebrate species in both **(A)** the Pacific Northwest and **(B)** New Zealand.

Predators can have dramatic effects on communities

As you might imagine, given the reduced growth, reproduction, and survival of prey, predators (including parasites) can have dramatic effects not only on populations but also on communities.

One classic experiment, conducted by Robert Paine in the mid-1960s, showed the strong role that sea star predation has on rocky shore communities of the Pacific Northwest of North America. Paine noticed that one species in the community, the mussel *Mytilus californianus*, dominated in certain zones of the intertidal when the density of its sea star predator (*Pisaster ochraceus*) was low. Through the continual removal of *Pisaster* from the rocky shoreline, Paine showed that sessile invertebrate species number declined as the mussel monopolized space in the absence of its predator. After nearly 3 years, the experimental plots had roughly half the number of sessile invertebrate species compared with the natural condition where *Pisaster* was present (**Figure 55.10A**). Subsequent experimental removal of sea stars on New Zealand rocky shores, which share no species in common with the Pacific Northwest, resulted in similar reductions in sessile invertebrates (**Figure 55.10B**). Paine's

research on the sea star–mussel interaction showed not only that predation can affect the number of species in a community but that the effect can be controlled by one or a few species, which Paine called **keystone species** (named after the single wedge-shaped stone in the center of an archway called the "keystone" that holds all the other stones in place). Keystone species have large effects, not because of their abundance but because of the important role they play in communities. You'll learn more about keystone species in Chapter 56.

Some of the most compelling examples of the effects of predation on communities come from human-caused events such as the extinction (or near extinction) of predators, as in the case of wolves living in Yellowstone National Park (see Key Concept 56.3), or from the introduction of novel predators into communities. As you saw in the opening of this chapter, the inadvertent introduction of lionfishes to the Atlantic Ocean from the Pacific Ocean has led marine ecologists to hypothesize that there may be drastic effects for reef communities.

55.2 recap

Predation is a fundamental species interaction because all animals must feed. Predator–prey interactions result in the evolution of a range of capture mechanisms (extremes of swift pursuer to inconspicuous ambusher) and avoidance mechanisms (including avoiding detection, chemical defenses, warning signals, mimicry, and behavior). Most parasites specialize on hosts, and most hosts have many species of parasites. Predator populations can cycle with their prey and can have dramatic effects on communities.

learning outcomes

You should be able to:

- Classify prey defenses against predators in terms of type (e.g., behavioral, chemical, morphological).
- Give examples of plant defenses against herbivores and of reciprocal adaptations of herbivores to these defenses.
- Describe variations in parasite–host relationships and infer the causes of these variations.
- Predict future population sizes, given information on current population numbers, in cycling predator–prey populations.

1. Given what you know about lionfish feeding behavior, describe some avoidance mechanisms that might evolve in small reef fish in the Atlantic, to avoid being eaten by lionfishes.

2. What characteristics of herbivory and parasitism are likely important in promoting specialization?

3. Suppose the number of grass species in an alpine meadow is maintained by herbivory by hares. By feeding on the dominant space-occupier, the hares allow less dominant grass species to thrive (similar to the sea star–mussel example in Figure 55.10). Now suppose lynx predators are introduced into the system. When, in the population cycle of lynx and hare (see Figure 55.9), would you predict the number of grass species to be highest? When would it be lowest?

Predator–prey interactions are intrinsically compelling; the idea of a lionfish "vacuuming up" tiny reef fishes naturally captures our attention. But at the same time, lionfishes are not the only predators of reef fishes; sharks, barracudas, and even turtles hunt small reef fishes, potentially reducing the food available for lionfishes. Whenever any shared resource is limiting, organisms may compete to varying degrees to get that resource. In the next section we will consider how competition influences the ecology and evolution of species that overlap in their use of a limiting resource.

▶ key concept 55.3   Competition Is a Negative Interaction in which Species Overlap in the Use of Some Limiting Resource

Virtually no species has exclusive access to any given set of resources. All species must compete with other species for at least some resources. **Resources** are simply the components of the environment—such as food, water, light, and space—that are required by all species. The full set of these resources, in part, determines the **niche** of a species, defined as the physical and biological conditions required for growth, reproduction, and survival. Thus a species' niche defines where it can and cannot live. Although a species might be physiologically able to live under a wide range of conditions, competitors may restrict its use of those resources. Thus every species has a **fundamental niche**, defined by its physiological capabilities, and a **realized niche**, defined by its interactions with other species.

focus your learning

- The fundamental niche determines where a species can physiologically live, but its realized niche depends on interactions with other species, especially competitors.
- Interference competition occurs when interacting species actively interfere with each other's access to limiting resources, while exploitation competition occurs when interacting species reduce the quantities of their shared but limited resources.
- Most species involved in competitive interactions show coexistence, which is maintained through a variety of ecological processes, including resource partitioning, environmental conditions, disturbance, and predation.

As you saw in Key Concept 54.1, *intra*specific competition for resources is the main reason why populations do not grow indefinitely, and instead eventually reach some limit defined by their carrying capacity. *Inter*specific competition—competition among different species—affects individuals in much the same way. At some point an essential resource, or set of resources, may be in such short supply that competition becomes a factor in a population's growth. Such a resource that is in short supply is referred to as a **limiting resource**. At one extreme, if a species can prevent another species from using essential resources, the inferior competitor may

become locally extinct, a result called **competitive exclusion**. In reality, though, most species show some sort of **competitive coexistence**, or the ability to coexist with one another despite sharing limiting resources. Competitive coexistence is maintained through a variety of ecological processes.

Assuming competitive coexistence occurs, competition can manifest itself either directly or indirectly. **Interference competition** occurs when one species directly interferes with or excludes another species' access to a limiting resource. Interference competition can take many forms, from physical exclusion to chemical warfare among the competitors. A graphic example involves two ant species: the desert ant *Conomyrma bicolor* and the honeypot ant *Myrmecocystus mexicanus*. These species occupy the same habitat type—arid areas containing little vegetation—and they feed on similar foods—the sugary excretions of aphids and other sap-feeding insects. When *C. bicolor* workers find the entrance of a honeypot ant nest, they pick up small stones in their mandibles, carry them to the rim of the nest opening, and drop them down the hole—up to 200 stones in a 5-minute interval. This activity stops the honeypot ants from foraging and temporarily limits competition for food with the desert ant.

Exploitation competition occurs when a limiting resource is available to all competitors but the outcome of the interaction depends on the relative efficiency with which each species uses the resource. Exploitation competition affects the availability of a resource for another species but not in an exclusionary way, as we see with interference competition. Exploitation competition may lead to coexistence, provided that the species relying on the same resource have ways to divide up, or partition, that resource. For example, in the American Southwest at least three species of bees consume the nectar of the shin dagger agave (*Agave schottii*). The three bee species differ in where and when they collect shin dagger nectar. Honey bees tend to forage in places with the greatest numbers of shin dagger flowers, bumblebees in places with intermediate numbers of flowers, and carpenter bees where flowers are few and far between. Honey bees also tend to be most active when nectar output is greatest. With their larger nests and greater numbers of offspring to support, honey bees require greater foraging efficiency and greater energy intake. Foraging sites that are not worth their time are left to the other bees.

The bee–shin dagger agave example is one in which closely related bee species compete for agave nectar. But competition is not limited to closely related species; it commonly occurs among distantly related species as well. Bats and hummingbirds also visit shin dagger agave for its nectar. Even humans harvest agave (albeit the leaves rather than the nectar) to be used as an alternative for sugar.

Another important aspect of competition is that it is often asymmetrical (unequal): one species can be more negatively affected by the interaction than the other. Clearly this is the case under conditions of competitive exclusion, when one species drives the other to local extinction. But even when species coexist under competition, competitive success forms a continuum, as you saw with red and gray squirrels in the United Kingdom (see Figure 55.2). The effects of competition can vary for many reasons. Morphological, physiological, or behavioral differences can affect a species' ability to obtain limiting resources. In addition, the magnitude of competitive

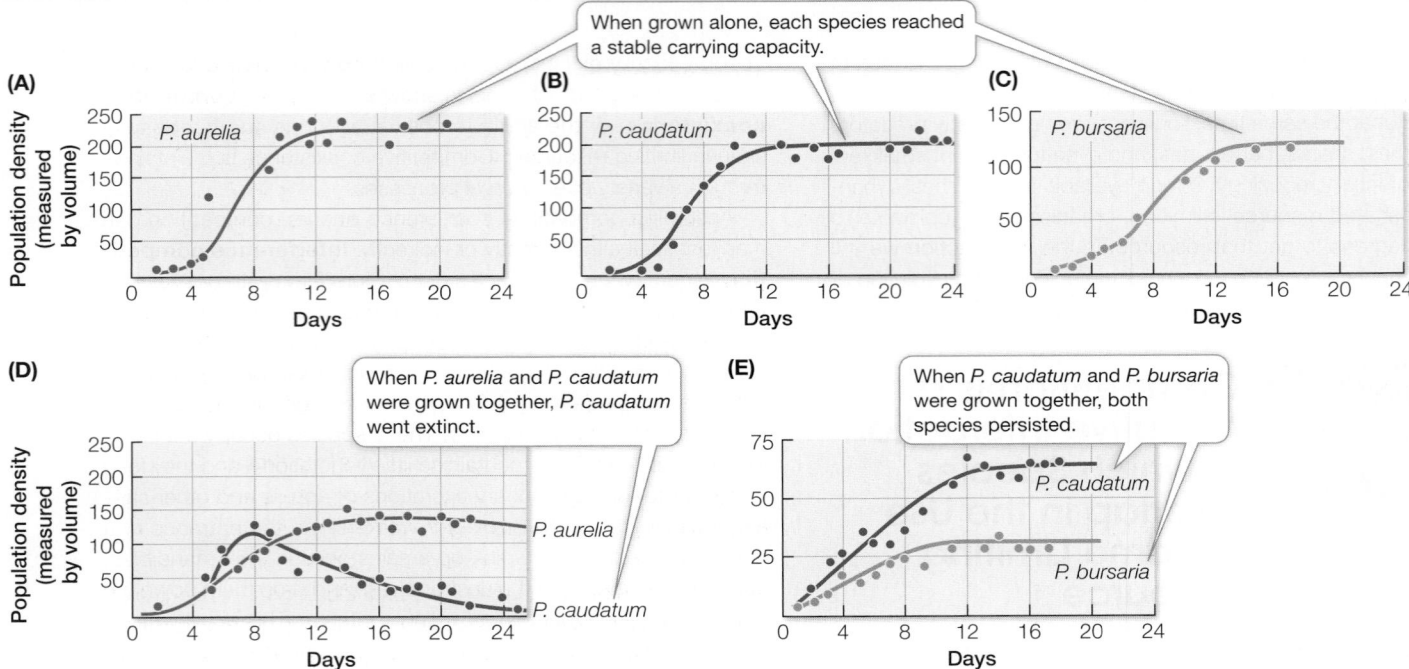

Figure 55.11 Competitive Exclusion or Coexistence? G. F. Gause grew three species of *Paramecium* in tubes filled with medium containing bacteria and yeast. **(A–C)** When grown alone, each species reached a stable carrying capacity. **(D, E)** When grown in pairs, *P. caudatum* faced two possible outcomes—extinction or coexistence—depending on whether *P. aurelia* or *P. bursaria* was present. *P. caudatum* and *P. aurelia* both fed mainly on floating bacteria, while *P. bursaria* fed mainly on yeast cells.

Q: Predict what would happen if *P. caudatum* and *P. bursaria* both fed mainly on yeast cells.

interactions can vary depending on environmental conditions and how they affect the competitive ability of individual species. We'll discuss how the outcome of competitive interactions can be modified by environmental factors below, but first let's consider the ways in which species can divide limiting resources to reduce their competitive effects on one another.

Resource partitioning allows species to coexist despite overlapping use of limiting resources

Even though the potential for intense competition is high, various ecological and evolutionary mechanisms reduce competition. As you saw with the bee–shin dagger agave example, one way is through **resource partitioning**, or the sharing of limiting resources by using them in different ways. Another example of resource partitioning comes from the classic experiments performed by G. F. Gause in the 1930s using *Paramecium*, a single-celled protist. Gause grew three species of *Paramecium* in test tubes filled with growth medium that contained bacteria and yeast for food. When grown alone, each species showed *logistic population growth** and reached a stable carrying capacity (**Figure 55.11A–C**). When pairs of species were grown together, however, they faced two possible outcomes: extinction or coexistence. *Paramecium caudatum* became extinct in the test tubes in the presence of *P. aurelia* but not when grown with *P. bursaria* (**Figure 55.11 D and E**). The difference in outcome, Gause suggested, was a consequence of *P. caudatum* and *P. aurelia* both competing for bacteria as a food source, while *P. bursaria*

avoided competition by eating the other food source, yeast. Thus *P. caudatum* and *P. bursaria* partitioned their food resource in the presence of one another and were able to coexist as a result.

***connect the concepts** Logistic population growth occurs when intraspecific competition causes population growth to slow down and eventually level off at some size known as the carrying capacity. Review Key Concept 54.2.

In some cases individuals within a species evolve different behaviors or morphologies depending on whether they are competing for resources with other species. Darwin remarked in *On the Origin of Species* that "Natural Selection leads to divergence of character; for more living beings can be supported on the same area the more they diverge in structure, habits, and constitutions." This "divergence of character" is referred to today as **character displacement**. On some of the islands of the Galápagos archipelago, for example, certain cactus species are pollinated exclusively by the small ground-finch (*Geospiza fuliginosa*), for which cactus nectar is an important food source (see Figure 22.8). On other islands, a carpenter bee (*Xylocopa darwinii*) competes with the finches for cactus nectar; the birds consequently feed more often on seeds and insects. On the islands where bees are absent, the nectar-feeding birds have smaller wingspans, presumably to make it easier for them to negotiate the flowers.

Sometimes organisms respond to competition by moving. The African wild dog (*Lycaon pictus*) (see Figure 58.17) is a carnivore that lives and forages in packs (groups of related individuals). Frequent vocalizations, called twitters, function to keep the pack together, but these acoustical signals also can alert the dogs' competitors, African lions (*Panthera leo*), to the presence of prey. Lions hearing the dogs' twitters can use them to locate dog packs and steal their kills. The dogs avoid competing with lions by selecting areas for their dens where the likelihood of being overheard by lions is low. The wild dog is considered a **fugitive species**—a species that leaves an otherwise

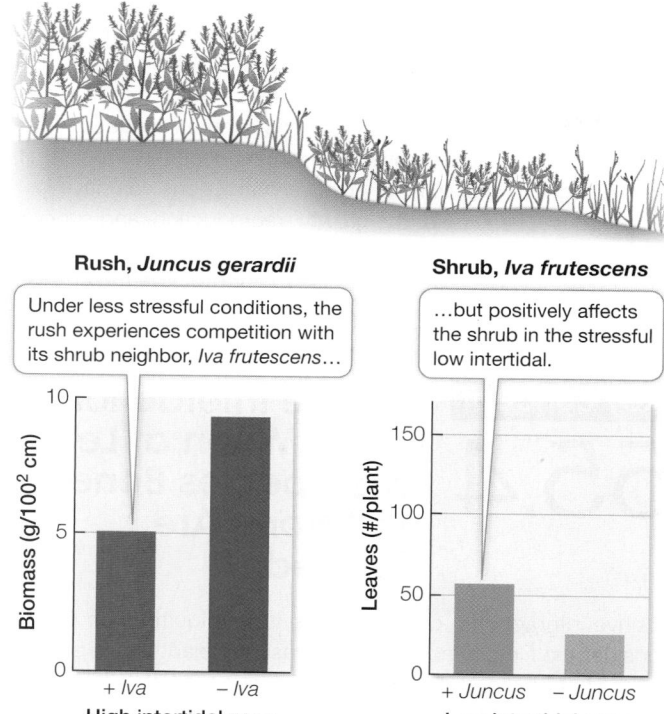

Rush, *Juncus gerardii*

Under less stressful conditions, the rush experiences competition with its shrub neighbor, *Iva frutescens*...

Shrub, *Iva frutescens*

...but positively affects the shrub in the stressful low intertidal.

Figure 55.12 Species Interactions in a Salt Marsh Plant species removal experiments showed that the shrub *Iva frutescens* competitively dominates the rush *Juncus gerardii* in the high intertidal where physical conditions are benign. But the interaction switches to a positive one in the low intertidal, where the tides flood on a daily basis and *Iva* grows better in the presence of *Juncus*.

Through a series of *Juncus* and *Iva* removal experiments in each tidal zone, the researchers found that in the relatively physically benign high intertidal, *Iva* is the competitive dominant, shading *Juncus* and causing it to have lower biomass (**Figure 55.12**). However in the low intertidal, where salinity and waterlogging from the tides are greater, *Iva* cannot compete with *Juncus* because it is physiologically intolerant of the physical conditions in this zone. In fact, the competitive relationship is turned into a positive one, in which *Iva* is dependent on *Juncus* to survive in the low intertidal. We'll discuss the role of positive interactions in the next section, but it is clear from examples such as this that the outcomes of species interactions depend on the particular circumstances, or context, under which they occur.

Competition can affect species' distributions

Competition can be important in determining where species are found. As you saw earlier, although species may be physiologically able to live under a wide range of conditions, competitors may restrict their use of resources to particular locations. An example of how competition can affect species distributions comes from the rocky intertidal zone of Scotland. Two species of barnacles, the acorn barnacle (*Semibalanus balanoides*) and Poll's stellate barnacle (*Chthamalus stellatus*), compete for space on the rocky shorelines of the North Atlantic Ocean. The planktonic larvae of both species settle in the intertidal zone and metamorphose into sessile adults. The smaller stellate barnacles generally live at higher levels in the intertidal zone, where they face longer periods of exposure and desiccation (drying out) than do acorn barnacles, which live at lower levels. There is little overlap between the areas occupied by adults of the two species (**Figure 55.13**). What explains their distinct distributions in the intertidal zone?

suitable habitat in order to avoid competition with another species.

The physical environment, disturbance, and predation can each alter the outcome of competition

The outcome of competition among species can be changed by a variety of factors, including the influence of the physical environment and processes such as disturbance and predation, which may injure or kill some individuals. These factors can cause a weakening or even a reversal of competitive outcomes between species if they negatively affect the competitively dominant species. By reducing the growth or survival of the competitively superior species, the physical environment, disturbance, or predation can allow the competitively inferior species a chance to acquire limiting resources, gain a foothold, and potentially increase its population size.

In an example from a Rhode Island salt marsh, Mark Bertness and colleague showed how competition could be altered by the effect of daily tides on the interaction between a rush, *Juncus gerardii*, and a shrub, *Iva frutescens*.

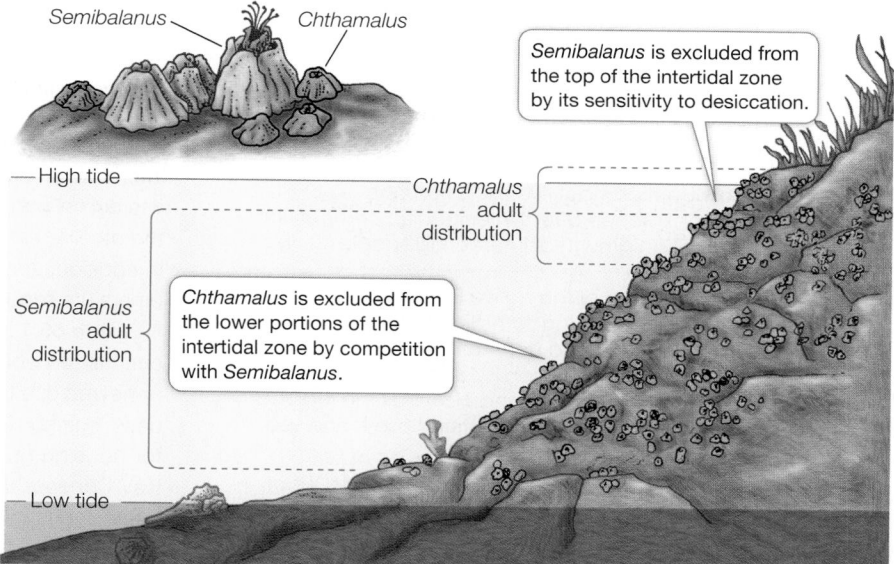

Semibalanus *Chthamalus*

Semibalanus is excluded from the top of the intertidal zone by its sensitivity to desiccation.

High tide

Chthamalus adult distribution

Semibalanus adult distribution

Chthamalus is excluded from the lower portions of the intertidal zone by competition with *Semibalanus*.

Low tide

Figure 55.13 Interspecific Competition Can Restrict a Species' Range Interspecific competition with acorn barnacles (*Semibalanus*) restricts stellate barnacles (*Chthamalus*) to a smaller portion of the intertidal zone than they could otherwise occupy. Larvae of both species settle throughout the intertidal, but at lower levels rock barnacles grow much faster and eliminate the stellate barnacle larvae. In the upper reaches of the intertidal, however, the greater susceptibility of acorn barnacles to desiccation (drying out) allows stellate barnacles to outcompete them. The two species can coexist in a small portion of the intertidal zone.

In a famous study conducted more than 50 years ago, Joseph Connell experimentally removed one or the other species of barnacle from its characteristic zone and observed the response of the remaining species. Stellate barnacle larvae normally settle in large numbers throughout much of the intertidal zone, including the lower levels where acorn barnacles are found (their fundamental niche), but they thrive at those lower levels only when acorn barnacles are not present (their realized niche). Connell found that the acorn barnacles grew so fast they smothered, crushed, or undercut the stellate barnacle juveniles. However, removing stellate barnacles from their spots higher in the intertidal zone did not lead to their replacement by acorn barnacles; the acorn barnacles are less tolerant of desiccation and failed to thrive there even when stellate barnacles were absent. The result of the competitive interaction between the two species is a distinctive pattern of intertidal zonation, with stellate barnacles restricted in their distribution by competition and acorn barnacles restricted in their distribution by their physiological limitations.

▶55.3 recap

Competition occurs when species overlap in their use of limiting resources, either through interference or exploitation. At one extreme, if a superior competitor prevents another species from using essential resources, the inferior species may be competitively excluded. In reality, most species show competitive coexistence, which is maintained through a variety of ecological processes, including resource partitioning, environmental conditions, disturbance, and predation. The distribution of a species may be affected by competition when another species restricts the use of limiting resources to a particular location.

learning outcomes

You should be able to:

- Contrast the concepts of fundamental niche and realized niche, and distinguish each with examples.
- Distinguish between interference competition and exploitation competition, and give examples of each.
- Describe how both competitive exclusion and competitive coexistence affect interspecific competition and the population growth of interacting species.

1. The fundamental niche of the native Eurasian red squirrel includes all of the United Kingdom. Since the introduction of the eastern gray squirrel, a non-native competitor, the Eurasian red squirrel's distribution is restricted to the northern regions of Scotland and Ireland. How has the realized niche of the red squirrel changed in response to the species' interactions with the gray squirrel?

2. Refer to Question 3 in the Key Concept 55.2 recap. What species interaction controls the coexistence of competing grass species? Explain how this interaction can promote species diversity.

3. Suppose you are sharing a milkshake with a friend. You each have a straw. Under scenario 1, you both drink from your straws, eventually consuming the entire milkshake together. Under scenario 2, your friend takes both straws and drinks the whole milkshake. Which scenario is an example of interference competition and which is an example of exploitation competition?

Prior to the 1990s, most ecologists believed that positive interactions were interesting, unusual phenomena, but of limited importance to populations and communities. Most of their focus was on the role of negative interactions such as predation and competition. But research over the last few decades shows that just like negative interactions, positive interactions can have important consequences for the growth, reproduction, and survival of species, especially under stressful conditions. Let's turn our attention to how positive interactions might promote the coexistence of species.

▶key concept 55.4 Positive Interactions Occur When at Least One Species Benefits and None Are Harmed

Positive interactions, or facilitations, are ubiquitous on Earth. Consider the simple fact that most vascular plants have beneficial pollinators, seed dispersers, and root-associated fungi and bacteria that are key to their survival. Consider all the species that provide essential habitat for other species. These positive interactions, while common, were often overlooked or underappreciated until recently.

focus your learning

- Positive interactions, including mutualisms and commensalisms, are ubiquitous on Earth and occur among all types of organisms.
- Conditions in stressful environments can limit species; positive interactions can ameliorate stressful conditions for affected species.
- Positive interactions have important effects on both population growth and community structure.

As mentioned in Key Concept 55.1, there are two types of positive interactions: mutualisms, in which both species benefit from the interaction, and commensalisms, in which one species benefits and the other is unaffected. Some positive interactions are symbiotic. In addition, some interactions can be **obligate** (necessary) while others are **facultative** (optional).

Table 55.1 lists mutualistic interactions that range from highly specialized obligate symbioses to highly generalized facultative nonsymbioses. Mutualistic interactions allow organisms to access limiting resources and often involve an exchange of food for housing or defense. Plants and their mycorrhizal fungi (see Key Concept 29.2 and Figure 29.9), corals and their photosynthetic endosymbionts (see Key Concept 26.4), and lichens formed from fungi and photosynthetic algae (see Key Concept 29.2 and Figure 29.9) all provide examples of obligate and/or facultative mutualistic interactions in which food is exchanged for housing and nutrients.

Some common mutualisms occur between sessile organisms—particularly flowering plants—and mobile animal species that can pollinate flowers or disperse their offspring. These interactions can be highly specialized or very general and facultative.

table **55.1** Examples of Mutualistic Interactions

Interaction	Obligate symbiosis	Facultative symbiosis	Obligate nonsymbiosis	Facultative nonsymbiosis
Plants–mycorrhizal fungi	x	x	x	x
Corals–algal endosymbionts	x	x		
Lichens (algae–fungi)	x			
Plants–insect pollinators			x	x
Plants–animal dispersers			x	x

For example, about three-fourths of the 250,000 flowering plant species on Earth require the transport of pollen by an animal partner. The benefit from the plant's perspective is clear: the animals move pollen from one plant to another and thereby promote sexual reproduction and thus genetic diversity. The most direct reward for pollinators is the pollen itself, which serves as food. Plant reproduction would not be served, however, if pollinators were to eat *all* of a plant's pollen; thus plants have evolved various adaptations to ensure that they benefit from the exchange. For example, some plants have two types of anthers: feeding anthers to produce pollen for pollinators, and fertilization anthers to produce pollen for reproduction.

Plants not only need to attract pollinators, but must also ensure that those pollinators carry their pollen to other members of the same species. Repeat visits by a pollinator to different individuals of a particular plant species increase the likelihood that the pollen will end up on the appropriate stigma; thus some plants have adaptations to encourage repeat visits by a given animal, while discouraging others. The nectar of tobacco flowers, for example, contains trace amounts of nicotine, an insecticidal neurotoxin. Many flower visitors, including hummingbirds, can ingest only tiny amounts of nicotine-laced nectar before moving on to other flowers. To other pollinators, however, nicotine may actually be addictive. Putting small amounts of a potentially addictive substance in nectar may be one way tobacco plants improve their odds of a repeat visit by the right pollinator species.

Many animals that eat fruits (called **frugivores**) provide a valuable service to the plants that produce those fruits by dispersing seeds. Seed dispersal by animals not only offers plants the advantages of delivery to potential germination sites away from the parent plant (described in Key Concept 37.1), but comes with the bonus of organic fertilizer for the seeds. Interactions between plants and frugivores, however, are not always reciprocal; in many cases, one party benefits more than the other. Whereas the frugivore is paid "in advance" for its transportation services, the seeds may never reach an appropriate destination for germination (your windshield, for example, will not do). From the plant's perspective, its partnership with frugivores requires a delicate balance between discouraging them from eating fruits before the seeds are capable of germinating and attracting them when the seeds are ready. In addition, the plant must protect the seeds from destruction in the frugivore's digestive tract and defend them against inappropriate consumers that would damage the seeds or fail to disperse them at all.

Positive interactions are more common in stressful environments

It is clear from the previous examples that positive interactions arise out of net positive benefits accrued by one or more species involved in the interaction. Over the last two decades studies have repeatedly shown that positive interactions, be they mutualisms or commensalisms, are more likely to occur in stressful environments such as deserts, salt marshes, and alpine communities. In these environments, where physical conditions can limit fitness, some species provide benefits for those species that lack mechanisms to deal with the stressful conditions.

A comprehensive study conducted by Ray Callaway and an international group of ecologists testing for the influence of positive interactions in alpine plant communities around the world illustrates the benefits of these interactions in stressful environments (**Figure 55.14**). The experiment was conducted on 11 mountain ranges and involved either removing neighboring plants around target plants or keeping them in place as controls (in total, 115 plant species were used). The experiment was conducted at both high-mountain elevations where temperatures are lower and at low-mountain elevations where temperatures are higher. A year later, the relative effect of the neighbor on the growth of the target plants was measured. Researchers found that target plants benefitted from their plant neighbors at high elevations simply as a consequence of the thermal protection that living close together provides under cold temperatures. In contrast, target plants competed with neighboring plants at low elevations, where temperatures are moderate and the same dense plant growth resulted in competition. Whether the observed positive interactions were mutualisms or commensalisms is unknown because the effects of the target species on the neighboring plant species were not measured. Nonetheless, in this and other studies (e.g., see Figure 55.12) it is clear that positive interactions can arise under a variety of stressful conditions.

Positive interactions can have dramatic effects on populations and communities

You have seen how positive interactions influence the physiology, behavior, and growth of species. Research has shown that positive interactions have effects at the population and community levels as well.

As you know from population biology, any interaction that affects the survival of individuals in a population can affect population growth. One example of the effects of a mutualism on population growth comes from an interaction between ants and acacia trees in Central America. In 1874, in Nicaragua, the naturalist Thomas Belt observed a peculiar interaction between bullhorn acacia trees (*Acacia cornigera*) and *Pseudomyrmex* ants. Bullhorn acacias get their common name from the enlarged, hollow thorns, in which the ants build nests. The trees also produce rewards for the ants, both in nectar-producing extrafloral structures and modified leaflet tips that are rich in oil and protein. These structures have no apparent purpose other than providing food for ants.

Belt suggested that the notoriously aggressive acacia ants defend the plants against herbivores in exchange for food and

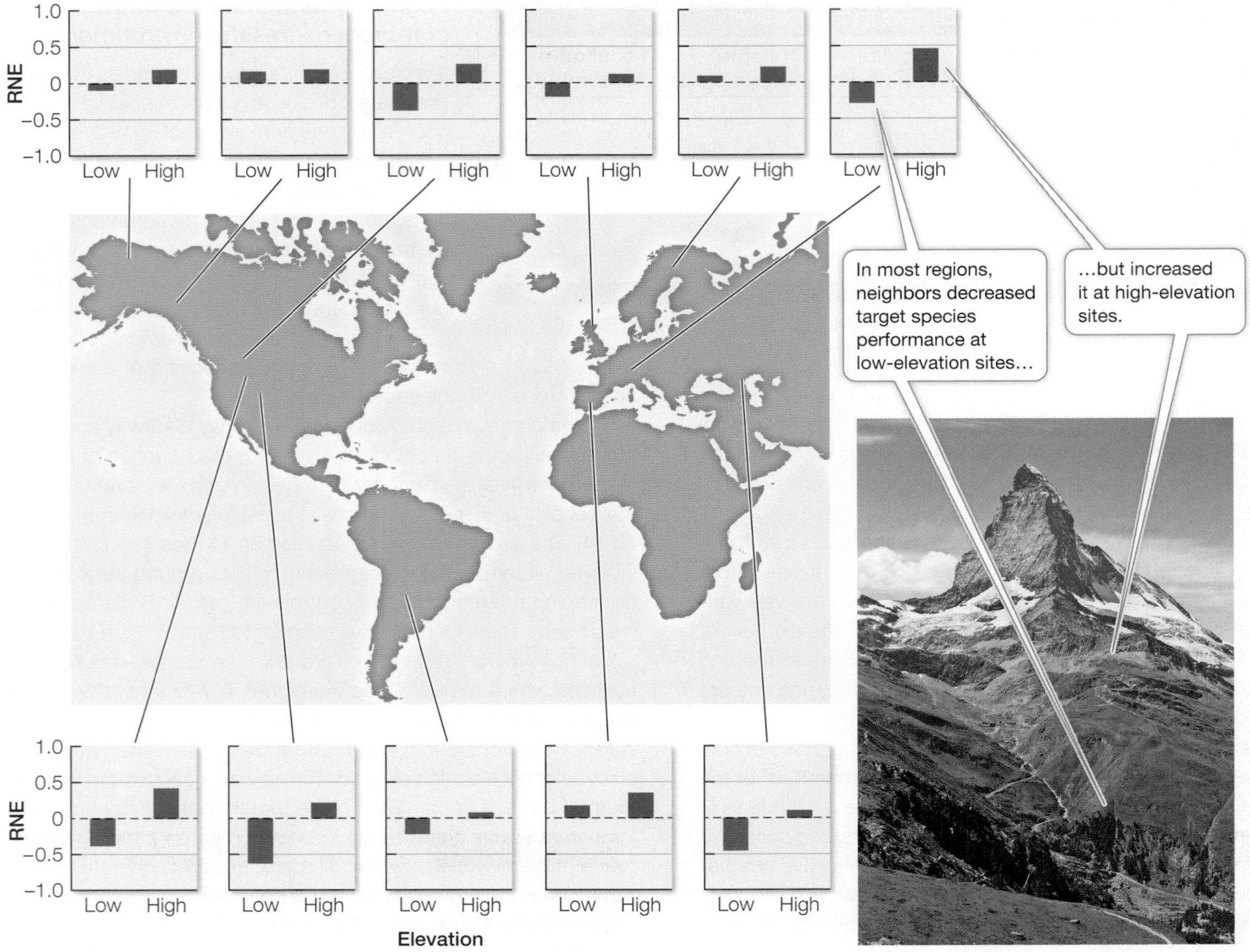

In most regions, neighbors decreased target species performance at low-elevation sites...

...but increased it at high-elevation sites.

Figure 55.14 Positive Plant Interactions Are More Common at Higher Elevations The species interactions of alpine plants were experimentally measured (using the relative neighbor effect [RNE], defined as the growth of the target plant species with neighboring plants present minus its growth with neighboring plants removed) in 11 mountain ranges worldwide. RNE values greater than zero (in blue) indicate that neighbors increased the growth of target species; RNE values less than zero (in red) indicate that neighbors decreased the growth of target species. Plants generally benefited from neighbors at high elevations where temperatures were colder and competed with them at lower elevations where temperatures were warmer.

shelter. This idea was tested by Daniel Janzen in 1966 when he removed ants from some acacias with insecticide, and showed that trees without ants suffered a reduction in growth and an increase in mortality (**Figure 55.15**). If a bullhorn acacia tree lacks an ant colony, the repeated loss of its leaves and growing tips to herbivores often kills the plant within 6–12 months. Since Janzen conducted his experiment, additional work on ants and acacias has revealed that ants do more than simply defend the plant against herbivorous enemies; they also clip weeds from around the base of the plants, presumably reducing competition. Thus the ant–acacia mutualism has considerable effects on the population size of each partner.

▶ Animation 55.1 **Mutualism**
www.Life11e.com/a55.1

Finally, possibly one of the most important types of positive interactions is that of species that provide habitat for other species and thus have widespread community effects. These dominant species mostly include plants such as trees or kelps but can include animals such as corals and sponges. For example, coral reefs are home to many species of fish that cannot survive outside the reef environment. Likewise, species that are forest specialists cannot survive outside the cool and shaded environment of trees. Most tree species facilitate many other types of species by simply providing sites for attachment, shelter from extreme environmental conditions, or hiding places from predators. Many of these species have no direct effect on the trees in which they live, and thus have a commensalism with the trees. In all these cases, these dominant plants and animals are the very foundation of the community, which would not exist without them.

experiment

Figure 55.15   Are Ants and Acacia Mutualists?

Original Paper: Janzen, D. H. 1966. Coevolution of mutualism between ants and acacias in Central America. *Evolution* 20: 249–275.

Bullhorn acacia trees (*Acacia cornigera*) have numerous structures that provide food and shelter for acacia ants (*Pseudomyrmex*). Daniel Janzen's experiments demonstrated that the trees benefit from their association with these ants and that the energy expended in growing ant-attractive structures is repaid with increased growth and survival.

HYPOTHESIS▶   *Acacia cornigera* trees deprived of their *Pseudomyrmex* ant populations will have lower growth and survival than trees populated by ant colonies.

METHOD

1. Define a population of *A. cornigera* trees; randomly designate some of them as untreated controls and the rest as experimental subjects.

2. Fumigate the experimental trees with insecticide to eliminate all *Pseudomyrmex* ants.

3. Apply Tanglefoot (a sticky material) to the base of the experimental trees to prevent ants from recolonizing them.

4. Record the growth and survival of the trees in both groups over a 10-month period.

RESULTS

After 10 months, control trees (with ants) had considerably higher growth and survival than did trees without ant populations.

CONCLUSION▶   *Pseudomyrmex* ants provide substantial fitness benefits to *Acacia cornigera* trees.

The "bull's horns" are enlarged, hollow thorns in which the ants build nests.

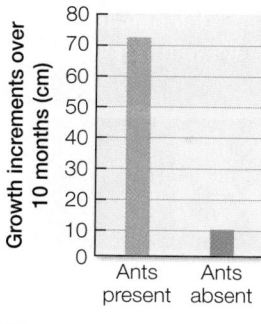

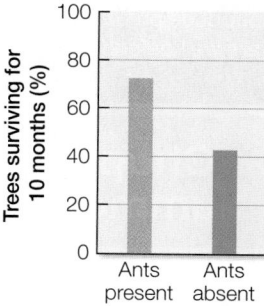

▶55.4 recap

Positive interactions, when at least one species benefits from an interaction and none are harmed, play a key role in enhancing coexistence among species. Positive interactions are more common in stressful environments, where physical conditions can limit the population growth of some species. Positive interactions can have dramatic effects on populations and communities, especially by providing habitat for other species.

learning outcomes

You should be able to:

• Give reasons why the types of mutualistic interactions, from generalized to specialized, and from obligate to facultative have evolved.

• Explain, with examples, why stressful environments are more likely to give rise to positive species interactions.

• List examples in which one species serves as habitat for other species. Explain whether each situation represents mutualism or commensalism, and justify your answer in each case.

---

1. Offer reasons for why insect pollinators and animal dispersers might evolve obligate versus facultative mutualistic relationships with plants.

2. Refer to Figure 55.13. Where in the rocky intertidal might *Semibalanus* benefit from positive interactions with other species? How about *Chthamalus*? Explain your answer.

3. Give an example of how species that provide habitat could affect the number of species in a community.

investigatinglife

Q&A Why are small coral-reef fish species so vulnerable to lionfish predation, and what can be done about it?

You have seen that this question is greatly informed by understanding the power that species interactions have on shaping the behavior, physiology, and morphology of interacting species. It is clear that the longer species are exposed to one another, the more likely it is that they will develop strategies that allow coexistence. When lionfishes were inadvertently introduced to coral reefs in the Atlantic Ocean, native coral-reef fishes were exposed to a simple but novel predator behavior for which they currently have no viable defense mechanisms. It is unclear whether native coral-reef fishes will evolve defensive mechanisms before coral-reef fish communities are irreparably damaged. One potential strategy to slow the invasion may be to introduce predators of lionfishes. Unfortunately, however, lionfishes are well defended by frilly fin rays with venomous spines and a zebralike barred coloration that provides crypsis. To make matters worse, populations of predators that might control lionfish populations (such as sharks) are drastically declining as a consequence of overfishing. Fortunately lionfishes are relatively easy to catch, and the future may lie in controlling the population through systematic harvesting by humans.

Future directions

Some people have argued that the best way to control lionfishes is through commercial fishing. Lionfishes are attracted to lobster traps and can also be easily speared. The only risk to harvesters is in being poked by the spines, but this can be avoided if the fish are handled with care. After some initial skepticism, restaurant owners are realizing that lionfish are very good eating; the meat has a light, buttery flavor. Efforts are now ramping up to create a market for this destructive invader. Lionfish and chips, anyone?

Chapter Summary 55

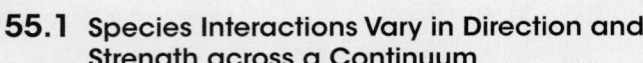

▶ 55.1 Species Interactions Vary in Direction and Strength across a Continuum

- Species vary in the effects they have on each other (positive, negative, neutral). Species interactions may involve feeding (**trophic**) effects and/or living in close association (**symbiosis**).

- The four main categories of species interactions are **predation** (predator kills and/or consumes its prey, or part of its prey), **competition** (at least two species use some of the same limiting resources and have negative effects on one another), **positive interactions**, or **facilitations** (at least one species benefits from the interaction and none are harmed), and **amensalism** (one species is harmed but the other is not affected). Review Focus: Key Figure 55.1, Activity 55.1

- Species interactions are not always clear-cut but instead fall along a continuum in which species vary in how strongly they affect one other. Review Figures 55.2, 55.3

- Some species interactions result in reciprocal evolutionary change over time, also known as **coevolution**. Review Activity 55.2

▶ 55.2 Predation Is a Trophic Interaction in which Predators Benefit and Prey Are Harmed

- Predator–prey interactions result in the evolution of a range of capture and avoidance mechanisms. Review Figures 55.4, 55.5, 55.6, Activity 55.3

- **Herbivory** (animals preying on plants) and **parasitism** (parasites feeding and living on or in hosts) are widespread but specialized interactions that usually do not result in the death of the prey. Review Figures 55.7, 55.8

- Population sizes of predators and their prey tend to cycle as a result of predator–prey interactions over time. Review Figure 55.9

- Some predators, called **keystone species**, can have dramatic effects on communities by preying on competitively dominant species. Review Figure 55.10

▶ 55.3 Competition Is a Negative Interaction in which Species Overlap in the Use of Some Limiting Resource

- **Interference competition** occurs when one species directly interferes with or prevents another species' access to a limiting resource. **Exploitation competition** occurs when all competitors have access to the limiting resource but the outcome depends on their relative efficiency at using the resource.

- **Competitive exclusion** occurs when a species prevents another species from using essential resources, which may cause the inferior competitor to become locally extinct. Review Figure 55.11D

- Most competing species show **competitive coexistence** through the processes of **resource partitioning**, disturbance, stress, and/or predation. Review Figures 55.11E, 55.12, 55.13

▶ 55.4 Positive Interactions Occur When at Least One Species Benefits and None Are Harmed

- The two types of positive interactions are **mutualisms**, in which both species benefit from the interaction, and **commensalisms**, in which one species benefits and the other is unaffected. Some positive interactions are symbiotic relationships. Some positive interactions are **obligate** (necessary), while others are **facultative** (optional). Review Table 55.1, Animation 55.1

- Positive interactions are more common in stressful environments, where physical conditions can limit the abundance and distribution of some species. Review Figure 55.14

- Positive interactions can have dramatic effects on populations and communities, especially by providing habitat for other species.

Go to **LearningCurve** (in **LaunchPad**) for dynamic quizzing that helps you solidify your understanding of this chapter. **LearningCurve** adapts to your responses, giving you the practice you need to master each key concept.

Apply What You've Learned

Review

55.1 Broad categories of species interactions are predation, competition, positive interactions, and amensalism.

55.1 Species interactions are often dynamic and asymmetrical; that is, species have unequal effects on one another.

Original Paper: Bartomeus, I., J. S. Ascher, D. Wagner, B. N. Danforth, S. Colla, S. Kornbluth et al. 2011. Climate-associated phenological advances in bee pollinators and bee-pollinated plants. *Proceedings of the National Academy of Sciences USA* 108: 20645–20649.

Rising global temperatures due to climate change affect the timing of natural events, including insect pollination and other species interactions. As temperatures rise, these events happen earlier. Not all species react to these changes in similar ways, and this can result in timing mismatches. For example, if the timing of insect emergence varies too much from that of flowering, pollination may not occur.

Scientists looked at 3,447 museum specimens of native bee species dating back to the 1870s. They chose 10 bee species that are known to range across North America, emerge in early spring, and pollinate both crops and wild plants. The scientists then compared the predicted pollination date for individual bees with flowering times of bee-pollinated spring plants measured in other studies. Dates of bee pollination activity and flowering times were correlated with temperature data in the same geographic areas.

Figure A shows scatter plots of the distribution of predicted pollination dates for individual bees between 1870 and 2010 (top) and the mean April air temperature between 1900 and 2010 (bottom). Predicted pollination date is the number of days between January 1 and the bee-collection date, a time in which bees were assumed to be engaging in pollination activity. Black lines indicate the trends for the entire time period, and red lines show the trends between 1970 and 2010. Statistical analysis revealed a strong positive correlation between predicted pollination dates and mean April air temperature. Between 1880 and 2010, bees advanced their pollination activity by 10.4 days; most of the advance (7.2 days, or 69 percent) has occurred since 1970.

Figure B shows comparisons of the change in days for bee pollination time and plant flowering time for two time intervals. The change in slope between the two dates (bees at left, plants at right) indicates the degree of difference between bee pollination time and plant flowering time for both time intervals. The interval indicates a mismatch of 4–6 days over the 118-year span of the longer study (1885–2003). For the more recent study (1971–1999), when mean April air temperature changed the most, the mismatch was 5 days over just 30 years.

Questions

1. Characterize the type of interaction occurring between plants and bees, and explain why it is necessary for the timing of bee emergence and plant flowering to be closely correlated.

Figure A

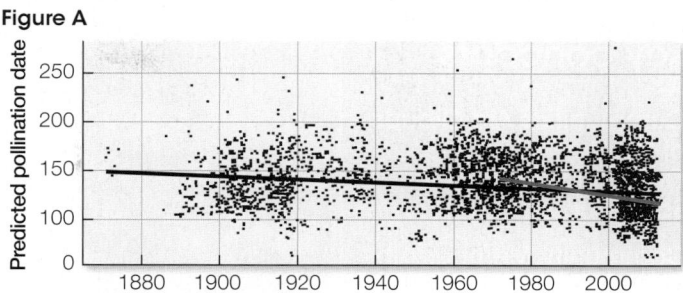

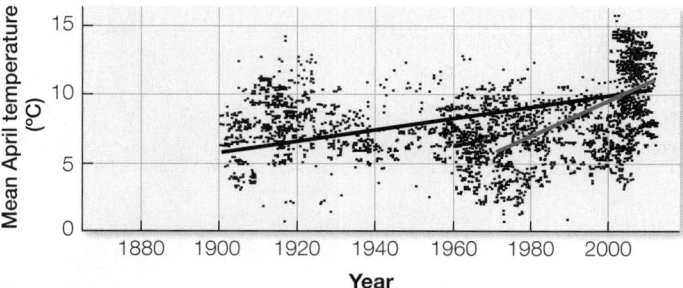

Figure B

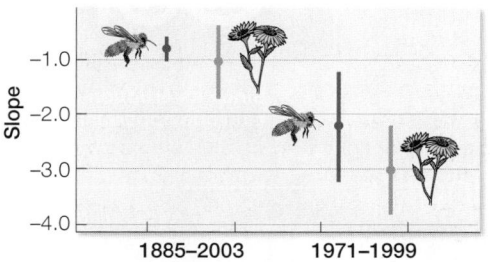

2. Explain and analyze the trends shown in the two plots of Figure A. What has happened over time to the trend in bee pollination date and the trend in mean April air temperature? Why are these two trends moving in opposite directions? What is the significance of the difference in slope of the entire time period versus the shorter time period after 1970?

3. Describe differences in the trends shown in the two studies in Figure B, and indicate which study best supports the hypothesis that climate warming is affecting plant–pollinator interactions.

4. The bees in these studies are generalist species; that is, they feed on (and pollinate) many species of plants, as compared with specialists, which associate with one or a few closely related species. Do you think specialist plant–pollinator interactions would be more or less affected by increasing temperatures? Explain your answer.

5. This is one of few studies conducted to date on the effect of climate change on plant–pollinator species interactions. How would you design a multiyear study to give more specific information about the presence (or absence) of timing mismatches between the emergence of insect pollinators and spring flowering of plants?

Go to **LaunchPad** for the eBook, LearningCurve, animations, activities, flashcards, and additional resources and assignments.

56

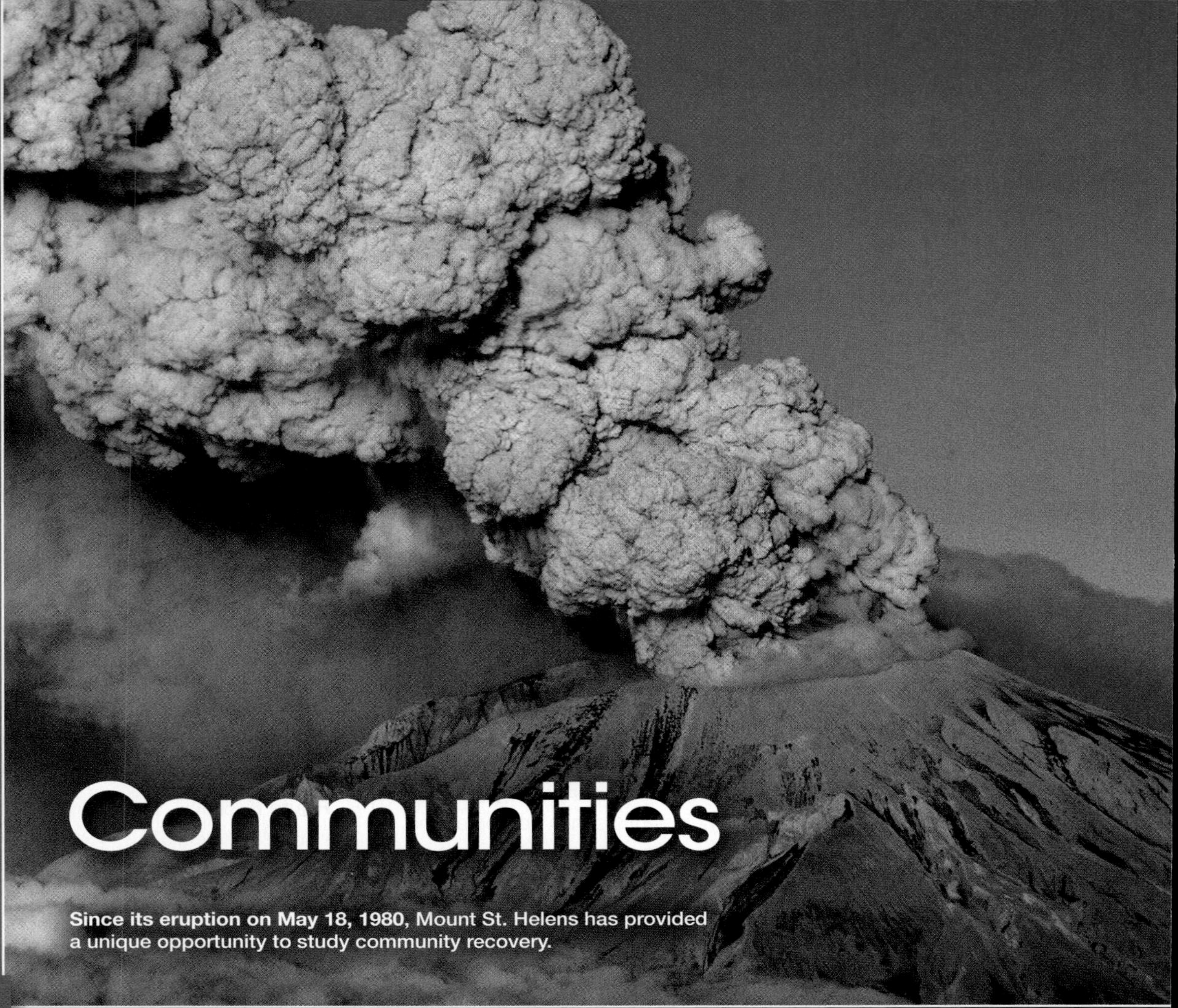

Communities

Since its eruption on May 18, 1980, Mount St. Helens has provided a unique opportunity to study community recovery.

▶ investigatinglife

Rising from the Ashes

When Mount St. Helens in Washington State erupted at 8:30 A.M. on May 18, 1980, almost all living things for miles around were obliterated in an instant. A huge magma-filled bulge that had been forming for months exploded, creating the largest avalanche in recorded history. Mud and rock flowed down Mount St. Helens and were deposited on alpine meadows, lush forests, and crystal-clear lakes and ponds, tens of meters deep in some areas. The bulk of the avalanche traveled to the North Fork Toutle River, where it scoured the entire valley, from floor to rim, with volcanic material. A huge pile of vegetation was deposited downstream. The blast also produced a cloud of hot air that burned forests to ash near the mountain, blew down trees over a large area, and left dead but standing trees for miles.

Volcanic eruptions are so rare that studying the aftereffects of the Mount St. Helens eruption gave ecologists an unprecedented view into how ecological communities respond to extreme natural catastrophes and the recovery processes they undergo. The event created brand-new environmental conditions to which colonizing species had to respond. At one extreme was the Pumice Plain, a large, gently sloping area close to the lava dome that had once

been covered in alpine meadows and forests. Pelted with hot, sterilizing pumice (a light and porous volcanic rock), it was now a harsh moonscape-like environment that lacked life and organic matter of any form. In the aptly named Blowdown Zone, covering most of the mountain's slopes, the destruction was less catastrophic. Here there was hope of a biological legacy buried beneath the piles of downed trees and vegetation covered in mud and ash. Beyond the Blowdown Zone, forests were covered in debris and ash that spewed from the mountain for months.

Shortly after the eruption, helicopters delivered the first scientists to the mountain. A few fortunate ecologists collected baseline data on the sequence of biological changes that began soon after the eruption. Now, more than 35 years later, hundreds of ecologists have studied the reemergence of life on Mount St. Helens. Much of what has been learned has been unexpected and has changed the way we view the resiliency and recovery of communities after catastrophic events.

 How have ecological communities on Mount St. Helens responded to the eruption, and what processes have been important to their recovery?

key concept 56.1   Communities Are Groups of Interacting Species Occurring Together in Space and Time

In ecological terms, a **community** is a group of species living together at the same place and time. Interactions among multiple species and their physical environment give communities their character and function. Although each species has unique interactions with the other species in its community (as you saw in the previous chapter), ecologists often find it useful to study the properties of the community as a whole.

focus your learning

- Ecologists may define communities based on taxonomy, resource usage (guilds), or functional groups.
- Ecologists may define communities in terms of food webs or interaction webs.
- Species diversity can be measured quantitatively.

The definition of a community given above is more theoretical than practical, in part because communities vary greatly in size and scope. For example, communities can be small and contained, such as that living within the purple pitcher plant (*Sarracenia purpurea*), a species common in North American wetlands (**Figure 56.1A**). Pitcher plants collect rainwater and are home to a thriving community of bacteria, protists, rotifers, and mosquito larvae. A community can also encompass a large geographic area, such as a desert (**Figure 56.1B**), where the borders of the community are not always well delineated. When borders become unclear, ecologists may designate boundaries somewhat arbitrarily, based on their ability to study the community.

Another practical problem in defining communities is determining the number of species present. Most communities contain thousands if not hundreds of thousands of species, from microscopic bacteria to towering trees. Some species are easily identified, but many others are not. Additionally, some species are present only at certain times of the year, for certain life stages, or tend to move among communities. Creating a species list for a community is a huge undertaking—one that is essentially impossible to complete, especially if all small or relatively unknown species are considered. Taxonomists have officially described about 1.8 million species on Earth, but we know from sampling studies of tropical insects and microorganisms that this number greatly underestimates the actual number, which could be 15 million or even more. For this reason, and because of the difficulty of studying many species at one time, ecologists often restrict their definition and study of communities to a particular subset of species.

Ecologists often use a subset of species to define communities

A community can be subdivided on the basis of taxonomic affinity—that is, by groups of organisms classified together because of evolutionary lineage (**Figure 56.2A**). For example, all the amphibians in a set of ponds on Mount St. Helens could be considered to be part of an amphibian community. A community can also be subdivided by resource usage. A **guild** is a group of species, often taxonomically distinct, that use similar resources (**Figure 56.2B**). For example, bats, birds, and bees all feed on pollen, forming a pollen-feeding guild. Yet another way to subdivide a community is by **functional group**—species that function in similar ways but may or may not use similar resources (**Figure 56.2C**). For example, nitrogen-fixing plants (legumes) can be placed in the same functional group.

Ecologists also organize communities into a **food web**, a representation of the trophic or energetic connections among species, as illustrated in **Figure 56.3** for Yellowstone National Park. Most communities contain so many species interacting in so many different ways that it is impossible to enumerate (or even identify) all of the links in a food web. Nevertheless, simplified food webs are useful in envisioning the kinds of interactions and sequence of energy flow occurring in a community. Food webs are generally divided by **trophic levels**, with each level including species that have similar ways of interacting and obtaining energy. Primary producers start the chain of trophic levels. At the next level are **primary consumers**—the herbivores that feed on primary producers. Organisms that eat herbivores, called **secondary consumers**, are the next trophic level. Those that eat secondary consumers are tertiary consumers, and so on. Some species, known as **omnivores**, feed on multiple trophic levels. For example, a bird might eat both insects and seeds. Finally, the waste products and dead bodies of organisms (known as **detritus**) provide another source of energy. Organisms that consume such materials are called **detritivores** or **decomposers** and serve an important function by returning nutrients back to the community through the uptake of primary producers. **Scavengers** eat larger dead organisms and are typically not considered detritivores.

Food web depictions can sometimes include nontrophic interactions such as competition and positive interactions (mutualisms and commensalisms). These so-called **interaction webs** provide a more realistic view of all the species interactions that might be

(A)

(B)

Figure 56.1 Defining Communities   Communities exist at a variety of spatial scales. The community within a single pitcher plant contains microorganisms and tiny invertebrates **(A)**, whereas a desert community might contain plants and animals over a large, hard-to-delineate, geographic area **(B)**.

(A) Taxonomic affinity

(B) Guild

(C) Functional group

Figure 56.2 Subsets of Species in Communities Ecologists often use subsets of species to define communities. **(A)** All the amphibian species in a community could be grouped together by taxonomic affinity. **(B)** All the species that use pollen as a resource form a guild. **(C)** All the legumes (e.g., vetch, clover, and peas) that have nitrogen-fixing bacteria could be placed in the same functional group. Species within functional groups may or may not use the same resources.

important in the community. Overall, the food web concept, regardless of whether it includes nontrophic interactions, is a useful visual representation of the important consumer relationships in a community. You will learn more about food webs later in this chapter and in Chapter 57.

 Media Clip 56.1 A Food Web in Africa
www.Life11e.com/mc56.1

Species diversity and composition are important descriptors of community structure

Communities vary widely in species number and **species composition** (or the kinds of species) present. A coral-reef community in the Great Barrier Reef, for example, has many more species of fish than one in the Caribbean. As you saw in Chapter 53, ecologists

have devoted considerable effort to measuring the variation in species diversity and species composition at multiple spatial and temporal scales (see Figure 53.12). Patterns at the community or local scale are important descriptors of what ecologists call **community structure**. Documenting community structure is an important first step in generating hypotheses about how communities work. Let's first consider the different ways that one metric of community structure—species diversity—is measured.

We have thus far used the term "species diversity" to describe the number of species in a sample, community, or region, but ecologists have a more technical definition. **Species diversity** is a measure that combines the number of species (**species richness**) and their relative abundances compared with that of other species (**species evenness**).

Trophic level

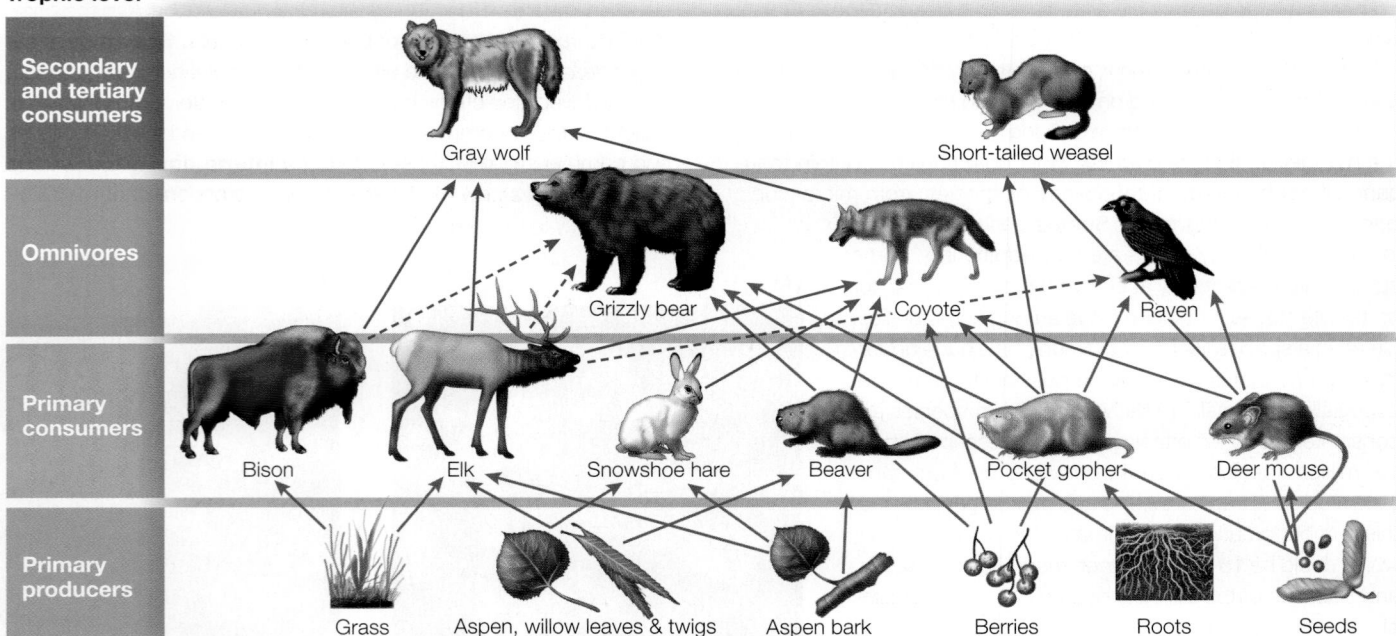

Figure 56.3 Food Webs Show Trophic Interactions in a Community This simplified food web for the meadows and forests of Yellowstone National Park includes only some vertebrates and the plants on which they depend. The arrows show trophic interactions and energetic connections. Herbivores whose sole source of food is plants (green arrows) are primary consumers. Carnivores that kill and

eat animals (purple arrows) are secondary and tertiary consumers. Omnivores such as grizzly bears, coyotes, and ravens eat both plant and animal tissues; ravens and grizzlies also eat carrion (dashed purple arrows), so these species are also scavengers.

 Activity 56.1 The Major Trophic Levels
www.Life11e.com/ac56.1

Pond A

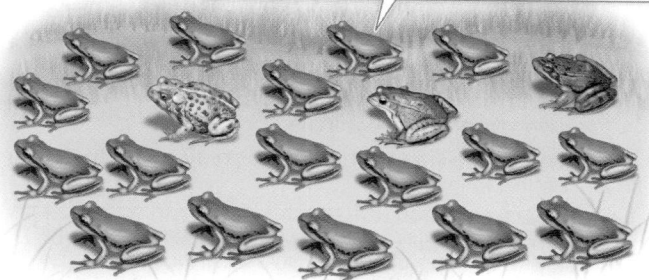

In pond A, the abundance of one species (Pacific tree frog) is high relative to the other species, so this community has low species evenness.

Pond B

In pond B, each species has the same abundance, so this community has high species evenness.

Figure 56.4 Species Richness and Evenness These two hypothetical amphibian pond communities have the same number of species (species richness) but different relative abundance of those species (species evenness).

Q: Which pond, A or B, has higher species diversity as measured using the Shannon index (see Table 56.1)? Explain.

 Activity 56.2 Measures of Species Diversity
www.Life11e.com/ac56.2

The contribution of both species richness and evenness to species diversity is illustrated in the following example from Mount St. Helens (**Figure 56.4**). Imagine that we collect samples of 20 individual amphibians in each of two different ponds on Mount St. Helens 3 years after its eruption. Our sample from Pond A contains 17 individuals of one species (Pacific tree frog) and only 1 individual of each of the other three species (an uneven distribution). Our sample from Pond B, however, contains 5 individuals of each of four amphibian species (an even distribution of individuals). Even though the species richness of the two communities is the same (four species), Pond A is less diverse because the less abundant species are encountered less frequently compared with the single most abundant species.

There are several ways to measure species diversity quantitatively. One common measure is the **Shannon index**,

$$H = -\sum_{i=1}^{s} p_i \ln(p_i)$$

where

H = the Shannon index value

p_i = the proportion of individuals found in the ith species

ln = the natural logarithm

s = the number of species in the community

The lowest possible value of H is zero; higher values represent greater species diversity. **Table 56.1** calculates the Shannon index for amphibians in the two Mount St. Helens pond communities shown in Figure 56.4. These calculations show that Pond A has the lower Shannon index value (H), confirming mathematically that this community has lower species diversity than Pond B. Given that both communities have the same species richness, the difference in species diversity is driven by the lower species evenness in Pond A.

How does the term "biodiversity" relate to species diversity? **Biodiversity** is used in reference to various levels of diversity, from genes to species to communities (**Figure 56.5**). Implicit in this term is the interconnectedness of different scales of diversity. For example, you saw in Chapter 20 that greater genetic diversity can lead to greater fitness of populations, which in turn can promote species diversity. The greater number of different types of communities can also lead to greater diversity at larger regional and continental scales, as you saw in Chapter 53.

table 56.1 Calculation of Species Diversity for Ponds A and B Using the Shannon Index

Pond A

Species	Abundance	Proportion (p_i)	ln (p_i)	p_i ln (p_i)
Pacific tree frog	17	0.85	-0.163	-0.139
Western toad	1	0.05	-2.996	-0.150
Cascades frog	1	0.05	-2.996	-0.150
Northern red-legged frog	1	0.05	-2.996	-0.150
Total	20	1.00		-0.589

Pond B

Species	Abundance	Proportion (p_i)	ln (p_i)	p_i ln (p_i)
Pacific tree frog	5	0.25	-1.386	-0.347
Western toad	5	0.25	-1.386	-0.347
Cascades frog	5	0.25	-1.386	-0.347
Northern red-legged frog	5	0.25	-1.386	-0.347
Total	20	1.00		-1.388

To calculate the **Shannon index** (H), the natural logarithm (ln) is applied to p_i for each species (i)...

...and then this value is multiplied by p_i once again.

All the values are summed for all the species in the community and multiplied by -1 to get H.

$$H = -\sum_{i=1}^{s} p_i \ln(p_i) = 0.589$$

Community B has higher species diversity than community A.

$$H = -\sum_{i=1}^{s} p_i \ln(p_i) = 1.388$$

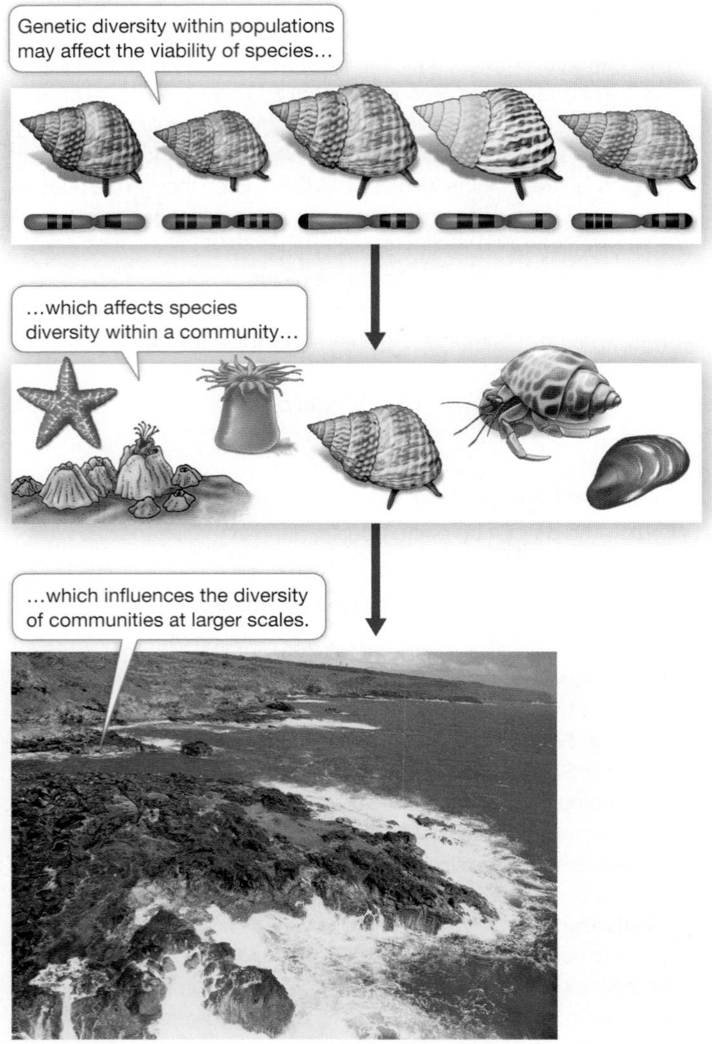

Genetic diversity within populations may affect the viability of species...

...which affects species diversity within a community...

...which influences the diversity of communities at larger scales.

Figure 56.5 Biodiversity Considers Multiple Spatial Scales Diversity can be measured at spatial scales that range from genes to species to communities. The term "biodiversity" encompasses diversity at all of these scales.

56.1 recap

A community is a group of species that occur together at the same place and time. Communities are often defined by a subset of species based on similar taxonomy, resource use (guild), ecological functions (functional group), or trophic and energetic connections (food webs). Species diversity and composition are descriptors of community structure. Species diversity is a measure that combines the number of species (species richness) and their relative abundances (species evenness). Biodiversity is a term that describes multiple scales of diversity, from genes to species to communities.

learning outcomes

You should be able to:

- Give examples of ways communities can be subdivided, and explain how these might also be used to understand communities.

56.1 recap

- Define food web, and analyze types of information it might provide about community structure.
- Use the equation for the Shannon index to determine the species diversity of a community, given appropriate data.
- Explain the significance of the value of *H* in the Shannon index.

---

1. Why are subsets of species used to define communities? What type of community subset was used to define the frogs and toads in ponds on Mount St. Helens?

2. What types of species interactions are missing from the food web depicted in Figure 56.3?

3. Refer to Table 56.1. Replace the abundance of amphibians in Pond A with the following values (Pacific tree frog = 6, western toad = 8, northern red-legged frog = 4, Cascades frog = 2), recalculate the values in the other three columns in the table accordingly, and then calculate a new Shannon index value. Which pond has the higher species diversity, Pond A or Pond B?

Measuring community structure prompts us to consider many interesting questions about communities. Why do communities differ in the number and composition of species? What processes determine community membership? In the next section we will explore how community membership is shaped by which individuals get to the community, deal with the environmental conditions there, and form interactions with other individuals.

▶ key concept 56.2 Community Membership Depends on Species Supply, Environmental Conditions, and Species Interactions

Suppose you are looking across the landscape of Mount St. Helens. You see lakes and ponds, alpine meadows, and forests. There is no doubt that these communities vary not only in their species composition but also their species richness. The lakes might be dominated by fishes, amphibians, and freshwater algae; the alpine meadows by various species of grasses, flowering plants, and rodents; and the forests by trees, shrubs, and birds. Even though some species might move from one community to another, the three communities would still be very different. How is it that species come together to form different communities?

focus your learning

- To become community members, species must be present in a region and able to colonize the community.
- Environmental conditions serve as an abiotic filter to determine which species colonize a community.
- Species already living in a community serve as a biotic filter, determining which colonizing species succeed or fail.

Community membership depends on three basic factors: (1) the regional species pool and dispersal ability (species supply), (2) physical and chemical environmental conditions (abiotic conditions), and (3) species interactions (biotic conditions). Each factor acts as a "filter" that excludes or includes species in particular communities (**Focus: Key Figure 56.6**). Let's consider each of these filters in more detail.

Community membership depends on species supply

In Chapter 53 we established that regional species pools provide the upper limits on the numbers and types of species that can live within communities (see Figure 53.12). An obvious first step to community membership is the presence of species at the regional level, and their ability to disperse to the community. For example, as communities started to recover from the Mount St. Helens eruption, they began to gain species supplied from the regional species pool of the Pacific Northwest of North America. Because they were able to disperse to the mountain, these species began to colonize the new and unique environments created in the aftermath of the devastation.

Maybe the best examples of how species supply affects community membership are seen in the introduction of novel species to communities. Humans have greatly expanded regional species pools, allowing new and potentially invasive species to enter communities. Chapter 55 described the catastrophic effects of the unintentional release of invasive lionfish in Atlantic Ocean and Caribbean coral-reef communities. The lionfish invasion demonstrates that "getting there" is a critical first step to community membership. All it took was the release of this predator, and its extraordinary ability to prey on reef fish, to dramatically change coral-reef communities.

Environmental conditions are critical to community membership

As you might expect, all species in a regional species pool do not end up becoming members of every community in that region. A species may make it to a community but fail to become a member because of unsuitable environmental conditions there. You might think of the influence of the environment as an "abiotic filter" that restricts species that are physiologically incapable of surviving in a given community. The lakes in the Mount St. Helens area, for example, have environmental attributes that accommodate fish, amphibians, and aquatic insects but not trees.

The physical environment can be a significant barrier to introduced non-native species. Consider, for example, marine organisms that are often introduced into foreign ports via ballast water (seawater that is pumped into and out of ballast tanks to stabilize large cargo-carrying ships). Ballast water picked up outside one port and then dumped in another can potentially release organisms (from bacteria to planktonic larvae to fish) that can colonize nearshore communities. While most ballast-water organisms do not survive the thermal, salinity, or light regimes of their new environment, a small percentage are not physiological constrained by such conditions and may gain a foothold, especially if they are introduced multiple times.

Community membership can shift with changing environmental conditions. For example, growing evidence indicates that climate change—in particular, rising temperatures—may improve conditions for some non-native species, increasing their ability to thrive.

focus: key figure

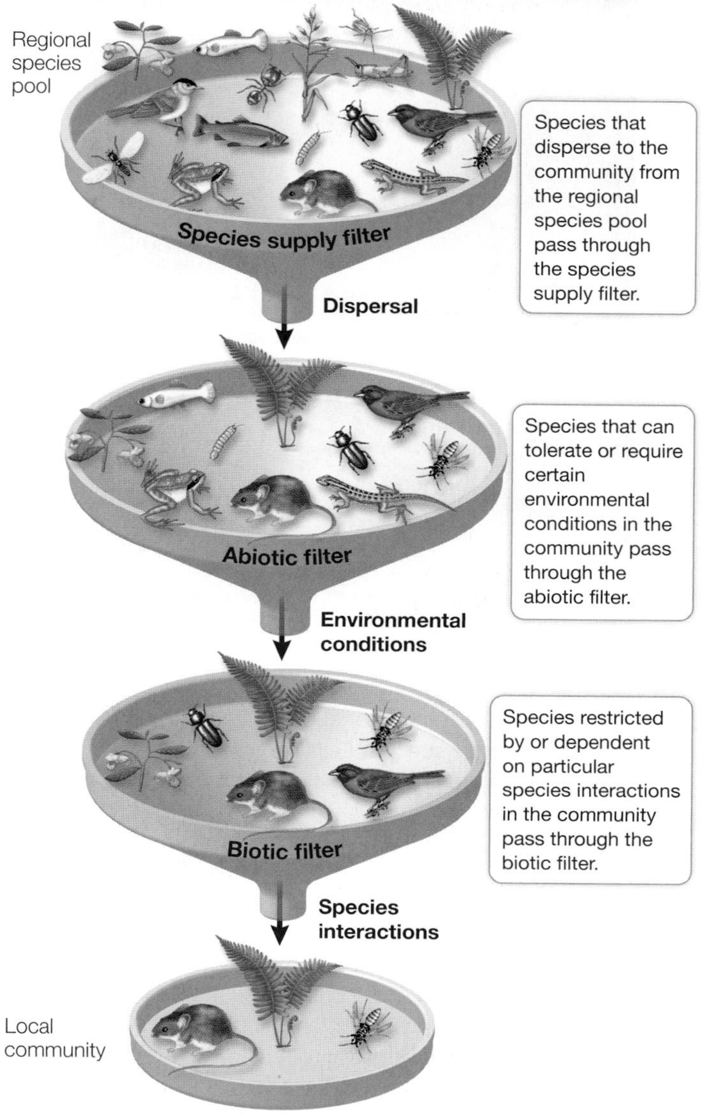

Figure 56.6 Filters Determine Community Membership Species members are included or excluded in a local community by passing through a set of "filters." Species are lost at each filter, so local communities contain a fraction of the species in the regional pool. In practice, all the filters work at the same time, rather than in series as the figure suggests.

Q: Would it make sense for the fish and frog species in the regional species pool to be present in the local community shown in the figure? Explain.

Jay Stachowicz and his colleagues found that the recruitment and growth of invasive marine invertebrates known as ascidians (sea squirts) in New England were dependent on warm winter and summer water temperatures (**Figure 56.7A**). Warmer temperatures gave the non-native organisms an earlier start in spring and increased the magnitude of their growth relative to that of native sea squirts. A similar story can be told for the European green crab (*Carcinus maenas*), whose recruitment to estuarine communities along the Pacific Northwest coast is dependent on warm water temperatures during El Niño conditions (**Figure 56.7B**).

(A) *Botryllus schlosseri*

(B) *Carcinus maenas*

Figure 56.7 Species Invasions Can Depend on Environmental Conditions **(A)** Invasive sea squirts in New England depend on warmer water temperatures to grow and outcompete native sea squirts. **(B)** The recruitment of European green crabs to Pacific Northwest estuaries depends on warm water conditions such as those during El Niño.

> **56.2 recap**

learning outcomes

You should be able to:

- Name the three factors that determine community membership, and evaluate their relative importance in given situations.
- Explain and illustrate with examples how environmental factors can act as a filter or barrier affecting community structure.
- Explain why studying introduced non-native species provides a unique test of community membership.

1. Refer to Figure 56.6. Which factor is most limiting to the species membership of the local community? Explain.
2. If you were trying to reduce the number of non-native species colonizing Mount St. Helens, what one best general management action would you employ?

The final requirement for species membership is the ability to coexist with other species. Let's consider the mediating effects of species interactions on the inclusion (or exclusion) of species in communities.

Resident species can restrict or promote community membership

Even if a species makes it to a community and is able to tolerate its environmental conditions, it must be able to live with all the other species in that community. It must be able to obtain food, survive without being eaten, and cope with competitors. Thus whether a species is successful in a new community will depend on the presence of resident species that either promote or restrict species membership.

The ability of resident species to exclude or slow the population growth of non-native species is called **biotic resistance**. While there are a handful of studies showing that competitors and herbivores can slow down the spread of non-native plant species, confirmed examples of biotic resistance are few. Identifying whether biotic resistance is common in nature may reflect the fact that many failed introductions of non-native species go undetected. Nevertheless, the evidence from the invasive species literature suggests that the influence of resident species may be a fairly porous filter when it comes to restricting species from, or promoting species to, communities.

> **56.2 recap**

Community membership is a consequence of regional species pools, environmental conditions, and species interactions. These three factors act as "filters" that control community membership. Biotic resistance—the process by which species interactions exclude or slow the population growth of colonizing species—is rarely documented.

Once a species is part of a community, how is it able to coexist with other species there? In the next section we will consider theories and examples that seek to explain how species interactions help form communities of varying composition and diversity.

> **key concept**
>
> # 56.3
>
> # Communities Are Complex Networks of Species Interactions That Vary in Strength and Direction

So far we have presented a rather static view of communities as groups of species residing in the same place at the same time. But communities are actually complex networks of species with connections that vary in their strength (i.e., magnitude) and direction (i.e., positive, negative, or neutral). Every species in a community experiences multiple direct and indirect interactions.

focus your learning

- Indirect interactions involving three or more species can affect species diversity and abundance in communities.
- Keystone, foundation, and ecosystem engineering species have major effects on community structure.
- The role of chance in maintaining community structure is poorly understood, but lottery, or neutral, models may help explain it.

Indirect interactions are important to community structure

A **direct interaction** is one that occurs between just two species. **Indirect interactions** occur when the direct relationships between two

(A) Direct interaction

A → B

This solid arrow represents a direct interaction.

(B) Indirect interaction

A → B → C

An indirect interaction between species A and C results when species B interacts directly with both species A and species C.

Figure 56.8 Direct and Indirect Species Interactions   (A) A direct interaction occurs between two species. **(B)** An indirect interaction (dashed arrow) occurs when the direct relationship between two species is mediated by a third species.

species are mediated by a third (or more) species (**Figure 56.8**). The addition of a third species to a two-species interaction has the possibility of dramatically changing the outcome of the original interaction.

Indirect interactions were observed by Charles Darwin in *On the Origin of Species* (1859). Having described the role of bees in flower pollination and seed production of plants in the region of England where he lived, Darwin hypothesized that the number of bees in the region was dependent on the number of field mice that prey on their combs and nests. Noting that mice, in turn, are eaten by cats, Darwin wrote, "Hence it is quite credible that the presence of a feline animal in large numbers in a district might determine, through the intervention first of mice and then of bees, the frequency of certain flowers in that district!"

Today the term "trophic cascade" would be applied to Darwin's observations. A **trophic cascade** occurs when the rate of consumption at one trophic level results in a change in species abundance or composition at lower trophic levels. For example, a carnivore eats an herbivore (a direct interaction) and decreases its abundance, causing an indirect positive effect on the primary producer eaten by the herbivore. One example of a trophic cascade is the indirect regulation of U.S. Rocky Mountain meadow and forest communities by wolves, which were reintroduced into Lamar Valley in Yellowstone National Park in 1995. William Ripple and his colleagues at Oregon State University have studied this system extensively.

The simplified food web in Yellowstone National Park depicted in Figure 56.3 shows that gray wolves in the park feed on elk, bison, and coyotes. Although they share some of these prey species with coyotes and grizzly bears, wolves exert particularly strong effects on the structure and dynamics of the community. We know this mostly because of how differently the community was structured in their absence, after wolves were hunted to extinction in Yellowstone in 1926. In the absence of wolves, the park service culled elk herds to prevent their populations from exploding. In 1968, in response to public pressure, the selective killing was stopped and the elk population rapidly increased (**Figure 56.9A**). The elk browsed aspen trees so intensely that the number of young trees recruited (added to the population) declined precipitously; once elk culling stopped, no new trees were recruited at all (**Figure 56.9B**). The elk also severely browsed streamside willows, with the result that beavers, which depend on willows for food, were mostly gone from Lamar Valley. In regions of the park where elk were absent, however, aspen and willow trees flourished. This observation suggested that the decline of the trees in Lamar Valley was indeed due to elk browsing rather than to climate conditions or other factors.

(A)

In the absence of wolves and culling, elk populations grew rapidly.

Number of elk (thousand) vs *Year*

Wolves eliminated (1926)   Elk culling suspended (1968)   Wolves restored (1995)

(B)

Wolves present │ Wolves absent

In the absence of wolves, browsing by elk prevented the recruitment of young aspens.

After wolves were restored, aspen forests began to regenerate.

Aspens recruited (%) vs *Aspen origination decade*

1900–1909 1910–1919 1920–1929 1930–1939 1940–1949 1950–1959 1960–1969 1970–1979 1980–1989 1990–1999

Wolves eliminated (1926)   Wolves restored (1995)

Figure 56.9 Wolves Initiated a Trophic Cascade in Yellowstone National Park   (A) Number of elk in Wyoming's Yellowstone National Park. **(B)** Aspen recruitment (new trees) in the presence and absence of wolves.

 Media Clip 56.2 **How Wolves Change Rivers** www.Life11e.com/mc56.2

In 1995, after wolves had been absent for 70 years, park managers reintroduced them to Yellowstone, and their population grew rapidly. The wolves preyed primarily on elk. The elk population of Lamar Valley dropped, and elk avoided the aspen groves, where they were especially vulnerable to wolf predation. Young aspen began to grow, willows regrew along streams, and the number of beaver colonies increased from one in 1996 to seven in 2003. Thus the presence or absence of a single predator influenced not only populations of its prey but also, indirectly, its prey's food resource and other species that depended on that resource. It is clear that wolves have a dramatic effect on the structure of meadow and forest communities through the trophic cascade they create.

Strongly interacting species often regulate community structure

The example of wolves in Yellowstone National Park offers clear evidence that species interactions vary considerably in their strength and direction, influencing the structure of communities. Some species have strong negative or positive effects on community structure, while others may have almost no effect at all. We can measure the **interaction strength**, or the effect of one species on the abundance of another species, by experimentally removing the "interactor" species and observing the response of the "target" species. If removing the interactor species causes a sharp decline in the population of the target species, we know that the interaction is a strong positive one. Alternatively, the interaction is strongly negative if the removal results in a large increase in the target species. If the removal has little or no effect on the abundance of the target species, then there is no interaction.

Some species can have strong negative effects on a community by excluding or diminishing populations of other species through competition for limiting resources. As discussed in Key Concept 55.3, competition is often asymmetrical: one species can be more negatively affected by the interaction than another. Asymmetrical competition leads either to competitive exclusion of the more negatively affected species (resulting in fewer species in the community than would be predicted by environmental conditions or dispersal alone) or to various mechanisms that promote coexistence, such as *resource partitioning.

---

* **connect the concept** As you saw with the bee–shin dagger agave example, resource partitioning, or the sharing of limiting resources by using them in different ways, is one way that species may reduce intense competition and coexist. See Key Concept 55.3.

Competition among species can also be mediated by factors that affect the dominant species' ability to acquire resources, so-called **resource-mediated coexistence**. For example, if events in the physical environment such as disturbance or stress affect the growth or survival of a dominant species, competitively inferior species may gain access to limiting resources. We will consider some specific examples of disturbance and stress in Key Concept 55.4. Here we consider how those effects can influence community structure.

The **intermediate disturbance hypothesis** describes how varying degrees of disturbance (an abiotic process that injures or kills individuals) affect species diversity in communities (**Figure 56.10**).

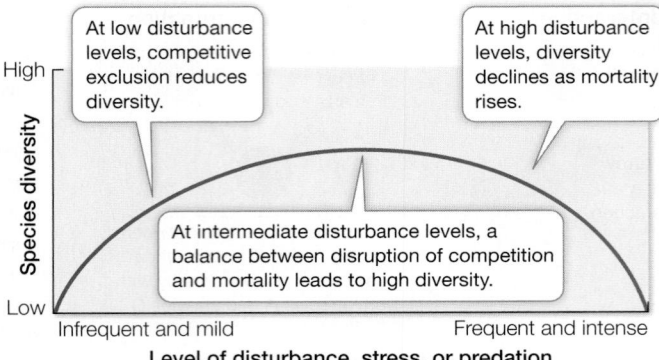

Figure 56.10 The Intermediate Disturbance Hypothesis Species diversity is expected to be greatest at intermediate levels of disturbance (an abiotic process that injures or kills individuals).

Joseph Connell, the author of the classic work on barnacle competition (see Key Concept 55.3), noted that the level of disturbance (both frequency and intensity) experienced by a community could have strong effects on its species diversity. He hypothesized that species diversity would be greatest in communities with intermediate disturbance and lowest in communities with low or high disturbance. At low levels of disturbance, competition from dominant species would lead to competitive exclusion of the inferior species, resulting in lower species diversity in the community. At the other extreme, a major disturbance would cause high mortality rates, leading to the extinction of some species in the community and resulting in lower species diversity. At intermediate levels of disturbance, species diversity would be greatest because (1) the influence of the dominant species is reduced by the disturbance, allowing the subordinate species to coexist, and (2) the mortality rates are not as high, so extinction events are less likely to occur.

Three other types of species that have large effects on community structure are considered below.

KEYSTONE SPECIES   As described in Key Concept 55.2, **keystone species** can have strong community-wide effects, not because of their size or abundance but because of the important role they play in communities (**Figure 56.11**). Keystone species mostly act by creating trophic cascades. For example, top predators such as wolves and sea stars are keystone species because they have a large influence over community species richness, abundance, and composition, even though keystone species may be small in size and their abundance is relatively low.

Sea otters (*Enhydra lutris*) that inhabit kelp forests on the west coast of North America illustrate how important keystone species can be in regulating community structure. Sea otters feed on sea urchins (*Strongylocentrotus* spp.), which feed on kelp, establishing a trophic cascade. Without predation by sea otters, sea urchin populations can explode, resulting in voracious consumption of kelp. Stretches of nearshore habitat that become nearly devoid of kelp beds, called "urchin barrens," can no longer support the diversity of life that is supported when kelp habitat is present.

FOUNDATION SPECIES   Some strongly interacting species are capable of providing habitat and food for other species. These **foundation species** have a large effect on communities as a consequence of their large size or great abundance (see Figure 56.11).

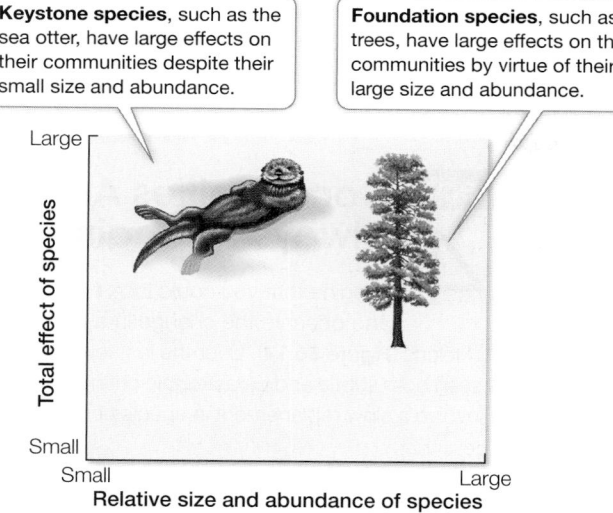

Keystone species, such as the sea otter, have large effects on their communities despite their small size and abundance.

Foundation species, such as trees, have large effects on their communities by virtue of their large size and abundance.

Figure 56.11 Keystone versus Foundation Species Species that have large effects on their communities do so by virtue of their relative size and abundance.

Trees are a good example of foundation species. For example, fig trees in tropical forests provide physical habitat for several thousand species of insects, non-avian reptiles, birds, and rodents in their canopy. Fig trees also serve as an important food source for species by producing fruits several times every year, especially at times when few, if any, other trees are fruiting. Dozens of frugivores depend on figs when no other fruits are present. Fig-eating animals include fruit bats, parrots, toucans, pigeons, flycatchers, trogons, orioles, rodents, howler monkeys, and even fish, which eat figs that fall into nearby streams. All of these animals represent prey for a diverse community of predators.

ECOSYSTEM ENGINEERING SPECIES Some species are specifically known for creating, modifying, or maintaining physical habitat for themselves and other species. These **ecosystem engineering species** can be either keystone or foundation species. These are species whose primary effect is to engineer the physical environment (rather than providing food), and they do not necessarily have great size or abundance in the community. For example, beavers create new habitats by cutting down (and killing) trees and using them to dam streams and create ponds and wetlands that provide habitat for species that would otherwise be unable to live in the area (**Figure 56.12**).

Species with similar effects on one another may coexist by chance

We have thus far discussed instances in which communities are composed of a few strongly interacting species and many weakly interacting species. But sometimes species in a community have relatively similar effects on one another. How is species diversity maintained when species vary little in their ability to obtain resources?

Over the last few decades, ecologists have proposed and tested so-called **lottery**, or **neutral**, **models** that emphasize the role of chance in the maintenance of species diversity. Like the name suggests, these models assume that when resources are made available, they are used at random by individuals of different species who happen to be in the "right place at the right time." As long as all individuals have similar chances of obtaining resources (or "winning the lottery"), and no clear advantage in population growth, then their presence in the community should be maintained by chance events that free up resources for individuals competing for those resources. If there is a large disparity among species in their ability to obtain resources, the dominant competitor will always have a greater chance of obtaining resources and will eventually monopolize them. So for lottery models to work, the element of equal chance for all individuals to obtain resources is paramount to coexistence.

The role of chance in maintaining species diversity, especially in unpredictable environments, has intuitive appeal. As long as individuals of species win the lottery every once in a while, they will continue to reproduce and grow their population. Lottery models have most often proven useful when applied to highly diverse communities such as coral-reef fishes and tropical rainforest trees, where hundreds of species overlap in their resource requirements (**Figure 56.13**).

Ecologists are pursuing various and sometimes conflicting hypotheses to explain why certain species coexist in space and time. It is unlikely

Minnesota

In 1940, beavers were nearly extinct in this region, and there were few wetlands (red).

By 1986, beavers had recolonized the region, and wetland area (red) had increased 13-fold.

1940   1961   1986

Figure 56.12 Beavers Are Ecosystem Engineers By damming streams, beavers created networks of different types of wetlands (shown in red) in a 45-square-kilometer watershed on Minnesota's Kabetogama Peninsula, thus increasing species diversity in the region.

Figure 56.13 Lottery Models Emphasize the Role of Chance
Highly diverse tropical rainforests have hundreds of tree species that overlap in their resource requirements. As long as all individuals have similar chances of obtaining those resources (or "winning the lottery"), and no clear advantage in population growth, then their presence in the community should be maintained by chance events.

that any one theory can provide a wholly satisfactory explanation of the mechanisms that drive structure in all communities.

56.3 recap

Communities are complex networks of direct and indirect species interactions that vary in strength and direction. Strongly interacting species such as keystone species, foundation species, and ecosystem engineering species often regulate community structure through their interactions with other species. In communities where species interactions are more equivalent, chance plays a greater role in determining community structure.

learning outcomes

You should be able to:

- Describe, analyze, and give examples of the importance of indirect interactions in forming and maintaining community structure.

- Explain the term "trophic cascade" as it relates to indirect interactions in communities.

- Define and compare keystone and foundation species, giving examples of each.

- Define a lottery, or neutral, model in community ecology, and describe its usefulness in explaining the chance occurrence of species in a community.

1. Suppose wolves were once again excluded from Yellowstone National Park. What direct and indirect interactions would change, and how would this affect aspen forests?

2. Make an argument for why beavers could be considered ecosystem engineering species and keystone species but not foundation species.

3. Using the lottery model as your guide, explain how species that depend on the same set of limiting resources can coexist.

When Mount St. Helens erupted in 1980, it was clear that the surrounding ecological communities would change forever. Change is a regular feature of all communities, but not all changes are as dramatic as those involving a volcanic eruption. In the next section we will consider how both subtle and catastrophic factors change communities over time.

key concept 56.4 Communities Are Always Changing

Imagine that you could look back in time and observe the change in a typical forest on Mount St. Helens (**Figure 56.14**). Over the last few decades, you would have seen both subtle and catastrophic changes. Subtle changes would involve a slow replacement in species composition

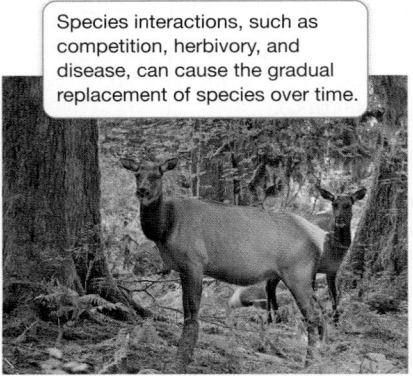

Species interactions, such as competition, herbivory, and disease, can cause the gradual replacement of species over time.

Changes in abiotic conditions, such as temperature or rainfall, can cause physiological stress, and eventually mortality.

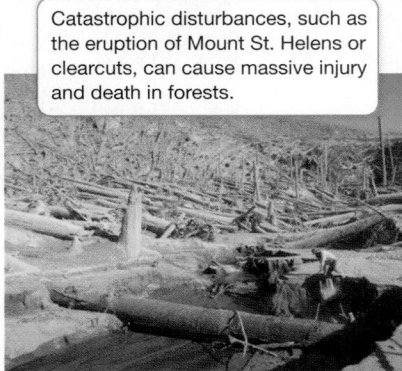

Catastrophic disturbances, such as the eruption of Mount St. Helens or clearcuts, can cause massive injury and death in forests.

Figure 56.14 Changes on Mount St. Helens Forest communities on Mount St. Helens have experienced significant changes over hundreds of years. The types of change have been both subtle and catastrophic.

as a consequence of competition, predation, and facilitation. More catastrophic changes would include volcanic eruptions and defor-estation, resulting in the destruction of some species and replace-ment with others over time. The community at any one time is the result of these various changes both natural and human-caused.

focus your learning

- Community succession is caused by factors that induce change, including disturbance and/or stress.
- Communities recover from a disturbance through the pro-cess of succession, which results in a climax community.
- Primary succession occurs after a catastrophic distur-bance whereas secondary succession occurs after a disturbance when most, but not all, organisms have been destroyed.
- Successional changes in a community are not necessarily predictable or repeatable, and may result in alternative states.

Change in communities can be caused by abiotic and biotic factors

How a community changes in species composition over time is known as **succession**. Succession is the result of both abiotic (physical and chemical) and biotic factors.

A **disturbance** is an abiotic event that physically or chemically injures or even kills some individuals, creating opportunities for other individuals to grow and/or reproduce. Some ecologists also consider biotic events, such as elephants trampling vegetation or beavers felling trees, to be disturbances. The magnitude of a disturbance can vary enormously. Many disturbances are limited to small areas—for example, a log carried by waves may crush algae and animals on a rocky shore. In contrast, hurricanes, forest fires, and volcanic eruptions can affect communities over hundreds or thousands of hectares. Although small-scale disturbances are far more frequent, the occasional large-scale event is often responsible for most of the changes in a community. For example, a single volcanic eruption on Mount St. Helens felled many more trees than multiple storm events ever did, even over decades.

Stress, in contrast to disturbance, occurs when some abiotic factor reduces the growth, reproduction, and/or the survival of some individuals. A common stress on the mountain communities of Mount St. Helens is the effect of temperature on the growth and reproduction of plants. As temperatures decrease with increasing elevation on the mountain, you might expect plants at higher alti-tudes to experience more stress than those at lower altitudes. But as illustrated in Figure 55.14, the stress that alpine plants experience at higher elevations can be tempered by the presence of neighboring plants. That is, some types of interactions can mitigate the effects of stress.

Biotic factors can influence community change as well. On Mount St. Helens, change might be driven by competition between different tree species, with the most dominant tree species surviving over time (see Figure 56.14). Herbivory by elk is another example of a species interaction that can initiate change. By preferentially brows-ing on certain species such as willow and aspen, elk can shift the trajectory of plant community succession from forest to meadow, as happened in Yellowstone National Park (see Figure 56.9).

Succession is a process of change in communities over time

Traditionally ecologists have viewed succession as a progression of stages in which species come and go until a climax commu-nity arises. The **climax community** is thought to be a stable as-semblage of species that experiences little change until an intense disturbance wipes out the community, sending it back to its initial stages again. As you will see later in this section, whether some communities reach a true stable end point is questionable. Let's consider the mechanics of succession in more detail.

Succession is probably most easily observed after a catastrophic disturbance kills all the organisms in a community, leaving an envi-ronment devoid or nearly devoid of life. This type of change is known as **primary succession**. Glaciers, volcanic activity, and in some cases floods or landslides cause disturbance that initiates primary succession. Primary succession by its very nature can be slow be-cause the early arrivals (known as **pioneer** or **early successional species**) must deal with extreme conditions. Many pioneer spe-cies deal with these conditions by employing certain *life history strategies or species interactions to their favor.

---

*connect the concept   Life history strategies are the lifetime patterns of growth, reproduction, and survival of a species (see Key Concept 54.3). Pioneer species tend to have life history strategies that maximize population growth (r-strategists).

One of the best-known examples of primary succession is that seen in plant communities in the wake of the retreat of glaciers in Glacier Bay, Alaska (**Figure 56.15A**). Captain George Vancouver first recorded the location of glacial ice there in 1794, while explor-ing the west coast of North America. Over the last 200 years, the glaciers have retreated up the bay, scraping the landscape down to bare rock and leaving a series of moraines—gravel deposits dropped where the glacial front was stationary for a number of years. No human observer was present to observe changes over the entire 200-year period, but ecologists have inferred the tempo-ral pattern of succession by studying the vegetation on moraines of different ages (**Figure 56.15B**). The youngest moraines, closest to the current glacial front, are populated with bacteria, fungi, and photosynthetic microorganisms that can support themselves on bare rock. Slightly older moraines farther from the glacial front are home to pioneer communities containing lichens, mosses, willows, and cottonwoods, which break down rocks and, when they die, decompose and contribute to the buildup of soil. Mosses and a few species of shallow-rooted shrubs in the genus *Dryas* eventually become established and contribute to soil building as they die and decompose. Still farther from the glacial front, suc-cessively older moraines have deeper soil layers that support shrubby willows and alder trees. Finally, a century after glacial retreat, a mature Sitka spruce forest dominates, fostering a di-verse array of forest species.

Nitrogen is virtually absent from glacial moraines, so the plants that grow best on recently formed moraines at Glacier Bay are *Dryas* and alders, both of which have nitrogen-fixing bacteria in nodules on their roots (see Figure 35.6B). Nitrogen fixation by these plants improves the soil so that spruce trees can grow (see Figure 56.15B).

(A)

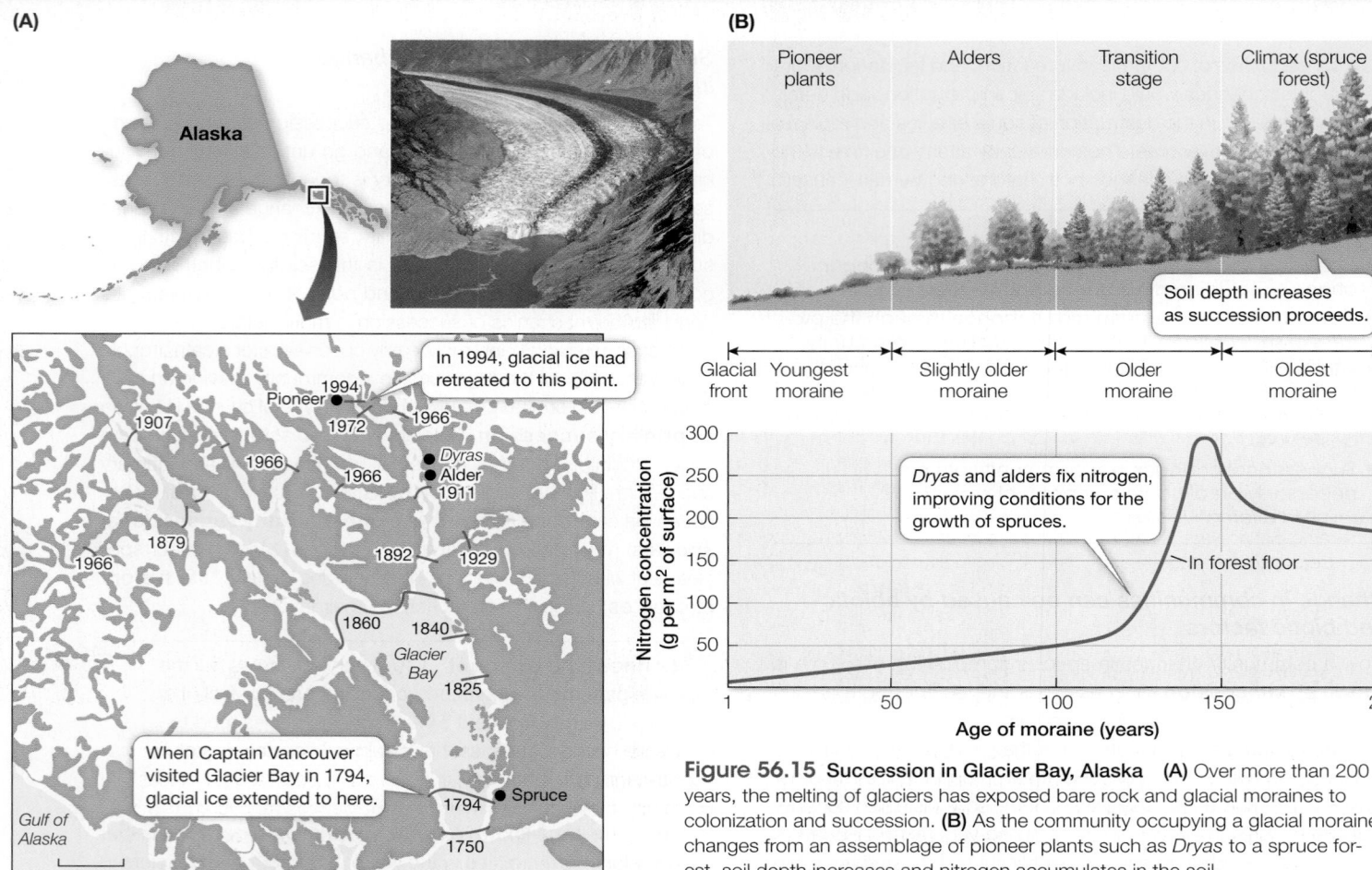

(B)

Figure 56.15 Succession in Glacier Bay, Alaska   **(A)** Over more than 200 years, the melting of glaciers has exposed bare rock and glacial moraines to colonization and succession. **(B)** As the community occupying a glacial moraine changes from an assemblage of pioneer plants such as *Dryas* to a spruce forest, soil depth increases and nitrogen accumulates in the soil.

Q: Based on the locations of the glaciers over time, indicate where the oldest and youngest communities (black dots) are located.

Animation 56.1 **Primary Succession on a Glacial Moraine**
www.Life11e.com/a56.1

Spruces then outcompete and displace the early colonists. If the local climate does not change dramatically, a climax community dominated by spruce trees is likely to persist for many centuries on old moraines at Glacier Bay.

On Mount St. Helens, ecologists have had a unique opportunity to study primary succession from its inception. As described in the opening story, the alpine communities surrounding the lava dome experienced the greatest destruction; they were the first to be hit by an avalanche of mud and hot water and then pelted by hot, sterilizing pumice. This created the so-called Pumice Plain, an area completely lacking life, including any traces of organic matter. The first plant species to arrive, a year after the eruption, was the dwarf lupine (*Lupinus lepidus*). Dwarf lupines, like *Dryas* and alders, utilized the nitrogen produced by bacterial symbionts in root nodules to colonize the sterile environment of the Pumice Plain. They subsequently facilitated other species of plants by trapping seeds and detritus and increasing the overall nutrient content of the soil. By providing safe sites for seedlings to become established, the dwarf lupines greatly increased the rate of primary succession on Mount St. Helens. Because of the early inroads made by dwarf lupine, researchers documented roughly 20 plant species living on the Pumice Plain 20 years after the eruption.

A surprising discovery about primary succession on Mount St. Helens was the important role of animals in directing change. For example, scientists observed that newly formed and isolated ponds and lakes were colonized by amphibians at a faster rate than one might predict given the surrounding harsh conditions. Ecologists discovered that frogs and salamanders were using tunnels created by the northern pocket gopher (*Thomomys talpoides*) as refuges when they made their way from pond to pond across the dry landscape (**Figure 56.16**). Pocket gophers successfully survived the eruption by living in tunnels under the mud. They also benefitted by the expansion of their preferred habitat—grassy meadows—after the eruption. Pocket gophers facilitated plant succession through their burrowing activities, which brought to the surface organic matter, seeds, and fungal spores buried deep under the pumice.

The other type of succession, known as **secondary succession**, involves the reestablishment of a community when most, but not all, organisms have been destroyed. Secondary succession is often initiated by human activities (such as clear cutting) as well as by natural disasters (such as storms and fires). Secondary succession is more common and it progresses more rapidly than primary succession. For example, even though the eruption of Mount St. Helens completely destroyed some areas near the blast zone, there were large areas where organisms survived and secondary succession took place.

experiment

Original Paper: Crisafulli, C. M., J. A. MacMahon and R. R. Parmenter. 2005. Small mammal survival and colonization on the Mount St. Helens Volcano: 1980–2002. Pp. 199–218 in V. H. Dale, F. J. Swanson and C. M. Crisafulli, eds. *Ecological Responses to the Eruption of Mount St. Helens*. New York: Springer.

Charles Crisafulli and his colleagues were some of the first ecologists to study the successional processes on Mount St. Helens after the eruption. They concentrated on the recovery of small mammals in three types of communities that the eruption had disturbed in different ways: a primary successional community named the Pumice Plain, which experienced complete destruction of all living things, and two secondary successional communities: the Blowdown Zone, in which trees fell and the area was covered in mud but some life remained underground, and the Tephra-fall Zone, where the intact forests and meadows were covered in volcanic debris. An undisturbed reference area, 21 kilometers away from the mountain, was used as a control. To compare the initial survival and subsequent recovery of small mammals in primary versus secondary successional communities, the researchers conducted extensive trapping between 1982 and 2000 at several locations on Mount St. Helens and in the reference area.

HYPOTHESIS▶ As a result of the variation in disturbance from the eruption, the primary successional community will have lower mammal species richness than the secondary successional communities, or the reference area, both initially and through time.

METHOD

1. Starting in 1982, several sites were established in the reference area and in three volcanically disturbed areas (Pumice Plain, Blowdown Zone, and Tephra-fall Zone). Trapping sessions were conducted at the sites up to four times a year, from June to October.

2. At each site, live traps baited with rolled oats and peanut butter were set up in multiple plots in a grid pattern. To capture northern pocket gophers, which remain underground, PVC pipe traps were constructed and placed in their burrows.

3. During a trapping session, the traps were checked each morning. Captured animals were identified, uniquely marked, weighed, examined to determine sex and reproductive conditions, and released.

(continued)

Ecologists compared primary and secondary succession on Mount St. Helens by following small mammals in the community. As we mentioned in the chapter opening, different Mount St. Helens communities experienced different levels of disturbance from the eruption. **Investigating Life: Rising from the Ashes** describes research by Charles Crisafulli and colleagues at the USDA Forest Service, showing that the recovery of small mammal populations on Mount St. Helens varied significantly between primary and secondary successional habitats. The Pumice Plain (primary successional habitat

completely devoid of life after the eruption) had the lowest species richness of small mammals, a pattern that persisted for at least 20 years. In contrast, small mammal populations recovered relatively quickly in the Blowdown Zone and the Tephra-fall Zone (secondary successional forest in which some life survived under downed trees and ash fall). In fact, within a few years, the number of small mammals in these two zones was similar to that in the undisturbed reference area 21 kilometers away. This comparison makes clear that the starting conditions characterizing secondary succession are critical in giving communities a "head start" in their recovery.

Both facilitation and inhibition influence succession

The progress of succession greatly depends on the activity of successive colonists, each of which modifies the environment to either facilitate or inhibit the colonization by other species. Facilitation is seen, for example, in the fixation of nitrogen by *Dryas* and alders that allows spruce trees to become established in Glacier Bay (see Figure 56.15B). At later stages of succession, though, competition by spruce trees for nitrogen, light, and water inhibits other tree species from establishing themselves. Likewise, dwarf lupines helped facilitate the establishment of later successional plants but, in turn, were inhibited by multiple insect herbivores, which were able to control the pace of primary succession.

Why do we see this interplay between facilitation and inhibition as succession progresses? In most post-disturbance successional sequences, the early pioneer species are exposed to physically challenging conditions that can be tolerated only with the help of other species. For example, in secondary succession, many of the first organisms to arrive on bare soil after a disturbance are detritivores, which process dead organic matter and release nutrients, thus facilitating the establishment of early colonizing plants. Early

Figure 56.16 Pocket Gophers to the Rescue The burrowing activity of northern pocket gophers, some of which survived the eruption underground, resulted in tunnels that amphibians used to move from one pond to another. It also brought organic matter, seeds, and fungal spores to the soil surface, creating microhabitats, like this one in the Pumice Plain, where plants can grow.

Q: Are pocket gophers keystone species, foundation species, and/or ecosystem engineering species?

experiment

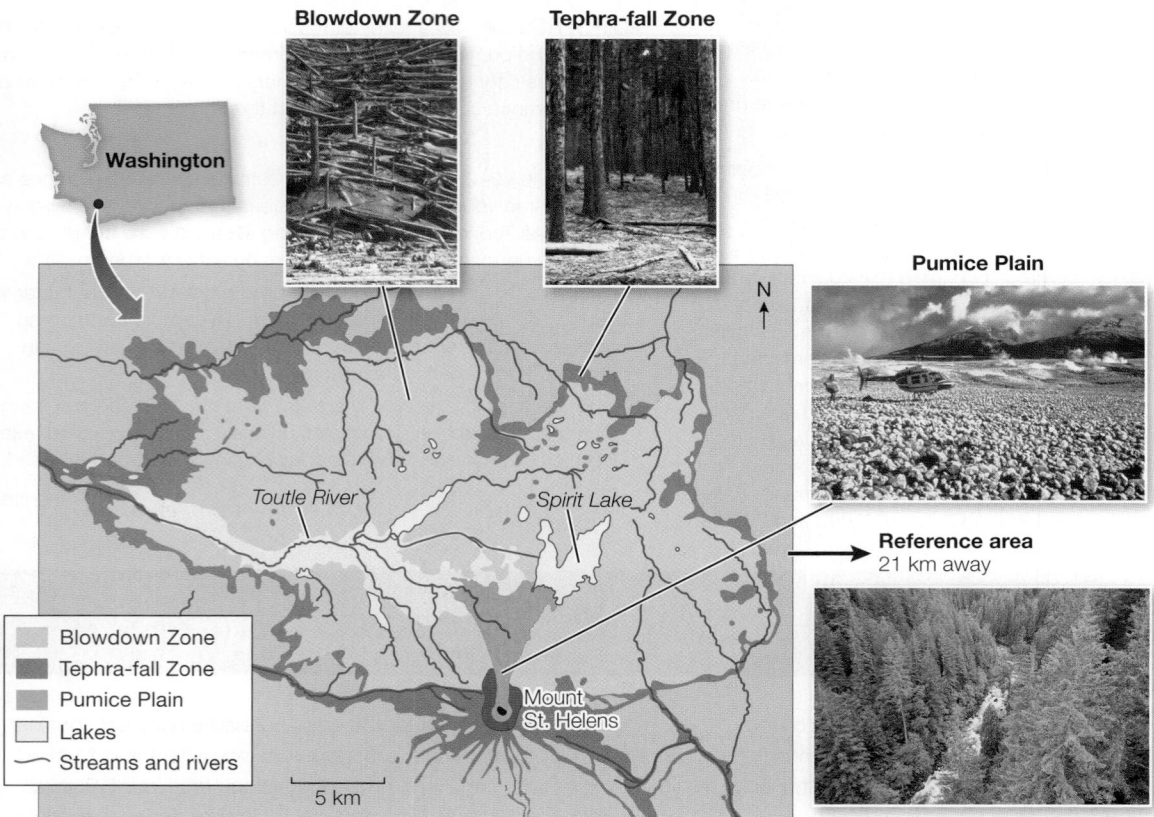

Blowdown Zone   **Tephra-fall Zone**

Washington

Pumice Plain

Toutle River   Spirit Lake

N ↑

Reference area
21 km away

- Blowdown Zone
- Tephra-fall Zone
- Pumice Plain
- Lakes
- Streams and rivers

Mount St. Helens

5 km

RESULTS

Early trapping efforts showed that a total of 17 species of small mammals survived the eruption with most located in the reference area. Over time, small mammal species richness increased in all the communities but proportionally more so in the Blowdown and Tephra-fall Zones (see graph). By 2000, the two secondary successional communities had species richness values roughly equivalent to those in the reference area. The Pumice Plain had only one year (1987) when more than one species of small mammal was present.

CONCLUSION▶   Numerous small mammal species survived and thrived in the post-eruption secondary successional communities of Mount St. Helens. In comparison, the primary successional community showed little recovery of small mammal richness over the 20-year period.

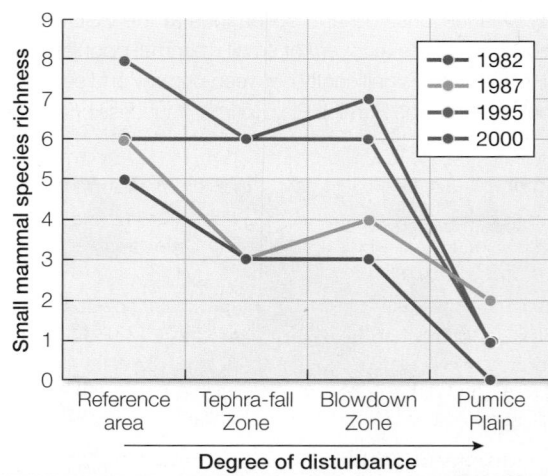

Small mammal species richness (y-axis, 0–9)

Legend:
- 1982
- 1987
- 1995
- 2000

x-axis categories: Reference area, Tephra-fall Zone, Blowdown Zone, Pumice Plain

Degree of disturbance →

successional species that can disperse to these habitats and tolerate and modify the abiotic environment will likely facilitate other species without those capabilities. As succession progresses, species that are less stress-tolerant but are larger, longer-lived, and competitively superior will likely come to dominate as the conditions improve, inhibiting earlier pioneer species. While there is typically a transition from early facilitation to later inhibition in most successional sequences,

as species diversity increases, a variety of both positive and negative interactions naturally arise, further shaping the pattern of succession.

Alternative successional pathways result in variations in community composition

The trajectory of succession in any community is not necessarily a repeatable and predictable process. Succession often occurs

work with the data

In addition to examining small mammal species richness, Crisafulli and his team were interested in the composition of those species and their abundance—data that could give them additional insights into the processes important to the species' recovery (or lack thereof). Below are capture data from 2000 on the small mammal species composition and proportional abundance in the four different post-eruption communities.

| Small mammal species | Proportion of individuals | | | |
	Pumice Plain	Blowdown Zone	Tephra-fall Zone	Reference area
Deer mouse (*Peromyscus maniculatus*)	1.00	0.20	0.25	0.10
Yellow-pine chipmunk (*Tamias amoenus*)	0	0.40	0	0
Cascade golden-mantled ground squirrel (*Spermophilus saturatus*)	0	0.05	0	0
Creeping vole (*Microtus oregoni*)	0	0.10	0	0
Shrew mole (*Neurotrichus gibbsii*)	0	0.05	0	0
Trowbridge's mole (*Sorex trowbridgii*)	0	0.10	0	0.05
Montane shrew (*Sorex monticolus*)	0	0.10	0.15	0.10
Southern red-backed vole (*Clethrionomys gapperi*)	0	0	0.45	0.65
Townsend's chipmunk (*Tamias townsendii*)	0	0	0.05	0
Ermine (*Mustela erminea*)	0	0	0.05	0
Northern flying squirrel (*Glaucomys sabrinus*)	0	0	0.05	0
Vagrant shrew (*Sorex vagrans*)	0	0	0	0.05
Northern water shrew (*Sorex palustris*)	0	0	0	0.05

QUESTIONS ▶

1. Using the data in the table above, calculate the species diversity of each of the post-eruption communities using the Shannon index (see Table 56.1). Which community has the lowest species diversity and which has the highest? How does this pattern compare with that of the species richness values for the four communities?

2. Plot species diversity versus the degree of disturbance from the eruption for the four communities. How well do the data fit the intermediate disturbance hypothesis (see Figure 56.10)? Explain.

3. Consider the presence or absence of particular species in the four communities. Which species is present in all the communities? What does that suggest about its life history?

4. The Blowdown Zone, Tephra-fall Zone, and reference area all have similar species richness (7, 6, and 6 species, respectively), but how do they compare in terms of species composition? Do they appear to be similar? If not, why do you think this is the case?

A similar **work with the data** exercise may be assigned in **LaunchPad**.

could be established after a similar disturbance? Ecologists use the term **alternative states** to refer to the different community assemblages that could develop at the same location under similar environmental conditions.

The theory behind alternative states can be visualized by imagining a landscape full of valleys, each of which represents a different community type or state (**Figure 56.17A**). Imagine, for example, valleys on the Pumice Plain of Mount St. Helens, each of which represents a community that might potentially form there. Impose upon this landscape changes in local conditions—such as the presence or absence of a particularly strongly interacting species—that might influence a community to migrate out of one valley (or state) and into another (**Figure 56.17B**). Intense grazing by herbivores could, for example, influence where dwarf lupine become established on the Pumice Plain. If lupine does not become established in a "new" valley, a different successional trajectory may be established, leading to an alternative state for the community. Now, what if conditions change again, and revert back to those of the initial state? Interestingly, even if original conditions are restored, the community may not revert to the initial state. If the change experienced by the original community was sufficiently large, the successional trajectory of the community can be irrevocably changed (**Figure 56.17C**). For example, removing herbivores from the system might not bring back dwarf lupine. The system is said to show **hysteresis** when the community is unable to shift back to its initial state even when the original conditions are restored.

Communities that have undergone alternative states, or what are sometimes called regime shifts, have become increasingly of interest as it becomes clear that human activities, such as habitat destruction, species introductions, and overharvesting, are shifting communities to alternative states. Once communities have shifted to a new regime it is often unclear whether the original community can be restored or whether hysteresis will

over long periods of time—longer than the life span of any one researcher—making it hard to know whether the changes in species composition are repeatable over time. For example, suppose dwarf lupines had not successfully colonized the Pumice Plain of Mount St. Helens because of intense herbivory. Might later successional stages, such as growth of Douglas fir forests, never develop? Possibly. What is the chance that different, alternative communities

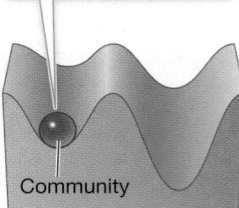

Stability occurs when the community resides within a state (valley).

Community

(A) Stability

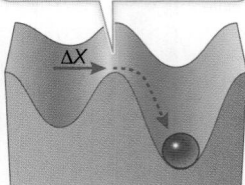

A change in some factor (ΔX) may cause the community to move to another state (valley).

ΔX

(B) Change

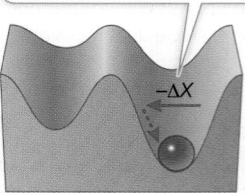

Reversal of the change ($-\Delta X$) may not result in a return to the original conditions if the initial shift was sufficiently large.

$-\Delta X$

(C) Hysteresis

Figure 56.17 Alternative States (A) A community is represented by a ball that moves within a landscape of community states (valleys). **(B)** In this analogy the valleys are shallower or deeper according to the magnitude of change (ΔX) needed to shift the community from one state to another. **(C)** Hysteresis occurs when reversal of the change ($-\Delta X$) does not return the community to its original state.

occur. Reintroducing wolves to Yellowstone National Park did successfully result in the recolonization of aspens as a consequence of elk population control (see Figure 56.9). But will protection of sea otters populations help reestablish kelp forests? Will Atlantic Ocean and Caribbean coral-reef fish communities recover if lionfish are eradicated there (see Chapter 55, Investigating Life: The Lionfishes King)? These are questions whose answers may be found in better understanding the factors that drive alternative states and the role that restoration plays in reversing the effects of those factors.

56.4 recap

Communities constantly undergo change in species composition over time as a result of abiotic and biotic factors, a process known as succession. Abiotic-induced changes to communities come in two forms: disturbance and stress. There are two types of succession: primary succession is the recovery of a community after a catastrophic disturbance kills all the organisms in it, and secondary succession is the reestablishment of a community when most, but not all, organisms have been destroyed. Both facilitation and inhibition influence succession. Communities under similar environmental conditions are sometimes subject to alternative successional pathways, leading to variations in community composition (alternative states).

learning outcomes

You should be able to:

- Distinguish between disturbance and stress, and give examples of each.
- Given specific examples, characterize biotic factors that might influence community structure, and describe their possible effects.
- Describe, with examples, the conditions required for primary succession to occur, including the time intervals involved.
- Describe, with examples, the concept of alternative states, or regime shifts, in a community, including how such states might arise.

1. What is the difference between disturbance and stress?
2. What are some biotic factors that induce change in communities?
3. True or false? "Primary succession is controlled only by facilitation." Explain your answer.
4. How does an alternative state, or regime shift, occur? Give an example from the chapter of such a shift.

We have focused thus far on the dynamic nature of community structure and on the factors on which community membership depends. In the next section we will consider whether the successful functioning of communities depends on the number of species they contain. In other words, how does species richness affect community productivity and stability?

▶ key concept 56.5 Relationships between Species Diversity and Community Function Are Often Positive

As more and more species have either become extinct or are threatened with extinction, ecologists are increasingly interested in understanding what a loss in species diversity means for communities and the way in which they function. Let's next consider research designed to understand how species diversity affects communities as well as agricultural crops.

focus your learning

- Community function can be measured by stability, which is associated with species richness.
- Polycultures are used to test the relationship between diversity, productivity, and stability.

Community function describes the way a community works and can be measured using metrics such as plant productivity, soil fertility, or water availability. Another measure of how well a community functions is its resistance to (or recovery from) disturbance, known as its **stability**. Ecologists have hypothesized that communities with more species function better than communities with fewer species. The argument is that no two species in a community uses resources in exactly the same way such that a community with more species will show more complete and efficient resource use. Also, should environmental conditions change, a species-rich community is more likely to contain some species that can persist under the new conditions. Thus a species-rich community should be more stable—that is, less likely to change over time in either productivity or species composition—than a species poor community. Let's examine these assumptions with some experiments.

Species diversity is associated with productivity and stability

To test the hypothesis that species-rich communities are more stable than species-poor communities, David Tilman and his colleagues at the University of Minnesota cleared 120 outdoor plots, in which they planted grasses in mixtures ranging from 1 to 24 grass species. At the end of each growing season, they measured total plant cover (a measure of grass biomass, and thus of net primary production) and the population densities of all the grasses in each plot. Over a period of 11 years, which included a serious drought, the plots with more species were more productive (**Figure 56.18A**), and their productivity was less variable from year to year. These findings were consistent with the hypothesis that species richness promotes productivity and keeps productivity stable. Moreover, in the plots with greater species richness, soil nitrogen was used more efficiently (**Figure 56.18B**). However, the population densities of *individual* species in the plots were not stable over the years (regardless of a plot's species richness) because different species performed better during drought years and wet years. In other words, higher species richness increased the stability of plant production in the plots, but not the stability of their species composition.

Researchers continue to debate whether species diversity is responsible for maintaining stability or is simply correlated with stability. This question is important because many of the alterations that humans have made in the structure of natural communities have reduced their species richness, and many of these human-altered communities—notably agricultural communities—are notoriously unstable.

Diversity, productivity, and stability differ between natural and managed communities

Although ecologists have been debating the relationships among species diversity, productivity, and stability for only a few decades, humans have been experimenting with those relationships, albeit inadvertently, for millennia—since plants were domesticated and agriculture was invented. Since the dawn of agriculture, crops have been susceptible to diseases and insect outbreaks: massive (often sudden) increases in populations of species that destroy or damage crops.

The practice of growing crops as **monocultures**—plantings of a single crop species—is one reason why managed agricultural communities are particularly unstable. Most farmers have little tolerance for the presence of any potential competitors for their crops and actively eliminate weeds (and the herbivore species that live with them) from their fields. Thus a typical agricultural community has very low species diversity. So the answer to the question of whether diversity causes or is merely correlated with stability may be sought in modern farming practices. The predisposition of agricultural communities to play host to outbreaks may well result from human influences on community structure.

For the last 20 years, ecologists have been using traditional subsistence agricultural plots as experimental models for testing the relationships between diversity and stability. Throughout the world, many farmers with small land holdings grow **polycultures** or multiple crops on the same plot. In Costa Rica, farmers often grow corn together with sweet potato. Such corn–sweet potato dicultures contain fewer sweet potato pests and more parasitoid wasps (which feed on those pests) than do sweet potato monocultures. Wasps feed on the corn pollen, and the tall corn plants act as a structural barrier, shade plant, and source of disruptive chemical signals that interfere with the ability of the sweet potato pests to find their host plants.

Recently, such applications of community ecology have been paying dividends. Although monoculture is overwhelmingly the

(A) Plant cover

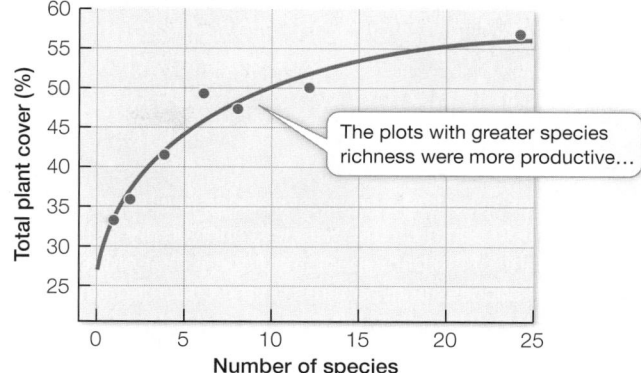

(B) Efficiency of nitrogen use

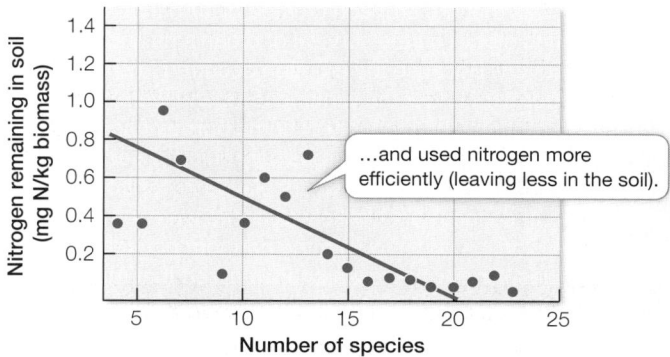

Figure 56.18 Species Richness Enhances Community Productivity David Tilman and his colleagues cultivated a total of 120 grassland plots, containing from 1 to 24 grassland species, for 11 years. **(A)** Total plant cover (a measure of grass biomass, and thus of net primary production). **(B)** The amount of nitrogen remaining in the soil is a measure of resource use efficiency.

dominant agricultural practice, polycultures are under development for agricultural production systems as varied as carp and shrimp farming, vermicomposting (raising worms for compost), and biofuel feedstock production.

56.5 recap

Studies show that the relationships between species diversity and community function (measured as productivity and stability) are often positive. Species diversity enhances productivity because species use resources in slightly different ways; mixtures of species tend to result in a more complete use of limiting resources. Species diversity is hypothesized to enhance stability (a measure of resistance to, or recovery from, disturbance) because the more species a community contains, the more likely some species can persist under variability. These species diversity–community function relationships can be used in agricultural settings to increase crop productivity.

learning outcome

You should be able to:

• Describe how and why growing polycultures can increase agricultural productivity.

1. Why do Costa Rican farmers often grow multiple crops on the same plot, and why might this be a good strategy under drought conditions?

Q&A: How have ecological communities on Mount St. Helens responded to the eruption, and what processes have been important to their recovery?

A similar question was asked by a group of ecologists who convened a week-long field camp on the twentieth anniversary of the eruption in 2000. Many of the participants had spent their entire careers studying the successional patterns on the disturbed landscapes of the mountain. When they left, they agreed to write a book, and 5 years later *Ecological Responses to the 1980 Eruption of Mount St. Helens* was published. The book offers unique insights into the successional processes that communities undergo. One important discovery was that the eruption created disturbances that varied in their effects both on the survival of organisms and their subsequent recovery. Because the eruption occurred in spring, many species were dormant under the winter snows and were able to survive the initial eruption. The recovery of some dormant species such as small mammals has been surprisingly quick. A second important discovery was the role that survivors played in controlling the pace and pattern of succession. Unlikely alliances were formed that hastened succession, as you saw with the northern pocket gopher and amphibians. Finally, the researchers realized that multiple mechanisms were responsible for the pattern of succession on Mount St. Helens. Dispersal, abiotic conditions, and species interactions all played a role in the successional process. The variety of ways in which species responded, and the resulting community assemblages, surprised the ecologists lucky enough to have studied the rise of these communities from the ashes.

Future directions

Despite decades of data and discoveries, research on Mount St. Helens has just begun. Will communities there follow paths of succession that are predictable and repeatable? Or will they form alternative states that are highly dependent on their historical legacies? Unfortunately, succession is a long process spanning many lifetimes of scientists. Steps have been taken to continue the "succession" of research on Mount St. Helens. One of the first steps was to secure Mount St. Helens, in 1982, as a National Volcanic Monument dedicated to research, recreation, and education. This designation, along with coordinated research funding, has set the foundation for long-term study sites and data collection and preservation. In addition, the early "pioneer" scientists have cultivated a strong collaborative culture that brings together young scientists who study different levels of biological organization—from molecules to ecosystems. The collaboration is renewed periodically with "science pulse" camps— week-long events attended by dozens of scientists from around the world who gather benchmark data on a diversity of plants, animals, and microbes. Around the campfire at night, the scientists share observations and recruit the next generation of those who will continue the scientific legacy and storytelling into the future.

Chapter Summary 56

▶ 56.1 Communities Are Groups of Interacting Species Occurring Together in Space and Time

- Communities vary in size and scope, and it is often impractical to define their boundaries. Ecologists often focus on a subset of species in a community based on similarities in taxonomy, resource use (**guilds**), function (**functional groups**), or trophic and energetic connections (**food webs**). Review Figure 56.2

- Food webs can be divided into **trophic levels** based on how groups of species interact and obtain energy. Food webs identify primary producers, **primary consumers** (herbivores), **secondary consumers** (carnivores), and so on. **Omnivores** feed on multiple trophic levels, and **detritivores** feed on the waste products and dead bodies of organisms. Review Figure 56.3, Activity 56.1

- **Species diversity** and composition are important descriptors of **community structure**. The **Shannon index** is a measure of species diversity that combines the number of species (**species richness**) and their relative abundances compared with other species (**species evenness**). Review Figure 56.4, Table 56.1, Activity 56.2

- **Biodiversity** is a term used to span multiple scales of diversity, from genes to species to communities. Review Figure 56.5

▶ 56.2 Community Membership Depends on Species Supply, Environmental Conditions, and Species Interactions.

- The three basic factors responsible for community membership are (1) the regional species pool and dispersal ability (species supply), (2) physical and chemical environmental conditions (abiotic conditions), and (3) species interactions (biotic conditions). These three factors act as "filters" that exclude (or include) species in communities. Review Focus: Key Figure 56.6

▶ 56.3 Communities Are Complex Networks of Species Interactions That Vary in Strength and Direction

- **Direct interactions** occur between two species. **Indirect interactions** occur when the direct relationships between two species are mediated by a third (or more) species. Review Figure 56.8

- A **trophic cascade** occurs when the rate of consumption at one trophic level results in a change in species abundance or composition at lower trophic levels. Review Figure 56.9

- Some species have especially strong negative effects on community structure, resulting in resource partitioning or **resource-mediated coexistence**. The **intermediate disturbance hypothesis** proposes that disturbance promotes species diversity, in part by reducing the negative effects of dominant competitive species. Review Figure 56.10

- **Keystone species** influence community structure disproportionally to their size or abundance, producing trophic cascades. **Foundation species**, by providing food and habitat for other species, have a large effect on communities as a consequence of their large size or great abundance. **Ecosystem engineering species** affect community structure by creating, modifying, or maintaining physical habitat for themselves and other species irrespective of their size or abundance. Review Figures 56.11, 56.12

(continued)

- In communities where species interactions are more equivalent, **lottery**, or **neutral**, **models** propose that as long as all individuals have similar chances of obtaining resources, and no clear advantage in population growth, then their presence in the community should be maintained by chance events that free up resources for competing individuals.

56.4 Communities Are Always Changing

- Communities constantly undergo change in species composition over time as a result of abiotic and biotic factors, a process known as **succession**. Disturbance and stress are two significant forms of abiotic-induced community change. A **disturbance** is an abiotic event that physically or chemically injures or kills individuals, whereas **stress** occurs when some abiotic factor reduces the growth, reproduction, or ultimately survival of individuals. Review Figure 56.14

- There are two types of succession: **primary succession** is the recovery of a community after a catastrophic disturbance kills all the organisms in it, and **secondary succession** is the reestablishment of a community when most, but not all, organisms have been destroyed. Review Figure 56.15, Animation 56.1, Investigating Life: Rising from the Ashes

- Facilitation and inhibition influence primary and secondary succession.

- The outcome of succession is not always predictable; alternative successional pathways, or regime shifts, lead to different community assemblages under similar environmental conditions. **Hysteresis** occurs when a community is unable to shift back to its original state even when the original conditions are restored. Review Figure 56.17

56.5 Relationships between Species Diversity and Community Function Are Often Positive

- Studies show that the relationships between species diversity and **community function** (measured as productivity and stability) are often positive. **Stability** is a measure of resistance to, or recovery from, disturbance. Review Figure 56.18

- Species diversity–community function relationships can be used in agricultural settings to increase crop productivity.

Go to **LearningCurve** (in **LaunchPad**) for dynamic quizzing that helps you solidify your understanding of this chapter. **LearningCurve** adapts to your responses, giving you the practice you need to master each key concept.

Apply What You've Learned

Review

56.3 Indirect interactions involving three or more species can affect species diversity and abundance in communities.

56.3 Keystone, foundation, and ecosystem engineering species have major effects on community structure.

56.4 Communities recover from a disturbance through the process of succession, which results in a climax community.

56.4 Successional changes in a community are not necessarily predictable or repeatable, and may result in alternative states.

56.5 Community function can be measured by stability, which is associated with species richness.

Original Paper: Alvarez-Filip, L., J. P. Carricart-Ganivet, G. Horta-Puga and R. Iglesias-Prieto. 2013. Shifts in coral-assemblage composition do not ensure persistence of reef functionality. *Scientific Reports* 3: 3486.

The calcium carbonate ($CaCO_3$) skeletons of stony corals form the basis for coral-reef community structure. As they build the reef, coral skeletons increase habitats for other reef species. Coral communities consist of coral species with two basic functional attributes. Coral species with high reproductive rates and relatively fast growth are more tolerant of environmental change, but these species tend to be small and do little to increase reef size or number of habitats. Massive corals that contribute the most to reef size and number of habitats are slower growing and are less tolerant of environmental change.

Corals are undergoing serious declines because of disease, pollution, and the effects of temperature and ocean acidification related to climate change. These declines can result in rapid changes in the composition of coral communities. As the pace of climate change escalates, changes in coral communities will likely increase.

Researchers developed simplified models consisting of four coral genera from the Caribbean to study the ability of coral reefs to retain functionality with changing species composition (**Figure A**). They looked at reef rugosity, which is a measure of reef complexity and habitat diversity, and calcification rate, which measures how fast coral skeletons are produced. Rugosity is calculated as a ratio of the contour of the reef relative to its length. A value of 1.00 equals a flat reef; larger values represent higher rugosity.

The table shows reef rugosity and estimated calcification rate for the four coral genera. **Figure B** shows how reef rugosity (dotted lines) and calcification rate (solid lines) would change in coral communities through time in two models: (*i*) steady coral cover decline from 45 to 10 percent (red lines) and (*ii*) steady coral recovery from 10 to 45 percent (blue lines).

Coral genus	Mean reef rugosity	Estimated mean calcification rate (kg $CaCO_3/m^2/yr$)
Acropora	3.33	22.30
Orbicella	1.87	13.80
Porites	1.49	6.12
Agaricia	1.52	2.43

Figure A

Acropora

Agaricia

Porites

Orbicella

Figure B

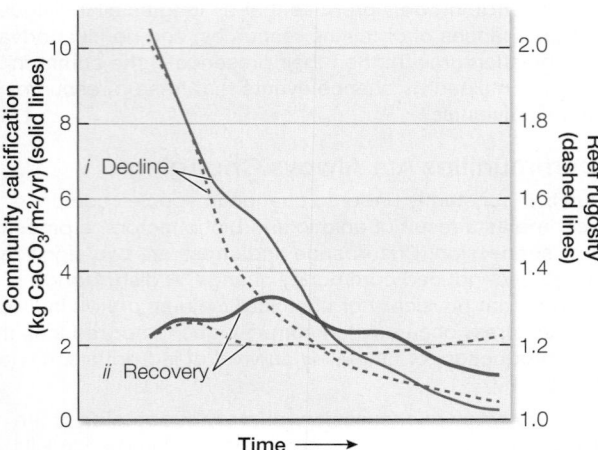

Questions

1. Based on their functions in the reef community, are corals keystone, foundation, or ecosystem engineering species? Explain your answer.

2. According to the table, which coral genera would be most important in building the coral reef and providing habitats for other species? Which genera would be least important? Why?

3. In the graph, the red lines (model *i*) show how reef rugosity and calcification rate would change if coral cover declined from 45 to 10 percent. Which coral genera are likely declining most rapidly, and why? How and why would the loss of these genera be likely to affect reef function and community biodiversity?

4. According to the graph, does model *ii* (blue lines), which shows how reef rugosity and calcification rate would change if coral cover increased from 10 to 45 percent, indicate that the reef is regaining its original functionality? On what do you base this conclusion? What factors might be affecting the reef's ability to regain its functionality?

5. Are the two model coral-reef communities illustrated here undergoing succession? On what do you base your conclusion? Using the concept of alternative states, explain possible differences in coral-reef community structure as the reef recovers, and consider how such differences might affect the reef's function.

Go to **LaunchPad** for the eBook, LearningCurve, animations, activities, flashcards, and additional resources and assignments.

57

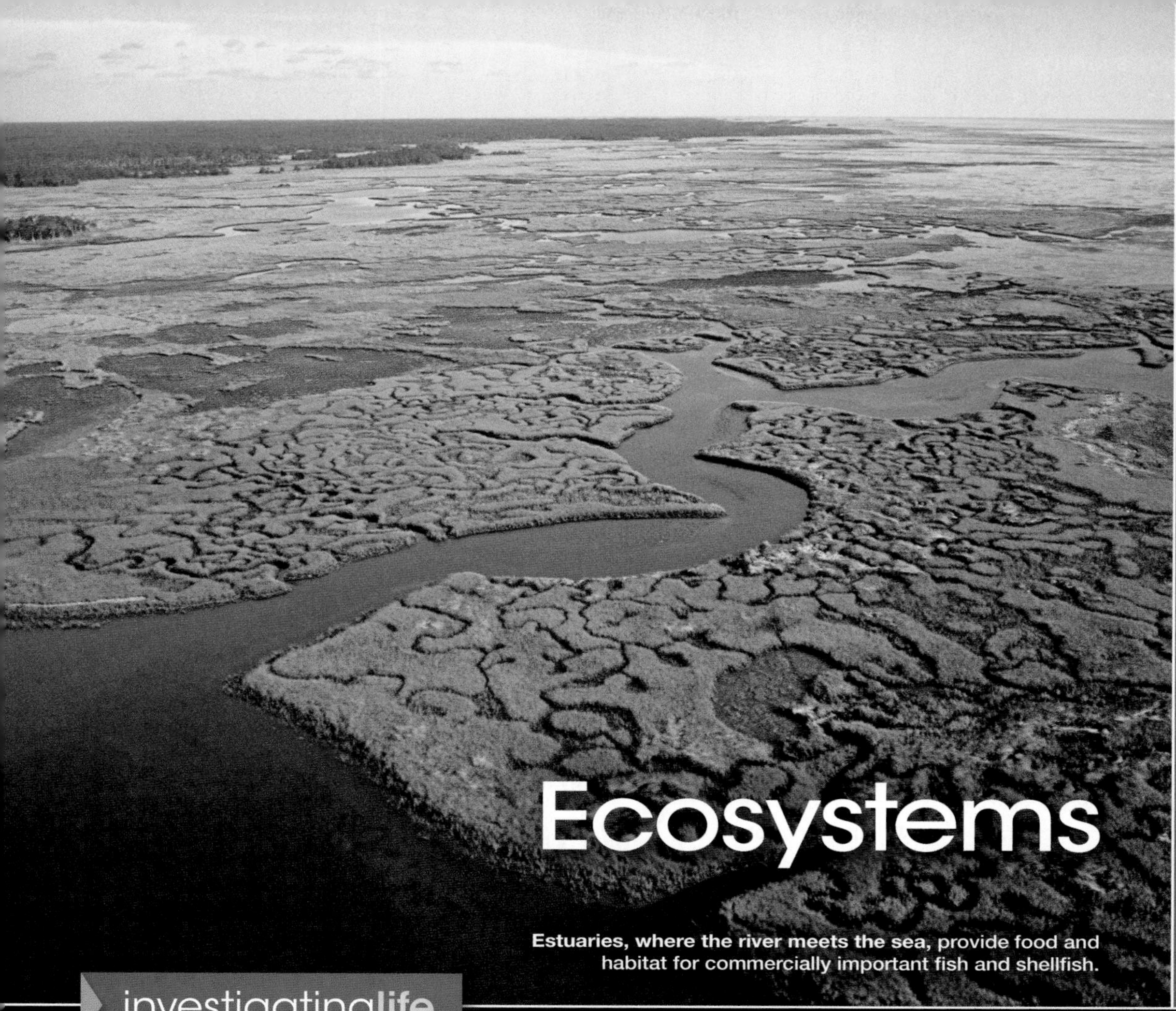

Ecosystems

Estuaries, where the river meets the sea, provide food and habitat for commercially important fish and shellfish.

investigating**life**

Food Webs in an Acidic and Warming Ocean

There is no dispute that the burning of fossil fuels over the last two centuries has increased Earth's atmospheric CO_2 concentrations. Estimates show that oceans are absorbing about 48 percent of all atmospheric CO_2. Marine primary producers use some of the human-caused CO_2 in photosynthesis, but the remainder reacts chemically with seawater, lowering its pH and causing oceans to become more acidic. Ocean acidification can have negative effects on calcifying organisms such as corals, mollusks, and crustaceans, which rely on calcium carbonate for the accretion and maintenance of their external shells. But the negative effects of increasing CO_2 and acidification are not universal. For example, primary producers such as phytoplankton, algae, and sea grasses are known to increase their production under elevated CO_2. Increasing photosynthesis increases O_2 production, which can potentially counteract growing acidity in surrounding seawater. Research is under way on the effects of ocean acidification on single species, but studies on how ocean acidity affects marine ecosystems and food webs are also needed. Moreover, along with the acidification of oceans, water temperatures are rising, potentially creating multiple stressors for organisms.

Estuaries are places where the dual effects of ocean acidification and temperature on primary producers and their food webs may be particularly relevant. Among the most productive ecosystems on Earth, estuaries support a diverse array of marine organisms, including commercially important fish species. Because estuaries are at the interface between land and sea, they have been highly affected by humans through intense overfishing, habitat destruction, and pollution from rivers. Estuaries also experience warm water temperatures, a condition that could intensify with global warming. But estuaries may also be places where the effects of ocean acidification could be mitigated if primary producers such as macroalgae and sea grasses are able to sequester CO_2 and provide more food and habitat for higher trophic levels.

Research on estuaries was conducted by a group of researchers on the western coast of Sweden. The group focused on a food web that includes algae primary producers and invertebrate consumers and exposed this food web to different conditions of CO_2 and temperature. What they found suggests that ocean acidification and temperature affect food webs in complex ways that may be hard to predict.

 How will food webs respond to the multiple effects of ocean acidification and warming in marine ecosystems?

key concept
57.1 Ecosystem Science Considers How Energy and Nutrients Flow through Biotic and Abiotic Environments

In Chapter 56 we looked at communities and the roles that species play in shaping their structure and function. In this chapter we will consider how energy flows and nutrients cycle through communities, a branch of ecology known as **ecosystem science**.

focus your learning

- Ecosystems include all the organisms in a given area and their physical and chemical environment.
- Energy sources in ecosystems result from the processes of photosynthesis, chemosynthesis, and metabolism; energy is transferred via primary and secondary production.
- Autotrophs and heterotrophs obtain required nutrients in different ways.

An **ecosystem** is all the organisms in a given area and the physical and chemical environment in which they live. The term was first coined by A. G. Tansley, a plant ecologist who was interested in how the flow of energy and chemicals was influenced by both the biotic and abiotic components of the environment.

Just like communities, ecosystems occupy a wide range of spatial scales, making them hard to delineate precisely (see Key Concept 56.1). For that reason, ecologists often define an ecosystem using subsets of species and components of the physical and chemical environment based on the questions they are investigating. For example, a soil ecosystem might include only the bacteria that transform nitrogen into forms plants can use. Or an estuarine ecosystem might be confined to the cycling of carbon through a food web of algae and their consumers, as you saw in the opening story.

Energy flowing through ecosystems originates with sunlight and inorganic and organic compounds

The flow of energy through ecosystems comes in a variety of forms (**Focus: Key Figure 57.1**). The most obvious source is the sun, which delivers radiant energy that warms Earth and provides energy for photosynthesis. Autotrophs such as plants, algae, and photosynthetic bacteria capture that radiant energy and fix it into organic compounds that make up the tissues of organisms (see Figure 57.1A). Energy can also come from the oxidation of inorganic nutrients

by archaea and bacteria through the process of **chemosynthesis**, which involves using energy from inorganic compounds to take up CO_2 and produce carbohydrates (see p. 545). Chemosynthesis typically occurs in ecosystems that lack sunlight and have high concentrations of particular inorganic compounds, such as hydrothermal vents, hot springs, and soils. These two forms of energy capture and carbon fixation are both involved in a process known as primary production. Ecologists use the term "primary productivity" to indicate the rate of primary production over time.

A third source of energy in an ecosystem is the product of primary production—the organic compounds synthesized through

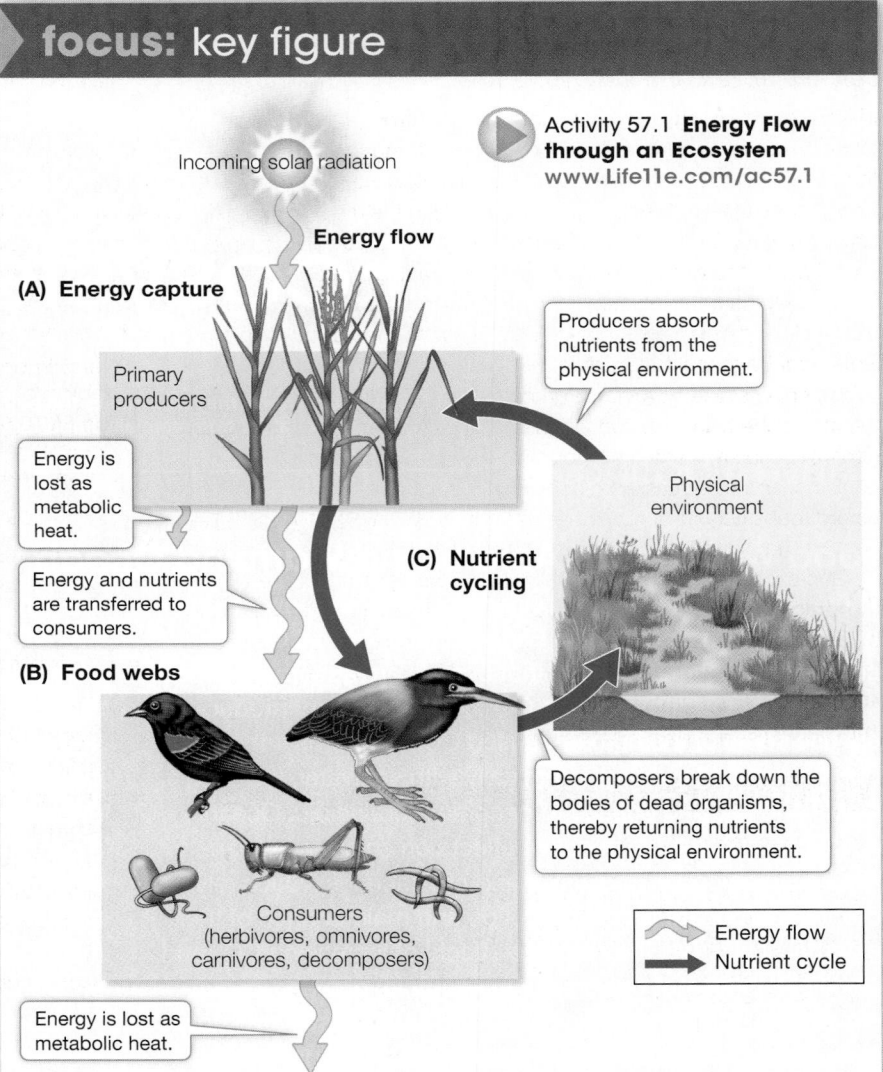

focus: key figure

Activity 57.1 **Energy Flow through an Ecosystem** www.Life11e.com/ac57.1

Figure 57.1 Energy Flows and Chemical Nutrients Cycle through Ecosystems (A) The flow of energy through ecosystems comes mostly from the sun, which delivers radiant energy for photosynthesis by primary producers. (B) Primary producers then provide energy to organisms at higher trophic levels in food webs, but some of that energy is lost as metabolic heat. (C) Nutrients continually cycle between living organisms and the physical components of ecosystems.

Q: Why does energy "flow" but nutrients "cycle" through ecosystems?

photosynthesis and chemosynthesis—that is made available for organisms at higher trophic levels in food webs (see Figure 57.1B). This source of energy, known as **secondary production**, is contained within the tissues of organisms and is used to power the metabolism of heterotrophs, or the consumers within food webs, which include herbivores, carnivores, omnivores, parasites and pathogens, and eventually decomposers. Metabolism involves converting the organic compounds in food into useable energy—primarily ATP—by processes such as glycolysis (see Key Concept 9.1). The main byproducts of metabolism are inorganic compounds, CO_2, and water. The energy generated by metabolism to support the "work" that organisms do in their daily life is eventually lost from the ecosystem in the form of metabolic heat.

Nutrients cycling through ecosystems originate in soil, water, and the atmosphere

In contrast to energy, which is ultimately lost from an ecosystem as heat, nutrients continually cycle between living organisms and the abiotic components of ecosystems (see Figure 57.1C). Organisms require 30–40 different nutrients, including the macronutrients carbon, hydrogen, nitrogen, calcium, phosphorus, potassium, and magnesium, and micronutrients such as iron, manganese, zinc, and copper. All the chemical components in the bodies of organisms ultimately originate from soil, water, or the atmosphere, but organisms acquire these materials in many different ways. Autotrophs such as plants take up certain elements directly from soil, water, and the atmosphere and incorporate them in their tissues. Heterotrophs generally acquire elements by consuming the biomass of other organisms, and then reassemble them via chemical reactions. Some heterotrophs acquire some elements from mutualistic microbes that convert them into forms that are usable by their hosts.

Gas exchange by living organisms returns certain elements to the atmosphere as gases. After organisms die, the materials in their bodies become detritus and are broken down by decomposers into simpler biochemical components, which then can be used by autotrophs once again. Elements not used by autotrophs can accumulate in soil, water, or sediments.

57.1 recap

An ecosystem is all the organisms in a given area and their physical and chemical environment. Autotrophs such as plants, algae, and photosynthetic bacteria capture sunlight and fix it into organic compounds (primary production). Energy is also derived through chemosynthesis: the oxidation of inorganic nutrients by archaea and bacteria. Energy captured through primary production is in turn available for organisms at higher trophic levels in food webs (secondary production). Nutrients continually cycle between living organisms and the abiotic environment.

learning outcomes

You should be able to:

• Compare different ways of defining an ecosystem, and contrast ecosystems with communities.

• List sources of energy in ecosystems, and associate each source with the processes that produce them.

57.1 recap

1. An ecologist measures the number of species in an ecosystem. Is she conducting ecosystem science? Explain.

2. Describe the two ways that energy and carbon are converted into primary production.

For the remainder of this chapter we will consider in more detail how energy flows and nutrients cycle through ecosystems as depicted in Figure 57.1. We will start with primary production, where energy and nutrients are first captured in ecosystems.

key concept 57.2 Energy and Nutrients in Ecosystems Are First Captured by Primary Producers

The overwhelming majority of the primary production on Earth comes from photosynthesis. Plants, algae, and bacteria fix carbon by capturing energy from the sun (or inorganic nutrients, in the case of chemosynthesis) and using that energy to convert carbon dioxide into carbon compounds that can then be used by higher trophic levels.

focus your learning

• Net primary production (NPP) varies by latitude and type of biome.

• The FACE (*free air CO_2 enrichment*) experiment studied the response of plants to elevated atmospheric CO_2 concentrations.

• Primary production in aquatic systems is limited by light (which is limited by depth) and by the availability of different nutrients, depending on the type of system.

Net primary production is the amount of carbon remaining in plants after respiration

Fixed carbon, be it from photosynthesis or chemosynthesis, is the currency used to measure primary production. The total amount of carbon fixed by primary producers in an ecosystem is called **gross primary production** (**GPP**). Not all GPP becomes living plant matter, or biomass, because primary producers use some of that energy for their own respiration and other metabolic processes. Thus **net primary production** (**NPP**), the amount of biomass incorporated into the tissues of primary producers after respiration, is the most common metric used to describe an ecosystem's carbon fixation. This relationship is represented mathematically as

$$NPP = GPP - respiration$$

The carbon not used in respiration is then allocated to such important processes as growth, reproduction, and survival in the face of physical stress, herbivory, or diseases (see Chapter 38).

Patterns of primary production vary with latitude and ecosystem type

Large-scale geographic variation in primary production is one of the most obvious ecological patterns on Earth. Only recently, however, have ecologists been able to reliably quantify patterns of productivity at global scales. Satellite-based optical sensing techniques now

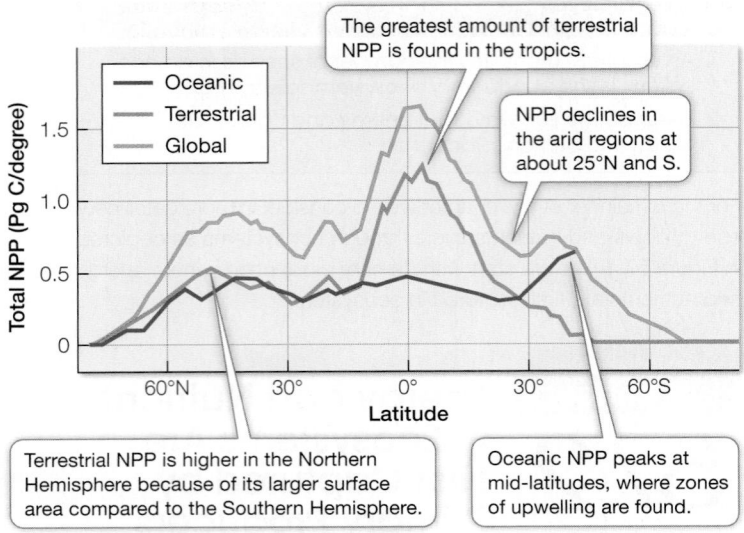

The greatest amount of terrestrial NPP is found in the tropics.

NPP declines in the arid regions at about 25°N and S.

Terrestrial NPP is higher in the Northern Hemisphere because of its larger surface area compared to the Southern Hemisphere.

Oceanic NPP peaks at mid-latitudes, where zones of upwelling are found.

Figure 57.2 Estimates of Total Oceanic, Terrestrial, and Global NPP by Latitude On land, NPP is highest in the tropics and declines at higher latitudes. In the oceans, NPP does not vary much with latitude except at mid-latitudes, especially in the Southern Hemisphere. (1 petagram or Pg = 10^{15} g).

help scientists detect the amount of primary production on land and in the ocean at global scales. How do terrestrial and oceanic ecosystems compare, and how is NPP distributed across the globe?

Estimates of global NPP show that the total amount of carbon taken up by terrestrial ecosystems (52%) is slightly higher than that taken up by the oceans (48%), even though oceans make up a greater percentage (70%) of Earth's surface. This difference arises because terrestrial ecosystems have a much higher per-area NPP (426 g C/m²/year) than do oceans (140 g C/m²/year).

Satellite imaging has also shown that NPP is distributed differently on land than in the ocean. NPP on land varies dramatically with latitude, whereas oceans show much less latitudinal variation (**Figure 57.2**). On land, NPP is highest in the tropics and declines at higher latitudes. Arid regions such as those in deserts (30° latitude) and at the poles (90° latitude) show extremely low NPP (**Figure 57.3A**). NPP rises slightly at mid-latitudes, especially in the Northern Hemisphere, where land area is larger than in the Southern Hemisphere (see Figure 57.2). In the oceans, NPP is highest at mid-latitudes and along coastal zones, especially those that experience upwelling (**Figure 57.3B**). NPP is also higher in the Southern Hemisphere, where ocean area is larger than in the Northern Hemisphere (see Figure 57.2).

Remote sensing has also revealed variation in NPP among different ecosystem types or biomes. Tropical rainforest ecosystems have the second-highest average and net NPP, even though they make up only a little over 3 percent of Earth's surface (**Figure 57.4**). In comparison, the open ocean ecosystems have a low average NPP, despite the large area over which they occur. Some marine ecosystems such as estuaries and algal beds and coral reefs have some of the highest average net primary productivity, but their small geographic areas make their contributions to global NPP modest. It is important to note, however, that the primary producers in these ecosystems contribute substantially to local and regional NPP and thus are critical to marine food webs.

(A)

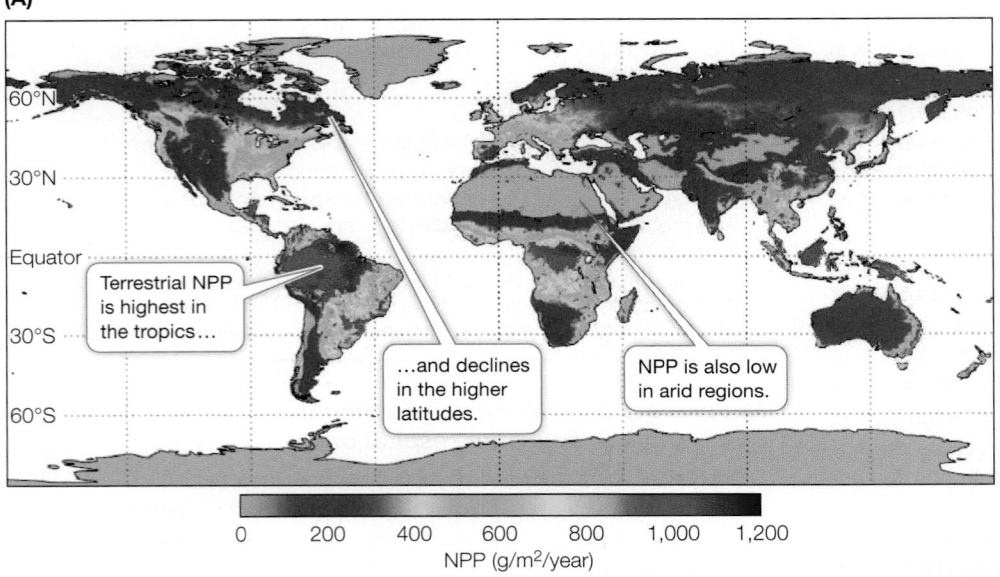

Terrestrial NPP is highest in the tropics...

...and declines in the higher latitudes.

NPP is also low in arid regions.

NPP (g/m²/year)

(B)

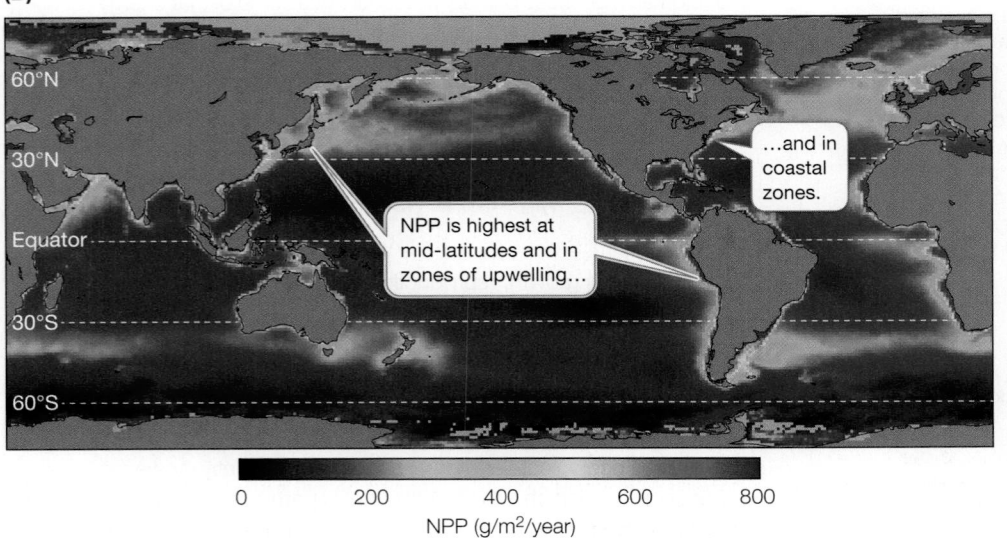

...and in coastal zones.

NPP is highest at mid-latitudes and in zones of upwelling...

NPP (g/m²/year)

Figure 57.3 Geographic Variation in Terrestrial and Oceanic NPP Maps of NPP based on satellite sensor data. **(A)** Terrestrial NPP is highest in the tropics and lowest in arid regions and at the poles. Gray areas represent unvegetated land masses, including deserts and ice caps. **(B)** Oceanic NPP is highest at mid-latitudes and in coastal and upwelling zones.

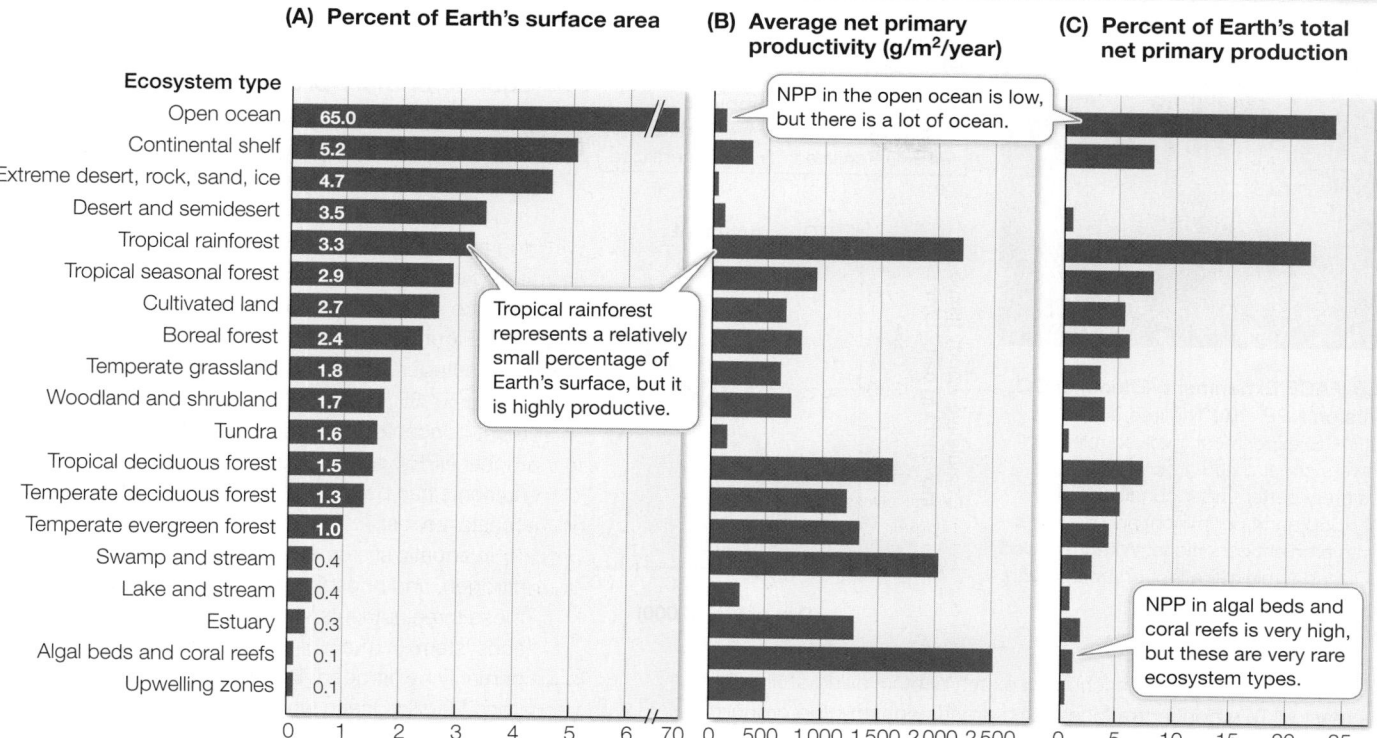

Figure 57.4 Ecosystems Differ in NPP The contribution of different ecosystem types to global NPP can be measured by **(A)** their geographic extent and **(B)** their average net primary productivity. **(C)** Combining these two measures gives the proportional contribution of each ecosystem type to Earth's total NPP.

Not surprisingly, the patterns of primary production that we see on land versus ocean, and among ecosystem types, can be explained by separate factors that control primary production in terrestrial versus aquatic systems. We will now consider these factors in more detail—first on land, and then in oceans and lakes.

Rainfall and temperature largely control terrestrial primary production

The strong latitudinal patterns in terrestrial NPP reflect variation in global climate, particularly rainfall and temperature, across latitudes and seasons. NPP increases as average annual rainfall increases up to a maximum (approximately 240 cm per year), but then declines in most ecosystems (**Figure 57.5A**). The decrease in NPP with high rainfall can occur because of sunlight being reduced by cloud cover over long periods of time or as a consequence of nutrient loss or flooding. NPP also increases with average annual temperature

Figure 57.5 Global Patterns of Terrestrial NPP Correlate with Climate The plots show the relationships between NPP for different terrestrial ecosystems and **(A)** precipitation and **(B)** temperature of those ecosystems worldwide. (1 megagram or Mg = 10^6 g.)

(**Figure 57.5B**). Thus it makes sense that close to the equator, where temperatures and rainfall are high throughout the year, conditions are ideal for primary producers. In deserts, even though temperatures can be high, plant growth is limited by a lack of rainfall. At higher latitudes or elevations, even though moisture is generally available, NPP is low because it is relatively cold much of the year and the growing season is shorter than at lower latitudes or elevations.

In addition to being affected by global climate, terrestrial NPP is also affected by nutrient availability and atmospheric CO_2 concentrations. How these two factors affect primary production is of special interest given the increasing use of artificial fertilizers and rising atmospheric CO_2 concentrations from the burning of fossil fuels (see Key Concept 57.4). A looming question is whether plants, by taking up CO_2, might be able to offset increasing CO_2 emissions. An ambitious experiment exploring this possibility was conducted in 1996.

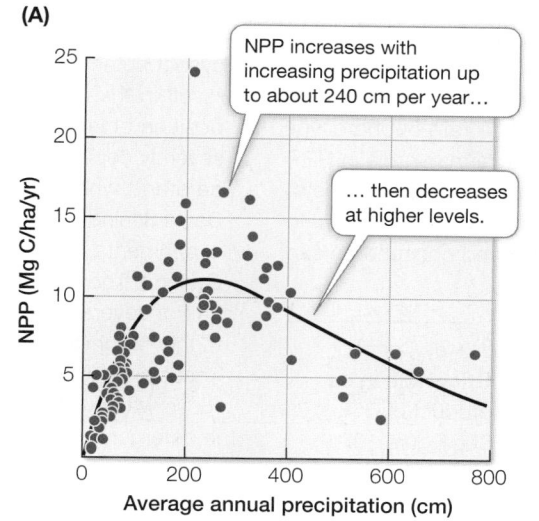

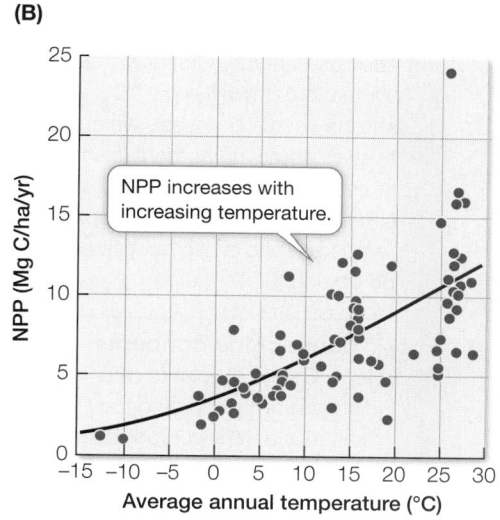

(A)

Figure 57.6 FACE Experiment: Effects of CO$_2$ and Nutrients on NPP **(A)** The free-air CO$_2$ enrichment (FACE) experiment involved releasing CO$_2$ from towers constructed around pine trees. **(B)** The trees grew better under conditions of elevated CO$_2$ (550 parts per million) and nutrients compared with ambient conditions. Without added nutrients, CO$_2$ had a minimal effect on tree growth.

(B)

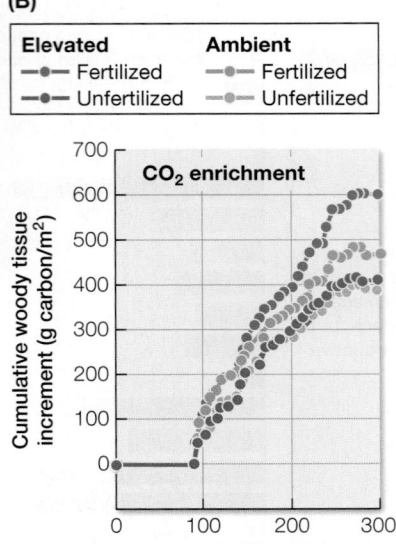

The FACE (*free-air CO$_2$ enrichment*) experiment involved artificially increasing CO$_2$ concentrations within arenas constructed around pine trees in North Carolina (**Figure 57.6A**). To test whether nutrients were a limiting factor, nitrogen was added to half the plots. By 2000, trees showed a slight increase in NPP (measured as woody tissue biomass) in the elevated CO$_2$ plots compared with the ambient plots (**Figure 57.6B**). The increase in NPP, however, was significant only in the fertilized plots, suggesting that nutrients are a limiting factor in the assimilation of CO$_2$ by plants. Since the initial FACE experiment, more than 12 similar large-scale experiments have been conducted in a variety of ecosystems, including grasslands and agricultural crops. The results generally confirm those found in North Carolina, but there are differences. Trees are generally more responsive than herbaceous species to elevated CO$_2$ concentrations, and grain crop yields increase far less than anticipated.

The FACE experiments remind us that terrestrial ecosystems can vary widely in their NPP, as you saw in Figure 57.4. Clearly some of this variability is a simple consequence of global and regional climate promoting the formation of certain types of ecosystems and biomes, as is illustrated by Walter climate diagrams. But the types of plant species present can also determine the NPP of an ecosystem. For a given set of environmental conditions, plant species can vary widely in their NPP depending on their photosynthetic pathway (e.g., ***C$_3$ versus C$_4$ plants**, or trees versus shrubs versus grasses), which regulates their rate of photosynthesis. Moreover, nutrient limitation can play a major role in NPP in certain ecosystems. For example, in lowland tropical forests, NPP is often limited by phosphorus, calcium, and potassium. In temperate and arctic ecosystems, nitrogen and phosphorus can be limiting.

***connect the concepts** As discussed in Key Concept 10.4, C$_3$ and C$_4$ plants differ in how they fix CO$_2$ and can be distinguished according to whether the first product of CO$_2$ fixation is a three- or four-carbon molecule. See Figure 10.16.

Light and nutrients combine to control aquatic primary production

Chapter 53 described how Earth's topography affects water depth, creating gradients in light, temperature, water pressure, and nutrient availability. The main primary producers in aquatic ecosystems—phytoplankton and algae—are most limited by light and nutrients along continental margins (coastal zones) and in the surface waters of oceans and lakes (photic zones) (see Figure 53.9C and D). Oceanic NPP peaks at mid-latitudes and along the continental margins (see Figure 57.3B) because these are regions where nutrient concentrations from ocean upwelling are high. Higher NPP is associated with estuaries as well, where nutrients can be "outwelled" from watershed runoff to coastal zones.

In aquatic systems, NPP is limited primarily by iron, nitrogen, and phosphorus. The relative importance of these three nutrients varies depending on the aquatic ecosystem. In much of the open ocean, NPP is limited primarily by nitrogen. But in some locations, particularly the equatorial Pacific Ocean where excess nitrogen has been detected in surface waters, NPP by phytoplankton may be more limited by iron. In a series of experiments in the mid-1990s, biological oceanographers added iron sulfite to the surface waters off the Galápagos Islands and in the Southern Ocean to test the hypothesis that oceanic NPP, and thus the potential of the ocean to take up CO$_2$ and combat global warming and ocean acidification (see p.1239), was limited primarily by iron. The experiments showed that NPP was as much as four- to tenfold higher in the plume of water that received the iron additions. While the experiments show that iron limitation is a major factor in reducing the ability of the ocean to absorb CO$_2$ from the atmosphere, large-scale fertilization of the oceans is impractical at best, and may be somewhat counteracted by respiration of CO$_2$ by zooplankton and bacteria feeding on phytoplankton.

In lake ecosystems, NPP is limited primarily by phosphorus. The effect of phosphorus on lake NPP was first studied in connection with the declining water quality of lakes in North America and Europe. Wastewater and sewage were implicated in a dramatic increase in **eutrophication**, a process of ecosystem change initiated by an increase in nutrients consumed by phytoplankton, cyanobacteria, and algae in aquatic systems. Eutrophication can result in the explosive growth of algae (algal blooms). Decomposition of large volumes of dead algae by bacteria can in turn severely deplete oxygen needed by aquatic life (hypoxia). To understand which nutrients were most important to the blooms, David Schindler and colleagues conducted whole lake fertilization experiments, starting in 1969, in a set of small lakes in Ontario, Canada. They added nitrogen, carbon, and phosphorus to all or half of several lakes and found that phosphorus was the major trigger for phytoplankton and cyanobacteria blooms (**Figure 57.7**). These experiments set in motion a campaign to reduce phosphorus in wastewater by recommending the banning of phosphates in detergents and stripping them chemically in sewage treatment plants.

This section, which was treated with a combination of phosphorus, nitrogen, and carbon, experienced a massive bloom of cyanobacteria.

A divider separates the two treatment areas.

This section was treated only with carbon and nitrogen, and remained clear.

Figure 57.7 Lake NPP Responds to Phosphorus Fertilization
Experiments show that the addition of phosphorus to a lake causes massive phytoplankton and cyanobacteria blooms.

57.2 recap

Fixed carbon, be it from photosynthesis or chemosynthesis, is used to measure primary production. Net primary production is the amount of biomass incorporated into the tissues of primary producers after respiration, and it varies with latitude and ecosystem type. On land, NPP is highest in the tropics where temperatures are high year-round and rainfall is abundant. In the oceans, NPP peaks along continental margins and in areas of upwelling, where the main primary producers—phytoplankton and algae—have access to light and nutrients. Nutrient runoff from agriculture and wastewater can cause eutrophication in aquatic systems, resulting in algal blooms and deadly low oxygen conditions (hypoxia).

learning outcomes

You should be able to:

- Analyze, with examples, the variation in NPP by latitude and biome in terrestrial and ocean ecosystems.

- Describe the FACE experiment and evaluate, with examples, how results of such experiments might shed light on the relationship between atmospheric CO_2 concentrations and plant production.

- Describe the process, causes, and effects of eutrophication in aquatic ecosystems, and compare the importance of various nutrients in this process.

1. Consider Figure 57.4. How does the average net primary productivity differ between the open ocean and algal beds and coral reefs? Why, then, is the percent of Earth's NPP so much higher for the open ocean?

57.2 recap

2. Atmospheric CO_2 concentrations are rising at unprecedented rates. Is it reasonable to assume that primary producers on land and in the oceans will be able to appreciably lower global CO_2 concentrations? What evidence is there to support your answer?

3. Describe the process of eutrophication and name the nutrient most responsible for this process in lake systems.

Once the energy captured by primary producers is fixed as NPP, consumption by heterotrophs transfers a fraction of that energy to higher trophic levels. Next we will consider in more detail the factors important to the movement and loss of energy between trophic levels.

▶ key concept
57.3
Food Webs Transfer Energy and Nutrients from Primary Producers to Consumers

You know from Chapter 56 that food webs can be constructed to represent the interactions and energy transfer that occur between different species and trophic levels. For example, the food web presented in Figure 56.3 shows the connections among primary producers and consumers in Yellowstone National Park. While the arrows indicate the trophic connections among species, they do not show how much energy is actually being transferred up the food web and converted into secondary production. We will now consider the factors important to secondary production.

focus your learning

- Production efficiency varies based on whether the consumer is an ectotherm or an endotherm, and on the quantity and quality of food available.
- Trophic efficiency varies among trophic levels and among ecosystems.
- Omnivory and the number of trophic levels affect the transfer of energy through food webs.
- Several factors may limit the number of trophic levels in a food web.

You know from Key Concept 8.1 that during energy transfer, some energy is lost as "unusable" to the system (second law of thermodynamics) and thus is unavailable to do work. For food webs, the consequence of this loss is that only part of the primary production consumed by heterotrophs is converted into heterotroph biomass. What factors are important to this movement and loss of energy from primary to secondary production? **Net secondary production**, or the amount of biomass obtained from the consumption of other organisms, depends on how much plant tissue is consumed (**consumption efficiency**), how much of the consumed food can actually be digested versus released as feces and urine (**assimilation efficiency**), and how much of the digested food is used in metabolic activities and released as CO_2 through respiration versus

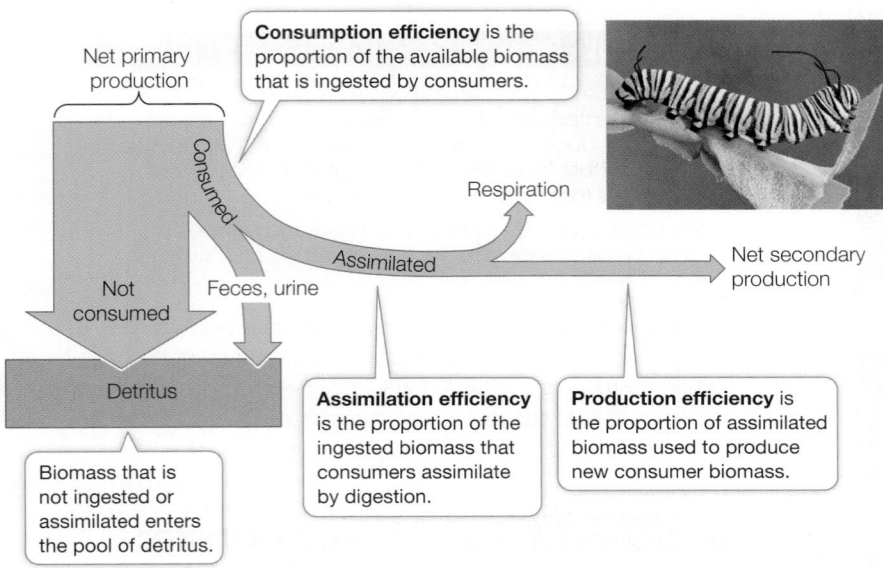

Consumption efficiency is the proportion of the available biomass that is ingested by consumers.

Net primary production

Consumed

Respiration

Assimilated

Net secondary production

Not consumed

Feces, urine

Detritus

Assimilation efficiency is the proportion of the ingested biomass that consumers assimilate by digestion.

Production efficiency is the proportion of assimilated biomass used to produce new consumer biomass.

Biomass that is not ingested or assimilated enters the pool of detritus.

Figure 57.8 Energy Flow and Net Secondary Production in an Herbivore   The proportion of energy transferred between a caterpillar and its plant food depends on efficiencies of consumption, assimilation, and production.

Q: Why do you think herbivores have lower production efficiencies than carnivores?

Consumer group	Production efficiency (%)
table 57.1 Production Efficiencies of Consumers	
Endotherms	
Birds	1.3
Small mammals	1.5
Large mammals	3.1
Ectotherms	
Fishes and social insects	9.8
Nonsocial insects	40.7
Herbivores	38.8
Detritivores	47.0
Carnivores	55.6
Non-insect invertebrates	25.0
Herbivores	20.9
Detritivores	36.2
Carnivores	27.6

Source: Data from Humphreys 1979.

stored as biomass (**production efficiency**) (**Figure 57.8**). We can think of production efficiency as the percentage of energy stored in assimilated food that is used to produce new biomass.

Consumers can vary dramatically in production efficiency. For example, endotherms have much lower production efficiencies than ectotherms (**Table 57.1**). Because endotherms have to maintain high body temperatures, they have higher metabolic rates than ectotherms, and thus have less energy left over to devote to growth and reproduction. Moreover, endotherms vary in production efficiency depending on body size and metabolism. Larger mammals have lower metabolic rates and higher production efficiencies compared with smaller mammals and birds, whose greater surface area-to-biomass ratios result in more heat loss and higher metabolic rates (see Key Concept 39.5).

Because production efficiency affects the growth and reproduction of consumers, changes in the quality and quantity of food can lead to population declines. Research on Steller sea lions in the Gulf of Alaska and the Aleutian Islands offers a case in point. Sea lion populations declined by 85 percent over a 25-year period starting in the early 1970s. Surveys of sea lion individuals showed that they were smaller and had lower birth rates compared with individuals before the decline, suggesting that food was a limiting factor. Prior to the decline, sea lions mostly fed on herring, rich in fats, along with lesser amounts of cod and pollock. But when herring populations declined in the 1970s, the sea lion diet consisted mostly of cod and pollock. The energy per gram of biomass of pollock and cod is roughly half that of herring, so sea lions would have needed to eat twice as much pollock and herring to make up for the loss in herring calories. While plenty of fish were available, consumption of a greater percentage of inferior food represented a decline in production efficiency, which in turn translated into a decline in sea lion populations.

The amount of energy transferred within food webs depends on trophic efficiency

In addition to understanding how energy flows between individual consumers, we can also consider how it flows between trophic levels. **Trophic efficiency** is a measure of the amount of energy at one trophic level divided by the amount of energy at the trophic level immediately below it. Trophic efficiency varies among ecosystems as well as trophic levels. Pyramid diagrams such as those in **Figure 57.9** illustrate the proportions of energy transferred from each trophic level to the next, making it possible to compare energy flow in different ecosystems. Pyramid diagrams can also be used to illustrate the amount of biomass found at each trophic level. As seen in Figure 57.9A and B, terrestrial ecosystems progressively lose energy from one trophic level to the next through trophic inefficiencies, especially between the primary producer and primary consumer levels. Note also that terrestrial ecosystems support less biomass at higher trophic levels than at lower trophic levels. Forest ecosystems have lower trophic efficiency than grasslands because much of the biomass in forests is in the form of wood, most of which is unavailable to primary consumers.

In aquatic systems, trophic efficiencies are much higher. Here pyramid diagrams also show a progressive loss of energy at higher trophic levels, as in terrestrial systems, but the corresponding biomass pyramids are inverted (see Figure 57.9C). Why might this be? The phytoplankton that are the primary producers in aquatic ecosystems grow and reproduce much more rapidly than do the zooplankton and small fishes that consume them, so their smaller biomass, with its rapid rate of primary production, can actually support a larger biomass of primary consumers. In part, this difference between terrestrial and aquatic systems explains why land is green but oceans are blue. Terrestrial ecosystems have primary production that far exceeds rates of herbivory, while the opposite is true for large portions of the ocean. Overall, estimates suggest

(A) Forest ecosystem

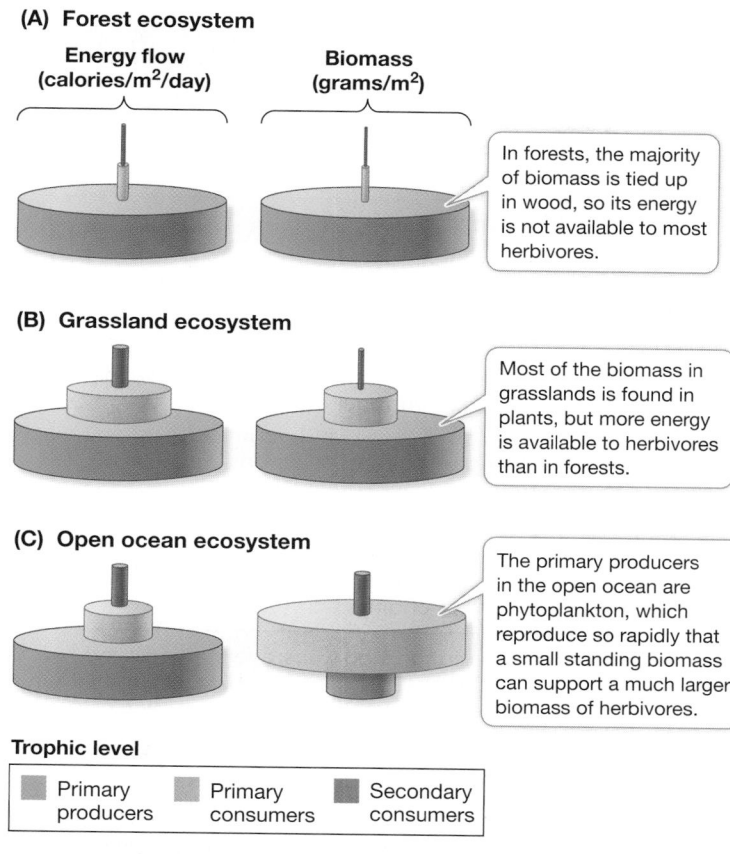

Energy flow (calories/m²/day)

Biomass (grams/m²)

In forests, the majority of biomass is tied up in wood, so its energy is not available to most herbivores.

(B) Grassland ecosystem

Most of the biomass in grasslands is found in plants, but more energy is available to herbivores than in forests.

(C) Open ocean ecosystem

The primary producers in the open ocean are phytoplankton, which reproduce so rapidly that a small standing biomass can support a much larger biomass of herbivores.

Trophic level

■ Primary producers   ■ Primary consumers   ■ Secondary consumers

Figure 57.9 Pyramid Diagrams Show Energy Flow and Biomass in Food Webs Pyramid diagrams allow ecologists to compare patterns of energy flow and biomass in food webs. **(A)** Forest ecosystems have lower trophic efficiency than **(B)** grasslands because much of the biomass in forests is wood, and thus unavailable to most primary consumers. **(C)** In open oceans, the rapid rate of primary production can support a larger biomass of primary consumers.

that, on average, only 13 percent of terrestrial biomass is consumed by primary consumers, compared with 35 percent in aquatic ecosystems.

Food webs are controlled by bottom-up and top-down forces

The control of energy flow in an ecosystem occurs in two ways. First, as discussed in Key Concept 57.2, the amount of energy entering a food web can depend on the NPP of that system, which depends on such factors as rainfall, temperature, light, and nutrients. The more NPP entering the system, the more energy that is passed up the food web to consumers. This flow of energy is referred to as "bottom-up" control because it comes from the "bottom," or primary producer trophic level **(Figure 57.10A)**. The other type of control on energy flow can come from consumers at higher trophic levels, who regulate the amount of NPP by feeding on lower trophic levels. This control is referred to as "top-down" control because it comes from the "top," or the secondary consumer trophic level.

All food webs are controlled by both bottom-up and top-down pathways, but the relative importance of that control can vary depending on the food web in question. For example, food webs that experience a trophic cascade driven by a keystone predator, as you saw in the examples of wolves in Yellowstone National Park (see Figure 56.9) and sea otters in kelp forests, are dominated by top-down control (see Key Concept 56.3). In those two cases, it was the top predator that controlled the prevalence of primary producers (aspens and kelp, respectively), rather than limiting resources per se.

There are also clear examples of bottom-up control in food webs. For example, in a study of pelagic ecosystems off the coast of western North America, researchers found a strong positive correlation between phytoplankton abundance, zooplankton abundance, and fish production (measured from fisheries-catch data). Phytoplankton

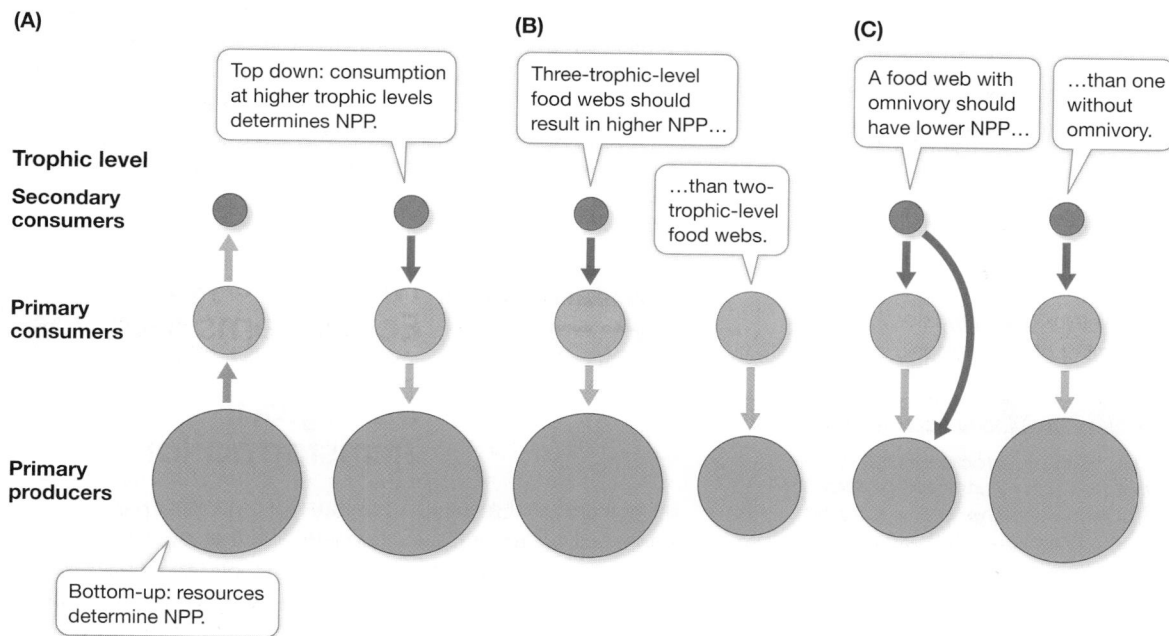

(A)

Top down: consumption at higher trophic levels determines NPP.

(B)

Three-trophic-level food webs should result in higher NPP...

...than two-trophic-level food webs.

(C)

A food web with omnivory should have lower NPP...

...than one without omnivory.

Trophic level

Secondary consumers

Primary consumers

Primary producers

Bottom-up: resources determine NPP.

Figure 57.10 Bottom-up and Top-down Control of Food Webs **(A)** Production in an ecosystem can be viewed as mostly being controlled by limiting resources (bottom-up) or by consumers at higher trophic levels (top-down). **(B)** A three-level food web, in which the secondary consumer affects the primary consumer's abundance, has higher NPP (represented by the size of the green circle) than a two- or four-level food web. **(C)** A three-level food web with omnivory has lower NPP than a food web without omnivory.

abundance in this region is largely controlled by ocean upwelling, which brings nutrients to the light-saturated surface waters for use by phytoplankton. The intensity of upwelling varies along the coast, thus creating "hot spots" of productivity that are reflected in the fish populations of the region.

The number of trophic levels can control the flow of energy through food webs

You know from the previous section that changes in the abundance of organisms at one trophic level can influence the energy flow at multiple trophic levels. You might imagine that the number of trophic levels in a food web could be important in the magnitude of that influence. Indeed, all things being equal, a three-level food web, in which the secondary consumer affects the primary consumer's abundance, should have higher NPP than a two- or four-level food web in which primary consumers have less secondary consumer control (**Figure 57.10B**). Omnivory, or feeding on more than one trophic level, can also change the way energy is transferred in food webs by essentially "collapsing" trophic levels on one another. For example, a three-level food web with a secondary consumer feeding on both the primary consumer and primary producer should have lower NPP than a three-level food web without omnivory (**Figure 57.10C**). That's because even though the secondary consumer indirectly benefits the primary producer by feeding at the primary consumer level, it partially negates this effect by directly feeding on the primary producer as well.

While the number of trophic levels in foods webs can vary, most food webs have three or four trophic levels, with the rare exception of five or more in some systems. What factors might be important in limiting the number of trophic levels in food webs? Several factors have been hypothesized to be important:

- *The amount of NPP entering the system*: This hypothesis suggests that bottom-up factors can influence trophic structure. Given that a large amount of energy is lost from one trophic level to the next (see Figure 57.9), the number of trophic levels will be limited by the amount of energy that can sustain populations at higher trophic levels.

- *The amount of disturbance*: This hypothesis suggests that longer food webs are less likely to recover from disturbances than shorter food webs, which can presumably reassemble more quickly given the fewer trophic levels, and species, involved.

- *The evolutionary constraints on top predators*: This hypothesis focuses on the idea that no organisms on Earth today, with the exception of humans, have evolved to capture apex predators such as birds of prey, sharks, killer whales, and polar bears. With evolutionary constraints maintaining top predators at the pinnacle of food webs, the number of higher trophic levels is limited.

57.3 recap

Once energy captured by primary producers is fixed as NPP, consumption by heterotrophs transfers a fraction of that energy to higher trophic levels (net secondary production). Terrestrial systems have lower trophic efficiencies than aquatic systems, where the rapid rate of primary production supports greater secondary production. The flow of energy in food webs is influenced by the number of trophic levels and the relative contribution of bottom-up

and top-down forces. Food webs are typically restricted to three or four trophic levels because of energy limitations at higher trophic levels, the effects of disturbance, and/or evolutionary constraints on top predators.

learning outcomes

You should be able to:

- Compare and contrast, with examples, the production efficiencies of ectotherms and endotherms.
- Analyze why trophic efficiencies are higher in aquatic than in terrestrial ecosystems, and explain how this affects biomass throughout the food web.
- Analyze, both qualitatively and quantitatively, the effect that length of a food chain has on NPP in an ecosystem.
- Explain how omnivory can change the transfer of energy in a food web.
- Provide and explain hypotheses relating to control of the length of food chains.

1. Suppose there are two food webs, one in a forest, the other in a lake. All else being equal, which food web would have the higher trophic efficiency, and why? Comparing different species, which would have the higher production efficiency, a bear or a crayfish? Explain your answer.

2. Now compare two food webs in different lakes. One has three trophic levels, and the other has four. All else being equal, which would have the higher NPP? Why? If you added omnivory to the top level of the food web with four trophic levels, how would it differ in NPP compared with the food web without omnivory?

3. Compare a tropical rainforest food web that has five trophic levels with a food web in a desert having three trophic levels. What hypothesis might best explain the difference in trophic level length between the two ecosystems?

We've considered how energy flows from primary producers to consumers, and the implications of that flow for food webs. Besides energy, a key component of the metabolism and growth of organisms is nutrients (including water). Earlier in the chapter you read that nutrients continually cycle between living organisms and the abiotic environment via nutrient cycling (see Figure 57.1C). We will consider nutrient cycling next.

key concept 57.4 Nutrient Cycling in Ecosystems Involves Chemical and Biological Transformations

All nutrients that make up the tissues of organisms originate from either soil, water, or the atmosphere, and then cycle between organisms and their environment through (1) primary and secondary production and (2) decomposition (**Figure 57.11**). We considered production in the last two sections, but not decomposition. **Decomposition** occurs when detritus, or dead organisms and their

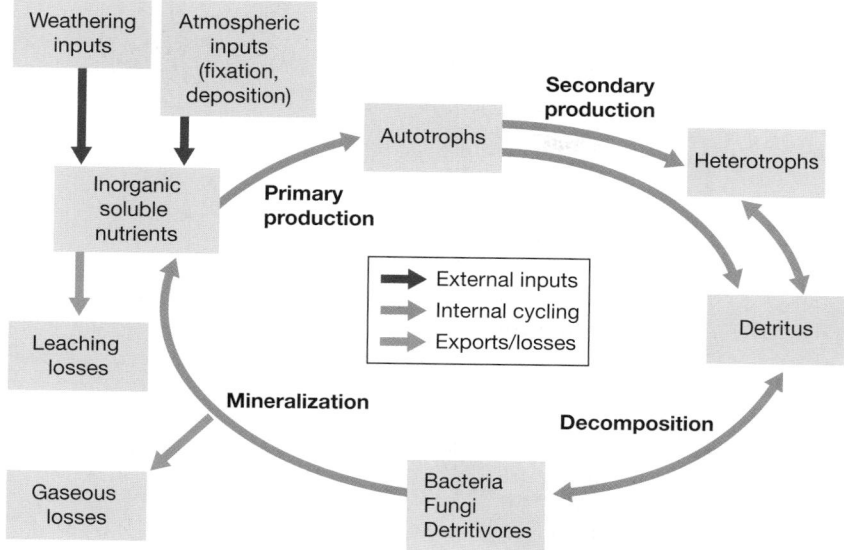

Figure 57.11 Nutrient Cycles A generalized nutrient cycle, showing the movement of nutrients among biotic and abiotic compartments of an ecosystem and the potential pathways for inputs and losses.

waste products, are broken down by bacteria and fungi into energy and nutrients. Decomposition releases nutrients in a soluble form that can then be taken up by primary producers and microorganisms. Detritus can come in many forms, including plants and animals, but is mostly made up of plant or algal matter. Detritus must first be broken down into smaller fragments by detritivores such as worms and arthropods before it can undergo mineralization (see Figure 57.11). **Mineralization** is the process by which animals, bacteria, and fungi convert organic compounds into inorganic soluble nutrients. Along the way, some of the nutrients may be lost to the system as gas to the atmosphere or minerals to ground water. The nutrient cycle is complete when the inorganic soluble nutrients produced through mineralization are used once again by primary producers or bacteria.

focus your learning

- Decomposition and mineralization are necessary to complete nutrient cycling in ecosystems.
- The average residence time of water molecules in the various pools of the hydrologic cycle ranges from one week to thousands of years.
- Fossil-fuel burning by humans is affecting Earth's carbon cycle, changing its climate and oceans.
- Nitrogen, an abundant gas in Earth's atmosphere, is cycled primarily by microbial activity.
- Most phosphorus is stored in rocks and deep-sea sediments, but it cycles rapidly through organisms and often limits plant growth.
- Most sulfur is pooled in rocks and ocean sediments, but cycling produces different compounds on land and in oceans.

The movement of nutrients through food webs from uptake to decomposition—that is, through the biotic components of ecosystems—occurs primarily on a local scale. In contrast, abiotic processes can move nutrients far beyond the boundaries of the local ecosystem. Each of the chemical elements that organisms use in large quantities

cycles in a distinctive way through the globe. Because geological, chemical, and biological processes are all important in moving materials around the planet, the pattern of movement of an element is called its **biogeochemical cycle**. The cycles are considered in terms of pools—the amount of elements within certain parts of the abiotic and biotic environment—and fluxes, or the rates of movement between pools. We will consider the biogeochemical cycling of water, carbon, nitrogen, phosphorus, and sulfur at the global scale because of their functional role in biological systems and their roles as pollutants in the environment.

Water cycles rapidly around the globe

We start our discussion of global biogeochemical cycles with the water cycle—known as the **hydrologic cycle**—not only because water serves as an essential element for organisms but because it plays an important role in transporting nutrients between compartments (**Figure 57.12**). Energy from the sun drives the hydrologic cycle by evaporating water from the vast surfaces of the oceans. Most of the water is returned to the oceans as rainfall, but a smaller fraction is transported over land. Ocean water vapor combines with terrestrial water vapor from soils, lakes and rivers, and plants as a consequence of *evapotranspiration, and falls as precipitation over land. Terrestrial precipitation eventually returns to the oceans via streams, coastal runoff, and groundwater flows. More than half of this volume of water is carried back to the oceans by Earth's four largest rivers: the Amazon in South America, the Nile in Africa, the Mississippi in North America, and the Yangtze in Asia.

---

*connect the concept Evapotranspiration is the evaporative transfer of heat and water from the surfaces of plants into the atmosphere, which reduces air temperature and increases moisture, as described in Key Concept 53.3 and Figure 53.10.

Despite their relatively small volume, rivers play a disproportionate role in the hydrologic cycle because the average residence time of a water molecule in rivers is only a few years. By comparison, the average residence time of a water molecule in lakes ranges from a few years to centuries. The larger the lake, the longer the residence time; the residence time for water in the top portion of Lake Superior, for example, is 1,500–2,000 years, and the water at the bottom of this massive lake cycles over much longer time scales. In the oceans the average residence time of a water molecule is about 3,000 years. Other pools of water include glaciers (with residence times of 20–100 years), seasonal snow cover (a few months), and soil moisture (1–2 months). The average residence time of water in the bodies of organisms is particularly brief, averaging just under a week.

Although large amounts of groundwater are present in underground pools called **aquifers**, this water has a long residence time underground and plays only a small role in the hydrologic cycle. In some places, however, aquifers are being depleted because humans are using groundwater more rapidly than it can be replaced, primarily for irrigation. On the North China Plain, depletion of shallow aquifers

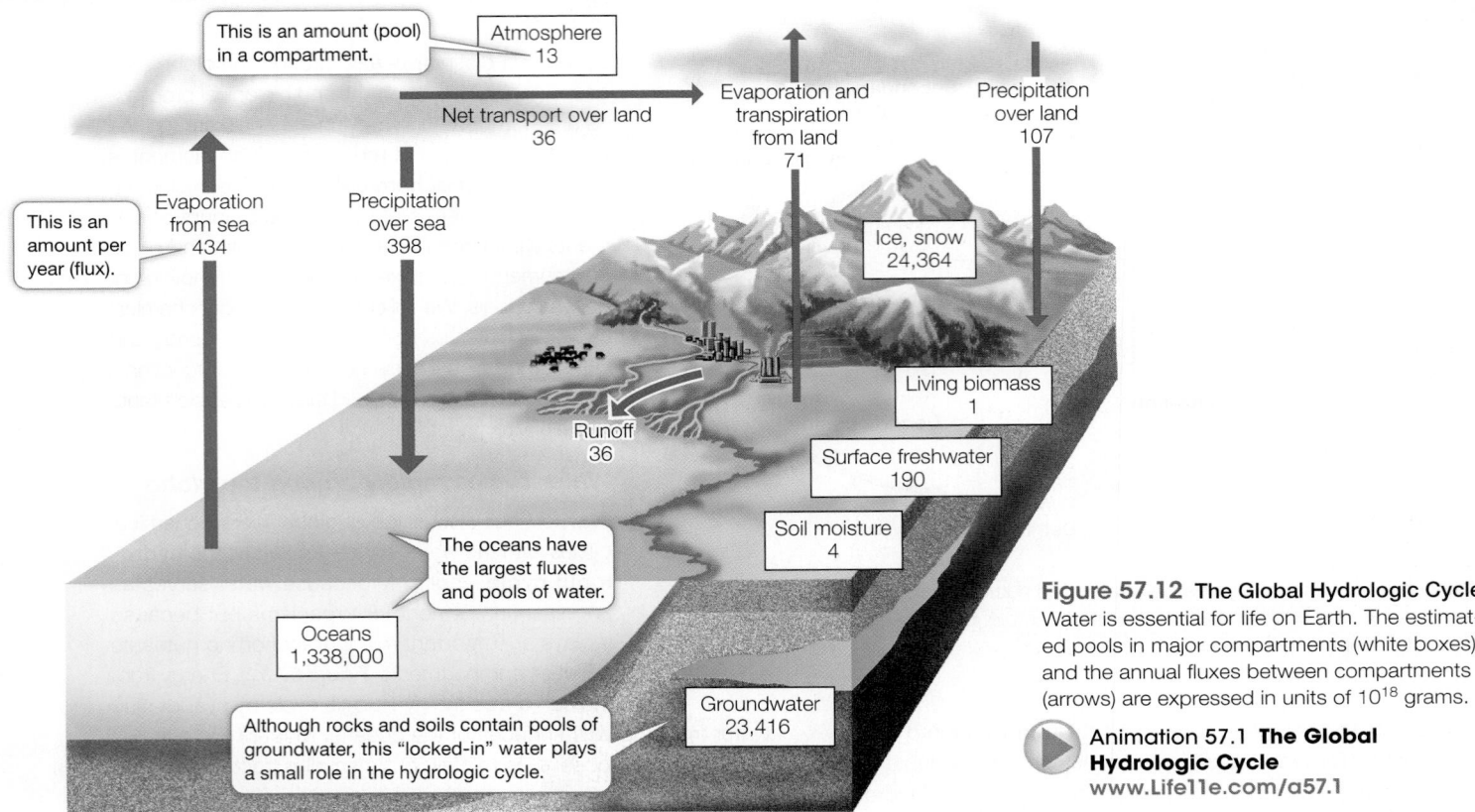

This is an amount (pool) in a compartment.

Atmosphere
13

Net transport over land
36

Evaporation and transpiration from land
71

Precipitation over land
107

This is an amount per year (flux).

Evaporation from sea
434

Precipitation over sea
398

Ice, snow
24,364

Living biomass
1

Runoff
36

Surface freshwater
190

Soil moisture
4

The oceans have the largest fluxes and pools of water.

Oceans
1,338,000

Although rocks and soils contain pools of groundwater, this "locked-in" water plays a small role in the hydrologic cycle.

Groundwater
23,416

Figure 57.12 The Global Hydrologic Cycle
Water is essential for life on Earth. The estimated pools in major compartments (white boxes) and the annual fluxes between compartments (arrows) are expressed in units of 10^{18} grams.

Animation 57.1 **The Global Hydrologic Cycle**
www.Life11e.com/a57.1

is forcing people to sink wells more than 1,000 meters deep to reach groundwater. If current water consumption patterns continue, it has been estimated that by 2025 at least 48 percent of the current world population will live in areas with inadequate water supplies.

The carbon cycle is being altered by human activities, resulting in climate change

All of the important macromolecules that make up living organisms contain carbon, and much of the energy that organisms use to fuel their metabolic activities is stored in carbon-containing (organic) compounds. On land, biological processes move carbon directly between organisms and the atmosphere as terrestrial organisms take up carbon during photosynthesis and return it to the atmosphere through respiration and metabolism (**Figure 57.13**). In contrast, carbon dioxide moves into ocean waters from the atmosphere primarily by simple diffusion at the ocean surface; this dissolved CO_2 is the source of the carbon used by marine primary producers. Even taken together, however, the amounts of carbon in the atmosphere, in soils, and in living and dead organisms are dwarfed by the vast quantities of carbon stored in terrestrial rocks, in fossil fuels, in marine sediments, and in seawater in the form of carbonate ions (CO_3^{-2}) or bicarbonate ions (HCO_3^-) (see Figure 57.13).

Over the span of life's existence on Earth, quantities of carbon were removed from active cycling when organisms died in large numbers and were buried in sediments lacking oxygen. In such anaerobic environments, with few detritivores to reduce organic carbon to CO_2, organic molecules accumulate and are eventually transformed into deposits of oil, natural gas, coal, or peat—the **fossil fuels** that modern humans use as a combustible source of energy. Humans have discovered and used these fossil fuels at

ever-increasing rates during the past 150 years. As a result, CO_2, one of the final products of burning fossil fuels, is being released into the atmosphere faster than it is absorbed in the oceans or incorporated into terrestrial biomass. Measurements of CO_2 on the top of Mauna Loa, Hawaii, show that, even though atmospheric CO_2 varies seasonally because of the change in NPP, it has steadily increased over the last 55 years of data collection (**Figure 57.14**).

ATMOSPHERIC CO_2 AND GLOBAL CLIMATE CHANGE Carbon dioxide is a greenhouse gas, so we would expect increasing atmospheric CO_2 concentrations to trap more heat in Earth's atmosphere and raise temperatures at Earth's surface (see Figure 53.2). What evidence do we have that this is occurring? Measurements of gases in air trapped in the Antarctic and Greenland ice caps show that temperatures have been warmer when atmospheric CO_2 concentrations have been higher and cooler when they have been lower (**Figure 57.15**). For example, the atmospheric CO_2 concentration was very low at the end of the last glaciation (180 parts per million [ppm]), 18,000 years ago, when temperatures were much colder than they are today. In contrast, during a warm interval between 11,000 years ago and the start of the Industrial Revolution (1750), the concentration of CO_2 in Earth's atmosphere was between 260 and 280 ppm. Today atmospheric CO_2 concentration is just over 400 ppm (see Figure 57.14), the highest recorded in the last 800,000 years. The increase in CO_2 (and two other greenhouse gases, CH_4 and N_2O) from the burning of fossil fuels has resulted in a roughly 1°C (1.8°F) rise in global temperatures compared with those from 1981 to 2010 (a comparison known as a "temperature anomaly").

How global climates and ecosystems will change in response to this rapid warming is a subject of intense investigation. Global

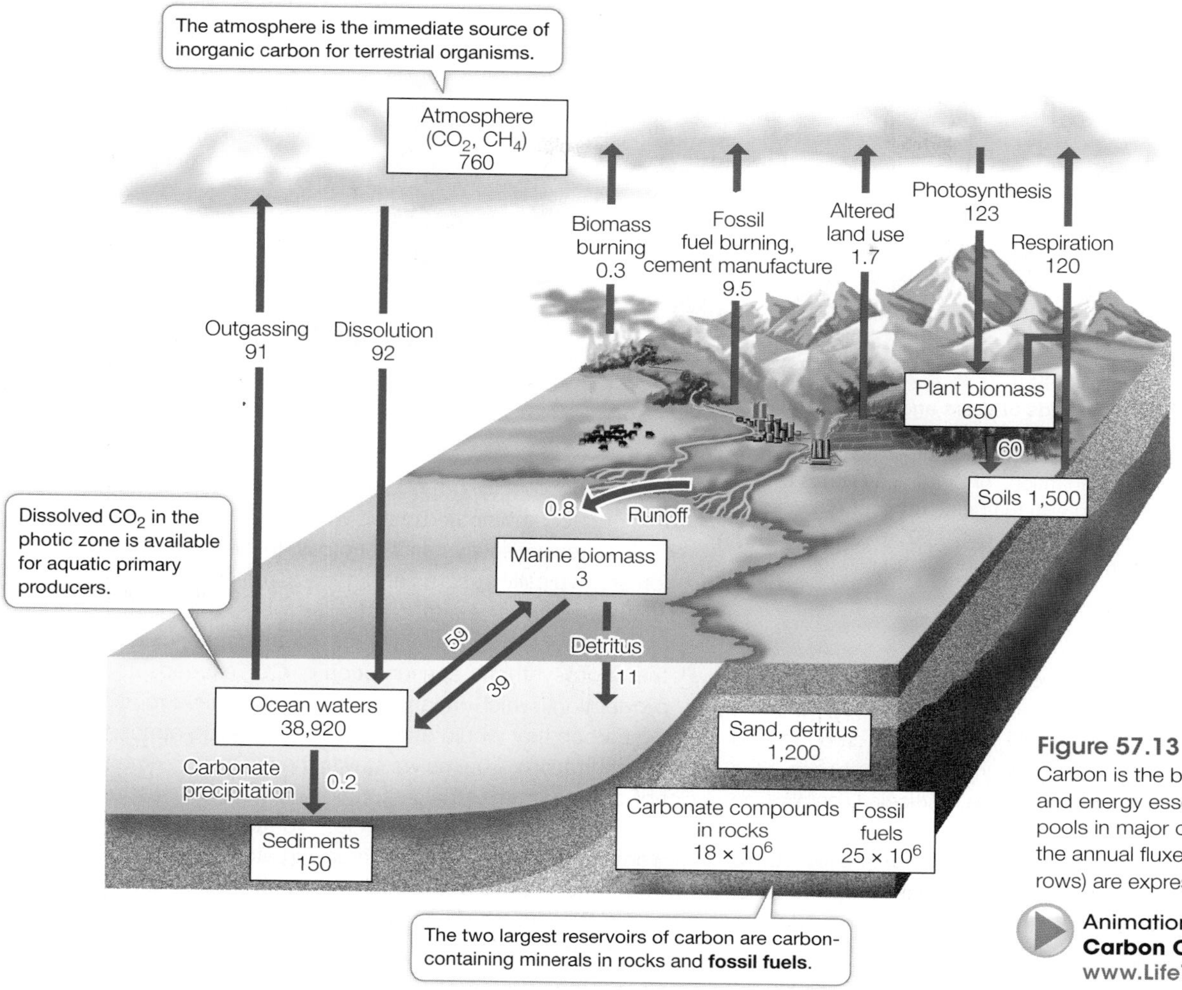

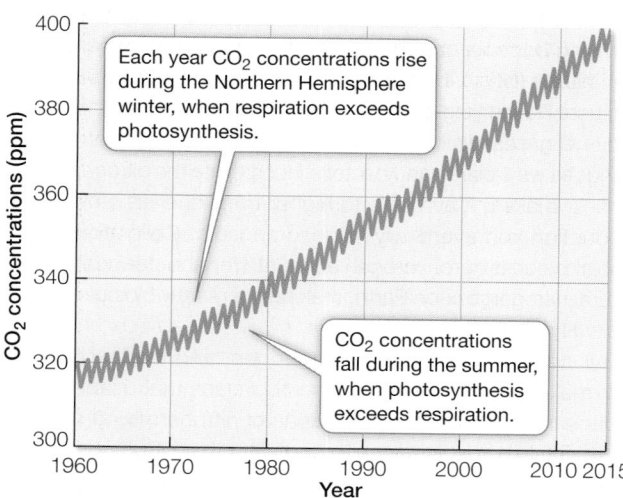

The atmosphere is the immediate source of inorganic carbon for terrestrial organisms.
Atmosphere (CO_2, CH_4) 760
Biomass burning 0.3
Fossil fuel burning, cement manufacture 9.5
Altered land use 1.7
Photosynthesis 123
Respiration 120
Outgassing 91
Dissolution 92
Plant biomass 650
60
Soils 1,500
0.8
Runoff
Dissolved CO_2 in the photic zone is available for aquatic primary producers.
Marine biomass 3
59
39
Detritus 11
Ocean waters 38,920
Carbonate precipitation 0.2
Sediments 150
Sand, detritus 1,200
Carbonate compounds in rocks 18×10^6
Fossil fuels 25×10^6
The two largest reservoirs of carbon are carbon-containing minerals in rocks and **fossil fuels**.

Figure 57.13 The Global Carbon Cycle Carbon is the basis of the organic molecules and energy essential for life. The estimated pools in major compartments (white boxes) and the annual fluxes between compartments (arrows) are expressed in units of 10^{15} grams.

Animation 57.2 The Global Carbon Cycle
www.Life11e.com/a57.2

warming has already resulted in the shrinking of Arctic sea ice, which is currently the lowest ever recorded. As temperatures continue to rise and glacial ice melts, sea level rises (because of both thermal expansion of ocean waters and the addition of glacial meltwater), increasing flooding of coastal cities and agricultural lands, especially during extreme storm events. Nearly one-third of the world's population lives in coastal regions, which make up only 4 percent of Earth's total land area.

Each year CO_2 concentrations rise during the Northern Hemisphere winter, when respiration exceeds photosynthesis.
CO_2 concentrations fall during the summer, when photosynthesis exceeds respiration.
CO_2 concentrations (ppm)
400
380
360
340
320
300
1960
1970
1980
1990
2000
2010 2015
Year

Figure 57.14 Atmospheric CO_2 Concentrations Are Increasing Carbon dioxide concentrations have been recorded since 1960 on top of Mauna Loa, Hawaii, far from most sources of human-generated CO_2 emissions. Although concentrations vary seasonally, the trend has been consistently upward.

Even though atmospheric CO_2 is increasing as the result of burning fossil fuels, less than half of the CO_2 released into the atmosphere by human activities remains in the atmosphere. Where does the rest of the CO_2 wind up? Much of it is absorbed by the oceans in inorganic forms. Over decades to centuries, the oceans exert a large influence over atmospheric CO_2 concentrations. Of the CO_2 absorbed by the ocean, some is used in photosynthesis by phytoplankton in the surface waters. These organisms remove dissolved CO_2 from water, thereby increasing the rate at which atmospheric CO_2 is absorbed by surface waters, but as you saw earlier, there are limits to this absorption given nutrient limitation in vast parts of the open ocean. In addition, many marine organisms (including clams, oysters, corals, and planktonic foraminiferans) incorporate carbon in their shells and other structures in the form of calcium carbonate ($CaCO_3$), which is synthesized by combining bicarbonate ions (HCO_3^-) and calcium ions (Ca^{2+}) dissolved in seawater. When these organisms die, those shells and their embedded carbon sink to the ocean floor.

Today's oceans absorb millions of tons of CO_2 from the atmosphere each day—more than at any time during the past 20 million years. As a result, water near the ocean surface is becoming more

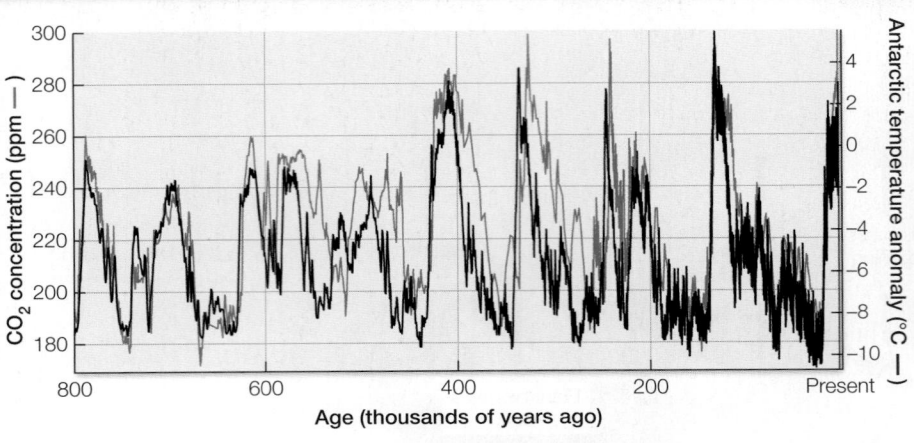

Figure 57.15 Higher Atmospheric CO$_2$ Concentrations Correlate with Warmer Temperatures Atmospheric concentrations of CO$_2$ (measured in air bubbles trapped in Antarctic ice) have varied with temperatures over Antarctica (estimated with oxygen isotope analysis) for more than 800,000 years. "Temperature anomaly" is the change in Antarctic temperatures compared to modern temperatures.

acidic. As CO$_2$ concentrations in the atmosphere rise, more of the gas diffuses into the water at the ocean surface, where it reacts with water to form carbonic acid (H$_2$CO$_3$). As levels of carbonic acid rise, the pH of seawater drops. This increase in acidity can have negative effects on many marine organisms, particularly corals. The combination of decreasing pH and increasing water temperature to which corals are being exposed results in *coral bleaching and sometimes the death of corals. Because so many other reef species depend on corals and the structure they provide, an entire reef community can collapse if its corals fail to thrive.

---

*connect the concept   Coral bleaching occurs when corals are exposed to warm and/or acidic waters, causing expulsion of their photosynthetic algal symbionts, as described in Chapter 1. These symbionts can sometimes be reacquired, as described in Investigating Life: Can Corals Reacquire Dinoflagellates Lost to Bleaching in Chapter 26.

But as you read in the opening story, the effects of CO$_2$ and ocean acidification vary depending on the organisms or ecosystems considered. **Investigating Life: Food Webs in an Acidic and Warming Ocean** describes research in Swedish by marine ecologists interested in the effects of ocean acidification and warming on an ecologically critical but predominately primary-producer-dominated ecosystem: estuaries. In this system, the researchers focused on benthic microalgae, an important food source for a diversity of sediment-dwelling herbivores, including crustaceans, snails, and worms. Benthic microalgae indirectly benefit from higher trophic levels, particularly omnivores that feed on macroalgae competitors and herbivores of the microalgae. Using experiments the researchers discovered that the biomass of benthic microalgae did not appreciably change under the elevated CO$_2$ and warming manipulations. As it turned out, the consumers mediated the effects of ocean acidification and warming on benthic microalgae, suggesting that consumers, or other strongly interacting species in food webs,

have the potential to modulate the effects of climate change in complicated and unpredictable ways.

In terrestrial ecosystems, photosynthesis, principally in tropical forests, typically absorbs about the same amount of carbon that is released by terrestrial plants, microbes, and animals through metabolism. In more recent times, as atmospheric CO$_2$ increases, the photosynthetic consumption of CO$_2$ exceeds its metabolic production, which means Earth's terrestrial vegetation is storing carbon that would otherwise be increasing atmospheric CO$_2$ concentrations—but as we saw in the FACE experiment, we cannot count on terrestrial vegetation to store the vast amounts of excess CO$_2$ that human activities produce. Furthermore, climate warming (another result of increasing atmospheric CO$_2$ concentrations, as you have already seen) increases plant metabolism and is thus likely to increase the flux of CO$_2$ from vegetation into the atmosphere.

The nitrogen cycle is dominated by biotic processes

Nitrogen gas is the most abundant gas in Earth's atmosphere, but most organisms cannot use nitrogen in its gaseous form. Nitrogen enters the system as atmospheric N$_2$ and is fixed by bacteria into ammonia (NH$_3$) (**Figure 57.16**). Ammonia is rapidly transformed into ammonium (NH$_4^+$), which can then be used by plants and bacteria. Nitrifying bacteria can transform ammonium to nitrite (NO$_2^-$) and then nitrate (NO$_3^-$), both forms of nitrogen that plants and bacteria can use. Denitrifying bacteria take nitrate and convert it back to N$_2$ and N$_2$O gases, which are then released back into the atmosphere. Along the way, plants and bacteria incorporate the nitrogen into their tissues, making it available to higher trophic levels as secondary production and eventually to decomposers. Collectively this microbial processing of nitrogen accounts for about 95 percent of all natural nitrogen flux on Earth, making it a mostly biologically driven cycle (**Figure 57.17**).

All living organisms require nitrogen, and the inability of the vast majority of organisms to use N$_2$ means that usable nitrogen is often in short supply. Populations of nitrogen-fixing organisms rarely increase in abundance to the extent that nitrogen is no longer limiting, because the end products of their nitrogen fixation are rapidly lost from ecosystems (ammonia by vaporization and denitrification; and nitrate, which is highly water-soluble, by leaching).

Human activities that fix nitrogen, such as the manufacture of artificial fertilizers, have had large and unanticipated effects on the

▶ experiment

Original Paper: Alsterberg, C., J. S. Eklöf, L. Gamfeldt, J. N. Haven-hand and K. Sundbäck. 2013. Consumers mediate the effects of experimental ocean acidification and warming on primary producers. *Proceedings of the National Academy of Sciences USA* 110: 8603–8608.

Christian Alsterberg and his colleagues considered how the food web in an estuarine ecosystem on the western coast of Sweden was influenced by ocean acidification and warming. The researchers chose to focus on benthic microalgae (single-celled algae) and its interactions with macroalgae and consumers. In this system, omnivores (i.e., a guild of medium-sized crustaceans and snails that feed on two trophic levels) affect the productivity of benthic microalgae in two ways: (1) they increase the light available to microalgae by feeding on macroalgae, and (2) they prey on herbivores of microalgae (i.e., a guild of small crustaceans, snails, and worms). To estimate the importance of ocean acidification and increased warming in a multi-species food web, the researchers conducted an experiment on the estuarine community in which CO_2, temperature, and omnivores were manipulated.

HYPOTHESIS▶ Increasing CO_2 and temperature will increase the biomass of macroalgae competitors, causing a decline in the biomass of benthic microalgae.

METHOD

1. Fill 30-liter buckets with seawater and add microalgae, macroalgae, and herbivores.
2. Apply three treatments:
   a. two levels of seawater temperature (ambient and increased)
   b. two levels of CO_2 (ambient and increased)
   c. two levels of omnivores (present and absent)
3. At the end of a 5-week period over the summer, measure the biomass of benthic microalgae (as Chl *a* concentrations), macroalgae, and herbivores.

RESULTS

To simplify, the results from the following treatments are presented: ambient CO_2 and temperature and increased CO_2 and temperature, each with and without omnivores.

The researchers found that benthic microalgae biomass did not change under increased CO_2 and temperature. This was true regardless of whether omnivores were present. However, the removal of omnivores did cause a decline in benthic microalgae under ambient conditions, as would be predicted from the food web shown above.

Omnivore removal caused an increase in the biomass of macroalgae competitors and herbivores, which then led to a decline in benthic microalgae. Interestingly, even though there was no effect of increased CO_2 and temperature on microalgae biomass, the results showed that both macroalgae and herbivores increased in biomass under these conditions compared with ambient conditions. This was especially the case when omnivores were removed.

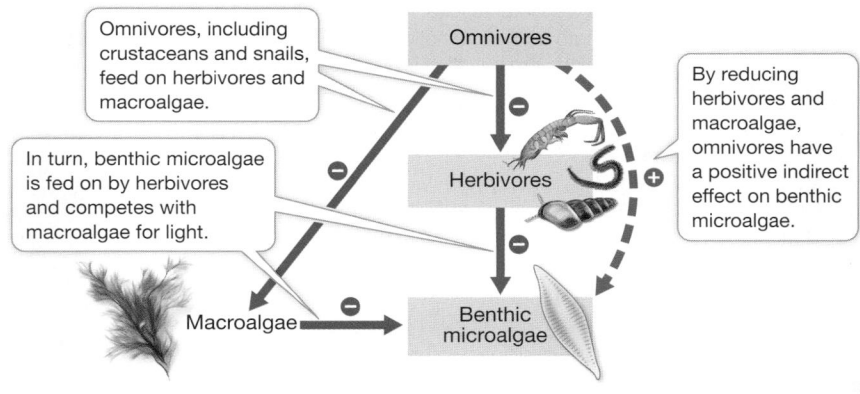

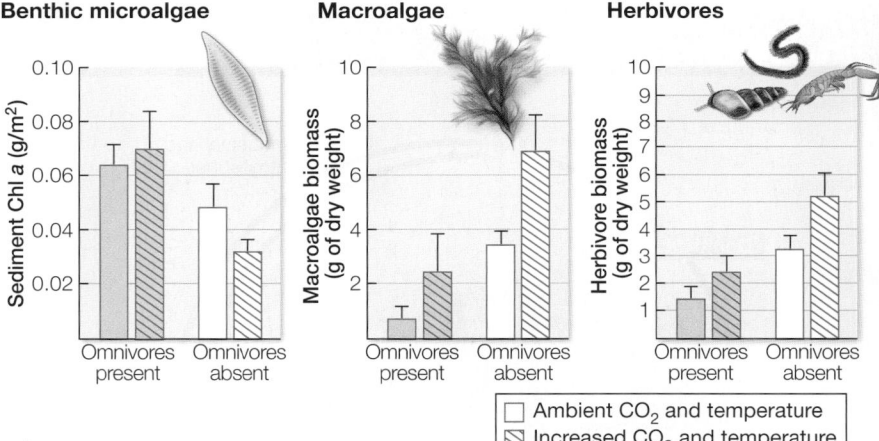

CONCLUSION▶ What can you conclude about the effects of increased CO_2 and temperature on benthic microalgae in this estuarine food web? Explore this question in the following work with the data.

investigatinglife work with the data follows on next page.

---

nitrogen cycle. The extensive use of artificial fertilizers on agricultural crops, coupled with the burning of fossil fuels (which generates nitric oxide and nitrogen dioxide), has resulted in total nitrogen fixation by humans being nearly equal to global natural nitrogen fixation (see Figure 57.17). This human-generated nitrogen flux has been increasing over the past half-century and is expected to continue to increase.

One consequence of artificial fertilizer production is an increase in the greenhouse gas nitrous oxide (N_2O), resulting in more tropospheric ozone (O_3) and smog. Trophospheric ozone is itself a greenhouse gas and thus can contribute to global climate change.

Some of the nitrous oxide that enters the atmosphere falls back to land in precipitation or as dry particles. This deposition of nitrogen from the atmosphere has increased dramatically during recent

work with the data

To interpret the Alsterberg et al. study, we consider why benthic microalgae biomass did not change under increased CO_2 and temperature within the context of the experimental manipulation of the food web.

QUESTIONS▶

1. Draw two food webs for this system, one with and one without the omnivores. Use arrows to indicate the direct (solid) and indirect (dashed) interactions and symbols to indicate negative (–) or positive (+) interactions.

2. Now, using data from the graphs in the experiment, modify the two food webs to show how increased CO_2 and temperature will affect the strength and direction of the food web interactions. Vary the width of the arrows to indicate the relative strength of the interactions, and use symbols to indicate whether the interactions are negative (–) or positive (+). Indicate whether benthic microalgae increase, decrease, or stay the same compared with ambient conditions.

3. Using the food webs you drew for the answer to Question 2 as your guide, explain why increased CO_2 and temperature had no effect on microalgae biomass compared with the ambient conditions.

4. Under increased CO_2 and temperature, what would happen to benthic microalgae biomass if the omnivores ceased feeding on macroalgae and just preyed on herbivores?

A similar **work with the data** exercise
may be assigned in **LaunchPad**.

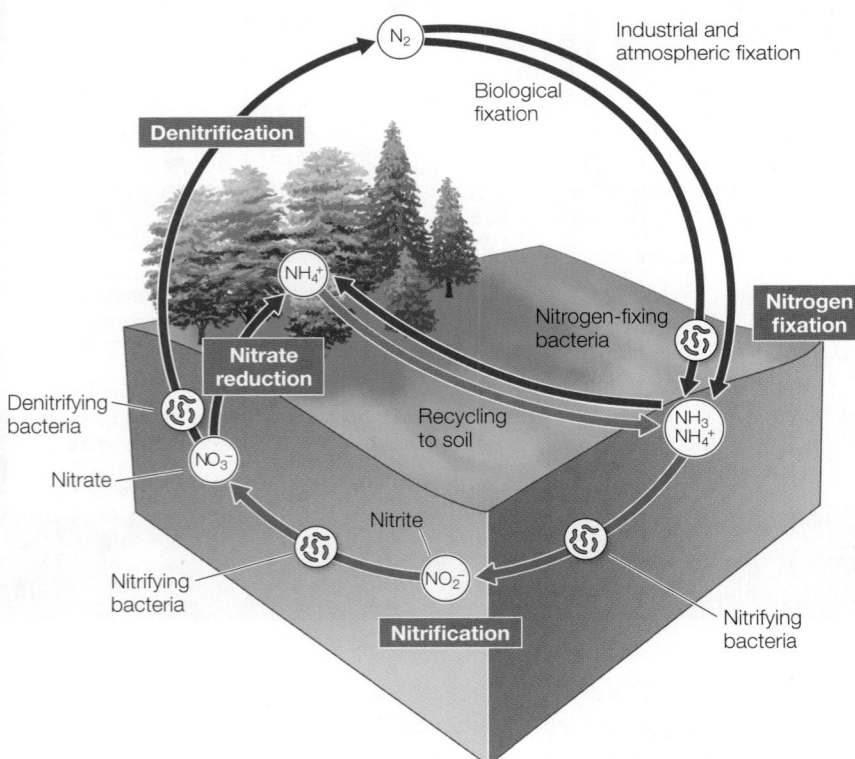

Figure 57.16 The Nitrogen Cycle Nitrogen fixation, nitrification, nitrate reduction, and denitrification are components of an essential chemical cycle that coverts atmospheric nitrogen gas into ammonium ions and nitrate ions—forms of nitrogen that can be taken up by plants—and returns N_2 to the atmosphere.

decades. Nitrogen deposition could affect the composition of terrestrial vegetation by favoring those plant species that are best adapted to take advantage of high nutrient levels, and which may then outcompete other species.

Another consequence of nitrogen fertilizers is eutrophication, a process we considered in Key Concept 57.2 for phosphorus additions in lake ecosystems. In this case, when more nitrogen fertilizer is applied to agricultural lands than can be taken up by the crops, the excess nitrogen moves out of the system in surface runoff, or downward into groundwater, and ultimately ends up in rivers, lakes, and oceans. So-called dead zones have formed near the mouth of the Mississippi River in the Gulf of Mexico as a result of the drainage from interior agricultural fields carrying high concentrations of nitrogen fertilizer.

 **Media Clip 57.1 Tracking Dead
Zones from Space**
www.Life11e.com/mc57.1

The global phosphorus cycle is dominated by geochemical processes

Phosphorus accounts for only about 0.1 percent of Earth's crust, but it is an essential nutrient for all life forms. It is a key component of cell membranes, DNA, RNA, and ATP. Unlike the other biogeochemical cycles discussed thus far, the phosphorus cycle does not have a significant atmospheric component, although some phosphorus is transported by dust particles. Most of Earth's phosphorus is in the form of phosphate in rocks and deep-sea sediments, and it cycles very slowly through the geological system because it requires sedimentary rock formation, uplift, and weathering. In contrast, phosphorus in organisms cycles rapidly, and it is often a limiting factor for their growth, particularly for plants. As you learned in Key Concept 57.2, runoff of phosphorus from fertilizers and detergents into aquatic systems can result in eutrophication.

The burning of fossil fuels affects the sulfur cycle

Sulfur is required by all organisms because it is critical in the production of proteins and lignin. Most of Earth's sulfur supply is locked up in rocks on land and as sulfate salts in deep-sea sediments, but sulfur moves to the atmosphere in a variety of ways. Roughly 10–20 percent of atmospheric sulfur is produced in volcanic eruptions. In terrestrial systems, when sulfur in the soil comes in contact with atmospheric oxygen, it is converted to sulfate salts, which can be taken up by plants and eventually incorporated into their tissue. This sulfur ultimately is returned to the atmosphere as H_2S via microbial decomposition. In marine systems, many phytoplankton and seaweeds manufacture large quantities of a sulfur-containing compound (dimethylsulfoniopropionate, or DMSP) to maintain their salt and water balance. When broken down, DMSP releases dimethyl sulfide (CH_3SCH_3), the principal odorant of rotting

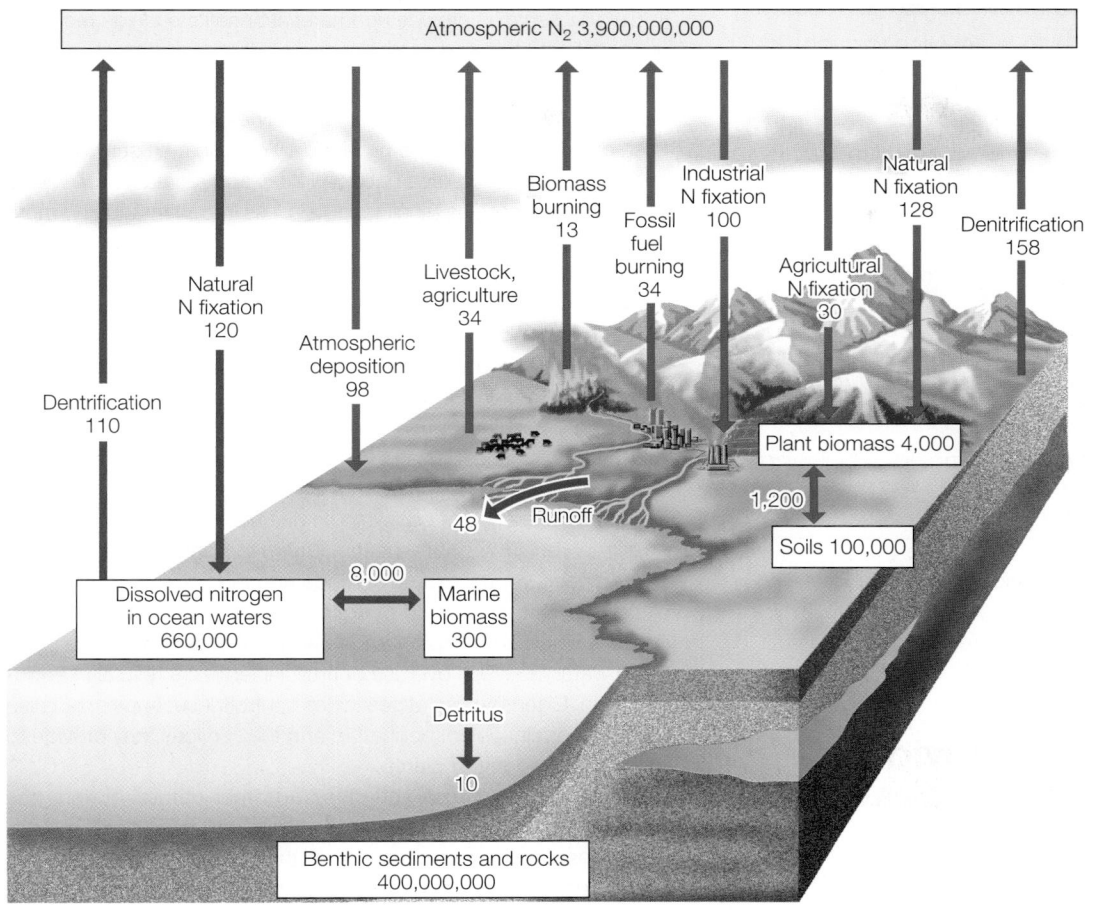

Atmospheric N_2 3,900,000,000

Biomass burning 13

Industrial N fixation 100

Natural N fixation 128

Denitrification 158

Fossil fuel burning 34

Agricultural N fixation 30

Natural N fixation 120

Livestock, agriculture 34

Atmospheric deposition 98

Dentrification 110

Plant biomass 4,000

1,200

Runoff 48

Soils 100,000

8,000

Dissolved nitrogen in ocean waters 660,000

Marine biomass 300

Detritus

10

Benthic sediments and rocks 400,000,000

Figure 57.17 The Global Nitrogen Cycle The largest pool of nitrogen is in the atmosphere in the form of nitrogen gas, N_2. Nitrogen cycles through the biosphere primarily via the processes of nitrogen fixation, which converts inorganic nitrogen to an organic form usable by plants, and denitrification, which returns N_2 to the atmosphere. The estimated pools in major compartments (white boxes) and the annual fluxes between compartments (arrows) are expressed in units of 10^{12} grams.

Q: How does the industrial fixation of nitrogen (as a percentage) compare with the other forms of fixation on land?

Animation 57.3 **The Global Nitrogen Cycle**
www.Life11e.com/a57.3

seaweed. Because the quantities of phytoplankton in the oceans are enormous, dimethyl sulfide production accounts for about half of the biotic component of the global sulfur cycle.

Atmospheric sulfur exists in gaseous and particulate forms and plays an important role in global climate. Cloud formation is dependent on small particles around which water can condense. Dimethyl sulfide is a major component of such particles, so increases in atmospheric sulfur concentrations increase cloud cover and reduce the amount of incoming solar radiation, affecting Earth's temperature. In addition, SO_2 and NO_2 produced by humans from the burning of coal and oil react with water molecules in the atmosphere to form sulfuric acid (H_2SO_4) and nitric acid (HNO_3), respectively. These acids can travel hundreds of kilometers in the atmosphere before they are released as **acid rain**. Acid rain can damage forests and lakes, affecting the food webs they support. Restriction on the emissions of sulfur in North America and Europe has resulted in significant reductions in acid rain. Research shows that once acidity is reduced, forest and aquatic systems can quickly recover. Acid rain remains a problem in some countries such as China and India, where rapid industrialization and the lack of restrictions create high sulfur emissions.

57.4 recap

Nutrient cycling in ecosystems and global biogeochemical cycles involves chemical and biological transformations through both production and decomposition. The hydrologic cycle is not only crucial in making water available to organisms but also in transporting nutrients within ecosystems. Carbon moves through biological systems and returns to the atmosphere through respiration. Burning of fossil fuels increases the carbon dioxide and sulfur in the atmosphere, which is implicated in global warming and acidic conditions in lakes and oceans. The nitrogen cycle is dominated by biotic transformations controlled by bacteria, whereas phosphorus and sulfur cycles are dominated by geochemical processes.

learning outcomes

You should be able to:

- Compare and contrast the roles of different water bodies (rivers, lakes, oceans, glaciers, underground aquifers) in the hydrologic cycle, including approximate residence times of water molecules.
- Describe and evaluate data on the effect of fossil-fuel burning on atmospheric CO_2.

(continued)

> **57.4 recap** (continued)

- List the important forms of nitrogen moving through the nitrogen cycle, and explain the importance of microbes in this cycle.
- Describe the basics of the phosphorus cycle, including speed of movement through various parts of the cycle.
- Explain the pools and fluxes of the sulfur cycle, and describe differences in the cycling of sulfur on land and in oceans.

1. The residence time of a water molecule is short in organisms, soil, and rivers and is much longer in lakes, glaciers, and oceans. Why?
2. Why is the concentration of CO_2 rising in the atmosphere and oceans? How does this rise in CO_2 affect climate and ocean chemistry?
3. Explain how the nitrogen cycle is dominated by biological processes whereas the phosphorus and sulfur cycles are dominated by geochemical processes.

As you have seen, biogeochemical cycles are intimately involved in how ecosystems function. Just as human alterations of those cycles are affecting ecosystems worldwide, the resulting changes could have important effects on humans as well.

key concept

57.5 Ecosystems Provide Important Services and Values to Humans

Although it seems obvious today that humans depend on ecosystems for survival, explicit recognition of the services they provide and their value is rather recent. Environmental writers introduced the idea of "natural capital" in the 1940s; it was in 1970 that ecosystems were first said to provide people with a variety of "goods and services." Here we consider more recent efforts to understand the importance of ecosystems to humans.

focus your learning

- Ecosystems provide four types of benefits to humans: provisioning, regulating, supporting, and cultural services.
- Alteration of ecosystems for human benefits involves trade-offs that sometimes result in deterioration of ecosystem services.
- To develop more sustainable use of ecosystems, the value of ecosystem services must be measured.

Ecosystems services are the benefits people obtain from ecosystems. Ecologists recognize four different categories of services depending on their role in the ecosystem:

1. *Provisioning services* provide products such as food, clean water, timber, and fiber.
2. *Regulating services* regulate events such as hurricanes, floods, disease outbreaks, and water and air quality.
3. *Supporting services* function in such processes as soil formation, carbon sequestration via net primary production, and nutrient cycling.

4. *Cultural services* provide nonmaterial benefits such as recreational activities and aesthetic and spiritual enrichment.

Most of these benefits are either irreplaceable, or the technology necessary to replace them is prohibitively expensive. For example, fresh drinking water can be provided by desalinating seawater, but only at great cost. The aesthetic, spiritual, and recreational benefits of ecosystems are less tangible, but no less important, and no more easily replaced.

Although humans have been altering ecosystems for millennia, the pace and scope of human alterations to ecosystem services have increased considerably in the past century. For example, the 2005 Millennium Ecosystem Assessment estimated that approximately 60 percent of the ecosystem services identified in its study are being degraded or used unsustainably, including fresh water, wild-caught fisheries, air and water purification, and the regulation of regional and local climate, natural hazards, and pests.

Human alteration of ecosystems has had many positive effects on human health and prosperity, but it necessarily involves trade-offs. Agriculture, for example, feeds and employs huge numbers of people. But the spread of agriculture into marginal lands may degrade soils and compromise the ability of ecosystems to provide clean water, as when overuse of artificial fertilizers results in eutrophication. Extensive use of pesticides controls insect pests, but also reduces populations of pollinators and the services they provide to both crops and native plants.

Similarly, the loss of wetlands and other natural buffers has reduced the ability of ecosystems to regulate flooding and other natural hazards. The damage from the tsunami that hit Indonesia and other Southeast Asian countries in December 2004 was greater in many places than it would have been had the mangrove forests that protect the coast not been cut down and converted to shrimp aquaculture farms. Hurricane Katrina, which struck the U.S. Gulf Coast less than a year later, would not have caused as much flooding in New Orleans had the wetlands surrounding the city been intact. Katrina's devastating effects were due in part to a situation that had been developing for decades.

New Orleans is located on the Mississippi River delta. Much of the city lies below sea level, buffered by dams and levees constructed by the Army Corps of Engineers. The upstream dams that protect New Orleans from flooding also prevent the river from depositing the sediments that have sustained the surrounding delta wetlands for centuries. Oil and natural gas producers have cut thousands of small canals through those wetlands in order to lay pipelines and install drilling rigs, and the extraction of oil and gas from beneath the land has caused it to sink. Increased dredging of shipping lanes and rising sea levels have contributed to a rise in salinity, killing off many of the great cypress tree swamps. These extensive alterations resulted in the loss of more than 80 percent (1.2 million acres) of the delta wetlands between 1930 and 2005. By the time Katrina made landfall, those wetlands could no longer protect New Orleans from flooding. Storm surges raced along the paths carved by canals and shipping lanes to breach the levees, inundating much of the city.

The value of ecosystem services can be measured

Practices that allow us to conserve or enhance ecosystems so as to benefit from specific ecosystem goods and services without compromising others are referred to as **sustainable**. To develop

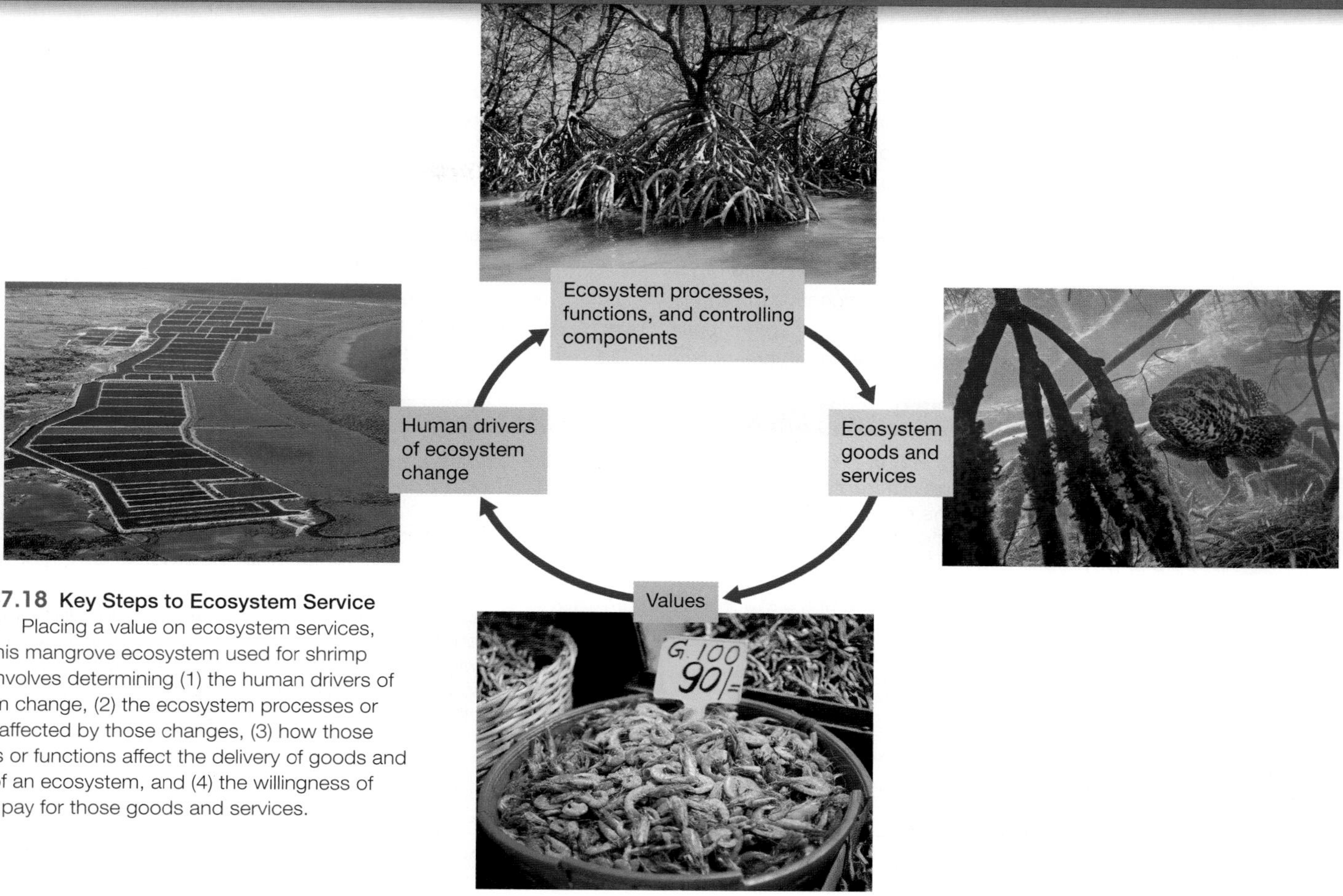

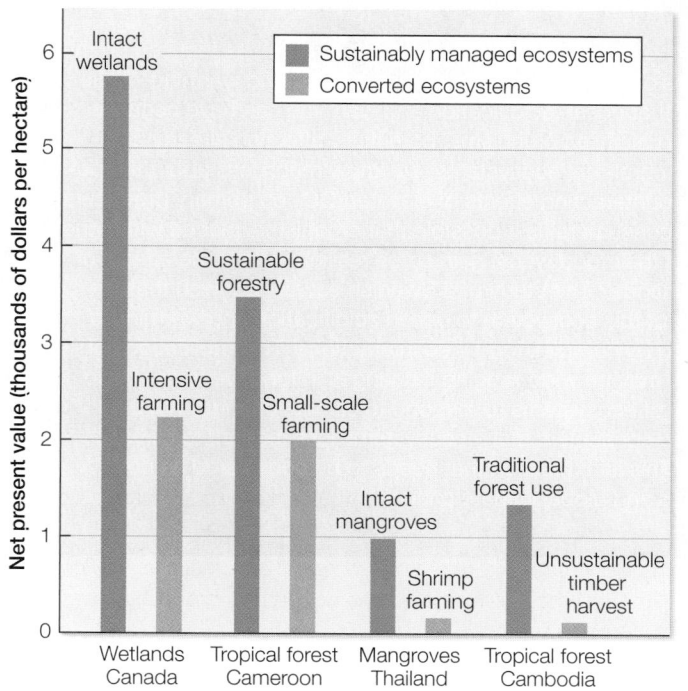

Ecosystem processes, functions, and controlling components

Human drivers of ecosystem change

Ecosystem goods and services

Values

Figure 57.18 Key Steps to Ecosystem Service Valuation Placing a value on ecosystem services, such as this mangrove ecosystem used for shrimp farming, involves determining (1) the human drivers of ecosystem change, (2) the ecosystem processes or functions affected by those changes, (3) how those processes or functions affect the delivery of goods and services of an ecosystem, and (4) the willingness of people to pay for those goods and services.

more sustainable practices, there has been growing interest in determining the economic value of ecosystem services. **Ecosystem values** are measures of how important ecosystem services are to people—that is, what they are worth. Placing a value on ecosystem services involves estimating the amount we are willing to pay to preserve or enhance ecosystem services. Ecosystem service valuation involves determining (1) the human drivers of ecosystem change, (2) the ecosystem processes or functions that are affected by those changes, (3) how those processes or functions influence all the goods and services of an ecosystem, and then (4) using a variety of economic and social evaluation tools to value those goods and services (**Figure 57.18**). Valuing ecosystem services can be difficult, because while it relatively easy to place a value on certain goods such as fish or timber, it is harder to value nonmarket services such as the existence of an endangered species or the ability to walk in an old-growth forest. Much of ecosystem service valuation involves determining through public surveys and census data the willingness of people to pay for such nonmarket services.

For those ecosystems in which services have been valued, in many cases the total economic value of a sustainably managed ecosystem is higher than that of a converted or intensively exploited ecosystem (**Figure 57.19**). For example, the value of coastal

Figure 57.19 The Economic Value of Sustainability Managed Ecosystems Many types of ecosystems are able to provide more goods and services, and thus have a higher value, when they are sustainably managed than when they are completely converted to human use.

Net present value (thousands of dollars per hectare)

- Sustainably managed ecosystems
- Converted ecosystems

Intact wetlands

Sustainable forestry

Intensive farming

Small-scale farming

Traditional forest use

Intact mangroves

Shrimp farming

Unsustainable timber harvest

Wetlands Canada   Tropical forest Cameroon   Mangroves Thailand   Tropical forest Cambodia

ecosystems such as mangroves can be driven largely by the coastal protection that mangroves provide from extreme storm or tsunami events, even though the short-term monetary gains from converting mangroves to shrimp farms can be substantial.

An important aspect of valuing ecosystem services is changing the impression that ecosystem services are "public goods" with no market value. People who do not stand to profit from the services provided by ecosystems have no incentive to pay for them, whereas individuals who convert ecosystems reap great economic benefits. Government action may be needed to create incentives that encourage sustainable ecosystem management. Raising public awareness is essential to the implementation of sustainable management programs. Most people do not realize the long-term value of ecosystem goods and services or understand how human activities affect the functioning of ecosystems. Maintaining and enhancing ecosystem goods and services that have no established market value is especially difficult. Perhaps one of the most difficult ecosystem services to maintain and value is that of biodiversity. The final chapter of this book is devoted to this important topic.

▶ 57.5 recap

Ecosystems provide benefits to humans in the form of goods and services. Ecologists recognize different categories of services depending on their role in the ecosystem. Ecosystem service valuation involves determining the market value or the willingness of people to pay for the services.

learning outcomes

You should be able to:

- List the four types of ecosystem services, explain the meaning of each, and give examples.
- Explain, with examples, the types of trade-offs that occur when humans change ecosystems for their own benefit.
- Define sustainability, and analyze its relationship to ecosystem services.

1. Mangroves provide coastal protection from extreme waves and winds, raw materials such as timber and fibers, erosion control, water purification, habitat for fisheries, carbon sequestration, and tourism, recreation, and education. Place each of these services into the four types of benefit categories.

2. Suppose you want to maximize the coastal protection service of mangroves with shrimp aquaculture farming (which requires removing the mangrove trees to create ponds). What sustainable solution might you devise to maximize the value of this ecosystem?

▶ investigatinglife

How will food webs respond to the multiple effects of ocean acidification and warming in marine ecosystems?

Much of the ocean acidification research to date has focused on the responses of single species to increasing CO_2 and warming. While we know that ocean acidification and warming can have negative effects on calcifying organisms, potentially affecting already compromised coral-reef ecosystems, we know much less about the potentially counteracting effects of elevated CO_2 on primary producers and their consumers. Is it possible that organisms, placed within the context of their food webs, gain some resilience from the effects of climate change? The study of a Swedish estuary showed that the response of this ecosystem to ocean acidification and warming is modulated by complex food web interactions. The implications of this study may be wide-ranging if changes in food webs decrease the ability of ecosystems to respond to climate change. Will the loss of top consumers make ecosystems less resilient to climate change? This is a question that needs to be considered as both food webs and global climate change.

Future directions

While the Swedish estuary experiment shows that a consumer can affect an ecosystem's response to climate change, climate change can also affect a consumer's role in an ecosystem. A good example is the recent dramatic shift in polar bear and orca whale observations in the Canadian Arctic. Sea ice in the Arctic is melting at unprecedented rates as the result of global warming. Polar bears need this sea ice to stalk their prey, mainly seals, but have been driven to inland hunting grounds with the loss of ice. Orca whales, by contrast, are unable to hunt their preferred prey, other species of whales, in ice-covered waters. As ice continues to melt, orcas have been spotted in areas of the Arctic where they have never been seen before. Some people have suggested that this shift from seal predation to whale predation could have important ecosystem-level effects as the food webs shift in response to the decline of sea ice.

Chapter Summary 57

57.1 Ecosystem Science Considers How Energy and Nutrients Flow through Biotic and Abiotic Environments

- An **ecosystem** includes all the organisms in a given area and their physical and chemical environment.
- Energy flowing through ecosystems originates with sunlight and inorganic and organic compounds. It involves the processes of photosynthesis, **chemosynthesis**, and **secondary production**. Review Focus: Key Figure 57.1, Activity 57.1
- Energy is lost from ecosystems as metabolic heat.
- Nutrients originate in soil, water, and the atmosphere and cycle through ecosystems via the processes of production and decomposition.

57.2 Energy and Nutrients in Ecosystems Are First Captured by Primary Producers

- **Net primary production** (**NPP**) is the amount of carbon remaining in plants after respiration, and varies with latitude and ecosystem type. Review Figures 57.2, 57.4
- Terrestrial NPP is highest in the tropics where temperatures are high year-round and rainfall is abundant. Review Figures 57.3A, 57.5
- Oceanic NPP varies little with latitude but peaks at mid-latitudes and in coastal zones and areas of upwelling where the main primary producers—phytoplankton and algae—have access to light and nutrients. Review Figure 57.3B
- In aquatic systems, nutrient runoff from agriculture and wastewater can cause **eutrophication**, resulting in algal blooms and low oxygen conditions (hypoxia) for aquatic life.

57.3 Food Webs Transfer Energy and Nutrients from Primary Producers to Consumers

- Once energy captured by primary producers is fixed as NPP, consumption by heterotrophs transfers a fraction of that energy to higher trophic levels, creating **net secondary production**. Review Figure 57.1
- Organisms differ in their **production efficiency**, or the percentage of energy stored in assimilated food that is used to produce new biomass. Review Figure 57.8, Table 57.1
- The amount of energy transferred within food webs depends on **trophic efficiency**, or the amount of energy used at one trophic level divided by the amount used at the trophic level below it. Review Figure 57.9
- The flow of energy in food webs is influenced by the number of trophic levels and the relative contribution of bottom-up (NPP passing up the food web) and top-down (consumers regulating NPP) forces. Review Figure 57.10A

- Most food webs have just three or four trophic levels because of energy limitations at higher trophic levels, the effects of disturbance, or the evolutionary constraints on top predators.

57.4 Nutrient Cycling in Ecosystems Involves Chemical and Biological Transformations

- Nutrients move through the biotic parts of ecosystems via production and **decomposition**. Review Figure 57.11
- At a global scale, nutrients move through geological, atmospheric, and biological pools, making up **biogeochemical cycles**.
- Water cycles rapidly through the globe as part of the hydrologic cycle, transporting nutrients and serving as an essential requirement for organisms. Review Figure 57.12, Animation 57.1
- Carbon moves through biological systems and returns to the atmosphere through respiration. Review Figure 57.13, Animation 57.2
- Burning of fossil fuels increases the carbon dioxide and sulfur in the atmosphere, causing global warming and acidic conditions in oceans and lakes. Review Figures 57.14, 57.15, Investigating Life: Food Webs in an Acidic and Warming Ocean
- The nitrogen cycle is dominated by biotic transformations controlled by bacteria, whereas the phosphorus and sulfur cycles are dominated by geochemical processes. Review Figures 57.16, 57.17, Animation 57.3

57.5 Ecosystems Provide Important Services and Values to Humans

- **Ecosystem services** benefit humans and can be categorized as those that provide provisioning, regulating, supporting, or cultural services.
- Most ecosystem services are either irreplaceable or prohibitively expensive to replace.
- **Ecosystem values** are measures of how important ecosystem services are to people—that is, what they are worth. Review Figure 57.18
- Ecosystem services can be valued by determining the market value or the willingness of people to pay for the services. Review Figure 57.19

See Activity 57.2 for a concept review of this chapter.

> Go to **LearningCurve** (in **LaunchPad**) for dynamic quizzing that helps you solidify your understanding of this chapter. **LearningCurve** adapts to your responses, giving you the practice you need to master each key concept.

▶ Apply What You've Learned

Review

57.1 Autotrophs and heterotrophs obtain required nutrients in different ways.

57.2 Primary production in aquatic systems is limited by light (which is limited by depth) and by the availability of different nutrients, depending on the type of system.

57.3 Omnivory and the number of trophic levels affect the transfer of energy through food webs.

57.4 Decomposition and mineralization are necessary to complete nutrient cycling in ecosystems.

57.5 Alteration of ecosystems for human benefits involves trade-offs that sometimes result in deterioration of ecosystem services.

57.5 To develop more sustainable use of ecosystems, the value of ecosystem services must be measured.

Original Paper: Slater, M. J. and A. G. Carton. 2009. Effect of sea cucumber (*Australostichopus mollis*) grazing on coastal sediments impacted by mussel farm deposition. *Marine Pollution Bulletin* 58: 1123–1129.

Some coastal marine ecosystems receive excess organic matter from aquaculture farms. This can concentrate organic matter in the sediment and decrease oxygen, leading to hypoxia. Scientists believe that benthic organisms, such as sea cucumbers, might help mitigate this potentially detrimental effect. Sea cucumbers feed on sediments containing bacteria, benthic marine algae, and the feces of cultured species; thus they recycle organic matter and alter nutrient fluxes in the ecosystem. They also disturb sediments, causing some amount of mixing with the water column above.

Researchers studied the effects of sea cucumber grazing on the chemistry of sediments produced by cultured mussels, filter feeders on phytoplankton. The researchers set up 16 experimental tanks with flowing filtered seawater. Three tanks were maintained as controls, with no grazing, and one tank was kept as a water filtration control. Each of the other 12 tanks housed a single sea cucumber, which was fed a specific amount of mussel-farm sediment daily. The experiment was run for four grazing periods that lasted 1, 2, 4, and 8 weeks. Chemical tests were run on the sediments at the end of each grazing period.

The graph shows the results of tests for total organic carbon (TOC). TOC measures all organic carbon including live and dead matter and not including CO_2 released from metabolism. It is sometimes used to measure organic pollution from sewage or agriculture.

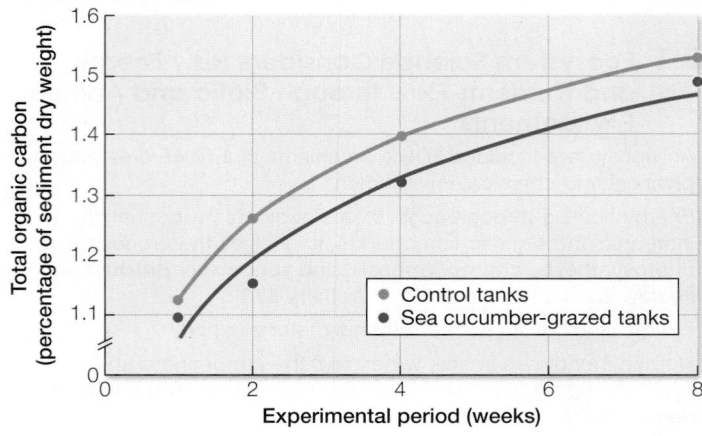

Questions

1. How would you describe the position of sea cucumbers in the marine benthic food web? How might their position affect net primary production (NPP) through this ecosystem?

2. Based on the graph, what can you conclude about the effect of sea cucumbers on the amount of TOC in sediments? How is this effect related to nutrient cycling?

3. Why do you think TOC (see graph) continues to increase over the course of the experiment? How does this increase relate to the activities of sea cucumbers?

4. A major problem in aquaculture operations such as mussel farming is hypoxia, or dangerously low levels of oxygen in the water. Given the information on TOC in this experiment, how do you think oxygen levels likely changed over the course of the experiment? What factors would cause these changes, and how does the presence of sea cucumbers affect the likelihood of hypoxia?

5. An aquaculture manager cultures mussels in a coastal farm. He has a serious pollution problem, due to the waste his mussels release into the water, and the resulting hypoxia. He is considering whether to introduce sea cucumbers into his farm. What ecological trade-offs occurred when the mussel farm was first developed? Would the addition of sea cucumbers into the system be considered an ecosystem service? Explain your answer.

Go to **LaunchPad** for the eBook, LearningCurve, animations, activities, flashcards, and additional resources and assignments.

A Changing Biosphere

A small population of the harlequin toad (*Atelopus varius*), a species in Costa Rica that was presumed extinct in 1996, was found in a mountainous protected area in 2003.

▶ investigatinglife

Fatal Fungus Final Fate for Frogs?

Most of us have at some point in our life encountered a frog in a local pond or park. Maybe more than almost any other wild animals, amphibians are commonly encountered by human beings. But since the late 1980s, ecologists have documented a sharp global decline in amphibian species— more than in any other major taxonomic group on Earth. Some regions show greater declines than others, but since 1980, 33 species of amphibians have become extinct, with an additional 115 species "missing in action" (not seen since that time) and 848 species in danger of extinction. A study by the International Union for Conservation of Nature (IUCN) showed that amphibians are more threatened than either birds or mammals, with 31 percent of amphibian species (1,994) being globally threatened and 37 percent (2,394) experiencing some type of population decline.

The decline has been particularly worrisome to scientists for a few key reasons: the sudden and global nature of the decline, the undisturbed nature of the areas where some the declines have occurred (for example, mountainous rainforest preserves in Costa Rica, where 40% of the amphibians have become extinct), and the reputation of amphibians as "biological indicators" of environmental conditions. Amphibians have permeable skin, live their lives in water and on land,

and typically do not move long distances, factors that make them particularly susceptible to changes in their immediate environment. The IUCN study showed that most amphibian declines were caused by habitat loss, habitat degradation, and overexploitation, but for 48 percent of the rapidly declining species, there was no clear explanation. What factors might be contributing to the rapid decline of these species?

Over the last two decades, a highly lethal and infectious skin disease called chytridiomycosis, caused by a fungal pathogen (*Batrachochytrium dendrobatidis*), has been linked to mass mortalities of amphibians, especially in the regions where amphibian declines have been most dramatic. The disease was first discovered in 1993 in dead frogs in Australia and then seemed to appear suddenly in the Americas, Africa, Europe, and New Zealand. However, it is not clear how or why the fungal disease has emerged as a threat, and whether humans have contributed to its spread. An increasing number of studies are looking for reasons why this deadly disease has emerged and how it might be prevented.

QA Is the most recent dramatic decline of amphibians the result of a novel fungal pathogen, and if so, what can be done about it?

58.1
Human Activities Are Changing the Biosphere, Resulting in Biodiversity Loss

Throughout Part Ten we have described the dramatic ways in which the biosphere is changing as a result of human activities. The causes of these changes have been varied, and include habitat destruction and degradation, overharvesting of species, species invasions and emerging diseases, and now climate change. One of the largest ecological consequences of these changes has been the rapid decline of biodiversity, or the loss of diversity at genetic, population, species, ecosystem, and global scales.

focus your learning

- Various factors decrease effective population size, eventually leading to extinction.
- Biodiversity connects species at all levels; its loss affects population, metapopulation, species, ecosystem, and global levels.
- It is very difficult to track and predict extinctions.

Biodiversity has great value to human society

The escalating loss of Earth's biodiversity is of great concern for many reasons:

- *Humans depend on thousands of species and their ecosystems for ecosystem goods and services.* As you learned in Key Concept 57.5, species and ecosystems provide humans with important goods (e.g., food, timber, and fiber) and services (e.g., regulating and supporting services such as coastal protection, water and air quality, soil formation, and carbon sequestration).

- *Humans derive enormous psychological benefits, including aesthetic pleasure, from interacting with other organisms.* These aesthetic benefits are the reason we surround ourselves with parks and recreate in the wilderness.

- *Living in ways that cause the extinction of other species raises ethical issues.* Loss of biodiversity is of concern to anyone who believes humans have an ethical obligation to the natural world.

- *Extinctions deprive the public and scientific community of opportunities to study and understand ecological relationships among organisms.* The more species that are lost, the more difficult it will be to understand the biosphere as it exists today.

Recognizing the importance of biodiversity propels us to develop strategies for protecting and managing it and has spawned the development of **conservation biology**—an integrative scientific discipline that relies on principles of ecology, economics, social science, and policy to protect and manage Earth's biodiversity. As you will see, we will refer to many ecological concepts from other chapters to establish, first, how and why biodiversity is being lost

Human-caused habitat destruction, overharvesting, species invasions and emerging diseases, and climate change cause the loss of individuals...

...leading to smaller and less genetically diverse populations, which are more susceptible to extinction (see Figure 58.2). The fewer populations there are, the smaller the metapopulations and the greater the risk of species extinction.

Once species are lost, species diversity can decline if those species play critical roles in the ecosystem.

Declines in species diversity in ecosystems lead to losses at regional, continental...

...and global scales.

Figure 58.1 Biodiversity Loss Is Interconnected across Scales   Human activities have caused the rapid decline of biodiversity, or the loss of diversity at genetic, population, species, ecosystem, and global scales. Implicit in biodiversity loss is the concept that the loss of diversity at one scale affects the loss of diversity at other scales (see Figure 56.5).

and, second, the role of conservation biology in protecting against and managing its further loss.

Diversity loss at one scale affects diversity loss at other scales

As you have seen in Key Concept 56.1, biodiversity is central to the interconnectedness of various levels of biological organization, from genes to species to communities. Thus the loss of diversity at one scale affects the loss of diversity at other scales (**Figure 58.1**). For example, when human-caused activities reduce the number of individuals in a population, they reduce what is called the **effective population size**, or the number of individuals that can contribute offspring to the next generation (**Focus: Key Figure 58.2**), resulting in further population size decline, or what has been termed an "extinction vortex." How does the effective population size continue to decline? First, small populations are more likely to experience *genetic drift and inbreeding, which can lead to reduced genetic diversity and **inbreeding depression**. Second, when population sizes are small, there is also a greater chance of **demographic stochasticity**, or fluctuations in population size as the result of random differences among individuals in reproduction and survival. For example, a small population might have, by chance, fewer females than males, resulting in fewer offspring than expected if the sex ratio were reversed. Finally, external mortality events such as extreme weather or habitat destruction can have devastating effects on small populations by increasing their chance of extinction even if only a small number of individuals are killed.

---

*connect the concepts   As described in Key Concept 12.3, inbreeding depression is a state of reduced biological fitness in a population arising from mating among close relatives that tend to have the same recessive, sometimes deleterious, alleles. Key Concept 21.2 describes how in small populations, genetic drift—random changes in allele frequencies from one generation to the next—may produce large changes in allele frequencies over time. Harmful alleles may increase in frequency, and rare advantageous alleles may be lost.

Biodiversity loss is also connected at higher metapopulation, species, and ecosystem levels (see Figure 58.1). For example, you know from Key Concept 54.4 that the larger the metapopulation, the greater the chance that dispersal among source populations can rescue sink populations from extinction. As more local populations become extinct, and as more metapopulations are compromised, there is a greater risk that species will become extinct as well. Unfortunately, if some of those species play critical roles in their ecosystems—if they are, for example, keystone or foundation species—the possibility exists that other species will be threatened with extinction as well. As species diversity declines in ecosystems, biodiversity loss manifests itself at larger regional, continental, and even global scales (see Figure 58.1).

There are many examples of how a series of unfortunate events at different levels of biodiversity has led to the extinction of species (or subspecies, in the case below). Consider the heath hen (*Tympanuchus cupido cupido*), a large bird in the grouse family (**Figure 58.3A**). Heath hens were common from Maine to Virginia prior to the 1800s but were hunted extensively for food. By 1830 only one population, on the island of Martha's Vineyard, Massachusetts, was left. By 1908 only 50 birds remained, and a reserve was established to recover the population, which eventually grew to several thousand birds. Unfortunately, in 1916 a fire and a series of environmental factors that included a hard winter, an influx of predators, and a poultry disease caused the population size to decline dramatically. By 1928 demographic stochasticity and inbreeding depression set in, leaving only 13 individuals—2 females and 11 males. Four years later, the heath hen was extinct.

Species diversity is being lost at unprecedented rates

The loss of the heath hen and other extirpations caused by humans (see Figure 58.3) remind us of the finality inherent in species

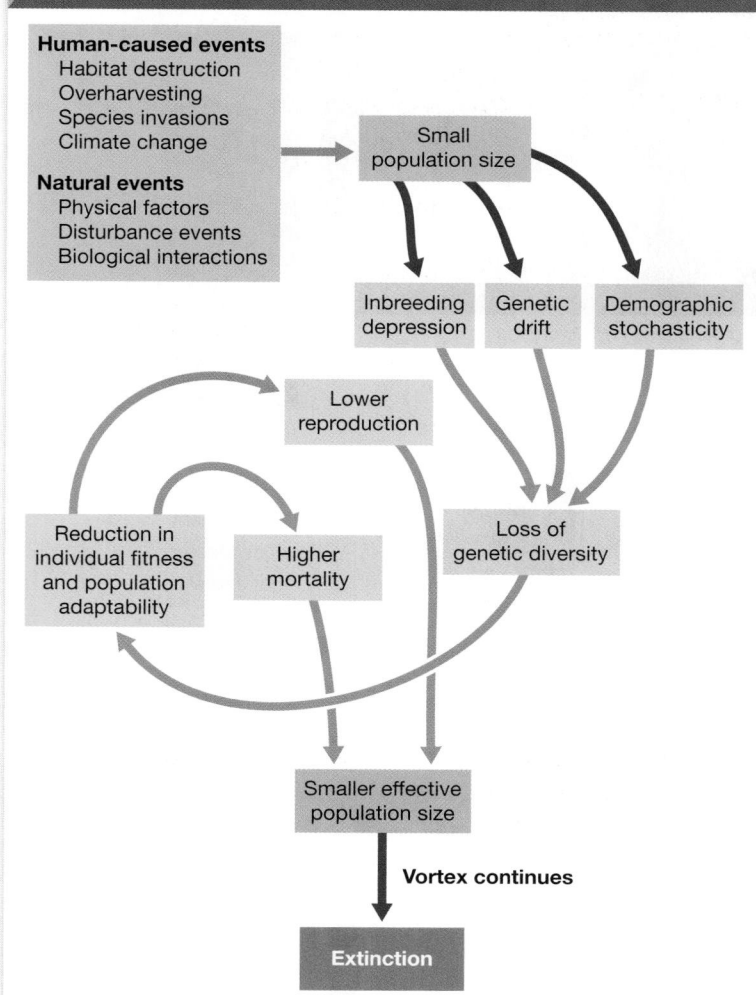

focus: key figure

Figure 58.2 Extinction Vortex   Human-caused and natural events can reduce the effective population size (number of individuals that can contribute offspring to the next generation) of species, eventually leading to population and species extinctions.

Q: Explain how the chytrid fungal pathogen described in the opening story could cause the extinction of a species such as the harlequin frog.

extinctions. However, over the scope of Earth's history extinction has been a constant theme; most of the species that have lived on Earth are extinct today. Extinctions have occurred throughout Earth's history at what is referred to as a "background" rate as changes in environmental conditions have favored some species and negatively affected others. But the rate of extinctions taking place today rivals those of the five great mass extinction events, which were the result of cataclysmic natural disturbances rather than human activities (see Table 24.1 and Figure 24.2). One estimate, from renowned evolutionary biologist Edward O. Wilson suggests that Earth is losing some 30,000 species per year, putting us in the midst of a sixth mass extinction event.

This mass extinction event has been ongoing for some time. For example, research suggests that when humans first arrived in North America from Siberia about 14,000 years ago, they

(A)

Tympanuchus cupido cupido

Tarsius pumilus

(B)

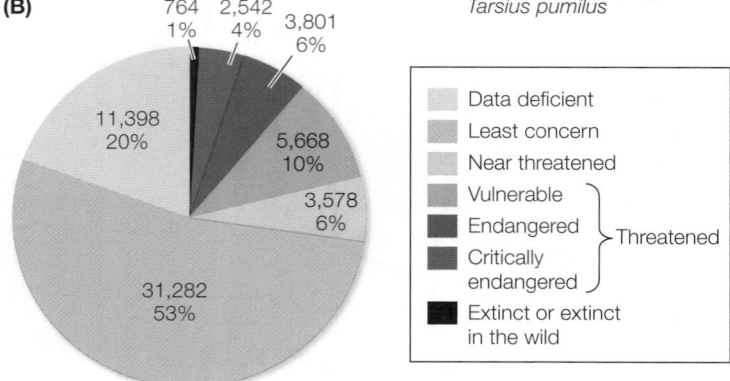

764 1%
2,542 4%
3,801 6%
11,398 20%
5,668 10%
3,578 6%
31,282 53%

Legend:
- Data deficient
- Least concern
- Near threatened
- Vulnerable
- Endangered } Threatened
- Critically endangered
- Extinct or extinct in the wild

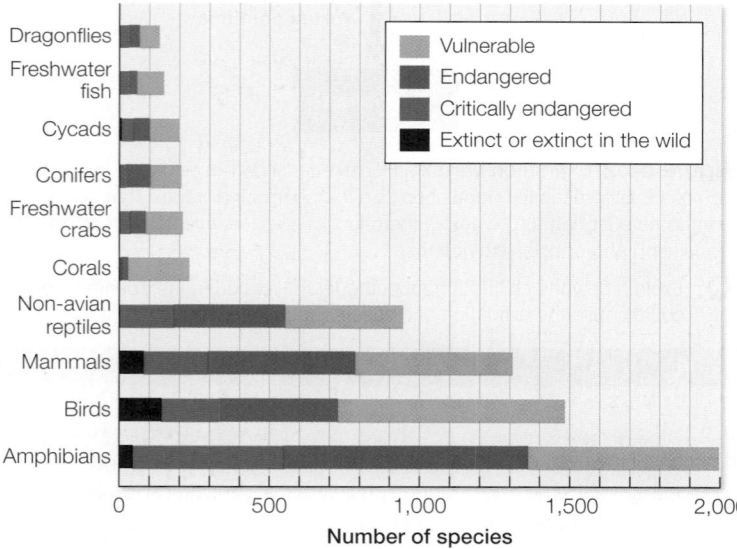

Bar graph categories (top to bottom): Dragonflies, Freshwater fish, Cycads, Conifers, Freshwater crabs, Corals, Non-avian reptiles, Mammals, Birds, Amphibians

Legend:
- Vulnerable
- Endangered
- Critically endangered
- Extinct or extinct in the wild

X-axis: Number of species — 0, 500, 1,000, 1,500, 2,000

Figure 58.3 Species Extinctions: Going, Going ... **(A)** The heath hen (left) became extinct in 1932. In 2008 the pygmy tarsier (right) was discovered in an Indonesian national park after having been presumed extinct for 85 years. **(B)** The pie chart shows the breakdown by extinction risk category of 59,033 species assessed by the International Union for Conservation of Nature. The bar graph shows the number of species in the various extinction risk categories (three of which are together termed "threatened") in taxonomic groups that have been comprehensively assessed.

Q: What percentage of species are extinct versus threatened with extinction? Of those species, which taxonomic group has experienced the most extinction and which is most at risk of extinction?

encountered a diverse and spectacular group of large mammals, including saber-toothed cats, dire wolves, mammoths, mastodons, giant ground sloths, and giant beavers. Most of this "megafauna" became extinct within a few thousand years after humans arrived. Although several hypotheses have been advanced to account for the rapid and simultaneous disappearance of so many large animals, overhunting by humans is the most likely explanation. Losses of megafauna coinciding with the arrival of humans have been documented worldwide, most notably in Australia, New Zealand, and Madagascar. Tracking and predicting extinctions are, however, difficult for several reasons:

- *We do not know how many species live on Earth today.* Many species that are likely to become extinct in the near future have not been named and described by scientists. Insects provide a case in point: although more than 1 million species have been described (see Key Concept 31.4), estimates of the number of species yet to be discovered range from 2 million to more than 50 million. Even for larger organisms, our understanding of diversity is incomplete. For example, worldwide, an annual inventory of newly described species reported 19,232 species discovered in 2009 alone; this list included 9,738 insects, 2,184 plants, 1,360 fungi, 71 mammals, and 7 birds.

- *The distributions of most described species, particularly those that are small, reclusive, and rare, are poorly known.* One tiny North American true bug, *Corixidea major* (so rare it has no common name), had been found in only one location near Clarksville, Tennessee, until entomologists using lights to collect insects at night discovered it in Virginia and Florida, extending the species' known range by more than 1,000 kilometers.

- *It is difficult to determine whether a species is truly extinct.* Rarely is the death of the last surviving member of a species recorded with certainty, as it was in the case of the last passenger pigeon (*Ectopistes migratorius*), a female named Martha, that died in the Cincinnati Zoo on September 1, 1914. The status of rare, reclusive species with poorly known life histories is much more difficult to determine, as has been the case with the pygmy tarsier (*Tarsius pumilus*) (see Figure 58.3A). This tiny primate, weighing less than 60 grams, was thought to have become extinct in its native cloud forests on the island of Sulawesi in Indonesia. In 2008—85 years after the last reported sighting of a living *T. pumilus*—a research team discovered individuals of this species living in one of the island's national parks.

- *We rarely know all of the interactions among species.* The loss of one species—be it as a member of a food web or as habitat for species—can place other species at risk. How many species are at risk as a result of their interactions with other species is hard to characterize and often requires detailed ecological knowledge.

 Media Clip 58.1 **New Species Found in the Twenty-First Century** www.Life11e.com/mc58.1

We can predict the effects of human activities on biodiversity

Despite gaps in our understanding of biodiversity, methods exist for estimating the extinction probability of species, taking into account their population sizes, genetic variation, life history traits, and ecology. The IUCN has published categories that define the risk of extinction of species. Species in imminent danger of extinction in all or most of their range are classified as "critically endangered" or "endangered"; those believed to be susceptible to extinction in the near future are classified as "vulnerable." Biologists consider species in any one of these three categories to be "threatened" (**Figure 58.3B**).

Rarity in and of itself is not always a cause for concern. Some species may be specialized for rare and unusual habitats and not be particularly abundant by nature. But as you saw earlier, "newly rare" species are of concern because large and rapid reductions in population sizes can lead to smaller effective population sizes through a variety of mechanisms (see Figure 58.2).

Certain aspects of species' life histories can be important in predicting the ability of species to recover from declines (see Key Concept 54.3). In fishes and mammals, for example, one of the best predictors of extinction risk is age at maturity, a life history trait that influences the rate of reproduction. Ecological requirements can also influence the ability of species to recover from rapid declines. Species with specialized habitat or dietary requirements, for example, are more likely to become extinct than species with more generalized requirements.

Ecologists can apply the principles of the species–area relationship and the theory of island biogeography (see Key Concept 53.5) to predict the effects of habitat loss on species—a major cause of extinction today. By measuring the rate at which species richness decreases with decreasing habitat patch size, they can estimate how many species that live in and depend on that habitat may be lost. We examined the connection between habitat area and species loss in Investigating Life: The Largest Experiment on Earth in Chapter 53.

▶ 58.1 recap

Biodiversity has great value to human society in the form of goods and services. Human activities have resulted in a rapid loss of biodiversity, a decline detectable at genetic, population, species, ecosystem, and global scales. Species extinctions can occur when human and natural events reduce effective population sizes and cause inbreeding depression, genetic drift, and demographic stochasticity. Methods exist for estimating the extinction probability of species that take into account their population sizes, genetic variation, life history traits, and ecology.

learning outcomes

You should be able to:

- List major factors causing declines in biodiversity.
- Explain the concept of effective population size; use examples to explain how small population sizes can lead to population and species extinctions.
- Discuss, with examples, why it is difficult to determine the extinction rates of species.

▶ 58.1 recap

1. Define biodiversity loss, and list four reasons for its loss.
2. Using the extinction vortex concept as your guide, describe how the heath hen became extinct in 1932.
3. Why is it difficult to determine how many species are extinct or threatened with extinction?

We've said that human activities have played a major role in the loss of biodiversity. In the next section we will describe the types of human activities that most threaten biodiversity.

▶ key concept

58.2 Most Biodiversity Loss to Date Is Caused by Habitat Loss and Degradation

Global Biodiversity Outlook 4, a report published by the United Nations in 2014, identified four principal pressures on biodiversity: (1) habitat loss and degradation, (2) overharvesting of biological resources, (3) species invasions and emerging diseases, and (4) climate change. Here we consider each of these pressures separately while keeping in mind that the threats to biodiversity are often not mutually exclusive.

focus your learning

- Human-caused habitat loss and degradation are the major causes of biodiversity loss.
- Overharvesting of single species for resources has been and still is an important cause of extinction or species decline.
- Introduction of non-native species has increased exponentially in the last 200 years, and some introduced species become invasive, usually with negative consequences.
- Climate change is affecting species through changes in distribution, timing of life history events, and decreased growth and reproduction.

Habitat loss and degradation endanger species

Most scientists agree that, up to this point in human history, **habitat loss** (reduction in habitat quantity) and **habitat degradation** (reduction in habitat quality) have been the major culprits in biodiversity loss. Our human footprint is large—it is estimated that we have transformed 50%–60% of the land surface, primarily for agriculture, timber, and livestock grazing. The building of infrastructure, such as cities, transportation corridors, and utilities (including hydroelectric dams) has altered another 2%–3% of Earth's surface.

As humans increasingly dominate the planet, the transformations they are causing affect whole biomes and the species dependent on them. For example, the current rate of loss of tropical rainforest—Earth's most species-rich biome—is about 2 percent of the remaining forest each year due to the increasing demands of a rapidly expanding human population for forest products and

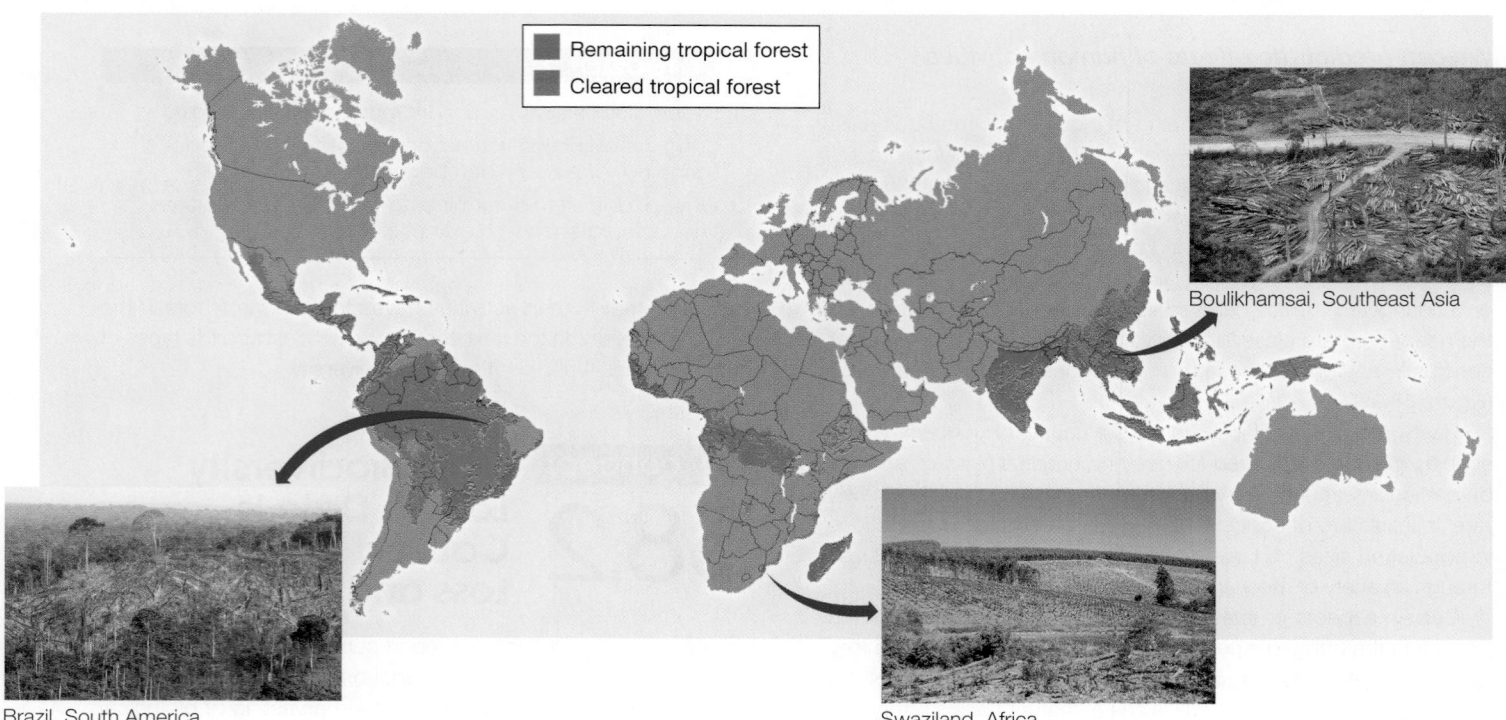

Boulikhamsai, Southeast Asia

Brazil, South America

Swaziland, Africa

Figure 58.4 The Disappearing Tropical Forest   Rainforests, the most abundant tropical forests, have long been recognized as centers of biodiversity that harbor vast numbers of species (see rainforest biome, p. 1154). Since the 1950s, tropical forests have been cleared for agriculture, highways, timber resources, and other needs of an exploding human population.

cleared agricultural land. Most of the rainforests of Asia have already been reduced to small fragments, with the only extensive remaining tracts being found on the islands of New Guinea and, to a much lesser extent, Borneo (**Figure 58.4**). As of 2016, the greatest amount of tropical rainforest deforestation had taken place in Brazil. If the current rate of loss continues, millions of rainforest species could become extinct before the end of this century.

Physical destruction of habitat, as when tropical rainforests are cut down or wetlands are drained and converted to agricultural use, has significant effects on the distributions and abundances of species. Reductions in suitable habitat have contributed to the extinction of thousands of species. As the remaining habitat gets divided into smaller and smaller fragments, it can become further degraded by edge effects (as you saw in Investigating Life: The Largest Experiment on Earth in Chapter 53 and Animation 53.4, Edge Effects). Recall that as habitat fragments become smaller, proportionally more land is exposed to edge effects (see Key Concept 53.5). The physical conditions at the edges of habitats often are more similar to those of the new habitat than to the original habitat, and this can be physically stressful for species acclimated to the original habitat. In addition, species from surrounding habitats can colonize the edges, where they may compete with or prey on the species living in the fragment.

One effect of forest fragmentation in much of North America has been an increase in the abundance of the brown-headed cowbird (*Molothrus ater*), a brood parasite that lays its eggs in the nests of other bird species. Nestlings of a brood parasite are raised by the host parents, to the detriment of their own young (**Figure 58.5A**). Historically, cowbirds followed bison and other grazing mammals, feeding on insects kicked up by the herds; thus their eggs were laid primarily in nests of grassland host species. Forest fragmentation has opened up new opportunities for the cowbirds, which can now lay their eggs in the nests of forest birds in habitat edges, affecting nestling mortality and fledgling success (**Figure 58.5B**). Fragmented forests, with relatively more edge than intact forest, thus favor the proliferation of cowbirds at the expense of forest birds.

Pollution is another cause of habitat loss and degradation. The negative effects of acid rain (see Key Concept 57.4), for example, have greatly affected lake and forest ecosystems. Among the most troublesome toxic pollutants in ecosystems today are heavy metal waste products of mining and manufacturing, and synthetic organic chemicals (pesticides) released into the environment to control pests. Multiple studies have implicated a variety of pesticides in the decline of amphibian species, particularly in areas with intense agriculture such as California and the midwestern United States.

A relatively hidden form of pollution comes from nondegradable plastic garbage in open ocean habitats (**Figure 58.6**). One garbage patch in the middle of the Pacific Ocean is estimated to be the size of Texas. Plastic is broken up into smaller pieces and when ingested by marine birds, mammals, or fish can be a choking hazard or disrupt endocrine functions such as reproduction, neural development, and immune function. Plastic or abandoned fishing nets can also entangle marine organisms, including marine mammals, resulting in death.

Overharvesting has driven many species to extinction and changed food webs

Overharvesting occurs when single species, or groups of similar species, are harvested for human use. Overharvesting for such

Figure 58.5 Cowbird Parasitism at Forest Edges **(A)** Cowbirds are nest parasites—they lay their eggs (black speckled) in the nests of host bird species and depend on them to raise their young (note the large cowbird chick). **(B)** At forest edges, host birds have higher nestling loss and lower fledging success as a result of cowbird parasitism.

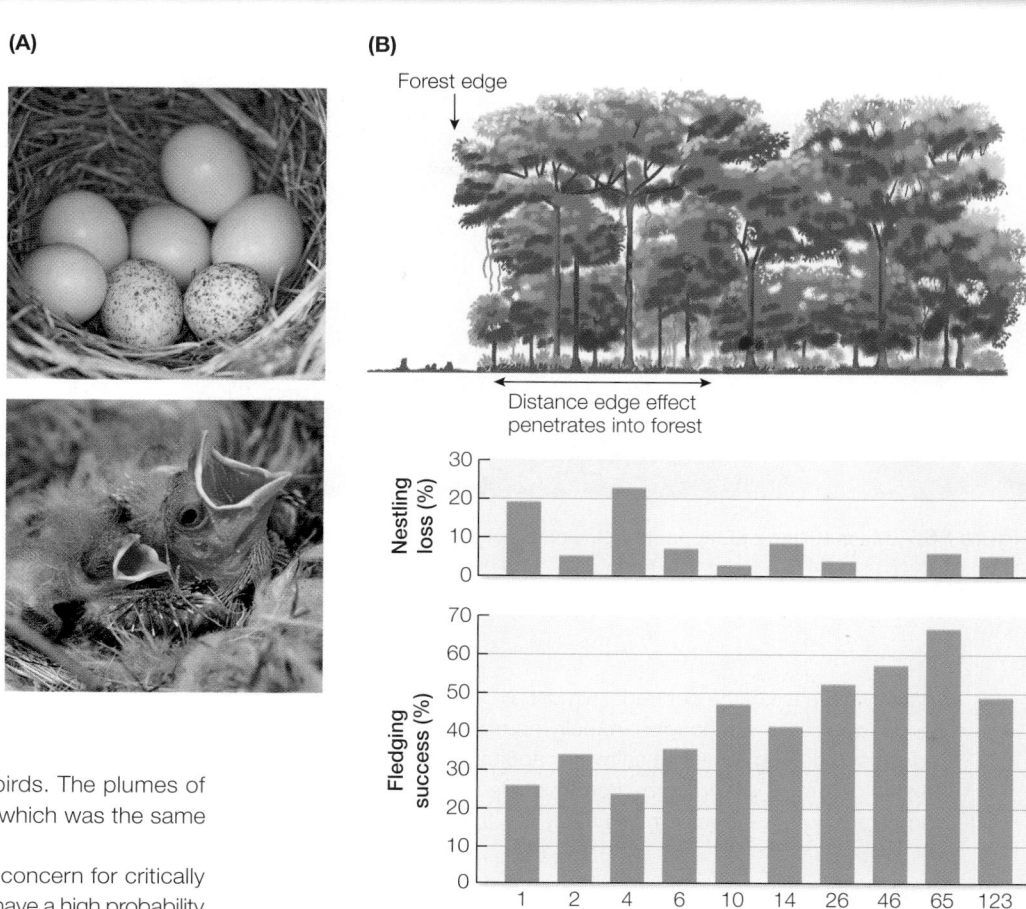

resources as food, clothing, ornamentation, pets, and medicines was once the most important and rapid cause of species extinction. For example, it has been estimated that overharvesting is responsible for roughly 40 percent of the approximately 150 bird species that have become extinct since the 1500s. Maybe one of the most egregious examples of overharvesting of birds came from "plume hunting" in the late nineteenth century. Feathers were harvested for ornamentation on hats and clothing—during the heyday, 5 million birds were killed each year, including 95 percent of Florida's shorebirds. The plumes of white egrets sold for $32 per ounce in 1915, which was the same price as gold.

Even today, overharvesting is still of great concern for critically endangered species that are not yet extinct but have a high probability of becoming so within the century, particularly as their habitat declines or is further degraded. Recently there has been an alarming uptick in illegal poaching of what might be considered charismatic megafauna, or species with widespread popular appeal. One example is the illegal poaching of elephants and rhinoceroses in much of Africa and Asia because of markets for ivory tusks and horns. A recent study estimated that in central Africa alone, regional elephant population sizes have declined by 64 percent because of illegal killing within the last decade (**Figure 58.7**). On a continental scale, the study showed that between 2010 and 2012, 40,000 elephants were killed illegally, a rate of 8 percent per year.

Massive and lucrative international trade in exotic pets such as birds, non-avian reptiles, frogs, mammals, aquarium fishes and corals, and ornamental plants threatens many species. It is estimated that 2–5 million birds and 2–3 million non-avian reptiles are captured live each year. In addition, it has been estimated that more than 1 billion ornamental fish comprising more than 4,000 freshwater and 1,400 marine species are traded internationally each year. For example, the Banggai cardinalfish (*Pterapogon kauderni*) is on the brink of extinction

Figure 58.6 Plastic Ocean Large amounts of plastic garbage have accumulated in the North Pacific Gyre, degrading the pelagic environment. Inset is a photo of tiny bits of plastic comingled with a sample of water taken from the vicinity of the North Pacific Gyre garbage patch.

Figure 58.7 Mass Killing of Elephants One of the largest known mass elephant killing events took place in Bouba Njida National Park, Cameroon, in 2012. Poachers killed more than 300 elephants with AK-47s and grenades.

endangered species but can unintentionally lead to disease transmission or species invasions, as you saw with the lionfishes example in Chapter 55.

Even though we tend to think of the oceans as boundless, they too are affected by overharvesting. Burgeoning human populations in need of food are placing unprecedented pressure on species harvested from the wild. Humans have fished for at least 40,000 years, but in recent centuries innovations in technology and increasing demand have accelerated overharvesting. A recent comprehensive study of fisheries found that nearly 14 percent of fished taxa collapsed between 1950 and 2007 (**Figure 58.8A**). The effects of fish harvesting vary widely across the globe, in part because restrictions in some locations have been implemented to encourage the recovery of overfished species (**Figure 58.8B**). Top predators such as sharks, bluefin tuna, and groupers, which are preferentially harvested, are at particular risk of extinction. These long-lived species show slow population recovery, and because they are keystone species, their losses are likely to have more lasting ecosystem ramifications.

What are some of the consequences of single-species overharvesting, particularly of species that are long-lived or top predators? First, species such as elephants and rhinoceroses that are characterized by slow population growth show slower recovery. Blue whales are at greater risk of extinction than are humpbacks,

entirely because of the pet trade; almost a million of these critically endangered fish are caught annually near Sulawesi, Indonesia, to satisfy the demand from saltwater aquarium enthusiasts. The exotic pet trade is worrisome from multiple aspects—harvesting and transporting species not only cause the decline of rare or

(A)

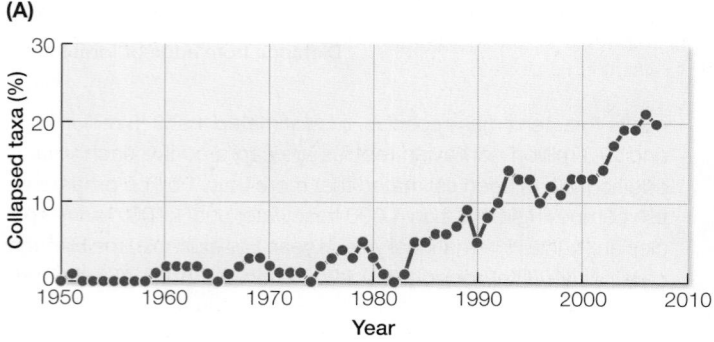

(B)

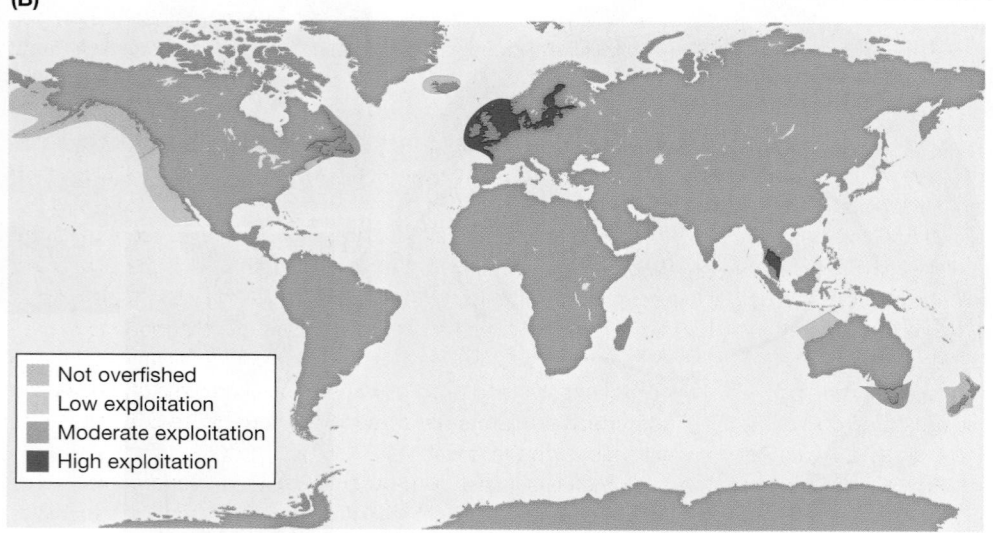

- Not overfished
- Low exploitation
- Moderate exploitation
- High exploitation

Figure 58.8 Overharvesting Has Caused Fisheries Collapses, but Some Recovery Is Occurring
(A) The number of collapsed fished taxa (estimated as a decline of biomass of the fished taxa to less than 10% of their unfished biomass) increased steadily from 1950 to 2007. **(B)** Worldwide status of exploited fish populations.

Q: Using the map, indicate which regions of the world are experiencing the highest fisheries exploitation.

even though both species experienced chronic overharvesting in the nineteenth century. You saw in Key Concept 54.4 that while humpback whales have been recovering relatively quickly with the cessation of hunting, blue whale populations have lagged behind because of their lower reproductive rates. Recovery of populations of the black rockfish (*Sebastes melanops*), which we introduced in Key Concept 54.4, has been extremely slow because older, larger females were preferentially harvested, and the remaining smaller females produce fewer and smaller larvae.

Second, consider what happens when an overharvested species plays an important keystone or foundation role in its community. A recent study by William Ripple and colleagues considered the status and ecological consequences of the decline of 31 of the world's largest carnivores. Most of the species studied have experienced substantial population declines over the past two centuries as the result of overharvesting, persecution (culling because of conflicts with humans), and habitat loss. Using data from several studies, the researchers showed that when top large carnivores declined, trophic cascades were interrupted, dramatically changing the abundance of herbivore and primary producer species (**Figure 58.9**). This research suggests that large carnivores are necessary to maintain biodiversity and ecosystem function in some ecosystems.

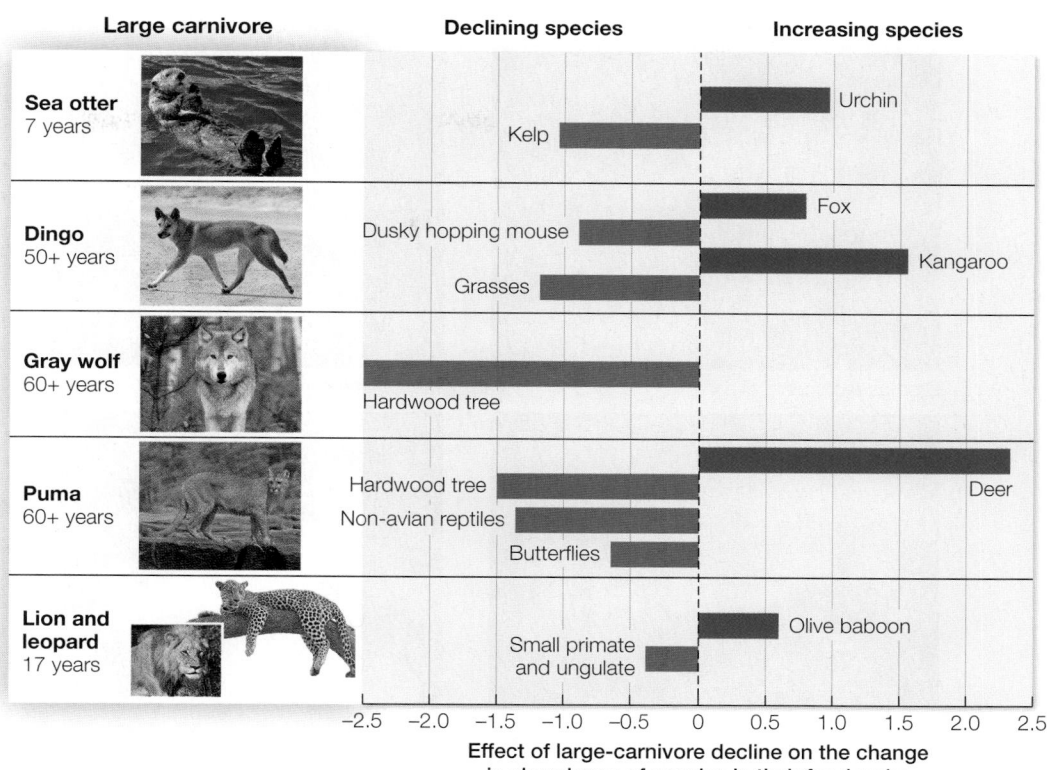

Figure 58.9 Large Carnivores Are Declining, Changing Food Webs Examples of the effects of large-carnivore decline on the change in abundance of species in their food webs. The blue bars represent the direct effects and the red bars represent the indirect effects. The number of years refers to the time since the carnivore species ceased to exist at the study site.

Invasive predators, competitors, and pathogens threaten many species

Deliberate or inadvertent introductions of non-native species have increased exponentially over the last 200 years. It is estimated that 50,000 species have been introduced to the United States alone. Many of these species were intentionally introduced to provide food, natural resources, or other important services for humans. But roughly 10 percent of these species have become **invasive**—that is, they reproduce rapidly, spread widely, and have mostly negative effects on the native species or ecosystems of the region. As you saw in Key Concept 56.2, once an introduced species arrives in a new community, a successful invasion usually involves multiple introduction events and the ability to cope with the physical environment and coexist with other species in the community.

While arguments can be made for the positive side of some species invasions, most invasions have had negative effects on species and ecosystems. For example, estimates suggest that 400 of the 958 species (42%) listed under the U.S. Endangered Species Act are at risk primarily because of predation, competition, or disease transmission by invasive species. Invasive species also can have

dramatic effects on the functions and services of ecosystems, affecting fire cycles, water availability, and coastal sedimentation (**Figure 58.10**). Invasive species have even changed the genetic diversity of native species through hybridization events with wild populations (e.g., non-native rainbow trout and native cutthroat trout in Montana). The total cost estimate for dealing with undesirable invasive species in the United States alone was roughly $120 billion per year.

Non-native species are spread in a variety of ways. Over the past 400 years, Europeans colonizing new continents have deliberately introduced plants and animals to reconstruct their once familiar surroundings. Many of these introductions have had disastrous effects on native flora and fauna. In Australia the introduction of European rabbits and foxes for sport hunting and of dogs and cats as pets has led to the extermination of nearly half the small- to medium-sized native marsupials over the last 100 years. As you saw in Key Concept 54.4, the use of biological control, or the deliberate introduction of species to control other invasive species, can sometimes cause even greater problems. Examples include the introduction of small predators such as mongoose to Hawaii to control rabbits and stoat (a type of weasel) to New Zealand to control rats. Instead of curbing the population sizes of rabbits and rats, these predators have been the major cause of bird species extinctions on these islands.

Sometimes invasions have occurred out of ignorance in what people believe are unlikely events of releasing non-native species

(A) Changes in fire cycle

Bromus tectorum

(B) Changes in hydrology

Tamarix ramosissima

(C) Changes in sedimentation

Spartina anglica

Figure 58.10 Species Invasions Can Change Ecosystem Functions and Services   Three examples of the effects that invasive species can have: **(A)** cheatgrass increases the fire cycle in sagebrush habitat, **(B)** salt cedar draws down water in desert streams, and **(C)** English cordgrass leads to sedimentation, transforming coastal mudflats into salt marshes.

(as was the case with the lionfishes described in Chapter 55). It can be difficult for people to imagine that species that are desirable and attractive in their place of origin can "go rogue" in a new region. Some of today's most noxious weeds were deliberately transported and planted in new places for their beauty, fragrance, or culinary value.

A more inadvertent mechanism of introduction comes from species that stow away during transportation. For example, you saw how non-native species can be transported in ballast water in Key Concept 56.2. The brown tree snake (*Boiga irregularis*) arrived on Guam in air cargo shortly after World War II and today can be found at densities up to 5,000 individuals per square kilometer. The snake has exterminated 15 species of land birds, including 3 found only on Guam.

Introduced pathogens have also wreaked havoc among native species, as exemplified by the fungal pathogen *Batrachochytrium dendrobatidis*, or *Bd* for short, which is implicated in the dramatic worldwide declines of amphibians over the last 20 years. As described in the story opening this chapter, the disease may be responsible for the large percentage of amphibian declines recorded globally and for which there has been no clear explanation (the others being a result of some combination of habitat loss, overharvesting, exposure to UV radiation, or pollution). Such "rapidly declining" events have taken place on all the major continents and occur within well-protected areas such as Yosemite National Park (USA), Monteverde Cloud Forest Reserve (Costa Rica), and Eungella National Park (Australia). Most of the mass mortality events in which *Bd* has been implicated have not been observed because

they happen so rapidly. Generally, researchers return to a site to find that population sizes are drastically reduced, or that some species are missing altogether, and discover the fungus present in some remaining individuals.

Researchers studying amphibian declines are interested in whether *Bd* is a novel and highly virulent pathogen that was recently introduced into wild populations ("novel pathogen hypothesis") or if *Bd* existed prior to the declines but emerged as a serious pathogen as a result of human-caused environmental changes to climate or UV radiation ("emerging endemic hypothesis"). Although *Bd* seems to thrive under fairly specific environmental conditions (high-elevation streams with permanent water sources and temperatures between 12°C and 27°C), most of the evidence now suggests that *Bd* is a non-native pathogen likely introduced through the global trade in amphibians for food, for use as laboratory animals, or for use as pets or display animals. **Investigating Life: Exploring the Decline of Central American Frogs** describes research by Karen Lips and colleagues on the 20-year decline of four species of frogs in the genus *Atelopus* at multiple sites in Costa Rica and Panama. Their data suggest that the *Bd* pathogen traveled as an "invasion front" from infected to uninfected populations over time and that, based on microscopic tissue analyses of frogs collected prior to the declines, *Bd* was probably never present in this region of the world until its introduction in the 1980s. Of the four *Atelopus* species considered in this study, three species (*A. chiriquiensis*, *A. senex*, and *A. zeteki*) are likely extinct, and one species (*A. varius*; photo at chapter opening), thought to be extinct in 1996, likely exists only in a small population near Quepos, Costa Rica.

experiment

Original Paper: Lips, K., J. Diffendorf, J. Mendelson and M. Sears. 2008. Riding the wave: Reconciling the roles of disease and climate change in amphibian decline. *PLoS Biology* 6: e72.

Karen Lips and her colleagues studied the spread of the fungal pathogen *Batrachochytrium dendrobatidis* (*Bd*), implicated in the dramatic declines of frogs in the genus *Atelopus* seen in the moist and mountainous regions of lower Central America (Costa Rica and Panama) starting in the late 1980s. They compiled biogeographic reports of *Atelopus* declines and conducted microscopic tissue analyses to test whether *Bd* was a non-native pathogen species recently introduced to Central America or an existing disease that emerged as the result of changes in climate or the environment.

HYPOTHESIS▶ Amphibian declines in Central America are caused by the non-native species invasion of the *Bd* pathogen rather than an environmentally induced emergence of *Bd* as an existing pathogen.

METHOD

1. The researchers mapped the location and approximate year of decline as a result of *Bd* infection for populations of four species of *Atelopus* at numerous sites in Costa Rica and Panama using data from published studies, the Global Amphibian Assessment, and their own fieldwork.

2. The distance of *Bd* spread was plotted against the years since *Bd* first caused a decline in *Atelopus* populations.

3. Sixty-four frogs, collected before 1987 in Monteverde, Costa Rica, were examined for *Bd* to determine if the pathogen was present prior to the decline.

RESULTS

1. The *Bd* pathogen was not found in frogs collected prior to 1987.

2. The map of *Atelopus* decline from *Bd*'s first appearance in 1987 to nearly 20 years later in 2006 shows that *Bd* spread from Costa Rica to Panama in a wavelike pattern from northwest to southeast.

3. There is a positive linear relationship between *Bd* travel distance and the years since *Bd* infection was detected (see inset on map).

CONCLUSION▶ The analyses of the prevalence of *Bd* in lower Central America over nearly 20 years support a classic pattern of a novel invasive disease spread across naive populations rather than the environmentally induced emergence of an already existing disease.

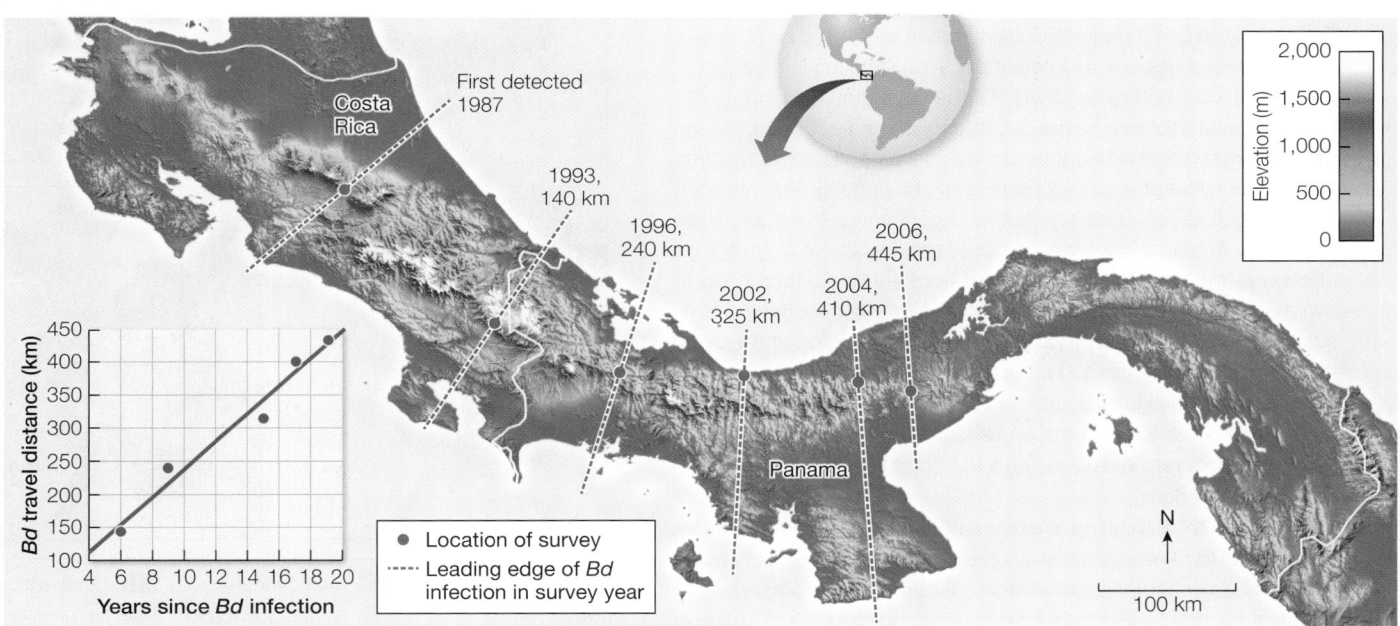

work with the data

The researchers were able to combine the distributional data of *Bd* infection from lower Central America with a much larger data set from South America to determine the percentage of *Atelopus* species that became extinct as a function of elevation. They found the following frog species extinctions with elevation:

Elevation (m)	Species extinctions (%)
0–199	30
200–1,000	85
1,001–4,000	100

QUESTIONS▶

1. The data from the results portion of the experiment can be used to calculate the rate of spread (km/year) of *Bd* by dividing the distance between pairs of locations (distance interval) by the number of years between the dates of decline. Calculate the rate of spread of the *Bd* pathogen between pairs of locations of frog declines (time interval). Is the rate of *Bd* spread similar among locations, or does it differ?

2. Calculate the average rate of spread for the *Bd* pathogen in lower Central America. Predict how far and in what direction *Bd* could travel by 2016 (10 years).

3. Using the map elevations in the results section of the experiment, and the species extinction data above, is it likely that the *Bd* pathogen will spread, and cause species extinctions, in the direction and distance estimated in your answer to Question 2? If so, what is your best estimate of the percentage of species that would become extinct? Explain.

A similar **work with the data** exercise may be assigned in **LaunchPad**.

Species and ecosystems are already being affected by climate change

As you saw in Key Concept 57.4, human-generated emissions of greenhouse gases are contributing to global climate warming, sea level rise, increased storminess, and ocean acidification, all of which are likely to become increasingly important causes of habitat loss and degradation, and ultimately species extinctions. Across North America, for example, average annual temperatures are predicted to increase by 2°C to 5°C by the end of the twenty-first century. If the climate warms to that extent, model projections show that the average temperature found at any given location in North America today could shift 500–800 kilometers to the north. Those species that cannot adapt to the warmer climate will have to shift their geographic ranges to stay within their physiological and ecological ranges. The shift in distributional range that organisms experience could result in habitat loss and fragmentation, especially if appropriate habitat does not shift as well or is lost altogether.

For example, as the globe warms, sea level is rising as a result of melting polar glaciers and warming seas. The Intergovernmental Panel on Climate Change (IPCC) estimates that the current rate of global sea level rise is 3 millimeters per year, resulting in a projected rise of 0.2–0.5 meters by 2100. As the seas slowly inundate the coast, coastal ecosystems will become increasingly flooded, putting them at great risk for habitat loss. Likewise, as polar regions warm, sea ice ecosystems are disappearing at an alarming rate, affecting species such as polar bears and seals dependent on that habitat (see the Future directions discussion in Chapter 57). Global warming has also been implicated in causing more frequent and extreme storm events, as seas warm and winds increase in intensity. Extreme storms such as Hurricane Sandy in 2012 are causing flooding and erosion hazards for both human and natural coastal communities. Finally, as discussed in Investigating Life: Food Webs in an Acidic and Warming Ocean in Chapter 57, as the ocean absorbs more atmospheric CO_2, seas become more acidic, leading to a host of individual species and ecosystem changes.

Scientists are beginning to explore how species and ecosystems will respond to climate change by predicting how it may affect organisms and looking for ways to mitigate those effects. Their research activities include analyses of past climate changes and studies of sites currently undergoing rapid climate change. It would be helpful to know, for example, how rapidly species responded to the end of the most recent ice age. Which species did and did not keep pace with the warming climate and rising seas? How much, and in what ways, do past ecological communities differ from those of today as a result of changes in climate?

Species that can disperse easily, such as birds, insects, and fish that can move considerable distances, may be able to shift their ranges as rapidly as the climate changes, provided they can find appropriate habitats. However, the ranges of other species, particularly plants, are likely to shift more slowly. For example, after the glaciers started to retreat in North America about 18,000 years ago, the ranges of plant communities slowly shifted northward (**Figure 58.11**). In addition, roughly 12,000 years ago, some novel, or what are termed "no analog" plant communities, formed under unique climate conditions that do not exist today. Thus it may be

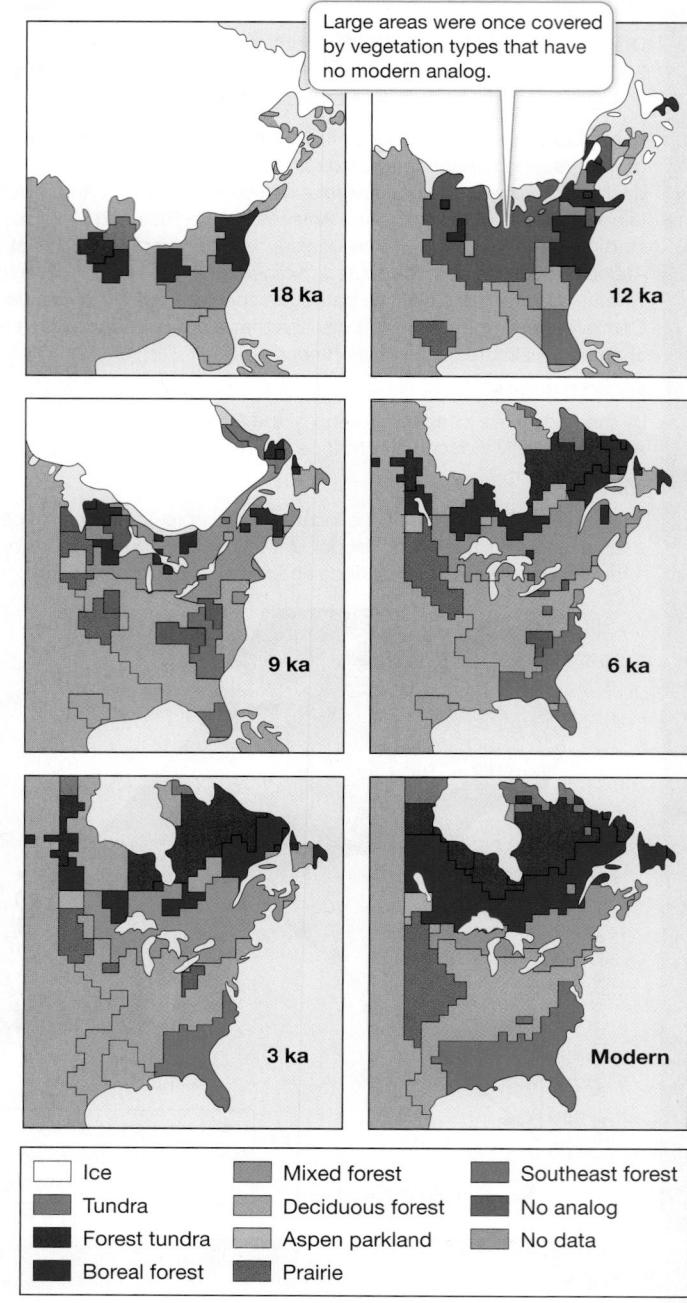

☐ Ice	■ Mixed forest	■ Southeast forest
■ Tundra	■ Deciduous forest	■ No analog
■ Forest tundra	■ Aspen parkland	■ No data
■ Boreal forest	■ Prairie	

Figure 58.11 Past Changes in Plant Communities Due to Climate Vegetation types in eastern North America have changed since the last glacial maximum, 18,000 years ago (ka, thousands of years before present). Vegetation composition was determined from pollen preserved in sediments.

Q: After the glaciers first started to retreat 18,000 years ago, what has been the pattern of plant community change with latitude? What were the "no analog" plant communities, and how did they form?

that as climate rapidly changes, unique combinations of species will come together to form similar novel communities.

Modern-day scientific observations have shown that for a wide variety of organisms, latitudinal and elevational distributions are

table 58.1 Recent Latitudinal and Elevational Range Shifts That Likely Result from Climate Change

Community or taxon	Location	Observed changes	Climate link
Tree line	Europe, New Zealand	Elevational shifts from lower to higher mountain zones	Increased air temperature
Alpine plants	European alps	Elevational shifts of 1–4 m per decade	Increased air temperature
Zooplankton, intertidal invertebrate, and fish communities	California coast, North Atlantic Ocean	Increasing abundance of warm-water species	Increased ocean temperature
39 butterfly species	North America, Europe	Northward range shifts up to 200 km over 27 years	Increased air temperature
Lowland birds	Costa Rica	Elevational shifts from lower to higher mountain zones	Decreased dry-season mist frequency
12 bird species	Britain	Northward range shifts up to 19 km over 20 years	Increased winter air temperature
Red fox (*Vulpes vulpes*), Arctic fox (*Alopex lagopus*)	Canada	Northward range shift of red fox and simultaneous southward range shift of Arctic fox	Increased air temperature

shifting in ways that are consistent with climate change (**Table 58.1**). For example, a study that considered the range shifts in alpine plants in the European alps showed that plant species have been moving to higher elevations consistent with increasing temperature. Researchers compared current plant community species richness with historical data from the eighteenth and nineteenth centuries. They found that alpine plant species richness increased over time, suggesting that plants were moving from lower elevations to higher elevations. Likewise, observations of the distribution of nonmigratory butterfly species in Europe and North America showed that of the 39 species examined, 63 percent had moved their ranges northward, while only 3 percent had moved southward.

Along with the shifts in distribution, evidence is mounting that important life history events are also occurring earlier in the spring (**Table 58.2**). The observations suggest that increases in temperature are triggering species to breed or migrate earlier than they have in the past few decades. There can also be physiological changes that reduce growth and reproduction. For example, since the mid-1980s the average minimum nightly temperature at La Selva Biological Station in the Caribbean lowlands of Costa Rica has increased from about 20°C to 22°C. On warmer nights, trees use more of their energy reserves to maintain themselves. As a result, even this small rise in temperature has reduced the average growth rate of six different tree species by about 20 percent.

Finally, climate change has the potential to cause species extinctions, although to date, none have been directly and definitively linked to this cause. However, as you saw in Figure 58.2, effects of climate change such as habitat loss or changes in life history or physiology could lead to lower effective population sizes, increased population extinctions, and eventually species extinctions.

58.2 recap

Biodiversity loss is the result of four major, and potentially interacting, factors. First, habitat loss and degradation (including pollution) have been primary culprits contributing to the extinction of tens of thousands of species. Second, overharvesting for food, clothing, ornamentation, pets, and medicines has eased for some species, but others are still under threat. Third, deliberate or inadvertent introduction of non-native species has increased exponentially, with some becoming invasive—meaning they reproduce rapidly, spread widely, and have mostly negative effects on native species. Finally, some species are already affected by climate change through distributional changes, timing of life history events, and decreased growth and reproduction and will continue to be impacted as climate change progresses.

learning outcomes

You should be able to:

- Describe, with examples, how human activities have resulted in habitat loss, fragmentation, and degradation.

- Give examples of situations where birds, mammals, and exotic animals have been overharvested, and describe and analyze the ecological implications of these examples.

- Describe and analyze methods by which species can invade or be introduced into new regions, and infer some possible results of these introductions.

- Correlate changes in climate with changes in species or ecosystems, giving evidence for the correlation.

(continued)

table 58.2 Recent Advances in the Timing of Life History Events That Likely Result from Climate Change

Taxon	Location	Observed changes	Period
Numerous plant species	Europe	Earlier flowering and leaf unfolding by 1.4–3.1 days per decade	Past 30–48 years
	North America	Earlier flowering and leaf unfolding by 1.2–2.0 days per decade	Past 35–63 years
18 butterfly species	United Kingdom	Earlier appearance by 2.8–3.2 days per decade	Past 23 years
Amphibians	United Kingdom	Earlier breeding	Past 25 years
Numerous bird species	Europe, North America	Earlier spring migration by 1.3–4.4 days per decade and earlier breeding by 1.9–4.8 days per decade	Past 30–60 years

> **58.2 recap** (continued)

1. How does habitat fragmentation contribute to the extinction of species?
2. Refer to Figure 56.9. Describe how the loss of wolves has likely resulted in a decrease in hardwood trees in Yellowstone National Park (see Figure 58.9).
3. Give three reasons why invasive species can have negative effects on biodiversity and ecosystems.
4. Suppose you have ecological data for a species of butterfly over the last 50 years. The data show that this species' range has moved north by 100 km and that its emergence occurs 4 days earlier than 50 years ago. Do these data suggest the butterfly species is responding to climate warming? Explain.

A major objective of understanding how and why biodiversity loss is occurring is to determine how conservation and management strategies might help preserve species. In the next section we will consider some positive steps that are being taken to recover and protect biodiversity.

> **key concept**

58.3 Protecting Biodiversity Requires Conservation and Management Strategies

As you have seen multiple times in this chapter, human activities play an integral role in biodiversity loss. Consequently, it falls to humans to develop, strengthen, and defend policies that conserve biodiversity.

focus your learning

- Humans are establishing protected areas to preserve biodiversity by restricting habitat loss and degradation.
- Coupled human–natural systems are a necessary part of protecting biodiversity.
- Captive breeding programs are temporary measures to maintain endangered species while threats in their natural environment are removed.

Biodiversity conservation integrates interests from ecological, institutional, and socioeconomic contexts into the decision-making process (**Figure 58.12**). Ecologists use scientific theory, empirical data, and tools from a variety of disciplines to help inform the protection and management of endangered and threatened species and ecosystems. They identify the factors that present risks to species and ecosystems and provide that information to managers, landowners, politicians, lawyers, nonprofit agencies, and the general public so that appropriate decisions can be made and action plans can be devised. Partnerships that incorporate these multiple contexts and stakeholders often result in conservation plans with the most support. Here we consider some conservation actions designed to protect and manage biodiversity in more detail.

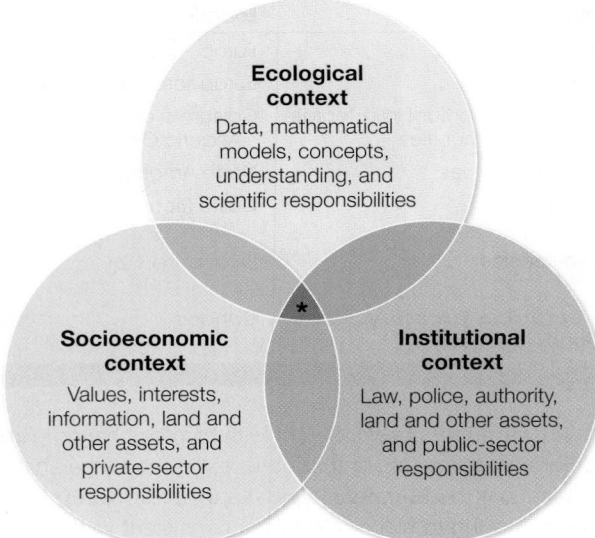

Figure 58.12 Humans Are an Integral Part of Protecting and Managing Biodiversity Protecting and managing biodiversity involves integrating interests from ecological, institutional, and socioeconomic contexts. Partnerships that incorporate these multiple contexts (*) often result in conservation plans with the most support.

Protected areas preserve habitat and curtail biodiversity loss

The establishment of **protected areas**, in which habitat loss or degradation is restricted or prohibited, is probably the most important component of efforts to conserve and manage biodiversity. Protected areas allow populations of multiple species to maintain themselves in the preserved habitat and may also serve as nurseries from which individuals can disperse into unprotected areas, replenishing populations that might otherwise become extinct. Without proper habitat, the long-term preservation of biodiversity will be nearly impossible.

Deciding which areas to protect usually involves a complex array of decisions by stakeholders. From an ecological perspective, two criteria need to be met: (1) the candidate habitat must support viable populations of the species it is meant to protect, and (2) the original ecosystem functions and services of the candidate habitat must be mostly intact.

DESIGNING PROTECTED AREAS   The first element in nature-preserve design involves identifying a large area that is relatively undisturbed and can serve as the core of the protected area (**Figure 58.13A**). This core natural area should allow populations of endangered species to maintain themselves and potentially serve as source individuals for populations outside the core area. This typically means that the core area should be large and compact with as little edge habitat as possible, to reduce edge effects.

The second element in designing a protected area involves buffer zones around the core area (**Figure 58.13B**). Buffer zones have some features required by the species of concern but involve less stringent controls on land use. For example, buffer zones allow some

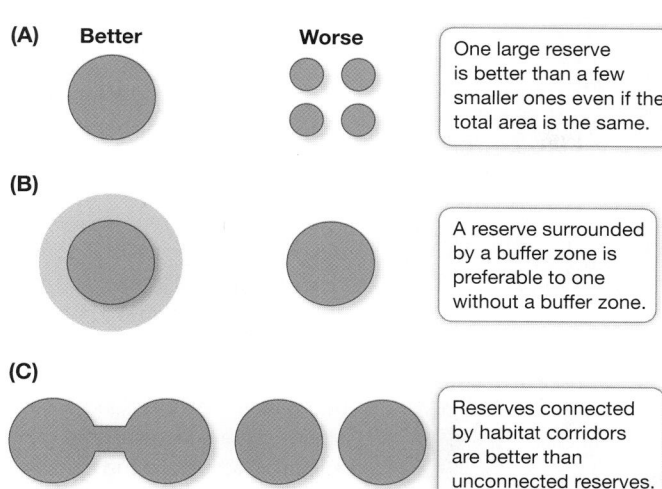

(A) Better / Worse

One large reserve is better than a few smaller ones even if the total area is the same.

(B)

A reserve surrounded by a buffer zone is preferable to one without a buffer zone.

(C)

Reserves connected by habitat corridors are better than unconnected reserves.

Figure 58.13 Nature Reserve Designs Based on Ecological Principles Some spatial configurations are better than others when designing reserves for fostering biodiversity.

Figure 58.14 Habitat Corridors: Passageways to Recovery Grizzly bears and other wildlife can use this overpass to travel safely over the Trans-Canada Highway and gain access to more protected area in Banff National Park, Canada.

resource extraction (e.g., timber, food, or medicines) or recreation while maintaining some of the original habitat.

The third element in nature-preserve design involves habitat connectivity (**Figure 58.13C**), which keeps populations from becoming isolated from the greater metapopulation and thus subject to extinction (see Key Concept 54.4). Traveling from one protected habitat to another requires habitat corridors, or patches that connect blocks of suitable habitat. Insight into the importance of corridors has led to new regional conservation initiatives, among the most notable of which is the Yellowstone to Yukon Conservation Initiative. This joint Canada–United States nonprofit organization has as its goal the sustainable preservation of the mountain ecosystem extending from Yellowstone National Park in the United States to Yukon, Canada. This stretch of land, the largest intact ecosystem of its kind on the planet, contains high-quality habitat for many of North America's most imperiled animals, including grizzly bears, gray wolves, lynx, and native fish. The initiative works with landowners to find sustainable ways of preserving high-quality, well-connected wildlife habitat in the region. For example, Banff National Park has 40 human-made corridors for wildlife that cross an 83-kilometer stretch of the Trans-Canada Highway (**Figure 58.14**). Managing the entire region in this way will not only provide passageways to habitat for these species, but will also provide room for their populations to shift in response to global climate change.

COUPLED HUMAN–NATURAL SYSTEMS Establishing protected areas is an essential component of efforts to maintain biodiversity, but this action alone is insufficient to stem global biodiversity loss. The extensive landscapes in which people live and extract resources should be integrated into biodiversity conservation efforts. Even urban and suburban areas can contribute to the conservation of species and habitats. The practice of encouraging biodiversity and sustainability in systems where humans and nature are intricately linked is known as **coupled human–natural system ecology**.

Research into coupled human–natural systems builds on the disciplines of ecology, physical sciences, social sciences, and economics. It is based on the principle that most ecosystem services are provided locally, and that people are motivated to work to protect their local interests. In practice, encouraging linkages between humans and nature comes in many forms. For example, the National Wildlife Federation has established a successful program in which people petition to have their backyards certified as wildlife-friendly. Criteria for certification include planting shrubs that provide food for birds and refraining from applying pesticides. In the case of coastal developments exposed to flooding hazards during extreme storms or sea level rise, humans are encouraging the protection and restoration of coastal wetlands and dunes, so-called green infrastructure, which provide important ecosystem services such as coastal protection, biodiversity, and recreation. Peregrine falcons (*Falco peregrinus*), once endangered by pesticide use, are thriving in urban settings where tall buildings mimic the cliffs that peregrines naturally nest on, and non-native pigeons are a main prey item. Green roofs, which are roofs covered with soil and native vegetation, are a way of creating parks and habitat, cooling buildings, and dealing with storm-water runoff, all important ecosystem services to humans. France passed a law in 2015 that requires all new buildings constructed in commercial areas to be partially covered by green roofs or solar panels.

Even some industrial sites can support biodiversity. The Turkey Point power plant in southern Florida uses large amounts of water to cool its generating units. To cool the heated water before discharging it, the Florida Power & Light Company dug a system of 38 canals that covers 2,428 hectares. These cooling canals are separated by low-lying berms that support a variety of native and non-native plants. Red mangroves grow along the edges of the canals. Today they support a thriving population of American crocodiles, a highly endangered species. Crocodiles living in the canals yield about 10 percent of all young crocodiles born in the United States.

(A)

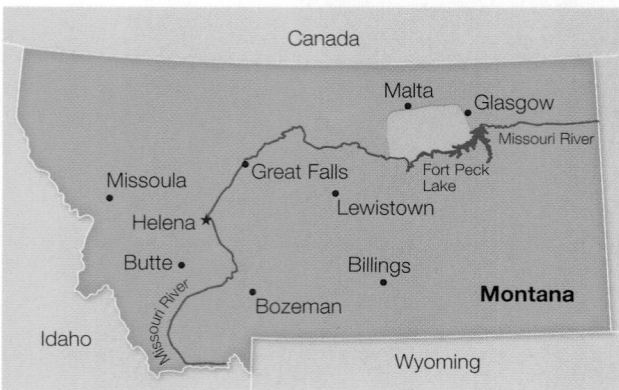

(B)

(C) *Bison bison*

Figure 58.15 Restoring a North American Prairie
(A) A major prairie restoration project (yellow area) is under way north of the Missouri River in Montana. **(B)** Restoration of prairie ecosystems can be as simple as removing fences that restrict large herds of native grazers such as bison. **(C)** The first bison were reintroduced to the area in 2005.

Degraded ecosystems can be restored

When a species is endangered as a consequence of habitat degradation rather than outright habitat loss, protecting the species may require restoring the habitat to a more natural state. Without human assistance, many degraded ecosystems recover very slowly, if at all. **Restoration ecology** involves renewing degraded ecosystems by active human intervention. Ideally, restoration involves reestablishing the original structure and function of native ecosystems through such techniques as active removal of pollutants or non-native species, revegetation to reestablish habitat and food resources, reintroduction of native species, and reestablishment of hydrological processes or disturbance regimes. For example, many species depend on particular patterns of disturbance, such as fire or windstorms, to maintain their populations (see Key Concept 56.4).

Recognition of the need for periodic disturbance to maintain healthy ecosystems is a relatively new dimension of conservation biology. We know, for example, that many plant species require periodic fires for successful establishment and survival, but for many years the official policy of the U.S. Forest Service, symbolized by the iconic mascot Smokey Bear, was to suppress all forest fires. Today, however, controlled burning is common, particularly in western North America. In order to use fire as an ecosystem management tool, it is important to know the historical pattern of fires in an area, which can be determined in part by studies of the annual growth rings and fire scars of trees. A schedule of controlled burning that recreates the historical pattern can reduce forest floor litter, avoiding a buildup of fuel that can lead to intense, tree-killing canopy fires.

Sometimes restoration involves using species that act as ecosystem engineers to help restore the original ecosystem. For example, as Key Concepts 56.3 and Figure 56.12 described, the recolonization of beavers, once hunted to near extinction on Minnesota's Kabetogama Peninsula, resulted in more wetlands and higher species diversity as a consequence of the beavers' dam-building activities. You've seen also how vegetation such as grasses can be planted to facilitate sediment capture, accretion, and ultimately the restoration of sea-grass beds, salt marshes, and sand dunes for habitat and coastal protection.

A large-scale terrestrial grassland restoration project—coined the "American Serengeti"—is under way in northeastern Montana. The goal of the project, which is run by the World Wildlife Fund and American Prairie Reserve in cooperation with public land managers and several other private conservation organizations, is to restore and conserve the native prairie and its fauna over a 1.2-million-hectare area near the Missouri River (**Figure 58.15**). When Lewis and Clark mapped this region 200 years ago, they saw large herds of bison, elk, deer, and pronghorn as well as abundant populations of their predators. But by the middle of the twentieth century, most North American prairies had been converted to cropland or were heavily grazed by domestic livestock, reducing the herds of large mammals to small remnant populations. Most of these populations are too small to maintain their genetic diversity or to function in their original ecological roles. However, the species have survived, so opportunities exist to reintroduce them if their habitat can be restored.

This ambitious restoration and protection project is feasible for three reasons. First, the private land in the area is owned by a small number of ranchers, each of whom owns extensive grazing leases on public lands administered by either U.S. federal agencies or the State of Montana. Second, most of the land has never been plowed, so native vegetation may recover rapidly when grazing pressures are reduced. Third, the area's human population is decreasing. Ranchers are aging, and some of their children are leaving for careers in urban settings. Once free-ranging herds of several thousand bison and large numbers of elk—along with their predators (wolves)—have been established, nature-minded tourists are expected to flock to the area to view the wildlife spectacle. Over the long term, the restored ecosystem should deliver major economic benefits to the region.

Captive breeding programs can maintain a few species

A few of the world's endangered species can be maintained in captivity while the external threats to their persistence are reduced or removed. However, captive propagation is only a temporary measure that buys time to deal with those threats. Zoos, aquariums, and botanical gardens do not have enough space to maintain adequate populations of more than a small fraction of Earth's endangered and threatened species. Nonetheless, captive propagation can play an important role by maintaining species and their genetic diversity during critical periods, providing a source of individuals for reintroduction into the wild, and raising public awareness of threatened and endangered species.

The California condor, North America's largest bird, survives today only because of captive propagation (**Figure 58.16**). Two centuries ago, condors ranged from British Columbia to northern Mexico, but by 1978 the wild population was plunging toward extinction. Many of the birds, which are scavengers, had died from ingesting animal carcasses containing lead shot or bullets. To save the condor from certain extinction, the Endangered Species Act was evoked and biologists captured all the remaining condors—only 22 individuals—and initiated a captive breeding program in 1983.

The first captive-bred birds were released in the mountains north of Los Angeles in 1992. Since that time, there have also been releases in northern Arizona and Baja California. Today captive-bred birds use the same roosting sites, bathing pools, and mountain ridges that their wild-born predecessors did. In 2003 a wild-born chick fledged in the wild for the first time in more than two decades. By 2014 the number of condors living in the wild had reached 228, with another 193 living in captivity. Most of the major threats to condor survival, including power lines, pesticides, and museum collectors, have been mitigated. Lead poisoning is still a problem, but as of July 1, 2008, under the Ridley–Tree Condor Preservation Act, California hunters are required to use non-lead bullets when hunting in the condor's range. Passage of this legislation marks a change in public attitudes from the days when cattle ranchers, in the mistaken belief that the condors killed livestock, opposed their reintroduction into the wild.

Ending trade is crucial to saving some species

Most endangered species cannot survive any further reductions in their breeding populations, so it is important to prevent their exploitation. The legal mechanism for prohibiting trade in these species or their products is an international agreement called the Convention on International Trade in Endangered Species (CITES). CITES currently prohibits international trade in items such as whale meat, rhinoceros horns, and many species of parrots, orchids, and others.

The recent history of elephant poaching for ivory illustrates how complex it can be to prevent hunting of endangered species. CITES instituted a ban on international trade in African elephant ivory in 1989, but demand for ivory remains strong, especially in Asia. As a result, as we described above, poaching of elephants continues in the forests of central and east Africa, where the animals are threatened. However, some countries, including Malawi and Zambia, have so many elephants that government officials kill them to control populations and prevent the animals from damaging crops. This excess of elephants led to the controversial decision to sanction sales of ivory from Namibia, Botswana, Zimbabwe, and South Africa in 2008, the first such sales in close to a decade. More than 100 tons of elephant tusks—the equivalent of 20,000 dead elephants—were auctioned off to authorized buyers from China and Japan and generated $15 million for elephant conservation efforts. Although the 2008 sales were monitored by CITES, concerns remain that the flood of legal ivory will be intermingled with poached ivory.

Species invasions must be controlled or prevented

The best way to reduce the damage caused by invasive species is to prevent their introduction or eradicate them earlier in their establishment phase. Given the tremendous volume of global trade,

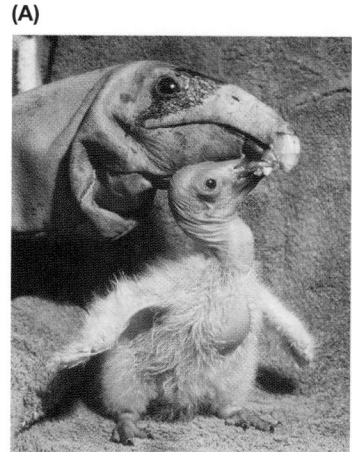

(A)

(B) *Gymnogyps californianus*

Figure 58.16 California Condors Make a Comeback (A) California condors raised in captivity are fed by humans wearing hand puppets so that the birds will not imprint on their human captors and will be able to survive in the wild. (B) Numbered wing tags allow conservation biologists to identify and track released adult condors. The survival of North America's largest bird species depends on this captive propagation project.

curtailing it might seem impossible, but some promising strategies do exist. For example, transoceanic transport of invasive species in ballast water could be largely eliminated by the simple procedure of deoxygenating ballast water before it is pumped out. This practice not only kills most organisms in the water but also extends the life of ballast tanks—an economic benefit to shippers.

In 1996 the U.S. Congress responded to concerns about ballast water with legislative action. After years of wrangling, in 2012 the U.S. Coast Guard amended its regulations on managing ballast water to set standards for "the allowable concentration of living organisms in ballast water discharged from ships in waters of the United States." Despite the adoption of these strict standards for protecting U.S. waterways, transport of invasive aquatic organisms in ballast water is an international problem whose potential solutions continue to run up against political and economic barriers.

Biodiversity can be conserved as a consequence of its economic value

As we've discussed previously, biodiversity can benefit humans in profound ways, providing important goods and services. You learned in Key Concept 57.5 that ecosystem goods and services can be valued by using a variety of methods to determine what people are willing to pay for them. Some goods such as timber or products from the animal trade are relatively easy to value, but nonmarket services such as the existence of an endangered species or the ability to visit a unique ecosystem are much harder to value. The following examples provide a sense of the nonmarket benefits of conserving rare or endangered species.

WILD DOGS AND ECOTOURISM Environmentally responsible travel to natural areas, the proceeds of which support conservation efforts and the economic well-being of the local communities, is a major source of income for many developing nations. For example, tourists visiting Africa often express interest in seeing wild dogs (*Lycaon pictus*) (**Figure 58.17**). However, diseases such as

Figure 58.17 Tourists Are Willing to Pay to See Wildlife
The wild dog (*Lycaon pictus*) is the second-most endangered carnivore in Africa. Tourists come to Kruger National Park in South Africa to catch a glimpse of this rare species.

rabies and canine distemper, along with habitat loss, road kills, deliberate extermination due to a perceived threat to livestock, and many other factors have decimated wild dog populations, making this the second-most endangered carnivore in Africa. (Another canid, the Ethiopian wolf, *Canis simensis*, is first.) South Africa is home to about 400 of Africa's remaining 5,000 wild dogs, most of which live in Kruger National Park. Their endangered status has piqued tourist interest in these charismatic animals; a survey of visitors to South Africa revealed that nearly three-fourths of them would be willing to pay an extra U.S. $12 for the opportunity to see wild dogs. Conservation biologists are working with lodge owners and ranchers elsewhere in South Africa and in Kenya to encourage them to reestablish wild dogs in areas from which they have disappeared.

POLLINATORS AND COFFEE   Taylor Ricketts and colleagues at Stanford University assessed the economic value of the pollination services provided by the bees that live in, and depend on, tropical forest patches adjacent to a coffee plantation in Costa Rica. They found that coffee production was highest at the sites that were closest to forest patches. They also hand-pollinated some coffee plants to show that the difference in production was a result of pollination services rather than other environmental conditions. The investigators calculated that the value of pollination services to the plantation on which the experiments were carried out was about $60,000 per year, more than the current conservation incentive payments offered to landowners to preserve forest patches.

▶ 58.3 recap

Ecologists use scientific theory, empirical data, and tools from a variety of disciplines to help inform socioeconomic and institutional sectors in the protection and management of biodiversity. To achieve these biodiversity conservation goals, various strategies are employed, including creating protected areas, restoring degraded habitat, breeding endangered species in captivity, ending trade of endangered species, controlling species invasions, and making biodiversity a marketable commodity.

learning outcomes

You should be able to:

• List guiding principles behind establishing protected areas and explain the reason for each principle.
• Discuss the benefits and limitations of captive breeding programs as a method for preserving endangered species.

1. Given the principles of reserve design and an understanding of how global warming can affect the distribution of species, design the best possible but realistic protected area for the conservation of terrestrial biodiversity today and in the future.
2. Using the example of the California condor, describe how species recovery is dependent on the integration of ecological information, socioeconomic interests, and institutional authority.

▶ investigatinglife

 **Is the most recent dramatic decline of amphibians the result of a novel fungal pathogen, and if so, what can be done about it?**

The research to date suggests that amphibian declines are caused by multiple human-instigated factors, including habitat loss and pollution, overharvesting, and even exposure to UV radiation, but that the final fatal blow may be the unintended introduction of the novel chytrid fungal pathogen *Batrachochytrium dendrobatidis*, or *Bd*. Population declines and species extinctions of amphibians have occurred at such alarming rates, and in areas that are considered undisturbed, that it can be hard to know how the *Bd* invasion can be stopped or realistically treated in the wild. Despite these hurdles,

pathogen's destructive effects. The first and foremost recommendation is to reduce the volume of amphibian trade and ensure that amphibians transported long distances are required by law to undergo disease testing and quarantine procedures. Second, improvements in diagnostics and knowledge of the disease are needed in both wild and captive settings. Recently, a method using PCR has been developed to definitively identify the fungus in the water and on the skin of potentially infected animals. Third, the risk of spread can be minimized by restricting the movement of captive amphibians, especially to the wild, and by disinfecting footwear and collecting equipment when working in aquatic environments. A variety of strategies that integrate ecological research, socioeconomic interests, and institutional authority will likely be required to save the remaining amphibian species that are susceptible to this highly fatal disease.

investigatinglife

Future directions

Although it is tragic that the harlequin frog (*Atelopus varius*), once ubiquitous in the mountains of Costa Rica, and the golden frog (*Atelopus zeteki*), the national symbol of Panama, could disappear so fast, hope still remains. An active area of research involves understanding why some amphibian species, such as the American bullfrog and African clawed frog, are more resistant to *Bd* while other species are highly susceptible. Some mechanisms that could explain species resistance to the disease are (1) the presence of specific types of bacteria or antimicrobial compounds that discourage *Bd* growth, (2) inherent genetic resistance to *Bd*, and (3) certain less virulent strains of *Bd*. More research will be needed to understand the complexities of the disease. In the meantime, captive breeding programs are under way for both the harlequin frog and golden frog so that they might one day be reintroduced into the Costa Rican and Panamanian nature preserves designed to protect them.

Chapter Summary 58

58.1 Human Activities Are Changing the Biosphere, Resulting in Biodiversity Loss

- Biodiversity has great value to human society in the form of goods and services, but human activities have caused its rapid decline at genetic, population, species, ecosystem, and global scales. **Review Figure 58.1**

- Human-caused events coupled with natural events can reduce a species' **effective population size** (number of individuals that can contribute offspring to the next generation), leading to population and species extinctions. Ever-smaller effective population sizes are affected by inbreeding depression, genetic drift, and **demographic stochasticity**. **Review Focus: Key Figure 58.2**

- Humans are causing biodiversity loss at unprecedented rates, rivaling that of the five previous mass extinction events, which were the result of cataclysmic natural disasters.

- We can estimate the extinction probability of species by taking into account their population sizes, genetic variation, life history traits, and ecology. **Review Figure 58.3**

58.2 Most Biodiversity Loss to Date Is Caused by Habitat Loss and Degradation

- **Habitat loss** (reduction in habitat quantity) and **habitat degradation** (reduction in habitat quality) have been the major causes of biodiversity loss. Estimates are that humans have transformed 50%–60% of the land surface, primarily for agriculture, timber, and livestock grazing. **Review Figure 58.4**

- Overharvesting of species for food, clothing, ornamentation, pets, and medicines was once the most important and rapid cause of species extinction and is still a concern for some species if conservation measures are not enacted. **Review Figures 58.7, 58.9**

- Deliberate or inadvertent introductions of non-native species have increased exponentially over the last 200 years, with roughly 10 percent of those species becoming invasive.

- **Invasive** species can endanger native species through predation, competition, and disease. They can cause changes to ecosystem functions or genetic diversity through hybridization with wild populations. **Review Figure 58.10, Investigating Life: Exploring the Decline of Central American Frogs**

- Human-generated emissions of greenhouse gases are contributing to global climate warming, sea level rise, increased storminess, and ocean acidification, all of which are becoming increasingly important causes of biodiversity loss.

- Scientific evidence shows that some species have been, or are currently being, affected by climate change through distributional changes, timing of life history events, and decreased growth and reproduction. **Review Figure 58.11, Tables 58.1, 58.2, Animation 58.1**

58.3 Protecting Biodiversity Requires Conservation and Management Strategies

- Ecologists use scientific theory, empirical data, and tools from a variety of disciplines to help inform socioeconomic and institutional sectors in the protection and management of biodiversity. **Review Figure 58.12**

- Creating **protected areas** and restoring degraded habitat can curtail biodiversity loss. Creating viable protected areas requires establishing core natural areas, buffer zones surrounding them, and habitat corridors connecting them. **Review Figures 58.13, 58.14**

- **Restoration ecology** involves renewing degraded ecosystems by reestablishing their original structure and function through active removal of pollutants or non-native species, revegetation to reestablish habitat and food resources, reintroduction of native species, or reestablishment of important processes such as disturbance. **Review Figure 58.15**

- Captive breeding programs and ending trade can help conserve and manage a few highly threatened or charismatic species. **Review Figure 58.16**

- The best way to reduce the damage caused by invasive species is to either prevent their introduction through trade restrictions or eradicate them in their early establishment phase.

- Biodiversity can be conserved because of the economic value of the goods and services it provides. **Review Figure 58.17**

See Activity 58.1 for a concept review of this chapter.

Go to **LearningCurve** (in **LaunchPad**) for dynamic quizzing that helps you solidify your understanding of this chapter. **LearningCurve** adapts to your responses, giving you the practice you need to master each key concept.

▷ Apply What You've Learned

Review

58.1 Various factors decrease effective population size, eventually leading to extinction.

58.1 Biodiversity connects species at all levels; its loss affects population, metapopulation, species, ecosystem, and global levels.

58.2 Human-caused habitat loss and degradation are the major causes of biodiversity loss.

58.3 Humans are establishing protected areas to preserve biodiversity by restricting habitat loss or degradation.

58.3 Coupled human–natural systems are a necessary part of protecting biodiversity.

Original Paper: Vickers, T. W. et al. 2015. Survival and mortality of pumas (*Puma concolor*) in a fragmented, urbanizing landscape. *PLoS One* 10(7): e0131490.

Habitat destruction and exploitation as a result of human activities have led to the extinction of many species. One species that is now endangered is the large cat called the puma or cougar (*Puma concolor*). Threats to pumas in southern California include habitat loss, habitat fragmentation, and conflicts with humans, which cause population declines, isolation, and ultimately smaller gene pools for the puma.

Two small puma populations occur in southern California between Los Angeles and San Diego. Earlier studies showed that pumas in both populations had very low genetic diversity, and little or no breeding occurred across populations. The populations are separated by urban and suburban areas, including highways, forming a barrier between the Santa Ana Mountains in the west (the more developed area) and the Peninsular Range to the east. Occasionally, however, pumas have been found to travel through agricultural areas, especially in the Peninsular Range.

Scientists fitted pumas in these populations with radio collars to track them and generate data on survivorship and causes of mortality, with the hope of conserving the populations (see figure). Between 2001 and 2013 they captured, marked, and monitored 74 animals in the two populations. Thirty-one marked animals came from the Santa Ana Mountains and 43 from the Peninsular Range. During the course of the study, 36 of the marked animals died. The table shows the causes of mortality in this sample.

	Number of individuals that died (% of population[s])		
Cause of mortality	Both populations combined	Santa Ana Mountains	Peninsular range
Vehicle collision	10 (28)	6 (46)	4 (19)
Legal hunting	6 (17)	0 (0)	6 (26)
Illegal hunting	4 (11)	3 (23)	1 (4)
Suspected disease	4 (11)	1 (8)	3 (13)
Confirmed disease	2 (5)	0 (0)	2 (9)
Fire	2 (5)	1 (8)	1 (4)
Killed for public safety	1 (3)	0 (0)	1 (4)
Killed by other pumas	1 (3)	0 (0)	1 (4)
Capture-related	1 (3)	0 (0)	1 (4)
Unknown	5 (14)	2 (15)	3 (13)
TOTAL	36	13	23

Questions

1. Among the known mortality causes (some are unknown), what are the major causes of death of pumas in these two populations?

2. Using the concept of effective population size, what would the likely effect of reducing puma population sizes be in this region of California?

3. Discuss how maintaining the puma populations in this region contributes to the overall biodiversity of California.

4. Based on the information provided in the text and table, should conservation methods for maintaining and protecting puma populations concentrate on the Santa Ana population, the Peninsular Range population, or both? Explain your answer.

5. How could the concept of coupled human–natural systems be employed to help maintain the puma populations?

Go to **LaunchPad** for the eBook, LearningCurve, animations, activities, flashcards, and additional resources and assignments.

APPENDIX A   The Tree of Life

Phylogeny is the organizing principle of modern biological taxonomy. A guiding principle of modern phylogeny is monophyly. A monophyletic group is considered to be one that contains an ancestral lineage and all of its descendants. Any such group can be extracted from a phylogenetic tree with a single cut.

The trees shown here provide a guide to the relationships among the major groups of extant (living) organisms in the tree of life as we have presented them throughout this book. The position of the branching "splits" indicates the relative branching order of the lineages of life, but the time scale is not meant to be uniform. In addition, the groups appearing at the branch tips do not necessarily carry equal phylogenetic "weight." For example, the ginkgo [78] is indeed at the apex of its lineage; this gymnosperm group consists of a single living species. In contrast, a phylogeny of the eudicots [86] could continue on from this point to fill many more trees the size of this one.

The glossary entries that follow are informal descriptions of some major features of the organisms described in Part Seven of this book. Each entry gives the group's common name, followed by the formal scientific name of the group (in parentheses). Numbers in square brackets reference the location of the respective groups on the tree.

It is sometimes convenient to use an informal name to refer to a collection of organisms that are not monophyletic but nonetheless all share (or all lack) some common attribute. We call these "convenience terms"; such groups are indicated in these entries by quotation marks, and we do not give them formal scientific names. Examples include "prokaryotes," "protists," and "algae." Note that these groups cannot be removed with a single cut; they represent a collection of distantly related groups that appear in different parts of the tree. We also use quotation marks here to designate two groups of fungi that are not believed to be monophyletic.

Go to **LaunchPad** for interactive versions of these trees, with links to photos, distribution maps, species lists, and identification keys.

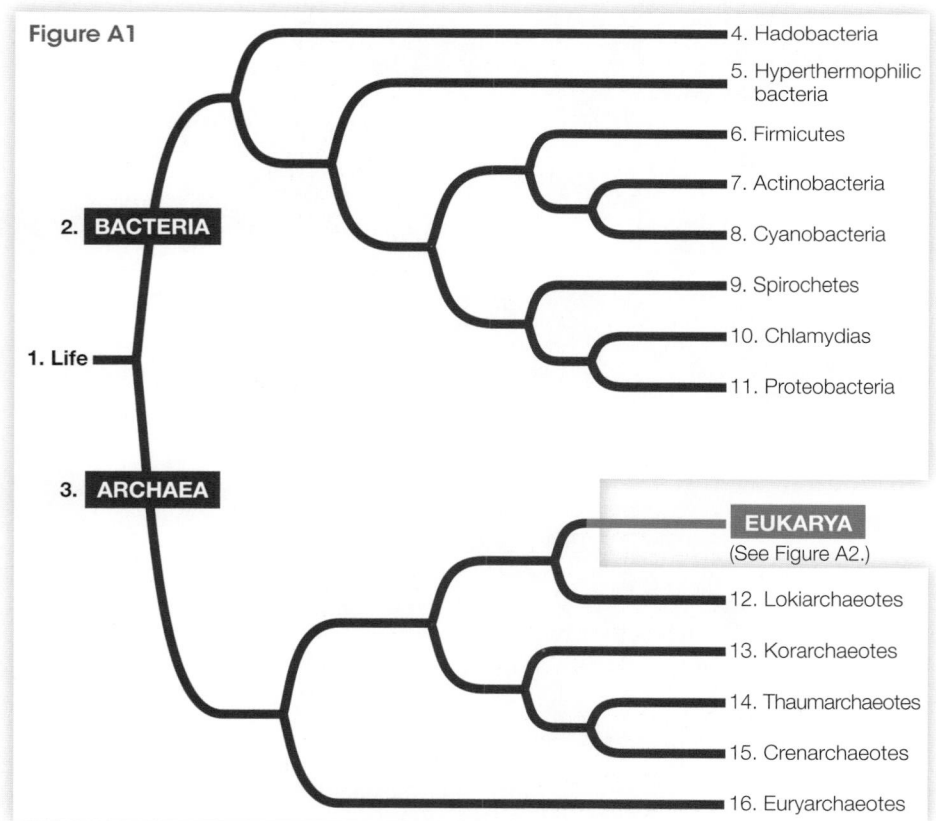

Figure A1

2. **BACTERIA**

4. Hadobacteria
5. Hyperthermophilic bacteria
6. Firmicutes
7. Actinobacteria
8. Cyanobacteria
9. Spirochetes
10. Chlamydias
11. Proteobacteria

1. Life

3. **ARCHAEA**

EUKARYA
(See Figure A2.)

12. Lokiarchaeotes
13. Korarchaeotes
14. Thaumarchaeotes
15. Crenarchaeotes
16. Euryarchaeotes

Figures A1 and A2   The Tree of Life   In this rendition, the prokaryotic lineages are shown in Figure A1 and the eukaryotic lineages are in Figure A2. Among the eukaryotes, the microbial lineages are shown in teal. The eukaryote groups that contain major radiations of large multicellular organisms are shown in brown (brown algae), green (Plantae), orange (fungi), and red (animals).

See the **Interactive Tree of Life**
www.Life11e.com/tree

Figure A2

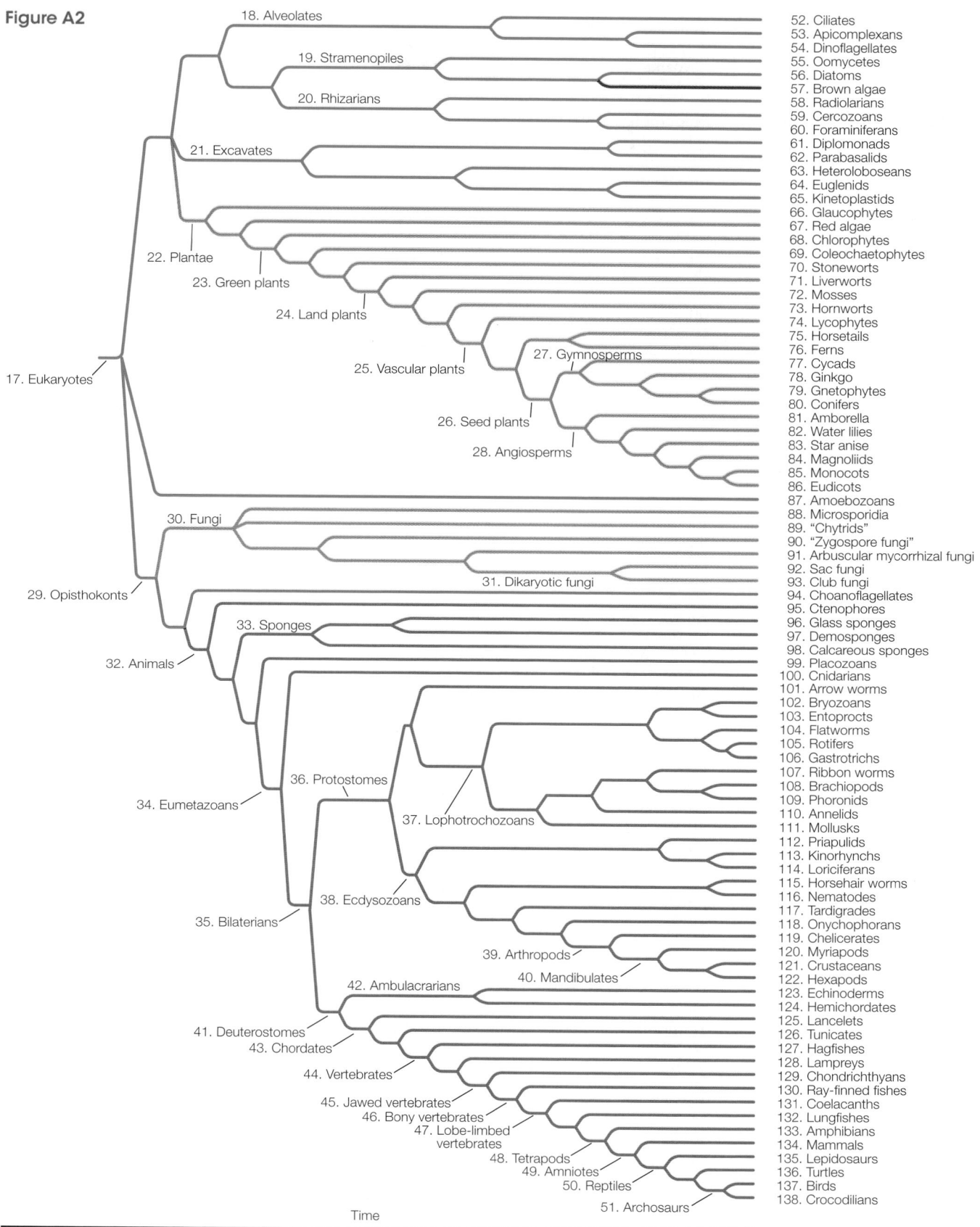

18. Alveolates
19. Stramenopiles
20. Rhizarians
21. Excavates
22. Plantae
23. Green plants
24. Land plants
25. Vascular plants
26. Seed plants
27. Gymnosperms
28. Angiosperms
17. Eukaryotes
30. Fungi
31. Dikaryotic fungi
29. Opisthokonts
33. Sponges
32. Animals
34. Eumetazoans
35. Bilaterians
36. Protostomes
37. Lophotrochozoans
38. Ecdysozoans
39. Arthropods
40. Mandibulates
42. Ambulacrarians
41. Deuterostomes
43. Chordates
44. Vertebrates
45. Jawed vertebrates
46. Bony vertebrates
47. Lobe-limbed vertebrates
48. Tetrapods
49. Amniotes
50. Reptiles
51. Archosaurs

Time

52. Ciliates
53. Apicomplexans
54. Dinoflagellates
55. Oomycetes
56. Diatoms
57. Brown algae
58. Radiolarians
59. Cercozoans
60. Foraminiferans
61. Diplomonads
62. Parabasalids
63. Heteroloboseans
64. Euglenids
65. Kinetoplastids
66. Glaucophytes
67. Red algae
68. Chlorophytes
69. Coleochaetophytes
70. Stoneworts
71. Liverworts
72. Mosses
73. Hornworts
74. Lycophytes
75. Horsetails
76. Ferns
77. Cycads
78. Ginkgo
79. Gnetophytes
80. Conifers
81. Amborella
82. Water lilies
83. Star anise
84. Magnoliids
85. Monocots
86. Eudicots
87. Amoebozoans
88. Microsporidia
89. "Chytrids"
90. "Zygospore fungi"
91. Arbuscular mycorrhizal fungi
92. Sac fungi
93. Club fungi
94. Choanoflagellates
95. Ctenophores
96. Glass sponges
97. Demosponges
98. Calcareous sponges
99. Placozoans
100. Cnidarians
101. Arrow worms
102. Bryozoans
103. Entoprocts
104. Flatworms
105. Rotifers
106. Gastrotrichs
107. Ribbon worms
108. Brachiopods
109. Phoronids
110. Annelids
111. Mollusks
112. Priapulids
113. Kinorhynchs
114. Loriciferans
115. Horsehair worms
116. Nematodes
117. Tardigrades
118. Onychophorans
119. Chelicerates
120. Myriapods
121. Crustaceans
122. Hexapods
123. Echinoderms
124. Hemichordates
125. Lancelets
126. Tunicates
127. Hagfishes
128. Lampreys
129. Chondrichthyans
130. Ray-finned fishes
131. Coelacanths
132. Lungfishes
133. Amphibians
134. Mammals
135. Lepidosaurs
136. Turtles
137. Birds
138. Crocodilians

A

acorn worms (*Enteropneusta*)   Benthic marine hemichordates [124] with an acorn-shaped proboscis, a short collar (neck), and a long trunk.

actinobacteria (*Actinobacteria*) [7]   Gram-positive bacteria [2] with genomes that have a high ratio of G-C to A-T nucleotide base pairs.

"algae"   Convenience term encompassing various distantly related groups of aquatic, photosynthetic eukaryotes [17].

alveolates (*Alveolata*) [18]   Unicellular eukaryotes with a layer of flattened vesicles (alveoli) supporting the plasma membrane. Major groups include the dinoflagellates [54], apicomplexans [53], and ciliates [52].

amborella (*Amborella*) [81]   An understory shrub or small tree found only on the South Pacific island of New Caledonia. Thought to be the sister group of the remaining living angiosperms [28].

ambulacrarians (*Ambulacraria*) [41]   The echinoderms [123] and hemichordates [124].

amniotes (*Amniota*) [49]   Mammals, reptiles, and their extinct close relatives. Characterized by many adaptations to terrestrial life, including an amniotic egg (with a unique set of membranes—the amnion, chorion, and allantois), a water-repellant epidermis (with epidermal scales, hair, or feathers), and, in males, a penis that allows internal fertilization.

amoebozoans (*Amoebozoa*) [87]   A group of eukaryotes [17] that use lobe-shaped pseudopods for locomotion and to engulf food. Major amoebozoan groups include the loboseans, plasmodial slime molds, and cellular slime molds.

amphibians (*Amphibia*) [133]   Tetrapods [48] with glandular skin that lacks epidermal scales, feathers, or hair. Many amphibian species undergo a complete metamorphosis from an aquatic larval form to a terrestrial adult form, although direct development is also common. Major amphibian groups include frogs and toads (anurans), salamanders, and caecilians.

amphipods (*Amphipoda*)   Small crustaceans [121] that are abundant in many marine and freshwater habitats. They are important herbi-vores, scavengers, and micropredators, and are an important food source for many aquatic organisms.

angiosperms (*Anthophyta* or *Magnoliophyta*) [28]   The flowering plants. Major angiosperm groups include the monocots [85], eudicots [86], and magnoliids [84].

animals (*Animalia* or *Metazoa*) [32]   Multicellular heterotrophic eukaryotes. The majority of animals are bilaterians [35]. Other groups of animals include the sponges [33], ctenophores [95], placozoans [99], and cnidarians [100]. The closest living relatives of the animals are the choanoflagellates [94].

annelids (*Annelida*) [110]   Segmented worms, including earthworms, leeches, and polychaetes. One of the major groups of lophotrochozoans [37].

anthozoans (*Anthozoa*)   One of the major groups of cnidarians [100]. Includes the sea anemones, sea pens, and corals.

anurans (*Anura*)   Comprising the frogs and toads, this is the largest group of living amphibians [133]. They are tail-less, with a shortened vertebral column and elongate hind legs modified for jumping. Many species have an aquatic larval form known as a tadpole.

apicomplexans (*Apicomplexa*) [53]   Parasitic alveolates [18] characterized by the possession of an apical complex at some stage in the life cycle.

arachnids (*Arachnida*)   Chelicerates [119] with a body divided into two parts: a cephalothorax that bears six pairs of appendages (four pairs of which are usually used as legs) and an abdomen that bears the genital opening. Familiar arachnids include spiders, scorpions, mites and ticks, and harvestmen.

arbuscular mycorrhizal fungi (*Glomeromycota*) [91]   A group of fungi [30] that associate with plant roots in a close symbiotic relationship.

archaeans (*Archaea*) [3]   One of the two primary divisions of life [1]. The term is often used only for the prokaryotic archaeans, which are unicellular organisms that lack a nucleus and lack peptidoglycan in the cell wall. Recent studies indicate that eukaryotes [17] are most closely related to the lokiarchaeotes [12] among the prokaryotic archaeans.

archosaurs (*Archosauria*) [51]   A group of reptiles [50] that includes dinosaurs and crocodilians [138]. Most dinosaur groups became extinct at the end of the Cretaceous; birds [137] are the only surviving dinosaurs.

arrow worms (*Chaetognatha*) [101]   Small planktonic or benthic predatory marine worms with fins and a pair of hooked, prey-grasping spines on each side of the head.

arthropods (*Arthropoda*)   The largest group of ecdysozoans [38]. Arthropods are characterized by a stiff exoskeleton, segmented bodies, and jointed appendages. Includes the chelicerates [119], myriapods [120], crustaceans [121], and hexapods (insects and their relatives) [122].

ascidians (*Ascidiacea*)   "Sea squirts"; the largest group of tunicates [126]. They are sessile (as adults), marine, saclike filter feeders.

B

bacteria (*Bacteria* or *Eubacteria*) [2]   Unicellular organisms lacking a nucleus, possessing distinctive ribosomes and initiator tRNA, and generally containing peptidoglycan in the cell wall. Different bacterial groups are distinguished primarily on nucleotide sequence data. One of the two primary divisions of life [1].

barnacles (*Cirripedia*)   Crustaceans [121] that undergo two metamorphoses—first from a feeding planktonic larva to a nonfeeding swimming larva, and then to a sessile adult that forms a "shell" composed of four to eight plates cemented to a hard substrate.

bilaterians (*Bilateria*) [35]   Those animal groups characterized by bilateral symmetry and three distinct tissue types (endoderm, ectoderm, and mesoderm). Includes the protostomes [36] and deuterostomes [41].

birds (*Aves*) [137]   Feathered, flying (or secondarily flightless) tetrapods [48].

bivalves (*Bivalvia*)   Major mollusk [111] group; clams and mussels. Bivalves typically have two similar hinged shells that are each asymmetrical across the midline.

bony vertebrates (*Osteichthyes*) [46]   Vertebrates [44] in which the skeleton is usually ossified to form bone. Includes the ray-finned fishes [130], coelacanths [131], lungfishes [132], and tetrapods [48].

brachiopods (*Brachiopoda*) [108]   Lophotrochozoans [37] with two similar hinged shells that are each symmetrical across the midline. Superficially resemble bivalve mollusks, except for the shell symmetry.

brittle stars (*Ophiuroidea*)   Echinoderms [123] with five long, whip-like arms radiating from a distinct central disk that contains the reproductive and digestive organs.

brown algae (*Phaeophyta*) [57]   Multicellular, almost exclusively marine stramenopiles [19] generally containing the pigment fucoxanthin as well as chlorophylls *a* and *c* in their chloroplasts.

bryozoans (*Ectoprocta* or *Bryozoa*) [102]   A group of marine and freshwater lophotrochozoans [37] that live in colonies attached to substrates; also known as ectoprocts or moss animals. They are the sister group of entoprocts.

C

caecilians (*Gymnophiona*)   A group of burrowing or aquatic amphibians [133]. They are elongate, legless, with a short tail (or none at all), reduced eyes covered with skin or bone, and a pair of sensory tentacles on the head.

calcareous sponges (*Calcarea*) [98]   Filter-feeding marine sponges with spicules composed of calcium carbonate.

cellular slime molds (*Dictyostelida*)   Amoebozoans [87] in which individual amoebas aggregate under stress to form a multicellular pseudoplasmodium.

cephalochordates (*Cephalochordata*) [125]   *See* lancelets.

cephalopods (*Cephalopoda*)   Active, predatory mollusks [111] in which the molluscan foot has been modified into muscular hydrostatic arms or tentacles. Includes octopuses, squids, and nautiluses.

cercozoans (*Cercozoa*) [59]   Unicellular eukaryotes [17] that feed by means of threadlike pseudopods. Group together with foraminiferans [60] and radiolarians [58] to comprise the rhizarians [20].

charophytes (*Charales*) [70]   *See* stoneworts.

chelicerates (*Chelicerata*) [119]   A major group of arthropods [39] with pointed appendages (chelicerae) used to grasp food (as opposed to the chewing mandibles of most other arthropods). Includes the arachnids, horseshoe crabs, pycnogonids, and extinct sea scorpions.

chimaeras (*Holocephali*)   A group of bottom-dwelling, marine, scaleless chondrichthyan fishes [129] with large, permanent, grinding tooth plates (rather than the replaceable teeth found in other chondrichthyans).

chitons (*Polyplacophora*)   Flattened, slow-moving mollusks [111] with a dorsal protective calcareous covering made up of eight articulating plates.

chlamydias (*Chlamydiae*) [10]   A group of very small Gram-negative bacteria; they live as intracellular parasites of other organisms.

chlorophytes (*Chlorophyta*) [68]   The most abundant and diverse group of green algae, including freshwater, marine, and terrestrial forms; some are unicellular, others colonial, and still others multicellular. Chlorophytes use chlorophylls *a* and *c* in their photosynthesis.

choanoflagellates (*Choanozoa*) [94]   Unicellular eukaryotes [17] with a single flagellum surrounded by a collar. Most are sessile, some are colonial. The closest living relatives of the animals [32].

chondrichthyans (*Chondrichthyes*) [129]   One of the two main groups of jawed vertebrates [45]; includes sharks, rays, and chimaeras. They have cartilaginous skeletons and paired fins.

chordates (*Chordata*) [43]   One of the two major groups of deuterostomes [41], characterized by the presence (at some point in development) of a notochord, a hollow dorsal nerve cord, and a post-anal tail. Includes the lancelets [125], tunicates [126], and vertebrates [44].

"chytrids" [89]   Convenience term used for a paraphyletic group of mostly aquatic, microscopic fungi [30] with flagellated gametes. Some exhibit alternation of generations.

ciliates (*Ciliophora*) [52]   Alveolates [18] with numerous cilia and two types of nuclei (micronuclei and macronuclei).

clitellates (*Clitellata*)   Annelids [110] with gonads contained in a swelling (called a clitellum) toward the head of the animal. Includes earthworms (oligochaetes) and leeches.

club fungi (*Basidiomycota*) [93]   Fungi [30] that, if multicellular, bear the products of meiosis on club-shaped basidia and possess a long-lasting dikaryotic stage. Some are unicellular.

club mosses (*Lycopodiophyta*) [74]   Vascular plants [25] characterized by microphylls. *See* lycophytes.

cnidarians (*Cnidaria*) [100]   Aquatic, mostly marine eumetazoans [34] with specialized stinging organelles (nematocysts) used for prey capture and defense, and a blind gastrovascular cavity. The sister group of the bilaterians [35].

coelacanths (*Actinista*) [131]   A group of marine lobe-limbed vertebrates [47] that was diverse from the Middle Devonian to the Cretaceous, but is now known from just two living species. The pectoral and anal fins are on fleshy stalks supported by skeletal elements, so they are also called lobe-finned fishes.

coleochaetophytes (*Coleochaetales*) [69]   Multicellular green algae characterized by flattened growth form composed of thin-walled cells. Thought to be the sister-group to the stoneworts [70] plus land plants [24].

conifers (*Pinophyta* or *Coniferophyta*) [80]   Cone-bearing, woody seed plants [26].

copepods (*Copepoda*)   Small, abundant crustaceans [121] found in marine, freshwater, or wet terrestrial habitats. They have a single eye, long antennae, and a body shaped like a teardrop.

craniates (*Craniata*)   Some biologist exclude the hagfishes [127] from the vertebrates [44], and use the term craniates to refer to the two groups combined.

crenarchaeotes (*Crenarchaeota*) [15]   A major and diverse group of prokaryotic archaeans [3], defined on the basis of rRNA base sequences. Many are extremophiles (inhabit extreme environments), but the group may also be the most abundant archaeans in the marine environment.

crinoids (*Crinoidea*)   Echinoderms [123] with a mouth surrounded by feeding arms, and a U-shaped gut with the mouth next to the anus. They attach to the substratum by a stalk or are free-swimming. Crinoids were abundant in the middle and late Paleozoic, but only a few hundred species have survived to the present. Includes the sea lilies and feather stars.

crocodilians (*Crocodylia*) [138]   A group of large, predatory, aquatic archosaurs [51]. The closest living relatives of birds [137]. Includes alligators, caimans, crocodiles, and gharials.

crustaceans (*Crustacea*) [121]   Major group of marine, freshwater, and terrestrial arthropods [39] with a head, thorax, and abdomen (although the head and thorax may be fused), covered with a thick exoskeleton, and with two-part appendages. Crustaceans undergo metamorphosis from a nauplius larva. Includes decapods, isopods, krill, barnacles, amphipods, copepods, and ostracods.

ctenophores (*Ctenophora*) [95]   Radially symmetrical, gelatinous marine animals [32], with eight rows of fused plates of cilia (called ctenes).

cyanobacteria (*Cyanobacteria*) [8]   A group of unicellular, colonial, or filamentous bacteria that conduct photosynthesis using chlorophyll *a*.

cycads (*Cycadophyta*) [77]   Palmlike gymnosperms with large, compound leaves.

cyclostomes (*Cyclostomata*)   This term refers to the possibly monophyletic group of lampreys [128] and hagfishes [127]. Molecular data support this group, but morphological data suggest that lampreys are more closely related to jawed vertebrates [45] than to hagfishes.

D

decapods (*Decapoda*)   A group of marine, freshwater, and semiterrestrial crustaceans [121] in which five of the eight pairs of thoracic appendages function as legs (the other three pairs, called maxillipeds, function as mouthparts). Includes crabs, lobsters, crayfishes, and shrimps.

demosponges (*Demospongiae*) [97]   The largest of the three groups of sponges [33], accounting for 90 percent of all sponge species. Demosponges have spicules made of silica, spongin fiber (a protein), or both.

deuterostomes (*Deuterostomia*) [41]   One of the two major groups of bilaterians [35], in which the mouth forms at the opposite end of the embryo from the blastopore in early development (contrast with protostomes). Includes the ambulacrarians [42] and chordates [43].

diatoms (*Bacillariophyta*) [56]   Unicellular, photosynthetic stramenopiles [19] with glassy cell walls in two parts.

dikaryotic fungi (*Dikarya*) [31]   A group of fungi [30] in which two genetically different haploid nuclei coexist and divide within the same hypha; includes club fungi [93] and sac fungi [92].

dinoflagellates (*Dinoflagellata*) [54]   A group of alveolates [18] usually possessing two flagella, one in an equatorial groove and the other in a longitudinal groove; many are photosynthetic.

dinosaurs (*Dinosauria*)   A group of archosaurs [51] that includes birds [137] as well as many extinct groups from the Mesozoic era. Extinct Mesozoic dinosaurs included some of the largest terrestrial vertebrates that have ever lived. Informally, many people use the term to refer only to the extinct Mesozoic species.

diplomonads (*Diplomonadida*) [61]   A group of eukaryotes [17] lacking mitochondria; most have two nuclei, each with four associated flagella.

dry-nosed primates (*Haplorhini*)   A group of primate mammals [134] that includes apes, monkeys, and tarsiers.

E

ecdysozoans (*Ecdysozoa*) [38]   One of the two major groups of protostomes [36], characterized by periodic molting of their exoskeletons. Nematodes [116] and arthropods [39] are the largest ecdysozoan groups.

echinoderms (*Echinodermata*) [123]   A major group of marine deuterostomes [41] with fivefold radial symmetry (at some stage of life) and an endoskeleton made of calcified plates and spines. Includes sea stars, crinoids, sea urchins, sea cucumbers, and brittle stars.

elasmobranchs (*Elasmobranchii*)   The largest group of chondrichthyan fishes [129]. Includes sharks, skates, and rays. In contrast to the other group of living chondrichthyans (the chimaeras), they have replaceable teeth.

embryophytes   *See* land plants [24].

entoprocts (*Entoprocta*) [103]   A group of marine and freshwater lophotrochozoans [37] that live as single individuals or in colonies attached to substrates. They are the sister group of bryozoans, from which they differ in having both their mouth and anus inside the lophophore (the anus is outside the lophophore in bryozoans).

eudicots (*Eudicotyledones*) [86]   A group of angiosperms [28] with pollen grains possessing three openings. Typically with two cotyledons, net-veined leaves, taproots, and floral organs typically in multiples of four or five.

euglenids (*Euglenida*) [64]   Flagellate excavates characterized by a pellicle composed of spiraling strips of protein under the plasma membrane; the mitochondria have disk-shaped cristae. Some are photosynthetic.

eukaryotes (*Eukarya*) [17]   Organisms made up of one or more complex cells in which the genetic material is contained in nuclei. Contrast with archaeans [3] and bacteria [2].

eumetazoans (*Eumetazoa*) [34]   A group of animals [32] characterized by body symmetry, a gut, a nervous system, specialized types of cell junctions, and well-organized tissues in distinct cell layers.

euphyllophytes (*Euphyllophyta*)   The group of vascular plants [25] that is sister to the lycophytes [74] and which includes all plants with megaphylls.

euryarchaeotes (*Euryachaeota*) [16]   A major group of prokaryotic archaeans [3], diagnosed on the basis of rRNA sequences. Includes many methanogens, extreme halophiles, and thermophiles.

eutherians (*Eutheria*)   A group of viviparous mammals [134], eutherians are well developed at birth (contrast to prototherians and marsupials, the other two groups of mammals). Most familiar mammals outside the Australian and South American regions are eutherians (see Table 33.1).

excavates (*Excavata*) [21]   Diverse group of unicellular, flagellate eukaryotes, many of which possess a feeding groove; some lack mitochondria.

F

ferns (*Pteridopsida* or *Polypodiopsida*) [76]   Vascular plants [25] usually possessing large, frond-like leaves and possessing thin-walled sporangia.

firmicutes (*Firmicutes*) [6]   Mostly Gram-positive bacteria [2] with genomes that have a low ratio of G-C to A-T nucleotide base pairs.

flatworms (*Platyhelminthes*) [104]   A group of dorsoventrally flattened and generally elongate soft-bodied lophotrochozoans [37]. May be free-living or parasitic, found in marine, freshwater, or damp terrestrial environments. Major flatworm groups include the tapeworms, flukes, monogeneans, and turbellarians.

flowering plants   See angiosperms [28].

flukes (*Trematoda*)   A group of wormlike parasitic flatworms [104] with complex life cycles that involve several different host species. May be paraphyletic with respect to tapeworms.

foraminiferans (*Foraminifera*) [60]   Amoeboid organisms with fine, branched pseudopods that form a food-trapping net. Most produce external shells of calcium carbonate.

fungi (*Fungi*) [30]   Eukaryotic heterotrophs with absorptive nutrition based on extracellular digestion; cell walls contain chitin. Major fungal groups include the microsporidia [88], "chytrids" [89], "zygospore fungi" [90], arbuscular mycorrhizal fungi [91], sac fungi [92], and club fungi [93].

G

gastropods (*Gastropoda*)   The largest group of mollusks [111]. Gastropods possess a well-defined head with two or four sensory tentacles (often terminating in eyes) and a ventral foot. Most species have a single coiled or spiraled shell.

Common in marine, freshwater, and terrestrial environments.

gastrotrichs (*Gastrotricha*) [106]   Tiny (0.06–3.0 mm), elongate acoelomate lophotrochozoans [37] that are covered in cilia. They live in marine, freshwater, and wet terrestrial habitats. They are simultaneous hermaphrodites.

ginkgo (*Ginkgophyta*) [78]   A gymnosperm [27] group with only one living species. The ginkgo seed is surrounded by a fleshy tissue not derived from an ovary wall and hence not a fruit.

glass sponges (*Hexactinellida*) [96]   Sponges [33] with a skeleton composed of four- and/or six-pointed spicules made of silica.

glaucophytes (*Glaucophyta*) [66]   Unicellular freshwater algae with chloroplasts containing traces of peptidoglycan, the characteristic cell wall material of bacteria.

gnathostomes (*Gnathostomata*)   See jawed vertebrates [45].

gnetophytes (*Gnetophyta*) [79]   A gymnosperm [27] group with three very different lineages; all have wood with vessels, unlike other gymnosperms.

green plants (*Viridiplantae*) [23]   Organisms with chlorophylls *a* and *b*, cellulose-containing cell walls, starch as a carbohydrate storage product, and chloroplasts surrounded by two membranes.

gymnosperms (*Gymnospermae*) [27]   Seed plants [26] with seeds "naked" (i.e., not enclosed in carpels). Probably monophyletic, but status still in doubt. Includes the conifers [80], gnetophytes [79], ginkgo [78], and cycads [77].

H

hadobacteria (*Hadobacteria*) [4]   A group of extremophilic bacteria [2] that includes the genera *Deinococcus* and *Thermus*.

hagfishes (*Myxini*) [127]   Elongate, slimy-skinned vertebrates [44] with three small accessory hearts, a partial cranium, and no stomach or paired fins. See also craniata; cyclostomes.

hemichordates (*Hemichordata*) [124]   One of the two primary groups of ambulacrarians [42]; marine wormlike organisms with a three-part body plan.

heteroloboseans (*Heterolobosea*) [63]   Colorless excavates [21] that can transform among amoeboid, flagellate, and encysted stages.

hexapods (*Hexapoda*) [122]   Major group of arthropods [39] characterized by a reduction (from the ancestral arthropod condition) to six walking appendages, and the consolidation of three body segments to form a thorax. Includes insects and their relatives (see Table 23.2).

hornworts (*Anthocerophyta*) [73]   Nonvascular plants with sporophytes that grow from the base. Cells contain a single large, platelike chloroplast.

horsehair worms (*Nematomorpha*) [115]   A group of very thin, elongate, wormlike freshwater ecdysozoans [38]. Largely nonfeeding as adults, they are parasites of insects and crayfish as larvae.

horseshoe crabs (*Xiphosura*)   Marine chelicerates [119] with a large outer shell in three parts:

a carapace, an abdomen, and a tail-like telson. There are only five living species, but many additional species are known from fossils.

horsetails (*Sphenophyta* or *Equisetophyta*) [75]   Vascular plants [25] with reduced megaphylls in whorls.

hydrozoans (*Hydrozoa*)   A group of cnidarians [100]. Most species go through both polyp and medusa stages, although one stage or the other is eliminated in some species.

hyperthermophilic bacteria [5]   A group of thermophilic bacteria [2] that live in volcanic vents, hot springs, and in underground oil reservoirs; includes the genera *Aquifex* and *Thermotoga*.

I

insects (*Insecta*)   The largest group within the hexapods [122]. Insects are characterized by exposed mouthparts and one pair of antennae containing a sensory receptor called a Johnston's organ. Most have two pairs of wings as adults. There are more described species of insects than all other groups of life [1] combined, and many species remain to be discovered. The major insect groups are described in Table 23.2.

"invertebrates"   Convenience term encompassing any animal [32] that is not a vertebrate [44].

isopods (*Isopoda*)   Crustaceans [121] characterized by a compact head, unstalked compound eyes, and mouthparts consisting of four pairs of appendages. Isopods are abundant and widespread in salt, fresh, and brackish water, although some species (the sow bugs) are terrestrial.

J

jawed vertebrates (*Gnathostomata*) [45]   A major group of vertebrates [44] with jawed mouths. Includes chondrichthyans [129], ray-finned fishes [130], and lobe-limbed vertebrates [47].

K

kinetoplastids (*Kinetoplastida*) [65]   Unicellular, flagellate organisms characterized by the presence in their single mitochondrion of a kinetoplast (a structure containing multiple, circular DNA molecules).

kinorhynchs (*Kinorhyncha*) [113]   Small (< 1 mm) marine ecdysozoans [38] with bodies in 13 segments and a retractable proboscis.

korarchaeotes (*Korarchaeota*)   A group of prokaryotic archaeans [3] known only by evidence from nucleic acids derived from hot springs.

krill (*Euphausiacea*)   A group of shrimplike marine crustaceans [121] that are important components of the zooplankton.

L

lampreys (*Petromyzontiformes*) [128]   Elongate, eel-like vertebrates [44] that often have rasping and sucking disks for mouths.

lancelets (*Cephalochordata*) [125]   A group of weakly swimming, eel-like benthic marine chordates [43].

land plants (*Embryophyta*) [24]   Plants with embryos that develop within protective structures; also called embryophytes. Both sporophytes and gametophytes of land plants are multicellular. Land plants possess a cuticle. Major groups are the liverworts [71], mosses [72], hornworts [73], and vascular plants [25].

larvaceans (*Larvacea*)   Solitary, planktonic tunicates [126] that retain both notochords and nerve cords throughout their lives.

lepidosaurs (*Lepidosauria*) [135]   Reptiles [50] with overlapping scales. Includes tuataras and squamates (lizards, snakes, and amphisbaenians).

life (*Life*) [1]   The monophyletic group that includes all known living organisms. Characterized by a nucleic-acid based genetic system (DNA or RNA), metabolism, and cellular structure. Some parasitic forms, such as viruses, have secondarily lost some of these features and rely on the cellular environment of their host.

liverworts (*Hepatophyta*) [71]   Nonvascular plants lacking stomata; stalk of sporophyte elongates along its entire length.

lobe-limbed vertebrates (*Sarcopterygii*) [47]   One of the two major groups of bony vertebrates [46], characterized by jointed appendages (paired fins or limbs).

loboseans (*Lobosea*)   A group of unicellular amoebozoans [87]; includes the most familiar amoebas (e.g., *Amoeba proteus*).

lokiarchaeotes (*Lokiarchaeota*) [12]   A group of prokaryotic archaeans [3] known from deep-sea hydrothermal vents, Among the known prokaryotic archaeans, they are the closest relatives of the eukaryotes [17].

"lophophorates"   Convenience term used to describe several groups of lophotrochozoans [37] that have a feeding structure called a lophophore (a circular or U-shaped ridge around the mouth that bears one or two rows of ciliated, hollow tentacles). Not a monophyletic group.

lophotrochozoans (*Lophotrochozoa*) [37]   One of the two main groups of protostomes [36]. This group is morphologically diverse, and is supported primarily on information from gene sequences. Includes bryozoans and entoprocts [103], flatworms [104], rotifers and gastrotrichs [106], ribbon worms [107], brachiopods [108], phoronids [109], annelids [110], and mollusks [111].

loriciferans (*Loricifera*) [114]   Small (< 1 mm) ecdysozoans [38] with bodies in four parts, covered with six plates.

lungfishes (*Dipnoi*) [132]   A group of aquatic lobe-limbed vertebrates [47] that are the closest living relatives of the tetrapods [48]. They have a modified swim bladder used to absorb oxygen from air, so some species can survive the temporary drying of their habitat.

lycophytes (*Lycopodiophyta*) [74]   Vascular plants [25] characterized by microphylls; includes club mosses, spike mosses, and quillworts.

M

magnoliids (*Magnoliidae*) [84]   A major group of angiosperms [28] possessing two cotyledons and pollen grains with a single opening. The group is defined primarily by nucleotide sequence data; it is

more closely related to the eudicots and monocots than to three other small angiosperm groups.

mammals (*Mammalia*) [134]   A group of tetrapods [48] with hair covering all or part of their skin; -females produce milk to feed their developing young. Includes the prototherians, marsupials, and eutherians.

mandibulates (*Mandibulata*) [40]   Arthropods [39] that include mandibles as mouth parts. Includes myriapods [120], crustaceans [121], and hexapods [122].

marsupials (*Marsupialia*)   Mammals [134] in which the female typically has a marsupium (a pouch for rearing young, which are born at an extremely early stage in development). Includes such familiar mammals as opossums, koalas, and kangaroos.

metazoans (*Metazoa*)   See animals [32].

microbial eukaryotes   See "protists."

microsporidia (*Microsporidia*) [88]   A group of parasitic unicellular fungi [30] that lack mitochondria and have walls that contain chitin.

mollusks (*Mollusca*) [111]   One of the major groups of lophotrochozoans [37], mollusks have bodies composed of a foot, a mantle (which often secretes a hard, calcareous shell), and a visceral mass. Includes monoplacophorans, chitons, bivalves, gastropods, and cephalopods.

monilophytes (*Monilophyta*)   A group of vascular plants [25], sister to the seed plants [26], characterized by overtopping and possession of megaphylls; includes the horsetails [75] and ferns [76].

monocots (*Monocotyledones*) [85]   Angiosperms [28] characterized by possession of a single cotyledon, usually parallel leaf veins, a fibrous root system, pollen grains with a single opening, and floral organs usually in multiples of three.

monogeneans (*Monogenea*)   A group of ectoparasitic flatworms [104].

monoplacophorans (*Monoplacophora*)   Mollusks [111] with segmented body parts and a single, thin, flat, rounded, bilateral shell.

mosses (*Bryophyta*) [72]   Nonvascular plants with true stomata and erect, "leafy" gametophytes; sporophytes elongate by apical cell division.

moss animals   See bryozoans [102].

mycoplasmas   Firmicutes [6] that lack cell walls and are among the smallest known cellular organisms.

myriapods (*Myriapoda*) [120]   Arthropods [39] characterized by an elongate, segmented trunk with many legs. Includes centipedes and millipedes.

N

nanoarchaeotes (*Nanoarchaeota*)   A group of extremely small, thermophilic, prokaryotic archaeans [3] with a much-reduced genome. The only described example can survive only when attached to a host organism.

nematodes (*Nematoda*) [116]   A very large group of elongate, unsegmented ecdysozoans [38] with thick, multilayer cuticles. They are

among the most abundant and diverse animals, although most species have not yet been described. Include free-living predators and scavengers, as well as parasites of most species of land plants [24] and animals [32].

neognaths (*Neognathae*)   The main group of birds [137], including all living species except the ostrich, emu, rheas, kiwis, cassowaries, and tinamous. *See* palaeognaths.

neopterans (*Neoptera*)   The largest group of flying (winged) hexapods [122], recognizable by the fact that they can fold their wings out of the way upon landing.

O

oligochaetes (*Oligochaeta*)   Annelid [110] group whose members lack parapodia, eyes, and anterior tentacles, and have few setae. Earthworms are the most familiar oligochaetes.

onychophorans (*Onychophora*) [118]   Elongate, segmented ecdysozoans [38] with many pairs of soft, unjointed, claw-bearing legs. Also known as velvet worms.

oomycetes (*Oomycota*) [55]   Water molds and relatives; absorptive heterotrophs with nutrient-absorbing, filamentous hyphae.

opisthokonts (*Opisthokonta*) [29]   A group of eukaryotes [17] in which the flagellum on motile cells, if present, is posterior. The opisthokonts include the fungi [30], animals [32], and choanoflagellates [94].

ostracods (*Ostracoda*)   Marine and freshwater crustaceans [121] that are laterally compressed and protected by two clamlike calcareous or chitinous shells.

P

palaeognaths (*Palaeognathae*)   A group of secondarily flightless or weakly flying birds [137]. Includes the flightless ostrich, emu, rheas, kiwis, and cassowaries, and the weakly flying tinamous.

parabasalids (*Parabasalia*) [62]   A group of unicellular eukaryotes [17] that lack mitochondria; they possess flagella in clusters near the anterior of the cell.

phoronids (*Phoronida*) [109]   A small group of sessile, wormlike marine lophotrochozoans [37] that secrete chitinous tubes and feed using a lophophore.

placoderms (*Placodermi*)   An extinct group of jawed vertebrates [45] that lacked teeth. Placoderms were the dominant predators in Devonian oceans.

placozoans (*Placozoa*) [99]   A poorly known group of structurally simple, asymmetrical, flattened, transparent animals found in coastal marine tropical and subtropical seas. Most evidence suggests that placozoans are secondarily simplified eumetazoans [34].

Plantae (*Plantae*) [22]   The most broadly defined plant group, corresponding to the group that descended from the primary endosymbiotic event that gave rise to chloroplasts. Includes green plants [23] as well as glaucophytes [66] and red algae [67]. In most parts of this book, we use the word **plant** as synonymous with land plant [24], a more restrictive definition.

plasmodial slime molds (*Myxogastrida*) Amoebozoans [87] that in their feeding stage consist of a coenocyte called a plasmodium.

pogonophorans (*Pogonophora*) Deep-sea annelids [110] that lack a mouth or digestive tract; they feed by taking up dissolved organic matter, facilitated by endosymbiotic bacteria in a specialized organ (the trophosome).

polychaetes (*Polychaeta*) A group of mostly marine annelids [110] with one or more pairs of eyes and one or more pairs of feeding tentacles; parapodia and setae extend from most body segments. May be paraphyletic with respect to the clitellates.

priapulids (*Priapulida*) [112] A small group of cylindrical, unsegmented, wormlike marine ecdysozoans [38] that takes its name from its phallic appearance.

primates (*Primates*) A group of mammals [134] that includes apes, monkeys, tarsiers, lemurs, and lorises.

"prokaryotes" Not a monophyletic group; as commonly used, includes the bacteria [2] and prokaryotic archaeans [3]. A term of convenience encompassing all cellular organisms that are not eukaryotes.

progymnosperms (*Progymnospermophyta*) An extinct group of seedless, vascular plants [25] that were among the first large, woody trees; lived in the Middle Devonian and Early Carboniferous periods.

proteobacteria (*Proteobacteria*) [11] A large and extremely diverse group of Gram-negative bacteria that includes many pathogens, nitrogen fixers, and photosynthesizers. Includes the alpha, beta, gamma, delta, and epsilon proteobacteria.

"protists" This term of convenience is used to encompass a large number of distinct and distantly related groups of eukaryotes, many but far from all of which are microbial and unicellular. Essentially a "catch-all" term for any eukaryote group not contained within the land plants [24], fungi [30], or animals [32].

protostomes (*Protostomia*) [36] One of the two major groups of bilaterians [35]. In protostomes, the mouth typically forms from the blastopore (if present) in early development (contrast with deuterostomes). The major protostome groups are the lophotrochozoans [37] and ecdysozoans [38].

prototherians (*Prototheria*) A mostly extinct group of mammals [134], common during the Cretaceous and early Cenozoic. The five living species—four echidnas and the duck-billed platypus—are the only extant egg-laying mammals.

pterobranchs (*Pterobranchia*) A small group of sedentary marine hemichordates [124] that live in tubes secreted by the proboscis. They have one to nine pairs of arms, each bearing long tentacles that capture prey and function in gas exchange.

pterygotes (*Pterygota*) Flying (winged) insects, the largest group of hexapods [122].

pycnogonids (*Pycnogonida*) Treated in this book as a group of chelicerates [119], but sometimes considered an independent group of arthropods [39]. Pycnogonids have reduced bodies and very long, slender legs. Also called sea spiders.

R

radiolarians (*Radiolaria*) [58] Amoeboid organisms with needlelike pseudopods supported by microtubules. Most have glassy internal skeletons.

ray-finned fishes (*Actinopterygii*) [130] A highly diverse group of freshwater and marine bony vertebrates [46]. They have reduced swim bladders that often function as hydrostatic organs and fins supported by soft rays (lepidotrichia). Includes most familiar fishes.

red algae (*Rhodophyta*) [67] Mostly multicellular, marine and freshwater algae characterized by the presence of phycoerythrin in their chloroplasts.

reptiles (*Reptilia*) [50] One of the two major groups of extant amniotes [49], supported on the basis of similar skull structure and gene sequences. The term "reptiles" traditionally excluded the birds [137], but the resulting group is then clearly paraphyletic. As used in this book, the reptiles include turtles [136], lepidosaurs [135], birds [137], and crocodilians [138].

rhizarians (*Rhizaria*) [20] Mostly amoeboid unicellular eukaryotes with pseudopods, many with external or internal shells. Includes the foraminiferans [60], cercozoans [59], and radiolarians [58].

rhyniophytes (*Rhyniophyta*) A group of early vascular plants [25] that appeared in the Silurian and became extinct in the Devonian. Possessed dichotomously branching stems with terminal sporangia but no true leaves or roots.

ribbon worms (*Nemertea*) [107] A group of unsegmented lophotrochozoans [37] with an eversible proboscis used to capture prey. Mostly marine, but some species live in fresh water or on land.

rotifers (*Rotifera*) [105] Tiny (< 0.5 mm) lophotrochozoans [37] with a pseudocoelomic body cavity that functions as a hydrostatic organ, and a ciliated feeding organ called the corona that surrounds the head. Rotifers live in freshwater and wet terrestrial habitats.

roundworms (*Nematoda*) [116] *See* nematodes.

S

sac fungi (*Ascomycota*) [92] Fungi that bear the products of meiosis within sacs (asci) if the organism is multicellular. Some are unicellular.

salamanders (*Caudata*) A group of amphibians [133] with distinct tails in both larvae and adults and limbs set at right angles to the body.

salps *See* thaliaceans.

sarcopterygians (*Sarcopterygii*) [47] *See* lobe-limbed vertebrates.

scyphozoans (*Scyphozoa*) Marine cnidarians [100] in which the medusa stage dominates the life cycle. Commonly known as jellyfish.

sea cucumbers (*Holothuroidea*) Echinoderms [123] with an elongate, cucumber-shaped body and leathery skin. They are scavengers on the ocean floor.

sea spiders *See* pycnogonids.

sea squirts *See* ascidians.

sea stars (*Asteroidea*) Echinoderms [123] with five (or more) fleshy "arms" radiating from an indistinct central disk. Also called starfishes.

sea urchins (*Echinoidea*) Echinoderms [123] with a test (shell) that is covered in spines. Most are globular in shape, although some groups (such as the sand dollars) are flattened.

"seed ferns" A paraphyletic group of loosely related, extinct seed plants that flourished in the Devonian and Carboniferous. Characterized by large, frondlike leaves that bore seeds.

seed plants (*Spermatophyta*) [26] Heterosporous vascular plants [25] that produce seeds; most produce wood; branching is axillary (not dichotomous). The major seed plant groups are gymnosperms [27] and angiosperms [28].

simians (*Simiiformes*) A group of primate mammals [134] that consists of apes and monkeys.

sow bugs *See* isopods.

spirochetes (*Spirochaetes*) [9] Motile, Gram-negative bacteria with a helically coiled structure and characterized by axial filaments.

sponges (*Porifera*) [33] A group of relatively asymmetric, filter-feeding animals that lack a gut or nervous system and generally lack differentiated tissues. Includes glass sponges [96], demosponges [97], and calcareous sponges [98].

springtails (*Collembola*) Wingless hexapods [122] with springing structures on the third and fourth segments of their bodies. Springtails are extremely abundant in some environments (especially in soil, leaf litter, and vegetation).

squamates (*Squamata*) The major group of lepidosaurs [135], characterized by the possession of movable quadrate bones (which allow the upper jaw to move independently of the rest of the skull) and hemipenes (a paired set of eversible penises, or penes) in males. Includes the lizards (a paraphyletic group), snakes, and amphisbaenians.

staphylococci (*Staphylococcus*) Firmicutes [6] that are abundant on the surface of the human body; can cause skin diseases and respiratory, intestinal, and wound infections.

star anise (*Austrobaileyales*) [83] A group of woody angiosperms [28] thought to be the sister-group of the clade of flowering plants that includes eudicots [86], monocots [85], and magnoliids [84].

starfish (*Asteroidea*) *See* sea stars.

stoneworts (*Charales*) [70] Multicellular green algae with branching, apical growth and plasmodesmata between adjacent cells. The closest living relatives of the land plants [24], they retain the egg in the parent organism.

stramenopiles (*Heterokonta* or *Stramenopila*) [19] Organisms having, at some stage in their life cycle, two unequal flagella, the longer possessing rows of tubular hairs. Chloroplasts, when present, surrounded by four membranes. Major stramenopile groups include the brown algae [57], diatoms [56], and oomycetes [55].

streptophytes (*Streptophyta*)   All green plants [23] other than chlorophytes [68].

T

tapeworms (*Cestoda*)   Parasitic flatworms [104] that live in the digestive tracts of vertebrates as adults, and usually in various other species of animals as juveniles.

tardigrades (*Tardigrada*) [117]   Small (< 0.5 mm) ecdysozoans [38] with fleshy, unjointed legs and no circulatory or gas exchange organs. They live in marine sands, in temporary freshwater pools, and on the water films of plants. Also called water bears.

tetrapods (*Tetrapoda*) [48]   The major group of lobe-limbed vertebrates [47]; includes the amphibians [133] and the amniotes [49]. Named for the presence of four jointed limbs (although limbs have been secondarily reduced or lost completely in several tetrapod groups).

thaliaceans (*Thaliacea*)   A group of solitary or colonial planktonic marine tunicates [126]. Also called salps.

thaumarchaeotes (*Thaumarchaeota*) [14]   A group of prokaryotic archaeans [3] known only from hot environments; they oxidize ammonia and are thought to play an important role in the nitrogen cycle.

therians (*Theria*)   Mammals [134] characterized by viviparity (live birth). Includes eutherians and marsupials.

theropods (*Theropoda*)   Archosaurs [51] with bipedal stance, hollow bones, a furcula ("wishbone"), elongated metatarsals with three-fingered feet, and a pelvis that points backwards. Includes many well-known extinct dinosaurs (such as *Tyrannosaurus rex*), as well as the living birds [137].

tracheophytes   *See* vascular plants [25].

trilobites (*Trilobita*)   An extinct group of arthropods [39] related to the chelicerates [119]. Trilobites flourished from the Cambrian through the Permian.

tuataras (*Rhyncocephalia*)   A group of lepidosaurs [135] known mostly from fossils; there is only one living tuatara species. The quadrate bone of the upper jaw is fixed firmly to the skull. Sister group of the squamates.

tunicates (*Tunicata*) [126]   A group of chordates [43] that are mostly saclike filter feeders as adults, with motile larval stages that resemble tadpoles.

turbellarians (*Turbellaria*)   A group of free-living, generally carnivorous flatworms [104]. Their monophyly is questionable.

turtles (*Testudines*) [136]   A group of reptiles [50] with a bony carapace (upper shell) and plastron (lower shell) that encase the body in a fashion unique among the vertebrates.

U

urochordates (*Tunicata*) [126]   *See* tunicates.

V

vascular plants (*Tracheophyta*) [25]   Plants with xylem and phloem. Major groups include the lycophytes [74] and euphyllophytes.

vertebrates (*Vertebrata*) [44]   The largest group of chordates [43], characterized by a rigid endoskeleton supported by the vertebral column and an anterior skull encasing a brain. Includes hagfishes [127], lampreys [128], and the jawed vertebrates [45], although some biologists exclude the hagfishes from this group. *See also* craniates.

W

water bears   *See* tardigrades.

water lilies (*Nymphaeaceae*) [82]   A group of aquatic, freshwater angiosperms [28] that are rooted in soil in shallow water, with round floating leaves and flowers that extend above the water's surface. They are the sister-group to most of the remaining flowering plants, with the exception of the genus *Amborella* [81].

wet-nosed primates (*Strepsirrhini*)   A group of primate mammals [134] that includes lemurs and lorises.

X

xenoturbellids (*Xenoturbellida*)   Wormlike organisms that feed on or parasitize marine mollusks. Relationships uncertain; thought to be a sister group of either ambulacrarians [42] or bilaterians [35].

Y

"yeasts"   Convenience term for several distantly related groups of unicellular fungi [30].

Z

"zygospore fungi" (*Zygomycota,* if monophyletic) [90]   A convenience term for a probably paraphyletic group of fungi [30] in which hyphae of differing mating types conjugate to form a zygosporangium.

APPENDIX B   Making Sense of Data: A Statistics Primer

This appendix is designed help you understand the analysis of biological data in a statistical context. We outline major concepts involved in the collection and analysis of data, and we present basic statistical analyses that will help you interpret, understand, and complete the Work with the Data and Apply What You've Learned problems throughout this book. Understanding these concepts will also help you understand and interpret scientific studies in general. We present formulas for some common statistical tests as examples, but the main purpose of this appendix is to help you understand the purpose and reasoning behind these tests. Once you understand the basis of an analysis, you may wish to use one of many free, online websites for conducting the tests (such as http://vassarstats.net).

Why Do We Do Statistics?

ALMOST EVERYTHING VARIES   We live in a variable world, but within the variation we see among biological organisms there are predictable patterns. We use statistics to find and analyze these patterns. Consider any group of common things in nature—all people aged 22, all the cells in your liver, or all the blades of grass in your yard. Although they will have many similar characteristics, they will also have important differences. Men aged 22 tend to be taller than women aged 22, but of course not every man will be taller than every woman in this age group.

Natural variation can make it difficult to find general patterns. For example, scientists have determined that smoking increases the risk of developing lung cancer. But we know that not all smokers will develop lung cancer and not all nonsmokers will remain cancer-free. If we compare just one smoker with just one nonsmoker, we may end up drawing the wrong conclusion. So how did scientists discover this general pattern? How many smokers and nonsmokers did they examine before they felt confident about the risk of smoking?

Statistics helps us find and describe general patterns in nature, and draw conclusions from those patterns.

AVOIDING FALSE POSITIVES AND FALSE NEGATIVES   When a woman takes a pregnancy test, there is some chance that it will be positive even if she is not pregnant, and there is some chance that it will be negative even if she is pregnant. We call these kinds of mistakes "false positives" and "false negatives."

Doing science is a bit like taking a medical test. We observe patterns in the world, and we try to draw conclusions about how the world works from those observations. Sometimes our observations lead us to draw the wrong conclusions. We might conclude that a phenomenon occurs, when it actually does not; or we might conclude that a phenomenon does not occur, when it actually does.

For example, planet Earth has been warming over the past century (see Key Concept 57.4). Ecologists are interested in whether plant and animal populations have been affected by global warming. If we have long-term information about the locations of species and about temperatures in certain areas, we can determine whether shifts in species distributions coincide with temperature changes. Such information, however, can be very complicated. Without proper statistical methods, one may not be able to detect the true impact of temperature, or instead may think a pattern exists when it does not.

Statistics helps us avoid drawing the wrong conclusions.

How Does Statistics Help Us Understand the Natural World?

Statistics is essential to scientific discovery. Most biological studies involve five basic steps, each of which requires statistics:

- **Step 1: Choose an Experimental Design**
  Clearly define the scientific question and the methods necessary to tackle the question.

- **Step 2: Collect Data**
  Gather information about the natural world through observations and experiments.

- **Step 3: Organize and Visualize the Data**
  Use tables, graphs, and other useful representations to gain intuition about the data.

- **Step 4: Summarize the Data**
  Summarize the data with a few key statistical calculations.

- **Step 5: Make Inferences from the Data**
  Use statistical methods to draw general conclusions from the data about the world and the ways it works.

Step 1: Choose an Experimental Design

We make observations and conduct experiments to gain knowledge about the world. Scientists come up with scientific ideas based on prior research and their own observations. These ideas start as a question such as "Does smoking cause cancer?" From prior experience, scientists then propose possible answers to the question in the form of hypotheses such as "Smoking increases the risk of cancer." Experimental design then involves devising comparisons that can test predictions of the hypotheses. For example, if we are interested in evaluating the hypothesis that smoking increases cancer risk, we might decide to compare the incidence of cancer in nonsmokers and smokers. Our hypothesis would predict that the cancer incidence is higher in smokers than in nonsmokers. There are various ways we could make that comparison. We could compare the smoking history of people newly diagnosed with cancer with that of people whose tests came out negative for cancer. Alternatively, we could assess the current smoking habits of a sample of people and see whether those who smoke more heavily are more likely to develop cancer in, say, 5 years. Statistics provides us with tools for assessing which approach will provide us with an answer with smaller costs in time and effort in data collection and analysis.

> *We use statistics to guide us in planning exactly how to make our comparisons—what kinds of data and how many observations we will need to collect.*

Step 2: Collect Data

TAKING SAMPLES   When biologists gather information about the natural world, they typically collect representative pieces of information, called **data** or observations. For example, when evaluating the efficacy of a candidate drug for medulloblastoma brain cancer, scientists may test the drug on tens or hundreds of patients, and then draw conclusions about its efficacy for all patients with these tumors. Similarly, scientists studying the relationship between body weight and clutch size (number of eggs) for female spiders of a particular species may examine tens to hundreds of spiders to make their conclusions.

We use the expression "sampling from a population" to describe this general method of taking representative pieces of information from the system under investigation (**Figure B1**). The pieces of information together make up a **sample** of the larger system, or **population**. In the cancer therapy example, each observation was the change in a patient's tumor size 6 months after initiating treatment, and the population of interest was all individuals with medulloblastoma tumors. In the spider example, each observation was a pair of measurements—body weight and clutch size—for a single female spider, and the population of interest was all female spiders of this species.

Sampling is a matter of necessity, not laziness. We cannot hope (and would not want) to collect and weigh *all* of the female spiders of the species of interest on Earth! Instead, we use statistics to determine how many spiders we must collect to confidently infer something about the general population and then use statistics again to make such inferences.

DATA COME IN ALL SHAPES AND SIZES   In statistics we use the word "variable" to mean a measurable characteristic of an individual component of a system. Some variables are on a numerical scale, such as the daily high temperature or the clutch size of a spider. We call these **quantitative variables**. Quantitative variables that take on only whole number values (such as spider clutch size) are called **discrete variables**, whereas variables that can also take on a fractional value (such as temperature) are called **continuous variables**.

Other variables take categories as values, such as a human blood type (A, B, AB, or O) or an ant caste (queen, worker, or male). We call these **categorical variables**. Categorical variables with a natural ordering, such as a final grade in Introductory Biology (A, B, C, D, or F), are called **ordinal variables**.

Each class of variables comes with its own set of statistical methods. We will introduce a few common methods in this appendix that will help you work on the problems presented in this book, but you should consult a biostatistics textbook for more advanced tests and analyses for other data sets and problems.

Step 3: Organize and Visualize the Data

Your data consist of a series of values of the variable or variables of interest, each from a separate observation. For example, **Table B1** shows the weight and length of 34 fish (*Abramis brama*) from Lake Laengelmavesi in Finland. From this list of numbers, it's hard to get a sense of how big

table **B1**	Weight and Length Measurements for a Sample of *Abramis brama* from Lake Laengelmavesi	
Individual number	Fish weight (g)	Fish length (cm)
1	242	30.0
2	290	31.2
3	340	31.1
4	363	33.5
5	390	35.0
6	430	34.0
7	450	34.7
8	450	35.1
9	475	36.2
10	500	34.5
11	500	36.2
12	500	36.2
13	500	36.4
14	575	38.7
15	600	37.3
16	600	37.2
17	610	38.5
18	620	39.2
19	650	38.6
20	680	39.7
21	685	39.5
22	700	37.2
23	700	38.3
24	700	40.6
25	714	40.6
26	720	40.9
27	725	40.5
28	850	41.5
29	920	42.6
30	925	44.0
31	950	45.9
32	955	44.1
33	975	45.3
34	1,000	41.6

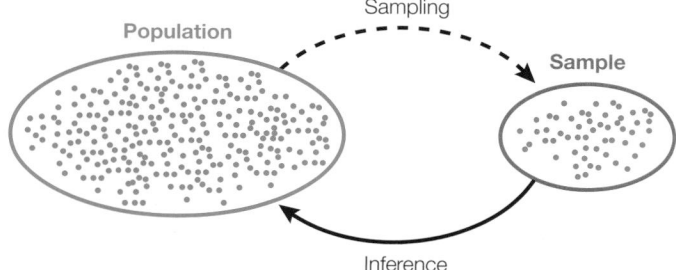

Figure B1 Sampling from a Population   Biologists take representative samples from a population, use descriptive statistics to characterize their samples, and then use inferential statistics to draw conclusions about the original population.

table **B2** Summary of Fish Weights of *Abramis brama* from Lake Laengelmavesi		
Weight (g)	Frequency	Relative frequency
201–300	2	0.06
301–400	3	0.09
401–500	8	0.24
501–600	3	0.09
601–700	8	0.24
701–800	3	0.09
801–900	1	0.03
901–1,000	6	0.18
Total	34	1.0

table **B3** Poinsettia Colors		
Color	Frequency	Proportion
Red	108	0.59
Pink	34	0.19
White	40	0.22
Total	182	1.0

the fish in the lake are, or how variable they are in size. It's much easier to gain intuition about your data if you organize them. One way to do this is to group (or bin) your data into **classes**, and count up the number of observations that fall into each class. The result is a **frequency distribution**. **Table B2** shows the fish weight data as a frequency distribution. For each 100-gram weight class, Table B2 shows the number, or frequency, of observations in that weight class, as well as the relative frequency (proportion of the total) of observations in that weight class. Notice that the data take up much less space when organized in this fashion. Also notice that we can now see that most of the fish fall in the middle of the weight range, with relatively few very small or very large fish.

 Media Clip B1 Interpreting Frequency Distributions
www.Life11e.com/mcB1

It is even easier to visualize the frequency distribution of fish weights if we graph them in the form of a **histogram** such as the one in **Figure B2**. When grouping quantitative data, it is necessary to decide how many

classes to include. It is often useful to look at multiple histograms before deciding which grouping offers the best representation of the data.

Frequency distributions are also useful ways of summarizing categorical data. **Table B3** shows a frequency distribution of the colors of 182 poinsettia plants (red, pink, or white) resulting from an experimental cross between two parent plants. Notice that, as with the fish example, the table is a much more compact way to present the data than a list of 182 color observations would be. For categorical data, the possible values of the variable are the categories themselves, and the frequencies are the number of observations in each category. We can visualize frequency distributions of categorical data like this by constructing a **bar chart**. The heights of the bars indicate the number of observations in each category (**Figure B3**). Another way to display the same data is in a **pie chart**, which shows the proportion of each category represented like pieces of a pie (**Figure B4**).

Sometimes we wish to compare two quantitative variables. For example, the researchers at Lake Laengelmavesi investigated the relationship between fish weight and length, from the data presented in Table B1. We can visualize this relationship using a **scatter plot**, in which the combination of the weight and length of each fish is represented as a single point (**Figure B5**). These two variables have a **positive relationship** since the slope of a line drawn through the points is positive. As the length of a fish increases, its weight tends to increase in an approximately linear manner.

Tables and graphs are critical to interpreting and communicating data, and thus should be as self-contained and understandable as possible. Their content should be easily understood simply by looking at them.

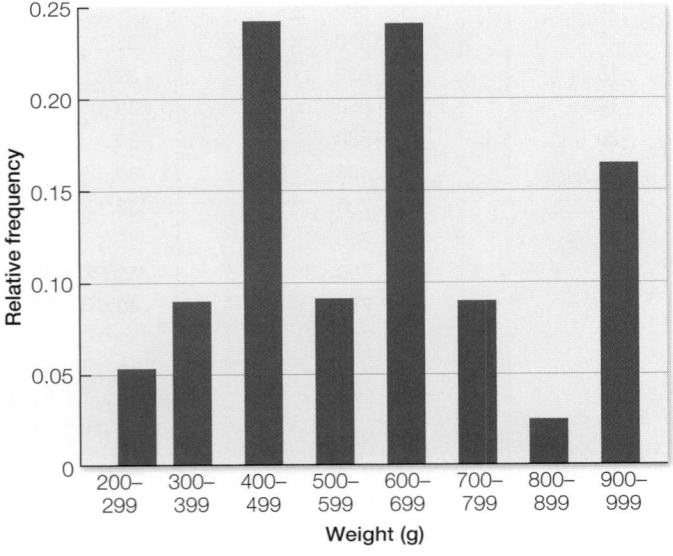

Figure B2 Histograms Depict Frequency Distributions of Quantitative Data   This histogram shows the relative frequency of different weight classes of fish (*Abramis brama*).

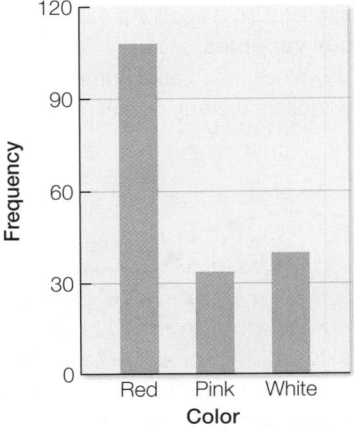

Figure B3 Bar Charts Compare Categorical Data   This bar chart shows the frequency of three poinsettia colors that result from an experimental cross.

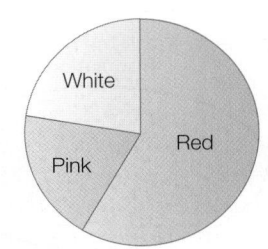

Figure B4 Pie Charts Show Proportions of Categories   This pie chart shows the proportions of the three poinsettia colors presented in Table B3.

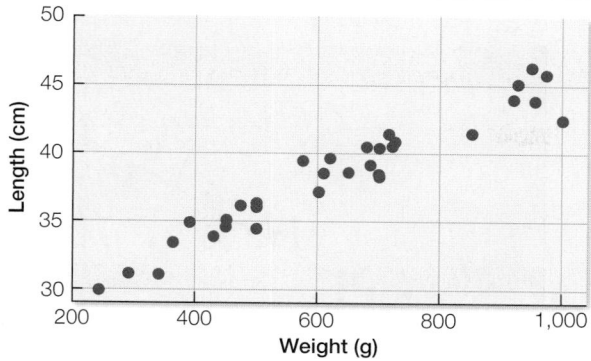

Figure B5 **Scatter Plots Contrast Two Variables**   Scatter plot of *Abramis brama* weights and lengths (measured from nose to end of tail). These two variables have a positive relationship since the slope of a line drawn through the points is positive.

Axes, captions, and units should be clearly labeled, statistical terms should be defined, and appropriate groupings should be used when tabulating or graphing quantitative data.

Step 4: Summarize the Data

A **statistic** is a numerical quantity calculated from data, whereas **descriptive statistics** are quantities that describe general patterns in data. Descriptive statistics allow us to make straightforward comparisons among different data sets and concisely communicate basic features of our data.

DESCRIBING CATEGORICAL DATA   For categorical variables, we typically use proportions to describe our data. That is, we construct tables containing the proportions of observations in each category. For example, the third column in Table B3 provides the proportions of poinsettia plants in each color category, and the pie chart in Figure B4 provides a visual representation of those proportions.

DESCRIBING QUANTITATIVE DATA   For quantitative data, we often start by calculating **measures of center**, quantities that roughly tell us where the center of our data lies. There are three commonly used measures of center:

- The **mean**, or average value, of our sample is simply the sum of all the values in the sample divided by the number of observations in our sample (**Figure B6**).
- The **median** is the value at which there are equal numbers of smaller and larger observations.
- The **mode** is the most frequent value in the sample.

It is often just as important to quantify the variation in the data as it is to calculate its center. There are several statistics that tell us how much the values differ from one another. We call these **measures of dispersion**. The easiest one to understand and calculate is the **range**, which is simply the largest value in the sample minus the smallest value. The most commonly used measure of dispersion is the **standard deviation**, which calculates the extent to which the data are spread out from the mean. A deviation is the difference between an observation and the mean of the sample. The standard deviation is a measure of how far the average observation in the sample is from the sample mean. Two samples can have the same range, but very different standard deviations if observations in

Figure B6 **Descriptive Statistics for Quantitative Data**
Below are the equations used to calculate the descriptive statistics we discuss in this appendix. You can calculate these statistics yourself, or use free internet resources to help you make your calculations.

Notation:

$x_1, x_2, x_3, \ldots x_n$ are the n observations of variable X in the sample.

$\sum_{i=1}^{n} x_i = x_1 + x_2 + x_3, \ldots + x_n$ is the sum of all of the observations. (The Greek letter sigma, Σ, is used to denote "sum of.")

In regression, the independent variable is X, and the dependent variable is Y. b_0 is the vertical intercept of a regression line. b_1 is the slope of a regression line.

Equations

1. Mean: $$\bar{X} = \frac{\sum_{i=1}^{n} x_i}{n}$$

2. Standard deviation: $$s = \sqrt{\frac{\sum (x_i - \bar{x})^2}{n-1}}$$

3. Correlation coefficient: $$r = \frac{\sum (x_i - \bar{x})(y_i - \bar{y})}{\sqrt{\sum (x_i - \bar{x})^2 (y_i - \bar{y})^2}}$$

4. Least-squares regression line: $Y = b_0 + b_1 X$

   where $$b_1 = \frac{\sum (x_i - \bar{x})(y_i - \bar{y})}{\sum (x_i - \bar{x})^2} \text{ and } b_0 = \bar{y} - b_1 \bar{x}$$

5. Standard error of the mean: $$SE_{\bar{x}} = \frac{s}{\sqrt{n}}$$

one are clustered closer to the mean than in the other. In **Figure B7**, for example, sample 1 has a smaller standard deviation than does sample 2, even though the two samples have the same means and ranges.

To demonstrate these descriptive statistics, let's return to the Lake Laengelmavesi study (see the data in Table B1). The mean weight of the 34 fish (see equation 1 in Figure B6) is:

$$\bar{X} = \frac{\text{sum of the weight of all fish in sample}}{\text{number of fish in sample}} = \frac{21{,}284 \text{ g}}{34} = 626 \text{ g}$$

Since there is an even number of observations in the sample, then the median weight is the value halfway between the two middle values:

$$\frac{610 \text{ g} + 620 \text{ g}}{2} = 615 \text{ g}$$

The mode of the sample is 500 g, which appears four times. The standard deviation (see equation 2 in Figure B6) is:

$$s = \sqrt{\frac{\sum (x_i - \bar{x})^2}{n-1}} = 206.6 \text{ g}$$

and the range is 1,000 g – 242 g = 758 g.

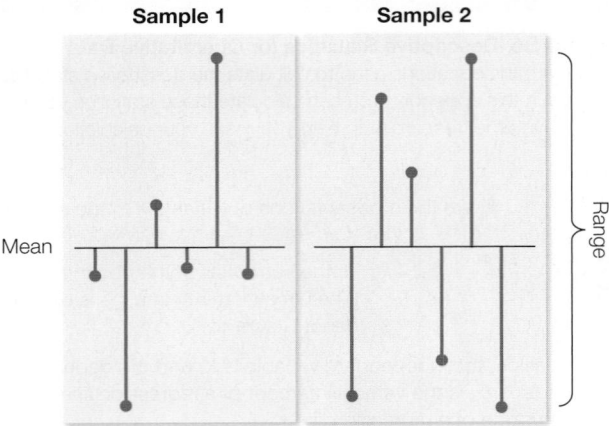

Figure B7 Measures of Dispersion Two samples with the same mean (black horizontal lines) and range. Red lines show the deviations of each observation from the mean. Samples with large deviations have large standard deviations. Sample 1 has a smaller standard deviation than sample 2.

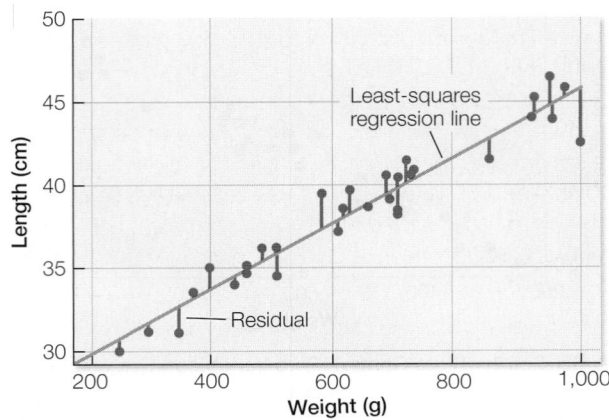

Figure B9 Linear Regression Estimates the Typical Relationship between Two Variables Least-squares linear regression line for *Abramis brama* weights and lengths (measured from nose to end of tail). The regression line is given by the equation $Y = 26.1 + 0.02X$. It is the line that minimizes the sum of the squares of the residuals.

DESCRIBING THE RELATIONSHIP BETWEEN TWO QUANTITATIVE VARIABLES Biologists are often interested in understanding the relationship between two different quantitative variables: How does the height of an organism relate to its weight? How does air pollution relate to the prevalence of asthma? How does lichen abundance relate to levels of air pollution? Recall that scatter plots visually represent such relationships.

We can quantify the strength of the relationship between two quantitative variables using a single value called the Pearson product–moment **correlation coefficient** (see equation 3 in Figure B6). This statistic ranges between −1 and 1 and tells us how closely the points in a scatter plot conform to a straight line. A negative correlation coefficient indicates that one variable decreases as the other increases; a positive correlation coefficient indicates that the two variables increase together; and a correlation coefficient of zero indicates that there is no linear relationship between the two variables (**Figure B8**).

One must always keep in mind that *correlation does not mean causation*. Two variables can be closely related without one causing the other. For example, the number of cavities in a child's mouth correlates positively with the size of his or her feet. Clearly cavities do not enhance foot growth; nor does foot growth cause tooth decay. Instead the correlation exists because both quantities tend to increase with age.

Intuitively, the straight line that tracks the cluster of points on a scatter plot tells us something about the typical relationship between the two variables. Statisticians do not, however, simply eyeball the data and draw a line by hand. They often use a method called least-squares **linear regression** to fit a straight line to the data (see equation 4 in Figure B6). This method calculates the line that minimizes the overall vertical distances between the points in the scatter plot and the line itself. These distances are called **residuals** (**Figure B9**). Two parameters describe the regression line: b_0 (the vertical intercept of the line, or the expected value of variable Y when $X = 0$), and b_1 (the slope of the line, or how much values of Y are expected to change with changes in values of X).

Step 5: Make Inferences from the Data

Data analysis often culminates with statistical inference—an attempt to draw general conclusions about the system under investigation—the larger population—from observations of a representative sample (see Figure B1). When we test a new medulloblastoma brain cancer drug on ten patients, we do not simply want to know the fate of those ten individuals; rather, we hope to predict the drug's efficacy on the much larger population of all medulloblastoma patients.

STATISTICAL HYPOTHESES Once we have collected our data, we want to evaluate whether or not they fit the predictions of our hypothesis. For example, we want to know whether or not cancer incidence is greater in our sample of smokers than in our sample of nonsmokers, whether or not clutch size increases with spider body weight in our sample of spiders, or whether or not growth of a sample of fertilized plants is greater than that of a sample of unfertilized plants.

Before making statistical inferences from data, we must formalize our "whether or not" question into a pair of opposing hypotheses. We start with the "or not" hypothesis, which we call the **null hypothesis** (denoted H_0) because it often is the hypothesis that there is no difference between sample means, no correlation between variables in the sample, or no difference between our sample and an expected frequency distribution.

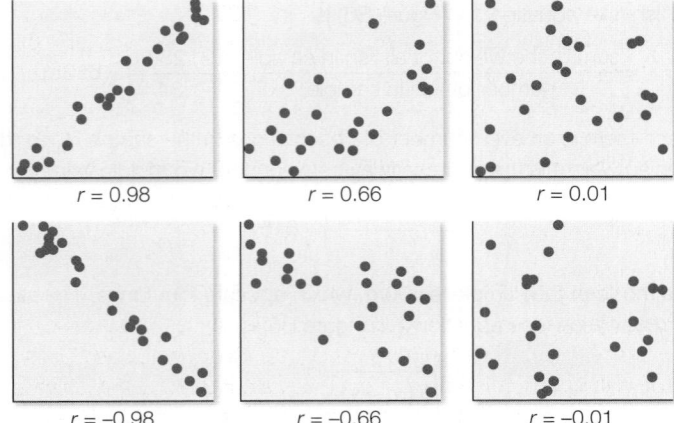

Figure B8 Correlation Coefficients The correlation coefficient (*r*) indicates both the strength and the direction of the relationship.

The **alternative hypothesis** (denoted H_A) is that there is a difference in means, that there is a correlation between variables, or that the sample distribution does differ from the expected one.

Suppose, for example, we would like to know whether or not a new vaccine is more effective than an existing vaccine at immunizing children against influenza. We have measured flu incidence in a group of children who received the new vaccine and want to compare it with flu incidence in a group of children who received the old vaccine. Our statistical hypotheses would be as follows:

H_0: Flu incidence was the same in both groups of children.

H_A: Flu incidence was different between the two groups of children.

In the next few sections we will discuss how we decide when to reject the null hypothesis in favor of the alternative hypothesis.

JUMPING TO THE WRONG CONCLUSIONS There are two ways that a statistical test can go wrong (**Figure B10**). We can reject the null hypothesis when it is actually true (**Type I error**), or we can accept the null hypothesis when it is actually false (**Type II error**). These kinds of errors are analogous to false positives and false negatives in medical testing, respectively. If we mistakenly reject the null hypothesis when it is actually true, then we falsely endorse the incorrect hypothesis. If we are unable to reject the null hypothesis when it is actually false, then we fail to realize a yet undiscovered truth.

Suppose we would like to know whether there are more females than males in a population of 10,000 individuals. To determine the makeup of the population, we choose 20 individuals randomly and record their sex. Our null hypothesis is that there are not more females than males; and our alternative hypothesis is that there are. The following scenarios illustrate the possible mistakes we might make:

- *Scenario 1*: The population actually has 40% females and 60% males. Although our random sample of 20 people is likely to be dominated by males, it is certainly possible that, by chance, we will end up choosing more females than males. If this occurs, and we mistakenly reject the null hypothesis (that there are not more females than males), then we make a Type I error.

- *Scenario 2*: The population actually has 60% females and 40% males. If, by chance, we end up with a majority of males in our sample and thus fail to reject the null hypothesis, then we make a Type II error.

Fortunately, statistics has been developed precisely to avoid these kinds of errors and inform us about the reliability of our conclusions.

Our conclusion	The real world	
	Null hypothesis true (*not more females*)	Null hypothesis false (*more females*)
Null hypothesis true (*not more females*)	√	Type II error (*false negative*)
Null hypothesis false (*more females*)	Type I error (*false positive*)	√

Figure B10 Two Types of Errors Possible outcomes of a statistical test. Statistical inference can result in correct and incorrect conclusions about the population of interest.

The methods are based on calculating the probabilities of different possible outcomes. Although you may have heard or even used the word "probability" on multiple occasions, it is important that you understand its mathematical meaning. A **probability** is a numerical quantity that expresses the likelihood of some event. It ranges between zero and 1; zero means there is no chance the event will occur, and 1 means the event is guaranteed to occur. This only makes sense if there is an element of chance, that is, if it is possible the event will occur and possible that it will not occur. For example, when we flip a fair coin, it will land on heads with probability 0.5 and land on tails with probability 0.5. When we select individuals randomly from a population with 60% females and 40% males, we will encounter a female with probability 0.6 and a male with probability 0.4.

Probability plays a very important role in statistics. To draw conclusions about the real world (the population) from our sample, we first calculate the probability of obtaining our sample if the null hypothesis is true. Specifically, statistical inference is based on answering the following question:

Suppose the null hypothesis is true. What is the probability that a random sample would, by chance, differ from the null hypothesis as much as our sample differs from the null hypothesis?

If our sample is highly improbable under the null hypothesis, then we rule it out in favor of our alternative hypothesis. If, instead, our sample has a reasonable probability of occurring under the null hypothesis, then we conclude that our data are consistent with the null hypothesis and we do not reject it.

Returning to the sex ratio example, let's consider two new scenarios:

- *Scenario 3*: Suppose we want to infer whether or not females constitute the majority of the population (our alternative hypothesis) based on a random sample containing 12 females and 8 males. We would calculate the probability that a random sample of 20 people includes at least 12 females assuming that the population, in fact, has a 50:50 sex ratio (our null hypothesis). This probability is 0.13, which is too high to rule out the null hypothesis.

- *Scenario 4*: Suppose now that our sample contains 17 females and 3 males. If our population is truly evenly divided, then this sample is much less likely than the sample in scenario 3. The probability of such an extreme sample is 0.0002, and would lead us to rule out the null hypothesis and conclude that there are more females than males.

This agrees with our intuition. When choosing 20 people randomly from an evenly divided population, we would be surprised if almost all of them were female, but would not be surprised at all if we ended up with a few more females than males (or a few more males than females). Exactly how many females do we need in our sample before we can confidently infer that they make up the majority of the population? And how confident are we when we reach that conclusion? Statistics allows us to answer these questions precisely.

STATISTICAL SIGNIFICANCE: AVOIDING FALSE POSITIVES Whenever we test hypotheses, we calculate the probability just discussed, and refer to this value as the ***P*-value** of our test. Specifically, the *P*-value is the probability of getting data as extreme as our data (just by chance) if the null hypothesis is in fact true. In other words, it is the likelihood that chance alone would produce data that differ from the null hypothesis as much as our data differ from the null hypothesis. How we measure the difference between our data and the null hypothesis depends on the kind of data in our sample (categorical or quantitative) and the nature of the null hypothesis (assertions about proportions, single variables, multiple variables, differences between variables, correlations between variables, etc.).

For many statistical tests, *P*-values can be calculated mathematically. One option is to quantify the extent to which the data depart from the null hypothesis and then use look-up tables (available in most statistics textbooks or on the internet) to find the probability that chance alone would produce a difference of that magnitude. Most scientists, however, find *P*-values primarily by using statistical software rather than hand calculations combined with look-up tables. Regardless of the technology, the most important steps of the statistical analysis are still left to the researcher: constructing appropriate null and alternative hypotheses, choosing the correct statistical test, and drawing correct conclusions.

After we calculate a *P*-value from our data, we have to decide whether it is small enough to conclude that our data are inconsistent with the null hypothesis. This is decided by comparing the *P*-value to a threshold called the **significance level**, which is often chosen even before making any calculations. We reject the null hypothesis only when the *P*-value is less than or equal to the significance level, denoted α. This ensures that, if the null hypothesis is true, we have at most a probability α of accidentally rejecting it. Therefore the lower the value of α, the less likely we are to make a Type I error (see the lower left cell of Figure B10). The most commonly used significance level is $\alpha = 0.05$, which limits the probability of a Type I error to 5%.

If our statistical test yields a *P*-value that is less than our significance level α, then we conclude that the effect described by our alternative hypothesis is statistically significant at the level α and we reject the null hypothesis. If our *P*-value is greater than α, then we conclude that we are unable to reject the null hypothesis. In this case, we do not actually reject the alternative hypothesis, rather we conclude that we do not yet have enough evidence to support it.

POWER: AVOIDING FALSE NEGATIVES   The **power** of a statistical test is the probability that we will correctly reject the null hypothesis when it is false (see the lower right cell of Figure B10). Therefore the higher the power of the test, the less likely we are to make a Type II error (see the upper right cell of Figure B10). The power of a test can be calculated, and such calculations can be used to improve your methodology. Generally, there are several steps that can be taken to increase power and thereby avoid false negatives:

- **Decrease the significance level**, α. The higher the value of α, the harder it is to reject the null hypothesis, even if it is actually false.
- **Increase the sample size**. The more data one has, the more likely one is to find evidence against the null hypothesis, if it is actually false.
- **Decrease variability in the sample**. The more variation there is in the sample, the harder it is to discern a clear effect (the alternative hypothesis) when it actually exists.

It is always a good idea to design your experiment to reduce any variability that may obscure the pattern you seek to detect. For example, it is possible that the chance of a child contracting influenza varies depending on whether he or she lives in a crowded (e.g., urban) environment or one that is less so (e.g., rural). To reduce variability, a scientist might choose

to test a new influenza vaccine only on children from one environment or the other. After you have minimized such extraneous variation, you can use power calculations to choose the right combination of α and sample size to reduce the risks of Type I and Type II errors to desirable levels.

There is a trade-off between Type I and Type II errors: as α increases, the risk of a Type I error decreases but the risk of a Type II error increases. As discussed above, scientists tend to be more concerned about Type I than Type II errors. That is, they believe it is worse to mistakenly believe a false hypothesis than it is to fail to make a new discovery. Thus they prefer to use low values of α. However, there are many real-world scenarios in which it would be worse to make a Type II error than a Type I error. For example, suppose a new cold medication is being tested for dangerous (life-threatening) side effects. The null hypothesis is that there are no such side effects. A Type II error might lead regulatory agencies to approve a harmful medication that could cost human lives. In contrast, a Type I error would simply mean one less cold medication among the many that already line pharmacy shelves. In such cases, policymakers take steps to avoid a Type II error, even if, in doing so, they increase the risk of a Type I error.

STATISTICAL INFERENCE WITH QUANTITATIVE DATA   Statistics that describe patterns in our samples are used to estimate properties of the larger population. Earlier we calculated the mean weight of a sample of *Abramis brama* in Lake Laengelmavesi, which provided us with an estimate of the mean weight of all the *Abramis brama* in the lake. But how close is our estimate to the true value in the larger population? Our estimate from the sample is unlikely to exactly equal the true population value. For example, our sample of *Abramis brama* may, by chance, have included an excess of large individuals. In this case, our sample would overestimate the true mean weight of fish in the population.

The **standard error** of a sample statistic (such as the mean) is a measure of how close it is likely to be to the true population value. The **standard error of the mean**, for example, provides an estimate of how far we might expect a sample mean to deviate from the true population mean. It is a function of how much individual observations vary within samples (the standard deviation) and the size of the sample (*n*). Standard errors increase with the degree of variation within samples, and they decrease with the number of observations in a sample (because large samples provide better estimates about the underlying population than do small samples). For our sample of 34 *Abramis brama*, we would calculate the standard error of the mean using equation 5 in Figure B6:

$$SE_{\bar{x}} = \frac{s}{\sqrt{n}} = \frac{206.6\,g}{\sqrt{34}} = 35.4\,g$$

The standard error of a statistic is related to the **confidence interval**— a range around the sample statistic that has a specified probability of including the true population value. The formula we use to calculate confidence intervals depends on the characteristics of the data, and the particular statistic. For many types of continuous data, the bounds of the **95% confidence interval** of the mean can be calculated by taking the mean and adding and subtracting 1.96 times the standard error of the mean. Consider our sample of 34 *Abramis brama*, for example. We would say that the mean of 626 g has a 95% confidence interval from 556.6 g to 695.4 g (626 ± 69.4 g). If all our assumptions about our data have been met, we would expect the true population mean to fall in this confidence interval 95% of the time.

Researchers typically use graphs or tables to report sample statistics (such as the mean) as well as some measure of confidence in them (such as their standard error or a confidence interval). This book is full of examples. If you understand the concepts of sample statistics, standard errors, and confidence intervals, you can see for yourself the major patterns in the data, without waiting for the authors to tell you what they are. For example, if samples from two groups have 95% confidence intervals of the mean that do not overlap, then you can conclude that it is unlikely that the groups have the same true mean.

 Activity B1 Standard Deviations, Standard Errors, and Confidence Intervals Simulation www.Life11e.com/acB1

Biologists conduct statistical tests to obtain more precise estimates of the probability of observing a given difference between samples if the null hypothesis that there is no difference in the populations is true. The appropriate test depends on the nature of the data and the experimental design. For example, we might want to calculate the probability that the mean weights of two different fish species in Lake Laengelmavesi, *Abramis brama* and *Leusiscus idus*, are the same. A simple method for comparing the means of two groups is the *t*-test, described in **Figure B11**. We looked earlier at data for *Abramis brama*; the researchers who collected these data also collected weights for six individuals of *Leusiscus idus*: 270, 270, 306, 540, 800, and 1,000 g. We begin by stating our hypotheses and choosing a significance level:

> ## research tools
>
> **Figure B11 The t-test**   What is the *t*-test? It is a standard method for assessing whether the means of two groups are statistically different from each other.
>
> **Step 1:** State the null and alternative *hypotheses*:
> H_0: The two populations have the same mean.
> H_A: The two populations have different means.
>
> **Step 2:** Choose a significance level, α, to limit the risk of a Type I error.
>
> **Step 3:** Calculate the *test statistic*: $t_s = \dfrac{\bar{y}_1 - \bar{y}_2}{\sqrt{\dfrac{s_1^2}{n_1} + \dfrac{s_2^2}{n_2}}}$
>
>   *Notation*: y_1 and y_2 are the sample means; s_1 and s_2 are the sample standard deviations; and n_1 and n_2 are the sample sizes.
>
> **Step 4:** Use the test statistic to assess whether the data are consistent with the null hypothesis:
>
>   Calculate the *P-value* (P) using statistical software or by hand using statistical tables.
>
> **Step 5:** Draw conclusions from the test:
>
>   If P ≤ α, then reject H_0, and conclude that the population distribution is significantly different.
>
>   If $P > \alpha$, then we do not have sufficient evidence to conclude that the means differ.

H_0: *Abramis brama* and *Leusiscus idus* have the same mean weight.

H_A: *Abramis brama* and *Leusiscus idus* have different mean weights.

$$\alpha = 0.05$$

The test statistic is calculated using the means, standard deviations, and sizes of the two samples:

$$t_s = \frac{626 - 531}{\sqrt{\dfrac{207^2}{34} + \dfrac{310^2}{6}}} = 0.724$$

We can use statistical software or one of the free statistical sites on the internet to find that the *P*-value for this result is $P = 0.497$. Since P is considerably greater than α, we fail to reject the null hypothesis and conclude that our study does not provide evidence that the two species have different mean weights.

You may want to consult an introductory statistics textbook to learn more about confidence intervals, *t*-tests, and other basic statistical tests for quantitative data.

STATISTICAL INFERENCE WITH CATEGORICAL DATA   With categorical data, we often wish to ask whether the frequencies of observations in different categories are consistent with a hypothesized frequency distribution. We can use a **chi-square goodness-of-fit** test to answer this question.

Figure B12 outlines the steps of a chi-square goodness-of-fit-test. As an example, consider the data described in Table B3. Many plant species have simple Mendelian genetic systems in which parent plants produce progeny with three different colors of flowers in a ratio of 2:1:1. However, a botanist believes that these particular poinsettia plants have a different genetic system that does not produce a 2:1:1 ratio of red, pink, and white plants. A chi-square goodness-of-fit test can be used to assess whether or not the data are consistent with this ratio, and thus whether or not this simple genetic explanation is valid. We start by stating our hypotheses and significance level:

H_0: The progeny of this type of cross have the following probabilities of each flower color:

Pr{Red} = 0.50, Pr{Pink} = 0.25, Pr{White} = 0.25

H_A: At least one of the probabilities of H_0 is incorrect.

$$\alpha = 0.05$$

We next use the probabilities in H_0 and the sample size to calculate the expected frequencies:

	Red	Pink	White
Observed	108	34	40
Expected	(.50)(182) = 91	(.25)(182) = 45.5	(.25)(182) = 45.5

Based on these quantities, we calculate the chi-square test statistic:

$$\chi_s^2 = \sum_{i=1}^{C} \frac{(O_i - E_i)^2}{E_i} = \frac{(108 - 91)^2}{91} + \frac{(34 - 45.5)^2}{45.5} + \frac{(40 - 45.5)^2}{45.5} = 6.747$$

research tools

Figure B12 The Chi-Square Goodness-of-Fit Test   What is the chi-square goodness-of-fit test? It is a standard method for assessing whether a sample came from a population with a specific distribution.

Step 1: State the null and alternative *hypotheses*:
H_0: The population has the specified distribution.
H_A: The population does not have the specified distribution.

Step 2: Choose a significance level, α, to limit the risk of a Type I error.

Step 3: Determine the *observed frequency* and *expected frequency* for each category:

The observed frequency of a category is simply the number of observations in the sample of that type.

The expected frequency of a category is the probability of the category specified in H_0 multiplied by the overall sample size.

Step 4: Calculate the *test statistic*: $\chi_s^2 = \sum_{i=1}^{C} \frac{(O_i - E_i)^2}{E_i}$

Notation: C is the total number of categories, O_i is the observed frequency of category i, and E_1 is the expected frequency of category i.

Step 5: Use the test statistic to assess whether the data are consistent with the null hypothesis:

Calculate the *P-value* (P) using statistical software or by hand using statistical tables.

Step 6: Draw conclusions from the test:

If $P \leq \alpha$, then reject H_0, and conclude that the population distribution is significantly different than the distribution specified by H_0.

If $P > \alpha$, then we do not have sufficient evidence to conclude that population has a different distribution.

We find the P-value for this result to be $P = 0.034$ using statistical software. Since P is less than α, we reject the null hypothesis and conclude that the botanist is likely correct: the plant color patterns are not explained by the simple Mendelian genetic model under consideration.

This introduction is meant to provide only a brief introduction to the concepts of statistical analysis, with a few example tests. **Figure B13** summarizes some of the commonly used statistical tests that you may encounter in biological studies.

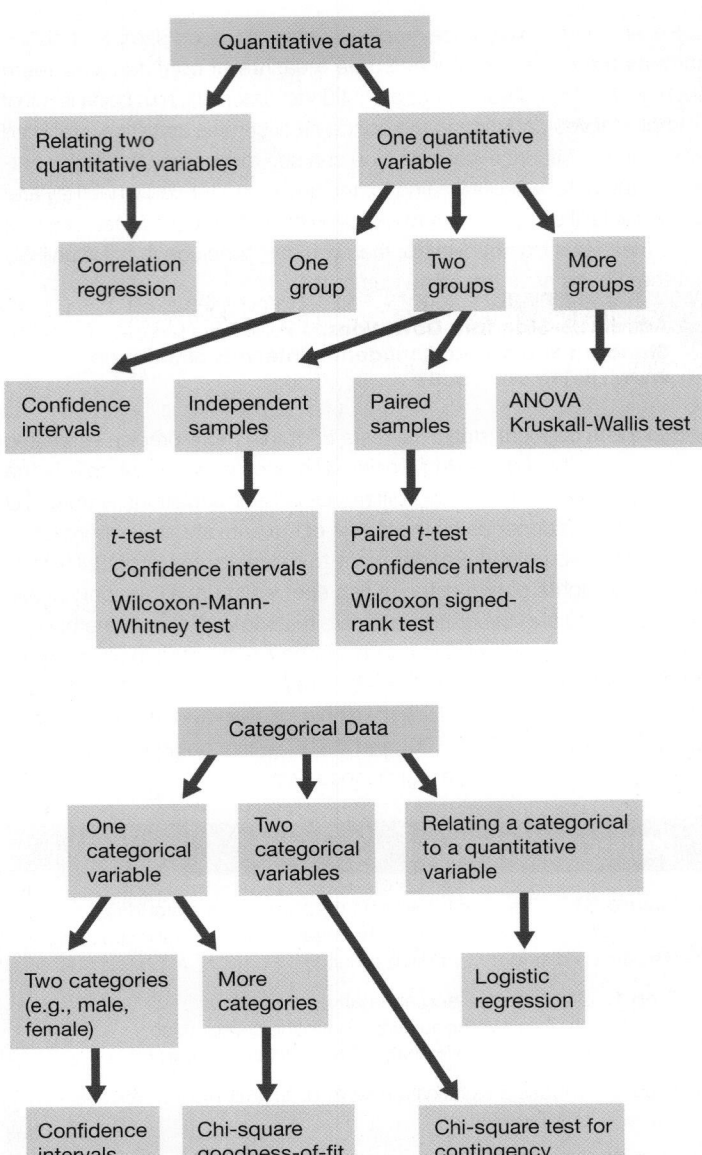

Figure B13 Some Common Methods of Statistical Inference
This flowchart shows some of the commonly used methods of statistical inference for different combinations of data. Detailed descriptions of these methods can be found in most introductory biostatistics textbooks.

APPENDIX C Some Measurements Used in Biology

Measures of	Unit	Equivalents	Metric → English conversion
Length	meter (m)	base unit	1 m = 39.37 inches = 3.28 feet = 1.196 yards
	kilometer (km)	1 km = 1000 (10^3) m	1 km = 0.62 miles
	centimeter (cm)	1 cm = 0.01 (10^{-2}) m	1 cm = 0.39 inches
	millimeter (mm)	1 mm = 0.1 cm = 10^{-3} m	1 mm = 0.039 inches
	micrometer (μm)	1 μm = 0.001 mm = 10^{-6} m	
	nanometer (nm)	1 nm = 0.001 μm = 10^{-9} m	
Area	square meter (m^2)	base unit	1 m^2 = 1.196 square yards
	hectare (ha)	1 ha = 10,000 m^2	1 ha = 2.47 acres
Volume	liter (L)	base unit	1 L = 1.06 quarts
	milliliter (mL)	1 mL = 0.001 L = 10^{-3} L	1 mL = 0.034 fluid ounces
	microliter (μL)	1 μL = 0.001 mL = 10^{-6} L	
Mass	gram (g)	base unit	1 g = 0.035 ounces
	kilogram (kg)	1 kg = 1000 g	1 kg = 2.20 pounds
	metric ton (mt)	1 mt = 1000 kg	1 mt = 2,200 pounds = 1.10 ton
	milligram (mg)	1 mg = 0.001 g = 10^{-3} g	
	microgram (μg)	1 μg = 0.001 mg = 10^{-6} g	
Temperature	degree Celsius (°C)	base unit	°C = (°F − 32)/1.8
			0°C = 32°F (water freezes)
			100°C = 212°F (water boils)
			20°C = 68°F ("room temperature")
			37°C = 98.6°F (human internal body temperature)
	Kelvin (K)*	K = °C − 273	0 K = −460°F
Energy	joule (J)		1 J ≈ 0.24 calorie = 0.00024 kilocalorie[†]

*0 K (−273°C) is "absolute zero," a temperature at which molecular oscillations approach 0—that is, the point at which motion all but stops.

[†]A *calorie* is the amount of heat necessary to raise the temperature of 1 gram of water 1°C. The *kilocalorie*, or nutritionist's calorie, is what we commonly think of as a calorie in terms of food.

ANSWERS TO ALL IN-CHAPTER QUESTIONS

Chapter 1

RECAP 1.1

1. The origin of photosynthesis resulted in a gradual build-up of oxygen in the Earth's atmosphere. Before the Earth had an oxygen-rich atmosphere, UV radiation on the Earth's surface was so intense that it killed any organisms on land; life could only survive if it was protected from UV radiation by water. But as O_2 accumulated in the atmosphere, O2 molecules began reacting with one another to form ozone (O_3). A layer of ozone gradually built up in the high atmosphere, and by about 500 million years ago it was sufficient to block enough UV radiation that life could colonize land.

2. Among the list of common characteristics of life on Earth, there are some features that we might expect to be general to any origin of life (such as the need to extract energy from the environment and use it to do work), but other features that we expect to be unique to each origin of life. For example, although another origin of life might use a genetic information system of some kind, there is no reason that we would expect the details of how that system functioned to be the same. Another origin of life might well use something other than DNA, or use different nucleotides to make up DNA. Moreover, even if another origin of life used the same basic building blocks to make DNA, the genetic code that specified which combinations of nucleotides encode which amino acids in proteins would be expected to be different. Furthermore, it is unlikely that the same set of amino acids would be used to construct proteins. It is the commonalities of these details across life on Earth that allow us to conclude that all life on Earth has a single common origin.

3. Fish typically have eyes that are used for sight, and eyes obviously require light to function. In a normal surface population, any mutation that disrupts eye-sight would be strongly selected against, because fish with non-functional eyes would be at a disadvantage compared to fish with functional eyes. Such mutations occur, but they are quickly eliminated from a surface population through natural selection. But in a cave population, there is no selection for eyesight, so such mutations are not selected against, and would accumulate in a cave population. In the absence of selection for eye function, we would expect gradual loss of a complex structure like an eye. In fact, many species of cave organisms do gradually evolve eyelessness over time. But how do they compensate for the loss of sight? Usually, there is strong selection for other sensory systems that do not require light, such as cells that detect small vibrations, or chemical scents. The important point is that in the dark environment, selection conditions change, and some features are expected to be lost, and others gained, because different features are favored under natural selection in the new environment.

4. Biologists use quantitative measures of similarities and differences between specimens to establish the relatedness of different species. The more similar, the more likely they have a recent common ancestor and the more different, the more likely their common ancestor is more remote in evolutionary time. Knowledge about phylogenetic relationships is also obtained from the fossil record and more recently from genomic analysis.

RECAP 1.2

1. A hypothesis is a proposed explanation for an observation or a phenomenon. An experiment is a rigorous test of that proposed explanation.

2. In a controlled experiment, all variables are held constant while one variable is manipulated to determine its effect on the system being studied.

3. When questions involve systems in which variables cannot be controlled, which is the case with many questions in natural systems, a comparative approach can be used to establish correlations between the variable of interest and its possible effects.

4. Since all life is related through evolution, different species use similar molecular, biochemical, cellular, physiological, and even behavioral components, systems, and mechanisms. Therefore what is learned from one species is likely to be applicable to other species.

RECAP 1.3

1. Modern biology is used to improve agricultural species—both plants and animals. An example is the genetic improvements of the plants that produce food grains. Biology is also used to treat diseases in agricultural species (e.g., antibiotics given to food animals) and to yield food plants resistant to herbicides.

2. The use of antibiotics creates a situation of artificial selection analogous to natural selection. Any disease organism that survives the antibiotic because of its genetic makeup will transmit that trait to its offspring. As the antibiotic continues to be used, the disease organisms with the resistance trait will multiply, resulting in decreased efficacy of the drug.

3. The Investigating Life thread in this chapter makes evident that global warming presents an enormous threat to reef-building corals, which support a high level of diversity in the marine ecosystem. Another notable example is the plight of the polar bear, which depends on sea ice to be able hunt its prey—seals. With the decrease in sea ice in the Arctic, the hunting areas and therefore hunting abilities of the polar bear have become more and more limited.

FIGURE QUESTIONS

Figure 1.9 This tree shows that the last common ancestor for fungi and animals was much more recent than the last common ancestor for fungi and plants. Fungi are therefore more closely related to animals than to plants.

Figure 1.15 The data reveal that the eastern- and western-breeding populations of bluefin tuna share the same mid-Atlantic feeding ground. Since the western-breeding population is commonly found east of the boundary line, the imposition of restrictions west of the boundary line would not have been effective in protecting the western-breeding population.

WORK WITH THE DATA, P. 14

1.

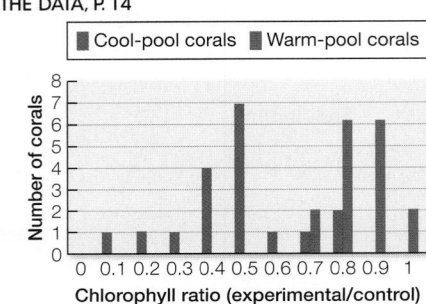

2. For the cool-pool corals, the probability of $H_0 = (0.5)17 = 0.00000762939$. Therefore we can safely reject H_0 and conclude that there is indeed a significant effect of heat stress on coral bleaching (at $P < 0.00001$). In other words, if the null hypothesis were true and there were no real effect of heat stress on coral bleaching, we would expect to see this many chlorophyll ratios below 1.0 fewer than 1 time out of 100,000 trials. For the warm-pool corals, there are seven observed values < 1, and none > 1. In this case, the probability of $H_0 = (0.5)7 = 0.0078125$. Therefore we can again reject H_0 and conclude that there is a significant effect of heat stress on coral bleaching in warm-pool corals as well (this time at $P < 0.01$).

3. The results of the randomization trials will differ depending on how well the cards are shuffled and how many replicates are compared, but the probability of finding a difference as great as 0.35 (the observed difference) in truly randomized samples of the two groups is very low ($P < 0.001$). Therefore we can again reject the null hypothesis and conclude that the effects of heat stress on coral bleaching are indeed higher in the corals from cool pools than in those from warm pools.

4. The different distributions of chlorophyll ratios for the cool- and warm-pool corals indicate that although both populations suffer bleaching as a consequence of heat stress, the populations from the cool pools are more sensitive. This suggests that corals from warmer environments might replace those from cooler environments under long-term conditions of global warming.

APPLY WHAT YOU'VE LEARNED

1. The results show that a population is more likely to go extinct if a deleterious environmental change is sudden and less likely to go extinct if the change is gradual. The complete loss of a species is more likely to happen if a deleterious change in the environment occurs rapidly and wipes out most of the populations of that species. Even if a few individuals survive, they may not be present in large enough numbers to effectively reproduce and maintain a viable population. However, if the environmental change is gradual, the population has time to allow surviving members to reproduce and enhance the proportion of individuals with the ability to survive, which allows the population to undergo adaptation to the change

2. The investigation shows how populations of living things evolve in response to changing environments. It models the process of natural selection that operates on populations and results in evolution. Environmental change and natural selection have helped shape the characteristics of organisms throughout the history of life on Earth.

3. All living things on Earth arose from a common ancestor and share certain characteristics. For example, all living things contain DNA, which encodes the information that passes from generation to generation. DNA is affected by mutation, which produces variation within a population. All living things are also affected by natural selection, which results in adaptation of populations and species over time. Because all living things possess DNA and are affected by natural selection, the results of the investigation are expected to be broadly applicable to life.

4. This was a controlled study because all of the cells were treated the same way (all variables were controlled) except for one condition that was varied. The varied condition was the amount of rifampicin added to the culture medium. (EXAMPLE) A comparative study could track changes in characteristics of a population of fish (or other species) cut in two by the building of a dam or other human-made construction. In a case like this, the study would have to show that the dam separated the original population into distinct groups that were isolated from each other and exposed to different environmental conditions. The study could then compare the characteristics of the separated groups of fish after a period of time to observe how the groups change in response to their respective environments

5. Yes. The research organization could use the results of the investigation as an example to show that any population of bacteria likely has some proportion of individuals with genetic mutations that allow a degree of resistance to an antibiotic. Over time, selection for resistance will result in an increasing proportion of individual bacteria in that population that can grow in the presence of the antibiotic. Therefore, an antibiotic effective at killing bacteria does not remain effective over time as the bacterial populations evolve resistance to the antibiotic. This means that new antibiotics need to be developed periodically to take the place of older ones that lose their effectiveness as bacteria develop resistance to them. Using this reasoning, the research organization can show the need for a continuous pipeline of new antibiotics to deal with the problem of bacterial resistance.

Chapter 2

RECAP 2.1

1. Elements with the same number of valence electrons are placed in the same vertical groups in the periodic table. For example, the electrons in sodium occupy the following shells (in order from the nucleus outward): 1s2, 2s2, 2p6, 3s1. Having one electron in the valence shell places sodium in Group 1. Lithium and potassium, which likewise have one electron in the valence shell, are in Group 1 as well.

2. An atom is stable when its eight electrons occupy its outermost shell (with the exception of hydrogen, which is stable with two electrons). In bonding, atoms either share electrons with, or donate to or receive electrons from, another atom to achieve stability.

3. The human body has the same elemental composition as Earth's crust but in very different proportions.

RECAP 2.2

1. In a covalent bond, the shared electrons between two atoms are actually part of both atoms. They are close together, and it takes a lot of energy to break them apart. In an ionic attraction, the electrons are transferred and the two atoms are relatively far apart, so it takes less energy to break an ionic attraction.

2. a. nonpolar; polar; polar; nonpolar;

   b. hydrophobic; hydrophilic; hydrophilic; hydrophobic

   c. C=O: δ^- at O; O—P: δ^- at O;

3. This is an example of van der Waals forces, which act over a short distance and do not involve polarity.

RECAP 2.3

1. $X\,C_6H_{12}O_6 + X\,O_2 \rightarrow X\,CO_2 + X\,H_2O$

2. In the burning of propane, there is a chemical transformation. Some chemical energy is released as heat and light.

RECAP 2.4

1. A solution contains a solvent (e.g., water) and a substance dissolved in the solvent (e.g., table salt, NaCl). Water is the medium of life because most molecules in living organisms either dissolve in water or interact with it.

2. A 1 molar (1 M) solution has a concentration of 1 mole per liter, which means it has 6.02×10^{23} molecules per liter. A 10^{-8} M solution has 6.02×10^{15} molecules per liter. If there is 10^{-6} liters, the number of molecules is 6.02×10^9.

3. This reaction increases the H^+ concentration in the blood and thereby decreases the pH. To increase the pH, and thereby decrease the H^+, the reverse reaction could be used as a buffer:

$$H^+ + HCO_3^- \rightarrow CO_2 + H_2O$$

WORK WITH THE DATA

1. The average local ratios differ from the all-countries average ratio. To test the hypothesis that the ratios are significantly different, a chi-square test could be used.

2. A graph of the data shows an increasing $^{13}C{:}^{12}C$ ratio as latitude away from the equator increases. This implies that different plants occur in different regions. Animal feeds used locally tend to be composed of plants grown locally.

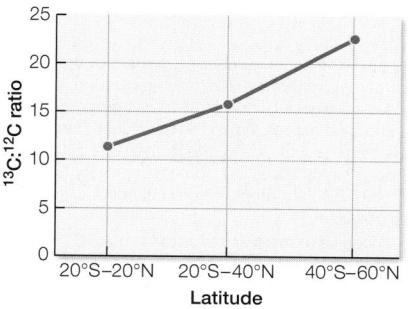

FIGURE QUESTIONS

Figure 2.3 The technique could be used to monitor therapy. As therapy progresses and the patient improves, one would expect that the brain region involved would show more activity.

Figure 2.6

Bohr model

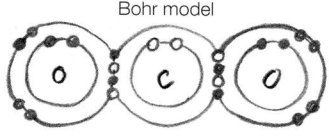

Structural formula

Ball-and-stick model

Space-filling model

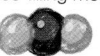

Figure 2.8 Ca (atomic number 20) has two valence electrons. Cl (atomic number 17) has seven valence electrons. So combining, two electrons are lost from one Ca atom to two atoms of Cl: formula $CaCl_2$.

Figure 2.9 At the chemical level, evaporation eliminates water for hydration of the ions, so the ions are no longer separated. At the physical level, the salt no longer dissolves and crystals form.

Figure 2.10 Heat will break hydrogen bonds. Because these bonds affect the interactions of some chemical groups with other groups at different locations in the molecule, the three-dimensional structure of the molecule will change.

APPLY WHAT YOU'VE LEARNED

1.

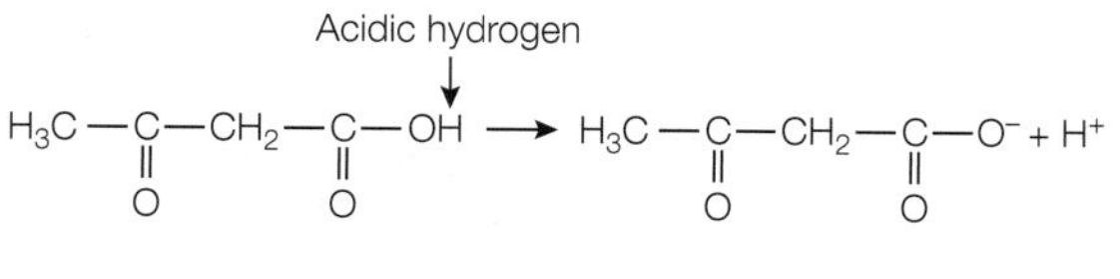

Acidic hydrogen

$$H_3C-\underset{\underset{O}{\|}}{C}-CH_2-\underset{\underset{O}{\|}}{C}-OH \longrightarrow H_3C-\underset{\underset{O}{\|}}{C}-CH_2-\underset{\underset{O}{\|}}{C}-O^- + H^+$$

2.

$$HCO_3^- + H^+ \rightleftharpoons H_2CO_3$$

The pH would not change. As H^+ was added due to ionization of acetoacetic acid, it would push the equilibrium of the above reaction to the right, as shown below.

$$HCO_3^- + H^+ \longrightarrow H_2CO_3$$

added H^+ reacts with HCO_3^- to shift equilibrium to the right

That is, bicarbonate ions would attach to the added H^+, so the overall free H^+ would not change.

3. Eventually the bicarbonate buffer system would be overwhelmed. It would no longer be able to take up the additional H^+ ions building up in the blood and shuttle them into carbonic acid. At that point, H^+ ions would accumulate, which would cause the blood pH to drop. The body would not be able to maintain a constant pH and its cells would be exposed to lower than normal pH conditions in the blood. In this case, the person would become severely ill because the cells would be exposed to an abnormal environment that then disrupts normal cell functions.

4. Acetone does not contain an acidic hydrogen, so it would not be expected to affect any acid or base buffering systems in the blood.

5. Four test possibilities are: insulin level (lower than normal), blood glucose level (elevated), blood pH (lower than normal), and bicarbonate level (lower than normal).

Chapter 3

RECAP 3.1

1. Structural isomers have the same number of atoms, but they are arranged differently. *Cis-trans* isomers typically have two double-bonded carbon atoms, and each of the two carbons is covalently bonded to two different groups as well (e.g., to –H, –CH$_3$, or another attached group). When the two substituents are oriented in the same plane relative to the double bond, the isomers are *cis*. When the two groups are oriented in opposing directions, the isomers are *trans*. Optical isomers have carbon atoms with four different atoms or groups bonded to them, and this makes two mirror images possible.

2. Mannose and galactose have the same atomic formula, C$_6$H$_{12}$O$_6$, but the arrangement of atoms is different. Compare carbons 2 and 4. These sugars have the hydroxyl (–OH) functional group. Its polarity helps the sugars dissolve in water. The –OH group also can participate in bonding the sugar to other molecules through condensation reactions.

3. H-A-OH + H-B-OH → H-A-B-OH + H$_2$O
   H-A-B-OH + H-C-OH → H-A-B-C-OH + H$_2$O

RECAP 3.2

1. When one amino acid (such as lysine) is replaced by another, the primary structure of the protein is altered. The change could affect tertiary structure if the protein folds differently as a result of electrostatic attractions between charged amino acids (+ to –). In this example, a negatively charged amino acid (aspartic acid) has taken the place of a positively charged one (lysine), and this may prevent correct folding, particularly if a negatively charged amino acid elsewhere in the polypeptide chain is involved in folding (it is attracted to a + amino acid). The same forces might be at work in the interaction of separate chains for quaternary structure.

2. The observations support explanation "a." Glycine is small and nonpolar. Glutamic acid and arginine are larger and polar (charged). Serine and alanine are small: the protein retains its shape. But serine is polar (it has –OH as its R group), and that does not affect the structure. Valine is larger and nonpolar, and this affects shape. So the issue is size.

3. See Figure 3.10. Heat breaks hydrogen bonds and other weak interactions that maintain protein shape. Disulfide bonds also are required for normal protein shape. Styling and perms partially denature keratin, then renature the protein in a new shape. Your investigation might involve measuring keratin protein structure of hair before and after disrupting hydrogen bonds and disulfide bonds.

RECAP 3.3

1. Starch and glycogen have many carbon atoms bonded covalently to one another. These covalent bonds require energy for formation and release energy when broken.

2. The many hydrogen bonds between polysaccharide chains of cellulose hold them together, and the covalent bonds within the chains are strong individually.

RECAP 3.4

1. The phospholipids would aggregate, with the polar heads facing in and the tails facing out—the reverse of what happens in membranes. This would probably be aggregates, rather than a bilayer.

2. Fats are solid at room temperature and have single bonds (saturated) between fatty acid carbon atoms (saturated). Oils are liquid at room temperature and have double bonds between some fatty acid carbon atoms (unsaturated).

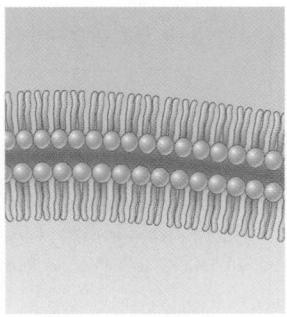

3. Steroids and some vitamins are classified as lipids because they are composed largely of carbon atoms linked together with substituent hydrogen atoms and are insoluble in water and soluble in nonpolar solvents.

WORK WITH THE DATA, P. 44

1. The genetically engineered silk was better than the native spider silk. It was thicker, required more force to break, and resisted strain better than native spider silk.

2. Fiber thickness is reported because it is one indicator of fiber strength.

3. The *t*-test could determine this.

4. Spider silk is stronger than Kevlar or steel and weighs less.

WORK WITH THE DATA, P. 52

1. Disulfide bonds began forming almost immediately after reoxidation began. Enzyme activity began appearing 100 minutes after reoxidation began. There are two reasons for the delay between the beginning of disulfide bond formation and the reappearance of enzyme activity. First, there are four disulfide bonds in the protein, all of which have to re-form before enzyme activity is restored. In other words, the first disulfide bonds to form aren't sufficient to restore activity, so there is a lag before activity reappears. Second, there are other chemical interactions, such as hydrogen bonding and hydrophobic interactions that occur after the protein has initially folded due to disulfide bond formation and which are also necessary for enzyme activity.

2. The absorption peak for the native protein was at about 278 nm; the peak for the reduced (denatured) protein was at about 275 nm. Reoxidation resulted in a return to the native spectrum. Under the denaturation conditions of these experiments, as long as the primary structure of RNase A is retained, the proper environmental conditions will result in a return to the native structure and a fully functional molecule.

FIGURE QUESTIONS

Figure 3.7   Primary structure will be unaffected, because it is held together by strong covalent bonds.

Figure 3.9   Regions in lysozyme that face the outside (water) environment are hydrophilic, whereas those on the inside are generally hydrophobic.

Figure 3.12   Weak interactions shown within or between molecules require a relatively low amount of energy to break them apart.

Figure 3.14   Heat shock proteins protect cellular proteins from being denatured and possibly broken down. This can occur not only when proteins are exposed to heat, but in any chemical situation where their structure is altered (e.g., a change in pH).

APPLY WHAT YOU'VE LEARNED

1. The forces involved are noncovalent and involve mainly hydrophobic and van der Waals interactions, since the R groups tend to be hydrophobic groups (Val, Ile, Phe, Gly). The R groups of these amino acids could stick out from the insulin polypeptide backbone so as to interact with its target proteins to form these noncovalent interactions.

2. These amino acids are invariant across all vertebrate species shown in the table, suggesting that they are essential in maintaining insulin's biological activity.

3. Certain changes are allowed as long as the overall tertiary structure of the protein is not changed significantly. Only amino acids that are similar in structure and properties would be able to function in a similar way at any position.

4. Cysteine residues are likely to be essential because they form disulfide bridges, linking the two chains together and stabilizing the protein's tertiary structure. No other amino acid can fulfill this role. We could compare these amino acids across the same vertebrate species shown above to see whether any have other amino acids at the cysteine locations.

Chapter 4

RECAP 4.1

1. Purines contain two nitrogen–carbon rings, while pyrimidines have one ring. The double helix of DNA has uniform width because a purine on one strand is always opposite a pyrimidine on the other strand.

2. While DNA molecules are similar in diameter and configuration, their base sequences are different. Differences in base sequence provide the informational content of DNA.

3. The number of possible 25-unit sequences of four nucleotides is 4^{25}, a large number indeed. Because of internal base pairing of the single strand (as in RNA), many folded configurations are possible, which allows specific binding to target molecules.

RECAP 4.2

1. The presence of O$_2$ in the atmosphere produces an oxidizing condition that prevents the reduction reactions observed in the Miller–Urey experiment

2. If microbes survived heat, the initial part of Pasteur's experiment might begin with microbes already present. They would grow in both the open and closed flasks. To get the results that Pasteur did, his flasks must not have contained such microbes. An answer for the proposed experiment on heat-stable microbes might be to inactivate them using reagents, such as mercaptoethanol, that destroy proteins.

3. A suggested experiment might be to dry the samples after the Miller–Urey experiment (allowing condensation reactions—polymerization) and then apply energy in the form of heat. This condition might have existed in volcanic rock on early Earth.

RECAP 4.3

1. A hallmark of living systems is the ability to reproduce, and this occurs from preexisting organisms. The instructions for producing an identical organism must be passed on to the offspring. This implies informational molecules. In living systems, chemical changes constantly occur, but in ordinary chemistry they are too slow to benefit the organism. So catalysts are needed to speed up the reactions.

2. A challenge in developing theories for the origin of life was the need for both a molecule that could carry information and a molecule that could act as a catalyst.

Without either, life as we know it could not exist. That an informational molecule, RNA, could also act as a catalyst solved this challenge; ribozymes, also known as catalytic RNA or RNAzyme, are RNA molecules that are capable of catalyzing specific biochemical reactions.

3. Most catalysts in living systems are proteins. But the polymerization of amino acids into proteins that are catalytic must have happened before the protein catalysts were initially formed. Having an RNA, that perhaps was originally informational, act as the catalyst for protein formation solves this "chicken–egg" issue.

RECAP 4.4

1. The cell membrane forms a compartment in which the chemicals needed for the reactions of life can be concentrated. The membrane creates a suitable internal chemical environment for the organism that is distinct from that of the environment.

2. First, examine rocks that are more than 3 billion years old. Then look at slices of rocks under microscopes for objects that look like cells or chains of cells. Finally, chemically analyze the rocks for chemical signatures for life, such as a carbon isotope ratio resulting from photosynthesis.

WORK WITH THE DATA, P. 74

1. Plot the data. In both experiments with unheated soil, there was a modest time-dependent increase in ^{14}C-labeled gases. These data are consistent with living organisms converting the molecules supplied into gases. The amount of gas produced was small in comparison with lab-based experiments: average radioactivity was 10,000 cpm, while potential was 257,000 cpm. Efficiency of production of gases was 10,000/257,000 = 0.038 = 4%.

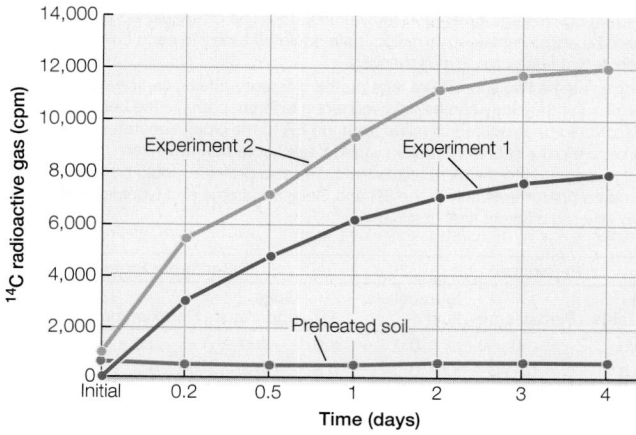

2. The graph of the data shows no increase in ^{14}C gases. Heat destroys hydrogen bonds in proteins and nucleic acids. Again, these data are consistent with living organisms having produced ^{14}C gases.

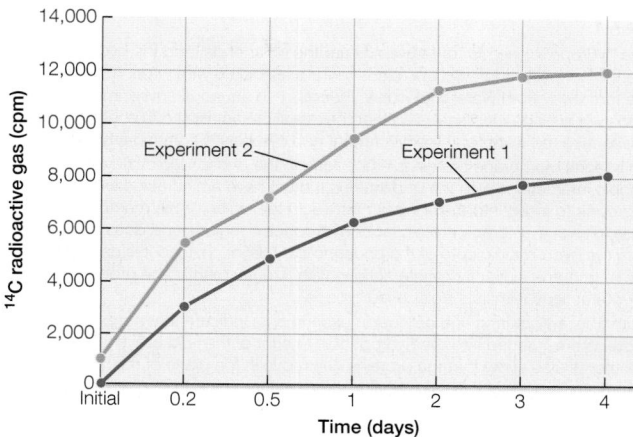

3. The data for hematite are similar to the data for Martian soil. So while the Martian soil data are consistent with life, they are also consistent with non-living soil components.

FIGURE QUESTIONS

Figure 4.3 If a folded RNA molecule were heated, hydrogen bonds between bases in the RNA would break and the molecule would assume a random shape, losing its specific shape.

Figure 4.4 Hydrogen bonds

Figure 4.5 There must be specific information in DNA sequences that signal their transcription. These specific sequences must bind to proteins that are involved in transcription.

APPLY WHAT YOU'VE LEARNED

1. The ratio of purines (A + G) to pyrimidines (C + T) is always one-to-one. This pattern is observed because of the double helix structure and base pairing between the two strands making up the double helix. There is always one purine on one strand and a pyrimidine that pairs with it on the complementary strand.

DNA	A	G	Purines	C	T	Pyrimidines	Ratio purines to pyrimidines
Herring sperm	27.8	22.2	50	22.6	27.5	50.1	1.00
Rat bone marrow	28.6	21.4	50	21.5	28.4	49.9	1.00
Human sperm	30.7	19.3	50	18.8	31.2	50	1.00
E. coli	26	24.9	50.9	25.2	23.9	49.1	1.04
Yeast	31.3	18.7	50	17.1	32.9	50	1.00

2. The ratio of purines (A + G) to pyrimidines (C + U) ranges from 0.87 to 1.24, with lots of variation in between. Therefore there is no constant pattern in this ratio in RNA across many species. This indicates that the number of purines and pyrimidines varies within an RNA strand, which we know to be single-stranded.

RNA	A	G	Purines	C	U	Pyrimidines	Ratio purines to pyrimidines
Rat liver	19.2	28.5	47.7	27.5	24.8	52.3	0.91
Carp muscle	16.4	34.4	50.8	31.1	18.1	49.2	1.03
Yeast	25.1	30.2	55.3	20.1	24.6	44.7	1.24
Rabbit liver	19.7	26.8	46.5	25.8	27.6	53.4	0.87
Cat brain	21.6	31.8	53.4	26.0	20.6	46.6	1.15

3. The difference in ratios of purines to pyrimidines in DNA and RNA across species highlights the double-stranded nature of DNA and the single-stranded nature of RNA. Only in the double-stranded structure would you have a constant ratio of purines to pyrimidines because they are paired in a one-to-one ratio. In single-stranded RNA, there is no requirement for pairing purines and pyrimidines, and the variability in their content reflects differences in the genetic sequences of the strands.

4. Only E. coli has about equal AT and GC content. Human sperm and yeast have more AT than GC content, and rat bone marrow and herring sperm have more GC than AT content.

DNA	A	G	C	T	A+T	G+C
Herring sperm	27.8	22.2	22.6	27.5	55.3	44.8
Rat bone marrow	28.6	21.4	21.5	28.4	57	42.9
Human sperm	30.7	19.3	18.8	31.2	61.9	38.1
E. coli	26.0	24.9	25.2	23.9	49.9	50.1
Yeast	31.3	18.7	17.1	32.9	64.2	35.8

5. Herring sperm and rat bone marrow cells have similar AT and GC content. Their genetic makeups are determined by the sequences of bases in DNA, so even though they have similar overall base content, they each have unique sequences of all bases—A, T, G, and C—that encode the genes within the DNA.

Chapter 5

RECAP 5.1

1. Life is continuous: the cell theory explains that this comes from cell continuity. Life is complex chemically: the cell theory states that this chemistry occurs within cells. Life evolves and changes over time: the cell theory states that evolution occurs in cells as they change genetically and pass on new characteristics to the next generation of cells.

2. The membranes that enclose cells and organelles create compartments. The cell membrane allows the chemical reactions of life to occur away from the surrounding environment; in particular, the cell membrane permits the internal environment to have a distinct composition and concentration. Organelles' membranes permit specialized functions and keep harmful substances away from the rest of the cell (cytoplasm).

RECAP 5.2

1. The cell wall is composed of a large complex of polysaccharides and peptides. It supports the cell and gives it shape. The capsule is a slimy polysaccharide layer that encloses the cell walls of some prokaryotes and keep cells from drying out. It may be used to attach to other cells. The flagellum is composed of strands of a large protein called flagellin and is attached to the cell wall and membrane by a motor complex. It is used for cell movement.

2. The nucleoid consists of DNA and proteins. Cytoplasm contains all the other small and large molecules for life, dissolved or suspended in water.

RECAP 5.3

1. Prokaryotic cells are smaller than eukaryotic cells and lack organelles.
2. (a) Four membranes: two in the chloroplast and two in the mitochondrion. (b) Two membranes: the lysosomal membrane and the cell membrane (via vesicle; the molecules do not themselves cross any membranes). (c) No membranes: ribosomes do not have membranes. However, if the ribosomes were associated with the endoplasmic reticulum (ER), the answer would be two membranes: one into the ER and another out of the ER.
3. Microtubules line the long axons of nerve cells, where they act as tracks for vesicles that carry substances down the neuron. Without microtubules, the contents of these vesicles cannot be delivered to their destination, which can result in nerve problems. Depolymerization of microtubules can interfere with cell divisions because microtubules are a key part of the mitotic spindle that is used to move chromosomes during cell division.
4. For a lysosomal enzyme, the pathway would be ribosome → interior of ER → Golgi apparatus → Golgi vesicles → lysosome.
5. For an extracellular protein (animal cell), the pathway would be ribosome → interior of ER → Golgi apparatus → Golgi vesicles → cell membrane → extracellular region.

RECAP 5.4

1. Initially, there is a digestion of molecules that attach the cell membrane to the extracellular matrix. Then enzymes are produced and secreted from the cell that digests the macromolecules of the extracellular matrix. The cell migrates by microfilaments.
2. A cell wall contains cellulose fibers which provide strong support yet allow a plant to have some flexibility to bend. The cell wall structure provides an extra barrier that helps protect the cell from infection, yet cell walls in some cells have pores that allow the flow of materials in and out of the cell.
3. Chemical analysis could identify a plant cell by the presence of cellulose in the extracellular matrix and an animal cell by the presence of collagen in the extracellular matrix. These two compounds are distinct to the two different types of cells.

RECAP 5.5

1. The cell membrane could fold inward and then pinch off, forming vesicles for the Golgi apparatus and flat sheets for the ER.
2. The endosymbiotic theory states that one cell engulfed another cell. Over time the engulfed cell lost some DNA and the ability to perform independently of the host cell. Thus the engulfed cell became dependent on the host cell for its survival.

WORK WITH THE DATA, P. 99

1. The tannins appear in the chloroplast thylakoids and then are transferred to the vacuole by vesicles that enclose the tannins.
2. The staining for chlorophyll and tannins is in the same place. Since chlorophyll is in the chloroplast, this is consistent with a chloroplast origin for tannins, as described in the answer to Question 1.
3. Staining and chemical analysis both showed tannins, as well as chlorophyll, in the lower fraction of organelles, so this is probably a chloroplast fraction that contains tannins.

WORK WITH THE DATA, P. 103

1. The reasoning behind these experiments was as follows: if microfilaments are essential for cell movement, movement should not occur in the presence of cytochalasin B; if microtubules are essential for cell movement, movement should not occur in colchicine; if cell movement requires the synthesis of new proteins, movement should not occur in cycloheximide; and if cell movement requires energy, movement should not occur in dinitrophenol. The latter three experiments were important controls to rule out involvement of the three processes involved. The cycloheximide + cytochalasin B and the dinitrophenol + cytochalasin B controls demonstrated that the drugs had independent effects on the cells.
2. The experiment with cytochalasin B implicated microfilaments in cell movement. The experiment with colchicine ruled out microtubules. The experiments with cycloheximide showed that new protein synthesis is not necessary for cell movement. The experiments with dinitrophenol showed that cell movement does not require new inputs of energy.

FIGURE QUESTIONS

Figure 5.9   Assembly of the ER from the nuclear envelope; transport between Golgi apparatus cisternae; transport from ER to Golgi apparatus; endocytosis; exocytosis; transport from Golgi apparatus to cell membrane

Figure 5.11   Cells with high energy requirements, such as muscle cells, would have a lot of mitochondria.

Figure 5.18   Nexin links cause cilia and flagella to bend when microtubule doublets try to slide past one another. Absence of nexin would result in reduced flagella and ciliary function. This is called immotile cilia syndrome.

APPLY WHAT YOU'VE LEARNED

1. The results show that the protein begins its path in the endoplasmic reticulum, passes through the Golgi apparatus, and ends up in the cell membrane.

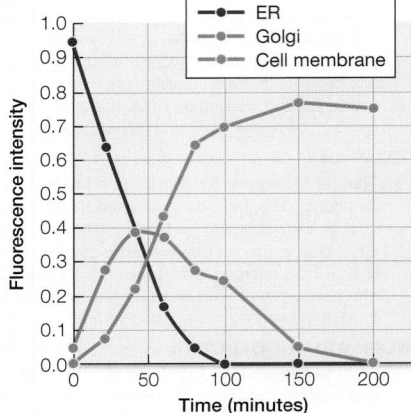

2. An inhibitor of protein synthesis would have prevented the fusion protein from being synthesized. Very little fluorescence (if any) would have been observed, and no fluorescence would have been observed moving into the Golgi apparatus or cell membrane.
3. The researchers would not have learned anything new because the cell fractionation method would have provided the same information about changes in subcellular locations of fluorescence over time. However, the results should have confirmed what was found using microscopy, which can provide additional weight of evidence for making claims about the path of the protein.
4. The researchers would have to use two different fluorescent tags that glow with different colors (e.g., green and red) so they could be distinguished from each other. Then the equipment used to collect data could be tuned to each color, and data collected at each color simultaneously.
5. As long as the two fluorescent tags on the different proteins can be distinguished from each other, they can be followed along different paths in the cell. The peroxidase would be expected to move from the ER to the Golgi apparatus to the peroxisome, while the secretory protein would be expected to move from the ER to the Golgi apparatus to the extracellular fluid. The two proteins would be found in the same compartments only in the ER and Golgi apparatus and would part ways as they entered different final destinations.

Time (min)	Peroxidase				Secretory Protein			
	ER	Golgi apparatus	Peroxisome	Extracellular fluid	ER	Golgi apparatus	Peroxisome	Extracellular fluid
0	1.0	0.0	0.0	0.0	1.0	0.0	0.0	0.0
10	0.6	0.3	0.1	0.0	0.6	0.3	0.0	0.1
20	0.4	0.4	0.2	0.0	0.4	0.4	0.0	0.2
50	0.2	0.3	0.5	0.0	0.2	0.3	0.0	0.5
100	0.1	0.2	0.7	0.0	0.1	0.2	0.0	0.7
200	0.0	0.0	1.0	0.0	0.0	0.0	0.0	1.0

Chapter 6

RECAP 6.1

1. The hydrophilic "heads" of fatty acids are the polar ends, and the hydrophobic "tails" are the nonpolar ends. So the heads tend to associate with water molecules, and the tails away from water molecules. Placed in an aqueous environment, the fatty acids will arrange themselves so that their tails interact with one another while their heads face the water of the environment and cytoplasm, forming a bilayer.
2. An integral membrane protein is embedded in the phospholipid bilayer by hydrophobic interactions with the lipid interior. It must have amino acids with hydrophobic R groups to insert into the nonpolar fatty acid tail region of the membrane bilayer. A peripheral membrane protein lacks hydrophobic regions and does not interact with the hydrophobic core of the phospholipid bilayer. Instead it is usually bound to the membrane indirectly by interactions with integral membrane proteins or directly by interactions with lipid polar head groups.
3. Both freeze-fracturing and cell fusion experiments indicate that nonpolar membrane proteins are inserted into the hydrophobic interior of the lipid bilayer. Cell fusion experiments also show that the proteins can move in the plane of the membrane.
4. To measure membrane fluidity, label a small amount of a lipid or protein with a dye and allow it to incorporate into the membrane of a cancer cell and a noncancer cell. This may make a localized labeled spot on the cells. The localized region will be seen to diffuse over the cells over time. In the cancer cell, this rate of diffusion may be faster.

RECAP 6.2

1. Enzymes called glycosidases cut the bonds between carbohydrates and other molecules. Obtain cells that are bound together (e.g., sponges) and then separate them. Then treat the cells with glycosidases to remove carbohydrates. This should block cell recognition and adhesion.
2. Plasmodesmata are most similar to gap junctions, because both have a membrane-lined channel.

RECAP 6.3

1. The properties that affect diffusion across a membrane are size and mass (smaller is faster), electric charge (less polar is faster), and concentration gradient (the higher the gradient, the faster the rate of diffusion).

2. If blood is hypotonic, water will enter red blood cells, causing them to swell and perhaps burst. (This is called hemolytic anemia.)

3. The hydrophobic cell membrane lipids are relatively impermeable to charged ions such as K^+. A channel will allow diffusion of K^+ out of the cell until equilibrium is reached, with equal concentrations inside and outside the cell.

RECAP 6.4

1. Substances tend to diffuse and reach equilibrium, where they are evenly distributed in the environment. Energy is needed to overcome this natural tendency.

2. An antiporter transports two substances in opposite directions. In the case of the Na^+–K^+ pump, Na^+ is transported across the cell membrane outward, and K^+ is transported across the cell membrane inward.

3. In primary active transport, ATP hydrolysis supplies the energy needed for transport against a concentration gradient. In secondary active transport, the energy comes from a gradient set up by a different (secondary) active transport.

4. Both active transport and facilitated diffusion speed up transport across a membrane by using a protein that binds to the substance transported. However, active transport is an energy-requiring process that transports a substance against its concentration gradient and diffusion tendency. By contrast, facilitated diffusion does not require energy and transports a substance along its concentration gradient and in conjunction with its tendency to diffuse.

RECAP 6.5

1. Phagocytosis involves a large cell membrane–derived vesicle forming around macromolecules. Pinocytosis involves smaller vesicles forming around water and its dissolved solutes.

2. In receptor-mediated endocytosis, a molecule binds to a cell membrane protein receptor and causes the membrane to form a vesicle around the molecule and receptor. The vesicle with its contents enters the cell and usually fuses with a lysosome. An example is the endocytosis of lipoprotein particles from blood into liver cells.

3. Diatom wall components move from the Golgi apparatus to the cell wall by exocytosis.

WORK WITH THE DATA, P. 122

1. The mRNA-injected oocytes swelled because of osmotic uptake of water. At 4 minutes, the mRNA-injected cells had taken up so much water that they burst. The control cells did not take up excess water and therefore stayed intact.

2. Water permeability increased with more mRNA injected, presumably because there was more aquaporin in the membranes that had more mRNA. The relationship could be evaluated statistically by linear regression.

3. The data on the mRNA-injected oocytes for mercuric chloride alone showed reduced water permeability, indicating that a protein was involved. Adding mercaptoethanol restored water permeability. There was not much water permeability in the controls without added mRNA, and mercuric chloride and mercaptoethanol had no effect on this.

FIGURE QUESTIONS

Figure 6.1   Hydrophobic interactions keep some proteins embedded whereas ionic attractions keep others on the surface.

Figure 6.10   Overfertilizing makes the soil water hypertonic to the interior of the plant root cells. Water leaves the plant roots by osmosis, moving toward the hypertonic medium. Water in the plant organs also travels to the roots, where it leaves by osmosis. Because cellular water is important in maintaining turgor pressure, the loss of water resulting from overfertilization causes the plant to wilt.

Figure 6.15   Blocking the Na^+–K^+ pump will result in equal concentrations of Na^+ inside and outside the cell. Generally, this will increase the cellular Na^+. The absence of a Na^+ gradient will mean there is no potential energy to drive glucose uptake into the cell, so the glucose concentration will be reduced.

APPLY WHAT YOU'VE LEARNED

1. When measured at a constant temperature, the cell membranes show a trend toward greater fluidity with decreasing temperature of a species' native environment. The least fluid membranes are found in animals that experience the highest temperatures (rats with body temperatures of 37°C). Fluidity is slightly greater in desert pupfish at 34°C and increases even more in goldfish raised at 25°C. This trend continues with increasing fluidity in goldfish raised at 5°C. The greatest fluidity is observed in arctic sculpin raised in extremely cold temperatures (0°C).

2. The data suggest that the ratio of saturated to unsaturated fatty acids in phospholipids making up the membrane also influences its fluidity.

3. Saturated fatty acids can pack together more tightly than unsaturated fatty acids, reducing membrane fluidity. Shifting to a greater proportion of unsaturated fatty acids would be appropriate as temperatures decrease, because molecular motion decreases as temperature decreases. At very low temperatures, the membrane needs more unsaturated fatty acids to reduce tight packing and maintain fluidity. At very high temperatures, the membrane can accommodate tight-packing saturated fatty acids because molecular motion is greater.

4. From the fluorescence value, the graph indicates that the cell membranes of the animal must have been raised at a temperature of about 15°C. This temperature falls between data collected for goldfish at 5°C and 25°C. This means that the ratio of saturated to unsaturated fatty acids in this animal might be expected to fall somewhere between 0.66 and 0.82 based on the data provided in the table, and probably close to 0.70.

Chapter 7

RECAP 7.1

1. Autocrine signaling would be important for a cell to maintain a specialized role. For example, a cell might receive a signal to specialize and form a tissue of many cells. The first cell would self-stimulate to grow and divide to form the tissue in response to self-signaling.

2. Hormones are delivered to target-cell receptors through the circulatory system. In animals, this is blood; in plants, it is the vascular system.

3. Specificity for signal response occurs because only certain cells make the receptor for a given signal.

RECAP 7.2

1. The dissociation constant is the ratio of the rate constant for dissociation of a ligand and its receptor to the rate constant for binding. The higher the K_D, the less likely it is that a signal will bind to its receptor, and the less likely it is that a cell will respond to a signal at particular concentrations of ligand and receptor.

2. The chemical nature of receptor–ligand bonding is shape and the presence of noncovalent interactions, such as hydrophobic and ionic interactions. This provides specificity. Specificity is important to ensure an appropriate biological response to each signal.

3. Cytoplasmic receptors lack nonpolar regions that allow insertion into the cell membrane. Their exteriors are polar, so they can interact with water in the cytoplasm. While membrane receptors stay in the membrane, cytoplasmic receptors may be translocated to the nucleus after binding to their ligand.

RECAP 7.3

1. Amplification occurs because each activated ras molecule catalyzes the phosphorylation, and therefore activation, of many molecules of Raf, each of which in turn activates many molecules of MEK, etc. The general principle is one enzyme → many molecules of its substrate.

2. Different cells can have different target molecules to which cAMP binds, and these targets can have different activities and functions. Binding of cAMP changes the structure of its target (e.g., tertiary structure of a protein).

3. Regulation of cascades occurs in three ways: by differing concentration of a second messenger (e.g., NO is a gas that breaks down readily), by differences in concentration of a single component of the pathway, and by enzymes that change target molecules (e.g., phosphatases that remove phosphate groups on proteins in a cascade).

RECAP 7.4

1. Oxytocin is involved with both G protein–coupled and ion channel responses. It binds to a G protein–coupled receptor (see Figure 7.7), resulting in signal transduction through the IL_3–DAG pathway, which releases Ca^{2+} into the cytoplasm via ion channels. In the brain, the effect of Ca^{2+} is to indirectly stimulate nerve cell activity by the opening of ion channels for Na^+.

2. Ion channel opening is a rapid response to signaling. Enzyme modification is also a relatively rapid response mechanism. Changes in gene expression involve many steps, so cell signaling via transcription factors is slower.

RECAP 7.5

1. *Characteristics of direct communication by cell junctions*: The size of signal molecules is limited by the size of openings between cells. It is not specific. It is fast. There can be cytoplasmic connection between cells.

   Characteristics of receptor-mediated communication: The signal molecules can be larger. It is specific. It is slower. There is no direct cytoplasmic connection.

   The advantage of direct communication over chemical signaling is that direct communication enables a rapid, coordinated response of many cells.

2. From single-celled *Chlamydomonas* evolved larger and larger groups of cells with communicating junctions (*Gonium*, *Pandorina*, *Eudorina*). Even larger groups had some differentiation (*Pleodorina* and *Volvox*).

3. Experiments might involve applying a solution containing the antibody to the upper part of the *Hydra* body. The antibody would block diffusion of the signal molecule from the apex to the upper body and—if the hypothesis is correct—would allow a bud to form in the upper body. A sham experiment, in which the solution without antibody is applied, would be a control. In this case, a bud would not form in the upper body.

WORK WITH THE DATA, P. 143

1. There was a higher level of trust in the donors who received oxytocin than in those who did not. The difference was about 15 percent. A *t*-test could be used to assess the difference for statistical significance.

2. There was no difference in investment between oxytocin-treated and untreated people when they were told exactly how much to invest. This indicates that oxytocin does not just induce people to take a greater risk, but when considered with the results of the first experiment, oxytocin increases risk when there is a social interaction between people involving trust.

FIGURE QUESTIONS

Figure 7.3   As with the binding of enzyme to substrate, the binding of caffeine and adenosine is noncovalent. Both substances bind to a specific site on the receptor by their shape and by interactions, including hydrophobic interactions (the rings).

Figure 7.10   Raf activity is an early event in the protein kinase cascade that stimulates cell division in kidney cancer. Blockage of raf prevents phosphorylation of MEK, and all the subsequent steps in the protein kinase cascade. Cell division is reduced and the tumor shrinks.

Figure 7.17   Direct communication between cells allows them to rapidly share signals, which travel from one cell to another in a group. This can result in common activities for a group of cells, which is important for tissues.

APPLY WHAT YOU'VE LEARNED

1. The figure shows that different amounts of the membrane protein OmpC are produced and inserted into the membrane, depending on environmental solute concentration.

2. The membrane protein EnvZ is a protein kinase that phosphorylates itself and then phosphorylates OmpR in the cytoplasm. This sequential phosphorylation could explain how signal transduction occurs in this signal pathway.

$$EnvZ + ATP \rightarrow EnvZ\text{-}P + ADP$$
$$EnvZ\text{-}P + OmpR \rightarrow EnvZ + OmpR\text{-}P$$

3. The numbers of molecules of EnvZ are much smaller than the numbers of molecules of OmpR in the cytoplasm, suggesting that a smaller number of the membrane proteins act to stimulate a larger number of cytoplasmic proteins that carry the signal to the next point in the signal transduction pathway.

Chapter 8

RECAP 8.1

1. Exergonic reactions release free energy because the energy of the reactants is greater than that of the products. The reverse is true for endergonic reactions, which require an input of energy. ΔG is the free energy change of a reaction—products minus reactants. A positive ΔG means that the reaction is endergonic and requires energy, while a negative ΔG indicates that a reaction is exergonic and releases energy.

2. Endergonic reactions require the input of energy to create more ordered molecules. The second law of thermodynamics states that order tends to increase in the universe. Endergonic reactions are coupled in time and space with exergonic reactions, which increase the disorder and release the energy needed for the endergonic reactions to proceed. Overall, organisms need to take in energy from their environment continually to maintain these reactions.

RECAP 8.2

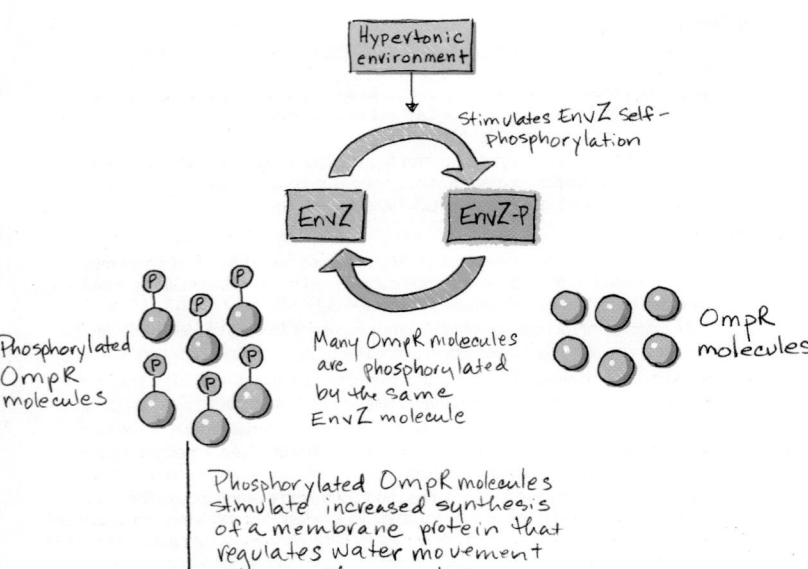

1. ATP has terminal phosphate groups that repel one another, so it takes energy to form a bond between them. Some of this energy is stored as potential energy. Also, hydrolysis of phosphate releases energy because the free energy of the P~O bond is higher than the free energy of the OH bond formed by hydrolysis.

2. a. The reaction is endergonic (positive ΔG).
   b. ATP hydrolysis can drive the reaction ($\Delta G = -7.3$ kcal/mol).

RECAP 8.3

1. Enzymes have a three-dimensional (tertiary) structure with an active site in which the substrate fits. Chemical groups at the active site also bind to the substrate non-covalently.

2. While an enzyme-catalyzed reaction proceeds more rapidly to its equilibrium than an uncatalyzed reaction, the actual equilibrium point is unaffected.

3. a. The presence of water may prevent O_2 from reaching the enzyme.
   b. Boiling denatures proteins, so polyphenol oxidase is irreversibly altered by boiling, and its active site is destroyed.

RECAP 8.4

1. As part of the catalysis, an enzyme can add or remove H^+ in a substrate, which may facilitate reaction with the substrate. The course of these H^+ reactions involves acidic or basic R groups on amino acids at the active site. Metal ions of enzymes may gain or lose electrons during the reaction, donating them temporarily to the substrate.

2. Coenzymes are small and nonprotein; enzymes are large proteins or RNA. Coenzymes do not catalyze reactions; enzymes are catalysts. Coenzymes add or remove chemical groups (e.g., H^+) and are permanently changed as a result of the reaction; enzymes are not permanently changed by participating in a reaction.

3. At the beginning, the reaction rate will increase linearly with added enzyme since the enzyme will be saturated with substrate even after many additions. Thus the reaction rate will increase proportionally with the amount of enzyme added, but only until the enzyme is no longer saturated with substrate. At this point the enzyme will be present in excess with respect to substrate, and additional enzyme will have no additional effect. The curve will level off to a constant maximum rate when this point is reached.

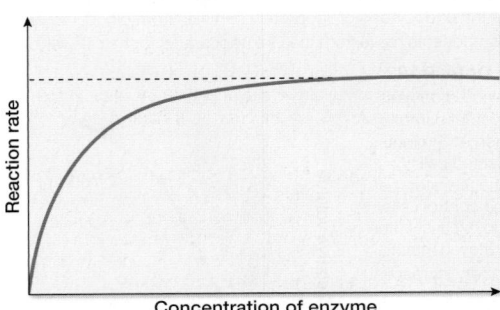

RECAP 8.5

1. Feedback inhibition occurs in a biochemical pathway when that pathway's end product can act as an inhibitor of the enzyme that catalyzes the first step in the pathway. In a systems diagram, each node represents an enzymatic transformation. Feedback inhibition can cross multiple pathways, thereby allowing the changing concentration of one molecule to affect several pathways that lead to its synthesis.

2. See Figure 8.17. A competitive inhibitor binds to the active site of the enzyme and shifts the equilibrium to enzyme molecules in the active form.

3. To determine whether catalase has an allosteric or a nonallosteric mechanism, perform an experiment with varying amounts of substrate and plot the rate of catalase-versus-substrate concentration. An S-shaped curve will indicate an allosteric mechanism. A hyperbolic curve will indicate a nonallosteric enzyme. To determine if a pollutant is a competitive or a noncompetitive inhibitor, add the pollutant to the catalase to lower the rate of reaction, then add increasing amounts of substrate. A competitive inhibitor will be removed from the active site and the rate of reaction will increase. A noncompetitive inhibitor will not allow the rate to increase as more substrate is added.

WORK WITH THE DATA, P. 165

1.

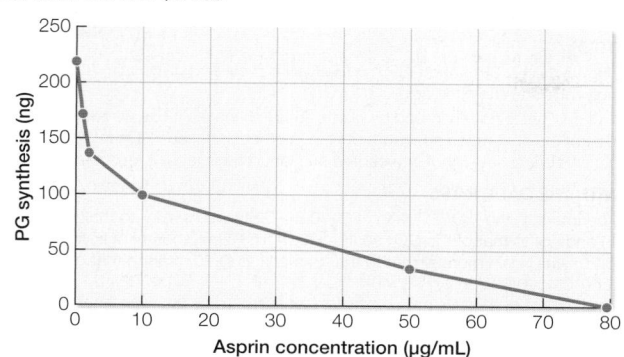

2. There is an inverse relationship between enzyme activity (PG synthesis) and aspirin concentration, so aspirin is an inhibitor of the reaction in lung tissue homogenates. Because aspirin also inhibited PG synthesis in human platelets, we can conclude that inhibition occurs in different tissues of different organisms.

3. In all three individuals, aspirin given in vivo inhibited PG synthesis by platelets. This generalizes the results from the test tube experiment to a living organism.

FIGURE QUESTIONS

Figure 8.7   The synthesis of ATP has a ΔG of +7.3 kcal/mol. So any reaction driving this synthesis must have a ΔG of at least –7.3 kcal/mol. Because of the second law of thermodynamics, the transfer of energy is not efficient (entropy increases), so the ΔG of the driving reaction would be more than +7.3 kcal/mol.

Figure 8.8   Yes. In this case, activation energy is needed to make the substrate(s) able to take up energy and then the reaction proceeds.

Figure 8.12   No. Typically, the tertiary structure of a protein is due to weak non-covalent forces such as hydrophobic interactions, ionic attractions, and hydrogen bonds. Binding of substrate alters these forces but not covalent bonds.

Figure 8.19   Lysosomal enzymes have an amino acid sequence and therefore tertiary structure that exposes their active site at pH 4.8. At pH 7.2, the tertiary structure does not expose the active site, so the enzyme is inactive.

APPLY WHAT YOU'VE LEARNED

1. The molecules are luciferin, oxygen, ATP, and magnesium ion. Luciferin reacts with oxygen to undergo chemical change that involves emission of light as this change takes place. In order to drive this reaction, phosphate bonds in ATP are hydrolyzed, which releases chemical energy. Some of this energy is then released in the form of light as luciferin and oxygen react. Magnesium ion is needed as a cofactor of the enzyme that catalyzes reactions involving the other molecules.

2. The firefly takes in food as a source of chemical energy. As the food is digested, some of the chemical energy in the bonds of the food molecules is stored in phosphate bonds of ATP as ADP is phosphorylated to form ATP. This chemical energy is released when ATP is hydrolyzed during the luciferin reaction, with some of the energy released in the form of light.

3.

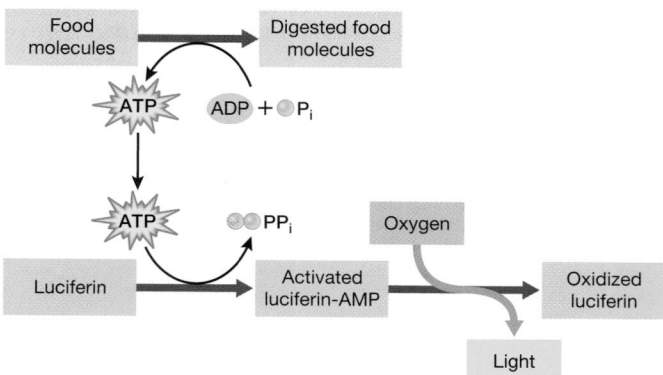

4. The enzyme's catalytic activity is abolished when the enzyme is heated. The catalytic activity is also affected by changes in pH. The effect of heating is likely due to disruption of the enzyme's tertiary structure by high heat; most proteins cannot maintain their active shape if they are exposed to high heat. The effect of changing pH is likely due to changes in ionization of basic amino acid side chains of lysine and arginine or acidic side chains of aspartic acid and glutamic acid. Changes in ionization in these side chains can affect binding of the substrate or interactions between groups within the enzyme that stabilize its three-dimensional structure.

5. This is an example of induced fit, which the enzyme uses to orient substrates with one another and with amino acid side chains of the enzyme that are important in its catalytic mechanism. This shape change of the enzyme excludes water from the active site, ensuring that ATP will not be hydrolyzed by water and will react only with luciferin to transfer AMP and release pyrophosphate, PP_i.

Chapter 9

RECAP 9.1

1. A complex chemical transformation occurs in a series of separate reactions that form a metabolic pathway. Each reaction is catalyzed by a specific enzyme. Many metabolic pathways are similar in all organisms, from bacteria to humans. In eukaryotes, many metabolic pathways are compartmentalized, with certain reactions occurring inside specific organelles. Some key enzymes in each metabolic pathway can be inhibited or activated to alter the rate of the pathway.

2. NAD^+ is a reducing agent (it accepts electrons), and O_2 is an oxidizing agent (it donates electrons).

3. a. It is an oxidation (removal of H from C2 and C3 of succinate).

   b. It is exergonic, because it is an oxidation.

   c. It requires the redox coenzyme NAD or FAD.

RECAP 9.2

1. In glycolysis, the net yield is 2 ATP per glucose and 1 NADH per glucose.

2. Pyruvate oxidation produces the two-carbon molecule acetyl CoA, which is activated and then participates as a first substrate in the citric acid cycle.

3. The citric acid cycle regenerates the four-carbon acceptor molecule for the next acetyl CoA from pyruvate. Glycolysis does not regenerate a starting material.

RECAP 9.3

1. A series of electron carriers in the inner mitochondrial membrane transports electrons by reduction–oxidation. As the electrons are added to each carrier, protons are transported via the carrier into the intermembrane space.

2. The experiment in Figure 9.9 shows that in the absence of electron transport, a gradient of protons across the membrane is sufficient to produce ATP if the ATP synthase is present in the membrane. The experiment in Figure 9.10 shows directly that ATP synthase can carry protons from a gradient, harnessing the potential energy from the gradient to make ATP.

3. If cytochrome c remains reduced and cannot accept electrons, the electron transport (respiratory) chain stays reduced and NADH and $FADH_2$ remain reduced. This prevents oxidation reactions in the citric acid cycle and pyruvate oxidations, so pyruvate cannot be converted to acetyl CoA. Instead, pyruvate is converted to lactic acid, regenerating some NAD that can be used so that glycolysis can continue. Because the electron transport chain is not working, no proton gradient is set up in the mitochondria, and ATP is not made by oxidative phosphorylation.

4. If antimycin A were present, it would make no difference to the results of the experiment, since an artificial proton gradient was already set up.

RECAP 9.4

1. NAD^+ is an electron acceptor for reactions in glycolysis and pyruvate oxidation. If it is not present in its oxidized form, the oxidation of these substrates in the two reactions does not occur, and the pathways stop.

2. Fermentation replenishes NAD^+ by oxidizing NADH during chemical oxidations forming lactate or ethanol from pyruvate.

RECAP 9.5

1. *A catabolic interconversion of a lipid*: Fatty acids are catabolized to acetyl CoA. This enters the citric acid cycle and is oxidized to CO_2, providing reduction energy to form NADH and FAHD. *An anabolic interconversion of a protein*: Amino acids are converted to intermediates in glycolysis and the citric acid cycle. These intermediates can be used in anabolism to form glucose, fats, or other amino acids to make proteins.

2. Phosphofructokinase acts early in glycolysis and is a focal point for turning off or activating the rest of the energy pathways. It is turned off by later products (ATP, citrate) and turned on when energy is needed (ADP).

3. Some amino acids are converted to intermediates of glycolysis. Once they enter glycolysis, these intermediates are further metabolized to a glycolytic intermediate that can be converted to glycerol, which is incorporated into triglycerides. Glycolysis and pyruvate oxidation produce acetyl CoA, which is converted to fatty acids and incorporated into lipids. Glucose is converted in glycolysis to acetyl CoA, which is then converted to fatty acids as above.

WORK WITH THE DATA, P. 183

1. Both groups of mice gained weight as they aged, but the mutant mice gained less weight than the normal mice.

2. The amount of food eaten and the levels of exercise were the same for both groups. So the lower weight gain in the mutant mice was not due to food intake or exercise.

3. UCP1 levels were much higher in the mutant mice than in the normal mice. The correlation between higher UCP1 and lower weight gain indicates that uncoupling of mitochondrial ATP synthesis from electron transport in brown fat may be responsible for lowered weight gain.

FIGURE QUESTIONS

Figure 9.3 The H atom comes from the oxidation of a substrate.

Figure 9.7 −20 kcal/mol

Figure 9.13 DNA can be an energy source. When DNA is hydrolyzed, the nucleotides can be metabolized into intermediates in the citric acid cycle. The intermediates are then oxidized and the energy released to reduced coenzymes, which release their energy when oxidized in the mitochondria to form ATP. However, because it is the genetic material and must be preserved, DNA is protected from hydrolysis after it is made by being sequestered in the cell nucleus.

APPLY WHAT YOU'VE LEARNED

1. The body first uses glucose supplied from the last meal, then begins to break down glycogen. Once all of the glycogen is gone, the body makes glucose through gluconeogenesis using molecules present in the body.

2. Mammals break down proteins and use amino acids to make glucose through gluconeogenesis. Mammals also use glycerol from triglycerides, but they do not use fatty acids in this way. Fatty acids do not stimulate gluconeogenesis

3. The person will lose muscle mass because proteins in muscle cells will be broken down to make glucose via gluconeogenesis. This is not advisable because this will weaken the person's body and put him or her in a state of low body fitness. People generally do not realize this because they think that dieting will cause loss of fat rather than muscle.

4.

$$
\begin{array}{ccc}
\mathrm{CH_3} & & \mathrm{CH_3} \\
| & & | \\
\mathrm{CH-NH_3^+} & \longrightarrow & \mathrm{C{=}O} \\
| & & | \\
\mathrm{COO^-} & & \mathrm{COO^-} \\
\textbf{Alanine} & & \textbf{Pyruvate}
\end{array}
$$

Any of the carbon atoms in alanine could be labeled with carbon-14 to trace it through the gluconeogenesis pathway, since only the NH_3 group is removed from the alanine carbon chain.

Chapter 10

RECAP 10.1

1. The light reactions convert light energy into chemical energy in the form of ATP and NADPH. In the light-independent reactions, ATP and NADPH power the fixation of CO_2 to form carbohydrate. NADP and ADP are regenerated.

2. Researchers used the isotope ^{18}O to show that the O_2 produced during photosynthesis came from water labeled with ^{18}O and not from carbon dioxide labeled with ^{18}O.

RECAP 10.2

1. A pigment molecule tends to lose the absorbed energy of a photon by returning to ground state and emitting the energy as light or heat. Alternatively, the energy can be transferred as an excited electron from the pigment to another molecule, reducing that molecule.

2. An absorption spectrum plots the extent of absorption by pigments (y axis) versus the wavelength of light to which the pigments are exposed (x axis). An action spectrum also plots wavelengths of light, but in this case the y axis is a biological activity (e.g., photosynthesis).

3. See Figure 10.8. In cyclic electron transport, ATP is produced chemiosmotically by electron transport in the thylakoid membrane.

RECAP 10.3

1. Researchers exposed algae to CO_2 labeled with ^{14}C for varying times and looked to see where the ^{14}C label ended up. After a very short exposure time of 3 seconds, all of the label was found in 3-phosphoglycerate (3PG). After a longer exposure time of 30 seconds, the label could be found in more compounds, suggesting that 3PG is the initial product that is then used for synthesis of additional molecules. See Figure 10.10.

   a. The light reactions produce ATP and NADPH, which are used in CO_2 fixation reactions (Calvin cycle).

   b. Light-induced pH changes in the stroma activate some Calvin cycle enzymes.

   c. Light-induced electron transport reduces disulfide bridges in four of the Calvin cycle enzymes, thereby activating them.

2. In the dark, photosynthetic electron transport stops at photosystem II → reduced PQ (plastoquinone). Initially the chlorophylls in light-harvesting complexes remain reduced, so reaction center chlorophylls remain reduced and thus photosystem II remains reduced. In the dark, the Calvin cycle stops at the reduction phase, which requires NADH. No RuBP is regenerated, so there is no rubisco activity. The initial reactions are no oxidation of photosystem I and no reduction of NADP to NADPH.

RECAP 10.4

1. In C_4 plants, CO_2 is initially fixed in the leaf mesophyll cells but is then transferred (as a four-carbon molecule) to the bundle sheath cells, where decarboxylation reactions release CO_2 for use in the Calvin cycle. The bundle sheath cells are located in the interior of the leaf where less atmospheric O_2 can reach them than reaches cells near the surface of the leaf.

2. In CAM plants, CO_2 is initially fixed into a four-carbon compound (malate) at night when it is cooler and water loss is minimized, and the stomata open. During the day, when the stomata close to reduce water loss, the accumulated malate is transferred

from the vacuole to the chloroplasts, where its decarboxylation supplies the CO_2 for the Calvin cycle and the light reactions supply the necessary ATP and NADPH.

RECAP 10.5

1. See Figure 10.17. 3PG and G3P link the Calvin cycle and glycolysis. Acetyl CoA links glycolysis to the citric acid cycle.

2. Most of sunlight is not absorbed by plants. This is because it has wavelengths that are not part of the absorption spectrum of plant pigments; because it is reflected back to space; and because it is not intercepted by plant organs (e.g., it reaches the ground).

WORK WITH THE DATA, P. 195

1. In Experiment 1, the $^{18}O/^{16}O$ ratio of O_2 (0.84–0.86) was similar to that of the H_2O (0.85) and not to that of the CO_2 sources (0.2–0.61). In Experiment 2, the ratio in O_2 (0.20) was again more similar to that of the H_2O (0.20) than to that of the CO_2 sources (0.40–0.50).

2. The source of the oxygen atoms in O_2 is H_2O.

WORK WITH THE DATA, P. 203

1.

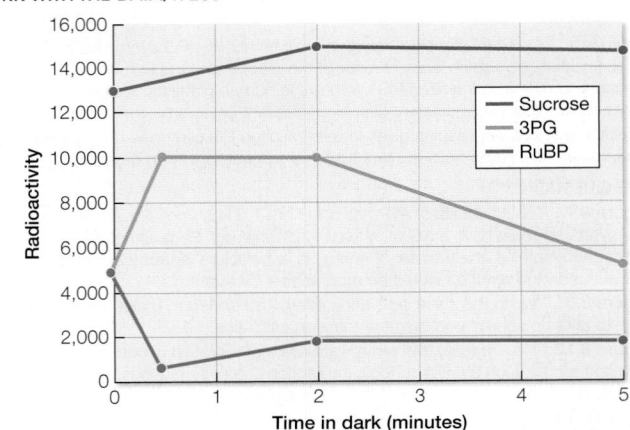

The data show an initial rise of 3PG (in first 30 sec) because rubisco is initially active and can catalyze the reaction of CO_2 with RuBP to produce 3PG. Between 30 sec and 2 min, the amount of 3PG levels off as rubisco becomes inactive in the dark. After 2 min, the amount of 3PG falls as it enters other pathways (see Figure 10.18).

2. The level of RuBP went down in the dark initially because it was consumed in the reaction catalyzed by rubisco.

FIGURE QUESTIONS

Figure 10.2 Reduction occurs in the stroma of the chloroplast and the reducing agent is NADPH.

Figure 10.4 In phycobilins, yellow light absorbs at a shorter wavelength (540 nm) that is more energetic than the longer wavelength (660 nm) at which chlorophyll absorbs. This means that the energy transfer from phycobilins to chlorophyll is thermodynamically favored (higher to lower energy).

Figure 10.7 The herbicide rather than NADP reductase would accept electrons from noncyclic photosystem I. NADPH would not be formed. This would severely reduce the transfer of solar energy to chemical energy in the light-requiring reaction system.

APPLY WHAT YOU'VE LEARNED

1. Light stimulates the light reactions to occur, which then stimulates the Calvin cycle. Also, light-induced changes activate Calvin cycle enzymes.

2. Shade-tolerant plants have much lower maximum rates of photosynthesis even when raised in the same light conditions as shade-intolerant plants and exposed to the same light intensities. This suggests that either there are fewer light-harvesting centers and fewer Calvin cycle enzymes in shade-tolerant plants, and/or that the Calvin cycle enzymes of shade-tolerant plants are adapted to function at lower rates than those of shade-intolerant plants.

3. Both shade-tolerant and shade-intolerant plants develop more chlorophyll when raised under greater degrees of shade. This helps compensate for the low light levels to allow the plants to collect greater amounts of light.

4. Specific leaf surface areas are larger in shade-tolerant plants. This allows the light-harvesting centers to be spread out as much as possible so that they can capture maximum light under shade conditions. This helps compensate for the low light levels to allow the plants to collect enough light to carry out photosynthesis.

5. The shade-intolerant plants will have a much greater fluctuation in photosynthesis rate than the shade-tolerant plants. This prediction is made based on the much larger range of photosynthesis rate data in the graph of the shade-intolerant species compared with that of the shade-tolerant species.

Chapter 11

RECAP 11.1

1. A common signal for the initiation of binary fission is adequate nutrients.

2. If the cell divided before DNA was fully replicated, each new cell would not receive a full complement of the genetic material, DNA.

3. In eukaryotes there is more DNA, the DNA is in numerous molecules, the DNA molecules are much larger, and the DNA is located in a separate cell compartment, the nucleus.

RECAP 11.2

1. See Figure 11.3.

2. The Cdk's are made throughout the cell cycle. However, their protein kinase active site is not available for target substrates for cell cycle activities unless a particular cyclin molecule binds to the Cdk. It is the cyclins that are made, bind to Cdk's to activate, and then break down. The transient nature of the cyclins controls each Cdk activation in sequence at the cell cycle control points.

3. Growth factors are proteins made by cells that can travel to other cells, or act on the cells that make them, usually to stimulate cell division. Growth factors bind to specific receptors on target cells, setting off signal transduction inside the cells. This can lead to gene expression for cyclins, for example, and the cell cycle is stimulated.

4. P16 blocks the interaction of cyclin and its Cdk that act at the G1–S boundary. If there is more p16 in older people, it may mean that their cell cycle is blocked and they cannot repair damaged tissues by cell replacement.

RECAP 11.3

1. A chromosome is a DNA molecule in the cell, and when the cell is in mitosis the chromosome is complexed with proteins to produce a visible, condensed structure. A chromatid is a DNA molecule complexed with proteins that is the product of S phase DNA replication. Chromatids are in pairs (the two products of replication) and lie attached to one another via the centromere until anaphase of mitosis. A daughter chromosome appears in an anaphase of mitosis, and was formerly a chromatid but has now separated from its partner as it migrated to the spindle pole.

2. During interphase, DNA is somewhat condensed by histone proteins into nucleosomes, and these fold over one another to form chromatin fibers. During prophase, the fibers attach as loops to proteins, and these in turn loop extensively to form the chromosome.

3. Chromosomes are attached to spindle microtubules, and molecular motors on the microtubules move the chromosomes along. In addition, spindle microtubules shorten from the poles, and this causes the attached chromosomes to move to the poles. Taxol prevents these processes, and so inhibits cell division. The drug also inhibits the division of normal cells that enter M phase.

4. In plant cells, a cell plate forms from Golgi vesicles and this makes a new cell wall to separate the two daughter cells. Cell membranes grow below the cell walls. In animal cells, a "purse string" of microfilaments contracts to pinch off the cell membrane, and the cells separate.

5. A nonfunctional cohesion would not allow close attachment of chromatids during cell division, and there would not be a centromere. The two chromatids would be separate and there would be no organization for kinetochore attachment. There would be ineffective segregation of one chromatid of a pair to each daughter cell.

RECAP 11.4

1. Take a dividing cell and examine the chromosomes. If each chromosome differs in size and centromere placement from the others, the plant sample is haploid. If there are two copies of each chromosome, the plant is diploid.

2. Fertilization involves the union of two haploid gametes, each with one set of chromosomes, made from meiosis. So the fertilized cell is diploid.

3. All sexual life cycles have some haploid and diploid cells, fertilization, and meiosis.

4. See Figure 11.14.

RECAP 11.5

1. In crossing over, there is an exchange of some genetic material between non-identical chromosomes of a pair. So the resulting chromosomes carry new combinations of genes, which can be passed on to offspring in a gamete and fertilization. In independent assortment, it is random which chromosome of a homologous pair ends up in a particular gamete. So different gametes will usually have a different set of chromosomes; that is, chromosome 1 from the father, chromosome 2 from the mother, and so on. Fertilization therefore results in diploids, each of which has a different set of chromosomes.

2. *Prophase I meiosis*: Chromosomes have chromatid pairs attached, and the two homologs are lined up gene for gene beside one another.
Prophase mitosis: Chromosomes have attached chromatid pairs but are not lined up beside one another.
Anaphase I meiosis: Homologous chromosomes, each with two attached chromatids, separate and move to the poles.
Anaphase of mitosis: Chromatids separate and become single daughter chromosomes and move to the poles.

3. In the formation of male gametes, the X and Y chromosomes fail to separate in meiosis I anaphase. At the end of meiosis II, there will be two types of gametes: half without an X or Y, and half with X and Y. If the latter fertilizes a normal egg with a single X chromosome, the offspring will be XXY. However, if there is a similar nondisjunction in the formation of female gametes, there will be eggs with two X chromosomes. If the XX egg is fertilized by a normal, Y-containing sperm, the offspring will be XXY.

4. Polyploidy refers to an extra set or more of chromosomes (e.g., 3n instead of 2n). It arises because of failure of all chromosomes to disjoin in meiosis in gamete formation, resulting, for example, in a diploid gamete.

RECAP 11.6

1. Signals for necrosis include great damage to a cell or starvation.

2. Apoptosis is necessary when cells are damaged and could be mutated, and when there are too many cells for organ structure.

3. Apoptosis is regulated by signals such as hormones and DNA damage, and a signal transduction pathway that results in apoptotic processes.

RECAP 11.7

1. Normal cells: have control over cell division; stay in tissue and do not migrate. Malignant tumor cells: lose control over cell division; migrate to other places in the body. Benign tumor cells: lose control over cell division and then at some point stop dividing; do not migrate to other parts of the body.

2. In normal cells, oncogene products are not active or are made in low amounts, while in cancer cells oncogene products are made in larger amounts or mutated forms, and these act to stimulate the cell cycle. In normal cells, tumor suppressor gene products are made and are active at blocking the cell cycle. In cancer cells, tumor suppressor gene products are either mutated to be nonfunctional or are not made, and in either case are not active in blocking cell division, so cells divide and form tumors.

3. Cancer cells are not synchronous in the cell cycle. At a given point in time, some are in G1, some in S, and so on. So targeting all the phases might be better than targeting just one.

WORK WITH THE DATA, P. 217

1. Labeling of the G1 nuclei in the G1/S cells was mostly complete by 16 hours.

2. The G1 control showed when DNA replication would normally occur in G1 nuclei. The G1/G1 control showed that the fusion process itself did not stimulate DNA replication; G1 cells had to be fused to S cells for DNA replication to be stimulated. Nuclei of the G1 and G1/G1 cells did not start to become labeled until about 8 hours after fusion, because these cells had to go all the way through G1 before entering S and replicating their DNA. By contrast, labeling of the G1 nuclei in the G1/S cells began soon after fusion.

3. G2 cells are further into the cell cycle than S phase cells are. It took several hours for the S phase cells to pass through S and G2 to begin mitosis.

4. The timing of mitosis in the hybrid S/G2 cells was similar to that in the unfused S cells and the S/S hybrids. This result indicates that G2 cells cannot stimulate mitosis of nuclei that are still in S phase.

FIGURE QUESTIONS

Figure 11.3 6 pg

Figure 11.21 Apoptosis may be a way to eliminate cells that might develop mutations causing the cell to be harmful to the organism, such as a normal cell turning into a cancer cell. In addition, as an organism develops, organs must be a defined size and shape. Apoptosis may eliminate excess cells that would make an organ too big or misshapen.

Figure 11.25 The cell cycle treatments affect the cell cycles of all dividing cells in the body, not just those of the tumor cells. By contrast, targeted drugs affect altered proteins present only in tumor cells. The side effects of the general cell cycle drugs would be on organs and systems that rely on dividing cells. For example, blood cells undergo apoptosis after a period in the bloodstream and must be replaced by dividing cells; if division in these cells is blocked by an anticancer drug, the patient may develop side effects such as poor immunity (too few white blood cells) and anemia (too few red blood cells).

APPLY WHAT YOU'VE LEARNED

1. Cells would halt at various points in the cell cycle:
Extract 1: cells would accumulate in G2.
Extract 2: cell would accumulate in M.
Extract 3: cells would accumulate in G1.
Extract 4: cells would accumulate in S.

2. Different cells in a tumor may have different errors in cell cycle control. One cell may respond to a G1 inhibitor and another cell to an S phase inhibitor.

3. The researcher could isolate compounds from each cell extract and test them individually in the same tests carried out with the extract in order to locate the specific molecule that acts as an inhibitor in each case. These molecules can be analyzed to determine their chemical structures in order to synthesize them and carry out additional tests, including animal tests, to ensure that they are safe.

4. Only one protein is overexpressed in cancer cells compared with non-cancer cells, and that protein is Cdk1, which functions at the checkpoint during mitosis. The plant extract that inhibits the cyclin B–Cdk1 complex was extract #2, and so extract #2 would be the best choice for treating this tumor.

5. Researchers would be interested in the plant extract that halted phosphatase activity but not the extract that halted kinase activity. Phosphatase activates Cdk, which leads to progression through the checkpoint. The goal of an anticancer drug is to stop progression through the checkpoint so cell division stops. Therefore it would be desirable to inhibit the phosphatase that activates Cdk. It would not be desirable to inhibit the kinase, which is already doing what is necessary to inhibit Cdk.

Chapter 12

RECAP 12.1

1. The F_1 generation showed only one phenotype and it was a parental phenotype; blending predicted that the F_1 would show an intermediate phenotype, not like either parent. The F_2 generation also showed the phenotype of the other parent, not seen in the F_1; blending predicted that the phenotype would disappear.

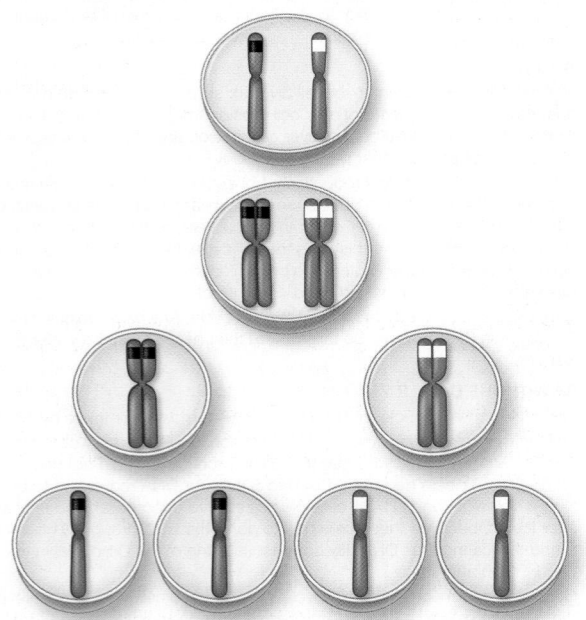

2.

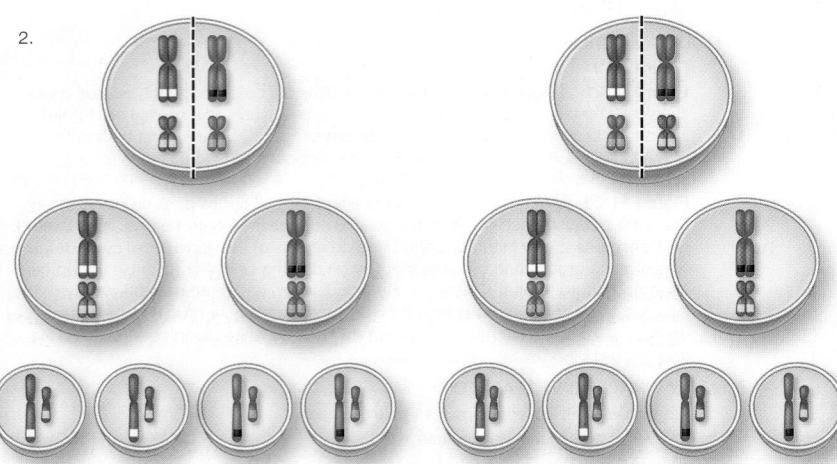

3. $BB \times bb$; $bb \times bb$; $Bb \times bb$; $Bb \times Bb$
4. a. Autosomal dominant
   b. ¼

RECAP 12.2

1. In codominance, in a heterozygote both alleles are evident in the phenotype. In incomplete dominance, in a heterozygote an intermediate phenotype between the parents is seen.
2. In blood type AB, both the A and B alleles are expressed. On the red blood cell surface, two types of proteins are seen.
3. Yellow, blue, and white in a 1:2:1 ratio.

RECAP 12.3

1. Penetrance is the proportion of individuals with a certain genotype that actually show the phenotype associated with that genotype; that is, what proportion of pea plants with "Tall" genotype are tall? Expressivity is the extent to which a genotype actually shows its associated phenotype; that is, are all the pea plants with the "Tall" genotype equally tall, or is there a gradation of height?
2. Qualitative variation involves phenotypes, such as colors and textures, that cannot be easy measured; for example, green and yellow peas. Quantitative variations are phenotypes that can be measured; for example, tall and short pea plants.
3. 75%

RECAP 12.4

1. In genetic linkage, alleles on the same chromosome are generally carried together when a gamete is formed. That is, if in a heterozygote $AaBb$ the A and B are on the same chromosome (and a and b on the other chromosome), gametes will be AB or ab.

2.

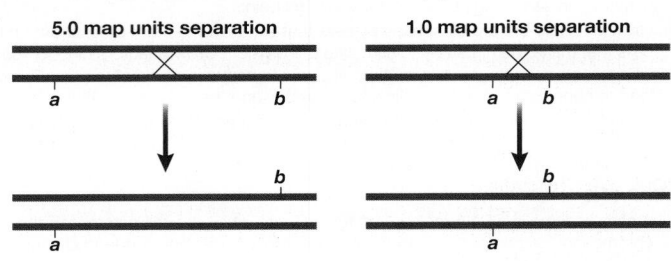

3.

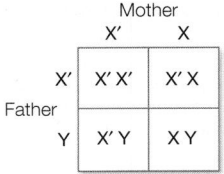

4. A sex-linked gene is carried on the X chromosome. In a genetic cross involving an organism with the XY system for sex differentiation, a sex-linked gene usually has no counterpart on the Y chromosome. So in an XY male, all X-linked alleles, dominant or recessive, are expressed in the phenotype.
5. The cross $RRYY \times rryy$ produces $RrYy$ (round, yellow) F_1 offspring. If the seed shape and seed color genes were linked with no recombination between them, the F_2 would also be all $RrYy$. A distance of 10 map units between two genes means that on average 10% of the F_2 offspring will have recombinant phenotypes, in this case round green (5%) and wrinkled yellow (5%).

   The cross in Figure 12.19 is $BbVgvg$ (gray, normal) $\times bbvgvg$ (black, vestigial). If there were no linkage between the genes, then the gray, normal parent would produce four types of gametes: BVg, bVg, Bvg, and bvg. When these combine with the bvg gametes produced by the other parent, four types of offspring in a 1:1:1:1 ratio will result: $BbVgvg$ (gray, normal), $bbVgvg$ (black, normal), $Bbvgvg$ (gray, vestigial), and $bbvgvg$ (black, vestigial).

   The body color (G/g) and wing size (A/a) genes are linked; eye color (R/r) is unlinked to the other two genes. The distance between the linked genes is 18.5 units.

RECAP 12.5

1. Because the gene is carried on mitochondrial DNA, it is passed through the mother only. Thus if the women does not have the disease but her husband does, their child will not be affected. However, if the woman has the disease but her husband does not, their child will have the disease.

RECAP 12.6

1. Mating and recombination allows genes to be exchanged between bacteria and results in a genetically diverse population of these organisms. Because they are haploid, any new gene combinations will be expressed in the recombined offspring.
2. A plasmid that integrates into the host chromosome will be stable when the cell divides and therefore any genes on the plasmid will be expressed. In addition, if the bacterium mates, the genes on the plasmid may be transferred to another cell.
3. Plasmids can carry genes that confer antibiotic resistance. When a bacterium carry a plasmid mates with one that does not, the plasmid replicates and transfers one of the copies to the recipient bacterium. Now there are two bacteria with antibiotic resistance.

WORK WITH THE DATA, P. 243

1. The chi-square goodness-of-fit test is explained in Appendix B, on pp. 1285–1286. The formula for calculating chi-squared is given in Research Tools Figure B12.

Cross	Chi-squared	P-value
1	0.33	0.57
2	0.69	0.40
3	0.02	0.8
4	0.24	0.59
5	1.52	0.21
6	0.05	1
7	0.19	0.66
8	1.41	0.23
9	0.19	0.66
10	0.35	0.55

All the individual crosses have *P*-values greater than 0.05, so no significant difference from a 3:1 ratio.

2. Combined data: Chi-squared 0.11, *P* = 0.74. In this case, similar *P*-values were obtained both for the individual crosses (small sample size) and for the combined dataset (large sample size).

WORK WITH THE DATA, P. 257

1. Using a chi-squared test with 3 degrees of freedom the chi-squared value is 429.96 and the *P*-value is less than 0.0001. This indicates a highly significant deviation from the expected 9:3:3:1 ratio.

2. The genes are linked, because there more parental type phenotypes and fewer recombinant (non-parental) phenotypes in the test cross progeny than would be expected for unlinked genes.

The map distance is:

$$[(578 + 307) / (578 + 307 + 1413 + 1117)] \times 100 = 26 \text{ map units}$$

These data are different from the data shown in Figure 12.17, because the parental types are different: black normal and gray vestigial in this cross, and gray normal and black vestigial in Figure 12.17.

3. *BBVgVg* and *bbvgvg*

FIGURE QUESTIONS

Figure 12.2 7

Figure 12.11 Yellow: *BBee* and Brown: *bbEE*.

Figure 12.18 Heterozygous, $X^R X^r$. The female parent passed the X^r chromosome to the offspring, which had a Y chromosome from the male parent.

APPLY WHAT YOU'VE LEARNED

1. F_1 genotype: 100% *DdBb*

   F_1 phenotype: 100% long legs, gray body

   F_2 genotypes: 1:2:1 ratio 25% *DDbb*:50% *DdBb*:25% *ddBB*

   F_2 phenotypes: 25% long legs, black body; 50% long legs, gray body; 25% short legs, gray body

2. F_1 genotype: 100% *BbCc*

   F_1 phenotype: 100% gray body, straight wings

   F_2 genotypes: 1:2:1 ratio 25% *BBcc*:50% *BbCc*:25% *bbCC*

   F_2 phenotypes: 25% gray body, curved wings; 50% gray body, straight wings; 25% black body, straight wings

3. In the F_2 generation of the *DDbb* × *ddBB* cross, you would also find the short legs, black body phenotype (*ddbb* genotype), which would be a new phenotype not observed earlier.

   In the second case, the F_2 generation of the *BBcc* × *bbCC* cross, you would find the black body, curved wings phenotype (*bbcc* genotype), which would be a new phenotype not observed earlier.

   The *BBcc* × *bbCC* cross would produce the larger frequency of recombinants because the distance between the two linked loci (*B* and *C*) is longer (75.5 – 48.5= 27.0 map units) than the distance between the two linked loci (*D* and *B*) (48.5 – 31.0=17.5 map units). There is a greater probability of crossing over with a longer distance between linked loci.

4.

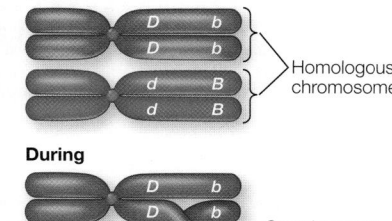

Before

Homologous chromosomes

During

Crossing over

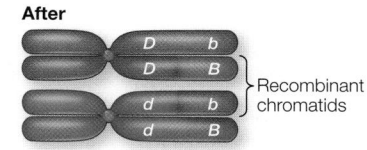

After

Recombinant chromatids

Chapter 13

RECAP 13.1

1. DNA was located in the eukaryotic cell nucleus, where chromosomes carrying genes were located. The amount of DNA was the same in somatic cells of an organism, and halved in the products of meiosis, as expected by the genetic material. Different species had different amounts of DNA, just as they seemingly had different numbers of genes.

2. Avery and his colleagues performed genetic experiments using cell extracts of one strain of bacteria (A) that could change a recipient strain (B) that was genetically different into the extracted strain (A). When they pretreated the extract of the donor strain with enzymes that hydrolyzed proteins, transformation still occurred. So proteins in the donor extract were not responsible for genetic transformation.

3. Bacteriophage T2 has only two types of molecules, DNA and protein. So labeling one or the other could indicate which got into a host cell to cause genetic changes.

4. The proposed experiments might use S strain pneumococcus and transform R strain, as in Figure 13.1. Incubate separate batches of S strain bacteria in ^{32}P or ^{35}S. Make cell-free extracts of the S strains. Incubate with R cells and look for their transformation to the S phenotype. Then check to see if there is ^{32}P or ^{35}S label in the newly transformed cells. It would be expected that only ^{32}P label (DNA) would enter the cells.

RECAP 13.2

1. X-ray diffraction indicated that DNA is double-stranded and twisted into a helix. There were indications that the bases were inside the helix, and the sugars and phosphates on the outside. Base composition data from many organisms showed that the percentages of the purine A = those of the pyrimidine T, and that the percentages of the purine G = those of the pyrimidine C. This suggested that A might be opposite T on the inside of the double helix, and G opposite C. When Watson and Crick built molecular models with the atoms and bonds of polynucleotide strands, the base pairing was confirmed, as the A-T and G-C pairs fit nicely together.

2. The double-stranded structure is essential in the replication of DNA, as the opposite strands can each act as a template for a new strand, so that two new identical strands are made. This is key in the replication of the genetic material when cells divide. The two strands can unravel at places, exposing the bases in the inside for gene expression. This is important because genes must be expressed for the phenotype.

3. The bases in DNA expose chemical groupings that can interact with groups on proteins. These include polar groups (e.g., C=O) on the bases that can attract oppositely polar groups (e.g., NH_2) on proteins, as well as form hydrogen bonds with groups on proteins.

RECAP 13.3

1. The bacteria were prelabeled with both strands of DNA with heavy ^{15}N. After one round of replication in light ^{14}N, three models for DNA replication had different predictions:

   a. Conservative: There would be original DNA (all heavy) and new DNA (all light) in equal amounts. This did not occur.

   b. Dispersive: There would be light, heavy, and intermediate DNA in no fixed proportions. This did not occur.

   c. Semiconservative: There would be only intermediate DNA, with one light and one heavy strand. This did occur.

2. The five proteins involved in DNA replication are DNA helicase (unwinds the double helix), single-strand binding proteins (stabilize and keep apart the two strands in unwound regions), primase (binds to DNA to make a short primer), DNA polymerase (adds nucleotides to a growing chain), and DNA ligase (seals up nicks in DNA, due to its lagging strand replication in short pieces).

3. The special sequences at the ends of chromosomes are called telomeres, and they can be replicated if necessary. After DNA replication, the primer regions at the 3′ ends of the long DNA in chromwosomes are removed. This shortens the DNA and makes it unstable. In some cells, such as gamete-producing cells and cancer cells, the telomeric sequences are recognized by an enzyme complex called telomerase, which catalyzes the replication of any lost telomeric sequences and keeps the DNA at its original length.

4. DNA replication adds new nucleotides to the 3′ end of DNA, where there is an —OH group on the sugar at the 3′ position. If there is no —OH group, there cannot be a condensation reaction and formation of a bond to the next nucleotide, so replication stops.

RECAP 13.4

1. If there were a mutation in S phase before meiosis (e.g., a T changed to a C), the replicated strand would be mispaired; e.g.,

   AAGGTT ⟶ AAGGCT
   TTCCAA   TTCCAA

   If this mispairing was not recognized and repaired, a gamete formed might have the mutation:

   AAGGCT
   TTCCAA

And after the cell containing the mutation divided, one of the daughters would have the mutation and the other would retain the normal sequence.

Top strand replicated:	AAGGCT
	TTAAGA Mutation
Bottom strand replicated:	AAGGTT
	TTCCAA Normal

2. If the colon cancer cell had a mutation in a tumor suppressor gene, it might not be repaired. This would lead to unregulated cell division.

RECAP 13.5

1. Primers in PCR bind to a short region of DNA and allow elongation of that strand of DNA when DNA polymerase is added along with nucleotide.
2. If a primer has a species-specific sequence—that is, a sequence of nucleotides complementary to a sequence unique to a certain organism—PCR will amplify a target DNA extracted from a field sample that has the unique sequence. This amplification will show that the organism with that sequence is present

WORK WITH THE DATA, P. 277

1. These data fit the semiconservative model of DNA replication because the heavy strands were templates for new light strands; after one round of replication, all the DNA had one heavy (original) strand and one light (newly made) strand, and so was intermediate in weight.

Generation	Percent heavy DNA	Percent intermediate DNA	Percent light DNA
1	0	100	0
2	0	50	50
3	0	25	75
4	0	12.5	87.5

2. After seven generations there would be about 1.5 percent intermediate DNA and 98.5 percent light DNA.
3. In the first generation, the bands would be the same as in Figure B: all intermediate. In the second generation, 1/2 would be intermediate and 1/2 heavy. In the third generation, 1/4 would be intermediate and 3/4 heavy. In the fourth generation, 1/8 would be intermediate and 7/8 heavy.

Generations

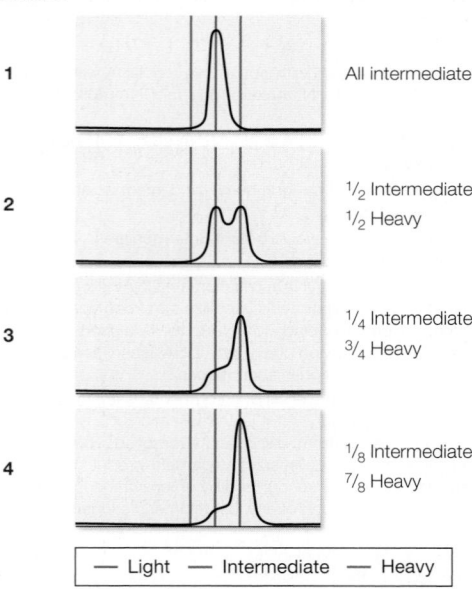

1	All intermediate
2	1/2 Intermediate 1/2 Heavy
3	1/4 Intermediate 3/4 Heavy
4	1/8 Intermediate 7/8 Heavy

— Light — Intermediate — Heavy

4. In a conservative model, the first generation would be 1/2 heavy, 1/2 light; the second generation 1/4 heavy, 3/4 light; the third generation 1/8 heavy, 7/8 light; and the fourth generation 1/16 heavy, 15/16 light. In a dispersive model, the first generation would be all intermediate; the second generation all half-way between intermediate and light; the third generation all half-way between the second generation peak and light; and the fourth generation all half-way between the third generation peak and light.

Generations   Conservative   Dispersive

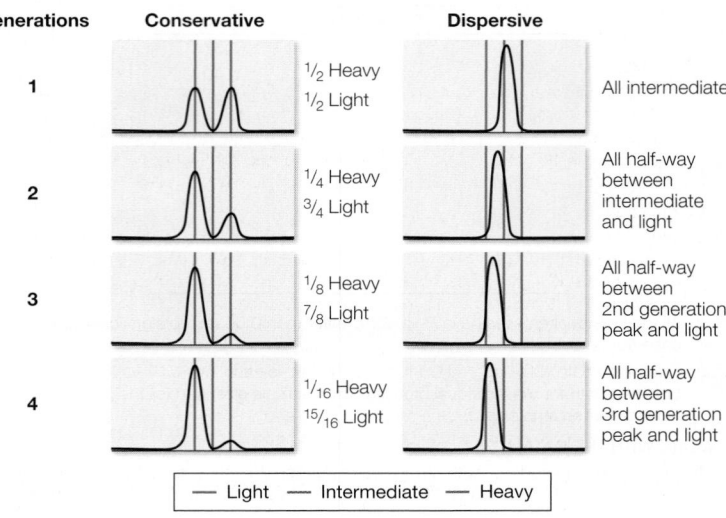

Generations	Conservative	Dispersive
1	1/2 Heavy 1/2 Light	All intermediate
2	1/4 Heavy 3/4 Light	All half-way between intermediate and light
3	1/8 Heavy 7/8 Light	All half-way between 2nd generation peak and light
4	1/16 Heavy 15/16 Light	All half-way between 3rd generation peak and light

— Light — Intermediate — Heavy

FIGURE QUESTIONS

Figure 13.6   Hydrogen bonds occur between the bases on opposite strands, within base pairs. Covalent bonds occur between the atoms that make up nucleotides and between the nucleotides in a DNA strand. van der Waals forces occur between the flat bases that stack on top of each other within the double helix, stabilizing them in the stacking.

Figure 13.13   On opposite sides of the origin, the leading strand will be the lagging strand, and vice-versa.

Figure 13.17   Shortened telomeres are unstable. Also, without telomerase there is unreplicated DNA. For these two reasons, gamete-forming cells express telomerase.

Figure 13.19   Both PCR primers in the test tube and DNA replication primers in the cell act to begin DNA replication. However, PCR primers can be complementary to any DNA strand, whereas DNA replication primers bind only to the origin of replication.

APPLY WHAT YOU'VE LEARNED

1. It will take about 280 hours to complete the synthesis of the shortest DNA strand from one starting point. This is much longer than the 8 hours that the cell has to complete DNA replication, so no one can claim that DNA replication involves just one starting point. The results of the calculation suggest that multiple starting points must be involved in replicating a strand of DNA
2. The results suggest that multiple starting points for synthesis are present in a single DNA molecule. The number of starting points corresponds to the number of radioactive bands. This result is consistent with the hypothesis made based on the calculations in Question 1.
3. The results suggest that DNA synthesis is bidirectional. This is shown by the presence of two dark patches running 180 degrees from each starting point of synthesis, followed by lighter patches indicating a period of dilution during the change from radioactive to nonradioactive thymidine. If DNA synthesis were unidirectional, we would see single dark patches followed by lighter patches to either the left or the right but not in both directions.
4.

Radioactivity   Origin of replication

Chapter 14

RECAP 14.1

1. A model organism is easy to grow and manipulate in the laboratory and has representative characteristic(s) common to a larger group of organisms (e.g., a mouse for mammals).
2. Srb, Horowitz, Beadle, and Tatum treated wild-type *Neurospora* with X rays to cause mutations. Strains that could not grow without the addition of an amino acid (arginine) were isolated. The strains were separately incubated without arginine but with intermediates along the proposed biochemical pathway that makes arginine. Growth in one of the intermediates meant that the mutation for that strain must occur in a gene encoding an enzyme that acts before that substance in the pathway. In this way, the collection of mutant strains led to identification of the intermediates in the pathway and their order, and ultimately to the enzyme that acts to convert each intermediate into the next.

3. Enzymes: 4 → 2 → 3 → 1 → 5
   Compounds: C → F → E → D → G → T

RECAP 14.2

1. The central dogma states that DNA is transcribed to RNA, which gets translated into protein, a process that is unidirectional.

2. In retroviruses, the genome is RNA. To replicate, the RNA is converted to a DNA copy, which is then transcribed into RNA. This violated the original central dogma in that the DNA-to-RNA conversion was thought to be unidirectional.

RECAP 14.3

1. If the code were just single letters (A or T or G or C), each "letter" would translate to one amino acid (41). But there are 20 amino acids, so a single-letter code could not unambiguously identify all amino acids. A doublet code would create 16 possible codons (42), not enough to uniquely identify all 20 amino acids. A triplet code using four letters has 64 unique possibilities (43), more than enough for 20 amino acids and stop codons. The code is redundant (there is more than one codon per amino acid) but not ambiguous (a codon does not stand for more than one amino acid).

2. RNA polymerase binds to DNA at the promoter. The DNA is unwound to expose the bases. The enzyme has binding sites for substrates, the ribonucleoside triphosphates. The enzyme then adds nucleotides to a growing chain by complementary base pairing to template DNA.

3. DNA must be replicated exactly. Any error in DNA in a gene that encodes a protein will result in an error in the RNA that is transcribed from that DNA region. This could result in a different codon and therefore a different amino acid at that location in the protein. The protein's function may change. RNA is made in many copies. So an error in an RNA could result in an error in the protein translated from it, but since there are many more normal copies of that RNA, there would be plenty of the normal protein for normal function.

RECAP 14.4

1. At the 5′ end, a "cap" of modified GTP is added. This facilitates binding of mRNA to the ribosome and protects the mRNA from hydrolysis by ribonucleases. At the 3′ end a poly A tail is added, with 100–300 A nucleotides. This assists in export of mRNA from the nucleus.

2. snRNP particles bind to consensus sequences at the 5′ and 3′ splice sites on pre-mRNA in the nucleus. The two ends approach each other, mediated by the snRNPs. mRNA is cut first at the 5′ end and then at the 3′ end, releasing the intron. The two exons are then joined.

3. 192 amino acids would ne encoded by 576 nucleotides. Add start and stop codons for a total of 582 base pairs in DNA. The DNA has extra base pairs for the promoter and terminator of transcription, introns, and a sequence for mRNA binding to the ribosome.

RECAP 14.5

1. rRNA acts as a scaffold for proteins to make the ribosome structure, with binding sites for tRNA. An rRNA has a nucleotide sequence region complementary to a region on mRNA so the two RNAs can bind and begin translation. An rRNA acts as the catalyst for peptide bond formation.

2. a. 3′–TACGGGCCCAATTCTTAAAATTTTACT-5′
   b. The bottom strand is transcribed: It has sequences that are transcribed into start (AUG) and stop (UGA) codons in RNA
   c. mRNA: AUGCCCGGGUUAAGAUAUUUUAAAUGA
      Polypeptide: Met-Pro-Gly-Leu-Arg-Tyr-Phe-Lys

3. A polysome is formed when more than one ribosome is bound to mRNA at the same time. This can occur because ribosomes move along mRNA is a 5′-to-3′ direction, translating as they go, much like a cafeteria line. Polysomes allow more proteins to be made at a given time from an mRNA.

RECAP 14.6

1. Signal sequences are translated to regions in a protein that bind to recognition molecules and/or receptors associated with a particular destination in the cell.

2. Post-translational modifications are made to specific amino acids at particular locations on a protein. The locations are determined by the three-dimensional shape of the protein, allowing an enzyme for the modification to bind to the amino acid at that location. So overall, the amino acid sequence of the protein determines which amino acids will be modified posttranslationally, and the amino acid sequence is determined by the genetic code and DNA sequence for the gene.

WORK WITH THE DATA, P. 291

1. 34105: gene a   33442: gene b   36703: gene c.
2. The mutant strains may have had some residual enzyme activity, allowing for growth.
3. Arginine is part of proteins and is essential for the tertiary structure of proteins. Replacing it with other amino acids is not possible.
4. The double mutant cannot do the reactions ornithine → citrulline and citrulline → arginine. So adding ornithine or citrulline does not allow growth, since arginine is not produced.

WORK WITH THE DATA, P. 297

1. Minus poly U mRNA: Charged tRNA does not bind to the codon at the ribosome, so no polypeptide can be made.

   Minus ribosomes: There are no locations for adjacent charged tRNAs to bind and also no peptidyl synthetase to catalyze peptide bond formation.

   Minus ATP: tRNA cannot get charged with amino acids, and mRNA cannot translocate along the ribosome. No polypeptides are made.

   Plus RNase: mRNA is destroyed. See Minus poly U mRNA, above.

   Plus DNase: All the components for polypeptide synthesis are present.

   Radioactive glycine instead of phenylalanine: Charged glycine tRNA binds to mRNA codon GGU, GGA, GGG, or GGC. These codons are not present in the poly U mRNA, so no radioactive polypeptides are made.

   Mixture of 19 radioactive amino acids minus phenylalanine: Charged tRNAs are made, but their codons are not present, so no polypeptides are made.

2. The data show that RNA (poly U) was essential for protein synthesis, since its absence resulted in no protein synthesis.

3. According to the genetic code (see Figure 14.5), UUU is the codon for phenylalanine. The other amino acids were not added to protein in the poly U experiment because only the UUU codon was present.

FIGURE QUESTIONS

Figure 14.2 Prokaryotic cells lack a nucleus, so transcription and translation are not spatially separated.

Figure 14.4 Both RNA synthesis and DNA replication require a polymerase enzyme that binds to DNA; the DNA must unwind to expose the bases; the new polymer is complementary to a template strand; and the substrates are nucleoside triphosphates.

Figure 14.12 The chemical forces that hold the molecules of the ribosome together include hydrogen bonds, ionic attractions, and hydrophobic interactions. These can be disrupted by heat or detergent.

Figure 14.15 If there is no stop codon, translation continues because there are more nucleotides at the end of mRNA past the stop codon location. The protein is not properly released from the ribosome.

Figure 14.17 A protein with no "address" stays in the cytoplasm.

APPLY WHAT YOU'VE LEARNED

1. Leu-1 is a mutation in the protein catalyzing step C. The product of this step (α-ketoisocaproate) is the first in the pathway able to overcome blockage in this mutant; therefore, Leu-1 must be a mutation in the enzyme at step C. Leu-2 is a mutation in the protein catalyzing step B. The product of this step (3-isopropylmalate) is the first in the pathway able to overcome blockage in this mutant; therefore, Leu-2 must be a mutation in the enzyme at step B.

2. The mutant cells carrying the Leu-1 mutation are deficient in enzyme C but carry the wild type enzyme B. The mutant cells carrying the Leu-2 mutation are deficient in enzyme B but carry the wild type enzyme C. Therefore, when the two cells fuse, the diploid cell contains one copy of wild type enzyme B and one copy of wild type enzyme C, which restores the wild type phenotype.

3. The table summarizes the predicted enzyme activities:

Enzyme	Predicted enzyme activity		
	Leu-1 (haploid)	Leu-2 (haploid)	Fused cells (diploid) Leu-1, Leu-2
A	Wild type	Wild type	Wild type
B	Wild type	None	Wild type
C	None	Wild type	Wild type
D	Wild type	Wild type	Wild type

4. One mutant strain is affected in the gene encoding the α subunit and the other mutant strain is affected in the gene encoding the β subunit. Thus, haploid cells cannot produce a functioning $\alpha_2\beta_2$ enzyme because these cells have only one copy of each gene, one of which is normal and one of which is not. Diploid cells, however, have two copies of each gene with one normal gene for α supplied by one of the mating pair and one normal gene for β, supplied by the other of the mating pair. The presence of a copy of a normal gene for each subunit allows the diploid cells to produce normal polypeptide chains of α and β, which associate to form a functional $\alpha_2\beta_2$ enzyme.

Chapter 15

RECAP 15.1

1. Point mutations that cause phenotypic changes could have resulted in a different amino acid in the encoded protein that consequently changes a protein's function; changed a promoter so a gene's expression is significantly altered; or created a stop codon that terminates expression prematurely, resulting in a shorter nonfunctional protein. Point mutations that are phenotypically silent may arise in codons where redundancy ensures no amino acid change; cause codon changes that result in amino acid changes that are not significant to protein function; or occur in noncoding regions of DNA, such as introns.

2. See Figure 15.4. Deletions are missing part of a chromosome; duplications have an extra copy (or copies) of a chromosomal region; in inversions, a chromosome region is out of sequence; and in translocations, a piece of one chromosome breaks off and attaches to another chromosome.

3. Spontaneous mutation occurs without an external agent causing it. Example: tautomeric shift of A, so that at DNA replication A base-pairs not with T but with C. In an induced mutation, an environmental agent changes DNA. Example: nitrous acid deaminates C to U, so that at DNA replication, instead of C pairing with G, it is U pairing with A.

4. C can be methylated to 5-methylcytosine. When deaminated spontaneously or by a mutagen, this base forms T. This is a normal base and is not removed by DNA repair. Other base changes are repaired.

RECAP 15.2

1. a. The amino acid sequence produced by the mutant allele is Leu-Ile-Ser-Ile-Ala. This is a missense mutation.
   b. The mutation replaces proline with serine. Proline is a nonpolar amino acid that is usually part of bends or loops in a protein; serine is a polar amino acid with a smaller side chain. The mutation is likely to affect enzyme activity because it is likely to affect protein structure.

2. Simply having a particular genetic mutation is not sufficient to lead to cancer. There are other genetic and environmental factors that may be involved in developing breast cancer. For example, a person can have a mutation in a different gene for DNA repair, such that it takes over the functions that are lost by the *BRCA* mutation.

3. The symptoms occur only if there are a large number of repeats of the CGC sequence in the promoter region of the *FMR1* gene, so that they disrupt expression of the gene. Family members may carry the mutation but not show symptoms because the mutation contains a low number of repeats rather than a high number.

RECAP 15.3

1. Direct DNA sequencing of the cystic fibrosis gene could be done. A person who is a carrier will test positive for both the normal and the mutant alleles.

2. Mapping a disease-causing mutation can be done by linkage analysis. A polymorphic DNA marker such as an STR can be linked to the occurrence of a disease in many patients. This means that the marker must lie on the chromosome near the mutant disease-causing gene. DNA sequencing can then isolate the gene involved. From the gene sequence or genetic technology, the protein encoded by the gene can then be isolated and its function described. So genotype precedes phenotype.

RECAP 15.4

1. DNA analysis can be done on any tissue at any time in the life cycle of an individual. In addition, heterozygotes can be detected. Phenotype analysis by enzyme activity requires gene expression in an accessible tissue at a certain time and place. In many cases, heterozygotes cannot be detected.

2. A patient's DNA could be tested to see if it hybridizes to sequences of the mutant β-globin gene. If it did, this would mean that the patient carries the mutation, and no further sequencing of the patient's DNA would need to be done to verify this.

RECAP 15.5

1. Metabolic inhibitors block important chemical transformations in cancer cells. An inhibitor may either block the accumulation of a harmful substance or block cancer-specific transformations to harmful substances.

2. In vivo gene therapy inserts the wild-type form of a gene that is mutated or abnormally expressed in tissues of a person with a genetic disease or other disease. Typically a virus is used to deliver the gene, and the DNA either inserts into the host chromosome or stays outside the cell nucleus in a virus that does not replicate. An example is adding a gene for glutamate decarboxylase to the brains of patients with a neurotransmitter deficiency in Parkinson's disease (see Figure 15.19).

3. a. The mutation that leads to PKU is rare in the human population; most people do not have the harmful allele, and the highest probability is that the father is homozygous normal. Because the mother has PKU (she is homozygous mutant), the developing fetus is heterozygous.
   b. High levels of phenylalanine cause brain damage. If the mother's phenylalanine levels were too high, the baby would be born with brain problems.
   c. The woman should be on a phenylalanine-restricted diet.

WORK WITH THE DATA, P. 329

1. In the three families, breast cancer occurred in two-thirds of the patients early in life, suggesting that the cancer was hereditary.

2. All three mutations were present only in the breast cancer patients. In Families A and B, the two point mutations may have caused codon mutations and a different amino acid in each case in the BRAC1 protein, causing it to be nonfunctional. In Family C, the deletion was 11 base pairs, which resulted in a frame-shift mutation and meant that the codons after that mutation were read incorrectly. This caused massive changes in amino acid sequence and a nonfunctional BRCA1 protein.

3. *BRCA1* is active in breast, ovary, and thymus tissues. So mutations would also be expressed there. Since *BRCA1* is involved in DNA repair, all three of these tissues would experience poor repair, leading to additional mutations that can cause cancer.

FIGURE QUESTIONS

Figure 15.1 Silent mutation, because the genetic code is redundant (many mutations do not change the amino acid translated) and because many amino acids in a protein are not essential for the activity of the protein (e.g., do not affect the active site).

Figure 15.3 Chromosomal mutations can be detected by staining dividing cells with dyes specific for each chromosome. Stained chromosomes can then be identified, and missing pieces or translocated pieces can be observed. Inversions can be detected by a special method called banding, whereby dyes on chromosomes produce banding patterns instead of colors. In this case, a reversal of bands can be seen.

Figure 15.13 The advantage of a genetic ID would be to predict future propensities for diseases and possibly act to prevent them, and to identify a person in an accident, war, or crime (this is already done with soldiers and people in federal prisons). A disadvantage would be that in the wrong hands there could be an invasion of privacy. Genetic markers would need to be used with caution as they are not necessarily predictive; environmental factors often play a role in disease as well. While a federal law in the United States prohibits using genetic data to discriminate against people applying for insurance and employment, the concern remains that genetic data may nevertheless influence decision makers in these and other arenas.

Figure 15.14 Chromosome linkage analysis involves actual crosses and is done by analyzing the results of recombination between alleles. DNA linkage analysis is done on single chromosomes by retrospective analysis of mating. Chromosome linkage analysis examines phenotypes, whereas DNA linkage analysis examines genotypes (DNA).

APPLY WHAT YOU'VE LEARNED

1. Person 1 is heterozygous, with one normal allele and one mutant allele. The normal allele translates to the amino acids: Pro-Trp-Thr-Gln-Arg-Phe. The mutant allele has a stop codon at position 4 in the mRNA (a nonsense mutation): Pro-Trp-Thr-(stop), and the polypeptide is short, only 38 amino acids instead of 146. So this globin would not be functional.
   Person 2 is heterozygous, with one normal and one mutant allele. In this case, the mutant allele is a deletion of the first T in the second codon. This causes a frame shift: Pro-Gly-Pro-Arg-Gly-Ser…, and the resulting polypeptide has a very different sequence and is most likely nonfunctional.

2. One example is a mutation that affects the promoter such that RNA polymerase cannot bind to initiate transcription. Another example is a mutation that affects mRNA splicing sites in the DNA that leads to abnormal deletions or insertions in the mature mRNA transcript.

3. The couple has a 1 in 4 chance of producing a child who is homozygous with two normal alleles, a 1 in 2 chance of producing a child who is heterozygous with one normal allele and one mutant allele, and a 1 in 4 chance of producing a child who is homozygous with two mutant alleles. A child with two mutant alleles would suffer severe anemia and would have to receive regular blood transfusions throughout life.

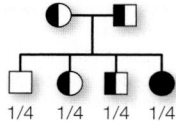

1/4   1/4   1/4   1/4

Chapter 16

RECAP 16.1

1. The cell could reduce the transcription of the Xase gene, hydrolyze Xase mRNA, prevent Xase RNA translation, hydrolyze Xase protein, or inhibit Xase protein.

2. In the presence of lactose, the promoter is exposed and RNA polymerase binds to begin transcription. In the absence of lactose, the promoter is occupied by the repressor and there is no transcription.

3. Sigma factors direct RNA polymerase to bind to recognition sequences at certain promoters. If several promoters of genes with related functions (e.g., heat shock response) have the same recognition sequence, transcription of those genes will occur at the same time and rate.

4. To keep a constant, low-level expression of repressor protein, the regulatory gene would have to have an inefficient promoter, and synthesis of the repressor would be constitutive.

RECAP 16.2

1. General transcription factors bind to the promoter and to RNA polymerase in a complex to direct RNA polymerase to the promoter to initiate transcription and locally denature DNA so that the template strand is available for base pairing during RNA synthesis. Specific transcription factors bind to specific promoters or promoters with recognition sequences. Other transcription factors bind to enhancer sequences that can be far from the actual promoter and induce DNA to bend to attract the rest of the initiation complex for transcription.

2. Proteins such as transcription factors fit into the DNA double helix by structural motifs, and their amino acids may form hydrogen bonds with bases on the interior

of the double helix. The sequences of amino acids (proteins) and of bases (DNA) are specific so that only certain proteins bind to certain DNA sequences.

3. In response to a signal from the environment or within the cell, a specific transcription factor is made and/or translocated into the nucleus where it binds to a recognition sequence in DNA at the promoters of target genes. This binding attracts the initiation complex for transcription, so multiple genes are activated.

RECAP 16.3

1. In the bacteriophage: Phage DNA is injected into the host cell. Early phase genes are transcribed at phage promoters by using DNA sequences similar to those of the host cell (positive regulation). This leads to early protein that binds to host promoters to shut them down (negative regulation). Other early proteins lead host RNA polymerase to transcribe middle and late phage genes (positive regulation).

   In HIV: HIV RNA is injected into the host cell. HIV reverse transcriptase is activated to make cDNA and integrase to splice cDNA into the host chromosome (positive regulation). Later, host RNA polymerase binds to HIV promoters to make HIV mRNAs (positive regulation). HIV tat protein acts as an anti-terminator for the transcription of HIV genes integrated into the host genome.

2. HIV infection, cDNA formation and integration would occur as normal. Viral genes would be expressed and a large precursor protein would be made. But it would not be cut into separate viral proteins. HIVE particles would not be packaged or released from the cell.

RECAP 16.4

1. Histone proteins are positively charged and bind to DNA, generally blocking transcription. Acetylation of histones neutralizes the positive charge and thus the histones do not bind to DNA as tightly, which opens up the chromatin structure for transcription. By contrast, histone deacetylation removes acetyl groups, restoring positive charges on histones so transcription is repressed

2. See the opening story and Investigating Life: Gene Expression and Behavior (p. 351). The behavioral environment appears to alter DNA epigenetically by increasing DNA methylation at promoters in genes in brain tissues, thus changing the rate of gene transcription.

3. X chromosome inactivation is shown in Figure 16.16. The *Xist* gene on the X chromosome is transcribed to make a short RNA that binds to the rest of the X chromosome, inhibiting transcription of the other genes. Chromosome proteins bind to the inactive X chromosome, causing heterochromatin to form and inhibiting gene expression. X chromosome inactivation is believed to occur in order to balance the expression of X-linked genes between males (XY) and females (XX) since the Y chromosome does not usually contain X-linked genes.

4. You could sequence the relevant genes of colorectal cancer cells and look for mutations that lead to aberrant function, then isolate the proteins involved and determine that their functions are indeed abnormal. To show epigenetic silencing, you might sequence the promoters of the genes and look for epigenetic changes (e.g., cytosine methylation, which would be increased if there were transcriptional silencing). Then you could examine the tumor cells to see if the active proteins were there but in small amounts.

RECAP 16.5

1. miRNAs and siRNAs can bind by base pairing to target mRNAs and prevent their translation because tRNA cannot bind; or the inhibitor RNAs can bind to pre-mRNA in the nucleus and lead to its hydrolysis by RNase; or the inhibitor RNAs can bind to DNA at the transcription site and block RNA polymerase from working for transcription.

2. mRNA can fold back on itself by hydrogen bonding of complementary bases, forming looped structures. These structures can bind to proteins that then inhibit translation at the ribosome.

3. The proteasome binds to proteins that are targeted with ubiquitin for breakdown. Within the proteasome are proteases that hydrolyze targeted proteins.

4. On average, a human gene can form at least four different mRNAs by alternative splicing. Each of these mRNAs is translated to a unique protein. So the number of different proteins is much greater than the number of genes.

5. miRNAs targeted to activated oncogenes will block the translation of target mRNAs that would make proteins that otherwise stimulate cell division.

WORK WITH THE DATA, P. 351

1. a. mRNA was measured in the heads because gene expression in the brain determines whether a honey bee will be a worker or a queen.

   b. The level of inhibition of DNMT mRNA was about 60 percent after 48 h. This is good but not perfect inhibition. So some DNMT mRNA probably remained.

2. The reduction of DNA methylation was about 20 percent. So some 5-methylcytosine remained.

3. In the controls, about 23 percent of the larvae developed into queens. But in the larvae that had reduced DNA methylation, about 72 percent developed into queens. This is a remarkable shift given that DNMT mRNA and DNA methylation were reduced but not as much. There must be a threshold for methylation, above which gene expression is affected.

FIGURE QUESTIONS

Figure 16.1   Yes. There can be multiple binding sites for regulatory proteins at the promoter region.

Figure 16.7   An advantage of the nucleus is compartmentation—the separation of transcription/processing and translation. This allows for finer regulation of gene expression. Also, the nucleus is protected from nucleases in the cytoplasm, which might hydrolyze mRNA and reduce its lifetime as it is made.

Figure 16.13   Lysogeny is a term used for bacteriophage to describe the integration of the viral genome into the host cell for a period of time. Although the term is not used for eukaryotic viruses, HIV does integrate its genome as cDNA and so could be called lysogenic.

Figure 16.14   5-methylcytosine mutations (deamination) are not repaired. So locations with 5-methylcytosine will tend to accumulate mutations and their potential to regulate transcription will vary.

Figure 16.17   7

APPLY WHAT YOU'VE LEARNED

1. An inhibitor of DNA methylation restores tumor suppressor activity in the cancer cells. The same effect is observed with an inhibitor of histone deacetylation. These results suggest that both DNA methylation and histone deacetylation operate to shut down expression of genes that suppress tumor growth.

2. The data show that inhibitors of both histone deacetylation and DNA methylation act together with greater effect than either inhibitor alone. This suggests that some gene expression is inhibited by DNA methylation, some by histone deacetylation, and some by both processes. In this last group of genes, both inhibitors must be required to release the genes from being silenced.

3. Gene 1 can be reactivated by either inhibitor, but Gene 2 can be reactivated only by a combination of the two inhibitors. For Gene 1, the extent of DNA methylation and histone deacetylation may not be great. In this case, either DNA demethylation or histone acetylation may tip the balance just enough to alter the nucleosome to allow access of RNA polymerase to bind to the promoter and express the gene. In the case of Gene 2, RNA polymerase may be prevented from binding to the promoter for two reasons: the promoter site could be heavily methylated and histones could also be deacetylated, thus preventing loosening of the nucleosome to expose the promoter region. Therefore, for Gene 2 both DNA methylation and histone deacetylation must be inhibited for expression to be restored.

Chapter 17

RECAP 17.1

1. Before sequencing, a long DNA molecule is cut into many fragments. These cuts are random (i.e., they are made at different random locations). The overlaps are important in arranging the final sequence. If there were only one type of cut, aligning the final sequence would not be possible. The sequences of the fragments are arranged by computer using the overlaps as a guide.

2. Open reading frames are recognized by the presence of a promoter sequence, triplet start and stop codons, transcription termination sequence, and recognition sequences at the beginning and end of introns. The reading frame gives the promoter sequence (signal for when and where the gene is expressed) and the sequence of amino acids of the protein (an indication of its function).

3. Comparative genomics is the study of DNA sequences across different organisms. Similar sequences detected in different DNA samples may help reveal the function of a protein (when a sequence matches that of a known protein in another organism), the identity of a species known only by its DNA (metagenomics), and evolutionary relationships. The gene that influences the relative sizes of dogs, for example, was identified by comparative genomics.

RECAP 17.2

1. The following characteristics of prokaryotic genomes are essential for organisms that exist in a rapidly changing environment and must adapt rapidly: (1) Small genomes can show large changes with a few mutations. (2) Plasmids allow individual cells to share genomic information rapidly. (3) Transposons shuffle gene locations and therefore the rate of gene expression, and they cause mutations that make evolution possible.

2. Transposons either splice out of the host chromosome directly and move to a different location, or they make a DNA copy that is inserted at a different location, leaving the original transposon in place.

3. One could take a sample of belly-button "lint" and extract DNA from it, then do PCR to amplify sequences that are prokaryotic and identify the species present. If this were done for many people, there would presumably be a unique set of species abundances for each person.

RECAP 17.3

1. Genes for tissue formation are present in C. elegans but not in yeast.

2. Gene families allow for evolutionary "tinkering." If there are several copies of a gene in the genome, mutations in one copy of the gene may temper a harmful effect of the wild-type copies of the gene. If the mutations are beneficial, they will be selected for.

3. In some tissues, such as the egg cell, there is a need for massive protein synthesis (after fertilization) and therefore a need for a lot of ribosomes. The rate of transcription of the rRNA genes is not sufficient to meet the cell's need for ribosomes, so there are multiple copies of the gene.

4. The four plants exist in different environments. For example, corn grows in drier and cooler regions than rice. Mutations of the basic plant genome allowed plants that carried them to adapt to the different environment to survive and reproduce.

RECAP 17.4

1. Each human gene has several introns. Alternative splicing of pre-mRNA from a single gene can result in different mRNAs and therefore different proteins.

2. The human genome sequence has haplotypes that are linked to differential sensitivity to a drug. If the patient's DNA is isolated and amplified, a classification with regard to the haplotypes can be made, and whether the patient will respond to a particular drug can be determined.

RECAP 17.5

1. The proteome is analyzed by identifying large numbers of proteins using two-dimensional gel electrophoresis and by identifying fragments of proteins by mass spectrometry.

2. Patterns of metabolites correlate with different physiological states. The simplest example of why it would be useful to have a database of primary and secondary metabolites involved in human metabolism might be the high level of blood glucose that is associated with diabetes. Most diseases are more complex than this. Diagnosis may be helped by a pattern of metabolites in the metabolome.

WORK WITH THE DATA, P. 363

1. There was a 4.4 percent change in DNA sequence between the two cat genomes in 10.8 million years. This indicates a rate of change of 0.4 percent per million years. By comparison, the rate of change between the human and gorilla genomes was 0.6 percent per million years, or 50 percent faster.

2. a. 14,425 gene families are shared by all the mammalian genomes examined.

   b. 103 are unique to the tiger and domestic cat genomes; 231 are unique to the human and mouse genomes.

   c. Over 90 percent of the mammalian genome is common to all mammals. Relatively few gene families are unique to each kind of mammal examined.

3. The tiger is a hunter, and its genome reflects this, with genes for smell, signaling and digestion.

FIGURE QUESTIONS

Figure 17.4   A retrovirus infects a cell and makes a cDNA copy of its genome. The cDNA is inserted into the host cell by the action of viral integrase. The cDNA is carried along with the host chromosome as the cell divides. Through recombination, the cDNA adds a gene for excision. It also adds adjacent genes and is able to move about the genome.

Figure 17.6   No. Protein-coding genes represent only a small fraction of eukaryotic genomes. Increasing size could include genes that do not encode proteins as well as repetitive sequences.

Figure 17.10   In both processes a protein structure is moving along a nucleic acid. In the polysome, the ribosome moves along mRNA; in rRNA synthesis, RNA polymerase moves along DNA. In both processes a polymer product is made: in the polysome the product is a polypeptide, and in the rRNA the product is RNA. The two processes are similar also in having a "cafeteria" system in which multiple polymers are made at the same time.

APPLY WHAT YOU'VE LEARNED

1. For SNP 5689, individuals with genotype *AA* have the highest concentrations (and thus the slowest metabolism) of Calm. Individuals with *GG* have the lowest concentrations (fastest metabolism) of Calm. Heterozygotes are intermediate. For SNP 8835, *AA* homozygotes have the lowest concentrations (fastest metabolism), *CC* homozygotes have the highest concentrations (slowest metabolism), and heterozygotes are intermediate. For SNP 11286, the heterozygotes have the highest concentrations (slowest metabolism). The two homozygotes have roughly similar concentrations and thus similar metabolisms.

2. Because *AA* individuals at this SNP have a slower metabolism of the drug than do *GG* individuals, the *AA* individuals should receive a lower dose.

3. *AA* homozygotes have the slowest metabolism and *GG* homozygotes have the fastest. Thus one would expect that the enzyme encoded by the *A* allele of the gene would be less active than the enzyme encoded by the *G* allele.

4. *AA* homozygotes have the fastest metabolism and *CC* homozygotes have the slowest. Thus one would expect that the inhibitor encoded by the *A* allele of the gene would be less active than the inhibitor encoded by the *C* allele.

5. Give the treatment mice the drug and leave others as the control. Prepare liver tissue from the treatment and the control mice. Use two-dimensional gel electrophoresis and mass spectrometry to identify differences in the proteins from the livers of treatment and control mice.

Chapter 18

RECAP 18.1

1. The sequences for recognition by a restriction enzyme often occur near one another on the opposite strands of DNA. For example,

   5′….GAATTC….3′

   3′….CTTAAG….5′

   The top and bottom strands have the same 5′-to 3′-sequence. A restriction enzyme recognizes this and cuts between the G and A:

   5′…G AATTC….3′

   3′….CTTAA   G….5′

   Note that the cuts are staggered, not directly opposite each other.

2. DNA from two genetically different strains of bacteria, each carrying a different antibiotic resistance gene, was cut with the restriction enzyme *Eco*RI and then spliced together using DNA ligase.

RECAP 18.2

1. GFP reporter: all cells alive. Antibiotic resistance: most cells dead and only a selection alive. GFP reporter: need only a UV lamp to see the reporter gene product. Antibiotic resistance: need to alter growth medium and wait until nonselected cells are dead to identify cells with the vector.

2. The sequence would be: d, e, a, b, c.

RECAP 18.3

1. a. This involves tissue-specific gene expression, so use RT-PCR, which is a "snapshot" of the mRNAs in the cell at a given time and can identify the amylase mRNA.

   b. This calls for genomic analysis, so use a genomic library and identity the amylase gene in the strains.

2. All cellular RNA is isolated and cDNA copies of the RNAs are made using reverse transcriptase. Primers for DNA replication are made for the two ends of the DNA for the specific gene whose expression is to be studied. Then a PCR procedure is carried out using these primers. The degree of amplification from these primers is proportional to the specific cDNA target, which in turn is proportional to the amount of the specific RNA in the original cell extract.

3. Synthetic DNA can be used to make a probe for hybridization to identify a genotype, as in genetic screening for human diseases. Mutations in the DNA can be used to detect mutations in patients.

4. Both tissues express some identical genes that are involved in cellular activities, such as energy metabolism (e.g., genes encoding enzymes in the citric acid cycle) and informational functions (e.g., making rRNA). The tissues will differ in tissue-specific mRNAs. For example, the brain makes proteins needed for electrical activity and communication between cells, while the pancreas makes the hormone insulin. But note that all tissues have the same DNA sequences—all of them.

RECAP 18.4

1. Antisense RNA and siRNA act at to prevent translation by binding to mRNA by base pairing and leading to its breakdown in the cytoplasm.

2. Germinate pea seeds and extract RNA from the embryos at different times. Make cDNA from the mRNA's and hybridize the cDNA's to a library of genes from the entire plant genome. The extent of hybridization to the genes will indicate differences in which genes are expressed.

3. CRISPR has the additional advantage of specifically changing the nucleotides in genes, that is, introducing mutations or reversions.

RECAP 18.5

1. Expression vectors have restriction site(s) for insertion of genes, a DNA replication origin, a tissue-specific active promoter for the proposed host cell, and often a sequence for signaling the protein to be sent to the extracellular medium for isolation in the lab.

2. Biotechnology is faster (one instead of many generations of breeding) and specific (only a single gene is introduced, instead of many during selection for a complex characteristic).

3. Because the bacterium *E. coli* normally lives in the human intestine, there was a danger that lab strains resistant to multiple antibiotics might infect people. Furthermore, there may have been concern that the genes of antibiotic-resistant strains might recombine with the genes of normal strains, resulting in the development of multiple-antibiotic-resistant strains. Another concern was that bacteria could be used to harbor genes encoding toxins as biological warfare agents. To alleviate these concerns, the host *E. coli* strains were developed with multiple mutations so that they would require a special lab environment to survive. For example, mutations in genes for the synthesis for vital metabolites were introduced so the bacteria would need to be supplied with metabolites that the human intestine does not provide.

WORK WITH THE DATA, P. 393

1.

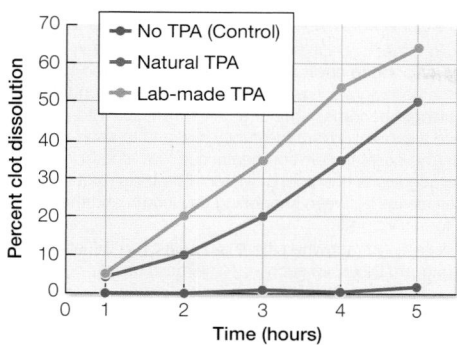

With no TPA there was no clot dissolution, but with TPA there was dissolution. Lab-made TPA was somewhat superior to natural TPA both in terms of the rate of clot dissolution and the final percentage of dissolution.

2. Again, lab-made TPA was somewhat superior to natural TPA in dissolving clots in the rabbits. A comparison could be made between the final percent of dissolution using a t-test.

FIGURE QUESTIONS

Figure 18.2 DNA ligase is needed for the formation of strong covalent bonds to link the DNAs.

Figure 18.4 Cells with GFP reporter plasmids are not killed. This means that the cell or organism with the plasmid can be grown directly, without killing other cells. Also, it means that live organism experiments, following the fates of cells with recombinant DNA over time, are possible.

Figure 18.10 DNA sequences that encode posttranslational modifications of the enzyme would be necessary, specifically the signal sequence that is involved in targeting the enzyme to lysosomes.

Figure 18.13 Synthetic cells could be programmed to make natural molecules with new properties such as antibiotics with altered structures and targets, or plastics that are biodegradable, or fuels.

APPLY WHAT YOU'VE LEARNED

1. The daughters all are homozygous for the transgene. Under Mendelian expectations, the daughters should be heterozygous.

2. The male offspring inherit their X chromosome from their mother. The female parent in this cross lacked the transgene.

3. Because the F_1 females are homozygous for the transgene, they should produce male offspring that all have the transgene, regardless of their mate's genetic makeup. Males have only one copy of the X chromosome. These males should then produce daughters that are all homozygous for the transgene. The cycle would continue, yielding more and more flies with the transgene.

4. Additional transgenes that inhibit malaria transmission would be needed. In actual practice, there are genes that cause the mosquitoes to be resistant to *Plasmodium*. Thus the mosquitoes would still bite humans, but the likelihood that they would transmit malaria would be reduced.

5. Because the technology is very powerful and CRISPR can theoretically be used in just about any organism, there is a legitimate concern about accidental release of the construct. Laboratory safety protocols are vitally important. The researchers have used stringent containment procedures in accordance with approval from their institutions.

Chapter 19

RECAP 19.1

1. Transplantation experiments with amphibian embryos showed that in the early amphibian embryo, cell fate is not yet determined and the environment in which a cell is placed can result in determination. In the later embryo, cell fate has been determined and the environment has no effect.

2. Stem cells in adult body tissues are typically multipotent, while stem cells in the embryo are typically pluripotent.

3. One could analyze mRNA in egg cells, in the parent differentiated cells, and in the reprogrammed cells. This could be done by reverse transcriptase PCR or by gene expression arrays.

4. In animals, cloning has been used to reproduce valuable organisms and increase populations of rare species. In humans, reproductive cloning could be used by couples who want offspring that are genetically related to one or both of them but who cannot produce eggs or sperm. It could allow a lesbian couple to have children that are genetically related to them or gay male couples to have genetically related offspring (using a surrogate mother to carry the child). It could allow a couple to have a child who is genetically the same as a child who is dying.

RECAP 19.2

1. Unequal distribution of a cytoplasmic component in the fertilized egg can result in only one of two daughter cells receiving the component after cell division. This asymmetry can be maintained by the cytoskeleton. It can mean that only one of the two cells will express certain genes, resulting in a difference in cell fate determination.

2. The anchor cell would not be able to send a differentiation signal to the overlying surface tissue, so the latter would not develop into the vulva.

3. The *p21* gene would be activated and the cell cycle would be blocked; in the presence of other factors, muscle cells would form.

RECAP 19.3

1. Organ identity genes specify different organs of the flower in plants. They are identified by loss-of-function mutations (mutations that result in a missing organ) or gain-of-function mutations (a promoter for the gene specifying one organ can be artificially spliced to a gene specifying a different organ, causing the latter gene to specify the first organ).

   Which particular organ will be generated depends not just on the expression of a single organ identity gene but on the combination of the expression of other genes. For instance, the whorl that generates petals (whorl 2) and the whorl that generates stamens (whorl 3) both express class B genes. The difference between these whorls lies in the other organ identity genes that they also express, with whorl 2 also expressing class A genes and whorl 3 also expressing class C genes.

2. A mutation that caused expression of class A genes instead of class C genes would lead to an AB combination instead of AC, and petals would develop instead of stamens.

3. All neuronal precursors might undergo apoptosis, and no neurons would form. The Hunchback protein gradient would not form properly, and the embryo would not establish its anterior–posterior axis.

RECAP 19.4

1. Substituting the fruit fly genetic switch for eye formation in the mouse embryo for the mouse genetic switch still resulted in eye formation in the mouse, and vice versa. This indicates that the genetic switches for development are conserved in evolution.

2. BMP4 is involved with cartilage development.

3. The observations are consistent with the hypothesis that there has been selection in some human populations for mutations of the enhancer that controls expression of the glycoprotein in red blood cells. This genetic change would be expected to have a selective advantage in human populations that are exposed to malaria at high levels, because the mutation confers greater resistance to malaria in humans who carry it.

RECAP 19.5

1. The developmental gene *Dll* controls the formation of legs in arthropods. Insects have three pairs of legs on their three thoracic segments, whereas centipedes have many legs on both thoracic and abdominal segments. In insects, the expression of *Dll* is repressed by the gene product of *Ubx*. *Ubx* is expressed in the abdominal segments of all arthropods, but it has different effects in different species. In centipedes, *Ubx* is co-expressed with *Dll* to promote the formation of legs. During the evolution of insects, a mutation in the *Ubx* gene sequence resulted in a modified Ubx protein that represses *Dll* expression in abdominal segments.

2. Marine sticklebacks have bony structures that protect them from predatory marine fish. In freshwater sticklebacks, these structures are not prominent. The differences are due to the expression of a developmental regulatory gene, *Pitx1*, which encodes a transcription factor normally expressed in regions of the developing embryo that in marine sticklebacks form the head, trunk, tail, and pelvis. However, in several freshwater stickleback populations, the *Pitx1* gene is no longer expressed in the pelvis, and spines do not develop. This same change in regulatory gene expression has evolved to produce similar phenotypic changes in several independent populations and is an example of parallel evolution.

FIGURE QUESTIONS

Figure 19.8 Yes, If the cell on the right of the figure had fewer receptors but the same exposure to inducer, it might have less activation of transcription.

Figure 19.16 If *Ubx* was expressed in all segments, the expression of the wing-forming gene would be inhibited and halteres would form.

Figure 19.19 If Gremlin was blocked in the developing feet of duck embryos, there would not be apoptosis between the digits and so chickenlike feet would form.

Figure 19.20 *Drosophila* with the centipede *Ubx* gene would grow legs from abdominal segments.

WORK WITH DATA, P. 404

1. In an RCT, the only difference between treated and control subjects is the response to the variable under investigation. Genetic and other differences between experimental and control groups are assumed to be randomized and equal between the groups.

2. The difference in the score for lameness in the stem cell group at 60 days was −1.11. This is greater than the difference in the experimental group (−0.33). So there

was more improvement in the stem cell group. A statistical test to show the significance of the differences would be the *t*-test.

3. Plot with Excel. The beneficial effect of stem cell treatment was evident after 30 days.

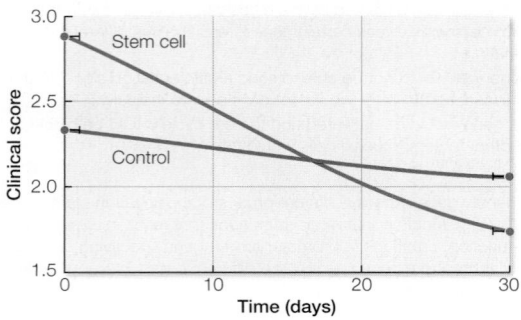

APPLY WHAT YOU'VE LEARNED

1. The two species differ with respect to how long the transcription factor expression remains high, with the duration of high expression being longer in Species 1. Thus this difference in timing is an example of heterochrony. A reasonable hypothesis is that the morphological difference between the species arises from differential timing of the transcription factor's expression; specifically, a longer period of high expression (as in Species 1) leads to longer bristles.

2. A direct test of the hypothesis would involve experimental manipulation of the transcription levels in the two species. Increasing the duration of high expression of hairball in Species 2 should result in a bristle size more like that of Species 1. Decreasing the duration of high expression of hairball in Species 1 should result in a bristle size more like that of Species 2.

3. The differences in the presence of bristles in the other segments are likely due to the expression of hairball in those segments. Given that hypothesis, the prediction is that expressing hairball appropriately in those other segments should result in bristles in those segments in Species 1 and 2. This is an example of heterotopy.

4. You could express the Hox gene from Species 3 in Species 1 and Species 2 and eliminate the expression of their normal Hox gene. If the differences in the Hox gene are responsible for the bristle pattern difference, then the flies of Species 1 and Species 2 with Species 3 Hox expression should have a bristle pattern like that of Species 3. Similarly, expressing the Species 1 Hox gene in Species 3 should result in a Species 1 bristle pattern in Species 3.

Chapter 20

RECAP 20.1

1. In science, the word "theory" does not mean just a guess or an untested idea. Instead, theory refers to a well-tested body of knowledge that explains the facts that we observe in the natural world. The millions of observations biologists make of living and fossil organisms every year demonstrate the factual basis of evolution. We can observe changes over time in the extensive fossil record of the Earth, just as we can observe the process of evolution at work in living natural populations and in controlled laboratory experiments. Observing that evolution occurs, however, does not by itself explain how evolution occurs. "Evolutionary theory" refers to the body of knowledge about the processes of evolution and our models of how those processes work. For example, using mathematical modeling, a biologist can show that we expect all biological populations of a finite size to evolve by genetic drift. That does not mean, however, that other processes are not resulting in evolutionary change as well. Charles Darwin's major contribution was to argue that the process of natural selection was a major factor in the evolution of populations of living organisms over time. That idea has been tested many thousands of times by many thousands of biologists since Darwin's time, and natural selection has been repeatedly shown to be important in the evolutionary change that biologists have observed, and continue to observe, across life on Earth.

2. Although most individual bacteria would die upon exposure to an antibiotic, bacteria that could survive short-term exposure would multiply rapidly after the antibiotic treatment ended. Over time, the population of bacteria would evolve resistance to the antibiotic, as any mutations that allowed survival would increase in frequency. The full treatment is judged to be effective against virtually all the bacteria in the population. If no bacteria survive the full course of the antibiotic, then the population cannot evolve resistance. If the treatment is stopped short, then there is an increased likelihood that some of the bacteria (those with the greatest antibiotic resistance) will survive, and the population of bacteria will evolve increased antibiotic resistance.

3. Humans select traits in domestic plant and animal populations based on our interest in the trait, rather than on how it affects the natural reproductive rate or survivorship of the organisms. Many of the traits selected by humans would not be advantageous in wild populations. For example, humans have selected many cattle breeds for high body fat and high body weight. These traits result in large calves, which in turn result in calving difficulties for cows. Ranchers often have to assist in the

birth of such calves, because the calf (and likely its mother) would often die without such assistance. In a natural population, there would be selection for smaller calf size and birth weight, which would increase the successful reproductive rate and survivorship.

4. Behaviors can respond to environmental cues that are predictive of future conditions, and these behaviors can be selected for if they are under genetic control. For example, day length becomes shorter as we move closer to winter, so individual mammals have a survival advantage if they respond to shortening days by going into hibernation. In this case, the environmental cue (day length) is predictive of future environmental conditions (the cold of winter). The traits exist in the present because these associations (as between shortening day length and the approach of winter) have existed for a long time.

5. Natural selection cannot act when there is no effect on the effective reproductive rate of the organism. Diseases such as Alzheimer's usually occur long after the reproductive years have passed. As long as the disease does not affect the relative likelihood of the survival of the affected person's offspring (as a result of reduced parental care, for example), we would not expect natural selection to lead to any reduction in Alzheimer's disease in human populations.

RECAP 20.2

1. Mutations provide the genetic variation on which all the other evolutionary processes act.

2. Neutral traits (traits that confer neither an advantage nor a disadvantage) are free to increase or decrease in a populations by drift alone. In small populations, chance effects have a large role. If the difference in fitness between individuals is relatively small, and there are few individuals in a population, then which individuals survive and reproduce is likely to be a result of factors that do not relate directly to the fitness of the organism. Under these conditions, even some mildly deleterious traits are expected to become fixed in the population over short periods of time.

3. Self-fertilization reduces the frequency of heterozygotes but does not change the allele frequencies in a population. By contrast, sexual selection (nonrandom mating in preference for a particular phenotype) produces a directional change in the population. Hence, the population evolves.

4. A small population might be under strong selection for a particular trait that is favored locally. However, if there is extensive gene flow from neighboring populations where the particular trait is not favored, then selection for the trait in question will be overwhelmed by the gene flow from the surrounding populations.

RECAP 20.3

1. a. Frequency of allele *a*: 0.60; of allele *A*: 0.40;

   b. Frequency of genotype *aa*: 0.40; of genotype *Aa*: 0.40; of genotype *AA*: 0.20;

   c. Expected frequency of genotype *aa*: 0.36; of genotype *Aa*: 0.48; of genotype *AA*: 0.16

2. $2pq = 2(0.2)(0.8) = 0.32$

   Since the observed frequency of heterozygotes is lower than the frequency predicted by Hardy–Weinberg expectations, any of the following are reasonable explanations of this pattern:

   • Toads from two or more adjacent subpopulations could be breeding in the same pond. The frequencies of alleles in each population may differ, resulting in high frequencies of different alleles in each subpopulation. In other words, the assumption of no gene flow has been violated.

   • Toads may not be breeding randomly within the population. For example, if closely related toads are more likely to mate with each other than with distantly related toads, then fewer heterozygotes would be expected in population (a form of inbreeding).

   • Heterozygous individuals may be at a disadvantage in the population (e.g., fewer heterozygotes may survive to adulthood). This would represent a violation of the Hardy–Weinberg assumption of no selection.

3. Presumably, the observed population is large enough to calculate that the low frequency of heterozygotes is significant. However, if the present generation of toads was produced by a small number of parents in the previous generation, then random effects of a small population size could explain the observed pattern.

RECAP 20.4

1. Stabilizing selection results in a reduction in variation within the population and an increase in frequency of the modal phenotype. Directional selection results in a change in the modal phenotype of the population in one direction. Disruptive selection results in a population with a bimodal distribution of phenotypes.

2. As shown in Figure 20.13, mortality increases with both smaller and larger birth weight compared to the optimal birth weight of about 7.5 pounds. Significantly smaller babies are more likely to be premature or undernourished, and thus less likely to survive. Significantly larger babies may cause difficulties in delivery, putting both the baby and mother at greater risk and stress during childbirth. Thus, babies close to the modal birth weight are most likely to survive and grow into adults, thereby increasing the relative frequency of alleles for intermediate birth size in the population.

3. Almost any phenotype of an organism that differs markedly from its close relatives is likely the result of directional selection. Among the many thousands of possible

examples, consider the long necks of giraffes, the long trucks of elephants, the great body size of whales, the large brain of humans, the great height of giant sequoias, and the large floating leaves of water lilies. Directional selection obviously can result in reduction in size as well, so also consider the tiny body size of hummingbirds compared to other birds, or the tiny leaf size of duck weed compared to other flowering plants. Undoubtedly, you can think of many other examples.

RECAP 20.5

1. The heterozygotes between the sickle-cell and normal alleles may have an advantage in defense against malaria. The normal allele may be able to function normally, whereas the sickle-cell allele provides protection against the malaria parasite.

2. A large population of humans could be tested for infection with malaria, and the frequency of infection, and the consequences of the infection, could be compared in people who do versus do not carry the sickle cell gene as heterozygotes. If the heterozygous individuals have an advantage in malaria resistance, we would expect them to exhibit lower infection rates or reduced effects of infection.

3. High genetic variation leads to more opportunities for the presence of beneficial alleles or traits. If the environment suddenly changes, and beneficial traits are already present in the population, then selection can increase the frequency of those traits and the population can rapidly evolve. However, high genetic variation does not guarantee that appropriate beneficial alleles will already be present in the population.

RECAP 20.6

1. Insects may be physiologically constrained by their system of respiration, which may not be able to support as large a body as that of birds. This is an example of an historical constraint that limits insect body size.

2. Anything that resulted in sudden, widespread environmental change would be likely to affect many species simultaneously. Examples include a meteorite or comet impact, widespread volcanic activity, or sudden climatic changes.

WORK WITH THE DATA, P. 431

1. Four groups of moths show clear indications of long wing tails, and each of these groups is closely related to moths that lack wing tails. Therefore the trait must have evolved in parallel at least four times. The long wing tails are present in 11 species on the tree, but some of these species are closely related to one another, and so the trait could have evolved in the common ancestor of each closely related species cluster. Therefore the trait appears to have originated about four times independently.

2. Although the long wing tails appear to have evolved four times independently, there is evidence that directional selection continued in many species as the trait evolved. For example, notice the parallel changes from green to yellow to red (indicating increasing length of wing tails) in many of the species.

WORK WITH THE DATA, P. 441

1. For *C. philodice*, 43.2% of all viable males are heterozygous, so 56.8% must be homozygous. To get expected numbers of heterozygous and homozygous mating males, we multiple the expected proportions (from all viable males) by the total number of mating males sampled. Therefore we expect to see (0.432)(50) = 21.6 heterozygous mating males, and (0.568)(50) = 28.4 homozygous mating males. If we repeat the same calculations for *C. eurytheme*, we expect (under the given assumption) to see (0.478)(59) = 28.2 heterozygous mating males and (0.522)(59) = 30.8 homozygous mating males.

2. Chi-square calculations for *C. philodice*:

Genotype	Expected (E)	Observed (O)	O − E	(O − E)²	(O − E)²/E
Heterozygotes	21.6	31	9.4	88.36	4.091
Homozygotes	28.4	19	−9.4	88.36	3.111

The sum of the last column gives the chi-square test statistic: 7.202. Since this value is greater than the critical value ($P = 0.05$) of 3.841, the observed results are significantly different from the expectations at $P < 0.05$. In other words, we can reject the null hypothesis and conclude that the proportions of each genotype (heterozygotes and homozygotes) of mating males are significantly different from the proportions of these genotypes seen among all viable males in *C. philodice*.

Chi-square calculations for *C. eurytheme*:

Genotype	Expected (E)	Observed (O)	O − E	(O − E)²	(O − E)²/E
Heterozygotes	28.2	45	16.8	282.17	10.005
Homozygotes	30.8	14	−16.8	282.17	9.162

The sum of the last column gives the chi-square test statistic: 19.167. Since this value is greater than the critical value ($P = 0.05$) of 3.841, the observed results are significantly different from the expectations at $P < 0.05$. In other words, we can reject the null hypothesis and conclude that the proportions of each genotype (heterozygotes and homozygotes) of mating males are significantly different from the proportions of these genotypes seen among all viable males in *C. eurytheme*.

3. $(0.75)^{16}$ is approximately 0.01002. Therefore the investigators would need to analyze at least 16 larvae from each batch of eggs to judge the genotype of the father with 99 percent certainty. An easy way to find this answer is to multiply 0.75 × 0.75, and then multiply the answer by 0.75, and continue until the result is approximately 0.01. Keep track of the number of times you multiply by 0.75 to find the appropriate sample size.

FIGURE QUESTIONS

Figure 20.6 Over time, sexual reproduction led to new combinations of the existing genes. Many different genes affect bristle number, so the selection for high (and low) bristle numbers led to selection for new genetic combinations that produced high (or low) bristle numbers. In addition, there was selection for any new mutations at any genes that led to high (or low) bristle numbers.

Figure 20.12 If the mean value of a trait in a population is below the optimal value for that trait, then the mean will evolve to be closer to the optimum through directional selection. But, as the population mean reaches the optimum, then the population will experience stabilizing selection (with selection against individuals that are both below and above the population mean).

Figure 20.20 Although TTX resistance is an advantage within the range of the toxic newt, the alleles that confer TTX resistance also lead to slower movement by the snakes (a disadvantage). There is therefore a trade-off in the benefits versus the cost of TTX resistance. In the range of the newt, the advantage outweighs the cost; outside the range of the newt, the cost outweighs the benefit. Thus there is selection for TTX resistance where the snakes occur with the newt, but selection against TTX resistance elsewhere.

APPLY WHAT YOU'VE LEARNED

1. *A. carolinensis* from islands with introduced *A. sagrei* have significantly larger toepads with more lamellae compared to lizards from islands without *A. sagrei*. As the *A. sagrei* were only introduced to the islands in 1995, these differences in foot structure appear to have arisen since that time. The fact that the toepads have evolved so quickly indicates that there is strong selection for larger toepads with more lamellae in lizard populations on the invaded islands.

2. The common garden experiment confirms that the observed differences have a genetic basis, and are not due to different expression of the same genes on the two sets of islands. If the lizards raised in the common garden experiment had not shown the same level of differences that were observed in the wild populations, then the observed changes could not be attributed to evolution, which refers to genetic changes in populations over time.

3. The most important evolutionary process in this example is selection. Given that tree-top *Anolis* typically have larger toepads compared to ground-dwelling species, it is logical that lizards with larger toepads and more lamellae are more likely to survive and reproduce in tree tops (compared to lizards with smaller toepads and fewer lamellae). By living longer, lizards with larger toepads will produce more offspring over time, and so their genes will be increasingly represented in subsequent generations.

4. In this case, individuals of *A. carolinensis* with smaller toepads and fewer lamellae would have an advantage on the ground and on low perches on the invaded islands. Therefore, we would predict that the evolutionary change would occur in the opposite direction; the average toepads would become smaller on the invaded islands compared to un-invaded islands.
   - Selection for larger toepads appears to be occurring.
   - Mutation is certainly occurring in the populations (as it does in all species), although it likely has a very small effect at this time scale.
   - The populations on each island are likely to be fairly small, so drift is occurring.
   - Gene flow among populations is unknown, but it is likely to be low, since new invasions of islands appear to be rare.
   - Non-random mating among the lizards may be occurring, although there is no evidence of this process described in the experiment.

5. The other four processes of evolution (mutation, drift, gene flow, and nonrandom mating) are likely occurring as well, although they are less important to this example:
   - The variation in toepad size and number of lamellae would not exist without mutations in the genes that produce these structures, so mutation was critical for introducing genetic variation into the populations. But mutations occur very slowly, so very few new mutations that affect toepad size would be expected over 15 years.
   - Genetic drift is certainly occurring on the islands, because the populations on each island are limited. However, drift would not produce a consistent directional effect in toepad size across islands, and so it cannot account for the consistently larger toepads in *A. carolinensis* on invaded islands.
   - There was no attempt to measure gene flow among the islands in this experiment. But the fact that *A. sagrei* only occurred on islands where it was introduced suggests that inter-island movement of lizards is low and not likely a major factor in the study.
   - Lizards might choose mates based on their toepad size (non-random mating), which would affect the distribution of toepad size in the population, producing more variation on which selection could then act. However, there is no evidence for this process in the described experiment.

Chapter 21

RECAP 21.1

1. Phylogenetic trees can represent any process in which biological lineages diverge, such as speciation (in which case the tree will depict the evolutionary relationships among species), viral replication (in which case the tree will depict the evolutionary relationships among different viral lineages), or gene duplication (in which case the tree will depict the evolutionary relationships among genes). In most cases in this book, phylogenetic trees are used to depict the evolutionary relationships among species or higher groups of organisms.

2. Because homologous characters are similar as a result of their common descent. Similarities that result from convergent evolution (the wings of birds and insects, for example) can be misleading about evolutionary relationships if they are mistaken as homologous characters.

3. Selection for similar environmental conditions often leads to convergence in traits. For example, fish and dolphins both have fins and are similar in body shape because there is strong selection for these traits in an aquatic environment. But these traits have evolved independently in the two groups. Biologists can usually detect such homoplasies because they conflict with a large number of homologous traits in the groups that are similar as a result of their recent shared ancestry. Biologists can minimize homoplasies using the principle of parsimony.

RECAP 21.2

1. The pine tree and the sunflower.
2. The fern.
3. True roots and vascular cells.
4. No, because each trait can be placed along a single branch in the tree.

RECAP 21.3

1. The West Nile Virus in the United States appears to be most closely related to a strain of the virus isolated in Israel. A reasonable hypothesis is that the virus emerged in Africa in the 1930s and subsequently moved into Asia and Europe, probably multiple times. In the late 1990s, a strain of the virus from Israel appears to have been transported to New York, perhaps carried by mosquitoes on an airplane or in a cargo shipment. Once in the United States, the virus spread quickly in native bird populations across North America.

2. The rate of divergence is represented by the slope of the regression line. Since the regression line goes approximately through the origin (0, 0), the slope is approximately 7.5% change in *cyt b* per 4 my, or 1.875% change per million years.

RECAP 21.4

1. Classifications One and Three (the group Reptilia is paraphyletic if the birds are excluded).
2. Classification Three (the group Homotheria is polyphyletic).
3. Classification Two.

WORK WITH THE DATA, P. 457

1.

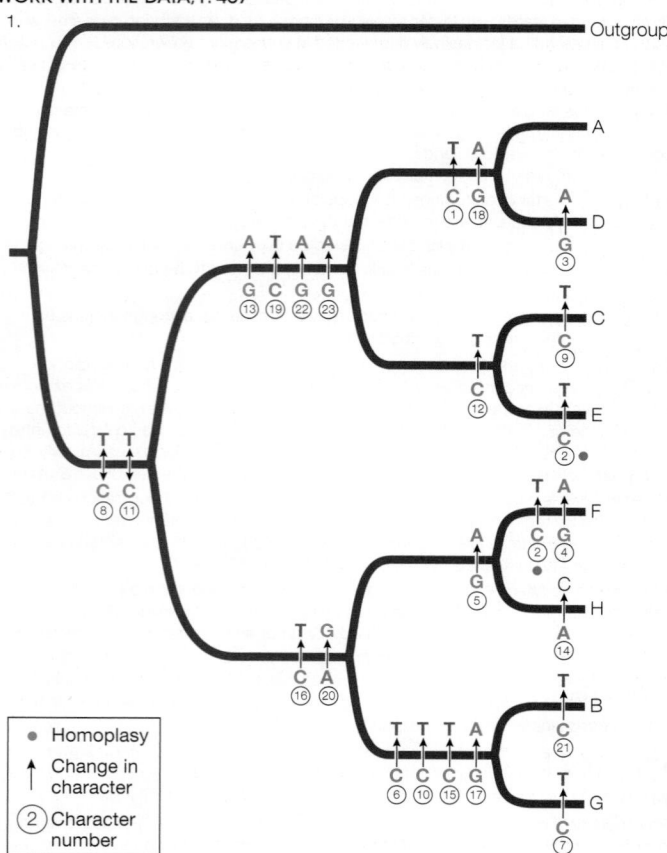

- ● Homoplasy
- ↑ Change in character
- ② Character number

2. Ancestor of A and D: TCGGGCCCCCCCCAACCGATACAA
   Ancestor of C and E: CCGGGCCCCCCTAACCGGTACAA
   Ancestor of F and H: CCGGACCCCCCCGACTGGCGCGG
   Ancestor of B and G: CCGGGTCCCTCCGATTAGCGCGG
   Ancestor of A, C, D, and E: CCGGGCCCCCCCAACCGGTACAA
   Ancestor of B, F, G, and H: CCGGGCCCCCCCGACTGGCGCGG
   Ancestor of A, B, C, D, E, F, G, and H: CCGGGCCCCCCCGACCGGCACGG

3. The reconstructed phylogeny requires 23 transitions and only 1 transversion. The viruses were grown in the presence of a mutagen, and the mutagen used in this experiment results predominantly in transitions.

FIGURE QUESTIONS

Figure 21.8 If the HIV sequences from the victim had been found to be more closely related to those of any other HIV-positive individuals in the local community than they were to those of the physician's patient, then the phylogeny would be inconsistent with the alleged crime, and the results could have been used to exonerate the physician.

Figure 21.12 The expected branch length would be about 0.09 substitutions per nucleotide.

APPLY WHAT YOU'VE LEARNED

1. The individual labeled CC01 is consistent with being the source of the infection, because that is the only individual with HIV sequences that are more closely related to HIV sequences in all of the other individuals, rather than to sequences sampled only within a single individual. Viruses from CC01 are found across the tree, which is consistent with CC01 transmitting viruses over time to the other individuals. (In the trial, CC01 was revealed to be the defendant in the case).

2. The viral sequences sampled from each of the other individuals in the cluster are monophyletic, indicating a single origin, with no further transmission to other individuals. Thus, the tree is inconsistent with CC02–CC08 being the original source of infection in this cluster.

3. The outgroup was needed to root the tree, which in this case identifies the direction of the transmission of events. Inclusion of viruses from outside the epidemiological cluster also tests the hypothesis that all viral sequences in this cluster are derived from a single ancestral virus.

Chapter 22

RECAP 22.1

1. Biologists are interested in many aspects of species, and their emphases differ depending on the questions they are asking about species. Biologists who are interested in the processes that led to lineage-splitting emphasize the mechanisms that allow lineages to diverge and to remain isolated from one another after divergence (i.e., how species come into existence and remain distinct from one another). Biologists who are mostly concerned with identifying species and understanding their distribution in time and space are more likely to emphasize historical aspects and how we can tell species apart (i.e., how we recognize species in nature). Still other biologists may be interested primarily in how species co-exist in communities, and these biologists are likely to emphasize the ecological role of species (how do species differ in their adaptations and niches?). Each of these emphases leads different biologists to emphasize different attributes of species.

2. Sexual lineages must have a substantial degree of reproductive isolation or they cannot remain distinct from one another. So even if a biologist is primarily using a morphological species concept, the morphological differences between species cannot be maintained in time and space unless the two species are largely reproductively isolated from each other. Obviously, reproductive isolation is central to the biological species concept. The many versions of lineage species concepts conceptualize species as distinct lineages in the tree of life. If these lineages are to remain distinct from one another through time, they must be substantially reproductively isolated from one another. Thus, reproductive isolation is important (either directly or indirectly) to all of the various species concepts.

3. The biological species concept limits its scope to "actually or potentially interbreeding natural populations." Clearly, asexual organisms do not form such populations. Nonetheless, they do form lineages that exist in time and space, and these lineages may be maintained by other factors, such as available niche space. Most biologists who emphasize the biological species concept are interested primarily in the mechanisms that lead to reproductive isolation and its maintenance over time. The lineage concepts of species are not in conflict with the biological species concept, but they represent a more inclusive way to think about all species, regardless of their sexuality. The lineage concepts also allow biologists to study species across geological time (where reproductive interactions are not possible).

RECAP 22.2

1. If the only difference between the diverging lineages is at a single locus, then both of the new alleles must be functional when they interact with the products of other gene loci (in both lineages). Any genetic incompatibility produced by these new alleles would be expected to affect the parental lineages as well. There are much greater numbers of possible incompatibilities across different gene loci. Rather than two deleterious changes at the same locus (one in each lineage), the

Dobzhansky–Muller model allows neutral changes at any pair of loci whose products interact. It is the negative interaction of these products in the hybrid between the two lineages that results in genetic incompatibility.

2. Segregation doesn't occur normally, because chromosome arms are linked differently on the different species, resulting in duplications and deletions in meiosis. If two different fusions of chromosomes occur in two different lineages, then the resulting chromosomes cannot pair normally in meiosis in the hybrids. If you attempt to diagram meiosis in the hybrid that would result from a cross of the divergent lineages in Figure 22.4 (see below), you will see that homologous pairings require parts of different chromosomes to align with one another. These chromosomes will then be pulled in two different directions as the cell divides in meiosis I, resulting either in a likely failure of the cell to divide or in an uneven distribution of the chromosome arms in the two daughter cells. Production of normal cells with an even distribution of the various chromosomes' arms is impossible, so the hybrid organisms will not produce functional gametes.

3. A likely possibility is that the incompatible alleles have not yet become fixed in the various strains, so only some combinations of crosses result in genetic incompatibility.

RECAP 22.3

1. Polyploids (e.g., tetraploids) can be produced in the first generation. Hybrids between diploids and tetraploids may be produced in the second generation, forming triploid individuals, but meiosis in the triploids results in gametes with unbalanced chromosome numbers, which results in sterile offspring. An F_1 tetraploid is able to self-fertilize or mate with another tetraploid and produce fertile offspring. Thus, the polyploidy organism is reproductively isolated from the diploid species, making it a different species.

2. Species that arise in allopatry initially occur in separate, but usually adjacent, ranges (e.g., see Figure 22.6). Therefore, we would expect many closely related species to exhibit this same pattern. The ranges of highly mobile species are more likely to change over time, so the pattern should be strongest among relatively sedentary species. Biologists have confirmed these predictions across many diverse groups of organisms.

3. a.

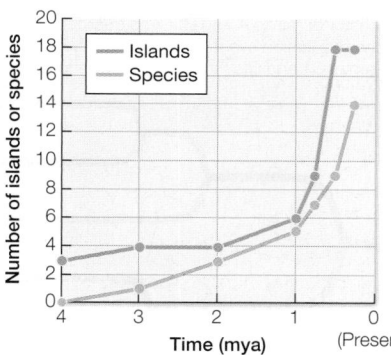

b. Yes, because the curve for the number of species lags behind the curve for the number of islands; but the two curves exhibit very similar changes in slope through time. As new islands arise, new opportunities for speciation also arise. The number of species at any one time is always just below the number of distinct islands.

c. There are currently 18 islands in the archipelago and only 14 species (as recognized by some biologists). This suggests that there are still opportunities for additional speciation by geographic isolation. Based on our graph from question 3a, we expect populations of species that occur on two or more islands to diverge into distinct species over time. To test this hypothesis, we could collect samples of each population and examine genetic divergence between the samples. Significant genetic divergence between the populations on different islands suggests that the distance between the islands is a significant barrier to gene flow, so the populations are expected to diverge into distinct species over time (see Key Concept 22.2). Indeed, as we noted in the text discussing Darwin's finches, some biologists recognize the genetically divergent (but morphologically similar) species on different islands as distinct species, which is why they recognize as many as 18 different species of Darwin's finches.)

RECAP 22.4

1. a.

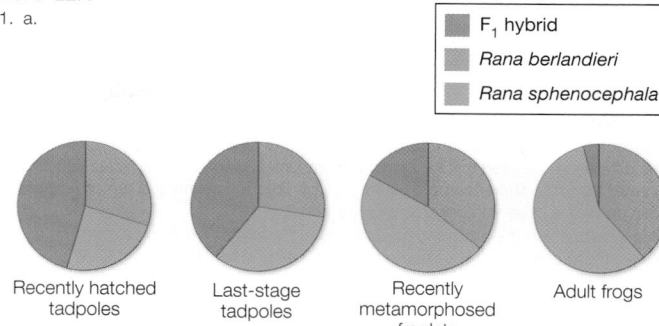

Recently hatched tadpoles   Last-stage tadpoles   Recently metamorphosed froglets   Adult frogs

d. Hybrids represent a smaller proportion of the samples at each successive life stage, indicating lower hybrid survivorship compared to either parental type. Postzygotic isolating mechanisms that are consistent with this pattern include low hybrid zygote viability and (for the adult frogs) low hybrid adult viability.

e. Selection against hybridization would favor non-overlapping breeding seasons between the two species. Over time, more individuals that result from offset breeding seasons would be expected to survive (since no hybrids are produced in these cases). Assuming there is a genetic component that influences when a frog will breed, the populations would be expected to shift to produce non-overlapping breeding seasons (as seen in allopatric versus sympatric populations of these two species; see Figure 22.11). As overlap between breeding seasons is reduced, we would expect to see fewer hybrids at every life stage.

2. There are many possible designs of experiments that might prove informative. Here is an example of one that would examine the effect of flower position on pollinator attraction: Take one species of flower and divide the flowers into two groups. Position each flower to be either upright or pendant, then record the number and type of pollinators that are attracted to flowers in each group. Test to see if the differences between the two groups are statistically significant.

3. There is gene flow into the hybrid zone, but the hybrid zone acts as a "genetic sink." The hybrids are less fit compared to either parental type, so there is little gene flow out of the hybrid zone. An expanding hybrid zone would require hybrids to move out of the hybrid zone and compete successfully against individuals of each of the two parental species.

RECAP 22.5

1. If a pollinator specializes on one particular flower type, that will reduce opportunities for cross-fertilization among different flower types. Any polymorphism in flowers can thus lead to rapid genetic divergence between individuals that express each flower type. A similar process occurs with sexual selection in animals. If some females prefer to mate with males that have a certain trait and other females prefer a different male trait, then the population can quickly lead to genetic isolation and differentiation in the two groups.

2. A new immigrant to an isolated area may encounter no competition for a particular feeding strategy or life style. Under these conditions, there are many "open niches" that can be co-opted by the invading species, leading to a rapid increase in speciation.

WORK WITH THE DATA, P 479

1. For red flowers, the proportion of seeds from hybrid matings is 0.1298 (27/208). The 95% confidence limits of this proportion are approximately 0.09–0.18.

   For pink flowers, the proportion of seeds from hybrid matings is 0.3813 (53/139). The 95% confidence limits of this proportion are approximately 0.30–0.47.

2. The null hypothesis is that there is no significant difference in the proportion of hybrid seeds produced from red versus pink flowers. This hypothesis is rejected; the probability that the null hypothesis is true is $P < 0.0002$.

3. Scientific studies often raise new questions that can be addressed by follow-up experiments. Repeating this study in several different sites of sympatry and allopatry would address the unlikely possibility that red flowers are favored at the study site because of local conditions that have nothing to do with the presence of *P. cuspidata* (such as the local abundance of pollinators that prefer red flowers). A possible control experiment would involve removing all *P. cuspidata* from an area of sympatry and testing to see if the absence of *P. cuspidata* eliminates the selective advantage of red-flowered *P. drummondii*.

FIGURE QUESTIONS

Figure 22.5 Genetically distant pairs of species almost always develop Dobzhansky–Muller incompatibilities, because there are many more divergent genes that are likely to function poorly together. But Dobzhansky–Muller incompatibilities can develop with divergence at just two genes, so some otherwise genetically similar pairs of species can become reproductively isolated very quickly.

Figure 22.6 The glaciers eliminated most of the highlands that formerly connected the two areas, so there is now little appropriate habitat that would allow the differentiated species to interact. But if the interactions were possible, it is likely that the hybrids would exhibit reduced fitness (as explained by the Dobzhansky–Muller model), and there would be selection for prezygotic isolating mechanisms that would minimize hybridization.

Figure 22.11 Overlap in breeding seasons in areas of sympatry results in increased hybridization. The hybrids have reduced fitness, so there is selection in favor of frogs that breed at times when hybridization is less likely.

APPLY WHAT YOU'VE LEARNED

1. Hydrocarbon 2 shows a pattern consistent with reinforcement. Under reinforcement, the difference between the sympatric populations and the other species should be greater than the difference between the allopatric populations and the other species. This pattern is observed only for hydrocarbon 2.

2. Hydrocarbon 2 in the allopatric populations of *D. serrata* would be expected to evolve to be more like the sympatric populations and less like *D. birchii*. Thus, its concentration would be expected to decrease in the lab populations. The concentrations of the other hydrocarbons would be expected to remain about the same, as they do not appear to be undergoing reinforcement.

3. The frequency of hybridization should decrease. The artificial selection is operating on the hydrocarbon because hybridization with the other species is disfavored. As a result, fewer hybridizations should take place.

4. The level of postzygotic reproductive isolation should not change, as postzygotic isolation generally does not evolve via direct selection.

Chapter 23

RECAP 23.1

1.

Sequence *a* AATGCAGGGTATA—CG
Sequence *b* ATT—CAGGGTATA—CC
Sequence *c* ATTGCAGCGTATAACC
Sequence *d* ATTGCAGGGTATA—CG

Sequence

		a	*b*	*c*	*d*
Sequence	*a*		3	4	1
	b	12		3	2
	c	12	12		3
	d	14	13	13	

2. A simple count of nucleotide differences misses multiple substitutions, including coincident substitutions, parallel substitutions, and back substitutions (see Figure 23.2). Evolutionary models can be used to estimate the number of multiple substitutions that have occurred, based on the number of actual differences that are observed. For example, the second nucleotide of Sequences *a* and *b* from Question 1 exhibits a single difference between an A and T. Although a single substitution from a T to an A in Sequence *a* could explain this difference, it is also possible that there were two substitutions (e.g., first a substitution from T to C, then another from C to A). A simple count of differences misses such multiple substitutions.

RECAP 23.2

1. This problem can be investigated by sequencing and comparing the genes for opsins in surface-dwelling (eyed) and cave-dwelling (eyeless) crayfish. If the genes of the eyeless species are no longer under any selection, we would expect to observe a similar rate of synonymous and nonsynonymous substitutions in the genes. If there has been strong selection for a new function (something other than vision), we would expect a higher rate of nonsynonymous substitutions compared with synonymous substitutions (indicating positive selection). We would compare the rates observed in the eyeless species to the rates seen in the surface-dwelling (eyed) species. In the surface-dwelling species, we would expect to see a higher rate of synonymous compared with nonsynonymous substitutions, which is expected under purifying selection.

2. A given neutral mutation will arise more often in a large population than in a small population, but any mutation that does arise is more likely to be fixed in a small population than in a large population. These two influences of population size exactly cancel each other out, so that overall, the rate of fixation of neutral mutations depends only on the mutation rate and is independent of population size. In a diploid population of size *N* and a neutral mutation rate μ per gamete per generation at a locus, the number of new mutations would be, on average, $2N\mu$, because $2N$ gene copies are available to mutate. The probability that a given mutation will be fixed by drift alone is its frequency, which equals $1/(2N)$ for a newly arisen mutation. We can multiply these two terms to get the rate of fixation of neutral mutations (*m*) in a given population of *N* diploid individuals:

$$m = 2N\mu \frac{1}{2N}$$

But we can simplify this equation by canceling out $2N$ on the right side of the equation:

$$m = \cancel{2N}\mu \frac{1}{\cancel{2N}}$$

This just leaves $m = \mu$. In words, the rate of fixation of neutral alleles is equal to the mutation rate of neutral alleles and independent of population size.

3. a. Codon numbers 12, 15, and 61 are likely to be evolving under positive selection for change because each of these codons has experienced a higher rate of nonsynonymous substitutions (which give rise to amino acid replacements) than synonymous substitutions.

   b. Codon numbers 80, 137, 156, and 226 are likely evolving under purifying selection, as the vast majority of changes at these codons are synonymous substitutions, which do not result in amino acid replacements. Substitutions that result in amino acid changes (nonsynonymous substitutions) undoubtedly occur but are usually selected against in the population. Codon number 165 has experienced similar numbers of synonymous and nonsynonymous substitutions. There are approximately three times as many possible nonsynonymous substitutions as there are synonymous substitutions. Therefore the number of synonymous substitutions is slightly higher than expected if the rates of each type of substitution are equal. Codon 165 may be evolving under weak purifying selection; it is the codon that is closest to neutral among the codons shown in the table.

4. Example hypothesis 1: Genome size varies primarily because noncoding DNA can effect changes in gene expression, and many species have evolved largely through changes in gene expression.

   Example hypothesis 2: Genome size varies primarily as a function of population size. Species with small population sizes may accumulate large amounts of noncoding DNA because of weak selection against accumulation of slightly deleterious "junk" DNA in small populations.

   Contrasting the two hypotheses: If hypothesis 1 explains most of the variation in genome size among organisms, then we would expect genome size to be largely independent of population size, which is not the case. For example, Figure 23.9 indicates that the species with the largest genomes, like the lungfish, generally have much smaller population sizes than the species with the smallest genomes, like the *E. coli* or yeast. However, both hypotheses may explain some of the variation in genome size, and careful, controlled experiments would be needed to determine how much of the variation in genome size is explained by the effects of selection (hypothesis 1) versus population size (hypothesis 2).

RECAP 23.3

1. New genes can add new functions. For example, lateral transfer of a gene that confers antibiotic resistance would provide a huge advantage to bacteria that are subjected to antibiotics.

2. The tree indicates that three gene duplication events and one gene loss occurred. In the tree below, the three gene duplication events are marked in blue, and the gene loss is marked in red. Gene duplication event 1 marked the duplication between gene C and the ancestor of genes A and B. Gene duplication event 2 resulted in genes A and B. Gene duplication event 3 happened in the ancestor of humans and resulted in human genes C1 and C2. Finally, there was a loss of gene B in gorillas (event 4 on the tree).

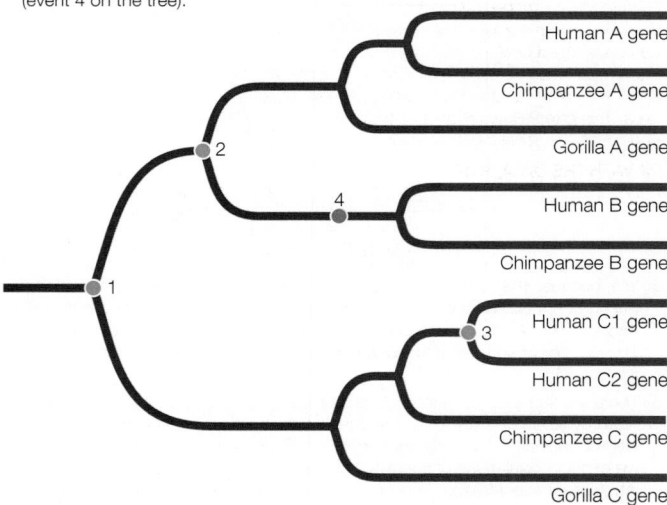

Human A gene
Chimpanzee A gene
Gorilla A gene
Human B gene
Chimpanzee B gene
Human C1 gene
Human C2 gene
Chimpanzee C gene
Gorilla C gene

3. Gene duplication provides the raw material for much of evolution. A duplicated gene is free to diverge from its ancestral function, since the original copy of the gene can continue to provide the original function.

4. Concerted evolution refers to highly repeated gene families where all the copies evolve together so that all the copies maintain a high degree of similarity. Two processes can produce concerted evolution: unequal crossing over and biased gene conversion. The two processes differ as follows:

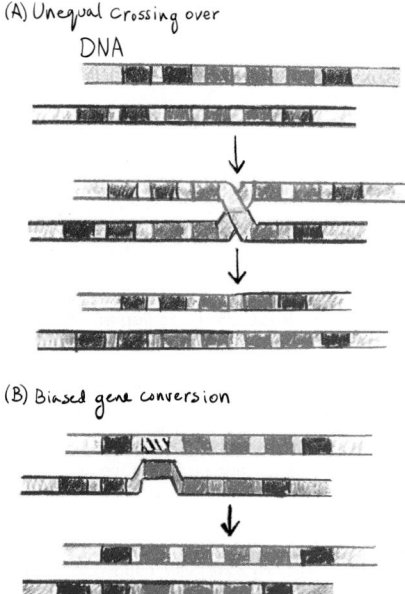

RECAP 23.4

1. Comparisons of evolutionary changes in proteins, made across species in which these proteins differ in function, provide insight into how change in protein sequence relates to change in protein function. For example, changes across many different genes in many different species of puffer fish show which changes are related to the evolution of tetrodotoxin resistance in sodium channel genes. These studies help us understand the function of sodium channels, which helps us understand the genetic basis of various neurological pathologies.

2. Starting with a large pool of random sequences, in vitro evolution involves rounds of selection and mutation to produce new molecules with specific functions. This corresponds exactly to the processes of mutation and natural selection in natural populations. The only difference is that the selection and mutational processes are controlled by humans in a laboratory setting, to produce a molecule with a desired function.

3. Most emerging diseases are caused by living, evolving organisms. Evolutionary comparisons of DNA sequences from these pathogens allow quick identification and often reveal the source of the disease (from a particular host species or geographical area). Previously unidentified pathogens can be compared with their closest relatives. This often allows biologists to make predictions about the best strategies for controlling the spread of the disease.

WORK WITH THE DATA, P. 494

1.

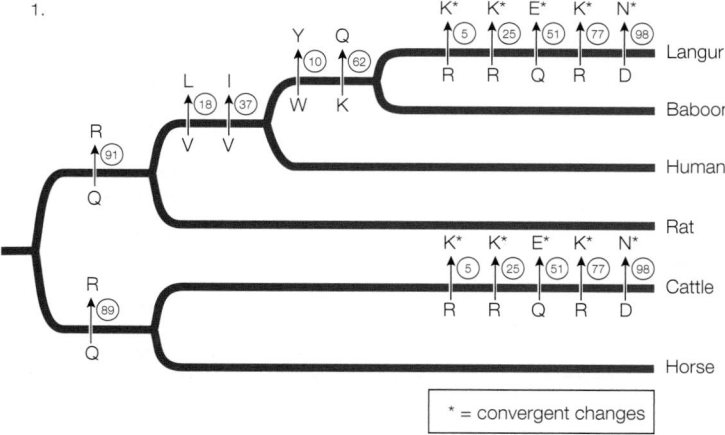

*= convergent changes

Positions 14, 21, 50, 75, and 87 show convergence in amino acids between cattle and langur.

2. Position 17 is convergent between cattle and the ancestor of langurs and baboons.

3. No, there are no other convergent changes. The fact that almost all convergent events occur between cattle and langurs supports the hypothesis that these convergent changes are related to the convergence in function associated with a shift to foregut fermentation.

WORK WITH THE DATA, P. 503

1. Birds appear to be the original source of influenza viruses that now cause flu in the various mammal species (horses, pigs, and humans). The bird lineages extend to the base of the tree, and appear to be closely related to each lineage of influenza virus that occurs in a mammal.

2. The tree shows five cross-species transmissions: two from birds to horses, two from birds to pigs, and one from birds to humans. The human flu viruses are most closely related to viruses in pigs (hence the common name "swine flu"), and it is possible that the virus was transferred from birds to pigs and then from pigs to humans in a short period of time late in the second decade of the 1900s.

3. In 1918, a new strain of influenza had just entered human populations (as well as pig populations). Humans' immune systems would not have had any prior experience with this new strain of influenza, and so would not have been prepared for an effective defense.

4. Earlier flu cases were likely caused by other cross-species transfers of different strains of influenza into human populations. Those flu strains appear to have been eliminated from human populations, and so were not available for sampling in this study (and therefore do not appear on the tree).

FIGURE QUESTIONS

Figure 23.10 Approximately 100 million years ago.

Figure 23.12 Two gene duplication events were restricted to the zebrafish lineage: the event that gave rise to the zebrafish *En1a* and *En1b* genes, and the event that gave rise to the zebrafish *En2a* and *En2b* genes.

APPLY WHAT YOU'VE LEARNED

1.

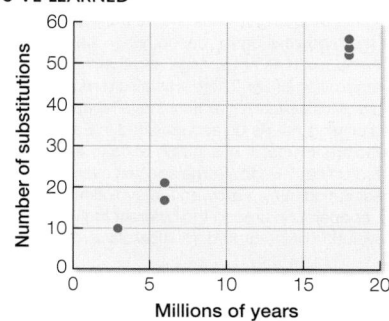

There are, on average, 54 synonymous substitutions between lineages that diverged 18 million years ago (36 million years of evolution altogether). Thus, there have been about 1.5 synonymous substitutions per million years for the entire exon. Given five generations per year and 600 synonymous sites, the rate is 3×10^{-7} synonymous substitutions per generation for the entire exon and 5×10^{-10} synonymous substitutions per site.

2. The per-site synonymous substitution rate is 5×10^{-10} (from Answer 1 above). If all synonymous substitutions are neutral, the substitution rate should equal the mutation rate.

3. Based on 3×10^{-7} synonymous substitutions per generation for the entire exon, 270 synonymous substitutions are expected (90,000,000 years × 2 lineages × 5 generations/year × 0.0000003 synonymous substitutions/generation). This is much higher than the 174 observed synonymous differences between these two species. A likely explanation for the discrepancy is that not all of the actual substitutions are observed, because of multiple, coincident, parallel, and back substitutions.

4. There are, on average, 18 nonsynonymous substitutions between *D. ananassae* and the other three species. Given 36 million years of evolution, 2000 nonsynonymous sites, and five generations per year, the per-generation, per-site nonsynonymous substitution rate is 5×10^{-11}. This is one-tenth the synonymous rate of 5×10^{-10} calculated in Question 1. This much-lower nonsynonymous rate suggests strong purifying selection.

5. The biologist is probably incorrect, and the gene is likely to still be a functional gene and not a pseudogene. Because they have lost function, pseudogenes are unlikely to be affected by selection. Thus, nonsynonymous sites of pseudogenes should have substitution rates about the same as those of synonymous sites, which is not what is observed.

Chapter 24

RECAP 24.1

1.

2.

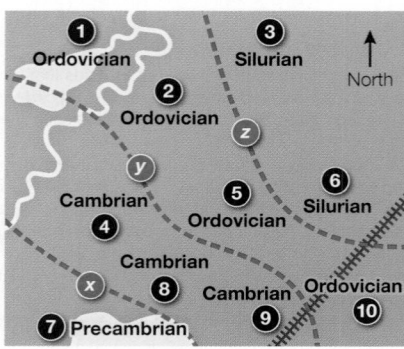

3. Site *z* falls near the predicted border between the Ordovician and Silurian periods on the map shown in the answer to Question 2. Therefore it would be the best site to examine to refine the border between these two periods.

RECAP 24.2

1. Climate refers to long-term average expectations of weather at a given location; weather refers to the daily events. Even if the long-term expectations of climate are of gradual warming, we still expect occasional winter blizzards at a given location away from the tropics. A long-term increase in the average temperature of 1°C represents a major climate change, but we would still expect a wide range of temperature conditions throughout the year at a given location.

2. Several major environmental changes occurred near the end of the Precambrian. This was the end of the "snowball Earth" period, when life was likely confined to a few relatively warm areas, such as hot springs and thermal vents. Warming of Earth resulted in rapid diversification of life. There was also a rapid accumulation of oxygen in the atmosphere. Multicellular organisms have higher metabolic requirements than do single-celled organisms. As O_2 accumulated in Earth's atmosphere from the earliest photosynthetic organisms, it was initially toxic to the single-celled, anaerobic organisms that existed on early Earth. Some species evolved to tolerate and then use O_2 for aerobic respiration, which allowed the evolution of large colonies of organisms. Eventually, cooperation among the different individuals in the multicellular colonies led to the evolution of integrated, multicellular organisms.

RECAP 24.3

1. A major factor was the evolution of two genes in legumes that allowed these plants to use atmospheric nitrogen directly by forming symbioses with a few species of nitrogen-fixing bacteria. This change dramatically increased the amount of nitrogen available for terrestrial plant growth.

2. The Archean and Proterozoic eons lasted about 1.3 billion and 2.0 billion years, respectively, compared with the 0.542 billion years of the Phanerozoic eon. Yet most of the tree depicts events in the much shorter Phanerozoic eon, since most of the major groups of life shown on the tree diverged during this time. If the tree depicted time proportionally, the branches at the base of the tree would need to be much longer. However, if this change were made, the divergence of the eukaryotes would be so compressed that their relationships could not be shown clearly on one page.

WORK WITH THE DATA, P. 516

1. Since $W = S(1 + R)^N$, then $W = (1 \text{ g})(1.0001)^{50,000} = 148.38$ grams.

2. The Permian lasted 48 million years (299 mya to 251.0 mya); therefore 50,000 years represents about one-thousandth of this time, or 0.1 percent.

3. Given that even a very modest increase in body size of 0.01 percent per generation is enough to result in an increase in body mass from 1 gram to more than 148 grams in just 50,000 years; that 50,000 years represents only about one-thousandth of the Permian; and that the experiment shows that increases in atmospheric oxygen can result in much more rapid evolution of insect body size than calculated above, it seems reasonable to conclude that the increased oxygen concentrations during the Permian were sufficient to account for the evolution of giant dragonflies.

WORK WITH THE DATA P. 517

1. First, plot the data from the table:

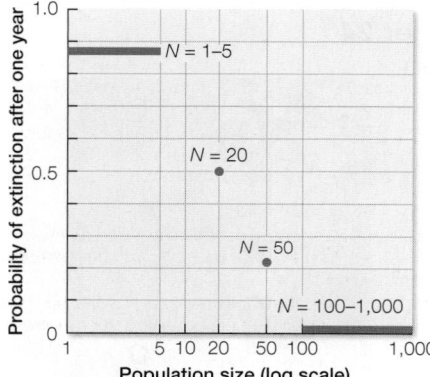

Then approximate an extinction curve from the plotted points:

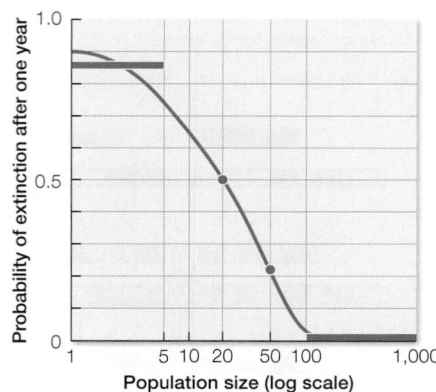

2. The survivorship after 2 years would clearly be lower than after 1 year. After 1 year, about half the populations of size 20 would go extinct. To calculate the number that would go extinct after 2 years, you would need to know the distribution of population sizes among the half of the populations that did not go extinct in the first year. If we assume that the average size of the surviving populations was still 20 after 1 year, then we would expect another half of the remaining populations to go extinct in the second year (for a 2-year extinction rate of 75 percent of the original populations).

FIGURE QUESTIONS

Figure 24.3 There is a spreading zone in the middle of the Pacific Ocean where new lithospheric crust is forming. This pushes the oceanic plates on either side to the east and west, forming subduction zones where oceanic plates meet continental plates on both sides of the Pacific Ocean.

Figure 24.4 When Earth cools, more of its water is tied up in ice on land, in the form of glaciers. Thus there is less water in the oceans, so sea levels are lower.

Figure 24.8 Ash and gases produced by massive volcanic eruptions near the end of the Permian blocked sunlight and resulted in rapid cooling of Earth. There was massive die-off of the lush Permian forests, and the decay of these forests used up much of the atmospheric oxygen. In addition, the loss of these photosynthetic organisms meant that less oxygen was being produced to enter the atmosphere.

APPLY WHAT YOU'VE LEARNED

1. The Ediacaran fossils would have been discovered in older sedimentary strata lying below younger strata where Cambrian-type shelly biota fossils were found. There may have been some overlap where both fossil types were found. The time line provides evidence for this answer, since it indicates that Ediacaran fossils are older than Cambrian-type shelly biota and that there was a short period during which both were present simultaneously.

2. The Ediacaran fauna evolved only after levels of atmospheric oxygen were high enough to support multicellular life forms. Therefore one factor that favored the success of the Ediacaran was the increase in atmospheric oxygen that occurred at this time. Another factor was the ending of glaciation events, which occurred as Earth warmed. The resulting loss of ice cover allowed sunlight to penetrate seawater more effectively, which spurred increased rates of photosynthesis. This further increased oxygen concentrations in aquatic environments. Warmer temperatures in general increased reaction rates, including those in life forms to speed their growth and reproduction; this also favored success of the Ediacaran fauna at that time in Earth's history.

3. The physical environment remained favorable for the evolution of multicellular organisms, since the Cambrian-type shelly biota also arose during this time period. Therefore it was not the physical environment that led to the extinction of the Ediacaran fauna. Most likely, the Ediacaran animals were preyed on by other organisms. They would have been easy prey since they were sessile and large and lacked protective exoskeletons. The Cambrian-type shelly biota were better protected. Their exoskeletons and motility made them much less vulnerable to predation, and thus to extinction.

4. The statement suggests that the lack of fossil discovery does not mean that organisms producing those fossils did not exist. The lack of discovery of fossils may just mean that fossils exist but have not yet been found, or that the organisms died without leaving fossils, or that fossils were formed but have degraded over time. We have to keep open the possibility that the Ediacaran biota lived before or beyond the time period for which we have fossil evidence at present. However, until such evidence is found, the available evidence supports the hypothesis that the Ediacaran fauna arose and went extinct in the time period shown in the diagram.

5. Suppose a geological formation is dug up and surveyed for the presence of fossils. If Ediacaran fossils are found in the same strata as fossils of organism X, one could conclude that organism X existed in the same time period as the Ediacaran biota.

This would date the existence of organism X to the same time period as the Ediacaran biota. This would be possible because fossils located in the same strata were deposited in sediment at the same time. The dating of the strata would be possible because the Ediacaran biota existed in a fairly narrow time window. This would all change, however, if new evidence placed Ediacaran biota at an earlier or later time than what scientists have observed so far.

Chapter 25

RECAP 25.1

1. Before DNA sequencing was developed, biologists usually grouped all prokaryotes together, emphasizing the differences between eukaryotes and prokaryotes. But the characteristics that differentiated these two "groups" (such as the presence of a nucleus, or the various organelles) are all derived features of eukaryotes. When gene sequences were compared across all of life (especially the ribosomal RNA genes, which are easily compared across all living organisms), it became clear that the diversity among "prokaryotes" was far greater than the diversity among eukaryotes. Indeed, it soon became apparent that the ribosomal RNA genes of prokaryotic archaea were actually more similar to the ribosomal genes of eukaryotes than they were to the ribosomal RNA genes of bacteria, and Archaea was proposed as a domain separate from Bacteria. As biologists studied Archaea in more detail, they discovered that many other features were also distinct from Bacteria.

2.

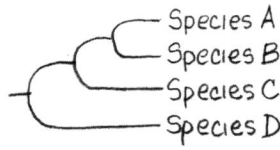

Expected tree based on gene x:
— Species A
— Species B
— Species C
— Species D

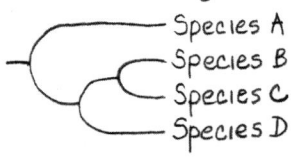

Expected tree based on consensus of non-transferred genes:
— Species A
— Species B
— Species C
— Species D

3. The genomes of eukaryotes contain mostly genes that are more closely related to prokaryotic archaea than to bacteria. This leads biologists to conclude that an ancestral prokaryotic archaean lineage first acquired the features that are characteristic of eukaryotes (such as the nucleus). These early eukaryotes then acquired an endosymbiotic proteobacterium, which was eventually incorporated as the eukaryote mitochondrion. Later, one group of eukaryotes acquired another endosymbiont, this time an early cyanobacterium, which became the chloroplast of photosynthetic eukaryotes.

RECAP 25.2

1. Similarities between any two groups shown in the table can occur for three reasons: (1) the similar feature between two groups is ancestral to life, and the third group has a derived character; (2) the similar feature is a derived character, indicative of a shared ancestry (such as between prokaryotic archaea and their eukaryotic relatives), and the third group (bacteria) has the ancestral condition; or (3) the two endosymbioses of bacteria within eukaryotes may have led to similarities between bacteria and eukaryotes. An example of the first category is that the two prokaryotic groups lack a nucleus, whereas eukaryotes have a nucleus (a derived character). The similarity of the RNA polymerases of prokaryotic archaea and eukaryotes is likely an example of the second category (a derived feature that indicates an evolutionary relationship between these two groups). A possible example of the third category is the ester-linked membrane lipids of bacteria and eukaryotes, since the prokaryotic archaea have ether-linked membrane lipids.

2. All organisms that are alive today are descendants of a common ancestor of life. Many changes have occurred in all these different species, including prokaryotes as well as eukaryotes. Many prokaryotes have undergone radical changes that allow them to live in extreme environments or in unusual ways. Numerous examples could be used, such as the following: One group of euryarchaeotes, the extreme halophiles (salt lovers), live exclusively in very salty environments, and some can live in lakes with pH values as high as 11.5. They have many adaptations that allow them to live in these environments, and so they are more "derived" in this respect than is any eukaryote. As another example, several groups of bacteria, such as chlamydias, are highly derived intercellular parasites. They have many derived changes that allow them to have very small cells than can live within the cells of other species. Thus the terms "primitive" and "derived" make sense only with respect to a particular feature of the organism, and no living organisms can be considered to be primitive overall (i.e., the common ancestor of life was very different from all organisms that are alive today).

RECAP 25.3

1. Biofilms represent a complex community of microbial organisms, living together in a gel-like polysaccharide matrix. Many free-living prokaryotes that come into contact with a solid surface bind to the surface and secrete a sticky polysaccharide, which protects the cells. These cells may then secrete signal molecules that attract other microorganisms to the matrix, which develops into a complex community of organisms over time. This community may become highly resistant to attack and become very difficult to remove. Biofilms are of considerable interest to human health and industry, as they may form on any solid surface (e.g., teeth, contact lenses, artificial joints, the inside of pipes), resulting in decay. Biofilms were also critical to the evolution of many multicellular communities, such as stromatolites.

2. Bacteria are essential for healthy digestion, and only a very few types of bacteria are pathogens. Humans use some of the metabolic products, such as vitamins B_{12} and K, that are produced by bacteria that live in the large intestine. Large communities of bacteria also line human intestines with a dense biofilm, which aids nutrient transfer from the gut into the human body. Thus our gut microbiome is essential to human health.

3. Nitrogen-fixing prokaryotes convert nitrogen from the atmosphere into a chemical form that is usable by living organisms. In addition, denitrifying bacteria release organic nitrogen back into the atmosphere as nitrogen gas, keeping nitrogen from eventually leaching into the oceans. Thus these two groups of bacteria (nitrogen fixers and denitrifiers) allow the cycling of nitrogen and make life on land possible.

RECAP 25.4

1. The greatly reduced genomes of most viruses provide few sequences that can be compared with other organisms. Also, the genomes of many viruses evolve very quickly, making comparisons even more difficult. The reduced nature of viruses provides few morphological clues to their relationships, and their tiny size means they do not produce fossils. Viruses have evolved many times throughout the history of life, so they are related to organisms across the tree of life. For all of these reasons, it is often difficult to place viruses precisely on the tree of life.

2. Many viruses probably represent escaped components of cellular organisms that now evolve independently of their hosts. Other viruses are likely to represent highly reduced, parasitic organisms that evolved from cellular ancestors but lost their cellular structures as they became independent of their cellular hosts.

3. Viruses called bacteriophages, or "phages" for short, infect and kill bacterial cells. Phage therapy, first developed during World War I, involves applying these phages to kill pathogenic bacteria. Phage therapy was largely replaced by the use antibiotics in the 1930s and 1940s. With the increase in evolution of bacterial resistance to antibiotics, however, phage therapy is once again an active area of research.

WORK WITH THE DATA, P. 538

1.

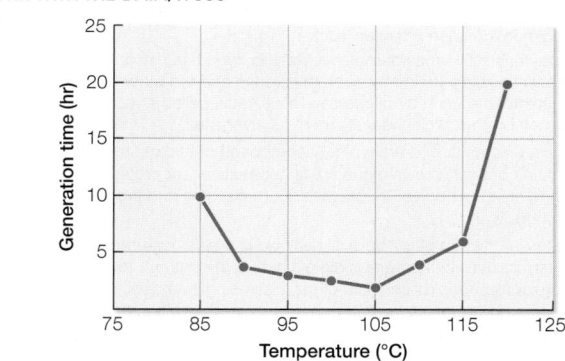

2. 105°C, where growth is fastest (generation time is lowest).

3. Simply lower the temperature of these cell cultures to below 120°C and look for cell growth.

FIGURE QUESTIONS

Figure 25.4   Lateral gene transfers are most likely to occur when there is a selective advantage to the transfer. One species may receive genes from many others, but the chances are smaller that a substantial portion of the genome of one species will be transferred into another. In contrast, the stable core of genes controls critical metabolic functions of the cell, so all of these genes are expected to be inherited from ancestors to descendants together.

Figure 25.24   Parasitic bacteria can depend on the gene products of their hosts, so some of their genes can be lost without loss of function. Small genomes allow smaller cell size, and small size is beneficial for a parasite that lives within other cells.

APPLY WHAT YOU'VE LEARNED

1. As a denitrifier, *Paracoccus denitrificans* would be most appropriate to reduce nitrate concentrations. Its thermal range would also be appropriate under most environmental conditions.

2. As an obligate anaerobe, *Clostridium novyi* would be appropriate to use in hypoxic conditions. Its thermal range is also suitable for mammalian body temperatures.

3. *Trichodesmium thiebautii*. Because the addition of ammonium temporarily increased the productivity of the ecosystem, one can infer that the ecosystem is nitrogen-limited. As a nitrogen fixer, *T. thiebautii* converts atmospheric nitrogen into a form other organisms can use.

4. *Clostridium novyi*. Vancomycin targets bacteria with thick peptidoglycan-filled cell walls. That is the defining characteristic of Gram-positive bacteria, and *C. novyi* is the only Gram-positive bacterium in the table.

Chapter 26

RECAP 26.1

1. Loss of the stiff prokaryotic cell wall allowed infolding of the cell membrane, greatly increasing the surface area of the cell without increasing its volume. These infoldings also allowed for greater structure within the cell and eventually led to the evolution of the endoplasmic reticulum.

2. When atmospheric oxygen concentrations were low early in Earth's history, organisms could use only anaerobic respiration. Increased availability of atmospheric oxygen allowed the evolution of aerobic respiration, which proceeds more rapidly and harvests energy more efficiently than anaerobic respiration does. But as cells increased in size, they would decrease in surface area-to-volume ratio unless the cell membrane had infoldings. Increased levels of atmospheric oxygen also led to the endosymbiosis of an aerobic bacterium (the ancestor of the eukaryotic mitochondrion), which facilitated the evolution of aerobic eukaryotes.

3. The three rRNA genes of corn are not one another's closest relatives because the nuclear, mitochondrial, and chloroplast genomes have different origins, and the relationships shown in the tree reconstruct the endosymbiotic events that gave rise to mitochondria and chloroplasts.

4. The mitochondrial genome of corn is more closely related to the genome of *E. coli* than it is to the nuclear genome of corn because the mitochondria were derived from an endosymbiosis with a proteobacterium. Likewise, the chloroplast genome of corn is more closely related to the genome of *Chlorobium* than it is to the nuclear genome of corn because the chloroplasts were derived from an endosymbiosis with a cyanobacterium.

5. The term "protists" is used, for convenience, to refer to a diverse group of distantly related, polyphyletic microbial eukaryotes. Some protists are more closely related to each of the major groups of multicellular eukaryotes (including plants, fungi, and animals) than they are to one another.

RECAP 26.2

1. a. Foraminiferans have external shells of calcium carbonate, whereas radiolarians have long, stiff pseudopods and radial symmetry. The external shells of foraminiferans and the internal skeletons of radiolarians are both important components of ocean sediments and sedimentary rocks.

   b. Ciliates are covered with numerous hairlike cilia, whereas dinoflagellates generally have two flagella (one in an equatorial groove and the other in a longitudinal groove). Both ciliates and dinoflagellates have sacs, called alveoli, just beneath their cell membranes, which identify them as alveolates.

   c. Diatoms are unicellular and are typically composed of two nested plates (like a petri dish). In contrast, brown algae are large, multicellular organisms composed of branched elements or leaflike growths. Both diatoms and brown algae are photosynthetic.

   d. The vegetative unit of a plasmodial slime mold is a plasmodium: a wall-less mass of cytoplasm containing numerous diploid nuclei. In contrast, the vegetative unit of cellular slime molds consists of separate, single amoeboid cells. In both groups, when environmental conditions become unfavorable, the vegetative units form fruiting structures.

2. Most eukaryotes were unicellular until the beginning of the Cambrian. The rapid diversification and increased size of multicellular eukaryotes (especially animals) near the beginning of the Cambrian led to greatly increased chances of fossilization.

3. Examples; other answers are possible.

   a. Alveolates: Dinoflagellates can be toxic (e.g., they can poison mollusks during red tides); the apicomplexan *Plasmodium* is the causative agent of malaria.

   b. Stramenopiles: Some brown algae are harvested and used by humans as an emulsifier in ice cream, cosmetics, and other products.

   c. Excavates: The parasitic *Giardia lamblia*, a diplomonad, causes the intestinal disease giardiasis (an infection common among hikers and campers who drink spring or stream water in recreational areas). Kinetoplastids include several medically important species of pathogenic trypanosomes that cause diseases such as sleeping sickness, Chagas disease, and leishmaniasis.

RECAP 26.3

1. Two paramecia exchange genetic material during conjugation, but there is no increase in the number of individuals. As a result, there is sex (genetic recombination) but no reproduction.

2. The sex that occurs in conjugation allows the removal of deleterious mutations through genetic recombination. Without conjugation, paramecia can only reproduce clonally, which is not viable over long periods of time.

3. Alternation of generations involves alternation of *multicellular*, *independent* haploid and diploid stages. Sperm and eggs are unicellular and are not independent organisms.

RECAP 26.4

1. The female mosquitoes transmit the *Plasmodium* parasites to vertebrate hosts. The mosquitoes take up *Plasmodium* along with blood from an infected host and transmit the parasite to new, previously uninfected hosts when they feed on the blood of the new hosts.

2. Red tides are massive blooms of free-living protists (often dinoflagellates) that occur in oceans (and sometimes freshwater bodies) around the world. Many dinoflagellates involved in red tides produce toxins that kill many other species, especially fish and other vertebrates. Dinoflagellates are important symbionts of corals; when environmental conditions change, the dinoflagellates die or are ejected from the host corals, resulting in bleaching. The corals may take up new dinoflagellates, or they may become weakened and perish.

3. Diatoms are important as primary producers in many ecosystems. Of more direct importance to humans, fossilized oils from diatoms are the primary source of petroleum and natural gas. In addition, the remains of diatom skeletons produce diatomaceous earth, which is used for insulation, filtration, polishing, and as an insecticide.

WORK WITH THE DATA, P. 569

1. No. The new strain (B211 in this experiment) was taken up by 92 percent of the coral colonies after bleaching (and by 100% of the surviving colonies), even though at least 58 percent of the experimental colonies retained some of their original symbionts after bleaching. This shows that bleached coral colonies are likely to take up strains of *Symbiodinium* that are available in the environment, even if some of their original symbionts survive the bleaching.

2. No. Between week 3 and week 6 after bleaching, an additional 17 percent of the experimental colonies died despite having acquired new strains of *Symbiodinium*.

3. This observation suggests that the detection assays for *Symbiodinium* are not sensitive enough to detect very small levels of the symbionts. Presumably, the non-B211 strains were present in some of the experimental colonies in week 3, but at such low levels that they were not detected. These symbionts then recovered in 8 percent of the colonies and were at detectable levels by week 6. Another possibility is that contamination of the experimental samples with non-B211 symbionts took place between weeks 3 and 6 of the experiment. Neither possibility challenges the conclusion that corals can acquire new strains of *Symbiodinium* from their environment after bleaching events.

FIGURE QUESTIONS

Figure 26.1 DNA sequences of genes that are found in all genomes (such as the ribosomal RNA genes) can be used to connect organelles to their endosymbiotic ancestors. If the organelles did not evolve through endosymbiosis, then we would expect homologous copies of organellar and nuclear genes from the same species to cluster in a phylogenetic tree. However, this is not the case. Instead, phylogenetic analyses show that the ribosomal RNA genes of mitochondria in eukaryotes are more closely related to homologous genes from proteobacteria than to eukaryotes, and the ribosomal RNA genes of chloroplasts in eukaryotes are more closely related to homologous genes from cyanobacteria than to eukaryotes. These analyses provide convincing evidence in support of the endosymbiotic origins hypothesis.

Figure 26.19 Conjugation is considered sex without reproduction because genetic material is exchanged (sex) but no new *Paramecium* are produced.

APPLY WHAT YOU'VE LEARNED

1. All of the apicomplexans shown here have a parasitic lifestyle. Therefore this lifestyle was likely present in the common ancestor of the apicomplexans. The hosts of the three apicomplexans differ, but they are all mammals. Thus it is reasonable to infer that the common ancestor of these species likely had a mammalian host.

2. No, all of the dinoflagellates are equally distantly related to the apicomplexans, so there is no reason to favor one genus over another in choosing an outgroup.

3. The genus *Alexandrium* is more closely related to *Pfiesteria* than to any of the other genera, so this genus is a good choice for comparison, as the two genera share the most recent common ancestor. Therefore comparing *Alexandrium* to *Pfiesteria* would help the investigator determine which attributes evolved in *Pfiesteria*, and which were already present in the common ancestor of *Alexandrium* and *Pfiesteria*.

4. Answers will vary. One example is the morphological similarity of loboseans and heteroloboseans, despite these groups not being closely related.

Chapter 27

RECAP 27.1

1. The broadest use of the term "plant" refers to eukaryotes that descended from the ancestor with the first primary endosymbiosis of chloroplasts: glaucophytes, red algae, the various lineages of green algae, and land plants. Another use of the term refers to green plants (those that contain chlorophyll *b*, or the green algae and land plants). The most common use of "plant" refers only to land plants

2. "Algae" is a convenience term that is used to refer to many distantly related, photosynthetic, aquatic organisms. This includes many groups of eukaryotes outside the Plantae.

WORK WITH DATA, P. 574

1.

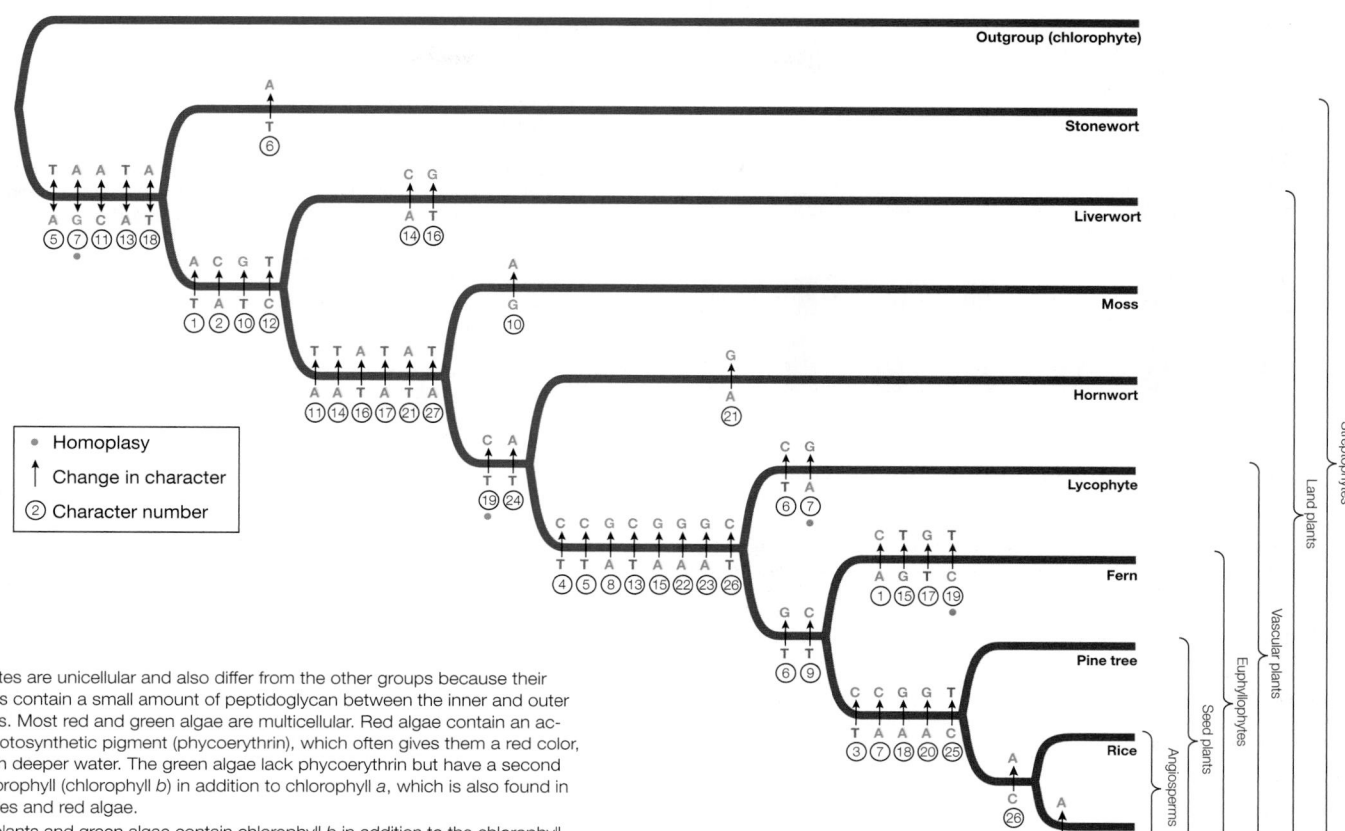

3. Glaucophytes are unicellular and also differ from the other groups because their chloroplasts contain a small amount of peptidoglycan between the inner and outer membranes. Most red and green algae are multicellular. Red algae contain an accessory photosynthetic pigment (phycoerythrin), which often gives them a red color, especially in deeper water. The green algae lack phycoerythrin but have a second kind of chlorophyll (chlorophyll *b*) in addition to chlorophyll *a*, which is also found in glaucophytes and red algae.

4. Both land plants and green algae contain chlorophyll *b* in addition to the chlorophyll *a* also found in glaucophytes and red algae. Land plants are more closely related to some green algae lineages than to others. The green algae most closely related to the land plants are the stoneworts; these two groups share protected embryos, a cuticle, a multicellular sporophyte, the presence of gametangia, and thick-walled spores.

RECAP 27.2

1. Alternation of generations is a life cycle in which a multicellular diploid sporophyte generation produces spores by meiosis, alternating with a multicellular haploid gametophyte generation that produces gametes by mitosis. Humans (and other animals) do not have alternating sexual and asexual multicellular stages. They undergo meiosis to produce sex cells and use mitosis for growth and maintenance, not for reproduction.

2. Terrestrial adaptations found in all major lineages of land plants (including liverworts, mosses, and hornworts) include the cuticle, which provides a coating of waxy lipids that retards water loss; gametangia (multicellular organs that enclose plant gametes), which prevent gametes from drying out; protected embryos; pigments that provide protection against ultraviolet radiation; thick-walled spores; and a mutually beneficial association with fungi that promotes nutrient uptake from the soil. Stomata, which are small closable openings in leaves and stems that are used to regulate gas exchange, are a terrestrial adaptation that is found in all land plants except liverworts.

RECAP 27.3

1. Xylem conducts water and minerals from the soil to aerial parts of the plant. Xylem also provides structural support. Phloem conducts the products of photosynthesis from sites of production to sites of storage within the plant.

2. The earliest rootlike structures were rhizomes—horizontal portions of stem that bear water-absorbing filaments called rhizoids. Rhizomes are retained in some modern plants, such as liverworts and mosses, and are also found in some vascular plants, such as ferns and lycophytes. Ferns and lycophytes, along with other vascular plants, also have true roots, a specialized organ for extracting water and minerals from the soil. Leaves may have originated twice among the vascular plants. The simple microphylls of lycophytes may have arisen from sterile sporangia. In contrast, the larger and more complex megaphylls, or true leaves, may have arisen from the flattening of a portion of a branching stem system that exhibited overtopping growth.

3. Heterospory refers to the presence of two distinct types of spores, one that develops into a female gametophyte and the other into a male gametophyte. One advantage of heterospory is that it promotes outcrossing among individuals. The fact that heterospory has evolved multiple times from homospory provides evidence that the heterosporous condition is advantageous.

2. Changes for each branch are indicated on the tree diagram provided with the answer to Question 1. Note that there are alternative (equally parsimonious) reconstructions for character numbers 6, 14, and 16.

3. Characters 7 and 19 exhibit homoplasy.

4. These groups are labeled in the figure provided as the answer to Question 1.

WORK WITH DATA, P. 576

1. The 95% confidence intervals for the means can be calculated by multiplying the standard errors of the means by 1.96, and then adding and subtracting this product from the respective mean.

The 95% confidence limits for each measure of chlorophyll *a* concentration (mg/L) in the various growth media are as follows:

Growth medium only	Growth medium + propionate	Growth medium + acetate	Growth medium + citrate	Growth medium + sucrose	Growth medium + glucose
10.98– 15.84	1.66–1.78	7.79–11.01	14.10– 16.42	15.94– 19.32	29.84– 44.58

Chlorophyll *a* concentrations in growth medium + propionate are clearly lower than in growth medium alone, as the 95% confidence intervals do not overlap. There is almost no overlap between the growth medium only and growth medium + acetate conditions, with the added acetate showing lower chlorophyll *a* concentrations. There is substantial overlap in the confidence intervals for growth medium alone and growth medium + citrate, suggesting no significant difference in those two conditions. The other two conditions, with sucrose or glucose added to the growth medium, result in significantly higher chlorophyll *a* concentrations than does growth medium alone.

2.

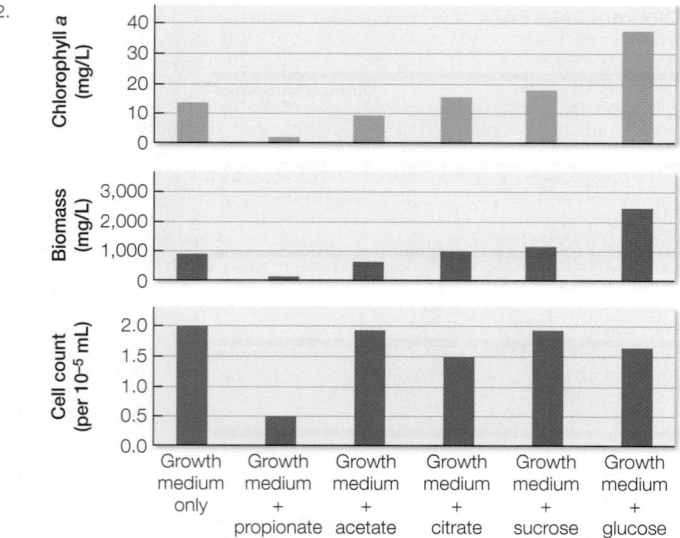

The measures of chlorophyll a concentration and biomass are strongly correlated, whereas cell counts are not as strongly correlated with these other measures. The reason is that favorable growth conditions produce large cells (with greater biomass) that contain proportionally more chlorophyll a, whereas poor growth conditions produce small cells that contain proportionally less chlorophyll a. Although poor growth conditions sometimes result in fewer cells than favorable growth conditions, the growth conditions have a greater effect on cell size than on cell numbers.

FIGURE QUESTIONS

Figure 27.1 The terms "algae" and "nonvascular plants" are both convenience terms, as neither of those groups forms a monophyletic group.

Figure 27.6 In plants, the haploid portion of the life cycle involves a multicellular organism, whereas in animals it involves only single-celled gametes.

Investigating Life Experiment The algae grow best in 50% wastewater and 50% growth medium.

APPLY WHAT YOU'VE LEARNED

1. The last common ancestor of seed plants and ferns lived in the mid-Devonian about 390 million years ago.
2. Spikemosses did not exist until the late Devonian. Spikemosses were not yet present when the last common ancestor of seed plants and ferns lived and thus fossils of spikemosses would not be expected in the same stratum with fossils of the common ancestors of ferns and seed plants. Because the common ancestor of ferns and seed plants lived before spikemosses did, we would expect fossils of that ancestor to be in a lower stratum.
3. Yes, rhyniophytes existed from the late Silurian through much of the Devonian, and survived for tens of millions of years after the last common ancestor of seed plants and ferns lived. So it would not be surprising if fossils of these two lineages were in the same stratum.
4. The last common ancestor of embryophytes lived in the Ordovician about 470 million years ago. This lineage was the first group of plants to live out of water and colonize land.

Chapter 28

RECAP 28.1

1. Pollen grains are highly resistant to dehydration and can be carried by wind or pollinators. Once a pollen grain lands on a stigma of a flower, it can grow a pollen tube, from which it releases sperm directly to the megagametophyte.
2. Seeds protect the embryo from excessive drying and from predators. Seeds can remain in a resting state in soil, often for many years, until conditions become favorable for growth. Many seeds also aid in dispersal by wind, water, or animals. When growth begins, the seed provides the developing sporophyte with the food reserves it needs.
3. Wood provided structural stability for the stems of plants and allowed them to reach great heights. The forests of the world are able to exist because of this evolutionary innovation.

RECAP 28.2

1. Some plants require fire for seeds to be released. For example, fire is needed to open the tightly closed cones and release the seeds of some pines, which are then able to germinate.
2. The megagametophyte is the female gametophyte, or multicellular haploid tissue contained within the ovary of all seed plants. The pollen grain is the multicellular male gametophyte, which produces the sperm that fertilize the megagametophyte.

3. These fleshy cones serve to attract animals, much as the fruits of flowering plants do. The animals eat the fleshy cones and the seeds they contain, and the seeds pass through the digestive tracts of the animals and are dispersed by their feces.

RECAP 28.3

1. One sperm produces the diploid zygote, and another produces the (usually) triploid endosperm that will provide nourishment for the developing embryo.
2. Pollination refers to the arrival of pollen on a stigma. After the pollen produces a pollen tube, it releases sperm to the megagametophyte, resulting in fertilization.
3. Fruits that attract animals to a food resource often have a fleshy, edible ovary. Animals gather or eat the fruit, and the seeds may pass through the digestive tract alive and germinate after they are deposited in the animal's stool. In some cases, such as tree nuts, the seeds are themselves edible, and the fruit serves as a hard, protective covering. Animals such as squirrels gather the nuts, eat some, and store others for later. Some of the stored nuts then germinate into new plants. In contrast, some seeds have sticky surfaces that adhere to animals that brush by the fruiting plant. If you have ever had to pull burrs off your clothes or your pet, you have pulled out fruits of plants that are being dispersed in this manner.
4. Flowers are the site of pollination and fertilization and, in many species, serve to attract pollinators. Fruits may provide protection or means of dispersal for the seeds; often, the fruits are attractive to an animal that then disperses the seeds in its feces. The seeds protect the developing embryos until conditions are favorable for germination.

RECAP 28.4

1. There are many possible answers, some of which are shown below.

Product	Plant Source	Medical Application
Atropine	Belladonna	Dilate pupils for eye examination
Bromelain	Pineapple stem	Control tissue inflammation
Digitalin	Foxglove	Strengthen heart muscle contraction
Ephedrine	Ephedra	Ease nasal congestion
Menthol	Japanese mint	Relieve coughing
Morphine	Opium poppy	Relieve pain
Quinine	Cinchona bark	Treat malaria
Taxol	Pacific yew	Treat ovarian and breast cancers
Tubocurarine	Curare plant	Muscle relaxant (used in surgery)
Vincristine	Periwinkle	Treat leukemia and lymphoma

2. Among the possible answers: Plants produce oxygen and remove carbon dioxide from the atmosphere, and they play important roles in forming soils and renewing soil fertility. Plant roots help hold soil in place, providing protection against erosion by wind and water. Plants also moderate the local climate in various ways, such as by increasing humidity, providing shade, and blocking wind.
3. Grasses produce the grain crops (including rice, wheat, and corn) that account for a large percentage of the calories in human diets. Grasses are also the basis for feeding grazing animals, such as cattle, which provide humans with meat and dairy products.

WORK WITH THE DATA, P. 596

1.

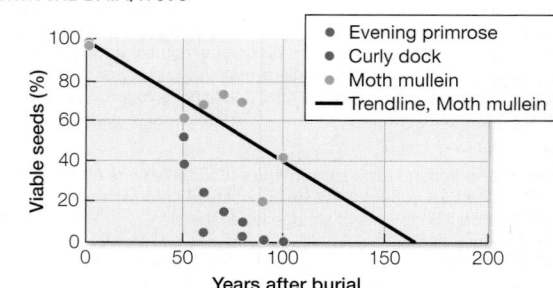

2. One approach to this problem is to calculate a linear trend line for survivorship of *Verbascum blattaria* seeds by calculating a linear regression line (see Appendix B) and then projecting it forward in time to the point at which it intersects zero percent survival. The resulting regression equation is $y = 102.09 - 0.62x$. The graph plotted in answering Question 1 uses this approach and predicts that the last *Verbascum blattaria* seeds should germinate in about year 165 of the experiment (set y to 0, and solve for x using the regression equation; the result is $x = 164.7$ years). This approach assumes a linear decline in viability of the seeds. It may be more reasonable to assume an exponential decay in seed viability (similar to radioactive decay; see

Figure 24.1). If seeds decay exponentially, then we would expect some low level of survivorship of *Verbascum blattaria* seeds well beyond year 165.

3. At least four factors are related to seed survivorship:

   a. Size of the seed: larger seeds have more food reserves (endosperm).

   b. Thickness of the seed coat: thicker seed coats provide better protection of the seed.

   c. Density of the seed coat: tougher seed coats provide better protection of the seed.

   d. Level of dormancy of the embryos: deeper dormancy results in longer survivorship.

FIGURE QUESTIONS

Figure 28.2   The haploid gametophyte stage of the moss is the most prominent stage, and the stage we recognize most readily as a moss. The haploid stage of ferns, in contrast, is very inconspicuous, so we usually see the larger diploid sporophyte. In flowering plants, the haploid gametophytes are rarely seen, except as pollen.

Figure 28.17   Although rodents destroy the seeds of nut-bearing plants when they eat them, these plants produce far more nuts than rodents can immediately eat. Rodents gather extra nuts and stash them for later consumption in or around their nests, often burying them in soil in the process. Many of these extra nuts then germinate before they are consumed by the rodents.

APPLY WHAT YOU'VE LEARNED

1. The bright red color of *M. cardinalis* flowers is consistent with of the expectations of a bird-pollinated species. The nectar guides (stripes on the petals) of *M. lewisii* flowers are consistent with the expectations of a bee-pollinated species.

2. The corolla of *M. cardinalis* is narrow and tubelike, while that of M. lewisii is broader and flatter. The narrow tubular corolla of *M. cardinalis* is an adaptation that accommodates the long beaks and tongues of its hummingbird pollinators, while restricting pollination by other species.

3. Variation at the *yup* locus did not significantly affect visitation by hummingbirds, but bees were much less likely to visit the *CC* genotype (red) flowers.

4. Although birds showed a slightly higher preference for red flowers over pink flowers, the difference shown here is not statistically significant. However, bees visit red flowers at a significantly lower rate than pink flowers. It may be that the primary function of red flowers is to discourage visitations by bees so that pollination is restricted to hummingbirds.

Chapter 29

RECAP 29.1

1. Unicellular fungi absorb nutrients directly from their environment, but multicellular fungi use structures known as hyphae for absorption. In both cases, fungi secrete digestive enzymes into the environment and then absorb the breakdown products through their cell membranes. The hyphae of multicellular fungi are long, thin networks that can penetrate soil and decaying or living organic material.

2. The large surface area-to-volume ratio allows rapid absorption over a wide area, as the hyphae can be immediately adjacent to the food source. However, this can also lead to rapid water loss and drying, which is why fungi tend to be restricted to (or at least thrive best in) moist environments.

RECAP 29.2

1. Fungi break down organic matter, especially the cellulose and lignin of plant walls, thus returning the carbon in dead plants back into a form that can be used again by other organisms. Fungi also break down the keratin of animals, which is resistant to bacterial decay. Without fungi, these materials would build up in the environment and be a global sink for carbon.

2. Lichens represent a mutualistic association between a fungus and a photosynthetic partner (a unicellular alga and/or a cyanobacterium). The fungus obtains fixed carbon (organic compounds) from its photosynthetic partner and provides it with minerals and water. Both the fungus and the photosynthetic partner benefit and can grow in harsh environments (such as on rocks or the bark of trees) where neither organism would survive well, if at all, on its own.

3. As with lichen associations, mycorrhizal fungi form a mutualistic partnership with a photosynthetic species, in this case a multicellular plant. Many plants depend on these mycorrhizal associations for survival. The plant provides organic compounds (the products of photosynthesis) to the fungus, and the fungus provides minerals and water from the soil to the plant. The high surface area-to-volume ratio of the fungal hyphae greatly increases the plant's ability to absorb water and minerals. Also, the hyphae act like a sponge, holding water around the roots

RECAP 29.3

1. In mating types, there is no phenotypic distinction as exists between male and female sexes, and there are often more than two mating types. Mating types are morphologically indistinct, but within each species, each mating type is genetically different. Individuals within a species may not mate with another of the same mating type. In contrast, male and female sexes are defined by the relative size of their gametes. (Females have large gametes, and males have small gametes; reproduction requires the combination of a male and a female gamete

2. Microsporidia are among the smallest eukaryotes known. They are obligate parasites on animals. The host cell is penetrated by a polar tube through which the contents of the spore are injected into the host. This sporoplasm then replicates within the host cell and produces new infective spores. In some insects, once an individual is infected, the microsporidian parasite may be transmitted from parent to offspring.

3. Most chytrids possess flagellated gametes and reproduce in water (or at least in wet environments, such as moist soil).

4. The zygospore is the diploid phase of the zygospore fungi life cycle. It has a thick, multilayer cell wall and can serve as a resting stage that may remain dormant for months until environmental conditions are favorable for growth. At that time, its diploid nuclei undergo meiosis, and the zygospore produces a stalked sporangiophore, which bears from one to many sporangia. Each sporangium contains haploid spores, which are the products of meiosis.

5. The haploid ascospores of sac fungi are contained in sacs known as asci. Many species of sac fungi bear these asci on cup-shaped fruiting bodies known as ascoma. In contrast to these cup-shaped fruiting bodies, the fruiting bodies of club fungi include the more familiar mushrooms, puffballs, and bracket fungi, which are called basidiomata. These basidiomata bear the basidia, in which fertilization and meiosis take place to produce the haploid basidiospores.

RECAP 29.4

1. Some fungi are eaten directly (including mushrooms and a few lichens), but fungi play a bigger role in the production of bread (from baker's yeast), cheese (fungi give many cheeses their distinctive flavors), and drinks (through fermentation of yeasts). In addition, fungi enhance and facilitate the growth of many crop plants.

2. The museum lichen collections represent an indirect record of past air quality that can be used in areas where no direct measurements of air quality were originally taken. In addition, lichen diversity surveys represent a fast and inexpensive way to monitor air quality across an environment.

3. Site 5 shows the highest diversity and density of lichens and so is probably farthest from the city center. Site 4 is next, followed by Site 1, then Site 3, and finally Site 2. In addition to distance from the city center and prevailing wind direction, other predictive factors could include distance to point-pollution sources (such as factories or power plants) and distance to major highways (a source of pollution from automobile exhaust). Other answers are also possible; it is important for such studies to control for factors such as the species of tree examined and the exposure of the branches to similar light and humidity conditions.

4. By using selective tree harvest, rather than clear-cutting, forests may be managed to retain largely intact communities of many organisms, including mycorrhizal fungi. If some of the original forest trees are retained, replacement of harvested trees is greatly facilitated by the continued presence of viable mycorrhizal fungal communities.

WORK WITH THE DATA, P. 631

1.

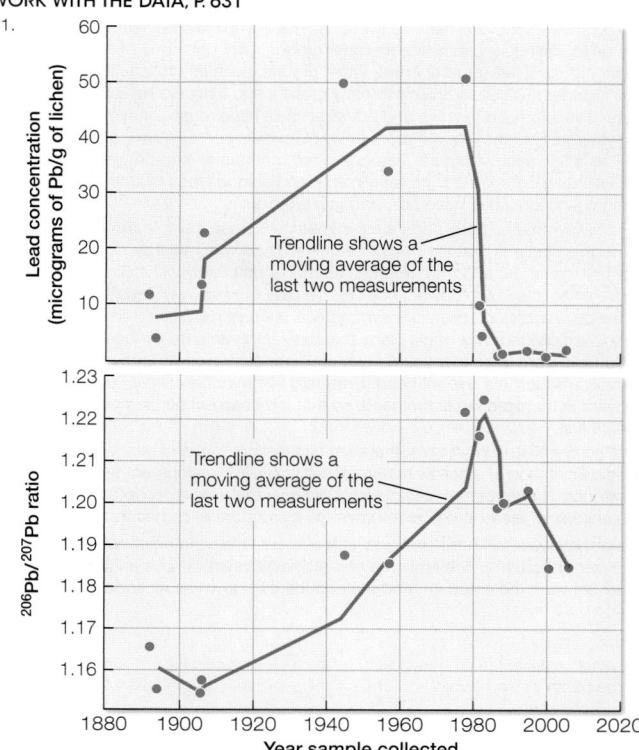

2. Yes, the levels of environmental lead contamination and the ratios of lead isotopes are consistent with the identified sources. The closure of the lead smelter was offset by the use of leaded gasoline, but there was a marked drop-off of environmental lead contamination after the elimination of both of these sources. The trend lines show a peak in environmental lead contamination around the 1960s to 1980s, and then a rapid drop after the end of leaded gasoline use in the early 1980s. The measured ratios of $^{206}Pb/^{207}Pb$ are consistent with the suspected sources of lead contamination through time. One way to extend the analysis would be to analyze additional lichen samples collected during the relatively poorly sampled period between 1907 and 1978 (from which period only two samples are included). Note, however, that such an analysis would depend on the existence of appropriate samples, such as museum collections made during this time.

FIGURE QUESTIONS

Figure 29.15   The two gametes differ in size; a male gamete is defined as the smaller of the two gametes, and the female is the larger gamete.

Figure 29.17   In the dikaryotic state, each cell contains a separate haploid nucleus from each parent. In the diploid state, the two nuclei are fused.

APPLY WHAT YOU'VE LEARNED

1. There is a strong linear relationship between the negative effects of exposure to garlic mustard on a species and its dependence on its fungal mutualists, as the plants that are most dependent on the mutualism are most negatively affected by exposure to garlic mustard. This is consistent with garlic mustard reducing growth by interfering with the mycorrhizal mutualists.

2. Trees from mature forests are most dependent on arbuscular mycorrhizal fungi. A reasonable explanation for this finding is that trees in mature forests rely heavily on drawing nutrients and water from their roots, the site of action for arbuscular mycorrhizal fungi. Plants that live on forest edges and particularly those that live in disturbed areas are less reliant on the mutualists than are those that live in mature forests. This may be due to edge species being less reliant on gathering water and nutrients via their roots.

3. Plants with active arbuscular mycorrhizal fungal mutualists are more resilient to drought and temperature extremes than those without. Thus the loss of the fungal mutualists by exposure to garlic mustard may lead to the plants being less resilient to drought and temperature extremes.

   The plants with arbuscular mycorrhizal fungi should be able to withstand extremes in climate better than those without. Loss of fungal mutualists via the action of garlic mustard should lead to greater negative consequences for the plants as a result of climate change.

Chapter 30

RECAP 30.1

1. Most animals are multicellular heterotrophs with internal digestion, muscular systems that allow movement, and nervous systems. This combination of features generally allows us to distinguish animals from other groups, although none of these features is diagnostic (by itself) for all animals. Other groups (such as protists, fungi, and plants) include multicellular species; many protists and fungi are heterotrophs; some protists have internal digestion; and not all animals have muscular systems and nervous systems. Evidence for the monophyly of animals comes from gene sequences, as well as a few microstructural features: a common set of extracellular matrix molecules, including collagen and proteoglycans; and unique types of junctions between cells (tight junctions, desmosomes, and gap junctions).

2. a. In radial symmetry, body parts are symmetrical across multiple planes that run through a single axis at the body's center. Animals with radial symmetry have no front or rear ends, and they are often sessile or drift freely with currents. If they move under their own power, they can typically move slowly equally well in any direction. In contrast, bilaterally symmetrical animals have mirror-image right and left halves divided by a single plane that runs along an anterior–posterior midline. They have front ends that usually contain a concentration of sensory systems and nervous tissues in a distinct head. Bilaterally symmetrical animals usually move forward in the direction of the head, so that the head encounters new environments first.

   b. Among the bilaterian animals, there are two distinct forms of gastrulation—the initial indentation of a hollow sphere of cells early in development that forms the blastopore. In protostomes the blastopore eventually develops into the mouth of the animal, whereas in deuterostomes the blastopore becomes the anus.

   c. Diploblastic animals have embryos with two cell layers (an outer ectoderm and an inner endoderm). The embryos of triploblastic animals have an additional cell layer between the ectoderm and the endoderm, known as mesoderm.

RECAP 30.2

1. Bilaterally symmetrical organisms have an anterior and a posterior end. As the animal moves through the environment, its anterior end encounters potential food or predators first. It is therefore advantageous for the sensory organs and central nervous system to be concentrated at the anterior end.

2. The body cavities of many animals function as hydrostatic skeletons. As the muscles that surround a cavity contract, the fluids must shift to another part of the cavity. In this way, animals can extend parts of their bodies and move specific body parts. Segmentation allows specialization of the body parts, and soft-bodied animals can change the shape of each part independently, thereby increasing the precision of movement. In animals with a hard external skeleton (such as the arthropods), segmentation and the accompanying appendages (controlled by muscles attached to the exoskeleton) allow even greater specialization of movement. Some arthropod appendages are used for walking, swimming, and even flying. The central nervous system is used to sense the environment (including food, appropriate temperatures, and potential predators) and to coordinate movement.

   Acoelomate animals lack a body cavity enclosed in mesoderm. Pseudocoelomate animals have a body cavity enclosed in mesoderm; this body cavity contains the gut and internal organs composed of endoderm, but these latter organs are not lined with mesoderm. Coelomate animals have a body cavity that is enclosed in mesoderm, and the internal organs are also lined with mesoderm.

RECAP 30.3

1. Filter feeders filter water or air and trap small food particles they contain. These particles may include small animals, but the filter feeder does not typically actively chase and feed on individual prey. Predators, in contrast, actively seek out and feed on other individual animals, killing them in the process. Parasites also feed on other animals, usually without killing them; a parasite may reside inside another animal or feed on its parts from the outside (such as a mosquito or a tick does).

2. Herbivores must digest relatively fibrous, tough plant material. So they usually need a longer gut (compared with carnivores) or a digestive system that permits fermentation of plant material. They usually need mouthparts that allow chewing of leaves or sucking of plant fluids. A predator needs to be able to move quickly enough to catch its prey and needs adaptations for subduing the prey (such as teeth, jaws, claws, venom, constriction, etc.).

RECAP 30.4

1. In a multicellular organism, every cell has the same genotype, and the cells are all physiologically interconnected and interdependent to form one functioning organism. In a colonial species, the individuals are more loosely integrated with one another, and individuals that make up a colony may also be able to exist independently of one another. In some cases, different individuals in a colony have different genotypes, although some colonies may be composed of clonally duplicated organisms that function as an integrated whole. Every individual in a colony is typically multicellular, unlike the single cells in a multicellular individual.

2. Typically the characteristics of an animal in one life stage are beneficial under certain conditions but detrimental under others. A change that improves one characteristic, such as a thicker shell for protection against predation, usually comes at a cost of some kind, such as less mobility. In a life cycle, an animal may evolve to produce more eggs, but that usually comes at a cost of the investment of resources in each egg, which leads to lower survivorship of the average offspring. For this reason, there are limits on the changes that can occur in the evolution of life cycles.

RECAP 30.5

1. First, phylogenetic analysis showed that ctenophores were the sister group of other animals. Ctenophores have nervous systems, but some other animals (such as sponges and placozoans) do not. Therefore either nervous systems evolved once in the ancestor of all animals and were subsequently lost in sponges and placozoans, or nervous systems evolved separately in ctenophores and in other animals. Second, analysis of whole genomes showed that the genes involved in nervous systems have been independently duplicated and specialized in ctenophores, cnidarians, and bilaterians, thereby suggesting that the nerve nets of ctenophores and cnidarians, and the centralized nervous systems of bilaterians, each evolved independently.

2. The group of organisms called "animals" represents a specific monophyletic group of multicellular organisms on the tree of life. Sponges retain many of the ancestral features of animal relatives (such as choanoflagellates), and split from the other animals before the evolution of complex organ systems. Placozoans, by contrast, may have evolved from ancestors with distinct organ systems, but they appear to have lost these systems and become secondarily simplified.

3. Placement of glass microscope slides (or other smooth substrates for placozoan attachment) in warm tropical waters often results in colonization by placozoans. The glass slides can be suspended in water in survey areas, then later retrieved and examined for the presence of placozoans.

WORK WITH DATA, P. 650

1.

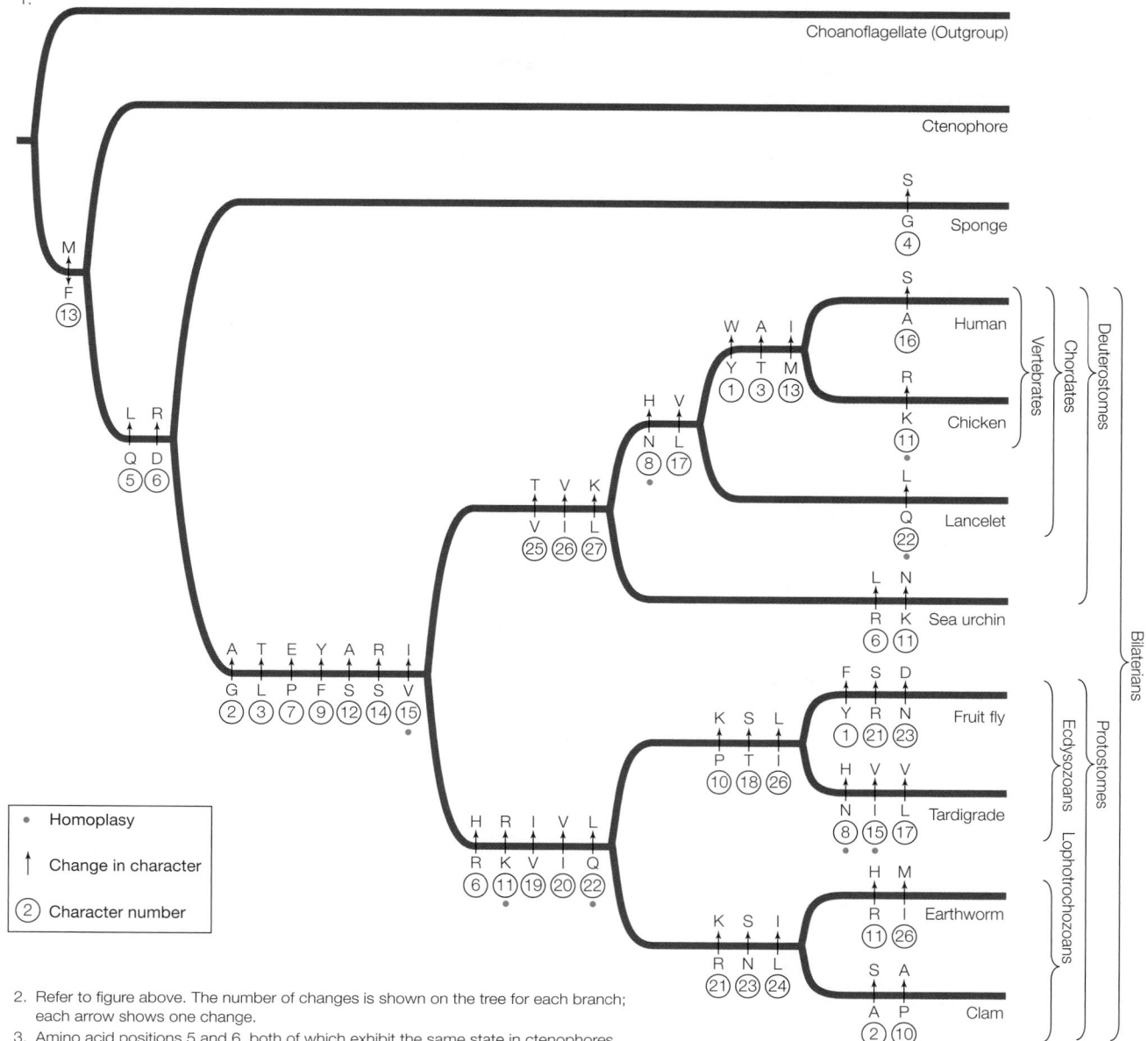

2. Refer to figure above. The number of changes is shown on the tree for each branch; each arrow shows one change.

3. Amino acid positions 5 and 6, both of which exhibit the same state in ctenophores and the outgroup, but a derived state in the remaining animals, support the ctenophores as the sister group of the remaining animals.

4. Each of these groups in marked on the tree above (answer to Question 1).

FIGURE QUESTIONS

Figure 30.1   Nervous systems are shown evolving three times: nerve nets in ctenophores and in cnidarians, and central nervous systems in bilaterians.

Figure 30.14   Producing many small eggs allows a species to rapidly expand its population size under favorable environmental conditions. However, when conditions are less favorable, the low parental investment in each egg means that very few, if any, offspring are likely to survive. In contrast, producing one large egg with high parental investment greatly increases the likelihood of each offspring surviving, but many fewer offspring can be produced, so populations cannot change rapidly to take advantage of temporarily favorable conditions.

APPLY WHAT YOU'VE LEARNED

1. Based on these data, the chemical disturbance of CsCl induced reverse development as compared with the control population. Increasing concentrations of CsCl resulted in stronger reverse development, as evidenced by the higher percentages of animals in polyp and reducing medusa stages in response to increasing concentrations of CsCl. Reverse development allows for some cnidarians to survive periods that are not suitable for their survival and reproduction as medusae.

2. The medusa form is motile and sexual, while the polyp form is sessile and asexual. Disadvantages associated with reverse development include decreased outbreeding opportunities, decreased dispersal, and investment in reverse development that could otherwise be devoted to maintenance or reproduction as a sexual medusa. These negative trade-offs are tolerated because reverse development allows survival in an environment that is not suitable for medusae.

3. Yes, you would expect the two populations to look different. Polyps promote survival and maintenance of local populations because they can reproduce without a mate, but they limit the ability to disperse. Medusae may enhance dispersal probability and/or population expansion and increase genetic diversity, but medusae are more energetically expensive, and developing and maintaining reproductive structures can be costly. So when conditions are good (high food abundance), it is favorable to be a sexual, motile medusa, and when conditions are poor (low food abundance), it is favorable to be an asexual, sessile polyp.

Chapter 31

RECAP 31.1

1. A soft body covering maximizes flexibility and body conformation, allows movement into confined spaces, and permits direct exchange of gases and fluids with the environment. Movement of soft-bodied organisms is largely through the actions of

muscles acting on a hydrostatic skeleton. A hard body covering, in contrast, provides greater protection from predation and dehydration and provides a structure to anchor muscles that attach to appendages, which can assist in movement and feeding. Respiration in a hard-bodied organism must be accomplished through external openings and movement of gases inside the body.

2. Their jointed appendages and rigid exoskeletons provide support for walking on land. The exoskeletons also provide protection against dehydration, which is important in a terrestrial environment.

RECAP 31.2

1. The dorsoventrally flattened body of flatworms ensures that each cell is near a body surface to allow gas exchange directly with the surface. In addition, the highly branched gut increases the surface area for transfer of nutrients to the nearby cells.

2. Most annelids have a thin, permeable body wall that serves as a surface for gas exchange. Annelids rapidly lose water across this body wall if they are removed from moist environments.

3. The basic body organization of a mollusk includes a muscular foot, a visceral mass, and a mantle that typically secretes a hard, calcareous shell. The primary modifications of the major groups of mollusks have involved the foot and the shell. Over the course of evolution, the foot has been modified as a crawling structure (as in gastropods), a burrowing structure (as in bivalves), a clinging structure (as in chitons), or a sensory and feeding structure (as in the arms and tentacles of cephalopods). The shell has been modified to form a series of flexible but protective plates (as in chitons), a hinged pair of valves (as in the burrowing bivalves), a greatly reduced structure for internal support (as in many cephalopods), or a mobile, spiraling chamber to protect the visceral mass (as in gastropods).

RECAP 31.3

1. In many wormlike ecdysozoans, the cuticle is relatively thin and flexible. This provides only modest body protection and support but allows the exchange of water, gases, and minerals across the body surface. Species in which the cuticle is thin are typically restricted to moist environments. In contrast, most arthropods have a thicker, more rigid cuticle that protects against dehydration and predation and provides support for muscle attachment, which allows for colonization of drier environments.

2. Many nematodes act as scavengers in the soil and are important in decomposition and soil formation (which is critical to agriculture). The nematode *C. elegans* is an important model organism that is widely used by geneticists and developmental biologists. Other species of nematodes parasitize humans, causing diseases such as trichinosis and elephantiasis.

RECAP 31.4

1. The segmented bodies of arthropods, with rigid exoskeletons and jointed appendages, provide support for walking, swimming, and flying, so arthropods are well suited to life in many different environments. Their exoskeletons also provide protection against dehydration and predation.

2. Incomplete metamorphosis involves a series of gradual changes among instars. Complete metamorphosis involves a dramatic morphological change between two developmental stages, as between caterpillars and butterflies

3. One factor contributing to the success of insects is that flight gives insects greater access to plants. Many insect species are specialists on one or a few plant species, and plant diversity is far greater on land and in freshwater environments than in the oceans. Although some insects live in fresh water for part or all of their life cycles, these freshwater environments are closely associated with surrounding terrestrial environments. Crustaceans have been much more successful in the oceans than have insects, and crustaceans may simply outcompete insects in marine environments.

WORK WITH THE DATA, P. 678

1. Estimates:
   a. Number of host-specific beetles species in the forest canopy = (Number of beetles specific to *L. seemannii*) × (Number of species of canopy trees) = (163) × (70) = 11,410.
   b. Number of non-host-specific beetles species in the forest canopy = 1,200 − 163 = 1,037.
   c. Number of beetle species on the forest floor = one-third of the number of species in the canopy (75% of beetles are in the canopy, and 25% are on the forest floor). Based on (a) and (b), there are 11,410 + 1,037 = 12,447 species in a hectare of canopy. Therefore we can estimate 12,447/3 = 4,149 species of beetles on a hectare of forest floor.
   d. Number of species of all insects other than beetles = 1.5 times the number of beetles (40% of insects are beetles, so 60% of insects are non-beetles). From (a), (b), and (c), we can estimate 11,410 + 1,037 + 4,149 = 16,596 beetles in a hectare of Panamanian forest.

   Therefore the number of insects other than beetles in a hectare of forest = (1.5) × (16,596) = 24,894.

   This gives a total estimate for the number of insect species in an average hectare of Panamanian forest of 16,596 + 24,894 = 41,490.

2. a. The estimate for the number of host-specific tropical canopy insects = (50,000) × (163) = 8,150,000.

   b. Add 1 million for generalist and temperate canopy beetles: total beetles = 8,150,000 + 1,000,000 = 9,500,000.
   c. As in Question 1, beetle species on the forest floor = one-third of the number of species in the canopy. Therefore the number of species of ground beetles = 9,500,000/3 = 3,166,667 species.

   As in Question 1, species of all insects other than beetles = 1.5 times the number of beetles. Our worldwide estimate for number of beetle species is 9,500,000 + 3,166,667 = 12,666,667. Therefore we estimate non-beetle species diversity at (12,666,667) × (1.5) = 19,000,000.

   d. Summing the number of species of beetles and non-beetles, we get an estimate of worldwide insect species diversity of 12,666,667 + 19,000,000 = 31,666,667 insect species.

   Note: Many biologists have debated whether Erwin's assumptions are reasonable. Clearly, each estimate is highly dependent on how representative *L. seemanii* is as a tropical forest tree. If the average tropical forest tree has many fewer host-specific beetle species than does *L. seemanii*, then these estimates would be inflated. Likewise, an overestimate of the number of tropical forest trees, or of the percentage of ground-dwelling beetles, or of the percentage of all insects that are not beetles, would lead to further inflation of the estimates. In addition, species diversity of beetles may be higher in Panama than in other areas of the tropics. However, any of these estimates could be underestimates as well.

   Only about 1 million species of insects have been described by biologists worldwide to date (see Table 31.2). All entomologists agree that many more species of insects remain to be discovered, and many new species are discovered and described every year. Most entomologists currently think that Erwin's estimates were high. Each of Erwin's assumptions is now being tested; these tests require extensive work on additional species of trees, additional groups of insects, and in additional areas of the world.

FIGURE QUESTIONS

Figure 31.24 Both are three-part body plans, divided into head, thorax, and abdomen. Both crustaceans and insects have antennae and feeding appendages on the head. However, crustaceans have additional appendages growing from both the thorax and the abdomen, whereas in insects, additional appendages are limited to the thorax. In addition, crustaceans have more than three pairs of limbs on the thorax, whereas insects have three pairs of limbs on the thorax and, in most groups, two pairs of wings.

Figure 31.28 The uppermost dorsal branch on the ancestral arthropod limb is thought to have functioned in gas exchange. The *Hox* gene expression data suggest that this structure may have been homologous to the insect wing.

APPLY WHAT YOU'VE LEARNED

1. The graph shows tracheal density of the insects' legs (red line) and total body (black line) plotted against body length. Researchers wanted to understand the upper limits of tracheal density as a possible limitation for insect size. The tracheal density of the leg is more limiting because it reaches the upper limit threshold at a smaller body size.

2. Insects take in oxygen through pores called spiracles, which open through the exoskeleton to the outside of the body. Oxygen exchange occurs in the tracheal network, an extensive system of tubes that branch into every part of the body. All living cells in the animal are within micrometers of the tracheal network. In contrast, humans and other vertebrates transport oxygen and carbon dioxide in blood and through a network of arteries, veins, and capillaries.

3. Based on the graph, one would predict that the largest living beetle could reach a body length of about 15 cm (with 95% confidence limits of about 12–22 cm). This is the approximate length at which the upper limit of tracheal leg density is reached. The largest living beetle (*T. giganteus*) is slightly bigger than the regression line would predict, but it is within the confidence limits of the prediction.

Chapter 32

RECAP 32.1

1. Deuterostomes have radial cleavage, development of the blastopore into an anus (with formation of the mouth at the opposite end of the embryo from the blastopore), and development of the coelom from mesodermal pockets that bud off from the cavity of the gastrula. However, radial cleavage is thought to be the ancestral condition for all bilaterians and not unique to deuterostomes, as it is also found among some protostomes. The development of the blastopore into an anus is unique to deuterostomes, but it may also represent the ancestral condition for bilaterians.

2. DNA sequences of many different genes provide the strongest support for the monophyly of deuterostomes.

3. The three major groups of deuterostomes are echinoderms, hemichordates, and chordates. Sea stars are echinoderms, acorn worms are hemichordates, and humans, rattlesnakes, and tunas are chordates.

RECAP 32.2

1. Echinoderm larvae have bilateral symmetry, whereas the adults have pentaradial symmetry.

2. Sea lilies and sea urchins use their mucus-covered tube feet to filter-feed by catching passing food particles, including phytoplankton. Sea cucumbers have anterior tube feet that are modified into large, sticky tentacles that can be protruded from the

mouth to capture food, then withdrawn to wipe the food into the mouth. Many sea stars use their tube feet to grasp and pull open bivalves, then push their stomach into the bivalves to digest them. Like some echinoderms, hemichordates are filter-feeders, but they lack tube feet. Instead, they capture their food on a large, mucus-covered proboscis and then move the mucus and food by cilia into the mouth.

RECAP 32.3

1. Chordates are characterized by a dorsal hollow nerve cord, a post-anal tail, and a notochord. Vertebrates have an anterior skull that encloses a large brain, an internal skeleton supported by the vertebral column, internal organs that are suspended in the coelom, and a well-developed circulatory system driven by a ventral heart.

2. Hagfishes have a weak circulatory system with three accessory hearts (rather than a single, large heart). They have only a partial skull and a simpler brain than other vertebrates, and they lack separate, jointed vertebrae in their skeletal system. Because hagfishes lack fundamental traits shared by all other vertebrates, some biologists consider hagfishes to be the sister group of vertebrates, rather than a member of the vertebrates. Another possibility is that these traits in hagfishes have been secondarily lost or simplified.

3. The four appendages common to most vertebrates are the two pectoral appendages and the two pelvic appendages. In most swimming vertebrates, these appendages function as fins. They are commonly used for propulsion (especially the pectoral fins) but are also used for steering, stabilization, and manipulation of the body position in water. Among tetrapods, the appendages are often modified into limbs used for walking, running, jumping, burrowing, climbing, grasping, and manipulating objects. There have been several reversals to finlike limbs in aquatic tetrapods (several times among amphibians, turtles, birds, and mammals, for example). The pectoral limbs of tetrapods were modified into wings for powered flight in at least three different lineages (birds, bats, and the extinct pterosaurs). The limbs have been modified for gliding as well (in fishes, amphibians, lizards, and mammals). One or both pairs of appendages have been lost or greatly reduced in many groups of fish, amphibians, reptiles (including birds), and mammals. Some well-known examples of limb reduction or loss include the completely legless caecilians and snakes, the loss of external hindlimbs in whales and manatees, and the greatly reduced forelimbs of flightless birds.

RECAP 32.4

1. Amphibians exchange gases and fluids through their permeable skin, which makes them highly vulnerable to many environmental toxins and some pathogens (especially an introduced chytrid fungus). Many species of amphibians have a biphasic life cycle, so they are vulnerable to habitat degradation and loss of both aquatic and terrestrial environments. Most amphibians do not move long distances, so they do not easily move into new habitats when their local environment is destroyed. For these reasons, they are also sensitive to rapid climate changes. Many species of amphibians have highly specialized habitat requirements and live in very restrictive ranges. Habitat loss or change within these restricted ranges often results in extinction.

2. The major mammal lineages rapidly diversified after the mass extinction event at the Cretaceous–Tertiary boundary. The extinction of many large lineages of dinosaurs may have opened up opportunities for mammalian diversification. By this time, the major continents had separated by continental drift, so there were independent radiations of mammals in Laurasia, Africa, South America, Europe, and southeastern Asia.

3. Fossil remains of extinct theropod dinosaurs show that many features once thought to be restricted to birds, such as feathers, actually evolved much earlier among the theropods. Other typical "bird" morphological features, such as air-filled bones and a furcula (wishbone), are also typical of the larger group of theropods. Among living reptiles, DNA sequence analyses clearly unite birds with the crocodilians (the other living archosaurs). The combined evidence from many sources that birds are a surviving group of theropod dinosaurs is now overwhelming.

4. Hair evolved in the ancestor of mammals; feathers evolved among theropod dinosaurs (seen today among the birds). Among the living tetrapods, birds and mammals are endothermic. Hair and feathers provide body insulation for mammals and birds, respectively. Without these forms of insulation, the maintenance of metabolic body heat would be difficult. Fossil evidence shows that many extinct theropod dinosaurs also had feathers, so many paleobiologists predict that they were endothermic as well. Endothermy would also be expected in large, active predators—a description that fits our current view of many theropod dinosaurs.

RECAP 32.5

1. New World monkeys are all arboreal, and most have a long, prehensile tail that they use to grasp branches in climbing. Many Old World monkeys are more terrestrial, and they all lack a prehensile tail.

2. Changes in a regulatory gene that affects skull development resulted in delayed somatic development in humans (neoteny). This led to the evolution of adult human skulls that are shaped more like those of juveniles (with relatively large brains and small jaws). Thus the human skull increased in size without changing as much in shape as occurs in our closest relatives, the chimpanzees.

WORK WITH THE DATA, P. 700

1.

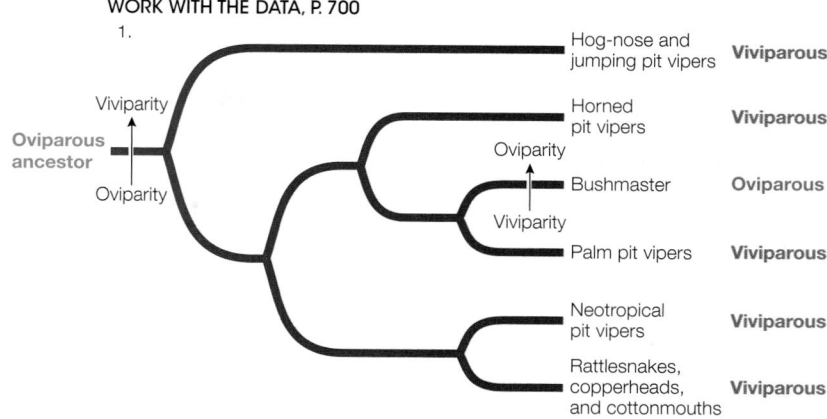

The minimum number of changes is two, one of which is a reversal from viviparity to oviparity.

2.

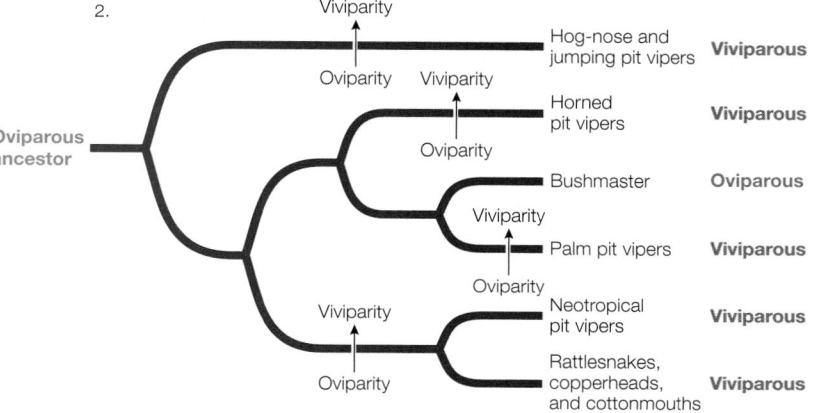

Four changes from oviparity to viviparity are required, or two additional transitions in reproductive mode compared with the answer to Question 1.

FIGURE QUESTIONS

Figure 32.17   Most adult frogs are insectivorous, but insects can be very limited during early spring or the beginning of wet seasons. By reproducing in the water, frogs can take advantage of an abundant seasonal resource (algal blooms in temporary water) as a food source for their offspring. This allows the developing tadpoles to specialize on an abundant resource until they are large enough to feed on insects.

Figure 32.19   In the amniote egg, nutrition is provided by the yolk sac, and the allantois functions in gas exchange and waste collection. The chorion holds all the various membranes together with the embryo, and also functions in gas exchange. The amnion provides structural support for the developing embryo. In viviparous species, the placenta provides nutrition, as well as gas and waste exchange, through the maternal circulatory system via the umbilical cord. Note that the yolk sac and allantois are involved in the makeup of the umbilical cord (retaining their roles in nutrition, gas exchange, and waste exchange), and the amnion and chorion retain their functions of structural support.

APPLY WHAT YOU'VE LEARNED

1. In *Sceloporus* lizards, there is a general trend toward viviparity occurring in colder (blue lineage) environments, with a few exceptions, lending support for the cold-climate hypothesis. Viviparity is advantageous in cold environments because mothers can behaviorally maintain higher temperatures for developing embryos. In addition, increased maternal investment leads to larger offspring among viviparous lizards, so the young are more likely to survive the shorter growing season of cold environments.

2. Advantages of viviparity: It can keep eggs safe. Behavioral thermoregulation can keep embryos at an optimal temperature for development. Increased maternal investment results in larger offspring that are more likely to survive the shorter growing seasons of cooler environments.

   Disadvantages of viviparity: If the mother dies during the gestation, all the eggs die with her. Each female produces fewer offspring. Females are encumbered with the developing embryos, which can make escape from predators harder.

   Advantages of oviparity: Females are freed from the weight of eggs, so they can reproduce more often and escape faster in response to predation attempts. They can also produce larger clutches of offspring.

Disadvantages of oviparity: Developmental temperature of eggs is dependent on the nest site, which limits the temperature of development in colder environments. Females cannot continue to contribute resources to developing eggs, which results in smaller offspring.

3. In mammals, the evolution of viviparity occurred only once. In squamates, viviparity is estimated to have originated *more than* 100 times, so there is a larger sample size of independent events for testing hypotheses. The transition in mammals occurred *more than* 100 million years ago. Many of the transitions to viviparity occurred much more recently in squamates (see the *Sceloporus* phylogeny), making it easier to reconstruct the probable environmental conditions of the time.

4. (Answers will vary.) An appropriate study design should test or demonstrate that (1) females of viviparous species behaviorally thermoregulate, and internal body temperature is different from nest temperature of oviparous species, and (2) differences in incubation temperature affect the fitness of offspring. An example study design was used by Ji et al. (2007), who submitted females of the common sun skink (*Mabuya multifasciata*) to five different thermal regimes throughout their gestation period and compared female body temperatures across the thermal regimes and to those of nongravid females. The researchers also measured phenotypic and performance traits of the offspring to assess the relationship between fitness and developmental temperature.

Chapter 33

RECAP 33.1

1. Plant organs, such as leaves and roots, have structures that allow maximal acquisition of gases and nutrients. Also, plants can grow throughout their lifetimes, enabling them to respond to environmental cues. A plant can redirect its growth to exploit opportunities in its immediate environment.

2. Plants grow throughout their lifetimes; animal organs often stop growing at adulthood. Plant cells are totipotent; in animals, only the zygote and perhaps the earliest embryonic cells are totipotent. Plant cells divide unevenly in cytokinesis to orient organ development; in animal development, cell movements are prominent.

3. Apical–basal patterns develop in plants by unequal cell division leading to differentiation between daughter cells destined for the tip, or apex, of the plant and those destined for its base in the early embryo. Radial patterns develop by cell differentiation, whereby the embryo is first a sphere and then a cylinder.

4. Without apical–basal polarity, there will be just a massive globular embryo, and root and shoot apices will fail to develop. So these organs will could arise at any place in the developing plant, or not arise at all.

RECAP 33.2

1.

Tissue	Structure	Function
Parenchyma	Thin-walled, large vacuoles	Photosynthesis in leaves, storage in roots
Collenchyma	Elongated, thick-walled at corners	Flexible support in petioles and stems
Sclerenchyma	Thick-walled, bundles	Support of stems, occurs in fruits

2. Collenchyma has primary walls at the corners of cells and is somewhat flexible; sclerenchyma has thick, less-flexible secondary walls.

3. When tracheids die, pits in their walls remain and allow water to flow between the cells. Vessel elements are long tubes, laid end-to-end, that die and form a continuous tube.

RECAP 33.3

1. An apical meristem has stem cells that continuously divide. Some of the daughter cells form new organs (e.g., leaves), but others remain to divide further as part of the undifferentiated pool.

2. Root and shoot apical meristems give rise to primary meristems, which are responsible for primary growth of plants and form all plant tissues. There are three primary meristems: (1) protoderm forms dermal tissues; (2) ground meristem forms ground tissues; and (3) procambium forms vascular tissues.

3. The root apical meristem forms the root cap and root primary meristem. The latter forms the dermal, ground, and vascular tissues of the root. The root grows by cell expansion in a region just above the root apical meristem.

4. Primary: apical meristem; secondary: lateral meristem (vascular cambium) and lateral meristem (cork cambium)

Primary: Xylem, phloem, parenchyma, dermal tissue; secondary: secondary xylem and phloem, cork

Primary: growth in length; secondary: growth in thickness

RECAP 33.4

1. Seeds of wild relatives of crop plants have genes that may have been selected against during plant domestication. These genes might encode phenotypes that adapt the plants to their wild environment but were not valuable to farmers at the time of domestication. But the environment can change, and the genes that were selected against can again become valuable. An example might be a changing climate where roots that grow deep would be advantageous. The cultivated crop may lack genes for this phenotype, but genes for root growth might still be present in a wild relative.

WORK WITH THE DATA, P. 729

1. CG appears to be highest in the upper leaves and shoot, where it is made. The lower leaves and stems contain less CG, indicating that CG is transported there on the way to other plant organs.

2. The fifth leaf accumulated CG, which was not transported down the plant vascular system because of girdling. There was some residual accumulation in the sixth leaf that had probably arrived before the girdling. The stem below the sixth leaf had very low CG, indicating the transport was blocked.

FIGURE QUESTIONS

Figure 33.3 The plane of cell division is controlled by two factors: the deposit of cell wall material by the Golgi apparatus and the orientation of microtubules that in turn orient cellulose microfibrils.

Figure 33.5 There is an asymmetry in the cell division forming basal and apical cells. In animal cells, such asymmetry is set up by uneven distribution by cytoplasmic granules that act as signals for cell fate determination.

Figure 33.6 The leaf cells contain chloroplasts and carry out photosynthesis.

Figure 33.12 Pericycle cells are differentiated yet totipotent, and can dedifferentiate to give rise to all cell types of the root.

APPLY WHAT YOU'VE LEARNED

1. The cells are incubated in radioactive thymidine for 2 hours, during which dividing cells take up the radioactive substance and incorporate it into their DNA. After the cells are removed and placed in nonradioactive thymidine, dividing cells take up the nonradioactive substance. Thus cells dividing more than 2 hours after the start of the experiment will not be labeled. As cells formed in the first 2 hours begin to elongate and mature, the cells are pushed up the root by the new cells being made. As cells move from the zone of division to the zone of elongation, the number of labeled cells in the zone of division decreases. As cells move from the zone of elongation to maturation, the number of labeled cells in the zone of elongation decreases.

2. The number of labeled cells increases first in the zone of division, then in the zone of elongation, and finally in the zone of maturation. This indicates that root cells are added at the bottom by cell division. The cells then grow and elongate the root, then finally differentiate into the various tissue types that make up the mature root. Many cells are present in the zone of division, as the initials divide. It takes time for these new daughter cells to grow and move up the root, as they are pushed by newly formed cells below them.

3. Initials in specific locations grow up from their original locations in the meristem, with the initials forming epidermal cells located on the outside, those forming the cortex inside this, and those forming transport tissues nearest the center. These regions form the protoderm, ground tissue, and procambium, which are the precursors that differentiate into all types of root tissue. As they grow upward without migrating, the tissues naturally orient into a cylinder with several layers of cells.

4. Cells in the QC divide very slowly; in this experiment, a single cell was dividing during the first 2 hours (when labeled thymidine was present), but no more division occurred after that time. There could very well have been no division during these 2 hours, as previous experiments show that corn QC cells divide only once every 170 hours. Also, the QC contains very few cells (the exact number varies by species). This indicates that it contributes only a few cells to the growing root. These few cells might replenish initial cells lost through damage or aging, without causing excess growth of root tissue. However, this experiment does not show how (or whether) cells move from the QC to the initials.

Chapter 34

RECAP 34.1

1. Plant cells swell when they take up water. The cell wall resists this swelling, causing an increase in pressure within the cell. This results in more rigid forms of the plant organs and also reduces the tendency of the plant cells to take in more water.

2. Aquaporins are proteins that form membrane channels, allowing the passage of water via osmosis. They occur in the cell membrane and vacuolar membrane. Their presence and numbers affect the rate of water movement across the membrane, from a region of higher water potential to a region of lower water potential. One would expect increased expression of aquaporins in the vacuolar membrane as the cell expands, allowing increased uptake of water and turgor pressure.

3. The apoplast lies outside the cell membrane and consists of the cell wall and intercellular spaces. The symplast is the cytoplasm of cells that can be considered as a continuous compartment if cells are connected by plasmodesmata. Water moves rapidly through the apoplast and more slowly through the symplast.

RECAP 34.2

1. A tree was cut near its base, and the upper part of the cut stem was placed in a poisonous solution. The poison rose in the tree, killing cells as it went. This indicated that the fluid could rise without the need for the root.

2. a. Yes. The difference in water potential between the soil and the leaf (1.7 MPa) is enough to overcome gravity and draw water to the top of the tree.

   b. No. If the soil water potential decreased to −1.0 MPa, it would be more negative than inside the root cells and water would leave the roots (and enter the soil).

   c. If all the stomata closed, the leaf water potential would not be as negative. This in turn would make the xylem water potential less negative, and so on down to the roots. This would make the difference between the leaf water potential and the root water potential insufficient for water to flow from the roots to the leaves (toward a more negative water potential).

3. Transpiration evaporates water from cell walls in the leaves. This must occur because it begins the process. The increase in negative pressure potential resulting from transpiration draws more water into the cell walls and begins to exert tension (the second part of the process) on the entire water column within the xylem. Cohesion draws water molecules together and makes them "stick." It prevents water in the column from breaking, thus losing tension and failing to rise.

4. The great difference between water pressure potential in the soil and the air is sufficient to pull water upward through the xylem, as water evaporates into the air. This is evidence for transpiration. The continuous column of water moving upward through the xylem is evidence for cohesion. Cut stems show high negative pressure potentials, showing that the xylem is under considerable tension.

RECAP 34.3

1. In sunlight, an H^+ pump in the cell membrane of guard cells pumps H^+ out of the cells. This sets up an electrochemical gradient across the guard cell membrane and K^+ enters the cells. This in turn sets up an osmotic gradient and water enters the guard cells, causing them to become turgid and a gap to appear between adjacent guard cells. This gap is an open stoma. In darkness, the proton pump becomes less active, K^+ diffuses passively out of the cell, and the stoma closes.

2. On a hot, dry day, mesophyll cells lose water and this results in stomatal closure. Stomata remain open on cooler, humid days.

RECAP 34.4

1. A source is an organ that produces more carbohydrates that it requires. Sources can be photosynthetic (e.g., leaves) or storage organs with starch (e.g., seeds). A sink is an organ that does not produce enough carbohydrates for its own use and must import them from a source. Examples of sinks include roots and flowers.

2. Source cells load sucrose into sieve tubes of the phloem, decreasing their water potential. Water from adjacent xylem vessels enters the sieve tubes. This results in pressure inside the tubes, so the sap flows. At the sink, sucrose is removed, so the water potential inside the tubes increases. Water leaves to enter the xylem. The combination of these two events causes pressure flow.

WORK WITH THE DATA, P. 742

1. There was increased water-use efficiency (biomass accumulation divided by transpiration) in plants overexpressing *HARDY* because there was decreased transpiration (line 2) and increased photosynthesis–carbon fixation (line 3).

2. Use a *t*-test for paired samples.

FIGURE QUESTIONS

Figure 34.3 A hypertonic environment (a higher concentration of solutes outside than inside root cells) will result in osmosis. Plants wilt when water leaves root cells, reducing turgor pressure.

Figure 34.5 Tight junction

Figure 34.7 With high humidity, there would be reduced transpiration from the leaf. This would in turn reduce water pull from the leaf veins and in turn reduce water from the xylem and root uptake.

APPLY WHAT YOU'VE LEARNED

1. The increase in cell volume indicated that TIP may increase water permeability of the cell membrane and is therefore an aquaporin. The unrelated mRNA condition showed no increased cell volume, so the volume increase was not just due to any mRNA. The uninjected cells were controls to show that there was no natural increase when cells were placed in an environment that had increased water potential relative to the cell interior. This confirms low osmotic permeability of the frog oocytes.

2. The oocytes burst because there was a continuing uptake of water due to the plant aquaporin. No, plant cells injected with TIP mRNA would not have burst. In plant cells, the cell wall exerts turgor pressure potential that stops the cell from bursting.

3. The table shows direct measurements of increased osmotic water permeability in oocytes injected with TIP mRNA. This indicates that the increase in size shown in the figure, due to the presence of TIP in the oocyte membrane, was due to increased water flow, and in this case water flowed into the cell because the oocytes were in an environment of higher water potential (hypotonic).

4. In the apoplastic pathway, the PIPs would be most prominent in the stele and in the endodermis, where the Casparian strip prevents apoplastic movement of water. In the symplastic pathway, the PIPs would also be found in epidermal and cortical cells.

5. Use a specific stain that can visualize only PIP. Elongating cells would have increased staining in their cell membranes for PIP and in their tonoplasts for TIP.

Chapter 35

RECAP 35.1

1. Solutions that were thought to be pure were not, as they contained trace amounts of elements that could not be detected at the time. Recently developed chemical methods made it possible to detect essential elements that were not previously recognized.

2. Yellow leaves in a young plant may be the result of an iron or a sulfur deficiency; yellowness in older leaves suggests a nitrogen deficiency.

RECAP 35.2

1. Plants express specialized transporter molecules to move nutrients in soil water across root cell membranes. The more of a certain transporter that is made, the more its nutrient moves into the plant.

2. Plants can grow toward nutrients in the soil and orient stems and leaves to maximize exposure to light and air for photosynthesis.

RECAP 35.3

1. Heavy irrigation after a prolonged dry period may produce runoff of topsoil (the A horizon) and leaching of ions (especially anions) into the subsoil, making fewer nutrients available to plant roots. Converting land use from virgin deciduous forest to crops will change the composition of living organisms in the soil, as many organisms that live in association with tree roots will disappear. The soil structure and texture will also change, because roots will no longer be present to hold the soil together and make air spaces. The soil chemistry will change, because crops take up nutrients from the soils and the nutrients are removed from the system when the crops are harvested.

2. Cation exchange frees ions bound to soil particles into the soil solution, where they can enter plant roots.

3. There are no differences between organic and inorganic nitrogen fertilizers in terms of plant nutrition; both enter the plant root as nitrate NO_3^-.

RECAP 35.4

1. In both cases, the plants supply the other organism with photosynthate (e.g., sugars). Mycorrhizae supply phosphorus; bacteria in nodules supply fixed nitrogen.

2. Numerous species fix nitrogen. Loss of one species might allow populations of other species to expand and take on additional nitrogen fixation. Loss of all nitrogen-fixing species would mean that only abiotic methods (e.g., industrial methods) could be used for nitrogen fixation. Loss of nitrogen-fixing species would likely reduce overall nitrogen in the soil, meaning less would be available for plant growth.

3. The corn crop depletes the soil of nitrate. The soybeans do not require nitrate, as they have nitrogen-fixing nodules. When corn is rotated with soybeans, the soil becomes replenished with fixed nitrogen via free-living bacteria.

RECAP 35.5

1. Carnivorous plants capture animals, digest their proteins, and absorb the amino acids. The primary nutrient they acquire is nitrogen.

2. The experiment with mutant *Arabidopsis* suggests that *Arabidopsis* uses either its own or exogenous strigolactones for growth regulation and has the appropriate receptor and response mechanisms. This reinforces the idea that an ancient mechanism to attract beneficial microbes is also used for modern plant growth regulation. Or the reverse might be true: the original function of strigolactone might have been as a plant hormone, and its role in plant–microbe interactions might have evolved later.

3. Holoparasitic plants can gain reduced carbon through association with hosts, so the genes encoding photosynthesis functions are not under selection pressure, because having them would not confer any survival and reproductive advantage for the parasites. So any mutation that renders such a photosynthesis gene nonfunctional will not be deleterious.

WORK WITH THE DATA, P. 759

1.

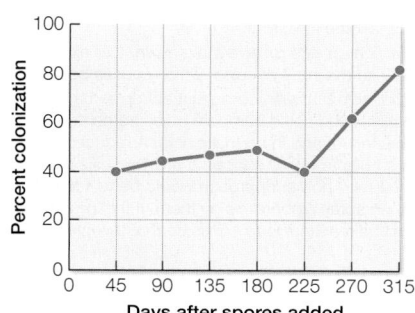

Root colonization was maximal 315 days after spores were added. The lag time was probably due to the need for the spores to germinate and grow into hyphae that could colonize the cassava roots.

2. a. Cassava root crop production was just as good with fungal spores alone (38 g) as with fertilizer alone (35 g). Adding spores to fertilizer had a positive effect (43 g).

b. A *t*-test for paired samples could be used to test for significance in differences for spores-treated and untreated experiments.

FIGURE QUESTIONS

Figure 35.5 Negatively charged ions (anions) can bind to certain soil particles that are positively charged and undergo ion exchange. But this does not occur at the pH of typical soils. For most soils with negatively charged particles, anions can leach out of the soil rapidly unless taken up by the plant roots.

Figure 35.8 O_2 is a strong oxidizing agent, and could attract electrons (and hydrogen atoms) that are needed for nitrogenase. These electrons come from a strong reducing agent.

APPLY WHAT YOU'VE LEARNED

1. Yes, some of the data support the researchers' conclusion that nickel is an essential micronutrient. Specifically, the germination data presented in the graph show that seed germination rate is related to the nickel quantity present in the seed. Seeds produced from plants grown under extremely low nickel concentrations have significantly lower germination rates than seeds produced from plants grown in higher concentrations of nickel. Because barley plants have high rates of germination failure without nickel, they cannot complete their life cycle in the absence of this nutrient. Over three generations, the plants were gradually depleted of nickel. This makes nickel an essential nutrient.

2. The data show that the mass of seeds produced by a plant is not affected by lack of nickel availability to the plant. This means that production of seeds within a parent plant occurs through processes that are not dependent on nickel. This conclusion does not contradict the conclusion from Question 1, which states that germination of seeds is dependent on nickel. Seed production within a parent plant and germination of the seed once it leaves the parent plant occur by two separate processes, only one of which is dependent on nickel.

3. Because soil is a complex mixture of organic and inorganic matter, it is more difficult to control the nutritional content of soil than it is to make a liquid growth medium with a defined and well-controlled content. The researchers would not have been able to easily remove nickel from soil and verify its nickel content. The results from using hydroponic growth conditions can be extrapolated to growth in soil because barley is able to adequately obtain the nutrients it needs from both hydroponic solutions and soil. Therefore there is no reason to think that different results would be observed using soil.

Chapter 36

RECAP 36.1

1. Fire produces ash, which enriches the soil with plant nutrients. A seed that germinates as a result of fire could have an advantage in such nutrient-rich soil. Such seeds might have a selective advantage in habitats such as forests, where periodic fires are natural events.

2. Both hormone receptors and photoreceptors are proteins that change when bound to a signal and set off a signal transduction response in a cell. Hormones are chemical in nature (such as small molecules, proteins, and steroids), whereas photoreceptors change when light binds to them.

3. Seeds would be treated with a mutagen and then planted. Those seeds that germinated immediately might have a mutation that affects dormancy. These mutant plants could be compared with wild-type plants to isolate the genes involved.

RECAP 36.2

1. An inhibitor of mRNA translation would block the gibberellin-induced synthesis of hydrolases. A proteasome inhibitor would block the hydrolysis of the repressor that normally is broken down when gibberellin affects cells. So gibberellin signal transduction would be inhibited and there would be no increase in hydrolase synthesis. In the case of auxin and coleoptiles, without inhibitors, there would be cell expansion and coleoptile growth. These effects would not occur in the presence of mRNA translation and proteasome inhibitors.

2. Cells away from the lighted side receive more auxin. Cell elongation on the shaded side is accelerated with the input of auxin, which causes bending toward the light.

3. To test for the relationship between corn stunt spiroplasma infection and suspected inhibition of gibberellins, you could compare measurements of gibberellins in plants infected with the bacterium and in normal plants. A reduction of gibberellins in the spiroplasma-infected plants would support the hypothesis that the infection inhibits synthesis of gibberellins. Another approach would be to infect normal plants with the spiroplasma and then spray gibberellins on them. If this process reversed the stunt phenotype, it would be reasonable to conclude that the infection reduced synthesis of gibberellins.

RECAP 36.3

1. Cytokinins stimulate axillary buds to grow into branches, which would make the plant more bushy. Auxins help maintain apical dominance, which prevents branching. Thus this plant most likely has a low ratio of auxin to cytokinin.

2. The charcoal in the bag absorbs ethylene gas, which is released by ripening fruits. The lack of ethylene prevents over-ripening and decay.

3. Both ethylene and brassinosteroids promote leaf senescence, so either or both of these hormones are most likely active. Cytokinins delay leaf senescence, so they are most likely inactive in this plant.

RECAP 36.4

1. An action spectrum for the bending of coleoptiles toward light shows that blue light is most effective. Genetic screens for blue-light insensitivity have shown that there is a blue-light receptor present on normal plants that is not present in plants unable to respond to blue light.

2. Red light causes phytochrome to convert from the P_r to the P_{fr} form. The latter is active in promoting seed germination. Far-red light reduces P_{fr} by causing it to covert to P_r, which is inactive in promoting seed germination.

WORK WITH THE DATA, P. 772

1. Light was shone from the right of the observer.

2. These data indicate that the tip is necessary for bending to light (if the tip was cut off, the plant did not bend). The fact that cutting the very apex (1.27 mm) of the tip allowed some bending indicates that there was not significant injury and the plant was still functional.

3. The data indicate that exposure of the tip to light is necessary for the bending response. The six coleoptiles that bent slightly even though they were covered possibly were ineffectually covered—that is, they might have had some light leakage, like that noted for the coleoptiles where the ink on the tubes was cracked.

WORK WITH THE DATA, P. 781

1. a. A high number means that more light energy at that wavelength is needed for the response (seed germination) than a low number.

   b.

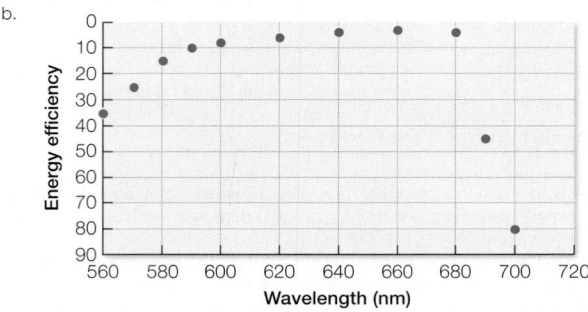

   The highest efficiency was in the red part of the visible spectrum (600–680 nm).

2. a. The photoreceptor has two forms that reversibly interconvert: red light stimulates the formation of the active receptor, and far-red light stimulates the conversion of the active receptor to its inactive form.

   b. A small amount of phytochrome was in the P_{fr} state in the dark.

FIGURE QUESTIONS

Figure 36.2   You could collect many seeds from strains of pea plants in nature. Do not treat these seeds with a mutagen, but allow them to germinate in the dark and then expose them to ethylene.

Figure 36.5   To stimulate the breakdown of starch, seeds are imbibed in water and mashed up.

Figure 36.9   Proton pumps are present in mitochondria and chloroplasts, where they are involved in the production of a proton gradient in chemiosmotic synthesis of ATP.

Figure 36.10   Only ubiquitin and proteasome are the same. The other molecules are specific to the particular hormone. For example, there is a specific receptor for auxin and a different receptor for gibberellin.

APPLY WHAT YOU'VE LEARNED

1. Acting alone, *phyC* enhances leaf area, whereas *phyA*, *phyB*, and *phyD* do not.

2. *Arabidopsis* seeds would be treated with a mutagen and planted. Those plants with reduced leaf size would be isolated and grown to examine for *phyC* mutations.

3. The leaf area phenotypes produced by these three conditions are equal. All three groups have leaf areas about one-third smaller than that of the WT group. This suggests that (1) the *phyA* and *phyC* genes have the same effect on controlling leaf area (these results show the effect of the mutant, or nonfunctional gene, so the functional gene should increase leaf area by one-third); (2) either gene will work with equal effect in the absence of the other (shown by *phyAD* and *phyCD*); and (3) having both genes dysfunctional does not increase the effect (shown by *phyACD*).

4. A deficiency in *phyB* causes a very long leaf (a long petiole) and a very small leaf area (small leaf). This suggests that, if this gene were functional, it would control leaf growth by greatly inhibiting petiole length and greatly increasing leaf area. This would produce a leaf with a large surface area for photosynthesis, which could be highly adaptive. *PhyA* and *phyC* have the same effect on leaf area as *phyB*, but to a lesser extent. *PhyA* and *phyC* also inhibit petiole length, with *phyA* having the greater effect and *phyC* adding slightly to this effect. It is likely that, in plants in nature, the control of these two leaf features results from the ratio of the various phytochromes present.

Chapter 37

RECAP 37.1

1. The pollen tube is required to deliver the sperm cell from the stigma of the plant to the egg nucleus in the embryo sac, which is located in the ovary at the bottom of the style. Failure of the pollen tube to grow properly and reach the egg nucleus will result in failure of reproduction, including failure of seed and fruit formation.

2. In the megagametophyte, one of the original nuclei becomes an egg cell; the remainder form the protective and nutritive embryo sac. In the microgametophyte, the eight original cells from meiosis become four two-celled pollen grains, each containing a sperm cell as well as a tube cell, which forms the pollen tube. The female structure is larger and contains nutrients for the developing embryo. The pollen grains are tiny and numerous, making movement easier and pollination more likely.

3. In triploid cells undergoing meiosis, there cannot be pairing of homologous chromosomes in meiosis I, so meiosis I is abnormal and functional gametes do not form. A fruit is formed from the ovary wall of the flower. Seedless grapes are probably propagated by cuttings (vegetative reproduction).

4. All fruits protect the seed and aid in seed dispersal, but the variety of fruits means they can fulfill these functions in many different ways, depending on locations and environmental factors. Some seeds, such as many tree seeds, fall to the ground and remain near their parent. Others are much smaller and have structures (wings, burs) that enable their dispersal by wind or animals. In this way, they can move long distances, increasing their range and chances of survival.

RECAP 37.2

1. a. The mutation stabilized the CO protein.

   b. The mutation caused nonfunction of the FD protein.

   c. The mutation increased expression of the FLC protein.

   d. The mutation caused constitutive expression of the CO protein.

2. Long-day plants typically bloom when days are long and nights are short, as in the summer. But in a greenhouse setting, plants can be tricked into responding as though the nights are short by turning on the lights for a short time in the middle of the night (night interruption lighting). This breaks up the long night, so the plants act as though the night is short—that is, they begin to flower.

3. Examples of evidence include:
   • Spinach will not flower if the shoot is masked to imitate short days but will flower if the bud (and not the shoot) is masked.
   • If a photoperiodically induced leaf is immediately removed from a plant, flowering will not occur. But if the leaf remains on the plant for several hours (allowing time for the substance to move to the bud), flowering will occur.
   • In grafting experiments, if only one plant (or plant part) is induced to flower, the other will flower as well.

RECAP 37.3

1. Asexual reproduction is rapid, and it conserves the genetic adaptations of a plant. However, in asexual reproduction genetic homogeneity makes the offspring and population derived from it susceptible to a changing physical or biological environment, since the phenotypes in the population are all the same. Sexual reproduction produces genetic diversity in offspring and the population derived from it, and the phenotypic diversity makes the population adaptive in a changing environment. However, sexual reproduction is a slower process than asexual reproduction.

2. Asexual reproduction can occur without the need for pollination, producing many plants quickly. Many grow by means of stolons or rhizomes, which make them well adapted for poor or eroded soils, such as those on hillsides, or as pioneer plants in unstable soils such as sand dunes. Asexual reproduction is also adaptive in deserts, where it may be difficult for seedlings to become established.

3. Several approaches might be taken, such as a genetic screen for meiotic cells that do not separate chromosomes at anaphase I, or a search for proteins (and then their genes) that bind to SWII protein.

WORK WITH THE DATA, P. 796

1. a. The plants were grown in short-night conditions so that flowering would not be induced by having a long night. The only condition that would affect flowering would be the brief light during a long night.

   b. The data show that a 700-nm wavelength of light was effective at inducing flowering whereas a 680-nm wavelength inhibited flowering, indicating that the receptor is phytochrome. The far-red light for 8 min at 700 nm converted phytochrome to its active form, P_r.

2. During the day there is more red than far-red light, and the P_r form of phytochrome gets converted to the P_{fr} form. At night the P_{fr} converts back to P_r. This takes time; in the "No light" column, the data show that it took 8.5 hours for P_r to accumulate to sufficient concentration to promote flowering. This was delayed if red light (660 nm) was administered at the beginning of the dark period, as this converted even more P_r to P_{fr}. However, if far-red light (720 nm) was administered at the start of the dark period, the P_{fr} that had accumulated during the day got converted to the active form, P_r, which allowed flowering to occur after a shorter period in the dark (7 hours).

WORK WITH THE DATA, P. 797

1. The plants with their leaves removed did not flower. Therefore the leaf senses photoperiod and must be present for flowering to occur.

2. The receptor is in the leaf.

3. The environmental signal (photoperiod) is received only by the single treated leaf. The signal sets in motion a signaling pathway that ends in the bud at the shoot apex, where flowering is initiated. The data imply that an inducer of flowering is

produced in leaves and is able to travel through the plant from the leaf to the shoot apex.

FIGURE QUESTIONS

Figure 37.3 Pollen tube growth requires loosening of the cell wall (hydrolase enzymes), deposition of new cell wall material (synthesis of cellulose and matrix polysaccharides), expansion of the cell membrane and cell contents (involvement of the Golgi apparatus), and cell expansion (driven by osmosis from the vacuole).

APPLY WHAT YOU'VE LEARNED

1. The *gi* mutant inhibits or greatly delays flowering in the pea plant.

2. Yes, the evidence suggests that the *gi* mutant is unable to produce the floral chemical. When only the *gi* mutant is present (*gi/gi*), flowering never occurs. But whenever the WT is present, flowering does occur. This suggests that only the WT contains the floral chemical and that it is able to move in both directions (from stock to scion, or vice versa), causing the signal to reach the apex of the plant and stimulate flowering.

3. Under long-day conditions, the WT gene is expressed. It stimulates production of the FT protein that acts as a floral signal, and the plant flowers after 15–20 vegetative nodes have formed. However, the *gigas* gene is a null, or inactive, gene; it does not stimulate production of floral signal. In cases where the *gigas* gene is expressed, the plant remains vegetative; it either does not flower at all or flowers only after a long delay.

Chapter 38

RECAP 38.1

1.
Avr2Avr3	Healthy	Healthy	Diseased
Avr1Avr4	Healthy	Healthy	Healthy

2. Both general and specific immunity involve similar signaling pathways. For general immunity, the researcher could look for PAMPs on the pathogen, and a weaker response. For specific immunity, the researcher could look for gene-for-gene resistance (*Avr* genes) and programmed cell death and the formation of a necrotic lesion at the site of pathogen invasion.

3. Salicylate initiates systemic resistance. Application of this molecule at the site of a wound would trigger the production of protective PR proteins against further infection. This would be a long-term effect.

RECAP 38.2

1. It indicates that latex is important in defense against herbivores such as slugs.

2. Because nicotine is a secondary metabolite that is always present in tobacco leaves, it is a constitutive defense that can deter many herbivores. Induced responses would be triggered by herbivores that are not affected by the nicotine.

3. a. This is an induced defense, as the frass contains an elicitor.

   b. Chemicals in the frass act as elicitors that trigger signal transduction pathways. One pathway involves changes in electric potential of the cell membrane, which is transmitted via the symplast to the whole plant. Another pathway triggers production of reactive oxygen species, which act as signaling molecules in pathways that lead to change in gene expression. Yet another pathway leads to production of jasmonate and other hormones, which trigger systemic defenses, and may even travel to neighboring plants to stimulate their defense responses.

4. Jasmonates are made in response to an elicitor from herbivory. Jasmonates can induce the synthesis of volatile molecules, which diffuse through the air to other plants or organs in the same plant.

RECAP 38.3

1. a. The effects of reduced rainfall could include dehydration and osmotic stress. Genetic responses might include alterations in leaf anatomy, with a thicker cuticle to reduce evaporation; a more extensive root system to obtain water; and accumulation of solutes in the roots, which would reduce root water potential and result in more water uptake in dry soils.

   b. Flooding reduces the amount of O_2 available to the plants and results in reduced respiration. Genetic responses might include increased production of pneumatophores or aerenchyma to supply air to submerged plant tissues.

2. She could add genes encoding proteins that produce salt glands or genes that encode membrane proteins that store excess salt in the vacuole. Sequestering salt in the vacuole keeps it away from the rest of the cell, where it can harm chemicals and reactions. (See Key Concept 18.5 for salt-tolerant plants.)

3. a. The lack of a plant cover would allow the accumulation of toxic ions in the soil.

   b. The genes involved might encode proteins that keep the toxic ions away from the plant cells (e.g., store the ions in glands) or away from the plant cells' cytoplasm (e.g., store the ions in vacuoles).

WORK WITH THE DATA, P. 809

1. Wild-type *Sr33* conferred rust resistance. Either a deletion mutant of *Sr33* (bar E5) or a point mutation of *Sr33* (bars E6–E9) resulted in rust susceptibility. Therefore *Sr33* can confer rust resistance.

2. Fielder with high-expressing *Sr33* was rust-resistant, while Fielder without *Sr33* was rust-susceptible. This reinforces the conclusion in the earlier experiments that *Sr33* confers resistance to wheat rust.

3. *Sr35* confers resistance in this relative of cultivated wheat.

4. *Sr35* also confers resistance in bread wheat. Therefore there are two genes that can confer resistance.

FIGURE QUESTIONS

Figure 38.1 Many human diseases are caused by pathogens. Here we suggest just one of many correct answers for each type of pathogen.

Virus: Influenza

Bacterium: Tuberculosis

Fungus: Candidiasis

Nematode worm: Hookworm

Figure 38.2 Both plant organs are exposed to all types of pathogens, although their identities may be different. So both root and leaf cells exposed to the environment would express the defense molecules.

Figure 38.3 Recognition is by shape (like a lock and key) and by chemical interactions between the molecules, such as hydrophobic interactions and ionic interactions.

Figure 38.12 LEA proteins might include chaperonins that bind to target proteins and stabilize them so that the targeted proteins remain in their active state, and other proteins that stabilize membranes and keep them from drying out.

APPLY WHAT YOU'VE LEARNED

1. The wound alone causes the release of a large amount of nicotine. Caterpillar herbivory (which includes the production of FACs) causes slightly less nicotine to be released. This suggests that FACs might have a slight inhibiting effect on nicotine release. Nicotine is a constitutive response; it is present naturally in the tobacco plant and does not have to be induced as a result of gene expression.

2. Jasmonic acid production is stimulated by the production of elicitors when herbivory occurs. Jasmonic acid then causes changes in gene expression that lead to the production of secondary metabolites, such as protein inhibitors (see Figure 38.5). These protein inhibitors limit the effectiveness of the herbivore. Jasmonic acid differs from nicotine in that it does not directly affect the herbivore but stimulates production of another substance that has the effect (i.e., jasmonic acid is an intermediate substance).

3. The burst of jasmonic acid production is much greater with the addition of FACs from the caterpillars. This stronger response suggests that FACs are elicitors that *Nicotiana* must recognize to trigger genetically controlled defense responses in the plant.

4. Jasmonic acid triggers induced responses that limit hornworm herbivory in the tobacco plant. Ethylene suppresses nicotine production, making it easier for hornworm herbivory to occur. These are essentially opposite reactions. This could represent an example of the "evolutionary arms race" between plant and predator.

Chapter 39

RECAP 39.1

1. Cells are limited in size by whether their surface area is sufficient to meet their metabolic needs by diffusion and transport of nutrients and wastes across the cell membrane.

2. The major advantage of being multicellular is that every cell does not have to provide for all of its metabolic needs. Different cells can become specialized to perform specific functions that contribute to the survival and health of the whole organism. With that specialization comes greater efficiency. Also, multicellularity avoids the problem of the decreasing surface area-to-volume ratio that accompanies an increase in size of the organism. A third advantage is that bigger organisms can more easily avoid predation and in turn prey on smaller organisms.

3. The gut is a good example of how multiple tissue types contribute to the structure and function of an organ. Epithelial tissues provide the boundary between the gut and its lumen where food is digested, and also provide transport mechanisms for the uptake of nutrients. Muscle tissues provide the motility necessary to move the gut contents from mouth to anus. Nervous tissues control motility of the gut and also provide sensory information used to control gut functions. Connective tissues give structural integrity to the gut, and one connective tissue—blood—carries nutrients absorbed from the gut to all the other cells of the body.

RECAP 39.2

1. Maintenance of the internal environment is the primary focus of physiology of multicellular animals because the internal environment serves the needs of all the cells of the body. Cell types are specialized to contribute something to the maintenance of the internal environment, but no cell can provide for all of its own needs and therefore cannot exist alone.

2. Negative feedback is information about the regulated variable that can be compared to a set point to generate an error signal. The error signal is relayed to effector organs to bring the regulated variable back to the set point. Positive feedback is information about the regulated variable that can be converted to command signals that push the regulated variable farther from its set point. Feedforward information changes the set point in anticipation of an imbalance so that the regulated variable can remain stable as conditions change.

3. You would inject known volumes (V_1) of mannitol and Evans blue at known concentrations (C_1) into the blood of a subject. After a sufficient amount of time for the injected solutions to circulate around the body, you would take a blood sample and measure the concentrations of Evans blue or mannitol in the blood plasma (C_2). You can then use the formula given for calculating V_2 for mannitol, which will be the extracellular volume, and V_2 for Evans blue, which will be the blood plasma volume. Subtracting the blood plasma volume from the ECF volume will give the interstitial fluid volume.

RECAP 39.3

1. Fluctuations in body temperature can disrupt complex physiological processes because different steps or components of those processes are likely to have slightly different Q_{10} values. As a result, a temperature change can alter the integration of those steps or components.

2.

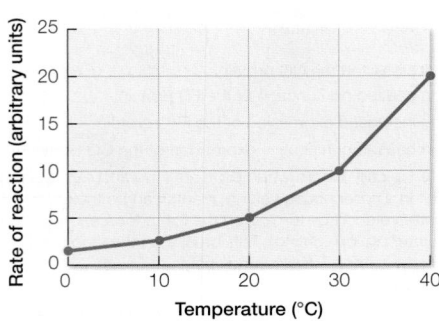

3. Isozymes can have different temperature optima, so expressing different isozymes that have temperature optima that match seasonal changes in the climate can result in seasonal acclimatization.

RECAP 39.4

1. In nature, the ectotherm can use behavioral adaptations such as basking and seeking shade to prevent significant changes in its body temperature. In the laboratory, it does not have the opportunity to bring those adaptations into play.

2. The animal gains heat from radiation, convection, or conduction if the environment is warmer than the surface temperature of the animal, and the animal loses heat if the environment is cooler than the surface temperature of the animal. Evaporation can only result in a loss of heat from the animal, and the magnitude of the loss depends on the amount of water that is evaporated on a heat exchange surface such as skin or respiratory passages.

3. In fishes, the blood leaving the heart goes to the gills, where it comes into temperature equilibrium with the surrounding water. In most fish, that blood flows down the center of the fish through a large vessel leading to lots of smaller vessels that perfuse the tissues of the fish, thus cooling those tissues. In "hot" fishes, the major blood flow from the gills is through peripheral vessels that give rise to smaller segmental vessels going into the muscle tissue. The vessels carrying blood in (arteries) run parallel and close to the vessels carrying blood out (veins). The veins carry heat produced in the muscle, but because of this countercurrent flow of blood in the arteries and veins, the heat from the venous blood warms the arterial blood flowing into the muscle, thus decreasing the loss of heat from the muscle tissue. Countercurrent heat exchange in "hot" fishes acts to keep the heat generated by the powerful swimming muscles in the muscle tissue.

RECAP 39.5

1. Within the thermoneutral zone, metabolism can remain at basal levels because the regulation of body temperature can be achieved by passive means such as changing blood flow to the skin and changing posture. However, as environmental temperature falls, a point will be reached when active thermoregulation is required, resulting in an increase in metabolism. This point is the lower critical temperature. Conversely, as the environment warms, a point will be reached when increasing blood flow to the skin and changing posture will not prevent a rise in core temperature, and active mechanisms such as sweating or panting are required, resulting in a rise in metabolism. This point is the upper critical temperature.

2. The segment of the metabolic rate curve in Figure 39.16 that falls below the lower critical temperature (low end of the thermoneutral zone) represents the amount of heat production necessary to balance the amount of heat loss to the environment. That segment of the metabolic rate curve can be described by the equation shown. If there were no metabolic functions other than balancing heat loss, the heat loss and therefore the MR would be zero when $T_b - T_a$ equaled zero. Thus this MR line below the basal metabolic rate level projects to a T_a that is equal to T_b.

In the equation, the constant K stands for thermal conductance, which is the inverse of insulation. If the curve represented a desert versus a similar sized arctic animal, the curve would have a higher slope (K). The lower critical temperature would be higher for the desert versus the arctic species.

3. Heat is lost from the surface of the body, so a bigger, stockier animal will have a lower surface area-to-volume ratio than a smaller, slimmer animal. Thus, for closely related species, selection has favored larger, stockier variants in the colder latitudes.

4. Feedforward information serves to change set points in anticipation of an upcoming stressor. For example, exposure to a cold environment can elevate the set point for metabolic heat production, so that the increased heat loss from the body to the cold environment is balanced without requiring a drop in core body temperature to trigger the response.

5. The mammalian thermostat is described as a proportional thermostat with an adjustable set point because the drive it generates to effector organs is proportional to the difference between hypothalamic temperature and a threshold hypothalamic temperature for activation of that response. The descriptor of adjustable set point refers to the role of feedforward information in altering the hypothalamic temperature thresholds for different thermoregulatory responses.

WORK WITH THE DATA, P. 832

1. Standardizing results to percent increases in treatments 1 and 2, averaging and calculating standard deviations for those two groups, comparing the two groups to determine if there is a possible order effect through use of a paired t-test.

Subject	Percent increase, experiment 1 Control group (no cooling)	Percent increase, treatment 2 Treatment group (palmar cooling)
1	35.29	11.11
2	24.00	27.59
3	4.55	5.80
4	28.30	12.50
5	25.86	35.44
6	3.28	14.93
7	2.70	76.74
8	6.00	28.57
9	−5.26	28.95
	Treatment group	Control group
10	39.39	7.87
11	11.49	5.62
12	30.00	8.75
13	32.31	39.44
14	57.89	26.19
15	42.62	21.59
16	53.33	18.67
17	48.98	1.35
Mean	25.93	21.83
Standard deviation	18.80	17.53

The P-value (0.058) shows there is no significant difference between the two data sets and therefore there cannot be an order effect.

2. The standardized values (percent increases) are sorted according to treatment—palmar cooling or no cooling. Means and standard deviations are calculated and a paired t-test is applied to the data.

Subject	Palmar cooling	No cooling
1	11.11	35.29
2	27.59	24.00
3	5.80	4.55
4	12.50	28.30
5	35.44	25.86
6	14.93	3.28
7	76.74	2.70
8	28.57	6.00
9	28.95	−5.26
10	39.39	7.87
11	11.49	5.62
12	30.00	8.75
13	32.31	39.44
14	57.89	26.19
15	42.62	21.59
16	53.33	18.67
17	48.98	1.35
Mean	32.80	14.95
Standard deviation	18.54	12.88

The P value of < 0.008 is highly significant, indicating that the increases during the palmar cooling treatments were greater than the increases during the no cooling treatments.

3. The hypothesis that extracting excess heat from the body will increase the capacity of an individual's muscles to do work is supported by these data. Normalizing the data to percent increases and then testing for order effects showed that there was no effect from whether the cooling or the control treatment came first. Testing for treatment effect, however, yielded a highly significant benefit to the cooling treatment.

FIGURE QUESTIONS

Figure 39.2 The muscles are connected to the bones and the bones are connected to each other by connective tissue. Bone is connective tissue, and without bones, the muscles would not be able to exert forces on the environment. Blood is also a connective tissue and supplies the muscles with oxygen and removes wastes and heat. Epithelial tissues line the blood vessels. Neural tissues carry the motor commands to control the contractions of the muscles.

Figure 39.8 The fall of the mouse's temperature at cold environmental temperatures and its rise at high environmental temperatures indicates the limitations on the mouse's ability to regulate its body temperature. At high environmental temperatures, the mouse's metabolic rate rises because active heat loss (e.g., sweating, panting, or licking) requires energy.

Figure 39.12 The fact that both Arctic and tropical mammals fall on this same curve means that BMR has not evolved as an adaptation to climate. If the BMR of arctic mammals was elevated to counter increased heat loss to the environment, the data points for arctic mammals should fall above the curve and the data points for tropical mammals should fall below the curve.

APPLY WHAT YOU'VE LEARNED

1. The group that reached a body temperature of 30°C in a significantly shorter time was the +2°C group. The starting temperature did have an effect on the rate of rewarming, which was 4.9°C/hr for the +2°C group and 3.8°C/hr for the −12°C group.

2. When an animal's body temperature is constant, there is a balance between heat entering the body and heat leaving the body.

$$\text{heat}_{in} = \text{heat}_{out}$$
$$\text{metabolism} + R_{abs} = R_{out} + \text{convection} + \text{conduction} + \text{evaporation}$$

Therefore for body temperature to rise, heat_{in} has to be greater than heat_{out}. For both groups of squirrels in this experiment, heat_{in} can only be coming from a single source, metabolism, since neither group has any source of radiant heat to absorb. However, the squirrels in the −12°C group encounter a much lower ambient temperature. Although they experience little conductive, convective, or evaporative heat loss when sequestered in a closed chamber, the −12°C group could experience greater radiative heat loss, which is a function of the difference between an animal's surface temperature and the ambient temperature. In addition, the −12°C group is starting the rewarming process at a lower T_b. Thus more metabolic heat must be generated by the −12°C group to reach a T_b of 30°C thus requiring a greater total metabolic output and a longer time.

3. The two primary generators of metabolic heat in mammals are shivering and nonshivering heat production. Shivering is the rapid cycling of contractions of antagonistic muscles that do no work on the environment, thus all of the energy released from ATP during this process is released as heat. Nonshivering heat production comes from brown fat. This tissue contains many mitochondria and a good blood supply (thus the brown color in comparison to white fat). The brown fat mitochondria contain a protein, thermogenin that uncouples the movement of protons from the production of ATP. Thus the fat substrate is metabolized, producing heat without producing ATP. These two mechanisms produce the heat to raise the body temperatures of the squirrels.

Chapter 40

RECAP 40.1

1. Epinephrine is a derivative of an amino acid, transported in solution in the blood, and acts on cell surface receptors to trigger an intracellular signaling cascade that produces an immediate response. Estrogen is a steroid hormone that is transported in the blood in combination with a protein carrier. Since it is lipid soluble, it diffuses into target cells to bind to an intracellular receptor that is then translocated to the nucleus and alters gene expression. Responses to the hormone are mediated through the products of that gene expression, all of which takes more time than the actions of epinephrine.

2. The presence of the same hormone in two species is not an indication of closeness of their evolutionary relationships because hormone structures are highly conserved even as their functions change. Therefore distantly related species can share the same hormones.

3. A single hormone can have different effects in the same species depending on the types of cells that express its receptors and also on the nature of the receptors themselves, which can activate different signal transduction pathways.

RECAP 40.2

1. During development, the posterior pituitary is derived from an outpocketing of the brain and the anterior pituitary is derived from an outpocketing of the embryonic mouth tissue. Thus the posterior pituitary is neural tissue, and it is nerve cells that produce and release the posterior pituitary hormones, which are neurohormones. The anterior pituitary comes from gut epithelium, and the secretory cells develop from gut epithelial cells.

2. Hypothalamic neurons release tiny quantities of releasing or release-inhibiting hormones that diffuse into portal blood vessels between the hypothalamus and the anterior pituitary. These hypothalamic hormones leave the portal capillaries in the anterior pituitary and diffuse to anterior pituitary cells that have receptors for those particular release or release-inhibiting hormones. Those anterior pituitary cells respond by changing the release of their hormones, which include four tropic hormones (thyrotropin, adrenocorticotropin, and the gonadotropins LH and FSH) and growth hormone.

3. Negative feedback in the hypothalamic control of endocrine function involves the inhibitory effect of the end hormone on the hypothalamic production of the hypothalamic releasing (or release-inhibiting) hormone that controls the anterior pituitary cells that produce the tropic hormones controlling the cells or gland that produce and release the end hormone.

RECAP 40.3

1. Decapitation of *Rhodnius* prevented molting when done 1 hour after feeding but not when done 1 week after feeding because of the time necessary for PTTH to be released and diffuse to the prothoracic gland in sufficient amounts to stimulate the release of ecdysone.

2. The female external phenotype is the default pathway in humans. That pathway is altered in normal XY individuals because of genes on the Y chromosome that stimulate the development of testes rather than ovaries. The testes produce the male steroid sex hormone that stimulates development of the male external phenotype. However, if an individual lacks receptors for the male steroid sex hormone, then it has no effect and the female phenotype develops.

3. In the prepubertal individual, sex steroids are produced in low amounts, but these low amounts provide sufficient negative feedback to the hypothalamus to inhibit the production of GnRH. As a result, the release of LH and FSH from the anterior pituitary is low. However, at the time of puberty, sensitivity of the hypothalamus to the circulating sex steroids is reduced, releasing the negative feedback control. Therefore production and release of GnRH as well as LH and FSH go up, stimulating the changes associated with puberty.

RECAP 40.4

1. A major cause of hypothyroidism is lack of iodine. Without adequate iodine, the thyroid cannot secrete active T_3 and T_4. As a result there is no negative feedback on the production of TRH and TSH. The elevated TSH causes increased production of thyroglobulin and growth of the thyroid gland. A major cause of hyperthyroidism is production of an antibody to the TSH receptor that binds to the receptor and keeps it activated. The result is continuous stimulation of thyroglobulin production. Even though the individual produces active T_3 and T_4, they cannot exert negative feedback on the hypothalamic–pituitary–thyroid system.

2. Vitamins are required in the diet because they are not produced in the body. Since vitamin D is produced in the body by the action of UV light on a precursor in the skin, converting it to calciferol, it is not a vitamin. Calciferol is converted to the active form calcitriol by enzymes in the liver and kidneys. Calcitriol (vitamin D) promotes the uptake of calcium from the gut and the kidneys, thereby raising blood calcium levels.

3. Insulin controls the rate of glucose uptake by cells by stimulating the cells to insert glucose transporters (passive transporters) into the cell membrane from a pool of those receptors in the cytoplasm. The glucose transporters in the cell membrane increase the permeability of the cell to glucose, which can then diffuse into the cell as long as its concentration in the extracellular fluid is high enough.

4. Stress stimulates the release of cortisol from the adrenal cortex. Cortisol helps the body deal with short-term stress by reducing cellular uptake of glucose, elevating blood pressure, inhibiting activity of the gut and the reproductive system, and even inhibiting the activity of the immune system. In the short run, these responses make more energy available to deal with the immediate stressor. Negative feedback from cortisol, acting through the hypothalamus and the pituitary gland normally shuts off the cortisol response to short-term stress. If, however, stress becomes chronic, as in a difficult job situation, cortisol levels remain high and the resulting higher blood pressure can lead to heart disease, the impaired immune system can lead to infections and diseases, the impaired digestion can lead to ulcers, and the impaired glucose metabolism can lead to diabetes.

WORK WITH THE DATA, P. 853

1.

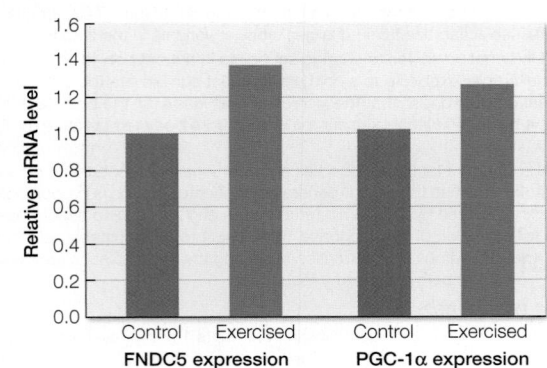

2. To determine if exercise increases BDNF levels in the hippocampus, you would assay BDNF levels in the hippocampus of the exercised and non-exercised mice, obtain average values and standard deviations, and do a *t*-test to determine the probability that the values for the two groups are the same. If *P* is < 0.05, the values for the two groups are significantly different.

3. To determine if increased expression of FNDC5 in the hippocampus is sufficient to stimulate an increase in BDNF expression, you would culture some hippocampal cells in media that stimulates FNDC5 expression and some hippocampal cells in media that suppresses FNDC5 expression. You would then measure the levels of BDNF in those cultures. A *t*-test would reveal the probability that FNDC5 had no effect on BDNF expression. If those probabilities are < 0.05, you would conclude that differences in FNDC5 expression are sufficient to influence BDNF expression.

WORK WITH THE DATA, P. 862

1. In the experiment in which irisin was injected into the mice kept on a high-fat diet, it was important to measure UCP1 expression as well as betatrophin because a relationship between irisin and UCP1 had been established earlier in exercise experiments. Therefore the measurement of UCP1 in that experiment was a positive control for activity of the injected irisin.

2. The mice were placed on a high-fat diet to model the condition of obesity and pre-diabetic insulin resistance.

3. The data support the hypothesis that irisin could reduce obesity and lower the risk for type II diabetes for the following reasons. Using *t*-tests confirms that the two groups of mice prior to irisin application did not differ in body mass. However, the irisin-treated mice showed a significant (*P* < 0.05) reduction in body mass, whereas the control mice showed no significant change in body mass. The irisin-treated mice had significantly lower fasting insulin levels and blood glucose levels than the control mice did, also indicating a lowered risk for type II diabetes.

FIGURE QUESTIONS

Figure 40.7 These results suggest that cells in the hypothalamus are producing a MSH release inhibiting hormone.

Figure 40.9 A mutation resulting in nonfunctional androgen receptors would result in XY individuals with female phenotype including external sex organs and secondary sexual characteristics.

APPLY WHAT YOU'VE LEARNED

1. The epinephrine in the irrigation fluid is affecting the dilator muscle fibers. When comparing the two types of surgeries, those that used the epinephrine-containing solution dilated the pupil past the 5-mm diameter and kept it dilated for a longer period of time than those surgeries that did not use the epinephrine-containing irrigation fluid. Since dilator muscle contractions are what cause the pupil to enlarge, these must be the muscle fibers affected by epinephrine.

2. Although both the constrictor muscle fibers and dilator muscle fibers are smooth muscle tissue, only the dilator muscle fibers contain a corresponding receptor protein on their cell surface that will bind to epinephrine. Thus the dilator fibers are target cells for epinephrine while the constrictor fibers are not. Once the epinephrine binds to the receptor protein on the dilator fibers, it begins a signaling cascade within the muscle fibers that promotes strong and prolonged contractions within the fibers. This produces the larger than 5-mm diameter pupil seen in the surgeries that used epinephrine-containing irrigation fluid.

3. Epinephrine affects the cardiovascular system by increasing the heart rate and blood pressure. It causes arterioles in the digestive system and skin to contract but causes the arterioles in the skeletal muscles to dilate. It causes the liver to break down glycogen into glucose and stimulates adipose tissue to break down fat into fatty acids. These different effects are stimulated by the same hormone binding to receptors on the surface of each type of cell. The reason the responses are so varied is because of what happens inside each cell type after the receptor has been activated. Each cell will have a different signaling cascade that results in different intracellular

processes. The difference in processes is determined by the type of cell (cardiac muscle vs. smooth muscle vs. hepatocytes vs. adipocytes, etc.) and by the types of intracellular signaling molecules that are created.

4. To elicit a pupillary dilation from the constrictor fibers, these muscle fibers must relax beyond their normal resting state. This is the effect of atropine. Since atropine inhibits the interaction of acetylcholine with its receptor, this must mean that acetylcholine is the neurotransmitter that stimulates constrictor fibers to contract and reduce the pupil size. It also means that there is normally a baseline level of acetylcholine that keeps the constrictor fibers somewhat contracted. Thus the autonomic nervous system controls pupil diameter by releasing both epinephrine and acetylcholine—epinephrine to cause dilator fibers to contract and enlarge the pupil, and acetylcholine to cause constrictor fibers to contract and decrease the pupil size.

Chapter 41

RECAP 41.1

1. Innate immunity is nonspecific, acts rapidly, and recognizes broad classes of molecules. Adaptive immunity is specific, acts more slowly, and is longer lasting.

2. The TLR pathway is involved in innate immunity to bacterial infections. Bacteria have molecules that act as pathogen-associated molecular patterns (PAMPs), and these normally activate the TLR pathway which results in white blood cells producing cytokines and other defensive molecules. If this does not occur, people will be especially susceptible to bacterial infections

3. The genomes of both the insect and the human would have genes for innate immunity, such as barriers and the TLR pathway, but only the human would have genes for adaptive immunity: antibodies, T cell receptors, and T and B cells.

RECAP 41.2

1. The physical barrier of skin prevents infection. Mucus traps and removes bacteria so they cannot infect. Lysozyme hydrolyzes bacterial cell walls. Defensins insert into the bacterial cell membranes, rendering them leaky. Once below the skin, the bacteria come in contact with complement proteins, which provoke phagocytosis, and phagocytes that ingest and destroy the bacteria.

2. The bacterium has an arrangement of atoms on its molecules called PAMPs. These are recognized by receptors on the human body cells, and they initiate a response. Dust particles lack PAMPs.

3. The splinter initiates the inflammatory response. See Figure 41.5.

RECAP 41.3

1. Humoral immune response: Free virus binds to B cells already making antibodies that react with viral antigens (recognition). These B cells are selected to divide and form a clone. The clone produces more antibodies against the virus. These antibodies bind to free virus in the bloodstream, and phagocytes ingest and hydrolyze the virus–antibody complexes.

Cellular immune response: Virus infects cells in the respiratory tract. Some of these cells break down the viral proteins and present fragments on their surfaces (antigen presentation). T cells with a T cell receptor that can bind to the viral fragment bind to the antigen-presenting cells. These T cells stimulate the formation of cytotoxic T cells that can then kill the virus-infected cells.

2. The older people had memory cells that could make antibodies to bind to the 1918 flu strain. Since the 1918 flu strain was similar to the 2009 virus, these people were able to mount a strong immune response to the 2009 virus.

3. Vaccines are inactive antigens that still provoke an immune response. Vaccination promotes a proliferation of clones that produce antibodies (B cells) or T cell receptors (T cells) that bind to the injected viral antigen. Some memory cells remain. These will be needed to mount a massive immune response when the virus (with antigen) infects later.

RECAP 41.4

1. See Figure 41.8. The antigen-binding site of an antibody has heavy and light chains in a unique three-dimensional configuration that binds a particular antigenic determinant. This is similar to an enzyme active site that binds a substrate. In both cases, binding is noncovalent. A major difference is in the result of binding: an antigen does not change its covalent structure when it binds to an antibody, whereas a substrate does change covalently when it binds to an active site.

2. Both immunoglobulins and T cell receptors have constant and variable protein regions, bind antigens, and have great variability in primary structure. T cell receptors are membrane proteins of T cells. Immunoglobulins can be either membrane proteins of B cells or secreted proteins in the blood.

3. There are thousands of different enzymes in an individual but potentially millions of different specific antibodies. Every cell in an animal has the genetic information for all enzymes. Each immunoglobulin, however, is derived from a unique gene (produced by DNA rearrangements) in a B cell or a clone.

RECAP 41.5

1. By inhibiting T cell development, cyclosporine blocks T_H cell binding to transplanted cells. This stops cellular immunity from occurring, because T_H cell binding releases cytokines to attract T_C cells that would kill the transplanted cells.

2. Inside the infected cell, the virus is broken down into fragments, and some peptide fragments are displayed on the cell surface along with MHC protein (class I). A

cytotoxic T cell displaying a T cell receptor that is specific for the peptide on the target cell binds to the target cell and initiates the cellular immune response.

3. Increased Treg activity would inhibit the cellular immune response, so the tumor would evade the adaptive immune system.

RECAP 41.6

1. An antigen in peanuts binds to a B cell displaying an anti-peanut protein antibody. This causes production of a clone of plasma cells that initially make IgG and then switch to IgE. The IgE binds to mast cells, which release histamine, causing symptoms of distress

2. Desensitization involves giving the person a small amount of peanut protein that provokes IgG synthesis but not IgE, so mast cells are not stimulated.

3. The anti-CTLA4 treatment for cancer removes inhibition of the T cell cytotoxic response to self antigens. This causes T cells to bind to and kill cancer cells. A side effect might be generalized autoimmunity.

4. Experiments might involve testing vaccinated people for neutralizing antibodies against HIV (humoral immunity) and looking for T cell activity against HIV-infected cells (cellular immunity).

WORK WITH THE DATA, P. 878

1. While some people had low titers and other had high titers of anti-Russian flu antibodies, the general trend over time was for the titers to increase. As these people were exposed to the typical flu strains year after year, their anti-flu virus responses, including to the Russian flu, increased. This indicates that there were some general antibodies that bind to all flu strains.

2. There was little change in the antibody titers to CMV. Note that the y axis is much lower quantitatively in this case than the y axis for anti-flu titers. CMV has epitopes unrelated to the flu virus, so anti-CMV antibodies will not be made when a person is exposed to flu. The data indicate that the increase in flu virus strain immunity was flu virus–specific.

3. The antibodies that developed were mostly broadly reactive. These antibodies could be used to develop a vaccine that might work on old as well as new strains of flu virus.

FIGURE QUESTIONS

Figure 41.5   Antihistamines block mast cells from making histamine. Blood vessels at a site of injury do not dilate or become leaky, so phagocytes are not attracted to the site and fluid does not enter the damaged tissue. In other words, inflammation does not occur in response to bacterial infection.

Figure 41.6   A reduction in T_H cells negatively affects both adaptive cellular and humoral immunity. With fewer T_H cells, immune response to an HIV infection is weak, both in terms of antibodies generated against the virus and cellular immune responses generated against virus-infected cells.

Figure 41.7   Small numbers of previously unexposed B cells are constantly differentiating to make antibody against Ebola and any other possible antigen. If there is no Ebola infection, these cells do not form clones in clonal selection and die off. But other previously unexposed B cells making anti-Ebola antibodies take their place.

APPLY WHAT YOU'VE LEARNED

1. By 3 weeks after the initial treatment, ADA and deoxyadenosine levels were normal. Weekly injections were probably needed because ADA gets broken down in blood serum.

2. Matched sibling and matched family donors had almost the same survival percentage: 87% and 88%, respectively. Matched unrelated donors were not nearly as successful, with only 67% survival. However, even this was better than the mismatched family donors, which had only 43% survival. The lowest survival of all was shown by mismatched unrelated donors: 29%. The survivorship of untreated patients might be similar to that with mismatched unrelated donors, but perhaps with even lower survival.

3. In gene therapy, hematopoietic stem cells are the cells that receive the genetic modification. Thus these are the only cells that will be able to produce and maintain normal levels of ADA. However, all cells produce ADA, because all cells undergo a certain level of DNA metabolism. ERT could keep other body cells from reaching dangerously high levels of deoxyadenosine.

Chapter 42

RECAP 42.1

1. Budding and regeneration begin with somatic (body) cells of the parent, whereas parthenogenesis begins with a germ (sex) cell, an egg that is not fertilized.

2. Global climate change is more likely to have a negative impact on asexually reproducing species, because they have less genetic variability to facilitate the evolution of adaptive changes that meet the new challenges imposed by climate change.

3. Flora must have reproduced through parthenogenesis, an ability that would be adaptive for an island-dwelling species should a weather event eliminate all potential mates or cause a female to be relocated to another island with no potential mate. If the offspring of a parthenogenic reproductive event are male, they would be able to mate with their mother to initiate a new population of sexually reproducing individuals.

RECAP 42.2

1. Human oocytes remain in prophase of the first meiotic division much longer than spermatocytes, and therefore the opportunity for crossing over is much greater.

2. Without cytoplasmic bridges between the spermatids of mammals, all of the spermatids would not receive gene products of the X chromosome that seem to be necessary for sperm survival. Thus all surviving sperm would lack a Y chromosome, and the offspring of that individual would all be female.

3. The egg is covered by protective layers that prevent sperm from contacting the egg cell membrane. The egg emits species-specific chemical attractants that stimulate sperm of its species to swim toward the egg. The protective coat of the egg also contains species-specific signals that stimulate the acrosomal reaction of the sperm, releasing enzymes that digest a path through the protective coat. The sperm head is coated with molecules (bindin) that have receptors on the inner protective coat of the egg, and when those molecules bind their receptors, the sperm contacts and fuses with the egg cell membrane. That event stimulates the blocks to polyspermy (fast and slow) that prevent additional sperm from fusing with the egg cell membrane.

4. a. The possibility of finding a mate is low.

   b. Sequential hermaphroditism would be advantageous to avoid inbreeding in a species that produces large numbers of offspring that remain in the same area.

   c. Sequential hermaphroditism would also be advantageous in a population where single dominant females control essential resources such as territory. Male offspring in that population would have the chance to reproduce, but females would not until the dominant female was removed. At that time a male could transition to female and take the place of the single dominant female in the group.

RECAP 42.3

1. Semen contains fibrinogen secreted by the seminal vesicles. The prostate secretes a clotting enzyme that causes the fibrinogen to polymerize and form a clot. The prostate also secretes the enzyme fibrinolysin, which eventually breaks down the clot and liberates the sperm.

2. Sertoli cells nurture and promote the maturation of sperm, beginning with the primary spermatocyte stage. They provide a suitable environment and essential nutrients. Tight junctions between Sertoli cells protect the developing sperm from water-soluble toxic substances that might be circulating in the blood.

3. Sexual stimulation of the male causes nerve endings in the penis to release a neurotransmitter that stimulates endothelial cells in the blood vessels to release NO. NO diffuses into the muscle cells controlling the diameter of the penile arteries, causing them to relax through increased production of cGMP. The resulting increase in blood flow in those arteries compresses the veins carrying blood out of the penis, and it becomes engorged with blood (an erection). A phosphodiesterase breaks down cGMP, so a phosphodiesterase inhibitor prolongs the effects of cGMP production.

4. Since prepubertal testes are responsive to LH and FSH, the cause of their inactivity must either be low production and release of LH and FSH by pituitary gonadotropes or low production of GnRH by the hypothalamus.

RECAP 42.4

1. The events in the ovary that are responsible for the onset of menses are the degradation of the corpus luteum and the consequent decrease in production and release of estrogen and progesterone. However, if a blastocyst implants in the uterus, it produces human chorionic gonadotropin (hCG), which maintains the corpus luteum until the extraembryonic membranes take over the production of estrogen and progesterone.

2. The granulosa cells nurture the developing egg, providing it with nutrients, growth factors, and mRNAs. The granulosa cells also produce estrogen and progesterone in response to FSH. Estrogen is produced by the granulosa cells from testosterone that is produced in the outer cells of the follicle, the thecal cells. Estrogen stimulates the granulosa cells to produce FSH receptors, but at the same time estrogen exerts negative feedback on the production and release of FSH from the anterior pituitary. Since multiple follicles begin to mature in each menstrual cycle, only the one with the most FSH receptors fully matures as the estrogen levels rise and FSH levels fall.

3. Prior to the onset of labor, the contractility of the uterine muscle increases, and at the same time the growing fetus is causing more stretch of the uterine muscle. Stretch of the uterine muscle stimulates contractions, pressing the head of the fetus onto the cervix. That stimulates neural signals to the hypothalamus, causing it to release oxytocin from the posterior pituitary. Oxytocin increases the uterine contractions, which in turn puts more pressure on the cervix, resulting in even more oxytocin being released from the posterior pituitary—a positive feedback loop

RECAP 42.5

1. The condom is the only contraceptive technology that can protect against STDs because if used properly it protects each partner from contact with the body fluids of the other

2. The pill delivers a high dose of progesterone that exerts negative feedback on the hypothalamic–pituitary axis to inhibit the production and release of LH and FSH, which are essential for the maturation and release of an egg from the ovary. RU-486 is an antiprogestive drug. During the luteal stage of the menstrual cycle, high levels of progesterone from the corpus luteum maintain the endometrium of the uterus in a receptive condition to receive a blastocyst. Blocking that action of progesterone can prevent implantation of the blastocyst or promote sloughing of the endometrium.

3. IVF can be used to prevent transmission of a genetic disease because cells can be taken from the early embryo in culture and analyzed for the suspected defective gene. Only embryos that do not carry the defective gene can then be selected for implantation into the uterus of the mother.

WORK WITH THE DATA, P. 909

1.

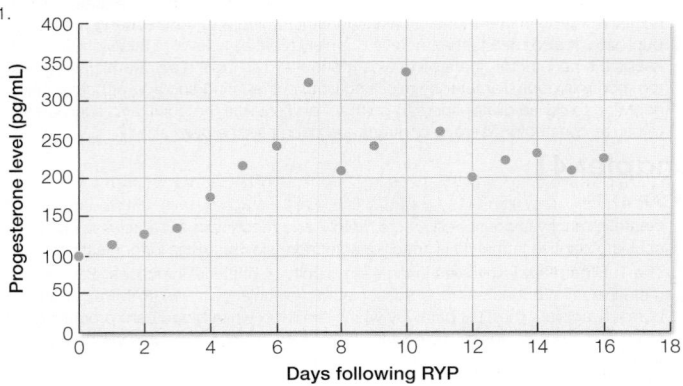

The data for this group were mixed, with minor peaks in mean PRL for the group occurring on days 7 and 10, indicating that some of the wallabies had broken diapause over that period of time. These rises occurred later than measured in the saline-treated (control) wallabies, but sooner than in the daily-treated wallabies. Thus the 48-hr PRL treatments were marginally effective in preventing the breaking of diapause.

FIGURE QUESTIONS

Figure 42.3 The selective advantage that could accrue to a bisexual species that becomes unisexual is that each member of the population could produce offspring. A disadvantage would be loss of genetic diversity.

Figure 42.6 If two sperm fertilized an egg, there would be two sets of male chromosomes in the egg, which would disrupt the subsequent cell divisions.

Figure 42.9 An enlarged prostate gland could squeeze the urethra shut and prevent or impair urination.

Figure 42.13 Hormonal contraceptive pills simulate the second half of the ovarian cycle, the luteal phase when estrogen and progesterone levels are high. These two hormones exert negative feedback that inhibits the release of gonadotropin, LH, and FSH which are essential for maturation of the egg and ovulation.

APPLY WHAT YOU'VE LEARNED

1. Following the mating behavior between the male and group A female koalas, the female koala's pituitary gland released a surge of LH into the bloodstream. This surge in LH stimulated ovulation. An egg was fertilized by the male's sperm, beginning a period of pregnancy that lasted about 35 days. No LH increases or pregnancies occurred in the Group B females. This experiment eliminates the possibility that the signal for ovulation could be a stimulus, such as a pheromone, from close proximity to a male koala.

2. The results from groups A and C show that neither artificial insemination procedures without semen nor tactile stimulation of the urogenital sinus stimulate LH surges or ovulation. However, AI with semen or artificial stimulation followed by semen injection both result in LH surges and ovulation. These results lead to the hypothesis that some component of the male semen is both the necessary and sufficient stimulus for the LH surge and subsequent ovulation. Clearly tactile stimulation alone or injection of a liquid volume into the urogenital sinus did not induce ovulation.

3. A possible selective advantage of reflex ovulation is that the female can choose the optimal time and the optimal mate for reproduction. If she has mature eggs available to ovulate, but she is not sexually receptive unless environmental conditions are good and a potential mate is at hand, reflex ovulation can optimize her chances of reproductive success.

Chapter 43

RECAP 43.1

1. The sperm contributes the centriole to the embryo, and the centriole is the origin of the microtubules of the primary cilia, which serve signaling functions.

2. The unfertilized frog egg has an animal (upper) and a vegetal (lower) hemisphere. The outer cytoplasm of the animal hemisphere is pigmented, but the vegetal hemisphere cytoplasm is not. The sperm binds to the animal hemisphere and stimulates rotation of the outer (cortical) cytoplasm toward the site of sperm entry. This rotation creates a band of more lightly pigmented cytoplasm opposite the site of sperm entry.

3. In the unfertilized frog egg, β-catenin and its degrading enzyme GSK-3 are homogeneously distributed. With sperm entry and cortical rearrangement, vesicles containing a GSK-inhibiting protein are translocated from the vegetal pole to the site opposite sperm entry. That protein regionally prevents the degradation of β-catenin on the dorsal side of the blastula.

RECAP 43.2

1. Complete cleavage occurs in eggs that have little yolk. In these eggs the early cleavage furrows can divide the egg completely. In eggs with a lot of yolk, the cleavage furrows cannot penetrate completely and the blastula forms as a disc of cells that sits on top of the yolk mass.

2. Cleavage in mammals is very slow, allowing enough time for gene expression to occur. In sea urchins and frogs, cleavage occurs very rapidly, precluding gene expression between cell divisions.

3. Mammals have regulative development, meaning that each cell of the blastula is not irreversibly determined, so if the blastula is divided in half, identical twins can develop. Since development is mosaic in most invertebrates, if the blastula is divided, each part will be lacking certain elements and will not be able to complete development.

4. Germ cells are determined at the time of early cell divisions of the embryo that take place during formation of the blastula. These presumptive germ cells will not differentiate until the gonads develop and they migrate into those gonads. The origin of skin cells is ectoderm that is determined during a later developmental stage—the gastrula—and these ectodermal cells differentiate rapidly to form the skin of the embryo.

RECAP 43.3

1. Gastrulation in both the sea urchin and the frog proceeds with an involution forming a blastopore. In the sea urchin, the cells involuting are similarly determined, but in the frog, tissue interactions differ according to location around the blastopore, and the involuting cells become differentially determined.

2. The criterion of necessity means that a component of a process must be present for that process to occur. Spemann's experiment, constricting the fertilized egg so that the gray crescent was only in one daughter cell or in both daughter cells, showed that the gray crescent was necessary for embryo development. The criterion of sufficiency means that a component of a process can alone cause the process to occur. In the Spemann–Mangold experiments, sufficiency was demonstrated by showing that when the dorsal lip of the blastopore was transplanted to another location on the blastula, it stimulated another axis of development.

3. The first involuting cells of the primary organizer express the transcription factor Goosecoid, and as those cells move anteriorly, Goosecoid suppresses the expression of certain transcription factors in neighboring tissues that have to be suppressed so that organs appropriate for the head region can develop. Cells that leave the dorsal lip of the blastopore later express different transcription factors that also have suppressive action on expression of other transcription factors in neighboring tissues, enabling the induction of region-appropriate organs.

4. Gastrulation in reptiles and amphibians differs because the blastula of amphibians is a sphere, but the equivalent stage of reptilian development is a flat sheet of cells, the blastodisc. In amphibians, the cell movements of gastrulation take place through a hole in the blastula (blastopore), while in reptilian blastodiscs, a longitudinal slit (primitive streak) forms and the cell movements of gastrulation take place through that slit. In amphibians, the ingress of cells through the blastopore creates the archenteron (primitive gut). No archenteron forms in the reptilian embryo, but the endodermal and mesodermal cells that pass through the primitive streak migrate forward to become gut and other structures.

RECAP 43.4

1. The ectoderm over the notochord thickens on either side of the midline to form ridges aligned on an anterior–posterior axis. These ridges grow until they meet on the midline, forming the neural tube. The most lateral cells of these ridges break free to form neural crest cells, while still more lateral ectoderm comes together and merges to cover the neural tube.

2. A somite is a segmental and bilateral block of mesodermal cells that gives rise to vertebrae, ribs, trunk muscles, and limbs.

3. Vertebrates have groups of Hox genes arranged in linear order on multiple chromosomes. These genes are expressed along the anterior–posterior axis in the same sequence as they occur on the chromosomes. As a result, each anterior–posterior region of the embryo receives different combinations of Hox gene products in different amounts, producing a large combinatorial mix of signaling molecules.

RECAP 43.5

1. The yolk sac and allantois grow out from the hypoblast and consist of endoderm and mesoderm. The chorion and amnion develop from ectoderm and mesoderm.

2. The blastocyst divides into a trophoblast group and a hypoblast group of cells. The trophoblast interacts with the uterine lining to embed the embryo into the endometrium and to send out projections to make contact with maternal blood vessels. Cells of the hypoblast grow out to form what in birds would be the yolk sac, but there is no yolk. These hypoblast cells and the trophoblast cells form the placenta.

3. The first trimester is a time of great sensitivity of the embryo to environmental risks because this is the time when many signaling cascades are setting up subsequent processes of determination, differentiation, and development. Therefore the earlier any necessary lifestyle or environmental changes are made, the better.

WORK WITH THE DATA, P. 929

1.

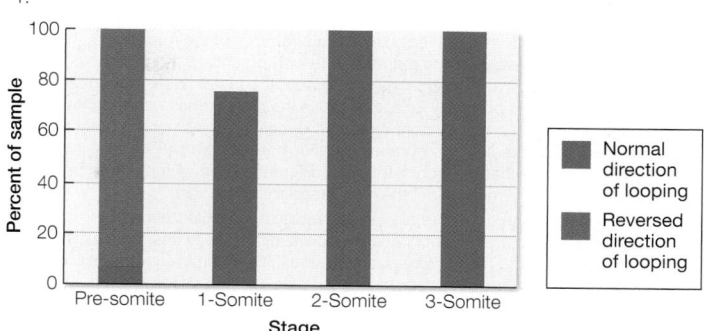

2. The experiment did not include left flow because we know left flow is the normal stimulus and that it is only disrupted by a fast right flow. This experiment was aimed at determining the sensitive period for the ability of that disruptive stimulus to alter the normal developmental pattern.

3. The data from embryos of different developmental stages support the conclusion that there is a sensitive phase of development where the left–right symmetry of the embryo is determined, and that it is between the presomite and 1-somite stages.

4. The results on 1-, 2-, and 3-somite wild-type embryos exposed to fast right flow show that the left–right asymmetry was determined at an earlier stage and that the embryo is no longer sensitive to direction of flow over the node.

FIGURE QUESTIONS

Figure 43.7   Gastrulation involves the involution of vegetal pole cells over the lip of the blastopore to form the archenteron thus forming the lining of the gut.

Figure 43.8   Epiboly spreads the animal pole cells over the entire embryo and results in the formation of ectoderm, the cells that form the skin and structures associated with the skin.

Figure 43.13   Hormonal contraceptive pills simulate the second half of the ovarian cycle, the luteal phase when estrogen and progesterone levels are high. These two hormones exert negative feedback that inhibits the release of gonadotropin, LH, and FSH which are essential for maturation of the egg and ovulation.

APPLY WHAT YOU'VE LEARNED

1. No, the symptoms are not consistent with vitamin A deficiency. The baby's defects are in mesenchyme tissues. A deficiency of vitamin A would most likely lead to defects of tissues derived from the ectoderm. In fact, the defects exhibited are consistent with an excess of vitamin A.

2. The ectoderm and the mesoderm express different genes. This results in mesoderm-derived cells being sensitive to excesses of vitamin A and ectoderm-derived cells being sensitive to too little vitamin A.

3. Individuals with this variant may be expected to be less susceptible to valproic acid. Because the variant enhances progression through G1 phase, the effect of valproic acid impeding this progression should be mitigated.

4. Even for women who have not been exposed to valproic acid, supplements of folic acid have been shown to reduce the frequency of neural tube defects. Given that valproic acid exposure increases risk for these defects, additional supplements of folic acid may counteract the effect of valproic acid. The extent to which additional folic acid could help would depend in part on whether and how these chemicals are involved in the same developmental processes. Valproic acid may have effects early in pregnancy—perhaps even before the woman knows she is pregnant—because neural tube formation is affected by processes that occur early in pregnancy.

5. Because deficiencies of folic acid increase the risk of neural tube defects, the efficiency of the receptor should affect this risk. Individuals who have receptor proteins that are more efficient in bringing folic acid to the cell should have greater cellular availability of folic acid and thus be at lower risk for neural tube defects compared with individuals who have less efficient receptor proteins.

Chapter 44

RECAP 44.1

1. a. Refer to Figure 44.1.
   b. At a chemical synapse, a chemical signal crosses the synaptic space between neurons. A neurotransmitter released by the presynaptic nerve endings diffuses across the synaptic space to bind to receptors on the postsynaptic membrane.

2. A neuron can receive both excitatory and inhibitory input at multiple synapses located throughout its dendrites and on the cell body. Summation of inputs determines whether the postsynaptic neuron becomes sufficiently depolarized to initiate an action potential.

3. Astrocytes can take up neurotransmitters that have been released into the synapse and thereby control the postsynaptic response. Astrocytes can also directly release neurotransmitters that bind to receptors to affect the excitability of a neuron.

RECAP 44.2

1. The Nernst equation calculates the membrane potential that will exist across a membrane due to the movement of a specific ion across that membrane, given the concentration difference of that ion on the two sides of the membrane. The dominant ion responsible for the membrane resting potential is K^+, but the potential calculated from the K^+ concentrations by the Nernst equation does not equal the measured membrane potential. This is because the membrane is also slightly permeable to other ions, and they contribute to the measured membrane potential. The Goldman equation predicts membrane potential more accurately because it takes into account all ions that have concentration differences across the membrane as well as the relative permeability of the membrane to those ions.

2. Inhibitory inputs to the dendrites of a neuron cause hyperpolarization of the dendrite membrane, and excitatory inputs cause depolarization. These changes in membrane potential spread to the neuron's cell body. The resulting membrane potential of the cell body is a graded membrane potential because it is always reflecting the sum of the dendritic inputs. That graded membrane potential spreads to the base of the axon, which fires action potentials if its membrane potential reaches threshold for its electrically gated Na^+ channels. Therefore the rate of action potential generation is a function of the graded membrane potential of the neuron's cell body, hence an integration of all of the dendritic inputs.

3. When the voltage-gated Na^+ channels open in a patch of membrane, that patch depolarizes. That local depolarization spreads by electrical conduction to neighboring regions of membrane, bringing them to threshold and thereby causing their electrically gated Na^+ channels to open and regenerating that action potential. This process continues down the length of the axon.

4. If the axons of the motor neurons that stimulate the muscle are demyelinated, the conduction velocity will be decreased and reflected in a longer delay for the muscle to respond to the electrical stimulation.

RECAP 44.3

1. The motor end plate contains chemically gated Na^+ channels and voltage-gated Na^+ channels. The chemically gated channels respond to the neurotransmitter acetylcholine to depolarize the motor end plate. The voltage-gated channels respond to that local depolarization by generating action potentials that spread to the adjacent muscle cell membrane.

2. The action of a neurotransmitter depends on the receptor types in the postsynaptic membrane, and they can initiate excitatory or inhibitory responses in the postsynaptic neuron.

3. Asphyxiation is the inability to breathe. Breathing requires regular activity in the motor neurons controlling the respiratory muscles. The inhibition of acetylcholinesterase by sarin results in the accumulation of acetylcholine in the synapses between the respiratory motor neurons and the respiratory muscles, reducing the ability of those muscles to relax between breaths.

4. Electrical synapses are fast but do not integrate information well. Electrical synaptic input does not allow temporal summation of inputs, as electrical synapses require a large area of contact between pre- and postsynaptic cells for effective transmission thus limiting the numbers of synapses that can be formed between two neurons. Finally, electrical synapses cannot provide inhibitory input.

RECAP 44.4

1. Ganglia are collections of nerve cell bodies. They tend to be concentrated in the anterior region of many invertebrates because that is where large numbers of sense organs are located.

2. See Figure 44.14.

3. In the knee-jerk reflex the same stimulus can cause contraction of a muscle through an excitatory synapse and relaxation of the antagonist muscle because of an inhibitory interneuron.

4. The part of the vertebrate brain that increases the most in going from fish to reptiles to mammals is the most anterior part of the brain, the cerebrum.

WORK WITH THE DATA, P. 945

1. Plugging the values in the table into the equation gives:

Substituting $2.3RT/F$ log for RT/F ln:

Assuming T = room temperature, $2.3RT/F$ = 58 (see Figure 44.5).

So,

$V_m = 58 \log 0.1118$

$V_m = -55.2$ mV

If you use the Nernst equation:

K^+: $E_K = 58 \log (5/140) = -84$ mV

Na^+: $E_{Na} = 58 \log (145/10) = +67$ mV

Cl^-: $E_{Cl} = 58 \log (110/20) = +43$ mV

Clearly the membrane potential is not due to K^+ alone. The membrane permeabilities to Na^+ and to Cl^- have a slight depolarizing influence on the resting potential of the mammalian neuron.

WORK WITH THE DATA, P. 954

1. During the dark phase the WT mice had significant discrimination scores during testing, indicating that they learned to recognize the training object. The DS mice did not have significant discrimination scores and thus did not learn. These were the same results as obtained during the light phase.

2. When the experiment was done during the dark phase, the results for the saline-treated mice showed the same results as during the light phase—the WT animals learned to recognize the training object, and the DS mice did not. The results for the drug-treated mice, however, were different than they were in the light-phase experiment. PTZ treatment during the dark phase did not result in an improved ability of the DS animals to learn to recognize the training object. Thus the conclusion is that the drug has to be given during the light phase to have an effect.

FIGURE QUESTIONS

Figure 44.2 The Purkinje cell has the most dendrites and therefore is likely to receive more inputs than the pyramidal cell, and certainly more than the retinal cell.

Figure 44.11 You would expect either Cl^- channels or K^+ channels to be on the postsynaptic membrane.

Figure 44.15 The pig has a proportionally larger olfactory bulb than the human—the other mammal in this comparison—so its olfactory sensory sensitivity is especially well developed.

APPLY WHAT YOU'VE LEARNED

1. The 1-second light stimulation of the hypocretin neurons caused depolarization of their membrane potentials and a large increase in the rate at which they fired action potentials. These results support the concept that ChR2 is a Na^+ channel and that light increases the Na^+ permeability of that channel, depolarizing the membrane and bringing the membrane potential to above the threshold potential for firing of APs.

2. The fact that the firing of the neurons followed precisely the trains of very short bursts of light (15 ms) indicates that the APs of the neurons were direct responses to the light.

3. The light stimulation of hypocretinergic neurons expressing ChR2 in sleeping mice shortened the latency to awakening. The same light stimulation had no effect on the mice not expressing ChR2. Above a frequency of 1 flash per second, the awakening response was not altered by the frequency of stimulation.

4. The fact that a blocker of the hypocretin receptor eliminated the awakening effect of the light stimulation supports the conclusion that the effect was mediated by the release of hypocretin by the stimulated neurons.

Chapter 45

RECAP 45.1

1. Mechanical stimulation of the stretch receptor dendrites by stretch of the muscle they innervate causes graded depolarization of the stretch receptor neuron cell body—the receptor potential. The receptor potential spreads to the axon hillock, and if it depolarizes the membrane of the axon hillock to threshold, a Na^+ action potential is fired in the axon.

2. The interpretation of an action potential depends on its target neuron in the brain. If the action potential arrives in a nerve of the optic tract projecting to the visual cortex, it is interpreted as light, but if a similar action potential arrives in an olfactory nerve projecting to the olfactory bulb, it is interpreted as smell.

3. The molecules responsible for the distinctive flavor of peppers are received through chemoreceptors that activate a G-protein signaling cascade that open cation channels in the receptor membrane. Capsaicin, however, directly activates a TRP protein that also serves as a thermosensor, and that protein itself is a cation channel that opens when activated by its specific stimuli.

4. Adaptation is the property of a sensory system to stop responding to a constant level of receptor stimulation. Adaptation is important when it is of value to recognize when stimulation changes but it is not necessary to have continuous information. An example is the activation of skin tactile sensors by your clothing. Adaptation would not be advantageous when the information is critical, such as pain sensations, or when constant levels of information are essential, such as from the sensory neurons that signal postural muscle tone.

RECAP 45.2

1. The odorant receptor neurons (ORNs) that share the same receptors project to the same glomerular cells in the olfactory bulb. Thus the identification of a particular odorant depends on the pattern of connectivity of the ORNs that are activated.

2. An olfactory system signals intensity of a stimulus—whiff versus repulsive odor—by the frequency of action potentials it generates. The discrimination between a skunk and a skunk cabbage depends on the mixture of odorant molecules.

3. For salt, extracellular fluids including saliva have a high concentration of NaCl, therefore a lower sensitivity is appropriate for sensing concentrations above that constitutive level. For sweet, the selective pressure for its evolution was the ability to discriminate between high and low energy foods. A low sensitivity motivates the selection of higher energy foods. Bitter is a characteristic of many protective, hence poisonous, compounds that evolved in plants to discourage predation, so a high sensitivity helps avoid poisoning.

4. With only five classes of taste receptors we can sense a great variety of tastes because (1) some classes have multiple receptor genes; (2) a substance may trigger multiple receptor classes and therefore create a mixed signal; and (3) information from taste receptors is integrated with information from olfactory receptors as well as thermoreceptors (TRP channels) and tactile receptors.

RECAP 45.3

1. Different mechanosensor cell types enable responses to different aspects of touch such as sharpness, texture, pressure, vibration, and itch. Different rates of adaptation of touch receptors make it possible to discriminate between stimuli that are relevant and those that aren't. Slowly adapting mechanosensors such as those in postural muscles provide continuous information. Rapidly adapting mechanosensors provide information about changing conditions and also improve spatial and temporal sensory ability.

2. The activity of a muscle spindle stretch receptor increases the activity in the motor neuron to that muscle. The functional significance of this property of muscle spindle stretch receptors is that it enables continuous adjustments to changes in load, such as when you are holding a glass that is being filled. The activity in the Golgi tendon organ inhibits the activity in the muscle creating the stretch of that Golgi tendon organ. The functional significance is the prevention of damage to the muscle and tendons through the generation of too much tension, such as when you are attempting to pick up objects that are too heavy.

3. Different frequencies of sound pressure waves cause different frequencies of flexion of the tympanic membrane, and these movements are transmitted and amplified by the ossicles of the middle ear into vibrations of the oval window membrane of the fluid-filled inner ear. Those vibrations create pressure waves in the fluid of the inner ear. That fluid surrounds the basilar membrane in the vestibular and tympanic canals. The membrane grades from thick at the proximal end of the canal to thin at the distal end. Pressure waves of different frequencies in the fluid of the tympanic canal cause the basilar membrane to vibrate in different locations. The hair cells are on the basilar membrane, and therefore different sets of hair cells are activated by pressure waves of different frequencies.

RECAP 45.4

1. Ommatidia are excellent at detecting movement as moving objects in a visual field switch neighboring ommatidia "on" and "off." However, since each ommatidium is receiving light from only a small but discrete portion of the total visual field, only a pixilated (low-resolution) image can be formed.

2. When photons excite rhodopsin, its conformation changes and that activates a G protein that activates a phosphodiesterase that converts cGMP to GMP. The decrease in cGMP levels causes the cGMP gated Na^+ channels to close, reducing the dark current. Na^+ is also pumped out of the proximal end of the photoreceptor cell and its membrane potential hyperpolarizes.

3. Rods and the various cones have different spectral sensitivities because they have slightly different opsins. The structure of the opsin determines which wavelengths of light a cell will absorb and how it will activate its associated 11-cis-retinal.

4. You cannot discriminate color in the periphery of your visual field because the cone cells are concentrated in the fovea, which receives light only from the center of the visual field. Rods do not distinguish color and are more abundant in the peripheral retina.

5. Two reasons why vision is impaired when you come from a brightly to a dimly lit environment are that bright light bleaches many rhodopsin molecules, and the amacrine cells have to readjust the range of brightness sensitivity of the retina.

WORK WITH THE DATA, P. 963

1. The expression of TRPA1 in the python is much greater in the TG than in the DRGs. In contrast, the expression of TRPA1 in the rat snake is low and not very different in the TG and the DRGs. There are no remarkable differences in the expression of TRPV1 in the TG versus the DRGs in either the rat snake or the python.

2. Since the expression of TRPA1 is dramatically higher in the TG of the pit snake but not in the TG of the non-pit snake, these data support the conclusion that the TRPA1 channel plays a role in the function of the pit organ.

3. If expression of the TRPA1 channel confers on a gene expression model system a temperature sensitivity in the range characteristic of the pit organ, that result would be strong evidence that the TRPA1 channel is the IR sensor in the pit organ.

FIGURE QUESTIONS

Figure 45.4 *Bombykol* has the function of attracting mates from a distance, so it has to diffuse widely. A territory-marking pheromone serves to indicate a specific place and therefore should not diffuse widely, so it would have a lower diffusion coefficient.

Figure 45.9 Several mechanisms can enable bilaterally placed ears to detect directionality of a sound: (1) The sound will be louder the closer the ear is to the source, and also the head creates a sound shadow for sounds coming from the side. (2) The sound will reach the closer ear sooner than the more distant ear. (3) If the sound is not coming from directly in front or in back, there will be a phase mismatch between the sound waves reaching the two ears.

Figure 45.13 Floaters in the eye would have to be in the vitreous humor.

APPLY WHAT YOU'VE LEARNED

1. According to the data, moths with disrupted tymbal organs displayed the same types of evasive maneuvers as moths with intact tymbal organs. There is no statistical difference in the evasive maneuvers made by silenced versus clicking moths. This finding demonstrates that the moths with the disrupted tymbal organs are still capable of "hearing" the echolocation signals coming from predatory bats and of responding with evasive maneuvers. If a significant difference had been seen in Figure A, this could indicate that tymbal organ disruption impeded a moth's ability to either hear the bat signals or respond with an evasive behavior.

2. According to the data, moths with intact tymbal organs were captured significantly less often than moths with disrupted tymbal organs. This is true for all silenced moths no matter what type of evasive maneuver they used to escape bat predation. If evasive maneuvers alone were responsible for moth escapes, we would not expect the number of captures between clicking and silenced moths to be significantly different. The fact that they are significantly different supports the hypothesis that the clicking of the tymbal organ serves as an echolocation-jamming device, which when paired with evasive maneuvers results in fewer captures.

3. Different frequency sound waves cause flexion of the basilar membrane at different locations because the basilar membrane varies in thickness and stiffness. At the base it is most stiff and thick and is only flexed by high frequency pressure waves in the cochlear canals. At the apical end it is thinner and less stiff and is flexed by low frequency pressure waves in the cochlear canal. The bat can hear higher frequency sounds than humans can and therefore their basilar membranes must be thicker and stiffer at the basal end than are the basilar membranes of humans.

Chapter 46

RECAP 46.1

1. a. The conscious sensory (afferent) stimulus of seeing a venomous snake prompts voluntary commands in the CNS to the muscles in the legs (efferent output) to run from the danger.

   b. The conscious sensory stimulus of the aroma of freshly baked bread can stimulate the autonomic responses of salivation and increased stomach activity.

   c. The unconscious stimulus of a fall in blood pressure can cause an involuntary command to increase heart rate.

2. Failure of the anterior region of the neural tube to close has the greatest effect on the telencephalon and results in lack of development of the cerebrum.

3. Areas of association cortex involved in reading and language are situated between the visual area in the occipital cortex and the auditory area in the temporal cortex. Reading integrates interpretation of visual images with the auditory patterns of spoken language.

4. Areas of the body that are shown by the two-point spatial discrimination test to be highly sensitive to touch are represented by large areas of somatosensory cortex.

5. The comparison of cortical capacity of humans with that of other mammals cannot be based just on brain size, because the human cortex is highly convoluted, giving the cortex a larger area. A higher percentage of the human cortex is devoted to association functions (association cortex) that increase cortical capacity.

RECAP 46.2

1. A knife wound to the left side of the neck could sever the sympathetic chain of ganglia on that side. This could break the connection between the preganglionic and postganglionic sympathetic neurons that innervate the pupil of the left eye and cause it to dilate.

2. A small spot of light on the retina can illuminate just the on-center of the receptive field of a ganglion cell and therefore activate it maximally. A larger spot of light would illuminate receptor cells around the on-center that have an inhibitory influence, and therefore decrease the activation caused by illuminating the center of that receptive field.

3. If the optic chiasm were cut right on the midline, the axons from the ganglion cells in the medial halves of the retina would not cross over to the opposite sides of the brain. As a result, the left eye would not see things in the left visual field and the right eye would not see things in the right visual field. Without input from both eyes, there would be no binocular cells and therefore no depth perception

RECAP 46.3

1. Paralysis of skeletal muscles during REM sleep prevents the possibility that an individual would act out his or her dreams while in the dreaming state.

2. A person with a severed corpus callosum would not be able to describe in words an object in his left visual field, but he could draw it or point to a similar object in a picture.

3. Since H.M. was capable of immediate memory but could not retain longer-term declarative memories, we can conclude that the hippocampus is necessary for acquiring new declarative experiences and consolidating them into long-term memory.

4. Humans, great apes, elephants, and some marine mammals have an expanded insular lobe, which is involved in integrating physiological information from throughout the body, as well as social and emotional information, and may therefore generate a sense of self-awareness.

WORK WITH THE DATA, P. 986

1. The probability of this sequence out of a random firing of nine cells is first a function of the probability of that one combination from all of the combinations of cells that could constitute six elements of the nine possible. That would be $1/(9 \times 8 \times 7 \times 6 \times 5 \times 4) = 1/60{,}480 = 1.6 \times 10^{-5}$. Then, for the sequence, what is the probability that it would be in the correct order? There is one possible correct order of those six elements in $6 \times 5 \times 4 \times 3 \times 2 \times 1$ possible combinations or 1 in 720, thus the

probability would be 1.4×10^{-3}. So the probability of recording those specific six elements out of nine in the correct order would be 2.24×10^{-8}.

2. The probability of recording four elements in the sequence in the proper order would be: $1/(9 \times 8 \times 7 \times 6) = 1/4,698 = 2.1 \times 10^{-4}$, and the probability of getting those four elements in the proper order would be $1/24$. Thus the probability of recording those specific four elements out of nine in the correct order would be $2.1 \times 10^{-4} \times 0.042 = 0.88 \times 10^{-3}$.

3. Both of these recordings have a probability of < 0.01, so they would be considered significant low-probability events.

4. The time frame of the sequence recording during sleep is about five times faster than the occurrence of that sequence during wakefulness when the rat is running the maze.

FIGURE QUESTIONS

Figure 46.1   Hormones reach the CNS through the circulation. Neurohormones are secreted by neurons into the interstitial fluid, from which they can diffuse locally or enter the blood.

Figure 46.8   When an individual is stressed such as when speaking in public, the sympathetic nervous system is activated (fight-or-flight response), and that inhibits the activity of the salivary glands.

Figure 46.9   The blind spot in each eye corresponds to the location on the retina where the axons of the ganglion cells leave the retina to form the optic nerve. There are no photoreceptors at those locations.

Figure 46.10   The right and the left visual fields would be narrower and there would be no depth perception.

APPLY WHAT YOU'VE LEARNED

1. The subjects who were well rested gave correct answers to working-memory tests at least 95 percent of the time. Sleep-deprived subjects were almost 10 percent more likely to give incorrect answers.

2. Compared with well-rested subjects, sleep-deprived subjects took approximately 50 milliseconds longer to answer the questions, which could reflect decreased alertness. The results could also be due to an effect of sleep deprivation on cognitive processing time.

3. The left side of the cerebral cortex is responsible for processing and understanding language. Since the tests involved letter recognition and sequencing, this is likely to be the area most activated by the tests.

4. Sleep deprivation caused a 20–30 percent decrease in parietal lobe activity but only a 10–15 percent decline in prefrontal lobe activity. Testing this conclusion would require a statistical evaluation of the differences between activities in the two brain areas as a function of the sleep-deprivation treatment.

Chapter 47

RECAP 47.1

1. A sarcomere is bounded by the Z lines and consists of overlapping actin and myosin filaments. The actin filaments are anchored on the Z lines and do not extend all the way to the middle of the sarcomere. The myosin filaments are anchored on the M band at the center of the sarcomere but do not extend all the way to the Z lines. Thus there are regions where the actin and myosin filaments do not overlap (i.e., the I band and the H zone). When the muscle, hence the sarcomere, contracts, the myosin filaments move along the actin filaments toward the Z lines. As a result, the I band gets narrower. At the same time, the actin filaments move closer to the M band, so the H zone gets narrower.

2. One subunit of troponin anchors it to tropomyosin filaments. One subunit of troponin binds to the actin filament. One subunit binds Ca^{2+}. When Ca^{2+} binds to the troponin subunit, it causes a conformational change in the troponin, resulting in the tropomyosin filament pulling away from the myosin-binding site of the actin. Actin–myosin binding results in contraction of the sarcomere.

3. Rigor mortis is due to the loss of ATP that is required to break actin-myosin cross bridges. With the cessation of breathing and circulation at death, muscles do not receive oxygen and glucose from the blood and production of ATP through oxidative metabolism ceases. However, some ATP can continue to be produced by anaerobic metabolism, and the muscles with the most glycogen will be able to sustain anaerobic metabolism longer

4. By destroying acetylcholinesterase, Malathione prevents the breakdown of acetylcholine released at the motor end plates, and as a result there is continuous and extreme activation of the muscle cell motor endplates. Action potentials continue to be fired in the muscle cell membranes resulting in tonic release of Ca^{2+} into the sarcoplasm and sustained activation of actin/myosin cross bridge formation.

5. Gap junctions enable sheets of cardiac or smooth muscle to contract as a unit. That makes it possible for cardiac muscle cells to work together to pump blood or smooth muscle cells to exert a unified function such moving food through the gut. The individual control over skeletal muscle fibers enables fine control over complex movements.

RECAP 47.2

1. A skeletal muscle is made up of many muscle fibers. Each muscle fiber is innervated by one motor neuron constituting a motor unit, but one motor neuron may form synapses with multiple muscle fibers. The nervous system can alter the frequency

of action potentials in one motor unit, and it can increase the number of motor units activated in the same muscle.

2. Postural muscles must remain continuously contracted for long periods of time, but they are not generally used for quick, powerful movements.

3. The energy source for the sprint comes primarily from preformed ATP and creatine phosphate along with some contributions from glycolysis. These sources of ATP are rapidly mobilized. The longer 10-kilometer run requires production of ATP through oxidative metabolism, which requires O_2 transport and many more enzymatic reactions. It is therefore a slower source of ATP that cannot sustain as high a workload.

RECAP 47.3

1. Arthropods have external skeletons and therefore cannot grow without shedding (molting) their exoskeletons.

2. Since astronauts are losing bone mass, it is reasonable to hypothesize that the activity of their osteoclasts has increased, or the activity of their osteoblasts has decreased. It is possible that these changes are being induced by altered activity of their osteocytes in response to lack of weight-bearing stresses on their bones.

3. If the joint has a large force arm relative to the load arm, it can generate great pressures. If the force arm is short relative to the load arm, the end of the load arm can move over a large distance very quickly but cannot exert much pressure. Thus a larger force arm:load arm ratio for the jaw joint enables the jaws to apply great pressure over a small distance, whereas the smaller force arm:load arm ratio for the elbow enables the lower arm to move quickly over a larger distance but does not enable it to apply great pressure.

WORK WITH THE DATA, P. 1011

1.

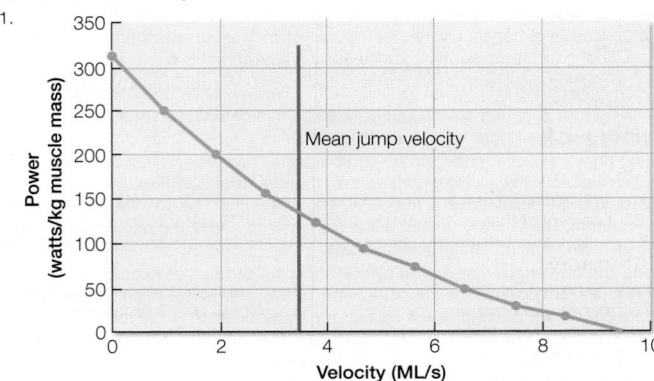

2. The jump takes 50 ms (see Figure B), and the shortening in terms of muscle lengths is $7.5\ mm/33.6\ mm = 0.22\ ML$. So, velocity of muscle contraction in ML/sec = $0.22/.05\ sec = 4.4$

3. The power generated by the jumping muscle is maximal at the observed mean jumping velocity.

WORK WITH THE DATA, P. 1013

1. If the workout capacity goes up with cooling, that suggests that the heat extraction is enabling more work, or in other words, decreasing muscle fatigue that otherwise limits workout capacity.

2. Cooling increased the work capacity for each workout in comparison to the previous workout, and over time, the rate of change of capacity was much greater with cooling than without. This difference in rate of change indicates that increasing work capacity (Figure A) leads to increased rate of conditioning (Figure B).

FIGURE QUESTIONS

Figure 47.4   If a motor neuron innervates only one or a few fibers, it is likely that those fibers are used in fine movements. If a single motor neuron innervates many muscle fibers of a muscle, it will be able to command strong, forceful movements.

Figure 47.6   ATP is necessary for skeletal muscle contraction to: (1) establish the ion gradient across the cell membranes, (2) break the actin–myosin bonds and "recock" the myosin heads, and (3) pump Ca^{2+} from the sarcoplasm into the sarcoplasmic reticulum.

Figure 47.11   A champion weight lifter is likely to have muscles with a high proportion of fast-twitch fibers.

Figure 47.13   The slopes of the curves are influenced by the time course of the mechanisms supplying the ATP supporting the activity. Because the first 10–20 seconds of high-level muscle work are supported by preformed ATP and CP, no further production of ATP and CP is involved. The production of ATP by glycolysis can produce ATP rapidly but not in large amounts, so the speed of performance goes down in the first minute of intense exercise even though the production of ATP by oxidative metabolism is gradually increasing. Oxidative metabolism is a more efficient and sustainable producer of ATP, but it involves many more biochemical reactions than glycolysis, and cannot proceed as rapidly. Also, it is limited by the delivery of O_2 to the mitochondria, so the speed of sustained performance goes down further.

Figure 47.21   The knee joint is a class 3 lever.

APPLY WHAT YOU'VE LEARNED

1. In both groups muscle glycogen fell to about 25 percent of its resting value by the point of fatigue. There were no significant differences ($P > 0.05$) in muscle glycogen at any time point in the two groups. Thus over 3 hours both groups depleted their muscle glycogen at about the same rate. Blood glucose in the placebo group fell steadily after the first hour, but blood glucose did not fall in the glucose-fed group, and blood glucose levels were significantly different in the two groups after the first hour. Both groups had similar levels of carbohydrate oxidation for the first 2 hours, after which the placebo group had declining carbohydrate oxidation levels that were significantly different from those of the glucose-fed group.

2. The difference between the two groups in time to fatigue was significant, and therefore the data show that carbohydrate feeding delayed muscle fatigue. Carbohydrate feeding also maintained blood glucose levels and the rate of carbohydrate oxidation. In addition, the data show that up until hour 3, this effect does did not involve a slowing of muscle glycogen depletion.

3. For the glucose-fed group, there was no significant decline in muscle glycogen between hour 3 and 4, indicating that other sources of fuel were being metabolized once muscle glycogen was depleted. A possible explanation for these effects of carbohydrate feeding is that it enabled the subjects to support more of their metabolic work with carbohydrates. We can hypothesize that the placebo group had to compensate sooner with fuel sources other than carbohydrates.

4. The body uses a mixture of carbohydrate and fats as fuel for ATP production via oxidative metabolism. Fats are slower to be mobilized than carbohydrates. The blood FFA data indicate that the placebo group began to depend more on FFA metabolism after 1 hour of exercise, but the glucose-fed group did not increase utilization of FFAs until between 3 and 4 hours of exercise. At that time the subject depended less on muscle glycogen. Overall, the carbohydrate feeding extended the time that the subject could exercise without increasing utilization of FFAs

5. Suggest eating small snacks composed of simple sugars periodically during runs to prolong the time to fatigue.

Chapter 48

RECAP 48.1

1. At sea level, assuming a normal atmospheric pressure of 760 mm Hg, the P_{O2} will be 20.9 percent of that, or 159 mm Hg. Therefore the P_{O2} at 2,000 meters will be 127 mm Hg. When breathing air at two times the atmospheric pressure, the P_{O2} will be 318 mm Hg.

2. Flatworms have a flat, thin body form, so all cells of the body are close enough to the surrounding environment that O_2 and CO_2 can diffuse directly between the cells and the environment. The two components of Fick's law reflected in these anatomical features are high surface area for diffusion, and minimal path length.

3. As water temperature increases, the O_2 content of water decreases (O_2 solubility decreases with increase in temperature), but since the fish's body temperature is at equilibrium with the water temperature, its metabolism increases (Q_{10} effect). Therefore the fish has to ventilate its gills more to satisfy its O_2 needs, and the work of gill ventilation increases O_2 demand even more

4. In going from sea level to high altitude, the partial pressure of O_2 in the inhaled air decreases, so the partial pressure gradient causing O_2 to diffuse into the body goes down. Since the partial pressure of CO_2 is extremely low at sea level, it does not decrease significantly at high altitude. Therefore the partial pressure gradient for CO_2 diffusing out of the body does not change.

RECAP 48.2

1. If ventilation (water) and perfusion (blood) of the gills are in the same direction (concurrent flow), then the concentration gradient of O_2 across the gills will gradually equilibriate somewhere between the maximum O_2 concentration of the water coming in and the O_2 concentration of the blood coming in. In countercurrent flow, it is possible for the O_2 concentration of the blood leaving the gills to be almost as great as the O_2 concentration of the water entering the gills

2. An important feature of avian lungs is the parabronchi, which enable air to travel unidirectionally through the lungs. When a bird inhales, the incoming air goes to the posterior air sacs. That fresh air flows into the parabronchi during the next exhalation. The subsequent inhalation causes air in the lungs to flow into the anterior air sacs while fresh air flows into the posterior air sacs. The subsequent exhalation empties the anterior air sacs to the environment while the air in the posterior air sacs enters the lungs. Thus the through-pass of the parabronchi and the bellows action of the air sacs result in continuous, unidirectional flow of air through the avian lungs.

3. The amount of He in the system remains the same but becomes equally distributed in the spirometer and lungs. So the initial amount of He is 0.05×30 L = 1.5 L. The final amount of He is still 1.5 L, but it is distributed over 30 L + the person's FRV. Thus,

    1.5 L = 0.046 × (30 L + FRV)

    1. L = 1.38 L + 0.046FRV

    0.12 L = 0.046FRV

    2.61 L = FRV

RECAP 48.3

1. Some premature infants may be born before cells in their alveoli have started to secrete surfactant. Without surfactant, inflation of the lungs requires extra effort to overcome surface tension in the fluid lining the alveoli

2. There is always tension between the pleural membranes because of the outward pull of the chest wall and the inward pull of the lung tissue. This tension keeps the lungs partially inflated even between breaths. A puncture of the chest wall allows air to enter space between the pleural membranes, equilibrating its pressure with atmospheric pressure. As a result, there is no longer a force maintaining partial inflation of the lungs.

3. The tension between the pleural membranes is maximal at the peak of inhalation because it is maintaining maximum distension of the lung tissue, but once the lungs are filled, air flow ceases and the alveolar pressure is the same as atmospheric pressure.

RECAP 48.4

1. Because of mixing with dead-space air, the maximum P_{O2} in the alveoli is about 100 mm Hg, and hemoglobin can fully saturate at that P_{O2}. The fact that mixed venous blood normally returns to the heart 75 percent saturated means that if the blood flows through any tissue that is very active metabolically, it has a reserve of O_2 that it can off-load to satisfy that local high demand.

2. When the concentration of 1,3-BPG in stored red blood cells decreases, so does the concentration of 2,3-BPG that can be generated by conversion from 1,3-BPG. As a result, the hemoglobin–O_2 binding/dissociation curve shifts to the left. Therefore the O_2 is more tightly bound to the hemoglobin and is not released for use by the respiring cells.

3. The conversion of CO_2 to H_2CO_3 and then to HCO_3^- is a reversible reaction depending on the concentrations of reactants and products. In the respiring tissues the P_{CO2} is high and drives the reaction in the red blood cells toward HCO_3^-, which moves into the blood plasma in exchange for Cl^-. In the alveoli, the P_{CO2} is low, so this entire suite of reactions is driven in reverse, meaning that the diffusion of CO_2 out of the red blood cells causes HCO_3^- to be moved back into the red blood cells to be converted back to H_2CO_3 and then to CO_2.

RECAP 48.5

1. During tidal breathing, a regular cyclical pattern of firing of respiratory motor neurons in the dorsal respiratory group of the medulla drives muscle contractions of the diaphragm. The need for increased respiratory gas exchange is first met with an increase in the rate of that oscillatory pattern of firing in the dorsal respiratory group, but as the need increases, the ventral respiratory group is recruited and those neurons drive contractions of the intercostal muscles, increasing the volume of air exchanged with each breath.

2. When you go to high altitude, the diffusion gradient for O_2 decreases as atmospheric pressure decreases. However, the diffusion gradient for CO_2 does not change; ambient CO_2 is close to 0 at sea level and also at high altitude. Therefore breathing to satisfy O_2 needs results in a greater loss of CO_2 than at sea level. Since CO_2 is the major stimulus for respiration, respiration slows or ceases until the P_{CO2} returns to a level that stimulates breathing.

3. Because the carotid and aortic bodies have high metabolic rates, they are compromised by a decrease in their O_2 supply, whether it is because of decreased blood flow or decreased P_{O2} in the blood.

WORK WITH THE DATA, P. 1030

1. Since helium is not absorbed by the respiratory system, the amount of helium remains the same after breathing from the air reservoir as before, but it is distributed in both the reservoir air and the functional residual volume (FRV) of the patient. Using the helium dilution method, we can calculate the FRV.

    0.050×30 L = 0.044 (30 L + FRV)

    1.5 L = 1.32 L + 0.044 × FRV

    (1.5 − 1.32)/0.044 = FRV = 4.1 L

    Subtracting the expiratory reserve volume of 1.5 L gives a residual volume of 2.6 L.

2. The patient is always short of breath because the P_{O2} in his alveoli is seriously diminished by the stale air in his very large residual volume as well as in his increased FRV during tidal breathing.

WORK WITH THE DATA, P. 1037

1. O_2 available in venous blood:

    Total venous blood = 355 kg × 0.148 × 0.66 = 34.7 kg

    Total venous hemoglobin = 34.7 kg × 0.58 × 0.24 = 4.8 kg

    Total venous O_2 available = 4.8 kg × 1.34 L/kg × 0 .9 = 5.8 L

2. O_2 available in myoglobin:

    Total amount of myoglobin: 355 kg × 0.33 × 44g/kg = 5.2 kg

    O_2 bound to myoglobin: 5.2 kg × 1.34 L/kg = 7 L

3. Total O_2 reserves at beginning of dive:

    2.5 L + 5.8 L + 7.0 L = 15.3 L

4. Maximum sleep episode = total O_2 reserves divided by the diving metabolic rate. So, 5.3 L divided by the rate of O_2 consumption which for a long dive would be

0.0035 L/kg min × 355 kg. So, 15.3L/ 1.24 L/min = 12.3 min, which corresponds well with the observed maximum sleep episode of 13.5 min.

5. Assuming the overall dive metabolic rate of 4.5 mL O_2/min kg, the maximum dive time would be 15.31/(0.0045 mL O_2/min × 355 kg) = 10 min, which is much less than the observed dive times.

FIGURE QUESTIONS

Figure 48.6 If the flow rate of the water were increased, or the flow rate of the blood were decreased, it would be possible to maximally saturate the blood with a concurrent exchanger. This exchange would be less efficient than countercurrent exchange, however.

Figure 48.10 If there are 23 branchings of the airways, each giving rise to 2 new branches, the total number of terminal airways would be 2^{23}, or 8,388,608.

Figure 48.13 The hemoglobin would not fully saturate with O_2 at 100 mm Hg P_{O_2}.

APPLY WHAT YOU'VE LEARNED

1. In Figure A, in the 15 minutes that the spiracles were open, the intratracheal O_2 level increased from 19 to 142 mm Hg. In the 22–23 minutes in which the spiracles were closed, the intratracheal level of O_2 dropped from 142 to approximately 37.5 mm Hg. In the approximately 70 minutes that the pupae entered the fluttering phase, the intratracheal O_2 level fluctuated between 37.5 and 18.75 mm Hg. In Figure B, the first chamber's atmospheric O_2 concentration was 48 mm Hg, but the fluttering phase kept the intratracheal O_2 levels at a fairly consistent 36.75 mm Hg. In the second chamber, the atmospheric O_2 levels were 159 mm Hg. The fluttering phase allowed some fluctuation in the intratracheal O_2 levels, from a high of roughly 38.25 mm Hg to a low of roughly 36 mm Hg. In the third chamber, atmospheric levels of O_2 were almost double normal atmospheric O_2 levels, at 301.5 mm Hg. But the fluttering phase kept the intratracheal O_2 level at an almost constant 30 mm Hg.

2. When the tracheal tubes open, the atmospheric O_2 diffuses into the tubes, bringing the intratracheal O_2 level close to normal atmospheric O_2 levels (159 mm Hg). However, over the approximately 23 minutes that the spiracles are closed, the O_2 levels are slowly depleted as more O_2 is dissolved into the hemocoel and distributed to all of the cells to be used in aerobic respiration. We could predict that this O_2 level would continue to decline, approaching zero, if it were not for the fluttering phase. The fluttering phase appears to allow CO_2 to escape into the atmosphere while also allowing small amounts of O_2 to enter the tracheal tubes and maintain O_2 levels at a fairly constant level.

3. A delicate balance must be maintained between intratracheal O_2 levels and atmospheric O_2 levels. When atmospheric levels of O_2 are normal (middle graph in Figure B), the fluttering phase allows CO_2 release and replacement of O_2 to seek a fairly constant level of O_2 within the trachea. In a low-oxygen environment (first graph in Figure B), the fluttering phase is increased, compared with a normal atmospheric O_2 level. Beginning at a lower level of O_2 at the time of spiracle closure would make it necessary to increase the amount of fluttering to keep O_2 levels at an appropriate intratracheal concentration for adequate aerobic respiration. However, when atmospheric O_2 levels are abnormally high, the fluttering phase is greatly diminished (third graph in Figure B). Opening the spiracles as often as what is seen under normal atmospheric O_2 levels could expose cells to higher than normal O_2 levels. These levels could prove detrimental to insect cells. Therefore decreasing the amount of fluttering would be one way to prevent too much O_2 from entering the tracheal tubes and reaching a toxic cellular level.

Chapter 49

RECAP 49.1

1. Circulatory systems supply exercising muscle cells with oxygen and nutrients. They also take CO_2 and heat out of muscles.

2. A sponge has water channels throughout its tissues, which means that the external medium can circulate close to all of the sponge's cells, where exchanges of nutrients, oxygen, and wastes take place.

3. Hemolymph is the extracellular fluid in animals with an open circulatory system. In animals with a closed circulatory system, blood plasma is the extracellular fluid contained in the heart and blood vessels, and interstitial fluid is the extracellular fluid outside the circulatory system. Blood contains cellular elements in addition to plasma.

4. The fight-or-flight response involves increasing blood flow to tissues that are necessary for action, such as the skeletal muscles, and decreasing blood flow to tissues that are not necessary for action, such as the gut. A closed circulatory system can increase blood flow by increasing pressure, and it can also direct that blood flow to critical tissues by changing the resistance in the vessels leading to those tissues.

RECAP 49.2

1. Deoxygenated blood returns from the systemic circulation in veins that converge on the sinus venous that leads to the atrium. From the atrium, blood flows into the single, muscular ventricle. When the ventricle contracts, blood under pressure flows into the elastic bulbus arteriosus. The elastic recoil of the bulbus arteriosus maintains a continuous blood flow through the gills where respiratory gas exchange occurs. The oxygenated blood from the gills flows through the systemic circuit.

2. The direction of O_2 diffusion across the gill membranes depends on the P_{O_2} gradient between the blood and the water flowing across the gills. By taking air into its

lungs, the fish can maintain a higher P_{O_2} in its blood than in the surrounding water. As a result, O_2 diffuses from the blood to the water, increasing the P_{O_2} in the water surrounding the eggs.

3. Ectothermic reptiles have a left and a right aorta. Except for the crocodilians, the ventricles are not completely divided and thus blood from the right ventricle can flow either into the right aorta or the pulmonary aorta. When the animal is resting or submerged and not breathing, the resistance in the lung circuit is high and blood from both ventricles flows into the aortas. When the animal is breathing, resistance in the lung circuit goes down, and blood from the right ventricle flows into the lung circuit while blood from the left ventricle flows into the aortas. In crocodilians there is complete separation of the ventricles and the right aorta opens into the right ventricle. But, just where the aortas leave the heart there is a connection between them. When the animal is breathing and resistance in the lung circuit is low, back pressure from the left aorta closes the valve between the right ventricle and right aorta so all blood from the right ventricle flows to the lungs. When resistance in the lung circuit is high, then pressure in the right ventricle is high enough to open the right aortic valve and blood from the right ventricle flows in the right aorta to the systemic circuit.

4. At birth, when the newborn starts to breathe, the resistance in the pulmonary circuit falls. If the ductus arteriosus does not close, the higher pressure in the left ventricle will pump blood into the pulmonary circuit through the open (patent) ductus arteriosus. This will cause the pulmonary circuit to become congested and the systemic circuit to be deprived of oxygenated blood.

RECAP 49.3

1. Aortic valve stenosis would decrease the flow of blood into the aorta. Consequences would include decreased pressure in the aorta and major arteries and decreased blood flow to tissues, causing fatigue and shortness of breath. Pressure in the left ventricle would increase, as would its work load. Blood could back up in the pulmonary circuit, causing pulmonary hypertension. Pulmonary valve stenosis would decrease blood flow to the lungs and therefore decrease delivery of oxygenated blood to the left heart, causing fatigue and shortness of breath. The pressure in the right ventricle would increase, as would pressure in the venous return vessels.

2. Contraction of the myocardium depends on the presence of Ca^{2+} ions in the sarcoplasm. Increasing Ca^{2+} in the sarcoplasm prolongs the duration of the Ca^{2+} pulse resulting from the cardiac action potential. Thus the heart beats more strongly, contractions last longer, and cardiac output increases

3. Stenosis of the aortic valve should produce a heart murmur during systole, as that is when pressure in the left ventricle is pushing blood through the aortic valve. Prolapse of the aortic valve should create a heart murmur during diastole, as that is when greater pressure in the aorta is pushing blood back into the ventricle through the prolapsed aortic valve.

RECAP 49.4

1. The partial pressure of O_2 is lower at high altitudes; therefore the athletes experience hypoxia and acclimate to the high altitude environment by producing more red blood cells. As a result, when they compete at lower altitudes, they have greater aerobic capacity.

2. Blood clotting is a massive event involving large numbers of cells and signals, yet it can be initiated by a small wound that exposes collagen fibers. Platelets that come into contact with those fibers are activated. They become sticky and stick to other platelets, which also become activated. The activated platelets release chemical signals that trigger a long line of clotting reactions, and with each step, the reaction grows. Thus a small event triggers a larger and larger number of events leading to clotting—a cascade of events.

3. Arterioles are called resistance vessels because of the smooth muscles in their walls that can constrict the vessels and increase their resistance. Changing the resistance of different arterioles allows blood flow to be directed to the tissues and organs that need it the most. Veins are called capacitance vessels because their thin walls with many elastic fibers can stretch to hold larger volumes of blood. When the body is at rest, all of the blood does not have to circulate to serve the body's needs, so a large proportion of it can be held in reserve in capacitance vessels.

4. The blood proteins are responsible for the osmotic pressure that draws interstitial fluid back into the capillaries. If the blood proteins are metabolized, the blood osmotic pressure drops and more plasma that leaves the capillaries due to hydrostatic pressure remains in the interstitial spaces. Thus interstitial fluid accumulates in the abdominal cavity and the extremities.

RECAP 49.5

1. Figure 49.19 shows that a drop in blood flow to a tissue due to a fall in MAP results in autoregulatory responses that decrease the resistance to flow in that tissue. Any decreased resistance in the systemic circuit will cause a drop in MAP, further depriving that tissue of O_2 and nutrients and increasing the autoregulatory response. With this positive feedback loop, the autoregulatory response can worsen the problem that initiated it.

2. The sympathetic division of the ANS stimulates vasoconstriction in tissues not essential for a fight-or-flight response, such as the digestive system and the skin. This increase in total peripheral resistance results in a rise in MAP. The sympathetic innervation of the heart causes it to beat faster and more strongly, also contributing to

a rise in MAP. The parasympathetic division of the ANS has opposite effects, slowing the heart and decreasing MAP.

3. Baroreceptors are firing at a rate midpoint in their range when blood pressure is normal. Therefore they have the potential to either decrease or increase their firing rate, depending on whether MAP falls or rises. They therefore contribute to regulatory responses to both rises and falls in MAP.

4. Autoregulatory mechanisms operate at the local level to dilate or constrict arterioles in response to local changes in the chemical environment. These changes in arteriole diameter affect the amount of blood reaching local tissues. As these changes occur, they have an effect on the body's blood pressure, which is detected by baroreceptors that elicit neural and hormonal signals that cause widespread cardiovascular changes throughout the body.

WORK WITH THE DATA, P. 1055

1. Comparing mutant mice not CsA treated, but treated with or without RNAi, the left ventricular wall thickness (LVWT) of the controls was 0.96 ± 0.12 and that of the RNAi-treated mice was 0.70 ± .088. A t-test for the probability that these two mean LVWTs should be the same gives $P = 0.0025$, indicating that they are significantly different: the RNAi-treated animals had significantly lower LVWT.

2. Comparing mutant mice that did receive CsA but did or did not receive RNAi treatment, the LVWT of the control mice was 1.66 ± 0.22 and that of the RNAi mice was 0.85 ± .06. A t-test calculates $P < 0.001$, indicating that these two mean LVWTs are significantly different.

FIGURE QUESTIONS

Figure 49.4 If these curves were for the right ventricle, the volume curve would be the same, but the pressure curves during systole would be lower.

Figure 49.9 The breadth of the ventricular muscle action potential is important because it determines the duration of systole and therefore the time available for the emptying of the heart. More complete emptying means greater stroke volume.

Figure 49.10 An occasional block of the AV node could result in a lengthening of the interval between the P and R waves (the PR interval) on the ECG and possibly a missed beat. A long-lasting block of the AV node could result in complete dissociation between the atrial rhythm and the ventricular rhythm.

APPLY WHAT YOU'VE LEARNED

1.

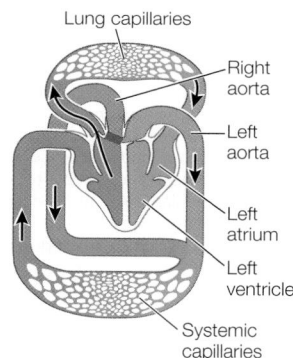

Lung capillaries
Right aorta
Left aorta
Left atrium
Left ventricle
Systemic capillaries

Blood from the right ventricle is blocked from exiting the right aorta by the surgery. Blood from the right ventricle can pass only through the pulmonary artery. This change converted the crocodilian heart into a heart similar to that of a bird, because a bird heart obligates the blood to go from the right ventricle to the lungs with no opportunity for bypass. In a bird, blood goes from the right ventricle to the lungs to the left atrium to the left ventricle to the body and then back to the right atrium and right ventricle again. In a crocodilian, the blood can go from the right ventricle to either the lungs or the body.

2. When an alligator is diving and spending time underwater, it is not actively breathing. In these instances (which can extend for long periods of time), the animal does not need to expend energy to pump blood through the lungs. An adaptive advantage is the ability to conserve energy by not pumping blood to the lungs when there is no benefit to be gained from doing so.

3. If bypassing the pulmonary circuit conserves energy in these animals, an indication might be that growth rates of control animals would have been higher than those of experimental animals, since control animals would have more energy available to divert to growth. However, if we require a P value < 0.01 to conclude that there is a significant difference between the control and experimental groups for both sedentary and exercised animals, there are no significant differences in body mass or body length. This indicates that growth rates between control and experimental animals were similar, suggesting that the pulmonary bypass capability confers no adaptive advantage in terms of growth of juvenile alligators, even when they are stressed by exercise. Therefore these data alone do not allow us to say that the shunt affords a significant conservation of energy related to growth.

4. An investigator could compare the duration of breath-hold dives in experimental and control animals. This indicator would provide a measure of the time animals can

spend not breathing, which would be an advantage in their ability to survive underwater for extended periods of time. Animals with greater dive times could evade predators more easily and also gain greater predatory advantages themselves by their ability to lie in wait for prey for extended times underwater.

A second indicator could be the amount of food consumed and waste produced by experimental and control animals under otherwise identical conditions. This indicator could be used to compute calories burned and would support the hypothesis if experimental animals burned more calories than control animals, since experimental animals would need more energy to go about their daily activities.

A third indicator would be the number of eggs produced by female alligators in experimental and control groups. To support the hypothesis, the results from this study would have to show that experimental females produce fewer eggs over a defined time period than control females. The quantity of eggs is related to the amount of energy the animals have available to put toward reproduction, and loss of the shunt may cause experimental animals to have less energy to expend in producing eggs.

Chapter 50

RECAP 50.1

1. When gray whales are migrating and on their wintering grounds, their energy source is their stored fat (blubber). Fat has a high energy content and low water content and therefore is ideal for long-term storage of energy with minimal mass. It also contributes to the buoyancy of the whale body.

2. Vitamin A is fat-soluble and can accumulate in the body. Vitamin C is water-soluble and rapidly excreted in the urine, so a toxic level of it cannot accumulate.

3. The micronutrient iron is largely recycled in the body, but due to menstrual blood loss, premenopausal women do lose iron each month.

4. If proteins were added to a solution delivered into the blood, it would stimulate an immune response in the patient to foreign proteins. Therefore it is essential to supply all amino acids in such a mix.

RECAP 50.2

1. Herbivores must spend much time feeding because their food has low energy content and requires considerable processing both mechanically (chewing) and chemically (digestion).

2. All of the digestive enzymes act through hydrolysis of their substrates.

3. Antibiotic therapy can greatly diminish or alter the gut microbiota and therefore result in altered digestion.

4. The symptoms are due to damage to the villi. Damaged villi are less effective at absorbing the products of digestion and therefore contribute to undernutrition and even lack of micronutrients such as iron. Unabsorbed digestive products pull water into the gut, resulting in diarrhea, and also support microbial metabolism, leading to bloating. Damaged villi compromise the surface area available for digestion, affecting fat absorption as well as absorption of nutrients that require transport. The undigested fat results in fatty stools.

RECAP 50.3

1. The tongue pushes the chewed food to the back of the mouth stimulating the swallowing reflex that through activation of many muscles pushes the food into the esophagus. Stretching of the smooth muscle in the esophagus stimulates contraction of that smooth muscle pushing the food toward the stomach. The directionality of this movement is facilitated by the enteric nervous system causing the smooth muscle ahead of the bolus of food to relax. When the bolus of food reaches the esophageal sphincter, that anticipatory wave of relaxation opens the sphincter allowing the food to enter the stomach. The coordination of smooth muscle contraction and relaxation creates the waves of peristalsis that generally moves the food in the esophagus towards the stomach even in the absence of gravity.

2. Stomach acid is produced by chief cells in the gastric pits of the stomach. In these cells, carbonic anhydrase catalyzes the hydration of CO_2 to produce H_2CO_3, which dissociates into HCO_3^- and H^+. The H^+ is transported across the pit lumen side of the cell in exchange for K^+. The HCO_3^- is exchanged across the opposite end of the cell into the interstitial fluid in exchange for Cl^- ions. The excess K^+ in the cell leaks out the luminal end and is pumped back in via the H^+/K^+ exchanger. Thus the concentration of H^+ in the stomach lumen and the separation of HCO_3^- in the interstitial fluid depend on the anatomical integrity of the stomach wall. If that integrity is destroyed, the H^+ cannot be separated from the HCO_3^- and there will be no pH change.

3. Bile emulsifies fats in the diet, creating tiny micelles that present a large surface area for the action of water-soluble lipases. Bile prevents the fatty micelles from coalescing into larger fat globules with a smaller surface area. Equivalent molecules are not required in the lymph and blood because the fats in lymph and blood are incorporated into lipoproteins that are coated with water-soluble proteins.

4. The gut microbiota is a significant source of nutrition for ruminants because their microbiota grow on the ingested food in the rumen and reticulum before the semi-digested food is exposed to the HCl and digestive enzymes of the stomach and small intestine. In humans, abundant microbiota are present in the small and large intestine and therefore are not killed by the stomach acid and digested in the upper region of the small intestine, where they could otherwise serve as a major source of nutrition.

5. The ruminant does not produce the cellulose hydrolyzing enzymes necessary to digest the plant materials it eats. The rumen and reticulum have cultures of microorganisms that produce cellulases and break down the plant matter. The resulting fermenting mixture of plant matter and microorganisms moves into the omasum, where water is reabsorbed. From the omasum, the mass of semi-digested plant matter and the associated microorganisms move to the abomasum (true stomach) that secretes HCl, which aids digestion and kills the microorganisms that are an important component of the animal's nutrition.

6. If even a small amount of pepsinogen is activated by hydrolytic cleavage to produce the active enzyme pepsin, that pepsin will act on additional pepsinogen to release more pepsin molecules creating an autocatalytic cascade. This is an example of positive feedback in which a product of a reaction (pepsin) stimulates still more reaction (pepsinogen → pepsin), which amplifies formation of the product of that reaction (pepsin).

RECAP 50.4

1. Introducing nutrient solutions into the jejunum would not stimulate release of secretin or CCK, therefore there would not be bile secretion to emulsify large complex lipids and there would not be pancreatic enzymes to digest complex carbohydrates and proteins. Thus, the jejunal formula would have to consist of medium to short chain fatty acids and partially hydrolyzed carbohydrates and proteins.

2. The three classes of lipoproteins differ in their relative compositions of protein, triglycerides, and cholesterol. The high-density lipoproteins have the lowest percentage of triglycerides and deliver cholesterol from tissues to the liver for excretion as bile. Low-density lipoproteins have less protein and triglyceride and more cholesterol. These particles have transferred most of their triglycerides to adipose and other cells and are left with a lot of cholesterol that can be deposited into the walls of arteries. Very low-density lipoproteins consist mostly of triglycerides, which they transfer to adipose cells, resulting in the production of low-density lipoproteins.

3. In both muscle and liver, insulin promotes the uptake of glucose and its incorporation into glycogen. When insulin levels fall, the effects are different in liver and muscle, in that in liver the reduced levels activate glucose phosphatase. This makes it possible for the glucose produced by the breakdown of glycogen to be released into the interstitial fluid. This process does not occur in muscle, so the glucose is trapped in the muscle.

4. In the experiments in which the lateral hypothalamus or the ventromedial hypothalamus was lesioned and subsequent changes in body mass measured, the animals either gained or lost a large amount of mass, but they eventually plateaued at a new level, indicating the continued ability to regulate, but at a different level.

5. Pyruvate and lactate produced by muscles working anaerobically enter the circulation and are taken up by the liver, where they are converted to glucose by processes of gluconeogenesis. This glucose can then return to the blood and support further glycolysis by the muscles.

6. Rats that have a mutation that eliminates their production of leptin eat more and become obese. If these rats are joined parabiotically with normal rats that produce leptin, the obese rats eat less and lose mass.

WORK WITH THE DATA, P. 1078

1. The average fasting blood glucose level was 11.00 mM for the lard-fed mice and 10.3 mM for the fish-oil-fed mice. These values are significantly different at the $P < 0.05$ level according to a two-tailed, non-paired t-test, (these calculations can be done in Excel).

2. The average fasting blood insulin level was 3.96 ng/mL for the lard-fed mice and 1.09 ng/mL for the fish-oil-fed mice. A t-test, two-tailed, non-paired showed that $P < 0.0001$ which is highly significant.

3. The time course of the response to an insulin challenge can be plotted. T-tests can be used to see which points are significantly different.

   The 30-minute values are not significantly different, but the rest of the time points are: 60 minutes, $P < 0.005$; 90 and 120 minutes, $P < 0.001$.

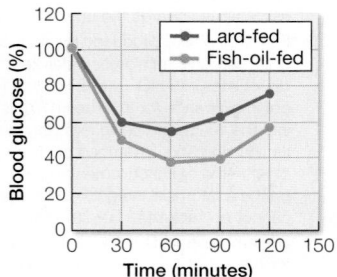

4. The lard-fed mice were less responsive to insulin and had a slightly higher fasting blood glucose level. This suggests an impairment of their carbohydrate metabolism. The fish-oil-fed mice had a more robust response to the insulin challenge.

FIGURE QUESTIONS

Figure 50.14 Without a gallbladder, the release of bile into the small intestine is continuous and slow, meaning a large amount of fat passing through cannot be optimally emulsified and digested. As a result it passes through the small intestine without being absorbed and produces fatty stool and diarrhea.

Figure 50.15 The position of the rumen and reticulum before the abomasum allows large populations of gut microbiota to grow and break down the ingested plant material before passing into the true stomach, where they are killed by stomach acid.

Figure 50.16 The fact that CCK and secretin can slow the stomach yet stimulate the pancreas means that these two hormones can regulate the passage of food through the digestive tract at a rate that enables complete digestion.

Figure 50.17 The uptake of glucose by cells of the nervous system does not depend on hormonal stimulation—it depends only on the concentration difference of glucose between the interstitial fluid and the cell interior. All of the hormone actions in this figure refer to mechanisms aimed at maintaining a constant blood glucose concentration, and this is the variable that guarantees adequate glucose supply for the nervous system.

APPLY WHAT YOU'VE LEARNED

1. The higher triglyceride levels in winter indicate a greater dependence on fat metabolism during winter, and this is supported by the decrease in winter of lactate, a product of glycolysis (carbohydrate metabolism), and the increase in succinate, which is a citric acid cycle intermediate. Since both carbohydrate and triglyceride metabolites enter the citric acid cycle as acetyl CoA, these data support the idea of an increase in dependence on fat metabolism during winter.

2. The data show significant changes in three highly abundant phyla and no change in one highly abundant phylum. Thus we can conclude that there are major seasonal changes in the gut microbiome. These changes could be due simply to the presence or absence of food in the gut over a substantial period of time, or the changes could play a role in the different metabolic states of the bears in summer and winter.

3. The data indicate that the summer microbiota promote greater mass and fat gain than winter microbiota. Two hypotheses that could explain this difference are that (1) some signal from the summer microbiota stimulates greater food intake, and (2) the summer microbiota is more efficient at facilitating digestion when there is high nutrient flow through the gut.

4. The data suggest that a seasonal change in the gut microbiome of the bears supports and may promote increased food intake and digestive efficiency in the summer months. The change from the summer to the winter gut microbiome may result from some microbial species being able to survive a long duration of fasting, or the change in microbiome may have a functional significance in altering energy metabolism to promote storage in the summer and conservation in the winter.

Chapter 51

RECAP 51.1

1. A freshwater vertebrate has to osmoregulate by conserving salts and excreting water. Having a closed circulatory system makes it possible to use pressure to filter the extracellular fluid to excrete water without the loss of salts and other solutes.

2. An animal that lives in the intertidal may be isolated in tidal pools that can be highly diluted by rain or highly concentrated by evaporation. *Artemia* may survive by osmoconforming over a wide range of salinities, but it cannot survive at the extremes of fresh water or saturated salt solutions without being able to osmoregulate. *Artemia* uses active transport of ions across its gills to regulate hypertonically in very dilute environments and to regulate hypotonically in highly concentrated environments.

3. Bundles of secretory tubules drain into the salt gland ducts. The cells of those tubules actively transport Cl⁻ ions into the lumen. Na⁺ follows, then water flushes the contents of the tubules into the duct.

RECAP 51.2

1. Ammonia is toxic and must be either excreted rapidly or detoxified. Ammonia is highly soluble in water and diffuses rapidly, so in fish, ammonia can diffuse into the blood and then be lost to the environment by continuous diffusion across the gill membranes. Because mammals are tidal breathers, the ammonia in their blood would have to reach higher levels to be lost by diffusion across alveolar membranes, so mammals detoxify ammonia by converting it into urea or uric acid.

2. Uric acid is considerably less soluble than urea. It precipitates out of solution in humans, in whom the total osmolarity of extracellular fluid is 300 mosm/L and therefore the contribution of uric acid to that concentration is considerably less than 300 mosm/L. Sharks and rays could not produce an extracellular osmolarity greater than that of seawater by retaining uric acid.

3. Most terrestrial animals have to conserve water. The advantage of excreting uric acid is that it can be precipitated out of solution in the excretory system, enabling the reabsorption, and therefore conservation, of water.

RECAP 51.3

1. Freshwater flatworms have to deal with a constant influx of water into their extracellular fluids by osmosis. The flame cells make it possible to filter that extracellular fluid into the protonephridia without loss of solutes, and the protonephridia can further process the filtrate by secretion and reabsorption to increase elimination of wastes and conserve solutes.

2. The blood plasma is filtered in capillaries to produce the coelomic fluid, and the two are similar in composition. The coelomic fluid is processed in the metanephridia through processes of tubular secretion and reabsorption to become the urine that is excreted from the body.

3. The insect excretory system eliminates nitrogenous waste as uric acid. By actively reabsorbing Na⁺ and K⁺ in the hindgut and rectum, the insect concentrates the waste fluid to the point that uric acid precipitates out of solution thus lowering the

osmolarity of the contents of the hindgut and rectum and enabling more reabsorption of water due the active reabsorption of Na$^+$ and K$^+$.

RECAP 51.4

1. To move into more arid terrestrial habitats, vertebrates had to minimize their water loss and their need to have access to standing water to practice external fertilization. Whereas both amphibians and reptiles evolved various adaptations to decrease evaporative water loss from their skins, only reptiles gained independence from needing standing water to reproduce, by the evolution of the amniote egg. However, the amniote egg requires fertilization before it is encased in a protective shell, and thus the evolution of internal fertilization paralleled the evolution of the amniote egg.

2. The afferent arterioles bring blood into the glomerulus at high pressure to support filtration. The efferent arterioles carry the remaining blood from the glomerulus to the capillaries serving the renal tubules, where secretion and reabsorption of solutes occur.

3. The anatomical components of the glomerular filtration mechanism are: fenestrations in the glomerular capillary walls, the basal lamina of the capillary endothelial cells, and the slits between the processes of the podocytes of Bowman's capsule.

4. The composition of the urine is made different from the composition of the blood by processes of selective active reabsorption and secretion by the renal tubules.

RECAP 51.5

1. The highest concentration of mitochondria would be in the renal cortex because of all of the active transport mechanisms in the convoluted tubules. The lowest concentration would be at the bottom of the renal medulla where there are only thin regions of the loops of Henle. The intermediate concentration would be in the upper level of the renal medulla because of the active transporters of Na$^+$ in the thick ascending limb of the loop of Henle.

2. The thick ascending limb of the loop of Henle actively transports Na$^+$ from the tubular fluid to the extracellular fluid, and Cl$^-$ follows. The increase in NaCl in the extracellular fluid of the medulla pulls water out of the descending limb osmotically, increasing the concentration of the tubular fluid. Since the ascending limb is not permeable to water, the more concentrated tubular fluid cannot pull water into the ascending limb. When it reaches the thick ascending limb, more NaCl is pumped out into the extracellular space. Through this mechanism, the concentration of the extracellular fluid in the renal medulla is maintained higher than the blood osmolarity.

3. The desert rat would have longer loops of Henle and therefore a longer renal medulla. The desert rat has evolved under selective pressure to conserve water. One way of doing that is to produce a more concentrated urine. The longer the loops of Henle, the greater the concentration gradient that can be created in the renal medulla, so the desert rat should have longer loops of Henle.

4. Bicarbonate in the blood is filtered into the tubular fluid, but the H$^+$ ions transported into the tubular fluid by the tubule cells combine with that bicarbonate to produce H$_2$CO$_3$ (carbonic acid), which dissociates into H$_2$O and CO$_2$. The CO$_2$ diffuses into the tubule cells, where its hydration to H$_2$CO$_3$ is catalyzed by carbonic anhydrase. The resulting HCO$_3$$^-$ is transported across the basal end of the tubule cells and into the interstitial fluid.

RECAP 51.6

1. The afferent and efferent arterioles regulate GFR by changing their resistance to flow. When blood pressure rises, the afferent arterioles respond by constricting and increasing their resistance. When blood pressure falls, the efferent arterioles respond by constricting and increasing their resistance. Constriction of the afferent arterioles decreases GFR, and constriction of the efferent arterioles increases GFR. Constriction of the afferent arterioles is due to the general property of smooth muscle – when stretched, it responds by contracting.

With a decrease in blood pressure and therefore a decrease in GFR, the feedback information is a fall in solute concentration in the distal tubule. In response, the macula densa cells signal the afferent arterioles to dilate and cause the efferent arterioles to constrict by releasing renin, which initiates the conversion of angiotensinogen to angiotensin, which in turn stimulates the efferent arterioles to constrict.

2. The secretion of renin and the resulting production of active angiotensin are triggered by a fall in blood pressure and therefore a fall in GFR. One action of the renin–angiotensin–aldosterone system is to increase water reabsorption from the kidney to maintain blood volume. To reabsorb water, it is necessary to increase the reabsorption of Na$^+$ to create the osmotic gradient that will pull water from the tubule fluid into the tubule cells. However, if the exchanger moved Na$^+$ and K$^+$ one for one, it would not create an osmotic gradient

3. The chicken broth has the same osmolarity as the body fluids, so its absorption from the gut would not increase the blood osmolarity, but it would increase blood volume and blood pressure. The increase in blood pressure would stimulate the carotid baroreceptors, resulting in an inhibition of ADH release. As a result of decreased ADH, the kidney would reabsorb less water and therefore produce more urine. In addition, the increased blood volume would stimulate the release of atrial natriuretic peptide, causing increased secretion of Na$^+$ and therefore increased loss of water, thus reducing fluid volume, blood volume, and blood pressure.

WORK WITH THE DATA, P. 1104

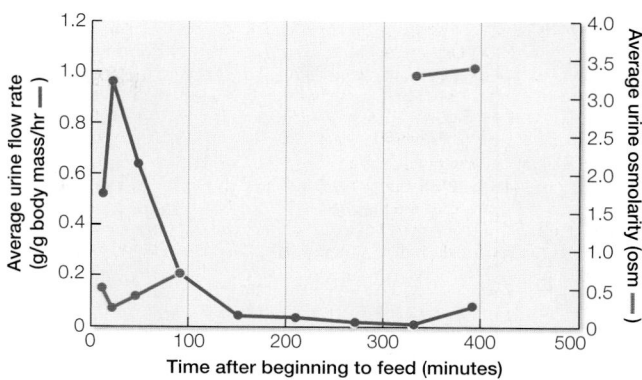

1. The average urine flow rate over the first hour is 0.18 g/g body mass/hr and for a 30-g bat would be 5.4 g. Blood intake = 0.60 × 30 g = 18 g. Therefore the percent of the ingested volume excreted in the first hour would be 30%.

2. The maximum concentrating factor in this experiment is 3,400 mosm/300 mosm = 11.3.

3. Extrapolating the missing data by connecting the two curves with a straight line would not be reasonable because the urine flow rate falls close to its lowest level at 2.5 hours (150 minutes). It is reasonable to expect that the urine osmolarity would be close to maximum at this time.

FIGURE QUESTIONS

Figure 51.4   For the freshwater planarian, the excretory product would be hypotonic to the interstitial fluid because the animal has to be excreting excess water that is entering its body by osmosis.

Figure 51.6   Na$^+$ and K$^+$ are transported into the Malpighian tubule so that water follows and flushes the contents of the tubule into the gut. They are transported out of the hindgut and rectum to create a concentration gradient for osmotic reabsorption of water from the gut, made possible by the precipitation of uric acid.

Figure 51.9   The blood flowing down the vasa recta always has a slightly lower osmolarity than the interstitial fluid at the same level because it is flowing and the equilibration is not instantaneous. The opposite is true for the ascending limb of the vasa recta, where the blood always has a slightly higher osmolarity than the surrounding interstitial fluid. As a result, the blood in the ascending limb absorbs water osmotically and carries some excess salt and water out of the renal medulla.

Figure 51.10   Blocking the Na$^+$ transport out of the thick ascending limb of the loop of Henle would result in a loss or a decrease (depending on the extent of the drug action) of the concentration gradient in the medulla. As a result, the tubular fluid in the collecting duct could not be concentrated, so urine flow would increase and there would be increased loss of water.

Figure 51.16   The vampire bat has long loops of Henle like the desert gerbil.

APPLY WHAT YOU'VE LEARNED

1. Yes, the RMT is a good predictor of the ability of a species to concentrate its urine.

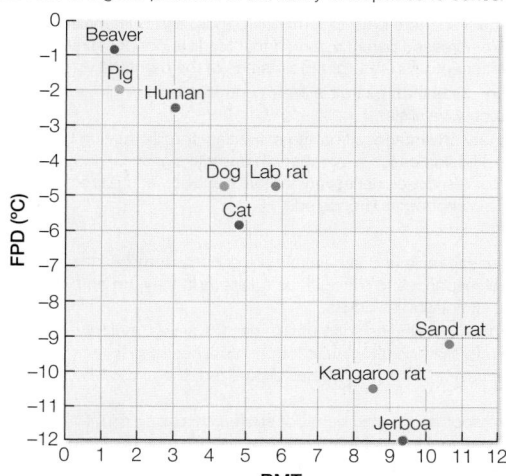

2. Although several species with low percentages of long loops cannot produce highly concentrated urine, some with high percentages of long loops have only moderate capacity for producing concentrated urine and three of the species that produce the most concentrated urine have low percentages of long loops. If you calculate

a correlation coefficient for these data, it is 0.44, and with a sample size of 7, the *p* value is 0.32, so this is not a significant correlation.

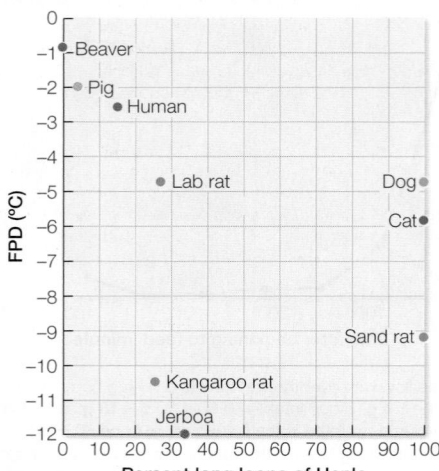

3. RMT is a much better predictor of an animal's ability to produce concentrated urine. As you can see in the percent of long loops of Henle versus FPD plot, animals such as the dog, cat, and sand rat, which have 100 percent long loops of Henle, were unable to produce the high concentrations of urine seen in the kangaroo rat and jerboa. By contrast, in the RMT versus FPD plot, animals with a high RMT, such as the sand rat, kangaroo rat, and jerboa, were able to produce higher concentrated urine than animals with a lower RMT.

Chapter 52

RECAP 52.1

1. Fixed action patterns are genetically determined, adaptive behaviors that are usually activated by specific stimuli normally encountered in nature. Pavlovian conditioned reflexes are physiological processes triggered by learned, non-natural stimuli. Behaviors resulting from operant conditioning are random behaviors without physiological or adaptive significance; they are activated by non-natural stimuli that the animal has learned to associate with receiving a reward when expressing the behavior.

2. A genetically determined behavior might not be expressed in a deprivation experiment if the animal is not in the appropriate developmental stage or physiological condition, and if the necessary stimuli for the behavior are absent.

3. Proximate causes of mating-display behavior include physiological reproductive condition and presence of stimuli from a sexually receptive conspecific. The ultimate cause of mating-display behavior is reproductive success—how many of one's offspring live and mature to contribute to the next generation.

RECAP 52.2

1. If the expression of vasopressin and oxytocin genes is increased in montane voles, the voles should be more likely to form pair bonds, and both males and females should exhibit increased parental care. If that result is not observed, it could be due to the lack of receptors in the right brain areas of the montane vole.

2. You would conclude that the trait is likely controlled by a single gene with a dominant and a recessive allele.

3. The evolutionary advantages of having a species-specific mating behavior controlled by a gene expression cascade are that: (1) a single signal can induce a complex combination of responses necessary for that behavior, and (2) temporal coordination of all of those components is ensured.

RECAP 52.3

1. If neonatal genetic male and female rats are gonadectomized, they both develop female sexual responses to estrogen as adults, and they are both unresponsive to testosterone stimulation as adults.

2. The songbird will not sing its song until it reaches sexual maturity, and at that time it has to learn to sing the template it formed as a nestling. To learn, it must practice and hear itself sing. Once its song has crystalized, the bird will be able to sing it even if deafened.

3. Variability in a species-specific behavior such as birdsong might be adaptive if it enables an individual to establish identity with a different local population or to create a new song feature that enhances its attractiveness to potential mates. Variability in this same species-specific behavior could be maladaptive if it detracts from individual recognition. For example, in a situation in which an individual is returning to its nesting grounds to mate with a previously established mate or to feed offspring, it is important to be able to establish individual identity.

RECAP 52.4

1. The benefit of the marsh-nesting site is that it reduces the risk of predation. The cost is that there is greater competition for those sites and less food availability. Feeding

in upland areas opens up a greater potential food source, and it does not involve the cost of defending a feeding territory. A risk, however, may be exposure to predation and the need to leave the nest unguarded during feeding.

2. The cost of being a dominant male cichlid is the energetic expenditure involved in territory defense and courtship of females. Also, there are opportunity costs in having less time to feed. The benefit is the likelihood of attracting a mate and fertilizing her eggs. The nondominant male does not incur the costs of territory defense and display and benefits from being able to achieve some reproductive success.

3. Female elephant seals have to have access to a beach to give birth and rear their young, and males can guarantee that access by excluding other males from their piece of beach. Thus the major focus of male elephant seal territorial behavior is exerting physical dominance over other males, and natural selection has optimized the males' ability to do so. Female elephant seals have no choice over who their "beachmaster" is. In contrast, female prairie-chickens do not need the territory of a male to rear their young, so they are able to simply choose the male they find most attractive. Therefore natural selection in prairie-chickens has favored male displays that are directed at females rather than at other males.

4. One source of selective pressure would be minimization of opportunity costs; therefore the amount of calories obtainable per unit of foraging time would be a critical variable. Another selective pressure is the danger of predation; the relative risk of exposure to predators in the open versus protected environment would be another critical variable.

RECAP 52.5

1. A free-running circadian rhythm is one that is not entrained to a regularly occurring signal, such as sunrise and sunset, but is expressing an endogenous periodicity. When a time cue such as a regular light–dark cycle is imposed on a free-running rhythm, that cue will either phase advance or phase delay the circadian clock to bring it into phase with the environmental light–dark cycle.

2. The direction to the sun can be used to tell compass directions if you know the time of day. A time-compensated solar compass uses an endogenous timing mechanism to compensate for the movement of the sun's position in the sky. Thus, if a honey bee finds food to the south at noon, its waggle dance on the honeycomb in the hive will be straight up, but as time passes into the afternoon, the direction of the waggle dance will have to be more and more to the left of straight-up to indicate that the direction to the food source is south.

3. The most efficient forms of inter-individual communication are tactile communication, which can be highly specific to the individual, and visual communication, which can convey considerable specific information in a very short time to an individual that is focused on the sender. Sound is an efficient form of communication for a broader population because it permeates the whole environment. Olfaction can be good for either individual or for population communication, depending on the volatility of the odorant.

RECAP 52.6

1. Selection will favor an individual helping its parents raise offspring rather than breeding itself when the chances of its own successful breeding are very low. Raising several brothers or sisters contributes more to an individual's inclusive fitness than not raising one's own offspring.

2. Because of the haplodiploid mechanism of sex determination in hymenopterans, sisters share 75 percent of their genes, but parents and offspring share only 50 percent. The unit of reproductive success in eusocial insects is the colony formed around a queen. By helping her mother raise sisters, it is likely that a worker will share 75 percent of her genes with a new queen, but if the new queen were one of her offspring, the worker would share only 50 percent of her genes.

3. The main risk of mobbing a potential predator is the possibility of being attacked and injured by the predator. A benefit of mobbing is that the predator is distracted from focusing on specific prey, such as the eggs or nestlings of the mobbers. The mobbing also warns others in the population of the presence of a predator. Of course, the major benefit is that the predator is driven away. Hypotheses about individuals taking part in mobbing would include that they are likely to be related, that they are more likely to participate if they have nestlings, and that monogamous males are more likely to join in mobbing than are promiscuous males.

WORK WITH THE DATA, P. 1124

1. If the song provides information about mate recognition, the female should show a strong preference for her mate's directed song over the directed song of other males.

2. Standard deviations are 5.9 and 19.3. The *P*-value is <0.001. Yes, the females clearly demonstrate recognition of and attraction to their mate's directed song versus the directed song of an unfamiliar male.

3. These data suggest that the directed song provides information about individual identity and plays a role in pair bonding of the male and female.

4. Undirected song may be an opportunity for the male to experiment with and practice slight variations in his song, but it may also represent a lower level of motivation to perform.

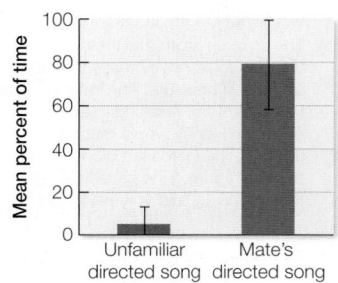

FIGURE QUESTIONS

Figure 52.2 It is unlikely that specific recognition of a complex stimulus could be genetically programmed.

Figure 52.11 The extended activity bouts of the mouse in the second panel represent the portion of the animal's circadian rhythm that corresponds to what its circadian clock is predicting as night. Therefore the light pulse in this experiment would be similar to seeing light early in the night. This is similar to what our experience would be after traveling from east to west—sunset would come later than our circadian clock would predict. In other words, we would get a light stimulus in the early portion of our circadian night. The result would be a delay of the circadian rhythm to bring it in line with the local light–dark cycle.

APPLY WHAT YOU'VE LEARNED

1. When females have no larvae or old larvae that no longer need parental care, they invest their energy in producing new broods. However, during the short period when larvae are dependent on parental investment ("new larvae"), the female fertility rate is low. Thus females seem to direct their investment of time and energy into the activities most likely to result in reproductive success.

2. Question 3 of Tinbergen applies. It asks about how a behavior influences survival and reproductive success, and therefore, it deals with ultimate causes of behavior. This study looked into the egg-laying behavior of female beetles under different circumstances. The results can be used to support the hypothesis that female beetles shift their energy expenditures to maximize their reproductive success. This is an ultimate cause of behavior since it deals with evolutionary pressures to maximize reproductive success.

3. The data indicate a correlation between increases in juvenile hormone and suppression of egg-laying behavior. The data also indicate a correlation between increases in methyl geranate and suppression of copulation behavior. These correlations suggest that hormone and pheromone changes may be underlying physiological mechanisms that influence the shift in egg-laying behavior and sexual behavior to parental care behavior in burying beetles.

4. Tinbergen's question about immediate causation 1 applies, and it deals with proximate causes of behavior. The studies were aimed at elucidating the underlying biological mechanisms that produce the observed changes in behaviors. The results can be used to support the hypothesis that expression of different behaviors comes about as the result of changing levels of hormones and pheromones produced by the female. This is a proximate cause of behavior since it deals with an immediate physiological mechanism that determines how an individual is behaving at a particular time.

5. A cost of the parental care behavior is the opportunity lost for mating and egg laying that could be done instead of caring for young larvae. This means that there are times when the beetles do not actively reproduce, and so fewer offspring are produced over time. A benefit is the protection and nurturing of young larvae, which improves the likelihood that they will survive to carry on the next generation.

Chapter 53

RECAP 53.1

1. Ecology is the study of the interrelationships among organisms and the physical environment, while environmentalism is the use of that ecological knowledge (as well as social and ethical concerns) to inform personal and public policy related to stewardship of the natural world. Ecology is a useful scientific enterprise because it can allow us to better anticipate how our actions may have negative consequences for ourselves and the natural world.

2. You could conduct observational surveys in the field to look for disease prevalence in live and dead sea stars. If multiple individuals were surveyed at multiple sites, then you might be able to make inferences of the prevalence of the disease at population and metapopulation scales. You could conduct an experiment in which you expose sea stars to the suspected pathogen to determine if they contract and die from the disease. Finally, you could create a population model to simulate how death from the disease would affect the populations of sea stars and compare the results with your observational data.

RECAP 53.2

1. No. A month of extremely hot weather does not necessarily tell us that Earth's climate is warming. Climate is the long-term average of atmospheric conditions over years to millennia.

2. Because Earth is a sphere, the intensity of solar radiation hitting its surface varies with latitude. At high latitudes, incoming solar energy comes in at an angle and has to travel through more of the atmosphere than at the equator, making it less intense than at the equator, where sunlight strikes the surface perpendicularly. The consequence of this latitudinal variation in energy input is a major difference in temperature and precipitation at every 30° latitude. At the equator, climate is tropical with warm and wet conditions. At 30° N and S, climate is warm and dry, producing desert conditions. At 60° N and S, climate is temperate with cold and wet conditions. At the poles, climate is cold and dry.

3. If you were sailing from New York to the United Kingdon (west to east), you would want to use the Westerlies and the Gulf Stream to your advantage. Thus you would

want to sail in a northeastern direction across the Atlantic Ocean. If you were sailing from San Francisco to Japan, you would want to take advantage of the NE Trade Winds and the North Equatorial Current in the Pacific Ocean.

4. Earth's axis is tilted, changing the amount of sunlight a particular region receives over the course of a year as Earth orbits the sun. When the Northern Hemisphere is tilted toward the sun, it experiences summer conditions, but that means that the Southern Hemisphere is tilted away from the sun and experiences winter conditions.

RECAP 53.3

1. Mountain ranges and valleys can create local and regional differences in temperature and precipitation due to their effects on atmospheric circulation. They also influence the flow of fresh water in rivers and lakes. The topography of the seafloor creates variations in water depth, which affects light penetration, water temperature, water pressure, and water movement (i.e., currents, waves, and tides). These changes in physical conditions create different zones containing characteristic biota, which are similar to terrestrial biomes on land.

2. Cities are warmer because concrete, asphalt, and the dark roofs of buildings absorb heat from solar radiation and radiate that heat in the evening hours. Cities

Year	Age (x)	Number of female birds (N_x)	Number of female offspring $(N_{x \text{ offspring}})$	Survivorship (l_x)	Fecundity (m_x)	Net reproductive rate $(l_x m_x)$	Generation time $(x\, l_x\, m_x)$
2012	0	100	0	0	0	0	0
2013	1	50	75	0.50	1.50	0.75	0.75
2014	2	40	80	0.40	2.00	0.80	1.60
2015	3	30	60	0.30	2.00	0.60	1.80
						$R_0 = 2.15$	4.15

also produce more heat through the burning of fossil fuels from cars, factories, and buildings. The temperature differential between cities and the surrounding countryside can create "country breezes," air movement generated as rising warm air from the city is replaced by cold surface air from surrounding rural areas. Parks can serve to cool the surrounding air through the evapotranspiration of trees and other vegetation.

RECAP 53.4

1. Biomes are characterized using information about the growth forms of their dominant plants, which reflect the evolution of those plants under annual patterns of temperature and precipitation. Tropical rainforests are found in equatorial regions where conditions are consistently wet and warm. The dominant vegetation is tropical trees, with up to 500 species per square kilometer. Temperate evergreen forests, by contrast, grow at middle to high latitudes, where winters are mild and wet and summers are cool and dry. The dominant trees are a handful of conifer species. The two forest types are similar in that they thrive under wet conditions and do not lose their leaves or needles on a seasonal basis.

2. Wallace saw dramatically different terrestrial species inhabiting the adjacent islands of Bali and Lombok, even though they were separated by 24 kilometers and had similar physical environments. This observation led him to hypothesize that different biota had been separated by a barrier (in this case, a deep channel), which kept them apart over evolutionary time. Wallace surmised that biogeographic regions are the result of the isolation that species experienced by residing on different continents or islands.

3. The fact that ice sheets extended into temperate regions, where species diversity is currently lower than in the tropics, better supports the species diversification time hypothesis. This hypothesis proposes that the amount of time in which speciation has taken place is greater in the tropics, where severe climate conditions did not slow down speciation or increase extinction. Thus the tropics have accumulated more species over time simply because of the lack of dramatic changes in climate.

RECAP 53.5

1. The premise of island biogeography is that the number of species on an island represents a balance between the immigration rate and the extinction rate. A far-away island will have fewer immigrants than an island close to the species pool. In addition, a smaller island will have fewer resources, and thus sustain fewer species and have higher extinction rates. Larger islands provide greater resources and can sustain larger populations (which tend to have lower extinction rates than small populations).

2. Habitat fragmentation essentially creates isolated islands of habitat surrounded by unsuitable habitat. The smaller and more isolated the habitat fragment, the fewer species that can be supported within the habitat and the less likely it is that species can move to more suitable fragments. In addition, the fragment can be exposed to edge effects—hazards outside the habitat fragment that effectively reduce its size. Leaving habitat fragments that are large enough to sustain species can reduce the effects of habitat fragmentation or by creating corridors that species can use to move safely from one fragment to another.

WORK WITH THE DATA, P. 1164

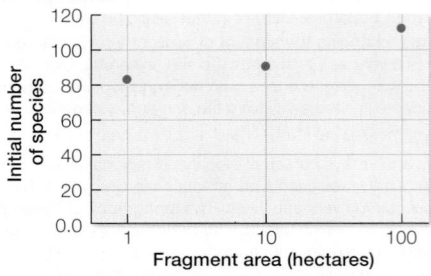

Yes, there is a positive relationship between number of bird species and size of the fragment area.

3. The scaling factor shows that to increase the t_{50} (i.e., the time it takes fragments to lose half their species) by 10-fold, the fragment area would need to increase 1,000-fold. Thus, if t_{50} for 1 hectare = 5 years, the fragment would need to be 1,000 hectares to ensure that half the species would remain after 50 years. The fragment would need to be 10,000 hectares to ensure that half the species would remain after 100 years.

4. Yes, because the average area would need to be 10,000 hectares to ensure that half the species would remain after 100 years.

FIGURE QUESTIONS

Figure 53.2   An increase in greenhouse gases increases the flux of infrared radiation back to Earth's surface and has a warming effect on climate.

Figure 53.9   Temperature inversions occur when there are large temperature extremes that produce afternoon heat and morning cold. A valley concentrates heat from the sun during the day, and as the heat rises, it forms a layer of warm air (an inversion layer). The inversion layer traps cold, dense, and moisture-laden air that then descends into the valley overnight, producing fog.

Upwelling occurs when prevailing winds blow parallel to the coastline, displacing surface water away from the shoreline and thus allowing deep, cold water to rise to the surface. This cold water can cool moist air on the coast, creating morning fog.

Figure 53.12   All the species that are limited to a region are part of the regional species pool. Thus the regional species pool can influence species diversity and composition from the regional to local scales.

Figure 53.13   The northern hemisphere has more biome types. This is likely a consequence of the northern hemisphere having much more land area than the southern hemisphere, creating the opportunity for a diversity of temperature and rainfall regimes.

Figure 53.14   One would expect speciation to increase as land masses separated because species would become reproductively isolated from one another, thus increasing the chance that they could follow different evolutionary trajectories. The separation of species in this way is known as vicariance.

APPLY WHAT YOU'VE LEARNED

1. According to Figure B, the ability of reef fish species to respond to loss of corals is inversely proportional to the number of coral species present. The results show that fish from more diverse areas (PNG) were more sensitive to changes in the number of coral species than were fish from lower diversity areas (GBR, FP). This tendency for fish species to decline under lower coral diversity situations is a consequence of the fact that more diverse reefs contain more specialized fish species (species that depend for food and/or shelter on one or a very few coral species). These results suggest that if managers want to protect the most fish species possible, they should focus their efforts on the most diverse coral reefs where these specializations occur.

2. The PNG and GBR sites are relatively close together. Thus it would be easier for organisms, such as fish and coral larvae, to move between and colonize these sites. These colonization events could serve to increase species diversity at each site. The FP site is very isolated, and the probability of new species from the other sites reaching this area is much lower.

3. The PNG site is very near the equator, so it has a warm, stable environment. One theory that could explain species diversity as it relates to latitude is the species diversification time hypothesis. Equatorial species have had a very long time to evolve, unaffected by conditions (such as glaciation) that caused extinctions at higher latitudes. The other sites are slightly farther from the equator, which might make them slightly less likely to reach high levels of diversity. Another explanation might be the theory of island biogeography and the related concept of species–area relationship. That is, higher species diversity is associated with greater land area, and isolated islands will show decreasing numbers of species with increasing distance from a source of new species. Although both PNG and GBR are islands, both sites are close to other islands and to a large landmass (Australia), making them likely to be colonized by species from these areas. The FP site is much more isolated and therefore much less likely to be colonized.

4. As climate change progresses, coral reefs are likely to degrade and show local extinctions of coral species. The experiment shows that reefs with higher coral diversity are likely to have greater declines in fish species number as well. Thus, if climate change continues on its present course and coral reefs continue to decline in the future, the reefs that remain will likely have lower fish species diversity. Because areas with high coral diversity have more specialized fish species, these species will be most at risk. PNG—and to a lesser extent GBR—will likely lose specialized fish species. FP, which is more isolated and has a lower coral diversity, will likely lose fewer (or no) fish species.

Chapter 54

RECAP 54.1

1.

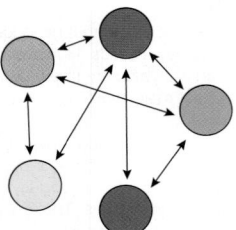

Yes, the five populations of humpback whales make up one metapopulation on their summer feeding grounds. All populations except the yellow and blue populations overlap in their summer feeding areas and thus have the potential to exchange individuals. The yellow and blue populations overlap with the other three populations, however, so they are still part of the metapopulation.

2. Intraspecific competition for food resources is likely responsible for fluctuations in the aphid population size over time. Aphid populations undergo boom-and-bust cycles triggered by their population densities. Predation keeps these fluctuations in check by controlling the density of aphids before they reach boom-and-bust conditions.

3. Approaches used to estimate population sizes include full censuses, surveys using quadrats or transects, mark–recapture methods, and DNA analyses.

4. The most appropriate method would probably be mark–recapture. Quadrats and transects are appropriate only for plants and sessile animals. In a large area, or in water, it is impractical or impossible to do a full census. For animals that can move, the best approach is to collect a sample of animals, mark them, and then recapture them and count the marked organisms. This allows the researcher to calculate a reasonable estimate of the population.

RECAP 54.2

1. $N_t = N_0 + (B - D) + (I - E)$

N_t = population size at time t; N_0 = population size at time 0; B = number of individuals born between time 0 and time t; D = number of individuals that died between time 0 and time t; I = number of individuals that immigrated between time 0 and time t; E = number of individuals that emigrated between time 0 and time t.

2.

Year	Age (x)	Number of female birds (N_x)	Number of female offspring (N_x offspring)	Survivorship (l_x)	Fecundity (m_x)	Net reproductive rate ($l_x\,m_x$)	Generation time ($x\,l_x\,m_x$)
2012	0	100	0	0	0	0	0
2013	1	50		0.50	1.50	0.75	0.75
2014	2	40	80	0.40	2.00	0.80	1.60
2015	3	30	60	0.30	2.00	0.60	1.80
						$R_0 = 2.15$	4.15

a. $R_0 = 2.15$; $G = 4.15/2.15 = 1.93$; $r = \ln(2.15)/1.93 = 0.40$.

b. $N_t = (120)\,e^{\,0.40 \times 20} = (120)(2,982) = 357,720$.

c. $N_t = 300/(1 + [(300 - 120)/120]\,e^{-0.40 \times 20}) = 300/1.0005 = 300$.

3.

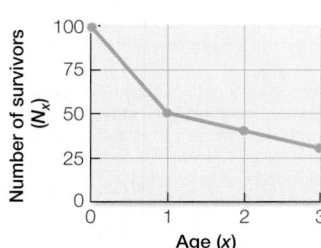

This cohort has a type II survivorship curve, which is typical for birds, fish, and plants.

4. Density-dependent factors controlling this cohort could include limiting resources that decrease population growth under high densities as a result of intraspecific

competition; and predators or pathogens that can have a differential effect on the mortality of individuals in high population densities. Density-independent factors controlling this cohort could include extreme cold or an exceptionally strong hurricane that kills a large number of individuals in the population.

RECAP 54.3

1. a. Life expectancy has increased steadily over the past 175 years, suggesting that advances in food and health security (i.e., environmental factors) have allowed humans to live longer. Changes in genetically determined life span would occur over evolutionary time.

   b. The differences in life span for citizens of Japan versus Angola are also mostly the result of environmental factors. Citizens of Japan generally have a better quality of life than those of Angola, due to Angola's high rates of HIV infection and civil unrest.

2. There are life history trade-offs because constraints on optimality include genetic variation for evolution to act on as well as the mediating effects of the physical and biological environments. Examples of trade-offs are allocation of resources, high growth versus high reproduction, and high reproduction versus survival.

3. *Lobelia telekii* reproduces just once, produces many seeds, and has a short life span, which falls on the side of an *r*-strategist. *L. keniensis* reproduces multiple times, produces fewer but larger seeds, and lives longer, suggesting a *K*-strategist life history. In terms of the guppy example, those living in high-predation streams have an *r*-strategist life history whereas those living in low-predation streams have a more *K*-strategist life history.

RECAP 54.4

1. Species that have a *K*-strategist life history live longer, reproduce at a larger size, and produce fewer offspring over a longer period of time compared to *r*-strategists. As a result, recovery may require managing the number or size of the individuals within the population so that they live long enough, and/or reach a large enough size, to successfully reproduce at a rate that increases their population size over time.

2. a. Some factors that affect the metapopulation size of the Edith's checkerspot butterfly in the San Francisco Bay area include the number and size of serpentine rock outcrops with the host plants of the butterfly, the effects of climate on the populations, the dispersal rates between the populations, and the number of source and sink populations.

   b. With the extinction of the Morgan Hill population, the metapopulation faces likely extinction. Large populations, such as Morgan Hill, serve as sources that recolonize smaller populations after temporary local extinctions. Without such source populations, continuation of the metapopulation over the long term is much less likely.

WORK WITH THE DATA, P. 1171

1. The equation in Figure 54.5A states that $N = (n_1 \times n_2)/M$. In words, the estimated total number of ticks (population size N) equals the total number of individuals captured, marked, and released in the first sample ($n_1 = 180$) times the total number of individuals captured in the second sample ($n_2 = 33$) divided by the number of marked individuals recaptured in the second sample ($M = 8$). Thus N (estimated population of the sampled lawn) = $(180 \times 33)/8 = 5{,}940/8 = 742.5$ adult ticks.

2. Calculate this by dividing 742.5 (the estimated number of ticks in the lawn population, per Question 1) by 700 (number of m²) = approximately 1.06 ticks per square meter.

3. This study was conducted in order to evaluate the risk that residents of this suburban neighborhood have of encountering the tick vectors of Lyme disease in their own yards; a high risk of encounter means that there is also likely to be a high risk of contracting the disease. A density of slightly more than one tick per square meter of lawn suggests that residents of this community are indeed likely to encounter ticks and, accordingly, have a high probability of contracting Lyme disease if they spend time outdoors on their lawns.

WORK WITH THE DATA, P. 1174

1. $N_t = (7.4 \text{ billion}) \, e^{0.0118 \times 85} = (7.4)(2.73) = 20.2$ billion.

2. $N_t = (10 \text{ billion}) e^{0.005 \times 20} = (10)(1.01) = 10.1$ billion.

3. 2015 data: $N_t = 12 /[1 + ([12 - 7.4]/12) e^{-0.0118 \times 85}] = 12/1.14 = 10.5$ billion.
   2080 data: $N_t = 12 /[1 + ([12 - 10]/12) e^{-0.005 \times 20}] = 12/1.15 = 10.4$ billion.

4. Reducing the population growth rate will reduce the population size by 2100 slightly more than reducing the carrying capacity will. But either parameter would be important to reduce population size.

FIGURE QUESTIONS

Figure 54.6 The blue curve ($r = 1$) has a faster growth rate, and thus a steeper slope, than the red curve ($r = 0.25$).

Figure 54.8 Exponential growth occurs when the rate of change in population size is multiplicative but constant over time. Logistic growth occurs when resources are limited and intraspecific competition slows growth of the population to its maximum size (carrying capacity).

Figure 54.9 Humans have a type I survivorship because they have high overall survivorship through adulthood but a steep decline later in life. Species with this type of survivorship curve typically have low reproduction rates but provide parental care to their offspring, which reduces the risk of death in early stages of development.

Figure 54.12 Yes, the life history strategies of the guppy populations are likely genetically determined. When the two populations were provided with unlimited food

and lacked exposure to predation, they did not change their strategies.

Figure 54.14 Yes, the rate of change is predicted to increase. From 1840 to 1940, female life expectancy increased by 25 years (from 45 years in 1840 to 70 years in 1940), thus the rate of change is 55% ((25/45) ×100%) over 100 years. From 1940 to 2040, female life expectancy is predicted to increase by 25 years (from 70 years in 1940 to 95 years in 2040), thus the rate of change is 36% ((25/70) × 100%) over 100 years. So over the same amount of time (100 years), life expectancy increased faster in the first time interval (1840–1940) than it is predicted to do in the second time interval (1940–2040).

APPLY WHAT YOU'VE LEARNED

1. Individual age values are shown in the table below. They are calculated by multiplying l_x by m_x for each age class. The net reproductive rate for the population is calculated as $R_0 = \text{sum} \, (l_x m_x)$, or $0 + 0.63 + 0.49 + 0.31 + 0.17 + 0.09 + 0.06 + 0.03 + 0.02 + 0.01 + 0 = 1.81$. This represents the mean number of offspring produced by each reproductive individual in the cohort during her lifetime.

Age (x)	0	1	2	3	4	5	6	7	8	9	10
$x \, l_x m_x$	0	0.63	0.49	0.31	0.17	0.09	0.06	0.03	0.02	0.01	0

2. The population is increasing. Because the life table adjusts for deaths at each year in the life cycle, death rates are built into the life table. If the value of R_0 is greater than 1 (as this one is), the population is increasing.

3. The generation time is calculated by first multiplying the age class (x) by the reproductive rate for that age class ($l_x m_x$) to get $x \, l_x m_x$. The values are summed as $G = \text{sum} \, (x \, l_x m_x)$, or $0 + 0.63 + 0.98 + 0.93 + 0.68 + 0.45 + 0.36 + 0.21 + 0.16 + 0.09 + 0 = 4.49$.

Age (x)	0	1	2	3	4	5	6	7	8	9	10
$x \, l_x m_x$	0	0.63	0.98	0.93	0.68	0.45	0.36	0.21	0.16	0.09	0

4. The per capita rate of increase is calculated as $r = (\ln R_0)/G$. Thus $r = (\ln 1.81)/4.49 = 0.132$.

5. Sharpnose shark populations are more likely to have a higher r value than are populations of temperate sharks. Sharpnose sharks mature and reproduce early, so they produce offspring more rapidly than larger sharks, even though they have a shorter life span. Also, sharpnose sharks are small and not deliberately fished but are part of bycatch. Larger, temperate species, because of their slow growth, late maturity, and limited offspring, reproduce more slowly. If fishing pressure becomes intense, temperate sharks could be overfished.

Chapter 55

RECAP 55.1

1. (a) Mutualism (b) Amensalism (c) Parasitism (d) Competition

2. The conditions that can modify or change a two species interaction include physical conditions, biological processes such as dispersal, and other interacting species.

3. No, it is unlikely that the lionfish prey in the Atlantic have had enough time to adjust in an evolutionary way to their novel predator. This conclusion is supported by the data, which show that some reefs in the Atlantic have suffered a 65 percent decline in small-bodied coral-reef fish.

RECAP 55.2

1. Perhaps the simplest behavior a reef fish might adopt—or that might evolve in reef fish—would be to avoid headfirst capture by the lionfish. This might involve adopting a way to detect a lionfish's presence and fleeing the encounter before the predator could use jets to create a headfirst capture. Or after being hit by a jet of water, the prey could adopt unpredictable swimming movements to avoid headfirst capture.

2. Both herbivory and parasitism typically involve a symbiosis in which the herbivore or parasite is smaller than, and may live on, the plant or host. This close relationship naturally leads to species evolving specialized mechanisms that counter their effects on one another. In addition, the prey of herbivores are plants, which are immobile, and thus may also evolve special mechanisms in response to the potential for intense herbivory.

3. The number of grass species would be highest when the hare population is highest and lowest when the hare population is lowest.

RECAP 55.3

1. A realized niche is defined by a species' interactions with other species. The red squirrel's realized niche has become smaller because of the competition with the gray squirrel.

2. Herbivory by the hares is the interaction that allows the grass species to coexist with one another. By grazing on the dominant grass species, the hares allow more subordinate grass species to thrive.

3. Scenario 1 is an example of exploitation competition because you and your friend are sharing the milkshake resource. Scenario 2 is an example of interference competition because your friend is excluding you from the milkshake resource

RECAP 55.4

1. An obligate mutualism would evolve when the two species involved in the interaction benefit from each other more than from other partners. A facultative mutualism would

evolve when the two species involved in the interaction benefit from partnerships with multiple species. For example, if multiple species can deliver pollen or seeds to their desired locations, then facultative relationships will be more likely to evolve.

2. Both species of barnacles are more likely to benefit from positive interactions under the more stressful conditions of the high intertidal where desiccation occurs.

3. Examples include trees, coral reefs, and kelp forests, where the habitat provided by these species facilitates many other species that depend on them.

WORK WITH THE DATA, P. 1193

1. The researchers used both methods to collect data for two different purposes. In the open aquarium trials, they wanted to observe the behavior of the lionfish and its prey in a relatively natural setting to see what strategies the lionfish used to capture and eat the prey. They learned that the lionfish would capture the prey when the prey turned and oriented its head toward the head of the lionfish. This headfirst orientation allowed the lionfish to more easily capture its prey.

   In the container trials, the researchers wanted to observe the behavior of the lionfish, particularly the production of water-jet pulses from its mouth, in a setting where the lionfish could see, but not capture or eat, the prey. In this way the researchers could measure the number of water jets produced and the maximum distance a water jet could travel, to get an idea of what behaviors a lionfish might employ when its prey was hard to catch (in this case, because the prey was in an inaccessible container).

2. Given that fish prey in the Atlantic Ocean are more naive to lionfish predatory behavior than fish prey in the Pacific Ocean, it would make sense that lionfishes would need to produce fewer and closer jets of water to capture their fish prey in the Atlantic Ocean compared with the Pacific Ocean. The Pacific Ocean fish should be harder to catch and thus require more water jets produced from a greater distance from the prey.

3. The field observations suggest that Pacific Ocean lionfishes need to resort to the water jet–blowing behavior to catch their prey more often than Atlantic Ocean lionfishes do. This pattern could be explained by the hypothesis that fish prey in the Atlantic Ocean are more naive to lionfishes and thus easier to catch. Atlantic Ocean fish prey are less likely to take appropriate evasive actions or require the use of the water jet–blowing behavior to confuse or disorient them. In addition, it is likely that the production of water jets is metabolically costly, and that lionfishes would use this behavior only when the cost of producing it is outweighed by the advantages it confers. Presumably this would be more likely to be the case for prey in the Pacific Ocean than for prey in the Atlantic Ocean.

FIGURE QUESTIONS

Figure 55.1 Yes. In the buffalo photo, trophic interactions include parasitism by the ticks, predation by the birds both of ticks on the buffalo and insects in the grass, and herbivory by the buffalo. Symbioses include the parasitic ticks on the buffalo. In the wolf and bear photo, the wolves are involved in a trophic interaction.

Figure 55.2 Effects of species 1 on species 2

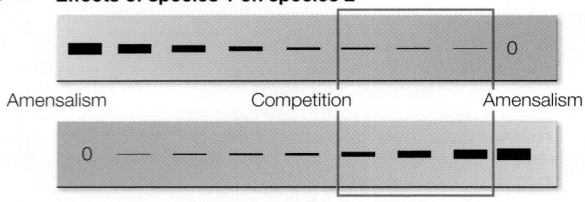

Figure 55.9 The peak number of lynx usually occurs after the peak number of hares. One reason this might occur is that as hare abundance rises, lynx have more prey and thus can produce more offspring. However, these offspring are not born immediately, so the rise in lynx numbers lags behind the rise in hare numbers.

Figure 55.11 If both *P. caudatum* and *P. bursaria* fed on yeast, one of the species would likely become extinct.

APPLY WHAT YOU'VE LEARNED

1. Pollination is a positive interaction, or facilitation. Specifically, it is a mutualism, meaning both partners benefit. Bees obtain food (pollen and nectar) from plants. Plants rely on bees for pollination, which is necessary for successful reproduction. For a successful pollination interaction, bees must emerge at the same time that plants flower.

2. Over time, the trend in bee pollination date is slightly downward, indicating that the bees were pollinating slightly earlier in the year. The mean April temperature showed an upward trend, indicating that temperatures rose over the time period. The opposing directions of the trends indicate that as temperatures rose, bees were pollinating earlier in the year. The steeper slope after 1970 on both graphs suggests that temperature has increased faster since 1970. Statistical analysis bears this out; 69 percent of the bees' advance in pollination activity has occurred since 1970.

3. The recent study (1971–1999) shows plant flowering advancing faster than bee pollination dates, and thus best supports the hypothesis that climate warming is affecting plant-pollinator interactions by creating mismatches in the timing of these events.

4. Specialist plant–pollinator interactions are more likely to be affected by timing mismatches caused by rising temperatures. Generalist species (such as those in these studies) can rely on many species for food or pollination interactions; if one or several species pairs show timing mismatches, other species will likely be present

to participate in the interaction. Specialist pairs, which depend on only one or a few species, are far more susceptible to the effects of timing mismatches. If timing mismatches occur, there may be no species available to interact with.

5. There are many possible designs, but more precise information would require analysis of specific plant–pollinator pairs, rather than the generalized groups in the current studies. The study would require seasonal measurements of plant flowering dates and pollinator emergence and pollination activity dates (as opposed to collection dates) for several years, correlated directly with climate data. Ideally studies should focus on pairs for which background information is already available on past emergence and flowering dates. Good candidates would include pollinators of important crop plants, such as apple trees or tomatoes.

Chapter 56

RECAP 56.1

1. Subsets of species are used to describe communities because it is impractical to account for every species in a community. There are at least two reasons for this: (1) communities can vary in spatial and temporal scale, making it hard to designate boundaries around them, and (2) it is impossible to identify every species in a community because many are not described. For example, many bacteria, small microbes, and small invertebrates are hard to see, let alone identify.

   The community subset used for frogs and toads on Mount St. Helens was a taxonomic one—they all belong to an amphibian community.

2. Nontrophic interactions, such as competition and facilitation, are not depicted in the food web.

3. Pond A

	Abundance	Proportion (p_i)	ln (p_i)	p_i ln(p_i)
Pacific tree frog	6	0.3	−1.20	−0.36
Western toad	8	0.4	−0.92	−0.37
Northern red-legged frog	4	0.2	−1.61	−0.32
Cascades frog	2	0.1	−2.30	−0.23
H				1.28

The new Shannon index value for pond A is H = 1.28. Pond B, with H = 1.39, still has slightly higher species diversity than Pond A.

RECAP 56.2

1. The most important factor limiting species membership in the local community is dispersal and immigration. Seven species did not pass through the species supply filter, while 4 species and 3 species, respectively, did not pass through the abiotic and biotic filters.

2. Given that the species supply filter has the most potential to exclude species from a local community, the best management action would be to exclude, as much as possible, non-native species from entering the mountain in the first place.

RECAP 56.3

1. Elk would be released from the direct interaction of predation by wolves and would increase in abundance. The increase in elk would result in more grazing pressure on aspens, causing them to decline. Thus, fewer wolves would have indirect effects for aspen. With fewer aspens, there would be other indirect effects of the lack of wolf predation: fewer snowshoe hare, beaver, and other rodents that depend on aspens for food. Fewer small herbivores could affect other carnivores in the system such as coyote, raven, and short-tailed weasels.

2. Beavers most closely fit the ecosystem engineering species definition because they are able to create, modify, and/or maintain physical habitat by cutting down (and killing) trees and using them to dam streams and create ponds and wetlands that provide habitat for themselves and other species. Beavers might also be considered a keystone species because their effect is large relative to their size or abundance, but keystone species are thought to mostly act through food webs by creating trophic cascades. Beavers would not be considered foundation species because foundation species have large effects on the communities as a consequence of their large size or great abundance.

3. The lottery model poses that in communities where species use similar limiting resources and have similar effects on one another, the element of equal chance for all individuals to obtain those resources determines coexistence. The model assumes that when resources are made available, they are used at random by individuals of different species that happen to be in the "right place at the right time." As long as all individuals have similar chances of obtaining resources (or "winning the lottery") and no clear advantage in population growth, then their presence in the community should be maintained by chance events that free up resources for individuals competing for those resources.

RECAP 56.4

1. Disturbance is an abiotic event that may physically or chemically injure or even kill some individuals, creating opportunities for other individuals to grow and/or reproduce. For example, hurricanes, forest fires, and volcanic eruptions are disturbances. Stress occurs when some abiotic factor reduces the growth, reproduction, and ultimately survival of some individuals. For example, drought or extreme cold can cause stress in organisms.

2. Some biotic factors that induce change in communities include species interactions, such as predation, competition, or disease. Others include physical damage by organisms, such as trampling or digging.

3. False. Primary succession is controlled by facilitative and inhibitory processes acting together over time. The progress of succession depends on the early colonists, each of which can facilitate and inhibit other colonizing species. For example, in Glacier Bay *Dryas* and alders allowed spruce trees to become established, but later in succession, competition from spruce trees led to the decline of early successional species.

4. An alternative state, or regime shift, occurs when a disturbance or stress causes a community to follow a different successional trajectory, leading to an alternative community state. Examples include the extirpation and then reintroduction of wolves to aspen forests of Yellowstone National Park, the presence or absence of sea otters in kelp forests along the west coast of North America, and the effects of beavers on wetlands in Minnesota.

RECAP 56.5

1. Costa Rican farmers often grow multiple crops together (such as corn and sweet potatoes) to reduce the number of insect pests that attack the plants. For example, parasitoid wasps are attracted to corn pollen and attack sweet potato pests, thus benefitting the sweet potato crop. Likewise, the corn plants receive pollination services from the wasps.

   Growing polycultures could be a good strategy under drought conditions if each species performs differently under low water conditions. For example, if one species performs better than another, overall productivity could still be high compared with a condition in which only one species, or monoculture, that did not respond favorably to drought.

WORK WITH THE DATA, P. 1221

1.

The species diversity data for small mammals seem to fit the intermediate disturbance hypothesis well. As the degree of disturbance experienced by the four community types increases, so does species diversity of small mammals in those communities, up to a point—the Blowdown Zone. Species diversity then declines under the extreme conditions of the Pumice Plain.

The pattern of species diversity seen in small mammals in the Tephra-fall and Blowdown Zones could be determined by the variety of different habitats and resources that became available to them in the secondary successional communities that developed after the eruption. Compared with the reference area, these two communities likely foster more species and at a higher relative abundance because of these new resources. In the primary successional community of the Pumice Plain, though, the habitats and resources are not nearly as diverse or abundant and thus cannot support more than one small species.

3. The deer mouse (*Peromyscus maniculatus*) is the only species present in all four communities. This suggests that the deer mouse has a life history that allows it to live in primary, secondary, and climax successional communities. It is likely able to disperse widely, growth quickly, and reproduce often—all characteristics of an early successional, *r*-strategist species. The deer mouse is also likely to be an opportunistic and generalist species, living in a variety of habitats and feeding on a variety of food items.

4. The Tephra-fall and reference area communities have more species in common than the Blowdown Zone, but all three communities have their own characteristic small mammal species assemblages. Given that each community is represented by a different successional stage, with vegetation characteristic of secondary and climax successional communities, it makes sense that the species composition of the small mammals inhabiting those communities will reflect these differences.

Small mammal species	Pumice Plain			Blowdown Zone			Tephra-fall Zone			Reference		
	Proportion (p_i)	ln (p_i)	p_i ln(p_i)	Proportion (p_i)	ln (p_i)	p_i ln(p_i)	Proportion (p_i)	ln (p_i)	p_i ln(p_i)	Proportion (p_i)	ln (p_i)	p_i ln(p_i)
Deer mouse (*Peromyscus maniculatus*)	1.00	0	0	0.20	−1.61	−0.32	0.25	−1.39	−0.35	0.10	−2.30	−0.23
Yellow-pine chipmunk (*Tamias amoenus*)	0			0.40	−0.92	−0.37	0			0		
Cascade golden-mantled ground squirrel (*Spermophilus saturatus*)	0			0.05	−3.00	−0.15	0			0		
Creeping vole (*Microtus oregoni*)	0			0.10	−2.30	−0.23	0			0		
Shrew mole (*Neurotrichus gibbsii*)	0			0.05	−3.00	−0.15	0			0		
Trowbridge's mole (*Sorex trowbridgii*)	0			0.10	−2.30	−0.23	0			0.05	−3.00	−0.15
Montane shrew (*Sorex monticolus*)	0			0.10	−2.30	−0.23	0.15	−1.90	−0.28	0.10	−2.30	−0.23
Southern red-backed vole (*Clethrionomys gapperi*)	0			0			0.45	−0.80	−0.36	0.65	−0.43	−0.28
Townsend's chipmunk (*Tamias townsendii*)	0			0			0.05	−3.00	−0.15	0		
Ermine (*Mustela erminea*)	0			0			0.05	−3.00	−0.15	0		
Northern flying squirrel (*Glaucomys sabrinus*)	0			0			0.05	−3.00	−0.15	0		
Vagrant shrew (*Sorex vagrans*)	0			0			0			0.05	−3.00	−0.15
Northern water shrew (*Sorex palustris*)	0			0			0			0.05	−3.00	−0.15
H		0			1.68			1.44			1.04	

The Pumice Plain had the lowest species diversity (*H* = 0) and the Blowdown Zone had the highest species diversity (*H* = 1.68). The Pumice Plain had the lowest species richness (1 species) compared to the Blowdown Zone, which had the highest species richness (7 species). The Reference and the Tephra-fall Zone both had the same species richness (6 species).

2.

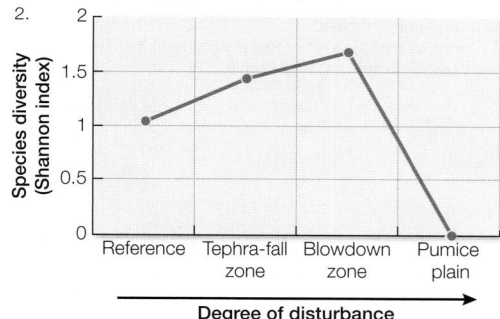

FIGURE QUESTIONS

Figure 56.4 Pond B has higher species diversity (*H* = 1.388) than Pond A (*H* = 0.589), because even though both ponds have the same species richness, Pond B has greater species evenness.

Figure 56.6 No, they would not be found in the local community depicted in the figure. The species in the local community are all terrestrial species, and thus fish and frogs, which are aquatic, would be excluded from the community by the abiotic filter.

Figure 56.15 The oldest communities are located in the areas that have been exposed the longest since glacial retreat, such as the mouth of the bay, where succession has proceeded for 200 years. As the glaciers melt and retreat up the bay, the communities become younger and younger, with the youngest pioneer community closest to the glacial front.

Figure 56.16 Pocket gophers most closely fit the ecosystem engineering species definition because they are able to create, modify, and/or maintain physical habitat for themselves and other species through their burrowing activities. They might also be considered a keystone species because their effect is large relative to their size and abundance. However, keystone species are thought to act mostly through food webs by creating trophic cascades.

APPLY WHAT YOU'VE LEARNED

1. Corals are foundation species because their effects on the coral-reef community are due to their relatively large size and abundance. They are also ecosystem engineering species; as they grow, their skeletons physically modify the environment, increasing habitat for other reef species.

2. The genera *Acropora* and *Orbicella* would be most important in building reef structure, and *Porites* and *Agaricia* would be least important. *Acropora* and *Orbicella* have the highest rugosity, indicating greater complexity, which results in a diversity of reef habitats. They also have the highest calcification rates, indicating that they add reef structure more rapidly than the other two genera.

3. Because of the rapid decline in both reef rugosity and rate of calcification, the genera *Acropora* and *Orbicella* are likely declining most rapidly. These are the genera most responsible for the large size and complexity of the reef. The loss of these genera would affect reef function by reducing the number of habitats and thus the number and abundance of many species in the reef community.

4. The reef does not appear to be regaining its original functionality, at least in the time span of the model. Although the coral cover increases from 10 to 45 percent, neither rugosity nor calcification rate shows a corresponding increase. Rugosity varies only slightly over the time span, ending with a slight upward trend. Calcification rate increases near the middle of the time period before showing a continuous downward trend. The low levels of both factors, compared with their very high levels in the original healthy reef (seen at the start of model *i*), indicate a much lower functionality. The low functionality may be a matter of timing; a reef's function can be compromised when foundation species are lost (as shown in model *i*), but because the foundation species (massive corals such as *Acropora* and *Orbicella*) are replaced slowly, it takes much longer for the reef to recover. The lower functionality probably indicates that foundation species are being replaced by smaller species such as *Porites* and *Agaricia*, both of which have lower rugosity and calcification rates. Their effect may be significant in the short term because of their rapid reproductive rate, but less than that of *Acropora* and *Orbicella* because they do not contribute significantly to reef building.

5. Yes, both model coral-reef communities are undergoing succession. For model *ii*, the recovery could be considered secondary succession, given that it is restoring its structure after a major disturbance (loss of most of its coral structure). According to the concept of alternative states, under similar environmental conditions, succession can result in different assemblages of organisms. That is, the new community does not return to its original state. In model *ii*, the coral community does not reach the same level of rugosity or calcification as the starting conditions of model *i*, indicating that either different species are present and/or their abundance has changed. The lower levels of rugosity and calcification will result in fewer habitats and lower biodiversity. Given more time, as the reef continues to undergo succession, more massive, slower-growing corals may again predominate, and the reef may be able to support higher species diversity. But the possibility that the coral-reef communities show hysteresis, or the inability to recover to the original community, is also a possibility if the effects of climate change intervene.

Chapter 57

RECAP 57.1

1. Not necessarily. Ecosystem science is the study of how energy flows and nutrients cycle through the biotic and abiotic environment. The number of species is a descriptor of the biotic environment but does not give any information about the energy flow or nutrient cycling of the ecosystem.

2. Energy and carbon can be converted into primary production through photosynthesis and chemosynthesis. Photosynthesis involves the capture of radiant energy by plants, algae, and photosynthetic bacteria to be used to fix CO_2 into organic compounds. Chemosynthesis involves the capture of energy from inorganic compounds by archaea and bacteria to be used to fix CO_2 into organic compounds. Chemosynthesis typically occurs in ecosystems such as hydrothermal vents, hot springs, and soils that lack sunlight and have high concentrations of particular inorganic compounds.

RECAP 57.2

1. The open ocean has very low average net primary productivity (~100 g/m^2/yr), compared with algal beds and coral reefs (2,500 g/m^2/yr). However, because there is so much more open ocean (nearly 70% of Earth's surface) than there are algal beds and coral reefs (only 0.1% of Earth's surface), open ocean accounts for a much higher percent of Earth's NPP.

2. No, primary producers will not be able to lower global CO_2 appreciably. CO_2-enrichment experiments show that primary producers can increase their NPP under increased CO_2 concentration, but there is a limit. That limit is set by other factors, such as nutrients, light, and water. In addition, the effects may be somewhat counteracted by respiration of CO_2 by consumers feeding on primary producers.

3. Eutrophication can, but does not always, result in the explosive growth of algae, which can severely deplete oxygen (hypoxia) when they decompose. The level of hypoxia is driven by the amount of algae subject to decomposition and the level of oxygen available in the system.

RECAP 57.3

1. The lake food web would have the higher trophic efficiency. Forest ecosystems have lower trophic efficiency than aquatic systems because much of the NPP in forests is in the form of wood and is unavailable to consumers. In contrast, the lake food web is based on phytoplankton and algae, which are more easily converted to secondary production. Estimates suggest that, on average, only 13 percent of terrestrial biomass is consumed by herbivores, compared with 35 percent in aquatic ecosystems. The crayfish, which is an ectotherm, would have a higher trophic efficiency than the bear, which is an endotherm. Endotherms maintain higher metabolic rates than ectotherms and thus have less energy left over to devote to growth and reproduction.

2. The lake with three trophic levels would have the higher NPP. A three-level food web in which the top carnivore affects an herbivore's abundance should have higher NPP than a two- or four-level food web in which herbivores have less predator control. Omnivory can also change the way energy is transferred in food webs by essentially "collapsing" trophic levels on one another. A four level food web with a top consumer feeding on both the herbivore and the primary producer should have lower NPP than a four-level food web without omnivory. Even though the top consumer indirectly benefits the primary producer by feeding at the herbivore level, it partially negates this effect by directly feeding on the primary producer level as well.

3. One hypothesis, which focuses on the amount of NPP entering an ecosystem, suggests that bottom-up factors can influence the number of trophic levels in an ecosystem. Because tropical rainforests have higher NPP than deserts (see Figure 57.4), they may support more trophic levels if the amount of energy that can sustain populations at higher trophic levels is greater.

RECAP 57.4

1. The residence time of water depends on the rate at which it moves from one pool to another. Organisms, soil, and rivers all have short water residence times because they have relatively small volumes of water compared with lakes, glaciers, and oceans, and thus the rate of water movement from one pool to the next is much faster.

2. The concentration of CO_2 is rising in the atmosphere and oceans because of fossil-fuel burning. Today atmospheric CO_2 concentration is just over 400 parts per million, the highest recorded in the last 800,000 years. CO_2 is a greenhouse gas and, along with other greenhouse gases emitted by fossil-fuel burning, has caused a rise in Earth's temperature. The increase in CO_2 has resulted in a roughly 1°C (1.8°F) rise in global temperatures compared with those from 1981 to 2010.
Slightly less than half of the CO_2 emitted by burning of fossil fuels has been absorbed by the oceans. CO_2 reacts with water to form carbonic acid (H_2CO_3). As levels of carbonic acid rise, the pH of seawater drops. This increase in acidity can have negative effects on many marine organisms that have calcium carbonate skeletons.

3. Nitrogen enters the biotic system as atmospheric N_2 and is fixed by bacteria into ammonia (NH_3). Ammonia is rapidly transformed into ammonium (NH_4^+), which can then be used by plants and bacteria. Nitrifying bacteria can transform ammonium to nitrate (NO_3^-), another form of nitrogen that plants and bacteria can use. Denitrifying bacteria take nitrate and convert it back to N_2 and N_2O gases, which are then released back into the atmosphere. Collectively, this microbial processing of nitrogen is very fast and accounts for about 95 percent of all natural nitrogen flux on Earth, making it a mostly biologically driven cycle.
Unlike nitrogen, phosphorus and sulfur are found in rocks and deep-sea sediments. They cycle very slowly through the geologic system because they require sedimentary rock formation, uplift, and weathering. Once they reach organisms, they rapidly cycle through the biological component of the ecosystem.

RECAP 57.5

1. Timber and fibers are provisioning services; coastal protection and erosion control are regulating services; water purification, habitat for fisheries, and carbon sequestration are supporting services; and tourism, recreation, and education are cultural services.

2. You could leave the mangrove ecosystem intact at the edge of the shoreline so that coastal protection would not be compromised. You could place shrimp farms inland behind these mangrove buffers, where the loss of forest would not have an appreciable effect on protection in extreme storms or tsunamis.

WORK WITH THE DATA, P. 1242

1.

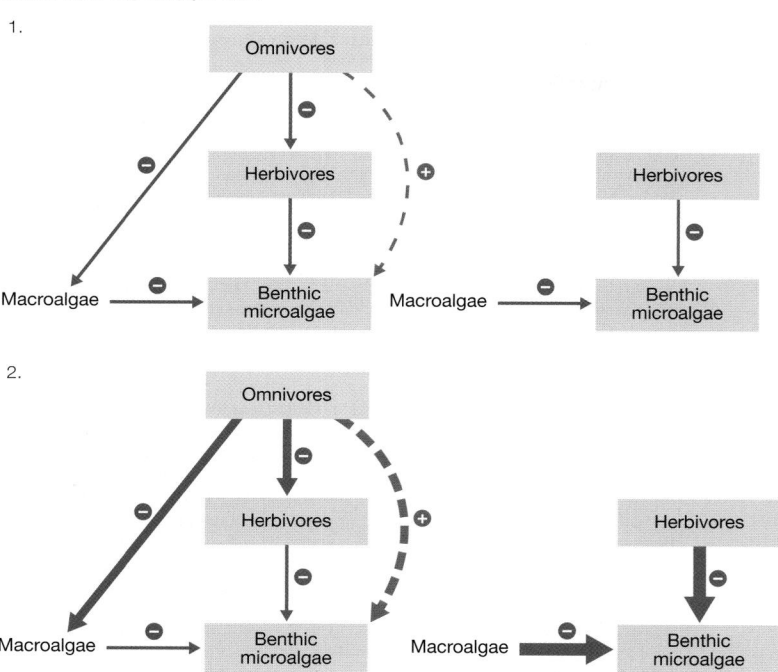

2.

Under increased CO_2 and temperature, macroalgae and herbivores increase in biomass. When omnivores are present, this greater biomass will be utilized by omnivores, resulting in a stronger negative interaction between them and their food sources. The stronger negative interaction will lead to a stronger indirect positive effect for benthic microalgae, which will not change in their biomass as a result. However, when omnivores are absent, macroalgae and herbivores will not be kept in check, and their negative interactions with benthic microalgae will grow even stronger, resulting in the decline of benthic microalgae.

3. The food webs and supporting data in the graphs show that with increased CO_2 and temperature, macroalgae and herbivores both increased in biomass, causing negative effects for benthic microalgae (the biomass declined). However, these negative effects for benthic microalgae did not manifest themselves when omnivores were present. By keeping macroalgae and herbivores in check under increased CO_2 and temperature, omnivores indirectly benefited benthic microalgae and allowed them to maintain the same biomass as that seen under ambient conditions.

4. If omnivores ceased feeding on macroalgae, benthic microalgae would decline in biomass because their competitors would increase. But the decline in biomass would be less than if these top consumers were removed altogether. The omnivores would still feed on the herbivores, creating a trophic cascade that would benefit benthic microalgae.

FIGURE QUESTIONS

Figure 57.1   Energy is said to flow through an ecosystem because it moves from the original source (either the sun or inorganic nutrients) to primary producers and consumers, and then is lost as metabolic heat. Nutrients are said to cycle because they continually move between living organisms and the abiotic components of an ecosystem.

Figure 57.8   The consumption and assimilation efficiencies are lower for herbivores because primary producers have structural and chemical defenses that make them harder to consume and digest than animals.

Figure 57.17   Industrial fixation of nitrogen accounts for 30 Tg of the 80 Tg total nitrogen fixed, or 37.5 percent.

APPLY WHAT YOU'VE LEARNED

1. Sea cucumbers feed on bacteria, benthic marine algae, and decaying organic matter such as feces. Thus they are omnivores, feeding from several levels in the food web. They are also decomposers because they feed on decaying organic matter. Their effect on NPP is hard to predict. By feeding on both primary producers and consumers, sea cucumbers may have little effect on NPP. Bacteria are also decomposers, so sea cucumbers may be feeding on a competitor.

2. According to the graph, TOC increases steadily over the 8-week period, but it increases significantly less in sea cucumber–grazed tanks. This indicates that grazing

by sea cucumbers decreases TOC by a relatively constant amount over time. The decrease in TOC is evidence that organic carbon is being broken down, or recycled, into smaller inorganic compounds, which become available as nutrients for consumers in the ecosystem. In addition, some of the TOC is lost as CO_2 to the system through metabolism and respiration. Thus sea cucumbers have a noticeable effect on nutrient cycling in sediments.

3. There are at least two reasons why TOC continues to increase. First, a specific amount of mussel feces containing TOC is added each day (mirroring, albeit experimentally, what would happen in a mussel farm). Second, as this organic waste is added, bacteria increase as they feed on and decompose the waste. Both waste (dead organic matter) and bacterial biomass (live organic matter) are included in the TOC value. The experiment shows that TOC begins to level off as the experiment proceeds. For the control treatment (black circles) this likely means that, after an initial rise in bacterial population size, the bacteria begin to suffer food limitation. This is also shown in the sea cucumber treatment (clear circles), but with the added effect of sea cucumber grazing on TOC, which includes both waste and bacteria. As sea cucumbers feed, they break down and mineralize the TOC. However, the single cucumber used per tank in these trials is insufficient to cause a significant decline in TOC. It is likely that a larger sea cucumber population would further control TOC; if that population were large enough, it could solve the hypoxia problem.

4. As TOC levels increase, bacterial levels also increase, as bacteria use TOC as a food source and decompose it. Increased bacterial activity uses up oxygen, often resulting in hypoxia. In this experiment, oxygen levels would be decreasing as TOC increases; oxygen would decrease less as TOC begins to level off. The presence of sea cucumbers would make hypoxia less likely, because they would break down TOC and bacteria. Thus as TOC levels decline, oxygen levels will increase. But in this particular experiment, there is only one sea cucumber per test aquarium. This is not sufficient to break down most of the TOC; therefore oxygen levels are still likely to be too low for a healthy ecosystem.

5. The aquaculture manager chose to produce mussels in a natural coastal environment. This resulted in food for humans, but in the process he risked polluting the coastal environment. This is a trade-off between an economic good (food) and an environmental good (a clean, unpolluted coastal environment). The addition of sea cucumbers would definitely be considered an ecosystem service. It uses a natural method (sea cucumbers, a species already present in the ecosystem) to solve a pollution problem, instead of resorting to chemical or physical removal of the TOC, which might cause further pollution and disruption of the environment.

Chapter 58

RECAP 58.1

1. Biodiversity loss is the loss of diversity at genetic, population, species, ecosystem, and global scales. The four major causes of biodiversity loss are habitat destruction (reduction in habitat quantity) and degradation (reduction in habitat quality); overharvesting of species; species invasions and emerging diseases; and climate change.

2. The heath hen's population sizes were reduced by hunting until only one very small population remained on the island of Martha's Vineyard, Massachusetts, in 1830. By 1908 only 50 birds remained. Although the population recovered to several thousand birds once a reserve was established for them, human-caused and natural factors led to a precipitous decline. The reduction in size caused inbreeding depression and demographic stochasticity and reduced the effective population size even more. By 1928 there were 2 females and 11 males, not enough individuals or genetic diversity to save the heath hen from extinction.

3. There are multiple reasons why it is hard to determine how many species are extinct or threatened with extinction. First, we do not know how many species live on Earth today because many have not been described. Some of those undescribed species may have already become extinct. Second, many described species are small, reclusive, and rare, and thus hard to keep track of. Third, it is difficult to determine whether a species is truly extinct, especially if it is rare. For example, there have been cases in which species were thought to be extinct but were later discovered to still be extant. Fourth, we rarely know all the interactions among species. It is hard to determine if the loss of one species will lead to the loss of others unless we have detailed ecological knowledge of all the species affected.

RECAP 58.2

1. The more fragmented a habitat is, the smaller and more isolated the fragments are and the greater the edge effects are. Fragmentation reduces the population sizes of species and isolates them from one another, reducing their ability to disperse among the greater metapopulation. Once population sizes are small and isolated, they become susceptible to extinction vortex processes (see Figure 58.2).

2. Gray wolves have been shown to create a trophic cascade that benefits hardwood trees such as aspens. Research by William Ripple and colleagues in Yellowstone National Park found that gray wolves, by feeding on elk, released aspen seedlings from herbivory, thus creating mature aspen forests. After wolves were hunted to

extinction in Yellowstone in 1926, the number of aspens declined precipitously because of intensive elk grazing, as is shown in Figure 56.9.

3. Three reasons why invasive species can have negative effects on biodiversity and ecosystems are (1) they can threaten native species with extinction through predation, competition, and disease transmission; (2) they can change the functions and services of ecosystems by affecting fire cycles, water availability, and sedimentation; and (3) they can change the genetic diversity of native species through hybridization events.

4. Yes, these data strongly suggest that climate warming is causing the butterfly species to shift its distribution to cooler, more optimal temperature conditions farther north. Likewise, emergence time is occurring earlier, when temperature conditions are optimal for that stage of the species' life history.

RECAP 58.3

1. Reserve design relies on three important principles: first, identify a large enough area that is relatively undisturbed and can serve as the core for protection; second, include a buffer zone around the core area that has some features required by the species of concern but is less restrictive to human use; and third, have habitat connectivity that keeps populations from becoming isolated from the greater metapopulation and thus subject to extinction. Taking into account global warming in terrestrial systems, the protected area should have similar protected areas, potentially connected by corridors of land, to the north (or south in the southern hemisphere) that could serve as a refuge from warming temperatures. An example is the Yellowstone to Yukon Conservation Initiative discussed in the chapter.

2. The California condor declined in its native range (British Columbia to Mexico) because of lead poisoning from carcasses containing lead shot, electrocution from power lines, the effects of pesticides on eggshell thinning, and hunting (ecological information). To save the California condor from extinction, the Endangered Species Act (institutional authority) was invoked and the species was removed from the wild in 1983 and subsequently bred in captivity. Although the breeding program was successful, ecological information indicated that birds could not be released into the wild until the threats that caused their decline were mitigated. In 2008, the Ridley–Tree Condor Preservation Act was passed to require California hunters to use non-lead bullets when hunting in the condor's range. Along with passage of the act, public awareness was raised for hunters and cattle ranchers who mistakenly believed that condors killed livestock (socioeconomic interests).

WORK WITH THE DATA, P. 1259

1.

Distance interval (km)	Time interval (years)	Rate of spread (km/yr)
0–140 = 140	1987–1993 = 6	140/6 = 23
140–240 = 100	1993–1996 = 3	100/3 = 33
240–325 = 85	1996–2002 = 6	85/6 = 14
325–410 = 85	2002–2004 = 2	85/2 = 43
410–445 = 35	2004–2006 = 2	35/2 = 18

The rates of spread do differ. They range from 14 to 43 km/year depending on the location.

2. The average rate of *Bd* spread is 26 km/year. Assuming this average, the pathogen could spread 260 km east and south by 2016 (26 km/yr × 10 years).

3. The majority of the habitat east and south of the last location where *Bd* was detected is mostly low elevation (0–199 meters), although there is a thin stretch of mountain habitat in the north. Given that far fewer frog species have become extinct because of *Bd* at elevations of 199 meters or less (30%), it seems less likely that the pathogen will be successful in this region of Panama. Based on the elevation data, the species extinction might reach 30%.

FIGURE QUESTIONS

Figure 58.2 The chytrid fungal pathogen leads to high mortality of frogs, thus lowering their population size. As population size declines, there is greater chance of inbreeding depression, genetic drift, and demographic stochasticity, which decrease

genetic diversity. As genetic diversity declines, individual fitness declines, leading to lower reproduction and higher mortality in the population. This causes the population to continue to maintain a lower effective population size, which can ultimately lead to extinction at the population and species level.

Figure 58.3 Of the 59,033 species categorized by the IUCN, 1 percent are extinct and 20 percent are threatened. Birds have suffered the most extinctions (~150 species), but amphibians are most at risk of extinction (~1,900 species).

Figure 58.8 The highest fisheries exploitation is occurring off northern Europe in the northeast Atlantic Ocean and off Thailand and Vietnam in the South China Sea. Moderate exploitation is occurring in the northwest Atlantic Ocean off New England and Canada, as well as off the southern tip of Australia.

Figure 58.11 After the glaciers first started to retreat in North America about 18,000 years ago, the ranges of plant communities shifted northward and expanded considerably. Roughly 12,000 years ago, "no analog" plant communities, unlike any plant assemblage found today, formed under the unique climate conditions of the time. It may be that as climate changes, unique combinations of species will come together to form similar "no analog" communities.

APPLY WHAT YOU'VE LEARNED

1. Vehicle collisions and hunting, both legal and illegal, are the two known major causes of puma deaths in these populations. Suspected and confirmed diseases together rank as the third most common cause. The two major causes result from human actions, particularly loss of habitat and/or habitat degradation, which forces pumas into urban and suburban areas. Disease is presumably a natural cause.

2. As effective population size decreases, a population enters an "extinction vortex," decreasing more every year and becoming more likely to go extinct. The pumas are living in an increasingly urban area and are at risk from vehicle collisions, hunting (both legal and illegal), and other urban-based factors, in addition to natural risks. As their population size declines, their genetic diversity also declines, making successful reproduction less likely. These factors suggest that, without some kind of intervention, pumas in these two populations will soon go extinct (with the Santa Ana population dying out first).

3. Maintaining any population and saving it from loss or extinction contributes to biodiversity, because all species and populations are part of biodiversity. Each lost population represents the loss of a strand in the web of life, which is connected to other strands (species) in the community. The puma, as a top predator, contributes to control of its prey populations, which in turn affects lower levels in the food chain, down to producers. Loss of the puma could therefore affect community structure at several levels. This in turn will limit the effectiveness of ecosystem functions and the ecological services (both practical and aesthetic) provided by California ecosystems.

4. Both populations are very small, so it makes sense to maximize the size of the gene pool by attempting to conserve both populations. However, finding an appropriate protected area might skew the protection toward one population: the less urbanized Peninsular Range population. If a protected area is identified in the Peninsular Range, habitat corridors could be provided to connect this population with the Santa Ana population, thus extending protection to both populations.

5. In a coupled human–natural system, humans in urban and suburban areas are enlisted to increase biodiversity in their own habitats. For small species (such as pollinators), people can plant native gardens or even window boxes. This is more difficult, but still possible, for large top predators, such as pumas, with large geographic ranges. If conservationists establish a protected area for pumas in the Peninsular Range, a fairly large buffer zone would need to be established around it. The buffer zone would involve agricultural areas, and farmers would need to help maintain it. This would involve an extensive education program, since many farmers see pumas as "the enemy." (In fact, the legal hunting that causes many puma deaths results from pumas feeding on livestock after being forced out of their natural habitat.) Farmers could be encouraged to establish edge habitats on their properties, allowing pumas safe passage to wilder areas. Coupled human–natural systems friendly to pumas would be more difficult to establish in urban areas, but wildlife corridors—particularly those allowing safe passage across highways—would be one obvious example.

GLOSSARY

3′ end (3 prime) The end of a DNA or RNA strand that has a free hydroxyl group at the 3′ carbon of the sugar (deoxyribose or ribose).

5′ cap (5 prime) Methylated guanine nucleotide bonded to the 5′ end of eukaryotic mRNA.

5′ end The end of a DNA or RNA strand that has a free phosphate group at the 5′ carbon of the sugar (deoxyribose or ribose).

A

A horizon *See* topsoil.

abiotic (a′ bye ah tick) [Gk. *a*: not + *bios*: life] Nonliving components of the environment including its physical and chemical characteristics. (Contrast with biotic.)

abomasum (a bo mey′ sum) The fourth of the four chambers of the stomach in ruminants; the true stomach.

abortion Any termination of pregnancy, whether induced or natural (in which case it is called a spontaneous abortion), that occurs after a fertilized egg is successfully implanted in the uterus.

abscisic acid (ABA) (ab sighs′ ik) A plant growth substance with growth-inhibiting action. Causes stomata to close; involved in a plant's response to salt and drought stress.

abscission (ab sizh′ un) [L. *abscissio*: break off] The process by which leaves, petals, and fruits separate from a plant.

aboral The opposite side of the body from the mouth.

absorption spectrum A graph of light absorption versus wavelength of light; shows how much light is absorbed at each wavelength.

absorptive heterotrophs Organisms (primarily fungi) that feed by **absorptive heterotrophy**, i.e., by secreting digestive enzymes into the environment to break down large food molecules, then absorbing the breakdown products.

absorptive state State in which food is in the gut and nutrients are being absorbed. (Contrast with postabsorptive state.)

acquired immune deficiency syndrome (AIDS) Decline in immune system function due to depletion of T_H cells and macrophages infected with HIV.

accessory fruits A fruit in which some of the flesh develops from tissue exterior to the carpel.

accessory sex organs Anatomical structures that allow transfer of sperm from male to female for internal fertilization. (Contrast with primary sex organs.)

acclimation Refers to increased tolerance for environmental extremes (e.g., extreme cold) after prior exposure to them.

acclimatization Refers to intrinsic seasonal adjustments in the "set points" of an animal's physiological functioning (e.g., metabolic rate).

acetyl coenzyme A (acetyl CoA) A compound that reacts with oxaloacetate to produce citrate at the beginning of the citric acid cycle; a key metabolic intermediate in the formation of many compounds.

acetylcholine (ACh) A neurotransmitter that carries information across vertebrate neuromuscular junctions and some other synapses. It is then broken down by the enzyme acetylcholinesterase (AChE).

acid [L. *acidus*: sharp, sour] A substance that can release a proton in solution. (Contrast with base.)

acid growth hypothesis The hypothesis that auxin increases proton pumping, thereby lowering the pH of the cell wall and activating enzymes that loosen polysaccharides. Proposed to explain auxin-induced cell expansion in plants.

acid rain Rain that has a lower pH than normal as a result of acid-forming precursor molecules introduced into the atmosphere by human activities.

acoelomate An animal that does not have a coelom.

acrosome (a′ krow soam) [Gk. *akros*: highest + *soma*: body] The structure at the forward tip of an animal sperm, which is the first to fuse with the egg membrane and enter the egg cell.

ACTH *See* corticotropin.

actin [Gk. *aktis*: ray] A protein that makes up the cytoskeletal microfilaments in eukaryotic cells and is one of the two contractile proteins in muscle. (*See also* myosin.)

action potentials (APs) Generated by neurons, these are electrical signals that transmit information via waves of depolarization or hyperpolarization of the cell membrane.

action spectrum A graph of a biological process versus light wavelength; shows which wavelengths are involved in the process.

activation energy (E_a) The energy barrier that blocks the tendency for a chemical reaction to occur.

activation gate The property of a voltage-gated Na^+ ion channel that increases its permeability to Na^+ ions when the membrane is depolarized.

activator A transcription factor that stimulates transcription when it binds to a gene's promoter. (Contrast with repressor.)

active site The region on the surface of an enzyme or ribozyme where the substrate binds, and where catalysis occurs.

active transport The energy-dependent transport of a substance across a biological membrane against a concentration gradient—that is, from a region of low concentration (of that substance) to one of high concentration. (*See also* primary active transport, secondary active transport. Contrast with facilitated diffusion, passive transport.)

adaptation (a dap tay′ shun) (1) In evolutionary biology, a particular structure, physiological process, or behavior that makes an organism better able to survive and reproduce. Also, the evolutionary process that leads to the development or persistence of such a trait. (2) In sensory neurophysiology, a sensory cell's loss of sensitivity as a result of repeated stimulation.

adaptive defenses One of the two general types of defenses against pathogens. Involves antibody proteins and other proteins that recognize, bind to, and aid in the destruction of specific viruses and bacteria. Present only in vertebrate animals. (Contrast with innate defenses.)

adaptive radiation A series of evolutionary events that results in an array (radiation) of related species that live in a variety of environments, differing in the characteristics each uses to exploit those environments.

addition rule The probability of one of two or more events that are mutually exclusive is the sum of their independent probabilities.

adenine (A) (a′ den een) A nitrogen-containing base found in nucleic acids, ATP, NAD, and other compounds.

adeno-associated virus A small, non-pathogenic virus with a single-strand DNA genome, that can be used for gene therapy in humans and does not integrate unto the host chromosomes.

ADH *See* vasopressin.

adhesion Binding of one cell or substance to another.

adrenal cortex The outer region of the adrenal gland that produces and secretes steroid hormones.

adrenal gland (a dree′ nal) [L. *ad*: toward + *renes*: kidneys] An endocrine gland located near the kidneys of vertebrates, consisting of two parts, the adrenal cortex and adrenal medulla.

adrenal medulla Tissue that makes up the core of the adrenal gland and secretes epinephrine and norepinephrine.

adrenergic receptors G protein-coupled receptor proteins that bind to the hormones epinephrine and norepinephrine, triggering specific responses in the target cells.

adrenocorticotropic hormone (ACTH) *See* corticotropin.

adventitious roots (ad ven ti' shus) [L. *adventitius*: arriving from outside] Roots originating from the stem at ground level or below; typical of the fibrous root system of monocots.

aerenchyma In plants, parenchymal tissue containing air spaces.

aerobic (air oh' bic) [Gk. *aer*: air + *bios*: life] In the presence of oxygen; requiring or using oxygen (as in **aerobic metabolism**). (Contrast with anaerobic.)

aerotolerant anaerobe A prokaryote that does not require an oxygen atmosphere for survival, but which can live while exposed to O_2.

afferent (af' ur unt) [L. *ad*: toward + *ferre*: to carry] Carrying to, as in neurons that carry impulses to the central nervous system (**afferent neurons**), or a blood vessel that carries blood to a structure (**afferent arterioles**). (Contrast with efferent.)

after-hyperpolarization A fall in membrane potential below normal resting level following an action potential.

aggregate fruit A fruit that develops from several carpels from the same flower.

agonist A chemical substance (e.g., a neurotransmitter) that elicits a specific response in a cell or tissue. (Contrast with antagonist.)

AIDS *See* acquired immune deficiency syndrome.

air sacs Structures in the respiratory system of birds that receive inhaled air; they keep fresh air flowing unidirectionally through the lungs, but are not themselves gas exchange surfaces.

albedo effect (al bee' doh) [L. *albus*: white] The solar radiation reflected by the surface of Earth.

alcoholic fermentation *See* fermentation.

aldosterone (al dohs' ter own) A steroid hormone produced in the adrenal cortex of mammals. Promotes secretion of potassium and reabsorption of sodium in the kidney.

aleurone layer (al yu' rown) In some seeds, a tissue that lies beneath the seed coat and surrounds the endosperm. Secretes digestive enzymes that break down macromolecules stored in the endosperm.

allantoic membrane In animal development, an outgrowth of extraembryonic endoderm plus adjacent mesoderm that forms the allantois, a saclike structure that stores metabolic wastes produced by the embryo.

allantois (al' lun toh is) [Gk. *allant*: sausage] An extraembryonic membrane enclosing a sausage-shaped sac that stores the embryo's nitrogenous wastes.

Allee effect An increase in the population growth rate (*r*) as the population density decreases.

allele (a leel') [Gk. *allos*: other] The alternate form of a genetic character found at a given locus on a chromosome.

allergic reaction [Ger. *allergie*: altered] An overreaction of the immune system to amounts of an antigen that do not affect most people; often involves IgE antibodies.

allopatric speciation (al' lo pat' rick) [Gk. *allos*: other + *patria*: homeland] The formation of two species from one when reproductive isolation occurs because of the interposition of (or crossing of) a physical geographic barrier such as a river. Also called geographic speciation. (Contrast with sympatric speciation.)

allopolyploidy The possession of more than two chromosome sets that are derived from more than one species.

allosteric regulation (al lo steer' ik) [Gk. *allos*: other + *stereos*: structure] Regulation of the activity of a protein (usually an enzyme) by the binding of an effector molecule to a site other than the active site.

alpha diversity Species diversity at the local or community scale. (Compare with beta diversity, gamma diversity.)

α (alpha) helix A prevalent type of secondary protein structure; a right-handed spiral.

alternation of generations The succession of multicellular haploid and diploid phases in some sexually reproducing organisms, notably plants.

alternative hypothesis In statistical inference, the hypothesis that contrasts with the null hypothesis; typically the hypothesis of primary interest.

alternative splicing A process for generating different mature mRNAs from a single gene by splicing together different sets of exons during RNA processing.

alternative states Different community assemblages that develop at the same location under similar environmental conditions.

altricial Born or hatched in a relatively undeveloped state that requires care and feeding by the parents.

altruistic Pertaining to behavior that benefits other individuals at a cost to the individual who performs it.

alveolus (al ve' o lus) (plural: alveoli) [L. *alveus*: cavity] A small, baglike cavity, especially the blind sacs of the lung.

amacrine cells Cells in the retina that communicate laterally between ganglion and bipolar cells.

amensalism A type of interaction in which one species is harmed and the other is unaffected. (Contrast with competition, commensalism, mutualism.)

amine hormones Small hormone molecules synthesized from single amino acids (e.g., thyroxine and epinephrine).

amino acid An organic compound containing both NH_2 and COOH groups. Proteins are polymers of amino acids.

ammonia NH_3, the most common nitrogenous waste. (Compare with urea, uric acid.)

ammonotelic (am moan' o teel' ic) [Gk. *telos*: end] Pertaining to an organism in which the final product of breakdown of nitrogen-containing

compounds (primarily proteins) is ammonia. (Contrast with ureotelic, uricotelic.)

amnion (am' nee on) The fluid-filled sac within which the embryos of reptiles (including birds) and mammals develop.

amniote egg A shelled egg surrounding four extraembryonic membranes and embryo-nourishing yolk. This evolutionary adaptation permitted mammals and reptiles to live and reproduce in drier environments than can most amphibians.

amniotic fluid The fluid in the amniotic sac that contains mammalian and reptilian embryos.

amphipathic (am' fi path' ic) [Gk. *amphi*: both + *pathos*: emotion] Of a molecule, having both hydrophilic and hydrophobic regions.

amplitude The magnitude of change over the course of a regular cycle.

amygdala A component of the limbic system that is involved in fear and the memory of fearful experiences.

anabolic reaction (an uh bah' lik) [Gk. *ana*: upward + *ballein*: to throw] A synthetic reaction in which simple molecules are linked to form more complex ones; requires an input of energy and captures it in the chemical bonds that are formed. (Contrast with catabolic reaction.)

anaerobic (an ur row' bic) [Gk. *an*: not + *aer*: air + *bios*: life] Occurring without the use of molecular oxygen, O_2. (Contrast with aerobic.)

anaerobic respiration A metabolic pathway in which glucose is catabolized in the absence of oxygen.

analogous In biology, refers to similar structures found in two or more groups of organisms, but with separate, independent evolutionary origins. An example is the wings of bats and the wings of birds, which evolved independently as wings from wingless ancestors.

ancestral trait The trait originally present in the ancestor of a given group; may be retained or changed in the descendants of that ancestor.

androgen (an' dro jen) Any of the several male sex steroids (most notably testosterone).

aneuploidy (an' you ploy dee) A condition in which one or more chromosomes or pieces of chromosomes are either lacking or present in excess.

angiosperms Flowering plants; one of the two major groups of living seed plants. (*See also* gymnosperms.)

angiotensin (an' jee oh ten' sin) A peptide hormone that raises blood pressure by causing peripheral vessels to constrict. Also maintains glomerular filtration by constricting efferent vessels and stimulates thirst and the release of aldosterone.

angular gyrus A part of the human brain believed to be essential for integrating spoken and written language.

animal hemisphere The metabolically active upper portion of some animal eggs, zygotes, and embryos; does not contain the dense nutrient yolk. (Contrast with vegetal hemisphere.)

animal physiology The study of animal form and function.

animal pole   The point at the end of the developing egg where yolk is least concentrated.

anion   (an' eye on) [Gk. *ana*: upward] A negatively charged ion. (Contrast with cation.)

annual   A plant whose life cycle is completed in one growing season. (Contrast with biennial, perennial.)

antagonist   A biochemical substance (e.g., a drug) that blocks the normal action of another biochemical substance. (Contrast with agonist.)

anterior pituitary   The portion of the vertebrate pituitary gland that derives from gut epithelium. Produces trophic hormones.

anterior   Toward or pertaining to the tip or headward region of the body axis. (Contrast with posterior.)

anther   (an' thur) [Gk. *anthos*: flower] A pollen-bearing portion of the stamen of a flower.

antheridium   (an' thur id' ee um) [Gk. *antheros*: blooming] The multicellular structure that produces the sperm in nonvascular land plants and ferns.

antibody   One of the myriad proteins produced by the immune system that specifically binds to a foreign substance in blood or other tissue fluids and initiates its removal from the body.

anticodon   The three nucleotides in transfer RNA that pair with a complementary triplet (a codon) in messenger RNA.

antidiuretic hormone (ADH)   *See* vasopressin.

antigen   (an' ti jun) Any substance that stimulates the production of an antibody or antibodies in the body of a vertebrate.

antigen presentation   In cellular immunity, a process in which a cell ingests and digests an antigen, and then exposes fragments of that antigen to the outside of the cell, bound to proteins in the cell's cell membrane.

antigenic determinant   The specific region of an antigen that is recognized and bound by a specific antibody. Also called an epitope.

antiparallel   Pertaining to molecular orientation in which a molecule or parts of a molecule have opposing directions.

antiporter   A membrane transport protein that moves one substance in one direction and another in the opposite direction. (Contrast with symporter, uniporter.)

antisense RNA   A single-stranded RNA molecule complementary to, and thus targeted against, an mRNA of interest to block its translation.

anus   (a' nus) An opening through which solid digestive wastes are expelled, located at the posterior end of a tubular gut.

aorta   (a or' tah) [Gk. *aorte*: aorta] The main trunk of the arteries leading to the systemic (as opposed to the pulmonary) circulation.

aortic body   A chemosensor in the aorta that senses a decrease in blood supply or a dramatic decrease in partial pressure of oxygen in the blood.

aortic valve   A one-way valve between the left ventricle of the heart and the aorta that prevents backflow of blood into the ventricle when it relaxes.

aphasia   A deficit in the ability to use or understand words.

apical dominance   In plants, inhibition by the apical bud of the growth of axillary buds.

apical hook   A form taken by the stems of many eudicot seedlings that protects the delicate shoot apex while the stem grows through the soil.

apical meristem   The meristem at the tip of a shoot or root; responsible for a plant's primary growth.

apomixis   (ap oh mix' is) [Gk. *apo*: away from + *mixis*: sexual intercourse] The asexual production of seeds.

apoplast   (ap' oh plast) In plants, the continuous meshwork of cell walls and extracellular spaces through which material can pass without crossing a plasma membrane. (Contrast with symplast.)

apoptosis   (ap uh toh' sis) A series of genetically programmed events leading to cell death.

appendix   A small blind-end sac that is attached to the initial segment of the colon.

aquaporin   A transport protein in plant and animal cell membranes through which water passes in osmosis.

aquifer   A large pool of groundwater.

archegonium   (ar' ke go' nee um) The multicellular structure that produces eggs in nonvascular land plants, ferns, and gymnosperms.

archenteron   (ark en' ter on) [Gk. *archos*: first + *enteron*: bowel] The earliest primordial animal digestive tract.

arcuate nucleus   A group of neurons in the hypothalamus that produce and secrete many neuroendocrines and are involved in regulation of appetite.

arteriole   A small blood vessel arising from an artery that feeds blood into a capillary bed.

artery   A muscular blood vessel carrying oxygenated blood away from the heart to other parts of the body. (Contrast with vein.)

artificial insemination   An infertility treatment that involves the artificial introduction of sperm into the woman's reproductive tract.

ascospores   Haploid spores contained with the sexual reproductive structure (the ascus) of sac fungi.

ascus   (ass' cus) (plural: asci) [Gk. *askos*: bladder] In sac fungi, the club-shaped sporangium within which spores (ascospores) are produced by meiosis.

assimilation efficiency   The proportion of ingested biomass that consumers assimilate by digestion.

assisted reproductive technologies (ARTs)   Any of several procedures that remove unfertilized eggs from the ovary, combine them with sperm outside the body, and then place fertilized eggs or egg–sperm mixtures in the appropriate location in a female's reproductive tract for development.

association cortex   In the vertebrate brain, the portion of the cortex involved in higher-order information processing, so named because it integrates, or associates, information from different sensory modalities and from memory.

asthenosphere   (ass thenn' o sphere) [Gk. *asthenes*: weak] The viscous, malleable (changeable) layer of Earth's mantle. It is overlain by the solid lithospheric plates.

astrocyte   [Gk. *astron*: star] A type of glial cell that contributes to the blood–brain barrier by surrounding the smallest, most permeable blood vessels in the brain.

atherosclerosis   (ath' er oh sklair oh' sis) [Gk. *athero*: gruel, porridge + *skleros*: hard] A disease of the lining of the arteries characterized by fatty, cholesterol-rich deposits in the walls of the arteries. When fibroblasts infiltrate these deposits and calcium precipitates in them, the disease becomes **arteriosclerosis**, or "hardening of the arteries."

atmosphere   A thin layer of gases surrounding Earth composed of nitrogen gas, oxygen, argon, carbon dioxide, and traces of hydrogen, helium, ozone, methane, and many other gases.

atom   [Gk. *atomos*: indivisible] The smallest unit of a chemical element. Consists of a nucleus and one or more electrons.

atomic number   The number of protons in the nucleus of an atom; also equals the number of electrons around the neutral atom. Determines the chemical properties of the atom.

atomic weight   The average of the mass numbers of a representative sample of atoms of an element, with all the isotopes in their normally occurring proportions. Also called atomic mass.

ATP (adenosine triphosphate)   An energy-storage compound containing adenine, ribose, and three phosphate groups. When it is formed from ADP, useful energy is stored; when it is broken down (to ADP or AMP), energy is released to drive endergonic reactions.

ATP synthase   An integral membrane protein that couples the transport of protons with the formation of ATP.

atrial natriuretic peptide (ANP)   A hormone released by the atrial muscle fibers of the heart when they are overly stretched, which decreases reabsorption of sodium by the kidney and thus blood volume.

atrioventricular node   A modified node of cardiac muscle that organizes the action potentials that control contraction of the ventricles.

atrioventricular (AV) valve   A heart valve between an atrium and a ventricle.

atrium   (a' tree um) [L. *atrium*: central hall] An internal chamber. In the hearts of vertebrates, the thin-walled chamber(s) entered by blood on its way to the ventricle(s). Also, the outer ear.

autocatalysis   [Gk. *autos*: self + *kata*: to break down] A positive feedback process in which an activated enzyme acts on other inactive molecules of the same enzyme to activate them.

autocrine   A chemical signal that binds to and affects the cell that makes it. (Contrast with paracrine.)

autoimmunity   An immune response by an organism to its own molecules or cells.

autonomic nervous system (ANS)   The portion of the peripheral nervous system that controls such involuntary functions as those of guts and glands. Also called the involuntary nervous system.

autonomous specification   Determination of cell fates due to informational molecules originally present in the egg.

autophagy   The programmed destruction of a cell's components.

autopolyploidy   The possession of more than two entire chromosome sets that are derived from a single species.

autoregulatory mechanisms   In mammalian circulatory systems, local control of blood flow through capillary beds by constriction or dilation of incoming arterioles in response to local metabolite concentrations.

autosome   Any chromosome (in a eukaryote) other than a sex chromosome.

autotroph   (au' tow trowf') [Gk. *autos*: self + *trophe*: food] An organism that is capable of living exclusively on inorganic materials, water, and some energy source such as sunlight (photoautotrophs) or chemically reduced matter (see chemoautotrophs). (Contrast with heterotroph.)

auxin   (awk' sin) [Gk. *auxein*: to grow] In plants, a substance (the most common being indoleacetic acid) that regulates growth and various aspects of development.

avirulence (*Avr*) genes   Genes in a pathogen that may trigger defenses in plants. (*See* gene-for-gene concept.)

Avogadro's number   The number of atoms or molecules in a mole (weighed out in grams) of a substance, calculated to be 6.023×10^{23}.

axillary bud   A bud that forms in the angle (axil) where a leaf meets a stem.

axon   [Gk. axle] The process (branching structure) of a neuron that conducts action potentials away from the cell body. (*See also* dendrites.)

axon hillock   The junction between an axon and the neuron's cell body; where action potentials are generated.

axon terminal   The end portion of an axon, which passes action potentials to another cell. Axon terminals can form synapses and release neurotransmitter.

B

B cell   A type of lymphocyte involved in the humoral immune response of vertebrates. Upon recognizing an antigenic determinant, a B cell develops into a plasma cell, which secretes an antibody. (Contrast with T cell.)

B horizon   See subsoil.

bacillus   (bah sil' us) [L: little rod] Any of various rod-shaped bacteria.

bacterial conjugation   See conjugation.

bacteriophage   (bak teer' ee o fayj) [Gk. *bakterion*: little rod + *phagein*: to eat] Any of a group of viruses that infect bacteria. Also called phage.

bacteroids   Nitrogen-fixing organelles that develop from endosymbiotic bacteria.

bar chart   A figure that displays frequency distributions of categorical data using bar lengths to represent relative frequency.

bark   All tissues external to the vascular cambium of a plant.

baroreceptor   [Gk. *baros*: weight] A pressure-sensing cell or organ. Sometimes called a stress receptor.

basal metabolic rate (BMR)   The minimum rate of energy turnover in an awake (but resting) bird or mammal that is not expending energy for thermoregulation.

base   (1) A substance that can accept a hydrogen ion in solution. (Contrast with acid.) (2) In nucleic acids, the purine or pyrimidine that is attached to each sugar in the sugar–phosphate backbone.

base pair (bp)   In double-stranded DNA, a pair of nucleotides formed by the complementary base pairing of a purine on one strand and a pyrimidine on the other. (*See* complementary base pairing.)

basidioma   (plural: basidiomata) A fruiting structure produced by club fungi.

basidium   (bass id' ee yum) In club fungi, the characteristic sporangium in which four **basidiospores** are formed by meiosis and then borne externally before being shed.

Batesian mimicry   The convergence in appearance of an edible species (mimic) with an unpalatable species (model).

behavioral ecology   An evolutionary approach to the study of animal behavior that studies how behaviors are adaptive in different environmental conditions.

behaviorism   One of two classical approaches to the study of proximate causes of animal behavior, derived from the discoveries of Ivan Pavlov and focused on laboratory studies. (Compare with ethology.)

benign   A tumor that grows to a limited extent and does not spread to other parts of the body. (Contrast with malignant.)

beta diversity   The change in species number and composition, or turnover of species, across a landscape. (Compare with alpha diversity, gamma diversity.)

β (beta) pleated sheet   A type of protein secondary structure; results from hydrogen bonding between polypeptide regions running antiparallel to each other.

biased gene conversion   A mechanism of concerted evolution in which a DNA repair system appears biased in favor of using particular nucleotide sequences as templates for repair, resulting in the rapid spread of the favored sequence across all copies of the gene. (*See* concerted evolution.)

bicuspid valve   The left AV valve. So named because it has two leaves.

biennial   A plant whose life cycle includes vegetative growth in the first year and flowering and senescence in the second year. (Contrast with annual, perennial.)

bilateral symmetry   The condition in which only the right and left sides of an organism, divided by a single plane through the midline, are mirror images of each other.

bilayer   A structure that is two layers in thickness. In biology, most often refers to the phospholipid bilayer of membranes. (*See* phospholipid bilayer.)

bile   A secretion of the liver made up of bile salts synthesized from cholesterol, various phospholipids, and bilirubin (the breakdown product of hemoglobin). Emulsifies fats in the small intestine.

binary fission   Reproduction of a prokaryote by division of a cell into two comparable progeny cells.

bindin   Species specific recognition molecule on the sperm acrosomal process.

binocular vision   Overlapping visual fields of an animal's two eyes; allows the animal to see in three dimensions.

binomial nomenclature   [Gk. two names] A taxonomic naming system in which each species is given a binomial, a genus name followed by a species name.

biodiversity   The diversity of important ecological entities that span multiple spatial scales, from genes to populations to species to communities.

biofilm   A community of microorganisms embedded in a polysaccharide matrix, forming a highly resistant coating on almost any moist surface.

biogeochemical cycle   Movement of inorganic elements such as water, nitrogen, phosphorus, and carbon through living organisms and the physical environment.

biogeographic region   Distinct regions of biota characterized by different species that occur at the continental-scale. (Contrast with biome.)

biogeography   The scientific study of the distribution and diversity of organisms across Earth.

bioinformatics   The use of computers and/or mathematics to analyze complex biological information, such as DNA sequences.

biological control   The use of natural enemies (predators, parasites, or pathogens) to reduce the population density of an economically damaging (pest) species.

biological species concept   The definition of a species as a group of actually or potentially interbreeding natural populations that are reproductively isolated from other such groups. (Contrast with lineage species concept; morphological species concept.)

biology   [Gk. *bios*: life + *logos*: study] The scientific study of living things.

biome   (bye' ome) A major division of the ecological communities of Earth, characterized primarily by distinctive vegetation and the annual patterns of temperature and precipitation. A given biogeographic region contains many different biomes.

biosphere   (bye' oh sphere) All regions of Earth (terrestrial and aquatic) and Earth's atmosphere in which organisms can live.

biota   (bye oh' tah) All of the organisms—animals, plants, fungi, and microorganisms—found in a given area. (Contrast with flora, fauna.)

biotechnology   The use of living cells or organisms to produce materials useful to humans.

biotic   (bye ah' tick) [Gk. *bios*: life] Living components of the environment such as organisms. (Contrast with abiotic.)

biotic resistance   The ability of resident species to exclude or slow the population growth of non-native species.

bipedal locomotion   Terrestrial locomotion by means of two rear limbs or legs.

bipolar cells   Cells in the retina that communicate between photoreceptors and ganglion cells.

blastocoel   (blass' toe seal) [Gk. *blastos*: sprout + *koilos*: hollow] The central, hollow cavity of a blastula.

blastocyst   (blass' toe cist) An early embryo formed by the first divisions of the fertilized egg (zygote). In mammals, a hollow ball of cells.

blastoderm   A layer of undifferentiated cells that are a very early stage of development prior to the specification of the embryonic axis.

blastodisc   (blass' toe disk) An embryo that forms as a disc of cells on the surface of a large yolk mass; comparable to a blastula, but occurring in animals such as birds and reptiles, in which the massive yolk restricts complete cleavage.

blastomere   Any of the cells produced by the early divisions of a fertilized animal egg.

blastopore   The opening created by the invagination of the vegetal pole during gastrulation of animal embryos.

blastula   (blass' chu luh) An early stage of the animal embryo; in many species, a hollow sphere of cells surrounding a central cavity, the blastocoel. (Contrast with blastodisc.)

block to polyspermy   Any of several responses to entry of a sperm into an egg that prevent more than one sperm from entering the egg.

blood clotting   A cascade of events involving platelets and circulating proteins (clotting factors) that seals damaged blood vessels.

blood plasma   (plaz' muh) The liquid portion of blood, in which blood cells and other particulates are suspended.

blood–brain barrier   The selective impermeability of blood vessels in the brain that prevents most chemicals from diffusing from the blood into the brain.

blue-light receptors   Pigments in plants that absorb blue light (400–500 nm). These pigments mediate many plant responses including phototropism, stomatal movements, and expression of some genes.

BMR   *See* basal metabolic rate.

body cavity   An internal, fluid-filled space, enclosed by mesoderm in triploblastic animals.

body plan   The general structure of an animal, the arrangement of its organ systems, and the integrated functioning of its parts.

Bohr effect   A shift in the O_2-binding curve of hemoglobin in response to excess H^+ ions such that the hemoglobin releases more O_2 in tissues where pH is low.

bone   A rigid component of vertebrate skeletal systems that contains an extracellular matrix of insoluble calcium phosphate crystals as well as collagen fibers.

Bowman's capsule   An elaboration of the renal tubule, composed of podocytes, that surrounds and collects the filtrate from the glomerulus.

brain   The centralized integrative center of a nervous system.

brainstem   The portion of the vertebrate brain between the spinal cord and the forebrain, made up of the medulla, pons, and midbrain.

brassinosteroids   Plant steroid hormones that mediate light effects promoting the elongation of stems and pollen tubes.

Broca's area   A portion of the human brain essential for speech. Located in the frontal lobe just in front of the primary motor cortex.

bronchioles   The smallest airways in a vertebrate lung, branching off the bronchi.

bronchus   (plural: bronchi) The major airway(s) branching off the trachea into the vertebrate lung.

brown fat   In mammals, fat tissue that is specialized to produce heat. It has many mitochondria and capillaries, and a protein that uncouples oxidative phosphorylation.

bud   A protuberance on a stem that contains an undeveloped leaf, shoot, or flower.

budding   Asexual reproduction in which a more or less complete new organism grows from the body of the parent organism, eventually detaching itself.

buffer   A substance that can transiently accept or release hydrogen ions and thereby resist changes in pH.

bulbourethral glands   Secretory structures of the human male reproductive system that produce a small volume of an alkaline, mucoid secretion that helps neutralize acidity in the urethra and lubricate it to facilitate the passage of semen.

bulbus arteriosus   The last chamber of the fish heart.

bulk flow   The movement of a solution from a region of higher pressure potential to a region of lower pressure potential.

bulk transport   *See* solvent drag.

bundle of His   Fibers of modified cardiac muscle that conduct action potentials from the atria to the ventricular muscle mass.

bundle sheath cell   Part of a tissue that surrounds the veins of plants.

C

C horizon   *See* parent rock.

C_3 plants   Plants that produce 3PG as the first stable product of carbon fixation in photosynthesis and use ribulose bisphosphate as a CO_2 receptor.

C_4 plants   Plants that produce oxaloacetate as the first stable product of carbon fixation in photosynthesis and use phosphoenolpyruvate as a CO_2 acceptor. C_4 plants also perform the reactions of C_3 photosynthesis.

calcitonin   Hormone produced by the thyroid gland; lowers blood calcium and promotes bone formation. (Contrast with parathyroid hormone.)

calcitriol   A hormone derived from vitamin D whose actions include stimulating the cells of the digestive tract to absorb calcium from ingested food.

calorie   [L. *calor*: heat] The amount of heat required to raise the temperature of 1 gram of water by 1°C. Physiologists commonly use the kilocalorie (kcal) as a unit of measure (1 kcal = 1,000 calories). Nutritionists also use the **kilocalorie**, but refer to it as the **Calorie** (**Cal**, capital C).

Calvin cycle   The stage of photosynthesis in which CO_2 reacts with RuBP to form 3PG, 3PG is reduced to a sugar, and RuBP is regenerated, while other products are released to the rest of the plant. Also known as the Calvin–Benson cycle.

CAM   *See* crassulacean acid metabolism.

Cambrian explosion   The rapid diversification of multicellular life that took place during the Cambrian period.

cAMP (cyclic AMP)   A compound formed from ATP that acts as a second messenger.

cancellous bone   A type of bone with numerous internal cavities that make it appear spongy, although it is rigid. (Contrast with compact bone.)

capillaries   [L. *capillaris*: hair] Very small tubes, especially the smallest blood-carrying vessels of animals between the termination of the arteries and the beginnings of the veins. **Capillary beds** are networks of capillaries where materials are exchanged between the blood and the interstitial fluid.

capsid   The outer shell of a virus that encloses its nucleic acid.

capsule   In some prokaryotes, a dense layer of polymers that surrounds the cell wall.

carbohydrases   Digestive enzymes that hydrolyze carbohydrates.

carbohydrates   Organic compounds containing carbon, hydrogen, and oxygen in the ratio 1:2:1 (i.e., with the general formula $C_nH_{2n}O_n$). Common examples are sugars, starch, and cellulose.

carbon skeleton   The chains or rings of carbon atoms that form the structural basis of organic molecules. Other atoms or functional groups are attached to the carbon atoms.

carbonic anhydrase (CA)   An enzyme that catalyzes the hydration of CO_2 to produce carbonic acid.

carboxylase   An enzyme that catalyzes the addition of carboxyl functional groups (O=C—OH) to a substrate.

cardiac cycle   Contraction of the two atria of the heart, followed by contraction of the two ventricles and then relaxation.

cardiac muscle   A type of muscle tissue that makes up, and is responsible for the beating of, the heart. Characterized by branching cells with single nuclei and a striated (striped) appearance. (Contrast with smooth muscle, skeletal muscle.)

cardiovascular system   [Gk. *kardia*: heart + L. *vasculum*: small vessel] The heart, blood, and vessels are of a circulatory system.

carnivore [L. *carn*: flesh + *vorare*: to devour] An organism that eats animal tissues. (Contrast with detritivore, herbivore, omnivore.)

carnivorous plant Plant capable of trapping and digesting animals to obtain nutrients.

carnivory An act of predation in which the predator and prey are both animals. (Contrast with herbivory, parasitism, or omnivory).

carotid body A chemosensor in the carotid artery that senses a decrease in blood supply or a dramatic decrease in partial pressure of oxygen in the blood.

carpel (kar' pel) [Gk. *karpos*: fruit] The organ of the flower that contains one or more ovules.

carrier proteins Proteins that bind another substance to transport it from one location to another; carriers bind small molecules and transport them across membranes.

carrying capacity (*K*) The maximum number of individuals in a population (i.e., maximum population size) that can be supported by the resources present in a given environment.

cartilage bone A type of bone that begins its development as a cartilaginous structure resembling the future mature bone, then gradually hardens into mature bone. (Contrast with membranous bone.)

cartilage In vertebrates, a tough connective tissue found in joints, the outer ear, and elsewhere. Forms the entire skeleton in some animal groups.

Casparian strip A band of cell wall containing suberin and lignin, found in the endodermis. Restricts the movement of water across the endodermis.

caspase One of a group of proteases that catalyze cleavage of target proteins and are active in apoptosis.

catabolic reaction (kat uh bah' lik) [Gk. *kata*: to break down + *ballein*: to throw] A synthetic reaction in which complex molecules are broken down into simpler ones and energy is released. (Contrast with anabolic reaction.)

catabolite repression In the presence of abundant glucose, the diminished synthesis of catabolic enzymes for other energy sources.

categorical variables Variables that take on qualitative categories as values, such as human blood types (A, AB, B, or O).

cation (cat' eye on) An ion with one or more positive charges. (Contrast with anion.)

cation exchange The process in soil by which cations bound to soil particles are released from the particles and made available to plant roots in soil water through the binding of hydrogen ions to the soil particles.

CCK *See* cholecystokinin.

Cdk *See* cyclin-dependent kinase.

cDNA *See* complementary DNA.

cDNA library A collection of complementary DNAs derived from mRNAs of a particular tissue at a particular time in the life cycle of an organism.

cecum (see' cum) [L. blind] A blind branch off the large intestine. In many nonruminant mammals, the cecum contains a colony of microorganisms that contribute to the digestion of food.

cell The simplest structural unit of a living organism. In multicellular organisms, many individual cells serve as the building blocks of tissues and organs.

cell adhesion The binding of one cell to another, often mediated by noncovalent forces.

cell body The portion of the neuron that contains the nucleus and gives rise to dendrites and an axon.

cell cycle checkpoints Points of transition between different phases of the cell cycle, which are regulated by cyclins and cyclin-dependent kinases (Cdk's).

cell cycle The stages through which a cell passes between one mitotic division and the next. Includes all stages of interphase and mitosis. (*See* mitosis.)

cell fate The type of cell that an undifferentiated cell in an embryo will become in the adult.

cell junctions Specialized structures associated with the plasma membranes of epithelial cells. Some contribute to cell adhesion, others to intercellular communication.

cell membrane The membrane that surrounds the cell, regulating the entry and exit of molecules and ions. Every cell has a cell membrane, and it is also called the plasma membrane.

cell potency In multicellular organisms, an undifferentiated cell's potential to become a cell of a specific type. (*See* multipotent, pluripotent, totipotent.)

cell recognition Binding of cells to one another mediated by membrane proteins or carbohydrates.

cell theory States that cells are the basic structural and physiological units of all living organisms, and that all cells come from preexisting cells.

cell wall A relatively rigid structure that encloses cells of plants, fungi, many protists, and most prokaryotes, and which gives these cells their shape and limits their expansion in hypotonic media.

cellular immune response Immune system response mediated by T cells and directed against parasites, fungi, intracellular viruses, and foreign tissues (grafts). (Contrast with humoral immune response.)

cellular respiration The catabolic pathways by which electrons are removed from various molecules and passed through intermediate electron carriers to O_2, generating H_2O and releasing energy.

cellular respiration The catabolic pathways by which electrons are removed from various molecules and passed through intermediate electron carriers to O_2, generating H_2O and releasing energy.

cellular specialization In multicellular organisms, the division of labor such that different cell types become responsible for different functions (e.g., reproduction or digestion) within the organism.

cellulose (sell' you lowss) A straight-chain polymer of glucose molecules, used by plants as a structural supporting material.

central dogma The premise that information flows from DNA to RNA to polypeptide (protein).

central nervous system (CNS) That portion of the nervous system that is the site of most information processing, storage, and retrieval; in vertebrates, the brain and spinal cord. (Contrast with peripheral nervous system.)

central vacuole In plant cells, a large organelle that stores the waste products of metabolism and maintains turgor.

centriole (sen' tree ole) A paired organelle that helps organize the microtubules in animal and protist cells during nuclear division.

centromere (sen' tro meer) [Gk. *centron*: center + *meros*: part] The region where sister chromatids join.

centrosome (sen' tro soam) The major microtubule organizing center of an animal cell.

cephalization (sef ah luh zay' shun) [Gk. *kephale*: head] The evolutionary trend toward increasing concentration of brain and sensory organs at the anterior end of the animal.

cerebellum (sair uh bell' um) [L. diminutive of *cerebrum*: brain] The brain region that controls muscular coordination; located at the anterior end of the hindbrain.

cerebral cortex The thin layer of gray matter (neuronal cell bodies) that overlies the cerebrum.

cerebral hemispheres The bilateral divisions of the cerebrum.

cerebrum (su ree' brum) [L. brain] The dorsal anterior portion of the forebrain, making up the largest part of the brain of mammals; the chief coordination center of the nervous system and the major information-processing areas of the vertebrate brain consists of two cerebral hemispheres.

cervix (sir' vix) [L. neck] The opening of the uterus into the vagina.

channel protein An integral membrane protein that forms an aqueous passageway across the membrane in which it is inserted and through which specific solutes may pass.

chaperone A protein that guards other proteins by counteracting molecular interactions that threaten their three-dimensional structure.

character displacement An evolutionary phenomenon in which species that compete for the same limiting resources diverge in morphology and/or behavior.

character In genetics, an observable feature, such as eye color. (Contrast with trait.)

chemical bond An attractive force stably linking two atoms.

chemical equilibrium *See* equilibrium.

chemical evolution The theory that life originated through the chemical transformation of inanimate substances.

chemical reaction The change in the composition or distribution of atoms of a substance with consequent alterations in properties.

chemical synapse Neural junction at which neurotransmitter molecules released from a presynaptic cell induce changes in a postsynaptic cell. (Contrast with electrical synapse.)

chemically gated channel A type of membrane channel that opens or closes depending on the presence or absence of a specific molecule that binds either to the channel protein itself or to a separate receptor that alters the three-dimensional shape of the channel protein.

chemiosmosis Formation of ATP in mitochondria and chloroplasts, resulting from a pumping of protons across a membrane (against a gradient of electrical charge and of pH), followed by the return of the protons through a protein channel with ATP synthase activity.

chemoautotroph An organism that obtains energy by oxidizing inorganic substances, using some of that energy to fix carbon. Also known as a chemolithotroph. (Contrast with chemoheterotroph, photoautotroph, photoheterotroph.)

chemoheterotroph An organism that must obtain both carbon and energy from organic substances. (Contrast with chemoautotroph, photoautotroph, photoheterotroph.)

chemoreceptor A receptor protein that binds to specific molecules (such as odorant molecules or pheromones) in the environment.

chemosynthesis The use of energy from inorganic chemical compounds by archaea and bacteria to fix CO_2 and produce carbohydrates using the Calvin cycle.

chi-square goodness-of-fit A statistical test used to assess whether the frequencies of observations in different categories are consistent with an hypothesized frequency distribution.

chiasma (kie az'muh) (plural: chiasmata) [Gk. cross] An X-shaped connection between paired homologous chromosomes in prophase I of meiosis. A chiasma is the visible manifestation of crossing over between homologous chromosomes.

chief cells One of three types of secretory cell found in the gastric pits of the stomach wall. Chief cells secrete the protein-digesting enzyme pepsin. (*See* parietal cells.)

chitin (kye'tin) [Gk. kiton: tunic] The characteristic tough but flexible organic component of the exoskeleton of arthropods, consisting of a complex, nitrogen-containing polysaccharide. Also found in cell walls of fungi.

chlorophyll (klor'o fill) [Gk. kloros: green + phyllon: leaf] Any of several green pigments associated with chloroplasts or with certain bacterial membranes; responsible for trapping light energy for photosynthesis.

chloroplast [Gk. kloros: green + plast: a particle] An organelle bounded by a double membrane containing the enzymes and pigments that perform photosynthesis. Chloroplasts occur only in eukaryotes.

choanocyte (ko'an uh site) The collared, flagellated feeding cells of sponges.

cholecystokinin (CCK) (ko'luh sis tuh kai'nin) A hormone produced and released by the lining of the duodenum when it is stimulated by undigested fats and proteins. It stimulates the gallbladder to release bile and slows stomach activity.

chordamesoderm Tissue of the vertebrate embryo that will form the notochord and somite mesoderm, and plays roles in organizing the embryonic axis.

chorion (kor'ee on) [Gk. khorion: afterbirth] The outermost of the membranes protecting mammal, bird, and reptile embryos; in mammals it forms part of the placenta.

chromatin The nucleic acid–protein complex that makes up eukaryotic chromosomes.

chromosomal mutation Loss of or changes in position/direction of a DNA segment on a chromosome.

chromosome (krome'o sowm) [Gk. kroma: color + soma: body] In bacteria and viruses, the DNA molecule that contains most or all of the genetic information of the cell or virus. In eukaryotes, a structure composed of DNA and proteins that bears part of the genetic information of the cell.

chylomicrons (ky low my'cron) Particles of lipid coated with protein, produced in the gut from dietary fats and secreted into the extracellular fluids.

chyme (kime) [Gk. kymus: juice] Created in the stomach; a mixture of ingested food with the digestive juices secreted by the salivary glands and the stomach lining.

cilia (sil'ee ah) (singular: cilium) [L. eyelashes] Hairlike organelle used for locomotion by many unicellular organisms and for moving water and mucus by many multicellular organisms. Generally shorter than flagella.

circadian rhythm (sir kade'ee an) [L. circa: approximately + dies: day] A rhythm of growth or activity that recurs about every 24 hours.

circulatory system A physiological system consisting of a muscular pump (heart), a fluid (blood or hemolymph), and a series of conduits (blood vessels) that transports materials around the body.

cis-trans isomers In molecules with a double bond (typically between two carbon items), identifies on which side of the double bond similar atoms or functional groups are found. If they are on the same side, the molecule is a *cis* isomer; in a *trans* isomer, similar atoms are on opposite sides of the double bond. (*See* isomers.)

clade [Gk. klados: branch] A monophyletic group made up of an ancestor and all of its descendants.

class A group of things defined by particular attributes.

class I MHC proteins Cell surface proteins that participate in the cellular immune response directed against virus-infected cells.

class II MHC proteins Cell surface proteins that participate in the cell–cell interactions (of T-helper cells, macrophages, and B cells) of the humoral immune response.

class switching Occurs when a B cell changes the immunoglobulin class it synthesizes (e.g., a B cell making IgM switches to making IgG).

cleavage The first few cell divisions of an animal zygote. (*See also* complete cleavage, incomplete cleavage.)

climate The long-term average atmospheric conditions (temperature, precipitation, humidity, wind direction and velocity) at a particular place over longer time (years to millennia). (Contrast with weather.)

climax community The final stage of succession; a community that is capable of perpetuating itself under local climatic and soil conditions and persists for a relatively long time.

clinal variation [Gk. klinein: to lean] Gradual change in the phenotype of a species over a geographic gradient.

clonal deletion Inactivation or destruction of lymphocyte clones that would produce immune reactions against the animal's own body.

clonal lineages Asexually reproduced groups of nearly identical organisms.

clonal selection Mechanism by which exposure to antigen results in the activation of selected T- or B-cell clones, resulting in an immune response.

clone [Gk. klon: twig, shoot] (1) Genetically identical cells or organisms produced from a common ancestor by asexual means. (2) To produce many identical copies of a DNA sequence by its introduction into, and subsequent asexual reproduction of, a cell or organism.

closed circulatory system Circulatory system in which the circulating fluid is contained within a continuous system of vessels. (Contrast with open circulatory system.)

clumped dispersion The spatial arrangement of individuals within a population, characterized by clumped spatial patterns. (Compare with random dispersion, regular (uniform) dispersion.)

CO (CONSTANS) Gene coding for a transcription factor that activates the synthesis of florigen (FT); involved in the induction of flowering.

co-repressor In the regulation of bacterial operons, a molecule that binds to the repressor, causing it to change shape and bind to the operator, thereby inhibiting transcription.

coastal zone The marine life zone that extends from the shoreline to the edge of the continental shelf. Characterized by relatively shallow, well-oxygenated water and variable temperatures and salinities.

coccus (kock'us) (plural: cocci) [Gk. kokkos: berry, pit] Any of various spherical or spheroidal bacteria.

cochlea (kock'lee uh) [Gk. kokhlos: snail] A spiral tube in the inner ear of vertebrates; it contains the sensory cells involved in hearing.

codominance A condition in which two alleles at a locus produce different phenotypic effects and both effects appear in heterozygotes.

codon Three nucleotides in messenger RNA that direct the placement of a particular amino acid into a polypeptide chain. (Contrast with anticodon.)

coelomate Possessing a coelom.

coenocytic (seen'a sit ik) [Gk. koinos: common + kytos: container] Referring to the condition, found in some fungal hyphae, of "cells" containing many nuclei but enclosed by a single plasma membrane. Results from nuclear division without cytokinesis.

coevolution Evolutionary process in which an adaptation in one species leads to the evolution of an adaptation in a species with which it interacts.

cohesion The tendency of molecules (or any substances) to stick together.

cohort life table A life table in which the fate of a group of individuals born during the same time frame (a cohort) is followed from birth to death.

cold-hardening A process by which plants can acclimate to cooler temperatures; requires repeated exposure to cool temperatures over many days.

coleoptile A sheath that surrounds and protects the shoot apical meristem and young primary leaves of a grass seedling as they move through the soil.

collagen [Gk. *kolla*: glue] A fibrous protein found extensively in bone and connective tissue.

collecting duct In vertebrates, a tubule that receives urine produced in the nephrons of the kidney and delivers that fluid to the ureter for excretion.

collenchyma (cull eng' kyma) [Gk. *kolla*: glue + *enchyma*: infusion] A type of plant cell, living at functional maturity, which lends flexible support by virtue of primary cell walls thickened at the corners. (Contrast with parenchyma, sclerenchyma.)

colon [Gk. *kolon*] The portion of the gut between the small intestine and the anus. Also called the large intestine.

commensalism [L. *com*: together + *mensa*: table] A type of positive interaction in which one species benefits from the interaction while the other is unaffected. (Contrast with amensalism, competition, mutualism.)

communication A signal from one organism (or cell) that alters the functioning or behavior of another organism (or cell).

community A group of species living together at the same place and time.

community function The way a community works as measured by metrics such as plant productivity or stability.

community structure The set of characteristics or patterns that describe a community, including the number, composition, and abundance of species.

compact bone A type of bone with a solid, hard structure. (Contrast with cancellous bone.)

companion cell In angiosperms, a specialized cell found adjacent to a sieve tube element.

comparative experiment Experimental design in which data from various unmanipulated samples or populations are compared, but in which variables are not controlled or even necessarily identified. (Contrast with controlled experiment.)

comparative genomics Computer-aided comparison of DNA sequences between different organisms to reveal genes with related functions.

comparator In a regulatory system, the mechanism for comparing a feedback variable to a set point to generate command signals to effector organs.

competition A nontrophic interaction in which two or more species overlap in the use of at least some of the same required limiting resources, negatively affecting their growth, reproduction, and/or survival. (Contrast with amensalism, commensalism, mutualism.)

competitive coexistence The ability of species to coexist with one another despite sharing limiting resources. (Contrast with competitive exclusion.)

competitive exclusion A result of competition between species for resources, in which one species completely eliminates the other from a given habitat. (Contrast with competitive coexistence.)

competitive inhibitor A nonsubstrate that binds to the active site of an enzyme and thereby inhibits binding of its substrate. (Contrast with noncompetitive inhibitor.)

complement system A group of eleven proteins that play a role in some reactions of the immune system. The complement proteins are not immunoglobulins.

complementary base pairing The AT (or AU), TA (or UA), CG, and GC pairing of bases in double-stranded DNA, in transcription, and between tRNA and mRNA.

complementary diet A mixture of foods that supplies all essential nutrients.

complementary DNA (cDNA) DNA formed by reverse transcriptase acting with an RNA template; essential intermediate in the reproduction of retroviruses; used as a tool in recombinant DNA technology; lacks introns.

complete cleavage Pattern of cleavage that occurs in eggs that have little yolk. Early cleavage furrows divide the egg completely, and the blastomeres are of similar size. (Contrast with incomplete cleavage.)

complete gut A gut with an entrance and a separate exit.

complete metamorphosis A change of state during the life cycle of an organism in which the body is almost completely rebuilt to produce an individual with a very different body form. Characteristic of insects such as butterflies, moths, beetles, ants, wasps, and flies.

complex ions Groups of covalently bonded atoms that carry an electric charge (e.g., NH_4^+, the ammonium ion).

complex life cycle In reference to parasitic species, a life cycle that requires more than one host to complete.

composite transposon Two transposable elements located near one another that transpose together and carry the intervening DNA sequence with them. (See transposable element.)

compound (1) A substance made up of atoms of more than one element. (2) Made up of many units, as in the **compound eyes** of arthropods.

concentration gradient A difference in concentration of an ion or other chemical substance from one location to another, often across a membrane. (See active transport, facilitated diffusion.)

concerted evolution The common evolution of a family of repeated genes, such that changes in one copy of the gene family are replicated in other copies of the gene family, and thus evolve "in concert." (See biased gene conversion, unequal crossing over.)

condensation reaction A chemical reaction in which two molecules become connected by a covalent bond and a molecule of water is released (AH + BOH → AB + H_2O.) (Contrast with hydrolysis reaction.)

conditional mutation A mutation that results in a characteristic phenotype only under certain environmental conditions.

conditioned reflex A form of associative learning first described by Ivan Pavlov, in which a natural response (such as salivation in response to food) becomes associated with a normally unrelated stimulus (such as the sound of a bell or a metronome).

conduction The transfer of heat from one object to another through direct contact.

cone cells In the vertebrate retina, photoreceptor cells responsible for color vision.

cone In conifers, a reproductive structure consisting of spore-bearing scales extending from a central axis. (Contrast with strobilus.)

confidence interval A numerical interval that is calculated to contain the true value of some parameter of interest at a stated probability level. For example, a **95% confidence interval** contains the true value of some parameter of interest in 95% of cases.

conidium (ko nid' ee um) (plural: conidia) [Gk. *konis*: dust] A type of haploid fungal spore borne at the tips of hyphae, not enclosed in sporangia.

conjugation (kon ju gay' shun) [L. *conjugare*: yoke together] (1) A process by which DNA is passed from one cell to another through a conjugation tube, as in bacteria. (2) A nonreproductive sexual process by which *Paramecium* and other ciliates exchange genetic material.

connective tissue A type of tissue that connects or surrounds other tissues; its cells are embedded in a collagen-containing matrix. One of the four major tissue types in multicellular animals, including cartilage, bone, blood, and fat.

connexon In a gap junction, a protein channel linking adjacent animal cells.

conservation biology An integrative scientific discipline that relies on principles of ecology, economics, social science, and policy to protect and manage Earth's biodiversity.

conspecifics Individuals of the same species.

constant region The portion of an immunoglobulin molecule whose amino acid composition determines its class and does not vary among immunoglobulins in that class. (Contrast with variable region.)

constitutive Always present; produced continually at a constant rate. (Contrast with inducible.)

constitutive proteins Proteins that an organism produces all the time, and at a relatively constant rate.

consumption efficiency The proportion of available biomass that is ingested by consumers.

continental drift The gradual movements of the world's continents that have occurred over billions of years.

continuous variables Variables that can take on a continuous range of values.

contraception Birth control methods that prevent fertilization or implantation (conception).

contractile vacuole (kon trak' tul) A specialized vacuole that collects excess water taken in by osmosis, then contracts to expel the water from the cell.

controlled experiment An experiment in which a sample is divided into groups whereby experimental groups are exposed to manipulations of an independent variable while one group serves as an untreated control. The data from the various groups are then compared to see if there are changes in a dependent variable as a result of the experimental manipulation. (Contrast with comparative experiment.)

controlled system A set of components in a physiological system that is controlled by commands from a regulatory system. (Contrast with regulatory system.)

convection The transfer of heat to or from a surface via a moving stream of air or fluid.

convergent evolution Independent evolution of similar features from different ancestral traits.

convergent extension Movements of cells in the sea urchin blastula that form the archenteron. These cells elongate, flatten, and interdigitate to form a tubular structure.

convolutions Foldings of the vertebrate brain's cerebral cortex into ridges called gyri and valleys called sulci. The level of cortical convolution increases taxonomically and is especially extensive in humans.

coprophagy The consumption of feces.

copulation Reproductive behavior that results in a male depositing sperm in the reproductive tract of a female.

coral bleaching The loss of color in coral colonies, usually because of the loss or decline of symbiotic zooxanthellae.

Coriolis effect The deflection of air or water as a result of differences in Earth's rotational speed at different latitudes.

cork cambium [L. *cambiare*: to exchange] In plants, a lateral meristem that produces secondary growth, mainly in the form of waxy-walled protective cells, including some of the cells that become bark.

cornea The clear, transparent tissue that covers the eye and allows light to pass through to the retina.

coronary artery An artery that serves the cardiac muscles.

coronary thrombosis A blood clot in a blood vessel of the heart.

corpora allata The insect endocrine gland that secretes juvenile hormone.

corpora cardiaca An insect endocrine gland that secretes PTTH.

corpus luteum (kor' pus loo' tee um) (plural: corpora lutea) [L. yellow body] A structure formed from a follicle after ovulation; produces hormones important to the maintenance of pregnancy.

correlation coefficient A measure of the strength of relationship between two quantitative variables, ranging from –1 (a perfect negative relationship) to 1 (a perfect positive relationship).

cortex [L. *cortex*: covering, rind] (1) In plants, the tissue between the epidermis and the vascular tissue of a stem or root. (2) In animals, the outer tissue of certain organs, such as the adrenal gland (adrenal cortex) and the brain (cerebral cortex).

corticosteroids Steroid hormones produced and released by the cortex of the adrenal gland.

corticotropin A tropic hormone produced by the anterior pituitary hormone that stimulates cortisol release from the adrenal cortex. Also called adrenocorticotropic hormone (ACTH).

corticotropin-releasing hormone (CRH) A hormone produced by the hypothalamus that controls the release of cortisol from the anterior pituitary.

cortisol A corticosteroid that mediates stress responses.

cost–benefit approach An approach to evolutionary studies that assumes an animal has a limited amount of time and energy to devote to each of its activities, and that each activity has fitness costs as well as benefits. (See also trade-off.)

cotyledon (kot' ul lee' dun) [Gk. *kotyledon*: hollow space] A "seed leaf." An embryonic organ that stores and digests reserve materials; may expand when seed germinates.

countercurrent flow An arrangement that promotes the maximum exchange of heat, or of a diffusible substance, between two fluids by having the fluids flow in opposite directions through parallel vessels close together.

countercurrent heat exchange In "hot" fishes, an adaptation of the circulatory system such that arterial blood flowing into the muscles is warmed by venous blood flowing out of the muscles, thereby conserving body heat by countercurrent exchange.

countercurrent multiplier The mechanism that increases the concentration of the interstitial fluid in the mammalian kidney through countercurrent flow in the loops of Henle and selective permeability and active transport of ions by segments of the loops of Henle.

coupled human–natural system ecology The practice of encouraging biodiversity and sustainability in systems where humans and nature are intricately linked.

coupled transporters Membrane proteins that transport two substances across a membrane, often with the transport of one driving the transport of another.

covalent bond Chemical bond based on the sharing of electrons between two atoms.

CpG islands DNA regions rich in C residues adjacent to G residues. Especially abundant in promoters, these regions are where methylation of cytosine usually occurs.

cranial nerves The 12 pairs of nerves that go directly from the brain to different parts of the body without going down the spinal cord.

crassulacean acid metabolism (CAM) A metabolic pathway enabling the plants that possess it to store carbon dioxide at night and then perform photosynthesis during the day with stomata closed.

creatine phosphate (CP) A phosphorylated creatine molecule that is abundant in skeletal muscle and can shuttle high energy phosphate from mitochondria to myofibrils.

CRISPR In archaea, clustered regularly interspaced short palindromic sequences in DNA that are adjacent to sequences from invading viruses. When the two sequences are transcribed, the RNA binds a complementary RNA that binds a nuclease, Cas9, that can cleave and inactivate the genome of the invading virus. This mechanism can be adapted for inactivating and mutating any gene.

critical period See sensitive period.

crop A simple food storage sac, the first of two stomachlike organs in many animals (including reptiles, earthworms, and various insects). (See also gizzard.)

crossing over The mechanism by which linked genes undergo recombination. In general, the term refers to the reciprocal exchange of corresponding segments between two homologous chromatids.

crosstalk Interactions between different signal transduction pathways.

crustose A growth form of organisms, such as lichens, in which the organism forms a thin, close, tight bond with the surface of a rock, tree, or other object.

crypsis [Gk. *kryptos*: hidden] The resemblance of an organism to some part of its environment, which helps it to escape detection by enemies.

cryptochromes [Gk. *kryptos*: hidden + *kroma*: color] Photoreceptors mediating some blue-light effects in plants and animals.

ctene (teen) [Gk. *cteis*: comb] In ctenophores, a comblike row of cilia-bearing plates. Ctenophores move by beating the cilia on their eight ctenes.

cumulus A thick gelatinous layer that protects a mammalian ovum.

cupula Gelatinous swelling in the semicircular canals of the vestibular system. A cupula encloses hair cell stereocilia that react to shifting fluid in the canal ducts.

cuticle (1) In plants, a waxy layer on the outer body surface that retards water loss. (2) In ecdysozoans, an outer body covering that provides protection and support and is periodically molted.

cyclic AMP See cAMP.

cyclic electron transport In photosynthetic light reactions, the flow of electrons that produces ATP but not NADPH or O_2.

cyclin A protein that activates a cyclin-dependent kinase, bringing about transitions in the cell cycle.

cyclin-dependent kinase (Cdk) A protein kinase whose target proteins are involved in transitions in the cell cycle and which is active only when complexed with additional protein subunits, called cyclins.

cytokine A regulatory protein made by immune system cells that affects other target cells in the immune system.

cytokinesis (sy' toe kine ee' sis) [Gk. *kytos*: container + *kinein*: to move] The division of the cytoplasm of a dividing cell. (Contrast with mitosis.)

cytokinin (sy' toe kine' in) A member of a class of plant growth substances that plays roles in senescence, cell division, and other phenomena.

cytoplasm The contents of the cell, excluding the nucleus.

cytoplasmic determinants In animal development, gene products whose spatial distribution may determine such things as embryonic axes.

cytoplasmic segregation The asymmetrical distribution of cytoplasmic determinants in a developing animal embryo.

cytoplasmic streaming The flow of cytoplasm around fungal and plant cells.

cytosine (C) (site' oh seen) A nitrogen-containing base found in DNA and RNA.

cytoskeleton The network of microtubules and microfilaments that gives a eukaryotic cell its shape and its capacity to arrange its organelles and to move.

cytotoxic T (T$_C$) cells Cells of the cellular immune system that recognize and directly eliminate virus-infected cells. (Contrast with T-helper cells.)

D

DAG *See* diacylglycerol.

daily torpor A daily reduction in metabolic rate and body temperature to conserve energy.

dalton (DA) A measure of mass for atoms equal to the atomic mass unit.

data Quantified observations about a system under study.

daughter chromosomes During mitosis, the separated chromatids from the beginning of anaphase onward.

dead space The lung volume that fails to be ventilated with fresh air (because the lungs are never completely emptied during exhalation).

declarative memory Memory of people, places, events, and things that can be consciously recalled and described.

decomposer *See* detritivore.

decomposition The physical and chemical breakdown of detritus by bacteria and detritivores, leading to the release of nutrients and energy.

deductive logic Logical thought process that starts with a premise believed to be true then predicts what facts would also have to be true to be compatible with that premise. (Contrast with inductive logic.)

defensin A type of protein made by phagocytes that kills bacteria and enveloped viruses by insertion into their plasma membranes.

deficiency disease A condition (e.g., scurvy and beriberi) caused by chronic lack of any essential nutrient.

delayed hypersensitivity Allergic response in which a T cell clone resulting from binding to an antigen-presenting cell releases of cytokines over a period of hours to days, resulting in such effects as inflammation and rash. (Contrast with immediate hypersensitivity.)

deletion A mutation resulting from the loss of a continuous segment of a gene or chromosome. Such mutations almost never revert to wild type. (Contrast with duplication, point mutation.)

demethylase An enzyme that catalyzes the removal of the methyl group from cytosine, reversing DNA methylation.

demographic stochasticity Fluctuations in population size as the result of random differences among individuals in reproduction and survival.

demography The study of how births, deaths, immigration, and emigration affect population changes over time.

denaturation Loss of activity of an enzyme or nucleic acid molecule as a result of structural changes induced by heat or other means.

dendrites [Gk. *dendron*: tree] Branching fibers (processes) of a neuron. Dendrites are usually relatively short compared with the axon, and commonly carry information to the neuronal cell body.

denitrifiers Bacteria that release nitrogen to the atmosphere as nitrogen gas (N_2).

density-dependent Factors such as limiting resources, predators, or pathogens that have a negative effect on population size as a consequence of the density of individuals. (Contrast with density-independent.)

density-independent Factors such as the physical environment that have a negative effect on population size independent of the density of the population. (Contrast with density-dependent.)

dentine The layer of dense bony material beneath the enamel of the tooth.

deoxyribose A five-carbon sugar found in nucleotides and DNA.

depolarization A change in the resting potential across a membrane so that the inside of the cell becomes less negative, or even positive, compared with the outside of the cell. (Contrast with hyperpolarization.)

deprivation experiments Raising an animal in an environment devoid of behavioral models from which it could learn a species specific behavior to see if the behavior is still expressed.

derived trait A trait that differs from the ancestral trait. (Contrast with synapomorphy.)

dermal tissue system The system of cells that forms the outer covering of a plant, consisting of epidermis in the young plant and periderm in a plant with extensive secondary growth. (Contrast with ground tissue system, vascular tissue system.)

descent with modification Darwin's premise that all species share a common ancestor and have diverged from one another gradually over time.

descriptive statistics Quantitative measures that describe general patterns in data.

desmosome (dez' mo sowm) [Gk. *desmos*: bond + *soma*: body] An adhering junction between animal cells.

desmotubule A membrane extension connecting the endoplasmic reticulum of two plant cells that traverses the plasmodesma.

determinate growth A growth pattern in which the growth of an organism or organ ceases when an adult state is reached; characteristic of most animals and some plant organs. (Contrast with indeterminate growth.)

determination In development, the process whereby the fate of an embryonic cell or group of cells (e.g., to become epidermal cells or neurons) is set (becomes **determined**).

detritivore (di try' ti vore) [L. *detritus*: worn away + *vorare*: to devour] An organism that obtains its energy from the dead bodies or waste products (**detritus**) of other organisms, releasing nutrients and energy. (Contrast with carnivore, herbivore, omnivore.)

development The process by which a multicellular organism, beginning with a single cell, goes through a series of changes, taking on the successive forms that characterize its life cycle.

diacylglycerol (DAG) In hormone action, the second messenger produced by hydrolytic removal of the head group of certain phospholipids.

diaphragm (dye' uh fram) [Gk. *diaphrassein*: barricade] (1) A sheet of muscle that separates the thoracic and abdominal cavities in mammals; responsible for breathing. (2) A method of birth control in which a sheet of rubber is fitted over the woman's cervix, blocking the entry of sperm.

diastole (dye ass' toll ee) [Gk. dilation] The portion of the cardiac cycle when the heart muscle relaxes. (Contrast with systole.)

dichotomous (dye cot' oh mus) [Gk. *dichot*: split in two + *tomia*: removed) A branching pattern in which the shoot divides at the apex producing two equivalent branches that subsequently never overlap.

diencephalon The portion of the vertebrate forebrain that develops into the thalamus and hypothalamus.

differentiation The process whereby originally similar cells follow different developmental pathways; the actual expression of determination.

diffusion Random movement of molecules or other particles, resulting in even distribution of the particles when no barriers are present. (Contrast with active transport, diffusion.)

digestive vacuole In protists, an organelle specialized for digesting food ingested by endocytosis.

dihybrid cross A mating in which the parents differ with respect to the alleles of two loci of interest.

dikaryon (di care' ee ahn) [Gk. *di*: two + *karyon*: kernel] A cell or organism carrying two genetically distinguishable nuclei. Common in fungi.

dioecious (die eesh' us) [Gk. *di*: two + *oikos*: house] Pertaining to organisms in which the two sexes are "housed" in two different individuals, so that eggs and sperm are not produced in the same individuals. Examples: humans, fruit flies, date palms. (Contrast with monoecious.)

diploblastic   Having a body derived from only two embryonic cell layers (ectoderm and endoderm, but no mesoderm) (Contrast with monoblastic, triploblastic.)

diploid   (dip' loid) [Gk. *diplos*: double] Having a chromosome complement consisting of two copies (homologs) of each chromosome. Designated 2n. (Contrast with haploid.)

direct development   Pattern of development (notably among insects) in which hatchlings look like miniature versions of adults. (Contrast with metamorphosis.)

direct fitness   That component of fitness resulting from an organism producing its own offspring. (Contrast with inclusive fitness.)

direct interaction   An interaction that occurs between two species, such as predation, competition, or a positive interaction. (Compare with indirect interaction.)

directional selection   Selection in which phenotypes at one extreme of the population distribution are favored. (Contrast with disruptive selection, stabilizing selection.)

disaccharide   A carbohydrate made up of two monosaccharides (simple sugars).

discoidal cleavage   In animal development, a type of incomplete cleavage that is common in fishes, reptiles, and birds, the eggs of which contain a dense yolk mass.

discrete variables   Quantitative variables that take on only whole number values.

dispersal   Movement of individuals into (immigration) or out of (emigration) an existing population.

dispersion patterns   The spatial arrangement of individuals in space within a population. (*See* regular, random, and clumped patterns.)

disruptive selection   Selection in which phenotypes at both extremes of the population distribution are favored. (Contrast with directional selection, stabilizing selection.)

dissociation constant (K_D)   A type of equilibrium constant measuring the tendency of a two substances that are bound together to separate into the two smaller components. In cell signaling, the two substances are a ligand and receptor.

distal convoluted tubule   The portion of a renal tubule from where it reaches the renal cortex, just past the loop of Henle to where it joins a collecting duct. (Compare with proximal convoluted tubule.)

disturbance   An abiotic event that physically or chemically injures or kills some individuals, creating opportunities for other individuals to grow and/or reproduce. (Contrast with stress.)

disulfide bridge   The covalent bond between two sulfur atoms (–S—S–) linking two molecules or remote parts of the same molecule.

DNA (deoxyribonucleic acid)   The fundamental hereditary material of all living organisms. In eukaryotes, stored primarily in the cell nucleus. A nucleic acid using deoxyribose rather than ribose.

DNA fingerprint   An individual's unique pattern of allele sequences, commonly short tandem repeats and single nucleotide polymorphisms.

DNA helicase   An enzyme that unwinds the double helix.

DNA ligase   Enzyme that unites broken DNA strands during replication and recombination.

DNA methyltransferase   An enzyme that catalyzes the methylation of DNA.

DNA microarray   A small glass or plastic square onto which thousands of single-stranded DNA sequences are fixed so that hybridization of cell-derived RNA or DNA to the target sequences can be performed.

DNA polymerase   Any of a group of enzymes that catalyze the formation of DNA strands from a DNA template.

DNA replication   The creation of a new strand of DNA in which DNA polymerase catalyzes the exact reproduction of an existing (template) strand of DNA.

DNA testing   In human genetics, the determination of genotype by analysis of DNA sequence.

domain   (1) An independent structural element within a protein. Encoded by recognizable nucleotide sequences, a domain often folds separately from the rest of the protein. Similar domains can appear in a variety of different proteins across phylogenetic groups (e.g., "homeobox domain," "calcium-binding domain"). (2) In phylogenetics, the three monophyletic branches of life (Bacteria, Archaea, and Eukarya).

dominance   In genetics, the ability of one allelic form of a gene to determine the phenotype of a heterozygous individual in which the homologous chromosomes carry both it and a different (recessive) allele. (Contrast with recessive.)

dormancy   A condition in which normal activity is suspended, as in some spores, seeds, and buds.

dorsal   [L. *dorsum*: back] Toward or pertaining to the back or upper surface. (Contrast with ventral.)

dorsal lip   In amphibian embryos, the dorsal segment of the blastopore. Also called the organizer, this region directs the development of nearby embryonic regions.

double fertilization   In angiosperms, a process in which the nuclei of two sperm fertilize one egg. One sperm's nucleus combines with the egg nucleus to produce a zygote, while the other combines with the same egg's two polar nuclei to produce the first cell of the triploid endosperm (the tissue that will nourish the growing plant embryo).

double helix   Refers to DNA and the (usually right-handed) coil configuration of two complementary, antiparallel strands.

duodenum   (do' uh dee' num) The beginning portion of the vertebrate small intestine. (*See also* ileum, jejunum.)

duplication   A mutation in which a segment of a chromosome is duplicated, often by the attachment of a segment lost from its homolog. (Contrast with deletion.)

E

early successional species   *See* pioneer species.

easterlies   Prevailing winds that blow from east to west at high latitudes. (Compare with westerlies.)

ecdysone   (eck die' sone) [Gk. *ek*: out of + *dyo*: to clothe] In insects, a hormone that induces molting.

ecology   [Gk. *oikos*: house] The scientific study of the interrelationships of organisms with their living (biotic) and nonliving (abiotic) environments.

ecosystem   (eek' oh sis tum) The organisms of a particular community together with the physical and chemical environment in which they live.

ecosystem engineering species   Species that create, modify, or maintain physical habitat for themselves and other species, and can be keystone or foundation species as well.

ecosystem science   A branch of ecology that considers how energy flows and nutrients cycle through communities.

ecosystem values   Measures of the economic and biophysical values of ecosystem services to humans.

ecosystems services   Benefits that humans obtain from ecosystems such as natural products, protective and regulating services, and recreational, aesthetic, and spiritual enrichment.

ectoderm   [Gk. *ektos*: outside + *derma*: skin] The outermost of the three embryonic germ layers first delineated during gastrulation. Gives rise to the skin, sense organs, and nervous system.

ectomycorrhizae   Mutualistic fungi that cover the roots of plants and assist in the uptake of water and minerals from the soil by the plant.

ectoparasite   A parasite that lives on the surface of its host organism. (Compare with endoparasite.)

ectotherm   [Gk. *ektos*: outside + *thermos*: heat] An animal that is dependent on external heat sources for regulating its body temperature (Contrast with endotherm.)

edge effects   Abiotic and biotic changes to the edge of a habitat brought about by habitat fragmentation.

effective population size   The number of individuals within a population that can contribute offspring to the next generation of the population.

effector   A component of a physiological system that responds to information by *effecting* changes (making change happen) in the internal environment; examples include muscles and the secretory cells of the digestive tract.

effector cells   In cellular immunity, B cells and T cells that attack an antigen, either by secreting antibodies that bind to the antigen or by releasing molecules that destroy any cell bearing the antigen.

effector mechanisms   Adaptations that enable an organism to alter its physiological or behavioral state.

effector protein   In cell signaling, a protein responsible for the cellular response to a signal transduction pathway.

efferent (ef' ur unt) [L. *ex*: out + *ferre*: to bear] Carrying outward or away from, as in neurons that carry impulses outward from the central to the peripheral nervous system (**efferent neurons**), or a blood vessel that carries blood away from a structure (**efferent arterioles**). (Contrast with afferent.)

egg In all sexually reproducing organisms, the female gamete; in birds, reptiles, and some other vertebrates, a structure within which early embryonic development occurs. (*See also* amniote egg, ovum.)

ejaculatory duct The duct that carries semen from the seminal vesicle to the urethra.

electrical synapse A type of synapse at which action potentials spread directly from presynaptic cell to postsynaptic cell. (Contrast with chemical synapse.)

electrocardiogram (ECG or EKG) A graphic recording of electrical potentials from the heart.

electrochemical gradient The concentration gradient of an ion across a membrane plus the voltage difference across that membrane.

electroencephalogram (EEG) A graphic recording of electrical potentials from the brain.

electromagnetic radiation A self-propagating wave that travels though space and has both electrical and magnetic properties.

electron A subatomic particle outside the nucleus carrying a negative charge and very little mass.

electron shell The region surrounding the atomic nucleus at a fixed energy level in which electrons orbit.

electronegativity The tendency of an atom to attract electrons when it occurs as part of a compound.

element A substance that cannot be converted to a simpler substance by ordinary chemical means.

elicitors Molecules associated with plant pathogens that attach to specialized receptors in plants to initiate a defensive response.

elongation (1) In molecular biology, the addition of monomers to make a longer RNA or protein during transcription or translation. (2) Growth of a plant axis or cell primarily in the longitudinal direction.

embolus (em' buh lus) [Gk. *embolos*: stopper] A circulating blood clot. Blockage of a blood vessel by an embolus or a bubble of gas is called an **embolism**. (Contrast with thrombus.)

embryo [Gk. *en*: within + *bryein*: to grow] A young animal, or young plant sporophyte, while it is still contained within a protective structure such as an egg, uterus, or seed.

embryo sac In angiosperms, the female gametophyte. Found within the ovule, it consists of eight or fewer cells, membrane bounded, but without cellulose walls between them.

embryonic stem cell (ESC) A pluripotent cell in the blastocyst.

emigration Dispersal of individuals out of an existing population. (Contrast with immigration.)

enamel The hard material, composed primarily of calcium phosphate, that covers the tooth.

endemic (en dem' ik) [Gk. *endemos*: native] A species that occurs in a particular location and nowhere else on Earth.

endergonic A chemical reaction in which the products have higher free energy than the reactants, thereby requiring free energy input to occur. (Contrast with exergonic.)

endocrine gland (en' doh krin) [Gk. *endo*: within + *krinein*: to separate] An aggregation of secretory cells that secretes hormones into the blood. The endocrine system consists of all endocrine cells and endocrine glands in the body that produce and release hormones. (Contrast with exocrine gland.)

endocrines Chemical signals released by cells that enter the blood and affect cells anywhere in the body that have appropriate receptors. Also known as hormones.

endocytosis A process by which liquids or solid particles are taken up by a cell through invagination of the plasma membrane. (Contrast with exocytosis.)

endoderm [Gk. *endo*: within + *derma*: skin] The innermost of the three embryonic germ layers delineated during gastrulation. Gives rise to the digestive and respiratory tracts and structures associated with them.

endodermis In plants, a specialized cell layer marking the inside of the cortex in roots and some stems. Frequently a barrier to free diffusion of solutes.

endogenous retroviruses Retroviruses that have been incorporated into the host's genome.

endomembrane system A system of intracellular membranes that exchange material with one another, consisting of the Golgi apparatus, endoplasmic reticulum, and lysosomes when present.

endometrium The epithelial lining of the uterus.

endoparasite A parasite that lives inside the body of its host organism. (Compare with ectoparasite.)

endophytic fungi Fungi that live within the aboveground portions of plants without causing obvious harm to the host plant.

endoplasmic reticulum (ER) [Gk. *endo*: within + L. *reticulum*: net] A system of membranous tubes and flattened sacs found in the cytoplasm of eukaryotes. Exists in two forms: rough ER, studded with ribosomes; and smooth ER, lacking ribosomes.

endorphins Molecules in the mammalian brain that act as neurotransmitters in pathways that control pain.

endoskeleton [Gk. *endo*: within + *skleros*: hard] An internal skeleton covered by other, soft body tissues. (Contrast with exoskeleton.)

endosperm [Gk. *endo*: within + *sperma*: seed] A specialized triploid seed tissue found only in angiosperms; contains stored nutrients for the developing embryo.

endospore [Gk. *endo*: within + *spora*: to sow] In some bacteria, a resting structure that can survive harsh environmental conditions.

endosymbiosis theory [Gk. *endo*: within + *sym*: together + *bios*: life] The theory that the eukaryotic cell evolved via the engulfing of one prokaryotic cell by another.

endotherm [Gk. *endo*: within + *thermos*: heat] An animal that can control its body temperature by the expenditure of its own metabolic energy. (Contrast with ectotherm.)

endotoxin A lipopolysaccharide that forms part of the outer membrane of certain Gram-negative bacteria that is released when the bacteria grow or lyse. (Contrast with exotoxin.)

energetic cost The difference between the energy an animal expends in performing a behavior and the energy it would have expended had it rested. (Compare with opportunity cost, risk cost.)

energy budget A quantitative description of all paths of energy exchange between an animal and its environment.

energy The capacity to do work or move matter against an opposing force. The capacity to accomplish change in physical and chemical systems.

enhancers Regulatory DNA sequences that bind transcription factors that either activate or increase the rate of transcription.

enkephalins Molecules in the mammalian brain that act as neurotransmitters in pathways that control pain.

enteric nervous system The nerve nets in the submucosa and between the smooth muscle layers of the vertebrate gut.

enthalpy (*H*) The total energy of a system.

entrain To advance or delay an organism's circadian clock each day so that it is in phase with the light-dark cycle of the organism's environment.

entropy (*S*) (en' tro pee) [Gk. *tropein*: to change] A measure of the degree of disorder in any system. Spontaneous reactions in a closed system are always accompanied by an increase in entropy.

enveloped virus A virus enclosed within a phospholipid membrane derived from its host cell.

environmental genomics Sequencing technique used when biologists are unable to work with the whole genome of a prokaryote species but instead examine individual genes collected from a random sample of the organism's environment.

environmentalism The use of ecological knowledge, along with economics, ethics, and many other considerations, to inform both personal decisions and public policy relating to stewardship of natural resources and ecosystems.

enzyme–substrate complex (ES) An intermediate in an enzyme-catalyzed reaction; consists of the enzyme bound to its substrate(s).

epiblast The upper or overlying portion of the avian blastula which is joined to the hypoblast at the margins of the blastodisc.

epiboly The movement of cells over the surface of the blastula toward the forming blastopore.

epidermis [Gk. *epi*: over + *derma*: skin] In plants and animals, the outermost cell layers. (Only one cell layer thick in plants.)

epididymis (epuh did' uh mus) [Gk. *epi*: over + *didymos*: testicle] Coiled tubules in the testes that store sperm and conduct sperm from the seminiferous tubules to the vas deferens.

epigenetics The scientific study of changes in the expression of a gene or set of genes that occur without change in the DNA sequence.

epiglottis The flap of tissue that covers the entrance of the larynx to prevent food and liquid from entering the trachea.

epinephrine (ep i nef' rin) [Gk. *epi*: over + *nephros*: kidney] The "fight or flight" hormone produced by the medulla of the adrenal gland; it also functions as a neurotransmitter. Also known as adrenaline.

epiphyseal plates Plates of cartilage near the ends of long bones where bone growth occurs up to adolescence when growth ceases.

epistasis Interaction between genes in which the presence of a particular allele of one gene determines whether another gene will be expressed.

epithelial tissue A type of animal tissue made up of sheets of cells that lines or covers organs, makes up tubules, and covers the surface of the body; one of the four major tissue types in multicellular animals.

epitope *See* antigenic determinant.

epoch A subdivision of a period in a geological time scale.

equilibrium Any state of balanced opposing forces and no net change.

ER *See* endoplasmic reticulum.

erection The state of the penis in which it is stiffened by engorgement with blood.

error signal In regulatory systems, any difference between the set point of the system and its current condition.

ERV *See* expiratory reserve volume.

erythrocyte (ur rith' row site) [Gk. *erythros*: red + *kytos*: container] A red blood cell.

erythropoietin (EPO) (ur rith' row poi' i tn) A hormone produced by the kidney in response to lack of oxygen that stimulates the production of red blood cells.

ESC *See* embryonic stem cell.

esophagus (i soff' i gus) [Gk. *oisophagos*: gullet] That part of the gut between the pharynx and the stomach.

essential amino acids Amino acids that an animal cannot synthesize for itself and must obtain from its food.

essential element A mineral nutrient required for normal growth and reproduction in plants and animals.

essential fatty acids Fatty acids that an animal cannot synthesize for itself and must obtain from its food.

ester linkage A condensation (water-releasing) reaction in which the carboxyl group of a fatty acid reacts with the hydroxyl group of an alcohol. Lipids, including most membrane lipids, are formed in this way. (Contrast with ether linkage.)

estivation (ess tuh vay' shun) [L. *aestivalis*: summer] A state of dormancy and hypometabolism that occurs during the summer;

usually a means of surviving drought and/or intense heat. (Contrast with hibernation.)

estrogen Any of several steroid sex hormones; produced chiefly by the ovaries in mammals.

estrus (es' trus) [L. *oestrus*: frenzy] The period of heat, or maximum sexual receptivity, in some female mammals. Ordinarily, estrus is also the time of release of eggs in the female.

estuary An aquatic habitat in which salt water and fresh water mix, as when a river meets the ocean. Includes such ecosystems as salt marshes, mangrove forests, mudflats, and seagrass beds.

ether linkage The linkage of two hydrocarbons by an oxygen atom (HC—O—CH). Ether linkages are characteristic of the membrane lipids of the Archaea. (Contrast with ester linkage.)

ethology [Gk. *ethos*: character + *logos*: study] An approach to the study of animal behavior that focuses on studying many species in natural environments and addresses questions about the evolution of behavior. (Compare with behaviorism.)

ethylene One of the plant growth hormones, the gas $H_2C{=}CH_2$. Involved in fruit ripening and other growth and developmental responses.

etiolated Process in flowering plants grown in the absence of light; characterized by long weak stems, smaller leaves due to longer internodes, and a pale yellow color.

euchromatin Diffuse, uncondensed chromatin. Contains active genes that will be transcribed into mRNA. (Contrast with heterochromatin.)

eudicots Angiosperms with two embryonic cotyledons. (*See also* monocots.)

eukaryotes (yew car' ree oats) [Gk. *eu*: true + *karyon*: kernel or nucleus] Organisms whose cells contain their genetic material inside a nucleus. Includes all life other than the viruses, prokaryotic archaea, and bacteria. (Contrast with prokaryotes.)

eusocial Pertaining to a social group that includes nonreproductive individuals, as in honey bees.

eustachian tube A connection between the middle ear and the throat that allows air pressure to equilibrate between the middle ear and the outside world. Also called the pharyngotympanic tube.

eutrophication (yoo trofe' ik ay' shun) [Gk. *eu*: truly + *trephein*: to flourish] A process in aquatic ecosystems initiated by a pulse of nutrients that results in an increase of algal growth that may lead to decomposition and depleted oxygen conditions (hypoxia).

evaporation The transition of water from the liquid to the gaseous phase.

evapotranspiration The evaporative transfer of heat and water from the surfaces of plants into the atmosphere, which reduces air temperature and increases moisture.

evolution Any gradual change. Most often refers to organic or Darwinian evolution, which is the genetic and resulting phenotypic change in populations of organisms from generation to generation. (*See* macroevolution, microevolution. Contrast with speciation.)

evolutionary developmental biology (evo-devo) The study of the interplay between

evolutionary and developmental processes, with a focus on the genetic changes that give rise to novel morphology. Key concepts of evo-devo include modularity, genetic toolkits, genetic switches, and heterochrony.

evolutionary radiation The proliferation of many species within a single evolutionary lineage.

evolutionary reversal The reappearance of an ancestral trait in a group that had previously acquired a derived trait.

evolutionary theory The understanding and application of the mechanisms of evolutionary change to biological problems.

ex vivo gene therapy Alteration of the human genome sequence in the cells removed from the body and then reimplanted into the body for medical benefit.

excision repair DNA repair mechanism that removes damaged DNA and replaces it with the appropriate nucleotide.

excretory systems In animals, organs that maintain the volume, solute concentration, and composition of the extracellular fluid by excreting water, solutes, and nitrogenous wastes in the form of urine.

executive functions Cognitive processes (e.g., attentional control, inhibitory control, reasoning, problem solving, planning) that allow goal-directed control of thought and behavior.

exergonic A chemical reaction in which the products of the reaction have lower free energy than the reactants, resulting in a release of free energy. (Contrast with endergonic.)

exocrine gland (eks' oh krin) [Gk. *exo*: outside + *krinein*: to separate] Any gland, such as a salivary gland, that secretes to the outside of the body or into the gut. (Contrast with endocrine gland.)

exocytosis A process by which a vesicle within a cell fuses with the plasma membrane and releases its contents to the outside. (Contrast with endocytosis.)

exon A portion of a DNA molecule, in eukaryotes, that codes for part of a polypeptide. (Contrast with intron.)

exoskeleton (eks' oh skel' e ton) [Gk. *exos*: outside + *skleros*: hard] A hard covering on the outside of the body to which muscles are attached. (Contrast with endoskeleton.)

exotoxin A highly toxic, usually soluble protein released by living, multiplying bacteria. (Contrast with endotoxin.)

expanding triplet repeat A three-base-pair sequence in a human gene that is unstable and can be repeated a few to hundreds of times. Often, the more the repeats, the less the activity of the gene involved. Expanding triplet repeats occur in some human diseases such as Huntington's disease and fragile-X syndrome.

expiratory reserve volume (ERV) The amount of air that can be forcefully exhaled beyond the normal tidal expiration. (Contrast with inspiratory reserve volume, tidal volume, vital capacity.)

exploitation competition Competition in which both species reduce the quantities of their shared but limited resources. (Contrast with interference competition.)

exponential growth   A pattern of population growth characterized by a multiplicative pattern in which, as the number of individuals increases, the number of new individuals added per unit time accelerates. (Contrast with logistic growth.)

expression vector   A DNA vector, such as a plasmid, that carries a DNA sequence for the expression of an inserted gene into mRNA and protein in a host cell.

expressivity   The degree to which a genotype is expressed in the phenotype; may be affected by the environment.

extensor   A muscle that extends an appendage. (Contrast with flexor.)

external fertilization   The release of gametes into the environment; typical of aquatic animals. Also called spawning. (Contrast with internal fertilization.)

external gills   Highly branched and folded extensions of the body surface that provide a large surface area for gas exchange with water; typical of larval amphibians and many larval insects.

extracellular fluid   In closed circulatory systems refers to the fluid in the circulatory system and the fluid outside it.

extracellular matrix   A material of heterogeneous composition surrounding cells and performing many functions including adhesion of cells.

extraembryonic membranes   Four membranes that support but are not part of the developing embryos of reptiles, birds, and mammals, defining these groups phylogenetically as amniotes. (See amnion, allantois, chorion, yolk sac.)

extreme halophiles   A group of euryarchaeotes that live exclusively in very salty environments.

extremophiles   Archaea and bacteria that live and thrive under conditions (e.g., extremely high temperatures) that would kill most organisms.

eye cups   Photosensory organs in flatworms; components of one of the simplest visual systems in animals.

F

F$_1$ The first filial generation; the immediate progeny of a parental (P) mating.

F$_2$ The second filial generation; the immediate progeny of a mating between members of the F$_1$ generation.

facilitated diffusion   Passive movement through a membrane involving a specific carrier protein; does not proceed against a concentration gradient. (Contrast with active transport, diffusion.)

facilitation   A positive interaction. See positive interaction.

facultative   A type of species interaction that is optional to at least one species involved in the interaction. (Compare with obligate.)

facultative anaerobe   A prokaryote that can shift its metabolism between anaerobic and aerobic modes depending on the presence or absence of O$_2$. Alternatively, facultative aerobe. (Compare with obligate anaerobe.)

facultative parasite   An organism that can parasitize other living organisms but is also capable of growing independently.

fast-twitch fibers   Skeletal muscle fibers that can generate high tension rapidly, but fatigue rapidly ("sprinter" fibers). Characterized by an abundance of enzymes of glycolysis. (Compare to slow-twitch fibers.)

fat   (1) A triglyceride that is solid at room temperature. (Contrast with oil.) (2) Adipose tissue, one type of connective tissue. (See brown fat.)

fate map   A diagram of the blastula showing which cells (blastomeres) are "fated" to contribute to specific tissues and organs in the mature body.

fatty acid   A molecule made up of a long nonpolar hydrocarbon chain and a polar carboxyl group. Found in many lipids.

fauna   (faw' nah) All the animals found in a given area. (Contrast with flora.)

FD (FLOWERING LOCUS D)   Gene coding for a transcription factor in the shoot apical meristem that binds to florigen; involved in the induction of flowering.

feather   A structure growing from the skin of theropods (living birds and their extinct relatives) that consists of a hollow shaft fringed with vanes of barbs. Functions to support flight, plumage display, and thermoregulation.

feces   [L. *faeces*: dregs] Waste excreted from the digestive system.

fecundity (m$_x$)   The average number of offspring produced by each individual (or female) within the population.

feedback   In regulatory systems, information about the relationship between the set point of the system and its current state. (Contrast with feedforward information.)

feedforward information   In regulatory systems, information that changes the set point of the system. (Contrast with feedback.)

fenestrations   Small holes in the walls of some capillaries.

fermentation   (fur men tay' shun) [L. *fermentum*: yeast] The anaerobic degradation of a substance such as glucose to smaller molecules such as lactic acid or alcohol with the extraction of energy.

fertilization   Union of gametes. Also known as syngamy.

fertilizer   Any of a number of substances added to soil to improve the soil's capacity to support plant growth. May be organic or inorganic.

fiber   In angiosperms, an elongated, tapering sclerenchyma cell, usually with a thick cell wall, that serves a support function in xylem. (See also muscle fiber.)

fibrin   A protein that polymerizes to form long threads that provide structure to a blood clot.

fibrinogen   A circulating protein that can be stimulated to fall out of solution and provide the structure for a blood clot.

fibrous root system   A root system typical of monocots composed of numerous thin adventitious roots that are all roughly equal in diameter. (Contrast with taproot system.)

Fick's law of diffusion   An equation that describes the factors that determine the rate of diffusion of a molecule from an area of higher concentration to an area of lower concentration.

fight-or-flight response   A rapid physiological response to a sudden threat mediated by the hormone epinephrine.

filament   In flowers, the part of a stamen that supports the anther.

filter feeder   An organism that feeds on organisms much smaller than itself that are suspended in water or air by means of a straining device. Also called suspension feeder.

first filial generation   See F$_1$.

first law of thermodynamics   The principle that energy can be neither created nor destroyed.

first polar body   The small daughter cell from the first meiotic division of a primary oocyte.

fitness   The contribution of a genotype or phenotype to the genetic composition of subsequent generations, relative to the contribution of other genotypes or phenotypes. (See also inclusive fitness.)

fixed action pattern   In ethology, a genetically determined behavior that is performed without learning, stereotypic (performed the same way each time), and not modifiable by learning.

fixed   In population genetics, the change in a gene pool where there are at least two alleles of a gene to a situation where only one allele remains.

flagellum   (fla jell' um) (plural: flagella) [L. *flagellum*: whip] Long, whiplike appendage that propels cells. Prokaryotic flagella differ sharply from those found in eukaryotes.

flexor   A muscle that flexes an appendage. (Contrast with extensor.)

flora   (flore' ah) All of the plants found in a given area. (Contrast with fauna.)

floral meristem   In angiosperms, a meristem that forms the floral organs (sepals, petals, stamens, and carpels).

floral organ identity genes   In angiosperms, genes that determine the fates of floral meristem cells; their expression is triggered by the products of meristem identity genes.

florigen   A plant hormone involved in the conversion of a vegetative shoot apex to a flower.

flower   The sexual structure of an angiosperm.

fluid feeder   An animal that feeds on fluids it extracts from the bodies of other organisms; examples include nectar-feeding birds and blood-sucking insects.

fluid mosaic model   A molecular model for the structure of biological membranes consisting of a fluid phospholipid bilayer in which suspended proteins are free to move in the plane of the bilayer.

foliose   Having a leafy growth form.

follicle   [L. *folliculus*: little bag] In female mammals, an immature egg surrounded by nutritive cells.

follicle-stimulating hormone (FSH)   A gonadotropin produced by the anterior pituitary.

food web A representation of the trophic or energetic connections among species within a community.

forebrain The region of the vertebrate brain that comprises the cerebrum, thalamus, and hypothalamus. (Compare with hindbrain, midbrain.)

fossil fuels Fuels, including oil, natural gas, coal, and peat, formed over geologic time from organic material buried in anaerobic sediments.

foundation species A species that has large, community-wide effects as a consequence of its large size and abundance, and its provision of habitat or food for other species.

founder effect Random changes in allele frequencies resulting from establishment of a population by a very small number of individuals.

fovea [L. *fovea*: a small pit] In the vertebrate retina, the area of most distinct vision.

frame-shift mutation The addition or deletion of a single or two adjacent nucleotides in a gene's sequence. Results in the misreading of mRNA during translation and the production of a nonfunctional protein. (Contrast with missense mutation, nonsense mutation, silent mutation.)

Frank–Starling law The principle that the stroke volume of the heart increases with increased return of blood to the heart.

free energy (*G*) Energy that is available for doing useful work, after allowance has been made for the increase or decrease of disorder.

free-running A descriptor of an endogenous rhythm that is not entrained by an environmental cue.

frequency distribution A figure that displays the frequency of different classes of data.

frequency-dependent selection Selection that changes in intensity with the proportion of individuals in a population having the trait.

frontal lobe The largest of the brain lobes in humans; involved with feeling and planning functions; includes the primary motor cortex.

frugivore [L. *frugis*: fruit + *vorare*: to devour] An animal that eats fruit.

fruit In angiosperms, a ripened and mature ovary (or group of ovaries) containing the seeds. Sometimes applied to reproductive structures of other groups of plants.

fruticose Having a shrubby growth form.

FSH *See* follicle-stimulating hormone.

FT (*FLOWERING LOCUS T*) Gene that codes for florigen, a small, diffusible protein involved in the induction of flowering.

fugitive species A species that leave an otherwise suitable habitat in order to avoid competition with another species.

full census A count of every individual in a population.

functional genomics The assignment of functional roles to the proteins encoded by genes identified by sequencing entire genomes.

functional group (1) A characteristic combination of atoms that contributes specific properties (such as charge or polarity) when attached to larger molecules (e.g., carboxyl group,

amino group). (2) A group of species that function in similar ways, whether or not they use the same resources. (Compare with guild.)

functional residual volume (FRV) The residual volume plus the expiratory reserve volume.

fundamental niche A species' niche as defined by its physiological capabilities. (Contrast with realized niche.)

G

G protein A membrane protein involved in signal transduction; characterized by binding GDP or GTP.

G protein–coupled receptors A class of receptors that change configuration upon ligand binding such that a G protein binding site is exposed on the cytoplasmic domain of the receptor, initiating a signal transduction pathway.

G0 A resting phase in which a cell is not preparing for the cell division cycle.

G1 In the cell cycle, the gap between the end of mitosis and the onset of the S phase.

G1-to-S transition In the cell cycle, the point at which G1 ends and the S phase begins.

G2 In the cell cycle, the gap between the S (synthesis) phase and the onset of mitosis.

gain-of-function mutation A mutation that results in a protein with a new function. (Contrast with loss-of-function mutation.)

gallbladder In the human digestive system, an organ in which bile is stored.

gametangium (gam uh tan' gee um) (plural: gametangia) [Gk. *gamos*: marriage + *angeion*: vessel] Any plant or fungal structure within which a gamete is formed.

gamete (gam' eet) [Gk. *gamete/gametes*: wife, husband] The mature sexual reproductive cell: the egg or the sperm.

gametogenesis (ga meet' oh jen' e sis) The specialized series of cellular divisions that leads to the production of gametes. Called oogenesis when referring to processes in the ovary and spermatogenesis in the testis.

gametophyte (ga meet' oh fyte) In plants and photosynthetic protists with alternation of generations, the multicellular haploid phase that produces the gametes. (Contrast with sporophyte.)

gamma diversity Species diversity within a geographic region; the regional species pool. (Compare with alpha diversity, beta diversity.)

ganglion (gang' glee un) (plural: ganglia) [Gk. lump] A cluster of neurons that have similar characteristics or function.

ganglion cells Cells at the front of the human retina that transmit information from the bipolar cells to the brain.

gap genes In *Drosophila* development, segmentation genes that define broad areas along the anterior–posterior axis of the early embryo. Part of a developmental cascade that includes maternal effect genes, pair rule genes, segment polarity genes, and Hox genes.

gap junction A 2.7-nanometer gap between cell membranes of two animal cells, spanned by

protein channels. Gap junctions allow chemical substances or electrical signals to pass from cell to cell.

gastric pits Deep infoldings in the walls of the stomach lined with secretory cells.

gastrin A hormone secreted by cells in the lower region of the stomach that stimulates the secretion of digestive juices as well as movements of the stomach.

gastrointestinal system *See* gut.

gastrovascular cavity Serving for both digestion (gastro) and circulation (vascular); in particular, the central cavity of the body of jellyfish and other cnidarians.

gastrovascular system A branching body cavity in Cnidaria and Platyhelminthes that has one opening to the environment and serves to digest food and distribute nutrients to the body.

gastrulation Development of a blastula into a gastrula. In embryonic development, the process by which a blastula is transformed by massive movements of cells into a gastrula, an embryo with three germ layers and distinct body axes.

gated channel A membrane protein that changes its three-dimensional shape, and therefore its ion conductance, in response to a stimulus. When open, it allows specific ions to move across the membrane.

gel electrophoresis (e lek' tro fo ree' sis) [L. *electrum*: amber + Gk. *phorein*: to bear] A technique for separating molecules (such as DNA fragments) from one another on the basis of their electric charges and molecular weights by applying an electric field to a gel.

gene [Gk. *genes*: to produce] A unit of heredity. Used here as the unit of genetic function which carries the information for a polypeptide or RNA.

gene duplication The generation of extra copies of a gene in a genome over evolutionary time. A mechanism by which genomes can acquire new functions.

gene expression The transcription and translation into a protein of the information (nucleotide sequence) contained in a gene.

gene family A set of similar genes derived from a single parent gene; need not be on the same chromosomes. The vertebrate globin genes constitute a classic example of a gene family.

gene pool All of the different alleles of all of the genes existing in all individuals of a population.

gene therapy Treatment of a genetic disease by providing patients with cells containing functioning alleles of the genes that are nonfunctional in their bodies.

gene tree A graphic representation of the evolutionary relationships of a single gene in different species or of the members of a gene family.

gene-for-gene concept In plants, a mechanism of resistance to pathogens in which resistance is triggered by the specific interaction of the products of a pathogen's *Avr* genes and a plant's *R* genes.

general transcription factors In eukaryotes, transcription factors that bind to the promoters of most protein-coding genes and are required for their expression. Distinct from transcription factors that have specific regulatory effects only at certain promoters or classes of promoters.

generation time (G)   The average age of the parents of all the offspring produced within a cohort.

genetic code   The set of instructions, in the form of nucleotide triplets, that translate a linear sequence of nucleotides in mRNA into a linear sequence of amino acids in a protein.

genetic drift   Changes in gene frequencies from generation to generation as a result of random (chance) processes.

genetic marker   (1) In gene cloning, a gene of identifiable phenotype that indicates the presence of another gene, DNA segment, or chromosome fragment. (2) In general, a DNA sequence such as a single nucleotide polymorphism whose presence is correlated with the presence of other linked genes on that chromosome.

genetic screen   A technique for identifying genes involved in a biological process of interest. Involves creating a large collection of randomly mutated organisms and identifying those individuals that are likely to have a defect in the pathway of interest. The mutated gene(s) in those individuals can then be isolated for further study.

genetic structure   The frequencies of different alleles at each locus and the frequencies of different genotypes in a Mendelian population.

genetic switches   Mechanisms that control how the genetic toolkit is used, such as promoters and the transcription factors that bind them. The signal cascades that converge on and operate these switches determine when and where genes will be turned on and off.

genetic toolkit   A set of developmental genes and proteins that is common to most animals and is hypothesized to be responsible for the evolution of their differing developmental pathways.

genetics   The scientific study of the structure, functioning, and inheritance of genes, the units of hereditary information.

genitalia   The reproductive organs, especially the external accessory sex organs.

genome   (jee′ nome) The complete DNA sequence for a particular organism or individual.

genome sequencing   Determination of the nucleotide base sequence of the entire genome of an organism.

genomic equivalence   All somatic cells in a complex organism have the entire genome of that organism.

genomic library   All of the cloned DNA fragments generated by the breakdown of genomic DNA into smaller segments.

genomics   The scientific study of entire sets of genes and their interactions.

genotype   (jean′ oh type) [Gk. *gen*: to produce + *typos*: impression] An exact description of the genetic constitution of an individual, either with respect to a single trait or with respect to a larger set of traits. (Contrast with phenotype.)

genus   (jean′ us) (plural: genera) [Gk. *genos*: stock, kind] A group of related, similar species recognized by taxonomists with a distinct name used in binomial nomenclature.

geographic range   The entire region within which a species occurs.

geological time scale   The division of Earth's history into ordered, named periods of time, based on major biotic and abiotic changes.

germ cell   [L. *germen*: to beget] A reproductive cell or gamete of a multicellular organism. (Contrast with somatic cell.)

germ layers   The three embryonic layers formed during gastrulation (ectoderm, mesoderm, and endoderm). Also called cell layers or tissue layers.

germ line gene therapy   Alteration of the human genome sequence in the gametes or cells giving rise to gametes for medical benefit. (Contrast with somatic cell gene therapy.)

germ line mutation   Mutation in a cell that produces gametes (i.e., a germ line cell). (Contrast with somatic mutation.)

germination   Sprouting of a seed or spore.

gestation   (jes tay′ shun) [L. *gestare*: to bear] The period during which the embryo of a mammal develops within the uterus. Also known as pregnancy.

GH   *See* growth hormone.

GHIH   *See* growth hormone inhibiting hormone.

ghrelin   A hormone produced and secreted by cells in the stomach that stimulates appetite.

gibberellin   (jib er el′ lin) A class of plant growth hormones playing roles in stem elongation, seed germination, flowering of certain plants, etc.

gizzard   (giz′ erd) [L. *gigeria*: cooked chicken parts] The second of two stomachlike organs in birds, other reptiles, earthworms, and various insects, that grinds up food, sometimes with the aid of fragments of stone. (*See also* crop.)

glans penis   The highly sensitive tip of the mammalian penis.

glia   (glee′ uh) [Gk. *glia*: glue] One of the two classes of neural cells (along with neurons, with which glia interact); glia do not typically conduct action potentials. Types of glia include astrocytes, oligodendrocytes, and Schwann cells.

glomerular filtration rate (GFR)   The rate at which the blood is filtered in the glomeruli of the kidney.

glomerulus   (glo mare′ yew lus) (plural: glomeruli) [L. *glomus*: ball] Sites in the kidney where blood filtration takes place. Each glomerulus consists of a knot of capillaries served by afferent and efferent arterioles.

glucagon   Hormone produced by alpha cells of the pancreatic islets of Langerhans. Glucagon stimulates the liver to break down glycogen and release glucose into the circulation.

gluconeogenesis   The biochemical synthesis of glucose from other substances, such as amino acids, lactate, and glycerol.

glucose   [Gk. *gleukos*: sugar, sweet wine] The most common monosaccharide; the monomer of the polysaccharides starch, glycogen, and cellulose.

glyceraldehyde 3-phosphate (G3P)   A phosphorylated three-carbon sugar; an intermediate in glycolysis and photosynthetic carbon fixation.

glycerol   (gliss′ er ole) A three-carbon alcohol with three hydroxyl groups; a component of phospholipids and triglycerides.

glycogen   (gly′ ko jen) [Gk. *glyk*: sweet] An energy storage polysaccharide found in animals and fungi; a branched-chain polymer of glucose, similar to starch.

glycolipid   A lipid to which sugars are attached.

glycolysis   (gly kol′ li sis) [Gk. *gleukos*: sugar + *lysis*: break apart] The enzymatic breakdown of glucose to pyruvic acid.

glycoprotein   A protein to which sugars are attached.

glycosidic linkage   Bond between carbohydrate (sugar) molecules through an intervening oxygen atom (–O–).

glycosylation   The addition of carbohydrates to another type of molecule, such as a protein.

glyoxysome   (gly ox′ ee soam) An organelle found in plants, in which stored lipids are converted to carbohydrates.

goiter   Swelling in the neck resulting from an enlarged thyroid gland.

Goldman equation   An equation that calculates membrane potential by taking into consideration the concentration differences of all ions on both sides of the membrane and the relative permeability of the membrane to those ions.

Golgi apparatus   (goal′ jee) A system of concentrically folded membranes found in the cytoplasm of eukaryotic cells; functions in secretion from the cell by exocytosis.

Golgi tendon organ   A mechanoreceptor found in tendons and ligaments; provides information about the force generated by a contracting muscle.

gonad   (go′ nad) [Gk. *gone*: seed] An organ that produces gametes in animals: either an ovary (female gonad) or testis (male gonad).

gonadotropin   A trophic hormone that stimulates the gonads.

gonadotropin-releasing hormone (GnRH)   Hormone produced by the hypothalamus that stimulates the anterior pituitary to secrete gonadotropins.

Gondwana   The large southern land mass that existed from the Cambrian (540 mya) to the Jurassic (138 mya). Present-day remnants are South America, Africa, India, Australia, and Antarctica.

graded membrane potential   Small local change in membrane potential caused by opening or closing of ion channels.

grafting   Artificial transplantation of tissue from one organism to another. In horticulture, the transfer of a bud or stem segment from one plant onto the root of another as a form of asexual reproduction.

Gram-negative bacteria   Bacteria that appear red when stained using the Gram-staining technique. These bacteria have an outer membrane outside the relatively thin peptidoglycan layer of the cell wall.

Gram-positive bacteria   Bacteria that appear blue to purple when stained using the

Gram-staining technique. These bacteria have an outer cell wall consisting of a thick layer of peptidoglycan.

Gram stain A differential purple stain useful in characterizing bacteria. The peptidoglycan-rich cell walls of gram-positive bacteria stain purple; cell walls of gram-negative bacteria generally stain orange.

granulosa cells Cells of the ovarian follicle of mammals that surround and nurture the oocyte.

gravitropism [L. *gravitas*: weight, force + Gk. *tropos*: to turn] A directed plant growth response to gravity.

gray crescent In frog development, a band of diffusely pigmented cytoplasm on the side of the egg opposite the site of sperm entry. Arises as a result of cytoplasmic rearrangements that establish the anterior–posterior axis of the zygote.

gray matter In the nervous system, tissue that is rich in neuronal cell bodies. (Contrast with white matter.)

greenhouse gases Gases in the atmosphere, such as carbon dioxide and methane, that are transparent to sunlight, but trap heat radiating from Earth's surface, causing heat to build up at Earth's surface.

gross primary production (GPP) The total amount of carbon fixed by primary producers in an ecosystem. (Compare with net primary production.)

ground meristem That part of an apical meristem that gives rise to the ground tissue system of the primary plant body.

ground tissue system Those parts of the plant body not included in the dermal or vascular tissue systems. Ground tissues function in storage, photosynthesis, and support.

growth An increase in the size of the body and its organs by cell division and cell expansion.

growth factor A chemical signal that stimulates cells to divide.

growth hormone (GH) A peptide hormone released by the anterior pituitary that stimulates many anabolic processes.

growth hormone inhibiting hormone (GHIH) Also called somatostatin, a peptide hormone produced in the hypothalamus that inhibits the activity of growth hormone secreting cells in the anterior pituitary.

guanine (G) (gwan' een) A nitrogen-containing base found in DNA, RNA, and GTP.

guard cells In plants, specialized, paired epidermal cells that surround and control the opening of a stoma (pore).

guild A group of species, often taxonomically distinct, that use the same resources. (Compare with functional group.)

gustation The sense of taste.

gut An animal's digestive tract.

gymnosperms Seed plants that do not produce flowers or fruits; one of the two major groups of living seed plants. (*See also* angiosperms.)

gyres Large-scale circular ocean currents caused by prevailing winds and Earth's rotation.

gyri (singular: gyrus) The ridges (folds) in the cerebral cortex.

H

habitat The particular environment in which an organism lives.

habitat degradation The conversion of an ecosystem by human activities that reduces habitat quality. (Compare with habitat loss.)

habitat loss The conversion of an ecosystem by human activities that reduces habitat quantity. (Compare with habitat degradation.)

hair cell A type of mechanoreceptor in animals. Detects sound waves and other forms of motion in air or water.

half-life The time required for half of a sample of a radioactive isotope to decay to its stable, nonradioactive form, or for a drug or other substance to reach half its initial dosage.

halophytes Organisms that can grow in high salinity.

Hamilton's rule The principle that, for an apparent altruistic behavior to be adaptive, the fitness benefit of that act to the recipient times the degree of relatedness of the performer and the recipient must be greater than the cost to the performer.

haplodiploidy A sex determination mechanism in which diploid individuals (which develop from fertilized eggs) are female and haploid individuals (which develop from unfertilized eggs) are male; typical of hymenopterans.

haploid (hap' loid) [Gk. *haploeides*: single] Having a chromosome complement consisting of just one copy of each chromosome; designated 1*n* or *n*. (Contrast with diploid.)

haplotype Linked nucleotide sequences that are usually inherited as a unit (as a "sentence" rather than as individual "words").

Hardy–Weinberg equilibrium In a sexually reproducing population, the allele frequency at a given locus that is not being acted on by agents of evolution; the conditions that would result in no evolution in a population.

haustorium (haw stor' ee um) (plural: haustoria) [L. haustus: draw up] A specialized hypha or other structure by which fungi and some parasitic plants draw nutrients from a host plant.

Haversian systems Units of organization in compact bone that reflect the action of intercommunicating osteoblasts.

hCG *See* human chorionic gonadotropin.

HDLs *See* high-density lipoproteins.

heart In circulatory systems, a muscular pump that moves extracellular fluid around the body.

heat of vaporization The energy that must be supplied to convert a molecule from a liquid to a gas at its boiling point.

heat shock proteins Chaperone proteins expressed in cells exposed to high or low temperatures or other forms of environmental stress.

hematocrit The proportion of the blood volume that consists of red blood cells.

hemiparasite A parasitic plant that can photosynthesize, but derives water and mineral nutrients from the living body of another plant. (Contrast with holoparasite.)

hemizygous (hem' ee zie' gus) [Gk. *hemi*: half + *zygotos*: joined] In a diploid organism, having only one allele for a given trait, typically the case for X-linked genes in male mammals and Z-linked genes in female birds. (Contrast with homozygous, heterozygous.)

hemocoel The primary body cavity of many invertebrates, which contains circulatory fluid.

hemoglobin (hee' mo glow bin) [Gk. *heaema*: blood + L. *globus*: globe] Oxygen-transporting protein found in the red blood cells of vertebrates (and found in some invertebrates).

hemolymph The extracellular fluid in closed circulatory systems.

Hensen's node In avian embryos, a structure at the anterior end of the primitive groove; determines the fates of cells passing over it during gastrulation.

hepatic portal vein The vein that begins in capillary beds in the gut and ends in capillary beds in the liver.

herbivore (ur' bi vore) [L. *herba*: plant + *vorare*: to devour] An animal that eats plant tissues. (Contrast with carnivore, detritivore, omnivore.)

herbivory (ur' bi vore) [L. *herba*: plant + *vorare*: to devour] An act of predation in which the predator is animal and the prey is a plant or algae. (Contrast with carnivory, parasitism, omnivory.)

hermaphroditism (her maf' row dite ism) The coexistence of both female and male sex organs in the same organism.

heterochromatin Densely packed, dark-staining chromatin; any genes it contains are usually not transcribed. (Contrast with euchromatin.)

heterochrony [Gk. different time] Alteration in the timing of developmental events, contributing to the evolution of different phenotypes in the adult. (Contrast with heterometry, heterotypy.)

heterocyst A large, thick-walled cell type in the filaments of certain cyanobacteria that performs nitrogen fixation.

heterometry [Gk. different measure] Alteration in the level of gene expression, and thus in the amount of protein produced, during development, contributing to the evolution of different phenotypes in the adult. (Contrast with heterochrony, heterotypy.)

heteromorphic (het' er oh more' fik) [Gk. different form] Having a different form or appearance, as two heteromorphic life stages of a plant. (Contrast with isomorphic.)

heterosis The superior fitness of heterozygous offspring as compared with that of their dissimilar homozygous parents. Also called hybrid vigor.

heterosporous (het' er os' por us) Producing two types of spores, one of which gives rise to a female megaspore and the other to a male microspore. (Contrast with homosporous.)

heterotherm An animal that regulates its body temperature at a constant level at some times but not others, such as a hibernator.

heterotopy   [Gk. different place] Spatial differences in gene expression during development, controlled by developmental regulatory genes and contributing to the evolution of distinctive adult phenotypes.

heterotroph   (het' er oh trof) [Gk. *heteros*: different + *trophe*: feed] An organism that requires preformed organic molecules as food. (Contrast with autotroph.)

heterotypy   [Gk. different kind] Alteration in a developmental regulatory gene itself rather than the expression of the genes it controls. (Contrast with heterochrony, heterometry.)

heterozygous   (het' er oh zie' gus) [Gk. *heteros*: different + *zygotos*: joined] In diploid organisms, having different alleles of a given gene on the pair of homologs carrying that gene. (Contrast with hemizygous, homozygous.)

hexose   [Gk. *hex*: six] A sugar containing six carbon atoms.

hibernation   [L. *hibernum*: winter] The state of inactivity of some animals during winter; marked by a drop in body temperature and metabolic rate. (Contrast with estivation.)

high-density lipoproteins (HDLs)   Lipoproteins that remove cholesterol from tissues and carry it to the liver; HDLs are the "good" lipoproteins associated with good cardiovascular health.

high-throughput sequencing   Rapid DNA sequencing on a micro scale in which many fragments of DNA are sequenced in parallel.

highly repetitive sequences   Short (less than 100 bp), nontranscribed DNA sequences, repeated thousands of times in tandem arrangements.

hindbrain   The region of the developing vertebrate brain that gives rise to the medulla, pons, and cerebellum. (Compare with forebrain, midbrain.)

hippocampus   [Gk. sea horse] A part of the forebrain that takes part in long-term memory formation.

histamine   (hiss' tah meen) A substance released by damaged tissue, or by mast cells in response to allergens. Histamine increases vascular permeability, leading to edema (swelling).

histogram   A figure that displays frequencies of classes of quantitative data binned by ranges of a particular variable.

HIV   *See* human immunodeficiency virus.

holometabolous   Undergoing complete metamorphosis.

holoparasite   A fully parasitic plant (i.e., one that does not perform photosynthesis). (Contrast with hemiparasite.)

homeobox   180-base-pair segment of DNA found in certain homeotic genes. A specific sequence within the homeobox—the **homeodomain**—regulates the expression of other genes and through this regulation controls large-scale developmental processes. (*See* homeotic genes.)

homeodomain   *See* homeobox.

homeostasis   (home' ee o sta' sis) [Gk. *homos*: same + *stasis*: position] The maintenance of a steady state, such as a constant temperature, by means of physiological or behavioral feedback responses.

homeotherms   Animals that regulate their body temperatures within a narrow range when confronted with a much greater range of environmental temperatures.

homeotic genes   Genes that act during development to determine the formation of an organ from a region of the embryo. (Compare with Hox genes.)

homeotic mutation   Mutation in a homeotic gene that results in the formation of a different organ than that normally made by a region of the embryo.

homing   In animal navigation, the ability to return to a nest site, burrow, or other specific location.

hominin   Lineages that includes modern humans (*Homo sapiens*) and their extinct ancestors (e.g., Australopithecines, *Homo erectus*).

homolog   (1) In cytogenetics, one of a pair (or larger set) of chromosomes having the same overall genetic composition and sequence. In diploid organisms, each chromosome inherited from one parent is matched by an identical (except for mutational changes) chromosome—its homolog—from the other parent. (2) In evolutionary biology, one of two or more features in different species that are similar by reason of descent from a common ancestor.

homologous pair   A pair of matching chromosomes made up of a chromosome from each of the two sets of chromosomes in a diploid organism.

homology   (ho mol' o jee) [Gk. *homologia*: of one mind, agreement] A similarity between two or more features that is due to inheritance from a common ancestor. The structures are said to be **homologous**, and each is a **homolog** of the others.

homoplasy   (home' uh play zee) [Gk. *homos*: same + *plastikos*: shape, mold] The presence in multiple groups of a trait that is not inherited from the common ancestor of those groups. Can result from convergent evolution, evolutionary reversal, or parallel evolution.

homosporous   Producing a single type of spore that gives rise to a single type of gametophyte, bearing both female and male reproductive organs. (Contrast with heterosporous.)

homotypic   Pertaining to adhesion of cells of the same type.

homozygous   (home' oh zie' gus) [Gk. *homos*: same + *zygotos*: joined] In diploid organisms, having identical alleles of a given gene on both homologous chromosomes. An individual may be a homozygote with respect to one gene and a heterozygote with respect to another. (Contrast with hemizygous, heterozygous.)

horizons   The horizontal layers of a soil profile, including the topsoil (A horizon), subsoil (B horizon) and parent rock or bedrock (C horizon).

horizontal cells   Cells in the retina that communicate laterally between photoreceptors and bipolar cells.

hormone   (hore' mone) [Gk. *hormon*: to excite, stimulate] A chemical signal produced in minute amounts at one site in a multicellular organism and transported to another site where it acts on target cells.

host   An organism that harbors a parasite or symbiont and provides it with nourishment.

Hox genes   Conserved homeotic genes found in vertebrates, *Drosophila*, and other animal groups. Hox genes contain the homeobox and specify pattern and axis formation in these animals.

human chorionic gonadotropin (hCG)   A hormone secreted by the placenta that sustains the corpus luteum and helps maintain pregnancy.

Human Genome Project   A publicly and privately funded research effort, successfully completed in 2003, to produce a complete DNA sequence for the entire human genome.

human immunodeficiency virus (HIV)   Human immunodeficiency virus, the retrovirus that causes acquired immune deficiency syndrome (AIDS).

humoral immune response   The response of the immune system mediated by B cells that produces circulating antibodies active against extracellular bacterial and viral infections. (Contrast with cellular immune response.)

humus   (hew' mus) The partly decomposed remains of plants and animals on the surface of a soil.

hybrid vigor   *See* heterosis.

hybrid zone   A region of overlap in the ranges of two closely related species where the species may hybridize.

hydrocarbon   A compound containing only carbon and hydrogen atoms.

hydrogen bond   A weak electrostatic bond which arises from the attraction between the slight positive charge on a hydrogen atom and a slight negative charge on a nearby oxygen or nitrogen atom.

hydrologic cycle   The movement of water from the oceans to the atmosphere, to the soil, rivers, and lakes, and back to the oceans.

hydrolysis reaction   (high drol' uh sis) [Gk. *hydro*: water + *lysis*: break apart] A chemical reaction that breaks a bond by inserting the components of water ($AB + H_2O \rightarrow AH + BOH$). (Contrast with condensation reaction.)

hydrophilic   (high dro fill' ik) [Gk. *hydro*: water + *philia*: love] Having an affinity for water. (Contrast with hydrophobic.)

hydrophobic   (high dro foe' bik) [Gk. *hydro*: water + *phobia*: fear] Having no affinity for water. Uncharged and nonpolar groups of atoms are hydrophobic. (Contrast with hydrophilic.)

hydrophobic interaction   A weak chemical interaction caused when molecules that are not attracted to water interact to exclude water.

hydroponic   Pertaining to a method of growing plants with their roots suspended in nutrient solutions instead of soil.

hydrostatic skeleton   A fluid-filled body cavity that transfers forces from one part of the body to another when acted on by surrounding muscles.

hyperaccumulators   Plant species that store large quantities of heavy metals such as arsenic, cadmium, nickel, aluminum, and zinc.

hyperpolarization A change in the resting potential across a membrane so that the inside of a cell becomes more negative compared with the outside of the cell. (Contrast with depolarization.)

hypersensitive response A defensive response of plants to microbial infection in which phytoalexins and pathogenesis-related proteins are produced and the infected tissue undergoes apoptosis to isolate the pathogen from the rest of the plant.

hypertonic Having a greater solute concentration. Said of one solution compared with another. (Contrast with hypotonic, isotonic.)

hypertonic regulation Maintaining body fluids at a greater solute concentration than the environment. (Contrast with hypotonic regulation.)

hypha (high' fuh) (plural: hyphae) [Gk. *hyphe*: web] In the fungi and oomycetes, any single filament.

hypoblast The lower tissue portion of the avian blastula which is joined to the epiblast at the margins of the blastodisc.

hypothalamus The part of the brain lying below the thalamus; it coordinates water balance, reproduction, temperature regulation, and metabolism.

hypothermia Below-normal body temperature.

hypothesis A tentative answer to a question, from which testable predictions can be generated. (Contrast with theory.)

hypotonic Having a lesser solute concentration. Said of one solution in comparing it to another. (Contrast with hypertonic, isotonic.)

hypotonic regulation Maintaining body fluids at a lesser solute concentration than the environment. (Contrast with hypertonic regulation.)

hypoxia A deficiency of oxygen.

hysteresis The inability of a community that has undergone change to shift back to the initial community type, even when the original conditions are restored.

I

igneous rocks Rocks solidified from lava or magma.

ileum The final segment of the small intestine. (*See also* duodenum, jejunum.)

imbibition Water uptake by a seed; first step in germination.

immediate hypersensitivity A rapid, extensive overreaction of the immune system against an allergen, resulting in the release of large amounts of histamine. (Contrast with delayed hypersensitivity.)

immediate memory A form of memory for events happening in the present that is almost perfectly photographic, but lasts only seconds. (Contrast with long-term memory, short-term memory.)

immigration Dispersal of individuals into an existing population. (Contrast with emigration.)

immunity [L. *immunis*: exempt from] In animals, the ability to avoid disease when invaded by a pathogen by deploying various defense mechanisms.

immunoglobulins A class of proteins containing a tetramer consisting of four polypeptide chains—two identical light chains and two identical heavy chains—held together by disulfide bonds; active as receptors and effectors in the immune system.

immunological memory The capacity to more rapidly and massively respond to a second exposure to an antigen than occurred on first exposure.

imperfect flower A flower lacking either functional stamens or functional carpels. (Contrast with perfect flower.)

implantation The process by which the early mammalian embryo becomes attached to and embedded in the lining of the uterus.

imprinting In animal behavior, a rapid form of learning in which an animal learns, during a brief critical period, to make a particular response (which is then maintained for life) to some object or other organism.

in vitro evolution A method based on natural molecular evolution that uses artificial selection in the laboratory to rapidly produce molecules with novel enzymatic and binding functions.

in vivo gene therapy Alteration of the human genome sequence in cells in the body for medical benefit.

inactivation gate The property of a voltage-gated Na$^+$ ion channel that decreases its permeability to Na$^+$ when the membrane is depolarized.

inbreeding Breeding of two closely genetically related individuals.

inbreeding depression Reduction of fitness in the offspring of inbreeding.

inclusive fitness The sum of an individual's genetic contribution to subsequent generations both via production of its own offspring and via its influence on the survival of relatives who are not direct descendants. (*See also* indirect fitness, kin selection. Contrast with direct fitness.)

incomplete cleavage A pattern of cleavage that occurs in many eggs that have a lot of yolk, in which the cleavage furrows do not penetrate all of it. (*See also* discoidal cleavage, superficial cleavage. Contrast with complete cleavage.)

incomplete dominance Condition in which the heterozygous phenotype is intermediate between the two homozygous phenotypes.

incomplete metamorphosis Insect development in which changes between instars are gradual. (Contrast with direct development, complete metamorphosis.)

independent assortment During meiosis, the random separation of genes carried on nonhomologous chromosomes into gametes so that inheritance of these genes is random. This principle was articulated by Mendel as his second law.

indeterminate growth An open-ended growth pattern in which an organism or organ continues to grow as long as it lives; characteristic of some animals and of plant shoots and roots. (Contrast with determinate growth.)

indirect fitness The component of inclusive fitness due to the reproductive success of related individuals times the percentage of shared genes between those individuals (e.g., 50% for a sibling).

indirect interaction An interaction in which the relationship between two species is mediated by a third (or more) species. (Compare with direct interaction.)

induced A process that occurs only under certain conditions or signals.

induced fit A change in the shape of an enzyme caused by binding to its substrate that exposes the active site of the enzyme.

induced mutation A mutation resulting from exposure to a mutagen from outside the cell. (Contrast with spontaneous mutation.)

induced pluripotent stem cells (iPS cells) Multipotent or pluripotent animal stem cells produced from differentiated cells in vitro by the addition of several genes that are expressed.

inducer (1) A compound that stimulates the synthesis of a protein. (2) In embryonic development, a substance that causes a group of target cells to differentiate in a particular way.

inducible Produced only in the presence of a particular compound or under particular circumstances. (Contrast with constitutive.)

inducible genes Genes that are expressed only when their products—**inducible proteins**—are needed.

induction In embryonic development, the process by which a factor produced and secreted by certain cells determines the fates of other cells.

inductive logic Involves making observations and then formulating one or more possible scenarios—hypotheses—that might explain those observations. (Contrast with deductive logic.)

inferior vena cava Large vein that returns deoxygenated blood to the right atrium from the lower body.

inflammation A nonspecific defense against pathogens; characterized by redness, swelling, pain, and increased temperature.

inflorescence A structure composed of several to many flowers.

inflorescence meristem A meristem that produces floral meristems as well as other small leafy structures (bracts).

ingroup In a phylogenetic study, the group of organisms of primary interest. (Contrast with outgroup.)

initials Cells that perpetuate plant meristems, comparable to animal stem cells. When an initial divides, one daughter cell develops into another initial, while the other differentiates into a more specialized cell.

initiation In molecular biology, the beginning of transcription or translation.

initiation complex In protein translation, a combination of a small ribosomal subunit, an mRNA molecule, and the tRNA charged with the first amino acid coded for by the mRNA; formed at the onset of translation.

initiation site The place within a promoter where transcription begins.

innate defenses In animals, one of two general types of defenses against pathogens. Nonspecific and present in most animals. (Contrast with adaptive defenses.)

inner cell mass   Derived from the mammalian blastula (blastocyst), the inner cell mass that will give rise to the yolk sac (via hypoblast) and embryo (via epiblast).

inorganic fertilizer   A chemical or combination of chemicals applied to soil or plants to make up for a plant nutrient deficiency. Often contains the macronutrients nitrogen, phosphorus, and potassium (N-P-K).

inositol trisphosphate (IP$_3$)   An intracellular second messenger derived from membrane phospholipids.

inspiratory reserve volume (IRV)   The amount of air that can be inhaled above the normal tidal inspiration. (Contrast with expiratory reserve volume, tidal volume, vital capacity.)

instar   (in' star) An immature stage of an insect between molts.

insular cortex   The outer layers of the insular lobe.

insular lobe   A region of the cerebrum that folds in between the temporal lobes and the overlying frontal and parietal lobes.

insulin   (in' su lin) [L. *insula*: island] A hormone synthesized in islet cells of the pancreas that promotes the conversion of glucose into the storage material, glycogen.

integral membrane proteins   Proteins that are at least partially embedded in the plasma membrane. (Contrast with peripheral membrane proteins.)

integrin   In animals, a transmembrane protein that mediates the attachment of epithelial cells to the extracellular matrix.

integument   [L. *integumentum*: covering] A protective surface structure. In gymnosperms and angiosperms, a layer of tissue around the ovule that will become the seed coat.

interaction strength   A measure of the effect of one species (the interactor) on the abundance of another species (the target species).

interaction webs   A concept that describes both the trophic (vertical) and non-trophic (horizontal; such as competition, mutualism, commensalism) interactions among the species in a food web.

intercalated discs   Structural features of cardiac muscle cells that connect cells together and provide electrical continuity through gap junctions.

intercostal muscles   Muscles between the ribs that can augment breathing movements by elevating and suppressing the rib cage.

interference competition   Competition in which one species actively interferes with another species' access to a limiting resource. (Contrast with exploitation competition.)

interference RNA (RNAi)   *See* RNA interference.

interferons   Glycoproteins produced by virus-infected animal cells; interferons increase the resistance of neighboring cells to the virus.

intermediate disturbance hypothesis   A hypothesis describing how varying degrees of

disturbance (or stress or predation) affects species diversity within communities. Species diversity is greatest at intermediate levels of disturbance because competitive exclusion at low levels of disturbance and mortality at high levels of disturbance reduces species diversity.

intermediate filaments   A part of the cytoskeleton that includes filaments intermediate in diameter between microtubules and microfilaments.

internal environment   In multicellular organisms, includes blood plasma and interstitial fluid (i.e., the extracellular fluids that surround the cells).

internal fertilization   The release of sperm into the female reproductive tract; typical of most terrestrial animals. (Contrast with external fertilization.)

internal gills   Gills enclosed in protective body cavities; typical of mollusks, arthropods, and fishes.

internal skeleton   *See* endoskeleton.

interneuron   A neuron that communicates information between two other neurons.

internode   The region between two nodes of a plant stem.

interphase   In the cell cycle, the period between successive nuclear divisions during which the chromosomes are diffuse and the nuclear envelope is intact. During interphase the cell is most active in transcribing and translating genetic information.

interspecific interactions   Interactions between members of two or more species.

interstitial fluid   Extracellular fluid that is not contained in the vessels of a circulatory system.

intertidal zone   A nearshore region of oceans that is periodically exposed to the air as the tides rise and fall.

intestine   The portion of the gut following the stomach, in which most digestion and absorption occurs.

intracellular receptor   A receptor that binds its ligand inside the cell. An example is the estrogen receptor in mammals.

intraspecific competition   Competition for shared limited resources by individuals of the same species.

intron   Portion of a gene within the coding region that is transcribed into pre-mRNA but is spliced out prior to translation. (Contrast with exon.)

invasive species   A non-native species that reproduces rapidly, spreads widely, and has mostly negative effects on the native species or ecosystems of the region to which it has been introduced.

inversion   A rare 180° reversal of the order of genes within a segment of a chromosome.

involution   Cell movements that occur during gastrulation of frog embryos, giving rise to the archenteron.

ion   (eye' on) [Gk. *ion*: wanderer] An electrically charged particle that forms when an atom gains or loses one or more electrons.

ion channel   An integral membrane protein that allows ions to diffuse across the membrane in which it is embedded.

ionic attraction   An electrostatic attraction between positively and negatively charged ions.

iris   (eye' ris) [Gk. *iris*: rainbow] The round, pigmented membrane that surrounds the pupil of the eye and adjusts its aperture to regulate the amount of light entering the eye.

islets of Langerhans   Clusters of hormone-producing cells in the pancreas.

isomers   Molecules consisting of the same numbers and kinds of atoms, but differing in the bonding patterns by which the atoms are held together.

isomorphic   (eye so more' fik) [Gk. *isos*: equal + *morphe*: form] Having the same form or appearance, as when the haploid and diploid life stages of an organism appear identical. (Contrast with heteromorphic.)

isotonic   Having the same solute concentration; said of two solutions. (Contrast with hypertonic, hypotonic.)

isotope   (eye' so tope) [Gk. *isos*: equal + *topos*: place] Isotopes of a given chemical element have the same number of protons in their nuclei (and thus are in the same position on the periodic table), but differ in the number of neutrons.

isozymes   Enzymes of an organism that have somewhat different amino acid sequences but catalyze the same reaction.

iteroparity   Reproducing multiple times in a lifetime. (Contrast with semelparity.)

J

jasmonic acid (jasmonate)   A plant hormone involved in triggering responses to pathogen attack as well as other processes.

jejunum   (jih jew' num) The middle division of the small intestine, where most absorption of nutrients occurs. (*See also* duodenum, ileum.)

joint   In skeletal systems, a junction between two or more bones.

juvenile hormone   In insects, a hormone maintaining larval growth and preventing maturation or pupation.

juxtacrine   A mode of cell communication where the cell producing a signal is in contact with the cell bearing the receptor for that signal.

K

K-strategist   A species whose life history strategy allows it to persist at or near the carrying capacity (K) of its environment. (Contrast with *r*-strategist.)

karyogamy   The fusion of nuclei of two cells. (Contrast with plasmogamy.)

karyotype   The number, forms, and types of chromosomes in a cell.

K_D   *See* dissociation constant.

keystone species   Species that have strong community-wide effects despite their small size and abundance.

kidneys   A pair of excretory organs in vertebrates.

kilocalorie (kcal)   *See* calorie.

kin selection   That component of inclusive fitness resulting from helping the survival of relatives containing the same alleles by descent from a common ancestor. (Contrast with direct fitness.)

kinetic energy (kuh-net' ik) [Gk. *kinetos*: moving] The energy associated with movement. (Contrast with potential energy.)

kinetochore (kuh net' oh core) Specialized structure on a centromere to which microtubules attach.

Koch's postulates A set of rules for establishing that a particular microorganism causes a particular disease.

L

lacteals The smallest vessels of the lymphatic system.

lactic acid fermentation A metabolic pathway in which glucose is catabolized in the absence of oxygen with the production of lactic acid.

lagging strand In DNA replication, the daughter strand that is synthesized in discontinuous stretches. (*See* Okazaki fragments. Contrast with leading strand.)

landscape A geographic area consisting of multiple ecosystems or habitats.

larva (plural: larvae) [L. *lares*: guiding spirits] An immature stage of any animal that differs dramatically in appearance from the adult.

larynx A structure between the pharynx and trachea that contains the vocal cords.

lateral gene transfer The transfer of genes from one species to another, common among bacteria and archaea.

lateral meristem Either of the two meristems, the vascular cambium and the cork cambium, that give rise to a plant's secondary growth.

lateral root A root extending outward from the taproot in a taproot system; typical of eudicots.

lateralization A phenomenon in humans in which language functions come to reside in one cerebral hemisphere, usually the left.

laticifers (luh tiss' uh furs) In some plants, elongated cells containing secondary plant products such as latex.

Laurasia The northernmost of the two large continents produced by the breakup of Pangaea.

law of independent assortment *See* independent assortment.

law of mass action The rate of a chemical reaction is directly proportional to the concentrations of the reacting substances.

law of segregation *See* segregation.

laws of thermodynamics [Gk. *thermos*: heat + *dynamis*: power] Laws derived from studies of the physical properties of energy and the ways energy interacts with matter. (*See also* first law of thermodynamics, second law of thermodynamics.)

LDLs *See* low-density lipoproteins.

LDP *See* long-day plant.

leaching In soils, a process by which mineral nutrients in upper soil horizons are dissolved in water and carried to deeper horizons, where they are unavailable to plant roots.

leading strand In DNA replication, the daughter strand that is synthesized continuously. (Contrast with lagging strand.)

leaf (plural: leaves) In plants, the chief organ of photosynthesis.

leaf primordium (plural: primordia) An outgrowth on the side of the shoot apical meristem that will eventually develop into a leaf.

leghemoglobin In nitrogen-fixing plants, an oxygen-carrying protein in the cytoplasm of nodule cells that transports enough oxygen to the nitrogen-fixing bacteria to support their respiration, while keeping free oxygen concentrations low enough to protect nitrogenase.

lek A display ground within which male animals compete for and defend small display areas as a means of demonstrating their territorial prowess and winning opportunities to mate.

lens In the vertebrate eye, a crystalline protein structure that makes fine adjustments in the focus of images falling on the retina.

lenticels Small porous regions on the bark of woody stems and roots in dicots.

leptin A hormone produced by fat cells that is believed to provide feedback information to the brain about the status of the body's fat reserves.

Leydig cells The cells between the seminiferous tubules of the testes that secrete testosterone.

LH *See* luteinizing hormone.

lichen (lie' kun) An organism resulting from the symbiotic association of a fungus and either a cyanobacterium or a unicellular alga.

life history The lifetime pattern of growth, reproduction, and survival for an average individual.

life history strategy The way in which individuals within and among species allocate resources to growth, reproduction, and survival based on genetic and environmental factors.

life table A summary of how survival and reproductive rates vary with the age, size, or sex of individuals within a population, and used to estimate population growth rate.

ligament A band of connective tissue linking two bones in a joint.

ligand (lig' and) Any molecule that binds to a receptor site of another (usually larger) molecule.

light reactions The initial phase of photosynthesis, in which light energy is converted into chemical energy.

light-harvesting complex In photosynthesis, a group of different molecules that cooperate to absorb light energy and transfer it to a reaction center. Also called antenna system.

light-independent reactions Reactions that follow the light reaction and use the energy captured in the light reactions to drive the reduction of CO_2 to form carbohydrates.

lignin A complex, hydrophobic polyphenolic polymer in plant cell walls that crosslinks other wall polymers, strengthening the walls, especially in wood.

limbic system A group of evolutionarily primitive structures in the vertebrate telencephalon that are involved in emotions, drives, instinctive behaviors, learning, and memory.

limiting resource The required resource whose supply (or lack thereof) most strongly influences the size of a population.

lineage A series of populations, species, or genes descended from a single ancestor over evolutionary time.

lineage species concept The definition of a species as a branch on the tree of life, which has a history that starts at a speciation event and ends either at extinction or at another speciation event. (Contrast with biological species concept, morphological species concept.)

linear regression A statistical method of fitting a straight line to describe the relationship between two variables in a scatter plot.

linkage analysis Genetic analysis in which a gene and its alleles are linked to an unknown gene and its alleles and the location of the latter is determined by its co-segregating with the former in genetic crosses.

lipases Digestive enzymes that break down fats.

lipid (lip' id) [Gk. *lipos*: fat] Nonpolar, hydrophobic molecules that include fats, oils, waxes, steroids, and the phospholipids that make up biological membranes.

lipid bilayer *See* phospholipid bilayer.

lipoproteins Lipids packaged inside a covering of protein so that they can be circulated in the blood.

liposome (lip' o zome) Spherical structure contained by a membrane of phospholipids. Can be used to deliver drugs to cells.

lithosphere (lith' o sphere) [Gk. *lithos*: strong] The crust of sold rock plates that overlays the viscous mantle of Earth. The movements of the lithosphere are the source of plate tectonics. (Contrast with asthenosphere.)

liver A large digestive gland. In vertebrates, it secretes bile and is involved in the formation of blood.

loading In phloem, the process of transferring the products of photosynthesis, usually sugars, into sieve tubes.

loam A type of soil consisting of a mixture of sand, silt, clay, and organic matter. One of the best soil types for agriculture.

locus (low' kus) (plural: loci, low' sigh) In genetics, a specific location on a chromosome. May be considered synonymous with gene.

logistic growth A pattern of population growth characterized by a multiplicative pattern that eventually levels off as density dependent processes slow growth at some carrying capacity (Contrast with exponential growth.)

long-day plant (LDP) A plant that requires long days (actually, short nights) in order to flower. (Compare to short-day plant.)

long-term depression (LTD) A long-lasting decrease in the responsiveness of a neuron resulting from a period of prior stimulation. (Contrast with long-term potentiation.)

long-term memory Information that is maintained in the brain and is accessible for long periods of time—hours to years. (Contrast with immediate memory, short-term memory.)

long-term potentiation (LTP)   A long-lasting increase in the responsiveness of a neuron resulting from a period of intense stimulation. (Contrast with long-term depression.)

loop of Henle   (hen' lee) Long, hairpin loop of the mammalian renal tubule that runs from the cortex down into the medulla and back to the cortex; creates a concentration gradient in the interstitial fluids in the medulla.

lophophore   A U-shaped fold of the body wall with hollow, ciliated tentacles that encircles the mouth of animals in several different groups. Used for filtering prey from the surrounding water.

loss-of-function mutation   A mutation that results in the loss of a functional protein. (Contrast with gain-of-function mutation.)

lottery model   A hypothesis proposing that species diversity in communities is maintained by a "lottery" in which resources, made available by the effects of disturbance, stress, or predation, are captured at random by individuals of different species. Also called neutral model.

low-density lipoproteins (LDLs)   Lipoproteins that transport cholesterol around the body for use in biosynthesis and for storage; LDLs are the "bad" lipoproteins associated with a high risk of cardiovascular disease.

lower critical temperature   The environmental temperature at which a resting endotherm must increase its metabolic rate to avoid a drop in its body temperature.

lumen   (loo' men) [L. *lumen*: light] The open cavity inside any tubular organ or structure, such as the gut or a renal tubule.

lung   An internal organ specialized for respiratory gas exchange with air.

luteinizing hormone (LH)   A gonadotropin produced by the anterior pituitary that stimulates the gonads to produce sex hormones.

lymph   [L. *lympha*: liquid] A fluid derived from blood and other tissues that accumulates in intercellular spaces throughout the body and is returned to the blood by the lymphatic system.

lymph node   A specialized structure in the vessels of the lymphatic system. Lymph nodes contain lymphocytes, which encounter and respond to foreign cells and molecules in the lymph as it passes through the vessels.

lymphatic duct   One of two large lymphatic vessels that drains lymph into a subclavian vein.

lymphocyte   One of the two major classes of white blood cells; includes T cells, B cells, and other cell types important in the immune system.

lysogeny   A form of viral replication in which the virus becomes incorporated into the host chromosome and remains inactive. Also called a lysogenic cycle. (Contrast with lytic cycle.)

lysosome   (lie' so soam) [Gk. *lysis*: break away + *soma*: body] A membrane-enclosed organelle originating from the Golgi apparatus and containing hydrolytic enzymes. (Contrast with secondary lysosome.)

lysozyme   (lie' so zyme) An enzyme in saliva, tears, and nasal secretions that hydrolyzes bacterial cell walls.

lytic cycle   A viral reproductive cycle in which the virus takes over a host cell's synthetic machinery to replicate itself, then bursts (lyses) the host cell, releasing the new viruses. (Contrast with lysogeny.)

M

macroevolution   [Gk. *makros*: large] Evolutionary changes occurring over long time spans and usually involving changes in many traits. (Contrast with microevolution.)

macroglia   Non-neuronal cells of the central and peripheral nervous system that provide homeostatic functions for neurons, modulate synaptic connections, provide myelin coverings for axons, and create the blood-brain or blood-spinal cord barrier.

macromolecule   A giant (molecular weight > 1,000) polymeric molecule. The macromolecules are the proteins, polysaccharides, and nucleic acids.

macronutrient   In plants, a mineral element required in concentrations of at least 1 milligram per gram of plant dry matter; in animals, a mineral element required in large amounts. (Contrast with micronutrient.)

macroparasites   Relatively large parasite species, such as arthropods and worms. (Compare with microparasites.)

MADS box   DNA-binding domain in many plant transcription factors that is active in development.

maintenance methylase   An enzyme that catalyzes the methylation of the new DNA strand when DNA is replicated.

major histocompatibility complex (MHC)   A complex of linked genes, with multiple alleles, that control a number of cell surface antigens that identify self and can lead to graft rejection.

malignant   Pertaining to a tumor that can grow indefinitely and/or spread from the original site of growth to other locations in the body. (Contrast with benign.)

malnutrition   A condition caused by lack of any essential nutrient.

Malpighian tubule   (mal pee' gy un) A type of protonephridium found in insects.

mandible   A crushing part of arthropod mouth parts, in mammals, the lower jaw bone, in birds the upper or lower part of the beak.

mantle   (1) In mollusks, a fold of tissue that covers the organs of the visceral mass and secretes the hard shell that is typical of many mollusks. (2) In geology, Earth's crust below the solid lithospheric plates.

mark–recapture method   A method of estimating population sizes of mobile organisms by capturing, marking, and releasing a sample of individuals, then capturing another sample at a later time.

mass   A measure of the quantity of matter present; the greater the mass, the greater the quantity of matter.

mass extinction event   A period of evolutionary history during which rates of extinction are much higher than during intervening times.

mass number   The sum of the number of protons and neutrons in an atom's nucleus.

mast cells   Cells, typically found in connective tissue, that release histamine in response to tissue damage.

maternal effect genes   Genes coding for morphogens that determine the polarity of the egg and larva in fruit flies. Part of a developmental cascade that includes gap genes, pair rule genes, segment polarity genes, and Hox genes.

mating   The joining of individuals for the purpose of reproduction.

mating type   A particular strain of a species that is incapable of sexual reproduction with another member of the same strain but capable of sexual reproduction with members of other strains of the same species.

maximum likelihood   A statistical method of determining which of two or more hypotheses (such as phylogenetic trees) best fit the observed data, given an explicit model of how the data were generated.

mean   The sum of all values in a sample divided by the number of observations in the sample.

measures of center   Quantities that describe various aspects of the center of a group of observations.

measures of dispersion   Measures that quantify the dispersion of observations in a sample of observations.

mechanically gated channel   A molecular channel that opens or closes in response to mechanical force applied to the plasma membrane in which it is inserted.

mechanoreceptor   A cell that is sensitive to physical movement and generates action potentials in response.

median   The value at which there are equal numbers of larger and smaller observations in a sample.

medulla   (meh dull' luh) (1) The inner, core region of an organ, as in the adrenal medulla (adrenal gland) or the renal medulla (kidneys). (2) The portion of the brainstem that connects to the spinal cord.

medusa   (plural: medusae) In cnidarians, a free-swimming, sexual life cycle stage shaped like a bell or an umbrella.

megagametophyte   In heterosporous plants, the female gametophyte; produces eggs. (Contrast with microgametophyte.)

megakaryocytes   A bone marrow cell that produces blood platelets.

megaphyll   The generally large leaf of a fern, horsetail, or seed plant, with several to many veins. (Contrast with microphyll.)

megasporangia   The plant structures that bear megaspores.

megaspore   [Gk. *megas*: large + *spora*: to sow] In plants, a haploid spore that produces a female gametophyte.

megastrobilus   In conifers, the female (seed-bearing) cone. (Contrast with microstrobilus.)

meiosis   (my oh' sis) [Gk. *meiosis*: diminution] Division of a diploid nucleus to produce four haploid daughter cells. The process consists of two successive nuclear divisions with only one cycle of chromosome replication. In meiosis I, homologous chromosomes separate but retain their chromatids. The second division meiosis II, is similar to mitosis, in which chromatids separate.

meiosis I   *See* meiosis.

meiosis II   *See* meiosis.

Meissner's corpuscles   Rapidly adapting skin mechanoreceptors sensitive to light touch and vibration. Found primarily in non-hairy skin; provide continuous information about changes in things touching the skin.

melanocyte-stimulating hormone (MSH)   A pituitary peptide hormone that stimulates melanocytes to produce the pigment melanin.

melatonin   A hormone released by the pineal gland. Involved in photoperiodicity and circadian rhythms.

membrane   A phospholipid bilayer forming a barrier that separates the internal contents of a cell from the nonbiological environment, or enclosing the organelles within a cell. The membrane regulates the molecular substances entering or leaving a cell or organelle.

membrane potential   The difference in electrical charge between the inside and the outside of a cell, caused by a difference in the distribution of ions.

membranous bone   A type of bone that develops by forming on a scaffold of connective tissue. (Contrast with cartilage bone.)

memory cells   Long-lived lymphocytes produced after exposure to antigen. They persist in the body and are able to mount a rapid response to subsequent exposures to the antigen.

menopause   In human females, the end of fertility and menstrual cycling.

menstrual cycle   A term for the uterine cycle of humans that refers to the monthly sloughing of the uterine lining, or menses.

menstruation   The process by which the endometrium breaks down, and the sloughed-off tissue, including blood, flows from the body.

meristem   [Gk. *meristos*: divided] Plant tissue made up of undifferentiated actively dividing cells.

meristem culture   A method for the asexual propagation of plants, in which pieces of shoot apical meristem are cultured to produce plantlets.

meristem identity genes   In angiosperms, a group of genes whose expression initiates flower formation, probably by switching meristem cells from a vegetative to a reproductive fate.

Merkel's discs   Slowly adapting skin mechanoreceptors that provide continuous information about anything touching the skin.

mesenchyme   (mez' en kyme) [Gk. *mesos*: middle + *enchyma*: infusion] Embryonic or unspecialized cells derived from the mesoderm.

mesoderm   [Gk. *mesos*: middle + *derma*: skin] The middle of the three embryonic germ layers first delineated during gastrulation. Gives rise to the skeleton, circulatory system, muscles, excretory system, and most of the reproductive system.

mesoglea   (mez' uh glee uh) [Gk. *mesos*: middle + *gloia*: glue] A thick, gelatinous noncellular layer that separates the two cellular tissue layers of ctenophores, cnidarians, and scyphozoans.

mesophyll   (mez' uh fill) [Gk. *mesos*: middle + *phyllon*: leaf] Chloroplast-containing, photosynthetic cells in the interior of leaves.

messenger RNA (mRNA)   Transcript of a region of one of the strands of DNA; carries information (as a sequence of codons) for the synthesis of one or more proteins.

metabolism   (meh tab' a lizm) [Gk. *metabole*: change] The sum total of the chemical reactions that occur in an organism, or some subset of that total (as in respiratory metabolism).

metabolome   The quantitative description of all the small molecules in a cell or organism.

metagenomics   The practice of analyzing DNA from environmental samples without isolating intact organisms.

metamorphosis   (met' a mor' fo sis) [Gk. *meta*: between + *morphe*: form, shape] A change occurring between one developmental stage and another, as for example from a tadpole to a frog. (*See* complete metamorphosis, incomplete metamorphosis.)

metanephridia   (singular: metanephridium) [Gk. *meta*: akin to) The paired excretory organs of annelids.

metapopulation   A group of geographically isolated populations linked together by dispersal.

metastasis   Spread of cells a tumor to regions of the body away from the primary tumor.

methanogens   Methane-producing microbes.

methylation   The addition of a methyl group (—CH_3) to a molecule.

MHC   *See* major histocompatibility complex.

MI   *See* myocardial infarction.

micelle   A particle of lipid covered with bile salts that is produced in the duodenum and facilitates digestion and absorption of lipids.

microbiome   The diverse community of bacteria that lives on or within the body of an organism and is essential to bodily function.

microbiota   The diverse communities of bacteria that live on or within the body.

microevolution   Evolutionary changes below the species level, affecting allele frequencies. (Contrast with macroevolution.)

microfilament   In eukaryotic cells, a fibrous structure made up of actin monomers. Microfilaments play roles in the cytoskeleton, in cell movement, and in muscle contraction.

microgametophyte   In heterosporous plants, the male gametophyte; produces sperm. (Contrast with megagametophyte.)

microglia   Glial cells that act as macrophages and mediators of inflammatory responses in the central nervous system.

micronutrient   In plants, a mineral element required in concentrations of less than 100 micrograms per gram of plant dry matter; in animals, a mineral element required in concentrations of less than 100 micrograms per day. (Contrast with macronutrient.)

microparasites   Parasite species too small to be seen with the naked eye, such as viruses, bacteria, protists, and fungi. (Compare with macroparasites.)

microphyll   A small leaf with a single vein, found in club mosses and their relatives. (Contrast with megaphyll.)

micropyle   (mike' roh pile) [Gk. *mikros*: small + *pylon*: gate] Opening in the integument(s) of a seed plant ovule through which pollen grows to reach the female gametophyte within.

microRNA (miRNA)   A small, noncoding RNA molecule, typically about 21 bases long, that binds to mRNA to inhibit its translation.

microsporangia   The plant structures that bear microspores.

microspore   [Gk. *mikros*: small + *spora*: to sow] In plants, a haploid spore that produces a male gametophyte.

microstrobilus   In conifers, male pollen-bearing cone. (Contrast with megastrobilus.)

microtubules   Tubular structures found in centrioles, spindle apparatus, cilia, flagella, and cytoskeleton of eukaryotic cells. These tubules play roles in the motion and maintenance of shape of eukaryotic cells.

microvilli   (singular: microvillus) Projections of epithelial cells, such as the cells lining the small intestine, that increase their surface area.

midbrain   One of the three regions of the vertebrate brain. Part of the brainstem, it serves as a relay station for sensory signals sent to the cerebral hemispheres. (Compare with forebrain, hindbrain.)

middle lamella   (la mell' ah) [L. *lamina*: thin sheet] A layer of polysaccharides that separates plant cells; a shared middle lamella lies outside the primary walls of the two cells.

migration   Roundtrip movement of individuals within a population as a consequence of seasonal variation in resources.

mineral nutrients   Inorganic ions required by organisms for normal growth and reproduction.

mineralization   The process by which animals, bacteria, and fungi convert organic compounds into inorganic soluble nutrients.

mismatch repair   A mechanism that scans DNA after it has been replicated and corrects any base-pairing mismatches.

missense mutation   A change in a gene's sequence that changes the amino acid at that site in the encoded protein. (Contrast with frame-shift mutation, nonsense mutation, silent mutation.)

missense substitution   A change in a gene from one nucleotide to another that results in a change to the corresponding amino acid in the encoded protein. (Contrast with synonymous substitution.)

mitochondria   (my' toe kon' dree uh) (singular: mitochondrion) [Gk. *mitos*: thread + *chondros*: grain] Energy-generating organelles in eukaryotic cells that contain the enzymes of the citric acid cycle, the respiratory chain, and oxidative phosphorylation.

mitosis   (my toe' sis) [Gk. *mitos*: thread] Nuclear division in eukaryotes leading to the formation of two daughter nuclei, each with a chromosome complement identical to that of the original nucleus.

mitosomes   Reduced structures derived from mitochondria found in some organisms.

mitral valve   *See* bicuspid valve. So named because it has a similar shape to a miter (a religious headdress).

mode   The most frequent value in a sample of observations.

model organisms   Also known as **model systems**, these include the small group of species that are the subject of extensive research. They are organisms that adapt well to laboratory situations and findings from experiments on them can apply across a broad range of species. Classic examples include white mice and the fruit fly *Drosophila*.

moderately repetitive sequences   DNA sequences repeated 10–1,000 times in the eukaryotic genome. They include the genes that code for rRNAs and tRNAs, as well as the DNA in telomeres.

molds   Sac fungi composed of filamentous hyphae that do not form large ascomata.

mole   A quantity of a compound whose weight in grams is numerically equal to its molecular weight expressed in atomic mass units. Avogadro's number of molecules: 6.023×10^{23} molecules.

molecular clock   The approximately constant rate of divergence of macromolecules from one another over evolutionary time; used to date past events in evolutionary history.

molecular evolution   The scientific study of the mechanisms and consequences of the evolution of macromolecules.

molecular weight   The sum of the atomic weights of the atoms in a molecule.

molecule   A chemical substance made up of two or more atoms joined by covalent bonds or ionic attractions.

molting   The process of shedding part or all of an outer covering, as the shedding of feathers by birds or of the entire exoskeleton by arthropods.

monoblastic   Having a body derived from a single embryonic cell layer. (Contrast with diploblastic, triploblastic.)

monocots   Angiosperms with a single embryonic cotyledon; one of the two largest clades of angiosperms. (*See also* eudicots.)

monoculture   In agriculture, a large-scale planting of a single crop species. (Contrast with polyculture).

monoecious   (mo nee' shus) [Gk. *mono*: one + *oikos*: house] Pertaining to organisms in which both sexes are "housed" in a single individual that produces both eggs and sperm. (In some plants, these are found in different flowers within the same plant.) Examples include corn, peas, earthworms, hydras. (Contrast with dioecious.)

monohybrid cross   A mating in which the parents differ with respect to the alleles of only one locus of interest.

monomer   [Gk. *mono*: one + *meros*: unit] A small molecule, two or more of which can be combined to form oligomers (consisting of a few monomers) or polymers (consisting of many monomers).

monophyletic   (mon' oh fih leht' ik) [Gk. *mono*: one + *phylon*: tribe] Pertaining to a group that consists of an ancestor and all of its descendants. (Contrast with paraphyletic, polyphyletic.)

monosaccharide   A simple sugar. Oligosaccharides and polysaccharides are made up of monosaccharides.

monosomic   Pertaining to an organism with one less than the normal diploid number of chromosomes.

morphogen   A diffusible substance whose concentration gradient determines a developmental pattern in embryonic animals and plants.

morphogenesis   (more' fo jen' e sis) [Gk. *morphe*: form + *genesis*: origin] The development of form; the overall consequence of determination, differentiation, and growth.

morphological species concept   The definition of a species as a group of individuals that look alike. (Contrast with biological species concept, lineage species concept.)

morphology   (more fol' o jee) [Gk. *morphe*: form + *logos*: study, discourse] The scientific study of organic form, including both its development and function.

mosaic development   Pattern of animal embryonic development in which each blastomere contributes a specific part of the adult body. (Contrast with regulative development.)

motile   (mo' tul) Able to move from one place to another. (Contrast with sessile.)

motor cortex   The region of the cerebral cortex that contains motor neurons that directly stimulate specific muscle fibers to contract.

motor end plate   The depression in the postsynaptic membrane of the neuromuscular junction where the terminals of the motor neuron sit.

motor proteins   Specialized proteins that use energy to change shape and move cells or structures within cells.

motor unit   A motor neuron and the muscle fibers it controls.

MSH   *See* melanocyte-stimulating hormone.

mucosa   The first layer of the gut that lines the luminal surface.

mucus   A viscous substance secreted by mucous membranes (e.g., mucosal epithelium). A barrier defense against pathogens in innate immunity in animals and a protective coating in many animal organ systems.

Muller's ratchet   The accumulation—"ratcheting up"—of deleterious mutations in the nonrecombining genomes of asexual species.

Müllerian mimicry   Convergence in appearance of two or more unpalatable species.

multifactorial   The interaction of many genes and proteins with one or more factors in the environment. For example, cancer is a disease with multifactorial causes.

multiple fruits   A fruit that develops from several carpels of multiple flowers.

multiplication rule   The probability of two or more independent events occurring together is the product of their independent probabilities.

multipotent   Having the ability to differentiate into a limited number of cell types. (Contrast with pluripotent, totipotent, unipotent.)

muscle fiber   A single muscle cell. In the case of skeletal muscle, a syncitial, multinucleate cell.

muscle spindles   Mechanoreceptors embedded in skeletal muscle that sense changes in the length of the muscle.

muscle tissue   Excitable tissue that can contract through the interactions of actin and myosin; one of the four major tissue types in multicellular animals. There are three types of muscle tissue: skeletal, smooth, and cardiac.

mutagen   (mute' ah jen) [L. *mutare*: change + Gk. *genesis*: source] Any agent (e.g., a chemical, radiation) that increases the mutation rate.

mutation   A change in the genetic material not caused by recombination.

mutualism   A type of positive interaction in which both species benefit from the interaction, sometimes in a highly dependent and symbiotic manner. (Contrast with amensalism, commensalism, competition.)

mycelium   (my seel' ee yum) [Gk. *mykes*: fungus] In the fungi, a mass of hyphae.

mycologists   Scientists who study fungi.

mycorrhiza   (my' ko rye' za) (plural: mycorrhizae) [Gk. *mykes*: fungus + *rhiza*: root] An association of the root of a plant with the mycelium of a fungus.

myelin   (my' a lin) Concentric layers of plasma membrane that form a sheath around some axons; myelin provides the axon with electrical insulation and increases the rate of transmission of action potentials.

myocardial infarction (MI)   Blockage of an artery that carries blood to the heart muscle; a "heart attack."

MyoD   The protein encoded by the *myoblast determining* gene. A transcription factor involved in the differentiation of myoblasts (muscle precursor cells).

myofibril   (my' oh fy' bril) [Gk. *mys*: muscle + L. *fibrilla*: small fiber] A polymeric unit of actin or myosin in a muscle.

myoglobin   (my' oh globe' in) [Gk. *mys*: muscle + L. *globus*: sphere] An oxygen-binding molecule found in muscle. Consists of a heme unit and a single globin chain; carries less oxygen than hemoglobin.

myosin   One of the two contractile proteins of muscle. (*See also* actin.)

myxamoeba   An independent, unicellular, haploid life stage of cellular slime molds.

N

natural killer cell   A type of lymphocyte that attacks virus-infected cells and some tumor cells as well as antibody-labeled target cells.

natural selection   The differential contribution of offspring to the next generation by various genetic types belonging to the same population. The mechanism of evolution proposed by Charles Darwin.

nauplius (naw' plee us) [Gk. *nauplios*: shellfish] A bilaterally symmetrical larval form typical of crustaceans.

necrosis (nec roh' sis) [Gk. *nekros*: death] Premature cell death caused by external agents such as toxins.

negative feedback In regulatory systems, information that decreases a regulatory response, returning the system to the set point. (Contrast with positive feedback.)

negative regulation A type of gene regulation in which a gene is normally transcribed, and the binding of a repressor protein to the promoter prevents transcription. (Contrast with positive regulation.)

negative-sense RNA A strand of RNA that is complementary to mRNA.

nematocyst (ne mat' o sist) [Gk. *nema*: thread + *kystis*: cell] An elaborate, threadlike structure produced by cells of jellyfishes and other cnidarians, used chiefly to paralyze and capture prey.

neoteny (knee ot' enny) [Gk. *neo*: new, recent + *tenein*: to extend] The retention of juvenile or larval traits by the fully developed adult organism.

nephron (nef' ron) [Gk. *nephros*: kidney] The functional unit of the kidney, consisting of a structure for receiving a filtrate of blood and a tubule that reabsorbs selected parts of the filtrate.

Nernst equation A mathematical statement that calculates the potential across a membrane permeable to a single type of ion that differs in concentration on the two sides of the membrane.

nerve A structure consisting of many neuronal axons and connective tissue.

nerve nets Diffuse, loosely connected aggregations of nervous tissues in certain non-bilatarian animals such as cnidarians.

nervous tissue Tissue specialized for processing and communicating information; one of the four major tissue types in multicellular animals.

net primary production (NPP) The amount of energy or biomass incorporated into the tissues of primary producers after respiration. (Compare with gross primary production.)

net reproductive rate (R_0) The mean number of offspring produced by an individual in a population during its lifetime.

net secondary production The amount of energy or biomass obtained from the consumption and assimilation of other organisms, taking into account respiration.

neural crest cells During vertebrate neurulation, cells that migrate outward from the neural plate and give rise to connections between the central nervous system and the rest of the body.

neural network An organized group of neurons that contains three functional categories of neurons—afferent neurons, interneurons, and efferent neurons—and is capable of processing information.

neural tube An early stage in the development of the vertebrate nervous system consisting of a hollow tube created by two opposing folds of the dorsal ectoderm along the anterior–posterior body axis.

neurohormone A chemical signal produced and released by neurons that subsequently acts as a hormone.

neuromuscular junction Synapse (point of contact) where a motor neuron axon stimulates a muscle fiber cell.

neuron (noor' on) [Gk. *neuron*: nerve] A nervous system cell that can generate and conduct action potentials along an axon to a synapse with another cell.

neurotransmitter A substance produced in and released by a neuron (the presynaptic cell) that diffuses across a synapse and excites or inhibits another cell (the postsynaptic cell).

neurulation Stage in vertebrate development during which the nervous system begins to form.

neutral allele An allele that does not alter the functioning of the proteins for which it codes.

neutral model *See* lottery model.

neutral theory A view of molecular evolution that postulates that most mutations do not affect the amino acid being coded for, and that such mutations accumulate in a population at rates driven by genetic drift and mutation rates.

neutron (new' tron) One of the three fundamental particles of matter (along with protons and electrons), with mass slightly larger than that of a proton and no electrical charge.

niche (nitch) [L. *nidus*: nest] The set of physical and biological conditions a species requires to grow, reproduce, and survive.

nitrate reduction The process by which nitrate (NO_3^-) is reduced to ammonia (NH_3).

nitrification The oxidation of ammonia (NH_3) to nitrate (NO_3^-) in soil and seawater, carried out by chemoautotrophic bacteria (nitrifiers).

nitrifiers Prokaryotes that can convert ammonium into nitrate.

nitrogen fixation Conversion of atmospheric nitrogen gas (N_2) into a more reactive and biologically useful form (ammonia), which makes nitrogen available to living things. Carried out by nitrogen fixers—bacteria, some of them free-living and others living within plant roots.

nitrogen fixers Organisms capable of fixing atmospheric N_2 to form NH_3.

nitrogenase An enzyme complex found in nitrogen-fixing bacteria that mediates the stepwise reduction of atmospheric N_2 to ammonia and which is strongly inhibited by oxygen.

node [L. *nodus*: knob, knot] (1) In plants, a (sometimes enlarged) point on a stem where a leaf is or was attached. (2) A region at the anterior end of the primitive grove—the site of ingression of cells during gastrulation. In mammals and birds it serves a function similar to the dorsal lip of the blastopore in amphibians.

node of Ranvier A gap in the myelin sheath covering an axon; the point where the axonal membrane can fire action potentials.

non-disjunction Failure of sister chromatids to separate in meiosis II or mitosis, or failure of homologous chromosomes to separate in meiosis I. Results in aneuploidy.

non-REM sleep A state of deep, restorative sleep characterized by high-amplitude slow waves in the EEG. (Contrast with REM sleep.)

noncompetitive inhibitor A nonsubstrate that inhibits the activity of an enzyme by binding to a site other than its active site. (Contrast with competitive inhibitor.)

noncyclic electron transport In photosynthesis, the flow of electrons that forms ATP, NADPH, and O_2.

nonpolar Having electric charges that are evenly balanced from one end to the other. (Contrast with polar.)

nonpolar covalent bond A covalent bond between atoms that has equal sharing of electrons.

nonsense mutation Change in a gene's sequence that prematurely terminates translation by changing one of its codons to a stop codon. (Contrast with frame-shift mutation, missense mutation, silent mutation.)

nonsense substitution A change in a gene from one nucleotide to another that prematurely terminates translation by changing one of its codons to a stop codon. (Contrast with synonymous substitution.)

nonsynonymous substitution A change in a gene from one nucleotide to another that changes the amino acid specified by the corresponding codon (i.e., AGC → AGA, or serine → arginine). (Contrast with synonymous substitution.)

nonvascular land plants All land plants that lack a vascular system.

norepinephrine A neurotransmitter found in the central nervous system and also at the postganglionic nerve endings of the sympathetic nervous system. Also called noradrenaline.

notochord (no' tow kord) [Gk. *notos*: back + *chorde*: string] A flexible rod of gelatinous material serving as a support in the embryos of all chordates and in the adults of tunicates and lancelets.

nuclear envelope A double membrane that encloses the cell nucleus.

nucleases Digestive enzymes that break down nucleic acids.

nucleic acid (new klay' ik) A polymer made up of nucleotides, specialized for the storage, transmission, and expression of genetic information. DNA and RNA are nucleic acids.

nucleic acid hybridization A technique in which a single-stranded nucleic acid probe is made that is complementary to, and binds to, a target sequence, either DNA or RNA. The resulting double-stranded molecule is a hybrid.

nucleoid (new' klee oid) The region that harbors the chromosomes of a prokaryotic cell. Unlike the eukaryotic nucleus, it is not bounded by a membrane.

nucleolus (new klee' oh lus) A small, generally spherical body found within the nucleus of eukaryotic cells. The site of synthesis of ribosomal RNA.

nucleoside   A nucleotide without the phosphate group; a nitrogenous base attached to a sugar.

nucleosome   A portion of a eukaryotic chromosome, consisting of part of the DNA molecule wrapped around a group of histone molecules, and held together by another type of histone molecule. The chromosome is made up of many nucleosomes.

nucleotide   The basic chemical unit in nucleic acids, consisting of a pentose sugar, a phosphate group, and a nitrogen-containing base.

nucleus   (new' klee us) [L. *nux*: kernel or nut] (1) In cells, the centrally located compartment of eukaryotic cells that is bounded by a double membrane and contains the chromosomes. (2) In the brain, an identifiable group of neurons that share common characteristics or functions.

null hypothesis   In statistics, the premise that any differences observed in an experiment are simply the result of random differences that arise from drawing two finite samples from the same population.

nutrient   A food substance; or, in the case of mineral nutrients, an inorganic element required for completion of the life cycle of an organism.

O

obligate   A type of species interaction that is necessary to at least one species involved in the interaction. (Compare with facultative.)

obligate aerobe   An organism that requires an oxygen atmosphere to live.

obligate anaerobe   An anaerobic prokaryote that cannot survive exposure to O_2. (Compare with facultative anaerobe.)

obligate parasites   Organisms that can only survive and grow in or on other living organisms, to the detriment of the host.

occipital lobe   One of the four lobes of the brain's cerebral hemisphere; processes visual information.

ocean zone   The marine life zone that extends from the coastal zone into open ocean. Characterized by deep water that varies in its oxygenation and relatively stable temperatures and salinities.

odorant   A molecule that can bind to an olfactory receptor.

oil   A triglyceride that is liquid at room temperature. (Contrast with fat.)

Okazaki fragments   Newly formed DNA making up the lagging strand in DNA replication. DNA ligase links Okazaki fragments together to give a continuous strand.

olfaction   (ole fak' shun) [L. *olfacere*: to smell] The sense of smell.

olfactory bulb   Structure in the vertebrate forebrain that receives and processes input from olfactory receptor neurons.

olfactory receptor neurons (ORNs)   Neurons with receptors for different odorants.

oligodendrocyte   A type of glial cell that myelinates axons in the central nervous system.

oligosaccharide   A polymer containing a small number of monosaccharides.

omasum   The third of the four chambers of the stomach in ruminants; concentrates food by water absorption before it enters the true stomach (abomasum).

ommatidia   [Gk. *omma*: eye] The units that make up the compound eye of some arthropods.

omnivore   [L. *omnis*: everything + *vorare*: to devour] An organism that feeds on multiple trophic levels; also an organism that feeds on animals and primary producers. (Contrast with carnivore, detritivore, herbivore.)

omnivory   An act of predation in which the predator is animal and the prey is an animal or a plant. (Contrast with carnivory, herbivory, parasitism.)

oncogene   [Gk. *onkos*: mass, tumor + *genes*: born] A gene that codes for a protein product that stimulates cell proliferation. Mutations in oncogenes that result in excessive cell proliferation can give rise to cancer.

one gene–one polypeptide   The idea, now known to be an oversimplification, that each gene in the genome encodes only a single polypeptide—that there is a one-to-one correspondence between genes and polypeptides.

one-gene, one-enzyme hypothesis   *See* one gene–one polypeptide.

one-gene, one-polypeptide relationship   *See* one gene–one polypeptide.

oogonium   (oh' eh go' nee um) (plural: oogonia) (1) In some algae and fungi, a cell in which an egg is produced. (2) In animals, the diploid progeny of a germ cell in females.

ootid   In oogenesis, the daughter cell of the second meiotic division that differentiates into the mature ovum.

open circulatory system   Circulatory system in which extracellular fluid leaves the vessels of the circulatory system, percolates between cells and through tissues, and then flows back into the circulatory system to be pumped out again. (Contrast with closed circulatory system.)

operant conditioning   Using a reward or a punishment to control the expression of a behavior by an animal in response to an unrelated stimulus.

operator   The region of an operon that acts as the binding site for the repressor.

operon   A genetic unit of transcription, typically consisting of several structural genes that are transcribed together; the operon contains at least two control regions: the promoter and the operator.

opportunity cost   The sum of the benefits an animal forfeits by not being able to perform some other behavior during the time when it is performing a given behavior. (Compare with energetic cost, risk cost.)

opsin   (op' sin) [Gk. *opsis*: sight] The protein portion of vertebrate visual pigments; associated with the pigment molecule 11-*cis*-retinal. (*See also* rhodopsin.)

optic chiasm   [Gk. *chiasma*: cross] Structure on the lower surface of the vertebrate brain where the two optic nerves come together.

optic nerve   The nerve that carries information from the retina of the eye to the brain.

optical isomers   Two molecular isomers that are mirror images of each other.

optimal foraging theory   The application of a cost–benefit approach to feeding behavior to identify the fitness value of feeding choices.

oral   [L. *os*: mouth] Pertaining to the mouth, or that part of the body that contains the mouth.

orbital   A region in space surrounding the atomic nucleus in which an electron is most likely to be found.

ordinal variables   Categorical variables with a natural ordering, such as the grades A, B, C, D, and F.

organ   [Gk. *organon*: tool] A body part, such as the heart, liver, brain, root, or leaf. Organs are composed of different tissues integrated to perform a distinct function. Organs, in turn, are integrated into organ systems.

organ identity genes   In angiosperms, genes that specify the different organs of the flower. (Compare with homeotic genes.)

organ of Corti   Structure in the inner ear that transforms mechanical forces produced from pressure waves ("sound waves") into action potentials that are sensed as sound.

organ system   An interrelated and integrated group of tissues and organs that work together in a physiological function.

organelle   (or gan el') Any of the membrane-enclosed structures within a eukaryotic cell. Examples include the nucleus, endoplasmic reticulum, and mitochondria.

organic fertilizers   Substances added to soil to improve the soil's fertility; derived from partially decomposed plant material (compost) or animal waste (manure).

organizer   Region of the early amphibian embryo that directs early embryonic development. Also known as the primary embryonic organizer.

organogenesis   The formation of organs and organ systems during development.

origin of replication (*ori*)   DNA sequence at which helicase unwinds the DNA double helix and DNA polymerase binds to initiate DNA replication.

ortholog   [Gk. *ortho*: true, direct] Homologous gene whose divergence can be traced to speciation events. (Contrast with paralog.)

osmoconformer   An aquatic animal that equilibrates the osmolarity of its extracellular fluid to be the same as that of the external environment. (Contrast with osmoregulator.)

osmolarity   The concentration of osmotically active particles in a solution.

osmoregulator   An aquatic animal that actively regulates the osmolarity of its extracellular fluid. (Contrast with osmoconformer.)

osmosis   (oz mo' sis) [Gk. *osmos*: to push] Movement of water across a differentially permeable membrane, from one region to another region where the water potential is more negative.

ossicles   Three bones in the middle ear that transmit vibrations from tympanum to the oval window of the inner ear. Individually named the malleus ("hammer"), incus ("anvil"), and stapes ("stirrup").

osteoblast (oss' tee oh blast) [Gk. *osteon*: bone + *blastos*: sprout] A cell that lays down the protein matrix of bone.

osteoclast (oss' tee oh clast) [Gk. *osteon*: bone + *klastos*: broken] A cell that dissolves bone.

osteocyte (oss' tee oh site) [Gk. *osteon*: bone + *kytos*: cell] An osteoblast that has become enclosed in lacunae within the bone it has built.

otoliths (oh' teliths) [Gk. *ous*: ear + *lithos*: stones] Crystals of calcium carbonate in the inner ear. Also called ear stones.

outgroup In phylogenetics, a group of organisms used as a point of reference for comparison with the groups of primary interest (the ingroup).

oval window The flexible membrane that, when moved by the bones of the middle ear, produces pressure waves in the inner ear.

ovarian cycle (oh var' ee un) In human females, the monthly cycle of events by which eggs and hormones are produced. (Contrast with uterine cycle).

ovary (oh' var ee) [L. *ovum*: egg] Any female organ, in plants or animals, that produces an egg.

overtopping Plant growth pattern in which one branch differentiates from and grows beyond the others.

oviduct In mammals, the tube serving to transport eggs to the uterus or to the outside of the body.

oviparity Reproduction in which eggs are released by the female and development is external to the mother's body. (Contrast with viviparity.)

ovoviviparity Pertaining to reproduction in which fertilized eggs develop and hatch within the mother's body but are not attached to the mother by means of a placenta.

ovulation Release of an egg from an ovary.

ovule (oh' vule) In plants, a structure comprising the megasporangium and the integument, which develops into a seed after fertilization.

ovum (oh' vum) (plural: ova) [L. egg] The female gamete.

oxaloacetate Four-carbon salt of an acid with the formula HOOC—CO—CO—COOH that is an intermediate in metabolic pathways such as the citric acid cycle.

oxidation (ox i day' shun) Relative loss of electrons in a chemical reaction; either outright removal to form an ion, or the sharing of electrons with substances having a greater affinity for them, such as oxygen. Most oxidations, including biological ones, are associated with the liberation of energy. (Contrast with reduction.)

oxidation–reduction (redox) reaction A reaction in which one substance transfers one or more electrons to another substance. (*See* oxidation, reduction.)

oxidative phosphorylation ATP formation in the mitochondrion, associated with flow of electrons through the respiratory chain.

oxygenase An enzyme that catalyzes the addition of oxygen to a substrate from O_2.

oxytocin A hormone released by the posterior pituitary that promotes social bonding.

P

***P*-value** The calculated probability of observing a given result by chance sampling, given the null hypothesis is true.

pacemaker cells Cardiac cells that can initiate action potentials without stimulation from the nervous system, allowing the heart to initiate its own contractions.

Pacinian corpuscles (puh sin' ee uhn) Encapsulated mechanoreceptors in deep skin sensitive to pressure and vibration.

pair rule genes In *Drosophila* (fruit fly) development, segmentation genes that divide the early embryo into units of two segments each. Part of a developmental cascade that includes maternal effect genes, gap genes, segment polarity genes, and Hox genes.

pancreas (pan' cree us) A gland located near the stomach of vertebrates that secretes digestive enzymes into the small intestine and releases insulin into the bloodstream.

parabronchi Passages in the lungs of birds through which air flows.

paracrine [Gk. *para*: near] Pertaining to a chemical signal, such as a hormone, that acts locally, near the site of its secretion. (Contrast with autocrine.)

parallel evolution The repeated evolution of similar traits, especially among closely related species; facilitated by conserved developmental genes.

paralog Homologous gene whose divergence can be traced to gene duplication events. (Contrast with ortholog.)

paraphyletic (par' a fih leht' ik) [Gk. *para*: beside + *phylon*: tribe] Pertaining to a group that consists of an ancestor and some, but not all, of its descendants. (Contrast with monophyletic, polyphyletic.)

parasite [Gk. *parasitos*: one who eats at another's table] An organism that kills and/or consumes parts of an organism much larger than itself (known as its host). Parasites sometimes, but not always, kill their host.

parasitic plants Plants that obtains some of their nutrition by growing on other plants.

parasitism An act of predation in which the predator (a parasite) lives symbiotically on or in the prey (its host) and consumes only certain tissues without necessarily killing the host. (Contrast with carnivory, herbivory, omnivory.)

parasympathetic nervous system The division of the autonomic nervous system that works in opposition to the sympathetic nervous system. (Contrast with sympathetic nervous system.)

parathyroid glands Four glands on the posterior surface of the thyroid gland that produce and release parathyroid hormone.

parathyroid hormone (PTH) A hormone secreted by the parathyroid glands that stimulates osteoclast activity and raises blood calcium levels. Also called parathormone. (Contrast with calcitonin.)

parenchyma (pair eng' kyma) A plant tissue composed of relatively unspecialized cells without secondary walls. (Contrast with collenchyma, sclerenchyma.)

parent rock The soil horizon consisting of the rock that is breaking down to form the soil. Also called bedrock, or the C horizon.

parental (P) generation The individuals that mate in a genetic cross. Their offspring are the first filial (F_1) generation.

parietal cells One of three types of secretory cell found in the gastric pits of the stomach wall. Parietal cells produce hydrochloric acid (HCl), creating an acidic environment that destroys many of the harmful microorganisms ingested with food. (*See* chief cells.)

parietal lobe One of four lobes of the cerebral hemisphere; processes complex stimuli and includes the primary somatosensory cortex.

parsimony principle Preferring the simplest among a set of plausible explanations of any phenomenon.

parthenocarpy Formation of fruit from a flower without fertilization.

parthenogenesis [Gk. *parthenos*: virgin] Production of an organism from an unfertilized egg.

partial pressures In a mixture of gasses at a specific pressure, the contribution of each gas to that total pressure.

passive transport Diffusion across a membrane; may or may not require a channel or carrier protein. (Contrast with active transport.)

patch clamping Technique for isolating a tiny patch of membrane to allow the study of ion movement through a particular channel.

pathogen (path' o jen) [Gk. *pathos*: suffering + *genesis*: source] An organism that causes disease.

pathogen associated molecular patterns (PAMPs) The molecules recognized by pattern recognition receptors.

pathogenesis-related (PR) protein A plant protein involved in systemic acquired resistance whose synthesis is induced in response to an infection by a pathogen such as a fungus.

pattern formation In animal embryonic development, the organization of differentiated tissues into specific structures such as wings.

pattern recognition receptors (PRRs) Proteins expressed by cells of the innate immune system in animals or plants that bind to molecules from pathogens containing pathogen associated molecular patterns (PAMPs).

PCR *See* polymerase chain reaction.

peat Partly decomposed vegetable matter, especially of sphagnum moss, that is typically formed in acidic bogs.

pedigree The pattern of transmission of a genetic trait within a family.

penetrance The proportion of individuals with a particular genotype that show the expected phenotype.

penis An accessory sex organ of male animals that enables the male to deposit sperm in the female's reproductive tract.

pentaradial symmetry   Symmetry in five or multiples of five; a feature of adult echinoderms.

pentose   [Gk. *penta*: five] A sugar containing five carbon atoms.

PEP carboxylase   The enzyme that combines carbon dioxide with PEP to form a 4-carbon dicarboxylic acid at the start of C_4 photosynthesis or of crassulacean acid metabolism (CAM).

peptidases   Digestive enzymes that break down peptides.

peptide hormones   Hormone molecules made up of chains of 2 to 50 amino acids (upper size arbitrary); encoded by genes and produced by translation.

peptide linkage   The bond between amino acids in a protein; formed between a carboxyl group and amino group (—CO—NH—) with the loss of water molecules.

peptidoglycan   The cell wall material of many bacteria, consisting of a single enormous molecule that surrounds the entire cell.

peptidyl transferase   A catalytic function of the large ribosomal subunit that consists of two reactions: breaking the bond between an amino acid and its tRNA in the P site, and forming a peptide bond between that amino acid and the amino acid attached to the tRNA in the A site.

per capita birth rate (*b*)   In population growth models, the average number of offspring that an individual produces over some time interval.

per capita death rate (*d*)   In population growth models, the average individual's chance of dying in some time interval.

per capita growth rate (*r*)   In population models, the average individual's contribution to total population growth rate. In mathematical terms, the difference between per capita birth rate (*b*) and per capita death rate (*d*) or (*b-d*).

perennial   (per ren' ee al) [L. *per*: throughout + *annus*: year] A plant that survives from year to year. (Contrast with annual, biennial.)

perfect flower   A flower with both stamens and carpels; a hermaphroditic flower. (Contrast with imperfect flower.)

pericycle   [Gk. *peri*: around + *kyklos*: ring or circle] In plant roots, tissue just within the endodermis, but outside of the root vascular tissue. Meristematic activity of pericycle cells produces lateral root primordia.

periderm   The outer tissue of the secondary plant body, consisting primarily of cork.

period   (1) A category in the geological time scale. (2) The duration of a single cycle in a cyclical event, such as a circadian rhythm.

periodic table   The arrangement of elements in order of increasing atomic number, with vertical groupings of similar valence electrons.

peripheral membrane proteins   Proteins associated with but not embedded within the plasma membrane. (Contrast with integral membrane proteins.)

peripheral nervous system (PNS)   The portion of the nervous system that transmits information to and from the central nervous system, consisting of neurons that extend or reside outside the brain or spinal cord and their supporting cells. (Contrast with central nervous system.)

peristalsis   (pair' i stall' sis) Wavelike muscular contractions proceeding along a tubular organ, propelling the contents along the tube.

peritoneum   The mesodermal lining of the body cavity in coelomate animals.

peritubular capillaries   Small blood vessels that surround the renal tubules.

peroxisome   An organelle that houses reactions in which toxic peroxides are formed and then converted to water.

petal   [Gk. *petalon*: spread out] In an angiosperm flower, a sterile modified leaf, nonphotosynthetic, frequently brightly colored, and often serving to attract pollinating insects.

P$_{fr}$ *See* phytochrome.

pH   The negative logarithm of the hydrogen ion concentration; a measure of the acidity of a solution. A solution with pH = 7 is said to be neutral; pH values higher than 7 characterize basic solutions, while acidic solutions have pH values less than 7.

phage therapy   The therapeutic use of bacteriophages to treat pathogenic bacterial infections.

phagocyte   [Gk. *phagein*: to eat + *kystos*: sac] One of two major classes of white blood cells; one of the nonspecific defenses of animals; ingests invading microorganisms by phagocytosis.

phagocytosis   The process by which a cell uses its cell membrane to enclose and engulf a macromolecule or particle in the extracellular environment.

pharmacogenomics   The study of how an individual's genetic makeup affects his or her response to drugs or other agents, with the goal of predicting the effectiveness of different treatment options.

pharming   The use of genetically modified animals to produce medically useful products in their milk.

pharynx   [Gk. throat] The part of the gut between the mouth and the esophagus.

phenotype   (fee' no type) [Gk. *phanein*: to show] The observable properties of an individual resulting from both genetic and environmental factors. (Contrast with genotype.)

pheromone   (feer' o mone) [Gk. *pheros*: carry + *hormon*: excite, arouse] A chemical substance used in communication between organisms of the same species.

phloem   (flo' um) [Gk. *phloos*: bark] In vascular plants, the vascular tissue that transports sugars and other solutes from sources to sinks.

phloem sap   Liquid contents on the phloem tubes.

phosphatidyl inositol-bisphosphate (PIP$_2$)   A phospholipid in the cell membrane that is involved in cell signaling after a ligand binds to a nearby receptor.

phosphodiester linkage   The connection in a nucleic acid strand, formed by linking two nucleotides.

phosphoenolpyruvate (PEP)   Phosphorylated salt of a three-carbon acid, formula OPO_3H_2—CO—CO—COOH that is an intermediate in metabolic pathways such as glycolysis.

phospholipid   A lipid containing a phosphate group; an important constituent of cellular membranes. (*See* lipid.)

phospholipid bilayer   The basic structural unit of biological membranes; a sheet of phospholipids two molecules thick in which the phospholipids are lined up with their hydrophobic "tails" packed tightly together and their hydrophilic, phosphate-containing "heads" facing outward. Also called lipid bilayer.

phosphorylation   Addition of a phosphate group.

photic zone   The life zone in lakes and oceans that is penetrated by light and therefore supports photosynthetic organisms.

photoautotroph   An organism that obtains energy from light and carbon from carbon dioxide. (Contrast with chemoautotroph, chemoheterotroph, photoheterotroph.)

photoheterotroph   An organism that obtains energy from light but must obtain its carbon from organic compounds. (Contrast with chemoautotroph, chemoheterotroph, photoautotroph.)

photomorphogenesis   In plants, a process by which physiological and developmental events are controlled by light.

photon   (foe' ton) [Gk. *photos*: light] A quantum of visible radiation; a "packet" of light energy.

photoperiodism   Control of an organism's physiological or behavioral responses by the length of the day or night (the **photoperiod**).

photophosphorylation   Mechanism for ATP formation in chloroplasts in which electron transport is coupled to the transport of hydrogen ions (protons, H$^+$) across the thylakoid membrane. (Compare with chemiosmosis.)

photoreceptors   (1) In plants, pigments that trigger a physiological response when they absorb a photon. (2) In animals, the sensory receptor cells that sense and respond to light energy. (*See* cone cells, rod cells.)

photorespiration   Light-driven uptake of oxygen and release of carbon dioxide, the carbon being derived from the early reactions of photosynthesis.

photosensitivity   Sensitivity to light.

photosynthates   Carbohydrate products of photosynthesis.

photosynthesis   (foe tow sin' the sis) [Gk. creating from light] Metabolic processes carried out by green plants and some microorganisms by which visible light is trapped and the energy used to synthesize compounds such as ATP and glucose.

photosynthetic lamellae   The internal membrane system of the cyanobacteria.

photosystem   [Gk. *phos*: light + *systema*: assembly] A light-harvesting complex in the chloroplast thylakoid composed of pigments and proteins. **Photosystem I** absorbs light at 700 nm, passing electrons to ferredoxin and from there to

NADPH. **Photosystem II** absorbs light at 680 nm and passes electrons to the electron transport chain in the chloroplast.

photosystem I *See* photosystem.

photosystem II *See* photosystem.

phototropin Photoreceptor proteins that mediate light reception in plants.

phototropism [Gk. *photos*: light + *trope*: turning] A directed plant growth response to light.

phycoerythrin A red protein pigment that occurs in red algae.

phylogenetic tree A graphic representation of these lines of evolutionary descent.

phylogeny (fy loj' e nee) [Gk. *phylon*: tribe, race + *genesis*: source] The evolutionary history of a particular group of organisms or their genes.

phytoalexins Substances toxic to pathogens, produced by plants in response to fungal or bacterial infection.

phytochrome (fy' tow krome) [Gk. *phyton*: plant + *chroma*: color] A plant pigment regulating a large number of developmental and other phenomena in plants. It has two isomers: P_r, which absorbs red light, and P_{fr}, which absorbs far red light. P_{fr} is the active form.

phytomers In plants, the repeating modules that compose a shoot, each consisting of one or more leaves, attached to the stem at a node; an internode; and one or more axillary buds.

phytoremediation A form of bioremediation that uses plants to clean up environmental pollution.

pie chart A circular figure that displays proportions of different classes of data in an observed sample.

pigment A substance that absorbs visible light.

piloting A form of navigation in which an animal finds its way by remembering landmarks in its environment.

pilus (plural: pili) A structure that links prokaryote cells at the beginning of conjugation.

pineal gland Gland located between the cerebral hemispheres that secretes melatonin.

pinocytosis Endocytosis by a cell of liquid containing dissolved substances.

pioneer species Species that arrive first to primary successional habitats and must deal with extreme conditions; they tend to have life history strategies that maximize population growth (see *r*-strategists). Also called early succession species.

pistil [L. *pistillum*: pestle] The structure of an angiosperm flower within which the ovules are borne. May consist of a single carpel, or of several carpels fused into a single structure. Usually differentiated into ovary, style, and stigma.

pith In plants, relatively unspecialized tissue found within a cylinder of vascular tissue.

pituitary gland A small gland attached to the base of the brain in vertebrates. Its hormones control the activities of other glands. Also known as the hypophysis.

placenta (pla sen' ta) The organ in female mammals that provides for the nourishment of the fetus and elimination of the fetal waste products.

plankton Aquatic organisms that float in the water column, dependent on currents and wind for movement. Plankton include many protists, some algae, and larval animals.

planula (plan' yew la) [L. *planum*: flat] A free-swimming, ciliated larval form typical of the cnidarians.

plaque (plack) [Fr. a metal plate or coin] (1) A circular clearing in a layer (lawn) of bacteria growing on the surface of a nutrient agar gel. (2) An accumulation of prokaryotic organisms on tooth enamel. Acids produced by these microorganisms cause tooth decay. (3) A region of arterial wall invaded by fibroblasts and fatty deposits.

plasma cell An antibody-secreting cell that develops from a B cell; the effector cell of the humoral immune system.

plasmid A DNA molecule distinct from the chromosome(s); that is, an extrachromosomal element; found in many bacteria. May replicate independently of the chromosome.

plasmodesmata (singular: plasmodesma) [Gk. *plassein*: to mold + *desmos*: band] Cytoplasmic strands connecting two adjacent plant cells.

plasmogamy The fusion of the cytoplasm of two cells. (Contrast with karyogamy.)

plate tectonics [Gk. *tekton*: builder] The scientific study of the structure and movements of Earth's lithospheric plates, which are the cause of continental drift.

platelet A membrane-bounded body without a nucleus, arising as a fragment of a cell in the bone marrow of mammals. Important to blood-clotting action.

pleiotropy (plee' a tro pee) [Gk. *pleion*: more] The determination of more than one character by a single gene.

pleural membrane [Gk. *pleuras*: rib, side] The membrane lining the outside of the lungs and the walls of the thoracic cavity. Inflammation of these membranes is a condition known as pleurisy.

pluripotent [L. *pluri*: many + *potens*: powerful] Having the ability to form all of the cells in the body. (Contrast with multipotent, totipotent, unipotent.)

pneumatophores Roots that grow upward out of the water or mud to reach the air and obtain oxygen for the root systems of trees that live in swampy or tidal habitats.

podocytes Cells of Bowman's capsule of the nephron that cover the capillaries of the glomerulus, forming filtration slits.

poikilotherms Animals whose body temperatures are largely dependent on environmental temperature.

point mutation A mutation that results from the gain, loss, or substitution of a single nucleotide. (Contrast with deletion, duplication.)

polar A molecule with separate and opposite electric charges at two ends, or poles; the water molecule (H_2O) is the most prevalent example. (Contrast with nonpolar.)

polar covalent bond A covalent bond in which the electrons are drawn to one nucleus more than the other, resulting in an unequal distribution of charge.

polar nuclei In angiosperms, the two nuclei in the central cell of the megagametophyte; following fertilization they give rise to the endosperm.

polarity (1) In chemistry, the property of unequal electron sharing in a covalent bond that defines a polar molecule. (2) In development, the difference between one end of an organism or structure and the other.

pole plasm Cytoplasm at one end of an insect egg that determines that the cells that form at that location will become germ cells.

pollen grain [L. *pollin*: fine flour] In seed plants, microscopic grains that contain the male gametophyte (microgametophyte) and gamete (microspore).

pollen tube A structure that develops from a pollen grain through which sperm are released into the megagametophyte.

pollination The process of transferring pollen from an anther to the stigma of a pistil in an angiosperm or from a strobilus to an ovule in a gymnosperm.

poly A tail A long sequence of adenine nucleotides (50–250) added after transcription to the 3' end of most eukaryotic mRNAs.

polyandry Mating system in which one female mates with multiple males. (Contrast with polygyny.)

polyculture In agriculture, a large scale planting of multiple crop species. (Contrast with monoculture.)

polygyny Mating system in which one male mates with multiple females. (Contrast with polyandry.)

polymer [Gk. *poly*: many + *meros*: unit] A large molecule made up of similar or identical subunits called monomers. (Contrast with monomer.)

polymerase chain reaction (PCR) An enzymatic technique for the rapid production of millions of copies of a particular stretch of DNA where only a small amount of the parent molecule is available.

polymorphic (pol' lee mor' fik) [Gk. *poly*: many + *morphe*: form, shape] Coexistence in a population of two or more distinct traits.

polyp (pah' lip) [Gk. *poly*: many + *pous*: foot] In cnidarians, a sessile, asexual life cycle stage.

polypeptide A large molecule made up of many amino acids joined by peptide linkages. Large polypeptides are called proteins.

polyphyletic (pol' lee fih leht' ik) [Gk. *poly*: many + *phylon*: tribe] Pertaining to a group that consists of multiple distantly related organisms, and does not include the common ancestor of the group. (Contrast with monophyletic, paraphyletic.)

polyploid (pol' lee ploid) Possessing more than two entire sets of chromosomes.

polyploidy (pol' lee ploid ee) The possession of more than two entire sets of chromosomes.

polyribosome (polysome) A complex consisting of a threadlike molecule of messenger RNA and several (or many) ribosomes. The ribosomes move along the mRNA, synthesizing polypeptide chains as they proceed.

polysaccharide A macromolecule composed of many monosaccharides (simple sugars). Common examples are cellulose and starch.

pons [L. *pons*: bridge] Region of the brainstem anterior to the medulla.

population (1) In biology, any group of organisms coexisting at the same time and in the same place and capable of interbreeding with one another. (2) In statistics, a study group of interest.

population bottleneck A period during which only a few individuals of a normally large population survive.

population density The number of individuals in a population per unit of area or volume.

population dynamics The patterns and processes of change in populations over space and time.

population growth rate The rate of change in population size over time.

population growth The change in population size over time.

population size The total number of individuals in a population.

portal blood vessels Blood vessels that begin and end in capillary beds.

positional information In development, the basis of the spatial sense that induces cells to differentiate as appropriate for their location within the developing organism; often comes in the form of a morphogen gradient.

positive cooperativity Occurs when a molecule can bind several ligands and each one that binds alters the conformation of the molecule so that it can bind the next ligand more easily. The binding of four molecules of O_2 by hemoglobin is an example of positive cooperativity.

positive feedback In regulatory systems, information that amplifies a regulatory response, increasing the deviation of the system from the set point. (Contrast with negative feedback.)

positive interaction Trophic or nontrophic interaction between two species in which one or both species benefit and neither is harmed. (*See also* facilitation, mutualism, commensalism.)

positive regulation A form of gene regulation in which a regulatory macromolecule is needed to turn on the transcription of a structural gene; in its absence, transcription will not occur. (Contrast with negative regulation.)

positive relationship A relationship in which two variables tend to vary among observations in the same direction.

positive selection Natural selection that acts to establish a trait that enhances survival in a population. (Contrast with purifying selection.)

postabsorptive state State in which no food remains in the gut and thus no nutrients are being absorbed. (Contrast with absorptive state.)

posterior Toward or pertaining to the rear. (Contrast with anterior.)

posterior pituitary A portion of the pituitary gland derived from neural tissue; involved in the storage and release of antidiuretic hormone and oxytocin.

postzygotic isolating mechanisms Barriers to the reproductive process that occur after the union of the nuclei of two gametes. (Contrast with prezygotic isolating mechanisms.)

potassium equilibrium potential (E_K) The negative membrane potential that balances the tendency for K^+ ions to diffuse out of the cell down its concentration gradient.

potential energy Energy not doing work but with the potential to do so, such as the energy stored in chemical bonds. (Contrast with kinetic energy.)

power In reference to statistical tests: refers to the probability of correctly rejecting a null hypothesis when it is false.

P_r *See* phytochrome.

pre-motor cortex The region of the frontal lobe of the cerebrum that is anterior to the primary motor cortex.

pre-mRNA (precursor mRNA) Initial gene transcript before it is modified to produce functional mRNA. Also known as the primary transcript.

Precambrian The first and longest period of geological time, during which life originated.

precapillary sphincter A cuff of smooth muscle that can shut off the blood flow to a capillary bed.

precocial Born or hatched in a relatively advanced state that allows a high degree of independence from the parents.

predation A trophic or feeding interaction in which an individual of one species (a predator) kills and/or consumes individuals (or parts of individuals) of another species (its prey).

predator An organism that kills and/or consumes individuals (or parts of individuals) of another species (its prey).

pressure flow model An effective model for phloem transport in angiosperms. It holds that sieve element transport is driven by an osmotically generated pressure gradient between source and sink.

pressure potential (Ψ_p) The hydrostatic pressure of an enclosed solution in excess of the surrounding atmospheric pressure. (Contrast with solute potential, water potential.)

prevailing winds The circulation of air (wind) that moves across the Earth's surface in a predominately single direction.

prey [L. *praeda*: booty] An organism killed and/or consumed by a predator as an energy source.

prezygotic isolating mechanisms Barriers to the reproductive process that occur before the union of the nuclei of two gametes (Contrast with postzygotic isolating mechanisms.)

primary active transport Active transport in which ATP is hydrolyzed, yielding the energy required to transport an ion or molecule against its concentration gradient. (Contrast with secondary active transport.)

primary bronchi The first airways that branch off of the trachea.

primary cell wall In plant cells, a structure that forms at the middle lamella after cytokinesis, made up of cellulose microfibrils, hemicelluloses, and pectins. (Contrast with secondary cell wall.)

primary consumer An organism (herbivore) that feeds on primary producers.

primary embryonic organizer *See* organizer.

primary endosymbiosis The engulfment of a cyanobacterium by a larger eukaryotic cell that gave rise to the first photosynthetic eukaryotes with chloroplasts.

primary growth In plants, growth that is characterized by the lengthening of roots and shoots and by the proliferation of new roots and shoots through branching. (Contrast with secondary growth.)

primary immune response The first response of the immune system to an antigen, involving recognition by lymphocytes and the production of effector cells and memory cells. (Contrast with secondary immune response.)

primary lysosomes A lysosome before fusion with an endosome.

primary meristem Meristem that produces the tissues of the primary plant body.

primary motor cortex *See* motor cortex.

primary oocyte (oh' eh site) [Gk. *oon*: egg + *kytos*: container] The diploid progeny of an oogonium. In many species, a primary oocyte enters prophase of the first meiotic division, then remains in developmental arrest for a long time before resuming meiosis to form a secondary oocyte and a polar body.

primary producer A photosynthetic or chemosynthetic organism that synthesizes complex organic molecules from simple inorganic ones.

primary sex determination Genetic determination of gametic sex, male or female.

primary sex organs Testes and ovaries.

primary somatosensory cortex *See* somatosensory cortex.

primary spermatocyte The diploid progeny of a spermatogonium; undergoes the first meiotic division to form secondary spermatocytes.

primary structure The specific sequence of amino acids in a protein.

primary succession Succession of ecological communities that begins in an area devoid of life, such as on recently exposed glacial till or lava flows. (Contrast with secondary succession.)

primase (prigh' maze) An enzyme that catalyzes the synthesis of a primer for DNA replication.

primer Strand of nucleic acid, usually RNA, that is the necessary starting material for the synthesis of a new DNA strand, which is synthesized from the 3' end of the primer.

primitive streak A midline ridge that forms in the avian blastula that will become the blastopore.

probability A numerical quantity that expresses the likelihood of an event occurring on a scale from 0 (no chance of the event) to 1 (certainty of the event).

probe A segment of single-stranded nucleic acid used to identify DNA molecules containing the complementary sequence.

proboscis An elongated nose or sucking mouthpart that is typically tubular and flexible.

procambium Primary meristem that produces the vascular tissue.

processive Pertaining to an enzyme that catalyzes many reactions each time it binds to a substrate, as DNA polymerase does during DNA replication.

production efficiency The proportion of assimilated biomass used to produce new consumer biomass or net secondary production.

products The molecules that result from the completion of a chemical reaction.

progesterone [L. *pro*: favoring + *gestare*: to bear] A female sex hormone that maintains pregnancy.

prokaryotes Unicellular organisms that do not have nuclei or other membrane-enclosed organelles. Includes Bacteria and Archaea. (Contrast with eukaryotes.)

promoter A DNA sequence to which RNA polymerase binds to initiate transcription.

proofreading During DNA replication, a mechanism that excises a base that is incorrectly inserted according to the template (e.g., an A opposite a template G) and inserts the correct base (e.g., C opposite template G).

prop roots Adventitious roots in some monocots that function as supports for the shoot.

prophage (pro′ fayj) The noninfectious units that are linked with the chromosomes of the host bacteria and multiply with them but do not cause dissolution of the cell. Prophage can later enter into the lytic phase to complete the virus life cycle.

prostaglandin Any one of a group of specialized lipids with hormone-like functions. It is not clear that they act at any considerable distance from the site of their production.

prostate gland In male humans, surrounds the urethra at its junction with the vas deferens; supplies an acid-neutralizing fluid to the semen.

proteases Digestive enzymes that break the bonds between adjacent amino acids in proteins.

proteasome In the eukaryotic cytoplasm, a huge protein structure that binds to and digests cellular proteins that have been tagged by ubiquitin.

protected areas An area in which habitat loss or degradation by human activity is restricted or prohibited.

protein (pro′ teen) [Gk. *protos*: first] Long-chain polymer of amino acids with twenty different common side chains. Occurs with its polymer chain extended in fibrous proteins, or coiled into a compact macromolecule in enzymes and other globular proteins. The component amino acids are encoded in the triplets of messenger RNA, and proteins are the products of genes.

protein hormones Large hormone molecules made up of chains of generally greater than 50 amino acids; encoded by genes and produced by translation.

protein kinase (kye′ nase) An enzyme that catalyzes the addition of a phosphate group from ATP to a target protein.

protein kinase cascade A series of reactions in response to a molecular signal, in which a series of protein kinases activate one another in sequence, amplifying the signal at each step.

proteoglycan A glycoprotein containing a protein core with attached long, linear carbohydrate chains.

proteolysis [protein + Gk. *lysis*: break apart] An enzymatic digestion of a protein or polypeptide.

proteome The complete set of proteins that can be made by an organism. Because of alternative splicing of pre-mRNA, the number of proteins that can be made is usually much larger than the number of protein-coding genes present in the organism's genome.

prothoracic gland A gland in insects that secretes the molting hormone ecdysone.

prothoracicotropic hormone (PTTH) An insect hormone that controls the prothoracic gland.

prothrombin The inactive form of thrombin, an enzyme involved in blood clotting.

protocells A self-organized sphere of lipids proposed as a stepping-stone to the origin of life.

protoderm Primary meristem that gives rise to the plant epidermis.

proton (pro′ ton) [Gk. *protos*: first, before] (1) A subatomic particle with a single positive charge. The number of protons in the nucleus of an atom determine its element. (2) A hydrogen ion, H⁺.

proton pump An active transport system that uses ATP energy to move hydrogen ions across a membrane, generating an electric potential.

proton-motive force Force generated across a membrane having two components: a chemical potential (difference in proton concentration) plus an electrical potential due to the electrostatic charge on the proton.

protonephridium (plural: protonephridia) [Gk. *proto*: before + *nephros*: kidney] The excretory organ of flatworms, made up of a tubule and a flame cell.

provirus Double-stranded DNA made by a virus that is integrated into the host's chromosome and contains promoters that are recognized by the host cell's transcription apparatus.

proximal convoluted tubule The initial segment of a renal tubule, closest to the glomerulus. (Compare with distal convoluted tubule.)

proximate causes The immediate genetic, physiological, neurological, and developmental mechanisms responsible for a behavior or morphology. (Contrast with ultimate causes.)

PRRs *See* pattern recognition receptors.

pseudocoelomate (soo′ do see′ low mate) [Gk. *pseudes*: false + *koiloma*: cavity] Having a body cavity, called a pseudocoel, consisting of a fluid-filled space in which many of the internal organs are suspended, but which is enclosed by mesoderm only on its outside.

pseudogene [Gk. *pseudes*: false] A DNA segment that is homologous to a functional gene but is not expressed because of changes to its sequence or changes to its location in the genome.

pseudoplasmodium An aggregation of myxamoebas that resembles a plasmodium.

PTH *See* parathyroid hormone.

pulmonary arteries In vertebrates, blood vessels that carry blood from the heart to the lungs.

pulmonary circuit The portion of the circulatory system by which blood is pumped from the heart to the lungs or gills for oxygenation and back to the heart for distribution. (Contrast with systemic circuit.)

pulmonary valve A one-way valve between the right ventricle of the heart and the pulmonary artery that prevents backflow of blood into the ventricle when it relaxes.

pulmonary veins Veins that return oxygenated blood from the lungs to the left atrium.

pulp cavity The space in the interior of the tooth that contains blood vessels, nerves, and dentine-producing cells.

Punnett square Method of predicting the results of a genetic cross by arranging the gametes of each parent at the edges of a square.

pupil The opening in the vertebrate eye through which light passes.

purifying selection The elimination by natural selection of detrimental characters from a population. (Contrast with positive selection.)

purine (pure′ een) One of the two types of nitrogenous bases in nucleic acids. Each of the purines—adenine and guanine—pairs with a specific pyrimidine.

Purkinje fibers (pur kin′ jee) Specialized heart muscle cells that conduct excitation throughout the ventricular muscle.

pyrimidine (pe rim′ a deen) One of the two types of nitrogenous bases in nucleic acids. Each of the pyrimidines—cytosine, thymine, and uracil—pairs with a specific purine.

pyrogen [Gk. *pry*, fire] Molecule that produces a rise in body temperature (fever); may be produced by an invading pathogen or by cells of the immune system in response to infection.

pyruvate The ionized form of pyruvic acid, a three-carbon acid; the end product of glycolysis and the raw material for the citric acid cycle.

Q

Q_{10} A value that compares the rate of a biochemical process or reaction over 10°C temperature ranges. A process that is not temperature-sensitive has a Q_{10} of 1; values of 2 or 3 mean the reaction speeds up as temperature increases.

quadrat A four sided measurement area of any size used to count the individuals within the area.

qualitative Based on observation of an unmeasured quality of a trait, as in brown vs. blue.

quantitative Based on numerical values obtained by measurement, as in quantitative data.

quantitative trait loci A set of genes determining a complex character (trait) that exhibits quantitative variation (variation in amount rather than in kind).

quantitative variables Variables that are measured by a quantity, as opposed to ones that differ only in a quality. (Contrast with categorical variables.)

quaternary structure The specific three-dimensional arrangement of protein subunits. (Contrast with primary, secondary, tertiary structure.)

quorum sensing The use of chemical communication signals to trigger density-linked activities such as biofilm formation in prokaryotes.

R

R group The distinguishing group of atoms of a particular amino acid; also known as a side chain.

r-strategist A species whose life history strategy allows for a high intrinsic rate of population increase (*r*). (Contrast with *K*-strategist.)

radial cleavage Embryonic development in which the planes of cell division are parallel and perpendicular to the animal-vegetal axis of the embryo. (Compare to spiral cleavage.)

radial symmetry The condition in which any two halves of a body are mirror images of each other, providing the cut passes through the center; a cylinder cut lengthwise down its center displays this form of symmetry.

radiation The transfer of heat from warmer objects to cooler ones via the exchange of infrared radiation. (*See also* electromagnetic radiation, evolutionary radiation.)

radicle An embryonic root.

radioactive decay Spontaneous disintegration of a substance with the emission of ionizing radiation.

radioisotope A radioactive isotope of an element. Examples are carbon-14 (^{14}C) and hydrogen-3, or tritium (^3H).

radiometric dating A method for determining the age of objects such as fossils and rocks based on the decay rates of radioactive isotopes.

radula A rasplike structure used for scraping food particles off a surface.

rain shadow The relatively dry area on the down-wind side of a mountain range.

random dispersion The spatial arrangement of individuals within a population, characterized by random spatial patterns. (Compare with clumped dispersion, regular (uniform) dispersion.)

range The largest minus the smallest observed value for a variable in a sample.

rapid eye movement (REM) sleep A sleep state characterized by vivid dreams, skeletal muscle relaxation, and rapid eye movements. (Contrast with non-REM sleep.)

reactant A chemical substance that enters into a chemical reaction with another substance.

reaction center A group of electron transfer proteins that receive energy from light-absorbing pigments and convert it to chemical energy by redox reactions.

realized niche A species' niche as defined by its interactions with other species. (Contrast with fundamental niche.)

receptive field The area of visual space that activates a particular cell in the visual system.

receptor potential The change in the resting potential of a sensory cell when it is stimulated.

receptor protein A protein that can bind to a specific molecule, or detect a specific stimulus, within the cell or in the cell's external environment.

receptor *See* receptor protein, sensor.

receptor-mediated endocytosis Endocytosis initiated by macromolecular binding to a specific membrane receptor.

recessive In genetics, an allele that does not determine phenotype in the presence of a dominant allele. (Contrast with dominance.)

recognition sequence *See* restriction site.

recombinant chromatid In meiosis, a chromatid after crossing over that contains part of a sister chromatid, typically having exchanged that part with the sister chromatid.

recombinant DNA A DNA molecule made in the laboratory that is derived from two or more genetic sources.

recombinant frequency The proportion of offspring of a genetic cross that have phenotypes different from the parental phenotypes due to crossing over between linked genes during gamete formation.

rectum The terminal portion of the gut, ending at the anus.

redox reaction A chemical reaction in which one reactant is oxidized (loses electrons) and the other is reduced (gains electrons). Short for reduction–oxidation reaction.

reduction Gain of electrons by a chemical reactant. (Contrast with oxidation.)

refractory period The time interval after an action potential during which another action potential cannot be elicited from an excitable membrane.

regeneration The development of a complete individual from a fragment of an organism.

regional species pool All the species that are limited to a geographic region; sometimes called gamma diversity.

regular (uniform) dispersion The spatial arrangement of individuals within a population, characterized by evenly spaced individuals. (Compare with clumped dispersion, random dispersion.)

regulated specification Determination of cell fates due to signals mostly received from neighboring cells.

regulative development A pattern of animal embryonic development in which the fates of the first blastomeres are not absolutely fixed. (Contrast with mosaic development.)

regulatory sequence A DNA sequence to which the protein product of a regulatory gene binds.

regulatory system A system that uses feedback information to maintain a physiological function or parameter at an optimal level. (Contrast with controlled system.)

regulatory T cells (Tregs) The class of T cells that mediates tolerance to self antigens.

reinforcement The evolution of enhanced reproductive isolation between populations due to natural selection for greater isolation.

releaser Sensory stimulus that triggers performance of a stereotyped behavior pattern.

renal [L. *renes*: kidneys] Relating to the kidneys.

renal artery The main artery supplying the kidney.

renal pyramids Tissue of the renal medulla that projects into and releases urine into the renal pelvis which is drained by the ureter.

renal tubule A structural unit of the kidney that collects filtrate from the blood, reabsorbs specific ions, nutrients, and water and returns them to the blood, and concentrates excess ions and waste products such as urea for excretion from the body.

renal vein The main vein carrying blood from the kidney.

renin An enzyme released from the kidneys in response to a drop in the glomerular filtration rate. Together with angiotensin converting enzyme, converts an inactive protein in the blood into angiotensin.

replication The duplication of genetic material.

replication fork A point at which a DNA molecule is replicating. The fork forms by the unwinding of the parent molecule.

replicon A region of DNA replicated from a single origin of replication.

reporter gene A genetic marker included in recombinant DNA to indicate the presence of the recombinant DNA in a host cell.

repressor A protein encoded by a regulatory gene that can bind to a promoter and prevent transcription of the associated gene. (Contrast with activator.)

reproductive isolation Condition in which two divergent populations are no longer exchanging genes. Can lead to speciation.

reproductive signal In cell biology, the internal or external signal that causes a cell to initiate its mitotic or meiotic cell division cycle.

residual volume (RV) In tidal ventilation, the dead space that remains in the lungs at the end of exhalation.

residuals The deviations, along the y-axis from the linear regression line, of individual observations in a bivariate scatter plot.

resistance (R) genes Plant genes that confer resistance to specific strains of pathogens.

resource Components of the environment such as food, water, light, and space that are required by all species.

resource partitioning A situation in which species share limiting resources by using them in different ways, thereby allowing them to coexist.

resource-mediated coexistence A type of coexistence among competitors in which factors such as disturbance, stress, or predation that act on the dominant competitor, allow inferior competitors access to limiting resources.

respiratory chain The terminal reactions of cellular respiration, in which electrons are passed from NAD or FAD, through a series of intermediate carriers, to molecular oxygen, with the concomitant production of ATP.

respiratory gases Oxygen (O_2) and carbon dioxide (CO_2); the gases that an animal must exchange between its internal body fluids and the outside medium (air or water).

resting potential The membrane potential of a living cell at rest. In cells at rest, the interior is negative to the exterior. (Contrast with action potential.)

restoration ecology The science and practice of restoring damaged or degraded ecosystems by active human intervention.

restriction digestion An enzymatic reaction in which a molecule of DNA is fully cleaved by a restriction enzyme.

restriction enzyme Any of a type of enzyme that cleaves double-stranded DNA at specific sites; extensively used in recombinant DNA technology. Also called a restriction endonuclease.

restriction point (R) The specific time during G1 of the cell cycle at which the cell becomes committed to undergo the rest of the cell cycle.

restriction site A specific DNA base sequence that is recognized and acted on by a restriction endonuclease.

reticular-activating system [L. *reticulum*: net] A central region of the vertebrate brainstem that includes complex fiber tracts conveying neural signals between the forebrain and the spinal cord, with collateral fibers to a variety of nuclei that are involved in autonomic functions, including arousal from sleep.

reticulation In a phylogenetic tree, a combining of lineages (through hybridization of gene capture, for example).

reticulum The second of the four chambers of the ruminant stomach. Along with the rumen, it is where food is partially digested with the assistance of gut bacteria.

retina (rett' in uh) [L. *rete*: net] The light-sensitive layer of cells in the vertebrate or cephalopod eye.

retrovirus An RNA virus that contains reverse transcriptase. Its RNA serves as a template for cDNA production, and the cDNA is integrated into a chromosome of the host cell.

reverse transcriptase An enzyme that catalyzes the production of DNA (cDNA), using RNA as a template; essential to the reproduction of retroviruses.

reverse transcription Production of DNA using an RNA template.

reversible reaction A chemical transformation that can occur in either direction, so that reactants become products and vice versa.

reversion mutation A second- or third-round mutation that reverts the DNA to its original sequence or to a new sequence that results in a non-mutant phenotype.

rhizoids (rye' zoids) [Gk. root] Hairlike extensions of cells in mosses, liverworts, and a few vascular plants that serve the same function as roots and root hairs in vascular plants. The term is also applied to branched, rootlike extensions of some fungi and algae.

rhizome (rye' zome) An underground stem (as opposed to a root) that runs horizontally beneath the ground.

rhodopsin A vertebrate visual pigment involved in transducing photons of light into changes in the membrane potential of certain photoreceptor cells.

ribose A five-carbon sugar in nucleotides and RNA.

ribosomal RNA (rRNA) Several species of RNA that are incorporated into the ribosome. Involved in peptide bond formation.

ribosome A small particle in the cell that is the site of protein synthesis.

ribozyme An RNA molecule with catalytic activity.

ribulose bisphosphate carboxylase/oxygenase *See* rubisco.

risk cost The increased chance of being injured or killed as a result of performing a behavior, compared to resting. (Compare with energetic cost, opportunity cost.)

RNA (ribonucleic acid) An often single-stranded nucleic acid whose nucleotides use ribose rather than deoxyribose and in which the base uracil replaces thymine found in DNA. Serves as genome from some viruses. (*See* ribosomal RNA, transfer RNA, messenger RNA, and ribozyme.)

RNA interference (RNAi) A mechanism for reducing mRNA translation whereby a double-stranded RNA, made by the cell or synthetically, is processed into a small, single-stranded RNA, whose binding to a target mRNA results in the latter's breakdown.

RNA polymerase An enzyme that catalyzes the formation of RNA from a DNA template.

RNA processing Modification of RNA primary transcripts, for example splicing out of introns.

RNA splicing The last stage of RNA processing in eukaryotes, in which the transcripts of introns are excised through the action of small nuclear ribonucleoprotein particles (snRNP).

rod cells Light-sensitive cells in the vertebrate retina; these sensory receptor cells are sensitive in extremely dim light and are responsible for dim light, black and white vision.

root The organ responsible for anchoring the plant in the soil, absorbing water and minerals, and producing certain hormones. Some roots are storage organs.

root apical meristem Undifferentiated tissue at the apex of the root that gives rise to the organs of the root.

root cap A thimble-shaped mass of cells, produced by the root apical meristem, that protects the meristem; the organ that perceives the gravitational stimulus in root gravitropism.

root hair A long, thin process from a root epidermal cell that absorbs water and minerals from the soil solution.

root nodule A swelling on the root of a leguminous plant, such as the soybean, that contains nitrogen-fixing Rhizobium bacteria.

root system The organ system that anchors a plant in place, absorbs water and

dissolved minerals, and may store products of photosynthesis from the shoot system.

rotational cleavage The form of complete cleavage that occurs in mammals. The first cleavage plain is parallel to the animal–vegetal axis and the two second cleavage planes are at right angles to each other.

rough endoplasmic reticulum (RER) The portion of the endoplasmic reticulum whose outer surface has attached ribosomes. (Contrast with smooth endoplasmic reticulum.)

round window A flexible membrane at the end of the lower canal of the cochlea in the human ear. (*See also* oval window.)

RT-PCR A laboratory manipulation of RNA in which the RNA is first incubated with reverse transcriptase (RT) to make cDNA and then the cDNA amplified for analysis by the polymerase chain reaction (PCR).

rubisco Contraction of ribulose bisphosphate carboxylase/oxygenase, the enzyme that combines carbon dioxide or oxygen with ribulose bisphosphate to catalyze the first step of photosynthetic carbon fixation or photorespiration, respectively.

Ruffini endings Slowly adapting stretch receptor in the skin.

rumen The first of the four chambers of the ruminant stomach. Along with the reticulum, it is where food is partially digested with the assistance of gut bacteria.

ruminant Herbivorous, cud-chewing mammals such as cows or sheep, characterized by a stomach that consists of four compartments: the rumen, reticulum, omasum, and abomasum.

S

S phase In the cell cycle, the stage of interphase during which DNA is replicated. (Contrast with G1 phase, G2 phase.)

salt glands Glands on the leaves of some halophytic plants that secrete salt, thereby ridding the plants of excess salt.

saltatory conduction [L. *saltare*: to jump] The rapid conduction of action potentials in myelinated axons; so called because action potentials appear to "jump" between nodes of Ranvier along the axon.

sample A set of observations made from a population.

saprobe [Gk. *sapros*: rotten] An organism (usually a bacterium or fungus) that obtains its carbon and energy by absorbing nutrients from dead organic matter. Also called a saprotroph.

sarcomere (sark' o meer) [Gk. *sark*: flesh + *meros*: unit] The contractile unit of a skeletal muscle.

sarcoplasm The cytoplasm of a muscle cell.

sarcoplasmic reticulum The endoplasmic reticulum of a muscle cell.

saturated fatty acid A fatty acid in which all the bonds between carbon atoms in the hydrocarbon chain are single bonds—that is, all the bonds are saturated with hydrogen atoms. (Contrast with unsaturated fatty acid.)

scatter plot   A figure that displays the values of observations for two variables along perpendicular axes.

scavenger   An animal that eats larger dead organisms but is not typically considered a detritivore.

Schwann cell   A type of glial cell that myelinates axons in the peripheral nervous system.

scion   In horticulture, the bud or stem from one plant that is grafted to a root or root-bearing stem of another plant (the stock).

sclera   The white outer layer of the eyeball continuous with the cornea.

sclereid   One of the principle types of cells in sclerenchyma.

scrotum   In most mammals, a pouch outside the body cavity that contains the testes.

SDP   *See* short-day plant.

second filial generation   *See* F_2.

second law of thermodynamics   The principle that when energy is converted from one form to another, some of that energy becomes unavailable for doing work.

second messenger   A compound, such as cAMP, that is released within a target cell after a hormone (the first messenger) has bound to a surface receptor on a cell; the second messenger triggers further reactions within the cell.

second polar body   In oogenesis, the daughter cell of the second meiotic division that subsequently degenerates. (*See also* ootid.)

secondary active transport   A form of active transport that does not use ATP as an energy source; rather, transport is coupled to ion diffusion down a concentration gradient established by primary active transport.

secondary bronchi   The airways branching off of the primary bronchi.

secondary cell wall   A thick, cellulosic structure internal to the primary cell wall formed in some plant cells after cell expansion stops. (Contrast with primary cell wall.)

secondary consumer   An organism (carnivore) that feeds on herbivores or primary consumers. (Contrast with primary consumer.)

secondary endosymbiosis   The engulfment of a photosynthetic eukaryote by another eukaryotic cell that gave rise to certain groups of photosynthetic eukaryotes (e.g., euglenids).

secondary growth   In plants, growth that contributes to an increase in girth. (Contrast with primary growth.)

secondary immune response   A rapid and intense response to a second or subsequent exposure to an antigen, initiated by memory cells. (Contrast with primary immune response.)

secondary lysosome   Membrane-enclosed organelle formed by the fusion of a primary lysosome with a phagosome, in which macromolecules taken up by phagocytosis are hydrolyzed into their monomers. (Contrast with lysosome.)

secondary metabolite   A compound synthesized by a plant that is not needed for basic cellular metabolism. Typically has an antiherbivore or antiparasite function.

secondary oocyte   In oogenesis, the daughter cell of the first meiotic division that receives almost all the cytoplasm. (*See also* first polar body.)

secondary production   Source of energy for organisms and ecosystems that is derived from the consumption of organic compounds produced by other organisms.

secondary sex characteristics   Outward expression of sex determination in organs and tissues, such a body hair and stature in humans.

secondary spermatocyte   One of the products of the first meiotic division of a primary spermatocyte.

secondary structure   Of a protein, localized regularities of structure, such as the α helix and the β pleated sheet. (Contrast with primary, tertiary, quaternary structure.)

secondary succession   The reestablishment of a community when most, but not all, organisms have been destroyed. (Contrast with primary succession.)

secretin   (si kreet' in) A peptide hormone secreted by the upper region of the small intestine when acidic chyme is present. Stimulates the pancreatic duct to secrete bicarbonate ions.

sedimentary rock   Rock formed by the accumulation of sediment grains on the bottom of a body of water. Often contain stratified fossils that allow geologists and biologists to date evolutionary events relative to each other.

seed   A fertilized, ripened ovule of a gymnosperm or angiosperm. Consists of the embryo, nutritive tissue, and a seed coat.

seedling   A plant that has just completed the process of germination.

segment polarity genes   In *Drosophila* (fruit fly) development, segmentation genes that determine the boundaries and anterior–posterior organization of individual segments. Part of a developmental cascade that includes maternal effect genes, gap genes, pair rule genes, and Hox genes.

segmentation   Division of an animal body into segments.

segmentation genes   Genes that determine the number and polarity of body segments.

segregation   In genetics, the separation of alleles, or of homologous chromosomes, from each other during meiosis so that each of the haploid daughter nuclei produced contains one or the other member of the pair found in the diploid parent cell, but never both. This principle was articulated by Mendel as his first law.

selectable marker   A gene, such as one encoding resistance to an antibiotic, that can be used to identify (select) cells that contain recombinant DNA from among a large population of untransformed cells.

selective permeability   Allowing certain substances to pass through while other substances are excluded; a characteristic of membranes.

self-incompatability   In plants, the possession of mechanisms that prevent self-fertilization.

semelparity   Reproducing only once in a lifetime. (Contrast with iteroparity.)

semen   (see' men) [L. *semin*: seed] The thick, whitish liquid produced by the male reproductive system in mammals, containing the sperm.

semicircular canals   Three canals in the human inner ear that form part of the vestibular system.

semiconservative replication   The way in which DNA is synthesized. Each of the two partner strands in a double helix acts as a template for a new partner strand. Hence, after replication, each double helix consists of one old and one new strand.

seminal vesicles   Male accessory sex glands that contribute most of the volume of the semen including mucus, fibrinogen, and fructose.

seminiferous tubules   The tubules within the testes within which sperm production occurs.

sensitive period   The life stage during which some particular type of learning must take place, or during which it occurs much more easily than at other times. Typical of song learning among birds. Also known as the critical period.

sensor   Cell that is responsive to a particular type of physical or chemical stimulation.

sensory mechanisms   The means by which internal or external stimuli are converted into neural signals.

sensory organs   Structures that convert types of internal or environmental energy into electrical signals used by the nervous system.

sensory system   A set of organs and tissues for detecting a stimulus; consists of sensory cells, the associated structures, and the neural networks that process the information.

sensory transduction   The transformation of environmental stimuli or information into neural signals.

sepal   (see' pul) [L. *sepalum*: covering] One of the outermost structures of the flower, usually protective in function and enclosing the rest of the flower in the bud stage.

septate   [L. wall] Divided, as by walls or partitions.

septum   (plural: septa) (1) A partition or cross-wall appearing in the hyphae of some fungi. (2) The bony structure dividing the nasal passages.

sequence alignment   A method of identifying homologous positions in DNA or amino acid sequences by pinpointing the locations of deletions and insertions that have occurred since two (or more) organisms have diverged from a common ancestor.

Sertoli cells   Cells in the seminiferous tubules of the testes that nurture the developing sperm.

sessile   (sess' ul) [L. *sedere*: to sit] Permanently attached; not able to move from one place to another. (Contrast with motile.)

set point   In a regulatory system, the threshold sensitivity to the feedback stimulus.

sex chromosome   In organisms with a chromosomal mechanism of sex determination, one of the chromosomes involved in sex determination (in humans and many other animals, these are the X and Y chromosomes).

sex pilus A thin connection between two bacteria through which genetic material passes during conjugation.

sex-linked Gene that is carried on a sex chromosome, such as the X chromosome in mammals.

sex-linked inheritance Pattern of inheritance characteristic of genes located on the sex chromosomes of organisms having a chromosomal mechanism for sex determination.

sexual reproduction Reproduction involving the union of gametes.

sexual selection Selection by one sex of characteristics in individuals of the opposite sex. Also, the favoring of characteristics in one sex as a result of competition among individuals of that sex for mates.

Shannon index The quantitative index most commonly used to describe species diversity that includes measures of both species richness and species evenness.

shoot apical meristem Undifferentiated tissue at the apex of the shoot that gives rise to the organs of the shoot.

shoot system In plants, the organ system consisting of the leaves, stem(s), and flowers.

short tandem repeats (STRs) Short (1–5 base pairs), moderately repetitive sequences of DNA. The number of copies of an STR at a particular location varies between individuals and is inherited.

short-day plant (SDP) A plant that flowers when nights are longer than a critical length specific for that plant's species. (Compare to long-day plant.)

short-term memory Information that is maintained in the brain and can be accessed for a short period of time—seconds to minutes. (Contrast with immediate memory, long-term memory.)

side chain *See* R group.

sieve tube element The characteristic cell of the phloem in angiosperms, which contains cytoplasm but relatively few organelles, and whose end walls (**sieve plates**) contain pores that form connections with neighboring cells.

sigma factor In prokaryotes, a protein that binds to RNA polymerase, allowing the complex to bind to and stimulate the transcription of a specific class of genes (e.g., those involved in sporulation).

signal sequence The sequence within a protein that directs the protein to a particular organelle.

signal transduction pathway The series of biochemical steps whereby a stimulus to a cell (such as a hormone or neurotransmitter binding to a receptor) is translated into a response of the cell.

signaling mechanisms The means of transferring information between cells, organs, or organisms.

significance level A particular threshold of making a Type I error (incorrectly rejecting a true null hypothesis) in a statistical test, selected a priori by the investigator.

silencer A gene sequence binding transcription factors that repress transcription. (Contrast with promoter.)

silent mutation A change in a gene's sequence that has no effect on the amino acid sequence of a protein either because it occurs in noncoding DNA or because it does not change the amino acid specified by the corresponding codon. (Contrast with frame-shift mutation, missense mutation, nonsense mutation.)

silent substitution *See* synonymous substitution.

similarity matrix A matrix used to compare the degree of divergence among pairs of objects. For molecular sequences, constructed by summing the number or percentage of nucleotides or amino acids that are identical in each pair of sequences.

simple diffusion Diffusion that does not involve a direct input of energy or assistance by carrier proteins.

simple fruit A fruit that develops from a single carpel in a single flower.

single nucleotide polymorphisms (SNPs) Inherited variations in a single nucleotide base in DNA that differ between individuals.

single-strand binding protein In DNA replication, a protein that binds to single strands of DNA after they have been separated from each other, keeping the two strands separate for replication.

sink In plants, any organ that imports the products of photosynthesis, such as roots, developing fruits, and immature leaves. (Contrast with source.)

sink populations Populations that receive more immigrants from other populations within a metapopulation than they produce.

sinoatrial node (sigh' no ay' tree al) [L. *sinus*: curve + *atrium*: chamber] The pacemaker of the mammalian heart.

sinus venosus The first chamber of the heart of fish that opens into the atrium.

siRNAs (small interfering RNAs) Short, double-stranded RNA molecules used in RNA interference.

sister chromatid Each of a pair of newly replicated chromatids.

sister clades Two clades that are each other's closest relatives.

sister species Two species that are each other's closest relatives.

skeletal muscle A type of muscle tissue characterized by multinucleated cells containing highly ordered arrangements of actin and myosin microfilaments. Also called striated muscle. (Contrast with cardiac muscle, smooth muscle.)

skeletal systems Organ systems that provide rigid supports—skeletons—against which muscles can pull to create directed movements. (*See also* endoskeleton, exoskeleton.)

sliding DNA clamp Protein complex that keeps DNA polymerase bound to DNA during replication.

sliding filament model Mechanism of muscle contraction based on the formation and breaking of crossbridges between actin and myosin filaments, causing the filaments to slide together.

slow-twitch fibers Skeletal muscle fibers specialized for sustained aerobic work; contain myoglobin and abundant mitochondria, and are well-supplied with blood vessels. Also called oxidative or red muscle fibers. (Compare to fast-twitch fibers.)

slow-wave sleep *See* non-REM sleep.

small interfering RNAs *See* siRNAs.

small intestine The portion of the gut between the stomach and the colon; consists of the duodenum, the jejunum, and the ileum.

smooth endoplasmic reticulum (SER) Portion of the endoplasmic reticulum that lacks ribosomes and has a tubular appearance. (Contrast with rough endoplasmic reticulum.)

smooth muscle Muscle tissue consisting of sheets of mononucleated cells innervated by the autonomic nervous system. (Contrast with cardiac muscle, skeletal muscle.)

sodium–potassium (Na⁺–K⁺) pump Anti-porter responsible for primary active transport; it pumps sodium ions out of the cell and potassium ions into the cell, both against their concentration gradients. Also called a sodium–potassium ATPase.

soil fertility The ability of a soil to sustain plant growth, in terms of the soil's concentration of available plant nutrients.

soil solution The aqueous portion of soil, from which plants take up dissolved mineral nutrients.

soils The upper layers of Earth's surface that support plant growth.

solute A substance that is dissolved in a liquid (solvent) to form a solution.

solute potential (Ψ_s) A property of any solution, resulting from its solute contents; it may be zero or have a negative value. The more negative the solute potential, the greater the tendency of the solution to take up water through a differentially permeable membrane. (Contrast with pressure potential, water potential.)

solution A liquid (the solvent) and its dissolved solutes.

solvent Liquid in which a substance (solute) is dissolved to form a solution.

solvent drag Transport mechanism in which solutes move across a tissue layer through intercellular spaces by a flow of water (solvent).

somatic cell [Gk. *soma*: body] All the cells of the body that are not specialized for reproduction. (Contrast with germ cell.)

somatic cell gene therapy Alteration of the human genome sequence in somatic cells for medical benefit. (Contrast with germ line gene therapy.)

somatic mutation Permanent genetic change in a somatic cell (as opposed to a germ cell, the egg or sperm). These mutations affect the individual only; they are not passed on to offspring. (Contrast with germ line mutation.)

somatosensory cortex An area of the parietal lobe that receives touch and pressure information from mechanoreceptors throughout the body;

neurons in this area are arranged according to the parts of the body with which they communicate.

somatostatin Peptide hormone made in the hypothalamus that inhibits the release of other hormones from the pituitary and intestine.

somite (so' might) One of the segments into which an embryo becomes divided longitudinally, leading to the eventual segmentation of the animal as illustrated by the spinal column, ribs, and associated muscles.

soredia (singular: soredium) Propagules of lichens consisting of one or a few photosynthetic cells bound by fungal hyphae.

sorus (plural: sori) A spore-producing structure on the underside of a fern frond.

source In plants, any organ that exports the products of photosynthesis in excess of its own needs, such as a mature leaf or storage organ. (Contrast with sink.)

source populations Populations that serve as a net source of individuals to other populations within a metapopulation.

spatial summation In the production or inhibition of action potentials in a postsynaptic cell, the interaction of depolarizations and hyperpolarizations produced at different sites on the postsynaptic cell. (Contrast with temporal summation.)

spawning See external fertilization.

speciation (spee' see ay' shun) The process of splitting one population into two populations that are reproductively isolated from one another. (Contrast with evolution.)

species (spee' sees) [L. kind] The base unit of taxonomic classification, consisting of an ancestor–descendant group of populations of evolutionarily closely related, similar organisms. The more narrowly defined "biological species" consists of individuals capable of interbreeding with each other but not with members of other species.

species composition The kinds of species present in a community.

species concepts The ways that biologists think about the category of species.

species diversity A measure that combines the number of species (species richness) in a community and their relative abundances compared with one another (species evenness).

species evenness The relative abundances of species compared to one another.

species richness The total number of species within an area.

species–area relationship The relationship between the size of an area and the numbers of species it supports.

specific heat The amount of energy that must be absorbed by a gram of a substance to raise its temperature by one degree centigrade. By convention, water is assigned a specific heat of one.

sperm [Gk. sperma: seed] The male gamete.

sperm cells See sperm.

spermatid One of the products of the second meiotic division of a primary spermatocyte; four

haploid spermatids, which remain connected by cytoplasmic bridges, are produced for each primary spermatocyte that enters meiosis.

spermatogonium (plural: spermatogonia) In animals, the diploid progeny of a germ cell in males.

spherical symmetry The simplest form of symmetry, in which body parts radiate out from a central point such that an infinite number of planes passing through that central point can divide the organism into similar halves.

sphincter (sfink' ter) [Gk. sphinkter: something that binds tightly] A ring of muscle that can close an orifice, for example, at the anus.

spicule [L. arrowhead] A hard, calcareous skeletal element typical of sponges.

spinal reflex The conversion of afferent to efferent information in the spinal cord without participation of the brain.

spindle apparatus [O.E. spindle: a short stick with tapered ends] Array of microtubules emanating from both poles of a dividing cell during mitosis and playing a role in the movement of chromosomes at nuclear division.

spiral cleavage A form of complete cleavage in which the cleavage planes are at oblique angles to the animal-vegetal axis.

spirillum Any of various spiral-shaped bacteria.

spleen Organ that serves as a reservoir for venous blood and eliminates old, damaged red blood cells from the circulation.

spontaneous generation The idea that under ordinary circumstances, living organisms can form from non-living materials.

spontaneous mutation A genetic change caused by internal cellular mechanisms, such as an error in DNA replication. (Contrast with induced mutation.)

sporangiophore A stalked reproductive structure produced by zygospore fungi that extends from a hypha and bears one or many sporangia.

sporangium (spor an' gee um) (plural: sporangia) [Gk. spora: seed + angeion: vessel or reservoir] In plants and fungi, any specialized stucture within which one or more spores are formed.

spore [Gk. spora: seed] (1) Any asexual reproductive cell capable of developing into an adult organism without gametic fusion. In plants, haploid spores develop into gametophytes, diploid spores into sporophytes. (2) In prokaryotes, a resistant cell capable of surviving unfavorable periods.

sporophyte (spor' o fyte) [Gk. spora: seed + phyton: plant] In plants and protists with alternation of generations, the diploid phase that produces the spores. (Contrast with gametophyte.)

sporulation The formation of spores, or resting stages.

stability The resistance of an ecological community to (or recovery from) a disturbance.

stabilizing selection Selection against the extreme phenotypes in a population, so that the

intermediate types are favored. (Contrast with directional selection, disruptive selection.)

stamen (stay' men) [L. stamen: thread] A male (pollen-producing) unit of a flower, usually composed of an anther, which bears the pollen, and a filament, which is a stalk supporting the anther.

standard deviation A measure of the spread of observations in a sample. See Appendix B for the mathematical formula.

standard error A measure of how close a sample statistic (such as the mean) is likely to be to the true population value.

standard error of the mean A measure of how close a sample mean is likely to be to the true population value. Calculated by dividing the standard deviation of a sample by the square root of the sample size.

standard free energy The energy associated with a chemical reaction that can be used to do work. The sum of the enthalpy and the product of temperature and entropy.

starch [O.E. stearc: stiff] A polymer of glucose; used by plants to store energy.

Starling's forces The two opposing forces responsible for water movement across capillary walls: blood pressure, which squeezes water and small solutes out of the capillaries, and osmotic pressure, which pulls water back into the capillaries.

start codon The mRNA triplet (AUG) that acts as a signal for the beginning of translation at the ribosome. (Contrast with stop codon.)

static life table A life table that records the survival and reproduction of individuals of different ages during a single time period.

statistic A numerical quantity calculated from data.

stele (steel) [Gk. pillar] The central cylinder of vascular tissue in a plant stem.

stem In plants, the organ that holds leaves and/or flowers and transports and distributes materials among the other organs of the plant.

stem cell In animals, an undifferentiated cell that is capable of continuous proliferation. A stem cell generates more stem cells and a large clone of differentiated progeny cells. (See also embryonic stem cell.)

stereocilia Fingerlike extensions of hair cell membranes whose bending initiates sound perception. (See hair cell.)

steroid hormones Steroid molecules that circulate in the blood and effect cells with appropriate receptors. Examples are testosterone and estrogen.

sticky ends On a piece of two-stranded DNA, short, complementary, one-stranded regions produced by the action of a restriction endonuclease. Sticky ends facilitate the joining of segments of DNA from different sources.

stigma [L. stigma: mark, brand] The part of the pistil at the apex of the style that is receptive to pollen, and on which pollen germinates.

stock In horticulture, the root or root-bearing stem to which a bud or piece of stem from another plant (the scion) is grafted.

stoma (plural: stomata) [Gk. *stoma*: mouth, opening] Small opening in the plant epidermis that permits gas exchange; bounded by a pair of guard cells whose osmotic status regulates the size of the opening.

stomach An organ that physically (and sometimes enzymatically) breaks down food, preparing it for digestion in the midgut.

stomatal crypt In plants, a sunken cavity below the leaf surface in which a stoma is sheltered from the drying effects of air currents.

stop codon Any of the three mRNA codons that signal the end of protein translation at the ribosome: UAG, UGA, UAA. (Contrast with start codon.)

STR *See* short tandem repeats.

stratigraphy The study of geological strata.

stratum (plural: strata) [L. *stratos*: layer] A layer of rock laid down at a particular time in the past.

stress An abiotic factor that reduces the growth, reproduction, and/or the survival of some individuals. (Contrast with disturbance.)

stretch receptor A modified muscle cell embedded in the connective tissue of a muscle that acts as a mechanoreceptor in response to stretching of that muscle.

strigolactones Signaling molecules produced by plant roots that attract the hyphae of mycorrhizal fungi.

strobilus (plural: strobili) One of several conelike structures in various groups of plants (including club mosses, horsetails, and conifers) associated with the production and dispersal of reproductive products.

stroke An embolism in an artery in the brain that causes the cells fed by that artery to die. The specific damage, such as memory loss, speech impairment, or paralysis, depends on the location of the blocked artery.

structural gene A gene that encodes the primary structure of a protein not involved in the regulation of gene expression.

structural isomers Molecules made up of the same kinds and numbers of atoms, in which the atoms are bonded differently.

structural motif A three-dimensional structural element that is part of a larger molecule. For example, there are four common motifs in DNA-binding proteins: helix-turn-helix, zinc finger, leucine zipper, and helix-loop-helix.

style [Gk. *stele*: pillar or column] In the angiosperm flower, a column of tissue extending from the tip of the ovary, and bearing the stigma or receptive surface for pollen at its apex.

subduction In plate tectonics, the movement of one lithospheric plate under another.

submucosa (sub mew koe′ sah) The tissue layer just under the epithelial lining of the lumen of the digestive tract.

subsoil The soil horizon lying below the topsoil and above the parent rock (bedrock); the zone of infiltration and accumulation of materials leached from the topsoil. Also called the B horizon.

substrate (sub′ strayte) (1) The molecule or molecules on which an enzyme exerts catalytic

action. (2) The base material on which a sessile organism lives.

succession The gradual, sequential series of changes in the species composition of an ecological community following a disturbance.

succulence In plants, possession of fleshy, water-storing leaves or stems; an adaptation to dry environments.

sulci (singular: sulcus) The valleys between the folds of the cerebral cortex.

superficial cleavage A variation of incomplete cleavage in which cycles of mitosis occur without cell division, producing a syncytium (a single cell with many nuclei).

superior vena cava Large vein that returns deoxygenated blood to the right atrium from the upper body.

suprachiasmatic nuclei (SCN) In mammals, two clusters of neurons just above the optic chiasm that act as the master circadian clock.

surface area-to-volume ratio For any cell, organism, or geometrical solid, the ratio of surface area to volume; this is an important factor in setting an upper limit on the size a cell or organism can attain.

surface tension The attractive intermolecular forces at the surface of liquid; an especially important property of water.

surfactant A substance that decreases the surface tension of a liquid. Lung surfactant, secreted by cells of the alveoli, is mostly phospholipid and decreases the amount of work necessary to inflate the lungs.

survivorship (l_x) The fraction of individuals that survive from birth to a given life stage or age.

survivorship curve A plot of the number of individuals surviving through each life stage (survivorship, l_x) for a hypothetical cohort, usually of 1,000 individuals; includes three general curves called type I, II, and III.

suspension feeders An organism that feeds on organisms much smaller than itself that are suspended in water or air by means of a straining device. Also called filter feeder.

suspensor In the embryos of seed plants, the stalk of cells that pushes the embryo into the endosperm and is a source of nutrient transport to the embryo.

sustainable Pertaining to the use and management of ecosystems in such a way that humans benefit over the long term from specific ecosystem goods and services without compromising others.

swim bladder An organ used primarily to regulate buoyancy in fish.

symbiosis (sim′ bee oh′ sis) [Gk. *sym*: together + *bios*: living] The living together of two or more species in a prolonged and intimate relationship.

symbiotic A relationship in which two or more organisms live in close association with one another.

symmetry Pertaining to an attribute of an animal body in which at least one plane can divide the body into similar, mirror-image halves. (*See* bilateral symmetry, radial symmetry.)

sympathetic nervous system The division of the autonomic nervous system that works in opposition to the parasympathetic nervous system. (Contrast with parasympathetic nervous system.)

sympatric speciation (sim pat′ rik) [Gk. *sym*: same + *patria*: homeland] Speciation due to reproductive isolation without any physical separation of the subpopulation. (Contrast with allopatric speciation.)

symplast The continuous meshwork of the interiors of living cells in the plant body, resulting from the presence of plasmodesmata. (Contrast with apoplast.)

symporter A membrane transport protein that carries two substances in the same direction. (Contrast with antiporter, uniporter.)

synapomorphy A trait that arose in the ancestor of a phylogenetic group and is present (sometimes in modified form) in all of its members, thus helping to delimit and identify that group. Also called a shared derived trait. (Contrast with derived trait.)

synapse (sin′ aps) [Gk. *syn*: together + *haptein*: to fasten] A specialized type of junction where a neuron meets its target cell (which can be another neuron or some other type of cell) and information in the form of neurotransmitter molecules is exchanged across a synaptic cleft.

synapsis (sin ap′ sis) The highly specific parallel alignment (pairing) of homologous chromosomes during the first division of meiosis.

synaptic cleft The space between the presynaptic cell and the postsynaptic cell in a chemical synapse.

syncytium A single cell with many nuclei produced as a result of superficial cleavage.

synonymous (silent) substitution A change of one nucleotide in a sequence to another when that change does not affect the amino acid specified (i.e., UUA → UUG, both specifying leucine). (Contrast with nonsynonymous substitution, missense substitution, nonsense substitution.)

systematics The scientific study of the diversity and relationships among organisms.

systemic acquired resistance A general resistance to many plant pathogens following infection by a single agent.

systemic circuit Portion of the circulatory system by which oxygenated blood from the lungs or gills is distributed throughout the rest of the body and returned to the heart. (Contrast with pulmonary circuit.)

systems biology The scientific study of an organism as an integrated and interacting system of genes, proteins, and biochemical reactions.

systole (sis′ tuh lee) [Gk. contraction] Contraction of a chamber of the heart, driving blood forward in the circulatory system. (Contrast with diastole.)

T

T cell A type of lymphocyte involved in the cellular immune response. The final stages of its development occur in the thymus gland. (*See also* cytotoxic T cell, T-helper cell. Contrast with B cell.)

T cell receptor A protein on the surface of a T cell that recognizes the antigenic determinant for which the cell is specific.

T tubules A system of tubules that runs throughout the cytoplasm of a muscle fiber, through which action potentials spread.

T-helper (T_H) cell Type of T cell that stimulates events in both the cellular and humoral immune responses by binding to the antigen on an antigen-presenting cell; target of the HIV-I virus, the agent of AIDS. (Contrast with cytotoxic T cells.)

taproot The largest plant root from which other roots sprout laterally.

taproot system A root system typical of eudicots consisting of a primary root (taproot) that extends downward by tip growth and outward by initiating lateral roots. (Contrast with fibrous root system.)

taste bud A structure in the epithelium of the tongue that includes a cluster of chemoreceptors innervated by sensory neurons.

TATA box An eight-base-pair sequence, found about 25 base pairs before the starting point for transcription in many eukaryotic promoters, that binds a transcription factor and thus helps initiate transcription.

taxon (plural: taxa) [Gk. *taxis*: put in order] A biological group (typically a species or a clade) that is given a name.

telencephalon The outer, surrounding structure of the embryonic vertebrate forebrain, which develops into the cerebrum.

telomerase An enzyme that catalyzes the addition of telomeric sequences lost from chromosomes during DNA replication.

telomeres (tee' lo merz) [Gk. *telos*: end + *meros*: units, segments] Repeated DNA sequences at the ends of eukaryotic chromosomes.

temperature inversions Foggy conditions created when cold air is trapped within a warm inversion layer that forms over valleys during warm days.

template A molecule or surface on which another molecule is synthesized in complementary fashion, as in the replication of DNA.

temporal lobe One of the four lobes of the cerebral hemisphere; receives and processes auditory and visual information; involved in recognizing, identifying, and naming objects.

temporal summation In the production or inhibition of action potentials in a postsynaptic cell, the interaction of depolarizations or hyperpolarizations produced by rapidly repeated stimulation of a single point on the postsynaptic cell. (Contrast with spatial summation.)

tendon A collagen-containing band of tissue that connects a muscle with a bone.

terminal bud Growing point at the end of a plant stem or branch.

termination In molecular biology, the end of transcription or translation.

territorial behavior Aggressive actions engaged in to defend a habitat or resource such that other animals are denied access.

tertiary endosymbiosis The mechanism by which some eukaryotes acquired the capacity for photosynthesis; for example, a dinoflagellate that apparently lost its chloroplast became photosynthetic by engulfing another protist that had acquired a chloroplast through secondary endosymbiosis.

tertiary structure In reference to a protein, the relative locations in three-dimensional space of all the atoms in the molecule. The overall shape of a protein. (Contrast with primary, quaternary, secondary, structures.)

test cross Mating of a dominant-phenotype individual (who may be either heterozygous or homozygous) with a homozygous-recessive individual.

testis (tes' tis) (plural: testes) [L. *testis*: witness] The male gonad; the organ that produces the male gametes.

tetanus [Gk. *tetanos*: stretched] (1) A state of sustained maximal muscular contraction caused by rapidly repeated stimulation. (2) In medicine, an often fatal disease ("lockjaw") caused by the bacterium *Clostridium tetani*.

tetrad [Gk. *tettares*: four] During prophase I of meiosis, the association of a pair of homologous chromosomes or four chromatids.

thalamus [Gk. *thalamos*: chamber] A region of the vertebrate forebrain; involved in integration of sensory input.

thallus The body of a fungus or plant that is not differentiated into stem and leaves and lacks true roots and a vascular system.

thecal cells Cells of the ovarian follicle of mammals that bound the follicle and secrete the hormone testosterone.

theory [Gk. *theoria*: analysis of facts] A far-reaching explanation of observed facts that is supported by such a wide body of evidence, with no significant contradictory evidence, that it is scientifically accepted as a factual framework. Examples are Newton's theory of gravity and Darwin's theory of evolution. (Contrast with hypothesis.)

theory of island biogeography A theory proposing that the number of species on an island (or in another geographically defined and isolated area) represents a balance, or equilibrium, between the rate at which species immigrate to the island and the rate at which resident species go extinct.

thermogenin Also called Uncoupling Protein 1 (UPC1). A protein in the mitochondria of brown fat that uncouples oxidative phosphorylation.

thermoneutral zone [Gk. *thermos*: temperature] The range of temperatures over which an endotherm does not have to expend extra energy to thermoregulate.

thermoregulatory adaptations Physiological and behavioral mechanisms for altering the heat content of the body.

thoracic cavity [Gk. *thorax*: breastplate] The portion of the mammalian body cavity bounded by the ribs, shoulders, and diaphragm. Contains the heart and the lungs.

thoracic duct The connection between the lymphatic system and the circulatory system.

threshold The level of depolarization that causes an electrically excitable membrane to fire an action potential.

thrombin An enzyme involved in blood clotting; cleaves fibrinogen to form fibrin.

thrombus (throm' bus) [Gk. *thrombos*: clot] A blood clot that forms within a blood vessel and remains attached to the wall of the vessel. (Contrast with embolus.)

thylakoid (thigh' la koid) [Gk. *thylakos*: sack or pouch] A flattened sac within a chloroplast. Thylakoid membranes contain all of the chlorophyll in a plant, in addition to the electron carriers of photophosphorylation. Thylakoids stack to form grana.

thymine (T) Nitrogen-containing base found in DNA.

thyroid gland [Gk. *thyreos*: door-shaped] A two-lobed gland in vertebrates. Produces the hormone thyroxine.

thyroid-stimulating hormone (TSH) Hormone produced by the anterior pituitary that stimulates the thyroid gland to produce and release thyroxine. Also called thyrotropin.

thyrotropin Hormone produced by the anterior pituitary that stimulates the thyroid gland to produce and release thyroxine. Also called thyroid-stimulating hormone (TSH).

thyrotropin-releasing hormone (TRH) Hormone produced by the hypothalamus that stimulates the anterior pituitary to release thyrotropin.

thyroxine Hormone produced by the thyroid gland; controls many metabolic processes.

tidal The bidirectional form of ventilation used by all vertebrates except birds; air enters and leaves the lungs by the same route.

tidal volume (TV) The volume of air inhaled and exhaled by a resting individual. (Contrast with expiratory reserve volume, tidal volume, vital capacity.)

tight junction A junction between epithelial cells in which there is no gap between adjacent cells.

tissue A group of similar cells organized into a functional unit; usually integrated with other tissues to form part of an organ.

titin A protein that holds bundles of myosin filaments in a centered position within the sarcomeres of muscle cells. The largest protein in the human body.

Toll-like receptors (TLRs) Membrane proteins of macrophages and dendritic cells with extracellular domains that recognize molecules from microbial pathogens.

tonoplast The membrane of the plant central vacuole.

topography The shape and surface features of Earth, in the form of mountains, valleys, ocean basins, etc.

topsoil The uppermost soil horizon; contains most of the organic matter of soil, but may be depleted of most mineral nutrients by leaching. Also called the A horizon.

totipotent [L. *toto*: whole, entire + *potens*: powerful] Possessing all the genetic information and other capacities necessary to form an entire individual. (Contrast with multipotent, pluripotent, unipotent.)

trachea (tray' kee ah) [Gk. *trakhoia*: tube] A tube that carries air to the bronchi of the lungs of vertebrates. When plural (**tracheae**), refers to the major airways of insects.

tracheary element Either of two types of xylem cells—tracheids and vessel elements—that undergo apoptosis before assuming their transport function.

tracheid (tray' kee id) A type of tracheary element found in the xylem of nearly all vascular plants, characterized by tapering ends and walls that are pitted but not perforated. (Contrast with vessel element.)

trade winds Prevailing winds that blow from east to west across the tropics.

trade-off The relationship between the fitness benefits conferred by an adaptation and the fitness costs it imposes. For an adaptation to be favored by natural selection, the benefits must exceed the costs.

trait In genetics, a specific form of a character: eye color is a character; brown eyes and blue eyes are traits. (Contrast with character.)

transcription The synthesis of RNA using one strand of DNA as a template.

transcription factors Proteins that assemble on a eukaryotic chromosome, allowing RNA polymerase II to perform transcription.

transect A linear measurement of any length used to count individuals along the stretch of the line.

transfection Insertion of recombinant DNA into animal cells.

transfer RNA (tRNA) A family of folded RNA molecules. Each tRNA carries a specific amino acid and anticodon that will pair with the complementary codon in mRNA during translation.

transformation (1) A mechanism for transfer of genetic information in bacteria in which pure DNA from a bacterium of one genotype is taken in through the cell surface of a bacterium of a different genotype and incorporated into the chromosome of the recipient cell. (2) Insertion of recombinant DNA into a host cell.

transgenic Containing recombinant DNA incorporated into the genetic material.

transition In genetics, a mutation such that a purine base is converted to the other purine (e.g., adenine to guanine) or a pyrimidine to the other pyrimidine (e.g., C to T).

transition state In an enzyme-catalyzed reaction, the reactive condition of the substrate after there has been sufficient input of energy (activation energy) to initiate the reaction.

transition-state intermediate Unstable molecule in a reaction which is formed during the reaction when reactants change into products.

translation The synthesis of a protein (polypeptide). Takes place on ribosomes, using the information encoded in messenger RNA.

translocation (1) In genetics, a rare mutational event that moves a portion of a chromosome to a new location, generally on a nonhomologous

chromosome. (2) In vascular plants, movement of solutes in the phloem.

transmembrane domain A protein region rich in hydrophobic amino acids that spans the phospholipid bilayer.

transmembrane protein An integral membrane protein that spans the phospholipid bilayer.

transpiration [L. *spirare*: to breathe] The evaporation of water from plant leaves and stem, driven by heat from the sun, and providing the motive force to raise water (plus mineral nutrients) from the roots.

transposable element (transposon) A segment of DNA that can move to, or give rise to copies at, another locus on the same or a different chromosome.

transversion A mutation that changes a purine to a pyrimidine or vice versa.

tree of life A term that encompasses the evolutionary history of all life, or a graphic representation of that history.

TRH *See* thyrotropin-releasing hormone.

tricuspid valve The right AV valve. So named because it has three leaves.

triglyceride A simple lipid in which three fatty acids are combined with one molecule of glycerol.

trimesters The three stages of human pregnancy, approximately 3 months each in length.

tripartite synapse The idea that a synapse includes not only the pre- and postsynaptic neurons involved but also encompasses many connections with glial cells called astrocytes.

triploblastic Having a body derived from three embryonic cell layers (ectoderm, mesoderm, and endoderm). (Contrast with diploblastic, monoblastic.)

triploid (3*n*) cell A cell containing three sets of an organism's chromosomes.

trisomic Containing three rather than two members of a chromosome pair.

trochophore (troke' o fore) [Gk. *trochos*: wheel + *phoreus*: bearer] A radially symmetrical larval form typical of annelids and mollusks, distinguished by a wheel-like band of cilia around the middle.

trophic [Gk. *trophes*: nourishment] The feeding relationship between two or more species.

trophic cascade A series of trophic interactions within a food web in which the rate of consumption at one trophic level results in the change in species abundance or composition at lower trophic levels.

trophic efficiency A measure of the amount of energy or biomass at one trophic level divided by the amount of energy at the trophic level immediately below it.

trophic level A subdivision of a food web that includes species that have similar ways of interacting and obtaining energy.

trophoblast [Gk *trophes*: nourishment + *blastos*: sprout] At the 32-cell stage of mammalian development, the outer group of cells that will

become part of the placenta and thus nourish the growing embryo. (Contrast with inner cell mass.)

tropic hormones Hormones produced by the anterior pituitary that control the secretion of hormones by other endocrine glands.

tropomyosin [troe poe my' oh sin] One of the three protein components of an actin filament; controls the interactions of actin and myosin necessary for muscle contraction.

troponin One of the three components of an actin filament; binds to actin, tropomyosin, and Ca^{2+}.

trypsin A protein-digesting enzyme. Secreted by the pancreas in its inactive form (trypsinogen), it becomes active in the duodenum of the small intestine.

TSH *See* thyroid-stimulating hormone.

tube feet A unique feature of echinoderms; extensions of the water vascular system, which functions in gas exchange, locomotion, and feeding.

tubulin A protein that polymerizes to form microtubules.

tumor [L. *tumor*: a swollen mass] A disorganized mass of cells. Malignant tumors spread to other parts of the body.

tumor necrosis factor A family of cytokines (growth factors) that causes cell death and is involved in inflammation.

tumor suppressor A gene that codes for a protein product that inhibits cell proliferation; inactive in cancer cells. (Contrast with oncogene.)

turgid Swollen or distended. In plants, a cell is turgid when its water potential is lower than that of the surrounding environment.

turgor pressure [L. *turgidus*: swollen] The force exerted by a plant cell against the cell wall when it has taken up water by osmosis. (*See* pressure potential.)

twitch A muscle fiber's minimum unit of contraction, stimulated by a single action potential.

tympanic membrane [Gk. *tympanum*: drum] The eardrum.

Type I error The incorrect rejection of a true null hypothesis.

type I survivorship curve A survivorship curve in which individuals experience high overall survivorship through adulthood but a steep decline late in life. (Compare with type II survivorship curve, type III survivorship curve.)

Type II error The incorrect acceptance of a false null hypothesis.

type II survivorship curves A survivorship curve in which individuals experience a constant chance of surviving at all ages. (Compare with type I survivorship curve, type III survivorship curve.)

type III survivorship curve A survivorship curve in which individuals experience low survivorship early in life and higher survivorship once they reach maturity. (Compare with type I survivorship curve, type II survivorship curve.)

U

ubiquitin A small protein that is covalently linked to other cellular proteins identified for breakdown by the proteosome.

ultimate causes In ethology, the evolutionary processes that produce an animal's capacity and tendency to behave in particular ways. (Contrast with proximate causes.)

uncompetitive inhibitor An enzyme inhibitor that binds to the enzyme-substrate complex.

undershoot *See* after-hyperpolarization.

unequal crossing over When a highly repeated gene sequence becomes displaced in alignment during meiotic crossing over, so that one chromosome receives many copies of the sequence while the second chromosome receives fewer copies. One of the mechanisms of concerted evolution. (*See also* biased gene conversion.)

uniporter [L. *unus*: one + *portal*: doorway] A membrane transport protein that carries a single substance in one direction. (Contrast with antiporter, symporter.)

unipotent An undifferentiated cell that is capable of becoming only one type of mature cell. (Contrast with totipotent, multipotent, pluripotent.)

unloading In phloem, the process of transferring sugars and other molecules in phloem to adjacent plant tissues.

unsaturated fatty acid A fatty acid whose hydrocarbon chain contains one or more double bonds. (Contrast with saturated fatty acid.)

upper critical temperature The environmental temperature at which a resting endotherm must actively increase its heat loss to avoid a rise in its body temperature.

upwelling A process by which offshore winds in combination with the Coriolis effect push warmer surface water away from the shore, allowing cold, nutrient-rich water from deeper waters to rise to the surface.

uracil (U) A pyrimidine base found in nucleotides of RNA.

urea (you ree' uh) A compound that is the main form of nitrogen excreted by many animals, including mammals. (Compare with ammonia, uric acid.)

ureotelic Pertaining to an organism in which the final product of the breakdown of nitrogen-containing compounds (primarily proteins) is urea. (Contrast with ammonotelic, uricotelic.)

ureter (your' uh tur) Long duct leading from the vertebrate kidney to the urinary bladder or the cloaca.

urethra (you ree' thra) In most mammals, the canal through which urine is discharged from the bladder and which serves as the genital duct in males.

uric acid A compound that serves as the main excreted form of nitrogen in some animals, particularly those which must conserve water, such as birds, insects, and reptiles. (Compare with ammonia, urea.)

uricotelic Pertaining to an organism in which the final product of the breakdown of nitrogen-containing compounds (primarily proteins) is uric acid. (Contrast with ammonotelic, ureotelic.)

urinary bladder A structure in which urine is stored until it can be excreted to the outside of the body.

urine (you' rin) In vertebrates, the fluid waste product containing the toxic nitrogenous by-products of protein and nucleic acid metabolism.

uterine cycle In human females, the monthly cycle of events by which the endometrium is prepared for the arrival of a blastocyst. (Contrast with ovarian cycle).

uterus (yoo' ter us) [L. *utero*: womb] A specialized portion of the female reproductive tract in mammals that receives the fertilized egg and nurtures the embryo in its early development. Also called the womb.

V

vaccination Injection of virus or bacteria or their proteins into the body, to induce immunity. The injected material is usually attenuated (weakened) before injection and is called a vaccine.

vacuole (vac' yew ole) Membrane-enclosed organelle in plant cells that can function for storage, water concentration for turgor, or hydrolysis of stored macromolecules.

vagina (vuh jine' uh) [L. sheath] In female animals, the entry to the reproductive tract.

valence shell The outermost energy shell of an atom, containing the valence electrons involved in chemical reactivity.

van der Waals forces Weak attractions between atoms resulting from the interaction of the electrons of one atom with the nucleus of another. This type of attraction is about one-fourth as strong as a hydrogen bond.

variable region The portion of an immunoglobulin molecule or T cell receptor that includes the antigen-binding site and is responsible for its antigen specificity. (Contrast with constant region.)

vas deferens (plural: vasa deferentia) Duct that transfers sperm from the epididymis to the urethra.

vasa recta Blood vessels that parallel the loops of Henle and the collecting ducts in the renal medulla of the kidney.

vascular bundle In vascular plants, a strand of vascular tissue, including xylem and phloem as well as thick-walled fibers.

vascular cambium (kam' bee um) [L. *cambiare*: to exchange] In plants, a lateral meristem that gives rise to secondary xylem and phloem.

vascular tissue system The transport system of a vascular plant, consisting primarily of xylem and phloem. (Contrast with dermal tissue system, ground tissue system.)

vasopressin A hormone that promotes water reabsorption by the kidney. Produced by neurons in the hypothalamus and released from nerve terminals in the posterior pituitary. Also called antidiuretic hormone or ADH.

vector (1) An agent, such as an insect, that carries a pathogen affecting another species. (2) A plasmid or virus that carries an inserted piece of DNA into a bacterium for cloning purposes in recombinant DNA technology.

vegetal hemisphere The lower portion of some animal eggs, zygotes, and embryos, in which the dense nutrient yolk settles. The vegetal pole is to the very bottom of the egg or embryo. (Contrast with animal hemisphere.)

vegetal pole The point at the end of the developing egg where yolk is most concentrated.

vegetative Nonreproductive, nonflowering, or asexual.

vegetative meristem An apical meristem that produces leaves.

vegetative reproduction Asexual reproduction through the modification of stems, leaves, or roots.

vein [L. *vena*: channel] A blood vessel that returns blood to the heart. (Contrast with artery.)

ventral [L. *venter*: belly, womb] Toward or pertaining to the belly or lower side. (Contrast with dorsal.)

ventricle A muscular heart chamber that pumps blood through the lungs or through the body.

venule A small blood vessel draining a capillary bed that joins others of its kind to form a vein. (Contrast with arteriole.)

vernalization [L. *vernalis*: spring] Events occurring during a required chilling period, leading eventually to flowering.

vertebral column [L. *vertere*: to turn] The jointed, dorsal column that is the primary support structure of vertebrates.

very low-density lipoproteins (VLDLs) Lipoproteins that consist mainly of triglyceride fats, which they transport to fat cells in adipose tissues throughout the body; associated with excessive fat deposition and high risk for cardiovascular disease.

vessel element A type of tracheary element with perforated end walls; found only in angiosperms. (Contrast with tracheid.)

vestibular system (ves tib' yew lar) [L. *vestibulum*: an enclosed passage] Structures within the inner ear that sense changes in position or momentum of the head, affecting balance and motor skills.

vestibule A central part of the inner ear involved in balance.

vicariance (vye care' ee uns) [L. *vicus*: change] The evolutionary separation of species due to a barrier that results in the geographic isolation of species that once were connected to one another.

villi (vil' lus) (singular: villus) [L. *villus*: shaggy hair or beard] Hairlike projections from a membrane; for example, from many gut walls.

virion (veer' e on) The virus particle, the minimum unit capable of infecting a cell.

virus Any of a group of ultramicroscopic particles constructed of nucleic acid and protein (and, sometimes, lipid) that require living cells in order to reproduce. Viruses evolved multiple times from different cellular species.

visceral mass The bulk of the digestive, reproductive, excretory, and respiratory systems of mollusks, contained within the shell.

vital capacity (VC)   The maximum capacity for air exchange in one breath; the sum of the tidal volume and the inspiratory and expiratory reserve volumes.

vitamin   [L. *vita*: life] An organic compound that an organism cannot synthesize, but nevertheless requires in small quantities for normal growth and metabolism.

vitelline envelope   The inner, proteinaceous protective layer of a sea urchin egg.

viviparity   (vye vi par′ uh tee) Reproduction in which fertilization of the egg and development of the embryo occur inside the mother's body. (Contrast with oviparity.)

vivipary   Premature germination in plants.

voltage   A measure of the difference in electrical charge between two points.

voltage-gated channel   A type of gated channel that opens or closes when a certain voltage exists across the membrane in which it is inserted.

vomeronasal organ (VNO)   Chemosensory structure embedded in the nasal epithelium of amphibians, reptiles, and many mammals. Often specialized for detecting pheromones.

W

waggle dance   A behavior of honeybees that communicates the direction and distance to a food source or a location for a new colony.

Walter climate diagram   A graphic technique that plots temperature and precipitation data to visualize a growing season for plants (those months when average temperatures are above freezing and average precipitation is sufficient for plant growth).

warning signal   A visual or acoustical defense mechanism used by prey species to advertise to predators that they may be toxic.

water potential (Ψ, psi)   In osmosis, the tendency for a system (a cell or solution) to take up water from pure water through a differentially permeable membrane.

water vascular system   In echinoderms, a network of water-filled canals that functions in gas exchange, locomotion, and feeding.

wavelength   The distance between successive peaks of a wave train, such as electromagnetic radiation.

weather   The short-term state of atmospheric conditions (temperature, precipitation, humidity, wind direction and velocity) at a particular place and time. (Contrast with climate.)

weathered   The mechanical and chemical processes by which rocks are broken down into soil particles.

Wernicke's area   A region in the temporal lobe of the human brain that is involved with the sensory aspects of language.

westerlies   Prevailing winds that blow from west to east across the mid-latitudes. (Compare with easterlies.)

white blood cells   Cells in the blood plasma that play defensive roles in the immune system. Also called leukocytes.

white matter   In the central nervous system, tissue that is rich in axons. (Contrast with gray matter.)

wild type   Geneticists' term for standard or reference type. Deviants from this standard, even if the deviants are found in the wild, are usually referred to as mutant. (Note that this terminology is not usually applied to human genes.)

wood   Secondary xylem tissue.

X

xerophyte   (zee′ row fyte) [Gk. *xerox*: dry + *phyton*: plant] A plant adapted to an environment with limited water supply.

xylem   (zy′ lum) [Gk. *xylon*: wood] In vascular plants, the tissue that conducts water and minerals; xylem consists, in various plants, of tracheids, vessel elements, fibers, and other highly specialized cells.

xylem sap   The liquid contents of the xylem.

xylem vessels   Continuous tubes that conduct water in plants.

Y

yeasts   Microscopic fungi consisting of single oval cells that reproduce by budding.

yolk   [M.E. *yolke*: yellow] The stored food material in animal eggs, rich in protein and lipids.

yolk sac   In reptiles, birds, and mammals, the extraembryonic membrane that forms from the endoderm of the hypoblast; it encloses and digests the yolk.

Z

zeatin   A cytokinin originally purified from corn kernels.

zeaxanthin   A blue-light receptor involved in the opening of plant stomata.

zona pellucida   A jellylike substance that surrounds the mammalian ovum when it is released from the ovary.

zone of cell division   The apical and primary meristems of a plant root; the source of all cells of the root's primary tissues.

zone of cell elongation   The part of a plant root, generally above the zone of cell division, where cells are expanding (growing), primarily in the longitudinal direction.

zone of maturation   The part of a plant root, generally above the zone of cell elongation, where cells are differentiating.

zygospore   Multinucleate, diploid cell that is a resting stage in the life cycle of zygospore fungi.

zygote   (zye′ gote) [Gk. *zygotos*: yoked] The cell created by the union of two gametes, in which the gamete nuclei are also fused. The earliest stage of the diploid generation.

zymogen   The inactive precursor of a digestive enzyme; secreted into the lumen of the gut, where a protease cleaves it to form the active enzyme.

ILLUSTRATION CREDITS

98 *Grapes*: © Julia Lelija/Shutterstock. Page 98 *Electron microscope*: © iStock.com/Savas Keskiner. Page 98 *Light microscope*: © MaZiKab/ Shutterstock. Page 98 *Tannin images*: From Brillouet et al., 2013. *Annals of Botany* 112: 1003. Page 99 Figures A, B: From Brillouet et al., 2013. *Annals of Botany* 112: 1003. Page 109: © Charles Mazel/Visuals Unlimited, Inc.

Chapter 6 *Opener*: © Russell Kightley/ Science Source. *inset*: © AF archive/Alamy Stock Photo. 6.2: After L. Stryer, 1981. *Biochemistry*, 2nd Ed., W. H. Freeman. 6.4: © D. W. Fawcett/ Science Source. 6.7A: Courtesy of D. S. Friend, U. California, San Francisco. 6.7B: Courtesy of Darcy E. Kelly, U. Washington. 6.7C: Courtesy of C. Peracchia. 6.10A *top*: © Stanley Flegler/Visuals Unlimited, Inc. 6.10A *bottom*: © Ed Reschke/Getty Images. 6.10B *top*: © David M. Phillips/Science Source. 6.10B *bottom*: © Ed Reschke/Getty Images. 6.10C *top*: © David M. Phillips/Science Source. 6.10C *bottom*: © Ed Reschke/Getty Images. 6.17: From M. M. Perry, 1979. *J. Cell Sci.* 39: 26. Page 122: From G. M. Preston et al., 1992. *Science* 256: 385. Page 128: © blickwinkel/Alamy Stock Photo.

Chapter 7 *Opener*: © Wim van Egmond/Visuals Unlimited, Inc. 7.3A: Data from PDB 3EML. V. P. Jaakola et al., 2008. *Science* 322: 1211. 7.3B: © iStock.com/Georgii Dolgykh. 7.12: © Stephen A. Stricker, courtesy of Molecular Probes, Inc. 7.17: Courtesy of David Kirk. Page 147: © iStock.com/ micro_photo. Page 149 *Figure A*: Courtesy of Janice Haney Carr/CDC. Page 149 *Figure B*: From Forst et al., 1989. *Proceedings of the National Academy of Sciences* 86: 6052.

Chapter 8 *Opener*: © George H.H. Huey/ Corbis. *inset*: © TH Foto/Alamy Stock Photo. 8.1: © Stephen Dalton/Minden Pictures/Getty Images. 8.9: Data from PDB 148L. Kuroki et al., 1993. *Science* 262: 2030. 8.11A: Data from PDB 1AL6. B. Schwartz et al., 1997. 8.11B: Data from PDB 1BB6. V. B. Vollan et al., 1999. *Acta Crystallogr. D. Biol. Crystallogr.* 55: 60. 8.11C: Data from PDB 1AB9. N. H. Yennawar, H. P. Yennawar, & G. K. Farber, 1994. *Biochemistry* 33: 7326. 8.12: Data from PDB 1IG8 (P. R. Kuser et al., 2000. *J. Biol. Chem.* 275: 20814) and 1BDG (A. M. Mulichak et al., 1998 *Nat. Struct. Biol.* 5: 555). 8.15: Data from PDB 1PTH. P. J. Loll et al., 1995. *Nat. Struct. Biol.* 2: 637. 8.18B: © joloei/Shutterstock. Page 171: © B. Mete Uz/Alamy Stock Photo.

Chapter 9 *Opener*: © Don W. Fawcett/ Science Source. *inset*: © Poulsons Photography/ Shutterstock. 9.8: From Y. H. Ko et al., 2003. *J. Biol. Chem.* 278: 12305. Courtesy of P. Pedersen. 9.11A: © sportpoint/Shutterstock. 9.11B: © volkova natalia/Shutterstock. 9.14: © iStock.com/ Ana Abejon.

Chapter 10 *Opener*: © Silken Photos/Alamy Stock Photo. 10.1: © Andrew Syred/SPL/Science Source. 10.10A: Courtesy of Lawrence Berkeley National Laboratory. 10.14A, 10.16A: © E. H. Newcomb & S. E. Frederick/Biological Photo Service. 10.18: © Aflo Foto Agency/Alamy Stock Photo. Table 10.1 *Rice*: © iStock.com/detshana. Table 10.1 *Maize*: © iStock.com/Ginginna. Table 10.1 *Cactus*: © iStock.com/Dan Eckert.

Chapter 11 *Opener*: © Steve Gschmeissner/ Science Source. 11.1A: © Dr. Stanley Flegler/ Visuals Unlimited, Inc. 11.1B: © Biodisc/Visuals Unlimited, Inc. 11.1C: © Stefan Sollfors/Alamy Stock Photo. 11.2B: © John J. Cardamone Jr./ Biological Photo Service. 11.7 *Chromosome*: © Biophoto Associates/Science Source. 11.7 *Nucleus*: © D. W. Fawcett/Science Source. 11.8: © Biophoto Associates/Science Source. 11.9: © Nasser Rusan. 11.10B: © Conly L. Rieder/ Biological Photo Service. 11.12A: © Dr. David Phillips/Visuals Unlimited, Inc. 11.12B: © B. A. Palevitz, E. H. Newcomb/Biological Photo Service. 11.13: © Egmont Strigl/Alamy Stock Photo. 11.14 *left*: © Andrew Syred/SPL/Science Source. 11.14 *center*: David McIntyre. 11.14 *right*: Courtesy of Andrew D. Sinauer. 11.15: © C. A. Hasenkampf/ Biological Photo Service. 11.16: Courtesy of J. Kezer. 11.20: Courtesy of Dr. Thomas Ried and Dr. Evelin Schröck, NIH. 11.21A: © Dr. Gopal Murti/ Visuals Unlimited, Inc. 11.22: © Dennis Kunkel Microscopy, Inc.

Chapter 12 *Opener*: © Picture Partners/ Alamy Stock Photo. 12.8 *Dark*: © iStock.com/ Marina Golskaya. 12.8 *Chinchilla*: © iStock.com/ purelook. 12.8 *Point*: David McIntyre. 12.8 *Albino*: © ZTS/Shutterstock. 12.11: Courtesy of Madison, Hannah, and Walnut. 12.12: Courtesy of the Plant and Soil Sciences eLibrary (http://plantandsoil.unl. edu); used with permission from the Institute of Agriculture and Natural Resources at the University of Nebraska. 12.13: © Mark Taylor/Naturepl. com. 12.14: © Peter Morenus/U. of Connecticut. 12.21A: © Dr. L. Caro/Science Source.

Chapter 13 *Opener*: © Biology Pics/Science Source. 13.3: © Lee D. Simon/Science Source. 13.5 *Franklin*: © Science Source. 13.6A: © A. Barrington Brown/Science Source. 13.6B: Data from S. Arnott & D. W. Hukins, 1972. *Biochem. Biophys. Res. Commun.* 47(6): 1504. 13.12A: Data from PDB 1SKW. Y. Li et al., 2001. *Nat. Struct. Mol. Biol.* 11: 784. 13.17B: © Dr. Peter Lansdorp/ Visuals Unlimited, Inc. Page 287: From D. M. Prescott and P. L. Kuempel, 1972. *Proceedings of the National Academy of Sciences* 69: 2842.

Chapter 14 *Opener*: Courtesy of the National Institute of Allergy and Infectious Diseases (NIAID). 14.1B: Courtesy of Eric Kalkman, University of Glasgow. 14.3: Data from PDB 1MSW. Y. W. Yin & T. A. Steitz, 2002. *Science* 298: 1387. 14.6B: From D. C. Tiemeier et al., 1978. *Cell* 14: 237. 14.10: Data from PDB 1EHZ. H. Shi & P. B. Moore, 2000. *RNA* 6: 1091. 14.12: Data from PDB 1GIX and 1G1Y. M. M. Yusupov et al., 2001. *Science* 292: 883. 14.16B: Courtesy of J. E. Edström and *EMBO J*.

Chapter 15 *Opener*: © The Florida collection/ Alamy Stock Photo. *inset*: © Armando Gallo/ Corbis. 15.8: From C. Harrison et al., 1983. *J. Med. Genet.* 20: 280. 15.10B: © David M. Martin, M.D./SPL/Science Source. 15.12: © Philippe Plailly/Science Source. 15.13B: U.S. Army photo. 15.15: © Simon Fraser/Science Source. Page 314: © Stanley Flegler/Visuals Unlimited, Inc.

Chapter 16 *Opener*: © imageBROKER/Alamy Stock Photo. *inset*: © Oleksii Sergieiev/Alamy Stock Photo. 16.11B: © Lee D. Simon/Science

Source. 16.16A: Courtesy of Irina Solovei, U. Munich (LMU), Germany. Page 352: © Dr. Donald Fawcett/Visuals Unlimited, Inc.

Chapter 17 *Opener*: © Shotshop GmbH/Alamy Stock Photo. 17.10B: Courtesy of O. L. Miller, Jr. 17.16 *left*: © kostudio/Shutterstock. 17.16 *right*: © Bruce Stotesbury/PostMedia News/Zuma Press. Page 368: © Mediscan/Corbis. Page 369: © Sinclair Stammers/Science Source. Page 370 *Drosophila*: David McIntyre. Page 370 *Arabidopsis*: David McIntyre (from seed provided by Tobias Baskin).

Chapter 18 *Opener*: © Erik Isakson/Getty Images. 18.3: © Dr. Jack Bostrack/Visuals Unlimited, Inc. 18.4: © iStock.com/Stephen Sewell. 18.11: Courtesy of the Golden Rice Humanitarian Board, www.goldenrice.org. 18.12: Courtesy of Eduardo Blumwald. 18.13: Courtesy of Tom Deerinck and Mark Ellisman of the National Center for Microscopy and Imaging Research at the University of California at San Diego. Page 394: © Dr. George Chapman/Visuals Unlimited, Inc. Page 398: From V. M. Gantz and E. Bier, 2015. *Science* 348: 442.

Chapter 19 *Opener*: © Dr. Yorgos Nikas/ Science Source. *inset*: Courtesy of VetStem Biopharma, www.vetstem.com. 19.4 *inset*: From I. Wilmut et al., 1997. *Nature* 385: 810. 19.7A: From J. E. Sulston & H. R. Horvitz, 1977. *Dev. Bio.* 56: 100. 19.11A: David McIntyre. 19.12A: From A. Ephrussi and D. St. Johnston, 2004. *Cell* 116: 143. 19.12B: Courtesy of Ruth Lehmann. 19.12C *left*: From E. A. Wimmer, 2012. *Science* 287: 2476. 19.12C *right*: From D. Tautz, 1988. *Nature* 332: 284. 19.13B: Courtesy of C. Rushlow and M. Levine. 19.13C: Courtesy of T. Karr. 19.13D: Courtesy of S. Carroll and S. Paddock. 19.16: © blickwinkel/Alamy Stock Photo. 19.18: © Bone Clones, www.boneclones.com. 19.19: Courtesy of J. Hurle and E. Laufer. 19.20 *Cladogram*: After R. Galant & S. Carroll, 2002. *Nature* 415: 910. 19.20 *Insect*: © Stockbyte/PictureQuest. 19.20 *Centipede*: © Burke/Triolo/Brand X Pictures/ PictureQuest. 19.21: Courtesy of Mike Shapiro and David Kingsley. Page 411, Page 412 *top*: From Meyerowitz 2002; courtesy of J. L. Riechmann. Page 412 *bottom*: Courtesy of D. Daily and W. Sullivan. Page 414: Courtesy of F. R. Turner, Indiana U. Page 416: © David Scharf/Science Source.

Chapter 20 *Opener*: © Fritz Pölking/Corbis. *inset*: © Michael Durham/Minden Pictures/Getty Images. 20.1 *H.M.S. Beagle*: Painting by Ronald Dean, reproduced by permission of the artist and Richard Johnson, Esquire. 20.1 *Darwin*: © The Art Gallery Collection/Alamy Stock Photo. 20.5A: © Luis César Tejo/Shutterstock. 20.5B–D: David McIntyre. 20.8: © Richard Du Toit/Minden Pictures/ Getty Images. 20.14: Courtesy of David Hillis. 20.19A: © Norbert Probst/imagebroker/Corbis. 20.19B: © Oliver Lucanus/NiS/Minden Pictures/ Getty Images. 20.20A: Courtesy of Edmund D. Brodie, Jr. Page 447, Figure A: © Fred LaBounty/ Alamy Stock Photo. Page 447, Figure B: © blickwinkel/Alamy Stock Photo. Page 447, Figure C: Courtesy of Yoel Stuart.

INDEX